AA004521

12th International Fatigue Congress (FATIGUE 2018)

MATEC Web of Conferences Volume 165 (2018)

Poitiers Futuroscope, France
27 May – 1 June 2018

Part 1 of 2

Editor:

Gilbert Henaff

ISBN: 978-1-5108-6386-6

Printed from e-media with permission by:

Curran Associates, Inc.
57 Morehouse Lane
Red Hook, NY 12571

Some format issues inherent in the e-media version may also appear in this print version.

Printed by Curran Associates, Inc. (2018)

For additional information, please contact EDP Sciences – Web of Conferences at the address below.

EDP Sciences – Web of Conferences
17, Avenue du Hoggar
Parc d'Activité de Courtabœuf
BP 112
F-91944 Les Ulis Cedex A
France

Phone: +33 (0) 1 69 18 75 75
Fax: +33 (0) 1 69 28 84 91

contact-edps@webofconferences.org

Additional copies of this publication are available from:

Curran Associates, Inc.
57 Morehouse Lane
Red Hook, NY 12571 USA
Phone: 845-758-0400
Fax: 845-758-2633
Email: curran@proceedings.com
Web: www.proceedings.com

TABLE OF CONTENTS

PART 1

PART 2

edp sciences

MATEC Web of Conferences

Statement of Peer review

In submitting conference proceedings to MATEC Web of Conferences, I certify to the Publisher that I adhere to the **Policy on Publishing Integrity** of the journal in order to safeguard good scientific practice in publishing.

1. All articles have been subjected to peer review administered by the proceedings editors.
2. Reviews have been conducted by expert referees, who have been requested to provide unbiased and constructive comments aimed, whenever possible, at improving the work.
3. Proceedings editors have taken all reasonable steps to ensure the quality of the materials they publish and their decision to accept or reject a paper for publication has been based only on the merits of the work and the relevance to the journal.

Title, date and place of the conference

12th International Fatigue Congress FATIGUE 2018

27 May - 1st June 2018

Poitiers Futuroscope, France

Proceedings editor(s):

Prof. Gilbert HÉNAFF

Date and editor's signature

30 March 2018

17, avenue du Hoggar - PA de Courtabœuf – BP 112 - 91944 Les Ulis Cedex A (France)
Tél. : 33 (0)1 69 18 75 75 - Fax : 33(0)1 69 07 45 17 – www.edpsciences.org

LCF and crack growth: recent results obtained by DIC

Luca Patriarca, *Stefano* Foletti, and *Stefano* Beretta*

Politecnico di Milano, Department of Mechanical Engineering, via La Masa 1, 20156, Milano

Abstract. Nowadays, several components for aerospace and energy applications (i.e., turbine disks, off-shore pipelines, and combustor chambers) are designed following a safe-life approach in the low cycle fatigue (LCF) regime. As also in the LCF regime it is worth adopting a damage tolerance design, fatigue life is assessed considering a crack propagating from the first load cycle, starting from an initial defect present in the component's most-stressed region or from a microstructural feature. In this frame, fatigue crack growth is described considering LCF conditions, where the driving force at the crack tip can be described with the ΔJ_{eff} parameter. The paper aims to give an overview of the potential applications and perspectives for the adoption of digital image correlation (DIC) measurements to study LCF nucleation and propagation. A visual technique based on the DIC displacement fields was developed to measure crack opening and closing levels in the presence of large strains. The technique is applied to different materials (steels, AlSi10Mg, nickel-based super-alloys). In addition, further applications of the DIC strain measurements are revised for the study of the nucleation stage and the crack tip process zone.

1 Introduction

Design approaches that consider the propagation of cracks are still currently under development as the request from industry (for example aerospace and energy) demands the application of the damage tolerance approach [1–3]. In this design perspective, a crack is supposed to nucleate from the first loading cycle in regions of the component where high stresses and strains develop [4]. Designers are then forced to consider the presence of propagating cracks in regions where the low cycle fatigue (LCF) design is considered. The overall reliability of the component is evaluated by introducing a series of safety factors, calculated following semi-probabilistic approaches, which are employed to consider the natural scatter of the external applied loads [5]. In such conditions, fatigue properties can be conservatively predicted, assuming the presence of small shallow cracks, with a depth corresponding to the detection limit of non-destructive techniques (NDT) [6].

Nowadays, advanced measurements techniques are available to characterize the mechanical behavior of metallic alloys. In particular, the digital image correlation (DIC) technique is currently under crescent interest from the engineering mechanical community [7,8]. The reasons for this success are different: (i) it is non-contact, (ii) it requires a (relatively) cheap set-up, (ii) it can be performed at different magnifications (providing the user can define a proper speckle pattern); (iv) it can be used to perform real-time full field measurements. DIC requires a specific specimen preparation aiming to define a target area where a proper speckle pattern has to be produced. The peculiarity of the DIC technique is that it can be used at different magnifications providing analyses that range from the macroscale (on the order of millimeters and centimeters) to the micro scale (microns).

DIC measurements are currently adopted in several applications for the mechanical characterization of materials along with the classical measurements techniques (extensometers, strain gauges, etc.). The adoption of DIC can be particularly beneficial for the characterization of the fatigue process as it allows to detect the early damage accumulation preceding the crack nucleation and, successively, to study the crack propagation stages. As an example, Abuzaid and co-workers shown that strain heterogeneities can be measured along specific microstructural features and correlated with the nucleation of cracks [9]. Displacement fields were used to calculate opening and closing levels in the high cycle fatigue (HCF) regime [10] and for calculating important fracture mechanics parameters, e.g. the stress intensity factor, [11].

In this work we discuss the application of DIC to LCF, an area where there are few applications in literature [12,13]. In the first part of the paper we discuss the crack closure phenomena, with particular emphasis on a technique based on DIC virtual extensometers which enables to detect the opening and closing point of a crack under LCF conditions. These results are successively validated by a second technique based on virtual strain gauges which originates from an experimental evidence presented in the work of Vormwald and co-workers, [14]. In the second part of the paper, we mainly focus on a brief discussion on the adoption of DIC strain measurements to detect early damage development during LCF loadings.

* Corresponding author: stefano.beretta@polimi.it

To summarize, this study aims to provide a general overview of the potential applications of visual measurements for deepening the study of LCF behavior. Specifically, we present LCF analyses on a structural steel used in railway transportation, an AlSi10Mg alloy produced by additive manufacturing (AM) and a coarse-grained Ni-based super alloy. Prospectively, this represents a quite-large set of experimental observations which covers different industrial applications.

2 DIC set-up and correlation strategy

The DIC technique is based on the correlation of images of a target area captured before and during specimen deformation. The specimen surface is required to be properly prepared to produce a light-intensity pattern when observed under an optical microscope. The DIC set-up adopted in this study is composed of: (i) a 2 megapixel Allied vision Manta G201B CDD digital camera; (ii) a set of lens produced by Optem with a final magnification ranging between 10x and 65x; (iii) a set of micro-stages to position the microscope; (iv) a Schott ACE I EKE rim light fiber optic illuminator. The final resolution of the DIC measurements performed in this study ranges between 1 and 2 μm/px.

The specimen geometry adopted was in all cases the classical cylindrical shape with the cross-section diameter that depends on the specific material under investigation. The LCF specimens were successively machined by electro-discharge machining to obtain two symmetric flattened surfaces required to perform the DIC measurements (a schematic of the geometry is included in the inset of Fig. 1). In the case of crack propagation analysis, one or more artificial defects are machined in the flattened surfaces. For the crack nucleation tests the flattened surfaces were properly prepared on the basis of the microstructural investigation analysis required. Etching and electron-back scatter diffraction (EBSD) were used to define the grain boundary and grain orientation maps, X-ray computed tomography to detect manufacturing defects as, for example, the porosity induced by lack of fusion in AM alloys. These analyses provided an initial scan of the target surface, successively DIC measurements were correlated and overlapped to this microstructural information to correlate the specific strain heterogeneities and quantify the effect of the microstructural features.

All the experiments presented in this study were conducted under a servo-hydraulic MTS loading frame in strain control. The frequency of the cyclic loads was set to 0.5Hz. Depending on the specific target of the experiment, the images were manually or automatically captured. At a fixed lens magnification, the dimension of the area under investigation cannot be changed, the region of interest (ROI) is then fixed before starting the test. The test can be interrupted at a specific number of cycles or at a specific load (or deformation) and multiple images are manually captured to measure the displacement and strain fields for a large area (bigger than the single ROI). When the image acquisition is performed automatically, a video is acquired and successively un-zipped. In this case, the LCF cycle is performed at lower frequency and a number of 250 images are the generally stored for one cycle and successively correlated. For this type of acquisition, the microscope cannot be moved and the ROI is fixed to a specific specimen surface position (e.g. the crack tip).

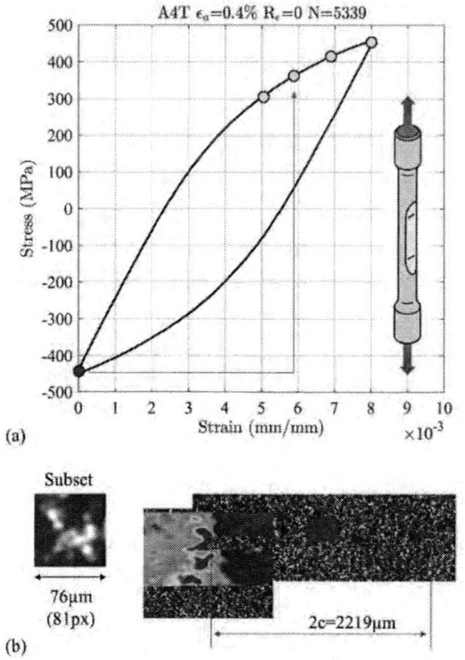

Fig. 1. DIC acquisition strategy: (a) the images are captured for the entire LCF cycle and successively correlated using the image at minimum stress as the reference one; (b) the subset size and the positioning of the DIC strain fields to locate the crack tip and calculate the crack length.

The DIC acquisition strategy for a general LCF cycle is schematically depicted in Fig. 1. The subset selected for the correlation is typically 81px, but depending on the specific magnification required and the speckle pattern designed, smaller subsets were selected. For LCF crack propagation tests the crack tip was located using the strain fields calculated correlating the reference image taken at the minimum load (red dot in Fig. 1) with the images captured during the LCF cycle (yellow dots). The displacement and strain maps calculated with this DIC strategy refer to the incremental quantities which respect to the reference image captured in the same loading cycle. This technique was used to detect the opening and closing stresses for a LCF propagation test and the local strain ranges for a LCF nucleation test.

3 DIC results and analyses

In this section, the application of DIC displacement measurements is detailed for the calculation of the crack opening and closing stresses under LCF conditions. The most significant results are summarized for three different alloys: an A4T railway steel, an AlSi10Mg produced by SLM and a coarse-grained super-alloy. The first material is used to explain the experimental technique to calculate opening and closing levels.

3.1 Application to A4T railway steel

The analysis of the crack loading loop by means of DIC displacement fields is performed by the virtual extensometer technique. A sequence of virtual extensometers is positioned along the crack plane behind and after the crack tip as shown in Fig. 2a. The distance between the extensometers is 5px which corresponds to approximately 4.5μm considering the typical resolutions adopted for these measurements. The reading of each extensometer is plotted as a function of the applied stress in Fig. 2b. This plot provides an approximate crack profile change with loading. The crack tip is indicated at the nominal position of $d = 0$.

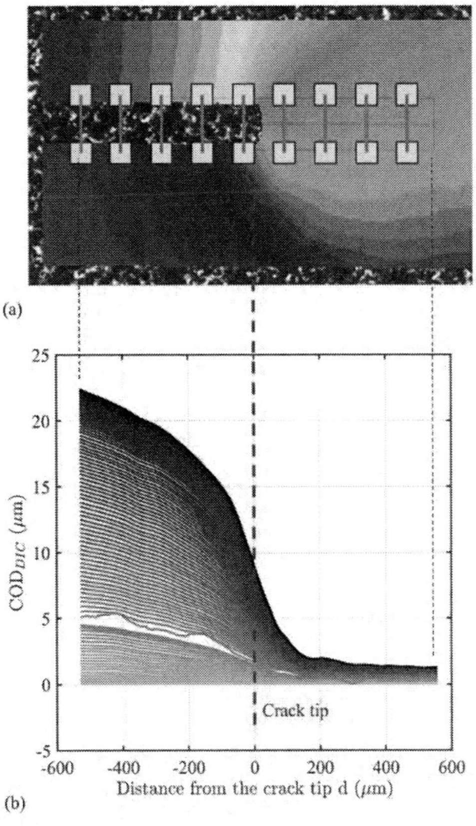

Fig. 2. Virtual extensometer readings from DIC displacement fields.

The opening stress is determined separately for each virtual extensometer reading defined as the crack opening displacement (COD_{DIC}), [10]. From the minimum load is observed an initial linear relation between the remote stress and the COD_{DIC} that originates from the linear elastic behaviour of the material in between the extensometer gauge length. In fact, the virtual extensometer cannot be positioned exactly on the crack surfaces at nominal zero distance.

After removing the initial linear part of the data set, the $COD_{DIC,pl}$-stress plot displays an initial vertical region (Fig. 3). One possible definition of the opening is provided in [15]. The opening point is defined as the intersection between the plot with a vertical line displaced by 1.5% of the maximum $COD_{DIC,pl}$.

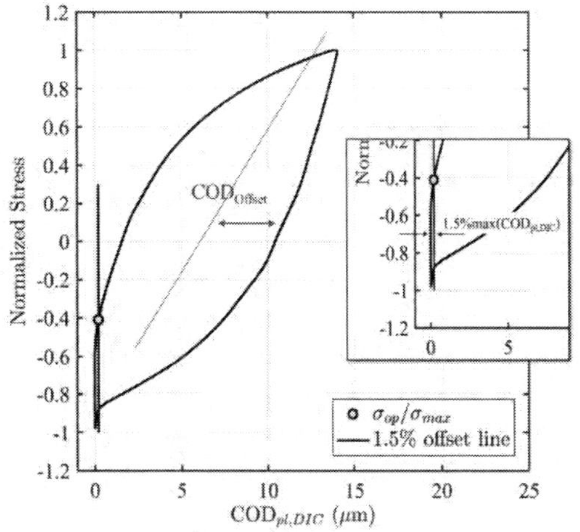

Fig. 3. The COD_{DIC}-stress plot is analysed according the 1.5% offset method to locate the opening point.

The closing point is defined according the procedure for HCF conditions suggested in [16], herein adapted for LCF. The data are linearly fitted considering the upper part of the $COD_{DIC,pl}$-stress plot where the crack is considered to be open according an equation of the type $COD_{DIC,pl}=m\cdot\sigma+q$. The COD_{offset} (Fig. 3) is then calculated for each data point according Eq. 1.

$$COD_{offset} = COD_{DIC,pl} - (m\cdot\sigma+q) \quad (1)$$

Plotting the COD_{offset} versus stress results in a drop-shaped diagram as shown in Fig. 4a. The advantage of this representation is that it can be clearly defined the inflection point of the unloading branch which is considered the closing point of the crack tip. According to the interpretation provided in [16], the inflection point represent the point where it is observed a change in the compliance that results from unloading a plastically deformed material to the onset of crack surface contact which implies a different compliance.

An additional interpretation is also provided with the schematic drawn in Fig. 4b. During unloading, and after the crack surfaces in between the virtual extensometers get in contact, there should be observed a dominant linear relation between the COD_{DIC} and the applied load. This is observed in Fig. 3 and in Fig. 4a in correspondence of the nose of the plot. Accordingly, the closing stress precedes this nose and is consistently defined according the inflection point definition.

https://doi.org/10.1051/matecconf/201816501001

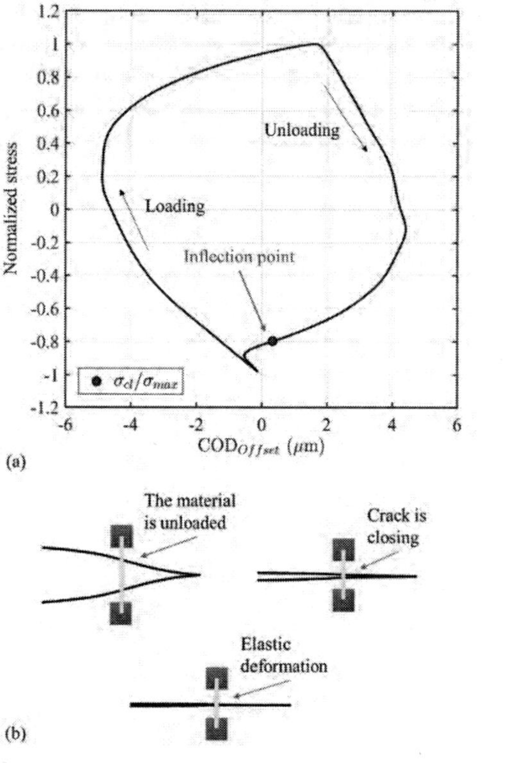

Fig. 4. Determination of the closing stress from COD$_{offset}$ analysis: (a) the inflection point is determined during unloading; (b) schematic of the closing stage.

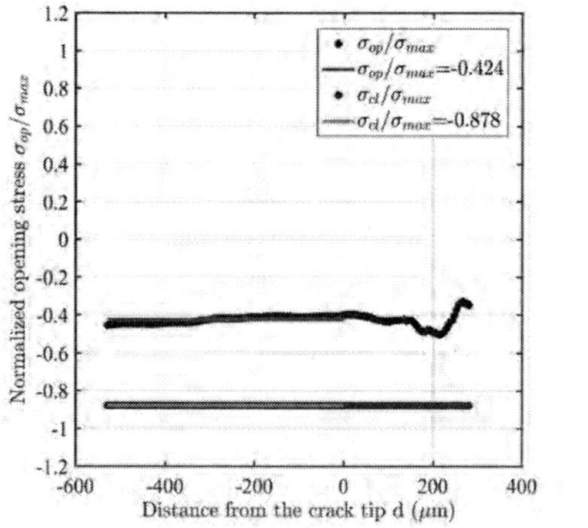

Fig. 5. Opening and closing stress measurements for different virtual extensometer position behind and ached the crack tip.

The measurements of the opening and closing stresses are then repeated for all the virtual extensometers introduced to provide a robust measurement. Fig. 5 summarizes the opening and closing levels calculation performed for the specific loading cycle and crack length. The opening and closing stresses are calculated averaging the values calculated using the virtual extensometers behind the crack tip. Noticeably, the opening stress values show a large scatter considering virtual extensometer ahead the crack tip. The number of virtual extensometers to be considered behind the crack tip is an aspect which is still under investigation. Due to the large deformations in LCF conditions, the opening stress depends on the position of the virtual extensometer. The opening stress measured decreases moving farther from the crack tip. Depending on the magnification adopted in the DIC set-up, this effect has to be taken into account as it might produce over-estimation of the driving force parameter related to the opening stress level.

Following the experimental approach of Vormwald [14], local strain measurements behind the crack tip are measured by virtual strain gauges and used to detect the opening and closing stresses [13]. Fig. 8a shows one example of virtual strain gauge defined on the DIC strain fields. The blu rectangle defines the area on the DIC strain field where the local axial strain values are extracted and averaged. These strain measurements are then plotted with the bulk hysteresis cycle (Fig. 8b).

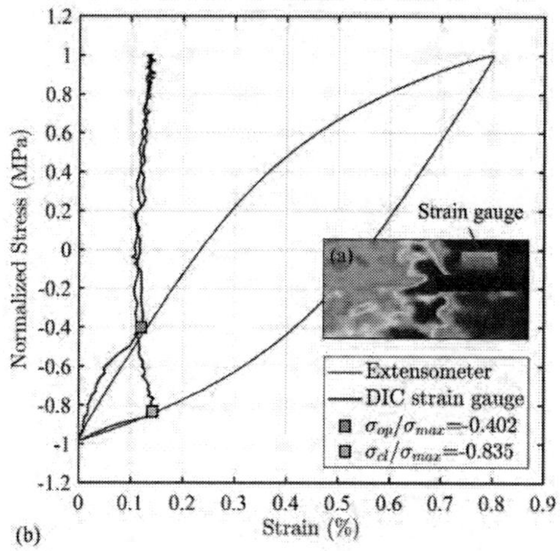

Fig. 8. Virtual strain gauge measurements for the opening and closing stresses: (a) the DIC virtual strain gauge positioned behind the crack tip; (b) the hysteresis cycle measured with the extensometer is compared with the hysteresis loop from the extensometer.

The material behaviour remains the same until the crack is closed, while after crack opening the crack surfaces in proximity of the crack tip release the elastic stress. The virtual strain gauge detects the elastic stress release displaying a vertical step in the hysteresis loop indicating that the measured strain remains constant. Once the crack surfaces are subjected to reverse loading the reading of the virtual strain gauge restores to the bulk hysteresis loop. The opening and closing levels measured with the virtual strain gauge confirms the measurements performed with the virtual extensometers (Fig. 5 and Fig. 8).

3.2 Application to AlSi10Mg (AM)

The aforementioned experimental technique was applied to study LCF crack propagation for an AlSi10Mg alloy produced by AM. For this material, in the current state of the manufacturing process (AM), lack of fusions and pores cannot yet be completely avoided. Design of these alloys has to deal with a crack propagation assessment [17].

The specimens were produced by means of a selective laser melting process (Fig. 6a). A semi-circular artificial defect with 200μm radius was introduced to vehiculate the crack nucleation. Three strain ranges at $R_\varepsilon=-1$ were selected which span from the elastic dominant behaviour ($\Delta\varepsilon=0.52\%$) to large plastic hysteresis loops ($\Delta\varepsilon=0.76\%$). The crack length was detected using DIC strain measurements as depicted in Fig. 6a. Since the axial strains are calculated correlating the images captured in one single cycle, the development of the plastic wake along the crack path is not captured within these strain maps. However, they allow to precisely locate the crack tip position and track the crack propagation path and the crack propagation rates.

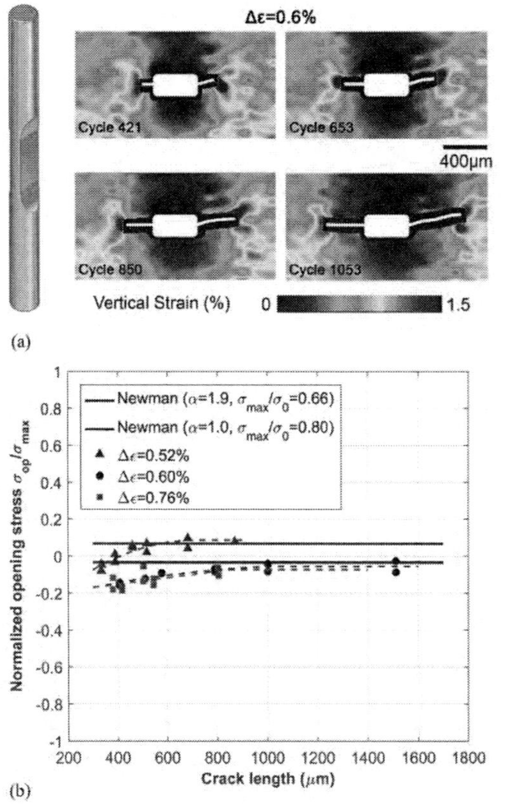

Fig. 6. Analysis of LCF crack propagation in ALSi10Mg alloy produced by AM: (a) The DIC strain fields determines the crack tip position at fixed test interruptions; (b) the normalized opening stress is shown to stabilize with crack propagation for different strain ranges [18].

The opening stresses were calculated for different crack lengths (fig.6b). The normalized opening stress was observed to increase with the number of cycles until an asymptotic value was reached. A prediction with the Newman's model [19] is also provided showing a good agreement between the stabilized opening stress levels and the ones predicted by the model.

Under fatigue loadings, local strain heterogeneities arise as a precursor of crack nucleation. In particular, microstructural features as manufacturing defects, grain boundaries, inclusions, etc can act as a local stress riser and determine the occurrence of localized plastic deformations. In the last years, a growing number of research works are based on the correlation between the detection of microstructural features and the characterization of the local strain fields that quantify the irreversible processes occurring at the micro-scale. One of the pioneering work in this field was performed on an Hastelloy X by Abuzaid and co-workers in [9]. This work demonstrated that local DIC measurements can be linked with microstructural data (grain orientations from EBSD) to quantify the specific effect of the grain boundaries in the slip transmission process. The same experimental approach was herein applied to LCF crack nucleation experiment for detecting damage development for a AlSi10Mg alloy and, in the next section, for a coarse-grained Ni-based super-alloy.

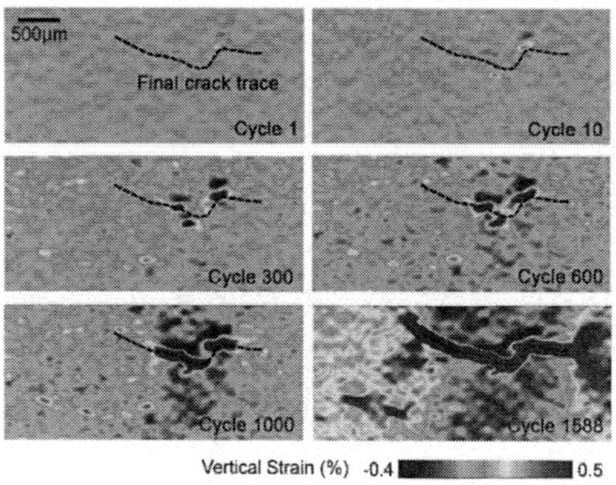

Fig. 10. Crack nucleation from pre-existing AM defects: the DIC strain localisations show the early crack nucleation from the first loading cycles in correspondence of a subsurface defect [18].

A set of cylindrical specimens for LCF tests were produced with selective laser melting process to investigate the crack nucleation for the AlSi10Mg alloy. The specimens were initially analysed by means of X-ray computed tomography to detect the porosity position and size. At a fixed deformation range, in LCF conditions, there exists a critical pore dimension that determines crack nucleation. In this condition, multiple specimen locations can enhance damage initiation as multiple pores are larger than the threshold. DIC strain measurements were then performed in the area where the largest superficial defect was identified. Fig. 10 summarizes the axial strain maps measured for one

AlSi10Mg specimen tested at R_ε=-1 and $\Delta\varepsilon$=0.72%. The DIC measurements clearly show the early strain accumulation in a region corresponding to the specimen failure.

This study demonstrated that the fatigue life of the AlSi10Mg alloy produced by additive manufacturing can be predicted based only on a purely crack propagation model also in LCF regime as the number of cycles required to nucleate the crack is negligible compared to the specimen life. This supports the life-prediction model developed in [18].

3.3 Application to a large-grained Ni-based super-alloy

A third alloy was investigated to analyse the closure levels for more extreme microstructures. The selected alloy is a coarse-grained Ni-based super-alloy tested in LCF conditions [20]. The test temperatures were 25°C for the specimens S1 and S2, 650°C for the specimens S3 and S4 (Fig. 7). In these experiments, the crack path was observed to be strongly affected by the local microstructure. In addition, the elastic modulus and the yield stress depend on the specific specimen tested. Specimens S1, S2 and S4 were tested under the same strain range of $\Delta\varepsilon$=0.8%. The hysteresis loop shapes, so the level of plasticity, change with the local microstructure in the gauge section of the specimens. For this class of alloys, it is then mandatory to analyse and quantify the large scatter in the mechanical behaviour under the same loading conditions. In Fig. 7 the opening stress values are compared with the prediction performed with the Newman's model. As evident from the diagrams, the comparison with the Newman's model is arguable, especially for loading conditions that differ from the pure elastic material behaviour.

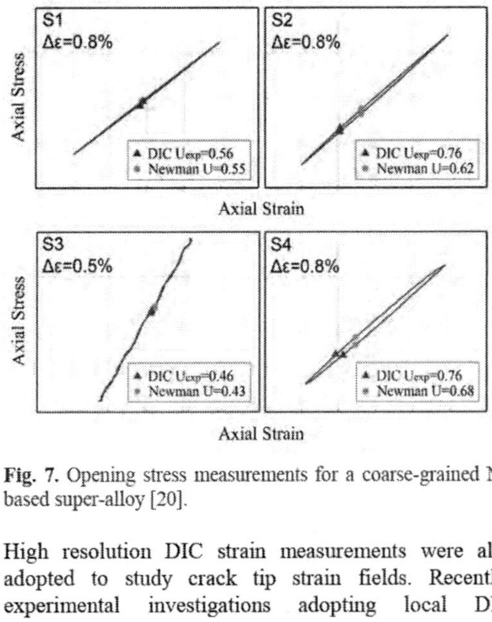

Fig. 7. Opening stress measurements for a coarse-grained Ni-based super-alloy [20].

High resolution DIC strain measurements were also adopted to study crack tip strain fields. Recently, experimental investigations adopting local DIC measurements were performed on a Ni-based single crystal alloy [21–23], and also on shape memory single crystals [24]. These studies highlight the importance of DIC measurements to quantify the effect of the microstructure (i.e. the crystal orientation) on the crack tip process zone. In a previous work, we reported the strain fields at the crack tip for a coarse-grained Ni-based super-alloy [20]. Fig. 9 shows the main results of this study: the development of the crack was monitored for different specimens subjected to LCF conditions from an artificial defect. Due to the large grains, the crack tip process zone was observed to depend on the local grain orientation. The strain localisations were then used to index the slip systems which are activated depending on the grain orientation, presence of grain boundaries and orientation of the crack plane.

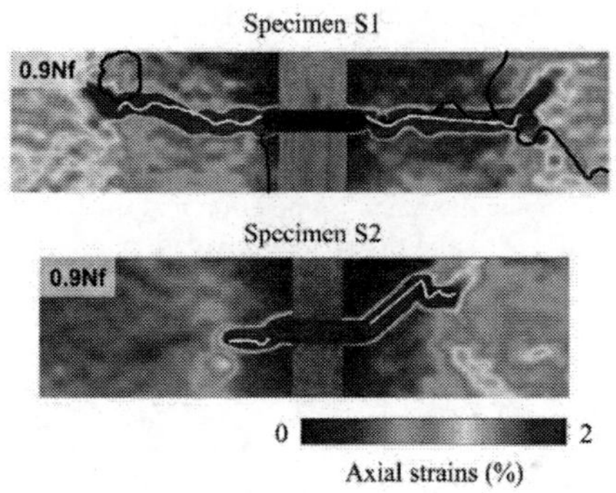

Fig. 9. Crack tip strain fields for a coarse-grained Ni-based super-alloy. The strain localisations were used to index the active slip systems [20].

For coarse-grained cast alloys, the local microstructure can promote damage initiation as a competitive mechanism with the defects originated during the manufacturing process. Specific orientations of grains contained in grain clusters can promote very high localized strains as shown, for example, in Fig. 11a.

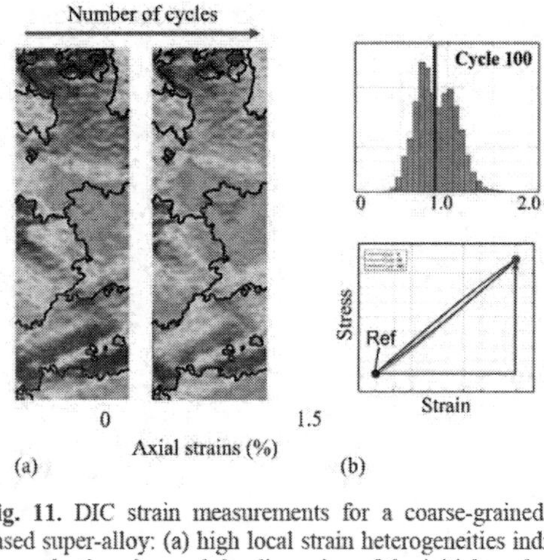

Fig. 11. DIC strain measurements for a coarse-grained Ni-based super-alloy: (a) high local strain heterogeneities indicate the nucleation sites and the dimension of the initial cracks; (b) correlation strategy and histogram of the strain values in the DIC area (from a private study).

The strain maps show the axial strain along the loading direction after correlation the images captured at the minimum load with the images captured at maximum load. In this case, for the specific correlation strategy, the reported strains refer to the range displayed during one LCF cycle, and not the absolute strain accumulation during the specimen life. The values of the axial strains slightly decrease with the number of cycles. These measurements provide an important tool to evaluate the most detrimental grain boundaries which can promote crack nucleation.

5 Final remarks

This paper presented an overview of the potential applications of DIC measurements for material characterization under LCF conditions. The measurements were used to analyse both the nucleation and crack propagation phenomena, for different material processes (cast versus additive manufacturing) and for different grain sizes. The results support the following conclusions:

- DIC virtual extensometer technique is nowadays a robust method to detect opening and closing stress levels in LCF regime;
- Classical models for predicting crack opening and closing levels are limited for fine grained-microstructures, while for large grained alloys further enhancements are required;
- Crack tip strain fields improve the level of information into the crack growth mechanisms;
- High resolution DIC measurements enable to detect the heterogeneous strains which anticipate crack nucleation and quantify the local irreversibility of the damage process.

References

[1] Beretta S, Foletti S, Madia M, Cavalleri E. Structural integrity assessment of turbine discs in presence of potential defects: probabilistic analysis and implementation. Fatigue Fract Eng Mater Struct 2015;38:1042–55.

[2] Leverant GR, McClung RC, Millwater HR, Enright MP. A new tool for design and certification of aircraft turbine rotors. ASME Turbo Expo 2002 Power Land, Sea, Air, American Society of Mechanical Engineers; 2002, p. 699–704.

[3] McClung RC, Enright MP, Millwater HR, Leverant GR, Hudak SJ. A software framework for probabilistic fatigue life assessment of gas turbine engine rotors. J ASTM Int 2004;1:1–16.

[4] Murakami Y, Miller KJ. What is fatigue damage? A view point from the observation of low cycle fatigue process. Int J Fatigue 2005;27:991–1005.

[5] Beretta S, Foletti S, Rusconi E, Riva A, Socie D. A log-normal format for failure probability under LCF: Concept, validation and definition of design curve. Int J Fatigue 2016;82:2–11.

[6] Cristea ME, Beretta S, Altamura A. Fatigue limit assessment on seamless tubes in presence of inhomogeneities: Small crack model vs. full scale testing experiments. Int J Fatigue 2012;41:150–7.

[7] Chu TC, Ranson WF, Sutton MA. Applications of digital-image-correlation techniques to experimental mechanics. Exp Mech 1985;25:232–44. doi:10.1007/BF02325092.

[8] Schreier H, Orteu JJ, Sutton MA. Image Correlation for Shape, Motion and Deformation Measurements: Basic Concepts, Theory and Applications. 2009. doi:10.1007/978-0-387-78747-3.

[9] Abuzaid WZ, Sangid MD, Carroll JD, Sehitoglu H, Lambros J. Slip transfer and plastic strain accumulation across grain boundaries in Hastelloy X. J Mech Phys Solids 2012;60:1201–20.

[10] J. Carroll, C. Efstathiou, J. Lambros, H. Sehitoglu, B. Hauber, S. Spottswood RC. Investigation of fatigue crack closure using multiscale image correlation experiments. Eng Fract Mech 2009;76:2384–2398. doi:10.1016/j.engfracmech.2009.08.002.

[11] McNeill SR, Peters WH, Sutton MA. Estimation of stress intensity factor by digital image correlation. Eng Fract Mech 1987;28:101–12.

[12] Hos Y, Freire JLF, Vormwald M. Measurements of strain fields around crack tips under proportional and non-proportional mixed-mode fatigue loading. Int J Fatigue 2015;89:87–98. doi:10.1016/j.ijfatigue.2016.01.018.

[13] Rabbolini S, Beretta S, Foletti S, Cristea ME. Crack closure effects during low cycle fatigue propagation in line pipe steel: An analysis with digital image correlation. Eng Fract Mech 2015;148:441–56. doi:10.1016/j.engfracmech.2015.07.070.

[14] Vormwald M, Seeger T. The Consequences of Short Crack Closure on Fatigue Crack Growth under Variable Amplitude Loading. Fatigue Fract Eng Mater Struct 1991;14:205–25.

[15] Pommier S, Bompard P. Bauschinger effect of alloys and plasticity-induced crack closure: a finite element analysis. Fatigue Fract Eng Mater Struct 2000;23:129–40.

[16] Chen D, Nisitani H. Analytical and experimental study of crack closure behavior based on an S-shaped unloading curve. Mech. fatigue crack Clos., ASTM International; 1988.

[17] Romano S, Brückner-Foit A, Brandão A, Gumpinger J, Ghidini T, Beretta S. Fatigue properties of AlSi10Mg obtained by additive manufacturing: Defect-based modelling and prediction of fatigue strength. Eng Fract Mech

2017.

[18] Romano S, Patriarca L, Foletti S, Beretta S. LCF behaviour and a comprehensive life prediction model for AlSi10Mg obtained by SLM. Submitt to Int J Fatigue 2018.

[19] Newman JJC. A Crack Opening Stress Equation for Fatigue Crack Growth. Int J Fract 1984;24:R131–5.

[20] Patriarca L, Beretta S, Foletti S, Monti S, Vacchieri E. Crack closure in LCF regime for a coarse-grained Ni-based superalloy tested at high temperatures. Eighth Int. Conf. Low Cycle Fatigue, Dresden: 2017.

[21] Rabbolini S, Pataky GJ, Sehitoglu H, Beretta S. Fatigue crack growth in Haynes 230 single crystals: an analysis with digital image correlation. Fatigue Fract Eng Mater Struct 2015;38:583–96.

[22] Rabbolini S, Luccarelli PG, Beretta S, Foletti S, Sehitoglu H. Near-tip closure and cyclic plasticity in Ni-based single crystals. Int J Fatigue 2016;89:53–65.

[23] Beretta S, Foletti S, Rabbolini S, Sehitoglu H. Fatigue Crack Propagation in Haynes 230: A Comparison between Single and Polycrystal Crack Closure Levels. Solid State Phenom., vol. 258, Trans Tech Publ; 2017, p. 243–8.

[24] Wu Y, Ojha A, Patriarca L, Sehitoglu H. Fatigue crack growth fundamentals in shape memory alloys. Shape Mem Superelasticity 2015;1:18–40.

Thermomechanical fatigue – Mechanism-based considerations on the challenge of life assessment

Hans-Jürgen Christ[1,*]

[1]Institut für Werkstofftechnik, Universität Siegen, Paul-Bonatz-Str. 9-11, D-57076 Siegen, Germany

Abstract. The combination of cyclic mechanical and cyclic thermal loading leads to thermomechanical fatigue (TMF) which is considered to be the primary life-limiting factor for engineering components in many high-temperature applications. Extensive low-cycle fatigue (LCF) data, which is traditionally used for design purposes, has been generated isothermally on various high-temperature materials, and thus, it is tempting to try to predict TMF life based mainly on isothermal LCF data. In this contribution, studies on different metallic structural high-temperature materials, which have mainly been carried out in the author's laboratory, are reviewed addressing the question, in which way and to which extent a reliable, unerring and robust TMF life assessment is possible on the basis of isothermally obtained fatigue life data. It is shown by means of examples that a sound TMF life prediction first of all requires a detailed mechanistic understanding of the isothermal cyclic stress-strain response and the relevant damage mechanisms. Furthermore, the TMF-specific peculiarities in both the non-isothermal cyclic stress-strain behaviour and the non-isothermal damage evolution process must be known. If all these requirements are fulfilled and reflected in the TMF life assessment methodology applied, a reasonable predictive accuracy can be attained.

1 Introduction

Thermomechanical fatigue (TMF) is the primary life-limiting factor for engineering components in many high-temperature applications. For example, during the transient regimes of start-up and shut-down operations, blades in gas turbines are submitted to low-cycle TMF due to the temperature gradients generated caused by temperature variations often in combination with internal air cooling (see Fig. 1). Depending on the location of the volume element considered and the cooling situation, various types of strain-temperature phasing can arise. In laboratory testing most commonly the two extreme cases, in-phase (IP) and out-of-phase (OP) TMF, are studied.

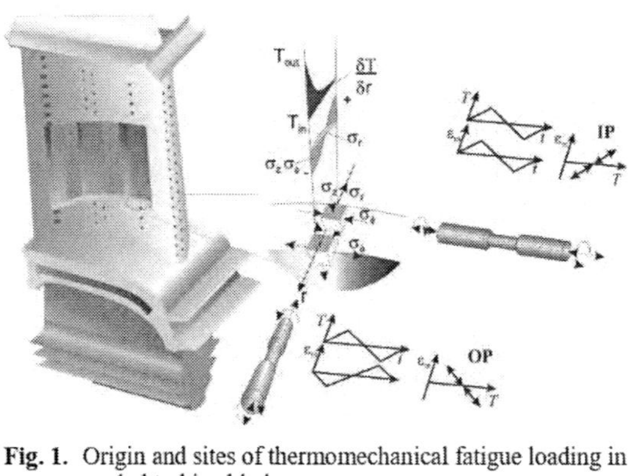

Fig. 1. Origin and sites of thermomechanical fatigue loading in cooled turbine blades.

Due to the life-reducing effect of TMF it is important to investigate the TMF behaviour of engineering materials in order to model thermomechanical fatigue life correctly. On the other hand, for many structural materials extensive, isothermally determined low-cycle fatigue (LCF) data has been generated at various temperatures, and thus, it would be very efficient, if TMF life could be predicted based largely on isothermal LCF data. However, it is well known that depending on the actual loading conditions damage can evolve differently [1,2], and life prediction may well be non-conservative if the relevant damage mechanisms are not captured accurately. Still, studies that assess the predictive capabilities of TMF life models [1-10] are rare. Part of this results from the large number of model parameters that need to be determined for the more advanced life models. Given this scenario, the objective of the present study is to answer the question, whether TMF life can in principle be considered to be predictable from isothermally obtained data. Furthermore, the requirements with respect to material behaviour and knowledge base are considered which have to be fulfilled to allow for the development of TMF life models that are robust, easily implemented, and do not require sophisticated testing.

This paper is based on the results which have been obtained in various studies on the TMF behaviour of structural metallic materials in the author's laboratory during more than two decades. Selected results are presented in order to illustrate exemplarily the broad spectrum of phenomena. After showing the strong effect of temperature on the isothermal cyclic stress-strain behaviour and the damage mechanisms, some peculiarities resulting from the varying temperature of TMF are illustrated. Then it is documented that TMF life prediction is very difficult, if cyclic plasticity is low.

* Corresponding author: hans-juergen.christ@uni-siegen.de

Brittle behaviour often means that the stage of fatigue crack initiation dominates fatigue life. Then the contribution of the stage of crack propagation is less important leading to an almost unpredictable behaviour. Finally, the need of a close correlation of the relevant damage mechanism and the suitable damage parameter is verified.

2 Materials and particular experimental details

This study reports on the behaviour of four different materials which were investigated in the framework of Ph.D. works [11-15]. An AISI 304 L-type stainless steel was used, as a large amount of data generated on this material in isothermal LCF tests, creep-fatigue tests and TMF tests are available.

The second material chosen was X8019+12.5SiCp that is a dispersion-strengthened SiC particulate-reinforced aluminium alloy (matrix composition: Al-8Fe-4Ce, in wt.%). Within the matrix, fine and thermally very stable dispersoids with a high volume fraction of 0.23 are present. The dispersoids provide for excellent creep properties at elevated temperatures. The reinforcement with 12.5 vol.% SiC particulates increases both the elastic modulus and the fatigue crack growth threshold of the alloy.

As the third material the high-temperature titanium alloy IMI 834 was applied which is a near-α Ti alloy of nominal composition Ti-5.8Al-4.0Sn-3.5Zr-0.7Nb-0.5Mo-0.35Si-0.06C (in wt.%). All data used in this study was obtained on the bimodal microstructure which is designed to yield an optimum compromise between creep and fatigue performance.

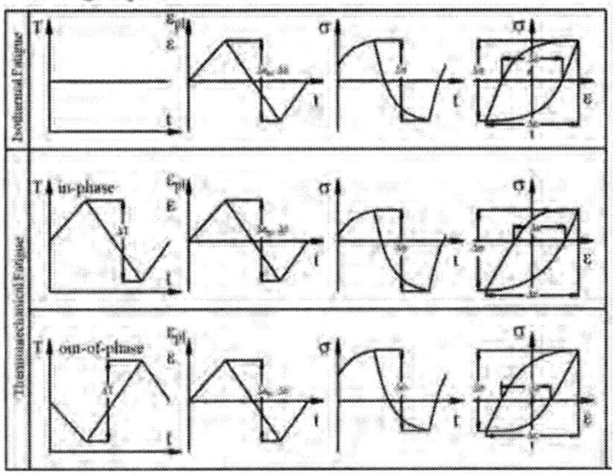

Fig. 2. Types of laboratory tests used in this study.

Different third generation near-γ TiAl alloys were also used in this study. The alloy γ-PX (similar to TNB-V5) was produced by Plansee AG, Austria with a nominal composition of Ti-47Al-5.1Nb-0.25C-0.4B (at.%). The material was subjected to a heat treatment such that a duplex microstructure was attained. The γ-alloy Ti-45Al-8Nb-0.2C (at.%), referred to as TNB-V2, has been developed by *Helmholtz Zentrum Geesthacht*. The heat treatment applied resulted in a nearly lamellar microstructure.

The Ductile to Brittle Transition Temperature (DBTT), which is a significant quantity for TiAl alloys, was determined for the alloys considered to be about 750°C. The transition from brittle to ductile behaviour manifests itself in a strong increase in fracture ductility when exceeding the temperature of 750°C because of the onset of dislocation climb.

In all cases, the isothermal fatigue (IF) and TMF tests evaluated (see Fig. 2) were selected such that a wide range of test parameters were covered. The TMF tests were of in-phase and out-of-phase type, *i.e.* maximum test temperature coincided with maximum strain (in tension) or minimum strain (in compression), respectively. All tests except of those on the γ-TiAl alloy were conducted in true-plastic-strain control, and a triangular wave form was used for the command signal. Because of the brittleness of the TiAl alloy, total-strain control had to be applied in this case

Fig. 3. Testing system used for high-temperature fatigue testing (left). View into the vacuum chamber (right) showing the induction coil, a ribbon-type thermocouple and a water-cooled extensometer.

A servohydaulic test system of type MTS 810 was applied for both the isothermal and thermomechanical fatigue tests. As can be seen in Fig. 3, this system is equipped with a vacuum chamber enabling tests in vacuum at a pressure below 10^{-5} mbar. These vacuum tests were carried out as reference tests in addition to the standard tests in air, in order to separate the environmental effect on fatigue life. The influence of environment can be very strong as depicted in Fig. 4 representing the ratio of the number of cycles until failure N_f in vacuum to N_f in air versus the mean temperature of OP tests at different total strain amplitudes $\Delta\varepsilon/2$ carried out on TiAl-based alloys. It is clearly seen that this ratio increases strongly with increasing mean temperature reaching values of up to 60.

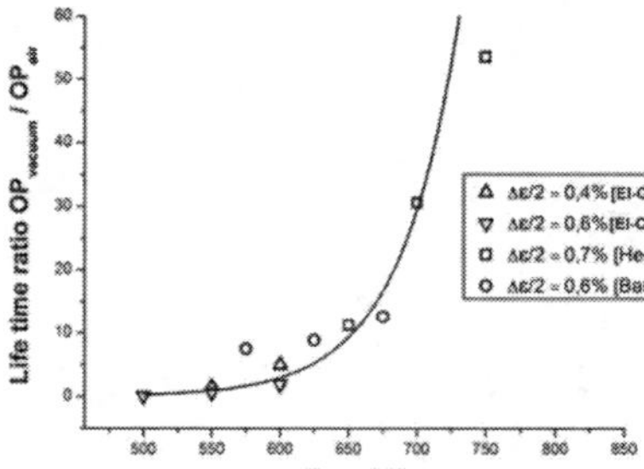

Fig. 4. Representation of the effect of environment (air or vacuum) on the TMF life for OP tests on TiAl-based alloys.

3 Results and discussion

3.1. Isothermal fatigue mechanisms

It is well-known and has been studied in detail that the cyclic deformation mechanism depends strongly on temperature applied in IF. Figure 5 tries to summarize some of these effects. While at low temperature dislocations motion is dominated by gliding, dislocation climb gains importance with increasing temperature. Most of the metallic materials show a strong interaction of point defects with moving dislocations giving rise to dynamic strain aging (DSA). The cyclic deformation response of a material can be affected via dislocation/microstructure interactions by microstructural changes resulting from phenomena such as deformation-induced transformation (e.g., martensite formation in metastable austenitic stainless steels), mechanical twinning (e.g., in intermetallic phases), deformation-induced precipitation (e.g., in solution-annealed alloys) and coarsening (e.g., in precipitation-strengthened materials).

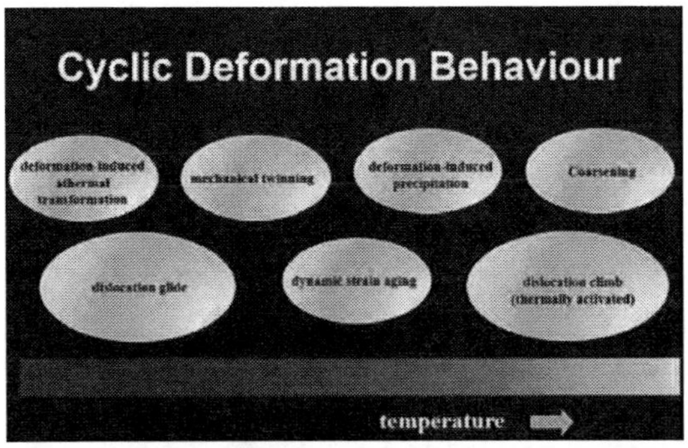

Fig. 5. Examples for the effect of temperature on the cyclic stress-strain behaviour.

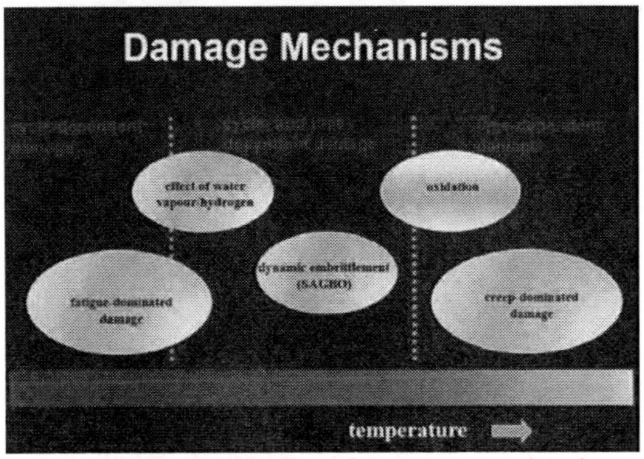

Fig. 6. Examples for the effect of temperature on the relevant damage mechanism.

A similar strong dependence on temperature exists for the dominant damage mechanism which is responsible for isothermal fatigue life of a material. Figure 6 represents a small selection of possible mechanisms. With increasing temperature a shift from cycle-dependent damage to time-dependent damage occurs. The last one is mainly a consequence of environmental effects and creep. Some materials are known to be prone to specific damage contributions such as hydrogen effects in the case of titanium alloys at moderate temperatures or dynamic embrittlement, often also termed stress-assisted grain boundary oxidation (SAGBO), of polycrystalline Ni-base superalloys.

As a prominent example for a temperature influence, the effect of DSA on cyclic stress-strain response and fatigue life N_f of AISI 304L is shown in Fig. 7 [16]. DSA leads to an increase of the maximum and saturation stress amplitude in an intermediate temperature range observed in tests carried out at a constant plastic strain amplitude of 0.5%. As a direct consequence of this stress amplitude raise the cyclic life drops in the corresponding temperature range, before creep damage reduces N_f further (see Fig. 8).

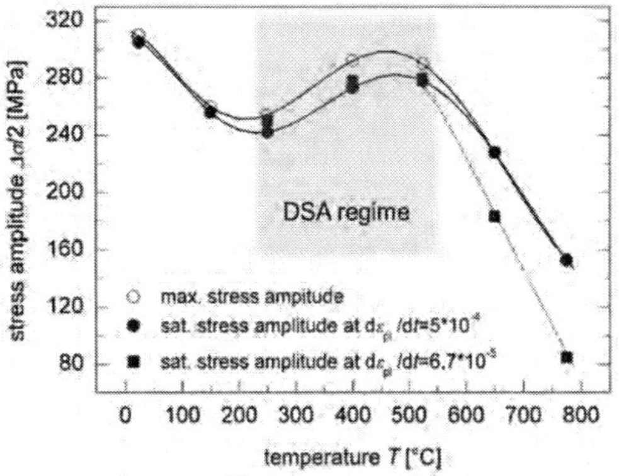

Fig. 7. Influence of DSA on the maximum and saturation stress amplitude of AISI 304L cyclically loaded at a plastic strain amplitude of 0.5%.

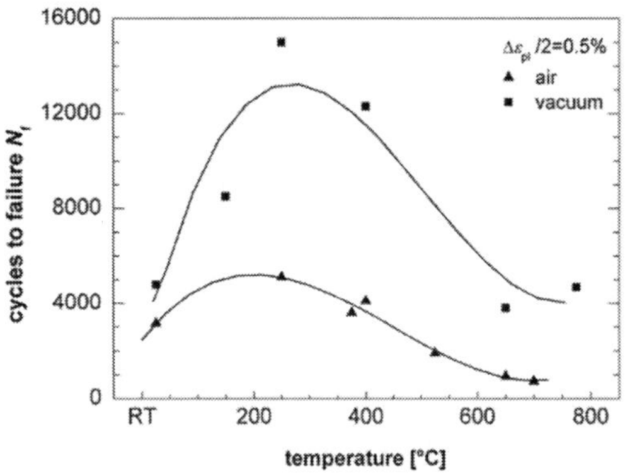

Fig. 8. Influence of DSA on fatigue life of AISI 304L obtained in air and in vacuum at a plastic strain amplitude of 0.5%.

The reason for the strong change in the cyclic stress-strain response of the austenitic stainless steel is the impeding effect of point defects on the dislocation mobility in the intermediate temperature range. This interaction of dislocations with point defects manifests itself in a tremendous change in the dislocation arrangement, as shown in the TEM micrographs of Fig. 9. While below the DSA temperature range, dislocations form cells, which is a wavy-slip-type structure typically of high plastic strain amplitudes, DSA leads to a planar dislocation arrangement. At sufficiently high temperature, the effect of DSA vanishes and a cell arrangement with a tendency to form subgrains prevails.

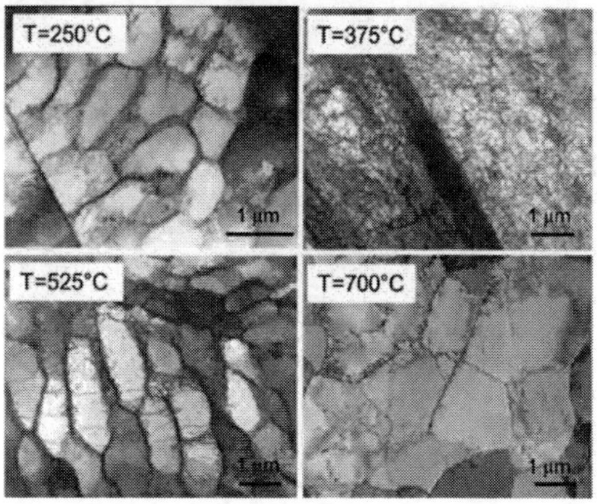

Fig. 9. Change of the dislocation arrangement with temperature: wavy arrangements above and below the temperature range of DSA, planar arrangement in the DSA range.

At temperatures above 650°C and/or slow cyclic plastic strain rates, austenitic stainless steels are prone to a change of the fatigue crack propagation from a transgranular to a intergranular type (see Fig. 10). This transition leads to a strong increase of the crack growth rate.

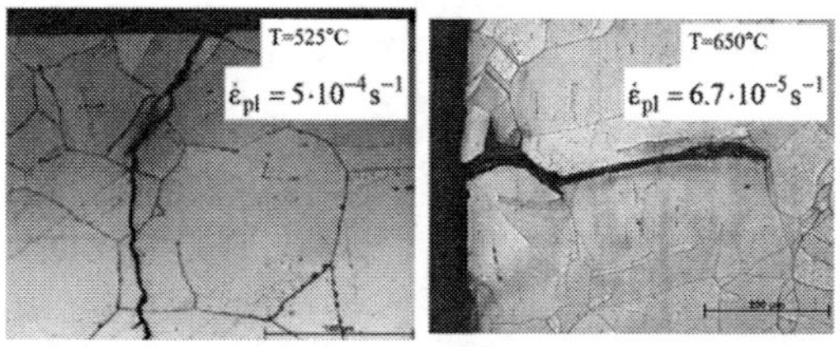

Fig. 10. Change from transgranular to intergranular fatigue crack propagation in AISI 304L austenitic steel.

Two examples for microstructural instabilities affecting strongly the cyclic stress-strain response are shown in Fig. 11 und Fig. 12. Figure 11 refers to the near-α Ti alloy IMI 834. At temperature exceeding about 600°C the boundaries in the colonies of the lamellar grains of the bimodal microstructure disintegrate leaving behind silicides which mark the former boundaries.

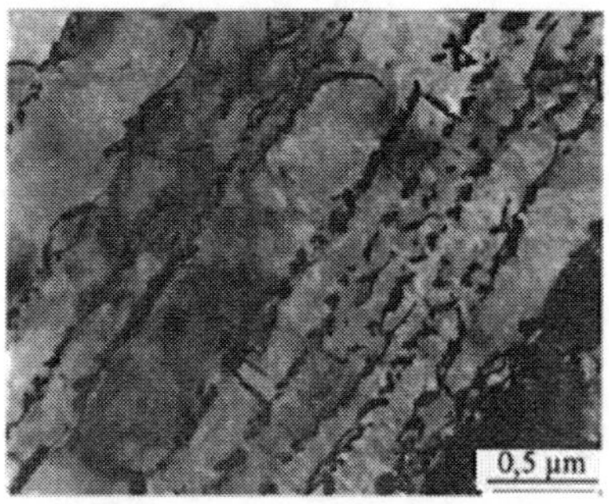

Fig. 11. Disintegration of lamellar boundaries in IMI 834

In γ-TiAl alloys the microstructural stability is limited to about 800°C. Above this temperature the lamellar structure is destroyed by a kind of dynamic recrystallization (see Fig. 12) reducing the strength strongly.

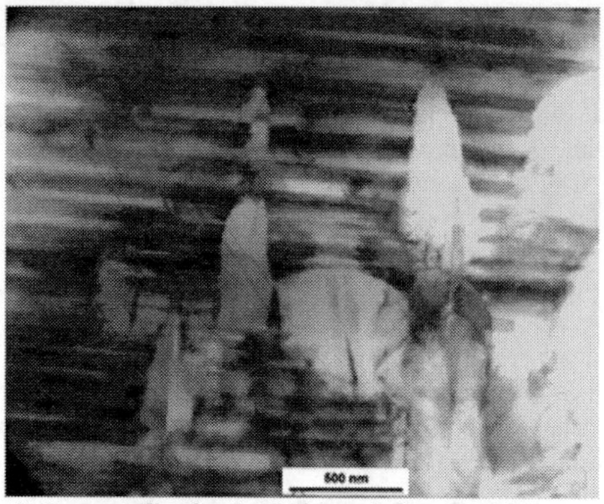

Fig. 12. Recrystallization in TiAl-based alloy TNB-V2.

Figures 13 und 14 [8] show typical examples for crack initiation sites as observed on IMI 834 after cyclic deformation at low and high temperature, respectively. Up to a testing temperature of about 400°C, most cracks are formed within the primary α grains (Fig. 13). At higher magnification, it becomes apparent that cracks initiate along planar slip bands.

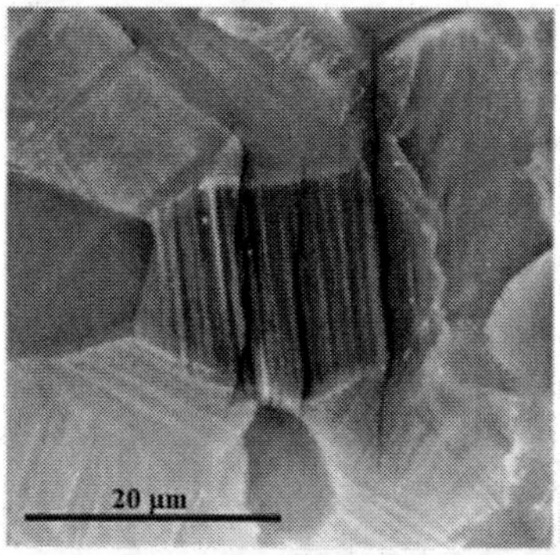

Fig. 13. Crack initiation in IMI 834, isothermal fatigue at 400°C.

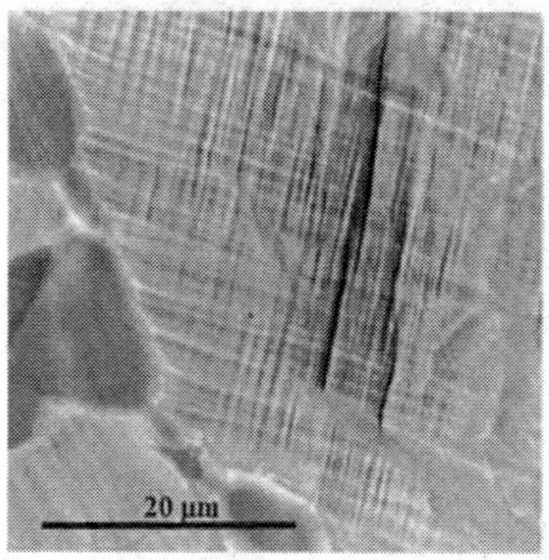

Fig. 14. Crack initiation in IMI 834, isothermal fatigue at 600°C.

The few fatigue cracks which were found in the primary α grains at T=600°C seem to become non-propagating as soon as they reach the grain boundary. The cracks now form in the transformed α grains (the prior β grains) and seem to grow unhindered (Fig. 14). A possible explanation for this change in the crack initiation mechanism is a transition in slip mode from planar to wavy which occurs at around 600°C.

However, this changes seems to be much less important than the change in the relevant environmental effect. Titanium alloys are known to be susceptible to hydrogen (e.g., resulting from the moisture in air) in particular in the intermediate temperature range. The detrimental effect of oxidation is more pronounced at high temperature, since the TiO_2 surface layer formed is no longer stable. Rather, the oxide is dissolved in the material leading to a brittle oxygen-enriched surface zone, which is referred to as α-case.

This temperature effect must be considered by an appropriate model concept as shown in Fig. 15. The straight line represents the correlation between the maximum stress established in cyclic saturation of plastic-strain-controlled fatigue tests performed at a plastic strain amplitude of 0.2% with the number of cycles to failure. The full data points refer to corresponding tests carried out in air. Modelling the effect of hydrogen explains only the fatigue life reduction at temperatures up to about 400°C. To explain the reduced life at higher temperatures oxidation has to be taken into account. The combination of both effects allows a very satisfactory fatigue life assessment (see [8] for model details).

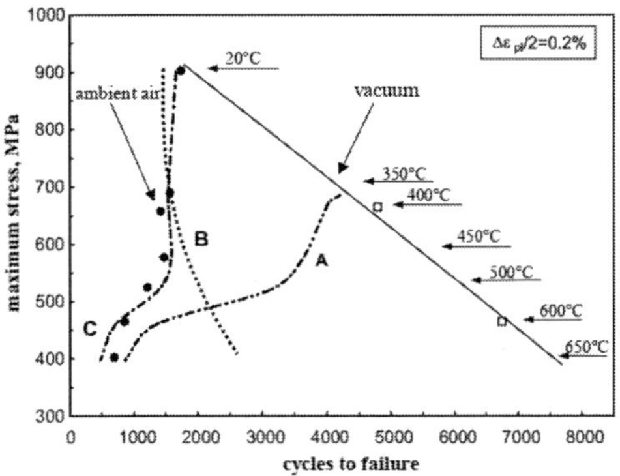

Fig. 15. Effect of oxidation (dashed line A) and hydrogen embrittlement (dotted line B) on cyclic life of IMI 834. The combination of both effects (line C) leads to a reasonable life assessment for isothermal conditions.

The considerations and supporting examples shown in this chapter clearly demonstrate that before dealing with TMF a sound understanding of the isothermal behaviour has to be developed. This behaviour forms the reference related to which the effects arising from the varying temperature of TMF can be identified. Hence, this sound basis obtained from extensive isothermal testing is a must for any rational consideration of TMF.

3.2 Peculiarities of the TMF behaviour

The strong effect of temperature on both the cyclic stress-strain behaviour and the relevant damage mechanism documents that already under isothermal conditions a high-temperature fatigue life assessment is very complex. The complexity increases immensely, if the temperature is continuously varied as is the case under TMF conditions.

A first important step in an identification of the TMF-specific behaviour is the knowledge of the so-called pseudoisothermal TMF (PI-TMF) behaviour. The PI-TMF behaviour predicts the cyclic stress-strain response which would exist, if the material behaves at each temperature in a TMF cycle identical to the corresponding isothermal case. A comparison of the PI-TMF behaviour with the experimentally determined behaviour (e.g., in form of the hysteresis loops) reveals the differences and provides valuable information on the TMF-specific mechanisms.

For the calculation of PI-TMF behaviour viscoplastic constitutive equations can be used. In the investigations carried out in the author's research group a multi-component behaviour approach is favoured, since all necessary model parameters can be determined from isothermal stress-strain hysteresis loops. The basic concept was proposed by Masing already in 1923 [17] assuming that a material behaves like a composite consisting of a parallel arrangement of ideally elastic-plastic elements which differ in yield stress. The distribution function of these microscopic yield stresses is a kind of a mathematical representation of the microstructure and can be determine by a double differentiation of one single stress-strain branch of a hysteresis loop (details can be found in ref. [18]).

The expansion of this multi-component approach to TMF is described in ref. [19] and requires the following procedure:

- The yield stress distribution function must be determined from isothermal tests at various temperatures.
- The distribution function must be interpolated regarding the temperature.
- The distribution function must then be transformed into a discrete distribution (e.g., 30 elements)
- The TMF loading situation must be determined in terms of $T(t)$ and $\varepsilon(t)$.
- The stress-strain relationship can then be calculated stepwise by
  - o stepwise changing temperature and strain in each element and
  - o calculating the total stress after each step from the individual stresses of all elements.

Figure 16 shows calculated hysteresis loops for two IP TMF tests with temperature ranges falling partially in the DSA regime in a representation of stress versus temperature. The calculated PI-TMF loops are compared with the corresponding experimental hysteresis loops. The course of the maximum and minimum saturation stress also plotted shows the DSA-induced increase in strength in the intermediate temperature range.

In this temperature regime of DSA, the extreme values of the hysteresis loops coincide, while outside this regime the experimental hystereses document a higher strength as observed in the isothermal tests and predicted by the PI-

TMF loops. The reason for this behaviour is illustrated in Fig. 17. TMF loading leads to typical dislocation arrangements which were not found for isothermal conditions (compare Fig. 9). These arrangements consist of dislocation-poor channels embedded in a wall structure (Fig. 17a) or a labyrinth-type arrangement (Fig. 17b).

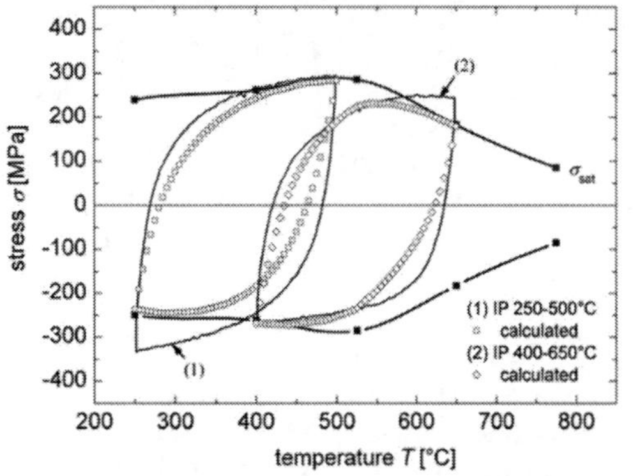

Fig. 16. Comparision of the isothermal and the athermal (TMF) stress strain response for AISI 304L.

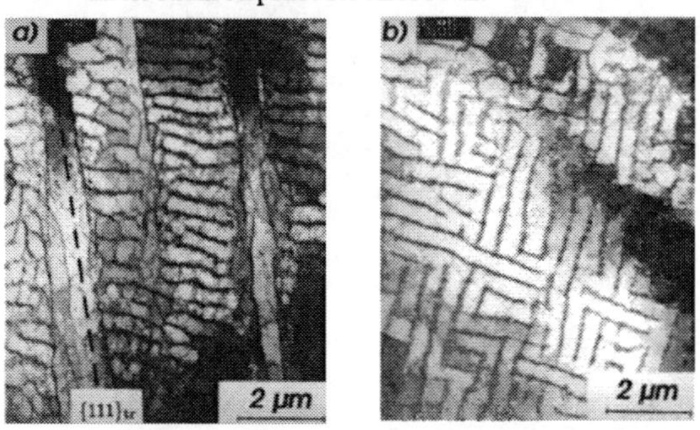

Fig. 17. Typical dislocation arrangements found exclusively after TMF loading: a) dislocation channels and walls, b) labyrinth-type arrangement.

Additional tests, where TEM samples were prepared after different numbers of loading cycles from specimens, which were fatigued under identical TMF conditions, indicate that despite the varying temperature a steady state condition is established in the microstructure. That means that the change from cycle to cycle is very small or is even negligible.

In order to understand the reason for the deviation of the TMF stress-strain response from the one of isothermal fatigue, tests were performed with block-wise constant temperature. Figure 18 shows the answer of AISI 304L to a sudden temperature change from 400°C to 650°C during fatigue testing at a constant plastic strain amplitude of 0.5%. As mentioned above the first temperature is in the DSA regime and a planar type of dislocation arrangement is formed. The increase to 650°C leads to a stress amplitude which is significantly higher than the corresponding stress amplitude of the isothermal test at this temperature (dashed line). Therefore, it can be concluded that the dislocation arrangement formed after the temperature change is strongly affected by the DSA prehistory. The consequence of this observation is that if TMF modelling deals with conditions where the DSA temperature range is passed through the required stress-strain data should be determined in temperature-change tests instead of IF testing (see details in [16]). Figure 19 compares such a calculated hysteresis loop with the corresponding experimental loop. The agreement is very reasonable documenting that the deviations depicted in Fig. 16 result from the TMF-specific dislocation arrangement.

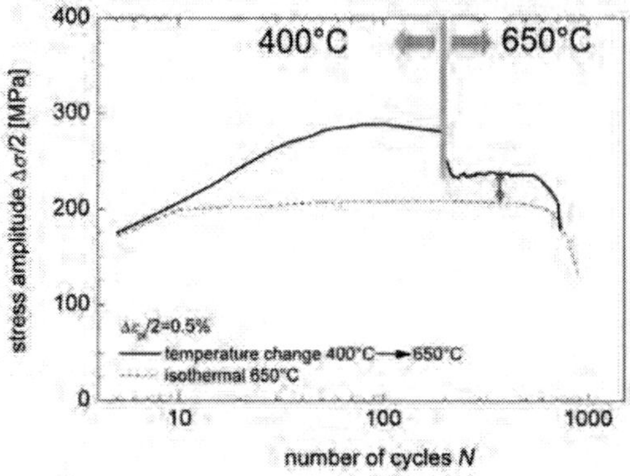

Fig. 18. Cyclic stress-strain response of AISI 304L to a change of temperature while keeping the other parameters constant.

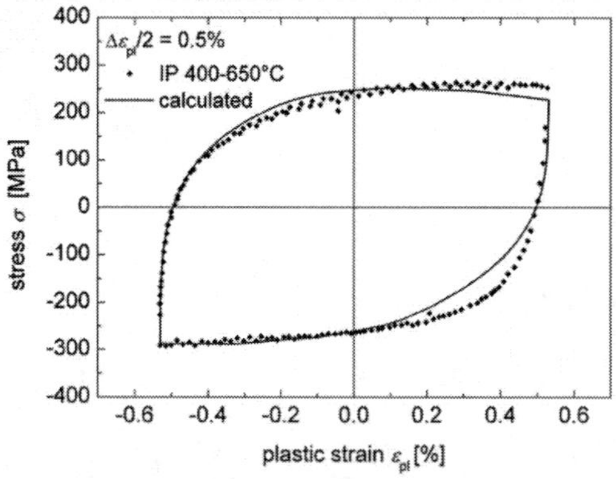

Fig. 19. Comparison of a calculated and an experimental TMF hysteresis loop. The calculation was done using the stress amplitude from temperature change tests.

The formation of a TMF-specific microstructure was also observed for the near-α Ti alloy IMI 834. This material shows in isothermal tests a transition from a planar dislocation arrangement to a wavy one at around 600°C. As a consequence, the dislocation arrangement becomes more wavy, if the maximum temperature applied in the cycle of a TMF test exceeds 600°C. A comparison of the microstructures formed in two TMF tests with a minimum temperature of 400°C and a maximum temperature of 600°C or 650°C is shown in Fig. 20.

The existence of a TMF-specific microstructure manifests itself in a different stress amplitude after the change from TMF to isothermal conditions [20]. Simply by stopping the temperature variation and holding the temperature constant, such a change can be carried out. Figure 21 shows a test, where the minimum temperature (400°C) of the TMF test was kept constant in the isothermal part of the fatigue experiment. A comparison with the corresponding isothermal test clearly shows that the TMF part establishes a stable dislocation arrangement that is harder at low temperatures than the corresponding isothermal microstructure.

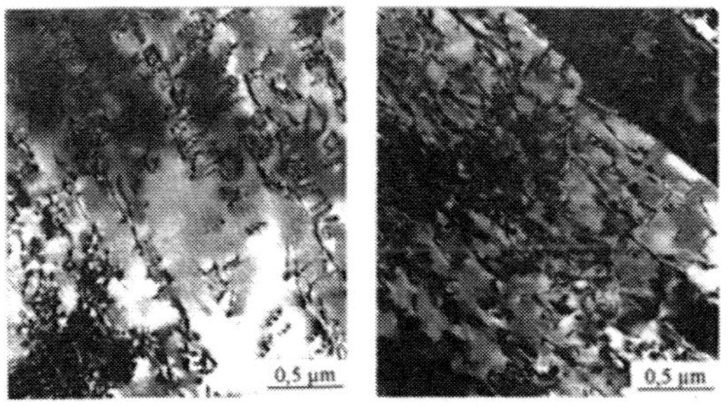

Fig. 20. Comparison of the dislocation arrangements in IMI 834 after TMF testing with a temperature range of 400°C-600°C (left) and 400°C-650°C (right). The slight increase of the maximum temperature of the TMF cycle promotes a wavier microstructure.

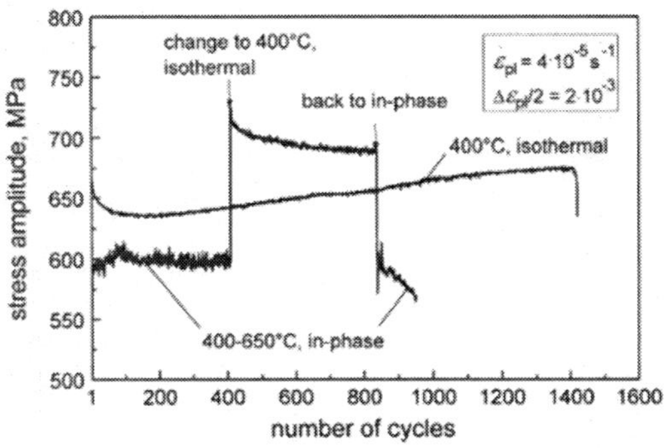

Fig. 21. Cyclic stress-strain response of a test carried out with a change from TMF to isothermal conditions. The higher stress amplitude in the isothermal part of the change test documents the higher strength of the dislocation arrangement established under TMF conditions.

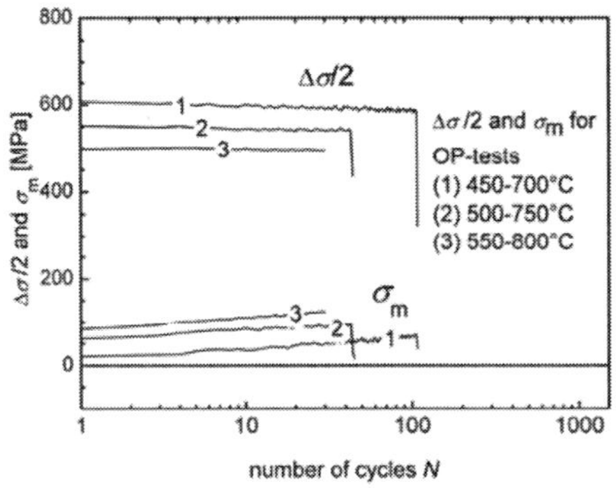

Fig. 22. Change of mean stress σ_m and stress amplitude $\Delta\sigma/2$ with number of loading cycles during OP-TMF testing of the TiAl alloy γ-PX.

In the case of the near-γ TiAl alloys a very stable microstructure was observed in the temperature range of interest. No significant differences between the isothermally formed dislocation arrangements and the corresponding TMF arrangements were detected. Consequently, the isothermal fatigue behaviour is characterized by a pronounced cyclic saturation. This holds also true under TMF conditions with respect to the stress amplitude. However, a mean stress develops continuously as seen in Fig. 22 for OP-TMF. The non-isothermal conditions lead to cyclic softening at high temperatures and (slight) cyclic hardening at low temperatures resulting in a mean stress in OP and IP testing which is positive (tensile) and negative (compressive), respectively. In particular, the positive mean stress developing during OP conditions strongly deteriorates TMF life as will be discussed in the next chapter.

As part of the answer to the basic question of TMF life predictability addressed in this paper, the conclusion from these (and many other existing) examples is that a thorough study of the specifics of TMF is indispensable in order to select the suitable approach for TMF life assessment. Hence, at least some TMF tests must be carried out, and the testing parameters should be chosen in such a way that they match the service conditions closely.

3.3 TMF life assessment and immanent brittleness

Most of the models reported in the literature, which are used for TMF life prediction, are connected in some way to the hysteresis loop or the deformation energy density. This holds true for many empirical approaches but also for most fracture mechanics concepts suitable for high temperatures. An application of these concepts is very problematic, if the material considered behaves brittle. Brittle behaviour is in this context synonymic to a negligible fraction of life spent in crack propagation, and hence the stage of crack initiation is more important. Unfortunately, crack initiation is a very complex process starting from an atomic level, and despite much research effort the process still detracts from a mechanism-based and quantitative description.

Fatigue life results under TMF loading conditions are compared with isothermal fatigue life data in Fig. 23 for the near-γ TiAl alloy γ-PX. TMF data is plotted as a function of the maximum temperature of the corresponding TMF cycle. The fatigue life of IP tests is approximately twice the life observed in isothermal LCF tests performed at the highest temperature of the TMF test. One reason for higher fatigue lives under IP conditions is the negative mean stress as discussed before. On the other hand, an oxide layer grows at high temperature on the specimen surface in the tension part. Apparently, the tension loading is not crucial for oxide scale cracking or spalling. For this reason, cracks initiate slowly.

In contrast, OP loading conditions reduce fatigue life tremendously. The fatigue life decreases at least by a factor of 5. The short TMF life in OP mode can be attributed to the accelerated oxidation-induced crack initiation during OP testing. The oxide layer mainly formed at high temperatures becomes brittle at lower temperatures. In the low temperature part of the cycle (tension) the oxide cracks easily and hence, promotes crack initiation. Once formed, cracks lead to failure quickly, because of the brittleness of the alloy at low temperatures (tension). This manifests itself in very different fracture surfaces of OP and IP TMF tests as shown in Fig. 24.

A comparison of the results of the OP tests in vacuum (Fig. 23) with isothermal air tests documents that the positive mean stress shown in Fig. 22 reduces cyclic life strongly. However, the effect of environment leads to a further tremendous life reduction.

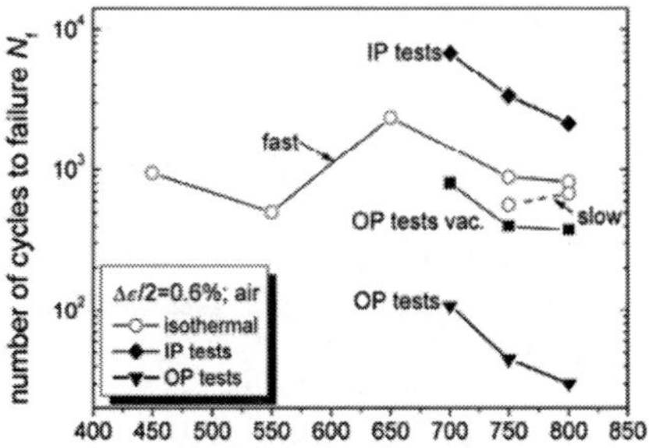

Fig. 23. Fatigue life data of the near-γ TiAl alloy γ-PX as a function of testing temperature and maximum temperature of isothermal and non-isothermal conditions, respectively.

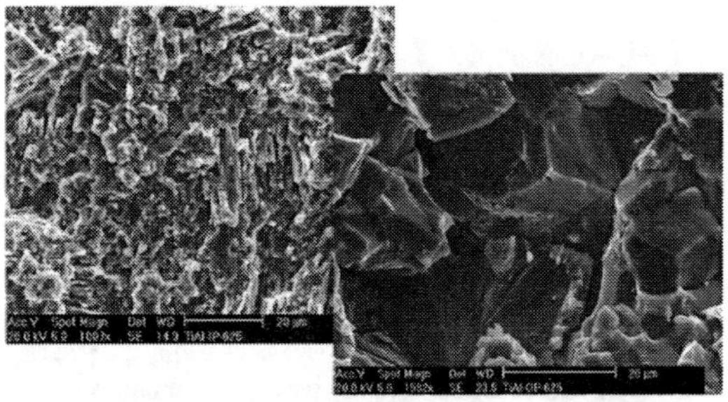

Fig. 24. Fracture surface from IP (left) and OP (right) TMF tests performed at a strain amplitude of 0.6% and a temperature range of 550°C-750°C.

The pronounced susceptibility of the cyclic life of the near-γ TiAl alloys to mean stresses demands the use of a damage parameter which explicitly considers this effect. The concept of Ostergren [21] provides a simple empirical, energy-based approach, which holds true for the isothermal tests. In order to consider the effect of oxidation, temperature and frequency, an expansion of the model is necessary, which can be obtained by implementing a combination of the frequency-modified Ostergren model [22] with the oxidation damage model of Antolovich [23].

If this concept is applied to describe TMF life on the basis of the isothermal data, the match is very poor indicating that TMF damage evolution deviates strongly from the isothermal behaviour (see Fig. 25). IP TMF life is conservatively predicted, while the experimentally observed OP TMF life is much shorter than expected from the assessment. Hence, as already stated in the beginning of this section, a sufficient ductility of the material considered is a fundamental requirement for a successful application of current TMF life prediction methodologies.

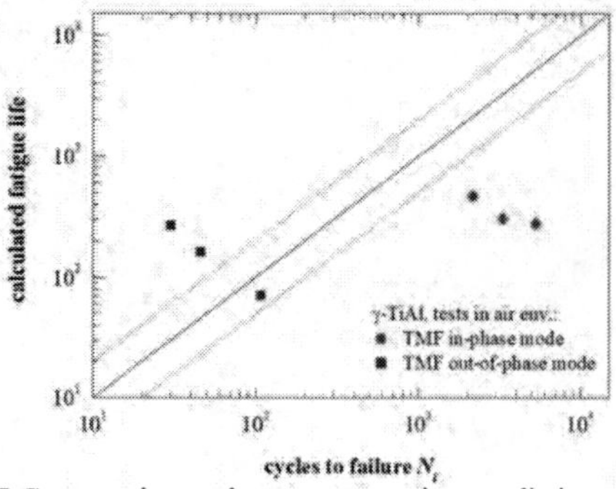

Fig. 25. Conservative and non-conservative prediction of IP TMF and OP TMF, respectively.

However, recently a new damage parameter could be developed and introduced, which seems to capture the strong effect of mean stress, oxidation and TMF phasing for the brittle TiAl-based alloys in such a way that the unified description of isothermal and TMF life works satisfying [24].

In this life assessment model, a damage parameter P was defined strongly emphasizing the influence of mean stress σ_m. This parameter P (similar to the one suggested by Buchholz in his Ph.D. thesis (Dresden 2012)) includes the temperature as a variable quantity and contains only one fitting parameter α:

$$P = \frac{1}{2}\left(\frac{\sigma_{tlrp}}{R_{p0.2}(T_{tlrp})} + \frac{\sigma_{clrp}}{R_{p0.2}(T_{clrp})}\right) + \left(\frac{\sigma_m}{R_{p0.2}(T_m)}\right)^{\alpha} \quad (1)$$

The subscripts *tlrp*, *clrp* and *m* correspond to the tensile load reversal point, the compression load reversal point and the mean value, respectively. All these parameters can be calculated from the above described multi-component model by means of the simulation of the hysteresis loop considering the changes deformation behaviour.

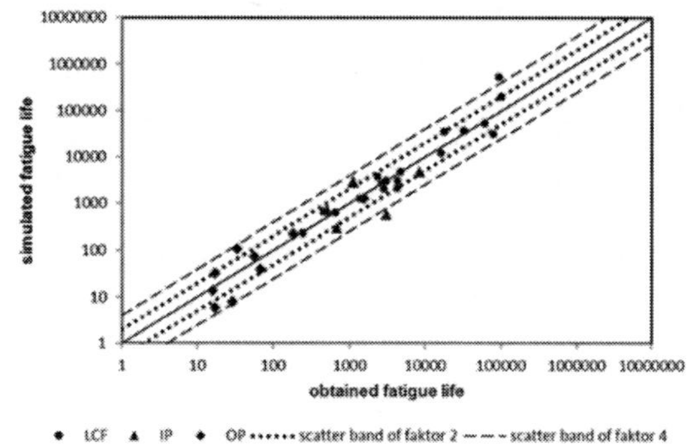

Fig. 26. Comparison of simulated and observed lifetimes for LCF, IP TMF and OP TMF.

Finally, the correlation between this damage parameter and the number of cycles until failure is given by a power law:

$$P = A \cdot N_f^{\,b} \quad (2)$$

Figure 26 shows a comparison between the simulated and the observed life for both LCF and TMF (IP and OP) conditions. Most data points lie within the scatter band of factor 2.

3.3 Successful mechanism-based TMF life prediction

From the previous sections it should have become clear that a successful development and application of a life prediction model for TMF conditions requires a thorough experimental characterization of both the isothermal and non-isothermal fatigue behaviour of the material considered and a mechanism-based and realistic theoretical approach with respect to damage evaluation. Unfortunately, each material has its specialities so that a generally applicable methodology cannot be expected. Rather an individual consideration which is laborious and time-consuming must be carried out. Selected examples are given in this section to shed some light on the required line of actions.

Fracture mechanics methods are a suitable means of describing fatigue life both for isothermal and thermomechanical conditions (see ref. [8] for details). However, the concepts must be selected carefully such that they relate closely to relevant damage mechanisms and microstructural processes. Hence, no fracture mechanics damage parameter is *a priori* qualified. Rather extensive testing in combination with detailed microstructural and fractographic studies must be carried out, before an appropriate concept can be selected. This approach is illustrated for the alloy X8019/12.5$_p$ in Figs. 27, 28 and 29.

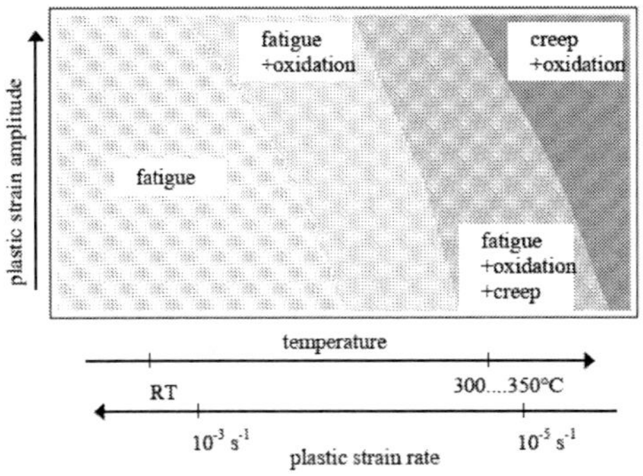

Fig. 27. Map of the damage regimes as a function of plastic strain amplitude (ordinate) and temperature or plastic strain rate (abscissa) for the particle-reinforced aluminium alloy X8019/12.5$_p$.

Figure 27 presents a map which shows the regimes of plastic strain amplitude (ordinate) and temperature or plastic strain rate (abscissa) where damage can be attributed to fatigue, creep, oxidation or combinations of these damage types. On the basis of such a map, which resulted from metallographic inspection of correspondingly tested samples, a suitable damage parameter can be chosen and applied to the respective loading parameter regime (Fig. 28). In the case of TMF, adaptation of these concepts to non-isothermal conditions requires even more consideration in order to deduce simplifications which enable the adaptation but are also reasonable from a mechanistic point of view.

In the case of the dispersoid-strengthend aluminum alloy X8019 in SiC particle-reinforced as well as non-reinforced condition the transformation of the cyclic stress-strain behaviour and the damage evolution from isothermal to non-isothermal conditions appears to be justified without any major adaptations. This is a consequence of a stable microstructure, which leads to a very strong dislocation-particle interaction irrespective of the testing mode (IF, IP-TMF or OP-TMF). Hence, no appreciable difference in the dislocation arrangement is observable. Moreover, the temperature range of the intended application of these alloys is small so that environmental effects are only of minor significance.

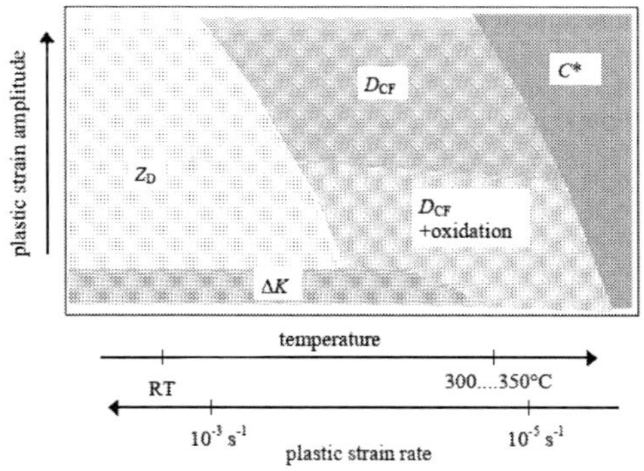

Fig. 28. Schematic map representing the regimes of the respective life prediction concepts deduced from Fig. 26 for X8019/12.5$_p$.

Figure 29 compares the results of the fracture-mechanics-based fatigue life assessment with the experimentally determined number of cycles to failure. The predictive capability is very satisfactory both for isothermal and thermomechanical conditions.

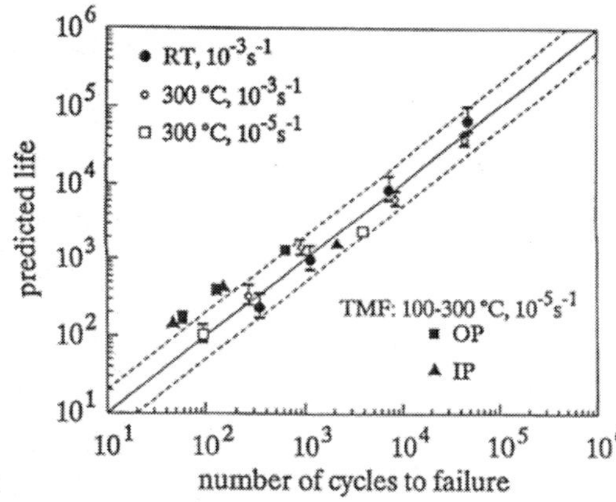

Fig. 29. Comparison of experimentally obtained and calculated IF and TMF lives of X8019 at plastic strain rates of 10^{-5} s^{-1} and 10^{-3} s^{-1}.

It has been described already above and depicted in Fig. 15 that the isothermal fatigue life of the near-α Ti alloy IMI 834 is strongly influenced by environmental effects. The fatigue lives in vacuum are reasonably well predicted with the assumption of pure fatigue. The curves labelled A and B in Fig. 15 represent the calculated results for dry air and humid argon environment, respectively. The linear combination of both environmental damage contributions yields satisfactory results over the whole temperature range (curve C).

The consideration of environmental effects on fatigue life in TMF is more complicated and demands additional assumptions. Hydrogen embrittlement was assumed to be negligible in the case of IP TMF as stresses are mostly compressive in the low-temperature part of the cycle. Consequently, oxidation was the only environmental effect considered for IP-TMF. By contrast, both environmental degradation mechanisms were accounted for to predict N_f of OP-TMF tests. Oxygen uptake was assumed to be not affected by the sign of the stress. Since high tensile stresses coincide with low temperature in OP-TMF, hydrogen embrittlement must be expected due to the reaction of the alloy with water vapour. Figure 30 compares the experimentally observed TMF lives of IP and OP tests at two values of the plastic strain amplitude with the results of the prediction calculation. As can be seen, all data points lie within a ±2 scatter band.

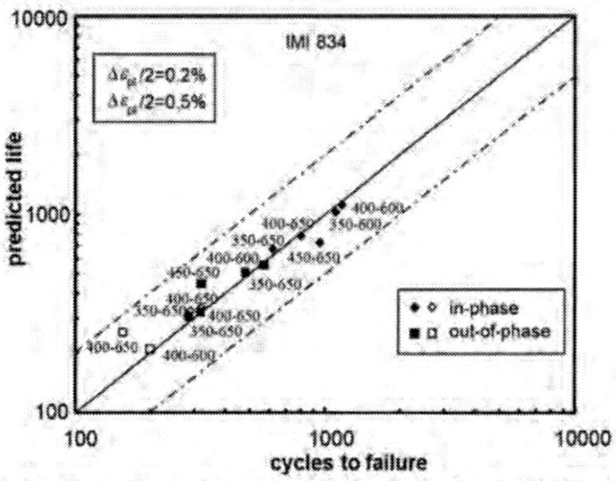

Fig. 30. Comparison of experimentally obtained and calculated lives for IMI 834 under TMF conditions at plastic strain amplitudes of 0.2% (full symbols) and 0.5% (open symbols). Temperature range of test is given next to symbol.

The behaviour auf the austenitic stainless steel of type AISI 304L has been described in detail above in terms of the temperature dependence of the isothermal stress-strain response and the TMF-specific implications. AISI 304L shows a cyclic stress-strain behaviour under TMF conditions that deviates characteristically from that observed under isothermal conditions. This phenomenon can be attributed to the occurrence of dynamic strain aging. The significance of environmental effects is moderate for the austenitic stainless steel. The main damage contribution through the interaction with the surrounding air as compared to vacuum is the preferential grain boundary oxidation leading to a premature crack initiation. This process is most important at high temperature and under conditions where the internally developing creep damage is negligible. Hence, OP loading with high maximum temperature promotes environmental effects. In principle, the influence of air is similar for isothermal and thermomechanical conditions so that isothermally obtained life data can be applied to TMF life assessment.

The multi-component model introduced in chapter 3.2 helps to identify the TMF-specific behaviour and can easily be expanded in such a way that these effects are considered. For this purpose, additional fatigue tests applying temperature changes were carried out to capture the effect of cyclic deformation in the temperature regime of dynamic strain aging.

For high maximum temperatures of the TMF cycle, superimposed dwell times or very low strain rates, creep and relaxation have to be taken into account as important contributions to the deformation of AISI 304L. By means of a simple relaxation law based on Norton's power law creep equation, the change of the yield strength resulting from relaxation was taken into account in each elementary volume leading to a very reasonable prediction of the stress-strain behaviour.

Three different approaches were used to predict TMF life of AISI304L solely from isothermal cyclic stress-strain and life data [10]. Here, a phenomenological energy-based approach and a damage mechanics concept [25] are briefly introduced, while a fracture mechanics method is described in ref. [26].

A well-known damage parameter for high-temperature cyclic loading conditions is the creep-fatigue damage parameter D_{CF} proposed by Riedel [27]. This parameter can be considered to describe the deformation energy density per loading cycle, and therefore the respective life assessment approach belongs to the class of phenomenological, energy–based life prediction concepts. The definition of D_{CF} as used in this study is given in Eq. (3).

$$D_{CF} = 2.9\frac{(\Delta\sigma)^2}{2E} + 1.9\,\Delta\sigma\left(\Delta\varepsilon_{pl} + \Delta\varepsilon_{cr}\right)\left(1 + \frac{\Delta\varepsilon_{cr}}{\Delta\varepsilon_{pl}}\right)^{(1+n)} \quad (3)$$

Here, n is the cyclic hardening exponent, and $\Delta\sigma$, $\Delta\varepsilon_{pl}$ and $\Delta\varepsilon_{cr}$ are the ranges of stress, plastic strain and creep strain per cycle. $\Delta\varepsilon_{cr}$ can either be determined by means of additional creep tests or by comparing hysteresis loops observed at different plastic strain rates. According to the last-mentioned method, the creep strain range is determined as the difference of the total inelastic strain range observed at the plastic strain rate of interest and the plastic strain range obtained during cycling at high strain rates (no creep contribution) applying the same stress range.

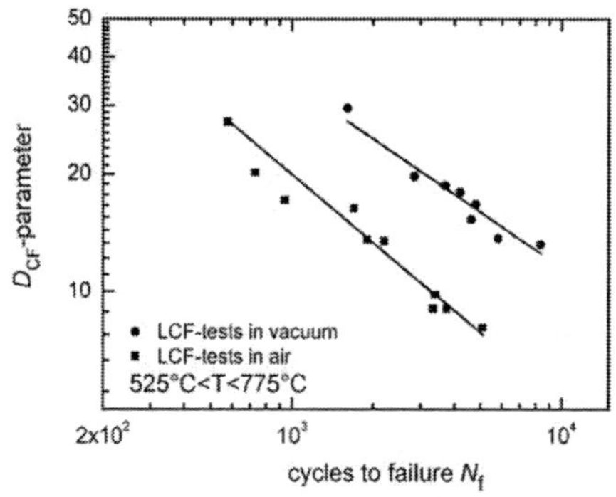

Fig. 31 Correlation between the number of cycles until failure of isothermal tests carried out in air and in vacuum and the corresponding value of the D_{cf} damage parameter.

Figure 31 shows that the parameter D_{CF} correlates nicely with the isothermal fatigue life data. However, two separate lines exist for the two environments considered. Clearly a simple power law describes the relation of D_{CF} with N_f in each case.

$$N_f = L\left(D_{CF}\right)^{-q} \qquad (4)$$

The constants L and q can directly be taken from the lines shown in Fig. 31 and their values contain implicitly the effect of environment on fatigue life.

The so-called *loop reversion method* after Nitta et al. [28,29] was used to determine the creep strain range for the IP-TMF cycles, while no creep contribution was assumed for OP cycling. Loop reversion means that the descending branch of the IP TMF hysteresis loop is mirrored, shifted to a common origin and compared with the ascending branch. The creep strain range is defined as the difference between the inelastic strain of the ascending branch at the stress of T_{max} and the plastic strain taken from the mirrored branch at the same stress (see Fig. 32).

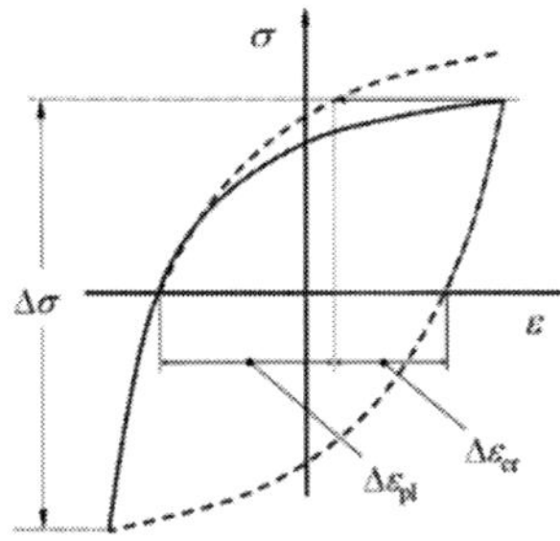

Fig. 32. Schematic representation of the "loop reversion method" used to determine the creep strain contribution of in-phase TMF loading.

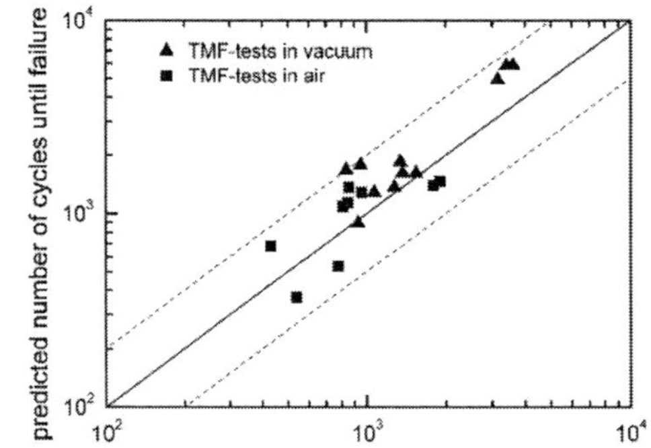

Fig. 33. Predicted TMF life calculated on the basis of the correlation between D_{cf} and N_f found for isothermal conditions (Fig. 31) in comparison to the experimentally observed TMF life data.

TMF life was calculated using the constants L and q from the isothermal tests. Figure 33 compares the predicted number of cycles until failure with the experimentally determined ones. Taking the complexity of TMF loading and the simplicity of the model used into account, the agreement must be considered as very satisfactory. All data points lie within a band of a maximum deviation of a factor of two (dashed lines in Fig. 33).

The damage mechanics model used is based on an equation for the increase of the damage state variable ω proposed by Lemaitre and Chaboche [30]:

$$\mathrm{d}\omega = \left\{ \frac{(1-\omega)^{-\lambda}}{(\lambda+1)\cdot N_c} + \frac{\left[1-(1-\omega)^{\beta+1}\right]^{\alpha}}{(\beta+1)\cdot N_f \cdot (1-\alpha)\cdot(1-\omega)^{\beta}} \right\} \cdot \mathrm{d}N \qquad (4)$$

The first term in braces represents the creep contribution to damage evolution, while the second term results from fatigue. λ denotes the creep damage tolerance factor and was determined from creep data according to

$$t_r = \frac{1}{\lambda+1} \cdot \frac{\sigma^{-n}}{A_0} \qquad (5)$$

using the time until rupture t_r under creep conditions at creep stress σ and the two parameters stress exponent n and coefficient A_0 of Norton's creep law.

N_c represents the life in terms of the number of cycles to failure as a consequence of creep damage and can be calculated for cyclic loading from the cycle period t_{cyc}, the apparent activation energy of steady state creep Q, the stress exponent n and the coefficient A_0:

$$\frac{1}{N_c} = (\lambda+1) \int_0^{t_{cyc}/2} \frac{\sigma(t)^n}{A_0} \cdot \exp\left(\frac{-Q}{RT}\right) \mathrm{d}t \qquad (6)$$

The constants α and β of the fatigue term in Eq. (4) were obtained from the isothermal fatigue life data, separately for air and vacuum [31]. Furthermore, the number of cycles to failure for plain fatigue N_f was calculated according to Ostergren [21]:

$$N_f = L_1 \cdot \left(\Delta\varepsilon_{pl} \cdot \sigma_{max}\right)^{\eta} \cdot \nu^{1-k} \qquad (7)$$

σ_{max} denotes the maximum stress and ν the test frequency. The constants L_1, η and k were deduced from the isothermal fatigue lives, separately for the tests performed in vacuum and air. The values, which were obtained, are L_1=1166, η=-1.86, and k=0.76 for vacuum [13] and L_1=3577, η=-1,04, and k=0,51 for air [11].

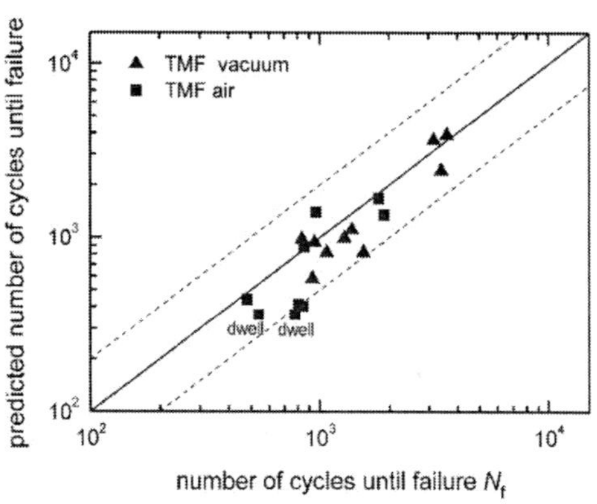

Fig. 34. Predicted TMF life calculated on the basis of the damage mechanics model according to Eqs. (4-7) using solely isothermal results in comparison to the experimentally observed TMF life data.

Figure 34 compares the experimentally observed TMF life data with the predicted lives. A very reasonable agreement can be stated irrespectively of plastic-strain/temperature phasing (IP or OP) for both the vacuum tests and air tests. It is worth noting that also the effect of superimposed dwell times (see data points labelled "dwell") leads to a conservative prediction.

3 Conclusions

From the results presented as examples to illustrate the variety of behavioural patterns of metallic structural materials under isothermal and non-isothermal conditions, the conclusion can be drawn that the following prerequisites should be fulfilled, in order to reach a reasonable predictive capability of TMF life assessement on the basis of mainly isothermal data:

- The mechanisms relevant for isothermal fatigue must be understood as a function of temperature in the temperature range of TMF loading.
- The specifics of the TMF behaviour both in terms of cyclic stress-strain response and damage evolution must be known and taken into account.
- The material considered should exhibit sufficient cyclic plasticity; otherwise the brittleness requires a particular and material-specific consideration.
- The method used for TMF life assessment must be selected on the basis of the relevant damage mechanisms.

Deutsche Forschungsgemeinschaft (DFG) is gratefully acknowledged for the financial support of the studies this article is based on.

References

1. H.L. Bernstein, in: *Low-cycle fatigue and life prediction, ASTM STP 770*, C. Amzallag, B.N. Leis, P. Rabbe (Eds.), American Society for Testing and Materials, Philadelphia, PA, 105 (1982)
2. F. Ellyin *Fatigue damage, crack growth and life prediction* (Chapman & Hall, London, 1997)
3. R. Danzer, *Lebensdauerprognose hochfester metallischer Werkstoffe im Bereich hoher Temperaturen* (Gebr. Borntraeger, Berlin, 1988)
4. H.J. Maier, R.G. Teteruk, H.-J. Christ, Materials at High Temperatures **19**, 9 (2002)
5. A. Nagesha, M. Valsan, R. Kannan, K. Bhanu Sankara Rao, V. Bauer, H.-J. Christ, Intern. Journal of Fatigue **21**, **636** (2009)
6. R. Zauter, F. Petry, H.-J. Christ, H. Mughrabi, in: *Thermomechanical behavior of materials, ASTM STP 1186*, H. Sehitoglu (Ed.), American Society for Testing and Materials, Philadelphia, PA, 70 (1993)
7. H.-J. Christ, A. Jung, H.J. Maier, R. Teteruk, SADHANA **28**, 147 (2003)
8. H.-J. Christ, R. Teteruk, A. Jung, H J. Maier, in: *Thermomechanical Fatigue Behavior of Materials, 4th Volume, ASTM-STP 1428*, A. McGaw, S. Kalluri, J. Bressers, S. D. Peteves (Eds.), American Society for Testing and Materials, West Conshohocken, PA, 145 (2003)
9. H.-J. Christ, Mater. Sci. Engng A, **A468-470**, 98 (2007)
10. V. Bauer, H.-J. Christ, *Intermetallics* **17**, 370 (2009)
11. V. Bauer, *Verhalten metallischer Konstruktionswerkstoffe unter thermomechanischer Belastung - Experimentelle Charakterisierung und modellmäßige Beschreibung* (Berichte aus der Werkstofftechnik, Shaker Verlag, Aachen, 2007)
12. P. Pototzky, *Thermomechanisches Ermüdungsverhalten der Hochtemperaturtitanlegierung IMI834* (Fortschritt-Berichte VDI, Reihe 5, Nr. 565, VDI Verlag, Düsseldorf, 1999)
13. R. Teteruk, *Modellierung der Lebensdauer bei thermomechanischer Ermüdungsbeanspruchung unter Berücksichtigung der relevanten Schädigungs-mechanismen* (Fortschritt-Berichte VDI, Reihe 5, Nr. 653, VDI Verlag, Düsseldorf, 2002)
14. A. Jung, *Einfluß einer Partikelverstärkung auf das Hochtemperaturermüdungsverhalten einer dispersions-gehärteten Aluminiumlegierung* (Fortschritt-Berichte VDI, Reihe 5, Nr. 600, VDI Verlag, Düsseldorf, 2000)
15. Ali El-Chaikh, *Charakterisierung der Schädigungs-mechanismen bei thermomechanischer Ermüdung einer hochfesten Titanaluminid-Legierung*, Doctorate Thesis, to be published in 2018
16. H.-J. Christ, V. Bauer, in: *Creep-Fatigue Interactions – Test Methods and Models, ASTM-STP 1539*, A. Saxena und B. Dogan (Eds..), ASTM International, West Conshohocken, PA, USA, 178 (2011)
17. G. Masing, Wissenschaftliche Veröffentlichungen aus dem Siemens-Konzern **3**, 119 (1923)
18. H.-J. Christ, *Wechselverformung von Metallen* (Springer Verlag, Berlin, Germany, 1991)
19. H.-J. Christ, V. Bauer, Comput. Mater. Sci. **57**, 59 (2012)
20. P. Pototzky, H.J. Maier, H.-J. Christ, Met. Mat. Trans. A **29A**, 2995 (1998)
21. W.J. Ostergren, J. Test. Evaluation **4**, 327 (1976)
22. W.J. Ostergren, E. Krempf, ASME Paper 78-PVP-63, ASME (1978)
23. S. D. Antolovich, S. Liu, R. Baur, Metall. Trans A **12**, 473 (1981)
24. H.-J. Christ, A. El-Chaikh, B. Wollny, in: *LCF8, Eighth International Conference on Low Cycle Fatigue*, T. Beck, E. Charkaluk (Eds.), DVM, Berlin, 87 (2017)
25. H.-J. Christ, V. Bauer, in: *Proc. of the Hael Mughrabi Honorary Symposium: Plasticity, Failure and Fatigue of Structural Materials – From Macro to Nano,* K.J. Hsia, M. Göken, T. Pollock, P.D. Portella, N.R. Moody (Eds.), TMS, Warrendale, 49 (2008)
26. V. Bauer, H.-J. Christ, in: *Proc. Fifth Intern. Conf. On Low Cycle Fatigue,* P.D. Portella, H. Sehitoglu, K. Hatanaka (Eds.), DVM, Berlin, 207 (2004)
27. H. Riedel, *Fracture at High Temperatures* (Springer Verlag, Berlin, 1987)
28. A. Nitta, K. Kuwabara, Current Japanese Materials Research **3**, 203 (1988)
29. A. Nitta, K. Kuwabara, T. Kitamura, *CRIEPI Report E282015*, Central Research Institute of Electric Power Industry, Tokyo (1983)
30. J. Lemaitre, J.L. Chaboche, *Mechanics of Solid Materials* (Cambridge University Press, Cambridge,1990)
31. F. Petry, H.-J. Christ, H. Mughrabi, in: *Microstructure and Mechanical Properties of Materials*, E. Teckhoff, O. Vöhringer (Eds.), DGM-Informationsgesellschft Verlag, Oberursel, 79 (1993)

Qualification of selective laser-melted Al alloys against fatigue damage by means of measurement and modeling techniques

Mustafa Awd[1*], *Jan* Johannsen[2], *Shafaqat* Siddique[1], *Claus* Emmelmann[2], and *Frank* Walther[1]

[1]Department of Materials Test Engineering (WPT), TU Dortmund University, D-44227, Dortmund, Germany
[2]Fraunhofer Research Institute for Additive Manufacturing Technologies (IAPT), D-21029 Hamburg, Germany

Abstract. Aluminum alloys processed through selective laser melting possess unique features of microstructure, defect morphology and mechanical properties. Constitution of fine cellular dendrites results from the high-cooling rate of the melt pool during the consolidation process. Investigation of the microstructure by scanning electron microscopy identifies supersaturation of Si particles as a secondary strengthening mechanism. On the contrary, platform heating that induces coarser microstructure leads to migration of Si particles from the Al matrix to the eutectic phase. As a result, tensile strength is reduced by ~3%, while fracture strain is increased by ~17%. Fine-grained structures exhibit a lower amount of plastic damage accumulation as well as delayed crack initiation as determined by the applied measurement techniques. Finite element models of the investigated configurations are obtained using scans of computed tomography under consideration of process-induced defects. Comparison of modeling and experimental results concluded that dominant fatigue damage mechanisms are related to the loading regime from low-cycle (LCF) to very-high-cycle fatigue (VHCF). Thus, process-inherent features of microstructure and porosity have different quantitative effects concerning the applied load. In VHCF, a material configuration with platform heating possesses an improved fatigue strength by ~33% at 1E9 cycles, concerning the material configuration without platform heating.

1 Introduction

Production of metallic components using layer-wise build-up in powder-bed-fusion processes, like selective laser melting (SLM), made the most significant progress from rapid prototyping into rapid manufacturing applications [1,2]. The process has seen a wide variety of alloy processing such as steels, titanium (Ti), aluminum (Al), and magnesium (Mg) alloys. An advantage emerges over electron beam melting (EBM) since thermal conditions during the latter limits its use to high melting temperature materials like Ti-6Al-4V or Ti-Al intermetallic [2].

Edwards and Ramulu found fatigue strength of as-built SLM Ti-6Al-4V inferior to the wrought alloy due to porosity and residual stresses [3]. Cain et al. reported that mechanical properties have a significant directional anisotropy. The observation was attributed to anisotropic microstructure with a strong gradient in the cooling direction. Effect of residual stress distribution in the as-built specimens was most significant in the fatigue crack growth rate and fracture toughness. Relaxation of residual stresses by post-process annealing reduced anisotropy and increased fracture toughness [1].

Kasperovich and Hausmann concluded that application of optimized scanning parameters with adequate energy density enhanced relative density which improved mechanical properties accordingly. The tensile strength of SLM Ti-6Al-4V in as-built condition was better than the wrought alloy while fracture strain was reduced. In comparison to hot-isostatic pressing (HIP), stress relief (SR) did not enhance high-cycle fatigue (HCF) strength since failure initiated from internal defects which persist after the treatment. In addition to the elimination of defects by HIP, ductility was recovered up to $19.4 \cdot 10^{-2}$ [4]. Wauthle et al. confirmed the same finding concerning HIP for lattice Ti-6Al-4V structures [5].

Leuders et al. reported a prolonged crack initiation phase in SLM Ti-6Al-4V, when porosity is relatively less, in HCF regime. The microstructure which is observed to be dominant α' increased crack initiation resistance signifycantly [6]. Wycisk et al. found that shot peening reduced fatigue limit by 15% since it subjected internal porosity to local tensile stresses. In HCF, the analysis reported failures from pores in the diameter range of 40 µm or higher [7]. After HIP, crack initiation at internal defects was no longer found. In completely tensile cyclic loading, crack initiation shifts from the surface to subsurface locations [8]. Linear-elastic fracture mechanics (LEFM) calculated fatigue lifetime reliably when long cracks existed but overestimated the lifetime in case of short cracks [9].

* Corresponding author: mustafa.awd@tu-dortmund.de

Brandl et al. attempted improving fatigue strength of AlSi10Mg using platform heating (PH) in the SLM process. The as-built microstructure consisted of an Al matrix where a Si network of equiaxed cells is embedded. In this study, PH did not increase fatigue strength until a peak hardening treatment is applied after processing [2]. Siddique et al. introduced micro-computed tomography (μ-CT) in combination with finite-element modeling (FEM) as means to pre-calculate fatigue lifetime for SLM AlSi12 alloy specimens. The analysis introduced stress-concentration factors for pores as a function of diameter and proximity to the surface. Given this, HIP of AlSi12 was assumed to enhance fatigue strength, but due to microstructure alteration and Si agglomeration, fatigue strength was significantly reduced [10]. Nevertheless, the fatigue strength of SLM AlSi12 in as-built condition was superior to cast alloy. Application of PH reduced mean fatigue strength as well as standard deviation. Since crack initiation from internal porosity was observed, reduction of porosity by PH increased fatigue strength reliability [11].

Siddique et al. improvised a stochastic approach for fatigue lifetime calculation based on fracture mechanics and the weakest-link theory. The lifetime calculation was found reasonable in low-cycle fatigue (LCF), but not in HCF. The study found that stress-concentration factors induced by surface roughness are much higher than stresses induced by internal porosity. The first reduced fatigue lifetime by one order of magnitude [12]. Hybrid Al structures with reasonable fatigue strength were possible within a hybrid manufacturing framework. After SR the resulting lifetime was better than that of pure SLM structures. The interface between SLM deposition and wrought alloy suffered from microstructure non-homogeneity as well as significant residual stresses [13]. Production of components from high-strength Al-Sc alloys by SLM was reported with a tensile strength of ~500 MPa. Considering fatigue, the strength was significantly reduced by the existence of significant internal porosity [14]. The latest conclusion envisages that if proper process window is developed, high-strength Al components for dynamic applications in the aerospace industry by SLM are feasible.

In this study, a novel plot-based Weibull model is presented for fatigue lifetime calculation of Al alloys. The model relies on real-time measurement and modeling techniques for dynamic loading qualification of SLM Al alloys. For two alloys, AlSi10Mg and AlSi12, the combined techniques are applied to conclude about processing and post-processing procedures during SLM process. The model is assessed by experimental fatigue data.

2 Materials and methods

2.1 Experimental testing

The specimens employed in this study were manufactured at Fraunhofer IAPT using SLM 250 HL. The maximum laser power of the system is 400 W based on a fiber laser source. Maximum volume capacity inside the building chamber is 250 × 250 × 280 mm^3 in 3D space. An inert atmosphere of argon gas was applied for shielding the melt tracks during build-up which was flowing from the left of the chamber to the right. The function of the shielding gas is to ensure purity of melt pools to prevent the formation of inclusions [11]. The scanning method is designed such that porosity and residual stresses are kept to a minimum. The factorial study is presented in Table 1. The experimental design aims at investigating the effect of platform heating (PH) at 200 °C on microstructure and eventually fatigue properties by a joint study of experi-mental testing and modeling procedure for two common compositions of cast Al alloys, AlSi10Mg and AlSi12 respectively.

Table 1. Experimental design.

Batch	A	B	C	D
Alloy	AlSi10Mg	AlSi10Mg	AlSi12	AlSi12
SR	no	no	yes	yes
PH	no	yes	no	yes

The as-built specimens of batches A and B were scanned in a μ-CT facility with the experimental setup being explained in detail in [14]. Similar results for batches C and D are presented in [12]. The relative densities for batches A and B were 99.96% and 99.94%, respectively. The effect of PH on reducing the number of pores was consistent with [12], but the relative density of the PH batch was less which disagrees with the previous observation regarding batches C and D. The volumetric defect distributions of batches A and B are presented in Fig. 1. It is observable that batch A developed defects of bigger volume than batch B while in the latter the smaller diameter porosity (< 120 μm) was more.

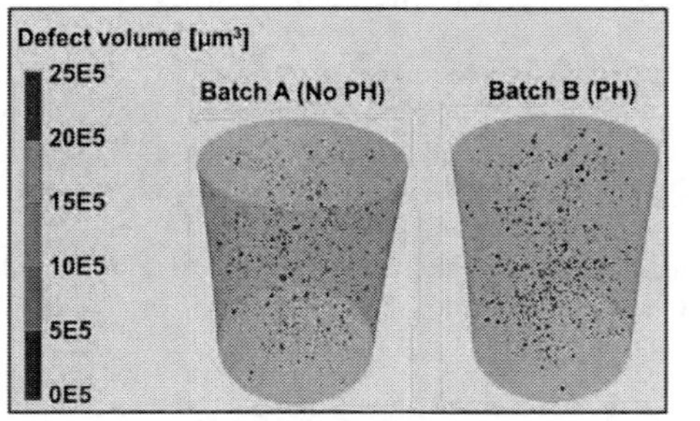

Fig. 1. Volumetric defect distribution from μ-CT of batches A and B.

The relationship between stress developed by porosity and their size was described in detail in [10]. The latter identified pore diameter and location as the most influential factors. The statement renders relative density of such minor difference irrelevant in comparison to the pore distribution within the volume as well as the existence of large critical pores. One more issue that interacts with this factor is the reduction of residual stresses through PH. The latter was also proved to induce microstructure coarsening [11]. The interaction between remnant porosity, residual stresses as well as microstructure resulting from PH is revealed in the mechanical quasi-static and cyclic experiments.

In-situ tensile testing in μ-CT was applied, in this study, to be compared with standard ex-situ tensile test for batch D. The aim was to separate porosity and microstructure related mechanisms and the identification of relaxation behavior and crazing effects during plastic deformation. The design of the in-situ experiment inside μ-CT can be envisaged in Fig. 2.

(a) (b)

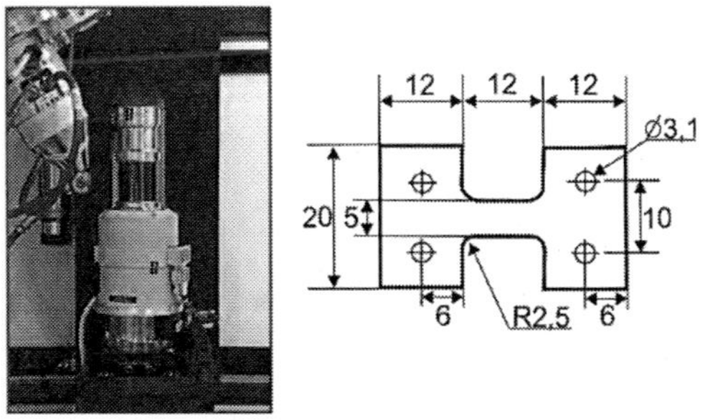

Fig. 2. Design of in-situ μ-CT tensile experiment: (a) mounted in-situ module; (b) specimen geometry.

Quasi-static tensile experiments were carried out on Instron 3369 with a 50 kN load cell using an extensometer with a gauge length of 10 mm. The specimen geometry is shown in Fig. 3a while stroke rate was fixed at 1 mm/min. The same specimen was used in fatigue experiments at Instron 8872 with 10 kN load cell. The experimental setup of fatigue tests is shown in Fig. 3b. Tensile testing was applied for bridging the relationship between microstructure, defects on one side and fatigue strength on the other through FEM and statistical modeling.

(a)

(b)

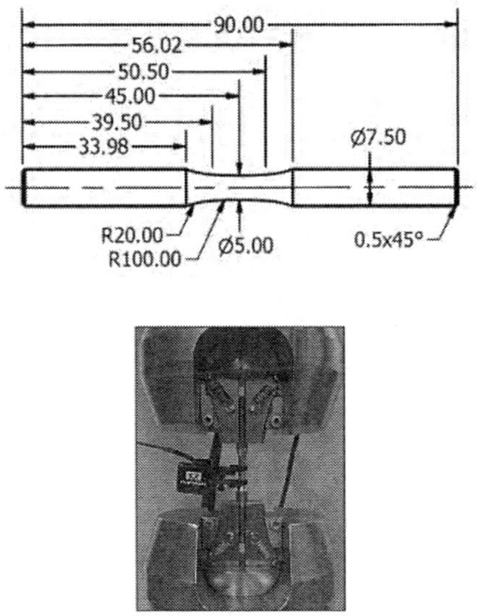

Fig. 3. Design of mechanical testing experiments: (a) specimen geometry; (b) test setup.

2.2 Modeling and simulation

Deformation behavior under quasi-static conditions was simulated in this study in Abaqus 6.14. The material law was the true stress-strain curve obtained in a tensile test. The deformation curves of batches A and B are shown in Fig. 4. It can be seen that batch B has higher values of Young's modulus, yield strength, ultimate tensile strength and fracture strain. The result agrees with [11] for fracture strain but disagrees with it in ultimate tensile strength. The cited work suggested that microstructure coarsening by PH increased ductility but reduced tensile strength.

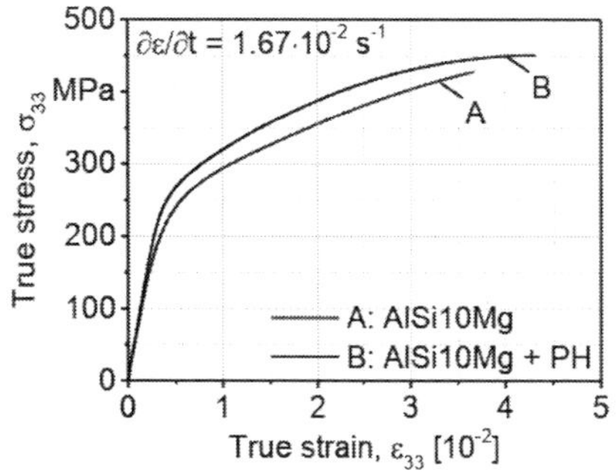

Fig. 4. Flow behavior under quasi-static conditions of batches A and B.

Deformation simulations in Abaqus 6.14 were carried out using the built-in material laws for isotropic and kinematic hardening. A comparison between both material laws as well as the experimental flow curve for batches A and B can be found in Fig. 5a and Fig. 5b, respectively.

(a)

(b)

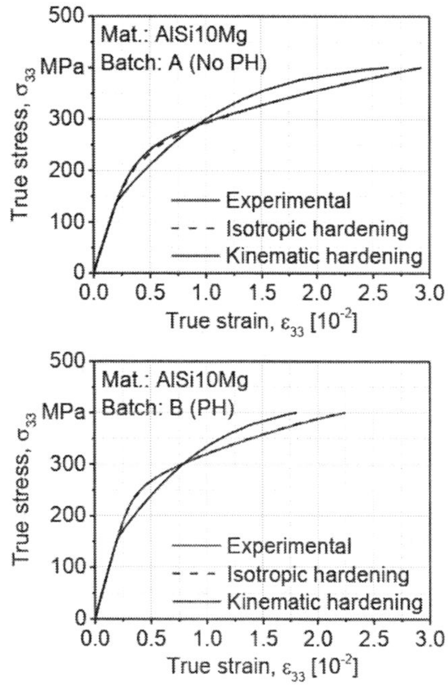

Fig. 5. Comparison between flow curves of the experimental test against simulated conditions under isotropic and kinematic hardening considerations for batches: (a) A (No PH); (b) B (PH).

The simulation was set to terminate when the cross-sectional stress of 400 MPa is developed. The agreement between the experimental flow curves and the isotropic hardening law is good while the discrepancy between both and the kinematic hardening law is clear. Kinematic hardening represents the evolution of peaks of hysteresis under cyclic loading capturing the Bauschinger effect. The finding yields kinematic hardening more representative when modeling fatigue strength of dispersion-strengthened Al alloys. The conclusion found agreement in [15].

3 Results and discussion

3.1 Mechanical properties

Siddique et al. reported tensile flow curve of PH AlSi12, here batch D, as seen in Fig. 6 [11]. A comparison between this ex-situ flow behavior against in-situ μ-CT test can be seen in the same figure. During the in-situ test, intermediate μ-CT scans are performed at steps 1-4. Each scan takes ~40 min at which the indicated force is kept acting on the specimen by the controller. The creeping effect on the stress-strain relation is clear in the figure. The result suggests that cold straining can enhance ductility of the microstructure at the expense of maximum strength. A by-product is to study relaxations under constant static loading as well as deformation of pores and crazing effects.

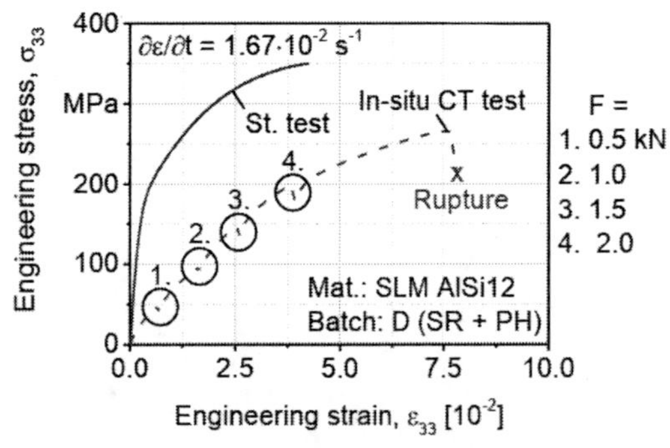

Fig. 6. Comparison between flow curves of the standard ex-situ tensile test and the in-situ μ-CT tensile test.

In comparison to quasi-static tensile test shown in Fig. 4, the comparison between cyclic properties in a continuous load increase test (LIT) is presented in Fig. 7. The principle of this test is explained in detail in [14]. Regarding ultimate tensile strength, batch B was stronger by ~4.65% in addition to better elasticity modulus. The difference was more pronounced in the cyclic properties since fracture stress was ~37.5% higher in favor of batch B. This promotes the assumption that micro- and macro-structure of this batch under PH was significantly tolerant to cyclic damage. Towards ~110 MPa, the damage accumulation profile was similar. Afterward, the damage profile of batch A stabilizes on a horizontal plateau until fracture at ~160 MPa. Otherwise, batch B adopts constant slope of plastic damage until ~170 MPa at which specimen stiffens until fracture.

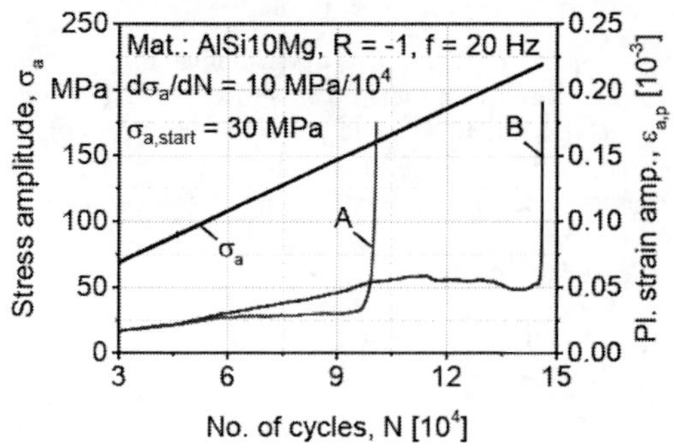

Fig. 7. Plastic strain amplitude reaction in a load increase test for batches: (a) A (No PH); (b) B (PH).

The results of the LIT indicated the response of the plastic damage could vary as a function of the instantaneous stress amplitude. To separate and confirm the mechanism, constant amplitude testing was applied for batches A and B at the critical stress ranges identified in Fig. 7. As shown in Fig. 8a that under 120 MPa specimen stiffens dynamically in the first ~500 cycles of the test until a stable stiffness is reached which persists until the end of the test. The footprint of this response is evident in Fig. 7. At 140 MPa, which is close to the fracture stress in Fig.7, the specimen softens from the beginning onwards until fracture.

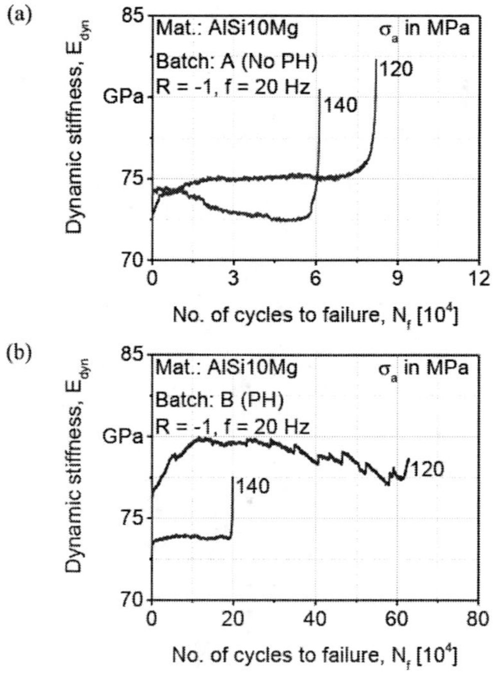

Fig. 8. Evolution of dynamic stiffness under constant amplitude cyclic loading for batches: (a) A (No PH); (b) B (PH).

In Fig. 8b, the profile at 120 MPa indicates sharp hardening and increase of dynamic stiffness from the beginning, until 10^4 cycles. Afterward, cycles of softening and abrupt hardening develops with overall softening of the specimen. The observation gives evidence about the dynamic healing of the microstructure which resists cyclic damage. The phenomenon is rather attributed to multi-scale cyclic damage features with evidence in the micro-structure to be found in a later study. At 140 MPa, the stiffness has a stable horizon since its near to the transition point in Fig. 7.

Ultrasonic fatigue tests at 20 kHz (10 kHz effective frequency due to pulse-pause-ratio) were held to identify the effect of PH on fatigue strength beyond 10^6 cycles. The testing methodology was explained in detail in [13]. In Fig. 9, the corresponding S-N curves for batches C and D are reported. Two main observations can be envisaged here. First is that fatigue strength of batch D being ~33% higher and its standard deviation is less in comparison to batch C. The second is consistent with LCF results reported in [11], but the first is not. The observation indicates alteration of damage mechanisms from LCF to VHCF with possibilities to tailor properties specifically for a given fatigue regime. The same analysis is planned for batches A and B.

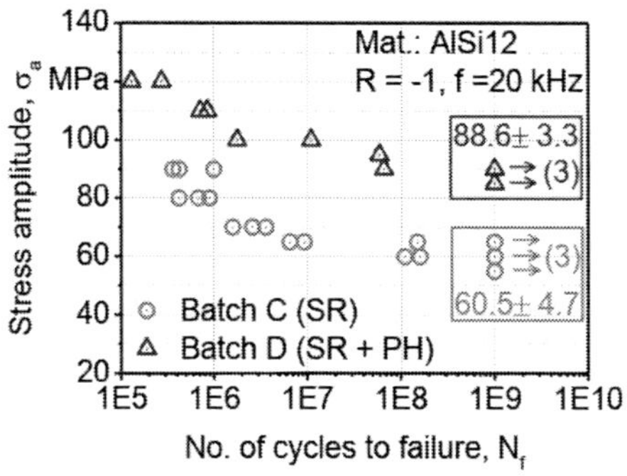

Fig. 9. Very-high-cycle fatigue strength of batches: (a) C (SR + No PH); (b) D (SR + PH).

3.2 Model-based correlation

After simulating stress evolution using the kinematic hardening law as shown in Fig. 5, based on μ-CT scans of Fig. 1, FEM results were post-processed in nCode fatigue lifetime calculation software. Element-wise cycles to failure were calculated based on fatigue cut-off criterion. The results were exported to OriginLab data analysis software to plot the Weibull distribution of lifetimes of individual elements. At two stress levels of 140 and 120 MPa, the lifetime distribution for batch A can be seen in Fig. 10. It can be seen from Fig.10a to Fig. 10b that the whole plot is shifting to the right as lifetime increases against lower stress amplitudes. The wide distribution of lifetimes from 10^2 to 10^{12} cycles is related to local stress state at the element.

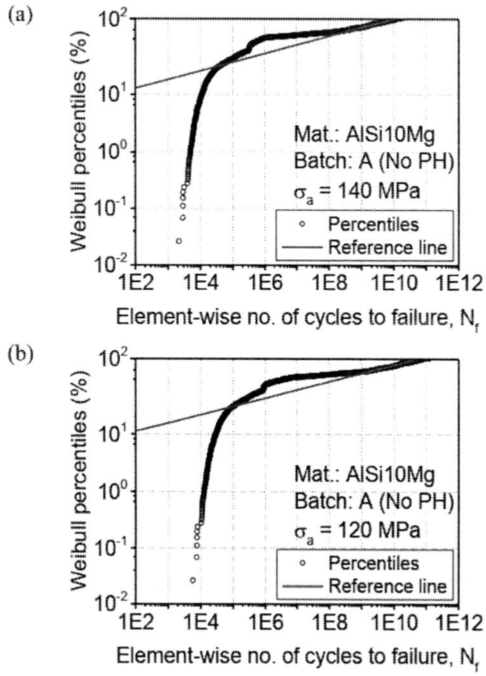

Fig. 10. Plot-based Weibull model of fatigue strength of batch A (No PH) at two stress amplitudes (MPa): (a) 140; (b) 120.

As shown in Fig. 10 for batch A and Fig. 11 for batch B, the intersection between Weibull percentiles and the reference line is the mean lifetime of the batch. A one-sided confidence interval of 90% is selected as the lower bound fatigue lifetime criterion. The rule means that the lower bound fatigue lifetime lies at 10% in Fig. 10 and 11. The upper bound is identified by the first point after the reference line where the slope of Weibull percentiles first becomes infinity. In this manner, fatigue strength characterization is held by identification of a population mean as well as lower and upper bounds as a safe-economic design interval. The approach follows the same concept presented in [14]. The latter was based on Monte-Carlo simulation and the load increase test with extensive computational effort. The approach presented in this work utilizes a flow curve from a tensile test for simulation of kinematic hardening behavior and hysteresis evolution in Abaqus. Post-processing of the element-wise stresses in nCode is followed by plot-based Weibull modeling.

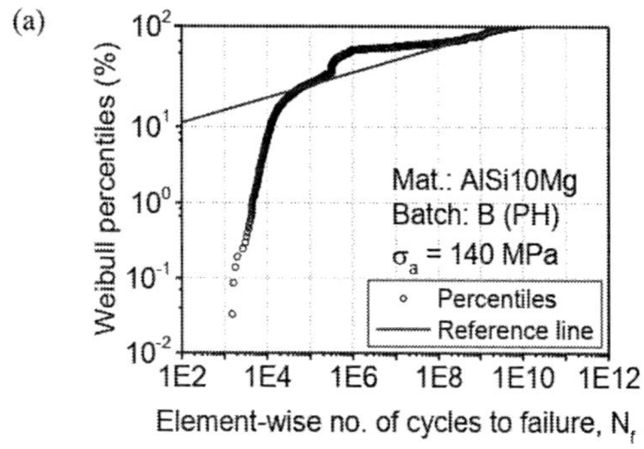

(b)

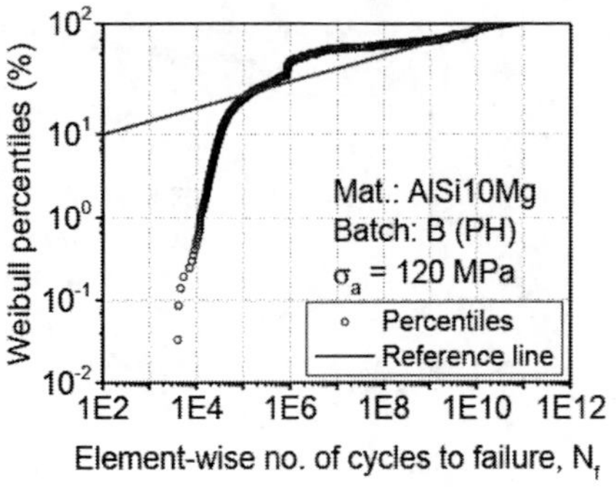

Fig. 11. Plot-based Weibull model of fatigue strength of batch B (PH) at two stress amplitudes (MPa): (a) 140; (b) 120.

A comparison between experimental and calculated fatigue lifetime is shown in Fig. 12 for batches A and B. In both batches, the simulation agrees well with experimental data from two perspectives. The first is that mean solid line passes near the mean value of the three validation experiments for each batch and each stress level. The second one is that experimental points are confined to the upper and lower bounds even at experimental points of extreme values. VHCF validation of the predicted lifetimes is planned in later studies. The speed of this procedure is seen as relevant in additive manufacturing since process flexibility allows for tailoring of properties with a wide variety. High-throughput characterization is thus necessary. Selection of proper process parameters to achieve a certain combination of properties with reliability is possible.

(a)

(b)

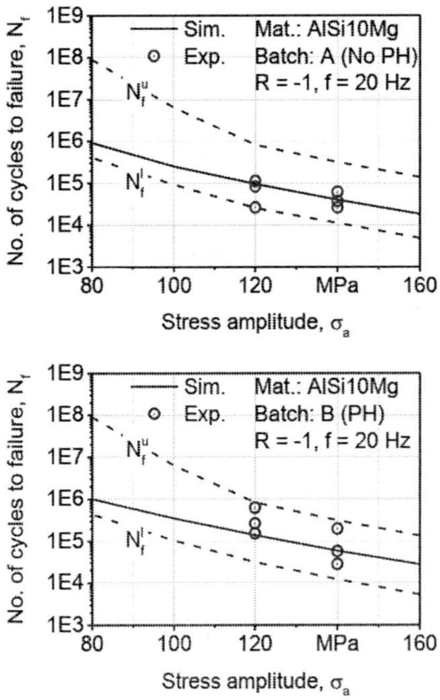

Fig. 12. Comparison between experimental and predicted fatigue strength of batches: (a) A (No PH); B (PH).

4 Conclusions and outlook

The current study aimed at investigating efficient testing strategies to qualify additively manufactured aluminum alloys. Two aluminum alloys, AlSi10Mg and AlSi12, were investigated with platform heating as the main factor. The latter factor is identified in the previous literature as a control parameter to reduce remnant porosity and coarsen microstructure [10,11]. In this study, it was found consistent that platform heating improved ductility, but tensile strength was not reduced in AlSi10Mg alloy. Consequently, LCF strength of platform heated specimens was better with superiority margin increasing as stress amplitude is reduced. Although the observation contradicts with reported data on the issue, that may be attributed to a change in scanning strategy. This offers an opportunity for a future investigation regarding residual stresses and microstructure control.

In AlSi12 alloy, platform heating improved VHCF strength significantly as determined by ultrasonic fatigue tests. The conclusion needs to be confirmed for AlSi10Mg and other additively manufactured alloys. In-situ tensile testing inside μ-CT presented the phenomenon of stress relaxation under constant static load. The phenomenon highlights cyclic creeping that might occur at stress ratios higher than zero. This open question also offers an opportunity for simulation of the effect and probing appropriate material laws to describe it.

The continuous load increase test was used in a previous study to perform an FE simulation of fatigue damage [14]. Based on a demanding computation algorithm of Monte-Carlo simulation, fatigue lifetimes were calculated. In this study, the load increase test was used to relate degradation in material's stiffness under constant amplitude to the wide-range fatigue stresses by monitoring plastic strain damage. Development of dynamic stiffness under constant amplitude was consistent with the plastic damage response in a load increase test. It was observed that the stiffness increased asymptotically towards fracture which can be used as failure criterion in a proper fatigue damage modeling scheme.

Built-in kinematic and isotropic hardening material laws of Abaqus were applied to capture cyclic hysteresis evolution and quasi-static deformation, respectively. According to findings of [15] that identified hardening behavior of dispersion-strengthened Al alloys as kinematic, the resulting flow curve was used to calculate fatigue lifetimes. The latter was computationally efficient as well as reasonably accurate. The result offered designers three levels of design safety by identification of mean population, lower and upper bounds of fatigue lifetime.

The plot-based Weibull model presented here used the probability distribution of element-wise stresses as failure criterion of a macroscale specimen. The finding represents a plasticity-based Weibull model as more accurate and practical in comparison to the approach based on fracture mechanics which was presented in [12]. The VHCF failure mechanism which is agreed to be a micromechanical problem can identify in such genre of modeling a potential opportunity to combine a crystal plasticity model with a weakest-link expression for identification of lifetimes required to develop a technical crack beyond microstructurally short ones.

The strategy behind the study is to develop efficiently and accurately presenting models of fatigue lifetime to qualify additively manufactured components of wider potential of tailored properties and functional grading. The field of potential application for the presented methodologies extends to multi-exposure SLM sections and intrinsic heat treatments which are very effective property control tools for allotropic alloys of the Al-Si and Ti-Al alloying systems.

Acknowledgment: The authors would like to thank the German Research Foundation for its financial support to the conduct of this research work in terms of a research grant entitled: "Mechanism-based understanding of functional grading focused on fatigue behavior of additively processed Ti-6Al-4V and Al-12Si alloys" (WA 1672/25-1; EM 95/2-1).

References

1. Cain, V.; Thijs, L.; van Humbeeck, J.; van Hooreweder, B.; Knutsen, R.: Crack propagation and fracture toughness of Ti6Al4V alloy produced by selective laser melting. Additive Manufacturing (2015) 5: 68-76.
2. Brandl, E.; Heckenberger, U.; Holzinger, V.; Buchbinder, D.: Additive manufactured AlSi10Mg samples using Selective Laser Melting (SLM): Microstructure, high cycle fatigue, and fracture behavior. Materials & Design (2012) 34: 159-169.
3. Edwards, P.; Ramulu, M.: Fatigue performance evaluation of selective laser melted Ti-6Al-4V. Materials Science and Engineering A (2014) 598: 327-337.
4. Kasperovich, G.; Hausmann, J.: Improvement of fatigue resistance and ductility of TiAl6V4 processed by selective laser melting. Journal of Materials Processing Technology (2015) 220: 202-214.
5. Wauthle, R.; Vrancken, B.; Beynaerts, B.; Jorissen, K.; Schrooten, J.; Kruth, J-P.; van Humbeeck, J.: Effects of build orientation and heat treatment on the microstructure and mechanical properties of selective laser melted Ti6Al4V lattice structures. Additive Manufacturing (2015) 5: 77-84.
6. Leuders, S.; Thöne, M.; Riemer, A.; Niendorf, T.; Tröster, T.; Richard, H.; Maier, H.: On the mechanical behavior of titanium alloy TiAl6V4 manufactured by selective laser melting: Fatigue resistance and crack growth performance. International Journal of Fatigue (2013) 48: 300-307.
7. Wycisk, E.; Emmelmann, C.; Siddique, S.; Walther, F.: High cycle fatigue (HCF) performance of Ti-6Al-4V alloy processed by selective laser melting. Advanced Materials Research (2013) 816-817: 134-139.
8. Wycisk, E.; Siddique, S.; Herzog, D.; Walther, F.; Emmelmann, C.: Fatigue performance of laser additive manufactured Ti-6Al-4V in very high cycle fatigue regime up to 10^9 cycles. Frontiers in Materials (2015) 2: 2117.
9. Wycisk, E.; Solbach, A.; Siddique, S.; Herzog, D.; Walther, F.; Emmelmann, C.: Effects of defects in laser additive manufactured Ti-6Al-4V on fatigue properties. Physics Procedia (2014) 56: 371-378.
10. Siddique, S.; Imran, M.; Rauer, M.; Kaloudis, M.; Wycisk, E.; Emmelmann, C.; Walther F.: Computed tomography for characterization of fatigue performance of selective laser melted parts. Materials & Design (2015) 83: 661-669.
11. Siddique, S.; Imran, M.; Wycisk, E.; Emmelmann, C.; Walther, F.: Influence of process-induced microstructure and imperfections on mechanical properties of AlSi12 processed by selective laser melting. Journal of Materials Processing Technology (2015) 221: 205-213.
12. Siddique, S.; Awd, M.; Tenkamp, J.; Walther, F.: Development of a stochastic approach for fatigue life prediction of AlSi12 alloy processed by selective laser melting. Engineering Failure Analysis (2017) 79: 34-50.
13. Siddique, S.; Awd, M.; Tenkamp, J.; Walther, F.: High and very high cycle fatigue failure mechanisms in selective laser melted aluminum alloys. Journal of Materials Research (2017) 32, 23: 4296-4304.
14. Awd, M.; Tenkamp, J.; Hirtler, M.; Siddique, S.; Bambach, M.; Walther, F.: Comparison of microstructure and mechanical properties of Scalmalloy® produced by selective laser melting and laser metal deposition. Materials (2018) 11, 1: 1-17.
15. Reynolds, A.; Lyons, J.: Isotropic and kinematic hardening in a dispersion-strengthened aluminum alloy. Metallurgical and Materials Transactions A (1997) 28, 5 : 1205-1211.

Influence of as-built surface on fatigue strength and notch sensitivity of Ti6Al4V alloy produced by DMLS

Gianni Nicoletto[1], *Radomila* Konečná[2], *Ludvík* Kunz[3], *Martin* Frkáň[2]

[1]University of Parma, Parma, Italy
[2]University of Žilina, Žilina, Slovakia
[3]Institute of Physics of Materials, Brno, Czech Republic

Abstract. Direct Metal Laser Sintering (DMLS) is one of the additive manufacturing technologies involving a metal powder bed and a laser source to obtain components of complex geometry. The fatigue life of an as-built DMLS part is negatively influenced by the rough surface compared to a machined surface. Since most DMLS components are expected to have also stress concentrations, here the combined effect on fatigue of a as-built surface and a geometrical notch in as-built DMLS Ti6Al4V was investigated using a new test methodology. Miniature specimens having three different directions with respect to build direction were fabricated and tested determining the directional fatigue notch sensitivity of as-built DMLS Ti6Al4V Optical inspection of notch surface quality and roughness measurements demonstrated a ranking of the directional notches that inversely correlated to the measured fatigue performance. The link among the curved surface orientation, the peculiar layer-wise additive manufacturing fabrication and the fatigue notch sensitivity is clarified.

1 Introduction

Direct Metal Laser Sintering (DMLS) is one of the additive manufacturing technologies suitable for the rapid processing of a metal powder bed to obtain components of complex geometry with minimal material and energy wastes, [1]. A range of important metal alloys, such as Ti alloys, Ni-based super alloys, Cr-Co alloys and Al-Si alloys, can be processed with a laser providing the concentrated thermal energy source that selectively fuses thin regions of a powder bed that quickly solidify. The process is repeated layer after layer until the entire part is built.

Whereas the effect of defects in materials prepared by DMLS has been widely discussed and the setting of optimal DMLS process parameters can eliminate their formation in a significant way, the problem of as-built surface roughness needs further investigation because in a part with a complex geometry, most of the surface area will neither be accessible for machining nor polishing and will remain in the as-built condition. When used in service the fatigue life of the as-built DMLS part is negatively influenced by the surface roughness typical of the process. The magnitude of the as-built surface roughness is dependent on melting energy, powder size distribution, the thickness of each built layer, [2].

The surface topography consists of multiple sharp radii of curvature that act as micro notches and lead to local stress concentrations, [3]. Therefore, fatigue crack initiation in as-built surface occurs at considerably lower stresses compared to a conventional machined surface.

The importance of the DMLS technology, the fact that Ti6Al4V is most commonly used material in the aerospace industry for its excellent overall properties and the lack of material allowables needed for part design have motivated the investigation of the fatigue properties of selective laser melted Ti6Al4V using smooth specimens with as-built surfaces, [4-9].

However, few metal additively manufactured components are expected to have simple geometries without any corners or radii that would act as stress concentrations. Therefore, the combined effect on fatigue of a rough as-built surface and a geometrical notch needs to be clarified to enable structural part qualification for service, [10]. Kahlin et al. originally reported the fatigue behavior of DMLS Ti6Al4V with as-built surfaces and quantified the notch fatigue factor for as-built tensile bar with a circumferential notch. The surface roughness itself was found to contribute most significantly by comparison of the fatigue strength of as-built and machined smooth specimens. The notch fatigue factor for combination of a geometrical notch and the rough surface was also determined in [10].

The loading axis of the notched specimens used in [10] was parallel to the build direction. Understanding the influence of fabrication orientation, process parameters and surface curvature on the effective as-built quality and its impact on fatigue strength is a key topic to be further explored.

* Corresponding author: gianni.nicoletto@unipr.it

This study focuses on the experimental investigation of the surface quality of as-built notches having different orientations with respect to the build direction and the resulting sensitivity of DMLS Ti6Al4V to notches and fatigue. The approach introduced in [11] based on the use of directional miniature specimens is adopted to determine the combined effect of an as-built notch and its fabrication direction with respect to build direction on the notch fatigue response.

2 Experimental details

Details of the material and its fabrication process, the peculiar specimen geometry and fatigue test methodology used, and the characterization techniques of the as-built surface roughness are now provided.

2.1 Material and specimen production

The Ti6Al4V ELI alloy powder used here was characterized by spherical powder particles of predominant diameter range from 25 to 45 μm. Specimen production was performed with the DMLS system EOSINT M290 (EOS GmbH, Germany) working with a layer thickness of 60 μm, a max laser power of 400 W and a chamber temperature of 80 °C. The contour and hatch strategy was adopted to melt the powder layer-by layer. Hatching direction was rotated from one layer to the next of a prescribed angle.

After fabrication, the following stress relief heat treatment was applied: heating up to 740 °C, exposure for 2 h and slow cooling to 540 °C all in a vacuum furnace followed by rapid cooling in Argon. The resulting quite isotropic structure was investigated elsewhere, [12]. The tensile properties of the heat treated DMLS Ti6Al4V were also reported elsewhere, [13].

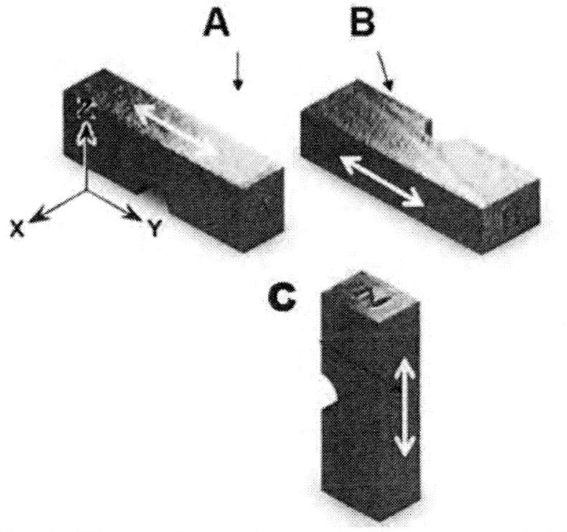

Fig. 1. Mini specimen orientations with respect to build direction. White arrow defines the applied stress direction.

A special miniature prismatic specimen geometry introduced in [14] was adopted to investigate the directional and geometrical notch effect on fatigue. The Fig. 1 shows the three different specimen directions with respect to build direction and their denomination. The size of the nominal section is 5 x 5 mm^2 and the length of the specimen is 22 mm. The prismatic specimen geometry is characterized by the presence of a semi-circular notch of 2 mm radius. The theoretical stress concentration factor was quantified in K_t = 1.63 using the finite element method. This is a mild notch effect but a sharp notch effect is reported elsewhere, [15].

Since the specimens can be oriented as desired with respect to the build direction, the anisotropic fatigue response can be readily investigated. Three sets of such specimens for a total of about 80 specimens in a single build were readily and cheaply produced. All the surfaces were left in the as-built condition to maximize the roughness effect on fatigue.

2.2 Surface roughness characterization

The optical characterization of the as-built notch quality was performed on polished and non-etched specimens with a light microscope Zeiss Neophot 32. Details of the notch root irregularities and initiated cracks were observed on etched (10% HF for 10 s) specimens.

The as-built roughness of the notch root on the length ~2 mm was measured on a Mitutoyo SJ 210 tester. The roughness parameters evaluated were R_a (arithmetical mean roughness value) and R_z (mean roughness depth), which are the most often used for roughness characterization in industry. The focus was the determination of the link between build direction of the notch and local roughness.

2.3 Fatigue test methodology

The three sets of mini specimens oriented in three different directions with respect to the build direction were tested in fatigue using a plane bending machine under a load ratio R = 0 with run-out fixed at 2x10^6 cycles.

The validation of the test methodology used here, [14], compared the plane bending test results of as-built Type C specimens and the rotating bending test results of as-built 6-mm-diameter specimens with longitudinal axis parallel to the build direction. The two sets of specimens were fabricated in DMLS Ti6Al4V. Since the two independent fatigue curves showed coherent trends, this new test methodology can be conveniently used,.

Since the miniature specimen geometry is characterized by a flat surface and a semicircular notch on opposite faces, Fig. 1, the tensile cyclic stress was applied either on the flat surface, see white arrows in Fig. 1, to obtain the un-notched reference fatigue behavior or on the opposite notch root to obtain the notch fatigue behavior as schematically shown in Fig. 2.

Thus the three set of specimens are used to determine the directional un-notched and notched fatigue performance of the as-built DMLS Ti6Al4V, [8].

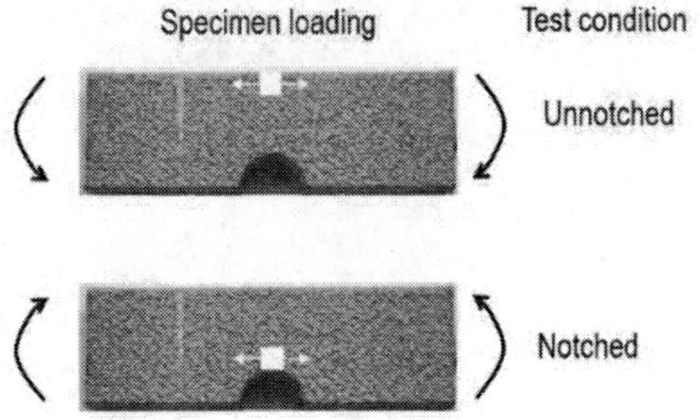

Fig. 2. Alternative cyclic bending loadings with R = 0 of the specimens and different conditions investigated.

3 Results and discussion

This section presents initially the notched fatigue behavior of the DMLS Ti6Al4V obtained with the directional mini specimens followed by the as-built surface characterization and roughness measurements at the notch root. Observations of fatigue crack initiation show the interaction of crack and the local surface conditions. A technology-based explanation of the directional notch fatigue behavior is finally introduced.

3.1 Notch fatigue behavior of DMLS Ti6Al4V

3.1.1 As-built notch fatigue behavior

Fig. 3 shows the combined directional and notch fatigue effect obtained in the three sets of the present mildly notched mini specimens. The maximum nominal stress of the bending load cycle with R = 0 is used as independent variable. Three different and well defined trends are determined for the different directions of fabrication.

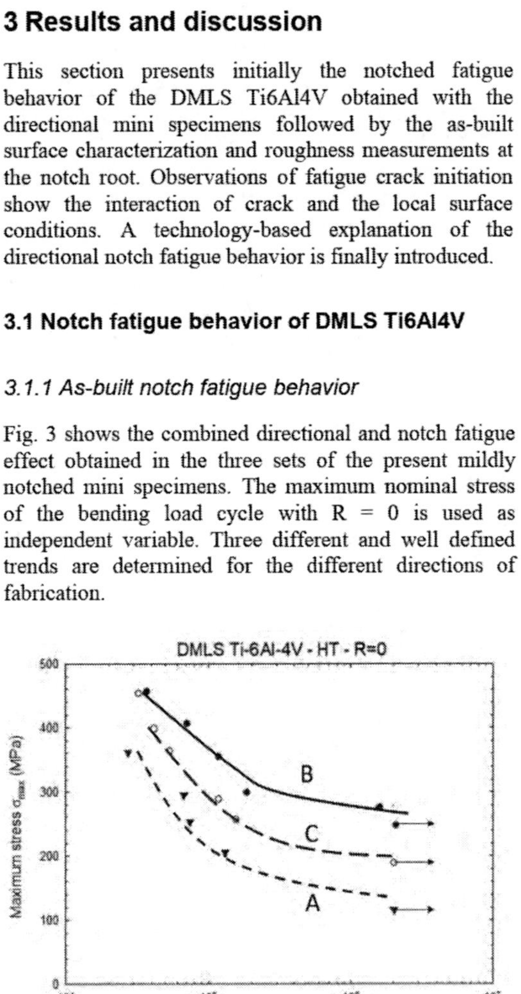

Fig. 3. Directional notch fatigue behavior for the same stress concentration factor.

Considering Fig. 1, the Type B specimen shows the best fatigue performance while the Type A specimen the worst fatigue performance. Type C specimens determine an intermediate behavior. The directional fatigue behavior of un-notched DMLS Ti6Al4V after the same stress relieving treatment used here was determined with the three sets of mini specimens and presented elsewhere, [8]. For the purpose of this study the data could be grouped in a unique scatter band with a mean representative high cycle fatigue strength of about 360 MPa at $2x10^6$ cycles. This value will be used to in the estimation of the notch fatigue factor in a subsequent section.

3.1.2 Notch fatigue factor

The classical definition of notch fatigue factor K_f is the ratio of the un-notched and of the notch high cycle fatigue strengths, see [16]. Typically, the surface finish considered depends on conventional fabrication processes (i.e. machining) and K_f quantifies only the geometrical notch effect on fatigue.

Here according to this approach, the un-notched and notched fatigue strength data to be examined are for as-built surfaces. Therefore, considering the reference un-notched fatigue strength of 360 MPa, [11], three distinct K_f values (i.e. one per specimen direction) are estimated from the present experiments on DMLS Ti6Al4V, namely $K_{f,A}$ = 2.6, $K_{f,B}$ = 1.4 and $K_{f,C}$ = 1.6 where the second subscript defines the notch direction with respect to build direction.

Since the geometrical stress concentration factor of the notched mini specimens is K_t = 1.63, the following more general result is reached for DMLS Ti6Al4V:

$$K_{f,A} > K_t; \qquad K_{f,B} < K_t \qquad K_{f,C} \approx K_t. \qquad (1)$$

The as-built notch effect in fatigue is apparently directional due to a technology-dependent effect superposed to the classical geometrical notch effect.

These findings are possibly the first of this kind and therefore they are not readily validated. Nonetheless, recent valuable fatigue data of [10] for with as-built un-notched and notched specimens made of DMLS Ti6Al4V defined a notch fatigue factor $K_{f.}$ = 2.4 for a sharper notch geometry (i.e. K_t = 2.5) and a different loading condition (i.e. cyclic tension, R = 0.1). Since the specimen axis of [10] was parallel to the build direction, as here is the case for Type C specimens, the result of [10] support the previous result that $K_{f,C} \approx K_t$. Further investigations on different powder bed fused metals are underway.

3.2 As-built notch surface quality

This section presents the characterization of the notch surfaces of the three specimen types of Fig. 1 using magnified optical microscopy and roughness measurements.

Fig. 4 to 6 show the magnified view of the round notches of the three specimen types along with the build direction. The images confirm the strong dependence of the surface quality in terms of roughness and adherence to the theoretical semi-circular geometry on the fabrication parameters.

Especially important for discussing the notch fatigue test results, it is the roughness of the notch root where the stress concentration is localized and crack initiation occurs. Inspections of Fig. 4 to 6 show qualitatively that the surface roughness of the notch is the lowest for the Type B specimen and the highest for Type A specimen. Type C specimen shows a roughness that depends on the local surface orientation with respect to the build direction.

Fig. 4. As-built surface roughness of Type A round notch.

Fig. 5. As-built surface roughness of Type B round notch (note that build direction is perpendicular to page).

Fig. 6. As-built surface roughness of Type C round notch.

Roughness measurements of parameters R_a and R_z were performed at the root of each type of notch and are presented in Table 1. Since the surfaces were curved, the measurement length was shorter than the length recommended by the standard.

Table 1. Notch root roughness vs fatigue specimen type.

Parameter	Type A	Type B	Type C
R_a [µm]	20.506 ± 3.6	8.531 ± 0.8	10.105 ± 0.6
R_z [µm]	107.599 ± 21.596	44.835 ± 3.3	54.599 ± 5.9

The average roughness measures R_a and R_z are however coherent with the visual inspection of Fig. 4 to 6. Furthermore, the data of Tab. 1 demonstrate that the surface roughness of the notch root of the three types of specimens having the same nominal semicircular geometry inversely correlates with the experimental fatigue ranking of Fig. 3. Type B has the lowest roughness and the highest notch fatigue strength while Type A has the highest roughness and the lowest notch fatigue strength. Type C has an intermediate roughness and a intermediate notch fatigue strength.

3.3 Fatigue crack initiation at notch root

The previous views of the notches gave an overall view of the surface quality in dependence with surface orientation. Now etched details of the notch roots for the different specimens of Fig. 7 to 9 show the interaction of local roughness and material structure and the fatigue crack mechanisms.

Fig. 7 shows the high roughness of the surface of a Type A specimen with evidence of partly melted particles and dross formation due to the down-skin orientation of this type notch. Sharp re-entrant notches are typically observed favoring multiple crack initiation observed. The influence of down-skin notch quality vs up-skin notch quality on fatigue behavior of DMLS Ti6Al4V has been discussed recently in [8]. Down-skin notches showed the worst fatigue performance.

Fig. 8 shows the notch root contour of Type B specimens which is much more regular than the previous Fig. 7. The shape is locally a circular arc and the stress concentration is expected to be well described by the K_t as in the case of a machined surface. The etched structure reveals the contour path and the equiaxed structure of the hatched internal area.

Finally, Fig. 9 shows the notch root contour of Type C specimens with a fatigue crack at the center. Globally, the roughness is intermediate compared to the previous Figs. 7 and 8. Interestingly, the upper notch arc shows an increasing roughness compared to the lower part. This is due to the influence of surface orientation with respect to the build direction.

Fig. 7. Magnified Type A notch root and fatigue cracking

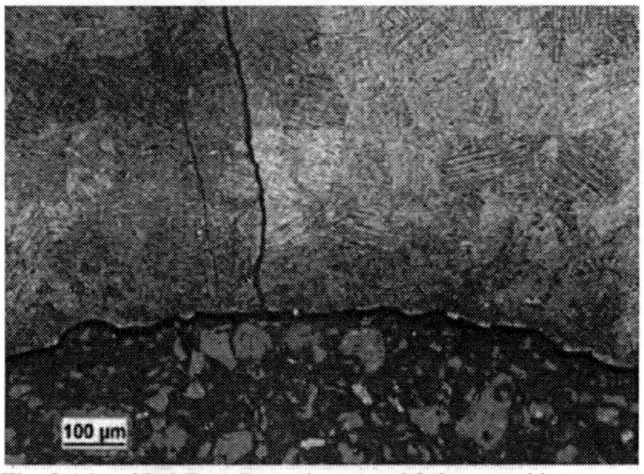

Fig. 8. Magnified Type B notch root and fatigue cracking

Fig. 9. Magnified Type C notch root and fatigue cracking

3.4 Directional layer-wise notch generation

The layer-by-layer notch generation is now schematically introduced in Fig. 10 to 12 to provide a link between the technology-dependent contribution to notch roughness and the directional fatigue strength of the three specimen types.

The notch root of Type A specimens, shown in Fig. 10, is affected very significantly by the step-wise approximation of circular geometry. The layer thickness is also expected to give an increasing negative contribution in fatigue. On the other hand, the notch root geometry of Type B specimens is not affected by the layer-wise fabrication, Fig. 11. The contour and hatch strategy typically used defines accurately the circular notch geometry by contouring. Finally, the notch root of Type C specimens is globally affected by the step-wise approximation of circular geometry but the notch root is affected the least, Fig. 12. In this case, however, the stress concentration is however acting perpendicular to the layers, which is normally considered the weakest direction.

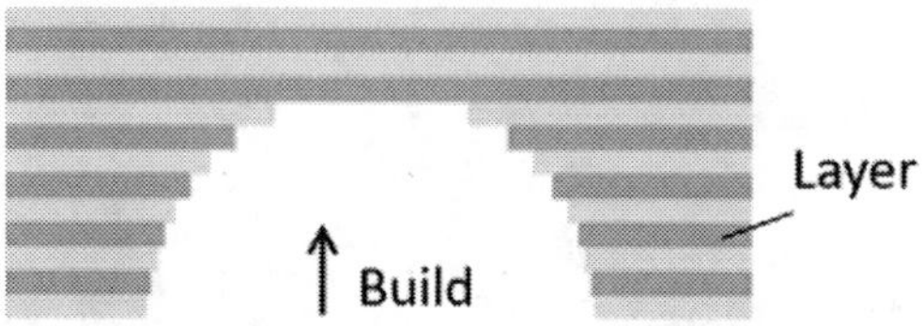

Fig. 10. Layer-wise generation of the Type A round notch.

Fig. 11. Contour and hatch strategy used for obtaining the Type B round notch (note that build direction is perpendicular to page).

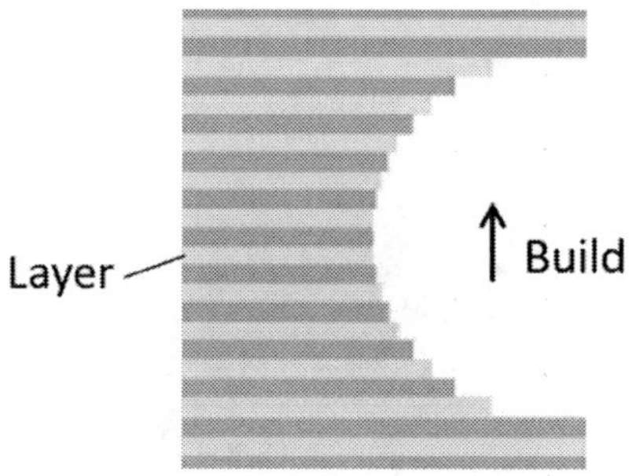

Fig. 12. Layer-wise generation of Type C round notch.

The schemes of Fig. 10 to 12 provide a link between curved surface fabrication direction and fabrication parameters and therefore the technological-dependent negative contribution to the geometrical notch effect quantified in Fig. 3.

4 Conclusions

This study was focused on the experimental investigation of fatigue notch sensitivity of as-built DMLS Ti6Al4V the presence of different orientations with respect to the build direction using a test methodology recently introduced, [14]. The main conclusions are the following:

- The test methodology using miniature fatigue specimens efficiently provided original insight into the dependence of the fatigue behavior on the directionality of the notch fabrication.
- Of the three directional specimens investigated and shown in Fig. 1, Type A showed the worst notch fatigue performance, Type B the best and Type C an intermediate notch fatigue performance.
- Optical inspection of the notch surface quality and roughness measurements demonstrated a ranking of the directional notches that inversely correlated to the measured fatigue performance.
- Three directional notch fatigue factors $K_{f,A}$, $K_{f,B}$, $K_{f,C}$ were originally estimated from the present data for the same K_t demonstrating a technological contribution in addition to the geometrical effect.
- A recent notch fatigue factor from the literature supported the present estimate for the same material and notch orientation.
- The link among the curved surface orientation, the peculiar layer-wise additive manufacturing fabrication and the fatigue notch sensitivity is schematically defined.

Part of the research was supported by the project Slovak VEGA grant No. 1/0685/2015. Specimen fabrication by the technological partner BEAM-IT, Fornovo Taro (Italy) is gratefully acknowledged.

References

1. W.E. Frazier, J. Mater. Eng. Perform., **23** (2014).
2. J. Kranz, D. Herzog, C. Emmelmann, J Laser Appl **27** (2015).
3. K.S. Chan, Surf Topogr Metrol Prop **3** (2015).
4. S. Leuders, M. Thöne, A. Riemer, T. Niendorf, T. Tröster, H. Richard, Int J Fatigue **48** (2013).
5. E. Wycisk, A. Solbach, S. Siddique, D. Herzog, F. Walther, C. Emmelmann, Phys. Proc, **56**, (2014).
6. P. Edwards M. Ramulu, Materials Science and Engineering: A, **598**, (2014)
7. T. M. Mower, M. J. Long, Materials Science and Engineering, **A651** (2016)
8. Greitemeier D, Palm F, Syassen F, Melz T. Int J Fatigue **94** (2017)
9. P. Li, D.H. Warner, A. Fatemi, N. Phan, Int J Fatigue, **85** (2016)
10. M. Kahlin, H. Ansell, J.J. Moverare, Int J Fatigue **101** (2017)
11. G. Nicoletto, Int J Fatigue, **106** (2018)
12. R. Konečná,, G. Nicoletto, A. Bača, L. Kunz, Materials Science Forum, **891** (2017).
13. R. Konečná, G. Nicoletto, S. Fintová, M. Frkáň, Structural Integrity Procedia **7** (2017)
14. G. Nicoletto, Int J Fatigue, **94** (2016)
15. M. Frkáň, G. Nicoletto, R. Konečná, Procs. XIX ICMFM, Porto, Portugal (2018) (in preparation)
16. R. Juvinall, K.M. Marshek, *Fundamentals of Machine Component Design*, New York: John Wiley, (2012)

https://doi.org/10.1051/matecconf/201816502003

Application of the small punch test to determine the fatigue properties of additive manufactured aerospace alloys

Robert Lancaster[1,*], *Henry* Illsley[1], *Spencer* Jeffs[1], *Roger* Hurst[1] and *Gavin* Baxter[2]

[1]Swansea University, Institute of Structural Materials, Bay Campus, Fabian Way, Swansea, SA1 8EN, United Kingdom

[2]Rolls-Royce plc, PO Box 31, Derby, DE24 8BJ, United Kingdom

Abstract. Additive layer manufacturing (ALM) processes are becoming increasingly prevalent in the aerospace industry as design engineers look to profit from the numerous advantages that these advanced techniques can offer. However, given the safety critical nature and arduous operating conditions to which these components will be exposed to whilst in service, it is essential that the mechanical properties of such structures are fully understood. Transient microstructures are a typical characteristic of ALM components and resulting from the thermal cycles that occur during the build operation. Those microstructures make any mechanical assessment an involved procedure when assessing the process variables for any given parameter set. A useful mechanical test technique is small-scale testing, in particular, the small punch (SP) test. SP testing is capable of localised sampling of a larger scale component and presents an attractive option to mechanically assess complex parts with representative geometries, that would not be possible using more conventional uniaxial test approaches. This paper will present the recent development of a small-scale testing methodology capable of inducing fatigue damage and a series of novel tests performed on different variants of Ti-6Al-4V.

1 Introduction

Additive layer manufacturing (ALM) is a rapidly growing technology receiving widespread attention from numerous industrial sectors. The process has evolved significantly over recent years and is capable of fabricating near net shape, fully dense structural parts with a high potential for geometrical optimisation. The emergence of ALM is linked to the significant benefits that the process can offer compared to more conventional manufacturing processes, such as forging or casting. These include considerable cost savings due to less material wastage, short lead times and improved buy-to-fly ratios in the gas turbine industry, whilst also being capable of forming highly intricate components that would not be possible with more traditional methods.

One such ALM technique is electron beam melting (EBM), a powder bed process where an electron beam melts metal powder on a retractable base plate to build a three-dimensional component layer by layer, typically within a near vacuum environment. Due to the inert atmosphere, materials which have a high affinity towards oxygen, such as titanium alloys, can be readily manufactured using this process. Previous research details the various aspects of EBM Ti alloys, particularly Ti-6Al-4V, including the presence of porosity, build artefacts, optimal parameter sets and powder characteristics [1], [2]. However, understanding the transient microstructures which are typically produced in an EBM build material, that can in turn lead to varying the mechanical properties across the component, is a fundamental research requirement. Furthermore, considering the multiple interactions of intra and inter build process variables on the integrity and consistency of the final structure, traditional laboratory scale test approaches are deemed unsuitable for mechanical characterisation as it is difficult to extract representative test specimens which comply to relevant standards. As such, small scale testing can offer a pragmatic alternative for initial assessment and verification.

Previous research has revealed that small scale mechanical testing approaches, in particular the small punch (SP) test, are now widely regarded as an established means of obtaining useful mechanical property information from limited material quantities. To date these methods have largely been adopted to characterise the creep [3], [4], tensile [5], [6] and fracture characteristics [7], [8] of numerous material systems [9], [10] from a range of industrial applications for sampling [11], [12]. This paper will now detail the research undertaken in realising a small-scale test methodology capable of characterising the localised fatigue properties of ALM built structures. A series of novel experiments are presented on Ti-6Al-4V materials fabricated through two different manufacturing processes.

2 Experimental methods

2.1. Materials

For this research, the study was carried out on two Ti-6Al-4V variants, manufactured using different processing routes: (i) a forged disk and (ii) Electron Beam Melting

* Corresponding author: r.j.lancaster@swansea.ac.uk

(EBM). The EBM material was built in a vertical orientation on an Arcam System configured with EBM Control 3.2 Service Pack software. The pre-alloyed plasma atomised Ti-6Al-4V powder feedstock contained spherical particles ranging from 45-100μm in diameter. The first layer of powder was deposited onto a 220 x 300 x 10 mm^3 thick stainless steel base plate that was pre-heated with the electron beam to approximately 730°C. Prior to melting each layer, the new layer of powder was preheated and sintered by scanning the beam over the cross-sectional geometry in an attempt to maintain a consistent build temperature of 730°C throughout and reduce the effects of charging of the powder during melting. Further information on this process can be found in [13], [14].

Figures 1 and 2 illustrate the microstructures for the two materials. The EBM material is shown in both the face parallel to the build direction and the face perpendicular to the build direction, highlighting the anisotropy in the material. When looking into the build direction, or the SP specimen surface, (Figure 1a, b) the microstructure displays an ordered lamellar structure consisting of an extremely fine acicular grain morphology. This results from the thermal characteristics of the EBM process where a small melt pool and rapid cooling results in a columnar prior-β grain structure with epitaxial β <100> crystal growth parallel to the build direction (Figure 1c, d). Here, the elongated grain widths were measured perpendicular to the length of the grain using the mean linear intercept (MLI) method. The forged variant (Figure 2) has an equiaxed bimodal structure, with an α volume fraction of approximately 60%.

Additional microstructural measurements characterising the typical grain sizes of the different Ti-6Al-4V variants are given in Table 1. In each material, the measurements are taken perpendicular to the textured grain's largest axis and represent an average from over 600 calculations from multiple locations within the sample.

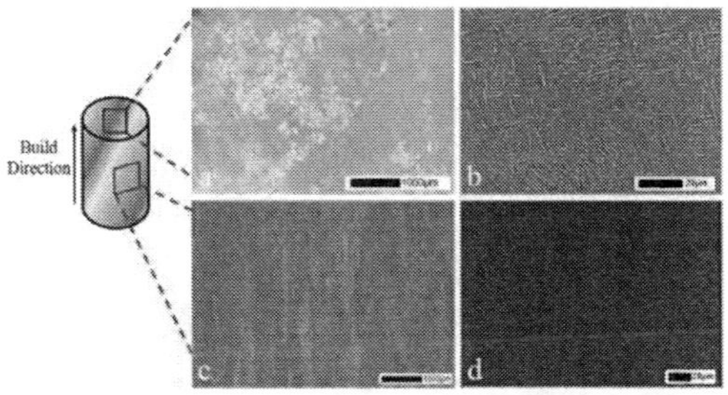

Figure 1. Microstructure of EBM Ti-6Al-4V (a) perpendicular to the build direction at low magnification; (b) perpendicular to the build direction at high magnification; (c) parallel to the build direction at low magnification and (d) parallel to the build direction at high magnification. The force is applied through the face perpendicular to the build direction.

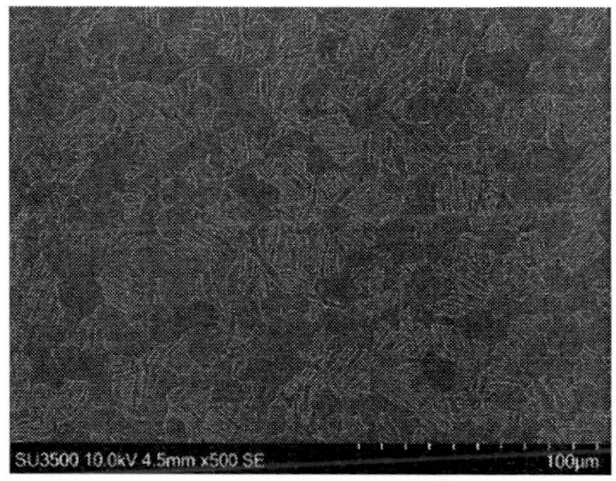

Figure 2. Microstructure of forged Ti-6Al-4V, illustrating the top surface of the SP specimen where the punch makes contact.

Table 1. Microstructural measurements of the two Ti-6Al-4V variants

Measurement	Forged	EBM
Mean Linear Intercept (1/μm)	0.0778	0.0041
Prior-β grain width (μm)	13	246
Transformed alpha width (μm)	1.6868	0.7867

2.2 Small punch specimens

SP specimens were extracted from the threaded stub ends of conventional tensile specimens for the two materials and were then turned down to a diameter of 9.5mm. The resulting cylinders were then sectioned into slices approximately 800μm in thickness before being ground down to the required thickness of 500μm ±5μm using successively finer grades of silicon carbide abrasive paper, the final grade being 1200 grit in accordance with the recommendations of the European Code of Practice for Small Punch testing (EUCoP) [15]. The thickness was then measured at 5 locations around the periphery and the centre of the specimen to ensure uniformity.

2.3 Experimental set up

A bespoke SP fatigue test capability has been developed at Swansea University to test miniature disc specimens at both ambient and elevated temperatures under relatively low loading conditions. The apparatus has been designed to be inserted into the load train of a universal, servo-actuated test machine, with a load capacity of ±5kN to ensure intimate control of the applied load. The test configuration consists of two Nimonic-90 punch indenters, one of which is positioned above the SP disc and the other located directly beneath. The two indenters maintain contact with the specimen throughout the test through a small residual force, which is introduced to eradicate any potential air gaps in the load train which may occur if the centre of the specimen begins to thin. The disc then vertically oscillates in movement to apply a fully reversed force to the material, replicating a fully reversed fatigue loading ratio of R=-1. To monitor the evolving deformation, a linear variable displacement transducer (LVDT) measures the displacement of the disc. These values are recorded in addition to the position and force measurements from the test machine.

In the test set up, as presented in Figure 3, the miniature disc sample is clamped within an upper and lower die which is positioned within an upper housing along with the upper indenter which is located above the specimen. The lower indenter is placed within the lower pull rod and is slowly raised until it makes contact with the bottom surface of the disc, from which it is then held in a stationary position under zero force.

For both the upper and lower indenters, the applied force is transferred to the specimen via a 2 mm diameter hemispherical ended Nimonic-90 punch, perpendicular to the top and bottom surfaces of the disc, through a receiving hole of 4 mm diameter with a 0.2 mm radius chamfer. These dimensions are in direct compliance to the directives detailed in the European CoP [15]. A schematic representation of this arrangement is displayed in Figure 3.

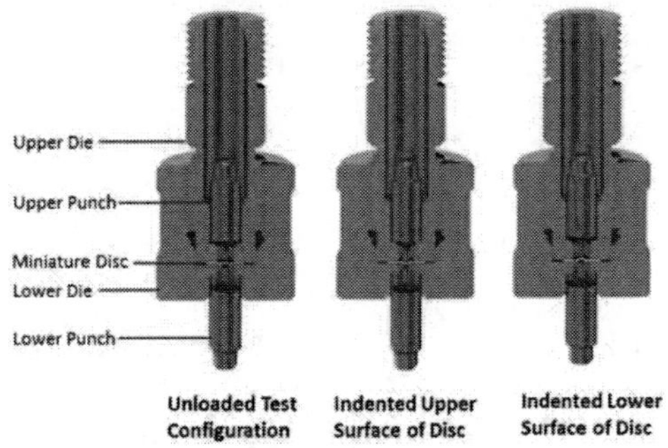

Figure 3. Representation of the small punch fatigue test configuration.

3 Results & Discussion

3.1 Small punch fatigue results

A series of SP fatigue tests were performed under various applied load ranges to generate representative force (F) – number of fatigue cycles to failure (N) curves, or an F-N curve (Figure 4) for each of the materials. For each of the experiments, an R=-1 waveform was adopted and was applied under a loading frequency of 1Hz using a sinusoidal waveform. Each test was interrupted once a displacement range of 250μm was achieved, from which a clearly defined crack was identifiable. At this point, the disc was considered fractured and the test was terminated.

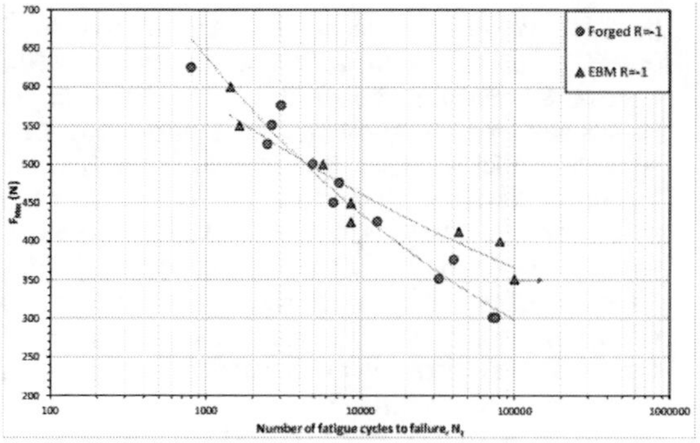

Figure 4. Small punch fatigue results (20°C, R=-1, 1Hz sinusoidal waveform).

The behaviour of the two materials (as given in Figure 4) is highly comparable to the response that would typically be expected from a uniaxial stress (S) – number of fatigue cycles to failure (N) curve, in that as the force range applied to the specimen is increased, the number of cycles required to produce failure decreases. Even though the two test methods appear to produce a similar S/F-N response, the deformation behaviour is quite different. In a SP test, whether the load is applied to replicate tensile, creep or fatigue damage, the miniature disc sample is subjected to an initial period of compressive loading, followed by a bending motion and finally a stage of tensile elongation where the membrane deforms and thins, all of which occur under a transient biaxial stress condition. This is in contrast to a uniaxial arrangement where the specimen simply elongates or deforms under an applied tensile load, whether the load is static (as is the case for tensile or creep) or dynamic (fatigue). Therefore, biaxiality needs to be considered when attempting to correlate SP fatigue results with uniaxial data, particularly for anisotropic or heavily textured materials such as ALM parts.

Figure 5 shows the evolution of displacement in each of the SP fatigue tests on EBM Ti-6Al-4V and how the top and bottom surface displacement values behave during the duration of each.

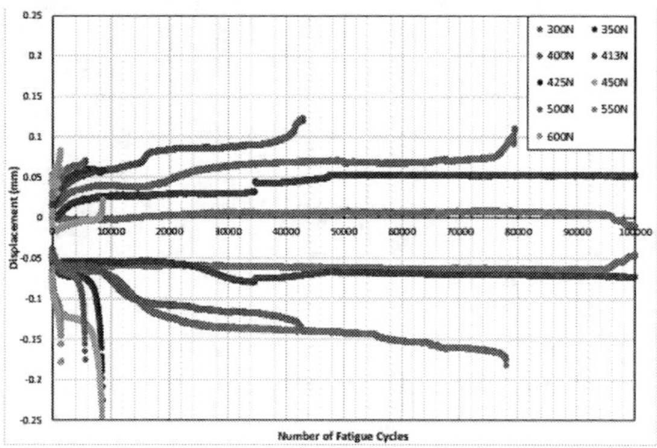

Figure 5. Evolution of upper and lower surface displacement behaviour for the small punch fatigue tests on EBM Ti-6Al-4V.

As can be seen, it appears that for the higher load tests (which produced the lowest lives), preferential deformation is more prevalent on the bottom surface of the disc, from the lower punch indenter penetrating upwards into the bottom surface of the material. This is a direct result of the experimental procedure itself, in which the load is initially exerted onto the lower punch and therefore, the bottom surface of the disc tends to deform at a slightly faster rate compared to the top surface. This pattern however, is not applicable for lower load tests since the magnitude of the applied load is not sufficiently high enough to introduce any significant plastic behaviour upon the initial loading cycles, similar to the yield strength when considering a uniaxial arrangement.

From these results, hysteresis force-displacement loops were generated for each test. Figure 6 presents the force-displacement loop responses for a SPF test performed on EBM Ti-6Al-4V under a maximum load of 400N. The figure shows the repeatability of the applied loading waveform through the entirety of the test where the maximum and minimum force values remain consistent throughout. Furthermore, the degree of plastic deformation, or displacement, in the disc can be seen to increase with an increasing number of fatigue cycles, replicating the expected behaviour of a component when subjected to a more traditional fatigue damage mechanism.

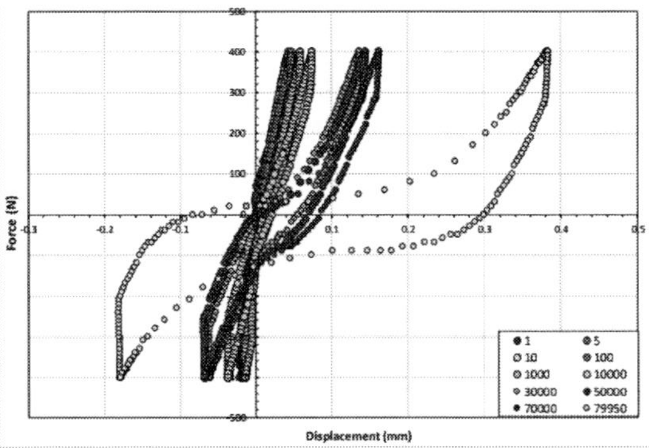

Figure 6. Hysteresis force-displacement loops for a SPF test performed on EBM Ti6Al-4V at a maximum force of 400N (20°C, R=-1, 1Hz sinusoidal waveform).

3.2 Fractography

Supporting fractography was gathered for each of the tested specimens to develop an understanding of the contrasting manners of failure in the two materials. Figure 7 presents a direct comparison for a SPF test performed on each material under an applied F_{Max} of 500N. The fracture behaviour in either case is quite similar, with a dominant 'star' type crack present, with the cracks propagating radially from a central location as the materials accumulate an increasing number of fatigue cycles, until the stretched membrane is unable to support the applied load and fractures.

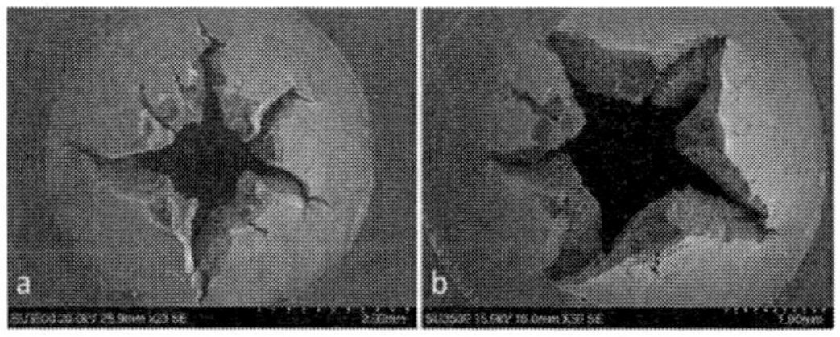

Figure 7. Low Magnification SEM Fracture surface images of (a) EBM and (b) Forged Ti-6Al-4V SPF tests performed at F_{Max}=500N.

Additional fractography is provided in Figure 8, highlighting some of the prominent features on each of the fracture surfaces at a higher magnification.

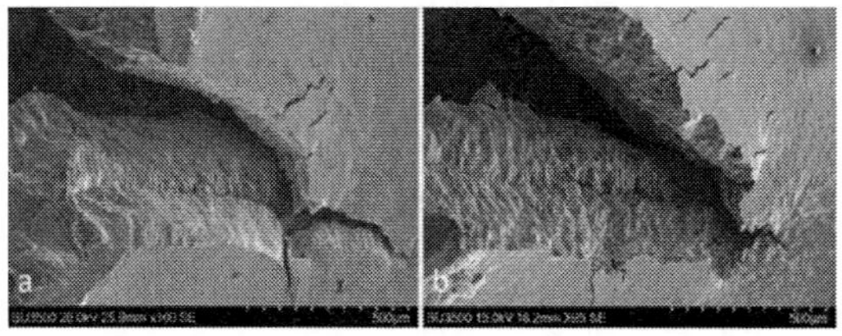

Figure 8. High Magnification SEM Fracture surface images of (a) EBM and (b) Forged Ti-6Al-4V SPF tests performed at F_{Max}=500N.

In both materials, the fracture characteristics are very similar, with secondary cracking clearly observed along the length of the fractured segments, the extent of which is highly comparable between the two.

When considering fatigue crack growth, the fracture surface of the material is usually associated with the presence of striations, a feature that provides adequate evidence that the component was subjected to a form of cyclic loading. Striations typically represent a single increment of crack growth that occur within a loading cycle through the plastic blunting process. However, as is commonly found in uniaxial fatigue tests where a compressive loading ratio is employed, striations were not seen due to crack closure events as the disc cycles through zero load in each fatigue cycle. As such, any striations that may have been present in the radial cracks have been simply worn away by the crack opening-closing process.

3.3 Crack path morphologies

Following fractographic analysis, the failed SP fatigue specimens were mounted, ground and polished back to analyse in further detail the mechanisms of crack growth through the microstructure in each material. Figure 9 presents images taken using backscattered electrons of the path of the crack through the two microstructures respectively. The crack behaviour through the EBM microstructure (Figure 9(a)) is complex with the crack travelling in a transgranular manner through the columnar prior β grains but the effect of the α laths is less clear. Overall the crack may be deemed to propagate in an inter-lath manner.

In the forged specimen (Figure 9(b)), the radial crack propagates in a transgranular fashion through the grains of the bimodal microstructure for the majority of the crack growth with some intergranular crack growth also present. This is in accordance with the behaviour seen in Figure 10, which shows the crack path through the microstructure of a forged Ti-6Al-4V uniaxial fatigue specimen tested with an R ratio of -1. The path of this crack is also largely mixed mode with aspects of both intergranular and transgranular crack propagation mechanisms present.

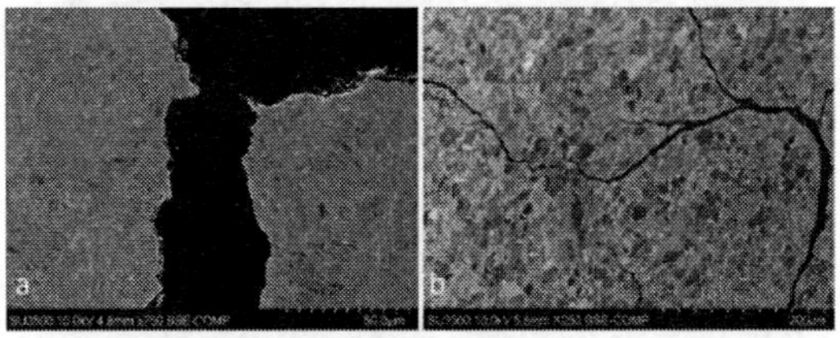

Figure 9. Crack path morphologies for (a) EBM and (b) Forged Ti-6Al-4V SPF test specimens.

Figure 10. Crack path morphology for a Forged Ti-6Al-4V uniaxial fatigue test.

4 Conclusions

This paper details the ongoing research at Swansea University in developing a bespoke SPF testing capability. A series of test results have been presented from novel SPF experiments on two Ti-6Al-4V variants, from which comparative force-number of fatigue cycles to failure curves were determined. The two materials showed a very similar response and the post-test fractography supports this trend. Additional information is generated in the form of representative force-displacement fatigue loops which reflect the deformation behaviour of the miniature disc specimen subjected to SP fatigue type loading. An investigation of the crack path morphologies from a SPF test compared to a uniaxial experiment have produced a similar mechanism of failure, signifying that the proposed test method shows real promise for material property evaluation of miniature disc specimens.

5 Future work

Further research is in progress with the authors now assessing the effects of a number of fatigue testing parameters including: alternative loading ratios, such as a zero to maximum cycle to investigate whether characteristic fatigue markings are present; loading frequencies to replicate low cycle fatigue conditions; different loading waveforms such as a trapezoid and how this compares to sinusoidal behaviour and also the impact of performing such experiments at elevated temperature and whether the material behaves in a similar manner as would be expected in a uniaxial regime.

Given that the results presented in this paper have been generated under load control conditions, the authors will also investigate the damage evolution of a small punch disc under a displacement controlled cycle, in an attempt to replicate strain-control type fatigue damage.

This additional testing will complement ongoing empirical and numerical modelling efforts to correlate SP fatigue results with uniaxial results and endeavour to understand the stress evolution of a SP disc that is subjected to such a complex form of biaxial loading. A series of interrupted tests will also be performed to aid the understanding of the onset of cracking when subjected to a fatigue type loading condition.

The current research was funded under the EPSRC Rolls-Royce Strategic Partnership in Structural Metallic Systems for Gas Turbines (grants EP/H500383/1 and EP/H022309/1). The provision of materials and supporting information from Rolls-Royce plc is gratefully acknowledged by the authors. Mechanical tests were performed at Swansea Materials Research and Testing Ltd. (SMaRT).

References

[1] S. Tammas-Williams, H. Zhao, F. Léonard, F. Derguti, I. Todd, and P. B. Prangnell, Mater. Charact., **102**, 47–61 (2015)

[2] L. E. Murr, E. V Esquivel, S. A. Quinones, S. M. Gaytan, M. I. Lopez, E. Y. Martinez, F. Medina, D. H. Hernandez, E. Martinez, J. L. Martinez, S. W. Stafford, D. K. Brown, T. Hoppe, W. Meyers, U. Lindhe, and R. B. Wicker, Mater. Charact., **60**, 2, 96–105 (2009)

[3] M. Bruchhausen, K. Turba, F. De Haan, P. Hähner, T. Austin, and Y. De Carlan, J. Nucl. Mater., **444**, 1–3, 283–291 (2014)

[4] S. P. Jeffs, R. J. Lancaster, and T. E. Garcia, *Mater. Sci. Eng. A*, **636**, 529–535 (2015)

[5] S. Davies, S. Jeffs, R. Lancaster, and G. Baxter, "High Temperature Deformation Mechanisms in a DLD Nickel Superalloy," *Materials (Basel).*, **10**, 5, 457 (2017)

[6] T. E. García, C. Rodríguez, F. J. Belzunce, and C. Suárez, J. Alloys Compd., **582**, 708–717 (2014)

[7] K. Turba, R. Hurst, and P. Hähner, Int. J. Press. Vessel. Pip., **111-112**, 155–161 (2013)

[8] M. Bruchhausen, S. Holmström, I. Simonovski, T. Austin, J. M. Lapetite, S. Ripplinger, and F. de Haan, Theor. Appl. Fract. Mech., **86**, 2–10 (2016)

[9] D. Andrés, R. Lacalle, and J. A. Álvarez, Mater. Des., **96**, 122–130 (2016)

[10] R. J. Lancaster, W. J. Harrison, and G. Norton, Mater. Sci. Eng. A, **626**, 263–274 (2015)

[11] R. C. Hurst, R. J. Lancaster, S. P. Jeffs, and M. R. Bache, Theor. Appl. Fract. Mech., **86**, 69–77 (2016)

[12] T. Izaki, T. Kobayashi, J. Kusumoto, and A. Kanaya, Int. J. Press. Vessel. Pip., **86**, 9, 637–642 (2009)

[13] X. Gong, T. Anderson, and K. Chou, Manuf. Rev., **1**, 2, 1–12 (2014)

[14] N. Hrabe and T. Quinn, Mater. Sci. Eng. A, **573**, 271–277 (2013)

[15] CEN Workshop Agreement CWA 15267, CEN Members National Standards Body, 1–69 (2007)

Influence of as-built surface and heat treatment on the fatigue resistance of Additively Layer Manufacturing (ALM) AlSi10Mg alloy.

Julius Noel Domfang Ngnekou[1,2*], Julien Nicolai[1], Yves Nadot[1] , Gilbert Henaff [1], Lionel Ridosz[2],

[1]Institut Pprime UPR CNRS 3346, Department of Physique and Mechanics of Materials, ENSMA-Université de Poitiers, 1 avenue Clément Ader, Téléport 2, 86960 Chasseneuil-Futuroscope, France
[2]Zodiac Aerospace, 61 rue Pierre Curie, 78370 Plaisir, France

Abstract: This work concerns the fatigue resistance of a AlSi10Mg material produced by additive manufacturing, and more precisely the competition between as built manufacturing surface and as-machined surface on the fatigue resistance. Samples were built by a powder-bed process with an EOS-M280 machine using standard in two configurations (0° and 90°) in order to evaluate the impact of building direction on fatigue life. The impact of as-built surface on fatigue behavior is quantified for each specimen configuration. A T6 heat treatment is performed on samples in order to evaluate the impact of microstructure on fatigue behavior. For each configurations, the S-N curves is determined in as-built and T6 materials with a load ratio R= -1. The fracture surfaces are carefully analyzed in order to determine the critical defect size for each sample. A Kitagawa type diagram representing the fatigue limit as a function of the defect size is derived from these measurements. All the results were compared to those obtained in as-machined samples.

1 Introduction

Additive Layer Manufacturing (ALM) is a breakthrough process compared to conventional processes such as casting, in particular to produce parts with a complex shape. In the aeronautic context, the introduction of such an innovative process implies a qualification process in which the understanding of the fatigue resistance is crucial [1–3]. Thus, the characterization of the microstructure inherited from the process is required in order to access the impact of a given microstructure on the fatigue behaviour. According to Domfang et al [4], the microstructure in the bulk of the non-heat treated ALM AlSi10Mg material can be described using five parameters: (i) defect population, (ii) melt-pools, (iii) dendritic structure, (iv) precipitation structure and (v) grains. It has been observed that melt-pools, dendritic structure and grains are strongly anisotropic in shape depending on the building direction. In addition the melt-pools and dendrites disappear after a T6 heat treatment [4]. On the other hand, the surfaces of near net-shape pieces are characterized by a high roughness that could impact the fatigue behavior. Since most of the ALM parts are partially machined, the aim of this study is to compare the fatigue behavior of as-machined samples to samples with a near net shape cross section also called as-built samples.

2 Samples fabrication

Specimens were fabricated by ALM method using a pre-alloyed powder with an average particle size of 30 µm. All specimens were constructed using a EOS INT-M290 machine with standard parameters. The layering is performed by means of a scraper. Two families of specimen are considered in this study and denominated P2 and P3. P2 specimens were machined from cylinders bars while P3 were manufactured with a near net-shape cross section. All specimens were constructed on a platform placed at 200°C. The materials thus manufactured were all subjected to a soft post-blast treatment to clean the surface. This cleaning was carried out using micro glass oxide marbles with a size of the order of 100 to 200 microns. These marbles were projected from about 10 cm to the specimen with a jet pressure of 4 bars. Before cutting specimen from the platform, a recommended stress relaxation treatment was performed during 2 hours at 210°C. P2 and P3 specimens were built along two directions, namely XY and Z as defined in Figure 1-a. P2 fatigue test specimens were machined (MA) from the bars in order to overcome the manufacturing gross surface integrity effects, namely the surface roughness and the presence of a population of defects in the subsurface due to edge lasing [5,6]. One bar corresponds to one specimen. No polishing was applied on the sample surfaces. On the other hand, for P3 fatigue specimen tests only threads were machined while

* Corresponding author: julius.domfang@ensma.com

the gauge length was left in the as-built condition. In the specific case of P3 XY specimens, the production supports were removed by means of grinding wheel. Some specimens have subsequently been subjected to a T6 heat treatment with conditions listed in Table 2, leading to a denomination of each individual fatigue specimen given in Table 3. This denomination used in this paper depends on the production number, the orientation and the application or not of a T6 treatment.

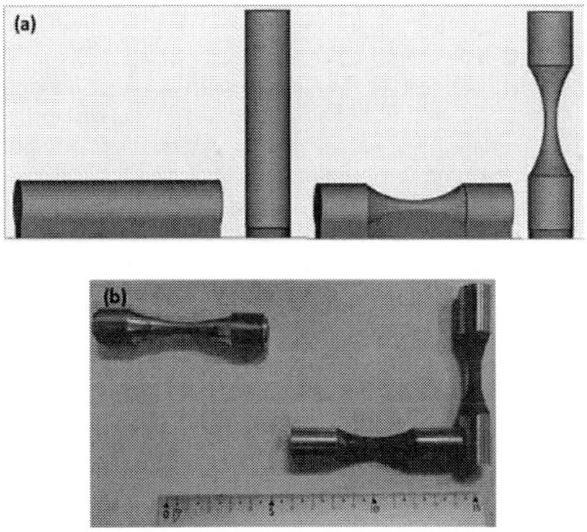

Figure 1:(a) manufactured parts from which samples where removed , (b) left: machined sample, right: near-net shape samples

Table 1: description of the T6 heat-treatment

Steps	Temperature (°C)	Duration (h)	Environment
Solid solutioning	540	8	Air
Quenching	20		Water
Tempering 1	20	24	Air
Tempering 2	160	10	Air

Table 2: sample designation

Production	Type	T6	Machine /As-Built	Designation
P2	XY	No	Machined (MA)	P2-XY-MA
		Yes		P2-XY-MA-T6
	Z	No	Machined (MA)	P2-Z-MA
		Yes		P2-Z-MA-T6
P3	XY	No	As-Built (AB)	P3-XY-AB
		Yes		P3-XY-AB-T6
	Z	No	As-Built (AB)	P3-Z-AB
		Yes		P3-Z-AB-T6

3 Experiments

An Energy Dispersive Spectrometry (EDS) analysis was carried out to quantify the chemical elements present in the alloy. The results reported in Table 3 show that the composition determined is in accordance with the standard composition. However, the non-detection of some elements such as iron means that the EDS technique is not able to quantify this particular component, especially when it is present in a small proportion and homogeneously distributed in the volume.

The fatigue limit were defined at one million cycles. A step by step method was used, as it is the only way to evaluate the fatigue limit in presence of natural defects. It is however assumed that no significant loading history effect is introduced by the loading steps applied prior to failure, as shown by Roy et al [7] in the case of a A356 cast Al-alloy. In order to plot a Kitagawa-type diagram, the fatigue limit is determined for each specimen after failure according to the following method:

- When the specimen fails during the first step, a Basquin equation is used to calculate the fatigue limit.

$$\sigma_{D-1}^{te} = A \times N_f^{\alpha}$$

- When the specimen fails after one or several loading steps, the fatigue limit is determined by a linear interpolation according to the following equation:

$$\sigma_{D-1}^{te} = \frac{N_f}{10^6} \times (\sigma_n - \sigma_{n-1}) + \sigma_{n-1}$$

Table 3: chemical composition of AlSi10Mg

	Al	Si	Mg
DIN EN 1706:2010	Bal.	9-11	0.2-0.45
W% EDX measure	Bal.	9.55	0.4

https://doi.org/10.1051/matecconf/201816502004

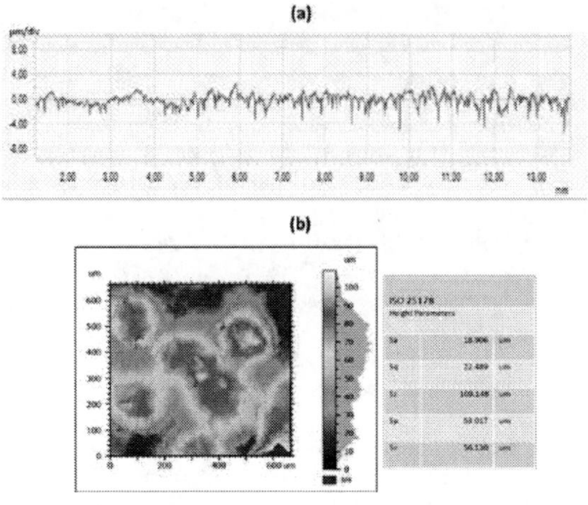

Figure 2: surface roughness characterization (a) machined samples (b) near-net shape sample using interferometric profilometer

4 Results and discussion

In Figure 3 the fatigue S-N curves of P2 and P3 machined or as-built, with or without heat treatment. For machined samples, two directions are study: XY and Z while for as-built samples the oblique samples are furthermore presented.. The black symbols correspond to the samples tested without T6 and open symbols correspond to heat treated samples. One point corresponds to one experimental failure of a given sample.. Even though the microstructure is strongly anisotropic [8–10], no difference in fatigue strength is noticed between samples machined along XY and Z directions. However they is a small impact of building strategy for as-built samples. However, in the case of as-built surfaces, the scatter in data exhibits only a subtle dependence on the building direction. Those differences can be related to the surface roughness. Figure 3 clearly shows that the as-built samples are less resistant to fatigue than the machined ones, even after a peak hardening T6 heat treatment. The fracture surface

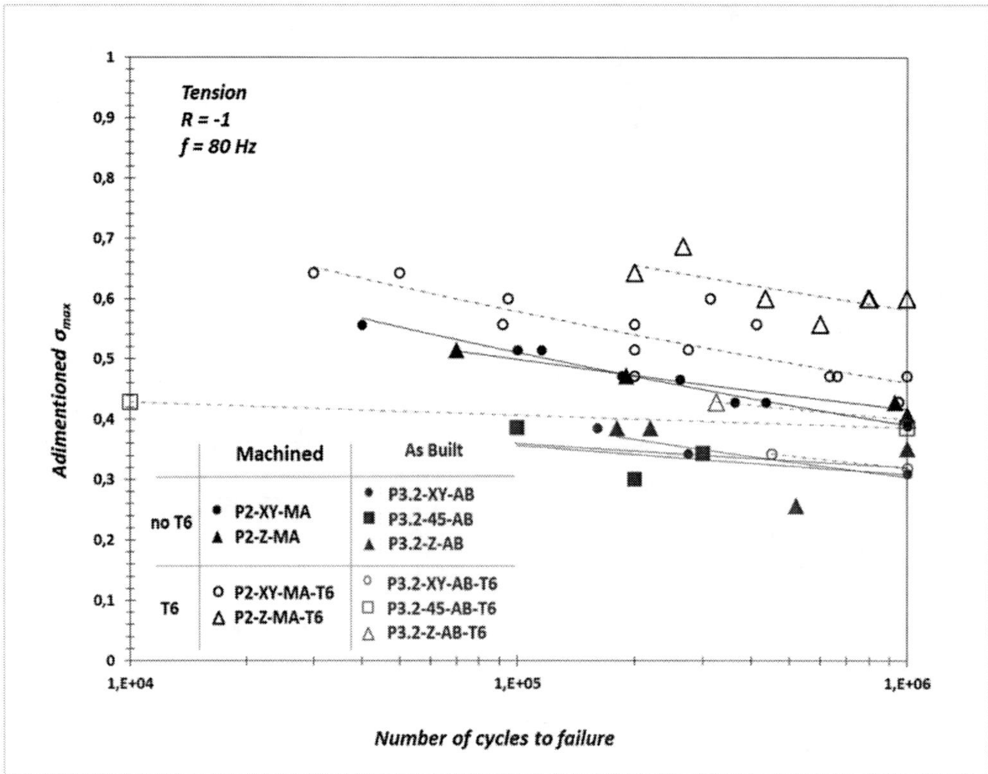

Figure 3: effect of building direction, defect and T6 heat-treatment on fatigue life.

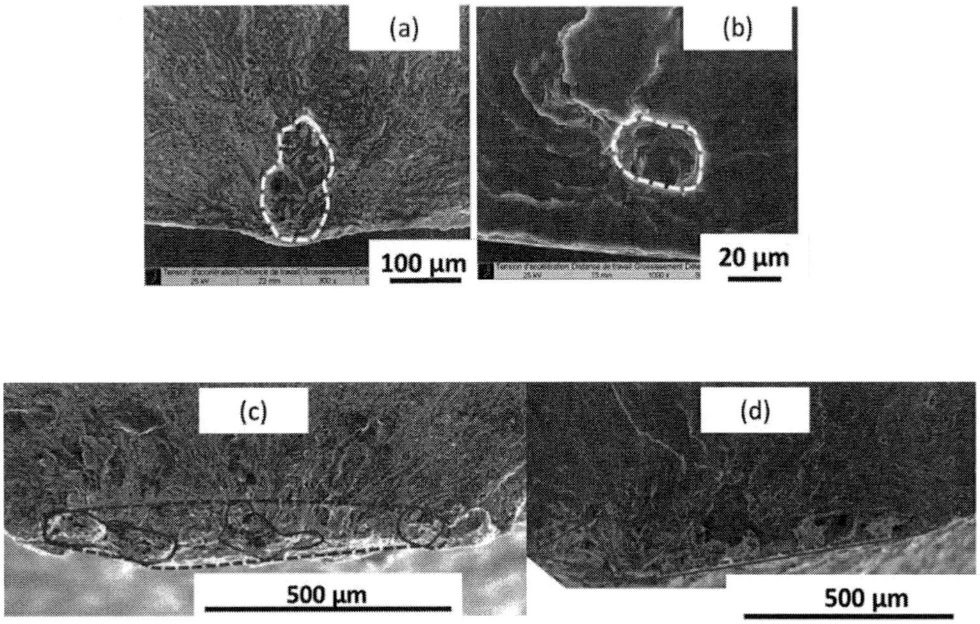

Figure 4: SEI images of initiation sites on fracture surfaces (a,b) P2-MA (c, d) P3-AB

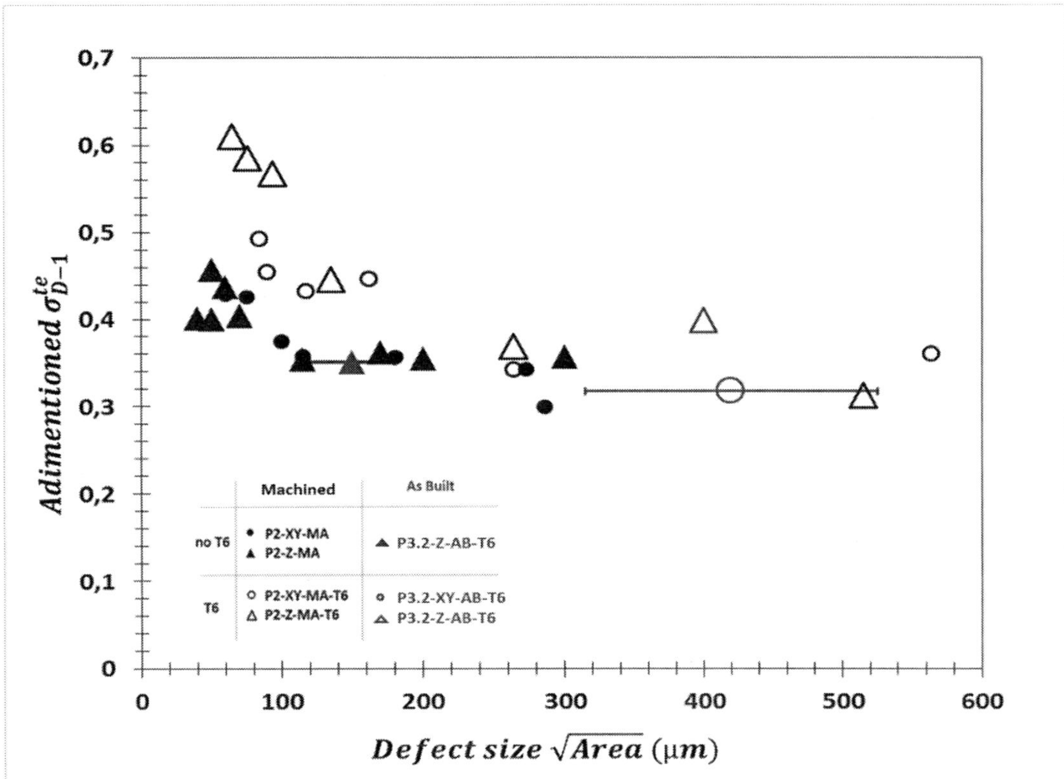

Figure 5: Kitagawa-type diagram showing the impact of heat treatment, specimen building orientation and surface roughness.

examination in Figure 4 furthermore suggests that the P3 production contains larger defects than P2. As it can be observed in Figure 2, that as-built samples (P3) are more rough than machined ones (P2) the fatigue drop is due at least to roughness and defect size. For each productions P2 and P3 specimens containing similar defects, an improvement in fatigue resistance is noticed after T6 heat treatment. Figure 5 exhibit the impact of defect size on the fatigue limit through the Kitagawa type diagram. In order to plot the Kitagawa diagram, the defect size was measure on fractures surfaces using Murakami parameter [11]. In this way, Figure 5 consolidates the SN curves interpretations in terms of impact of building direction and heat treatment. However it clear that the impact of HT seems to dwindle in presence of big defects. Same approach was carry out for surfaces as-built specimen. So, fracture surfaces of as-built failed samples was analysed and the defect size was quantified with and without T6 heat treatment. Fatigue limit was calculated and associated to a corresponding defect size. Results as compared to machined Kitagawa diagram on Figure 5. Even in the presence of a degraded surface and regardless the heat treatment, the results of the as-built specimens coincide with those of the machined specimens. Meaning that the defect size seems to be the first order parameter that controlled the fatigue limit.

5 Conclusion

This study addresses the fatigue behaviour of AlSi10Mg alloy elaborated by selective laser melting. The impact of as-built surface have been quantified and compared to machined surface. On the basis of the Kitagawa diagram established, it has been observed that the defect size is the first order parameter that controls the fatigue limit. For both machined and as-built samples, the T6 heat treatment improves the fatigue limit and reveal a slight building orientation effect.

Acknowledgement

The authors gratefully acknowledges funding provided by Zodiac Aerospace and French National Research and Technology Association (ANRT). This work has been partially supported by « Nouvelle Aquitaine » Region and by European Structural and Investment Funds (ERDF reference: P-2016-BAFE-94/95). Finally the authors would like to thanks Thierry Rouge-Carassat for his comment on a present manuscript.

References

1. S. Romano, A. Brandão, J. Gumpinger, M. Gschweitl, S.B. a, Mater. Des., **131**, 32–48, (2017).

2. L. Portolés, O. Jordá, L. Jordá, A. Uriondo, M. Esperon-Miguez, S. Perinpanayagam, J. Manuf. Syst., **41**, 65–75, (2016).

3. J. Mardaras, P. Emile, A. Santgerma, Procedia Struct. Integr., **7**, 109–115, (2017).

4. J.N.D. Ngnekou, G. Henaff, Y. Nadot, J. Nicolai, L. Ridosz, Procedia Eng., **213**, 79–88, (2018).

5. K. V. Yang, P. Rometsch, T. Jarvis, J. Rao, S. Cao, C. Davies, X. Wu, Mater. Sci. Eng. A, **712**, 166–174, (2018).

6. J. Damon, S. Dietrich, F. Vollert, J. Gibmeier, V. Schulze, Addit. Manuf., **20**, 77–89, (2018).

7. M. Iben Houria, Y. Nadot, R. Fathallah, M. Roy, D.M. Maijer, Int. J. Fatigue, **80**, 90–102, (2015).

8. J.N. Domfang Ngnekou, Y. Nadot, G. Henaff, J. Nicolai, L. Ridosz, Procedia Struct. Integr., **7**, 75–83, (2017).

9. N. Takata, H. Kodaira, K. Sekizawa, A. Suzuki, M. Kobashi, Mater. Sci. Eng. A, **A 704**, 218–228, (2017).

10. W. Li, S. Li, J. Liu, A. Zhang, Y. Zhou, Q. song Wei, C. Yan, Y. Shi, Mater. Sci. Eng., **A 663**, 116–125, (2016).

11. Y. Murakami, M. Endo, Int. J. Fatigue, **16**, 163–182, (1994).

Influence of defects on axial fatigue strength of maraging steel specimens produced by additive manufacturing

Daniele Rigon[1,*], *Giovanni* Meneghetti[1], *Michael* Görtler[2], *Daniele* Cozzi[2], *Wolfgang* Waldhauser [2] and *Manuele* Dabalà[1]

[1] University of Padova, Department of Industrial Engineering, via Venezia 1 – 35131 Padova, Italy
[2] Joanneum Research Forschungsgesellschaft mbH, Institute for Surface Technologies and Photonics, Niklasdorf Leobner Straße, 94 A-8712 Niklasdorf, Austria

Abstract. Nowadays many materials such as steels, aluminium and titanium alloys can be realised by powder bed solutions melting subsequently powder layers by means of a laser or electron beam (Laser Beam Melting – LBM and Electron Beam Melting – EBM). The microstructure realised by layer-by-layer solidification having high cooling rate cannot be considered isotropic. Therefore, the mechanical properties could be influenced by the building direction. Regarding maraging steel, the study of the influence of the building direction and the heat treatment on the static and axial fatigue strength has been investigated in a previous contribution. A large scatter of the fatigue test results was found because of the presence of detrimental surface and subsurface defects. The aim of this contribution is to present additional axial fatigue test results of maraging steel characterized by different build orientation and providing an analysis of the defects observed at the crack initiation area of the fracture surface.

1 Introduction

In PBF technologies, such as Selective Laser Melting SLM (or Direct Metal Laser Sintering DMLS) and Electon Beam Melting (EBM) layers of metal powder are subsequently melted by a laser or an electron beam source in order to obtain the part designed by a CAD software without the typical constraints of traditional manufacturing processes [1–4].

The main advantage of adopting an AM process is the possibility of producing components having customised and complex geometries. Indeed, from structural point of view, topological optimization is the key task in design for additive manufacturing for maximising the performance of the component for a given set of loading.

As reported in literature, the application of design for AM are currently reserved to aerospace, aeronautical and automotive fields because at the moment this new manufacturing process takes advantage in customized components with low production volumes due to the high process costs.

Nowadays, SLM process is relatively well developed in order to produce many alloys having comparable static mechanical properties with respect to wrought ones, as many contribution demonstrate [5–10].

Conversely, the critical point regarding the design of AM mechanical parts is the presence of defects which degrades the fatigue strength of the material [11,12]. Indeed, main causes of lower fatigue performance in AM parts were imputed to the surface finishing, the post heat treatment and the presence of defects currently unavoidable from the manufacturing process[13–15]. Typical defects that can be observed in SLM parts are incomplete fusion holes having irregular 3D shapes also known in literature as Lack Of Fusion (LOF). Other defects are pores with elliptical or spherical shape and crack-like defects. The formation mechanism of these defects is explained in [16].

For a design point of view, an accurate estimation of the fatigue limit, intended as the maximum stress that guarantee a fatigue life of a chosen number of cycles, cannot be done without taking into account the mechanics of the defects.

As well-known from [17], the fatigue strength is controlled by the defects having the maximum size and given the complex geometries of the LOF is appropriate to adopt the $\sqrt{area}$ parameter.

Therefore in some contribution [18], statistics of extreme values has been applied in order to estimate the maximum size of the defect starting from inspection by CT scanning.

The aim of this contribution is to add axial fatigue results of maraging steel produced by SLM to those previously published [19] and quantify the $\sqrt{area}$ of the defects present at the crack initiation zone of the fracture surface of the specimens by using a Scanning Electron Microscope (SEM).

2 Materials and methods

2.2. Material and specimen's geometry

* Corresponding author: daniele.rigon.1@phd.unipd.it

Table 1. Chemical composition of the powder adopted in the batch EOS and SISMA.

Batch	Fe (wt-%)	Ni (wt-%)	Co (wt-%)	Mo (wt-%)	Ti (wt-%)	Al (wt-%)	Cr (wt-%)	Cu (wt-%)	C (wt-%)	Mn (wt-%)	Si (wt-%)	P (wt-%)	S (wt-%)
EOS	balance	17-19	8.5-9.5	4.5-5.2	0.6-0.8	0.05-0.15	≤ 0.5	≤ 0.5	≤ 0.03	≤ 0.1	≤ 0.1	≤ 0.01	≤ 0.01
SISMA	balance	17-19	8.5-10	4.5-5.2	0.8-1.2	/	≤ 0.25	/	≤ 0.03	≤ 0.15	≤ 0.1	≤ 0.01	≤ 0.01

Two batches of cylindrical maraging steel specimens were considered in the contribution. The first batch (E) is produced by a EOSINT M280 system, while the second one was provided by SISMA srl, that uses SISMA MYSINT 100 system to produce components made of maraging steel and other alloys. The material under investigation is maraging steel or also commonly known as 18Ni 300. The chemical composition of powders and the process parameters adopted to produce fatigue testing samples are reported in Table 1 and 2, respectively.
The specimen's geometry designed for fatigue testing is shown in Fig. 1. For each batch the specimens were built with their axis oriented both at 0° and 90° with respect to the building direction that correspond to the translation of the platform which allows the deposition of the subsequent layer during the process. A schematic sketch of the orientation of the samples can be seen in Fig. 2. In addition, half of the total number of specimens were subjected to aging heat treatment for 6 hours at a temperature of 490°C and then air cooled.

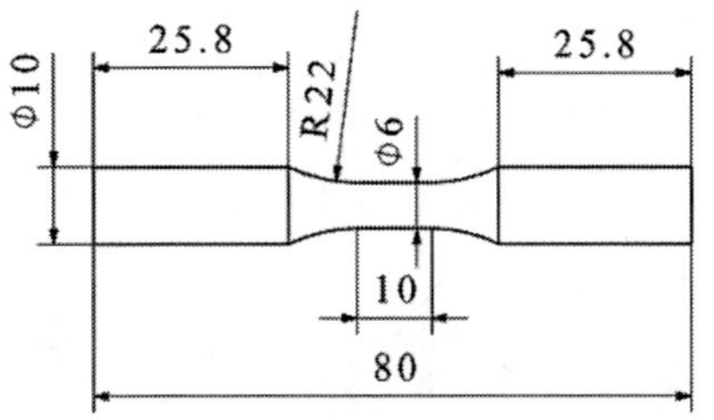

Fig. 1. Specimen's geometry.

For a clear comprehension of the results in the present contribution the nomenclature of the single test series is summarized in Table 3, where the prefix E and S are referred to EOS and SISMA respectively, 0° and 90° indicate the building direction and NT / T refer to Not heat-Treated and the heat-Treated condition, respectively.

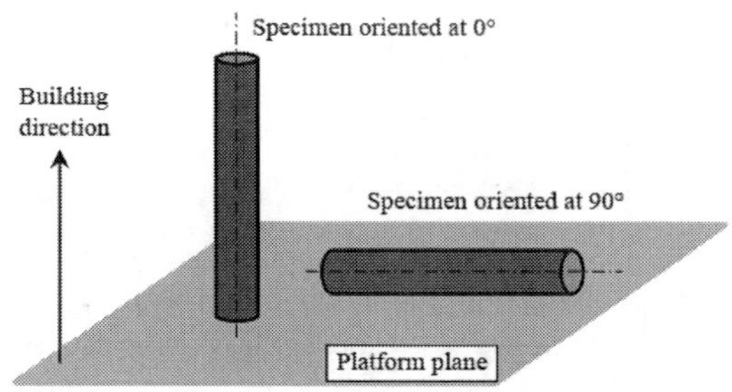

Fig. 2. Orientation of specimens' axis with respect to the building direction.

2.2. Testing protocol

Push pull, constant amplitude fatigue tests were carried out by using on the S batch and results were compared with the E batch previously tested [19]. All the specimens were polished by using progressively finer emery paper from grade 80 up to grade 800.

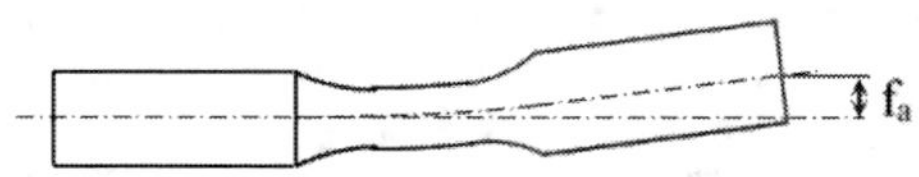

Fig.3. Deflection of the specimen's axis f_a observed in [19].

Since it is known that in additive components the residuals stress induced by rapid cooling of the melted zone might cause geometrical distortion (as demonstrated in [19]), the gross section of each specimens were turned in order to reduce misalignments. Before machining, deflection of the specimen's axis f_a (Fig. 3) was measured for all specimens.
The fatigue tests were carried out by using a servo-hydraulic SCHENCK HYDROPULS PSA 100 machine having a 100 kN load cell and equipped with a TRIO

Table 2. Process parameters adopted to manufacture maraging steel specimens for fatigue testing.

Batch	Laser power [W]	Layer thickness [µm]	Laser spot diameter [µm]	Laser scan rate [mm/s]	Powder dimension [µm]	Temperature of the platform [°C]	Scanning strategy
EOS	400	40	100	Not provided	60	40	Parallel vector
SISMA	120	20	55	500	15÷45	Not provided	Not provided

Table 3. Nomenclature of the test series.

Batch	N° of specimens	Building direction	Heat treatment
E_AD_0°_NT	12	0°	NT
E_AD_0°_T	12	0°	T
E_AD_90°_NT	12	90°	NT
E_AD_90°_T	12	90°	T
S_AD_0°_NT	10	0°	NT
S_AD_0°_T	8	0°	T
S_AD_90°_NT	10	90°	NT
S_AD_90°_T	10	90°	T

Sistemi RT3 digital controller. The load frequency was set in the range between 10 Hz and 30 Hz, depending on the applied load level.
After fatigue testing the fracture surfaces were observed by means of Scanning Electron Microscope (SEM) in order observed the cause of failure at the crack initiation point. In particular, since the LOFs are the most frequent and often more detrimental in terms of fatigue strength, the $\sqrt{\text{area}}$ were measured following the guide line reported in [17].

3 Measurements

3.1. Roughness

In Fig.4 is reported a histogram of the Ra mean values measured on the surface of 5 specimens per test series along the longitudinal direction. It can be seen that all test series present a Ra value approximately 0.5 μm.

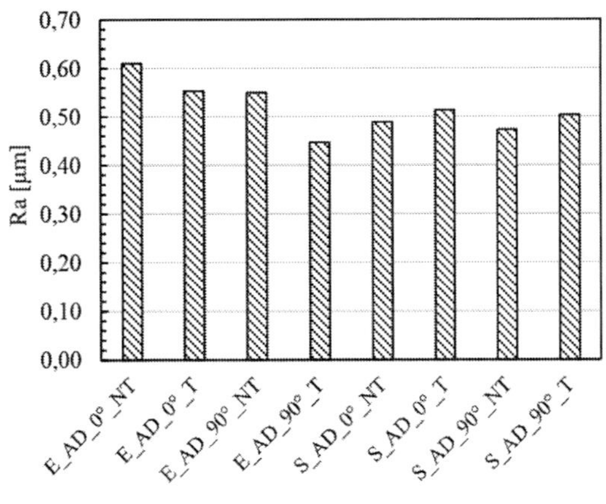

Fig. 4. Ra mean value for each test series.

3.2. Geometrical distortion

Geometrical distortion of each batch, in terms of deflection of the specimen's axis f_a, was measured, and the synthesis of the mean values is reported in Fig.5. In the previous work [19] deflection values were used for evaluating the mean stress that is caused by the machine gripping system. Contrary, in this contribution the gross section of specimens provided by SISMA srl, were turned in order to avoid mean stress state during fatigue testing, therefore the value of f_a is reasonable negligible.

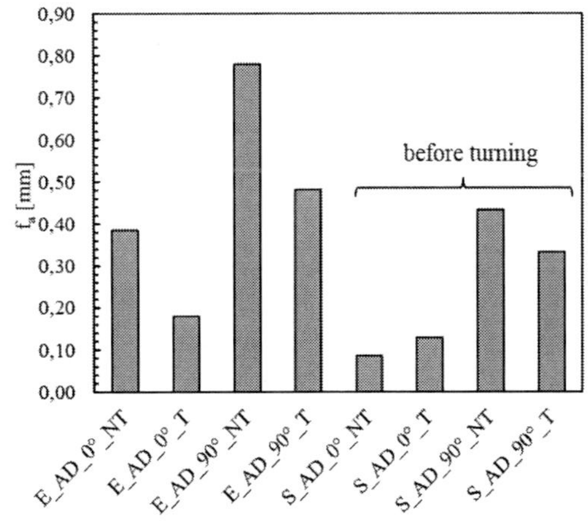

Fig. 5. Mean value of deflection f_a for each test series.

3.3. Hardness

Micro-hardness (HV) was measured in one specimen per test series and the results are reported in Fig.6. Due to the aging hardening heat treatment the hardness of the is doubled with respect to the NT series. This was not observed in S batch, where it seems that the aging treatment has not been effective.

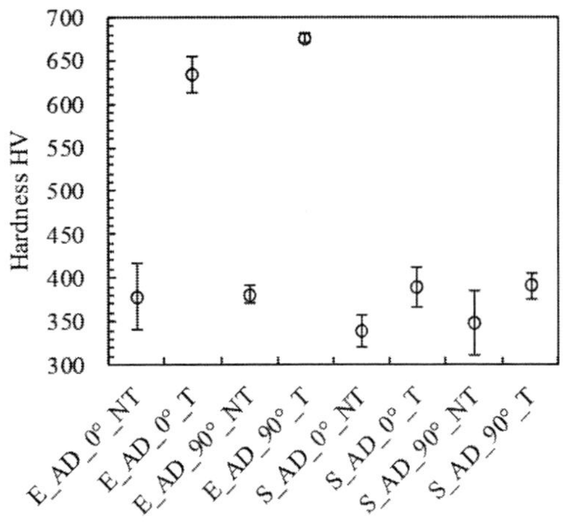

Fig. 6. Microhardness (HV) results for each test series.

4 Results

4.1. Fatigue test results

Fig. 7(a-d) show the fatigue test results in terms of nominal stress amplitude of the test series related to the

https://doi.org/10.1051/matecconf/201816502005

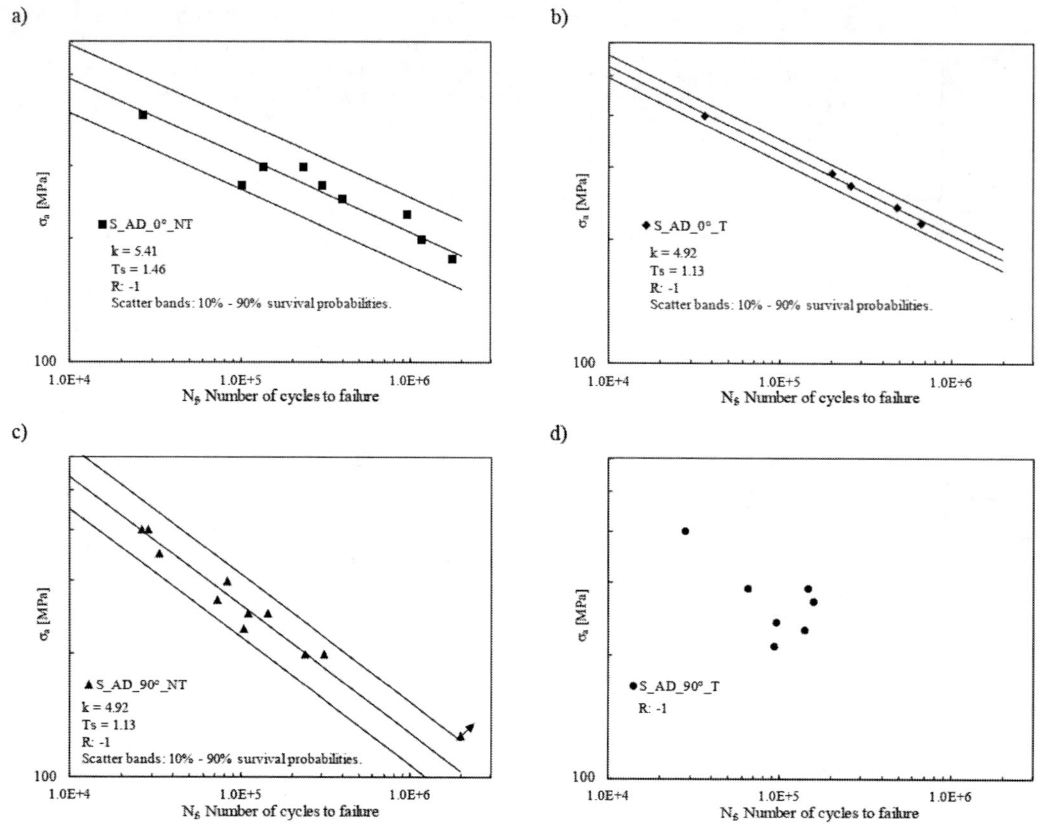

Fig. 7. Fatigue test results of the test series related to the (S) batch in terms of nominal stress amplitude.

(S) batch. The single scatter bands are referred to 10-90% probability of survival obtained from statistical analysis of the fatigue data.

Since the fatigue test results of the (E) batch were affected by mean stresses, for a comparison purposes both (S) batch and (E) batch results have been reported in Fig. 8 in terms of SWT parameter. As a result of turning, the SWT value of the specimens of the batch S is simply the applied nominal stress amplitude.

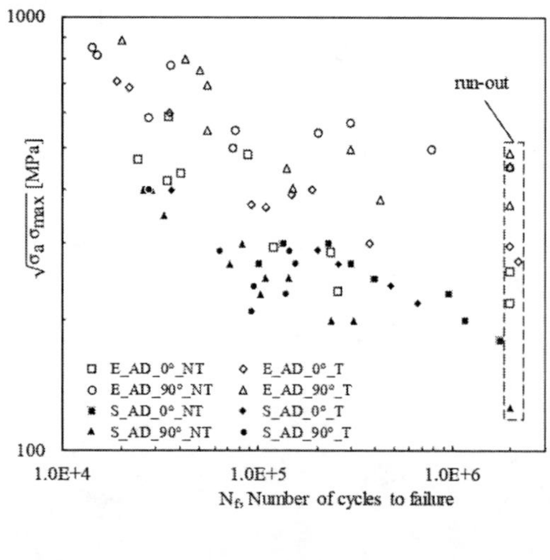

Fig. 8. Constant amplitude fatigue tests results of S-batch compared to E-batch already published in [X]. The S-batch results are considered without mean stress effect following the turning of the specimens' gross sections.

For all details of E batch results the reader is referred to [19].

4.2. Fractography

In Fig. 9 the available fracture surfaces of specimens 0° oriented at the crack initiation point are shown. In agreement with results from the literature of the recent past, the cause to failure in AM components are LOFs, and such defects present irregular shapes. As reported in Fig. 9, the effective area adopted to evaluate the $\sqrt{area}$ parameter is highlighted by a red dotted line in each image. To choose the effective areas, suggestions reported in [17] were followed.

5 Conclusions

In this contribution, an experimental analysis of the defects that affect the fatigue behaviour of two batches of maraging steel specimens produced by additive manufacturing has been reported.

https://doi.org/10.1051/matecconf/201816502005

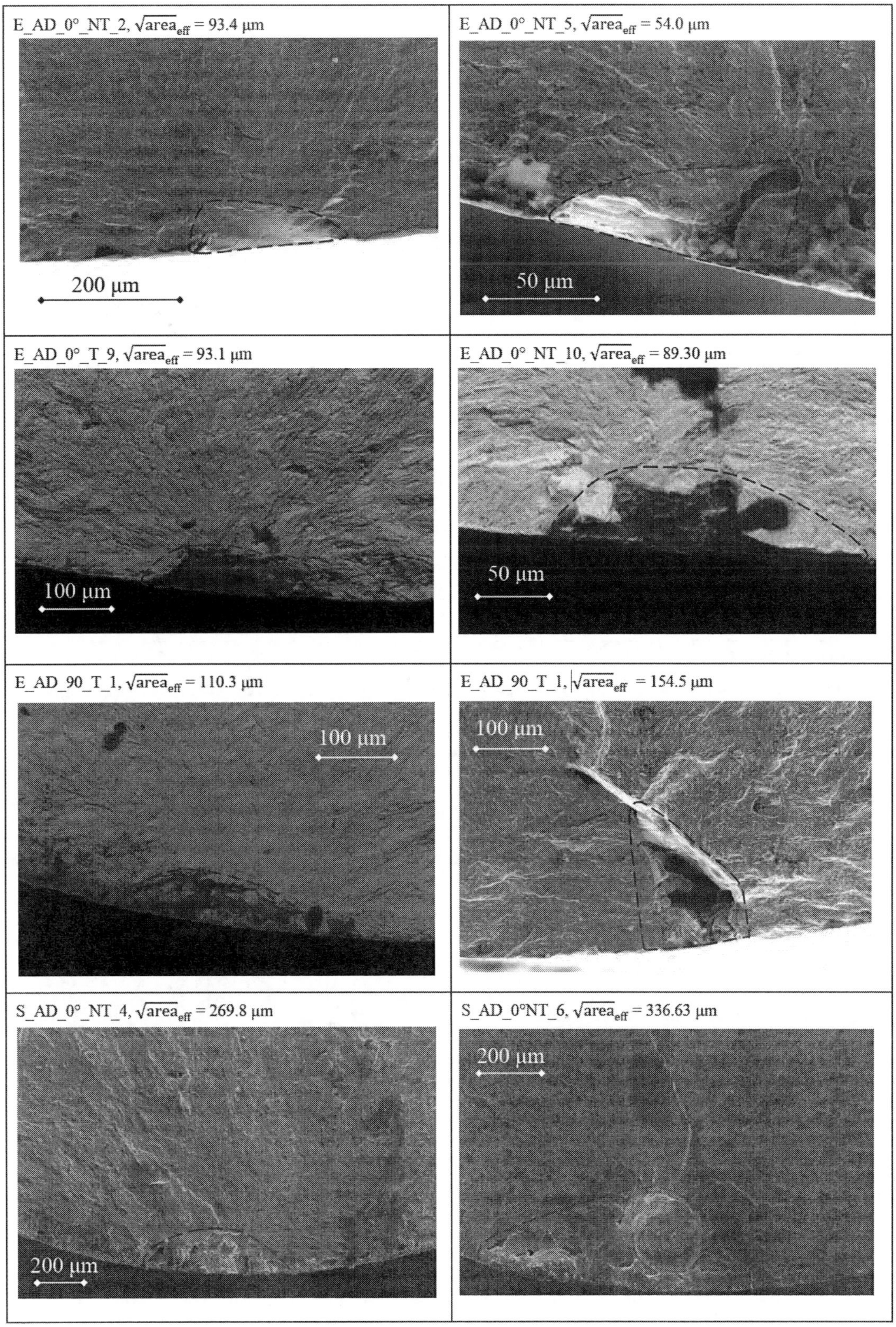

Fig 9. Examples of LOF at the crack initiation zone of the (E) and (S) batches. The $\sqrt{area}_{eff}$ was estimated from the effective area highlighted by red dotted line.

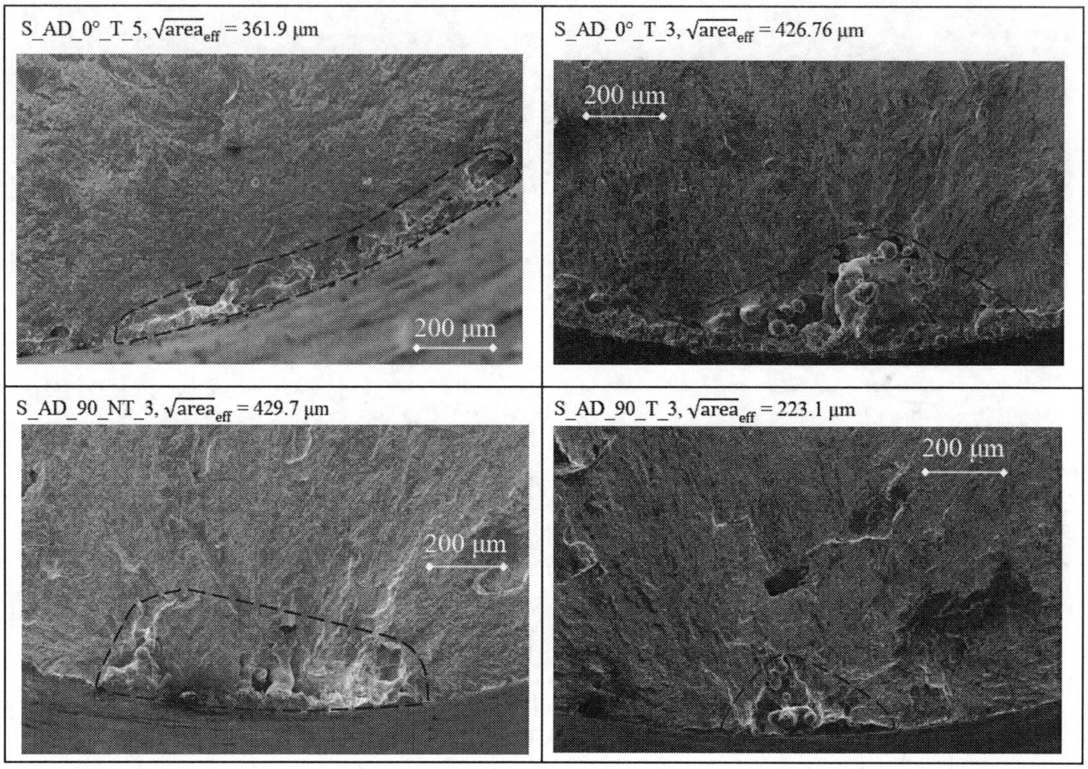

Figure 9 (continued).

Both batches are divided in four test series by taking into account the build orientation and heat treated/as-built condition (see Table 3).
As observed on the specimen related to the (E) batch, geometrical distortion caused by residual stresses also occurred on the (S) batch, with more relevance in specimens which were not heat treated (NT). While for the (E) batch tested in the previous work the effect of mean stress caused by geometrical distortion had to be taken into account in this contribution the specimens' gross sections were turned to avoid misalignments.

For comparing the two batches, fatigue test results were plotted in terms of SWT parameter. All the fatigue data related to (S) batch show a fatigue strength lower than those related to (E) batch. Focusing on the single test series of the (S) batch, the specimens oriented at 0°, both (NT) and (T), exhibit a fatigue strength from 45% to 60% higher than those oriented at 90° (fatigue strength evaluated at 1 million cycles) in disagreement with the results of the (E) test series [19]. It is worth highlighting that the values of the micro-hardness obtained on (S) specimens heat treated are not in agreement with the (E) ones and those reported in literature.

The inspections of the fracture surfaces have demonstrated that the fatigue failures of the AM components are mainly caused by the defects, which in literature are referred to as Lack Of Fusion (LOF). In agreement with the literature, this kind of defects present an irregular shape and are mainly located both at the surface and at the sub surface area of the component. In the examples reported in Fig. 9, it can be observed that the effective area of the defects related to the (E) batch are about 4 times smaller than those calculated for the (S) batch. In addition, at the crack initiation area, a cluster of small defects is often observed; one possible interpretation is to consider such cluster as a single defect having an effective area that circumscribes all defects, but such hypothesis has not been validated yet.

References

1. J.P. Kruth, B. Vandenbroucke, J. van Vaerenbergh, I. Naert, AFPR, S4, 2005.
2. L. Rännar, A. Glad, C. Gustafson, Rapid Prototyp. J. **13** (2007)
3. L.E. Murr, S.M. Gaytan, D.A. Ramirez, E. Martinez, J. Hernandez, K.N. Amato, P.W. Shindo, F.R. Medina, R.B. Wicker, J. Mater. Sci. Technol. **28** (2012).
4. J. Kranz, D. Herzog, C. Emmelmann, J. Laser Appl. **27** (2015).
5. H.D. Carlton, A. Haboub, G.F. Gallegos, D.Y. Parkinson, A.A. MacDowell, Mater. Sci. Eng. A. **651** (2016).
6. Z. Wang, T.A. Palmer, A.M. Beese, Acta Mater. **110** (2016).
7. D. Manfredi, F. Calignano, M. Krishnan, R. Canali, E. Paola, S. Biamino, D. Ugues, M. Pavese, P. Fino, Light Met. Alloy. Appl., InTech, (2014)

8. M.J. Donachie, Titanium: A Technical Guide, 2nd Edition, (2000).
9. W. Xu, M. Brandt, S. Sun, J. Elambasseril, Q. Liu, K. Latham, K. Xia, M. Qian, Acta Mater. **85** (2015).
10. J.J. Lewandowski, M. Seifi, Metal Additive Manufacturing: A Review of Mechanical Properties, (2016).
11. P. Li, D.H. Warner, A. Fatemi, N. Phan, Int. J. Fatigue. **85** (2016)
12. A. Yadollahi, N. Shamsaei, Int. J. Fatigue. **98** (2017).
13. R. Konečná, G. Nicoletto, L. Bubenko, S. Fintová, Eng. Fract. Mech. **108** (2013).
14. G. Nicoletto, S. Maisano, M. Antolotti, F. Dall'Aglio, Procedia Struct. Integr. **7** (2017).
15. G. Nicoletto, Procedia Struct. Integr. **7** (2017).
16. H. Gong, K. Rafi, H. Gu, T. Starr, B. Stucker, Addit. Manuf. **1** (2014)
17. Y. (Yukitaka) Murakami, Metal fatigue: effects of small defects and nonmetallic inclusions, Elsevier, (2002).
18. S. Romano, A. Brückner-Foit, A. Brandão, J. Gumpinger, T. Ghidini, S. Beretta, Eng. Fract. Mech. **187** (2017).
19. G. Meneghetti, D. Rigon, D. Cozzi, W. Waldhauser, M. Dabalà, Procedia Struct. Integr. **7** (2017).

Determination of the anisotropic fatigue behaviour of additively manufactured structures with short-time procedure $PhyBaL_{LIT}$

Bastian Blinn[1,*], *Marcus* Klein[2], and *Tilmann* Beck[1]

[1]Institute of Materials Science and Engineering, TU Kaiserslautern, 67663 Kaiserslautern, Germany
[2]State Materials Testing Institute Darmstadt (MPA), Chair and Institute for Materials Technology (IfW), Technische Universität Darmstadt, 64283 Darmstadt, Germany

Key Words: Additive Manufacturing, load increase tests, $PhyBaL_{LIT}$, fatigue behaviour, anisotropy, 316L

Abstract. Additive Manufacturing techniques provide completely new possibilities in component design and creation of innovative material structures. To utilize the whole potential of Additive Manufacturing, the microstructure, the mechanical properties and their interrelations as well as their relationship to the Additive Manufacturing process parameters are essential. Investigations of the fatigue behaviour of additively manufactured (AM-) metallic materials are still available in limited extent. However, as a prerequisite for efficient and reliable use of AM-components in safety relevant structures, sound knowledge of fatigue behaviour and properties of these structures is indispensable. A central aspect in Additive Manufacturing is the anisotropic mechanical behaviour under monotonic and cyclic loading in dependency on the building direction [1, 2]. In the present work, the microstructure and mechanical properties of Selective Laser Melted (SLM) as well as Laser Deposition Welded (LDW) AISI 316L stainless steel specimens are investigated with special focus on the influence of the building direction. The investigated specimens are built in horizontal and vertical direction, resulting in layer planes oriented parallel and perpendicular to the loading direction, respectively. The fatigue tests have been performed on a servohydraulic testing system with measurement of stress-strain-hysteresis as well as of plastic deformation induced changes in temperature and specific electrical resistance. S-N_f-curves in the HCF-regime of AM-specimens have been determined with the time and material efficient Physically Based Lifetime calculation procedure $PhyBaL_{LIT}$ [3]. Anisotropic fatigue behaviour of the different AM-specimens has been rated with load increase tests (LIT) and the usage of S-N_f-curves calculated by the $PhyBaL_{LIT}$ method.

1 Introduction

Within in the last decades, Additive Manufacturing underwent an extremely progressive development. This technology enables possibilities in manufacturing components far beyond the limits of conventional methods, such as casting or machining. Besides the possibility to realize complex structures, which can be optimized to their individual application, this technology opens the opportunity to locally modify materials properties of components during the manufacturing process [4].

To exploit the potential of this innovative manufacturing technology, it is necessary to investigate and understand the microstructure and the mechanical properties of additively manufactured (AM-) structures as well as the influence of the manufacturing parameters on materials properties.

One of the most important manufacturing parameters in Additive Manufacturing is the building direction, as already shown in many investigations of AM-materials, e.g. titanium alloys, aluminium alloys, nickel-base superalloys and stainless steels [1, 2, 5-9]. Especially the influence of the building direction on the monotonic mechanical properties have been investigated in a large extent [1, 2, 5, 6]. In these investigations, the building direction in which the layer planes are perpendicular oriented to the loading direction (vertical building direction) show lower tensile strength compared to specimens with layer planes oriented parallel to the loading direction (horizontal building direction).

Currently, AM-components are mostly used in applications with minor safety requirements. To use this technique in safety relevant structural parts it is indispensable to get sound knowledge about the fatigue behaviour of AM-materials. Investigations about fatigue properties are available in a limited extent and, hence, a gap of knowledge in this field has to be closed. The investigations of Mower et al. on AISI 316L stainless steel [2], Shamseaeia et al. on Ti6Al4V [8] and Yadollahi et al. on 17-4-PH [1] show, according to the monotonic properties of AM-materials, an anisotropic fatigue behaviour depending on the building direction, i.e. a significantly higher fatigue strength of horizontally built specimens compared to the vertical building direction.

The production of AM-structures is, compared to conventional manufacturing, still relatively time and cost intensive, resulting in high efforts for characterization of the fatigue behaviour of AM-materials by typically 20 to 30 constant amplitude tests (CATs) to determine one S-N_f curve. To reduce this effort, short-time procedures

* Corresponding author: blinn@mv.uni-kl.de

based on load increase tests (LITs) and Physically Based Lifetime calculation $PhyBaL_{LIT}$ [3] are considered especially useful in case of AM-structures.

In this work, the fatigue behaviour of Selective Laser Melted (SLM-) and Laser Deposition Welded (LDW-) specimens made of AISI 316L stainless steel built in horizontal (SLM-H, LDW-H) and vertical (SLM-V, LDW-V) direction were investigated. Additional to the results shown in Blinn et al. [10], LITs at a frequency of 2 Hz and lifetime calculations by the short-time procedure $PhyBaL_{LIT}$ [3] were conducted. Therefore additionally to the LITs, CATs in the HCF-regime were performed. Based on the cyclic deformation curves of one LIT and two CATs of the differently manufactured specimens, S-N_f-curves were calculated with the $PhyBaL_{LIT}$ method and verified by CATs. The anisotropic fatigue behaviour of AM-specimens were rated from the calculated S-N_f-curves and additionally by the results of LITs.

2 Materials

The specimens were manufactured from AISI 316L stainless steel with the chemical composition, determined by spectrophotometric analyses, given in Table 1. The LDW-specimens exhibit significantly higher contents of chromium and nickel and lower content of nitrogen. These differences in chemical composition significantly influence austenite stability [11], which can be rated with the M_{d30}-Temperatures determined from equation (1) according to Angel [12]. As shown in [10], austenite stability has to be taken in account for comparison of the mechanical properties of the differently manufactured specimens made of 316L, because lower austenite stability can result in higher deformation or stress induced transformation of the austenite into α'-martensite, which strongly influences the mechanical properties.

$$M_{d30,\,Angel} = 413 - 462(C + N) - 9.2Si - 8.1Mn - 13.7Cr - 9.5Ni - 18.5Mo \quad [12]\ (1)$$

Table 1. Chemical compositions and $M_{d30,\,Angel}$-temperatures of the different AM-specimens [10, 11]

Amount of alloying element in wt.-%	ASTM A 182		SLM	LDW
	Min	max		
C	--	0.03	0.02	0.03
N	--	0.10	0.08	0.10
Si	--	1.00	0.61	0.53
Mn	--	2.00	1.44	1.30
Cr	16.00	18.00	17.68	16.41
Ni	10.00	15.00	13.07	10.54
Mo	2.00	3.00	2.26	2.04
Fe	60.80	72.00	64.68	68.75
$M_{d30,\,Angel}$ in °C	61.8	-117.1	-58.3	-28.5

In the Additive Manufacturing processes blanks with a simple geometry were manufactured using the facilities and powders given in Table 2 [10, 11]. To focus the investigations on properties of the volume microstructures, and to exclude the influence of surface topography, resulting from the Additive Manufacturing process, blanks were by subsequently turned to the final specimen geometry given in Fig. 1 and mechanically polished in the gauge length.

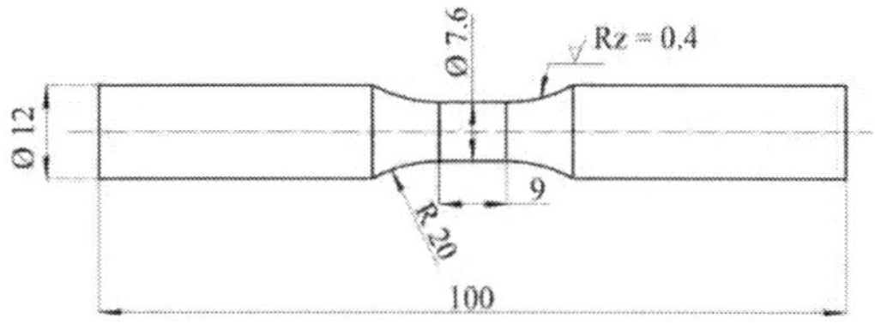

Fig. 1. Geometry of the fatigue specimens [10]

Table 2. Manufacturing parameters of the AM-blanks [10, 11]

AM process	SLM	LDW
Machine type	EOS M 290	DMG MORI LASERTEC 65 3D
Laser power in W	400	2000
Powder size in µm	25-45	50-150
Av. Layer thickness in µm	40	400
Dimensions of the blanks in mm	bars Ø14x102	cuboids 15x15x103

3 Experimental methods

For investigation of the microstructure of the differently manufactured specimens, light optical micrographs (LOMs) of etched cross sections of the shafts of the fatigue specimens were taken with a Leica DM 6000 M device. The SLM-specimens were etched using a V2A etchant, while LDW-specimens were etched with an Adler etchant, which results in better visualization of the grain structure in case of the higher chromium and nitrogen content of the LDW-specimens. Furthermore, scanning electron microscopy (SEM) observations were performed using a FEI Quanta 600 device.

The fatigue tests (LITs and CATs) were performed on a servo-hydraulic testing system with a stress ratio R=-1 and a frequency of f=2Hz. The LITs were started at a stress amplitude of $\sigma_{a,\ start}$=100 MPa. The stress amplitude was then increased stepwise by $\Delta\sigma_a$=20 MPa and a step length of 9 000 cycles. Additionally for the LDW-specimens, modified LITs with a step length of 4 500 cycles and an increase of stress amplitude of $\Delta\sigma_a$=10 MPa were conducted to get a higher resolution in materials properties in the HCF-regime.

During the fatigue tests, stress-strain-hysteresis loops, the plastic deformation induced change in temperature and in electrical resistance according to Biallas et al. [13] were measured continuously. The experimental setup is shown in Fig. 2. Strain measurements were conducted with a conventional extensometer. The temperature was measured with one thermocouple in the gauge length (T_1) and two reference thermocouples at the elastically loaded shafts (T_2 and T_3). The change in temperature was determined in according to equation (2), to compensate external influences on the specimen temperature. The change in temperature calculated this way results from the plastic deformation work dissipated in the gauge length volume of the specimen and, therewith, directly correlates with the area of the stress-strain-hysteresis loop in each cycle [13].

$$\Delta T = T_1 - \frac{T_2 + T_3}{2} \tag{2}$$

For Measurement of change in electrical resistance, a constant direct current of 8 A was applied to the specimen. The voltage drop was measured between two contact pins integrated in plastics clamps above and below the gauge length (see Fig. 2). The specific electrical resistance calculated from this according to Ohm's law is influenced by microstructural features such as dislocation density and dislocation structures, which makes this measurement sensitive to fatigue processes in the microstructure of the specimens [13].

Additionally, the content of the ferromagnetic fraction in the gauge length, was measured using a FERITSCOPE™ MP 30E device before and after tensile and fatigue tests, to determine and quantify the amount of transformation from paramagnetic austenite into ferromagnetic α'-martensite during monotonic and cyclic loading. This measurement is essential to determine the influence of the austenite stability on the mechanical properties.

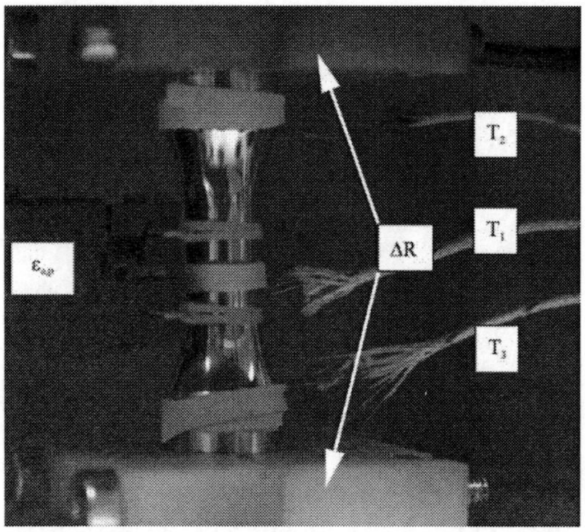

Fig. 2. Experimental setup of the fatigue tests [10]

4 Results and Discussion

4.1 Microstructure and monotonic properties

4.1.1 Microstructure

LOMs of the AM-specimens' microstructure show the layer boundaries along the building direction resulting from the additive manufacturing process (see Fig. 3 a, f and Fig. 4 a). The AM-microstructures also exhibit an elongation of grains along the building direction visible in LOMs and grain growth over the layer boundaries (see Fig. 3 a, f and Fig. 4 a), which is confirmed by electron backscatter diffraction (EBSD-) mappings (see Fig. 3 b and Fig. 4 b). This unique grain structure results from the directed heat flux during the additive manufacturing process. This leads to a grain aspect ratio significantly >1, which also can be seen in lower apparent grain size in cross sections (CS)/ longitudinal sections (LS) of vertically/ horizontally built specimens compared to CS/LS of horizontally/ vertically built specimens (see Fig. 3 and Fig. 4).

LDW-specimens exhibit, compared to SLM-specimens bigger grains, which is caused by higher layer thickness (see Fig. 3 a, f, Fig. 4 a and Table 2), and therefore lower cooling rates in the manufacturing process.

Fig. 3. LOMs of CS of a) SLM-H-, d) SLM-V-, LS of c) SLM-H- and f) SLM-V- and EBSD-mappings of CS of b) SLM-H- and e) SLM-V –specimens [10]

Besides their unique grain structure, AM-materials exhibit defects like pores or oxide inclusions. Porosity of the differently manufactured specimens were rated by means of LOMs (see Table 3). SLM-specimens exhibit significantly higher porosities compared to LDW-specimens, which is caused by smaller layer thickness [14] and lower laser power [15] resulting in lower penetration depth of the laser in SLM-process. This leads to a higher amount of failure binding (see Fig. 5 a) and therefore higher porosity in SLM-material. SLM-V specimens exhibit lower porosity compared to SLM-H-specimens, which may be caused by slight differences in manufacturing parameters and is not considered as a systematic difference between the microstructure of SLM-H- and SLM-V-specimens (comp. Fig. 3 a to f and c to d).

Additionally to pores, the LDW-specimens exhibit silicon manganese (SiMn-) oxide inclusions [11] resulting from a partially incomplete protecting gas flow during manufacturing [16]. These oxide inclusions are bigger than pores in LDW-specimens (see Fig. 5 b), implying that oxide inclusions have a higher influence on fatigue strength in LDW-specimens.

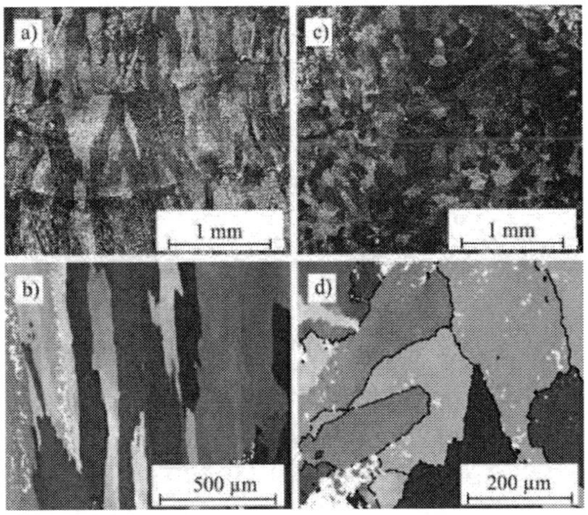

Fig. 4. LOMs of CS of a) LDW-H and c) LDW-V-, and EBSD-mappings of CS of b) LDW-H- and d) LDW-V–specimens [10]

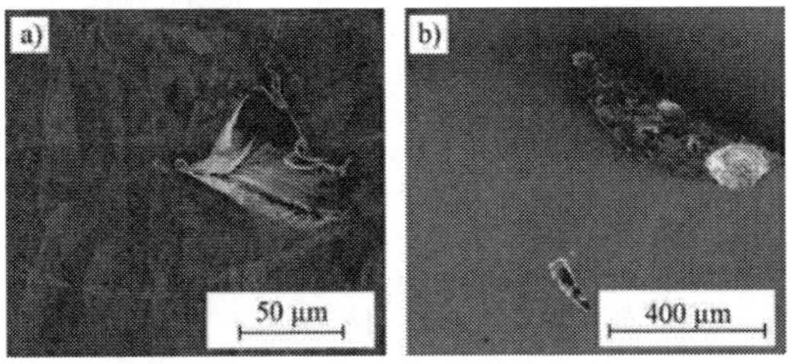

Fig. 5. SEM-image a) of a pore in SLM-material and b) of a pore and SiMn-oxide inclusion in LDW-material [10]

Table 3. Porosity and monotonic properties of AM-specimens [10]

Specimen	SLM-H	SLM-V	LDW-H	LDW-V
Porosity in %	1.54 ± 0.49	0.85 ± 0.27	0.03 ± 0.01	0.02 ± 0.01
HV 30/10	218	213	171	177
R_m in MPa	681 ± 7	612 ± 2	629 ± 7	564 ± 9
$R_{p0.2}$ in MPa	609 ± 43	490 ± 2	438 ± 40	322 ± 2
Young´s Modulus in GPa	167 ± 12	152 ± 7	172 ± 12	170 ± 10
A in %	28.9 ± 3.9	21.4 ± 1.6	37.8 ± 0.3	26.8 ± 4.9
$\Delta\xi$ in Fe-%	0.0	0.0	0.5	0.4

4.1.2 Monotonic properties

Results of tensile tests are given in Table 3 and show an anisotropic behaviour of AM-specimens in dependency of the building direction in accordance to [1, 2, 5, 6]. Horizontally built specimens show higher tensile strength, 0.2% yield stress and elongation at fracture compared to vertical building direction. In elastic properties, i.e. young´s modulus, no significant influence of building direction or manufacturing process occurs.

In contrast to tensile tests, Vickers (HV30) hardness measurements do not show anisotropic properties depending on the building direction. The SLM-specimens possess higher tensile and 0.2% yield stress compared to LDW-specimens, which correlates to the hardness values and results from higher cooling rates associated with smaller layer size in case of SLM (see Table 2). The elongation at fracture of LDW-specimens is higher compared to SLM-specimens with the same building direction. This correlates to $\Delta\xi$-measurements (see Table 3), which indicate lower austenite stability of the LDW-specimens (also see Table 1). This results show, that austenite-α'-martensite transformation and, hence, austenite stability influences the mechanical properties of the investigated AISI 316L variants.

4.2 Fatigue behaviour

4.2.1 Load increase tests (LITs)

The fatigue behaviour of the AM-specimens was determined by evaluation of three key results of LITs: The shape of cyclic deformation curves, the last load step of linear growth of materials reaction σ_l and the load step at failure σ_f. The cyclic deformation curves in LITs show in the first load steps a linear growth of the measured quantities. Afterwards, all specimens show in an individual load step a transition to significant plastic deformation, followed by progressive growth of materials reaction (see Fig. 6 c). The last load step before this transition is defined as σ_l and is used as reference for fatigue behaviour at lower stress amplitudes, while σ_f is used as reference at higher stress amplitudes.

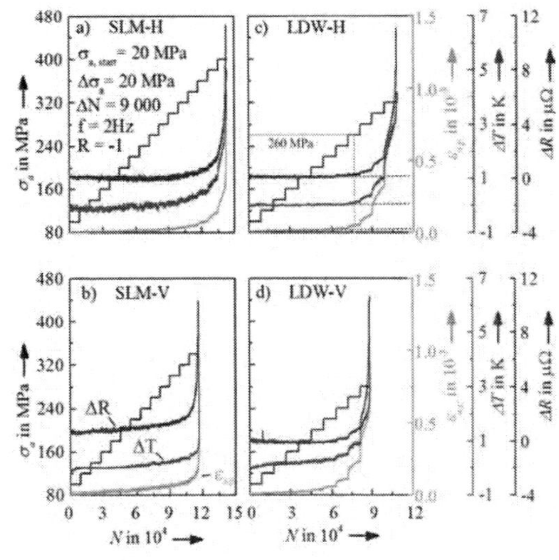

https://doi.org/10.1051/matecconf/201816502006

Fig. 6. Cyclic deformation behaviour of AM-specimens obtained by LITs with measurements of plastic strain amplitude ($\varepsilon_{a,p}$), change in temperature (ΔT) and electrical resistance (ΔR), and a schematic description in c) of determination of σ_l

Table 4. Results of σ_l, σ_f and $\Delta\xi$ for different AM-specimens in LITs

Specimen	SLM-H	SLM-V	LDW-H	LDW-V
σ_l in MPa	280	240	240	200
σ_f in MPa	400	360	340	280
$\Delta\xi$ in Fe-%	0.0	0.0	0.2	0.1

Note that the different measured quantities show identical load steps at first significant plastic deformation processes for each type of specimen.

The results of LITs given in Table 4 show an anisotropic fatigue behaviour of AM-specimens. Vertically built specimens show lower values of σ_l and σ_f and higher $\varepsilon_{a,p}$ at identical load amplitude compared to horizontal building direction and, hence, similar to monotonic properties, lower performance in LITs in accordance to results of [1, 2, 8].

In contrast to SLM-specimens, the LDW-specimens show an increase of magnetic fraction in the gauge length after LIT, which is caused by the lower austenite stability (see Table 1 and Table 4), resulting in austenite-α'-martensite-transformation. However by comparison of the same building direction, SLM-specimens show higher fatigue performance in LITs compared to LDW-specimens, identifiable by higher values of σ_f and σ_l. Moreover, the LDW-specimens show higher plastic deformation compared to SLM-specimens with the same building direction.

4.2.2 S-N$_f$-curves determined with PhyBaL$_{LIT}$

The fatigue strength of the different AM-specimens in the HCF-regime is compared based on results of CATs and S-N$_f$-curves calculated by the short-time procedure PhyBaL$_{LIT}$. In this approach, the slope of the S-N$_f$-curves is calculated based on the cyclic deformation behaviour. For this, Basquin's law [17] (see equation (3)) is combined with Morrow's correlation between the slope of S-N$_f$-curves in the HCF-regime, represented by the fatigue strength exponent b, and the cyclic hardening exponent n' of cyclic hardening curves (σ_a-$\varepsilon_{a,p}$-curves, see equation (5)) [18]. The cyclic hardening curves can be described by the power law given in equation (4) with the cyclic hardening coefficient K' and cyclic hardening exponent n'.

$$\sigma_a = \sigma'_f(2N_f)^b \qquad [17]\ (3)$$

$$\sigma_a = K' \cdot \varepsilon_{a,p}^{n'} \qquad [18]\ (4)$$

$$b = \frac{-n'}{5n'+1} \qquad [18]\ (5)$$

The PhyBaL$_{LIT}$ method allows calculating of an S-N$_f$-curve based on only three fatigue tests, which means a significant reduction of experimental effort: One load increase test is performed to characterize cyclic deformation behaviour under different load amplitudes with one specimen. Based on the LIT results, the stress amplitude for two different constant amplitude tests (CATs) is determined: One CAT is conducted at high stress amplitude below the load step of failure in the LIT and the other CAT is performed at a low stress amplitude, in which plastic deformation already occurs (see Fig. 7).

To determine the cyclic hardening curve, the mean values of $\varepsilon_{a,p}$ in all load steps of LIT between the stress levels of the two CATs are determined and plotted in a σ_a-$\varepsilon_{a,p}$-diagram (see Fig. 7). Note that in most cases the cyclic hardening curve directly taken from the LIT cannot be used for lifetime calculation, because the materials reaction in the load steps with higher stress amplitudes are influenced by previous fatigue processes at lower stress amplitudes and therefore do not represent the materials behaviour for constant amplitude fatigue loading. To increase the data density of the σ_a-$\varepsilon_{a,p}$-curves for constant amplitude loading, the mean $\varepsilon_{a,p}$ values of the single LIT load steps are then transferred by interpolation to constant amplitude values based on the values of $\varepsilon_{a,p}$ at the half of lifetime ($N_f/2$) of each CAT (see Fig. 7). Based on the resulting cyclic hardening curve, the cyclic hardening exponent n' and, with equation (5), the slope of S-N$_f$-curve b can be determined.

To determine the position of the S-N$_f$-curve, σ_a and N_f of the CAT at higher stress amplitude is inserted in equation (3) and the fatigue strength coefficient σ'_f of the S-N$_f$-curve is calculated. The CAT at higher stress amplitude is chosen for determination of σ'_f to minimize the influence of lifetime scatter, which is higher at lower stress amplitudes in HCF-regime.

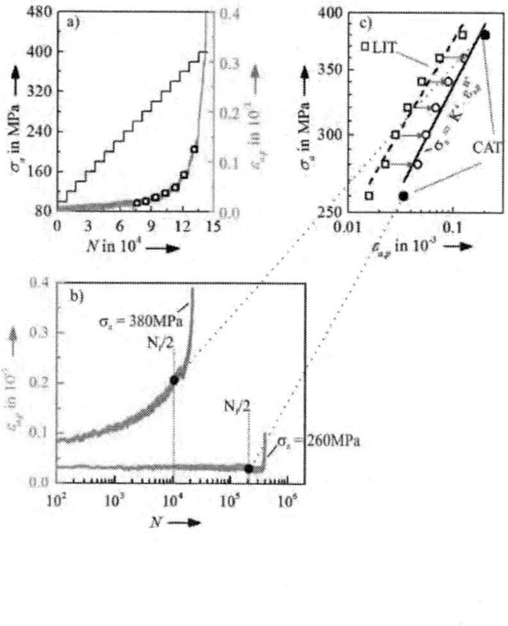

Fig. 7. Description of the short-time procedure $PhyBaL_{LIT}$ with results of $\varepsilon_{a,p}$ measurements of SLM-H-specimens

The cyclic deformation curves of the different AM-specimens used for calculation of the S-N_f-curves by the $PhyBaL_{LIT}$ method are given in Fig. 8 a, b, d and e . The test parameters of LITs for LDW-specimens differ slightly compared to those used in case of SLM-specimens: Due to the relatively low stress amplitude range between onset of first materials reaction and specimen failure, the CATs, resulting out of LITs in Fig. 6 c and d, lead to a cyclic hardening curve with a relatively small amount of data. Therefore, additional LITs were performed with a reduced step length of 4500 cycles and a load increase of only 10 MPa to enhance resolution at HCF- relevant stress amplitudes. These modified LITs were started for the LDW-V-specimens at a stress amplitude of $\sigma_{a,\ start}$=120 MPa and for LDW-L-specimens of $\sigma_{a,\ start}$=180 MPa.

The stress amplitudes of the "high stress" CATs for SLM-H- and LDW-V-specimens were determined as the last load step below σ_f. The SLM-V- and LDW-H-specimens fail in LITs after a low number of cycles in the last load step. Hence, to avoid influence of crack growth, a stress amplitude two steps below σ_f is determined for the "high stress" CAT. For CATs at lower stress amplitude, stress amplitudes with indications of microplastic deformation in the LIT, but below the transition area to significant plastic deformation were used.

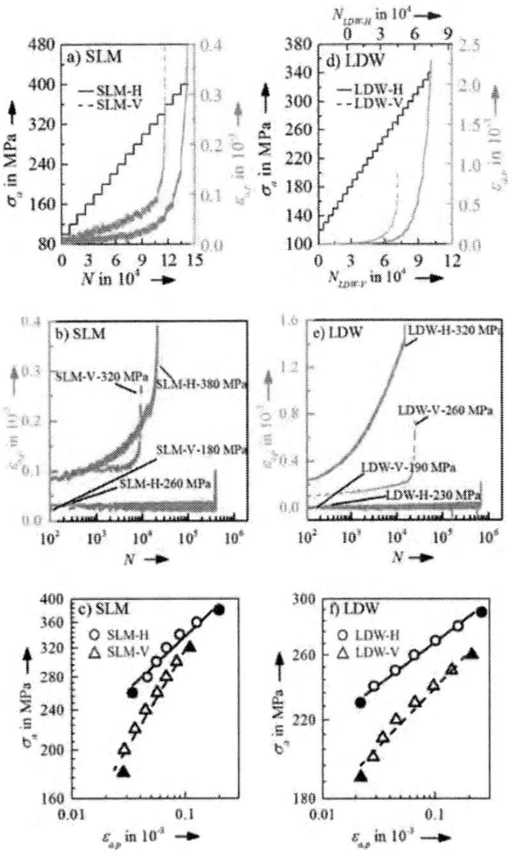

Fig. 8. Cyclic deformation curves with measurements of $\varepsilon_{a,p}$ in LIT in a) and d) and in CATs in b) and e) for $PhyBaL_{LIT}$ method and resulting σ_a-$\varepsilon_{a,p}$-curves in c) and f)

The resulting cyclic hardening curves show a high coefficient of determination R^2 and are hence considered appropriate for calculation of the cyclic hardening exponent n' (see Table 5). The S-N_f-curves calculated by $PhyBaL_{LIT}$ and shown in Fig. 9, are in good accordance with the data of additional CATs in the HCF-regime. Similar to the S-N_f-curves shown in [10], the S-N_f-curves for vertically built specimens show a slightly steeper slope compared to horizontal building direction (see Fig. 9 and Table 5). This anisotropic fatigue behaviour correlates to the results of LITs and CATs, which show for SLM- and LDW-specimens higher fatigue strength in horizontal building direction (see Fig. 6 and Fig. 9) in accordance to [1, 2, 8].

Comparing the different manufacturing processes in identical building directions, SLM-specimens show higher lifetime at high stress amplitudes but lower lifetime at low stress amplitudes compared to LDW-specimens, which can also be seen in the calculated slopes of S-N_f-curves (see Fig. 9 and Table 5). The flatter slope in S-N_f-curves of LDW-specimens is influenced by lower austenite stability of the LDW-material, leading to austenite-α'-martensite-transformation during CATs. This transformation of austenite into α'-martensite does not lead to cyclic hardening in cyclic deformation curves of LDW-specimens as can be seen in Fig. 8 and, hence, seems to be a locally restricted process in highly stressed regions at microstructural defects leading to a higher lifetime, especially at lower stress amplitudes (see Fig. 9) [19, 20]. These results show, that for characterisation of fatigue properties of the AISI 316L stainless steel, the austenite stability and, hence, chemical composition has to be taken in account.

Table 5. S-N_f-curve parameters in HCF-regime of AM-specimens calculated by means of PyBaL_{LIT} and CAT data for verification

Specimen	SLM-H	SLM-V	LDW-H	LDW-V
n'	0.220	0.392	0.082	0.135
B	-0.105	-0.132	-0.058	-0.081
σ'_f	1165	1180	627	622
R^2	0.96	0.96	0.99	0.98

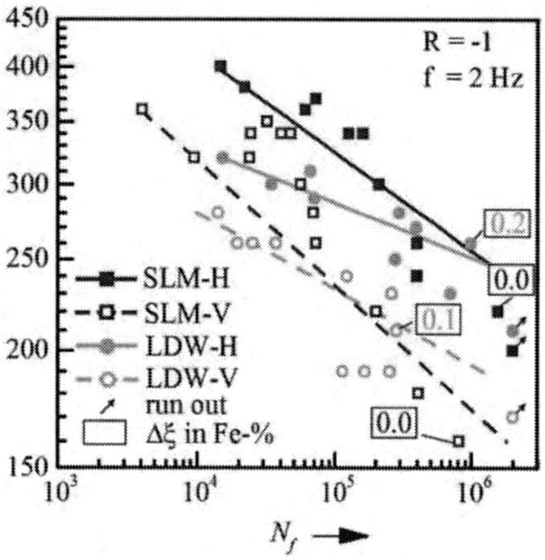

Fig. 9. S-N_f-curves calculated by means of PhyBaL_{LIT} and experimental data of CATs

The scatter in S-N_f-curves is caused by microstructural defects described in section 4.1.1 which correlates with investigations of Günther et al. [21] at Ti6Al4V and Mower et al. [2] at AISI 316L stainless steel. In SLM-specimens the size and distance of crack initiating pores to the surface affect the scatter in S-N_f-curves (see Fig. 10 a). In LDW-specimens, different locations, shape and size of oxide inclusions lead to scatter in lifetime while the relatively small and few pores in LDW-material are assumed not to influence fatigue lifetime significantly (see Fig. 10 b).

The SLM-V-specimens have lower porosity compared to SLM-H-specimens (see Table 3). However, horizontally built specimens show lower lifetime scatter (comp. [10]). In correlation to SLM-material, horizontally built specimens show lower scatter compared to vertical building direction in LDW-material. Hence, the horizontally built AM-specimens seem to have higher capability to reduce locally increased stresses at microstructural defects. This correlates to the higher plasticity at fracture in tensile tests (see Table 3) and higher plastic strain amplitude in last load steps in LIT of horizontally built specimens compared to vertical building direction respectively.

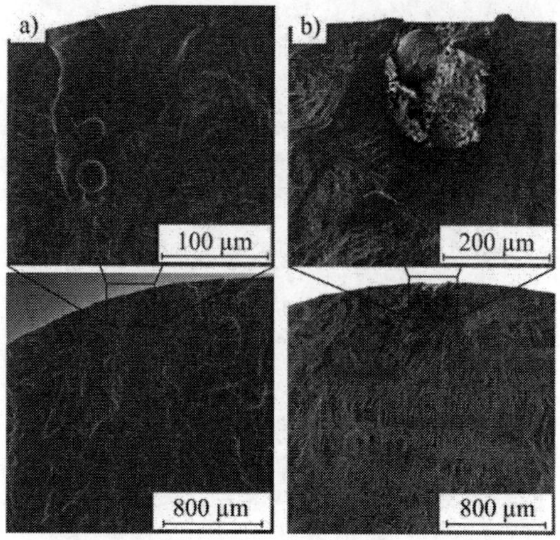

Fig. 10. Fracture surface of a) SLM-H-specimen (σ_a = 300MPa; N_f = 212 218) with crack initiation at a pore and b) LDW-H-specimen (σ_a = 260MPa; N_f = 982 098) with crack initiation at a SiMn-oxide inclusion [10]

5 Summary and conclusion

Specimens manufactured by Selective Laser Melting (SLM) and Laser Deposition Welding (LDW) made of AISI 316L stainless steel were investigated with regard to their microstructure and mechanical properties. Focus of the investigations was to determine the influence of the building direction on fatigue behaviour of these additively manufactured (AM-) specimens by means of short-time procedures, i.e. load increase tests (LITs) and PhyBaL_{LIT}.

The microstructure of AM-specimens exhibit grain elongation along the building direction and grain growth over the layer boundaries visible in light optical micrographs. Furthermore, the AM-materials exhibit microstructural inhomogeneities, i.e. pores in both AM-materials and SiMn-oxide inclusions in LDW-material. In SLM-material pores and in LDW-material oxide inclusions, which are bigger compared to the pores in LDW-materials, are determined to cause scatter in fatigue lifetime.

Monotonic properties of AM-specimens show anisotropic behaviour, i.e. higher elongation at fracture, tensile strength and 0.2% yield stress in horizontal building direction compared to vertically built specimens. Further, in the same building direction, the SLM-specimens show higher tensile strength and 0.2% yield stress compared to LDW-specimens and lower elongation at fracture, which correlates to higher amount of austenite-α'-martensite-transformation resulting from lower austenite stability in LDW-material.

The LITs show anisotropic fatigue behaviour of AM-specimens, which is confirmed by results of constant amplitude tests (CATs) and S-N_f-curves calculated by means of PhyBaL_{LIT}. The result of lifetime calculation are in good accordance to the CATs. Horizontally built specimens from both manufacturing processes show higher fatigue strength, lower amount of scatter and a slightly flatter slope of the S-N_f-curves.

By comparison of the fatigue behaviour of the different Additive Manufacturing processes for identical building direction, the LDW-specimens show higher lifetime at lower stress amplitudes and lower lifetime at higher stress amplitudes than SLM-specimens. This is influenced by austenite-α'-martensite-transformation in LDW-material, which leads to higher lifetime especially at lower stress amplitudes and shows, that austenite stability and, hence, chemical composition significantly influences the fatigue properties of AISI 316L stainless steel.

The results demonstrate, that the short-time procedures LIT and $PhyBaL_{LIT}$ can be used to determine the fatigue behaviour of AM-specimens. With LIT the anisotropic fatigue behaviour can be described very efficiently and, hence, this method can be used for investigations of the influence of variations in manufacturing parameters or post processing on anisotropic fatigue behaviour. $PhyBaL_{LIT}$ proved to be a viable method to determine fatigue strength in dependency on AM-process and building direction very efficiently.

The LIT cannot determine the influence of austenite-α'-martensite-transformation at lower stress amplitudes, because the low number of cycles in the LIT do not lead to deformation induced phase transformation at lower stress amplitudes. However phase transformation in CATs reduces the amount of local plastic deformation at microstructural defects, especially at lower stress amplitudes, leading to a flatter σ_a-$\varepsilon_{a,p}$-curve, which influences the slope of S-N_f-curve. Hence, with the $PhyBaL_{LIT}$ method the influence of phase transformation can be determined due to the additional information of CATs, if phase transformation is locally restricted at microstructural defects.

Acknowledgement

The research described in this paper was done in collaboration to the Institute for Manufacturing Technology and Production System at the University of Kaiserslautern and was financially supported by European Union's European Regional Development Fund (ERDF) and the Commercial Vehicle Cluster (CVC) Südwest.

References

1. A. Yadollahi, N. Shamsaei, S. M. Thompson, A. Elwany, L. Bian: Int J Fatigue, *Effects of building orientation and heat treatment on fatigue behavior of selective laser melted 17-4 PH stainless steel.* **94**: p. 218-235, (2017).
2. T. M. Mower, M. J. Long: Mater Sci Eng A, *Mechanical behavior of additive manufactured, powder-bed laser-fused materials.* **651**: p. 198-213, (2016).
3. H. S. Kramer, P. Starke, M. Klein, D. Eifler: Int. J. Fatigue, *Cyclic hardness test PHYBALCHT – Short-time procedure to evaluate fatigue properties of metallic materials.* **63**: p. 78-84, (2014).
4. S.R. Daniewicz, N. Shamsaei: Int J Fatigue, *An introduction to the fatigue and fracture behavior of additive manufactured parts.* **94**: p. 167, (2017).
5. J. J. Lewandowski, M. Seifi: Annu Rev Mater Res, *Metal Additive Manufacturing: A Review of Mechanical Properties.* **46**: p. 151-186, (2016).
6. R. Casati, J. Lemke, M. Vedani: J Mater Sci Technol: *Microstructure and Fracture Behavior of 316L Austenitic Stainless Steel Produced by Selective Laser Melting.* **32**: p. 738-744, (2016).
7. T. Bauer, K. Dawson, A.B. Spierings, K. Wegener: 26th Annual International Solid Freeform Fabrication (SFF) Symposium: *Microstructure and mechanical characterisation of SLM processed Haynes® 230®.* (2015)
8. N. Shamsaeia, A. Yadollahi, L. Bian, S. M. Thompson: ADDMA, *An overview of Direct Laser Deposition for additive manufacturing;Part II: Mechanical behavior, process parameter optimization andcontrol.* **8**: p. 12-35, (2015).
9. E. Brandl, C.Leyens, F. Palm: IOP Conf. Series: Materials Science and Engineering 26, *Mechanical properties of additive manufactured Ti-6Al-4V using wire and powder based processes.*, **26**, (2011).
10. B. Blinn, M. Klein, C. Gläßner, M. Smaga, J. C. Aurich, T. Beck: Metals, *Investigation of microstructure and fatigue behavior of additively manufactured AISI 316L stainless steel with regard to the influence of a heat treatment.* (2018) (accepted).
11. C. Gläßner, B. Blinn, M. Burkhart, M. Klein, T. Beck, J. C. Aurich: 7. WGP Jahreskongress, *Comparison of 316L test specimens manufactured by Selective Laser Melting, Laser Deposition Welding and Continuous Casting.* Aachen: Apprimus Verlag, (2017).
12. T. Angel: J Iron Steel Inst, *Formation of martensite in austenitic stainless steel - Effects of deformation, temperature and composition.* **177**: p. 165-174. (1954).
13. G. Biallas, A. Piotrowski, D. Eifler: FFEMS, *Cyclic stress-strain, stress-temperature and stress-electrical resistance response of NiCuMo allyed sintered steel.* **18**: p. 605-615, (1995).
14. A. B. Spierings, G. Levy: SFF Symposium, *Comparison of density of stainless steel 316L parts produced with selective laser melting using different powder grades.* (2009).
15. C. Kamath, B. El-dasher, G. F. Gallegos, W. E. King, A. Sisto: Int J Adv Manuf Technol, *Density of additively-manufactured, 316L SS*

parts using laser powder-bed fusion at powers up to 400 W. **74**: p. 65-78. (2014).

16. P. Ganesh , R. Kaul, G. Sasikala, H. Kumar, S. Venugopal, P. Tiwari, S. Rai, R. C. Prasad, L. M. Kukreja: Metallogr Microstruct Anal, *Fatigue Crack Propagation and Fracture Toughness of Laser Rapid Manufactured Structures of AISI 316L Stainless Steel.* **3**: p. 36-45, (2014).
17. O.H. Basquin: ASTM proceedings: *The exponential law on endurance tests.* **10**: p. 625-630, (1910).
18. J. Morrow: ASTM Int, *Cyclic plastic strain energy and fatigue of metals.* **STP 378**: p. 45-87, (1964).
19. A. Grigorescu, P.-M. Hilgendorff, M. Zimmermann, C.-P. Fritzen, H.-J. Christ: Adv Mat Res, *Effect of geometry and distribution of inclusions on the VHCF properties of a metastable austenitic stainless steel.* **891-892**: p. 440-445, (2014).
20. A.C. Grigorescu, P.-M. Hilgendorff, M. Zimmermann, C.-P. Fritzen, H.-J. Christ: Int J Fatigue, *Cyclic deformation behavior of austenitic Cr–Ni-steels in the VHCF regime: Part I – Experimental study.* **93**: p. 250-260, (2016).
21. J. Günther, D. Krewerth, T. Lippmann, S. Leuders, T. Tröster, A. Weidner, H. Biermann, T. Niendorf: Int J Fatigue, *Fatigue life of additively manufactured Ti–6Al–4V in the very high cycle fatigue regime.* **94**: p. 236-245, (2017).

The effect of texture on the low cycle fatigue property of Inconel 718 by selective laser melting

*Zhongjiao Zhou, Xu Hua, Changpeng Li, Guofeng Chen**

Materials & Manufacturing Qualification Group, Corporate Technology, Siemens Ltd., China, Beijing 100102, China

Abstract. In recent decades, additive manufacturing (AM) technology has shown its great advantages to produce end-use products with complex design and high-added value. However, the AM-specific characters, such as inherent material anomalies (porosity, lack of fusion defects, or inclusions), anisotropy, location-specific properties and residual stresses, prevent AM from widely adoption in safety-critical parts. Therefore, the damage tolerance assessment of AM parts is desperately necessary. In this study, the impact of residual stress and the induced texture (columnar/equiax grain structure) after different heat treatment on the low cycle fatigue (LCF) behavior of Inconel 718 fabricated through selective laser melting (SLM) is investigated. The results showed that the texture of AMed parts can be controlled by suitable heat treatment, based on the residual stress during AM processing acting as the drive force to recrystallization. For SLMed Inconel 718 samples with columnar grains, anisotropic LCF properties exist, while no obvious sensitivity to orientations is shown for samples with equiaxed grains. This work is significantly meaningful to speed up the design-to-product transformation of safety-critical AM parts and optimize the orientation of components for various applications.

1 Introduction

Additive manufacturing (AM) technique has already been known for more than 20 years, but it experiences considerable publicity and growth in the past few years due to its great advantages to produce end-use products with complex design and high-added value. Currently, most of successful cases of AM are for non-mission-critical applications, in which the microstructure and monotonic response of alloy are studied. However, to meet the full potential that AM has to offer, especially for safety-critical components (e.g., rotating parts, fracture-critical parts, etc.), the qualification considering creep and fatigue testing are necessary [1].

Inconel 718 is the most utilized superalloy for aerospace application, high temperature plants and cryogenic systems due to its low price/performance ratio, good formability and weldability properties [2]. Inconel 718 belongs to the class of precipitate strengthened Ni-base superalloys, with the primary strengthening phase being γ'' with a minority γ'□that may co-precipitate with the γ''. The good weldability makes Inconel 718 ideally suited for the investigation of SLM process [3].

During AM processing, due to the rapid heating of the upper surface by the fast moving heat source (laser in present study) and the relatively slow heat conduction, a steep temperature gradient develops. Then the expansion of the heated top layer is restricted by the underlying material, thus forming residual stress during processing [4]. If the residual stress is high enough, it may cause the distortion and even failure of AMed parts. In the case that the residual stress is relatively low, it may act as the drive force to recrystallization during the following heat treatment. The recrystallization will break the well-known columnar-grained structure of as-built AMed parts, thus influencing the mechanical properties.

Many work have been reported to control the texture for AMed parts [5-8]. Dinda et al. [5] were the first to report on the ability to steer the growth direction of grains through changing the direction of the laser heat source. Later, Dehoff et al. [6] demonstrated the premise of texture control within AM materials in a site-specific manner in the electron beam melting (EBM) process, through manipulation of the thermal conditions within the melt pool in the different regions, although not in a perfect manner. Recently, Helmer et al. [7, 8] have shown the ability to influence the grain structure in the EBM process for Inconel 718 through changing the parameters associated with the standard raster scan strategy utilized by the process.

However, based on the black box algorithm of the AM process, applying texture control over generic cross-sections within bulk structures in a controlled manner is extremely challenging. A relatively simple and easy to implement method is needed for texture control. In addition, the influence of various texture on the mechanical properties of AMed parts is not very clear.

As mentioned above, due to the nature of fast melting and solidification, residual stress is inevitable during AM processing. The goal of this work is to control texture by subsequent heat treatment, in which different degree of recrystallization occurs drived by

* Corresponding author: guofeng.chen@siemens.com

 https://doi.org/10.1051/matecconf/201816502007

residual stress. Afterwards, the influence of textures on low cycle fatigue properties of SLMed Inconel 718 is studied.

2 Materials and experimental

Commercial Inconel 718 powders with an average particle size of 20 to 50 μm were used as the feedstock materials in present experiments. The manufacturing of the samples was performed in an EOSINT M280 machine operated at a power of ~ 285 W with a scanning speed of 0.96 m/s, laser focus spot size of 70 μm, and laser thickness of 40 μm, under an argon atmosphere with an oxygen level of 0.01%. The scanning strategies was bidirectional scanning with a rotation of 90° between layers (Scan Strategy YX) that repeats the Y scan and X-scan (as shown in Fig. 1a).

Then, two heat treatment methods were applied to obtained different textures:

1. SA (Solution + double ageing): solution at 980 °C/1 h, AC + double aging at 760 °C/10 h, furnace cooling for 2 h to 650 °C and holding at 650 °C/8 h, AC
2. HA (homogenization + double ageing): homogenization at 1065 °C/1 h, AC + double aging at 760 °C/10 h, furnace cooling for 2 h to 650 °C and holding at 650 °C/8 h, AC.

To investigate the microstructure, the AMed samples were mechanically polished via a standard metallographic procedure using 1 μm diamond powder in the final stage of polishing. After final electropolishing the specimen was etched in a solution composed of 70 vol% H_3PO_4 and 30 vol% H_2O at 5 V at room temperature. The texture was taken using an electron backscatter diffraction (EBSD) system mounted on a field-emission scanning electron microscope (FE-SEM; Supra 35; Zeiss; Germany). An X-ray CT was utilized as non-destructive measuring technique for internal porosity detection due to its capability of providing a complete analysis of size, shape, volume and distribution of pores/defects within the entire analyzed volume.

After microstructural characterization, low cycle fatigue (LCF) experiments were performed. The LCF testing was performed on cylindrical dog-bone specimens, with printing orientation and size of LCF samples shown in Fig. 1. To examine the influence of anisotropy of the SLMed In718 material, the loading axis of the different samples was oriented parallel and transverse to the build direction. All LCF experiments were conducted under strain-controlled conditions, with R=-1 at 650 °C and strain amplitude of 0.40% (Fig. 1c).

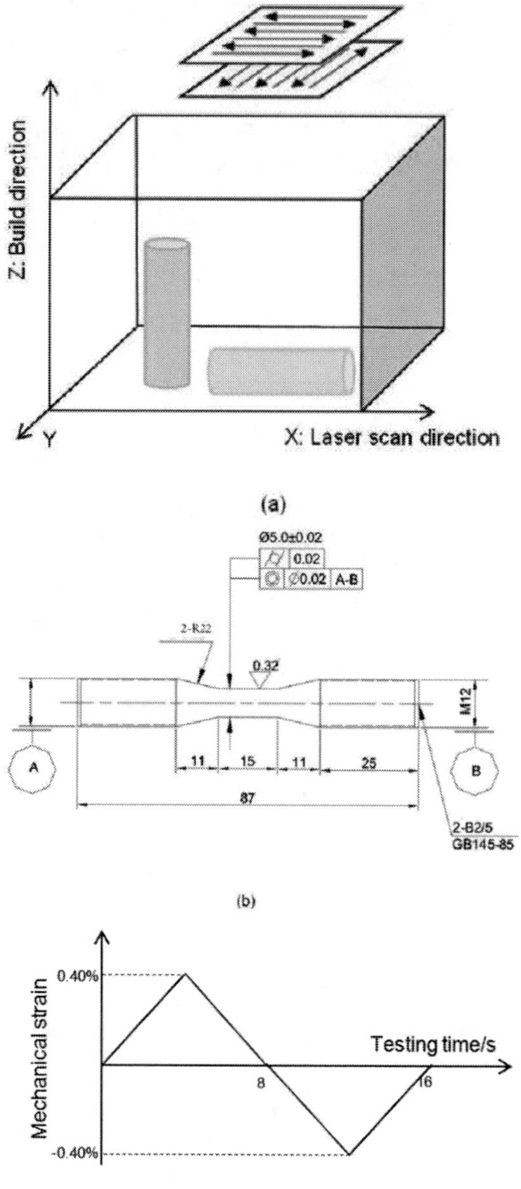

Fig. 1. The schematic of sample orientation (a), LCF testing sample (b) and the LCF testing condition (c).

3 Results and Discussion

The resultant textures of as-built SLMed Inconel 718 are illustrated in Fig. 2. Obviously, the standard linear heat source resulted in a columnar grain with orientation of <001> and <110> aligned parallel to the build direction (Fig. 2a). However, in the plane transverse to the building direction, equiaxed grains rather than columnar grains were observed (Fig2. b). Most of these

https://doi.org/10.1051/matecconf/201816502007

columnar grains showed 3 μm in widths and 18.5 mm in length (Fig2. c).

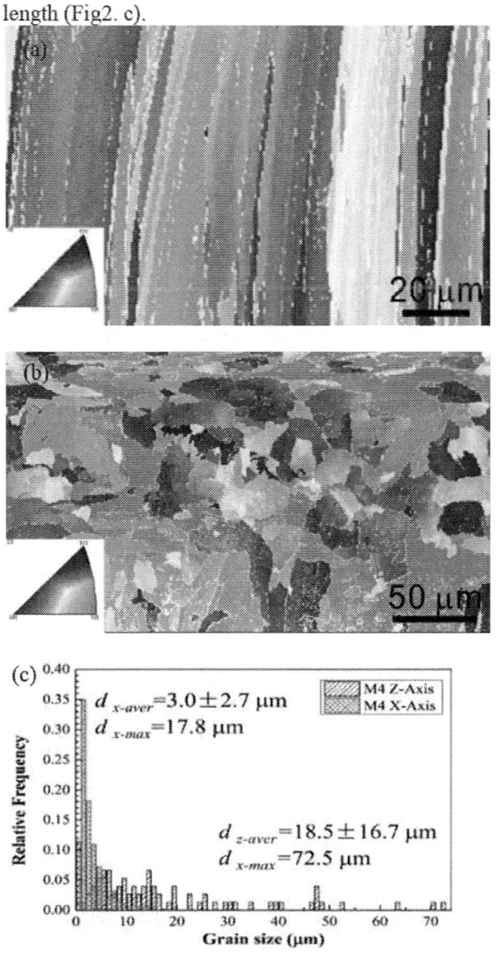

Fig. 2. EBSD analysis depicting the texture on the plane parallel to the building direction (a), on the plane transverse to the building direction (b) and the grain size distribution of columnar grains (c).

The pore distribution analyzed by X-ray CT is shown in Fig. 3. The density is about 99.85% with the largest pore with diameter 40 mm. three types of pores were observed: Large irregular pores caused by a lack of melting, shrinkage micropores due to a lack of feeding within interdendritic zones and spherical pores caused by trapped gas.

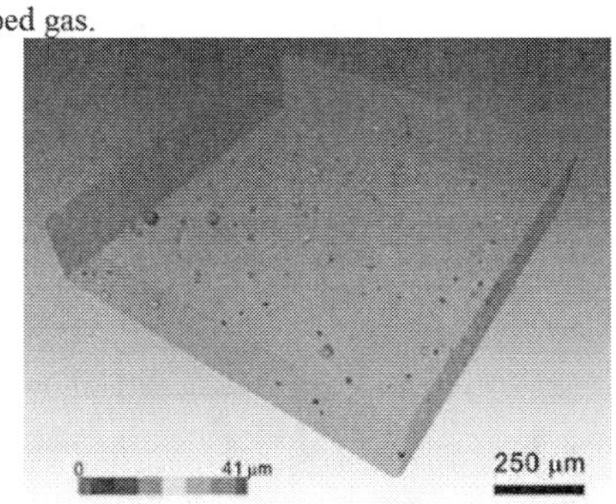

Fig. 3. The CT reconstructed 3D volume of SLMed In718 showing the pores/defects distribution

After heat treatment, different degree of recrystallization occurred which was drove by the residual stress during AM process, leading to different textures, as shown in Fig. 4. As for SA samples, only partly recrystallization occurred and the columnar grains still existed parallel to the build direction (Fig. 4a). However, as for the HA samples, fully recrystallization occurred and all the columnar grained had transformed to equiaxed grains, as shown in Fig. 4b.

Fig. 4. The EBSD results of SA sample (a) and HA sample (b).

The textures have a significant influence on the LCF life, as shown in Fig. 5. Obvious anisotropy exhibited for samples with columnar grains. Amongst the textures, the columnar grains oriented parallel to the build direction exhibited the highest life on average compared to the transverse columnar material. However, for samples with equiaxed grains, the samples parallel to the building direction showed comparable LCF life to those transverse to the building direction, namely the mechanical anisotropy nearly disappeared.

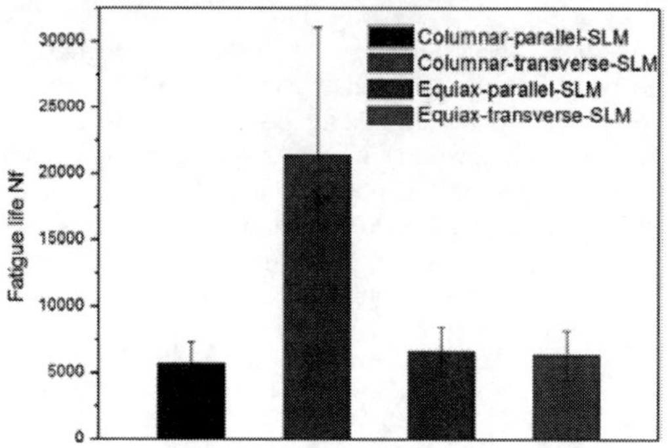

Fig. 5. The LCF life of samples with different textures

Based on the cyclic stress amplitude for each material orientation and texture, the material achieved stabilized hysteresis within the first few initial cycles as illustrated in Fig. 6. The HA sample, namely samples with equiaxed grains showed higher cyclic stress amplitude compared to the SA samples with columnar grain, which may be attributed to the cyclic Young Modulus in various texture orientation.

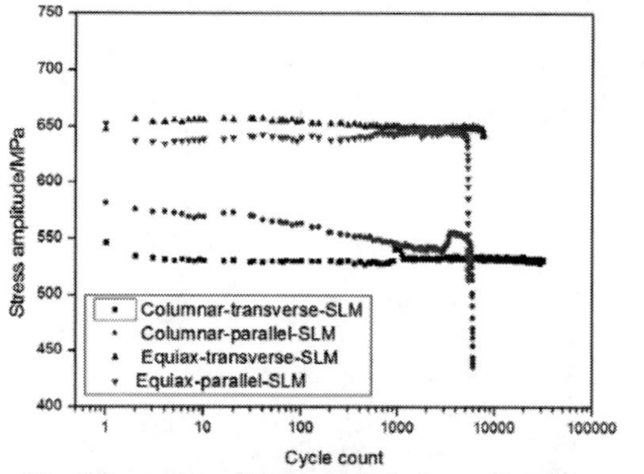

Fig. 6. Comparison of stress amplitude for samples with different textures.

Generally, various textures are desired for different applications based on the loading condition and property requirement. Just as investment casted turbine blades for aerospace application, the low Young Modulus <100> single crystal or columnar orientation is preferred to obtain lower stress parallel to the growth direction of blades, while equiaxed and small grains are needed for turbine disks to ensure longer lifetime. With the development of AM techniques, the texture control will be required; especially for AMed parts whose creep or fatigue properties are required.

The present work has illustrated the possibility to control texture of AMed parts by suitable heat treatment utilizing the residual stress during AM processing. In further study, the quantitative characterization of residual stress as well as the relationship between the residual stress and solution temperature need to be investigated, which will be significantly meaningful to speed up the design-to-product transformation of AMed parts and optimize the orientation of components for various applications.

4 Conclusion

In this work, the effect of textures and material anisotropy on the low cycle fatigue behavior of SLMed Inconel 718 was examined at 650 °C. The following conclusions can be obtained:

1. Based on the residual stress during AM processing acting as the drive force to recrystallization, the texture of AMed parts can be controlled by suitable heat treatment.
2. For SLMed Inconel 718 samples with columnar grains, anisotropic LCF properties exist and the columnar grains oriented parallel to the build direction exhibited the highest life.
3. For SLMed Inconel 718 samples with equiaxed grains, no obvious sensitivity to orientations is shown.

References

1. M. Seifi, A. Salem, J. Beuth, O. Harrysson, J.J. Lewandowski, JOM, **68**, 3 (2016)
2. C. Sims, The Superalloys, John Wiley and Sons, (1972)
3. P. Blackwell, J Mater Process Technol, **170**, 240 (2005)
4. B. Cheng, K. Chou, 25th Annual International Solid Freeform Fabrication Symposium - An Additive Manufacturing Conference, 1076 (2014)
5. G. Dinda, A. Dasgupta, J. Mazumder, Scr. Mater. **67**, 503 (2012).
6. R.R. Dehoff, M.M. Kirka, W.J. Sames, H. Bilheux, A.S. Tremsin, L.E. Lowe, S.S Babu, Mater. Sci. Technol. **31**, 931 (2015).
7. H.E. Helmer, C. Körner, R.F. Singer, J. Mater. Res. **29**, 1987 (2014).
8. H. Helmer, A. Bauereiß, R.F. Singer, C. Ko¨rner, Mater. Sci. Eng. A **668**, 180 (2016).

Influence of porosity on the fatigue behaviour of additively fabricated TA6V alloys

Viet-Duc Le[1,*], *Etienne* Pessard[1], *Franck* Morel[1], *François* Edy[2]

[1]LAMPA, Arts et métiers ParisTech, Centre d'Angers, France

[2] IRT Jules Verne, Bouguenais, France

Abstract. This work is focused on the influence of porosity when dealing with the fatigue behaviour of the TA6V alloys fabricated by the selective laser melting (SLM) process. The presence of porosity is one of the major issue facing additive manufacturing (AM) of metallic components subjected to fatigue loading. In order to study the effect of porosity on the fatigue behaviour, a vast experimental campaign has been undertaken. These materials are fabricated by the SLM process with different building directions (horizontal, vertical and diagonal) thanks to which specimen batches with different pore sizes are obtained. It is observed that despite a low porosity fraction (around 0.01%), the influence of pores on the fatigue strength of the materials is pronounced. A mapping approach is presented, which links the applied stress, the pore size and the fatigue life. This approach makes it possible to accurately characterize the effect of the porosity size, and also to construct Kitagawa-Takahashi diagrams from S-N data.

1 Introduction

During recent years, additive manufacturing (AM) technologies have continued to grow steadily due to industrial and academic developments. One of the main advantages of this technology is the capacity to fabricate complex geometries directly from a computer-aided design. Among several metal AM processes, Selective Laser Melting (SLM) is one of the most commonly used. In this process, the energy of the laser source is used point-wise to melt powder within a powder-bed layer. Upon completion of a single layer, recoating of a new powder-layer is done and subsequent melting is pursued to continue the manufacturing process to build the 3D geometry. As for the material, the TA6V (Ti6Al4V) titanium alloy is one of the most common alloy manufactured by the SLM process and has applications in many industries, in particular the aeronautic and the medical domains. The mechanical properties of TA6V alloy obtained by this process have been intensively studied in the literature [1] [2] [3] [4] [5] . For the monotonic tensile behaviour, it has been observed that the tensile strength of the AM TA6V is equivalent or even higher than the alloys obtained by conventional processes [6] [7] [8]. However, in terms of fatigue behaviour, the AM TA6V alloy shows significantly lower fatigue strength [9] [10] [11] . This fact is due principally to microstructural defects in the AM materials, in particular porosity. The impact of pore size on the fatigue behaviour of the AM TA6V alloys has been investigated in the literature [12] [13] [14]. However, to the best of the authors' knowledge, the effects of pore at the surface or in the volume or the effect of the pore geometry are rarely mentioned.

The principal aim of this paper is to contribute to the understanding of the impact of porosity on the fatigue behaviour of the TA6V alloys obtained by the SLM process. In order to achieve this aim, a vast experimental fatigue campaign for uniaxial loading mode (loading ratio R=0.1) has been undertaken. Five different specimen batches have been investigated. They were obtained using different AM building directions (horizontal, vertical and diagonal). All of the specimens were machined to avoid the effect of the as-built surface.

The fatigue behaviour and the associated fatigue crack initiation mechanisms are presented. The influence of porosity, including the effects of the surface/volume pores and of the pore geometry will be highlighted thanks to a mapping approach that links the applied stress, the pore size and the fatigue life. A first results of a fatigue model will also be discussed.

2 Material and experimental conditions

2.1 Material and specimen fabrication

The TA6V alloy used in this study is grade 5 ELI for which the referred chemical composition (source www.titane-services.fr) is shown in Table 1.

The specimen were fabricated by IRT Jules Verne by using a SLM®280 HL machine. The fabrication process has been optimized thanks to a vast experimental campaign in which several process parameters such as laser power, scan speed, hatch distance or focus were

* Corresponding author: viet-duc.le@ensam.eu

taken into consideration. Note that the process parameters were the same for all of the specimen batches investigated in the present work.

Table 1. Referred chemical compositions of the investigated TA6V alloy

Al	V	C	Fe	H	N	O
6.0	4.0	≤0.08	≤0.25	≤0.0125	≤0.05	≤0.13

For monotonic tensile tests, three specimen batches with 5 specimens per batch, referred to as 0°, 45° and 90°, corresponding to the building direction, were fabricated. Note that for the 0° configuration, the specimen axis is perpendicular to the building direction. The tensile specimens have a plate shape with a gauge section of approximately 6x2 mm².

For the fatigue tests, five specimen batches are investigated in this study, referred to as 0°-B1, 0°-B2, 45°-B1, 45°-B2, 90°. Table 2 summaries the fabricated fatigue specimen batches. Fig. 1 shows the 0° and 45° fatigue specimen batches.

Table 2 : Summary of the fabricated fatigue specimen batches

Building direction	Specimen surface	Batch Id	Number of specimens
0°	Machined	0°-B1	10
		0°-B2	10
45°	Machined	45°-B1	14
		45°-B2	6
90°	Machined	90°	20

After fabrication, all of the specimens were post heat treated by a usual heat treatment (annealed at 850 °C for 2 hours following by a slowly cooling within the furnace) to relax the residual stresses.

The machining of all specimens was done after the heat treatment. The Ra roughness of the machined surface is lower than 0.8. The fatigue specimen geometry (shown in Fig. 2) is given as per the ISO 2072:2010 standard and is the same for all batches. The position of all of the fatigue specimens on the plate was registered in order to investigate the possible effect of the specimen position on the fatigue strength of the specimens.

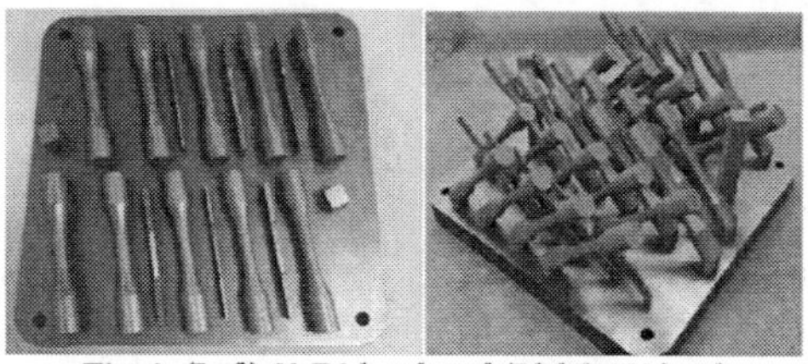

Fig. 1. (Left) 0°-B1 batch and (Right) 45° batch

2.2 Material and mechanical property characterization

2.2.1 Microstructure characterization

The microstructure was observed using a light optical ZEISS microscope on polished and chemically etched samples (by using the Kroll solution for 20 s). These samples were cut from the fatigue specimens after fatigue testing.

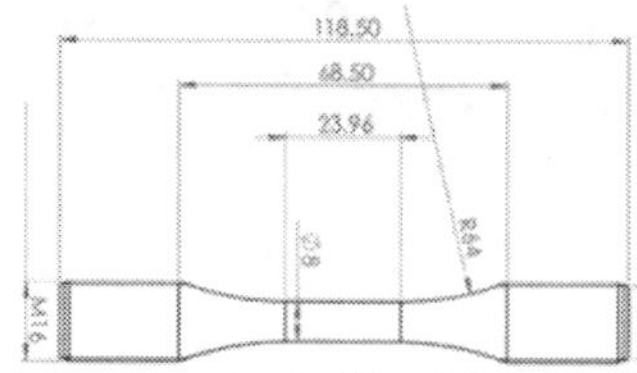

Fig. 2. Geometry for all of the fatigue specimens

2.2.2 Monotonic tensile testing

The monotonic tensile tests were done using an INSTRON servo-hydraulic machine. The strain was measured thanks to an extensometer with an initial length L_0=25 mm and a displacement range ▭L=+/- 2.5 mm. All of the tests were conducted with a constant strain rate of 0.005 min^{-1} at ambient temperature.

2.2.3 Fatigue testing

All of the fatigue tests presented below were carried out at ambient temperature and pressure in laboratory air. The fatigue tests were conducted with a constant stress amplitude with a loading ratio R=0.1 at a frequency of 20 Hz. A maximum fatigue life of $2x10^6$ cycles was used. The stopping criterion was chosen to be the complete rupture of the specimen. The run-out specimens that survived $2x10^6$ cycles were re-tested at a higher load. All of the fatigue failure surfaces were analysed on a ZEISS Gemini SEM.

3 Results and discussion

3.1 Microstructure and porosity characterization

It is well known in the literature that the microstructure of the TA6V alloys obtained by AM processes is columnar [15] with grains growing according to the building direction. In the present work, all of the specimens were fabricated and post-heat treated with the same parameters and heat treatment conditions. Therefore, the microstructures of all of the batches must be the same for all specimen orientations.

The microstructural characterization was realized on samples cut from the fatigue specimens after fatigue tests. The cut planes are perpendicular to the specimen axis in order to measure the pore size in the principal stress plane. It can be seen in Fig. 3. that the microstructure of the 0° specimen is columnar but this is not the case for the 45° and 90° specimens. In fact, for the 0° samples, the cut planes are parallel to the building direction. For 45° and 90° specimens, the observation planes are not parallel to the building direction.

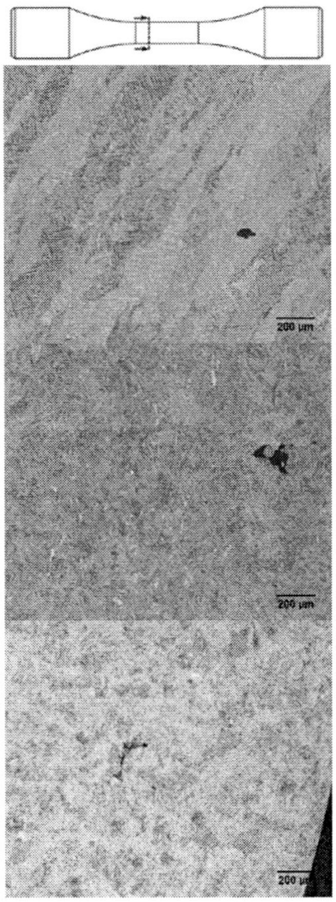

Fig. 3. Microstructures observed on polished samples from 0°-B1 batch (top), 45°-B1 batch (middle) and 90°-B1 batch (bottom). Observation planes are perpendicular to the specimen axis.

The porosity characterizations were conducted on 3 batches 0°-B1, 45°-B1 and 90°. The characterization methodology proposed by Murakami [16] was used. As per this method, the size of the biggest pore observed in each standard inspection area of size S_0 is measured. The resulting distribution corresponds to the extreme value distribution related to the area S_0. In the present work, the inspection area S_0 is equal to the area of the taken images, S_0=2.34 mm². The $\sqrt{area}$ of pores is used to define the pore size. For each batch, 300 observations were realized from samples cut from 3 fatigue specimens after fatigue tests. The three specimens were intentionally chosen so that their fatigue strengths are representative of the fatigue strength of the batch.

It is worth noting that almost all images contain no more than one pore so the porosity can also be estimated. Table 3 shows the porosity density of the three investigated batches.

Table 3. porosity density (surface measurement) of the three batches 0°-B1, 45°-B1 and 90°

Batch Id	Porosity (%)
0°-B1	0.003
45°-B1	0.013
90°	0.004

The pore size distributions for the three batches 0°-B1, 45°-B1 and 90° are shown in Fig. 4. The y-axis is presented by using the reduced variable, *Y*, defined as a function of the cumulative probability, *F*, given as the following:

$$Y = -\ln(-\ln F) \qquad (1)$$

It can be observed that the size of the pores observed for 0°-B1 specimens are the smallest while the pore sizes for the 45°-B1 are the biggest.

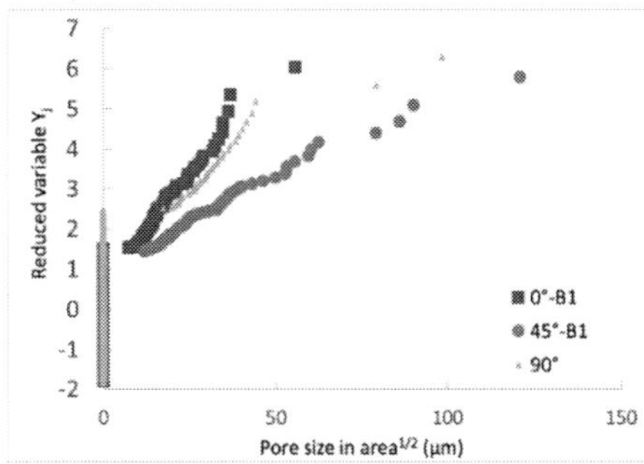

Fig. 4. Pore size distributions of the three batches 0°-B1, 45°-B1 and 90°-B1 observed on polished samples.

Finally, it is worth mentioning the anisotropy of the pore geometry. In the investigated alloys, it will be shown in the next section that the critical pores observed at the fatigue crack initiations are principally related to the lack of fusion. In the work of Kasperovich et al. [17], the authors showed, thanks to micro-tomography observations, that the pores relating to the lack of fusion have a very spread geometry, i.e. a large area but small thickness. The plane on which pores spread out is generally perpendicular to the building direction. Fig. 5 shows a pore observed by optical microscopy on a 0° specimen.

Fig. 6 shows a schema illustrating the anisotropies of the microstructure and of the pores and also the orientation of the observation planes for specimens of 3 building directions.

One of the questions in this work is to know which parameter is more important to consider in fatigue between the particular microstructure and the porosity population.

3.3 Monotonic tensile behaviour.

The comparison of the ultimate tensile strength (Rm), the yield strength (Rp0.2) and the fracture strain of the six investigated batches is shown in Fig. 7. The error bars correspond to the standard deviations of the determined properties.

It can be seen that the 45° specimens show the greatest yield strength and ultimate tensile strength and the 90° specimens have the lowest values.

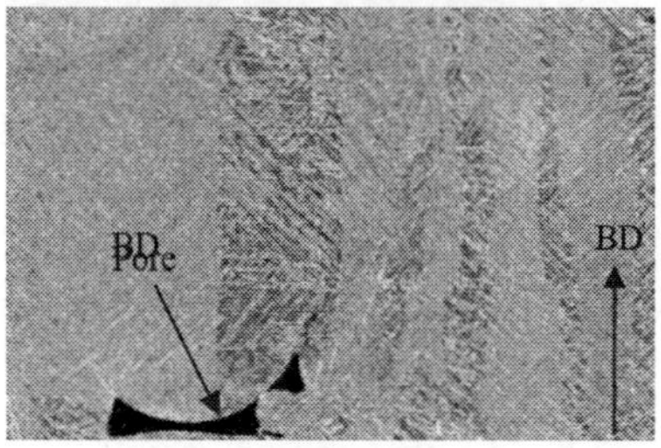

Fig. 5. A pore observed by optical microscopy on a 0° specimen in the present project.

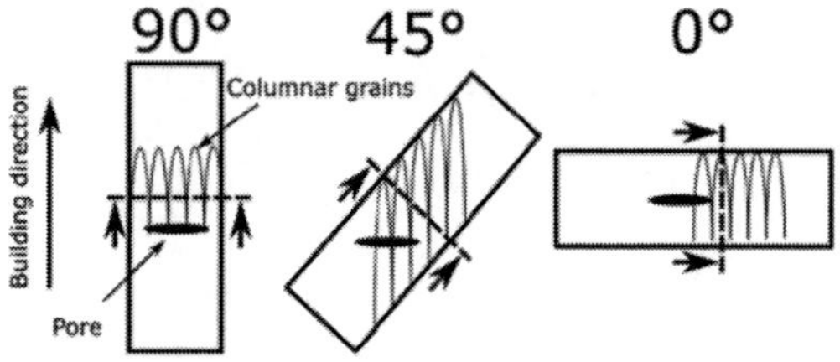

Fig. 6. Schema showing the orientation of pores and the observation planes (dashed lines) for 0°, 45° and 90° specimens.

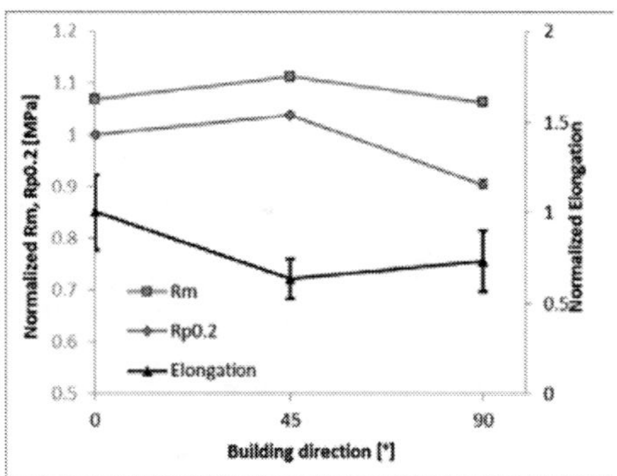

Fig. 7. Monotonic tensile properties for three investigated batches

3.4 Fatigue behaviour and fatigue crack initiation mechanisms

3.4.1 S-N curves

Fig. 8 shows the S-N curves for the 5 investigated batches: the top figure for 0° batches, the middle figure for 45° batches and the bottom figure for 90° batch. The regression lines correspond to the Stromeyer theoretical model [18] as given by the following equation.

$$\text{Log}\, N = A - B\, \text{Log}(\Delta\sigma - \Delta\sigma_0) \qquad (2)$$

Firstly, a large difference can be seen between the two 0° batches as well as between the two 45° batches, in spite of the fact that the same process parameters were used. The measurements of the fabrication conditions such as the oxygen level or the plate and chamber temperatures also showed that there was no notable difference between batches. For the moment, the authors cannot explain why there is such differences. It is assumed that the fabrication repeatability was not obtained in the present work.

Another observation that can be made for the 0° and 45° batches is that the S-N data scatter is higher for the 0°-B2 and 45°-B2 batches for which the fatigue strengths are better. For the 90° batch, it can be seen that the fatigue scatter is very high. It seems that there are two distinguished S-N curves: one for the 3 points indicated by the arrows and the other one is for the remaining 17 points. The Stromeyer curve shown was fitted to these 17 points.

The comparison of all of the specimen batches is shown in Fig. 9. It can be seen that generally, except the 3 points mentioned previously, the fatigue strength of the 90° batch is the lowest. For the 0° and 45° specimens, the difference between the 0°-B2 and 45°-B2 is very slight. The difference between the 0°-B1 and 45°-B1 is also small in the high cycle number regime (N_f>10^5 cycles) and more pronounced in the low cycle number regime.

It can also be seen that the higher fatigue strength of the 45°-B1 batch compared to the 90° batch is not in agreement with the pore size distributions measured on polished samples of these batches. It was shown in Fig. 4. that the pore size of the 45°-B1 specimens is larger than for the 90° specimens. One possible explanation is that the pore size distributions *measured on polished samples* are not representative of the real material porosity for all of the building directions. As mentioned in section 3.1, the critical pores in the investigated alloys have a very spread geometry in the planes perpendicular to the building direction. The authors assume that for this reason, the probability that the observation planes contain the biggest pores is small for the 90° samples despite a large number of observations (300 observations with an area S_0=2.34 mm² were realized on 3 specimens).

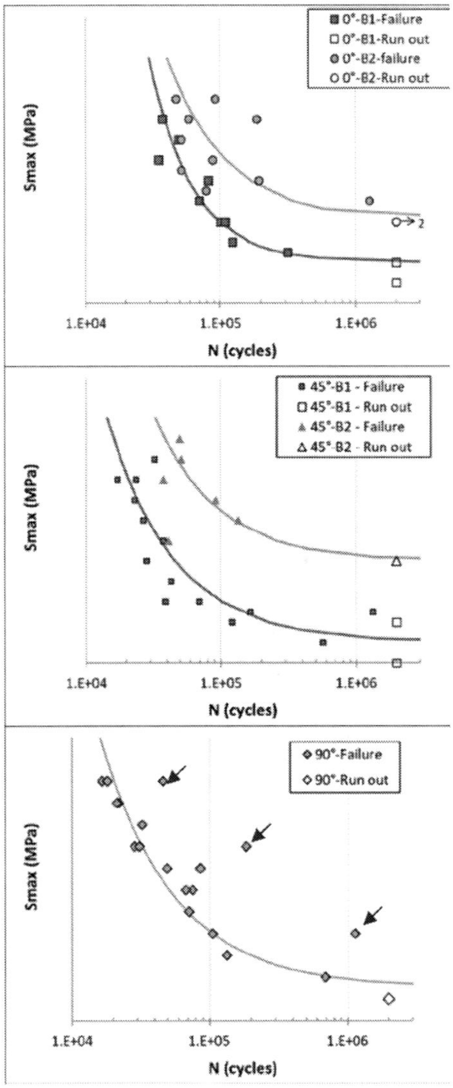

Fig. 8. Wöhler curves for all of specimen batches: 0° batches (top); 45° batches (middle) and 90° batch (bottom)

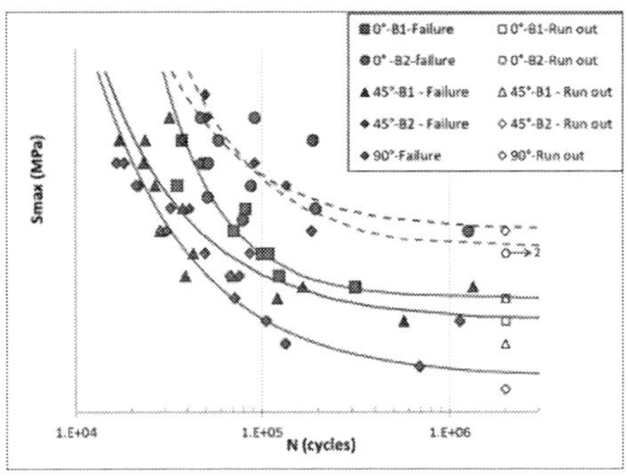

Fig. 9. Comparison of the Wöhler curve of all of the specimen batches

3.4.2 Crack initiation mechanisms

All of the fatigue cracks initiate from one site. Almost all the fatigue cracks initiate from a pore (observed on 59 specimens out of a total of 60). Among them, 4 cracks initiate from internal pores and the others initiate from surface pores. Only one specimen with crack initiation from the microstructure was observed. Fig. 10 shows crack initiations from surface pores observed on 0°, 45° and 90° specimens.

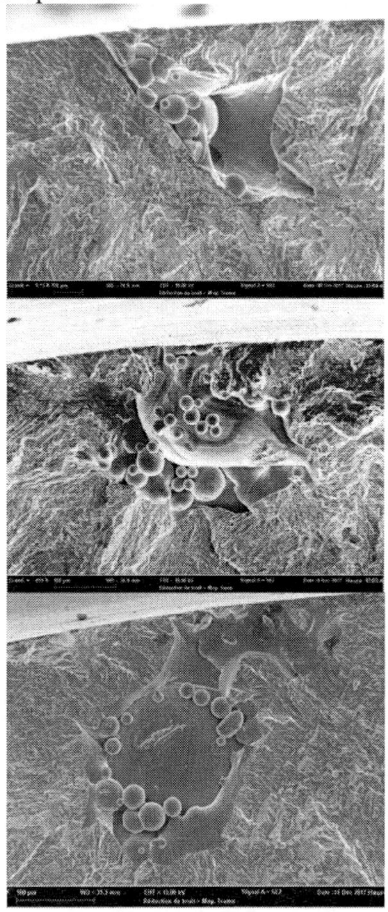

Fig. 10. A fatigue crack initiation from surface pore observed on a 0° specimen (Top), 45° specimen (Middle) and 90° specimen (Bottom).

3.5 Characterization of the effect of pore size on the fatigue behaviour

All of the fatigue failure surfaces were analysed by using a SEM in order to measure the $\sqrt{area}$ of the pores at the crack initiation sites. Fig. 11 shows the comparison of the pore size distribution of all of the specimen batches. The *normalized* fatigue strengths, $\sigma_{D,n}$, estimated thanks to the Stromeyer fitting curves at a fatigue life of $2x10^6$ cycles are also shown. It can be seen that the pore size distributions are generally in a good agreement with the fatigue strengths. In fact, it was shown that the 90°

 https://doi.org/10.1051/matecconf/201816502008

specimens have the lowest fatigue life. The 0°-B2 and 45°-B2 showed a similar fatigue S-N behaviour while the fatigue life of 0°-B1 batches is slightly higher than the 45°-B1 batch. Another observation could be made is that by comparing with the pore size distributions measured on polished samples for the three batches 0°-B1, 45°-B1 and 90° (Fig. 4), the size of the pores at the crack initiation sites is much larger. Also, a disagreement can be seen: for the distributions measured on polished samples, the size of the pores of the 45°-B1 specimens are the largest however for the measurements on the crack initiation sites, the pore size of the 90° specimens is the biggest. This disagreement was discussed in section 3.4.1.

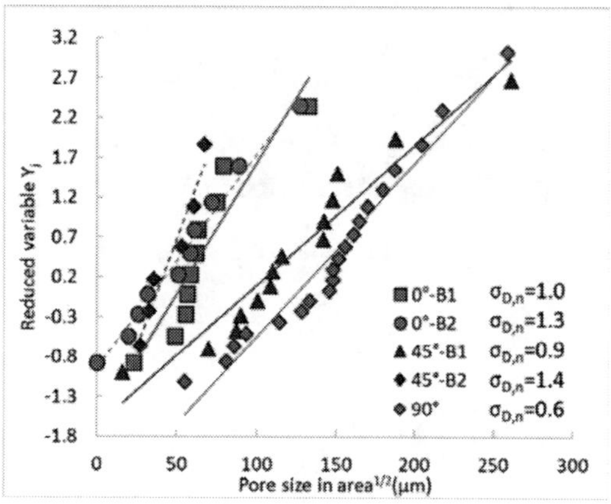

Fig. 11. Comparison of the pore size distributions measured on the crack initiation sites and the normalized fatigue strengths, $\sigma_{D,n}$, of the five investigated specimen batches.

Because of the large scatter related to the fatigue data as well as to the pore size distributions, the presentation of the pore size distributions is limited to a very coarse view of the effect of pore size. In order to better understand the pore size effect, a mapping that links the applied stress (S), the pore size measured at the crack initiation site ($\sqrt{area}$) and the fatigue life (N_f) is used and shown in Fig. 12 for the three building directions. A linear interpolation in terms of Log(N_f) was used to fill the maps. The Y-axis scale is the same for all of the three maps. The run-out specimens are also presented with a fatigue life of 2x10^6 cycles and with a pore size measured on the failure surface of the same specimen after re-tested at higher load.

Thanks to these diagrams, two interesting observations can be highlighted. Firstly, the arrows shown in the top and bottom maps of Fig. 12 (for the 0° and the 90° batches) indicate some "perturbation" points that show a better fatigue strength compared to others. The observations on the fatigue failure surfaces showed that these points correspond to the crack initiation from *internal* pore as shown in the top map of Fig. 13. Secondly, the arrows shown in the middle map of Fig. 12 indicate also three points with a slightly better fatigue life. These points relate to crack initiation from *spherical gas* pores (see Fig. 13-Bottom) that have a much more regular geometry than the pores associated with the lack of fusion or non-melted particles.

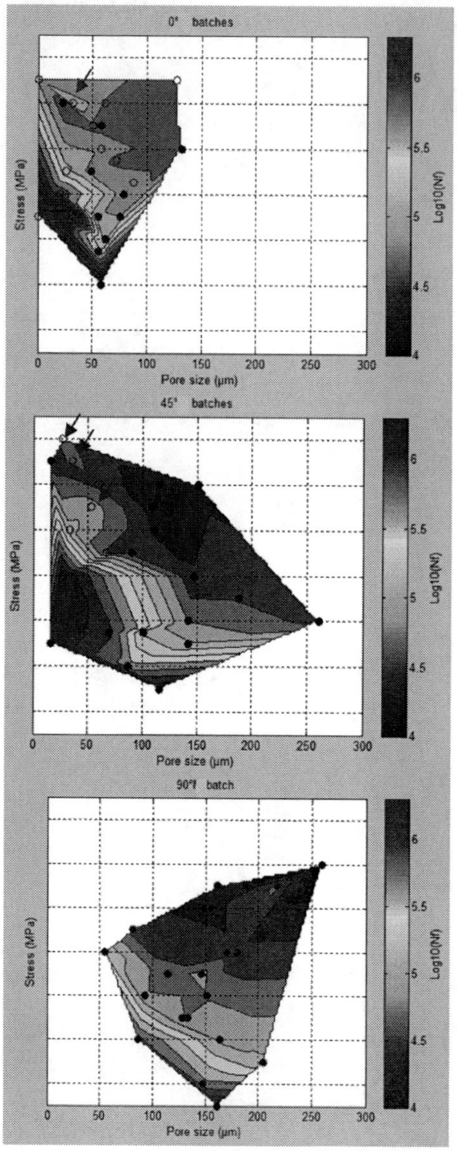

Fig. 12. S-$\sqrt{area}$ -N_f mapping for the 0° batches (top), 45° batches (middle) and 90° batch (bottom). For the 0° and 45° batches, the solid points correspond to the B1 batch and the open points for B2 batch.

The superposition of all of the fatigue data of the specimen batches without internal and spherical gas pores in Fig. 14 shows good agreement between the stress, the pore size and the fatigue life in spite of large scatter related to the S-N curves as well as to the pore size distributions.

The effect of the microstructure on the fatigue life does not seem pronounced in comparison with the pore size. As an example, two points enclosed by the white ellipse in Fig. 14 correspond two critical pores with a similar size and geometry. The fatigue life of the two specimens are close (67619 and 69350 cycles) despite one is a 45° specimen and the other is a 90° specimen.

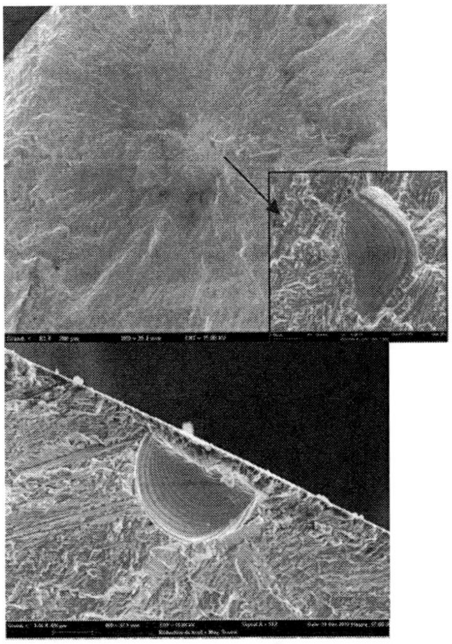

Fig. 13. Top - a fatigue crack initiation form an *internal* pore observed on a 0°-B1 specimen; Bottom – a fatigue crack initiation from a *spherical gas* pore observed on a 45°-B2 specimen

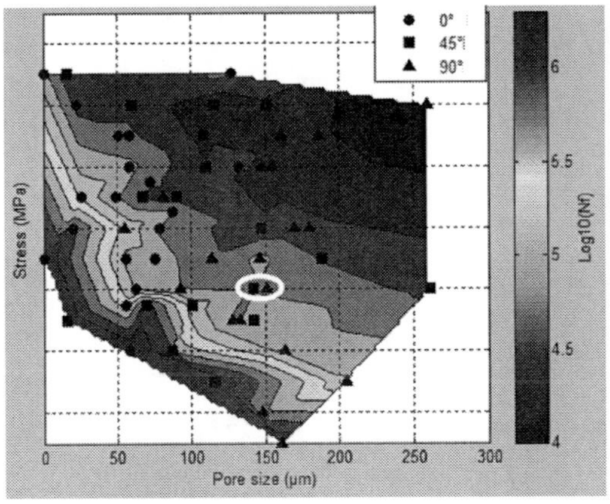

Fig. 14. S-$\sqrt{area}$ -N_f mapping for all of the specimen batches (circle: 0° batches, square: 45° batches and triangle: 90°) with neither internal pores nor spherical gas pores.

Another advantage of this S-$\sqrt{area}$ –N_f maps is that they can also be used as a pseudo-Kitagawa – Takahashi diagram. If we assumed that the applied stress at a fatigue life of 2×10^6 cycles is the fatigue strength of the material, then the dark red region in these diagrams can be considered as the Kitagawa – Takahashi diagram. In order to estimate the trend of the Kitagawa-Takahashi diagram of the investigated alloy, three fatigue criteria, the classical linear elastic fracture mechanic, El-Haddad model and the Murakami approach, were used. The equations of the three criteria are given as follows.

- The classical LEFM criterion:

$$\sigma_{D,a} = \frac{\Delta K_{th}}{2 \times 0.65 \times \sqrt{\pi \sqrt{area}}} \tag{3}$$

- El-Haddad approach:

$$\sigma_{D,a} = \frac{\Delta K_{th}}{2 \times 0.65 \times \sqrt{\pi(\sqrt{area} + \sqrt{area_0})}} \tag{4}$$

where $\sqrt{area_0} = \frac{1}{\pi}\left(\frac{\Delta K_{th}}{0.65 \times \Delta\sigma_{D0}}\right)^2$, $\Delta\sigma_{D0}$ is the fatigue strength of the porosity-free material.

- Murakami approach:

$$\sigma_{D,a} = \frac{1.43\,(\mathrm{Hv} + 120)}{\sqrt{area}^{1/6}} \times \left(\frac{1-R}{2}\right)^{\alpha} \tag{5}$$

where $\alpha = 0.226 + \mathrm{Hv} \times 10^{-4}$

The material constants ΔK_{th} and $\Delta\sigma_{D0}$ were taken from the literature. The material hardness was experimentally measured on a 90° sample (average value according to the sample height) and is supposed to be the same for all of the building directions. Fig. 15 shows on a log-log scale the three fatigue approaches and the experimental points for which the fatigue life is higher than $10^{5.5}$ cycles.

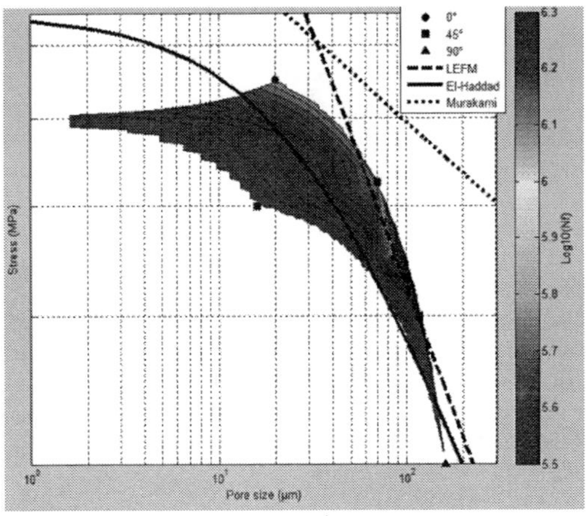

Fig. 15. S-$\sqrt{area}$ -N_f mapping for specimens of which the fatigue life $N_f > 10^{5.5}$ cycles. The dotted line relates to the Murakami approach, the dashed line corresponds to the classical LEFM fatigue criterion, the solid line for the El-Haddad model.

It can be seen that the El-Haddad model fits the experimental data better while the Murakami approach is very non-conservative. However, because of the limited number of experimental data, more fatigue tests in the high cycle fatigue regime need to be done.

4 Conclusion

In the present work, a vast experimental campaign was carried out to investigate the fatigue behaviour and the crack initiation mechanisms of the TA6V alloy obtained by the SLM process. Five fatigue specimen batches were fabricated with three building directions (0°, 45° and 90°). All of the specimens were machined in order to avoid the effect of the as-built surface. The microstructural and porosity characterizations showed high anisotropies in terms of microstructure and pore geometry. The pore size distributions were also measured by optical observations on polished samples for 0°, 45° and 90° batches. It could be established a link between the pore size distribution and the fatigue strength for 0° and 45° batches. For the 90° specimens the link seems not to be clear. The authors assumed that this is due to the high anisotropy of the pore geometry.

As for the fatigue behaviour, the 90° batch showed the lowest fatigue life while the difference between 0° and 45° specimens is not pronounced. A large difference was also shown in terms of the fatigue behaviour between the two 0° machined batches and between the two 45° machined batches even though the same process parameters were used. Also, high scatter related to the S-N curves was observed.

The characterization of the fatigue crack initiation mechanisms showed almost all of the fatigue cracks initiate from pores. In several observations, the pores associate with a large non-melted particle that could be formed by the balling effect.

The characterization of the pore size effect on the fatigue life was done thanks to a mapping approach that links the stress (S), the pore size ($\sqrt{area}$) and the fatigue life (N_f). Good agreement between these three parameters is shown for all of the investigated batches despite the large scatter in terms of S-N data and pore size. It was also highlighted the beneficial effects of the distance to the specimen surface (i.e. internal vs surface pores) and the pore geometry (i.e. spherical gas pores). This mapping approach can also be used as a pseudo Kitagawa-Takahashi diagram. Three fatigue criteria, classical LEFM, El-Haddad and Murakami criteria, were applied to investigate the trend of pore size – fatigue strength relationship of the investigated alloy. Among them, the El-Haddad model best fits the experimental data.

Acknowledgement: This study is part of the FATAL project managed by IRT Jules Verne (French Institute in Research and Technology in Advanced Manufacturing Technologies for Composite, Metallic and Hybrid Structures). The authors wish to associate the industrial and academic partners of this project; Respectively ACB, Arts et Métiers ParisTech, CNRS, DAHER, Ecole Centrale de Nantes, Europe Technologies, FIVES, Renault and University of Nantes.

References

1. S. Leuders, T. Lieneke, S. Lammers, T. Tröster et T. Niendorf, J. Mater. Res. , vol. **29**, pp. 1911-1919, 7 2014.
2. G. Kasperovich et J. Hausmann, J. Mater. Process. Technol. , vol. **220**, pp. 202-214, 6 2015.
3. P. Edwards et M. Ramulu, Mater. Sci. Eng., A , vol. **598**, pp. 327-337, 3 2014.
4. W. Xu, S. Sun, J. Elambasseril, Q. Liu, M. Brandt et M. Qian, JOM, vol. **67**, pp. 668-673, 1 2015.
5. H. Gong, K. Rafi, H. Gu, G. D. J. Ram, T. Starr et B. Stucker, Mater. Des. , vol. **86**, pp. 545-554, 12 2015.
6. L. Bian, S. M. Thompson et N. Shamsaei, JOM, vol. **67**, pp. 629-638, 2 2015.
7. H. K. Rafi, T. L. Starr et B. E. Stucker, The Int. J. Adv. Manuf. Technol. , vol. **69**, pp. 1299-1309, 6 2013.
8. L. Facchini, E. Magalini, P. Robotti, A. Molinari, S. Höges et K. Wissenbach, RAPID PROTOTYPING J , vol. **16**, pp. 450-459, 10 2010.
9. S. Leuders, S. Meiners, L. Wu, A. Taube, T. Tröster et T. Niendorf, J. Mater. Process. Technol., vol. **248**, pp. 130-142, 2017.
10. A. Fatemi, R. Molaei, S. Sharifimehr, N. Phan et N. Shamsaei, Int. J. Fatigue , vol. **100**, pp. 347-366, 7 2017.
11. A. Fatemi, R. Molaei, S. Sharifimehr, N. Shamsaei et N. Phan, Int. J. Fatigue , vol. **99**, pp. 187-201, 6 2017.
12. J. Günther, D. Krewerth, T. Lippmann, S. Leuders, T. Tröster, A. Weidner, H. Biermann et T. Niendorf, Int. J. Fatigue , vol. **94**, pp. 236-245, 1 2017.
13. D. Greitemeier, F. Palm, F. Syassen et T. Melz, Int. J. Fatigue , vol. **94**, pp. 211-217, 1 2017.
14. S. Beretta et S. Romano, Int. J. Fatigue , vol. **94**, pp. 178-191, 1 2017.
15. L. Thijs, F. Verhaeghe, T. Craeghs, J. V. Humbeeck et J.-P. Kruth, Acta Mater., vol. **58**, pp. 3303-3312, 5 2010.
16. Y. Murakami, Elsevier, 2002.
17. G. Kasperovich, J. Haubrich, J. Gussone et G. Requena, Mater. Des. , vol. **105**, pp. 160-170, 9 2016.
18. C. E. Stromeyer, *Proceedings of the Royal Society of London A: Mathematical*, Physical and Engineering Sciences, vol. **90**, pp. 411-425, 1914.
19. K. S. Chan, M. Koike, R. L. Mason et T. Okabe, Metall. Mater. Trans. A , vol. **44**, pp. 1010-1022, 10 2012.
20. R. Li, J. Liu, Y. Shi, L. Wang et W. Jiang, Int. J. Adv. Manuf. Technol. , vol. **59**, pp. 1025-1035, 8 2011.
21. R. Konecna, G. Nicoletto, S. Fintova et M. Frkan, *Procedia Struct. Integrity* , vol. **7**, pp. 92-100, 2017.
22. M. Benedetti, E. Torresani, M. Leoni, V. Fontanari, M. Bandini, C. Pederzolli et C. Potrich, J.

https://doi.org/10.1051/matecconf/201816502008

Mech. Behav. Biomed. Mater. , vol. **71**, pp. 295-306, 7 2017.

23. A. Townsend, N. Senin, L. Blunt, R. K. Leach et J. S. Taylor, Precis. Eng. , vol. **46**, pp. 34-47, 10 2016.

24. I. Koutiri, E. Pessard, P. Peyre, O. Amlou et T. D. Terris, J. Mater. Process. Technol. , vol. **255**, pp. 536-546, 2018.

Effect of microstructure on fatigue crack tip field of laser melting deposited Ti–5Al–5Mo–5V–1Cr–1Fe titanium alloy

Kai Wang [1], *Rui* Bao [1,*], *Siyuan* Ren [1], *Dong* Liu [2,3,4] and *Chuliang* Yan [1]

[1]Institute of Solid Mechanics, Beihang University (BUAA), Beijing 100191, China
[2]National Engineering Laboratory of Additive Manufacturing for Large Metallic Components (Beihang University), Beijing 100191, China
[3]Engineering Research Center of Ministry of Education on Laser Direct Manufacturing for Large Metallic Components, Beijing 100191, China
[4]School of Materials Science and Engineering, Beihang University, Beijing 100191, China

Abstract. In this paper, experimental and simulated investigations were carried out to understand the effect of microstructure on the fatigue crack tip strain field of laser melting deposited Ti–5Al–5Mo–5V–1Cr–1Fe titanium alloy. The mechanical properties of the laser melting deposited material corresponding to different microstructure zones were measured by digital image correlation method under in-situ static stretching, and the results indicate that the direction of primary α lamellas has great influence on the mechanical property of each grain. The crack tip strain fields considering the influence of material microstructure under constant amplitude conditions were measured, and the strain field of plastic zone has significant difference. Crystal plasticity finite element simulation of the static tensile and crack tip field considering the microstructure effect had been conducted and compare with experiment results. The simulation results are in good agreement with the test results in both the static tension strain field and crack tip strain field.

1 Introduction

Crack tip field is very important to the investigation of fatigue crack growth behavior of various materials [1-2]. Traditional fracture mechanics, which is based on material homogeneity and isotropy hypothesis, has greatly developed and solved many engineering practical problems in many years. As a technique of additive manufacturing, laser melting deposition (LMD) is a rapid solidification material additive layered manufacturing technology for building components from a computer-aided design model [3-4]. The macrostructure morphology of LMD titanium alloy is featured by fine equiaxed grains and long columnar grains alternately arrayed due to the high temperature gradient in the overlap zone and relatively low temperature gradient in the body zone of the molten pool [5-7]. And the mechanical properties present inhomogeneity and anisotropy affected by the special microstructure [8-13], which brought great challenges to the research of the crack tip field.

There are amounts researches focused on the effect of microstructure on the fatigue crack growth (FCG) behavior of additive manufactured titanium alloys. Lu [14] observed a periodic fluctuation in the crack growth rate of LMD Ti-6.5Al-3.5Mo-1.5Zr-0.3Si alloy under constant amplitude loading, which was resulted from the layered structure with heat-affected bands. Acceleration and retardation in fatigue crack growth rates are observed during microstructurally-small fatigue crack propagation of laser engineered net shaped Ti-6Al-4V alloy due to the blocking effects at prior β grain boundaries [15]. Xie et al. claimed that the FCG is dominated by the intrinsic crack resistance of the samples when $\Delta K \geq \Delta K_T$ [16]. Galarraga et al. found that increasing the stress ratio can reduce the influence of microstructural features on region II crack growth in the electron beam melted Ti-6Al-4V ELI alloy [17]. However, few studies focus on the effect of special microstructure on fatigue crack tip field of additive manufactured titanium alloys.

In this paper, experimental and simulated investigations were carried out to understand the effect of microstructure on fatigue crack tip field of LMD Ti–5Al–5Mo–5V–1Cr–1Fe titanium alloy. The mechanical properties of the material corresponding to different microstructure zones were measured by digital image correlation (DIC) method under in-situ static stretching. The crack tip field with the influence of microstructures were also measured and compared. Crystal plasticity finite element method (CPFEM) simulation of the static tensile and crack tip field considering the microstructure effect had been conducted and compare with experiment results.

2 Experiment and simulation

The Ti–5Al–5Mo–5V–1Cr–1Fe (Ti-55511) alloy, known as TC18 in China, is a typical near β titanium alloy with

* Corresponding author: rbao@buaa.edu.cn

high strength, high toughness and excellent fatigue property [18]. The specimens used in this study were cut from a 300mm × 100mm × 40mm LMD TC18 thick plate. The macrostructure morphology of LMD TC18 titanium alloy is a typical equiaxed and columnar grains alternately arrayed, as shown in Fig. 1a. The microstructure inside the grains is characterized by elongated primary α (α_p) lamellas orderly arranged, as shown in Fig. 1b. However, the directions of α_p lamellas in each grain are not same.

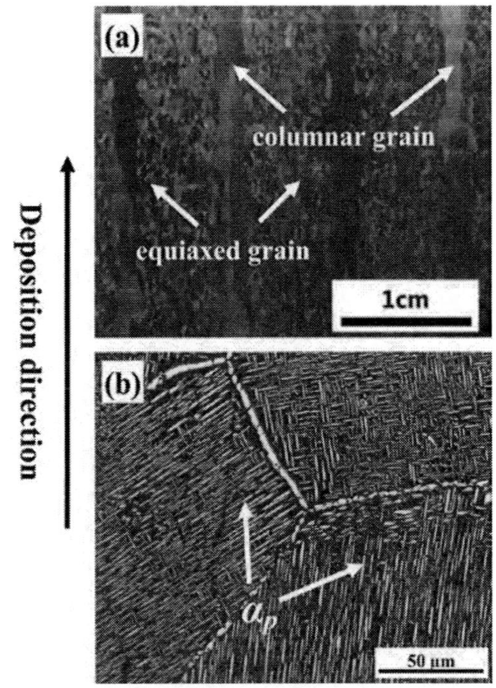

Fig. 1. The macrostructure (a) and microstructure (b) of LMD TC18 alloy.

Two kinds of specimens were used for the research: static tensile specimen to investigate the mechanical properties of various grains with different grain orientations, and single edge notched tensile (SENT) specimen to measure the crack tip strain field with different material characteristics. The nominal sizes of static tensile and SENT specimen used are shown in Fig. 2a and Fig. 2b, and the thickness of both specimens is approximately 1 mm. The notch direction of SENT specimen is parallel to the deposition direction. An in-situ stretching observation system, mainly consisted of ZEISS AxioScope metalloscope, CARE Measurement & Control Co IBTC-5000 in-situ tension testing machine and VIC-2D Digital Image Correlation software, as shown in Fig. 3, was used to measure the specimen deformation under static stretching and cyclic loading. Both surfaces of all specimens were sanded step by step with 300, 600, 800, 1200, 2000 sandpapers and then etched by a mixture of 1 ml HF, 6 ml HNO_3 and 100 ml H_2O to provide sufficient features as speckles for the DIC test and observe the metallographic structure meanwhile.

Static tensile test use a 100X total magnification with a field of view approximately 1.25×1.0mm. Displacement step-by-step loading was used during the in-situ static tensile test until destroyed, and take three photos at each step to avoid random errors caused by factors such as vibration. As for the test of the crack tip strain field measurement, the crack of SENT specimens should firstly be prefabricated to 3.0mm with a stress ratio 0.1, and the last step maximum load of gradually lower alternating load is 80MPa. Then use force loading with 0MPa, 0.1×80MPa and 80MPa three steps, and also take three photos at each level of load.

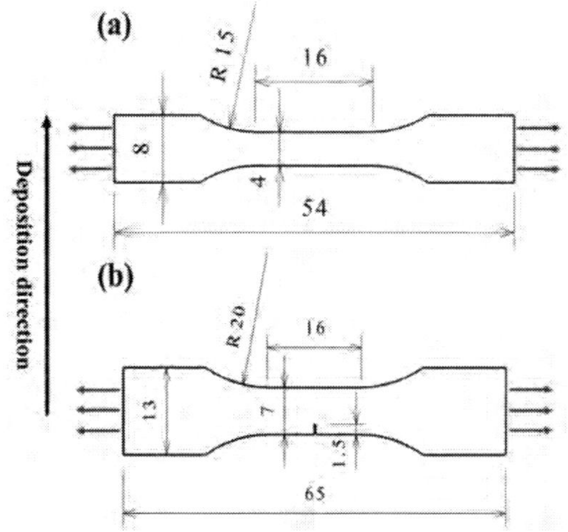

Fig. 2. Specimens nominal sizes: (a) static tensile specimen, (b) SENT specimen.

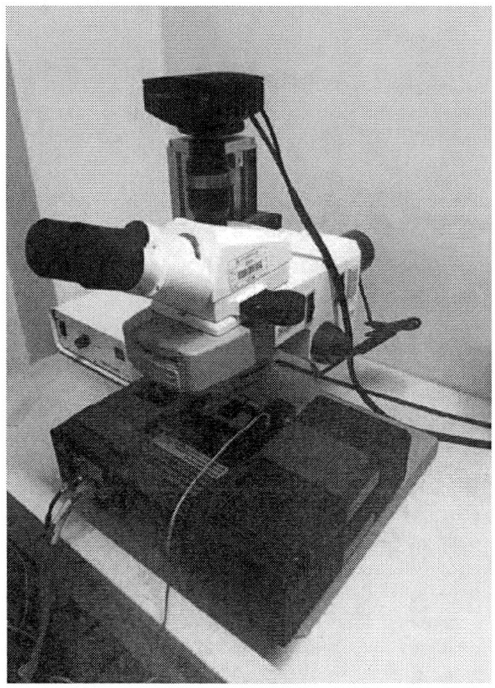

Fig. 3. In-situ stretching observation system

Crystal plasticity finite element method (CPFEM) was used to calculate static tensile and crack tip field considering the microstructure effect. A three-dimensional geometric model was established based on the Voronoi diagram principle as shown in Fig. 4a, and its seed points were optimized to make the grain size controllable. The orientation of each grain is random. The Python scripting language was used to build an optimized mesh model in ABAQUS, and a multi-step calculation was performed based on this model. The similar method was used to create a monocrystal model with crack as shown in Fig. 4b. Change the grain orientation in calculations to investigate its effect on the crack tip field. A force loading was used, and the loading direction has been marked in the figure.

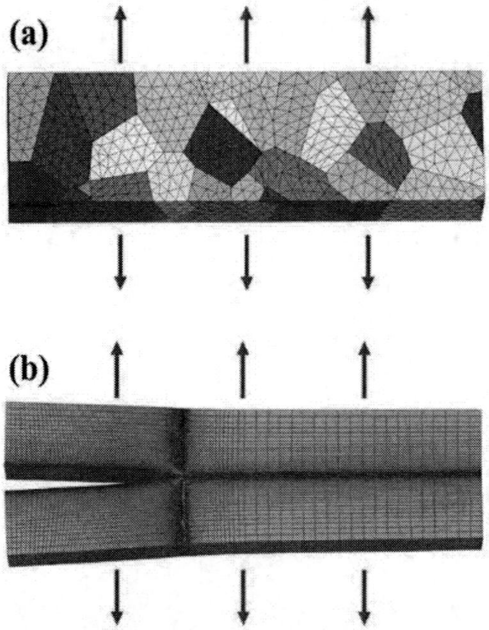

Fig. 4. CPFEM model: (a) static tensile polycrystalline model, (b) monocrystal model with crack.

3 Results and discussion

3.1 In-situ static stretching

The microstructure of measurement area is shown in Fig. 5a, in which the direction of loading and α_p lamellas are also marked. The corresponding stretch direction strain field of displacement loading size Δdis=1.0mm are shown in Fig. 5b. The results indicate that the great include angle of adjacent α_p lamellas direction would cause significant stress concentration at the grain boundary, as marked by circle in Fig. 5b, which is coincide with the effect of grain orientation.

In addition, the deformation of those grain with large angle α_p lamellas is observably lower, just as the purple area in Fig. 5b. The strain field of Δdis=0.4mm shown in Fig. 5c gives a better illustration of this problem. If take 0.2% as the yield standard, the whole specimen was already yield in Δdis=0.4mm. However, the tensile strain value of the area indicated by red is significantly greater than that of the purple area. It may due to the retardation of large angle α_p lamellas to the slip caused by maximum shear stress, the direction of which is ±45°. At the same time, there are many deformation bands along ±45° in Fig. 5b and Fig. 5c, which should also be caused by the maximum shear stress.

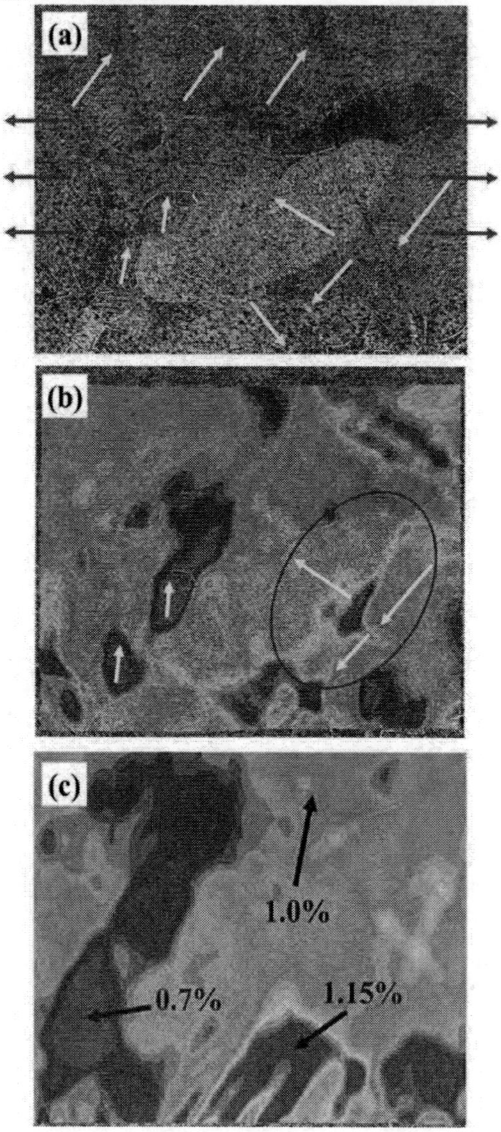

Fig. 5. (a) The microstructure of measure area; The strain field of stretch direction measured by DIC method: (b) displacement loading size Δdis=1.0mm, (c) Δdis=0.4mm.

Unlike plastic deformation, the elastic deformation of various microstructure is very close. In other words, the elastic property of different microstructure is similar. It may due to the totally different mechanism of elastic and plastic deformation. The elastic deformation of metal is caused by the length change of metal bond under normal

 https://doi.org/10.1051/matecconf/201816502009

stress. While the plastic deformation is achieved by dislocations and slips under shear stress. It also proves that the effect of α_p lamellas is mainly retard the dislocations and slips caused by shear stress.

The stretch direction strain field calculated by CPFEM is shown in Fig. 6. The results indicate that the deformation of those grains with random orientation is also different, which is similar with the experiment results.

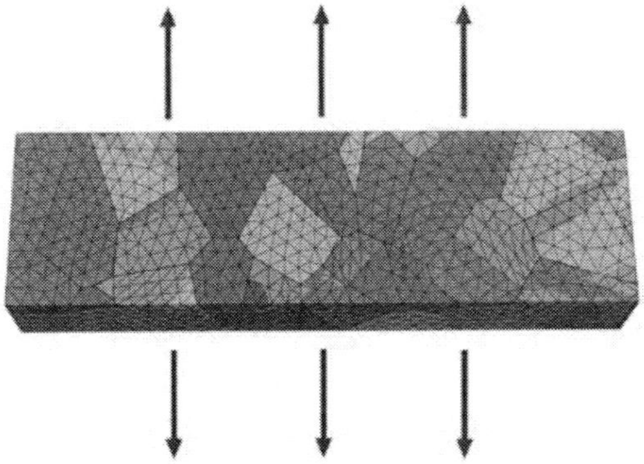

Fig. 6. The static tensile result of different grain orientation calculated by CPFEM.

3.2 Crack tip strain field

The crack tip strain fields measured by DIC method are shown in Fig. 7. The specimen was stretched along X direction and the crack is along Y, as the coordinate system shown in Fig. 7a. The strain field of X direction *exx* is shown in Fig. 7a, which is a typical butterfly shape. The Y direction strain *eyy* and shear strain *exy* are respectively shown in Fig. 7b and Fig. 7c.

As introduced above, the crack of several specimens with various microstructure were prefabricated to 3.0mm firstly. And three specimens with straight crack were chosen to investigate the effect of microstructure on the fatigue crack tip field. Considering the strain in X direction is much larger than Y direction, only exx was chosen to compare. A straight line from the crack tip in the Y direction was chosen to compare, which is parallel to the crack and is the axis of symmetry of the crack tip field.

The relationship of X direction strain *exx* and the distance to the crack tip *r* along the line chosen above are shown in Fig.8. The three cracks with different microstructure in crack tip area measured by DIC method are shown in Fig. 8a. Obviously there is a significant difference in the plastic zone, which size is approximately $r_0 \leqq 0.10$mm, but small difference in elastic region. It may due to the difference in elastic property at different microstructure is small, but the plasticity property is likely to have significant difference. It is worth noting that the correct choice of crack tip position makes the *exx* of three cracks does not continue to increase significantly at the crack tip, but gradually converges to a certain value. It conforms to the inference of elastic-plastic fracture mechanics that the strain at the crack tip is limited, not infinity.

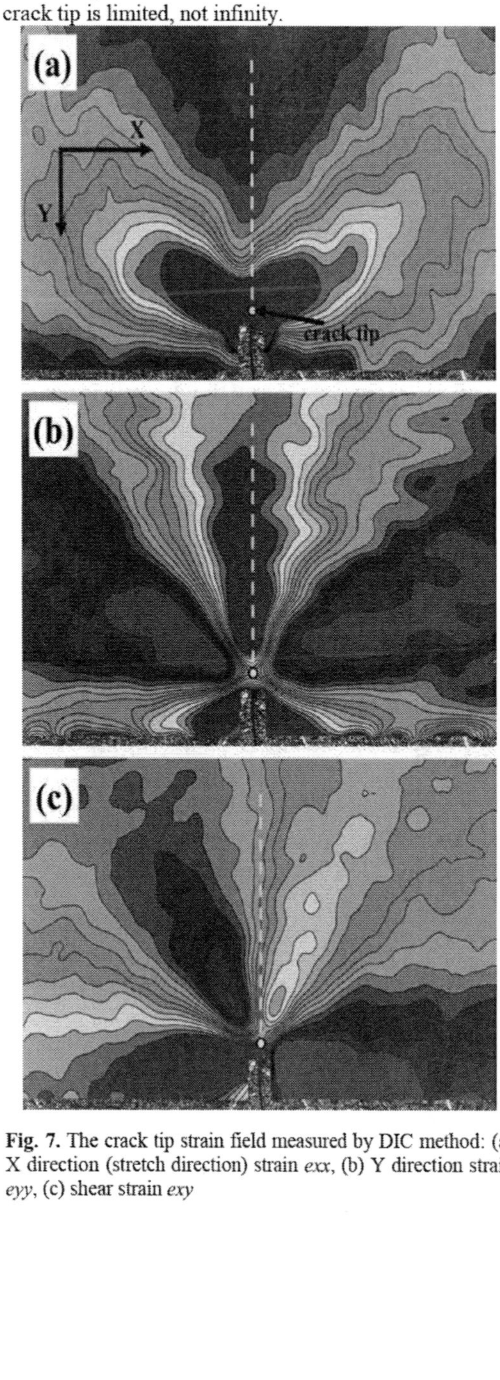

Fig. 7. The crack tip strain field measured by DIC method: (a) X direction (stretch direction) strain *exx*, (b) Y direction strain *eyy*, (c) shear strain *exy*

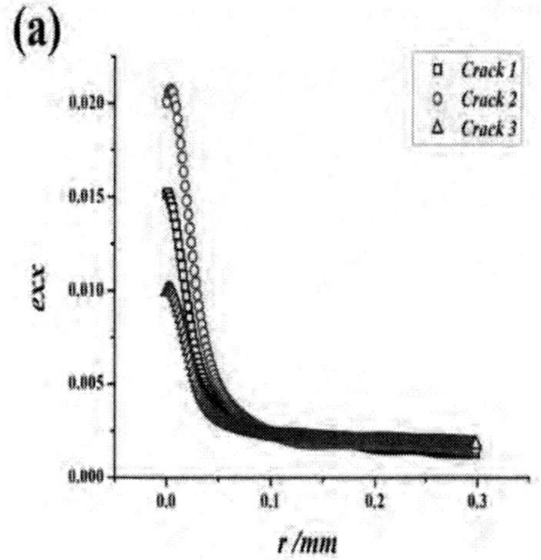

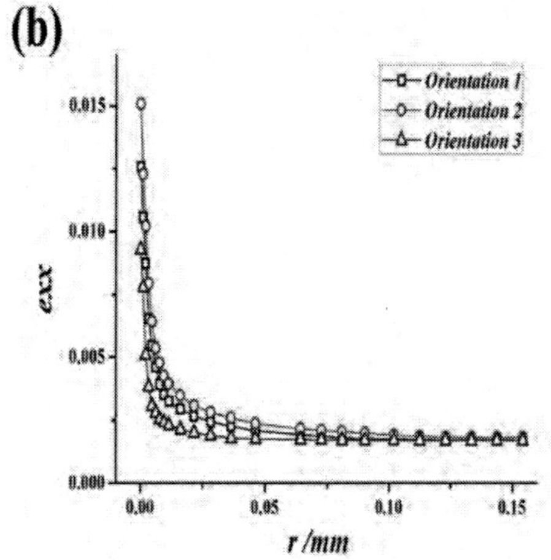

Fig. 8. The relationship of X direction strain *exx* and the distance to the crack tip *r* of: (a) three cracks with different microstructure in crack tip area measured by DIC method, (b) the same crack under different grain orientations calculated by CPFEM.

The results of three typical grain orientations calculated by CPFEM are shown in Fig. 8b. As the small size of the model, the load applied is smaller than test, thus the plastic zone is also smaller. However, the overall trend of *exx* with distance *r* is similar to the results of experiments. More importantly, the difference between the three orientations in the plastic zone is significantly greater than that in the elastic zone, which is confirm to the conclusion of experiments.

4 Conclusions

In this study, the effect of microstructure on the fatigue crack tip field of laser melting deposited Ti–5Al–5Mo–5V–1Cr–1Fe titanium alloy was investigated. The main conclusions from this study can be summarized as follows:

1) The direction of orderly arranged α_p lamellas in each grain has significant effect on its mechanical property of LMD TC18 titanium alloy. The great include angle of adjacent α_p lamellas direction would cause significant stress concentration at the grain boundary. And the deformation of those grain with α_p lamellas direction deviates significantly from $\pm 45°$ is observably lower.

2) The strain fields of plastic zone corresponding to cracks in different microstructure may have great difference, but the strain of elastic region is close. It may due to the difference in elastic property at different microstructure is small, but the plasticity property is likely to have significant difference.

3) CPFEM simulation results show good agreements with the test results on both static tension strain field and the crack tip strain field. The effect of the orientations of α_p lamellas on mechanical behaviors can be reflected by setting different grain orientations in CPFEM.

Acknowledgement

This work was supported by the National Natural Science Foundation of China (NO. 11672012)

References

1. J. M. Vasco-Olmo, F. A. Díaz, Eng. Fract. Mech. **166**, 82-96 (2016)
2. S. E. Ferreira, M. A. Meggiolaro, Int. J. Fatigue (2018)
3. H. P. Qu, H. M. Wang, Mater. Sci. Eng. A, **466**, 1–2 (2007)
4. G. K. Lewis, E. Schlienger, Mater. Des. **21**, 4 (2000)
5. C. M. Liu, X. J. Tian, H. B. Tang, H. M. Wang, J. Alloys Compd. **572**, 17-24 (2013)
6. Y. Zhu, D. Liu, X. Tian, H. Tang, H. Wang, Mater. Des, **56** 7 (2014)
7. G. C. Li, J. Li, X. J. Tian, X. Cheng, B. He, H. M. Wang, Mater. Sci. Eng. A, **684**, 233-238 (2017)
8. J. Alcisto, A. Enriquez, H. Garcia, S. Hinkson, T. Steelman, E. Silverman, et al. J. Mater. Eng. Perform. **20**, 2 (2011)
9. H. Galarraga, Lados, R. R. Dehoff, M. M. Kirka, P. Nandwana, Addit. Manuf. **10**, 47 (2016)
10. D. Agius, K. Kourousis, C. Wallbrink, T. Song,. Mater. Sci. Eng. A **701** (2017)
11. N. Shamsaei, A. Yadollahi, L. Bian, S. M. Thompson, Addit. Manuf. **8**, 12-35 (2015).
12. X. Shi, S. Ma, C. Liu, Q. Wu, J. Lu, Y. Liu, et al. Mater. Sci. Eng. A, **684**, 196-204. (2017)

13. Q. Chunlei, J.E. Nicholas, Adkins, M. Moataz Attallah, Mater. Sci. Eng. A, **578** 8 (2013)
14. S. Lu, R. Bao, K.Wang, D. Liu, Y. Wu, B. Fei, Mater. Sci. Eng. A, **690**, 378-386 (2017)
15. Y. Zhai, D. A. Lados, E. J. Brown, G. N. Vigilante, Int. J. Fatigue, **93**, 51-63 (2016)
16. Y. Xie, M. Gao, F. Wang, C. Zhang, K. Hao, H. Wang, et al, Mater. Sci. Eng. A, **709** (2017)
17. H. Galarraga, R. J. Warren, D. A. Lados, R. R. Dehoff, M. M. Kirka, Eng. Fract. Mech. 176 (2017)
18. C. M. Liu, H. M. Wang, X. J. Tian, H. B. Tang, D. Liu. Mater. Sci. Eng. A, **586**, 323-329 (2013)

Hydrogen effect on the fatigue behavior of LBM Inconel 718

Simon Puydebois[1, 2,3], *Abdelali* Oudriss[2], *Pierre* Bernard[1], *Laurent* Briottet[3], *Xavier* Feaugas[2]

[1] ArianeGroup, Forêt de Vernon, F-27200, Vernon, France.
[2] LaSIE, CNRS UMR 7356, Université de La Rochelle, F-17000, La Rochelle, France.
[3] Univ. Grenoble Alpes, CEA, LITEN, DTBH, F-38000 Grenoble, France.

Abstract. For several years, Inconel 718 made by Laser Beam Melting (LBM) has been used for components of the Ariane propulsion systems manufactured by ArianeGroup. In the aerospace field, many components of space engines are used under hydrogen environment. The risk of hydrogen embrittlement (HE) can be therefore a first order problem. Consequently, to improve the HE sensitivity of LBM Inconel 718, a systematic approach needs to be developed to characterize the microstructure at different scales and its interaction with hydrogen. This study addresses the impact of gaseous hydrogen on the material mechanical behavior under fatigue loadings. In a first step, the low cycle fatigue behavior under 300 bar of hydrogen gas has been evaluated with specimen loaded at a constant load ratio of R=0.1 and a frequency of 0.5 Hz. A reduction in the cycle number of fracture is shown. This reduction of fatigue life is a consequence of the impact of hydrogen damage processes. The impact of hydrogen is evaluated at the stages of crack initiation, crack propagation. These results are discussed in relation with the hydrogen embrittlement mechanisms and particularly in terms of hydrogen / plasticity interactions. To achieve this, the fracture surface morphology was first examined using scanning electron microscopy and second samples near the fracture surface were extracted using Focused-Ion Beam machining from regions containing striation. The main result observed is a reduction of the size of dislocation organization in relation with a decrease of the striation distance.

1 Introduction

This study will focus on the fatigue behavior under hydrogen pressure of a material used for the Ariane propulsion systems manufactured by ArianeGroup. This is an Inconel 718 nickel based super alloy made by laser beam melting process (LBM). The additive process (AM) is defined as the process of manufacturing objects by adding material, unlike traditional manufacturing methods such as machining (subtractive fabrication methods). This manufacturing technique has real potential in several industrial fields because of many advantages such as the reduction of the production cycle, more geometrical freedom, precise control of the mass of the parts, reduction of processing steps and flexibility of material selection.
One of the fuels used for propulsion of the upper and lower stages of the Ariane V launcher is hydrogen. Thus, some components are subjected to mechanical loading in a hydrogenated environment.
Since 1950, several studies have shown that hydrogen has a detrimental effect on the mechanical properties of many materials [1-4]. Several studies have been conducted to investigate the impact of hydrogen on Inconel 718 in conventional manufacturing [5-9]. The work of Bruchhausen *et al.* [5] revealed a reduction of the number of cycles to failure when the material is loaded under H_2 pressure. However, no study has been conducted to evaluate the impact of gaseous hydrogen on the mechanical behavior of Inconel 718 obtained by additive manufacturing method.
The aim of the present work is to evaluate the influence of H_2 on the mechanical properties and on the damage processes of this alloy submitted to low cycle fatigue loading. Firstly, a thorough microstructural analysis has been done to identify the various metallurgical heterogeneities.
Secondly, the mechanical behavior has been studied by low cycle fatigue loading under air and under H_2 gas at room temperature. The impact of hydrogen on the damage process will be discussed with regard to the different stages of damage (crack initiation, crack propagation, toughness). Each of these damage steps has been studied using observations under a scanning electron microscope.
The hydrogen / plasticity interactions during the crack growth process are discussed using samples extracted by Focused-Ion Beam machining from regions containing striation.

2 Experimental procedure

The microstructure was investigated at different scales using an optical Leica microscope DM6000M, a Scanning Electron Microscope (SEM Philips FEI Quanta 200FEG / ESEM) associated to an analysis EBSD and a Transmission Electron Microscope (TEM JOEL 2010 200 kV).

* Corresponding author: Puydebois.simon@live.fr

The first step of this metallurgical characterization sought to identify the inclusion state and the presence of porosity. Several studies [10-11] show the presence of porosity inherent to the additive manufacturing process and to the laser parameter (linear laser energy density, laser scanning speed, etc…). The surface preparation consists of mechanical polishing with SiC paper up to grade 4000, followed by OPS polishing on a felted polishing pad. The sample is then rinsed and cleaned in an ultrasonic bath with ethanol, and finally etched for 1 minute 30 seconds in a solution of 1/3 HCL and 2/3 HNO_3.

In a second step, Electron BackScatter Diffraction (EBSD) helps to identify several metallurgical parameters such as grain size, crystallographic texture, crystallographic grain orientations, grain nature, grain boundary fraction grain and boundary connectivity. The surface preparation prior to EBSD mapping consists of polishing up to grade 4000, diamond paste ¼ µm and vibrating polisher with 0.02µm SiO_2 for 3 hours. The incident electron acceleration voltage is set to 20 kV with a vacuum close to 10-5 mbar with a measuring step of 3 µm. TSL software analyzes OIM® 5 was used to analyze data from EBSD. These analyses were performed according to two characteristic directions of manufacturing.

The transmission electronic microscope sample is firstly polished with SiC paper up to grade 4000 with a thickness of 80-100µm. This sample is then thinned electrolytically with a Tenupol-5 thinning machine using an electrolyte composed of 450 ml of methanol, 50 ml of perchloric acid and 100 ml of 2-butoxyethanol. The operation is carried out at a temperature around - 25°C, under a potential of 10 V [12].

The mechanical properties of low cycle fatigue of Inconel 718 LBM were addressed in air, and under 300 bar of hydrogen at room temperature on axisymmetric fatigue sample. Given that the weakening effect of hydrogen is higher at room temperature [13], the test have been performed at room temperature in order to lead a conservative study. The gage length of each specimen was previously polished with a 4000 grade SiC paper in order to present an identical surface state. Before each fatigue test, four nitrogen rinsing / vacuum cycles are performed followed by a pressure rinsing with the testing gas (hydrogen of high purity 99.9999%) to decrease the level of impurity. These tests were conducted on a hydraulic testing machine equipped with an autoclave. The mechanical load is measured by a load cell located in the pressure chamber. The fatigue tests were carried out with a load ratio of 0.1, for five maximum stress values with a frequency of 0.5 Hz and a triangular signal. After rupture, an observation of the fracture surface was conducted by scanning electron microscopy.

3 Results

3.1 Microstructural analysis

3.1.1 Inclusion state and porosities

The studied material, Inconel 718 LBM, is mainly composed of 54% Ni, 18% Cr, 17% Fe, 5% Nb and 3% Mo. Following manufacturing, the material was first treated with an annealing treatment to obtain a fine and homogeneous microstructure, then with a second heat treatment to precipitate intermetallic phases (γ 'and γ").

Energy Dispersive X-ray Spectroscopy analyzes has highlighted the presence of three types of inclusion. The majority of the identified inclusions are aluminas. Titanium carbide and titanium nitride have also been found in smaller proportions. The size of these inclusions varies from 5 to 25 µm. Very few cavities are observed. The area ratio of inclusions and porosities is less than 0.05% of the total surface (fig. 1).

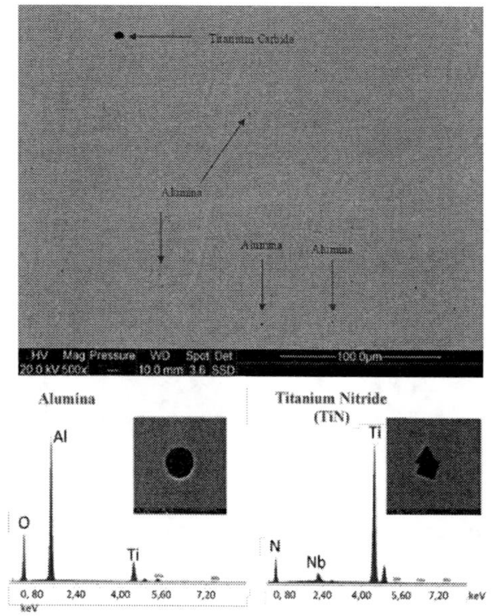

Fig. 1. A) SEM observation of porosity and inclusion B) EDS spectrum of alumina C) EDS spectrum of Titanium Nitride

The analysis of the microstructure by optical microscopy after etching and EBSD mapping revealed the grain size homogeneity and a random crystallographic texture in the material whatever the direction. Moreover, EBSD mapping reveals that the majority of the grain boundaries has an index of coincidence Σ below 29 (Fig. 2 B.). A lower scale was observed and identified using TEM, the presence of γ " precipitates in higher proportions and γ' and δ in lower proportions was observed. The morphology and size of these precipitates (Fig. 2 C, D, E and F)) has been quantified by TEM analysis.

The Electron BackScatter Diffraction scan shows a random crystallographic texture in the material whatever the manufacturing direction. However, it revealed the heterogeneity of grain size due to the manufacturing process. Figure 3 shows homogeneous grain size distribution along the X and Y axis with a mean grain size of 62.3 μm. The results along the Z axis show a double grain size distribution with mean values of 66 μm and 153 μm. This is the only observed effect of additive manufacturing on the microstructure of the studied material.

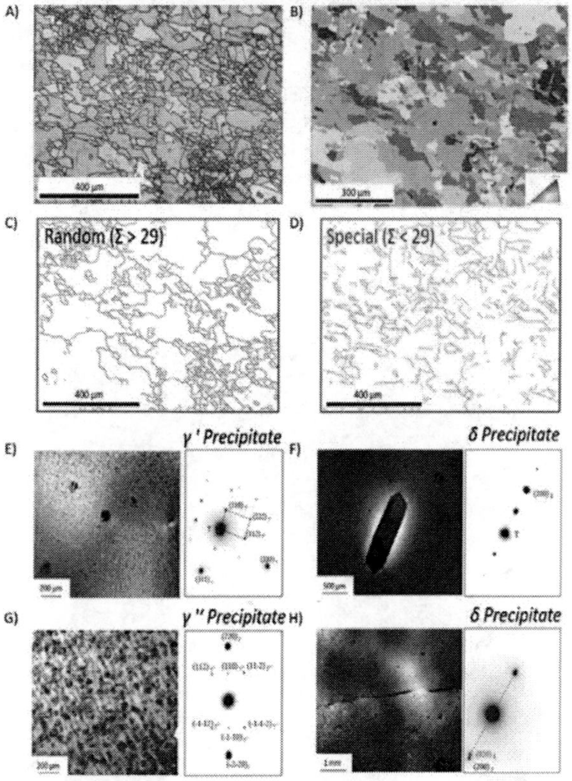

Fig. 2. A) EBSD grains mapping along plan 1; B) EBSD orientation mapping; C) EBSD Random grain boundaries mapping D) EBSD Special grain boundaries mapping E) Dark field image and diffraction pattern of the γ ' precipitated; F) Dark field image and diffraction pattern of the globular precipitate δ; G) dark field image and diffraction pattern of the γ " precipitated; H) bright field image and diffraction pattern of the acicular precipitate δ

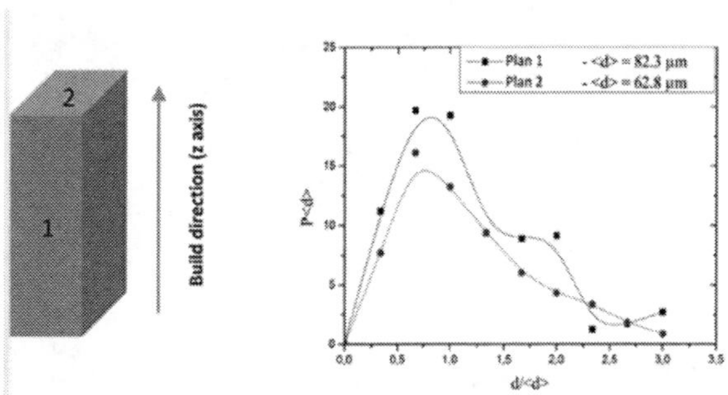

Fig. 3. Grain size distribution according to the two the normal plans to the additive manufacturing direction

3.2 Low cycle fatigue behavior under H_2 pressure

3.2.1 Mechanical results

Figure 4.a shows the S-N curve of Inconel 718 FA displaying the maximum stress versus the number of cycles to failure. First, a decrease is noticed under hydrogen gas compared to samples tested in air. This effect is lower when the stress amplitude is reduced. This reduction in the number of cycles under hydrogen breakdown ranges from a factor 2 for the lowest mechanical loading to a factor of 10 for the highest load.

The fatigue tests were launched after 30 minutes or one day under hydrogen pressure (marked by blue circles). The duration of the immersions before the tests has no effect on the total fatigue life time. The stress/strain fatigue curves shows a fatigue adaptation phenomenon in both conditions illustrated in figure 4.b by a progressive accumulation of plastic strain followed by apparent elastic cycles after stabilization. The stabilization of the plastic strain happens after 10 to 100 cycles of fatigue. The comparison of the two environmental conditions shows that there are no effect of hydrogen on the cumulative plastic strain at stabilization (fig. 4.B).

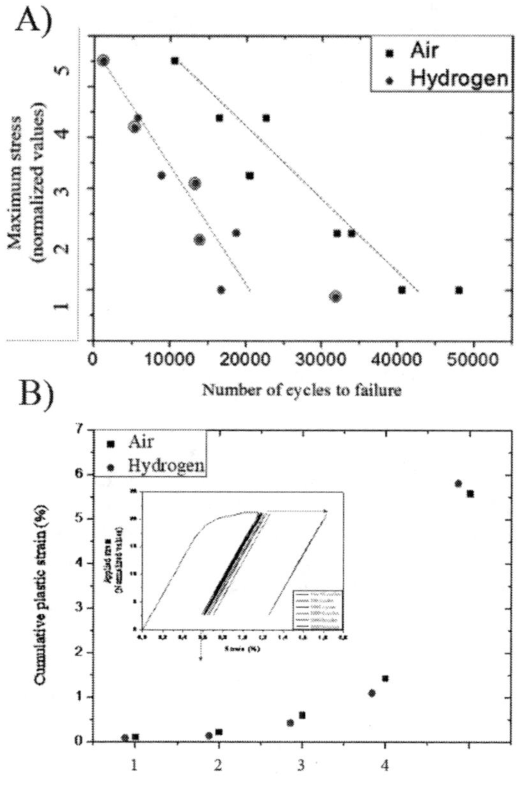

Fig. 4. A) S-N curve of Inconel 718 in air and in 300 bar H_2 B). Evolution of the stabilizing plastic deformation as a function of the level of loading for the two experimental conditions (illustration of the adaptation phenomenon)

 https://doi.org/10.1051/matecconf/201816502010

Figure 4.B shows that hydrogen has no macroscopic effect on the mechanical property of the material. In order to confirm the previous results, observations were made on a smaller scale in the core of the specimen after failure. For each level of stress, we observed the structure of dislocation after failure using a transmission electron microscope.

TEM observations show that under most tested conditions the dislocations structure is organized as a dislocations cell network (table 1). By contrast, the dislocations structure of the higher stress level under hydrogen (σ_5) presents slip bands. For this experimental condition the number of cycles to failure is low, close to 1000 cycles. In this case, the low density of dislocations did not lead to the formation of dislocations cells. In order to quantify the effect of hydrogen on the dislocation structure, a statistical study of the cells size was performed. For each condition, the average diameter was measured on a sample of 50 dislocation cells.

Table 1. Dislocation structure for different stress levels under air and under hydrogen

	Air	H
σ_1	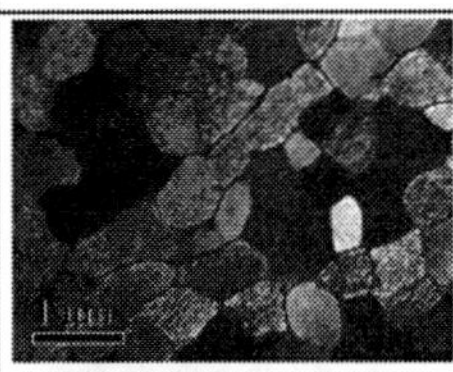	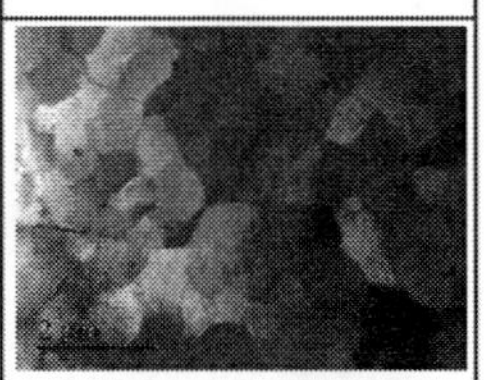
σ_2	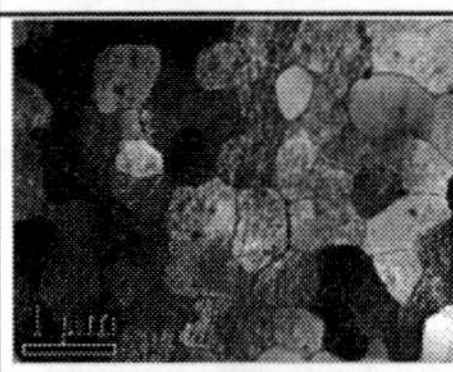	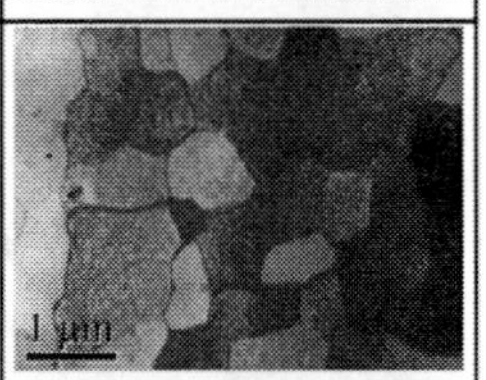
σ_3	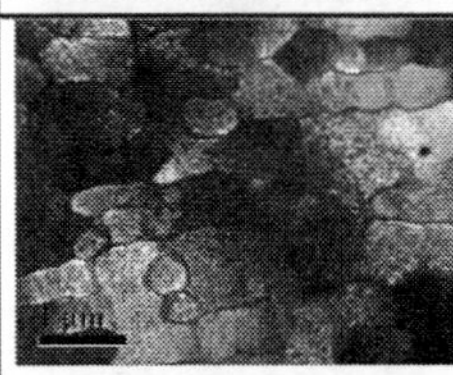	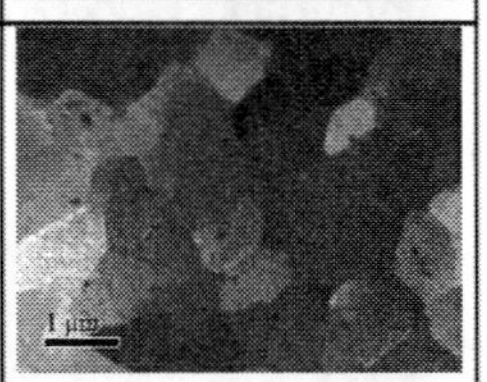
σ_4		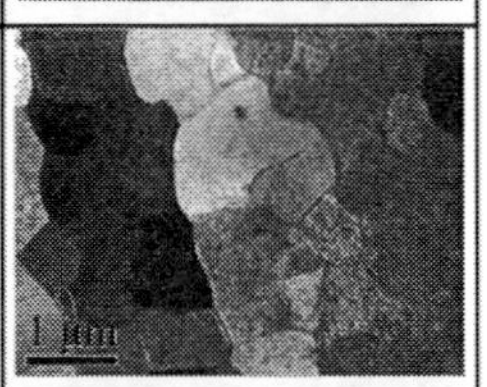
σ_5	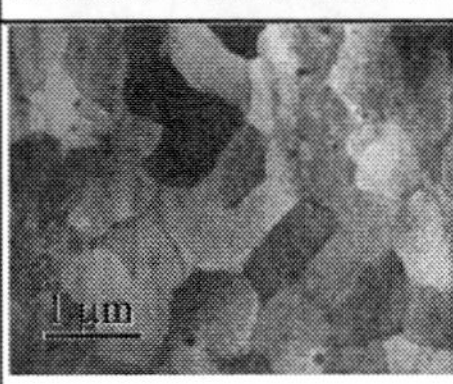	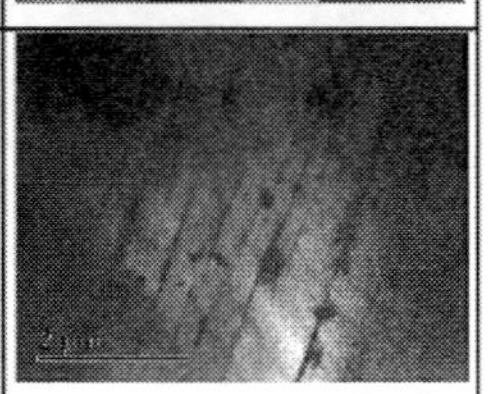

Figure 5.a shows a linear evolution of cells size with increasing stress. Moreover, a difference in the mean cells size according to whether the sample was tested under air or under hydrogen is identified. Finally, a relation between the dislocations cells size and the number of cycles to failure is shown (Figure 5.B). This evolution is directly explained by the increase of the dislocation density during cycling. The evolution of the average diameter of dislocation cells is described [14] as proportional to the evolution of the dislocation density:

$$d \approx Kc \cdot \rho_0^{-1/2} \quad (1)$$

with Kc a constant material, d the mean diameter of the cells, and ρ_0 the dislocation density.

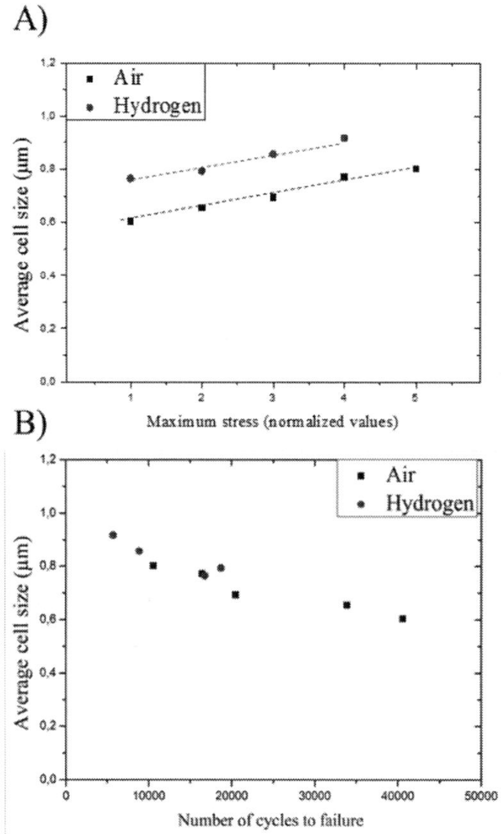

Fig. 5. A) Average cell size for different stress levels under both condition B) Average cell size vs. the number of cycles to failure.

The hydrogen effect shown in Figure 5.A is an indirect effect of cell size. Indeed, the evolution of the cell size is not due to the hydrogen/plasticity interaction in the core of the material. This effect is due to decrease of the number of cycles to failure under hydrogen. The low diffusion coefficient of hydrogen in the crystal lattice [15] and in the grain boundary [16] does not allow the hydrogen to reach the center of the specimen after one day under hydrogen. We will therefore question the effect of hydrogen on damage process on the surface.

3.2.2 Effect of hydrogen on damage processes

3.2.2.1 Initiation stage

The initiation stage was described in a previous paper [17] using SEM observations. In order to identify the effect of hydrogen on the first step of fatigue life, scanning electron microscope analysis were performed on the gage length surface of the specimen to observe crack initiation.

Under both conditions at failure, the specimen gage length exhibits the emergence of a sliding system, which is present in higher proportion for the tests performed under hydrogen.

Crack initiation takes place on dislocation slip bands under 300 bar of hydrogen (fig.6). The localization of crack initiations under air is not obvious to identify (fig.6).

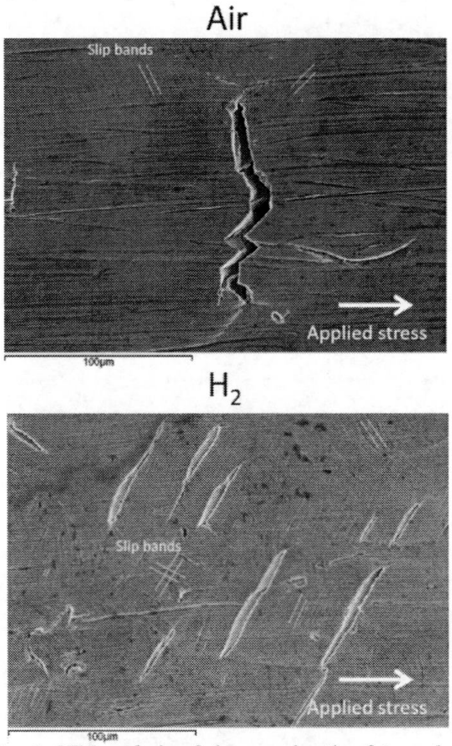

Fig. 6. SEM analysis of the gage length of Inconel 718 fatigue specimen in air and in 300 bar H_2 for $\sigma 2$

For each experimental condition, the density of surface cracks was obtained using 3D optical microscopy along the gage length of the samples.

The results (fig.7) show an increase of the number of cracks with the increasing maximal stress.

Moreover, hydrogen induces an increase in the cracks initiation density by a factor 2 to 8 and a localization of crack initiation close to the fracture surface.

The increase in slip band emergence and crack density under hydrogen is a proof of hydrogen/plasticity interaction with a decrease in cohesion energies at the interfaces (slip system).

Therefore, under hydrogen, crack initiation occurs at lower loads than under air.

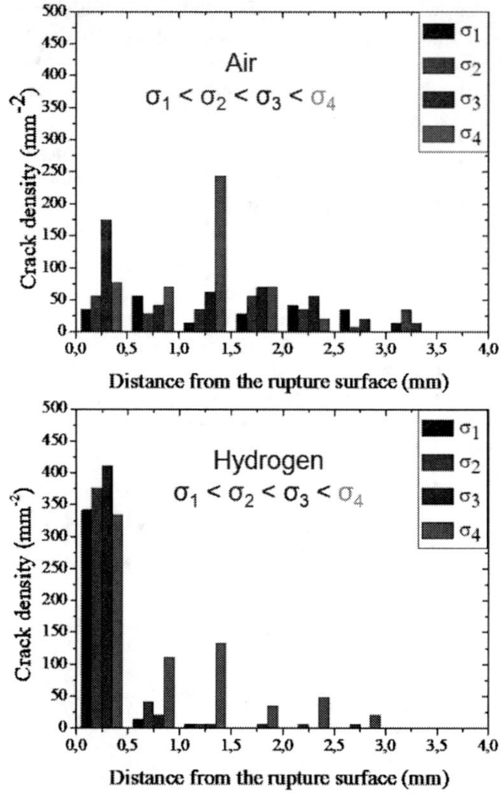

Fig. 7. Crack density as a function of the distance from the rupture surface under air and under hydrogen

3.2.2.2 Fatigue crack growth rate (FCGR) determination from LCF tests

In order to quantify the effect of hydrogen on crack growth rate, the following systematic study was carried out. The distance between striations has been measured and reported versus the distance to crack initiation as illustrated in figure 8.

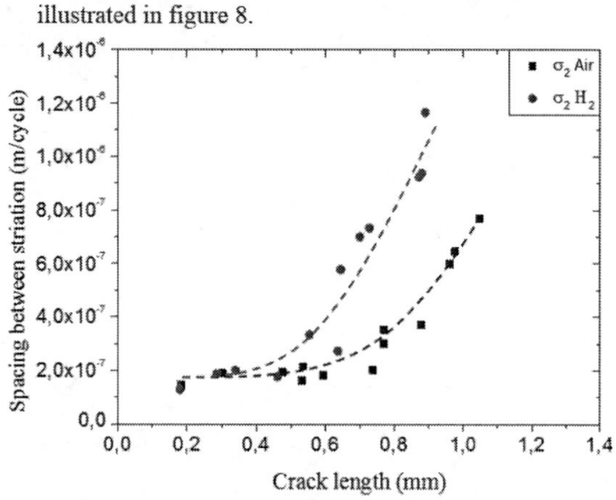

Fig. 8. Spacing between striation as a function of the crack length

The correlation between striations and mechanical cycles has been described by many authors as a 1: 1 correlation [18; 19; 20; 21]. Nedbal [22] and Ruckert [23] works respectively on AlCu4Mg1 and Al Alloy 7475-T7351 showed a good correlation between the macroscopic fatigue crack growth rate and the distance between striations for an inter-striation distance ranging between 0.1 and 1 µm.

Considering that one fatigue striation corresponds to one cycle, we can assimilate the distance between striations at the rate of crack propagation. Considering that the tests were performed under stress control, the crack length is directly related to the stress intensity factor

Figure 8 shows the evolution of crack growth rate as function of the crack length for the stress level σ_2. The FCGR is composed of two stages. These two propagation kinetics were also observed by Connors on 7075-T73 aluminum alloy [21].

These results also show an early increase of the crack growth rate in the presence of hydrogen. At this stress level, hydrogen does not appear to have a significant impact on the crack growth rate, for crack lengths lower than 450 µm,.

The effect of hydrogen tends to be higher for a higher crack length and hence for a higher stress intensity factor.

3.2.2.3 Hydrogen/plasticity interactions during fatigue crack growth

In order to question the impact of hydrogen on the crack growth rate by fatigue striations, thin sections near the fracture surface were extracted using Focused-Ion Beam machining from regions containing striation (fig.9). The purpose of this study is to observe the dislocations structure under the fracture surface at different crack lengths. This study was done for a given stress level σ_2.

Fig. 9. Illustration of a sampling area (a = 1100 µm under air)

Sampling was performed at crack initiation (150 µm), at an intermediate crack length corresponding to the transition of the two kinetics (900 µm under air and 450 µm under hydrogen), and at the crack tip before final rupture (1100 µm under air and 900 µm under hydrogen).

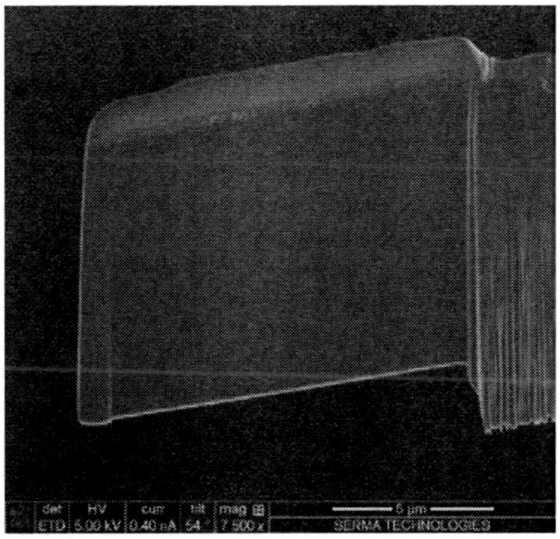

Fig. 10. Example of a FIB thin section (a = 900 µm under hydrogen)

The first observations show, in all conditions, the presence of slip bands identified by stereographic projection as being on the dense planes of type (111) and in dense directions of type <110> (figure 12).

A study of the interband spacing as function of the spacing between striations is presented in Figure 11. The inter-band distance was corrected by a disorientation factor obtained using stereographic projection.

For samples taken after rupture under air, each sliding band is associated with the presence of a fatigue striation. For samples taken under hydrogen, we have a 1: 1 correlation between slip band and fatigue striation for crack lengths lower than 450 µm. Which is highlighted by a cos (45°) slope between inter-bands spacing and inter-striations spacing (fig.11).

The sample taken at 900 µm and tested under hydrogen, shows a 1: 1 correlation between fatigue striations and dense primary slip bands (in white line in figure 13). However, on this sample we also observed the presence of micro-bands (in dash line in figure 13). This may reflect an increase in plastic activity in presence of hydrogen.

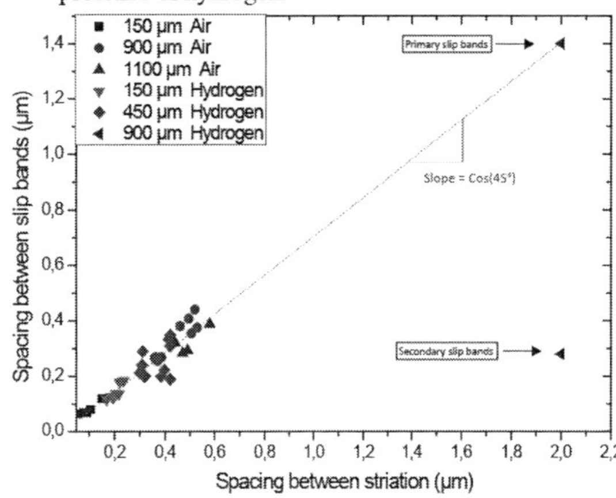

Fig. 11. Spacing between slip bands vs. spacing between striations

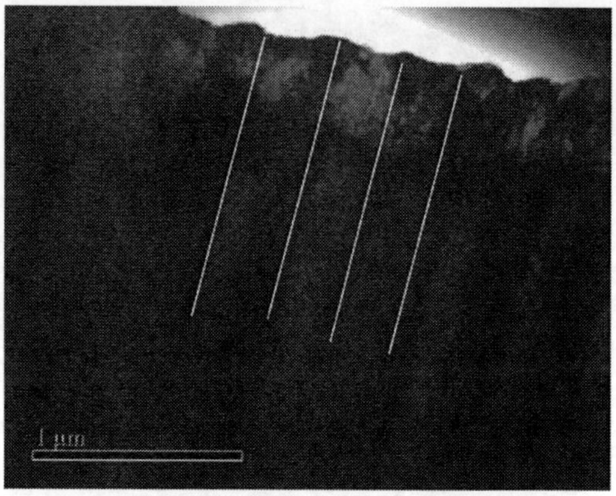

Fig. 12. TEM observations of the dislocation structure under rupture surface for a = 450 μm under hydrogen

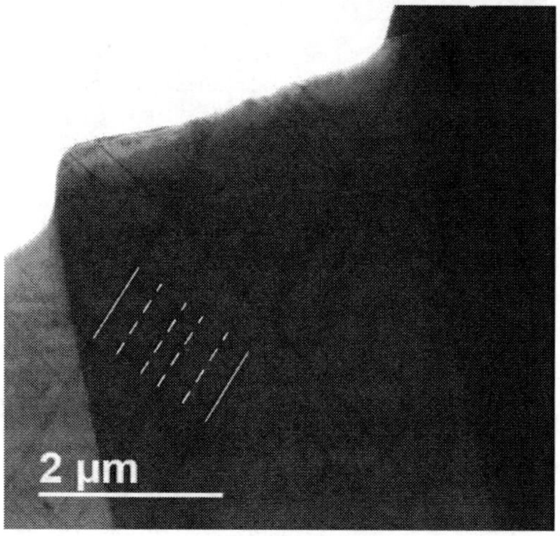

Fig. 11. TEM observations of the dislocation structure under rupture surface for a = 900 μm under hydrogen

4. Discussion

This study helped to identify the effect of the manufacturing process on the microstructure and the different effects of hydrogen on different stages of low cycle fatigue life. The interactions between hydrogen and microstructure can therefore be questioned:

The low cycle fatigue tests showed a decrease of the lifetime by a factor of 2 to 10. The effect of hydrogen increases with the increasing applied force. It is well known that hydrogen effect on lifetime increases with the increase of stress level and consequently with the plastic activity [24]. The time under hydrogen pressure before test has no effect on the number of cycles to failure. The effect of hydrogen on lifetime seems to be due to the presence of hydrogen on the surface of the specimen during loading.

Indeed, no macroscopic effect of hydrogen has been observed on the cumulated plastic deformation during the study of the ratcheting effect or on the dislocation structure in the tested specimen core. The size of the cells is directly related to the mechanical history of the material and more precisely to the number of cycles to failure. Hydrogen has no direct effect on the dislocation structure in the test specimen core. Therefore, the study was directed towards the surface phenomena and more precisely the interactions between hydrogen and the mechanism of damage.

SEM observations of gage length show a localization of crack initiation on sliding systems under hydrogen. In addition, an increase in the crack density was observed close to the rupture zone under hydrogen. This location of hydrogen crack density in areas of high plastic activity highlights the presence of interactions between hydrogen and plasticity. The debonding of the sliding system will be facilitated by the presence of hydrogen (HID).

The effect of hydrogen on crack growth rate has also been studied.

The determination of the crack growth rate in low cycle fatigue tests, by the measurement of striation spacing, shows an increase of the crack growth rate under hydrogen pressure.

Focused ion beam (FIB) technique was used to observe the dislocation structure beneath the fracture surface containing striations. These samples were performed to determine how hydrogen impacts the fatigue crack growth.

The observations of samples in air show the presence of sliding bands directly connected to a fatigue striation. This may imply that during each cycle, the activation of a sliding plane allows the blunting of the crack tip and its propagation. We observed the same phenomenon under hydrogen for short crack lengths (lower than 450 μm in this case). For these hydrogen crack length, no effect was observed on the inter-striation spacing and consequently on the crack propagation rate.

For larger crack lengths under hydrogen, we found the presence of sliding macrobands with a 1: 1 correlation with fatigue striation. The observations also show the presence of intermediate bands closer to each other. This secondary network of slip bands has lower dislocations density.

These observations suggest that for higher stress intensity factor under hydrogen a secondary slip network is activated. This can be explained by an increase in plastic activity in presence of hydrogen or/and by an increase in hydrogen/plasticity interactions with the increase in plastic activity.

Conclusion

After heat treatment, no effect of additive manufacturing on the microstructure was observed. The only observations that can be associated with additive manufacturing are a bimodal distribution of grain size according to the direction of manufacturing and a small proportion of porosities.

This work has shown a predominant effect of hydrogen/plasticity interaction on the crack initiation step and on the crack growth step. During these two stages of fatigue, the effect of hydrogen increases with the plastic activity. Hydrogen/plasticity interactions occur only at the material surface.

Acknowledgements

The authors wish to thank ArianeGroup for their financial support.

References

1. P. Cotterill; Progr Met Phys (1961), **9,** (4), pp 205–301.
2. H. Johnson, A.Troiano, Nature (1957),**179** (4563)
3. D. Williams, R. J Jaffee, Less Common Met, (1960), **2** (42)
4. W. Yeniscavich, R. Wolfe, R. Lieberman; J Nucl Mater (1959) ; **1** (3): pp 271–280.
5. M. Bruchhausen, Int. J. Fatigue, (2015), **70** p 137-145
6. J. Jewett, R. Walter, W. Chandler , Frohmberg R. NASA contractor report NASA, (1973) **CR-2163**; Rocketdyne;.
7. P. Hicks, C. Alstetter, Metall Trans A (1992); **23**A: pp237–249.
8. R. J. Walter and W. T. Chandler, (1973), NASA **CR-124410**
9. V. Frick, G.R. Janser, and J.A. Brown, Space Shuttle Materials (1971), **Vol. 3**, pp. 597- 608
10. J-P. Choi, G-H. Shin, S. Yang, D-Y. Yang, J-S. Lee , M. Brochu , J-H. Yu ,Powder Technology (2017), **310,** pp 60–66
11. Q. Jia, D. Gu ,Journal of Alloys and Compounds (2014), **585,** pp 713–721
12. L.Xiao, D.L. Chen, M.C.Chaturvedi, Scripta materialia (2005), **52**, pp 603-607
13. V. Frick, G.R. Janser, J.A. Brown, "In: Space Shuttle Materials (1971),**3** , pp.597-608,
14. D. Holt Journal of Applied Phys (1970), **41**
15. J. J.M. Jebaraj, D. J. Morrison, I. I. Suni, Corrosion Science (2014), **80**, pp 517–522
16. T. M. Harris, M. Latanision, Metallurgical Transactions A February 1991, Volume 22, Issue 2, pp 351–355
17. S. Puydebois, A. Oudriss, L. Briottet, X. Feaugas, P. Bernard, International Hydrogen Conference (2016) Jackson Lake (USA) pp 324-331
18. J. Au, J. S. Ke, ASTM STP733 (1980), p. 202
19. P.J.E. Forsyth and D. A. Ryder, ACTA Metallurgica (1961), **63**, pp 117-124
20. L. De Baglion de La Dufferie, Manuscript for PhD Ecole Nationale Supérieure de Mécanique et d'Aérotechnique de Poitiers, 2011.
21. W.C. Connors, Mater. Charact (1994). **33** (3) pp 245–253.
22. I. Nedbal, J. Siegl, J. Kunz, Advances Fracture Research (Proc. ICF 7). Eds. K.Salama et al. Vol.**5**. Oxford, Pergamon Press (1989), p. 3483-3491.
23. C.O.F.T. Ruckert, A.A. Messias Filho, W.W. Bose Filho, D. Spinelli, and J.R. Tarpani, Journal of Materials Engineering and Performance (2011), **20** (3), pp 381-389
24. J. A. Harris, M.C VanWanderham, (1973). Final Report to NAS8-26191, Pratt and Whitney Government Products Division.

The role of pores and microstructural heterogeneity on the tooth root fatigue strength of sintered spur gears

Matteo Benedetti[1], *Vigilio* Fontanari[1*], *Alberto* Molinari[1], *Pietro* Valcozzena[1], and *Wolfgang* Pahl[2]

[1]Department of Industrial Engineering, University of Trento, 38122 via Sommarive 9, Trento, Italy
[2]GKN Sinter Metals, 39031 via delle Fabbriche 5, Brunico (BZ), Italy

Abstract. The automotive industry employs a considerable amount of sintered parts, mainly as transmission and engine components. Gears are the parts that mostly benefit, in terms of cost saving, from the near net shape P/M technology. However, the porosity along with the heterogeneous microstructure can detrimentally affect the mechanical behaviour, especially the fatigue strength. The possibility of increasing sintered density up to 90% and more, the use of high strength alloys, as well as post sintering treatments have been extensively investigated obtaining consistent increases in the fatigue strength. The present study focuses on the effects of porosity and microstructure on tooth root bending fatigue of small module spur gears. The aim is to investigate the synergistic contribution of pore morphology and microstructure heterogeneity to the initiation of fatigue cracks and to the following crack paths. High density parts produced by high strength pre-alloyed powders were studied. Part of the specimens was case-hardened to obtain a martensitic/bainitic microstructure in the surface layer. Bending fatigue tests up to a fatigue endurance of three million cycles were performed. A careful fractographic analysis was conducted. The obtained results were discussed using the fracture mechanics approach of Murakami, considering the pores as pre-existing defects, whose propagation strongly depends on the microstructural heterogeneity.

1 INTRODUCTION

The structural parts produced by powder metallurgy (P/M) represent a competitive solution, alternative to traditional production, in many industrial sectors, including the automotive industry that employs a considerable portion of the total production of P/M parts.

The progressive improvement of the production processes has led to the design and fabrication of structural parts able to withstand high stress levels. The gears represent one of these applications, where maximum advantages in terms of competitiveness and quality can be achieved by using a near net shape technique such as P/M [1-5].

The research efforts have always been devoted to setting up processes able to increase the relative density of the sintered parts up to 90% and more; in addition, the formulation of new high strength pre-alloyed powders and the development of surface modification techniques have strongly improved the 'in service' performances as well as the quality of the final product. Despite this positive trend, some crucial aspects still remain open. One of these is the comprehension of the synergistic effect of porosity and microstructure in the damage process of the material when subjected to 'in service' loads [6-10]. This aspect turns out to be particularly critical for structural pieces that work under cyclic loading conditions and are therefore prone to fatigue damage. It can be observed that, for relatively low sintered density, the damage usually occurs very early, located at the necks between powder particles, which are the weakest point of the microstructure. A widespread damage occurs and the final failure is often caused by the coalescence of numerous defects previously initiated [11]. By increasing the sintered density to high values, even up to near-full density, the triggering of the damage is delayed and the pores can be regarded as local defects of different criticality, able to concentrate the stresses and to act as internal notches or, in certain situations, even as cracks [5, 12 to 15]. In this case the fatigue response is dictated by the most critical defect, which is generally an irregularly shaped pore located in the regions of maximum stress.

The microstructural heterogeneity can play a very important role in the damage localization and in the definition of a preferential crack path [5,8]. Particularly critical are the soft phases that usually envelop the critical pores and represent the microstructure of first propagation of the fatigue crack.

In some papers [6-8], the pores are treated as pre-existing cracks and a fracture mechanics approach is adopted to predict the propagation conditions of crack-like defects. The Murakami criterion [16], initially proposed for determining the fatigue limit of steels containing inclusions, can be very promising among the fracture mechanics based methods. In particular, in its most general formulation this approach envisages estimating the fatigue limit of the material with the

* Corresponding author: vigilio.fontanari@unitn.it

https://doi.org/10.1051/matecconf/201816502011

following equation, that includes also the influence of the load ratio R.

$$\sigma_{lim} = C \cdot \frac{HV+120}{(\sqrt{area})^{1/6}} \cdot \left(\frac{1-R}{2}\right)^{\alpha} \quad (1)$$

The Vickers hardness HV accounts for the effect of the microstructure, whereas the term '*area*' contains the information on the defect geometry, i.e. the dimension of the defect projected on the critical plane of crack propagation, characterized by the maximum normal stress (opening mode I). Finally, C and α are empirical parameters related respectively to the position of the pore with respect to the surface and to the material microstructure. In particular C = 1.41 for surface defects, C = 1.56 for internal defects, while the exponent $\alpha = 0.226 + HV \cdot 10^{-4}$.

Bergmark [6] used the Murakami model for sintered products by introducing a further corrective factor that indirectly takes into account the effect of porosity in terms of variation of the elastic modulus with respect to that of the fully dense material.

A critical aspect for the application of the Murakami model concerns the definition of the critical pore and the identification of the hardness that has to be included in the formula.

In the context of this research topic, the present work is aimed at investigating the tooth root fatigue behaviour of small module spur gears produced by P/M technology, starting from different pre-alloyed powders, compacted up to high green density and consolidated adopting different sintering conditions. The principal objective is to explore the synergic effect of pore structure and microstructure on the fatigue behaviour. For this reason, the fatigue response is discussed starting from an in-depth fractographic analysis aimed at identifying the role of porosity and microstructure in the initiation and propagation of cracks. The Murakami model is applied for distinguishing the synergistic role of porosity and heterogeneous microstructure in defining the fatigue limit of the material.

2 EXPERIMENTAL ACTIVITY

The experimental activity was carried out on the P/M spur gears shown in Fig. 1, whose main constructive parameters are: number of teeth N= 20, module m = 0.71 mm and pressure angle θ =20°.

The tooth root bending fatigue strength for three types of material is investigated. The reference standard is ISO 6336 [17] for the calculation of the load capacity of spur and helical gears.

The gears are produced using pre-alloyed as well as diffusion bondend powders, whose nominal compositions are summarized in Table 1.

In the blending phase, carbon in the form of graphite is added in a quantity equal to 0.25% wt for A85Mo05, and 0.15% wt both for DHP and DDH.

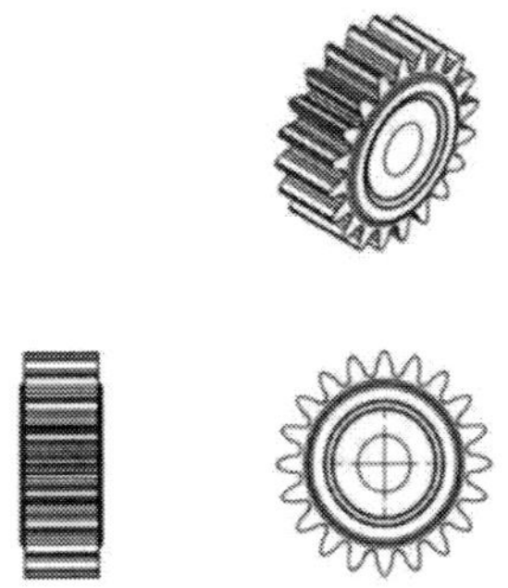

Fig. 1. Geometry of the P/M spur gears.

Table 1. Nominal compositions of the powders

Powder label	Composition (%wt)	
A85Mo05	0.85 Mo, Fe bal.	pre-alloyed
DHP	1.5 Mo, 2.0 Cu, 4.0 Ni, Fe bal	diffusion bonded
DDH	1.5 Mo, 2.0 Cu, Fe bal	diffusion bonded

The gears were produced by GKN Sinter Metals. For A85Mo05 a conventional sintering cycle was conducted in a continuous furnace at a temperature of 1120 °C for 30 min in an atmosphere of endogas (A85Mo05); DDH and DHP underwent to a sinter-hardening cycle that provides forced cooling from the temperature of 1120 ° C down to 300 °C, followed by stress relieving in a separate oven at 180 ° C for 1 h. The A85Mo05 gears were case hardened (austenitizing at 860°C with oil quenching) after sintering to improve the mechanical behaviour of the surface layer.

2.1 Microstructure and porosity

The density was measured using the Archimedes method according to ASTM B311, which permits a quantitative evaluation of the total and open porosity. The porosity was also studied by image analysis using an optical microscope, in order to identify the pores morphology and also their spatial distribution in the gear body. In particular, local measurements were carried out at the tooth root, as well as inside the tooth and in the gear body. Table 2 lists the mean porosity values calculated according to ASTM B311, while table 3 reports the results of the image analysis measurements. It can be observed that porosity has a non-homogeneous distribution into the gear volume.

Table 2. mean density, total porosity and open porosity following ASTM B311.

	A85Mo05	DHP	DDH
Density (g cm^{-3})	7.55	7.35	7.41
Total porosity(%)	4.09	7.56	6.36
Open porosity(%)	0.60	1.39	0.96

Table 3. mean porosity values in different regions of the gear measured using image analysis on metallographic sections.

Porosity %	A85Mo05	DHP	DDH
Tooth body	7.13	8.85	7.63
Tooth root	6.29	7.99	6.20
gear body	3.04	7.15	4.17

The A85Mo05 specimens contains irregularly shaped pores, not uniformly distributed in the gear volume: the remarkable difference in pores content between the tooth and the gear body is evident (table 3). While A85Mo05 exhibits the lowest total porosity among the three variants, its pore content in the critical tooth root region is comparable with other materials. Concerning the microstructure (fig. 2), a surface layer with martensitic microstructure and limited presence of bainite and retained austenite can be observed. The bulk is instead characterized by lower bainite and martensite.

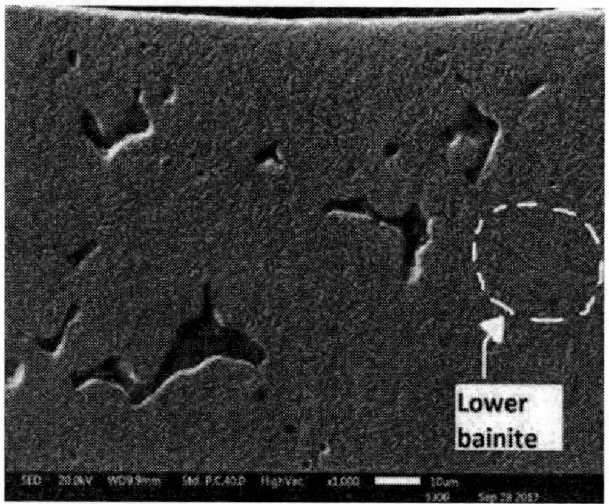

Fig. 2. Microstructure of A85Mo05

The DHP steel is characterized by a more homogeneous pore distribution in the gear volume. The microstructure is heterogeneous (fig. 3) with austenite (white areas), lower bainite (light gray areas) and martensite (dark gray areas). The distribution of the alloy elements is inhomogeneous in the diffusion bonded powders and during sintering there is no complete diffusion of Ni and Cu into the base powders (1.5% Mo Prealloyed iron). The Ni-rich austenite is predominantly located around the pores [18]. The martensite is richer in Cu, while the base powder, poor in Ni and Cu, is predominantly bainitic.

The DDH steel shows a non-uniform pore distribution, comparable to A85Mo05. The microstructure (fig. 4) is predominantly martensitic in the surface layer with some evidences of lower bainite, while the upper bainite prevails in the bulk. Also in this case the pre-alloyed powders do not presents a homogeneous distribution of Cu, consequently areas richer in Cu are still present after sintering and assume a martensitic microstructure upon cooling.

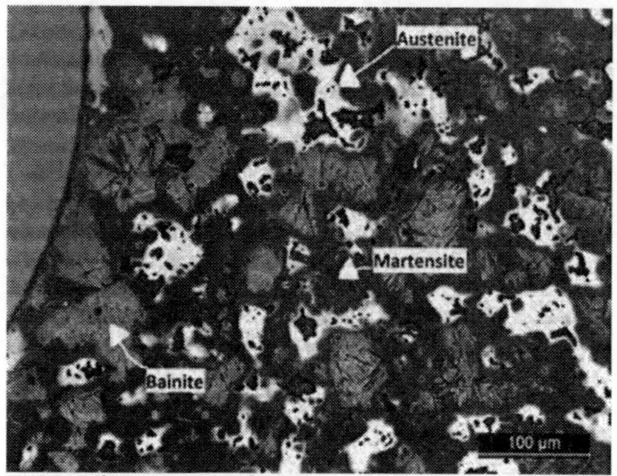

Fig. 3. Microstructure of DHP

Fig. 4. Microstructure of DDH

Due to the post sintering treatments, all the materials are characterized by a surface layer with higher hardness compared to the bulk one. The A85Mo05 alloy was modified by case-hardening, while the other two material variants are produced by sinter-hardening, undergoing a controlled cooling from the sintering temperature down to 300 °C, whose effects are more evident in the surface layer. The microhardness profiles ($HV_{0.05}$) are plotted in figure 5: the surface hardening extends to a depth of nearly 0.5 mm.

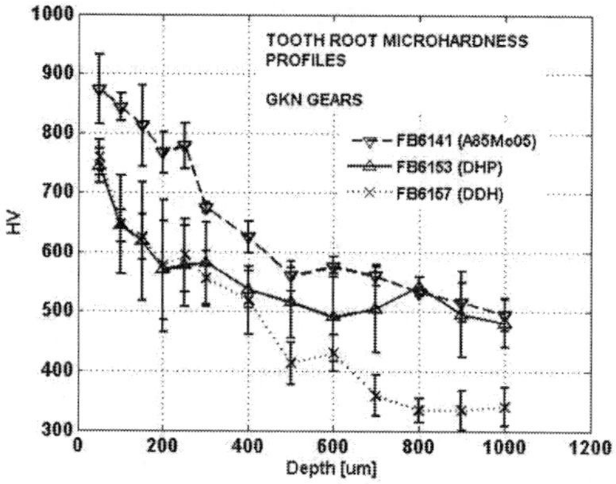

Fig. 5. Microhardness profiles in the surface layer

2.1 Fatigue testing

The fatigue tests were conducted with a resonant Rumul Mikroton 20 kN testing machine operating at frequencies of about 150 Hz. The testing configuration shown in fig. 6 provides for the simultaneous gripping of two teethwith intespaciong of five teeth.

Unfortunately, due to the very small module of the teeth, this condition cannot be precisely reproduced by the testing configuration. For this reason, the stress state in the tooth root region is evaluated with a finite element analysis, that allows to model the effective testing configuration: In this way stress at the tooth root can be correctly calculated. 20-node brick element were used to build up the FE model. The sub-modelling technique was adopted to obtain a refined evaluation of the local stress. The FEM sub-model developed for the analysis is shown in fig. 7. The FE convergence analysis was performed by reproducing the ISO 6336 loading configuration and comparing the FE tooth root bending stress with that calculated using the ISO6336 formula.

Fig. 6. Tooth bending fatigue testing configuration

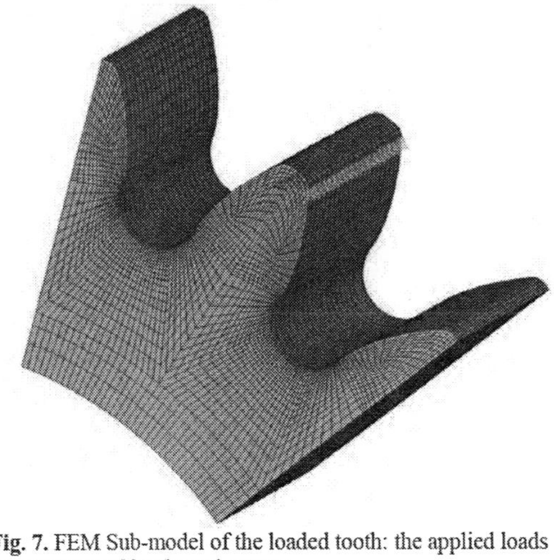

Fig. 7. FEM Sub-model of the loaded tooth: the applied loads are represented by the red arrows.

A load ratio R of 0.1 was adopted. A fatigue life interval between $3\cdot10^4$ and $3\cdot10^6$ cycles was explored. The fatigue strength at $3\cdot10^6$ cycles was determined using the staircase method performing at least 15 tests. Specimens surviving the $3\cdot10^6$ cycles are considered as runouts. The choice of $3\cdot10^6$ cycles is dictated by the ISO6336 standard. In fact, for this fatigue live, an endurance factor equal to one is assumed in the design formula.

The fatigue curve in the finite life regime was evaluated using the Basquin law reported by Eqn. 2. A least-square regression of the results of 15-20 tests, distributed on different stress levels was carried out:

$$\sigma_a \cdot N^{1/m} = C \qquad (2)$$

In this formula σ_a represents the stress amplitude and N is the number of cycles to failure. The dispersion of the results is calculated using the central standard deviation of the data, assuming uniform scatter for different fatigue lives. A measure of the results scattering is expressed by the ratio between the stress amplitude corresponding to 90% and 10% probability of failure:

$$T_{\sigma,10:90} = \frac{\sigma_{a.90}}{\sigma_{a.10}} \qquad (3)$$

The evidence of a knee in the fatigue curve at fatigue lives lower than about $2\cdot10^6$ cycles may be indicative of the presence of a fatigue limit, the confirmation thereof would however require investigating the response to longer lifetimes. In the figures 8-10 the results of the bending fatigue tests and the P_{10}, P_{50} and P_{90} fatigue curves are plotted.

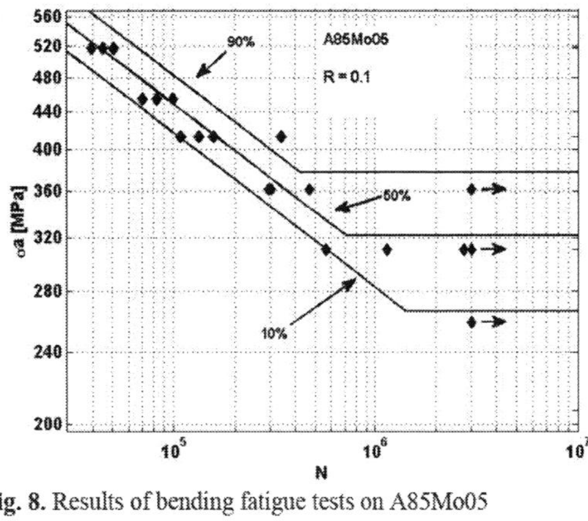

Fig. 8. Results of bending fatigue tests on A85Mo05

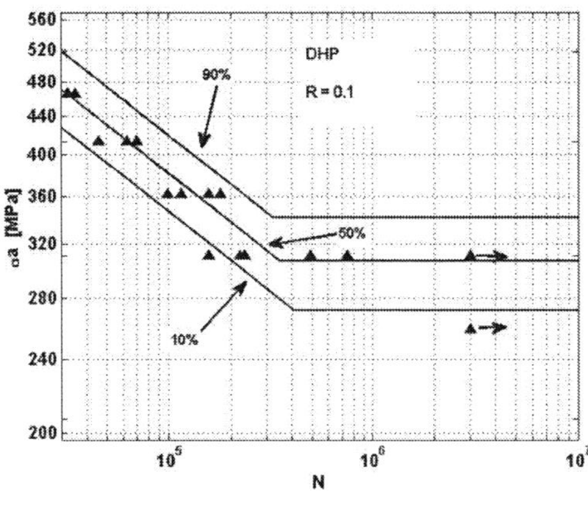

https://doi.org/10.1051/matecconf/201816502011

Fig. 9. Results of bending fatigue tests on DHP

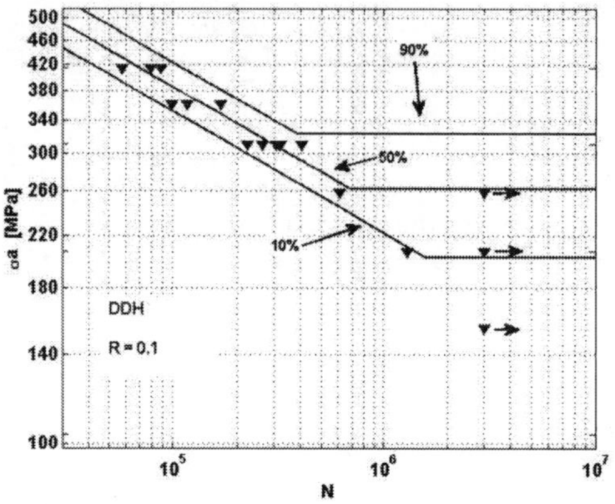

Fig. 10. Results of bending fatigue tests on DDH

The principal parameters that describe the fatigue behaviour for the three materials are collected in table 4.

Table 4. Characteristic parameters of the fatigue curves.

	$\sigma_{a,3\cdot10^6}$ (P_{50})	$T_{\sigma,10:90}$	C	m
A85Mo05	321 (MPa)	1.42	3163 (MPa)	5.90
DHP	307	1.26	2858	5.71
DDH	262	1.59	3846	5.00

The obtained results are compared with results published in [3-5] for similar microstructures with different porosity contents. In general, the literature data show a progressive but not particularly marked decrease in the fatigue strength with density for relative densities higher than 90%, whereas a very marked worsening occurs at lower densities.

2.2 Fractographic analysis

The fractographic study was carried out both by metallographic observations and by electron scanning microscopy (SEM) analysis. In the first case, the main objective is to study the preferential crack path, whereby the tooth is sectioned with a plane orthogonal to the gear axis, polished and prepared with the conventional metallographic procedure. The SEM analysis, on the other hand, is aimed at identifying the crack initiation sites and at evaluating the size of the critical pores.

The representative images for the three material variants are shown in the following figures. The crack path for A85Mo05 (Fig. 11a) is not very serrated: a limited evidence of deviations due to the presence of porosity can be observed. In fig. 11b a surface pore is highlighted which has probably played the role of crack initiation site. In the same region some evidences of intergranular fracture have been observed, that is probably produced by grain boundary embrittlement due to cementite precipitation after carbon enrichment during case hardening.

The fracture path for DHP (Fig. 12) and for DHH (Fig. 13) is much more influenced by the presence of pores that define preferential propagation directions and determined a very serrated path

a)

b)

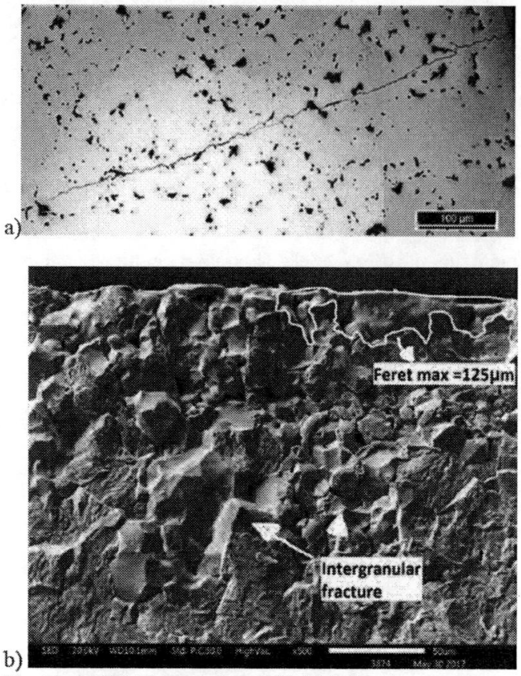

Fig. 11. A85Mo05: a) crack path, b) initiation site

a)

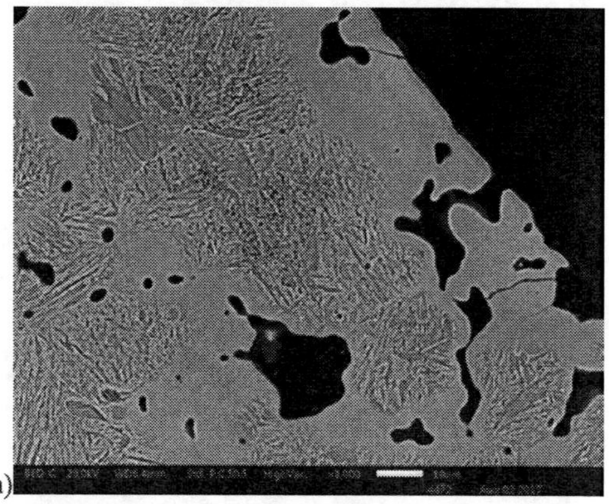

b)

Fig. 12. DHP: a) crack path, b) initiation site

a)

b)

Fig. 13. DDH: a) crack path, b) initiation site

For both material variants the SEM analysis allows to identify pores with very articulated morphology close to the surface, which can be reasonably considered as initiation sites for the fatigue crack. Some evidences of multiple crack initiation followed by crack coalescence have been observed.

Both for DDH and DHP, small cracks nucleated at pores proximal to the tooth surface (Fig. 14) have been observed. This suggests the occurrence of a first stage of diffused damage with multiple crack nucleation, followed by the onset of a main crack that can connect several cracks in its propagation path. The cracks not involved in this process of coalescence stop growing as that shown in figure 14. This observation agrees with the crack propagation behaviour reported in the technical literature [5,8,12,13].

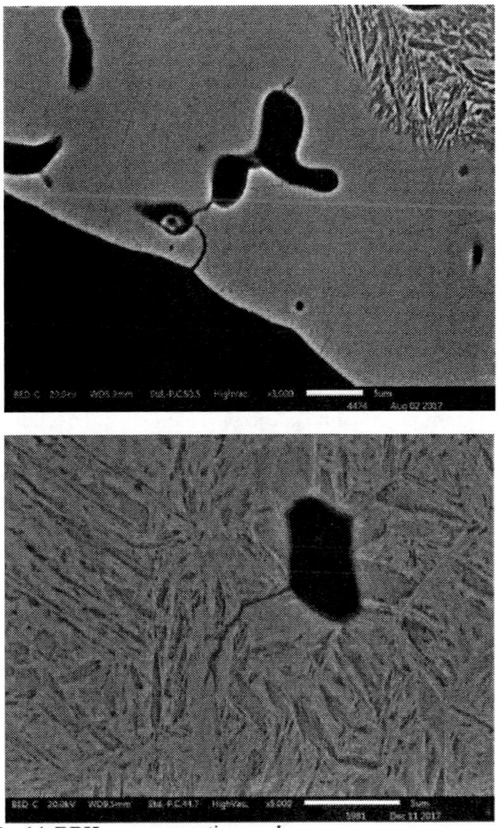

Fig. 14. DDH: non propagating cracks

An extensive analysis of the most critical pores on the fracture surface was carried out to estimate the characteristic size of the initial defect to be used in the Murakami formula. To calculate the area of the critical pores the maximum Feret diameter is used, the evaluation of which is schematically described in figure 15. The area to be inserted in equation 1 is that of the circle defined by the maximum Feret diameter, that inscribes the pore.

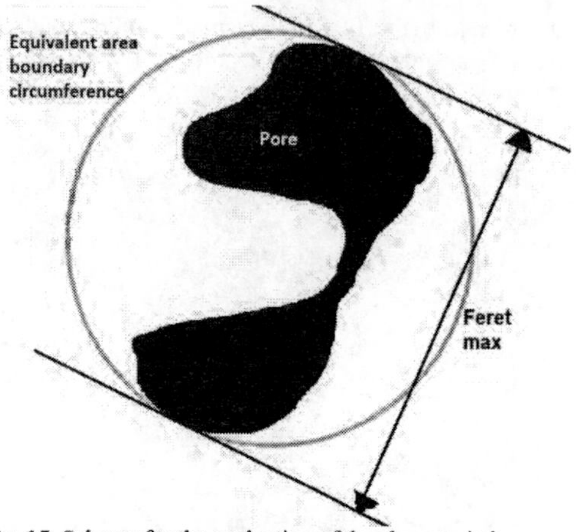

Fig. 15. Scheme for the evaluation of the characteristic dimension of the pore, defined by the Feret diameter

For the evaluation of the fatigue strength using the Murakami model it is necessary to provide the hardness value representative of the material microstructure. The microstructural heterogeneity produced by the non-uniform distribution of the alloying elements can play an important role. The significant differences between the bulk microstructure and the phase surrounding the pores were previously highlighted (fig. 2). To shed light on the role of the local microstructure, the estimate of the fatigue strength using the Murakami model has been realized both considering the average hardness of the material and the local hardness taken in the surroundings of the pores, which characterizes the microstructural phase in which the crack initiation occurs. The results are collected in tables 5 and 6 and compared with the experimental values.

Table 5. Fatigue strength estimated with the Murakami model considering the average microstructural hardness

	HV mean	$\sqrt{area}$ (μm)	$\sigma_{l,\,Mur}$ (MPa)	$\sigma_{l,\,exp}$ (MPa)	Δ (%)
A85Mo05	873	111	524	321	+63
DHP	744	152	437	307	+42
DDH	758	136	452	262	+73

Table 6. Fatigue strength estimated with the Murakami model considering the local hardness.

	HV local	$\sqrt{area}$ (μm)	$\sigma_{l,\,Mur}$ (MPa)	$\sigma_{l,\,exp}$ (MPa)	Δ %
A85Mo05	515	111	345	321	+7
DHP	325	152	233	307	-24
DDH	378	136	264	262	+1

A consistent improvement in the estimation of the fatigue strength is obtained using the local hardness in the Murakami formula. This result confirms, even considering the approximations introduced for the evaluation of the critical defect size, the considerable role played by the microstructure in the neighbourhood of the critical pores. From another point of view, it is worth noticing that the initiation phase, spent by the cracks propagating trough the microstructural phases enveloping the pores, takes a large part of the entire fatigue life.

Moreover, a plausible explanation of the different errors obtained for the three variants can be given. As far as A85Mo05 is concerned, the overestimation may be due to the fact that the grain boundary embrittlement occurring in the case hardened layer is not taken into account by the Murakami model. On the contrary, for the sinter-hardened DHP material, characterized by soft austenite surrounding the critical pores, an underestimation of the fatigue strength is obtained. This can be due by the partial or total transformation of the retained austenite into martensite with consequent local increase in volume, that can produce crack closure slowing propagation speed or even leading to arrest the fatigue crack. These hypotheses need to be confirmed with a more in-depth study.

3 Conclusions

The tooth root bending fatigue behaviours for three P/M material variants characterized by high density and high strength microstructures were studied. The main conclusions can be listed as follows:

1) the fatigue curves have a marked knee that can suggest to consider the fatigue strength at $3 \cdot 10^6$ cycles as a reasonable estimate of the fatigue limit. However, the scatter of the fatigue results is quite large and therefore additional tests will be needed to explore longer fatigue lives.
2) by studying the porosity distribution and morphology as well as the inhomogeneous microstructure after sintering characterized by evident differences in hardness, the factors that most affect the fatigue response can be evaluated.
3) The SEM analysis of the fracture surfaces allows to identify the critical pores and to advance the hypothesis that the main crack is formed as a result of the coalescence of multiple cracks nucleated in the most stressed region.
4) For the evaluation of the critical pore it is necessary to conduct the SEM fractographic analysis, as the metallographic analysis does not allow to clearly identify the pores criticality. The pore area identified by SEM observation on the fracture surface approaches that of the crack surface projected on the critical plane, as required by the Murakami model
5) The representative parameter of the pore criticality can be the maximum Feret diameter, which identifies the circumference inscribing the serrated perimeter of the pore
6) The application of the Murakami method allows to highlight the remarkable importance of the microstructure, which acts synergistically with the pore in the damage localization

References

1. C.M. Sonsino CM., Powder Metall., 33(3), 235–245, (1990)
2. P.K Jones, K. Buckley-Golder, R. Lawcock et al., Int J Powd Metall., 33(3), 37–44, (1997).
3. L. Forden, S. Bengtsson, M. Bergström,. Powder Metall., 48(1):10–12, (2005)
4. S. Dizdar, P. Johansson, Proceedings of EURO PM2007, Toulouse, France; 1-7, (2007).
5. M. Benedetti, C. Menapace, Powder Metallurgy, 60 (2), 149-156, (2017)
6. A. Bergmark, Powder Metallurgy Progress, 5(3), 131-137, (2005)
7. M. Andersson, Powder Metallurgy Progress, 11 (1-2), 21-3,1 (2011)
8. G. Straffelini, V. Fontanari, A. Hafez, M. Benedetti, Powder Metallurgy, 52(4), 298-303, (2009)
9. J.M. Torralba, L. Esteban, E. Bernardo, et al., Powder Metall. 57(5):357–364, (2014)
10. H. D'Armas, L. llanes, J. Penafiel, J. Bas, M. Anglada, Mater Sci Eng A.;277:291–296. (2000)
11. G. Straffelini, M. Benedetti, V. Fontanari, Materials and Design, 61, 101–108, (2014)
12. M. Kabatova, E. Dudrova, A. S. Wronski, Fat. Fract. Engng Mat., 32, 214–222, (2009)
13. E. Dudrova, M. Kabatova, Powder Metall. 59(2):148–167, (2016)
14. Danninger H, Xu C, Khatibi G, B Weiss, B Lindqvist, Powder Metall. 55(5):378–387, (2012)
15. X. Deng, G.Piotrowski, N.Chawla, K.S. Narasimhan, Mat. Sc. Engng. A, 491(1-2), 28-38, (2008)
16. Y. Murakami, Y.: *Metal Fatigue: effects of small defects and nonmetallic inclusions*. Oxford (UK) Elsevier Science Ltd., (2002)
17. ISO 6336-1:2006, Calculation of load capacity of spur and helical gears. (2006)
18. F. Bernier F, P. Plamondon, J-P. Bailon, G. L'Esperance, Powder Metall 2011;54:559–65.

The effect of corrosive environment on fatigue life and on mean stress sensitivity factor

Marta Morgantini[1,*], *Donald* MacKenzie[1], *Yevgen* Gorash[1], and *Ralph* van Rijswick[2]

[1]Mechanical and Aerospace Department, University of Strathclyde, 16 Richmond St, G1 1QX, Glasgow, United Kingdom
[2]WEIR Minerals B.V., Egtenrayseweg 9, 5928 PH Venlo, The Netherlands

Abstract. An experimental investigation of the effect of corrosive environment on the fatigue life of circular cross section specimens made from structural low carbon steel S355 is presented in this paper. Uniaxial load controlled fatigue tests in the regime of 10^5 to 10^7cycles were performed in two different environments: at room temperature in air (without corrosion) and in-situ fresh water solution environment. Various approaches to evaluate mean stress effect on stress-life behaviour in inert environment are reviewed and used to estimate the mean stress sensitivity factors both for air and corrosive environment. Mean stress sensitivity factor is calculated based on experimental results and predicted using the different approaches. Moreover, a comparison between Haigh Diagram predicted by these models and experimental results is proposed and the Walker approach is found to be the one that best fits experimental data.

1 Introduction

Corrosion fatigue is a common failure mechanism where mechanical parts or components are subjected to cycling loads in corrosive and aggressive environment. Corrosion fatigue failure is therefore an essential design consideration in many industrial fields, where pumps components work under the influence of alternating load due to cyclic pressure in a corrosive environment that degrades material surface. Many studies have been conducted to understand how corrosive environment influences fatigue life of steel [1] and how fatigue strength of material decreases. Usually, sea water condition are tested and many data are available in literature relevant to this condition. However, other corrosive environment less aggressive have received less attention in research and, therefore, less data are available in literature. This study, because of application requirements, considers corrosion fatigue behaviour in fresh water environment.

The effect of mean stress on fatigue life is a well-established research subject. Several models have been proposed to evaluate its effect on fatigue limit and more generally on fatigue strength [2, 3, 4, 5, and 6]. Many comparison of these approaches have been done to understand which of them better predicted the material behaviour and many experimental data have been used and compared, included low alloy steel [7]. However, all of these models are design without taking into account the effect of corrosive environments. Recently, some researcher have presented the effect of mean stress on crack propagation in corrosive environment, particularly for low alloy steel [8], however in this study tests based on stress-life approach are presented and therefore a fracture mechanics analysis is not taken into account. The aim of this study is understand qualitatively which effect has corrosion on mean stress sensitivity factor and if any of the existing approach can be used, as a starting point, to predict mean stress effects on corrosion fatigue life.

2 Experimental Procedure

Material used for tests in this research work is a structural forged low carbon steel, S355J2G3+N. Chemical composition and mechanical property are already discussed in previous work [9]. Tests in air and in corrosive environment were carried out using 6 mm diameter circular cross section specimens with tangentially blended fillets between the test section and the test ends, according to ASTM E466-07 standard [10]. Roughness of the gauge length surface of each specimens were calculated through the arithmetic mean based on four measurements. Values of roughness of all population of specimens were included between 0.095 and 0.130 micron. Both fatigue tests in air and in fresh water environment were performed under axial loading and load controlled with servo-hydraulic fatigue testing system Instron 8801 and Instron 8802.

Fatigue tests in air were conducted at room temperature and at a fixed frequency of 15 Hz. When the specimen achieved $6x10^6$ number of cycles without being broken, the test was stopped and the result saved as 'run out'. Two conditions of stress ratio R were used for tests in air: fully reversed load R=-1 and positive mean stress R=0.

* Corresponding author: marta.morgantini@strath.ac.uk

To simulate the synchronous effect of corrosion and fatigue phenomena, a corrosion cell was developed and home-made produced in the AMRL laboratory, compatible with both Instron 8801 and Instron 8802. The cell is designed in-built to the specimen and it allows a continuous flow of corrosive solution during the application of cycling loads and, consequently, it makes possible the corrosion of central part of specimens. In fact the corrosive cell encases the sample central part and leaves its grip parts free. The duct inlet of the corrosion cell is connected through a plastic pipe to a water pump of nominal flow of 600 l/h placed in a glass tank to permit the circulation of solution during the test. Similarly, the duct outlet is connected to the tank to allow the water solution to return in the tank to be oxygenated by an air pump of nominal capability of 100 l/h. Thanks of this basic circuit, in-situ corrosion fatigue tests were possible.

Characteristics of aqueous solution used in all corrosion fatigue tests are shown in Table 1.

Three condition of stress ratio were tested in fresh water corrosive environment: fully reversed load R=-1 and two positive mean stress R=0 and R=0.5.All tests were carried out at fixed frequency of 10 Hz. A work presented by Thierry Palin-Lu et al [11], devoted to the effect of corrosion on the gigacycle fatigue strength of similar low alloy steels, shows a continuous decrease of fatigue strength with number of cycles. For this reason, all tests were left to run till a rupture occurred (specimen divided in two parts), because a fatigue limit was not expected in this conditions.

Table 1. Characteristics of aqueous solution environment.

Temperature [°C]	Salinity [ppm]	pH	Conductivity [mS]	Flow [l/h]
25±1	824 NaCl	6.9	2.0±0.2	190

3 Experimental Results

Experimental data obtained from tests conducted in air, for both conditions of stress ratio, are plotted in a log-log scale S-N curve and in Fig. 1. Interpolation of data is presented in the S-N curves based on the best fits of experimental data. Data trends can be expressed using Basquin's law, up to $2x10^6$ number of cycles, in the form:

$$\sigma_a = A \times N^B \quad (1)$$

Under positive mean stress fatigue strength decreases and the change of slope of interpolated data appears at a lower number of cycles, as typical behaviour of alloy steel. Data obtained in air are considered as the reference for evaluation of the effect of corrosion on the fatigue strength and they are used to calculate the mean stress sensitivity factor in air.

Results obtained in corrosive environment in the three stress ratio configuration are plotted in a log-log scale S-N diagram together with results obtained in air. All experimental results and data interpolations are shown in Fig. 2. For all the stress ratio configuration, the interpolation derived from experimental data can be express using Basquin's law, as for air data. The main difference is that in air, the validity of the equation is limited to a certain region, in corrosive environment is it valid for all the region taken into account in this work (10^5 to 10^7).

The effect of corrosive environment on the fatigue strength in the region of low number of cycles is less detrimental than in the higher number of cycles region: this is due to corrosion time depending. In fact, since tests were carried out at 10 Hz, to achieve 10^5 cycles the sample was corroded less than 3 hours during the effect of fatigue load; in this amount of time even if specimens surface started to be corroded, this effect was not so intense and therefore the fatigue strength is not reduced too much. At higher number of cycles, the effect of corrosion is therefore more intense and fatigue strength at $4x10^6$ is reduced by a factor of 2.

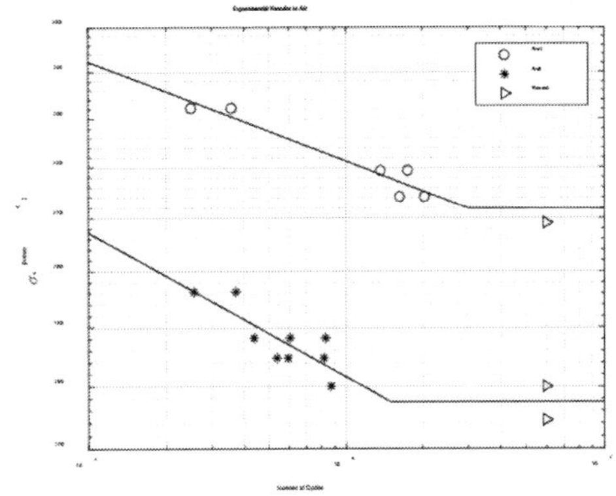

Fig. 1: S-N Curves in Air

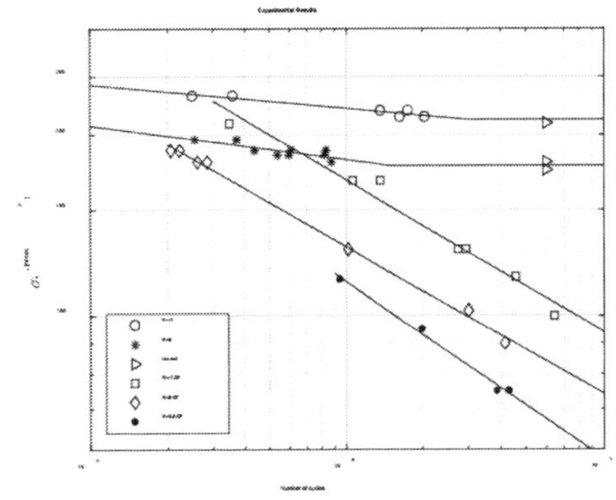

Fig. 2: Experimental S-N curve

Fatigue strength decreases constantly, following Basquin's law and consequently a fatigue limit in corrosive environment is not occurring, at least in the region up to 10^7. This behaviour is in line with results found in literature, for similar material and environment [12].

S-N curves show a reduction of fatigue strength in relation of increased positive mean stress in fresh water environment, as well as in air.

4 Mean stress analysis

To assess the mean stress effect on fatigue life and corrosion fatigue life the mean stress sensitivity factor model developed by Schutz and used by FKM Guideline [13] is taken into account. The sensitivity factor M_σ is define in (2):

$$M_\sigma = \frac{\sigma_{a,R=-1}}{\sigma_{a,R=0}} - 1 \qquad (2)$$

Where $\sigma_{a,R=-1}$ is the stress amplitude for a fixed number of cycles obtained from fully reversed loading tests and $\sigma_{a,R=0}$ is the stress amplitude for the same number of cycles obtained from test at positive mean stress R=0.

Moreover, four stress-life approaches for fatigue in air, already developed in literature to predict the effect of mean stress in fatigue life in air, are assessed in this work to evaluate how they correspond to experimental data in air and in corrosive environment. The selection of the following four methods is based on a previous analysis published by N. E. Dowling [14]. The approaches and relative equations are: Modified Goodman relation (3), Geber parabola relation (4), Smith-Watson-Topper (SWT) relation (5) and Walker relation (6).

$$\frac{\sigma_a}{\sigma_{eq}} + \frac{\sigma_m}{\sigma_u} = 1 \qquad (3)$$

$$\frac{\sigma_a}{\sigma_{eq}} + \left(\frac{\sigma_m}{\sigma_u}\right)^2 = 1 \qquad (4)$$

$$\sigma_a = \frac{\sigma_{eq}}{\sqrt{\frac{2}{1-R}}} \qquad (5)$$

$$\sigma_a = \frac{\sigma_{eq}}{\left(\frac{2}{1-R}\right)^{1-\gamma}} \qquad (6)$$

Where σ_a is the stress amplitude, σ_m is the mean stress, R is the stress ratio, σ_u is the ultimate strength, σ_{eq} is the completely reversed stress expected to cause the same life as the actual combination of amplitude and mean stress (σ_m, σ_a) and γ is a mathematical parameter introduced by Walker to allow a superior accuracy.

From each equation the expression of stress amplitude σ_a is acquired and it is used in the definition of the mean stress sensitivity factor to evaluate the general dependency of this factor on material characteristics. Trends are shown in Fig. 3.Mean stress sensitivity factors as function of Equation (3) and (4), shown a dependency on the ratio of the equivalent stress and the ultimate tensile stress. It means that under these approaches the sensitivity to the mean stress is depends on the material ultimate strength, that is different for each material, and also it changes with the number of cycles: material is more sensible to mean stress at lower number of cycles, when consequently the equivalent stress is higher. Mean stress sensitivity factors obtained using σ_a from equation (5) and (6) is plotted as a function of γ. The factor based on Smith-Watson-Topper equation depends only on stress ratio, so for a fixed R value, the sensitivity of all material is the same and it doesn't even change for number of cycles. The sensitivity related to Walker approach results to be constant with the number of cycles, as SWT, but it changes with material by the parameter γ.

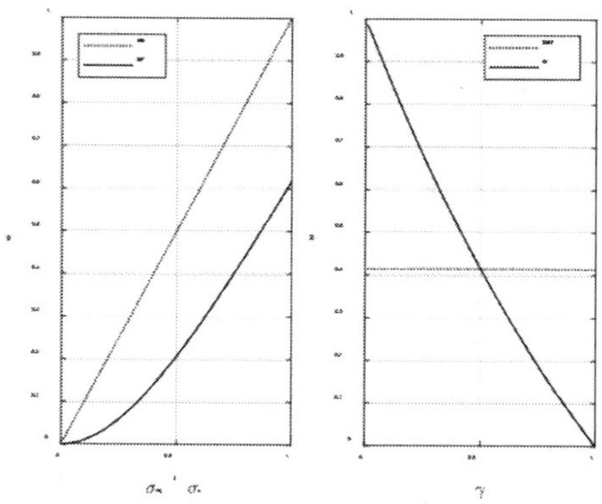

Fig. 3. Mean stress sensitivity factors trends

Using value obtained from experimental tests of fatigue strength at R=-1 in air at 5x10^6 cycles, values of $\sigma_{a,R=0}$ from each approaches are determined and the corresponding numerical value of mean stress sensitivity factor, M_σ, calculated. Data employed in each equation are report in Table 2 .

The value for parameter γ, used in equation (6) was calculated based on the relation (7), proposed by N.E. Dowling [15]:

$$\gamma = -0.0002 * \sigma_u + 0.8818 \qquad (7)$$

Table 2. Data employed in equations

	$\sigma_{eq}=\sigma_{a,R=-1}$ [MPa]	σ_u [MPa]	R	γ
Goodman	212	500	0	-
Geber	212	500	0	-
SWT	212	-	0	-
Walker	212	-	0	0.7818

Table 3. Mean stress sensitivity factor results

	$\sigma_{a,R=0}$ [MPa]	M_σ
Goodman	149	0.4240

Geber	183	0.1556
SWT	150	0.4142
Walker	182	0.1633
KFM	-	0.08
Exp. Data	178	0.1910

In Table 3 are reported amplitude stress at R=0 and mean stress sensitivity factor obtained from the solution of equations (3) to (6), mean stress sensitivity factor suggested by FKM Guideline for low carbon steel S355 and mean stress value calculated from experimental data in air.

The same methodology is employed to plot in a Haigh Diagram the trends of each approach, again based on the equivalent stress extracted from experimental tests. In this case, amplitude and mean are evaluated not only for R=0 but for all combination of stress ratio in order to cover the positive mean stress quarter of the Haig Diagram. The graph is shown in Fig. 4.

The mean stress sensitivity factor that better estimates the experimental results is the one obtained using Walker equation. The same results is also shown in the Haigh Diagram, where Walker equation is the best representative for experimental data. SWT and Goodman approaches appear to be too conservative for this material. The FKM mean stress sensitivity factor, that is 0.08, is, from the other side, the less conservative approach in the prediction of mean stress effect in air.

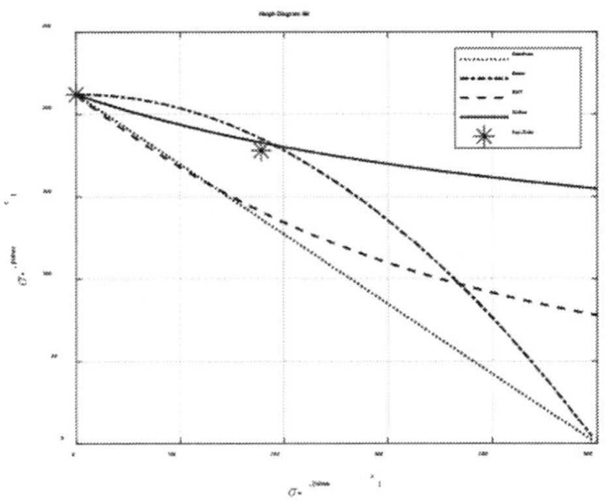

Fig. 4. Haigh Diagram in air

The mean stress sensitivity factor and Haigh Diagram are evaluated also in corrosive environment, using the same approach that has been implemented for data in air. Since a fatigue limit doesn't exist in corrosive environment, as shown in S-N curves, mean stress sensitivity factors are evaluated for two different point of the specimen life: at 10^6 cycles and 10^7 cycles.

Data employed to evaluated amplitude stress from equation (3) to (6) are reported in Table 4. Ultimate strength, R ratio value and parameter γ are the same employed in air.

Table 4. Data employed in corrosive environment

	$\sigma_{eq}=\sigma_{a,R=-1}$ 10^6 cycles [MPa]	$\sigma_{eq}=\sigma_{a,R=-1}$ 10^7 cycles [MPa]
Goodman	165.5	94.1
Geber	165.5	94.1
SWT	165.5	94.1
Walker	165.5	94.1

Table 5. Sensitivity factors in corrosive environment

	M_σ 10^6	M_σ 10^7
Goodman	0.3360	0.1882
Geber	0.1024	0.0342
SWT	0.4142	0.4142
Walker	0.1633	0.1633
Exp. Data	0.2923	0.2673

Mean stress sensitivity factors calculated from the four approaches and from experimental data are reported in

Table 5. Results obtained from experimental data, at both conditions of number of cycles, show that material is more affected by mean stress in corrosive environment rather than in air, in fact the sensitivity factor increases of 0.1. Results obtained using SWT approach and Walker approach are constant both for the environment and for the number of cycles because they are not able to take into account number of cycles effect. Differently, results obtained from Goodman and Geber approach, since they depend on equivalent stress, they change both with environment condition and number of cycles.

A different relation to calculate the parameter γ is proposed, based on the best fit of experimental results. Equation is shown in (8) and it is a function of the number of cycles. In fact the main difference between equation (7) and (8) is the dependency of equation (8) to number of cycles that, in corrosive environment, is an important factor since the strength of material continuously decreases. Moreover relation proposed in this work is limited to low carbon steel material and it can not help to find the value of γ of different materials, as equation proposed by Dowling.

$$\gamma = -2\times10^{-9}\times N + 0.6734 \qquad (8)$$

Mean stress sensitivity factors obtained by Walker approach are calculated again at 106 and 107 cycles based on the new expression of γ and results are reported in Table 6. In this case, the factor approaches better the behaviour shown by experimental data.

Table 6. Mean stress sensitivity factors from Walker approach

	M_σ 10^6	M_σ 10^7
Walker	0.2558	0.2716

Trends of relation between stress amplitude and mean stress obtained from the four approaches are plot in Haigh Diagram. Results at 10^6 cycles are shown in Fig. 5. Walker relation has been plotted twice, both with γ constant and γ function of number of cycles. Haigh Diagram at 10^7 cycles is shown in Fig. 6.

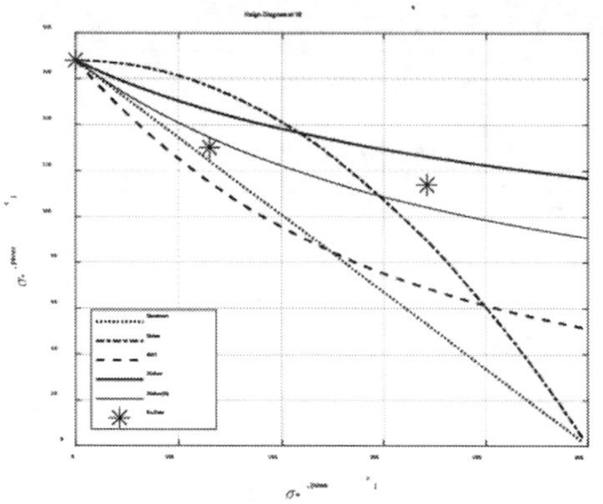

Fig. 5. Haigh Diagram at 10^6 cycles

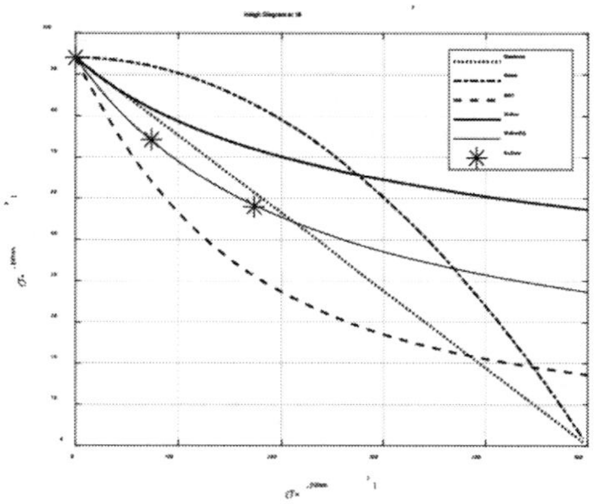

Fig. 6. Haigh Diagram at 10^7 cycles

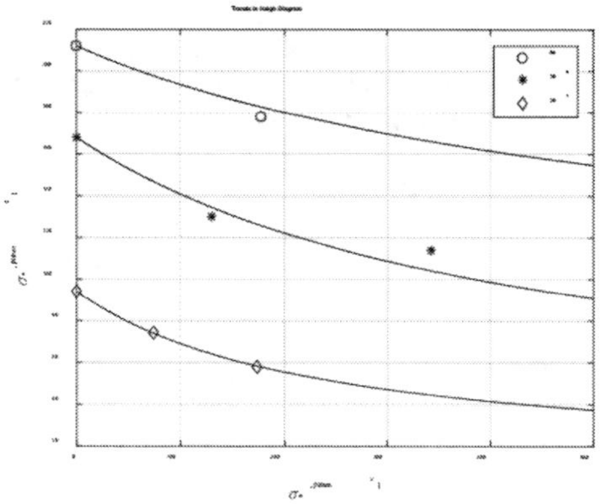

Fig. 7. Comparison of experimental data

Haigh Diagrams in corrosive environment show that the Walker approach, using γ function of number of cycles, is the relation than better approach experimental data. This approach is particular interesting for high positive mean stress ($R \geq 0.3$): whereas other approaches show a severe decrease of stress amplitude, Walker approach in this area is more gradual.

Fig. 7 shows experimental data in air, in corrosive environment at 10^6 and in corrosive environment at 10^7 plotted in a Haigh Diagram. The allowable combination of mean and amplitude stress decreases with corrosive environment and in this environment mean stress affects more the fatigue strength.

5. Conclusion

This work presents an experimental investigation of the effect of fresh water environment on the fatigue strength and the effect of stress ration on corrosion fatigue life of a structural low carbon steel.

The experimental results show that the effect of corrosive environment is to reduce the fatigue strength of the material: deterioration is higher at lower level of stress amplitude but it occurs even a lower number of cycles. At $4x10^6$ number of cycles the fatigue strength is reduced by a factor of two compared to fatigue strength in air. Furthermore, a fatigue limit is not observed in the region up to 10^7 number of cycles in all of three stress ratio condition presented in this work. Further investigation, with test at higher number of cycles, is required to better understand if a fatigue limit could occur in this non-aggressive corrosive environment.

The effects of mean stress on fatigue life and corrosion fatigue life are evaluated using the mean stress sensitivity factor and four predictive models well known in literature for the evaluation of fatigue in air. It is found that mean stress sensitivity factor in corrosive environment increases of 0.1 compared to the one in air: in corrosive environment in fact the material is more affected by mean stress. Among the approaches taken into account, Walker method is the one that better fits data in air. This prediction is based on a parameter γ which value is based on a relation proposed by Dowling. In corrosive environment it has found that the mean stress sensitivity factor varies depending in number of cycles. In this environment a new relation for γ is proposed based on a best fit of experimental data. Using the new relation, function of the life of carbon steel, Walker equation fits better experimental results plotted in a Haigh Diagram. The main limitation of this equation is that is based on experimental data related to a specific material in a certain environment. A generalization of this relation is not therefore possible for other materials.

References

1. P. P. Milella. Fatigue and Corrosion in Metals, *Springer* 767-804, (2013).
2. J. Goodman, Mechanics Applied to Engineering, *Longman,* 631-636 (1919).
3. J. O. Smith, *U.o.Illinois,* Bulletin **334**, (1942).
4. K. N. Smith, P. Watson, T. H. Topper, *J. of Metals, ASTM,* Vol. 5 n **4**, 767-778, (1970).
5. J. Morrow, Fatigue Design Handbook, Sect. **3.2**, AE-4, (1968).
6. K. Walker, ASTM STP 462, 1-14, (1970).

7. T. Wehener, A. Fatemi, *Int. J. Fatigue*, 13 n **3**, 241-248, (1991).

8. O. Adedipe, F. Brennan, A. Kolios, *Marine Structures*, **42,** 115-136, (2015).

9. M. Morgantini, D. MacKenzie, T. Comlekci, R. Van Rijswick, *Procedia Engineering*, vol **213,** (2018).

10. ASTM E466-07 , *ASTM International,* (2007).

11. T. Paulin-Luc, R. Perez-Mora, C. Bathias, G. Dominiguez, P.C. Paris, J. M. Arana, Eng. Fracture Mechanics, **77**, 1953-1962, (2010).

12. R. Pérez-Mora, T. Palin-Luc, C, Bathias, P. C. Paris, *Int. J. of Fatigue*, **74**,156-165, (2015)

13. E. Haibach, FKM Guideline: Analytical Strength Assessment of components in Mechanical Engineering, 5th Rev. (2003).

14. N. E. Dowling, *Proceeding of Fatigue 2004*, 2004-01-2227 (2004).

15. N. E. Dowling, C. A. Calhoun, A. Arcari, *FFRMS*, **32**, 163-179 ,(2009).

Cyclic plasticity and low cycle fatigue damage characterisation of thermally simulated X100Q heat affected zone

Ronan J. Devaney[1,4,*], *Heiner* Oesterlin[3], *Padraic E.* O'Donoghue[2,4], and *Sean B.* Leen[1,4]

[1]Mechanical Engineering, College of Engineering and Informatics, NUI Galway, Ireland.
[2]Civil Engineering, College of Engineering and Informatics, NUI Galway, Ireland.
[3]Fraunhofer Institute for Mechanics of Materials IWM, Freiburg, Germany.
[4]Ryan Institute for Environmental, Marine and Energy Research, NUI Galway, Ireland.

Abstract. This paper presents the cyclic plasticity and low cycle fatigue (LCF) damage characterisation of thermally simulated heat affected zone (HAZ) for API 5L X100Q weldments. Microstructures representative of the HAZ for two cooling rates are generated using a Gleeble thermomechanical simulator for manufacture of strain-controlled cyclic plasticity test specimens. The simulated HAZ specimens are subjected to a strain controlled test programme which examines the cyclic effects of strain-range and the tensile response at room temperature. A modified version of the Chaboche rate independent plasticity model, which accounts for early stage damage is implemented to characterise the cyclic plasticity response, including isotropic and kinematic hardening effects. The constitutive parameters are fitted to experimental data using an optimisation procedure developed within a MATLAB code. The measured response of the simulated HAZ specimens is compared to that of the X100Q parent material (PM), and the simulated HAZ is shown to share the early stage fatigue damage behaviour of the PM, but exhibits significantly a higher yield and cyclic strength.

1 Introduction

Steel catenary risers (SCRs) are pipelines used to conduct hydrocarbons or injection fluids between a production facility and a subsea wellhead. SCRs are subjected to heave caused by floating production facilities, vortex induced vibrations and cyclic pressurisation, making fatigue performance a primary concern. Weight reduction is also a key requirement for deep and ultra-deepwater SCRs. The use of next generation high strength steels facilitates a reduction in riser weight and installation time through the use of lower wall thickness pipe sections.

Welded connections are of particular concern for fatigue failure in SCRs [1]. The thermal cycles experienced during the welding process alter the microstructure and thus the constitutive and fatigue performance of the PM adjacent to the weld. The material and geometrical mismatch resultant from the welding process causes the welded connection to act as a fatigue hot-spot [2].

Mechanical characterisation of the HAZ is difficult due to its size and continuously changing nature from the fusion line to the unaffected PM. Ringinen et al. [3] used Gleeble thermomechanical simulation to generate HAZ material from the candidate SCR material X100. The microstructure and impact strength of the simulated HAZ material were compared to HAZ material obtained from welded joints, and the simulated HAZ material was shown to exhibit a representative microstructure, and an impact strength within the spread of the experimentally observed values for the HAZ material obtained from welded X100 joints.

Guiying and co-workers also used Gleeble thermomechanical simulation to generate HAZ specimens for an X100 grade steel, with peak temperatures and cooling rates representative of the coarse grain HAZ (CGHAZ), fine grain HAZ (FGHAZ) and intercritical HAZ (ICHAZ) [4]. The specimens were microstructurally examined and subjected to a tensile and impact test programme. The simulated ICHAZ was shown to exhibit a significantly reduced impact toughness relative to the PM and the other HAZ specimens. The tensile and yield strength of the simulated FGHAZ was lowest when compared to the PM and the other thermally transformed zones.

However, influence of the varied HAZ microstructure on the cyclic plasticity and fatigue performance has not been considered. This paper presents an experimental programme to characterise the tensile and LCF behaviour of X100Q PM and HAZ. Gleeble thermomechanical simulation is used to generate simulated HAZ specimens, with a peak temperature and cooling rates representative of the FGHAZ in the as-welded condition; post-weld heat treatment is typically avoided for offshore SCR welding due to the additional time required and associated costs. The simulated HAZ material is then used in a mechanical test programme, consisting of strain controlled room temperature tensile and LCF tests. The test programme results are used to fit a modified version of the Chaboche

* Corresponding author: r.devaney1@nuigalway.ie

rate independent plasticity model using an optimisation procedure developed within a MATLAB code. The simulated HAZ specimens are shown to exhibit superior cyclic and tensile strength when compared to the parent material.

2 Methodology

2.1. Experimental

2.1.1 HAZ thermal simulation

The as-received X100Q material is in the form of two 406.4 mm OD, 25 mm WT 2 m long seam welded pipe sections. The chemical composition supplied by the plate manufacturer is shown in Table 1.

Table 1. Chemical composition of the X100Q steel (%wt).

C	Si	Mn	P	S	Al	B
0.14	0.21	1.17	0.008	0.001	0.084	0.0025
Cr	Cu	Mo	N	Nb	Ni	Ti
0.32	0.01	0.3	0.004	0.027	0.02	0.005

Thermal simulation of X100Q HAZ was conducted using a Gleeble 3150 thermomechanical simulator at Fraunhofer IWM, Freiburg, Germany. Cylindrical specimens 120 mm long with a 10 mm diameter were manufactured from the as-received material for HAZ thermal simulation as shown in Figure 1. A 16 mm long gauge length of uniform thermal history was calibrated for each cooling rate by using a specimen with three type K thermocouples spot-welded, one at the centre of the specimen, one offset 5 mm from the centre of the specimen and a third thermocouple offset 8 mm from the centre of the specimen. Temperature control was achieved through a closed loop feedback system based on the temperature at the central thermocouple and the specimens were resistance heated and cooled using a combination of conduction through water cooled copper grips and air blasting.

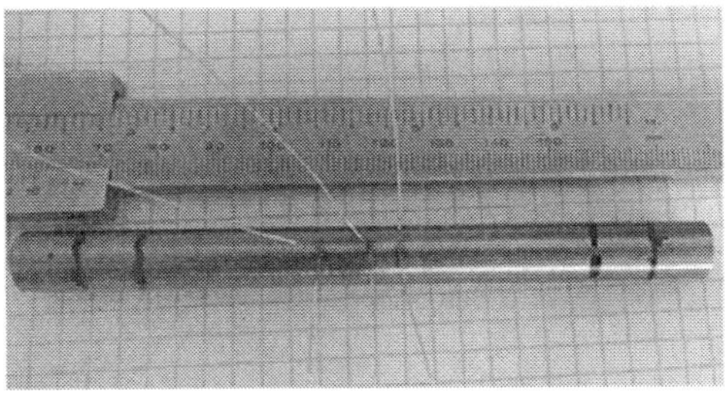

Figure 1. An X100Q specimen with type K thermocouples spot welded for Gleeble thermomechanical simulation.

Two thermal cycles in the expected range for the FGHAZ were identified for the thermomechanical simulation of HAZ specimens. For both cycles the specimens were heated at a rate of 100°C/s above the Ac_3 temperature of 847°C to a peak of 950°C. The peak temperature was maintained for 0.5 seconds before the specimens were cooled to 110°C at rates of 10°C/s and 30°C/s.

2.1.2 Microstructural analysis

Microscopy specimens were manufactured from the as-received material and the post-test Gleeble specimens, mounted in epoxy resin, then ground and polished to a 0.06 µm finish using a diamond suspension and etched using Vilella's reagent. Microscopy was performed using an Olympus BX51M optical microscope at NUI Galway. Optical micrographs of the PM and two simulated HAZ microstructures are shown in Figure 2. The microstructure of the PM consisted of lower bainite with visible prior austenite grain boundaries (PAGBs) within which are sheaves of bainite consisting of bainitic ferrite (BF) laths separated by carbides. There was a visible reduction in grain size for the 10°C/s simulated HAZ which exhibited an upper bainite microstructure also with visible BF laths and PAGBs. The 30°C/s simulated HAZ displayed the greatest deviation from the PM, with a granular bainite microstructure consisting of equiaxed BF and retained austenite (RA).

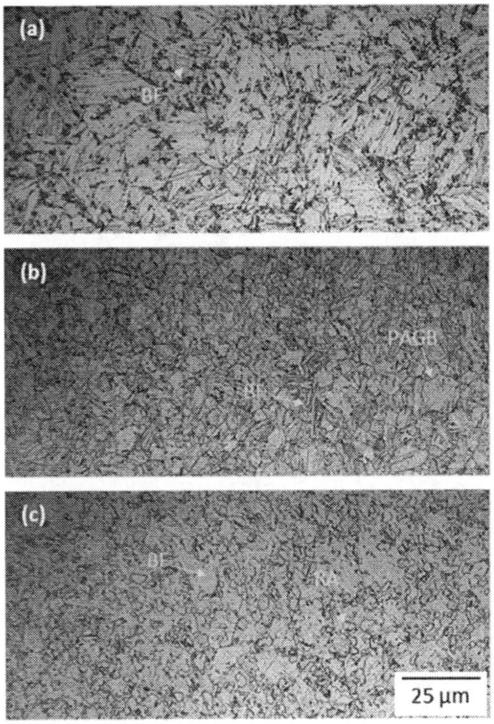

Figure 2. Optical micrographs of (a) X100Q PM, (b) 10°C/s simulated HAZ and (c) 30°C/s simulated HAZ. Microstructural features are labelled in yellow.

2.1.3 Tensile testing

Tensile specimens were machined from the Gleeble simulated HAZ specimens, with a gauge length of 36 mm, a diameter of Ø6 mm and an average surface roughness of 0.2 μm on the gauge length. The specimens were machined with M10 threaded ends.

2.1.4 Cyclic testing

LCF specimens were prepared from the as-received X100Q and the Gleeble simulated HAZ specimens. The PM LCF specimen were machined to a gauge length and diameter of 16 mm and Ø6.5 mm respectively, while the HAZ specimen are machined to a gauge length and diameter of 12 mm and Ø7 mm respectively. All LCF specimens are polished to an average surface roughness of 0.2 μm on the gauge length.

2.2 Modelling

2.2.1 Constitutive model

A damage mechanics methodology is used during constitutive modelling, where damage is modelled as an effective reduction in load resisting area due to the formation and coalescence of micro voids within the material. The experimentally observed early stage damage in X100Q PM highlighted by Devaney et al. [5] is accounted for by including the effects of damage when modelling the constitutive response of the materials. The damage accumulation is experimentally measured throughout each LCF test using the following equation

$$D_N = 1 - \frac{E_N}{E_0} \tag{1}$$

where D_N is the level of damage at cycle N, E_0 is the initial elastic modulus and E_N is the unloading modulus at cycle N. Using this definition of damage and the assumption of isotropic damage, the effective stress is defined as

$$\tilde{\sigma} = \frac{\sigma S}{S - S_D} = \frac{\sigma}{1-D} = E\varepsilon^e \tag{2}$$

where $\tilde{\sigma}$ is effective stress, S is the area of load resisting material on a plane perpendicular to the applied load, S_D is the area of voids a plane perpendicular to the applied load and ε^e is elastic strain. The yield function for the Chaboche time independent nonlinear kinematic-isotropic hardening (NLKIH) plasticity model with damage is

$$f = \left(\frac{\boldsymbol{\sigma}}{1-D} - \boldsymbol{\chi}\right)_{eq} - r - k \tag{3}$$

where f is the yield function, $\boldsymbol{\sigma}$ is the stress tensor, $\boldsymbol{\chi}$ is the kinematic back stress tensor, eq denotes von Mises equivalent stress, r is the isotropic hardening/softening and k is the cyclic yield stress. The increment in plastic strain is calculated as follows

$$\dot{\boldsymbol{\varepsilon}}^p = \dot{\lambda}\frac{df}{d\boldsymbol{\sigma}} = \frac{3}{2}\frac{\dot{\lambda}}{1-D}\frac{\left(\frac{\boldsymbol{\sigma}}{1-D} - \boldsymbol{\chi}\right)_{dev}}{\left(\frac{\boldsymbol{\sigma}}{1-D} - \boldsymbol{\chi}\right)_{eq}} \tag{4}$$

where $\dot{\boldsymbol{\varepsilon}}^p$ is the increment in the plastic strain tensor, $\dot{\lambda}$ is the plastic multiplier, and dev denotes deviatoric stress. The Armstrong-Frederick nonlinear kinematic hardening evolution with damage for each back stress is as follows

$$\dot{\boldsymbol{\chi}}_i = \frac{2}{3}C_i\dot{\boldsymbol{\varepsilon}}^p(1-D) - \gamma_i\boldsymbol{\chi}_i\dot{p} \tag{5}$$

where $\dot{\boldsymbol{\chi}}_i$ is the increment in back stress tensor, $\boldsymbol{\chi}_i$ is the back stress tensor, C_i are hardening moduli and γ_i are recall parameters, also $\boldsymbol{\chi} = \sum_{i=1}^{3}\boldsymbol{\chi}_i$. The increment in Chaboche nonlinear isotropic kinematic hardening/softening is given as

$$\dot{r}_i = b_i Q_i (1-D)\dot{p} - b_i r_i \dot{p} \tag{6}$$

where $\dot{r}_i$ is the isotropic hardening/softening increment, r_i is the level of isotropic hardening/softening, Q_i is the saturated hardening/softening stress and b_i controls the rate of hardening/softening, also $r = \sum_{i=1}^{2} r_i$.

3 Results

3.1. Experimental

3.1.1 HAZ thermal simulation

Typical examples of the agreement shown between the chosen simulated HAZ thermal cycles and the thermal cycles measured from the central thermocouple on the Gleeble specimen for both the 10°C/s cooling rate and the 30°C/s are shown in Figure 3. The effect of the latent heat released during the bainite transformation is visible in the 30°C/s cooling curve. From these results the average Ar3 temperature for the 30°C/s and 10°C/s cooling rates can be estimated as 380°C and 413°C respectively.

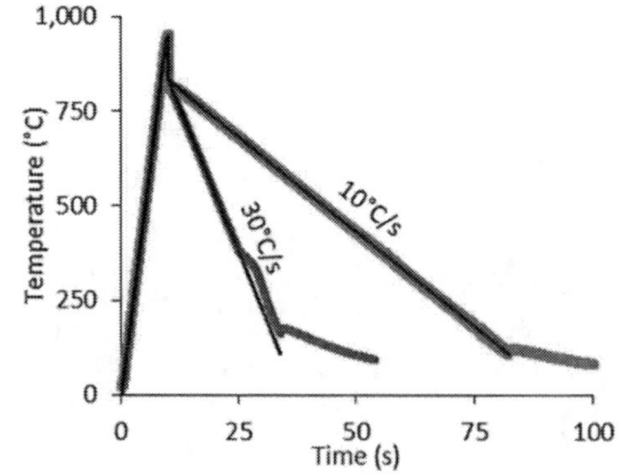

Figure 3. A comparison between the desired thermal cycles and the thermal cycles measured from the central thermocouple achieved during Gleeble thermal simulation of HAZ material.

 https://doi.org/10.1051/matecconf/201816503002

3.1.2 Tensile testing

The stress-strain tensile response of both the 10°C/s and the 30°C/s simulated HAZ specimen is shown in Figure 4, and an image of a post-test tensile specimen is shown in Figure 5. On either side of the homogeneous HAZ region, an interface zone (IF) exists in which there is a gradation in microstructure (and presumably therefore mechanical behaviour) from those of the HAZ to those of the PM. In act, this is the observed location of significant necking and tensile failure for both specimen. Otherwise, it can be seen that the higher cooling rate leads to a harder response and a higher ductility. This can perhaps be attributed to the more refined microstructure of this case shown in Figure 2 (c). However, this requires further investigation, for example, using the through-process modelling methodology [6] under development within the present Science Foundation Ireland (SFI) project "Multi-scale, through-process characterisation for next-generation welded connections" (MECHANNNICS), in order to quantitatively assess the effects of the gradation in microstructure in the IF zone. Real joints will inevitably contain such IF regions also and, indeed, these are common locations of failure, which has motivated this work.

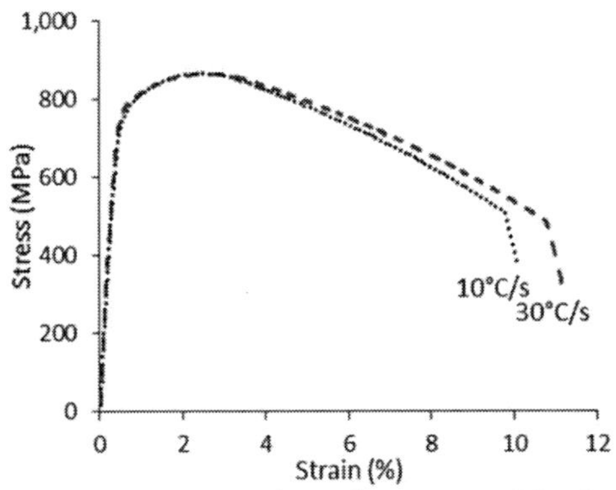

Figure 4. The stress-strain tensile response of the simulated HAZ specimens.

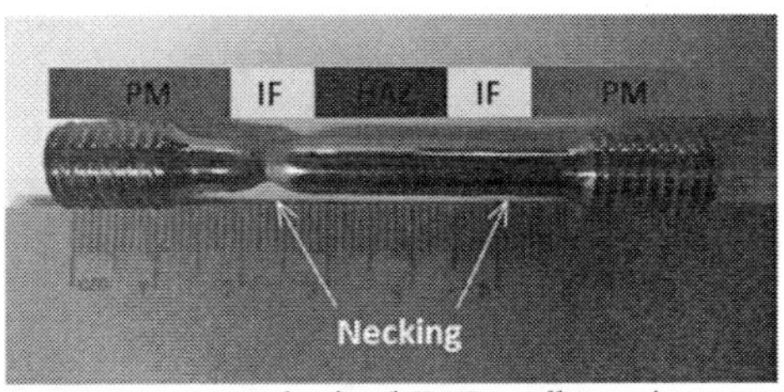

Figure 5. A post-test simulated HAZ tensile specimen tensile specimen, showing necking at the IF regions. The unaffected PM, IF zone and thermally calibrated region of simulated HAZ is shown schematically.

3.1.3 Cyclic testing

A comparison of the measured stress-strain response of the PM and the 10°C/s simulated HAZ for the 1st cycle of a $\Delta\varepsilon = 1\%$ strain controlled test is shown in Figure 6. The onset of yield for the simulated HAZ specimen occurs at approximately 100 MPa higher than for the PM, and the stress amplitude of the simulated HAZ is about 146 MPa or 18% greater than the PM over the 1st cycle. There is also a significant contrasting response in terms of hardening-softening behaviour of the HAZ and PM materials; the simulated HAZ cyclically hardens within the first cycle, by approximately 15 MPa, whereas the PM softens by approximately 5 MPa.

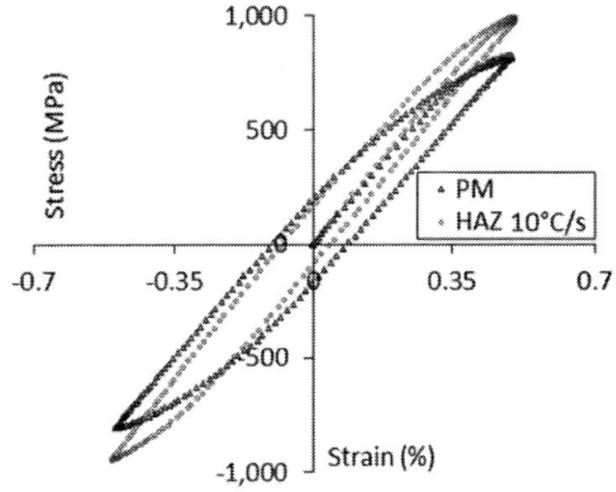

Figure 6. A comparison between the first cycle stress-strain responses of the 10°C/s HAZ and the PM at a 1% strain range.

The measured damage evolution of the 10°C/s simulated HAZ for a $\Delta\varepsilon = 1\%$ test is compared to that of the PM in Figure 7. The early stage damage for each test was observed through the degradation in unloading modulus as shown in Eq 1. The same general trend in damage evolution can be seen for the initial cycles of both the PM and the HAZ, with a rapid rise for the first couple of cycles before exhibiting stabilisation. The 10°C/s simulated HAZ LCF specimen failed through buckling in the IF zone after 7 cycles for the $\Delta\varepsilon = 1\%$ test.

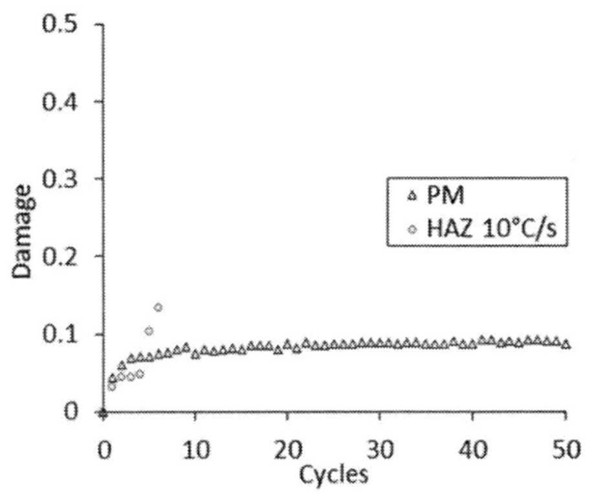

Figure 7. A comparison between the experimentally measured damage evolutions for the 10°C/s HAZ and the PM at a 1% strain range.

Figure 8 shows a comparison between the first cycle stress-strain responses of the simulated HAZ and the PM for the $\Delta\varepsilon = 0.8\%$ test. The stress amplitude for the first cycle of the $\Delta\varepsilon = 0.8\%$ simulated HAZ test is 85 MPa greater than the PM. Both materials exhibit hardening over the first cycle, but the extent of hardening in the

https://doi.org/10.1051/matecconf/201816503002

simulated PM is almost negligible in comparison to that of the 10°C/s simulated HAZ. A comparison of the measured damage evolutions for the 10°C/s simulated HAZ and the PM for the $\Delta\varepsilon = 0.8\%$ test is shown in Figure 9. The damage level in the simulated HAZ matches that of the PM at a value of approximately 0.05 for the majority of life, then increases slightly in the cycles preceding failure.

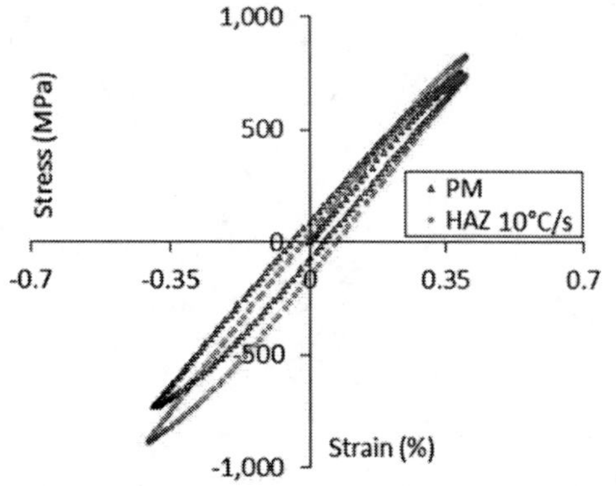

Figure 8. A comparison between the first cycle stress-strain responses of the 10°C/s HAZ and the PM at a 0.8% strain range.

For the $\Delta\varepsilon = 0.6\%$ test the stress-strain response for both the 10°C/s simulated HAZ and the PM is predominantly elastic. As a result, the stress amplitude of the simulated HAZ is just 37 MPa greater than the PM for the first cycle, as shown in Figure 10, and both the simulated HAZ and the PM show a low levels of hardening, with the extent of hardening greater in the simulated HAZ specimen.

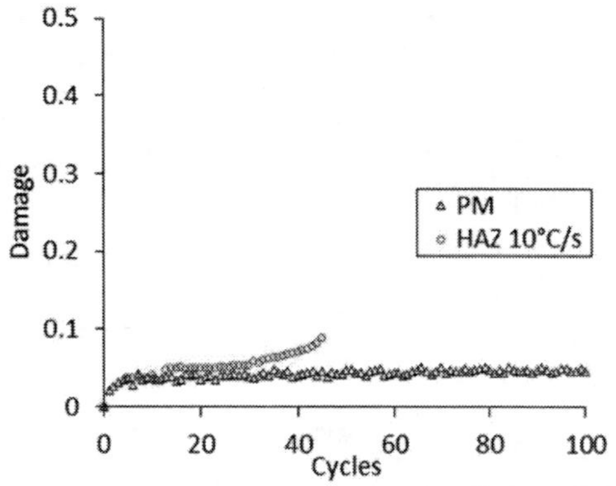

Figure 9. A comparison between the experimentally measured damage evolutions for the 10°C/s HAZ and the PM at a 0.8% strain range.

Figure 11 shows a comparison of the measured damage evolutions for the 10°C/s simulated HAZ and the PM for the $\Delta\varepsilon = 0.6\%$ test. Both materials follow the same trend of a rapid rise in damage within the first few cycles followed by stabilisation, but the stabilised level of early stage damage in the 10°C/s simulated HAZ prior to the onset of failure is greater than that of the PM.

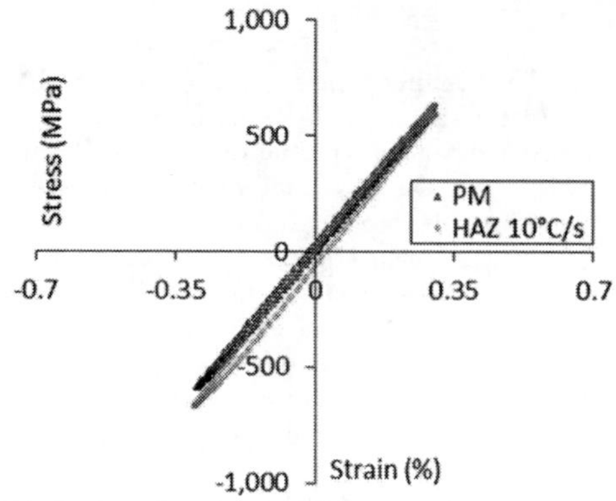

Figure 10. A comparison between the first cycle stress-strain responses of the 10°C/s HAZ and the PM at a 0.6% strain range.

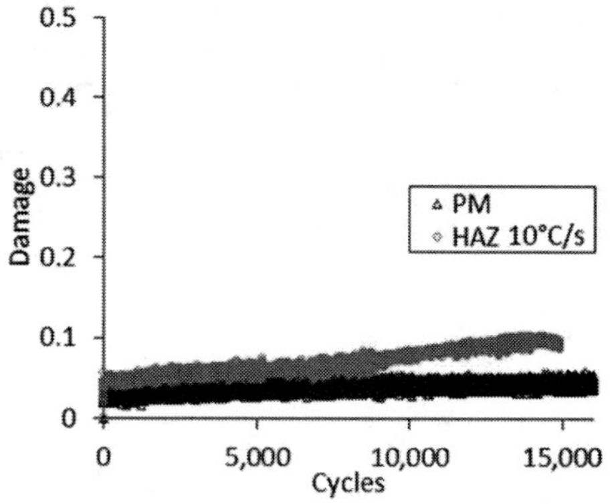

Figure 11. A comparison between the experimentally measured damage evolutions for the 10°C/s HAZ and the PM at a 0.6% strain range.

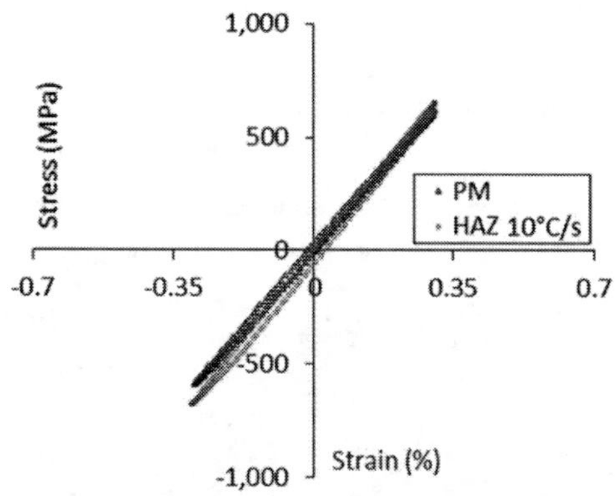

Figure 12. A comparison between the first cycle stress-strain responses of the 30°C/s HAZ and the PM at a 0.6% strain range.

Figure 12 shows a comparison of the first cycle stress-strain responses for the 30°C/s simulated HAZ and the PM. The stress amplitude of the 30°C/s simulated HAZ is approximately 50 MPa greater than the PM and 14 MPa greater than the 10°C/s simulated HAZ, indicating that the higher cooling rate has resulted in the formation of a harder HAZ material.

The increased hardness and therefore reduction in cyclic plasticity for the 30°C/s simulated HAZ specimen is also indicated by the lower level of stabilised early stage damage exhibited for the $\Delta\varepsilon = 0.6\%$ test when compared to the PM or the 10°C/s simulated HAZ, as shown in Figure 13. This suggests that the early stage damage observed for the PM and simulated HAZ is a function of cyclic plasticity, which has long been recognised as the primary mechanism in fatigue failure [7].

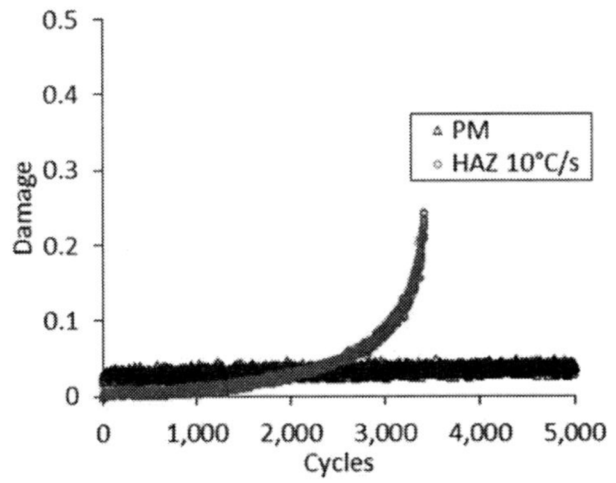

Figure 13. A comparison between the experimentally measured damage evolutions for the 30°C/s HAZ and the PM at a 0.6% strain range.

During cyclic testing of the simulated HAZ specimens it was noted that the significant change in the mechanical response of heat treated gauge length resulted in a concentration of deformation in the IF zone between the thermally calibrated region of simulated HAZ and the unaffected PM. Similar to the tensile specimens, cyclic failure is also generally found to occur in the IF zones, as shown schematically in Figure 15. Figure 15 (a) shows an example of a buckling failure in the IF zone and Figure 15 (b) shows a fracture failure in the IF zone.

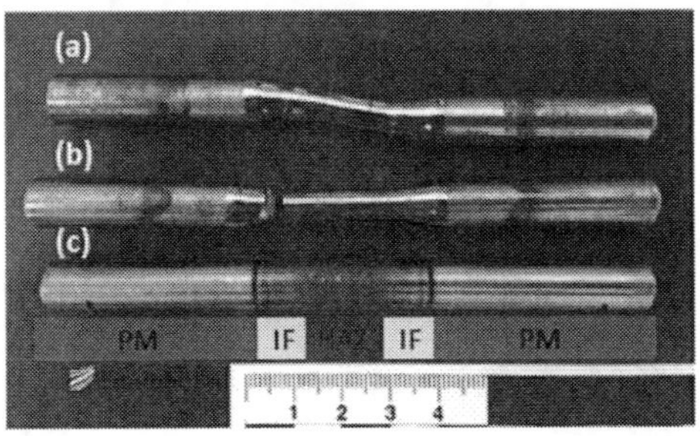

Figure 14. (a) Buckled simulated HAZ LCF test specimen. (b) Simulated HAZ LCF test specimen with failure outside the heat treated gauge length. (c) Gleeble simulated HAZ specimen post-test. The unaffected PM, IF zone and thermally calibrated region of simulated HAZ is shown schematically.

3.2 Modelling

3.2.1 Constitutive model

A typical example of the fit to the first cycle and stabilised stress-strain response achieved using the Chaboche NLKIH constitutive model with parameter optimisation accounting for damage is shown in Figures 15 and 16 for the $\Delta\varepsilon$=1% 10°C/s simulated HAZ specimen. For both cycles the predicted stress-range of the model is within 4.5% of the experimental observations.

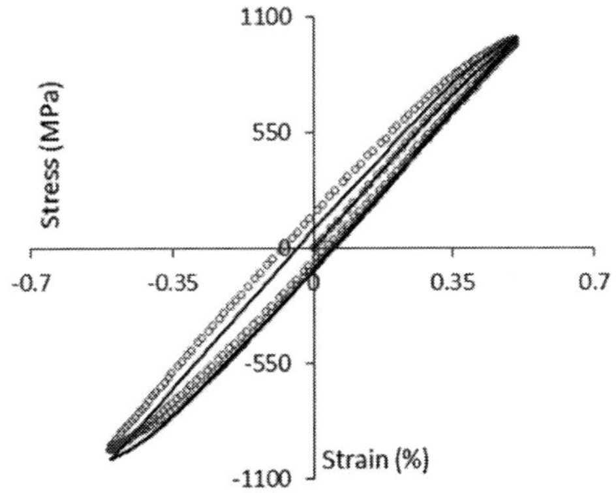

Figure 15. The fit between the model (black line) and the experimental (open symbols) stress-strain response for the first cycle of the $\Delta\varepsilon$=1% 10°C/s simulated HAZ specimen test.

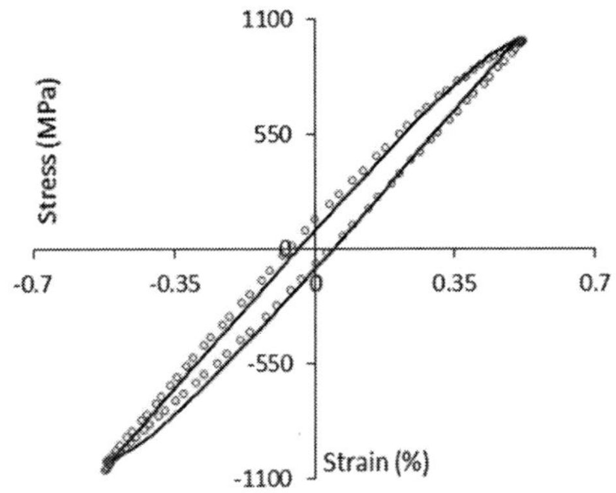

Figure 16. The fit between the model (black line) and the experimental (open symbols) stress-strain response for a stabilised cycle of the $\Delta\varepsilon$=1% 10°C/s simulated HAZ specimen test.

Table 2 shows the NLKIH parameters identified for the X100Q PM and 10°C/s simulated HAZ from the parameter optimisation process accounting for damage, as described by the authors in [5]. The increased hardness of the simulated HAZ material can be seen from the increase in the cyclic yield stress, while the difference in the cyclic response is most apparent when comparing the significant level of isotropic softening exhibited by the PM to the low level of cyclic hardening displayed by the 10°C/s simulated HAZ.

Table 2. The optimised NLKIH parameters for the X100Q PM and 10°C/s simulated HAZ.

	PM	10°C/s HAZ
E (GPa)	209.8	217.8
k (MPa)	279.5	420.6
$b_{1,2}$	0.0104 0.3912	4050 21335
$Q_{1,2}$ (MPa)	-181.8 -68.2	50 20
$C_{1,2,3}$ (GPa)	1200.7 360 94.7	2342.9 364.6 10.7
$\gamma_{1,2,3}$	4562.3 2858.7 32.7	6478.5 3180.5 32

4 Discussion

The two distinct microstructures and related sets of mechanical properties obtained from the PM during the HAZ simulation test programme highlight the importance of controlling cooling rate for welding in SCRs, where it is typically desirable to avoid the requirement for post weld heat treatments due to the high costs associated with offshore installation resources. The bainitic transformation temperatures observed through the liberation of latent heat from the simulated HAZ specimen were below the value of 518°C reported by Wang et al. [8] for an X100Q steel. It is likely that this difference is partially caused by the slower cooling rate of 5°C/s as it has been shown by Qiao and co-workers[9] and indicated by the results reported here, that the transformation temperature decreases with an increase in cooling rate. The increased accuracy of measurements taking using dilatometry will also result in an earlier detection of transformation temperature.

The present work has permitted identification of the cyclic plasticity and early stage LCF damage of thermally simulated FGHAZ material. This is a key step forward for modelling and design of multi-material welded X100Q connections where HAZ behaviour is a key challenge, due to the typically small size-scale of HAZ, and the associated difficulty of manufacturing tensile and fatigue test specimens. As mentioned above, the present work is part of a larger programme of research (the MECHANNICS project, funded by SFI), which is developing multi-scale, through-process predictive methodologies for welded materials and joints. Hence, a key next step for X100Q, is the application of this through-process methodology developed in the MECHANNICS project, e.g. see Mac Ardghail et al. [6], to the present X100Q, specifically for the IF zone. This will facilitate more detailed analysis of the failure behaviour of the microstructure-gradient IF zones, which are so important for fatigue failure, as also demonstrated here. This issue has been most documented for "Type IV cracking" high temperature creep failures in 9Cr power plant steel welds [10]. The concentration of failures in this region is due to its susceptibility for partial transformation to austenite and over-tempering of the PM as a result of the thermal cycles experienced.

From the cyclic plasticity constitutive behaviour obtained for the 10°C/s simulated HAZ specimen for the initial and stabilised cycles, it was possible to demonstrate that the material shared the early stage damage behaviour of the PM which has been attributed to the formation of micro cracks within the material as discussed by the authors in [5].

5 Conclusions

Two simulated HAZ microstructures representative of the FGHAZ in X100Q welds have been generated using Gleeble thermomechanical simulation. On microstructural examination the 10°C/s simulated HAZ exhibited an upper bainite microstructure, while the 30°C/s simulated HAZ showed a granular bainite microstructure, in contrast to the lower bainite microstructure of the PM.

The simulated HAZ specimens were subjected to a room temperature LCF and tensile test programme where the constitutive, cyclic plasticity and early stage damage behaviour was observed. During the LCF and tensile test programme a soft material region was identified at the interface of the simulated HAZ material and the parent metal in which all failures occurred. The softening of the material in this region is attributed to a combination of partial transformation to austenite and over tempering of the parent material.

The combined application of the NLKIH constitutive model with damage and MATLAB parameter optimisation process which accounts for early stage experimental damage model allows identification from the experimentally observed cyclic plasticity response of the 10°C/s simulated HAZ. Future work will focus on quantitative analysis and investigation of the IF zone, for example, using the through-process methodology of the MECHANNICS project.

This publication has emanated from research conducted with the financial support of Science Foundation Ireland as part of the MECHANNICS joint project between NUI Galway and University of Limerick under Grant Number SFI/14/IA/2604.

The authors would also like to thank Eva Augenstein of Fraunhofer IWM.

References

[1] M. Abelanet, D. Karunakaran, R. Jones, S. Eyssautier, and P. White, "Deepwater steel catenary risers require attention to fatigue factors," *Offshore magazine*, no. 10. Oct-2013.

[2] R. J. Devaney, P. E. O'Donoghue, and S. B. Leen, "Global and local fatigue analysis of X100 & X60 steel catenary riser girth welds," *J. Struct. Integr. Maint.*, vol. 2, no. 3, pp. 181–189, Jul. 2017.

[3] D. A. Ringinen, A. V Chastukhin, G. E. Khadeev, L. I. Efron, and P. P. Stepanov, "A WELDABILITY STUDY OF STEEL OF STRENGTH CLASS X100," *Metallurgist*, vol. 57, no. 12, pp. 68–74, 2013.

[4] Q. Guiying, Z. Ming, G. U. Yu, H. A. N. Xiulin, W. Xu, X. Furen, and L. Bo, "Mechanical Properties of the Heat Affected Zone of an X100 Grade Pipeline Steel," in *Advances in Materials and Processing Technologies XV*, 2014, pp. 795–802.

[5] R. J. Devaney, P. E. O'Donoghue, and S. B. Leen, "Experimental characterisation and computational modelling for cyclic elastic-plastic constitutive behaviour and fatigue damage of X100Q for steel catenary risers," *Int. J. Fatigue*, Under review, 2018.

[6] P. Mac Ardghail, N. Harrison, and S. B. Leen, "A through-process, thermomechanical model for predicting welding-induced microstructure evolution and post-weld high temperature fatigue response," *Int. J. Fatigue*, In press, 2018.

[7] J. Morrow, "Cyclic Plastic Strain Energy and Fatigue of Metals," in *Internal Friction, Damping, and Cyclic Plasticity*, 100 Barr Harbor Drive, PO Box C700, West Conshohocken, PA 19428-2959: ASTM International, 1965, pp. 45–87.

[8] H. Wang, F. Wang, G. Shi, Y. Sun, J. Liu, Q. Wang, and F. Zhang, "Effect of quenching temperature on microstructure and yield strength of Q-T-treated X100Q bainitic steel," *Mater. Res. Express*, no. 475, pp. 560–564, Mar. 2018.

[9] Z. X. Qiao, Y. C. Liu, L. M. Yu, and Z. M. Gao, "Formation mechanism of granular bainite in a 30CrNi3MoV steel," *J. Alloys Compd.*, vol. 475, no. 1–2, pp. 560–564, 2009.

[10] J. A. Francis, W. Mazur, and H. K. D. H. Bhadeshia, "Review Type IV cracking in ferritic power plant steels," *Mater. Sci. Technol.*, vol. 22, no. 12, pp. 1387–1395, Dec. 2006.

A Non-Linear Model for Corrosion Fatigue Lifetime Based on Continuum Damage Mechanics

Jie Zhang [1,2], *Stijn* Hertelé [2], and *Wim* De Waele [1]

[1]SIM vzw, Technologiepark 935, BE-9052 Zwijnaarde, Belgium
[2]Soete Laboratory, Department of EEMMeCS, Ghent University, Ghent, Belgium

Abstract. Many structures, such as oil platforms and wind turbines, are built in the marine environment. These structures do not merely suffer from variable cyclic loading due to wind, waves and current, but also from corrosion. Their interaction can give rise to corrosion fatigue, thus reducing the service life and integrity of the structure. The research community witnesses a challenge to identify damage mechanisms of combined fatigue loading and corrosion, and to link this to the lifetime prediction of offshore structures. In this paper, a non-linear corrosion fatigue model is proposed to describe damage accumulation based on continuum damage mechanics. Fatigue endurance limit, load frequency and corrosion rate are included in the model as basic parameters which influence the interaction between fatigue and corrosion. A sequential load effect is revealed via integration of the non-linear accumulation of damage. Parametric studies are conducted to show the abilities of the model. Preliminary simulation results agree to a strong extent with experimental data in the form of corrosion fatigue S-N curves. Nonetheless, deviations are observed at lower life times and these are prone to further investigation. In the future, parameters will be calibrated and further validation experiments will be performed.

1 Introduction

Offshore structures, such as supporting structures of offshore renewable energy applications, experience millions of cyclic loads due to wind, waves and current. Fatigue failure can occur when accumulated damage (propagation of an initiated fatigue crack) reaches the critical tolerance value of the structures. Offshore structures are also exposed to harsh environmental conditions during their operational life (1). The synergistic interaction between a corrosive environment and cyclic loading may amplify the development of fatigue damage, and is referred to as corrosion fatigue. As a result, materials deteriorate more severely (2, 3) which – if unaccounted for – may lead to catastrophes involving loss of lives. Therefore, it is important to develop a suitable model to predict the lifetime of the structure, and to instruct inspection and maintenance properly and timely.

It is reported that pits easily form and grow on the surface of materials in a corrosive environment (4-6). When this happens, changes in the surface morphology introduce stress concentrations which facilitate fatigue crack initiation. Therefore, dependent on the time and degree of corrosion, the formation of pits leads to a significant reduction of the fatigue resistance of metals (7, 8). However, it is difficult to satisfactorily model the entire physical mechanism due to the complexity of the corrosion fatigue process, which includes the stage of pit formation, pit growth, fatigue crack initiation and its propagation.

This study proposes a new model, by extending CDM approach by combining evolution model of fatigue limit. The effect of corrosion in translated into its associated influence on surface roughness. Damage of corrosion fatigue is discretized into fatigue damage in different pre-corroded conditions. The capability of mode to predict load interaction sequences effects are showed by a two-level loading study.

2 Theoretical model

The corrosion fatigue model proposed in this study is based on non-linear continuum damage mechanics (CDM), which is suitable for making engineering predictions about the initiation, propagation, and fracture of materials without resorting to a microscopic description that would be too complex for practical engineering analysis (9). The CDM method was first introduced by Kachanov (10) based on thermodynamic principle and. The proximity to local failure is defined as a cumulative damage variable. The rate of damage increase is based on state variables, which can be measured by macroscopic properties, Material degradation laws represent the effects of damage on the stiffness and remaining life of the material that is damaging as a result of thermomechanical load and ageing (11). CDM approaches describing fatigue are based on the generic equation shown below:

* Corresponding author: author@e-mail.org

$$dD = f\left(\sigma_{mean}, \sigma_a, D, MC\right) dn \qquad (2\text{-}1)$$

where dD is the fatigue damage increment per cycle dn, σ_a and σ_{mean} are alternating stress and mean stress, respectively; MC is material property constant.

For uniaxial loading, Chaboche et al. (12)proposed the following damage rate equation for high cycle fatigue:

$$dD = \left[1-(1-D)^{1+\beta}\right]^{\alpha}\left[\frac{\sigma_a}{M(1-D)}\right]^{\beta} dn \qquad (2\text{-}2)$$

where β is a fixed material parameter representing, and M and α are functions themselves of applied load parameters. Concretely, M describes the dependency between mean stress and the fatigue limit. A linear dependency in terms of stress amplitude is commonly used:

$$M\left(\sigma_{mean}\right) = M_0\left(1-b\sigma_{mean}\right) \qquad (2\text{-}3)$$

with, M_0 and b additional material parameters. Then, α describes and is expressed as follows:

$$\alpha = 1-\frac{1}{H}\left\langle\frac{\sigma_a-\sigma_f\left(1-b\sigma_{mean}\right)}{\sigma_u-\sigma_a}\right\rangle \qquad (2\text{-}4)$$

Where the symbol <> represents the Heaviside step function (<x>=x if x>0 and <x>=0 if x<0); σ_u and σ_f are ultimate tensile stress and fatigue limit; H is a material constant.

Local failure of material (associated with crack initiation at that point) is often assumed to occur when D = 1. However, since CDM models like the above are often calibrated on the basis of S-N curves which express the total lifetime of a material (including crack initiation and propagation), researchers have also assumed crack initiation to occur at a critical damage value D_c smaller than 1. In such case, D_c is to be considered as an additional model parameter.

For the case of multiaxial loading, Chaudonneret(13) extended Chaboche work and proposed the multiaxial fatigue model for proportional and non-proportional loadings.

$$dD = \left[1-(1-D)^{1+\beta}\right]^{\alpha(\sigma_{mean}, A_{II})}\left[\frac{A_{II}}{M_0\left(1-3b\sigma_{H,mean}\right)(1-D)}\right]^{\beta} dn \qquad (2\text{-}5)$$

$$\alpha = 1-\frac{1}{H}\left\langle\frac{A_{II}-\sigma_f\left(1-3b\sigma_{H,mean}\right)}{\sigma_u-A_{II}}\right\rangle \qquad (2\text{-}6)$$

Where A_{II} is the amplitude of the octahedral shear stress, and $\sigma_{H,mean}$ is the mean value of hydrostatic pressure during the cycle. A_{II} and $\sigma_{H,mean}$ can be expressed as follows:

$$A_{II} = \frac{1}{2}\sqrt{\frac{3}{2}\left(\sigma_{ij,\max}-\sigma_{ij,\min}\right)^2} \qquad (2\text{-}7)$$

$$\sigma_{mean}^{H} = \frac{1}{6}\left[\sigma_{kk,\max}+\sigma_{kk,\min}\right] \qquad (2\text{-}8)$$

Where $\sigma_{ij,max}$ and $\sigma_{ij,min}$ are the maximum and minimum stress values of ij component of the deviator during a cycle respectively.

The basic idea of the model proposed in this study is to discretize fatigue damage into different corrosion conditions. For reasons of simplicity, the D_c value equals to 1 in this study. The corrosion effect is represented by a time dependent reduction of the fatigue limit. Notably, ultimate tensile strength, elongation and elastic modulus of engineering metals (steels, aluminium and magnesium alloy) insignificantly decrease when immersed in a certain corrosive environment for different time span (14-16). Therefore, the ultimate tensile strength is assumed as a constant value in the proposed model. The effect of corrosion on fatigue properties, on the other hand, is far more influential. It can be translated into S-N curves obtained from pre-corroded specimens as described in related researches (17-20). A time-dependent interpolation between S-N curves in absence and presence of different degrees of corrosion then allows to describe the coupled effect of corrosion and fatigue.

In the proposed model, the reduction of fatigue limit due to corrosion is expressed in terms of the associated increase in roughness. The latter can be expressed in various ways, such as the amplitude parameters R_a and R_q, height parameters R_p and R_z, spacing parameters H_{Sc} and S_m and hybrid parameters Δ_q and λ_q(21). We adopt R_a as a roughness characteristic in the model. R_a is the arithmetic average of the absolute values of the profile height deviations from the mean line, recorded within a certain evaluation length (22), as shown in Eq.):

$$R_a = \frac{1}{n}\sum_{i=1}^{n}\left|y_i\right| \quad \text{or} \quad R_a = \frac{1}{L}\int_0^L\left|Y(x)\right|dx \qquad (2\text{-}9)$$

where y is the depth (or peak) from the mean line (see Figure 1, right). Therefore, the maximum depth of a corrosion pit serves as an upper bound of roughness R_a.

In general terms, fatigue limit is expressed in terms of roughness and a certain reference roughness, as shown in Eq. (2-10):

$$\sigma_f^* = f\left(R_a, \sigma_f\right) \qquad (2\text{-}10)$$

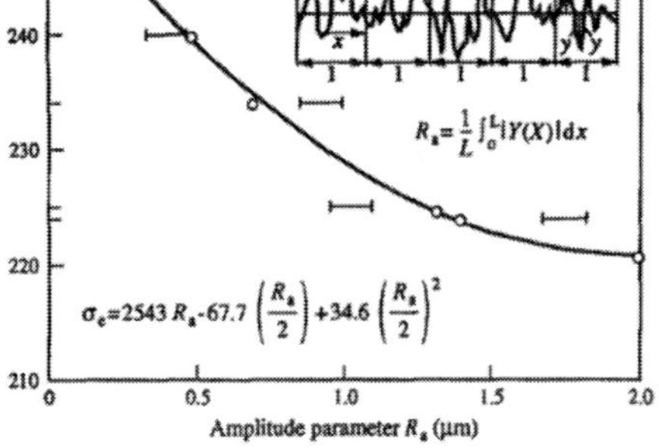

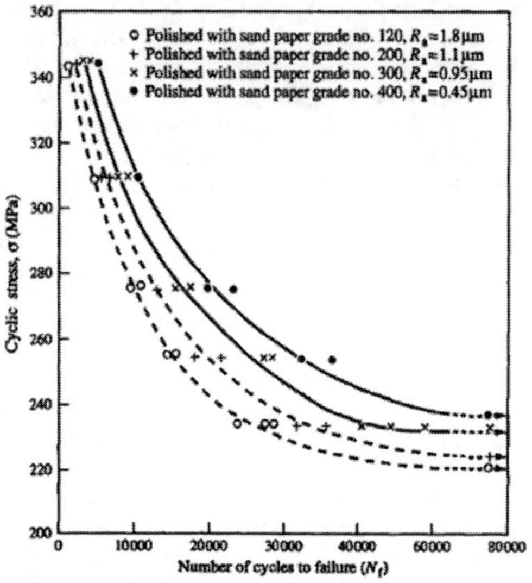

Figure 1 effect of surface roughness on number of cycles to failure(left) and effect of R_a on fatigue limit(right) (21)

The development of corrosion pit size depends on electrochemical conditions. Based on abundant studies data collected in (23, 24) , the time evolution of pit depth follows a power law, as shown in Eq.):

$$d = gt^h \quad (2\text{-}11)$$

where d is the depth of the corrosion pit; t is time; g and h are constants which are entirely dependent on material and electrochemical parameters.

According to Rybalka`s study(25), the depth of growing pit on 20Kh13 steel in 0.01 m NaCl solution can be described as a function of time by $h \sim t^{0.5}$. A similar observation was made by Zhang et al. (26), who summarized various test data and plotted pit depth versus the square root of time, as shown in Figure 2. Cavanaugh(27) used Weibull function and neural network method characterized the pit depth and found its kinetics followed a $t^{1/3}$ relation. Sriraman(28) similarly proposes that depth is considered proportional to the cube root of time by $d = Bt^{1/3}$.

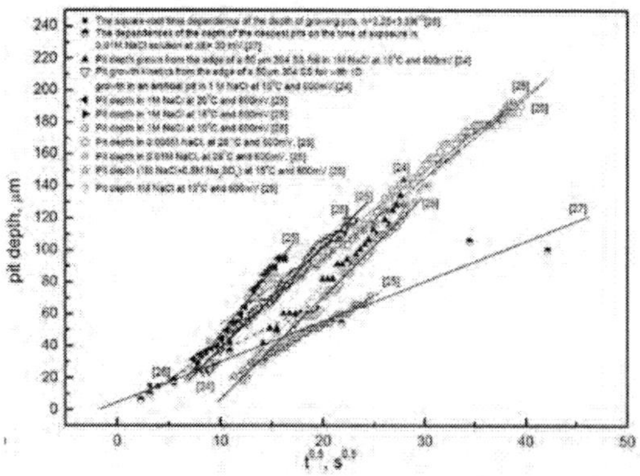

Figure 2 corrosion pit depth vs $t^{0.5}$ (26)

According to Eqs. (2-9) to (2-11) it can be concluded that the evolution of fatigue limit in corrosion environment can be a function of time and parameters from material and corrosion.

$$\sigma_f^* = F\left(\sigma_f, t, \mathrm{MC}\right) \quad (2\text{-}12)$$

To couple the effect of corrosion with fatigue, t can be linked to the number of fatigue cycles by means of the cycle frequency f, namely $t = n/f$.

Substituting fatigue limit σ_f in damage model by σ_f^*, the corrosion fatigue damage rate can be expressed as follows:

$$\frac{dD_{cf}}{dn} = power\left[1-\left(1-D_{cf}\right)^{1+\beta}, 1-\frac{1}{H}\left\langle\frac{A_{II}-\sigma_f^*\left(1-3b\sigma_{H,mean}\right)}{\sigma_u - A_{II}}\right\rangle\right]g\left[\frac{A_{II}}{M_0\left(1-3b\sigma_{H,mean}\right)\left(1-D_{cf}\right)}\right]^{\beta}$$

(2-13)

The model proposed in this paper contains a few parameters which are material properties due complexities in different physical contexts, and those parameters con be calibrated via matrix of experiments.

3 Results and discussion

Based on a hardnend and tempered steel, 30NiCoMoV12, reported in literature (29), where an S-N curve is provided with fatigue limit equal to 391MPa. This curve could be reasonably approximated using the basic Chaboche model, when adopting the following model parameters:

$$\alpha = 1 - 12.48\left\langle\frac{\sigma_a - 391}{1035 - \sigma_a}\right\rangle^{0.434}$$

Then, an example of virtual corrosion effect was applied by implementing the model proposed in this study. The hypothetical data for relationship between time and roughness, roughness and fatigue limit are listed in Table 1.

Table 1 virtual data of corrosion time, roughness and fatigue limit

corrosion time/min	0	6	12	18
surface roughness/μm	0.45	0.7	1.1	1.8
fatigue limit/MPa	391	250	160	50

In Figure 3, blue and black curves are corrosion fatigue SN curves predicted by the model proposed in frequency 1 Hz and 10 Hz respectively. Comparing to the red curve, the two predicted curves indicate a reduced fatigue resistance, accelerated initiation at higher stresses and elimination of the fatigue limit. Generally, the results agree with the trend observed experimentally

(30). Additionally, the model successfully predicts how a lower loading frequency leads to a lower fatigue resistance, as more corrosion damage develops during each cycle.

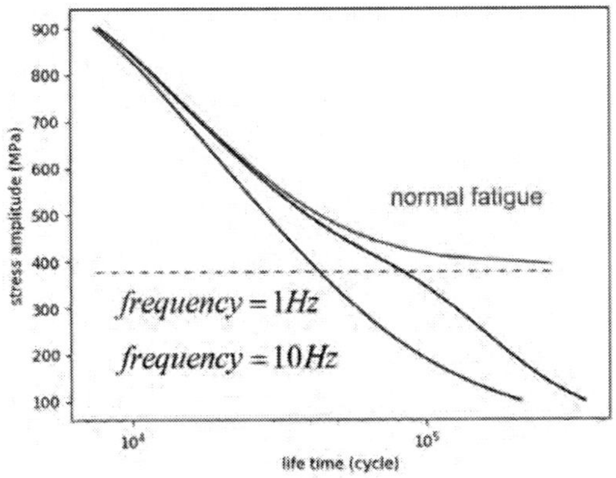

Figure 3 corrosion fatigue SN curves predicted in different load frequency in same corrosion condition

This model reveals non-linearity instinctively through differential equation of damage evolution rate, it can describe and predict load interaction sequence effect in the multi-level loading conditions. Two load levels at 450MPa and 550MPa have been applied as high-low and low-high sequence. The predicted data are plotted in Figure 4, which indicates the non-linearity of damage accumulation by the model proposed. By plotting two loads switching at damage value equal to 0.6, Figure 5 illustrates one case of predicted evolution of corrosion fatigue damage accumulation by proposed model. The left in Figure 5 shows the total life is shortened comparably when the load sequence starting from low to high, vice versa, the total life is prolonged when load sequence is from high to low.

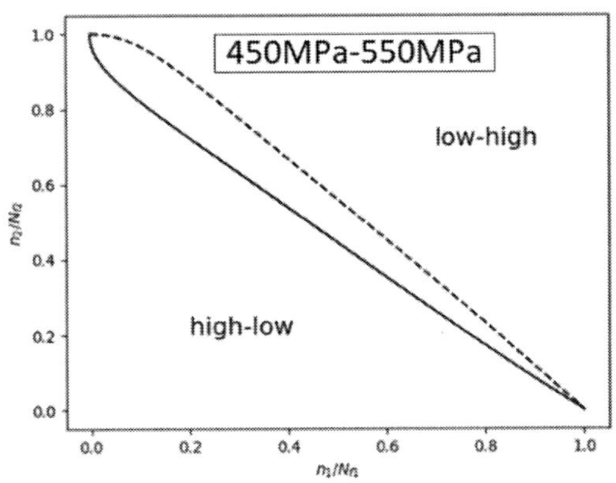

Figure 4 Prediction of sequence effect of two-level loading

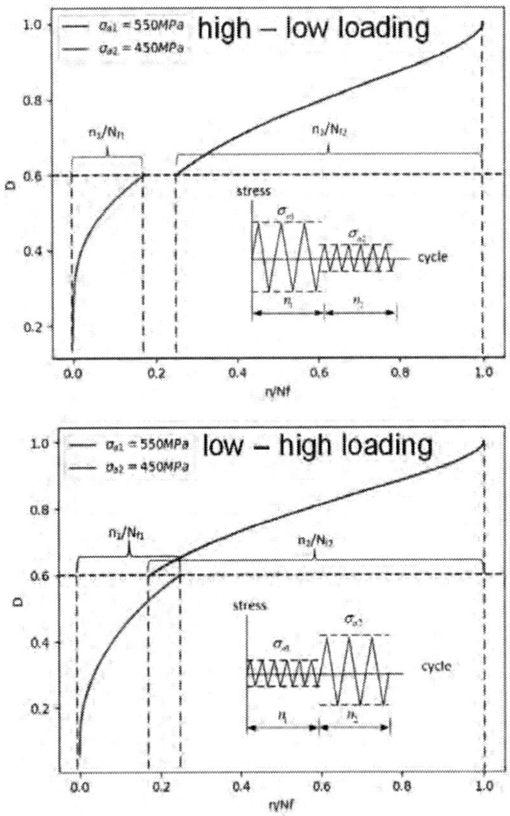

Figure 5 An example of predicted evolution of damage accumulation in high-low and low-high load sequence

4 Conclusions

This study proposes an innovative corrosion fatigue model based on non-linear continuum damage mechanics, which includes the corrosion effect by a time dependent degradation of the fatigue resistance curve. The numerical simulation shows that the model is capable to predict the corrosion effect and load sequence effect, and it successfully describes an interaction between corrosion and fatigue. This paper has focused on the methodology description and a study of the model capabilities. A validation study is currently being carried out, to evaluate the predictive ability of the model based on well-chosen calibration tests.

The authors gratefully acknowledge the financial support via MaDurOS program from SIM Flanders (Strategic Initiative Materials) and Vlaio (Flemish Agency for innovation and Entrepreneurship).

References

1. Adedipe O, Brennan F, Kolios Renewable and Sustainable Energy Reviews. 2016;
2. Marcus P. Corrosion Mechanisms in Theory and Practice, Third Edition: Taylor & Francis; 2011.

3. Magnin T, Delafosse D. CORROSION TECHNOLOGY-NEW YORK AND BASEL-. 2002;17
4. Rajasankar J, Iyer NR.. 2006;73(5)
5. Rokhlin SI, Kim JY, Nagy H, Zoofan B. 1999;62(4)
6. Dolley, Lee, Wei. 2000;23(7)
7. Havn T, Maritime A, Osvoll H. 2002/1/1/. NACE: NACE International.
8. Wang QY, Pidaparti RM, Palakal MJ. AIAA Journal. 2001;39(2)
9. Ebara R. Materials Science and Engineering: A. 2007;468-470
10. Krajcinovic D. Damage mechanics. Mechanics of Materials. 1989;8(2–3).
11. Kachanov LM. Rupture time under creep conditions. International journal of fracture. 1999;97(1-4)
12. Struik LCE. TU Delft, Delft University of Technology; 1977.
13. Chaboche JL, Lesne PM. Fatigue & Fracture of Engineering Materials & Structures. 1988;11(1)
14. Chaudonneret M. Journal of Engineering Materials and Technology. 1993;115(4)
15. Fu S, Gao H, Chen G, Gao L, Chen X. Materials Science and Engineering: A. 2014;593
16. Obert B, Ngo K, Hashemi J, Ekwaro-Osire S, Sivam TP. Journal of Materials Engineering and Performance. 2000;9(4)
17. Hou Y, Lei D, Li S, Yang W, Li C-Q. International Journal of Corrosion. 2016
18. Hu W, Shen Q, Zhang M, Meng Q, Zhang X. International Journal of Damage Mechanics. 2012;21(8)
19. Huang Y, Ye X, Hu B, Chen L. International Journal of Fatigue. 2016
20. Abdulstaar M, Mhaede M, Wagner L.. Advanced Engineering Materials. 2013;15(10)
21. Wang Z-b, Liu W-t, He X-f. Chinese Journal of Aeronautics. 2004;17(4)
22. Bayoumi MR, Abdellatif AK. 1995;51(5)
23. Institute ANS, Engineers ASoM. American Society of Mechanical Engineers; 2009
24. Harlow DG, Wei RP. Engineering Fracture Mechanics. 1998;59(3)
25. Kondo Y. CORROSION. 1989;45(1)
26. Rybalka KV, Beketaeva LA, Shaldaev VS, Kasparova LV, Davydov AD.. Russian Journal of Electrochemistry. 2009;45(11)
27. Zhang X-Y, Li S-X, Liang R, editors. ICF13; 2013.
28. Cavanaugh MK, Buchheit RG, Birbilis N. Corrosion Science. 2010;52(9)
29. Sriraman MR, Pidaparti RM. Journal of Aircraft. 2009;46(4)
30. Dattoma V, Giancane S, Nobile R, Panella FW. International Journal of Fatigue. 2006;28(2)
31. Devereux OF, McEvily AJ, Staehle RW, The University of Connecticut, Storrs, Connecticut: National Association of Corrosion Engineers; 1972.

Hydrogen-assisted fatigue crack growth in ferritic steels – a fractographic study

Di Wan[1,*], *Antonio* Alvaro[2], *Vigdis* Olden[2] and *Afrooz* Barnoush[1]

[1]Department of Mechanical and Industrial Engineering, Norwegian University of Science and Technology, Richard Birkelands vei 2B, 7491 Trondheim, Norway
[2]SINTEF Materials and Chemistry, 7456 Trondheim, Norway

Abstract. Fatigue crack growth (FCG) behavior of a Fe-3wt.%Si ferritic alloy under different environmental conditions using in-situ electrochemical (cathodic) hydrogen (H) charging has been investigated. Three frequencies have been applied. Results clearly show that the FCG rate increased by a factor spanning from 20 to 1000 times, depending on the loading frequencies, when compared to the reference test in air. Lower frequency leads to higher FCG rate. A comprehensive fractographic analysis was carried out: the area fraction of different fracture surface features was measured and taken into statistical analysis. Based on these investigations, the possible mechanisms of H-enhanced FCG are discussed. Similar tests in high-pressure H gas from other studies were also compared and discussed. These results give a preliminary understanding of H effect in fatigue crack propagation procedure in ferritic alloys.

1 Introduction

It is generally acknowledged that engineering structures are normally experiencing different loads such as static load, cyclic load, impact load or a combination of several loads from the above-mentioned category. And a very common condition is cyclic loading combining a static loading. Researches regarding the fracture behavior from repeated loading have been published over a century and continue increasing. According to the data from Battele Laboratory under contract to United States Government agencies in 1983, the costs of fractures and efforts to prevent fracture amounted to 119 billion US dollars per year, which occupied a considerable part of the gross national product at that time [1]. It can be expected that this number becomes different today due to a faster development in industry. The classical fatigue crack growth behavior can be simplified by a logarithmic linear relation, which is called the Paris' law (or Paris-Erdogan law), as shown in Eq. (1):

$$da/dN = C{\cdot}\Delta K^m \quad (1)$$

where a is the crack length, N is the number of cycles and da/dN is the crack growth rate, describing the infinitesimal crack length per increasing number of load cycles. C and m on the right-hand side are constants depending on material and test conditions. ΔK is the range of the stress intensity factor during fatigue loading, and is defined as Eq. (2):

$$\Delta K=K_{max}-K_{min} \quad (2)$$

where K_{max} and K_{min} are the maximum and minimum value of the stress intensity factor, respectively.

This relation is favored by many engineers since by taking a simple logarithmic operation, the crack growth rate and the stress intensity factor range can be treated as a linear relation, and it's easier for engineers to predict how fast the crack can grow without repairing it.

A factor that degrades the mechanical behavior of materials is the environmental effect, especially the environment containing hydrogen (H), in which the materials can become more brittle than in other cases. This phenomenon is named hydrogen embrittlement (HE) and has been studied since the first publication by Johnson [2] in 1875. The HE effect of metallic materials has been intensively discussed over the past century. It draws engineers' attention since it can cause catastrophic failure of metallic structures that serve in a H-containing environment. Several different mechanisms have been developed, and the most common ones are hydrogen-enhanced localized plasticity (HELP) [3-8], hydrogen-enhanced decohesion (HEDE) [9-12], adsorption-induced dislocation emission (AIDE) [13], hydrogen-enhanced vacancy production [14, 15], hydrogen-induced phase transformation [16-18] etc. These mechanisms describe the HE effect in terms of different aspects ranging from chemical bonding up to microstructure level. A systematic review has been established by Murakami et al. [19] regarding the H effect on the fatigue crack propagation in steels.

For most oil and gas industries in the arctic region, the metallic structures will serve in a H-containing environment (e.g. ocean) with a continuous vibrational loading. Hence the fatigue properties with environmental

* Corresponding author: di.wan@ntnu.no

effect of those structures are of significant importance. The purpose of this study is to study the mechanism in the degradation of structural materials that serve in the condition with cyclic loading and H environment and try to correlate the physical mechanism with engineering aspects. For this purpose, Fatigue Crack Growth Rate (FCGR) tests were done on a ferritic alloy with large grain size in different environments. The FCGR data were analyzed and correlated with the classical Paris' law. The results showed a strong effect of promoting crack propagation in H environment by a factor spanning from 20 to 1000 times. Scanning Electron Microscopy (SEM) characterization was done on the fractured specimens. The reasons for the degradation were discussed.

2 Experimental

2.1. Materials

The materials used in this study is a Fe-3wt.%Si alloy with pure ferritic structure. The grain size of the as-received material is about 300 μm. The chemical composition of this material is shown in Table 1.

Compact Tension (CT) specimens according to the ASTM E647 standard were cut from the raw material by Electric Discharge Machining (EDM). The geometry is shown in Figure 1.

Table 1. Chemical composition of the investigated material.

Element	C	Si	Mn	P
wt.%	0.018	3.000	0.055	0.008
Element	Cu	Al	Ti	Nb
wt.%	0.013	0.015	0.001	0.002
Element	S	Cr	Ni	Mo
wt.%	0.003	0.010	0.006	0.003
Element	V	B	Zr	Fe
wt.%	0.001	0.0002	0.0010	Bal.

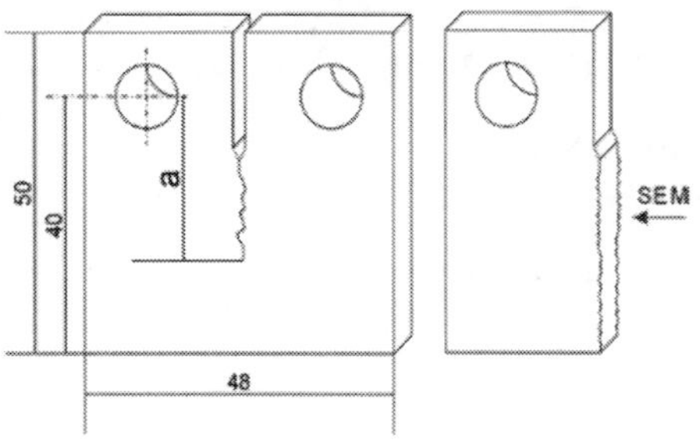

Figure 1. Schematic of the CT specimen (thickness = 8, unit: mm). SEM characterization was done on the fractured half-specimen.

2.2 Fatigue crack growth rate test

The CT specimens were first pre-cracked by fatigue test by the same procedure as described in Ref.[20]. The final pre-cracks normally have a length ranging between 1 to 3 mm. For the FCGR test, an initial stress intensity factor range (ΔK) corresponding to a 2.5% increase from the last ΔK value of the pre-crack test was applied. The fatigue crack growth rates and the crack length were recorded via an alternate current potential drop (AC-PD) device. The tests were done at room temperature in lab air and in an electrochemical cell. The frequencies were chosen to be 0.1 Hz, 1 Hz and 10 Hz and the load ratio R was kept 0.5 during the test to reduce the crack closure effect.

For H-charging, the electrolyte was chosen to be a neutral 0.1M Na_2SO_4 solution. The specimen was in-situ cathodically charged with a constant potential of -1400 mV_{SCE}. Multimeters were used during the charging procedure, to make sure the whole circuit runs as designed.

After tests, the specimens were put into liquid nitrogen and broken into two parts. The half-specimens were used for further characterization.

2.3 Characterization

For characterization, a Quanta 650 environmental scanning electron microscope (ESEM, Thermo Fisher Scientific Inc., USA) was used. The ESEM was operated at an accelerating voltage of 20 kV. To investigate the dislocation structure induced during the FCG test, electron channeling contrast imaging (ECCI) was performed on the cross-section of the fractured specimens. A solid-state four-quadrant backscatter electron (BSE) detector was used, and the contrast was controlled to an optimum state for dislocation imaging.

3 Results

3.1. FCGR results

The FCGR data is shown in Figure 2. It can be seen from the data that when H was applied, the crack propagation rate increased significantly. The enhancement depends strongly on the test frequency. When the frequency decreases, the crack propagation becomes faster. With the highest frequency (10 Hz) in the present study, the FCGR increased by about 20 times with respect to the test in lab air, while the lowest frequency (0.1 Hz) made the FCGR increase by about 1000 times. These data show a relatively good linear relation and were fitted by the Paris' law. The fitted parameters are shown in Table 2.

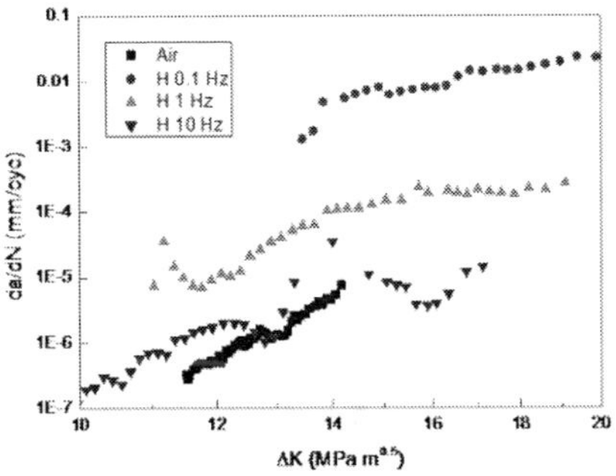

Figure 2. *da/dN* vs *ΔK* data for the studied specimens.

Table 2. Fitted parameters based on the Paris' law.

Test condition	*C'*	*m'*
Air, R=0.5, f=10 Hz	3E-17	7.890
H, R=0.5, f=10 Hz	3E-15	7.898
H, R=0.5, f=1 Hz	5E-13	6.986
H, R=0.5, f=0.1 Hz	7E-9	5.048

3.2. Fractography

Fractography was taken on all half-specimens after the FCGR test. The fracture features were statistically analyzed. As shown in Figure 3a, the fracture surface of the specimens can be divided into three different regions, namely the EDM region, pre-crack region and the FCG region. It should be noted that the crack front was not necessarily a straight line due to non-uniform distribution of the stress field ahead of the crack tip. To avoid ambiguity between the pre-crack region and the FCG region, 1 mm was skipped from the measured starting place, and 0.6 mm was skipped from the edges through the thickness direction. As a result, only the highlighted part in Figure 3a was considered in the statistics. For fracture features, normally three distinct features can be identified, namely transgranular (TG) type, intergranular (IG) type and the so-called "quasi-cleavage" ("QC") type, as shown in Figure 3b-d. The area percentage of each fracture mode was analyzed on all H-charged specimens from the fractography, and the result is shown in Figure 4. The specimen tested in air showed only TG fracture and a minor percentage of IG fracture over the tested *ΔK* range. It should be noted that the name of "QC" fracture is somehow ambiguous, but it has been such named in many literature regarding crack growth with H environment (e.g. [8, 15, 21, 22]) and we just keep the convention. It can be seen from the results that H mainly changes the relative amount between TG and "QC" fracture modes. When H is charged to the specimen, "QC" fracture starts to happen to the material. When the frequency decreases, the "QC" percentage increases and the TG percentage decreases, with minor change in the IG percentage.

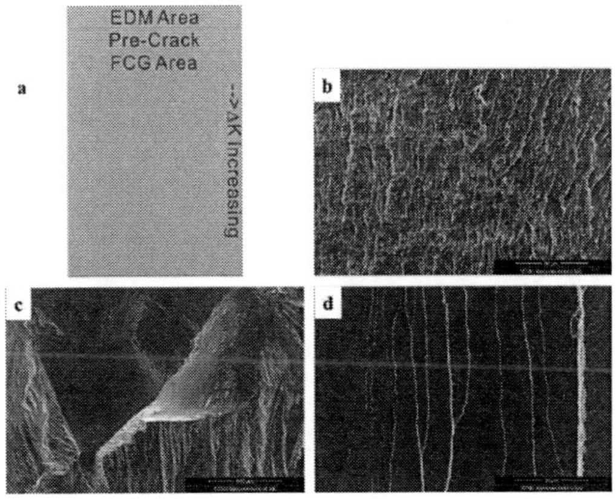

Figure 3. a. schematic description of the statistical analysis; b. typical TG fracture; c. typical IG fracture; d. typical "QC" fracture.

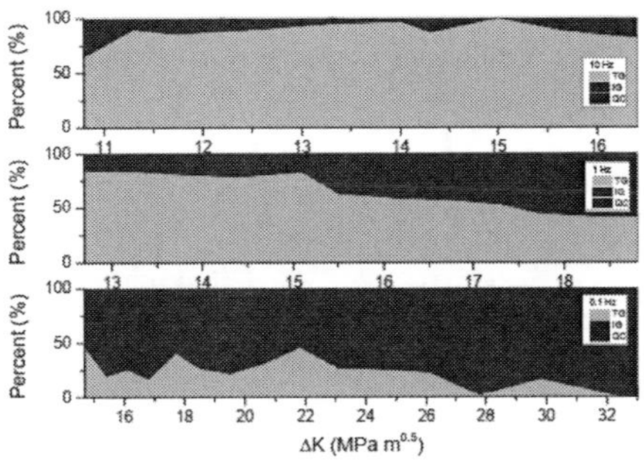

Figure 4. Fracture mode distribution with respect to *ΔK* level.

Images with high magnification were taken regarding each fracture feature on specimens tested in all conditions. The results are shown in the following figures.

In the TG fracture zone (as shown in Figure 5), the structures are similar in H-free and H-charged cases. Small and fine striations that are almost vertical to the global crack growth direction could be observed over the fracture surface. Tear edges are commonly observed, and the striations are normally vertical to the direction of the tear edges.

Figure 6 shows the magnified IG fracture surface. The figures show clearly fracture following grain boundaries (GBs), leaving a smooth surface from the boundaries without visible striations. There is not much different between the H-free and all the H-charged conditions. It is worth noting that the IG fracture type occupies only a small fraction according to the statistics in Figure 4.

Figure 7 shows the "QC" fracture in all conditions. Sparsely distributed brittle-like striations that are almost perpendicular to the global FCG direction could be observed on the surfaces. In the H-charged cases the fracture left sharp edges on the surface, and the striations were faceted in all H-cases. River-lines are visible on the fracture surfaces, and the lines are more or less parallel

https://doi.org/10.1051/matecconf/201816503004

to the global FCG direction. It is important to be pointed out that the "QC" feature could not be observed in the air case. Figure 7a shows the most similar feature in air as "QC" in H-cases, but the striations are not faceted and the tear ridges are not as sharp as the H-cases. The "QC" feature is characteristic from the H influence.

3.3. ECCI characterization

Preliminary ECCI characterization was performed on the cross-section of the fractured specimen to investigate the dislocation structures below the fracture surface. The result is shown in Figure 8. The global FCG direction is from left to right, but local variations could be observed. Plastic zones can be identified by the bright contrast due to a high dislocation density. The bright contrast above the yellow dash-line in Figure 8a comes from Fe-coating. Away from the plastic zone, sparsely distributed dislocations were observed in the H-free case, while in the H-charged case, dislocations were better organized into vein-like structures. This is the preliminary result and further systematic and detailed work is ongoing.

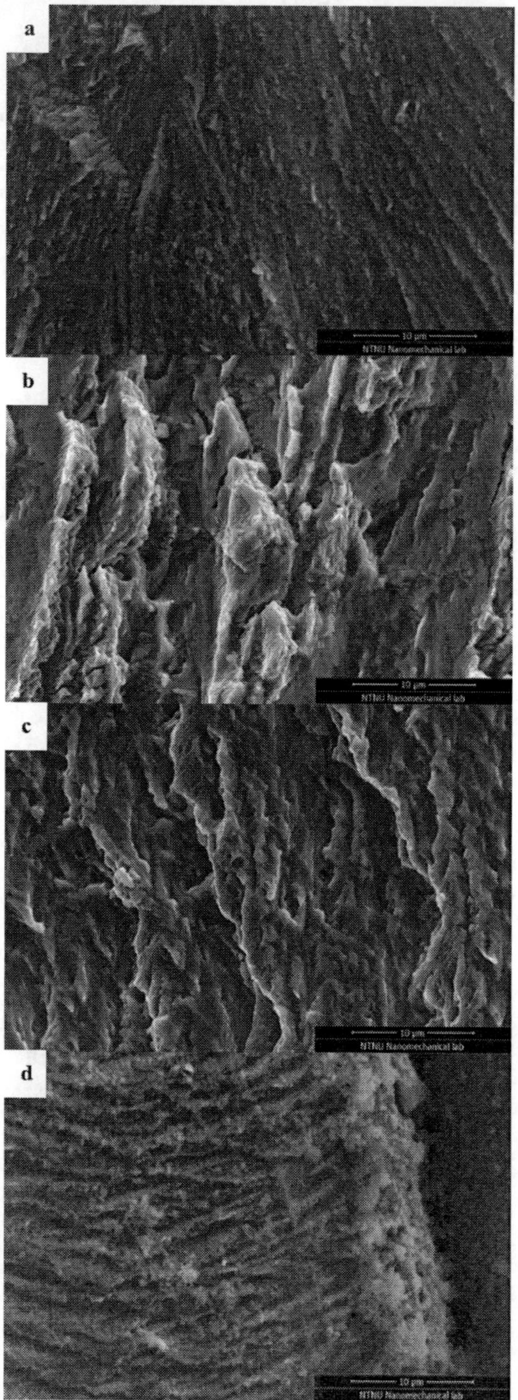

Figure 5. Magnified images showing structure in TG fracture region (a. in Air; b. in H, f = 10 Hz; c. in H, f = 1 Hz; d. in H, f = 0.1 Hz; global FCG direction from top to bottom for each case)

https://doi.org/10.1051/matecconf/201816503004

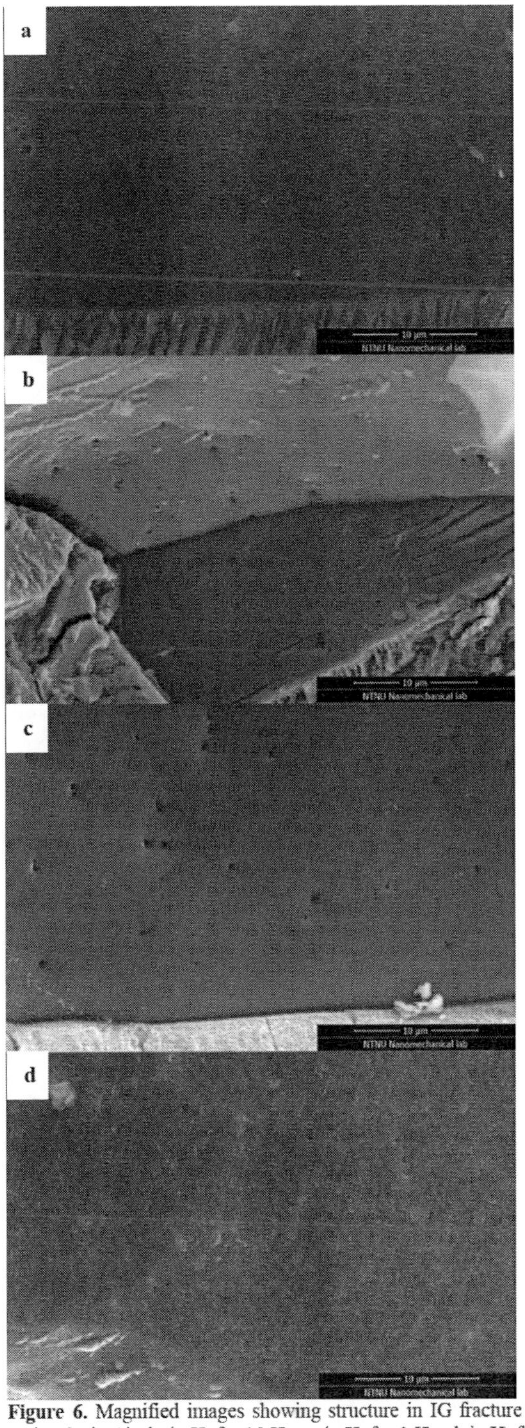

Figure 6. Magnified images showing structure in IG fracture region (a. in Air; b. in H, f = 10 Hz; c. in H, f = 1 Hz; d. in H, f = 0.1 Hz; global FCG direction from top to bottom for each case)

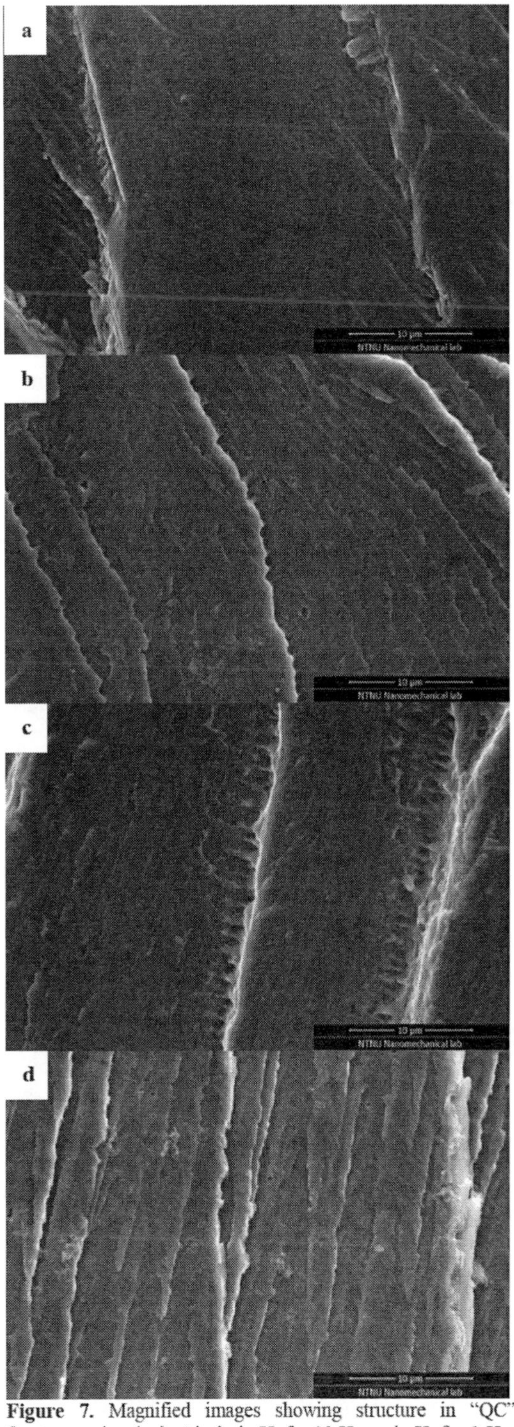

Figure 7. Magnified images showing structure in "QC" fracture region (a. in Air; b. in H, f = 10 Hz; c. in H, f = 1 Hz; d. in H, f = 0.1 Hz; global FCG direction from top to bottom for each case)

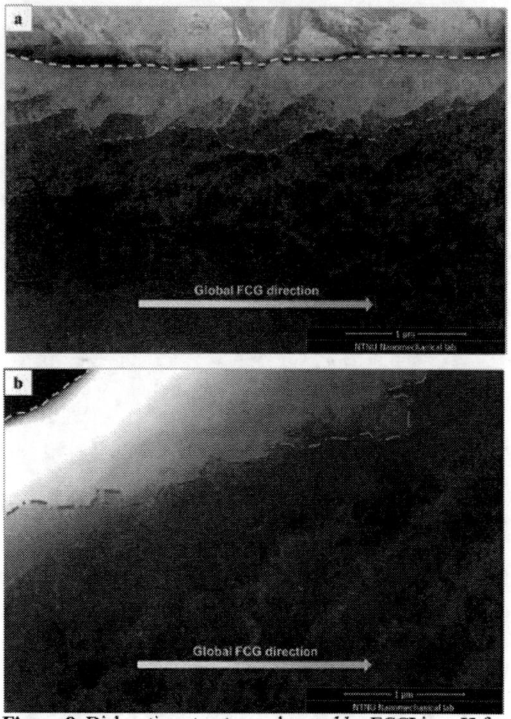

Figure 8. Dislocation structures observed by ECCI in a. H-free case and b. H-charged case from the region fractured at similar ΔK level. The yellow dash-line is the fracture surface and the light blue broken-line is the indicated plastic zone under the fracture surface.

4 Discussion

4.1 Paris' relation

It is clearly shown that the measured crack growth data can be properly correlated with the Paris' law. This relation can describe the crack growth procedure in a mathematically simple and linear relation, also when aggressive environment applies. These data show that the Paris' law is validated for these test conditions. However, the local variation in different grains or different fracture features can give error in the estimation. A systematic verification at the microscale level is ongoing. This will be analyzed and discussed in a future work by the present authors.

4.2 H effect on FCG procedure

It is clearly shown in the crack growth data that when H was charged, the crack growth rate was increased up to 1000 times in comparison with the reference test in air. The test frequency also played a significant role in the crack growth rate enhancement. It could be inferred that the H degradation is a time-dependent process. When the frequency decreases, H has more time to invade the lattice and deteriorate the mechanical behavior, globally makes the crack grows faster. Chen et al. [23] also suggested that the H-assisted cracking is time-dependent in Fe-3%Si single crystals.

From the statistics in Figure 4, the frequency also influences the fracture mode distribution. A higher frequency leads to more TG fracture and a lower frequency gives more "QC" fracture, with the IG fraction almost keeps unchanged. The possible reasons might be that the "QC" fracture was transformed from the TG fracture by the effect from H.

Matsuoka et al. [24, 25] has explained the formation of this fracture feature in terms of reduced plasticity expansion in the crack-tip region. Based on their model, these striations are the results from slip localization, which is compatible with the HELP mechanism [3, 4]. They suggested that when the cyclic load was exerted to the specimen, the stress would be localized in the vicinity of the crack-tip, and the H atoms would be attracted by the highly localized tensile stress field. The localized H concentration made the crack propagation much easier even at a lower stress level during unloading procedure. Consequently, the distances between striations are larger in the H-charged specimens than the one tested in air. Similar features have also been observed in polycrystalline pure iron [22] and Fe-Si single crystals [26, 27] with H influence.

Another key point from Figure 4 is that when ΔK level increases, there is always a tendency of increasing "QC" fraction and decreasing TG fraction. This is the most significant for 1 Hz, but a clear tendency is also seen for 0.1 Hz and 10 Hz. Since "QC" feature is characteristic for the H influence, we can assume the fraction of "QC" fracture is a factor that can give an estimation of the HE level. Since the HE happens generally in high strength materials, it is suggested that the combination of high stress level and high H concentration makes the embrittlement possible. During FCGR tests, there is always a stress concentrated zone ahead of the crack-tip, which can also attract H atoms and enhance the local H concentration.

4.3 Dislocation structure

The ECCI results shown in Figure 8 are some preliminary results from dislocation investigations. The regions of interest came from a similar ΔK level during the FCGR tests. The H-case shows highly organized vein-like dislocation structure while the H-free case does not show similar feature. This could be the effect from H that dislocations were localized along some specific crystallographic directions, while impeded in other directions. As a result, concentration to some organized structures. However, there are some points still remain unclear such as the grain orientations, the accommodation from nearby grains, the local stress field, etc. Further investigations are ongoing and will be discussed more in detail in a future work by the present authors.

Nevertheless, some proof could be found from literature. For example, Wang et al. [8] did a similar work on pure iron with the help of scanning transmission electron microscope (STEM). It has been explained in this work that the attachment of H to dislocations or other obstacles will change the stress field associated with crystallographic defects, and consequently enhancing the dislocation motion preferred in some specific directions while suppressing in other directions. As a result, the observed dislocation structure should be more organized in the H-charged case than that from the H-free case.

5 Conclusions

In this study, the fatigue crack growth behavior of a ferritic Fe-3wt.%Si alloy was tested both in lab air and with in-situ cathodic H-charging. Post-mortem fractography analysis was done through SEM. The main results can be summarized in the following points:

- The FCGR is increased by in-situ cathodic H-charging by a number ranging from 20 to 1000 times in comparison with a reference test in air;
- The higher crack growth rate is attributed to the increasing proportion of the "QC" type fracture, and it depends on the loading frequency and ΔK level;
- IG fracture occupies only a small part in comparison with TG type and "QC" type. GBs are not preferred crack propagation path;
- The preliminary investigations on dislocation structures show difference in H-free and H-charged cases, and more detailed systematic work is ongoing.

This work is financially supported by the Research Council of Norway through the HyF-Lex project (Petromaks 2 Programme, Project Number: 244068/E30).

References

[1] P. Peralta, C. Laird, Fatigue of Metals, in: D.E. Laughlin, K. Hono (Eds.), *Physical Metallurgy (Fifth Edition)*, **vol 2**, Elsevier, Amsterdam, (2014), pp. 1765-1880.
[2] W.H. Johnson, P R SOC LONDON **23** (1875) 168-179.
[3] C.D. Beachem, METALL MATER TRANS B **3** (1972) 441-455.
[4] H.K. Birnbaum, P. Sofronis, MAT SCI ENG A **176** (1994) 191-202.
[5] I.M. Robertson, ENG FRACT MECH **64** (1999) 649-673.
[6] I.M. Robertson, H.K. Birnbaum, ACTA METALL **34** (1986) 353-366.
[7] I.M. Robertson, H.K. Birnbaum, P. Sofronis, Hydrogen Effects on Plasticity, **vol 15**, (2009), pp. 249-293.
[8] S. Wang, A. Nagao, P. Sofronis, I.M. Robertson, ACTA MATER **144** (2018) 164-176.
[9] W.W. Gerberich, P.G. Marsh, J.W. Hoehn, Hydrogen Induced Cracking Mechanisms - Are There Critical Experiments?, in: A.W. Thompson, N.R. Moody (Eds.), *Hydrogen Effects in Materials*, Minerals, Metals & Materials Society (TMS), Warrendale, Pennsylvania, USA, (1996), pp. 539-554.
[10] R.A. Oriani, CORROSION **43** (1987) 390-397.
[11] R.A. Oriani, P.H. Josephic, ACTA METALL **25** (1977) 979-988.
[12] R.P. Gangloff, Environment-Induced Cracking of Materials **1**, (2008) 141-165.
[13] S.P. Lynch, ACTA METALL **36** (1988) 2639-2661.
[14] Y. Fukai, J ALLOY COMPD **356-357** (2003) 263-269.
[15] M. Nagumo, MAT SCI ENG **20** (2013) 940-950.
[16] H.K. Birnbaum, J LESS COMMON METALL **104** (1984) 31-41.
[17] N. Narita, C.J. Altstetter, H.K. Birnbaum, METALL TRANS A **13** (1982) 1355-1365.
[18] D.S. Shih, I.M. Robertson, H.K. Birnbaum, ACTA METALL **36** (1988) 111-124.
[19] Y. Murakami, R.O. Ritchie, Effects of hydrogen on fatigue-crack propagation in steels, *Gaseous Hydrogen Embrittlement of Materials in Energy Technologies*, **vol 2**, Woodhead Publishing, (2012), pp. 379-417.
[20] A. Alvaro, O.M. Akselsen, X. Ren, P.-A. Kane, Fatigue Properties of a 420 MPa Structural Steel at Low Temperature, The 25th International Ocean and Polar Engineering Conference, **vol 4**, Kona, Big Island, Hawaii, USA, 2015, pp. 331-337.
[21] Y. Ogawa, H. Matsunaga, J. Yamabe, M. Yoshikawa, S. Matsuoka, I J FATIGUE **103** (2017) 223-233.
[22] Y. Ogawa, D. Birenis, H. Matsunaga, A. Thøgersen, Ø. Prytz, O. Takakuwa, J. Yamabe, SCR MATER **140** (2017) 13-17.
[23] X. Chen, W.W. Gerberich, METALL TRANS A **22** (2012) 59-70.
[24] S. Matsuoka, H. Tanaka, N. Homma, Y. Murakami, I J FRACT **168** (2010) 101-112.
[25] S. Matsuoka, N. Tsutsumi, Y. Murakami, T JAP SOC MECH ENG A **74** (2008) 1528-1537.
[26] H. Vehoff, P. Neumann, ACTA METALL **28** (1980) 265-272.
[27] H. Vehoff, W. Rothe, ACTA METALL **31** (1983) 1781-1793.

A frequency dependent embrittling effect of high pressure hydrogen in a 17-4 PH martensitic stainless steel

Jean-Gabriel Sezgin[1,*], *Junichiro* Yamabe[1,2,3]
[1]AIST-Kyushu University Hydrogen Materials Laboratory (HydroMate), National Institute of Advanced Industrial Science and Technology (AIST), West Zone 1,4F, D-409 Ito Campus, Kyushu University, 744 Moto-oka, Nishi-ku, Fukuoka, 819-0395 Japan
[2]International Research Center for Hydrogen Energy, Kyushu University, 744 Moto-oka, Nishi-ku, Fukuoka, 819-0395 Japan
[3]Research Center for Hydrogen Industrial Use and Storage (HYDROGENIUS), Kyushu University, 744 Moto-oka, Nishi-ku, Fukuoka, 819-0395 Japan

Abstract. The effects of hydrogen on tensile and fatigue-life properties of 17-4PH H1150 steel have been investigated by using a smooth, round-bar specimen for tensile tests and circumferentially-notched specimen for fatigue-life tests. The specimens were precharged by an exposure to 35-100 MPa hydrogen gas at 270°C for 200 h. For the 100 MPa hydrogen exposure, the steel showed a significant degradation in ductility loss, translated by a relative reduction in area, RRA, of 0.31. The fatigue-life test of the present notch specimen (stress concentration factor of 6.6) reflects the fatigue crack growth (FCG) for long cracks. The fatigue limit of the non-charged and H-charged notched specimens, defined by the threshold of non-propagation for long cracks, was not affected by hydrogen. At a higher stress amplitude, the H-charged specimen showed a significant FCG acceleration ratio compared to the non-charged specimen. Although, an upper bound of the FCG acceleration seemed to exist, this ratio was approximately 100. The fracture surface of the H-charged specimen was covered with quasi-cleavage (QC) at a lower stress amplitude and with a mixture of QC and intergranular (IG) facets at higher stress amplitudes. It has been suggested that a cycle-dependent crack growth accompanied by QC occurs at a lower stress amplitude, whereas a mixture of cycle-dependent crack growth (accompanied by QC) and time-dependent crack growth (accompanied by IG) occurs otherwise. This mixture justifies the 100 times FCG acceleration ratio in spite of the existence of the upper bound.

1 Introduction

Hydrogen has been recently considered to be a sustainable carrier in order to tackle the future energetic challenges. However, a limited number of steels is available to manufacture such systems due to hydrogen compatibility issues. To enhance cost-effectiveness (*e.g.* of FCV), substantial efforts have been assigned to clarify the degradation of mechanical and metallurgical properties directly caused by hydrogen. This phenomenon is usually called Hydrogen Embrittlement (HE) [1]–[3]. Depending on the supply of hydrogen, the application and the nature of the alloy, various mechanisms are available in the literature [4]–[8], however, this phenomenon has not been deeply understood and a precise mechanism has not been established yet.

From the viewpoint of strength design for components, the utilization of high-strength steels is relevant to combine weight reduction and high performance. However, extensive experimental results demonstrate that the high-strength steel with ultimate tensile strength (UTS) exceeding approximately 1 GPa is considerably or severely susceptible to HE [9]–[18]. In addition, it has also been reported that an upper bound of fatigue crack growth (FCG) ratio by hydrogen (30 times acceleration) exists only for low- or medium-strength steels with UTS < 900 MPa [18]. This fact suggests that the HE mechanism of high-strength steels may be different from the mechanism observed on low- and medium-strength ones. Data available for some steels such as JIS-SCM435 and JIS-SNCM439 with UTS < 900 MPa [18] or the SAE52100 (JIS-SUJ2) with UTS of 1900 MPa [19] support this assumption. Fig. 1 shows a schematic illustration of the test frequency dependence of the hydrogen-enhanced FCG acceleration ratio of these steels. For the low-alloy steels with UTS < 900 MPa, an upper bound of the hydrogen-enhanced FCG acceleration ratio exists and the fatigue crack presents a cycle-dependent growth. This fracture surface is covered with quasi-cleavage (QC) and the acceleration is explained by the hydrogen-induced successive crack growth (HISCG) model [20] based on localized slip deformations enhanced by hydrogen. In contrast, for the high-strength steel with a UTS of 1.9 GPa, no upper bound of the hydrogen-enhanced FCG acceleration ratio exists and this fatigue crack shows a time-dependent growth. This fracture surface is covered with intergranular (IG) surface and the acceleration is explained by the hydrogen-enhanced deformation twin model [19].

In the current paper, the material of interest is a 17-4PH stainless steel with an UTS of around 1 GPa. It is well-known that the 17-4PH steel is susceptible to HE [21]. The objective of this paper is to clarify the effect of hydrogen on the threshold behaviour for long crack, the FCG acceleration of the steel and then, propose an appropriate mechanism to explain the hydrogen-enhanced FCG acceleration.

* Corresponding author: sezgin.jean-gabriel@aist.go.jp

https://doi.org/10.1051/matecconf/201816503005

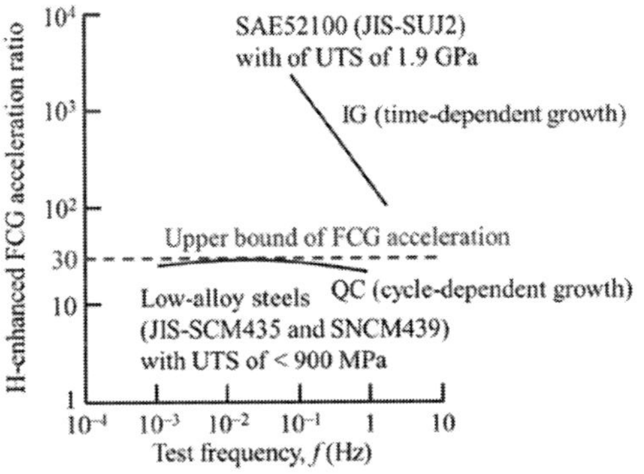

Fig. 1. Test frequency dependence of hydrogen-enhanced FCG acceleration for various steels [18], [19]

2 Material characterization

2.1. Metallurgical aspects

The material used in this paper is a 17-4PH stainless steel. The chemical composition is given in Table 1. To adjust its mechanical properties, the steel is heat-treated in two steps. The first step is a solution treatment at 1040°C for 1h with water quenching. The second step is a precipitation hardening treatment at 620°C for 2 h with air cooling. This heat-treated steel is referred to as a 17-4PH H1150 steel.

Table 1. Chemical composition of the 17-4 PH steel in mass%

	C	Si	Mn	P	S
17-4 PH	*0.04*	*0.31*	*0.87*	*0.034*	*0.004*
	Cu	**Ni**	**Cr**	**Nb**	
	3.3	*4.24*	*15.57*	*0.34*	

After heat treatment, the obtained microstructure is presented on Fig. 2-a). This SEM micrograph is obtained by using the backscattered electron (BSE) signals and suggests a martensitic microstructure. This point is confirmed by the electron backscatter diffraction (EBSD) map given in Fig. 2-b). This inverse pole figure (IPF) map clearly shows the characteristic lath of a martensitic microstructure. During martensitic transformation, in case of a high cooling rate, some residual austenite may remain in the alloy, influencing therefore the repartition of hydrogen in solution. The proportion of residual austenite in martensitic structures is quite important for two reasons. The first one is the high content of hydrogen in these areas because of the high solubility. The second one is the metastability of this phase. The phase map in Fig. 2-c) shows the proportion of FCC iron (in red) contained in the BCC structure (in green). The proportion of residual austenite has been measured and is below 4%. This fact disproves any hypotheses relying on an effect of strain induced reversion of martensite.

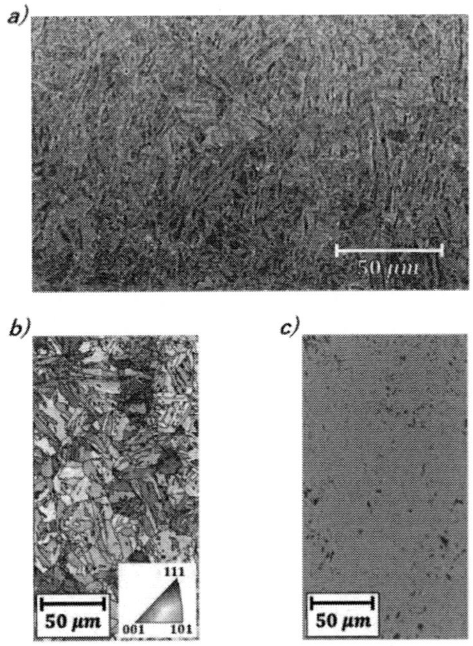

Fig. 2. Microstructure of 17-4PH H1150: (a) BSE image, (b) IPF map, and (c) phase map. Maps of (b) and (c) were obtained by EBSD

2.2 Sample preparation and test procedures

The three types of specimen used in this study are illustrated in Fig. 3. The first geometry used for tensile tests is a smooth, round-bar specimen, as shown on Fig. 3-a). The tensile tests of non-charged and H-charged specimens were performed in air at room temperature at a displacement rate of 1 mm/min.

The fatigue-life tests were performed on a circumferentially-notched specimen as shown in Fig. 3-b). The specimen has been designed in accordance with the ASTM G142-98 standard [22]. The net section diameter is 6.4 mm at notch root with a nominal root radius specified at 0.083 mm, while the gross section diameter is 12.8 mm. The stress concentration factor (K_t) for this geometry is about 6.6. The tests of non-charged and H-charged specimens were conducted in air at room temperature at a stress ratio, R, of –1 and under test frequencies ranging from 0.001 to 1 Hz. The existing literature [23] has demonstrated that the fatigue life of this notched specimen can be predicted from FCG rate obtained with an ordinary compact-tension (CT) specimen. Thus, the fatigue-life test of the present notched specimen evaluates the FCG behavior of 17-4PH H1150.

After fatigue-life tests, the hydrogen content is measured from the specimens described in Fig. 3-c). These specimens are processed from fatigue-life ones.

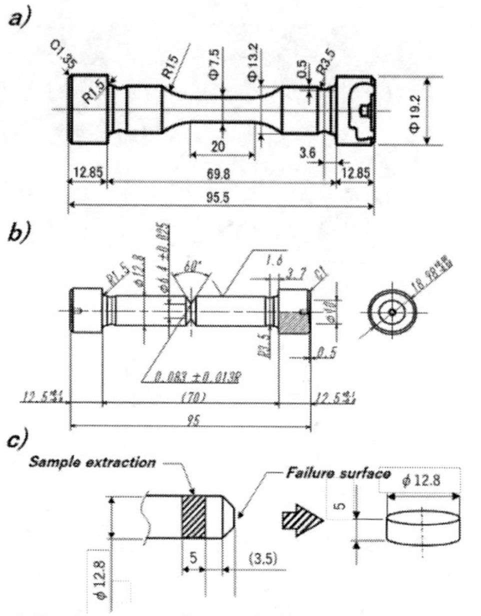

Fig. 3. Specimen geometries used in (a) tensile tests, (b) in fatigue-life tests and (c) in hydrogen-content measurements

2.3 Hydrogen charging and hydrogen-content measurements

Before the tensile and fatigue-life tests, some specimens were exposed to 35 MPa or 100 MPa hydrogen gas at 270°C for 200 h. According to previous investigations, the present exposure condition ensures the uniform distribution of hydrogen content in the specimen. During the fatigue-life tests, hydrogen desorption from the specimens may occur; thus, the hydrogen content after fatigue-life tests were measured using thermal desorption analysis (TDA) under a rising temperature at a heating rate of 100°C/h. The samples used for the TDA measurements were extracted far enough from the failure area, at 5 mm of the net section, and have a diameter of 12.8 mm and a height of 5 mm (Fig. 3-c).

3 Tensile properties

The nominal stress-strain curves are provided in Fig. 4, and show that barely no effects of hydrogen on the elastic behavior as well as the UTS are observable. The values of UTS were 1013 MPa (non-charged), 1016 MPa (35 MPa H-charged), and 1026 MPa (100 MPa H-charged). In contrast, the plastic behavior was strongly affected by hydrogen. The elongation at failure of the non-charged specimen was 37% and the ratio in elongation (compared to the non-charged specimen was 0.43 for the 35 MPa H-charged specimen and 0.38 for the 100 MPa H-charged specimen. In addition, the ductility loss is usually quantified in terms of relative reduction in area (RRA). The values of RRA were 0.43 for 35 MPa H-charged specimen and 0.31 for 100 MPa H-charged specimen, respectively. From a regulatory point of view [24], this material is therefore strongly affected by hydrogen.

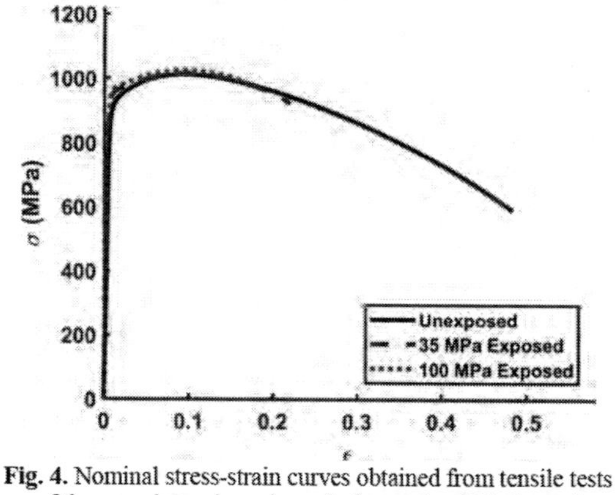

Fig. 4. Nominal stress-strain curves obtained from tensile tests of the non-charged specimen (reference) and H-charged specimens

Fig. 5 shows the fracture surfaces of the tensile specimens. The non-charged specimens failed by an ordinary cup-and-cone ductile failure. In contrast, the H-charged specimen showed a brittle failure, accompanied by a mixture of QC and IG surfaces.

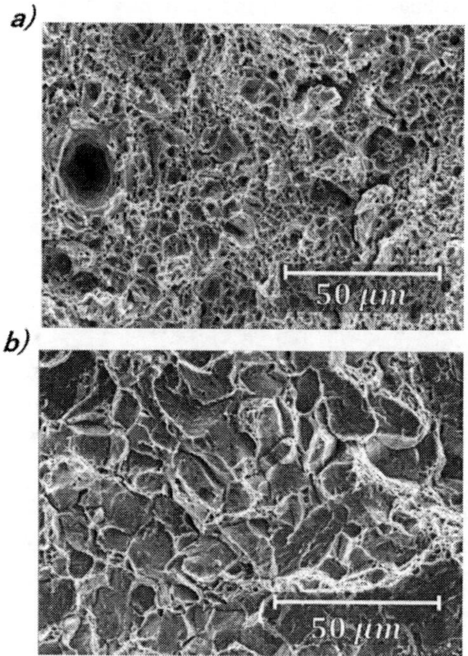

Fig. 5. Fracture surface morphologies of the non-charged tensile specimen (a) and the H-charged tensile specimen (b) by SEM

4 Fatigue-life properties

4.1 Residual hydrogen content

Because solute hydrogen strongly affects mechanical properties of a material, it is then worthy to measure the hydrogen content in the specimen. Fig. 6

presents the residual hydrogen content of the H-charged specimens after the fatigue-life tests in air at room temperature. As a reference, residual hydrogen content of the specimens exposed to 10 MPa and 35 MPa hydrogen gas at 270°C are also plotted. Regardless the hydrogen-gas pressures, the hydrogen desorption from the H-charged specimens is hardly observable. In addition, the hydrogen content, C_H, is higher in case of high pressure of hydrogen gas since the content follows the Sieverts law, *i.e.*, $C_H = SF^{1/2}$, where S is the solubility of hydrogen and F is the fugacity.

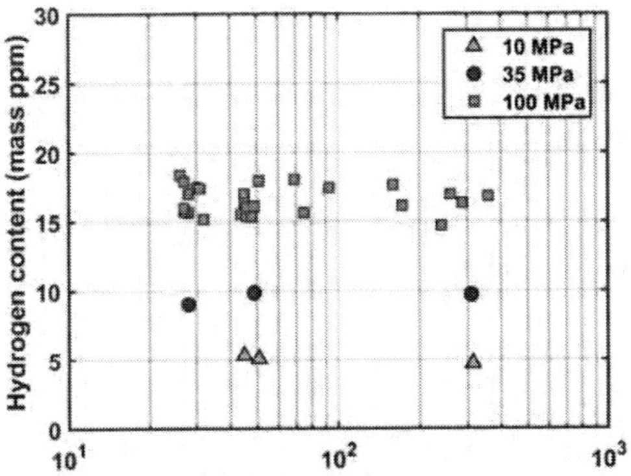

Fig. 6. Residual hydrogen content of H-charged specimens after fatigue-life tests

4.2 *S-N* and threshold properties

Fig. 7 shows the *S-N* curves of the non-charged and H-charged specimens tested in air at room temperature under R = -1 and f = 0.01–10 Hz. The data related to the non-charged specimens are represented in light blue, whereas the data referring to the H-charged specimens are represented in red. The fatigue-life of the H-charged specimens is remarkably degraded compared to the non-charged ones in the low-cycle regime. In parallel, within the high-cycle regime, the fatigue-life reduction of the H-charged specimen is limited. In addition, the fatigue limit σ_W, herein defined as the non-failure maximal stress amplitude for 1×10^7 cycles, is equal in case of non-charged and H-charged specimens. This value is σ_W = 110 MPa.

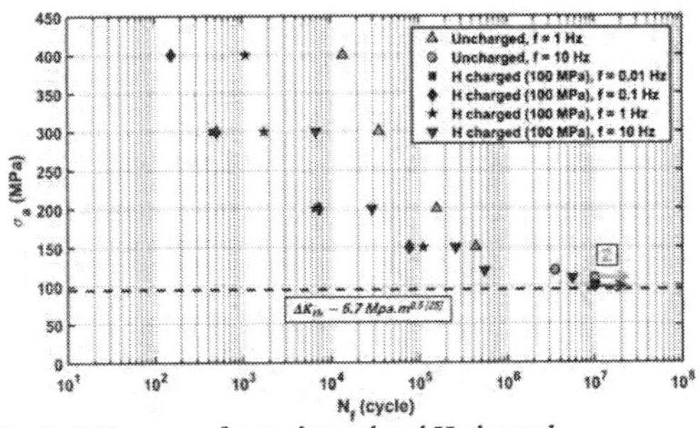

Fig. 7. *S-N* curves of non-charged and H-charged circumferentially-notched specimens tested in air at room temperature under R = -1 and various test frequencies

Fig. 8 presents optical micrographs of the cross section at the fatigue limit for the non-charged and the H-charged specimens. Fatigue cracks emanating from the notch root are observed on both specimens. However, these cracks are not observable on the non-failed specimens tested at a stress amplitude of 105 MPa. Although further investigations are needed, these cracks seem to be non-propagating cracks. The fatigue limit of the circumferentially-notched specimens is therefore determined by the threshold of non-propagation of fatigue cracks (different from the threshold for crack initiation). The dashed black line in Fig. 7 represents the predicted fatigue limit based on the threshold stress intensity factor range for long cracks at R = -1, ΔK_{th} found to be 6.7 MPa.m$^{1/2}$ for 17-4PH steels [25]. To calculate the corresponding stress amplitude, the expression (1) of the stress intensity factor in mode I has been used:

$$\Delta K_I = \Delta\sigma\,\sqrt{\pi b}\,F(\xi) = 2\sigma_a\,\sqrt{\pi b}\,F(\xi) \qquad (1)$$

$$F(\xi) = \frac{1}{2}\left(1 + \frac{1}{2}\xi + \frac{3}{8}\xi^2 - 0.363\xi^3 + 0.731\xi^4\right)\sqrt{1-\xi}$$

where ξ is the ratio of the net diameter, b, to the gross diameter, R. Although the heat-treatment conditions differ from the literature, the experimental fatigue limit of the circumferentially-notched specimen is reasonably consistent with the predicted fatigue limit based on the threshold value for long crack.

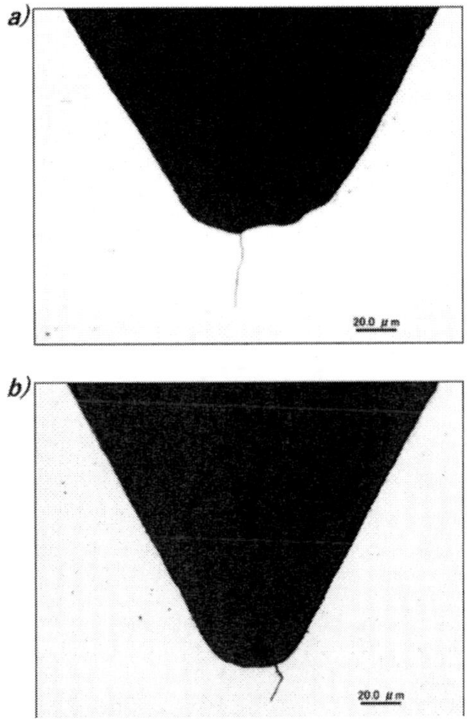

Fig. 8. Non-propagating cracks emanating from the notch root of non-charged and H-charged specimens at the fatigue-limit

condition. The stress amplitude of both specimens is equal (σ_a = 110 MPa)

4.3 FCG acceleration under various combination of stress amplitudes and test frequencies

The experimental result in Fig. 7 shows a strong decrease of the fatigue life due to the presence of hydrogen. This decrease is about one order of magnitude depending on the testing frequency. This graph also shows that the embrittling effect of hydrogen is emphasized for lower frequencies and high amplitudes. In contrast, for lower amplitude, the effect of hydrogen is minimized. In order to quantify the impact of hydrogen on the fatigue life, the ratio of the fatigue-life for the non-charged specimen to the H-charged specimen, N_f / N_{f_H}, has been introduced. This parameter represents the ratio of the hydrogen-enhanced FCG acceleration; therefore, N_f / N_{f_H} is hereinafter referred to as the FCG acceleration ratio.

Fig. 9 presents the evolution of the FCG acceleration ratio as a function of the test frequency for several stress amplitudes (stress ratio R = -1). Except for the stress amplitude of 400 MPa, an upper bound of the FCG acceleration seems to exist, although the acceleration ratio is quantitatively different. The FCG acceleration is promoted at a higher stress amplitude, and its value is approximately 100 times higher in case of σ_a = 300 MPa. As shown in Fig. 1, the hydrogen-enhanced FCG acceleration ratio based on the HISCG model is below 30. Thus, another mechanism, different from the HISCG model, may contribute to the observed FCG acceleration ratio of 100, although an upper bound of the FCG acceleration seems to exist.

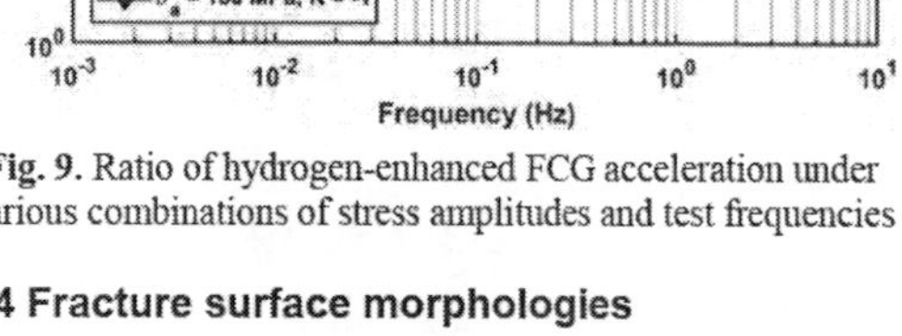

Fig. 9. Ratio of hydrogen-enhanced FCG acceleration under various combinations of stress amplitudes and test frequencies

4.4 Fracture surface morphologies

Fracture surface morphologies of non-charged and H-charged specimens have been investigated by SEM imagery. Fig. 10 shows the fracture surface of the non-charged specimen at σ_a = 120 MPa and f = 10 Hz (high-cycle regime), whereas Fig. 11 shows the fracture surface obtained at σ_a = 400 MPa at f = 1 Hz (low-cycle regime). Both surfaces indicate typical fatigue striations or microstructure-dependent surface.

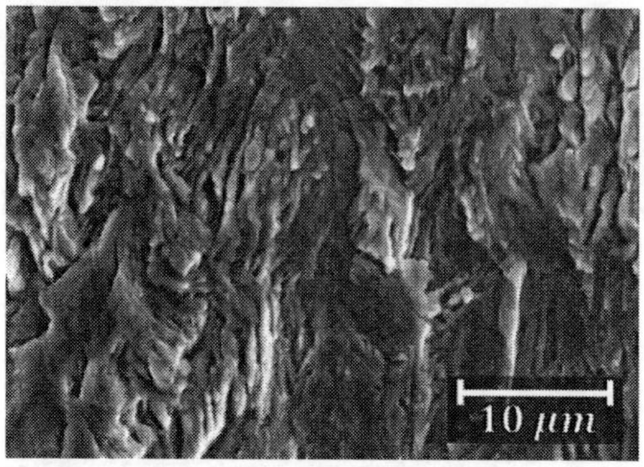

Fig. 10. Fracture surface morphology of the non-charged specimen at σ_a = 120 MPa and f = 10 Hz

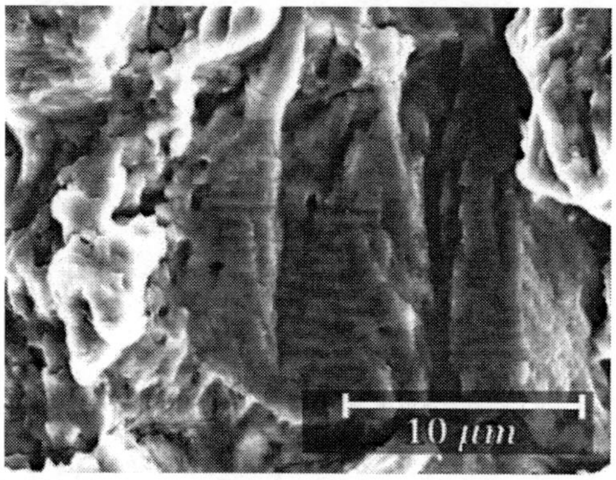

Fig. 11. Fracture surface morphology of the non-charged specimen at σ_a = 400 MPa and f = 1 Hz

Fig. 12 presents the fracture surface of the H-charged specimen tested at σ_a = 120 MPa and f = 10 Hz. The fracture surface is different from the non-charged specimen and is covered with QC surface. The SEM micrograph presented in Fig. 13 is related to the H-charged specimen tested at σ_a = 400 MPa and f = 1 Hz. For this condition, the fracture surface is covered with QC and IG surfaces. Fig. 14 also presents the fracture surface of the H-charged specimen at σ_a = 300 MPa and f = 0.01 Hz. In this case, the FCG acceleration of the H-charged specimen is approximately 100 times higher compared to the non-charged specimen. The corresponding SEM micrograph reveals both QC and IG surfaces.

Fractographic observations have shown that the non-charged specimens usually fail by fatigue striations. In contrast, the introduction of hydrogen has led to a brittle fracture accompanied by QC and IG surfaces. At higher stress amplitudes, the fracture surface is covered with QC and IG surfaces, whereas at lower stress amplitude, the fracture surface is only covered with QC surface.

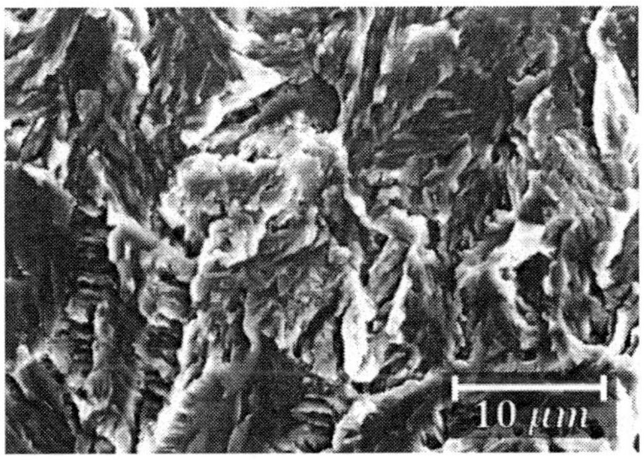

Fig. 12. Fracture surface morphology of the H-charged specimen at σ_a = 120 MPa and f = 10 Hz

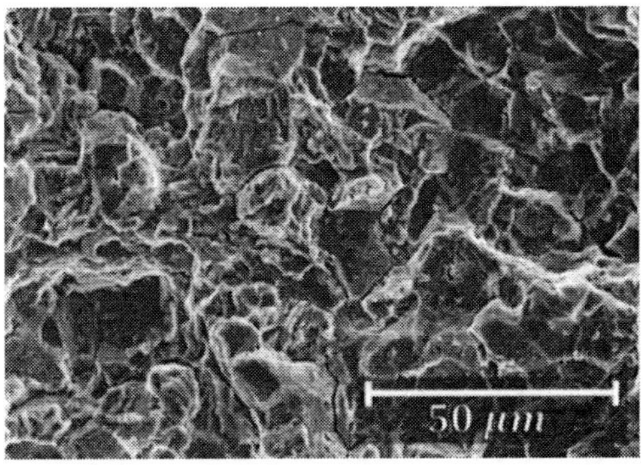

Fig. 13. Fracture surface morphology of the H-charged specimen at σ_a = 400 MPa and f = 1 Hz

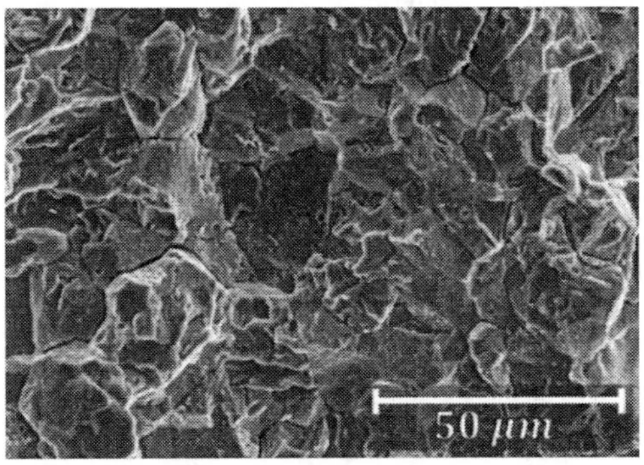

Fig. 14 Fracture surface morphology of the H-charged specimen at σ_a = 300 MPa and f = 0.01 Hz

5 Discussion

As shown in Fig. 9, the FCG acceleration ratio at a higher stress amplitude is about 100, and an upper bound for the FCG acceleration ratio seems to exist. According to the HISCG model, the fatigue crack growth of the H-charged specimen is at most 30. Thus, another mechanism, different from the HISCG model, may contribute to reach the experimental FCG acceleration ratio of 100.

For a lower stress amplitude, the FCG acceleration ratio of the H-charged specimens has an upper bound and the actual value is lower than 10. Furthermore, as shown in Fig. 12, the fracture surface was only covered with QC surface. In consequence, it is considered that such failure surface results from the HISCG model.

In contrast, for a higher stress amplitude, as known in Fig. 13 and Fig. 14, the fracture surface was covered with QC and IG surfaces. It means that both time-dependent crack growth (leading to IG areas) and cycle-dependent crack growth (leading to QC areas, result of the HISCG mechanism) occur. It is therefore proposed that the resulting mechanism is issued from the superposition of both elementary mechanisms as expressed in (2).

$$\frac{N_f}{N_{fH}} = \alpha\left(\frac{N_f}{N_{fH}}\right)_{QC} + (1-\alpha)\left(\frac{N_f}{N_{fH}}\right)_{IG} \quad (2)$$

where (N_f/N_{f_H}) is the acceleration ratio considering only the mechanism i and α is the surface fraction which have failed by QC. Fig. 15 shows a schematic illustration of the frequency dependence of FCG acceleration ratio of H-charged specimens at higher and lower stress amplitudes. In case of both QC and IG failure, the resulting FCG acceleration ratio of the H-charged specimens appears to result from the superposition of a cycle-dependent crack growth (based on the HISCG mechanism) and a time-dependent crack growth. Consequently, a larger surface fraction of IG leads to an enhanced contribution of the time-dependent crack growth.

In the current study, the fraction of IG area represents approximately 20% of the failure surface for the σ_a = 300 MPa condition (measurement performed at on a larger surface than Fig. 14). Therefore, the contribution of the cycle-dependent crack growth is higher than the time-dependent crack growth. As a result, it is considered that at a higher stress amplitude, the H-charged specimens present an upper bound for the FCG acceleration ratio. However, this ratio is approximately 100, which is considerably higher than the ratio expected based on the HISCG (~30). Similar tendencies have been observed on low-alloy steels (JIS-SCM435 and JIS-SNCM439) with UTS > 900 MPa [18].

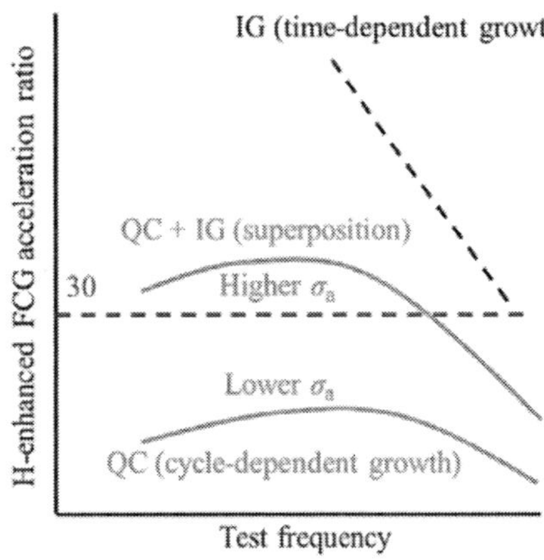

Fig. 15. Schematic illustration of test frequency dependence of FCG acceleration ratio of the H-charged specimens at higher and lower stress amplitudes

6 Conclusion

This paper has investigated the effect of hydrogen on threshold and fatigue crack growth (FCG) properties of the 17-4PH H1150 stainless steel under a wide range of test frequencies. Smooth round-bar specimens for tensile tests and circumferentially-notched specimens for fatigue-life tests have been used and exposed to 100 MPa hydrogen gas at 270°C for 200 h. The tensile and fatigue-life tests of non-charged and H-charged specimens have been performed in air at room temperature. The following points have been put in evidence:

1. This steel is highly sensitive to hydrogen since the relative reduction in area (RRA) was found to be 0.43 and 0.31 in respect with the 35 MPa and 100 MPa H-charging conditions (prior to the tensile tests).
2. For H-charged specimens, hydrogen has not significantly desorbed during the fatigue-life tests and is almost constant.
3. The fatigue limit of the circumferentially-notched specimens has been determined considering the threshold for non-propagation of fatigue cracks. The tests on the non-charged and H-charged specimen have revealed that the fatigue limit is not affected by hydrogen.
4. At higher stress amplitudes, the H-charged specimen presented a significant FCG acceleration ratio compared to the non-charged specimen. Although an upper bound of the hydrogen-enhanced FCG acceleration ratio seems to exist, this ratio was approximately 100. In contrast, the hydrogen-enhanced FCG acceleration ratio at lower stress amplitudes is lower than 10.
5. The H-charged specimens showed a fracture surface covered with quasi-cleavage (QC) at lower stress amplitudes and a mixture of QC and intergranular (IG) at higher stress amplitudes.
6. It has been suggested that the crack growth is cycle-dependent (accompanied by QC) at a lower stress amplitude. Otherwise, a mixture of cycle-dependent crack growth (QC) and time-dependent crack growth (accompanied by IG areas) seems to take place.

References

[1] A. M. Brass and J. Chene, "Influence of deformation on the hydrogen behavior in iron and nickel base alloys: A review of experimental data," *Mater. Sci. Engng A*, vol. 242, pp. 210–221, 1998.

[2] J. P. Hirth, "Effects of hydrogen on the properties of iron and steel," *Metall. Trans. A*, vol. 11, no. 6, pp. 861–890, 1980.

[3] R. P. Gangloff and B. P. Somerday, *Gaseous hydrogen embrittlement of materials in energy technologies: mechanisms, modelling and future developments*. Elsevier, 2012.

[4] L. Coudreuse, *Fragilisation par l'hydrogène et corrosion sous contrainte ; Corrosion sous contrainte – phénoménologie et mécanismes*, Ed. D. Des. 1992.

[5] Q. Liu, Q. Zhou, J. Venezuela, M. Zhang, J. Wang, and A. Atrens, "A review of the influence of hydrogen on the mechanical properties of DP, TRIP, and TWIP advanced high-strength steels for auto construction," *Corros. Rev.*, vol. 34, no. 3, pp. 127–152, 2016.

[6] Q. Liu and A. Atrens, "A critical review of the influence of hydrogen on the mechanical properties of medium-strength steels," *Corros. Rev.*, vol. 31, no. 3–6, pp. 85–103, 2013.

[7] J. Venezuela, Q. Liu, M. Zhang, Q. Zhou, and A. Atrens, "A review of hydrogen embrittlement of martensitic advanced high-strength steels," *Corros. Rev.*, vol. 34, no. 3, pp. 153–186, 2016.

[8] S. Ramamurthy and A. Atrens, "Stress corrosion cracking of high-strength steels," *Corros. Rev.*, vol. 31, no. 1, pp. 1–31, 2013.

[9] R. J. Walter and W. T. Chandler, "Effects of high-pressure hydrogen on storage vessel materials," 1968.

[10] G. Sandoz, "A unified theory for some effects of hydrogen source, alloying elements, and potential on crack growth in martensitic AISI 4340 steel," *Metall. Trans.*, vol. 3, no. May, pp. 1169–1176, 1972.

[11] W. T. Chandler and R. J. Walter, "Testing to determine the effect of high-pressure hydrogen environments on the mechanical properties of metals," in *Hydrogen embrittlement testing*, ASTM International, 1974.

[12] W. T. Chandler and R. J. Walter, "Hydrogen-environment embrittlement of metals and its control," 1975.

[13] J. Toplosky and R. O. Ritchie, "On the influence of gaseous hydrogen in decelerating fatigue crack growth rates in ultrahigh strength steels," *Scr. Metall.*, vol. 15, no. 8, pp. 905–908, 1981.

[14] N. Bandyopadhyay, J. Kameda, and C. J. Mcmahon, "Hydrogen-induced cracking in 4340-type steel: Effects of composition, yield strength, and H2 pressure," *Metall. Trans. A*, vol. 14, no. 4, pp. 881–888, 1983.

[15] R. P. Gangloff, "Hydrogen assisted cracking of high strength alloys," *Compr. Struct. Integr.*, vol. 6, pp. 31–101, 2003.

[16] C. San Marchi and B. P. Somerday, "Technical Reference On Hydrogen Compatibility Of Materials," *Sand2008-1163*, no. code 4001, p. 292, 2012.

[17] J. Yamabe, H. Itoga, T. Awane, T. Matsuo, H. Matsunaga, and S. Matsuoka, "Pressure Cycle Testing of Cr – Mo Steel Pressure Vessels Subjected to Gaseous Hydrogen," vol. 138, no. February, pp. 1–13, 2016.

[18] S. Matsuoka, H. Matsunaga, J. Yamabe, S. Hamada, and T. Iijima, "Various strength

properties of SCM435 and SNCM439 low-alloy steels in 115 MPa hydrogen gas and proposal of design guideline," *Trans. JSME (in Japanese)*, vol. 83, no. 854, pp. 17-264-17–00264, 2017.
[19] J. Yamabe, T. Matsumoto, S. Matsuoka, and Y. Murakami, "A new mechanism in hydrogen-enhanced fatigue crack growth behavior of a 1900-MPa-class high-strength steel," *Int. J. Fract.*, vol. 177, no. 2, pp. 141–162, 2012.
[20] S. Matsuoka, J. Yamabe, and H. Matsuoka, "Criterial for determinating hydrogen compatiblity and the mechanisms for hydrogen-assisted, surface crack growth in austenitic stainless steels," *Eng. Fract. Mech.*, vol. 153, no. 3, pp. 103–127, 2016.
[21] L. W. Tsay, W. C. Lee, R. K. Shiue, and J. K. Wu, "Notch tensile properties of laser-surface-annealed 17-4 PH stainless steel in hydrogen-related environments," *Corros. Sci.*, vol. 44, no. 9, pp. 2101–2118, 2002.
[22] A. Standard, "G142-98" Standard test method for determination of susceptibility of metals to embrittlement in hydrogen containing environments at high pressure, high temperature, or both," *ASTM Int. West Conshohocken, PA*, vol. 4, 2004.
[23] J. Yamabe *et al.*, "Qualification of chromium-molybdenum steel based on the safety factor multiplier method in CHMC1-2014," *Int. J. Hydrogen Energy*, vol. 40, no. 1, pp. 719–728, 2015.
[24] "ANSI/CSA CHMC 1 - 2014: Test Method for Evaluating Material Compatibility in Compressed Hydrogen Applications - Phase I - Metals," *CSA Gr.*, 2014.
[25] B. M. Schönbauer, K. Yanase, and M. Endo, "The influence of various types of small defects on the fatigue limit of precipitation-hardened 17-4PH stainless steel," *Theor. Appl. Fract. Mech.*, vol. 87, pp. 35–49, 2017.

Influence of gaseous hydrogen on plastic strain in vicinity of fatigue crack tip in Armco pure iron

Tomoki Shinko[1], *Gilbert* Hénaff[1,*], *Damien* Halm[1] and *Guillaume* Benoit[1]

[1]Pprime Institute, Department of physics and mechanics of materials, ISAE-ENSMA, 86961 Futuroscope Chasseneuil Cedex, France

Abstract. A multi-scale characterisation of the crack tip plasticity has been investigated in a fatigue crack propagation under gaseous hydrogen at gas pressure of 35 MPa in a commercially pure iron, Armco iron. The dislocation structure beneath a fracture surface was observed by a Transmission Electron Microscopy (TEM), and the cyclic and monotonic plastic zones were evaluated by an Out-of-Plane Displacement (OPD) measurement. By the TEM observation in a non-accelerated regime (ΔK = 11 MPa×m$^{1/2}$), a dislocation cell structure was observed even in the brittle intergranular fracture in hydrogen. This result indicates a certain amount of plastic strain is introduced into the grains in front of an intergranular crack in hydrogen, and this may explain the mechanism of hydrogen-induced intergranular fatigue crack propagation. On the other hand, in an accelerated regime (ΔK = 18 and 20 MPa×m$^{1/2}$), a distribution of scattered dislocation tangles without any cell or vein structure was observed in hydrogen. Besides, the inhibition of the cyclic plasticity near the crack path in hydrogen was confirmed by the OPD measurement. These results are clear evidences of hydrogen-induced localization of cyclic plasticity in the vicinity of a crack tip, which suggests a mechanism model of hydrogen-enhanced fatigue crack growth based on the plasticity localization.

1 Introduction

Hydrogen is expected as one of the most desired energy carriers in the future infrastructures of a sustainable society. On the other hand, it is known that gaseous hydrogen environment may result in a significant degradation of mechanical properties in metallic materials which is commonly known as "hydrogen embrittlement" [1–3]. Especially, the degradation of fatigue resistance due to hydrogen environment is a crucial problem for the industrial usage. Many theories have been proposed to explain the mechanism of hydrogen embrittlement. In particular, the following three theories are today considered as the most important ones: Hydrogen-Enhanced DEcohesion (HEDE) [4], Hydrogen-Enhanced Localized Plasticity (HELP) [5] and Adsorption-Induced Dislocation Emission (AIDE) [6]. The HEDE mechanism supposes that the hydrogen atoms adsorbed at a crack tip or segregated at the grain boundaries weaken interatomic bonds and cause an atomic decohesion. This leads to a brittle fracture such as a cleavage fracture and an intergranular decohesion. In the HELP mechanism, the solute hydrogen atoms in front of the crack tip enhance and localize a dislocation activity. This promotes a nucleation and coalescence of micro voids, resulting in a crack propagation with less blunting and less macroscopic ductility of the material. The fracture surface often exhibits small, shallow dimples. The AIDE mechanism is based on the facilitation of dislocation nucleation due to adsorbed hydrogen. In this mechanism, adsorbed hydrogen on the surface and the sub-surface weakens interatomic bonds and promotes the dislocation nucleation at a crack tip which enhances the local plastic strain near the crack tip. The principle of FCG enhancement by the AIDE is similar with the HELP mechanism, however the main difference is that the AIDE involves the sub-surface effect of adsorbed hydrogen while the HELP is based on the solute hydrogen in the material.

In the previous researches by Pprime institute [7,8], it has been confirmed that the Fatigue Crack Growth Rates (FCGRs) in Armco iron under 35 MPa of gaseous hydrogen environment are accelerated up to 100 times higher than the one in air, as shown in Fig. 1. And also, the FCGRs in hydrogen exhibit different tendencies depending on the stress intensity factor range ΔK. These tendencies can be decomposed into following three distinct regimes: (a) a non-accelerated regime (ΔK < 12 MPa×m$^{1/2}$) where FCGR is not accelerated and is almost the same as in air; (b) a transition regime (12< ΔK < 14 MPa×m$^{1/2}$) where FCGR is sharply accelerated; and (c) an accelerated regime (14 MPa×m$^{1/2}$ < ΔK) where the slope of FCGR becomes the same level as in air. It has also been confirmed that the hydrogen-induced FCG enhancement depends on the loading frequency *f*. As well as the FCGR is affected by hydrogen, the fracture mode is also clearly changed due to hydrogen. In the non-accelerated regime, the fracture mode is brittle intergranular fracture in hydrogen while a ductile transgranular quasi-cleavage (QC) fracture takes place in air. In the transition regime, fracture mode in hydrogen

* Corresponding author: gilbert.henaff@ensma.fr

gradually changes to a transgranular QC fracture. Subsequently in the accelerated regime, the fracture surface becomes fully transgranular QC fracture with course and flat striations (i.e. brittle striations).

In this study, in order to clarify the mechanism of the hydrogen-induced FCG enhancement in Armco iron, the influence of hydrogen on microscopic and macroscopic plasticity at a crack tip has been investigated by means of the following two methods: (1) a microscopic plasticity near the crack tip has been investigated by a Transmission Electron Microscopy (TEM) of dislocation structure immediate beneath fracture surface; and (2) macroscopic cyclic and monotonic plasticity in front of the crack tip has been evaluated by a measurement of lateral shrinkage around the fatigue crack path (i.e., Out-of-Plane Displacement, OPD) by using an interferometric confocal microscope.

2 Material and testing method

The material used in this study is a commercially pure iron, Armco iron (AK Steel Corp.). The chemical composition of Armco iron is shown in Table 1. An annealing heat treatment at 1073 K (800 °C) for 60 minutes was applied in order to relax a residual stress and homogenize the ferrite microstructure. Figure 1 shows an Electron Back-Scattered Diffraction (EBSD) mapping of the heat-treated Armco iron. The mean grain size after heat treatment is approximately 90 µm. The mechanical properties determined after heat treatment are: yield strength σ_y = 170 MPa and maximum elongation A = 25 %.

Specimen preparation and fatigue crack propagation tests were performed by following the ASTM E647 standard. CT specimens with a width W = 40 mm and a thickness B = 8 mm were machined in the longitudinal-transverse orientation against hot-rolling direction. The specimen surface was finished by diamond abrasives with a diameter of 1 µm.

Fatigue crack propagation tests were conducted under loading control at room temperature under various atmospheres: vacuum (5×10^{-10} Pa), laboratory air, 3.5 MPa of gaseous nitrogen and 35 MPa of gaseous hydrogen. The tests were performed by using Hycomat test bench [9] developed by Pprime institute. It consists of an autoclave assembled to a servo-hydraulic machine, which has a capacity to perform mechanical tests under a maximum gas pressure of 40 MPa at the highest temperature of 150 °C (423 K). The various loading frequencies of 0.2, 2 and 20 Hz were applied to investigate its influence on FCGR. Crack length was measured by optical microscope and a Direct Current Potential Drop (DCPD) technique during the tests. The stress ratio R (= K_{min}/K_{max}) was 0.1, where K_{min} and K_{max} are the minimum and maximum stress intensity factors, respectively.

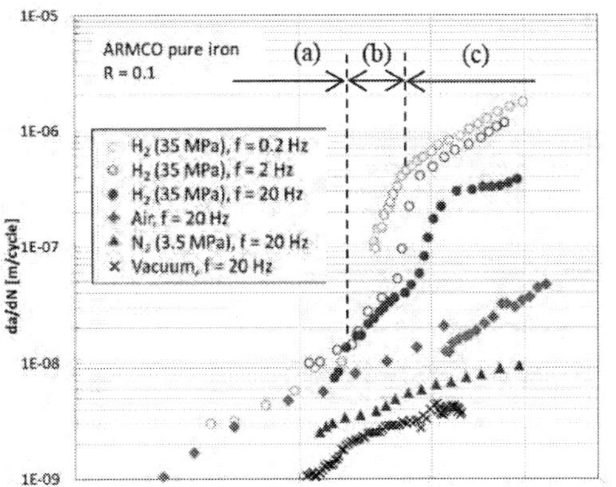

Fig. 1. FCGRs (= da/dN) vs ΔK in Armco iron [8]. FCGR curves in hydrogen can be separated as three distinct regime: (a) a non-accelerated regime, (b) a transition regime and (c) an accelerated regime.

3 Result

3.1. TEM observation of dislocation structure immediately beneath fracture surface

To investigate the microscopic plasticity immediate vicinity of a crack tip, the dislocation structure in the material within 10 µm depth from fracture surface has been observed by TEM. The thin foils for TEM observation were prepared from the fracture surfaces by using a Focused Ion Beam (FIB, Helios NanoLab G3, FEI Company). The thickness of foil was approximately 100 nm. Each TEM foil presented in this study consists of one grain, as confirmed by a transmission EBSD analysis on the foils. Figure 3 shows the SEM images of fracture surfaces and the dark-field TEM images of cross-section immediately beneath the fracture surface at ΔK = 11 MPa×m$^{1/2}$ at f = 20 Hz in air (Fig. 3 (a)) and at f = 2 Hz in 35 MPa of gaseous hydrogen (Fig. 3 (b)). This ΔK value corresponds to the non-accelerated regime. A rectangular part with different contrast in the center of some SEM

Table 1. Chemical composition of Armco iron (mass %).

Element	C	Mn	P	S	Cu	N	Al
Content	0.001	0.050	0.003	0.003	0.009	0.004	0.005
Element	Cr	Mo	Ni	Sn	Si	Fe	
Content	0.015	0.002	0.14	0.002	0.004	Balance	

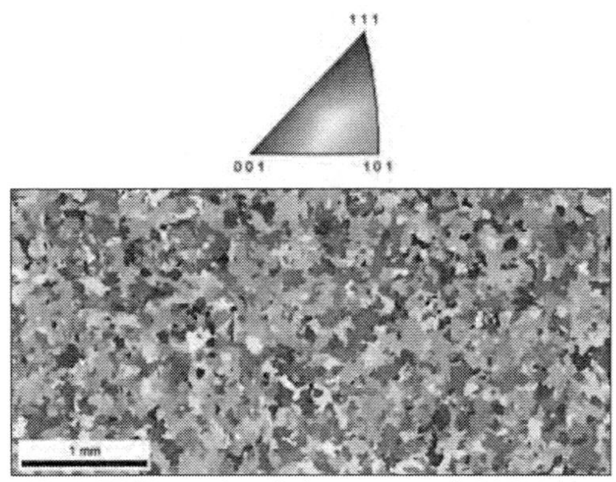

Fig. 2. EBSD mapping of microstructure of ARMCO iron used in this study.

images is an area covered by platinum layer to protect the material from the FIB milling.

The fracture surface in air (Fig. 3 (a-1)) represented a typical transgranular QC fracture. This fracture mode in air continued until the unstable rupture at the end of the test. In the TEM image (Fig. 3 (a-2)), clear dislocation walls in the inside of material can be observed. Besides, it can be observed that the dislocation walls form a cell structure, in other words sub-grains, with a diameter of few μm. On the other hand, one can see that the fracture surfaces in hydrogen (Fig. 3 (b-1) and (b-3)) are fully covered by intergranular facets evidencing an intergranular fracture. And also, some secondary intergranular cracks and stripe-like plastic markings on the surface of intergranular facets have been observed as shown in Fig. 3 (b-3) and (b-4). The presence of the plastic markings suggests that a certain amount of plastic strain is introduced to the grains in front of the crack tip. Both TEM image of the dislocation structure in hydrogen (Fig. 3 (b-2) and (b-5)) shows a similar dislocation cell structure as observed in air. The diameter of dislocation cells is approximately 1 μm which is slightly smaller than the one in air.

Figure 4 shows the same type of SEM and TEM observations in case of ΔK = 20 MPa×m$^{1/2}$ in the accelerated regime. It has been confirmed that the fracture surface showing the transgranular QC fracture (Fig. 4 (a-1)) and the dislocation structure showing the cell structure (Fig. 4 (a-2)) are not changed from the ones observed in non-accelerated regime (Fig. 3 (a)). On the other hand, as shown in Fig. 4 (b-1) and (b-4), the fracture surfaces at ΔK = 20 MPa×m$^{1/2}$ in hydrogen represented a transgranular QC fracture with brittle striations. Figure 4 (b-2) and (b-5) show the entire TEM images of the foils, and Fig. 4 (b-3) and (b-4) are the magnified images of the areas indicated by the rectangles

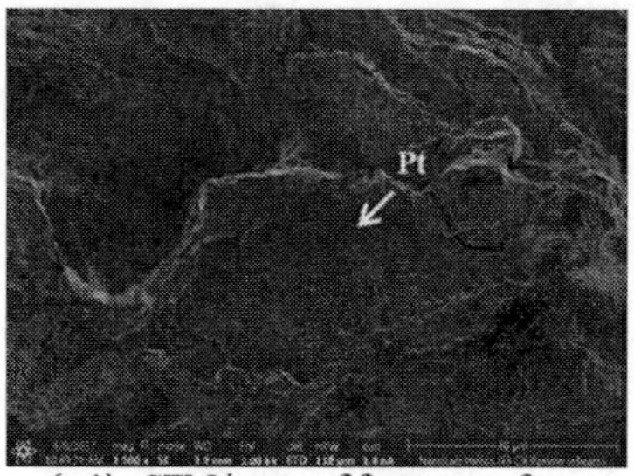

(a-1) SEM image of fracture surface.

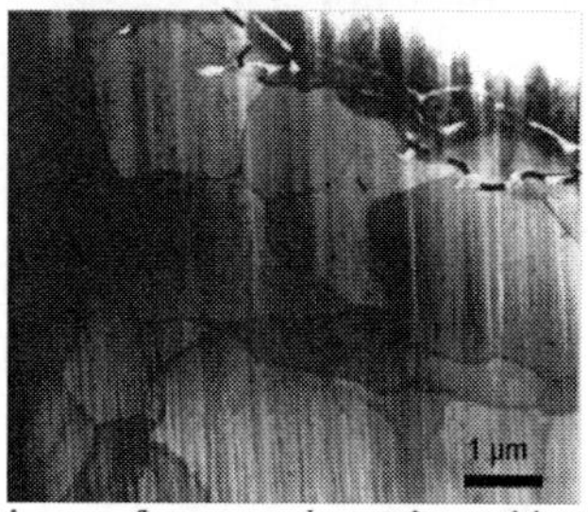

(a-2) TEM image of cross-section at the position indicated by the red line in (a-1).

(a) In air, f = 20 Hz.

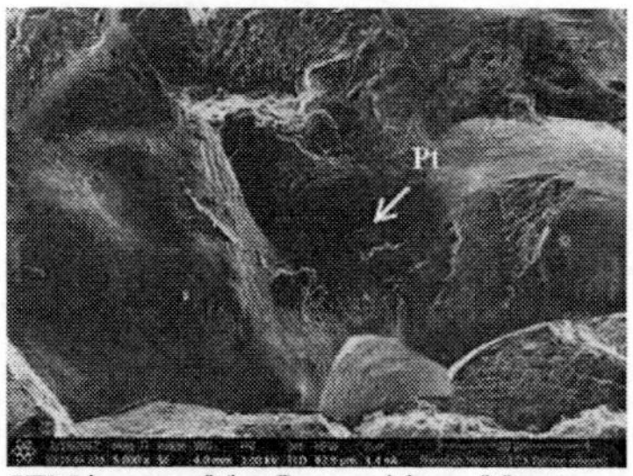

(b-1) SEM image of the first position of fracture surface.

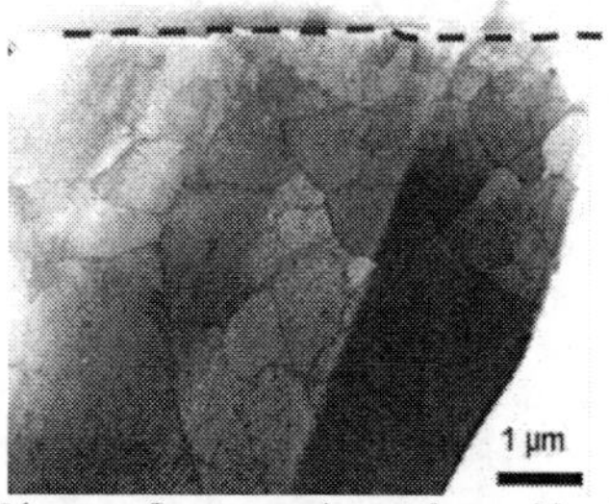

(b-2) TEM image of cross-section at the position indicated by the red line in (b-1).

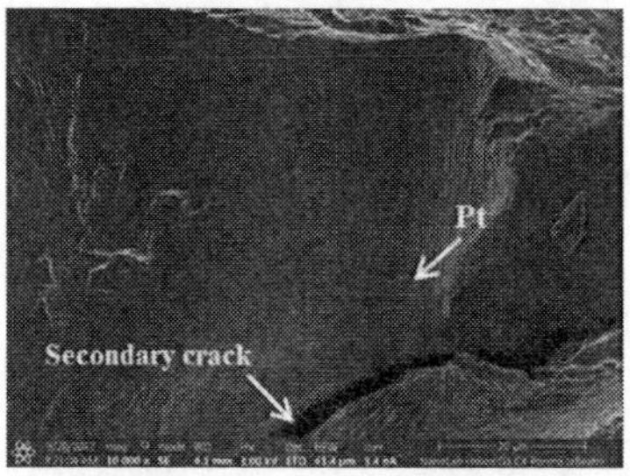

(b-3) SEM image of the second position of fracture surface.

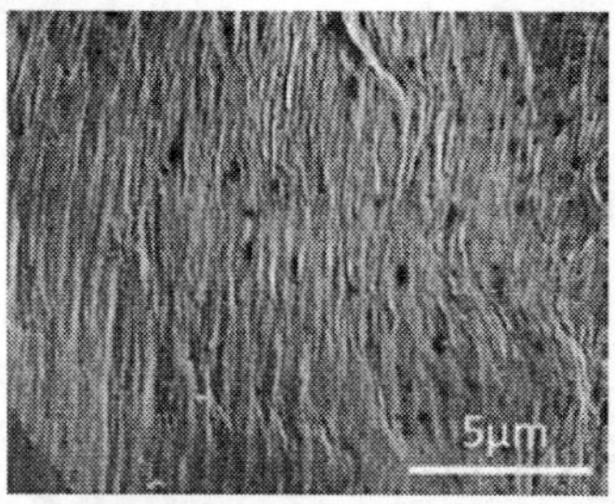

(b-4) Higher magnified SEM image of stripe-like plastic markings in the area indicated by the red rectangle in (b-3).

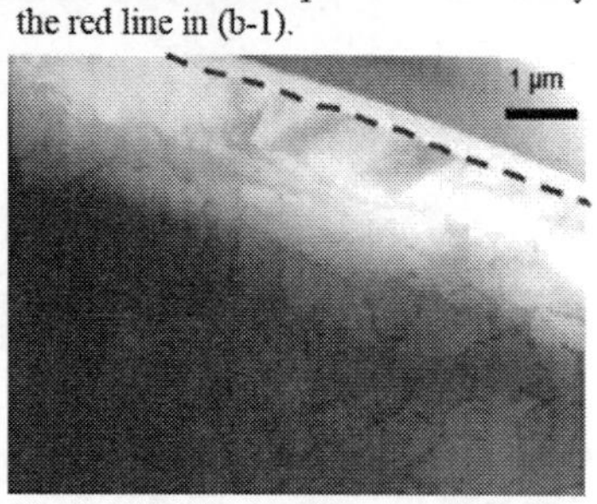

(b-5) TEM image of cross-section at the position indicated by the red line in (b-3).

(b) In 35 MPa of gaseous hydrogen, f = 2 Hz.

Fig. 3. SEM images of fracture surfaces and TEM images of cross-section immediately beneath fracture surface at ΔK = 11 MPa×m$^{1/2}$ in air, f = 20 Hz and in gaseous hydrogen (35 MPa), f = 2 Hz. Crack propagation direction is from left to right in the images. Red dotted lines in TEM images show the position of fracture surface.

in Fig. 4 (b-2) and (b-5). The red arrows in the SEM and TEM images in hydrogen indicate the position of the striations crossing the TEM-observed area. In these TEM images in hydrogen, the dislocation walls do not form any cell or vein structure, but are disconnected in short pieces and randomly dispersed. In addition, one can notice that there are long dislocation walls starting from the positions of striations, more precisely, one starting from the striation in the center of Fig. 4 (b-2) and another one starting from the striation in the left side of Fig. 4 (b-5) and (b-6), even though there is no other long dislocation wall observed. The dislocation wall in Fig. 4 (b-2) is separated into two and winding, while another one in Fig. 4 (b-5) and (b-6) is straight and angled approximately 45 degrees from the fracture surface.

3.2. Measurement of Out-of-Plane Displacement (OPD) along crack path

In order to investigate the macroscopic plasticity (cyclic and monotonic plasticity) in vicinity of the crack tip, the OPD has been measured by using an interferometric confocal microscope (Talysurf CCI6000, Taylor Hobson).

The surface relief curves measured at ΔK = 12 and 18 MPa$\times$m$^{1/2}$ in various testing conditions are shown in Fig. 5. By comparing Fig. 5 (a) and (b), one can confirm that the displacement and the width of OPD increase with the increment of ΔK from 12 to 18 MPa$\times$m$^{1/2}$. The black dotted line shows the half of monotonic plastic zone size $r_p/2$. The monotonic plastic zone size in plane

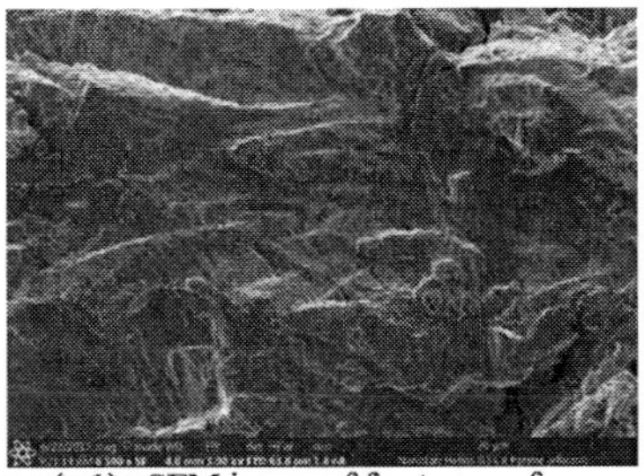

(a-1) SEM image of fracture surface.

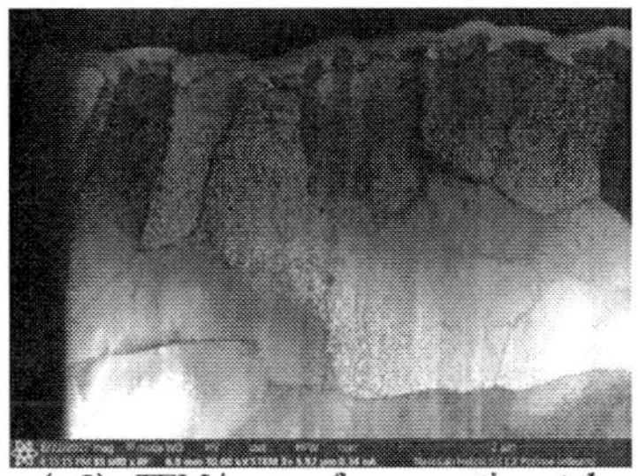

(a-2) TEM image of cross-section at the position indicated by the red line in (a-1).

(a) In air (f = 20 Hz).

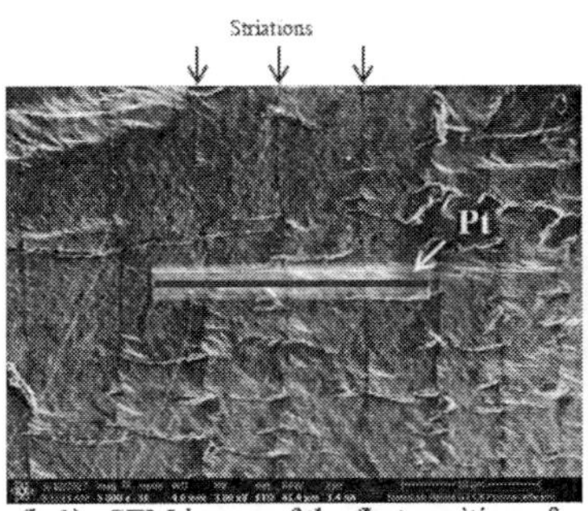

(b-1) SEM image of the first position of fracture surface.

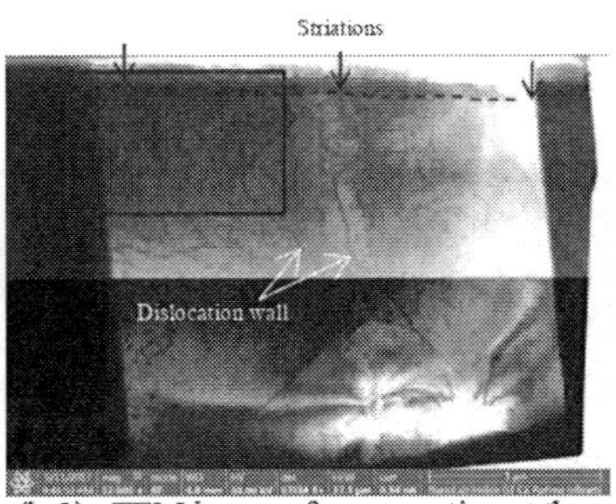

(b-2) TEM image of cross-section at the position indicated by the red line in (b-1).

(b-3) Higher magnified TEM image of the area indicated by the red rectangle in (b-2).

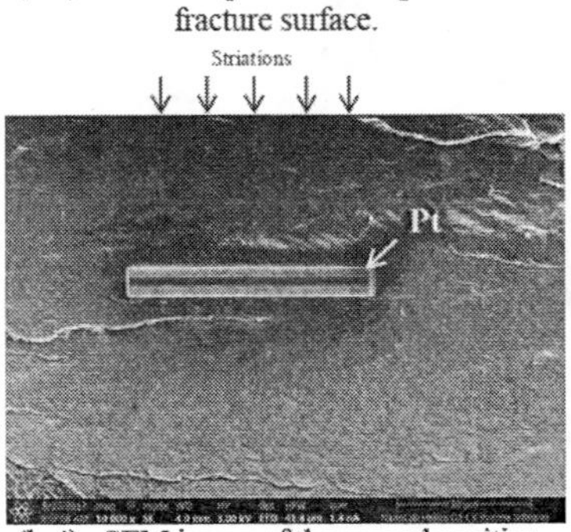

(b-4) SEM image of the second position of fracture surface.

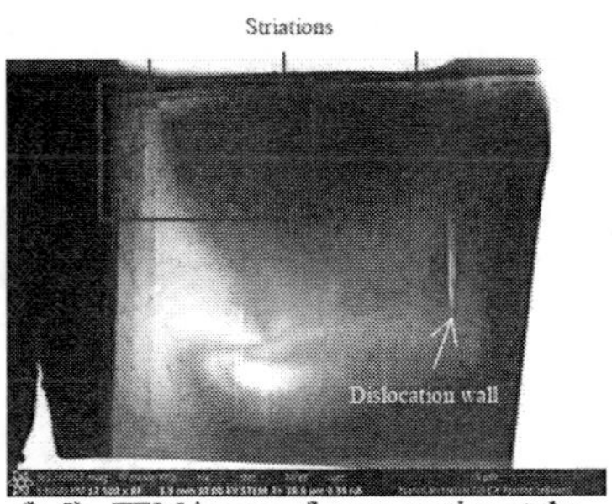

(b-5) TEM image of cross-section at the position indicated by the red line in (b-4).

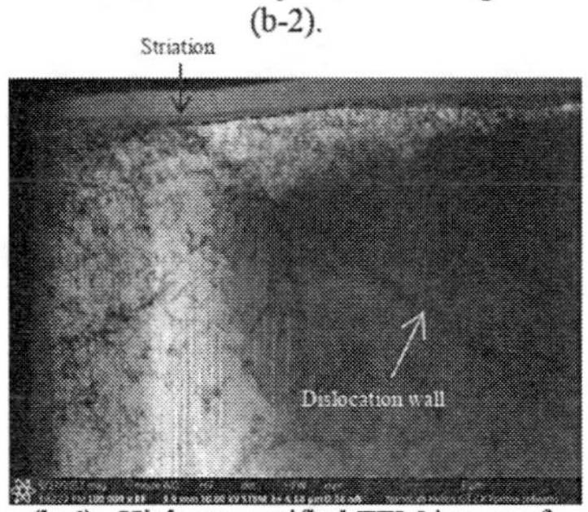

(b-6) Higher magnified TEM image of the area indicated by the red rectangle in (b-5).

(b) In 35 MPa of gaseous hydrogen (f = 2 Hz).

Fig. 4. SEM images of fracture surfaces and TEM images of cross-section immediately beneath fracture surface at ΔK = 20 MPa$\times$m$^{1/2}$ in air (f = 20 Hz) and in 35 MPa of gaseous hydrogen (f = 2 Hz). Crack propagation direction is from left to right in all images. Red dotted lines show the location of fracture suface. Red arrows show the position of striations crossing the TEM-observed area.

stress condition r_p was calculated by the following equation [10]:

$$r_p = \frac{1}{\pi}\left(\frac{K_{max}}{\sigma_y}\right)^2. \quad (1)$$

The values of $r_p/2$ are obtained as 0.85 mm at ΔK = 12 MPa×m$^{1/2}$ and 1.92 mm at ΔK = 18 MPa×m$^{1/2}$, which are close to the positions of the beginning of OPD. Therefore, the observed OPD is considered as a result of monotonic plasticity around the crack tip. In a comparison of the surface relief curves, no clear influence of environment or testing conditions on the OPD was confirmed except for a quite large fluctuation in the surface relief curves of each specimen. This result denotes that hydrogen does not influence on the large scale of plastic strain in a monotonic plastic zone.

Figure 6 shows the surface topographic images of side-surface near the crack path at ΔK = 12 and 18 MPa×m$^{1/2}$ in nitrogen and in hydrogen (at f = 20 Hz). Color gradation indicates surface height. Central vertical blue part is a fatigue crack path. At ΔK = 12 MPa×m$^{1/2}$, the surface topographies in both environments (Fig. 6 (a-1) and (b-1)) are smooth. On the other hand, at ΔK = 18

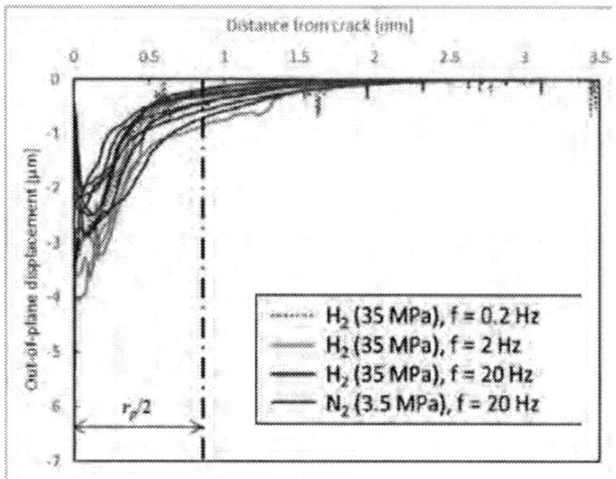

(a) ΔK = 12 MPa×m$^{1/2}$

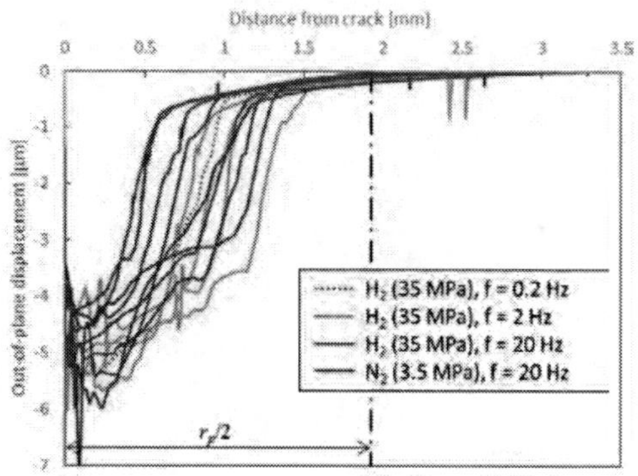

(b) ΔK = 18 MPa×m$^{1/2}$

Fig. 5. Surface relief curves presenting the Out-of-Plane Displacement (OPD) near crack rips at ΔK = 12 and 18 MPa×m$^{1/2}$ in various testing conditions. The width of half monotonic plastic zone size $r_p/2$ is indicated to compare with the OPD width.

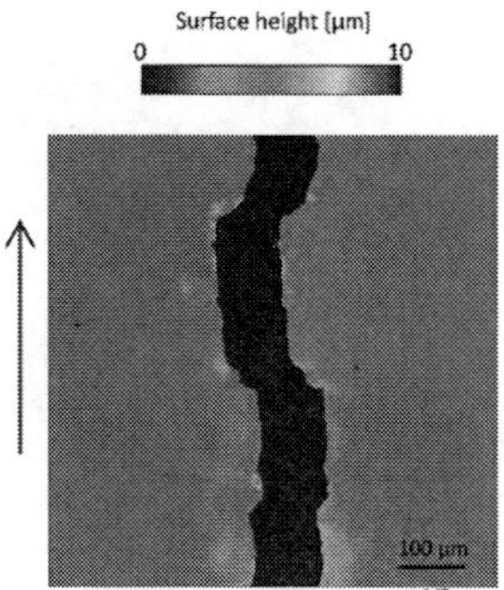

(a-1) ΔK = 12 MPa×m$^{1/2}$

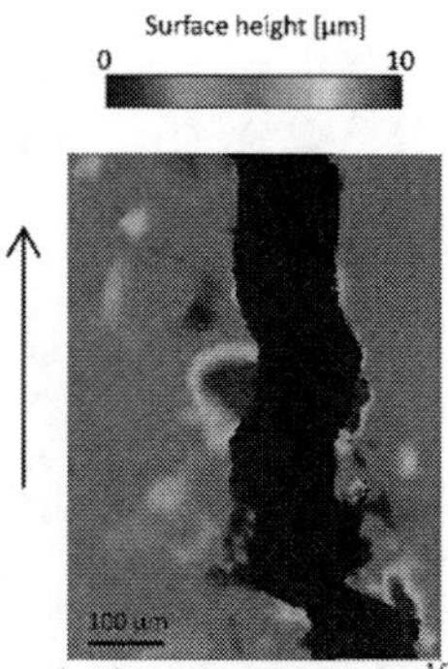

(a-2) ΔK = 18 MPa×m$^{1/2}$

(a) In nitrogen, f = 20 Hz

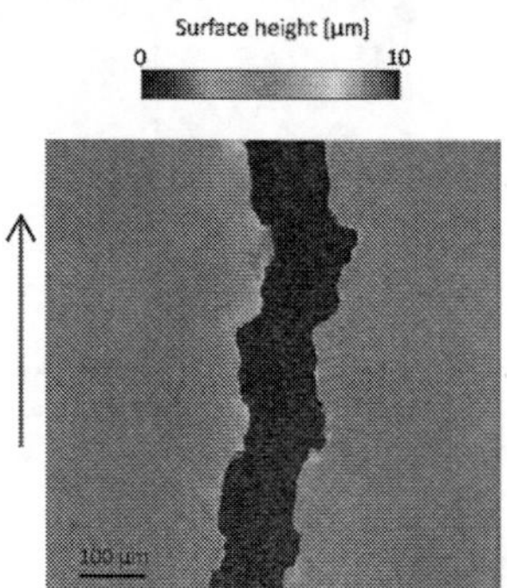

(b-1) ΔK = 12 MPa×m$^{1/2}$

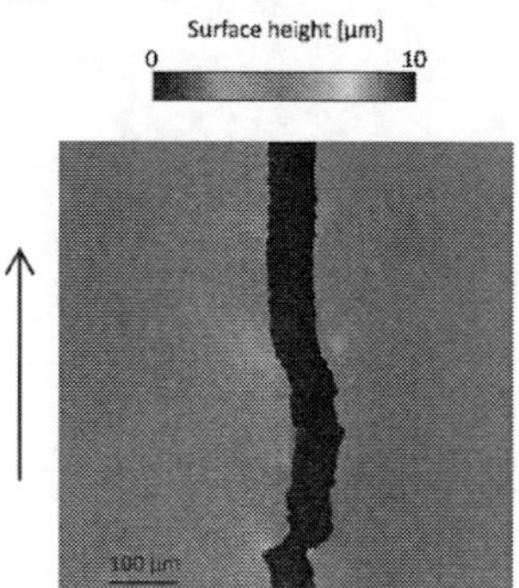

(b-2) ΔK = 18 MPa×m$^{1/2}$

(b) In hydrogen, f = 20 Hz

Fig. 6. High magnification images of surface topography near crack rips in nitrogen (3.5 MPa, f = 20 Hz) and in hydrogen (35 MPa, f = 20 Hz). The arrows indicate the direction of crack propagation.

MPa×m$^{1/2}$, the surface topographic image in hydrogen (Fig. 6 (b-2)) shows no significant surface unevenness as same as at ΔK = 12 MPa×m$^{1/2}$, meanwhile a heterogeneous surface unevenness over a range of few hundreds of μm from the crack path is clearly observed in nitrogen (Fig. 6 (a-2)). According to Rice [11], the cyclic plastic zone size Δr_p in plane stress condition can be estimated by following equation:

$$\Delta r_p = \frac{1}{\pi}\left(\frac{\Delta K}{2\sigma_y}\right)^2 \quad (2)$$

The value of half cyclic plastic zone size $\Delta r_p/2$ is calculated as 389 μm at ΔK = 18 MPa×m$^{1/2}$. Because the surface unevenness observed in Fig. 6 (a-2) spreads out within 100-200 μm distance from the crack path which is close to the calculated value of $\Delta r_p/2$, the surface unevenness observed in Fig. 6 (a-2) is likely a result of cyclic plasticity in front of the crack tip. Therefore, from the results of OPD measurement, it can be concluded that the presence of hydrogen inhibits the cyclic plasticity at the crack tip in the accelerated regime, while hydrogen does not influence on the monotonic plasticity.

4 Discussion

4.1. Fracture mechanism in non-accelerated regime

From the obtained results of TEM observation in the section 3.1, it has been revealed that the dislocation cell structure is formed in the grains beneath intergranular facets in the non-accelerated regime in hydrogen. The presence of dislocation cell structure means that a certain amount of plastic strain is introduced to grains even though a macroscopic feature of fracture surface represents a brittle intergranular fracture. This introduced plastic strain in the grains has been also confirmed by the presence of slip markings on intergranular facets as mentioned in the section 3.1. A similar dislocation cell structure beneath intergranular fracture surface in hydrogen has also been confirmed in the tensile tests in a pure iron by Wang et al. [12] and in a pure nickel by Martin et al. [13]. The diameter of dislocation cells observed in their studies is 200 - 500 nm which is slightly smaller than the one in this study. In addition, Nishikawa et al. [14] has performed an in-situ observation of intergranular crack propagation in a low carbon steel under gaseous hydrogen, and confirmed the formation of slip bands in grains in front of intergranular crack. This result also demonstrates the presence of plastic strain introduced near the crack tip of the intergranular crack in hydrogen.

In order to explain the mechanism of hydrogen-induced intergranular fracture, Wang et al. [12] proposed "a hydrogen-induced plasticity-mediated decohesion process". Their model is based on the interaction between dislocation-dragged hydrogen atoms and grain boundaries. In their model, the solute hydrogen atoms in the vicinity of crack tip promote a slip activity in the grains. The enhanced mobile dislocations in slip planes transport and redistribute the hydrogen atoms to grain boundaries. Consequently, the highly accumulated hydrogen atoms at grain boundaries weaken cohesive energy of grain boundaries by the HEDE mechanism, and result in a grain boundary decohesion.

4.2. Fracture mechanism in accelerated regime

Another noteworthy result of the TEM observation is that the observed dislocation structure in the accelerated regime in hydrogen was a randomly distributed dislocation tangles without any cell or vein structure. This type of dislocation structure implies that the fracture at the crack tip occurred before the plastic strain leached to the sufficient level to form the dislocation cell structure, hence the fracture mode in hydrogen became more brittle-like crack propagation with less plasticity at the crack tip compared with the one in air. This agrees well with the suppression of cyclic plasticity at a crack tip by hydrogen observed in the OPD measurement and the presence of brittle striation on the fracture surface in the accelerated regime.

A series of researches by Kyushu university has also investigated the dislocation structure in the vicinity of the fatigue crack tip in the accelerated regime in a pure iron [15], a low carbon steel [16] and a Fe-Si alloy[17–19]. In these studies, the one in a pure iron by Ogawa et al. [15] has to be especially noted here. They conducted a crack propagation test in a pure iron and performed a Grain Reference Orientation Deviation (GROD) analysis and a TEM observation of the cross-section beneath fracture surface in the accelerated regime (at ΔK = 17 MPa×m$^{1/2}$). Their GROD analysis showed a great reduction of plastic strain distribution within few hundreds μm from a crack path in hydrogen, although there was a deviation of plastic strain distribution along a crack path showing a slightly-deformed part and an almost non-deformed part. Their TEM observations in the non-deformed part presented the dislocation structure similar to the one in this study, a scattered dislocation tangles distribution. On the other hand, the dislocation cell structure was observed within limited depth (< 2 μm) in a slightly-deformed part. Their obtained results of the GROD analysis and the TEM observation are in a good agreement with the ones of the OPD measurement and the TEM observation in this study, respectively.

It has to be mentioned that a clear modification of OPD around the crack path due to hydrogen has been observed in a 15-5PH steel by Bilotta et al. [20]. In a 15-5PH steel, it has been confirmed that hydrogen enhances maximum OPD of monotonic plastic zone without change in the OPD width, namely, increases the slope of OPD curve. This result is different from the one obtained in this study since no influence of hydrogen on the OPD of monotonic plastic zone is observed in Armco iron. However, the size of OPD of monotonic plastic zone in a 15-5PH is few hundreds μm which is the same order of size as the cyclic plastic zone in Armco iron clearly influenced by hydrogen. Therefore, this suggests that hydrogen affects only the plasticity within a certain distance from the crack tip. Another possible explanation is a difference in crack propagation mechanism. Bilotta et al. also confirmed that the hydrogen-enhanced FCG in a 15-5PH steel is governed by K_{max} rather than by ΔK. By considering this fact, the influence of hydrogen on plasticity at crack tip and a mechanism of consequent FCG enhancement might be different between high strength steels like a 15-5PH steel and ductile steels or a pure iron.

Murakami et al. [21] and Matsuoka et al. [22] have proposed a model of FCG acceleration mechanism as called a "hydrogen-enhanced successive fatigue crack growth (HESFCG)" model based on a hydrogen-induced localization of plasticity. In this model, a steep gradient of hydrogen content causes a localization the plasticity at the crack tip and prevents a crack tip blunting. As a consequence, a crack tip becomes sharp and a crack advances continuously during a rising part of cyclic loading, resulting in the increase of crack advancement per one cycle and the difference in the fracture morphology. The experimental results obtained in this study basically agree with the HESFCG model. Based on the HESFCG model, it is elicited that an accelerated crack advancement does not exceed a maximum value of a Stretched Zone Width (SZW_{max}) defined as: $SZW_{max} = 95\,(1-\nu^2)(K_{max}/E)^2$ [22], where ν is Poisson's ratio and E is an elastic modulus. In the case of Armco iron, ν is 0.3 and E is 207 GPa, hence SZW_{max} is calculated as 9.96×10^{-7} m at $\Delta K = 20$ MPa×m$^{1/2}$. It can be confirmed in Fig. 1 that this value of SZW_{max} is slightly lower but fairly close to the FCGR at $f = 0.2$ Hz in hydrogen of 35 MPa. This fact predicts that this FCGR value already reaches the upper limit of acceleration.

5 Conclusion

In this study, a multi-scale analysis of the plasticity ahead of crack tip in hydrogen-affected fatigue crack propagation has been performed in a commercially pure iron, Armco iron. The main conclusions obtained in this study are shown below.

(1) In a non-accelerated regime, the dislocation structure developed immediately beneath fracture surface consists of a cell structure with the cell diameter of few μm, identical to the one observed in air. This indicates that a certain amount of plastic strain is introduced into grains even in a brittle intergranular fracture. The plastic strain and slip activity in front of a crack tip may explain a mechanism of hydrogen-induced intergranular fracture.

(2) In an accelerated regime, the dislocation structure near the crack tip is a scattered dislocation tangles without any cell or vein structure.

(3) From the results of the OPD measurement, it has been clarified that hydrogen localizes cyclic plasticity ahead of a crack tip in an accelerated regime, on the other hand, hydrogen does not affect monotonic plasticity.

(4) Based on the experimental results mentioned in (2) and (3), the hydrogen-enhanced FCG in the accelerated regime can be understood by a model based on the localization of cyclic plasticity due to hydrogen.

Acknowledgement

Funding from the Poitou-Charentes region (G. Bilotta and T. Shinko PhD thesis grants) is gratefully acknowledged. This work pertains to the French Government program "Investissements d'Avenir" (LABEX INTERACTIFS, reference ANR-11-LABX-0017-01).

References

1. G. Hénaff, G. Odemar, A. Tonneau-Morel, Int. J. Fatigue, **29**, 1927–1940, (2007).
2. Y. Murakami, S. Matsuoka, Eng. Fract. Mech., **77**, 1926–1940, (2010).
3. S.P. Lynch, Metall. Mater. Trans. A, **44**, 1209–1229, (2012).
4. R.A. Oriani, P.H. Josephic, Scr. Metall., **6**, 681–688, (1972).
5. H.K. Birnbaum, P. Sofronis, Mater. Sci. Eng. A, **176**, 191–202, (1994).
6. S.P. Lynch, N.E. Ryan, Proc. Second Int. Congr. Hydrog. Met., 369–376, (1977).
7. G. Bilotta, G. Hénaff, G. Benoit, C. Moriconi, D. Halm, M. Arzaghi, ASME 2015 Press. Vessel. Pip. Conf., 1–8, (2015).
8. G. Bilotta, M. Arzaghi, G. Hénaff, G. Benoit, D. Halm, Vol. 6B Mater. Fabr., V06BT06A026, (2016).
9. Z. Sun, C. Moriconi, G. Benoit, D. Halm, G. Henaff, Metall. Mater. Trans. A, **44**, 1320–1330, (2012).
10. J. Schijve, *Fatigue of Structures and Materials*, Second edi, (2009).
11. J.R. Rice, Fatigue Crack Propag., **ASTM STP 4**, 247–309, (1967).
12. S. Wang, M.L. Martin, P. Sofronis, S. Ohnuki, N. Hashimoto, I.M. Robertson, Acta Mater., **69**, 275–282, (2014).
13. M.L. Martin, B.P. Somerday, R.O. Ritchie, P. Sofronis, I.M. Robertson, Acta Mater., **60**, 2739–2745, (2012).
14. H. Nishikawa, Y. Oda, H. Noguchi, J. Solid Mech. Mater. Eng., **5**, 263–278, (2011).
15. Y. Ogawa, D. Birenis, H. Matsunaga, A. Thøgersen, Ø. Prytz, O. Takakuwa, J. Yamabe, Scr. Mater., **140**, 13–17, (2017).
16. H. Nishikawa, Y. Oda, Y. Takahashi, H. Noguchi, J. Solid Mech. Mater. Eng., **5**, 179–190, (2011).
17. Y. Takahashi, M. Tanaka, K. Higashida, H. Noguchi, Scr. Mater., **61**, 145–148, (2009).
18. Y. Takahashi, J. Sakamoto, M. Tanaka, K. Higashida, H. Noguchi, Scr. Mater., **64**, 157–160, (2011).
19. Y. Takahashi, J. Sakamoto, M. Tanaka, K.

Higashida, H. Noguchi, Scr. Mater., **64**, 721–724, (2011).
20. G. Bilotta, G. Henaff, D. Halm, M. Arzaghi, Int. J. Hydrogen Energy, **42**, 10568–10578, (2017).
21. Y. Murakami, T. Kanezaki, Y. Mine, **39**, 1327–1339, (2008).
22. S. Matsuoka, N. Tsutsumi, Y. Murakami, Trans. Japan Soc. Mech. Eng. Ser. A (in Japanese), **74**, 1528–1537, (2008).

Fatigue Properties of Ultra-Fine Grain Austenitic Stainless Steel and the Effect of Hydrogen

Masanobu Kubota[1,*], *Arnaud* Macadre[1], *Koichi* Mori[2] and *Ryo* Mori[2]

[1]International Institute for Carbon-Neutral Energy Research (WPI-I2CNER), Kyushu University, 744 Motooka, Nishi-ku, Fukuoka, Japan
[2]Graduate School of Mechanical Engineering, Kyushu University, 744 Motooka, Nishiku, Fukuoka, Japan

Abstract. The fatigue properties of ultra-fine grain austenitic steel (UFG16-10), which has a 1 μm average grain size, were studied as part of the project aimed at the development of high-strength low-cost stainless steels for hydrogen service. The fatigue properties of the UFG16-10 were compared with that of a coarse grain material with the same chemical composition (CG16-10) and two kinds of commercial steels, JIS SUS316 and JIS SUH660. The fatigue strength of the UFG16-10 was 2.8 times higher than that of the CG16-10. The effect of hydrogen on the fatigue limit of the UFG16-10 was not significant. However, the fatigue life of the UFG16-10 was reduced by hydrogen in the short life regime. In the fatigue crack growth test, the UFG16-10 showed a good crack growth resistance that was equivalent to that of the SUH660 and significantly higher than that of the SUS316. However, the crack growth rate was significantly accelerated by hydrogen. The cause of the hydrogen-assisted fatigue crack growth of the UFG16-10 was transformation of the microstructure at the crack tip from austenite to strain-induced martensite. This was also the cause of the reduced fatigue life of the hydrogen-charged UFG16-10.

1 Introduction

The studies of ultra-fine grained materials produced by ECAP [1], HPT [2], ARB [3], Warm tempforming [4], HCR [5], etc., are actively being advanced. An austenitic stainless steel, which has a grain size less than 2 μm, has been commercialized since 2013 [6]. Regarding the research on the fatigue properties of the ultra-fine grain materials, a unique fatigue mechanism was found such that the recrystallization at the crack tip during crack propagation occurs in low-melting-point materials such as copper [1, 8] and a magnesium alloy [7]. This resulted in no improvement in the fatigue strength due to the grain growth, whereas the static strength was significantly improved by the grain refinement. On the other hand, a significant increase in the fatigue strength is obtained in carbon steels [9-11] and stainless steels [5, 12] by the grain refinement. Grain refinement using a friction machining tool was applied to the specimen surface, then improved fatigue strength was obtained [13]

However, the mechanism that causes an increase in the fatigue strength in the ultra-fine grain steels remains unclear, for instance, the fatigue strength increases with the degree of grain refinement treatment, but it then decreases with further treatment [9]. In addition, a negative effect on the improvement of the fatigue strength that the crack growth resistance of a low carbon steel is reduced by the grain refinement was also reported [15]. Specifically, the contribution of the fact that "grains are small" to the improved fatigue strength is not sufficiently understood, since there are many other factors which can influence the fatigue behavior such as an increase in the initial dislocation density, phase transformation during grain refinement process, formation of micro-defects, change in distribution of inclusions, etc. Further studies are needed in order to elucidate the fatigue mechanism in the ultra-fine grain materials.

According to the recent studies on hydrogen-assisted fracture related to fuel cell vehicles and hydrogen refuelling stations, 316L austenitic stainless steel was not susceptible to hydrogen during slow strain rate testing and fatigue crack growth test in high-pressure hydrogen [16]. However, the yield strength was relatively low. On the other hand, Cr-Mo steels had a higher yield strength, but hydrogen significantly reduced the ductility [17] and accelerated the crack growth [18]. Therefore, development of a new alloy that have both high-strength and hydrogen compatibility is required.

In this context that grain refinement is a promising method to develop high-strength material and austenitic stainless steels are hydrogen compatible but low strength, the fatigue properties of an austenitic stainless steel having a 1 μm average grain size and the effect of hydrogen were studied.

2 Materials

* Corresponding author: Kubota.masanobu.304@m.kyushu-u.ac.jp

The ultra-fine grain austenitic stainless steel used in this study was produced by a thermo-mechanical treatment which had been developed by Takaki et al. [19, 20]. Figure 1 shows the conditions of the thermo-mechanical treatment. The grain refinement was achieved by reverse transformation of the strain-induced martensite to austenite following a cold rolling with a 90 % reduction rate. Therefore, the chemical composition that was optimized for this grain-refinement process was for a semi-stable austenitic stainless steel that can easily transform to strain-induced martensite during the cold rolling [19]. Table 1 shows the chemical composition of the material. The ultra-fine grain austenitic stainless steel used in this study is mainly consisted of 16 mass% Cr, 10 mass% Ni and Fe (balance). Therefore, the material was labelled as UFG16-10 in this paper.

The grain size was controlled by the temperature during the reverse transformation from strain-induced martensite to austenite. A coarse grain 16Cr-10Ni steel was also produced from the same ingot as the UFG16-10, and was used as the reference material. The average grain size was 21 μm. This material was labeled as CG16-10. In addition, two commercial steels, JIS SUS316 and JIS SUH660, were also used to compare the fatigue properties with the UFG16-10. The SUS316 had Cr and Ni contents similar to that of the 16Cr-10Ni steel. The SUH660, which was an alloy similar to A286, had approximately the same ultimate tensile strength (UTS) as that of the UFG16-10. The heat treatment was solution heat treatment by heating at 1303 K for 10.8 ksec followed by rapid cooling for the SUS316, and precipitation hardening at 833 K for 3.6 ksec followed by air cooling for the SUH660. The chemical composition of the reference materials is also shown in Table 1.

The microstructure of the materials is shown in Fig. 2. Since it was hard to observe 1 μm grains by an optical microscope, the microstructure was observed by electron back scatter diffraction patterns (EBSD). The average grain size of the test materials was 1 μm for the UFG16-10, 21 μm for the CG16-10, 76 μm for the SUS316, and 10 μm for the SUH660.

The mechanical properties of the materials are shown in Table 2. The yield strength and the ultimate tensile strength of the 16Cr-10Ni steel are significantly improved by the grain refinement. For more details of the tensile properties of the UFG16-10, Ref. [21] is available. The UTS of the CG16-10 was comparable to that of the SUS316. The UTS of the SUH660 is the same level as that of the UFG16-10. However, the elongation and reduction of area were significantly higher in the UFG16-10 than in the SUH660. The relation between grain refinement and ductility change depends on the material, grain refinement process and post heat treatment. There are examples of

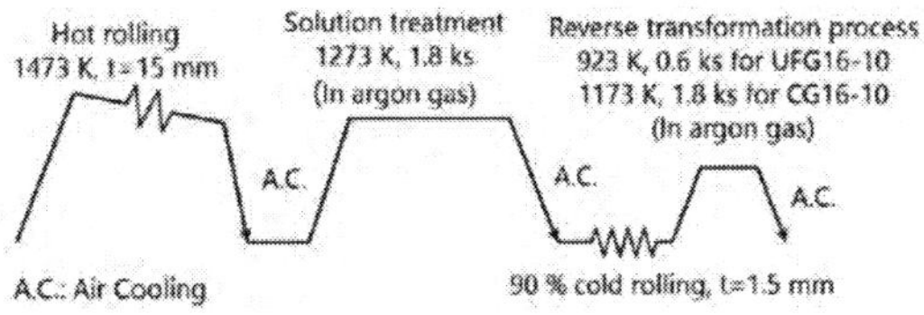

Fig. 1. Thermo-mechanical treatment for production of ultra-fine grain austenitic steel developed by Takaki et al.

Table 1. Chemical composition of the materials (mass%).

Material	C	Mn	Si	P	S	Ni	Cr	Mo	V	Al	Ti	B	N
UFG16-10	0.002	0.08	0.002	0.003	0.002	10.1	16.4	-	-	-	-	-	0.032
CG16-10													
SUS316	0.04	1.30	0.029	0.0036	0.0027	10.01	16.80	2.02	-	-	-	-	-
SUH660	0.03	0.38	0.12	0.017	0.001	24.18	13.75	1.10	0.23	0.15	2.28	0.003	-

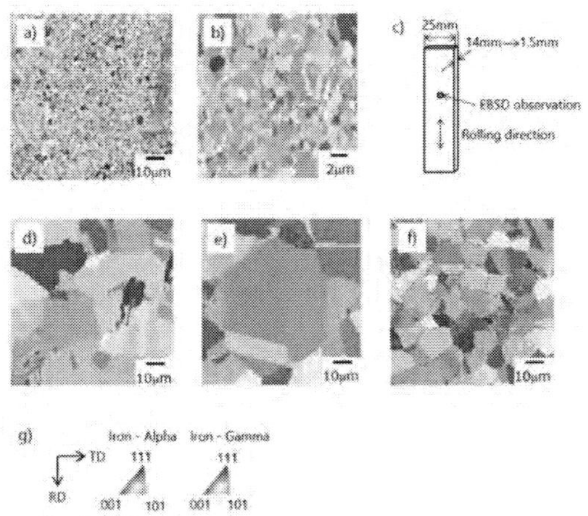

Fig. 2. Microstructure of the ultra-fine grain austenitic steel and reference materials observed by EBSD. The figures are crystal orientation map. a) UFG16-10. The average grain size was 1 μm. b) Magnified image of a). c) Observed area for the UFG16-10. d) CG16-10. The average grain size was 21 μm. e) SUS316. The average grain size was 76 μm. f) SUH660. The average grain size was 10 μm. g) Crystal orientation.

Table 2. Mechanical properties of the materials

Material	Yield strength σ_Y, $\sigma_{0.2}$ (MPa)	UTS σ_B (MPa)	Elongation δ (%)	Reduction of area ϕ (%)	HV1
UFG16-10	585	793	50.0	76.1	295
CG16-10	156	629	66.2	84.0	145
SUS316	242	610	81.8	81.3	163
SUH660	495	827	35.0	46.0	260

both the reduction in ductility [1-3, 5, 8, 12] and increase in ductility [4, 7, 9, 10]. In this study, the grain refinement was effective in improving the tensile strength without a drastic degradation of the ductility.

The effect of hydrogen was characterized by hydrogen-charged specimens. The specimens were exposed to 10 MPa hydrogen gas at 543 K for 72 h. The hydrogen concentrations of the materials measured by a thermal desorption spectrometry (TDS) are shown in Fig.

3. For the measurement, several hydrogen-charged samples with different thicknesses were prepared. As shown in the graph, the hydrogen concentration of the samples was constant regardless of the thickness of the samples. The thickness of the thickest chips was the same as that of the fatigue test specimen. Therefore, it is reasonable to consider that the hydrogen in the fatigue test specimen was uniformly distributed throughout the thickness.

The hydrogen concentration was approximately 22.1 mass ppm for the UFG16-10 and 22.4 mass ppm for the CG16-10. According to the TDS result, there was almost no effect of the grain size on the hydrogen content, although there is the result that grain refinement increased the amount of absorbed hydrogen [22]. Since the volume fraction of the grain boundary rapidly increases with a decrease in the grain size in the region where the grains are finer than 1 μm [23], it can be considered that the volume fraction of the grain boundary in the UFG16-10 was not very different from that in the CG16-10. This is the reason for that the hydrogen contents were almost the same for the UFG16-10 and the CG16-10.

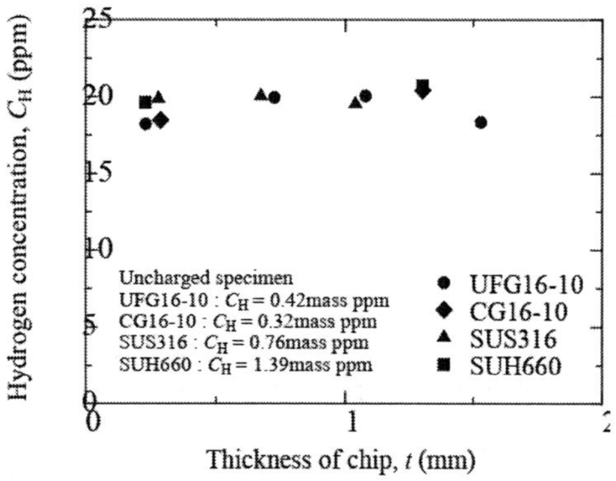

Fig. 3. Hydrogen concentration of the hydrogen charged samples having different thicknesses

3 Effect of hydrogen on fatigue strength

The *S-N* curves were obtained by the specimen shown in Fig. 4. The thickness of the specimen was 1.5 mm. Small specimens are often used in the study of ultra-fine grain materials [2, 10], however the size of the ultra-fine grain steel used in this study was bulk size. The specimen had blunt notches, which stress concentration factor was 1.47, to prevent buckling of the specimen during the compression loading and to avoid heating of the specimen during the fatigue loading. It was confirmed that there was no significant increase in the temperature of the specimen during the fatigue test. The surface of the notch root was finished by buffing.

The fatigue tests were carried out under a tension-compression loading with a stress ratio, R, of -1 at a loading frequency, f, of 15 Hz at room temperature in laboratory air.

Figure 5 shows the *S-N* curves of the UFG16-10 and the reference materials. The stress amplitude in the graph is the nominal value at the notch root. The fatigue limit of the UFG16-10 was 2.8 times higher than that of the CG16-10. The grain refinement was very effective in improving not only the static strength [21] but also the fatigue strength of the 16Cr-10Ni steel. Furthermore, the UFG16-10 had a higher fatigue strength than the SUH660 despite the fact that the ultimate tensile strength of the SUH660 is slightly higher than that of the UFG16-10.

The fatigue strength of the CG16-10 was very similar to that of the SUS316. This is reasonable when considering the similar tensile strengths of these materials and the empirical relationship between the ultimate tensile strength and fatigue limit.

The effect of the hydrogen on the fatigue strength is also shown in Fig. 5. The reduction in the fatigue limit of the UFG16-10 by the hydrogen charge was not significant. For austenitic stainless steels, work-hardening is frequently used to improve mechanical properties. However, in some cases, the work-hardened austenitic stainless steels severely suffer from hydrogen-assisted fatigue failure [24, 25]. The result of this study clearly demonstrated the significant potential of the UFG steel to be a high-strength austenitic steel for hydrogen service systems use.

On the other hand, the fatigue life of the UFG16-10 in the short life region was significantly reduced by the hydrogen. In the short life region, crack growth is dominant rather than crack initiation, the crack growth behavior of the materials were investigated.

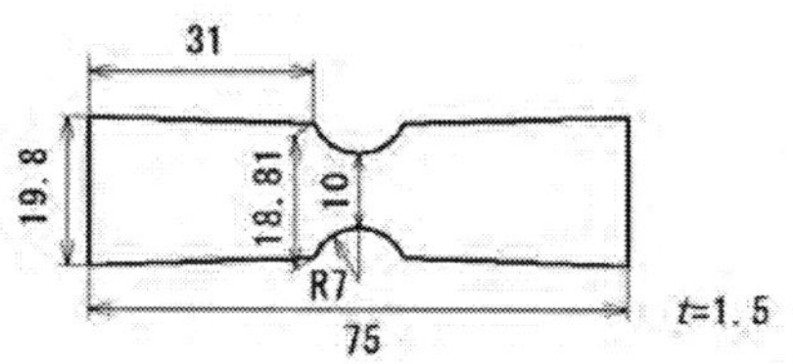

Fig. 4. Fatigue test specimen for *S-N* test

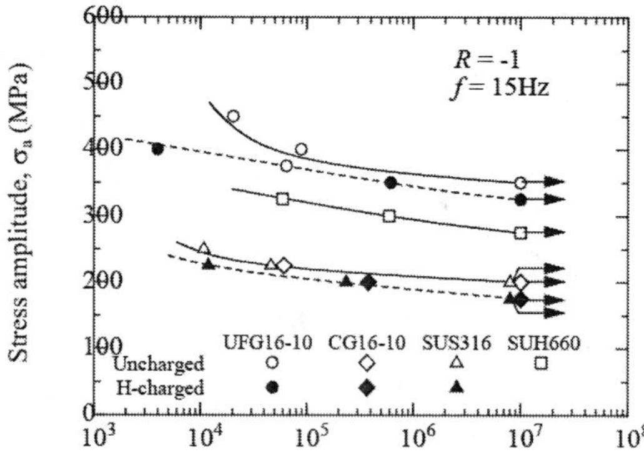

Fig. 5. Fatigue strength of the UFG16-10 and the effect of hydrogen

4 Effect of hydrogen on fatigue crack growth

The crack growth test was carried out using a CT specimen shown in Fig. 6. According to the material processing method shown in Fig. 1, there is no limitation of the thickness of the UFG16-10. However, for our experiment, the maximum thickness of the UFG16-10 specimen was 1.5 mm because of the capacity of the rolling mill we have.

Therefore, the thickness of the CT specimen was also 1.5 mm. Even so, plain strain condition was achieved in the UFG16-10 and the SUH660 within the stress intensity factor range that the experiment was done because the yield strengths of both materials were sufficiently high. For the SUS316, the plain stress condition was not achieved. Since crack growth resistance in the plain stress condition is higher than that in the plain strain condition [26], it should be noted that the crack growth resistance of the SUS316 in this study was apparently high.

The crack growth test was performed at f = 5 Hz and 0.5 Hz with R = 0.1 at room temperature in laboratory air. The slower loading frequency was adopted by considering the experimental fact that the effect of hydrogen on the fatigue crack growth behavior was enhanced with a decrease of the loading frequency [27]. The crack length was measured by an unloading elastic compliance method using a back face strain gauge.

Figure 7 shows the results of the crack growth test. For the uncharged materials, the crack growth rate of the UFG16-10 was remarkably low compared to that of the SUS316, and it was equivalent to that of the SUH660. It is presumed that this higher crack growth resistance contributed to the significantly improved fatigue strength of the UFG16-10.

Regarding crack growth resistance, there is a report that grain refinement has a negative impact on the crack growth resistance [15]. The mechanism that achieved the excellent crack growth resistance in the UFG16-10 was discussed in the separated paper [28]. In brief, the crack tip opening displacement (CTOD) was reduced by the small grains as shown in Fig. 8. According to the established crack growth theory, crack growth rate is proportional to the CTOD [29]. The reason why the CTOD was reduced by the small grain was the reduction of stress concentration at the grain boundary ahead of the crack tip due to piling up of dislocations and consequent reduction in the plastic deformation at the crack tip. The CTOD is proportional to the square of the stress intensity factor, and the plastic zone size at the crack tip is also proportional to the square of the stress intensity factor [30]. Then, the CTOD is proportional to the plastic zone size.

For the hydrogen charged materials, the crack growth rate of the UFG16-10 was accelerated by hydrogen. It was consistent with the significant reduction in the fatigue life of the UFG16-10 in the short fatigue life regime due to hydrogen shown in Fig. 5. The acceleration became more significant with the decrease in the loading frequency.

The accelerated fatigue crack growth was also found in the SUS316. On the other hand, the SUH660 did not suffer from hydrogen-assisted crack growth. The acceleration of the crack growth by hydrogen was grater in the order of the UFG16-10, SUS316 and SUH660. According to the past study, stability of austenite was closely related to the hydrogen-assisted fatigue crack growth [16]. Nickel equivalent that is an index of austenite stability that was calculated by Oshima's equation [31] was 18.3 for the UFG16-10, 20.2 for the SUS316 and 32.0 for the SUH660. The fact that the order of the nickel equivalent values was consistent with that of the degree of the acceleration of the crack growth implied the accelerated crack growth of the UFG16-10 was caused by the transformation of microstructure to strain-induced martensite at the crack tip during the crack growth.

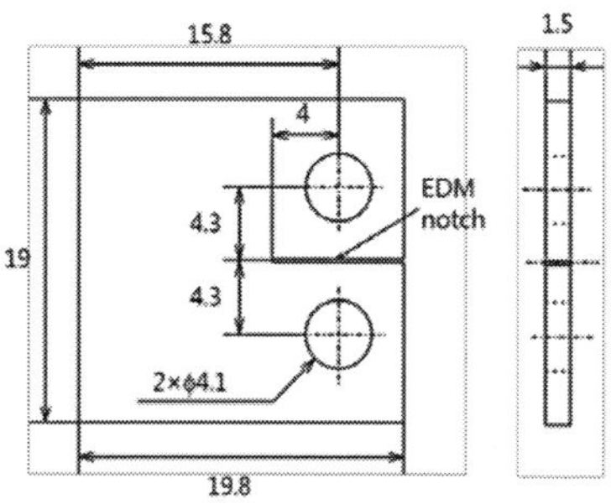

Fig. 6. CT specimen for fatigue crack growth test

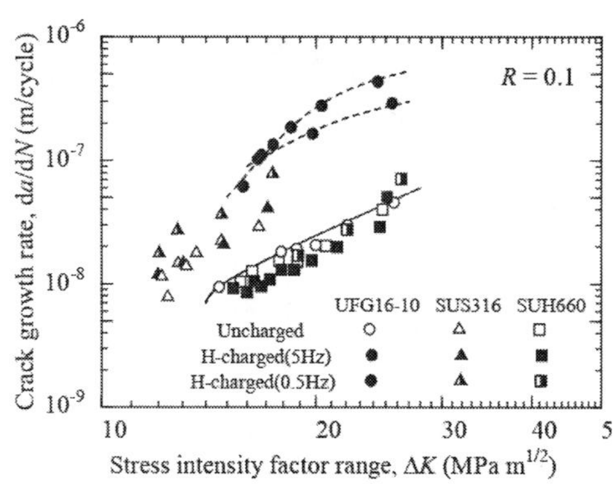

Fig. 7. Results of crack growth test and the effects of hydrogen and loading frequency

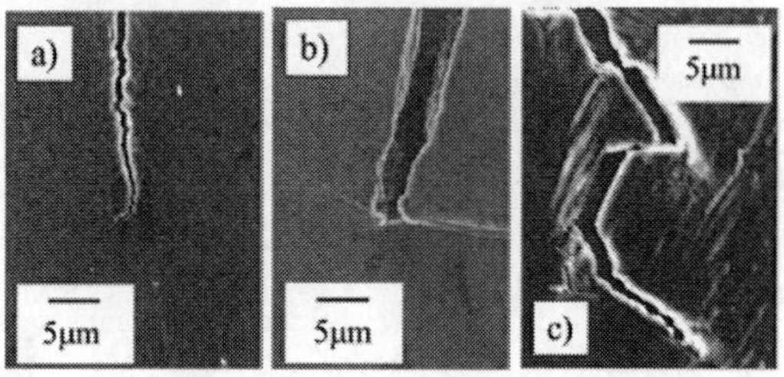

Fig. 8. Crack tip opening displacement at similar ΔK. a) UFG16-10. $\Delta K = 22.1$ MPa m$^{1/2}$. b) SUS316. $\Delta K = 19.3$ MPa m$^{1/2}$. c) SUH660. $\Delta K = 23.1$ MPa m$^{1/2}$.

Figure 9 shows the fracture surface of the UFG16-10. The fracture surface of the uncharged material showed a ductile feature and transgranular crack path. In the case that the grain refinement reduced the crack growth resistance, the fracture surface showed brittle feature [18]. Therefore, the good fatigue crack growth behavior of the uncharged UFG16-10 could be achieved by maintaining ductile crack growth.

The fracture surface of the hydrogen-charged material exhibited a granular pattern. The morphology of the fracture surface was clearly different. As shown in Fig. 9 c), the unit size of the granular pattern was almost equivalent to the grain size. It revealed the crack growth mechanism was changed to an intergranular cracking. Cracking at twin boundary might also be mixed. It was obvious that the change in the crack growth mechanism due to hydrogen caused the accelerated crack growth in the hydrogen charged UFG16-10.

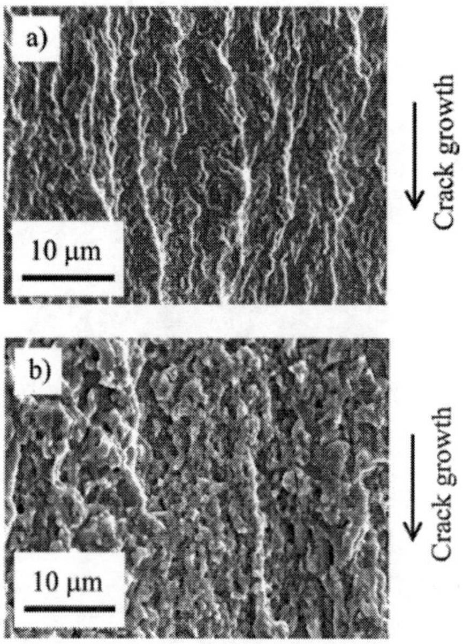

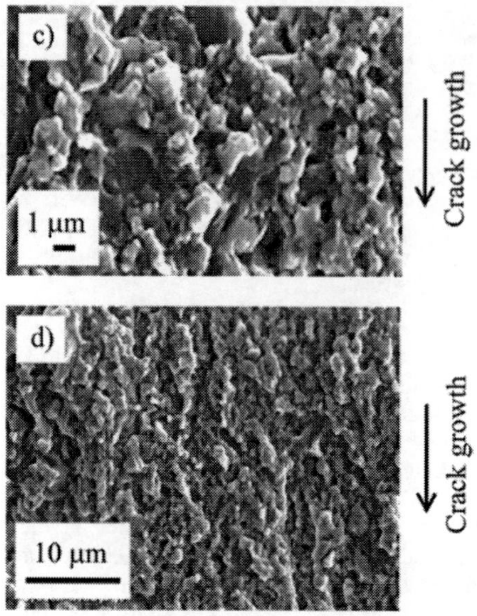

Fig. 9. Fracture surfaces of the UFG16-10 observed at $\Delta K = 20$ MPa m$^{1/2}$. a) Uncharged. b) Hydrogen charged, $f = 5$ Hz. c) Enlarged view of b). d) Hydrogen charged, $f = 0.5$ Hz.

5 Effect of hydrogen on crack initiation

The fatigue failure process can be divided into two stages, crack initiation and crack propagation. Thus, the effect of hydrogen on the crack initiation behavior of the UFG16-10 was also investigated.

The fatigue test for the observation of crack initiation was carried out under a tension and compression loading at $f = 5$ Hz with $R = -1$ at room temperature in laboratory air. Figure 10 shows the specimen. The specimen has blunt notches to facilitate crack observation. The notch root was observed by a surface replication technique. The notch root was finished by buffing. Since the effect of hydrogen on the fatigue life of the UFG16-10 was significant in the short life regime as shown in Fig. 5, the crack initiation behavior was observed in the short life regime.

The crack initiation behavior of the UFG16-10 is shown in Fig. 11. The fatigue crack originated from a non-metallic inclusion. The reproducibility was confirmed by multiple specimens. The size of the inclusion was approximately 2 μm. It was very small compared with the sizes of the inclusions that lead crack initiation in other commercial high-strength steels [32]. In addition, this type of crack initiation is found in the high-strength steels having the UTS more than 1.2 GPa or the hardness more than HV400 [33]. Therefore, it might be considered special mechanisms due to small grains to the crack initiation of the UFG16-10.

Figure 12 shows the crack initiation behavior of the hydrogen-charged UFG16-10. The fatigue life of this hydrogen-charged specimen was significantly shorter than that of the uncharged specimen shown in Fig. 11.

https://doi.org/10.1051/matecconf/201816503007

Similar to the uncgarged material, the crack originated from an nonetallic inclusion.

Figure 13 shows the crack length as a function of the number of cycles. Whereas the fatigue life was significantly reduced by hydrogen, the crack initiation lifves of the hydrogen-charged and uncharged materials were not much different. In the subsequent crack propagation of the crack length reached approximartely 10 μm, the small crack of the hydrogen-charged material grew fatser than that of the uncharged one. The results of the observations of the crack initiation and small crack propagation showed that hydrogen had no significant effect on the crack initiatin of the UFG16-10 and accelerated the small crack propagation. As shown in Fig. 7, the propagation of the large crack was also accelelated by hydrogen. Therefore, the reduced fatigue life of the hydrogen-charged UFG16-10 was due to acceleration of the crack growth by hydrogen.

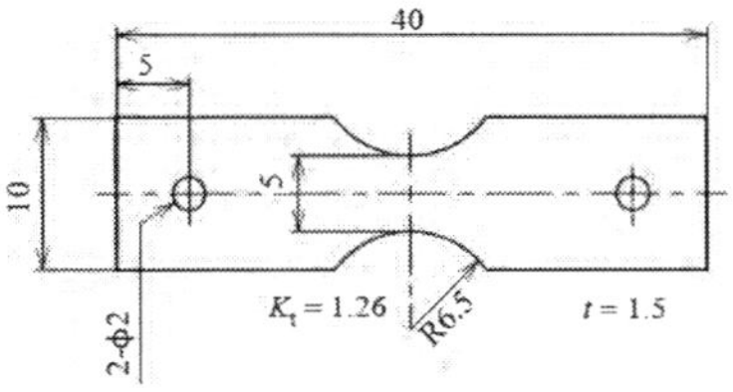

Fig. 10. Fatigue test specimen for identification of crack origin

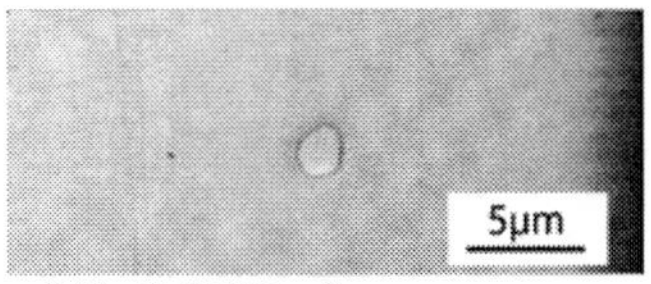

a) $N = 0, N/N_f = 0$

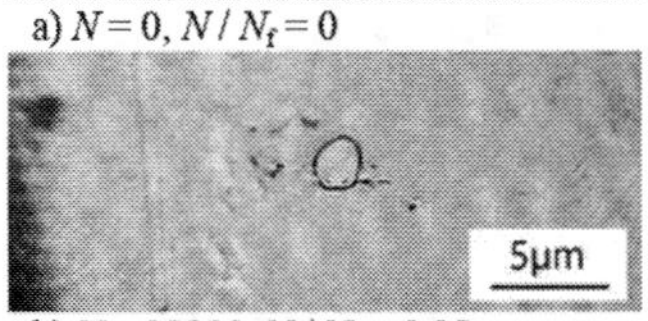

b) $N = 30000, N/N_f = 0.28$

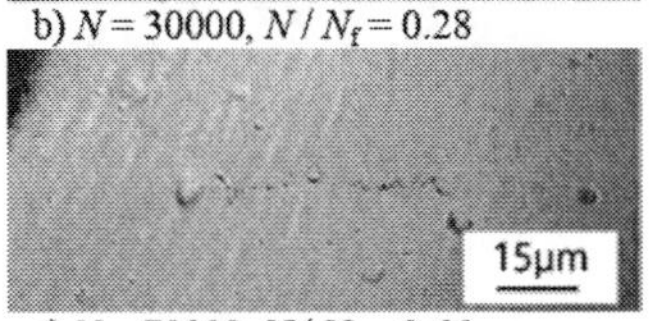

c) $N = 70000, N/N_f = 0.66$

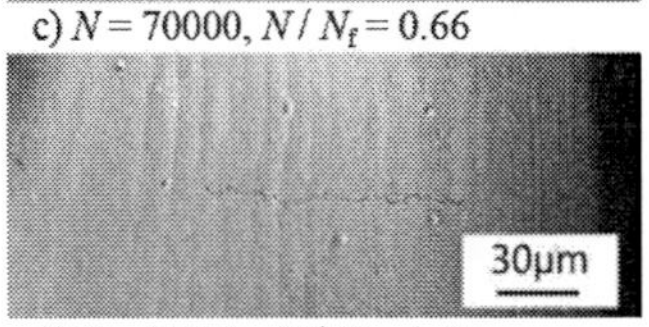

d) $N = 80000, N/N_f = 0.75$

Fig. 11. Crack initiation of the uncharged UFG16-10. The stress amplitude at the notch root was 660 MPa. The fatigue life was 106500 cycles.

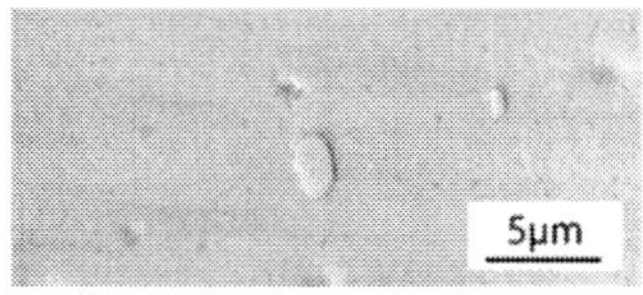

a) $N = 0, N/N_f = 0$

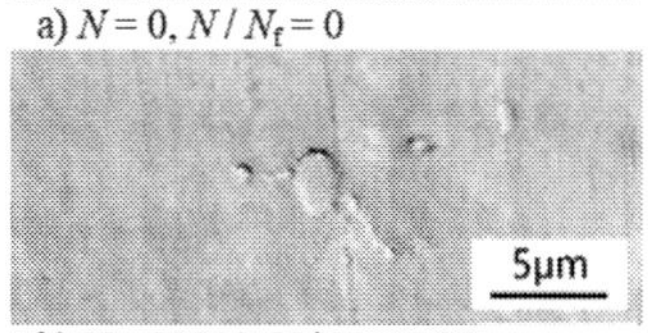

b) $N = 20000, N/N_f = 0.24$

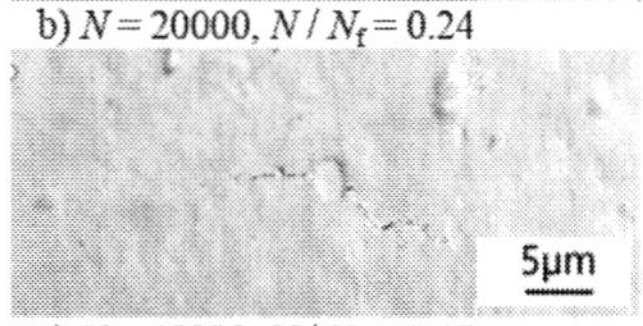

c) $N = 40000, N/N_f = 0.47$

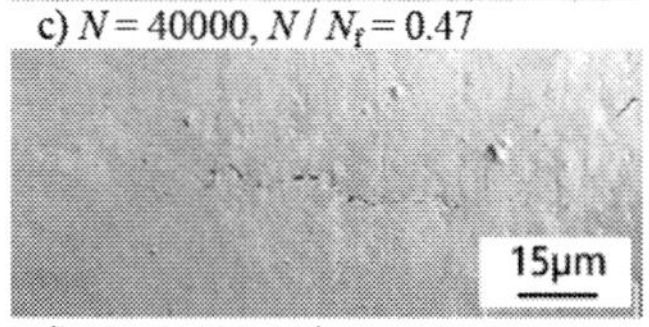

d) $N = 60000, N/N_f = 0.72$

Fig. 12. Crack initiation of the hydrogen-charged UFG16-10. The nominal stress amplitude at the notch root was 523 MPa. The fatigue life was 83500 cycles.

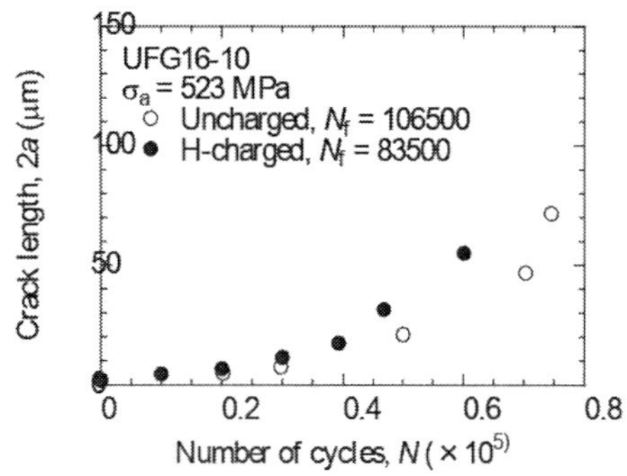

Fig. 13. Effect of hydrogen on the crack initiation and the small crack propagation of the UFG16-10.

6 Conclusion

The effect of hydrogen on the fatigue properties of the ultra-fine grain austenitic stainless steel (UFG16-10) was studied.

(1) The fatigue limit of the uncharged UFG16-10 was 2.8 times higher than that of the CG16-10. The fatigue limit of the UFG16-10 was reduced by hydrogen, however, the reduction was no so significant.

(2) The fatigue life of the UFG16-10 was significantly reduced by hydrogen.

(3) The crack growth resistance of the uncharged UFG16-10 was significantly higher than that of SUS316 and equivalent to that of the SUH660.

However, the severe hydrogen-assisted crack growth occurred in the UFG16-10.

(4) The crack origin was a non-metallic inclusion for both hydrogen-charged and uncharged specimens. There was no significant change in the crack initiation life due to hydrogen.

(5) The cause of the reduced fatigue life of the hydrogen-charged specimen was the accelerated crack growth by hydrogen.

The ultra-fine grain austenitic stainless steel used in this study was provided by Professor Setuo Takaki at the Research Center for Steel, Kyushu University. All members in the Professor Takaki's laboratory provided us excellent support for this research. This study was also supported by the World Premier International Research Center Initiative (WPI), MEXT, Japan. The International Institute for Carbon-Neutral Energy Research (WPI-I2CNER) is supported by the World Premier International Research Center Initiative (WPI), MEXT, Japan.

References

1. M. Goto, N. Teshima, S. Z. Han, K. Euh, T. Yakushiji T, S. S. Kim, J. Lee, Engng Fract. Mech. **110**, 218 (2013)
2. J. Horky, G. Khatibi, B. Weiss, M. J. Zehetbauer, J Alloys and Compounds **509** (1), 323 (2011)
3. H. Kitahara, S. Matsushita, M. Tsushida, S. Ando, Materials Transactions **54** (4), 528 (2013)
4. N. Nagashima, Y. Furuya, Y. Kimura, T. Inoue, Trans. JSME **A78**-790, 923 (2012)
5. S. Brochet, A. B. Quintin, J. B. Vogt, J. C. Glez, J. D. Mithieux, Proc. ECF-16, Alexandroupolis, Greece, 2006.
6. Nippon steel and Sumitomo metal, Press release 2013, http://www.nssmc.com/news/20131203_100.html, Refered on March 23th, 2018
7. F. Akbaripanah, F. Fereshteh-Saniee, R. Mahmudi, H. K. Kim, Mater. Sci. Engng **A565**, 308 (2013)
8. S. Z. Han, M. Goto, C. Lim, C. J. Kim, S. Kim, Alloys and Compounds **434-435**, 304 (2007)
9. J. C. Pang, M. X. Yang, G. Yang, S. D. Wua, S. X. Li, Z. F. Zhang, Mater. Sci. Engng **A553**, 157 (2012)
10. M. Okayasu, K. Sato, M. Mizuno, D. Y. Hwang, D. H. Shin, Int J Fatigue **30**, 1358 (2008)
11. C. Ruffing, E. Kerscher, Adv. Mater. Res. **891-892**, 428 (2014)
12. H. Ueno, K. Kakihata, Y. Kaneko, S. Hashimoto, A. Vinogradov, Acta Materialia **59**, 7060 (2011)
13. K. Suzuki, T. Shimizu, Y. Todaka, K. Koujima, T. Yakushiji, M. Nakajima, J Japan Soc. Abrasive Technology **56**-2, 112 (2012)
14. J. H. Kim, S. K. Hwang, Y. T. Im, I. H. Son, C. M. Bae, Mater. Sci. Engng **A552**, 316 (2012)
15. H. K. Kim, M. I. Choi, C. S. Chung, D. H. Shin, Mater. Sci. Engng **A340**, 243 (2003)
16. H. Itoga, T. Matsuo, A. Orita, H. Matsunaga, S. Matsuoka, Trans JSME **A79**-808, 1726 (2013)
17. H. Matsunaga, M. Yoshikawa, R. Kondo, J. Yamabe, S. Matsuoka, Hydrogen Energy **40**, 5739 (2015)
18. T. Miyomoto, T. Matsuo, N. Kobayashi, Y. Mukaie, S. Matsuoka, Trans. JSME **A78**-788, 531 (2012)
19. K. Tomimura, S. Takaki, Y. Tokunaga, Tetsu-to-Hagane **74**-8, 1649 (1988)
20. S. Takaki, S. Tanimoto, K. Tomimura, Y. Tokunaga, Tetsu-to-Hagane **74**-6, 1052 (1988)
21. A. Macadre, K. Tsuboi, N. Nakada, T. Tsuchiyama, S. Takaki, Procedia Mater. Sci. **3**, 350 (2014)
22. M. Noda, S. Ishida, K. Funami, J Japan Inst. Light Metals **59**-1, 13 (2009)
23. N. Tsuji, Tetsu-to-Hagane **88**-7, 359 (2002)
24. E. Takeuchi, M. Hayakawa, N. Nagashima, Y. Furuya, S. Matsuoka, Trans. JSME **A73**-736, 1335 (2007)
25. Y. Kondo, M. Kubota, K. Ohguma, K. Shimada, Trans. JSME **A73**-736, 1351 (2007)
26. K. Tokaji, Z. Ando, K. Nagae, Trans. JSME **52**-475, 655 (1986)
27. M. Yoshikawa, T. Matsuo, N. Tsutsumi, H. Matsunaga, S. Matsuoka, Trans. JSME **80**-817, SMM0254 (2014)
28. M. Kubota, M. Arnaud, K. Mori and R. Mori, ISIJ internatioal, under review
29. R. Pippan, C. Zelger, E. Gach, C. Bichler and H. Weinhandl, FFEMS **34**, 1 (2010)
30. N. E. Frost, K. J. Narsh, L. P. Pook, Metal fatigue, Dover Publications, Inc., 1999.
31. T. Oshima, Y. Habara, K. Kuroda, Tetsu-to-Hagane **93**-8, 544 (2007)
32. Y. Murakami, Metal fatigue : Effect of small defect and nonmetallic inclusions, Elsevier, 2002
33. M. Saito, K. Ito, Trans. JSSE, **30**, 11 (1985)

Fatigue Crack Growth Behaviour of High Strength Ferritic Steels in High Pressure Hydrogen

Ashok Saxena[1,*] and *Kevin* Nibur[2]

[1]Department of Mechanical Engineering, University of Arkansas, Fayetteville, AR, 72701, USA
[2]Hy Performance Materials Testing, LLC, Bend, OR 97701, USA

Abstract. The design of safe and low-cost, high-pressure hydrogen storage systems are a critical need for harnessing clean power but must consider the propensity of hydrogen to accelerate fatigue crack growth rates in the construction materials. Design of safe pressure vessels needs robust models for predicting crack growth rates and how they are affected by variables such as loading frequency, load ratios, hydrogen pressure, gaseous impurities, temperature, and material variability. In this study, fatigue crack growth rates were measured in the liner material in 10 MPa gaseous hydrogen at various load ratios, R, in the range $-1 \leq R \leq 0.2$. The effects of varying loading frequency were investigated, and the results were pooled with those from literature for similar alloys tested in 103 MPa gaseous hydrogen pressure. The differences in crack growth rates between H_2 pressures of 10 to 103 MPa as well as the effects of frequency on the environment assisted crack growth rates were assessed. Loading frequency effects tend to saturate at frequencies of 1 Hz and less. H_2 pressure effects appear to saturate at pressures of 45MPa, while load ratio effects are not significant for $-1 \leq R \leq 0.2$ but become important for $R \geq 0.2$.

1. INTRODUCTION

Safe and low-cost, high-pressure hydrogen storage systems are a critical need for refueling stations for fuel-cell powered vehicles, for back-up power in residential and office buildings, and for fork-lifts in warehouses. Also, excess energy during high generation periods from renewable sources such as sun and wind can be stored in the form of compressed hydrogen that can be made available to power fuel cells during the low generation periods. H_2 powered vehicles can reduce harmful pollutants in large metropolitan areas.

Type I, steel pressure vessels that have been in use in the industry for decades to store hydrogen on ground are constructed from seamless high strength low alloy steels, but are restricted to maximum pressures of 450 bar. This is due to hardenability considerations that limit the maximum permissible wall thickness of the vessel. For onboard vessels, their weight also becomes a limiting factor. Therefore, the challenge is to continue to use steel tanks for storing hydrogen that have been time-tested, but limit the maximum stresses in the interior of the vessel via good design to levels that have been successfully used in the past.

Calculations show that the maximum operating pressure of commercially available steel pressure vessels can be approximately doubled by incorporating a hoop wrap using high strength wires with tensile strength of 2 to 4 GPa [1,2]. The wires are held together by epoxy that otherwise does not have a strucrural role. These designs also include subjecting the vessels to hydro-pressures that are about twice the maximum design pressure during manufacturing to produce plastic strains in the liner while maintaining elastic behavior in the high strength steel wire wrap [1,2]. This process is known as autofrettage and the pressure used is called the autofrettage pressure. Upon removal of the autofrettage pressure, compressive stresses are locked into the wall of the steel liner. This, along with effective transfer of stresses from the steel liner on to the wire wrap, reduces the maximum stress in the liner to safe levels at which these vessels have decades of service history with an excellent safety record.

The American Society for Mechanical Engineers (ASME) has developed codes [3,4] for designing pressure vessels for storing hydrogen but these codes are conservative because of gaps in our ability to confidently model the degradation kinetics of hydrogen embrittlement in these steels. There has been progress in our understanding of mechanisms of hydrogen embrittlement in the crack tip process zone [5] but not sufficient to allow development of robust models for predicting crack growth rates and how they are affected by variables such as loading frequency, load ratios, hydrogen pressure, gaseous impurities, temperature, and material variability. This need has been partially met by empirical studies conducted by Sandia National Laboratory and by Japanese and European research programs[6-8].

* Corresponding author: asaxena@uark.edu

However, many more such studies must be conducted to meet the overall need of the designers of these vessels. For example, the acceleration of fatigue crack growth rates have been quantified for tension only loading conditions, but autofrettaged pressure vessels experience compression to tension load cycles during service [9].

2. MATERIALS AND EXPERIMENTAL PROCEDURE

Metal liners with outside diameter of 406 mm and a nominal wall thickness of 31.75 mm that were made from SA 372 (or A372) Grade J, Class 70 steel were procured for testing. The test specimens for generating fatigue crack growth rate data were extracted from one of these liners. The material chemistry and all mechanical properties met the standard requirements for this grade of material. The yield strength was 770.75 MPa, the UTS was 884.5 MPa and percent elongation was 21.0%. The microstructure consisted of the typical tempered martensite resulting from an oil quenched tempered low alloy, medium carbon steel.

Single edge crack tension, SEC(T) specimens, Fig. 1, were machined from curved panels removed from the liner such that the crack orientation was circumferential – radial (C-R). This orientation matches with the potential crack growth direction, axial-radial, in the pressure vessel. The specimens and their sizes were especially designed to test material that was directly taken from the liner and were capable of being tested under tension-compression loading. For more details on the specimen and its design, refer to reference [10]. The radial direction was along the crack plane and the loading direction was circumferential. A few 50.8 mm C(T) specimens were also machined. All SEC(T) specimens were labelled WT-xx and the C(T) specimens were labelled WT-C(T)-xx.

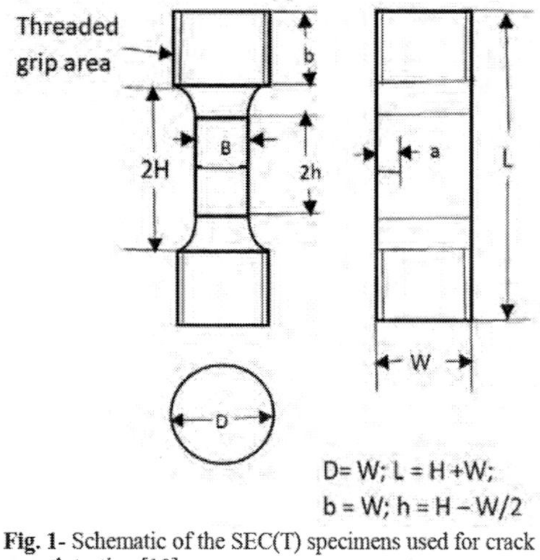

Fig. 1- Schematic of the SEC(T) specimens used for crack growth testing [10]

3. RESULTS AND DISCUSSION

In this section, the experimental results from this study and those from previously published studies are used to explore the kinetics of environment assisted cracking in A372 Grade J, Class 70 tempered martensitic steels for use in hydrogen storage applications. Figure 2 shows the effects of 10 MPa, H_2 on the FCGR behavior of our test material at a load ratio of 0.2 compared to the same behavior in air and in helium (He) environment. The results show that the crack growth rates are higher in air than in He and substantially higher in H_2 at 10 MPa. These differences underscore the importance of susceptibility of this material to environment assisted cracking and the need to fully understand it. It is also important to systematically explore the effects of variables such as load ratio, R, loading frequency, ν, and hydrogen pressure on the FCGR behavior.

3.1 Effects of Load Ratio on FCGR Behavior

Fatigue crack growth rate tests on SEC(T) specimens were conducted under negative load ratios of -1.0 and -0.5 as well as positive load ratios of 0.1 and 0.2. Positive load ratio data from these tests were compared with data from prior studies conducted by Sandia [11,12] as shown in Fig. 3. The ASME trend line [3] for FCGR behavior in air is also provided for reference. Fatigue crack growth rate data obtained on SEC(T) specimens from this study and Sandia's data obtained on C(T) specimens appear to be identical at load ratios of 0.1 and a H_2 pressure of 10 MPa. At R = 0.2 our data was at a pressure of 10 MPa while the Sandia data was at 103 MPa; the FCGR behavior at the higher pressure does show a distinct tendency for higher crack growth rates. The crack growth rates at R=0.5 are higher than the crack growth rates at R=0.1 and 0.2 even after reconciling the differences in pressure.

Figure 4 shows a comparison between FCGR behavior at negative load ratios of -0.5 and -1.0, and the positive load ratios of R=0.1 and 0.2. All data in this figure are from the current study and have been generated at a H_2 pressure of 10 MPa. The data show that there is no systematic difference between the FCGR behavior between R values of -0.5 and -1.0. This is entirely expected because negative loads do not contribute to the value of ΔK and if there are any plasticity induced crack closure effects that influence the value of the effective ΔK, they should have saturated for R =-0.5. We note that the R= 0.1 data also blends with the trends for R = -0.5 and -1.0 data. One of the tests at R=0.2 yielded crack growth rates that are very comparable to the rates for R = -1.0, -0.5, and 0.1 while the other showed a somewhat higher rate. This indicates that at R =0.2, the FCGR behavior of A372 Grade J Class 70 steels begins to

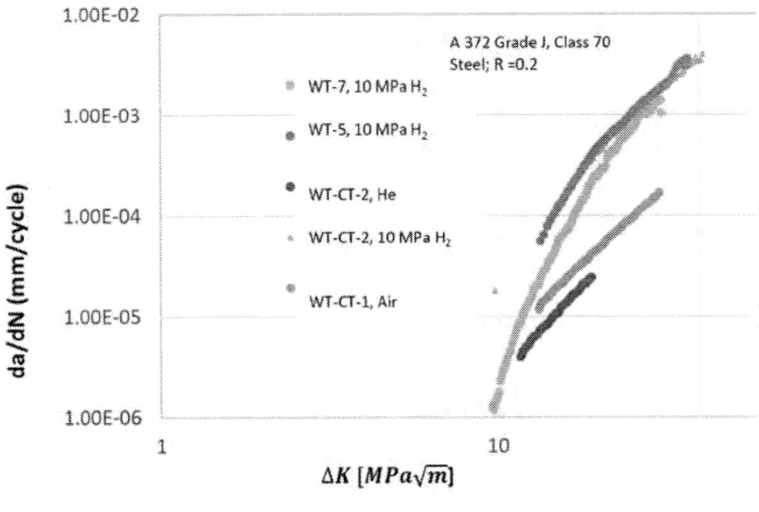

Fig. 2- FCGR behavior of A 372 Grade J, Class 70 steels in He, air, and H_2 at a load ratio, R = 0.2.

become sensitive to R in H_2 environment. For R < 0.2, and continuing into negative R values, the FCGR behavior is not sensitive to R even in H_2 environment. However, in comparison with the crack growth rates in air, the rates in H_2 are consistently higher. For $\Delta K > 20$ MPa$\sqrt{m}$, the differences between air and H_2 are an order of magnitude or more but for $\Delta K < 10$ MPa$\sqrt{m}$, the differences seem smaller.

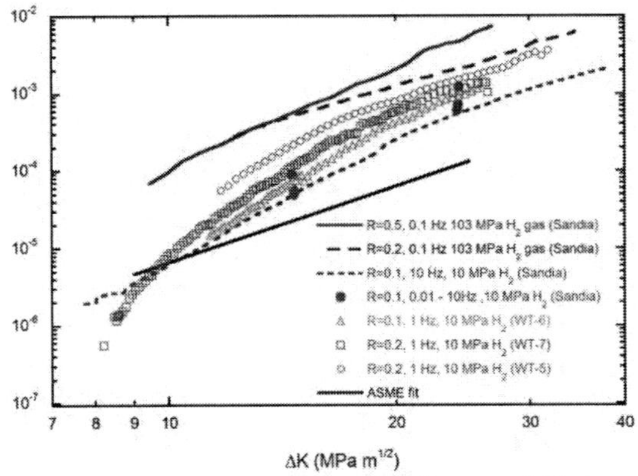

Fig. 3- Fatigue crack growth rates of the liner material in H_2 environment at positive load ratios. The data referred to as Sandia data is from references [11,12]

Figure 5 shows a plot of FCGR for values of ΔK of 10, 20, and 30 MPa$\sqrt{m}$ for R values of -1.0, 0.5, 0.1, 0.2, and 0.5. This plot confirms that the dependence of FCGR on R in high pressure H_2 begins to accelerate at R = 0.2. An unanswered question is whether for R > 0.5, the FCGR continues to increase or if it saturates to a plateau level once R reaches 0.5. Some recent work shows some effects of load ratio even when it increases from 0.5 to 0.7[13]. From applications point of view, this is a critical question that must be addressed in future studies. Also, as mentioned before, the data for R≤ 0.2 was obtained at a H_2 pressure of 10 MPa from this study, and at R =0.5, the test pressure was 103 MPa [11,12]. The da/dN for R =0.5 is higher than for lower load ratios but since it was at higher H_2 pressure, we are unable to separate the effects of R from those due to pressure.

Figure 6 shows all the available data from our study for -1.0 ≤ R ≤ 0.2 to characterize scatter and produce trends for use in design. The figure also shows the 95% confidence upper bound trend and the mean behavior from the data. The R =0.5 and R = 0.2 trend from Sandia data at 103 MPa H_2 pressure and the ASME trend in air are also included for comparison. The fatigue crack growth rate constants are given in an earlier publication [9]

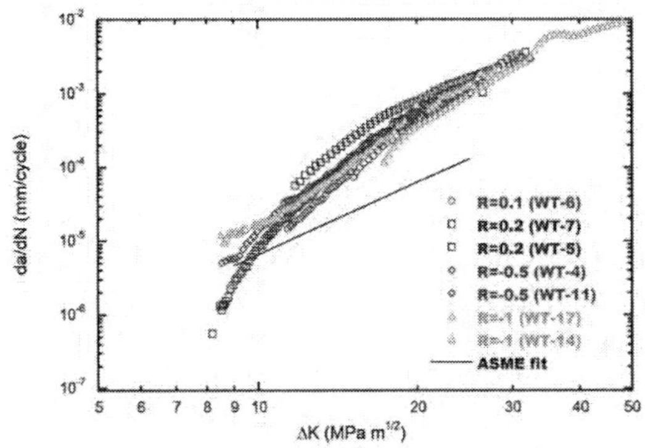

Fig. 4- Fatigue crack growth rate in the liner material for -1.0 ≤ R ≤ 0.2 in 10 MPa hydrogen. The trend labelled ASME fit is for air environment.

* Corresponding author: asaxena@uark.edu

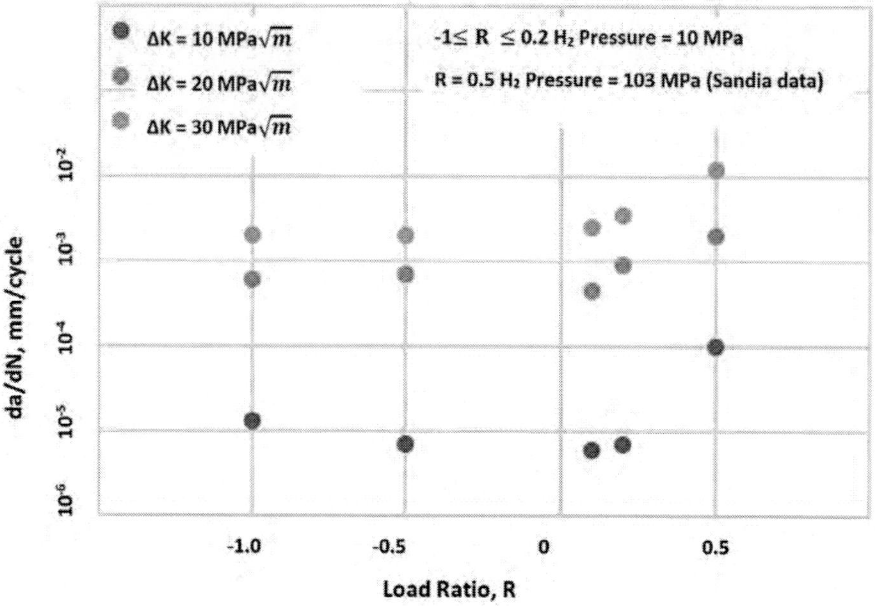

Fig. 5- FCGR behavior of A372 Grade L Class 70 as a function of load ratio R for three values of ΔK

3.2 Effects of Loading Frequency:

The effects of loading frequency were investigated in previous studies on A372 Grade J Class 70 steels [12] in the range of 10 to 0.001 Hz at R =0.1 and H_2 pressures of 10, 45, and 103 MPa for ΔK levels of 15 and 24 MP$\sqrt{m}$. As the loading frequency decreased, the crack growth rates increased modestly. The largest increase in crack growth rates were observed when the loading frequency decreased from 10 Hz to 1 Hz, by approximately a factor of 3. Between 1 Hz and 0.001 Hz, the increase in crack growth rates were small, especially in relation to random scatter in the data. Also, the effects of frequency appeared to be stronger at H_2 pressure of 10 MPa than at 45 and 103 MPa. The results from our studies are shown in Fig.7. Fatigue crack growth rates at a H_2 pressure of 10 MPa in the range of $0.01 \leq \nu \leq 10$ Hz at R = 0.1 and R =-1.0 are shown for four values of ΔK that were nominally 30, 25, 15, and 9 MPa$\sqrt{m}$. The actual ΔK values varied from the nominal values and are noted on Fig. 7. These data clearly show only a small effect of frequencies ranging from 1 Hz to 0.01 Hz on the crack growth rates at all ΔK levels. It is thus concluded that tests conducted at frequencies of 1 Hz or slower yield crack growth rates that realistically include the effects of H_2 pressure on the FCGR behavior. This conclusion is of practical importance when choosing the loading conditions for generating design relevant data.

3.3 Effects of Hydrogen Pressure on FCGR Behavior:

The effect of H_2 pressure on the fatigue crack growth behavior was investigated by Somerday et al. [11, 12] so their results are included here. It is well known that the effects of, load ratio, loading frequency, and pressure can be synergistic so it is impossible to totally isolate the effects of one variable from that of the others. However, it is possible to get some sense of the relative importance of these variables from experimental trends. Figure 8 shows the effect of varying H_2 pressure on the FCGR behavior of A 372 Grade J, Class 70 steels in which the frequencies have been varied by 4 orders of magnitude. Varying frequency over 4 orders of magnitude produces a difference within a factor of 2 to 3 at most in crack growth rates as also reported in Fig.7. By comparison, increasing H_2 pressure from 10 MPa to 45 MPa causes an increase in the FCGR by as much as by a factor of 4. But, further increases in FCGR due to increases in pressure from 45 MPa to 100 MPa were small. Note that the 100 MPa crack growth data in Fig. 8 was for R = 0.2 compared to R = 0.1 condition for the other tests. Generation of additional data is necessary to document this trend.

Figure 9 shows the FCGR behavior of A372 Grade J Class 70 steel at ΔK levels of 10, 20, and 25 MPa$\sqrt{m}$ for various H_2 pressures. The rates are presented as average actual rates and in a normalized form in which the actual rates are normalized by the rates in air at the corresponding ΔK value. Differences in crack growth

* Corresponding author: asaxena@uark.edu

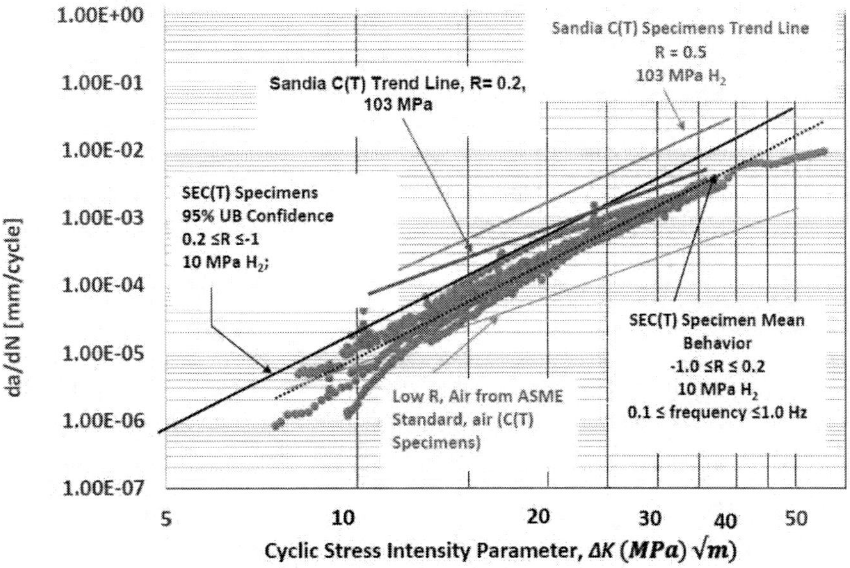

Fig. 6- All FCGR data produced as part of this study for $-1 \leq R \leq 0.2$ on 10 MPa H_2 using SEC(T) specimens and comparison with Sandia data from C(T) specimens at H_2 pressure of 103 MPa for R values of 0.5 and 0.2 [9].

rates due to pressure seem to be smaller for ΔK values in the range of 10 MPa $\sqrt{m}$ than at higher ΔK values of 20 and 25 MPa$\sqrt{m}$. More data at H_2 pressure of 100 MPa is needed to fully confirm this trend. Figure 9 also shows how the effects of increasing H_2 pressure saturate beyond 45 MPa at the various ΔK levels. This should also be explored with more data in the future to confirm (or not) this trend.

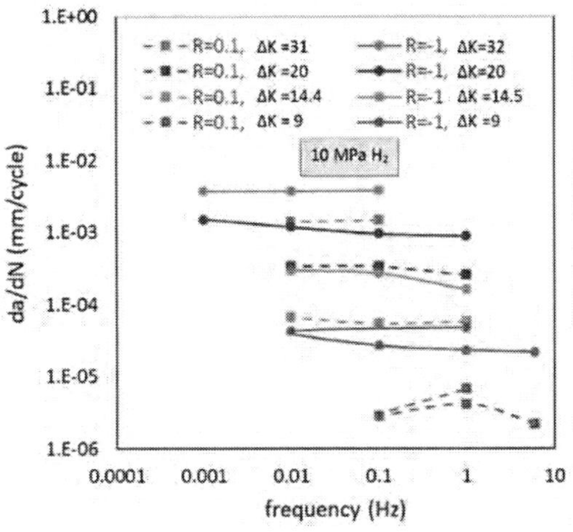

Fig. 7- Effect of loading frequency on the fatigue crack growth rates in 10 MPa H_2 pressure at R = -1.0 and at R = 0.1 at four ΔK levels for A372 Grade J Class 70 steel [9].

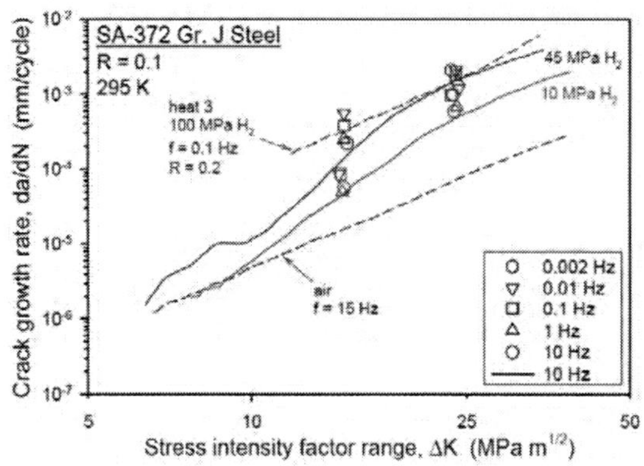

Fig. 8- FCGR behavior of A372 Grade J, Class 70 steel at various H_2 pressures for R =0.1. Reproduced from reference [17] with permission.

4. Summary

As mentioned previously, the effects of load ratio, loading frequency, and pressure are complex and synergistic in nature. However, a picture that appears to be evolving based on the evidence presented in this paper subject to additional experimental confirmation is as follows:

- FCGR behavior in high pressure H_2 gaseous environment at negative load ratios of -1.0 and -0.5 are comparable to those at load ratios of 0.1 and 0.2 if the negative portion of the loading cycle is not included in the estimation of ΔK. This clearly underscores the benefits of the autofrettage process in enhancing the fatigue crack growth life of hydrogen storage vessels. The benefits

* Corresponding author: asaxena@uark.edu

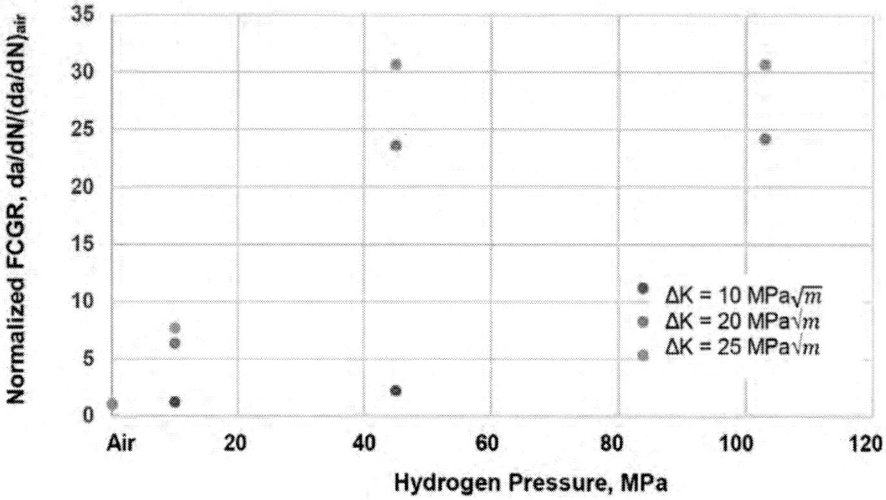

Fig. 9- FCGR behavior in H_2 gas environment at various pressures normalized by the corresponding crack growth rates in air to highlight the extent of embrittlement [9]

accrue from a reduced ΔK level as well as through a reduction in load ratio. The load ratio effects in high pressure gaseous H_2 appear to become significant for $R > 0.2$ and especially in the pressure range of 100 MPa.

The loading frequency is a secondary variable in determining the crack growth rates in H_2 environment and decreasing it does increase crack growth rates in the range of 10 to 1 Hz but it changes marginally when decreased beyond 1 Hz. Thus, $0.1\text{Hz} \leq \nu \leq 1\text{Hz}$ is a suitable range for testing while producing data relevant for design of pressure vessels for storing H_2.

The available data show that FCGR behavior in gaseous H_2 do increase with increasing pressure. The magnitude of this effect appears to be highest in going from air environment to 10 MPa H_2 pressure and appears to saturate at 45 MPa level. Additional systematic studies to accurately quantify the effects of pressure are recommended.

5. References

1. A. Saxena, A. Prakash, K.A. Nibur, and I. Miller, "Considerations of the Effects of H2 in the Design of Type II Storage Vessels Built for Fatigue Resistance", Proceedings of the International Hydrogen Conference, B.P. Somerday and P. Sofronis editors, ASME-861387, 2017, pp 382-.389.
2. A. Prakash, G.R. Sharp, B.T. Deeken, W.J. Head, W.H. Thomson, "Steel Wrapped Pressure Vessel" US Patent US 9,266,642- B2 issued Feb. 2016.
3. ASME Section VIII, Division 3, Code 2004 Edition, "Alternate Rules for Construction of High Pressure Vessels", ASME New York 10010.
4. M.D. Rana, G.B. Rawls, J.R. Sims, E. Upitis, , "Technical Basis and Application of New Rules on Fracture Control of High Pressure Hydrogen Vessel in ASME Section VIII Division 3, PVP 2007-2603, 2007, ASME PVP Conference, San Antonio, Texas.
5. R.P. Gangloff "H_2-Enhanced Deformation and Fracture in the Crack Tip Plastic Zone", Proceedings of the International Hydrogen Conference, B.P. Somerday and P. Sofronis editors, ASME-861387, 2017, pp.1-35.
6. K.A. Nibur, and B.P. Somerday, "Fracture and Fatigue Test Methods in Hydrogen Gas" Gaseous Hydrogen Embrittlement of Materials in Energy Technologies, R.P. Gangloff and B.P. Somderday, eds., Woodhead Publishing Ltd., Cambridge, Vol. 1, 2012, pp.195-236.
7. B.P. Somerday, P. Bortot, and J. Felbaum, "Optimizing Measurement of Fatigue Crack Growth Relationships for Cr-Mo Pressure Vessel Steels in Hydrogen Gas", PVP2015-45424, Proc. ASME PVP Conference, July 19-23, 2015, Boston, MA.
8. K. A. Nibur, B.P. Somerday, C. San Marchi, J.W. Foulk, M. Dadfarina, P. Sofronis, "The Relationship Between Crack Tip Strain and Subcritical Cracking Thresholds in Steels in High Pressure Hydrogen Gas" Met. Transactions A, Vol. 44, 2013, pp248-269.
9. A. Saxena, K.A. Nibur, and A. Prakash, "Applications of Fracture Mechanics in Assessing Integrity of Hydrogen Storage Systems" Engineering Fracture Mechanics, 187, 2018, pp 368-380.
10. A. Saxena, F. Bassi, K.A. Nibur, and J.C. Newman, Jr., "On Single-Edge-Crack Specimen Tension-Compression Fatigue Crack Growth Testing", Engineering Fracture Mechanics, 176, 2017, pp 343-350.

* Corresponding author: asaxena@uark.edu

11. B. P. Someday, P. Berlot, J. Felbaum, "Optimizing Measurement of Fatigue Crack Growth Relationships for Cr-Mo Pressure Vessel Steels in Hydrogen Gas", PVP-2015-45424, Proceedings of the ASME 2015 Pressure Vessel and the Piping Division Conference, July 19-23, 2015, Boston, MA
12. B.P. Somerday, C. San Marchi, and K. A. Nibur, "Measurement of Fatigue Crack Growth Rates in SA372 Grade J Steel in 100 MPA Hydrogen Gas Following Article KD-10, PVP 3013-97455, Proceedings of ASME 2013 Pressure Vessel and the Piping Division Conference, July 14-28, 2013, Paris, France.
13. San Marchi, C., Bortot, P., Felbaum, J., Wada, Y. and Ronevich, J.A. "Fatigue and Fracture of High-Hardenability Steels for Thick-walled Hydrogen Pressure Vessels" Proceedings of the International Conference on hydrogen Safety, Sept 11-13, 2017, Hamburg Germany.

Effect of hydrogen on fatigue crack growth of quenched and tempered CrMo(V) steels

Luis Borja Peral[1,*], *Sergio* Blasón[1], *Alfredo* Zafra[1], *Cristina* Rodríguez[1], and *Javier* Belzunce[1]

[1] Polytechnic Engineering School, University of Oviedo, campus universitario, 33203 Gijón, Spain

Abstract. In order to select the most appropriate steel to deal with pressurized hydrogen during long times, the fatigue crack propagation rate of quenched and tempered 2.25Cr1Mo and 2.25Cr1Mo0.3V steel grades was evaluated by means of tests performed on thermally pre-charged specimens in a hydrogen reactor at 195 bar and 450ºC during 21 hours. Cylindrical samples to measure the hydrogen content and their desorption kinetics at room temperature and compact tensile specimens to determine the fatigue crack growth rate were used. Finally, scanning electronic microscopy was used in the study of fracture surfaces. Using the aforementioned pre-charging conditions, significant amounts of hydrogen were introduced, being much larger in the 2.25Cr1Mo0.3V steel grade (vanadium carbides provide strong hydrogen tramps). Regarding fatigue tests, the fatigue crack growth rate was increased notably due to the presence of hydrogen in the 2.25Cr1Mo grades for frequencies lower than 10 Hz. On the other hand, the presence of vanadium carbides has significantly improved the fatigue crack propagation performance in the presence of internal hydrogen.

1 Introduction

Vessels used to store and transport hydrogen must be able to provide a safe service during long periods of time in direct contact with gaseous hydrogen under high internal pressure, being so essential to ensure good resistance to hydrogen embrittlement. Generally, quenched and tempered steels alloyed with chrome, chrome-molybdenum or chrome-molybdenum-vanadium are used in these industrial installations. Nevertheless, in order to reduce the vessel wall thickness, high yield strength steels are preferred, although it is well known that these steels are more sensitive to hydrogen embrittlement [1]. This type of components, which are also subjected to important cyclic loads, are used to work under high pressures of hydrogen (up to 100 MPa) and hydrogen absorption and diffusion into the steel microstructure might take place during the working period. The combination of all these factors, acting simultaneously, might give rise to hydrogen embrittlement processes, therefore decreasing the mechanical properties of these steel grades. For this reason, it is important to know, how hydrogen diffuses and accumulates in the process zone existing ahead of any existent defect until reaching a critical concentration which could trigger the failure of the component. One important task for material scientists and engineers is the development of materials and systems able to withstanding the effects of cyclic loading in hydrogen environments. Complete and detailed studies quantifying the influence of hydrogen on fatigue crack growth rate are still scarce, as most studies have mainly examined the influence of hydrogen on tensile properties. Nevertheless, in order to produce components able to perform satisfactorily in services for up to 15 years under high hydrogen pressures, it is an urgent need to produce reliable data on the fatigue behaviour of candidate materials in hydrogen environments [2-3].

In order to simulate the interaction between hydrogen atoms and the steel microstructure, there are two basic ways to analyse the effects of hydrogen in the mechanical properties of the steels: testing in high-pressure hydrogen gas (external hydrogen) and testing in air after pre-charging the specimens in a hydrogenated medium (internal hydrogen). In this second case, specimens can be pre-charged at high temperature from a gaseous atmosphere or electrochemically from an aqueous solution at room temperature. In this work, pre-charging gaseous hydrogen at high temperature was used. In this way, given the exponential variation of the diffusion coefficient and solubility with temperature, the kinetics of hydrogen entry into the steel can be significantly accelerated and, after a few hours, a similar concentration of hydrogen to that obtained after years of service at room temperature can be reached.

* Corresponding author: luisborjapm@gmail.com

Nevertheless, the largest drawback of this technique is the necessity to carry out the mechanical tests just after the hydrogen pre-charge in order to limit their desorption.

Compact specimens of quenched and tempered CrMo and CrMoV steels were thermally pre-charged and fatigue crack propagation rates were determined in this work. The influence of frequency on fatigue crack growth rate was also studied. Finally, scanning electronic microscopy (SEM) was used to analyse failure micromechanisms.

2 Experimental procedure

2.1 Materials and heat treatments

Two different low-alloyed ferritic steels from the Cr-Mo family have been selected in this study, with and without vanadium. The chemical composition of 2.25Cr1Mo and 2.25Cr1Mo0.3V steels is shown in Table 1.

Table 1. Chemical composition (weight %).

Steel	C	Mn	Si	Cr	Mo	V
2.25Cr1Mo	0.14	0.56	0.16	2.23	1	-
2.25Cr1Mo0.3V	0.15	0.52	0.09	2.27	1	0.3

2.25Cr1Mo and 2.25Cr1Mo0.3V steels were studied in the as-received (AR) condition (see Table 2). In addition, 2.25Cr1Mo steel was austenitized at 940°C for 30 min, quenched in water and tempered at 600°C for two hours. In this way, due to the applied heat treatment, two steel grades with the same chemical composition were obtained in the case of the 2.25Cr1Mo steel. The sequence of heat treatments and the nomenclature of all the studied grades (based on the tempering temperature) are shown in Table 2.

Table 2. Applied heat treatments.

Steel grade	Heat treatment sequences
2.25Cr1Mo_690	AR: 940°C/3h+ water quench + 690°C/30h tempered
2.25Cr1Mo_600	940°C/30min+water quench+600°C/2h tempered
2.25Cr1Mo0.3V_650	AR: 925°C/90min+ water quench + 650°C/2h tempered

Different specimens were machined to perform the different tests: small cylindrical samples to measure hydrogen contents and the kinetics of hydrogen egression at room temperature (RT) and also compact tensile specimen for fatigue crack growth rate determination.

2.2 Hydrogen pre-charging

All the specimens were pre-charged with gaseous hydrogen in a high-pressure hydrogen reactor. In order to make sure that the specimens were saturated with hydrogen, the following conditions were applied: 21 hours at 450°C under a pressure of 19.5 MPa of pure hydrogen. After this maintenance time, a cooling phase of 1h, until attaining a temperature of 85°C (keeping the hydrogen pressure at 19.5MPa to minimize hydrogen egress) was always used. Afterwards, the hydrogen pre-charged specimens were removed from the reactor and rapidly immersed in liquid nitrogen (-196°C), where they were kept until the moment of testing, in order to avoid hydrogen loss.

2.3 Measurement of hydrogen content

The amount of hydrogen absorbed in the different steel grades and the hydrogen desorption curves at room temperature were determined. With this purpose, cylindrical samples with a diameter of 10 mm and a length of 30 mm were used (≈20g).
The hydrogen content was measured in a LECO DH603 analyzer, which is able to measure hydrogen concentrations between 0.1 and 2500 ppm. The measuring principle of the equipment is based on the difference of thermal conductivity between a reference gas flow of pure nitrogen and a secondary flow carrying nitrogen and the hydrogen thermally extracted from the specimen. This conductivity difference is measured by means of a Wheatstone bridge and converted into an electrical signal that is interpreted by the equipment software, which finally gives the sample hydrogen concentration in ''ppm''.
The procedure to obtain the desorption curves was the following: all the samples were removed from the liquid nitrogen at the same time, and exposed in air at room temperature. Then, the hydrogen concentration of the samples was measured at different time intervals. Before starting the measure, each sample was cleaned in an ultrasonic bath with acetone for 5 minutes, and then carefully dried using cold air. The analysis to determine hydrogen concentration consists on keeping the sample at 1100°C during approximately 400 seconds in the aforementioned LECO DH603 analyzer.

2.4 Fatigue crack growth tests

Fatigue crack growth rate determination was performed using compact tensile, CT, specimens, with a width of 48 mm and a thickness of 10 mm (see Figure 1). The specimens were always fatigue pre-cracked before pre-charging to obtain an initial crack length versus width ratio, a/W=0.2, using an R ratio of 0.1 and following the ASTM E647 standard [4]. The initial applied ΔK was about 35 $MPa\sqrt{m}$, and always a R ratio of 0.1 was used in these tests performed at RT. Crack length was continuously monitored with the aid of a CTOD extensometer and a nominal frequency of 10 Hz was employed. Moreover, in the case of the hydrogen pre-charged specimens, frequencies of 1 and 0.1 Hz were also used in order to evaluate the influence of this parameter in HE. The curves representing crack growth rate (da/dN) versus stress intensity factor range (ΔK) of uncharged and hydrogen pre-charged specimens were subsequently compared.

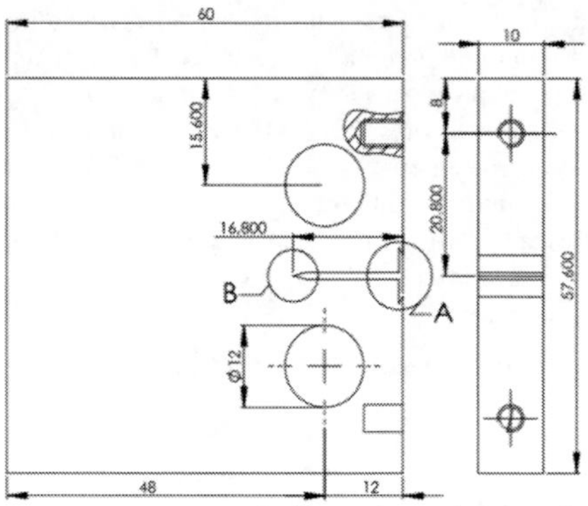

Fig. 1. Geometry and dimensions (mm) of CT specimens.

2.5 Observation of microstructures and fracture surfaces

The steel microstructures were observed in a scanning electron microscope (SEM JEOL-JSM5600) using an acceleration voltage of 20kV (the samples were previously ground and finally polished with diamond paste of 1μm and etched with Nital). The fracture surfaces of the tested specimens were also carefully examined using the same technique.

3 Results

3.1 Microstructure

The resultant microstructures of 2.25Cr1Mo and 2.25Cr1Mo0.3V steels after the heat treatments described in Table 2 are shown in Figure 2.

(a)

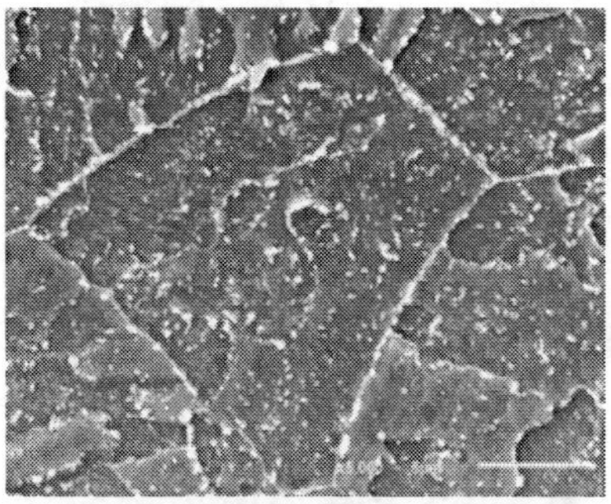

(b)

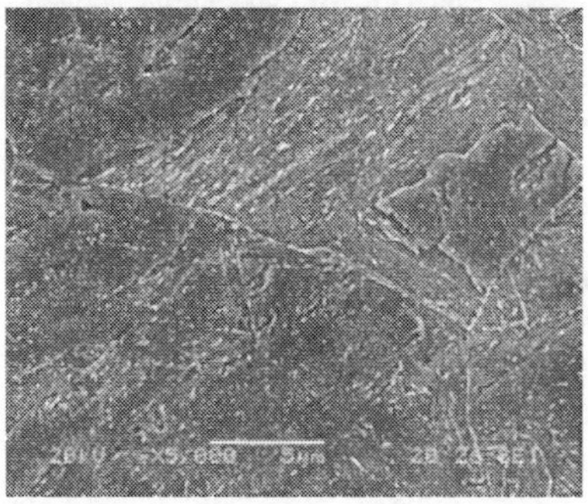

(c)

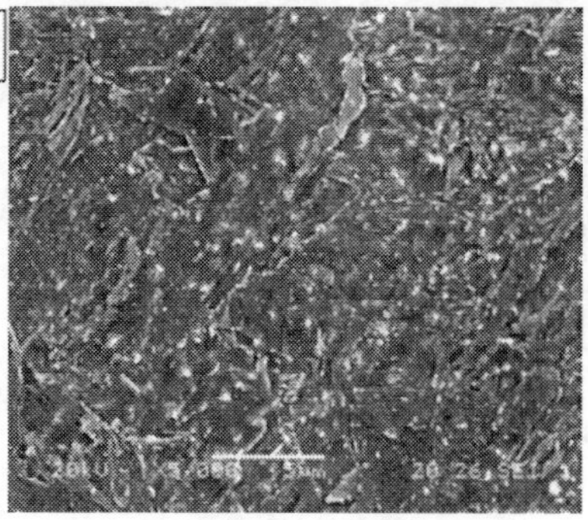

Fig. 2. Microstructure of heat treated steels (x5000). 2.25Cr1Mo_690 (a). 2.25Cr1Mo_600 (b) and 2.25Cr1Mo0.3V_650 (c).

All these microstructures correspond to tempered martensite. Microstructural differences between the steels and heat treatments are also noticed in the hardness (Brinell hardness) and tensile properties shown in Table 3. σ_y is the yield strength; σ_u, the ultimate tensile strength; and e the total elongation.

Table 3. Hardness and tensile properties of the different steel grades.

Grado de acero	HB	σ_y (MPa)	σ_u (MPa)	e (%)
CrMo_690	170	430	580	27
CrMo_600	285	761	887	19.5
CrMoV_650	262	667	829	20

3.2 Hydrogen pre-charging

Hydrogen solubility in ferritic steels at high temperature, considering that traps are not active at these temperatures, as reported by Pillot [5], can be calculated using Equations (1) for CrMo steels and (2) for CrMoV steel grades.

$$C_H = 0.152 \cdot exp(-27079/R \cdot T) \cdot \sqrt{P} \quad (1)$$

$$C_H = 9.23 \cdot 10^{-3} \cdot exp(-6943/R \cdot T) \cdot \sqrt{P} \quad (2)$$

Where C_H represents the hydrogen content expressed in 'mass ppm', R is the gas constant (8.31 J/mol·K), T is the temperature (K) and P the hydrogen pressure (Pa). Applying these equations to our charging conditions, hydrogen contents of 7.5 and 13 ppm were respectively introduced into the CrMo and CrMoV steel samples. It may be mentioned that significant hydrogen loss takes place in the course of the cooling phase (1 hour) until reaching the reactor sample extraction temperature of 85°C.

3.3 Hydrogen desorption curves

The hydrogen desorption curves obtained at room temperature with each steel grade are shown in Figure 3. These figures represent the hydrogen content evolution (ppm) versus the exposure time in air (hours) at room temperature. The initial hydrogen content, C_{H0}, corresponds to the first point of the desorption curve (hydrogen content at the end of the cooling phase following thermal pre-charge). The residual hydrogen, C_{Hf}, is the hydrogen strongly trapped in the steel microstructure, hydrogen content after a long exposure at room temperature, being the diffusible hydrogen, the amount that is able to get out from traps and diffuse out of the steel when enough time is available, the difference between these two values (C_{H0} - C_{Hf}).

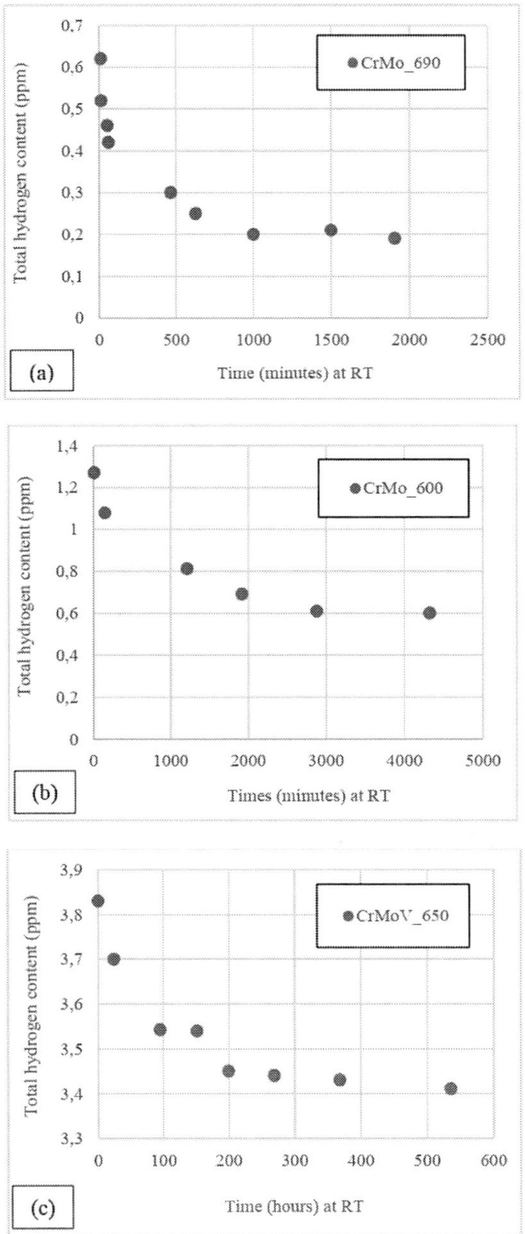

Fig. 3. Hydrogen desorption curves. CrMo_690 (a) CrMo_600 (b) and CrMoV_650 (c).

In line with the heat treatments previously applied and regarding the 2.25Cr1Mo steel, when the tempering temperature increases, the initial and residual hydrogen contents decrease, as this microstructure has lower density of traps and strong traps. 2.25Cr1Mo steels always desorbed hydrogen much faster than V-added steel, despite having less initial content (0.60-1.30 ppm respectively) than the vanadium steel (3.80 ppm). For example, all diffusible hydrogen diffused out from the V-free steels in about twenty-four hours but only 3 % diffused out from 0.3V after the same time. It was previously demonstrated that fine precipitated V carbides are very strong hydrogen traps [5, 6] and this explains the high values of initial and residual hydrogen contents measured in this steel grade. Table 4 shows the results obtained with the three steel grades.

Table 4. Initial (C_{H0}), residual (C_{Hf}) and diffusible (C_{H0}-C_{Hf}) hydrogen contents measured in the different steel grades.

Grade of steel	C_{H0} [ppm]	C_{Hf} [ppm]	C_{H0}-C_{Hf} ppm	%
2.25Cr1Mo_690	0.60	0.20	0.40	67
2.25Cr1Mo_600	1.30	0.60	0.70	54
2.25Cr1Mo0.3V_650	3.80	3.40	0.40	10

3.4 Hydrogen diffusion coefficient estimation

Data adjustment of the experimentally measured hydrogen desorption curves was used to estimate the apparent hydrogen diffusion coefficient of the three steel grades. We performed an axysimetric bi-dimensional diffusion analysis for the cylindrical samples using a commercial FEM software (Abaqus Cae). The hydrogen initial distribution, after the cooling phase (from 450 to 85°C performed in the high pressure reactor) for the CrMo_690 steel is shown in Figure 4. This hydrogen distribution corresponds to the first point of the desorption curve in Fig. 3 (a), where an average hydrogen content of 0.6 ppm was experimentally measured, although the specimen surface has the residual hydrogen content (0.2 ppm) and hydrogen contents as high as 0.9 ppm are still present in the centre of the specimen. This hydrogen distribution was taken as the first point (boundary condition) to analyse the desorption curve of Fig. 3(a).

The Fick diffusion law ($J= -D_{app}\nabla C_L$, being J the hydrogen flux, D_{app} the apparent diffusion coefficient, and C_L represents the lattice hydrogen concentration) was applied to a cylinder with a radius of 5 mm using the apparent diffusion coefficient, D_{app}, as variable, and the hydrogen evolution with time at RT was calculated. The diffusion coefficient which better adjusted the experimental data was taken as the apparent diffusion coefficient of the steel. The D_{app} value estimated for the CrMo_690 steel grade was $2.4 \cdot 10^{-9}$ m^2/s and the adjustment to the experimental data is shown in Figure 5.

Table 5 shows the apparent diffusion coefficient calculated on the three steel grades. As tempering temperature of the CrMo steel increases, the diffusion coefficient significantly increases, although the variation of the diffusion coefficient is even greater for the CrMoV steel, which has a relatively very low hydrogen diffusion coefficient.

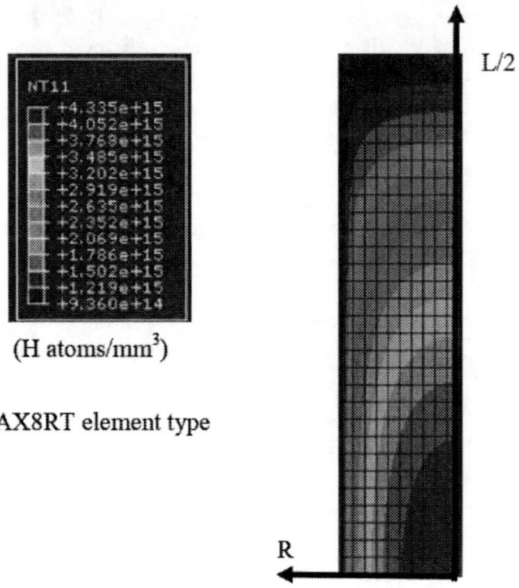

Fig. 4. Hydrogen initial distribution in the CrMo_690 steel (C_H=0.6ppm) with R=5 mm and L=30mm.

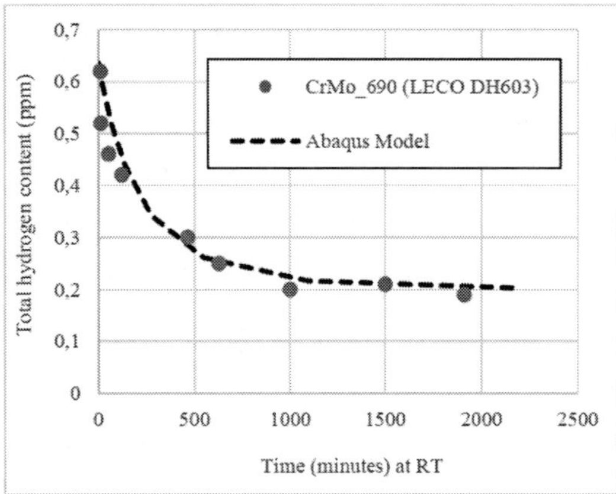

Fig. 5. CrMo_690 grade of steel. Adjustment of the apparent diffusion coefficient.

The apparent hydrogen diffusion coefficients of the three studied steels are shown in Table 5.

Table 5. Hydrogen diffusion coefficients (D_{app}).

Grade of steel	C_{H0} [ppm]	C_{Hf} [ppm]	D_{app} [m^2/s]
2.25Cr1Mo_690	0.60	0.20	$2.4 \cdot 10^{-9}$
2.25Cr1Mo_600	1.30	0.60	$8.0 \cdot 10^{-10}$
2.25Cr1Mo0.3V_650	3.80	3.40	$8.0\ 10^{-11}$

The hydrogen diffusion coefficient measured in V-added steel was one or two orders of magnitude lower than that obtained in the V-free steels (depending on tempering temperature). This result corroborates previous ones, in which vanadium carbides were seen to be very strong hydrogen microstructural traps [5,6].

3.5 Fatigue crack growth curves

The curves representing the fatigue crack growth rate (da/dN) vs the stress intensity factor range (ΔK) obtained with the three steel grades are shown as a function of the test frequency in Figure 6. The acceleration effect in the fatigue crack growth rate was only noted when the test frequency was lower than 10 Hz and this effect was more important under the lowest test frequency (0.1 Hz), as it would be expected according to previous works [2,3,7].

On the other hand, the ratio between hydrogen crack growth rate $(da/dN)_H$ measured with hydrogen pre-charged specimens under the lowest frequency (0.1 Hz) and the crack growth rate obtained with uncharged specimens $(da/dN)_{NoH}$ was plotted as a function of the applied ΔK in Figure 7. This factor is an indicator of the hydrogen embrittlement susceptibility under cyclic loads. It is worth to note that the aforementioned acceleration ratio is the largest for low ΔK values and frequencies. For the other hand, crack growth acceleration under internal hydrogen was much lower in the V-added steel, although the internal hydrogen present in this steel was the largest.

3.6 Fracture surfaces

The fracture surfaces of the fatigued CT specimens were analysed under the scanning electron microscope. As mentioned above, the most relevant differences were noted at the lowest test frequencies (0.1Hz) and for the lowest ΔK values. Fig. 8a shows the fatigue surface of the uncharged CrMo_690 steel grade (the most hydrogen embrittled grade under cyclic loads), where typical fatigue striations are clearly seen. On the contrary, in the fatigue surface of the hydrogen pre-charged specimens tested at 0.1 Hz, striations dissapeared and PRHIC (plasticity related hydrogen induced cracking) micromechanisms (white circles, Fig. 8b) with some small areas of brittle intergranular (IG) fractures (white arrows, Fig. 8b) are visible [8]. In this case, hydrogen accumulation in front of the crack, where the hydrostatic stress attains a maximum value, promotes internal interface decohesion (block and packet martensite boundaries and prior austenite grain boundaries).

Regarding now the 2.25Cr1MoV_650 grade (the less sensible studied grade to hydrogen embrittlement), the hydrogen precharged specimens tested under a frequency of 0.1 Hz showed a similar failure appearence to that observed in the uncharged specimen, being also crack propagation rates very close in both cases (Fig. 6c).

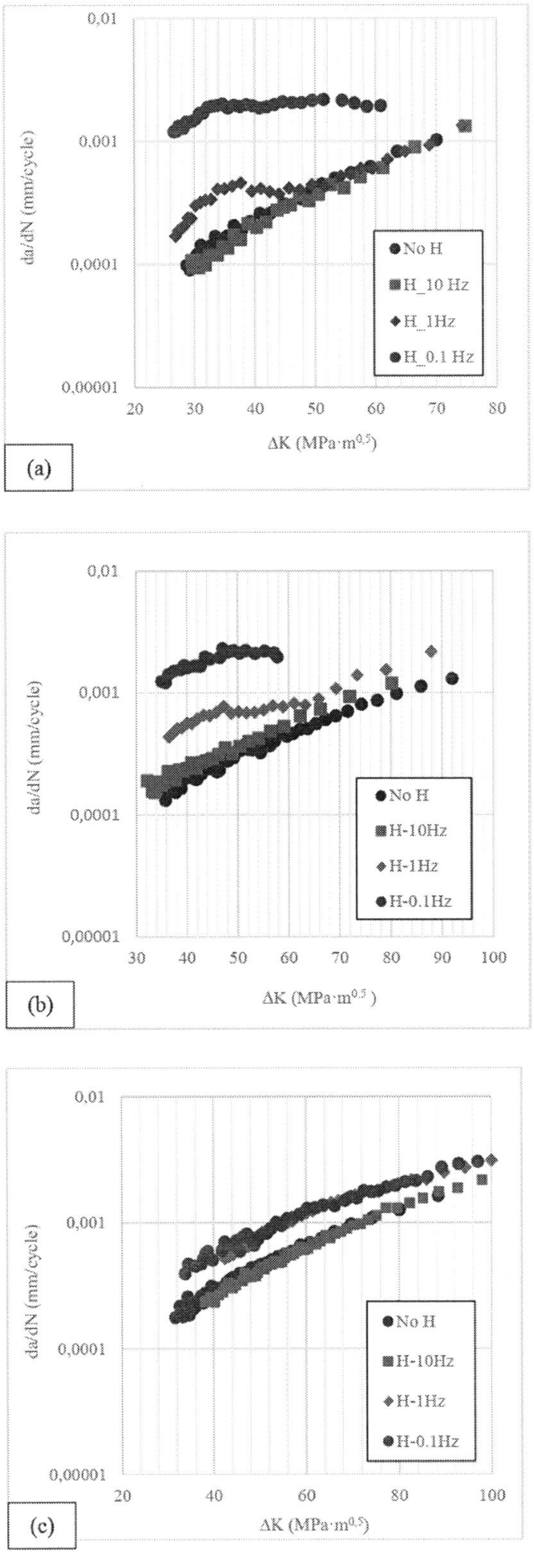

Fig. 6. Fatigue crack growth rates versus ΔK curves CrMo_690 (a), CrMo_600 (b) and CrMoV_650 (c).

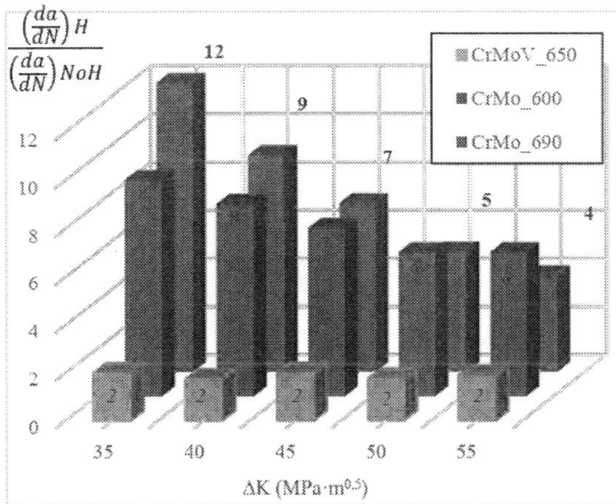

Fig. 7. Fatigue crack growth acceleration factor due to internal hydrogen. R=0.1 and f=0.1 Hz.

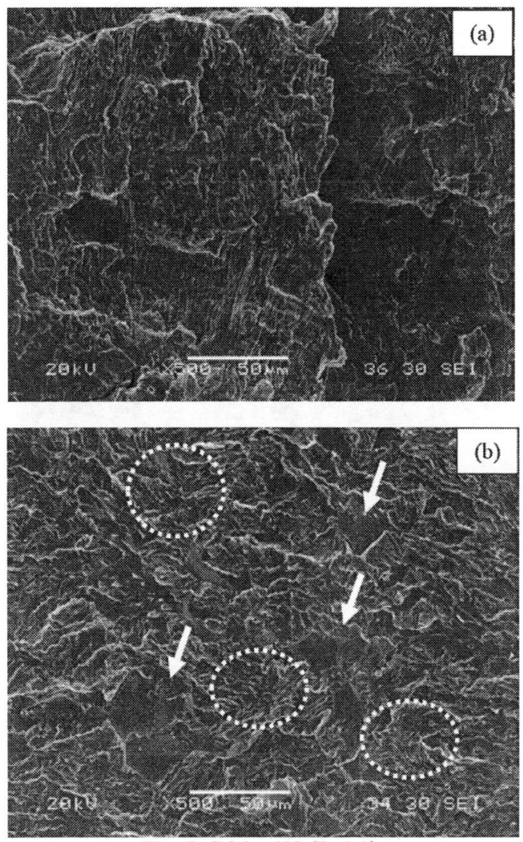

Fig. 8. CrMo_690 (R=0.1).
f=10Hz, uncharged, ΔK ≈40 MPa·$\sqrt{m}$ (a).
f=0.1Hz, hydrogen pre-charged, ΔK ≈40 MPa·$\sqrt{m}$ (b).

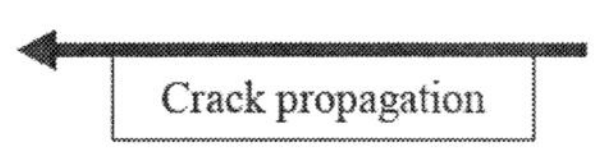

This steel grade (V-added steel) did not exhibit any trace of intergranular or PRHIC fractures. The most remarkable difference between the uncharged and precharged specimen was the presence of secondary cracks due to the presence of hydrogen (white arrows, Fig.9.b)

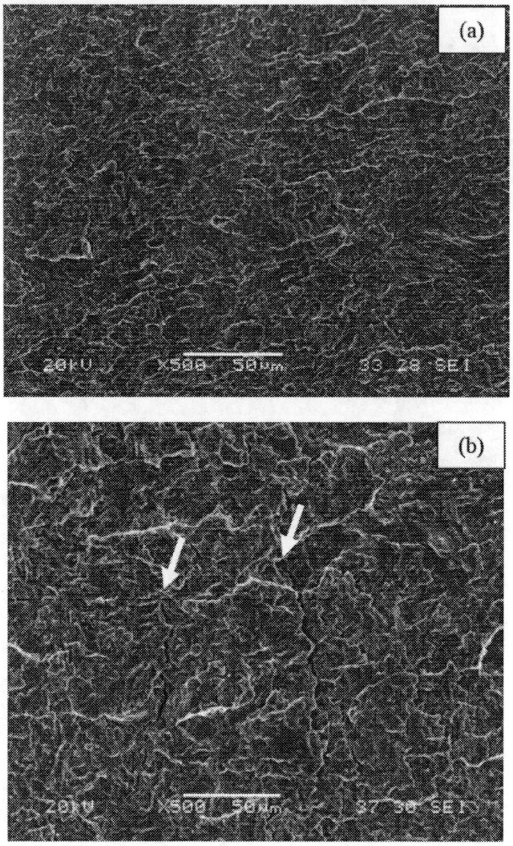

Fig. 9. CrMoV_650 (R=0.1).
f=10Hz, uncharged, ΔK ≈40 MPa·$\sqrt{m}$ (a).
f=0.1Hz, hydrogen pre-charged, ΔK ≈40 MPa·$\sqrt{m}$ (b).

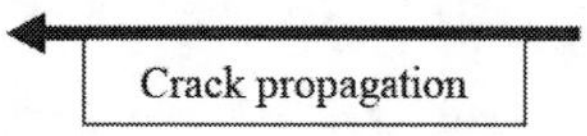

4 Discussion

In order to discuss the fatigue results obtained in this work, it is necessary to remember that when a mechanical load is applied on a hydrogen precharged CT specimen, hydrogen located in the crack tip surroundings will diffuse up to the zone of maximum triaxiality (maximum hydrostatic stress) existing just ahead of the crack tip, where it will accumulate, giving rise to the embrittlement phenomenon (hydrogen enhanced descohesion, HEDE, and/or hydrogen enhanced localized plasticity, HELP mechanisms). In this way, the influence of the test frequency on the fatigue crack growth rate will depend on the time that hydrogen has to move in each fatigue load cycle to attain a critical concentration in the aforementioned process region (see Figure 10).

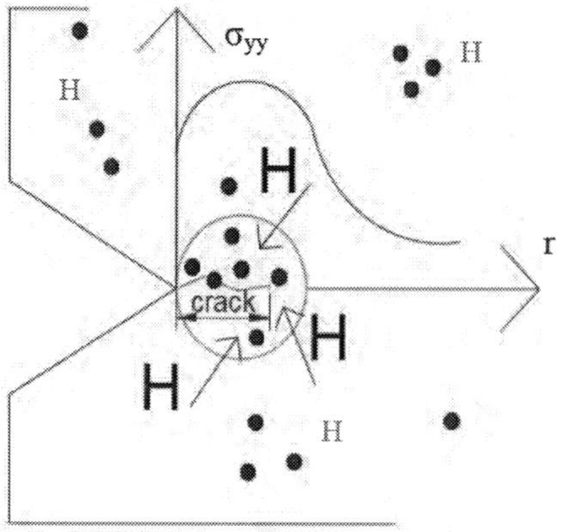

Fig. 10. Crack extension in the process zone ahead of an opened crack tip.

So, the hydrogen crack growth acceleration factor observed under low fatigue frequencies, lower than 10 Hz (see figures 6 and 7) might be explained from the effective diffusion distance, x, available for hydrogen movement along the fatigue cycle, that can be estimated from Fick´s second law, equation 3.

$$x = \sqrt{Dt} \qquad \text{(equation 3)}$$

D being the apparent hydrogen diffusion coefficient in the steel and t, the time available in each fatigue cycle (t = 1/f). Table 6 shows the hydrogen effective diffusion distances calculated for each steel grade as a function of the applied test frequency.

Table 6. Effective diffusion distance (μm).

Acero	D_{app} (m^2/s)	10 Hz	1 Hz	0.1 Hz
		0.1 s/cycle	1 s/cycle	10 s/cycle
CrMo_690	$2.4\cdot10^{-9}$	15	49	155
CrMo_600	$8\cdot10^{-10}$	9	28	89
CrMoV_650	$8\ 10^{-11}$	3	9	28

In this way, when the test frequency is high (10Hz), fatigue cycle load time is low, and consequently, the hydrogen diffusion distances are also very low. Hence, hydrogen accumulation in the crack tip region is barely significant and the fatigue crack growth rate doesn´t differ from the one determined without hydrogen in air (see Figure 6). Nevertheless, when the test frequency decreases, hydrogen effective diffusion distances increase significantly, reaching values, which allow important hydrogen accumulation in the crack tip region. In this case, hydrogen diffusion would be coupled with the crack growth rate, hydrogen atoms will diffuse in and out of the process zone in each one of the fatigue cycles, enabling the existance of the so-called embrittled process zone due to hydrogen accumulation. This fact would explain the increase in the crack growth rate observed in the CrMo steels cycled at low frequencies. By the contrary, hydrogen effect was notably lower in the CrMoV steel. It is worth noting that the effective diffusion distances in the V-added steel tested at 0.1 Hz were nearly similar to that observed in the CrMo steel tempered at 690ºC and tested at 10 Hz. As mentioned above, V-added steel has the capacity to strongly retain hydrogen due to the existance of submicrometric vanadium carbides, reducing the apparent diffusion coefficient and, consequently the effective disffusion distances, even under quite low frequencies.

It is also important to note that from a certain ΔK value (or a certain crack propagation rate, da/dN), a kind of ''plateau '' in the crack propagation rate was detected. A plateau like this one is typically referred in the scientific literature when a phenomenon of stress corrosion cracking (SCC) or hydrogen assisted cracking (HAC) is present [9]. Under these circumstances if the applied stress intensity factor in the fatigue cycle exceeds a threshold value, K_{ISCC}, the crack extends at a constant growth rate independent of the applied ΔK.

Another explanation for this phenomenon could be based on the progressive loss of hydrogen during the fatigue test. For example, based on the desorption curve present in Figure 3, the fatigue CT specimen of CrMo_690 steel grade after 10 hours testing at 1 Hz, would have already lost all its diffusible hydrogen and it should be expected that the fatigue crack growth rate curve would superpose with the one determined without hydrogen (Figure 6a). Nevertheless, when the fatigue test frequency decreased to 0.1 Hz, significant diffusible hydrogen still remains in the fatigue specimen at the end of the fatigue test and, consequently, fatigue crack growth rate is still higher than the crack propagation rate measured in the uncharged specimen (without hydrogen). Finally, the hydrogen effect in the crack growth rate of the CrMoV steel grade was always barely important, as hydrogen diffusion coeficient and effective diffusion distances were both very low.

5 Conclusions

The presence of internal hydrogen causes an increase of the crack propagation rate in the CrMo steel grades and this effect is larger as the test frequency applied in the fatigue test is reduced. This behaviour can be explained assuming that hydrogen embrittlement is controlled by the diffusion of hydrogen atoms toward the process zone existing in the the advancing crack tip. As frequency decreases, hydrogen diffusion distances in the fatigue cycle increase and hydrogen concentration in the process zone can attain a critical value for the decohesion of intenal boundaries, martensite blocks and packets or prior austenitic grains.

The CrMoV steel has a uniform dispersion of submicrometric vanadium carbides which act as strong traps for hydrogen, significantly reducing the apparent diffusion coefficient and the effective diffusion distances in the fatigue cycle. The fatigue behaviour of this steel grade in presence of internal hydrogen was much improved and the effect of frequency was in this case nearly negligeable.

Finally, hydrogen pre-charged CT specimens gradually lose hydrogen during the fatigue test. Consequently, it is important to control the total duration of the test, in order to avoid excessive hydrogen loss.

The authors would like to thank the Spanish Ministry of Economy and Competitivity for the support received for the development of the research project MAT2014-58738-C3 (SAFEHIDROSTEEL) and L. Borja Peral the financial aid given by FEDER and the Asturias government, through the project FC-15-GRUPIN14-001.

References

[1] Gangloff R.P., Sommerday B.P. Edit., Gaseous hydrogen embrittlement of materials in energy technologies, *Woodhead Publishing*, (2012).

[2] Murakami Y., Matsuoka S., Effect of hydrogen on fatigue crack growth of metals, *Engineering Fracture Mechanics*, **77** (2010) 1926-1940.

[3] Murakami Y., Effects of hydrogen in metal fatigue, *Proceedings of the international hydrogen energy development forum*, Fukuoka, Japan (2007) 96-105.

[4] ASTM E647, Standard test method for measurement of fatigue crack growth rates, Annual Book of ASTM Standards, **Vol. 03.01**, (2015).

[5] S. Pillot, L. Coudreuse, Hydrogen-induced disbonding and embrittlement of steels used in petrochemical refining, *Woodhead Pub. Ltd.*, (2012).

[6] Lee T., "Effects of Vanadium Carbides on Hydrogen Embrittlement of Tempered Martensitic Steel", *Met. Mater. Int.*, **Vol. 22**, No 3 (2016), pp. 364-372.

[7] A.Sciuccati et al, ``Effect of hydrogen and low temperature on fatigue crack growth of pipelines steels'', *Engineering Fracture Mechanics* (2013) 10-25.

[8] Nibur K.A., "The Relationship Between Crack-tip Strain and Subcritical Cranking Tresholds for Steels in High-Pressure Hydrogen Gas", *The Minerals, Metals & Materials Society and ASM International*, (2012).

[9] R.H. Jones, Edit., Stress-corrosion cracking. Materials performance and evaluation, *ASM International*, (2017).

Hydrogen-assisted fatigue crack propagation in a pure BCC iron. Part II: Accelerated regime manifested by quasi-cleavage fracture at relatively high stress intensity range values

Domas Birenis[1,*], *Yuhei* Ogawa[2,3], *Hisao* Matsunaga[4,5,6], *Osamu* Takakuwa[5], *Junichiro* Yamabe[5,6,7], *Øystein* Prytz[1], and *Annett* Thøgersen[8]

[1]Department of Physics, Centre for Materials Science and Nanotechnology, University of Oslo, Blindern, NO-0316 Oslo, Norway
[2]Graduate School of Engineering, Kyushu University, 744 Motooka, Nishi-ku, Fukuoka 819-0395, Japan
[3]Research Fellow of the Japan Society for the Promotion of Science, 744 Motooka, Nishi-ku, Fukuoka 819-0395, Japan
[4]Department of Mechanical Engineering, Kyushu University, 744 Motooka, Nishi-ku, Fukuoka 819-0395, Japan
[5]Research Center for Hydrogen Industrial Use and Storage (HYDROGENIUS), Kyushu University, 744 Motooka, Nishi-ku, Fukuoka 819-0395, Japan
[6]International Institute for Carbon-Neutral Energy Research (I2CNER), Kyushu University, 744 Motooka, Nishi-ku, Fukuoka 819-0395, Japan
[7]AIST-Kyushu University Hydrogen Materials Laboratory (HydroMate), 744 Motooka, Nishi-ku, Fukuoka 819-0395, Japan
[8]SINTEF Materials and Chemistry PO Box 124 - Blindern, NO-0314 Oslo, Norway

Abstract. Hydrogen effect on fatigue performance at relatively high values of stress intensity factor range, ΔK, of pure BCC iron has been studied with a combination of various electron microscopy techniques. Hydrogen-assisted fatigue crack growth rate is manifested by a change of fracture features at the fracture surface from ductile transgranular in air to quasi-cleavage in hydrogen gas. Grain reference orientation deviation (GROD) analysis has shown a dramatic suppression of plastic deformation around the crack wake in samples fatigued in hydrogen. These results were verified by preparing site-specific specimens from different fracture features by using Focused Ion Beam (FIB) technique and observing them with Transmission Electron Microscope (TEM). The FIB lamella taken from the sample fatigued in air was decorated with dislocation cell structure indicating high amount of plasticity, while the lamella taken from the quasi-cleavage surface of the sample fatigued in hydrogen revealed a distribution of dislocation tangles which corresponds to smaller plastic strain amplitude involved at the point of fracture. These results show that a combination of critical hydrogen concentration and critical stress during fatigue crack growth at high ΔK values triggers cleavage-like fracture due to reduction of cohesive force between matrix atoms.

1 Introduction

Hydrogen has the potential to become one of the major energy carriers, but to become a feasible fuel alternative it has to close the gap against the traditional energy sources in terms of cost. One fraction of the high total cost of hydrogen fuel comes from the structural materials used in hydrogen industry. Industry is actively looking for cheaper materials and low-alloy or carbon steels with body-centered cubic (BCC) lattice structure are among the best contenders. Unfortunately, the understanding of hydrogen effect on these types of steels is still lacking [1] and the performance under cyclic loading in hydrogen environment is one of the most important properties which needs to be studied in detail for a successful realisation of BCC steels in hydrogen energy industry.

For steels in corrosive environment a stable crack growth region described by the Paris law in fatigue crack growth (FCG) curve usually shows a dual stage behavior. At relatively low ΔK values (in this article we refer to it as Stage I) the fatigue crack growth rate (FCGR) seems almost unaffected by environment, while at higher ΔK values (Stage II) the FCGR transitions into accelerated regime where the crack grows at higher rate. Such behavior is generally interpreted as a superposition of sustained load stress-corrosion cracking and pure mechanical fatigue [2]. Similar trends are applicable for fatigue crack growth in steels exposed to hydrogen gas [3]. According to fracture surface examinations the hydrogen-assisted cracking in high-strength steels manifests as a brittle intergranular fracture [4,5], while low or medium-strength steels subjected to hydrogen-assisted fatigue crack growth (HAFCG) show transgranular quasi-cleavage (QC) fracture accompanied by brittle-like striations on the fracture surfaces [6,7]. These observations resulted in several proposed models for explanation of the QC fracture formation

* Corresponding author: domas.birenis@fys.uio.no

mechanism [6,8,9], which are based on the most common theories for HE, such as hydrogen-enhanced decohesion (HEDE) [10,11], hydrogen-enhanced localized plasticity (HELP) [12,13], hydrogen-enhanced stabilization of lattice defects [8,14], or the combination of them. However, these models still lack direct and critical supporting evidence, mostly due to the fact that most of the experiments have been performed on materials with complex microstructures and several alloying elements. These factors make the interpretation difficult and thus, the analysis of the hydrogen effect on the fracture processes is further complicated.

In this study, in order to minimize complexities arising from microstructure and material impurity we chose to study pure Fe, which is the simplest system representing BCC steels. FCG data was acquired for specimens fatigued in air and in hydrogen gas at different pressures and then fracture surfaces, fracture paths and dislocation structures around the crack wake were investigated by using scanning electron microscopy (SEM) and transmission electron microscopy (TEM). Even though pure Fe in air demonstrated fully ductile transgranular fracture throughout the measured ΔK range, failure of the specimen in hydrogen was attributed to two essentially different types of fracture: intergranular (IG) at relatively low ΔK values and quasi-cleavage (QC) at relatively high ΔK values. This difference in fracture surfaces marked a two stage FCGR behavior with a different level of HAFCG acceleration and this article deals only with high ΔK regime (Stage II), while in Part I of this series of articles the Stage I regime is analyzed in detail [15]. The Grain Reference Orientation Deviation (GROD) analysis has shown a significant plastic deformation reduction in the samples fatigued in hydrogen in comparison with the ones fatigued in air. In addition, these results were verified by using TEM observations of thin foils extracted directly from QC fracture features. Namely, the foil extracted from sample fatigued in air was decorated with a dislocation cell structure which is a typical feature for heavily plastically deformed iron. On the other hand, in case of hydrogen the foils possessed a distribution of individual dislocations and dislocation tangles indicating a low plastic strain deformation and crack was confirmed to propagate along {100} type cleavage plane. The combination of these results implies that hydrogen-induced cleavage cracking, resulting in a significant reduction of plastic deformation, was the main contributor to the HAFCG in stage II regime.

2 Experimental

The material and experimental methods were the same as described in Part I of this article series. Additionally, ΔK-constant tests were conducted at $\Delta K = 17$ MPa • $m^{1/2}$ in air, 0.7 MPa and 90 MPa hydrogen. Also, a misorientation angle-dispersed plastic deformation distribution (MADPDD) technique was used to quantify plastic deformation. This technique uses the raw data from the GROD maps to plot the number of pixels versus misorientation angle, resulting in a MADPDD curve.

3 Results and discussion

1.1. Fatigue crack growth data

The FCG curves obtained from ΔP-constant tests in air, nitrogen and different pressures of hydrogen gas are illustrated in Fig. 1. The crack growth curve is linear in logarithmic scale for air and nitrogen throughout the tested ΔK range, but in the case of hydrogen gas, there is an obvious distinction between Stage I (relatively low ΔK regime) and Stage II (higher ΔK regime) FCG. At increasing hydrogen pressure the FCG acceleration in Stage II almost does not change and shows a difference of approximately 30 times when compared to FCG in air. On the other hand, the transition point between the two regimes is strongly affected by the hydrogen pressure.

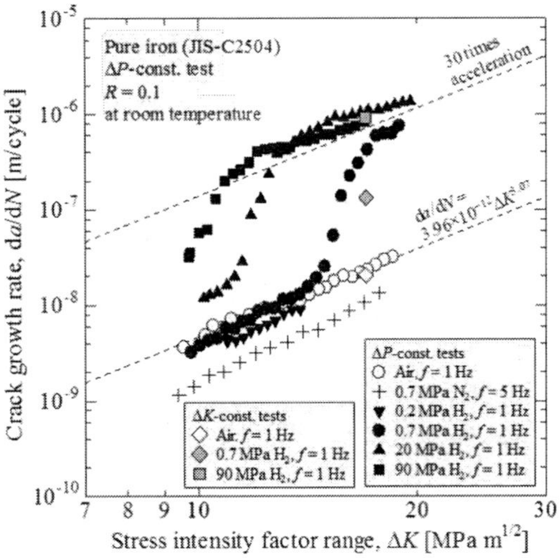

Fig. 1. FCG curves of pure Fe measured in laboratory air, 0.2 MPa nitrogen and 0.2, 0.7, 20, 90 MPa hydrogen gas.

Namely, increasing the hydrogen pressure made the transitional stress intensity factor range, ΔK^T, from Stage I to II to shift to lower values. The trends shown in Fig. 1 are common in other BCC materials [16].

1.2. Fracture surface morphology

Fracture surfaces of samples fatigued in air, 0.7 MPa, 20 MPa, and 90 MPa hydrogen at different ΔK values are represented in Fig. 2. Fracture surfaces of specimens fatigued in air were covered only by ductile features (striation-like stripe patterns) throughout the tested ΔK range and fracture was fully transgranular (TG). On the other hand, the fracture surfaces of the specimens fatigued in hydrogen can be distinguished into two different regions: ductile transgranular fracture with a varying fraction of IG fracture was observed in regions corresponding to Stage I (Fig. 2c, e) and QC features

https://doi.org/10.1051/matecconf/201816503010

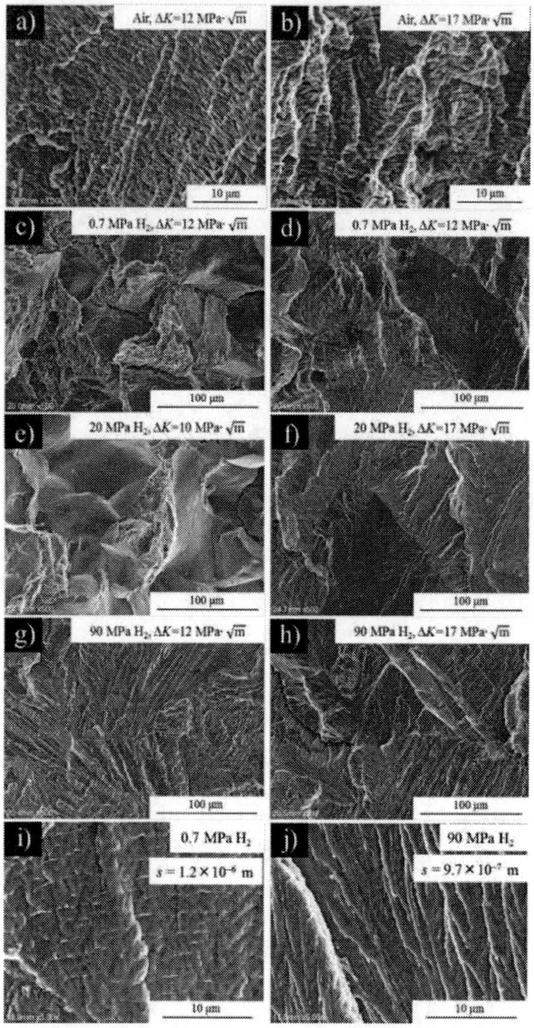

Fig. 2. SEM images of the fracture surfaces of the specimens subjected to ΔP-constant tests in (a, b) air, (c, d) 0.7 MPa hydrogen, and (e, f) 20 MPa hydrogen, (g, h) 90 MPa hydrogen at different ΔK values. (c, e) and (d, f, g, h) correspond to stage I and II crack growth regime respectively. High magnification images show striation spacing in (i) 0.7 MPa hydrogen and (j) 90 MPa hydrogen gas.

decorated by brittle-like striations dominated the surfaces corresponding to Stage II (Fig. 2d, f, g, h) FCG. In addition, it is confirmed that the spacing between brittle-like striations on QC fracture surfaces (Fig. 2i, j) match with the macroscopic FCG rate, which indicates that these striations form on a cycle-by-cycle basis. Fraction of intergranular fracture in Stage I increased at increasing hydrogen pressures. A similar transition from IG to QC fracturing between Stages I and II has also been reported in some pipeline steels [17,18] tested in hydrogen gas.

A summarized overview of fracture feature distribution across the tested ΔK ranges is represented in Fig. 3. The onset of the appearance of QC features on the fracture surface coincides very well with the transition point of FCGR from Stage I to Stage II regime – the ΔK^T value. In addition, the fraction of QC features reaches close to 100% surface coverage at the point where the transition ends and FCG curve becomes parallel to the one in air, but at higher level. Therefore, FCG acceleration enhancement was controlled by the coverage of QC fracture, and as it reaches 100% there is no further enhancement by increasing hydrogen pressures. On the other hand, higher hydrogen pressures caused a shift of ΔK^T to the lower values implying that a combination of critical stress and hydrogen concentration is responsible for triggering QC type fracture.

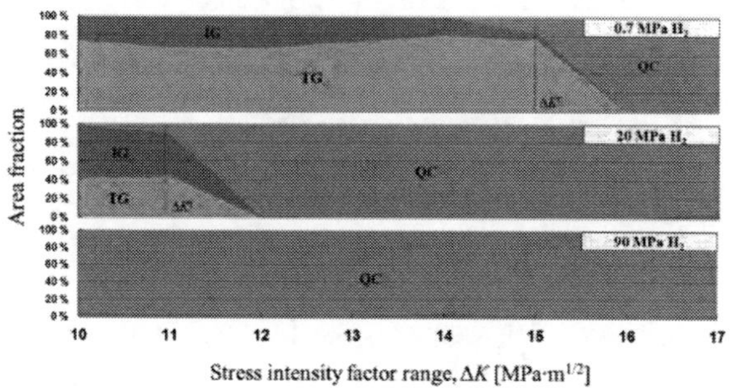

Fig. 3. Statistical overview of the fraction of fracture surface features at different ΔK values.

1.3. Plastic deformation and dislocation structure analyses by EBSD and ECC imaging

In order to reveal the path of the crack growth we acquired EBSD maps at the crack wake area and to visualize plastic deformation GROD analyses on the specimens fatigued at ΔK = 17 MPa·m$^{1/2}$ (Fig. 4) were performed. As visible from the illustrations the crack path in air (Fig. 4a) is purely transgranular, wavy and branches in several locations. In comparison, the crack path in case of hydrogen (Fig. 4b, c) is also transgranural, but much more straight with path direction changing mostly across the grain boundaries. The plasticity analysis by using GROD maps on Stage II FCG (Fig. 4d, e, f) showed significant differences in samples fatigued in air and hydrogen gas. To evaluate these results in a more quantitative way we display a MADPDD graphs corresponding to the GROD maps in Fig 5. In case of air the sample was strongly plastically deformed up to 200 μm away from the crack (Fig. 4d) and strain induced grain misorientation reached more than 40° (Fig. 5). However, in case of 0.7 MPa hydrogen the amount of plasticity was largely reduced and confined to the areas where the crack path was changing direction, while regions where the crack was running straight through the grains the plastic deformation was minimal. These observations are even more pronounced in the sample fatigued at 90 MPa hydrogen. As visible from Fig. 5 the maximum misorientation angle is drastically reduced for higher hydrogen pressures indicating that plasticity is not just localized as suggested in previous studies [9,19], but also reduced. Interestingly enough, the plastic deformation differences

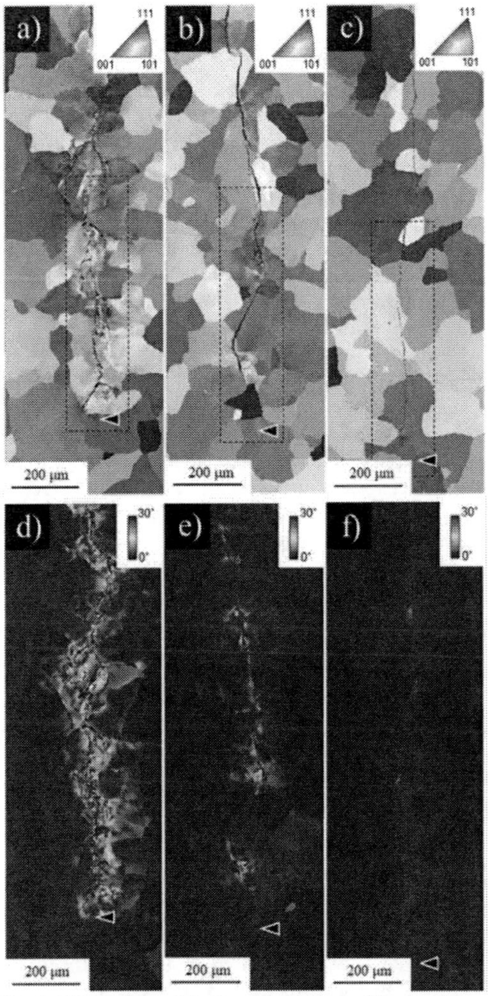

Fig. 4. SEM images of the fracture surfaces of the specimens subjected to ΔP-constant tests in (a, b) air, (c, d) 0.7 MPa hydrogen, and (e, f) 90 MPa hydrogen at different ΔK values. (c, e) and (d, f) correspond to stage I and II crack growth regime respectively. The crack growth directions are bottom to top.

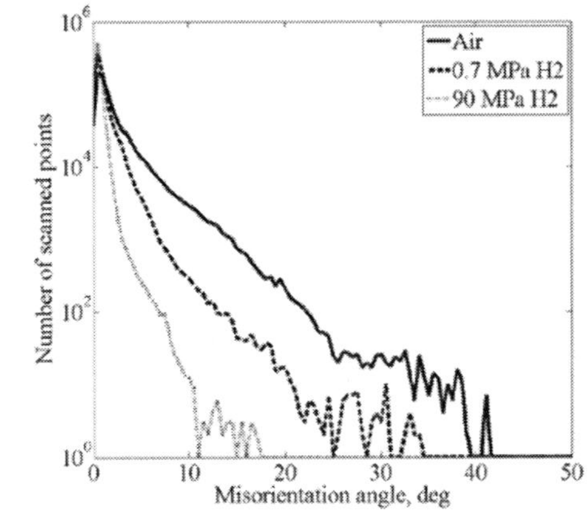

Fig. 5. MADPDD graph obtained from the GROD maps indicated in Fig. 4d, e, f.

between 0.7 MPa and 90 MPa hydrogen do not have very large influence on FCG (Fig. 1) confirming our previous assertion that it is the type of fracture, which has major influence on FCG, while the amount of plasticity plays a secondary role.

The ECC images of the plastic deformation induced substructures formed in the regions marked by black dashed rectangles in Fig. 4a-c are shown in Fig. 6. In the sample fatigued in air a heavy plastic deformation resulted in a formation of fine sub-grains close to the crack and deformed regions further away from the crack (Fig. 6a). In contrast, the sample fatigued in 0.7 MPa hydrogen gas demonstrated a much less distorted microstructure with sub-grain formation only in the most deformed regions (Fig. 6b). In case of 90 MPa the microstructure is almost free of distortions and no sub-grain formation was detected. {100} type plane traces determined by EBSD analysis are indicated on two grains in Fig. 6b, c implying the crack growing along {100} type cleavage plane. However, EBSD analysis provides only a 2 dimensional picture of the lateral surface, and it is not known how the crack plane is oriented deeper into the sample. Therefore, additional examination is needed to assure this assumption.

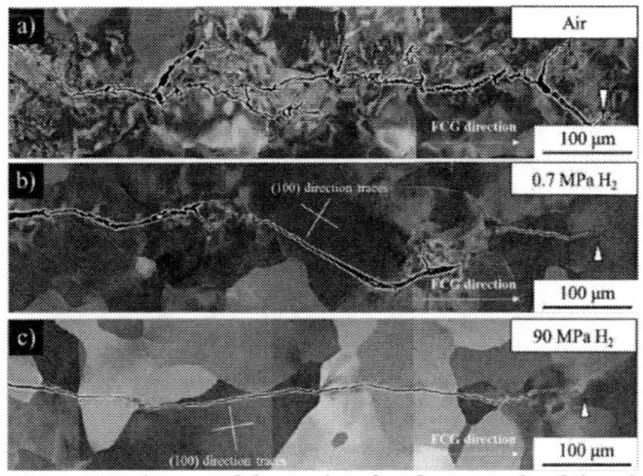

Fig. 6. ECC images of the crack wakes from samples fatigued in a) air, b) 0.7 MPa H_2 and c) 90 MPa H_2. The captured areas correspond to the regions marked by black dashed rectangles in Fig. 4a-c. The black arrowheads indicate crack tip positions, while yellow arrowheads show the fracture path along cleavage plane.

1.4. Dislocation structures characterized by TEM

Thin-foil samples for TEM were directly extracted from the ductile striations formed in air and QC regions formed in hydrogen gas using FIB, and the dislocation structure immediately beneath the fracture surfaces were examined by dark field-scanning TEM (DF-STEM) observation and are represented in Fig. 7. Dislocation cells were seen to dominate the dislocation structure in the sample fatigued in air (Fig. 8a) with a uniform cell size distribution. However, below the QC fracture surface formed in hydrogen gas (Fig. 8b, c), only dislocation tangles and individual dislocations were observed. According to the past studies [20–22], dislocation evolution is directly linked to the plastic strain amplitude applied to the material. At initial stages, individual dislocations and dislocation tangles are formed, and further plastic deformation forces dislocations to redistribute into veins, walls and finally cell structure. Applying these findings to our work, we can state that samples fatigued in air sustained high plastic strain amplitude, while samples fatigued in hydrogen suffered from much lower plastic strain amplitude. In addition, due to the fact that FIB lamellas were extracted from the regions, where fracture surface was perpendicular to the plane of lamella, we can recover the information about the fracture plane by using selected area diffraction in TEM. The diffraction patterns acquired in TEM indicate that in samples fatigued in hydrogen the QC fracture plane corresponds very well with the {100} cleavage plane. These results strongly support the statement that hydrogen heavily suppresses plasticity evolution and promotes cleavage fracture when the hydrogen concentration and stress level ahead of the crack tip reaches critical values.

In a study on low-carbon steel by Nishikawa et al. [8], the proposed model for brittle striation formation during fatigue in hydrogen consists of three subsequent processes: i) crack tip opening; ii) microscopic defect nucleation ahead of the crack tip due to hydrogen-enhanced stabilization of point defects [14,23], iii) final crack growth by a coalescence of the main crack with the primary defects due to ductile tearing. A crucial detail in this model is that crack growth progresses by ductile tearing, meaning that crack wake should contain significant plastic deformation and fracture surface should be dimpled on microscopic scale. However, they did not indicate the direct evidence of these dimple features ahead of the crack tip, and our findings of underdeveloped dislocation structures and relatively flat fracture surface between the striations contradict this model. Additionally, the preferential fracture path along {100} cleavage plane observed in our study cannot be explained according to the ductile tearing process.

According to Troiano and Oriani [10,11], when hydrogen atoms enter the material they reduce the cohesive force between matrix atoms and activate the fracture under lower stress level than when hydrogen is absent. This model fits very well with our observations. Another interesting idea was introduced recently by Hajilou et al. [24]. The authors found similar plasticity suppression during the bending test using micro-sized cantilever of Fe-3%Si alloy under in-situ electrochemical hydrogen-charging. According to the authors, large amount of hydrogen is attracted to the crack tip hydrostatic stress field and locks dislocations from motion, thereby leading to brittle type fracture. On a microscopic scale, a direct molecular dynamics calculation conducted by Song and Curtin also supported cleavage cracking in pure iron, due to the inhibited dislocation emission from crack tip source caused by hydrogen aggregation [25]. To some extent, all of these hypotheses can be applied to the present case to describe cleavage-like fracture. However, it is important to note that the dislocation density in the specimens fatigued in hydrogen gas is not negligible, and the cleavage fracture may be supported by the presence of these dislocations for example by assisting transport [26,27] or trapping [28,29] of hydrogen.

As a result, we propose the following sequence for fatigue crack growth in hydrogen atmosphere during one loading cycle: i) as the cycle starts, the crack tip is being blunted by alternating slip mechanism and dislocations are being emitted from the crack tip, which may result in the final dislocation tangle structure visible in post-mortem samples (Fig. 7b, c); ii) a maximum hydrostatic stress field is established just ahead of the main crack tip [30], which lowers the chemical potential and raises hydrogen concentration in this region; iii) the combination of critical stress and critical hydrogen concentration causes a primary cleavage micro crack formation ahead of the crack tip; iv) the main crack and the primary cleavage micro crack coalesce and propel the main crack forward until it is arrested. Striation spacing match with macroscopic FCGR indicates that the brittle striations on QC are formed on a cycle-by-cycle basis, as was also confirmed in some conventional BCC steels [8,31]. This fact demonstrates that the macroscopic behavior of HAFCG in BCC iron is still controlled by a stable manner, even though on microscopic scale the fracture event is catastrophic.

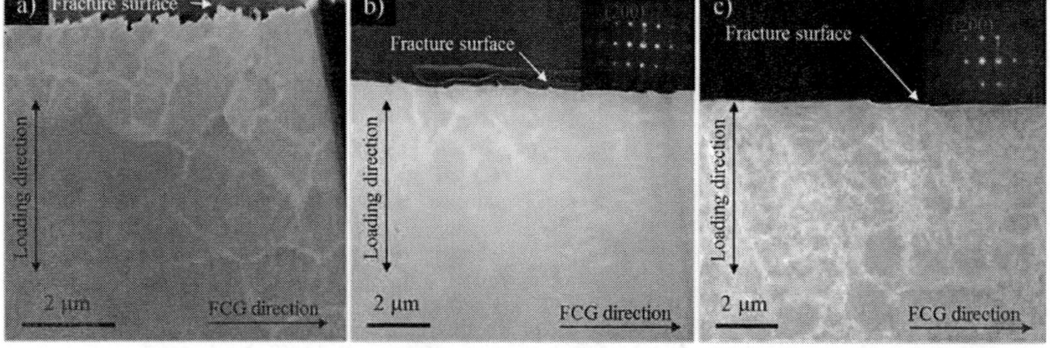

Air | 0.7 MPa H_2, ΔK=17 MPa·m$^{1/2}$ | 90 MPa H_2, ΔK=17 MPa·m$^{1/2}$

Fig. 7. Dislocation structures immediatly beneath the fracture surfaces of samples fatigued in a) air, b) 0.7 MPa H_2, and c) 90 MPa H_2. The SADPs acquired from [110] zone axis are indicated in the insets.

4 Conclusions

1. Hydrogen gas enhanced FCG acceleration in pure iron above critical stress intensity factor range, ΔK^T, by ~30 times and this enhancement was independent of hydrogen pressure. The ΔK^T value was shifted to lower values by increasing hydrogen pressure.
2. The FCG acceleration beyond ΔK^T manifested by the formation of hydrogen induced quasi-cleavage (QC) fracture and below ΔK^T by intergranular (IG) fracture in combination with ductile transgranural features.
3. According to EBSD and ECCI analyses the plastic deformation and dislocation structure evolution around the crack wake were considerably reduced in case of hydrogen when the fracture surfaces are characterized by QC type features.
4. TEM micrographs showed dislocation tangle structure in hydrogen gas and dislocation cell structure in air, which were attributed to deformation at low plastic strain and high plastic strain amplitudes respectively. In addition, electron diffraction revealed the fact that the fracture in hydrogen was along {100} crystal planes, implying that hydrogen-induced cleavage cracking was the main contributor to the HAFCG.

This work was supported by JSPS KAKENHI Grant Number JP16H04238 and JP16J02960. The Research Council of Norway is acknowledged for its support through the Norwegian Center for Transmission Electron Microscopy, NORTEM (197405/F50). This study also forms part of the "HIPP" project from the PETROMAKS2 program, funded by the Research Council of Norway [Grant Number: 102006899].

5 References

[1] M. Dadfarnia, A. Nagao, S. Wang, M. L. Martin, B. P. Somerday, and P. Sofronis, Int J Fract **196**, 223 (2015).
[2] S. Suresh and R. O. Ritchie, Met. Sci. **16**, 529 (1982).
[3] C. San Marchi and B. P. Somerday, *Technical Reference for Hydrogen Compatibility of Materials* (Albuquerque, 2012).
[4] R. P. Wei and D. Landes, Mater. Res. Stand. **12**, 45 (1972).
[5] J. Yamabe, T. Matsumoto, S. Matsuoka, and Y. Murakami, Int. J. Fract. **177**, 141 (2012).
[6] T. J. Marrow, P. J. Cotterill, and J. E. King, Acta Metall. Mater. **40**, 2059 (1992).
[7] Y. Ogawa, D. Birenis, H. Matsunaga, A. Thøgersen, Ø. Prytz, O. Takakuwa, and J. Yamabe, Scr. Mater. **140**, 13 (2017).
[8] H. Nishikawa, Y. Oda, and H. Noguchi, J. Solid Mech. Mater. Eng. **5**, 370 (2011).
[9] S. Matsuoka, H. Tanaka, N. Homma, and Y. Murakami, Int. J. Fract. **168**, 101 (2011).
[10] R. A. Oriani, Annu. Rev. Mater. Sci. **8**, 327 (1978).
[11] A. R. Troiano, Trans. ASME **52**, 54 (1960).
[12] C. D. Beachem, Metall. Trans. **3**, 441 (1972).
[13] H. K. Birnbaum, *Mechanisms of Hydrogen Related Fracture of Metals* (1989).
[14] M. Nagumo, ISIJ Int. **41**, 590 (2001).
[15] Y. Ogawa, D. Birenis, H. Matsunaga, O. Takakuwa, J. Yamabe, Ø. Prytz, and A. Thøgersen, in *12th Int. Fatigue Congr.* (2018).
[16] J. Yamabe, M. Yoshikawa, H. Matsunaga, and S. Matsuoka, Procedia Struct. Integr. **2**, 525 (2016).
[17] B. P. Somerday, P. Sofronis, K. A. Nibur, C. San Marchi, and R. Kirchheim, Acta Mater. **61**, 6153 (2013).
[18] R. L. Amaro, N. Rustagi, K. O. Findley, E. S. Drexler, and A. J. Slifka, Int. J. Fatigue **59**, 262 (2014).
[19] Y. Murakami, T. Kanezaki, Y. Mine, and S. Matsuoka, Metall. Mater. Trans. A Phys. Metall. Mater. Sci. **39 A**, 1327 (2008).
[20] H. Mughrabi, K. Herz, and X. Stark, Int. J. Fract. **17**, 193 (1981).
[21] M. S. Pham and S. R. Holdsworth, Metall. Mater. Trans. A **45**, 738 (2013).
[22] J. Polák, *Fatigue of Steels* (Elsevier B.V., 2007).
[23] K. Takai, H. Shoda, H. Suzuki, and M. Nagumo, Acta Mater. **56**, 5158 (2008).
[24] T. Hajilou, Y. Deng, B. R. Rogne, N. Kheradmand, and A. Barnoush, Scr. Mater. **132**, 17 (2017).
[25] J. Song and W. A. Curtin, Nat. Mater. **12**, 145 (2012).
[26] J. Tien, A. W. Thompson, I. M. Bernstein, and R. J. Richards, Metall. Trans. A **7**, 821 (1976).
[27] M. Dadfarnia, A. Nagao, S. Wang, M. L. Martin, B. P. Somerday, and P. Sofronis, Int. J. Fract. **196**, 223 (2015).
[28] R. A. Oriani, Acta Metall. **18**, 147 (1970).
[29] J. P. Hirth, Metall. Trans. A **11**, 861 (1980).
[30] M.-J. Lii, X.-F. Chen, Y. Katz, and W. W. Gerberich, Acta Metall. Mater. **38**, 2435 (1990).
[31] M. Yoshikawa, T. Matsuo, N. Tsutsumi, H. Matsunaga, and S. Matsuoka, Trans. JSME (in Japanese) **80**, SMM0254 (2014).

Hydrogen-assisted fatigue crack propagation in a pure BCC iron. Part I: Intergranular crack propagation at relatively low stress intensities

Yuhei Ogawa[1,2*], *Domas* Birenis[3], *Hisao* Matsunaga[4,5,6], *Osamu* Takakuwa[4,5], *Junichiro* Yamabe[7,8], *Øystein* Prytz[3] and *Annett* Thøgersen[9]

[1]Graduate School of Engineering, Kyushu University, 744 Motooka, Nishi-ku, Fukuoka 819-0395, Japan
[2]Research Fellow of the Japan Society for the Promotion of Science, 744 Motooka, Nishi-ku, Fukuoka 819-0395, Japan
[3]Department of Physics, Centre for Materials Science and Nanotechnology, University of Oslo, PO Box 1048 - Blindern, NO-0316 Oslo, Norway
[4]Department of Mechanical Engineering, Kyushu University, 744 Motooka, Nishi-ku, Fukuoka 819-0395, Japan
[5]Research Center for Hydrogen Industrial Use and Storage (HYDROGENIUS), Kyushu University, 744 Motooka, Nishi-ku, Fukuoka 819-0395, Japan
[6]International Institute for Carbon-Neutral Energy Research (I2CNER), Kyushu University,744 Motooka, Nishi-ku, Fukuoka 819-0395, Japan
[7]AIST-Kyushu University Hydrogen Materials Laboratory (HydroMate),744 Motooka, Nishi-ku, Fukuoka 819-0395, Japan
[8]Department of Mechanical Engineering, Fukuoka University, 8-19-1 Nanakuma, Jonan-ku, Fukuoka 814-0180, Japan
[9]SINTEF Materials and Chemistry, PO Box 124 - Blindern, NO-0314 Oslo, Norway

Abstract. The role of hydrogen on intergranular (IG) fracture in hydrogen-assisted fatigue crack growth (HAFCG) of a pure iron at low stress intensity was discussed in terms of the microscopic deformation structures near crack propagation paths. The main cause of IG fracture was assumed to be the hydrogen-enhanced dislocation structure evolution and subsequent microvoids formation along the grain boundaries. Additionally, the impact of such IG cracking on the macroscopic FCG rate was evaluated according to the dependency of IG fracture propensity on the hydrogen gas pressure. It was first demonstrated that the increased hydrogen pressure results in the larger area fraction of IG and corresponding faster FCG rate. Moreover, gaseous hydrogen environment also had a positive influence on the FCG rate due to the absence of oxygen and water vapor. The macroscopic crack propagation rate was controlled by the competition process of said positive and negative effects.

1 Introduction

For the storage and transportation of compressed gaseous hydrogen in the forthcoming hydrogen energy-based society, hydrogen-assisted fatigue crack growth (HAFCG) in structural steels is of significant concern with regards to the safe design of high-pressure components such as pressure vessels or pipelines [1,2]. For elucidating the detailed fatigue crack growth (FCG) behaviour of iron and steels with ferritic, *i.e.* body-centered cubic (BCC) crystal structure under the presence of hydrogen, we conducted FCG tests of a commercially pure iron in gaseous hydrogen environment with various pressures at room temperature. Our experiments revealed that the FCG behavior in gaseous hydrogen environment was characterized by two distinctive parts, *i.e.* the regime with no or slight acceleration of FCG compared to in air at relatively low stress intensities, and the highly accelerated regime at higher stress intensities, both of which will be referred to in the following manuscript as Stage I and II regimes, respectively. Additionally, hydrogen significantly altered the fracture surface morphologies in both Stage I and II FCG regimes. Whereas the large FCG acceleration in Stage II regime was accompanied by transgranular (TG) quasi-cleavage (QC) type fracture surfaces, brittle appearance intergranular (IG) features were observed in Stage I, contrary to the fact that there was no significant change of macroscopic FCG rate in this FCG regime. The main focus of this paper is the underlying mechanism dominating such hydrogen-induced IG crack propagation in Stage I FCG regime and its role on the macroscopic FCG acceleration. The details of Stage II HAFCG will be explained in the second part (Part II) of this series of papers presented by Birenis et al. [3].
While several researchers have also reported similar alteration of fracture features in hydrogen gas in the past, *i.e.* appearance of IG and QC with regards to low carbon or low alloy commercial steels [1,4,5], most of the interest had been particularly focused on the Stage II FCG with QC type fracture [5–7]. However, because the large part of fatigue life in practical components is

* Corresponding author: yuhei.ogawa.bz@gmail.com

generally accounted for by the crack propagation under low stress intensity, the Stage I FCG accompanied by IG fracture should not be depreciated when attempting to evaluate the overall HAFCG phenomena.
In this study, a combination work of various observation techniques including electron backscattered diffraction (EBSD), electron channelling contrast imaging (ECCI) and transmission electron microscopy (TEM) were applied to the crack propagation paths in hydrogen gas. The role of hydrogen in triggering the fracture along grain boundaries (GB) is then discussed based on the amount of plastic deformation and dislocation structures underneath the IG fracture surfaces.

2 Material and Experiments

A hot-rolled commercially pure iron (JIS-C2504) plate with the chemical composition of 0.001%C-0.07%Mn-0.010%P-0.003%S was used in this study. The yield stress, σ_y and ultimate tensile strength, σ_B, measured in air were 133 and 252 MPa, respectively. For the FCG tests, compact-tension (CT) specimens with a width, W, of 50.8 mm and a thickness, B, of 10.0 mm were cut out from the L-T orientation of the hot-rolled plate.
The FCG tests were conducted in air, 0.7 MPa nitrogen gas (purity of > 99.999%) and hydrogen gas with the pressures of 0.2, 0.7 and 20 MPa (purity of > 99.999%) at room temperature, according to the ASTM-E647 standard [8]. The constant load range condition (ΔP-constant tests) was used for obtaining the FCG curves in each testing environment. On the other hand, the tests were also conducted under constant stress intensity factor range, ΔK (ΔK-constant tests, $\Delta K = 12$ MPa·m$^{1/2}$) in air and 0.7 MPa hydrogen for analyzing the crack propagation behavior under a specific ΔK value. The crack length during the FCG tests were measured by unloading the elastic compliance method with the compliance measured with a clip-on gauge. The load ratio, R of 0.1 was chosen for all the experiments, and the testing frequency was 1 Hz for tests in air and hydrogen, and 5 Hz for the test in nitrogen gas.
The specimens subjected to ΔP-constant tests were broken into two parts by additional fatigue loading, and then used for the observation of the fracture surfaces via scanning electron microscope (SEM). On the other hand, ΔK-constant tests were halted before fracture and the specimens were cut along mid-thickness sections, then the lateral surfaces were polished with colloidal SiO_2 for the analyses via EBSD and ECCI. EBSD analysis was conducted by using a Schottky type field-emission (FE) SEM operated at 15 kV, and the electron beam step size of 0.1 ~ 1.0 μm was used for data collection. ECCI was also performed by a FE-SEM with the acceleration voltage of 15 kV. Based on crystallographic orientation maps obtained via EBSD, we evaluated the amount of plastic deformation around the crack propagation paths by using two types of strain analyses, *i.e.* grain reference orientation deviation (GROD) and kernel average misorientation (KAM). Site-specific TEM samples were extracted directly from the fracture surfaces of the ΔK-constant test specimens using a focused ion beam (FIB) machine. The samples were observed using a field-emission-TEM, operated at 300 kV in dark-field scanning TEM (DF-STEM) mode.

3 Results and discussion

3.1 Macroscopic FCG behavior

The relationships between FCG rate, da/dN and ΔK (FCG curve) measured in each testing environment are presented in Fig. 1. The FCG curves in air and nitrogen were linear throughout the tested ΔK range, and the FCG rate in nitrogen was approximately 70% slower than in air. On the other hand, clear distinction of Stage I and II crack propagation is visible on the FCG curves in hydrogen gas, demonstrating two important key findings. Firstly, the transitional stress intensities dividing Stage I and II, ΔK^T shifted to lower values as the pressure of hydrogen gas was increased. Secondly, although the FCG rate at Stage I in 0.7 MPa hydrogen was equivalent with that in air, acceleration or deceleration of FCG was confirmed in Stage I at higher and lower hydrogen gas pressures. Namely, 20 MPa hydrogen resulted in 2 ~ 3 times accelerated FCG in Stage I, whereas in 0.2 MPa hydrogen, it became slightly slower. In addition to the results of ΔP-constant tests, the FCG rates measured via ΔK-constant tests in air and 0.7 MPa hydrogen were plotted in the same diagram. No detectable difference between the FCG rates in air and 0.7 MPa hydrogen was found in ΔK-constant tests, and the values of FCG rates were almost consistent with those measured in ΔP-constant tests at corresponding ΔK values. Contrary to Stage I regime where hydrogen had only a mild influence on the macroscopic FCG rates, dramatic acceleration of FCG was confirmed in Stage II with the acceleration rate approximately 30 times compared to the FCG rate in air. Although, the ΔK^T point had strong dependence on the hydrogen gas pressures, the FCG rates after the transition showed almost the same values in 0.7 ~ 90 MPa hydrogen gas, and were lying on the line corresponding to ~30 times acceleration.

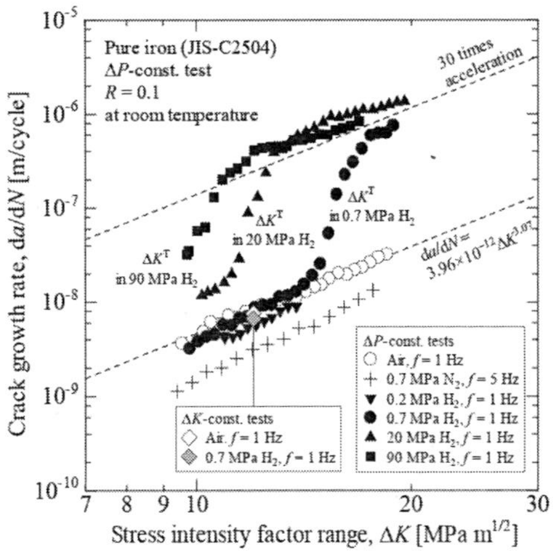

Fig. 1 Relationship between FCG rate, da/dN, and stress intensity factor range, ΔK, measured in each environment.

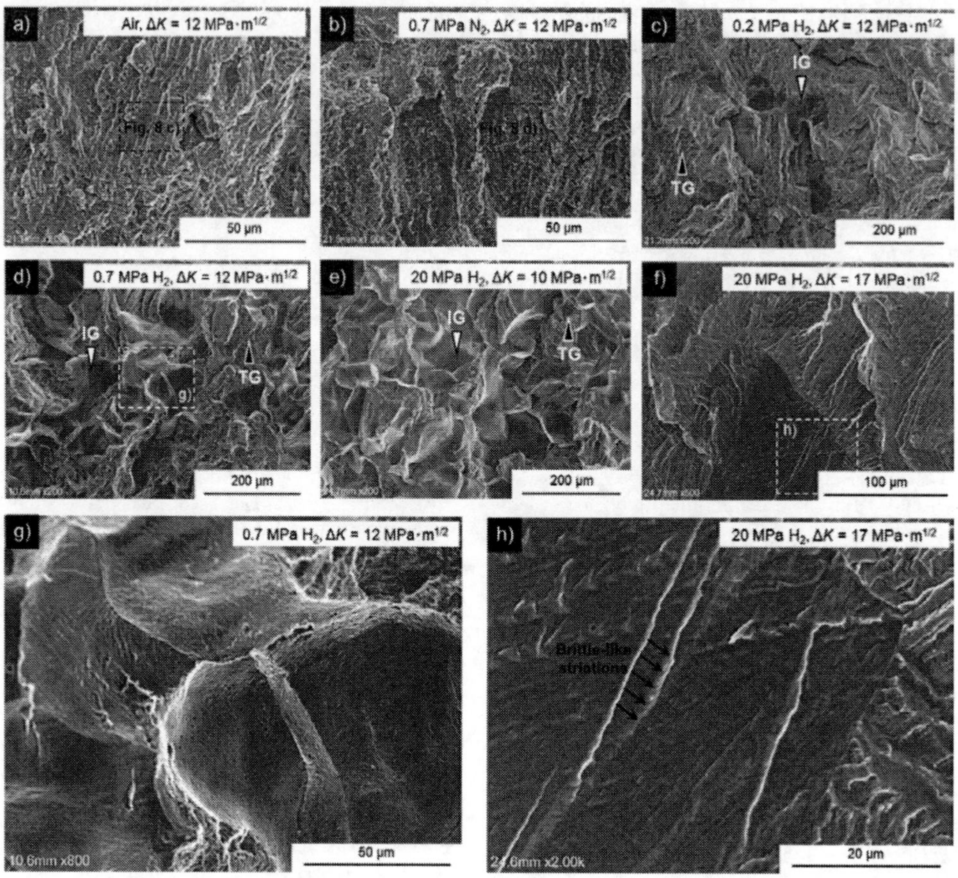

Fig. 2 SEM micrographs of the fracture surfaces formed in each environmental condition with the corresponding ΔK values of $10 \sim 17$ MPa·m$^{1/2}$. g) and h) are the magnified views of the regions surrounded by white dotted lines in d) and f) respectively. The crack growth directions are from bottom to top.

3.2 Fracture surface morphologies

Figs. 2 (a) ~ (e) present the SEM micrographs of the fracture surfaces corresponding to $\Delta K = 10 \sim 12$ MPa·m$^{1/2}$ in air, nitrogen and 0.2 ~ 20 MPa hydrogen. In the case of air, clear TG striation-like stripe patterns appeared on the whole fracture surface (Fig. 2 (a)), whereas the striations in the case of nitrogen gas became blurred in appearance (Fig. 2 (b)). The spacing of those TG striations was 1 or 2 orders of magnitude larger than the measured FCG rate at the corresponding ΔK value, which is typical for steels tested under relatively low stress intensity conditions [9,10]. Conversely, the fracture surfaces in hydrogen gas were covered by brittle-like appearance IG facets (Figs. 2 (c) ~ (e)) mixed with the ductile TG fracture features. Moreover, the formation of the IG fracture surface seemed to be more pronounced at higher pressure hydrogen (20 MPa) when we compared Figs. 2 (c) ~ (e) to each other. Figs. 2 (f) and (h) show the fracture surface formed in 20 MPa hydrogen gas at the ΔK value of 17 MPa·m$^{1/2}$, *i.e.* Stage II FCG regime. In this case, IG facets which were observed in the lower ΔK (Fig. 2 (e)) completely disappeared, and the fracture surface was covered by flat QC planes accompanied by brittle-like fatigue striations (Fig. 2 (h)), as observed in the previous papers regarding commercial steels [5–7].

In order to correlate the propensity of IG fracture and the macroscopic FCG rates in Stage I, we examined the area fractions of IG facets on the fracture surfaces and plotted them as the function of ΔK. The results of those examinations are presented in Fig. 3. In this diagram, each plot was constructed according to the three SEM pictures with the magnification of 200 times, which were taken from the positions on the fracture surfaces corresponding to the target ΔK values. The area fraction of IG reached almost zero at the onset of Stage II crack growth in 0.7 and 20 MPa hydrogen, where the fracture morphology was abruptly replaced by QC type features as shown in Fig. 2 (f). However, the more important finding here is that the fraction of IG fracture surface clearly increased with the increment of hydrogen gas pressure, meaning the macroscopic FCG rate in Stage I was controlled by the propensity of the IG cracking. Specifically, the higher IG fracture ratio resulted in the more accelerated FCG rate.

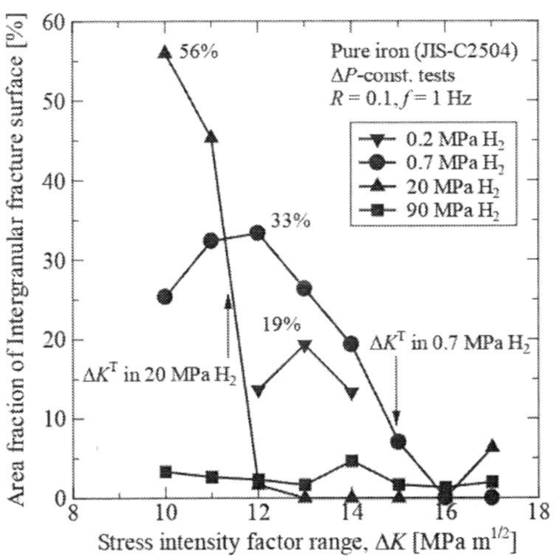

Fig. 3 Area fraction of the IG fracture surface in 0.2, 0.7 and 20 MPa hydrogen gas as the function of stress intensity factor range, ΔK.

3.3 Dislocation structure underneath the hydrogen-induced IG fracture surface

Generally, IG fracture surface is one of the representative fractographic features in the hydrogen-assisted cracking of high-strength steels [11–13]. In such cases, the crack propagation process is dominated by a relatively brittle manner both in macroscopic and microscopic scale, therefore the FCG rate can be dramatically accelerated even at intermediate testing frequencies such as $f = 1$ Hz. On the other hand, only a slight acceleration (2 ~ 3 times compared to in air) was detected in case of pure iron (Fig. 1), although similar brittle appearance IG facets were clearly observed on the fracture surfaces (Figs. 2 (c) ~ (f)), implying that the microscopic mechanism dominating the IG crack propagation in Stage I FCG in pure iron was not identical with the one in high-strength steels. In Fig. 2 (g), a higher magnification image of the IG fracture surface formed in 0.7 MPa hydrogen is presented. Even though the IG facets seemed brittle in a macroscopic view (Figs. 2 (c) ~ (e)), numerous trenches and uneven, wavy surfaces can be seen on microscopic scale, indicating the contribution of some plasticity on the fracture process. To elucidate whether this unevenness was the result of dislocation activity or not, a thin-foil sample was directly extracted from the IG fracture surface on the ΔK-constant-tested sample in 0.7 MPa hydrogen, and was then visualized using TEM. The dark-field STEM micrograph of the dislocation structure beneath the IG fracture surface in 0.7 MPa hydrogen at $\Delta K = 12$ MPa·m$^{1/2}$ is indicated in Fig. 4 (b) together with a SEM picture of the location for the TEM sample extraction (Fig. 4 (a)). The dislocations were organized into cellular structures with clear and thin cell walls, indicating that the final separation of the grain boundary was mediated by a high level of plastic strain.

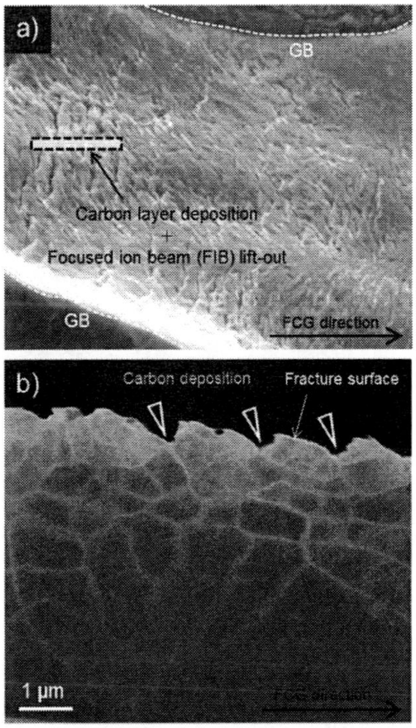

Fig. 4 (a) SEM micrograph of the IG fracture surface formed in 0.7 MPa hydrogen at $\Delta K = 12$ MPa·m$^{1/2}$ from which the thin-foil sample was extracted. (b) Dark field STEM image of the dislocation structure immediately beneath the IG fracture surface shown in (a).

3.4 Deformation structure analyses around the crack propagation paths via EBSD and ECCI

For the quantitative evaluation of the amount of plastic deformation around the crack propagation paths in air and hydrogen in Stage I FCG, the specimens subjected to ΔK-constant tests in air and 0.7 MPa hydrogen gas were cut along mid-thickness sections and the surfaces were then observed via EBSD. Figs. 5 (a) and (b) present the crystallographic orientation maps along the mid-thickness fracture paths in air and 0.7 MPa hydrogen at $\Delta K = 12$ MPa·m$^{1/2}$. In these EBSD analyses, the beam step size of 1.0 μm was used. The crack propagation in air was almost fully TG, on the other hand, in 0.7 MPa hydrogen, some parts of the fracture path were IG (indicated by arrowheads), as was confirmed in the fracture surface examination (Fig. 2 (d)). In Figs. 5 (c) ~ (f), the GROD and KAM maps which were constructed from Figs. 5 (a) and (b) are presented. Significant extent of plastic deformation can be seen around the fracture path in the case of air (Figs. 5 (c) and (e)). Likewise, severe plasticity was also detected in hydrogen both around the TG and IG fractured regions, which was in accordance with the observation result using TEM (Fig. 4 (b)). Fig. 6 shows the histograms of the GROD and KAM values constructed from the raw EBSD-scanned data indicated in Figs. 5 (c) ~ (f). According to this diagram, both of the GROD and KAM values in 0.7 MPa hydrogen gas surprisingly showed slightly higher

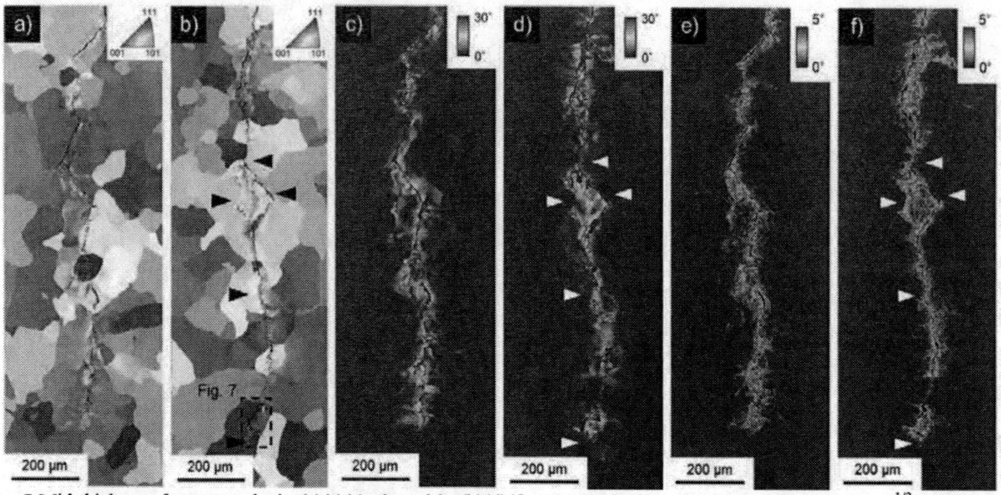

Fig. 5 Mid-thickness fracture paths in (a)(c)(e) air and in (b)(d)(f) 0.7 MPa hydrogen gas at ΔK = 12 MPa·m$^{1/2}$ analyzed via EBSD. (a)(b), (c)(d) and (d)(e) are the crystallographic orientation maps, GROD maps and KAM maps respectively. The IG fracture paths observed in hydrogen gas are marked with black and white arrowheads in (b), (d) and (f). The crack growth directions are from top to bottom.

misorientation angles than in air, meaning the crack propagation process in hydrogen was accompanied by more intense plastic deformation. In the second part (Part II) of this series of papers, it will be showed that hydrogen has the role to suppress the plasticity expansion around the crack tip region and enhances the fracture along {100} type cleavage plane in Stage II FCG regime, resulting in the formation of QC type features (c.f. Fig. 3 (f)) [3]. However, it is worth noting that the above-noted results imply the completely opposite role of hydrogen in Stage I FCG regime. Specifically, when the stress intensity level is below the critical stress intensity to trigger cleavage-like fracture along {100} plane ($\Delta K < \Delta K^T$), hydrogen accelerates the crack-tip plasticity rather than suppressing it.

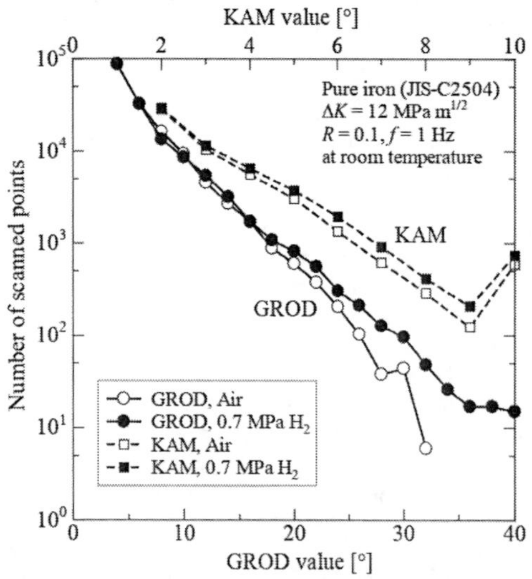

Fig. 6 Histograms of the GROD and KAM values measured in the EBSD analyses shown in Fig. 5.

Our present result regarding Stage I, *i.e.* the enhancement of plasticity under the presence of hydrogen, is consistent with the results reported by Wang et al. [14], who found that the size of dislocation cells in hydrogen-charged pure nickel subjected to high-pressure torsion (HPT) became smaller compared to non-charged material. More recently, they also examined the dislocation structures immediately beneath the IG fracture surface formed in the FCG test of a structural carbon steel tested in 40 MPa hydrogen gas, and reported that the dislocations showed smaller cell diameters compared to the cells formed beneath ductile striations in air [15]. Throughout the series of publications, they explained such alteration of dislocation behavior based on the framework of hydrogen-enhanced localized plasticity (HELP) hypothesis [16–18]. According to HELP assertion, hydrogen distributed around the dislocation stress field reduces the elastic interaction of dislocations with other obstacles, thereby enhancing their mobility and velocity [17,18], as well as hydrogen in normal interstitial lattice sites also reduces the nucleation energy of dislocations [19,20]. Wang et al. assumed that this role of hydrogen, *i.e.* easing dislocation generation and mobility, can result in the finer and more evolved dislocation structures under the presence of hydrogen [14,15], which seems to be applicable also in the case of severely strained region ahead of the fatigue crack tip in pure iron.

Fig. 7 shows the high-mag. images around the crack tip region in 0.7 MPa hydrogen obtained via EBSD (beam step size of 0.1 µm) and ECCI. In this region, the crack propagated originally through TG path (indicated by upward arrows in Figs. 7 (a) ~ (c)) and then changed its path along the GB (indicated by downward arrows in Figs. 7 (a) ~ (c)). As shown in the low-mag. observation (Fig. 5), the plastic strain visualization via EBSD (Figs. 7 (b) and (c)) demonstrated that severe plastic deformation was operated in the adjacent grains

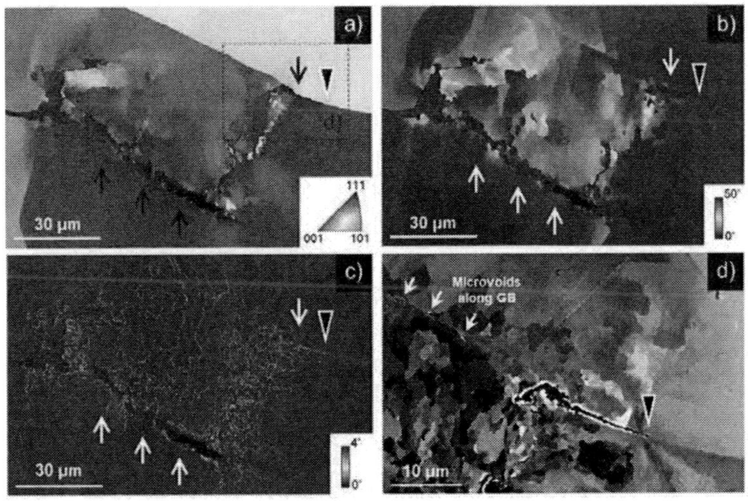

Fig. 7 High-mag. observation around the tip of fatigue crack propagated in 0.7 MPa hydrogen gas at $\Delta K = 12$ MPa·m$^{1/2}$. (a), (b) and (c) are the crystal orientation, GROD and KAM maps respectively analyzed via EBSD. (d) is the ECC image of the region surrounded by black dotted line in (a). The crack growth directions are from left to right.

surrounding the IG crack propagation path, also demonstrating the clear contrast changes in the corresponding ECC image (Fig. 7 (d)). However, it is worth noting that some microvoids were generated along the GB, where the evolved dislocation structure bumped into it as shown in Fig. 7 (d) with white arrows. Combining all the above, we can assume the following sequences for IG cracking during Stage I HAFCG:

1. Firstly, numerous dislocations are emitted from the tip of transgranularly propagating fatigue crack, and severe plastic strain ahead of the crack tip organizes them into cellular structure. As the cell structure evolution proceeds, more dislocations are absorbed into the cell walls and annihilate each other, which makes the cell walls thinner and results in the formation of sub-grain boundaries. Moreover, both of the dislocation emission [19–21] and evolution [14,15,22] processes are accelerated by hydrogen. These modifications of dislocation behavior raise higher misorientation angles in GROD and KAM analyses of the sample tested in hydrogen gas (Fig. 6).

2. Secondly, if some GBs are located near the propagating crack tip region, dislocations pile-up onto the GBs, leading to the local stress concentration and formation of some lattice defects such as vacancies due to the dislocations-GB interactions, *i.e.* disordering of the GB structure. It is commonly accepted that hydrogen in metals potentially stabilizes the point defects and enhances the generation of non-thermal equilibrium superabundant vacancies as reported by Nagumo et al. [23,24], demonstrating a possibility that the agglomeration of those stabilized vacancies can result in the nucleation of microvoids along the GBs. Recently, Lawrence et al. investigated the density of vacancies and vacancy clusters in hydrogen pre-charged single crystalline and polycrystalline pure nickel after tensile deformation by means of thermal desorption spectroscopy (TDA) and positron annihilation spectroscopy (PAS), and reported that polycrystalline material showed higher peak temperature in TDS and longer positron lifetime in PAS [25]. Their results strongly suggest that the GB itself – or the volumes adjacent to them – are the preferential sites for hydrogen-induced vacancies or vacancy cluster formation, rather than grain interior, which is in accordance with our present observation results (Fig. 7).

3. Finally, when sufficient numbers of GB microvoids are formed, their coalescence trigger the final separation of the GBs, and propel the crack tip ahead.

Referring to the TEM micrograph shown in Fig. 4 (b), we can observe the asperity formed on the IG fracture surface, which may be caused by the microvoids formation and coalescence processes. Furthermore, each pit of the troughs accurately corresponds to the positions at which the dislocation cell walls intersect with the fracture surface (indicated by black arrowheads), further supporting our assumption that dislocations-GB interaction is the main reason in triggering the microvoiding near GB regions. On the other hand, a slightly different mechanism has been proposed by Wang et al. [26] and others [16,22] to explain the IG fracture of hydrogen-charged pure nickel and pure iron subjected to monotonic tensile tests. In their assumption, the combined effect of HELP and the hydrogen-enhanced decohesion (HEDE) model [27] play a crucial role in causing the fracture along GBs. According to the HELP framework, the hydrogen atoms – which distribute inside the grains – are transported by moving dislocations [28,29] and then deposited into GBs in conjunction with the dislocations piling-up onto GBs. This process achieves the local stress concentration and hydrogen agglomeration around GBs, which, theoretically, is enough to satisfy the critical condition to operate atomistic decohesion [16,22,26]. Additionally, they also found that the development level of the dislocation structure underneath the IG fracture surface far exceeds the one under the absence of hydrogen at an equivalent strain level [22,26], as per the recent investigation conducted by Wang et al. [14,15]. Martin et al. suggested that the work hardening associated with such hydrogen-enhanced plasticity evolution has a role in assisting GB separation by rising the local stress level to exceed the ideal GB cohesive force [22]. Likewise, in our study, the higher GROD and KAM values in hydrogen demonstrated similar enhancement of plasticity, possibly resulting in the further hardening of the matrix near the crack tip region. Specifically, both the microvoid coalescence model and HELP-mediated HEDE model proposed by Martin et al. [22] may have contributed to the GB separation in Stage I, although the latter was not directly confirmed in the present study.

3.5 Comparison of the TG fracture morphology formed in air, nitrogen and hydrogen gas

If the IG fracture in HAFCG of pure iron is dominated by the aforementioned two mechanisms, *i.e.* GB microvoiding model and the model by Martin et al. [22], the propensity of IG cracking should show the dependency on the hydrogen concentration inside the material. This hypothesis can reasonably explain the dependency of IG fracture ratio and corresponding FCG rate on the testing hydrogen gas pressure as indicated in Fig. 3. Namely, higher hydrogen gas pressure, *i.e.* higher hydrogen concentration in the material, increased the propensity for IG fracture which resulted in an accelerated FCG rate in 20 MPa hydrogen gas. However, one query for such an assumption is the slightly slower FCG rate measured in 0.2 MPa hydrogen gas compared to that in air (Fig. 1). When attempting to explain this mysterious phenomenon, we have to consider both positive and negative effects of gaseous hydrogen environment for the FCG processes. Here, the positive effect represents the extremely low content of impurities, such as oxygen or water vapor in the hydrogen gas used for the experiments. It is understood that these components have a strong influence on the FCG rates of steels or aluminum alloys, and there already exist numerous studies which mention their deleterious aspect on the FCG behavior [30,31]. Additionally, it was also reported that the presence or absence of oxygen and water vapor cause a significant change of fatigue striation morphology on the fracture surface [30]. In general, the fatigue striations in non-aggressive environments are formed through the repetition of normal slip-off in loading processes and reverse-slip during unloading processes [32]. McEvily et al. found that the striations of Type304 stainless steel in a vacuum environment become more blurred in appearance compared to those formed in air, and discussed these differences with reference to the crack blunting process [30]. Specifically, the absence of adsorption of oxygen or water vapor on the bare surface in crack tip facilitates larger crack blunting and vanishes the border of striations formed in prior loading cycles. Likewise, in our study, these two aspects, *i.e.* crack deceleration (Fig. 1) and vanishing of clear striations (Fig. 2 (b)) were observed in the case of crack propagation in 0.7 MPa nitrogen gas. For comparison, in Fig. 8, the detailed appearance of the fracture surfaces formed in air, 0.7 MPa nitrogen and 0.7 MPa hydrogen gas at ΔK = 12 MPa·m$^{1/2}$, are presented. The high magnification image in Fig. 8 (b) indicates the region surrounded by white dotted lines in the overview image in Fig. 8 (a), and Figs. 8 (c) and (d) are the details of the areas surrounded by black dotted lines in Figs. 2 (a) and (b), respectively. When comparing Figs. 8 (b) and (d), the TG part of the fracture surface in 0.7 MPa hydrogen (Fig. 8 (b)) shows an obvious similarity to the fracture surface formed in 0.7 MPa nitrogen gas (Fig. 8 (d)); however, they prove to be different from the TG striations formed in air (Fig. 8 (c)). This similarity of fracture surface appearance strongly suggests that the TG crack path in hydrogen gas at Stage I has a role in decelerating the FCG rate compared to in air, whereas IG fracture surface enhances the faster crack propagation. In summary, at lower pressure (0.2 MPa), the positive effect exceeded the negative effect, while such effects balanced out in the case of 0.7 MPa hydrogen.

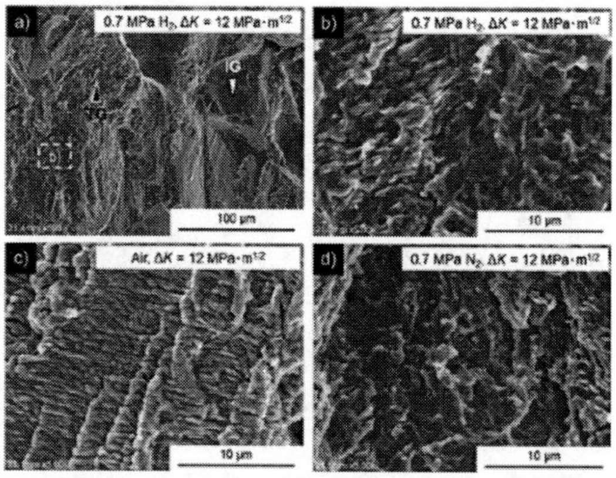

Fig. 8 SEM micrographs of the fracture surfaces formed in (a)(b) 0.7 MPa hydrogen gas, (c) air and (d) 0.2 MPa nitrogen gas at the corresponding ΔK value of 12 MPa·m$^{1/2}$. (b) shows the magnified view of the area surrounded by a white dotted line in (a). (c) and (d) correspond to the regions surrounded by black dotted lines in Fig. 2 (a) and (b), respectively. The crack growth directions are from bottom to top.

4 Conclusions

1. FCG behavior in hydrogen gas was clearly divided into two distinctive regimes, *i.e.* relatively low stress intensity regime with no or slight acceleration of FCG rate compared to in air (Stage I), and higher stress intensity regime with dramatic acceleration of FCG (Stage II). Fracture surface observation revealed that hydrogen triggers the fracture along grain boundaries (GB) in Stage I FCG, whereas in Stage II, the fracture was characterized by a quasi-cleavage (QC) type feature.

2. The FCG rates measured in Stage I showed considerable dependency on the testing hydrogen gas pressures. Namely, high pressure, *e.g.* 20 MPa resulted in 2 ~ 3 times acceleration and low pressure, *e.g.* 0.2 MPa resulted in a slightly slower FCG compared to that

in air. While at intermediate pressure, *e.g.* 0.7 MPa, the FCG rate was almost equivalent to that measured in air.

3. The area fraction of IG fracture surface in Stage I is also dependent on the testing hydrogen gas pressure as is the case with the measured FCG rate. This result indicated that increased hydrogen gas pressure, i.e. hydrogen content inside the material, enhances the propensity for IG cracking and subsequent FCG acceleration in Stage I.

4. TEM and EBSD analyses demonstrated significant plasticity around the IG fracture paths in hydrogen gas, meaning that the IG fracture in Stage I FCG is not just a brittle type fracture. Additionally, ECCI observation revealed the microvoiding along GBs near the propagating crack tip, indicating that hydrogen-modified dislocation behavior, and its interaction with GBs, plays a crucial role in triggering the IG cracking.

5. In low pressure hydrogen, *e.g.* 0.2 MPa, and 0.7 MPa nitrogen gas, the measured FCG rates showed considerably slower values compared to those in air. Even though hydrogen facilitates GB cracking on some parts on the fracture surface, other parts of the fracture surface in hydrogen, i.e. TG region, exhibited a definite similarity with the fracture surface formed in nitrogen gas. This result implies that TG crack paths in hydrogen have a role in decelerating FCG rates due to the absence of oxygen and water vapor as is the case in nitrogen gas. The macroscopic FCG rate in hydrogen gas is determined by the competition process of this positive aspect (TG cracking) and negative aspect (IG cracking).

This work was supported by JSPS KAKENHI Grant Number JP16H04238 and JP16J02960. The Research Council of Norway is acknowledged for its support through the Norwegian Center for Transmission Electron Microscopy, NORTEM (197405/F50). This study also forms part of the "HIPP" project from the PETROMAKS2 program, funded by the Research Council of Norway [Grant Number: 102006899].

References

[1] H.J. Cialone, J.H. Holbrook, Metall. Trans. A **16** (1985) 115–122.

[2] B.P. Somerday, P. Sofronis, K.A. Nibur, C. San Marchi, R. Kirchheim, Acta Mater. **61** (2013) 6153–6170.

[3] D. Birenis, Y. Ogawa, H. Matsunaga, O. TakakuwaJ. Yamabe, Ø. Prytz, A. Thøgersen, to be presented in 12th International Fatigue Congress (Fatigue 2018).

[4] A.J. Slifka, E.S. Drexler, N.E. Nanninga, Y.S. Levy, J.D. McColskey, R.L. Amaro, A.E. Stevenson, Corros. Sci. **78** (2014) 313–321.

[5] Y. Ogawa, H. Matsunaga, J. Yamabe, M. Yoshikawa, S. Matsuoka, Int. J. Fatigue **103** (2017) 223–233.

[6] Y. Takahashi, H. Nishikawa, Y. Oda, H. Noguchi, Mater. Lett. **64** (2010) 2416–2419.

[7] T.J. Marrow, P.J. Cotterill, J.E. King, Acta Metall. Mater. **40** (1992) 2059–2068.

[8] ASTM, 2010, **E647-08-1**. West Conshohocken: ASTM International.

[9] H. Cai, A.J. McEvily, Mater. Sci. Eng. A **314** (2001).

[10] A.J. McEvily, H. Matsunaga, Sci. Iran. B Mech. Eng. **17** (2010) 75–82.

[11] Z. Sun, G. Benoit, C. Moriconi, F. Hamon, D. Halm, F. Hamon, G. Hénaff, Int. J. Hydrogen Energy **36** (2011) 8641–8644.

[12] J. Yamabe, T. Matsumoto, S. Matsuoka, Y. Murakami, Int. J. Fract. **177** (2012) 141–162.

[13] C.J. McMahon, Eng. Fract. Mech. **68** (2001) 773–788.

[14] S. Wang, A. Nagao, K. Edalati, Z. Horita, I.M. Robertson, Acta Mater. **135** (2017) 96–102.

[15] S. Wang, A. Nagao, P. Sofronis, I.M. Robertson, Acta Mater. **144** (2018) 164–176.

[16] I.M. Robertson, P. Sofronis, A. Nagao, M.L. Martin, S. Wang, D.W. Gross, K.E. Nygren, Metall. Mater. Trans. B **46** (2015).

[17] I.M. Robertson, Eng. Fract. Mech. **68** (2001) 671–692.

[18] H.K. Birnbaum, P. Sofronis, Mater. Sci. Eng. A **176** (1994) 191–202.

[19] A. Barnoush, H. Vehoff, Acta Mater. **58** (2010) 5274–5285.

[20] A. Barnoush, H. Vehoff, Scr. Mater. **55** (2006) 195–198.

[21] S. Lynch, Corros. Rev. **30** (2012) 105–123.

[22] M.L. Martin, B.P. Somerday, R.O. Ritchie, P. Sofronis, I.M. Robertson, Acta Mater. **60** (2012).

[23] M. Nagumo, Mater. Sci. Technol. **20** (2004) 940–950.

[24] K. Takai, H. Shoda, H. Suzuki, M. Nagumo, Acta Mater. **56** (2008) 5158–5167.

[25] S.K. Lawrence, Y. Yagodzinskyy, H. Hänninen, E. Korhonen, F. Tuomisto, Z.D. Harris, B.P. Somerday, Acta Mater. **128** (2017) 218–226.

[26] S. Wang, M.L. Martin, P. Sofronis, S. Ohnuki, N. Hashimoto, I.M. Robertson, Acta Mater. **69** (2014) 275–282.

[27] R.A. Oriani, P.H. Josephic, Acta Metall. **22** (1974) 1065–1074.

[28] J. Tien, A.W. Thompson, I.M. Bernstein, R.J. Richards, Metall. Trans. A **7** (1976) 821–829.

[29] M. Dadfarnia, M.L. Martin, A. Nagao, P. Sofronis, I.M. Robertson, J. Mech. Phys. Solids **78** (2014) 511–525.

[30] A.J. McEvily, J.L. Gonzalez Velazquez, Metall. Trans. A **23** (1992) 2211–2221.

[31] N.M. Grinberg, Int. J. Fatigue **4** (1982) 83–95.

[32] R. Pippan, C. Zelger, E. Gach, C. Bichler, H. Weinhandl, Fatigue Fract. Eng. Mater. Struct. **34** (2011).

Fatigue behavior of 316L austenitic stainless steel in air and LWR environment with and without mean stress

Wen Chen[1,*], *Philippe* Spätig[1], and *Hans-Peter* Seifert[1]

[1]Laboratoray for Nuclear Materials, Nuclear Energy and Safety Research Division, Paul Scherrer Institute, 5232 Villigen-PSI, Switzerland

Abstract. The fatigue life design curves in nuclear codes are generally derived from uniaxial strain-controlled fatigue test results. Evidently, the test conditions are very different from the actual components loading context, which involves much more complex thermo-mechanical loading including mean stress, static load holding time and variation in water chemistry, etc. In this work, the mean stress and environmental effects on fatigue life of 316L austenitic stainless steel in air and light water reactor (LWR) environment were studied using hollow fatigue specimens and testing under load-controlled condition. Both positive (+50 MPa) and negative (-20 MPa) mean stresses showed beneficial effect on fatigue life in LWR environment and in air. This is tentatively attributed to mean stress enhanced cyclic hardening, which leads to smaller strain response at the same loading force. -20 MPa mean stress was found to increase fatigue limit, whereas the effect of +50 MPa mean stress on fatigue limit is still unclear. The preliminary results illustrate that the environmental reduction of fatigue life is amplified in load-controlled fatigue tests with tensile mean stress.

1 Introduction

Fatigue is a common mode of failure of structural materials. Austenitic stainless steels (SSs) are extensively used as structural materials in aggressive environments, such as in pressure-boundary components in the primary reactor coolant circuit of light water reactors (LWR). These components are in contact with high-temperature water and subjected to complex thermo-mechanical loading conditions during service [1]. Design against fatigue of primary pressure boundary components is usually based on Section III, Division 1 of the ASME Boiler and Pressure Vessel Code (hereafter called ASME Code). It relies on the use of fatigue curves and endurance limits, derived mainly from strain-controlled low-cycle fatigue (LCF) tests with small, smooth specimens in air at room temperature, which do not explicitly consider the possible effects of LWR environments [1]. The ASME III design fatigue curves are based on best-fit "mean" curves of the test data, which were reduced by a factor of 2 on stress/strain range or 20 on cycles (whichever is most conservative). The design margins of 2 and 20 on stress/strain and cycles, respectively, in the ASME Code design curve were intended to cover the effects of some variables (e.g., surface finish/roughness, material variability, load sequence, size effects and atmosphere) that can influence the fatigue life of components, but were not actually investigated in the tests, which provided the original data for the curves. They did not intend to cover the effects of LWR environments. Conservatism in the ASME Code fatigue evaluations may arise from (a) the fatigue evaluation procedures and/or (b) the fatigue design curves. The sources of conservatism in the procedures include the use of design transients that are significantly more severe than those experienced in service, conservative grouping of transients, and use of simplified elastic-plastic analyses that result in higher stresses/strains. However, the ASME Code permits new and improved approaches to fatigue evaluations (e.g. finite-element analyses) that can significantly decrease the conservatism in the current fatigue evaluation procedures. Uncertainties on conservatism remain due to the significant differences between the original "crude" fatigue design and real, complex fatigue loading and water chemistry conditions of large components with complex geometry in service. The original fatigue design ignores environmental effects and is derived from laboratory fatigue testing with small standard samples.

Significant environmental reduction of fatigue lives were observed in laboratory investigations under simulated LWR coolant conditions in recent years (Figure 1) [2-6] (namely environmental-assisted fatigue (EAF)). Under certain environmental and loading conditions, fatigue lives of SSs can be lower than in the fatigue design curve (Figure 1). Therefore, the margins in the ASME code may be less conservative than originally intended. The decrease in fatigue life of SS in high-temperature water depends on strain rate, dissolved oxygen (DO) level in water, and temperature and is usually more pronounced in hydrogenated pressurized water reactor (PWR) and boiling water reactor/hydrogen

* Corresponding author: wen.chen@psi.ch

https://doi.org/10.1051/matecconf/201816503012

water chemistry (BWR/HWC) conditions at low corrosion potentials (low DO) than in oxidizing boiling water reactor/normal water chemistry (BWR/NWC) conditions (high DO). The fatigue life is reduced significantly when three threshold conditions are satisfied simultaneously, i.e., when the applied strain amplitude and service temperature are above a minimum threshold level (≈ 0.15% and 150°C), and the loading strain rate is below 0.4 %/s. Environmental effects are absent or moderate, e.g., less than a factor of 2 decrease in life, when any one of the threshold conditions is not satisfied.

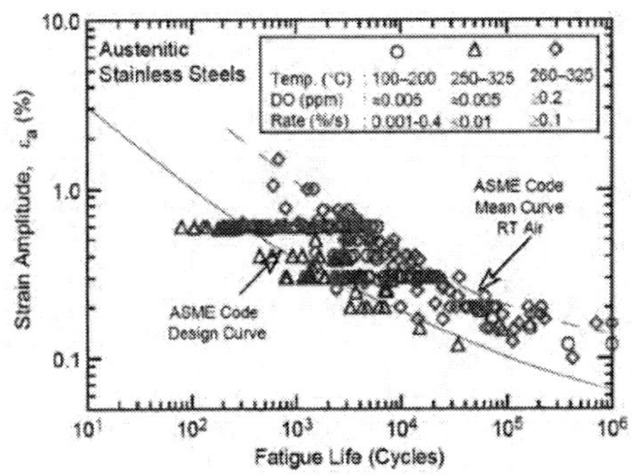

Figure 1. Comparison of fatigue test data in LWR environments with the ASME Code fatigue mean and design curves [7].

Meanwhile, based on laboratory investigations, different proposals [7, 8], such as Environmental Factors Approach, US NRC Regulatory Guide 1.207 and NUREG/CR-6909 or other national equivalents (JSME in Japan), were established for incorporating environmental effects into the fatigue design procedure of the ASME Code. These procedures are related to some relevant uncertainties and potential undue conservatism and their practical application is rather complex. Thus, they have found little acceptance in the nuclear industry so far and new international efforts (EPRI EAF, INCEFA+) were started to close some important knowledge gaps. These gaps are related to the effects of means stress, surface conditions, long static load hold periods, multiaxial loading, load history or changing temperature and strain rates.

It is well known that mean stress can affect fatigue life with tensile and compressive mean stress being detrimental and beneficial, respectively. In pressurized piping, mean stress may arise from, e.g., internal pressure, weld residual stress, dead weight loading or static temperature gradients. Mean stress effects on EAF were not investigated so far. If at all, mean stress is taken into account in the ASME Code through a modified Goodman correction, which shows that positive tensile mean stress has a detrimental effect on fatigue life. Solomon, on the other hand, observed an increase in fatigue life with positive tensile mean stress in load-controlled LCF tests with SSs [9].

The goal of the present work is to evaluate the magnitude of mean stress effects in air and high-temperature water and the adequacy or inadequacy of the typical mean stress corrections for EAF. For this purpose, load- and strain-controlled fatigue tests were performed with hollow specimens in air and hydrogenated high-temperature water and the first preliminary results of these experiments are discussed in this paper.

2 Materials and Experimental Procedure

2.1 Material and specimens fabrication

The material was provided by Sandvik as a tube of 316L low-carbon austenitic SS, with a 219.1 mm outer diameter, 19.5 mm wall thickness, and 130 cm in length. The AISI 316L steel was produced following ASTM A-511-12 and ASTM A-312-15 technical requirement. The material composition is given in Table 1. The material was delivered in heat treated solution annealed (non-sensitized) and quenched condition. It has a homogeneous texture-free grain structure (Figure 2) with equiaxed austenite grains and a high share of twin boundaries (31%) and a mean grain size of 53 µm (ASTM grain size 5.2).

Table 1. Chemical composition of the investigated austenitic stainless steel (in wt.%).

Steel	C	Si	Mn	P	S	Cr	Mo	Ni
316L	0.011	0.56	1.77	0.031	0.024	17.20	2.02	11.14

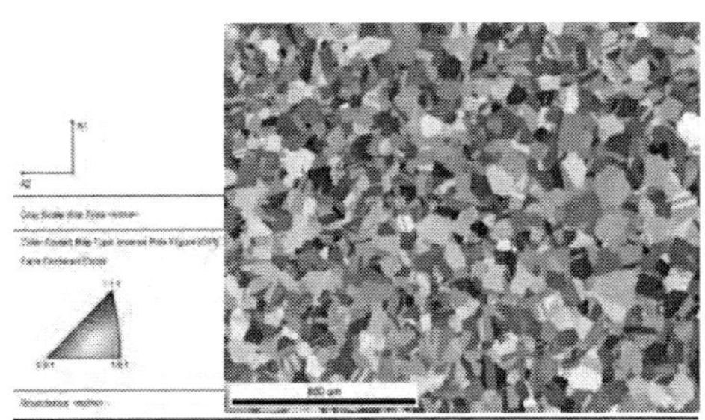

Figure 2. Grain structure (EBSD) of original material.

The mechanical behavior was characterized by tensile tests (DIN 50125, 10×50 mm cylindrical B specimens) at room temperature, 100, 200 and 300°C and a nominal strain rate of 0.1 s^{-1} (Figure 3). The measured 0.2% yield stresses decrease from 260 MPa at 20°C to 150 MPa at 300°C (Figure 4). In the same temperature range, the material provider reported elastic Young modulus decreases from 200 to 179 GPa (Figure 4).

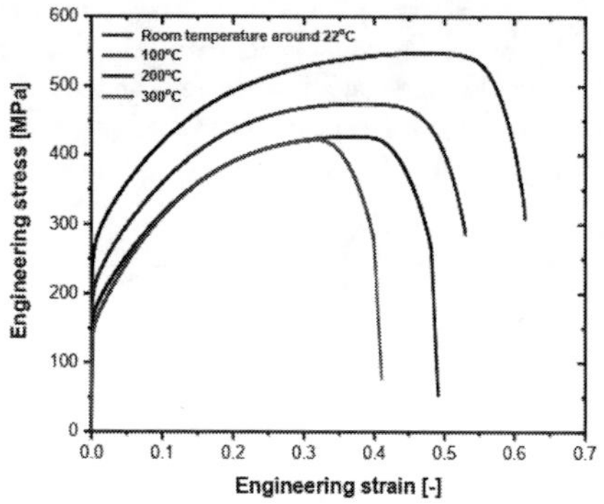

Figure 3. Engineering stress-strain curves of 316L SS tested at room temperature, 100, 200 and 300°C.

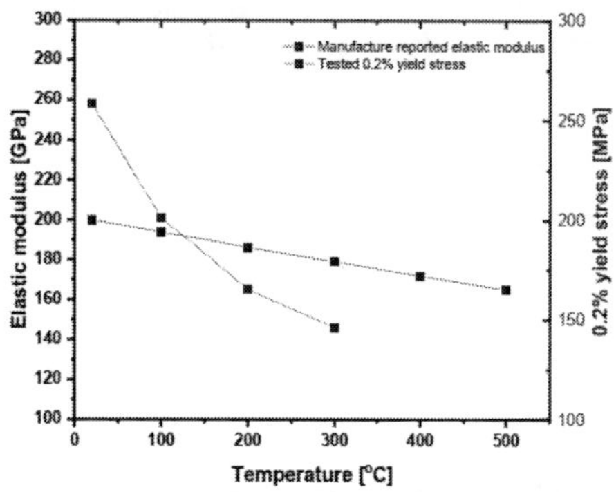

Figure 4. Measured 0.2% yield stress at 22 to 300°C and the material provider reported elastic modulus from 22 to 500°C.

Hollow specimens with an outer diameter of 10 mm and wall thickness of 2.5 mm and 2 mm were fabricated for fatigue tests in water and in air (Figure 5). The gauge length/tensile axis of the hollow specimens is parallel to the longitudinal direction of the tube. The inner and outer surface roughness of the specimens was 0.41 μm in Ra / 3.57 μm in RzD and 0.15 μm in Ra / 1 μm in RzD, respectively. The grinding scratches of the outer surface are parallel to the tensile axis and scratches of the inner surface slightly tilted to the tensile axis. The scratches on the inner surface may act as local stress raiser and thus affect fatigue initiation. However, the crack planes after fatigue tests were found perpendicular to the loading axis and scratches indicating that their influence on fatigue life is limited.

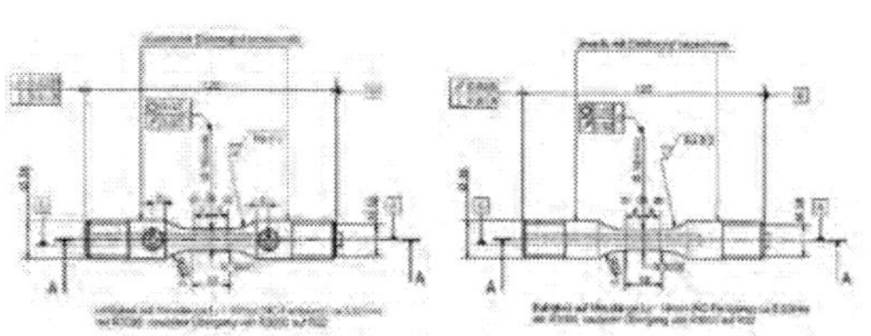

Figure 5. Hollow specimens used for the tests in high-temperature water (left) and in air (right).

2.2 Fatigue test facilities and experimental conditions

The facility for fatigue tests in high-temperature water (Figure 6) consists of an Instron 8862 electro-mechanical fatigue test machine and a water loop that allows thermo-mechanical loading under BWR or PWR water chemistry conditions. A detailed description of the facility is described in [10]. The fatigue tests were conducted at 288°C and pressurized water was characterized by high-purity, hydrogenated water with 150 ppb dissolved hydrogen (BWR/HWC). The conductivity in the high-purity inlet and outlet water were around 0.055 μS/cm (impurity content < 1 ppb). The specimen was heated by the high-temperature water flowing through the hollow specimen with minimal axial and through-wall temperature gradients. The internal pressure was 200 bars. Fatigue tests in air were performed with a Schenck RMC 100 type electro-mechanical machine. High-temperature air environment was sustained by an ATS series 2961 oven with three heating cells and a EUROTHERM 2704 type thermal controller.

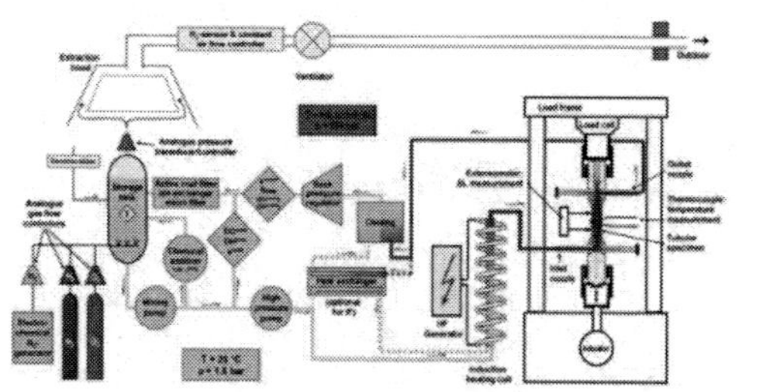

Figure 6. Schematic of facility for fatigue test in high-temperature water [10].

The tests were run both in load- and strain-controlled modes using a sinusoidal waveform at a frequency of 0.17 s^{-1}, mainly in the LCF region ($N_f \leq 10^5$). The load-controlled fatigue tests were typically performed with stress amplitudes larger than 170 MPa at 288°C, which induced elasto-plastic deformation. For EAF study, the strain amplitudes of interest were in the range of 0.1 to 0.8%. This corresponds roughly to stress amplitude in load-controlled experiment between 170 and 250 MPa at 288°C. In load-controlled fatigue tests, the strain was measured by a calibrated MTS 635.53F-30 type extensometer for high-temperature water (HTW) condition and an Epsilon model 3648-010M-025-ST extensometer for high-temperature air (HTA) condition. In strain-controlled fatigue tests, a Sandner Sensor EXA15-1u type extensometer was applied for more precise strain control. The extensometer for testing in high-temperature air was liquid cooled by Huber Minichiller 300. A stepwise ramping loading scenario was implemented to avoid over loading during the first cycles. For tests in high-temperature water, the specimens were stepwise loaded with amplitudes of +1 kN, -2 kN, +3 kN, -4 kN and coming analogous force levels each for single cycle to the desired amplitude. 1 kN corresponds to 16.98 MPa stress for HTW specimen.

Same stress step, namely 16.98 MPa, was used to start tests in HTA to ensure same loading history was imposed on the specimens in both environmental conditions.

3 Results and Discussion

3.1 Fatigue tests

Fatigue tests were conducted without mean stress and with mean stresses (-20, -10, 10, +50 MPa) in water and in air at 288°C under load-controlled mode. Strain-controlled fatigue tests were performed for comparison with load-controlled test results. The stress-life (S-N) results in water condition and in air condition are plotted in Figure 7 and Figure 8 respectively. For strain-controlled tests, the calculated average stress amplitude was considered for S-N plotting. For the tests in water environment, fatigue life was determined when a water leakage was detected. For the tests in air condition, fatigue life was defined when the measured elongation was larger than 1.0 mm or when the specimen totally broke into two parts. Optical and scanning electron microscopy observations showed that, the cracks initiated on the inner surface for tests in water and in air. In Figure 7 and Figure 8 the curves were fitted with Langer equation (1) :

$$\sigma_a = B(N_f)^{-b} + \sigma_{fs} \quad (1)$$

where σ_a stands for stress amplitude, N_f is fatigue life, σ_{fs} for fatigue strength, B and b are fitting constants to experimental data. The Langer equation fits well the load-controlled test results in Figure 7, where a fixed value of the exponent b was chosen equal to 0.9:

In 288°C BWR/HWC:

σ_m = 0 MPa: $\sigma_a = 72112.17(N_f)^{-0.9} + 170$ (2)

σ_m = +50 MPa: $\sigma_a = 144312.26\ (N_f)^{-0.9} + 170$ (3)

σ_m = -20 MPa: $\sigma_a = 166395.38\ (N_f)^{-0.9} + 194.52$ (4)

From Figure 7, we conclude that -20 MPa mean stress rises up the fatigue limit with respect to that without mean stress. However, +50 MPa mean stress does not show obvious effect on fatigue limit. While tests in air are still in progress, the available data are currently sufficient to show some trends. Only tests with $N_f < 10^5$ are presented, so that the determination of the fatigue limit (namely for $N_f > 10^6$) could not be determined. From the preliminary results, the fatigue tests with zero mean stress in air condition follow the Langer equation well.

In 288°C air:

σ_m = 0 MPa: $\sigma_a = 105919.57\ (N_f)^{-0.9} + 162$ (5)

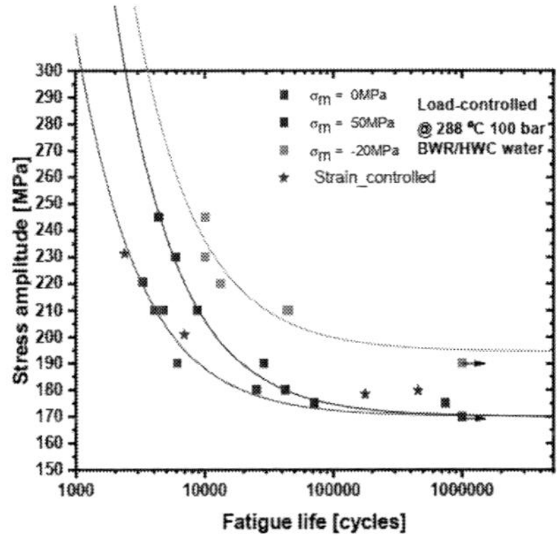

Figure 7. S-N of fatigue tests conducted under load-controlled with different mean stresses in 288°C BWR/HWC; Strain-controlled tests performed in 288°C BWR/HWC plotted with green star.

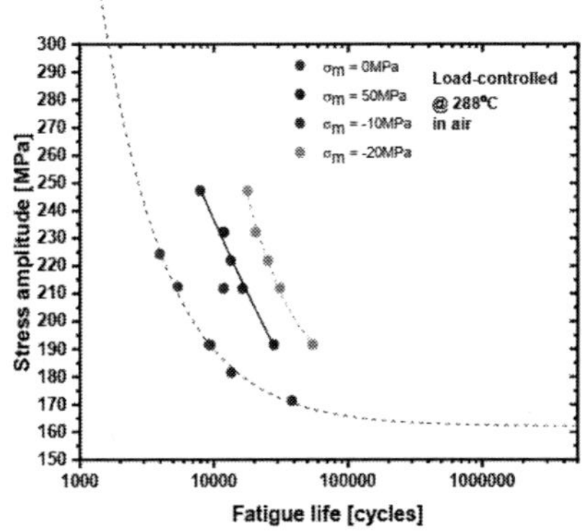

Figure 8. S-N of fatigues tests conducted under load-controlled with different mean stresses in 288°C air condition.

In 288°C air and BWR/HWC conditions, the most salient results show that both negative mean stress and positive mean stress increase fatigue life when comparing with the tests without mean stress. In air condition, more tests in the HCF region are necessary to determine the fatigue limits precisely and provide sufficient data points for a reliable fitting. However, it was still possible to estimate the fatigue life difference of the obtained results with different mean stress: for +50 MPa mean stress, the fatigue life increases by a factor of 2-3.5 and for -20 MPa mean stress the increase is about 5.7-6.5 comparing with the tests without mean stress in 288°C air. This is clearly a non-standard behavior, where one generally observes a detrimental effect of positive mean stress and beneficial one of negative mean stress on fatigue life. Solomon also reported that 100 MPa mean stress extends fatigue life on 304L stainless steel in air and PWR water at 150°C and 300°C [9, 11]. More recently, Spätig et al. reported that +50 MPa increase the fatigue life of 316L stainless steel based on load-controlled tests with different mean stresses. However, the effect of a smaller mean stress of 20 MPa or 10 MPa was quite modest and within the usual fatigue life scatter of the test without mean stress [4].

The strain-controlled test results are plotted together with the load-controlled results in Figure 7 for comparison. The calculated average stress amplitudes of strain-controlled tests are globally consistent with the S-N fitting curve at 288°C BWR/HWC without mean stress, even though different stress/strain distribution and microstructures evolution may be developed in these two loading control modes.

The strain-life (ε-N) data in 288°C BWR/HWC and 288°C air is plotted in Figure 9 and Figure 10 respectively. The strain amplitude in vertical axis is the average strain amplitude value ($\overline{\varepsilon_a} = \overline{\Delta\varepsilon}/2$) over whole fatigue life of load-controlled tests.

The test data are fitted with Equation (6):

$$\overline{\varepsilon_a} = A(N_f)^{-B} + C \quad (6)$$

The fitted curves are:

In 288°C BWR/HWC:

$$\overline{\varepsilon_a} = 2.61(N_f)^{-0.773} + 9.527\times10^{-4} \quad (7)$$

In 288°C air:

$$\overline{\varepsilon_a} = 1.29(N_f)^{-0.654} + 9.527\times10^{-4} \quad (8)$$

For some tests, the average strain amplitude is not presented in Figure 10 because the strain was not measured. We assumed however that the material has same fatigue strain limit in 288°C water and 288°C air. Thus, the fatigue strain limit (namely the constant C) in high-temperature water was considered for fitting the test data in air.

In Figure 9 and Figure 10, the test data with $N_f <$ 1000 cycles fit the strain-life model (Equation (6)) well. However, when $N_f >$ 1000 cycles, mean stress slightly deflects the test points away from the fitting curves. Kujawski [12] also found similar phenomenon when he tried to correlate experimental data of 7075-T651 Al with positive and negative mean stress.

Drawing firm conclusions on mean stress influence on average strain vs. fatigue life relationship requires more tests, both in high-temperature water and high-temperature air. However, the sole stress or average strain parameter seems to be insufficient to correlate fatigue tests data with mean stress. An approach using a parameter accounting simultaneously for stress amplitude and strain amplitude is certainly more appropriate to correlate the data with each other. For instance, correlating the data using Smith-Watson-Topper (SWT) parameter or a modified SWT parameter has already been shown to be a powerful method to correct the effect of mean stress/strain on non-symmetric cyclic loading in SSs. A modified SWT parameter correlation will be presented in a separated paper. In load-controlled fatigue test of SSs, which have FCC structure, mean stress hardens the material which is reflected by decreasing strain amplitude in comparison to tests with zero mean stress. More detail of mean stress hardening effect is addressed in the section 3.2.

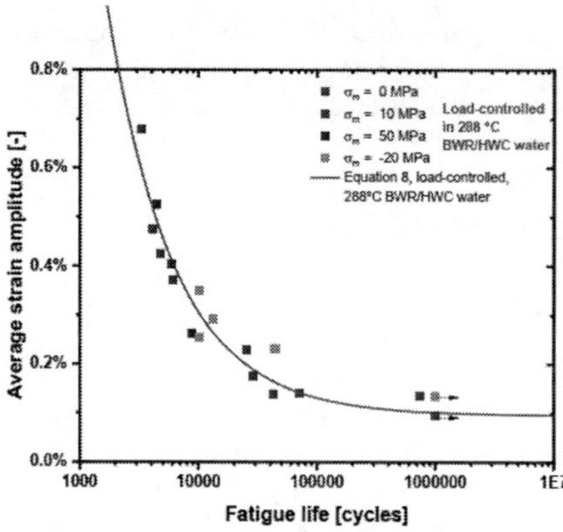

Figure 9. Strain-life of fatigue tests in 288°C BWR/HWC under load-controlled condition, the strain amplitude is the averaged value over whole fatigue life.

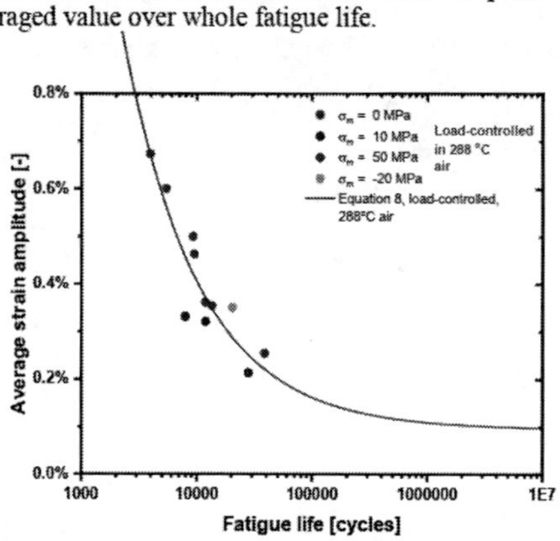

Figure 10. Strain-life of fatigue tests in 288°C air under load-controlled condition, the strain amplitude is the averaged value over whole fatigue life.

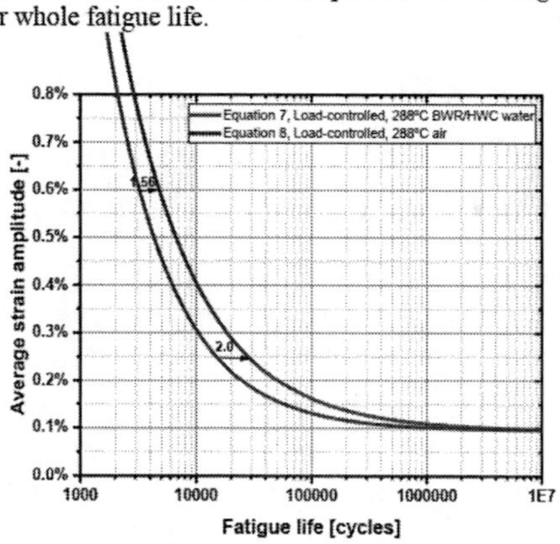

Figure 11. Fitting curves of load-controlled test data in 288°C BWR/HWC water and in 288°C air respectively based on Equation (6).

The fitted strain-life curves in water and in air conditions are plotted together in Figure 11. The F_{en} factor ($N_f^{288°C\ air} / N_f^{288°C\ BWR/HWC}$) is roughly determined to be 1.56-2.0 in the LCF region, where the average strain amplitude is larger than 0.25%. This is consistent with JSME Code's claim that F_{en} is constant for all strain amplitude larger than 0.11%. In JSME Code, no environmental effect is involved when the strain amplitude is equal or less than 0.11%. However, Kamaya [13] recently reported that the reduction of fatigue life of 316L steel in PWR water environment is not significant when the strain amplitudes are 0.25% and 0.22 %. He also observed that F_{en} varies with strain range. He concludes that larger strain range induces bigger environment detrimental effect and result in larger F_{en} value. Nevertheless, JSME Code, Kamaya's finding and most environmental detrimental effect study are based on strain-controlled tests. Their conclusions should be carefully translated to load-controlled cases. In general, reactor environment does not shift the fatigue strain limit in strain-controlled tests. Whether load-controlled tests present a similar phenomenon remains to be confirmed with more tests in HCF regime.

3.2 Means stress effect

In Figure 7 and Figure 8, both positive and negative mean stresses clearly indicate a beneficial effect on fatigue life. This effect is tentatively attributed to mean stress hardening effect, which leads to reduce strain. The strain vs. cycle relationship of several selected tests condition, namely σ_a = 190 MPa, σ_m = 0, +50, -20 MPa in both environments, is presented in Figure 12 and Figure 13. For the test without mean stress, the material firstly undergoes hardening, then softening and finally follows relatively stable amplitude for most of the fatigue life. This is similar to the behavior in strain-controlled cyclic loading. The tests with mean stress show smaller initial strain than the tests without mean stress in both environmental conditions. This may stems from a larger plastic strain hardening for the tests with mean stress, which occurs during the stepwise ramping loading cycles. The initial strain of the test with +50 MPa mean stress is smaller than the one of the test with -20 MPa mean stress. This trend continues until around 20000 cycles, where the strain amplitude of -20 MPa mean stress test becomes lower than the strain amplitude of +50 MPa mean stress test, reflecting a stronger secondary hardening with -20 MPa mean stress. The secondary hardening initiated around 1000 cycles in the test with -20 MPa mean stress.

The average strain amplitudes of -20 MPa mean stress tests are calculated to be the smallest of 0.134% (in water) & 0.234% (in air), following those of +50 MPa mean stress tests with 0.176% (in water) & 0.214% (in air). The average strain amplitudes of tests without mean stress are 0.371% (in water) & 0.5% (in air). The strain evolution in water is analogous with the one in air. With σ_a = 190 MPa, the tests without and with +50 MPa mean stress have approximately a similar fatigue life in water and in air conditions. However, the fatigue life of -20 MPa mean stress tests is much higher in water than in air. This may attribute to relatively higher strain amplitude in air than in water. When strain amplitude is below 0.2% for -20 MPa mean stress test, environmental effect is limited. In addition, fatigue life is extremely sensitive to strain range change in low strain amplitude region, namely where the strain amplitude close to the fatigue limit. These reasons may lead to higher fatigue life in water than in air for tests with -20 MPa mean stress. However, the intrinsic scatter in fatigue also has to be considered. Let's us mention that higher strain amplitude in air than in water is found in the tests with relatively low stress amplitude (such as 190 MPa).

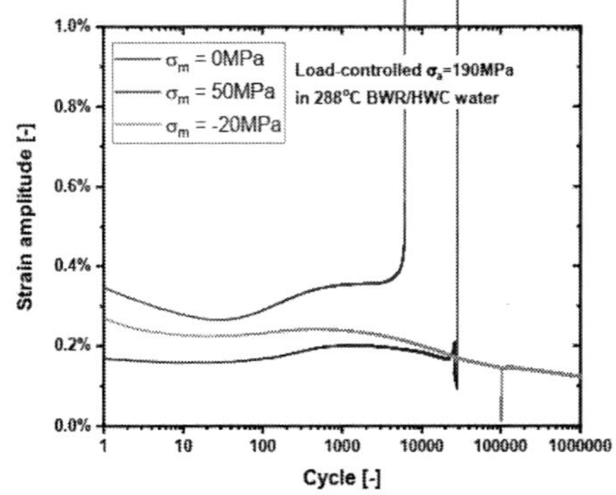

Figure 12. Strain amplitude vs. cycle of fatigue tests with σ_a = 190 MPa, σ_m = 0, +50, -20 MPa in 288°C BWR/HWC.

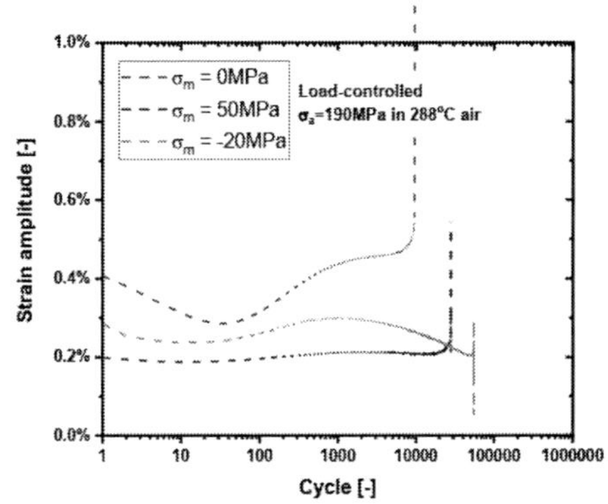

Figure 13. Strain amplitude vs. cycle of fatigue tests with σ_a = 190 MPa, σ_m = 0, +50, -20 MPa in 288°C air.

Comparing Figure 12 with Figure 13, tests in air have systematically higher starting strain than tests in water for all mean stress levels. A plausible reason is the effect of the internal pressure in the specimen tested in water. A pressure in the range 100-200 bar pressurized hollow specimen induces an asymmetric strain response in load-controlled condition [4]. More FEM stress/strain state computation is necessary to better understand the stronger hardening in water.

The physical mechanisms of cyclic hardening, softening and stabilization in strain-controlled of austenitic steels have been studied via microstructures characterization in the last decades [14-17]. However, the physical processes in load-controlled condition are not thoroughly understood, especially the mechanism of cyclic secondary hardening. In Spätig et al dislocation investigation [18], the secondary hardening is accompanied with formation of corduroy structures, which is consistent with nano-scale dislocation loops and defects free channels. This kind structure is supposed to acts as obstacles for mobile dislocations and thus strengthens the material.

3.3 Environmental effect

As mentioned in the introduction section, when temperature, strain amplitude and strain rate are simultaneously larger than critical values (≈150°C, 0.15 %, and 0.4%/s respectively), the water environment significantly assists fatigue damage. The environmental correction factor F_{en} factor was originally defined as the ratio of the fatigue life in room temperature air to the fatigue life in water reactor environment in strain-controlled conditions [19]. F_{en} factor depends on temperature, strain-rate, and dissolved oxygen content for austenitic stainless steels [1, 3, 20, 21]. In this paper, the environmental factor in load-controlled condition is defined analogously as $N_f^{288°C\ air}/N_f^{288°C\ water}$.

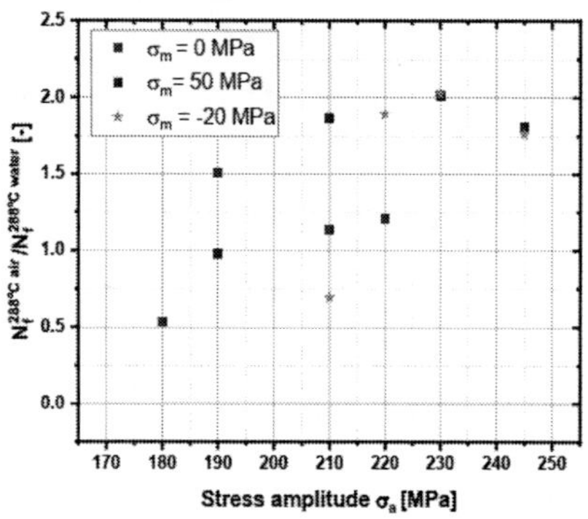

Figure 14. $N_f^{288°C\ air}/N_f^{288°C\ water}$ vs. σ_a under different mean stress of 0, +50, -20 MPa.

$N_f^{288°C\ air}/N_f^{288°C\ water}$ calculated from the experimental N_f is plotted in Figure 14 for the different mean stresses. For σ_m = 0 MPa, high-temperature water environment starts to take effect around σ_a = 185 MPa, where the corresponding average strain amplitude ($\overline{\varepsilon_a}$) is about 0.3%. For σ_m= +50 MPa, environmental effects seem to happen at a higher σ_a of about 195 MPa corresponding to $\overline{\varepsilon_a}$ = 0.2%. Finally, for σ_m=-20 MPa, environmental effects appear around σ_a = 215 MPa, where the strain amplitude ($\overline{\varepsilon_a}$) is around 0.26%. In load-controlled condition, -20MPa mean stress increases the stress amplitude threshold for environmental effect. When $N_f^{288°C\ air}/N_f^{288°C\ water} > 1$, the results with +50 MPa and -20 MPa mean stress shown in Figure 14 suggest larger environmental factors than those without mean stress. Positive mean stress amplifies environmental influence was commonly reported. However, the exact reason for the observed non-standard phenomenon originated from -20 MPa mean stress is unclear.

4 Conclusions

In this work, a series of load-controlled fatigue tests were conducted to investigate the mean stress and light water reactor environment effect on fatigue. From the presented results, it clearly show both negative mean stress (-20 MPa) and positive mean stress (+50 MPa) increase fatigue life in air or in water conditions. -20 MPa mean stress improves the fatigue life more notably than +50 MPa. The stress-life and strain-life data were fitted using Langer equation. -20 MPa mean stress increases the fatigue limit obviously, whereas +50 MPa mean stress effect on fatigue limit is unclear.

In an average strain $\overline{\varepsilon_a}$ versus fatigue life N_f representation, the F_{en} factor appears practically independent of average strain amplitude.

In stress-life representation, both negative and positive mean stress seems to enhance the environment detrimental effect. The exact reason is still unclear. The reactor water environment effects on crack initiation and crack propagation need to be investigate.

The beneficial effect of tensile mean stress is attributed to a primary and secondary cyclic hardening effect, which lead to a lower strain response under same stress amplitude. However, the physical mechanism of load-controlled cyclic hardening is not thoroughly understood. Microstructures investigation is necessary for mean stress hardening effect understanding.

Acknowledgement

This work was conducted within the frame of SAFE research project. The financial support of the Swiss Federal Nuclear Inspectorate (ENSI) is gratefully acknowledged. The technical assistance and contribution to the mechanical testing of R. Schwenold at Paul Scherrer Institute is appreciated.

References

1. D. B. Magnus Dahlberg, "Fatigue margins for austenitic stainless steels in ASME boiler and pressure vessel code- a literature study," (Swedish Radiation Safety Authority, Sweden, 2012).

2. S. M. Takao NAKAMURA, Environmental Fatigue Evaluation in PLM Activities of PWR plant. E-Journal of Advanced Maintenance **2**, 82-100 (2010).

3. M. Higuchi, T. Nakamura, Y. Sugie, Development of an Environmental Fatigue Evaluation Method for Nuclear Power Plants in JSME Code. Journal of Environment and Engineering **6**, 452-468 (2011).

4. P. Spätig, H. P. Seifert, 17th Int. Conference on Environmental Degradation of Materials in Nuclear Systems - Water Reactors (TMS, Ottawa, 2015).

5. H. P. Seifert, S. Ritter, H. J. Leber, Corrosion fatigue crack growth behaviour of austenitic stainless steels under light water reactor conditions. Corrosion Science **55**, 61-75 (2012).

6. H. P. Seifert, S. Ritter, H. J. Leber, Corrosion fatigue initiation and short crack growth behaviour of austenitic stainless steels under light water

reactor conditions. Corrosion Science **59**, 20-34 (2012).

7. O. C. a. G. L. Stevens, "Effect of LWR Coolant Environments on the Fatigue Life of Reactor Materials," (Argonne National Laboratory, Argonne, 2014).

8. U. S. N. R. C. O. o. N. R. Research, Regulatory Guide 1.207 [electronic resource] : guidelines for evaluating fatigue analyses incorporating the life reduction of metal components due to the effects of the light-water reactor environment for new reactors. (U.S. Nuclear Regulatory Commission, Office of Nuclear Regulatory Research, [Washington, D.C.], 2007).

9. H. D. Solomon, C. Amzallag, A. J. Vallee, R. E. De Lair, ASME 2005 Pressure Vessels and Piping Conference (ASME, Colorado, 2005).

10. H. J. Leber, S. Ritter, H.-P. Seifert, Thermo-Mechanical and Isothermal Low-Cycle Fatigue Behavior of Type 316L Stainless Steel in High-Temperature Water and Air. CORROSION **69**, 1012-1023 (2013).

11. H. D. Solomon, C. Amzallag, A. J. Vallee, R. E. De Lair, 15th International Conference on Environmental Degration (TMS, Colorado, 2011).

12. D. Kujawski, A deviatoric version of the SWT parameter. International Journal of Fatigue **67**, 95-102 (2014).

13. M. Kamaya, Influence of strain range on fatigue life reduction of stainless steel in PWR primary water. Fatigue & Fracture of Engineering Materials & Structures **40**, 2194-2203 (2017).

14. D. Kuhlmann-Wilsdorf, C. Laird, Dislocation behavior in fatigue. Materials Science and Engineering **27**, 137-156 (1977).

15. M.-S. Pham, S. R. Holdsworth, Evolution of Relationships Between Dislocation Microstructures and Internal Stresses of AISI 316L During Cyclic Loading at 293 K and 573 K (20 °C and 300 °C). Metallurgical and Materials Transactions A **45**, 738-751 (2013).

16. J. Nellessen, S. Sandlöbes, D. Raabe, Effects of strain amplitude, cycle number and orientation on low cycle fatigue microstructures in austenitic stainless steel studied by electron channelling contrast imaging. Acta Materialia **87**, 86-99 (2015).

17. M. Heczko, J. Polák, T. Kruml, Microstructure and dislocation arrangements in Sanicro 25 steel fatigued at ambient and elevated temperatures. Materials Science and Engineering: A **680**, 168-181 (2017).

18. M. Heczko, P. Spätig, H. P. Seifert, T. Kruml, Correlation between Dislocation Structures and Mechanical Fatigue Response of 316L Austenitic Steel Loaded with and without Mean Stress at High Temperature in Air and Water Environment. Solid State Phenomena **258**, 534-537 (2017).

19. F. H. E. de Haan-de Wilde, F. J. Blom, Overview of International Implementation of Environmental Fatigue (Update). V01AT01A015 (2017).

20. L. Midmore, "Environmentally Assisted Fatigue Gap Analysis and Roadmap for Future Research," (ELECTRIC POWER RESEARCH INSTITUTE, California, 2011).

21. O. K. Chopra, W. J. Shack, A Review of the Effects of Coolant Environments on the Fatigue Life of LWR Structural Materials. Journal of Pressure Vessel Technology 131, 021409-021409-021421 (2009).

Evolution Study of the Surface States of Low Carbon Microalloyed Steel before and after Corrosion in NaCl Solution

Faouzia Benkafada[1,*], *Djahida* Kerdoud[2], and *Ali* Bouchoucha[1]

[1]University of Constantine 1, Department of Mechanical Engineering, Constantine, Algeria
[2]University Centre Abdelhafid Boussouf, Mila, Algeria

Abstract. Surface topography is of paramount interest in science, technology, and industry. In this work investigation of the evolution of the surface with multiparameter characterization was carried out using optical profiler. Surface profile is measured two-dimensionally applying Gaussian filter of cutoff 0.8 mm. R_a, R_q, R_{sk}, R_{ku}, R_p, R_v and R_z values according to the standard ISO 4287 were measured. Following ISO 25178, 3D roughness parameters: S_a, S_q, S_{sk}, S_{ku}, S_p and S_v values were reported. The appearance of the texture of our microalloyed steel has been indicated by evaluating the texture aspect S_{tr}. Analysis of Abott-Firestone curves and related parameters S_{mr} and S_{dc} was presented. Some significant characterization parameters such as surface hardness, weight loss and corrosion rate were analysed and compared.

1 Introduction

Currently, the micro-alloyed steels have emerged as an important class of materials and drawn lots of research focus. They are called micro-alloyed steels because they contain only a small number of alloying elements such as: vanadium, niobium or titanium [1-2]. In addition, they have a ferritic matrix with extremely fine grained structure due to the effects of the alloying elements [3]. Consequently, micro-alloyed steels are generally characterized by high strength and ductility resulting from the fine grains and ferrite matrix. They are thus widely used in the oil and gas industry due to their beneficial mechanical properties.
In recent years, there is a growing awareness of the necessity to evaluate the surface quality. The surface roughness is recognized as being one of the most important properties. A wide range of roughness parameters can be used to describe a given surface [4-5]. The most common parameters are calculated from line profiles according to the ISO 4287 standard [6], but due to an increased use of 3D profilers a set of complimentary area roughness parameters have been defined in the ISO 25178 standard [7].

In the present study, an effort has been made to determine the surface properties of micro-alloyed steels by analysing the change in several 3D parameters. In addition, microstructural evaluation, surface hardness measurements, weight loss, and corrosion rate were performed and considered in order to assess the suitability of such steels for application to gas transmission pipelines.

2 Experimental details

2.1 Materials

The steel used in this study is a low carbon ferritic steel with a ferrite-pearlite fine-grained microstructure. In delivery condition, the steel manifests approx. 86.64 ferrite and 13.35 pearlite, with a measured average grain size of ASTM 8 (see Table 1) [8].

Table 1. Composition of as-received steel.

Steel	Ferrite	Pearlite	ASTM grain
Vol%	86.64	13.35	8

The chemical composition is detailed in Table 2 [8].

Table 2. Chemical composition (in wt.%) of the steel used in the experiment.

Steel	C	Mn	Si	P	S	Cr	Cu	Ni	F
Wt.%	0.111	0.955	0.175	0.005	0.022	0.037	0.293	0.013	98.3

2.2 Metallographic preparation

In order to perform the gravimetric measurements, pipeline tests of Ø12 diameter were machined into a rectangular form of 25 mm × 15 mm × 4.45 mm. Specimens were subjected to a mechanical polishing at a speed of 500 rpm using the silicon carbide abrasive papers sequentially of up to 1000 grades, the polished surface was then thoroughly rinsed with distilled water and degreased in acetone before used and dried with an electric drying machine. All experiments were carried out under thermostatic conditions 25°C (±0.1°C).

* Corresponding author: fbenkafadar@yahoo.fr

https://doi.org/10.1051/matecconf/201816503013

2.3 X-ray diffraction

The XRD analysis was performed using X'Pert3 Powder Diffractometer. Diffraction patterns were recorded at 45 kV and 40 mA in 0.013°/s detector's movement speed, the scanning was carried out with a step length of 3.35° (2θ) and a step time of 29.07 s. Flat diffracted beam pyrolitic graphite monochromator was used to remove fluorescent X-rays. It becomes very relevant when analyzing iron compounds with CuKα radiation. Noise that emerges because of the fluorescence (without monochromator) would be so large that only the most intensive diffraction maximums (e.g. from ferrite) would appear in the diffraction pattern. The diffraction patterns were recorded automatically by a data acquisition system. The peaks obtained were identified with those available in PDF-2 data base.

2.4 Surface roughness measurements

An optical profilometer, AltiSurf® 500, was used to characterize the roughness of the sample. Using the software, Phenix V2, cartographies of the disc surfaces were done with 5x5 mm dimensions. An acquisition with steps of 2 micron, a speed of 400 micron/sec and a double frequency 400/800 Hz. Roughness profiles of a 4.9 mm length were extracted from the diagonal of the cartographies and their parameters (R_a, R_z, etc.) were calculated with Altimap topography XT software. Optical images were captured by a CCD camera and then recorded.

2.5 Weight loss measurements

The samples were tinted with anti-rust paint all over the sides except around one facade that is to be buried in the ground at pH from 7 to 9. After weighing with sensitive electronic balance, samples were immersed in 10 ml 0.9% NaCl solutions. Weight loss of metal samples was noted at every 24 h time interval. The experiments were carried out in duplicate and the average values were reported.
After the completion of the test, the corrosion rate is calculated by the following equation:

$$V = \Delta M / S.\ T\ (mg/h.\ cm^2) \quad (1)$$

Where:

V : corrosion rate ($mg/h.cm^2$)
S : the total area of specimen (cm^2)
T : time of immersion (h)
$\Delta M = M_i - M_f$ (mg)
ΔM : the average weight loss (mg)
M_i : initial mass (mg)
M_f : final mass (mg)

2.6. Hardness testing

Vickers hardness of metallographic samples were evaluated using a digital Vickers microhardness and Wolpert UH930 Universal Hardness testing machines. Tests were carried out in three different positions on the polished surface of the samples and the average Vickers hardness value was reported.

3 Results and discussion

3.1 X-ray diffraction analysis

X-ray diffraction patterns (Fig. 1) of the samples, that were used in microscopic analysis, are quite similar, and all visible very sharp diffraction peaks, that have interplanar spacings *d*(110) = 0.2027 nm; *d*(200) = 0.14336 nm and *d*(211) = 0.1171 nm, belong to ferrite (α-Fe) cubic lattice. There are no other diffraction peaks, also typical to carbide phase. This confirmed the literature data that carbide phase in the steel could be identifiable by XRD only if there are 5 or more percents of it.

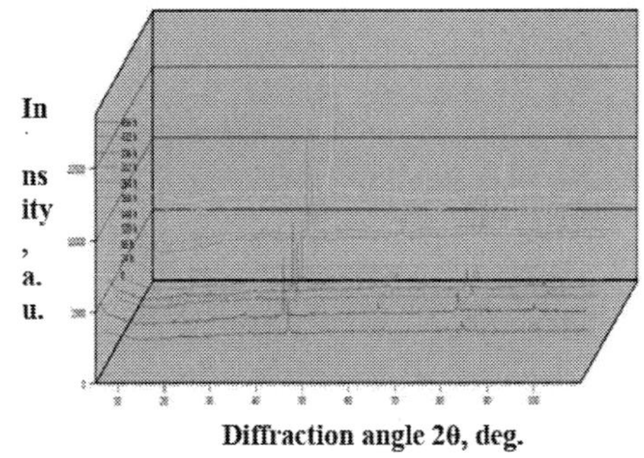

Fig. 1. X-ray diffraction patterns of the steel. Curves : 1 - 0 h, 2 - 24 h, 3 - 96 h, 4 - 120 h, 5 - 144 h, 6 - 168 h, 7- 264 h, 8 - 312 h, 9 - 336 h, 10 - 432 h, 11 - 456 h.

It can be seen that there were not any significant changes in the amount of each phase formed, and as the time exposure is extended, the intensity of the peaks corresponding to ferrite increases slightly, this indicates that the interplanar space was smaller.

3.2 Optical profiler analysis

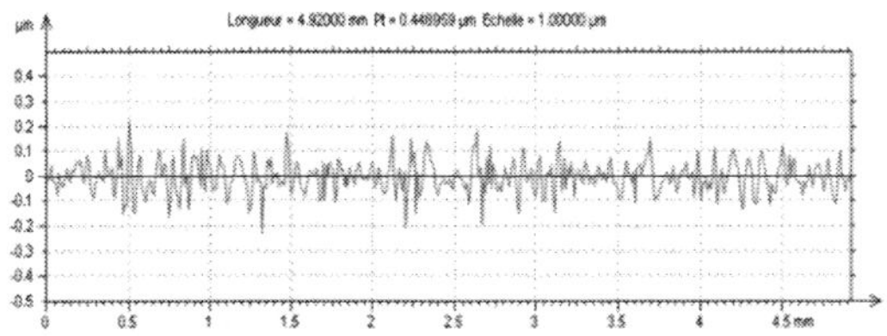

Fig. 2. 2D roughness profile of sample obtained after 144 h of immersion into NaCl solution.

Using a simple 2D profile, as shown in Fig. 2, may not provide a complete picture of the sample surface. This profile can provide accurate height, width and roughness data, but may miss detailed surface features that can be measured by adding 3D analysis capabilities.
According to ISO 4287, 2D amplitude parameters values are shown in Table 3.

For sample immersed for 144 h, it is important to note that the sum of R_P and R_V is equal to R_Z, which is lower than the value R_t. This result is in good agreement with the literature given by the ISO 4287 standard.

Table 3. 2D amplitude parameters.

ISO 4287		
2D amplitude parameters		
R_p	0.152645	µm
R_v	0.140306	µm
R_z	0.292951	µm
R_c	0.143695	µm
R_t	0.403858	µm
R_a	0.0446722	µm
R_q	0.0563578	µm
R_{sk}	-0.021757	
R_{ku}	3.47164	
Functional parameters – Roughness profile		
R_{mr}	100.000	%
R_{dc}	0.0907388	µm

With 3D capabilities, an entire area can be mapped, as shown in Fig. 3. This enables visual inspection of defects and uniformity.

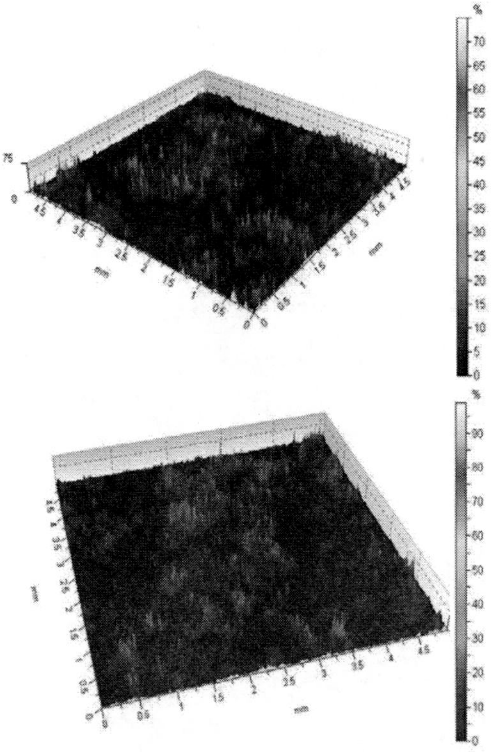

Fig. 3. 3D surface roughness image of samples obtained after 144 h and 168 h of immersion into NaCl solution, respectively.

3D roughness images for the patterned samples showed shiner blue colour after corrosion, where small surface pits and spikes could appear depending on the two basic parameters, immersion time into NaCl solution and localized area of corrosion. Other colours show deep point like grooves.

Following ISO 25178, 3D surface roughness parameters are indicated in Table 4.

Table 4. 3D height parameters.

ISO 25178		
Height parameters		
S_q	0.835091	µm
S_{sk}	0.315758	
S_{ku}	10.8696	
S_p	10.4655	µm
S_v	9.70796	µm
S_z	20.1734	µm
S_a	0.539842	µm

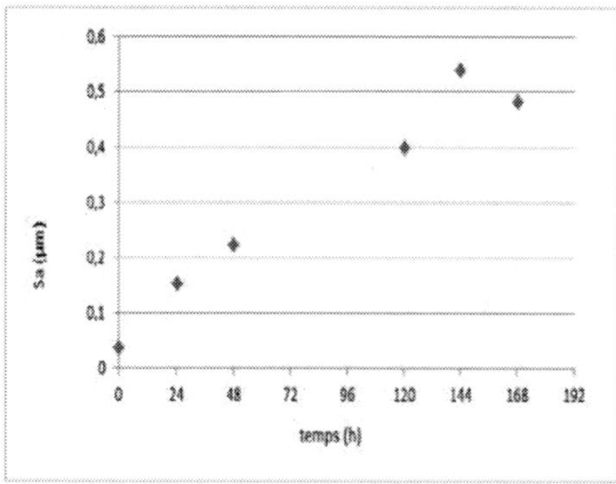

Fig. 4. Variation of arithmetic mean height S_a with immersion time into NaCl.

It is observed from Fig. 4 that an increase in arithmetic mean height S_a is accompanied by an enhancement of the corrosion. Interestingly, this trend is similar to that in the variation of corrosion rate with corrosion period (Fig. 8). It can be concluded that the corrosion rate increased with an increase of surface roughness. The reason of this phenomenon is that by increasing the roughness the active sites increased and deeper grooves played an active role as suitable places for localized corrosion. The depth of the valleys that influenced the diffusion of active ions during NaCl corrosion.

By contrast, in a smooth surfaces weak points are less activate than on rough surface and the formation of stable passive film on smooth surfaces is more likely to occur.

A bearing ratio analysis was performed on both types of samples as seen in Fig. 5. A bearing ratio curve indicates the percentage of a surface that falls above/below a particular depth. These curves quantified the percentage of valley area that tended to lead to corrosion.

Skewness and valley depth were found to correlate well with the tendency toward corrosion. Furthermore, a ratio of parameters derived from the bearing area analysis proved an excelent indicator of the tendency to corrode.

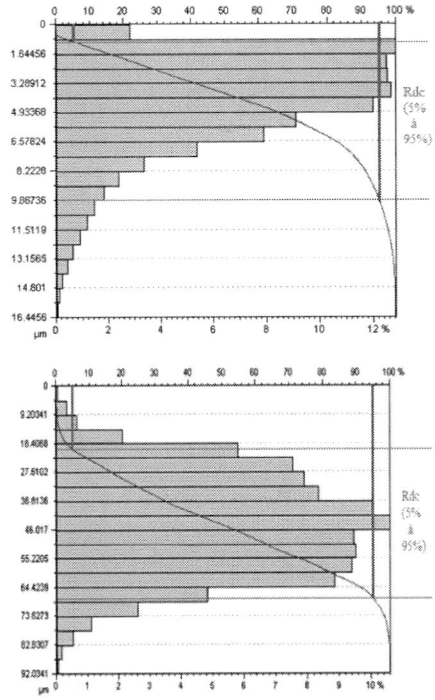

Fig. 5. Abott-Firestone curves before and after 144 h of immersion into NaCl, respectively.

Table 5. The bearing ratio values.

	Before corrosion	After 144 h of immersion into NaCl
5 % to 95% (µm)	9.04508	48.31795

From these data prsented in Table 5, it was determined that the deeper valley structure tended to hold processing NaCl solutions, allowing rusting to occur.

Surface anisotropy is also observed with the 3D roughness parameter S_{tr} ratio of texture aspect. The value can lie in the interval from 0 to 1. It can be concluded that the appearance of the surface texture after 144 h of immersion into NaCl solution is of the anisotropic type while the steel as-received is essentially isotropic before corrosion.

3.3 Hardness evaluation

Table 6. Vickers micro-hardness results.

As-received (HV)	151.0	161.6	156.4
After immersion into NaCl (HV)	66.8	78.3	81.46

After calculation, the average micro-hardness with a load of 4.903 N is found:

- Initial condition (as-received) : 156.33 HV
- After immersion into NaCl : 75.52 HV

Micro-hardness values have changed slightly from 151 to 156.4 and there is a decrease in the hardness of the surface layer after immersion into NaCl solution. It can be attributed to ferrite phase, it is expected that some ferrite zone have been formed in the microstructure of the steel. This can be related to the role of Si element (0.175%). Presence of Si element as ferrite stabilizer in the composition of the steel.

3.4 Weight loss and corrosion rate measurements

The results obtained for the variation of weight loss and corrosion rate with exposure time, with varied concentrations of NaCl solution are presented in Table 7.

Table 7. Weight loss and corrosion rate variations with immersion period and NaCl concentration.

Time (h)	Mass (g)	Weight loss (mg)	NaCl Concentration (mg/ml)	Corrosion rate (mg/h.cm^2)
0	11.67104	0	0	0
24	11.66680	4.24	10	0.04711
48	11.66027	10.77	20	0.05983
144	11.63080	40.24	40	0.07452
168	11.62938	41.66	50	0.06612

It must be pointed out that in the initial condition (steel as-received) no weight loss was seen. However, there is a significant variance between the weight losses of the samples tested in different chloride concentrations that has noteworthy effect on the weight loss of micro-alloyed steel.

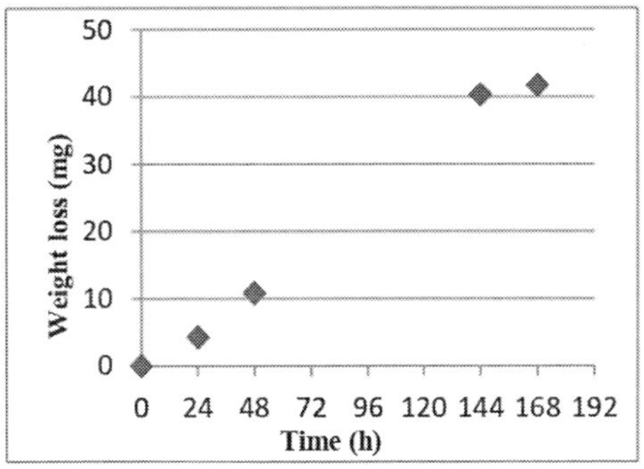

Fig. 6. Variation of weight loss with time exposure.

It is seen from the table that the samples lose weight with increasing exposure time. The observed increase in cumulative weight loss of the samples at constant temperature agrees with the observation made that corrosion rate increases with increasing immersion period time in NaCl solution, while the observed variation in the weight loss of the sample may be attributed to difference in the samples grain size [9]. It was equally revealed that concentration of NaCl affects rate of corrosion and posited that the bigger the grains of material, the more are its resistance to corrosion [10]. Hence, the as-received sample with the biggest grain size distribution revealed less weight loss as compared to the samples immersed in NaCl solution.

The NaCl medium with 10 mg/ml concentration had the least corrosion effect of the immersed samples by the end of the 24th hour achieving a weight loss value of 4.24 mg. At this value it could be considered to be fairly protective. The NaCl medium with 50 mg/ml concentration recorded the highest weight loss with a value of 41.66 mg showed that the NaCl exhibited very good corrosion at the end of the 168th hour. Additionally, it was further observed that the exposure time had greater significant effect on the corrosion rate.

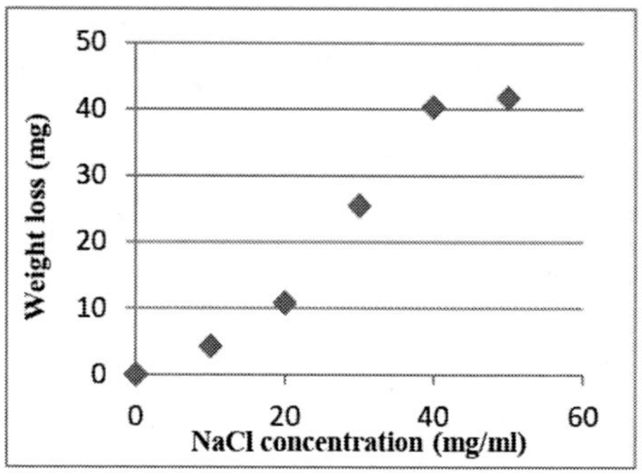

Fig. 7. Effect of NaCl concentration on weight loss.

Ferritic steels having a softer matrix can be easily and rapidly corroded by corrosion reactions depending on the time. On the other hand, hardness of the steel matrix provides good chemical and mechanical properties. Chloride ions induce localized corrosion on carbon steels and can as well lead to the formation of significant amount of corrosion products on steels [11].

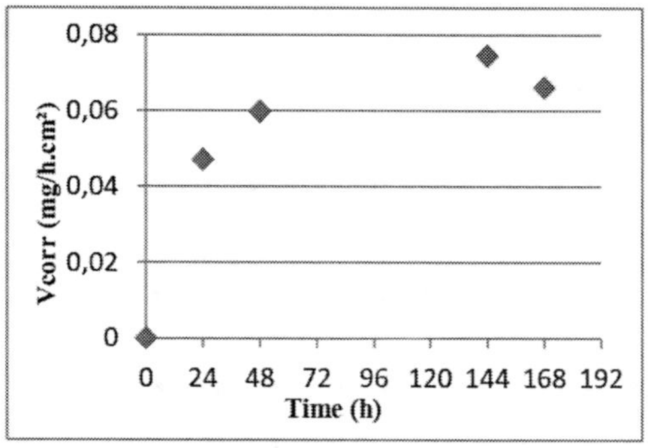

Fig. 8. Variation of corrosion rate with immersion time into NaCl.

Our results show that the corrosion rate of the steel increases sharply and then decreases to stabilize at a value close to 0.07012 ($mg/h.cm^2$) due to the presence of a passive film. Most corrosion-resistant metals and alloys are in the passive state [12], which confirms our results.

Due to the presence of the passive film, the passive dissolution which corresponds to a certain potential is slower than the active dissolution. It depends, inter alia, on the properties of the film, in particular its solubility in the electrolyte. During passivation, in other words during the transition from the active to the passive state, the dissolution rate decreases [13], which is in good agreement with our results.

The increasing degradation of micro-alloyed steel due to chloride as detected in this study compares well with outcomes on corrosion performance of carbon steel. However, micro-alloyed steel may be supposed to be basically well-suited for gas transmission pipelines application since its corrosion rates are characteristically low.

4 Conclusions

Based on the experimental results, the following conclusions can be drawn from the present investigation:

i. The XRD technique enables to identify the phases formed and the intensity of those peaks defines the amount of the steel phases. The microstructure of the samples consists of fine ferrite. The amount of ferrite reduced with hardness due to corrosion in NaCl solution.

ii. Weight loss of the samples is determined as a function of immersion period into NaCl solution. The results show that the microstructure and micro-hardness at different exposure time directly affect the corrosion rate, as the micro-hardness decreases the corrosion rate increases. The samples displayed higher corrosion rate. The presence of a passive film provides a higher protection characteristic.

iii. Sodium chloride concentration of up to 50 mg/ml increases the weight loss. Chloride further increases the corrosion rate. In the absence of chloride, no weight loss and corrosion rate were observed.

iv. Many parameters have been established regarding the measurement and assessment of surface roughness. Surface parameters have been shown to exert a great influence on corrosion and corrosion rate where the amplitude parameters have the most significant effect.

v. Micro-alloyed steel is supposed to be well-suited for gas transmission pipelines application due to its very low corrosion rate.

References

[1] M. Korchynsky, Proc. Conf. on Microalloying Union Carbide Corp., New York, (1977) 75.

[2] Yakubtsov I.A., Poruks P. and Boyd J.D. (2008) Microstructure and mechanical properties of bainitic low carbon high strength plate steels. Mater Sci Eng A, 480, 109-116.

[3] Seikh EM, Seikh AH. Effects of grain refinement on the corrosion behavior of micro-alloyed steel in sulfuric acid solutions. Int J Electrochem Sci (2012); 7:7567-78.

[4] F. Szigeti, Gy. Varga, G. Dezső, L. Péter, A. Százvai: Forgácsolási paraméterek felületi érdességre gyakorolt hatása környezetkímélő megmunkálásoknál, MTÉAR (Műszaki Tudomány az Észak-Alföldi Régióban) 2010 konferencia, Nyíregyháza, 2010 Május 19. kiadvány, p.193-198. (ISBN 978-963-7064-24-1, ISBN 978-963-7064-23-4).

[5] F. Szigeti, Gy. Varga, G. Dezső: Experimental investigation on roughness of drilled surfaces resulted from environmentally conscious machining, International Journal of Engineering, VIII. (2010) 3. pp.: 313-317. ISSN: 1584-2073.

[6] ISO 4287 – Geometrical product specifications (GPS) - Surface texture : Profile method - Terms, definitions and surface texture parameters (1997).

[7] Geometrical product specifications (GPS) - Surface texture: Areal - Part 2: Terms, definitions and surface texture parameters (ISO 25178-2:2012).

[8] Seminar IAP 24; "Corrosion et Protection des Canalisations", 28-03-2013.

[9] D. T. Oloruntoba, O. O. Oluwole, and E. O. Oguntade, "Comparative study of corrosion behaviour of galvanized steel and coated Al 3103 roofing sheets in carbonate and chloride environments," Materials and Design, vol. 30, no. 4, pp. 1371-1376, (2009).

[10] G. H. Aydogdu and M. K. Aydinol, "Determination of susceptibility to intergranular corrosion and electrochemical reactivation behaviour of AISI 316L type stainless steel," Corrosion Science, vol. 48, no. 11, pp. 3565-3583, (2006).

[11] Roman J, Vera R, Bagnara M, Carvajal AM, Aperador W. Effect of chloride ions on the corrosion of galvanized steel embedded in concrete prepared with cements of different composition. Int J Electrochem Sci (2014); 9:580-92.

[12] H. Bentrah, Master dissertation, University of Biskra, 2007.

[13] K. Mohamed Lakhdar, Master dissertation, "détérioration des canalisations sous l'effet des différents facteurs de la corrosion", University of Biskra, 2014.

Fatigue life enhancement of cast Mg alloy by surface modification in cold spray process

*S.K. Shaha, S.B. Dayani, H. Jahed**

Department of Mechanical & Mechatronics Engineering, University of Waterloo, 200 University Ave W, Waterloo, ON N2L 3G1, CANADA

Abstract. To improve the fatigue life of Mg alloy, high strength AA7075 spherical powder was deposited on AZ31B samples in nitrogen carrier gas environment at 400°C temperature in cold spray process followed by electrostatic painting with zinc phosphate. The fully reverse four-point rotating-bending fatigue tests were conducted on the coated and uncoated samples in different environmental conditions. It is seen that the cold sprayed AA7075 improved the yield strength of the AZ31B cast alloy. Similarly, a significant fatigue enchantment was observed in the coated samples, reaching a fatigue strength of 90 MPa compared to as-cast fatigue strength of 70 MPa at 10^7 cycles when tested in air, and fatigue strength of 80 MPa under a 3.5%NaCl testing environment. The SEM analysis at the interface of the tensile tested sample exhibited the interfacial fracture followed by delamination of the coating. Similarly, delamination of coating was merely detected in the fatigue fracture sample tested in the air, while the sample tested in the corrosive environment showed pits which allowed the solution to penetrate in the AZ31B substrate results delamination and premature failure. However, the presence of e-paint forms a passive layer which is hindering the pit formation and extending the fatigue life.

1 Introduction

Cold spray (CS) or gas dynamic cold spray, also known as supersonic particle deposition, is a high-efficient solid-state coating method and powder consolidation process [1]. As seen in Fig. 1, CS uses high-pressure carrier gas (usually nitrogen or helium) heated electrically, to obtain the very high velocity of the metal powders passing through a de Laval nozzle above the critical velocity to adhere particles. The most predominant bonding mechanism in CS is considered to "adiabatic shear instability". When the particles moving at the critical velocity to impact the substrate, a strong pressure field in spherical shape propagates at the interface of the substrate and particle at the point of contact. Because of the pressure field, shear load is produced which causes plastic deformation of the material laterally and creates localized shear straining. The shear straining under critical circumstances exhibits adiabatic shear instability results thermal softening. The adiabatic shear instability forms a viscous flow at outward direction leading to the temperatures near to material's melting temperature, which basically forms a bonding between the coating and substrate [2, 3]. So, a combination of mechanical interlocking along with the metallurgical bonding creates recrystallization at high strained coating/substrate interfaces is the main bonding mechanism in CS [4]. This system shows significant benefits over other coating systems. It can form mixtures of metallic and non-metallic particulates to create a coating. This is suitable for processing of heat sensitive materials as there would be very less heat input with no "heat-affected zone". Most importantly, CS coatings shows low porosity with high deposition efficiency [5, 6]. The CS process can be applied to dimensional restoration, coatings to improve the corrosion and wear-resistant, and repairing of structural components [7].

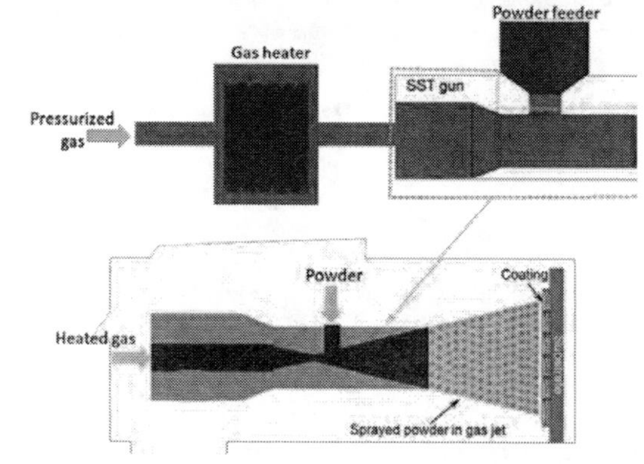

Fig. 1. A schematic illustration shows the cold spray system to deposit powder using Supersonic Spray Technology (SST).

Magnesium (Mg) alloys are the lightest metallic material, which is attractive to automotive, aerospace and electronics for light weighting trades [8-11]. Mg alloys are positioned in the anodic region of galvanic

* Corresponding author: hjahed@uwaterloo.ca

series. In general, the least noble of metals are typically corroding in the galvanic cell. Thus, Mg alloys are susceptible to form oxides, chlorides, and sulfides when in the harsh environment which limits its extensive applications. Studies reported that the fatigue and corrosion properties of Mg alloys can be improved by surface modification in CS method [12-18] by employing different coating materials, which has higher strength and better corrosion resistance. Aluminum and its alloys (AA) have shown to be an excellent coating material for coating the surface of the Mg alloy parts [19]. When AA comes in contact with air a thin aluminum oxide layer is made which protects the substrate from penetration of the oxygen which in turn results better oxidation resistance. A number of research work has been reported on the surface modification of Mg alloy substrate in CS process using AA as coating materials [12-18]. Tao et al. [20, 21] investigated the corrosion behaviour of AZ91D Mg alloy and concluded that the main cause for Mg shortening is its corrosion fatigue life. They concluded that the corrosion performance of the AZ91D can be significantly enhanced by surface modification in CS process using pure aluminium [20, 21]. Diab et al. [13] studied the fatigue behavior of pure aluminium coatings on extruded AZ31B Mg alloy in corrosive environment and reported a marginal improvement in fatigue performance. Other research also confirmed that the surface modification of extruded AZ31B alloy in CS process using pure Al increased the fatigue strength by 9% [22]. It is seen that surface modification in CS process using pure aluminium cannot meet the industrials requirements. Our recent studies [10] established that the fatigue performance of the cast AZ31B Mg alloys in different environmental conditions increased up to 25% by depositing AA7075 powder. However, it was seen that the CS sample tested in corrosion environment exhibited localized corrosion due to the presence of residual stress, which caused the premature failure results shorter the fatigue life. In practice, most of the case, the automotive parts are coated with inorganic top coat, which basically improve the corrosion resistance of the outer surface. However, there was no study on the effect of surface modification in CS process with addition of top-coat on the fatigue performance of the cast AZ31B Mg alloy. Therefore, in the present study, the feasibility of CS process on the surface modification to improve the corrosion fatigue performance was investigated. Here, the corrosion fatigue performance of the CS samples were investigated for different environmental conditions and the obtained results were discussed.

2 Experimental Details

In this study, the surface of the as-cast AZ31B Mg alloy was modified using powder of AA-based wrought AA7075 having higher strength and longer fatigue life than the AZ31B. As seen in Fig. 2, the shape of particles of the powder was spherical with an average size of 34 μm. The surface modification was performed at the Fatigue and Stress Analysis laboratory of the University of Waterloo, Waterloo, Canada, using the Supersonic Spray Technologies (SST) Series P CS system manufactured by Centerline Ltd. First, dog bone shape fatigue samples were extracted from a 300-mm diameter billet of as-cast AZ31B following ASTM: E8/E8M-11. Then, the sample surfaces were polished up to 600 grid SiC emery paper and rinse with acetone. Details about the coating parameters can be seen in [10]. As a top-coat, zinc phosphate was employed in e-painting process on the surface of the uncoated AZ31B and the sample coated with AA 7075.

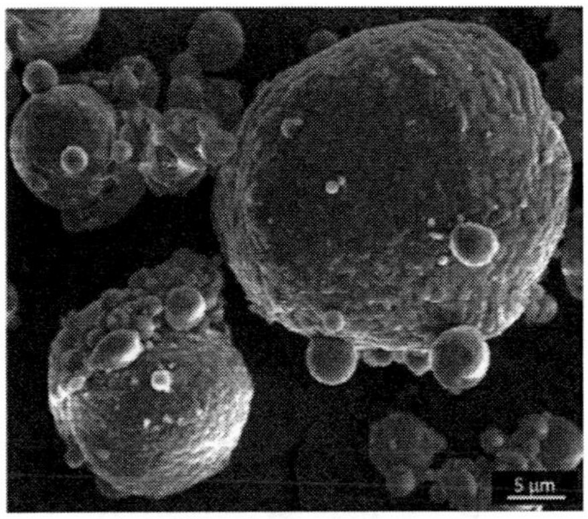

Fig. 2. SEM image shows the morphology of the AA7075 powder used as coating material.

Coated dog-bone shape specimens with a gauge length of 25 mm and a diameter of 7 mm were used for tensile and fatigue tests following ASTM E8. The tensile tests were performed on the coated and uncoated samples. A fully reversed (R=−1) load-controlled four-point bending fatigue tests were conducted using an Instron RR Moore machine. In this study, four types of samples were used for fatigue test. As seen in Fig. 3, they are: (1) uncoated, (2) e-painted only, (3) CS AA7075 coated only, and (4) CS AA7075 followed by e-painted (CS+EP). The tests were conducted at different stress amplitudes between 40 to 145 MPa to cover low and high cycle regime.

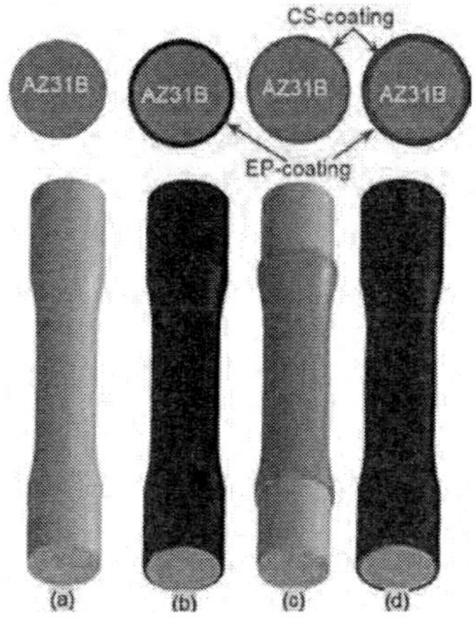

Fig. 3. Schematic of AZ31B fatigue samples used in this study: (a) uncoated, (b) e-paint, (c) coated with AA7075 (CS), and (d) coated with AA7075 followed by e-paint

A custom-made chamber was installed to the fatigue machine to perform the corrosion fatigue test (Fig. 4). From the top of the chamber, a pipe was inserted to flow the 3.5% NaCl solution onto the specimen gauge section. A constant flow rate of 40 ml/min during testing was maintained to make a continuous film of solution on sample. For all fatigue teste, a constant frequency of 30 Hz was maintained, and tests were stopped when samples break into two parts or reached at 10^7 cycles, which is considered as a run-out. At least two samples were tested for each condition. The tensile and fatigue fracture samples were then characterized using SEM.

Fig. 4. An experimental setup for corrosion-fatigue tests showing the NaCl solution chamber around the specimen.

3 Results and Discussion

3.1 Coating quality

Fig. 5 shows the obtained topography of the AZ31B substrate and the AA7075 coated sample in as-deposited and after grinding conditions. The root means square average roughness (R_a) values as typical surface characteristics measurement is considered as surface roughness. The measured R_a of the substrate (Fig. 5a) is much lower than the coated sample (Fig.5b, c). The measured R_a value for the substrate is 0.37 µm. Compared to the substrate, the average R_a value for the AA7075 CS coating surface is 25 times higher and close to 9.61 µm. The deviation of R_a is due to differences in the step-over distance, the standoff distance, and AA7075 powder size. It is also well known that the CS process exhibits rough surface finish due to powder consolidation process. The surface roughness is a critical issue in fatigue testing. As the Ra value of the as-coated sample was very high, surface roughness was reduced by grinding to 2.61 for tensile and fatigue tests.

3.2 Behavior of coating in tensile loading

The engineering stress-engineering strain plot of the surface modified, and unmodified conditions tested at room temperature is shown in Fig. 6. In general, the ductility and ultimate strength in both conditions are about to the same. However, a significant increase in yield strength (YS) of the surface modified samples was observed. The obtained results show that the as-cast AZ31B sample has a YS of 87 MPa, an ultimate tensile strength (UTS) of 196 MPa and an elongation of 12% during quasi-static tensile loading as seen in Fig. 6.

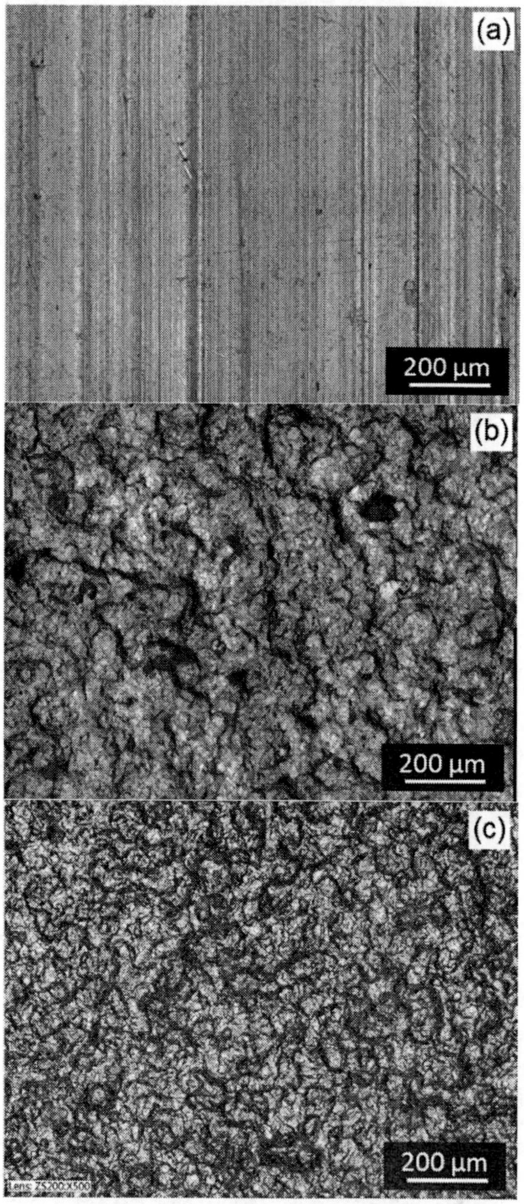

Fig. 5. Topography shows the surface morphology of the (a) uncoated AZ31B and (b) AZ31B substrate coated with Al7075 in as-coated condition and (c) after polishing.

In contrast, the surface modified samples show significantly higher YS values of 110 MPa where the coating surface was delaminated and cracked (marked by red arrow). The obtained higher YS was basically due the combination of peening effect, covering of porosity and higher strength of coating materials as mentioned in [7, 23].

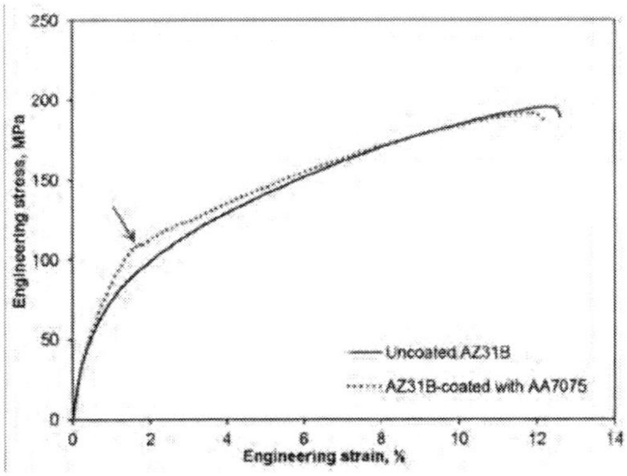

Fig. 6. Tensile results showing a comparison between the uncoated and coated AZ31B cast alloy.

There were many studies reported the tensile fracture surface of the AZ31B alloys; e.g., [10]. However, there was no information in the literature on how the CS coating behaves during tensile loading. Thus, in this study, the coating fracture and coating/substrate were investigated using SEM. As seen in the Fig. 7a, the spherical powder was plastically deformed and elongated due to the tensile load. At the same time, a number of secondary cracks in the coatings were identified. This is an indication of strong bonding between the particles.

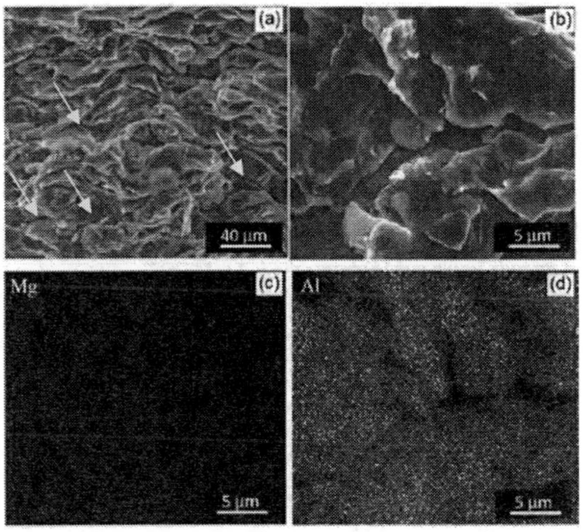

Fig. 7. SEM micrographs shows tensile fracture surfaces in the AA7075 coating (a) cross-section of the coating and (b) at the coating-substrate interface with corresponding EDX image mapping in (c) and (d).

The SEM image at the interface along with the EDX elemental distribution of the Mg and Al are presented in Fig. 7 (b-d). It is seen that the concentration of the AA is much higher than the Mg which indicates that the coating was delaminated and separated from the substrate inside the coating. Dayani et al. [7] reported that cast AZ31B coated with AA7075 shows higher bonding strength with cohesion type failure in bonding strength test. However, due to the different Young's modulus of the coating/substrate and higher bonding strength between them, delamination was observed in the coating, which basically led to the failure at slightly lower UTS.

3.3 Behavior of coating on cyclic loading

Fig. 8 illustrated the *S-N* curves of the coated and uncoated samples tested in air and corrosive environment of 3.5% NaCl solution at the different stress amplitudes. For comparison purposes, and to show the improvement in fatigue life in the corrosive environment, some earlier fatigue test results are also added to the figure. In particular, the *S-N* curves from fatigue testing of AZ31B extrusion surface modified with CS coated by pure aluminium tested in similar environmental conditions [13, 22] are also presented in the same figure. Generally, the samples modified with AA7075 alloy coating followed by e-paint obtained longer fatigue life related to the others. The fatigue strength of EP on substrate and CS+EP samples increased significantly under corrosive environment, even higher than the as-coated samples. At the same time, the CS+EP sample shows better fatigue behavior compared to the CS pure-Al on AZ31B extrusion in 3.5% NaCl solution [13, 22].

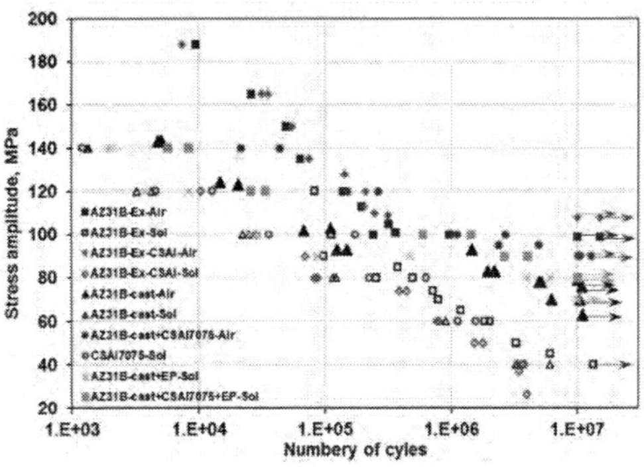

Fig. 8. A comparison of fatigue life of coated and uncoated AZ31B substrate in different environmental conditions.

Unlike the as-cast and as-coated sample fatigue behavior of EP on cast AZ31B and EP on AA7075 coating exhibits a plateau in *S-N* curve at the stress amplitudes of 50 MPa and 80 MPa, respectively. By contrast, the CS sample displays underprivileged fatigue behavior in the corrosive environment. The shorter fatigue life can be correlated to the presence of residual stress in the CS coating resulting stress corrosion cracking [6, 7]. Alternatively, the improvement of fatigue life in CS+EP can be associated with the corrosion properties. As of our previous study [24], the presence of EP increased the corrosion resistance significantly. Thus, the CS+EP samples obtained higher corrosion resistance results comparatively a reduction of pit formation which delaying the nucleation of cracks and extending the fatigue life. However, the induced residual stress during surface modification in CS process obtained a positive effect on the fatigue performance enhancement [25]. Compared to the unmodified as-cast and extruded AZ31B samples tested in all conditions, the fatigue performance of the surface modified AZ31B cast is improved remarkably in both, air and corrosive environment.

Fig. 9 illustrates SEM images of the fatigue fracture surfaces of CS and CS+EP samples tested at the stress amplitudes of 100 MPa in overall view and magnified views. It is seen that the there was no delamination visible at the interface of the CS sample tested in the air. As seen in the Fig.9 (a, c), the cracks were nucleated from the casting defects and propagate to the substrate. When the cracks reached at a certain length finally tensile like fracture (Fig. 7a) occurred in the coating. In contrast, the CS+EP, with a life around 3 million cycles, shows localized cavity (yellow arrows in Fig 9d) which permitted the NaCl solution to enter into the cavity and react with the substrate at the interface. Then, the little corrosion pits grow further and created a large cavity which basically causes a delamination of the coating. However, the presence of EP, delaying the formation of pits and prolongs the fatigue life.

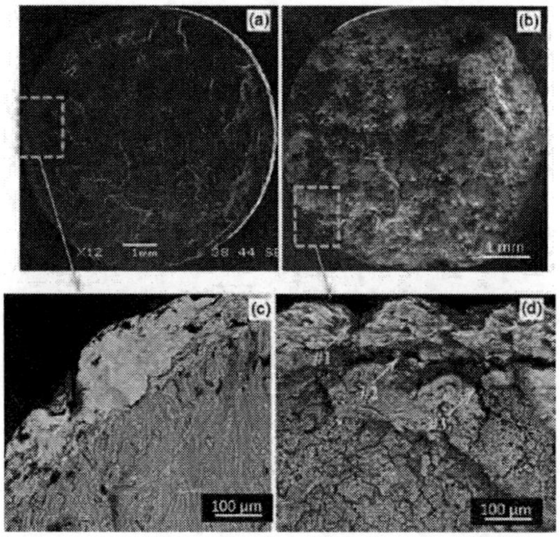

Fig. 9. SEM images show the overall (a, b) and magnified views (c, d) of fatigue fracture surfaces for the AZ31B magnesium alloy coated with AA7075 alloy samples tested at total stress amplitudes of 100 MPa in air (a, c) and 3.5% NaCl solution (b, d).

Note: #1 - location of pits to enter solution, #2 – delamination of coating due to corrosion, and #3-crack propagated into the matrix.

The enhancement of fatigue life is related to several factors which includes (i) AA7075 powder high strength, (ii) induced compressive residual stresses and (iii) high bonding strength between coating (AA7075) and substrate (AZ31B cast). As seen in [26], the AA7075 alloys possess a higher fatigue performance than the as-cast AZ31B. So, the surface cracking in the case of CS samples is deferred until high amount of stresses are imposed. The probability of development of sub-surface cracks in cast AZ31B which rupture at lower stresses is removed in CS samples as any premature cracks in the cast AZ31B were showed not to be progressive cracks. The higher impact velocity and temperature of the carrier gas of CS powder on the substrate induced severe plastic deformation resulting in adiabatic shear instability as discussed earlier [14, 25, 27]. Further, the thermal expansion coefficient (TEC) between coating (AA7075) and substrate (AZ31B) significantly influences the residual stress formation. During the surface modification in CS process, the temperature of the carrier gas was 400^0C. The higher value of TEC of coating as compared to substrate creates a higher level of mismatch at the interface which causes a higher magnitude of compressive residual stress. It was shown that the residual stress due to the TEC mismatch, and the high impact of CS particles on the as-cast AZ31B sample was about to −122 MPa in the CS AA7075 coatings [10]. Due to the presence of higher amount of compressive residual stress inhibited or delayed the initiation of cracks results improvement in the fatigue life in CS as-cast AZ31B. Moreover, the surface modification in CS process shielded the materials defects like casting porosity which essentially postponed the nucleation of cracks, resulting in longer fatigue life [10].

4. CONCLUSIONS

In summary, the surface modification in CS process significantly improved the yield strength of cast AZ31B Mg alloy in tensile loading. The fatigue life of the CS AZ31B cast Mg alloy is always higher than the uncoated/ unmodified samples in both, air and corrosive environment. However, only the CS process cannot achieve the infinite fatigue life. An efficient top coating is necessary. Thus, this study advocate that the surface modification with addition of top coating can improve the fatigue life significantly.

The authors would like to gratefully acknowledge the financial support of the Natural Sciences and Engineering Research Council of Canada (NSERC), Automotive Partnership Canada (APC) program with contributions from Multimatic Technical Centre, Ford Motor Company, and Centerline Windsor.

References

1. Moridi, S. M. Hassani-Gangaraj, M. Guagliano, M. Dao, Surf. Eng. **30,** 369 (2014).
2. T. Hussain, D. G. McCartney, P. H. Shipway, D. Zhang, J. Therm. Spray. Techn. **18**, 364 (2009).
3. T. Schmidt, F. Gärtner, H. Assadi, H. Kreye, Acta Mater. **54**, 729 (2006).
4. H. Assadi, F. Gärtner, T. Stoltenhoff, H. Kreye, Acta Mater., **51**, 4379 (2003).

5. S. K. Shaha, B. Marzbanrad, H. Jahed, *TMS 2018 147th Annual Meeting & Exhibition Supplemental Proceedings*, 2018.
6. S. K. Shaha, S. B. Dayani, H. Jahed, *TMS 2018 147th Annual Meeting & Exhibition Supplemental Proceedings*, 2018.
7. S. B. Dayani, S. K. Shaha, R. Ghelichi, J. F. Wang, and H. Jahed, Surf. Coatings Technol. **337**, 150 (2018).
8. S. M. H. Karparvarfard, S. K. Shaha, S. B. Behravesh, H. Jahed, B. W. Williams, J. Mater. Sci. Technol. **33**, 907 (2017).
9. S. M. H. Karparvarfard, S. K. Shaha, A. Hadadzadeh, S. B. Behravesh, H. Jahed, M. A. Wells, B. Williams, *Magnesium Technology 2017*.
10. D. Toscano, S. K. Shaha, B. Behravesh, H. Jahed, B. Williams, J. Mater. Eng. Perform. **26**, 3090 (2017).
11. D. Toscano, S. K. Shaha, B. Behravesh, H. Jahed, B. Williams, Mater. Sci. . Eng. A **706**, 342 (2017).
12. H. Bu, M. Yandouzi, C. Lu, D. MacDonald, B. Jodoin, Surf. Coatings Technol. **207**, 155 (2012).
13. M. Diab, X. Pang, H. Jahed, Surf. Coatings Technol. **309**, 423 (2017).
14. R. Ghelichi, D. MacDonald, S. Bagherifard, H. Jahed, M. Guagliano, B. Jodoin, Acta Mater. **60**, 6555 (2012).
15. M. R. Rokni, C. A. Widener, G. A. Crawford, M. K. West, Mater. Sci. Eng. A **625**, 19 (2015).
16. G. Shayegan, H. Mahmoudi, R. Ghelichi, J. Villafuerte, J. Wang, M. Guagliano, H. Jahed, Mater. Des. **60**, 72 (2014).
17. K. Spencer, D. M. Fabijanic, M. X. Zhang, Surf. Coatings Technol. **204**, 336 (2009).
18. Q. Wang, D. Qiu, Y. Xiong, N. Birbilis, M. X. Zhang, Appl. Surf. Sci. **289**, 366 (2014).
19. S. M. Hassani-Gangaraj, A. Moridi, M. Guagliano, Surf. Eng. **31**, 803 (2015).
20. Y. Tao, T. Xiong, C. Sun, L. Kong, X. Cui, T. Li, Corros. Sci. **52**, 3191 (2010).
21. Y. Tao, T. Xiong, C. Sun, H. Jin, H. Du, T. Li, Appl. Surf. Sci. **256**, 261 (2009).
22. E. Kalatehmollaei, H. Mahmoudi-Asl, H. Jahed, Int. J. Fatigue **64**, 33 (2014).
23. B. Marzbanrad, H. Jahed, E. Toyserkani, Mater. Des. **138**, 90 (2018).
24. S. K. Shaha, S. B. Dayani, H. Jahed, Int. J. Fatigue, (under review), 2018.
25. H. Jahed R. Ghelichi, *Modern Cold Spray : Materials, Process, and Applications* 2015.
26. T. S. Shih, Y.W. Chiu. Appl. Surf. Sci. **264**, 410 (2013).
27. R. Ghelichi, S. Bagherifard, D. Mac, M. Brochu, H. Jahed, B. Jodoin and M. Guagliano, Int. J. Fatigue, **65**, 51 (2014).

Fatigue behaviour of AA6061-T6 alloys in the corrosive environment

Ngoc Vu Nguyen[1], and *Peifeng* Li[2,*]

[1]School of Mechanical and Aerospace Engineering, Nanyang Technological University, Singapore
[2]School of Engineering, University of Glasgow, Glasgow G12 8QQ, UK

Abstract. The combined effects of corrosion and fatigue are known to be hazardous threats to structural integrity of aluminium alloys that are being extensively used in marine applications. This work investigated the fatigue crack initiation and growth behaviour of AA6061-T6 alloys in 3.5 wt% NaCl simulated seawater using scanning electron microscope and electron backscatter diffraction characterisation techniques. It was found that the fatigue resistance of AA6061-T6 is drastically downgraded when subjected to the corrosive environment of 3.5 wt% NaCl solution. High stress concentration at both sides of a pit mouth in conjunction with attacked grain boundaries facilitates fatigue crack nucleation, while the presence of hydrogen formed by corrosion reactions causes crack tip embrittlement and thus increases crack growth rate. Fractographic analysis reveals that there is a change in fatigue crack growth mechanism of AA6061-T6 alloys tested in the NaCl solution. At short crack length, the crack develops transgranularly along crystallographic planes due to hydrogen-enhanced decohesion process. Further crack growth is dominated by adsorption induced dislocation emission process, resulting in the mixed mode of intergranular and transgranular crack growth.

1 Introduction

Corrosion fatigue is the response of a material to alternating stresses in a corrosive environment by the initiation and propagation of cracks. It has been known as a dominant factor on the reduction of service life of marine structures and other equipment used in the offshore environment. Pitting corrosion combined with fatigue is the complex phenomenon involving various processes and mechanisms [1-4]. The pitting process causes local discontinuities on the surface of a structure and enhances the nucleation as well as propagation processes of small fatigue cracks. Due to the presence of these local discontinuities, short cracks can initiate and develop even below the threshold range of stress intensity factor determined for long crack growth [5]. In a corrosive environment, hydrogen is generated by cathodic corrosion reactions and then adsorbed at crack tips. There is broad consensus that degradation of crack growth resistance is caused by crack tip embrittlement due to the presence of hydrogen species.

Of the many proposed mechanisms, three most widely accepted interstitial mechanisms are: hydrogen-enhanced localised plasticity (HELP), hydrogen-enhanced decohesion (HEDE) and adsorption-induced dislocation emission (AIDE). The HELP mechanism is supported by the observations that solute hydrogen facilitates dislocation motion ahead of crack tips by forming an enhanced plastic zone size [6, 7]. The fracture process exhibits plasticity phenomena such as reduction of material yield strength and micro-void coalescences rather than embrittlement. However, this mechanism is only valid in a certain range of temperatures and strain rates. Contradictory results have been observed at elevated temperatures and higher strain rates. The second mechanism HEDE assumes that hydrogen adsorption weakens interatomic bonds at crack tips, resulting in decohesion (tensile separation) along cleavage planes or grain boundaries [8-10]. Therefore, the fracture surface may exhibit either brittle intergranular facets or transgranular cleavage. This mechanism is supported by brittle characteristics similarly observed in solid-metal induced embrittlement and metal-vapour induced embrittlement where decohesion occurs in the absence of considerable local deformation. The third mechanism AIDE proposed by Lynch [11] is based on fractographic observations such as extensive localised slips or nanoscopic dimples, indicating that crack growth occurs by localised slip or void coalescence processes. The AIDE mechanism assumes that adsorption of embrittling species weakens interatomic bonds and thereby enhances dislocation movements ahead of crack tips.

Fundamental understanding of hydrogen embrittlement mechanisms has been appreciably achieved thanks to advances in improved theoretic tools as well as high resolution experimental instruments. However, the embrittlement mechanisms and kinetics

* Corresponding author: peifeng.li@glasgow.ac.uk

involved are not fully resolved and are still controversial disputes as they cannot be generally applied to all materials in a particular environment such as the combination of corrosion and cyclic stresses.

This study investigated the effect of pitting corrosion on fatigue behaviour of AA6061-T6 alloys with the aim of having a better understanding of fatigue crack initiation and propagation mechanisms of aluminium alloys in a corrosive environment.

2 Experimental procedure

2.1 Materials and specimens

The material investigated in this study was aluminium alloys AA6061-T6 in the form of rolled sheet with a thickness of 4.06 mm. Dog bone fatigue specimens were fabricated in the rolling-transverse orientations with the rolling direction for cyclic loading. The dimension of the specimens was designed according to ASTM E466 standard requirements (Fig. 1). Prior to fatigue testing, the specimen surfaces were ground with the subsequent use of 800 and 1200 grit silicon carbide paper to ensure the surface roughness R_a less than 0.15 μm.

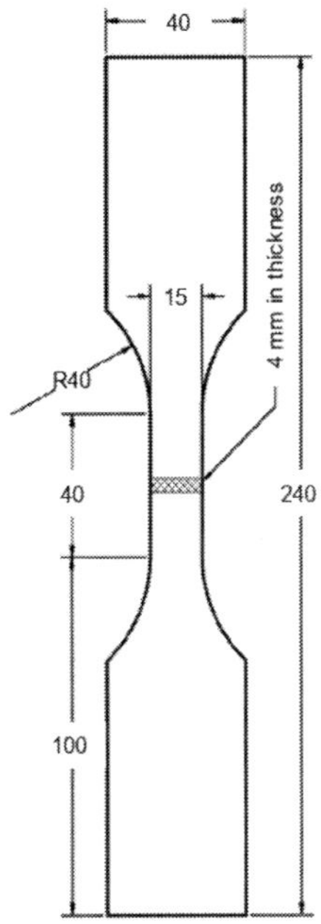

Fig. 1. Geometry and dimensions of the fatigue specimen. (unit: mm)

2.2 Corrosion fatigue testing

Fatigue experiments were conducted on the specimens under fully reversed tension-compression loading (stress ratio $R = -1$) at room temperature with the load frequency $f = 5$ Hz in an MTS 810 (MTS Systems Corp., USA) servo-hydraulic universal testing machine. At each stress level, at least three tests were repeated and the average number of cycles to failure was calculated to determine the *S-N* curve (the cyclic stress amplitude versus the number of cycles to failure) of the material. The fatigue tests were performed in a corrosive environment as well as in the air atmosphere to analyse the effect of corrosion on fatigue behaviour of aluminium alloys.

Fig. 2 shows the overall setup of fatigue testing of an aluminium alloy specimen in an environmental chamber in the MTS 810 machine. The fatigue specimen especially the gauge portion was immersed into the environmental chamber that consisted of 3.5 wt% sodium chloride (NaCl) simulated seawater. The NaCl solution was circulated through the environmental chamber by a peristaltic pump to provide continuous aeration and replenishment of the corrosive solution. The pH of this synthetic seawater solution was adjusted to 8.2, and the specimens were exposed to the solution environment for one hour before applying the cyclic load.

Immediately after the corrosion fatigue failure, the specimens were taken out of the chamber and cleaned to preclude corrosion processes from obscuring fine details of the fracture surfaces. Fatigue tests of some specimens in the corrosive environment were interrupted prior to failure and examined by optical microscope to identify critical cracks. Once a critical fatigue crack was identified, the specimen portion containing this crack was shortly cut, polished by the colloidal silica 0.04 μm suspension for two minutes and then etched by Keller's reagent for 30 seconds to reveal the microstructure for examination. Fracture features of these specimens were examined in a field emission scanning electron microscope (FE-SEM) JEOL 7600 (JEOL Ltd., Japan) equipped with an electron backscatter diffraction (EBSD) detector.

 https://doi.org/10.1051/matecconf/201816503015

Fig. 2. Corrosion fatigue testing setup in the MTS 810 servo-hydraulic machine.

3 Results and discussion

3.1 Fatigue *S-N* curves

The fatigue lives of AA6061-T6 alloy specimens tested in the air atmosphere and in the 3.5 wt% NaCl solution were measured at different stress levels. Fig. 3 illustrates the plot of the average number of cycles to failure (N_f) versus the applied cyclic stress amplitude (σ_a) on a logarithmic scale. As aluminium does not exhibit a typical fatigue endurance limit, there is a continuous reduction of fatigue life as the applied stress increases. The relation between the fatigue life and the stress amplitude can be quantitatively described using a power equation:

$$\sigma_a N_f^m = C \tag{1}$$

where m and C are the material constants determined from the fatigue experiments. The *S-N* curves (Fig. 3) were fitted to estimate the constants $m = 0.1018$ and $C = 580$ MPa for the specimens tested in the air while $m = 0.2538$ and $C = 2357$ MPa for the specimens tested in the 3.5 wt% NaCl solution.

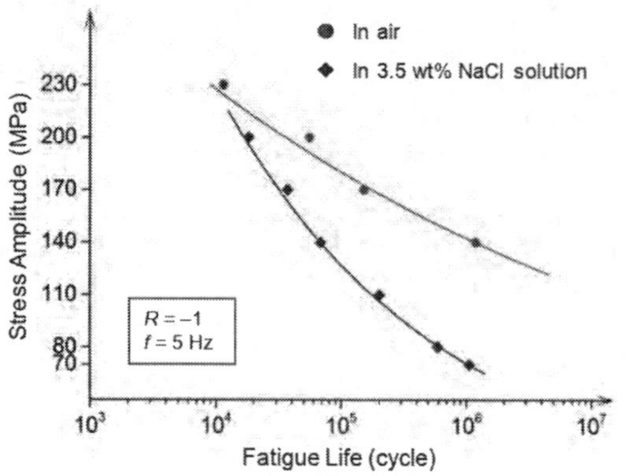

Fig. 3. Fatigue *S-N* curves of AA6061-T6 alloys tested in the air atmosphere and in the corrosive environment of 3.5 wt% NaCl solution.

The fatigue resistance of AA6061-T6 alloys is significantly downgraded in the NaCl solution. For the fatigue life of more than one million cycles, the allowable fatigue strength in the 3.5 wt% NaCl solution decreases by approximately 80 MPa compared to the strength in the air atmosphere. At a higher stress amplitude of approximately 140 MPa, specimens subjected to fatigue loading in the air can sustain more than one million cycles. The damage ($N_{f,NaCl}/N_{f,air}$) at a specific applied stress is defined as the ratio of the average fatigue life tested in the NaCl solution ($N_{f,NaCl}$) to that tested in the air ($N_{f,air}$). The damage ratio at three different stress levels 200 MPa, 170 MPa and 140 MPa were calculated to be 3.06, 4.08and 17.38 respectively. The increase of fatigue damage ratio in association with reduced stress amplitude indicates that the effect of corrosion environment on fatigue behaviour is more significant at lower stress amplitude as there would be more time for corrosion reactions and hydrogen embrittlement processes to occur in materials.

3.2 Fractographic observations

The effect of corrosion environment on fatigue crack initiation and propagation was elucidated using the SEM examiniations on the fracture features. Fig. 4 shows the fracture surface of a fatigue specimen failed in the 3.5 wt% NaCl solution environment. Fatigue cracks nucleate from both sides of the pit mouth forming multiple crack fronts. The crack initiation site is a typical crystallographic pit. As these crack fronts further propagate, they meet and merge with each other, leaving a tearing ridge on the fracture surface (Fig. 4(a)).

Two stages of short crack can be distinguished by the detailed fractographic features as shown in Fig. 4(b–d). In the vicinity of the crack origin, the fracture surface appears to be flat and crack propagates transgranularly along crystallographic planes, forming cleavage patterns (Fig. 4(b)). As the crack grows to a critical length, extensive localised slips and fatigue striations can be observed on the fracture surface instead of cleavage features (Fig. 4(c, d). The different fractographic features indicate that there is a change in growth mechanism of short fatigue cracks in aluminium alloy when subjected to a corrosive environment.

3.3 Mechanisms of fatigue crack initiation and propagation

The higher crack growth rate and the resultant reduction of fatigue life of aluminium alloys in the corrosive environment can be attributed to the presence of hydrogen formed by cathodic corrosion reactions. The mechanism of pitting formation is based on the dissolution of aluminium in the anode:

$$Al \rightarrow Al^{3+} + 3e^{-} \quad (2)$$

and several cathodic reactions, which could be the oxygen reduction:

$$O_2 + 2H_2O + 4e^{-} \rightarrow 4OH^{-} \quad (3)$$

or hydrogen development in neutral or acidic environments:

$$2H^{+} + 2e^{-} \rightarrow H_2 \quad (4)$$

$$2H_2O + 2e^{-} \rightarrow H_2 + 2OH^{-} \quad (5)$$

The presence of hydrogen causes crack tip embrittlement and escalates fatigue crack growth rate. Moreover, the cohesive bonding at crack tips can be weakened due to the high concentration of chloride in the simulated seawater. Fracture surface of corrosion fatigue specimens exhibits brittle characteristics in comparison with extensive ductile striations as in the case of fatigue samples tested in the air. It is generally accepted that hydrogen adsorption weakens interatomic bonds at crack tips and thus accelerates crack advancement.

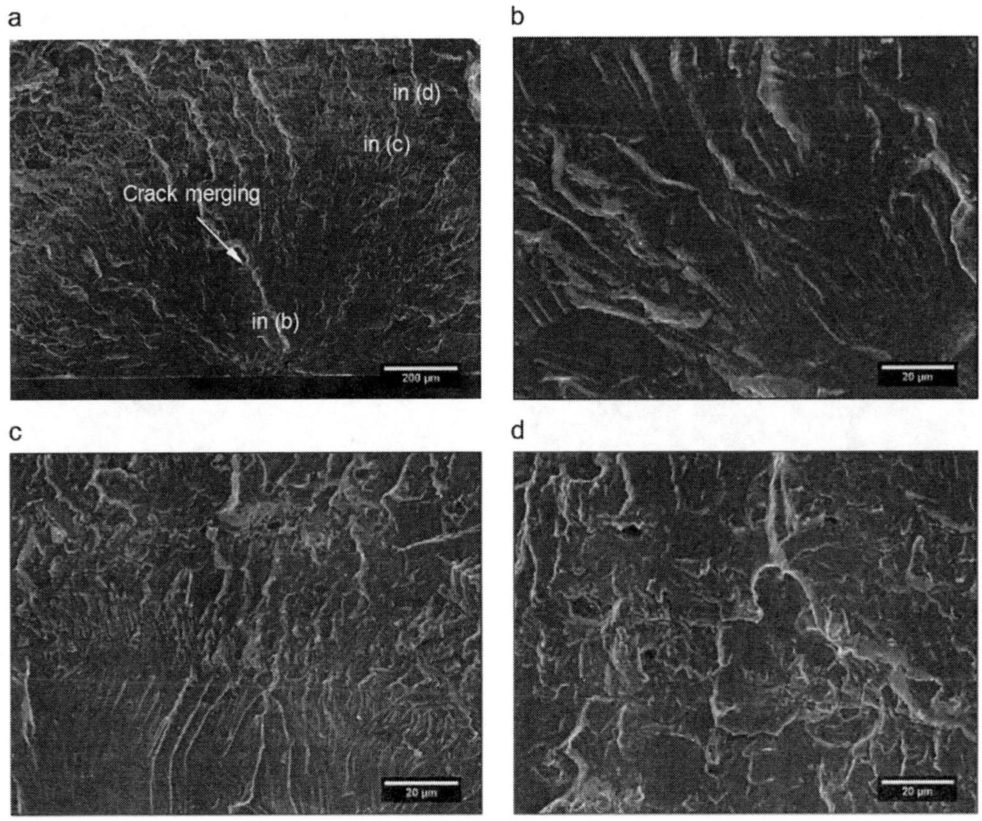

Fig. 4. SEM images of the fracture surface of a fatigue specimen tested in the NaCl solution: (a) overview, (b–d) details of selected fracture regions as indicated in (a).

However, the speed of hydrogen diffusion and adsorption could be responsible for the variation in crack growth mechanisms at different crack lengths. At short crack length where the crack growth rate is low, there is sufficient time for corrosion reactions to produce hydrogen and diffuse it to the crack front. Hydrogen is adsorbed at crack tips and weakens interatomic bonds such that decohesion occurs by tensile separation of interatomic bonds on low index crystallographic planes, i.e., the HEDE process. Capillary transport of hydrogen is limited as it requires a crack tip with significant radius. The plastic zone size ahead of the crack tip is relatively small at low stress intensity region, suggesting that the crack tip radius is too tiny to favour capillary flow.

As the crack grows further to a certain length corresponding to higher stress intensity, there would not be enough time for hydrogen diffusion at higher crack growth rate. Therefore, the crack tip is not embrittled yet, and most of dislocation activities caused by fatigue loading would produce crack tip blunting rather than cracking advancing because the plastic zone size is enhanced at higher stress intensity. This crack tip blunting process facilitates capillary transport of embrittling agents, hydrogen in this case, to the crack front. Capillary transport of hydrogen is dominant over surface diffusion at this stage of crack growth and thereby initiates the AIDE process. This process involves adsorption of embrittling species at crack tips and stimulating dislocation emission. This implies that most of dislocation induced by fatigue loading would advance fatigue crack on suitable slip systems. Therefore, extensive localised slips and fatigue striations occur at growing crack length as shown in Fig. 5.

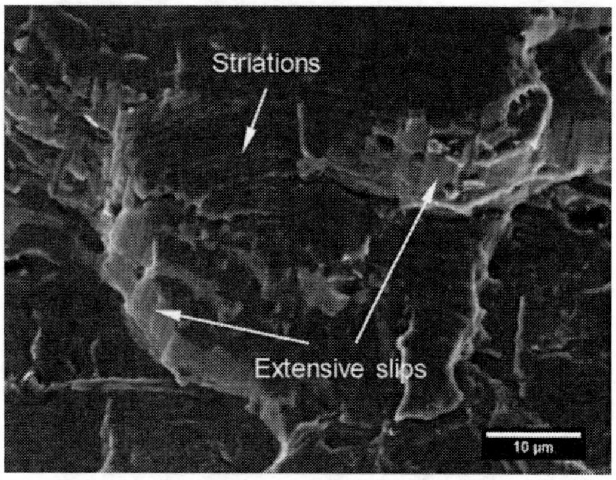

Fig. 5. SEM image of the fracture surface showing extensive localised slips and fatigue striations at growing crack length.

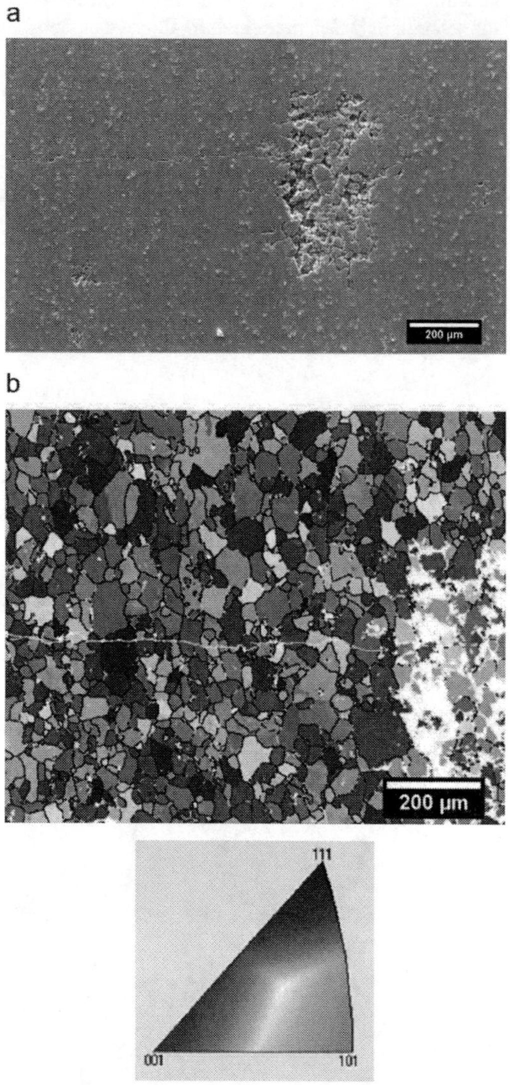

Fig. 6. (a) SEM image and (b) EBSD orientation map of a pit that initiates the fatigue crack.

The corrosion pit from which fatigue cracks initiate can be identified from the AA6061-T6 specimen subjected to interrupted fatigue in the corrosive environment (Fig. 6). The cracks initiate from both sides of the crystallographic pit mouth which are highest stress concentration sites [12]. Grain boundaries surrounding the corrosion pit are attacked due to anodic reactions of corrosion processes, which create a local acidic environment with high chloride concentration. This local acidic environment leads to a precipitate free zone, which facilitates intergranular corrosion at pit walls. Furthermore, fatigue loading can break down the protective aluminium hydroxide layer such that the aggressive environment can easily penetrate into the metal. High stress concentration at both sides of the pit mouth in conjunction with attacked grain boundaries favours fatigue crack nucleation and substantially reduces fatigue life as crack initiation can account for up to 80 percent of the total corrosion fatigue life. It should be noted that intergranular corrosion only occurs around the perimeter of crystallographic pits within the zone that local pitting potential is modified by corrosion reactions. Outside the corroded area, grain boundaries are not attacked, and the occasional occurrence of intergranular fatigue crack growth (Fig. 6(b)) does not involve intergranular corrosion but is associated with decohesion of grain boundaries due to the adsorption of hydrogen.

4 Conclusions

The fatigue performance of AA6061-T6 alloys is significantly downgraded in the 3.5 wt% NaCl corrosive solution. Fatigue crack nucleation and propagation are affected by the corrosive environment. High stress concentration at both sides of the pit mouth in conjunction with attacked grain boundaries facilitates fatigue crack initiation. The presence of hydrogen formed by corrosion reactions causes crack tip embrittlement and intensifies the crack growth rate. The detailed fractographic analysis reveals that the crack growth can be distinguished by two stages. Near the crack origin, the crack propagates transgranularly along crystallographic planes due to the HEDE process. Further crack growth is dominated by the AIDE process, thus resulting in a mixed mode of intergranular and transgranular crack growth.

References

1. S.S. Singh, T.J. Stannard, X.H. Xiao, N. Chawla, In situ x-ray microtomography of stress corrosion cracking and corrosion fatigue in aluminum alloys, JOM, **69**, 1404-1414 (2017).
2. M.N. Ilman, N.A. Triwibowo, A. Wahyudianto, M.R. Muslih, Environmentally assisted fatigue behaviour of stress relieved metal inert gas (MIG) AA5083 welds in 3.5% NaCl solution, Int. J. Fatigue, **100**, 285-295 (2017).
3. J.J. Schubbe, S.H. Bolstad, S. Reyes, Fatigue crack growth behavior of aerospace and ship-grade aluminum repaired with composite patches in a corrosive environment, Compos. Struct., **144**, 44-56 (2016).
4. Z.Q. Hua, S.S. Wu, Tensile, fatigue, and corrosion fatigue behavior of high performance die cast aluminum alloy, J. Mater. Res., **30**, 833-840 (2015).
5. J. Newman, E.P. Phillips, M. Swain, Fatigue-life prediction methodology using small-crack theory, Int. J. Fatigue, **21**, 109-119, (1999).
6. H.K. Birnbaum, P. Sofronis, Hydrogen-enhanced localized plasticity—a mechanism for hydrogen-related fracture, Mater. Sci. Eng. A, **176**, 191-202 (1994).
7. I. Robertson, The effect of hydrogen on dislocation dynamics, Eng. Fract. Mech., **68**, 671-692, (2001).
8. A.R. Troiano, The role of hydrogen and other interstitials in the mechanical behavior of metals, Trans. ASM, **52**, 54-80 (1960).
9. R. Oriani, Hydrogen embrittlement of steels, Annu. Rev. Mater. Sci., **8**, 327-357 (1978).
10. R. Oriani, A mechanistic theory of hydrogen embrittlement of steels, Berichte der Bunsengesellschaft für physikalische Chemie, **76**, 848-857 (1972).
11. S. Lynch, Environmentally assisted cracking: overview of evidence for an adsorption-induced localised-slip process, Acta Metall., **36**, 2639-2661 (1988).
12. Y.F. Huang, C. Wei, L.J. Chen, P.F. Li, Quantitative correlation between geometric parameters and stress concentration of corrosion pits, Eng. Fail. Anal., **44**, 168-178 (2014).

Behaviour of short and long cracks in air and in liquid metal in T91 steel

Jean-Bernard Vogt[1,*], *Carla* Carle, *Jérémie* Bouquerel[1], and *Ingrid* Proriol Serre[1]
[1]Univ. Lille, CNRS, INRA, ENSCL, UMR 8207 - UMET - Unité Matériaux et Transformations, F-59000 Lille, France

Abstract. The low cycle fatigue behaviour of a 9Cr1MoNbV martensitic steel has been investigated at 350°C in air and in lead-bismuth eutectic (LBE). Total strain controlled tests were performed from $\Delta\varepsilon_t$ = 0.40% to 1.2%. The material exhibited a pronounced cyclic softening in both environments. LBE reduced the fatigue resistance. LBE accelerated the formation of the long crack by promoting the growth of the first short cracks. A clear change in propagation mode was observed. In air, ductile fatigue striations were observed while in LBE a brittle fracture decorated by voluminous and largely spaced striations were visible. From EBSD analysis, it was concluded that in LBE, the long crack advanced quickly by repeated and discontinuous cleavage.

1 Introduction

T91 martensitic steel is a modified grade of the 9Cr1Mo steel family. It exhibits excellent combination of strength and ductility which can be achieved by control of the heat treatment. With the addition of Nb and V, T91 is particularly chosen for application where creep resistance is required. It is therefore employed for structural components in all industrial fields working at high temperatures such as refineries or power plants. With an amount of chromium less than 10.5 wt %, the alloy cannot be classified as stainless steel and is not really designed to operate in corrosive environments as can do the 12Cr steels.

T91 steel has been considered as one of the materials to be possibly employed for components of accelerator driven systems (ADS) or of Gen IV nuclear reactors. These systems employ liquid metals as spallation target or as coolant. Even if the ductility of T91 is somewhat appreciable, it can be decreased by irradiation e.g. [1] or by long term exposure at high temperature e.g. [2]. Besides these operating conditions that affect the bulk of the material, damage initiation by the liquid metal environment may occur as it is the case for aqueous environments e.g. [3]. Liquid metals do not involve similar corrosion mechanism as aqueous environments. Electrochemical corrosion is not a process that occurs when metallic alloys are in contact with a liquid metal.

A very amazing response that can exhibit a ductile material stressed in liquid metal is the so called liquid metal embrittlement (LME) or liquid metal assisted damage (LMAD). It has been shown by several authors working in the field of materials in contact with lead bismuth eutectic (LBE) that T91 is especially sensitive. For T91, the ductile to brittle transition in LBE at 300°C is strongly influenced by the heat treatment and especially the tempering temperature [4].

The objective of the paper is to study the effect of LBE on the mechanical response and on the behaviour of short and long cracks under cyclic loading by using low cycle fatigue tests (LCF).

2 Materials

The chemical composition of the studied T91 satisfies the standard. Expressed in wt%, it contains 0.1 C, **8.99 Cr**, **0.89 Mo**, **0.06 Nb**, **0.21 V**, 0.38 Mn, 0.22 Si, 0.11 Ni,
0. 0146 Al, 0.06 Cu, 0. 0034 Ti, 0.01 W, 0.021 P, 0. 0004 S, Fe bal.

The material 91 was heat treated at 1100°C for 15 minutes followed by water cooling and a subsequent tempering for 45 minutes at 770°C and then air cooling.

The microstructure is a fully tempered martensite (Figure 1) with an average hardness value of 220 HV.

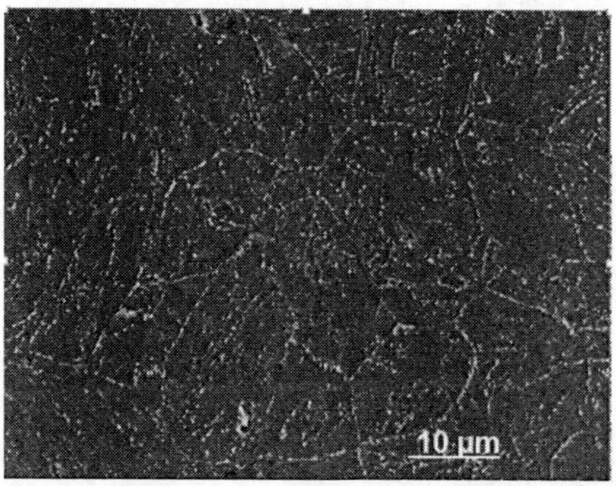

* Corresponding author: jean-bernard.vogt@univ-lille1.fr

Fig. 1. SEM micrograph of the material before fatigue testing

The chemical composition of the liquid metallic alloy is 45 wt% lead and 55 wt% bismuth (LBE).

3 Low cycle fatigue details

A tank and samples specifically designed to perform LCF tests in LBE were machined. The specimens were smooth and cylindrical with a gauge length of 10 mm and a gage diameter of 6 mm. In order to avoid effects due to the roughness of the surface and to residual stresses developed during the machining, the sample surface was carefully electro polished. The fatigue tests were carried out using a servo-hydraulic Schenk machine with a load capacity of 250 kN under total axial strain control $\Delta\varepsilon_t$ ranging from 0.40% to 1.2%. A fully push pull mode (R_ε = -1), a triangular waveform and a constant strain rate of 4.10^{-3} s^{-1} were used. All the tests were carried out at 350°C in air and in LBE.

The specimen was immersed into a small reservoir containing the LBE (Figure 2) which could not accept the extensometer to be attached onto the specimen. Therefore, the strain control was performed by using a strain gage extensometer fixed outside of the specimen.

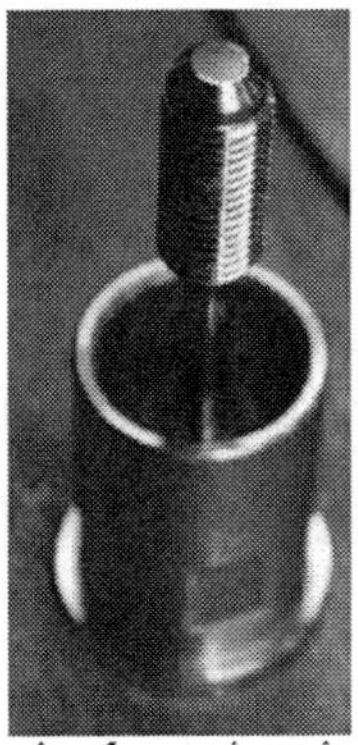

Fig. 2. Picture showing the specimen in the LBE reservoir

The fatigue life N_f, is defined as the number of cycles needed for a 25% drop in the tensile stress taking as a reference the (pseudo) stabilized hysteresis loop

4 Results

4.1 Stress response to strain cycling

Except for the test performed at $\Delta\varepsilon_t$ = 0.4%, the material cycled at 350°C presents a strong tendency to softening at the beginning of cycling and then at a moderate rate (see Figure 3). Then the fall of the stress is more moderate. Finally a marked decrease of the stress amplitude occurs that is related to the propagation of a macroscopic crack into the bulk just before the final failure.

Except for the final decrease in the stress amplitude which is affected by the liquid metal, the same overall behaviour is observed when cycling in the LBE environment. Note that the amount and duration of the initial softening clearly depends on the applied strain range only.

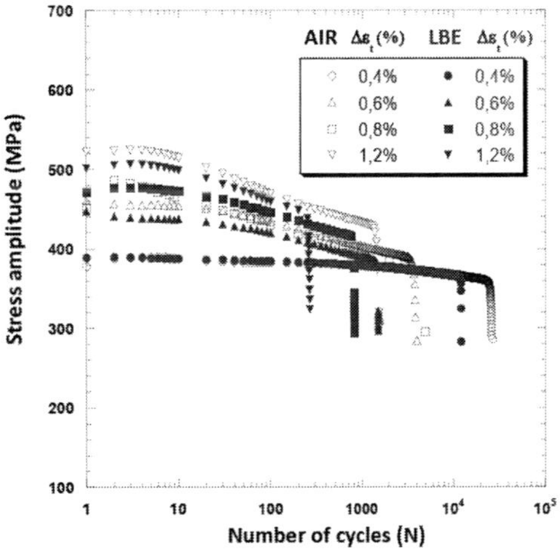

Fig. 3. Evolution of the stress amplitude with the number of cycles of T91 steel fatigued in air and LBE at 350°C

4.2 Effect of LBE on fatigue crack initiation resistance

The evolution of the total strain range as a function of the number of cycles to failure is plotted Fig. 4.

The main effect of LBE is to reduce the fatigue resistance in comparison with tests performed in air. Such an influence of a liquid metal on fatigue resistance has been reported for the Manet II martensitic (Fe-10Cr-0.8Mn-0.2V-0.1C) in LBE at 260°C [5] and in T91 at 300°C [6] and 350°C [7].

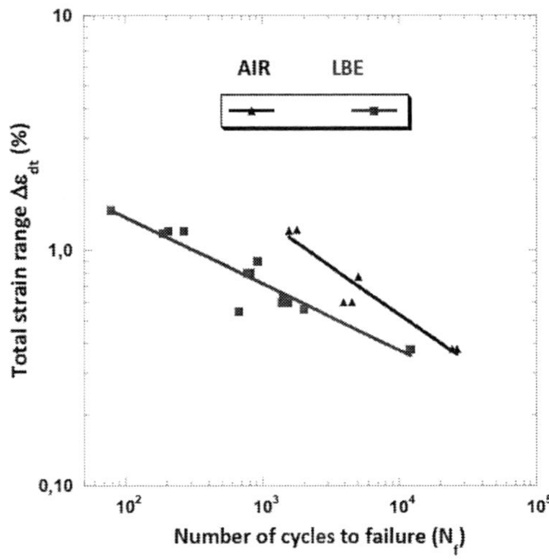

Fig. 4. Evolution of the total strain range with the number of cycles to failure of T91 steel fatigued in air and LBE at 350°C

4.3 Effect of LBE on fatigue crack propagation resistance

Normally, the study of fatigue crack propagation is performed by using pre-cracked specimens in order to connect the fatigue crack velocity da/dN to the cyclic stress intensity factor ΔK. This could not have been performed. However, the final part (the last 10 percents of the fatigue life) of the curves stress amplitude-number of cycles was analysed for that.

For each test, $\Delta N_{5\%-25\%} = N_{25\%} - N_{5\%}$ was calculated where $\Delta N_{25\%}$ and $\Delta N_{5\%}$ are the number of cycles correspond to a 25% drop in the tensile stress and to a 5% drop in the tensile stress respectively.

The result is reported in figure 5.

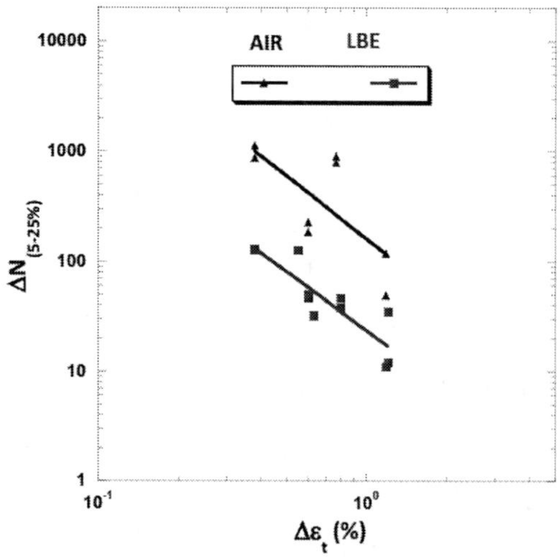

Fig. 5. Evolution of the number of cycles for propagation of the long crack versus the total strain range of T91 steel fatigued in air and LBE at 350°C

4.4 Short crack analysis

Fatigue specimen were cut transversally in order to appreciate the number and the depth of the short cracks. After cutting, the specimen were polished to have a mirror surface and then were etched with the Villela etchant. After fatigue in air, it was possible to observe on the transverse cross sections a lot of short cracks of different lengths (Figure 6).

After fatigue in LBE, the number of short cracks was considerably reduced so that it can be considered that only the main crack was visible. (Figure 7).

The strain range had a noticeable effect on the density of short cracks in air, i.e. the density increased with the strain range, but not for the tests performed in LBE.

Fig. 6. Observation of short cracks in T91 steel fatigued in air at 350° C and $\Delta\varepsilon_t$= 1.2%

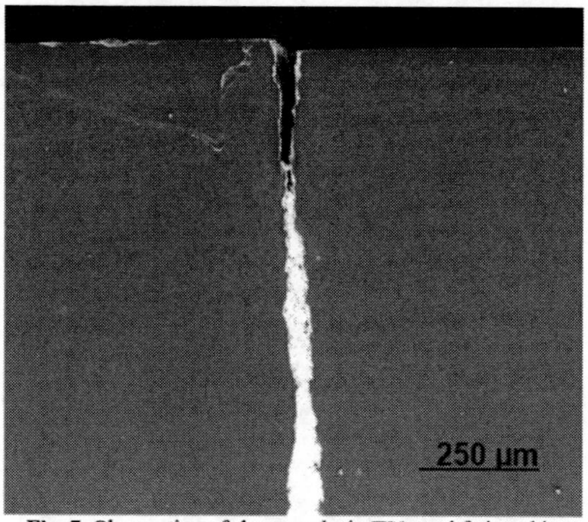

Fig. 7. Observation of short cracks in T91 steel fatigued in LBE at 350° C and $\Delta\varepsilon_t$= 0.6%

4.5 Long crack analysis

The long crack morphology was first observed directly on the gage length of the specimen.

After fatigue in air, the main crack was accompanied by a rough zone. The rough aspect tended to vanish gradually the observation left the crack tip. However, even at 1 mm from the crack tip, the plastic deformation was still visible.

After fatigue in LBE, the crack path appeared more serrated without any roughness on the lips.

The most striking difference between cycling in air and in LBE was the fracture surface morphology. In air, in agreement with the short crack distribution, the fracture surface contained multiple crack initiation sites from which several cracks propagate by fatigue in the bulk and a final fracture due to monotonic loading. Fatigue striations could be observed by SEM. The fracture surfaces of specimen cycled in LBE are totally different. First, the fracture surface was very flat and contained a single initiation site from which the crack

https://doi.org/10.1051/matecconf/201816503016

propagated by cyclic loading. The final fracture occurred by pure tensile loading. On the fracture surface related to cyclic loading, it was possible to see at very low magnification, and even with the naked eye, curved striations (Figure 8). For a given strain range, the distance between two striations increased as long as the crack advanced in the bulk.

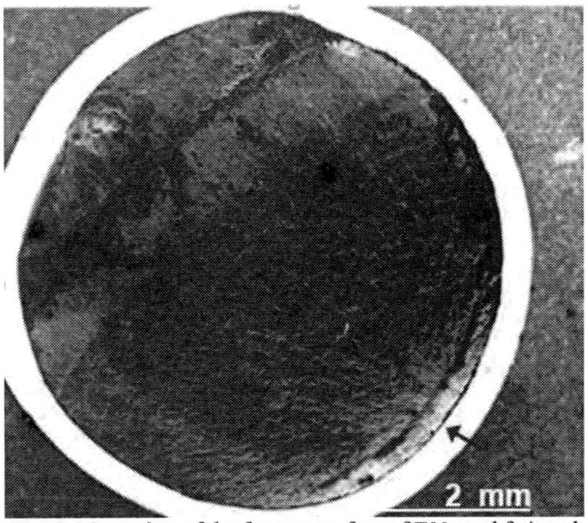

Fig. 8. Macroview of the fracture surface of T91 steel fatigued in LBE at 350° C and $\Delta\varepsilon_t$= 1.2% (the red arrow indicates crack initiation site)

At higher magnification, SEM observations pointed out transgranular brittle fracture (Figure 9).

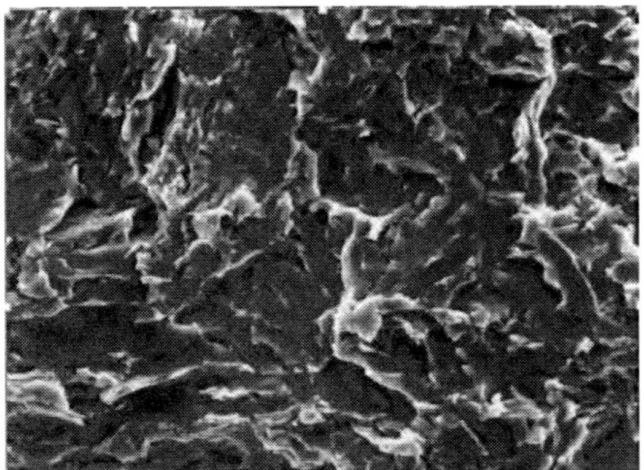

Fig. 9. SEM observation of the fracture surface between two striations of T91 steel fatigued in LBE at 350° C and $\Delta\varepsilon_t$= 1.2% showing brittle fracture

SEM-EBSD analysis were performed on specimen fatigued in air and in LBE at $\Delta\varepsilon_t$ = 1.2%. When the long crack was supposed to propagate in the bulk, then the test was stopped, the specimen removed and cut along the loading direction. Specific preparation was performed to obtain high quality EBSD patterns and further processing. EBSD patterns were acquired far away from the crack to have a reference state, on the crack lips very close to the external surface, at 2.5 mm from the external surface and very near the crack tip.

For the specimen fatigued in air, the inverse pole figures (IPF) revealed that the crack lips were decorated by a continuous raw of misoriented fine grains in the very first micrometers of the material adjacent to the fracture surface (Figure 10). The grains were highly misoriented and had a grain size ranging from 0.25 µm to 0.5µm.

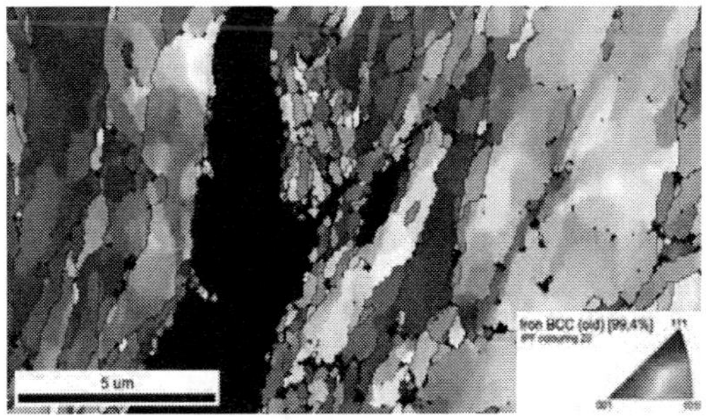

Fig. 10: IPF acquired near the crack tip in T91 fatigued in air at 350° C and $\Delta\varepsilon_t$= 1.2%

This was not observed for the specimen failed in LBE except at some spots where small groups of misoriented fine grains similar to those observed in the specimen fatigued in air could be observed (Figure 11).

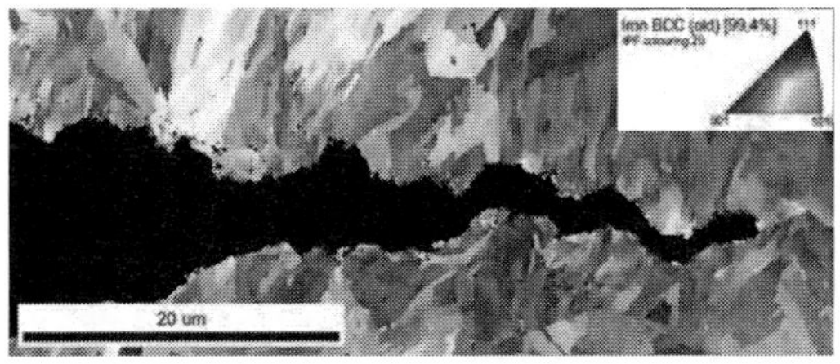

Fig. 11: IPF acquired near the crack tip in T91 fatigued in LBE at 350° C and $\Delta\varepsilon_t$= 1.2%

5 Discussion

The present investigation performed on T91 shows a conventional response typical of martensitic steel i.e. a cyclic softening. This is in accordance with the microstructure resulting from the heat treatment which produced and arranged a high density of dislocations. Annihilation can explain the softening and since it is a bulk effect, there is no reason why the presence of LBE should modifies the response. Therefore, in air as well in LBE, slip bands can emerge at the surface and intrusion-extrusion pairs can form. Intrusions are preferential sites for short crack initiation and their developments are sensitive to the presence of the environment.

In air, the very first short cracks with a dimension of a grain size are hindered by the presence of grain boundaries (including a misorientation effect between the grains) which can be overcome after a given number of cycles. The crack extension proceeds by

crystallographic growth and may again be delayed by other grain boundaries as its length reached three or four grain sizes. Longer microcracks (up to ten grain sizes typically) can eventually form by the coalescence of the shorter cracks and finally, only a very few of them propagated into the bulk perpendicularly to the stress axis.
In LBE, the absence of short cracks is a clear indication that the liquid metal provided a driving force to the very first short cracks to overcome the structural barriers.

The presence of fatigue striation in specimen fatigued in air is a clear indication that the crack propagated in a ductile manner. The very small distance between striations suggests that the crack extension is slow and involves plastic deformation during tensile and compressive phase of the loading signal. The striations may be the result of successive blunting and resharping due to plastic deformation at the crack tip. In LBE, the large distance between striations indicates that the crack advanced rapidly. And this is in good agreement with the observations of brittle transgranular fracture. The modification of fracture mode between fatigue in air and fatigue in LBE is a simple appearance of LME. There are several explanations and models for LME. In our situation, the brittle fracture theory proposed by Stoloff and Johnson[8] & Westwood and Kamdar [9] and based on the weakening of inter-atomic bonds by the liquid metal atoms at the crack sounds well. LME occurrence requires an actual wetting between atoms of the liquid metal and the steel. This can be not encountered if the crack tip is not wetted due to capillarity. However, in our case, EDX analysis of the crack tip after fatigue in LBE clearly showed that the liquid metal reached the very crack tip. Another reason for the non wetting is the presence of an oxide film. In the present experiments, LBE was oxygen saturated which means that even fresh surface formed in the LBE, dissolved oxygen can react with the fresh steel to form an oxide film. However, the stability of the oxide film is disturbed due to very localized plasticity. And this allows atoms to adsorb the steel and to promote a liquid metal assisted crack propagation.

Finally, the EBSD analysis revealed a very strong and localized activity of cyclic plasticity at tip of a crack propagating in air in which recrystallization operated and gave rise to new fine grains. The absence of such fine grains in large portion of the crack lip in specimen fatigued in LBE is well in agreement with brittle fracture. The presence of fine grains at some spots of the lips may be explained by the temporary arrest of the brittle crack due to temporary non-wetting of the surface by the atoms of the liquid metal.

6 Conclusion

The low cycle fatigue behaviour of the T91 martensitic steel has been investigated at 350°C in air and in LBE. LBE accelerates the formation of the long crack by promoting the growth of the first short cracks. In LBE, the long crack advanced quickly by repeated and discontinuous cleavage.

Acknowledgements

The work was partly funded by the European Atomic Energy Community's (Euratom) Seventh Framework Programme FP7/2007-2013 under Grant agreement no. 604862 (MatISSE project)

References

1. Y. Dai, J. Henry, Z. Tong, X. Averty, J. Malaplate, B. Long, J. Nuc. Mat., **415**, 306, (2011)
2. J. Blach, L. Falat, P. Ševc, Eng. Failure Analysis, **16**, 1397 (2009)
3. M. El May, N. Saintier, O. Devos, A. Rozinoer, Procedia Engineering, **133**, 528 (2015)
4. I. Serre, J.-B. Vogt, J. Nucl. Mater., **376**, 330, (2008)
5. D. Kalkhof, M. Grosse, J Nucl Mater, **318**, 143, (2003)
6. A. Verleene, J.-B. Vogt, I. Serre, A. Legris, Int. J. Fatigue, **28**, 843 (2006)
7. X. Gong, P. Marmy, B. Verlinden, M. Wevers, M. Seefeldt, Corros. Sci. **94** 377 (2014)
8. N. S. Stoloff, T. L. Johnston, Acta. Met., **11**, 251 (1963)
9. A.R.C. Westwood, M.H. Kamdar, Philos Mag **8**, 787,(1963)

https://doi.org/10.1051/matecconf/201816503017

A finite element and experimental analysis of durability tested springs

Ying Wang[1], *Constantinos* Soutis[1,*], and *Lorenzo* Gagliardi[2]

[1]Faculty of Science and Engineering, The University of Manchester, James Light Building, Manchester, M13NJ, UK
[2]Jaguar Land Rover, Warwick, CV35 0RR, UK

Abstract. Ageing of vehicles has become a major concern as the vehicles reach the end of their original design life. Corrosion is part of long-term ageing and forms one of the critical degradation processes affecting the mechanical properties, such as stiffness, affecting the structural integrity and life of key components. In this study, a finite element and experimental analysis was carried out to investigate the response of corroded springs in relation to the morphology of a corroded surface. An FE model was created using ABAQUS that correlates the stiffness changes with ageing. The analysis of corrosion-induced stress is then performed to determine the stress fields within the corrosion region. The durability tested springs were analysed with a microscope to identify the morphology of corrosion pits; the measured pattern was adopted in the subsequent FE simulations. A spring test machine was employed to perform the mechanical tests. The load-deflection behaviour of durability tested springs was recorded and then was used to validate the FE simulations.

1 Introduction

1.1 Background

Ageing of vehicles has become a major concern as the vehicles reach the end of their original design life. Corrosion is part of long-term ageing and forms one of the critical degradation processes affecting the mechanical properties, such as stiffness, affecting the structural integrity and life of key components. When localised corrosion occurs, stiffness and strength reduction can be difficult to predict because of the complicated effects of uneven surfaces and uneven material properties on the stress fields. Significant work is being accomplished by the automotive industry in this area, with emphasis on widespread fatigue damage [1-6]. The present study, however, considers long-term ageing in the context of corrosion and its effect on the system behaviour. The purpose of this study is to develop a full predictive FE model for suspension springs that can simulate and predict the dynamic behaviour of corroded springs.

Fig. 1 shows an example of the mechanical response of a suspension spring under static loading. The force-displacement relationship shows that the spring stiffness (or spring rate) is a variable, bilinear rate that is approximated, as indicated by k_1 and k_2 (slope of the curve). The parameters k_1 and k_2 are referred to as static stiffness here.

The key dynamic characteristic of the suspension spring is the dynamic stiffness at prescribed preloads, amplitudes and frequencies. Basically, the stiffness is considered to be a design variable where an increased/decreased stiffness can affect the driveability and the driving experience in many different ways. Changing the stiffness and predicting the consequences of it is a complex and expensive process, and the result can in some cases be very hard to predict, especially in the presence of surface corrosion defects. It is important to understand the mechanical response of automotive helical springs containing surface corrosion.

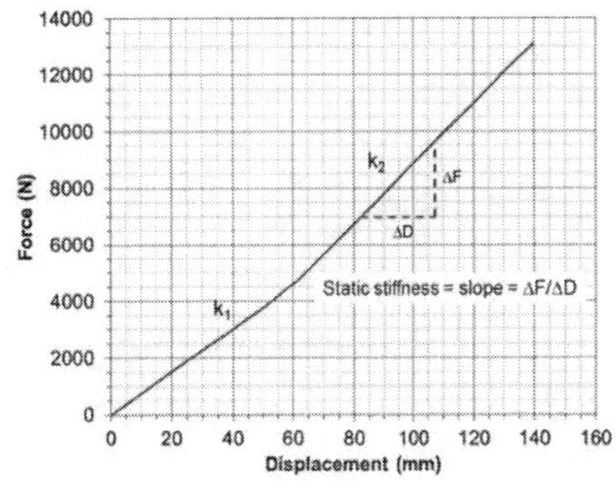

Fig. 1. Force-displacement curve of a suspension spring

Dynamic behaviour of isotropic and homogeneous helical springs has been widely studied in the past decades [1, 7-12], there are no in depth studies on the dynamic behaviour of a corroded suspension spring. For the corroded spring, the design formulas for helical compression springs are no longer suitable. In general, the geometry and loading encountered in three-

* Corresponding author: constantinos.soutis@manchester.ac.uk

 https://doi.org/10.1051/matecconf/201816503017

dimensional corrosion problems is too complex for the stresses to be solved analytically. The stress calculation is further complicated because it is a function of the position along the wire, corrosion defect size and shape. Therefore, in this study we will use current available numerical analysis tools to perform the stress analysis.

In the present study, a nonlinear dynamic analysis is carried out to investigate the dynamic response of corroded springs in relation to corrosion depth, degree of surface corrosion and distributions. An FE model is created using ABAQUS that is used to correlate the stiffness changes of the corroded spring with corrosion depth and the degree of surface corrosion (DSC). The analysis of corrosion-induced stress is then performed.

1.2 Dynamic stiffness

The complex stiffness for a spring that is exposed to a sinusoidal displacement $x_i(t)$ with the corresponding output force $f_o(t)$ is described as

$$K^* = \frac{F_o^*}{X_i} = \frac{F_o}{X_i}cos\delta + j\frac{F_o}{X_i}sin\delta \quad (1)$$

The displacement and the output force are given by $x_i(t) = X_i e^{j\omega t}$ and $f_o(t) = F_o e^{j(\omega t+\delta)}$, where $F_o^* = F_o e^{j\omega t}$. X_i is the displacement amplitude, F_o is the force amplitude and δ is the phase angle between the input displacement and the output force, and ω is the input frequency.

A more general form of equation (1) can be expressed as:

$$K^* = K' + jK''$$

where

$$K' = \frac{F_o}{X_i}cos\delta \text{ and } K'' = \frac{F_o}{X_i}sin\delta$$

The dynamic stiffness K_d is the magnitude of K^*

$$K_d = \sqrt{(K')^2 + (K'')^2} = \frac{F_o}{X_i} \quad (2)$$

2 Geometry of corrosion defects

2.1 Corrosion geometry and distribution

Fig. 2(a) shows an example of the durability tested spring that contains several corrosion defects. It is difficult to generate an actual profile of corrosion that can be implemented in the FE simulation. For simplicity, the corroded surface is assumed to be smooth with a uniform corrosion depth as shown in Fig. 2(b). Therefore, the corrosion defect can be described in terms of the parameters (α, β, D, r and R) shown in Fig. 2(c). The angle α is used to describe the region that is covered by the corrosion defect in the wire's cross-section ($0 \le \alpha \le 360^0$), β is the angle covered by the corrosion defect in the coil's section ($0 \le \beta \le 180^0$), D is the mean diameter of the corresponding coil, r is the radius of corroded surfaces and R is the radius of the spring wire. The depth of corrosion (d) is given by:

$$d = R - r \quad (3)$$

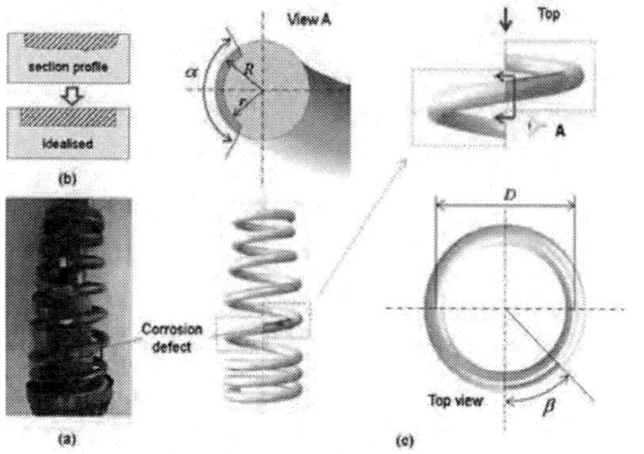

Fig. 2. Schematic showing: (a) the tested spring containing several corrosion defects; (b) the real and idealised surface profile of the section; (c) the geometry of the corrosion defect describing its associated geometrical parameters.

Fig. 3 demonstrates a serious of typical corrosion spatial distributions. For the single coil corrosion (e.g. coil 5), two patterns were selected for the modelling process: case A (1 in coil 5) and case B (2 in coil 5). The reason for choosing coil 5 is because the maximum stress is often seen in this coil. For the multi-coil corrosion case (coils 2 to 7), two further patterns were selected: case C (1 in coils 2-7) and case D (2 in coils 2-7). The meaning of these patterns is that, the 1 in coil 5 represents that 1 corrosion defect exists in coil 5; the 2 in coil 5 represent that 2 corrosion defects exist in coil 5; the 1 in coils 2-7 is that 1 corrosion defect exists in each coil for the coils 2 to 7 and the 2 in coils 2-7 is that 2 corrosion defects exist in each coil for the coils 2 to 7.

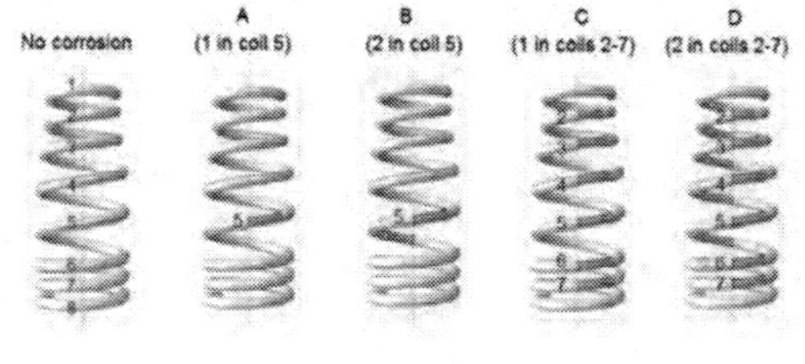

Fig. 3. Corrosion spatial distributions: A (1 in coil 5); B (2 in coil 5); C (1 in coils 2-7) and D (2 in coils 2-7).

2.2 Degree of surface corrosion

The degree of surface corrosion (DSC) was used to correlate the dynamic stiffness change of a corroded spring with corrosion surface area. The DSC is defined as the ratio (as a percentage) of the area of the corroded surfaces to the original surface area of the spring and is given by the following expression:

$$DSC = \frac{1}{A_0}\sum_{i=1}^{n} A_i \times 100\ \% \quad (4)$$

where n is the number of corrosion defects, A_i is the surface area of the ith corrosion defect, A_0 is the original surface area. $A_0 = 2\pi R l_t$, R and l_t are the radius and length of spring wire. In the FE model, the size of each corrosion defect can be obtained by changing angles (α or/and β) to get different combinations.

3 FE modelling

3.1 Description of FE model

The FE model was created using ABAQUS to simulate the dynamic behaviour of a helical compression spring. The spring CAD drawing (stp format) provided by JLR was imported into ABAQUS and then this imported file was edited using ABAQUS CAE to create the FE model of a corroded spring. The geometric parameters are determined in terms of the corrosion geometry model described above. Different corrosion geometries are selected to study its effect on the dynamic behaviour.

An 8-node brick element (C3D8) was used in the model. Approximately 65000-80000 elements were used in total, depending on the depth of corrosion and its distribution on the surface. Fig. 4 demonstrates the overview of the mesh and detailed view of a mesh at corrosion defect regions, respectively.

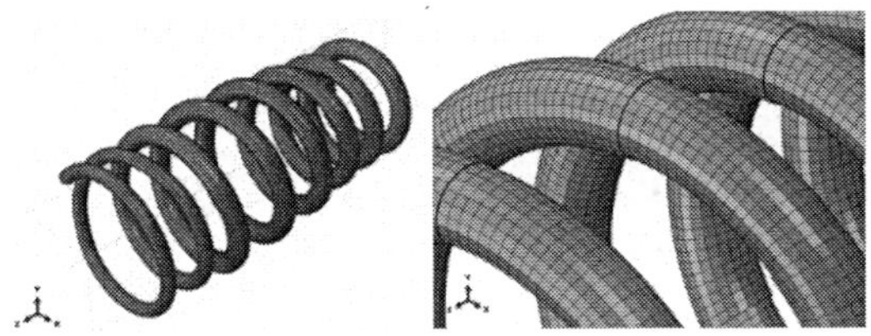

Fig. 4. Diagrams showing overview of mesh and detailed view of a mesh at the corrosion regions.

The spring material is a high-strength martensitic steel (54SiCr6). The isotropic material model was adopted; the elastic modulus and Poisson's ratios are assumed to be equal to 200 GPa and 0.28, respectively. The yield stress is 1680 MPa and is used in the FE model. For the elastic-plastic deformation, the true stress (σ) - true strain (ε) relationship of the spring material, provided by supplier, can be expressed as follow.

$$\sigma = 2430\varepsilon^{0.08} \ (0.0085 \leq \varepsilon \leq 0.6) \qquad (5)$$

3.2 Boundary conditions

The boundary condition for the FE model is shown in Fig. 5, in which one end of the spring is fixed and the load is applied from the other end. The load was applied using a "dummy node" which was specified at the axis of the coils. Every node in the contact region at the top of the coil is constrained to have the same displacement as the dummy node, via a linear multi-point constraint EQUATION available in ABAQUS [13].

The prescribed displacement describes the movement of the spring which occurs during driving. A static preload (e.g. 77 mm displacement) was applied first and then a dynamic load was superimposed from the other end to record the dynamic response for each frequency applied (see Fig. 5).

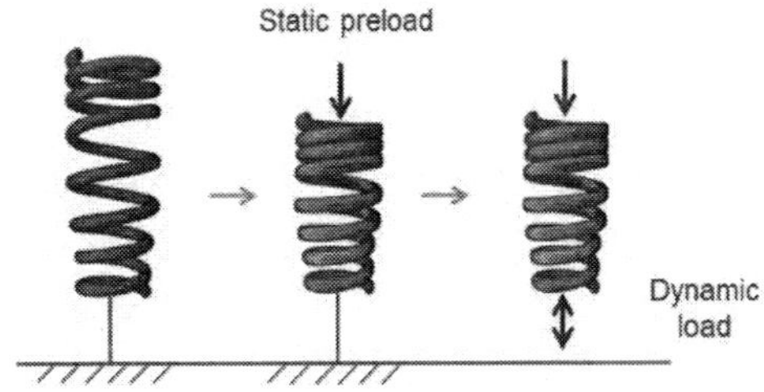

Fig. 5. Boundary conditions applied in the FE model.

The dynamical simulations were carried out in several steps, assigned with an individual displacement amplitude curve with a corresponding frequency, see Fig. 6. A sinusoidal displacement was assumed with the displacement amplitude of 5 mm. The increment size was adjusted for each step, since a higher frequency requires a smaller time step size in order to achieve high accuracy and for being able to capture the peaks in the time response. Fig. 7 shows an FE simulation example of the reaction force at the corresponding frequency, which indicates that the force increases with increased frequency.

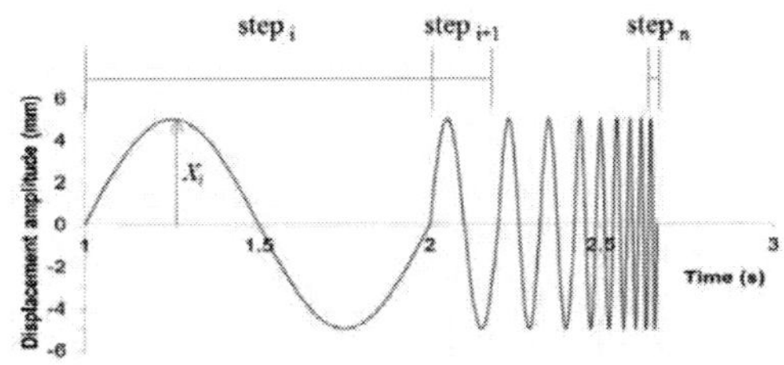

Fig. 6. Frequency step

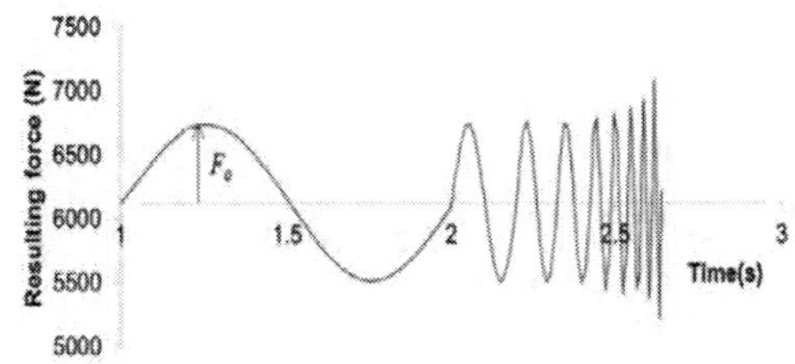

Fig. 7. Reaction force at corresponding frequency

Surface to surface contact is defined when the surfaces of the wire come into contact with each other. At this point simple Coulomb friction is introduced between coils. The coil contact is modelled using a hard contact-penalty algorithm provided by the ABAQUS/Standard finite element program. The frictional coefficient for surfaces in contact is assumed to be 0.5.

4 Dynamic stiffness changes due to corrosion

4.1 Influence of corrosion depth

Consider single coil corrosion (Case A) and let $\alpha = 360°$ and $\beta = 45°$. Dynamic stiffness is plotted against frequency for different corrosion depth shown in Fig. 8. As can be seen from the figure, the dynamic stiffness is moderately increased with an increase in frequency, if the frequency is below approximately 8 Hz. The increase in dynamic stiffness becomes larger when the frequency is more than 10 Hz. For other cases (Case B, C and D), the dynamic stiffness changes show a similar pattern but the magnitude of the stiffness is slightly different.

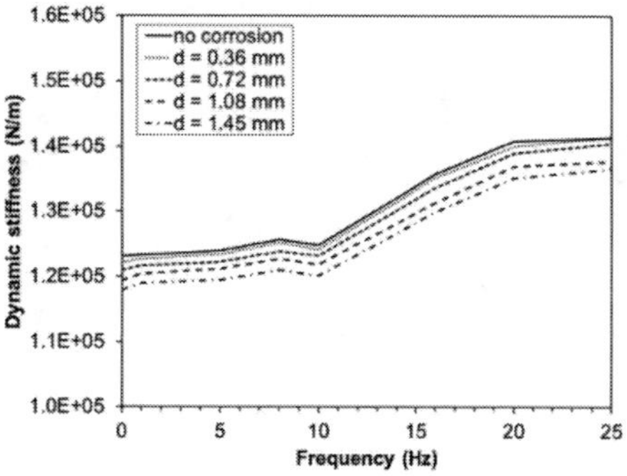

Fig. 8. Stiffness against frequency at different corrosion depth.

The dynamic stiffness at different frequencies is re-plotted against corrosion depth, along with a regression fit, shown in Fig. 9. The R^2 value for the regression lines is 0.996 for all cases. It can be seen that the dynamic stiffness is decreased as the corrosion depth increases for all the range of frequencies. The static stiffness (the case of frequency = 0) is also plotted, indicated by blue line.

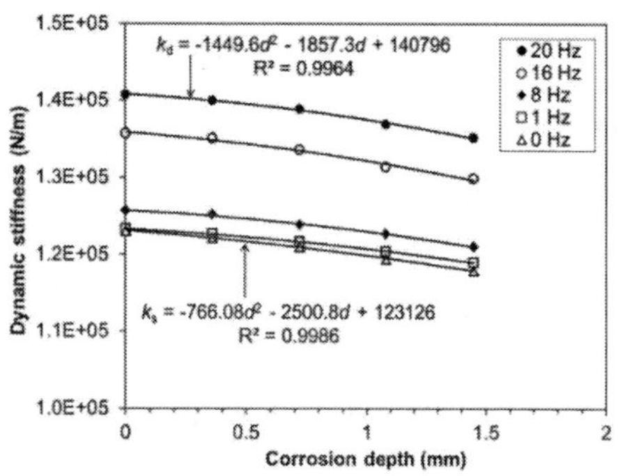

Fig. 9. Stiffness against corrosion depth at different frequencies

4.2 Influence of DSC

The dynamic stiffness is plotted against the degree of surface corrosion (DSC), along with a regression fit for each, shown in Fig. 10. The R^2 value for the regression lines is 0.999 for all. Two regression equations, for frequency = 0 Hz (static) and frequency=25 Hz, are shown in the figure, which can be used to predict the dynamic stiffness of the spring with different DSC. The trend for all the cases is very similar compared to the static. The only difference is the magnitude of the dynamic stiffness, depending on the frequency

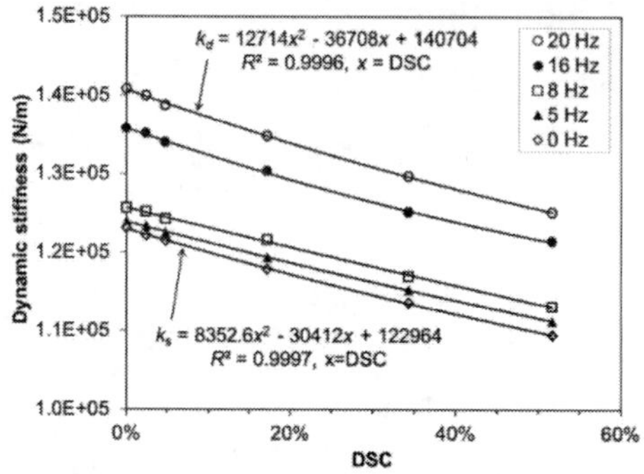

Fig. 10. Dynamic stiffness against DSC at different frequency.

5 Analysis of corrosion-induced stress

5.1 Uniform corrosion

The uniform corrosion observed on the tested spring was idealised as a smooth surface with a uniform corrosion depth in order to perform the FE analysis, as described in Section 2 (see Fig. 2). The stress distribution on the surface is significantly changed due to the corrosion defects. Fig. 11 shows an example of stress distributions in the presence of corrosion defects.

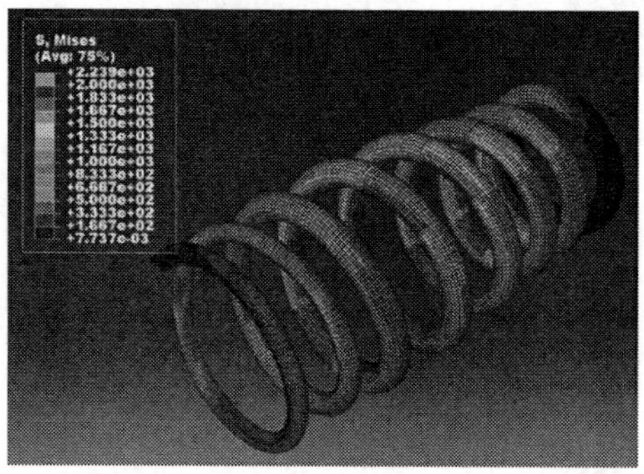

Fig. 11. Stress contours of the spring with surface corrosion (d = 0.36 mm, α = 360° and β = 45°).

Fig. 12 shows the history of von Mises stress at different frequency steps. The stress was measured at the location indicated by a symbol '+'. As can be seen from the figure, the stress is moderately increased with an increase in frequency while the corrosion depth shows a significant effect on the stress.

The maximum von Mises stresses at each frequency are plotted against corrosion depth in Fig. 13. This indicates that the von Mises stress is increased with the increase of corrosion depth.

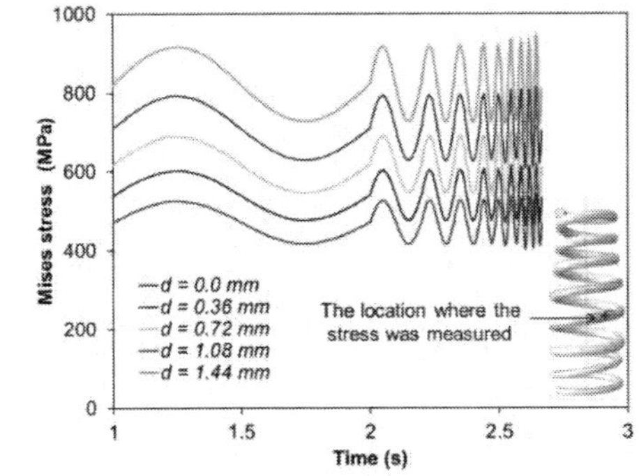

Fig. 12. Von Mises stress history plot at different corrosion depth (Case A, $\alpha = 360°$, $\beta = 45°$)

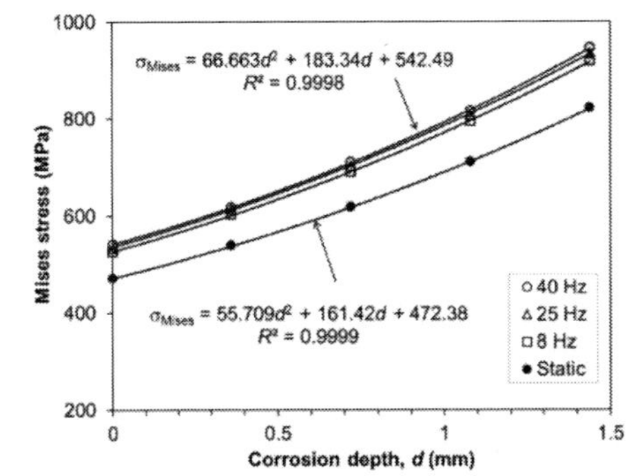

Fig. 13. Plot of maximum stresses at each frequency against corrosion depth.

Fig. 14 shows the effect of the degree of surface corrosion (DSC) on the magnitude and distribution of stresses. The von Mises stress was measured at the same position for each case, as indicated in the previous diagram (Figure 12). It is assumed that α is 360° and corrosion depth is 0.36 mm. The angle β is variable so that different DSC values can be obtained. As can be seen from the figure, the von Mises stress is gradually decreased with an increase in DSC.

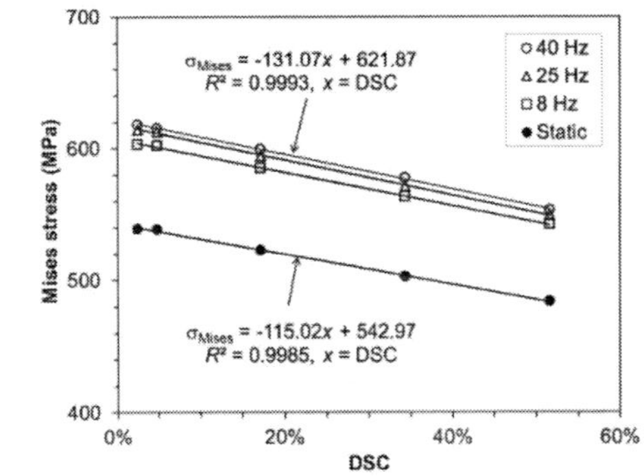

Fig. 14. Von Mises stress against DSC at different frequency.

5.2 Pitting corrosion

Fig. 15 shows the morphology of a corroded surface of the durability tested spring. The corrosion pits are randomly distributed within the corrosion region and the radius is between 0.1 and 0.3mm, approximately. The distributions of these pits are idealised into two patterns and then pattern 1 was adopted in the subsequent FE simulations.

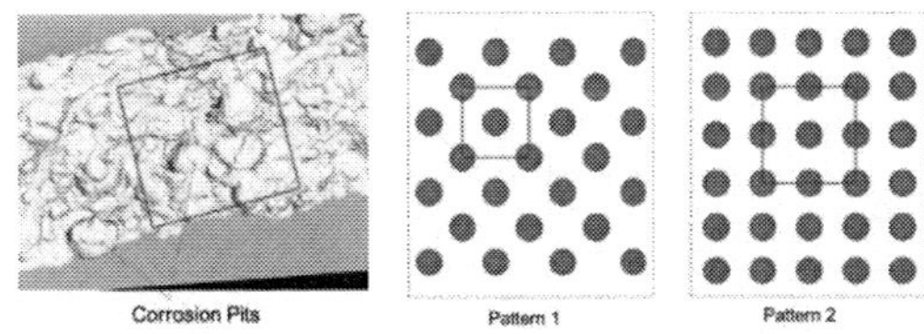

Fig. 15. The morphology of corroded surface observed on tested springs and two idealised distribution patterns.

A multiscale FE model was developed to determine the stress on these corrosion pits. Fig. 16 demonstrates how the FE model is built up. A full model is set up first; the sub-model is extracted from the full model using ABAQUS sub-model function (¼ of coil, 8×8×6mm and 2×2×1.5mm sub-models are selected) and then the hemispherical pits are added to the smallest one. An 8-node linear brick element (C3D8) is adopted in the full spring model and all sub-models. A 10-node quadratic tetrahedron element (C3D10) was used in the model with the pits (e.g. 2 pits, 8 pits and 32 pits model).

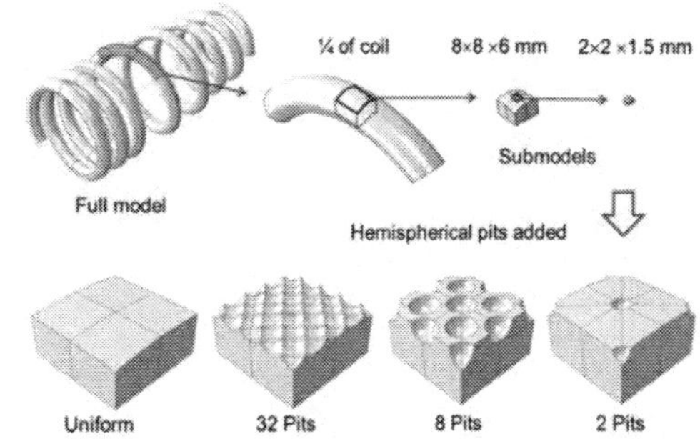

Fig. 16. Schematic showing the development of multiscale FE model for a spring with corrosion pits.

The FE simulation result is shown in Fig. 17, where the von Mises stress is measured at the bottom of the central pit, frequency is at 25 Hz, the static preload is 6000N and displacement amplitude is 5mm. The static (frequency=0) is also plotted to compare. As can be seen from the diagram, the worst case is 2 pits with the highest von Mises stress of 1368 MPa. The stress is decreased as the number of pits increases. The stress is 1087 MPa for 8 pits, 942 MPa for 32 pits and 748 MPa for uniform, respectively. The stress contours for each case are shown in Fig. 18.

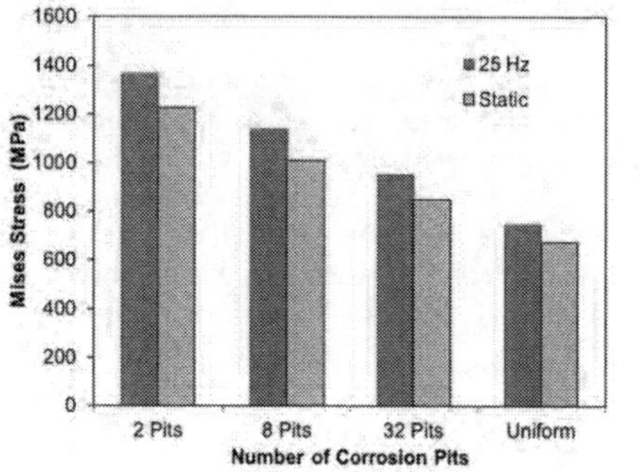

Fig. 17. Von Mises stress at the bottom of corrosion pits for the frequency of 25 Hz.

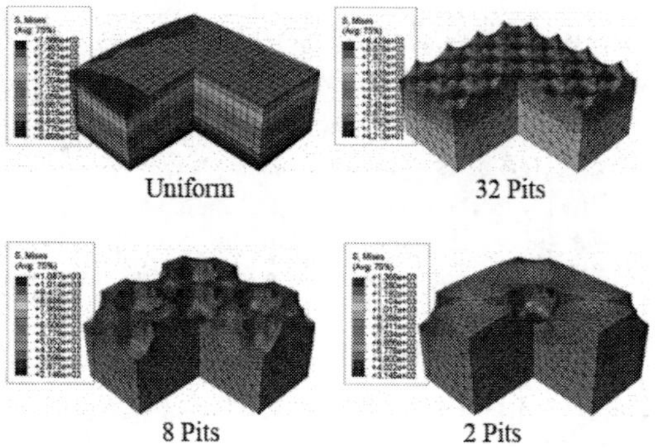

Fig. 18. Von Mises stress contours for 25 Hz.

5.3 Stress distribution with the coils

The stresses were measured at the points 2, 3, 4 and 5 on the surface (indicated by a symbol '+'), as shown in Fig. 19. The measured maximum von Mises stress at each corresponding frequency is plotted against frequency to determine the effect of frequency. As can be seen from the figure, the stress almost stays in the same level until the frequency reaches 20 Hz, but afterwards the stress at the point 5 (coil 5) is slightly increased with an increased frequency while the stress at other points (coils 2-4) is decreased as the frequency increases.

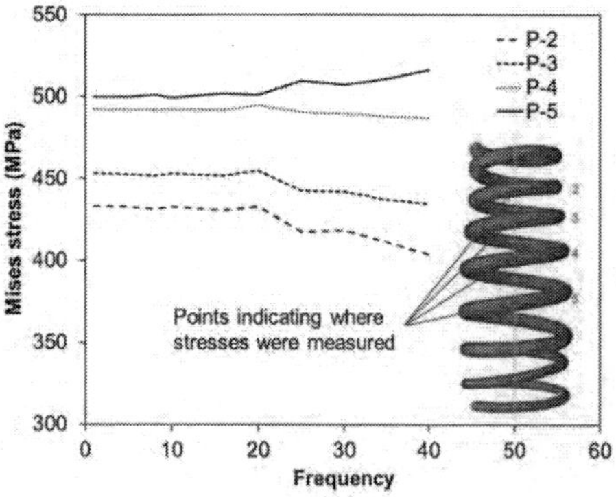

Fig. 19 Mises against frequency for each measured point.

6 Experimental measurements

Tests were conducted using a spring testing machine to validate the FE simulation results. The tests were defined by a series of steps, each step containing load, loading rate and data recording. The computer is connected via the serial port to the control interface. Test data is stored on the hard disc. The tests were performed at a room temperature.

The durability tested springs were analysed with a microscope to identify the surface condition. Small and shallow pits were found in few locations and the majority of the surface was in a very good condition. Therefore, the overall condition of the durability tested springs can be regarded as a nearly new spring.

Fig. 20 shows a comparison of FE simulations and experimental tests. It can be seen from the diagram, the mechanical responses from the FE simulation is very close to that from the experimental measurement. These observations confirmed the validity of the FE model.

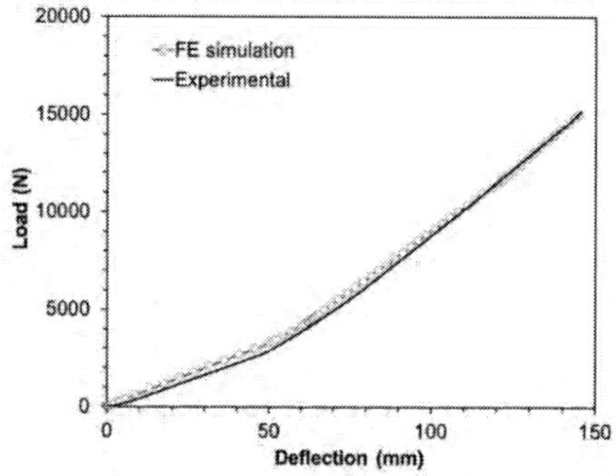

Fig. 20. Comparison of FE simulation and compression test.

7 Concluding remarks

A predictive FE model was developed which can simulate the dynamic response of springs with various corrosion defects. The model can be used to determine the stress/strain distribution in the presence of corrosion defects.

The dynamic characteristic for an uncorroded spring is similar to that of the corroded spring for the frequencies studied here. For example, the trend of dynamic stiffness changes for all selected cases is very similar; the only difference is the magnitude of the dynamic stiffness, depending on the frequency.

Dynamic stiffness (K_d) is increased with an increased frequency for all cases while the K_d is decreased with the increase of either corrosion depth or DSC.

The surface stress is significantly influenced by the morphology of corroded surfaces, corrosion depth and frequency. The stress on the corroded surface increases with the increase of corrosion depth.

Acknowledgements

This work was supported by Jaguar Land Rover and the UK-EPSRC grant EP/L025752/1 as part of the jointly funded Programme for Simulation Innovation.

References

1. A.M. Wahl, *Mechanical Springs*. (1963)
2. Y. Prawoto, M. Ikeda, S.K. Manville, and A. Nishikawa, Eng Fail Anal. **15**(8):1155–1174 (2008)
3. B. Pyttel, I. Brunner, B. Kaiser, C. Berger, and M. Mahendran, Int J Fatigue. **60**:101–109 (2014)
4. M.A. Zaccone, Practical Failure Analysis. **1**(3):51-62 (2001)
5. Y. Yamada, *Materials for Springs*. (2007)
6. M.T. Todinov, Int J Mech Sci. **41**:357-370 (1999)
7. L.E. Beckera, G.G. Chassieb, and W.L. Cleghornb, Int J Mech Sci. **44**:825–841 (2002)
8. C.L. Dym, J Mech Design. **131**(7):071004:1-5 (2009)
9. J.E. Mottershead, International Journal of Mechanical Sciences. **22**(5):267-283 (1980)
10. D. Pearson, J Mech Eng Sci. **24**(4):163-171 (1982)
11. M. Takata, F. Dammak, S. Abid, and M. Haddar, J Mech Mater Struct. **3**(4):641-658 (2008)
12. V. Yıldırım, Int J Mech Sci. **115-116**:280-298 (2016)
13. ABAQUS Aanalysis User's Mannul (2016)

MATEC Web of Conferences 165, 04001 (2018) https://doi.org/10.1051/matecconf/201816504001
FATIGUE 2018

Fatigue Performance of the Novel Titanium Alloy Timetal 407

William Davey[1], *Martin* Bache[1,*], *Helen* Davies[1] and *Matthew* Thomas[2]

[1]Institute of Structural Materials, College of Engineering, Bay Campus, Swansea University, Swansea, SA1 8EN, UK
[2]Timet UK, Holdford Road, Birmingham, B6 7BJ, UK

Abstract. Timetal 407 (Ti-407) is a novel titanium alloy formulated as a lower strength, more malleable alloy offering a range of cost reduction opportunities compared with Ti-6-4 (Ti-6Al-4V). An investigation of the room temperature, high cycle and low cycle fatigue properties of Ti-407 is presented. The effect of thermo-mechanical processing on microstructure is characterised and the fatigue properties of a microstructure containing 30-40% primary alpha volume fraction are presented and discussed. The Ti-407 results are compared with data generated from Ti-6-4 processed to provide a similar microstructure, to demonstrate both superior HCF endurance strength and ductility of the former.

1 Introduction

Titanium alloys have been extensively used within the aero-engine sector since the 1950s. The high strength to weight ratio at low to intermediate service temperatures, good corrosion resistance and low density relative to that of other suitable metals, allow the production of lightweight and durable components with applications ranging from the large-scale fan blades and supporting discs at the front of the engine through to relatively small vanes and blades in the high pressure compressor section. The well-established Ti-6-4 (Ti-6Al-4V), the 'workhorse titanium alloy', is still used for a large number of components within the engine, in part due to the range of microstructural forms produced via thermo-mechanical processing that can be tailored to specific applications. Ti-6-4 accounts for 50% of the total titanium production[1]. This alloy delivers an auspicious balance of mechanical properties including good fatigue performance with intermediate fracture toughness and moderately high tensile strength[2]. Over the years much work has been performed to optimise the manufacturing and production of Ti-6-4 components. However, economic competition is forcing more cost-effective production leading to the development of bespoke alloys, better suited for specific components in terms of both mechanical performance and ease of manufacture.

In general, the machinability of titanium and its alloys can be considered to be difficult. Early practitioners in the 1950s believed that "machining of titanium and its alloys would always be a problem, no matter what techniques are employed to transform this metal into chips[3]". More recent workers in the 1980s, when referring to the same issues, stated that "this is still true in so far as cutting tool materials are concerned[4]." The low thermal conductivity of titanium increases the temperature at the tool/workpiece interface, decreasing the tool life. When machining Ti-6-4, about 80% of the heat generated is conducted into the tool and is concentrated on the cutting edge and tool face. Additionally, titanium is highly chemically reactive. This results in the material welding to the cutting tool during machining leading to significantly shorter tool lives. The high strength of titanium which is maintained at elevated temperatures and the relatively low elastic modulus decreases further the machinability. For these reasons, the improvements in machining of titanium alloys has fallen behind the progress seen by other materials in this area[5]. Other alloys such as steels and nickel-based superalloys have benefited from the advancement in the development of cutting tools. This progress includes the use of coated carbides, ceramics, cubic boron nitride and polycrystalline diamond[6]. These cannot be used with titanium, however, because of a strict set of quality criteria driven by the aerospace and automotive industries. The tools need a high coefficient of thermal conductivity, good thermal resistance, high hardness, wear durability and hot hardness. Hence, tungsten carbide tools are used for the vast majority of titanium tooling processes.

TIMET is constantly addressing the need for new bespoke titanium alloys. The focus of the present study was placed on one such alloy, Ti-407. Relative to Ti-6-4, one of the main benefits and most important design features offered by Ti-407 is a significant improvement in ductility, in turn offering superior manufacturability and increased machining speeds with lower contact forces. During machining trials, Ti-407 displayed more than twice the tool life of the baseline Ti-6-4 alloy[7]. Compared to Ti-6-4 the low flow stress, greater malleability and wide process window of Ti-407 should allow it to be processed with fewer reheats, while exhibiting less surface cracking and giving a consistently good surface finish. Optimised Ti-407 manufacturing processes should favour nearer net shape giving higher yields and requiring less machining to finished component. Ti-407 has a slightly higher beta content, making it denser than Ti-6-4. However, by employing suitable designs, components can still be manufactured from Ti-407 as a weight neutral alloy.

Of particular interest, Ti-407 can replace Ti-6-4 in applications where the key design criteria are energy absorption during fracture and HCF endurance limit. The

[*] Corresponding author: m.r.bache@swansea.ac.uk

most relevant applications within the gas turbine are, therefore, the engine fan and compressor casings. Due to its relatively low strength, Ti-407 may not be able to replace all Ti-6-4 components, but promisingly, Ti-407 demonstration casings have shown 2.5 times the lateral expansion and absorbed more than twice the impact energy when compared to Ti-6-4[7].

2 Experimental Methods

2.1 Ti-407 production and sampling

The present study utilised Ti-407 alloy (Ti-0.85Al-3.9V-0.15O-0.25Si-0.25Fe) in the form of "pancake" forgings processed by TIMET. Billet sections were cut into six, 60° degree segments. These were then forged into the pancake form which is displayed in Figure 1, from which the mechanical test specimen blanks were extracted from the central regions.

Fig. 1. A pancake forging and specimen blank extraction.

Heat treatment trials were conducted in order to establish a solution treatment temperature that would generate a bimodal microstructure with approximately 30-40% primary alpha (α_p) volume fraction. This goal was based on results from previous research[8] which suggested that this specific volume fraction provided the optimum fatigue performance. Increasing content has been shown to increase tensile plasticity[9], refine the microstructure and limit the formation of areas of common orientation [10]. Whilst α_p within a lamellar matrix does offer these advantages, the percentage volume must be carefully restricted since beyond a critical volume, ductility ceases to improve and alloy element partitioning takes place. This degrades the basic strength of the lamellar constituent of the microstructure, therefore resulting in a decrease in fatigue life.

A SNOL 1.8Kw electric furnace was employed for microstructure trials, with two calibrated N type thermocouples in contact with the specimens (15x15mm cubes). The temperatures were constantly monitored throughout the duration of the heat treatments using calibrated Fluke thermometers. Samples were subjected to three different solution heat treatments, each for a two- hour period;

- 820°C producing 50% α_p volume fraction
- 840°C producing 30% α_p volume fraction
- 860°C producing 20% α_p volume fraction.

The resultant microstructures are displayed in Figure 2 along with image analysis based measurements of the α_p volume fractions.

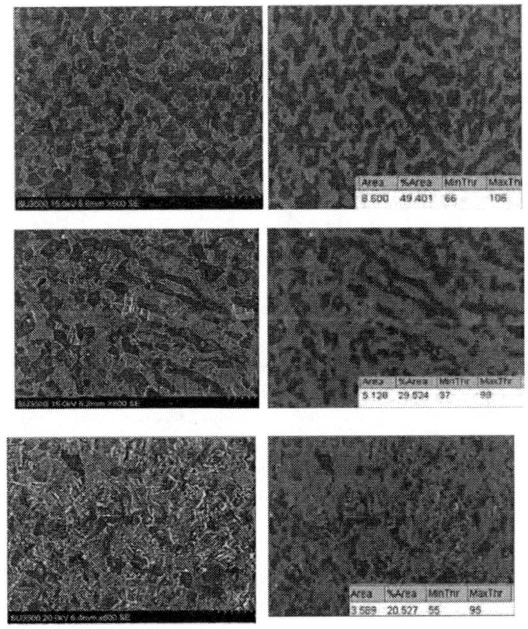

Fig. 2. Microstructures generated under the three solution heat treatments.

All mechanical specimen blanks were subsequently heat treated at 840°C for two hours, followed by an eight-hour aging treatment at 500°C. The 30% volume fraction of α_p grains demonstrated an average α_p grain diameter of 10 to 15 microns. The blanks were then machined using CNC turning operations into plain cylindrical specimens suitable for tensile and fatigue experiments, Figure 3 and Figure 4. A non-contacting shadowgraph was employed to measure the co-axiality of the test pieces and the critical dimensions of the parallel gauge section according to the guidance of aerospace based fatigue test standards [11].

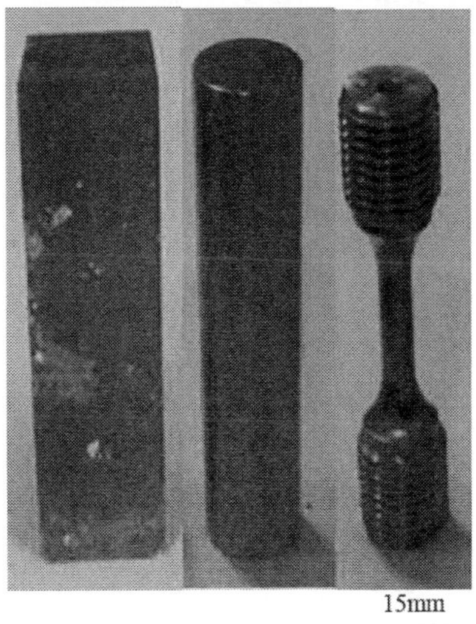

Fig. 3. Examples of post heat treated blank, CNC machined cylinder and final test piece.

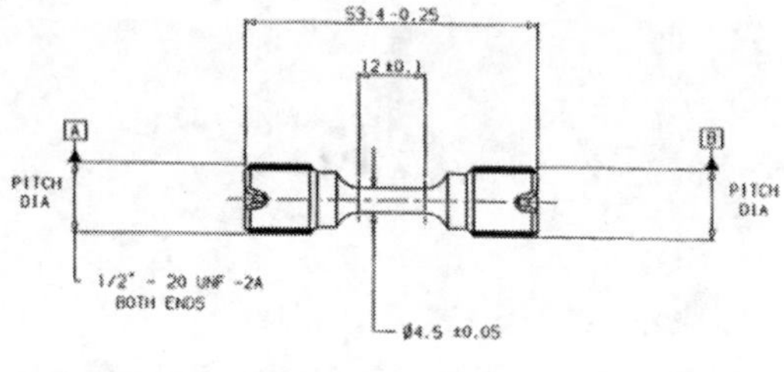

Fig. 4. Mechanical specimen geometry.

2.2 Ti-6-4 production and sampling

In order to provide a comparison between HCF performance, Ti-6-4 blanks were extracted from similar forged pancake material stock. The same method was employed to determine a heat treatment temperature that would produce a α_p volume fraction of approximately 30%. Two temperatures were trialled, again with the samples subject to a 2 hour solution treatment:

- 950°C producing ~30% α_p volume fraction.
- 960°C producing ~40% α_p ha volume fraction

The resultant microstructures are displayed in Figure 5 along with image analysis based on measurements of the α_p volume fraction.

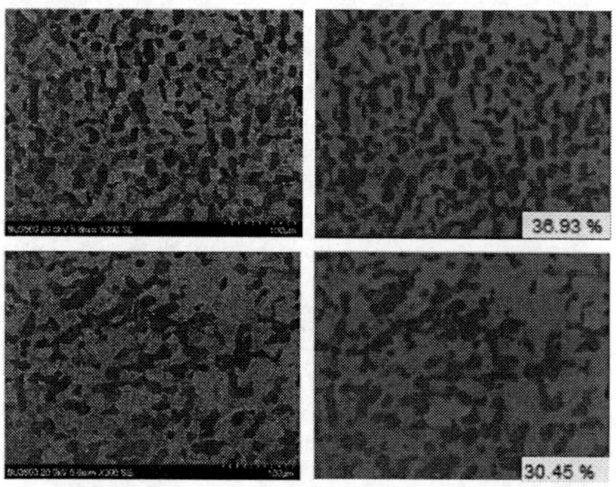

Fig. 5. Ti-6-4 microstructures generated from the two different heat treatments.

It is emphasised that the 950°C heat treatment we ultimately employed provides a bespoke microstructure, i.e. non-standard for typical aero-engine casing applications. Instead, this offered an academic comparison with Ti-407 processed to the same 30% α_p content.

2.3 HCF and LCF Testing

High cycle fatigue (HCF) tests were conducted on an Amsler vibrophore resonance machine at a load ratio of R=0 under room temperature, using a sinusoidal waveform with a frequency of 100Hz. Low cycle fatigue (LCF) tests, again under R=0 at room temperature, employing a 15 cycle per minute trapezoidal waveform comprising one second linear rise and fall ramps with a one second hold at both peak and minimum load, were performed on various servo-hydraulic test rigs. All test machine load cells were calibrated to BS EN ISO 7500-1:2015[12]. In all fatigue tests, complete rupture of the specimen was induced (i.e. no "run-outs" were generated).

3 Results

3.1 Tensile Testing

Key mechanical properties of the Ti-407 pancake material were measured through tensile testing. Following the guidelines of BS EN ISO 2002-1:2005 [13], a strain rate of 0.002/min was applied to define yield, increasing to a rate of 0.01/min to induce fracture. The tensile results are summarised in Table 1.

Table 1. Tensile data for Ti-407.

Attribute	Ti-407
Young's modulus (GPa)	114
Tensile strength (MPa)	764
Yield strength (MPa)	648
Ductility (%)	18

3.2 Fatigue Data and Fractography

The combined fatigue data for Ti-407 generated under LCF and HCF conditions are plotted in Figure 6. Least squares best fit trend lines are superimposed to the individual sets of fatigue data. The range of maximum cyclic stresses assessed under both regimes was identical (with individual tests performed between 500 and 650MPa). This helped to emphasise a significantly stronger response under HCF conditions relative to LCF performance. Figure 7 compares HCF performance between Ti-407 and the non-standard Ti-6-4.

The individual LCF and HCF Ti-407 fatigue fracture surfaces displayed a mixture of surface and subsurface initiation sites, see Figure 8. The sub-surface initiated failures were easily identified by the circular shaped, bright halo formed during sub-surface crack propagation, effectively within a vacuum environment. The fracture surface surrounding the halo, formed after the crack broke through at the surface and now open to oxidation, took a dull appearance. Quasi-cleavage facets were invariably revealed at the centre of the halo marking the precise site of fatigue crack initiation. These relatively flat, facetted regions were orientated near perpendicular to the applied tensile stress axis and were lustrous to the naked eye. Under higher magnification scanning electron microscope (SEM) inspection, the facets were formed of a cluster of elongated features representing the earliest stages of fracture through neighbouring α_p grains, with regions of a more ductile appearance between representing failure of the juxtaposed transformed α/β microstructure, see Figure 9.

3.3 Sub-fracture micro-texture analysis

After noting the predominance of quasi-cleavage facets at the epi-centre of sub-surface crack initiation sites, metallographic sections were prepared immediately below the fracture surface of selected specimens. The polished plane was prepared approximately 5mm sub-fracture, using standard grinding and etch techniques. Electron back scatter diffraction (EBSD) measurements were taken to generate crystallographic orientation maps extending across the entire specimen gauge section, exemplified by Figure 10. This represents a montage of 70 individual small area scans of approximately 500 x 500 microns conducted with a step size between 2-3 microns. Figure 3.7 shows the inverse pole figures (IPF) relating to this section. A moderate basal plane texture was evident with a preferred basal pole orientation of x3.6 random. The -12-10 and 01-10 maps also indicate a preferential axis of rotation of individual α_p grains.

4. Discussion

From the perspective of ranking fatigue performance, Ti-407 demonstrated a significantly stronger fatigue strength compared to the non-standard form of Ti-6-4 under HCF loading, Figure 7. Therefore, together with superior ductility, machinability and manufacturing properties Ti-407 could be viewed as a promising alternative alloy for aero-engine containment structures. A single solution heat treatment was adopted for the Ti-407 specimens tested during the present study, providing a α_p volume fraction of 30%. The heat treatment trials demonstrated that α_p content can vary across a relatively limited process window. Therefore it would be prudent to extend our fatigue assessment to different variants of the alloy to consider the sensitivity of this mechanical performance to microstructure.

The relatively high HCF strength of Ti-407 is a promising outcome for what is recognised as a comparatively low static strength alloy. Since fatigue mechanisms in general are intimately controlled by the onset of cyclic plasticity, a future investigation into the cyclic yield performance through strain controlled fatigue testing is envisaged. Specifically, with HCF performance in mind, any sensitivity to machining operations and surface finish should also be considered, although it is recognised that the sub-surface failures generated amongst the current data set suggest that these factors may not be controlling crack initiation.

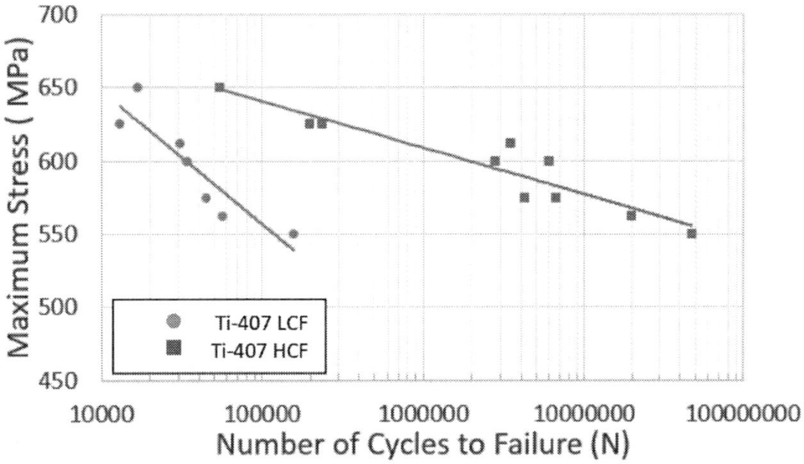

Fig. 6. LCF and HCF data for Ti-407.

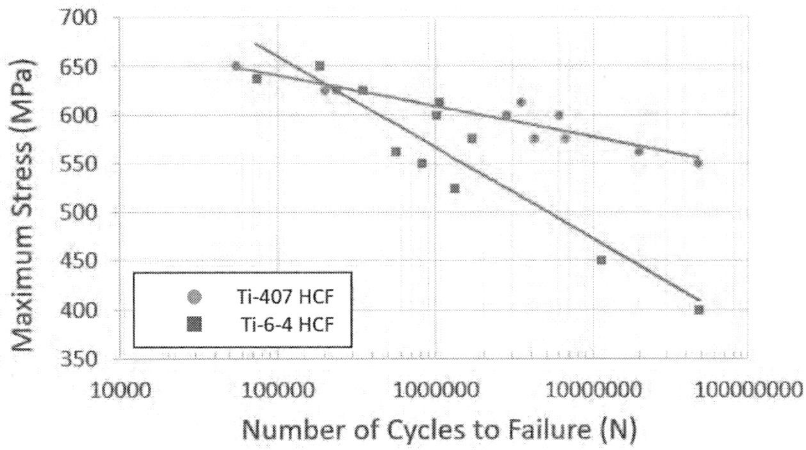

Fig. 7. HCF comparison of Ti-6-4 and Ti-407.

Fig. 8. Examples of surface (top) and sub-surface (bottom) crack initiation sites in Ti-407.

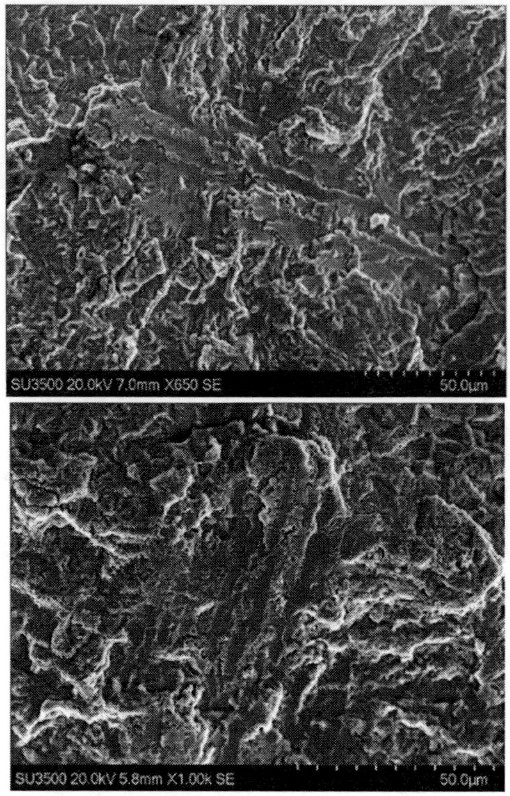

Fig. 9. Sub-surface facet clusters revealing elongated ap α grains exposed on the fracture surface.

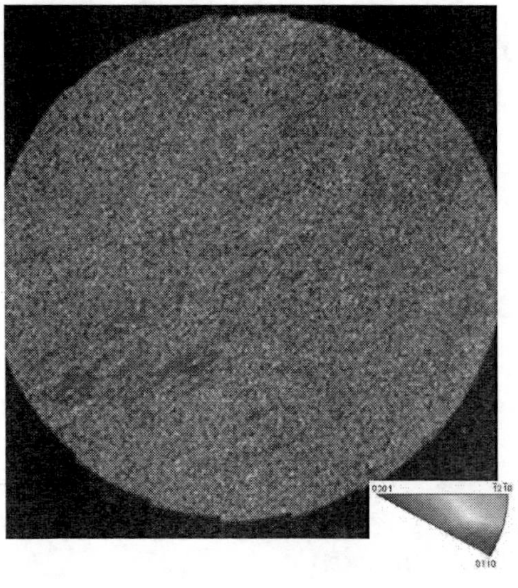

Fig. 10. EBSD orientation map of a sub-fracture section.

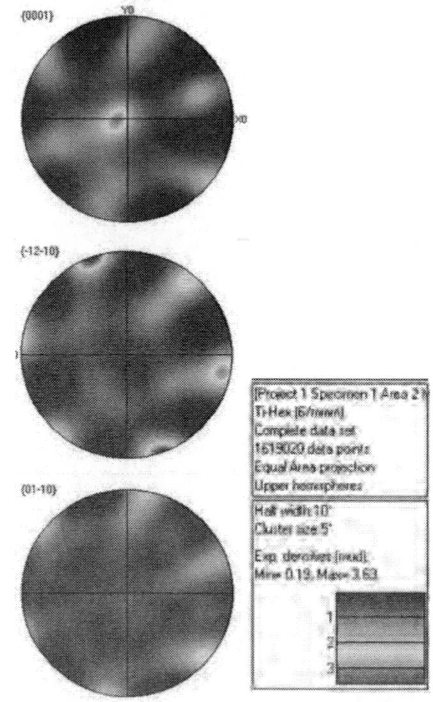

Fig. 11. Pole figures associated with Figure 10.

* Corresponding author: m.r.bache@swansea.ac.uk

Viewing the fatigue behaviour of Ti-407 in isolation (Figure 6), it was interesting to note the greater strength measured under HCF conditions and to consider this in the context of the time-dependent sensitivity of the α/β titanium class of alloys in general. Bache has previously demonstrated that fatigue tests performed on the compressor disc alloy Ti-834 at HCF frequencies induced a stronger response when compared to an LCF based 15cpm waveform[14]. That study went on to assess Ti-834 under a commonly employed two minute dwell waveform and showed forged variants of the alloy were susceptible to "cold dwell". In light of the current investigation on Ti-407, the one second dwell at peak load during the 15cpm LCF experiments appears to induce attributes of a similar dwell response. Although dwell sensitive behaviour would not be considered as a design criterion for containment casings, these results on Ti-407 present an academic interest that will be pursued further in the future.

The appearance and form of the quasi-cleavage faceting in the Ti-407 LCF and HCF fatigue specimens also draws comparisons to Ti-834[15]. At the heart of the initiation sites, facets were discernible associated with individual, elongated α_p grains, which together with multiple neighbouring facets created a cluster. The EBSD characterisation of the microstructural section extracted immediately below the fracture surface of a selected specimen (Figure 3.5), indicated regions shaded in red where α_p grains are all preferentially orientated with their basal plane orthogonal to the applied tensile stress axis. EBSD measurements were also achieved directly from some of the individual facets exposed on the fracture surface. Despite the relatively small α_p grain diameters at less than 15 μm, the regions of common crystallographic orientation provide a much larger "macrozone" or effective structural unit to control deformation and crack initiation mechanisms. Invoking the Stroh model [16] and the subsequent Evans-Bache model[14,17] which applied this theory to the understanding of fatigue failure in a wide selection of α/β titanium alloys, the macrozone with basal orientations perpendicular to the tensile axis would act as a "hard region". The adjacent regions containing a collection of grains with more random orientation would be more favourable for slip deformation thus acting as the neighbouring "soft regions". A process of stress redistribution induces the initial stages of crack initiation within the hard region, eventually observed on the fracture surface as the lustrous cluster of facets lying perpendicular to the tensile stress axis.

The empirically derived Evans-Bache model was later supported by crystal plasticity modelling reported by Zheng et al [18]. Their results suggested that compared to standard cyclic loading, facet formation and crack initiation was encouraged under dwell conditions due to a large increase in the dislocation density within the grains favourably orientated for time-dependent slip. The dwell at peak stress also enables dislocations to escape obstacles over time, resulting in higher intensity dislocation pile-ups at the boundaries between hard and soft grains. In turn this generates higher localised stresses in the hard grains which, in combination with the bulk applied tensile stresses, help the facets to initiate on the basal planes.

The principles behind the Evans-Bache model, but applied to regions or macrozones of common orientation, are represented in the schematic of Figure 12.

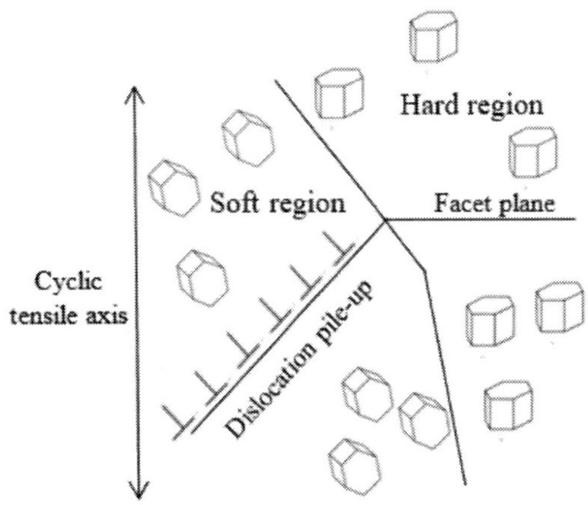

Fig. 12. Adaptation of the Evans-Bache model to incorporate hard and soft regions made up of multiple grains with common basal and random orientations respectively.

It should be emphasised that the material employed for the present study still remains to be optimised for different thermo-mechanical processing routes. Therefore, the microstructural condition produced via the relatively simple billet to pancake forging would not prevail in engineering components. This is particularly true for thin walled containment structures that would probably utilise ring-rolling techniques. The general micro-texture and presence or otherwise of macrozones could be different in the component material variants and so further fatigue assessments will be required.

5. Conclusions

The following conclusions were drawn from the present programme of research:

- The novel alloy Ti-407 could offer a cost-effective replacement for Ti-6-4 engine fan and compressor casings due to its ease of manufacture.
- Ti-407 offers superior HCF fatigue strength despite a relatively low UTS.
- HCF strength is greater than that measured under LCF loading, suggesting Ti-407 fatigue behaviour may be subject to dwell sensitivity.
- Quasi-cleavage facets were invariably noted at crack initiation sites, including examples of sub-surface failures.
- Clusters of facets revealed the underlying microstructure, including elongated α_p grains with transformed product between.
- EBSD defined regions of grains with common basal plane orientation.

6. Acknowledgements

The current research was funded under the EPSRC Rolls-Royce Strategic Partnership in Structural Metallic Systems for Gas Turbines (grants EP/H500383/1 and EP/H022309/1). The provision of materials and supporting information from TIMET UK is gratefully acknowledged. Mechanical tests were performed at Swansea Materials Research and Testing Ltd. (SMaRT).

7. References

1. R. Wood and R. Favor, *Titanium Alloys Handbook*, MCIC-HB-002, Battelle (1972)
2. www.timet.com. [Accessed: 31-Oct-2017].
3. H. Siekmann, Tool Eng., **34**, no. 1, pp. 78–82, (1995)
4. R. Komanduri and W. Reed, Wear, **92**, no. 1, pp. 113–123, (1983)
5. E.O. Ezugwu and Z.M. Wang, *J.* Mater. Process. Technol., **68**, no. 3, pp. 262–274, (1997)
6. M. Rahman, Y.S. Wong and A.R. Zareena, JSME International Journal Series C, **46**, no. 1. pp. 107–115, (2003)
7. S. James, Y. Kosaka, R. Thomas and P. Garratt, *13th International conference on titanium,* Eds. Venkatesh et al, pp.721-725, TMS, (2016)
8. G.Q. Wu, C.L. Shi, W. Sha, A.X. Sha and H.R. Jiang, Mater. Des., **46**, pp. 668–674, (2013)
9. G. Lütjering, Mater. Sci. Eng. A, **243**, no. 1–2, pp. 32–45, (1998)
10. M.T. Whittaker, W.J. Evans, R. Lancaster, W. Harrison and P.S. Webster, Int. J. Fatigue, **31**, no. 11–12, pp. 2022–2030, (2009)
11. J. Krautkräm et al, J. Appl. Phys., **58**, no. 3, pp. 6737–6741, (2006)
12. BS EN ISO 7500-1: *Metallic materials. Verification of static uniaxial testing machines. Tension/compression testing machines. Verification and calibration of the force-measuring system* (2004).
13. BS EN 2002:1, *Metallic Materials - Test Methods: Tensile testing at ambient temperature* (2005)
14. M.R. Bache, Int. J. Fatigue, **25**, pp. 1079–1087, 2003.
15. L. Germain and M.R. Bache, Ti 2007 Science and Technology, **2,** Eds Ninomi et al, Japan Institute of Metals 2007 pp. 953-956 (2015)
16. A. Stroh, *Proc. R. Scociety London A Mathmatical,* Phys. Eng. Sci., **223**, no. 1154, pp. 404–414, (1954)
17. W.J. Evans and M.R. Bache, Titanium '92 Science and Technology, **3**, Eds Froes and Caplan pp.1693-1700, TMS, (1993)
18. Z. Zheng, D.S. Balint and F.P.E. Dunne, Int. J. Plast., **87**, pp. 15–31, (2016)

* Corresponding author: m.r.bache@swansea.ac.uk

A cyclic slip irreversibility based model for fatigue crack initiation of nickel base alloys

Hsin Shen Ho[1,*], *Marion* Risbet[2], and *Xavier* Feaugas[3]

[1]School of Mechanical Engineering, Zhengzhou University, 450001 Zhengzhou, China
[2]UMR CNRS 7337, Université de Technologie de Compiègne, 60205 Compiegne cedex, France
[3]UMR CNRS 7356, Université de La Rochelle, 17000 La Rochelle cedex, France

Abstract. The opportunity to define a microscopic law of fatigue crack initiation using Manson-Coffin law formulated in terms of cyclic slip irreversibility deduced from AFM measurements is discussed for a polycrystalline superalloy with different grain sizes and precipitate sizes. The results show that the modified Manson-Coffin law, relating cyclic slip irreversibility parameter to fatigue crack initiation life, is sustained through a two-parameter power law: ε_f' and c. The analysis suggests that the exponent c-value can be related to the degree of plastic strain incompatibility between grains, and the cumulative irreversible cyclic plastic strain to crack initiation is a relevant damage indicator for crack initiation. Consequently, our approach allows giving a physical base of engineering law.

1 Introduction

Fatigue damage of materials that exhibits flow localization is very often related to a surface phenomenon [1]. In these materials, the cyclic slip irreversibility, which is a phenomenon resulted from the slip displacements caused by the movement of dislocations in the forward part of the fatigue loading cycle that are not fully recovered in the reverse part of the fatigue loading cycle, is a primary source of fatigue crack initiation [2]. Cyclic slip irreversibility may arise from quite different physical mechanisms such as irreversible dislocation displacements resulted from cutting of precipitates and surface roughening caused by to-and-fro glide of dislocations, as recapitulated in Refs. [3-4]. Some other important factors known to promote cyclic slip irreversibility are the local microstructure and environment [5-6].

Cyclic slip irreversibility is difficult to be analyzed quantitatively [7]. Thereby, there exist many different forms of characterizing the cyclic slip irreversibility in which they are given numerical values, as summarized in Ref. [3]. Though, the cyclic slip irreversibility defined as the fraction of microstructurally irreversible cyclic plastic strain with respect to the total cyclic plastic strain is found to be one of the most relevant microstructural and life-determining fatigue damage parameter [7-8]. Some works have recently demonstrated the opportunity to use atomic force microscopy to measure slip band emergence and consequently the degree of reversibility during cyclic loading [3,5,8]. In low-cycle fatigue tests, a global and generic cumulative damage rule known as the Manson-Coffin (MC) law is generally used to express the relationship between total fatigue life and plastic strain amplitude [9-10]. Thereby, three questions arise: (i) whether the MC law can be formulated in terms of cyclic slip irreversibility, (ii) is the MC law changing if the early stage of damage is considered as fatigue crack initiation life, and (iii) what are the factors that influence the correlation between the two empirical parameters (ductility coefficient and power exponential value) dominating the microscopic law of fatigue crack initiation?

Thus, the aims of the present paper are firstly, to test the validity of MC rule formulated in terms of cyclic slip irreversibility for fatigue crack initiation of a Ni-base superalloy with different grain sizes and precipitate sizes, where cyclic slip irreversibility is a result of repeated irreversible cutting of the coherent γ' precipitate particle during cycling, and secondly, to understand the nature and/or physical content of the two empirical parameters of the modified MC rule.

2 Material and methods

2.1. Material

The target material is a Ni-base superalloy (Waspaloy) strengthened by spherical precipitates (γ', Ni_3Al), with the following chemical composition (wt.%): 1.80 Al; 3.33 Ti; 12.64 Co; 19.45 Cr; 7.31 Mo; balance Ni. The average grain size (*D*) and the average precipitate size (*d*) of the as-received state of Waspaloy are 64 μm and 42 nm, respectively. A two-step heat treatment (solutionizing and aging treatments) is carried out on the as-received state to modify *D* and *d* so as to obtain specimens with well-controlled metallurgical states, i.e.

* Corresponding author: hsinshen.ho@zzu.edu.cn

different grain sizes of the matrix γ with a constant γ' precipitate size and inversely (see Ref. [5] for details on heat treatment steps). In the present study, eight classes of specimens are selected with the following average grain size (D) and average precipitate size (d): (30 µm, 7 nm); (50 µm, 10 nm); (50 µm, 20 nm); (50 µm, 30 nm); (50 µm, 40 nm); (100 µm, 10 nm); (100 µm, 20 nm) and (200 µm, 20 nm) (*c.f.* Table 1 indicating the stereological elements, e.g. grain size (D) and precipitate size (d), but also the values of the other parameters, e.g. fatigue crack initiation life (N_i), mean extrusion height ($\langle h \rangle$), band thickness (w) and mean inter-band spacing (I_B), of the specimens tested at three different levels of plastic strain amplitude (ε_p)).

2.2 Methods

The specimens are mechanically polished, then chemically-etched using standard metallographic techniques, as detailed in Ref. [5].

Interrupted low-cycle fatigue tests are conducted at ambient temperature using an Instron servohydraulic machine on hourglass-shaped cylindrical specimens. The fatigue tests are performed on controlled plastic strain amplitude (ε_p) of 0.05 %, 0.3 % and 0.9 % with a fully reversed triangular wave form at a frequency of 0.05 Hz until the nucleation of first cracks. Cracks are considered as nucleated when the length of a crack is equivalent to the dimension of a grain.

During fatigue testing, the fatigue hysteresis loops are recorded in order to track the cyclic behavior and to calculate the internal stresses. The cyclic stress amplitude (σ_a) is composed of effective stress (σ_{eff}) and backstress (X). The backstress (X), which is to be discussed further, can thus be expressed as:

$$X = \sigma_a - \sigma_{eff} \text{ with } \sigma_{eff} = \frac{1}{2}(\sigma_a - \sigma_r) + \frac{1}{2}\sigma^* \quad (1)$$

where σ^* is the thermal part of the flow stress and σ_r denotes the reverse yield stress, which are measured from the experimental hysteresis loops with a plastic strain offset of $(4 \pm 2) \times 10^{-5}$.

The surface states are characterized by scanning electron microscope (SEM) and atomic force microscope (AFM), using respectively a (FEG) Zeiss Sigma SEM and a Veeco Dimension 3100 AFM at room temperature. For each specimen, about 8 damaged grains and 8 undamaged grains are investigated. And for each grain, a few regions of 10 µm × 10 µm are examined. Each region extracted using the topography-imaging mode contains around 20 slip bands. The surface roughness, representing the arisen surface marking from material surface, is then analyzed using the experimental procedure developed in Refs. [5,8]. In addition to the surface analysis, the bulk material is examined at room temperature using JEOL JEM2011 transmission electron microscope (TEM) operating at 200 kV.

Table 1. Metallurgical states of investigated specimens.

Plastic strain amplitude (ε_p)	Grain size (D, µm)	Precipitate size (d, nm)	Fatigue crack initiation life (N_i, cycle)	Mean extrusion height ($\langle h \rangle$, nm)	Band thickness (w, nm)	Mean inter-band spacing (I_B, µm)
0.0005	30	7	33500	50	80	8
0.003	30	7	1500	50	80	2.5
	50	10	1100	98	121	1.75
	50	20	800	107	140	2.17
	50	30	700	109	146	1.93
	50	40	650	116	171	2.42
	100	10	950	97	130	1.7
	100	20	700	98	150	2.11
0.009	30	7	320	50	80	1.0
	50	40	65	40		1.5
	100	10	196	69		0.8

3 Microstructure-based reformulation of the Manson-Coffin fatigue crack initiation life

If the Manson-Coffin-type law is assumed to be valid for damage initiation in polycrystals Waspaloy, this law relates the plastic strain amplitude (ε_p) to the cycle number of crack initiation (N_i) through a two-parameter power law [9-10]:

$$\varepsilon_p = \varepsilon_f'(2N_i)^c \quad (2)$$

where ε'_f and c (< 0) are the fatigue ductility coefficient and exponent, respectively.

Inspired by earlier works of Mughrabi and Risbet et *al.*, which propose to formulate a life law in terms of microstructural parameters instead of the function of macroscopic plastic strain amplitude endured by the specimen during fatigue test, the plastic strain irreversibility (p_{irr}^{cum}) is introduced in the present work as a microstructural parameter to explicitly take into account the fatigue damage resulted from irreversible plastic strain [7-8,11]. The parameter p_{irr}^{cum} is defined as the fraction of the microstructurally irreversible cyclic plastic strain with respect to the total cyclic plastic strain.

$$p_{irr}^{cum} = \frac{\gamma_{irr,pl,loc}^{cum}}{\gamma_{pl,loc}^{cum}} \quad (3)$$

where $\gamma_{irr,pl,loc}^{cum}$ is the cumulative irreversible cyclic plastic strain, expressed as:

$$\gamma_{irr,pl,loc}^{cum} = \frac{h}{D} \quad (4)$$

where h is the mean extrusion height and D is the average grain size, and $\gamma_{pl,loc}^{cum}$ is the local plastic strain, representing the total plastic strain accumulated in a slip band during N fatigue cycles. $\gamma_{pl,loc}^{cum}$ is expressed as a function of the volume fraction of slip bands (f_v), which itself depends on inter-band spacing (l_B) and volume fraction of one slip band (f_1):

$$\gamma_{pl,loc}^{cum} = \sum_{k=1}^{N} \gamma_{pl,loc} = \left(\sum_{k=1}^{N} l_B(N,\varepsilon_p)\right)\frac{4M\varepsilon_p}{Df_1} = \gamma_{pl,loc}^{cum}(N,\varepsilon_p) \quad (5)$$

where M is the Taylor factor and N is the number of cycles. In shearable-precipitated alloy, the plastic strain is supported by multiplication of slip bands, and thus l_B is dependent on ε_p. By combining the expressions for $\gamma_{irr,pl,loc}^{cum}$ in Eq. (4) and $\gamma_{pl,loc}^{cum}$ in Eq. (5), p_{irr}^{cum} can be expressed as a function of N:

$$p_{irr}^{cum} = \frac{hf_1}{Nl_B(N_i,\varepsilon_p)\varepsilon_p 4M} = p_{irr}^{cum}(N,\varepsilon_p) \quad (6)$$

4 Experimental validations

4.1 Empirical relationship between plastic strain irreversibility and fatigue crack initiation life

Fig. 1. Variation of the fraction of the microstructurally irreversible cyclic plastic strain (p_{irr}^{cum}) as a function of the number of cycles (N_i) for different metallurgical states tested under three different levels of plastic strain amplitude (ε_p).

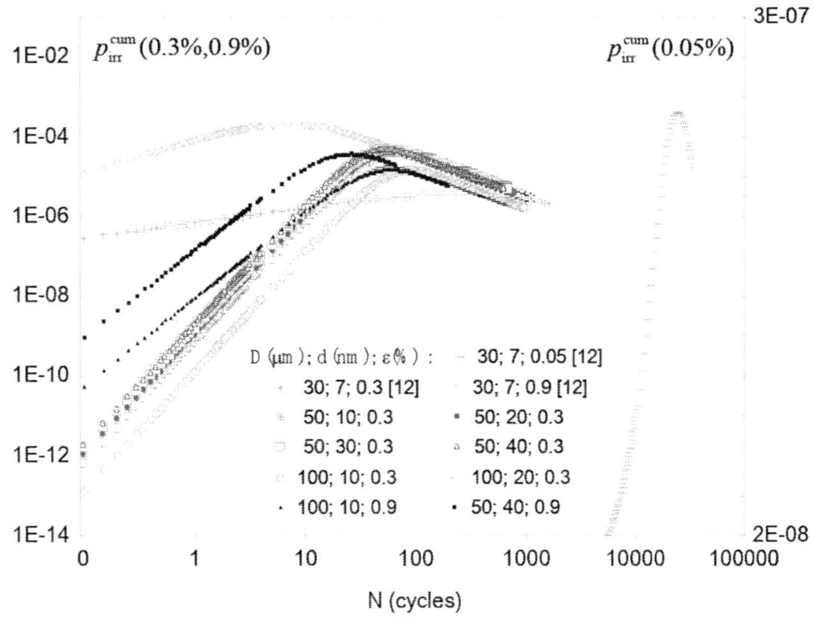

Eq. (6) shows that p_{irr}^{cum} is dependent on N and our results reveal that p_{irr}^{cum} first increases and then decreases with increasing N, as exhibited in Fig. 1. Figure 1 shows that the functional dependency between p_{irr}^{cum} and N remains valid for all the specimens investigated in the present study, in coherence with earlier results obtained with the same alloy of only one metallurgical state tested under three different levels of plastic strain ε_p [8].

At the fatigue crack initiation phase ($N = N_i$), $p_{irr}^{cum}(N_i)$ is found to be dependent on ε_p (and hence also on N_i), as is exhibited in Fig. 2. It is shown in this figure that $p_{irr}^{cum}(N_i)$ increases when ε_p increases (or when N_i decreases). The variation of $p_{irr}^{cum}(N_i)$ with ε_p (and N_i) fits well with the results reported in the literature [8].

Fig. 2. Variation of plastic strain irreversibility ($p_{irr}^{cum}(N_i)$) as a function of plastic strain amplitude (ε_p) for different grain sizes (D).

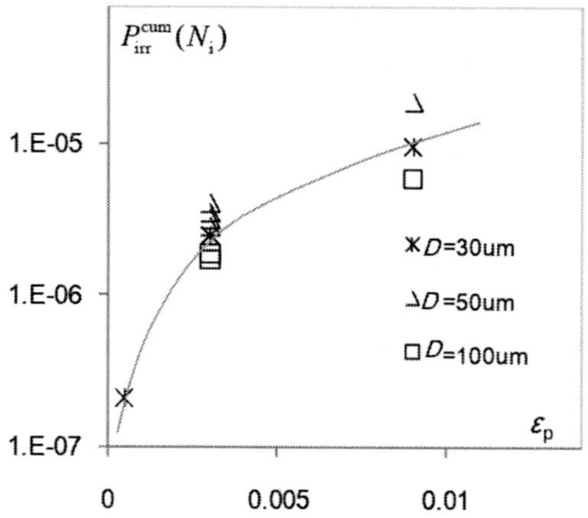

The Manson-Coffin relation in Eq. (2) formulated in terms of cyclic slip irreversibility for $N = N_i$ expressed in Eq. (6) can be rewritten in modified form as:

$$p_{irr}^{cum}(N_i)I_B(N_i)D = \frac{\langle h\rangle f_1 D}{2^{c+1}2M\varepsilon_f'} \times N_i^{-(c+1)}$$

$$= \text{constant} \times N_i^{-(c+1)} \quad (7)$$

Eq. (7) implies that the product $p_{irr}^{cum}(N_i)I_B(N_i)D$ is linked with the fatigue crack initiation life (N_i) in a well-defined manner. Thereby, the variations of the product $p_{irr}^{cum}(N_i)I_B(N_i)D$ with N_i determined experimentally from AFM observations are assessed for three different grain sizes. In Fig. 3, the following expressions are found:

$$p_{irr}^{cum}(N_i)I_B(N_i)D = 3\times10^{-15}N_i^{-0.38},$$
$$\text{for} D \approx 30\ \mu\text{m} \quad (8)$$

$$p_{irr}^{cum}(N_i)I_B(N_i)D = 2\times10^{-14}N_i^{-0.57},$$
$$\text{for} D \approx 50\ \mu\text{m} \quad (9)$$

$$p_{irr}^{cum}(N_i)I_B(N_i)D = 2\times10^{-15}N_i^{-0.25},$$
$$\text{for } D \approx 100\ \mu\text{m} \quad (10)$$

Fig. 3. Evolution of the product $p_{irr}^{cum}(N_i)I_B(N_i)D$ with cycle number to crack initiation (N_i).

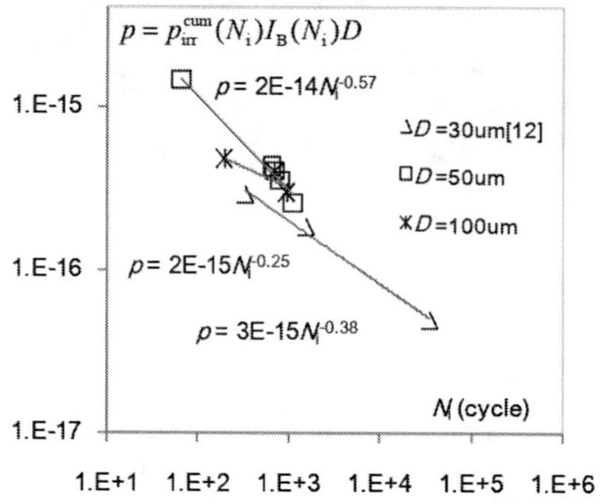

The data plotted in Fig. 3 and the relationships shown in Eqs. (8) - (10) suggest that the product $p_{irr}^{cum}(N_i)I_B(N_i)D$ is empirically connected to the fatigue crack initiation life (N_i) by a power law in the form of:

$$p \approx \alpha(2N_i)^{-\beta} \quad (11)$$

Eq. (11) seems to be rather similar to Eq. (7). Thus, from a comparison of Eqs. (7) and (11), it follows that:

$$p = p_{irr}^{cum}(N_i)I_B(N_i)D \quad (12)$$

and

$$\alpha = \frac{\langle h\rangle f_1 D}{2^{c+1}2M\varepsilon_f'} = \text{constant} \quad (13)$$

and

$$\beta = c + 1 \quad (14)$$

Morrow has suggested that for total fatigue life, c is ranging from − 0.5 to − 0.6 [12]. Yet, lower c-values are found for fatigue crack initiation life [13]. Thereby, the exponents β in Eqs. (8) - (10) are examined to verify if they are reasonably connected to c + 1. With the data obtained from the literature, one expects that $\beta = c + 1 < 0.5$ for the early-stage fatigue damage. Within the limit, the exponents β in Eqs. (8) - (10), found to be equivalent to 0.38 for $D \approx 30$ µm, 0.57 for $D \approx 50$ µm and 0.25 for $D \approx 100$ µm, are in fairly good agreement with expectation.

4.2 Correlation between fatigue ductility coefficient and fatigue ductility exponent

Subsequently, the two coefficients of the Manson-Coffin fatigue crack initiation law denoted as ε'_f and c are deduced from Eqs. (13) - (14) by considering the following experimental data: $\langle h\rangle = 100 \times 10^{-9}$ m, $f_1 = 3.3 \times 10^{-3}$ ($D \approx 30$ µm) or $f_1 = 2.0 \times 10^{-3}$ ($D \approx 50$ µm) or $f_1 = 1.0 \times 10^{-3}$ ($D \approx 100$ µm), and $M = 3$. The relationship between the two empirical parameters of MC law are further assessed.

In the literature, it is shown by correlation that ε'_f and c can be connected through a linear log (ε'_f) − c relationship [13]. Thus, an log (ε'_f) = log (ε'_p) − $c$log ($2N'_i$) type correlation between ε'_f and c is assessed (*c.f.* Fig. 4); the coefficients ε'_p and N'_i, reported to be the coordinates of a fixed point at varying c, would have potential to be related to the fatigue damage [14]. Fig. 4 exhibits that the log (ε'_f) value is decreasing with increasing c-value, implying that the supposed correlation between the logarithm of ε'_f and c is rather strong. Finally, with the correlation coefficient value (R) of 0.948, the correlation can be regarded to be rather viable.

Fig. 4. Correlation between log fatigue ductility coefficient (ε'_f) *vs*. fatigue ductility exponent (c).

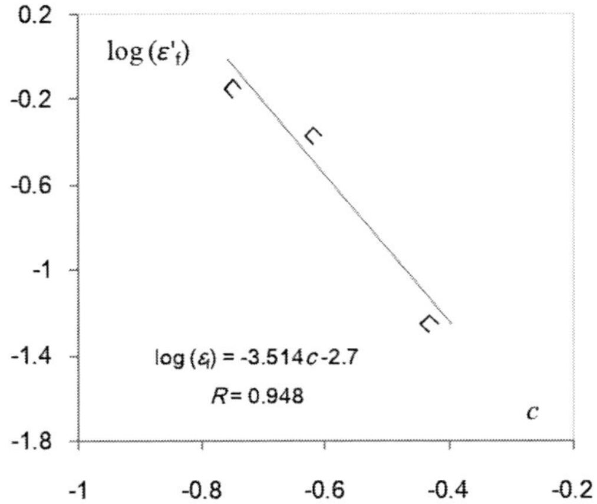

In an attempt to examine the factors influencing the correlation between ε'_f and c plotted in Fig. 4, the behavior of the fatigue ductility exponent (c) is assessed as a function of the grain size (D) for fatigue damage initiation in polycrystal Waspaloy. Our results show that c exhibits a decreasing tendency with increasing D. Thus, to understand the behavior of c *vs*. D, an approach based on an overall analysis of the impact of internal stresses on the occurrence of early fatigue damage is used in the following. It is worth noting that the initiation of cracks along slip bands is very often argued to be a consequence of the local stress concentration due to strain localization [11]. Thereby, to explore the possible contribution of the stress-strain relationship on the crack initiation process, the average value of the long-range internal stress (backstress X) estimated at the crack initiation phase is examined.

According to Feaugas et *al.*, the backstress (X) stemming from the long-range interaction can be decomposed into two distinct components called intragranular backstress (X_{intra}) and intergranular backstress (X_{inter}):

$$X = X_{\text{intra}} + X_{\text{inter}} \tag{15}$$

where X_{intra} is related to strain incompatibility between grains, which involves the crystal or grain boundary structure, while X_{inter} is a consequence of the heterogeneity of the dislocation spatial distribution [15]. The results show that the backstress (X) is inversely proportional to the grain size (D). Our analyses then show that the dependency of the backstress (X) on the grain size (D) is mainly related to the intergranular long-range internal stresses (X_{inter}) created by the plastic strain incompatibility between grains, in agreement with previous results reported in the literature [16]. In Fig. 5, X_{inter} is found to be a decreasing function of the grain size D and thus, the functional relationship between X_{inter} and D can be formalized using the following phenomenological law with D in micrometers:

$$X_{\text{inter}} = 920 \times D^{-1/2} \tag{16}$$

This finding suggests that the elastic-energy stored inside the slip bands can easily be relaxed when the grain size increases due to the decrease in the plastic strain incompatibility resulting from the slip transmission across the grain boundary, which implies that the grain of large-grain-sized polycrystals can be considered as pseudo-monocrystal.

Fig. 5. Evolution of intergranular long-range internal stresses (X_{inter}) as a function of grain size (D).

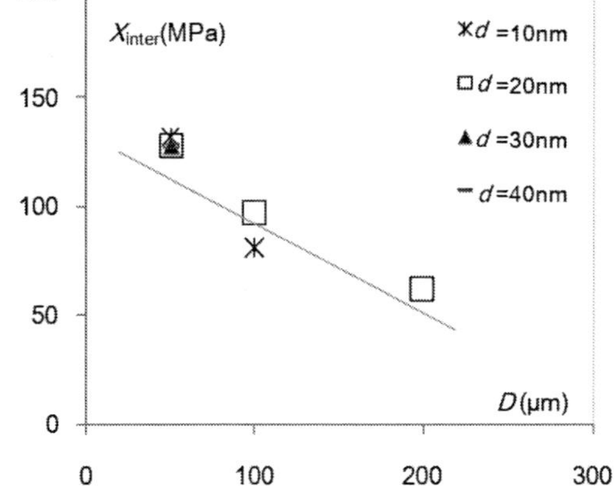

4.3 The cumulative irreversible cyclic plastic strain to fatigue crack initiation

Ho et *al.* showed that there exists a critical value of cumulative irreversible cyclic plastic strain to crack initiation ($\gamma^{cum}_{irr,pl,loc}$) for fatigued polycrystal Waspaloy and they found that fatigue damage is initiated when $\gamma^{cum}_{irr,pl,loc}$ is equal to $1.3 \pm 0.3 \times 10^{-3}$ [5]. It is therefore important to evaluate this quantity based on the modified Manson-Coffin fatigue crack initiation law in Eq. (7). Rewriting Eq. (4) by considering Eq. (13), we obtain:

$$\gamma^{cum}_{irr,pl,loc} = \frac{\alpha 2 M \varepsilon'_f 2^{c+1}}{D^2 f_1} \tag{17}$$

Evaluation of a critical cumulative irreversible cyclic plastic strain to crack initiation ($\gamma^{cum}_{irr,pl,loc}$) via Eq. (17) for the grain sizes of 30 μm, 50 μm and 100 μm yields very similar values of 3.3 × 10^{-3}, 2.0 × 10^{-3} and 1.0 × 10^{-3}, respectively. They are of the same order of magnitude as the critical value of 1.3 ± 0.3 × 10^{-3} computed directly from Eq. (4). However, the critical value of cumulative irreversible cyclic plastic strain to crack initiation for fatigued polycrystals Waspaloy is found to be much smaller than the critical value of cumulative irreversible cyclic plastic strain to failure, estimated to lie in the range of ~10 to ~30, for fatigued α-iron polycrystals and Cu-31% Zn (α-brass) single crystal [4]. In this sense, it can be expected that the characteristic critical value of cumulative irreversible cyclic plastic strain for a specific damage type in a particular material is rather unique, as assumed by the modified Manson-Coffin fatigue crack initiation law [5]. Thereby, the computed critical value of cumulative irreversible cyclic plastic strain equal to 1.3 ± 0.3 × 10^{-3} can thus be considered as a constant for the early stages of fatigue damage in polycrystal Waspaloy.

5 Conclusions

In summary, a microstructure-based reformulation of the modified Manson-Coffin law has been discussed for fatigued Ni-based superalloy, Waspaloy. The experimental results suggest the existence of a power-law relationship between cyclic slip irreversibility parameter and fatigue crack initiation life, and show that the two-parameter of the Manson-Coffin law can be related through a linear log (ε'_f) - c relationship. The analyses suggest that higher exponent c-value signifies higher sensitivity to the plastic strain incompatibility between grains and that fatigue crack initiation life can be controlled by limiting the value of cumulative irreversible cyclic plastic strain to crack initiation.

References

1. S. Suresh, Fatigue of Materials. second ed., Cambridge University Press, Cambridge (1998)
2. U. Essmann, U. Gosele U, H. Mughrabi, Philos. Mag. A **44**, 405-426 (1981)
3. A. Shyam, W.W. Milligan, Acta Mater. **53**, 835-844 (2005)
4. H. Mughrabi, Acta Mater. **61**, 1197-1203 (2013)
5. H.S. Ho, M. Risbet, X. Feaugas, Scr. Mater. **65**, 998-1001 (2011)
6. P. Zhang, Z.J. Zhang, L.L. Li, Z.F. Zhang, Scr. Mater. **66**, 854-859 (2012)
7. H. Mughrabi, Metall. Mater. Trans. A **40**, 1257-1279 (2009)
8. M. Risbet, X. Feaugas, Scr. Mater. **49**, 533-538 (2003)
9. L.F. Coffin, Trans. ASME **76**, 931-950 (1954)
10. S.S. Manson, NACA Rep. **1170**, 317-350 (1954)
11. M. Risbet, X. Feaugas, Eng. Fract. Mech. **75**, 3511-3519 (2008)
12. J.D. Morrow, ASTM STP 378. Philadelphia, PA: ASTM (1965)
13. H. Alush, Y. Katz, M.B. Maros, L. Toth, J Mater. Process. Technol. **157-158**, 16-22 (2004)
14. B. Tribonod, Int. J Fatigue **81**, 143-147 (2015)
15. X. Feaugas, H. Haddou, Metall. Mater. Trans. A **34**, 2329-2340 (2003)
16. X. Feaugas, Acta Mater. **47**, 3617-3632 (1999)

Peridynamic Simulation of Fatigue Crack Growth Behaviour with the Effect of Microstructure

Binchao Liu[1], *Rui* Bao[1,*], *Yamei* Niu[1], *Songsong* Lu[1], and *Kai* Wang[1]

[1] Institute of Solid Mechanics, Beihang University, China

Abstract. The purpose of this paper is to explore the influences of microstructures on crack growth behaviour in 2324-T39 aluminum alloy based on peridynamic(PD) theory. The microelastic bond-based peridynamic constitutive is modified as microplastic to describe the plasticity of aluminum alloys. A new method to establish polycrystalline models based on metallographs is adopted, and grains are reflected in simulations by setting transgranular and intergranular pairwise force in the corresponding bonds. Two kinds of microstructures are modeled according to metallographs, and a special kind of crack branch resulted from the link-up of the secondary crack with the main crack and the growth of the branched crack is successfully captured. The PD simulations reveal that microstructure orientation characteristics have an impact on crack propagation paths and crack growth modes, and it is easier for the secondary-crack resulted macroscopic crack branching to appear if grain boundaries locate not too close to the leading crack tip but within the crack tip plastic zone. The numerical results are verified by experiments and fractographic analysis.

1 Introduction

According to achievements in fracture mechanics, a through-thickness middle crack in an isotropic thin plate subjected to remote tension-dominated loading should propagate perpendicular to the load direction, i.e. Mode I. Under the above conditions, however, phenomenon such as crack kinking, bifurcation and branching are observed in the experiments, and it is discovered that material microstructure is one of the affecting factors [1, 2]. For metals with small grains, the influence of microstructure is embodied in microscale and inconspicuous in macroscale; for relatively larger grains or grains with apparent orientation, however, its influence may result in the macroscale differences of crack growth behaviors. Such impacts are observed even in traditional rolled aluminum alloys if their microstructures exhibit apparent features. For instance, Bao [3] observed a link-up crack branching in AA2324-T39 and AA7050-T7451, and Lu [4] discovered that the branching is affected by different microstructures of aluminum alloys.

Studying the crack growth behavior under the influence of microstructures has a great significance on damage tolerance design and evaluation of crack growth life. On one hand, researches on such impacts rely on experiments, which can manifest special phenomena and uncover general rules; on the other hand, numerical simulations based on experimental phenomena also play an important role with improving computing power and simulation technologies, for it largely reduces experiment costs as well as reveals key parameters under complex circumstances. Several methods such as cohesive zone model(CZM), extended finite element(XFEM) and molecular dynamics(MD), have been proposed to simulate crack growth and branching behaviour with varying success, but they suffer from various problems in modelling the fracture problems and the influence of material microstructure. For the finite element based methods, the fundamental difficulties lie in the assumption of classical continuum theory that the body remains continuous as it deforms. However, once a crack initiates and propagates, which means discontinuities appear, such assumption is violated. From the standpoint of mathematics, the classical continuum theory is formulated using spatial partial differential equations and these spatial derivatives lose their meanings at discontinuities where cracks occur [5]. For MD, however, time and spatial scale are limited to microscale, making it applicable to explain mechanism from molecular level but incompetent to predict crack growth behaviors in practical structures [6].

Peridynamics (PD), proposed by Silling in 2000 [7], has aroused intense scholarly interest since its birth and becomes a promising method to study crack propagation. It attempts to unify the mechanics of continuous and discontinuous media within one framework. The theory replaces the spatial derivatives with integrals that are valid regardless of the existence of a discontinuity, so that the peridynamic governing equations are applicable at fractures and external crack growth criteria are not necessary. Therefore, complex phenomena such as crack nucleation, crack branching, coalition of multiple cracks and crack arrest occur spontaneously, which sheds a new light on simulating crack growth in materials with complex orientation or microstructures [5].

* Corresponding author: rbao@buaa.edu.cn

In this paper, the original bond-based peridynamic (PD) theory is briefly introduced firstly. In order to simulate mechanic behaviors of aluminum alloy, the microelastic bond-based peridynamic constitutive is then extended to be microplastic. The models with artificial grains are established to make a proper difference in pairwise force between transgranular and intergranular bonds. Further, grain models according to metallographs of aluminum alloy (AA) 2324-T39 are established to observe crack growth behavior, and compared with test results.

2 Peridynamic constitutive

2.1 Original PMB model in Peridynamics

The original prototype microelastic brittle(PMB) material model, which is the first bond-based peridynamic model, was proposed by Silling in 2000 [7]. As shown in Fig. 1, assuming that an object occupies a certain spatial domain R in the reference configuration at time t, there exists a pairwise force f between one material point x and any other material point x' located inside its peridynamic horizon δ,

$$f\left(x,x',u\left(x,t\right),u\left(x',t\right),t\right) \tag{1}$$

where $u(x,t)$ is the displacement of material point x. As a contrast, there are no such interactions for material points out of the peridynamic horizon, which is similar to the theory of molecular dynamics.

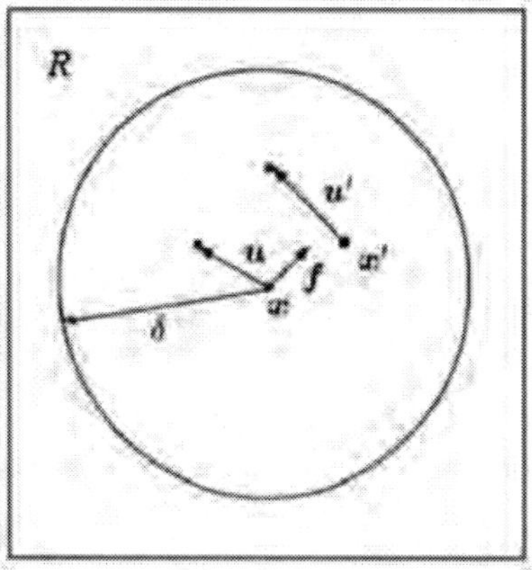

Fig. 1. The diagrammatic drawing of pairwise bond force between material points in original PMB.

According to the Newton's Second Law, the peridynamic governing equation can be written as

$$\rho(x)\ddot{u}(x,t)=\int_H f\left(x,x',u\left(x,t\right),u\left(x',t\right),t\right)dV'+b\left(x,t\right) \tag{2}$$

where $\rho(\mathrm{x})$ denotes the material density, $b(x,t)$ denotes a prescribed body force, and H denotes the neighborhood of material point x,

$$H=H\left(x,t\right)=\left\{x'\in R,\left\|x'-x\right\|\le\delta\right\} \tag{3}$$

In the following discussion, the relative position of two material points in the reference configuration is denoted as ξ, and their relative displacement is denoted as η,

$$\xi=x'-x,\eta=u\left(x',t\right)-u\left(x,t\right) \tag{4}$$

thus $\xi+\eta$ denotes current relative position between the particles.

It can be seen that all the mechanic responses are included in the peridynamic bond force f, which is also the key in establishing peridynamic models. In the bond-based peridynamics, pairwise force between two material points in elastic materials can be expressed as

$$f=cs\mu \tag{5}$$

where bond constant c is the peridynamic material parameter,

$$c=\frac{18K}{\pi\delta^4} \tag{6}$$

where K denotes the bulk modulus and s the relative stretch defined as

$$s=\frac{\left|\xi+\eta\right|-\left|\xi\right|}{\xi} \tag{7}$$

and μ is a parameter to judge whether the bond connecting two particles is broken,

$$\mu\left(x,\xi,t\right)=\begin{cases}1, s<s_0\\0, s>s_0\end{cases} \tag{8}$$

where s_0 denotes critical stretch. When relative stretch exceeds the critical stretch s_0, the bond is broken and two particles no longer interact with each other. Further on, the damage of material point x is defined as the broken bonds' proportion of all bonds connected to material point x,

$$\varphi\left(x,t\right)=1-\frac{\int_H \mu\left(x,\xi,t\right)dV_\xi}{\int_H dV_\xi} \tag{9}$$

when it reachs certain value φ_0, it is regarded that crack occurs on this matrial points.

2.2 Constitutive extension

So far only elastic materials have been discussed for the bond-based model as shown in Fig. 2(a). However, many metal materials like aluminum alloy display strong and apparent plastic behavior in mechanical response, thus plasticity must be taken into consideration in peridynamic constitutive establishment. Fig. 2 (b) shows that plasticity can be incorporated into microelastic bond-based model by treating each bond as though it has its own yield stress [8]. This bond-based material model is called microplastic, and the yield force density f_{yield} or the yield stretch s_{yield} can be calibrated according to the measured material yield stress. In order to simulate elastoplastic mechanic behaviors of aluminum alloy, this microplastic bond-based peridynamic model is adopted in this paper.

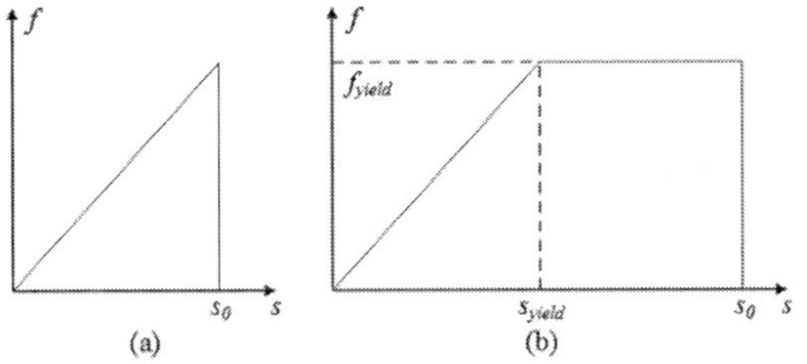

Fig. 2. The micro-elastic PMB model; (b) The micro-plastic model of extended PMB.

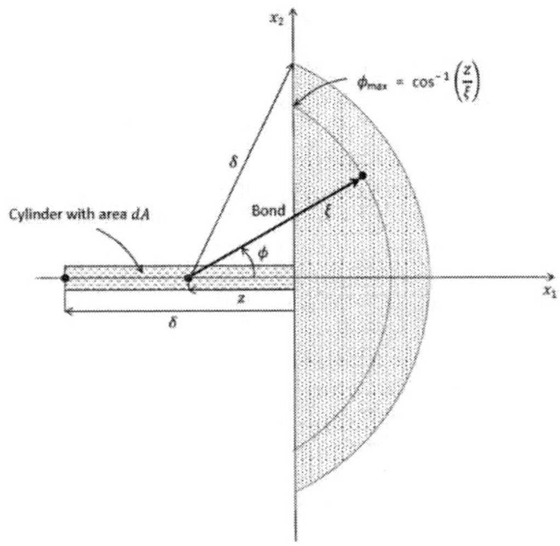

Fig. 3. Computation of force per unit area from bond force densities.

Assuming conditions of uniaxial strain in $x1$ direction shown in Fig. 3, the yield force density can be calibrated to the material yield stress as

$$\sigma_y = 2\pi\int_0^{\delta}\int_0^{\xi}\int_0^{\cos^{-1}\frac{z}{\xi}}\left(f_{yield}\cos\phi\right)\xi^2\sin\phi\, d\phi\, dz\, d\xi = \frac{\pi\delta^4 f_{yield}}{6} \quad (10)$$

where σ_y is the normal stress in uniaxial strain after all the bonds have yielded. However, it is obvious that bonds do not all yield at the same value of bulk strain, for bonds in x_1 direction are bound to yield earlier than others, and the initial yield normal stress σ_0 can be derived similarly as

$$\sigma_0 = \frac{\pi c\delta^4\varepsilon_{11}}{10} \quad (11)$$

where ε_{11} is the bulk normal strain component. On the other hand, this material starts to yield when $cs = f_{yield}$, hence

$$\varepsilon_{11} = \frac{f_{yield}}{c} \quad (12)$$

thus we arrive at

$$\sigma_0 = \frac{3\sigma_y}{5} \quad (13)$$

which indicates that the material displays elastic response before the normal stress reaches initial yield stress σ_0 , and then experiences a strain hardening response until all bonds yield. Such response is similar to the stress-strain curve of aluminum alloys. The yield stretch is finally obtained by combining (6), (11) and (13):

$$s_{yield} = \frac{\sigma_y}{3K} \quad (14)$$

The stress-strain curve given by updated microplastic PMB is shown as red dash line in Fig. 4. Compared with the blue solid line given by primer microelastic PMB, the updated PMB constitutive displays obvious yielding process, which is to some extent consistent with experiment data of AA2324-T39. Therefore, the improved constitutive is suitable to describe the mechanical response of aluminum alloy.

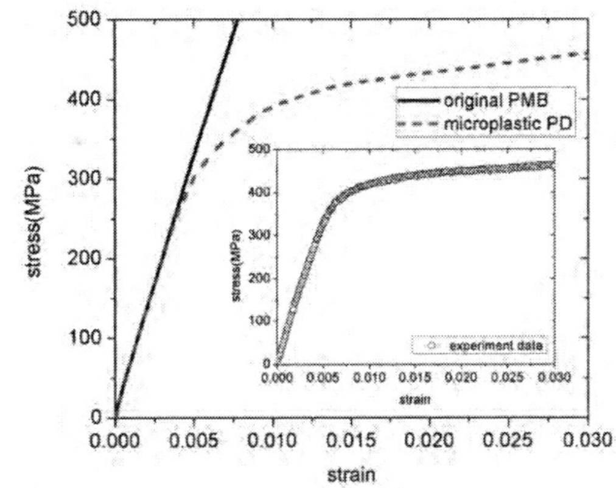

Fig. 4. Compared with primer microelastic PMB, the stress-strain curve of updated microplastic PMB gives obvious yielding process, which is more consistent with experiment data and suitable to describe mechanical response of aluminum alloy.

3 Models and parameters

In many studies concerning material microstructure, Voronoi tessellation method is adopted to establish polycrystalline models. In this study, however, a new modeling method using metallographs is adopted to exactly reflect the unique grain characteristics in materials. Metallograph of AA2324-T39 is respectively shown in Fig. 5(a), from which can be seen that AA2324-T39 microstructure displays apparent directional feature with its size in longitudinal orientation (L-orientation) larger than transverse orientation (T-orientation). According to different gray level in different region in metallographs, grains can be distinguished. Then we paint grains in different colors and obtain a picture with grains distinguished by a color characteristic value, RGB for example, as shown in Fig. 5 (b). The picture is afterwards imported into MATLAB for discretization and the spacial coordinates of the particles can be obtained, during which the picture magnification as well as pixel size must be taken into consideration. Meanwhile, the RGB map of picture is also read by MATLAB. If the RGB values of two particles equal with each other, the bond connecting them is of transgranular type and its pairwise force is computed as f_{trans}; otherwise, the bond is of intergranular type and its pairwise force is computed as f_{inter}.

Following the methods above, two plates with different microstructures according to the metallograph, namely L-T oriented AA2324-T39 and T-L oriented AA2324-T39, are established in the simulations as shown in Fig. 5 (c) and (d). The length, width and thickness of the model are respectively 5mm, 2.25 mm and 150μm, with a pre-crack of 0.8 mm. The spacing between material points in PD is set as 25 μm, which is considered to be small enough in comparison to the average grain size. To take the different transgranular and intergranular interaction into consideration, bond force in transgranular bond and intergranular bond is set different. The transgranular bond force is computed according to Eq. (6), and $f_{inter} / f_{trans} = 0.5$ is adopted.

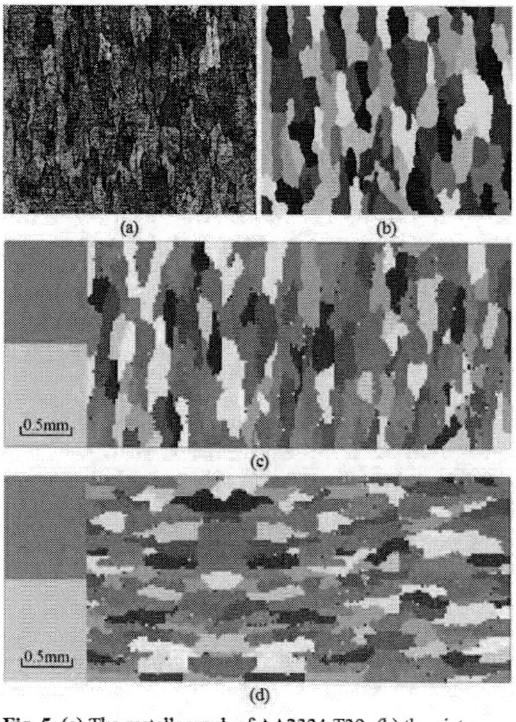

Fig. 5. (a) The metallograph of AA2324-T39; (b) the picture with painted grains; (c) the model with metallograph grains of L-T oriented AA2324-T39 and (d) the model with metallograph grains of T-L oriented AA2324-T39.

Numerical simulations are conducted by the peri-module of Large-scale Atomic Molecular Massively Parallel Simulation (PDLAMMPS), and material parameters are inputted according to mechanical properties of AA2324-T39 obtained from previous experiments [9], where the Young's modulus is E=74GPa, the material density is ρ=2730kg•m^{-3}, the ultimate strength is σ_b=530MPa and the yield strength is $\sigma_{0.2}$=480MPa. Here, the slight distinctions between elastic modulus observed in experiments of L-T oriented AA2324-T39 and T-L oriented AA2324-T39 are ignored. Considering both the computational accuracy, stability and convergence requirement, the horizon is set as 3.03 times the spacing of PD material points, where accordingly the critical damage value is about φ_0=0.35. The timestep is set as 2×10^{-9}s, which is small enough for the computational stability. To be consistent with the Kmax-level in the experiments, a remote tension stress of 150MPa is applied. The yield stretch is set as 0.0036 according to Eq. (14), and the critical stretch is set as 0.0065.

4 Branching in crack paths

4.1 Branching in crack paths

Crack growth processes of L-T oriented AA2324-T39 are shown in Fig. 6, which indicate that the appearance of secondary crack, the link-up of the secondary crack with the main crack and the growth of the branched crack can be successfully captured in PD simulations. The crack trajectory and the branching process agree quite well with those observed in the experiments shown in Fig. 7, and one obvious change of crack growth rate in branching process is marked by rectangle in Fig. 9(a). The whole branching process can be summarized and explained as follows:

① Firstly, before the secondary crack appears, the leading crack tip front is obstructed by a grain boundary perpendicular to the direction of crack propagation (Fig. 6, 116000 steps; Fig. 9(a), point A). The crack slows down or even stops growing because the bonding cohesion inside grains is stronger than the grains boundary so that crack needs more energy to grow through the grain.

② Secondly, the secondary crack appears on the boundary nearby the crack tip (Fig. 6, 177000 steps; Fig. 9(a), point B). The cohesion between grains is relatively weaker than inside, so it is easier for damage to accumulate here and thus for secondary crack to generate.

③ At last, as the secondary crack grows, it is connected to the leading crack along the grain boundary which prevents the growth of leading crack (Fig. 6, 228000 steps; Fig. 9(a), point C).

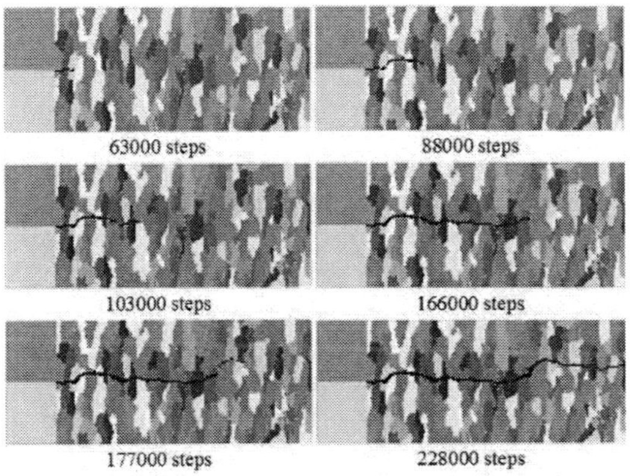

Fig. 6. The PD simulation of crack growth process in L-T oriented AA2324-T39.

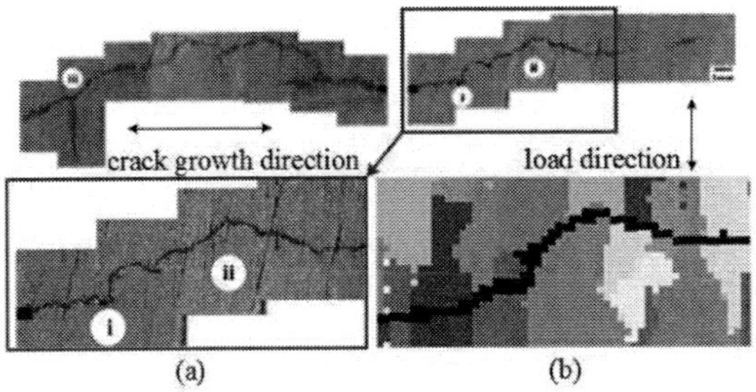

Fig. 7. (a) The experimental results in [4] and (b) the PD simulation results of crack propagation path in L-T oriented AA2324T39 which is a part of FIG. 6.

Crack growth processes of T-L oriented AA2324-T39 are shown in Fig. 8, which are similar to those of L-T orientated AA2324-T39. However, the location of the secondary cracks is much nearer to the main crack tip, which leads to smaller deflection of the whole crack path. The reason is that the secondary crack appears on grain boundaries and the dimension of the grains in T direction is relatively smaller than that in L direction. This phenomenon matches the experimental observation.

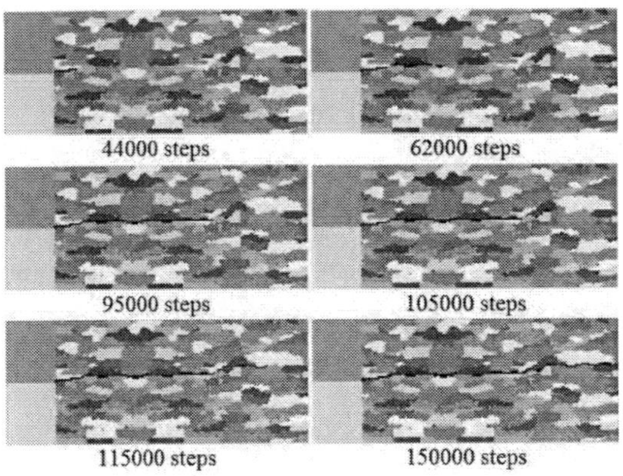

Fig. 8. The PD simulation of crack growth process in T-L oriented AA2324-T39.

It is worth noting that there are also some differences between the crack growth mode of L-T and T-L oriented AA2324-T39. For the L-T orientation, the leading crack tends to display transgranular propagation in most cases, and only display intergranular propagation when the secondary crack just appears. For the T-L orientation, the crack always displays intergranular propagation. Moreover, the observation that crack propagates faster in T-L oriented AA2324-T39 than L-T oriented AA2324-T39, which is shown in Fig. 9(c), accords well with the experiment results shown in Fig. 9(d). In fact, transgranular propagation mode needs and consumes more energy, and it is natural that cracks grow slower than intergranular propagation mode. Thus the differences of propagation mode and growth rate complement each other. The black rectangles in Fig. 9(a) indicates the moments when obvious crack branches occur.

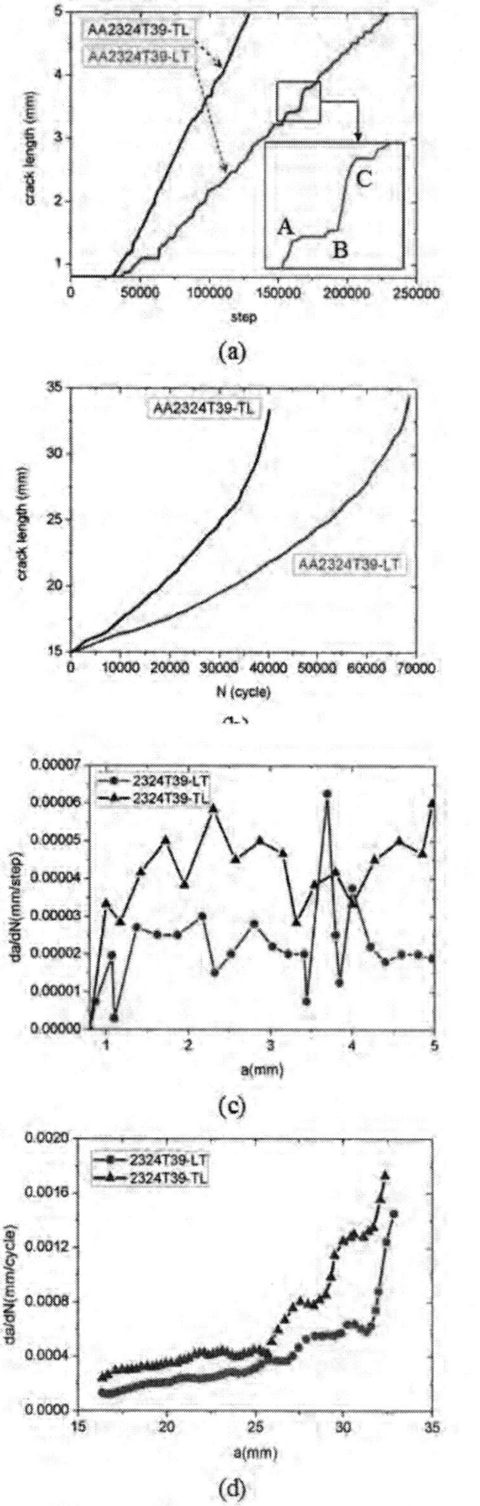

Fig. 9. The crack length obtained from (a) PD simulations and (b) experiments; and the crack growth rate obtained from (c) PD simulations and (d) experiments.

4.2 Secondary crack location

The mechanism of secondary crack has been studied through fracture surface analysis and finite element simulations in previous studies [4,10]. It has been pointed out the surface secondary crack which results in macroscopic crack branching appears within the leading crack plastic zone.

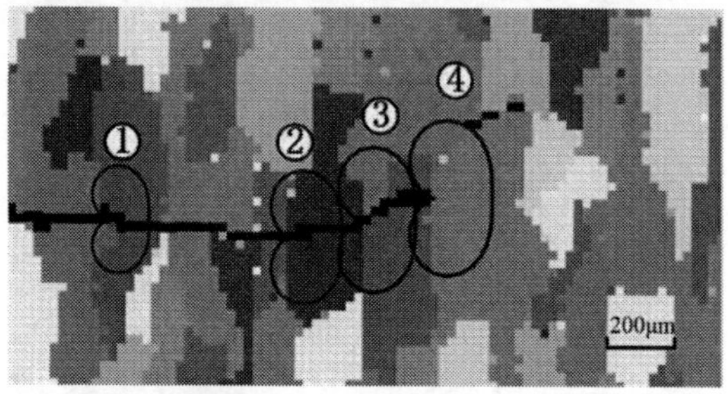

Fig. 10. The relationship of the crack tip plastic zone and the location of the second crack.

As mentioned above, the crack branching similar to the experimental results has been observed in PD simulations, and the relationship between the position of secondary crack and crack tip plastic zone is verified in PD as follows. Four corresponding plastic zones of leading crack are marked in Fig. 10, including two place ③④ where intergranular crack appear and other two places ①② with no intergranular cracks. The plastic zones of leading crack in plane stress cases are estimated by Eq. (15), which is the curve equation of the crack tip plasticity boundaries in the condition of small scale yielding under Mises criterion. It can be concluded from Fig. 10 that the surface secondary crack appears along the grain boundary within the scope of leading crack plastic zone, as marked by ③ and ④. Under the circumstances where there are no grain boundaries located in the crack tip plastic zone of the leading crack, as marked by ① and ②, it is rare for the secondary cracks to initiate.

$$r_p = \frac{K_I^2}{2\pi\sigma_{0.2}^2}\cos^2\frac{\theta}{2}\left[1+3\sin^2\frac{\theta}{2}\right] \qquad (15)$$

It is widely acknowledged that the deformation of material is relatively larger in the crack tip plastic area, thus the damage resulted from dislocations is also larger, and the secondary crack is easier to initiate at weaker points such as holes or inclusions. In PD simulations, the grain boundary is set to be relatively weak, so the secondary crack is inclined to appear on the grain boundary alongside the propagation direction in crack tip plastic area. Further, it is of greater probabilities for the link-up resulted crack branching to appear when the grain boundary locates from proper distances to the crack tip in plastic area. If the grain boundary is too close to the crack tip, the secondary crack on the grain boundary will be directly connected to the leading crack, and the crack branching is not obvious to observe. If there are no grain boundaries located in the crack tip plastic zone, on the contrary, it is likely for the leading crack to propagate transgranularly and therefore of little probabilities for the secondary crack to appear.

5 Conclusions

According to the PD simulation in effects of grain size and orientations on crack paths, it can be concluded that:

① The crack branching process and crack path trajectories in AA2324-T39 resulted from the link-up of secondary crack with the main crack can be reasonably captures by PD method.

② The crack path in AA2324-T39 can be reasonably described in PD simulations by setting distinct pairwise force in transgranular and intergranular bonds.

③ According to the PD simulations, the whole crack path in L-T oriented AA2324-T39 is more circuitous than that observed in T-L oriented AA2324-T39. These results indicate that the microstructure grain size and orientation characteristics has an influence on the crack growth paths.

④ In PD simulations, it is observed in L-T orientated AA2324-T39 model that the leading crack displays generally transgranular propagation, and the secondary crack which appears on the grain boundary usually displays intergranular propagation. As for T-L oriented AA2324-T39, however, the cracks always display intergranular propagation. The results indicate that the microstructure grain size and orientation characteristics also has an impact on the crack growth mode.

⑤ It is easier for the secondary-crack resulted macroscopic crack branching to appear if grain boundaries locate not too close to the leading crack tip but within the crack tip plastic zone.

References

1. J. Toribio, B. González, J.C. Matos, Eng. Fract. Mech. **77**, 11 (2010).
2. M. Benedetti, M. Beghini, V. Fontanari, B. Monelli, Int. J. Fatigue, **31** 11 (2009)
3. R. Bao, X. Zhang, Int. J. Fatigue, **32** 7 (2010)
4. S. Lu, R. Bao, T.Zhang, B. Fei,. Int. J. Fatigue, **92**, 459-469, (2016)
5. A. Agwai, I. Guven, E. Madenci, Int. J. Fract., **171** 1 (2011)
6. B. N. Cox, H. Gao, D. Gross, D. Rittel, J. Mech. Phys. Solids, **53** 3 2005.
7. S. A. Silling, J. Mech. Phys. Solids, **48** 1 (2000)
8. F. Bobaru, J. T. Foster, P. H. Geubelle, S. A. Silling,. *Handbook of peridynamic modeling* (Crc Press, Boca Raton, 2016)
9. T. Zhang, B. Rui, B. Fei, Int. J. Fatigue, **58** 1 (2014)
10. T. Zhang, B. Rui, S. S. Lu, B. Fei, Int. J. Fatigue, **82**, 602-615, (2016)

Effect of local anisotropy on fatigue crack initiation in a coarse grained nickel-base superalloy

Benedikt Engel[1*], *Tilmann Beck*[1], *Nadine Moch*[2], *Hanno Gottschalk*[2] and *Sebastian Schmitz*[3]
[1]Institute of Materials Science and Engineering, University of Kaiserslautern, Germany
[2]Bergische Universität Wuppertal, Gaußstraße 20, 42119 Wuppertal, Germany
[3]Siemens AG, Huttenstraße 12, 10553 Berlin, Germany

Abstract. In the present work, theoretical approaches, based on grain orientation dependent Young's modulus and Schmid factor are used to describe the influence of local grain orientation on crack initiation behaviour of the coarse grained nickel base superalloy René80. Especially for strongly anisotropic crystal structures with large grain size, such as the investigated material, the local elastic properties must be taken into account for assessment of fatigue crack initiation. With an extension of Schmid's law, the resulting shear stress amplitude, which triggers local cyclic plastic deformation, can be calculated depending on local Young's modulus and Schmid factor. A Monte Carlo simulation with 100,000 samples shows how random grain orientation affects these parameters. Furthermore, the product of Young's modulus and Schmid factor (called E·m) is used as a parameter to determine how grain orientation influences resulting shear stress amplitude for given total strain amplitude. In addition to the theoretical work using that approach, this model is also validated using isothermal LCF experiments by determining local grain orientation influence on the crack initiation site using SEM-EBSD analyses.

1 Introduction

Due to the increase in the share of regenerative energy in electricity production and the associated decrease in use of fossil sources, i.e. coal and natural gas, it is necessary to compensate the gap between current electricity demand and available renewables [1]. In order to react quickly to these gaps, gas turbines are used, which can be cold started and ramped up to maximum power in short time. While conventional power plants, such as coal or nuclear power plants, require considerable time to generate maximum output from cold start or idle conditions, modern gas turbine power plants, even in the 300 MW class, can achieve this in less than 30 minutes [2]. Hence, frequent starts and stops as well as load changes result in thermally induced low-cycle-fatigue (LCF) loadings of the turbine blades made of nickelbase superalloys, which can become more critical for component life than creep loads. Therefore, a fundamental understanding of the crack initiation and fatigue process of these materials becomes increasingly important. Current lifetime prediction models use classic deterministic approaches considering high safety factors and hence, less efficiency is achieved in turbine design than possible in case of optimal usage of the Ni-alloys' fatigue strength. Recent design approaches therefore use probabilistic models for lifetime calculation with improved material models. For this purpose, a significantly higher number of influencing parameters must be taken into account to describe fatigue crack initiation behaviour. In this context, the local grain orientation at the crack initiation site is of particular interest. Due to the large grain size of conventionally cast Ni-base turbine blades, the regions of the component undergoing highest cyclic stress only contain few grains. The random orientation of these grains leads to locally different elastic and plastic material properties significantly influencing the crack initiation behaviour. Related to Schmid's law for single crystals or single grains, the resulting shear stress in the slip systems depends on the applied normal stress as well as the orientation dependent Schmid factor. If the local shear stress exceeds the critical resulting shear stress (CRSS), dislocation slip becomes effective and, in case of cyclic loading, a fatigue crack may be initiated. Moreover, at a given integral normal stress, e.g. in a uniaxially loaded test piece, the local normal stress to be taken into account in Schmid's law also depends on local elastic grain stiffness, which again is a function of grain orientation. Nickel-based materials have a high anisotropy factor of more than 2.5. Hence, Young's modulus can vary from 130 GPa to 330 GPa [3,4] depending on grain orientation, which has a significant influence on the resulting shear stresses in the slip systems.

Strain-controlled isothermal LCF tests at conventionally cast Ni-Superalloys typically show large fatigue lifetime scatter up to a factor of 10, even in the LCF regime [5–9]. A major reason for this are orientation dependent resolved shear stresses in the randomly oriented coarse grains. Previous work has shown the influence of the Schmid factor on lifetime behaviour by calculating a resulting shear stress – lifetime diagram by multiplying the integral stress amplitude measured at half lifetime in strain controlled LCF-tests with the Schmid factor of the crack initiation grain. The resulting CRSS vs. Number of cycles to crack initiation diagram showed clearly reduced scatter compared with the representation total strain amplitude vs. number of cycles to failure [5,6,10]. However, these results and EBSD measurements also

* Corresponding author: engel@mv.uni-kl.de

show that fatigue cracks are not necessarily initiated in slip systems with the highest Schmid factor at the specimen surface [10]. Therefore, crack initiation prediction only based on the Schmid factor is obviously not covering all major influences on fatigue of coarse grained polycrystalline Ni-base superalloys. Research on HCF of Ni-base superalloys with grain sizes in the μm range has already shown that both, Young's modulus and Schmid factor including their interrelation via grain orientation must be considered to predict the crack initiating grains. In this context, Pollock et al. showed that crack initiation in René88DT, manufactured by powder metallurgy, is dependent on the occurrence of twins in addition to Schmid factor and Young's modulus. For this material cracks often initiated in grains with slip systems oriented parallel to twin boundaries [11–13].

2 Investigated Material

The investigated material is the polycrystalline Ni-base superalloy René80 frequently used for gas turbine blades [13–15]. Typical operation temperatures are between 760 °C and 982 °C. The material contains approximately 30 % - 35 vol.-% of cuboidal primary γ'- particles with an average size of 0.4 μm, measured by scanning electron microscopy, semi-coherently embedded in the fcc γ – matrix. In addition, spherical secondary γ' is observed within the nickel matrix. The specimen were delivered as 150 mm long casted bars, with diameters of 20 mm. The chemical composition measured by the manufacturer is given in Table1.

Table 1. Chemical composition of René80 (in weight %)

Element	*Composition*
Ni	balance
Cr	14.04
Co	9,48
Ti	5.08
Mo	4.03
W	4.02
Al	2.93
C	0.17
B	0.0151
Zr	0.011

LCF-tests at 850 °C were carried out using a 100 kN servo-hydraulic testing system (MTS type 810) with inductive heating (Hüttinger TIG 5/300). For temperature control a Eurotherm 2604 device was used combined with a ribbon-type thermocouple (type K) attached in the middle of the gauge-length. All isothermal LCF tests were performed in total strain control with fully reversed loading. A 12mm MTS high temperature extensometer was used for strain measurement. The temperature gradient within the cylindrical centre section of the sample (length 18mm, diameter 7mm), measured with three ribbon-type thermocouples, does not exceed ± 8°C within a distance of +/- 5 mm from the middle of the gauge length.

3 Description of anisotropic material behaviour

3.1 Elastic Anisotropy

Due to the crystalline structure of metallic materials, elastic properties are in most cases direction dependent within a single grain. A non-textured polycrystal is composed of statistically randomly oriented grains, separated from each other by grain boundaries. If a sufficiently high number of randomly oriented grains exists in the gauge length of a specimen or within technical component, the anisotropic elastic properties of the individual grains are averaged over the high grain number and so-called quasi-isotropic material behaviour can be assumed in good approximation. Hooke's law can then be used in its simple form representing a linear correlation between stress and strain for isotropic material behaviour. This assumption becomes questionable if only a small number of grains in a specimen gauge length or component are to be taken into account, especially in case of materials with high elastic anisotropy factor. An isotropic approach would then result in a large scatter of mechanical properties including fatigue strength and, hence, local grain properties need to be taken into account. The following section shows how the elastic properties of a crystallite can be determined for any given orientation.

If a general linear elastic material law is considered, the two second order stress and strain tensors are combined with a fourth rank stiffness tensor as shown in equation 3.1, known as generalized Hooke's law [14].

$$\varepsilon_{ij} = S_{ijkl}\sigma_{kl} \qquad (3.1)$$

For a completely anisotropic material, the stiffness matrix has 81 independent components. For a cubic lattice, three independent stiffness components S_{11} ,S_{12}, S_{44} exist due to rotational planes in the cubic system and symmetry of stress and strain tensor. Hence, using equation 3.2, an anisotropic cubic material can be described completely.

$$S_{ij} = \begin{pmatrix} S_{11} & S_{12} & S_{12} & & & \\ S_{12} & S_{11} & S_{12} & & & \\ S_{12} & S_{12} & S_{11} & & & \\ & & & S_{44} & & \\ & & & & S_{44} & \\ & & & & & S_{44} \end{pmatrix} \qquad (3.2)$$

To determine the stiffness as a function of grain orientation, the stiffness tensor must be rotated four times with a rotational matrix U (eq. 3.4) [15–17].

$$S'_{ijkl} = U_{ig} U_{jh} S_{ghmn}\, U_{km} U_{ln} \tag{3.3}$$

U represents a rotational matrix containing the rotation angles around the default crystal coordinate axes. A common convention for this is the Bungee notation with the Euler angles φ_1, φ_2 and θ. A detailed description of the Bungee notation and the Euler angles can be found in [17]. Each rotation can be described with its own rotational matrix and multiplication of all three rotational matrices results in the complete rotational matrix U (3.4), with s,c for sine and cosine.

$$U = \begin{bmatrix} c(\varphi1)\,c(\varphi2) - s(\varphi1)\,c(\theta)\,s(\varphi2) & -c(\varphi1)\,s(\varphi2) - s(\varphi1)\,c(\theta)\,c(\varphi2) & s(\theta)\,s(\varphi1) \\ s(\varphi1)\,c(\theta) + c(\varphi1)\,c(\theta)\,s(\varphi2) & -s(\varphi1)\,s(\theta) + c(\varphi1)\,c(\varphi2) & -s(\theta)\,c(\varphi1) \\ s(\theta)\,s(\varphi2) & s(\theta)\,c(\varphi2) & c(\theta) \end{bmatrix} \tag{3.4}$$

Equations commonly used in literature allow an analytical calculation of effective Young's Modulus in normal stress direction [4,17–19]. The equation used here consists of the three independent stiffness components S_{11}, S_{12} and S_{44} combined with an orientation factor r.

$$\frac{1}{E_{hkl}} = S_{11} - 2(S_{11} - S_{12} - \frac{1}{2} S_{44}) r_{\{hkl\}} \tag{3.5}$$

The orientation factor r can be expressed with the Euler angles φ_1, φ_2 and θ by the following equations for stresses in z-direction.

$$r_{\{hkl\}} = g_{13}^2 g_{23}^2 + g_{23}^2 g_{33}^2 + g_{33}^2 g_{13}^2 \tag{3.6}$$

with

$$g_{13} = \sin\theta \sin\varphi_2,\; g_{23} = \cos\varphi_2 \sin\theta,\; g_{33} = \cos\theta \tag{3.7}$$

Figure 1 shows the geometric dependency of orientation and r factor in a pole figure for cubic materials.

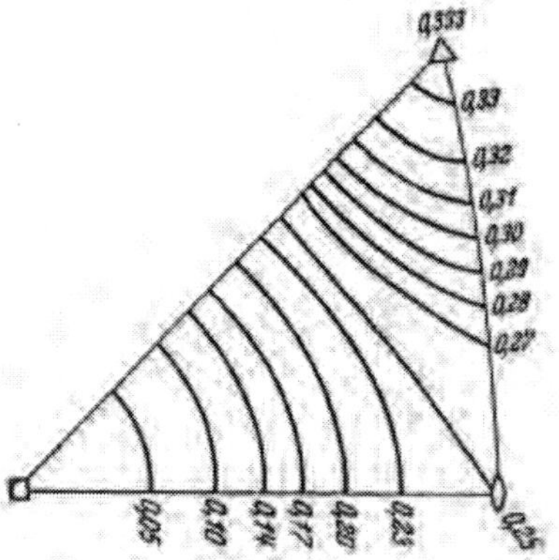

Fig. 1. Dependence of grain orientation on the geometrical r-factor [20].

Equation 3.6 corresponds with the bottom line of the rotational matrix U and shows the independence of φ_1 on the calculation of the Young's modulus due to its invariant rotation around the stress axis z.

3.2 Plastic Anisotropy

As a prerequisite of plastic deformation, it is assumed that the shear stress in the slip systems resulting from the applied stress must exceed a critical shear stress, which can be considered as a material-specific threshold value. To calculate the resulting shear stress a more generalized, Schmid-law is used [21–24].

$$\tau_{res} = s_{ij} \cdot \sigma \cdot n_i \tag{3.8}$$

Equation 3.8 shows the calculation of resulting shear stress where σ represents a given stress tensor, s_{ij} the normalised slip plane direction vector and n_i the normalised slip plane normal vector. With this equation, the resulting shear stress can be calculated based on the stress tensor in each slip system (or in every other crystallographic system of interest). For a rotation of the grain or elementary cell, the slip systems also rotates. For this purpose, the plane and directional vectors are rotated with the rotation matrix U mentioned in equation 3.4. Furthermore, the value of σ_{33} is set to 1 for uniaxial loading (in z direction) to calculate the Schmid factor. Plastic deformation can be expected to occur first in slip systems with the highest resulting shear stress, i.e. the highest Schmid factor ($\max(\tau_{res})$).

$$\tau_{res} = U \cdot n_i \cdot \sigma \cdot U \cdot s_{ij} \tag{3.9}$$

$$\text{and } \sigma = \begin{bmatrix} 0 & 0 & 0 \\ 0 & 0 & 0 \\ 0 & 0 & 1 \end{bmatrix} \tag{3.10}$$

3.3 Combination of Schmid factor and Young's modulus

To illustrate the correlation between Schmid factor and Young's modulus, a single grain in stress controlled loading is considered. The material is assumed to be isotropic and uniaxial load is applied. In this case the resulting shear stress amplitude can be determined using Schmid's law.

$$\tau_{a,res} = \sigma_a \cdot m \tag{3.11}$$

In the next step, the load condition is changed to a total strain control. Assuming approximately elastic material behaviour, the stress amplitude σ can be estimated conservatively from Hooke's law and be inserted into Eq. 3.11:

$$\tau_{a,res} = \varepsilon_{a,t} \cdot E \cdot m \tag{3.12}$$

Equation 3.12 clarifies that, in case of approximately elastic behaviour, the resulting shear stress in each considered grain only depends on the product of Young's modulus and Schmid factor. And as shown above, both, Schmid factor and Young's modulus depend on the same grain orientation while $\varepsilon_{a,t}$ is constant for strain controlled tests. Note that equation 3.12 is exactly valid only until the resulting shear stress exceeds the critical shear stress within the slip system. From this point on, plastic deformation occurs which is no longer associated with the linear assumption of Hooke's law and the resulting shear stress is overestimated by eq. 3.12.

4 Generation of random orientation

To investigate the correlation of Young's modulus and Schmid factor depending on grain orientation, a Monte Carlo Simulation with 100.000 samples was performed. To generate random orientations a normal vector placed on the [001] plane of a face-centered cubic cell is taken, located in the middle of a three-dimensional coordinate system. By rotating the cubic cell, the vector is oriented to different points on the surrounding spherical surface. With sufficiently high amount of random orientations, the whole surface is sufficiently covered and each point on this sphere representing an orientation of the cubic cell. Due invariant rotation around z-axis, because of uniaxial loading, the sphere can be parameterized with two Euler angles. The assumption of a uniform distribution of these angles leads to an accumulation of points at the corresponding poles, as the sphere and its projection in figure 2 a) shows.

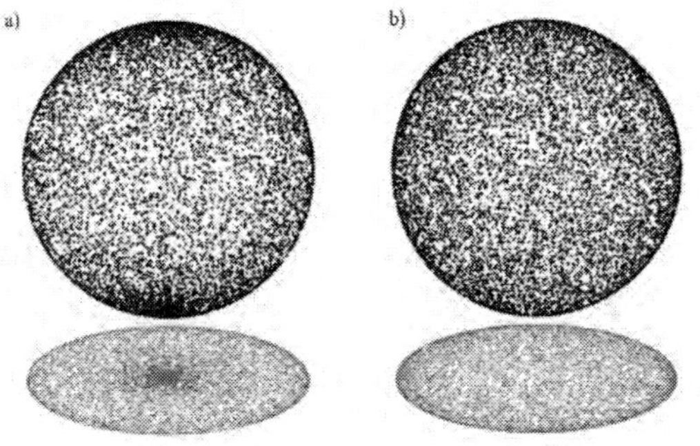

Fig. 2. 10.000 randomly distributed points on a sphere. a) Uniform distribution of the parameters b) homogeneous distribution achieved by Haar measure.

The distribution in a), with increased orientations in [001]- directions, would lead to misleading results in frequency distributions of schmid factor and Youngs modulus. To achieve a homogeneous distribution, the three dimensional Haar measure is projected onto the two dimensional sphere. Based on this distribution, which is shown in figure 2b), 100.000 random SO3 matrices were generated to simulate a statistically random orientation of the grain.

5 Results and Discussion

5.1 Monte Carlo Simulation

Due to lack of available materials data for René80, the calculations were performed instead based on data of INC738 LC, which is very similar in chemical composition and γ/γ' structure. The following Monte Carlo simulation for the distribution of Young's modulus was made for a single INC738 LC grain at 25°C and 850°C, respectively. The elastic constants for a wide temperature range were taken from [26]. For 25 °C the values are $S_{11} = 0.00797$, $S_{12} = -0.00323$ and $S_{44} = 0.00787$. The stiffness constants used in equation 1.5 and the Euler angles for the rotation factor were extracted from the rotational matrices according to equation 1.4. Figure 3 summarizes the results as a histogram.

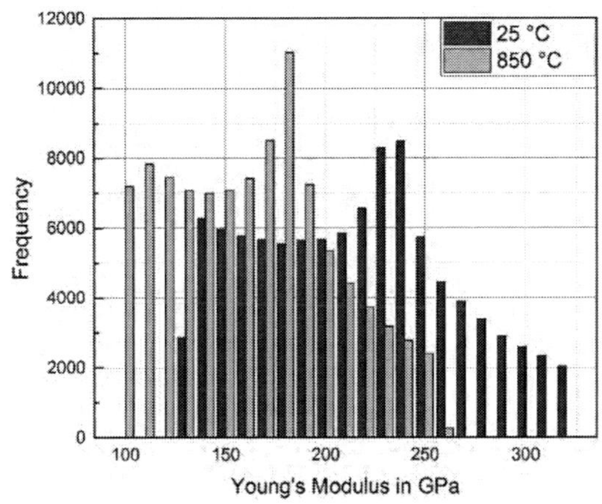

Fig. 3. Result of the Monte Carlo Simulation for 100.000 random orientations. INC738 at 25 °C and 850 °C.

Due to elastic anisotropy, Young's modulus ranges from 130 GPa up to 330 GPa at 25°C with a maximum density at 240 GPa representing the high anisotropy factor. The mean value is 213 GPa. A second simulation for 850 °C shows a similar distribution, but shifted to lower values due the temperature influence on Young's modulus. The elastic constants for INC738 LC at 850 °C are determined by linear interpolation of the values from [26] between 800 °C and 900 °C, resulting in $S_{11} = 0.010963$, $S_{12} = 0.00457$ -and $S_{44} = 0.010122$. Notably, the range of Young's Modulus is reduced compared 25°C leading to higher frequencies in the distribution at low Young's moduli. The maximum density of Young's moduli for 850 °C is found to occur at values 180 GPa, with a more pronounced peak in frequency compared to room temperature. The mean value is 158 GPa. The occurrence of increasing frequencies with increasing temperature can be explained by figure 4. For orientations close to [111] Young's Modulus decreases by 80 GPa from 25 °C to 900°C. For the same temperatures the Young's moduli near [100] orientation change only by 38 GPa. As a result, the distribution of Young's Moduli is "compressed" at higher temperatures, leading to a higher number density in the frequency distribution.

https://doi.org/10.1051/matecconf/201816504004

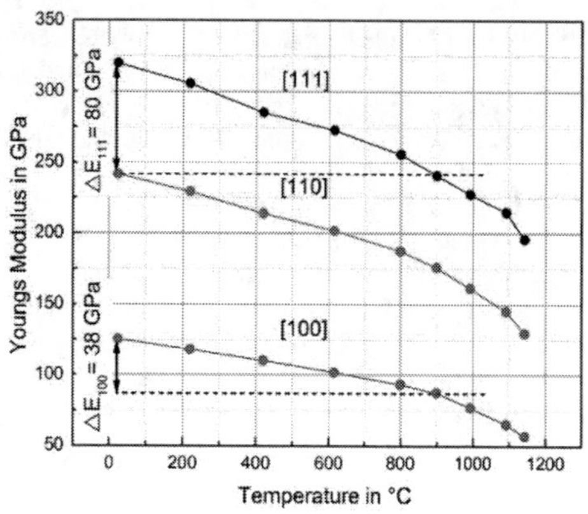

Fig. 4. Temperature Dependence of Young's modulus in selected lattice directions of INC738 LC calculated from data according to [7].

Opposed to Young's modulus, the Schmid factors for the considered slip systems are temperature independent because they result entirely from geometric considerations, taking only the crystallographic structure of the material and the vectors of the slip systems into account. The face centred crystallographic structure of nickel base superalloy leads to 12 primary slip systems of the type {111}[110]. To determine the highest Schmid factor, eq. 2.2 is used to calculate the Schmid factors for all slip systems in dependence of the rotational matrices, while only the highest is taken into account in the histogram in Fig. 5.

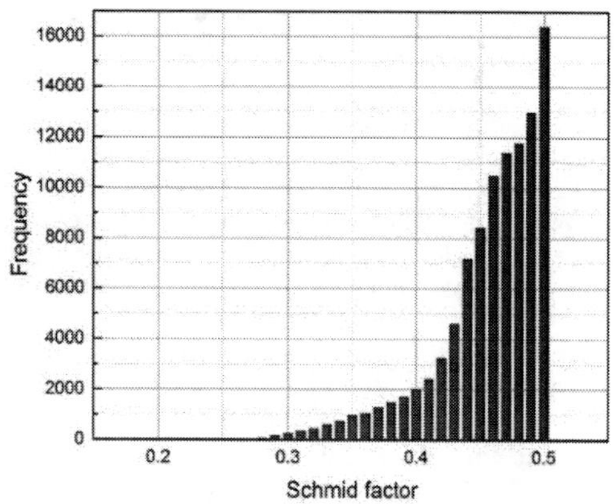

Fig. 5. Schmid factor distribution for 100.000 random grain orientations of a face centred cubic cell.

It can be seen that the frequencies in the histogram increase up to the maximum Schmid factor of 0.5. The reason for this is the large number of slip systems, due to which, for random orientation, a high probability exists for at least one of the twelve slip systems to feature a high Schmid factor. Only a few orientations show low maximum Schmid factors. These orientations correspond to those where the load axis direction approximately corresponds to the [111] directions of the elementary cell. The Average Schmid factor of 100.000 random orientations is 0.452 with a standard deviation of 0.0413.

5.1 E·m - Modell

The resulting shear stress in each slip system of a uniaxial loaded, anisotropic grain only depends on the product of Young`s modulus and Schmid factor, in the following referred to as E·m. Figure 6 shows E plotted against the Schmid factor as a result of a Monte Carlo simulation for 25°C in one diagram. The product E·m is indicated by a greyscale coding, while each point corresponds to a single, randomly chosen orientation.

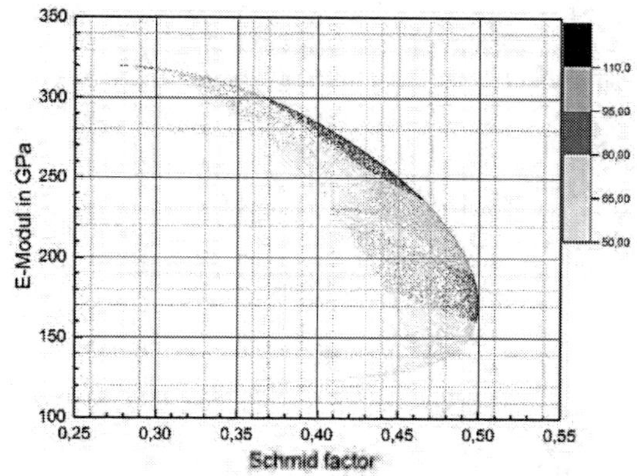

Fig. 6. Distribution of Schmid factor and Young`s modulus calculated for IN738 LC at T = 25°C with 100.000 random grain orientation

The distribution (Fig.6) shows highest shear stresses, i.e. highest E·m values for orientations with Schmid factors between 0.37-0.45 and high Young's moduli, from 240 GPa up to 300 GPa. As a result of the frequency distributions discussed above, these combinations occur less frequently while orientations with high Schmid factors up to 0.5 appear more often. It should be noted that these orientations correlate with low Young's moduli of 130 GPa - 200 GPa. A second simulation at 850 °C show that in spite of a different distribution of Young's moduli (see Fig. 3) the E·m distribution remains similar as for room temperature. Highest shear stresses occur for orientations with Schmid factors from 0.37 – 0.45 and corresponding Young's moduli of 190 GPa – 240 GPa.

Further visualisation of the behaviour of Young's modulus and Schmid factor depending on grain orientation is possible as a three-dimensional body (Fig. 7). In the centre of a considered sphere is a cubic cell. As described above, for random orientations of the cell, the [001] vector is directed to the sphere surface in dependence of its rotation. To demonstrate the influence of E·m in dependence of grain orientation, the [001] vector is rotated with the relevant rotational matrix and multiplied with the value of E·m. The distance from the centre to a point on the three-dimensional body's surface represents the E·m value for the randomly chosen orientations described in section 4, in case of uniaxial loading in [001]-direction. For better illustration, only one half of the axisymmetric body is plotted in Fig. 7.

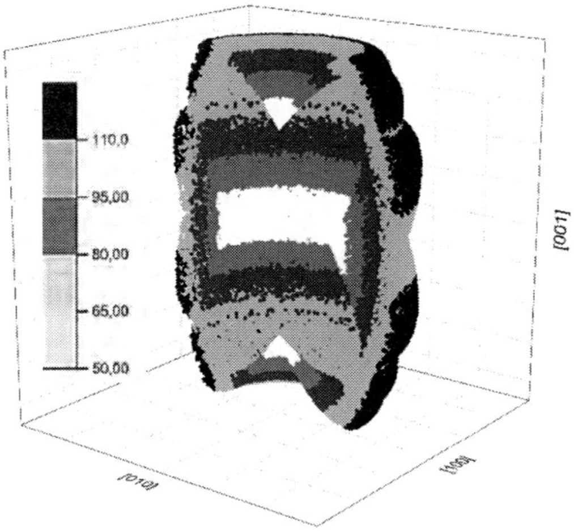

Fig. 7. Three dimensional representation of orientation dependence of E·m for uniaxial stress in [001] direction. Calculated for 200.000 orientations of INC738 LC at 25°C.

Besides visualizing the grain orientation dependency of E·m, Fig. 7 underlines the excellent properties of [001] directed single crystalline (SX) or directionally solidified (DS) Ni-base alloys: Orientations near [001] lead to lowest Young's moduli combined with moderate Schmid factors which results in a global minimum of E·m. Moreover, it becomes apparent that even small manufacturing induced changes in crystal orientation cause a steep increase in E·m, which detrimentally effect e.g. thermo-mechanical fatigue behaviour of turbine blades manufactured from SX and DS Ni-Base alloys, respectively.

5.2 Experimental Validation of E·m Modell

To validate the theoretically determined correlation between E·m and fatigue crack initiation, cracked specimen from isothermal LCF experiments were investigated. Because of the relatively low total strain amplitudes, single crack initiation within individual grains was observed [9] and, according to the nearly closed stress strain hysteresis loops recorded during the tests, globally elastic behaviour can be assumed in very good approximation. Due formation of a single crack, it can be assumed that only one grain has exceeded the critical shear stress during the test and plastic deformation occurred on local scale. This leads to the formation of persistent slip bands in the relevant grain, inducing extrusions and intrusions at the adjacent surface. The resulting micro notch effect leads to an increase in local stress and finally to crack initiation [10,27,28]. To investigate the grain orientation in the vicinity of the crack origin via EBSD, the specimen were cut longitudinaly, grinded, mechanically and electrolytically polished. Figure 8 shows a longitudinal cut through the region around the crack initiation grain (grain 1) of two exemplarily chosen specimens.

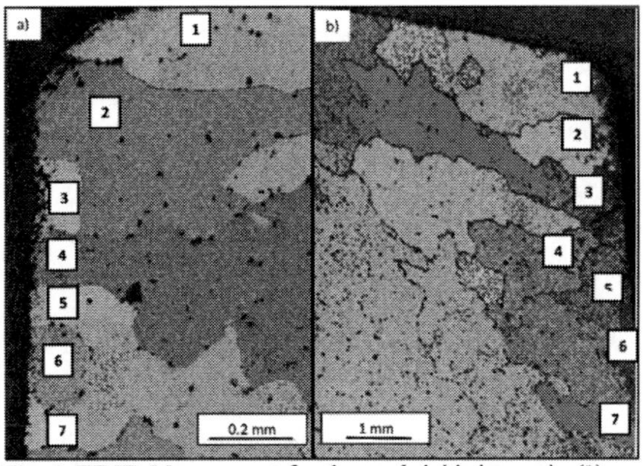

Fig. 8. EBSD Measurement for the crack initiation grain (1) and surrounding grains.

Table 2. Calculated Young's modul (for T = 850 °C), Schmid factors and E·m values.

Grain Nr. a)	Young's Moduli	Schmid factor	E·m
1	152	0.48	73.2
2	95	0.46	44.2
3	97	0.46	45.8
4	95	0.45	42.8
5	97	0.46	45.2
6	102	0.48	49.1
7	102	0.47	48.7

Grain Nr. b)	Young's Moduli	Schmid factor	E·m
1	167	0.45	76.48
2	173	0.46	81.61
3	178	0.43	77.60
4	106	0.48	52.05
5	96	0.46	44.95
6	128	0.49	63.74
7	143	0.49	70.47

The calculation of local Young's moduli and Schmid factors (Table 2) for the same specimens was carried out using the measured Euler angles φ_1, φ_2 and θ. In specimen a), both, grain 1 and 6 feature the highest Schmid factors of m = 0.48. If crack initiation would be assumed to occur in grains with high Schmid factors, the crack could also have started in grain 6. However, grain 1 provides a significantly higher Young's modulus which leads to higher E·m values and therefore to local plastic deformation and crack initiation in grain 1.

Considering sample b), it becomes apparent that fatigue failure did not start in the grain with highest E·m, but, the crack initiation grain 1) and the surrounding grains show higher values of E·m than the other grains. Thus crack initiation starts in this case at a grain agglomerate with high E·m values. Other feature significantly higher Schmid factors (Grain 6, 7) but lower shear stresses are induced due to the lower Young's moduli.

6 Concluding remarks

It could be shown that especially for elastic anisotropic materials such as nickel-base superalloys, that, besides the influence of the Schmid factor investigated in earlier studies [10, 21, 23], local elastic stiffness has an important impact on crack initiation. If homogenous strain is assumed in the considered region, the shear stress in the slip system depends linearly on the product of Schmid factor and Young's modulus. As soon as this resulting shear stress exceeds a critical value, plastic deformation, or, in case of cyclic loading, fatigue crack initiation may be initiated. A Monte Carlo simulation with 100.000 samples showed the frequency distributions of Young's modulus as well as Schmid factor and revealed orientations with Schmid factors in the range 0.37-0.45 and Young's moduli of about 240 GPa up to 300 GPa to be most critical at room temperature. These orientations induce highest shear stresses at a given uniaxial strain but occur less frequently. Orientations with maximum Schmid factor occur much more frequently but induce significant lower shear stresses within the slip systems because of lower Young's moduli. A similar behaviour was shown for T = 850 °C. The influence of both, Young's modulus and Schmid factor on the crack initiation site in coarse grained Ni-base superalloys was experimentally proven for isothermal total strain controlled fatigue tests with single crack initiation. EBSD measurements of cracked LCF specimens showed crack initiation in grains with orientations featuring intermediate Schmid factors and high Young's moduli and therefore high E·m or crack initiation in grain accumulation with high E·m. With knowledge of the elastic constants and the investigated crystal system, the E·m model can be applied to several materials. However, it should be noted that the presented failure model in its actual state only applies to uniaxial loadings and globally homogenous strain distribution is assumed. Ongoing research deals with the extension of the E·m model to multiaxial loading conditions as well as with FEM-based consideration of stress/strain inhomogeneities caused by interaction between differently oriented grains.

The investigations were conducted as part of the joint research programme COOREFLEX TURBO the frame of AG Turbo. The work was supported by the Bundesministerium für Wirtschaft und Energie (BMWi) as per resolution of the German Federal Parliament under grant number 03ET7041K. The authors gratefully acknowledge AG Turbo and Siemens AG for their support and permission to publish this paper. The responsibility for the content lies solely with its authors.

References

1. R. Kehlhofer, R. Bachmann, H. Nielson et al., Combined-cycle gas and steam turbine power plants. 2. Edition (2018)
2. P. Walsh and P. Fletcher, eds., *Gas Turbine Performance*: *Starting*, Blackwell Science Ltd, Oxford (2008)
3. L. Zhang, P. Yan, M. Zhao et al., *Tensile Anisotropy of a Single Crystal Superalloy* (Pacific Rim International Congress on Advanced Materials and Processing, 2013)
4. C. Dresbach, U. Dressler, J. Gussone et al., Mat.-wiss. u. Werkstofftech., **45**, (2014)
5. D. Kulawinski, *Biaxial-planare isotherme und thermo-mechanische Ermüdung an polykristallinen Nickelbasis-Superlegierungen* (2015)
6. B. Buchholz, *Entwicklung eines Werkstoffmodells für isotherme und anisotherme Ermüdung der Nickelbasislegierung René80* (2012)
7. H. T. Pang, and A. S. Reed P, Mater. Sci. Eng., A, **448**, 67 (2007)
8. K. Schneider, H. Arnim, and H. W. Grünling, Thin Solid Films, **84**, 36 (1981)
9. B. Engel, T. Beck, and S. Schmitz, *High Temperature Low-Cycle-Fatigue Of The Ni-Base Superalloy René80* (LCF8 - Eighth International Conference on Low Cycle Fatigue, Dresden, 2017)
10. T. Seibel, *Einfluss der Probengröße und der Kornorientierung auf die Lebensdauer einer polykristallinen Ni-Basislegierung bei LCF-Beanspruchung* (2014)
11. J. C. Stinville, W. C. Lenthe, M. P. Echlin et al., Int. J. Fatigue, **208**, 221 (2017)
12. J. C. Stinville, W. C. Lenthe, J. Miao et al., Acta Mater., **103**, 461 (2016)
13. J. Miao, T. M. Pollock, and J. Wayne, Acta Mater., **57**, 5964 (2009)
14. J. Rösler, H. Harders, and M. Bäker, *Mechanisches Verhalten der Werkstoffe* (2016)
15. O. Rand and V. Rovenski, *Analytical Methods in Anisotropc Elasticity* (2005)
16. C. Chou and N. Pagano, *Elasticity - Tensor, Dyadic and Engineering Approaches* (1967)
17. R. Newham, *Properties Of Materials - Anisotropy, Symmetry, Structure* (2005)
18. W. A. Wooster, *A Text-Book On Crystal Physics* (1949)
19. C. R. Reed, *The Superalloys - Fundamentals and Applications* (2006)
20. E. Schmid and W. Boas, *Kristallplastizität* (1935)
21. H. Gottschalk, S. Schmitz, T. Seibel et al., Mat.-wiss. u. Werkstofftech, **46**, 156 (2015)
22. S. Schmitz, T. Seibel, T. Beck et al., Comp. Mater. Sci., vol. **79**, 584 (2013)
23. P. Gumbsch and R. Pippan, *Multiscale Modelling of Plasticity and Fracture by Menas of Dislocation Mechanics*: *CISM Courses and Lectures Vol. 522*, (2010)

24. L. Bittner, H. Gottschalk, M. Gröger et al., *Modeling, Minimizing and Managing the Risk of Fatigue for Mechanical Components* (Stochastic Geometric Mechanics, Lausanne, Switzerland, 2016)
25. E. W. Weisstein, *CRC Concise Encyclopedia Of Mathematics* (2003)
26. U. Bayerlein, and H.G. Sockel, Superalloys, 695 (1992)
27. A. Weidner, R. Beyer, C. Blochwitz et al., Mater. Sci. Eng., A, **435-436**, 546 (2006)
28. D. Leidermark, J. Moverare, M. Segersäll et al., Procedia Engineer, **10**, 619 (2011)

The setup and application of the multi-scale in-situ test system for fatigue damage analysis

Yi Shi[1,*], *Xiaoguang* Yang[1], *Guolei* Miao[2] and *Duoqi* Shi[1]

[1]Beihang University, School of Energy and Power Engineering, 100191 Beijing, China
[2]Chengdu Holy Industry& Commerce Corp.LTD, 611936 Sichuan Province, China

Abstract. This essays aims at introducing the setup for the multi-scale in-situ test system which is applied for the fatigue crack initiation test. The setup of the experiment system is first introduced, including the image capturing system, optical path system, image acquisition and storage system and the three-axis mobile platform. Then the preparation of micro speckle and the corresponding technique for spatial adjustment are improved to realize the DIC measurement in micro scale. Finally three experiments from macro-meso scale to macro scale: fatigue initiation test, the observation of micro crack and the fatigue crack growth rate in macro scale were conducted to verify the application of the system. The test result can indicate the location of crack initiation, the crack growth rate and the evolution of displacement/strain field, etc.

1 Introduction

Fatigue is the main cause of failure of engineering structures and components. The fatigue damage of metal is composed of multiple process which is generally divided into the following stages: 1) the accumulated damage caused by the change in microstructure; 2) the nucleation of micro crack; 3) the crack growth of micro crack and formation of main crack; 4) the crack growth; 5) fracture [1]. Therefore fatigue failure is a multi-stage continuous process. The fatigue life is generally analyzed via the post-test fractography after the fatigue tests. However, this can only analyze the existing results instead of revealing the middle process. Actually understanding the intermediate process of fatigue damage will be helpful for analyzing the underlying failure mechanism and identifying key factors.

By now many researchers have applied different methods to study various stages of fatigue damage process. For example, the surface replica is used for the study of small crack growth[2-4]; the compliace method and potential drop method are applied to monitor the length of long crack[5-7]; the in-situ SEM methods are used to investigate the initation of crack [8-9]. However, all these methods have thier limitations. More specifically, the process of surface replica is complicated and inefficient. The specimen should be notched to ensure the crack can be captured inthe view field of SEM. Additionally, some methods can only acquire the length of long cracks. Therefore, a new experimental setup which can realize the in-situ multi-measurement based on the hydraulic fatigue testing machine is proposed.

2 Experiment system

The experiment system is composed of four sub-systems, including 1) image capturing system, 2) optical path system, 3) image acquisition and storage system and 4) three-axis mobile platform (**Fig. 1**). This system cooperates with the axial servo-hydraulic testing machine instead of tensile stage to realize the in-situ observation and measurement of related parameters in the loading process. It is widely accepted that the performance of cyclic loading from hydraulic testing machine is better than the tensile stage.

2.1 Image capturing system

The image capturing system includes optical lens and camera components. To ensure the resolution for the meso and micro scale, the selection of CCD camera and sensor should be careful. In this system, the German manufactured high resolution CCD camera named AVT-GT2450 is applied, with 2/3 inch sensor and 2448×2050 pixels resolution. The optical lens can achieve high resolution lens with 1.16-66.63 times zoom, with 0.17-9.49mm (diagonal length) working view field, which is situable for the meso scale. For the measurement in the macro scale, the present lens can be replaced with bilateral telecentric lens and combined with CCD camera [10]

* Corresponding author: shiyi_sepe@buaa.edu.cn

2.2 Optical Path System

The coaxial optical path system is applied in this system. Compared with the traditional ring LED light source, the coaxial optical path system could offer more uniform light which can improve the accuracy and reproducibility of machine vision. Therefore it is more convenient for the observation of surface information such as foreign object, crack and scratches. Meanwhile, to satisfy the light intensity requirements as well as the stable light source for the experiment, the high-intensity LED cold light source with fiber is applied. This light source can not only produce less heat under the working condition, but also provide long-lasting light.

2.3 Image Acquisition and Storage System

The Stream Pix software is selected for the image storage and acquisition. This software has capacity of high-frequency image acquisition and continuous or fixed-interval image storage.

2.4 Three-axis mobile platform

The main functions of the three-axis mobile platform includes 1) support the whole image capture system; 2) to reduce the vibration from other objects; 3) to adjust the focal length of lens. This platform is made up of three parts: 1) vibration isolation base platform; 2) three-axis coarse adjustment mobile platform; 3) three-axis fine adjustment mobile platform. More specifically, the main function of the three-axis adjustment mobile platform is to adjust the focal length of lens. For the observation and measurement in meso scale, the lens should be set at large zoom. However, with the increase in zoom rate, the depth of filed will decreases. If the resolution of fine-adjustment is low, it will be difficult to find the appropriate focal length for the lens. In addition, images are likely to be influenced significantly by small vibration. Therefore, it is necessary to apply good vibration isolation system.

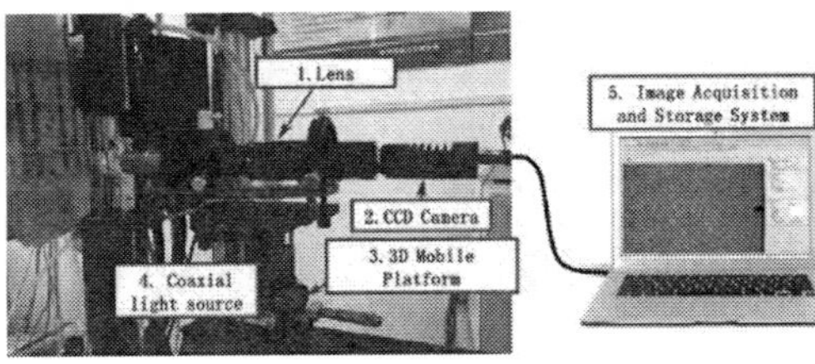

Fig. 1. The setup of the in-situ system

3. Digital image correlation (DIC) method

3.1 DIC basic theory

Digital image correlation method, as known as the digital speckle correlation method is different from the traditional contacting measurement such as extensometer, which is based on the image processing and numerical calculation and is a kind of non-contact whole-field deformation optical measurement method [11]. The basic principle is to calculate the corresponding displacement filed and strain field by analyzing (speckle or metal surface texture) deformation and displacement of the information carrier in the deformed images and unreformed images, shown in **Fig. 2**. The process of DIC method is divided into two parts: 1) matching the acquired deformed images and initial images; 2) calculating the displacement field in the region of interest and calculate the strain field DIC through the differential displacement field. In the DIC algorithm, the images are matched by the grey value in the subset. In other words, the correlation function is used to analyze the pre-deformation and post-deformation values and match tem based on the maximum correlation value in the process of DIC image. If the grey value of the deformed carrier is changed during the test, it will cause the DIC algorithm fail to further calculate the displacement field or deformation field. Therefore the quality of speckle is a key problem in the realization of DIC technique. It is required that the speckles should have the random grey value distribution. If the gradation is concentrated in a range, it will make it difficult to match in the operation.

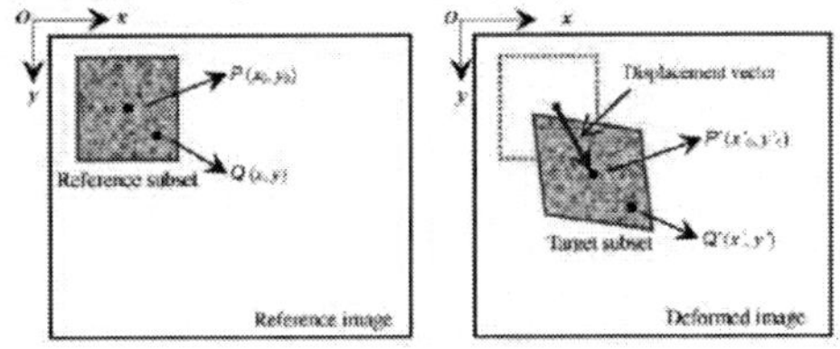

Fig. 2. DIC algorithm schematic

3.2 Preparation for the micro speckle

The speckle preparation is the key to achieve accurate measurement of DIC and the production of speckles for small scales it is important to the crack initiation test. In this paper, the micro speckles were produced by spraying method. The black graphite powder with a particle diameter of 10μm was used as the "paint". The graphite powder was sprayed with an airbrush and attached to the surface of specimen by means of electrostatic adsorption. This speckle preparation method is convenient, however to obtain high quality speckle the preparation parameter should be adjusted. For example the pressure of the airbrush, the spray angle and distance. To ensure the resolution of DIC test, the quality of the prepared speckles need to be evaluated. The global parameter estimation method based on average gray gradient of speckle images proposed by Pan Bing[12-13] was applied. The average gray gradient of this speckle is 15.46, and the speckle image and grey value distribution in shown in **Fig. 3**.

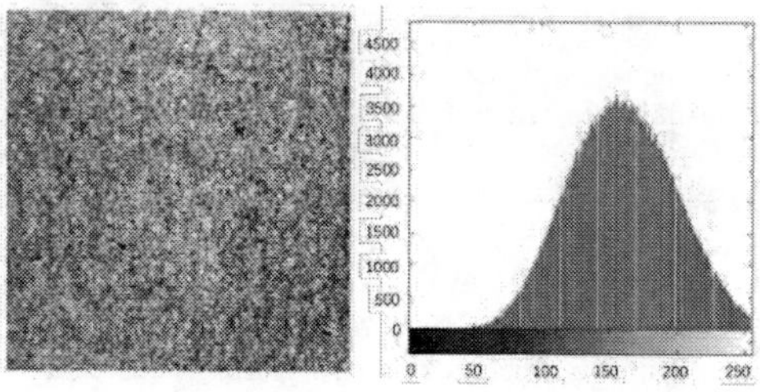

Fig. 3. The image of speckle made by graphit powder and its gray level distribution histogram

3.3 Spatial adjustment technique for the platform

Besides the quality of deformed carrier, the adjustment of the 3D mobile platform also makes difference to the test results. For the 2D-DIC test, since the out-of-plane displacement has a detrimental effect on the result, it must be ensured that the optical axis of the camera lens is in a spatial vertical state with the sample surface, shown in **Fig. 4**. The detailed steps are given below: 1) To ensure that the optical axis of the lens and the loading axis of the fatigue testing machine are in a plane perpendicular state, the pitch of the 3D mobile platform must be adjusted and the lens should be adjusted to a horizontal level with a level instrument; 2) to ensure the spatial vertical state of the optical axis and test specimen surface, the specimen should be rotated for adjustment. In addition to the empirical operation, a little load can be given (at the premise of not influencing the test result) to conduct and compare the displacement field distribution to the theoretical one. If the results is consistent, the lens are well perpendicular to the sample surface.

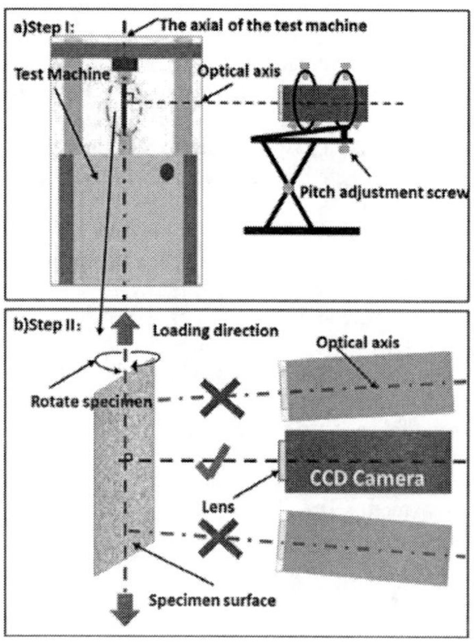

Fig. 4. The schematic of test equipment setting requirements and steps

4. Application for the Testing in Macro-meso Scale

4.1 The observation of crack initiation test

The purpose of this test is to realize the observation of crack initiation at room temperature combined with DIC method. Generally, the position of the crack initiation is difficult to locate at early stage by naked eyes because these cracks are too small and stochastic as well. However, with the help of contour calculated by DIC method, it can indicate the position due to the discontinuity at the surface.

4.1.1 Test specimen and procedure

GH536 is a kind of Chinese made nickel based superalloys which is applied in aero-engine flame tube of combustion chamber due to its excellent mechanical performance at elevated temperature. Its average grain size is approximately 50 micro meter. The grain size should be taken into consideration when prepare the speckle and design the gauge section. To identify the deformation between grains, there should be certain amount of speckles in one grain. In addition, the size of gauge section is related to the magnification rate. Lower magnification rate results in large pixel point thus there will be less speckles in one grain. In this test, the gauge section is 2mm×2mm(**Fig. 5**) and the magnification rate of lens is 7 times.
Before the preparation for the micro speckle, the specimen was polished by abrasive paper in the order from 800-3000 to the mirror surface. The purpose of this procedure is to eliminate the scratches which will exert negative effect on the strain field measurement. Finally, the U-STAR GP-300 airbrush with 0.5mm caliber was applied to manufacture the speckles.

4.1.2 Test Condition

The triangular wave with maximum load 3.0kN was applied in the test. The loading ratio was set at 0.1 with frequency 3Hz. The test was conducted at room temperature and controlled by loading. The test was interrupted for image capturing. In other words, a complete cycle of image information was captured at regular intervals (**Fig. 6**).In the capture progress, the loading rate was 0.1kN/s.

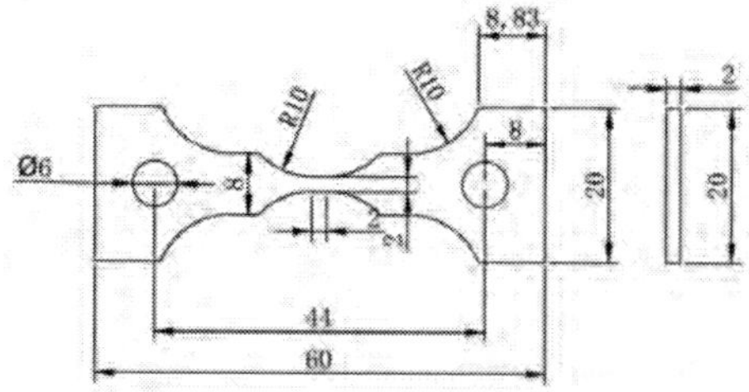

Fig. 5. The schematic of the test specimen

4.1.3 Result and analysis

When DIC analysis software is used to analyze the deformation field in the process of fatigue damage evolution, the types of strain fields obtained are different depending on the selection of reference pictures and deformation pictures. If the image of the specimen at the minimum loading in the first cycle is chosen as the reference image and other images at the peak loading is treated as deformed image, the evolution of the total strain field at the specimen surface in the fatigue progress can be obtained and analyzed (**Fig. 7**). On the other hand, if images taken at the minimum loading in each cycle were chosen as deformed image, the residual strain filed could be obtained (**Fig. 8**).

1) The analysis of total strain field

The calculation of total strain field is useful for locate the initiation of micro cracks. The physical size of micro crack is small and it is usually hard to find until it reaches hundred micro meter. Although the sprayed speckles on the surface seem to add difficulty in observation, the application of DIC method indicates the micro cracks. Since the DIC algorithm is based on the correlation between the deformed image and reference image. If the subset fails to find the corresponding subset in the deformed image due to the discontinuity on the surface caused by the micro cracks, there will be no calculation for this subset. In other words, this area will be 'faded' in the contour images. In addition, the value of strain can be calculated based on the DIC algorithm mentioned before. Therefore, in the process of analysis, the location of potential crack can be identified by the color of contour. The results of this test is shown in **Fig. 9**.In early stage, for example **Fig. 9**(b)-(c), six points were chosen from higher strain area. Meanwhile the strain value of each point was extracted. The diagram of evolution of total maximum strain was also plotted **Fig. 10**. Among the six points, before 40000 cycles point B took the first place. However, point C which was the smallest one initially exceeded point B after 40000 cycles. A significant increase can be observed for all points at approximately 4000 cycles except point F. Point F first stabilized at 0.02 before 4000 cycles and then grew less significantly than other points to 0.022. The contour at 46135 cycles is shown in **Fig. 11**. Micro cracks can be observed by the 'missing color' area. Meanwhile, the strain value of circled points is higher than other area, which further indicate these points are likely to be the initiation places. For point F which appeared higher strain value than point A, C, D, E at the beginning, no sign of initiation can be found,**Fig. 9**(g)-(h). On one hand, the strain value is obviously lower compared with other five points. On the other hand, there is no discontinuity in the contour in the point F area. Therefore, the application of DIC method is effective for identifying the initiation places. The value of strain also makes difference. From this test, some area like point F exhibits higher strain rate at first but does not initiate earlier than others. In this view filed, there were several micro cracks at the surfaces. If the advanced microscope is applied for this specimen, only one or even none cracks could fall into this view field and thus losing important information of the evolution. The specimen fractured under another 1919 cycles after 46135 cycles. This also means there are only little time left for obersevation when the micro cracks can be found by naked eyes.

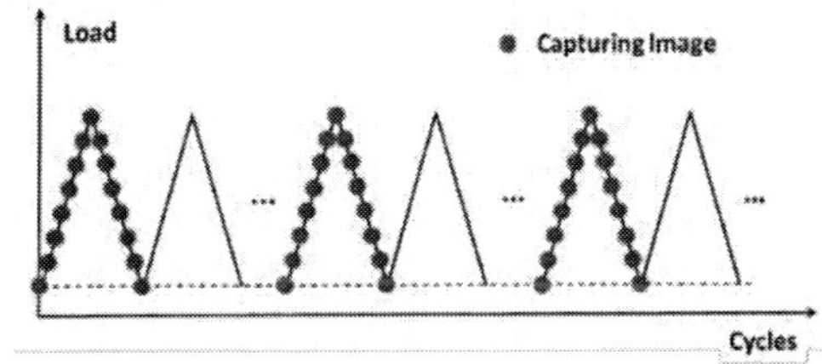

Fig. 6. Image information acqusition spectrum during loading

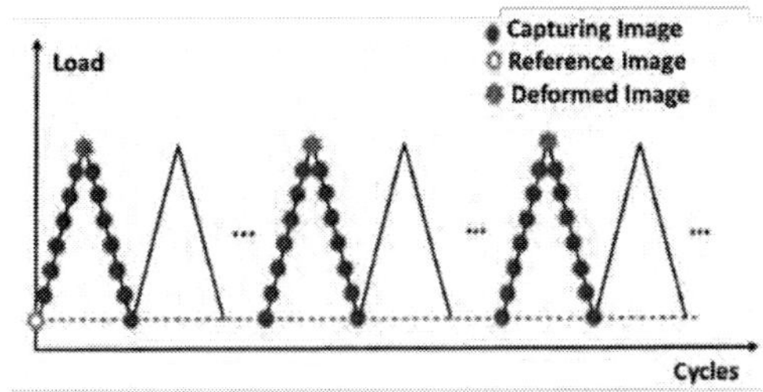

Fig. 7. The schematic of calculation for the maximum total strain by DIC anaylsis

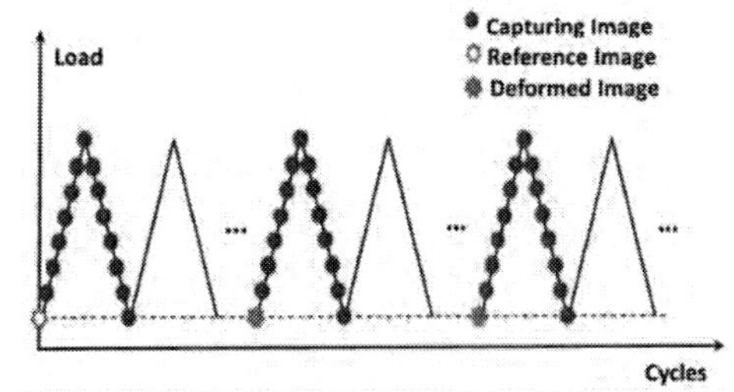

Fig. 8. The schematic of calculation for the residual strain by DIC analysis

https://doi.org/10.1051/matecconf/201816504005

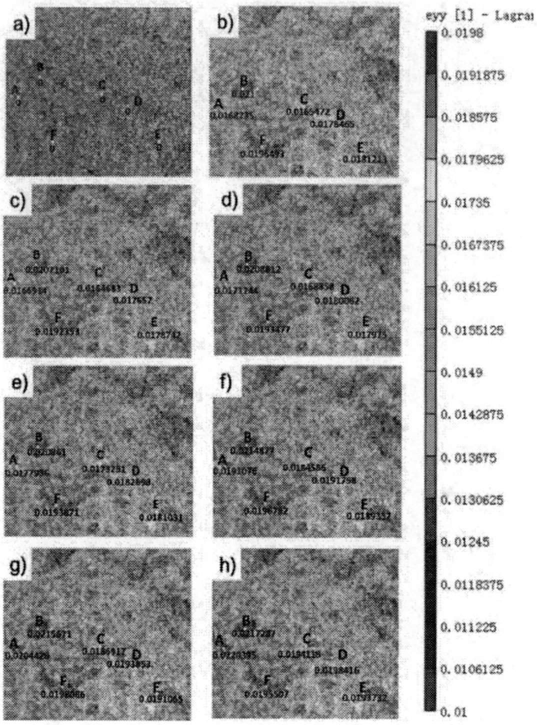

Fig. 9. The contour of the maximum total strain evolution in different cycles a) unloading; b) 3002 cycles; c) 13004 cycles; d) 18005 cycles; e)23006 cycles; f) 29727 cycles; g) 30727 cycles

2) The analysis of residual strain field

The residual strain can be calculated by DIC as mentioned before. The correlation method is a kind of subtraction. In other words it compares the difference between unloading at different cycles and thus indicates the residual strain field. By extracting the value from images, the curve of the accumulated residual strain is shown in **Fig. 12** Similar to the curve of total strain evolution, an obvious increase can be observed near 40000 cycles. The accumulated strain first stabilized at 1.6% and then increased slightly to 1.7%. However, after 40000 cycles, it soared to 2.6% with another 5000 cycles. The curve also reflect that initiation stage consumes most fatigue life.

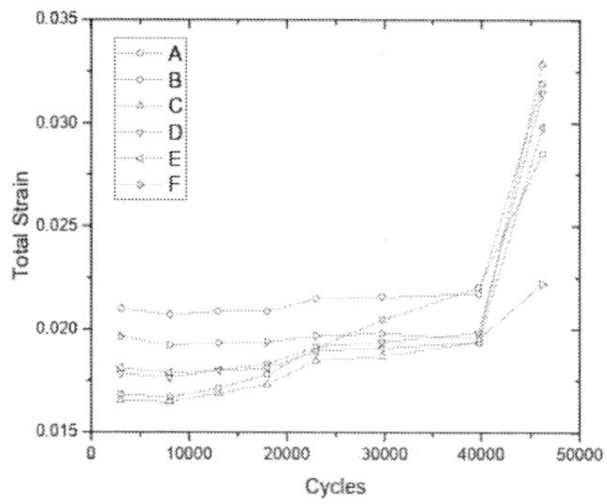

Fig. 10. The total maxmium strain evolution of six points

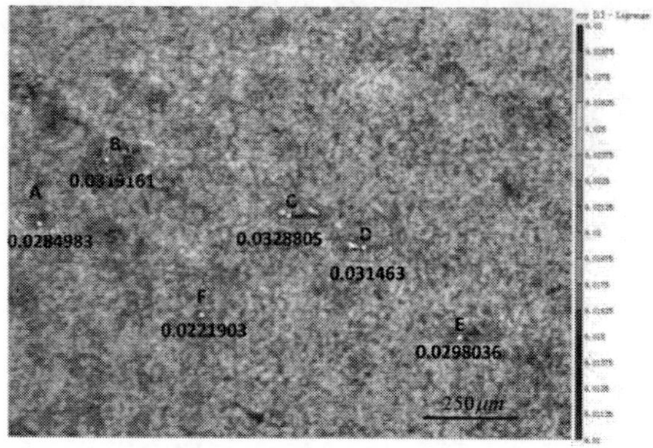

Fig. 11. The contour of total strain at 46135 cycles

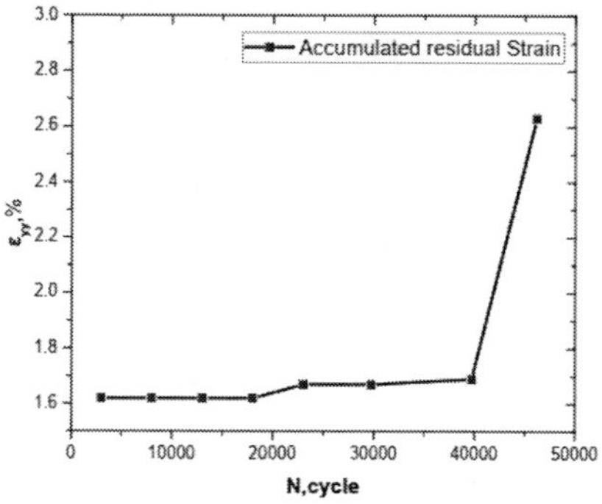

Fig. 12. The cruve of the accumulated residual strain

4.2 The observation of small crack growth

As is mentioned before, surface replica and some in-situ observation instruments are applied in investigation of small crack growth. For the setup of this in-situ system, it is also effective for this purpose.

4.2.1 Test specimen and procedure

The material of the specimen is a Chinese made nickel based powder metallurgy superalloys named FGH96. The geometry of the specimen is shown in **Fig. 13**, and the notch is done by EDM. In order to characterize the grain feature, the surface of the specimen was corroded by the etchant solution: copper chloride 5g+ethyl alcohol 100ml+ hydrochloric acid 100ml. The grain character and twin grains can be observed from our in-situ setup **Fig. 14**. The test condition is similar to the previous one but the maximum loading in this case is 4kN.

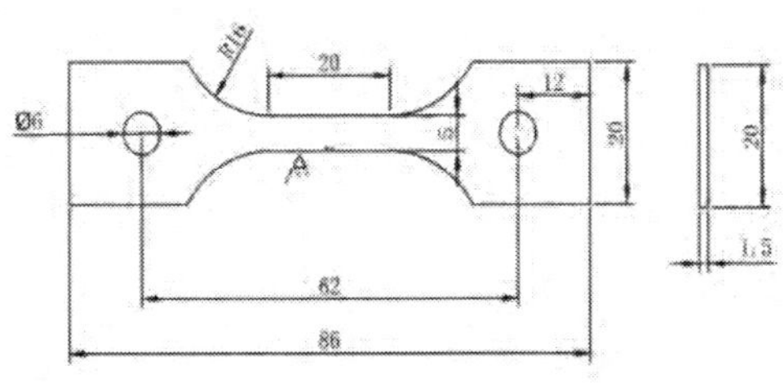

Fig. 13. The schematic of the test specimen

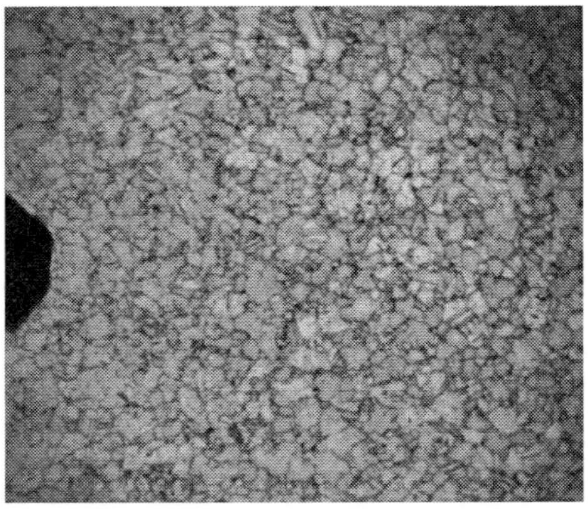

Fig. 14. The image of the corroded specimen captured by the setup

4.2.2Result and analysis

The crack imitation mode and path can be clearly captured. In **Fig. 15**, no crack can be observed before the 5002nd cycle. Later, at the 14341st cycle, a micro crack can be detected in the grain near the bottom of the notch. The initiation mode is identified as inner-grain initiation mode. In the process of crack growth, intergranular and transgranular changes alternately (**Fig. 16**). The crack path first grew in the 45 degree from the loading direction, and the direction changes continuously instead of being perpendicular. This is mainly due to the unevenly distributed grain size and orientation.The 45° defelection also indicates the inital failure was casued by the maximum shear stress.

According to the definition of initiation, microstructure small crack, physically small crack and long crack, the crack growth can be divided into four stages. The physically small crack is usually defined as the plastic area is larger than the average microstructure size and is smaller than 1mm. For FGH96, the average grain size is 23.8μm measured by EBSD in our previous work [14] . Therefore, the minimum size of the physically small crack should be 3-5 grains size. The percentage of each stage accounting for is shown in **Fig. 17**.The initiation stage accounts almost half part of the total life. Meanwhile, the small crack stage contributed approximately 50%. Only 4% life belongs to the long crack growth. Therefore, the stage of small crack cannot be neglected. Future efforts will be put on the study of physically small crack growth. The relationship between cycle number and crack length is given in **Fig. 18.** The crack growth rate at the transition point between the PSC and MSC increased significantly while the rate in the PSC decreased to 27500 cycle and then increased evidently until the LC.

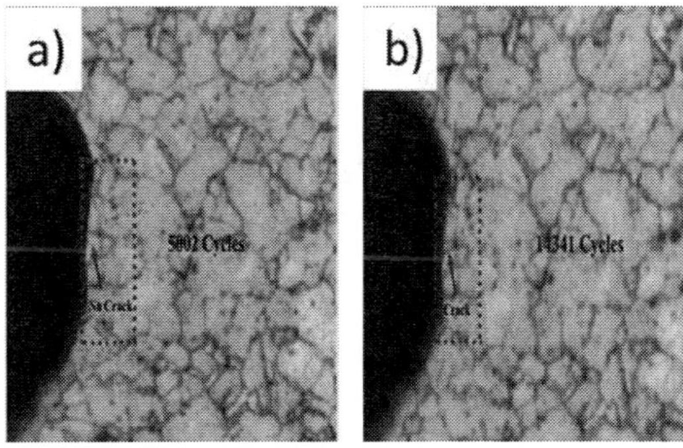

Fig. 15 Initiation in one grain a) the 5002th cycle ; b) the 14341st cycle

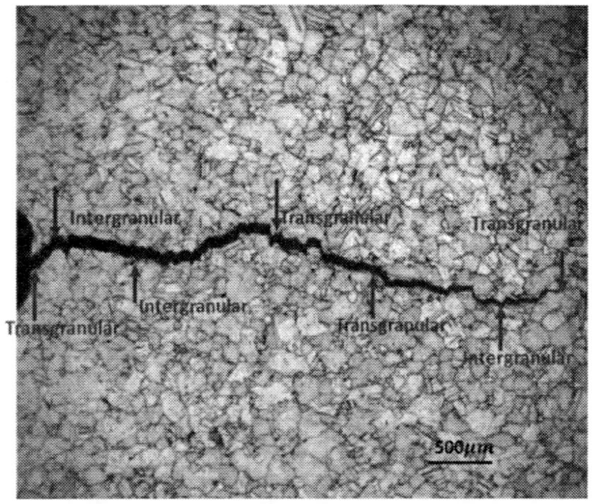

Fig. 16. The crack growth path (the 32869th cycle)

5. Application for the Testing in macro Scale

This multi-scale in-situ test system also has the capability of monitoring the long crack growth in macro scale. The deformation of the surface is also indicated by the DIC method.

5.1 Test specimen and procedure

The single edge notch tension (SENT) specimen was applied in this test, shown in **Fig. 19**. The specimen was designed based on the E647 standard[15]. Compared with the traditional compact tension specimen, the SENT has a wider area for observation. The surface of the SENT was polished by the abrasive paper to the mirror state. To enhance the image contrast, white paint was used as the background and black paint was chosen as the deformation carrier. The maximum loading was 5kN, with the R-0.1 at the frequency 2Hz. The capture method is similar to the previous mentioned tests.

 https://doi.org/10.1051/matecconf/201816504005

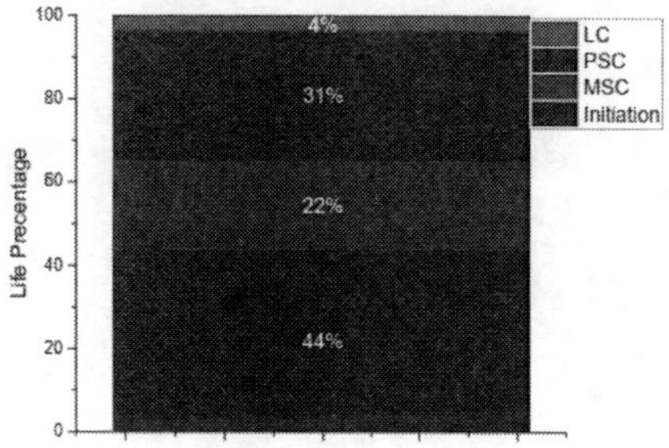

Fig. 17. The percentage of each stage of the taotal life

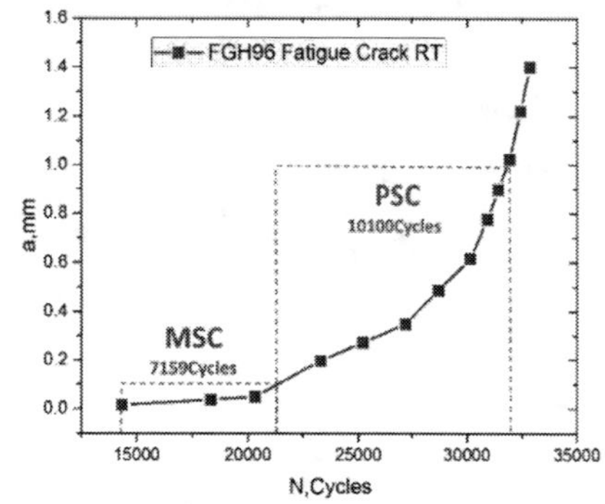

Fig. 18. The relationship between the crack legnth and cycles

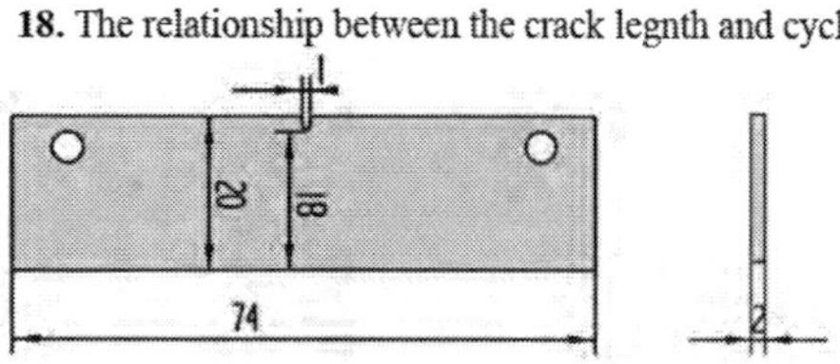

Fig. 19. The geometry of the SENT specimen

5.2 Result and analysis

The fatigue crack length can be obtained by the image captured by this system and the stress intensity can be calculated by the formula provided by the E647. The fatigue crack growth starts from 10^{-5} mm/cycle to 10^{-3} mm/cycle and the stress intensity falls between 30 to 65 $MPa \cdot m^{1/2}$. In the medium and higher level of stress intensity factor, the data points is linear in the logarithm scale, shown in **Fig. 20** which is consistent with the Paris law.

The strain can be calculated by DIC method which is similar to the method described beforehand. The evolution of strain field is shown in **Fig. 21**. The shape of the plastic zone at the crack tip is consistent with the plastic shape of the von Mises principle in the plane stress condition. Under the cylic loading, the strain at the crack tip increased significantly. In addition, the contour of the area along the crack seems faded due to the discontinuity which was discussed before. By extracting the maximum strain, the relationship between the strain and stress intensity factor is linear, shown in figure **Fig. 22**.

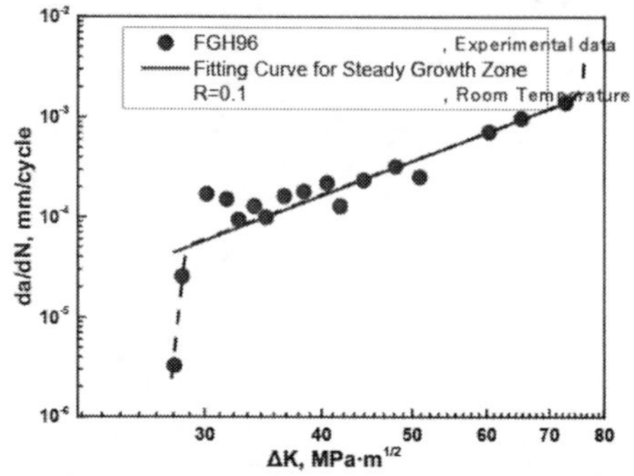

Fig. 20. The fatigue crack growth rate of FGH96

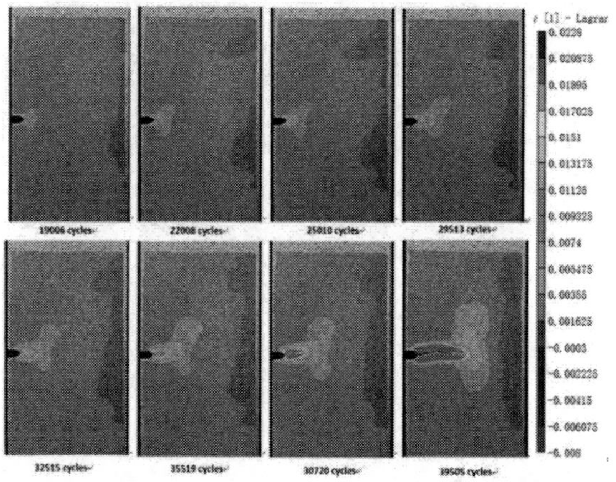

Fig. 21. The evolution of maximum strain of SENT specimen

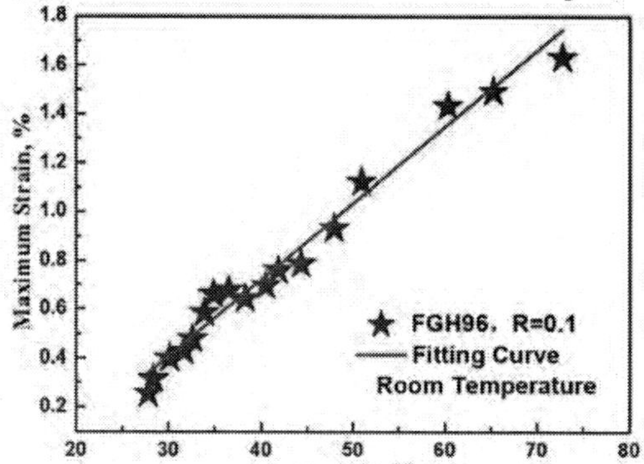

Fig. 22. The relationship between the maximum strain and stress intensity factor

The authors would like to thank the sponsorship by National Program on Key Basic Research Project (973 Program) (No. 2015CB057401).

References

1. J.Schijve, *Fatigue of Structures and Materials* (2001).
2. M.Goto,D.Knowles, Eng. Fract. Mech. **60(1)**, 1-18 (1998).
3. J.Newman,S.Willard,S.Smith,R.Piascik, Eng. Fract. Mech., **76(7)**, 898-910 (2009).
4. K.Zhang,X.Wu,C.Davies, Int. J. Fatigue., **104** (2017).
5. R.Bao, H.Yang, J.Zhang, L.Peng, B.Fei, Int. J. Fatigue., **47(1)**, 189-195 (2013)
6. X.He,Y.Zhang,H.Shi,J.Gu,C.Li,K.Kai, Mater. Sci. Eng. A, **618** 153-160 (2014)
7. S.Na,D.Yoon,J.Kim,H.Kim,D.Kim, Int. J. Fatigue., **101** 27-35 (2017)
8. H.Pang,P. Reed. Mater. Sci. Eng. A, **448(1)** 67-79 (2007)
9. R.Jiang,F.Pierron,S.Octaviani,P.Reed, Mater.Sci.Eng. A, (2017).
10. B.Pan, Acta Opt. Sin., **33(4)** (2013)
11. H.Gao,H.Liu,Z.Qi,H.Zheng, Ordnance Mater. Sci. Eng. (2016).
12. B.Pan,Z.Lu,H.Xie.Optics & Lasers in Eng. **48(4)** 469-477 (2010)
13. B.Pan,D.Wu,Y.Xia, J. Exp. Mech. **25(2)** 120-129 (2010)
14. G.Miao,X.Yang,D.Shi, Mater. Sci. Eng. A, **668** 66-72 (2016)
15. ASTM International. *Standard test method for measurement of fatigue crack growth rates E647-15* (2015)

3D characterization of the fracture mechanisms of a Fe-rich Al-Si-Cu alloy

Angelika Brueckner-Foit[1*], *Inigo* Bacaicoa[1], *Martin* Luetje[1], *Marcel* Wicke[1], *Andreas* Geisert[2], and *Martin* Fehlbier[2]

[1]Institute for Materials Engineering, University of Kassel, D-34125 Kassel, Germany
[2]Chair of Foundry Technology, University of Kassel, Kurt-Wolters-Strasse 3, D-34109 Kassel, Germany

Abstract: The effect of the defect size and morphology on the fatigue damage evolution was analysed in a recycled Al-Si-Cu alloy by micro-computed tomography and scanning electron microscopy. Fatigue tests were performed and the different crack initiation scenarios were characterized and classified. The interaction between shrinkage and gas pores was the key crack initiation mechanism and the ß-Al5FeSi particles did not play any role in the crack initiation phase. However, crack path analysis indicated that there is a certain amount of crack advance by brittle fracture of the β-phase particles. This is in accordance with the findings of tensile tests in which the ductility depended strongly on the iron content.

1 Introduction

Cast Al-Si-Cu alloys are extensively used in the aeronautic and automotive industry as they provide with excellent mechanical properties, castability and recycling possibilities [1,2]. The use of the recycled grade alloy as base material implies a significant reduction of CO_2 emission and process costs since the manufacturing of recycled aluminum requires only 5% of the energy in comparison with the primary alloy [3]. Nevertheless, the use of recycled aluminum for structural components can lead to the premature failure of the components if the stress state includes certain amount of cycling loading. This is the fundamental idea motivating the investigations carried out in recent years about the characterization of cast defects in recycled Al-Si alloys [4,5] and their influence on the fatigue damage [6,7].

The presence of iron in the cast melt causes the formation of brittle iron rich intermetallic inclusions that promotes the formation of shrinkage pores and reduces the tensile properties of the alloy [8]. Previous studies report that the iron inclusions do not have a direct effect on crack initiation since cracks tend to initiate from large shrinkage pores or small interacting gas pores [9]. However, the influence of the Fe-rich particles on the damage evolution for different defect scenarios requires further investigation in order to ensure the safety and reliability of structural parts. Moreover, non-equilibrium heat treatments can fragment and dissolve the iron rich compounds, and hence significantly improve the ultimate strength and the ductility of the alloy [10].

The effect of heat treatment on the crack initiation is limited due to the presence of large shrinkage pores but its influence on the crack propagation mechanisms is yet unknown.

In this study, fatigue tests were performed and the fracture surfaces were analyzed by scanning electron microscopy and micro-computed tomography in order to characterize the effect of the interaction between the Fe-rich phase and the various pores on the fatigue damage evolution. The effect of the Fe content is studied by systematically varying the iron content of the alloy and relating the damage mechanism to the three-dimensional morphology and disposition of defect structures obtained by computed-tomography scans and in the scanning electron microscope.

2. Material and experimental methods

Table 1 shows the chemical composition of the Fe-rich near-to-eutectic Al-Si-Cu alloy used in the present work. The Fe-content was varied adding different amounts of iron in the cast melt, as shown in table 1. The specimens were cast by gravity die casting at 760°C in a steel mold pre-heated at 350°C achieving a fixed cooling rate with a certain dispersion. The specimens were die cast without additional machining in order to consider the effect of the casting skin, since it has been recently reported that the casting skin can insulate the casting defects from the air environment and promote sub-surface crack initiation [9].

The Mn:Fe relationship is 0.28, which can promote partial substitution from ß-Al_5FeSi to the less detrimental α-$Al_{15}(Fe,Mn)_3Si_2$ [11], although a significant amount of ß-phase is formed.

* Corresponding author: a.brueckner-foit@uni-kassel.de

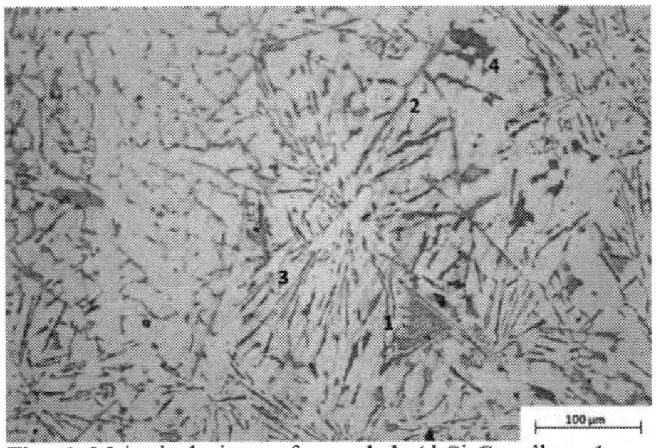

Fig. 1 Main inclusions of recycled Al-Si-Cu alloy: 1- α-$Al_{15}(FeMn)_3Si_2$, 2- β-Al_5FeSi, 3- Euectic silicon, 4- $CuAl_2$

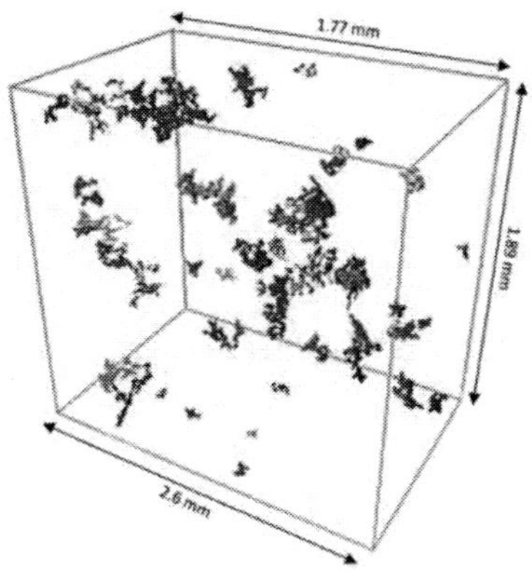

Fig. 2 3D reconstruction of the typical shrinkage pores in the studied alloy.

Fig. 3 3D reconstruction of the typical gas pores in the studied alloy.

The studied alloy is characterized by the eutectic silicon particles, Fe-rich inclusions and Cu-rich compounds, which can be observed in Fig. 1. The Chinese-script-like α-phase and the platelet-like ß-phase are the main iron rich inclusions present in the alloy. The brittleness of ß-Al_5FeSi plates and their tendency to form large clusters of particles significantly reduce the ductility and the ultimate strength of the alloy [10]. Some specimens were subjected to heat treatment at a solution temperature of 525°C and solution time 2 hours in order to investigate the influence of heat treatments on the microstructure and the fatigue damage evolution of the material. Fig. 2 shows volume elements with a high concentration of shrinkage and Fig. 3 the typical gas pores formed during the cooling phase of the casting process. Volume elements with different porosity levels were obtained as a result of the variability of the casting parameters.

Specimens with different Fe-content were tensile tested in a MTS E45 testing machine at a strain rate of 1.5 10-4 s-1. The fatigue tests were performed at different stress levels at R=-1 and 50 Hz and the fracture surfaces were subsequently analyzed by scanning electron microscopy and micro-computed-tomography with the aim of classifying the different fracture damage scenarios.

A high resolution lab computer tomograph (Zeiss Xradia Versa 520) was used for three-dimensional measurements of the casting defects at 70 KV and a power of 7 W with the 0.4X objective, achieving a resolution of 3.84 µm/voxel for the pores segmentation and 2.89 µm/voxel for Fe-rich inclusions. The image analysis was carried out with the rendering software Avizo [12], in which the casting pores were segmented by adjusting grey scale after applying the non-local means filter, whereas the Fe-rich inclusions were segmented manually.

Table 1 Chemical compositions of the Al-Si-Cu alloy with different Fe-additions (wt. %).

Alloy	Si	Cu	Mg	Fe	Zn	Mn	Al
A	12.96	1.52	0.68	**0.6**	0.48	0.17	rest
B	12.87	1.64	0.58	**1.37**	0.47	0.23	rest
C	12.99	1.8	0.43	**1.85**	0.51	0.3	rest
D	12.62	1.45	0.51	**2.28**	0.45	0.26	rest

3. Experimental results

3.1 Fatigue behaviour

Fig. 4 shows the results of the fatigue tests performed with the previously described specimens. The large scatter in the fatigue life at each stress amplitude level indicates the strong influence of the casting defects on the fatigue as well as the high variability of the morphology of defects associated with this casting process. An exhaustive analysis of the fracture surfaces both by scanning electron microscopy and micro-computed tomography revealed that there are distinct failure behaviors related to different fatal defects.

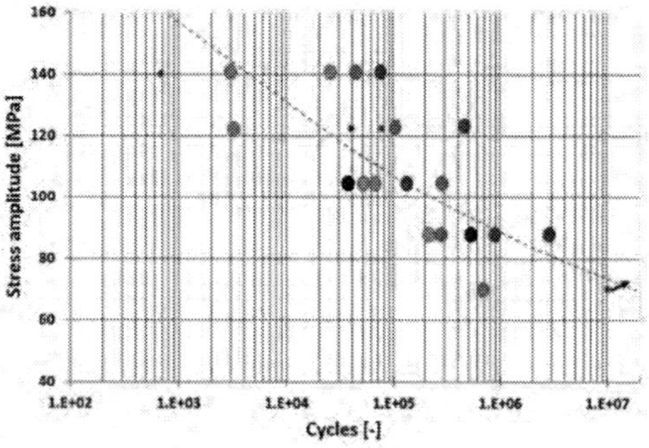

Fig. 4 Fatigue curve of die cast specimens; classification according to fracture origin; red circles: gas pores, green circles: interior and subsurface shrinkage pores, blue circles: interaction between gas pores and shrinkage pores, black circles: crack initiation from MgO inclusions on surface; small symbols: surface damage or non-identified fracture origin.

Cracks tend to initiate from gas pores located close to the surface in the cases in which large shrinkage pores are absent and the gas pores are located close to the surface (see Fig. 5). The variability of the cooling rate during the casting process increased the dispersion of the size of shrinkage pores. Under unfavorable conditions, specimens were cast with giant shrinkage pores of around 1 mm (see Fig. 7).

Nevertheless, the drop on lifetime was not dramatic in these cases and a substantial amount of internal flaw extension occurred prior to failure.

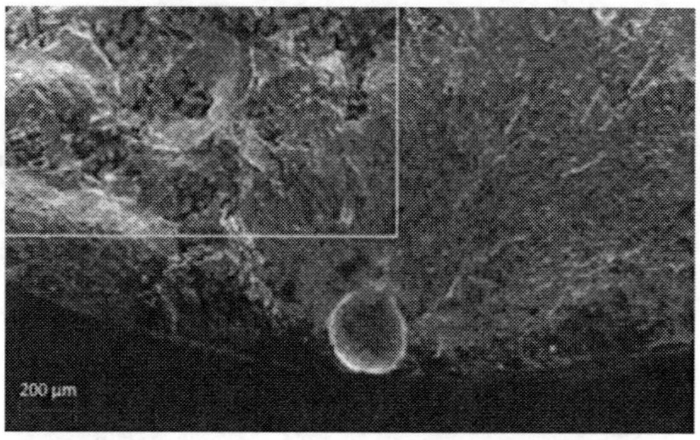

Fig. 5 Fracture surface of specimen failed at 140 MPa, 74,283 load cycles.

Fig. 6 Sample containing typical gas and shrinkage pores.

Interaction between defects was the most common scenario, what led to the premature failure of the specimens. This phenomenon can be explained by the fact that gas pores tend to agglomerate around shrinkage pores in the interdendritic structures (see Fig. 6 and Fig. 7). Furthermore, multiple crack initiation is likely to happen as a result of the intricate spatial pattern of bifurcations and branches of typical shrinkage pores (see e.g. Fig. 2).

* Corresponding author: a.brueckner-foit@uni-kassel.de

 https://doi.org/10.1051/matecconf/201816504006

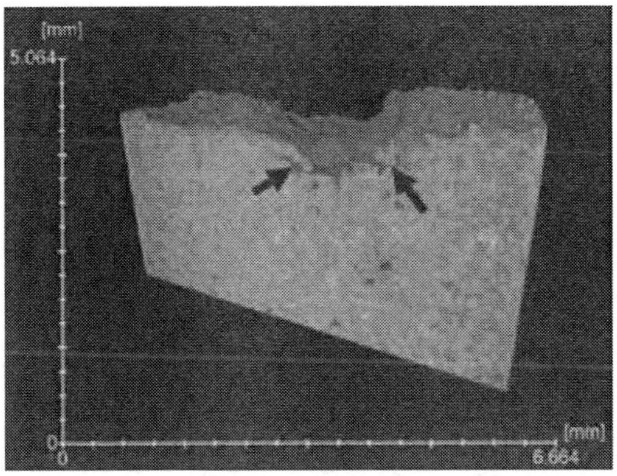

Fig. 7 Fracture surface of specimen failed at 140 MPa, 45,065 load cycles; large shrinkage pore with rough fracture surface and secondary cracks (blue arrows).

3.2 Effect of the Fe-rich inclusions

The ß-Al_5FeSi particles form large three-dimensional clusters (see Fig. 8), but they did not had a significant influence on crack initiation. These clusters are embedded flaws, as there are no iron inclusion in the surface or subsurface regions (see. Fig. 9). No direct influence of the ß-Al_5FeSi particles was observed in the SEM analysis of the fracture surfaces. This is in agreement with the fact that a higher concentration of these particles was observed in the inner part of the specimen; and therefore they have a limited effect on the crack initiation.

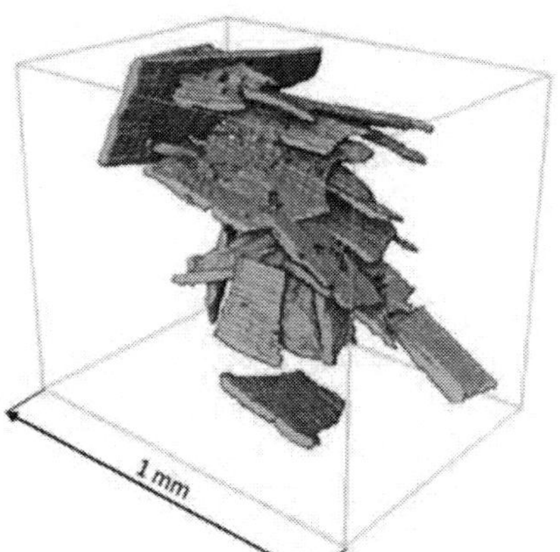

Fig. 8 3D reconstruction of a ß-Al_5FeSi inclusion cluster.

Agglomeration of complex clusters of inclusions distributed along the specimens can be observed in the 3D reconstructions of the ß-Al_5FeSi particles. Therefore, the inclusions that are depicted as individual needle-like particles in the 2D metallographic images may actually belong to a larger composite inclusion consisting of an agglomerate of plate-like particles.

The fracture profiles of the fatigue tests were examined on sections perpendicular to the fracture surface in the optical microscope and scanning electron microscope. This approach enables to examine if other microstructural features were involved in the evolution of the fatigue damage and if the crack development was driven by the fracture or the debonding of the ß-Al_5FeSi and the eutectic Si particles.

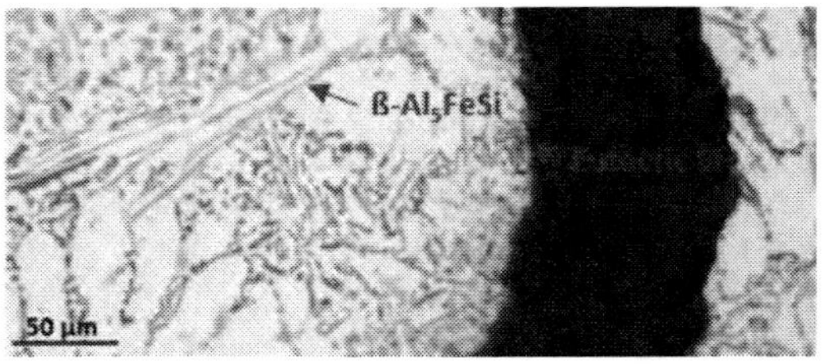

Fig. 9 Fracture surface cross section of a fatigue specimen at 122 MPa and 39,122 cycles.

Fig. 9 shows an example of the cross sectional fracture surface of fatigue specimen. It can be observed that a large cluster of ß-Al_5FeSi inclusions is located close to the crack path, and that the crack path follows the eutectic Si particles without breaking the cluster of the β-phase inclusions. This is related to the fact that the preferred orientation of these inclusions is perpendicular to the crack path in this particular case. However, cleavage fracture of a ß-Al_5FeSi particle can be observed in another area of the cross sectional fracture surface in the same fatigue specimen (see Fig. 10). Here, the plate-like morphology of the ß-Al_5FeSi inclusion causes the kinking of the crack path at the edges of the particle.

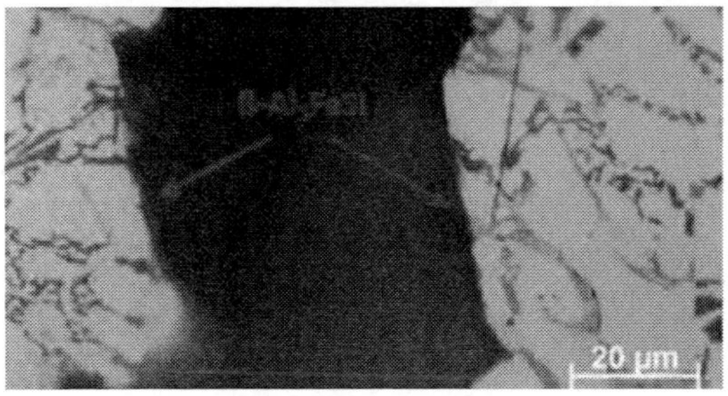

Fig. 10 Cleavage fracture of a ß-Al_5FeSi plate in the cross section of a fatigue specimen.

Fig.11 shows the kinking of the crack path as a result of cleavage fracture of hard intermetallic particles. It can be observed that the ß-Al_5FeSi platelets split right through the middle and drive the local direction of crack. The crack growths to the end of the ß-Al_5FeSi platelets, until it reaches the aluminum matrix that allows the relaxation at the crack tip by local plastic flow due to its ductility. It can also be observed Cu-rich particles located parallel have a similar effect on the crack path.

 https://doi.org/10.1051/matecconf/201816504006

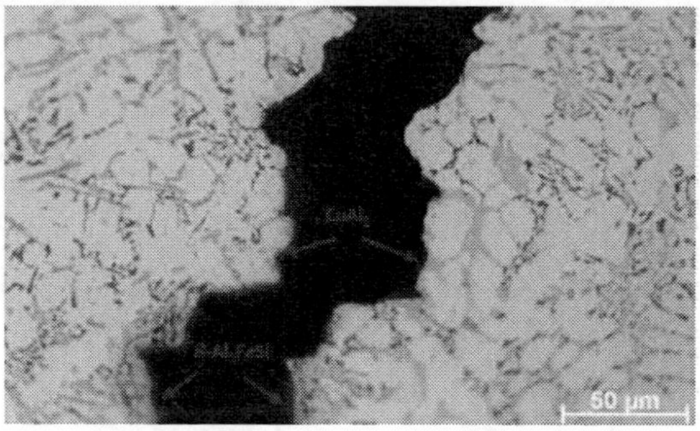

Fig. 11 Crack path in the cross sectional fracture surface of a fatigue specimen.

Fig. 12 shows the cross sectional fracture surface of a fatigue specimen in which the rupture of an entire ß-Al_5FeSi occurred. It can be observed that the crack propagated breaking various Fe-rich particles belonging to the same cluster. In this case the particles also broke through the middle at different levels of the cluster (see Fig. 8). The connection of the ß-Al_5FeSi plates in different plains leads to the kinking of the crack path. Besides the plate-like ß-Al_5FeSi particles, broken Chinese-script α-$Al_{15}(Fe,Mn)_3Si_2$ inclusions were also observed in the crack path at the cross sectional fracture surface of the fatigue specimen (see Fig. 13).

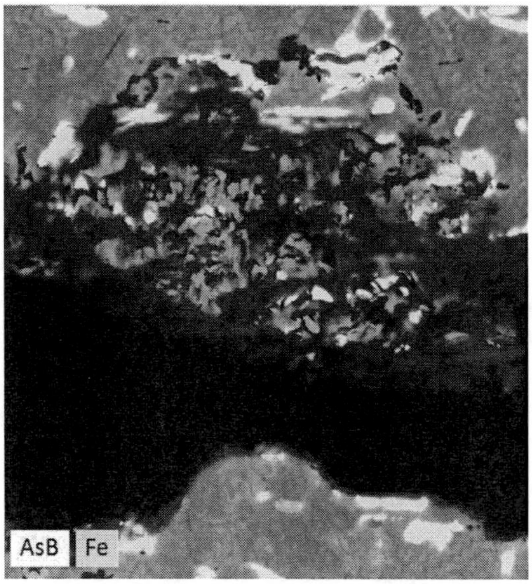

Fig. 12 SEM cross sectional fractograph of a fatigue specimen with a broken ß-Al_5FeSi cluster.

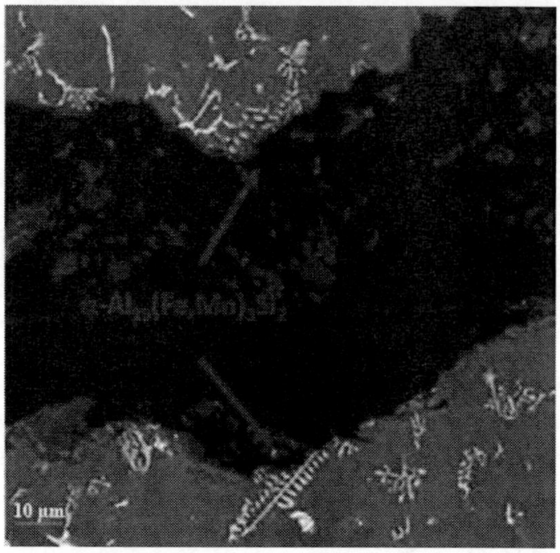

Fig. 13 Cleavage fracture of a Chinese-script like α-$Al_{15}(Fe,Mn)_3Si_2$.

3.3 Effect of heat treatment

The ß-Al_5FeSi did not have a direct influence on the crack initiation process since damage initiated from the big shrinkage pores or interacting gas pores, therefore the effect of heat treatment on fatigue life is limited. However, the non-equilibrium heat treatments with solution temperature at 525°C during 2h led to a significant level of fragmentation and dissolution of the ß-Al_5FeSi particles, what improved notably the ultimate strength and ductility of the alloy (see Fig.15). The addition of Fe in the alloy caused an exponential increase in the size of the ß-Al_5FeSi inclusions (see Fig. 12) that decreased the elongation at break of the alloy. Heat treatments also improved the mechanical properties of the modified alloy, although the effect diminished with increasing Fe content.

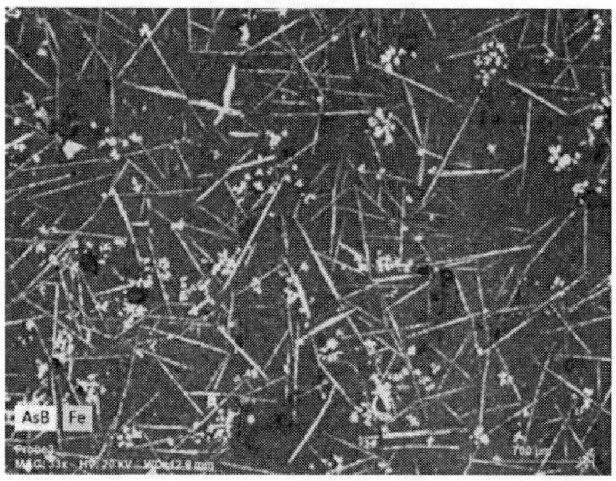

Fig. 13 Massive brittle ß-Al_5FeSi plates in the alloy with 2.3 Fe% (D)

The fracture surface analysis shows the cleavage fracture of massive ß-Al_5FeSi plates together with some fractured Cu-rich particles (see Fig. 13). The observation of the cross sectional fracture surfaces also reveals the dominating fracture mechanism. Fig. 14 shows two fractured ß-Al_5FeSi inclusions that have been split through the middle at the rupture of the specimen. The difference on the orientation of the broken plates causes the formation of sharp angles in the fracture surfaces.

Fig. 14 Cleavage fracture of a massive ß-Al_5FeSi inclusion in the alloy with 2.3% Fe content (D) together with some Cu-rich particles.

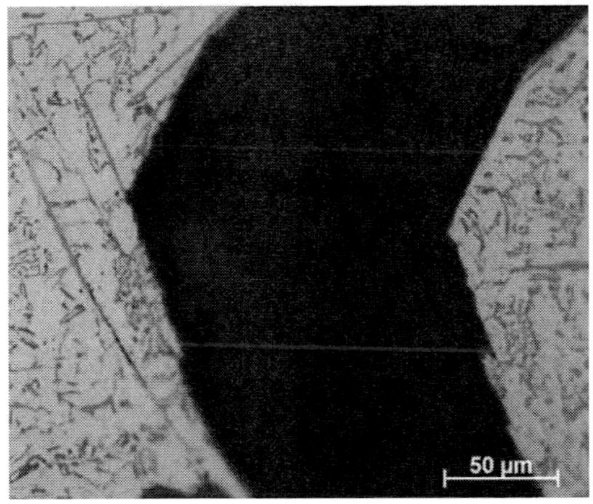

Fig. 15 Cross section of the fractured specimen with 2.3% Fe content.

In the heat-treated specimen with high Fe-content the main fracture mechanism was the cleavage of ß-Al_5FeSi inclusions that were not able to fragment and dissolve during the heat treatment. Nevertheless, it can be observed that the fracture mechanism was not fully dominated by the rupture of the Fe-rich inclusions, but the fracture of the eutectic Si particles (see Fig. 16).

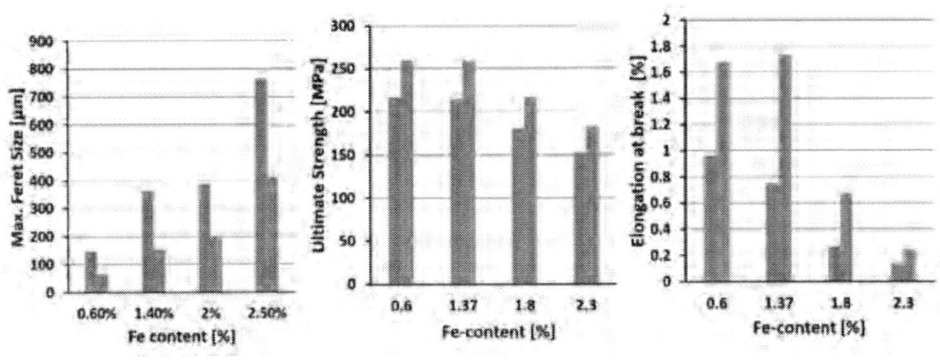

Fig. 16 Influence of the Fe content before (blue) and after (orange) heat treatment on: a) max. Feret size of the Fe-rich inclusions; b) ultimate strength (MPa); c) elongation at break (%).

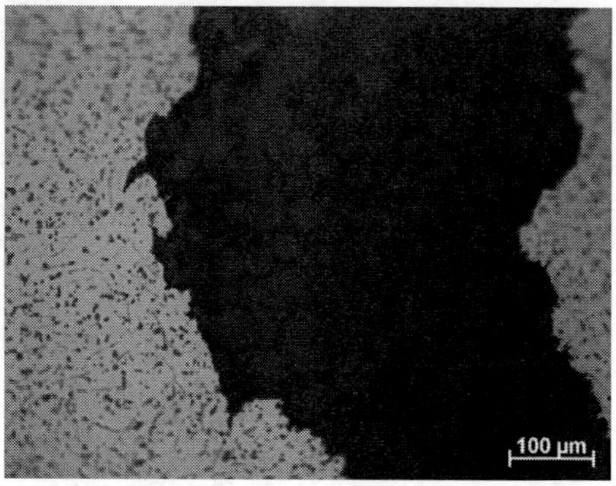

Fig. 17 Fracture surface (cross section) of the heat-treated specimen with 2.3% Fe.

4. Discussion

The previous results provide new insights into the understanding of the fatigue behavior of a recycled Al-Si-Cu alloy and the mechanism of the evolution of the fatigue damage. The 3D characterization of the fracture surfaces by micro-computed tomography and scanning electron microscopy shows that shrinkage pores and interacting gas pores are the main crack initiation sites of the alloy, what is in good agreement with other studies in this field [13]. Nevertheless, the analysis of non-machined specimens reveal that the casting skin has a significant influence in the fatigue behavior, since it insulates the material from the air environment and promotes internal crack initiation. The ß-Al_5FeSi play a moderate role in the evolution of damage due to the interaction of pores that drive the crack propagation. However, ß-Al_5FeSi inclusions located parallel to the crack path can break by splitting through the middle and causing some kinking in the crack path. The cracks propagate through these ß-Al_5FeSi particles until they reach the Al matrix, and stress relaxation occurs at the crack tip.

The strain localizations at conveniently oriented hard eutectic Si has a similar effect on the crack path. These findings imply that the strain localizations at ß-Al_5FeSi and eutectic Si particles could also drive the growth of other secondary microcracks initiated from shrinkage and gas pores and cause the coalescence at advanced stages of damage. Heat treatment did not affect the crack initiation mechanism since it is fully dominated by the interaction of shrinkage and gas pores, which remain after subjecting the specimens to heat treatment. However, the fragmentation and dissolution of the ß-Al_5FeSi particles (see Fig. 17) can alter the mechanism of the crack growth, particularly in the alloy with increased Fe content. This implies that there is a distinct effect of the Fe-content on the crack growth phase. However, this will not be seen in the lifetimes of fatigue specimens, as crack interaction of pore-initiated cracks competes with crack kinking caused by hard phases.

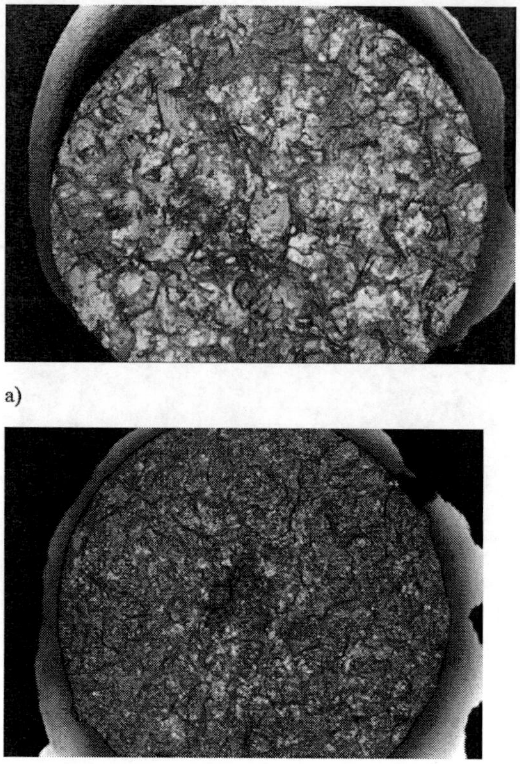

a)

b)

Fig. 17 a) Fracture surfaces of the tensile specimens with 1.85% Fe content before and b) after heat treatment.

5. Conclusions

The effect of the casting defects in a recycled Al-Si-Cu alloy on the fatigue damage evolution was analysed by micro-computed tomography and scanning electron microscopy. Based on the results, the following conclusions may be drawn:

1) A high dispersion in size and morphology of the defects was found due to the variation of the cooling rate during casting. Different defect cases led to various damage scenarios that were monitored and classified. The cast skin can insulated the crack propagation and enabled the crack initiation from the inner part of the material.

2) The interaction between shrinkage and gas pores located at the inter-dendritic spaces was the main cause of crack initiation. The hard particles such as eutectic Si particles and ß-Al_5FeSi plates did not play any role in the crack initiation process.

3) ß-Al_5FeSi plates and Cu-rich particles located parallel to the crack broke through the middle driving the crack propagation at some stages. However, this mechanism did not affect the lifetimes under fatigue loading.

4) The size of the ß-Al_5FeSi inclusions increased exponentially with the Fe-content, what caused a deep reduction of the ultimate strength and the ductility of the alloy.

5) Heat treatment partially fragmented and dissolved the ß-Al_5FeSi particles, although the effect is limited in the alloy with high Fe-content.

6. Acknowledgments

The authors would like to thank the Hessen State Ministry of Higher Education, Research and the Arts - Initiative for the Development of Scientific and Economic Excellence (LOEWE) - for financial support of the special research project "Safer Materials".

7. References

1. P.D. Lee, T.C. Lindley, C.W.J. Cheong, G.R. Davis, J. Elliot. Proceeding of COM, Vancouver, BC, 2003, p.457.

2. S.K. Das, J.A.S. Green. Aluminum industry and climate change assessment and responses. JOM (2010), 62 (2), 27-31.

3. EEA/OEA Recycling Division. Aluminum recycling in Europe. European Aluminum 2007, 1, 51.

4. N. Vanderess, E. Maire, A. Chabod, J-Y Buffière. Microtomographic study and finite element analysis of the porosity harmfulness in a cast aluminium alloy. Int J Fatigue (2011), 33, 1514–25.

5. I. Bacaicoa, M. Luetje, M. Wicke, A. Brueckner-Foit. Characterization of Casting Pores in Fe-rich Al-Si-Cu Alloys by Microtomography and Finite Element Analysis. Procedia Structural Integrity (2016) 2, 2269-2276.

6. L. Wang, N. Limodin, A. El Bartali, J.-F. Witz, R. Seghir, J.-Y. Buffiere, E. Charkuk. Influence of pores on crack initiation in monotonic tensile and cyclic loadings in lost foam casting A319 alloy by using 3D in-situ analysis. Mater. Sci. & Eng. A (2016), vol. 673, pp. 362-372.

7. I. Bacaicoa, M. Luetje, M. Wicke, A. Geisert, M. Fehlbier, A, Brueckner-Foit. Characterization of casting defects in Fe-rich Al-Si-Cu Alloys by Microtomography and Finite Element Analysis. Engng. Fract. Mech. (2017), 183, 159-169.

8. I. Bacaicoa, I., Luetje, M., Wicke, M., Geisert, A., Zeismann, F., Fehlbier, M., Brueckner-Foit, A. (2016). 3D Morphology of Al5FeSi inclusions in high Fe-content Al-Si-Cu Alloys. Procedia Structural Integrity, vol. 2. pp. 2269-2276.

9. M. Luetje, I. Bacaicoa, M. Wicke, A. Geisert, M. Fehlbier, A.Brueckner-Foit(2017). 3D characterization of fatigue damage mechanism in a cast aluminum alloy using X-ray tomography. International Journal of Fatigue, vol. 103, pp. 363-370.

10. I. Bacaicoa, P.K. Dwivedi, M. Luetje, F. Zesimann, A. Geisert, A. Brueckner-Foit, M. Fehlbier. Effect of non-equilibrium heat treatments on the microstructure and tensile prioperties of an Al-Si-Cu alloy. Materials Science and Engineering A (2016), vol. 673, 562-571.

11. P. Ashtari, H. Tezuka, T. Sato. Influence oft he Sr and Mn additions on intermetallic compound morphologies in Al-SI-Cu-Fe cast alloys. Mater. Trans. (2003), 44 (12), 2161-2174.

12. FEI Company. Avizo 3D Software. https://www.fei.com/software/amira-avizo (2016), 9.2.

13. Z. Li, N. Limodin, A. Tandjaoui, P. Quaegebeur, P. Osmond, D. Balloy. Influence of Sr, Fe and Mn content and casting process on the microstructures and mechanical properties of AlSi7Cu3 alloy .Mater. Sci. & Eng. A. (2017), vol. 24, pp. 286-297.

Changing Mechanisms of Surface Relief and the Damage Evaluation of Low Cycle Fatigued Austenitic Stainless Steel

Nao Fujimura[1,*], *Takashi* Nakamura[1], and *Kosuke* Takahashi[1]

[1]Hokkaido University, Division of Mechanical and Space Engineering, Faculty of Engineering, Kita 13, Nishi 8, Kita-ku, Sapporo, Hokkaido, Japan

Abstract. To quantitatively investigate the cause of the changes in arithmetic mean roughness R_a and arithmetic mean waviness W_a of austenitic stainless steel under low-cycle fatigue loading, precise observation focusing on persistent slip bands (PSBs) and crystal grain deformations was conducted on SUS316NG. During the fatigue tests, the specimen's surface topography was regularly measured using a laser microscope. The surface topographies were analysed by frequency analysis to separate the surface relief due to PSBs from that due to grain deformation. The height caused by PSBs and that by grain deformation were measured respectively. As a result, both of the heights rose with the increase of usage factor (*UF*). The amount of increase in the heights with respect to *UF* increased with strain range. The trend of development of both heights was similar with the trend of R_a and W_a. A comparison between R_a and the height caused by PSBs showed that these values strongly correlated with each other. A comparison between W_a and the height caused by grain deformation also showed that these values strongly correlated with each other. Consequently, the surface texture parameters R_a and W_a represent the changes in the heights of surface reliefs due to PSBs and grain deformation.

1 Introduction

When important industrial facilities are subjected to excessive cyclic loadings that lead to deformation, ensuring the structural health of the facilities requires precise damage evaluation from the view-point of material strength. In general, fatigue damage is often assessed using the linear cumulative damage law. However, it is necessary to focus on the physical damage process in order to evaluate the effects of excessive cyclic loadings on subsequent fatigue life more precisely.

The surface of metallic materials become rough during cyclic loading. If the changes in the surface topography can be related to the amount of physical damage, it may be possible to estimate the degree of fatigue damage from measuring surface topography. During cyclic loading, two mechanisms cause the surface roughening of metallic materials: (a) formation of persistent slip bands (PSBs) [1-5] and (b) deformation of crystal grains [6, 7]. In the PSBs, there are fine peaks and valleys, i.e., extrusions and intrusions caused by active slip systems. The waveforms of these fine reliefs caused by PSBs typically have a wavelength of around 1 μm. In contrast, the wavelengths of convex and concave structure due to grain deformation are several times larger than grain size [6, 7]. Thus, the wavelengths of surface relief due to grain deformation are much larger than those due to PSBs. Several studies have reported that frequency analysis (i.e., the wavelength difference) can be used to separate a PSBs-induced surface relief from a grain-deformation-induced one [8, 9].

In our previous study [10], we investigated the change in the surface topography under low-cycle fatigue loadings with constant strain range conditions. The surface relief due to PSBs were separated from those due to grain deformation using frequency analysis and the evolution of each relief was evaluated using two surface texture parameters: arithmetic mean roughness R_a and arithmetic mean waviness W_a. As a result, R_a and W_a increased with the increase of usage factor *UF*. The amount of increase rate in R_a and W_a with respect to *UF* changes with the strain range. Additionally, it suggested that the applied strain range and the degree of fatigue damage *UF* could be estimated by measuring R_a and W_a if the strain range condition is constant. However, it has not been quantitatively investigated whether the changes in R_a and W_a correspond to the evolution of two surface reliefs due to PSBs and grain deformation.

On the basis of the above background, this study conducted precise observation focusing on PSBs and grain deformations in SUS316NG to quantitatively investigate the cause of the changes in R_a and W_a under low-cycle fatigue loading. During the fatigue tests, the specimen's surface topography was regularly measured using a laser microscope. The surface topographies were analysed by frequency analysis to separate the surface relief due to PSBs from that due to crystal grain deformation. The height caused by PSBs and that by crystal grain deformation were measured respectively.

* Corresponding author: fujimura@eng.hokudai.ac.jp

To clarify the correspondence between the change mechanisms of surface relief and surface texture parameters, the heights of surface reliefs and surface texture parameters were compared and investigated.

2 Experimental procedures

2.1 Material and specimen

The material was a solution heat-treated SUS316NG (Nuclear Grade) austenitic stainless steel. The chemical components of the material were C: 0.01, Si: 0.40, Mn: 1.70, P: 0.013, S: 0.001, Ni: 12.09, Cr: 16.64, Mo: 2.48, N: 0.10 (mass%). Thermal treatment of the supplied material was undertaken at the temperature of 1050 degrees Celsius for 1 hour, followed by water quenching. The average grain size was 55 μm. The mechanical properties were: 0.2 proof stress and tensile strength were $\sigma_{0.2}$ = 252 MPa and σ_B = 580 MPa, respectively. An hourglass-shaped specimen with an R-part of 35 mm curvature radius and the minimum cross-section diameter of ϕ6 was used (Fig. 1). The specimen's surface was mirror finished by polishing with emery paper (grits from 120 to 2000) followed by buffing with diamond abrasives (particle size: 1 μm). To observe the deformation of each grain at the mirror-polished surface, a specimen etched in aqua regia was also prepared.

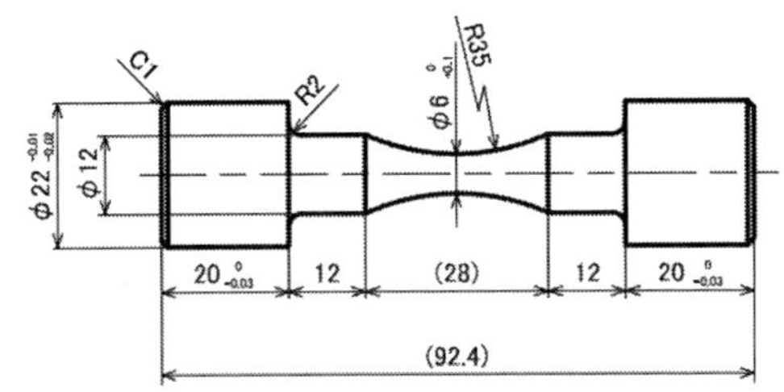

Fig. 1. Schematic of an hourglass-shaped specimen.

2.2 Fatigue testing method and conditions

Strain-controlled uniaxial push-pull fatigue tests were conducted using a servo hydraulic fatigue testing machine with a load capacity of 100 kN. Triangular loading with constant strain range was applied to specimens in ambient air at room temperature. Axial strain ε_{axis} was controlled by measuring the change in diameter at the specimen's minimum diameter part using an extensometer. The strain was calculated using the equation:

$$\varepsilon_{axis} = -2\ln(d/d_0) \quad (1)$$

where d_0 is the original diameter at the minimum diameter part, and d is the diameter during the fatigue tests. The constant strain range $\Delta\varepsilon$ were two conditions: $\Delta\varepsilon$ = 4% and 2%. The strain ratio and the strain rate were R_ε = -1 and 0.4 %/sec, respectively.

2.3 Measurement of surface topography

During the above-mentioned fatigue tests, the specimen's surface topography was regularly measured using a color 3D laser scanning microscope (VK-9700/9710 Generation II, KEYENCE). Cyclic loading was interrupted at arbitrary cycles, and then the specimen was removed from the testing machine for the surface topography measurement. The measurement interval was approximately 0.1 or 0.2 of the usage factor UF (=N/N_f), which represents the degree of fatigue damage as the consumption rate of life; N is the number of loading cycles, and N_f is the fatigue life. The fatigue life N_f at $\Delta\varepsilon$ = 4% and 2% were calculated using the following best-fit curve which were obtained from the fatigue test results in our previous study [11].

$$\Delta\varepsilon = 0.8771 \times N_f^{-0.5043} \quad (2)$$

3D surface images were taken at four to ten measurement points set on the circumference of the specimen's minimum diameter part. Two object lenses with different magnification were used for the observation of the surface reliefs due to PSBs and grain deformation: one's magnification was 150, and the other's was 20. The number of the high and low magnification images taken at a measurement point was six and one, respectively. The surface topography measurement conditions are summarized in Table 1. Different measurement ranges and resolutions were used for each surface relief because different objective wavelengths were used for the surface topographic images, as mentioned in the next subsection.

Table 1. Surface topography measurement conditions.

Measuring object of surface reliefs	PSBs	Grains
Object lens magnification	150	20
Axial measurement range [μm]	94	703
Circumference measurement range [μm]	70	528
XY resolution [μm/pixel]	0.092	0.688
Z resolution [μm/pixel]	0.02	0.02
Wavelength region of filtering [μm]	0.25 - 11.8	11.8 - 704

2.4 Image processing method

To separate the surface topography into the surface relief due to PSBs and that due to grain deformation respectively, the obtained 3D surface images were analysed by mean of two-dimensional fast Fourier transform (2DFFT) using the image analysis software (VK-Analyzer, KEYENCE).

In the analyses, low-pass and high-pass filters were applied to extract two wavelength range: one from λ_s = 0.25 μm to λ_c = 11.8 μm for the surface relief due to PSBs, one from λ_c = 11.8 μm to λ_f = 704 μm for that due to grain deformation (Table 1). The filters were brick-wall shaped. The λ_s and λ_f were the cut-off wavelengths to remove measurement noise and the specimen's shape error. These values, λ_s = 0.25 μm and λ_c = 11.8 μm,

https://doi.org/10.1051/matecconf/201816504007

were determined as they were the ones closest to the JIS-recommended value (λ_s = 0.25 μm in JIS B 0651: 2001; λ_f = 704 μm in JIS B 0610: 2001). The λ_c was the cut-off wavelength to separate the surface relief due to PSBs from that due to grain deformation. It was determined to be the shortest wavelength of the surface asperities caused by grain deformation. The shortest wavelength likely corresponds to the length of two adjacent small grains, which rotate opposite to each other. The smallest grain size was estimated to be about 6 μm from the grain size distribution of the supplied material, and 11.8 μm was chosen for λ_c as the value closest to twice the smallest grain size.

2.5 Height measurement of surface reliefs caused by PSBs and crystal grain deformation

The height of surface reliefs due to PSBs and that due to grain deformation were respectively measured as follows using the processed surface topographic images.

2.5.1 Height caused by PSBs

Fig. 2 shows the processed surface topographic image where PSBs and the surface relief formed on the mirror-polished specimen's surface were observed. Two measurement lines (light blue lines in (a)) were drawn perpendicular to PSBs (black lines in (a)), and then a profile curve at the intersecting point of the measurement line and PSBs (the red circle in (a)) was obtained as shown in (b). The height due to a PSB was defined as the vertical distance between the maximum peak height and the maximum valley depth in the profile curve of the PSB. Two height data were obtained per one PSB. Five to eight clearly visible PSBs were chosen in one surface topographic image, and the total number of the measured *h* was from ten to sixteen. Four surface topographic images, taken at different measurement points on the circumference, were used. The measurement interval was approximately 0.1 or 0.2 of *UF*. The *h* of the same surface relief on PSBs was measured when *UF* increased in the subsequent loading. The *h* at *UF* = 0 was 0 because PSBs had not appeared yet before cyclic loading.

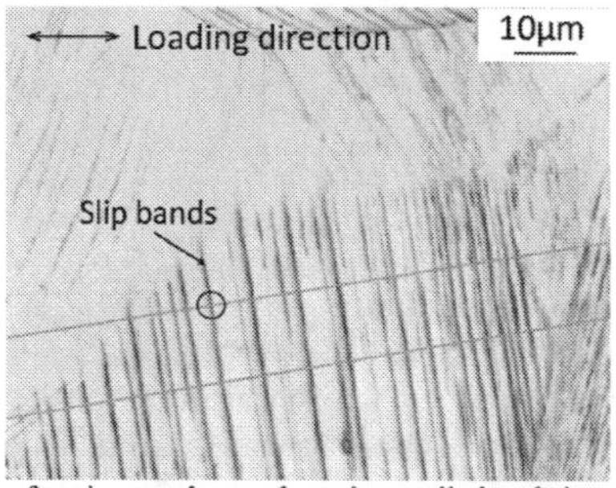

(a) Surface image observed persistent slip bands in grains.

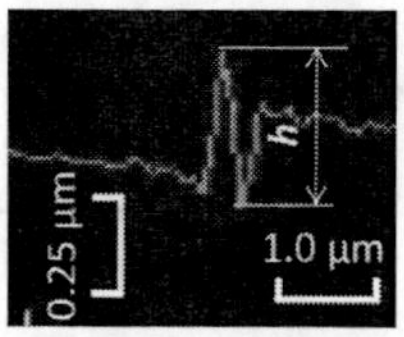

(b) Cross section of a PSB framed by the red circle in (a).

Fig. 2. Measuring method of the height caused by PSBs.

2.5.2 Height caused by crystal grain deformation

Fig. 3 shows the processed surface image where crystal grains deformation and the surface relief on the mirror-polished specimen's surface were observed. Five horizontal measurement lines (light blue lines in (a)) were drawn at an equal interval in the surface image. On the measurement lines, five measurement sections (red lines in (a)) were set at an equal interval: the section length was about twice the average grain size (110 μm). A profile curve at the measurement line in the section (the red dotted frame in (a)) was obtained as shown in (b). The height due to grain deformation was defined as the vertical distance between the maximum peak height and the maximum valley depth in the profile curve of the measurement section (the red line in (b)). Twenty five height data were obtained per one surface topographic image. Four surface topographic images, taken at different measurement points on the circumference, were used. The measurement interval was approximately 0.2 of *UF*. The *H* of the same surface relief on the section was measured when *UF* increased in the subsequent loading. The *H* at *UF* = 0 was 0 because grains had not been deformed yet before cyclic loading.

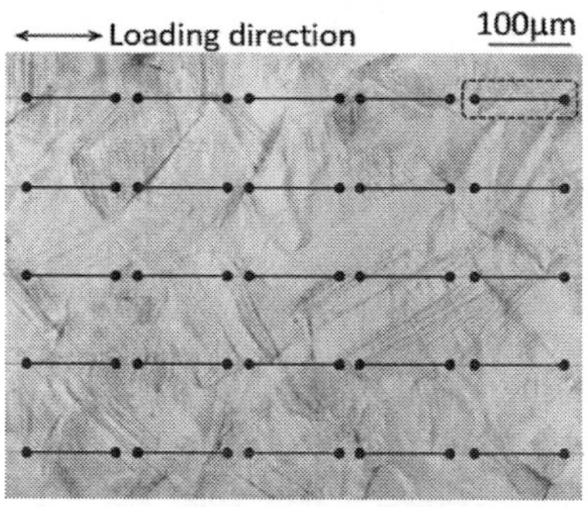

(a) Surface image observed crystal grain deformation.

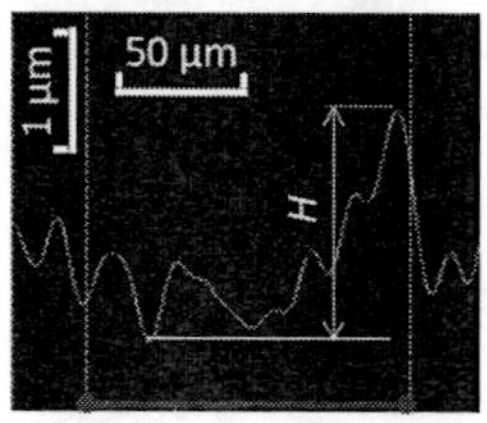

(b) Cross section of the area framed by a red dotted line in (a).

Fig. 3. Measuring method of the height caused by crystal grain deformation.

2.6 Calculation of surface texture parameters

Surface texture parameters R_a and W_a were measured to investigate the correlations between the parameters and the surface reliefs. Areal roughness and waviness parameters (arithmetic mean roughness R_a and waviness W_a) were determined from the processed surface topographic images, which were used for the height measurement of each surface relief. The area 5 μm inward from the edge of the measurement area was excluded for the calculations of R_a and W_a. If a crack observed in the measurement area, a region of a few μm surrounding a crack was also excluded. These processes were to prevent abnormal signals affecting the calculation of R_a and W_a. Fig. 4 shows the measurement results of surface texture parameters. As mentioned above, R_a and W_a increased with the increase of *UF*, and the amount of increase rate in R_a and W_a with respect to *UF* changes with the strain range.

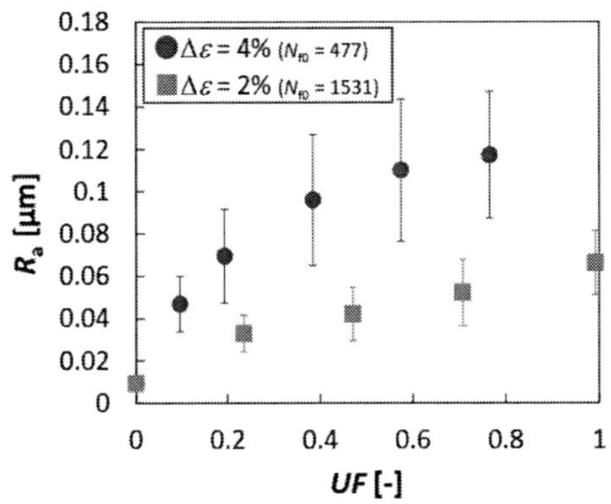

(a) Changes in R_a during cyclic loading.

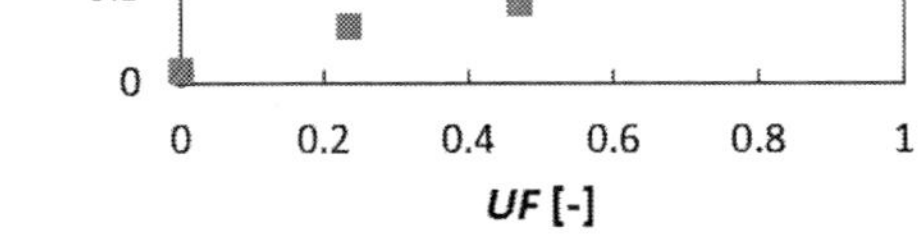

(b) Changes in W_a during cyclic loading.

Fig. 4. Measurement results of surface texture parameters.

3 Experimental results and discussion

3.1 Measurement result of *h* caused by PSBs

The measurement results of *h* are shown in Fig. 5. The vertical and horizontal axes represent *h* and *UF*, respectively. The circle and square marks indicate the average values of the measurement data at $\Delta\varepsilon = 4\%$ and 2%, and the error bars show the standard deviation. As shown in Fig. 5, the height *h* rose with the increase of *UF*. Focusing on the increase tendency of *h*, the rate of increase gradually decreases after $UF = 0.2$. The amount of increase in *h* with respect to *UF* increased with strain range: the *h* under $\Delta\varepsilon = 4\%$ was larger than that under $\Delta\varepsilon = 2\%$ at the same *UF*.

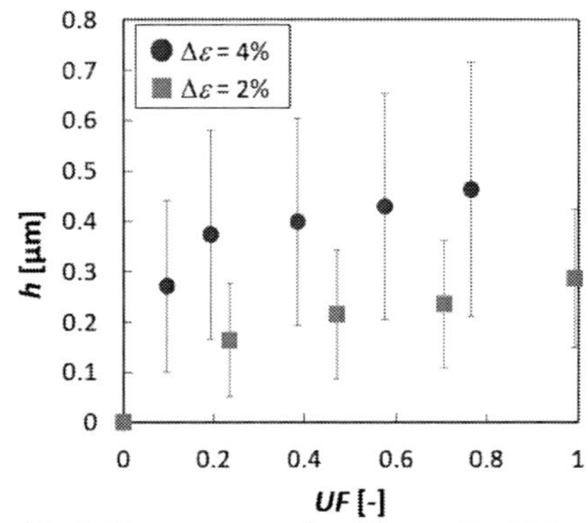

Fig. 5. Measurement results of *h* caused by PSBs.

3.2 Relationship between *h* and R_a

The trend of development of *h* (Fig. 5) was similar with the trend of R_a (Fig. 4(a)). To clarify the relationship between the height due to PSBs and the surface roughness parameter, *h* was compared with R_a. Fig. 6 shows the scatter diagram: the vertical and the horizontal axes represent R_a and *h*, respectively. Here, *h* was the average of the measured data per one surface topographic image, and R_a was determined from the image. The circle and square marks indicate the data at $\Delta\varepsilon = 4\%$ and 2%. The black dotted straight line is the regression line, passing through the origin, determined by applying the least-square method to all data on Fig. 6. As shown in Fig. 6, R_a increased with the increase of *h*. Focusing on the regression line, the coefficient of determination was $R^2 = 0.8629$, and the correlation coefficient was $R = 0.9289$. The values showed that the height *h* and surface roughness R_a strongly correlated with each other. Consequently, the surface roughness parameter R_a represents the change in the height of surface relief due to PSBs.

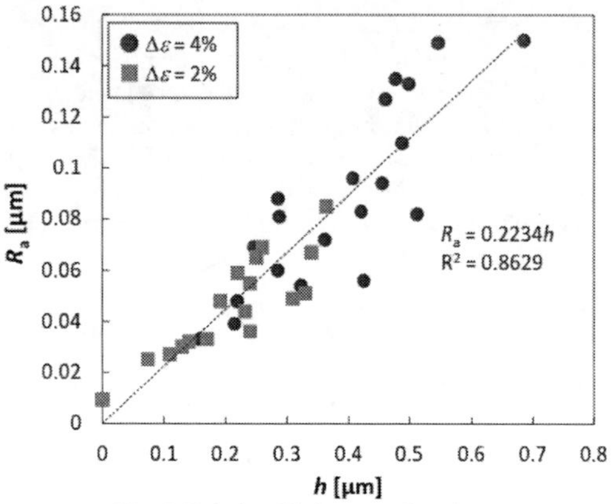

Fig. 6. Relationship between h and R_a.

3.3 Measurement results of *H* caused by crystal grain deformation

Figs. 7and 8 show the surface observation of the etched specimen before and after cyclic loading. Grains are visible on the etched surface before cyclic loading (Figs. 7(a) and 8(a)). Figs. 7(b), 7(c), 8(b) and 8(c) show the contour images of the surfaces fatigued under $\Delta\varepsilon$ = 4% and 2%: (b) at UF = 0.2 and (c) at UF = 0.6. Black lines traced the grain boundaries to define the correspondence between the position of surface relief and grains,. After applying cyclic loading until UF = 0.2, the convex parts (the red squares in Figs. 7(b) and 8(b)) and the concave parts (the blue dotted square in Figs. 7(b) and 8(b)) were observed on the surface. After more cyclic loading (Figs. 7(c) and 8(c)), these surface reliefs were developed: the convex part rose higher and the concave part sank deeper; however, their positions on the surface did not change. These convex and concave parts formed at near grain boundaries or within grains, and this observation showed that grain deformation caused the formation of these large surface reliefs.

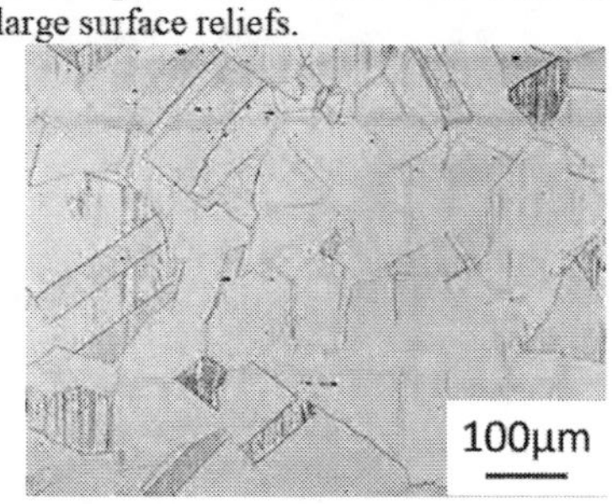

(a) Optical surface image at UF = 0.

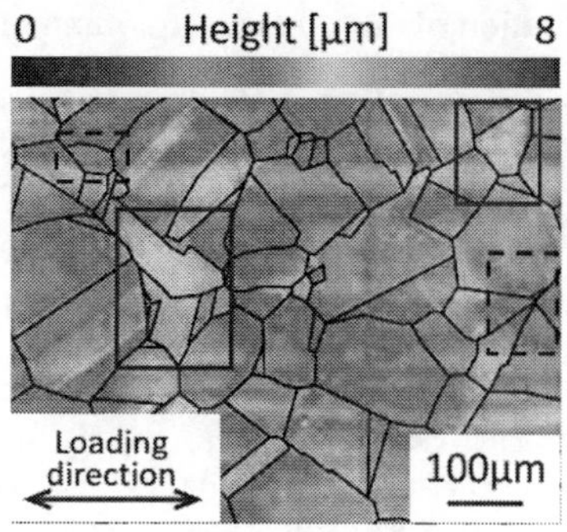

(b) Contour image of the surface fatigued at UF = 0.2.

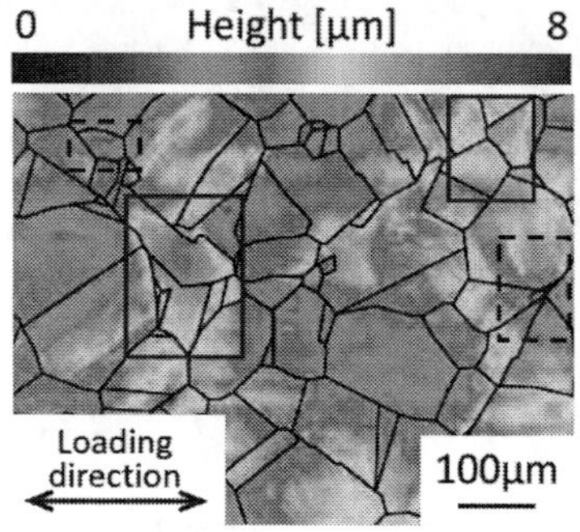

(c) Contour image of the surface fatigued at UF = 0.6.

Fig. 8. Surface observation of the specimen fatigued under $\Delta\varepsilon$ = 2%.

(a) Optical surface image at $UF = 0$.

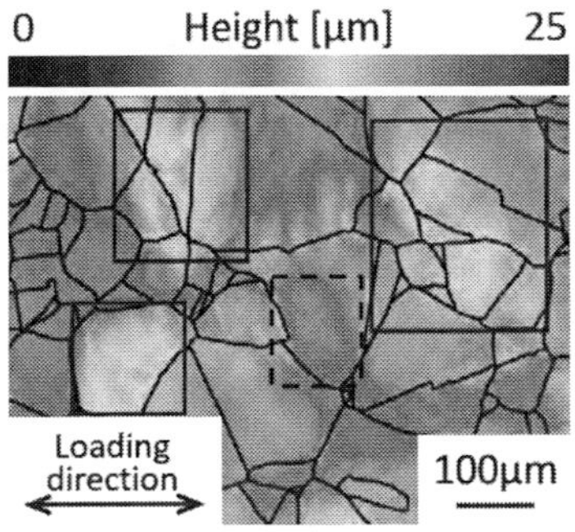

(b) Contour image of the surface fatigued at $UF = 0.2$.

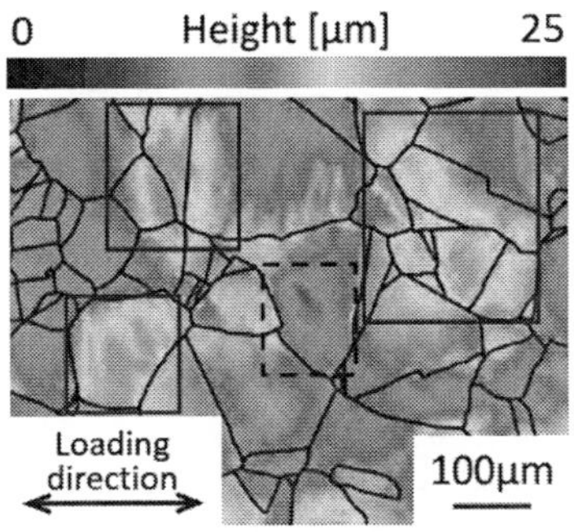

(c) Contour image of the surface fatigued at $UF = 0.6$.

Fig. 7. Surface observation of the specimen fatigued under $\Delta\varepsilon = 4\%$.

Fig. 9 shows the measurement results of the height H of the surface reliefs caused by grain deformation. The vertical and horizontal axes represent H and UF, respectively. The circle and square marks indicate the average values of the measurement data at $\Delta\varepsilon = 4\%$ and 2%, and the error bars show the standard deviation. As shown in Fig. 9, the height H rose with the increase of UF. The amount of increase in h with respect to UF increased with strain range.

3.4 Relationship between H and W_a

The trend of development of H (Fig. 9) was similar with the trend of W_a (Fig. 4(b)). To clarify the relationship between the height due to grain deformation and the surface waviness parameter, H was compared with W_a. Fig. 10 shows the scatter diagram: the vertical and horizontal axes represent W_a and H, respectively. Here, H

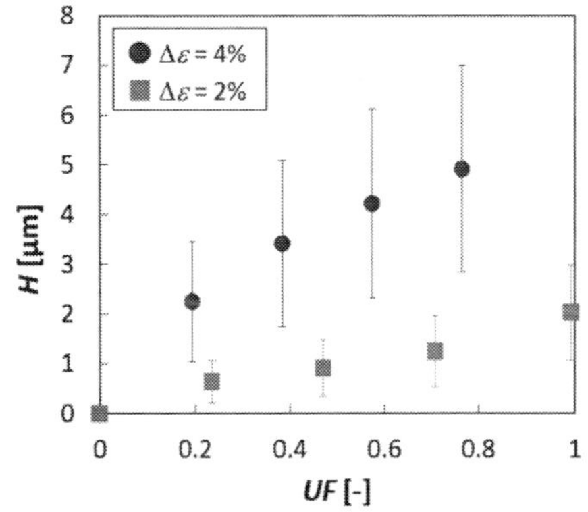

Fig. 9. Measurement results of H caused by grain deformation.

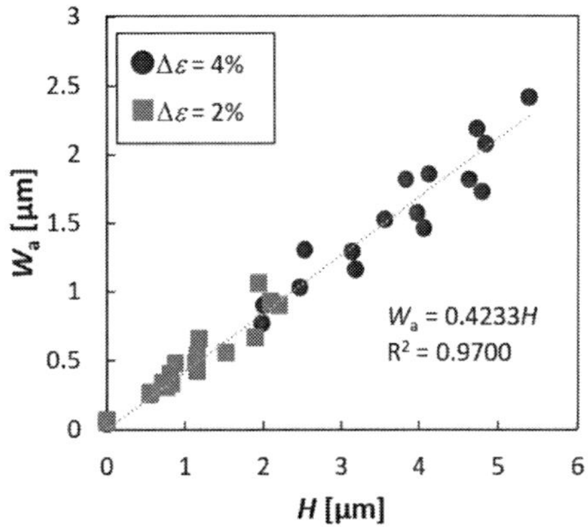

Fig. 10. Relationship between H and W_a.

was the average of the measured data per one surface topographic image, and W_a was determined from the image. The circle and square marks indicate the data at $\Delta\varepsilon = 4\%$ and 2%. The black dotted straight line is the regression line, passing through the origin, determined by applying the least-square method to all data on Fig. 10. As shown in Fig. 10, W_a increased with the increase of H. Focusing on the regression line, the coefficient of determination was $R^2 = 0.9700$, and the correlation coefficient was $R = 0.9849$. These values showed that the height H and surface waviness W_a strongly correlated with each other. Consequently, the surface waviness parameter W_a represents the change in the height of surface relief due to grain deformation.

4 Summary and conclusions

Precise surface observation focusing on the surface relief due to PSBs and crystal grain deformation were conducted on low-cycle fatigued austenitic stainless steel SUS316NG to quantitatively investigate the cause of the changes in surface texture parameters R_a and W_a. During fatigue tests, surface topography was regularly measured using a laser microscope. The height of surface reliefs caused by PSBs and grain deformation were measured, respectively. Surface topographies were analysed by frequency analysis to separate the surface relief due to PSBs from that due to grain deformation. The height caused by PSBs and that by grain deformation were measured respectively. The experiments led to the following conclusions:

1. Both heights of surface reliefs rose with the increase of *UF*. The amount of increase in the heights with respect to *UF* increased with strain range.
2. The trend of development of both heights was similar with the trend of R_a and W_a.
3. A comparison between R_a and the height caused by PSBs showed that these values strongly correlated with each other.
4. A comparison between W_a and the height caused by grain deformation also showed that these values strongly correlated with each other.
5. Surface texture parameters R_a and W_a represent the changes in the heights of surface reliefs due to PSBs and grain deformation.

This work was supported by JSPS Grant-in-Aid for Young Scientists (B) Number 17K14552.

References

1. A. Hunsche, P. Neumann, Acta Matallurgica, **34**, 2, 207-217 (1986).
2. P. Villechaise, L. Sabatier, J. C. Girard, Materials Science and Engineering, A**323**, 377-385 (2002).
3. J. Man, K. Obrtlík, J. Polák, Materials Science and Engineering. A**351**, 123-132 (2003).
4. J. Polák, J. Man, Materials Science and Engineering. A**596**, 15-24 (2014).
5. J. Polák, V. Mazánová, M. Heczko, R. Petráš, I. Kuběna, L. Casalena, J. Man, Engineering Fracture Mechanics, **185**, 46-60 (2017).
6. I. Shimizu, H. Terai, T. Nosho, T. Abe, Transactions of the Japan Society of Mechanical Engineerings. A64, 620, 1043-1051 (1998).
7. H. Ogawa, K. Hatanaka, Tansactions of the Japan Society of Mechanical Engineerings. A, **68**, 665, 57-64 (2002).
8. S. Nagaki, T. Abe, E. Ohta, K. Iguchi, Transactions of the Japan Society of Mechanical Engineerings. A, **57**, 536, 983-988 (1991).
9. Y. Kakimoto, S. Nagaki, S. Itagaki, T. Abe, Transactions of the Japan Society of Mechanical Engineerings. A, 59, 564, 1888-1894 (1993).
10. T. Shinko, T. Nakamura, N. Fujimura, Y. Nakata, Journal of the Society of Materials Science, Japan, **64**, 11, 910-917 (2015).
11. N. Fujimura, T. Nakamura, H. Oguma, Journal of Solid Mechanics and Materials Engineering, **7**, 3, 372-380 (2013).

Early damage and fatigue crack initiation at ambient and elevated temperatures in heat resistant austenitic steel

Jaroslav Polák[1,2*], *Veronika* Mazánová[1], *Roman* Petráš[1,2] and *Milan* Heczko[1]

[1]Institute of Physics of Materials, Žižkova 22, 61662 Brno, Czech Republic
[2]CEITEC IPM, Institute of Physics of Materials, Žižkova 22, 61662 Brno, Czech Republic

Abstract. Fatigue damage at ambient and at elevated temperature has been studied in advanced heat resistant austenitic steel. Cylindrical specimens were subjected to constant strain rate cycling and cycling with dwell in the tensile part of the cycle with different strain amplitudes. Surface relief evolution and initiation of surface and internal cracks were studied using high resolution SEM, FIB cutting, EBSD analysis, foil preparation and STEM imaging.

1 Introduction

Austenitic stainless steels are employed for construction of industrial equipment in a wide range of temperatures. Due to service loads and temperature variations they are subjected to fatigue damage. The nature of the fatigue damage depends substantially on the temperature. Recent studies at room temperature in 300 type steels [1-3] revealed that the cyclic strain localization is an important factor in the initiation and early growth of fatigue cracks. Cyclic plastic straining at room temperature is localized into persistent slip bands (PSBs) and repeated cyclic straining produces pronounced surface relief in the form of persistent slip markings (PSMs) at the locations where PSBs emerge on the surface [4]. PSMs consist of extrusions and intrusions. Preferentially intrusions represent crack-like defect from which fatigue cracks were proposed to initiate [4]. Alternative mechanism of crack initiation coming from the observations of the cracks in copper single and bicrystals [5,6] is based on the nucleation of the cracks on the grain boundary or on the PSB/matrix interface by direct condensation of vacancies.

Study of crack initiation in materials subjected to elastoplastic cyclic loading at high temperatures concentrated preferentially to the superalloys and austenitic steels [7-9]. The effect of oxidizing atmosphere and the influence of stress and strain relaxations resulting in the creep damage has been recognized. Nevertheless the role of individual damage mechanisms which contribute to the decrease of the fatigue life in high temperature environment has not been distinguished.

In this contribution experimental study of the typical mechanisms related to the initiation of fatigue cracks in advanced heat resistant austenitic stainless steel at ambient and at high temperatures will be presented.

2 Experimental

Highly-alloyed austenitic stainless steel Sanicro 25 which has good mechanical properties at elevated temperatures was studied. The material was supplied by the Sandvik, Sweden in the form of a cylindrical rod of 150 mm in diameter. The chemical composition of the material in wt.% was: 0.1 C, 22.5 Cr, 25.0 Ni, 3.6 W, 1.5 Co, 3.0 Cu, 0.5 Mn, 0.5 Nb, 0.23 N, 0.2 Si, balance Fe. Rough shape specimens were solution annealed at 1200 °C for one hour and cooled to room temperature in the air. The average grain size, found using the intercept method, was about 60 μm. Particles of Z-phase rich in Nb, Cr and N are found preferentially at grain boundaries but also in the interior of the grains. Some larger Z-phase particles are arranged in chains approximately parallel to the axis of the rod.

Cylindrical specimens for room temperature cyclic straining had 8 mm in diameter and 12 mm in gage length. Specimens cycled at high temperature were button end with the diameter of 6 mm and the gage length 15 mm. The axis of the specimen was parallel to the axis of the cylindrical rod. In the middle of the gage length of specimens cycled at room temperature a shallow notch of 0.4 mm in depth was ground. The area of the shallow notch was mechanically and electrolytically polished to facilitate the observation of crack initiation using optical microscopy (OM) and scanning electron microscopy (SEM). Also the whole gage length of cylindrical specimens subjected to high temperature cyclic loading was mechanically and electrolytically polished.

Specimens were subjected to symmetric tension-compression strain cycle at ambient and at elevated temperature in an electrohydraulic computer controlled MTS test system. The strain was controlled and

* Corresponding author: polak@ipm.cz

 https://doi.org/10.1051/matecconf/201816504008

measured using axial extensometer having the 8 mm gage length. Split resistance furnace with three zones was used to achieve high temperatures. At room temperature the constant strain rate was 5×10^{-3} s^{-1} and at temperature 700 °C the strain rate was 2×10^{-3} s^{-1}. Constant total strain amplitudes in the range of (2.5 – 7.0)$\times10^{-3}$ were applied. Stress and strain amplitudes were derived from the hysteresis loops during cycling. Plastic strain amplitude was determined as the half-width of the hysteresis loop at mean stress.

After cycling the specimen surface was inspected in detail using SEM-FEG Tescan Lyra3 XMU equipped with focused ion beam (FIB). FIB cuttings and SEM observations were used to reveal the shape of the extrusions, intrusions and starting cracks. In order to protect the shape of PSMs and cracks during FIB nanofabrication the surface of the specimen was in particular location first covered by a very thin sheet of platinum using electron deposition and later a thicker platinum layer was applied using ion deposition. FIB sectioning was performed perpendicular to the specimen surface and approximately perpendicular to the direction of PSMs. Final polishing of the cut was performed with low accelerating voltage of 5 kV.

Lamellae for TEM observations were prepared by fabricating craters from both sides of the future lamella. After extraction of the lamella from the specimen it was subsequently thinned repeatedly from both sides using low energy ions. Lamellae were then observed and characterized in STEM mode by JEOL JEM-2100F S/TEM equipped with FEG electron source and working at 200 kV.

For the observation of the internal damage of specimens cycled at high temperature the longitudinal cut along the whole gage length was performed. The cut was mechanically polished to obtain the perfectly smooth surface. Longitudinal cuts were observed in SEM and corresponding EBSD images of the selected areas were performed as well.

3 Results

3.1. Ambient temperature

Room temperature cyclic straining results in formation of distinct PSMs in individual grains of the material. Fig. 1 shows the surface of a typical grain in a specimen subjected to the constant strain amplitude 4×10^{-3} for 1850 cycles with the primary electron beam perpendicular to the surface. Low magnification picture shows large particles of Z-phase at grain boundaries and one larger grain with well-developed PSMs. One slip system is active producing several parallel PSMs. Some of them arise only in the centre of the grain and only a few of them reach one grain boundary. High magnification picture of the central part of the PSM shows wide PSM consisting of several parallel individual extrusions and intrusions. The exact 3-dimensional information about the shape of the PSMs is missing. The image documents the localization of the cyclic strain only into several PSBs producing surface relief in the form of PSMs while the rest of the grain (matrix) is deformed only elastically thus no relief on the surface corresponding to the matrix is produced.

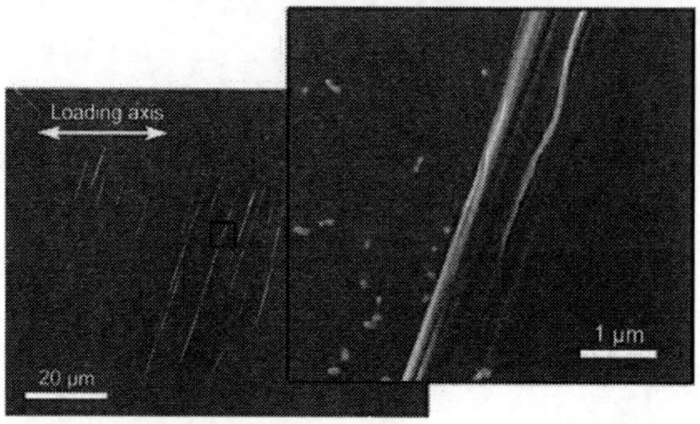

Fig. 1. Parallel PSMs in a grain after constant strain amplitude loading; $\varepsilon_a = 4\times10^{-3}$ for 1850 cycles.

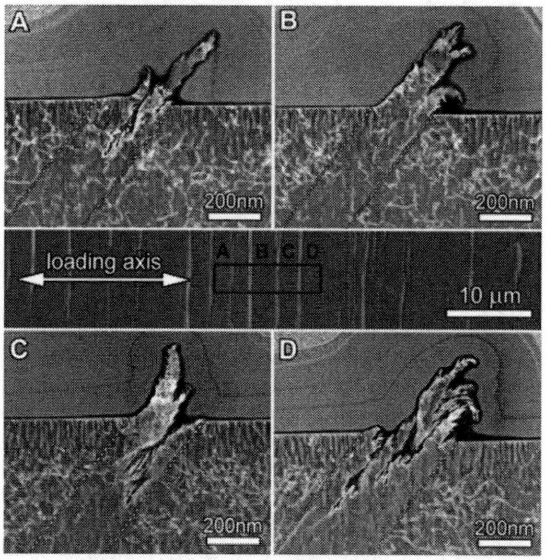

Fig. 2. Parallel PSMs in a grain of fatigued Sanicro25 steel and their profiles imaged using STEM of the foil prepared perpendicular to the surface. The surface was protected by electron (gray) and ion (not shown) deposition; dark field imaging.

3-dimensional information can be obtained using FIB sections perpendicular to the surface or fabricating thin lamella using FIB which can be later inspected in high resolution by TEM or STEM. Surface of a grain of the specimen cycled with constant strain amplitude 2.5×10^{-3} to fracture is shown in the central section of Fig. 2. Several parallel PSMs, perpendicular to the stress axis are shown at low magnification. Black rectangle denotes the position of the lamella which was produced perpendicular to the surface. The low angle annular dark field STEM images of the lamella showing four parallel PSMs denoted A, B, C, and D are shown above and below. The profiles of all PSMs and the internal dislocation structure in the nearest vicinity of PSMs are distinctly displayed. The directions of the prospective persistent slip bands (PSBs) in the material are indicated by dashed white lines. PSMs consist of extrusions and intrusions or sharp valleys between two neighbour extrusions. The direction of extrusion and intrusion growth corresponds to the trace of the primary slip plane and the trace of the primary Burgers vector. Thin extrusions are often bent and sometimes peculiar shapes are produced. Intrusions represent very sharp notches and due to high plastic strain amplitude in the PSBs fatigue cracks start from the tip of the intrusions. Starting cracks can be recognized from the PSMs A, C and D. PSM B consists only of two extrusions and no crack has been initiated. Closer examination reveals that from the sharp valley between two extrusions a crack has a tendency to arise.

Fig. 3. FIB section of two PSMs covered by platinum straps with a crack starting from one of them; inset shows the location of the cut.

The paths of secondary short cracks starting from the PSMs were followed using FIB cutting. Fig. 3 shows the FIB cut with two PSM profiles on the surface of a specimen cycled with strain amplitude 3.5×10^{-3} to fracture. The inset in the bottom right corner shows the position of the cut of two PSMs, each of them covered by the platinum layer. Both PSMs consist of extrusions and small intrusions. Only one intrusion/valley in the right PSM was deeper and from this intrusion/valley a fatigue crack developed and continued growing. The path of the fatigue crack initially followed primary slip plane but reaching approximately 10 μm the direction of its growth deviated from primary slip plane with a tendency to become perpendicular to the stress axis.

3.2. High temperature

Cyclic loading at the temperature of 700 °C leads to premature initiation of fatigue cracks. Fig. 4 shows typical surface of the specimens of Sanicro 25 steel after high temperature cyclic straining. Fig. 4a shows the specimen surface at 10% of fatigue life of a specimen subjected to constant strain rate cycling and Fig. 4b shows the surface of fractured specimen subjected to constant strain rate cycling with 10 min dwell in maximum tension in a cycle.

The surface of the specimen shown in Fig. 4a is oxidized with oxide more developed at grain boundaries in the form of oxide extrusions. Preferentially grain boundaries on the surface perpendicular to the stress axis are oxidized more markedly and the oxide extrusions are cracked (see the detail of cracked oxide extrusion in the black inset).

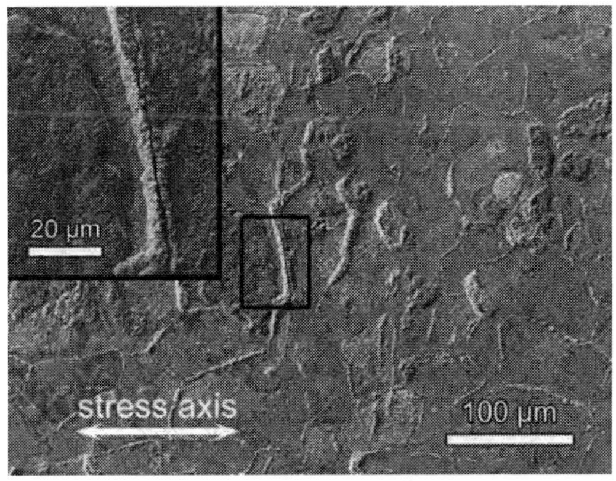

Fig. 4a. Surface of the specimen cyclically strained under constant strain amplitude loading at 700 °C, (a) constant strain rate, $\varepsilon_a = 3\times10^{-3}$, N= N$_f$/10.

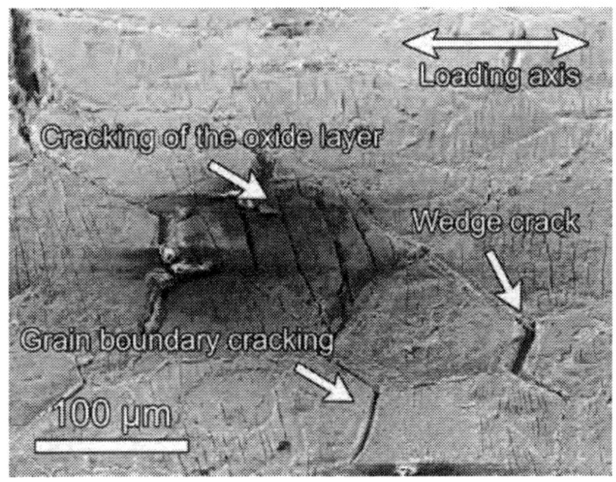

Fig. 4b. Surface of the specimen cyclically strained under constant strain amplitude loading at 700 °C, $\varepsilon_a = 4\times10^{-3}$, 10 min dwell in maximum tension, N= N$_f$.

Specimen subjected to cycling with 10 min dwell up to fracture has thicker layer of oxide than specimens cycled with constant strain rate. Very high density of cracks developed in the oxide layer both inside the grains and in the grain boundaries. Numerous cracks in the grain boundaries perpendicular to the stress axis became wide open and in some of them the sliding during dwell period lead to the formation of wedge cracks Fig. 4b shows high density of cracks in the thick oxide layer on the whole surface and also cracked grain boundary approximately perpendicular to the loading axis. Another grain boundary was subjected to sliding and wedge crack has formed.

The profile of the grain boundary cracks in cycling with constant strain rate and in cycling with 10 min dwells is shown using FIB cuts perpendicular to the

 https://doi.org/10.1051/matecconf/201816504008

surface in Figs 5a and 5b, respectively. In constant strain rate cycling (Fig. 5a) the oxide layer far from the grain boundary is thin but close to the grain boundary the oxidation is very intensive. The oxide intrusion goes deep in the grain boundary and due to higher volume of the oxide also characteristic oxide extrusion protrudes along the grain boundary. The thick oxide layer is cracked. The crack is shallow in the case of Fig. 5a.

Fig. 5a. FIB section of the surface cracks along the grain boundaries in the specimen cyclically strained under constant strain amplitude loading at 700 °C, constant strain rate, $\varepsilon_a = 3\times10^{-3}$, N= N_f/10.

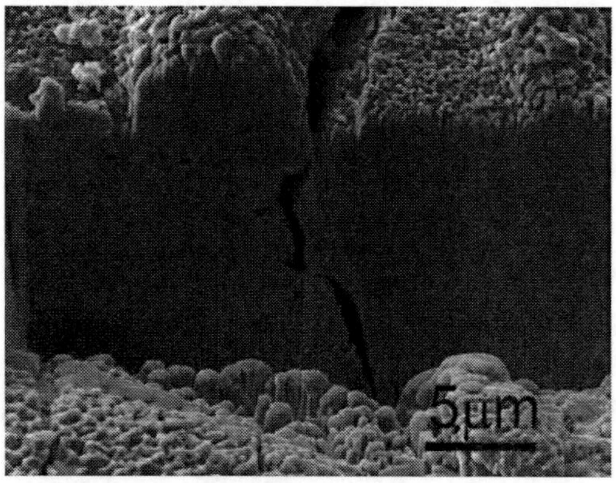

Fig. 5b. FIB section of the surface cracks along the grain boundaries in the specimen cyclically strained under constant strain amplitude loading at 700 °C with 10min dwell in maximum tension, $\varepsilon_a = 4\times10^{-3}$, N= N_f (secondary crack).

In cycling with dwells the thickness of the oxide layer on the surface of grains and even more the height of the oxide extrusion are substantially larger (Fig. 5b). The oxide intrusion along the grain boundary extends deep in the material. The oxide is cracked up to the tip of the oxide intrusion. Fig. 5 illustrates that the oxidation of the grain boundaries in cooperation with the cyclic loading which stimulates the cracking of the oxide intrusion contribute to the early initiation and growth of fatigue cracks.

In cyclic straining with dwells fatigue cracks initiate not only on the surface of the specimen but also in the interior. Fig. 6 shows the longitudinal cut of the specimen fatigued up to fracture with constant strain amplitude and 10 min dwell in the tensile maximum of the cycle. In addition to the principal macroscopic crack starting from the surface and a number of small surface cracks also several small cracks having the length of around 200 μm in the interior of the specimen has arisen. The interior cracks could be found in all locations on the surface of the longitudinal cut but higher density of the interior cracks was found close to the tip of the macroscopic crack. The origin of these cracks was studied by searching initial stages of their formation. Fig. 7 shows junction of three grains with number of small voids along the grain boundaries. This observation allows us to consider the mechanism of internal crack formation during elastoplastic cyclic loading with dwells. Cyclic loading produces high number of point defects, predominantly vacancies. In case of cyclic loading with dwells the dislocation density can substantially decrease during high temperature dwell. Vacancies have high probability to coalesce. Coalescence of vacancies along the grain boundaries and further condensation of the voids leads to the internal grain boundary cracks. The internal grain boundary cracks can be linked with the principal crack and contribute to the increase of the crack growth rate in cycling with dwells.

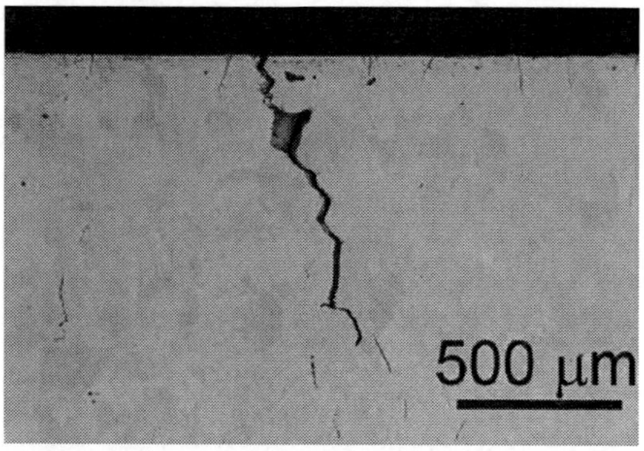

Fig. 6. Longitudinal cut of the specimen cyclically strained under constant strain amplitude at 700 °C with constant strain rate and 10min dwell in maximum tension, $\varepsilon_a = 4\times10^{-3}$, N= N_f.

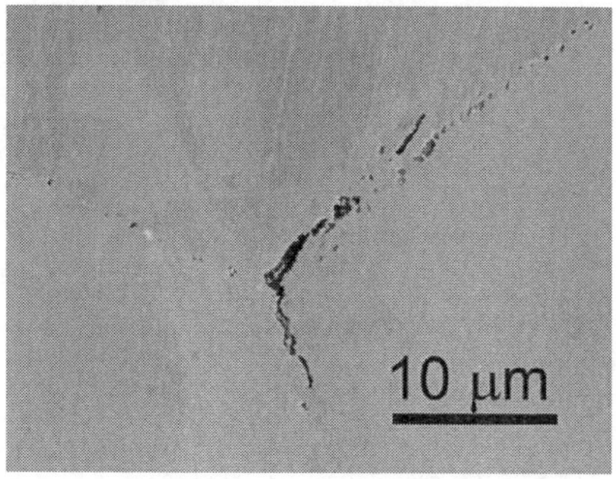

Fig. 7. Longitudinal cut of the specimen cyclically strained under constant strain amplitude loading at 700 °C with constant strain rate and 10min dwell in maximum tension, $\varepsilon_a = 4\times10^{-3}$, N= N_f.

Internal crack on the longitudinal cut of the specimen cycled with dwells in the tensile part of the cycle is shown in Fig. 8a. Corresponding EBSD image of the identical area displayed in Fig. 8a is shown in Fig. 8b. It is evident that internal crack follows grain boundaries. Moreover, it is not oxidized and therefore the hypothesis assuming that it originates due to production and coalescence of internal cavities on the grain boundaries is very probable.

(a) (b)

Fig. 8. Longitudinal section of the specimen subjected to constant strain rate and 10 min dwell in maximum tension showing internal crack, (a) SEM image, (b) EBSD image.

4 Discussion

Present results show clearly how the temperature influences fatigue crack initiation in high resistant austenitic steel. Two extreme cases were investigated i.e. room temperature and the highest projected temperature for this sort of austenitic stainless steel (700 °C). At room temperature in agreement with previous results [10-12] cyclic strain is highly localized. While the major volume of the grains undergoes only elastic straining the plastic strain amplitude in the bands of intensive cyclic slip (in PSBs) is high. As a result pronounced surface relief arises at locations where PSBs emerge on the surface of the grains and PSMs are produced. PSMs consist of extrusions and intrusions but observation of the surface using even high resolution SEM (see Fig. 1) or even AFM [1] cannot yield a true profile of the PSMs. This is due to very fine structure of the PSMs (see Fig. 2) and also due to the inclination of the primary slip plane to the surface. This inclination is in most cases close to 45 degrees. Therefore even very fine tips of the AFM cannot reproduce the true shape of the PSMs [1]. Either FIB cutting and SEM imaging or the production of the site-specific foil for TEM or STEM using FIB nanofabrication can yield the details of the extrusions and intrusions. In the center of Fig. 2 the rectangle indicates the location where the foil for transmission electron microscopy was produced using FIB machining. The foil was thinned and observed both in SEM and in STEM. The images in STEM give much more details and thus the profiles of all four PSMs are shown in Fig. 2. The images reveal complicated shapes of the PSMs in austenitic steel contrary to e.g. simple hill shapes in the grains of copper polycrystal [12]. PSM profiles are formed preferably by extrusions but parallel to extrusions (PSM C and D) or in between two extrusions (PSM A) crack-like intrusion appears. From sharp intrusions stage I cracks start and grow parallel to the primary slip plane. Fig. 3 shows the crack path of the stage I crack starting from one of two PSMs on the surface of the grain. Growing crack absorbed majority of applied plastic strain and as a result the other PSM did not produce an appreciable crack. The growing crack close to the surface followed primary slip plane and becoming longer its path has changed to approach a direction perpendicular to the stress axis.

Formation of the PSM, i.e. the growth of extrusions and formation of intrusions, have been recently treated quantitatively [4,13]. The driving force for extrusion and intrusion formation is the redistribution of matter between the PSB and the neighbor matrix. Continuous production of point defects during cyclic plastic straining in PSB and their systematic migration from the PSB to the matrix results in surplus of matter in PSB and deficiency of matter in a thin sheet close to the PSB/matrix interface. It provokes internal compression in PSB and internal tension in a thin sheet at the PSB/matrix interface. The relaxation of internal compressive stresses due to cyclic plastic straining leads to formation of extrusions in locations where PSBs emerge on the surface. At a later stage [13] relaxation of internal tensile stresses results in a parallel intrusion at the PSB/matrix interface. Originally this mechanism was derived for ladder-like dislocation structure of a PSB in copper. However, it can work in any PSB in which plastic strain amplitude is an order or two orders higher than plastic strain amplitude in the neighbor matrix.

Fatigue damage at the temperature of 700 °C is substantially affected by temperature. In cyclic loading at elevated temperature the oxidation of the surface is not homogeneous. In the interior of the grains a thin protective layer of the chromium oxide is formed that prevents further oxidation of the material. Preferential formation of the chromium oxides and nucleation of $M_{23}C_6$ chromium carbides at the grain boundaries [14] decreases the chromium content along the grain boundaries. As a result the grain boundaries are preferentially oxidized. An oxide extrusion and oxide intrusion arise at the grain boundaries. As soon as the oxide layer becomes thicker it is easily cracked in tensile part of the cycle. Preferentially the grain boundaries whose directions on the surface are approximately perpendicular to the loading axis become cracked. The cracking of the oxide extrusion and intrusion allows the access of oxygen to the crack tip. The alternating oxidation of the crack tip and cracking of the oxide layer is an effective mechanism of the early growth of initiated fatigue cracks in our austenitic steel. This mechanism is similar to that observed in high temperature cyclic straining of superalloys [15].

Introduction of dwells in tensile maximum of a cycle has pronounced effect on the formation of surface cracks but results also in the formation of internal cracks [16]. Initiation of internal cracks has been observed only in high temperature cycling with dwells [16]. Cyclic plastic straining produces point defects [4,13]. Due to high formation energy only a small fraction of produced point defects are interstitials. Since their migration energy is very low they disappear rapidly at edge dislocations. Vacancy type defects are produced in much higher concentration. They arise by annihilation of two opposite edge segments on neighbor slip planes or by non-conservative motion of the jogs on screw dislocations in the form of unite dipoles. The unite dipoles can disintegrate and vacancies, divacancies and trivacancies can migrate and interact. Fraction of vacancies is annihilated at edge dislocations. There is however high probability of an encounter of two vacancy type defects and formation of immobile nucleus of a void. This void can grow by absorbing other vacancies. Grain boundaries which are rich in Z-phase and $M_{23}C_6$ carbides are preferred sites for the formation of a number of small voids. Once the voids are present along the grain boundary they can be linked together and internal cracks are produced.

It was proven experimentally that the internal cracks start to be formed on the grain boundaries by formation of voids along the grain boundary (Fig. 7). The crack path of internal cracks following grain boundaries was corroborated by EBSD images of the area containing an internal crack (Fig. 8). In this regard it is interesting to note that some authors [5,6] propose a similar mechanism for initiation of fatigue cracks in room temperature cyclic straining without dwells. However the conditions in room temperature cyclic straining are less favorable for the accumulation of point defects since dislocation density is high and point defects created during cyclic plastic straining migrate preferentially to dislocations.

5 Conclusions

Constant amplitude cyclic straining of advanced austenitic stainless steel at room temperature and temperature 700 °C and subsequent analysis of the surface and internal fatigue damage led to the following conclusions

(i) High degree of cyclic lip localization is present in room temperature cyclic plastic straining.

(ii) Cyclic plastic strain is localized in PSBs and pronounced PSMs consisting of extrusions and intrusions are produced in locations where PSBs emerge on the surface.

(iii) Intrusions represent crack-like defects and true fatigue crack is initiated at its tip and continues to grow in stage I parallel to the primary slip plane.

(iv) High temperature cyclic straining results in preferential oxidation of the grain boundaries producing oxide extrusions and intrusions.

(v) Oxide extrusions and intrusions on the grain boundaries perpendicular to the stress axis on the surface are cracked and oxygen is admitted to the tip of the oxide intrusion. Repeated oxidation and cracking of the oxide leads to the early growth of the crack along the grain boundary.

(vi) Introduction of dwells in tensile maximum of the cycle results in grain boundary sliding and formation of wedge cracks and simultaneously to the formation of internal cracks in the whole volume of the cyclically deformed material. Internal cracks can link with the growing macroscopic crack and accelerate its growth.

The equipment used in the present project was provided by the Large Infrastructures for Research, Experimental Development and Innovation - project IPMinfra, LM2015069. The support by the Ministry of Education, Youth and Sports of the Czech Republic under the project CEITEC 2020 (LQ1601) is acknowledged.

References

1. J. Man, M. Valtr, M. Petrenec, J. Dluhoš, I. Kuběna, K. Obrtlík, J. Polák, Int. J. Fatigue **76**, 11 (2015)
2. L. Signor, P. Villechaise, T. Ghidossi, E. Lacoste, M. Gueguen, S. Courtin, Mater. Sci. Eng. A **649**, 239 (2016)
3. J.W. Pegues, M.D. Roach, N. Shamsaei, Mater. Sci. Eng. A **707**,657 (2017)
4. J. Polák, J. Man, Int. J. Fatigue **91**, 294 (2016)
5. L.L. Li, P. Zhang, Z.J. Zhang, H.F. Zhou, S.X. Qu, J.B. Yang, Z.F. Zhang, Acta Mater. **73**, 167 (2014)
6. H.W. Hoppel, P. Goik, C. Krechel, M. Goken, J Mater.Res. **32**(23), 4276, (2017)
7. S.D. Antolovich, E. Rosa, A. Pineau, Mater. Sci. Eng. **47**, 47 (1981)
8. J. Weiss and A. Pineau, Metall. Trans. A, **24**, 2247, (1993)
9. A. Pineau, S.D. Antolovich, Materials at High Temperatures **32**(3), 298 (2015)
10. J. Polák, R. Petráš, G.C. Chai, V. Škorík, Mater. Sci. Eng. A **658**, 221 (2016)
11. J. Polák, V. Mazánová, I. Kuběna, M. Heczko, J. Man, Metall Mater Trans A **47**, 1907 (2016)
12. J. Polák, V. Mazánová, M. Heczko, I. Kuběna, J. Man, Fatigue Fract. Eng. Mat. Str. **40**, 1101, (2017)
13. J. Polák, J. Man, Mater. Sci. Eng. A **596**, 15 (2014)
14. V. Mazánová, M. Heczko, J. Polák, Int. J. Fatigue (2018) (in print)
15. L. Viskari, H.M. Hörnqvist, K.L. Moore, Y. Cao, K. Stiller, Acta Mater **61**, 3630 (2013)
16. R. Petráš, J. Polák, Mater. Sci. Eng. A (2018) (in print)

Shear-mode Crack Initiation Behavior in the Martensitic and Bainitic Microstructures

Kentaro Wada[1,2,*], *Soma* Yoshimura[1], *Tomoko* Yamamoto[3], *Yoshihiro* Ohkomori[3] and *Hisao* Matsunaga[4,5,6]

[1]Guraduate School of Engineering, Kyushu University, 744 Motooka, Nishi-ku, Fukuoka 819-0395, Japan
[2]AIST-Kyushu University Hydrogen Materials Laboratory (HydroMate),744 Motooka, Nishi-ku, Fukuoka 819-0395, Japan
[3]Japan Casting & Forging Co., 46-59 Sakinohama Nakabaru Tobata-ku, Kitakyushu 804-8555, Japan
[4]Department of Mechanical Engineering, Kyushu University, 744 Motooka, Nishi-ku, Fukuoka 819-0395, Japan
[5]Research Center for Hydrogen Industrial Use and Storage (HYDROGENIUS), Kyushu University, 744 Motooka, Nishi-ku, Fukuoka 819-0395, Japan
[6]International Institute for Carbon-Neutral Energy Research (I2CNER), Kyushu University,744 Motooka, Nishi-ku, Fukuoka 819-0395, Japan

Abstract. Fully reversed torsional fatigue tests were conducted to elucidate the behaviour of shear-mode crack initiation and propagation in one martensitic and two bainitic steels. The relationship between the crack initiation site and microstructure was investigated by means of an electron backscatter diffraction (EBSD) technique. From the *S-N* diagram, two notable results were obtained: (i) the shear-mode crack was initiated on the prior austenitic grain boundary in martensitic steel, while in bainitic steels, the crack was initiated along the {110} plane; one of the slip planes of bcc metals, and (ii) the torsional fatigue limit of lower bainitic steel with finer grains was 60 MPa higher than that of upper bainitic steel with coarser grains even though the hardnesses were nearly equivalent. The mechanism determining the torsional fatigue strength in these steels is discussed from the viewpoint of microstructure morphology.

1 Introduction

Rolling contact fatigue (RCF) is one of the critical failure modes in the components with Hertzian contact surfaces such as ball bearings, back-up rolls and rail and wheel systems in trains. With the aim of improving the RCF strength, considerable efforts have been made especially in improving the cleanness of material [1, 2]. However, the improvement of cleanness does not result in the improvement of the strength in case that non-metallic inclusions no longer affect the crack initiation in RCF process [3, 4]. In other words, there is a threshold inclusion size at which the fracture origin changes from inclusions to microstructural features. Namely, to improve the RCF strength effectively, two factors should be considered: the threshold size of detrimental inclusion for each microstructure, and an alternative fracture origin when cleanness is below the threshold.

Several studies have focused on the transition point of the fracture origin from the defect to the matrix. For example, Murakami *et al.* [3] documented that the fatigue limit increases with a decrease in the size of defect (e.g. hole and inclusion). The defect smaller than a certain size is no longer detrimental to the strength, below which the reduction of defect size does not lead to the improvement of strength. Most of the existing studies have targeted the opening-mode (mode I) crack. By contrast, number of studies on shear-mode (modes II and III) crack is limited, though shear-mode crack growth plays a significant role in the RCF process [5-8]. Allegedly, such a transition of the crack initiation site also occurs in the shear-mode cracking; however, the phenomenon is still unclear. Therefore, the first purpose of this study is to elucidate the transition behaviour of the initiation site in the case of shear-mode cracking.

Elucidating potential initiation sites in the matrix – except the inclusions – are also important when the cleanness exceeds a certain point. For martensitic and bainitic steels, several researchers have pointed out in the case of uni-axial loading that crack initiation site is dependent on the material type and microstructure. For example, prior austenitic grain boundary in martensitic steels [9, 10], {110} slip planes or bainitic lath/block interface in bainitic steels [11] and martensite-austenite constituent [12] have been reported as crack initiation sites. However, it has not yet been clarified which microstructural feature dominates the shear-mode crack initiation in each case. To improve the RCF strength of materials, a better understanding of the role of microstructure on the shear-mode initiation is necessary; and is therefore the second purpose of this study.

In this study, in order to reveal the initiation behavior of shear-mode crack, torsional fatigue tests were conducted using a martensitic steel and the two types of bainitic steel with or without drill holes. In addition,

* Corresponding author: k.wada.007@s.kyushu-u.ac.jp

shear-mode crack initiation sites, except for the inclusion or defect, are discussed employing the EBSD technique.

2 Experimental

Three types of material were prepared. The first material is JIS-SKD61, called "back-up roll steel" (BR), whose chemical composition is 0.45C-5Cr-0.8Mo-0.3V. The second material is called "alloy steel for marine shaft" (MS), whose chemical composition is 0.4C-1Cr-0.5Mo. The BR was heat-treated at 1030 °C for 0.5 hours, then subsequently oil-quenched and tempered at 515 °C to obtain a martensitic microstructure. The MS was heat-treated to obtain two types of microstructure: upper bainite (MS-U) and lower bainite (MS-L). The microstructure and mechanical properties of each steel are shown in Fig. 1 and Table 1, respectively. In Figs. 1 (a), (c) and (e), crystal orientations are shown. In Figs. 1 (b), (d) and (f), red lines represent the high-angle grain boundaries on which the crystal misorientation is 15° or greater. The density of the high-angle grain boundary is higher in MS-L than in MS-U, meaning that MS-L has smaller grains. It is noted that MS-L occasionally includes coarse-grains as indicated by yellow arrows in Fig. 1 (f).

Table 2 Mechanical properties of martensitic steel JIS-SKD61 (BR), upper bainitic steels (MS-U) and lower bainitic steel (MS-L).

	BR	**MS-U**	**MS-L**
Vickers hardness, *HV*	560	329	334
0.2% proof stress, $\sigma_{0.2}$	1384 MPa	895 MPa	943 MPa
Tensile strength, σ_B	1823 MPa	1064 MPa	1059 MPa
Reduction in area, φ	33.6%	56.0%	66.0%

Cylindrical specimens were machined from

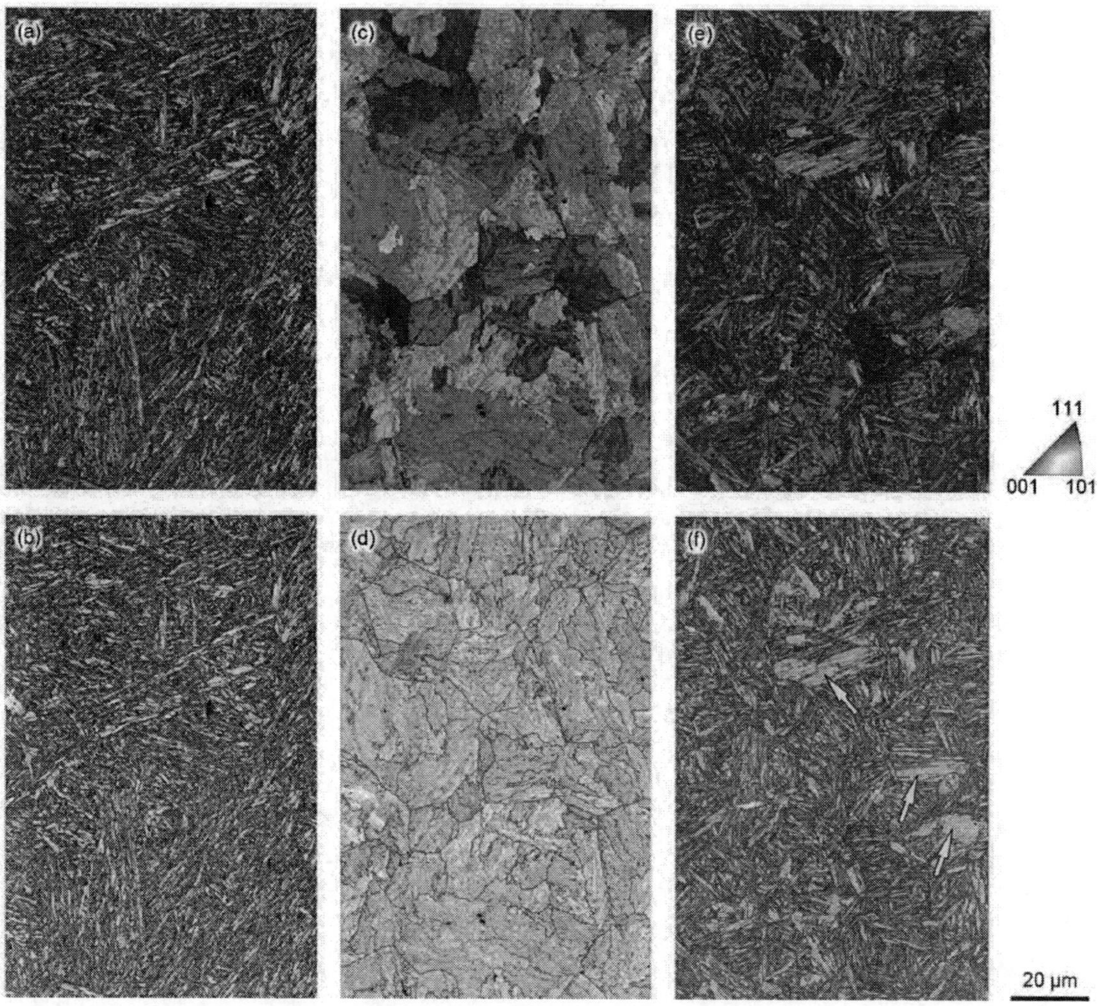

Fig. 1 Crystal orientation (a, c and e) and high-angle grain boundary traces (b, d and f) of martensitic steel SKD61 (a and b), upper bainitic steel (c and d) and lower bainitic steel (e and f). Yellow arrows indicate coarse grains.

these materials. The dimensions of the parallel part are 6.5 mm in diameter and 13 mm in length in BR specimens, and 5 mm in diameter and 21 mm in length in the MS specimens. The specimen surface was buffed with a diamond paste after emery polishing. In some BR specimens, a drill hole of 100 μm or 200 μm in diameter and depth was introduced onto the centre of parallel part.

Torsional fatigue tests were conducted at stress ratio, R, of −1 and test frequencies of 45 ~ 50 Hz. In the test of BR specimens, a static axial compression of 1000 MPa was superimposed on cyclic torsion to facilitate the crack growth in shear-mode direction by supressing the opening-mode crack branching. Fatigue tests were periodically halted to observe crack growth behavior by means of the plastic replica method.

The crystal orientation and misorientation angles of grain boundaries in each microstructure were analysed before and after the fatigue tests by using EBSD. Surface of all the samples was finished with buff polishing using colloidal silica. Step size for the EBSD analysis was 0.5 μm and the acceleration voltage was 15 kV.

3 Results and discussion

3.1. Results for back-up roll steel

In Fig. 2, *S-N* data of BR are presented. Black circles denote the results of smooth specimens, and red rectangle and green diamond denote those of holed specimens. It seems that both the holes of 100 μm and 200 μm led to no significant degradation of the fatigue life of BR. In the case of 100 μm hole, no crack initiated from the hole, whereas in the case of 200 μm, opening-mode cracks propagated from the hole in the directions oblique to the specimen axis, demonstrating that a defect smaller than 100 μm in size does not degrade the fatigue strength of BR.

Fig. 3 (a) shows a crack after the fatigue test at stress amplitude τ_a of 540 MPa. Considering the directions of crack branching, any point in the region surrounded by red broken line was assumed to be the crack initiation site. Fig. 3 (b) presents the crystal orientation around the crack initiation site. The microstructure around the crack can clearly be divided into two regions: the upper region, coupled with the colours blue, red and orange, and the lower region, coupled with the colours green, yellow and orange. The difference in colours represents the different prior austenite grains, meaning that the crack initiation site was the prior austenite grain boundary.

3.2. Results for alloy steel for marine shaft

Brue triangles and red reversed triangles in Fig. 2 show the *S-N* data of MS-U and MS-L, respectively. Two notable results were obtained. Firstly, the fatigue strength of MS-L had a large scatter. Secondly, the fatigue strength of MS-L was more than 60 MPa as high as that of MS-U. Fig. 4 shows the fracture surfaces of MS-L and MS-U, where no defect (e.g. inclusion) was found at crack initiation sites. The result inferred that the cracks were initiated in the matrix.

Fig. 5 (a) presents the crack growth behaviour in an MS-U specimen tested at τ_a = 280 MPa. Firstly, a fatigue crack was initiated and grew in shear-mode in the matrix, and subsequently branched into opening-mode. To investigate the initiation site of the shear-mode crack, crystal feature was investigated employing EBSD. Fig. 5 (b) shows the crystal orientation around the crack initiation site in MS-U, indicating that the shear-mode crack was initiated and propagated transgranulary. In Fig. 5 (c), red lines represent the high-angle grain boundaries (misorientation of 15° or greater), whereas blue lines represent the {110} planes. The result clearly exhibits that the shear-mode crack was initiated along {110} plane, which is one of the slip planes in bcc metals.

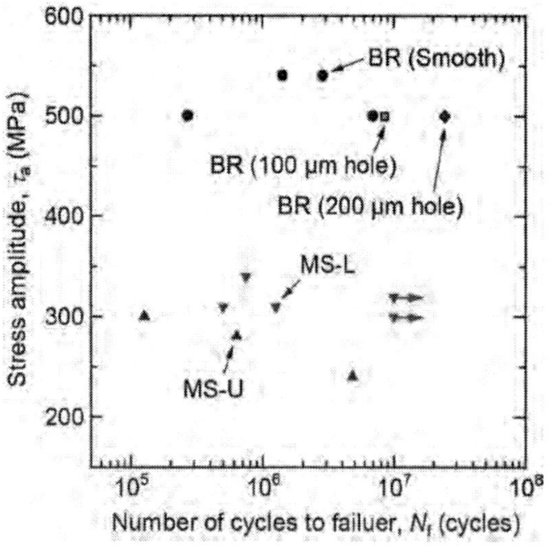

Fig. 2 *S-N* diagram of martensitic steel JIS-SKD61 (BR), upper bainitic steels (MS-U) and lower bainitic steel (MS-L).

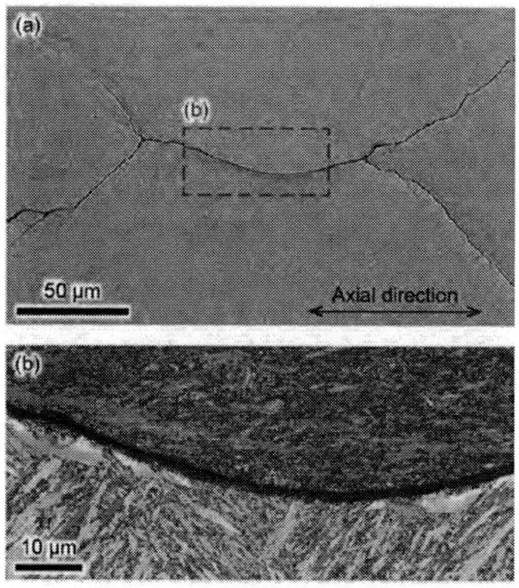

Fig. 3 Mopholorgy of shear-mode fatigue crack (a), and cristal orientaion mapping around the crack initiation site (b) on the JIS-SKD61 (BR) martensitic steel. The region surrounded by red broken line was assumed to be the crack initiation site.

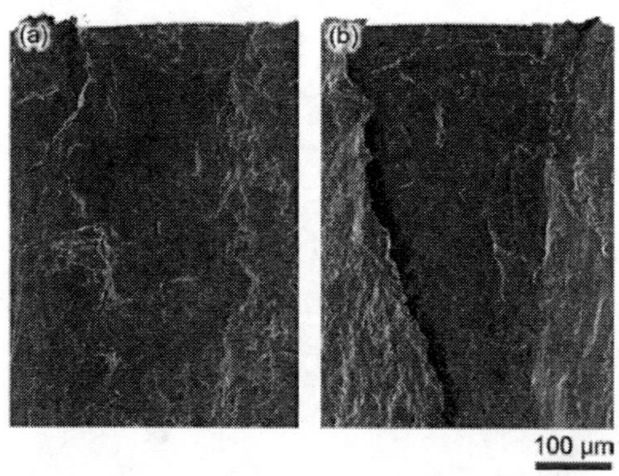

Fig. 4 Initiation site of shear-mode crack on fracture surface. MS-U tested at τ_a = 240 MPa (a), and MS-L tested at τ_a = 310 MPa.

In MS-L, non-propagating cracks were observed in unbroken specimens as demonstrated in Fig. 6 (a). Fig. 6 (b) presents the crystal orientation around the non-propagating crack, exhibiting that the shear-mode crack was transgranulary initiated also in MS-L. In Fig. 6 (c), the high-angle grain boundaries and {110} planes were presented, exhibiting that the shear-mode crack was initiated along the {110} plane, as was the case with MS-U. The microstructure around the crack was coarser than surroundings, and the crack was arrested in the region with finer grains.

3.3. Effect of microstructural morphology on shear-mode crack initiation

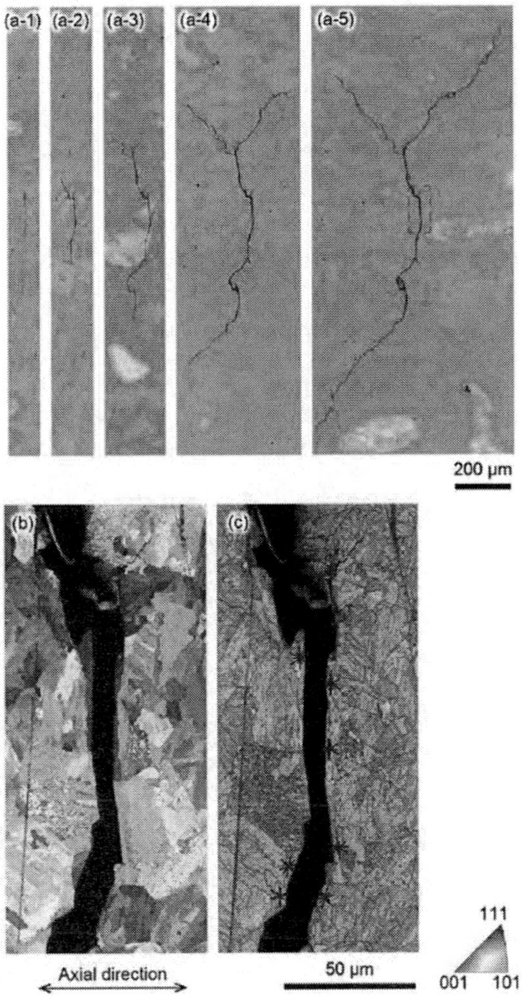

Fig. 5 Crack growth behavior on the MS-U upper bainitic steel tested at τ_a = 280 MPa (a-1~5), crack orientation mapping (b) and high-angle grain boundary traces (c) around the crack initiation site indicated by red broken line in (a-5). The applied cycles are 1.7×10^5 cycles in (a-1), 2.7×10^5 cycles in (a-2), 3.7×10^5 cycles in (a-3), 4.7×10^5 cycles in (a-4) and 5.2×10^5 cycles in (a-5).

As demonstrated above, the crack initiation site varied from intergranular to transgranular, i.e. a grain boundary in the martensitic steel and {110} slip planes in the bainitic steels. In the following, the reason for the difference is discussed.

Overall, the intergranular fracture occurs only in a limited number of cases. For example, it has been reported that intergranular fracture occurs by hydrogen

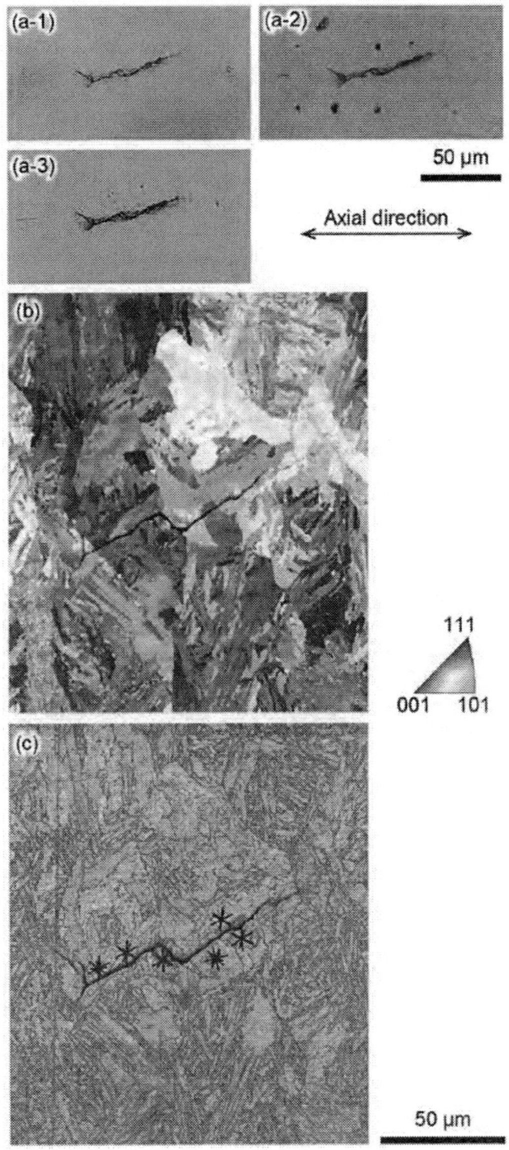

Fig. 6 Non-propagating crack observed in an MS-L specimen, tested at 320 MPa (a-1~3), crystal orientation mapping (b) and high-angle grain boundary traces (c) around the non-propagating crack. The applied cycles are 5.0×10^5 cycles in (a-1), 2.0×10^6 cycles in (a-2) and 1.0×10^7 cycles in (a-3).

embrittlement [13], tempering embrittlement [10, 14], intergranular segregation [15-17] and so on [18]. In addition, also in the case of cyclic shear loading under crack opening stress, intergranular cracking occurred [19, 20].

In BR having martensitic microstructure, it is assumed that an intergranular segregation occurred in the tempering process, weakening the grain boundary, in which the responsible elements in BR are P and S [16]. The existence of such an intergranular segregation should be investigated in the future to make ensure its role on the fracture process.

3.4. Mechanism determining torsional fatigue strength

It is well known that the fatigue limit of matrix obtained via uniaxial fatigue tests is well correlated with Vickers hardness [21]:

$$\sigma_w = 1.6HV \tag{1}$$

This correlation predicts that the fatigue limits of two types of MS steel with equivalent hardness ($HV \approx 330$) are the same under uniaxial loading. However, in fact, the torsional fatigue limit, τ_w, of MS-L was more than 60 MPa, as high as that of MS-U. The followings are cited as possible causes.

It has been reported that high-angle grain boundary plays a significant role as an obstacle for the crack growth along slip planes [11, 12]. As was shown in Fig. 1, MS-L had a higher density of the high-angle grain boundary than MS-U. This can contribute to the higher resistance against the shear-mode initiation and growth of microstructurally small crack.

In addition, the sizes of crack initiation units are different between MS-U and MS-L. As demonstrated above, shear-mode cracks were initiated along slip planes, meaning that one grain can act as the crack initiation unit. In MS-L, non-propagating cracks were observed at the fatigue limit. Namely, the fatigue limit of MS-L was determined by a crack growth threshold. In such a case, the grains, which are preferentially oriented to crack initiation, can act like an initial defect in the fatigue process. In MS-L, coarse-grained regions were occasionally observed (cf. yellow arrows in Fig. 1(f)), which were confirmed to become crack initiation sites as shown in Fig. 6. The result infers that such a large grain can preferentially initiate the shear-mode crack in MS-L. It is assumed that the scatter of the maximum size of the coarse-grain resulted in the scatter of the fatigue strength of MS-L (cf. Fig. 2), and the difference in the maximum grain size between MS-U and MS-L resulted in the difference in the fatigue limits, based on the weakest link hypothesis.

Conclusions

To elucidate the role of microstructure on shear-mode fatigue crack behaviour, torsional fatigue tests were conducted with three types of steels with a martensite, an upper bainite and a lower bainite. The following conclusions were obtained:

(1) In the martensitic steel JIS-SKD61, the hole of 100 μm in diameter and depth did not affect the fatigue process. Thus, it is assumed that the defect smaller than 100 μm in size has no effect on the fatigue strength.

(2) In the martensitic steel, shear-mode crack was initiated along the prior austenitic grain boundary, meaning that attention should be paid not only on increasing cleanness, but also on controlling the microstructure for improving the resistance of material against shear-mode cracking.

(3) In both the two types of bainitic steel, shear-mode cracks were initiated along {110} slip plane. Even in the lower bainitic steel with finer microstructure, the cracks were preferentially initiated from a coarse-grained region. The results revealed that controlling grain size is important to increase the torsional fatigue strength.

References

1 Y. Ohkomori, I. Kitagawa, K. Shinozuka, T. Toriyama, K. Matsuda, Y. Murakami, Tetsu-to-Hagane, **77**, 438 (1991)
2 Y. Ohkomori, C. Sakae, Y. Murakami, J. Soc. Mater. Sci., Jpn., **50**, 249 (2001)
3 Y. Murakami, M. Endo, Int. J. Fatigue, **16**, 163 (1994)
4 Y. Murakami, T. Endo, Int. J. Fatigue, **2**, 23 (1980)
5 A.V. Olver, Proc. Inst. Mech. Eng. P. J. Sport. Eng. Technol., **219**, 313 (2005)
6 P.E. Bold, M.W. Brown, R.J. Allen, Wear, **144**, 307 (1991)
7 H. Matsunaga, N. Shomura, S. Muramoto, M. Endo, Fatigue Fract. Eng. Mater. Struct., **34**, 72 (2011)
8 S. Beretta, M. Boniardi, M. Carboni, H. Desimone, Eng. Fail. Anal., **12**, 157 (2005)
9 M.-L. Zhu, F.-Z. Xuan, G.-Z. Wang, Mater. Sci. Eng. A, **515**, 85 (2009)
10 K.S. Ravichandran, E.S. Dwarakadasa, Metallurgical Transactions A, **21**, 3171 (1990)
11 M.C. Marinelli, I. Alvarez-Armas, U. Krupp, Mater. Sci. Eng. A, **684**, 254 (2017)
12 L. Lan, C. Qiu, D. Zhao, X. Gao, L. Du, Mater. Sci. Eng. A, **529**, 192 (2011)
13 C.J. McMahon, Eng. Fract. Mech., **68**, 773 (2001)
14 J.H. Hollomon, TRANSACTIONS OF THE AMERICAN SOCIETY FOR METALS, **36**, 473 (1946)
15 C.T. Liu, C.L. White, J.A. Horton, Acta Metall., **33**, 213 (1985)
16 J.R. Rice, J.-S. Wang, Mater. Sci. Eng. A, **107**, 23 (1989)
17 H. Erhart, H.J. Grabke, Met. Sci., **15**, 401 (1981)
18 C.L. Briant, S.K. Banerji, International Metals Reviews, **23**, 164 (1978)
19 S. Okazaki, K. Wada, H. Matsunaga, M. Endo, Eng. Fract. Mech., **174**, 127 (2017)
20. A. Otsuka, Y. Fujii, K. Maeda, Fatigue Fract. Eng. Mater. Struct., **27**, 203 (2004)

21. Y. Murakami, *Metal fatigue : effects of small defects and nonmetallic inclusions* (Elsevier, 2002)

Metastability and fatigue behavior of austenitic stainless steels

Marek Smaga[1*], *Annika Boemke, Tobias Daniel, Matthias W. Klein*[1]

[1]University of Kaiserslautern, Institute of Materials Science and Engineering, Kaiserslautern, Germany

Abstract This study presents the results of a detailed investigation of metastability and susceptibility to deformation induced α'-martensite formation of several austenitic steels (AISI 304, AISI 321, AISI 348 and two batches from AISI 347) in the solution-annealed state. Besides conventional characterization of metastability by calculating stacking-fault energy and threshold temperature (designated as M_S and M_{d30}), the present work introduced a new method for determining susceptibility to α'-martensite formation. The method was based on dynamically applied local plastic deformation and non-destructive micro-magnetic measurement of α'-martensite content. The parameter I_ξ was established, which correlated very well with the grade of α'-martensite formation during cyclic loading. The cyclic deformation and phase transformation behavior of cyclically loaded specimens from different metastable austenitic steels were investigated in total-strain and stress controlled fatigue tests with load ratio R = -1 at ambient temperature. The influence of the strain rate on the cyclic deformation and phase transformation behavior was also examined. During the fatigue tests, besides stress-strain hysteresis and temperature measurement, in situ micro-magnetic measurements were performed. Using the compressive measured data, the influence of plastic induced self-heating of the specimen and the strain rate on α'-martensite formation was analyzed.

1 Introduction

Fatigue is a process of mechanical failure resulting from the application of repeated cyclic loading with stresses mostly below stress limit determinates in monotonic tensile tests [1]. To enhance the monotonic and especially cyclic strength of metallic materials, microstructural modification is possible due to chemical alloying and heat treatment. Accordingly, crystallographically structured steels can be divided into two classes: (i) steels with a body-centered cubic (bcc) structure and (ii) steels with a face-centered cubic (fcc) structure [2]. The second class comprises austenitic steels, which were discovered to obtain a passivity (and, hence, "stainlessness"). Since the chromium contents of typical austenitic stainless steels exceed 16 wt%, their equilibrium microstructure at room temperature would be fully ferritic, if no other austenite-stabilizing alloying elements were added to the material. The elements most often used to obtain an austenitic microstructure are nickel, manganese, carbon and nitrogen [3]. However, changes in chemical composition not only influence the passivity of austenitic stainless steel but also significantly affect the metastability of the austenitic microstructure [4-7].

As is very well known, a huge number of austenitic stainless steels of the AISI 300 series exist in the metastable state [8-10]. These steels transform due to plastic deformation from paramagnetic austenite to a more stable microstructure, i.e. paramagnetic ε-martensite and/or ferromagnetic α'-martensite. The process of deformation induced phase transformation is presented in figure 1. Figure 1a shows a purely austenitic microstructure of AISI 304 in the solution-annealed state. Due to monotonic loading at ambient temperature up to ε = 8 % elongation, a deformation induced phase transformation from γ-austenite to α'-martensite occurred, which can be clearly observed in the micrograph presented in Fig. 1b. Further mechanical loading leads to an increase in the volume fraction of α'-martensite. The deformation-induced α'-martensite is clearly visible as dark needles embedded in the yellow-colored austenite grains (Fig. 1b and c). This clarifies that plastic deformation of metastable austenite at ambient temperature leads to significant changes in phase distribution from single-phase austenitic (Fig. 1a) to two-phase γ-austenite/α'-martensite microstructures (Fig. 1b and c).

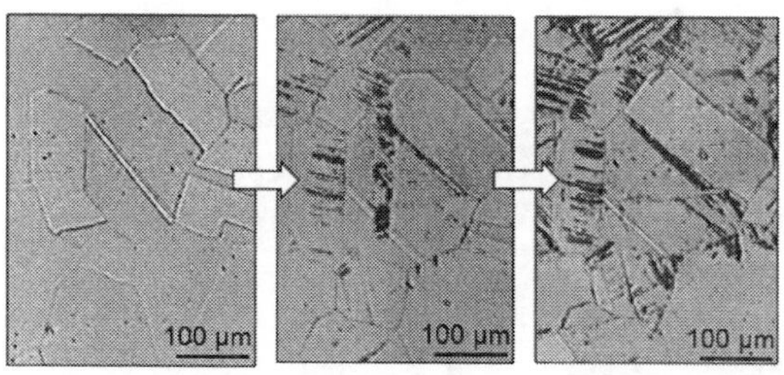

* Corresponding author: smaga@mv.uni-kl.de

Fig. 1. Deformation induced α′-martensite formation in metastable austenitic steel AISI 304 during tensile loading initial state (a), 8 % (b) and 12 % (c) elongation

Because, the fcc austenite has a lower strength compared to the bcc α'-martensite [11], it can be defined as the matrix and the α'-martensite as reinforcements of a "dynamically composite material" with a changing volume fraction and distribution of α'-martensite during mechanical monotonic and cyclic loading. Because α'-martensite formation leads to cyclic hardening, and this formation consequently influences significantly the fatigue life of metastable austenitic steels [11-22], it is important to have a reliable method for determining susceptibility to deformation induced α'-martensite formation. The susceptibility depends generally on two parameters: (i) the metastability of the initial austenitic microstructure, as a function of its chemical composition [4-7] and grain size [23], and (ii) the loading conditions such as load temperature [16, 19-20] and deformation rate [22].

In this study, the metastability of six austenitic stainless steels (AISI 304, AISI 321, AISI 348 and two batches from AISI 347) were characterized by calculating stacking-fault energy (SFE) and M_S and M_{d30} temperature. Further, a new method for determining susceptibility to deformation induced α'-martensite formation in metastable austenitic stainless steels was developed and applied to specimens based on dynamically applied local plastic deformation of metastable austenite and non-destructive micro-magnetic measurements. The influence of metastability and load conditions on deformation induced α'-martensite formation and cyclic deformation behavior were investigated and analyzed during total strain control fatigue tests.

2 Material and methods

2.1 Austenitic stainless steels

The chemical composition of investigated austenitic stainless steels is summarized in table 1. AISI 304 is a typical reference for so-called 18/10 stainless steels with 18 wt% Cr and 10 wt% Ni. AISI 321 is titanium-alloyed (% Ti ≥ 5 · % C) austenitic steel. Due to a higher affinity between carbon and titanium, compared to chromium, titanium carbides develop in the microstructure and arrest the carbon content. Hence, chromium carbide formation, which can result in chromium depletion at the grain boundaries, is depressed. Consequently, there can be an increase in intergranular corrosion resistance in AISI 321. A similar concept is used for niobium-alloyed (% Nb ≥ 10·% C) austenitic steels. In this case niobium carbides develop and arrest the carbon content, which is no longer in solution in austenite, for the formation of chromium carbides [3]. AISI 348 and 347 are niobium-alloyed steels and practically have the same chemical composition (nominally the difference is in the Nb content). From a one hand it can be seen, that three investigated type of austenitic stainless steels (AISI 304, 321 and 348/347) differed generally in corrosion resistance, but for the another hand chemical composition significantly influenced the metastability of the austenite and its susceptibility to forming α'-martensite even for the same type of austenitic steel (see section 3.1).

AISI	304	321	348	347a	347b
C	0.040	0.035	0.021	0.040	0.024
N	0.008	0.008	0.009	0.007	0.019
Ti	0.01	0.30	0.01	0.02	0.02
Nb	0.02	0.03	0.36	0.62	0.41
Cr	18.3	17.6	17.4	17.6	17.3
Ni	8.2	9.3	9.3	10.6	9.3
Mn	1.40	1.45	1.47	1.83	1.55
Mo	0.18	0.23	0.34	0.29	0.19
Si	0.36	0.56	0.54	0.41	0.63

Table 1. Chemical composition in weight %

Table 2 summarizes the monotonic mechanical properties, hardness, as well as volume fraction of α'-martensite after specimen failure for all the investigated austenitic stainless steels. The specimens were examined after solution-annealed heat treatment, taken from bars with a diameter of 20 mm (AISI 304, 321, 348) [16] and 25 mm (AISI 3457b) [11]. The following heat treatment parameters were used: 1050°C for 35 min followed by quenching in helium atmosphere. The AISI 347a specimen was taken from an original surge pipeline component with an outer diameter of 333 mm and wall thickness of 36 mm [17] and solution annealed at 1050°C / 10 min / $H_2$0.

Table 2. Mechanical properties, hardness and α'-martensite after specimen failure (ξ)

AISI	304	321	348	347a	347b
$R_{p0.2}$ in MPa	241	209	243	242	220
UTS in MPa	677	641	640	569	621
A in %	84	65	62	66	51
HV10	208	176	177	140	152
ξ in FE-%	9.4	23.4	25.3	4.4	33.2

2.2 Method for determining susceptibility to the formation of deformation induced α'-martensite

Susceptibility to the formation of deformation induced α'-martensite in austenitic stainless steels at constant temperature, e.g. ambient temperature (AT), depends on both chemical composition, therefore the stacking-fault energy (SFE), and the local microstructure state, such as grain size or dislocation density. The influence of chemical composition on austenite metastability has been intensively investigated in the literature [4-7] and several empirical equations are given to calculate SFE and M_S and M_{d30} temperature (see section 3.1). For calculations of the aforementioned metastability parameters, a homogenous distribution of chemical composition in the microstructure was assumed. But even after solution-annealing heat treatment, a local heterogeneity of chemical composition exists in the microstructure, which strongly influences the local formation of α'-martensite [7, 9-10]. Additionally, the same chemical composition can be obtained for austenitic stainless steels with higher dislocation density, e.g. after cold rolling or for austenitic microstructures with a comparatively smaller dislocation density e.g. after warm rolling. The ratio of cold hardening, before α'-martensite formation, is not included in the aforementioned parameters (M_{d30}, M_S, SFE) but significantly influences the stress-strain behavior and α'-martensite formation during monotonic and especially cyclic loading of metastable austenitic steels. Therefore, in order to determine the susceptibility of the formation of deformation induced α'-martensite in austenitic stainless steels an investigative method is required which includes both the influence of the chemical composition *and* the stress-strain state of the material. One possibility is to use a Poldi hardness tester (Fig. 2a) to make a local plastic deformation, and to perform micro-magnetic measurements by e.g. Feritscope™ sensor (Fig. 2c).

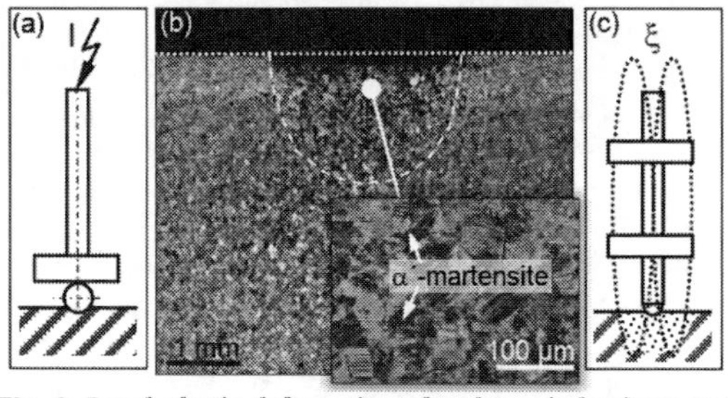

Fig. 2. Local plastic deformation after dynamic hardness test showed schematically in (a), optical micrograph with α'-martensite in metastable matrix after dynamic loading of AISI 348 (b) and Feritscope™ measurements principle (c)

The Poldi hardness tester was used to make a local plastic deformation in a flat specimen. Independently of the chemical based metastability of austenite (SFE) and the local microstructural state (grain size, dislocation density, etc.), the metastable austenite transformed more or less into stable α'-martensite by applied dynamic load at constant temperature e.g. ambient temperature. For the Poldi hardness test a special Brinell ball of 10 mm diameter was used resulting in relatively higher impact forces in comparison to static hardness tests. The plastic deformation occurred in the range of several millimeters (Fig. 2b). This range of plastic deformation is sufficiently high enough to average out the local inhomogeneity in chemical composition as well as in grain size, which occurred within the deformation zone. Hence the real local microstructural state (work-hardened or solution annealed) can be tested. After impact loading, micro magnetic measurements using a Feritscope™ sensor were performed. Because the α'-martensite is ferromagnetic, the volume fraction of α'-martensite could be nondestructively measured, which directly correlated with the susceptibility of deformation induced α'-martensite formation. The results from these tests are shown in section 3.1.

2.3 Fatigue setup

For the cycle-dependent recording of mechanical hysteresis loops, an extensometer (A) was fixed to the gauge length. The specimens were clamped in

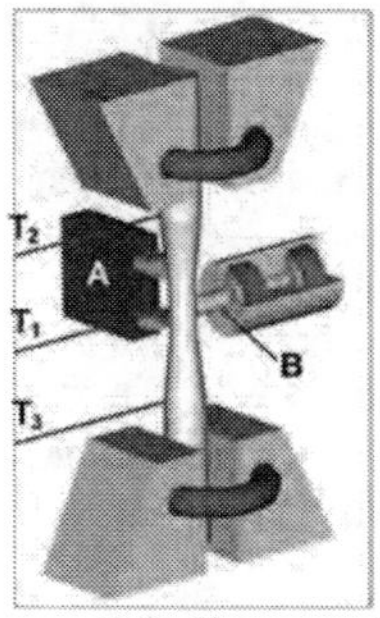

$$\Delta T = T_1 - 0.5 \cdot (T_2 + T_3)$$

A: extensometer
B: Feritscope™
T_1-T_3: thermocouples

Fig. 3. Experimental setup for fatigue tests using measurement methods

thermostatically-controlled wedge grips.

Three thermocouples T_1-T_3 were used for highly-accurate temperature measurements. The temperature changes ΔT were only caused by plastic deformation in the gauge length. The volume fraction of α'-martensite was detected in situ with a Feritscope™ sensor (B). During testing, the probe was attached to the specimen and energized by a low-frequency magnetic field in order to form a magnetic circuit. This field induced a voltage in the testing probe coil as a direct measurement of the permeability. The Feritscope™ magnetic fraction (ξ) is given in volume percent ferrite (FE%), without converting into α'-martensite content. In the literature, a linear correlation between FE% and α'-martensite content is reported [24]. To investigate the cyclic deformation behavior and the formation of deformation induced α'-martensite, specimens with a gauge length of 15 mm and diameter of 7.6 mm were used.

3 Results

3.1 Metastability of austenitic stainless steels and susceptibility to the formation of deformation induced α'-martensite

Metastabilty of metastable austenitic steels is mainly influenced by the chemical composition [4-7]. Hence, several empirically determined equations for calculating stacking-fault energy (SFE) and M_S and M_{d30} temperature thresholds have been developed to estimate the metastabilty of austenite. M_S temperature is the temperature at which spontaneous thermally induced austenite-martensite transformation without plastic deformation occurs [6]. The so-called M_{d30} temperature is used for plasticity-induced martensite formation. At M_{d30} temperature, by definition, 50% α'-martensite is formed as a result of 30% deformation [4, 5] Calculating the values of the SFE, M_S and M_{d30} temperatures, according to the following equations, provides table 3.

$M_{d30,Angel}$ in °C = 413 - 462· (C+N) - 13,7·Cr - 9,5·Ni - 8,1·Mn - 18,5·Mo - 9,2·Si [5]

$M_{S,Eichelmann}$ in °C = 1350 - 1665· (C+N) - 42·Cr - 61·Ni - 33·Mn - 28·Si [6]

SFE in mJ/m^2 = 25.7 + 2·Ni - 0.9·Cr - 1.2·Mn + 410·C - 77·N - 13·Si [7]

Table 3. SFE, M_S and M_{d30} temperature

AISI	304	321	348	347a	347b
M_{d30} in °C	12	43	48	25	46
M_S in °C	-171	-91	-66	-189	-81
SFE in mJ/m²	35	33	28	39	27

The M_{d30} temperature of the metastable austenitic steels AISI 304, 321, 348, 347a and 347b is shown in Fig. 4. Due to minimum M_{d30} temperature, AISI 304 was found to be more stable compared to titanium-alloyed AISI 321 and niobium-alloyed AISI 348 steels. The two batches of AISI 347 displayed a significant difference in metastability. AISI 347a had a small metastability, in the range of AISI 304, while AISI 347b had a higher one, in the range of AISI 348. Dependent of which parameter (SFE, M_S or M_{d30} temperatures) was used to determine the metastability of the specimens the differences between the types of austenitic stainless steels could be, more or less observed. It is important to note that the resulting temperatures were not absolute values. Their use was of a comparative kind, e.g. to compare steels or batches from different types of austenitic steels in terms of their susceptibility to austenite-martensite transformation. Hence, the method for determining susceptibility to the formation of deformation induced α'-martensite outlined in section 2.2 was used. Figure 4 summarizes the results from impact tests with subsequent Feritscope™ measurement. For each type/batch of austenitic stainless steels six different measurement points in the longitudinal section of the specimen were used. All sections before the impact loading showed ξ = 0.0 FE% due to mechanical and electrical polishing of the surface. The mean value of the α'-martensite measurement after impact loading was called I_ξ and is plotted together with the standard deviation in Fig. 4.

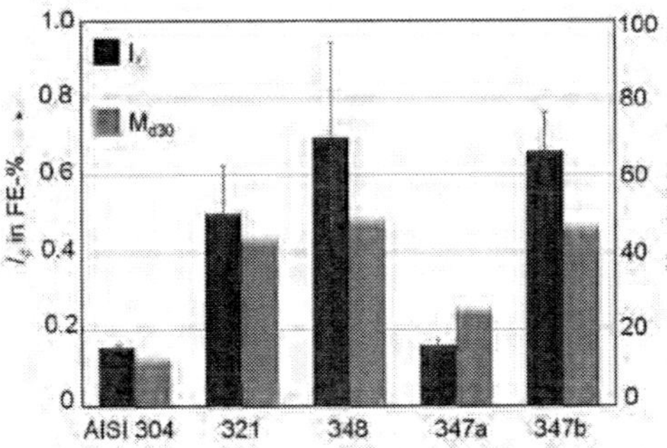

Fig. 4. M_{d30} temperatures as metastability parameter, and α'-martensite content (I_ξ) after impact tests as a susceptibility parameter to deformation induced α'-martensite formation of all investigated austenitic stainless steels

3.2 Influences of metastability on α'-martensite formation and cyclic deformation behavior

Fatigue behavior of metastable austenitic steels at temperatures up to the martensite deformation temperature is strongly influenced by cyclically induced α'-martensite formation. In respect of specimens AISI 304, AISI 321 and AISI 348 (tested under cyclic loading with total-strain amplitude $\varepsilon_{a,t}$ = 1 % and load frequency f = 0.2 Hz) the following results are presented: figure 5a shows the resulting stress amplitude (σ_a); figure 5b shows the development of ferromagnetic fraction of α'-martensite (ξ) measured in situ by Feritscope™, and figure 5c displays the change in temperature (ΔT) due to change in the strain energy density per specific volume, correlating with the area of stress-strain hysteresis.

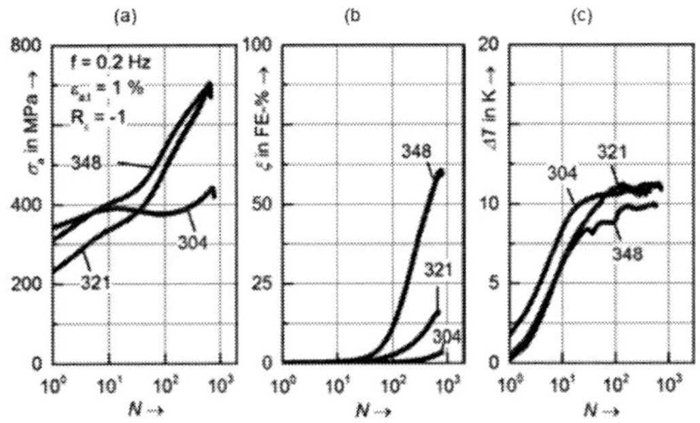

Fig. 5. σ_a-N (a), ξ-N (b) and ΔT-N (c) curves from total strain controlled fatigue tests, $\varepsilon_{a,t}$ = 1 %, f = 0.2Hz for AISI 304, AISI 321 and AISI 348

Clearly, the formation of α'-martensite led to pronounced cyclic hardening. The susceptibility to the formation of cyclic deformation induced α'-martensite correlated very well with metastability and susceptibility I_ξ parameters (see section 3.1). According to the I_ξ, M_{d30} and M_S temperatures as well as SFE of the investigated metastable austenite, the AISI 304 steel formed less α'-martensite during cyclic loading compared to the titanium-alloyed AISI 321 and the niobium-alloyed AISI 348 steels. After a material-dependent number of cycles N, the formation of α'-martensite commenced and increased continuously with increasing number of cycles up to specimen failure. Maximum value of α'-martensite (ξ = 60 FE%) was observed for AISI 348 at specimen failure. AISI 321 achieved ξ = 15 FE% at specimen failure but AISI 304 only ξ = 4 FE%. The σ_a -N curves (Fig. 5a) describe the cyclic deformation behavior of these three metastable austenitic stainless steels in detail. The stress amplitude of AISI 304 and AISI 348 commenced at a comparable value of σ_a while AISI 321 shows a lower value, which correlated with the monotonic determinate $R_{p0.2}$ limits (see table 2). Further cyclic loading led to cyclic hardening due to increased dislocation density [16] for all materials up to approximately ten cycles. While for the more metastable AISI 348 and AISI 321 steels a continuous increase in stress amplitude was observed, the less metastable AISI

304 showed a softening process with a decrease in stress amplitude, up to approximately 100 cycles. A secondary cyclic hardening in AISI 304 could only be detected when the α'-martensite formed. Cyclic hardening of the AISI 321 / AISI 348 steels resulted in a maximum stress amplitude σ_a= 700 / 710 MPa, which is higher than the tensile strength R_m = 641 / 640 MPa of these materials. The ΔT-N-curves (Fig 5c) show the change in the specimens' temperatures during cyclic loading. ΔT was determined as the difference between temperature in the middle of the specimen T_1 and mean value of the temperature measured with two further thermocouples (T_2 and T_3) fixed at the shafts of the specimens, where due to the comparatively thicker section only elastic deformation occurred (see Fig. 3). After the initial period (approximately ten cycles) with a significant increase of ΔT, the measured change in the specimens' temperature was in the range 9 K < ΔT < 12 K. For austenitic stainless steels this is a relatively small change in the temperature of the specimen (comp. Fig 5c with Fig. 8c), due to the relatively low frequency of f = 0.2 Hz. However, it is important to mention that for the development of α'-martensite not only is the change in temperature critical but so is the absolute temperature in the gauge length of the specimens. In these tests maximal temperatures were in the range 34 °C < T_{1max} < 37 °C.

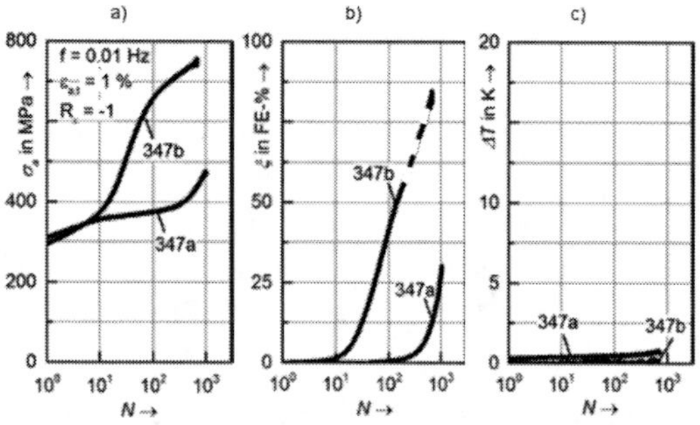

Fig. 6. σ_a-N (a), ξ-N (b) and ΔT-N (c) curves from total strain controlled fatigue tests, $\varepsilon_{a,t}$ = 1 %, f = 0.01 Hz, for AISI 347a and AISI 347b

To investigate the influence of metastability of nominally the same type of austenitic stainless steel, specimens from two batches of AISI 347 were examined. The specimens were cyclically loaded with total strain amplitude $\varepsilon_{a,t}$ = 1 % at very low frequency f = 0.01 Hz to keep their temperature at ambient level (see Fig. 6c) thereby eliminating the possible influence of temperature on α'-martensite formation. Figure 6 summarizes the results of these tests. Specimens from both batches showed a continuous cyclic hardening, however with different gradients after the first ten cycles. In the cycle range 10 < N < 100 AISI 347a showed only a slight increase of stress amplitude, while AISI 347b underwent a larger increase of σ_a. AISI 347b's increase in σ_a correlated directly with a significant development of α'-martensite (Fig. 6b) from 1 FE% to 41 FE%. As mentioned in section 2, ferromagnetic α'-martensite was measured in situ during fatigue tests by Feritscope™ sensor, for which readings above 60 FE% could be due to instrument linearity and calibration difficulties at high α'-martensite contents. Therefore the measured data in the range above 60 FE% were plotted as dashed lines in Fig. 6b. After a high rate of σ_a at cycles between N = 10 and N = 100, AISI 347b showed a decrease in the stress amplitude rate $d\sigma_a/dN$. However, further cyclic hardening took place which led to specimen failure at σ_a = 760 MPa, which is 120 MPa higher than the tensile strength of this material (table 2). With respect to the other batch, AISI 347a also showed evidence of a correlation between the development of α'-martensite and an increase in stress amplitude. However, the start of the α'-martensite formation occurred at a higher number of cycles in comparison to AISI 347b and the α'-martensite content was found to be, at the same number of cycles, significantly smaller with the maximum value at specimen failure ξ = 30 FE%. The results of these tests clearly depicted how substantial the differences can be between the formation of deformation induced α'-martensite in the same type of austenitic stainless steels due to differences in chemical composition and consequently the metastability of austenite. It should be noted that both batches had a chemical composition in the range of international standards for this steel.

3.3 Cyclic stress-strain behavior

Illustrating the hardening potential of metastable austenitic stainless steels due to cyclic induced α'-martensite formation, figure 7 displays the monotonic and cyclic stress-strain (CSS) curves from the low carbon steel AISI 4140 in two different heat treatment conditions [25] and the investigated AISI 347a.

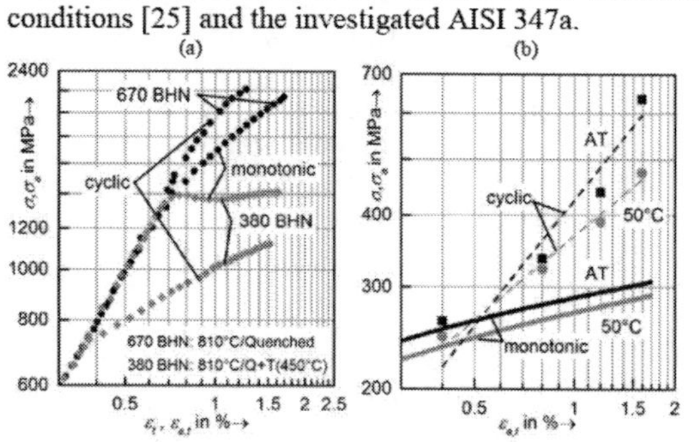

Fig. 7. Cyclic and monotonic stress-strain curves for AISI 4142 in tempered and in hardened state according [25] (a) and for AISI 347a at ambient temperature and at 50°C (b)

For AISI 4142, figure 7a shows that in quenched and tempered state, the cyclic stress strain curve is below the monotonic one and the material softens due to cyclic loading. This behavior can only be changed by heat treatment. The AISI 4142 in hardenend state reached stress amplitudes even above the monotonic stress-strain curves, which indicated a cyclic hardening process. However, the cyclic hardening of AISI 4142 in hardenend state led to stress amplitudes in the range of tensile strength. By comparison, in figure 7b, the metastable austenitic stainless steel AISI 347a shows a significant increase of stress amplitude during the cyclic loading due to α'-martensite formation. At ambient temperature the CSS curve is at lower total strain amplitudes, due to no pronounced α'-martensite formation, and below the monotonic stress-strain curve. But at higher $\varepsilon_{a,t}$ due to cyclic hardening, the CSS curve is above the monotonic stress-strain curve. Thus, the tensile strength of 569 MPa was exceeded by 120 MPa to a maximum stress level of 690 MPa. These results illustrate clearly a high potential in the cyclic hardening of metastable austenite in comparison to stable steels without deformation induced phase transformation. Because the cyclic loading of metastable austenite led to significant changes in the specimen temperature, the CSS curve of AISI 347a from tests at T = 50 °C is shown in addition to the CSS curve at ambient temperature (AT) in the Fig. 7b. Generally, comparable behavior with cyclic hardening can be seen for both temperatures, but the rate of cyclic hardening in the test at T = 50°C was lower than at AT. The difference in the stress level of the CSS curve at AT to CSS curve at T = 50 °C increased with the increasing total strain amplitude until a maximum difference of approximately 100 MPa for $\varepsilon_{a,t}$ = 1.6 %.

3.4 Influence of deformation rate and temperature on cyclic induced α′-martensite formation

Besides the metastability of austenite and the stress-strain state of microstructure, the strain rate also influences the formation of α'-marteniste. To investigate this phenomenon, total strain controlled fatigue tests with strain rate in the range $0.01\%\cdot s^{-1} < d\varepsilon_t/dt < 10\ \%\cdot s^{-1}$ were performed. Figure 8a shows the cyclic deformation behavior for different total-strain rates. The specimens from AISI 347a were cyclically loaded with constant total-strain amplitude $\varepsilon_{a,t}$ = 1.2% and R = -1.

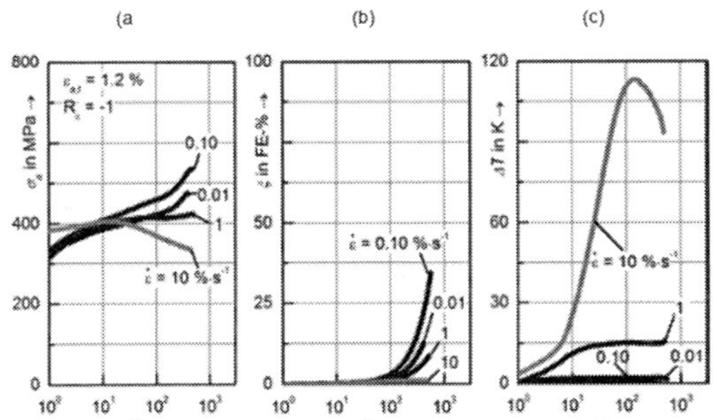

Fig. 8. σ_a-N (a), ξ-N (b) and ΔT-N (c) curves from total strain controlled fatigue tests, $\varepsilon_{a,t}$ = 1.2 %, $0.01\%\cdot s^{-1}<d\varepsilon_t/dt<10\ \%\cdot s^{-1}$, for AISI 347a

For the strain rate up to $d\varepsilon_t/dt = 1\%\cdot s^{-1}$, the cyclic deformation behavior was characterized by a continuous cyclic hardening until sample failure with different specificity. When the strain rate was increased from $0.01\%\cdot s^{-1}$ to $0.1\%\cdot s^{-1}$, higher stress amplitudes were measured. The ten-time increase in the strain rate led to an increase in the volume fraction of α'-martensite at the same number of cycles and after failure the specimen loaded with $d\varepsilon_t/dt = 0.01\%\cdot s^{-1}$ had ξ = 13 FE% and specimen loaded with $0.1\%\cdot s^{-1}$ had ξ = 34 FE% (Fig. 8b). It should be noted that the change in the temperature was in both tests very small, approximately ΔT ~ 1 K (Fig. 8c). A further increase in the strain rate to $1\%\cdot s^{-1}$ led to an increase in the sample temperature ΔT > 14 K. Due to the increase in the sample temperature, the α'-martensite formation set in later and reached a value of approx. ξ = 8 FE% at the end of the experiment. At the highest strain rate $d\varepsilon_t/dt = 10\%\cdot s^{-1}$, temperature changes ΔT > 100 K were achieved (Fig. 8c). The enormous increase in temperature after dislocation hardening resulted in cyclic softening of the material until sample failure with complete suppression of α'-martensite formation. By increasing the strain rate up to $0.1\%\cdot s^{-1}$, an increase in α'-martensite formation was observed. When the deformation rate was increased further, the formation of α'-martensite decreased again since it was not possible to suppress sample heating during the stress. Results from fatigue tests with the strain rate in the range $0.01\%\cdot s^{-1} < d\varepsilon_t/dt < 10\ \%\cdot s^{-1}$ clearly show that an increase of the strain rate led to increases in the formation rate of α'-martensite, but only if the specimens' temperature was kept approximately constant to ambient temperature (ΔT ~1K). When testing with a strain rate of $1\%\cdot s^{-1}$, $\varepsilon_{a,t}$ = 1.2 % and frequency f ≅ 0.2 Hz, a change in the temperature led to a comparatively reduced rate of α'-martensite formation. Therefore, the cyclically induced change in the specimen temperature has to be taken in to consideration by fatigue testing the metastable austenitic steels while significant changes in the formation of α'-martensite, and consequently cyclic deformation behavior, occurred. In the extreme case at higher strain rates or higher frequency, a completely different behavior was observed due to deactivated α'-martensite formation (see Fig. 8b at $d\varepsilon_t/dt < 10\ \%\cdot s^{-1}$ / f ≅ 2 Hz). In addition, as a function of metastability, small changes in the temperature can significantly influence the formation of α'-martensite and consequently cyclic deformation behavior. Especially for investigations of high cycle frequency (HCF) behavior up to the limit of number of cycles $N_L = 2\cdot 10^6$, a frequency higher than 2 Hz is required to obtain the ultimate number of cycles in a justifiable experimental time (e.g. to obtain N_L with f = 2 Hz, the test duration is 12 days). Hence, usually stress controlled fatigue tests are performed for investigations of the HCF behavior. Figures 9 and 10 illustrate how the test frequency of 2 Hz can influence the development of specimen temperature and consequently α′-martensite formation as well as cyclic deformation behavior.

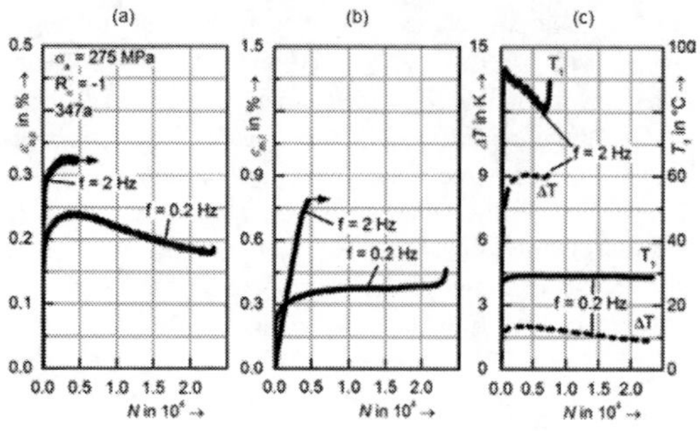

Fig. 9. $\varepsilon_{a,p}$-N (a), $\varepsilon_{a,p}$-N (b) and ΔT, T_1-N (c) curves from stress controlled fatigue tests, σ_a = 275 MPa, f = 2 and 0.2 Hz, for AISI 347a

At σ_a = 275 MPa, the specimen loaded with f = 2 Hz was tested at development temperatures T_1 (see Fig. 3) in the range of 100 °C. The development of plastic strain amplitude versus number of cycle ($\varepsilon_{a,p}$-N curve) showed cyclic softening at specimen failure at $N_f = 7 \cdot 10^3$. The same stress amplitude after cyclic softening led to a cyclic hardening process by cyclic loading with the frequency f = 0.2 Hz. The maximum measured temperature (T_1) in this test was 30 °C. However, to obtain the $N_L = 2 \cdot 10^6$ with f = 0.2 Hz, 115 days of test duration was needed, which is impractical for a systematic investigation of HCF behavior due to high resource expense at this frequency. Figure 10 shows selected stress-strain hysteresis for both test frequencies. It is clear to see that, the cyclic ratcheting occurred in the austenitic stainless steels by stress controlled fatigue test. The higher the test frequency, the higher the total middle strain (see. Fig. 9b) which led to a reduction of fatigue life.

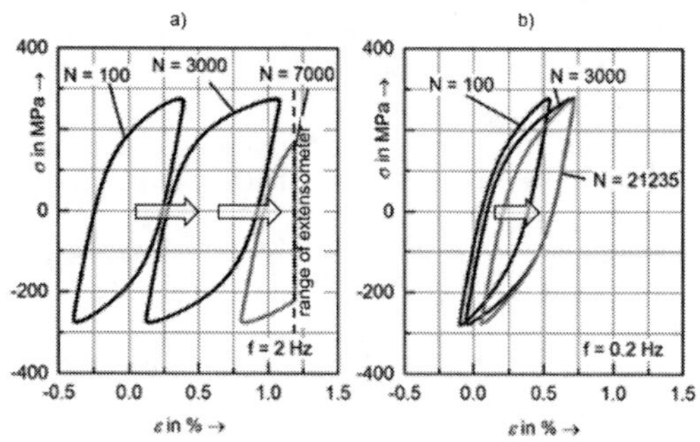

Fig. 10. Stress-strain hysteresis from stress controlled fatigue tests, σ_a = 275 MPa, f = 2 (a) and 0.2 Hz (b), for AISI 347a

4 Conclusion

Small changes in the chemical composition of austenitic stainless steels, including changes within batches of the same type, significantly influence the metastability of fcc austenite, which transforms by monotonic and/or cyclic plastic deformation in the more stable bcc martensitic phase. Hence, two-phase microstructures exist with a changing volume fraction and distribution of α′-martensite by mechanical loading. Consequently, a kind of "dynamically composite material" can develop, with α′-martensite as reinforcement embedded in the γ-austenitic matrix. As a result of cyclic plastic deformation, the α′-martensite forms, which significantly strengthens the material, and the stress amplitude range can achieve a strength higher than the tensile strength of the initial microstructure. The rate of α′-martensite formation depended on chemical composition, temperature and the stress-strain state of the microstructure. For determination of susceptibility to the formation of deformation-induced α′-martensite, a method based on dynamically applied local plastic deformation and micro-magnetic measurement was developed. The parameter established as I_ξ correlates very well with the grade of α′-martensite formation during cyclic loading. By investigating the cyclic deformation and transformation behavior of metastable austenitic stainless steels, the frequency or plastic deformation induced self-heating of the specimen has to be taken into account, because the metastability of austenitic microstructures directly depend on temperature.

5 Acknowledgement

The authors gratefully acknowledge financial support from Deutsche Forschungsgemeinschaft (DFG) as part of the Collaborative Research Center (SFB) 926, *Microscale Morphology of Component Surface.*

References

1. S. Suresh, *Fatigue of materials* (CUP, 2003)
2. J. Polák, *Fatigue of steels in Comprehensive Structural Integrity* (Elsevier, 2007)
3. P. Marshall, *Austenitic stainless steels* (Elsevier, 1984)
4. H. Becker, H. Brandis, W. Küppers, Techn. Berichte **12**, 35-54 (1986)
5. T. Angel, J. Iron Steel Inst, 165-174 (1954)
6. G.H. Eichelman, F.C. Hull, Trans. ASM **45**, 77-104 (1953)
7. S. Martin, O. Fabrichnaya, D. Rafaja, Mater. Letters **159**, 484-488 (2015)
8. J.K. Lai Leuk, C.H. Shek, K.H. Lo, *Stainless Steels: An Introduction and Their Recent Developments* (Bentham Science Publishers, 2012)
9. J. Man, I. Kubena, M. Smaga, O. Man, A. Järvenpää, A. Weidner, Z. Chlup, J. Polák, Proc. Struct. Integr. **2**, 2299-2306 (2016)
10. J. Man, M. Smaga, I. Kubena, D. Eifler and J. Polák, Eng. Fract. Mech. **185**, 139-159 (2017)
11. M. Smaga, R. Skorupski, D. Eifler, T. Beck, J. Mater. Res. **32**, 4452-4460 (2017)
12. M. Bayerlein, H.-J. Christ, H. Mughrabi, Mat. Sci. Eng. **A114**, L11-L16 (1989)

13. M. Smaga, F. Walther, D. Eifler, Mater. Sci. Eng. A **483-484**, 394-397 (2008)
14. C. Müller-Bollenhagen, M. Zimmermann, H.-J. Christ, Int J Fatigue **32**, 936-942 (2010)
15. M. Droste, C. Ullrich, M. Motylenk, M. Fleischer, A. Weidner, J. Freudenberger, D. Rafaja, H. Biermann, Int J Fatigue **106**, 143–152 (2018)
16. F. Hahnenberger, M. Smaga, D. Eifler, Int J Fatigue **69**, 36-48 (2014)
17. A. Sorich, M. Smaga, D. Eifler Ad. Mat. Res. **891-892**, 1231-1236 (2014)
18. C. Grigorescu, P. M. Hilgendorff, M. Zimmermann, C. P. Fritzen, H. J. Christ, Int J Fatigue, **93,** 250-271 (2016)
19. M. Smaga, F. Hahnenberger, A. Sorich, D. Eifler, Key Eng. Materials **465**, 439-442 (2011)
20. I. Nikitin, M. Besel, Int J Fatigue **30,** 2044-2049 (2008)
21. L. Vincent, J.C. Le Roux, S. Taheri, Int J Fatigue **38**, 84-91 (2012)
22. D.F. Pessoa, G. Kirchhoff, M. Zimmermann Int J Fatigue, **103**, 48-59 (2017)
23. A. Kisko, R.D.K. Misra, J. Talonen, L.P. Karjalainen Mater. Sci. Eng. A **578.** 408-416 (2013)
24. J. Talonen, P. Aspegren, P. Hänninen, Mater Sci Technol. **20**, 1506-1512 (2004)
25. R.W. Landgraf, *Cyclic deformation and fatigue behavior of hardened steels* (UIUC, 1968)

Effect of micro-notches on fatigue strength of electrodeposited nanocrystalline nickel thin films

Keisuke Tanaka[1,*], *Yuta* Murase[2], and *Hirohisa* Kimachi[2]

[1]Research Department, Nagoya Industrial Science Research Institute, Yotsuya-dori, Chikusa-ku, Nagoya, 464-0819, Japan
[2]Mechanical Engineering Department, Meijo University, Shiogamaguchi, Tempaku-ku, Nagoya, 468-8502, Japan

Abstract. The effect of micro-notches on the fatigue strength of nickel thin films was studied. Two types of thin films with 10 μm thickness were produced by electrodeposition using sulfamate solution without and with brightener: ultra-fine grained film (UFG) with the grain size of 384 nm and nano-crystalline grained film (NCG) with that of 17 nm. Micro-sized notches introduced by FIB had the width of 2 μm and various depths from 8 to 150μm. Fatigue tests were conducted under the stress ratio of 0.1. The fatigue strength decreased with increasing depth of notches. NCG had much higher strength than UFG compared at the same notch depth. Notches as small as 8μm did reduce the fatigue strength of both UFG and NCG. The fatigue limit was controlled by the initiation of cracks and no non-propagating crack was observed in specimens fatigued below the fatigue limit. A model of fictitious crack successfully predicted the reduction of the fatigue limit due to micro-notches. The characteristic crack length of NCG was much smaller than the UFG, while the fatigue strength of defect-free NCG was larger than that of UFG. SEM observation of fracture surfaces was conducted to reveal micromechanisms of fatigue crack initiation.

1 Introduction

Nickel thin films have been used as structural materials for MEMS (micro-electro-mechanical systems) devices made by LIGA process. To improve the fatigue strength of metallic thin films, nanocrystallization is one of the most promising technique. In our previous studies, free-standing nickel thin films with various grain sizes were produced by electrodeposition using sulfamate solution under constant current by changing brightener content [1-3]. The mean grain size measured by TEM ranged from 384 to 17 nm [4]. The fatigue strength was improved in finer-grained films, following the Hall-Petch relation. On the other hand, the resistance to fatigue crack propagation got lower for finer grain sizes [1-3,5], suggesting the increase of notch sensitivity by reducing the grain size. The effect of stress consentaration on the fatigue strength of nanocrystalline nickel need to be studied, because fatigue fracture may start at defects or notches in MEMS devices.

In this paper, the effect of micro-notches on fatigue strength reduction was studied by using two types of electrodeposited nickel thin films with the grain size of 384 and 17 nm.

2 Experimental procedure

Nickel thin films were produced by electrodeposition using sulfamate solution [1-3]. A polished stainless steel plate was used for a cathode and a pure nickel plate for an anode. Two deposition conditons were adapted : one was without brigtener (sodium allylsulfamate) and the other with brigtener of the amount of 2 g/L. The temperature of the bath was kept constant at 328 K for the former case and 313K for the latter using a hot-water circulating bath. The pH value of the solution was maintained at 3.7 to 4.2 by adding sulfamate acid. The solution was stirred by a magnetic stirrer to avoid pit formation. The current density was kept constant at 25 mA/cm^2 by using a constant current supply, and the deposition period was 28 min. The thickness of the thin film was around 10 μm. After deposition, thin films were removed from the cathode and subjected to the characterization of the tensile and fatigue properties as free-standing films.

The grain size was measured by TEM. The one without brigtener, UFG, had the grain size of 384 nm and the grain orientation of 100 fiber texture, while the other with brigtener, NCG, had that of 17 nm and randam grain orientation [4]. The tensile properties and

Table 1 Mechanical properties of UFG and NCG.

Material	Grain size d, nm	Youngs modulus E, GPa	Yield stress $\sigma_{0.2}$, MPa	Tensile strength σ_B, MPa	Fracture strain ε_f, %	Vickers hardness Hv
UFG	384	147	593	887	3.1	230
NCG	17	182	1363	2165	3.3	681

* Corresponding author: ktanaka@ccalumni.meijo-u.ac.jp

Vicker hardness of two films are summarized in Table 1. NCG with nanosized grains has much higher strength than UFG with ultra-fine grains. The Youngs modulus of UFG is lower than taht of NCG, because of 100 fiber texture.

Figure 1(a) shows smooth sepcimens (SM) made from electrodeposited films by buffing [2]. They have a small stress concetration factor of 1.043 at the centarl region. Micro-notches were introduced with FIB (Focused ion beam) at the the central region A as shown in Fig. 1(b). The notch width was 2 μm and the notch tip was square-shaped with a corner radius of 0.20 μm. The notch length was 8, 20, 50, 120, and 150 μm, and the specimens with these lengths are named MN8, MN20, MN120 and MN150, respectively.

Fatigue tests were conducted in a servo-electromagnetic fatigue testing machine (Shimadzu MMT-100N-10) under the stress ratio of R = 0.1 at a frequency of 20 Hz. The range of the gross stress in the minimum cross section (without micro-notch) was used to express the loading condition. The initiation and propagation of fatigue cracks was observed with a digital microscope during fatigue tests. After specimens were broken, SEM observation was conducted on the fatigue fracture surface and also the exact value of the specimen thickness was measured.

The stress intensity factor (SIF) for a single edge crack of length a in smooth specimens shown in Fig. 1(a) was calculated by the finite element method. The SIF value was expressed by

$$K = \sigma\sqrt{\pi a} \cdot F\left(a/W\right) \quad (1)$$

$$F(\xi) = 1.1697 - 0.2703\xi + 4.235\xi^2 - 6.049\xi^3 + 4.293\xi^4$$

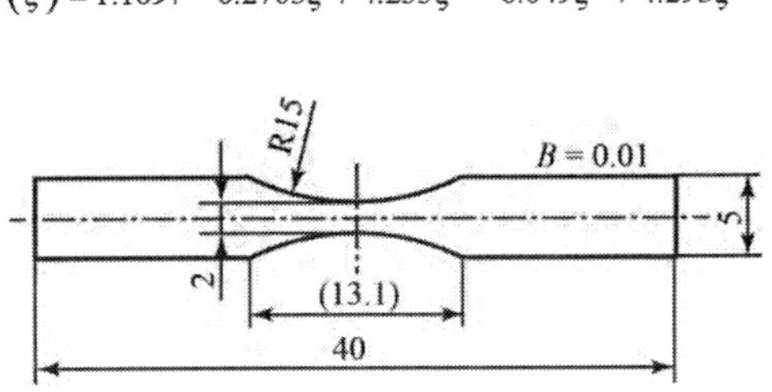

(a) Smooth specimen (SM).

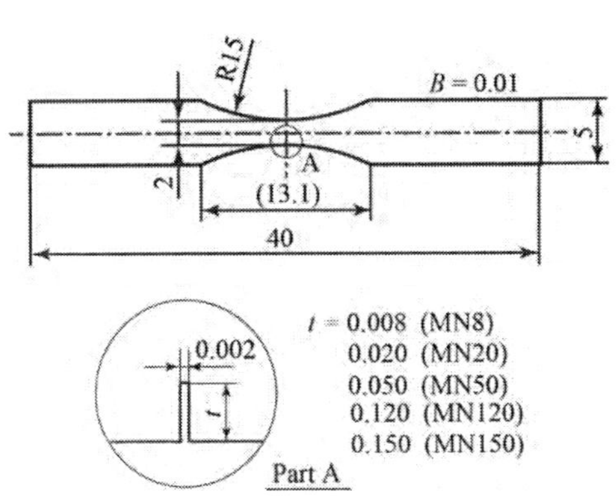

(b) Micro-notched specimen (MN).

Fig. 1 Smooth specimen and micro-notched specimen

where σ is the gross stress and W is the specimen width, and $\xi = a/W$. F is the correction factor. For a = 0, F = 1.1215×1.043 = 1.1697.

3 Experimental results

3.1. *S-N* relation

The *S-N* relation of UFG is shown in Fig. 2, where the gross stress range was taken as the ordinate. The data with the rightward arrow indicate the results for a specimen which was not fractured. The fatigue strength of smooth specimen (SM) was highest, and decreases with increasing notch depth.

Table 2 summarize the fatigue limit, $\Delta\sigma_w$, defined by the fatigue life of 10^7 cycles. The stress intensity factor was calculated from the fatigue limit by assuming the notch depth equal the crack depth a in Eq. (1). The equivalent crack length, a_{eq}, indicated in the table is the half crack length converted to an isolated crack which is calculated from

$$\Delta K_{th} = \Delta\sigma_w\sqrt{\pi a} \cdot F = \Delta\sigma_w\sqrt{\pi a_{eq}} \quad (2)$$

Therefore

$$a_{eq} = a \cdot F^2 \quad (3)$$

The ΔK_{th} value of UFG is not constant and decreases with decreasing the notch depth.

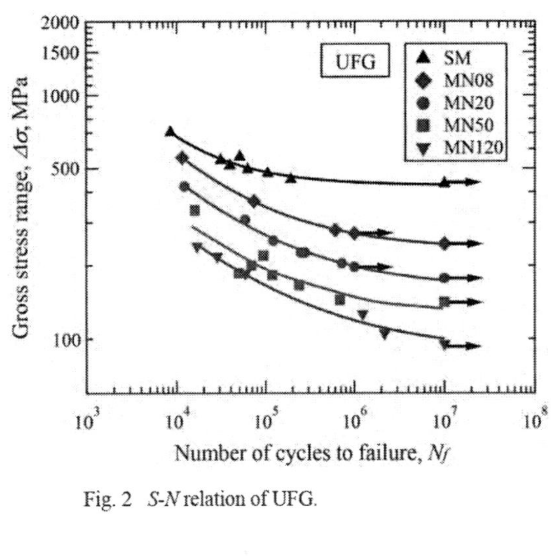

Fig. 2 *S-N* relation of UFG.

Table 2 Fatigue thresholds of UFG.

Specimen UFG	Notch length t, μm	Equivalent lenpth, a_{eq}, μm	Fatigue limit $\Delta\sigma_w$, MPa	SIF, ΔK_{th} MPam$^{1/2}$
SM	0	0	432	n/a
MN08	8	10.8	263	1.53
MN20	20	27.0	193	1.76
MN50	50	67.6	143	2.08
MN120	120	163.7	107	2.28

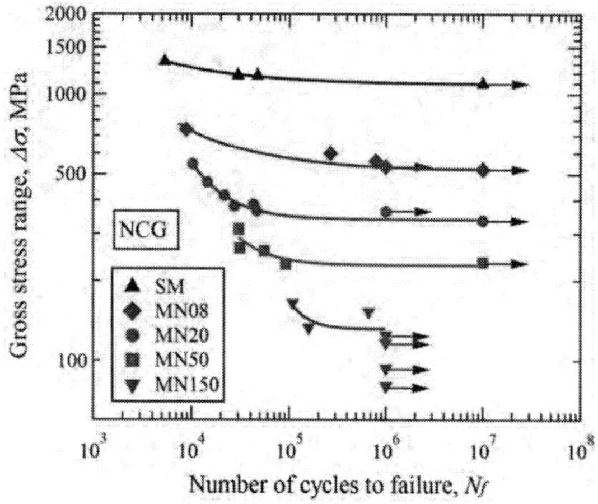

Fig. 3 *S-N* relation of NCG.

Table 3 Fatigue thresholds of NCG.

Specimen NCG	Notch length t, μm	Equivalent lenpth, a_{eq}, μm	Fatigue limit $\Delta\sigma_w$, MPa	SIF, ΔK_{th} MPam$^{1/2}$
SM	0	0	1096	n/a
MN08	8	10.8	541	3.16
MN20	20	27.0	359	3.30
MN50	50	67.6	232	3.39
MN150	150	206	129	3.27

Figure 3 shows the *S-N* relation of NCG. The fatigue strength of NCG is much higher than that of UFG, and decreases with increasing notch depth. *S-N* curves have a rather clear knee at stress cycles of about 10^5 to 10^6 cycles. Table 3 summarizes the fatigue limit at life of 10^7 cycles. The fatigue limit of NCG is much higher than that of UFG when compared at the same notch depth. The threshold stress intensity factor, ΔK_{th}, is nearly constant for all notch depths.

3.2 Fatigue crack initiation and propagation

The initiation and propagation of fatigue crack at the notch root was observed by video recording during fatigue tests. For fatigued specimens run out of 10^7, no non-propagating crack was observed near the notch root for both UFG and NCG. The fatigue limit given in Tables 2 and 3 corresponds to the threshold of crack initiation.

Figure 4 shows fatigue cracks observed by video recording. The arrow indicates the notch tip. The crack length is 50 μm for both UFG and NCG. The crack propagates straight perpendicular to the loading direction. The crack in NCG is very fine, while that in UFG is thick indicating the larger amount of plastic deformation associated with crack propagation. The minimum crack length that can be resolved by our video recording system was 10 μm. Therefore, the crack initiation life was defined as the number of cycles up to the detection of a 10 μm long crack. Crack propagation curves were obtained after the initiation of 10 μm crack. The rate was calculated by the second method.

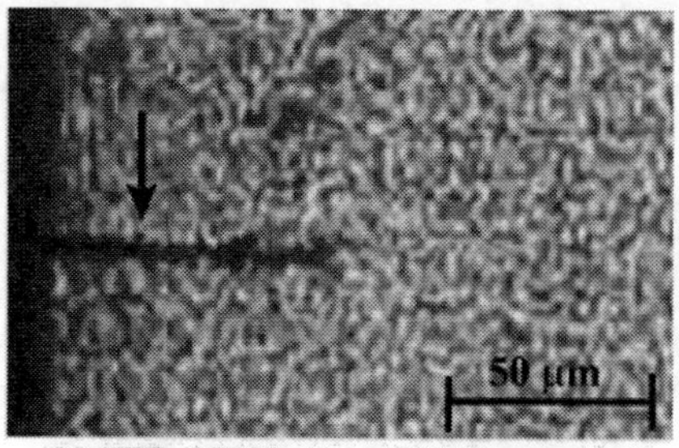

(a) UFG, MN20, $\Delta\sigma$ = 542 MPa, c = 50 μm N = 6202, N_f = 10992

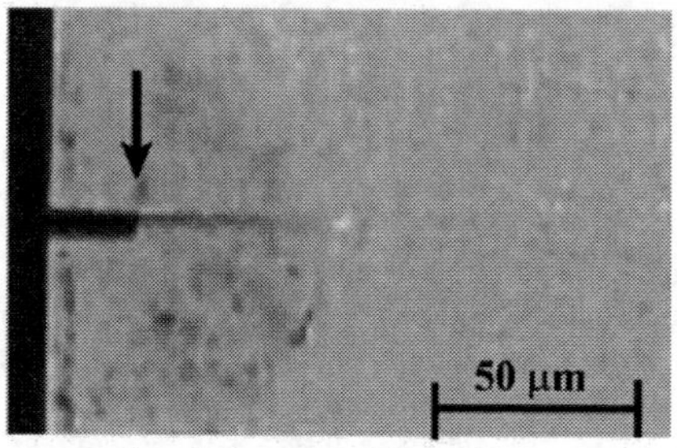

(b) NCG, MN20, $\Delta\sigma$ = 308 MPa, c = 50 μm N = 40627, N_f = 58131

Fig. 4. Video recorded micrographs of cracks in MN speciemns of UFG and NCG.

The relation between *da*/*dN* and the range of stress intensity factor, ΔK, for UFG and NCG is presented in Figs. 5 and 6, respectively, where ΔK is calculated by inserting the crack length plus the notch depth for a in Eq. (1). The open marks indicate the data obtained under small-scale yielding (SSY) conditions, while the solid marks are those under large-scale yielding conditions. The all data for NCG satisfy the SSY, and some of UFG at large K values do not.

The relation obtained for a long crack by load-shedding method is also shown with the small dots in the figures in the figure [6]. The threshold stress intensity factor for propagation of long cracks, $\Delta K_{th\infty}$, was 4.26 MPam$^{1/2}$ for UFG and 2.19 MPam$^{1/2}$ for NCG. The crack path of NCG was very straight and not accompanied by plastic deformation near crack tips. There might be no extrinsic shielding of the crack tip, the threshold value of 2.19MPam$^{1/2}$ is equal to the effective threshold. The threshold value of 4.26 MPam$^{1/2}$ for UFG is larger than that of NCG is mainly due to crack closure, increasing in proportion of the square root of the grain size [6].

For UFG shown in Fig. 5, the ΔK value of small cracks just after initiation is below the threshold of long crack, 4.26 MPam$^{1/2}$, and then the relation gradually merges to that of long cracks. There is no dip in the crack propagation relation, suggesting non-existence of crack arrest usually observed in sharply notched steel specimens [7]. On the other hand, for the case of NCG as shown in Fig. 6, the ΔK value of small cracks just after initiation is above the long crack threshold, 2.19 MPam$^{1/2}$, and then the relation between *da*/*dN* and ΔK for small cracks follows the relation of long cracks.

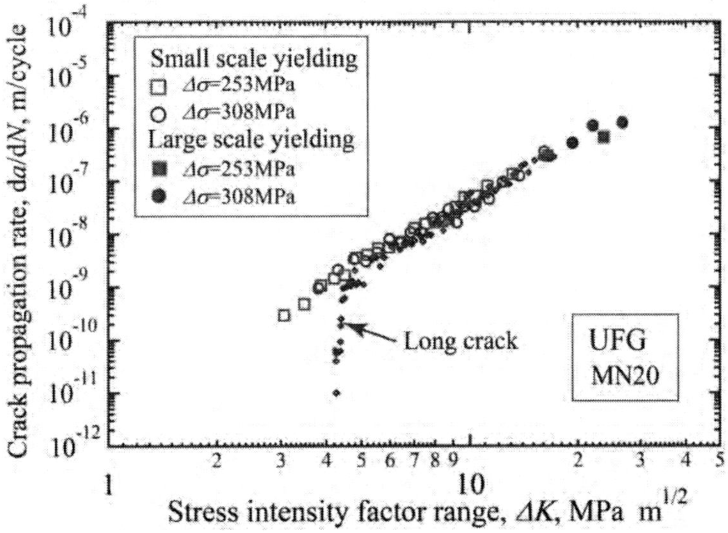

Fig. 5. Relation between fatigue crack propagation rate and stress intensity range of UFG.

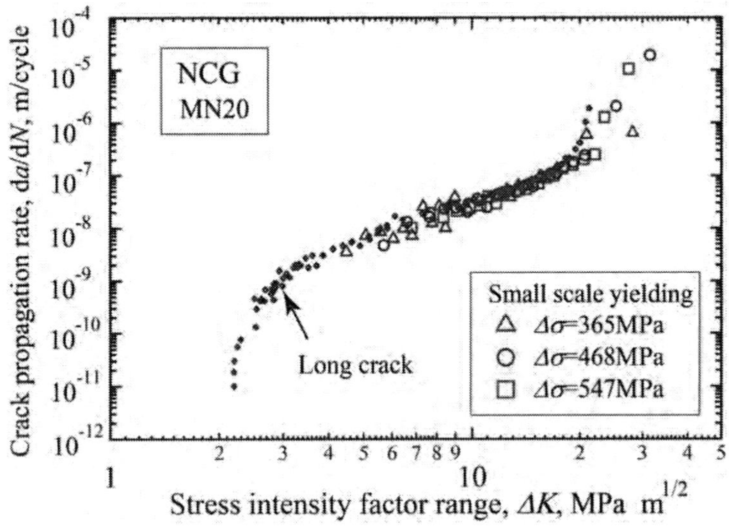

Fig. 6. Relation between fatigue crack propagation rate and stress intensity range of NCG.

3.3 SEM observation of fatigue fracture surfaces

SEM micrographs of fatigue fracture surfaces near the notch tip are shown in Fig. 7, where (a) is for MN50 of FG and (b) is for MN20 of NCG. For micro-notched specimens of UFG and NCG, there is no defect near the notch root. The fracture surface of UFG is rather rough and several parallel lines are seen across the thickness, corresponding the columnar grain structure made by electrodeposition. On the other hand, the fracture surface of NCG is very smooth and tiny granular patterns are observed. Cracks are expected to start at the central part of the notch root.

The fatigue fracture surface of SM specimens were also examined by SEM. Material defects were observed at the crack initiation site in SM specimens of NCG, while no defects were seen on the fracture surface of those of UFG. Figure 8 shows two examples of such defects seen on the fracture surface of SM specimens of NCG. Dotted lines in Fig. 8 indicate the boundary of defect areas. Defects made during electrodeposition gave rise to the crack initiation site, and reduced the fatigue strength of NCG even if SM specimens were supposed to be smooth or defect-free. Therefore, the true fatigue limit of defect-free NCG would be higher as discussed later.

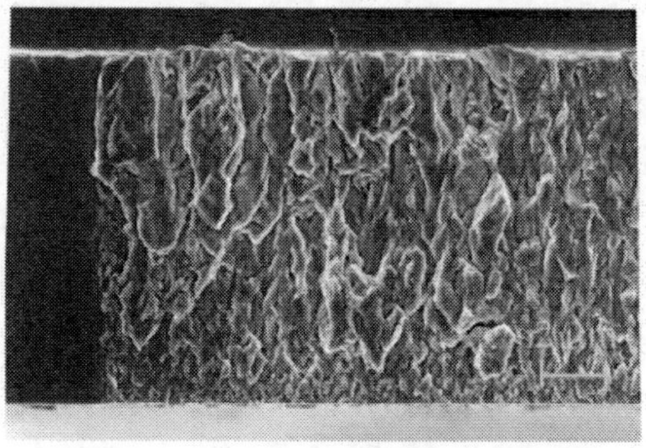

(a) UFG, MN50, $\Delta\sigma$ = 200 MPa, N_f = 68793

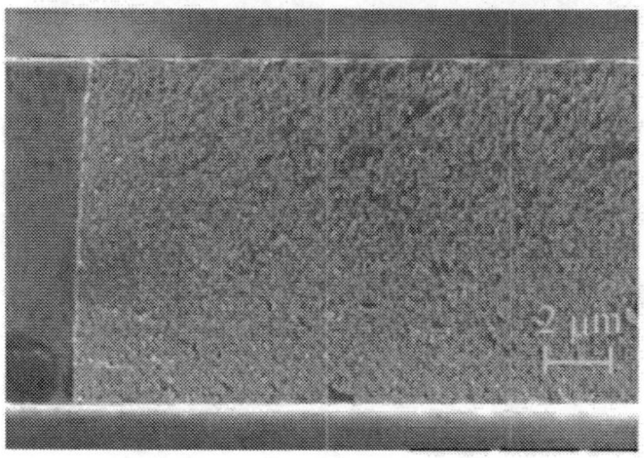

(b) NCG, MN20, $\Delta\sigma$ = 468 MPa, N_f = 14528

Fig.7 SEM micrographs of fatigue fracture surafce near the micro-notch root. The crack propagation rate is from left to right.

(a) NCG, SM, $\Delta\sigma$ = 1348MPa, N_f =5290

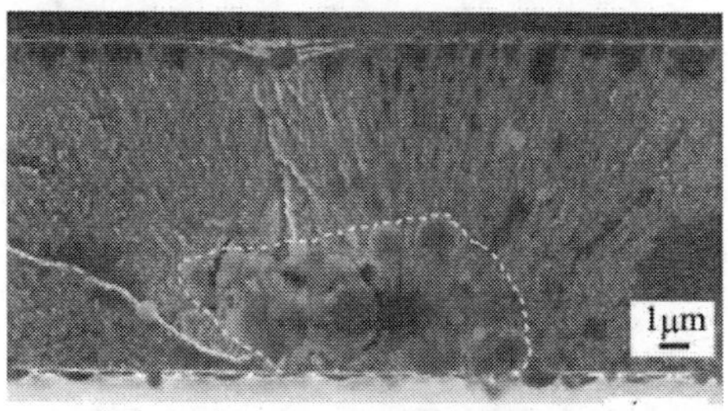

(b) NCG, SM, $\Delta\sigma$ = 1175MPa, N_f =47266

Fig.8 SEM micrographs of fatigue fracture surafce of SM specimens of NCG. The dotted area indictes the boundary of defects made by electrodepostion.

4 Prediction of fatigue strength reduction due to micronotchs.

4.1. Fictitious crack model

In sharply notched specimens, fatigue fracture is not controlled by the maximum stress alone, but by stress acting over a characteristic distance ahead of the notch tip. The fictitious crack model proposed by El Haddad et al [8] is a simple reconciliation for such microstructural influences. The fatigue fracture is assumed to be determined by the condition of the propagation of a fictitious crack existing ahead of the tip of pre-cracks or notches. This model was applied to the prediction of the effect of pre-crack length on the threshold of fatigue crack propagation by Tanaka et al [9]. The model is applied to the present experimental results. The essence of the model is described below.

For an isolated pre-crack with half length a in an infinite plate, the threshold stress, $\Delta\sigma_w$, is given by the following condition of crack extension:

$$\Delta K_{th\infty} = \Delta\sigma_w \sqrt{\pi\left(a + a_0\right)} \tag{3}$$

where $\Delta K_{th\infty}$ is the threshold stress intensity factor for long cracks, and a_0 is the length of the fictitious crack which is determined from the fatigue limit of defect-free specimens, $\Delta\sigma_{w0}$, as

$$a_0 = \left(\Delta K_{th\infty} / \Delta\sigma_{w0}\right)^2 / \pi \tag{4}$$

where $\Delta\sigma_{w0}$ is the fatigue limit of un-notched defect-free specimens. From Eqs. (3) and (4), the following equation is derived

$$\Delta\sigma_w / \Delta\sigma_{w0} = \left[a_0 / \left(a + a_0\right)\right]^{1/2} \tag{5}$$

The threshold SIF of short crack, ΔK_{th}, is

$$\Delta K_{th} / \Delta K_{th\infty} = \left[a / \left(a + a_0\right)\right]^{1/2} \tag{6}$$

The threshold condition of pre-cracked specimens gradually changes from $\Delta\sigma_w$ = $\Delta\sigma_{w0}$ for un-cracked specimens to ΔK_{th} = $\Delta K_{th\infty}$ for long pre-crack. The a_0 length is a parameter to distinguish short and long cracks.

To apply the fictitious crack model to single edge notches, the characteristic crack length is often determined by [7]

$$c_0 = \left(\Delta K_{th\infty} / 1.1215 \Delta\sigma_{w0}\right)^2 / \pi \tag{7}$$

The relation between c_0 and a_0 is

$$c_0 = a_0 / 1.1215^2 = 0.795 a_0 \tag{8}$$

4.2 Fatigue threshold of ultra-fine grained films

The characteristic crack length a_0 of UFG is determined from Eq. (4) by using the crack initiation threshold SIF value, ΔK_{th} = 2.28MPam$^{-1/2}$, of MN120 of UFG shown in Table 2 and the fatigue limit of SM specimen, $\Delta\sigma_{w0}$ = 432 MPa, multiplied by the stress concentration factor of 1.043. The multiplication is required to obtain the fatigue limit of un-notched specimens, $\Delta\sigma_{w0}^*$, because SM specimens have small stress concentration of 1.043. The value of $\Delta\sigma_{w0}^*$ is used for $\Delta\sigma_{w0}$ to calculate a_0 and c_0

values. Table 4 summarizes the values. They are about 20 times the grain size.

Figure 9 shows the relation between the threshold stress, $\Delta\sigma_w$, and the equivalent depth, t_{eq}, where the open marks indicate specimens not fractured and the solid marks indicate those fractured. The dotted line corresponds the relation of Eq. (5) predicted by using the parameters given in Table 4. The prediction agrees very well with the experimental data. Figure 10 shows the change of SIF with the notch length, where the dotted line indicates the prediction by Eq. (6). The decrease of ΔK_{th} for shorter notches is well predicted by Eq. (6).

Table 4 Charctersitic crack length of UFG and NCG.

Material	Threshold SIF ΔK_{th} MPam$^{1/2}$	Fatigue limt $\Delta\sigma_{w0}$ MPa	$\Delta\sigma^*_{w0}$ MPa	Characteristic crack length a_0 μm	c_0 μm
UFG	2.28	432	451	8.14	6.47
NCG	3.27	1096	1143	2.61	2.07

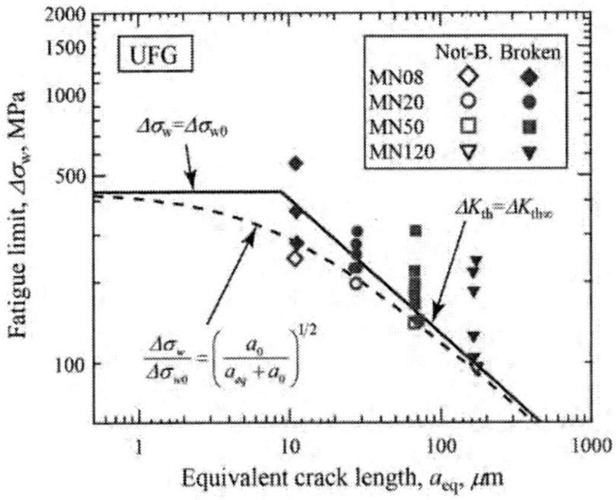

Fig. 9 Change of fatigue limit with equivalent crack length of UFG.

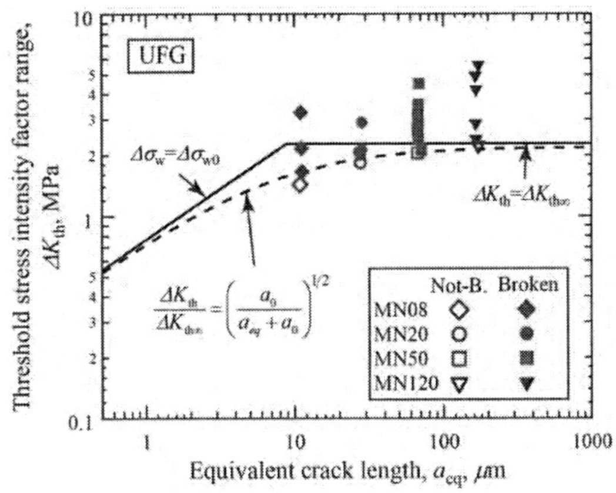

Fig. 10 Change of threshold SIF with equivalent crack size of UFG.

4.3 Fatigue threshold of nanocrystalline films

The crack initiation threshold SIF value, ΔK_{th}, of NCG shown in Table 3 is fairly constant, indicating the notch length of 8 μm means long enough to apply the constant SIF value for threshold. Using the fatigue limit, $\Delta\sigma_{w0}$ = 1096 MPa, of SM specimens multiplied by 1.043 and the ΔK_{th} of 3.27 MPam$^{-1/2}$, of MN150 of UFG, the a_0 value is calculated to be 2.61 μm as shown in Table 4. This value for NCG is about three times smaller than that of UFG, and about 150 times the grain size. A larger ratio of NCG partly comes from the fact that the fatigue limit of SM specimens is not equal to the fatigue limit of defect-free NCG as discussed later. Figures 11 and 12 show the change of the threshold stress and SIF with notch size. Both figures show a good agreement between experiments and the prediction.

The SIF value of the defect seen on the fracture surface of SM specimens was estimated from the defect area by the following Murakami's equation [11]

$$\Delta K = 0.65\Delta\sigma\sqrt{\pi\sqrt{area}} \quad (9)$$

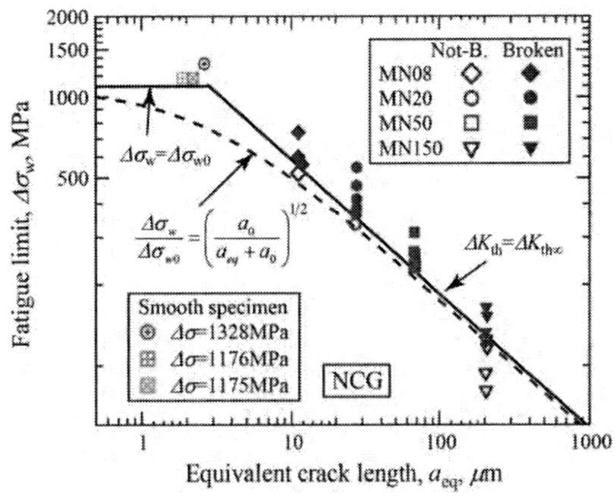

Fig. 11 Change of fatigue limit with equivalent crack length of NCG.

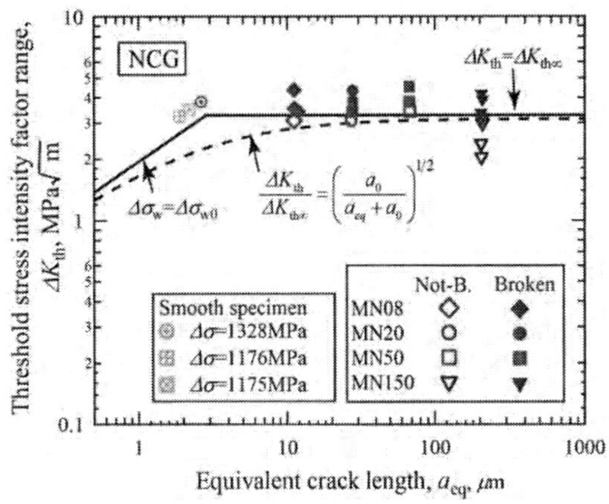

Fig. 12 Change of threshold SIF with equivalent crack length of NCG.

The equivalent crack length is defined by

$$\Delta K = \Delta\sigma\sqrt{\pi a_{eq}} \tag{10}$$

From the area of defects as shown in Fig. 8, the equivalent crack length, a_{eq}, is calculated by

$$a_{eq} = \left(\Delta K/\Delta\sigma\right)^2/\pi = 0.476\sqrt{area} \tag{11}$$

The applied stress range $\Delta\sigma$ is plotted as a function of a_{eq} in Figs. 11 and 12. As seen in Fig. 11, the ΔK value of SM specimens is on and above the constant ΔK_{th} line, meaning the constant ΔK criterion is applicable to micro-sized defects due to electrodeposition for NCG.

4.4 Discussion

The threshold SIF for crack initiation in micro-notched specimens is different from the threshold value of long crack propagation obtained by the load-shedding method in both UFG and NCG. For UFG, the crack initiation threshold, 2.28 $MPam^{-1/2}$, is smaller than the long crack propagation threshold 4.26 $MPam^{1/2}$. The threshold value of long cracks is the sum of crack initiation threshold plus the crack closure. The crack initiation threshold of micro-notches correspond the intrinsic threshold value. On the other hand, for NCG, the crack initiation threshold, 3.27 $MPam^{-1/2}$, is larger than the long crack propagation threshold 2.19 $MPam^{-1/2}$. In NCG, there is no crack closure operating, so 2.19 $MPam^{-1/2}$ is regarded as the intrinsic threshold value. The notch-tip shape might influence of initiation threshold.

In sharply notched specimens of ordinary micro-grain-sized steels, non-propagating cracks are observed near the notch root even when the applied stress is below the fatigue limit [7]. The non-propagating crack is caused by the increase of crack closure with crack extension, and its length is determined by the resistance curve method proposed by Tanaka et al [10]. In the present experiments, any non-propagation cracks are not observed in both UFG and NCG. In NCG, there is no crack closure at the threshold, so non-existence of non-propagating crack is expected. On the other hand, for UFG, the difference between the crack initiation threshold, 2.28 $MPam^{-1/2}$, and the long crack propagation threshold, 4.26 $MPam^{1/2}$, is the amount of crack closure, 1.98 $MPam^{1/2}$. A preliminary estimate of the length on non-propagation crack based on the resistance curve method is less than 10 μm. If non-propagating crack exist, its length is below the resolution limit of video recording. Further detailed observation by SEM is necessary.

The length of characteristic cracks, a_0, may represent the size of the process zone of fatigue crack initiation near notch tips. The a_0 value is 8.14 μm for UFG and 2.61 μm for NCG. It is larger for the coarser grain sized materials, UFG, than NCG. The ratio to the grain size is 20 for UFG to 150 for NCG. For ordinary micro-sized metals, a_0 is about 10 to 20 times the grain size [7]. SEM micrographs of the fracture surface near the notch tip presented in Fig. 7 do not show any area with distinct feature corresponding to the crack initiation process. In Fig. 7(a) of UFG, the size of a semi-elliptical region is seen at the centre of the notch root, which is smaller than a_0 or c_0. A large ratio of a_0/d_0 for NCG may suggest that nanosized grains behave as a group of several grains in the process of fatigue crack initiation. Since the grain size of NCG is very small, even a tiny radius of the notch tip may influence the crack initiation threshold.

The fictitious crack model is applied to predict the influence of the tip radius of notches on the fatigue threshold of NCG. Figure 13 illustrates the condition of crack initiation based on the fictitious crack model. The SIF value of the notch-tip crack at the threshold is given by

$$\Delta K_{th} = 1.1215 K_t \Delta\sigma_w \cdot \sqrt{\pi c_0} \cdot Y\left(c_0/\rho\right) \tag{12}$$

where $\Delta\sigma_w$ is the fatigue limit stress, K_t is the stress concentration factor, c_0 is the fictitious crack length, and Y is the correction factor for SIF. Y is a function of the ratio of crack length to the notch-tip radius ρ and is often expressed as [12]

$$Y\left(c_0/\rho\right) = 1/\sqrt{1+4.5\left(c_0/\rho\right)} \tag{13}$$

The term ($K_t\Delta\sigma_w$) means the maximum stress amplitude at the notch tip, which can be approximated by

$$\Delta\sigma_{max} = K_t\Delta\sigma_w = 2\Delta K_i/\sqrt{\pi\rho} \tag{14}$$

The ΔK_i value mean the threshold SIF for crack initiation. By substituting the threshold SIF of long cracks 2.19 $MPam^{-1/2}$ for ΔK_{th} and the initiation threshold of MN150, 3.27 $MPam^{-1/2}$, for ΔK_i in Eq. (12), the ratio of c_0 to ρ is obtained as

$$c_0/\rho = 0.15 \tag{15}$$

In the present experiment, the notch-tip radius is 1 μm, so c_0 is calculated as 0.15 μm. The value is about one fourteenth of the value presented in Table 4, and is nine times the grain size of NCG. It is within the range of the ratio for ordinary metals. If we calculate the fatigue limit of defect-free NCG using the following equation:

$$\Delta\sigma_{w0} = \Delta K_{th}/1.1215\sqrt{\pi c_0} = 2.19/1.1215\sqrt{\pi c_0} \tag{16}$$

The calculated value is 2.9 GPa, and is larger than the fatigue limit of SM specimens which have defects as the fatigue initiation site.

The influence of the tip radius of micro-notches on crack initiation will be the subject of the future study.

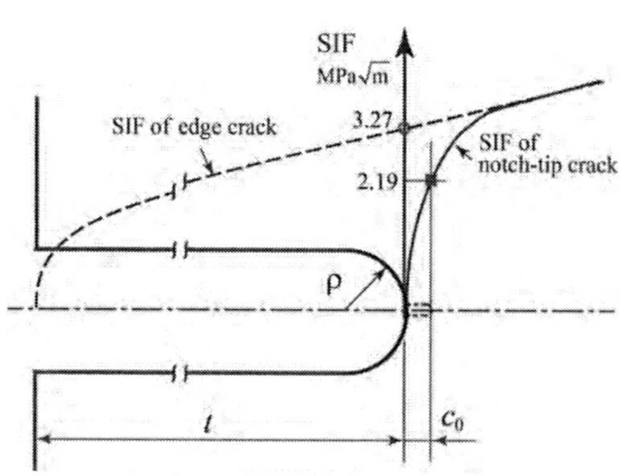

Fig. 13 SIF of fictitious crack at the notch root.

5 Conclusions

The effect of micro-notches on the fatigue strength of electrodeposited nickel thin films with ultra-fine grains (UFG) of 384 nm in length and nano-crystalline grains (NCG) of 17 nm were studied using crack-like notches made by FIB. Micro-notches are 2 μm wide with various depths ranging from 8 to 150 μm. The main results are summarized as follows:

(1) Micro-notches as small as 8 μm in length did reduce the fatigue strength of both UFG and NCG, and the amount of reduction was increased with increasing depth. The fatigue strength of NCG was larger than that of UFG when compared at the same length.
(2) The fatigue limit of micro-notched specimens was determined by the initiation of fatigue crack. No non-propagating crack was observed near the notch root of specimens fatigued just below the fatigue limit.
(3) The threshold value of SIF, ΔK_{th}, was decreased for short notches in UFG, while nearly constant in NCG.
(4) The fictitious crack model successfully predict the reduction of the fatigue strength as a function of notch depth. The length of fictitious crack, a_0, is smaller for NCG than UFG.

This work was supported by JSPS KAKENHI Grand Number 15H03896.

References

1. K. Tanaka, Y. Isokawa, H. Asano, H. KImachi, J. Soci. Mater. Sci. Jpn, **59**, 313 (2010).
2. K. Tanaka, M. Sakakibara, H. Kimachi, Procedia Engng, **10**, 542 (2011).
3. K. Tanaka, M. Sakakibara, H. Tanaka, S. Takeshita, H. Kimachi, J. Soci. Mater. Sci. Jpn, **61**, 946 (2012).
4. K. Tanaka, Y. Koike, K. Sano, H. Tanaka, S. Machiya, T. Shobu, H. Kimachi, J. Soci. Mater. Sci. Jpn, **64**, 528 (2015).
5. T. Hanlon, E. D. Tabachnikova, S. Suresh, Inter. J. Fatigue, **27**, 1147 (2005).
6. Y. Murase, H. Kobayashi, K. Tanaka, H. Kimachi, Proc. 33rd Symp. Fatigue, JSMS, CD-ROM No. 41 (2016).
7. Y. Akiniwa, K. Tanaka, Trans. Jpn Soci. Mech,. Eng. **A53**, 393 (1987).
8. M. H. El Haddad, K. N. Smith, T. H. Topper, J. Eng. Mater.Tech., Trans. ASME, **101**, 42 (1979).
9. K. Tanaka, Y. Nakai, M. Yamashita, Inter. J. Fracture, **17**, 519 (1981).
10. K. Tanaka, Y. Akiniwa, Eng. Fract. Mech., **30**, 963 (1988).
11. Y. Murakami, Engng. Frac. Mech., **16**, 373 (1992).
12. P. Lukas, M. Klesnil, Mater. Sci. Eng., **34**, 61(1978).

Mechanical Integrity of Steel Discs with Corrosion Pits

Jose Urbano[1,*], *Andrej* Golowin[2], *Stefan* Löhnert[3] and *Dieter* Bestle[1]

[1]Chair of Engineering Mechanics and Vehicle Dynamics, Brandenburg University of Technology, Siemens-Halske-Ring 14, 03046, Cottbus, Germany
[2]Rolls-Royce Deutschland Ltd & Co KG, Eschenweg 11, 15827 Blankenfelde-Mahlow, Germany
[3]Institute of Continuum Mechanics, Leibniz University Hannover, Appelstraße 11, 30167 Hannover, Germany

Abstract. Currently, prediction of crack initiation by corrosion pits is only possible by assuming regular geometrical shapes, such as semi-spheres or semi-ellipsoids. Moreover, typical fatigue life diagrams associate the crack initiation life with geometrical features, such as pit depth or aspect ratio, often leading to unsatisfactory correlations due to high pit shape variability and data scatter. In the context of blade-disc fixation in aero engine turbines, this limitation translates into highly conservative life estimations. Therefore, a new crack initiation predictor is formulated based on experimental testing and numerical analysis of 28 artificial corrosion pits. A low-cycle fatigue test campaign is conducted using three-point bending test specimens to simulate maximum takeoff operation conditions of the aero engine and the associated loading of the blade root designed as firtree. An artificial pit is located at the critical point of each test specimen, respectively. The prediction criterion is based on finite element analysis and is formulated as the lowest plastic strain of a plastic region with a certain volume in the corrosion pit. This reference volume is varied until an optimum correlation with experimental crack initiation life is obtained. The criterion shows a superior correlation with crack initiation life compared to pure geometrical parameters such as pit depth.

1 Introduction

Corrosion of critical parts is one of the key cost drivers in aerospace industry. Corrosion problems typically appear in the mature fleet and are considered as most difficult and expensive due to the potential size of the affected fleet and variety of factors influencing life prediction and fleet management.

In the design of steel discs for high-pressure turbines (HPT), corrosion criticality is evaluated by estimating the number of cycles required to initiate a crack by a corrosion pit (crack initiation life), and the number of cycles required for this crack to propagate and cause macroscopic fracture (crack propagation life). The crack propagation life is typically obtained from a linear elastic fracture analysis with an initial crack size equal to the depth of the corrosion pit [1–4]. The crack initiation life may also be estimated by crack propagation analysis until a specific engineering crack [5,6], crack growth rate or stress intensity factor is reached [1–4,7,8]. To overcome the high conservatism in such a crack initiation life calculation, current research is conducted on two major fronts.

The first focuses on formulating improved analytical models for estimating stress intensity factors using linear elastic fracture mechanics, see [8–10]. Recent efforts have been devoted to enhance the pit model by considering more specific shape factors, accurate stress concentration factors using finite element simulations, and multi-step crack growth hypotheses [3,4]. One of the major drawbacks of these approaches relies on the approximation of the pit shape as hemispherical or semi-elliptical. Although this is a necessary assumption for the development of analytical solutions, in reality the corrosion pit shape is rather irregular (Fig. 1), and it must be taken into account since it may have important implications on local strain concentrations.

Fig. 1. Corrosion pits at the firtree joint of an HPT disc.

The second research front deals with an improved micromechanical description of the crack initiation mechanisms using the continuum damage approach [11,12]. Damage models offer an analytical description of the mechanical interaction between pit growth and

* Corresponding author: urbanoca@b-tu.de

damage laws. Improved technologies such as accelerated 3D microstructure characterization may facilitate the applicability of such models in the future [13], however, due to the increased amount of model parameters, the industrial application of such approaches is still limited.

Instead of developing an analytical model for crack growth based on geometric simplifications, the present paper aims at formulating a crack initiation predictor which considers the real pit geometry and individual load level. The method is developed to predict the crack initiation life of a corroded high-pressure turbine disc of an aero engine. A high-pressure turbine disc is prone to corrosion pitting at the firtree joint, which is the location of highest mechanical loading (Fig. 1). Therefore, the mechanical integrity of the turbine disc may be evaluated by estimating the required number of cycles for crack initiation by a corrosion pit located at the firtree.

In practical terms, a crack initiation predictor is a criterion which exhibits a useful correlation with real crack initiation life. To formulate this predictor, an experimental test campaign with 28 simplified test specimens representing the disc firtree is carried out. Each test specimen contains an artificially created corrosion pit (hereafter referred as artificial pit), which imitates the shape of corrosion pits found in used HPT discs (hereafter referred as ex-service pit). From this experimental test campaign, crack initiation lives $N_{I,i}$ of all artificial pits are obtained. The surface scans of the artificial pits are used in a finite element analysis, and from the individual simulation results a predictor q_i is formulated for each artificial pit i. The goal is to maximize the correlation of the data set

$$(N_{I,i}, N_I(q_i)),\ i = 1, \ldots, 28, \tag{1}$$

between the experimentally found crack initiation life $N_{I,i}$ and the estimate $N_I(q_i)$ based on the numerical predictor q_i.

The paper is organized as follows: after describing the experimental test campaign, a finite element model of the test specimen will be discussed. A crack initiation predictor will be formulated using the representative-volume method, and validated with the data set (1). Finally, the found correlation will be applied to estimate the crack initiation life of ex-service pits.

2 Experimental test campaign

Since it is not possible to properly load and test corrosion pits on HPT discs directly, a test campaign based on a simplified mechanical model of the disc firtree was developed and a three-point bending test was chosen, see Fig. 2. The geometry of the specimen is designed as a rectangular plate with a V-notch such that the angle and depth of the V-notch correspond to the firtree lobe, Fig. 3. A population of 28 test specimens was manufactured using ex-service HPT discs for an ideal representation of material properties.

Each manufactured test specimen was sent to the National Physical Laboratory (NPL) at Teddington, UK, for artificial pit application. A corrosion experiment was

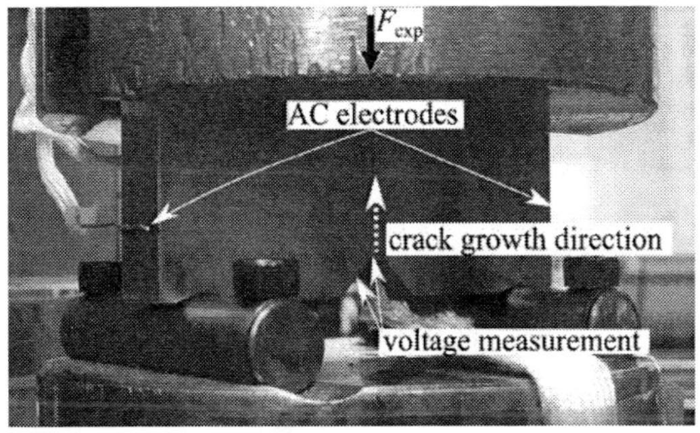

Fig. 2. Instrumentation of the test specimen.

designed to produce artificial pits with shapes and dimensions similar to those observed in ex-service discs. Each artificial pit is created using a droplet of saline solution on an electrified test specimen. Since the highest stress occurs at the bottom of the V-notch, it was decided to incubate the artificial pits at this location, see Fig. 3. The diameter of the droplet, the electrical current, and the exposure time where controlled to achieve the requested pit depth, as well as desired major and minor pit diameters. The 3D images of the artificial pits were compared to those taken from ex-service pits to ensure representativeness.

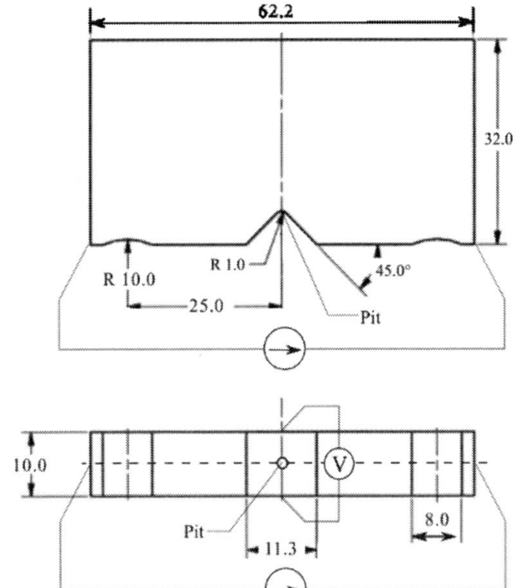

Fig. 3. Test specimen for three-point bending test (dimensions given in mm).

The pitted specimens were sent back to the Rolls-Royce Mechanical Test Operations Centre (MTOC) at Dahlewitz, Germany, for low-cycle fatigue testing. A single load cycle is defined by four loading stages: 1 s linearly increasing loading ramp to a maximum load F_{exp}, 1 s constant loading at F_{exp}, 1 s linearly decreasing loading ramp until 10% of F_{exp}, and 1 s constant loading at 10% of F_{exp}. The test cycles were repeated until failure by fracture and counted by N_T.

The maximum load F_{exp} is designed to achieve a desired stress field at the V-notch. It is based on the nominal load F_{nom} which produces the same stress field at the V-notch as the one in the firtree lobe during maximum takeoff. This loading is considered as the maximum loading during the complete flight cycle. Specimens with different pit depths were assigned to one of the four load levels

$$F_{\text{exp}} = \alpha F_{\text{nom}} \tag{2}$$
$$\text{where } \alpha \in \{95\%, 100\%, 105\%, 110\%\},$$

such that each of seven pit depth magnitudes was combined with each of the four load levels. The specimen is supported by two rollers on the lower side, see Fig. 2, and the contact between the loading piston and the upper side of the test specimen is a similar internal roller inside the piston with the same radius as the support rollers (not visible in Fig. 2, but see model in Fig. 6a).

The artificial pit intensives the nominal stress at the V-notch. Since this location experiences the highest bending stress, this is where crack initiation is expected. The crack growth follows the path of highest stress gradient, which in this case is perpendicular to the bending direction, see Fig. 2. The crack growth is estimated by inducing an alternating electric current across the test specimen and measuring the change of the electric potential v in the V-notch (see Fig. 2 for instrumentation location). A typical behavior of voltage v vs. number of test cycles N_T is shown in Fig. 4.

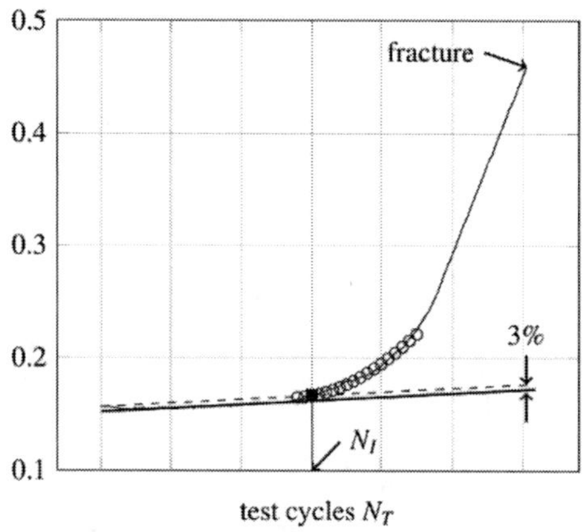

Fig. 4. Voltage at the test specimen vs. number of test cycles: raw signal (—), linear fit (—), 3% offset line (--), polynomial fit (○), and point of intersection (■).

In the beginning, the increase is linear and rather slow. Here, a linear crack growth may be assumed, and thus a direct proportional relation between measured voltage and crack length. After a specific amount of cycles, the crack growth is significantly accelerated. The amount of cycles up to this transition point is defined as the number of cycles for crack initiation N_I (also crack initiation life). The remaining cycles until specimen failure refer to crack propagation.

The identification of the cycles to crack initiation N_I is obtained by the following procedure: two sets of experimental data points are selected. The first set is taken at the initial stage of crack growth and a linear curve is fitted to this data set. Additionally, a second line with a 3% offset is defined. The second set of points is taken from the middle region of the test curve and a third-order polynomial is fitted to this data set. The point of intersection between the polynomial approximation and the 3% offset line is defined as number of cycles N_I to crack initiation. The range of the two data sets is based on experience from previous crack initiation measurements.

Traditionally, the crack initiation life is correlated with geometrical predictors, such as pit depth or aspect ratio. Such a correlation between the artificial pit depth D and the experimentally found value N_I for crack initiation for all 28 tested specimens is shown in Fig. 5. Despite the high scatter in the data, some relationship between crack initiation life and load level may be deduced: as the load level increases, crack initiation life becomes shorter. Further, the relevance of the pit depth seems to be more influential at lower load levels. For example, at 95% load level an increase in pit depth is generally associated with some shortening of crack initiation life, however, pit depth has almost no influence at 110% load level. Thus, this correlation picture of crack initiation life with pit depth has obvious limitations. First, a clear distinction of different load levels should be performed which, however, would decrease the population size and thus statistical significance. Furthermore, it must be considered that the pit shapes are extremely variable and irregular. Therefore, different crack initiation lives may be expected from different pits with same pit depth. In the following, a crack initiation predictor will be developed to overcome such limitations.

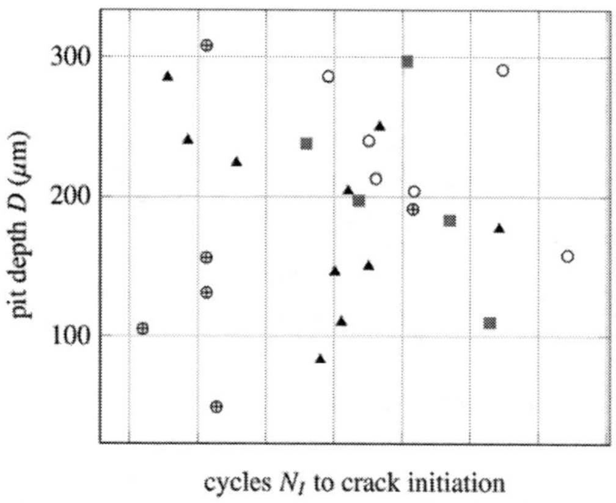

Fig. 5. Pit depth vs. crack initiation life for different load levels: 95% (○), 100% (■), 105% (▲) and 110% (⊕).

3 Finite element model of test specimen

Crack nucleation involves an interaction of elastic stress and plastic strain concentrations [3,6,14]. Thus, a crack predictor should consider the intrinsic mechanical loading at the pit to provide a more satisfactory correlation with crack initiation. Furthermore, a representative measure of the plastic strain concentration at the pit would make the desired predictor independent of the pit geometry. For these reasons, the goal is to formulate a crack initiation predictor based on the actual plastic strain concentration at the pit, which will be calculated from a finite element analysis.

A structural simulation of a corrosion pit using real surface measurements is possible by the following two steps: firstly, a geometry processing workflow in Siemens NX to consider the measured pit surface in the specimen CAD geometry; secondly, a plastic analysis in Abaqus CAE to obtain an accurate representation of the stress and strain field at the pit.

The geometry processing workflow prepares the surface scan to be applied to the model geometry. Each artificial pit was scanned with a high-precision laser-scanning microscope at NPL, and the surface scan was then processed in Siemens NX to result in an applicable CAD geometry. First, the surface scan was trimmed to the internal pit surface. Then, a smoothing was performed by fitting a B-spline surface to the trimmed surface scan, which facilitates the subsequent finite element meshing. Finally, the surface was positioned in a CAD geometry of the test specimen, and the material above the pit surface was removed using an extrusion procedure. This surface processing is similar to the one reported in [3]. It should be emphasized that the transition from the pit surface to the outer specimen surface should be kept as smooth as possible to avoid meshing errors. Fortunately, most of the high plastic strain regions are located at the inside of the pit, as it will be demonstrated later. Therefore, the relevant regions are barely affected by this rim smoothing.

The second step starts with the preparation of a finite element model, where the goal is to obtain a precise representation of the strain field at the pit surface. For this purpose, a global model is used for a coarse finite element analysis of the whole test specimen (Fig. 6a), and the pit region is then refined using two sequential submodels (Fig. 6b-c). The global model determines displacements of the test specimen considering load and fixation boundary conditions, but not the pit. Then, a first submodel transfers the obtained nodal displacements to the boundaries of a section of the specimen near the pit and calculates displacement with a refined mesh, where the pit is still not considered (Fig. 6b). Only the second sectional submodel in Fig. 6c has a sufficiently high mesh refinement to include the corrosion pit for obtaining the desired stress and strain information of the pit region. Of course, the section walls for the submodels must be positioned at locations where the strain field is not affected by the missing pit.

The global model in Fig. 6a considers the complete test assembly composed of test specimen, the two support cylinders, and an upper half-cylinder transmitting the piston load. The lower half of the support rollers is completely constrained, whereas the upper half has a friction contact with the specimen surface (friction coefficient $\mu = 0.4$). The contact between the upper half cylinder and the test specimen is simulated as a non-slip contact, and the uniform pressure on the upper cylinder matches the experimental load F_{exp}.

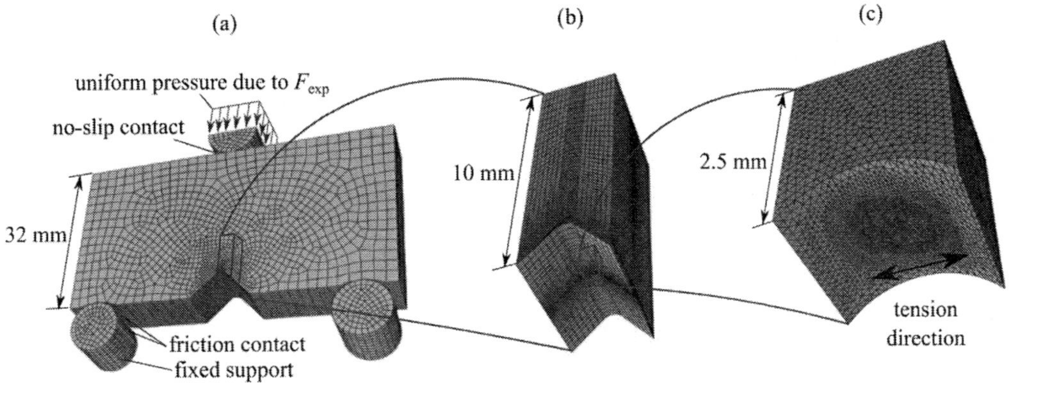

Fig. 6. Meshing of specimen-roller assembly: (a) global model, (b) first and (c) second submodel.

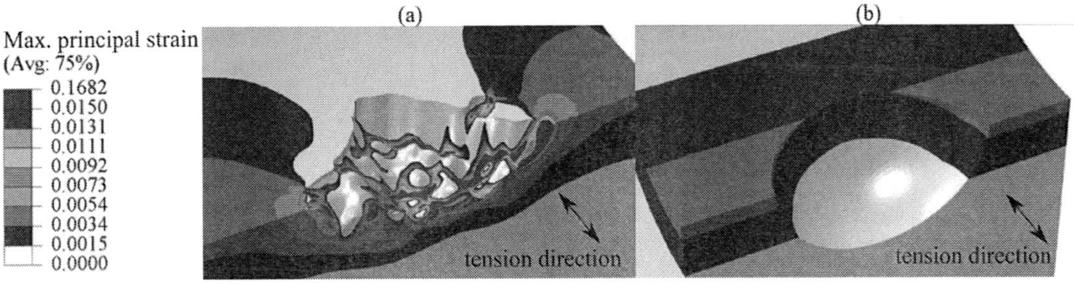

Fig. 7. Maximum plastic strain for an (a) ex-service and (b) idealized pit.

The static plastic analyses of the three models are conducted with Abaqus CAE. An isotropic plastic hardening material model based on the von Mises yield criterion is utilized for the test specimen. The stress-strain relationship at room temperature for the disc material (FV535 steel) is taken from [15]. Although the experiment involves cyclic loading, the simulation considers only the first loading ramp. This is consistent with the isotropic plastic hardening assumption, since no further incremental strain should be expected by simulating additional cycles.

The global model is meshed with second order hexahedral elements. The first submodel refines the region about the V-notch also with second order hexahedral elements. Finally, the second submodel is meshed using second order tetrahedral elements to provide the required geometric flexibility for modelling the pit surface. The average element size in the second submodel is 0.015 mm, and the mesh consists of approximately 180000 elements.

The geometry processing and the finite element analysis are applied to all 28 test specimens of the test campaign in Section 2 using individually measured pit surfaces and individual load levels to obtain the plastic strain values. This procedure is also suitable in Section 5, to simulate ex-service pits located in a HPT firtree.

In order to see the need of taking into account the real pit geometry, plastic strains are computed for one of the ex-service pits and compared to those of an idealized pit, which is defined as the smallest half-ellipsoid that completely encloses the ex-service pit, see Fig. 7. The latter is a classical simplification used in finite element modelling of corrosion pits, see [9,16]. The sectional view at the middle plane of the pit reveals that the ex-service pit has a totally different strain pattern than the idealized pit, although both have identical boundary conditions. Critical areas in Fig. 7a with high plastic strain occur mainly due to surface irregularities of the pit, especially at surface peaks and valleys. These high-strain regions tend to group into longitudinal bands perpendicular to the direction of tension, which has also been observed on previous finite element analyses of corrosion pits [3,9,16]. Regions with highest plastic strains most likely represent potential crack nucleation sites [3]. Therefore, beside plastic strain values, an additional outcome of the shown finite element analysis is the identification of potential crack nucleation locations.

4 Crack initiation prediction with the representative-volume method

Although the plastic strain provides insight in the location and severity of crack nucleation sites, a summarizing criterion for comparison of different pits is still absent. Therefore, the goal of the present section is to define a strain measure which is representative for the whole pit. In the following, the plastic strain will be associated with a characteristic volume, which is why this method will be called representative-volume method. The idea of the method is to identify the volume of the most critical plastic strain regions, such that its reference volume V_{ref} and associated plastic strain $\varepsilon_{\text{ref}}^{P}$ may be used for crack initiation life predictions which maximize the correlation with experimentally found results.

Let us define the index subset $\mathbb{D}$ as the collection of element indexes e whose maximum principal plastic strain $\varepsilon_{\max,e}^{P}$ exceeds a prescribed reference plastic strain $\varepsilon_{\text{ref}}^{P}$, i.e.,

$$\mathbb{D} = \{e: \varepsilon_{\max,e}^{P} \geq \varepsilon_{\text{ref}}^{P}\}. \tag{3}$$

The size $|\mathbb{D}|$ of this subset depends on the value of $\varepsilon_{\text{ref}}^{P}$, where a higher reference plastic strain will lead to a smaller subset $\mathbb{D}$. The sum of volumes V_e of all elements belonging to $\mathbb{D}$ is defined as the reference volume

$$V_{\text{ref}} = \sum_{e \in \mathbb{D}} V_e\,, \tag{4}$$

The relationship $V_{\text{ref}} = V_{\text{ref}}(\varepsilon_{\text{ref}}^{P})$ between reference volume and reference plastic strain is individual to every pit and may be graphically represented in a volumetric accumulation diagram as shown in Fig. 8. This diagram reveals the volumetric distribution of plastic strains in the pit, and may be obtained by the following procedure:

1. For each element e of the second submodel in Fig. 6c, compute the maximum principal plastic strain $\varepsilon_{\max,e}^{P}$ at the element centroid by finite element analysis.
2. Sort the values of the maximum plastic strain in ascending order to get an element index set

$$\mathbb{E} = \{e_1, \cdots, e_j, \cdots, e_{N_E}\} \tag{5}$$
$$\text{where } \varepsilon_{\max,e_{j+1}}^{P} \geq \varepsilon_{\max,e_j}^{P}\,,$$

   and N_E corresponds to the number of elements in the second submodel.
3. Calculate the reference volumes $V_{\text{ref},j}$ associated with occurring reference plastic strains $\varepsilon_{\text{ref},j}^{P}$ recursively as

$$\varepsilon_{\text{ref},j} = \varepsilon_{\max,e_j}^{P}\,, \tag{6}$$
$$V_{\text{ref},j} = V_{\text{ref},j+1} + V_{e_j}\,, \tag{7}$$
$$\text{where } e_j \in \mathbb{E},$$
$$V_{\text{ref},N_E} = V_{e_{N_E}}\,,$$
$$\varepsilon_{\text{ref},N_E} = \varepsilon_{\max,e_{N_E}}^{P}\,,$$
$$j = N_E - 1, N_E - 2, \ldots, 1.$$

The results (6) and (7) define the volumetric accumulation diagram for a specific pit as

$$V_{\text{ref},j} = V_{\text{ref}}(\varepsilon_{\text{ref},j}^{P})\,,\ j = 1, \cdots, N_E. \tag{8}$$

Additionally, a reference length L_{ref} associated with the reference volume V_{ref} may be introduced as

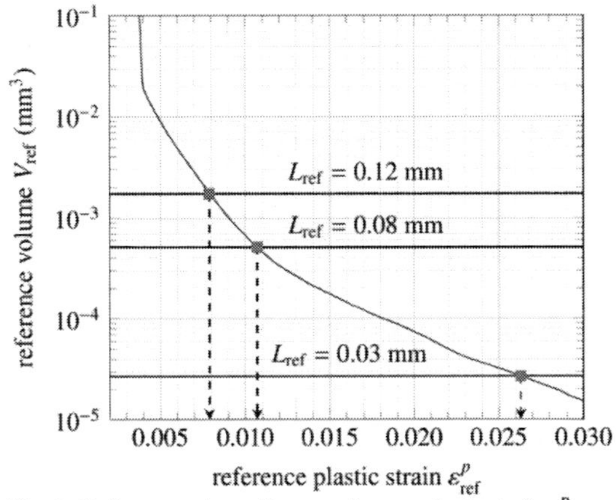

Fig. 8. Reference volume V_{ref} vs. reference plastic strain ε^P_{ref} for a specific artificial pit.

$$L_{ref} = \sqrt[3]{V_{ref}}\,. \tag{9}$$

The volumetric accumulation diagram in Fig. 8 associates a plastic strain ε^P_{ref} with the size of a plastic region V_{ref}. For instance, the reference lengths $L_{ref} = 0.12$ mm, 0.08 mm, and 0.03 mm are associated with the plastic strains $\varepsilon^P_{ref} = 0.0079$, 0.0107, and 0.02629. Obviously, larger element volumes are associated with lower plastic reference strains, Fig. 9.

Based on the assumption that high plastic strains will increase the probability of crack initiation, and thus shorten crack initiation life, we may now reproduce Fig 8 for all 28 artificial pits of the test campaign in Section 2. This is done by extracting plastic reference strains $\varepsilon^P_{ref,i}$ for prescribed values of L_{ref} and correlating them to the experimentally obtained crack initiation lives $N_{I,i}$, see Fig. 10. Obviously there is a strong correlation between these two quantities, which is why

$$q := \varepsilon^P_{ref}(V_{ref}) \quad \text{or} \quad q := \varepsilon^P_{ref}(L_{ref}) \tag{10}$$

may act as a predictor for estimating the crack initiation life N_I by

$$\ln N_I = -m \ln q + a. \tag{11}$$

An optimal choice for the slope $(-m)$ and y-intercept a may be found by a least-square fitting of the data $(q_i = \varepsilon^P_{ref,i}, N_{I,i})$, $i = 1, \ldots, 28$, to fulfil Eq. (11), i.e.,

$$\min_{m,a} \sum_{i=1}^{28} \left(\ln N_{I,i} + m \ln q_i - a\right)^2, \tag{12}$$

see solid lines in Fig. 10.

The fact that a rather general reference strain ε^P_{ref} corresponding to a reference volume V_{ref} of material act as a reliable predictor in a wide range (compare Figures 10a-c with L_{ref} ranging from 0.03 mm to 0.12 mm) indicates, that not only the singular point of maximum plastic strain ε^P_{max} is responsible for crack initiation. Instead, a whole region with plastic strains above some lower threshold $\varepsilon^P_{ref} < \varepsilon^P_{max}$ may be exposed to micro-cracks which then coalesce to a macroscopic crack determining the crack initiation life N_I.

The quality of the data fitting (12) may be quantified by the correlation coefficient

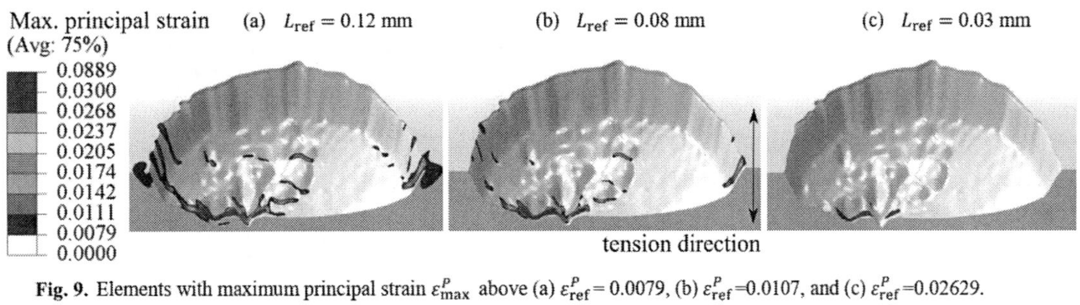

Fig. 9. Elements with maximum principal strain ε^P_{max} above (a) $\varepsilon^P_{ref} = 0.0079$, (b) $\varepsilon^P_{ref} = 0.0107$, and (c) $\varepsilon^P_{ref} = 0.02629$.

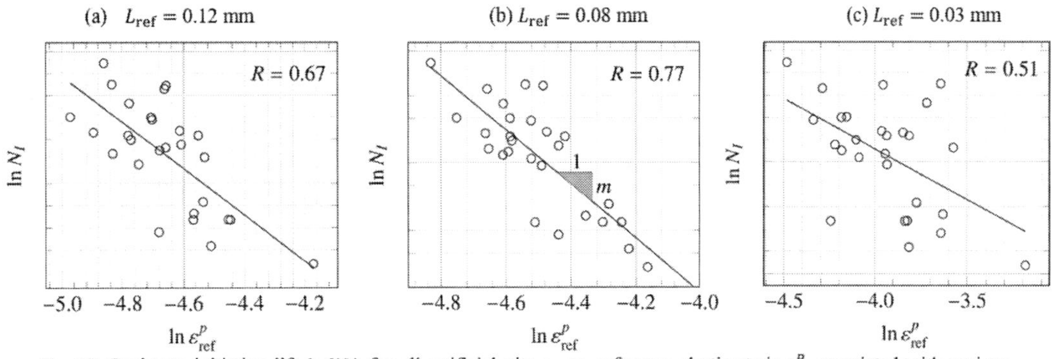

Fig. 10. Cycles to initiation life $\ln(N_I)$ for all artificial pits p_i vs. reference plastic strain ε^P_{ref} associated with various representative lengths.

$$R = \frac{n\sum N_{I,i}\widetilde{N}_{I,i} - (\sum N_{I,i})(\sum \widetilde{N}_{I,i})}{\sqrt{n\sum N_{I,i}^2 - (\sum N_{I,i})^2}\sqrt{n\sum \widetilde{N}_{I,i}^2 - (\sum \widetilde{N}_{I,i})^2}} \quad (13)$$

where $\widetilde{N}_{I,i} = \exp(-m \ln q_i + a)$.

The coefficient R is a statistical measure for the deviation between the experimentally obtained $(N_{I,i})$ and the predicted $(\widetilde{N}_{I,i})$ cycles to crack initiation. The coefficient delivers values in the range $-1 \leq R \leq 1$, where 1 indicates a perfect match.

According to Fig. 10, the strength of correlation depends on the reference length. A larger reference length L_{ref} = 0.12 mm supposes that rather low plastic strain values may influence crack initiation life, Fig. 10a, which, however, is inadequate. A small reference length L_{ref} = 0.03 mm corresponding to a small reference volume assesses only elements with the highest plastic strain peaks in the corrosion pits, Fig. 10c. The corresponding correlation is also unsatisfactory, since these highest strain values usually occur from mesh irregularities or sharp edges. Therefore, the representative length producing the best correlation has to be chosen carefully.

In order to identify the optimum reference length (or volume), several reference lengths have been fixed and the corresponding correlation coefficient for the artificial pits has been calculated, see Fig 11, where the grey markers correspond to the results in Fig.10. Obviously, the reference length L_{ref} = 0.08 mm delivers the highest value of the correlation coefficient, and thus the best correlation (which can also be seen in Fig. 10b). The optimal values of representative length, volume and plastic strain are denoted as L_R, V_R and ε_R, respectively.

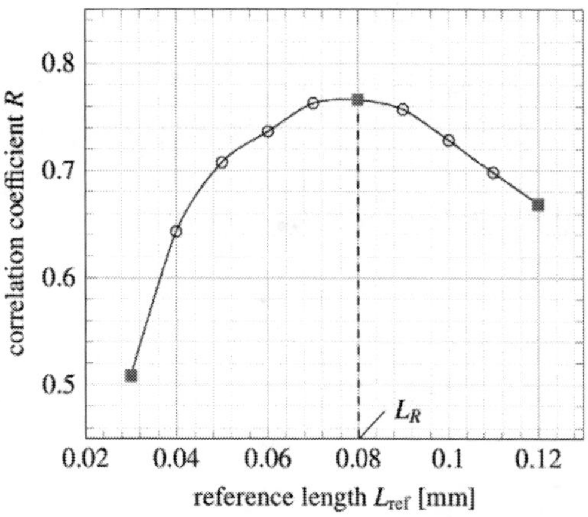

Fig. 11. Correlation coefficient R vs. characteristic length L_{ref} for tested specimens for test population.

A striking finding for application is that the optimum reference strain is rather robust against mesh refinement. Fig. 12 shows volumetric accumulation diagrams for one of the artificial pits (different from the artificial pit used in Fig. 8) using different mesh refinements. As expected, a higher mesh refinement may lead to a more precise description of regions with high plastic strain, (i.e., regions with low reference length L_{ref} < 0.03 mm), however, for the optimum reference length L_R = 0.08 mm nearly the same value ε_R is obtained from all finite element analyses. Therefore, an expensive mesh refinement of the pit surface is unnecessary for getting an optimal correlation with crack initiation life.

Throughout this work, the plastic strain (10) is used as a predictor for crack initiation life. The same procedure may also be based on stress values instead. However, since plastic strain is more sensitive than stress to load increments, a better correlation was observed by using the plastic strain as a crack initiation predictor.

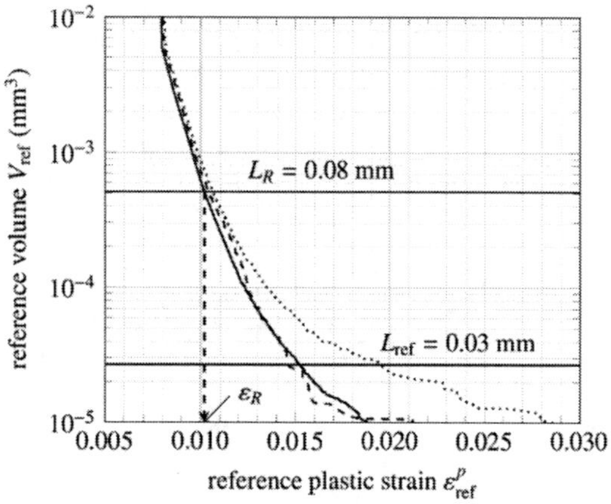

Fig. 12. Reference volume V_{ref} vs. reference plastic strain ε^p_{ref} obtained from different meshes with 83128 (--), 113936 (···), and 254637 (—) elements.

5 Application of the representative-volume method to ex-service pits

Finally, the proposed procedure is used to estimate the crack initiation life of six ex-service pits located at the firtree of a real HPT disc taken from an operated aero engine. Since the artificial pit population in Section 2 was designed to use similar loads and pit sizes as these ex-service pits, it is now possible to use the estimator (11) based on Fig. 10b also for ex-service pit life estimation. This means, that only the calculation of the representative strain of the ex-service pits is necessary.

According to Section 3, the surfaces of the ex-service pits have been scanned and smoothly approximated by a B-spline surface. They were then implemented in a finite element model to obtain the stress field and plastic strains. In the present case, the simulation load was set to 100% of the nominal load. Once the finite element problem is solved, a volumetric accumulation diagram similar to Fig. 8 may be drawn for each ex-service pit. The representative plastic strain $\varepsilon_{R,k}$ of each ex-service pit k is computed by using the optimum reference length L_R found in Section 4, and it is used as input $q_k = \varepsilon_{R,k}$ for the estimator (11) to obtain the associated crack initiation lives $N_{I,k}$, $k = 1, \ldots, 6$.

To compare the estimated crack initiation lives of ex-service pits with those of artificial pits, we consider the artificial pit data set at the optimum reference length,

Fig. 10b, and transfer it to a traditional fatigue life diagram in Fig.13, i.e., a non-logarithmic drawing with representative strain ε_R as ordinate and cycles to crack initiation N_I as abscissa. It is observed that the ex-service pits (grey markers) have a higher representative strain than most of the artificial pits. Therefore, a rather short crack initiation life may be expected from the estimating curve.

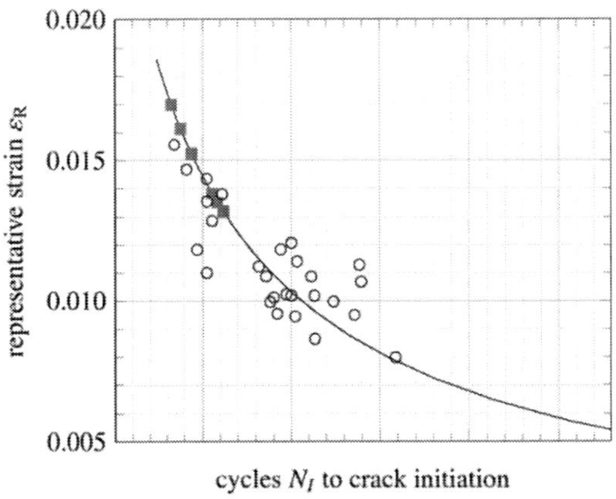

Fig. 13. Crack initiation life for ex-service pits (▪), determined with L_{ref} = 0.08 mm) and artificial pits (○).

6 Conclusions

Classical crack initiation predictors such as pit depth seem to offer only a low correlation with crack initiation life and are highly dependent on pit shape, as a test campaign on 28 artificially produced pits shows. Therefore, a new strategy called the representative-volume method is developed, which overcomes this limitation by formulating the representative strain as a predictor. The crack initiation life may be estimated by an exponential function trained by a population of 28 artificially pitted test specimens. The estimator is rather robust against different pit sizes and shapes, loading, and meshing of the finite element analysis. The proposed approach provides a valuable tool for corrosion pit characterization, which significantly reduces the conservatism of current methods and provides a platform for future technology development based on robotic visual inspection, automated damage measurement, and decision on damage repair or acceptance without human involvement.

This work has been carried out in collaboration with Rolls-Royce Deutschland. The first author is funded by ANTriebE (Adaptative Neuorientierung von Triebwerks-Entwurfssystemen) financed by the State of Brandenburg and the European Regional Development Fund. Rolls-Royce Deutschland's permission to publish this work is greatly acknowledged. The authors also thank the National Physical Laboratory for their assistance in the design, production and analysis of artificial corrosion pits.

References

1. B. M. Schönbauer, S. E. Stanzl-Tschegg, A. Perlega, R. N. Salzman, N. F. Rieger, S. Zhou, A. Turnbull, and D. Gandy, Int. J. Fatigue **65**, 33 (2014).
2. B. M. Schönbauer, A. Perlega, U. P. Karr, D. Gandy, and S. E. Stanzl-Tschegg, Int. J. Fatigue **76**, 19 (2015).
3. S. Xu and Y. Wang, Int. J. Fatigue **55**, 193 (2014).
4. S. I. Rokhlin, J. Y. Kim, H. Nagy, and B. Zoofan, Eng. Fract. Mech. **62**, 425 (1999).
5. A. Shekhter, B. R. Crawford, C. Loader, and W. Hu, Eng. Fail. Anal. **55**, 193 (2015).
6. D. L. McDowell, Mater. Sci. Eng. A **468–470**, 4 (2007).
7. P. Shi and S. Mahadevan, Eng. Fract. Mech. **68**, 1493 (2001).
8. M. Mao, X. Zhang, S. Tu, and F. Xuan, J. Aircr. **51**, 805 (2014).
9. M. Cerit, K. Genel, and S. Eksi, Eng. Fail. Anal. **16**, 2467 (2009).
10. A. Turnbull, L. N. McCartney, and S. Zhou, Scr. Mater. **54**, 575 (2006).
11. M. Amiri, A. Arcari, L. Airoldi, M. Naderi, and N. Iyyer, Corros. Sci. **98**, 678 (2015).
12. P. Hu, Q. Meng, W. Hu, F. Shen, Z. Zhan, and L. Sun, Corros. Sci. **113**, 78 (2016).
13. M. D. Sangid, Int. J. Fatigue **57**, 58 (2013).
14. J. T. Burns, J. M. Larsen, and R. P. Gangloff, Fatigue Fract. Eng. Mater. Struct. **34**, 745 (2011).
15. B. Erice, F. Gálvez, D. A. Cendón, and V. Sánchez-Gálvez, Eng. Fract. Mech. **79**, 1 (2012).
16. A. Turnbull, L. Wright, and L. Crocker, Corros. Sci. **52**, 1492 (2010).

Crack initiation prediction of additive manufactured ductile nickel-based superalloys

Thomas Lindström[1,*], *Robert* Eriksson[1], *Daniel* Ewest[1, 2], *Kjell* Simonsson[1], *Jan-Erik* Lundgren[2], and *Daniel* Leidermark[1]

[1]Division of Solid Mechanics, Linköping University, SE-58183 Linköping, Sweden
[2]Siemens Industrial Turbomachinery AB, SE-61283 Finspång, Sweden

Abstract. A model to predict crack initiation life of an additive manufactured nickel-based superalloy similar to Hastelloy X subjected to low-cycle fatigue loading at room temperature has been developed, taking material anisotropy into account. An anisotropic constitutive model based on the Hill yield criterion was developed, with linear kinematic hardening up to a saturation value of the back stress, above which the material behaves perfectly plastic. Low-cycle fatigue experiments has been performed on additive manufactured smooth bars with two different build orientations, with an angle of 0° and 90° relative to the building platform. A total of 20 experiments at room temperature were conducted with different strain ranges and *R*-values. To predict the crack initiation life of the specimens, a model based on the Smith-Watson-Topper (SWT) parameter has been established, where ten of the specimens were used to calibrate the initiation model, and the remaining specimens were used for validation. Using this model, the obtained crack initiation life agrees well with the experiments.

1 Introduction

Due to the increasing use of renewable energy on the energy market, usage of stationary gas turbines has changed and their role have more become to stabilize the power grid. As a consequence, the running profile of gas turbines will be more cycle-driven, and the components will be exposed to a more fatigue related loading which must be taken into consideration during the design [1]. In order to optimize gas turbine components with respect to efficiency, fuel flexibility and performance, additive manufacturing (AM) has been identified as a major technology due to the possibility to create more complex geometries, and to simplify assemblies of previously multiple components based on conventional manufacturing techniques.

Selective laser melting (SLM) is an AM technique where a thin layer of metallic powder is deposited on a substrate, on which a laser is melting the powder, creating one layer of solid material. A new layer of powder is then spread out and the procedure continues until a complete component is achieved. In SLM, the material will receive a small scale weld-like microstructure. The build platform is held at constant temperature during the process, causing rapid solidification of the material and high temperature gradients. These gradients affect the microstructure by generating elongated grains in the building direction which is the source of the anisotropic behaviour of the material. This has been observed in numerous studies for different alloys, c.f. [2-7]. Components for combustor applications in gas turbines is one example where AM has already been successfully adopted [8]. The use of AM can improve the performance of such components with respect to fuels of different nature and the manufacturing technique can be used to simplify repair and decrease costs. Combustor components in a gas turbine are exposed to high temperatures and large mechanical loads. To withstand the loading conditions, materials with satisfying properties for this working environment have to be used. Combustor components are therefore advantageously manufactured by ductile nickel-based superalloys, due to their oxidation resistance and strength at high temperature.

In order to fully understand how the manufacturing process affects the life of SLM ductile nickel-based superalloy components, accurate prediction models for crack initiation and propagation must be developed, taking material anisotropy into account. Due to the complex loadings of combustor components with cyclic thermal and mechanical loads, SLM manufactured components exposed to thermomechanical fatigue must be analysed and evaluated. In addition, for thermomechanical fatigue loads, the room temperature behaviour is of great importance, since crack growth can be present during e.g. an out-of-phase loading condition.

The aim of this paper is to evaluate the crack initiation behaviour and life of SLM manufactured smooth bars exposed to low-cycle fatigue (LCF) loadings at room temperature. The material used is an alloy similar to Hastelloy X, i.e. a ductile nickel-based superalloy with superior oxidation and high temperature strength. A total of 20 tests have been performed with different strain ranges and *R*-values using specimens manufactured in 0° and 90° direction relative to the building platform. A material model, that takes material anisotropy into

* Corresponding author: thomas.lindstrom@liu.se

account, has been developed to describe the material behaviour. Furthermore, a relation based on the Smith-Watson-Topper (SWT) parameter has been adopted to predict the initiation life of the specimens, where ten of the specimens have been used to calibrate the model, while the remaining ten have been used for validation.

2 Experiments

As mentioned previously, the investigated material used in the LCF experiment was a ductile nickel-based superalloy similar to Hastelloy X. To capture the anisotropic behaviour due to manufacturing, the specimens were built in two different orientations relative the building platform; 0° and 90°, as illustrated in Fig. 1.

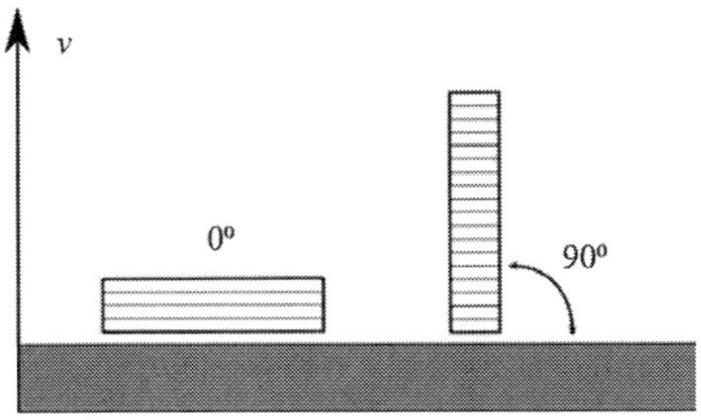

Fig. 1: Illustration of how the test specimens were manufactured, where v is the building direction. 0° specimen (left) manufactured with an angle of 0° to the build platform and 90° (right) manufactured with an angle of 90° from the build platform.

A total of 20 strain controlled LCF experiments were performed with different strain ranges for two R-values and building orientations; $R = 0$ and -1, and 0° and 90°, respectively. The data used in the experiments can be seen in Table 1. The crack initiation lives were taken as 5% load-drop from the trend line of the maximum stress in each cycle.

The specimens were manufactured as cylindrical smooth bars with optimized process parameters in an EOS M290 machine with building layers of 40 μm. As usual when manufacturing SLM components, the scan direction of the laser was rotated in the building plane after every built layer. Due to this rotation, the material properties in the plane parallel to the building platform becomes more or less uniform. The material is therefore assumed to be transversely isotropic, since the material properties in the building direction is different due to the elongated grain structure.

To avoid effects from the surface during fatigue testing, the as-machined diameter of the specimens were 16 mm, and then machined to an hourglass geometry by conventional techniques to the final diameter of 6.35 mm in the gauge section.

The gauge length of the extensometer for the experiments was 11.4 mm.

Table 1 : Data used in the experiments for all specimens.

	Specimen		Δε [%]
0°	S0:1	R=0	0.7
	S0:2		0.9
	S0:3		1.3
	S0:4		1.8
	S0:5		2.5
	S0:6	R=-1	0.7
	S0:7		0.9
	S0:8		1.1
	S0:9		1.5
	S0:10		2.0
90°	S90:1	R=0	0.6
	S90:2		0.8
	S90:3		1.0
	S90:4		1.3
	S90:5		2.6
	S90:6	R=-1	0.8
	S90:7		1.0
	S90:8		1.5
	S90:9		1.99
	S90:10		3.0

In Fig. 2, the applied strain range is plotted against the crack initiation lives obtained from the experiments for the 0° and the 90° specimens, where the crack initiation lives are normalized with the longest life obtained from experiments. As can be seen, the material shows anisotropic behaviour for LCF loading in room temperature, where the anisotropic crack initiation behaviour is most evident for large strain ranges.

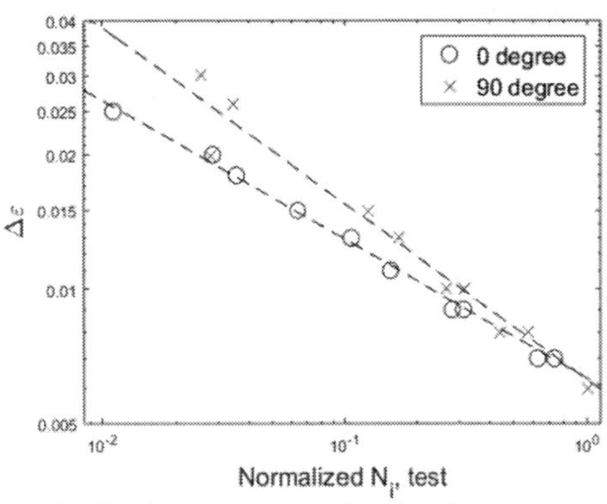

Fig. 2 : Nominal strain range, $\Delta\varepsilon$, plotted against normalized cycles to crack initiation, N_i, obtained from experiments.

As can be seen from Fig. 2, there is a difference between the two building orientations. One can also see that the 90° specimens have longer initiation life compared to 0° specimens, which agrees with several other studies on SLM nickel-based superalloy, see e.g. [2].

When studying the fracture surface of the specimens, the anisotropic behaviour is also evident. In Fig. 3, two specimens with the same loading condition but different building orientations are shown. Comparing the 0°

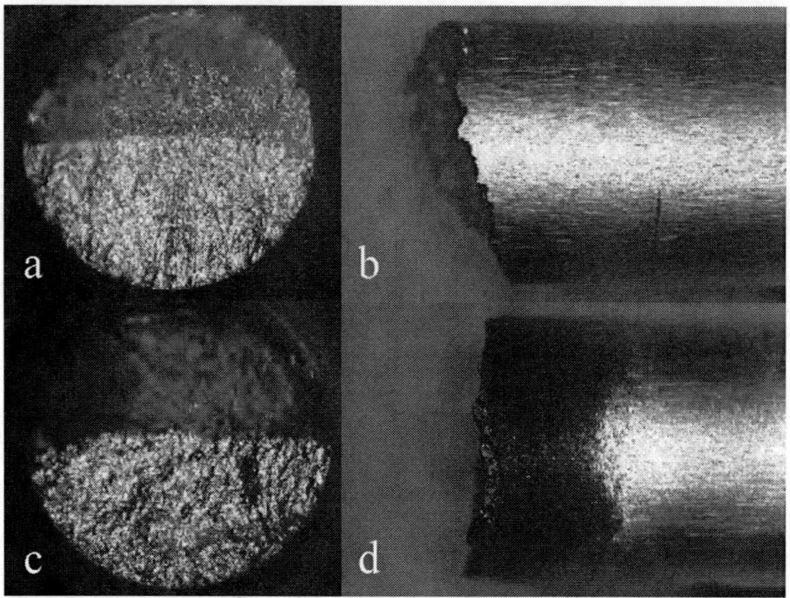

Fig. 3: Fracture appearance for specimen S0:3 a) shown from above, b) from the side, and for specimen S90:4 c) from above and d) from the side.

specimens with the 90° specimens, one can see that the former specimens have a much smoother fracture surface compared to the latter. Furthermore, by looking at the specimens from the side, the fracture of the 90° specimens seem to appear on a plane orthogonal to the loading direction, while the fracture plane for the 0° specimens are tilted w.r.t. the orthogonal plane to the loading.

3 Modelling

3.1 Constitutive modelling

In order to describe the crack initiation behaviour of the material, a transversely isotropic material model was developed, catching the elasto-plastic anisotropic behaviour in the context of small deformations. In the elastic region the stresses in a component subjected to applied strains, can be calculated as:

$$\sigma_{ij} = C^e_{ijkl}\varepsilon^e_{kl} \tag{1}$$

where C^e_{ijkl} is the fourth-order elastic stiffness tensor. Due to transversely isotropic material behaviour, the configuration of the elastic stiffness tensor is dependent on the material direction in the global coordinate system. The material direction is specified by the normal unit vector to the material plane in which the material properties are isotropic. In the later described FE model, the cylindrical bar is modelled with the axial direction in the z-direction in the global coordinate system. Therefore, according to Fig. 1, a 90° specimens corresponds to a material direction $\boldsymbol{v} = [0 \quad 0 \quad 1]^T$ and a 0° specimen corresponds to any unit vector in the x-y plane, here adopted as $\boldsymbol{v} = [0 \quad 1 \quad 0]^T$.

Since the constitutive relations for the material are dependent on the material direction, structural tensors have been used to describe the anisotropic material behaviour. For a transversely isotropic material, the structural tensor is calculated based on the material direction [9] as:

$$M_{ij} = v_i v_j \tag{2}$$

In the case of transversely isotropic materials, the elastic behaviour can be described with five independent elastic material parameters. Two of these, λ and μ_{TT}, describe the isotropic behaviour, while the three other constants, μ_{LT}, α and β, are used to describe the anisotropic behaviour. From this, the elastic stiffness tensor is calculated as:

$$\begin{aligned} C^e_{ijkl} = {} & \lambda(I_{ij}I_{kl}) + \mu_{TT}(I_{ik}I_{jl} + I_{il}I_{jk}) \\ & + \alpha(M_{ij}I_{kl} + I_{ij}M_{kl}) + \beta(M_{ij}M_{kl}) + \\ & (\mu_{LT} - \mu_{TT})(I_{ik}M_{jl} + M_{ik}I_{jl} + I_{il}M_{jk} + M_{il}I_{jk}) \end{aligned} \tag{3}$$

where I_{ij} is the second order unit tensor. For a 90° specimen, i.e. $\boldsymbol{v} = [0 \quad 0 \quad 1]^T$, the elastic stiffness tensor in Voigt notation becomes:

$$\boldsymbol{C}^e = \begin{bmatrix} \lambda + 2\mu_{TT} & \lambda & \lambda + \alpha & 0 & 0 & 0 \\ \lambda & \lambda + 2\mu_{TT} & \lambda + \alpha & 0 & 0 & 0 \\ \lambda + \alpha & \lambda + \alpha & \begin{matrix}\lambda + 2\alpha + \beta \\ +4\mu_{LT} - 2\mu_{TT}\end{matrix} & 0 & 0 & 0 \\ 0 & 0 & 0 & \mu_{TT} & 0 & 0 \\ 0 & 0 & 0 & 0 & \mu_{LT} & 0 \\ 0 & 0 & 0 & 0 & 0 & \mu_{LT} \end{bmatrix}$$

(4)

For elasto-plastic materials undergoing small deformations, the total mechanical strain can be decomposed into one elastic and one plastic part:

$$\varepsilon_{ij} = \varepsilon_{ij}^{e} + \varepsilon_{ij}^{p} \tag{5}$$

The elastic part of the strain tensor follows the relation described in Eq. (1), and the plastic part follows the associated flow rule as:

$$\mathrm{d}\varepsilon_{ij}^{p} = \mathrm{d}\lambda \frac{\partial f}{\partial \sigma_{ij}} \tag{6}$$

where $\mathrm{d}\lambda$ is the plastic multiplier and $\partial f/\partial \sigma_{ij}$ the direction of the plastic flow, where f describes the yield surface. Considering the anisotropic yield behaviour of the material, a Hill yield criterion is used where the parameters describe the different yield limits for the material in the different directions [10].
The yield function is expressed, in similarity with [1], as

$$f = \sqrt{\sigma_y^2 (s_{ij} - \alpha_{ij}) P_{ijkl} (s_{kl} - \alpha_{kl})} - \sigma_y = 0 \tag{7}$$

where s_{ij} is the deviatoric stress tensor, α_{ij} is the back stress tensor and P_{ijkl} is the fourth order anisotropic tensor describing the directional dependent yield behaviour of the material. The anisotropic tensor contains the Hill parameters, and is dependent on the material direction. Therefore, this anisotropic tensor is also built up with use of the structural tensor as:

$$\begin{aligned} P_{ijkl} = (F + 7G + 4N - 8L)(M_{ij}M_{kl}) \\ +(3G - N)(M_{ij}I_{kl} + I_{ij}M_{kl}) \\ +2(L - N)(M_{ik}I_{jl} + I_{ik}M_{jl}) + N(2I_{ik}I_{jl} - I_{il}I_{jk}) \\ -G(I_{ij}I_{kl}) + (N + 2L)(M_{il}I_{jk} + I_{il}M_{jk}) \end{aligned} \tag{8}$$

where $N = F + 2G$. For a 90° specimen, the anisotropic tensor in Voigt notation becomes:

$$\boldsymbol{P} = \begin{bmatrix} F+G & -G & -F & 0 & 0 & 0 \\ -G & F+G & -F & 0 & 0 & 0 \\ -F & -F & 2F & 0 & 0 & 0 \\ 0 & 0 & 0 & 2(F+2G) & 0 & 0 \\ 0 & 0 & 0 & 0 & 2L & 0 \\ 0 & 0 & 0 & 0 & 0 & 2L \end{bmatrix} \tag{9}$$

where F, G and L are the Hill parameters, defined as:

$$F = \frac{1}{2(\sigma_{LT,Y})^2} \tag{10}$$

$$G = \frac{1}{(\sigma_{TT,Y})^2} - F \tag{11}$$

$$L = \frac{1}{2(\tau_Y)^2} \tag{12}$$

where $\sigma_{LT,Y}$ is the yield stress in the direction orthogonal to the building platform, $\sigma_{TT,Y}$ is the yield stress in the direction parallel to the building direction, and τ_Y is the anisotropic shear yield stress, which is not of importance when only considering homogeneous uniaxial loading of 0° and 90° specimens.
In Eq. (7), the constant scalar σ_y is defined as:

$$\sigma_y = \left(\frac{3}{2(2F + G)}\right)^{1/2} \tag{13}$$

Note that, for $F = G$ and $L = 3G$, one ends up with the traditional von Mises yield criterion.
Furthermore, the direction of the plastic flow becomes:

$$\frac{\partial f}{\partial \sigma_{ij}} = \frac{\sigma_y^2 P_{ijkl}(s_{kl} - \alpha_{kl})}{\sqrt{\sigma_y^2 (s_{mn} - \alpha_{mn}) P_{mnpq} (s_{pq} - \alpha_{pq})}} \tag{14}$$

The material is assumed to follow a linear kinematic hardening behaviour, with a saturation value above which the material acts as a perfect plastic material. To capture the saturation of the back stress, the back stress is multiplied with a scalar function, H, which quenches the evolution of it at a certain saturation value.

$$\mathrm{d}\alpha_{ij} = \frac{2}{3} C^p \mathrm{d}\varepsilon_{ij}^{p} H(g_{sat}) \tag{15}$$

where C^p is the linear kinematic hardening coefficient, which is dependent on the material orientation. To take this into account, for a specimen with the axial direction and the applied load in the z-direction in the global coordinate system, C^p is calculated as:

$$C^p = M_{ij} A_{ijkl} M_{kl} \tag{16}$$

The fourth order tensor A_{ijkl} contains the kinematic hardening coefficients in each direction. When only considering 0° and 90° specimens, A_{ijkl} can, in Voigt notation, be written:

$$\boldsymbol{A} = \begin{bmatrix} C_{TT}^p & 0 & 0 & 0 & 0 & 0 \\ 0 & C_{TT}^p & 0 & 0 & 0 & 0 \\ 0 & 0 & C_{LT}^p & 0 & 0 & 0 \\ 0 & 0 & 0 & 0 & 0 & 0 \\ 0 & 0 & 0 & 0 & 0 & 0 \\ 0 & 0 & 0 & 0 & 0 & 0 \end{bmatrix} \tag{17}$$

where C_{TT}^p and C_{LT}^p are the linear kinematic hardening coefficient in the direction parallel and orthogonal to the building platform, respectively.
In Eq. (16), the structural tensor is used to pick out the kinematic hardening coefficient in the current direction.
In addition, in Eq. (15), H(*) is the Heaviside function and the expression inside the Heaviside, g_{sat}, accounts

https://doi.org/10.1051/matecconf/201816504013

for the saturation of the back stress tensor in the following way:

$$g_{sat} = \alpha_{sat} - \frac{\partial f}{\partial \sigma_{ij}} \alpha_{ij} \quad (18)$$

where α_{sat} is the magnitude of the saturation value of the back stress, which has been found to be linearly proportional to the kinematic hardening coefficient for this material. The saturation value of the back stress can therefore be calculated by scaling the kinematic hardening coefficient by the scalar w, according to:

$$\alpha_{sat} = wC^{p} \quad (19)$$

This constitutive model has similar features as the Ohno-Wang Model 1, further described in [11], but takes anisotropic material behaviour into account through the anisotropic tensor P_{ijkl}. Furthermore, the definition of the saturation value of the back stress differs between the above described model and the Ohno-Wang Model 1.
The material parameters have been extracted from stabilized hysteresis loops at mid-life of the specimens, and the parameters were taken as the mean of the values from the experiments for the two different build orientations.

3.2 Crack initiation model

From the experiments, it was observed that using the SWT parameter [12] to describe the initiation life, one can capture the differences between the two building orientation and establish a relation for the crack initiation life:

$$\sigma^{max}\varepsilon^{a} = a(N_i)^{b} \quad (20)$$

where the left-hand side of the expression is the SWT parameter, σ^{max} is the maximum stress, ε^{a} is the strain amplitude, N_i is the number of cycles to crack initiation and a, b are constants determined by calibration. The maximum stress and the strain amplitude is extracted at mid-life of the specimens. Fig. 4 shows the SWT parameter, normalized with the largest SWT parameter calculated from the experiments, against normalized cycles to crack initiation observed from LCF experiments.

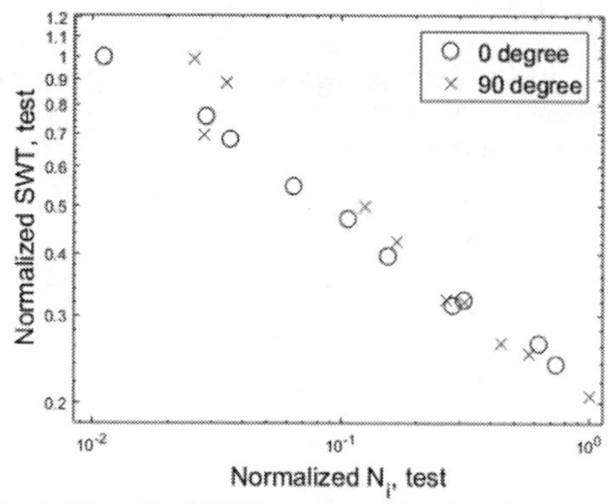

Fig. 4: Normalized SWT parameter plotted against normalized cycles to crack initiation, N_i, from the experiments.

4 Calibration and evaluation

Simulations of the LCF experiments were performed using the finite element program ABAQUS 6.12 [13], with the above described material description implemented as a user defined material model. Only the part of the specimen inside the extensometer region has been modelled. A total of 320 8-noded linear brick elements with reduced integration was used. The load was applied as a cyclic displacement in the axial direction on the top on the specimen and the bottom surface was fixed in the axial direction. The remaining boundary conditions were set so that the cross section of the specimen was allowed to contract. Since crack initiation is assumed to occur at the surface of the specimen, the stresses and strains were extracted from an element on the surface of the specimen. Furthermore, to avoid undesirable impact from the boundary conditions on the extracted result, the chosen element was picked as far away from the boundary as possible.
In the FE simulations, one loading cycle, with start and finish in tension has been simulated. The maximum stress was then extracted at the end of the load cycle.
In Fig. 5, a simulated hysteresis loop of specimen S0:5, c.f. Table 1, is shown. Comparing the maximum and minimum stresses for this simulated hysteresis with the values obtained from the experiment, it can be seen that the material model predicts these values with very good accuracy.

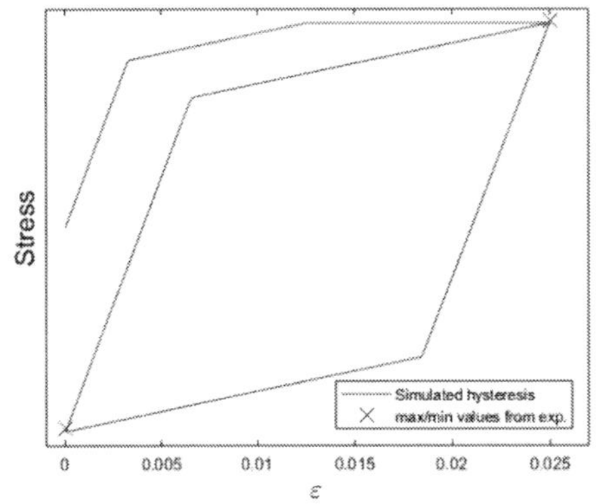

Fig. 5: Simulated hysteresis loop for specimen S0:5 compared to obtained maximum and minimum stress at mid-life from the experiment.

The 0° specimens (specimen S0:1-10 in Table 1) were used to calibrate the model. The SWT parameter of these specimens was extracted from the FE model and plotted against the crack initiation lives obtained from the experiments in a log-log-scale. The constants a and b from the crack initiation model, Eq. (20), were then evaluated by a curve fit of the SWT parameter and the crack initiation lives. In Fig. 6, the normalized crack initiation lives from the FE simulations of the 0° specimens are plotted against the normalized crack initiation lives obtained from the experiments. The scatter bands shows the standard deviation of the simulation results compared to experimental data.

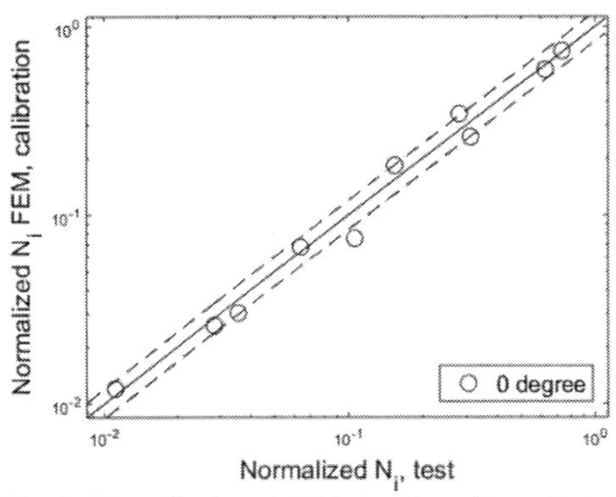

Fig. 6 : Normalized crack initiation lives, N_i, evaluated from the simulations for the calibrated specimens plotted against the normalized crack initiation lives from the experiments.

As a validation, the calibrated model was then used to predict the crack initiation lives of the 90° specimens (specimen S90:1-10 in Table 1). The results of this is shown in Fig. 7, where the predicted normalized crack initiation lives are plotted against the normalized lives obtained in the experiments. The scatter bands plotted in Fig. 7 are the standard deviation of the 0° specimens compared to the experiments, multiplied with a factor of 2.3.

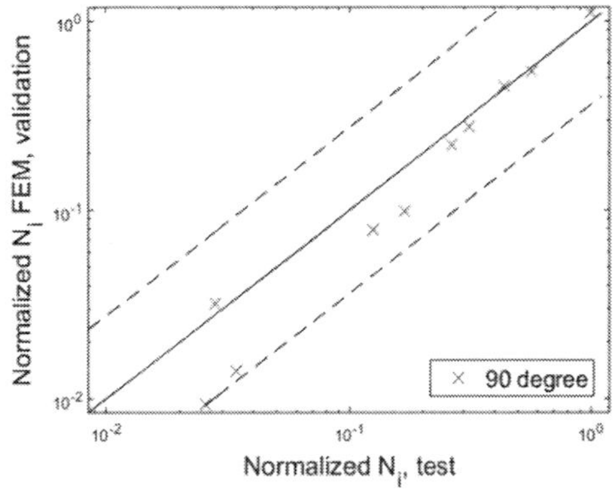

Fig. 7 : Normalized crack initiation lives, N_i, evaluated from simulations for the validated specimens plotted against the normalized crack initiation lives from the experiments.

The SWT parameter calculated with the FE model and the calculated SWT parameter extracted from the experiments are plotted against each other in Fig. 8. The largest deviation of the simulated SWT parameter compared to the SWT parameter calculated from experimental data is 10%, which indicates that the developed material model predicts the maximum stress during a mid-life load cycle with good agreement to experiments.

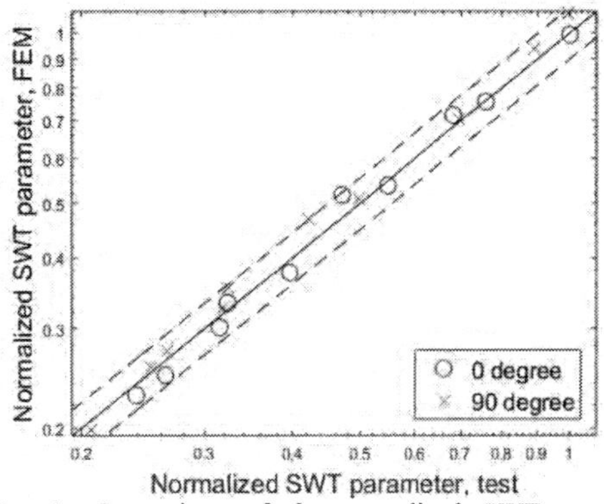

Fig. 8: Comparison of the normalized SWT parameter extracted from FE analyses and those calculated from experiments.

5 Discussion

The developed initiation model enables the prediction of the initiation life of both 0° and 90° specimens with the same relation with reasonable agreement to experiments. The crack initiation lives of all specimens used for validation are within 2.3 standard deviations of the specimens used for calibration, and the specimens with the largest applied strain ranges deviates most from the experimentally obtain lives. This indicates that there might be some further mechanism affecting the initiation life. Another thing that could affect the results is that all the material properties are taken as a mean value from experimental observations. An example of this is shown in Fig. 9, where the experimental elastic stiffness at mid-life of each specimen is plotted against the applied strain range. As can be seen, due to damage of the material, the elastic stiffness at mid-life decreases with increasing strain range.

In the presented crack initiation model, all the data are taken from mid-life of the component where a stable hysteresis is assumed. However, earlier loading cycles in the life of the components could possibly also affect the crack initiation life.

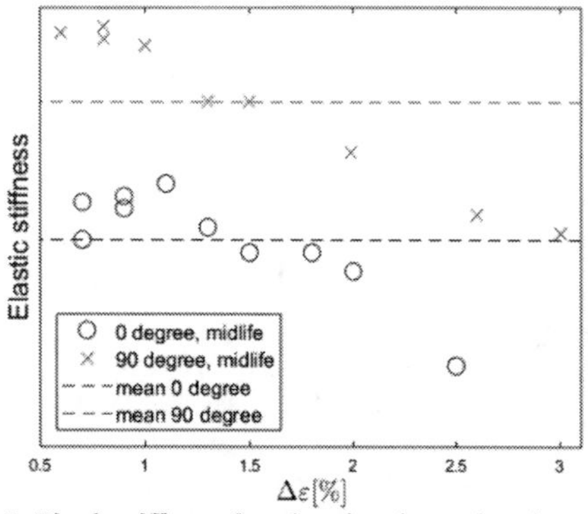

Fig. 9: Elastic stiffness plotted against the total strain range of the specimens at mid-life.

With the use of structural tensors to establish the constitutive relations for the material, an easy and straight forward approach to define the material orientation is obtained. The only input is the unit normal vector to the building platform on which the component is manufactured. The established material model can be used for any material orientation, not only those studied in this paper. This simplifies future studies on the material, when other material orientations might be in focus.

The developed material model has, similar to the Ohno-Wang Model 1, a linear kinematic hardening behaviour up to a saturation value of the back stress above which the material acts as a perfect plastic material. This material model predicts the maximum and minimum stress at a mid-life cycle with good agreement compared to experiments. Therefore, this model also predicts the crack initiation life accurately, since the only parameter needed from the FE simulations is the maximum stress when analyzing strain controlled smooth bars. Hence, it is not necessary in this case to use the complete hysteresis loop. However, when notched specimens are studied, a full hysteresis loop need to be considered, since local stresses and strains differ from global entities.

This study has been performed under the assumption that only the building orientation affects the material properties. However, other research, e.g. [14], have found that also the amount of material surrounding the melt pool during manufacturing can influence the material properties. This has not been taken into account in this work, but is worth having in mind for future studies.

From the fractography, a clear difference can be seen between the 0° specimens and the 90° specimens. The fracture surface of the 0° specimens is much smoother than on the 90° specimens, which could be due to the elongated grains in the loading direction for the 90° specimens. Furthermore, when studying the fracture surfaces from the side of the specimens, see Fig. 3b, the fracture of the 0° specimens does not seem to appear at the plane orthogonal to the loading, which could be explained by the anisotropic material properties in the plane orthogonal to the loading.

From Fig. 2 it can be observed that one of the 90° specimens with a strain range of 1.99% (specimen S90:9) collapses with the trend line of the 0° specimens and experiences almost similar life as the 90° 3% strain range specimen (specimen S90:10). Comparing the fracture surface of specimen S90:9 (Fig. 10b) with the one for specimen S90:8 (Fig. 10a), which have an applied strain range of 1.5%, it is not clear if there is a different governing fracture mechanisms between these. However, one can see a slight difference close to the specimen surface of the where specimen S90:9 have a more irregular fracture surface than specimen S90:8. Studying the fracture surface of specimen S90:10 (Fig. 10c), one can see a clear distinction compared to the other two specimens, indicating that at such large strain ranges, there is another governing fracture mechanism.

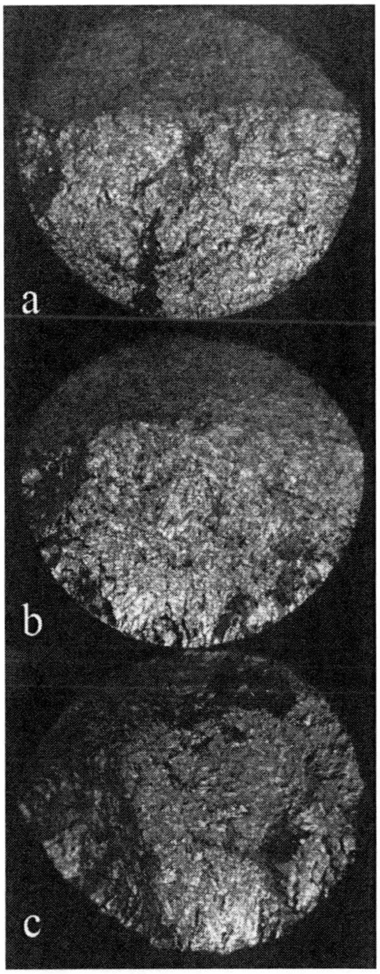

Fig. 10: Fracture surfaces of specimen a) S90:8 b) S90:9 and c) S90:10.

6 Conclusions

In this paper, a model for predicting crack initiation life for an SLM nickel-based superalloy subjected to LCF loading at room temperature has been established. The following conclusions can be drawn:

- A simple elasto-plastic constitutive model, with similar performance as the Ohno-Wang Model 1, that takes material anisotropy into account through structural tensor has been developed. This model can be used to simulate the elasto-plastic behaviour of the studied material with good agreement.
- The use of structural tensors for defining the transversely isotropic behaviour simplifies the modelling framework, as the building direction is a known entity from the manufacturing process.
- The adopted SWT model predicts the crack initiation lives in the two different building orientations and *R*- values with good correlation to the performed experiments.
- The studied material clearly exhibit anisotropic crack initiation behaviour for LCF loading at room temperature, which has been confirmed through both experiments and by fractography.

This research has been funded by the Swedish Energy Agency and Siemens Industrial Turbomachinery AB through "Turbines for Future Energy Systems" (Turbiner för framtidens energisystem), Grant No. 44112-1, the support of which is gratefully acknowledged. The authors would also like to thank Linnéa Tingö at Siemens Industrial Turbomachinery AB for the laboratory work.

References

[1] J. Zachary, ASME Turbo Expo, **3**, 607-619 (2012)
[2] H. Brodin, O. Andersson and S. Johansson, ASME Turbo Expo, **5A**, GT2013-95878 (2013)
[3] R. Konečná, G. Nicoletto, L. Kunz and A. Bača, Proc. Struct. Integr., **2**, 2381-2388 (2016)
[4] J. Suryawanshi, K. G. Prashanth and U. Ramamurty, J. Mater. Sci. Eng. A, **696**, 113-121 (2017)
[5] F. Wang, Int. J. Adv. Manuf. Tecnol., **58**, 545-551 (2012)
[6] A. Yadollahi, N. Shamsaei, S. M. Thompson, A. Elwany and L. Bian, Int. J. Fatigue, **94**, 218-235 (2017)
[7] D. Agius, K. I. Kourousis, C. Wallbrink and T. Song, J. Mater. Sci. Eng. A, **701**, 85-100 (2017)
[8] O. Andersson, A. Graichen, H. Brodin and V. Navrotsky, J. Eng. Gas Turbine Power, **139**, 031506 (2016)
[9] N. S. Ottosen and M. Ristinmaa, *The Mechanics of Constitutive Modeling* (2005)
[10] R. Hill, Proc. R. Soc. A, **193**, 281-297 (1948)
[11] N. Ohno and J.-D. Wang, Int. J. Plast., **9**, 375-390 (1993)
[12] K. N. Smith, T. Topper and P. Watson, J. Mater., **5**, 767-778 (1970)
[13] ABAQUS, *6.12 Documentation*, (2014).
[14] P. Kanagarajah, F. Brenne, T. Niendorf and H. J. Maier, J. Mater. Sci. Eng. A, **588**, 188-195 (2013)

Comparison of fatigue crack initiation behavior in different microstructures of TC21 titanium alloy

Changsheng Tan[1], Qiaoyan Sun[1], Lin Xiao[1], Yongqing Zhao[2], Jun Sun[1]*
1. State Key Laboratory for Mechanical Behavior of Materials, Xi'an Jiaotong University, Xi'an, Shaanxi, 710049, P.R. China
2. Northwest Research Institute of Nonferrous Metals, Xi'an, Shaanxi, 710016, P.R. China
**Corresponding author. Tel.: +86 29 82668614; fax: +86 29 82663453.*
E-mail address: qysun@mail.xjtu.edu.cn (Qiaoyan Sun).

Abstract. Cyclic heterogeneous deformation, slip characteristics and crack nucleation with different microstructures, such as bimodal microstructure (BM) and fine lamellar microstructure (FLM) in TC21 alloy (Ti-6Al-2Sn-2Zr-3Mo-1Cr-2Nb-0.1Si), were systematically investigated and analyzed during high cycle fatigue at room temperature. The results demonstrated that the FLM microstructure possesses higher high-cycle fatigue strength than those of the BM one. For BM, the heterogeneous plastic deformation existed within the different large primary α phase, such as equiaxed primary α and primary α lath. The cracks at interfaces and slip bands easily coalesce with each other to form large cracks in BM. However, the α laths with similar morphology and size (nanosize) distributed uniformly in FLM and could relatively deform homogeneously in micro-region, which delayed the initiation of the fatigue crack. Based on the electron-backscattered diffraction (EBSD) analysis, it found that the strain was nonuniformly distributed in BM, however, it is relatively homogeneous in FLM. Moreover, lots of straight cracks are parallel and along single intrusions within the β grain which delays the coalescence of cracks.

1 Introduction

Due to the high strength, high operating temperature, and corrosion resistance, titanium alloys have been applied widely to aerospace industries [1-3]. Based on the constitution of phase, the titanium alloy is generally classified into three groups: near α, α+β, and β titanium alloys. TC21 titanium alloy is a newα+β structural material with high strength and fracture toughness and has been successfully developed and applied in aircraft structures [4-5]. However, during those applications, the aircraft structure parts are mainly subjected to cyclic loading, especially for the high cycle fatigue loading [6]. It has been reported that fatigue failure is one of the major problems for the application of materials in transportation, manufacturing, medical devices and electronic industries [7].

It has been considered that fatigue life is determined by crack initiation because major of the fatigue life is spent on this region during the high cycle fatigue [8-9]. The crack initiation and coalescence in titanium alloy can be varied significantly by the microstructural parameters, such as phase dimensions, volume fraction, and morphology [10-11]. Many researchers have systematically investigated the influence of microstructures on the fatigue crack initiation behavior in titanium alloys [11-16]. It is found that the fatigue strength decreases in the order of bimodal microstructure, lamellar microstructure and equiaxed microstructure in Ti-6Al-4V alloy [16]. Huang et al [13-15] found that some small-scale heterogeneous microstructure regions (SHMR) in β grain interiors and the grain boundary (GB) α layer promote the crack initiation during fatigue loading in LM than BM which resulted in a lower fatigue strength for former one. While the crack initiation site depends on amount and size of primary α phase in bimodal microstructures, it can initiate either on the primary α/β transformation matrix or within the primary α phase [17]. Therefore, it is essential to reveal the relationship between micro-mechanism and fatigue life in titanium alloys.

During high cycle fatigue, the applied stress amplitude is often below the yield strength and still in the elastic region. However, it has been reported that it is not entirely elastic and the micro-plastic deformation occurs below the yield strength [18-20]. With further loading, localized plastic deformation was observed to initiate at individual grains with several preferred orientations which may result in crack formation [21]. This heterogeneous deformation within micro-regions

* Corresponding author: qysun@mail.xjtu.edu.cn

can lead to a fatal failure after high cycles [19, 22-23]. However, little information is available in the literature [10-11, 25] about the strain distribution and heterogeneous deformation with different microstructures during high cycle fatigue in TC21 titanium alloy.

Thus, in present work, the cyclic heterogeneous deformation, slip characteristics, crack nucleation and coalescence were systematically analyzed. Subsequently, the slip irreversibility and accumulative irreversible strain in BM and LM were calculated to explore the cyclic heterogeneous and localized deformation.

2 Material and experimental procedure

TC21 titanium alloy was produced by the Northwest Institute for Nonferrous Metal Research in China. The cylinder samples with 15 mm in diameter, 80 mm in height were obtained from the as-received TC21 disc. The heat treatments for bimodal microstructure and fine lamellar microstructure are listed in table 1.

Table 1. Heat treatments for the two microstructures

Structures	Temperature (°C)	Time (h)	Temperature (°C)	Time (h)
BM	900	1	600	4
FLM	970	1	600	4

* Both were air cooled

Tensile specimens with a diameter of 6 mm and a gage length of 30mm were obtained and tested on INSTRON 1195 Testing Machine. Fatigue specimens with an unnotched hourglass shaped specimen with threaded grip sections and 6 mm in diameter (the minimum gage sections) and 76 mm in length were performed on ZwickAmsler HFP 5100 with triangular waveforms at room temperature under a load ratio of R= -1.

Fractographic surface and cyclic deformation behavior after fatigue were observed using a field emission scanning electron microscope (FESEM) of HITACHI SU6600. The dislocation evolution was observed by transmission electron microscope (JEM-200CX) operating at 120 KV. Based on the electron-backscattered diffraction (EBSD), using a field emission gun SEM (Carl Zeiss Microscopy GmbH 73447, ZEISS) equipped with an HKL EBSD system, the strain distribution on the fatigue initiation site was investigated. The microstructure parameters were obtained by the Image Pro-Plus 6.0 software.

3 Results and discussions

3.1 Microstructure before fatigue

Fig. 1 shows the microstructures before fatigue. The bimodal microstructure consists of three different α morphology, including the equiaxed primary α (α_{p1}), the primary α laths (α_{p2}) and the secondary acicular α phase (α_s) in the β matrix (β_{trans}). The microstructural parameters for BM are listed in table 2. The average width of α phase in FLM is about 180nm.

Table 2. Microstructure parameters for the BM

Structures	Size	Volume fraction (%)	Aspect
α_{p1}	10.1 μm	25.4	1.7
α_{p2}	2.2 μm	13.1	6.3
α_s	100 nm		
β_{trans}		61.5	

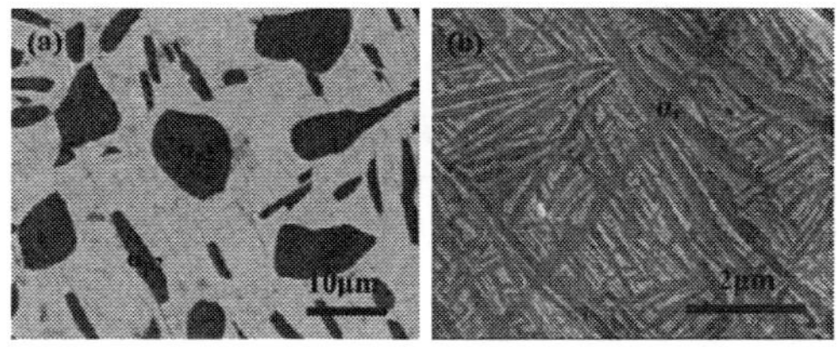

Fig. 1. Microstructures before fatigue: (a) BM; (b) FLM.

3.2 Mechanical properties

The mechanical properties for BM and FLM are listed in table 3. It can be seen that the FLM possesses lower ductility but higher yield and tensile strength than the BM one. The variations of the maximum stress with fatigue life for BM and FLM are compared. It indicates that FLM shows higher high cycle fatigue strength than BM at room temperature with a load ratio of -1. The fatigue strength for BM and FLM at 10^7 cycles with a stress ratio of minus one is approximately 490 MPa and 650 MPa, respectively.

Table 3. Tensile results for the two microstructures

Structures	σ_s (MPa)	σ_b (MPa)	ε(%)	Fatigue strength (MPa)
BM	1110	1240	12	490
FLM	1210	1280	2	650

The aforementioned empirical descriptions of fatigue life are for the fully reversed fatigue loading with the mean stress of zero. The mean stress level of the imposed fatigue cycle is known to play an important role in influencing the fatigue behavior of engineering materials [9]. The effect of mean stress on fatigue can also be represented in terms of constant-life diagrams which are often used as a design consideration in engineering applications [9]. In most cases, the popular Goodman model is used for the constant life diagrams [9,13, 26]. In general, the Goodman formula can be expressed as:

$$\sigma_a = \sigma_{-1} * (1 - \frac{\sigma_m}{\sigma_b}) \quad (1)$$

Where σ_a is the stress amplitude, σ_{-1} is the fatigue limit (R = -1) of the alloy, σ_m is the mean stress, and σ_b is the ultimate tensile strength.

Fig. 2 shows the Goodman diagrams of BM and LM which is constructed under 10^7 cycles at room temperature. The corresponding Goodman equations for BM (Eq. (2)) and LM (Eq. (3)) are as following:

$$\sigma_a = 490 * (1 - \frac{\sigma_m}{1200}) \quad (2)$$

$$\sigma_a = 650 * (1 - \frac{\sigma_m}{1280}) \quad (3)$$

It indicated from Fig. 2 that LM presented a better HCF strength than those of BM no matter what the load ratio is (R is from -1 to 1). This is consistent with the above high cycle fatigue results (R = -1) from table 3.

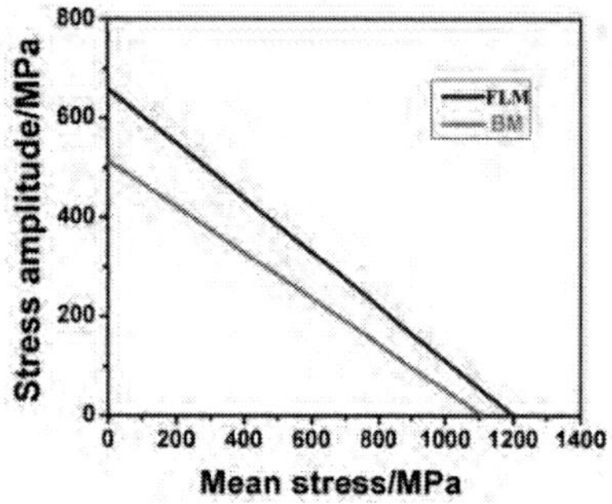

Fig. 2. Goodman diagrams of TC21 alloy with BM and FLM

3.3 The crack initiation and coalescence behavior

During the high cycle fatigue, most of the fatigue life is spent in the region of fatigue crack initiation. Therefore, it is essential to explore the relationship between fatigue crack nucleation and microstructures. The observation of second cracks and voids were performed near the site of the main crack, as described in details in Ref. [11]. Fig. 3 shows the fatigue crack initiation behavior of different microstructures. Most of the cracks nucleate at the primary α phase interface, including equiaxed and primary α phase in the BM at stress amplitude of 500 MPa, as shown in Fig. 3a and b. Therein, cracks at the primary α/β_{trans} interface tend to connect with each other along the interface (Fig. 3b). However, fatigue cracks mainly initiate either at the intrusions or intersection of two intrusions with various orientations on the rough surface in FLM at stress amplitude of 690 MPa, as shown in Fig. 3c and d.

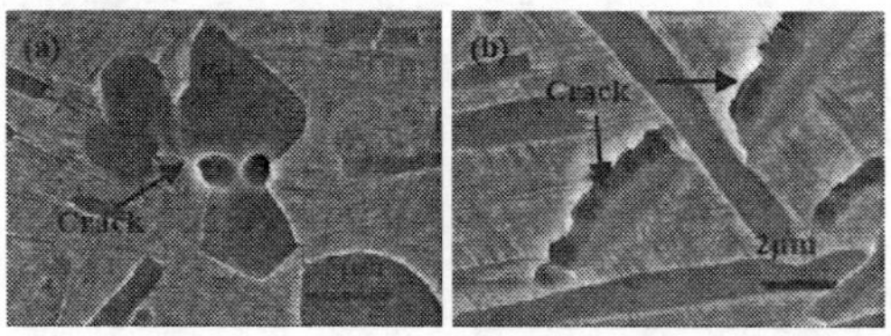

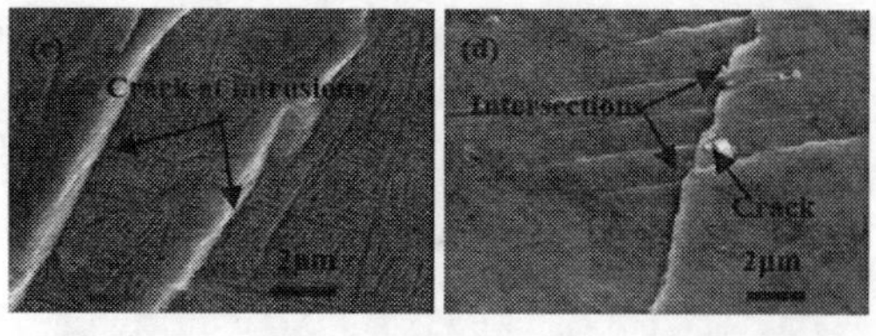

Fig. 3. Fatigue crack initiation behavior within different microstructures: (a-b) BM; and (c-d) FLM

The microcrack coalescence behavior is different from long cracks in titanium alloys [9, 11, 27-28]. Based on the observation of crack initiation behavior, it is found that crack easily nucleates at the primary α lath. Once initiated, it would connect with another crack to form large cracks, as shown in Fig. 4a. With further cycling, this crack propagates along the large α phase interface and finally transfers across the thin β lath, as seen in Fig. 4b. It has been reported that serious local deformation existed at the primary α lath, which leads to stress concentration at the interface [11, 25, 29].

However, fatigue crack mainly nucleates at the intrusions and propagates along the intrusions, as displayed in Fig. 4c. In most cases, these cracks are parallel to each other in the β grain boundaries interior (Fig. 4d). Little crack is found at the β grain boundaries in present work, which is different to the metastable titanium alloy [13, 15, 27]. Within the large β grain, the only one chance to concatenate with each other is at the intersection site along two directions of extrusions, as shown in Fig. 4d. Its low capability for microcrack coalescence in FLM at the initial stage may improve the initiation fatigue life.

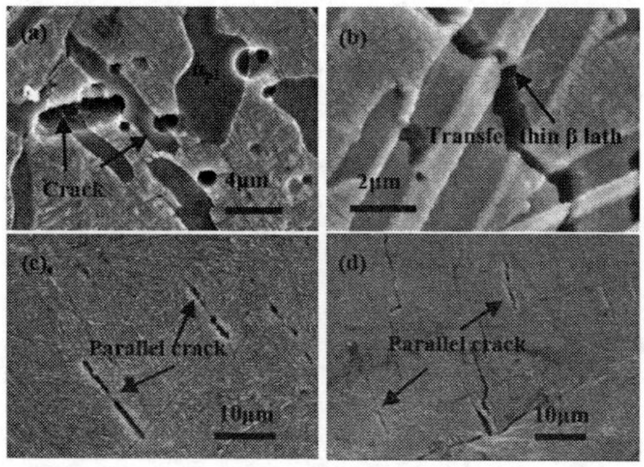

Fig. 4. Fatigue crack coalescence behavior within different microstructures: (a-b) BM; and (c-d) FLM

3.4 The dislocation evolution

Recently, Echlin et al [20] found that the localized plastic deformation was produced at some micro-regions in Ti-6Al-4V titanium alloys even at the elastic region. Additionally, it has been reported that the dislocation motion during each cyclic is not completely reversible and results in strain localization which leads to the crack formation [19, 22-24]. Thus, it is essential to investigate the dislocation structure during fatigue within various microstructures.

Fig. 5a-b illustrates various deformation processes for the primary α laths at stress amplitude of 560 MPa.

Some $(10\bar{1}0)$ primary slip bands occurred within α laths and interact with the interface (Fig. 5a), which leads to some slip steps and kinks. In addition to slip bands, some deformation twins have been activated to coordinate the plastic deformation in the primary α lath (Fig. 5b). For many metals, it has been reported that the main cyclic plastic deformation mechanism is dislocation motion, which is accomplished by slip or twinning in some cases [9-11,13,15]. Fig. 5c displays a number of $(10\bar{1}0)$ prismatic slip bands and dislocation arrays in the equiaxed primary α phase. Some dislocation debris and tangles exist in the β_{trans} matrix, as Fig. 5d shown.

Fig. 5. The dislocation structures in BM at stress amplitude of 560 MPa: (a) some prismatic slip bands in the primary α lath, (b) some deformation twins in primary α lath, (c) multi prismatic slip bands in equiaxed α phase, (d) dislocation pile-ups and tangles in the β_{trans} matrix

Fig. 6 shows the dislocation structures at different stress amplitudes in the FLM. Besides the dislocation pileups at the interface, easy slip transmission across the α/β interface is observed at 660 MPa, as indicated by the arrows in Fig. 6a. Many researchers have found that high ability of slip transmission is an efficient way to release the serious stress concentration at the interface, and improves the fatigue crack resistance [21, 25, 30]. The dislocation produces both in α and β phase (Fig. 6b), which indicates that a compatible deformation at the interface is induced [21, 25]. It has been reported that deformation twin would generate when the dislocation slip is not sufficient to coordinate plastic deformation [10-11, 13, 15]. Fig. 6c shows a lot of deformation twins in the secondary α phase with about 50 nm in widths at a high-stress amplitude of 705 MPa. Furthermore, abundant of secondary nano-twins with about 10nm in widths are observed within the deformation twins, as shown in Fig. 6d. It has been reported that nanotwins promote the back and forth motion of dislocations during cyclic loading in Cu [31], which improves the fatigue life. A relatively homogeneous cyclic deformation generates within micro-regions in FLM due to the easy slip transmission and plentiful nano-twins.

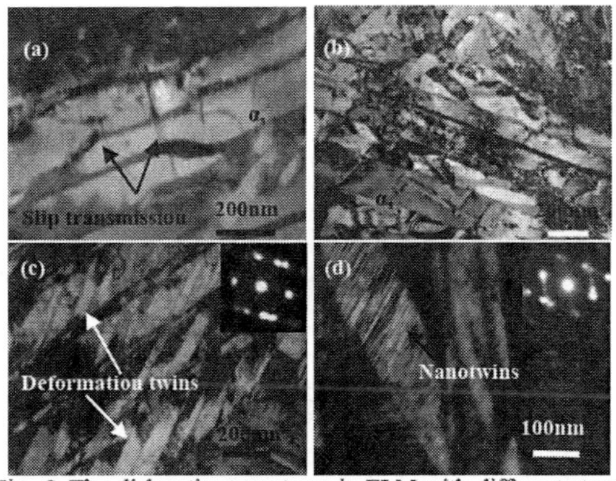

Fig. 6. The dislocation structures in FLM with different stress amplitude: (a) some slip transmission across the interface was observed at 660 MPa, (b) dislocation both in the α and β phase was observed at 690 MPa, (c) some deformation twins were induced at 705 MPa, (d) abundant of nano-deformation twins at 705 MPa

3.5 The non-uniform deformation

To compare the heterogeneous cyclic deformation, especially at the fatigue crack initiation site, the microstructure and strain distribution after fatigue were examined by the EBSD. Fig. 7 shows the process to prepare the fatigue samples for the EBSD observation. Firstly, the fatigue crack initiation site was marked and then removed the rough surface, as shown in Fig. 7a. Secondary, the surface was metallographically prepared with SiC and electrolytic polished for the SEM and EBSD observation (Fig. 7b).

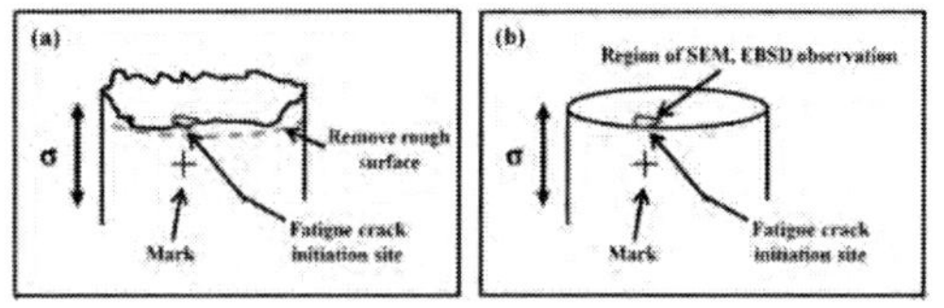

Fig. 7. The schematically illustration for SEM, EBSD observation near the fatigue crack initiation site: (a) a rough surface is removed, (b) regions of SEM, EBSD observation [10]

Fig. 8 shows the local misorientation and strain distribution at the fatigue crack initiation site both in BM and FLM. The large local misorientation tends to generate at the primary equiaxed α/β_{trans} interface, as shown in Fig. 8a. Fig. 8b indicates that the strain distribution is heterogeneous and easily localizes at the primary α lath. It has been reported that nonuniform deformation results in strain gradients and plastic interactions in the two-phase alloys [32-33]. Fatigue cracks tend to nucleate in this heterogeneous deformation region due to the high-stress concentration [11-13, 21, 33]. However, the local misorientation and strain distribution are relatively homogeneous in FLM than the BM one, as shown in Fig. 8c and d. Combined with the TEM results, it indicates that the cyclic deformation within micro-region is relatively homogeneous. Due to the uniform cyclic deformation and parallel cracks in FLM explain the better HCF strength than BM.

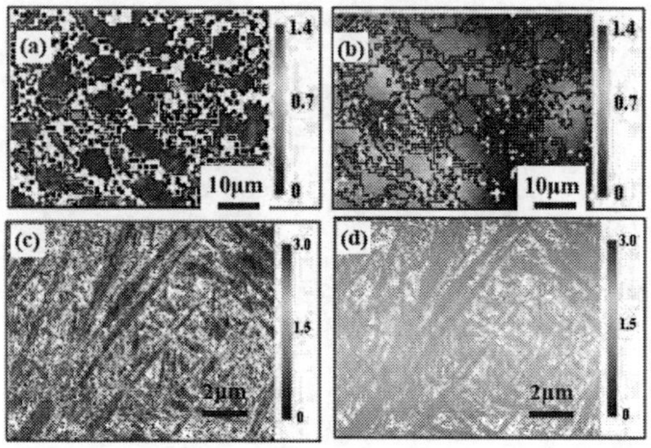

Fig. 8. The local misorientation and strain distribution near the fatigue crack initiation site within different microstructures: (a-b) BM, (a) the local misorientation, (b) strain distribution; (c-d) FLM, (c) the local misorientation, (d) strain distribution

4 Conclusions

Based on the above experimental results and analysis, the following conclusions can be drawn:

(1) High cycle fatigue strength was obtained in FLM than the BM one.

(2) The corresponding Goodman equations for BM and FLM are as following:

BM : $\sigma_a = 490*(1-\frac{\sigma_m}{1200})$;

FLM: $\sigma_a = 650*(1-\frac{\sigma_m}{1280})$.

(3) The heterogeneous plastic deformation existed within the large primary α phase. The cracks at interfaces and slip bands in the primary α lath easily coalesce with each other in BM. However, the α laths with similar morphology and size (nanosize) distributed uniformly in FLM. High ability of easy slip transmission across interface and abundant of nano-twins improves the fatigue initiation life.

(4) Based on the electron-backscattered diffraction (EBSD) analysis, it found that both the local misorientation and strain distribution were nonuniformly distributed in BM, however, it is relatively homogeneous within micro-region in FLM, which delayed the initiation of the fatigue crack.

Acknowledgments

This project was financially supported by the 973 Program of China (2014CB644003), the National Natural Science Foundation of China (51671158, 51471129, 51621063), and the 111 Project of China (B06025).

References

1. C. Leyens, M. Peters, *Titanium and titanium alloys: fundamentals and applications, John Wiley & Sons.* (2003)
2. D. Banerjee, J.C. Williams. Acta Mater. **61**(3) (2013)
3. G. Lütjering, J.C. Williams. *Titanium, Berlin: Springer.* (2003)
4. Y.Q. Zhao, G.J. Yang. Titanium Industry Progress. **23**(5) (2006)
5. X.N. Wang, Z.S. Zhu, L. Tong, Y. Zhou, X.H. Zhou, H.Q. Yu. Rare Metals Letters.**27** (2008)
6. B.A. Cowles. Int. J. Fracture. **80**(1996)
7. G.M. Castelluccio, W.D. Musinski, D.L. McDowell. Solid State Mater. Sci. **18** (2014).
8. A. Pineau, D.L. McDowell, E.P. Busso, S.D. Antolovich. Acta. Mater. **107** (2016)
9. S. Suresh. *Fatigue of materials, Cambridge university press.* (1998)
10. C.S. Tan, X.L. Li, Q.Y. Sun, L. Xiao, Y.Q. Zhao, J. Sun. Int. J. Fatigue. **75** (2015) 1-9
11. C.S. Tan, Q.Y. Sun, L. Xiao, Y.Q. Zhao, J. Sun. Mater. Sci. Eng. A. **711C** (2018)
12. K.S. Chan. Int. J. Fatigue. **32**(9) (2010)
13. C.W. Huang, Y.Q. Zhao, S.W. Xin, C.S. Tan, W. Zhou, Q. Li, W.D. Zeng. Int. J. Fatigue. **94** (2017)
14. C.W. Huang, Y.Q. Zhao, S.W. Xin, C.S. Tan, W. Zhou, Q. Li, W.D. Zeng. Mater. Sci. Eng. A. **682** (2017)
15. C.W. Huang, Y.Q. Zhao, S.W. Xin, W. Zhou, Q. Li, W.D. Zeng, C.S. Tan. J. Alloys Compd. **695** (2017)
16. G.Q. Wu, C.L. Shi, W. Sha, A.X. Sha, H.R. Jiang. Mater. Des. **46** (2013)
17. R.R. Boyer, J. Hall. Titanium'92: Sci Technol (1993)
18. P. Lukáš, M. Klesnil. Mater. Sci. Eng. A. **11**(1973)
19. H. Mughrabi. Philos. Trans. R. Soc. London. Ser. A. **373**(2038) (2015)
20. M.L.P. Echlin, J.C. Stinville, V.M. Miller, W.C. Lenthe, T.M. Pollock. Acta. Mater. **114** (2016)
21. W. Abuzaid, M.D. Sangid, J.D. Carroll, H. Sehitoglu, J. Lambros. J. Mech. Phys. Solids. **60**(6) (2012)
22. H. Mughrabi. Acta Mater. **61**(4) (2013)
23. H. Mughrabi. Metall. Mater. Trans. B. **40**(4) (2009)
24. H. Mughrabi. Fatigue. Fract. Eng. Mater. Struct. **22**(7) (1999)
25. C.S. Tan, Q.Y. Sun, L. Xiao, Y.Q. Zhao, J. Sun. J. Alloys. Compd. **724** (2017)
26. M.B. Bettaieb, A. Lenain, A.M. Habraken. Fatigue Fract. Eng. Mater. Struct. **36**(5) (2013)
27. R.K. Nalla, B.L. Boyce, J.P. Campbell, J.O. Peters. Metall. Mater. Trans. A **33** (13)(2002)
28. S.G. Ivanova, R.R. Biederman, R.D. Sisson. J. Mater. Eng. Perform. **11** (2) (2002).
29. C.S. Tan, Q.Y. Sun, L. Xiao, Y.Q. Zhao, J. Sun. Mater. Sci. Eng. A. (Minor review)
30. L.L. Li, Z.J. Zhang, P. Zhang, J.B. Yang, Z.F. Zhang. Acta Mater. **120** (2016)

31. Q.S. Pan, H.F. Zhou, Q.H. Lu, H.J. Gao, L. Lu. Nat. **551** (2017)
32. M.F. Ashby. Philos. Mag. **21**(170) (1970)
33. S. Ankem, H. Margolin, C.A. Greene, B.W. Neuberger, P.G. Oberson. Prog. Mater. Sci. **51**(5) (2006)

Effect of neighboring-microstructure on the rolling contact fatigue around non-metallic inclusion

Izumi Shimoji, Masahide Natori, Hideyuki Hidaka, and Yuji Miyamoto*

NSK Ltd., Core Technology R&D center, Fujisawa, Kanagawa, Japan

Abstract. It is well known that the flaking of a raceway causes the bearing failure, and the flaking mechanism depends on the lubrication condition. In the case of ideal lubrication condition with sufficient oil film thickness, the flaking originates from a non-metallic inclusion at the subsurface. It occurs through the formation of butterfly structure due to stress concentration around the inclusion. Therefore, the countermeasure of this type of flaking is to make cleaner steel by the reduction of inclusions. However, further reduction of inclusion imposes an extra challenge for the current steel making process. Therefore, in this paper, the effect of neighbouring microstructure on the butterfly structure is evaluated to propose a new life improvement method for rolling contact fatigue. It was suggested that it's important to eliminate strain localization near inclusion for improving bearing life.

1 Introduction

Rolling bearing is a machine element that supports the smooth rotation of a shaft, even under a heavy load and high speed rotation. The bearing ring and rolling element has a line or point contact, which induces a high compressive stress because of the small contact area. This causes the flaking of a raceway which limits the bearing life.

The bearing raceway can have flaking from various reasons, which mostly depend on the lubrication condition. Therefore, the bearings are recommended to be used under a good lubrication condition with a sufficient oil film thickness and clean oil. In this case, however, the flaking is caused by the crack originated from the subsurface inclusion. Hence, the life of the bearing is basically considered to be dependent on the size, position, and amount of nonmetallic inclusion.

The previous studies have shown that the fatigued microstructure neighboring to an inclusion is a butterfly-shaped structure [1,2]. The crack originates from the boundary between the butterfly structure and matrix, and propagate to the surface which results in flaking. The butterfly structure is formed by the microstructural changes occurred due to the nanocrystallization of matrix and decomposition of cementite under the Rolling-Contact Fatigue (RCF) stress [3]. It is believed that these microstructural changes is caused by the stress concentration around the inclusion. At present, the way to improve the bearing life is to use cleaner steel with reduced amount and smaller size of inclusion.

On the other hand, the cleanliness of steel is saturated to its maximum level and further reduction of inclusion imposes an extra challenge for the current steel making process. The author aimed to reduce the harmful effects of the inclusions by focusing on the reduction of butterfly structure formation. In order to achieve this, the factors affecting the formation of the butterfly structure are probed in detail.

The butterfly structure formation is suggested to be caused by the strain localization due to the stress concentration around the inclusion [4]. Therefore, a homogenous strain distribution in neighbouring microstructure can have a tendency to limit its formation. In order to inhibit the strain localization, average stress should be high enough to widely deform the surrounding regions of an inclusion. From this viewpoint, the increment of applied stress or decreasing the hardness of the material can have a possiblity to reduce the butterfly structure formation and consequently, can improve the bearing life.

In this research, we investigated the influence of the load condition and the initial hardness on the butterfly structure formation. The influence is evaluated by focusing on the strain distribution on the surrounding microstructure of inclusion in order to avoid the strain localization.

2 Experimental methods

The material used for this study was a bearing steel with a large amount of nonmetallic inclusions. A 6206 deep groove ball bearing was used, in which the hardness was adjusted for 735 HV and 600 HV by tempering at different conditions. The bearing was tested for RCF life under an ideal lubrication environment. The rotational speed was set to 3900 min^{-1}. The bearing was tested under the radial test load of 635 kgf and 1410 kgf. Therefore, there were three types of test conditions (a) standard test: 735 HV specimen with a radial test load Fr

* Corresponding author: shimoji-i@nsk.com

= 635 kg, (b) heavy load: 735 HV specimen with a radial test load Fr = 1410 kg and (c) soft microstructure: 600 HV specimen with a radial test load Fr = 1410 kgf.

The bearing test was stopped whenever there was flaking of inner or outer raceway. The test was also suspended at arbitrary time durations. The microstructural observation was done on the cross-section along the center of raceway by Optical Microscopy (OM) and Scanning Electron Microscopy (SEM). The sample was cut from the inner bearing ring and the preparation for observation was done by standard techniques of mounting, polishing and chemical etching. The butterfly structures were observed and the number of them was counted by OM on the cross-section area with 7 mm circumferential width. However, for the case of 600 HV specimen with the load of 1410 kgf, the observation area was increased to various cross sections to measure the sufficient number of butterfly structures.

The microstructure was further evaluated by EBSD/OIM analysis for measuring the strain values from the neighboring microstructure around the inclusion. The sample was prepared up to mirror finish by polishing with the colloidal silica. The Inverse Pole Figure (IPF) map and Image Quality (IQ) map were generated to evaluate the fatigued microstructure from the specimen.

3 Result and discussion

3.1 Influence of the radial test load on the formation of butterfly structures

This section, investigates the influence of an increase in the radial test load on the butterfly structure formation.

Figure 1 shows the frequency distribution of butterfly structures along the depth in the inner ring of a 6206 deep groove ball bearings with hardness of 735 HV. A standard radial test load of 635 kgf and a higher test load of 1410 kgf was applied for the RCF life test. The number of butterfly structures was measured at various depths from the raceway surface with 50 μm apart.

The formation of the butterfly structure is sometimes uncertain because of inclusion characteristics like size, position and the kinds which affect the surrounding stress state. This makes it difficult to compare the test results. To deal with this probabilistic behavior, the RCF tests were conducted for the various time duration. To evaluate this tendency the frequency of butterfly structure is summed up for the same depth for all test durations.

The dotted line in the figures shows a distribution of dynamic shear stress, which is calculated along the depth for each loading case by considering the Young's modulus of the material. The shear stress value is high for the higher test load of 1410 kgf such that the maximum shear stress value is about 1.3 times than that of 635 kgf. The position of maximum shear stress is slightly deeper by about 30 μm for 1410 kgf.

In case of 635 kgf test load, the frequency distribution of butterfly structure corresponds to the dynamic shear

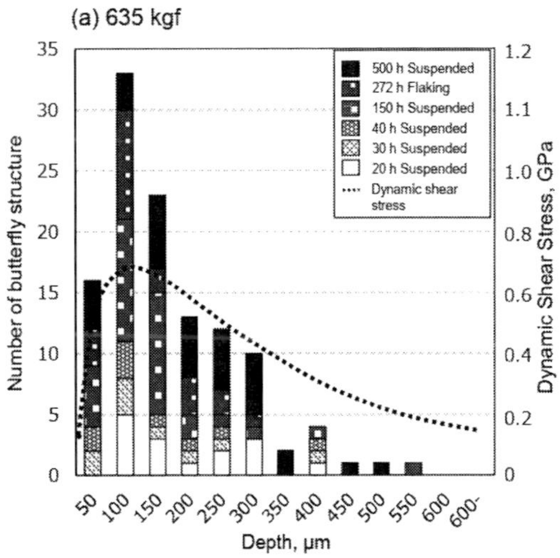

stress distribution. The higher frequency of butterfly

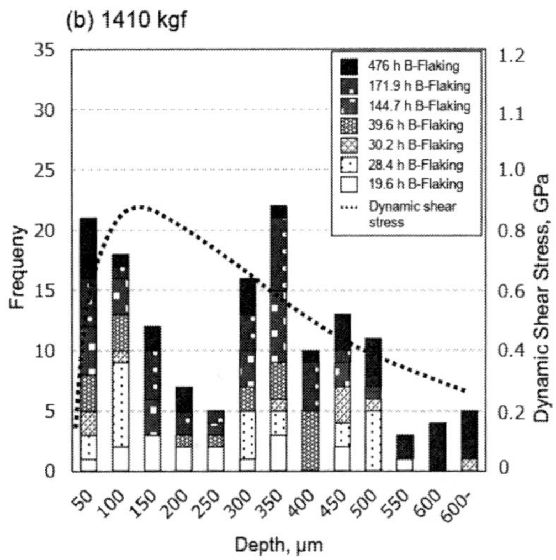

structure occurs at the depth of maximum shear stress. On the other hand, in case of 1410 kgf test load, the frequency distribution of butterfly structure does not correspond to

Fig.1. Frequency distribution of the butterfly structure formed along the depth in the inner ring of a 6206 deep groove ball bearing with hardness of 735 HV under RCF testing load of (a) 635 kgf and (b) 1410 kgf.

the dynamic shear stress.

Figure 2 shows the size distribution of butterfly structure along the depth of the inner ring. The size is described only by considering the length of the butterfly structure. The size distribution belongs to same butterfly structures of specimens, whose frequency is shown in fig. 1. In case of 635 kgf test load, the size distribution well corresponds to the dynamic shear stress. On the other hand, in case of 1410 kgf test load, the size of it decreases at the higher shear stress depth. These tendencies are almost similar to the frequency distribution in each load condition.

These results show that the generation of butterfly structure is reduced at the depth of the higher shear stress. This indicates that there is another factor except for the stress. To investigate this phenomenon, further

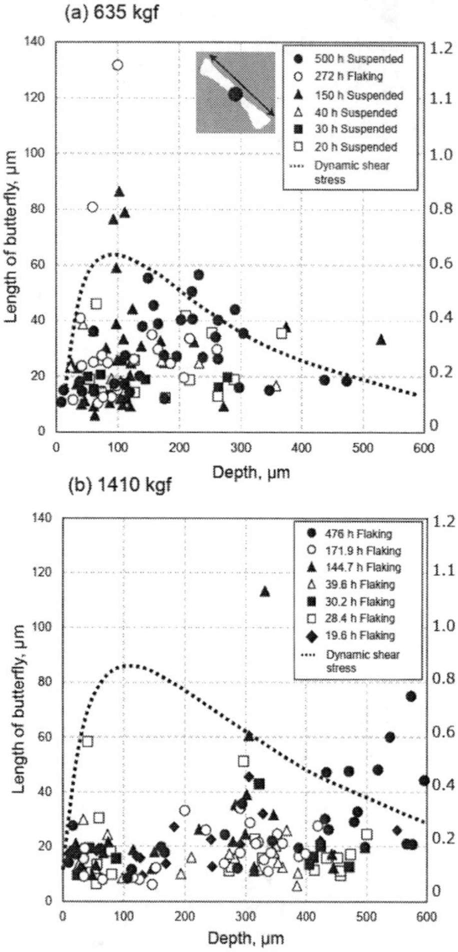

microstructural observations are made by OM and SEM. Figure 3 shows the images of microstructure obtained

Fig. 2. Distribution of length of the butterfly structure formed along the depth in the inner ring of a 6206 deep groove ball bearing with hardness of 735 HV under RCF testing load of (a) 635 kgf and (b) 1410 kgf.

by the OM of 735 HV specimen after the RCF life test with the load of 635 kgf and 1410 kgf for 20 h and 19.6 h, respectively. The top part of microstructure shows the inner ring raceway surface, whereas the subsurface region of microstructure with loading of 1410 kgf shows a darker contrast region known as a Dark Etching Constituent (DEC). The DEC extends from 50 to 300 μm depth from the surface. This is fatigued structure which indicates the presence of higher dynamic shear stress in the subsurface region [5].

Figure 4 shows the image of a microstructure obtained from the SEM of the DEC region shown in fig 3(b). The microstructure has slip-bands in a ferrite grain. Similar slip-bands were also found in the microstructure of the bearing specimen tested up to 500 h for the same test load.

Furthermore, the frequency of the butterfly structure was lower at the depth of 250 μm, as shown in fig. 1(b).

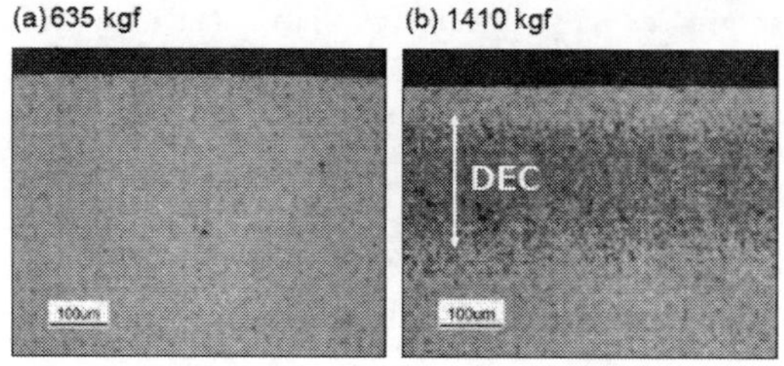

This part was observed by OM in the specimen which is RCF tested up to 476 h as shown in fig. 5. A large number of White Etching Constituent (WEC) was observed at this

Fig. 3. Microstructure of 735 HV specimen after RCF test for (a) 20 h at 635 kgf and (b) 19.6 h at 1410 kgf observed by optical microscopy. The DEC corresponds to the Dark Etching Constituent formed at subsurface region.

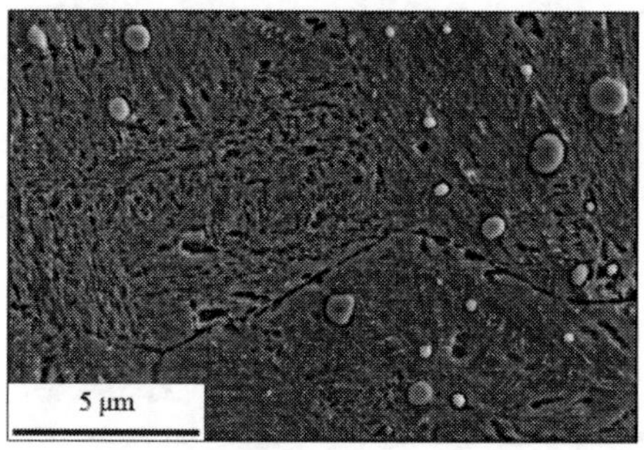

Fig.4. Microstructure of specimen after RCF test for 19.6 h at 1410 kgf observed by SEM. The microstructure corresponds to the DEC region formed at the subsurface.

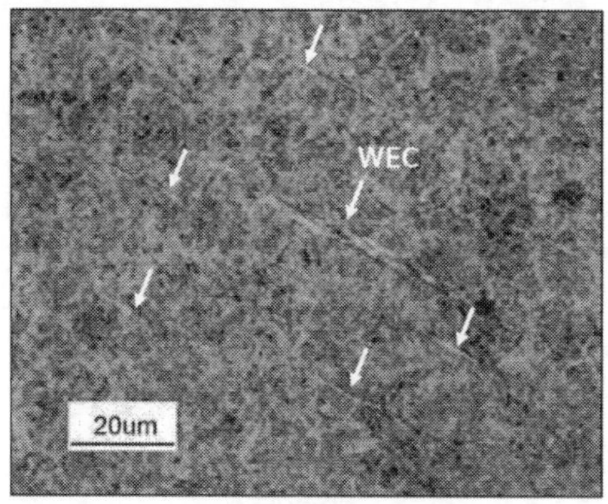

Fig. 5. Microstructure of 735 HV specimen after RCF test for 476 h at 1410 kgf observed by OM. The WEC corresponds to White Etching Constituent formed at subsurface region.

depth. It has been reported that the WEC structures are developed with in the DEC region [6] and the depth of 250 μm is the most developed microstructure area in this test condition of 1410 kgf test load.

A detailed quantitative analysis is done by EBSD/OIM to evaluate the microstructures in the region nearby of to a butterfly structure. Figure 6 shows the IPF and IQ map of 735 HV specimen after the RCF test for 20 h with 635 kgf test load. Fig. 6(a) and fig. 6(b) shows the butterfly structure formed nearby to an inclusion at the depth of 150 μm which belong to a subsurface region. The IPF map shows the color that indicates the grain orientation. The inside of butterfly structure was not indexed by EBSD due

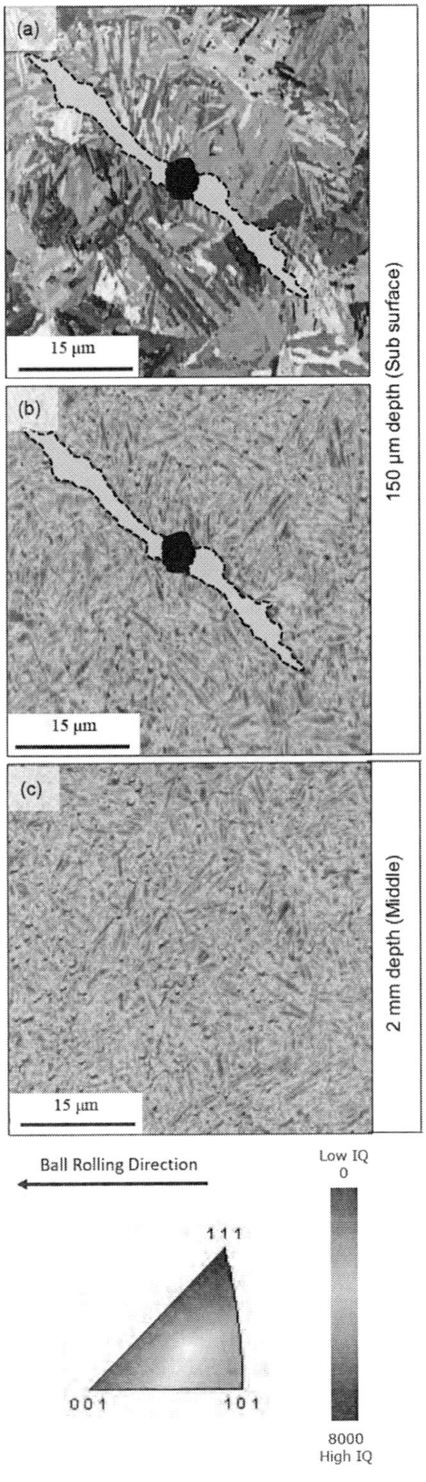

Fig. 6. The subsurface region IPF and IQ map shown at (a) and (b) respectively are for 735 HV specimen tested for 20 h at 635 kgf. The middle region IQ Map shown at (c) is for specimen tested for 19.6 h at 1410 kgf for reference.

to the nanocrystalline ferrite structure. There was a lack of sufficient data from these regions and therefore, it is highlighted by the gray area surrounded by dotted lines. The inclusion is highlighted with black color.

A method for visualizing the strain was established using the IQ values obtained from the EBSD/OIM analysis. The lower IQ value is caused by poor crystallinity, which corresponds to high strain expressed as red color, similarly, a higher IQ value corresponds to the lower strain expressed as blue color. Fig.6 (c) is the IQ map for 735 HV specimen at the depth of 2 mm which belong to an unfatigued region after the RCF test. The IQ map has green-yellow colored regions because of the presence of an initial strain in microstructure, which is generated during the heat treatment process with phase transformation.

In fig. 6(a), the lath microstructure outside the butterfly region was maintained even after RCF test. This also well corresponds to fig. 3 (a) where there is a no change of contrast in the subsurface region. The IQ distribution of the undeformed area in fig. 6(b) is also homogeneous, which is shown by the uniform green-yellow color distribution over the map. However, the stress field around the inclusion is symmetric according to the elastic stress theory, which doesn't match with the shape of butterfly structure which has wings inclined at 45° against to the rolling direction.

Moreover, it has been reported that the butterfly wing consists of nano-crystalline ferrite formed by the severe plastic deformation of the neighboring region of an inclusion [3]. The mechanism of the direction of butterfly wing has not been explained yet, but these results reconfirm that formation of butterfly is resulted from significant strain localization by stress concentration under RCF load.

Figure 7 shows the IPF and IQ map of 735 HV specimen after RCF test for 150 h at 1410 kgf. The observation was done at the depth of 101 μm which belong to a subsurface region. The figure shows a smaller butterfly structure nearby an inclusion as shown in fig. 2(b). Fig. 7(a), is an IPF map which shows that the surrounding lath microstructure is collapsed due to deformation. This also well corresponds to the subsurface DEC region, which shows a darker contrast in fig. 3 (b) and the slip-bands in fig. 4. Fig. 7(b) is an IQ map which shows a wide region of red color which belongs to the lower IQ value from a higher strain, distributed over the deformed microstructure.

From these results, it is suggested that the accumulation of strain in the neighboring region around the inclusion suppresses the formation of the butterfly structure.

3.2 Influence of the soft microstructure on the butterfly structure formation and bearing life

The previous section showed that the formation of butterfly structure can be suppressed through the elimination of the strain localization by increasing the RCF stress. This section investigates the effect of soft microstructure on the strain localization and butterfly structure formation. The soft microstructure has a

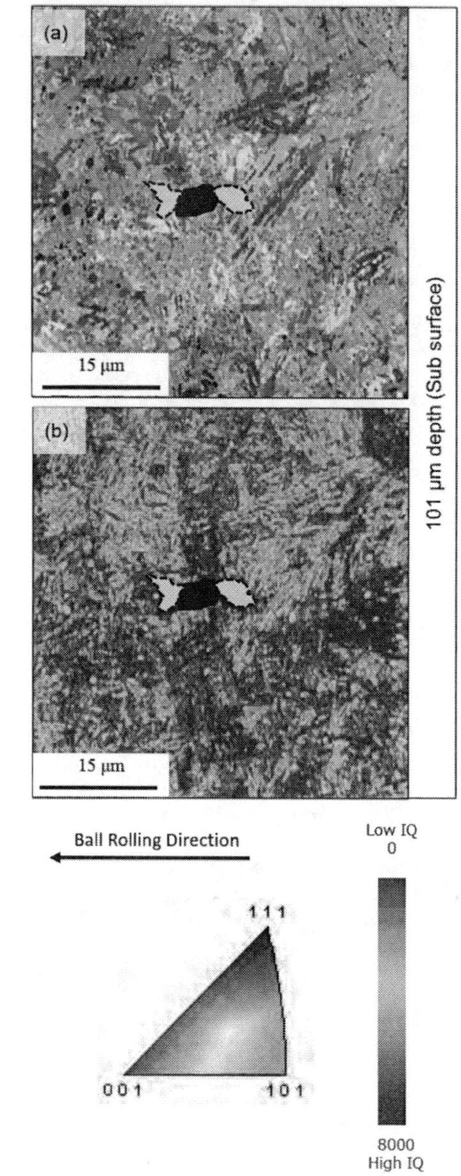

Fig. 7. The subsurface region IPF and IQ map shown at (a) and (b) respectively are for 735 HV specimen tested for 19.6 h at 1410 kgf.

relatively low hardness of 600 HV which is obtained by tempering.

Figure 8 shows the frequency distribution of butterfly structures along the depth of the inner ring of 600 HV specimen. The radial test load of 1410 kgf was applied for the RCF life test. The number of butterfly structures was measured at various depths from the raceway surface with 50 μm apart. The calculated dynamic shear stress was almost similar in the case of 600 HV and 735 HV for same radial load of 1410 kgf (fig. 1 (b)). The frequency of butterfly structure was relatively small. Almost, no butterfly structures were observed at the depth of maximum dynamic shear stress, even when the after the

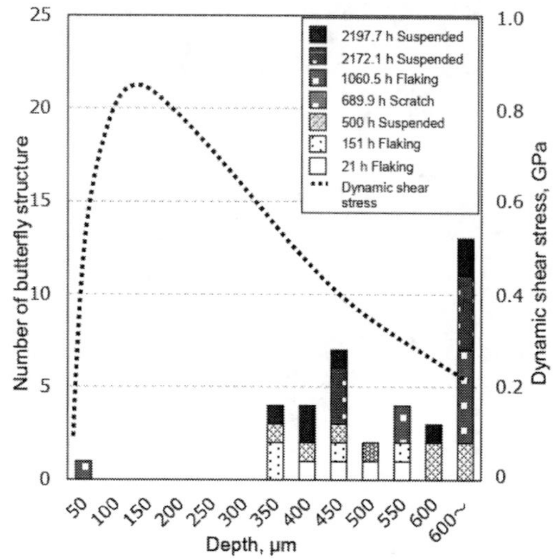

Fig. 8. Frequency distribution of the butterfly structure formed along the depth in the inner ring of a 6206 deep groove ball bearing with hardness of 600 HV under the load of 1410 kgf.

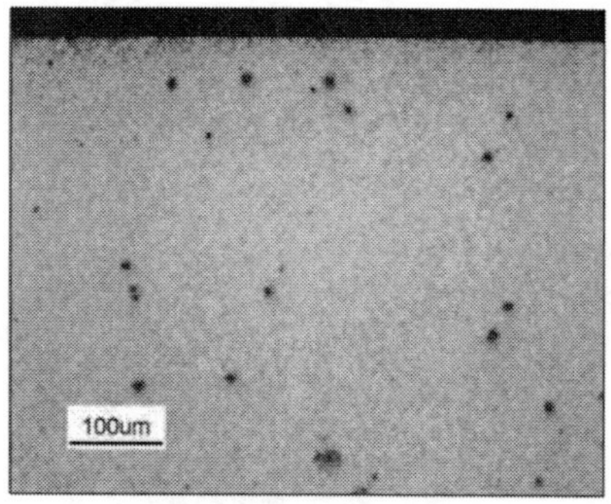

Fig. 9. Microstructure of specimen after bearing test for 20.8 h at 1410 kgf observed through OM.

observation area was increased for various cross-section. On the contrary, the frequency of butterfly structure was higher in the depth below 350 μm.

Figure 9 shows the OM image of the 600HV specimen RCF test at 1410 kgf for 20.8 h. Unlike 735 HV specimen, there was no DEC observed in the fatigued microstructure at subsurface region. This effect can be attributed to the difference of microstructure between 600 HV and 735 HV specimen. In order to show the amount of deformation in 600 HV specimen, a hardness measurement was done along its depth.

Figure 10 shows the hardness distribution along the depth of this 600 HV specimen. The hardness variation well corresponds to the shear stress distribution and the high hardness region is from 50-400 µm which doesn't has any butterfly structures as shown in fig. 8.

To further investigate this reduction of butterfly structure at subsurface region, the microstructure was evaluated by EBSD / OIM analysis.

Figure 11 shows the IQ maps of the subsurface region of 600 HV and 735 HV specimen tested at 1410 kgf.

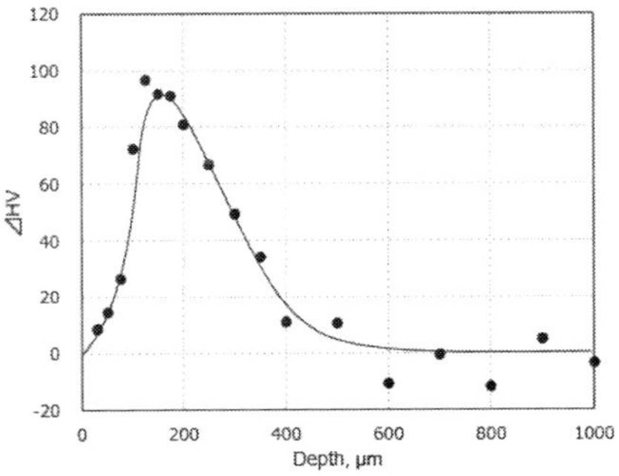

Fig. 10. The amount of change of the hardness at the each depth from the raceway of inner ring for 600HV specimen after RCF test in the test load 1410 kgf.

In case of 600 HV specimen, butterfly structures were absent around the inclusion in these subsurface regions (54 µm, 118 µm and 223 µm depth). Moreover, the surrounding area has a uniformly distributed lower IQ value shown by red color, this indicates the homogenous strain distribution in the microstructure in comparison of 735 HV specimen.

The strain is widely distributed all over the subsurface region of softer microstructure which eliminate the formation of the butterfly structure up to the depth of 350µm, as shown in fig. 8. This clearly shows that the homogenous strain distribution inhibit the formation of the butterfly structure near the inclusion.

Fig.12. Weibull plot for estimating the RCF life of 600 HV and 735 HV specimens with the test load of 1410 kgf.

In order to confirm the effect of reduction in the butterfly
structure on improving the bearing life, a life evaluation RCF test was done on 600 HV and 735 HV specimen.

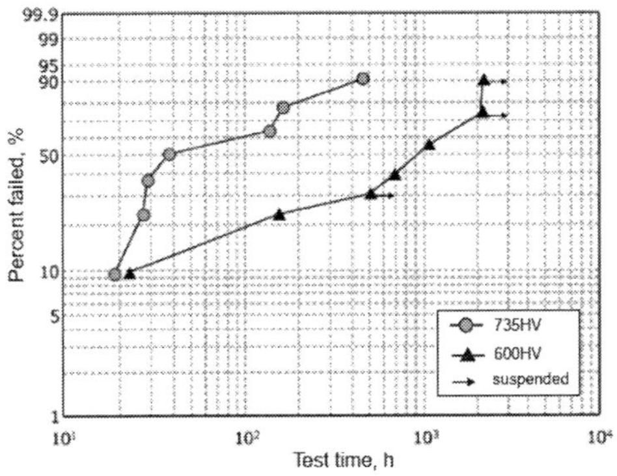

Figure 12 shows a Weibull plot for estimating the RCF life of 600 HV and 735 HV specimen with the test load of
1410 kgf. The Weibull plots show a similar L_{10} life, but a longer L_{50} life for 600 HV specimen. Both the specimen has a similar distribution of inclusion since they were prepared from same material ingot. Therefore, the reason for the L_{50} life improvement can be explained by the reduction of butterfly structure formation in the subsurface region as shown in fig. 8.

On the other hand, there was not a significant improvement of the L_{10} life. It was revealed that there are many surface cracks originated near the surface inclusions by SEM observation in RCF tested specimens

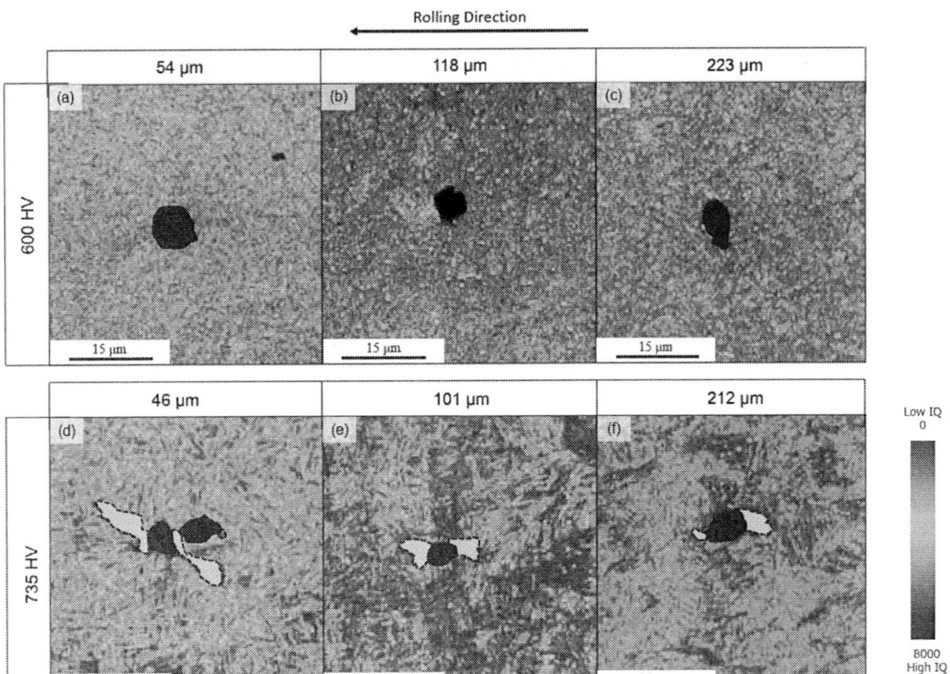

Fig.11 Subsurface region IQ map are from the depth of (a) 54 µm (b) 118 µm and (c) 223 µm for 600 HV specimen. Subsurface region IQ map are from the depth of (d) 46 µm (e) 101 µm and (f) 212 µm for of 735 HV specimen. The test load is 1410 kgf. The IQ maps are obtained from the OIM/EBSD analysis.

with flaking around L_{10} life. These results suggest that it is difficult to improve L_{10} life by reducing the formation of butterfly structure, because the initiation of crack from surface inclusion does not need the formation of the butterfly structure.

4 Conclusions

The formation of butterfly structure near the inclusion was evaluated from the viewpoint of strain distribution. The following conclusion were obtained.

1. The formation of butterfly structure depends on stress concentration around inclusion. However, this is also depends on deformation of neighboring microstructure around inclusion, which resulted from highly applied stress and reduction of hardness.
2. EBSD / OIM analysis is useful technique for evaluating damage of microstructure due to RCF in bearing. It is reveals that homogeneous strain distribution in the neighboring microstructure around the inclusion suppresses formation of butterfly structure.
3. The bearing life due to subsurface flaking shows an improvement by homogeneous strain distribution in the neighbouring microstructure around the inclusion.

5 References

1. M.H.Evans, *PhD Thesis* (nCATS, University of Southampton, 2013)
2. K. Sugino, K. Miyamoto, M. Nagumo, K. Aoki, ISIJ Int., **10** (1970) 98-111.
3. A. Grabulov and H. W. Zandbergen, *VHCF-4 Conference* (University of Michigan, 2007)
4. R. H. Vegter and J.T. Slycke, Journal of ASTM International, 7, 2 (2010)
5. J. Gegner, Tribology-Lubricants and Lubrication (2011)
6. N. Mitamura, H. Hidaka and S. Takaki, Material *Sicence Forum (Thermec'2007)*, **539-543**, 4255-4260 (2007)

Creep-fatigue interaction in heat resistant austenitic alloys

Hugo Wärner[1,*], *Mattias* Calmunger[1], *Guocai* Chai[1,2] and *Johan* Moverare[1]

[1]Department of Management and Engineering, Linköping University, 58183 Linköping, Sweden
[2]AB Sandvik Materials Technology R&D Center, 81181 Sandviken, Sweden

Abstract. This work includes an investigation of two commercial austenitic steels: UNS S21500 (Esshete 1250) and UNS S31035 (Sandvik Sanicro^{TM} 25). The materials were exposed to isothermal strain controlled fatigue with load controlled dwell time at maximum strain. The testing temperature used was 700°C and the test cycles were performed in tension. Mechanical test data were obtained and analysed in order to define creep-fatigue damage diagrams at failure for the investigated austenitic alloys. During the given conditions, Sanicro 25 showed superior creep-fatigue life, suffered less amount of creep elongation for the same amount of strain amplitude and dwell times compared to Esshete 1250. Both alloys showed creep-fatigue interaction damage for specific test configurations.

1 Introduction

Rising global energy consumption and the increase in emissions of greenhouse gases (e.g. CO_2) causing global warming, make need for more sustainable power generation [1, 2]. This could be accomplished by increasing the efficiency of biomass-fired, biomass co-firing and coal-fired power plants, which is achieved by increasing the steam temperature and steam pressure in the boiler sections [3–5]. By increasing the steam data to advanced-ultra super critical condition (A-USC) with temperature up to 700°C and pressures up to 36 MPa the efficiency can be increased to over 50% and the emission of CO_2 can be reduced by up to 10% compared to conditions with steam data at 25 MPa and 540°C (Super critical condition) [4, 6]. In addition, flexible generation of power is critical to compensate for the deficiencies of renewable power generation, such as solar and wind power, and this will increase the number of start-and stop cycles [3, 7, 8]. The change in operation conditions will increase the demands on the materials in the critical components of the power plants. Both cyclic condition and extended service in a long-term high temperature environment are operation conditions that such materials must withstand. Traditionally, ferritic and austenitic stainless steels are used for critical components of power plants, but with more demanding conditions other highly alloyed austenitic stainless steels could be more suitable for long-term service [6].

These new operating conditions involves a combined accumulation of cyclic damage and creep damage originated from start-up and shut-down and extended service at increased temperature and pressure. The interaction behaviour of creep and fatigue for a material is a complex and sometimes a competing process that depend on temperature, strain rate, dwell time, chemical composition and environment. Relevant investigations of the creep and fatigue process has been conducted for the austenitic 316 stainless steel by Hales [9] and Plumbridge [10]. The investigations reported four types of interaction cases where the type of initiation damage in the microstructure depend on if fatigue, creep or both fatigue and creep is the dominant process. For highly alloyed austenitic stainless steels, this has not been fully investigated and further work are needed to verify that the demands on safety for cyclic long-term usage is fulfilled.

This paper will present results from Creep-Low Cycle Fatigue (LCF) testing with alternating strain amplitudes and dwell times. Two austenitic alloys, candidates for critical components in the power generation industry, were investigated and the mechanical response of different creep-LCF conditions were analysed in order to determine the interaction behaviour and type of initiation damage.

2 Material and methodology

2.1 Austenitic alloys

The investigations involved two commercial austenitic alloys UNS S21500 (Esshete 1250) and UNS S31035 (Sandvik Sanicro^{TM} 25). Before mechanical testing Sanicro 25 were solution heat treated at 1220°C for 10 minutes and Esshete 1250 were solution heat treated at 1100°C for 15 minutes. Sandvik Materials Technology AB (SMT) provided and heat-treated the materials. The chemical composition of the materials in wt% is given in Table 1.

2.2 Methodology

The mechanical testing were performed using an MTS servo hydraulic testing machine equipped with an Instron 8800 control system. The strain was measured by an Instron 2632-055 extensometer and the isothermal condition, at 700°C, was controlled by an MTS 652.01 furnace. The $R\varepsilon$=0-type test cycle involved a strain

* Corresponding author: hugo.warner@liu.se

controlled part, dashed line in Figure 1, and a load controlled part, solid line in Figure 1. Strain ranges from 0.5-1.0% and dwell times, t_d, from 0-1800 seconds were tested. Setup and employment of the test as well as definition of plastic strain range, $\Delta\varepsilon_p$, stress range, $\Delta\sigma$, and total strain amplitude, $\Delta\varepsilon/2$, were conducted according to ASTM standard E2714-3 for Creep-Fatigue testing [9]. The number of cycles until failure, N_f, was defined as the point at which the amplitude stress, $\Delta\sigma/2$, decreases 10% below the tangent line constructed at the last point of zero curvature, as suggested by [11].

Table 1. Chemical composition (in wt%) of the austenitic alloys.

Material	C	Cr	Ni	W	Co	Cu	Mn	Nb	N	Si	V	Mo	Fe
Sanicro 25	0.1	22.5	25.0	3.6	1.5	3.0	0.5	0.5	0.23	0.2	-	-	Bal.
Esshete1250	0.1	15	9.5	-	-	-	6.3	1.0	-	0.5	0.3	1.0	Bal

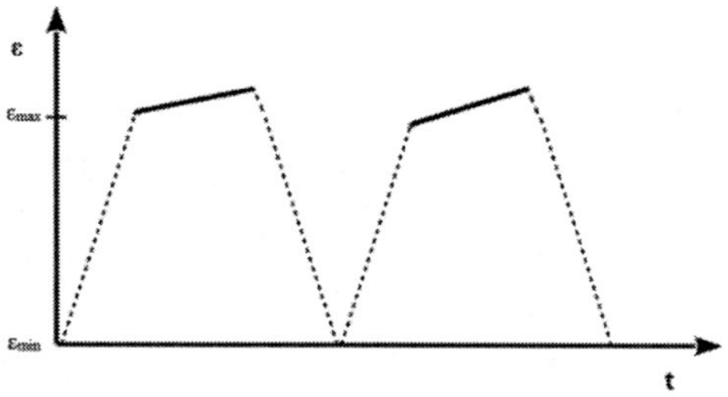

Fig. 1. Sketch of the Creep-LCF test cycle, strain (ε) versus time (t).

The creep-fatigue interaction diagrams were constructed similarly to those in "ASME boiler and pressure vessel code" [12] with the assumption that creep damage and fatigue damage may be evaluated separately and then merged to receive the total damage i.e. linear summation damage technique (Miners rule).

$$\sum(\Delta t/T_l)_k + \sum(n/N_l)_j \leq D \qquad (1)$$

This is displayed in equation (1).

Where D is the creep-fatigue damage factor, $(N_l)_j$ is the number of allowable cycles of type j, $(T_l)_k$ is the allowable time at stress level k, n is the actual number of cycles of type j and Δt is the actual time at stress level k.

The performed test used for construction of the diagrams can be viewed in Table 2. Reference data for creep rupture time at different stress levels were provided by SMT [13], however, the cyclic reference data were obtained from this study.

3 Results

3.1. Sanicro 25

In the stress versus cycles to failure diagram, Figure 2, at the beginning, Sanicro 25 shows cyclic hardening followed by cyclic softening after the maximum stress and until failure, for tests with different $\Delta\varepsilon/2$ and constant

Table 2. Test for the Creep-Fatigue interaction diagrams.

Test Nr	Material	$\Delta\varepsilon/2$ [%]	t_d [s]
1	Sanicro 25	0.25	0
2	Sanicro 25	0.25	300
3	Sanicro 25	0.375	600
4	Sanicro 25	0.25	600
5	Sanicro 25	0.25	1800
6	Esshete 1250	0.25	0
7	Esshete 1250	0.25	100
8	Esshete 1250	0.25	300
9	Esshete 1250	0.25	600
10	Esshete 1250	0.25	1800

t_d. Also higher strain amplitude yields shorter creep-fatigue life. In Figure 3 a) tests with constant $\Delta\varepsilon/2$ and different t_d, shows that longer t_d results in lower maximum stress and shorter creep-fatigue life. In Figure 3 b), the life is calculated in total test time instead of cycles to failure and plainly, tests with longer t_d have longer life.

The hysteresis loops in Figure 4 a) shows that with increasing number of cycles, maximum stress and elongation during t_d also increases. Although, in the same figure it is clear that the $\Delta\varepsilon_p$ decreases with increasing number of cycles. Comparing a cycle at the beginning of the test with a cycle at 90% of life, the reduction of $\Delta\varepsilon_p$ is 14.4%. At lower $\Delta\varepsilon/2$, displayed in Figure 4 b), Sanicro 25 show an earlier decrease of maximum stress than at higher $\Delta\varepsilon/2$ (Figure 4 a)) and $\Delta\varepsilon_p$ decreases with 44.4% compared with a cycle at the beginning of the test and a cycle at 90% of life.

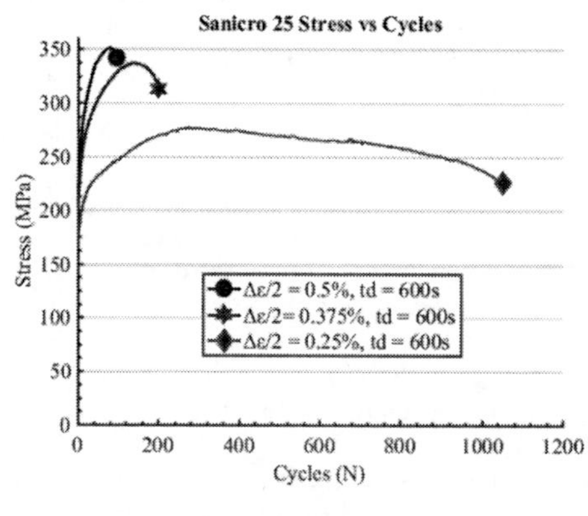

Fig 2. Maximum stress versus cycles to failure curve for Sanicro 25 tests with different strain amplitudes and constant dwell time.

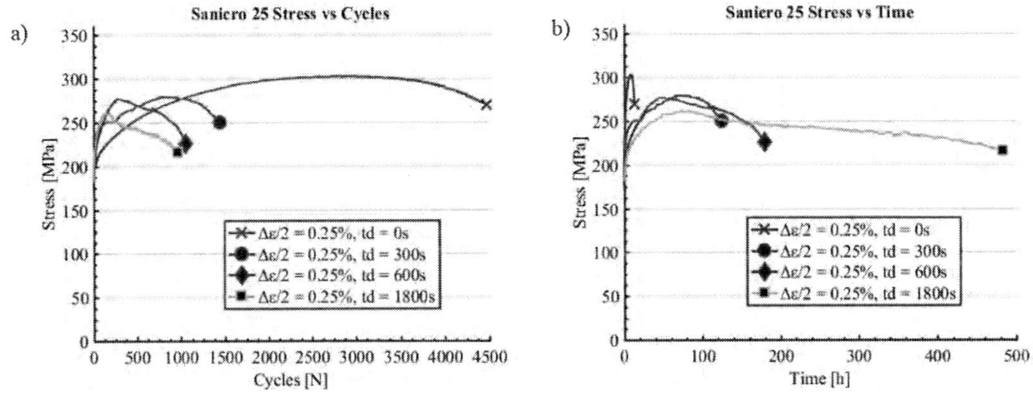

Fig 3. Sanicro 25 test result for tests with different dwell time and constant strain amplitudes, a) Maximum stress versus cycles to failure, b) Maximum stress versus time to failure.

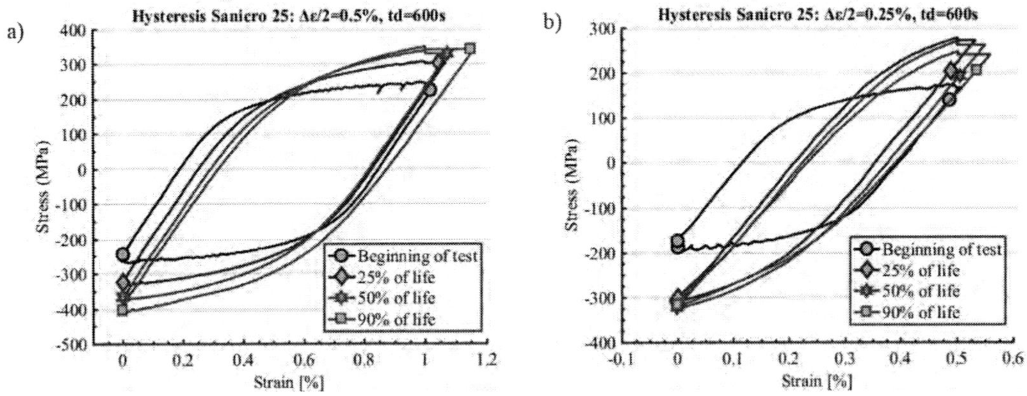

Fig 4. Sanicro 25 hysteresis curves for selected cycles, a) test with higher strain amplitude, b) test with lower strain amplitude.

3.2. Esshete 1250

In Figure 5, the mechanical response for Esshete 1250 can be observed. At the initial part of test life, for tests with lower $\Delta\varepsilon/2$ (0.25% and 0.375%), there are a rapid increase in cyclic hardening and then a cyclic softening after maximum stress. For the test with highest $\Delta\varepsilon/2$ (0.5%) instead of a reduction in hardening after the initial state to a steady-state with constant stress, the response is continuous cyclic hardening before failure. In Figure 6 a), tests with lower t_d (100s and 300s) have almost the same mechanical response and creep-fatigue life. This is also the case for tests with medium and long t_d (600s and 1800s). However, these similarities are not seen in Figure 6 b), were life is calculated in time.

The hysteresis loops in Figure 7, shows that with higher $\Delta\varepsilon/2$ (Figure 7 a)), the elongation during t_d increases with increasing number of cycles. Comparing a cycle at the beginning with a cycle at 90% of life, the increase in $\Delta\varepsilon_p$ is approximately 56%. In addition, the maximum stress increases from the beginning of the test to 25% of life, but then stay at the same level until failure. At lower $\Delta\varepsilon/2$ (Figure 7 b)) there are first an increase in maximum stress to 25% of life and then cyclic softening occurs until 90% of life. The $\Delta\varepsilon_p$ increases with increasing numbers of cycles, but the largest increase, 12%, is between a cycle at the beginning of the test to a cycle at 25% of life, instead of a cycle at 90% of life.

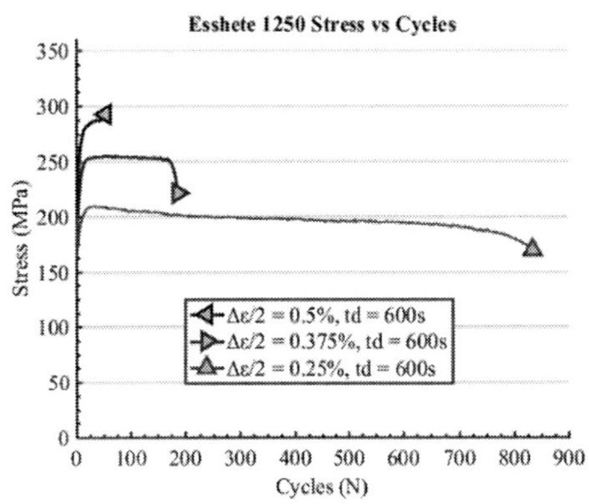

Fig 5. Maximum stress versus cycles to failure curve for

Esshete 1250 tests with different strain amplitudes and constant dwell time.

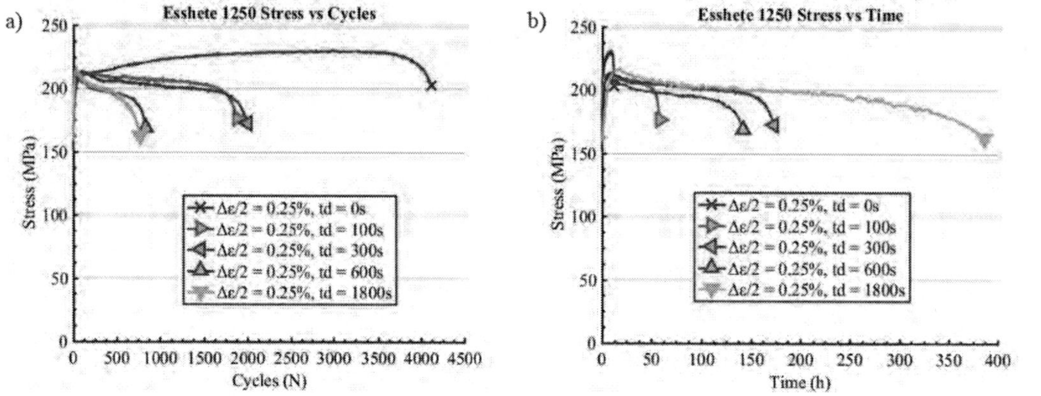

Fig 6. Esshete1250 test result for tests with different dwell time and same strain amplitudes, a) Maximum stress versus cycles to failure, b) Maximum stress versus time to failure.

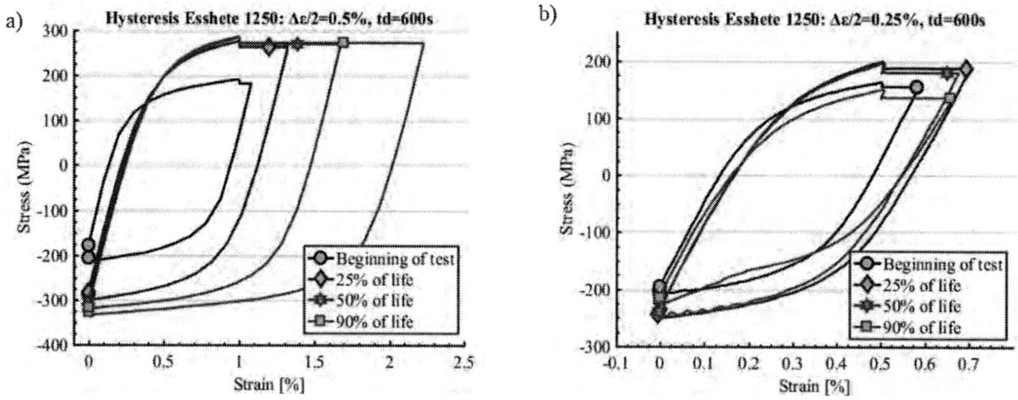

Fig 7. Hysteresis curves for selected cycles, a) test with high strain amplitude, b) test with lower strain amplitude.

3.3. Creep-fatigue Interaction diagram

In Figure 8, the linear summation of the creep damage, D_c, and the fatigue damage, D_f, are displayed for both Sanicro 25 and Esshete 1250. For Esshete 1250 (Triangular markers) the test with longest t_d (1800s), is situated (1.01) over the maximum total damage D=1 line and the test with t_d = 300s is close to the D=1 line. The other Esshete 1250 tests both exhibits total damage of approximately D=0.5.

The Sanicro 25 tests all exhibit more creep damage than fatigue damage. Comparing the two tests with t_d = 600s, the test with higher $\Delta\varepsilon/2$ showed more creep damage and total damage, D, but less amount of fatigue damage. The test with t_d = 300 seconds showed almost equal creep and fatigue damage and the test with longest t_d (1800s), showed an greater amount of creep damage situated close to the D=1 line.

4 Discussion

Comparing the different mechanical responses from the Creep-LCF testing of the two investigated austenitic alloys, there are differences in stress levels, N_f, $\Delta\varepsilon_p$,

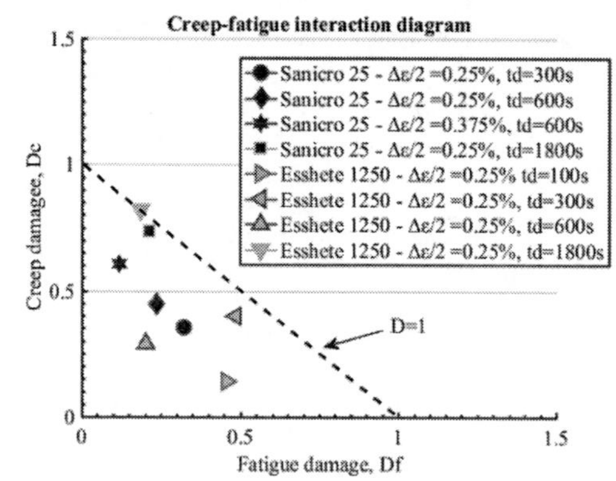

Fig 8. Creep-fatigue interaction diagram for both Sanicro 25 and Esshete 1250.

hardening and softening behaviour. The Sanicro 25 tests all showed increased N_f and higher stresses, compared to

Esshete 1250 tests with the same $\Delta\varepsilon/2$. This is evident in Figure 9, where stress at half life is plotted against N_f using the same dwell time (t_d = 600s).

Considering the hardening behaviour of Sanicro 25, it is evident in Figure 3 and 4 that there are a large increase in stress at the beginning of the tests. In addition, there are serrations of the curve at the beginning of the tests, in Figure 4. This is commonly related to dynamic strain aging (DSA) [14, 15]. DSA reduces the dislocation elimination rate and produce point defect concentration because of sufficient movement possibility of interstitial carbon and/or nitrogen atoms or carbon- and/or nitrogen-vacancy pairs or substitutional chromium atom and therefore increases the hardening behaviour which gives higher maximum stresses [14]. In Figure 7, this phenomenon is not seen for Esshete 1250 and is most likely due to the small amount of carbon and chromium content and the absence of nitrogen. This is probably associated with the reduction in $\Delta\varepsilon_p$ and the increase of maximum stress for the Sanicro 25 tests and the continuous increase of $\Delta\varepsilon_p$ and the stagnation of maximum stress for the Esshete 1250 tests.

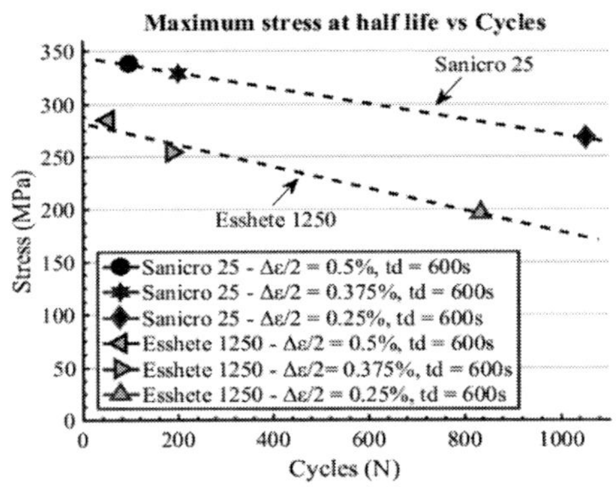

Fig 9. Maximum stress at half cycle life versus cycles to failure for the investigated austenitic alloys.

The tests with different t_d and constant $\Delta\varepsilon/2$ for Esshete 1250 (Figure 6), shows that in general the test with longer t_d and therefore more creep, exhibits shorter creep-fatigue life but longer life calculated in time. However, the test with t_d = 300 seconds show both, longer creep-fatigue life compared to the test with t_d = 100 seconds and longer life calculated in time compared to the test with t_d = 600 seconds. Consulting the interaction diagram in Figure 8, this test configuration is close to the maximum total damage D=1 line and is therefore not considered to suffer that much from interaction damage and have most likely a competitive damage process (case 2 in Figure 10) [16]. The two test with t_d = 100 seconds and t_d = 1800 seconds lies further at the fatigue part and the creep part of the diagram respectively, which is quite understandable given the difference in cycle configuration. In coherence with Hales and Plumbridge [9, 10], the damage interaction behaviour for the test with t_d = 100 seconds should be mostly fatigue dominated even though creep is still active. This should yield transgranular initiation followed by intergranular crack growth in an interactive manner (case 3 in Figure 10), because of the distance to the maximum damage D=1 line [16]. The test with t_d = 1800 seconds would most likely show both intergranular initiation and growth with some transgranular damage. This in a competitive manner because of the position close to the maximum damage line and therefore lack of interaction damage (case 2 in Figure 10). If the test with t_d = 600 seconds is considered in the interaction diagram, it has a central position with estimated high amount of interaction damage and should exhibit the transgranular crack damage from fatigue to grow into a intergranular progression caused by an additive process between fatigue and creep (case 3 in Figure 10). This interactive damage behaviour could also be contributed from internal triple point wedge cracks, grain boundary voids and initiations from matrix pore or inclusion according to [10].

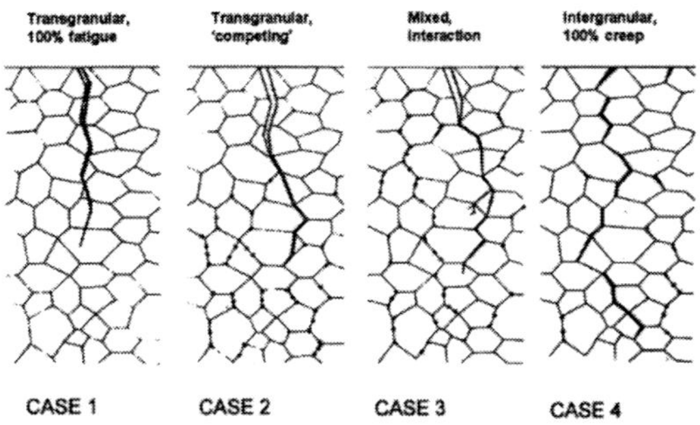

Fig 10. Different cases for creep-fatigue interaction damage, according to Hales [9] (Courtesy of publisher).

From findings of Hales [9], creep damage is more prominent at higher strain ranges. This coincides with the results in Figure 8, from Sanicro 25 test configurations with the same amount of dwell time (t_d = 600s) but different strain amplitudes i.e. the test with $\Delta\varepsilon/2$ = 0.375% and the test with $\Delta\varepsilon/2$ = 0.25%. In the figure, the test with the higher $\Delta\varepsilon/2$ (0.375%) exhibits a dominant part of creep damage and also less effect of interaction damage because of the position closer to the maximum interaction damage line compare to test 4. Further more, the test with lower $\Delta\varepsilon/2$ (0.25%) show less interaction damage compared to the Esshete 1250 test with identical test configuration which could be contributed by slightly more dominant creep damage process.

If the creep damage fraction is high ($D_c \geq 0.7$) in the Creep-fatigue interaction diagram (Figure 8) the interaction damage part becomes almost non-existent, for the two investigated austenitic alloys and the damage initiation mechanism should be close to case four in Figure 10, with intergranular growing cracks. When the tests with the highest amount of creep damage (t_d = 1800s) is compared for the alloys, the Esshete 1250 test has higher fraction of creep damage, even though it experience lower maximum stress and yield less amount of N_f. In this case, it is therefore reasonable to establish that Sanicro 25 has superior creep strength to Esshete

1250 and that it enhances the creep-fatigue life. According to Guocai et.al [13], the high creep strength at elevated temperatures for Sanicro 25 is because of interaction of dislocation and precipitates acts as pinning points for dislocation movement. Also nano-sized particles like $M_{23}C_6$ and Laves phase and smaller nano-precipitates such as copper rich- and MX particles were found to impede dislocation movement [13].

If the creep damage fraction is lower ($D_c \leq 0.7$), in the Creep-fatigue interaction diagram (Figure 8), the interaction damage influences the creep-fatigue life more and it is harder to predict the amount of interaction damage for a specific test cycle configuration. This is evident if the Esshete 1250 tests with $D_c \leq 0.7$ is considered in Figure 8. For Sanicro 25 tests, the interaction damage seems to stay close to 30% independent of the cycle configuration, but more tests and with different configurations need to be analysed before a general design trend can be established.

The obtained mechanical data from this study has been analysed using Hales and Plumbridges work [9, 10] to predict and estimate the type of initiation damage from creep-fatigue interaction behaviour. However, a microstructural investigation for the different tested specimens from this study should be conducted before a more comprehensive analysis and understanding of the damage mechanisms can be obtained.

5 Conclusions

The creep-fatigue interaction study for the two austenitic alloys contributed results, analysis of the mechanical response and estimated damage initiation mechanisms. From the discussed results, these conclusions are deduced:

- Sanicro 25 suffered higher maximum stresses but still shows superior creep-fatigue cycle life compared to Esshete 1250.
- Compared with Sanicro 25, Esshete 1250 shows more creep elongation for the same amount of strain amplitude and dwell times.
- Both alloys shows creep-fatigue interaction damage for most test configurations.

6 Acknowledgements

Present study was financially supported by AB Sandvik Materials Technology in Sweden and the Swedish Energy Agency through the Research Consortium of Materials Technology for Thermal Energy Processes, Grant No. KME-701. AFM Strategic Faculty Grant SFO-MAT-LiU#2009-00971 at Linköping University is also acknowledged.

References

[1] U.S. Department of Energy, "International Energy Outlook 2005," (2005).

[2] L. Trygg and S. Amir, "European perspective on absorption cooling in a combined heat and power system- A case study of energy utility and industries in Sweden," *Appl. Energy*, vol. **84**, no. 1319, (2007).

[3] R. Saidur, E. A. Abdelaziz, A. Demirbas, M. S. Hossain, and S. Mekhilef, "A review on biomass as a fuel for boilers," *Renew. Sustain. Energy Rev.*, vol. **15**, no. 5, pp. 2262–2289, Jun. (2011).

[4] A. G. Tumanovskii *et al.*, "Review of the coal-fired, over-supercritical and ultra-supercritical steam power plants," *Therm. Eng.*, vol. **64**, no. 2, pp. 83–96, Feb. (2017).

[5] J. YIN and Z. WU, "Corrosion Behavior of TP316L of Superheater in Biomass Boiler with Simulated Atmosphere and Deposit," *Chinese J. Chem. Eng.*, vol. **17**, no. 5, pp. 849–853, (2009).

[6] R. Blum and R. Vanstone, "Materials Development for Boilers and Steam Turbines Operating At 700 ° C," *Proc. 6th Int. Charles Parsons Turbine Conf. Dublin, Irel.*, pp. 498–510, (2003).

[7] Eurelectric, "Flexible generation: Backing up renewables," Eurelectric, Brussels, (2011).

[8] K. Dietrich, J. M. Latorre, L. Olmos, and A. Ramos, "The Role of Flexible Demands in Smart Energy Systems," in *Energy System*, pp. 79–97, (2013).

[9] R. Hales, "A Quantitative Metallographic Assessment of Structural Degradation of Type 316 Stainless Steel During Creep–Fatigue," *Fatigue Fract. Eng. Mater. Struct.*, vol. **3**, no. 4, pp. 339–356, (Oct. 1980).

[10] W. J. Plumbridge, "Uprating and life assessment under fatigue-creep conditions," *Int. J. Press. Vessel. Pip.*, vol. **59**, no. 1–3, pp. 119–129, (Jan. 1994).

[11] ASTM, "Creep-Fatigue Testing," *E2714 - 13*. ASTM International, pp. 1–15, (2014).

[12] R. Jetter, "Subsection NH—Class 1 Components in Elevated Temperature Service," in *Companion Guide to the ASME Boiler and Pressure Vessel Code, Volume 1, Third Edition*, Three Park Avenue New York, NY 10016-5990: ASME, pp. 409–445, (2007).

[13] G. Chai, M. Boström, M. Olaison, and U. Forsberg, "Creep and LCF Behaviors of Newly Developed Advanced Heat Resistant Austenitic Stainless Steel for A-USC," *Procedia Eng.*, vol. **55**, no. 55, pp. 232–239, (2013).

[14] L. H. De Almeida, I. Le May, and P. R. O. Emygdio, "Mechanistic modeling of dynamic strain aging in austenitic stainless steels," *Mater. Charact.*, vol. **41**, pp. 137–150, (1998).

[15] G. Chai, "Damage Mechanism of Low Cycle Fatigue in an Advanced Heat Resistant Austenitic Stainless Steel at High Temperature," *Procedia Mater. Sci.*, vol. **3**, pp. 1754–1759, (2014).

[16] R. P. Skelton and D. Gandy, "Creep – fatigue damage accumulation and interaction diagram based on metallographic interpretation of mechanisms," *Mater. High Temp.*, vol. **25**, no. 1, pp. 27–54, (Mar. 2008).

Low and high cycle fatigue of heat resistant steels and nickel based alloys in hydrogen for gas, steam turbines and generators applications

Alexander Balitskii [1, 2]*, *Jacek* Eliasz [1] and *Valentina* Balitska [3]

[1] West Pomeranian University of Technology, Szczecin, Poland
[2] Karpenko Physicomechanical Institute National Academy of Sciences, Lviv, Ukraine
[3] Lviv State University of Life Safety, Lviv, Ukraine

Abstract. It has been established that, at some region of hydrogen pressure and strain rate exists a maximum influence of hydrogen on the plasticity, low cycle fatigue and cyclic crack resistance of Ni-Co alloys and high nitrogen steels. The drop of plasticity of the dispersion-hardening materials within the temperature range of intense phase transformations is caused by the localization of strains on the grain boundaries due to the intense redistribution of alloying elements in the boundary regions. Moreover, the increase in plasticity observed at higher temperatures is caused both by partial coagulation of hardening phases and possible dissolution of small amounts of finely divided precipitations. The effect hydrogen on short-term strength and plasticity, high- and low-cycle durability of 15Cr12Ni2MoNMoWNb martensitic steel, 10Cr15Ni27Ti3W2BMo austenitic dispersion-hardened steel, heat resistant 3,5NiCrMoV rotor steel, 04Kh16Ni56Nb5Mo5TiAl and 05Kh19Ni55Nb2Mo9Al Ni-base superalloys in range of pressures 0–30 MPa and temperatures 293–1073 K was investigated. In the case of 15Cr12Ni2MoNMoWNb steel and 04Kh16Ni56Nb5Mo5TiAl alloy the dependence of low-cycle durability (N) and characteristics of plasticity (δ and φ) on the hydrogen pressure consists of two regions. In the first region (low pressures), the N, δ and φ abruptly drops, and in the second, the negative action of hydrogen becomes stable or decrease negligibility.

1 Introduction

The production of turbine and generator equipment requires a wide usage of dispersive hardened heat-resistant Fe-Ni, Ni-Cr, Ni-Co alloys and high nitrogen steels. Elevated temperature steels and alloys are exploited at high temperatures steam (up to 973 K), hydrogen containing gas mixtures (up to 1173 K). Therefore one of the most important requirements for such steels and alloys is their resistance to hydrogen degradation at very high temperatures. In other words their ability to keep high level of mechanical and fatigue properties under the action of hydrogen in wide range of exploitation parameters. At the same time, steels and alloys are known to be rather sensitive to hydrogen embrittlement [1-6].

Power equipment on NPP and advanced FPP, transport and energetic gas turbines requires their modernization due to increasing of working parameters up to the super high, extreme level. The production of future turbine equipment requires a wide usage of dispersive hardened heat-resistant Fe-Ni and Ni-Co alloys. Heat-resistant steels and alloys are exploited at high temperatures steam (up to 973 K), hydrogen containing gas mixtures (up to 1073...1173 K). Therefore one of the most important requirements for such steels and alloys is their resistance to high temperature, high pressure gaseous environments. In other words their ability to keep high level of mechanical properties under the action of hydrogen in wide range of exploitation parameters. The serviceability of structures in hydrogen is, as a rule, estimated according to the results of testing at room temperature and life extension procedures has base on such approaches. However, the operating conditions of the equipment for hydrogen power engineering include cyclic loading of the products in hydrogen in fairly broad temperature ranges. In most cases, the influence of gaseous hydrogen on mechanical properties weakens as temperature increases and, according to the presented in, the upper temperature of embrittlement under the analyzed conditions was equal to 573 K. At the same time, we reveal a significant decrease in the plasticity of heat-resistant nickel alloys in hydrogen under a pressure of 35 MPa at temperature 973 K. The aim of this investigation is to study the influence of high pressure hydrogen on the short-term strength and plasticity, low- and high-cycle durability of martensitic steel, austenitic dispersion-hardened steel, and Ni base superalloys at high temperatures.

2 Materials and experimental procedure

Static tensile tests were carried out on standard fivefold cylindrical specimens with working part diameter 5 mm by displacement rate $V = 6.7 \cdot 10^{-5}$ s^{-1} on the original equipment. This type specimens for determination high-cycle durability submitted to cyclic loading on scheme ''pure bending with rotation'' at fixed strain amplitude with frequency 50 Hz at room temperature. The low-cycle durability tests for pure strain-controlled sign-preserving bending was found for the strain amplitudes e=1.6% and a loading frequency of 0.5 Hz on polished plane specimens with a working part of 3x6x20 mm were carried out on typical equipment. We determined the influence of hydrogen under a pressure 30 MPa on the critical stress intensity factor (SIF) under a static load K_c. For this purpose 50x60x20 mm specimens with a preliminarily induced fatigue crack were loaded in off-centre tension in a high-pressure chamber. The cyclic crack resistance (CCR) characteristics were determined in three-point bending of 160x40x20 mm beam specimens at a frequency of loading of 20 Hz and a coefficient of cycle asymmetry R=0.22.

To determine the indicated mechanical characteristics in hydrogen, the working chambers were preliminarily evacuated, blownout with hydrogen, again evacuated, and filled with hydrogen up to a given pressure. At high temperatures, the specimens were held under the testing conditions for 30 min up to the attainment of thermal equilibrium. It has been established [1] that, at some values of hydrogen pressure and strain rate, which depend on the chemical compositions and structures of materials, maximum influence of hydrogen on the plasticity, low-cycle fatigue life, and static and cyclic crack resistance of martensitic steels and nickel alloys are achieved. In short-term tension, austenitic dispersion hardened steels are substantially embrittled by hydrogen after preliminary hydrogenation at elevated temperatures and upon attaining its content above 12 wppm, and the properties of hydrogenation specimens at room temperature in air and hydrogen are equal [1]. This is why we held a part of the specimens for 10 h in a hydrogen atmosphere under 623 K and 35 MPa. These regimes provide the hydrogenation of specimens to hydrogen contents of 3.2 wppm (15Cr12Ni2MoNMoWNb steel), 15 wppm (10Cr15Ni27Ti3W2BMo steel), 19 wppm (05Kh19Ni55Nb2Mo9Al alloy) and 20 wppm (04Kh16Ni56Nb5Mo5TiAl alloy). On the other hand during long term service in hydrogen or hydrogen containing gas structural material has been alloyed by hydrogen. For example stable high nitrogen steels P-900 (for retaining rings of NPP turbogeneraros): 7…20 ppm; dispersive hardened steels (Ni23Mo1,5Ti3), (Ni27Mo1,5W2): 8…25 ppm; Ni-Cr-Fe alloys (Ni55Cr19Fe12Mo9Nb2) (for distributive face of combustion chamber of gas turbine), (Ni63Mo6Ti3), (Ni56Mo6Nb4), (Ni42Fe36Cr14Nb3Mo2) (for inlet gas turbine nozzle), Ni-Co alloys (Ni56Co15Cr9W6Al5Mo4, Ni64Cr14Co10Mo5Al3Ti3) (for gas turbine rotor disks and blades), which strengthenid by γ' $(Ni)_3(Al,Ti)$ (stable up to 1373 K), γ''$(Ni)_3(Nb)$ (stable up to 1573 K) phases: 8…30 ppm.

Hydrogenated and non-hydrogenated specimens were tested in helium and hydrogen under different pressures. The hydrogen concentration (C_H) was determined with a LECO TCH 600 instrument.

3 Results and discussion

The martensitic steels contain about 10% of residual austenite after the optimal thermal treatment in the form of thin layers between the plates of martensite and on the boundaries of the former austenitic grains. The chemical composition, heat treatment regimes and mechanical properties of investigated alloys in helium and in hydrogen under a pressure of 35 MPa are given in [1]. The fatigue properties of 15Cr12Ni2MoNMoWNb martensitic steel, 04Kh16Ni56Nb5Mo5TiAl, 10Cr15Ni27Ti3W2BMo and 05Kh19Ni55Nb2Mo9Al dispersion hardened alloys were investigated. The alloying with niobium, vanadium, titanium, aluminium and boron leads to the formation of the carbide TiC, borides Me_3B_2, and intermetallics Ni_3(Al, Ti, Nb) in amounts up to 8% (10Cr15Ni27Ti3W2BMo steel and 05Kh19Ni55Nb2Mo9Al alloy) and 15–17% (04Kh16Ni56Nb5Mo5TiAl), which substantially increases their high-temperature strength and significantly influences the sensitivity to the action of hydrogen [1]. The γ' phase existed as $(Ni,Co)_3(Ti,Al)$ in these three alloys, added cobalt was preferentially partial to the γ and titanium was partial to the γ' phase. Meanwhile, increasing cobalt and titanium in the alloys an increase of the Co and Ti contents in the major γ' phases. For the other observed alloys some plate-shape phase was observed. Chemical analysis indicated that the plate-shape phase was rich in Ni and Ti, and was confirmed to be η phase (Ni_3Ti) by X-ray diffraction analysis.

In most cases, the influence of steam, gaseous hydrogen on mechanical properties, plasticity, crack resistance weakens as pressure, temperature increases and the upper temperature of embrittlement under the analyzed conditions is equal 573 K. At the same time, we reveal a significant decrease in the plasticity of heat-resistant nickel alloys in hydrogen under a pressure of 35 MPa at 1073 K. Maximum hydrogen effect on properties of Cr15Ni27Ti3W2BMo austenitic dispersion-hardened steel and Cr19Ni55Nb2Mo9Al alloy achieved on hydrogenated specimens at hydrogen pressure above 10 MPa. The comparison of the K_{th} threshold values of and the cyclic fracture toughness K_{fc} of 15Cr12Ni2MoNMoWNb and 10Cr15Ni27Ti3W2BMo steels has shown, that on the temperature dependence of K_{fc} and K_{th} of 15Cr12Ni2MoNMoWNb steel are absent the hydrogen embrittlement extremums. Hydrogen negative influence of the fatigue crack growth resistance characteristics of this steel, as a rule, essential and monotonic decrease with temperature increasing.

As the number of technology industries are increasingly used engine and vehicles with limited lifetime, due to the intensification of working parameters and decrease specific quantity of metal structures. In these structures allowed the emergence of strains that exceed the yield stress of materials, i.e., the frequent plastic (little cycles) deformation. Taking into account the impact of variables elastic–plastic loading, high temperatures and corrosive gas environments on the mechanical properties of materials is essential to ensure the reliability of some energy products and aerospace engineering. Especially important is to assess the degree of hydrogen embrittlement, which often leads to catastrophic failure of structures [1, 2].

It is well known, that on the stainless steel surface exists the spinel thin coating. But thermally-induced degradation-relaxation kinetics of such structures composed of spinel $Cu_{0.1}Ni_{0.1}Co_{1.6}Mn_{1.2}O_4$ ceramics are structural inhomogeneities [7] and surface active environment (including hydrogen) play a decisive role in non-exponential kinetics of negative relative resistance drift. Crossover from stretched- to compressed-exponential kinetics in spinel-metallic structures is mapped on free energy landscape of non-barrier multi-well system under strong perturbation from equilibrium, showing transition with a character downhill scenario resulting in faster than exponential decaying.

The effect of hydrogen on yield stress and ultimate strength of investigated materials is negligible. In the case of 15Cr12Ni2MoNMoWNb steel and 04Kh16Ni56- Nb5Mo5TiAl alloy the dependence of low-cycle durability (N) and characteristics plasticity (δ and φ) on the hydrogen pressure consists of two regions. In the first region (low pressures), the N, δ and φ abruptly drops, and in the second, the negative action of hydrogen becomes stable or decrease negligibility.

The action of hydrogen on the plasticity [1] and low-cycle durability of martensitic steels is identical: maximum at room temperature and decreases when temperature increases (Fig.1). At 673 K the mechanical properties of 10Cr15Ni27Ti3W2BMo steel specimens in helium and hydrogen is equal. As temperature increases from 293 to 473 K, the coefficient of influence of hydrogen on the low-cycle durability b of non-hydrogenated specimens of 10Cr15Ni27Ti3W2BMo steel decreases from 0.86 to 0.22 (Fig.1, curves 2, H). At the same, for hydrogenated specimens, it remains constant within the range 0.2–0.22 up to 773 K (Fig.1, curves 2, H + H). At 473 K, the additional influence of preliminarily dissolved hydrogen becomes negligible, i.e., for low-cycle fatigue, this temperature is sufficient for the hydrogenation of stable dispersion-hardening austenitic steels from the atmosphere of hydrogen. In tension, when the maximum tensile stresses are localized at much larger distances from the specimen surface, this effect is attained for 973 K [1]. The degree of embrittlement of 10Cr15Ni27Ti3W2BMo steel somewhat decreases at 873 K. However, for both loading modes, the temperature interval of significant hydrogen degradation for austenitic steels is much larger than for martensitic steels m15Cr12Ni2MoNMoWNb (Fig.1, curves 1). Analogically, in whole temperature diapason 293–1073 K the 04Kh16Ni56Nb5Mo5TiAl alloy is more sensitive to hydrogen embrittlement than 05Kh19Ni55Nb2Mo9Al alloy (Fig.1, curves 3 and 4). In helium ultimate strength, reduction of area w and number cycles to failure N of 04Kh16Ni56Nb5Mo5- TiAl alloy (17–18% intermetallics) are essentially high than properties of 05Kh19Ni55Nb2Mo9Al alloy (8% intermetallics) [1]. Because of the considerable degradation of 04Kh16Ni56Nb5Mo5TiAl alloy in hydrogen under the pressure 30 MPa for all indicated characteristics 05Kh19Ni55Nb2Mo9Al alloy is better.

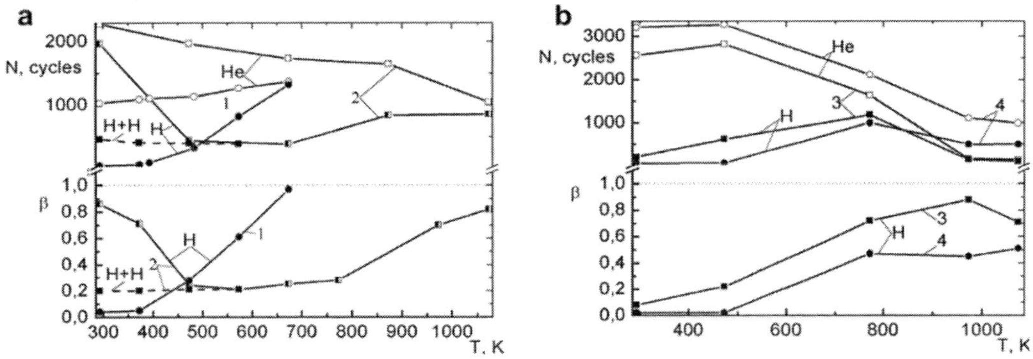

Fig. 1. Temperature dependences of number cycles to failure N (e=1.6%) in helium (He), hydrogen under the pressure 30 MPa (H) and hydrogen under the pressure 30 MPa after preliminary hydrogenation (35 MPa H2, 623 K,

10 h) (H + H) and coefficients of hydrogen influence b=N^H/N^He on 15Cr12Ni2MoNMoWNb (1), 10Cr15Ni27Ti3W2BMo (2) (a), 05Cr19Ni55Nb2Mo9Al (3) and 04Cr16Ni56Nb5Mo5TiAl (4) (b) alloys.

The additional effect of preliminary dissolved hydrogen on the properties of 15Cr12Ni2MoNMoWNb steel and 04Kh16Ni56Nb5Mo5TiAl alloy developed at hydrogen environment pressure least of 10 MPa. The parameters of loading and the modes of action of hydrogen for which the mechanical characteristics of the investigated alloys are minimum at 293 K can be formulated as follows: at the strain rate in short-term static tension in hydrogen V=6.7·10^{-5} s^{-1}; the frequency and amplitude of bending under the conditions of low-cycle fatigue ν=0.5 Hz and ε=1.6%, and the pressure of hydrogen must be higher than 10 MPa. In the case of 10Cr15Ni27Ti3W2BMo steel and 05Kh19Ni55Nb2Mo9Al alloy the low-cycle durability N, characteristics plasticity δ and φ decrease in whole hydrogen pressure range.

Preliminary dissolved hydrogen leads to a considerable additional decrease in the properties of this alloy. Maximum hydrogen effect achieved on hydrogenated specimens of 10Cr15Ni27Ti3W2BMo

Hydrogen essentially decreases number cycles to failure of 15Cr12Ni2MoNMoWNb martensitic steel in temperature range 293-623 K, 10Cr15Ni27Ti3W2BMo austenitic dispersion-hardened steel – in 293-973 K and Ni-base superalloys [8, 9] – in 293-1073 K. We compared the threshold values of ΔK_{th} and of the cyclic fracture toughness K_{fc} of 15Cr12Ni2MoNMoWNb and 10Cr15Ni27- Ti3W2BMo steels. On the temperature dependence of K_{fc} and ΔK_{th} of 15Cr12Ni2MoNMoWNb steel are absent the hydrogen embrittlement extremums [1]. Hydrogen negative influence of the fatigue crack growth resistance characteristics of this steel, as a rule, essential and monotonic decrease with temperature steel and 05Kh19Ni55Nb2Mo9Al alloy at hydrogen pressure above 10 MPa. Actually, the behaviour of different materials is different under high pressures. Martensitic steels and 04Kh16Ni56Nb5Mo5TiAl alloy have ultimate baric hydrogen resistance, i.e., the characteristic number of cycles to failure and plasticity in hydrogen which remains constant as the pressure of hydrogen increases further.

For 10Cr15Ni27Ti3W2BMo steel and 05Kh19Ni55Nb2Mo9Al alloy the limit is absent and their serviceability is a monotonically decreasing function of hydrogen pressure. The ultimate hydrogen embrittlement of these materials achieved only after preliminary high temperature hydrogenation. Hydrogen decreases high-cycle durability of martensitic steels and Ni-base alloys, the degree of hydrogen effect increase with strain amplitude decrease. The additional effect of preliminary dissolved hydrogen is negligible for 10Cr15Ni27Ti3W2BMo steel (Fig. 2a) and is absent for 05Kh19Ni55Nb2Mo9Al alloy (Fig. 2b).
increasing. More affected characteristics – threshold stress intensity factor ΔK_{th}. As the testing temperature rises, the values of ΔK_{th} of the 10Cr15Ni27Ti3W2BMo steel in helium and hydrogen increase, whereas the characteristic K_{fc} is practically independent of temperature in both atmospheres. The influence of hydrogen is maximum at 293 K.

On the whole, for 10Cr15Ni27Ti3W2BMo steel, at this frequency of loading and cycle asymmetry, which are essential during cyclic crack resistance testing, the influence of hydrogen is much weaker than its influence on the crack resistance under static loading (see Fig.1).

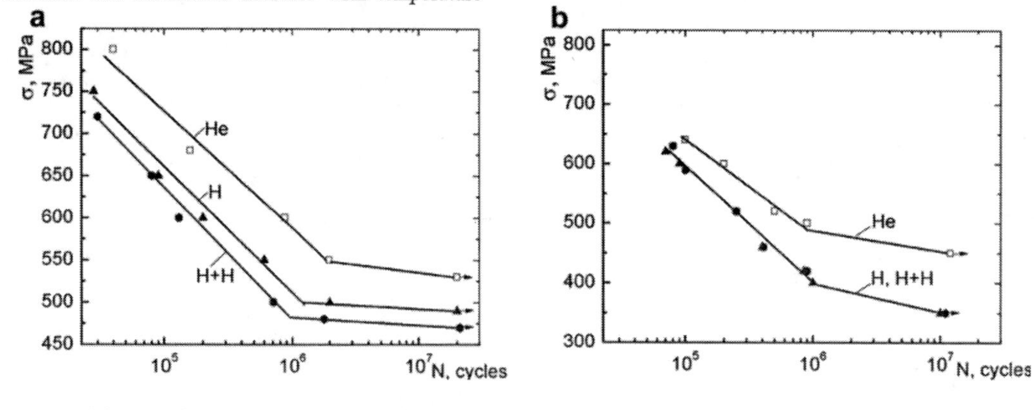

Fig. 2. High-cycle fatigue curves of 15Cr12Ni2MoNMoWNb (a) and 05Cr19Ni55Nb2Mo9Al (b) alloys in helium (He), hydrogen under the pressure 30 MPa (H) and hydrogen under the pressure 30 MPa after preliminary hydrogenation (35 MPa H2, 623 K, 10 h) (H + H) at 293 K.

There is the strong dependence of the threshold stress intensity factor K_{th} on the frequency of cyclic loading (Fig.3) for heat resistant 3,5NiCrMoV rotor steel [10, 11]. The tendency of monotonic decreasing of the K_{th} values with decreasing of frequency is observed both in air and hydrogen environment.

However this trend is significantly stronger for test in hydrogen. There is a strong dependence of fracture mechanics parameters on the values of yield stress of steels used for generators.

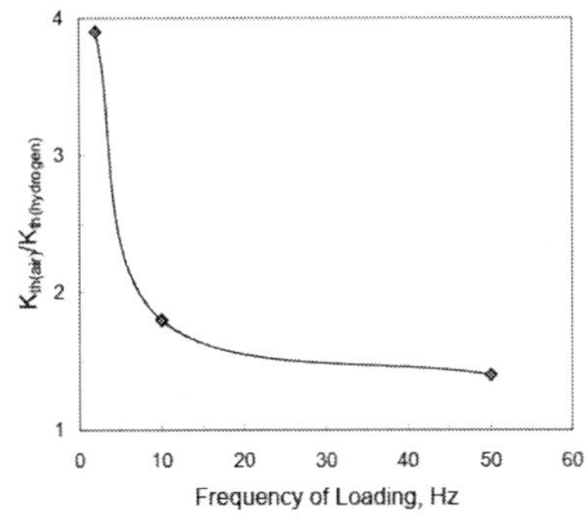

Fig.3. Ratio K_{th}(air)/K_{th}(hydrogen) versus frequency of loading for steel 38KhN3MFA.

The assessment of hydrogen effect on these parameters with comparison with air can be made from the following diagrams [2, 10, 11, 14] (Fig.4): $K_{IC}=F_1(\sigma_{0.2})$ or $K_{JC}=F^*_1(\sigma_{0.2})$, $\Delta K_{fc}=F_2(\sigma_{0.2})$, $\Delta K_{th}=F_3(\sigma_{0.2})$. Here: K_{IC} is a fracture toughness defined under static loading by standard method; K_{JC} is a static fracture toughness defined on a base of J-integral; ΔK_{fc} is a cyclic fracture toughness defined from a diagram of fatigue crack growth; ΔK_{th} is a range of threshold stress intensity factor, which can be also defined from fatigue crack growth diagram.

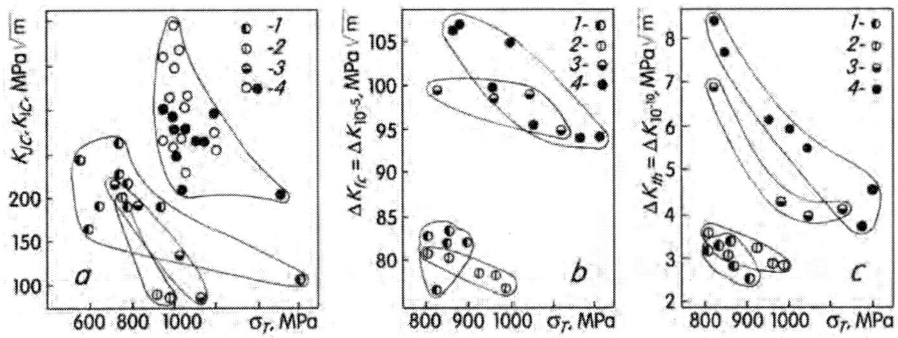

Fig.4. Diagrams $K_{IC}=F_1(\sigma_{0.2})$ (a), $K_{JC}=F^*_1(\sigma_{0.2})$(b), $\Delta K_{fc}=F_2(\sigma_{0.2})$, $\Delta K_{th}=F_3(\sigma_{0.2})$(c) for steels: 1 – 3,5NiCrMoV; 2 – 8Mn8Ni4Cr; 3 –18Mn4Cr; *4* – 18Mn18Cr on air (a) and after hydrogen saturation (b, c) (○ - K_{IC}, ● K_{JC}).

But it is necessary to take in to account the experimental results of the fatigue crack growth in specimens with rectangular cross sections which were weakened by sharp and blunt one-sided notches (for example the notch root ρ = 0.2; 5; 10 and 22.5 mm) [12, 13]. The tests were performed in the low- and high-cycle fatigue modes by imposing a constant value of the nominal load ratio R =–1; 0 and a moment amplitude Ma =15.84 N·m. All the results of fatigue tests must be analyzed in terms of the parameter ΔK including the influence of the notch.

4 Conclusions

The condition of hydrogen action in which the degradation of 15Cr12Ni2MoNMoWNb martensitic steel, 10Cr15Ni27Ti3W2BMo austenitic dispersion-hardened steel, heat resistant 38KhN3MFA rotor steel, 04Kh16Ni56Nb5Mo5TiAl and 05Kh19Ni55Nb2Mo9Al Ni-base superalloys reaches its limit was established. Plasticity characteristics and low-cycle durability of non-hydrogenated specimens of 15Cr12Ni2MoNMoWNb martensitic steel and 04Kh16Ni56Nb5Mo5TiAl alloy abruptly drops in 0–10 MPa pressure range and becomes stable or decrease negligibility as the pressure of hydrogen increases further. Maximum hydrogen effect on properties of 10Cr15Ni27Ti3W2BMo austenitic dispersion-hardened steel and 05Kh19Ni55Nb2Mo9Al alloy achieved on hydrogenated specimens at hydrogen pressure above 10 MPa. Hydrogen decreases high-cycle durability of martensitic steels and Ni-base alloys, the degree of hydrogen effect increase with strain amplitude decrease. The temperature range of essential hydrogen degradation is a structure-sensitive factor: 293–623 K in the case 15Cr12Ni2MoNMoWNb martensitic steel, 293–973 K for 10Cr15Ni27Ti3W2BMo austenitic dispersion-hardened steel and 293–1073 K for Ni-base alloys.

There is the strong dependence of the threshold stress intensity factor K_{th} on the frequency of cyclic loading for heat resistant 3,5NiCrMoV rotor steel. The tendency of monotonic decreasing of the K_{th} values with decreasing of frequency is observed both in air and hydrogen environment. There is a strong dependence of fracture mechanics parameters on the values of yield stress of steels used for hydrogen cooling generators.

References

1. A. Balitskii, V. Vytvytskyii, L. Ivaskevich, J. Eliasz, Int. J. Fatigue. **39**, 32 (2012)
2. A. Balitskii, European Commission fun. Int. Workshop, *Mater. Resist. to extreme cond. for future energy sys.*,21(2017)
3. A. Balitskii, S. Eriksson, G. Mascalzi, Fatigue-02: Proc.Eighth Int. Fatigue Congress (Stockholm), **5/5**, 3149 (2002)
4. A. I. Balitskii, V. I. Pokhmurskii, Fatigue-90: Proc. Fourth Int.Con. (Honolulu), **III/1** , 1839 (1990)
5. J. N. Vincent, L. Remy, Fatigue-84: Proc. 2nd Int. Con., **II**, 697 (1984)
6. J.-Y. Guédou, I. Augustins-Lecallier, L. Nazé, P. Caron, D. Locq, Superalloys – 2008, TMS (The Minerals, Metals & Materials Society), 21,2008
7. V. Balitska, O. Shpotyuk, M Brunner, I. Hadzaman. Chem. Phys. **501**, 121 (2018)
8. P. T. Newman, C. J. Beevers, Fatigue-84: Proc. 2nd Int. Con., **II**, 785 (1984)
9. T. L. Grobstein, E. Welsch, N. Panigrahi, J. D. McGervey, Fatigue-84: Proc. 2nd Int. Con., **II**, 1019 (1984)
10. A. Balitskii, Yu. Boichenko, A. Sosnin, N. Shokov. Met. Sci. Heat Treat Metals. **4**, 266 (1987)
11. A. Balitskii, V. Makarenko. Sov. Mat. Sci. **24**, 633 (1988)
12. D. Rozumek, S. Faszynka. Mat. Sci. **53**, 417 (2017)
13. Z. Marciniak, D. Rozumek, E. Macha. Intern. J. Fatigue. **30**, 800 (2008)
14. O.I. Balyts'kyi. Mat. Sci. **22**, 541 (2000)

Experimental and numerical investigation of strain distribution of notched lead fatigue test specimen

Audun Johanson[1,*], *Luigi Mario* Viespoli[2], *Bård* Nyhus[3], *Antonio* Alvaro[3], and *Filippo* Berto[2]

[1]Nexans Norway, Innspurten 9, 0663 Oslo, Norway
[2]Department of Mechanical and Industrial Engineering, Norwegian University of Science and Technology (NTNU), Norway
[3]Sintef Materials and Chemistry, Richard Birkelands vei 2B, 7031, Trondheim, NORWAY

Abstract. The work here presented focuses on the test methodology related to effect of stress concentrators in strain controlled structures. Cable sheathing as used in subsea power cables are investigated by cyclic fatigue testing, Digital Image Correlation (DIC) and 3D Finite Element Analysis. Focus is put on the strain distribution in conventional specimen geometries and under the presence of artificial notches. It is evident that standardized fatigue testing provides limited input to the final fatigue life of strain controlled power cable sheathing both accounting for expected and unintended stress concentrators. The limitations can be explained by measured strain distribution inherent in most fatigue test specimens. The use of DIC and 3D FEM provides valuable insight into both the theoretical and practical stress and strain distribution. This can help in understanding and overcoming geometrical test constraints, when compared to the actual component loading mode.

1 Background and Scope

Lead alloys sheathing is used as water barrier on high voltage subsea power cable. Subsea power cables are subjected to various cyclic loads during manufacturing, installation and operation. Cyclic loading of cable sheathing falls in the typical high cycle regime, but under global strain control in combination with non-linear and time dependent material response.

Cable sheathing design and manufacturing is associated with multiple discontinuities which intensify the local stress and strain field. Such discontinuities might be inherent design features such as intimate contact with various strips such as transversal steel armoring, or local defects associated with the sheathing extrusion process such as indigenous oxides or intermetallic, or exogenous particles. Studies of such effects and fatigue properties of power cable sheathing is limited and rarely provide suitable input for fatigue life calculations of power cable sheathing where appropriate loading mode and design features are taken into account [1].

Discontinuities in components subjected to mechanical loading will cause concentration(s) in the local stress and strain field, and shorten the fatigue life of the component. Such effect is dependent on the material in question, particularly inelastic and relaxation properties as well as work hardening. These are associated with lead alloys [2]. Otherwise said, both the monotonic and the cyclic behavior of lead alloys exhibit a significant strain rate dependence. Effects of discontinuities are therefore not equivalent at any strain rate.

Often lab scale testing regarding such effects are done by implementing a well-defined notch detail on the test specimen. Due to the aforementioned material characteristics, the interpretation of results thereof is non-trivial as the component loading mode with respect to the detail-of-interest is not necessarily reproduced in smaller scale. Additional complications arise from the fact that the lab specimen geometry introduce deformation incompatibility zones, such as specimen edges and fillet radiuses, that do not exist for the real component which put into question the direct lab testing results transferability to the real component case. In this case, both the distribution of strain and strain rate must be carefully accounted for to ensure the suitability of the lab scale results. The following work aims to investigate the strain distribution and fatigue failure of notched test specimens extracted from cable sheathing.

2 Material and experimental procedure

2.1. Materials

The test material is extracted from as- manufactured power cable where the tubular lead sheathing was extruded from raw material according to PB012K in EN 12548. The chemical composition is reported in Table 1. The procedure has been adopted to account for the presence of the defect in the lab specimen. A 0.5 x 0.5 x

* Corresponding author: audun.johanson@nexans.com

0.4 sheathing thickness notch was made by electro discharge machining (EDM) on the convex surface of the curved specimen (Fig. 1). The design of the notch/discontinuity has been made based a potential surface penetrating inclusion originating from the extrusion process. Due to the small thickness of the sheathing (3.3 mm), the sheathing curvature in the transversal section has been kept for later fatigue testing in order to prevent buckling in the compressive phase and thus increase the maximum achievable strain range, as well as limiting the induced work hardening when preparing the sample (Fig. 2).

Table 1. Chemical composition of lead alloy PB012K

Element	**Sn**	**Cd**
Weight %	0.175- 0.225	0.60- 0.90

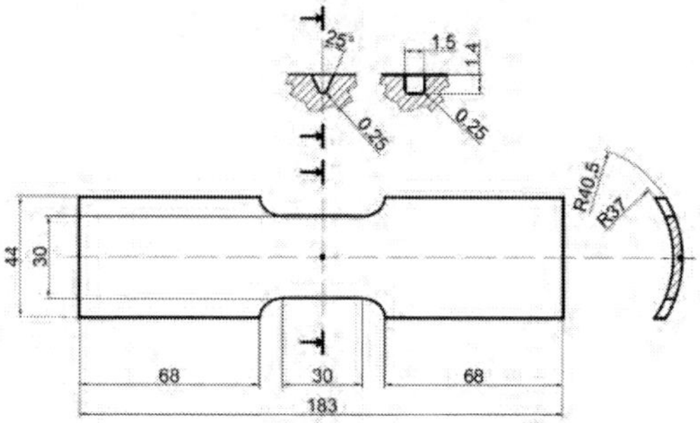

Fig. 1. Nominal dimensions of the specimens and of the notch detail.

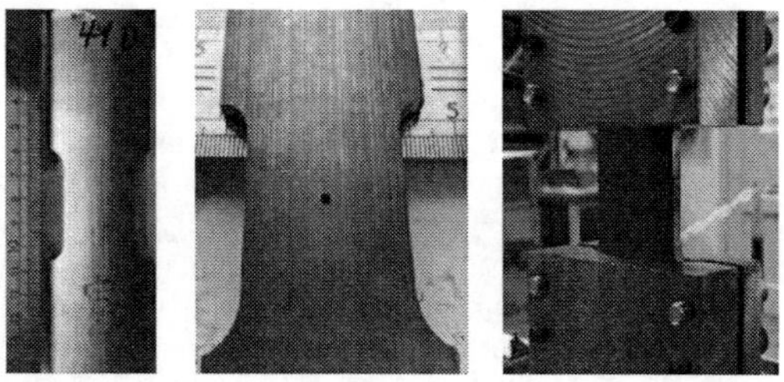

Fig. 2. The pictures show the sample un-notched (left), notched (middle) and in the custom-made, curved clamping system (right).

2.2. Fatigue test procedure

The effect of the discontinuity has been investigated by comparing cyclic tension-compression fatigue testing directed towards reversed bending of a power cables.

The tests have been run in position control, moving the head between two set points, which have been related to the strain range target values (0.15 % or 0.28 %) prior to the test controlled by a mechanical extensometer. The extensometer was removed after the set-point definition due to sample indents (which where polished away). The recorded positions obtained before each test, in average corresponded to 0.008 and 0.0041 mm for a strain of 0.28 % and 0.15 %, respectively, has been kept constant during the complete the test duration. An appreciable decrease in the nominal amplitude was found for notched specimens Fig. 3, suggesting the presence of the notch affects the global response of the specimen.

For two of the tests at higher strain amplitude, the extensometer was applied on both sides of the specimen in order to verify whether or not an out-of-plane deformation occurred. The fatigue tests were performed following a frequency stepping up (0.5, 1, 2 and then 5 or 10 Hz, according to the target frequency). During the test, load and displacement were recorded every 100 cycles and the number of cycles to failure was considered at the moment in which the tensile load fell below 0.2 kN, This was the highest lad value at which a through-thickness growing crack was visually detectable.

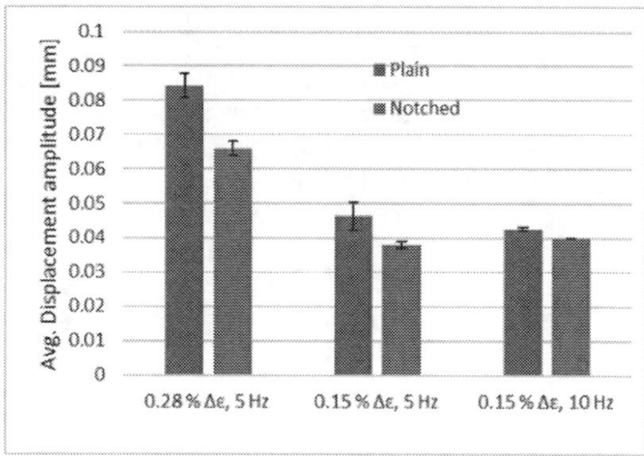

Fig. 3. Average displacement amplitude for plain and notched specimens.

2.2. Strain rate dependency

For lead alloys tested under displacement control, relaxation will cause a strong strain rate dependency. A series of tensile tests at various strain rates were conducted to identify the onset of appreciable strain rate dependency and thereby a common material law for finite element modelling. Tensile test where performed on dog- bone specimens (Fig. 4).

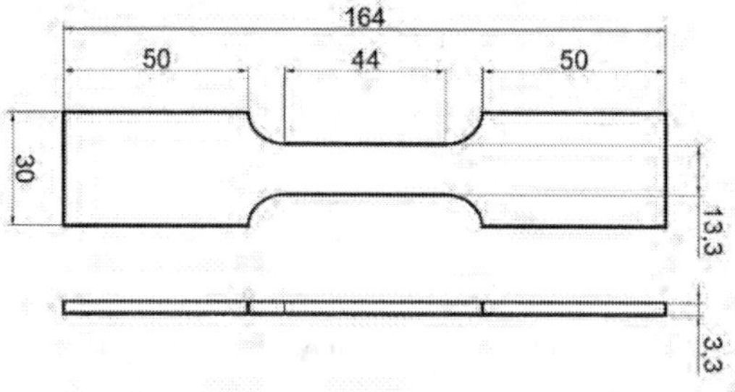

Fig. 4. Geometry of flat dog- bone tensile specimens.

The Stress-Strain curves obtained are shown in Fig. 5. It is evident how, for strain rates higher than 10^{-4} s^{-1}, the effect of creep at room temperature is negligible, especially in the range of interest for the fatigue testing (up to 0.28 % of nominal strain). Subsequently, the

Digital Image correlation (DIC) material characterization on the flat dog-bone samples in Fig. 4 and the DIC analysis on the curved samples in Fig. 7 , have been performed at a nominal strain rate of 10^{-3} s^{-1}.

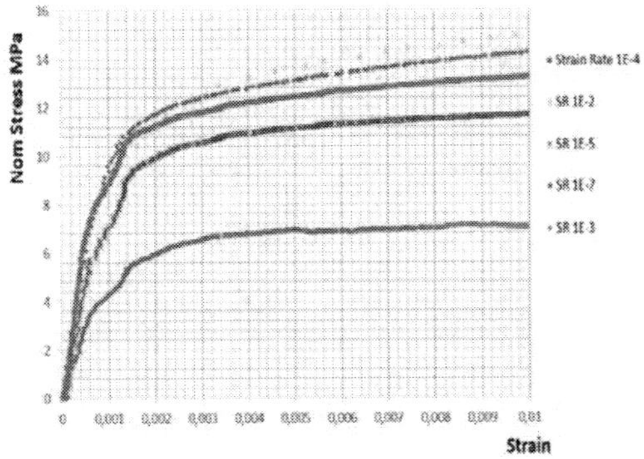

Fig. 5. Nominal Stress vs Strain curves for tensile dog-bone samples of lead alloy at different Nominal Strain Rates. At a rate higher than 10^{-4} s^{-1} the creeping effects are negligible within the cyclic strain range to-be investigated.

2.2. Strain distribution

The stress distribution in the specimen during the fatigue testing in relation to the failure location and final fatigue life from the fatigue testing is investigated by 3-dimensional finite element analysis (FEM) where monotonic tensile testing under DIC is used to validate the calculated strain distribution. Speckle pattern where applied on both flat dog-bone specimens and curved dog-bone specimens (Fig. 7) which were tested in a servo-hydraulic Zwick machine. A nominal strain rate of 10^{-3} s^{-1} where applied which assumes to constitute the lower bound strain rate below which relaxation/ creep effects takes effect (Fig. 5); in other words, above such strain rate value the material strain rate sensitivity is negligible.

Fig. 7. A frame of the DIC images acquired for the flat tensile dog-bone (left) and of the curved specimen, un-notched, of the same geometry used for the fatigue testing (right).

3 Experimental results

3.1. Fatigue tests

Fatigue testing have been carried out at frequencies 5 and 10 Hz: the test results are summarized in the plot in Fig. 6. Result at the higher strain range, i.e. 0.28 % show practically no difference in terms of fatigue life due to the presence of the notch. For the test performed at 0.15 % and 5 Hz, the presence of the notch infers a decrease in the average fatigue life of about 65 %, i.e. from an average value of 536,000 cycles to 181,000 cycles. However, by stepping up the frequency from 5 to 10 Hz, the detrimental effect induced by the presence of the notch seems to be reduced as in fact is rather the average fatigue life of the un-notched specimens to drop of about 40 %, from 536,000 to an average of 323,000 cycles. No significant difference is found for the notched specimens.

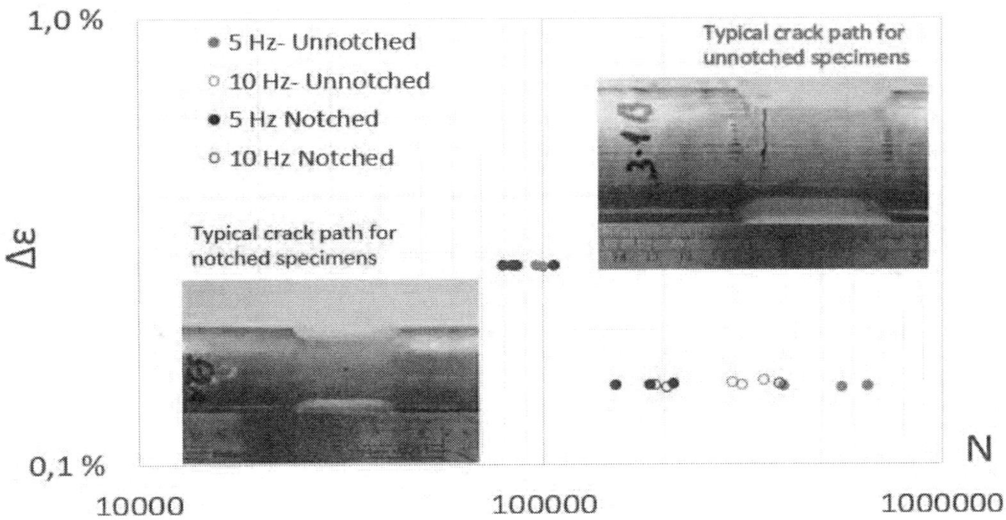

Fig. 6. Fatigue test results together with typical crack paths. It was found that for the lower strain range tests (0. 15 %) notched specimens, the failure consistently initiated from the specimen edge at the height/ position of the notch (Lower left picture). When the notch was not present, failure tended to initiate from the fillet edge. For 0.28 % strain range, both notched and unnotched tended to fail from the middle (notch- height) region.

https://doi.org/10.1051/matecconf/201816505003

Post mortem analysis of the specimens also reveals a difference when it comes to the crack location. In 7/ 9 un-notched specimens the crack initiated from the specimen fillet radius, 8/10 notched specimens failed by the crack initiating from the outer edge, but in correspondence of the notch height/ location (middle region of the sample). Note that 2/3 unnotched 0.28 % strain range samples failed from the middle edge position. i.e. the trend more evident for the 0.15 % strain range samples. This indicates that the presence of the notch induces a variation in the stress/strain field strong enough to alter the position of the initiation point but still the specimen edge constitutes the "weakest" point when it comes to either crack initiation or crack- initiation and growth. The investigation of this effect and the primary aim for continued DIC and FEM analysis reported in the following paragraphs.

3.2 Strain distribution

The strain distribution in the different regions of the specimen (Fig. 8) together with the resulting stress-strain plot obtained through post-processing of the DIC images.

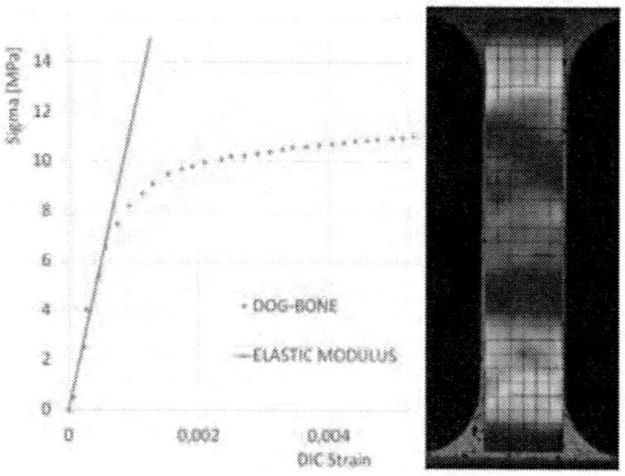

Fig. 8. Obtained tensile curve and strain distribution as acquired by DIC.

Tensile testing under DIC has permitted to record the true Stress-Strain characteristic of the alloy. This provides the material constitutive behaviour for the elastic-plastic FEM analysis on both the un-notched and the notched sample.

4 Finite element analysis

The aim of the analysis is to give basis for a better understanding of the influence of the notch on the failure dynamics and to eventually reproduce numerically the behavior of the alloy. The comparison between the DIC and FEM analysis is executed at the same nominal/global strain, i.e. measured from the displacement of the boundary divided by the free length in the FEM analysis.

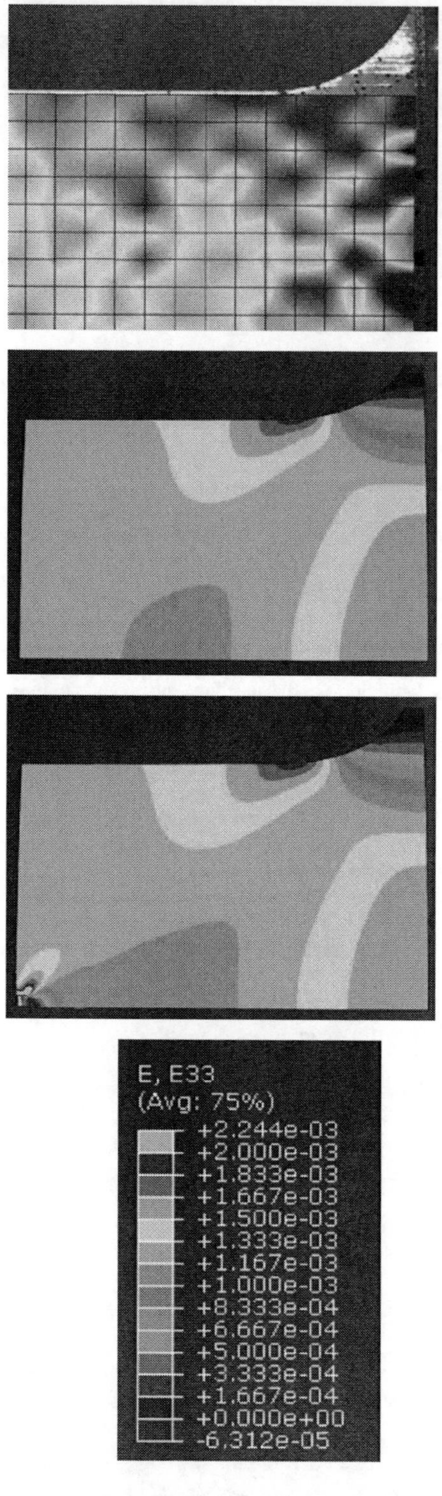

Fig. 9. Strain field along the longitudinal direction of a quarter of the un-notched specimen, DIC (top) and FEM (2. from the top). In figure (3. from the top), the same strain field for the notched specimen. The scale adopted for the strain is the same for all captures (lower). All strain field corresponds to a nominal strain of $1.1E^{-3}$.

For the test, the real border effects and eventual deformation of the sample close to the clamping area does not allow precise strain estimation. Therefore, the nominal strain is obtained from DIC on a vector which covers the full length of the sample. The evaluation of the equivalent nominal strain is then made by considering the FEM analysis. The results obtained show, at the same equivalent loading condition and in the same scale, that the FEM analysis is feasible to reproduce the strain field in magnitude and shape (Fig. 9, Fig. 10), with a concentration of the strain for the un-notched sample in correspondence of the fillet radius and the clamping area.

The results of the FEM analysis on the notched geometry shows a much higher strain, at comparable equivalent boundary displacement, which justifies the final fracture passing through the notch, but there is no evidence from the numerical stress field of the reason why fatigue fractures in the notched specimens have been observed to initiate on the edge of the sample, at the height of the notch, but not from the notch itself.

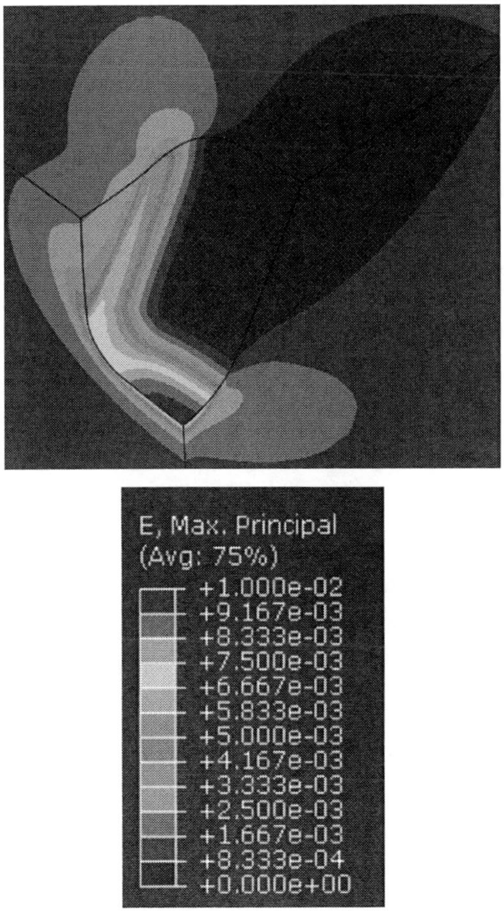

Fig. 10. Maximum principal strain in the notch, at the same nominal strain of the captures in Fig. 9. The scale shows a notable strain intensification due to the notch.

5 Conclusion and further work

It is evident from the fatigue test results that the impact of the notch is higher for low strain ranges. This relation appears to be dependent on test frequency. However, the imposed test frequencies imply a strain rate exceeding appreciable creep. Understanding of the impact of frequency, particularly taking into account the expected fatigue life reduction due to creep is critical and the subject of ongoing work. Location of failure initiation also supports a different effect of the notch for the two tested strain range levels, where high cyclic strain range fail from the middle region of the test specimens both with and without a notch. For lower cyclic strain range, failure only occur in the middle region when the notch is present. The Numerical calculations and tensile test experiments under DIC show that the strain concentration on specimen fillets is significantly higher than the nominal strain range and supports failure from this location. Numerical calculations show that the notched detail is associated with the highest strain concentration which justifies both the reduction in fatigue life and change in failure location. However, as the final crack do not tend to initiate at notch the test results appear inconsistent with the analysis.

As the lead alloy is expected to undergo irreversible deformation at the imposed strain ranges, hardening and local blunting can be mechanisms that cause re-distribution of the strain and differentiates the monotonic and cyclic strain fields thus explains a change in failure initiation. However: no evidence of blunting or major changes of the notch geometry has been observed after fatigue testing. Regardless, the material law used in the numerical study as well as the DIC strain is obtained from monotonic cases. The cyclic material response will differ. Ongoing work is put into the development of a dynamic material law and in obtaining the DIC recording to have a quantification of the strain distribution variation and development during the actual fatigue test. The inconsistently between local strain concentration and failure initiation can also be related to the crack initiation and growth dynamics- meaning that the strain field must not only favor crack initiation but continued growth.

Finally, it should be noted that failure initiation from the specimen edge is a test artifact with respect to the final application as such details are absent for tubular components. Crack initiation from the sample edge indicates that even if the relative effect of a notched-unnotched component is larger than for the notched - unnotched test specimen, the laboratory scale fatigue test result for both notched and unnotched samples are conservative compared to the real component. Despite of this, care should be taken in assessing fatigue life ductile components such as power cable sheathing, by standardized testing.

The work supports the use of numerical calculations coupled with experimental validations using DIC to both interpret fatigue results and design fatigue test specimens. This is critical for accurate fatigue life predications of component subjected to cyclic loading-particularly in the presence of local details and strain distribution.

6 Acknowledgment

The authors wish to thank NEXANS Norway and the Research Council of Norway for the financial support through ENERGIX Programme, Contract No.256367/E20.

References

1. International Lead Zinc Research Organization, Inc. (ILZRO), *"Program on Technology Innovation: Grain Growthin Lead Alloy Sheathin and its Influence on the Life of Lead- Sheathed Power Cables,"* EPRI, Palo Alto, Ca, (2007).
2. S. Guruswamy, *Engineering Properties and Applications of Lead Alloys*, New York: Marcel Dekker, (2000).

Modeling of creep-fatigue interaction effects on crack growth at elevated temperatures

Gabriel P. Potirniche[1*]

[1]Mechanical Engineering Department, University of Idaho, 875 Perimeter Drive MS 0902, Moscow, ID 83844-0902, USA

Abstract. The well-known load frequency effect on creep-fatigue crack growth is explained by the interactions between fatigue and creep loading and is quantified using the concept of plasticity-induced crack closure. It is shown that the hold time during creep loading affects crack growth rates during subsequent fatigue cycles. Longer hold times lead to lower crack-tip opening stresses and faster crack growth rates during fatigue loading. To model the impact of hold time on crack opening stresses during fatigue loading, a strip-yield model was developed for creep-fatigue crack growth. The strip-yield model computes crack-tip opening stresses, which determine the effective stress intensity factor range and crack growth rate during the fatigue portion of each loading cycle. Maximum stress intensity factor is used to compute the crack growth rate during the creep portion of each cycle. The proposed strip-yield model is used to compute creep-fatigue crack growth rates for several structural materials, i.e., an Astroloy, aluminium alloy 2650 and 316 stainless steel. The model predictions of crack growth rates compare well with published experimental data for these alloys. This model achieves reliable predictions of crack growth rates and life prediction on components subjected to creep-fatigue loading at elevated temperatures by considering loading interaction effects.

1 Introduction

Failure by crack growth during creep-fatigue loading is an important failure mode for structural components in service at elevated temperatures in power plants or in the aerospace industry. Creep-fatigue loading is a combination of a creep portion, in which the load is held constant at a certain level, and a fatigue portion, in which the load fluctuates between minimum and maximum values. The two types of loading can act synergistically on determining how fast cracks grow each creep-fatigue cycle. A common approach is to assume that total crack growth rate is the sum of growth rates during each portion of a loading cycle. Some models neglect the interaction effects of creep and fatigue loads on crack growth rate, while others consider these effects by modeling crack tip damage.

The objective of this paper is two-fold. First, it proposes a modeling approach for the influence of hold time during creep loading on the crack growth rate during subsequent fatigue cycles. Second, predictions of crack growth rates are presented for several alloys commonly used in structural applications at elevated temperatures. The approach presented in this study is denoted as Strip-Yield Model for Creep-Fatigue Crack Growth (SYM-CFCG). Creep-fatigue crack growth is predicted using a strip-yield methodology, devised to compute crack growth rates during the fatigue and creep portions of each loading cycle. Strip-yield methodology has been used successfully to predict fatigue crack growth rates in metallic components under constant and variable amplitude loading. This modeling approach is well suited to capture the effects of loading history on fatigue crack growth rates. In the present study, SYM-CFCG is used to capture loading history effects on crack growth by quantifying how creep loading affects crack growth rates during subsequent fatigue cycles.

Strip-yield modeling is a computational approach in which the crack plane, including both crack surface and uncracked ligament in front of the crack tip, is discretized with a certain number of one-dimensional elements of a given length and width. During crack growth under applied loading, new elements are generated with lengths computed according to the mechanics of crack opening and closing.

The proposed SYM-CFCG uses the weight function method to compute relevant fracture mechanics parameters related to crack growth, such as crack surface displacements and plastic and creep zones in front of the crack tip. SYM-CFCG also computes the plastic wake left behind the advancing crack tip and the contact stresses between crack surfaces during unloading. Moreover, SYM-CFCG computes crack-tip driving forces, such as effective stress intensity factor range ΔK_{eff}, and relevant creep fracture parameters C^*, $C(t)$ and $C(t)_{avg}$. Crack growth rates during fatigue loading are computed using ΔK_{eff} and during creep loading using K_{max} or $C(t)_{avg}$. Excellent predictions and correlations for several commonly used crack geometries are obtained. The geometries analyzed with SYM-CFCG were: middle-tension (MT), compact-tension (CT), single-edge notched tension (SENT) and double-edge notched tension (DENT) specimens. Materials considered in these simulations were a nickel-base superalloy, aluminum alloy 2650 and

* Correspondence: gabrielp@uidaho.edu

stainless steel 316 stainless. These materials are used in structural applications in power plants and the aerospace industry.

2 Modeling approach

2.1 Crack opening and closing

Total crack growth rate during a typical creep-fatigue cycle is commonly assumed as the sum of growth rates during loading and holding portions of that cycle. Figure 1 presents a typical creep-fatigue loading scheme used to measure crack growth rates, with the loading cycle as *A-B-C-D* and load hold time t_H.

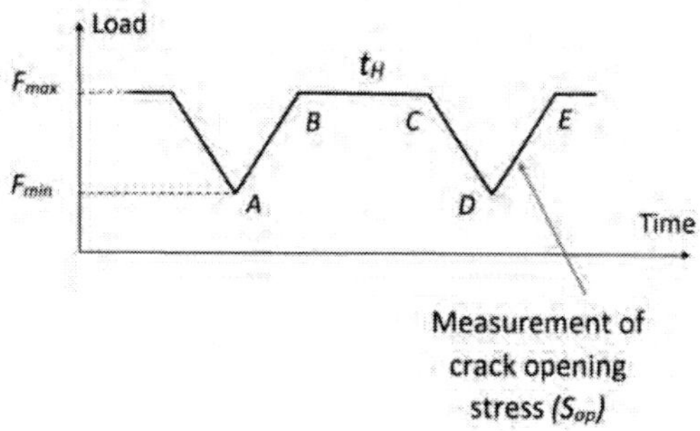

Fig.1. Typical creep-fatigue loading using in crack growth testing.

Crack growth rates during *A-B* are commonly predicted using ΔK_{eff} and during *B-C* using either K_{max} or $C(t)$ integral. Crack growth rate during fatigue loading (*A-B*) is highly dependent on crack-tip opening stress, which is influenced by plasticity-induced crack closure, although other closure mechanisms such as roughness- and oxide-induced crack closure may also play a role at elevated temperatures. In this study, the focus is exclusively on the plasticity-induced crack closure phenomenon.

The core hypothesis of this study, which attempts to explain the influence of creep loading on fatigue crack growth, is that the duration of hold time t_H affects the crack-tip opening stress during the subsequent cycle, and implicitly crack growth rate. This is expressed as

$$\left(\frac{da}{dN}\right)_{D-E} = f_1(S_{op}, \dots), \text{ and}$$

$$(S_{op})_{D-E} = f_2(t_H, \dots) \qquad (1)$$

where da/dN is crack growth rate per cycle, and f_1 and f_2 are some unknown functions. The goal of this study is to determine these dependencies represented by f_1 and f_2.

The above equation indicates that creep loading affects fatigue crack growth rate through hold time t_H. In general, for longer hold times t_H, opening stress S_{op} decreases, leading to faster crack growth rates during the fatigue portion of that cycle (*D-E*).

Figure 2 shows an interpretation of how the influence of hold time on the subsequent closing and opening of the crack operates. At maximum load (point *B* in Figure 1), the crack is fully open. During hold time t_H (*B-C*), the crack opening displacement increases as a result of stress relaxation and accumulation of creep strains in the near crack-tip region. Longer t_H values lead to larger crack-tip opening displacements. For $t_H = 0$ (i.e., pure fatigue loading) or small hold times, the additional crack opening displacement caused by creep deformation is negligible and, upon unloading, a portion of the crack closes, as illustrated in Figure 2. On the contrary, when t_H is large, the additional crack opening displacement caused by creep is significant, and upon unloading, the portion of the crack surface to experience closing at minimum load is shorter than that when $t_H = 0$ or t_H is small. For moderate to long hold times, crack surfaces may not even close at minimum load, and in this case the crack-tip opening stress will match minimum stress. In this situation, effective stress intensity factor range ΔK_{eff} is maximized, causing an acceleration of crack growth rate. A caveat to this interpretation is that hold times t_H cannot be too large, because the crack opening displacement at *C* becomes very large, causing significant crack-tip blunting and temporary deceleration of the crack growth rate as the crack tries to re-initiate and propagate in subsequent fatigue cycles. Crack tip blunting due to very large t_H is not accounted for in this study.

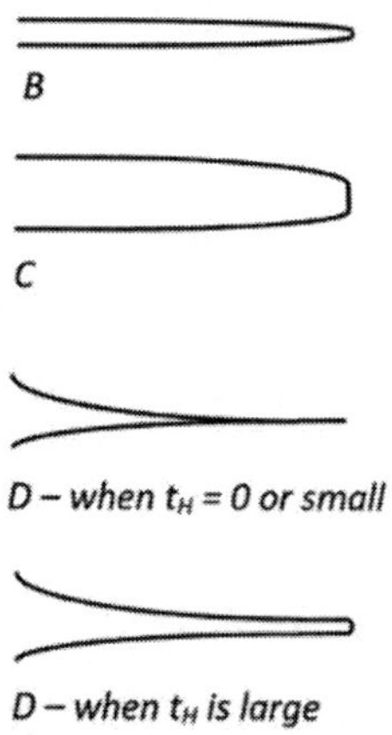

Fig. 2. Crack opening displacements at various points during the creep-fatigue loading cycle illustrated in Figure 1.

These considerations are supported by the finite element simulation results obtained by Sehitoglu and Sun [1] and Ramirez et al. [2]. Both studies showed that opening stresses decrease with longer hold times. Some differences between the two studies exist in terms of the magnitude of hold time impact on opening stresses, and they could be attributed to differences in the model setup and mesh refinement. While Sehitoglu and Sun used a two-dimensional small-scale deformation model and a mesh relatively coarse, Ramirez et al. performed a systematic study employing two- and three-dimensional models for several standard specimens and meshes with a high degree of refinement. Despite their quantitative differences, qualitatively both studies revealed that hold times play a major role in determining crack-tip opening stresses during creep-fatigue loading.

In addition to the above mentioned finite element efforts, the experimental results obtained by Piard et al. [3] of creep-fatigue crack growth in a nickel-base superalloy showed that longer hold times cause faster crack growth rates during fatigue loading in a creep-fatigue cycle. While Piard et al. measured and reported crack growth rates during loading and holding of individual creep-fatigue cycles, they did not measure crack-tip opening stresses. However, their experimental observation of cracks growing faster during loading when the prior hold time was longer could be explained by a decrease of crack-tip opening stresses with longer hold times. In section 3 of this paper, it will be demonstrated that such hypothesis is able to reliably explain their measured crack growth rates.

2.2 Crack-tip opening stress computation

The general framework of the proposed SYM-CFCG is similar to the approach published by the author for creep crack growth [4], however a significant improvement has been achieved in this study. While the strip-yield model in study [4] considered only monotonic loading during creep crack growth, currently proposed SYM-CFCG improves on the previous version by considering cyclic loads due to fatigue, and uses a procedure for computing crack-tip opening stresses due to the combined actions of plasticity and creep deformations at the crack tip. In this section the main highlights of this modeling approach are presented.

In an arbitrary specimen subjected to creep-fatigue loading, plastic and creep zones develop near the crack tip [5]. While the plastic zone develops instantaneously under applied load, the creep zone initiates and steadily grows under steady load. Longer t_H lead to larger creep zone sizes. During creep zone extension, the normal stress near the crack tip relaxes from its initial flow stress to lower values.

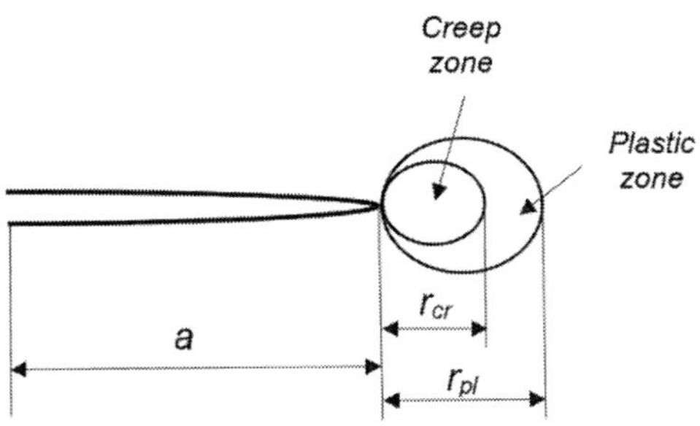

Fig. 3. Extent of plastic and creep deformation zones at the crack tip. Crack length is a.

Figure 3 shows the plastic and creep zone regions at the crack tip of a crack with length a. The size of plastic zone is r_{pl}, and of creep zone is r_{cr}. With progression of creep time, creep zone size r_{cr} increases.

Remotely applied stress is denoted as S, and it fluctuates between its minimum and maximum values, S_{min} and S_{max}, respectively. Normal stress in the plastic zone is assumed to have a uniform distribution and equal to $-\alpha\cdot\sigma_0$, where σ_0 is flow stress and α constraint factor. Flow stress σ_0 is the average of S_y and S_{ut}, the yield and ultimate strength, respectively. For plane stress $\alpha = 1$, and for plane strain $\alpha = 3$.

Plastic zone size at the crack tip is computed from the condition that total stress intensity factor at the fictitious crack tip $d = a + r_{pl}$ must be zero

$$K_S + K_{\sigma_0} = 0 \tag{2}$$

The two stress intensity factors in the equation (2) are computed using the weight function method, as presented by Wu and Carlsson [6]. All cracked specimens have been analyzed using the Wu and Carlsson weight functions. For exemple, the stress intensity factors in equation (2) for an MT specimen are the following

$$K_S = \frac{S\sqrt{W}}{\sqrt{\pi d}} \int_0^d \sum_{i=1}^{3} \beta_i(d) \left[1 - \left(\frac{x}{d}\right)^2\right]^{i-\frac{3}{2}} dx \tag{3}$$

$$K_{\sigma_0} = \frac{-\alpha\sigma_0\sqrt{W}}{\sqrt{\pi d}} \int_a^d \sum_{i=1}^{3} \beta_i(d) \left[1 - \left(\frac{x}{d}\right)^2\right]^{i-\frac{3}{2}} dx \tag{4}$$

where function β are defined in [6]. Formulas (3) and (4) are introduced in equation (2). The integrals are computed using a ten point Gauss quadrature, and equation (2) is solved for r_{pl} using the bisection method. Once plastic zone size r_{pl} is computed, it is kept constant throughout the entire creep loading of the particular cycle in which it is calculated.

Before the onset of crack growth, the initial crack plane is discretized using 10 elements for the plastic zone and 5 elements for the crack surface. The width of crack plane elements is the same as those presented in [4,7].

Initial element lengths are the values of crack surface displacements produced by remote loading S_{max}, local stress in the plastic zone $-\alpha\cdot\sigma_0$ and stress increment $\Delta\sigma_{cr}$ caused by creep relaxation at the crack tip. The normal stress caused by plastic deformation in the plastic zone is compressive, while stress relaxation caused by creep is tensile. Creep zone size is computed according to a formula presented by Riedel and Rice [8]. Stress relaxation $\Delta\sigma_{cr}$ is considered to occur not only inside the creep zone, but over a larger region ahead of the crack tip, and is computed by multiplying creep zone size with a correction factor, $\rho_{cr}\cdot r_{cr}$. For all simulations ρ_{cr} was chosen as 3.5. Further finite element simulations are planned to gain a better understanding on the extent of the near crack-tip region that experiences stress relaxation and redistribution during creep loading.

Riedel [5] showed that for a given creep hold time t_H, the crack-tip stress in the creep zone relaxes proportionally to $(t_H)^{1/(1+n)}$, where n is the stress exponent in Norton creep law. Using this result, the stress relaxation due to creep is computed as

$$\Delta\sigma_{cr} = \alpha\sigma_0 \left(\frac{t_0}{t_H}\right)^{\frac{1}{1+n}} \tag{5}$$

where t_0 is a time constant accounting for the nonlinear distribution of stresses in the creep zone. For all simulations, this time constant was chosen as $t_0 = 0.001$.

Element lengths at S_{max} were computed by superposition of three types of crack surface displacements, i.e., those caused by remote loading, flow stresses in the plastic zone and stress relaxation in the creep zone

$$L(x_i) = V_{S_{max}}(x_i) + \sum_{j=1}^{10}\left(-V_{\sigma_0}(x_i, x_j) + V_{\Delta\sigma_{cr}}(x_i, x_j)\right) \quad (6)$$

where x_i is element i coordinate, and x_j is the coordinate for element j that causes additional displacement at i because of plasticity and/or creep. Formulas for displacements V in equation (6) are computed using Wu and Carlsson weight functions.

Upon unloading at S_{min}, contact stresses between the two crack surfaces develop. They are computed following a procedure indicated by Newman [7] and considering the element lengths computed with equation (6). The equation for computing contact stresses at minimum load is

$$\sum_{j=1}^{10} V_{\sigma_j}(x_i, x_j) = V_{S_{min}}(x_i) - L(x_i) \quad (7)$$

From the above equation, element stresses σ_j are calculated. Those elements that exceed the flow stress in compression, taken here as $-\sigma_0$ are considered yielded in compression, and thus new elements lengths are computed using S_{min} instead of S_{max} in equation (6), and the newly computed contact stress distribution σ_j. As indicated by Newman [7], an element cannot exceed a stress of $\alpha \cdot \sigma_0$ in tension, or $-\sigma_0$ in compression.

After computing contact stresses at minimum load, during the next loading portion, crack-tip opening stress S_{op} is calculated using a procedure described by Daniewicz et al. [9].

2.3 Creep-fatigue crack growth laws

Total crack growth rate per cycle is the sum of the crack growth rates during fatigue and creep loading

$$\frac{da}{dN} = \left(\frac{da}{dN}\right)_{fat} + \left(\frac{da}{dN}\right)_{cr} \quad (8)$$

Fatigue crack growth rates are computed using effective stress intensity factor range ΔK_{eff}

$$\left(\frac{da}{dN}\right)_{fat} = C \cdot \Delta K_{eff}^{m} \quad (9)$$

where C and m are material constants, and $\Delta K_{eff} = K_{max} - K_{op}$, K_{max} is maximum stress intensity factor in the cycle, and K_{op} is opening stress intensity factor. For instance, K_{op} is calculated using opening stress S_{op} as follows

$$K_{op} = F \cdot S_{op}\sqrt{\pi a} \quad (10)$$

where the geometrical factor F depends on the type of testing specimen and crack length.

Creep crack growth rates are computed as

$$\left(\frac{da}{dt}\right)_{cr} = C' \cdot K_{max}^{m'} \quad (11)$$

where C' and m' are material constants for the creep crack growth power law. Researchers commonly use either K_{max} or $C(t)_{avg}$ integral to correlate crack growth rates for the creep portion of loading cycle. In this study, it was observed that K_{max} produces in general better results in terms of correlating crack growth rates compared to $C(t)_{avg}$. Equation (11) is presented in terms of time rate of crack growth, and it can be transformed to crack growth rate per cycle with the formula

$$\left(\frac{da}{dN}\right)_{cr} = \frac{1}{f}\left(\frac{da}{dt}\right)_{cr} \quad (12)$$

where f is the test frequency, with $f = 1/t_H$, when neglecting the loading and unloading times, which usually occur much faster than creep hold time.

Creep-fatigue interactions are modeled through the influence of load hold time on crack-tip opening stress. Accordingly, SYM-CFCG determines the variation of opening stress on creep loading. Other factors that determine crack-tip opening stresses are specimen type, ratio of applied stress to flow stress and ratio of crack length to specimen width.

3 Model verification

Verification of SYM-CFCG was done for several specimens, load ratios and constraint factors. Crack-tip opening stresses were computed and compared with published data in the case of fatigue loading ($t_H = 0$). Figure 4 shows the predicted opening stress from SYM-CFCG for MT, SENT and DENT specimens. These predictions are compared with data published by McClung [10] and Newman [11] for MT and SENT specimens.

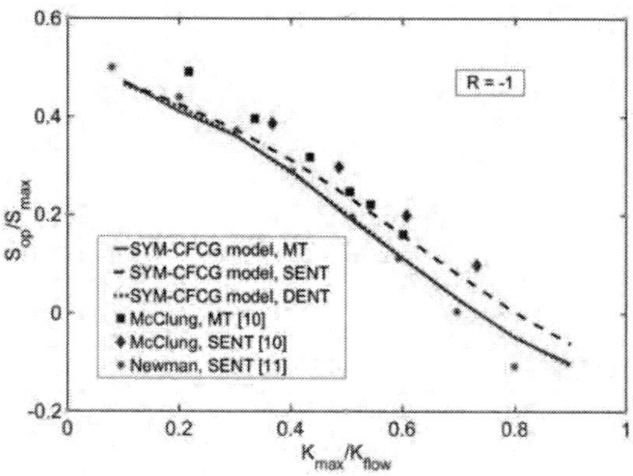

Fig. 4. Comparison of current model predictions (SYM-CFCG) with finite element results ([10]) and strip-yield modeling results ([11]) of crack-tip opening stresses in middle tension (MT), single-edge notched tension (SENT) and double-edge notched tension (DENT) specimens.

The model produces similar opening stresses with results from the McClung and Newman studies for these specimens, at a load ratio of $R = -1$, and for ratios of S_{max}/σ_0 between 0.1 and 0.9. Crack-tip opening stress is reported as normalized by maximum applied stress in the cycle.

Figure 5 shows a comparison of model predictions versus results of the Wang and Blom [12] study for CT specimens. SYM-CFCG predicts similar opening stresses for various load ratios between $R = -1$ and $R = 0.8$, and constraint factors of $\alpha = 1$ (plane stress), $\alpha = 2$ and $\alpha = 3$ (plane strain).

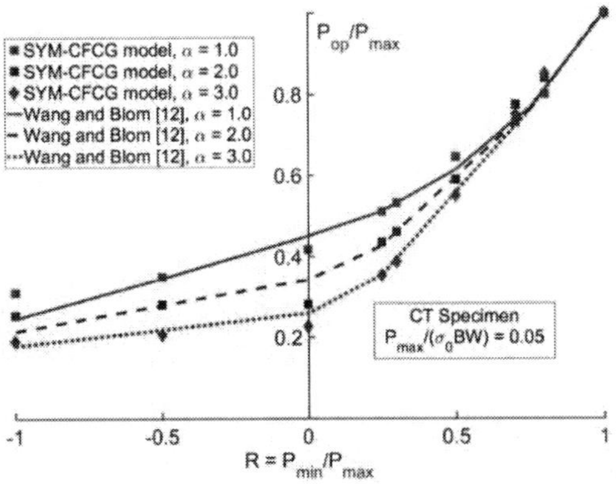

Fig. 5. Comparison of crack-tip opening stress predictions by the current model (SYM-CFCG) with results from literature [12] for the compact tension (CT) specimen.

Maximum applied load was selected according to Wang and Blom simulations as $P_{max}/(\sigma_0 BW) = 0.05$, where B is specimen thickness and W specimen width. Agreement between the current model predictions and published literature data is obtained in this case as well.

4 Results

In this section, SYM-CFCG predictions are presented and discussed for three materials, a nickel-base superalloy, aluminium alloy 2650 and 316 stainless steels. Experimental data used for comparison with model predictions were taken from published studies. For each case, the material constants, specimen geometry and loading were the same in SYM-CFCG simulations as those in the respective published studies, whenever data was available.

4.1 Creep-fatigue crack growth in an Astroloy

The first material for which SYM-CFCG simulation predictions are presented is a nickel-base superalloy, an Astroloy used at high temperatures for turbine engine disks. The experimental study of creep-fatigue crack growth was published by Piard et al. [3], and these simulations replicate their study. The Piard et al. data is perhaps one of the most relevant in terms of highlighting the interaction between creep and fatigue loading on crack growth rates. They measured the influence of hold time t_H on crack growth rates during fatigue loading $(da/dN)_{fat}$. Piard et al. showed in their experimental measurements that the longer the hold time, the faster cracks grow during the next loading from F_{min} to F_{max}.

SENT specimens were used in these simulations, similar to those in the Piard et al. tests. Ratio of initial crack length to specimen width was $a/W = 0.15$. Ratio of maximum stress to flow stress ratio used was $S_{max}/\sigma_0 = 0.06$, with $\sigma_0 = 910$ MPa. Material properties used in this study were: (i) Norton creep constants, $A = 2.8{\cdot}10\text{-}31$ s^{-1}MPa^{-n} and $n = 8.5$, (ii) Young's modulus at 750 °C is $E = 160$ GPa. Material constants used in fatigue crack growth equation (9) were $C = 6.47{\cdot}10^{-8}$ and $m = 3.1$, with crack growth rate measured in mm/cycle and ΔK in MPa·√m. The crack-tip creep zone correction factor used was $\rho_{cr} = 3.5$.

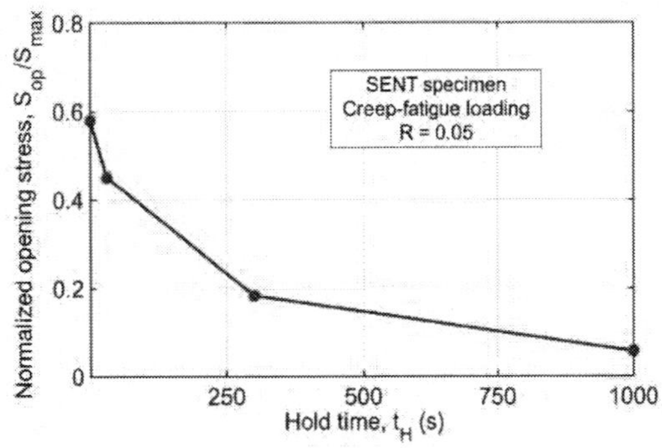

Fig. 6. Computed values of the crack-tip opening stress as a function of creep hold time for the Piard et al. [3] experiments.

Figure 6 shows the computed opening stresses using SYM-CFCG versus hold time for the four different hold times tested by Piard et al., i.e., 0s, 30s, 300s and 1000s. For pure fatigue loading (t_H = 0s), the normalized opening stress produced by SYM-CFCG simulations is $S_{op}/S_{max} = 0.58$. For a hold time of t_H = 30s, the opening stress decreases to $S_{op}/S_{max} = 0.45$. At a hold time of t_H = 300s, $S_{op}/S_{max} = 0.18$, while for the largest tested hold time of t_H = 1000s, the opening stress becomes almost equal to the minimum stress in the creep-fatigue cycle, $S_{op}/S_{max} = 0.057$, whereas $R = S_{min}/S_{max} = 0.05$). Crack-tip opening stresses decrease with increasing hold time per cycle, which in turn accelerates the crack growth rate during subsequent fatigue loading, according to equations (9) and (10).

Figure 7 presents the crack growth rates measured by Piard et al. and simulation predictions using SYM-CFCG. The comparison shows that the model predicts crack growth rates well during fatigue loading as a function of hold time. The hold time during creep loading varied from t_H = 30s to t_H = 1000s. The comparison also shows that for t_H = 0 (i.e. fatigue loading only, and no hold time), predictions match well experimental data for most of ΔK range.

SYM-CFCG can also predict the initial drop of crack growth rate curve. This decrease in growth rate is explained by the lack of plastic wake as crack starts growing. As the crack becomes longer, a plastic wake of stabilized thickness fully develops behind the crack tip, and crack growth rate stabilizes. Moreover, the model simulations also indicate a minimal drop in initial crack growth rate for tests that involved non-zero hold times. This is caused by diminished plastic wake influence on crack growth rates at longer hold times.

Figure 7 shows that there is some discrepancy between the model predictions and experimental data for t_H = 30s.

Some of this discrepancy can be explained considering that SYM-CFCG neglects the loading and unloading times, while in Piard et al each lasted 10s. A loading time of 10s may also have some effects on crack growth rates, and this is not taken into account in modeling simulations. The comparison between model simulations and experimental data at t_H = 300s is excellent, while some discrepancies are observed at t_H = 1000s. The imperfect match of experimental and simulation data at t_H = 1000s could be attributed in part to some other effects that were not included in the current model, such as severe damage in front of the crack tip at longer hold times.

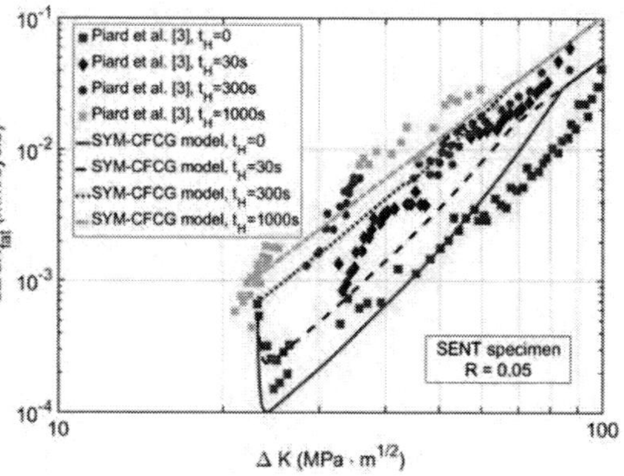

Fig. 7. Comparison of model predictions (SYM-CFCG) with experimental data ([3]) of crack growth rates in a single-edge notch tension (SENT) specimen of Astroloy at 750 °C.

The acceleration of crack growth rate computed by SYM-CFCG approach during loading as a function of hold time is due solely to the decrease in crack closure. It is possible that this effect could be accompanied by some additional damage phenomena that would accelerate the crack growth rate even further, especially for longer hold times.

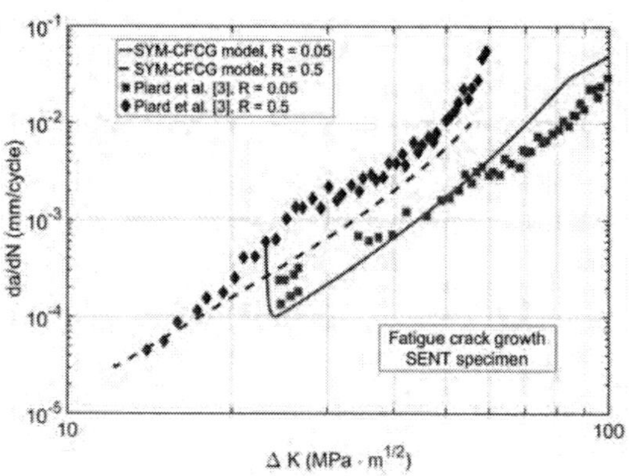

Fig. 8. Comparison between model predictions (SYM-CFCG) and experimental data ([3]) for fatigue crack growth in an Astroloy at 750 °C and two load ratios R.

Figure 8 shows the SYM-CFCG predictions and Piard et al. experimental results for fatigue crack growth at load ratios of $R = 0.05$ and $R = 0.5$. These simulations include only fatigue loading and unloading without any hold time. The model predicts well crack growth rates for both load ratios, based on computed crack-tip opening stresses.

4.2 Creep-fatigue crack growth in 2650 aluminum alloy

SYM-CFCG predictions of creep-fatigue crack growth rates in Cu-Mg aluminium alloy 2650 were compared with the experimental results of Henaff et al. [13,14]. Two testing temperatures were considered, 130 °C and 175 °C. Cu-Mg aluminium alloys are heavily used in the aerospace industry for fuselage and wing skins. Both simulations and Henaff et al. experiments were performed in CT specimens at a load ratio of R = 0.5. Material properties used in this study for alloy 2650 at 130 °C were: S_y = 372 MPa, S_{ut} = 412 MPa, E = 68 GPa, $A = 8.0\cdot10^{-53}$ $s^{-1}MPa^{-n}$, n = 17.7, $C = 1.10\cdot10^{-10}$, m = 3.84, $C' = 5.0\cdot10^{-9}$, m' = 2.35. At 175 °C the following material properties were used: S_y = 345 MPa, S_{ut} = 365 MPa, E = 65 GPa, $A = 4.08\cdot10^{-29}$ $s^{-1}MPa^{-n}$, n = 8.81, $C = 3.0\cdot10^{-10}$, m = 3.5, $C' = 2.0\cdot10^{-8}$, and m' = 2.35. For both temperatures, creep zone correction factor of ρ_{cr} = 3.5 was used.

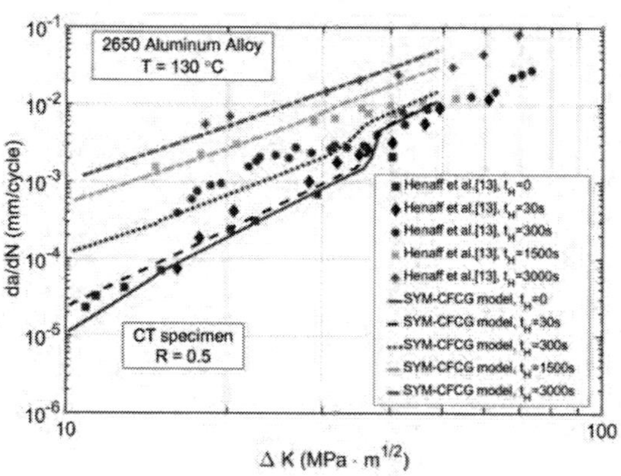

Fig. 9. Comparison between SYM-CFCG simulations and experimental test data ([13]) of creep-fatigue crack growth rates in aluminum alloy 2650 at 130 °C.

Figure 9 shows total crack growth rates for creep-fatigue tests performed at 130 °C in aluminium alloy 2650. The plot shows data for fatigue crack growth (t_H = 0) and creep-fatigue loading with hold times ranging from 30s to 3000s. Figure 10 shows results at 175 °C for fatigue and creep-fatigue loading with hold time up to t_H = 1500s. Both figures indicate good agreement between the SYM-CFCG simulations and experimental data of Henaff et al.

4.3 Creep-fatigue crack growth in 316 stainless steel

Predictions of creep-fatigue crack growth using SYM-CFCG were also performed for 316 stainless steel. This austenitic stainless steel has vast applicability in structural applications in power plants at elevated temperatures. Two relevant experimental studies in 316 stainless steel at 593 °C were used to compare with model predictions, namely the works of Michel and Smith [15] and Sadananda and Shahinian [16].

The study of Sadananda and Shahinian [16] was performed on 316 stainless steel in 20% cold worked condition. They used CT specimens with 5% side-grooving to maintain an even profile of the crack plane during growth and avoid crack tunneling. Material properties of this material at 593 °C were: S_y = 364 MPa, S_{ut} = 617 MPa, E = 153 GPa, A = 7.75·10^{-30} s^{-1}MPa^{-n}, n = 8.76, C = 3.0·10^{-5}, m = 1.2, C' = 6.5·10^{-4} and m' = 2.1.

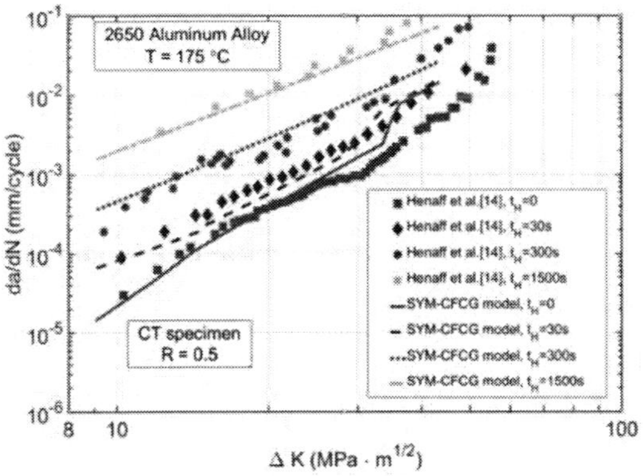

Fig. 10. Comparison between SYM-CFCG simulations and experimental test data ([14]) of creep-fatigue crack growth rates in aluminum alloy 2650 at 175 °C.

SYM-CFCG simulations were performed in a CT specimen with width W = 50 mm, un-notched thickness B = 12.7 mm, and notched thickness (due to side-grooving) B_N = 12.07 mm. Initial crack length was a_i = 12.5 mm. Fatigue loading cycle was applied with a maximum load F_{max} = 13.3 kN and load ratio R = 0.05. Two tests were performed to replicate the Sadananda and Shahinian experiments. One was a FCG test (t_H = 0) and another CFCG test with t_H = 60s. Creep crack zone correction coefficient was ρ_{cr} = 3.5. In addition to the tests in CT specimens, simulations using SENT specimens were also performed. These simulations were designed to replicate the experiments of Michel and Smith [15]. Their tests were performed on a SENT specimen geometry. SYM-CFCG simulations were performed on the same specimen, with width W = 63.5 mm, initial crack length a_i = 10 mm, un-notched thickness B = 12.7 mm and notched thickness B_N = 11.43 mm. Michel and Smith also performed experiments on 316 stainless steel in annealed condition. The material properties for the annealed condition at 593 °C were: S_y = 148 MPa, S_{ut} = 435 MPa, E = 153 GPa, A = 1.76·10^{-36} s^{-1}MPa^{-n}, n = 11.83, C = 1.6·10^{-5}, m = 1.5, C' = 1.5·10^{-7} and m' = 3.7. The loading conditions were F_{max} = 47 kN, R = 0, t_H = 0 and t_H = 60s.

Figures 11 and 12 show the comparison between SYM-CFCG simulations and experimental data from the two published studies [15,16] in 316 stainless steel at 593 °C in 20% cold-worked and annealed conditions. Both t_H = 0 and t_H = 60s simulations show excellent agreement between experiments and simulations. In Figure 11, the data of Sadananda and Shahinian shows some deviation from the modeling predictions for t_H = 60s loading, however the modeling results match well the experimental data of Michel and Smith [15] for similar hold times.

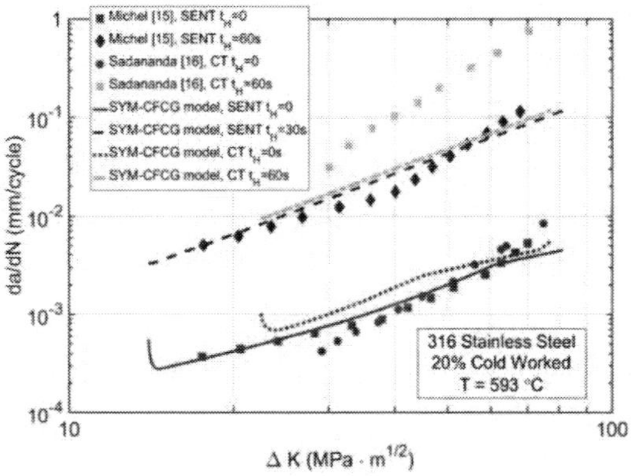

Fig. 11. Comparison between model simulations (SYM-CFCG) and experimental results ([15,16]) for creep-fatigue crack growth at 593 °C in 316 stainless steel 20% cold worked condition.

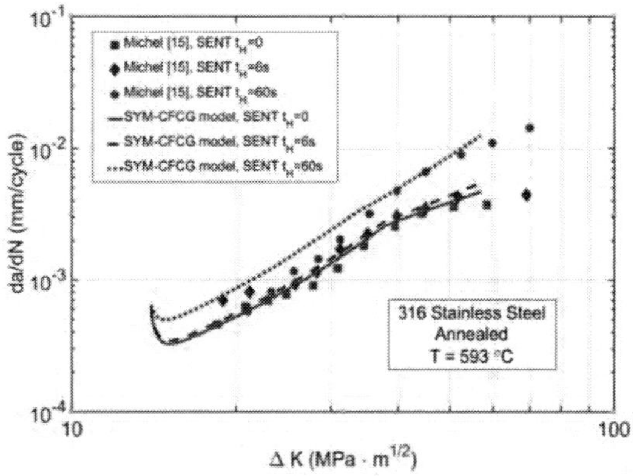

Fig. 12. Comparison between model simulations (SYM-CFCG) and experimental results ([15]) for creep-fatigue crack growth at 593 °C in 316 stainless steel annealed condition.

5 Conclusions

A strip-yield model was developed for creep-fatigue crack growth. The model takes into account both plastic and creep deformations at the crack tip, and the evolution of crack surface displacements during creep-fatigue loading. Load interaction effects on crack growth rates are accounted for by quantifying the influence of creep loading on crack-tip opening stresses due to plasticity-induced crack closure. The model identified that increased creep hold times reduce crack-tip opening stresses, and consequently, increase crack growth rates during fatigue loading. This conclusion is supported by previous finite element modeling results. The model was applied to simulate creep-fatigue crack growth in several structural alloys used at elevated temperatures. Model predictions compared well with published experimental data of crack growth rates in these alloys. The model can be used to perform reliable predictions of creep-fatigue lives in structural components used in power plants and the aerospace industry.

The funding support of this work by the U.S. Department of Energy under the Nuclear Energy University Program (Grant DE-NE0008443) work is gratefully acknowledged.

References

1. H. Sehitoglu, W. Sun, Eng Fract Mech **33**, 3 (1989)
2. J. Ramirez, G.P. Potirniche, H. Pugesek, N. Shaber, M. Taylor, R. Stephens, I. Charit., *12th International Fatigue Congress* (2018) (to be published)
3. A. Piard, G. Gamby, C.Carbou, J. Mendez, Eng Fr Mech, Eng Fract Mech **71** (2004)
4. G.P. Potirniche, *Fatigue and Fracture Mechanics: 38th Volume, STP1546* (2012)
5. H. Riedel, Fracture at High Temperatures (Springer-Verlag, 1986)
6. X.-R. Wu and A. J. Carlsson, Weight Functions and Stress Intensity Factor Solutions (Pergamon Press, 1991)
7. J.C. Newman, Jr., *Design of Fatigue and Fracture Resistant Structures, ASTM STP 761* (1982)
8. H. Riedel, J.R. Rice, *Fracture Mechanics: Twelfth Conference, ASTM STP 700* (1980)
9. S.R. Daniewicz, J.A. Colins, D.R. Houser, Fatigue **16** (1994)
10. R.M. McClung, Fatigue Fract. Engng Mater. Struct. **17**, 8 (1994)
11. J.C. Newman, Jr., *Fatigue 02: The Eight International Fatigue Congress* (2002)
12. G.S. Wang, A.F. Blom, Eng Fract Mech **40**, 4 (1991)
13. G. Henaff, F. Menan, G. Odemer, Eng Fract Mech **77** (2010)
14. G. Henaff, G. Odemer, G. Benoit, E. Koffi, B. Journet Int J Fatigue **31** (2009)
15. D.J. Michel, H.H. Smith, *ASME-MPC Symposium on Creep-Fatigue Interaction*, ASME (1976)
16. K Sadananda, P. Shahinian, Metall. Trans. A, **11A** (1980)

MATEC Web of Conferences **165**, 06001 (2018) https://doi.org/10.1051/matecconf/201816506001
FATIGUE 2018

Effects of thermomechanical history and environment on the fatigue behavior of (β)-Ti-Nb implant alloys

André Reck [1*], *Stefan* Pilz [1,3], *Ulrich* Thormann [2], *Volker* Alt [2], *Annett* Gebert [3], *Mariana* Calin [3], *Christian* Heiß [2], and *Martina* Zimmermann [1]

[1]Technical University Dresden, 01062 Dresden, Germany
[2]University Hospital Gießen-Marburg GmbH, 35385 Gießen, Germany
[3]Leibniz Institute for Solid State and Materials Research, 01062 Dresden, Germany

Abstract. This study examined the fatigue properties of a newly developed cast and thermomechanical processed (β)-Ti-40Nb alloy for a possible application as biomedical alloy due to exceptional low Young's modulus (64-73 GPa), high corrosion resistance and ductility (20-26%). Focusing on the influence of two microstructural states with fully recrystallized β-grain structure as well as an aged condition with nanometer-sized ω-precipitates, tension-compression fatigue tests (R=-1) were carried out under lab-air and showed significant differences depending on the β-phase stability under cyclic loading. Present ω-precipitates stabilized the β-phase against martensitic α'' phase transformations leading to an increased fatigue limit of 288 MPa compared to the recrystallized state (225 MPa), where mechanical polishing and subsequent cyclic loading led to formation of α''-phase due to the metastability of the β-phase. Additional studied commercially available (β)-Ti-45Nb alloy revealed slightly higher fatigue strength (300 MPa) and suggest a change in the dominating cyclic deformation mechanisms according to the sensitive dependence on the Nb-content. Further tests in simulated body fluid (SBF) at 37°C showed no decrease in fatigue strength due to the effect of corrosion and prove the excellent corrosion fatigue resistance of this alloy type under given test conditions.

1 Introduction

In the last two decades intensive research was carried out to develop new titanium alloys for the use as implant materials [1-4].

Focus of this research efforts were mostly ß-titanium alloys, since they show the lowest Young's modulus compared to α and (α+β) alloys while providing the possibility to a significant stiffness reduction due to tailored composition and thermomechanical treatment. Major advantage of these properties is the minimization of "stress shielding effects", which can lead to bone atrophy, impaired healing and consequently premature failure or loosening of implants due to the high difference of Young's moduli compared to the cortical bone (10-30 GPa) [1-2,5-8]. This is particularly important in case of implants for osteoporotic bones, which have a lower mass and density then the healthy bone and consequently an even more reduced stiffness corresponding to elastic modulus values <10 GPa [9].

The primary challenge is to maintain excellent static and cyclic strength while providing an alloy with lowest possible Young's modulus [1,10]. The fatigue behavior of new β-titanium alloys with low stiffness is therefore of central relevance, especially under the effects of corrosion [11-14]. Thus, many strengthening strategies involving diverse thermomechanical treatment strategies to positively influence the fatigue strength were in focus of research. Whereas work hardening for instance increases the static strength, fatigue strength remains mostly unaltered [15-24]. The most significant effect in case of TiNb-based alloys seems to be possible due to precipitation and grain boundary hardening in order to effectively hinder dislocations to move through the material and thus, positively influence the fatigue limit [16,18].

Another way to improve the fatigue strength is particle dispersion (e.g. yttrium oxide, titanium boride) in the matrix of these alloys or increasing the oxygen content [25-29].

Precipitation of hexagonal omega (ω)-phase is another promising approach and proved to be enhancing the fatigue strength of TNTZ [30]. Detrimental effects with higher volume content of ω-phase are an increase in Young's modulus as well as lower ductility due to the brittle nature of this phase. Consequently the content, distribution and size of the ω-precipitates have to be well under control with aging time and temperature depending on the exact alloy composition. Whereas TiNb-based β-titanium alloys without ω-phase show, determined by the exact alloy composition, the e/a

* Corresponding author: andre.reck@tu-dresden.de

relation and the bond order (Bo value)/d-electron-orbital energy level (Md value), typical stress-induced α''-phase formation as well as mechanical twinning, present isothermal or athermal ω-precipitations influence these features to great extent [31-33]. Moreover, diverse features like stress-induced ω-bands and "dislocation channels" depleted on ω-precipitations can be part of the deformation process, which was studied by Lai et al. in gum metal [34]. The additional sensitive dependence on the Nb-content and thus β-phase stability at room temperature [32,35] combined with changed deformation features made TiNb-based β-titanium alloys to very promising but also very complex implant materials, whose successful application is particularly determined by the fatigue properties and cyclic deformation behaviour.

Hence, this study aims at the identification of differences in S-N based fatigue properties of a binary (β)-Ti-40(wt%)Nb alloy with and without isothermal ω-precipitations. Comparisons will be drawn to a previous study focusing on the notch effect due to an industrial surface condition of the same alloy [36]. Furthermore commercially available (β)-Ti-45(wt%)Nb was studied to comprehend influences of Nb-content and first in-vitro tests are carried out to assess corrosion fatigue.

2 Experimental methods

2.1. Alloy processing and machining

All samples of Ti-40(wt%)Nb were produced through a two-step casting process consisting of arc melting of the pure elements cp2-Ti (99,7%) and Nb (99,9%) in highly purified argon followed by remelting via induction heating of the pre-alloys and casting into a cylindrical water-cooled Cu mold. Additional remelting steps were carried out to ensure the homogeneity of the β-titanium alloy.

Solution heat treatment (1273 K, 24 h) of the as-cast rods was carried out to resolve possible microsegregations followed by water quenching. In order to achieve a sheet form for further processing as well as grain refinement and work hardening, the rods were cold-rolled in several steps to a final thickness reduction of 93% (ϕ_{total} = 2.66). Recrystallization (1073 K, 20 min) was executed between cold-rolling and at the end of the thermomechanical treatment. (β)-Ti-40Nb samples with ω-phase were additionally aged at 573K for 4h to get a controlled precipitation of nanometer-sized ω-particles in the ß-matrix. Further details of the alloy processing can be found in our previous studies [16,35].

The investigated (β)-Ti-45(wt%)Nb alloy was bought in rod form with a diameter of 15 mm from MetSuisse Distribution AG and processed in the USA of ATI Specialty Alloys & Components.

In preparation for the fatigue tests all studied samples were furthermore machined according to the geometries in Fig.1. (β)-Ti-40Nb sheet samples were laser-cut in agreement with our previous study [36]. Recrystallized samples, afterwards called Ti-40Nb-Rk, and aged samples with omega ω-precipitations, afterwards called Ti-40Nb-Om4h were subsequently stepwise mechanically polished (Grit 2500) to remove the thin oxide layer from the heat treatment and to minimize surface roughness related notch influences and the heat affected zone (HAZ)) from the laser cutting [36]. (β)-Ti-45Nb rods were turned to the given geometry (Fig 1b) and also mechanically polished to Grit 2500.

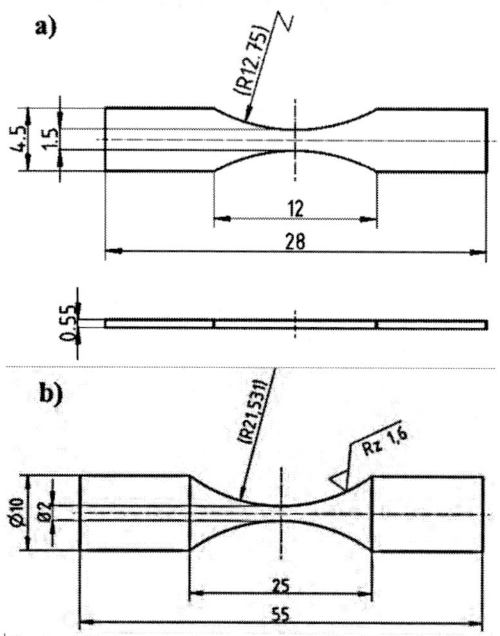

Fig. 1. a) Fatigue sample geometry (β)-Ti-40Nb; b) Fatigue sample geometry (β)-Ti-45Nb

2.2 Static and Fatigue tests

After processing and machining of the samples, uniaxial tensile tests on (β)-Ti-40Nb were conducted to assess fundamental effects in the static behavior due to aging. Known values of industrial (β)-Ti-45Nb were obtained from the datasheet of the manufacturer. Load-controlled tension-compression (R= -1) fatigue investigations were carried out subsequently.

A servo-electric test system (Thelkin) was used for the fatigue study. All tests at laboratory air were executed at 20 Hz after carefully adjusting the test system control parameters. These parameters regulate, depending on the stiffness and the strength of the samples, the vibration behaviour (optimal sinus curve) of the tested material.

To assess the fatigue range of the material load-increase tests were carried out followed by regular single-step tests on 3 to 4 stress levels each with 2 to 4 samples for each tested condition.

Most of the investigations were realized in lab-air under pressurized air cooling. The (β)-Ti-45Nb were additionally studied in simulated body fluid (SBF) to evaluate the possible influence of corrosion fatigue. The developed set-up is depicted in Fig.2. A built in heating plate holds the test temperature at a constant temperature of 310 K ± 2K (37°C). Ringer's solution consisting of a mixture of different chlorides (Tab.1) was used as SBF.

Sample mass was measured before and after the fatigue tests. Frequency adjustment to 10 Hz was necessary to obtain stable sinus-curve testing under in-vitro conditions.

Table 1. Composition of Ringer's solution for in-vitro tests

Component	NaCl	KCl	$CaCl_2$	$NaHCO_3$	Dest. H_2O
Content in g/l	8	0.2	0.2	1	-

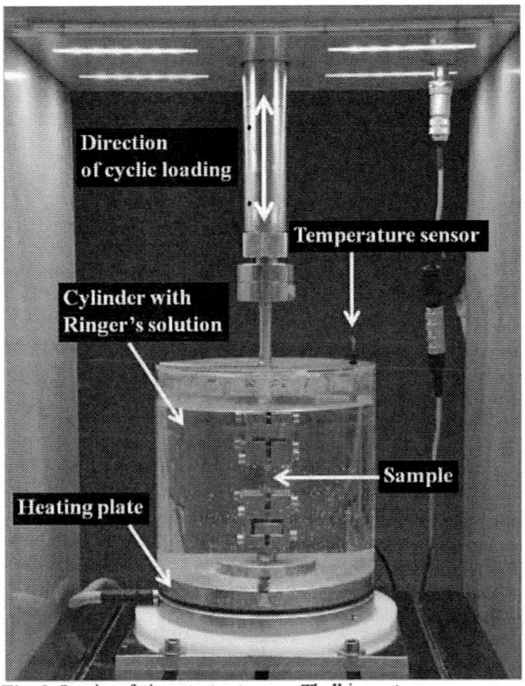

Fig. 2. In-vitro fatigue test set-up on Thelkin system

2.3 Analytical investigations

Before and after the fatigue tests different analytical methods were used.

The microstructure before cyclic loading was assessed with light and high-resolution electron microscopy. The high metastable character, especially of the recrystallized samples (Ti-40Nb_Rk) could be origin of possible phase transformation due to laser cutting or mechanical polishing.

XRD-measurements were executed after the processing of (β)-Ti-40Nb and TEM investigations were carried out to establish an understanding of the size and distribution of the ω-precipitations. After the fatigue tests detailed fractographic analysis was conducted to identify crack origin and important characteristics of fatigue failure.

3 Results and discussion

3.1 Microstructure and phase investigations

The microstructural condition before the fatigue tests but after the laser cutting and subsequently mechanical polishing steps of the Ti-40Nb_Rk samples is depicted in Fig. 3.

Compared to the initial microstructure immediately after the thermomechanical treatment with single (β)-grain structure, local phase transformation can clearly be identified. A higher magnification of the light microscopy investigations reveals formation of α''-phase during the preparation of the samples. Either local heat development during laser processing or applied residual compression stresses from the mechanical polishing could result in the activation of the phase transformation, due to the high metastable character of this alloy condition. Special relevance obtains this result considering the high importance of laser processing for implant materials in automatized industrial process chains [36].

The aged Ti-40Nb_Om4h samples are on the contrary free of α''-phase after the processing and preparation steps and consist only of β-phase and ω-precipitates with an intermediate size of around 10 nm, which are not present in the Ti-40Nb_Rk samples. Hence, it appears that the isothermal precipitates operate as stabilizer for the metastable β-matrix against phase transformations at a Nb-content of 40 wt%.

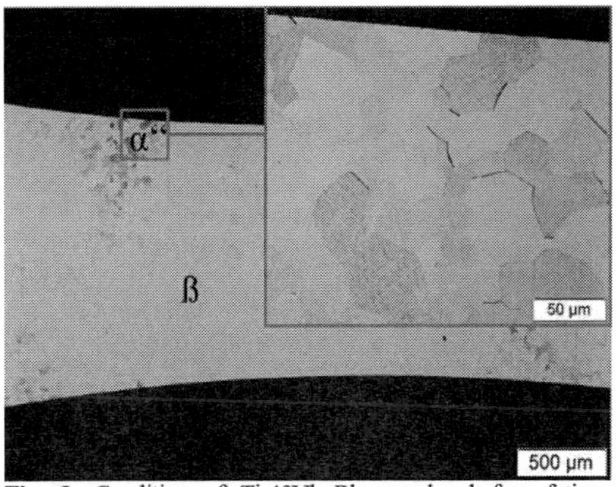

Fig. 3. Condition of Ti-40Nb_Rk samples before fatigue loading (light microscopy): Local stress-induced α''-phase transformation due to laser cutting and preparation

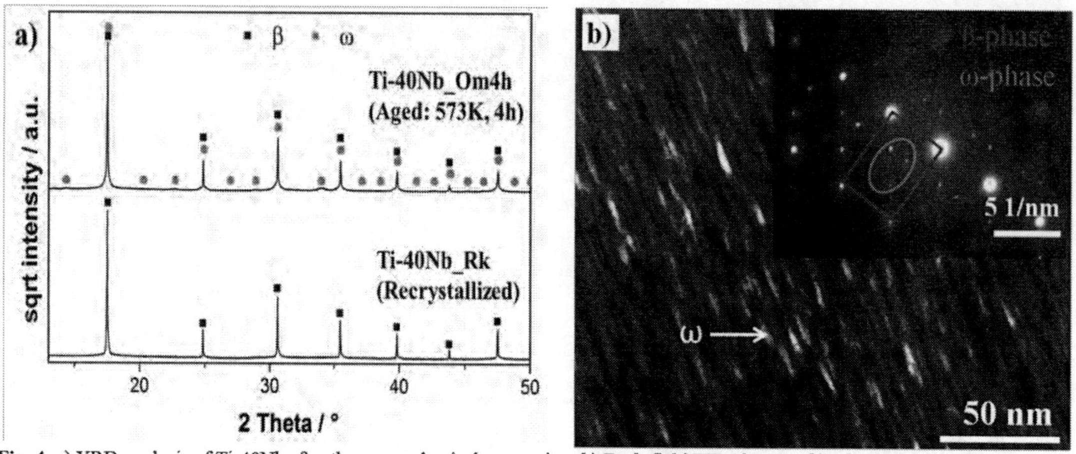

Fig. 4. a) XRD-analysis of Ti-40Nb after thermomechanical processing; b) Dark-field TEM image of isothermal (ω)-precipitates after aging (573 K, 4h) in the Ti-40Nb_Om4h samples

The major reason for the higher stability of the ß-phase is most likely a further reduction of the martensitic finish temperature (M_f) below room temperature due to the presence of the ω-phase, which normally is valid for binary Ti-Nb alloys over 42 wt% Nb [31]. In addition to the higher stability of the β-phase, the presence of the ω-phase also suppresses superelasticity and shape memory effects in the Ti-40Nb alloys, as could be observed in the recrystallized Ti-40Nb_Rk samples (Fig. 3) without ω-phase [31-32].

Results of the XRD-analysis as well as the TEM-investigation of the aged condition (Ti-40Nb_Om4h) are shown in Fig. 4. The XRD plot exhibits weak but distinct additional peaks of ω-phase. The close structural relationship of the ω- and the β-phase results in a high similarity of their diffraction patterns and an overlap of the β-reflections with reflections of the ω-phase. The slightly lower intensity of the β-peaks in the aged condition is however another effect due to the presence of the isothermal ω-precipitates and an indication for slight texture changes after the aging treatment [30].
The dark field TEM image (Fig. 4b) shows the homogeneous distribution and size of the hexagonal ω-phase. Distortion effects can thereby assigned to possible preparation effects caused by focused ion beam (FIB). Analyses reveals that the ω- phase is oriented to β after the common relationship $[111]_\beta$ // $[0001]_\omega$ and four variants are presented by four reflections in the TEM-diffraction pattern. One reflection is particularly strong, whereas the neighbour reflection is almost undetectable (marked in Fig. 4b), which is typically proof for the preferred precipitation of a distinct variant [37]. Similar results are found e.g. by Nakai et al. in aged TNTZ [30].

The commercially available (β)-Ti-45(wt%)Nb alloy shows compared to the thermomechanically treated (β)-Ti-40Nb alloy a single β-grain structure with linear segregations of α''-phase (not depicted). This feature is distributed alongside the studied rod and very likely caused due to the extrusion process during manufacturing.

From the standpoint of phase stability and the relevant Bo-Md-diagram for this binary alloy with 45 wt% Nb only β-phase should be present at room temperature [32,34]. However, an inhomogeneous alloy microstructure is present, which might influence the fatigue behaviour, e.g. crack initiation from microstructural barriers (e.g. α''-β phase or grain boundaries) or defects at the interior of the material [38].

3.2 Mechanical behavior

3.2.1 Static properties

The mechanical properties of the alloys studied under static load are summarized in Table 2. All alloys show similar low Young's moduli between 62-73 GPa, which is favourable to minimize "stress-shielding", but show very different strength and ductility values.

A significant precipitation hardening effect can be identified in the aged Ti-40Nb_Om4h samples due to the nucleation and growth of isothermal ω-precipitates compared to the recrystallized Ti-40Nb_Rk samples. Associated with that is on the one hand the slight increase in Young's modulus to 73 GPa and the decrease of ductility due to the brittle nature of ω. At the same time an increase in ultimate tensile strength up to 718 MPa can be observed.

Table 2. Static mechanical properties of the investigated alloys

Alloy	E / GPa	$\sigma_{0.2PS}$/ MPa	σ_{UTS}/ MPa	El / %
Ti-40Nb_Rk	64	315	498	26
Ti-40Nb_Om4h	73	639	718	19.7
Ti-45Nb [39]	62	475	503	34

An acceptable compromise between static strength and low Young's modulus can be achieved by careful adjustment of the aging parameters. In the given study this was the case for an aging for 4h at 573 K, thus limiting the size and volume fraction of the ω-phase. In addition, the typical "double-yielding" phenomenon caused by the stress-induced α''-transformation, which could be observed for the Ti-40Nb_Rk samples, is almost fully suppressed in the aged condition. The β-matrix is consequently effectively stabilized and strengthened with the presence of ω-phase in form of isothermal ω-precipitates.

The static strength of the commercially available (β)-Ti-45Nb alloy depicted in Table 2 shows an intermediate value with very high ductility and low Young's modulus, although direct comparison to the self-processed (β)-Ti-40Nb is limited mainly due to the influence of the different sample geometries. However, without the ω-phase the slightly higher Nb-content shifts this alloy to a more stable character, due to the sensitive dependence of the mechanical behaviour of binary Ti-Nb alloys on its chemical composition [31,34,40].

3.2.2 Fatigue results

The fatigue properties of the two studied Ti-40Nb conditions are depicted in Fig. 5.

Whereas the recrystallized state (Ti-40Nb_Rk) shows a tension-compression fatigue limit of σ_a= 225 MPa at 2x10^6 cycles, the aged condition with present ω-precipitations has a superior fatigue strength of around σ_a= 288 MPa. Since both heat treatment conditions were tested with the same mechanically polished surface and the same Nb-content of 40 wt% this is a clear indication of the effectiveness of the additional aging treatment. The precipitation hardening on the basis of the ω-phase proves to be an important factor regarding the successful application of Ti-40Nb as implant material. The improved mechanical properties of Ti-40Nb are in accordance with other studies on aged TNTZ and other dispersion strengthened TiNb-based alloys [28-30]. The relatively pronounced scatter of the fatigue results is caused by the inhomogeneous local formation of α''-phase in the Ti-40Nb_Rk samples (Fig. 3). Moreover, it has to be mentioned that due to the high degree of cold-rolling (ϕ_{total} = 2.66) and the subsequent heat treatment a minimal bending load due to a slight uneven surface of the Ti-40Nb_Om4h samples could not be completely ruled out.

Comparing the results presented in Fig. 3 with results of a previous study of Ti-40Nb-Rk samples with the same geometry and test conditions [36] but with laser cut surface a remarkable increase of fatigue strength from σ_a= 150 MPa (laser cut surface) to σ_a = 225 MPa (mechanically polished surface) is apparent. The effect introduced by the laser cut surface can be determined by a notch factor K_f of 1.5.

Fig. 6 shows the fatigue results for the (β)-Ti-45Nb alloy in lab-air as well as in Ringer's solution as simulated body fluid. The cyclic strength of both test conditions is comparable and no significant decrease due to the influence of corrosion is detectable leading to a fatigue limit of σ_a = 300 MPa at 2x10^6 cycles.

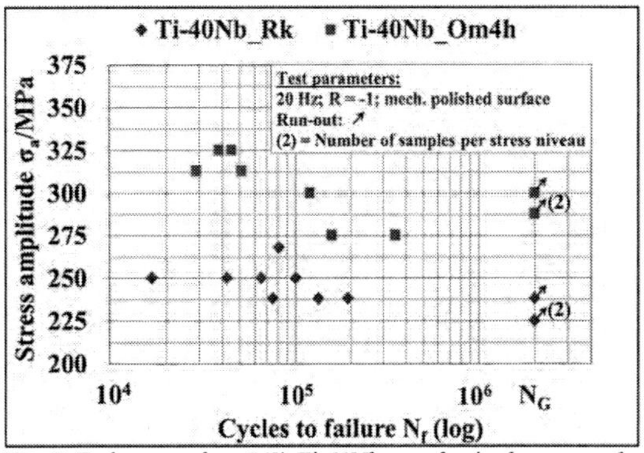

Fig. 5. Fatigue results of (β)-Ti-40Nb samples in the recrystallized (Ti-40Nb_Rk) and aged (Ti-40Nb_Om4h) condition

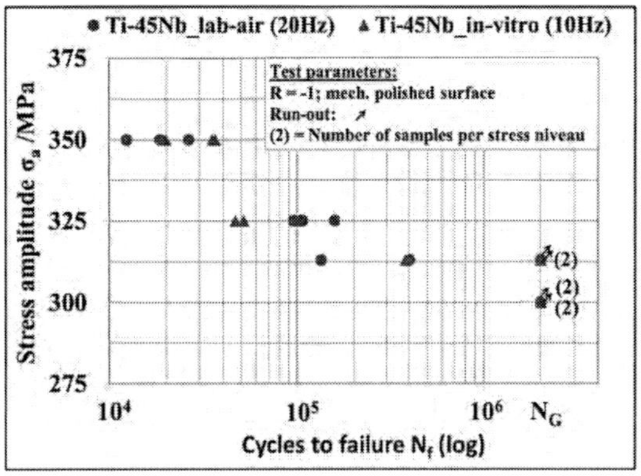

Fig. 6. Fatigue results of (β)-Ti-45Nb samples tested in lab-air and in-vitro in Ringer's solution

Mass difference before and after the fatigue tests did not give any evidence of corrosion effects. Visual inspection in SEM and light microscopy also showed that no corrosion developed during fatigue testing proving the already studied excellent corrosion resistance of (β)-Ti-45Nb [13, 41].

A direct comparison of (β)-Ti-45Nb with the recrystallized and aged condition of the (β)-Ti-40Nb alloy is restricted due to the difference in sample geometry, however a slightly superior fatigue strength of (β)-Ti-45Nb is undeniable, even considering the effect of precipi-tation hardening in Ti-40Nb_Om4h.

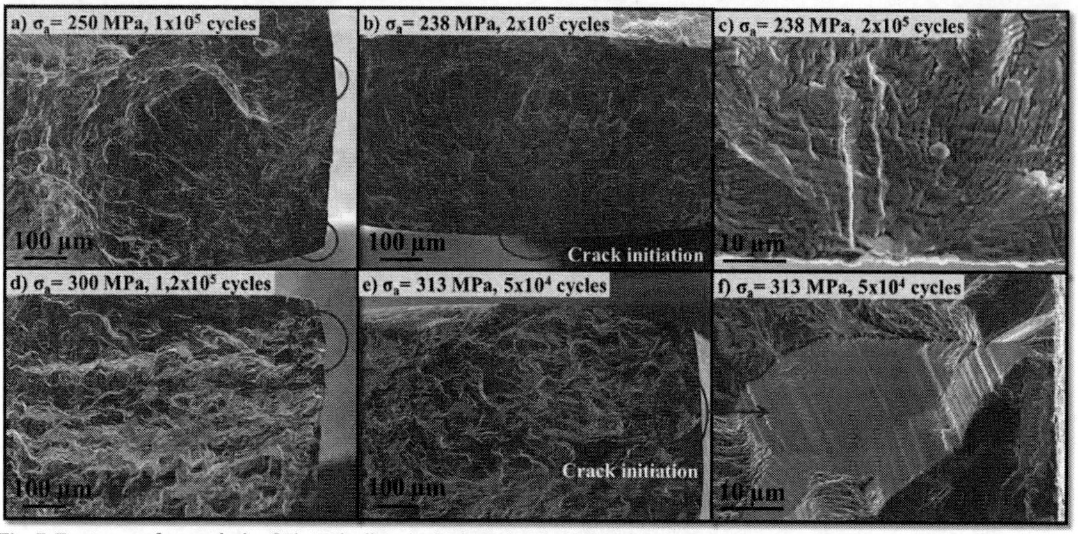

Fig. 7. Fracture surface analysis of Ti-40Nb alloy: (a-c) Ti-40Nb_Rk, (d-f) Ti-40Nb_Om4h

Comparing the studied binary Ti-Nb alloys with classical Ti-alloys such as Ti-6Al-4V, the cyclic strength of these renowned alloys cannot yet be reached.

However, in combination with its excellent biocompatibility and the very low Young's modulus the newly developed (β)-Ti-40Nb alloy, in particular in the aged condition, proves a high potential for future application as implant material [1].

3.2 Fracture surface analysis

All studied samples were subject of detailed fractographic analysis after the fatigue tests and all alloys and conditions show crack initiation at the surface. However, distinct differences in failure behaviour can be detected.

Fig. 7 depicts typical fracture surfaces of the (β)-Ti-40Nb alloy. The recrystallized condition (Ti-40Nb_Rk) shows transgranular cracks initiating from the sample surface. Higher stress amplitudes σ_a result in multiple crack initiation sites (Fig. 7a), whereas stress amplitudes below values of 250 MPa lead to single crack initiation. A detailed inspection reveals ductile fatigue fracture characteristics with striations in the area of fatigue crack propagation, while dimples indicate the ductile nature of the material in the overload fracture area. Cracks are growing radially outgoing from the initiation site. A feathered and faceted structure can be identified in higher magnifications, which appears to be a sign for high amounts of stress-induced α''-phase in this binary ß-titanium alloy.

Thus, the high instability of the β-phase at 40 wt% Nb seems to provide enough driving force to cause formation of localized α''-phase [31]. Whether this transformation is a result of the cyclic deformation or whether it was already present before fatigue testing has yet to be analyzed. In any case, the localized α''-phase in the near surface region acts as potential crack initiation site.

Compared with the widely studied static deformation behaviour and the expectable mechanisms after the Bo-Md diagram, slip and twinning should be dominating for this alloy group [32]. Kim et al. [42] studied the effects of superelasticity and shape memory emerging from a Nb-content of around 40 wt% and directly linked it with the double yielding behaviour and necessary minimal stresses below 200 MPa for α''-formation and reorientation. Hence, the cyclic stresses above this value in this study seem to initiate preferred transformation from β to α'' compared to deformation due to slip and twinning [31,42].

Fracture surfaces of the aged condition with ω-precipitations present in the β-matrix (Ti-40Nb_Om4h) are of intergranular appearance with cleavage structures instead of facets. This indicates a more brittle fatigue failure. The cleavage structures are more pronounced in the samples with high stress amplitudes σ_a. While cracks initiate from the sample surface in all cases, no multiple crack initiation sites were observed irrespective of the stress amplitude.

Fatigue crack paths are in many cases not perpendicularly oriented to the applied stress but rather in 35 to 45 degree following preferred crystallographic planes according to Schmidt's law.

The more brittle nature of fatigue failure in the aged samples are in good agreement with the superior proof and ultimate tensile strength as well as decreased ductility (Table 2). Since superelasticity and shape memory effects are almost fully suppressed by the ω-phase, cyclic deformation seems to be shifting towards dominating dislocation slip and possible twinning mechanisms. As a result initial crack formation (stage I) occurs along the plane of maximal shear stress [43].

The higher Nb-content in the additionally studied (β)-Ti-45Nb alloy results in fracture surfaces as depicted in Fig. 8. Like both conditions of the (β)-Ti-40Nb alloy fatigue cracks always initiated at the surface of the samples.

Higher magnification shows similar transgranular appearance compared to the fracture surface of the Ti-40Nb_Rk samples with α''-phase formation. Following the phase stability approach according to the electronic parameters of the Bo-Md diagram the higher Nb-content should inhibit stress-induced α''-phase formation and cause deformation dominated by slip bands [31,34].

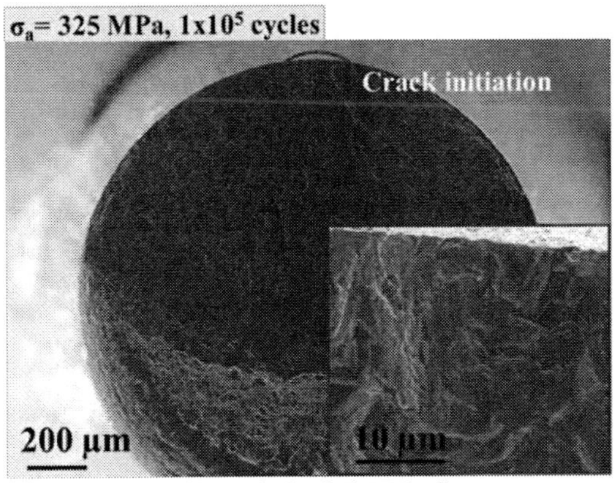

Fig. 8. Fracture surface analysis of Ti-45Nb alloy

Comparing the fracture surfaces, stress-induced α''-phase formation seems to be indeed an active cyclic deformation mechanism as well and must be subject of further investigations.

Regarding the influence of corrosion, no differences between the samples tested in air and in Ringer's solution are detectable, neither directly on the fracture surfaces nor at the radius area near the crack initiation site. This suggests that the natural forming oxide layer, which mainly consists of ductile Nb_2O_5 for Ti-Nb-based alloys instead of the more brittle TiO_2 of e.g. cp-titanium or Ti-6Al-4V, is effective enough to resist early corrosion fatigue failure [13-14,41].

4 Conclusions

Focus of the current study was the fatigue behavior of two binary (β)-Ti-Nb alloys and the influence of thermomechanical treatment as well as environment due to corrosion fatigue. A recrystallized and an aged condition with ω-precipitations of a newly developed Ti-40(wt%)Nb alloy were investigated and compared with results of industrial commercially available (β)-Ti-45(wt%)Nb alloy. The findings can be summarized as follows:

(1) The fatigue strength at 2x10^6 cycles of the aged (σ_a = 288 MPa) and the recrystallized condition (σ_a = 225 MPa) of the Ti-40Nb alloy differed significantly, caused by the contribution of precipitation hardening and the stabilizing character of the ω-phase against α''-phase formation.

(2) Aging of Ti-40Nb and the resulting less ductile deformation behaviour also became apparent in the fracture surfaces formed during fatigue loading. The recrystallized condition shows transgranular, more ductile failure perpendicular to the cyclic load direction, whereas the aged condition fails with a more brittle, intergranular characteristic alongside preferred crystallographic planes.

(3) The commercially available (β)-Ti-45Nb (σ_a = 300 MPa) alloy showed slightly superior fatigue strength compared to the aged (β)-Ti-40Nb alloy. Irrespective of the higher ß-phase stability from the higher Nb-content, the fracture surface analysis revealed similar characteristics as the aged condition of (β)-Ti-40Nb.

(4) No significant influence of corrosive environment on the fatigue strength could be observed during in-vitro testing, proving the excellent biocompatability of the tested binary (β)-Ti-Nb alloy.

Acknowledgement

The authors gratefully acknowledge the work and contribution of S. Volkmann for conducting the tests on Ti-45Nb and M. Kuczyk for assistance in the analytical investigations. This project was funded by the Deutsche Forschungsgemeinschaft (DFG) in the framework of SFB/Transregio 79, project M1.

References

1. M. Niinomi, J. Mech. Behav. Biomed. Mater. **1** (1), 30–42, (2008)
2. M. Niinomi and M. Nakai, Int. J. Biomater., (2011), doi:10.1155/2011/836587
3. Q. Chen, G. A. Thouas, Mat. Sci. & Eng.: R: Reports **87**, 1–57, (2015)
4. M. Geetha, A. K. Singh, R. Asokamani, A. K. Gogia, Prog. Mat. Sci. **54** (3), 397–425, (2009)
5. M. Niinomi, Mat. Sci. & Eng. A **243** (1-2), 231–236, (1998)
6. M. Niinomi, Y. Liu, M. Nakai, H. Liu, H. Li, Regen. Biomater. **3** (3), 173–185, (2016)
7. T. Ozaki, H. Matsumoto, S. Watanabe, S. Hanada, Mater. Trans. **45** (8), 2776 – 2779, (2004)
8. M. Long, H. Rack, Biomater. **19** (18), 1621–1639, (1998)
9. I. Jimenez-Palomar, A. Shipov, R. Shahar, A. H. Barber, Front. Mater. **2** (9), 1-13, (2015)
10. M. Abdel-Hady Gepreel, M. Niinomi, J. Mech. Behav. Biomed. Mater. **20**, 407–415, (2013)
11. C. Fleck, D. Eifler, Int. J. Fatigue **32** (6), 929–935, (2010)
12. R. A. Antunes, M. C. Lopes de Oliveira,, Acta Biomater. **8** (3), 937–962, (2012)
13. R. Godley, D. Starosvetsky, I. Gotman, J. Mat. Sci. **17**, 63–67, (2006)

14. F. Rubitschek, T. Niendorf, I. Karaman, H. J. Maier, J. Mech. Behav. Biomed. Mater. **5** (1), 181–192, (2012)
15. M. Marteleur, F. Sun, T. Gloriant, P. Vermaut, P. J. Jacques, F. Prima, Scri. Mater. **66** (10), 749–752, (2012)
16. A. Helth, S. Pilz, T. Kirsten, L. Giebeler, J. Freudenberger, M. Calin, J. Eckert, A. Gebert, J. Mech. Behav. Biomed. Mater. **65**, 137–150, (2017)
17. S. Bahl, S. Das, S. Suwas, K. Chatterjee, J. Mech. Behav. Biomed. Mater. **78**, 124–133, (2018)
18. S. Guo, J. Zhang, X. Cheng, X. Zhao, J. Alloys & Comp. **644**, 411–415, (2015)
19. S. Hanada, N. Masahashi, T. K. Jung, M. Miyake, Y. S. Sato, H. Kokawa, J. Mech. Behav. Biomed. Mater. **32**, 310–320, (2014)
20. C. Huang, Y. Zhao, S. Xin, C. Tan, W. Zhou, Q. Li, W. Zeng, Int. J. Fatigue **94**, 30–40, (2017)
21. C. Lan, Y. Wu, L. Guo, F. Chen, Mat. Sci & Eng.: A **690**, 170–176, (2017)
22. K. Ozaltin, W. Chrominski, M. Kulczyk, A. Panigrahi, J. Horky, M. Zehetbauer, M. Lewandowska, J. Mater. Sci. **49** (20), 6930–6936, (2014)
23. S. Ozan, J. Lin, Y. Li, Y. Zhang, K. Munir, H. Jiang, C. Wen, J. Mech. Behav. Biomed. Mater. **78**, 224–234, (2017)
24. J. Stráský, P. Harcuba, K. Václavová, K. Horváth, M. Landa, O. Srba, M. Janeček, J. Mech. Behav. Biomed. Mater. **71**, 329–336, (2017)
25. P. Majumdar, S. B. Singh, S. Dhara, M. Chakraborty, J. Mech. Behav. Biomed. Mater. **10**, 1–12, (2012)
26. K. Cho, M. Niinomi, M. Nakai, J. Hieda, R. Kanekiyo, Mater. Trans. **54** (10), 2000–2006, (2013)
27. H. Liu, M. Niinomi, M. Nakai, S. Obara, H. Fujii, Mat. Sci. & Eng.: A **704**, 10–17, (2017)
28. X. Song, M. Niinomi, M. Nakai, H. Tsutsumi, L. Wang, Mat. Sci. & Eng.: C **32** (3), 542–549, (2012)
29. X. Song, L. Wang, M. Niinomi, M. Nakai, Y. Liu, Mat. Sci. & Eng.: A **640**, 154–164, (2015)
30. M. Nakai, M. Niinomi, T. Oneda, Metall. & Mat. Trans. A **43** (1), 294–302, (2012)
31. M. Abdel-Hady, K. Hinoshita, M. Morinaga, Scri. Mater. **55** (5), 477–480, (2006)
32. D. Kent, G. Wang, M. Dargusch, J. Mech. Behav. Biomed. Mater **28**, 15–25, (2013)
33. M.J. Lai, C.C. Tasan, D. Raabe, Act. Mater. **100**, 290–300, (2015)
34. M. Morinaga, N. Yukawa, T. Maya, K. Sone, H. Adachi, Proc. **6th World Conf. Titan.**, France, 1601–1606 (1988)
35. J.J. Gutiérrez Moreno, M. Bönisch, N. T. Panagiotopoulos, M. Calin, D. G. Papageorgiou, A. Gebert, J. Eckert, G. A. Evangelakis, C. Lekka, J. Alloys & Comp. **696**, 481–489, (2017)
36. A. Reck, S. Pilz, M. Calin, A. Gebert, M. Zimmermann, Int. J. Fatigue **103**, 147–156, (2017)
37. T. Furuhara, T. Maki, T. Makino: J. Mater. Process. Technol. **117**, 318–23, (2001)
38. M. Delshadmanesh, G. Khatibi, M. Lederer, M. Zehetbauer, H. Danninger, SSP **258**, 337–340, (2016)
39. Company Certificate ATI 45Nb™ Alloy, Allegheny Technologies Incorporated (ATI), **Batch-No.TSC-522069**
40. S. Shin, C. Zhang, K.S. Vecchio, Mat. Sci. & Eng.: A **702**, 173–183, (2017)
41. P.F. Gostin, A. Helth, A. Voss, R. Sueptitz, M. Calin, J. Eckert, A. Gebert, J. Biomed. Mater. Res. **101B**, 269–278, (2013)
42. H.Y. Kim, S. Hashimoto, J. I. Kim, H. Hosoda, S. Miyazaki, Mater. Trans. **45** (7), 2443–2448, (2004)
43. S.A. Mantri, D. Choudhuri, T. Alam, V. Ageh, F. Sun, F. Prima, R. Banerjee, Scri. Mater. **130**, 69–73, (2017)

Fatigue life time modelling of Cu and Au fine wires

Golta Khatibi[1,*], *Ali* Mazloum-Nejadari[2], *Martin* Lederer[1], *Mitra* Delshadmanesh[1] and *Bernhard* Czerny[1]

[1]Christian Doppler Laboratory for Lifetime and Reliability of Interfaces in Complex Multi-Material Electronics, Institute for Chemical Technologies and Analytics, TU Wien, 1060 Vienna, Austria
[2]Infineon Technologies AG, Neubiberg, Germany

Abstract. In this study, the influence of microstructure on the cyclic behaviour and lifetime of Cu and Au wires with diameters of 25µm in the low and high cycle fatigue regimes was investigated. Low cycle fatigue (LCF) tests were conducted with a load ratio of 0.1 and a strain rate of ~2e-4. An ultrasonic resonance fatigue testing system working at 20 kHz was used to obtain lifetime curves under symmetrical loading conditions up to very high cycle regime (VHCF). In order to obtain a total fatigue life model covering the low to high cycle regime of the thin wires by considering the effects of mean stress, a four parameter lifetime model is proposed. The effect of testing frequency on high cycle fatigue data of Cu is discussed based on analysis of strain rate dependency of the tensile properties with the help of the material model proposed by Johnson and Cook.

1 Introduction

Fine metallic Cu and Au wires with diameters in the range of 15 µm to 75 are used for electrical connection between the semiconductor chip and the package in microelectronic devices. The wire bonded interconnects are most commonly made by thermosonic ball bonding technology [1]. During the operation, wire bonds are exposed to repetitive thermal and mechanical cyclic loading as a consequence of thermal extension and contraction of the materials in the package. Gradual accumulation of the plastic strain results in fatigue and final fracture of the wires and failure of the devices. During the thermosonic bonding, electronic flame-off process is applied to melt the tip of the wire by an electrical discharge and to form a free air ball (FAB) which is ultrasonically bonded to the substrate. The FAB process results in formation of a heat affected zone in the wire above the bonded area consisting of recrystallized larger grains. The inhomogeneity of the final microstructure along the wire bond loop depends on the processing condition and the microstructure of the wire prior to the bonding process [1]. The coarse grained heat affected area above the ball bond has been recognized as one of the vulnerable sites of the electronic packages [2]. Knowledge of static and cyclic behaviour of wires in the actual dimensions and microstructure is an important factor for reliable modelling and lifetime prediction of the packages. For automotive and high power electronics with required fault free performance of above 10 years, characterization of the fatigue properties from the low cycle (LCF) up to very high cycle regime (VHCF) is of high importance.

The influence of sample size and microstructure on the static mechanical properties of miniaturized materials has been extensively studied by using micro tensile-, compression tests or nano- indentation techniques. It was found that the number of grains in the sample volume or the ratio of thickness to grain size plays a major role on the mechanical response of small scaled materials [3]. Due to their high aspect ratio, investigation of the fatigue properties of thin wires requires especial experimental set-ups. Providing the perfect axial alignment and avoiding the buckling of the wire required special care. Thus fatigue studies on thin wires have been conducted mostly under bending or torsional modes [4,5] and to a very less extent in tension-tension or tension-compression loading modes [6-9]. Studies on the high cycle fatigue behaviour of thin micro wires are very scarce [9].

In the last decades, ultrasonic fatigue testing has been used as an established time saving method for investigation of VHCF behaviour of a wide range of materials. The effect of high strain rate / high frequency loading on the static and dynamic response of materials has been subject of experimental and theoretical studies [10,11]. The influence of strain rate and testing frequency on the tensile and fatigue properties of copper as a reference material has been studied more extensively with the results being partly controversial [10-15].

In the present study the influence of microstructure and mechanical properties of thin Au and Cu wires on their low and high cycle fatigue response was studied. The LCF stress controlled experiments were conducted at low strain rates (<1Hz). Using an ultrasonic resonance testing system working at 20 kHz allowed to perform fully reversed HCF experiments up to 1e9 loading cycles. Based on the experimental data and analytical approach, the effects of mean stress and the strain rate on the lifetime of Au and Cu micro wires is discussed. A lifetime model

* Corresponding author: golta.khatibi@tuwien.ac.at

https://doi.org/10.1051/matecconf/201816506002

for Au and Cu fine wires covering the low to high cycle regime is suggested.

2 Materials and Methods

Commercial high purity Cu and Au wires with diameters of 25μm which are used for interconnection in microelectronics devices were purchased from different suppliers. The microstructure of the wires was modified by annealing at 450°C for one hour in vacuum. The annealing conditions were selected based on a previous study with the aim to obtain a microstructure similar to that occurring in the heat affected zone of thermosonic wire bonds in microelectronic packages [2]. The wires are denoted as Cu-*i* and Au-*i* for the initial state and Cu-*a* and Au-*a* for the annealed condition.

2.1 Microstructural characterization

The microstructures of the Au and Cu wires in the initial condition (as purchased) are shown in figures 1a-b. The Au and Cu wires exhibited considerably distinct microstructures which were dependent on the processing parameters during the wire drawing. The microstructure and texture of fine Au and Cu wires used in electronic packages in dependency of processing parameters has been subject of a number of extensive investigations [16,17]. In our study, the Au-i wires show an inhomogeneous fine grained structure with a recrystallized region at the periphery and a strong fibre textured region in the core. The microstructure of Cu-i wires consists of smaller and larger recrystallized equiaxed grains with an average grain size of ~10μm. Annealing resulted in different degrees of recrystallization and grain coarsening in each material. A quasi-bamboo structured microstructure was obtained for Cu-*a* wires. The microstructure of the annealed Au-*a* samples remained multi-crystalline consisting of a few grains in the cross section of the wires.

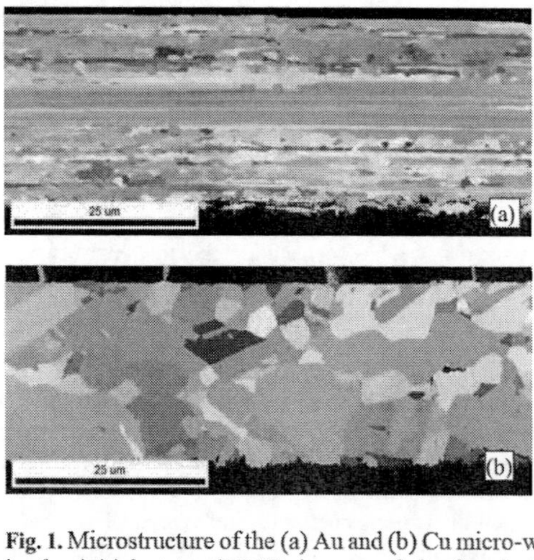

Fig. 1. Microstructure of the (a) Au and (b) Cu micro-wires in the initial state. (EBSD images of the longitudinal sections).

2.2 Tensile and fatigue tests

2.2.1 Tensile test and low cycle fatigue set-up

Tensile properties and Young's Modulus of the Cu wires were determined by using a micro-tensile machine equipped with a load cell with a capacity of 1N and accuracy of +/- 0.001N. A laser speckle extensometer with a strain resolution of about 10e-5 was used for contactless strain measurements. All tests were conducted at strain rate of 1.3e-4 s^{-1}.

Low cycle fatigue tests were conducted by using the same micro-tensile machine. Load controlled tensile-tensile low cycle fatigue tests were conducted up to about 2e4 cycles. The stress ratio (R= $\sigma_{min}/\sigma_{max}$) was ~ 0.11 - 0.12 to avoid possible buckling of the wires during the unloading cycles. The tests were started with a strain rate of 1.3e-4 s^{-1} which was increased to ~2e-4 s^{-1} after reaching the maximum stress. This testing speed is equivalent to a testing frequency of about 0.05 Hz. LCF curves were obtained from 20- 30 experiments for each sample series. Figure 2 shows the overlapping of the static and cyclic stress-displacement curves for the Cu-*i* fine grained wires. The observed ratcheting was a typical behaviour for the thin wires during the cyclic tension-tension loading (Fig. 2). All the tested wires showed a variation between the monotonic and cyclic fracture strain which was higher for annealed wires with a larger standard deviation of their tensile properties (Table 1) and the cyclic hardening effect.

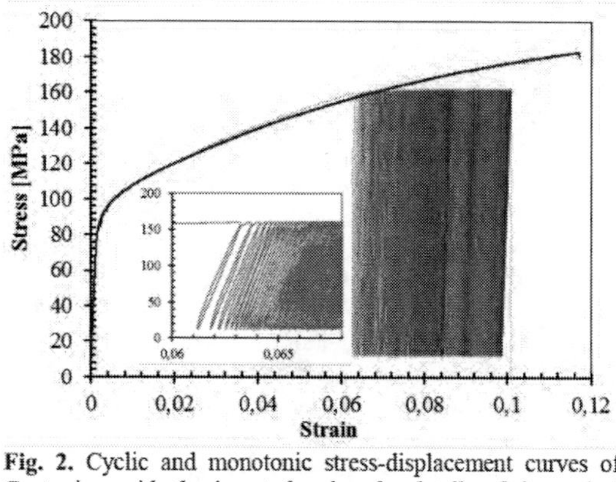

Fig. 2. Cyclic and monotonic stress-displacement curves of Cu-*i* wires with the insert showing the details of the typical ratcheting behaviour of thin wires subjected to tensile-tensile cyclic loading.

2.2.2 High cycle fatigue set-up

High cycle fatigue experiments were performed by using a standard ultrasonic resonance testing system working at 20 kHz which was adapted for testing of miniaturized samples [9]. The resonance system consists of a power supply, a piezoelectric transducer and a mechanical part which includes an acoustic horn. Further, the horn is connected to a sample holder with a resonance length ($\lambda/2$) of 128mm and a cross-section of 20x8 mm^2. During the loading, the mechanical part of the system is excited to longitudinal push-pull vibrations with a zero mean strain (i.e. R = -1). The distribution of displacement and strain varies in a sinusoidal manner along the system with the maximum of the strain being in the midsection of the holder. The cyclic strain amplitude is determined by using a strain gauge. Two grooves with a width and depth of about 0.3x0.3μm are machined into the both sides of the holder at this location. Fatigue tests are performed by placing and aligning the wires over the groove parallel to the loading direction and gluing them by using a cyanide-acrylate adhesive. Thus the gauge section of the wires across the groove is free-standing. The effective strain amplitude of the wires during the testing was estimated by means of finite element analysis and validated by using a miniaturized strain gauge which was applied over the groove resulting in a stress concentration value (k_t) of 1.3. Under the assumption of preliminarily elastic deformation in the high cycle regime, the cyclic stress amplitude can be obtained from the strain value measured with the strain gauge and the Young's modulus of the wires in accordance with Hooke's law. The loading cycles to failure (N_f) were determined by monitoring the surface of the wires with a high resolution video microscope during the entire testing. The tests were performed at room temperature with additional cooling of the specimen by pressurized dry air.

3 Results

3.1 Tensile properties

The stress-strain curves of the Au and Cu wires prior and subsequent to annealing are presented in figure 3 and the obtained mechanical properties are listed in Table 1. The curves show that the fine grained Au-*i* wires which exhibit the highest yield and tensile strength are highly prone to softening subsequent to annealing at 450°C. The ductility remains low with a value of about 4% for the initial and annealed wires. According to a study on similar wires [17] the high tensile strength of the cold drawn fine Au wires is related to the high percentage of <111> fibre component in the microstructure. By annealing at the temperature range of 400°C to 700°C the <100> recrystallization textures is predominantly formed resulting in softening and a decrease of the yield and ultimate tensile strength of the wire. Though annealing also results in a strong decrease of yield and tensile strength of the Cu wires, the wires show the same hardening behaviour.

Table 1. Mechanical properties of Cu and Au micro wires.

Wire material	Yield stress σ_y [MPa]	Tensile strength σ_u [MPa]	Fracture strain [%]	Young's Modulus E [GPa]
Cu -*i*	63± 4.5	178± 5	10.4 ± 2	95± 4
Cu -*a*	26 ± 5.3	112 ± 14	4.5± 2	78± 3
Au -*i*	130	209 ± 2	3.5 ± 1	75± 2
Au -*a*	31 ± 3	84 ± 5	4.5± 2	76 ± 5

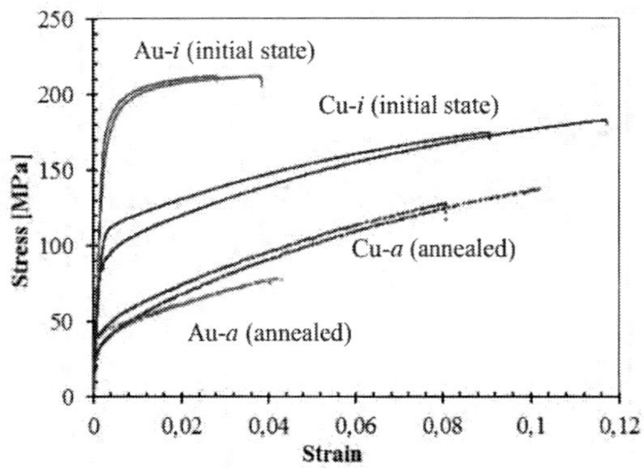

Fig. 3. Stress-strain behaviour of Cu and Au wires prior and after annealing.

3.1.1. Modelling of strain rate dependency

The strain rate dependency of the stress-strain response of the Cu wires was analysed by using the Johnson and Cook constitutive model [10]. The simulations were conducted by using

$$\sigma = \left(A + B \cdot \varepsilon_{pl}^{\ n}\right) \cdot \left(1 + C \cdot \ln \dot{\varepsilon}^*\right), \qquad (1)$$

where A, B, C, and n are material parameters, ε_{pl} is the equivalent plastic strain, $\dot{\varepsilon}^* = \frac{\dot{\varepsilon}}{\dot{\varepsilon}_0}$ is the dimensionless strain rate and $\dot{\varepsilon}_0$ is the reference strain rate. The temperature dependent term of the model could be neglected since all the experiments were performed at room temperature. The strain rate dependency of tensile response of Cu wires prior and after annealing for strain rates of 1e-4 s^{-1} (~ 0.05 Hz) and 32 s^{-1} (~ 20 kHz) is presented in figure 4 and the model parameters used for the numerical fits are given in table 2. The results show that at lower strain values the yield stress of Cu increases by a factor of $\left(1 + C \cdot \ln \dot{\varepsilon}^*\right) = 1.32$.

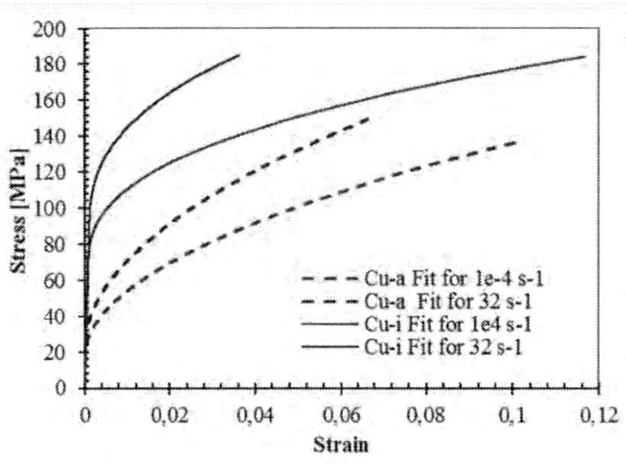

Fig. 4. Johnson and Cook model fits showing the strain rate dependency of Cu wires prior and after annealing.

Table 2. Material parameters for initial and annealed Cu micro wires. The value for C was adopted according to reference [10].

	A [MPa]	B [MPa]	C	n	$\dot{\varepsilon}_0$ s^{-1}
Cu -*i*	70	272	0.025	0.4	1e-4
Cu -*a*	20	388	0.025	0.52	1e-4

3.1 Fatigue life curves

The influence of microstructure and tensile properties on the low and high cycle fatigue life of Au and Cu micro-wires is presented in figures 5 and 6, respectively. The LCF curves are presented as a function of the stress amplitude against (N_f) up to 2e4 loading cycles (Fig. 5). The results of the fatigue experiments show a direct relationship between their cyclic and monotonic properties for all samples (Fig. 1, Table 1). The best fatigue performance was obtained for the fine grained Au-*i* wires with a tensile strength (σ_u) of 209 MPa while the lowest lifetime was observed for the annealed Au wires (σ_u = 84 MPa). For comparison reasons the experimental fatigue data were fitted by using a Basquin type equation,

$$\sigma_a = \sigma_f'.(N_f)^{-b} \tag{2}$$

as presented in figure 5. Using equation (2) the stress amplitude σ_a is expressed by a first order power law of loading cycles (N_f) with σ_f' being the fatigue stress coefficient and b being the fatigue stress exponent. The values of σ_f' are determined to be 112 MPa and 35 MPa for initial and annealed Au wires and 93 MPa and 53 MPa for initial and annealed Cu wires, respectively. These coefficients correlate very well to the corresponding $\sim\frac{\sigma_u}{2}$ values for each wire. The stress exponent values which indicate the slopes of the curves are found to be in the range of 0.01 to 0.04. The higher slope may indicate a stronger decrease of fatigue life at intermediate or higher loading cycles.

Previous studies on low cycle fatigue of thin Cu wires [6-7] and lines [8] with diameters /thicknesses in the range of 20μm to 95μm subjected to alternate tensile loading (R~0) reported a similar behaviour. The cyclic properties were found to be closely related to the monotonic tensile response and obey the Basquin law. The fatigue stress exponent (n) was found to be in the range of 0.01-0.07 and the fatigue strength coefficient correlates very well with the tensile strength of the material.

The high cycle fatigue life curves of the four wire materials with R=-1and f =20kHz are presented in figure 6. The trend of the S-N curves in the range of 5e5 to 1e9 cycles is similar to that found in LCF regime. The highest fatigue resistance is again obtained for Au-*i* wires with the highest yield and tensile strength. The lifetime curve lies between the stress amplitudes of 53 MPa (N_f = $1e6$) and 46 MPa which is considered as the fatigue limit at 1e9 cycles. The lifetime curve of Cu-i. wires which lies beneath that of Au-*i* exhibits a steeper slope with a fatigue limit of 25 MPa at 1e9. Annealing results in a strong reduction of the fatigue resistance of the wires which with a drop of about 50% is more severe for Au-*i* wires. The lifetime curve of Cu-*a* wires is slightly higher than that of Au-*a* wires and the fatigue limit of the both wire series is 15 -16 MPa at 1e9 cycles. In a previous study high cycle fatigue of bamboo structured thin wires of 10μm to 125 μm was investigated by using the same experimental set-up (R=-1, frequency 20 kHz) [9]. The authors found that the higher monotonic yield stress results in an improved fatigue resistance in the high cycle regime. This result complies well with the findings of the present work.

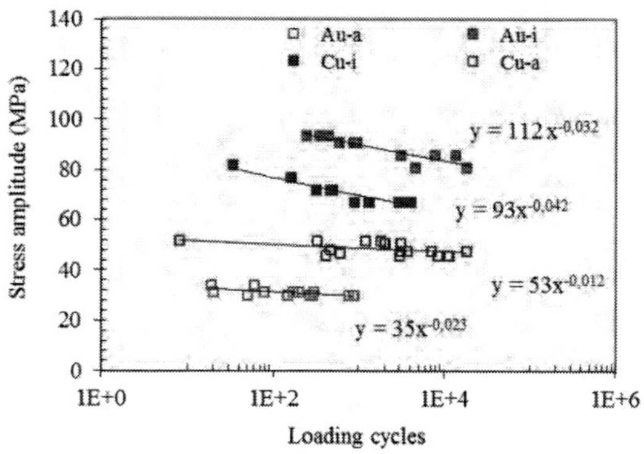

Fig. 5. Comparison of low cycle fatigue life curves of Au and Cu fine wires with different grain sizes in initial and annealed states.

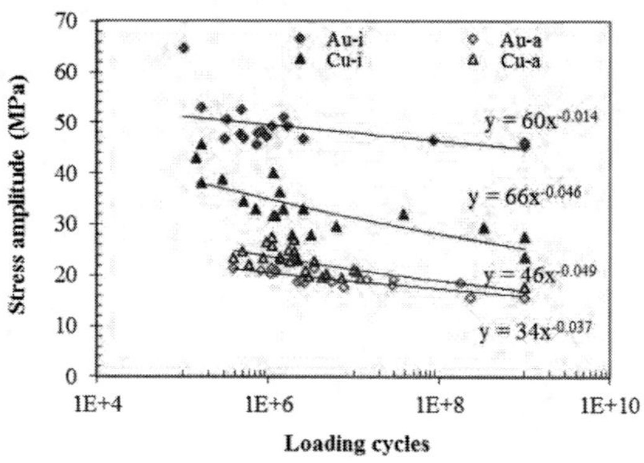

Fig. 6. Comparison of high cycle fatigue life curves of Au and Cu fine wires with different grain sizes in initial and annealed states.

3.1.1 Fracture surface observations

Figures 7 and 8 show the surface morphologies of the Cu wires next to the failure site indicating the effect of microstructure /grain size and the mean stress on the failure modes of the wires. The plastic deformation observed on the surface of the initial and annealed Cu wires subjected to tensile alternating loads resemble strongly to that caused by uniaxial tensile loading (Fig. 7a and 8a respectively). Failure occurs as a result of slip band formation, necking and rupture of the wires in a ductile manner. In the case of Cu-*a* quasi-bamboo structured wires, plastic deformation has caused formation of slip steps at the grain boundaries of the two grains which have been individually deformed until final rupture in a knife edge manner. The surface of Cu-*i* (7b) and Cu-*a* (8b) wires exhibit the typical deformation features which are observed on the surface of copper samples subjected to symmetrical cyclic loading. Strong surface roughening due to local slip band formation and deeper intergranular cracks near the ruptured area can be observed on the surface of a Cu-*i* wire (Fig. 7b). Figure 8b indicates that the slip bands have been developed mainly in the large grains with a preferred orientation with respect to the loading axis. Several elliptical shaped slip bands which formed in two large grains indicates a partly bamboo like structure in this region. Similar features were observed on the fatigue fracture surfaces of the Au wires.

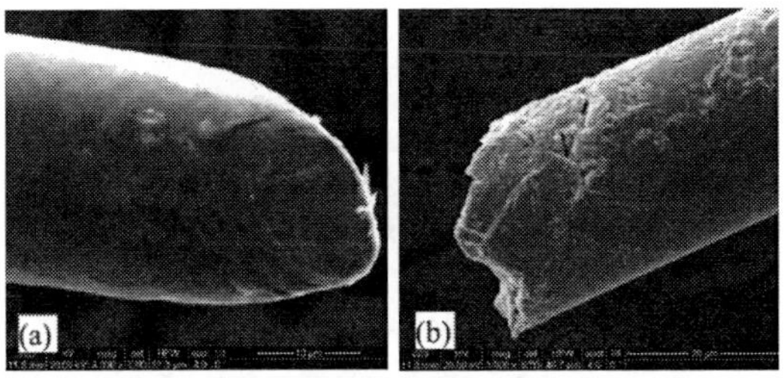

Fig. 7. Typical images of fatigue fracture surfaces of Cu-i wires after failure at low and high cycle regime

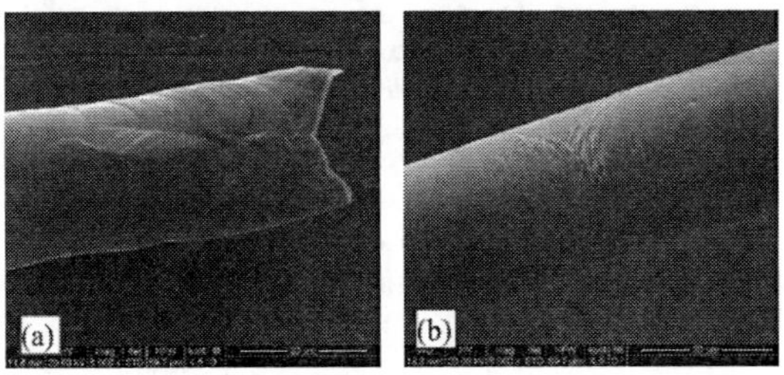

Fig. 8. Typical images of fatigue fracture surfaces of Cu-a wires after failure at low and high cycle regime

3.2 Fatigue life modelling

Establishment of lifetime curves covering the low to high cycle regime was conducted based on the experimental results and considering the effect of mean stress [18, 19]. For this purpose Gerber and Goodman methods were used for conversion of the maximum stress σ_{max} at low cycle fatigue tests to an effective stress amplitude $\sigma_e^{'}$ for symmetrical loading by the following equations:

$$\frac{\sigma_a}{\sigma_e^{'}} + \frac{\sigma_m}{\sigma_u} = 1 \tag{3}$$

$$\frac{\sigma_a}{\sigma_e^{'}} + \left(\frac{\sigma_m}{\sigma_u}\right)^2 = 1, \tag{4}$$

where σ_a , σ_m, and σ_u are the stress amplitude, the mean stress and the ultimate tensile stress in Goodman (3) and Gerber (4) equations, respectively. The results of transformation of the maximum stress amplitude to an equivalent symmetrical stress amplitude for the low cycle fatigue data for the annealed Au wires is shown in figure 9. The experimental stress amplitude values for low and high cycle fatigue are also included in this plot. It can be observed that the Gerber model provides a more conservative result.

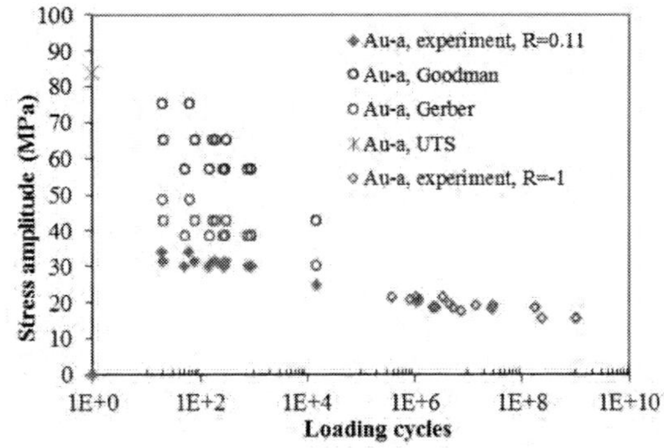

Fig. 9. Conversion of σ_{max} from LCF experiments to symmetrical stress amplitude values $\Delta\sigma/2$ (R=-1) based on using the Goodman and Gerber equations for Au-*i* wires.

The total fatigue life curves for Au and Cu wires for symmetrical loading conditions (R=-1) are presented in figures 10 and 11 based on the converted LCF data by using the Goodman equation and the obtained experimental HCF values. According to the Johnson and Cook material models proposed for the initial and annealed Cu wires (Table 2, Fig. 4), loading at a testing frequency of 20 kHz (~32 s^{-1}) results in an increase of the yield stress level of Cu by a factor of 1.32 at relative low strains. However this effect occurs after yielding of the material. In this study, the maximum stress amplitudes of the high cycle fatigue data in the whole loading range were found to be below the yield strength of all the wires. Thus an increase or shift of the experimentally determined lifetime curves due to testing at 20 kHz is not expected or it can be considered as negligible. Though the strain rate dependency of stress-strain response of Au wires have not been calculated in this work, concerning the HCF curves, similar arguments may also be applied. Detailed investigation of the influence of cyclic testing frequency on fatigue response of copper can be found in [13-15,20]. In order to establish lifetime models covering the low and high cycle fatigue range of the Au and Cu micro wires, a four parameter lifetime model originally proposed by

Weibull was used. The lifetime model is based on the equation (5):

$$S(N) = S_d + \frac{S_{\max} - S_d}{\exp\left\{\left(\frac{\log(N)}{\alpha}\right)^{\beta}\right\}}, \quad (5)$$

where S(N) is the stress amplitude, S_{max} is the ultimate tensile strength, S_d is the endurance limit of the material, N is the number of loading cycles to failure, α and β are material parameters. The parameters used for the model are given in Table 3 and the obtained lifetime models for Cu and Au wires prior and subsequent to annealing are presented in figures 10 and 11, respectively. The lifetime models for Au-*i* and Cu-*i* wires show a steep slope in the low and intermediate region resulting in a rather abrupt drop of fatigue resistance at higher loading cycles. The annealed Cu-*a* wires show a moderate decrease of fatigue resistance in the transition region from LCF to HCF regime. The fatigue curve of the Au-a wires with the weakest mechanical properties follows a flat trend in the whole range of lifetime.

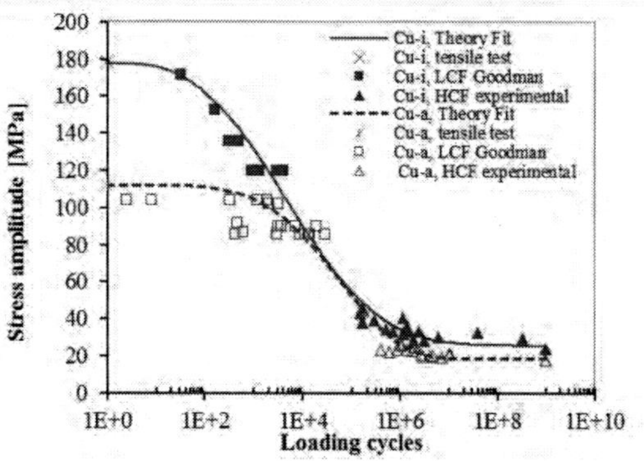

Fig. 10. Lifetime model for Cu fine wires prior and after annealing

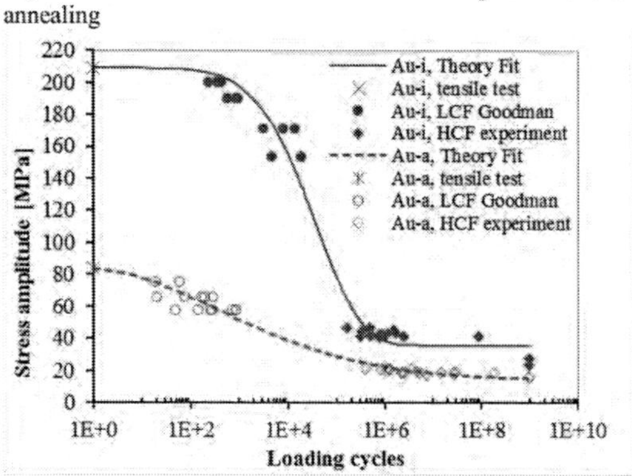

Fig. 11. Lifetime model for Au fine wires prior and after annealing

Table 3. Material parameters for lifetime models of the initial and annealed Cu and Au wires.

Wire material	S_d [MPa]	α	β
Cu-i	25.69	9.6916	2.9736
Cu-a	18.2	11.4741	4.8556
Au-i	35.71	10.8762	5.9965
Au-a	14.22	8.9872	1.789

The lifetime curves of the Au and Cu wires in the initial state are representative for the fatigue response of a typical wire bond in the region along the loop further from the bonded area, while the lifetime curves of the annealed wires estimate the fatigue performance of the heat affected zone above the ball bond [1,2].

4 Summary and Conclusion

In this study the influence of grain size, strain rate and loading mode on the tensile and fatigue properties of gold and copper micro wires used as interconnects in microelectronic packages have been investigated. Based on the experimental stress-strain curves and using the Johnson and Cook model, the strain rate dependency of material properties of Cu wires were assessed at strain rates of 1e-4 s^{-1} and 32 s^{-1}. LCF and HCF experiments were performed with R= 0.1 at ~ 0.05 Hz (1e-4 s^{-1}) and R=-1 at 20 kHz (~32 s^{-1}), respectively. The effect of mean stress was considered based on the Goodman equation which was used for conversion of the maximum stress of the LCF tests to an effective stress amplitude for symmetrical loading. The maximum stress (stress amplitude) during the HCF loading at 20 kHz remained below the yield strength of the material. Thus, based on the calculated strain rate material model for copper, the effect of high strain rate on the response to cyclic loading of the samples was considered to be negligible. Further experiments on the effect of testing frequency on the fatigue life of Cu and Au micro wires are going on. Finally, by using the experimental low and high cycle fatigue curves, total lifetime models for Au and Cu wires of 25 μm could be established. Depending on the expected service conditions, the proposed lifetime models may serve as a tool for a proper material choice for thermosonic ball bonds in electronic packages.

Acknowledgements: The financial support by the Austrian Federal Ministry for Digital and Economic Affairs and the National Foundation for Research, Technology and Development is gratefully acknowledged. The authors would like to thank A. Steiger Thirsfeld from USTEM-TU Wien for the EBSD images.

References

1. Copper Wire Bonding, P. S. Chauhan, A. Choubey, Z. W. Zhong, M. G. Pecht, Springer (2013)
2. A. Mazloum-Nejadari, G. Khatibi, B. Czerny, M. Lederer, J. Nicolics, L. Weiss, Microelectron. Reliab., 74 (2017) 147-154

3. B. Weiss, V. Gröger, G. Khatibi, A. Kotas, P. Zimprich, R. Stickler, B. Zagar, Sens. Actuators, A 3287, (2002) 1-11
4. A. Schreiner, J.T. Mortimer, T.P. Kicher, J Biomed Mater Res. 25 (1991) 589-608
5. Z. H. Zhao, C.Q. Feng, S. Zhao, M. Lei, Key Eng. Mater. 734 (2017) 176-184
6. R. Hofbeck, K. Hausmann, B. Ilschner, H.U. Künzi, Scripta Metallurgica (1986) 20 : 1601-1605
7. B. Yang, C. Motz, W. Grosinger, G. Dehm, Mater. Sci. Eng. A 515 (2009) 71–78
8. A. Wimmer, A. Leitner, T. Detzel, W. Robl, W. Heinz, R. Pippan, G. Dehm, Damage evolution during cyclic tension–tension loading of micron-sized Cu lines, Acta Mater. 67 (2014) 297–307
9. G. Khatibi, A. Betzwar-Kotas, V. Göger, B. Weiss, Fatigue Fract. Eng. Mater. Struct. 28 (2005) 723–73
10. G. R. Johnson, W. H. Cook, Proc. 7th Int Symp on Ballistics (1983)541–47
11. C. Laird, P. Charsley, in J.M Wells, O. Buck, L.D. Roth, J.K. Tien (Eds.), Ultrasonic Fatigue, The Metall. Soc. of AIME, Philadelphia (1982), pp. 187-205
12. J.S. Kim and H. Huh, Experimental Mechanics (2011) 51:845–855
13. S. Stanzl-Tschegg, Int. J. Fatigue, 105(2017) 86-96
14. P. Lukáš, L. Kunz, Z. Knésl, B. Weiss, R. Stickler, Mater. Sci. Eng. 70 (1985) 91-100
15. L. Kunz, in Copper alloys - Early applications and current performance - Enhancing processes, in: L. Collini (Ed.), InTech, Rijeka, Croatia, (2012) 93–126
16. J.H. Cho, A.D. Rollet, J.S. Cho, Y.J. Park, J.T. Moon, K.H. Oh, Metall. Mater. Trans. A, 37A (2006) 3085-3097
17. S.H. Kang, H.S. Jung, W.H. Bang, J.H. Cho, K.H. Oh *et al.,* Mater. Sci. Forum 1 (2005) 495-497
18. R. Eckert, C. Laird, J. Bassani, Mater. Sci. Eng. 91 (1987) 81-88
19. H. Hwangbo, J.H. Song, Mater. Sci. Eng. A 527 (2010) 2222-2232
20. H. Mayer, C. Laird, Mater. Sci. Eng. A187 (1994) 23-35

https://doi.org/10.1051/matecconf/201816506003

Tensile and fatigue assessments of brazed stainless steel joints using digital image correlation

Anke Schmiedt[1,*], *Simon* Jaquet[1], *Matthias* Manka[2], *Wolfgang* Tillmann[2], and *Frank* Walther[1]

[1]TU Dortmund University, Department of Materials Test Engineering (WPT), Baroper Str. 303, D-44227 Dortmund, Germany
[2]TU Dortmund University, Institute of Materials Engineering (LWT), Leonhard-Euler-Str. 2, D-44227 Dortmund, Germany

Abstract. For mechanical tests of brazed stainless steel joints, the local deformation behaviour within the small area of the brazing seam is a major concern, because local strains cannot be detected with standard mechanical extensometers. The current study allows a fundamental comprehension of the gauge length influence on the strain measurements of brazed joints with smooth and notch-containing surfaces, under quasi-static and cyclic loadings. Therefore, the optical measurement technique of digital image correlation (DIC) is used within tensile and fatigue tests of brazed AISI 304L/BAu-4 joints in an as-received and pre-corroded condition. A triggered image acquisition of the DIC system is successfully applied to evaluate the local ratcheting fatigue behaviour in the area of the brazing seam at a frequency of 10 Hz. The gauge length influence, analysed in the range of 0.5 to 12.5 mm, is more pronounced with increasing tensile and fatigue stresses and is significantly enhanced for notch-containing surfaces. Instrumented load increase tests with strain, electrical, magnetic and temperature measuring techniques have proven to be appropriate to estimate fatigue properties of the brazed joints with a deviation of 4%. Fatigue and corrosion fatigue damage mechanisms are evaluated by using scanning electron microscopy with secondary and back-scattered electron detectors.

1 Introduction

Brazing technologies are well established for an efficient and economic production of joined structures for numerous reliable applications, in particular for the automotive industry, chemical engineering and power generation. The potential and benefits of the joining technology, considering e.g. thermal stresses, material combinations, complex structures and process times, have already been extensively described [1]. Until now, several studies have been realised in order to investigate the mechanical properties as well as the damage and fracture behaviour of brazed joints under quasi-static and cyclic loadings [2]. For the evaluation of the strain measurements, deviating deformation behaviours of the brazing seam from the base material have to be considered. It is known, that local strain concentrations cannot be detected with standard extensometers and a significant influence of the gauge length is present.

The optical measurement technique of digital image correlation (DIC) was developed for a 3-dimensional, non-contact and full-field image analysis in order to generate deformation and strain maps across an entire material surface. Most of the investigations utilizing the DIC technique were performed with structures under static loads, aiming at a visualisation of the strain distribution [3]. Based on the local strain and stress concentrations, potential failure locations can be identified. An application of the optical technique, using standard digital cameras, for mechanical fatigue tests leads to a particular challenge, as common loading frequencies of 10 Hz already require a complex triggered image acquisition. Up to now, fatigue assessments using the DIC technique for the evaluation of the continuous cyclic deformations behaviours are relatively rare [3].

The high potential of the DIC technique to determine fatigue-induced damage evolutions in brazed steel joints was already presented [1]. Within this current study, the deformation and damage behaviour of AISI 304L/BAu-4 brazed stainless steel joints is analysed under quasi-static and fatigue loadings using a DIC system as well as standard mechanical extensometers for validation. A triggered image acquisition enables an evaluation of the continuous cyclic creep properties under fatigue loading with a frequency of 10 Hz. In addition to strain maps, various virtual strain gauges are placed post-test in the area of the brazing seam in order to demonstrate the influence of the gauge length of standard extensometers in mechanical tests of brazed specimens. For the fatigue assessment, time-efficient, instrumented stepwise load increase tests (LIT) are performed with the aim to determine the local strain distributions within the total stress range, leading to elastic and plastic material deformations. The applicability of the DIC technique for non-smooth surfaces is evaluated in tensile and fatigue tests of pre-corroded brazed specimens. For comparison, homogeneous base material specimens are analysed.

* Corresponding author: anke.schmiedt@tu-dortmund.de

2 Materials and experimental procedure

2.1 Materials

The austenitic stainless steel X2CrNi18-9 (AISI 304L; 1.4307), acc. to specification DIN EN 10088-3, is widely used in industry and was utilised as base material for the investigated brazed butt joints. The chemical composition, acc. to the test certificate, is listed in Tab. 1. Plastic deformations of the metastable austenite can lead to a phase transformation to ferromagnetic martensite. This fact was extensively investigated in terms of cyclic loadings for homogenous alloys in earlier studies [4-6]. The deformation-induced martensitic transformation, which is sensitive to the test temperature, strongly affects the cyclic deformation behaviour and the lifetime of metastable austenites. Hence, the self-heating of the material caused by cyclic loadings has to be taken into consideration. This effect was evaluated in anisothermal and isothermal fatigue tests at room temperature with and without cooling [5-7]. As a result of high temperatures due to self-heating, the martensitic transformation can be inhibited. In contrast, martensitic transformations in isothermal tests lead to a secondary cyclic hardening, which increases the lifetime [7].

A commercially available gold-based brazing alloy BAu-4 (Au 827), with the chemical composition Au82-Ni18, was applied as a 50 μm amorphous brazing foil between the base material prior to the brazing. The brazing process was carried out in a vacuum furnace with a vacuum of 10^{-5} to 10^{-6} mbar, using a brazing temperature of 1050 °C in conjunction with a short dwell time of 2 min, a heat treatment for 2 hrs at 950 °C and a subsequent free cooling down to room temperature. The microstructure of the brazed AISI 304L/BAu-4 joints in the as-received condition is shown in Fig. 1a.

Specimens for tensile and fatigue tests were produced from brazed cylindrical bars with the same specimen geometry, Fig. 2. The rejuvenated area was polished in order to obtain a surface quality of $R_z = 0.9$ μm and to ensure the consistency and reproducibility of the test results. In contrast to standardised tensile test specimens, the applied geometry shows a shorter gauge length of 12.5 mm. As a result, an increase of the elongation values during tensile tests is expected, if necking occurs. For the base material specimens, the AISI 304L cold drawn bars were heat-treated acc. to the temperature-time curve of the brazing process and manufactured as described in [8]. Before testing, a white speckle pattern was generated with an airbrush system on the black coated specimens for the DIC analysis.

Tab. 1. Chemical composition of AISI 304L base material acc. to test certification.

Alloying elements, wt.-%								
C	Cr	Ni	Mn	Si	P	S	N	Fe
0.018	18.23	8.06	1.05	0.42	0.035	0.027	0.074	bal.

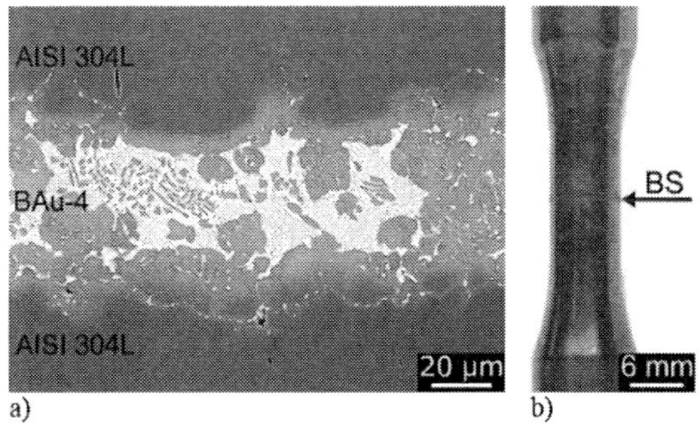

Fig. 1. a) Microstructure of the brazing seam (BS) in the as-received condition; b) Specimen after pre-corrosion.

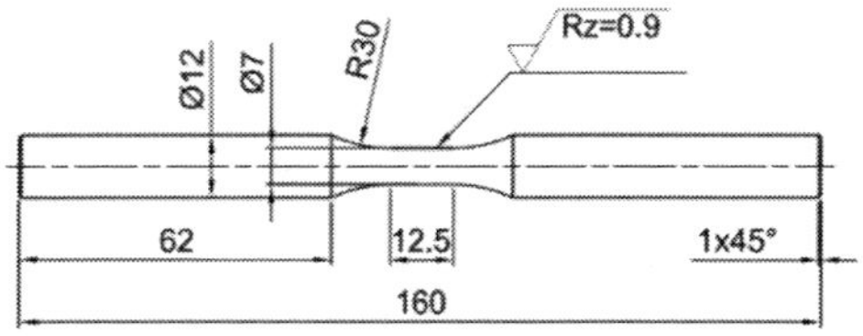

Fig. 2. Specimen geometry for tensile, fatigue and corrosion fatigue investigations (all dimensions in mm).

The prepared AISI 304L/BAu-4 and AISI 304L specimens were pre-corroded acc. to the VDA test 230-214 [9], which is established for exhaust gas-carrying components. Therefore, the DIC technique was evaluated for surfaces that are not smooth but notch-containing. The alternating corrosion exposure consists of daily periods of semi-immersion, drying and a steam phase in combination with a weekly heat ageing, conducted at 600 °C for the base material and at 400 °C for the brazed AISI 304L/BAu-4 joints. After 6 weeks of pre-corrosion, depositions of corrosion products can be detected on the specimen's surfaces. In the area of the brazing seam, the corrosive attack leads to a local circumferential metal dissolution, especially at the interface to the base material, Fig. 1b.

2.2 Experimental details

The deformation behaviour of AISI 304L/BAu-4 brazed stainless steel joints was analysed under quasi-static and cyclic loadings using a DIC system and standard mechanical extensometers. Tensile tests were performed with the base material and the brazed specimens prior to and after 6 weeks of pre-corrosion in order to evaluate the applicability of the DIC system for homogeneous and joint geometries as well as for smooth and non-smooth surfaces. For fatigue tests, brazed specimens of both conditions were used. Two specimens were tested for each material, condition and mechanical loading.

Quasi-static tensile tests were performed at a universal testing system (Shimadzu, F = 100 kN) with and without using an extensometer with a gauge length of 10 mm (GL10). Applying the extensometer allows to compare the DIC and standard strain measurements, while the complete gauge length of 10 mm can only be evaluated optically if the mounting springs of the extensometer do not obscure the view on the specimen's surface. In accordance with ISO 6892, the testing speed can be increased after the yield strength is determined. Up to the total strain ε_t in the range of 8 to 14%, tests with an extensometer were constantly controlled with a strain rate of $\dot{\varepsilon} = 0.00025\ s^{-1}$. For tests without the extensometer, the crosshead separation rate was constantly controlled by $v_c = 0.8$ mm/min, which correlates with an estimated strain rate over the parallel length $\dot{e}_{Lc} = 0.00025\ s^{-1}$. Subsequently, $v_c = 4.8$ mm/min was controlled until the failure of the specimen.

Fatigue tests were carried out using a servo hydraulic testing system (Instron, $F = \pm 100$ kN) and an extensometer with a gauge length of 12.5 mm (GL12.5), featured with a considerable measuring range of ±5 mm and allowing strain measurements of up to ±40%. The tests were performed by applying a stress ratio of $R = 0.1$ and a frequency of $f = 10$ Hz, using sinusoidal load time functions. In addition to constant amplitude tests (CAT) with a limited number of cycles $N = 2 \cdot 10^6$, a time- and source-efficient procedure, based on stepwise load increase tests, was applied, which is already established for a wide range of material classes and joints [10, 11]. With a constant stress ratio, the maximum stress was increased stepwise by $\Delta\sigma_{max} = 10$ MPa each $\Delta N = 10^4$ cycles, starting at a stress level of 50 MPa at which the material and joints are expected to be damage-free up to the level of specimen's failure.

In addition to the DIC system and the extensometer, the material response to the cyclic loading was continuously evaluated with electrical, magnetic and temperature measuring techniques. Therefore, the experimental setup for fatigue tests includes a crack growth monitor (Matelect) for alternating current (AC) potential drop measurements, a feritscope (Helmut Fischer) as well as high resolution thermal imaging cameras (InfraTec; Micro-Epsilon), Fig. 3. An alternating current power supply was fixed to the clamping jaws and the change in the deformation-induced electrical voltage ΔU, referring to the initial value at the test beginning, was recorded in the gauge length using two welded wires. Furthermore, the feritscope sensor was positioned at the base material close to the brazing seam in order to detect the change in magnetic portion $\Delta\zeta$ close to the surface, which can be correlated with the plasticity-induced phase transformation of the austenitic AISI 304L base material to ferromagnetic martensite.

The thermal imaging camera allows a visualisation of the temperature distribution until failure of the specimens with a temporal and spatial resolution. Differences down to 20 mK can be measured with the camera detector format of 640 × 512 pixels. As recommended in [6], three temperatures were evaluated after testing at the middle of the gauge length (T_1) and at the shafts of the specimen (T_2, T_3) for calculation of the change in temperature $\Delta T = T_1 - 0.5 \cdot (T_2 + T_3)$.

In addition to the tactile measurements, strain measurements within tensile and fatigue tests were performed, using 3D-digital image correlation (Limess) in order to determine the strain distribution at the specimen's surface until failure and to identify the local damage mechanisms for the brazed joints and base materials in different VDA pre-corrosion conditions. Using the DIC technique, the deformation state can be visualised with a temporal and spatial resolution. For fatigue investigations, specific modifications of the DIC system allow a frequency-dependent triggered image acquisition to continuously analyse the cyclic deformation behaviour during fatigue testing at a frequency of $f = 10$ Hz. Engineering tangential strains ε_{xx}, computed post-test in tensile direction for line elements (LE) with various lengths, show the influence of the gauge length in mechanical tests of brazed joints. Asymmetric cyclic loadings with mean stresses lead to a directional accumulation of inelastic strains, which are well known as cyclic ratcheting strains. For the evaluation of ratcheting strains, the total mean strain $\varepsilon_{m,t}$ and the total maximum strain $\varepsilon_{max,t}$, plotted as functions of the number of load cycles N, can be considered, respectively. In this context, the total maximum strain $\varepsilon_{max,t}$ is determined by $\varepsilon_{max,t} = \varepsilon_{m,t} + \varepsilon_{a,t}$ with the total strain amplitude $\varepsilon_{a,t}$. Within the current study, the ratcheting fatigue and cyclic deformation behaviour of brazed AISI 304L/BAu-4 joints were investigated based on $\varepsilon_{max,t}$, computed for various gauge lengths. Furthermore, the DIC results were compared with strain measurements of standard extensometers.

Within LIT, the transition from almost steady to significantly deviating material reactions at $\sigma_{max,e\ (LIT)}$ as well as the failure maximum stress $\sigma_{max,f\ (LIT)}$ have proven to be characteristic for the fatigue behaviour of various materials and joints [10, 11]. For metastable austenitic materials, a particular evaluation method of LIT was recommended in the literature [6]. It was shown that the high cycle fatigue strength corresponds well with the stress level, at which the cyclic hardening, due to the deformation-induced formation of martensite, leads for the first time to a temperature decrease [6]. Within the current study, the applicability of the evaluation method for brazed AISI 304L/BAu-4 joints in the as-received condition were examined based on load increase and constant amplitude tests. Further, LIT enable the assessment of the DIC technique for the complete stress range, leading to elastic as well as plastic deformations of the brazed joints. After testing, the fracture surfaces were analysed using scanning electron microscopy (SEM), with a secondary electron (SE) and a back-scattered electron (BSE) detector.

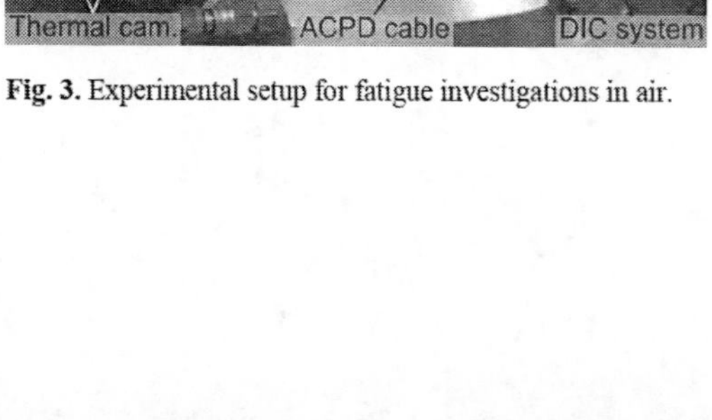

Fig. 3. Experimental setup for fatigue investigations in air.

3 Results and discussions

3.1 Tensile behaviour

Quasi-static tensile tests were performed with specimens of the AISI 304L base material (BM) and the brazed AISI 304L/BAu-4 joints (BJ) in the as-received (0w.) and 6 weeks pre-corroded (6w.) conditions to assess the DIC technique for homogeneous and joint geometries. According to the deformation state of a base material specimen, stresses below σ_{UTS} lead to homogeneous deformations within the rejuvenated area, Fig. 4. Close to failure at a nominal stress $\sigma_N = 400$ MPa, a local higher degree of deformation is detected in the necking area.

The DIC software is used to compute engineering tangential strains in tensile direction ε_{xx} for a 10 mm line element (LE10) in the centre of the specimen with the aim to evaluate the stress-strain curves [8]. As expected, the phenomenon of necking occurred within a major part of the measuring length. Thus, in comparison to standardised specimen geometries, an increase of strain values has to be considered. Comparable stress-strain curves are determined for the AISI 304L prior to and after 6 weeks of pre-corrosion, Fig. 5. Consequently, the long-time corrosive attack does not influence the quasi-static tensile behaviour of the base material, concerning strength as well as deformation properties. Two specimens were tested for both corrosion conditions and average values of the 0.2% yield strengths and ultimate tensile strengths with small standard deviations were determined with $\sigma_{YS} = 232 \pm 4$ MPa and $\sigma_{UTS} = 618 \pm 3$ MPa. [8]

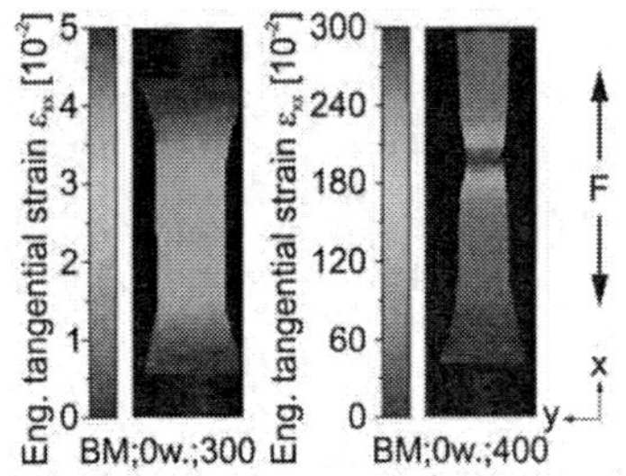

Fig. 4. DIC analysis for a tensile specimen of the base material prior to pre-corrosion at two different stress levels [8].

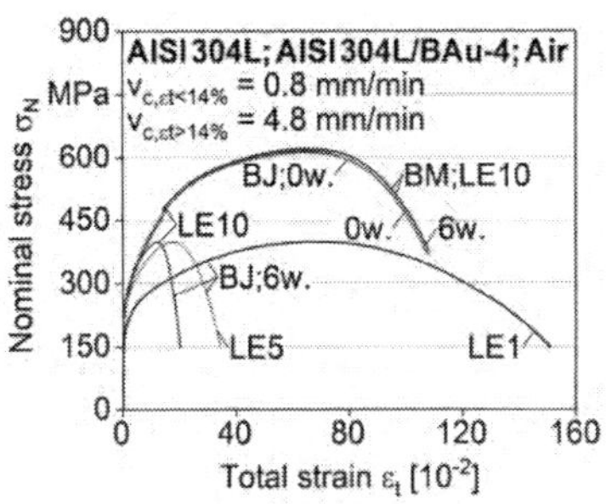

Fig. 5. Quasi-static stress-strain curves of the base material and the brazed joints prior to and after pre-corrosion.

For AISI 304L/BAu-4 brazed joints in the as-received and pre-corroded conditions, the DIC system allows to visualise local strain concentrations in the area of the brazing seam, Fig. 6. For the smooth surface of the as-received specimen, locally increased strain values can be barely identified for the nominal stress of 300 MPa, while a clear distinction between the base material and the brazing seam is possible after the σ_{UTS} is reached, Fig. 6. In addition to the qualitative evaluation, the engineering tangential strains ε_{xx} are computed with the DIC software for line elements with lengths of 0.5, 1, 3 and 5 mm (LE0.5, LE1, LE3, LE5), positioned at the brazing seam with the aim to evaluate the influence of the measurement length on the stress-strain curve. The maximum gauge length within these tests is limited to 5 mm due to the mounting springs of the extensometer, which was removed at $\varepsilon_t = 8\%$.

In contrast to the plastic deformation range, there is no effect of the gauge length on the elastic modulus, Fig. 7. Comparing the 0.5 and 5 mm line elements, the difference in the total strain $\Delta\varepsilon_t$ constantly increases with increasing stresses from e.g. $\Delta\varepsilon_t = 0.2\%$ at 300 MPa up to 5.0% at σ_{UTS}, Fig. 7. The total strains close to failure show the greatest difference of $\Delta\varepsilon_t = 11.8\%$. Consequently, the DIC system is well suited to evaluate local strain concentrations in the area of the AISI 304L/BAu-4 brazing seam for specimens with a smooth surface. The stress-strain curve, evaluated with the DIC system, is validated up to $\varepsilon_t = 8\%$ using an extensometer with a 10 mm gauge length (GL10), Fig. 7. Thus, mechanically determined strain values are well located within the range of the strain values, computed with the DIC system.

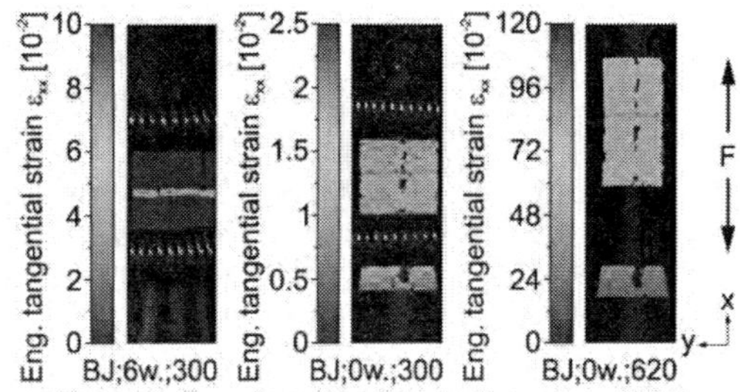

Fig. 6. DIC analysis for brazed tensile specimens prior to and after pre-corrosion.

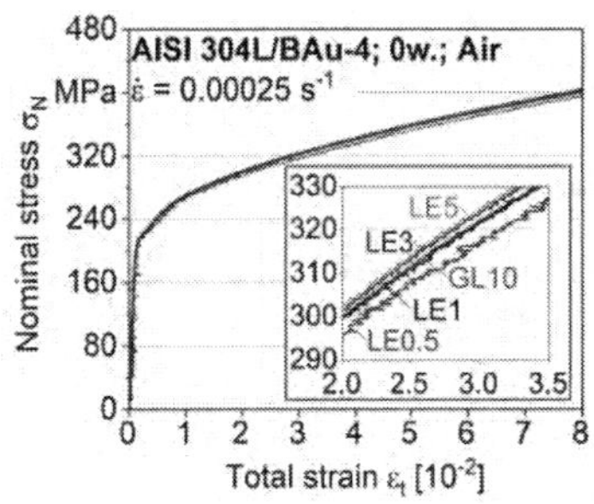

Fig. 7. Quasi-static stress-strain curves of the brazed joints prior to pre-corrosion.

Additional tensile tests of the brazed joints prior to and after pre-corrosion are performed without the extensometer in order to evaluate the total strains of a 10 mm line element (LE10) for direct comparison with the results of the base material, Fig. 5. Comparable stress-stain curves are determined for the base material and the brazed joints in the as-received condition and a significant degradation of the yield and ultimate tensile strengths due to the brazing process do not have to be considered. As the neck phenomenon is less distinct for the brazed specimens, the maximum total strain close to failure is reduced from 108% of the base material to 79%. For the brazed AISI 304L/BAu-4 joints, the VDA pre-corrosion for 6 weeks leads to a reduction of the σ_{UTS} down to 65% (σ_{UTS} = 399 MPa) and to a significant decrease of the total strain close to failure down to ε_t = 20%. Here, the applied nominal tensile stress σ_N is calculated based on the gross cross-sections.

The effect of the pre-corrosion on the deformation state of a brazed tensile specimen is presented for the nominal stress of 300 MPa, Fig. 5. The local strain increase at the brazing seam is more pronounced due to the corrosive attack. Thus, expected higher local degrees of deformation at notches under tensile stresses, especially due to splitting effects, are confirmed. For line elements with a length of 1 and 5 mm (LE1, LE5), the difference in the total strain of $\Delta\varepsilon_t$ = 49% at σ_{UTS} and of $\Delta\varepsilon_t$ = 117% close to failure were computed, Fig. 5. Due to the significant opening of the corrosion-dependent notch, an evaluation of the 0.5 mm line element was not possible. Consequently, a considerable influence of the gauge length on the strain values was determined with the DIC system for the pre-corroded brazed joints.

3.2 Fatigue behaviour

The material reactions within LIT of the brazed AISI 304L/BAu-4 specimens in the as-received and pre-corroded conditions were recorded by strain, electrical, magnetic, and temperature measuring techniques in order to evaluate the deformation and damage behaviour. The progression of the controlled maximum stress σ_{max} as well as the total strain amplitude $\varepsilon_{a,t}$, the total maximum strain $\varepsilon_{max,t}$, the loss energy density w, the change in AC voltage ΔU and the change in temperature ΔT are plotted as functions of the load cycles N for the as-received brazed joint, Fig. 8. The presented strain measurements were conducted with an extensometer with a gauge length of 12.5 mm (GL12.5).

Up to a stress level of approx. σ_{max} = 290 MPa at N = 24·10^4 cycles, values of $\varepsilon_{a,t}$ and $\varepsilon_{max,t}$ remain constant within the stress levels and increase linearly with increasing stress. Afterwards, $\varepsilon_{a,t}$ still shows a stepwise increase, while a more continuous growth is determined for the data of the total maximum strain $\varepsilon_{max,t}$. For stresses above 290 MPa and up to failure of the specimen at 47.4·10^4 cycles ($N_{f(LIT)}$), both curves show a pronounced progressive increase in combination with a degressive trend within the constant stress levels. At the failure maximum stress $\sigma_{max,f(LIT)}$ = 520 MPa, a considerable total maximum strain $\varepsilon_{max,t}$ of approx. 30% is reached. The determined directional accumulation of plastic strain is a result of an asymmetric cyclic loading with mean stresses and is well known as ratcheting fatigue. The ratcheting strains in form of the total maximum strain $\varepsilon_{max,t}$ and the total mean strain $\varepsilon_{m,t}$, respectively, are known to be more pronounced for elevated mean stresses and stress amplitudes [11]. The degressive trend of the cyclic deformation curves within the constant stress levels corresponds well with results in the literature on the AISI 304L austenite [12]. Within the first load cycles, the total maximum strain $\varepsilon_{max,t}$ significantly increases and is expected to be a result of the rearrangement of the dislocation structure, until it reaches a more stable configuration [12]. Subsequently, slightly increasing ratcheting rates are observed within the main part of the stress levels.

The measured values concerning the change in AC voltage ΔU remain at a constant level until 24·10^4 cycles and finally increase progressively until failure. Degressive curve progressions within the constantly controlled stress levels, as well as the trend throughout the entire fatigue life, correspond well with the total maximum strain $\varepsilon_{max,t}$. Thus, the electrical measuring technique is well suitable to characterise the cyclic deformation behaviour of brazed AISI 304L/BAu-4 joints to indicate damage-induced microstructural changes as well as to announce fatigue fractures at 50% of fatigue life.

The energy dissipation during the cyclic loading of the brazed joints at f = 10 Hz leads to a self-heating of the specimen with a maximum temperature in the gauge length T_1 of 41.6 °C and corresponds with a maximum change in temperature ΔT of 7.2 K at failure. A local temperature increase in the area of the brazed joint could not be determined with the high resolution thermal imaging camera, which allows to measure smallest temperature differences of 20 mK.

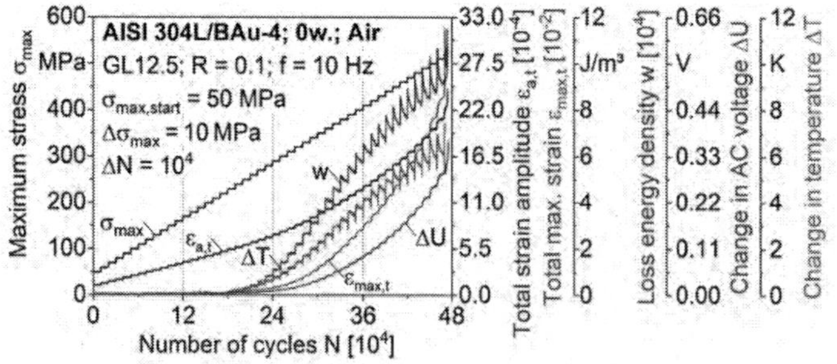

Fig. 8. Load increase test of a brazed specimen in the as-received condition.

The temperature of the specimen increases linearly until $20 \cdot 10^4$ cycles, followed by a progressively increasing trend, with respect to the average values of each stress level until failure. In contrast, a regressively decreasing temperature has to be considered for higher stresses, first detected at $\sigma_{max} = 380$ MPa. The decrease of the change in temperature is more pronounced with increasing stresses. In detail, the change in temperature $\Delta T = 7.0$ K, measured at the beginning of the stress level $\sigma_{max} = 510$ MPa at $N = 46.1 \cdot 10^4$ cycles, decreases to $\Delta T = 6.1$ K after 10^4 cycles, Fig. 9.

In order to evaluate the decreasing temperatures as an indicator for the energy dissipation during fatigue loading, the cyclic deformation behaviour of the brazed joints is evaluated by stress-strain hysteresis loops. The areas of the hysteresis loops are calculated with $w = \oint \sigma \, d\varepsilon$ [4] and the progression of the loss energy density w is plotted as a function of the load cycles N. In good agreement with the change in temperature ΔT, the loss energy density increases stepwise and linearly up to $20 \cdot 10^4$ cycles and, subsequently, progressively increases until failure, when considering the values at the beginning of the stress levels. For higher stresses, the loss energy density regressively decreases and therefore, indicates a cyclic hardening effect. The first decrease is detected at $\sigma_{max} = 350$ MPa, which is three stress levels below the first detected temperature decrease. In accordance with the temperature, the decrease is more pronounced with increasing stresses with a maximum change of $1.5 \cdot 10^4$ J/m^3 within the last completely tested stress level of $\sigma_{max} = 510$ MPa. The direct correlation of the temperature and the cyclic deformation behaviour was already explained with the fact that 95% of the loss energy density dissipates in heat [4].

The cycles hardening of metastable austenitic stainless steels, such as AISI 304L, was extensively investigated in the literature and is known to be mainly a result of a deformation-induced martensite formation [4-7, 12]. The enhanced strength of the martensitic phase leads to a significant change of the cyclic deformation behaviour and increases the fatigue lifetime [7]. The austenite stability depends on the chemical composition of the austenite, on the temperature as well as on the degree of plastic deformation [12]. Here, a maximum temperature change of $\Delta T \leq 20$K is expected to have no significant influence on the martensite formation [4].

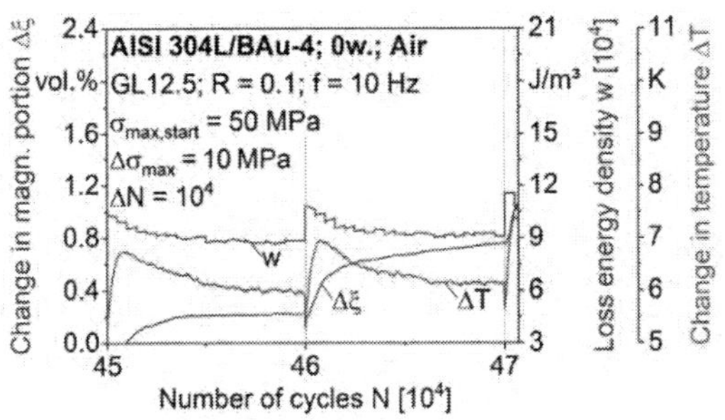

Fig. 9. Excerpt of the load increase test of a brazed specimen in the as-received condition.

Within LIT, a feritscope sensor was positioned at the base material close to the brazing seam to detect magnetic contents close to the surface, which indicate a deformation-induced formation of ferromagnetic martensite. Exemplarily, the change in magnetic portion $\Delta\zeta$ is presented for two stress levels of $\sigma_{max} = 500$ and 510 MPa, close to the failure of the specimen, Fig. 9. Therefore, a degressive trend of the change in magnetic portion $\Delta\zeta$ is detected for both stress levels with a maximum change of $\Delta\zeta = 0.8$ vol.%. The formation of martensite is expected to be the reason for a cyclic hardening effect, leading to a regressive trend of the loss energy density, which can be detected due to a regressive change in temperature. Thus, a combination of the electrical and temperature measurements is most suitable to characterise the cyclic deformation behaviour of brazed joints with ratcheting and cyclic hardening effects.

In the literature, a particular evaluation method of LIT was recommended for the characterisation of the fatigue behaviour of metastable austenitic materials [6]. It was shown that the high cycle fatigue strength corresponds well with the stress level, at which the deformation-induced formation of martensite and the temperature decrease due to the cyclic hardening occurs for the first time [6]. For brazed AISI 304L/BAu-4 joints of the current study, a temperature decrease in combination with a decreasing loss energy density w was first detected within the stress level of $\sigma_{max} = 380$ MPa. For validation, constant amplitude tests with a limited number of cycles of $N = 2 \cdot 10^6$ were performed, Fig. 10. The S-N curve is described in a double logarithmic scale acc. to Basquin with $\sigma_{max} = S_D \cdot (N_D/N_f)^{1/k}$ with the gradient k [8]. The fatigue strength at $2 \cdot 10^6$ cycles of 397 MPa and the gradient $k = 12.4$ are determined. Thus, the LIT combined with strain, temperature and electrical measurements are most suitable to estimate the fatigue strength of brazed AISI 304L/BAu-4 joints with a deviation of 4%.

The triggered DIC system was applied within LIT of the brazed specimens to evaluate the total maximum strain $\varepsilon_{max,t}$ as a function of the load cycles N. The engineering tangential strains in tensile direction ε_{xx} are computed for line elements with lengths in the range of 0.5 to 12.5 mm (LE0.5 to LE12.5) to evaluate the influence of the measuring length on the cyclic deformation and ratcheting fatigue behaviour, Fig. 11.

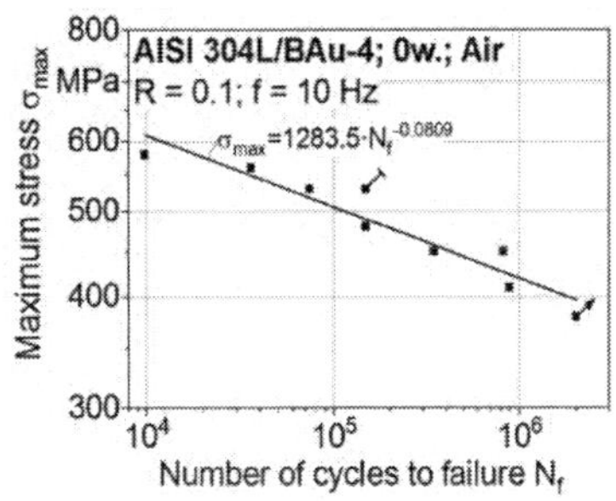

Fig. 10. Results of constant amplitude tests of brazed specimens in the as-received condition.

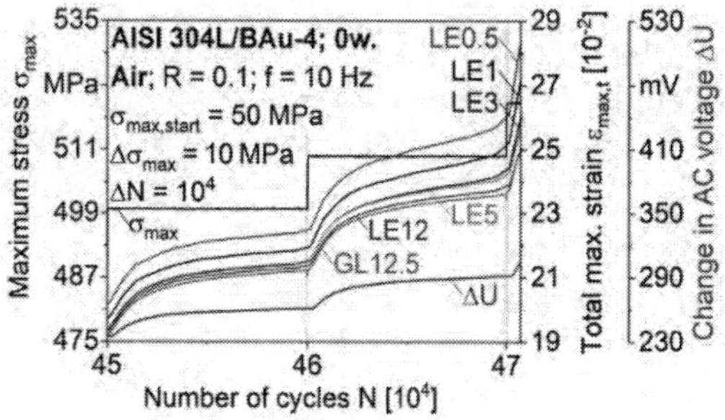

Fig. 11. Excerpt of the load increase test with a DIC analysis of a brazed specimen in the as-received condition.

When showing the total maximum strain $\varepsilon_{max,t}$ for the entire LIT for an as-received brazed specimen, a distinction between the values measured by the extensometer and computed by the DIC system is not possible. Therefore, an excerpt of the fatigue test results in the two stress levels before failure is illustrated, Fig. 11. Strain values, computed by the DIC system for a 12 mm line element (LE12), are in very good agreement with the results of the mechanical extensometer with a comparable gauge length (GL12.5). The comparison of the 0.5 and 5 mm line elements (LE0.5, LE5) for the two stress levels close to failure shows a maximum difference in strain values of $\Delta\varepsilon_{max,t} \approx 3\%$, Fig. 11. For lower stresses, the difference in strain values is less pronounced and deviations $\Delta\varepsilon_{max,t} < 0.5\%$ have to be considered for $\sigma_{max} < 450$ MPa.

The pre-corrosion leads to a significant influence on the fatigue behaviour with a reduction of the failure maximum stress down to $\sigma_{max,f(LIT)} = 220$ MPa (42%), Fig. 12. The forces are related to the gross cross-section of the brazed joints. For $\sigma_{max} = 180$ MPa, a transition from an almost steady to an increasing slope can be determined for the total maximum strain $\varepsilon_{max,t}$, computed for the various line elements. In accordance with the tensile test results, the influence of the gauge length on the strain values is more pronounced for notch-containing conditions than for the as-received specimens, when considering comparable strain levels. Here, a decrease of the DIC line element length from 5 to 0.5 mm leads to a difference of $\Delta\varepsilon_{max,t} \approx 3\%$ at $17 \cdot 10^4$ cycles, close to failure. Values of the 12.5 mm line element (LE12.5) are in good accordance with the results of the extensometer with the same gauge length (GL12.5).

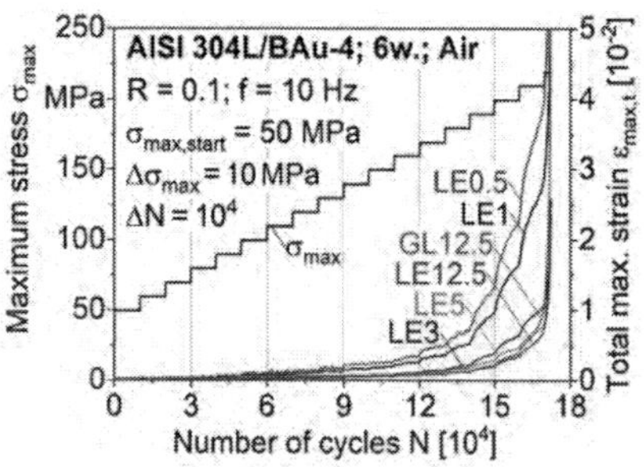

Fig. 12. LIT of a brazed specimen after pre-corrosion.

3.3 Fractographic investigations

Fracture surfaces of the brazed specimens were investigated using SEM in order to evaluate the fatigue and corrosion fatigue damage mechanisms. By using the BSE detector, the gold-based filler metal appears light grey in contrast to the dark grey stainless steel. On the top-view of a specimen in the as-received condition, the fatigue fracture area at the top of the figure can be distinguished from the forced fracture area, which shows pronounced deformation characteristics, Fig. 12a. The fatigue crack initiated at the surface of the specimen in the area of a geometrical imperfection of the brazing seam and exposes a free solidified surface of the filler metal at the fracture surface, Fig. 12b. After initiation, the crack propagated close to the interface between the base material and the filler metal, detected by BSE as spotted light and dark grey areas, and in the base material, respectively, Fig. 12d. In contrast, the residual forced fracture occurred in the centre of the brazing seam, Fig. 12c.

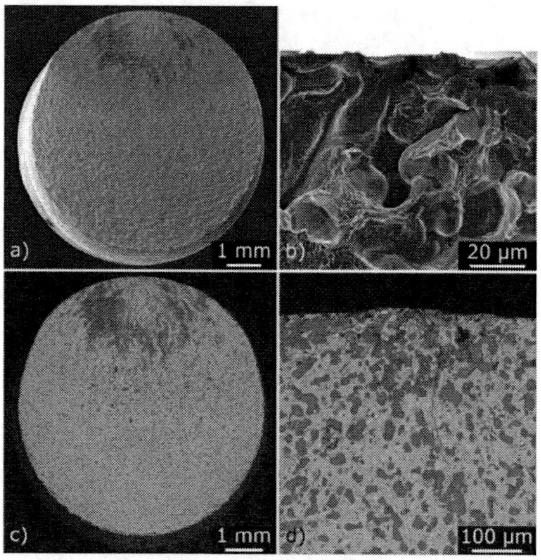

Fig. 13. Fracture surface of a brazed fatigue specimen in the as-received condition: Overview (a, c) and area of crack initiation (b, d); SEM with SE (a, b) and BSE (c, d).

Specimens, cyclically tested after 6 weeks of pre-corrosion, show a clearly visible corrosive attack, that progressed from the specimen's surface to the inner volume in the area of the brazing seam and significantly reduced the unaffected cross-section area, Fig. 14a. Within the circumferentially corroded region of the fracture surface, the filler metal can be distinguished from the two interfaces where the corrosive attack occurs locally. As the crack propagated alternatingly through one of the interfaces, the corroded base material below, as well as the corrosion products of the upper interface, can be identified, Fig. 14c.

Concerning the unaffected cross-section area, fracture characteristics comparable to the as-received specimens can be determined. Within the fatigue fracture area, which is positioned on the left side of Fig. 14b, the fatigue crack propagated preferentially through the base material, while the residual forced fracture occured in the centre of the brazing seam. In contrast to the crack initiation at one local imperfection close to the surface of the as-received specimens, a crack initiation zone within the circumferentially corroded interface leads to a minimum of six dark grey fatigue fracture areas that can be differentiated on the left side of the BSE image from the adjacent light grey residual forced fracture area in the centre of the brazing seam. A distinct transition from the forced fracture to the circumferentially corroded interface can be seen on the right side of the top-view.

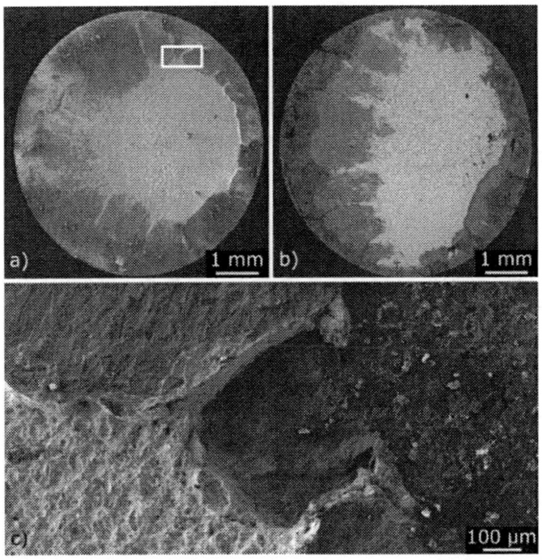

Fig. 14. Fracture surface of a pre-corroded brazed specimen after fatigue testing; SEM with SE (a, c) and BSE (b).

4 Conclusions

Pre-corrosions acc. to VDA 230-214 lead to a significant corrosive attack of brazed AISI 304L/BAu-4 joints, localised at the interface between the base material and the brazing seam. Compared to the as-received condition, the pre-corrosion leads to a reduction of the ultimate tensile strength σ_{UTS} and the failure maximum stress $\sigma_{max,f\,(LIT)}$ down to 65% and to 42%. In contrast, the pre-corrosion shows no influence on the tensile behaviour of the AISI 304L base material. In future studies, stress concentrations at the corrosion-dependent grooves of the brazed joints have to be evaluated.

LIT combined with strain, electrical, magnetic, and temperature measuring techniques have proven to be a most suitable approach to characterise the deformation and damage behaviour of brazed stainless steel joints. An evaluation method [6], developed for metastable austenitic materials, was successfully applied for the brazed specimens in order to estimate the fatigue strength with a deviation of 4%. Further, the LIT enables a time-efficient assessment of the DIC technique for brazed joints. An optimised triggered image acquisition was applied to investigate local ratcheting fatigue effects at the brazing seam and the gauge length influence.

It was shown that the DIC technique is well appropriate to characterise the local deformation behaviour of brazed stainless steel joints within tensile and fatigue tests. Visualised local strain concentrations in the area of the brazing seam are more pronounced with increasing mechanical stresses and are significantly enhanced due to notch-containing surfaces. In general, strain values, computed by the DIC system, are in good agreement with the results of mechanical extensometers.

For brazed AISI 304L/BAu-4 joints in the as-received condition, small differences in the strain values, computed for the gauge lengths of 0.5 and 5 mm with the DIC system, have to be considered. In tensile tests, the total strains at σ_{UTS} differ with $\Delta\varepsilon_t \approx 5\%$ and the total maximum strains within LIT close to failure differ with $\Delta\varepsilon_{max,t} \approx 3\%$. There is no effect of the gauge length on the elastic modulus. For tensile and fatigue tests of pre-corroded brazed joints, a considerable influence of the gauge length on the strain values was determined due to splitting effects at the notches. Although the current study allows a fundamental comprehension of local strain concentrations based on 0.5 mm gauge lengths, local deformations within the 50 µm brazing seam have to be evaluated in the future. Furthermore, the applicability of the DIC technique has to be evaluated for brazed joints with less pronounced deformation behaviours, such as AISI 304L/BNi-2 joints.

References

1. M. Koster, C. Kenel, C. Leinenbach, *ICF13* (2013).
2. C. Leinenbach, M. Koster, H.J. Schindler, J Mater Eng Perform **21**, 739 (2012).
3. D. Chen, S. Sun, J.M. Dulieu-Barton, Q. Li, W. Wang, Int J Fatigue **110**, 172 (2018).
4. M. Smaga, F. Walther, D. Eifler, Materials Science and Engineering: A **483**, 394 (2008).
5. T. Nebel, D. Eifler, Sadhana **28**, 187 (2003).
6. C. Müller-Bollenhagen, *Dissertation*, University Siegen (2010).
7. I. Nikitin, M. Besel. Int. J. Fat. **30**, 2044 (2008).
8. A. Schmiedt, M. Manka, T. Tillmann, F. Walther, Materialwiss Werkst, 10.1002/mawe201800007 (2018).
9. VDA Test Sheet 230-214, *Verband der Automobilindustrie* (2010).
10. F. Walther, Mater. Test. **56**, 519 (2014).
11. A. Schmiedt, M. Manka, T. Tillmann, F. Walther, Weld World, 10.1007/s40194-018-0557-y (2018).
12. F. Hahnenberger, M. Smaga, D. Eifler, Int J Fatigue **69**, 36 (2014).

Acknowledgements

The financial support of the German Research Foundation (DFG) is gratefully acknowledged (WA 1672/13, TI 343/96). The authors thank the Institute of Mechanics (IM) at TU Dortmund University for the provision of high resolution thermal camera.

Experimental and Numerical Study on Plastic Zone Variation ahead of Fatigue Crack Tip

Zhang Wei[1,*], *Zhou* Daoqing[1], and *Cai* Liang[1]

[1]School of Reliability and Systems Engineering, Beihang University, Beijing, 100191, China

Abstract. The plastic deformation ahead of crack tip is of great significance to analysis of the fatigue crack growth behaviour. Using the in-situ microscopy experiment technique, the variation of strain field in the vicinity of crack tip is investigated within load cycles at the small time scale. The contours of plastic zones are measured through the in-situ observation and digital image correlation (DIC). Finite element method (FEM) is also used to simulate the plasticity ahead of the crack tip. Furthermore, the numerical studies are extended to the single overload case to analyse the effect of large plastic zone on the subsequent crack growth. The evolution of residual stress is extracted by FEM simulation to explore the influence of plastic deformation before, during and after the single overload applied on the following crack propagation. Based on the FEM analysis, a model is proposed to approximate the size of the overload effect zone. Finally, some experimental data and numerical simulations are employed to validate this model.

1 INTRODUCTION

The plastic strain ahead of the crack tip has a significant effect on the fatigue crack propagation. The nonlinear loading interaction under the variable amplitude loading is usually considered to be determined by this local plastic deformation. Particularly, the retardation phenomenon caused by the overload is one typical and simple example. Furthermore, the constant amplitude loading with spike loads or overloads are very common in the engineering practice. Therefore, the studies on plastic zone ahead of fatigue crack tip under the overload condition has a great importance [1].

Irwin [2] make an assumption of small scale yielding ahead of crack tip, which makes it possible to analyse the metal material fatigue by using the linear elastic fracture mechanics theory. But this modification of small scale yielding is not used to explain the nonlinear load interaction at that time. Then, many studies are conducted to investigated the plasticity in the vicinity of crack tip [3][4], since it is considered as the key mechanism of the crack growth behaviour [5][6][7]. Dugdale and Barenblatt et al. also put forward a plastic zone estimation formula [8][9]. Irwin proposed a plastic zone size formulation, which could only calculate the plastic zone under the monotonic load [2]. Bannatine modified the Irwin method, which could estimate the plastic zone under cyclic loading. [10]. Suzuki determines the size of plastic zone by optical microscopy [11]. Paul simulates the size and shape of the monotonic and cyclic plastic zones by the numerical model [1].

In engineering practice, many structures are used under the variable amplitude loading and the variable loading induces the retardation which has influence on crack propagation. Therefore, many scholars studied the influence of retardation on crack growth. Sander et al. apply the overload testing on aluminum alloy compact tension specimens (CT) to obtain the relationship between the overload ratio and retardation effect [12]. Silva studies the fatigue crack growth retardation by overloads at negative stress ratios [13]. Kalnaus et al. investigate the fatigue crack propagation in AL6XN stainless steel to observe the obvious retardation phenomenon after the overload [14].

Plastic zone ahead of crack tip is usually used to analysis the nonlinear crack propagation under the overload condition. Willenborg model and wheeler model correlate the effective stress intensity factor range with the plastic zone size and predict the crack propagation after the overloading successfully [15][16]. Yuen and Taheri put forward a modified wheeler model, which can describe the retardation effect on crack propagation under the multiple overloads [17]. Additionally, the plasticity-induced closure is also a good alternative to deal with the crack growth under the overload. Zhang and Liu build up a relationship between the crack closure with the forward and reversed plastic zones [18]. Furthermore, Zhang and Jiang apply that closure model into the fatigue crack propagation prediction under variable amplitude loading [19][20].

However, there are some problems with the models and methods above. One is that the plastic zone ' s model proposed by Irwin, this model is too simple to consider the influence from the actual sharp. And this models calculate plastic zone size under the monotonic loading condition only and ignore the effect of the previous residual stress. Secondly, the scope of plasticity influence is assumed the distance from the origin to the

* Corresponding author: author@e-mail.org

position where the new plastic zone reaches the boundary of the big plastic zone induced by the overload, which is much smaller than experimental observation [21][22].

In this paper, the in-situ experiment and FEM simulation are employed to investigate the variation and influence of the plastic zone in the vicinity of fatigue crack tip under the constant amplitude loading with an overload. The residual stress extracted by simulation is used to study the effect of the plasticity induced by the overload on the subsequent crack growth.

2 Experiment and Simulation

2.1 Materials and specimen design

The material used in this study is aluminium alloy 2024-T3, which is a kind of hard aluminium with high strength and superior anti-fatigue property. The Young's modulus is 7.2GPa; the Poisson's ratio is 0.33; and the yield stress is 328MPa.

The specimen is a thin plate with a single-edge crack, whose width W=12mm, length L=76mm, thickness T=1mm, and the initial crack a=1.45mm, as shown by Fig 1.

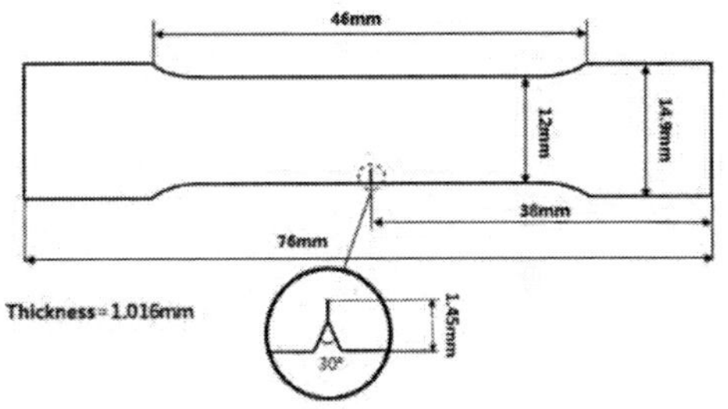

Fig. 1. The schematic illustration of specimen.

2.2 Fatigue testing and DIC procedure

The experimental system consists of a fatigue testing stage, an optical measuring microscope and a digital imaging processing system, as shown in Fig 2. Firstly, spray the matte paint onto the polished surface of the specimen with a very fine spray gun, and the speckles are required moderate size and evenly distributed around the crack tip, as shown in. Fig 3. After that, apply the cyclic loading with stress ratio R = 0 and maximum load $S_{max} = 1.5KN$ onto the specimen by the fatigue testing machine. The optical measuring microscope is used to trace the local strain field ahead of the crack tip. During the measurement process, the in-situ images at different loading levels are taken and collected at a very high speed.

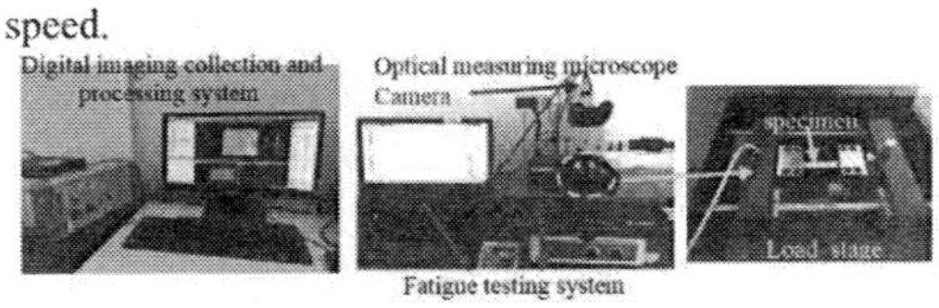

Fig. 2. The schematic illustration of experimental setup.

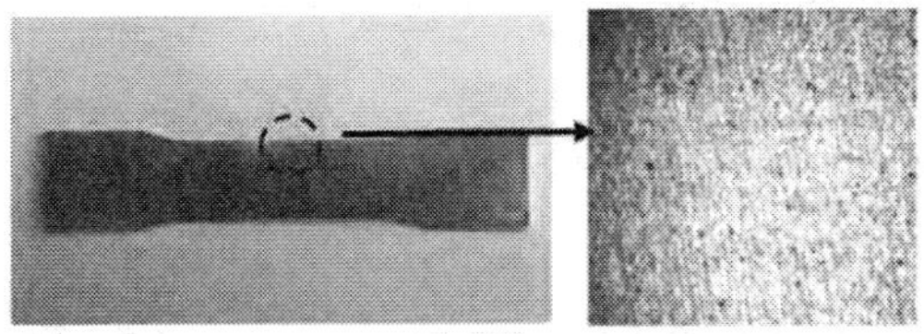

Fig. 3. The processed surface of the specimen.

The images are obtained, the variation of the strain ahead of crack tip can be analyzed through DIC (digital image correlation). The image feature points are recognized by DIC system, and the the strain field can be calculated based on the displacement field, as shown in Fig 4. Given the yield strain, the daimeter of the plastic zone can be measured. Repeat the above process at each loading level, variation of the plastic zone size can be traced.

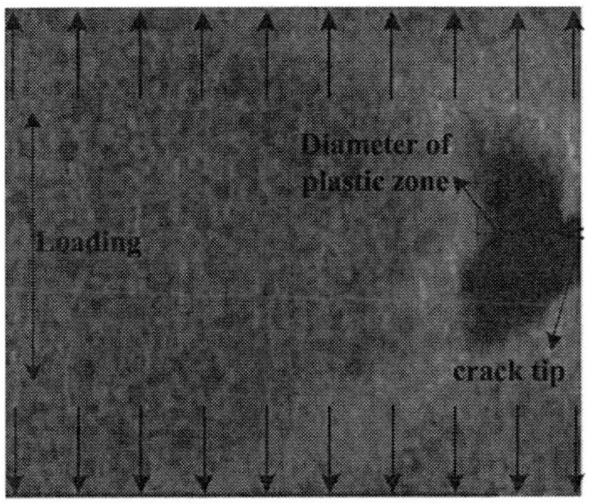

Fig.4. The relative strain field observed in the experiment

2.3 FEM modelling

Since the specimen is a very thin plate, it belongs to the plane stress problem. A 2D FEM model is built to simulate the variation of strain distribution in the vicinity of the crack tip. The specimen configuration, material properties, and loading/constraint conditions are kept equivalent to the real specimen in our experiment.

The constraints of the specimens are shown in Fig 5. Apply the constant loading with stress ratio R = 0 and maximum load $S_{max} = 1.5KN$. The variation of the stress-strain field and the diameter of plastic zone ahead of the crack tip are investigated. And the residual stress at the peak loads is computed to analyze the retardation effect region.

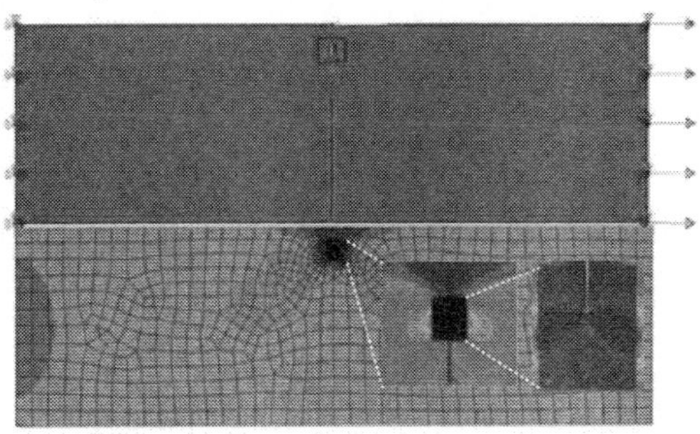

Fig. 5. The specimen's loading and restraint

https://doi.org/10.1051/matecconf/201816506004

2.4. Comparison between experiment and FEM model

The stress-strain fields ahead of the crack tip are shown in Fig 6. The upper-plots are the experiment measurements, while the down-plots are the simulation results. The diameters of plastic zone are shown in Fig 7. The simulation matches the testing data well, which can verify the effective of this simulatin approach.

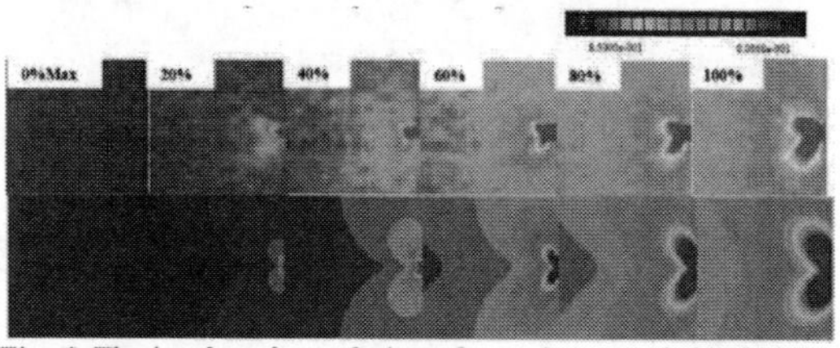

Fig. 6. The local strain evolution of experiment and simulation

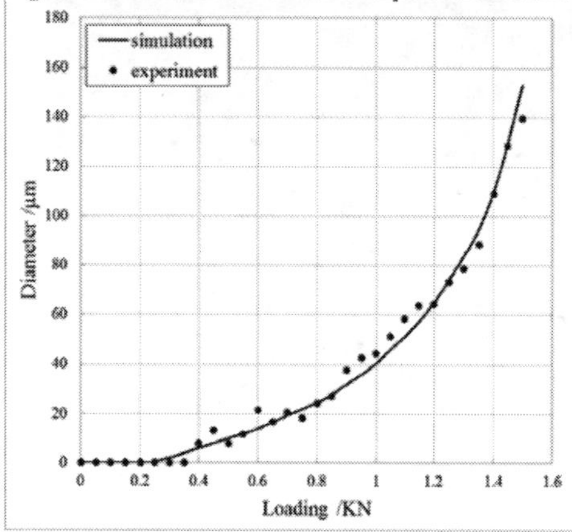

Fig. 7. The diameter of relative plastic zone.

3 Modelling and validation

3.1. Residual stress analysis

Through examining the influence of plasticity induced by the single overload on the subsequent crack growth, some experimental observations show that there are obvious differences between the effect region and its classical solution [23]. The classical models proposed by Irwin consider the shape of plastic zone is a circle, but the actual shape of plastic zone is observed as a butterfly shape by the expriment [24]. Therefore, the residual stress is used to investigate the effect region of plsticity.

The residual stress is obtained by the aforementioned simulation approach, which is used to analyze the variation of the plastic zone before, during and after a single overload. The schematic illustration of residual stress is shown in Fig 8.

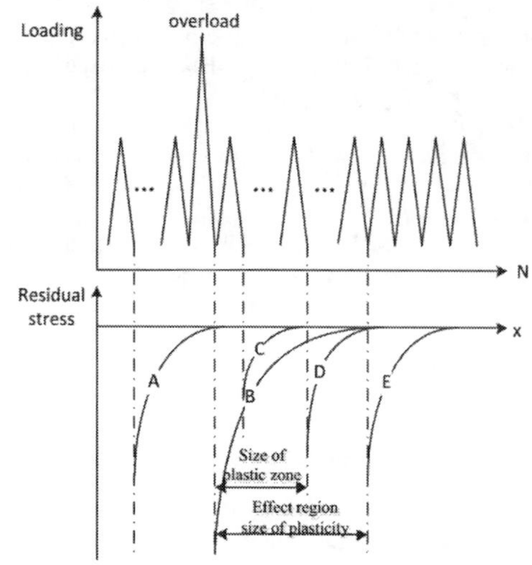

Fig. 8. The schematic illustration of residual stress.

Curve A represents the residual stress under the constant loading as the baseline. Once the overload is applied, residual stress changes greatly (the curve B). Afterwards, the residual stress is still affected by the overload (the curve C). Curve D is the residual stress at the boundary of the plastic zone calculated by the Irwin's model. The residual stress does not return back to the baseline. It indicates that this position is still in the effect region by the overload. Until the curve coincides with the curve E, the loading sequence effect vanishes entirely. The distance from the curve B to the curve D is the plastic zone size calculated by Irwin's model. The distance from curve B to the curve E is the effect region size of the overload.

Noroozi et.al point out that the distance from the origin to the first intersection point of residual stress curve and the X axis is the size of plastic zone [23]. The distance is defined as the criterion to judge if the residual stress has returned back to the baseline. The residual stress of every cycle is computed and compared with the baseline. By this way, the effect region of large plastic zone can be found by the comparison of the residual stresses.

Take the load condition with R=0 and r=1.2 as an example, the procedure is shown in Fig 9.

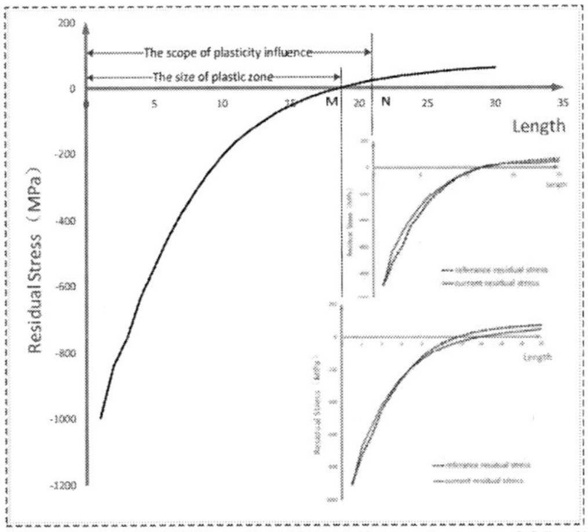

Fig. 9. Schematic illustration of the effect region (r=1.2&R=0)

Fig 9 shows the distribution of the residual stress along the crack direction. The point M is the boundary of plastic zone calculated by Irwin's model, where the residual stress has not returned back to the baseline. The residual stress at point N basically reverts to the original level. So the point N is considered as the boundary, at which the overload has no influence on the subsequent crack. The distance from the origin to the point M is the size of plastic zone. And the distance from the origin to the point N is the effect region size.

3.2. Modelling

For normalization, the ratio between the effect region size of the plastic zone and its classical solution is defined as φ, which is a function of overload ratio r and the stress ratio R. This equation is developed as following.

First of all, the overload ratio remains the same, the stress ratio is set as 0, 0.3, 0.5 and 0.7. The ratios between the effect region and its classical solution are calculated and fitted by the equation f(R). The fitting diagram under the loading of the stress ratios 0, 0.3, 0.5 and 0.7 is depicted below in Fig.10-13.

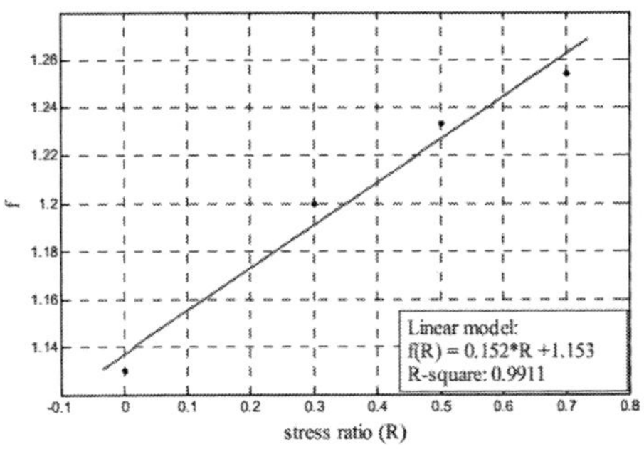

Fig. 10. The fitting diagram (R=0&r=1.2)

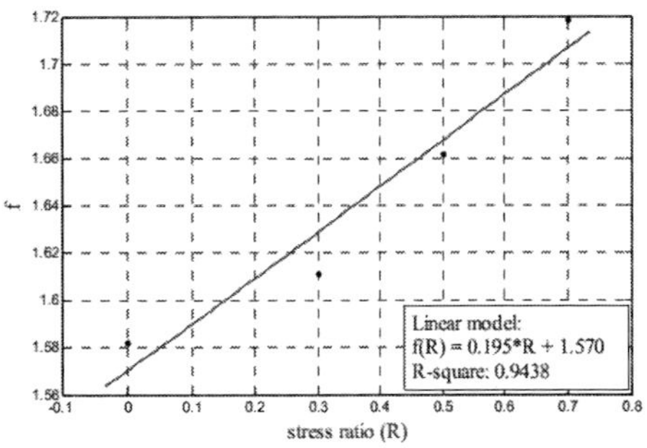

Fig. 11. The fitting diagram (R=0&r=1.5)

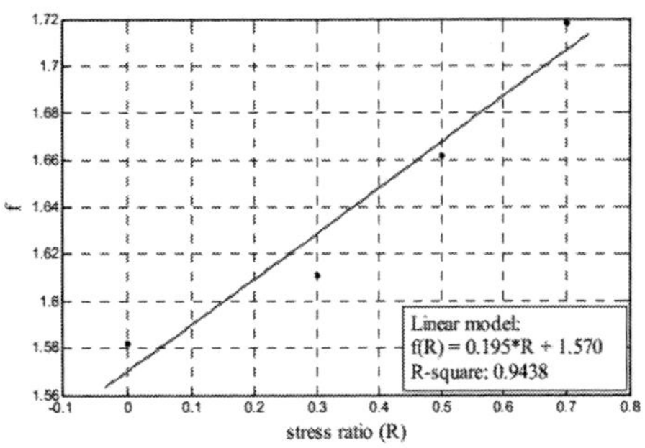

Fig. 12. The fitting diagram (R=0&r=1.8)

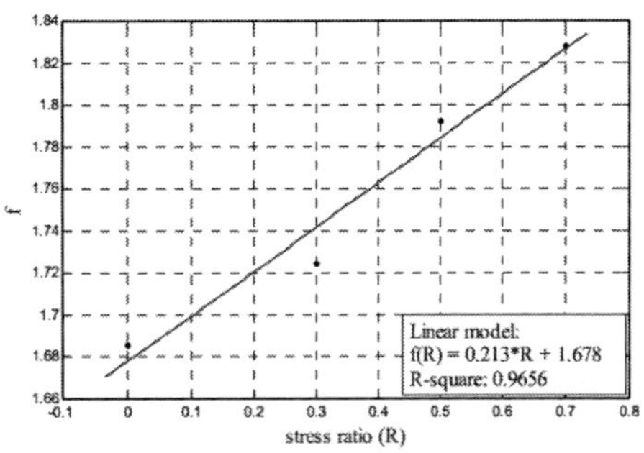

Fig. 13. The fitting diagram (R=0&r=2.0)

From the Fig 10-13, the relationship between them is linear. The fitting equation can be expressed as f(R)=A*R+B, where the values of A and B are related to the stress ratio and overload ratio.

Thus, A and B can be determined by equations of overload ratio. The value of A and B under different overload ratios are expressed as the fitting equation p(r) and q(r). And the fitting diagrams are demonstrated in Fig.14 and Fig.15.

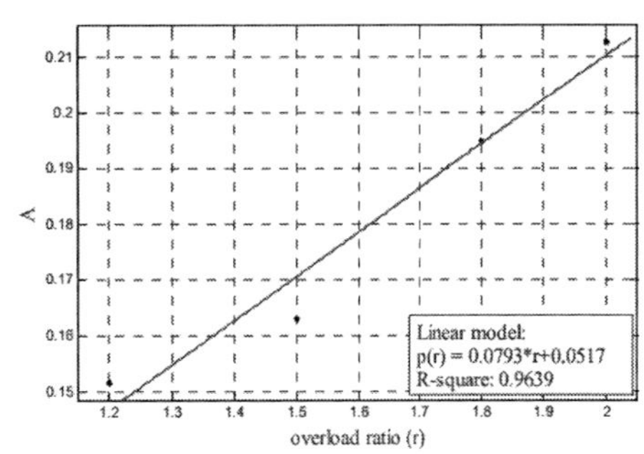

Fig. 14. The fitting diagram of parameter A

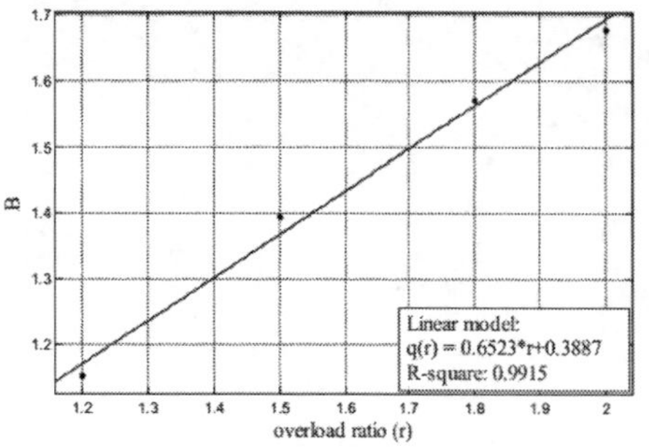

Fig. 15. The fitting diagram of parameter B

Substitute this result into the equation $\varphi(R,r)$, the equation can be rewritten as:

$$\varphi(R,r) = p(r) * R + q(r)$$
$$= (0.0793 * r + 0.0517) * R$$
$$+ (0.6523 * r + 0.3887)$$

This model can calculate the effect region size under tension-tension loading condition (the range of overload ratio is $1 \leqslant r \leqslant 2.3$) [25]. The value of $\varphi(R,r)$ is indirect proportion to the stress ratio and the overload ratio respectively. When the overload ratio is 1, this formulation describes the plastic effects under constant amplitude loading, which is apparently larger than the classical solution. It further manifests that the plastic zone shape has a noteworthy influence on the plasticity effect area. In addition, the overload ratio is a more sensitive factor of $\varphi(R,r)$, comparing with the stress ratio. Therefore, the plastic effect area under different overload ratio is investigated below.

3.3. Validation

Experiment

The experimental data of aluminum alloy D16 are used to validate the proposed model [21]. The initial center crack length is 20 mm. The mechanical properties of D16 in longitudinal (LT) orientation are as follows: σ_y =345 MPa, σ_{UTS} =483 MPa. The crack increment Δa under the overload effects is shown in Fig 16.

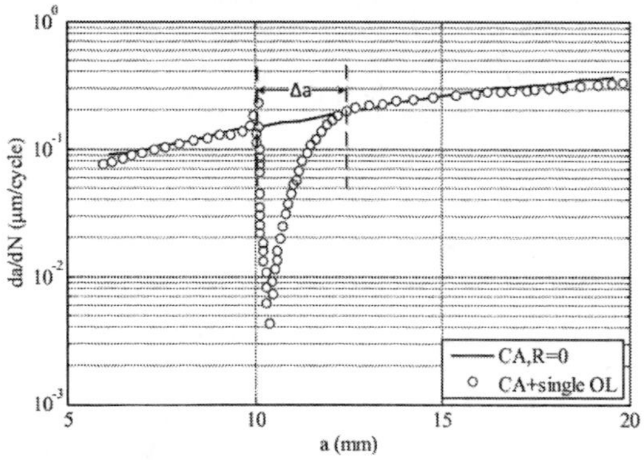

Fig. 16. The retardation after a single overload in D16

The crack length a is measured in Fig 15, and the details of the testing can be found in the paper [21]. The maximum stress intensity factor K_{max} is 22.89 MPa $* (m)^{\frac{1}{2}}$.

According to the Irwin's model [2], the plastic zone ahead of crack tip can be calculated:

$$L = (\frac{K_{max}}{\sigma_y})^2 * \frac{1}{\pi} = 14.86 * 10^{-4}\, m$$

The size of effect region is measured from the Fig 16:

$$d = 2.5 * 10^{-3} m$$

Then, the ratio between the effect region and its classical solution is obtained:

$$\rho = \frac{d}{L} = 1.6822$$

The ratio by the proposed model $\varphi(R,r)$ is:

$$\varphi(0,2) = 1.6933$$

It is noted that the proposed model can give reasonable predictions under the constant loading with overload.

Simulation

In addition, a simulation is conducted to further validate the aforementioned approach. The plastic effect region α is calculated by the proposed formulation of $\varphi(R,r)$. β is defined as the distance from the origin to the position where the residual stress return back to the baseline (the constant loading condition). The values of α and β are compared under different overload ratio cases.

This simulation example is considered as a single edge crack. Hence, the maximum stress intensity factor, K_{max}, is expressed as:

$$K_{max} = \sigma\sqrt{\pi a}[1.12 - 0.23 * \left(\frac{a}{b}\right) + 10.6 * \left(\frac{a}{b}\right)^2 - 21.7 * \left(\frac{a}{b}\right)^3 + 30.4 * \left(\frac{a}{b}\right)^4]$$

The size of plastic zone is calculated by Irwin's model [2], which can be written as:

$$L = (\frac{K_{max}}{\sigma_y})^2 * \frac{1}{\pi}$$

The diameter of effect region α is calculated as:

$$\alpha = \varphi(0.5,1.2) * L$$

Moreover, β is measured from the figures directly.

So the comparison between the distances α and β is obtained, and the relative error δ is expressed as:

$$\delta = \left|\frac{\alpha - \beta}{\alpha}\right|$$

The validations are shown in Fig 17 and Fig 18.

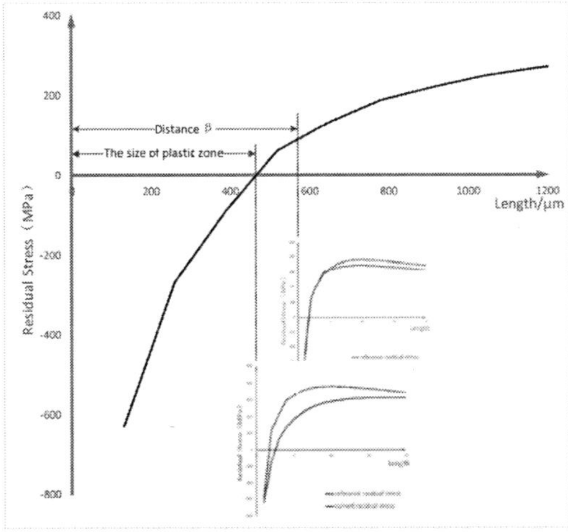

Fig. 17. Schematic illustration of effect region (r=1.2&R=0.5)

Under the loading of the R=0.5 and r=1.2, the parameters are calculated as following:

$$\alpha = \varphi(0.5,1.2) * L = 548.568\mu m$$

$$\beta = 580\mu m$$

$$\delta = \left|\frac{\alpha - \beta}{\alpha}\right| = 5.8\%$$

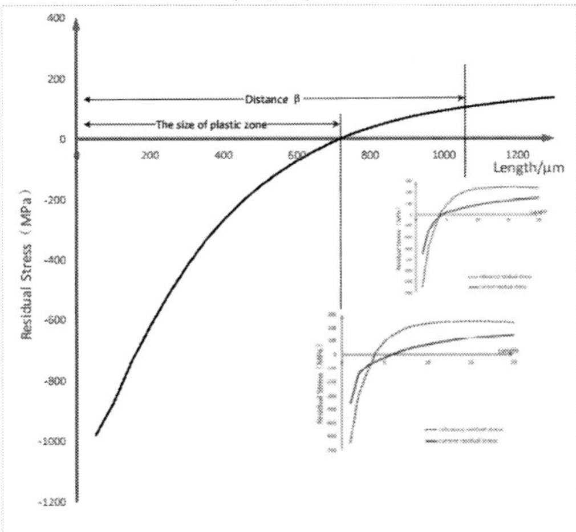

Fig. 18. Schematic illustration of effect region (r=1.5&R=0.3)

Under the loading condition of R=0.3 and r=1.5, the parameters are calculated as following:

$$\alpha = \varphi(0.3,1.5) * L = 976.57.57\mu m$$

$$\beta = 1040\mu m$$

$$\delta = \left|\frac{\alpha - \beta}{\alpha}\right| = 6.5\%$$

The process above proves that the proposed model matches the simulation well.

4 Conclusion

The in-situ optical microscopy testing is performed to measure the variation of strain distribution ahead of the crack tip under the constant amplitude loading. A FE model is established to simulate this process, and a very good agreement is received. Then, the FE model is employed to analyse the plasticity and residual stress caused by the overload around the crack tip. A new calculation model is proposed to estimate the retardation effect region under the constant amplitude loading with a single overload. Some conclusions can be drawn:

- The effect region size is larger than its classical solution.
- This model can calculate the size of retardation effect region under the constant loading and overload more accurately, which is validated by experiment and simulation.

In the future work, the experimental and numerical investigation will be extended to more complex loading case.

The research is financially supported by the National Natural Science Foundation of China (Grant No. 51405009).

References

1. S.K. Paul, E.F.M. **159**, 90-97(2016)
2. H. Tada, P.C. Pairs, G.R. Irwin, D.R.C., H.P. (1973)
3. W.Y. Maeng, M.H. Kim, J.N.M, **282**, 32-39(2000)
4. Y.Xiong, X.X. Hu, J. Katsuta, et al.,I.J.F, **30**, 67-73(2008)
5. D.L. Davidson, J. Lankford, T. Yokobori, et al., I.J.F, **12**, 579-585(1976)
6. G. Xin, H. Wang, X. Kang, et al., E.J.M.A.S., **29**, 739-745(2010)
7. R.A. Sousa, F.P. Figueiredo, F.F.E.M.S, (2017)
8. D.S. Dugdale, J.M.P.S, **8**, 100-104(1960)
9. G.I. Barenblatt, A.A.M, **7**, 55-129(1962)
10. J. Bannantine, P.H., **70**, 219-225(1990)
11. H. Suzuki, A.J. Mcevily, M.M.T.A, **10**, 475-481(1979)
12. M. Sander, H.A. Richard, I.J.F, **25**, 999-1005(2003)
13. F.S. Silva, I.J.F., **29**, 1757-1771(2007)
14. S. Kalnaus, F. Fan, A.K. Vasudervan, et al. E.F.M., **75**, 2002-2019(2008)
15. J. Willenborg, R.M. Engle, H.A. Wood, T.R.A.(1979)
16. O.E. Wheeler, J.B.E., **94**, 181-186(1972)
17. B.K.C. Yuen, F.Taheri, I.J.F., **28**, 1803-1819(2006)
18. W. Zhang, Y. Liu, I.J.F., **43**, 188-198(2012)
19. S. Jiang, W. Zhang, Z.L. Wang, J.V. **17**, 3650-3661(2015)
20. S. Jiang, W. Zhang, X. Li, et al., M.P.E, **4**, 1-9(2014)
21. J. Schijve, M. Skorupa, A. Skorupa, I.J.F., **26**, 1-15(2004)
22. T. Jian. A. Ping, et al., S.C.T.S., **55**, 673-683(2012)
23. A.H. Noroozi, G. Glinka, S. Lambert, E.F.M., 75, 188-206(2008)
24. M. Besel, E. Breitbarth, I.J.F.,**93**, 92-108(2016)
25. J.P. Gallagher, A.G.D.Y.Z.M, (1974)

Cyclic deformation behaviour and stability of grain-refined 301LN austenitic stainless structure

Jiří Man[1,*], *Antti* Järvenpää[2], *Matias* Jaskari[2], *Ivo* Kuběna[1], *Stanislava* Fintová[1], *Alice* Chlupová[1], *L. Pentti* Karjalainen[3], and *Jaroslav* Polák[1]

[1]Institute of Physics of Materials ASCR, Žižkova 22, 616 62 Brno, Czech Republic
[2]Kerttu Saalasti Institute, University of Oulu, Pajatie 5, 85500 Nivala, Finland
[3]Centre for Advanced Steels Research, University of Oulu, P.O. Box 4200, 90014 Oulu, Finland

Abstract. Low cycle fatigue (LCF) behaviour of metastable austenitic 301LN stainless steel with different grain sizes – coarse-grained (13 μm), fine-grained (1.4 μm) and ultrafine-grained (0.6 μm) – produced by reversion annealing after prior cold rolling was investigated. Fully symmetrical LCF tests with constant total strain amplitudes of 0.5% and 0.6% were performed at room temperature with a low constant strain rate of 2×10^{-3} s^{-1}. Microstructural changes in different positions within the gauge part of the specimens were examined by optical microscopy (polarized light) and electron backscatter diffraction (EBSD) technique; for quantitative assessment of the volume fraction of deformation induced martensite (DIM) a Feritscope FMP 30 was adopted. The cyclic stress-strain response and specific changes of hysteresis loop shapes in the very early stage of cycling are confronted with the character of DIM formation and its distribution in the whole volume of the material. A possible effect of strain rate (frequency of cycling) on the destabilization of austenitic structure during cyclic straining of materials with different grain sizes is highlighted.

1 Introduction

Grain refinement of metastable austenitic steels via so called reversion annealing after prior cold deformation attracts deserved interest of researches for many decades since it represents an effective tool for considerable improvement of tensile properties [1]. The choice of proper parameters of this thermomechanical treatment, namely the level of cold deformation and temperature and time of reversion annealing can lead up to ultrafine-grained (UFG) microstructure with an excellent combination of high strength and good ductility – see e.g. [2–12].

Whereas the behaviour of grain-refined austenitic steels during unidirectional tensile testing has been intensively studied (see above), to the behaviour of these materials during cyclic straining has been paid much less attention so far [10, 13–17]. Very limited number of papers [13,15–17] indicates positive effect of grain refinement on the fatigue life of austenitic steels cycled under low-cycle fatigue (LCF) conditions.

Järvenpää et al. [10] recently studied LCF behaviour and structure destabilization of 301LN steel reversion-annealed to different grain sizes ranging from 13 μm to 0.6 μm and cyclically strained with constant total strain amplitudes of 0.4% and 0.6%. They reported that the most stable structure during fatiguing with relatively high frequencies of 6 Hz and 1 Hz, was unexpectedly at both strain amplitudes the fine-grained structure with average grain size of 1.4 μm. The kinetics of deformation induced martensite (DIM) formation was found to deviate systematically from the other stainless steels [10].

The present work represents a continuation of our previous studies on the cyclic plasticity, stress-strain response and microstructural changes in grain refined 301LN steel cyclically strained under LCF conditions. To bring some new insight into the above unexpected behaviour strain controlled test with constant low strain rate of 2×10^{-3} s^{-1} was adopted. The cyclic stress-strain response and characteristic changes of hysteresis loop shape of 301LN steel with various grain sizes are confronted with microstructural changes documented in different locations of a sheet material both *in situ* during cycling and after interruption of cycling at the very early stage of fatigue life.

2 Experimental

A commercial austenitic 301LN stainless steel with the chemical composition (in wt.%) 0.025C, 0.53Si, 1.25Mn, 0.024P, 0.001S, 17.5Cr, 6.5Ni, 0.2Cu, 0.09Mo, 0.15N, and rest Fe was supplied by Outokumpu Stainless Oy (Tornio, Finland) in the form of 8 mm thick, hot rolled and annealed sheet. The sheet was subjected to the following thermo-mechanical treatment at the University of Oulu, Finland. Sheet material with initial fully austenitic structure was cold rolled (CR) with the

* Corresponding author: man@ipm.cz

Table 1. Conditions of RA, average grain size and tensile characteristics of various microstructural states of 301LN steel [10].

Microstructural state	Reversion annealing conditions		Average grain size	Tensile characteristics*		
	Temperature	Holding time		YS	UTS	ε_f
CGA1	1050 °C	200 s	13 μm	340 MPa	960 MPa	72%
FGA	900 °C	1 s	1.4 μm	550 MPa	1025 MPa	67%
UFGA	800 °C	1 s	0.6 μm	750 MPa	1100 MPa	57%

* YS = yield stress, UTS = ultimate tensile strength, ε_f = strain to fracture

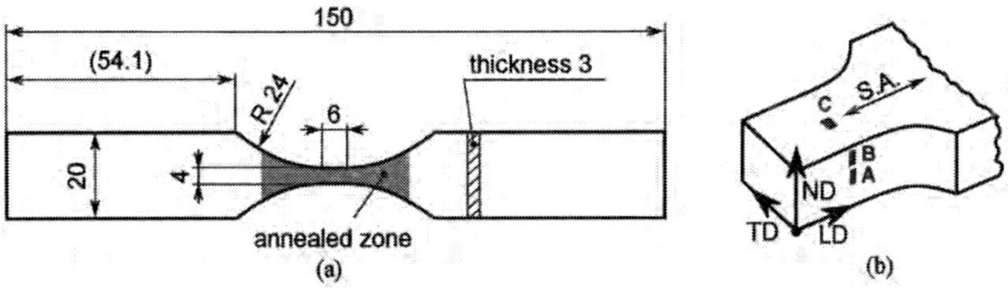

Fig. 1. Schema of specimen geometry (a) and specimen gauge part (b) with indicated positions of microstructural observations. All dimensions are in mm.

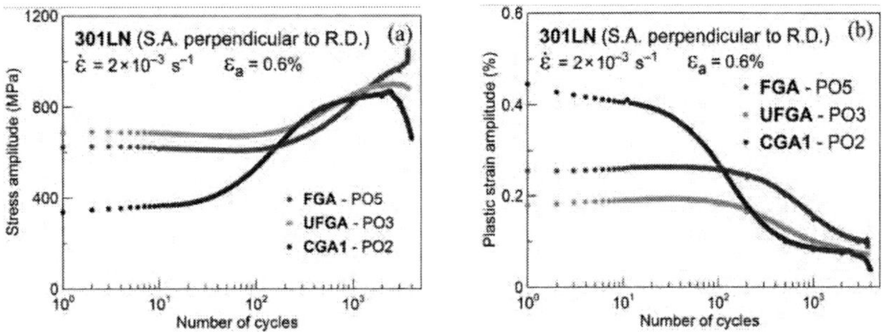

Fig. 2. Stress amplitude (a) and plastic strain amplitude (b) versus number of cycles in 301LN steel with three different grain sizes.

reduction of 63% to the thickness of 3 mm. Rectangular pieces 20×150 mm were cut from the CR sheet both perpendicular and parallel to the rolling direction. The pieces were reversion annealed with a high heating rate of 200 °C/s and cooling rate of 35 °C/s on a Gleeble 3800 thermo-mechanical simulator. The adoption of three different annealing conditions resulted in fully austenitic microstructures with different grain sizes, namely the coarse-grained (CGA1), fine-grained (FGA) and ultrafine-grained (UFGA) – see Table 1 where the basic tensile characteristics are also listed. More details on the thermomechanical treatment, characterization of microstructure including its behaviour during tensile tests can be found elsewhere [9, 10].

From the processed 3 mm thick sheets of individual materials the flat dog-bone shaped specimens having the gauge width and length of 4 mm and 6 mm, respectively, were machined by spark-erosion (see Fig. 1a). This geometry enabled to avoid the adoption of anti-buckling device used in our previous LCF tests [15, 16] which made possible to perform *in situ* Feritscope monitoring of α′-martensite formation on two perpendicular surfaces (positions A and C indicated in Fig. 1b). Before fatigue tests the central part of specimens was carefully polished mechanically and electrolytically.

LCF tests were conducted at room temperature in air in symmetrical push-pull cycle ($R\varepsilon = -1$) with constant total strain amplitudes of 0.5% and 0.6% to the different stages of fatigue life. Contrary to the recent study by Järvenpää el al. [10] the low strain rate of 2×10^{-3} s^{-1} was kept constant during all fatigue tests. The volume fraction of magnetic phase (α′-martensite) was assessed using a Feritscope Fischer MP30. The actual α′-martensite content was determined by multiplying the Feritscope reading by the correction factor of 1.7 [18]. During *in situ* measurements the measuring probe was attached to the specimen surface always after unloading from tension. After completion of fatigue tests the microstructural changes in different positions within the gauge part of specimens (see Fig. 1b) were examined by optical microscopy (polarized light and Beraha II colour etching) and EBSD (electron backscatter diffraction) technique in high-resolution SEM–FEG (LYRA 3 XMU from TESCAN) with a step size of 85 nm.

3 Results

3.1 Cyclic stress-strain response

During cycling with constant total strain amplitude of 0.6% the stress and plastic strain amplitude were evaluated from the recorded hysteresis loops and they are plotted vs. number of cycles for all grain sizes (CGA1, FGA and UFGA) in Fig. 2. The cyclic hardening/softening curves are similar to those observed

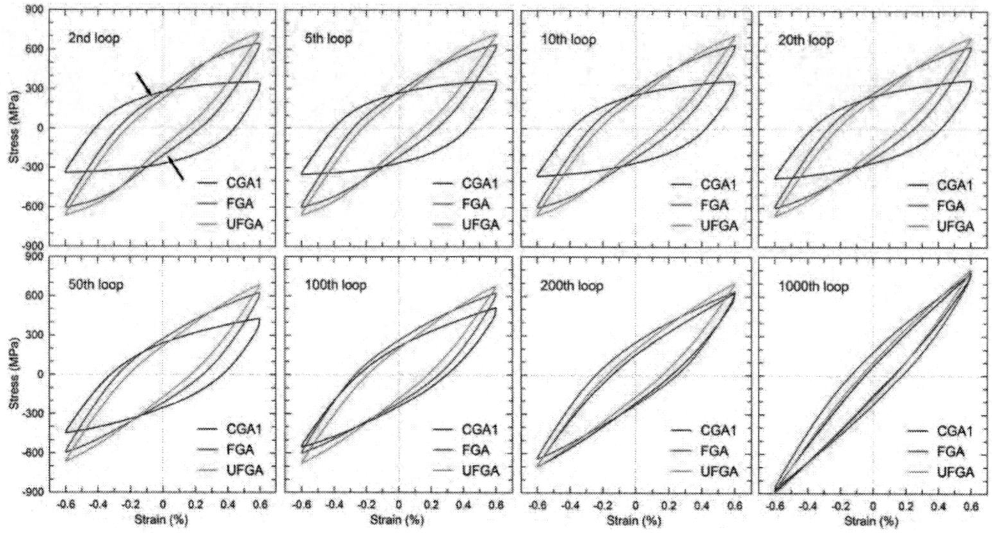

Fig. 3. Development of hysteresis loops in 301LN steel with different grain sizes during cyclic straining with ε_{at}=0.6%.

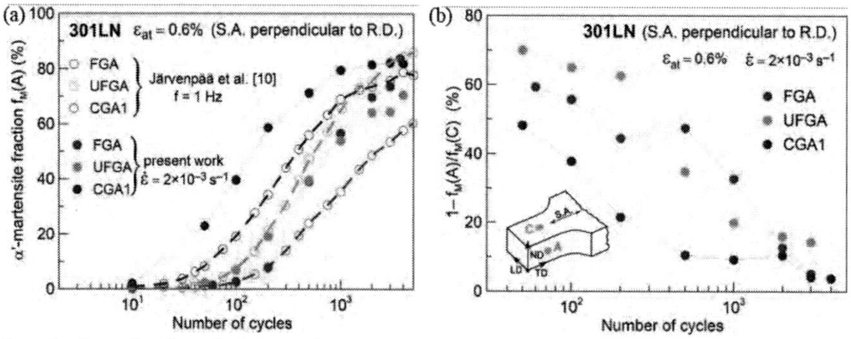

Fig. 4. Evolution of volume fraction of α′-martensite as assessed by Feritscope in position A (a) and relative difference in α′-martensite fraction measured in positions A and C (b) in the sheet of various microstructures of 301LN steel cyclically strained with total strain amplitude of 0.6%.

in our previous studies on the grain refined identical type 301LN austenitic stainless steel [10, 15, 16]. The stress amplitudes of FGA and UFGA steels at the start of cycling are considerably higher than that for coarse-grained counterpart (see Fig. 2a) which is in agreement with the differences in the yield stress (see Table 1).

The character of cyclic hardening/softening curves in 301LN steel of diverse grain sizes was different. Whereas only permanent hardening of different intensity during cycling was detected for CGA1 steel (cf. Figs. 2a and 2b)[1] FGA and UFGA steels showed more complicated behavior. Contrary to their coarse-grained counterpart they exhibit the initial stage of mild softening which is after approximately 100 cycles superseded by a strong cyclic hardening. As will be shown below the increase of stress amplitude in all microstructural states can be correlated with their structural stability of austenite.

During fatigue tests the stress-strain response within a cycle, i.e. hysteresis loops were recorded. While the coarse-grained material exhibits the usual smooth shape of hysteresis loops during cycling, some irregularities in the form of loop constriction are apparent at the early stages of cycling for both UFGA and FGA materials. They are highlighted by black arrows in Fig. 3 for the second loops. In agreement with our previous results [15] hysteresis loops are constricted in the region of low applied stress for both tensile and compression part of a loop. With increasing number of elapsed cycles the constrictions becomes less pronounced and after approximately 150 cycles it disappears completely and the shape of hysteresis loop becomes normal (see Fig. 3). Similar hysteresis loop changes at the early stage of fatigue life were documented recently for UFGA and FGA steels also for much higher frequencies adopted during cyclic straining [19].

[1] The sudden drop in stress amplitude clearly visible for CGA1 steel after 2 500 cycles resulted purely from the fatigue test layout due to Feritscope measurements on two perpendicular surfaces (see above) and did not reflect the true stress response of this steel. Since anti-buckling device was not adopted during fatigue tests the local bending of the specimens at the ends of specimen gauge length due to considerable hardening of the material during cycling became in later stages of fatigue life inevitable which resulted in virtual "cyclic softening".

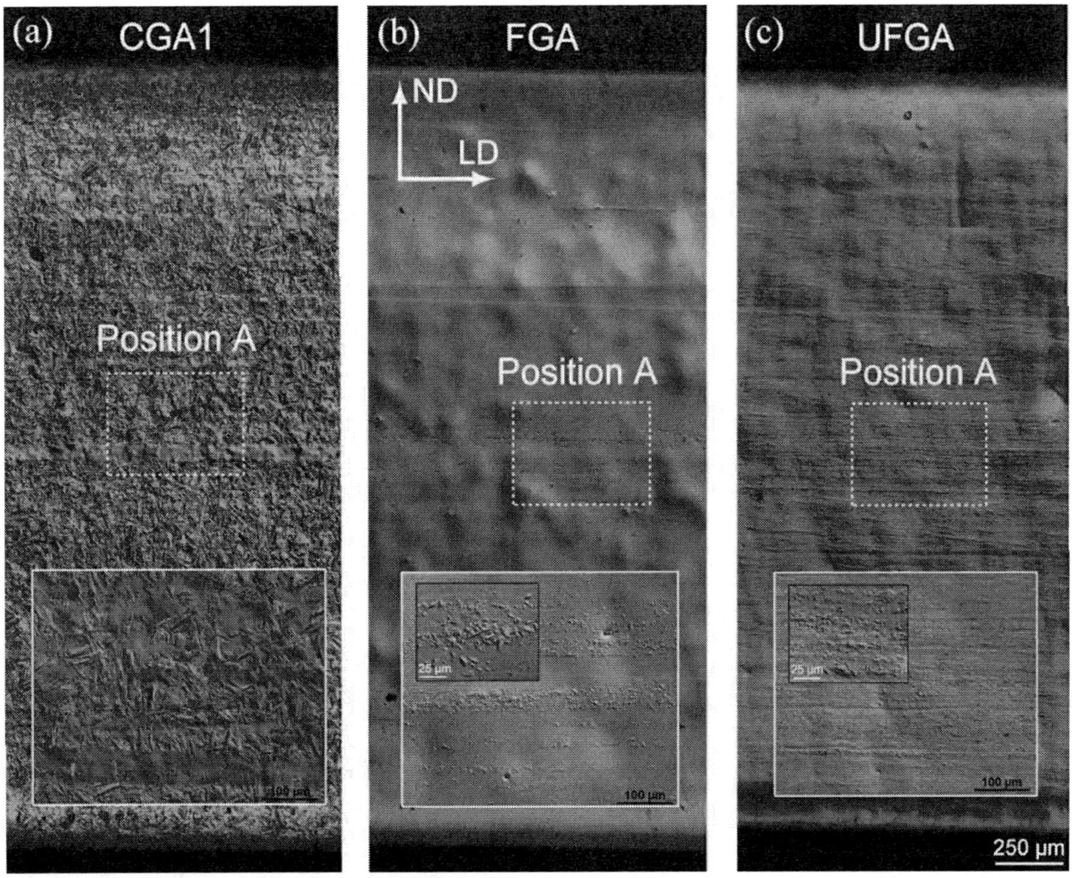

Fig. 5. Surface topography across the thickness of 301LN steel with different grain sizes fatigued with ε_{at} = 0.5% to very early stage of fatigue life (OM, polarized light). Stress axis is horizontal in all micrographs (c.f. Fig. 1b).

3.2 Deformation induced phase transformation

3.2.1 In situ monitoring of DIM formation during cyclic straining

The results of present *in situ* Feritscope monitoring of destabilization of initially fully austenitic structure, i.e. magnetic α'-martensite formation in the course of cycling of all three materials are shown in Fig. 4a together with results obtained recently for identical materials for cycling with the same total strain amplitude but with much higher frequency of 1 Hz [10]. Note that in both studies specimens were fabricated perpendicular to the rolling direction and the α'-martensite content was assessed by the Feritscope probe attached in the position A, i.e. in the middle of sheet thickness (see the inset in Fig. 4b).

As can be seen from Fig. 4a the kinetic curves of α'-martensite formation obtained in two laboratories (Oulu and Brno) exhibit similar, approximately sigmoidal shapes. In both laboratories the coarse-grained microstructure CGA1 was found to be the most metastable structure in the early stage of cycling. Irrespective of the small differences in the α'-martensite content in various microstructures during cycling which could be attributed, in addition to different applied strain rates, to larger cross-section area of the gauge part used by Järvepää et al. [10] a clear tendency to the similar α'-martensite content at the late stages of fatigue life is observed for all microstructures. The only exception represents the fine-grained structure cyclically strained with the frequency of 1 Hz which shows systematic deviation from the described behaviour and indicating thus higher stability of this microstructural state. However, as will be discussed later, for explanation of this discrepancy the actual temperature of the specimen gauge part and its increase during cycling should be also taken into account.

Another important result shows Fig. 4b in which relative difference in α'-martensite fraction measured in positions A and C, i.e. in the middle of the thickness and on the surface of a cyclically strained sheet respectively, is plotted vs. number of cycles. It is clear that in the early stage of cycling much higher content of α'-martensite in all three microstructural states is found in the middle of sheet thickness. This indicates the presence of some form of magnetic anisotropy across the sheet thickness. With continuing cycling, however, the relative

 https://doi.org/10.1051/matecconf/201816506005

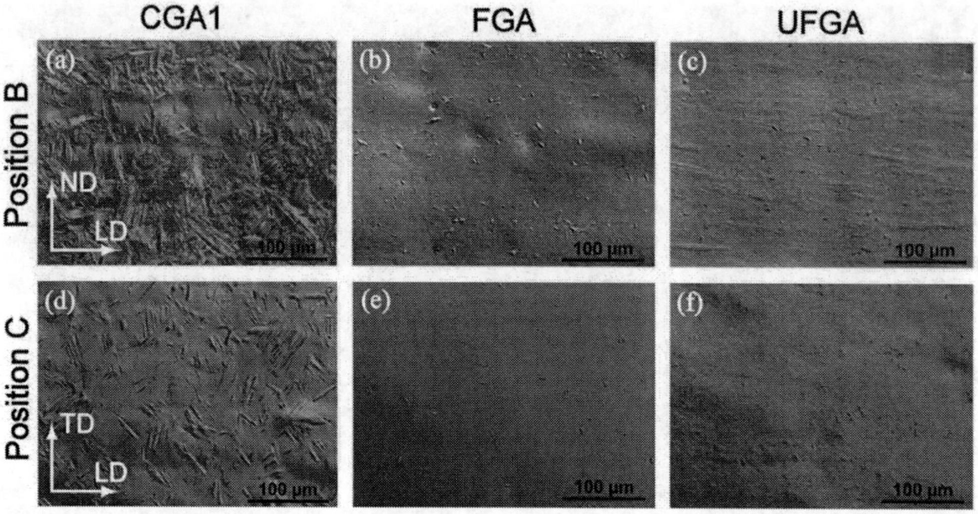

Fig. 6. Surface topography outside the central part of sheet thickness (position B) and on the sheet surface (position C) of 301LN steel with different grain sizes fatigued with ε_{at} = 0.5% to the very early stage of fatigue life (OM, polarized light). Stress axis is horizontal in all micrographs (c.f. Fig. 1b).

differences in α′-martensite content decrease gradually up to the value of 5% documented for FGA and CGA1 microstructures. Only UFGA microstructure still shows a higher difference of 15% at later stage of cycling.

3.2.2 Post mortem characterization of DIM in the very early stage of cyclic straining

To bring more insight into the specific features revealed for 301LN steel with various grain sizes in the early stage of cycling, namely the characteristic changes of hysteresis loop shapes and the presence of magnetic anisotropy across the sheet thickness, individual specimens were cyclically strained to the very early stage of fatigue life. Then the surface at different positions within the gauge part of specimens (both on the sheet surface and across the sheet thickness), carefully electrolytically polished prior fatigue tests, was documented by optical microscopy in polarized light and EBSD technique.

An example of the surface observations using polarized light – an extremely sensitive technique for visualization even of very fine slip features – for all microstructural states of 301LN steel is presented in Figs. 5 and 6. Fig. 5 shows the surface topography across the whole specimen thickness and the details of surface topography in the places denoted A, B and C, the position of which is schematically indicated in Fig. 1b, are shown in Figs. 5 and 6.

Fig. 5a witness that the density of slip markings generated in coarse-grained steel in the plane LD-ND after 150 cycles is relatively high. Even though it can locally fluctuate across the sheet thickness (see some featureless areas or even a band in the middle of sheet thickness in Figs. 5a and 6a), at macroscopic scale it remains approximately constant – cf. Figs. 5a and 6a. Contrary to this situation much lower density of slip

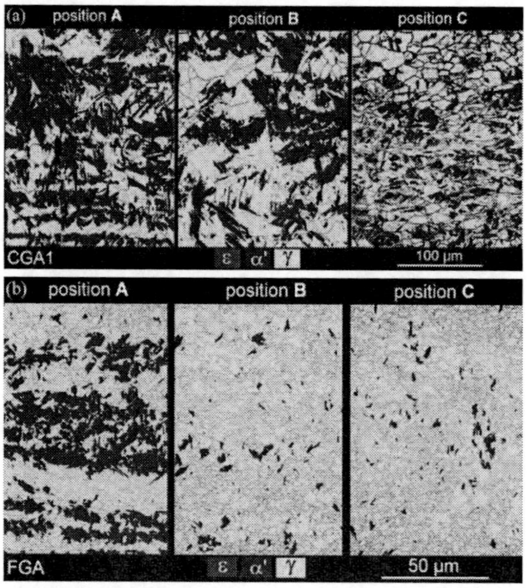

Fig. 7. Microstructure of coarse-grained (a) and fine-grained (b) 301LN steel fatigued with ε_{at} = 0.5% to the very early stage of fatigue life as revealed by EBSD at different positions indicated in Fig. 1b. Stress axis is horizontal in all micrographs.

markings has been revealed on the sheet surface, i.e. in the position C as apparent from Fig. 6d.

The surface observations of grain-refined structures yield considerably different picture on the distribution of slip markings within the volume of cyclically strained steels. The bands of high density of fine slip markings, much finer in comparison with those detected in CGA1 steel, separated by featureless areas were detected after 100 cycles in the central part of the sheet thickness (see details corresponding position A in the insets in Figs. 5b

Table 2. Results of quantitative assessment of microstructure of 301LN steel (FGA and CGA1 states) by EBSD and Feritscope at the very early stage of cyclic straining.

Microstructural state and fatigue conditions	**Position of EBSD or Feritscope measurement**	**Phase fraction (EBSD)** ε **(h.c.p.) [%]**	γ **(f.c.c.) [%]**	α' **(b.c.c.) [%]**	**Zero solution [%]**	**Vol. fraction of** α'**-martensite [%] (Feritscope)**
FGA (#PI 4)	A	0.47	60.25	29.31	9.97	3.37 ± 0.49
ε_{at} = 0.6%	B	0.38	90.48	3.40	5.74	not measured
N = 100 cycles	C	0.21	90.90	2.72	6.18	1.98 ± 0.29
CGA1 (#PI 1)	A	0.53	42.48	44.18	12.82	19.61 ± 0.36
ε_{at} = 0.5%	B	0.70	56.72	29.29	13.30	not measured
N = 150 cycles	C	0.97	60.39	12.32	26.33	12.01 ± 0.41

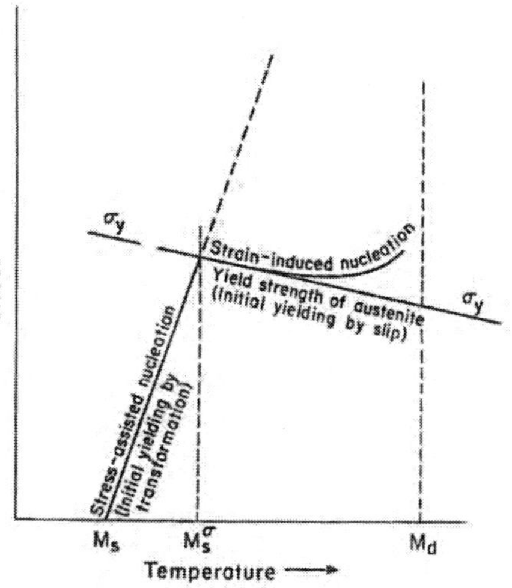

Fig. 8. Schematic representation of interrelationships between stress-assisted and strain-induced nucleation of α'-martensite in metastable austenitic alloys (after [23]).

and 5c). The thickness of individual bands can fluctuate along the gauge length. Even though some individual bands are discontinuous along the R.D., macroscopically the area of individual band occurrence is running along the whole gauge part of specimen; its thickness is around 500 μm for both FGA and UFGA structures. Outside the central part of sheet thickness (see microstructure in position B in Figs. 6b and 6c) the density of slip markings, approximately randomly distributed in the material, is considerably lower. The sheet surface is nearly featureless for both grain-refined structures (see Figs. 6e and 6f showing microstructure in the position C in FGA and UFGA steel, respectively).

The EBSD mapping with a step size of 85 nm in the above mentioned areas A, B and C was performed for all 301LN steel structures. Figs. 7a and 7b show the result obtained for CGA1 and FGA structures. Quantitative data on the EBSD phase analysis are together with the results of Feritscope measurements listed in Table 2.

Comparing EBSD phase maps with Figs. 5 and 6 a very good agreement can be seen between two different microscopic techniques. Nearly compact band-like areas of high density of α'-martensite are apparent in the central parts of sheet thickness of both CGA1 and FGA steel (see position A in Fig. 7a and 7b). Much lower density of α'-martensite distributed relatively randomly in the material is typical for the areas outside the central part of the sheet thickness (position B) and as well as for the sheet surface (position C) – see Fig. 7b. Situation in coarse-grained material seems to be more complex, cf. phase maps in Fig. 7a. Whereas at the position B the character of α'-martensite remains practically similar to that in the position A with lower α'-martensite density, numerous black dots (i.e. not resolved) in the phase map of sheet surface (place C) and corresponding presumably to α'-martensite indicate clearly a possible change in deformation mechanisms in the areas of higher stability of austenite.

Comparison of quantitative data on the α'-martensite content assessed at different positions of fatigued 301LN steels by the EBSD technique and Feritscope (see Table 2) shows that both methods reflect an inhomogeneous α'-martensite distribution across the sheet thickness in the very early stage of cycling. The local EBSD analysis indicates that the differences in α'-martensite fraction in the central parts of sheet thickness and the remaining parts of the sheet may be even much higher than those found by Feritscope (cf. quantitative data in the position A and positions B and C for both 301LN steel in Table 2).

4 Discussion

The experimental study on the destabilization of the initially fully austenitic structure during cyclic straining of 301LN steel grain-refined to three different grain sizes of 13 μm (CGA1), 1.4 μm (FGA) and 0.6 μm (UFGA) brought several very important results.

Both *in situ* Feritscope measurements during cyclic straining and *post mortem* EBSD analysis showed that the α'-martensite is not distributed homogeneously across the sheet thickness. The most metastable part in all grain-refined structures was always the central part of the sheet thickness. This inhomogeneity in the α'-martensite distribution resulting in a magnetic anisotropy is highest in the very early stage of fatigue tests and with continuing cycling gradually decreases (see Fig. 4b). This finding agrees with our previous results on cyclically strained grain-refined 301LN steels [10, 16, 20, 21] and it can be rationally reconciled with our recent studies on the stability of various product forms of wrought AISI 304-type [20, 22] and 316-type [20] austenitic stainless steels. Both studies demonstrated convincingly that that these Cr–Ni austenitic steels are never fully chemically homogeneous and furthermore due to the specific solidification behaviour and contemporary continuous casting practice the most metastable area of any wrought full-cross-section semi-product is always its central part. The local chemical analysis proved the presence of characteristic local variations in chemistry (and thus in the austenite stability) in the form of chemical banding and showed the validity of the above statements also in case of 301LN steel [16].

To understand the characteristic changes of hysteresis loop shape and the presence of cyclic softening at the early stages of cyclic straining of 301LN steel with FGA and UFGA structure the specific nature of α′-martensite formation has to be taken into account.

Olson and Cohen in their concept of deformation induced martensitic nucleation distinguish between two mechanisms (see Fig. 8) [23]), namely the stress-assisted nucleation occurring at temperatures between M_s and M_s^σ, and the strain-induced nucleation above M_s^σ. The character of α′-martensite formation during stress-assisted transformation is similar to that at usual spontaneous transformation below M_s at which the nucleation of α′-martensite plates at the specimen surface is accompanied by the characteristic surface topography [24, 25] and by a rapid local volume expansion of 2.57% [26]. These surface features are clearly apparent from the optical micrographs taken in polarized light for all microstructural states in Figs. 5 and 6.

In the early stage of cycling of FGA and UFGA structures the stress-assisted nucleation of α′-martensite is concentrated into thin bands the occurrence of which is restricted primarily only to the 500 μm thick central part of the sheet (see Figs. 5b and 5c) while the remaining parts of the sheet which are affected much less by the α′-martensite formation (see Figs. 6b,c and 6e,f). Due to the rapid volume expansion accompanying martensite formation the central part of fatigued sheet is thus in the state of internal compression stress and internal tensile stress is present in the remaining, much larger part of the material within the specimen gauge part. In the total strain controlled fatigue tests this situation will lead to the gradual development of small tensile mean stress and small increase of the hysteresis loop width, i.e. cyclic softening. As soon as the formation of stress-assisted α′-martensite bands ceases and a sufficient amount of α′-martensite will be generated also outside the most metastable central part of the sheet, the internal compression stress develops in the whole volume of the fatigued material and this will result in the start of intensive cyclic hardening. As is apparent from Fig. 2a and 2b the above described processes require in FGA and UFGA steels, cyclically strained with total strain amplitude of 0.6%, typically 100–150 cycles.

Contrary to the described behaviour, the situation in destabilization in coarse-grained CGA1 steel is much simpler. Due to the fact that the stress-assisted α′-martensite (presumably accompanied also by the strain-induced α′-martensite) is distributed more uniformly in

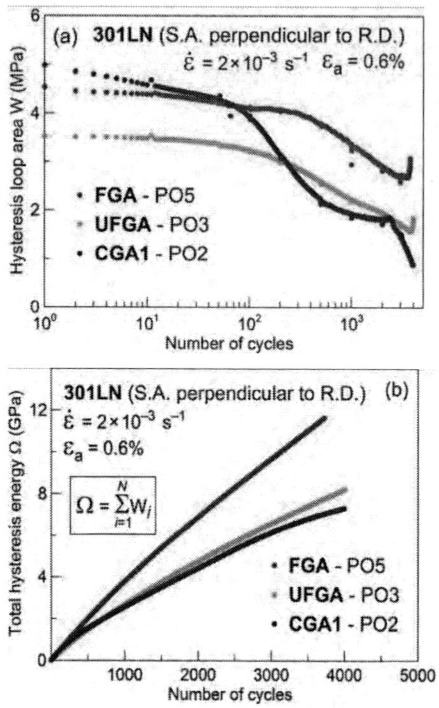

Fig. 9. Mean dissipated energy in a cycle (a) and total hysteresis energy (b) versus number of cycles in 301LN steel with different grain sizes.

the whole volume of material, the permanent internal compression stress will be present within the specimen gauge length. It results in the gradual development of the compression mean stress and permanent cyclic hardening as apparent from Fig. 2.

The presence of hysteresis loop constriction in the early stage of cycling of FGA and UFGA steels can be rationally explained by the nature of stress-assisted nucleation of α′-martensite in the form of thin bands. Due to this specific localized mode of α′-martensite formation accompanied by rapid volume expansion an extra-strain arises. In the total strain controlled test this event results in the drop of stress both in tensile and compression part of the hysteresis loop (at stress level well below macroscopic yield stress) and hysteresis loops thus become visually constricted as apparent from Fig. 3.

In order to explain much slower rate of α′-martensite formation in FGA structure fatigued with the frequency of 1 Hz contrary to the strain rate of 2×10^{-3} s^{-1} adopted in the present work the actual temperature of the specimen gauge part during cyclic straining should be taken into account. Since the related temperature measurements are still in progress only some energetic considerations will be discussed here.

Fig. 9 shows both the development of hysteresis loop area and total hysteresis energy for all microstructures cycled with low strain rate of 2×10^{-3} s^{-1} whose stress-strain response is shown in Fig. 2. The hysteresis loop area decreases gradually during the strain hardening in all structures, nevertheless as can be seen from Fig 2a the FGA structure due to higher cyclic plasticity (see also Fig. 3) has the highest hysteresis loop area during cycling. Since the fatigue process has a cumulative character much more important dependence is that shown in Fig. 9b. As can be seen from this dependence the total dissipated energy in FGA structure after several hundreds of cycles start to systematically deviate from those evaluated for CGA1 and UFGA structures. Since the great majority of hysteresis energy dissipates in heat [27] and taking into account a generally low thermal conductivity and especially low thermal diffusivity of austenitic stainless steels [28] it is feasible to expect a small but continuous increase of temperature due to adiabatic heating during cycling with the frequency of 1 Hz used by Järvenpää et al. [10]. Since the characteristic temperature M_{d30} is 23 °C for the present AISI 301LN steel [10], it is clear that even a small increment in temperature during fatigue tests can have an important impact on the stability of the structure and thus on the stress-strain response. However, as indicated recent results by Pegues et al. for AISI 304L austenitic steel with normal grain size [29], the effect of higher strain rates can have a beneficial effect on fatigue life in total strain controlled fatigue tests. Such behaviour should be, however, in the case grain-refined austenitic structures validated experimentally.

The support of the present work by the grant No. 13-32665S from GAČR and by the research project RVO 68081723 is gratefully acknowledged.

References

1. L.P. Karjalainen, T. Taulavuoiri, M. Sellman, A. Kyröläinen, Steel Res. Int. **79**, 404 (2008)
2. R.D.K. Misra, S. Nayak, P.K.C. Venkatasurya, V. Ramuni, M.C. Somani, L.P. Karjalainen, Metall. Mater. Trans. A **41**, 2162 (2010)
3. R.D.K. Misra, V.S.A. Challa, P.K.C. Venkatsurya, Y.F. Shen, M.C. Somani, L.P. Karjalainen, Acta Mater. **84**, 339 (2015)
4. B. Ravi Kumar, S. Sharma, B.P. Kashyap, N. Prabhu, Mater. Des. **68**, 63 (2015)
5. M. Shirdel, H. Mirzadeh, M.H. Parsa, Mater. Charact. **103**, 150 (2015)
6. P. Behjati, A. Kermanpur, L.P. Karjalainen, A. Järvenpää, M. Jaskari, H. Samaei Baghbadorani, A. Najafizadeh, A. Hamada, Mater. Sci. Eng. A **650** 119 (2016)
7. C. Celada-Casero, B.M. Huang, M.M. Aranda, J.-R. Yang, D. San Martin, Mater. Charact. **118**, 129 (2016)
8. D.M. Xu, G.Q. Li, X.L. Wan, R.L. Xiong, G. Xu, K.M. Wu, M.C. Somani, R.D.K. Misra, Mater. Sci. Eng. A **688**, 407 (2017)
9. A. Järvenpää, M. Jaskari, J. Man, L.P. Karjalainen, Mater. Charact. **127**, 12 (2017)
10. A. Järvenpää, M. Jaskari, J. Man, L.P. Karjalainen, Mater. Sci. Eng. A 703, 280 (2017)
11. A. Järvenpää, M. Jaskari, L.P. Karjalainen, Metals **8**, 109 (2018)
12. C. Lei, X. Li, X. Deng, Z. Wang, G. Wang, Mater. Sci. Eng. A **709**, 72 (2018)
13. J.-B. Vogt, A. Poulon, S. Brochet, J.-C. Glez, J.-D. Mithieux, in *Sixth Int. Conf. on Low Cycle Fatigue (LCF6)*, P.D. Portella, T. Beck and M. Okazaki, eds., pp. 97–102 (DVM e.V., Berlin, 2008)
14. A. Poulon, S. Brochet, J.-C. Glez, J.-D. Mithieux, J.-B. Vogt, Adv. Eng. Mater. **12**, 1041 (2010)
15. A. Chlupová, J. Man, I. Kuběna, J. Polák, L.P. Karjalainen, Proc. Eng. **74**, 147 (2014)
16. J. Man, A. Chlupová, I. Kuběna, T. Kruml, O. Man, A. Järvenpää, L.P. Karjalainen, J. Polák, in *Eighth Int. Conf. on Low Cycle Fatigue (LCF8)*, T. Beck and E. Charkaluk, eds., pp. 205–210 (DVM e.V., Berlin, 2017)
17. M. Droste, C. Ullrich, M. Motylenko, M. Fleischer, A. Weidner, J. Freudenberger, D. Rafaja, H. Biermann, Int. J Fatigue **106**, 143 (2018)
18. J. Talonen, P. Aspegren, H. Hänninen, Mater. Sci. Technol. **20**, 1506 (2004)
19. A. Järvenpää, M. Jaskari, L.P. Karjalainen, unpublished results (2017)
20. J. Man, I. Kuběna, M. Smaga, O. Man, A. Järvenpää, A. Weidner, Z. Chlup, J. Polák, Proc. Struct. Integr. **2**, 2299 (2016)
21. J. Man, A. Chlupová, I. Kuběna, M. Valtr, A. Järvenpää, T. Kruml, L.P. Karjalainen, J. Polák, Acta Mater. (to be published)
22. J. Man, M. Smaga, I. Kuběna, D. Eifler, J. Polák, Eng. Fract. Mech. **185**, 139 (2017)
23. G.B. Olson, M. Cohen, J. Less-Common Metals **28**, 107 (1972)
24. D.A. Porter, K.E. Easterling, *Phase transformations in metals and alloys* (Van Nostrand Reinhold (UK), 1986)
25. G.B. Olson, W.S. Owen, eds., *Martensite: A tribute to Morris Cohen* (ASM International, 1992)
26. P. Marshall, *Austenitic stainless steels: microstructure and mechanical properties* (Elsevier Applied Science Publishers, London and New York, 1984)
27. M. Smaga, F. Walther, D. Eifler, Mater. Sci. Eng. A **483–484**, 394 (2008)
28. L. Bodelot, L. Sabatier, E. Charkaluk, P. Dufrénoy, Mater. Sci. Eng. A **501**, 52 (2009)
29. J.W. Pegues, S. Shao, N. Shamsaei, J.A. Schneider, R.D. Moser, Fatigue Fract. Engng Mater. Struct. **40**, 2080 (2017)

Investigating the fatigue behavior of grain-oriented Fe-3%Si steel sheets using magnet-optical Kerr microscopy and micromagnetic multiparameter, microstructure and stress analysis

Shayan Deldar[1,*], *Marek* Smaga[1], and *Tilmann* Beck[1]

[1]University of Kaiserslautern, Institute of Materials Science and Engineering, Kaiserslautern, Germany

Abstract. Fatigue is considered as a reason for a significant number of mechanical failures of engineering materials. Conventionally, microstructural investigations along with stress-strain hysteresis measurements are performed to understand and characterize the fatigue behavior of metallic materials. Moreover, further physical data like temperature, electrical resistance and, in the case of ferromagnetic materials, magnetic properties can be used for a comprehensive characterization of fatigue process. The present work has employed Magneto-Optical Kerr Effect (MOKE) microscope and Micromagnetic Multiparameter, Microstructure and stress Analysis (3MA) system to illustrate magnetic domain structure and various intrinsic magnetic properties including magnetic Barkhausen noise (MBN) of the investigated material. In order to investigate the influence of the mechanical deformation processes on the magnetic parameters, samples were produced out of the grain-oriented electrical steel sheets and were subjected to a tensile test as well as a cyclic strain increase load test with R = 0 at ambient temperature.

1 Introduction

Although the magneto-mechanical behavior of ferromagnetic materials has been investigated for almost 150 years [1], a detailed understanding and visualization of the underlying phenomena caused by plastic deformation and fatigue process is yet to be elaborated. Electrical steels are considered as one of the conventional group of ferromagnetic materials well known to the magnetic community [2]. Their coarse magnetic domain structure, high magnetic flux density, and low iron loss have made this group of ferromagnetic materials suitable for usage in internal permanent magnet (IPM) motor cores [3, 4]. Developments in the automobile industry focusing on electric vehicles (EV) and hybrid electric vehicles (HEV) as well as efforts to increase the fuel efficiency of internal combustion engines resulted in a progressively increasing demand for research and development on this material class in the recent years [3]. Since the rotor core of an IPM motor is subjected to centrifugal forces which undergo fluctuations due to variations in the rotational speed of the motor during operation, the electrical steels used for this application not only should have acceptable mechanical properties, but also must maintain their magnetic characteristics in order to ensure a durable and efficient electrically powered system.

As a result, investigations on electrical steels recently directed to their magneto-mechanical behavior, fatigue properties, manufacturing technologies, and advanced magnetic measurements on the two main groups of electrical steels, grain-oriented and non-oriented [1, 5-11]. Perevertov et al. [1, 5] described magnetic domain structure and hysteresis loops of grain-oriented Fe-3%Si under the influence of tensile and compressive elastic stresses. Fatigue behavior of high grade non-oriented electrical steels was published by Liying et al. [9] focusing on crack characterization and fracture mechanisms as well as the Wöhler curve determined for a specific batch of non-oriented electrical steel (30WGP1600). Furthermore, Dehmani et al. [6, 7] investigated M300-35A non-oriented electrical steels taking into account the effect of the punching process and annealing as parts of the manufacturing procedures. The negative effect of punching and positive effects of mechanical polishing on the fatigue properties of non-oriented electrical steel has been demonstrated. The article also has focused on the microstructure and fracture surface morphology. One of the few works regarding the influence of the fatigue loading on the magnetic properties of electrical steels has been presented by Karthaus et al. [8] demonstrating the changes of the magnetic polarization hysteresis curve at different numbers of cycles as well as a comparison between fatigue tests with different maximum nominal stress amplitudes. In that study, fatigue-induced cracking of the specimen was not investigated.

In contrast to this, the present work aims to characterize the influences of mechanical strains on the magnetic properties of grain-oriented electrical steels

* Corresponding author: deldar@mv.uni-kl.de

with 3 wt.-% silicone content. The mechanical strain was applied in a stepwise strain controlled tensile. The correlation between mechanical loading, magnetic domain structure visualized by Magneto-Optical Kerr Effect (MOKE) and integral magnetic properties measured via a Micromagnetic Multiparameter, Microstructure and stress Analysis (3MA) system are discussed. A comprehensive analysis of the magnetic domain structure and magnetic Barkhausen noise (MBN) signal measured at the loaded and unloaded steps is presented.

2 Investigated material

The material investigated in this work is a "powercore" grain-oriented electrical steel. It holds a {110}<001> (Figure 1) texture along the rolling direction. This so-called Goss texture [5] is typical in grain-oriented electrical steel and effects magnetic anisotropy with large magnetic flux density in the rolling direction (RD). The chemical composition of the investigated material is shown in Table 1.

Table 1. The chemical composition of the investigated grain-oriented electrical steel.

Chemical composition	**C**	**Mn**	**Si**	**Fe**
In wt.-%	0.003	0.070	3.000	Bal.

The mentioned texture of the material is schematically illustrated in figure 1(a). tensile specimens were cut out of the steel sheet with the geometry shown in figure 1(b) and longitudinal axes oriented in RD. The mechanical stress and magnetic field are applied parallel to the rolling direction (RD).

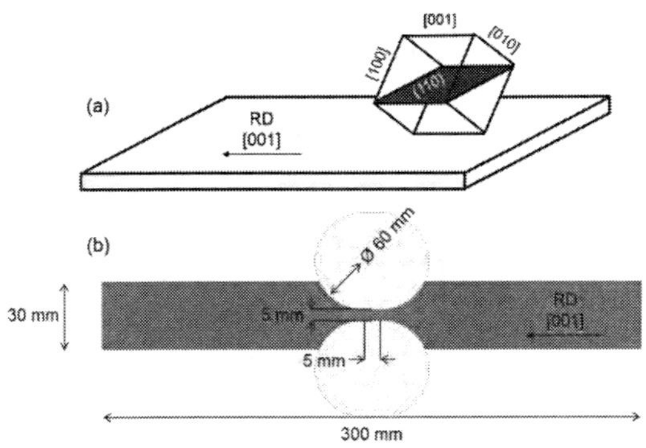

Fig. 1. Schematic of the Goss texture longitudinal to the rolling direction (a) and the specimen geometry (b).

The mechanical properties of the investigated material are summarized in table 2. In comparison to 30WGP1600 and M300-35A steels investigated respectively in [7, 9], the grain-oriented electrical steel studied in this work exhibits significantly lower yield stress and tensile strength. Note that the crystallographic texture of the grain-oriented electrical steel results in a variation of the mechanical properties with respect to the measurement direction while in case of non-oriented steels, the material has isotropic properties.

Table 2. Mechanical properties of the investigated grain-oriented electrical steel.

Description	**Value**
Thickness	0.27 mm
Yield stress	300 MPa
Tensile strength	340 MPa
Elongation to fracture	10 %
Hardness	175 – 195 HV0.1

3 Experimental setup

Before the experimental measurements, the samples were mechanically ground up to 4000 grit SiC paper in the deformation area of the specimen and subsequently polished on a cotton cloth in 3 µm diamond suspension.

3.1 Magneto-optical Kerr microscope and tensile test system for in-situ loading

For visualization of the magnetic domain structure, a magneto-optical Kerr microscope was employed in longitudinal mode. Figure 3 shows the experimental setup and a schematic overview of its components. Magneto-Optical Kerr Effect (MOKE) enables a regular polarization optical microscope to be used for magnetic domain and domain wall observation [12]. The working principle of the Kerr microscope is based on the application of linearly polarized light on the surface of the magnetic material. Due to the physics of Lorenz force when the linearly polarized light reflects from the surface of a ferromagnetic material, the plane of polarization is rotated because of the surface magnetization. The rotation of the polarization plane differs for the lights reflected from regions with different magnetizations. Therefore, the reflected light obtains independent polarization planes with respect to the regions from where they reflected. The reflected light with different polarization directions are the indicators of the magnetic domain contrast on the surface of the ferromagnetic material. To obtain a high signal to noise ratio and eliminate the negative effect of surface roughness of the specimen on the magnetic domain image quality, and to enhance image contrast differential imaging technique was used. In this method, two images are taken at two different magnetic stages of the hysteresis, one of which is usually the saturation state. The two images are then digitally subtracted from each other and as a result, topographic effects of the surface are eliminated and only the difference between the magnetic domain structures appears [2]. For the differential imaging to be successful, the two images must be at the same spatial position with high precision. In order to have a fixed position of the sample while differential imaging, a piezo stabilization system has been applied underneath the sample stage which immediately corrects the drifts which always occur due to the magnetic force on the ferromagnetic material under investigation.

In the schematic sketch, it is demonstrated that the illumination path of the light beam is applied to the surface at a certain angle (angle of incidence). This angle is necessary for the polarization rotation to occur. The larger the angle of incidence, the higher the raw image contrast. Since both, the illumination path and the reflection path (observation path), go through the objective lens of the Kerr microscope, one has to take into account for selection of the proper objective lens for MOKE application.

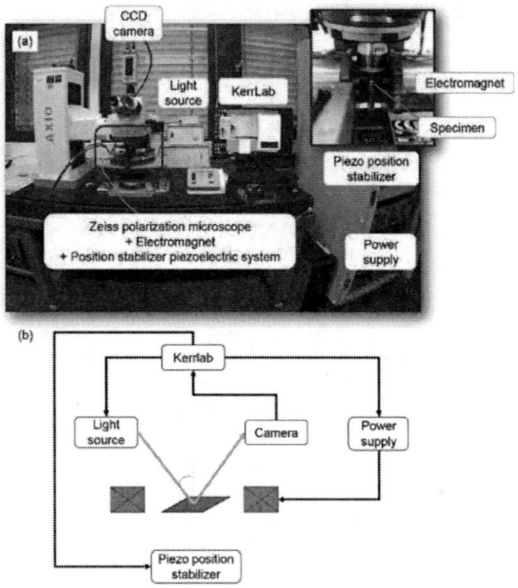

Fig. 2. Magneto-optical Kerr effect (MOKE) experimental setup (a) and a schematic sketch of the setup (b).

The tensile test system shown in figure 3 has been designed specifically for application as an in-situ tensile module for Kerr microscopy. The window placed in between the two electromagnets and the geometry of the electromagnets allow the objective lens to reach the working distance to the specimen. The electromagnets apply the magnetic field parallel to the loading direction. The setup is, in fact, capable of performing an elongation controlled tensile loading since the loading is done via turning the tension screw clockwise. The load cell then monitors the tension applied by the tension screw. As the device limits the force applied to a maximum of 1 kN the specimen geometry (figure 1(b)) was designed appropriately.

Because of the large size of the specimen, the preparation and polishing of the surface were challenging. Once the surface quality of the specimen in the deformation area is proper, magnetic domain observation during the application of the tension was possible.

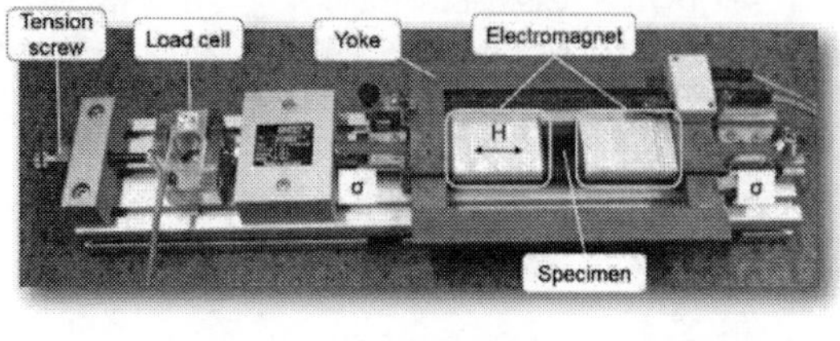

Fig. 3. The in-situ tensile test designed for Kerr microscope.

3.2 3MA sensor

Figure 4(a) shows a Micromagnetic Multiparameter, Microstructure and stress Analysis (3MA) system which is classified as a non-destructive evaluation device. The parameters obtained from this setup are used for online monitoring of e.g. machinery components and steel production processes. With this setup, only ferromagnetic materials can be characterized by their integral magnetic properties such as magnetic Barkhausen noise, incremental permeability, and harmonic analysis of the tangential magnetic field. As the magnetic properties experience variations under the influence of mechanical deformation, the 3MA system was used to record the corresponding magnetic behavior. [13, 14] In the present study, the focus is on the analysis of the magnetic Barkhausen noise signal obtained from the 3MA sensor.

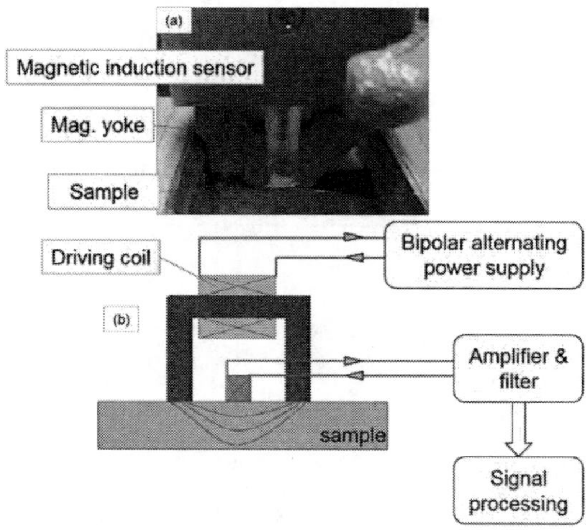

Fig. 4. Photograph of the 3MA sensor (a) and a schematic sketch of the 3MA system (b).

Magnetic Barkhausen noise is a physical property of magnetic materials which is a result of Bloch wall motion during the magnetization reversal process. A ferromagnetic material in the demagnetized state experiences a multi-domain magnetic microstructure, such that the material is, from the thermodynamical point of view, at its lowest total free energy [2]. As the material is applied to the external magnetic field, this multi-domain magnetic microstructure converts during a so-called magnetic reversal process to a single domain magnetization (saturation magnetization). The reversal process includes domain wall movement and rotation of magnetization. The domain wall movement mechanism is not a continuous process and, because in iron-based ferromagnetic materials Bloch walls are present, this mechanism is also called Bloch wall jumps [13]. The reason for these discontinuities in the domain wall movements is microstructural obstacles within the path of the domain walls by which the Bloch walls are pinned. Grain boundaries, dislocations, and precipitations are examples of the above mentioned microstructural obstacles. The pick-up coil (magnetic induction sensor in figure 4(b)) directly receives the Barkhausen noise and the data acquisition system converts it to an electrical voltage.

In order to induce an alternating magnetic field to the specimen, a bipolar alternating power supply is connected to a driving coil and the magnetic induction generated by the driving coil is transferred to the specimen by a high permeability magnetic core and the magnetic response of the specimen is read out by the pick-up coil. The signal is then amplified and analyzed.

4 Results and discussion

Figure 5(a) illustrates the grain structure of the investigated material which indicates an average grain size of 10 mm. Besides the crystallographic grain structure observable at the surface of the material, the coexistence of the magnetic domain structure in the undeformed state of the grain-oriented electrical steel and its interaction with the crystallographic orientation is pictured. Moreover, it is shown that because of the crystallographic texture of the material parallel to the rolling direction, a magnetocrystalline anisotropy exists in the material leading to the magnetization direction along the rolling direction. In the magnetic domain image, one can notice a disturbance of the domain structure in the grain on the left side of figure 5(b) compared to the grain on the right side of the image. The reason for such a magnetic structure difference is a small degree of misorientation in the left grain with respect to the right grain.

Figure 6(a) illustrates a cyclic strain increase load test performed by the in-situ tensile testing device (figure 3). The test with a strain ratio of $R_\varepsilon = 0$ was performed by loading and unloading the specimen repeatedly until the specimen failure. By each loading step, 1.7 % additional plastic strain was applied to the specimen and the measured force was recorded. After each loading step, the elastic strain was completely released by reducing the tensile force to zero. For each loading and unloading step in the plastic region (points c to g and c' to g'), longitudinal magnetic domain image was captured (figure 7) and maximum magnetic Barkhausen noise (MBN) signal within one hysteresis loop was extracted (figure 6(b)).

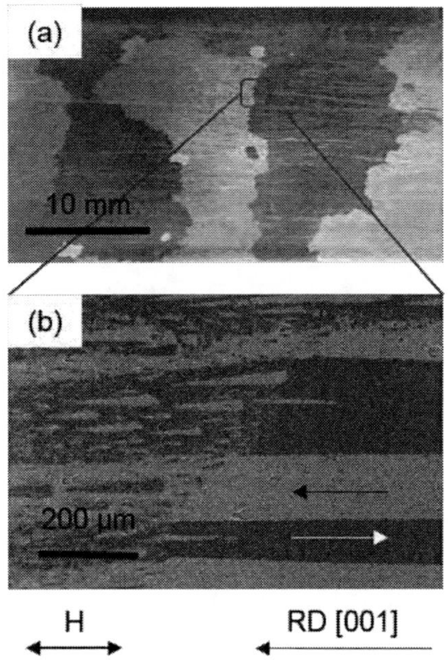

Fig. 5. Crystallographic microstructure (a) and magnetic domain structure (b) of the grain-oriented electrical steels at the initial state.

The blue and red lines in figure 6 indicate the loaded and unloaded condition in the strain increase load test. The dashed blue lines illustrate each loading/unloading step (for instance unloading from d to d' and loading from d' to d). Based on the behavior of the maximum MBN in the loaded condition (the blue line in figure 6(b)) the diagram was divided into 3 zones. At Zone 1 which was still in the elastic region, the maximum Barkhausen noise at the loaded state with respect to the initial state decreased. This means the stress-induced magnetic anisotropy helped the magnetization reversal by decreasing the energy required for domain wall unpinning as a driving force. At zone 2, where the plastic deformation started, the maximum MBN reached a minimum and began to rise at about 6 % total strain. The beginning of the plastic regime is where the dislocation density started to increase, in other words, more pinning sources were generated. Therefore, a contribution in favor of MBN increase is produced. Before zone 3, the sample has already experienced 4 % plastic deformation, therefore the mechanisms which resulted in MBN increase were already dominant. This argument is also valid when considering the maximum MBN curve at the unloaded condition (red curve in figure 6 (b)) where the MBN showed a steady increase throughout the strain increase load test because, in the unloaded condition, stress-induced effects did not assist the magnetization reversal by easing the domain wall movement.

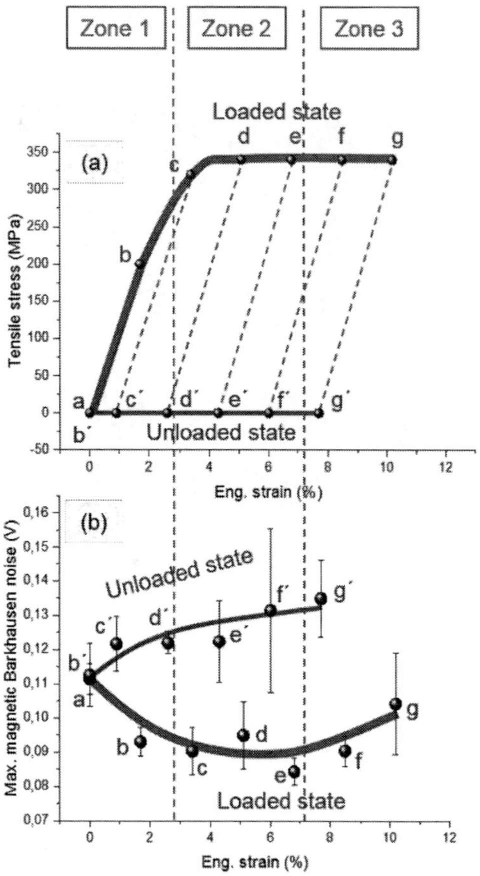

Fig. 6. Nominal stress vs. engineering strain curve of the cyclic strain increase load test performed by the in-situ testing device (a) and maximum magnetic Barkhausen noise measurement for each step in loaded and unloaded condition (b).

The MOKE measurements were done in the longitudinal mode. The applied magnetic field, the uniaxial magnetic anisotropy, the sensitivity direction (which is defined by the illumination path), the applied mechanical stress, and the rolling direction in figure 7 are all horizontally directed. The magnetic domain images in figure 7 indicate that domain refinement occurs by application of strain in both, loaded and unloaded condition. Furthermore, a comparison between the respective images in the loaded and unloaded condition in each step shows that the magnetic domain structure at the unloaded state tends to exhibit a higher domain wall density, i.e. finer domain structure. This correlates with figure 6(b) indicating higher maximum MBN in the unloaded (red curve) than in the loaded state (blue curve), i.e. the magnetic Barkhausen noise of the unloaded condition is always higher than that in the respective loaded condition.

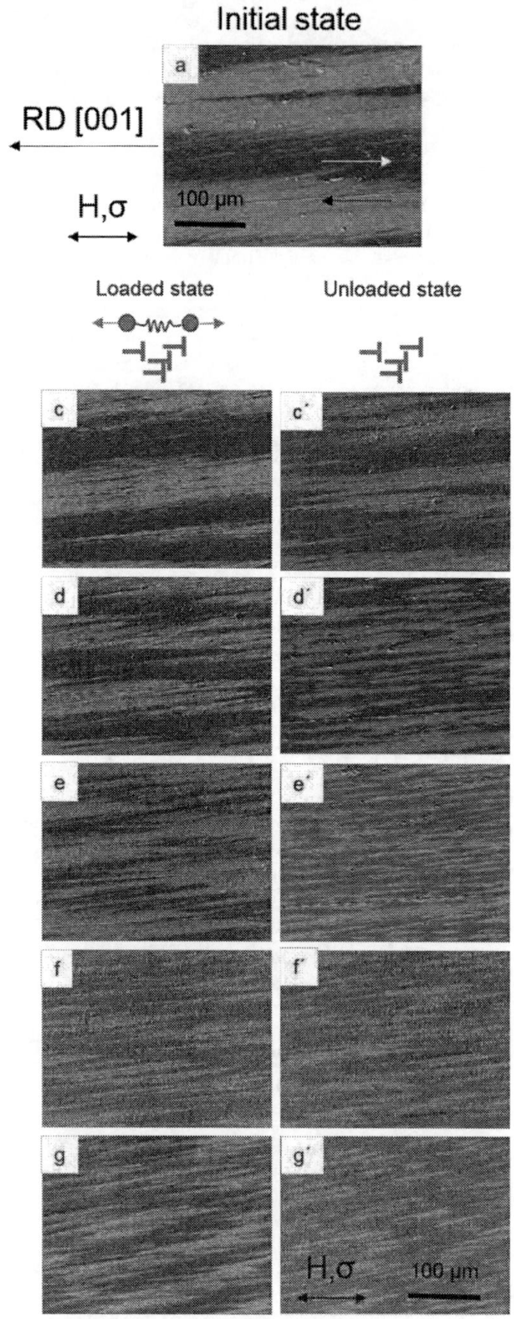

Fig. 7. Magnetic domain images at the states of the strain increase load test indicated in Figure 6.

5 Summary and conclusion

Cyclic strain increase tests with a strain ratio $R_{\varepsilon} = 0$ were performed to investigate the influence of the mechanical deformation processes on the magnetic domain structure and integral magnetic parameters. Magneto-Optical Kerr Effect (MOKE) microscopy and a Micromagnetic Multiparameter, Microstructure and stress Analysis (3MA) system were employed to illustrate magnetic domain structure and intrinsic magnetic properties including magnetic Barkhausen noise (MBN). The tensile test system by which the cyclic strain increase load test was performed was specifically designed for application as an in-situ tensile module for Kerr microscope. Presence of the stress-induced anisotropy acted as a driving force for domain wall movement which led to decrease of the maximum MBN in the elastic regime at the loaded condition. As the dislocation density increased during plastic deformation, the maximum MBN in the loaded state started to increase. The magnetic domains showed a refined structure with increasing plastic strain applied during cyclic strain increase load testing. The domain structures at the unloaded condition indicated higher domain wall densities compares to the loaded condition at the same plastic strain which is in agreement with the MBN curves.

Acknowledgment

This research was supported by German Research Foundation (DFG) within the Transregional Collaborative Research Center SFB/TRR 173 (Spin+X), project B08. We would also like to show our gratitude to Prof. Rudolf Schäfer for sharing his experiences concerning MOKE experimental setup.

References

1. O. Prevertov, R. Schäfer, J. Phys. D: Appl. Phys. **47**, 185001 (2014)
2. A. Hubert, R. Schäfer, *Magnetic Domains*, 3rd edition (Heidelberg: Springer) (1998)
3. Y. Oda, M. Kohno, A. Honda, JMMM **320**, 2430-2435 (2008)
4. I. Tanaka, H. Nitomi, K. Imanishi, K. Okamura, H. Yashiki, IEEE Transactions on Magnetics, **49**, 2297-3001 (2013)
5. O. Prevertov, R. Schäfer, J. Phys. D: Appl. Phys. **45**, 135001 (2012)
6. H. Dehmani, C. Brugger, T. Palin-Luc, C. Mareau, S. Koechlin, Procedia Engineering, **133**, 556-561 (2015)
7. H. Dehmani, C. Brugger, T. Palin-Luc, C. Mareau, S. Koechlin, Procedia Structural Integrity, **2**, 3256-3263 (2016)
8. J. Karthaus, S. Steentjes, D. Gröbel, K. Andreas, M. Merklein, K. Hameyer, Archives of Electrical Engineering, **66(2)**, 351-360 (2017)
9. D. Liying, Z. Guifeng, L. Jing, S. Wenmin, B. Yunjie, C. Zhaoyang, x. Huan, Journal of Wuhan University of Technology-Mater. Sci. Ed., **32**, 1329-1335 (2017)
10. M. Petrun, S. Steentjes, K. Hameyer, B. Polajzer, IEEE Transactions on Magnetics, **53**, 1-5 (2017)
11. N.J. Lewis, P.I. Anderson, Y. Gao, F. Robinson, JMMM, **452**, 495-501 (2018)
12. J. McCord, J. Phys. D: Appl. Phys. **48**, 333001 (2015)
13. G. Dobmann, *Physical Basics and Industrial Applications of 3MA – Micromagnetic Multiparameter Microstructure and Stress Analysis*, (IOS Press publishing house) (2007)
14. G. Hübschen, I. Alpeter, R. Tschuncky, H. Herrmann, *Materials Characterization Using Nondestructive Evaluation (NDE) Methods*, (Elsevier) (2016)

Study of the low cyclic behaviour of the IRIS alloy at high temperature

Soumaya NAANANI [1,2,*], *Muriel HANTCHERLI* [1], *Anis HOR* [2], *Catherine MABRU* [2], *Jean-Philippe MONCHOUX* [1] and *Alain COURET* [1]

[1] *CEMES, Université de Toulouse, CNRS, 29 rue Jeanne Marvig, 31055 Toulouse, France*
[2] *ICA, Université de Toulouse, CNRS, ISAE-SUPAERO, 10 avenue Edouard Belin - BP 54032 - 31055 Toulouse, France*

Abstract. This paper presents experimental results on low cycle fatigue (LCF) behaviour at high temperatures of the IRIS alloy (Ti-Al_{48}-W_2-$B_{0.08}$, at. %) densified by Spark Plasma Sintering (SPS). Cyclic stability is noticed, whatever the temperature and the total strain amplitude. Fracture occurs by mixed interlamellar and translamellar modes. Transmission electron microscopy (TEM) observations reveal that deformation is mainly due to ordinary dislocations and twins, and slip and cross-slip mechanisms have been clearly evidenced, without significant contribution of climb at 800°C. It has also been established that dislocation density depends on fatigue life time rather than on stress amplitude.

1 Introduction

Extensive investigations on the mechanical properties and processes have allowed commercial implementation of TiAl alloys in aerospace and automotive engines. Their main attractive properties are low density (4g/cm^3) and high strength at high temperature. They are used for instance in low pressure turbine blades of the new generation of aircraft engines, such as the GEnX by General Electric and the LEAP by Snecma-Safran, and in turbine wheels of new generations of turbochargers.
Giving these applications at elevated temperatures, the material is subjected to mechanical and thermal regimes, which generates a large amount of plastic deformation. Low cycle fatigue tests are then used to evaluate the lifetime of TiAl-based alloys, simulating service conditions. Fatigue properties of this class of materials were regularly studied since the pioneering work of Sastry and Lipsitt [1]. Most of these efforts were devoted to binary Ti–Al alloys [3-4] and later to alloys with Nb [5–8].
In this context, IRIS TiAl alloy (Ti-Al_{48}-W_2-$B_{0.08}$ at%) has been developed [8]. The innovation comes from the incorporation of tungsten (W) and boron (B) which allows respectively improving creep properties and preserving a good ductility. In addition, densifying the materials by spark plasma sintering technique proved interesting to favour fine microstructures. This process is based on powder metallurgy and allows very fast processing cycles, by applying a pulsed current and a uniaxial pressure on the powder. By heating the powder in the α+β field above the α+β+γ/ α+β line transformation of the phase diagram, a fine near-lamellar microstructure is obtained [10-11]. Mechanical properties [11] exhibited a very good compromise between elongation at room temperature (A = 1.6 %, when most TiAl alloys have a ductility below 1 %) and creep resistance (above 4000 h at 700°C under 300 MPa). The present study investigates the LCF behaviour and performance of IRIS at high temperatures and attempts to understand the cyclic deformation mechanisms.

2 Experimental procedure

2.1 Alloy composition and microstructure

The alloy investigated is the IRIS alloy with a composition of Ti-Al_{48}-W_2-$B_{0.08}$ (at%). Disks of 100 mm in diameter and 18 mm thick were densified by SPS in the FCT HPD-125 machine of the ICB Dijon in France. Temperature ramp was 100°C/min, and dwell temperature was 1375°C during 10min. At 1200°C, the temperature ramp was reduced to 25°C/min to avoid excessive overshoot. For these experiments, different dwell temperatures have been tested, in order to obtain the desired microstructure on more than ¾ of the diameter of the disks. Temperature was measured by a pyrometer at the center of the sample, 10 mm above the surface. The target microstructure (Figure 1) is formed by small lamellar colonies (35 to 45 µm in diameter), surrounded by γ grains (5 to 10 µm width) containing β_0 precipitates. Cylindrical specimens for the fatigue tests were then machined from SPS disks as illustrated in Fig 2.

* Corresponding author: Soumaya.Naanani@cemes.fr

https://doi.org/10.1051/matecconf/201816506007

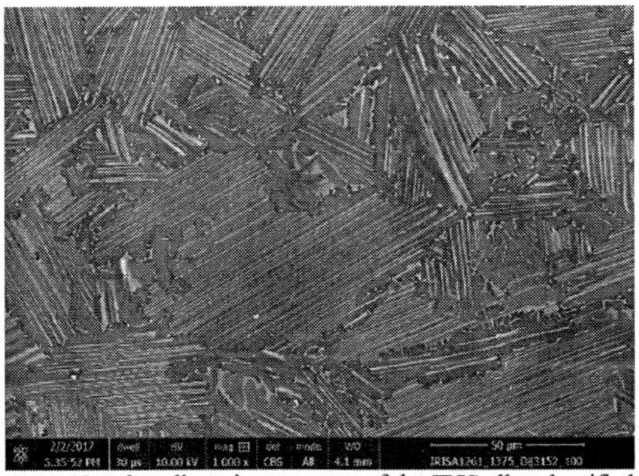

Fig.1. Near lamellar microstructure of the IRIS alloy densified by SPS.

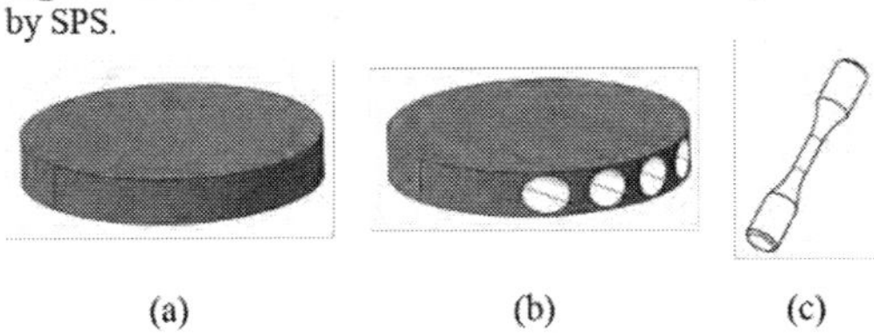

Fig. 2. (a) SPS disk of 100 mm in diameter and 15 mm in tickness of IRIS alloy with near lamellar microstructure. (b) Manufacturing of 4 cylinders per disk. (c) Machining of cylindrical fatigue specimen with: a gauge length of 14 mm, gauge diameter of 5 mm and a total length of 76 mm.

2.2 Low cycle fatigue test

Cylindrical specimen were mechanically polished in the gauge length with SiC papers in order to remove the surface irregularities due to the machining and minimise their effects on the fatigue life.
The low cycle fatigue tests were conducted on a MTS servohydraulic machine at a constant strain rate of 10^{-3} s^{-1} and at different total strain amplitudes ($\Delta\varepsilon_t/2$ = 0.3%, 0.45%, 0.6%) in symmetric tension compression cycle ($R_\varepsilon = \varepsilon_{min}/\ \varepsilon_{max} = -1$). The total strain amplitude was controlled by an extensometer directly placed on the gauge part of the specimens and which delivers a command signal adjusted in a triangular wave shape. All the tests were performed in air laboratory at three temperatures: 750, 800 and 850°C. Heating was ensured by an electrical resistance furnace providing a uniform temperature repartition on the specimen.

2.3 Fracture surfaces and microstructural observations

In order to identify and clarify failure mechanisms in IRIS, fracture surfaces were examined in a JEOL 2010 scanning electron microscope (SEM).
The cyclic deformed microstructures were analysed in a JEOL 2010 Transmission Electron Microscope (TEM) operating at 200 kV. Disks of 0.5 mm were cut perpendicularly to the tensile axis, close to the fracture surface and mechanically polished to a thickness of 20 µm and then electrolytically polished to electron transparency.

3 Results

3.1. Cyclic Stress Strain behaviour

Stress amplitudes as a function of the number of cycles at 800°C are presented for three different strain amplitudes in Fig.3. The cyclic behaviour is stable and no continuous hardening or softening is noticed. The fatigue life decreases as the total strain amplitude increases: from 5276 cycles for the lowest strain amplitude (0.3%) to 187 cycles for the highest strain amplitude (0.6%).

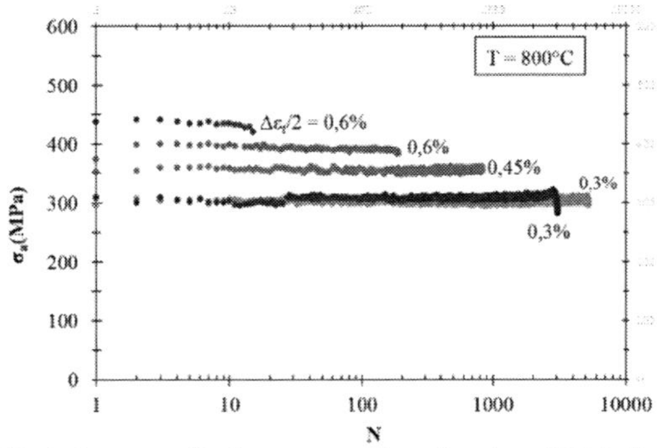

Fig.3. Stress amplitude versus number of cycles of IRIS alloy at 800°C.

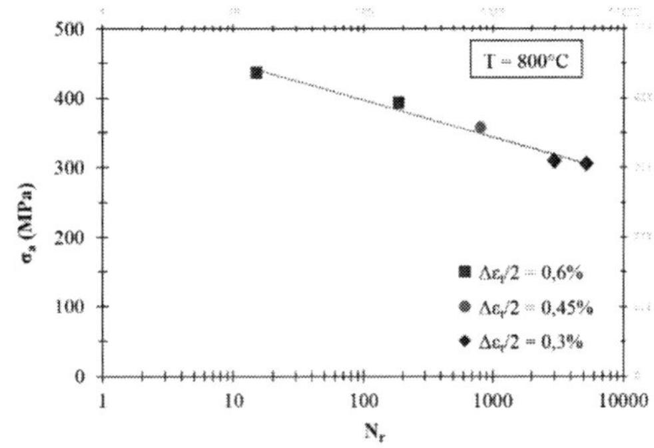

Fig.4. Stress amplitude versus number of cycles to failure for IRIS at 800°C.

An apparent dispersion of the number of cycles to failure for a given total strain amplitude is noticed (at $\Delta\varepsilon_t/2$ = 0.6% , N_f=15 and 187 cycles and at $\Delta\varepsilon_t/2$ = 0.3% , N_f= 5276 and 3071 cycles). However when plotting the stress amplitudes versus the number of cycles to failure for each fatigue test (Fig.4) all the results are on the same line. This indicates that the results are consistent and that fatigue life of IRIS can be easily predicted.

Tests at different temperatures were also performed to study the influence of the temperature on the fatigue life and the cyclic behaviour. Fig.5 shows stress amplitudes as a function of the number of cycles for the same total strain amplitude ($\Delta\varepsilon_t/2$ = 0.45%) at three different temperatures: 750°C, 800°C and 850°C.

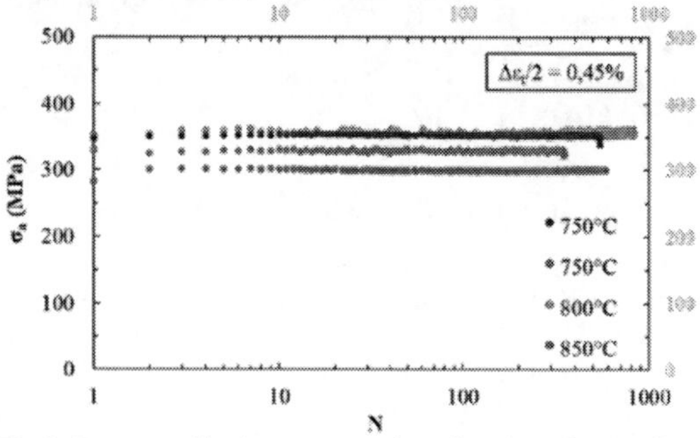

Fig.5. Stress amplitude versus number of cycles of IRIS alloy at high temperatures.

Cyclic behaviour is stable whatever the temperature. However, the fatigue life is lower at 750 and 850°C than at 800°C. This could be related to the brittle-ductile transition temperature, which occurs for this alloy between 700°C and 850°C as evidenced by Voisin et.al [11].

As specified in the introduction, literature provides many studies on LCF of TiAl alloys at various high temperatures and with various test conditions. It is thus difficult to strictly compare IRIS LCF behaviour with other TiAl alloys. However, focusing on a given temperature test, Table 1 presents some literature results of lamellar or near-lamellar TiAl alloys, obtained at 800°C with similar other test conditions (speed rates, strain amplitude).
Even if the microstructures and the speed rate of the LCF tests are not strictly the same, these results suggest that IRIS has higher fatigue lives than the TiAl alloys containing Nb at 800°C for the imposed strain amplitudes that has been investigated.

3.2. Fractures surface observations

All the fractures surfaces observations indicate a brittle fracture, typical of TiAl alloys [4]. Due to exposure of the fracture surfaces to air, an oxide layer has been formed (Fig.6).
For most of the fracture surfaces observed by optical and scanning electron microscope, the crack initiation site as well as the crack propagation regions are difficult to determine due to the rough surface and the absence of river patterns indicating the fatigue crack direction. The fracture surfaces analysed by SEM present a similar topography and the same fractures modes, regardless of the total strain amplitude applied. It consisted in a mixed mode comprising interlamellar and translamellar modes. These two modes are represented respectively by yellow and white arrows on figure 6. Secondary cracks along the lamellar colonies boundaries and interlamellar decohesion were also found (fig.7(a)). These fracture modes are typical of the TiAl alloys with lamellar microstructure [6][13-15].

However, the γ borders, which are important for the ductility of the alloy [11] are difficult to investigate: their size (5 to 10 µm width) and the irregular surface do not allow to detect their specific fracture mode.

3.3. Deformation mechanisms

The changes in the microstructure after fatigue failure were studied in specimens cycled at 800°C and for the extremes total strain amplitudes applied: 0.3% and 0.6%. In both specimens, the microscopic mechanisms have been investigated in the γ borders rather than in the lamellar colonies. The identification of the mechanisms is indeed more difficult in the latter case, due to perturbation of the dislocation morphologies by the lamellar interfaces. Globally, the presence of loops, twins, ordinary dislocations and debris are common to the two samples, but in a much higher proportion in the microstructure tested at 0.3%, as illustrated on figure 8. The microscopic mechanisms in the two samples are discussed in more details in what follows.

Table 1. LCF results on TiAl alloys at 800°C

Composition (at%)	Microstructure	$\dot{\varepsilon}$ (s^{-1})	$\Delta\varepsilon_t/2$ (%)	σ_a (MPa)	N_f
IRIS : Ti-Al_{48}-W_2-$B_{0.08}$	Near-lamellar	10^{-3}	0.3	304 ; 310	5276 ; 3071
			0.45	356	828
			0.6	392 ; 435	187 ; 15
Ti-Al_{46}-NB_2-Cr_2 [6]	Near-lamellar	2.10^{-3}	0.35	290	1046
	Full-lamellar	2.10^{-3}	0.35	300	1238
			0.45	340	205
TiAl-8%Nb [15]	Lamellar	2.10^{-3}	0.3	350	1500
			0.42	410	120

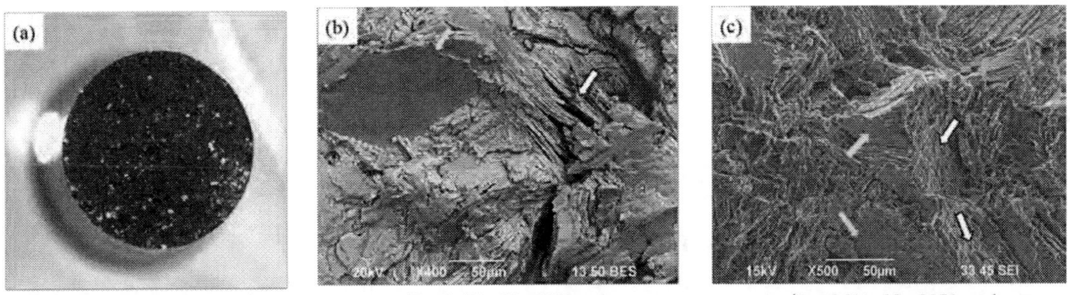

$\Delta\varepsilon_t/2$ = 0,6% ; N_f=187 cycles $\Delta\varepsilon_t/2$ = 0,6% ; N_f=187 cycles $\Delta\varepsilon_t/2$ = 0,3% ; N_f=3071 cycles

Fig.6. Fatigue fracture of specimens tested at 800°C. (a) General view (b) and (c) Yellow and white arrows indicates interlamellar and translamellar fracture respectively.

3.3.1 Deformation mechanisms in specimen: $\Delta\varepsilon_t/2$ = 0,6%; N_f =15 cycles

Figure 9 shows the activated deformation mechanisms in the γ borders cycled at high strain amplitude. The microstructure is mostly deformed by ordinary dislocations ($\vec{b}$ = 1/2 [$\bar{1}$10]) parallel to their screw direction (Figure 9. (a), (b)). These dislocations can be anchored at many pining points (Figure 9. (d)). Twinning is also observed in this specimen (Figure 9. (c)), but less commonly.

To identify moving planes of ordinary dislocations, tilt experiments were performed on several loops in the γ microstructure of this specimen. An example of these experiments is shown in figure 10. It consists in measuring the apparent width of the loop portion, at different inclination angles (represented respectively by W and θ in figure 10(b)). The variation of W according to θ corresponds to a sine curve (figure 10(c)), whose maximum (≈-18° in this case) represents the largest width of the portion loop. This identifies the moving plane, represented in the stereographic projection (figure 10. (b)) as a dotted line, and whose pole is marked as P. It can be seen that the moving plane is away of the Burgers vector $\vec{b}$ by only a few degrees, which is in the experimental errors of tilt experiments (±10°). Thus, it can be considered that P contains the loop, which indicates a glide mechanism.

Moreover, the plane P appears to be neither a common gliding plane one of the two possible (111) pure glide planes containing $\vec{b}$. So, the deformation mode activated here is cross slip. This procedure has been repeated for other ordinary loops and the deformation mechanisms found were either pure glide or cross glide.

3.3.2 Deformation mechanisms in specimen: $\Delta\varepsilon_t/2$ = 0,3%; N_f =5276 cycles

The same study was made on the specimen cycled at 0,3%. Dislocations density does not allow to perform many tilt experiments, it is indeed more difficult to find isolated loops. The microstructure revealed, as indicated above, a higher dislocation density than in the sample tested at $\Delta\varepsilon_t/2$ = 0.6%.

Figure 11 shows characteristic landscapes of deformed γ phase. γ borders can be deformed by ordinary dislocations ($\vec{b}_A$= 1/2 [$\bar{1}$10] and $\vec{b}_B$= 1/2 [110]) aligned in their Burgers vector (figure 11.(a)). However, unlike

* Corresponding author: Soumaya.Naanani@cemes.fr

 https://doi.org/10.1051/matecconf/201816506007

the previous case ($\Delta\varepsilon_t/2 = 0.3\%$), this is not the most frequent deformation mechanism.
The presence of twins is more important in this case (Figure 11(b) and (c)) and loops and debris are also present in a much higher proportion.

Figure 12 shows a tilt experiment of a loop identified as an ordinary dislocation ($\vec{b}$=1/2[$\bar{1}\bar{1}$0]). The stereographic projection shows that its Burgers vector does not lie in the loop plane (represented by a dotted line whose pole is marked as P on figure 12 (c)), this is therefore not a glide loop.

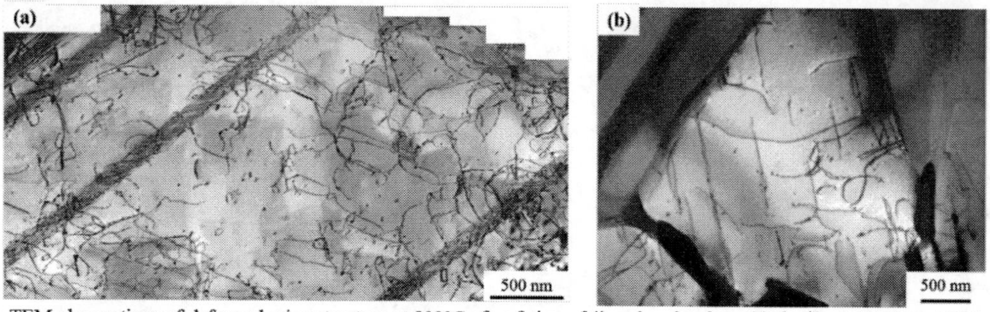

Fig.7. TEM observations of deformed microstructure at 800°C after fatigue failure in γ borders: (a) $\Delta\varepsilon_t/2$= 0,3%, N_f= 5276 cycles (b) $\Delta\varepsilon_t/2$= 0,6%, N_f = 15cycles

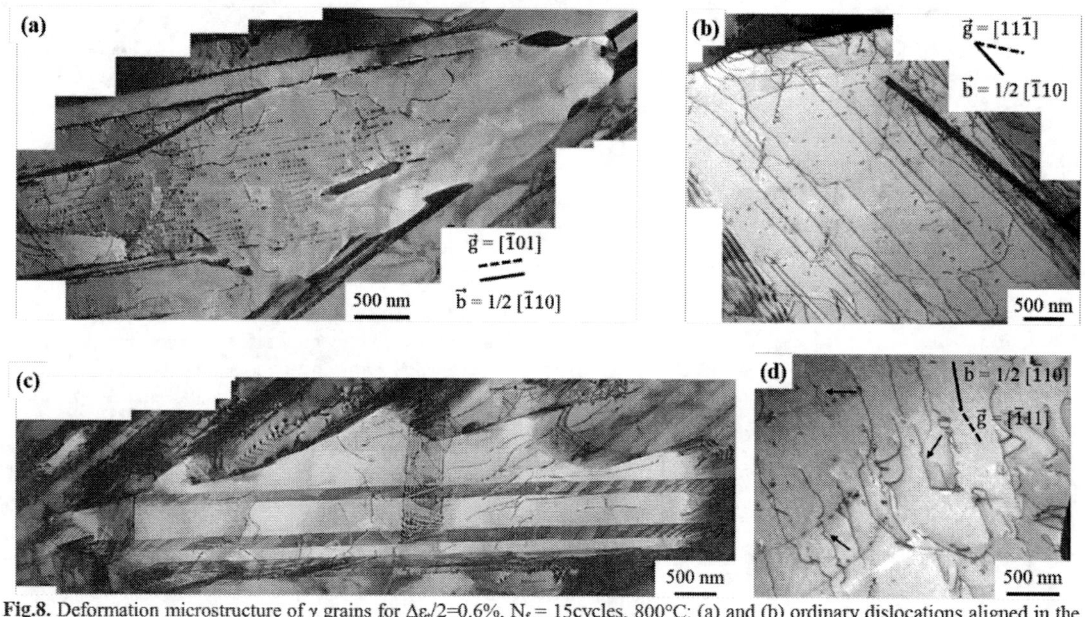

Fig.8. Deformation microstructure of γ grains for $\Delta\varepsilon_t/2$=0.6%, N_f = 15cycles, 800°C: (a) and (b) ordinary dislocations aligned in the screw direction. (c) Twins. (d) Pinning points

However this plane does not correspond to the pure climb plane [$\bar{2}\bar{2}$0] (represented in green on figure 12(c)), so the movement involved here is mixed climb.

Another tilt experiment on this microstructure indicated the presence of cross slip.

https://doi.org/10.1051/matecconf/201816506007

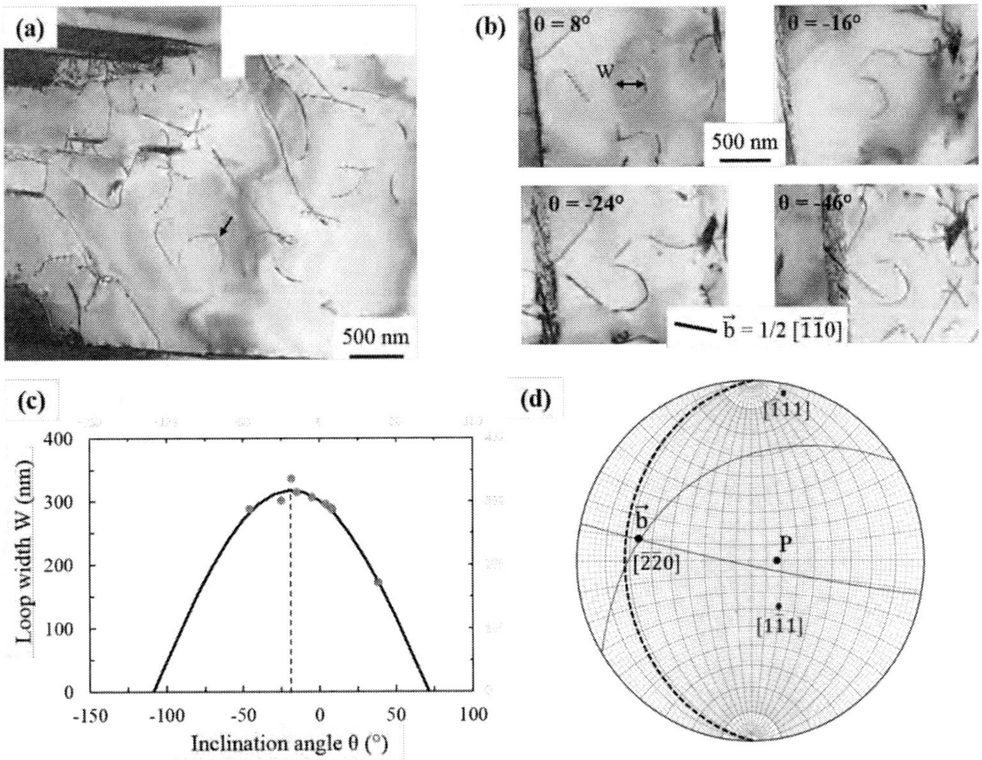

Fig.9. Tilt experiment on a loop portion in the sample tested at $\Delta\varepsilon_t/2 = 0.6\%$ ($N_f = 15$cycles). (a) General view of the γ grain that contains the loop. (b) Loop at different inclinations θ. (c) Variation of the loop width W according to the inclination angle θ. (d) Stereographic projection indicating the moving plane (dotted line), its pole (P), and two (111) planes (red lines).

* Corresponding author: Soumaya.Naanani@cemes.fr

Fig.10. Deformed γ grains for $\Delta\varepsilon_t/2 = 0.3\%$ ($N_f = 5276$ cycles, 800°C): (a) ordinary dislocations, (b) twins, (c) loops, twins, debris and ordinary dislocations.

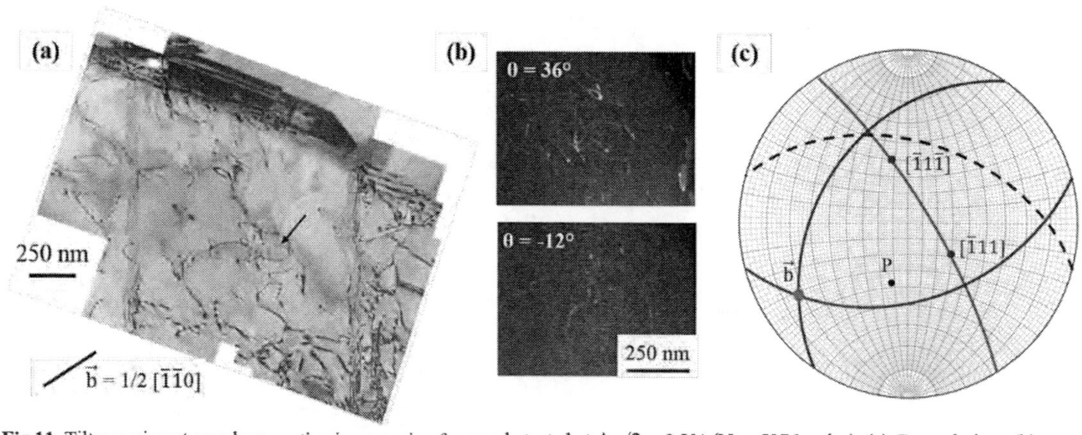

Fig.11. Tilt experiment on a loop portion in a γ grain of a sample tested at $\Delta\varepsilon_t/2 = 0.3\%$ ($N_f = 5276$cycles). (a) General view. (b) Loop at two inclinations θ. (d) Stereographic projection indicating the moving plane (dotted line), its pole (P), and two (111) planes (red lines)

4 Discussion

First results of investigation on LCF behaviour of IRIS at high temperatures (750, 800 and 850°C) are presented in this paper. The literature provides many fatigue tests results of TiAl alloys that are difficult to compare due to the various test conditions. When considering the few data available at 800°C with similar strain rates and strain amplitudes [6] [15], it appears that IRIS has a comparatively good lifetime.
Moreover the cyclic behaviour is very stable and no hardening or softening has been observed at the three temperatures. This feature is interesting in view of industrial use of the alloy, for which stable mechanical properties is an asset. The TiAl alloys with 2%Nb have a similar behaviour [6] [16] whereas TiAl-8%Nb present a noticeable softening at high temperatures [17] [15].

Temperature dependence appears rather weak, with only slight lifetime increase and decrease between 750°C-800°C, and between 800°C-850°C, respectively. This could be related to the brittle-ductile transition, which occurs around 800°C for this alloy [11]. Dependence on the imposed total strain amplitude follows a regular trend, indicating no abrupt change in damage mechanism.
Fracture follows a mixed interlamellar and translamellar propagation mode, which is usual in TiAl.

Study of the deformed microstructure at 800°C by TEM shows that the presence of ordinary dislocations, twins and loops are common to the two samples analysed. However, deformation at $\Delta\varepsilon_t/2$ = 0.6% is mainly controlled by glide of ordinary screw dislocations whereas at $\Delta\varepsilon_t/2$ = 0.3% there is no main deformation mechanism, several types of dislocations randomly organized are present. Unlike Sastry and Lipsitt [1], no braid structure has been observed.
TEM observations exhibit also that dislocation density is more important in the specimen deformed at $\Delta\varepsilon_t/2$ = 0.3% (N_f=5276 cycles and σ_a=304 MPa) than at $\Delta\varepsilon_t/2$ = 0.6% (N_f=15 cycles and σ_a=435 MPa). Density dislocation depends then on fatigue lifetime rather than on stress amplitude. This could be quantified by cumulative plastic strain ε_{pc} which corresponds to the sum of plastic strain of each cycle, and represents the forth and back movement of dislocations during the test. Table 2 shows that there is an important difference of this parameter between the two specimens: ε_{pc} is markedly more important for the specimens tested at $\Delta\varepsilon_t/2$ = 0.3 % (ε_{pc} = 491%) than at $\Delta\varepsilon_t/2$ = 0.6 % (ε_{pc} = 9%). The fact that the material can endure cumulated plastic strain as high as 491%, indicates clearly that most of the dislocation displacements are reversible during the tension and compression steps of the fatigue cycles, if not the specimen would be extremely deformed.

Table 2. Cumulative plastic strain ε_{pc} (%) of two specimen tested at 800°C.

$\Delta\varepsilon_t/2$(%)	σ_a (MPa)	N_f (cycles)	ε_{pc} (%)
0,3	310 MPa	3071	491
0,6	435 MPa	15	9

Furthermore during their reversible forth and back movement, the dislocations leave prismatic loops and debris, as observed in figures 9(b) and 11(c). However, all the movements are not completely reversible, and an increase in dislocation density occurs up to fracture. The high density of twins observed is compatible with reversibility of the motion of twins in tension and compression. Thus, the presence of twins probably explains also the rather low density of ordinary dislocations (lower than the density of dislocations which would be observed in samples deformed of 491%, without reversibility or recovery processes), most of the reversible deformation of the sample being accommodated by twinning. Finally, a high number of loops is observed. If they would be glide loops, they would very quickly annihilate due to very high line tension coming from their small diameter. Thus, they can only be prismatic and result from cross-slip processes [18]. At high temperatures, these loops would tend to vanish by climb mechanisms. Then, their stability is an approximate measure of the climb kinetics. In the study of Gloanec et.al [4], in situ TEM observations show that it takes about 25 minutes at 750°C for these loops to disappear. This shows that climb mechanisms are very slow in this temperature range, in reason of slow diffusion kinetics. The study of Kad and Fraser [19] shows that temperatures of 900°C are required to activate diffusion-controlled mechanisms for strain rates between 10^{-3} and 10^{-5} s^{-1}. However, some mixed-climb loops can form in the $\Delta\varepsilon_t/2$ = 0,3% experiments, as observed in Fig. 12. In this case, the duration of the experiments (18h) is high enough to allow growth of mixed-climb loops by diffusion processes. The contribution of this phenomenon to the total deformation, nevertheless, is believed to be negligible. Then, climb mechanisms are rather improbable in our experiments.

5 Conclusion

In this study, the IRIS alloy has been tested in low cycle fatigue, for temperatures ranges of 750-850°C and for total strain amplitudes $\Delta\varepsilon_t/2$ between 0.3 % and 0.6 %. The mechanical results and the microstructure characterizations by TEM lead to the following conclusions:

• The stress amplitude is remarkably constant during the tests, and the dependence of fatigue life on the strain amplitude follows a regular trend. Moreover, only weak temperature dependence is observed. These mechanical

* Corresponding author: Soumaya.Naanani@cemes.fr

properties are interesting in comparison with other alloys.

• Regular mixed interlamellar and translamellar propagation modes are observed, as usual in lamellar TiAl alloys.

• Dislocation density depends on fatigue lifetime rather than on stress amplitude.

• Deformation occurs mainly by ordinary dislocations and twins. The dislocation movements are found to be mainly reversible during the tension and compression steps, indeed extremely high dislocation densities are not observed, despite very elevated cumulative plastic strains (up to 491 %). Tilt experiments show slip and cross-slip are the preferred motion modes, climb being rather improbable.

This study has been conducted in the framework of the cooperative project "ALTIAERO" project of the "IDEX-ATS" program supported by the Université Fédérale – Toulouse Midi-Pyrénées, which is acknowledged.

References

1. S. M. . Sastry and H. . Lipsitt, Metall. Mater. Trans. A Phys. Metall. Mater. Sci., **8A**, 299–308, (1977).
2. H. Y. Yasuda, T. Nakano, and Y. Umakoshi, Philos. Mag. A, **71**, 127–138, (1995).
3. Y. Umakoshi, H. Y. Yasuda, and T. Nakano, Mater. Sci. Eng. A,**192–193**, no. PART 1, 511–517, (1995).
4. A. L. Gloanec, M. Jouiad, D. Bertheau, M. Grange, and G. Hénaff, Intermetallics, **15**, 520–531, (2007).
5. A. L. Gloanec, G. Henaff, M. Jouiad, D. Bertheau, P. Belaygue, and M. Grange, Scr. Mater., **52**, 107–111, (2005).
6. G. Malakondaiah and T. Nicholas, "High-temperature low-cycle fatigue of a gamma titanium aluminide alloy Ti-46Al-2Nb-2Cr," Metall. Mater. Trans. A, **27**, 2239–2251, (1996).
7. M. Petrenec, M. Šmíd, T. Kruml, K. Obrtlík, and J. Polák, "Low Cycle Fatigue of Cast γ-TiAl Based Alloys at High Temperature," Key Eng. Mater., **452–453**, 421–424, (2010).
8. T. Voisin, J. P. Monchoux, L. Durand, N. Karnatak, M. Thomas, and A. Couret, Adv. Eng. Mater., **17**, 1408–1413, (2015).
9. A. Couret, J. P. Monchoux, M. Thomas, and T. Voisin, "Procédé de fabrication d'une pièce en alliage en titane-aluminium," Patent WO2014199082 A1, (2013)
10. T. Voisin, J.-P. Monchoux, M. Perrut, and A. Couret, Intermetallics, **71**, 88–97, (2016).
11. T. Voisin, J. P. Monchoux, M. Thomas, C. Deshayes, and A. Couret, Metall. Mater. Trans. A Phys. Metall. Mater. Sci., **47**, 6097–6108, (2016).
12. K. S. Chan and D. S. Shih, Metall. Mater. Trans. A, **28**, 79–90, (1997).
13. B. D. Worth, J. M. Larsen, S. J. Balsone, and J. W. Jones, Mater. Trans. A, **28**, (1997).
14. S. J. Balsone, B. D. Worth, J. M. Larsen, J. W. Jones, M. Directorate, and A. Arbor, Scr. Mater, **32**, 1653–1658, (1995).
15. T. Kruml and K. Obrtlík, Int. J. Fatigue, **65**, 28–32, (2014).
16. F. O. R. Fischer, H. J. Maier, and H. Christ, *Influence of Environment , Mean Stress and Temperature on the Low-Cycle Fatigue Behaviour of a Near- γ Titanium Aluminide*, (2000).
17. F. Appel, T. K. Heckel, and H. J. Christ, Int. J. Fatigue, **32**, 792–798, (2010).
18. D. Caillard, M. Legros, and A. Couret, Philos. Mag., **93**, 203–221, (2013).
19. B. K. Kad and H. L. Fraser, Philos. Mag. A Phys. Condens. Matter, Struct. Defects Mech. Prop., **69**, 689–699, (1994).

Influence of the stacking fault energy and temperature on the prestrain memory effect of face centered cubic metal submitted to cyclic loadings

Clement Keller[1,*], *Gael* Marnier[2], *Wilson* Veloz[1], and *Lakhdar* Taleb[1]

[1]Groupe de Physique des Matériaux, INSA de Rouen, Université de Rouen, UMR CNRS 6634, avenue de l'Université, 76800 Saint-Etienne du Rouvray, France.
[2]ENSICAEN, bvd du Maréchal Juin, 14020 Caen, France.

Abstract. In this paper, the effect of a monotonic prehardening in tension on the subsequent cyclic stress strain curves is investigated as a function of the deformation mechanisms. To this aim, three materials are employed, copper for wavy dislocation slip, Ni20Cr alloy for planar slip and AISI 316L for intermediate deformation mechanisms. Samples were first prestrained in tension and then submitted to incremental strain cyclic tests at room temperature, and also 200°C for the 316L. Based on the analysis of the flow stress components (backstress and effective stress), the results show that all materials are sensitive to a prehardening depending on the prestrain level and cyclic test characteristics (amplitude and temperature).

1 Introduction

During the forming processes of metallic materials or accidental service overloads, strong plastic strains are introduced which may influence the following fatigue behavior in service of the metal parts. This link between the prior deformation (called PH in the following text) sequence and the mechanical behavior in fatigue has been widely studied during the past decades. This PH sequence generally induces softening depending on the material and the fatigue test.

For FCC materials, several researchers reported that a monotonic or cyclic prehardening sequence generally increases the stress level during the subsequent fatigue cycles compared to the stress obtained without PH, giving rise to a so-called "memory" effect. This effect seems to depend on the stacking fault energy (SFE) of the material and on the maximal strain amplitude. For Cu, it was reported that the flow stress for cycles with a given strain amplitude is not affected by previous cycles with larger strain amplitude [1]. In case of a tensile PH, the memory effect appears if the prestrain is larger than 20% [2–4]. For stainless steels of low SFE, Murakami et al. [5] described an increase in flow stress in fatigue if the samples were previously submitted to a PH with larger strain amplitude than during fatigue cycles. However it seems that this effect occurs if the prestrain amplitude is larger than 0.4%. For strong planar material such as CuAl or CuZn alloys, the prestrain cannot be recovered and the memory effect seems not to depend on the prior test nor on the subsequent fatigue test [1, 2, 4]. Following all these previous research, the dislocation slip characteristics clearly control the memory effect.

*e-mail: clement.keller@insa-rouen.fr

According to previous results published by the authors, up to three domains of memory effect can appear depending on the material, the prestrain level and cyclic stress amplitude [6]. The objective of this work is to try to explain the origin of these memory domains using experimental characterization of the memory effect for Cu, 316L and Ni20Cr with different dislocation slip tendencies with tests performed at room temperature and 200°C.

2 Material and experimental procedure

In order to investigate the role played by the dislocation slip tendency on the memory effect, three materials had been employed : pure copper (wavy slip), AISI 316L (intermediate between wavy and planar slip) and Ni20Cr alloy (planar slip). The samples of these materials were first submitted to different heat treatments to ensure fully recristallized microstructures with similar grain sizes around 50 μm for each material. The ratio of $\frac{\gamma_{sfe}}{\mu b^2}$ which characterizes the planarity of the dislocation slip (planar slip is linked to low value of this parameter) is illustrated for each material on table 1.

After this first heat treatment, specimens were submitted to a first tensile test. The strain levels of theses tensile tests were determined using the strain hardening stages for each material. For Cu, 316L and Ni20Cr, three different strain levels in tension were employed, one for each strain hardening stage (namely stages SI, SII, SIII). These stages, obtained usually by a derivative of the tensile curve, are illustrated figure 1 for copper. Cyclic behavior of theses samples has been then characterized using incremental strain amplitude test (total strain controlled) in order to obtain the cyclic stress strain curves (CSSC). For each PH strain level, the CSSC is then compared to the one of a virgin specimen. For each strain amplitude of the incremental test, the total number of cycles allows the quasi-saturation of the stress amplitude. Frequency has also been adapted in order to ensure similar strain rate for each strain amplitude. For 316L, in order to check the effect of temperature on the memory effect, similar CSSC are performed at 200°C using a three zones lamp furnace.

For the analysis of the mechanical behavior, the stress partition method which decomposes the stress into effective stress and backstress has been employed due to the already characterized link between these two stress components and the deformation mechanisms [7, 8].

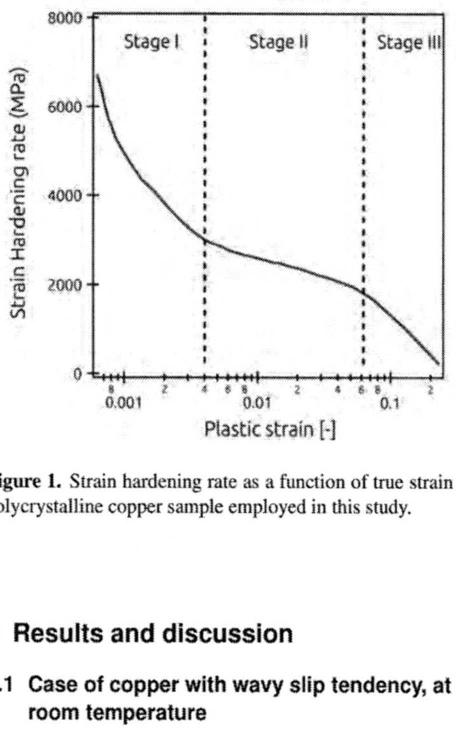

Figure 1. Strain hardening rate as a function of true strain for a polycrystalline copper sample employed in this study.

3 Results and discussion

3.1 Case of copper with wavy slip tendency, at room temperature

For copper, the figure 2(a) illustrates the CSSC of a virgin (VG) sample and three different prestrained specimens (one for each stage : 0,3% SI, 2 % SII and 10 % SIII). In agreement with the literature [2, 3], cyclic behavior of copper is sensitive to a prehardening depending on various factors. First, the overstress between the prestrain and virgin sample, which characterizes the memory effect, only appears if the prestrain is larger than a critical value. This critical value is related to the transition from the first to the second strain hardening stage in tension, related to the formation of heterogeneous dislocation structures such as tangles or walls [7]. Second, if the plastic strain amplitude of the cyclic test is larger than a second threshold value, then the overstress disapears and the cyclic mechanical behavior no longer depends on the first deformation step. Thus, two domains of memory effect are observed, one with positive overstress, for low cyclic plastic strain amplitude and a second one, without overstress for strain amplitudes larger than a critical value depending on the prestrain level.

Figures 2(b), (c) and (d) represent the evolution, associated with the CSSC of the backstress, the effective stress and mean stress respectively. For the low plastic strain amplitudes, for which the overstress is positive, the value of this parameter is largely linked to the backstress and slighly to the effective stress. Morever, this positive overstress appears simultaneously to a positive mean stress. This mean stress is associated to the backstress and begins to soften for a plastic strain amplitude close to 5.10^{-5}. This value, almost independent on the prestrain level for copper, is in agreement with the value reported by Polak et. al. for which softening occurs for a Cu virgin sample due to strain localization [9].

This beginning of stress softening for prestrained samples can be also characterized in term of ratio σ^+/σ^{PH} where σ^+ represents the maximal cyclic stress value at which softening begins and σ^{PH} represents the maximal stress value at the end of the prestrain step. For all pre-strained copper samples, this ratio is around 0.7, revealing that softening occurs for stress values lower than the yield stress inherited from the prehardening step. This value is larger than the 0.4 value reported for mild steel [10] and copper [11] both for stress controlled cyclic tests.

3.2 Case of 316L and Ni20Cr with planar slip tendency

3.2.1 At room temperature

For 316L and Ni20Cr alloys, similar tests as for copper have been performed. Due to the difference in strain hardening stages, prestrained values are different for the three materials. The two memory domains with positive overstress and zero overstress are also characterized for these two materials as illustrated in figure 3(a) for 316L and 3(b) for Ni20Cr.

Table 1 illustrates the characteristics of the positive overstress domain of memory effect for the three materials and different prestrain levels. In this table is presented, first, the threshold value of plastic strain amplitude, named $\frac{\Delta\varepsilon_p}{2}_{\text{soft}}$, which induces the softening of the mean stress (and the overall mechanical behavior). This value is then followed by the associated value of σ^+/σ^{PH}, already defined in the previous section. $\frac{\Delta\varepsilon_p}{2}_{P-0}$ characterizes the plastic strain amplitude corresponding to the transition between the positive and zero overstress and $\Delta\Sigma^{max}_{norm}$ represents the maximal flow stress increase for a prestrain sample compared to a virgin one. In order to be compared between the three materials, this value of stress increase

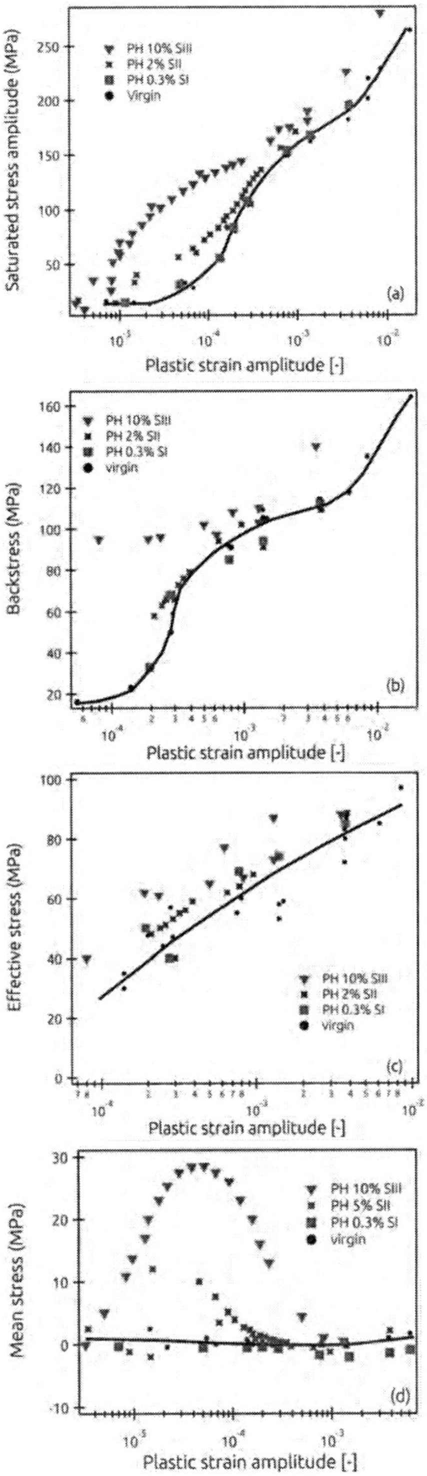

Figure 2. Cyclic stress-strain curves for polycristalline copper submitted to tensile prehardening : (a) saturated stress amplitude, (b) backstress, (c) effective stress and (d) mean stress.

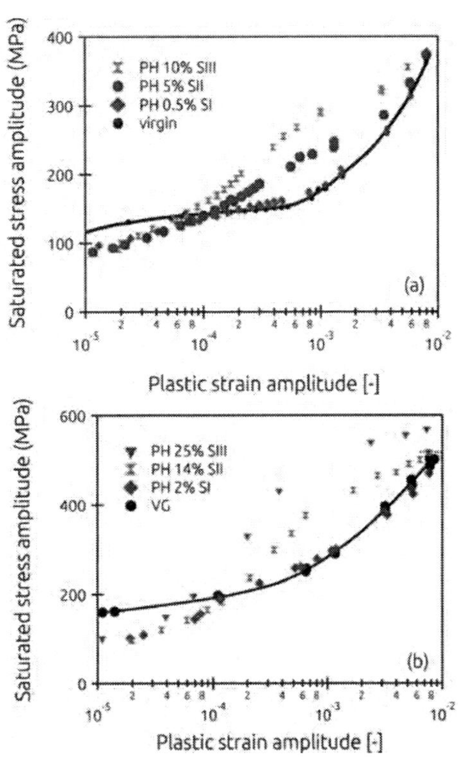

Figure 3. Cyclic stress strain curves for : (a) AISI 316L and (b) Ni20Cr alloy.

is normalised, for each material, by the stress level of the virgin sample (at the strain amplitude involving the maximal increase).

For 316L and Ni20Cr, the softening of mean stress starts for larger strain amplitude. Despite large difference in $\frac{\gamma_{sfe}}{\mu b^2}$ between Ni20Cr and 316L, the value for these two materials are very close. The comparison of the value of σ^+/σ^{PH} for the three materials illustrates the influence of the deformation mechanisms of the softening of the mechanical behavior. With the transition from wavy to planar slip, the σ^+/σ^{PH} ratio progressively shifts towards unity. For Ni20Cr, the softening of the mechanical behavior inherited from the prestrain only occurs if the loading surface is reached whereas, for 316L and copper, this softening occurs for maximal stress levels lower than the yield stress.

This evolution clearly highlights the role played by cross slip on the softening and the recovery of the mechanical behavior inherited from the prestrain. In case of Cu, the cross-slip is operative at very low strain amplitude as already characterized in literature [1] whereas for 316L, this cross slip needs larger strain amplitude and stress level to be operative. For Ni20Cr, due to the restricted activity of cross-slip, generalized plasticity is needed to transform the dislocation structures inherited from the monotonous prestrain into those characteristic of cyclic

Table 1. Summary of the characteristics of the positive overstress domain of memory effect for the three materials and different prestrain levels.

Material	$\frac{\gamma_{sfe}}{\mu b^2}$ (10^6) m^{-1}	Prestrain level	$\frac{\Delta\varepsilon_p}{2}_{soft}$	σ^+/σ^{PH}	$\frac{\Delta\varepsilon_p}{2}_{P-0}$	$\Delta\Sigma^{max}_{norm}$
		0.3%	-	-	-	-
Cu	16	2%	$4.5\ 10^{-5}$	0.69	$3.7\ 10^{-3}$	3.3
		10%	$5.5\ 10^{-5}$	0.67	$8.5\ 10^{-3}$	4.61
		0.5%	-	-	-	-
316L	6	5%	$2\ 10^{-4}$	0.84	$8\ 10^{-3}$	0.4
		10%	$3.9\ 10^{-4}$	0.845	$1\ 10^{-2}$	0.6
		2%	$2\ 10^{-4}$	0.99	$3\ 10^{-3}$	0.15
Ni20Cr	2.5	14%	$4\ 10^{-4}$	0.98	$8.7\ 10^{-3}$	0.47
		25%	$5\ 10^{-4}$	0.98	$1.3\ 10^{-2}$	0.83

loadings. The comparison of the last parameter of table 1 between the three materials illustrates the larger strain hardening capacity of the wavy slip material.

Compared to copper, a new domain is also characterized for 316L and Ni20Cr with a negative overstress for the lower plastic strain amplitudes. The analysis of the stress components reveals that this negative overstress is linked to the effective stress. Compared to copper which is significantly textured, 316L and Ni20Cr do not, in the unstrained configuration, exhibit any preferential grain orientations. For 316L and Ni20Cr, the first strain step involves a grain rotation towards (001) and (111) orientations, the larger the plastic strain, the larger the grain rotation. However, following literature, these two grain orientations involve larger stress amplitude for 316L [12]. As a consequence, this mechanism of grain rotation cannot explain the negative overstress. This overstress is then probably link to the existence, after the prestrain, of forest dislocations which can flip-flop easily for very low plastic strain amplitude during the fatigue test, reducing the stress amplitude [13]. The decrease of the threshold plastic strain amplitude between the negative and positive memory domain for larger prestrain levels in in agreement with this mechanisms. With the increase in monotonic prestrain, dislocations tend to form heterogeneous structures such as walls or cells with large dislocation density which, in turn, reduces their mobility to accommodate the cyclic plastic strain amplitude.

3.2.2 316L at 200°C

As illustrated in the previous section, the characteristics of the memory effect strongly depends on the cross-slip activity of the material. For 316L with intermediate strain mechanisms between wavy and planar slip, an increase in temperature should increase the cross-slip activity and, in turn, improve the softening capacity of the material. After a first monotonic prestrain at room temperature, the incremental cyclic tests are performed at 200°C. At this temperature, the ratio $\frac{\gamma_{sfe}}{\mu b^2}$ is very close to the one at Room Temperature - RT - (increase in γ_{sfe} [14] but decrease in μ [15]) but thermal energy should enhance cross slip activity without dynamic strain ageing.

In order to be compared, all the stress values at RT and 200°C were normalized by the shear modulus at each temperature (μ_{RT}=67 GPa and μ_{200}=60.5 GPa [15]). Figure 4 illustrates the normalized stress amplitude as a function of the plastic strain amplitude at RT and 200°C for a virgin, a prestrain in tension at 5% and prestrain in tension at 10% samples. A signicative difference in stress amplitude value appears between the CSSC of the virgin samples at RT and 200°C. This difference is larger at low plastic strain amplitude and seems to be due to the thermal activation of the short range interactions between dislocations and the microstructure.

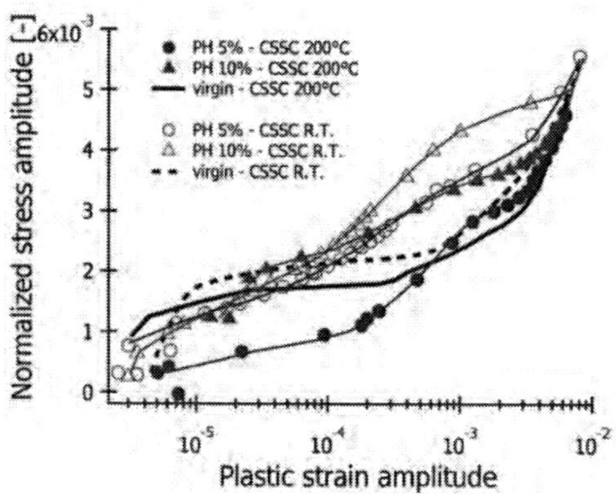

Figure 4. CSSC normalized by the shear modulus obtained at room temperature and 200°C for tensile prestrained 316L samples.

After a PH at 5% in tension, at 200°C the range of plastic strain amplitude with positive overstress is considerably reduced. Indeed, for this sample cyclically strained at 200°C, the transition plastic strain amplitude from negative to positive overstress is increased compared to RT (6.10^{-4} at 200°C vs 10^{-4} at RT) whereas the plastic strain amplitude inducing no memory effect is reduced (4.10^{-4} at 200°C vs 8.10^{-3} at RT).

Nevertheless, for PH a 10% in tension, the differences between the CSSC for the two investigated temperatures are weaker. Compared to RT, the elevation in temperature induces a lower plastic strain amplitude value at which the overstress vanished as for PH 5% and the overstress is significantly reduced as observed in figure 4.

These results show, first, as expected, that the increase in temperature enhances the crossslip activity which results in a reduction in overstresses and plastic strain amplitude at which this overstress is closed to 0. Second, the lower effect of the temperature for a prestrain of 10% compared to 5% shows that as soon as the third stage of prehardening has been reached, recovery of the dislocation structures inherited from the prehardening step is harder.

4 Conclusions

Using three materials Cu, 316L and Ni20Cr with different cross slip abilities, the investigation of the prehardening effect on the subsequent cyclic behavior highlights the following features :

- three domains of memory effect can be characterized (negative overstress, positive overstress, no overstress). The negative overstress domain is only observed material with planar slip tendency.

- the transition from positive to 0 overstress domains is linked to mechanical softening. This softening can appear for stress values lower than the yield stress for wavy slip material whereas for planar slip material softening occurs when yield stress is reached.

- an increase in temperature enhances cross slip activity and reduces the plastic strain amplitude range of the cyclic test for which positive overstress is observed.

The memory effect is then triggered by the cross-slip ability of the material which can be affected by the temperature test.

References

[1] C. Feltner, C. Laird, Acta metallurgica **15**, 1621 (1967)
[2] P. Lukáš, M. Klesnil, Materials Science and Engineering **11**, 345 (1973)
[3] H.J. Christ, G. Hoffmann, O. Öttinger, Materials Science and Engineering: A **201**, 1 (1995)
[4] C. Laird, J. Finney, A. Schwartzman, R. De la Veaux, Journal of Testing and Evaluation **3**, 435 (1975)
[5] S. Murakami, M. Kawai, Y. Ohmi, Journal of Engineering materials and Technology **111**, 278 (1989)
[6] G. Marnier, C. Keller, L. Taleb, International Journal of Plasticity **78**, 64 (2016)
[7] X. Feaugas, H. Haddou, Metallurgical Transactions **34A**, 2329 (2003)
[8] C. Keller, E. Hug, X. Feaugas, International Journal of Plasticity **27**, 635 (2011)
[9] J. Polák, K. Obrtlík, M. Hájek, A. Vašek, Materials Science and Engineering: A **151**, 19 (1992)
[10] H.F. Chai, C. Laird, Materials Science and Engineering **93**, 159 (1987)
[11] Q. Fan, H. Chai, M. Tong, Acta Metallurgica Sinica (English letters) **4**, 129 (2009)
[12] M. Mineur, P. Villechaise, J. Mendez, Materials Science and Engineering: A **286**, 257 (2000)
[13] L. Kunz, P. Lukáš, B. Weiss, D. Melisova, Materials Science and Engineering: A **314**, 1 (2001)
[14] L. Rémy, A. Pineau, B. Thomas, Materials Science and Engineering **36**, 47 (1978)
[15] M. Fukuhara, A. Sanpei, ISIJ International **33**, 508 (1993)

Fatigue life improvement of cast ZK60 Mg alloy through low temperature closed-die forging for automotive applications

SMH. Karparvarfard[1,*], *S.K. Shaha*[1], *S.B. Behravesh*[1], *H. Jahed*[1], and *B. Williams*[2]

[1]Deparment of Mechanical & Mechatronics Engineering, University of Waterloo, 200 University Ave W, Waterloo, ON N2L 3G1, Canada
[2]CanmetMATERIALS, Natural Resources Canada, 183 Longwood Road South, Hamilton, ON L8P 0A1, Canada

Abstract. The influence of low-temperature closed-die forging on the quasi-static and cyclic behaviour of as-cast ZK60 magnesium alloy was investigated. As-cast ZK60 billets were forged at a ram speed of 20 mm/sec and a temperature of 250 °C. While the yield strength of the starting alloy was 139 MPa, the forging process improved the yield strength significantly by ~68% to 234 MPa. Moreover, the stress-controlled push-pull fatigue tests at the stress amplitudes of 140 MPa to 180 MPa revealed that the fatigue life was enhanced by an order of magnitude. Microstructural analyses besides the texture measurements at different locations of the forged part manifested partial grain refinement and texture modification strengthening mechanisms. It is believed that the fatigue life improvement is achieved in the wake of the grain refinement and the subsequent material strengthening.

1 Introduction

Concerns over the adverse impact of fossil fuels have driven the global transportation sector to change their products' structure through the widespread adoption of lightweight materials. Magnesium (Mg), as the lightest commercial metal available in the industry, can play an integral role in the lightweighting of multi-material vehicle architecture [1]. However, to date, the usage of this metal has been restricted to the non-load bearing components such as seat frames and housing parts [2]. Moreover, the dominant manufacturing process is die-casting for its economic benefits [3], whilst casting brings about abundant of defects in the microstructure of alloys like porosities and inclusions, deteriorating its mechanical properties. As a result, wrought Mg alloys are gaining interests to be utilized in vehicles in load-bearing components [4]–[8].

Among different manufacturing techniques, forging is of particular interest not only because it can lead to enhanced mechanical properties, but also it can form components with complicated geometries [9][10]. The monotonic and cyclic behaviour of different open die- and semi-closed die forged Mg alloys have been investigated so far [11]–[19]. The forging temperature at these studies were quite high, i.e., ~450 °C. However, to expand the applications of forged Mg alloys in high volume production passenger cars, it is crucial to carry out the manufacturing process at a lower temperature, for it be cost-effective. In this study, we investigated the influence of low-temperature closed-die forging on the quasi-static and fatigue response of as-cast ZK60. The forging temperature in this study was 250 °C, which is less than that in previous studies, but also is high enough to achieve a crack-free material after the forging. Consequently, there is the need to better understand how the microstructure relates to the final mechanical properties, particularly the fatigue response, after low temperature forging.

2 Material and Experiments

Cylindrical billets of 63.5mm diameter and 65 mm length were machined from a commercial as-cast ZK60 Mg alloy ingot (5.8% Zn, 0.61% Zr, <0.3% total others, and Mg as balance- analytical weight percent). The billets were gradually heated to the temperature of 250 °C, and then transferred to a hydraulic press whose dies were also pre-heated to the same temperature. This was the lowest forging temperature for a crack-free sample [20]. The billet was forged into an I-beam shape [10] which allowed material flow in different directions at varying strain-rates. The forging was conducted at an initial ram speed of 20 mm/sec. To reduce friction between the billet and the die, a graphite lubricant was utilized. After forging, no heat treatment was employed, and the forged part was air- cooled to the room temperature. Figure 1 represents the forged part and the schematic locations of the dog-bone samples extracted from it. TD, LD, and FD (as shown in Figure 1) denote the radial, longitudinal, and forging directions, respectively.

* Corresponding author: smkarparvarfard@uwaterloo.ca

 https://doi.org/10.1051/matecconf/201816506009

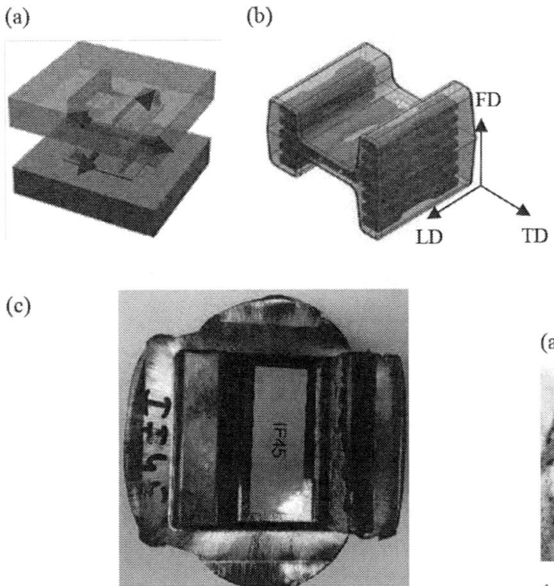

Fig. 1. (a) The schematic demonstration of the forging process of the as-cast billet; (b) the schematic representation of the final forged part and the location of dog-bone samples; (c) the photograph of the final closed-die forged part from the top view

Dog–bone samples were machined from different parts of the forged beam as shown schematically in Figure 1(b). Details about the geometry of the samples and the location of samples in the cast billet could be found in [11]. Quasi-static tests were carried out at the displacement rate of 1 mm/min. Strain was captured during these tests using ARAMIS 3D Digital Image Correlation (DIC) system with 5-megapixel resolution and the frame rate of 15 fps. Stress-controlled fully reversed ($R = -1$) push-pull fatigue tests were performed under standard laboratory conditions, using an Instron 8874 servo-hydraulic frame with a load capacity of ±25 KN at three different stress amplitudes, 140 MPa, 160 MPa, and 180 MPa. These stress amplitudes were chosen such that fatigue lives within the range of 10^3 to 10^5 cycles could be obtained. The test frequencies were 5 Hz, 10 Hz, and 15 Hz for the tests at 180 MPa, 160 MPa, and 140 MPa, respectively.

A Bruker D8 Discover X-ray diffractometer equipped with a 2D-detector using CuKα beam and a current of 40 mA was utilized to measure texture. The procedure for the microstructure analysis and texture measurement was the same as the one presented in [11]. Hardness testing was performed using a United True-Blue Rockwell tester.

3 Results and Discussion

3.1 Microstructure, texture, and hardness

The microstructure and texture of the as-cast ZK60 have already been detailed in [11]. The cast alloy exhibited an average grain size of 131±26 μm. Moreover, the texture measurements indicated that the c-axis of the hexagonal close-packed (HCP) crystal structure of the grains are randomly oriented. Figure 2 displays the typical optical microscope (OM) microstructure and the collected (0002) and $(10\bar{1}0)$ pole figures of the as-cast alloy.

For the microstructure and texture analyses of the forged part, samples were extracted from different locations on the mid-cross section of the I-beam. Figure 2 shows the microstructure of the forged material in the two flanges, and the web of the beam.

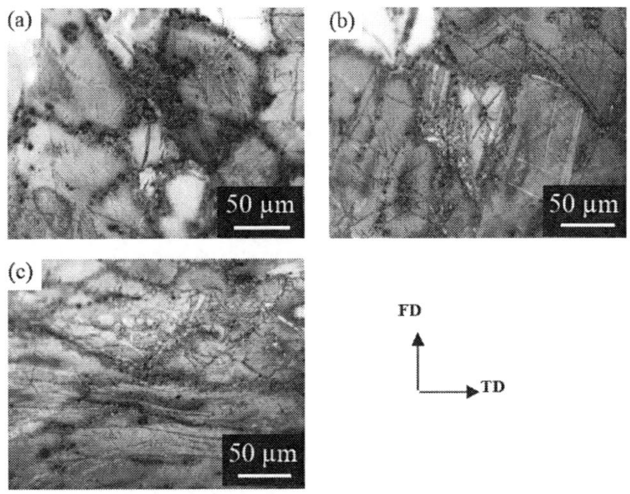

Fig. 2. The optical micrograph of forged ZK60 at (a) tall flange, (b) short flange, (c) and the web of the beam. The indicated directions apply to all figures.

According to Figure 2, the close-die forging brings about a bimodal grain structure in the forged alloy; i.e., elongated grains are surrounded by fine grains (2-5 μm). This indicates that partial dynamic recrystallization (DRX) is happening [21]. The incomplete DRX can stem from that the low processing temperature and high deformation rate inhibit a complete recrystallization happening [21]. Moreover, it is reported that rare-earth materials in ZK60 can impede dislocation movement by their pinning effect leading to a partial DRX [22]. Another significant aspect of microstructure is that qualitatively the volume fraction of fine recrystallized grains is higher in the web of the I-beam forged alloy (Figure 2.c). This affects the mechanical behavior of the forged alloy as will be discussed later.

Figure 3 depicts the calculated (0002) and $(10\bar{1}0)$ pole figures of the samples machined from the cross-section of the forged material at the abovementioned locations and the corner of the beam. To better understand the pole figures, the hexagonal close packed (HCP) grains of the crystallographic microstructure are shown schematically in Figure 3.

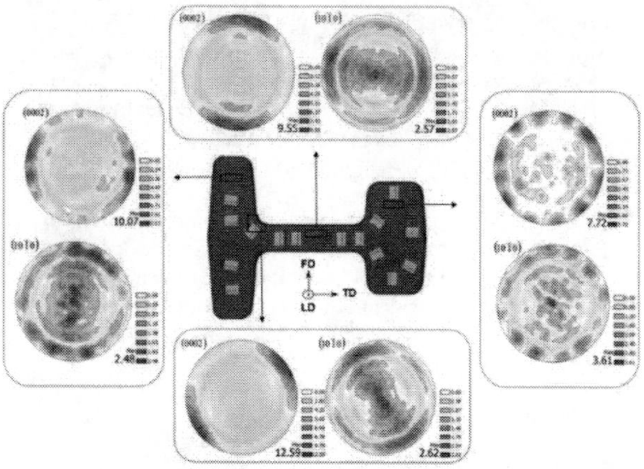

Fig. 3. (0002) and (10$\bar{1}$0) pole figures for the I-beam forged ZK60 Mg alloy at different locations

According to Figure 3, the plastic deformation as a result of the forging process develops a sharp basal texture in the forged ZK60 at different locations. This texture modification is because of the Rotational Dynamic Recrystallization (RDRX) that rotates the fine grains of the forged material [23]. Nevertheless, the orientation of the c-axis of the HCP grains changes locally. To be more specific, the die cavity provides a small space for the material flow within the web section, which causes severe compressive plastic deformation in the web of the I-beam forging. As a result, most of the HCP grains orient along the forging direction with the maximum pole intensity of 9.55 MRD. In contrast, in the shorter flange, more space is available in the die cavity for the material flow. Therefore, the level of compressive plastic deformation is less than the web and the HCP grains are more randomly distributed in the TD and FD, yet perpendicular to the LD direction. Also, in the taller and thinner flange, a texture with a higher intensity value of 10.07 MRD is induced, and the grains are mostly along the TD direction. The modified texture after the forging process affects the mechanical behaviour as will be discussed later.

Superficial hardness test was performed on the mid-cross section of the beam. Testing was carried out on 34 locations across the section in a 30-Ton scale. Figure 4 depicts the Rockwell hardness test results. It is noted that the hardness is roughly symmetric across the beam section. Also, the highest values were observed on the web of the section. This concurs with the microstructure results. It is believed that due to the geometry limitations of the die in the web area, plastic deformation is more severe in this region. Therefore, grains are finer, and material is harder within the web section.

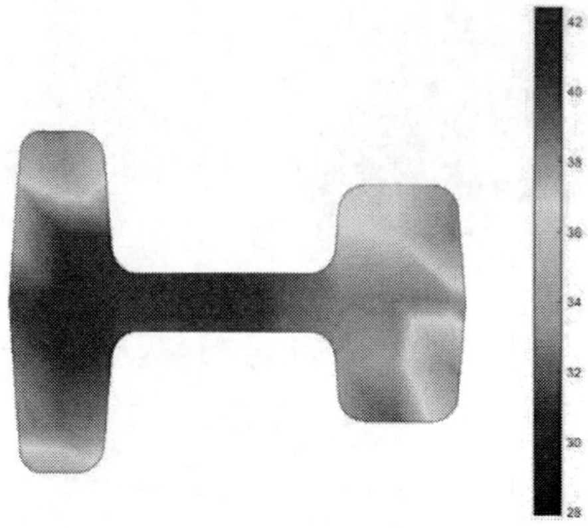

Fig. 4. Rockwell hardness measurement on the cross section of the forged ZK60 Mg alloy in a 30T scale

3.2 Quasi-static behaviour

Figure 5 displays the tensile quasi-static behaviour of as-cast and cast-forged ZK60 Mg alloy. For the sake of comparison, the effects of open-die forging at the temperature of 450 °C on the quasi-static behaviour of ZK60 and AZ31B Mg alloys which have already been published in literature are also represented in Figure 5 [11], [13].

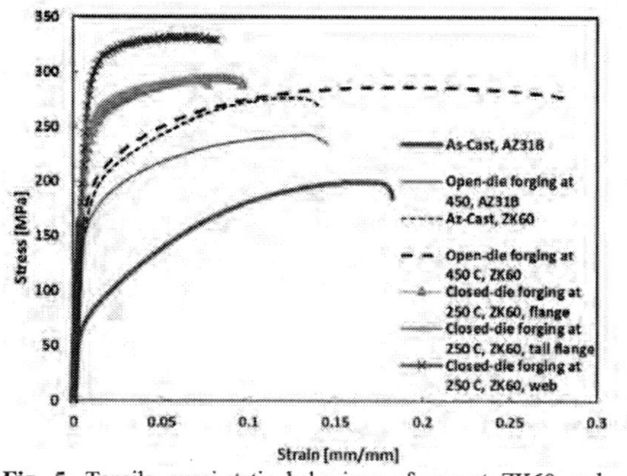

Fig. 5. Tensile quasi-static behaviour of as-cast ZK60 and closed-die forged ZK60 at 250 °C in the web and flanges of the I-beam

It is noted that the closed-die forging process promisingly enhanced the strength of the alloy comparing to the as-cast condition. While the yield strength of ZK60 cast was 139 MPa, the average yields in the two flanges and the web of the forged I-beam were improved to 218 MPa (~57% improvement) and 268 MPa (~93% improvement), respectively. In contrast, the ductility of the alloy was decreased from 15% to 9%. The average mechanical properties of as-cast, open-die forged [11], and closed-die forged ZK60 Mg alloy is summarized in Figure 6. In this graph, YS and FS

https://doi.org/10.1051/matecconf/201816506009

represent yield strength and fracture strain, respectively. It is believed that the underlying reason for the increase of yield strength and the decrease of ductility after the closed-die forging is the grain refining impact of the process (Figure 2). In fact, the forged alloy is exhibiting higher strength in the wake of its finer grains, thereby higher strength according to the Hall-Petch relation [24]. On the other hand, grain refinement yields higher density of grain boundaries that augments the impedance to the dislocation movement. It is reported that the lack of strain accommodating mechanisms as a result of the increase of barriers to dislocation movement after grain refinement reduces the ductility [25].

Lastly, higher strength is observed in the web of I-beam in comparison to its flanges. In fact, in this region more severe plastic deformation is occurring as a result of geometry constraints, therefore, grains are finer in the web (Figure 2); hence, the strength of this part is more than that in the I-beam flanges.

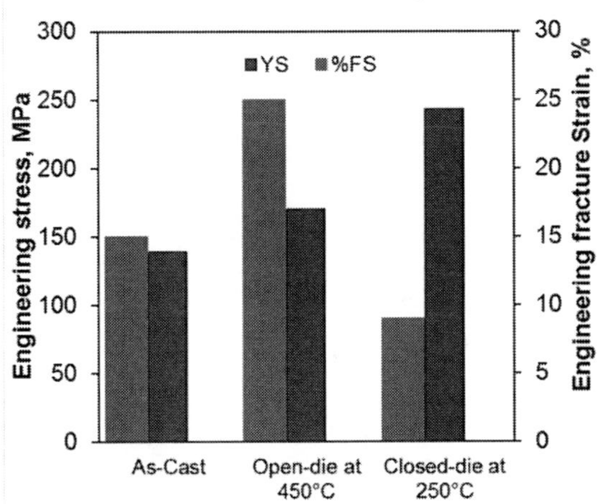

Fig. 6. The yield strength and fracture strain of as-cast, open-die forged [11], and closed-die forged ZK60 Mg alloy

3.3 Fatigue behaviour

Fatigue tests were performed under stress-controlled mode at three different stress amplitudes of 140, 160, and 180 MPa. Figure 7 demonstrates the S-N results for the as-cast and forged ZK60 Mg alloys. The scatter of the results for the forged condition is because the samples were extracted from different locations in the I-beam. However, the results revealed an order of magnitude improvement in the life at all the three load levels for the forged condition regardless of the specimen location. For instance, while the fatigue life of the as-cast alloy at the stress amplitude of 140 MPa was 6878 cycles, the average life for the closed-die forged samples was enhanced to 35812 cycles.

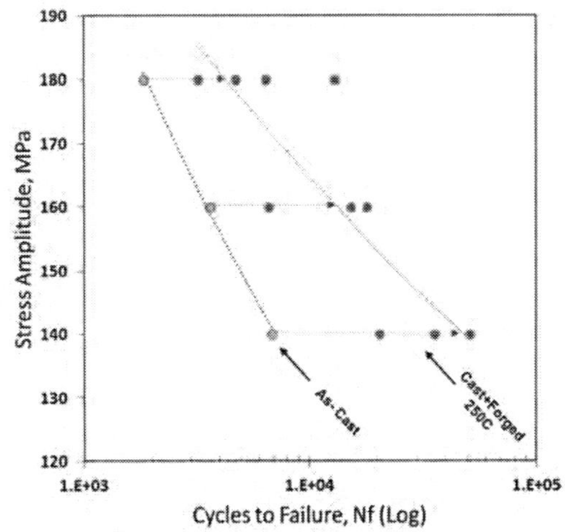

Fig. 8. Stress-controlled fully-reversed push-pull fatigue tests results of as-cast and cast-forged ZK60 Mg alloy

As discussed earlier, microstructural analysis besides texture measurement of the forged part at different locations manifested partial grain refinement and texture modification strengthening mechanisms. It is believed that the fatigue life improvement is achieved in the wake of the grain refinement and the subsequent material strengthening. Indeed, the forged alloy having higher strength exhibited less plastic deformation during the cyclic loading, thereby showing higher number of cycles to accumulate fatigue damage required for the crack initiation. In addition, the finer grains in the microstructure of the forged alloy created more obstacles against the crack propagation which retards the final failure [26]. In contrast, the cast alloy has lower grain boundary density with larger grains serving to increasing the crack propagation rate and lowering the fatigue life.

Furthermore, as was pointed out in section 3.1, the forged alloy showed a strong basal texture. The texture can bring about extension twinning happening once a compressive load is applied along the longitudinal direction of the I-beam. Upon unloading, some twin deformations are recovered, but some residual twins remain inside the microstructure [27]. These residual twins can affect the fatigue response of the alloy in two ways. First, the interactions between the twin-twin bands and also twin-dislocations can lead to the crack initiation causing premature failure [14][28]. Alternatively, some studies have suggested that the extension twinning can elongate the fatigue life in the light of crack growth retardation due to the roughness-induced crack closure [29]. The combination of the abovementioned competing factors leads to the enhanced fatigue response of the forged ZK60.

4 Conclusions

In the present study, the effects of closed-die forging on microstructure, texture, quasi-static, and cyclic behaviour of cast ZK60 Mg alloy was studied. From the above results and discussions, the following conclusions could be made:

- The forging process brought about grain refinement in the microstructure of the material. The grains in the web of the forged I-beam was finer, since more severe plastic deformation was occurring in that area due to the geometrical constrains.

- The forging process induced sharp basal texture in the crystallographic microstructure of the material so that the c-axes of the most hexagonal grains were perpendicular to the longitudinal direction. This texture modification can activate extension-twinning under high stress amplitudes.

- The yield strength of the closed-die forged alloy was higher than that of the as-cast and open-die forged alloys at higher temperature. The underlying reason for material strengthening was grain refinement and texture modification.

- Forging process improved the stress-controlled fatigue response, considerably. Indeed, at all stress amplitudes, the fatigue life of forged alloy was higher than that of as-cast alloy by an order of magnitude. Grain refinement in the forged part could significantly contribute to the fatigue life improvement by retarding the crack initiation and growth.

- The ultimate goal of this project is to develop knowledge for the adoption of the forged Mg alloys in the control arm of a massive production passenger car. It is believed that forging is a beneficial manufacturing process for ZK60 Mg alloy to be utilized in the automotive industry, as the strength and fatigue life of the alloy can be improved remarkably.

The authors would like to gratefully acknowledge the financial support of the Natural Sciences and Engineering Research Council of Canada, the Automotive Partnership Canada (APC) program. The authors would also like to thank Jonathan McKinley and Lucian Blaga from CanmentMATERIALS for performing the forging trials.

References

[1] B. L. Mordike and T. Ebert, "Magnesium Properties - applications - potential," *Mater. Sci. Eng. A*, vol. 302, no. 1, pp. 37–45, 2001.

[2] M. K. Kulekci, "Magnesium and its alloys applications in automotive industry," *Int. J. Adv. Manuf. Technol.*, vol. 39, no. 9–10, pp. 851–865, 2008.

[3] F. Pan, M. Yang, and X. Chen, "A Review on Casting Magnesium Alloys : Modification of Commercial Alloys and Development of New Alloys," *J. Mater. Sci. Technol.*, vol. 32, no. 12, pp. 1211–1221, 2016.

[4] H. Jahed and J. Albinmousa, "Multiaxial behaviour of wrought magnesium alloys – A review and suitability of energy-based fatigue life model," *Theor. Appl. Fract. Mech.*, vol. 73, pp. 97–108, 2014.

[5] J. Albinmousa, H. Jahed, and S. Lambert, "Cyclic behaviour of wrought magnesium alloy under multiaxial load," *Int. J. Fatigue*, vol. 33, no. 8, pp. 1127–1139, 2011.

[6] A. A. Roostaei and H. Jahed, "Role of loading direction on cyclic behaviour characteristics of AM30 extrusion and its fatigue damage modelling," *Mater. Sci. Eng. A*, vol. 670, pp. 26–40, 2016.

[7] A. A. Roostaei and H. Jahed, "Multiaxial cyclic behaviour and fatigue modelling of AM30 Mg alloy extrusion," *Int. J. Fatigue*, vol. 97, pp. 150–161, 2017.

[8] A. Hadadzadeh, M. A. Wells, S. Kumar, H. Jahed, and B. W. Williams, "Role of compression direction on recrystallization behavior and texture evolution during hot deformation of extruded ZK60 magnesium alloy," *J. Alloys Compd.*, vol. 702, pp. 274–289, 2017.

[9] M. Madaj, M. Greger, and V. Karas, "MAGNESIUM-ALLOY DIE FORGINGS FOR AUTOMOTIVE APPLICATIONS," *Mater. Tehnol.*, vol. 49, no. 2, pp. 267–273, 2015.

[10] G. Yu, "Forging Specimen Design for Magnesium Alloys," *(Master's Thesis), Retrieved from Univ. Waterloo Database*, 2016.

[11] S. M. H. Karparvarfard, S. K. Shaha, S. B. Behravesh, H. Jahed, and B. W. Williams, "Microstructure , texture and mechanical behavior characterization of hot forged cast ZK60 magnesium alloy," *J. Mater. Sci. Technol.*, 2017.

[12] S. M. H. Karparvarfard, S. K. Shaha, A. Hadadzadeh., S. B. Behravesh, H. Jahed, M. A.

Wells, and B. Williams, "Characterization of Semi-Closed Die-Forged ZK60 Mg Alloy Extrusion," *Magnes. Technol. 2017*, pp. 329–334, 2017.

[13] D. Toscano, S. . Shaha, S. B. Behravesh, H. Jahed, M. A. Wells, B. W. Williams, and J. McKinley, "Effect of Forging on Microstructure, Texture and Uniaxial Properties of Cast AZ31B Alloy," *J. Mater. Eng. Perform.*, vol. 26, no. 7, pp. 3090–3103, 2017.

[14] D. Toscano, S. K. Shaha, B. Behravesh, H. Jahed, and B. Williams, "Effect of forging on the low cycle fatigue behavior of cast AZ31B alloy," *Mater. Sci. Eng. A*, vol. 706, no. May, pp. 342–356, 2017.

[15] D. Toscano, S. K. Shaha, B. Behravesh, H. Jahed, M. Wells, B. Williams, and J. McKinley, "Effect of Forging on Microstructure, Texture and Compression Behavior of Extruded AZ31B," in *Proceedings of the 3rd Pan American Materials Congress*, Springer International Publishing, 2017, pp. 347–354.

[16] A. Gryguc, H. Jahed, B. Williams, and J. Mckinley, "MagForge – Mechanical Behaviour of Forged AZ31B Extruded Magnesium in Monotonic Compression," *Mater. Sci. Forum*, vol. 828–829, pp. 291–297, 2015.

[17] A. Gryguc, S. K. Shaha, S. B. Behravesh, H. Jahed, M. Wells, and B. Williams, "Compression Behaviour of Semi-closed Die Forged AZ80 Extrusion," *Charact. Miner. Met. Mater. 2017*, no. TMS Conference, Location: San Diego, CA, USA, Date: FEB,26-MAR,2, pp. 361–369, 2017.

[18] A. Gryguc, S. K. Shaha, S. B. Behravesh, H. Jahed, M. Wells, B. Williams, and X. Su, "Monotonic and cyclic behaviour of cast and cast-forged AZ80 Mg," *Int. J. Fatigue*, vol. 104, pp. 136–149, 2017.

[19] A. Gryguc, S.K. Shaha, H. Jahed, M. Wells, B. Williams, J. McKinley, "Tensile and fatigue behaviour of as-forged AZ31B extrusion," *Fract. Struct. Integr.*, vol. 38, pp. 251–258, 2016.

[20] N. Ogawa, M. Shiomi, and K. Osakada, "Forming limit of magnesium alloy at elevated temperatures for precision forging," *Int. J. Mach. Tools Manuf.*, vol. 42, no. 5, pp. 607–614, 2002.

[21] Y. Wu, H. Yan, S. Zhu, J. Chen, A. Liu, and X. Liu, "Flow behavior and microstructure of ZK60 magnesium alloy compressed at high strain rate," *Trans. Nonferrous Met. Soc. China*, vol. 24, no. 4, pp. 930–939, 2014.

[22] H. Watanabe, K. Moriwaki, T. Mukai, T. Ohsuna, K. Hiraga, and K. Higashi, "Materials Processing for Structural Stability in a ZK60 Magnesium Alloy," *Mater. Trans.*, vol. 44, no. 4, pp. 775–781, 2003.

[23] Y. Xu-yue, J. I. Ze-sheng, H. Miura, and T. Sakai, "Dynamic recrystallization and texture development during hot deformation of magnesium alloy AZ31," *Trans. Nonferrous Met. Soc. China*, vol. 19, no. 1, pp. 55–60, 2008.

[24] F. Yang, S. M. Yin, S. X. Li, and Z. F. Zhang, "Crack initiation mechanism of extruded AZ31 magnesium alloy in the very high cycle fatigue regime," *Mater. Sci. Eng. A*, vol. 491, pp. 131–136, 2008.

[25] C. M. Cepeda-Jimenez, J. M. Kolina-Aldareguia, and M. T. Perez-Prado, "Effect of grain size on slip activity in pure magnesium polycrystals," *Acta Mater.*, vol. 84, pp. 443–456, 2015.

[26] M. Tsushida, K. Shikada, H. Kitahara, S. Ando, and H. Tonda, "Relationship between Fatigue Strength and Grain Size in AZ31 Magnesium Alloys," *Mater. Trans.*, vol. 49, no. 5, pp. 1157–1161, 2008.

[27] L. Wu, A. Jain, D. W. Brown, G. M. Stoica, S. R. Agnew, B. Clausen, D. E. Fielden, and P. K. Liaw, "Twinning–detwinning behavior during the strain-controlled low-cycle fatigue testing of a wrought magnesium alloy, ZK60A," *Acta Mater.*, vol. 56, no. 4, pp. 688–695, 2008.

[28] H. Yu, Y. Xin, A. Chapuis, X. Huang, R. Xin, and Q. Liu, "The different effects of twin boundary and grain boundary on reducing tension-compression yield asymmetry of Mg alloys," *Nat. Publ. Gr.*, no. May, pp. 4–11, 2016.

[29] Y. J. Wu, R. Zhu, J. Tao, and W. Qing, "Role of twinning and slip in cyclic deformation of extruded Mg – 3 % Al – 1 % Zn alloys," *Scr. Mater.*, vol. 63, no. 11, pp. 1077–1080, 2010.

MATEC Web of Conferences **165**, 06010 (2018) https://doi.org/10.1051/matecconf/201816506010
FATIGUE 2018

Influence of the surface morphology on the cyclic deformation behavior of HSD® 600 steel

Matthias W. Klein[1*], *Marek* Smaga[1], and *Tilmann* Beck[1]

[1]Institute of Materials Science and Engineering, TUK , 67663 Kaiserslautern, Germany

Abstract. The presented research work investigates the fatigue properties in the low cycle fatigue (LCF) regime of the high manganese metastable austenitic High Strength and Ductility (HSD®) 600 TWIP steel dependent on its surface morphology. The steel features, according to its chemical composition following the alloying concept Mn-Al-Si-C and heat treatment, a fully austenitic microstructure that shows deformation induced twinning at ambient temperature. Due to this microstructural deformation mechanism, HSD® 600 steel has an outstanding combination of strength and formability. Besides monotonic deformation behavior, characterized by tensile tests, cyclic deformation behavior was investigated with varying the surface morphology of fatigue specimens. In order to create different surface morphologies, flat fatigue specimens were excised from larger sheets by waterjet-cutting. Depending on the surface morphology, further climb milling or up-climb milling in the gauge length was performed. The three investigated morphologies (as-received with rolling skin, climb milled and up-climb milled) differed in roughness, initial residual stresses and initial phase compositions. For all variants, total strain controlled fatigue tests with stepwise increasing load amplitudes as well as total strain controlled single step tests were performed in the low cycle fatigue regime with a load ratio of R_ε = -1 and a frequency of f = 0.2 Hz. Beside stress-strain hystereses, the changes in temperature ΔT and the magnetic properties ξ were measured. The magnetic properties directly correlate with the transformation from paramagnetic γ-austenite to ferromagnetic α'-martensite. The cyclic deformation behavior of the HSD® 600 steel in the LCF regime was characterized by cyclic softening until fracture at low total strain amplitudes but changed with increasing total strain amplitudes into initial cyclic hardening followed by cyclic softening. This initial cyclic hardening became more pronounced when the total strain amplitude increased. Furthermore, single step tests at lower total strain amplitudes showed a saturation state before fracture. A comparison between the monotonic and cyclic deformation behavior showed a significant difference of the stress levels at the same amounts of plastic deformation respectively. Nevertheless, the different surface morphologies led to different lifetimes at high total strain amplitudes but to similar lifetimes at lower total strain amplitudes.

1 Introduction

TWinning Induced Plasticity (TWIP) steels with the alloying concept Mn-Al-Si-C offer an outstanding combination of strength and formability due to deformation induced microstructural mechanisms like twinning [1]. These properties are needed for new design concepts for energy absorbing applications such as crash resistant components in the automotive industry or deep drawing of sheet materials [1].

TWIP steels typically have an austenitic microstructure at ambient temperature. The TWIP mechanism is characterized by the formation of deformation twins. These twins represent strong barriers for mobile dislocations and cannot be circumvented, and therefore further dislocation pile-up occurs. This ongoing process is called the dynamic Hall-Petch effect [2, 3]. The activation of deformation induced mechanisms depends on the Stacking Fault Energy (SFE) and consequently on the chemical composition of the material. High manganese steels with additions of aluminum and silicon achieve SFE < 60 mJ/m² [4].

Several investigations have characterized the cyclic deformation behavior of high manganese TWIP steels. Hamada et al. [5] showed that a short period of cyclic hardening is followed by cyclic softening and a saturation state until fracture. In addition Shao et al. [6] made similar observations and reported further that in the LCF regime the amount of the initial cyclic hardening increases with increasing total strain amplitude. At low and moderate total strain amplitudes ($\varepsilon_{a,t}$ < 1.0 %) the degree of cyclic softening is enhanced. Thus, the authors observed deformation induced twins beside the formation of dislocation cells with relatively small size. Both phenomena became more pronounced

* Corresponding author: mwklein@mv.uni-kl.de

by increasing the total strain amplitude. Moreover, Guo et al. [7] investigated the cyclic deformation behavior with varied total strain amplitudes from 0.2 % until 1.0 % of a high manganese TWIP steel. They showed that within the first 5 percent of lifetime, cyclic hardening occurs but the subsequent behavior is strongly dependent on the applied total strain amplitude. At total strain amplitudes of 1.0 % to 0.6 % the initial cyclic hardening merged into cyclic softening until fracture: At lower total strain amplitudes of 0.4 % and 0.3 % cyclic saturation was observed before fracture and at a total strain amplitude of 0.2 % the initial cyclic hardening remained until fracture. Furthermore, the cyclic yield strength, determined by stress-strain hystereses, exceeded the monotonic yield strength.

It is well known, that the surface morphology, consisting of residual stresses, phase composition and roughness, significantly influences fatigue properties by influencing the initial fatigue damage at the surface [8]. Therefore, in metastable austenitic steels, there are efforts to modify surface layers for improving fatigue lifetime by inducing compressive residual stresses, increasing the dislocation density, inducing of nanostructures and phase transformation from paramagnetic γ-austenite to paramagnetic ε-martensite and/or ferromagnetic α'-martensite [8, 9]. In addition, the development of surface morphologies during fatigue loading is also important. For example, relaxation and redistribution of residual stresses due to cyclic loading can be observed [10]. In the literature, there are many investigations about the influence of morphology on the fatigue properties of austenitic stainless steels with the alloying concept Fe-Cr-Ni-C, but hardly any on high manganese austenitic TWIP steels. Therefore, the aim of this investigation is to characterize the influence on fatigue properties in the LCF regime of a high manganese TWIP steel with different surface morphologies, created by different machining parameters. The initial surface morphologies were characterized by SEM, XRD and confocal microscopy methods before push-pull testing until fracture in the LCF regime.

2 Material and experimental setup

2.1 HSD® 600 steel

The investigated material was the HSD® 600 steel, delivered as flat sheets with a thickness of 5 mm in hot rolled state. The chemical composition is given in table 1.

Table 1. Chemical composition in weight %

Element	C	Mn	Al	Si
weight %	0.36	15.29	1.80	2.29

The material has a mean grain size of about 20 μm, determined including annealing twins. Note, that the grain size varies from relatively large grains up to 40 μm to small grains of about 10 μm (Fig. 1 a)). EBSD analysis shows no preferred grain orientation in this material. In addition, the hardness is 301±13, determined by Vickers indentation HV10.

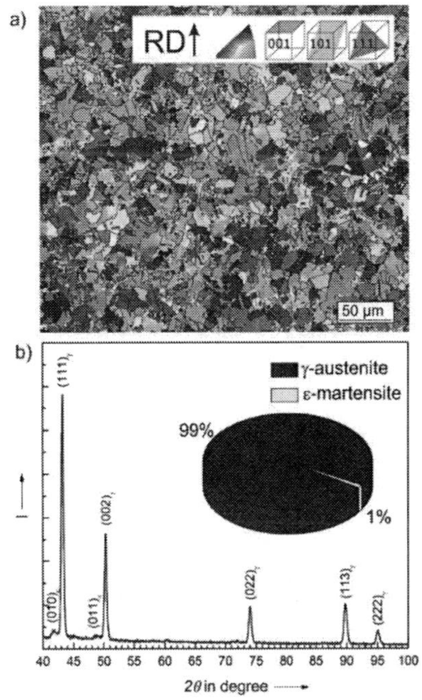

Fig. 1. Microstructure in longitudinal section of HSD® 600 steel at the initial state a) grain orientation in rolling direction (RD); b) x-ray diffraction pattern and phase composition

The initial state is nearly fully austenitic. Nevertheless, a small volume fraction of 1 wt % ε-martensite can be found by x-ray diffraction using the Rietveldt method [3] after wet grinding up to 1200 grit followed by electrolytic polishing (Fig 1 b)).

2.2 Experimental setup

The $\sin^2\psi$-method was used for determining residual stresses while x-ray diffraction patterns were measured with the PANalytical diffractometers MRD Pro and Empyrean operating at 40 kV and 40 mA using $CuK_{\alpha 1}$ radiation and a spot size of 3 x 3 mm. The Rietveldt-method was used for quantitative phase analysis. SEM investigations were carried out with a Quanta 600, FEI with an EBSD module hkl 5, Oxford instruments. The phase transformation from γ-austenite to α'-martensite during cyclic loading was measured in-situ using a magnetic sensor Feritscope® MP30E, Helmut Fischer. The ferromagnetic fraction (ξ) of the material is given in FE-% and correlates directly with the volume fraction of α'-martensite. Topographies were characterized by confocal microscopy using μSoft explorer, μSurf.

Monotonic loading of flat specimens was performed on the Zwick Z250 tensile test machine with a maximum force of 250 kN. The strain rate was $\dot{\varepsilon} = 2.5 \times 10^{-4}\ s^{-1}$

reaching the 0.2 % yield stress until $R_{p0.2}$ and was then increased to 6.7 x10^{-3} s^{-1}. The strain in loading direction was measured by an electromechanical extensometer, while strain in transverse direction was measured optically by a video extensometer.

To characterize the cyclic deformation behavior, single step tests and strain increase tests with flat fatigue specimens with a gauge length of 8 mm were performed. The tests were total strain controlled at ambient temperature (AT), at a frequency of f = 0.2 Hz and load ratio of R_ε = -1 in the LCF regime on a MTS 810 servohydraulic testing system with a maximum push-pull force of 100 kN. The load-time function was triangular. Besides stress-strain hystereses, changes in temperature T were calculated by equation (1) from temperatures measured by three thermocouples (Fig. 2).

$$\Delta T = T_1 - \frac{1}{2}(T_2+T_3) \qquad (1)$$

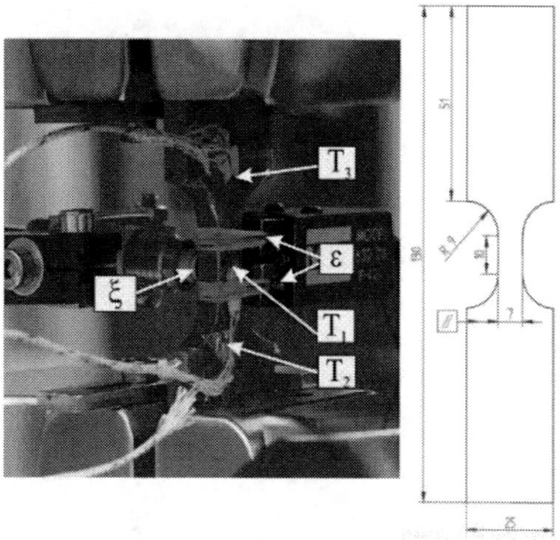

Fig. 2. Experimental setup for fatigue testing and specimen geometry (sample)

Due to machining, the specimens vary in thickness. Therefore, the gauge length of each specimen was measured before testing. Specimens with rolling skin morphology (i.e. from as-received sheets) had a thickness of 5 mm, while specimens with the climb milled and up-climb milled morphologies were 4.6 mm thick (see 2.3).

2.3 Investigated morphologies

Flat specimens were excised transverse to the rolling direction by waterjet-cutting and depending on the investigated surface morphology, further machining was performed before cyclic loading. Three different surface morphologies were investigated: as-received with a so called rolling skin of about 30-40 µm thickness formed during the manufacturing process (M_{RS}), after climb milling the gauge length (M_{CM}) and after up-climb milling the gauge length (M_{UCM}) (Fig. 3). For the machined morphologies M_{CM} and M_{UCM} the machining parameters, given in table 2, were identical, despite the rotation direction of the milling tool.

Table 2. Machining parameters for morphologies M_{CM} and M_{UCM}

tool coating	f_z	n	n_z
AlTiN	0.025	5091	4

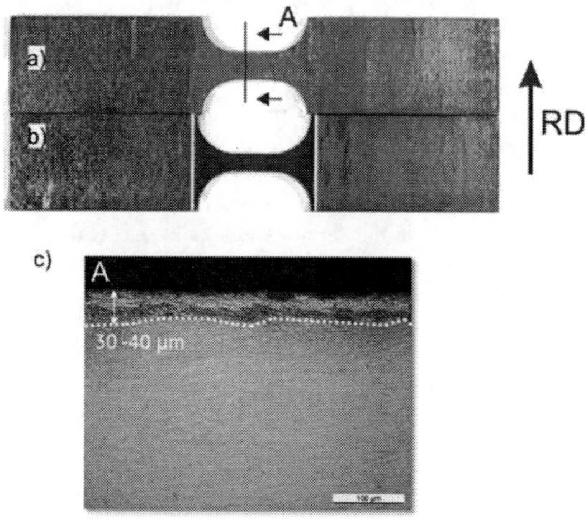

Fig. 3. Sample surface morphologies: a) as-received with rolling skin (M_{RS}); b) climb-milled (M_{CM}); c) cross section of M_{RS} with the thickness of the rolling skin

These four morphologies have different properties concerning roughness, initial residual stresses, phase composition and topography.

The topographies of the morphologies were quantified by the value R_z (Fig. 4). R_z is defined as the average distance between the highest peak and the lowest valley of the sampling length. The morphology M_{RS} showed the highest value of R_z due to no further machining after the specimens were cut. The climb milled specimen showed smoother surfaces than the up-climb milled surfaces, but on both surfaces the traces of the milling tool were visible.

Table 3. Roughness of the investigated surface morphologies

Morphology	M_{RS}	M_{CM}	M_{UCM}
R_z in µm	6.17	0.75	1.18

Due to machining process, the unevenness of the morphologies M_{CM} and M_{UCM} repeated periodically while the surface of M_{RS} was irregular.

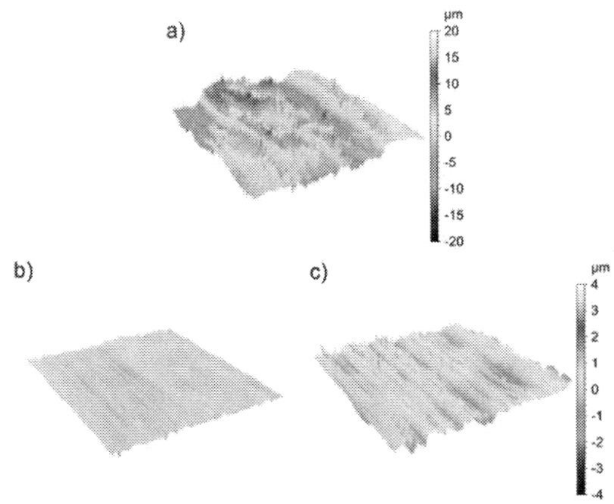

Fig. 4. Topographies of the investigated morphologies a) as-received with rolling skin (M_{RS}); b) climb-milled (M_{CM}); c) up-climb milled (M_{UCM})

Besides the roughness, the morphologies showed wide disparities in phase compositions and initial residual stresses (Fig. 5). A significant amount of α'-martensite was present on the surface of the morphology M_{RS}. The absence of α'-martensite after metallographic polishing indicated (Fig. 1), that this phase was present in the surface layer induced by the manufacturing process. The morphologies M_{CM} and M_{UCM} were austenitic, despite a small amount of ε-martensite of approximately 4 wt % in M_{CM} and 1 wt % in M_{UCM}.

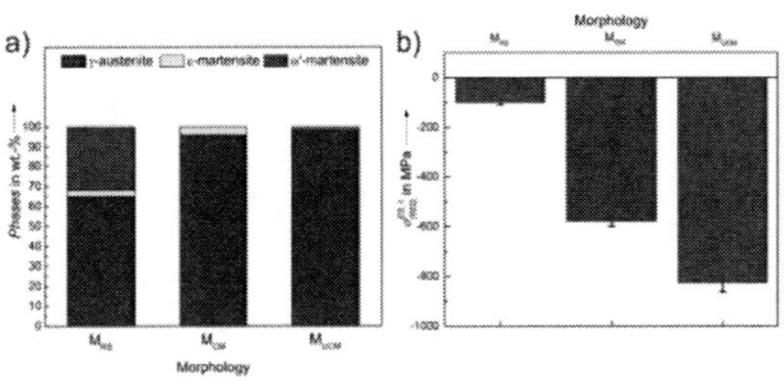

Fig. 5. Surface morphologies before loading: a) phase compositions; b) initial residual stresses

Each morphology showed compressive residual stress in the loading direction before loading, but the amount varied strongly. The morphology M_{UCM} showed the highest value of -800 MPa while morphology M_{CM} had lower values of -600 MPa. The milling process induced these compressive stresses. Due to no further machining after waterjet-cutting, the morphology M_{RS} showed significantly lower values of residual compressive stress of -100 MPa.

3 Results

3.1 Monotonic deformation behavior

For M_{RS} specimens, the monotonic deformation behavior was investigated transversely and longitudinally to the rolling direction (RD) (Fig. 6). The mechanical properties in both directions are given in table 2.

Table 4. Quasi-static properties longitudinal and transverse to rolling direction

	Longitudinal	**Transverse**
R_m in MPa	1028±10	1044±4
$R_{p0.2}$ in MPa	607±21	621±5
Elongation in %	53±2	51±1
Young's Modulus in GPa	169±7	180±1
Poisson's ratio	0.24±0.04	0.24±0.04

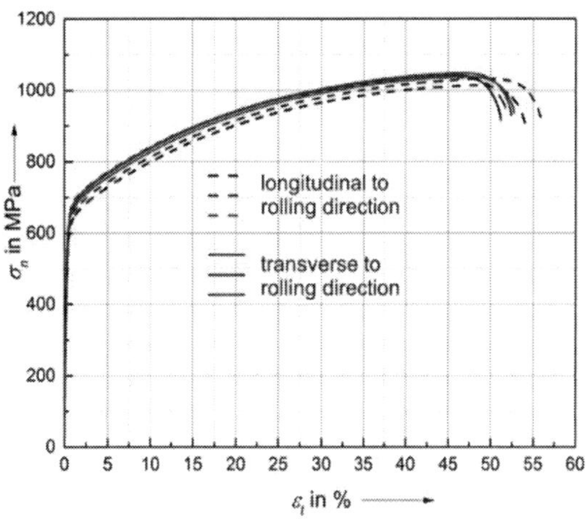

Fig. 6. Monotonic stress-strain curves of specimen taken longitudinally and transversely to rolling direction

The stress-strain curves showed a similar behavior of the material in both directions. Furthermore, the monotonic deformation behavior was characterized by a large amount of plastic deformation up to an elongation of 50 %, while tensile strengths of about 1000 MPa were reached. This demonstrates the aforementioned combination of strength and formability, which is typical for high manganese TWIP steels. No pronounced yield limit was noticeable.

3.2 Strain increase tests (SIT)

To characterize the cyclic deformation behavior at different total strain amplitudes, strain increase tests (SIT) were performed at specimens with the three different surface morphologies (Fig. 7). The total strain amplitude started at $\varepsilon_{a,t}$ = 0.2 % and was increased each 100 cycles by $\Delta\varepsilon_{a,t}$ = 0.05 %. For all morphologies in the first step, no cyclic hardening or softening was noticeable in the stress amplitude σ_a. From the second step, cyclic softening occurred, determined by decreasing stress amplitude σ_a. However, with increasing total strain amplitude $\varepsilon_{a,t}$ cyclic softening decreased until within the first cycles of one step cyclic hardening occurred (enlarged details of Fig. 7). Thus, the cyclic deformation behavior of all three morphologies depended on the applied total strain amplitude $\varepsilon_{a,t}$. This behavior was also noticeable in the changes in temperature because of its correlation with the area of stress-strain hystereses [11]. The temperature increased due to larger stress-strain hystereses while cyclic softening occurred in the first steps, and decreased in later steps while cyclic hardening was occurring.

Even though the behavior of the materials with three different surface morphologies is comparable in general, the surface morphologies led to different results of the strain increase tests. The specimens with morphology M_{RS} showed phase transformation from γ-austenite to α'-martensite, which was noticeable in changes to their magnetic properties while testing. The milled morphologies M_{CM} and M_{UCM} remained fully austenitic and no changes in magnetic properties occurred. Thus, the rolling skin of M_{RS} is metastable depending on its phases. Nevertheless, the morphology M_{RS} led to the lowest lifetime due to its significantly higher roughness.

The morphology M_{UCM} led to the highest lifetime, although the same applied total strain amplitudes led to lower stress amplitudes at the morphology M_{CM}. Note, that the described phenomena occurred significantly below the yield strength $R_{p0.2}$ which clearly differed from the results shown by the literature for high manganese austenites, and even for austenitic stainless steels [6, 7].

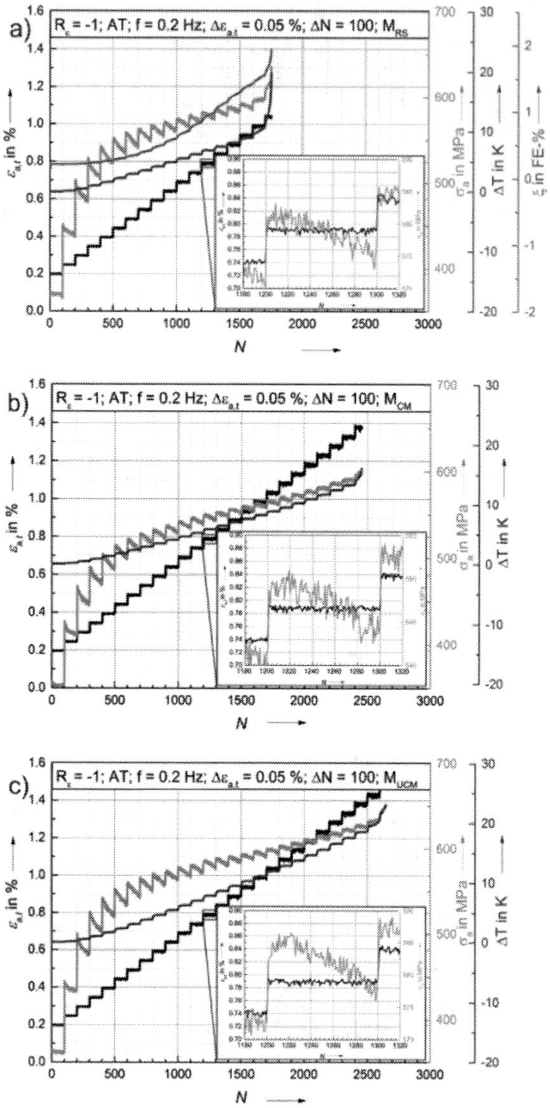

Fig. 7. Development of stress amplitude and temperature versus number of load cycles in strain increase tests (SIT): a) as-received with rolling skin with development of α'-martensite (M_{RS}); b) climb-milled (M_{CM}); c) up-climb milled (M_{UCM})

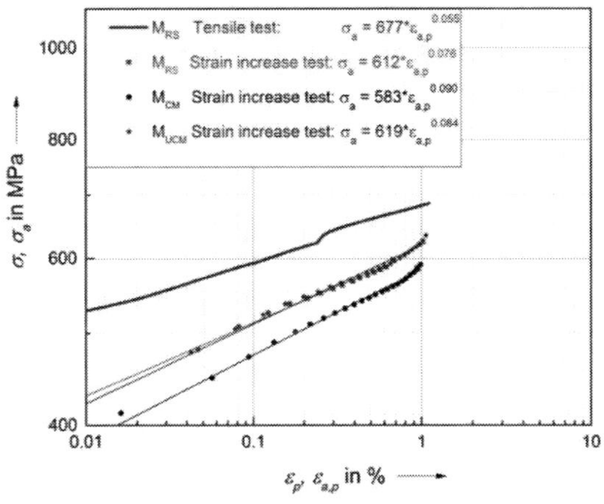

Fig. 8. Cyclic and monotonic stress-strain curves

Fig. 8 compares the monotonic stress-strain and cyclic stress-strain (CSS) curve estimated from strain increase tests (Fig. 7) and tensile tests (Fig. 6). For the CSS curve the average values of stress amplitude σ_a and plastic strain amplitude $\varepsilon_{a,p}$ of the last 25 cycles of each step of the strain increase tests are plotted against the values of stress σ and plastic strain ε_p of a tensile test, with loading direction transverse to the rolling direction. For every morphology, the stress level of cyclic loading was significantly lower than the stress level of monotonic loading. The cyclic yield strength is defined hereafter as the stress level at a plastic strain amplitude of $\varepsilon_{a,p}$ = 0.2 % to a comparable value to $R_{p0.2}$. At $\varepsilon_{a,p}$ = 0.2 % the cyclic yield strength was 505 MPa for morphology M_{CM} and 563 MPa for the morphologies M_{RS} and M_{UCM} respectively. Thus, the HSD® 600 steel softened due to cyclic loading. Therefore, a cyclic softening is also expected in LCF regime for single step tests. In addition, the stress levels of M_{RS} and M_{UCM} are almost congruent while M_{CM} shows a parallel, but slightly lower development of σ_a during the strain increase test.

3.3 Single step tests

Single step tests for all three surface morphologies at four different total strain amplitudes $\varepsilon_{a,t}$ = 0.4 %, 0.6 %, 1.0 % and 1.4 % were performed (Fig. 9). The results confirmed the cyclic deformation behavior shown by strain increase tests. With increasing total strain amplitude, the initial cyclic hardening became more pronounced. At total strain amplitudes of $\varepsilon_{a,t}$ = 1.0 % and $\varepsilon_{a,t}$ = 1.4 % cyclic hardening occurred within the first 8 or 10 cycles respectively. At total strain amplitudes of $\varepsilon_{a,t}$ = 0.6 % and $\varepsilon_{a,t}$ = 0.4 % the initial cyclic hardening occurred just within the first three cycles and was barely observed for M_{RS}. Furthermore for all three morphologies, the cyclic hardening merged to cyclic softening with increasing cycle number. At higher total strain amplitudes of $\varepsilon_{a,t}$ = 1.0 % and $\varepsilon_{a,t}$ = 1.4 % this cyclic softening remained until fracture. At lower total strain amplitudes of $\varepsilon_{a,t}$ = 0.6 % and $\varepsilon_{a,t}$ = 0.4 % the cyclic softening merged into saturation state before fracture, which became more pronounced with decreasing total strain amplitude.

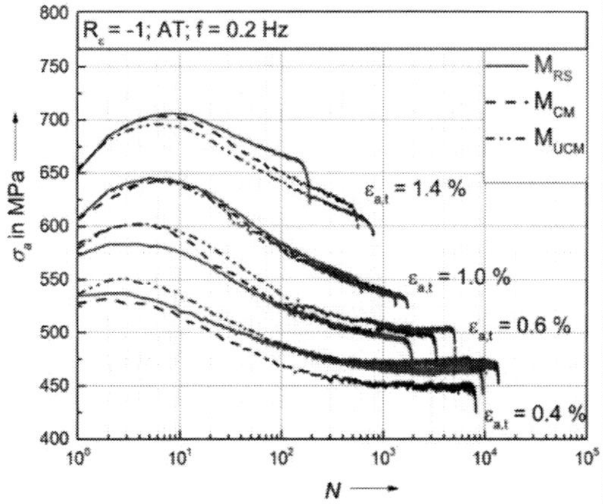

Fig. 9. Development of stress amplitude versus number of load cycles of the three investigated surface morphologies

As well as in strain increase tests, the morphology M_{RS} showed deformation induced phase transformation from γ-austenite to α'-martensite. Fig. 10 shows the cycle and load dependent development of α'-martensite. The

α'-martensite volume fraction increased continuously with the increasing number of cycles and reached higher values at higher loading amplitudes. Maximum values of α'-martensite in the range 0.51 FE-% < ξ < 1.33 FE-% were observed at specimen failure. The development of α'-martensite is comparable to the results from fatigue tests of metastable austenitic Cr-Ni stainless steel [9, 12].

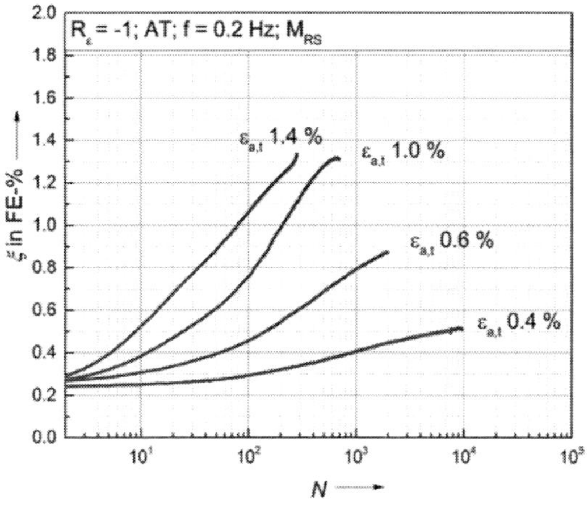

Fig. 10. Development of cyclic deformation induced α'-martensite (ξ) in M_{RS} during fatigue tests

The influence of the different surface morphologies on fatigue life in LCF regime is summarized in Fig. 11 by the total-strain based S-N curve.

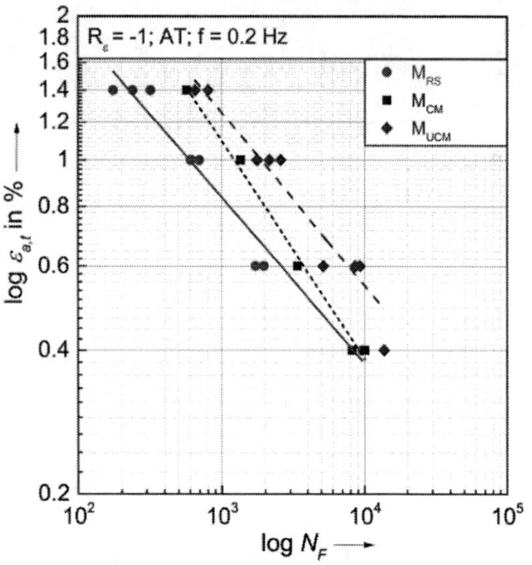

Fig. 11. S-N curve of specimens with different surface morphologies

The specimen with the highest lifetime at all loading amplitudes were M_{UCM}, while the specimen with morphology M_{RS} had the lowest lifetime. However at $\varepsilon_{a,t}$ = 0.4 % M_{RS} achieved a higher number of cycles to failure than M_{CM}, although the surface roughness was eight times higher (see. Fig. 4 and table 3). Therefore, the positive influence of α'-martensite formation on the surface during fatigue loading becomes noticeable but only at low total strain amplitude. For a better understanding of the influence of the surface morphologies at lower load amplitude further high cycle fatigue (HCF) and very high cycle fatigue (VHCF) tests will need to be performed.

4 Conclusion

The cyclic deformation behavior in the LCF regime as well as the monotonic deformation behavior of the TWIP HSD® 600 steel with three different surface morphologies were investigated by strain increase tests, single step tests and tensile tests respectively. The morphologies were as-received with rolling skin (M_{RS}), climb milled (M_{CM}) and up-climb milled (M_{UCM}). Machining was performed in the gauge length respectively. The three morphologies varied in initial residual stresses, roughness and initial phase composition. M_{RS} showed a significant amount of α'-martensite of 32 wt % while the morphologies M_{CM} and M_{UCM} were fully austenitic before loading. The roughest surface occurred in M_{RS} condition with a value of R_z of 6.17 µm, while the morphology M_{CM} showed a value of R_z of 0.75 and M_{UCM} of 1.18 µm. Every morphology had residual compressive stress before loading but the amount varied from -100 MPa for M_{RS} to -600 MPa for M_{CM} and -800 MPa for M_{UCM} respectively.

The strain increase tests as well as the single step tests showed a dependence of the deformation behavior on the applied total strain amplitude. With increasing strain amplitude, an initial cyclic hardening occurred and the higher the applied total strain amplitude the more pronounced was the initial cyclic hardening. This cyclic hardening merged into cyclic softening until fracture or, at lower total strain amplitudes, was followed by cyclic saturation. Nevertheless, the cyclic yield strength was significantly lower than the monotonic yield strength for all three surface morphologies.

Although the morphology M_{RS} had an initial amount of 32 wt % α'-martensite on its surface and showed further α'-martensite formation induced by cyclic loading, it led to the shortest lifetime due to its significantly higher roughness. The machined surfaces were comparable in their achieved lifetimes, but the climb milled morphology showed slightly lower stress levels at cyclic loading.

The authors gratefully acknowledge financial support from Deutsche Forschungsgemeinschaft (DFG) within the Collaborative Research Center (CRC) 926, the chair MT&S of Prof. Seewig for confocal microscopy and the use of µSoft explorer, and Salzgitter AG for providing the HSD® 600 steel.

References

1. G. Frommeyer, U. Brüx, P. Neumann, ISIJ Int. **43**, 438–446 (2003)
2. I. Gutierrez-Urrutia, D. Raabe, Scr. Mater. **66**, 992–996 (2012)
3. M. W. Klein, R. Skorupski, M. Smaga, T. Beck, Engineering Transaction, 69–75 (2017)
4. A. Saeed-Akbari, A. Schwedt, W. Bleck, Scr. Mater. **66** 1024–1029 (2012)
5. A. S. Hamada, L. P. Karjalainen, J. Puustinen, Mater. Sci. Eng., A **517**, 68–77 (2009)
6. C. W. Shao, P. Zhang, R. Liu, Z. J. Zhang, J. C. Pang, Z. F. Zhang, Acta Mater. **103**, 781–795 (2016)
7. P. Guo, L. Qian, J. Meng, F. Zhang, L. Li, Mater. Sci. Eng., A, **584**, 133–142 (2013)
8. R. Skorupski, M. Smaga, D. Eifler, Adv. Mater. Res. **891-892**, 464–469 (2014)
9. M. Smaga, R. Skorupski, D. Eifler, T. Beck, J. Mater. Res. **39**, 1–9 (2017)
10. W. Z. Zhuang, G. R. Halford, Int. J. Fatigue **23**, 31–37 (2001)
11. H. Knobbe, P. Starke, S. Hereñú, H.-J. Christ, D. Eifler, Int. J. Fatigue **80**, 81–89 (2015)
12. M. Smaga, F. Walther, D. Eifler, Mater. Sci. Eng., A **483-484**, 394–397 (2008)

Long-term durability of CFRP under fatigue loading for marine applications

Antoine Le Guen-Geffroy[1,*], *Pierre-Yves* Le Gac[1], *Malick* Diakhate[2], *Bertrand* Habert[3], and *Peter* Davies[1]

[1]Ifremer, Laboratoire de Comportement de Structures en Mer, Pointe du diable, 29280 Plouzané, France
[2]Univ. Bretagne Occidentale, IRDL-UMR CNRS 6027, F-29600 Morlaix, France
[3]Direction Générale de l'Armement, Boulogne-Billancourt, France

Abstract. Composite propellers are a major new development in the marine transport industry. The use of composite materials in seawater turbines is also of great interest to the marine renewable energy industry. Those systems present similarities being both constantly immersed in seawater, under fatigue loadings, of large dimensions, and they are often designed using Carbon Fibre Reinforced Polymer (CFRP). Previous studies have shown sensitivity of some composite materials to the seawater environment mainly in the quasi-static loading domain. However, investigations now need to be performed on the behaviour of CFRP under seawater environment and fatigue loadings. In this study, CFRP samples were aged in natural seawater until saturation, in order to investigate the influence of water absorption on the fatigue properties of the material. Results showed a small decrease in fatigue lifetime under tension but a more significant drop for sample subjected to four point flexural fatigue.

1 Introduction

Global warming is a priority for a large part of the scientific community and many research programs. The impact on the environment and animals has been largely shown [1, 2]. In order to limit global warming, simulations have shown the importance of limiting our production of greenhouse gases such as Nitrous Oxide (N_2O), Methane (CH_4) or the most produced and studied: Carbon Dioxide (CO_2) [3, 4].

As a response, in the engineering domain, work is being performed to produce carbon-free energies [5]. Current and most popular in-use technologies are wind turbines, hydro-power dams and solar panels. Various studies are also examining marine renewable energies encouraged by huge energy production potential. Among others, we can cite offshore wind-turbines, Ocean Thermal Energy Conversion (OTEC), wave energy and tidal and stream energy [6–8]. Stream energy converters consist, for the most part, of turbines immersed in water converting current energy into electrical energy [9]. These systems are often comparable to wind turbines in principle but the dimensions are not comparable. Due to the higher density of water compared to air, Grogan et al. showed the importance of using carbon fibres composites (CFRP) instead of glass fibre composites (GFRP) [10].

There is also activity within the marine transport and freight industry, engineers tend to reduce the gas emissions. As in the automotive and aeronautical industries, the reduction of weight and the improvement of propulsion efficiency allow reductions in gases emissions. An active current subject is the use of composite materials to manufacture the propellers. Theoretical studies have shown multiple advantages as weight reduction —almost 50%— and propulsion efficiency[10–15].

Both of these subjects concern the use of fibre-reinforced polymers to make rotating blades. Most of the design processes rely on either theoretical values or experimental values of unaged materials. However, these values tend not to be representative of material's properties throughout the system's lifetime. Indeed, composite materials have been shown to be sensitive to moisture and particularly to seawater [16–18]. For fatigue loadings, this could be characterized by a change in behaviour and the ultimate lifetime — if considered as existing [19, 20]— could be overestimated. This could result in early failures of the systems.

In this paper, investigations are carried out both on the unreinforced matrix resin and its carbon fibre reinforced composite. Changes in behaviour are studied in the unaged, saturated and dried states of the materials. Epoxy resin appears to be highly sensible to seawater but it does not appear as clearly in tensile fatigue loadings of composites, as it will be shown below.

2 Experimental

2.1 Materials

The composite material studied here has been chosen to be representative of what could be used to manufacture tidal turbines and marine propeller blades. As noted above, the material of interest is a carbon fibre reinforced epoxy.

The RTM light process was chosen to be the more representative process for the large-scale fabrication of propeller and turbine blades. Then, the same process was used

*e-mail: antoine.le.guen.geffroy@ifremer.fr

to fabricate experimental composite panels. Both composite panels and the mold were made by LoireTech. The two stacking sequences studied here were 10 plies unidirectional and [(0/90/0/90/0)s]. Fibre volume fraction was calculated using results from calcination made in a TGA apparatus and density measurements made in an helium pycnometer. Tests showed a fibre volume fraction of 50 % ± 4% depending on the different panels used in the study.

The fibre reinforcements are Torayca T700 unidirectional supplied as 300 g/m² layers. They have a tensile modulus of 230 GPa and an ultimate tensile strength of 4.9 GPa with 2.1 % strain at break.

The matrix, from Sicomin, is a DGEBA-DGEBF epoxy-based resin SR8100 with a slow hardener SD4772. This resin is made for RTM light process meaning that its contact with air has to be limited to prevent unwanted reaction of the resin.

Resin was vertically cast between two aluminium plates covered by PTFE tape. A three-edge frame separating the two plates assured the thickness. This allowed the resin to harden while limiting the surface in contact with the air.

Both composite and resin panels were cured at room temperature for 24 hours. Then a first post cure cycle was applied at 60 °C during 16 hours. The glass transition temperature (Tg) resulting was found to be at 69 ±1 °C. In order to obtain the maximum Tg, a second post cure was applied at 120 °C for 2 hours resulting in a Tg of 78 ±1 °C. During this second post curing, to prevent reaction of the polymer with the external environment, panels were vacuum bagged.

Water uptake and Single End Notch Bend (SENB) resin samples, and the composite panels, were cut with a water-cooled diamond saw according to the required sample's dimensions. Dog-bone resin samples were cut by an automated 2-axis CNC machine according to 1BA dimensions from ISO 527-2. The CNC machine also allows surface machining of the plates to reduce the thickness of the samples making the plate from 3 to 1 mm in thickness.

2.2 Fatigue apparatus

To investigate on the influence of seawater ageing on the mechanical properties, it is necessary to develop specific methods and test equipment. The goal is to be representative of what the material undergoes when it is in use. Therefore, standardized tests have been adapted in seawater environment.

The two fatigue tests performed were tensile-tensile and four point flexure. For each type of fatigue loading, at least three specimens were tested initially to break at quasi-static rates. Then 3 to 4 different load levels were fatigue tested aiming to have at least three specimens per load level. A constant load ratio of R = 0,1 was taken.

2.2.1 Tensile-Tensile tests

Tensile-tensile tests were performed on an MTS 250 kN hydraulic machine. Frequency was 1 Hz to limit self-heating of the sample as this can induce premature sample failures. As for the four-point flexure test apparatus, a specific apparatus was developed to allow tensile-tensile tests immersed in water. Previous studies tested successfully this apparatus for GFRP specimens. For CFRP, as strength and stress are much higher, the end conditions have to be modified. Tensile-tensile results presented here are for tests performed in air. A modified fixture is being developed and future works will investigate on the influence of immersion on tensile-tensile test performance. It should be noted that for GFRP the effect of wet ageing were considerably more significant than the influence of the test environment [21].

2.2.2 Four-point flexural tests

A specific test set-up developed at Ifremer allows immersion of four-point flexure samples in seawater. The standardized test from ASTM D6272 (one half support span and 32:1 span to depth ratio) was adapted and the apparatus was placed in an in-house developed seawater bath. This allows us to perform fatigue tests immersed in seawater. An aquarium heater regulates the temperature and an overflow system constantly recycles the water.

Control of the test is done by an hydraulic 25 kN Zwick Roell test machine (Figure 1). Tests were done at a frequency of 2 Hz.

Four point flexural loading was chosen because it involves a weak point of the composite; the compression response. The value of the maximum stress occurring in the middle -whether on top or on the bottom depending on the breakage- of the specimen is theoretically calculated by the equation from the standard as :

$$\sigma = \frac{3}{4} \times L \times \frac{P}{bd^2} \qquad (1)$$

In this equation, σ is the stress occurring on the specimen, L is the span value, P is the load and, b and d are respectively the width and the specimen's thickness. In our case, the support span L was $100mm$ and the specimens target dimensions were 120 by 25 by 3 millimetres.

Figure 1: Immersed four point bending apparatus

2.3 Ageing conditions and experimental method

To investigate the material's behaviour in marine environment, a common test is to follow its mass uptake while immersed in water. In this study, square coupons were immersed in constantly renewed natural seawater tanks developed at Ifremer. At regular intervals, samples are weighed after being dried on the surface with paper towel. Water uptake is expressed as the mass change in percent calculated as in equation 2. M(t) being the mass uptake value in % at time t, m_0 the initial mass and m_t the mass at time t (Figure 2 (a)).

$$M(t) = \frac{m_t - m_0}{m_0} \times 100 \qquad (2)$$

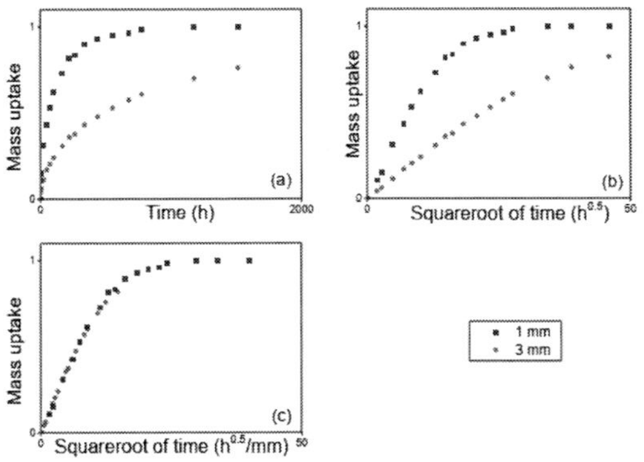

Figure 2: Illustration of Fick's law for normalized mass uptake of two resin samples of 1 and 3 mm thick in seawater at 40 °C

Water uptake is then plotted vs square root of time. The goal is to identify diffusion laws in particular the applicability of Fick's diffusion law. There are two major characteristics of Fick's diffusion law to validate its identification. The first one is that, when it is plotted, the curve is linear at the start and reaches a plateau (Figure 2 (b)). The second point is that when plotting the mass uptake for two different thicknesses of test samples, if square root of immersion time is divided by the thickness, the two curves superpose (Figure 2 (c)).

2.4 Acoustic Emission

Acoustic emission is a powerful tool to monitor damage mechanisms such as fibre rupture or matrix cracking [22–24] Other tools can monitor these phenomena, such as edge surface optical analysis or even tomography. Tomography is a powerful tool but it is expensive and difficult to use *in situ*. On the other hand, edge surface optical analysis presents good results and is less expensive than tomography but it only records the surface visible damages. Acoustic emission tend to be an intermediate tool for monitoring damage in the sample's entirety, is less expensive than the tomography and can be used *in situ*.

Under mechanical loading, cracking within the composite material generates transient elastic waves. A two channel AE system designed by MISTRAS group was used to record the waveforms generated during the laboratory fatigue tests. Two AE sensors (Micro 80 model) were used to also perform a linear localization of the acoustic sources (referred to as events).

Once the acoustic emission activity within the composite material is recorded, a post processing (cluster analysis) of the data is required to identify the acoustic signature of the different damage mechanisms. This classification aims mainly to identify the cracking mechanism (AE signature) that governs the damage evolution of the composite material.

A quasi-static load-unload tensile test was done starting from 30 % of its load to break and increasing load by 5 % (of the load to break) each time. This way we can identify acoustical signature of damage in the samples and re-use these criteria in fatigue tests to show damage evolution.

3 Results

3.1 Water Uptake

Composite coupons were aged for up to 6 months and neat resin ones for up to 3 months. Results showed a Fickian behaviour for both neat resin and composite.

An in-house developed routine, using Scilab's *leastsq* function, allowed the evaluation of diffusion coefficients (D) and maximal water uptake (M_∞). An example of experimental and fitting curve can be seen in figure 3.

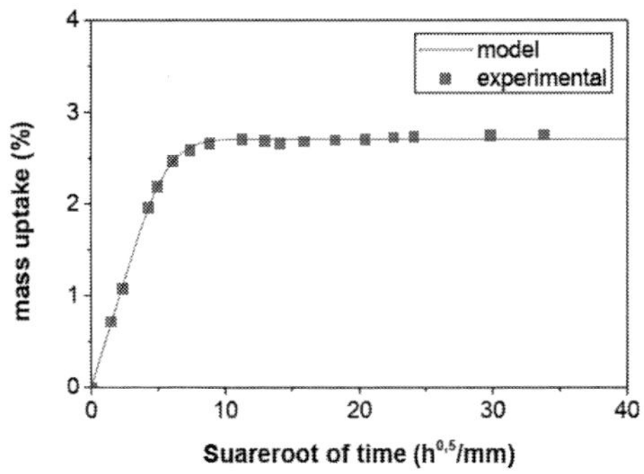

Figure 3: Fitted curve from Scilab routine on a neat resin sample at 60 °C in seawater

Results of corresponding D and M_∞ for the different conditions that are visible in Figure 4 are summed up in Table 1. Theoretical time-to-saturation t_∞ (in days) for a sample of 1 mm in thickness is calculated from analytical equation 3 extracting t, making $M_t = M_\infty$, and making the thickness $h = 10^{-3}$m.

$$t_\infty = \frac{M_t}{M_\infty} \times \frac{\pi}{16} \times \frac{h^2}{D} \qquad (3)$$

https://doi.org/10.1051/matecconf/201816507001

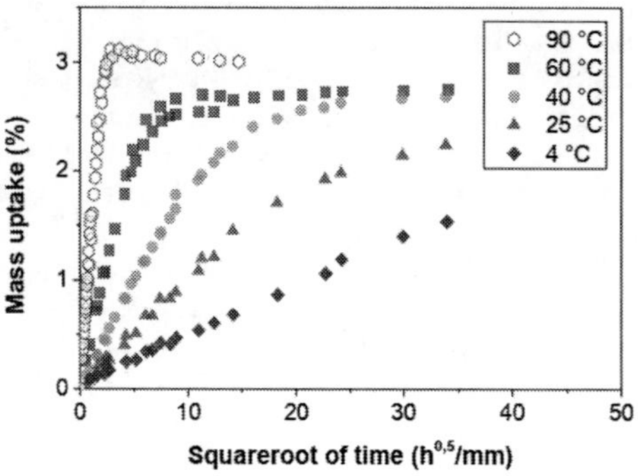

Figure 4: Neat resin seawater uptake for 1 and 3 mm thick samples at several temperatures

Table 1: Seawater uptake diffusion values on neat resin (* estimated values)

	D $(10^{-13}m^2/s)$	M_∞ (%)	T_{M_∞} (days)	Least Square
90 °C	110.50	3.06	0.8	0.27
60 °C	16.59	2.70	7.2	0.05
40 °C	2.81	2.61	42	0.29
25 °C	1.25	2.17	96*	0.09
4 °C	0.29	2.10*	419*	0.09

For the composite ageing, water uptake in the carbon fibres is often considered to only depend on the resin fraction. The mass to saturation is then inferior and can be calculated from equation 4 with $M_{\infty c}$ being the mass to saturation of the composite, V_r the mass content ratio of resin in the composite, in our case $V_r = 0.41$, and $M_{\infty r}$ the mass to saturation of the neat resin.

$$M_{\infty c} = V_r \times M_{\infty r} \quad (4)$$

Theoretical values of diffusion coefficients of the composite D_c can be calculated from various mixing factors found in the literature [25]. Here, we found that the Rayleigh factor presented better results for the samples that were saturated. The diffusion coefficient calculated with the Rayleigh factor is presented in equation 5 with V_f being the volume fibre ratio and D_r the resin diffusion coefficient.

$$D_c = D_r \times \frac{1 - V_f - 0.3058 \times V_f^4}{(1 + V_f - 0.3058 \times V_f^4) \times (1 - V_f)} \quad (5)$$

Comparison of fitted D and M_∞ of the composite and theoretical D_t and $M_{\infty t}$, calculated with equations 4 and 5, are shown in table 2.

We can see that theoretical values of mass to saturation underestimate the water content for the composite but overestimate the diffusion coefficient. This could be explained by the void content in the composite that comes during the manufacturing process. Fibre/resin interface could also induce errors to the prediction values.

Table 2: Seawater uptake diffusion values on composite(* estimated values)

	D $(10^{-13}m^2/s)$	D_t	M_∞ (%)	$M_{\infty t}$	Least Square
60 °C	9.71	11.22	1.36	1.10	0.14
40 °C	1.86	1.89	1.34	1.07	0.16
25 °C	0.29*	0.85	1.30*	0.89	0.02
4 °C	0.07*	0.13	1.30*	0.86	0.04

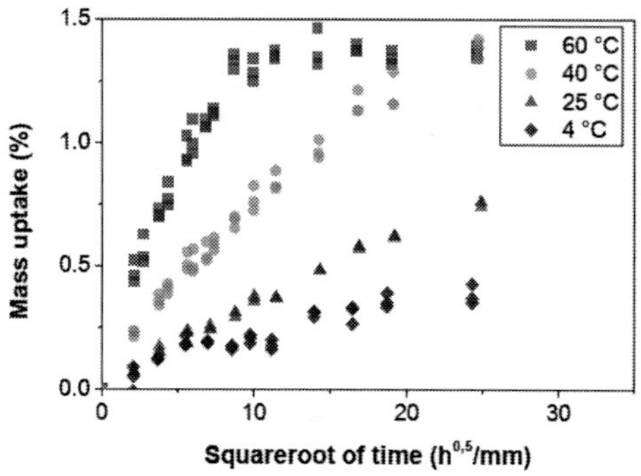

Figure 5: Water uptake in composite

3.2 quasi-static

3.2.1 Neat-resin

Neat resin specimens have been tested in three conditions: unaged, saturated and dried after being saturated. For the tensile tests, unaged and dried samples showed similarities in maximum strength and modulus. On the other hand, saturated samples showed a loss in maximum strength of almost 50 % and strain at break went from 10 % strain to over 25 % strain (Figure 6).

These results suggest that there is plasticization induced by seawater absorption on the polymer based on mechanical response and reversibility.

Fracture tests results showed the same tendency. Initial properties were quite brittle and saturated specimens were more ductile. Drying the specimens after being saturated showed reversibility of the behaviour.

3.2.2 Composite

Composite were also tested in quasi-static loading in order to obtain ultimate values for fatigue tests. Specimens were tested in tensile for various orientations, in Inter-Laminar Shear Stress (ILSS) accordingly to ISO 14130 and in four-point flexure (FPF). Although there was an apparent sensibility of the resin due to water absorption, the effect of water on the composite was limited. (See Table 3)

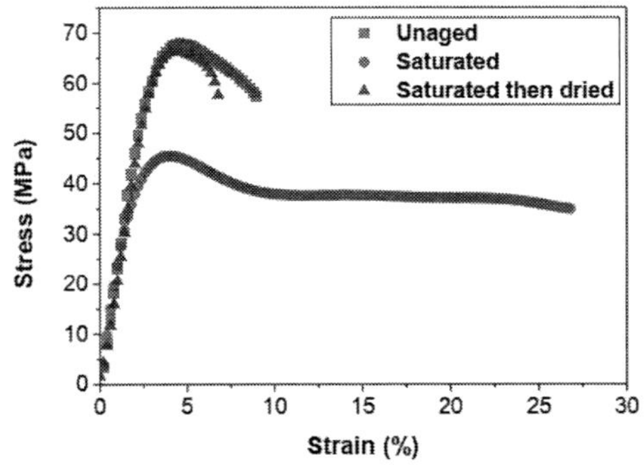

Figure 6: Tensile tests on unaged aged and dried neat resin

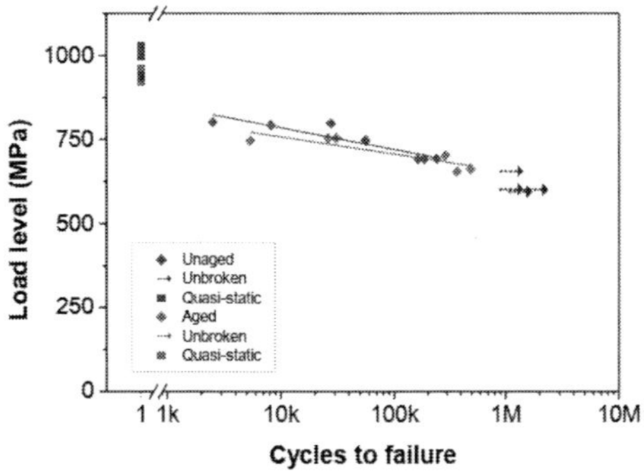

Figure 7: Tensile-Tensile results for unaged and aged 0/90° stack composite samples

Table 3: Quasi-static properties of composite

Property	Unaged value (MPa)	Aged value (MPa)	Loss/gain (%)
Tensile 0°	1190±78		
Tensile 0/90°	989±45	938±20	-5,4
Tensile 90°	42±2		
FPF 0/90°	740±46	681±17	-8
FPF 0°	849±29		
ILSS	40±3	35±3	-14,3

Figure 8: Example of fibre debonding on an unbroken fatigue sample

3.3 Fatigue

3.3.1 Tensile-Tensile

Tensile-tensile fatigue tests were only done on 0/90° stacking sequence. Samples were considered run out after 1 million cycles. When comparing fatigue properties in σ_{max} vs cycles to failure, saturated specimens showed almost equivalent performances. This phenomenon is to be considered knowing there is a 5% loss in ultimate tensile strength. Sensitivity to water absorption is then obvious but not as much as might have been expected given the 40% loss in tensile for the neat resin.

When considering the failure mode, it appeared that for the unaged and aged specimens, delamination of the outer ply is predominant. Figure 8 shows the delamination of an unbroken sample. The damaging process seems to be the same at all levels, with small debonds progressing along the specimen's length up to full delamination of the outer ply. We can see that debonding occurs in strips. For manufacturing purposes, unidirectional fibres are meshed every 20 mm by a small polyester weft strand to maintain fibres orientation. This induces plies to debond by strips independently.

3.3.2 Real time monitoring of damage evolution

As stated previously cracking within composite material generates acoustic activity. This latter has been analyzed with the aim of better understanding the failure mechanisms involved in the composite material damage. In a first step of the AE data post processing, an unsupervised classification method is performed since it can allows gathering the AE events into clusters (Figure 9), and thus evaluating the number of failure mechanisms within the composite material

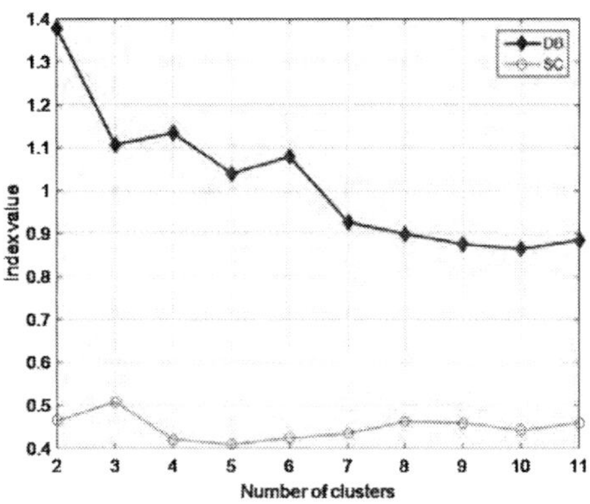

Figure 9: Evaluation of the optimal number of clusters

This unsupervised classification method, which used the k-means algorithm, has been successfully performed in previous studies to detect cracking mechanisms within

https://doi.org/10.1051/matecconf/201816507001

composite material [22, 26], and to monitor crack tip growth within wood material [27, 28]. A detailed description of this unsupervised method is available in these papers.

The AE data clustering aims, among other things, to propose the optimal number of clusters based on the values of two clustering evaluation indices: the Davies-Bouldin (DB) and the Silhouette Coefficient(SC). The optimal number of clusters leads to a highest value of SC index and a lowest value of DB index (Figure 9). With these values, the clusters are dense and well separated, which corresponds to the standard concept of a cluster. From Figure 9 it can be concluded that the optimal number of clusters is three. In a previous study [22] results of cluster analysis on AE data from infused composite specimens exhibited three cracking mechanisms.

Cluster analysis on AE data from composite specimens shows a typical result (Figure 10), which show the projection of the three clusters of AE data onto a 2D plot. Clusters boundaries are identified.

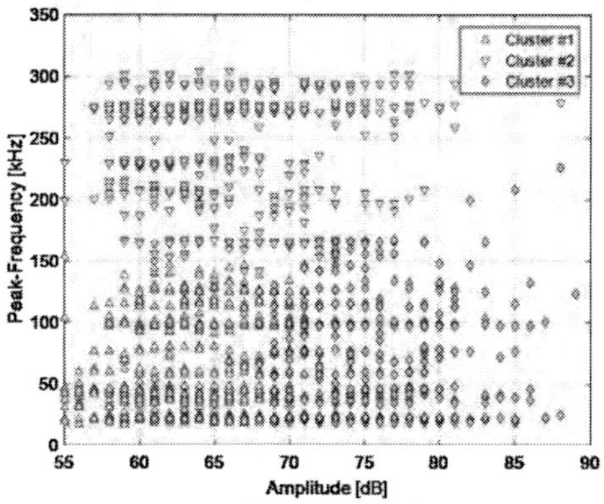

Figure 10: AE signatures of each damage mechanism cluster

Since this AE data clustering aims mainly to identify the cracking mechanism (AE signature) that governs the damage evolution of the composite material, AE events of Cluster 3 were selected, and the cumulative number of these events is compared with the evolution of the mechanical behaviour. This latter is analysed in terms of sample stiffness evolution during fatigue test (Figure 10).

On Figure 11 some changes in the slope of the cumulative number of events curve can be observed. These changes in slope can be detected in real time, and can be linked to damage evolution within the composite material. As a result, the change in slope can be selected as an experimental criterion to monitor damage evolution of the specimen.

3.3.3 Four point flexure

Four point flexural tests were performed on specimens in the unaged and aged states for up to 7 million cycles. As for tensile-tensile tests, there is a small loss in quasi-static

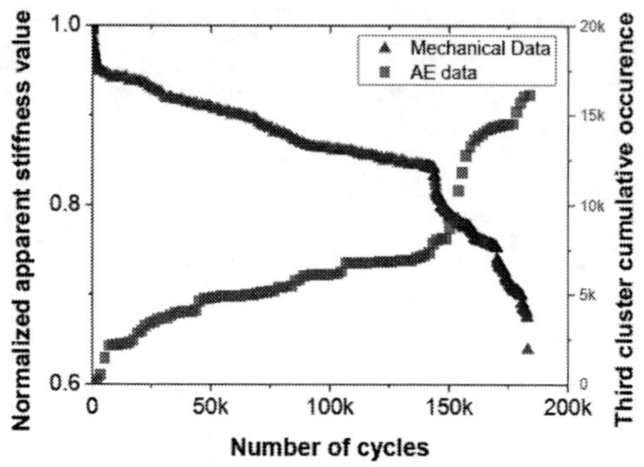

Figure 11: Evolution of AE events (cluster 3) during a fatigue test

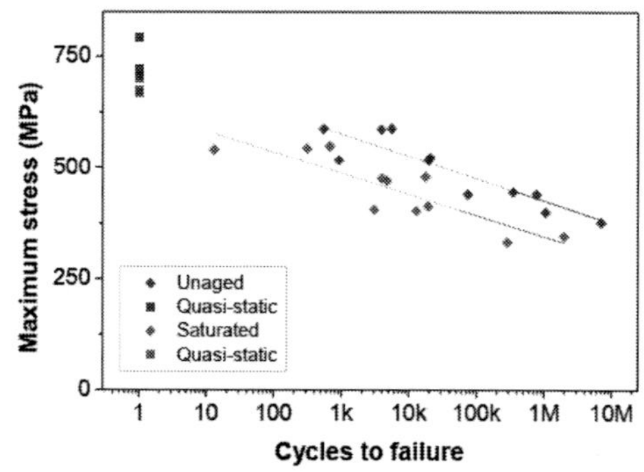

Figure 12: Four point flexure results for unaged and aged 0/90° stack composite samples

strength values. For the fatigue tests, the composite appears to be more sensitive to seawater with a loss of approximatively one decade of lifetime compared to the unaged stress values (Figure 12). It should be noted that for a maximum stress of 50% in the unaged state, the cycles to failure of a sample were found to be 7 million cycles. If the shift is considered as true, in the saturated state, at a stress/strength value of 50%, it should now fail at approximatively 700 000 cycles. This is partly confirmed by one of the samples tested at 50% in the aged state that failed at 1.9 million cycles. This must be taken into account for the dimensioning of structures, but today only the unaged fatigue values are considered. A réécrie avec les valeurs actuelles

As mentioned above, the four point flexure test sollicitates the composite in tension and in compression. As composites are weaker in compression than in tension, sample usually break in compression. however, for the tests performed here, both unaged and after ageing, the composite did not break in compression or in tension. The dominating mechanism is delamination as shown in figure 13.

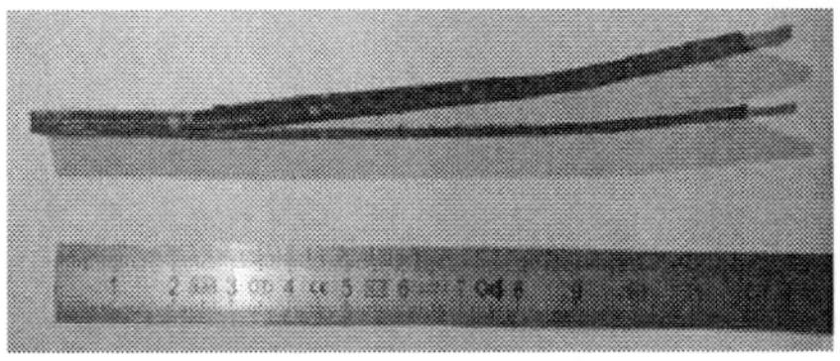

(a) Unaged

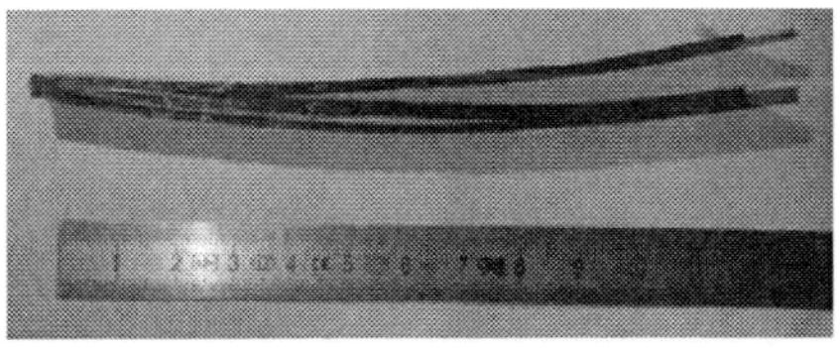

(b) aged

Figure 13: break faciès of two four point flexure samples

4 Conclusion

We showed in this study that CFRP composites are sensitive to a seawater environment, both under quasi-static and fatigue loading. The influence of the seawater environment on the fatigue properties is more marked for four point flexure fatigue results than for tension. A loss of one decade for the same stress values shows that dimensioning systems such as tidal turbine blades or marine propeller blades considering only the dried/unaged state values will not be conservative. It is therefore essential to take into account the impact of seawater environment and ageing on the properties of CFRP.

The out of plane properties, interlaminar shear strength and delamination in particular have shown to be the most sensitive to seawater. In tension, the outer surface showed debonded regions, while in four point flexure, the failure mode was clearly in interlaminar delamination. It seems important to conduct tests on the shear and delamination properties. Such properties can be quantified by standardized tests such as Double Cantilever Beam (CDB), Mixted Mode Bending (MMB) and, three and four Points Edge Notched bending tests (ENF & 4ENF). These tests are now underway. An interesting subject would be to compare results from similar composites but containing z-pins or with three dimensions woven fibres

It should also be noted that in this study, the specimens where only immersed until fully saturated. Long term immersion of sample could reveal even higher impact of seawater on mechanical properties.

Acoustic emission showed promising results that needs more investigations. One acoustical signature was found suitable for damage monitoring. This signature can be followed in real time and interrupt tensile tests could allow us to link the acoustic signal to the mechanical damaging by performing tomographies. This way, the acoustic signature could be linked to a specific damage mechanism in the composite.

Aknowledgment

Authors would like to thanks the DGA and Ifremer for financial support and FabHéli project supported by Loiretech, Meca and Naval Group for material supply.

References

[1] L. Hughes, Trends in Ecology & Evolution **15**, 56 (2000)

[2] T.L. Root, J.T. Price, K.R. Hall, S.H. Schneider, C. Rosenzweig, J.A. Pounds, Nature **421**, 57 (2003)

[3] P.M. Cox, R.A. Betts, C.D. Jones, S.A. Spall, I.J. Totterdell, Nature **408**, 184 (2000)

[4] M. Meinshausen, N. Meinshausen, W. Hare, S.C.B. Raper, K. Frieler, R. Knutti, D.J. Frame, M.R. Allen, Nature **458**, 1158 (2009)

[5] M.I. Hoffert, K. Caldeira, G. Benford, D.R. Criswell, C. Green, H. Herzog, A.K. Jain, H.S. Kheshgi, K.S. Lackner, J.S. Lewis et al., Science **298**, 981 (2002)

[6] R. Pelc, R.M. Fujita, Marine Policy **26**, 471 (2002)

[7] A.G. Borthwick, Engineering **2**, 69 (2016)

[8] A.F.d.O. Falcão, Renewable and Sustainable Energy Reviews **14**, 899 (2010)

[9] F.O. Rourke, F. Boyle, A. Reynolds, Renewable and Sustainable Energy Reviews **14**, 1026 (2010)

[10] D. Grogan, S. Leen, C. Kennedy, C. Ó Brádaigh, Renewable Energy **57**, 151 (2013)

[11] G. Marsh, Reinforced Plastics **48**, 34 (2004)

[12] C.C. Lin, Y.J. Lee, C.S. Hung, Composite Structures **89**, 206 (2009)

[13] M.R. Motley, Z. Liu, Y.L. Young, Composite Structures **90**, 304 (2009)

[14] J.P. Blasques, C. Berggreen, P. Andersen, Marine Structures **23**, 22 (2010)

[15] T. Taketani, K. Kimura, S. Ando, K. Yamamoto, *Study on Performance of a Ship Propeller Using a Composite Material*, in *The Third International Symposium on Marine Propulsors smp* (2013), Vol. 13, `http://www.marinepropulsors.com/proceedings/2013/11A.4.pdf`

[16] P. Davies, Y.D. Rajapakse, eds., *Durability of Composites in a Marine Environment 2*, Vol. 245 of *Solid Mechanics and Its Applications* (Springer International Publishing, Cham, 2018), ISBN 978-3-319-65144-6 978-3-319-65145-3, `http://link.springer.com/10.1007/978-3-319-65145-3`

[17] P. Davies, G. Germain, B. Gaurier, A. Boisseau, D. Perreux, Philosophical Transactions of the Royal Society A: Mathematical, Physical and Engineering Sciences **371**, 20120187 (2013)

[18] N. Tual, N. Carrere, P. Davies, T. Bonnemains, E. Lolive, Composites Part A: Applied Science and Manufacturing **78**, 380 (2015)

[19] A. Hosoi, N. Sato, Y. Kusumoto, K. Fujiwara, H. Kawada, International Journal of Fatigue **32**, 29 (2010)

[20] D. Backe, F. Balle, D. Eifler, Composites Science and Technology **106**, 93 (2015)

[21] A. Boisseau, P. Davies, F. Thiebaud, Applied Composite Materials **20**, 145 (2013)

[22] M. Diakhate, N. Tual, N. Carrere, P. Davies, *Cracking and Durability of Composites in a Marine Environment*, in *Challenges in Mechanics of Time Dependent Materials, Volume 2*, edited by B. Antoun, A. Arzoumanidis, H.J. Qi, M. Silberstein, A. Amirkhizi, J. Furmanski, H. Lu (Springer International Publishing, Cham, 2017), pp. 1–8, ISBN 978-3-319-41543-7

[23] L. Li, S.V. Lomov, X. Yan, V. Carvelli, Composite Structures **116**, 286 (2014)

[24] S. Chaki, W. Harizi, G. Bourse, M. Ourak, Composites Part A: Applied Science and Manufacturing **78**, 358 (2015)

[25] H.R. Dana, A. Perronnet, S. Fréour, P. Casari, F. Jacquemin, Journal of Composite Materials **47**, 1081 (2013)

[26] L. Li, S.V. Lomov, X. Yan, Composite Structures **123**, 45 (2015)

[27] M. Diakhate, E. Bastidas-Arteaga, R. Moutou Pitti, F. Schoefs, Engineering Fracture Mechanics **180**, 254 (2017)

[28] M. Diakhate, N. Angellier, R. Moutou Pitti, F. Dubois, Construction and Building Materials **156**, 911 (2017)

https://doi.org/10.1051/matecconf/201816507002

Effect of Thickness on Fatigue Crack Propagation in Injection Molded Short Carbon Fiber Reinforced Plastics

Kenichi Shimizu[1,*], *Yuya* Hasegawa[2] and *Keisuke* Tanaka[3]

[1]Department of Mechanical Engineering, Meijo University, Nagoya, Japan
[2]Graduate School, Department of Mechanical Engineering, Meijo University, Nagoya, Japan
[3]Department of Research, Nagoya Industrial Science Research Institute, Nagoya, Japan

Abstract. The influence of plate thickness on the fatigue crack propagation behavior was studied by using center-notched specimens which were cut from injection-molded plates of short carbon-fiber reinforced polyphenylene sulfide (PPS) at two fiber angles relative to the molding flow direction (MFD), i. e. θ=0 deg. (MD), 90 deg. (TD). The short carbon-fiber reinforced plastics (SCFRP) plates have three-layer structure where the fiber orientation is parallel to MFD in the shell layer and is nearly perpendicular in the core layer. The fraction of the core layer increases with increase in the plate thickness. In the relation between the crack propagation rate, da/dN, and stress intensity factor, ΔK, da/dN increases with increase in thickness for MD specimen. Conversely, da/dN decreases for TD specimen. The crack opening displacement along the crack was measured by using the digital image correlation (DIC) method. The measured crack opening displacement become larger with increase in the plate thickness for MD specimens. Contrary, measured values become smaller with increase in the plate thickness for TD specimen. The crack-tip-opening radius, $\Delta\rho$, was estimated from the parabolic approximation of the crack opening displacement distribution near the crack tip. The relationships between da/dN and $\Delta\rho$ for all specimens tend to merge into a unique relationship.

1 Introduction

Short-fiber reinforced plastics (SFRP) are expected to be used more widely in order to reduce the weight of vehicles such as automobiles. The injection molding process makes the production of SFRP components more economical and at higher rates. The mechanical properties of SFRP components are very much anisotropic, depending on the fiber orientation produced by injection molding. Since the applications of SFRP in fatigue-sensitive components are steadily increasing in automobile industries, the anisotropic fatigue properties should be assessed in relation to the fiber orientation.

The molded SCFRP plates have three-layer structure where the fiber orientation in the shell layer is parallel to molding flow direction (MFD) and that of the core layer is nearly perpendicular to MFD. The fiber orientation produced by injection molding has a big influence on the propagation rate and path of fatigue cracks [1-8]. Because the crack propagation rate perpendicular to aligned fibers is much slower than that parallel to fibers, the crack propagation rate is different between in the shell layer and in the core layer.

The thickness of the shell layer does not increase even if the plate thickness increases, i. e. the fraction of the core layer increases with increase in the plate thickness. Therefore, it is necessary to investigate the effect of thickness of the molded SCFRP plate on the crack propagation behavior.

In this study, the influence of plate thickness on the fatigue crack propagation behavior was discussed by using center-notched specimens which were cut from injection-molded plates of short carbon-fiber reinforced polyphenylene sulfide (PPS) at two fiber angles relative to the molding flow direction (MFD), i. e. θ=0 deg. (MD), 90 deg. (TD). The thickness of plates was 1, 2 and 3mm. And the crack opening displacement along the crack was measured by using the digital image correlation (DIC) method. The crack-tip-opening radius, $\Delta\rho$, was estimated from the parabolic approximation of the crack opening displacement distribution near the crack tip and the crack propagation rate in each thickness was estimated by using $\Delta\rho$.

2 Experimental procedure

2.1. Specimens

The experimental material is semi-crystalline brittle thermoplastics, PPS, reinforced with carbon fibers. The amount of fiber content was 30 wt%. Fatigue specimens were cut from an injection-molded plate (IMP) with the in-plane dimensions of 80×80 mm and the thickness of 1, 2 and 3mm. IMP have a three-layer structure where

* Corresponding author: kshimizu@meijo-u.ac.jp

two shell layers sandwich the core layer as shown schematically in Fig. 1(a). Figure 1(b) shows the cross section "A" observed by the optical microscope in IMP with the thickness of 1mm (IMP1), 2mm (IMP2) and 3mm (IMP3). In the shell layer, short fibers are parallel to the molding direction. In the core layer, fibers are transverse to the molding direction. The thickness of the shell layer does not increase even if the plate thickness increases, i. e. the fraction of the core layer increases with increase in the plate thickness. Table 1 shows the fraction of shell layer and core layer, f_s, and f_c, respectively.

Figure 2 shows the shape of test specimens which has a center notch of whole length 6 mm, the length of 80 mm, and the width of 20 mm. The region of length 15 mm was used for chucking to the testing machines through aluminum tabs. A fatigue crack was started from the initial notch, and extended in the direction as shown in Fig.1. The angle between the molding direction and the longitudinal direction of specimens was set to be 0 deg. (MD) and 90 deg. (TD).

The core layer may influence the crack propagation behavior of IMP. To evaluate its influence, specimens made of only skin layer were manufactured by thinning the 1mm thickness to 0.4 mm from the one side of plates only for cases of MD, TD. The layer was carefully removed by milling machines. The surface of the machined layer was polished by emery paper to eliminate damages due to machining and the single edge notch of length 2mm was machined. In the following, specimens made of skin-layer are called SLP.

Experimental values of anisotropic elastic constants of IMP and SLP are summarized in Table 2, where suffix 1 indicates the molding direction and 2 the perpendicular direction. The E_1 value decreases with increase in thickness. Conversely, the E_2 increases with increase in thickness. In short, the core layer of IMP reduced E_1 value and increased E_2 value.

Table 1. Fraction of shell layer and core layer to the thickness of the molded plate.

Specimen	Fraction of shell layer f_s (%)	Fraction of core layer f_c (%)
IMP1	86 (= 43 x 2)	14
IMP2	56 (= 28 x 2)	44
IMP3	40 (= 20 x 2)	60

Table 2. Elastic constants of injection-molded plate (IMP) and shell layer plate (SLP).

Specimen	E_1 (GPa)	ν_{12}	E_2 (GPa)	ν_{21}	G_{12} (GPa)
SLP	29.76	0.418	8.94	0.131	4.36
IMP1	28.01	0.379	12.63	0.174	4.73
IMP2	21.79	0.264	15.58	0.199	4.12
IMP3	14.97	0.196	18.68	0.229	3.72

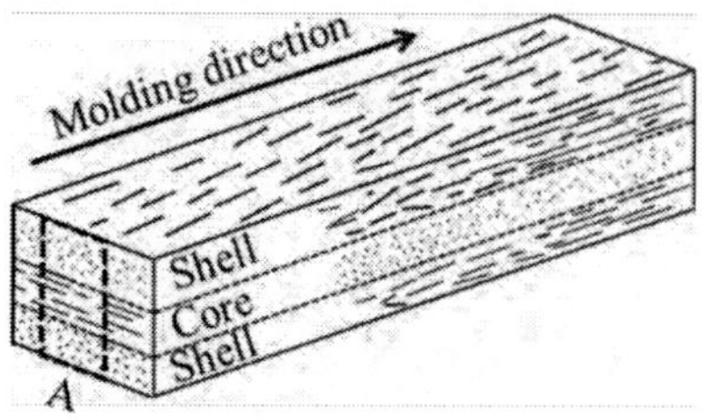

(a) Cross section observed by optical microscope

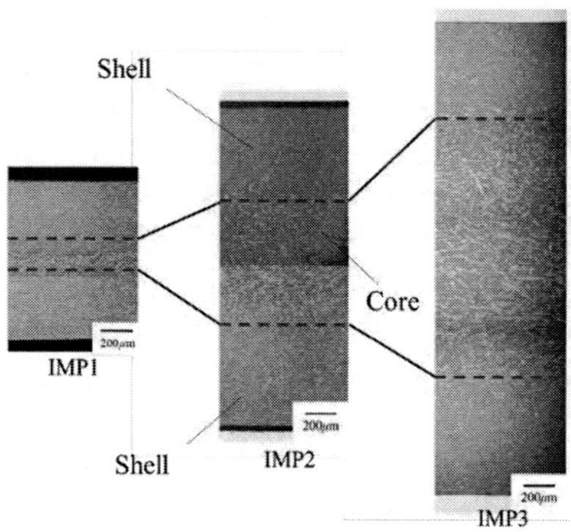

(b) Cross section observed by optical microscope

Fig. 1. Three-layer structure in injection-molded plate (IMP).

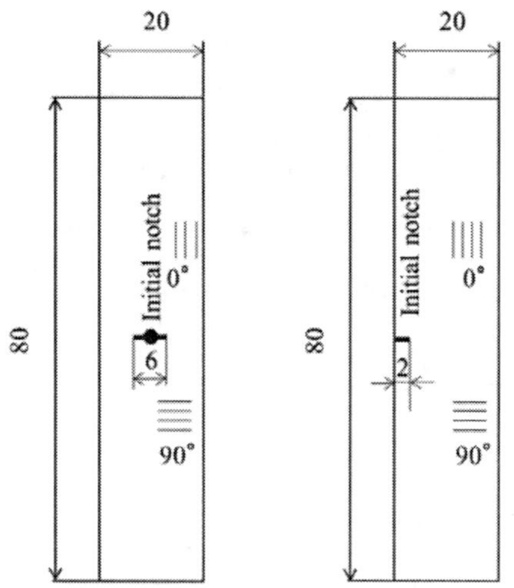

(a) IMP. (b) SLP.

Fig. 2. Specimen (in mm).

2.2 Fatigue crack propagation tests

Fatigue crack propagation tests were performed with a tension-compression electro-servo-hydraulic testing machine. Fatigue testing was done in air at room temperature under load-controlled conditions with the stress ratio R of 0.1. The waveform of the cyclic load was triangular and the frequency was 4 Hz. The crack length was measured with a video microscope. As the macroscopic fatigue crack propagation path was perpendicular to the loading axis, the crack length projected on the plane perpendicular to the loading axis was measured. The crack propagation rate, da/dN, is calculated from crack lengths a vs number of cycles by the secant method.

2.3 Fracture mechanics parameters

The energy release rate of crack propagation was calculated by the method of a modified crack closure integral (MCCI) of FEM [9,10]. For the case of crack growth under mode I in TD specimen, the energy release rate is converted to the stress intensity factor by using [11,12]

$$G = HK^2 \tag{1}$$

$$H = \sqrt{\frac{1}{2E_1E_2}}\left[\sqrt{\frac{E_1}{E_2}} + \frac{-2(\nu_{12}/E_1)+(1/G_{12})}{2(1/E_1)}\right]^{1/2} \tag{2}$$

For the MD specimen, H is calculated by replacing subscript, 1 and 2 in eq. (1). The H values for MD and TD of IMP and SLP are summarized in Table 3. For isotropic materials, we have $1/H$ equal to Young's modulus E and

$$G = K^2/E \tag{3}$$

The stress intensity factor for mode I, K_I, is expressed as

$$K = \sigma\sqrt{\pi a}\cdot F\left(a/W\right) \tag{4}$$

where $F(a/W)$ is a correction factor for the stress intensity factor.

Table 3. Stiffness parameter for MD and TD of IMP and SLP.

Specimen	MD (x10^{-3}/GPa)	TD (x10^{-3}/GPa)
SLP	52.2	95.2
IMP1	50.9	75.9
IMP2	61.3	72.4
IMP3	77.8	69.6

2.4 Measurement of crack opening displacement distribution

A pair of areas of 18x18μm rectangle (subset) was defined longitudinally at a gage length of 60μm containing a crack line in the image obtained from the digital microscope. Each subset was spaced at 9μm intervals near the crack tip and 110 μm other than that along a crack. From images around the fatigue crack under the maximum loading and the minimum loading, the displacements of each subset were evaluated by using digital image correlation method and the crack opening displacement distribution along the fatigue crack was obtained.

3 Experimental results and discussion

3.1. Fatigue crack propagation

Figure 3 shows optical micrographs of a fatigue crack in MD specimen. The crack propagation is blocked by fibers and the crack path is microscopically zigzag shaped and macroscopically straight perpendicular to the loading direction. But there is no clear difference depending on the thickness. Figure 4 shows a fatigue crack in TD specimen. The zigzag of the crack propagation path is smaller than that in MD specimen for each thickness of the plate.

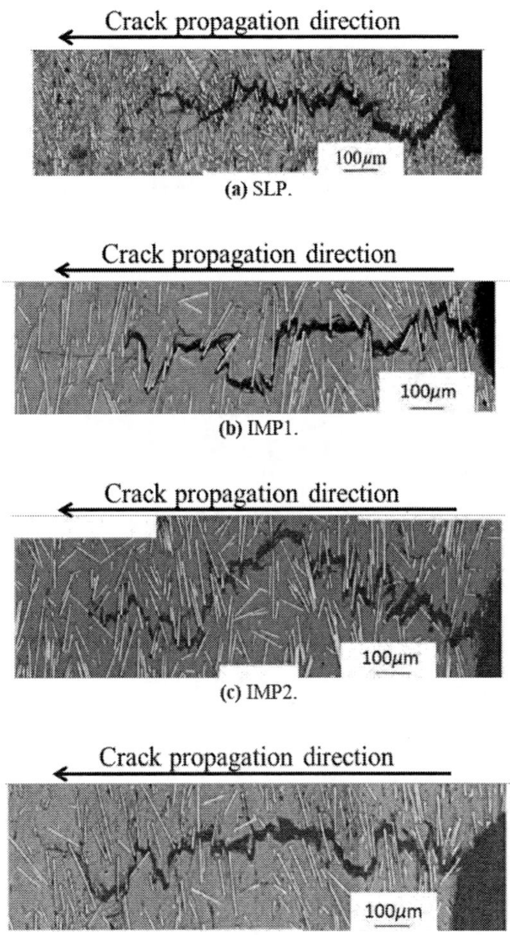

(a) SLP.
(b) IMP1.
(c) IMP2.
(d) IMP3.

Fig. 3. Optical micrographs of a fatigue crack in MD specimen.

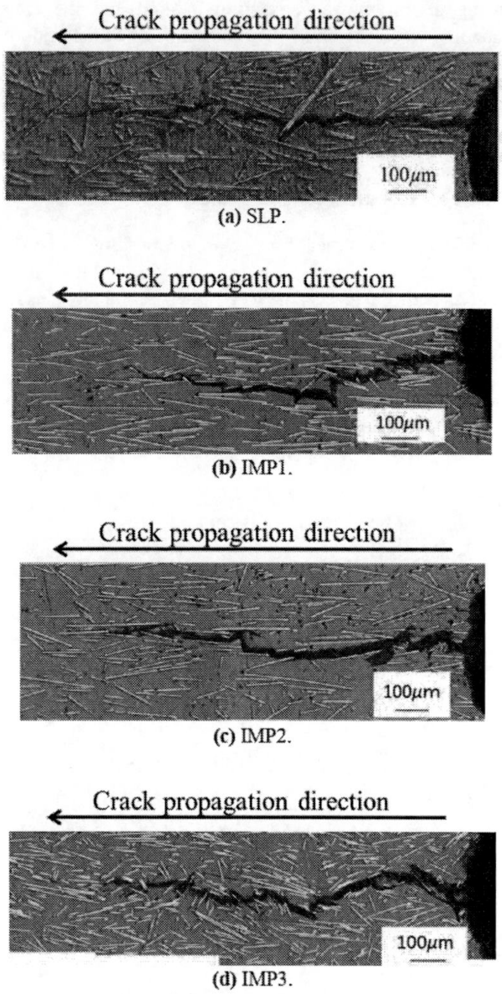

Fig. 4. Optical micrographs of a fatigue crack in TD specimen.

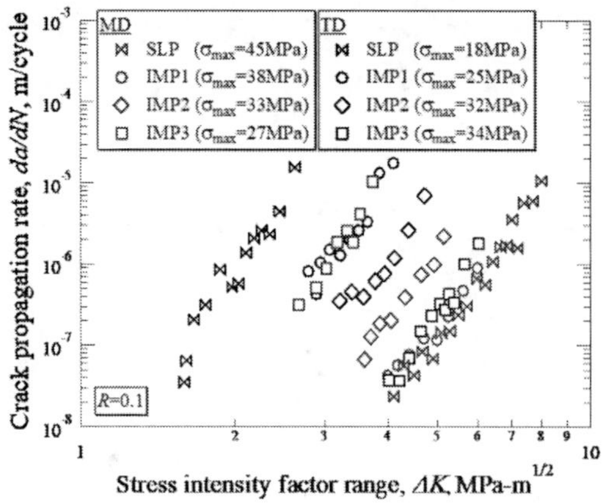

Fig. 5. Relation between crack propagation rate and stress intensity factor range.

The fatigue crack propagation rate, *da/dN*, is plotted against the range of stress intensity factor, *ΔK*, in Fig. 5. There is a linear relation between *da/dN* and *ΔK* in log-log diagram. In the relation between the crack propagation rate, *da/dN*, and stress intensity factor, *ΔK*, *da/dN* increases with increase in thickness for MD specimen including SLP. The resistance to crack propagation is improved by fiber reinforcement, and fibers aligned perpendicular to the crack growth direction block more severely crack propagation. Then, *da/dN* increases with increase in the thickness for the MD specimen because the ratio of shell layer decreases. Conversely, *da/dN* decreases with increase in thickness for TD specimen as the ratio of shell layer decreases Because the fiber is parallel to the crack propagation direction in the shell layer, the crack propagates faster than in the core layer for TD specimen.

3.2. Crack opening displacement

Figures 6 and 7 show the crack opening displacement obtained from the MD and TD specimen, respectively. Lines are the values calculated by Finite Element Method (FEM) analysis. Two-dimensional isoparametric four-node rectangular elements were used under the condition of plane stress. The orthotropic elastic constants used for FEM is given in Table 1. The influence of layered structure was ignored.

Measured values are considerably smaller than analytical values in SLP of MD specimen as shown in Fig. 6(a). Crack closure is occurred by the rough surface contact or the fiber bridging at the crack tip. Then measured values become larger with increase in the plate thickness as indicated in Fig. 6(b)~(d). On the other hand, measured values are little larger than analytical values in SLP of TD specimen as shown in Fig. 7(a). There is a possibility that the crack opening was promoted by delamination between the fiber and the matrix. Contrary to the MD specimen, measured values become smaller with increase in the plate thickness as indicated in Fig. 7(b)~(d).

3.3. Crack-tip-opening radius

For isotropic materials, the stress intensity factor divided by Young's modulus, *ΔK/E*, is often used to correlate the crack propagation data for different materials. This parameter can be interpreted to represent the range of the radius of crack-tip opening during one cycle [13]. The crack-tip opening radius is given by

$$\rho = (4/\pi)(K/E)^2 = (4/\pi)G/E \tag{5}$$

The range of the crack-tip-opening radius is

$$\begin{aligned}\Delta\rho &= \rho_{max} - \rho_{min} = (4/\pi E)\cdot(G_{max} - G_{min}) = (4/\pi E)\cdot\Delta G \\ &= (4/\pi)\cdot[(1+R)/(1-R)]\cdot(\Delta K/E)^2\end{aligned} \tag{6}$$

and the range of the energy release rate is defined by

$$\Delta G = G_{max} - G_{min} \tag{7}$$

where G_{max} and G_{min} are the energy release rate at the maximum and minimum stresses. For orthotropic materials, the crack-tip-opening radius is expressed as [12]

$$\rho = (4/\pi) H G \tag{8}$$

so the range of crack-tip-opening radius is given by

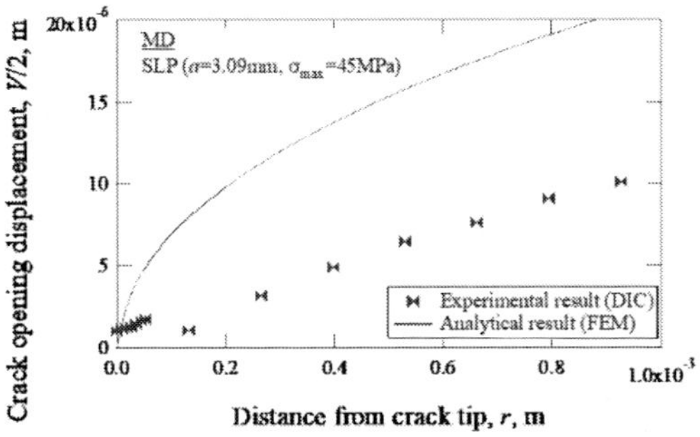

(a) SLP.

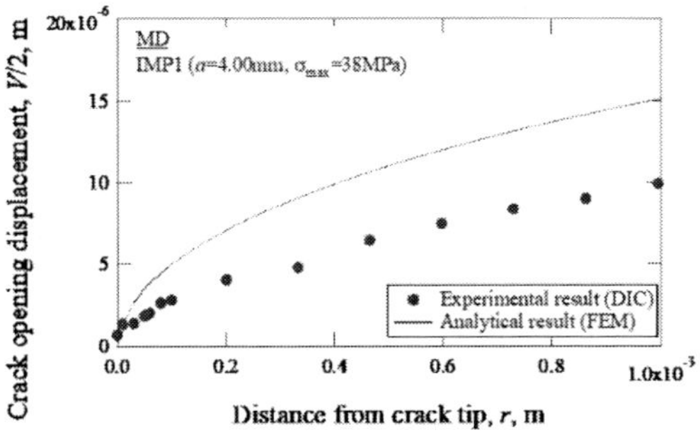

(b) IMP1.

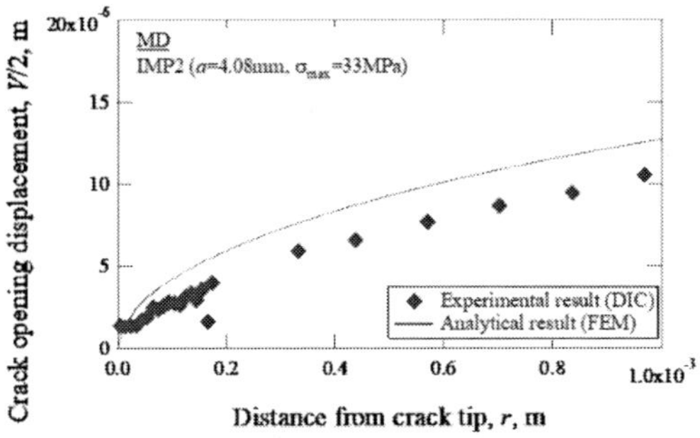

(c) IMP2.

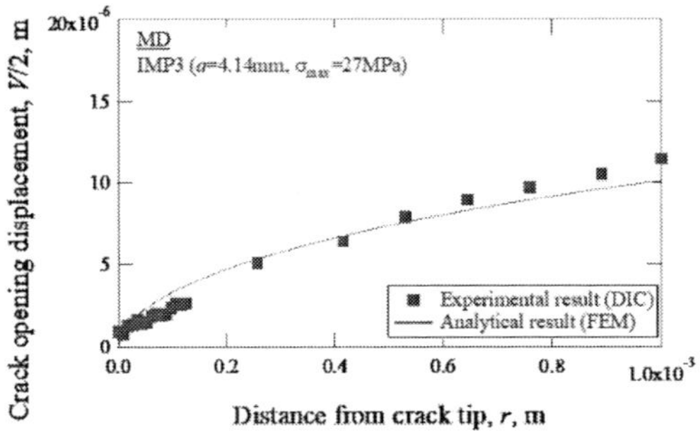

(d) IMP3.

Fig. 6. Crack opening displacement along a fatigue crack in MD specimen.

$$\Delta\rho = \rho_{max} - \rho_{min} = (4/\pi) H \Delta G \tag{9}$$

and the range of the energy release rate is defined by

$$\begin{aligned}\Delta G &= G_{max} - G_{min} = H\left(K_{max}^2 - K_{min}^2\right) \\ &= H\left[(1+R)/(1-R)\right](\Delta K)^2\end{aligned} \tag{10}$$

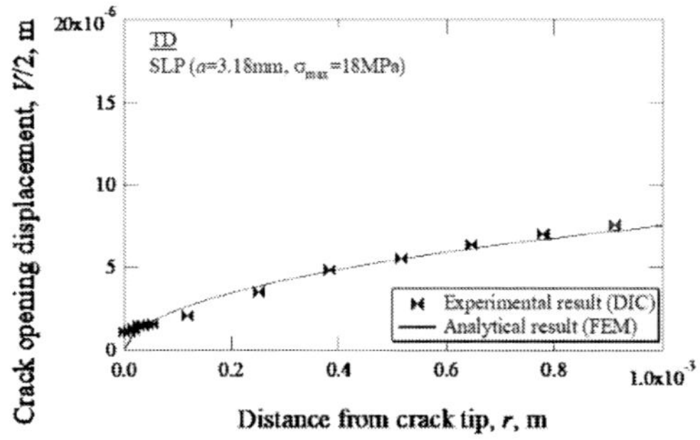

(a) SLP.

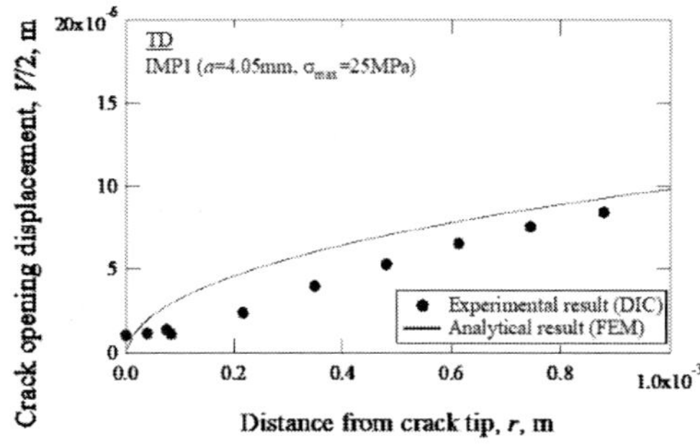

(b) IMP1.

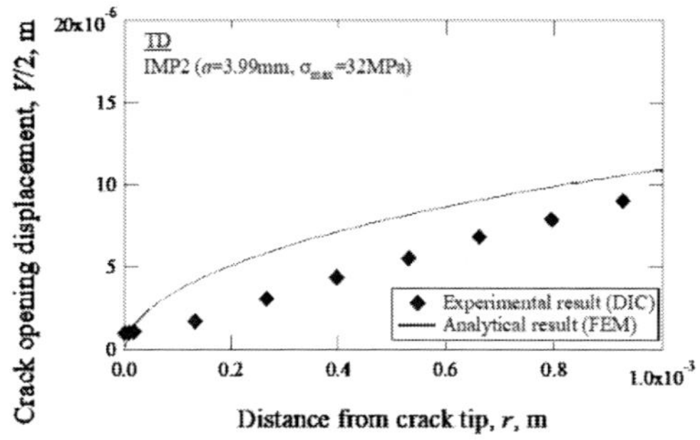

(c) IMP2.

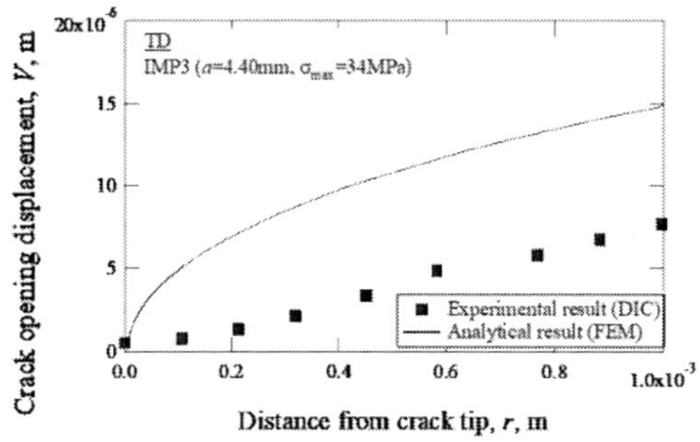

(d) IMP3.

Fig. 7. Crack opening displacement along a fatigue crack in TD specimen.

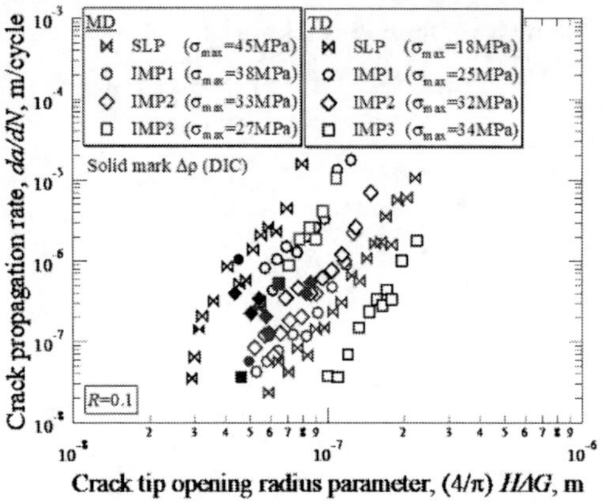

Fig. 8. Relation between crack propagation rate crack-tip-opening radius.

Table 4. Effective fraction of a crack-tip-opening radius.

Specimen	MD		TD	
	$U_{\Delta\rho}$	$(U_{\Delta\rho})_{est}$	$U_{\Delta\rho}$	$(U_{\Delta\rho})_{est}$
SLP	0.14	-	0.96	-
IMP1	0.35	0.26	0.49	0.84
IMP2	0.63	0.49	0.36	0.61
IMP3	0.84	0.64	0.10	0.46

The relation between da/dN and a crack-tip-opening radius, $\Delta\rho$ $(=(4/\pi)H\Delta G)$, is shown in Fig. 8. Open marks show the converted results from ΔK using eqs. (9) and (10) and solid marks are the result of parabola approximation from measured values. Using the data from x=0 to 6x10^{-3} mm in Figs. 6 and 7, the measured values are approximated by

$$v(x) = \sqrt{\frac{x}{\alpha}} \tag{11}$$

And $\Delta\rho$ is given by

$$\Delta\rho = \frac{1}{2\alpha} \tag{12}$$

Similarly, $\Delta\rho_{FEM}$ is able to obtain from the FEM analytical results in Figs. 6 and 7 and the data is equivalent to the open marks in Fig. 8.

Open marks for MD specimens come closer but the difference among TD specimens are remained similar to Fig. 5. On the other hand, solid marks for TD specimens come closer similarly to the MD specimens and the da/dN values in TD specimen is little higher than MD specimens when compared at the same value of $\Delta\rho$.

Effective fraction of a crack-tip-opening radius, $U_{\Delta\rho}$, is described as $U_{\Delta\rho} = \Delta\rho/\Delta\rho_{FEM}$. From eqs.(9) and (10), the following equation is established, $U_{\Delta\rho}=U^2$, in which U $(=\Delta K_{eff}/\Delta K)$ is defined as an effective faction of ΔK. For IMP specimens, $U_{\Delta\rho}$ is estimated from the values of SLP considering the fraction of 0 deg.(MD) and 90 deg.(TD) layer as $(U_{\Delta\rho})_{est}$. For example, $(U_{\Delta\rho})_{est}$ of IMP1_MD specimen is obtained by

$$(U_{\Delta\rho})_{est} = 0.14\times0.86 + 0.96\times0.14 = 0.26$$

Table 4 shows the values of $U_{\Delta\rho}$ and $(U_{\Delta\rho})_{est}$. The $U_{\Delta\rho}$ value is larger than $(U_{\Delta\rho})_{est}$ for MD specimens. And the difference increases with the increase in the thickness. It is considered that the crack length is longer in the core layer than in the shell layer and the difference becomes larger with the increase in the fraction of the core layer. Conversely, $U_{\Delta\rho}$ value is smaller than $(U_{\Delta\rho})_{est}$ for TD specimens. It is considered that the crack length is smaller in the core layer than in the shell layer and the difference becomes larger with the increase in the fraction of the core layer.

4 Conclusions

The influence of plate thickness on the fatigue crack propagation behavior was studied by using center-notched specimens which were cut from injection-molded plates of short carbon-fiber reinforced polyphenylene sulfide (PPS) at two fiber angles relative to the molding flow direction (MFD), i. e. 0 deg. (MD), 90 deg. (TD). The thickness of plates was 1, 2 and 3mm with three-layer structure where the fiber orientation in the shell layer is
parallel to MFD and that of the core layer is nearly perpendicular to MFD. Main results are as follows;

(1) The thickness of the shell layer does not increase even if the plate thickness increases, i. e. the fraction of the core layer increases with increase in the plate thickness.
(2) In the relation between the crack propagation rate, da/dN, and stress intensity factor, ΔK, da/dN increases with increase in thickness for MD specimen. Conversely, da/dN decreases with increase in thickness for TD specimen.
(3) The measured crack opening displacement is smaller than that analysed by FEM in shell layer plate (SLP) of MD specimen owing to the crack closure caused by the rough surface contact or the fiber bridging. On the other hand, measured values are little larger than analytical values in SLP of TD specimen. There is a possibility that the crack opening was promoted by delamination between the fiber and the matrix.
(4) The measured crack opening displacement become larger with increase in the plate thickness for MD specimens. Contrary, measured values become smaller with increase in the plate thickness for TD specimen.
(5) The crack-tip-opening radius, $\Delta\rho$, was estimated from the parabolic approximation of the crack opening displacement distribution near the crack tip. The relationships between da/dN and $\Delta\rho$ for all specimens tend to merge into a unique relationship

References

1. K. Friedrich, R. Walter, H. Voss, J. Karger-Kocsis, Composites, **17**, pp. 205-216 (1986)

2. M. G. Wyzgoski, G. E. Novak, Journal of Materials Science, **25**, pp. 4501-4510 (1990)
3. M. G. Wyzgoski, G. E. Novak, Journal of Materials Science, **26**, pp. 6314-6324 (1991)
4. W. J. Evans, D. H. Isaac, K. S. Saib, Composites Part A, **27**, pp. 547-554 (1996)
5. A. Pegoretti, T. Ricco, Composite Science and Technology, **59**, pp. 1055-1062 (1999)
6. A. Pegoretti, T. Ricco, Journal of Composite Materials, **34**, pp. 1009-1027 (2000)
7. A. Pegoretti, T. Ricco, Composites, Part A, **33**, pp. 1539-1547 (2002)
8. K. Tanaka, T, Kitano, N. Egami, Engineering Fracture Mechanics,**123**, pp. 44-58 (2014)
9. E. F. Rybicki, M. F. Kanninen, Engineering Fracture Mechanics, **9**, pp. 931-938 (1997)
10. R. Kruger, OCASE Report No. 2002-10, NASA/CR-2002-211628 (2002)
11. G. C. Sih, P. C. Paris, G. R. Irwin, International Journal of Fracture Mechanics, **1**, pp. 189-203 (1965)
12. G. C. Sih, H. Liebowitz, Fracture, Vol. 2 Mathematical Fundamentals, Academic Press, New York, pp. 67-190 (1968)
13. L. P. Pook, N. E. Frost, International Journal of Fracture Mechanics, **9**, pp. 53-61(1973)

Experimental monitoring of the self-heating properties of thermoplastic composite materials during tensile and cyclic tests

Laura Muller[1,*], Jean-Michel Roche[1], Antoine Hurmane[1], Catherine Peyrac[2], and Laurent Gornet[3]

[1]ONERA, The French Aerospace Lab, F-92322, Châtillon, France
[2]CETIM, 52 avenue Félix Louat, 60300 Senlis, France
[3]GeM, École Centrale de Nantes, 1 rue de la Noë, 44321 Nantes, France

Abstract. The sizing of industrial structures with a given material requires to understand completely its fatigue behaviour. Yet, the full estimation of the fatigue lifetime of a material requires time-consuming and expensive fatigue testing campaigns, and a huge number of samples. An alternative experimental procedure is based on the monitoring of the self-heating properties of the studied material. The aim is to correlate the fatigue limit with a change of the thermal behaviour of the material during the self-heating tests. The main advantage of these tests is that they need only a limited number of mechanical cycles, and require few samples to test. Ultimately, self-heating tests lead to an accelerated estimation of the fatigue behaviour. The purpose of the present paper is to validate this approach for a woven carbon/thermoplastic composite material.

1 Introduction

Over the years, composite materials are more and more used for primary structural components. This phenomenon is due to the high strength and stiffness to weight ratios of the composite materials. In these conditions, it has become very important to study their static and fatigue behaviour: indeed, structural components are designed with high safety factors, as the fatigue behaviour of the used composite material is not completely known and understood [1].

So far, the main method used in the industry to study the fatigue behavior of a given material is to build its Wöhler curve, also called S/N curve, which provides the number of cycles to failure (N) for each load level (S) (Fig. 1). The constitution of the Wöhler curve requires a large number of mechanical tests, including several samples per load level to avoid scattering, and until 10^7 cycles for the lowest load levels, meaning several months of test for the lowest load levels. Moreover, at these levels, an asymptote can appear, meaning that under this threshold value, the material will not prematurely break because of fatigue. This asymptote is called "fatigue limit" and is used to size structural components [2] (**Fig. 1**).

An alternative to the fatigue tests are the self-heating tests. This methodology is also based on cyclic loads, but instead of cycling each sample until failure for several given load values, several blocks with a limited number of cycles for each block are applied on the same sample, for increasing load values (**Fig. 2**).

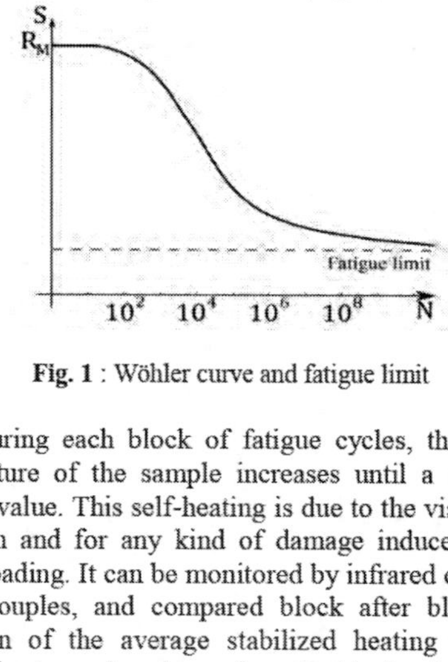

Fig. 1 : Wöhler curve and fatigue limit

During each block of fatigue cycles, the surface temperature of the sample increases until a stabilized heating value. This self-heating is due to the viscosity of the resin and for any kind of damage induced by the cyclic loading. It can be monitored by infrared camera or thermocouples, and compared block after block. The evolution of the average stabilized heating with the maximal stressed value of each block shows two tendencies: for the lowest load values, the stabilized heating is almost constant from one block to another. Then, after a particular load value, the stabilized heating strongly increases block after block. This change of the thermal behaviour may indicate an alteration of the viscous properties of the resin, and/or the initiation of fatigue damage (**Fig. 3**).

* Corresponding author: laura.muller@onera.fr

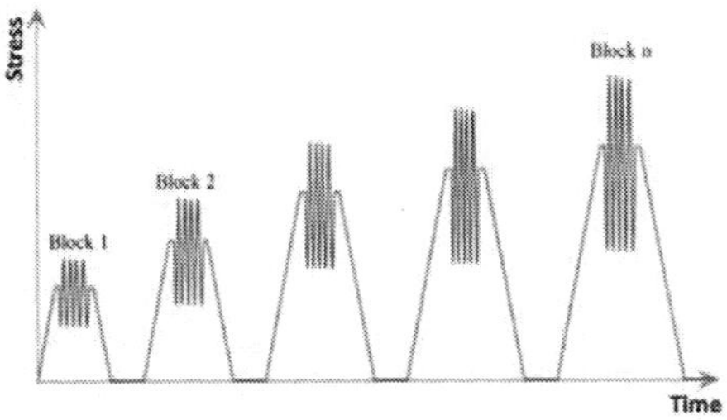

Fig. 2 : Schematic representation of self-heating tests

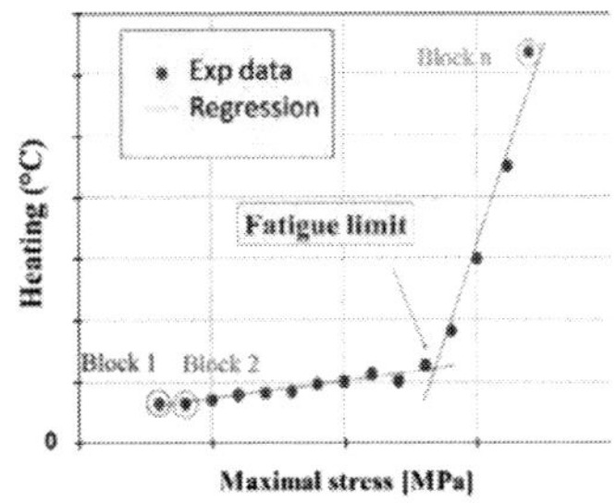

Fig. 3 : Self-heating curve

This approach has been proven efficient for metallic materials [3]–[7] and successfully verified for some composites [8], [9] : the specific load value between the two thermal tendencies during the self-heating tests is related to the fatigue limit determined on the Wöhler curve. It seems that the self-heating monitoring can indeed enable a reliable monitory of fatigue damage. The aim of this work is to apply this methodology to a woven carbon/thermoplastic composite material. Several experimental campaigns are carried out: a static test campaign to characterize the material's properties, conventional fatigue tests, and self-heating tests.

2 Quasi-static tensile test campaign

2.1 Material and conditioning

The studied material is a balanced plain woven carbon/thermoplastic composite. The thermoplastic matrix is a semi-crystalline polymer, the polyamide 6.6, also called PA66. This resin has the particularity to be very sensitive to moisture absorption : the polyamid can absorb 2% to 5% moisture, depending on the relative humidity of the specimen [10], [11]. This important sensitivity to moisture has a direct impact on the mechanical properties (like the Young modulus or the ultimate tensile strength) [8], [12]. To guarantee coherence of the experimental campaigns, samples have to be conditioned with a relative humidity (RH) controlled. In this study, the samples are conditioned at ambient temperature, with a RH fixed at 50%, to approach the ambient conditions. The conditioning is ensured by a desiccator, in which the conditioned air is generated by a saturate salted solution [13] and regularly controlled.

2.2 Quasi-static tensile tests

2.2.1 Experimental procedure

The quasi-static tensile tests are carried on a ZWICK testing machine, equipped with mechanic grips and a 150 kN load cell. The applied load speed is fixed to 2 MPa/s. Samples are rectangular (250x25mm²), along three directions: 0°, 90°, 45°. The temperature field and specimen displacement are monitored respectively by an infrared camera and optical camera for Digital Image Correlation (DIC). Their characteristics are detailed in [14].

2.2.2 Mechanical results

Mechanical characterization involves the determination of failure stress, Shear's and Young's modulus and Poisson ratio according to the 3 orthotropic directions of the material. Only warp and weft plane properties are measured. To do this, the samples are tested up to failure, and the measured displacement fields lead to calculated strain fields. The Young modulus E is defined as the ratio between the stress σ and the longitudinal strain ε_{longi} in the elastic domain of the material. Assuming that the material strength is equivalent to a laminate composite strength, the Young modulus is experimentally determined according to the EN 2561-91 norm, i.e. by calculating the slope of $\sigma(\varepsilon_{longi})$ on a stress range between 10% and 50% of failure stress (black curve in **Fig. 4**). The Poisson's ratio is defined as the ratio between the transverse ε_{transv} and the longitudinal ε_{longi} strain. It is determined on the slope $\varepsilon_{transv}(\varepsilon_{longi})$ with the same interval (10% to 50% of failure stress) as for Young's modulus.

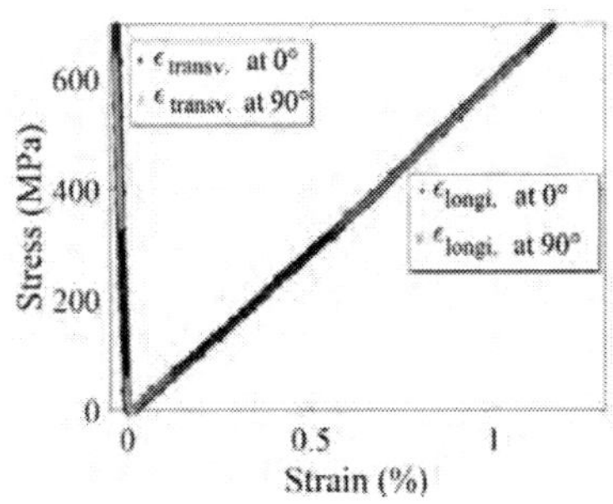

Fig. 4 : Stress-strain curve, with the longitudinal and transversal strain for sample at 0° and 90°, and the domains used for the evaluation of the Young modulus and the Poisson's ratio (black)

Table 1 summarizes the properties of each orientation by averaging the values obtained for each sample. As shown in **Fig. 4**, the properties at 0° and 90° are similar, due to the balanced nature of the woven composite.

* Corresponding author: laura.muller@onera.fr

Table 1: Elastic properties and ultimate tensile stress of the studied carbon T700/PA66 thermoplastic composite material

Orientation	Ultimate tensile stress	Young modulus	Poisson's ratio
0° (warp)	695 MPa	59 GPa	0.031
90° (weft)	595 MPa	58 GPa	0.033
45° (shear)	265 MPa		

2.3 Thermal behaviour

2.3.1 Theoretical background

Several authors [1], [15] have been studying the thermal behaviour of the material during a static tensile test. It is noted that temperature variations follow a three-step pattern:

- The first regime (I) corresponds to the elastic mechanical domain of the material, and the temperature linearly decreases with the loading stress.
- The second regime (II) corresponds to the anelastic mechanical behaviour, the temperature decreases non-linearly until a minimum.
- The third regime (III) is characterized by the non-linear increase of the temperature, due to viscosity and damage apparition, until the sample breaks.

The linear variation of temperature ΔT in the first domain (I) is related to the stress level increase $\Delta\sigma$ by thermoelastic effect [16], [17] :

$$\Delta T = -T_0 \, K_m \, \Delta\sigma \qquad (1)$$

T_0 is the initial average temperature of the sample, K_m is the thermoelastic coefficient defined by:

$$\mathrm{K_m} = \frac{\alpha}{\rho c_p} \qquad (2)$$

α is the thermal expansion coefficient, ρ the mass density and c_p the specific heat of the material. According to this law, the end of linearity of the temperature behaviour marks the end of the elastic domain. This change has been correlated with the appearance of micro-damage in [15], on thermoset composites.

Then, the transition from a drop to an increase temperature via a thermal minimum explains the transition from the second domain to the third domain. This transition was correlated with the appearance of macro damage in [15], which would generate enough heat to balance the cooling of the specimen.

This study will be interested in the threshold stress values associated with the transition from one step to another: the stress value related to the end of the elastic domain and the stress value related to the thermal minimum.

2.3.2 Results and discussion

In the case of a 0° specimen (and a 90° specimen also), the evolution of its average heating is in good agreement with the literature description, as shown in **Fig. 5**. To determine the two particular stress values mentioned before, the noise due to the infrared camera has to be filtered. To do this, a Gaussian filter is applied, as shown in **Fig. 6**. Once the thermal signal filtered, the $\ell 1$ trend filtering, related to "segmented regression", is applied : the variables (the thermal curve) are segmented into groups, and a regression analysis is performed on each segment [18], [19]. The thermal curve is interpolated as a succession of straight lines (**Fig. 7**), and the number of segments is defined by the $\ell 1$ trend filtering so that the correlation coefficient between the curve and the line is minimized. The accuracy of this minimization can be adjusted according to the $\ell 1$ parameter λ.

The junctions between two straight lines are highlighted by dots, **Fig. 7**. The two particular values that interest us are the one marking the end of the linear domain, i.e. the end of the first line; and the one corresponding with the minimal thermal value. In order to estimate a margin of error, a circular permutation of the thermal signal is applied, while maintaining the general trend of the curve: this is to redistribute the noise. Then, the $\ell 1$ trend-filtering is applied for each circular permutation step, and the new values of the junctions are saved. The process is repeated a hundred times, and the set of stored values for each junction will give us a margin of error (**Fig. 8**). The same procedure can is applied for a 90° specimen. The particular stress values are summarized in Table 2.

Table 2 : Particular stress values for 0° et 90° specimens

Orientation	End of linearity	Thermal minimum
Particular stress values at 0°	165 MPa [115 – 215] MPa	360 MPa [340 – 380] MPa
Particular stress values at 90°	195 MPa [187 – 205] MPa	347 MPa [340 – 255] MPa

The next step of the work consists in connecting these particular stress values to physical phenomena, especially to damage apparition. Two acoustic sensors were used to record each acoustic emission occurring during the test. The plotting of the cumulative energy as a function of the applied stress shows a starting threshold for damage, defined by the initiation of first acoustic emissions, as shown in **Fig. 9**. This threshold has been determined shortly before 400MPa for 0° specimens, corresponding to their thermal minimum. As the damage in a composite is assumed to be the same during a static or a fatigue tests, the starting threshold for damage in static might be a good approximation of the fatigue limit, which finally defines the low stress from which fatigue damage occurs. The rest of the work will consist in validating this theory by analysing the acoustic exploitation or microscopic scans, and to analyse the case at 45°, which present a totally different thermal behaviour.

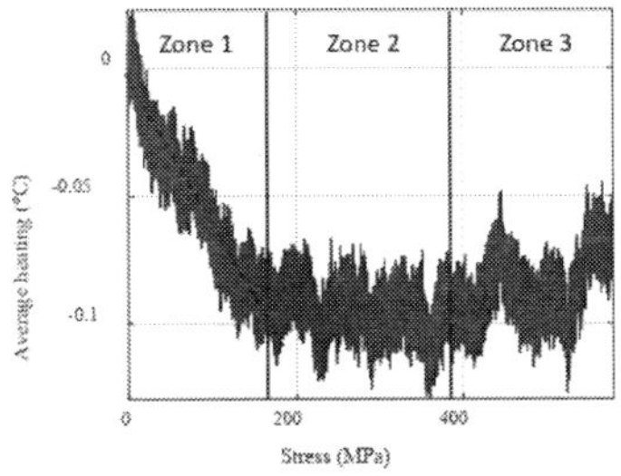

Fig. 5 : Evolution of the spatial average heating of the sample with the stress during a static test

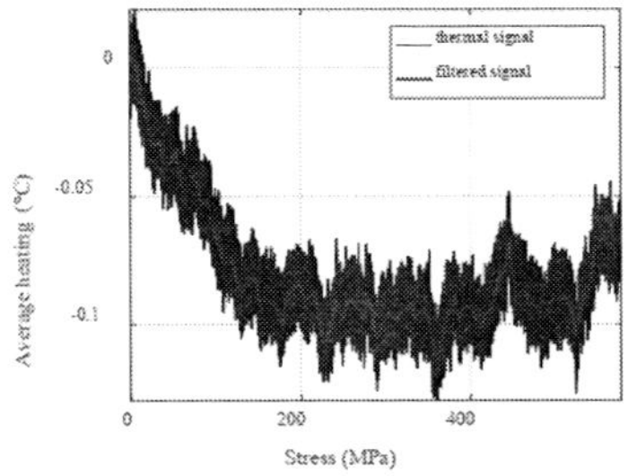

Fig. 6 : Filtering of the thermal signal

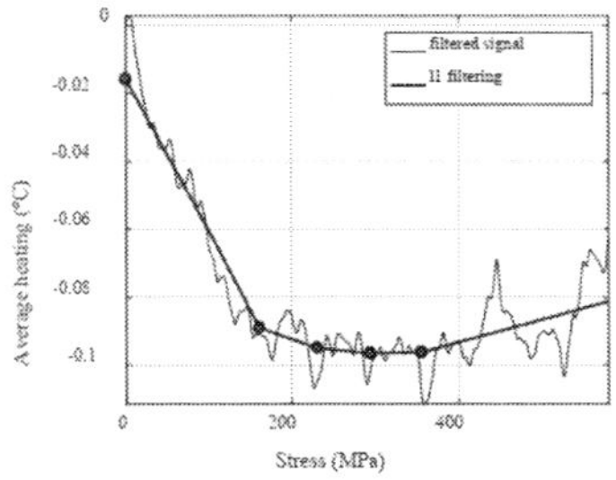

Fig. 7 : Application of $\ell 1$ trend filtering on the filtered signal

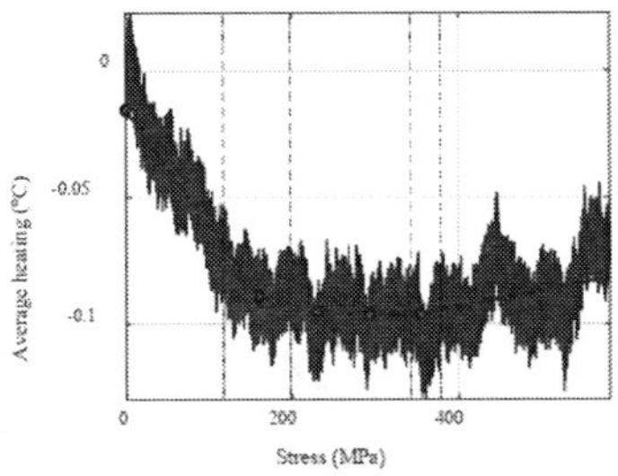

Fig. 8 : Incertitude range for the two particular stress values

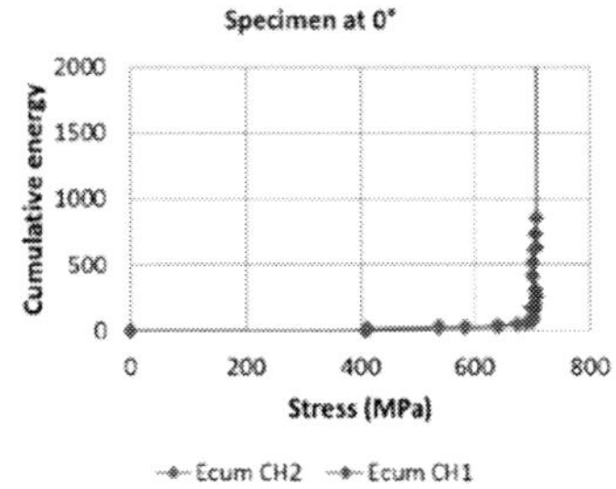

Fig. 9 : Cumulative energy measured by two acoustic sensors during a quasi-static tensile test, for a 0° specimen

3 Fatigue tests campaigns

Wöhler curves are traditionally used to determine the fatigue behaviour of a material, and lower stress levels will give the fatigue limit, used to size structures. The aim of the study is to test an alternative method (the self-heating monitoring approach) but a fatigue test campaign is necessary to have a reference. This campaign is carried out at CETIM. The Wöhler curves are built for the 0° and the 45° specimens.

3.1 Experimental setup

The fatigue tests are carried out on a MTS testing machine, equipped with hydraulic grips and a 25 kN load cell. The load ratio, defined by the ratio between the minimal and the maximal stress applied, is fixed at $R = 0.1$. Due to the high temperature rise of the matrix during the 45° tests, the Wöhler curve at 45° is realized with a cyclic load fixed at 5 Hz, and an air-cooling system keeps the specimen at 22°C. For the 0° tests, it is the fibers that are mainly under stress, the matrix does not heat up much, which makes it possible to carry out tests at 10 Hz which reduces the test time.

With rectangular specimens, the break occurs at the grips outlet, due to overloads, which disqualifies the test. To solve this problem and force the failure at the center of the sample (i.e. a valid rupture), the specimens are shaped with a dogbone geometry, following the recommendation of [20]. The dimensions are then: 20mm width at the centre against 25mm width in the grips, for a 200mm length. About thirty specimens are used for each orientation, to build the final curves.

3.2 Results and discussion

The obtained Wöhler curves are presented **Fig. 10** and **Fig. 11**. In both cases, the failure stress is recalculated from the measured thickness and the measured breaking width, in order to approach the real failure stress. To determine the fatigue limit, several models exist, listed by [21]. The Basquin model is mainly used to model the linear part, in a log-log scale. Stromeyer model include the asymptotic part in the Basquin model. Finally, the Bastenaire approximation, more complete, models the Wöhler curve from the highest to the lowest levels. Using the ESOPE software developed by Arcelor, these models are applied to the curves, the used formulas are summarized in Table 3, with A, B and C coefficients of the fit, N the number of cycles, S the maximal stress, and E the fatigue limit. The ESOPE software may plot the interpolation at 50% (the same probability to have an experimental dot to the left or right of the model), framed by the 10% and 90% interpolations.

Table 3 : Formulas of the main models used to fit the Wöhler curves

Model	Basquin	Stromeyer	Bastenaire
Formulas	$\log N = A.s + B$	$N = \frac{A}{s-E}$	$\mathrm{N} = \mathrm{A}.\frac{e^{-\left(\frac{s-E}{B}\right)^{C}}}{s-E}$

 https://doi.org/10.1051/matecconf/201816507003

For the study, it is the Bastenaire model that will be applied to both curves, in order to have both the fatigue behaviour of the material and the fatigue limit (**Fig. 12** and **Fig. 13**). The fatigue limits are determined around 500 MPa for the 0° direction and 180 MPa for the 45° direction.

For the 0° specimens, the fatigue limit at 0° is much higher than what could be assumed from the static tests. However, it can clearly be seen that the scattering is very important on the fatigue tests, and that the material does not seem to be subjected to fatigue damage: it has a brittle behaviour, and samples sharply break. In this case, the relevance of the use of Wöhler curve can be discussed. Indeed, in the case of composites, the damage propagation (i.e. fatigue behaviour) depends on the matrix behaviour, whereas the failure of the specimen, reported on the Wöhler curve, is led by the fiber's break. It seems that two separated phenomenon (fatigue and failure), depending on two different component, are mixed on the Wöhler curves. The advantages of self-heating tests are then twofold: it would not only reduce the cost of testing (compared to conventional fatigue tests), but this new methodology might also provide greater precision on the material's fatigue behaviour.

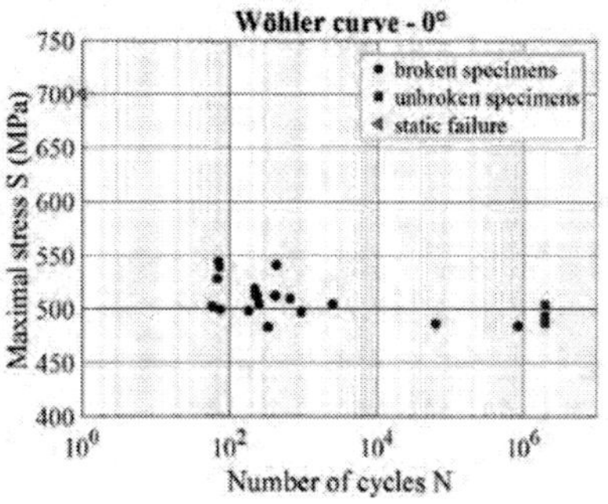

Fig. 10 : Wöhler curve at 0°

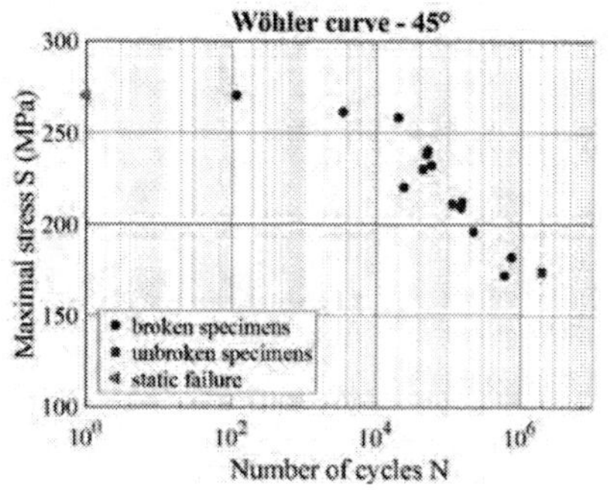

Fig. 11 : Wöhler curve at 45°

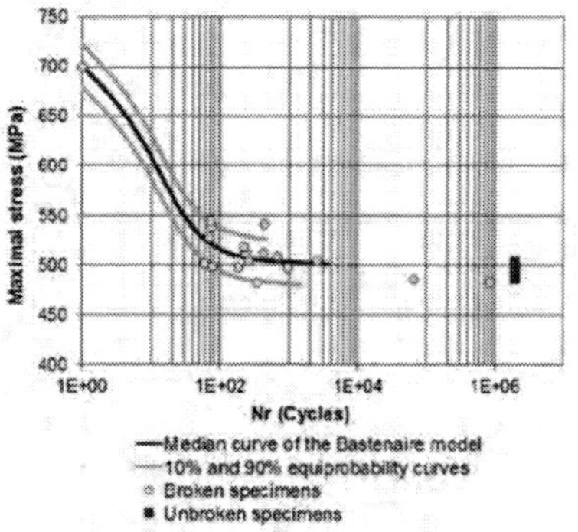

Fig. 12 : Wöhler curve at 0° approximated by Bastenaire model

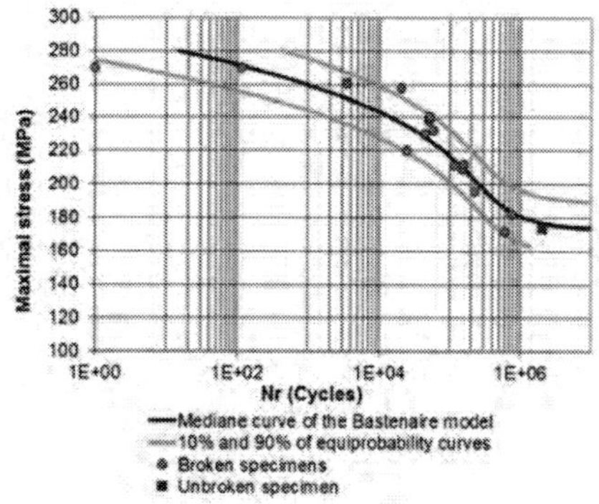

Fig. 13 : Wöhler curve at 45° approximated by Bastenaire model

4 Self-heating tests campaign

As shown before, in the case of some composite materials, the analysis of Wöhler curves might not be concluding and satisfying to determine the fatigue limit of a material. That's why the self-heating tests are investigated, to bring another point of view of fatigue behaviour, and hopefully more robust information.

4.1 Theoretical approach

During a cyclic mechanical loading, characterized by its pulsation $\omega = 2\pi f$ (with f the frequency of the solicitation), the global thermal signal can be written as a Fourier decomposition:

$$T(t) = T_{drift}(t) + \sum_{n=1}^{\infty} a_n \cos(n\omega t) + \sum_{n=1}^{\infty} b_n \sin(n\omega t) \quad (3)$$

With a_n and b_n the Fourier coefficients, t the time, T the temperature, T_{drift} the thermal drift of the signal.

In this study, only the first harmonic ($n = 1$) is considered. The development of (3) is detailed in [14], with the introduction of the heating $\theta(t) = T(t) - T(t = 0)$, of the amplitude A_1 and the phase ϕ_1 of the first harmonic, of the stabilized temperature rise θ_F and of the characteristic time parameter τ. Ultimately, the heating can be written as:

https://doi.org/10.1051/matecconf/201816507003

$$\theta(t) = \theta_{drift}(t) + A_1 \cos(\omega t + \phi_1) \quad (4)$$

With the first term developed as:

$$\theta_{drift}(t) = \theta_F(1 - e^{-\frac{t}{\tau}}) \quad (5)$$

Experimentally, the thermal drift is fitted by θ_{drift}, and the stabilized temperature rise θ_F is used as a thermal indicator for each load level during the self-heating tests. The oscillating part of the thermal signal corresponds to the first harmonic of the thermal signal, defined by its amplitude A_1 and its phase ϕ_1. This harmonic describes the thermoelastic behavior of the material.

The self-heating tests are carried out on a 100 kN-hydraulic machine, equipped with hydraulic cooled grips. To avoid any resonance effect between the frequency of the test and of the infrared camera (especially for the study of the harmonic), the frequency of the cyclic load is $f = 4.72\ Hz$. The stress ratio is the same as for the fatigue tests, i.e. $R = 0.1$. The tests are based on 12 blocks of cycles, 6,000 cycles per block. The displacement and thermal fields are respectively monitored by an optical camera dedicated to image correlation and an infrared camera (FLIR sc7000) similar to the one used for the static mechanical tests.

4.2 Experimental passive thermal monitoring at law IR camera frequency acquisition

The first approach, following the methodology described in the literature, is to monitor the temperature rise of the specimen with a low frequency acquisition (1 image per second), in order to analyse the thermal drift while minimizing the volume of recorded data. As shown in **Fig. 14** and **Fig. 15**, the stabilized temperature rise value θ_F is reached for different stress levels, and tends to increase with the stress level. The evolution of θ_F with the maximal stress level on each block leads to build the "self-heating curves", displayed in **Fig. 16** for three specimens: one at 0° and two at 90°. The stake here is to identify a change of thermal behavior, using two linear regressions, for a threshold stress value which will be compared to the fatigue limit, in order to verify if this thermal behavior change can be directly related to fatigue damage. The method used to identify the threshold value is the similar as for tensile tests: the best intersection between two linear regressions is kept, for several redistributions of the experimental data, and the several intersection data calculated produce a confidence interval.

For both orientations (0° and 90°), the self-heating curves show a definite variation of the thermal behavior. However, the precise and reliable identification of a threshold stress value associated with a clear change of heating tendency is not obvious. If the determined values for the 90° specimens seem reliable, the lack of data to define the second domain for the 0° specimen makes it impossible to properly estimate a stress threshold associated with a change of thermal behavior. Moreover, the values present a high scattering with a large confidence interval, as shown in Table 4. Despite the scattering, it would appear that the particular stress value of the self-heating curves is close to the particular stress value related to thermal minimum, on the thermal curve of quasi-static tensile tests. The exploitation of the self-heating tests has to be improved and completed with acoustic and mechanical data, to confirm this link between static tests and self-heating tests, and to better understand why the value obtained is below that the one obtained from the Wöhler curve.

Table 4 : Threshold stress values for each specimen, and their confidence range

Specimen	Particular stress value	Confidence interval
N°1 at 90°	280 MPa	[193 367] MPa
N°2 at 90°	320 MPa	[268 347] MPa

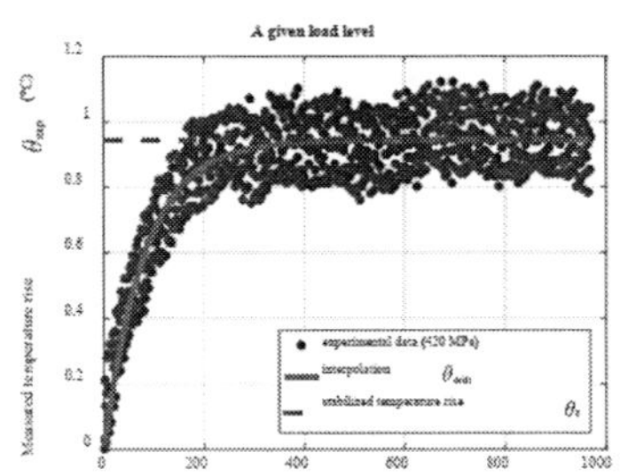

Fig. 14 : Measured self-heating θ_{exp} for a given stress level and estimation of the stabilized temperature rise θ_F. Thermal acquisition setup: 1 image per second. Composite at 90°

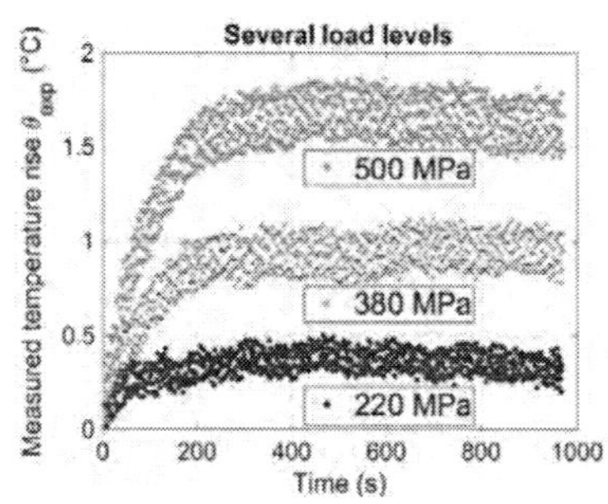

Fig. 15 : Variation of temperature rise between from one stress level to another. Composite at 90°

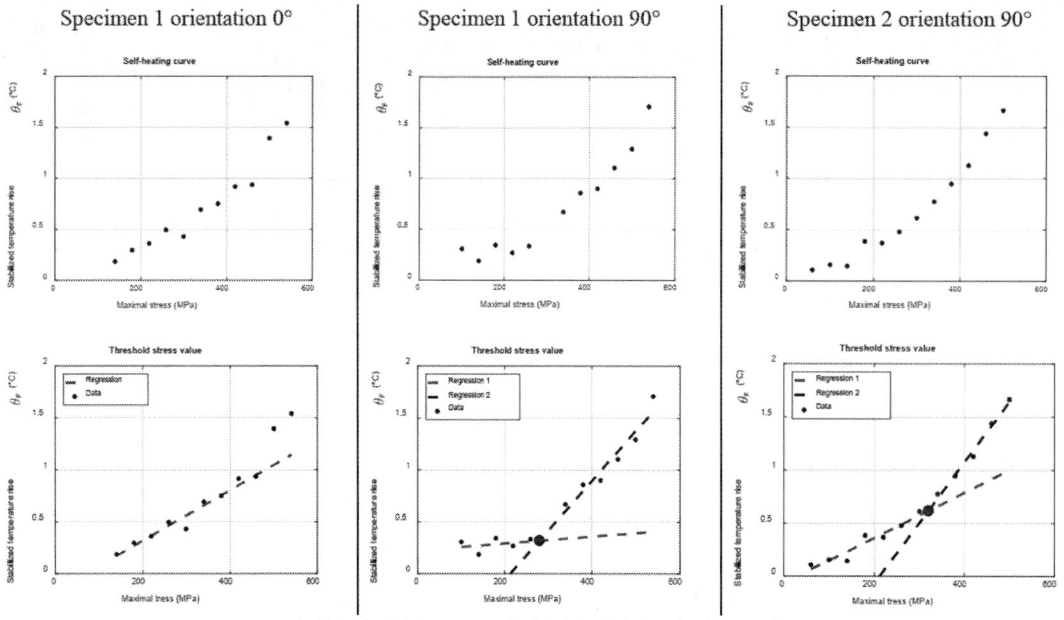

Fig. 16 : Self-heating curves and estimation of the threshold stress values

4.3 Experimental lock-in thermal monitoring at high IR camera frequency

The second chosen approach is to monitor the heating of the sample with a much higher frequency acquisition of 100 images per second in order to analyze the sinusoidal part of the thermal signal, in anti-phase with the mechanical signal (**Fig. 17**). This lock-in thermography approach had been used by Krapez *et al.* for metallic materials [22]. If this approach is validated, it has several advantages over the first one: the number of cycles per block may be shorten, and thus the environment has less influence on the result. Indeed, even if the thermal drift does not stabilize, is too small to be detected correctly, or is disturbed by an air current, the sinusoid component of the thermal signal will still be usable. For this study, to be able to compare results with "classic" self-heating tests, the applied mechanical load is the same as for the thermal drift study (12 blocks of cycles, 6,000 cycles per block, with the same load values). Thus the material is submittted to the same state of strain and temperature. For each block of cycles, the infrared camera is triggered to record 10 seconds (about 50 cycles), after 20 and 3,000 mechanical cycles.

Since the load ratio R is constant the stress amplitude grows from one block to another. According to the thermoelastic theory [16], the thermal amplitude A_1 is expected to increase proportionally as long as the material behavior remains in its elastic domain. But during the last blocks, a non-linearity of $A_1(\sigma_{max})$ is observed, using the same method of double linear regressions than for the thermal drift. This change appears at a threshold stress value (**Fig. 18**). Table 5 shows the threshold stress values estimated from the $A_1(\sigma_{max})$ curve, for two samples (one of each orientation). It appears that the threshold stress value is the same in both cases, with a relatively low scattering. This means that viscous phenomena and/or damage events altering the thermoelastic behaviour are activated at the same stress level. It also appears that this threshold stress value is close to the one associated with the thermal minimum during static tests, much more clearly than with the thermal drift study. So, even if the physical phenomenon of the non-linearity of the first harmonic are not yet understood (that will be the next step of this work), this approach seems at least as promising as the thermal drift study. As for the quasi static tensile tests, the case of the 45° specimen will be study separately, as the thermal behavior is different than for 0° and 90° specimens.

Table 5 : Threshold stress value assessed from the evolution of the first harmonic amplitude, and their confidence range

Specimen	Threshold stress value	Confidence interval
N°2 at 0°	321 MPa	[290 352] MPa
N°3 at 90°	320 MPa	[306 331] MPa

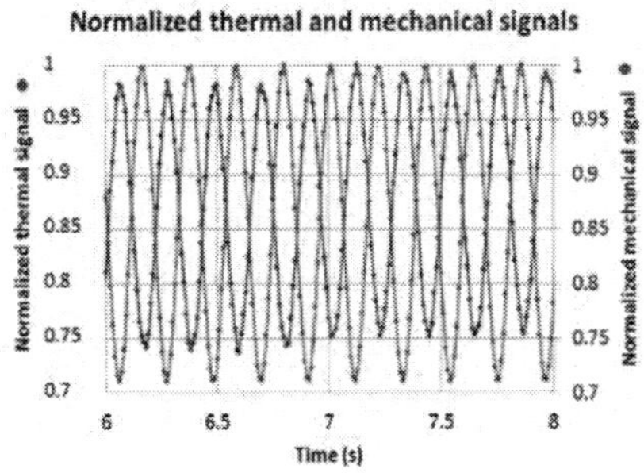

Fig. 17 : Anti-phase variations of the thermal and mechanical signals. Thermal acquisition setup: 100 images per second.

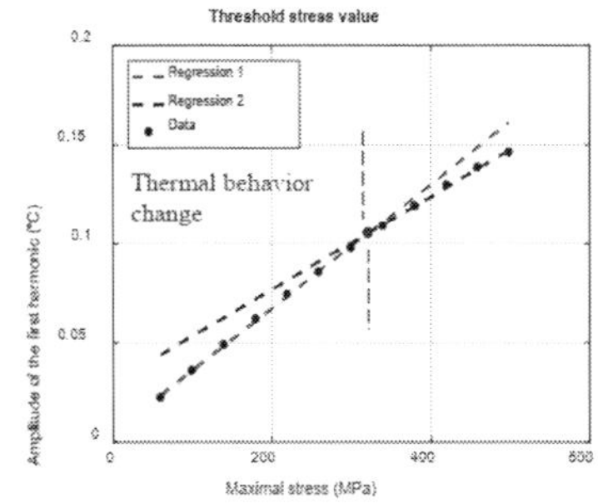

Fig. 18 : Threshold stress value of $A_1(\sigma_{maxi})$

5 Conclusion and outlooks

After checking that the self-heating evolutions described before for thermoset composites are still valid for woven carbon fiber thermoplastic matric composite materials in [14], the aim of this paper was to develop the exploitation of the several campaigns and to bring the fatigue tests as a comparison.

It appears that the particular threshold values highlighted during the quasi static tensile tests, (especially the one associated with the thermal minimum), and during the self-heating tests (the stabilized temperature rise determined from the thermal drift, the first harmonic determined on the sinusoid component), seem to be linked. New tests, as well as new methods of exploitation, are planned in order to better understand the origin of these thermal changes, and if they might be compared.

However, the constitution of the Wöhler curves did not make it possible to link these changes in thermal behaviour to a fatigue limit. The question will be: which method is most relevant to precisely understand the fatigue phenomena of woven composite material.

References

[1] C. Colombo, F. Libonati, F.Pezzani, A. Salerno, and L. Vergani, 'Fatigue behaviour of a GFRP laminate by thermographic measurements', *Procedia Eng.*, vol. 10, pp. 3518–3527, 2011.

[2] N. E. Dowling, *Mechanical Behavior of Material : Engineering methods for deformation, fracture, and fatigue*, International Edition. Pearson, 2013.

[3] M. P. Luong, 'Infrared thermographic scanning of fatigue in metals', *Nucl. Eng. Des.*, vol. 158, no. 2–3, pp. 363–376, 1995.

[4] G. La Rosa and A. Risitano, 'Thermographic methodology for rapid determination of the fatigue limit of materials and mechanical components', *Int. J. Fatigue*, vol. 22, no. 1, pp. 65–73, 2000.

[5] C. Doudard, S. Calloch, F. Hild, and S. Roux, 'Identification of heat source fields from infrared thermography: Determination of "self-heating" in a dual-phase steel by using a dog bone sample', *Mech. Mater.*, vol. 42, no. 1, pp. 55–62, Jan. 2010.

[6] R. Munier, C. Doudard, S. Calloch, and B. Weber, 'Identification of the micro-plasticity mechanisms at the origin of self-heating under cyclic loading with low stress amplitude', *Int. J. Fatigue*, vol. 103, pp. 122–135, Oct. 2017.

[7] M. Poncelet, C. Doudard, S. Calloch, F. Hild, B. Weber, and A. Galtier, 'Prediction of self-heating measurements under proportional and non-proportional multiaxial cyclic loadings', *Comptes Rendus Mécanique*, vol. 335, no. 2, pp. 81–86, Feb. 2007.

[8] L. Gornet, O. Westphal, A. Krasnobrizha, P. Rozycki, C. Peyrac, and F. Lefèbvre, 'Rapid determination of the high cycle fatigue limit of a woven carbon fibe thermoplastic matric', *JNC 19*, 2015.

[9] L. Jegou, Y. Marco, V. Le Saux, and S. Calloch, 'Fast prediction of the Wöhler curve from heat build-up measurements on Short Fiber Reinforced Plastic', *Int. J. Fatigue*, vol. 47, pp. 259–267, Feb. 2013.

[10] A. Malpot, F. Touchard, and S. Bergamo, 'Effect of relative humidity on mechanical properties of a woven thermoplastic composite for automotive application', *Polym. Test.*, vol. 48, pp. 160–168, 2015.

[11] M. Broudin *et al.*, 'Water diffusivity in PA66: Experimental characterization and modeling based on free volume theory', *Eur. Polym. J.*, vol. 67, pp. 326–334, Jun. 2015.

[12] A. Benaarbia, A. Chrysochoos, and G. Robert, 'Influence of relative humidity and loading frequency on the PA6.6 cyclic thermomechanical behavior: Part I. mechanical and thermal aspects', *Polym. Test.*, vol. 40, pp. 290–298, Dec. 2014.

[13] L. Greenspan, 'Humidity Fixed Points of Binary Saturated Aqueous Solutions', *Phys. Chem.*, vol. 81 A, 1977.

[14] L. Muller *et al.*, 'Experimental monitoring of the self-heating properties of thermoplastic composite materials', *Procedia Eng.*, vol. 213, pp. 183–191, 2018.

[15] L. Vergani, C. Colombo, and F. Libonati, 'A review of thermographic techniques for damage investigation in composites', *Frat. Ed Integrità Strutt.*, 2014.

[16] W.-T. Kim, M.-Y. Choi, Y.-H. Huh, and S.-J. Eom, 'Measurement of thermal stress and prediction of fatigue for STS using Lock-in thermography', *Proc 12th -PCNDT*, 2006.

[17] A. Salerno, A. Costa, and G. Fantoni, 'Calibration of the thermoelastic constants for quantitative thermoelastic stress analysis on composites', *Rev. Sci. Instrum.*, vol. 80, no. 3, p. 034904, Mar. 2009.

[18] S.-J. Kim, K. Koh, S. Boyd, and D. Gorinevsky, 'l1 Trend Filtering', *SIAM Rev.*, vol. 51, no. 2, pp. 339–360, May 2009.

[19] A. Moghtaderi, P. Borgnat, and P. Flandrin, 'Trend filtering : Empirical Mode Decompositions versus l1 and Hodrick-Prescott', *Adv. Adapt. Data Anal.*, vol. 03, no. 01n02, pp. 41–61, Apr. 2011.

[20] I. De Baere, W. Van Paepegem, C. Hochard, and J. Degrieck, 'On the tension–tension fatigue behaviour of a carbon reinforced thermoplastic part II: Evaluation of a dumbbell-shaped specimen', *Polym. Test.*, vol. 30, no. 6, pp. 663–672, Sep. 2011.

[21] E. Castillo and A. Fernández-Canteli, *A unified statistical methodology for modeling fatigue damage*. Springer Science & Business Media, 2009.

[22] J. C. Krapez, D. Pacou, and G. Gardette, 'Lock-in thermography and fatigue limit of metals', *Off. Natl. ETUDES Rech. Aerosp. ONERA-Publ.-TP*, no. 187, 2000.

* Corresponding author: laura.muller@onera.fr

Fatigue behaviour of fibre-reinforced composite T-joints

Ying Wang[1] and *Constantinos* Soutis[1,*]

[1]Faculty of Science and Engineering, The University of Manchester, James Light Building, Manchester, M13NJ, UK

Abstract. In this paper a study was carried out on the fatigue life of fibre-reinforced composite T-joints subjected to a tensile pull-out loading. The composite T-joints have been made of glass fabric infused with epoxy resin using a vacuum assisted resin transfer moulding technique. Methods such as the use of veil layers, tufting techniques and 3D weave have been employed to improve the interlaminar fracture toughness of the composite T-joints. All the tests were conducted in an Instron testing machine using a specially designed test fixture. Fatigue tests were performed in a load control mode with a stress ratio of $R = \sigma_{min}/\sigma_{max} = 0.1$. The cyclic loading pattern was a sinusoidal wave with a frequency of 6 Hz. The specimens were cycled at a series of constant maximum load values up to failure. Fatigue loads versus life data for each T-joint type were produced at various maximum applied loads. The 3D weave T-joints were found to have the best performance in both static and fatigue loading. Increasing the static properties increases fatigue life performance; the increasing rate in fatigue life is changed with the number of stress cycles. The location for the through-thickness reinforcement plays an important role in improving fatigue life of the T-joints. Fatigue life is significantly improved if the web is reinforced in through-thickness direction. A finite element (FE) failure model was also created using ABAQUS to determine the location where delamination is initiated and its subsequent propagation.

1 Introduction

T-joints are found in structural components such as skin-spar in wind turbine blades and aircraft wings and road vehicles. The wind-turbine blades are often made of composite materials due to their low density and excellent mechanical properties combination of strength, stiffness, fatigue performance and better corrosion resistance than metals [1]. Wind turbine blade failures, when in service are due to various complicated reasons but blade failures account for most accidents and damage to wind turbines [2]. Therefore, it is important to understand the way the composite T-joint fails and develop methods of improving fatigue life performance to enhance the durability of wind turbines.

Failure mode of a composite T-joint has been widely studied by many investigators [3-9]. The initiation of delamination in the composite T-joint is almost always initiated on the joint's corner due to the maximum stresses developed in this region. After crack initiation, the cracks may propagate along the flange-skin bond line only or along both the central line of the web and the flange-skin bond line, depending upon the T-joint reinforced configurations. Therefore, delamination crack propagation behaviour plays an important role in the control of fatigue failure of the T-joint.

To increase the resistance to delamination/crack propagation, through-thickness reinforcements are widely used to prevent the out-of-plane failures. Various techniques, such as stitching/tufting [7, 10-12], interlaminar veils [13, 14] , knitting [15], 3D weaving [7, 16, 17] and z-pining [18-20] have been developed to improve the properties in the through-thickness direction. Numerical approach such as finite element analysis was also used to model and simulate delamination behaviour in the composite T-joints by a number of investigators [11, 14, 18, 19, 21-24]. A large number of experimental investigations were performed on failure of composite T-joints with the aim to improve thorough-thickness properties.

Many studies [25-30] have been carried out on fatigue failure of composite materials over the past 20 years; a number of scientific and technical issues still need to be resolved, for example, predictive models for determining the strength of the T-joints with various reinforcement configurations have not been satisfactorily developed [16]. There are no in depth studies on fatigue behaviour of fibre-reinforced composite T-joints so for, and it is the subject of the paper.

In this study, a composite T-joint was manufactured as representative element of the blade structure. The composite T-joints have been made of glass fabric infused with epoxy resin using a vacuum assisted resin transfer moulding (RMT) technique. Methods such as the use of veil layers, tufting and 3D weaving techniques have been utilised to improve delamination resistance. A finite element model was created using ABAQUS that can be used to determine the initiation and propagation of delamination crack in the composite T-joints. Fatigue tests on different T-joint configurations were conducted in earlier work [31] and used to validate the FE model.

* Corresponding author: constantinos.soutis@manchester.ac.uk

Experimental measurements and FE analysis are presented and discussed in the following sections.

2 Experimental methods

2.1 Composite T-joints

The composite T-joints were made of glass fabric infused with epoxy resin using a vacuum-assisted RTM technique. The geometry and layup detail of the composite T-joint used as the baseline configuration in the study is shown in Fig. 1(a). The flange and skin parts consist of five layers of equal thickness arranged in a stacking sequence of $[45°/-45°]_5$ while the web consists of 10-layer $[45°/-45°]_{10}$ plate. The stiffener consists of 5-layer unidirectional (UD) $[0°]_5$, which is used to prevent large deflection that may not be representative of the larger structure. Various techniques such as veil layer, tufting and 3D weave were used to improve interlaminar fracture toughness and are listed in Table 1.

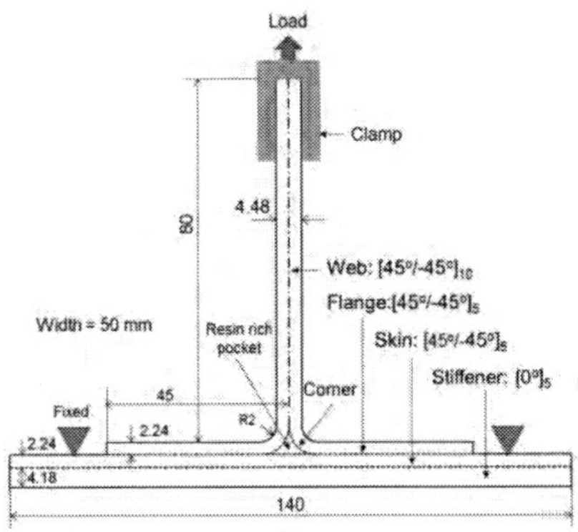

(a)

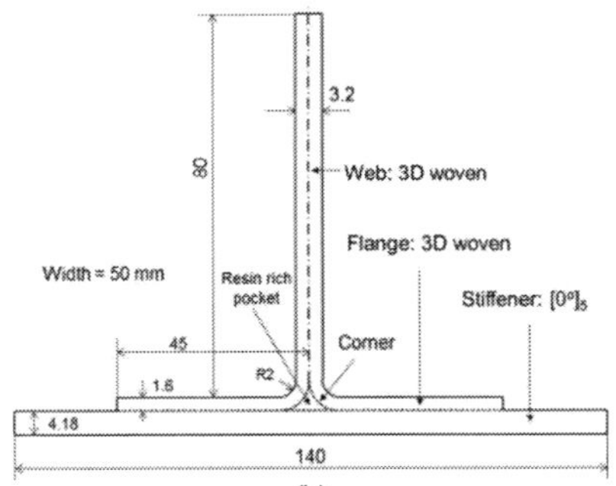

(b)

Fig. 1. Geometry and layup of composite T-joints used (all dimension in mm)

A non-woven fabric (polyamide/polyester veil) was added between the flange and the skin. The thickness of the veil is 0.055mm for polyester and 0.194mm for polyamide, respectively.

Tufting material used for out-of-plane stitching was a 120 tex Kevlar-29 thread with tenacity of 185-200 cN/tex. Stitch density was 10 mm in the length direction and 10 mm in the width direction.

Table 1. Test sample specifications

Type	Descriptions
Baseline	T-joint made from glass fibre and epoxy matrix (Fig. 1a)
PA veil (Polyamide veil)	Type Baseline T-joint + polyamide veil added between the flange and the skin (Fig. 1a)
PE veil (Polyester veil)	Type Baseline T-joint + polyester veil added between the flange and the skin (Fig. 1a)
Tufted	Type Baseline T-joint + Kevlar thread in through-thickness direction on flange-skin-stiffener (Fig. 1a)
3D-Layer to Layer	T-joint made from 3D Layer to Layer woven fabric (Fig.1b)
3D-Angle Interlock	T-joint made from 3D Angle Interlock woven fabric (Fig.1b)

3D woven fabrics, angle interlock and layer to layer, were selected to manufacture the composite T-joints. The fibre type used in the 3D weaves is E-glass. The structure was weaved as a flat panel and from halfway in width the middle layer was removed during the weave process and a gap in thickness produced. The woven panel was then opened up from this gap and the two flange sections of the joint were formed on each sides of the web shown in Fig. 1(b).

The resin is an epoxy based on Araldite LY 564, and the associated hardener used is a formulated amine based on XB 3486. Unidirectional E glass fibre (FGE664) was used for the stiffener (support plate) and ±45° glass fibre (FGE104) was used in the fabrication of the web, flange and skin of the T-joint. The areal density for [±45°] and UD fabrics is 609 and 594 g/m^2, respectively.

2.2 Test condition and procedure

The T-joint is subjected to tensile pull-out loading along the web. The skin part is considered to be supported near the two ends of the T-joint, as indicated in Fig. 1(a). The mechanical response of different types of T-joints was tested under tensile pull-out loading to obtain the ultimate failure load.

An Instron testing machine was employed to perform the mechanical tests at a loading rate of 2 mm/minute. Load and displacement values were recorded during tests. A criterion to terminate the test was set as either a 40% drop in the measured load or a total delamination length of 50 mm

Fatigue test was performed in a load control mode with a stress ratio of $R = \sigma_{min}/\sigma_{max} = 0.1$. The cyclic loading pattern was a sinusoidal wave at a frequency of 6 Hz while the thermal effect is not considered at this test frequency. The specimens were cycled at a series of constant maximum load values up to failure. Fatigue load levels were chosen with regards to maximum pull-out strength of samples obtained in the static tensile test. This allowed a complete knowledge of the stress-life relationship (*S-N* curves) of T-joints with different configurations. Constant fatigue load versus life data for each T- joint type were produced at various constant maximum applied loads. Failure criteria for fatigue tests were either crack length of 25 mm on each side of the web or one million fatigue cycles. Visual observations of each specimen were also undertaken at intervals during testing, and a record was made of any damage observed.

3 Experimental results

3.1 Tensile pull-out test

For fatigue testing it is necessary determine the ultimate tensile pull-out strength (UTS) under static loading. The test results were normalised on the base plate (flange-skin-stiffener) to remove the effect of thickness variation of the base plate. Taking into account the effect of bending, the maximum nominal stress occurring in the base plate is calculated using the beam bending theory [32] and the maximum nominal stress(σ) is given by

$$\sigma = \frac{3L}{2b}\left(\frac{P}{h^2}\right) \quad (1)$$

Here P is the applied load, L is the span of T-joint, h is the total thickness of the base section and b is the width of T-joint. When the span (L=100 mm in this case) and the width (b=50 mm) is fixed the magnitude of the stress occurred on the surface is inversely proportional to the square of the thickness (h).

The nominal stresses (σ) calculated by Equation (1) are plotted against the deflections (δ) and these σ-δ curves are shown in Fig. 2. It can be seen from the figure, the σ–δ behaviour is sensitive to the composite T-joint architectures. 3D woven T-joint always shows the best pull-out performance compared with other type T-joints. When the defection is less than a certain value, a linear relationship exists between the deflection and the applied stress produced by the load. The slope of the σ-δ curves is almost the same when the applied stress is less than around 65 MPa for the joints (Baseline, PA veil, PE veil and Tufted T-joints). In other words, the slope (or 'stiffness') appeared in the linear stage is not influenced by the veil layer or tufting. In the 3D woven T-joint, the slope of the σ–δ curves is almost the same as in that of the σ–δ curves from above type T-joints, although the slope of the σ–δ curve begins to decrease at a higher applied stress, e.g. around 210 MPa.

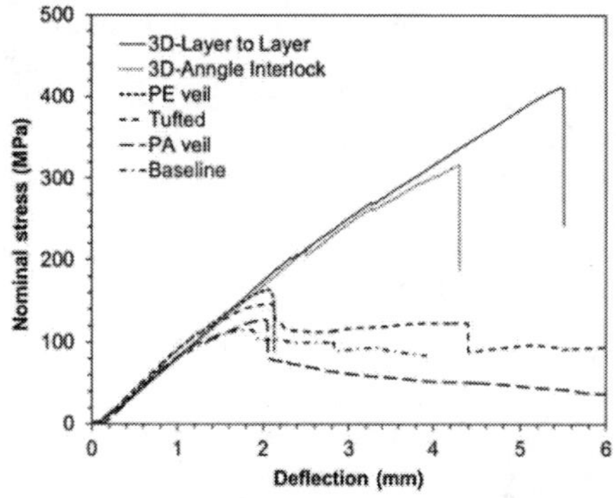

Fig. 2. The σ -δ curves of T-joints from tensile pull-out tests.

The reduction in the curve's slope is associated with initiation of damage developed in the T-joint and it continues until the maximum pull-out load is reached. The applied stress value for the reduction of the curve's slope depends upon the type of the composite T-joint.

The maximum pull-out loads were measured for all T-joints. At least five tests were performed for each type of T-joint. The 3D-layer to layer stitched T-joint shows the highest pull-out strength (417.5MPa) among all the tested specimens and its strength is increased by 254% when compared to the type Baseline specimen (117.8MPa). The second is the 3D-angle interlock specimen (319.8MPa). The third is the PE veil T-joint (169.6MPa) while the PA veil specimen is only 120.9 MPa. The Tufted specimen is 149.6MPa.

The test results confirm that when through-thickness reinforcements (3D woven joint) are used in the web the tensile pull-out strength is significantly increased in comparison with other type T-joints. The samples with the PE veils also show higher strength compared to Baseline type T-joint. The tensile pull-out strength in the tufted specimen is also improved, but much less compared to the 3D woven specimen in that the tufting yarns are applied only on the flange-skin-stiffener. These results emphasise the importance of the location where through-thickness reinforcements is applied. All these test results provide experimental evidence for the validation of the FE results, which are discussed in later section.

3.2 Fatigue behaviour

Fatigue test results in terms of applied stress and number of cycles to failure were plotted for each specimen to produce *S-N* curves for all type T-joints. Fig. 3 shows the fatigue test results of the Baseline T-joint. The solid line is the best regression fit of the data points with the R^2 value of 0.789; the dotted lines are 95% upper and lower confidence limits. As expected, increasing the fatigue load decreases the number of cycles to failure.

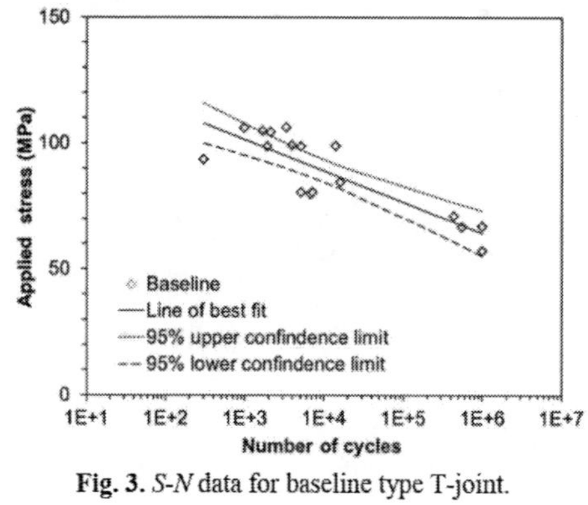

Fig. 3. *S-N* data for baseline type T-joint.

The *S-N* curve of the T-joint with a polyamide veil is shown in Fig. 4, along with a regression fit (solid line). The R^2 value is 0.727. A few data points are out of 95% confidence limit band. The result of fatigue tests from the PE veil T-joints is shown in Fig. 5. The R^2 value is 0.84.

Fatigue life performance of the tufted T-joints is shown in Fig. 6, where the solid line is the best fit and the dotted lines are 95% upper and lower confidence limits. The R^2 value is 0.853. As can be seen from the figure, some of the test data are located outside 95%

confidence limit band. The tufting technique improves delamination resistance resulting from the crack bridging by z-fibres, which reduces the driving force for crack growth. However, the localised fibre damage from the needle penetration and the resin rich pocket at tufted yarns lead to the reduction in fatigue life performance due to its direct interaction with the crack growth path.

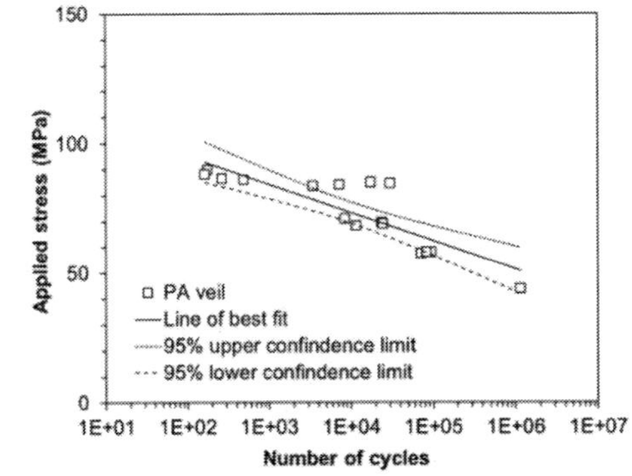

Fig. 4. *S-N* data for polyamide veil T-joint.

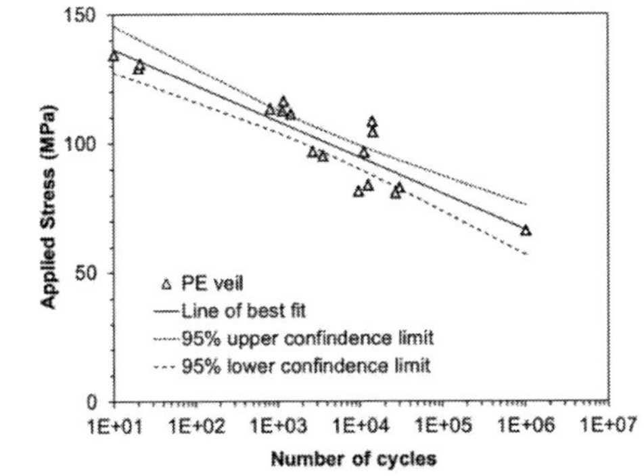

Fig. 5. *S-N* data for polyester veil T-joint

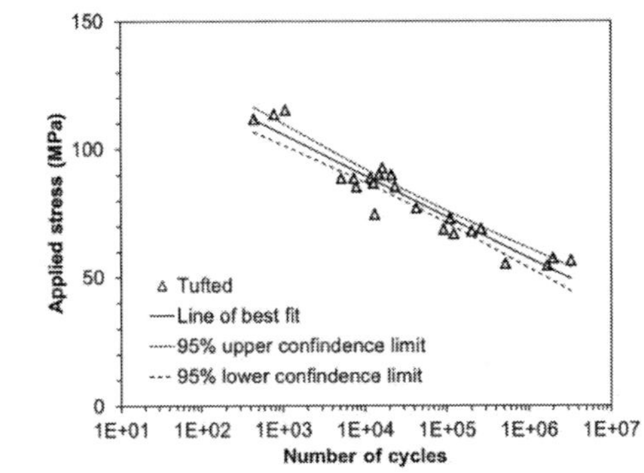

Fig. 6. *S-N* data for tufted t-joint

Fig. 7 shows fatigue data of the 3D woven specimen made of angle interlock woven structure. Due to limited availability of the 3D angle interlock fabric, the number of fatigue tested specimens was less compared to other structures; however, the low deviation shows the high accuracy of measured values with R^2 value of 0.976. Most of the test data are located in 95% confidence limit band as shown in Fig. 7.

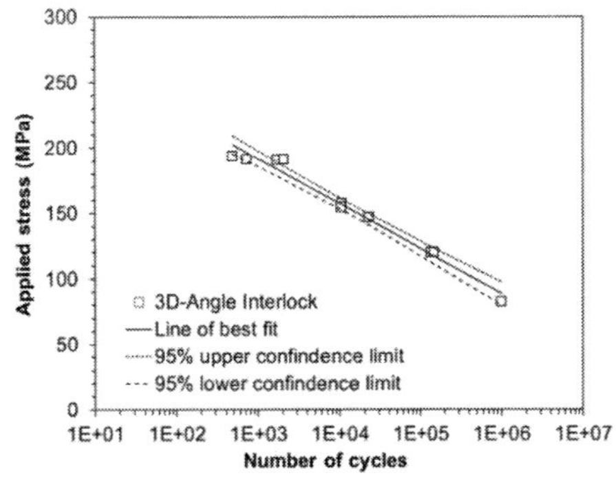

Fig. 7. *S-N* data for 3D-anagle interlock T-joint.

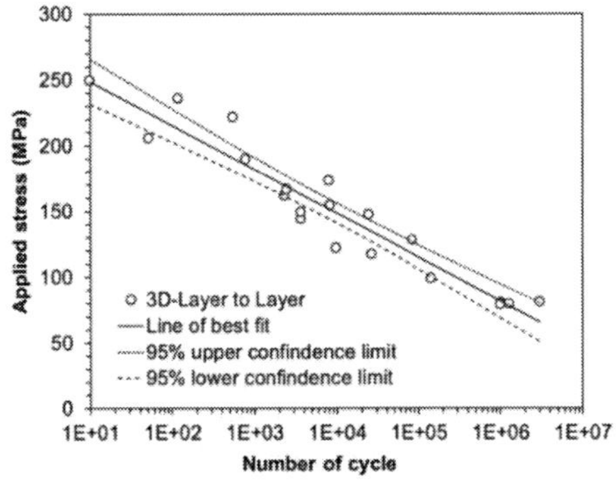

Fig. 8. *S-N* data for 3D-layer to layer stitched T-joint.

The test result of the 3D woven layer to layer stitched T-joint is shown in Fig. 8, along with a best regression fit (solid line). A number of specimens were tested in very low load levels (25% of tensile pull-out strength) to examine the possibility of failure of the joint in low applied stress and find the fatigue limit of the joint. Most of the tested specimens in this load level did not fail, even after 3×10^6 cycles, which was three times the test limit.

To clearly compare the fatigue behaviour of all types of T-joints, all six sets of data were plotted together shown in Fig. 9, along with a regression fit for each. The results from the static tests were also plotted at cycle number $n = 1$. As can be seen, all curves show the same trend. Increasing fatigue load leads to the decrease in the number of cycles to failure. Fatigue strength of the composite T-joints is significantly influenced by reinforcement configurations.

The best fatigue life performance is clearly seen in 3D- woven T-joint with almost 40-80% improvement in fatigue life, depending on the stress cycles. The use of inter-laminar veil is also able to increase T-joint fatigue strength, but not very significant. The tufting or stitching may cause localised fibre damage from the needle penetration and discontinuity at the stitching point; the resin rich pocket is created due to the presence of tuft yarns and therefore the micro-cracks are easily to be produced.

Taking to account the deviation of data and statistical analysis, it was found that there is no significant difference in fatigue performance for Baseline, PE veil

 https://doi.org/10.1051/matecconf/201816507004

and tufted T-joints as the 95% confidence limit bands of these data were overlapped quite significantly. However 3D woven T-joints show a good improvement in fatigue properties while the polyamide veil reduces fatigue properties of the PA veil T-joint.

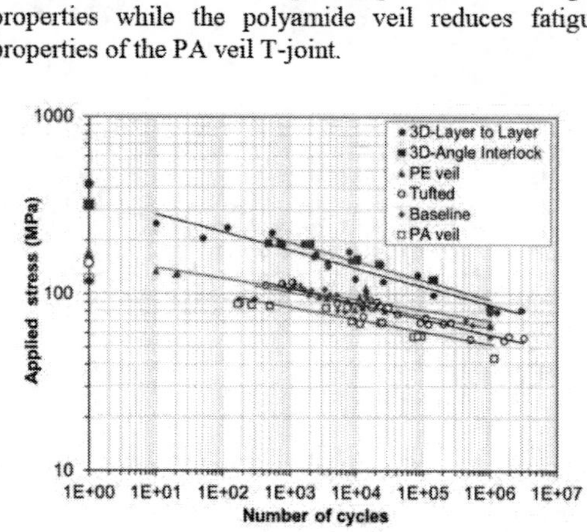

Fig. 9. Comparison of *S-N* curve for al type T-joint specimens.

Comparison of static and fatigue test results showed that they are not always in agreement. For instance 3D-layer to layer T-joint, which had the highest pull-out strength, did not perform well under fatigue loading and the 3D woven angle interlock structure performed best. However, when all test results were considered, a relatively good agreement between static and fatigue results was seen.

4 FE modelling of composite T-joints

4.1 T-joint FE model

The T-joint FE model was created using ABAQUS software [33]. The geometry of T-joints used for FE model is as the same as that used in the experimental tests (Figure 1a). Two types of local coordinate systems are used to describe the orientations of each ply. On the corner the cylindrical coordinate system is used. In this coordinate system, the 1, 2, and 3 directions refer to the radial (R), circumferential (T), and axial directions (Z), respectively. In the web, flange, skin and stiffener the rectangle coordinate system is used and the 1, 2, and 3 directions refer to the X, Y and Z, respectively. For both coordinate systems, the fibre direction is oriented along axis 2 while directions 1 and 3 correspond to thickness and width directions, respectively (see Fig. 10). Under this coordinate system the engineering constants for the composite material are listed in Table 2. The elastic modulus and Poisson's ratio for the epoxy resin are equal to 4 GPa and 0.34, respectively.

Table 2. Composite material properties

E_1 (GPa)	E_2 (GPa)	E_3 (GPa)	ν_{12}	ν_{13}	ν_{23}	G_{12} (GPa)	G_{13} (GPa)	G_{23} (GPa)
9	38	9	0.071	0.3	0.3	3.6	3.46	3.6

The FE simulations were run under displacement-controlled loading by applying an increasing tensile displacement along the web. To simulate the fixture roller bars, the skin surface is considered to be restricted in the vertical direction at both sides of the web with 50 mm distance from the centre. Three elements were used in the thickness direction and an eight-node linear brick element (C3D8R) was utilised. The detailed mesh of the model with the kinematic and loading boundary conditions is shown in Fig. 10.

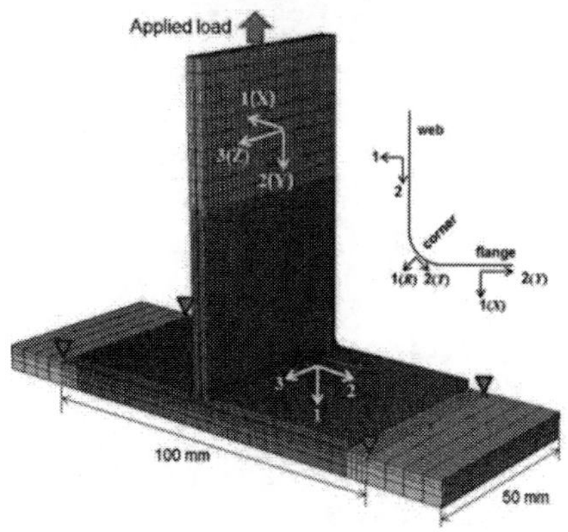

Fig. 10. Mesh, boundary conditions and local coordinate system of the T-joint FE model.

4.2 Modelling of delamination

Delamination is modelled using cohesive zone elements. Linear elastic traction-separation behaviour is used to model the interface where the delamination is expected to occur. The traction-separation model in ABAQUS assumes initially linear elastic behaviour followed by damage initiation and evolution.

The initiation of delamination damage can be estimated using the quadratic nominal stress criterion defined by the following expression [33]

$$\left\{\frac{\langle t_n\rangle}{t_n^0}\right\}^2+\left\{\frac{t_s}{t_s^0}\right\}^2+\left\{\frac{t_t}{t_t^0}\right\}^2=1 \qquad (2)$$

where t_n^0 is the interlaminar normal tensile strength, t_s^0 and t_t^0 are shear strengths in the two transverse directions.

Delamination crack growth is described using a linear softening fracture-based law, where a mixed mode fracture energy criterion is used. Failure under mixed-mode conditions is governed by a power law interaction of the energies required to cause failure in the individual modes. It is given by the following expression [33]

$$\left\{\frac{G_n}{G_n^c}\right\}^\alpha+\left\{\frac{G_s}{G_s^c}\right\}^\alpha+\left\{\frac{G_t}{G_t^c}\right\}^\alpha=1 \qquad (3)$$

where G_n^c is the normal strain energy release rate, G_s^c and G_t^c are shear strain energy release rates in the two transverse directions; α is assumed to be equal to 1, but could be 2. The effect of α value on the load-deflection curve was studied and the variation of α values shows less effect. Various fracture energy values are selected to investigate its effect on crack propagation. Selecting 'appropriate' stress-based or fracture-based remains a debating issue.

5 Numerical results

5.1 Stress distribution with T-joint geometry

The numerical analysis has confirmed that the maximum stresses are developed at the interface between the corner and resin (arc BC in Fig. 11(a)). The magnitude of these stresses is influenced by the parameters such as the radius of the corner (r) and the thickness of the plate (h).

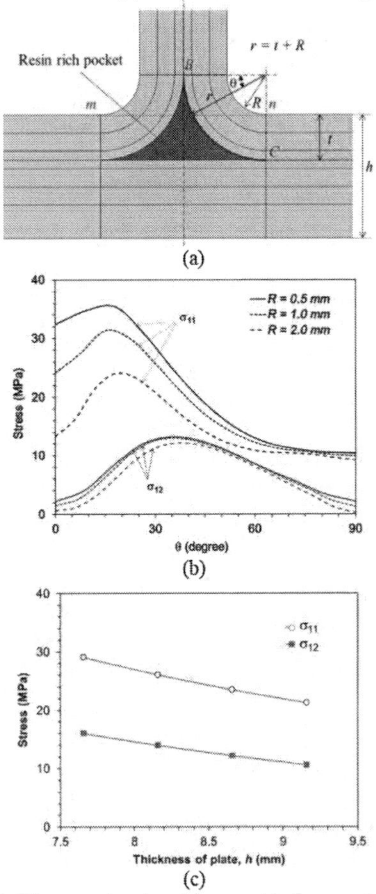

Fig. 11. Diagrams showing: (a) geometrical parameters of T-joint; (b) effect of corner radius on stress distribution along the arc BC; (c) effect of thickness on the maximum stress (σ_{11}).

The distribution of the stress (σ_{11}) is significantly influenced by the curvature of the corner as shown in Fig. 11(b). As can be seen from the figure, the maximum stress (σ_{11}) is increased by 40% when the radius R is reduced from 2 mm to 0.5 mm. However, there is no significant effect on the shear stress (σ_{12}) shown in Fig. 11(b), indicated by red colour lines.

The stress at the corner is sensitive to the thickness of the base plate. Fig. 11(c) demonstrates the relationship between the thickness and the maximum stresses occurred at the corner. For example, 11.3% increase (from 7.66 to 8.66mm) in the thickness (h) results in 19.1% decrease in stress σ_{11} and 23.8% decrease in stress σ_{12}, respectively.

5.2 Effect of through-thickness reinforcement in the web

When the web has through thickness reinforcements the tensile pull-out strength is significantly increased in both static and fatigue loading. With this reinforced structure, crack propagation was only along the bonded connection between skin and flange; no crack growth was seen in the web as shown in Fig. 12(a). The crack is inhabited to propagate in the web due to the nature of reinforcements.

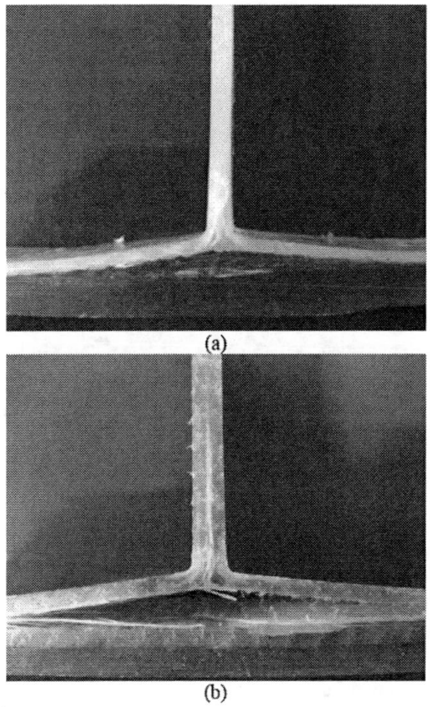

Fig. 12. (a) No crack growth occurs in the web for 3D woven T-joints and the crack grows only along the flange-stiffener bond line. (b) Crack growth is along the central line of web and the flange-skin bond line if no reinforced fibre in the web through thickness direction.

To model crack growth in the 3D woven T-joints, a tie constraint was imposed between pairs of surfaces (the central line of the web), instead of cohesive zone element, to simulate its structural feature. In this way, there is no crack propagation along the central line of the web and the crack propagation is along the flange-stiffener bonding line only. This approach reflects the experimental observations from the 3D woven T-joints.

The experimental observations showed that the crack propagation path was similar in both static and fatigue tested T-joints. The crack was initiated at the corner for both cases. For the T-joints with no through-thickness reinforcements in the web, after initial propagation in the web the crack propagation slowed down and then started to propagate along the interface between the flange and skin. This failure mode is applicable to Baseline, PA veil, PE veil and Tufted T-joints. Fig. 12(b) just shows an example of tested PE veil type T-joint.

FE simulation results are shown in Fig. 13. As can be seen from the figure, the maximum load to failure is significantly increased when the web is reinforced in the through thickness direction. The maximum load to failure is increased by 90%. This is fairly consistent with the experimental observations of tensile pull-out testing, although it remains difficult to accurately capture all the non-linearities and fluctuations seen in the measured load-deflection response.

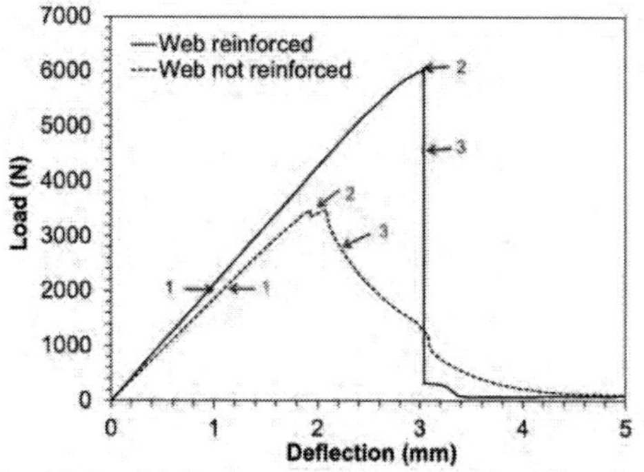

Fig. 13. Load-deflection curve shows that the failure load is significantly increased when the web has through-thickness reinforcements.

The simulated T-joint is framed at different phases of load to numerically investigate damage processes and delamination behaviour. Three levels of load are selected, indicated by the arrows in Fig. 13 (red: web reinforced and blue: web not reinforced). The simulated results are shown in Figures 14 and 15.

Comparison of these two cases reveals that the crack is initiated at the interface between the corner and resin as the high stress developed here. This is highly consistent with the experimental observations. After crack initiation, it continues to propagate in the web along the central line (Fig. 14(a)) when the web does not have through-thickness reinforcements while the crack is hardly to propagate in the web when through thickness reinforcement are applied (Fig. 15(a)).

The crack is propagated further with the increase of applied load. The simulated crack growth at the maximum load is shown in Fig. 14(b), where the web has no through-thickness reinforcements. Fig. 15(b) shows that the crack propagation is increased further along the bond line at the maximum applied load. After the maximum load the crack continues to grow with a big drop in the applied loads and the simulated images for these two cases are shown in Figs. 14(c) and 15(c), respectively.

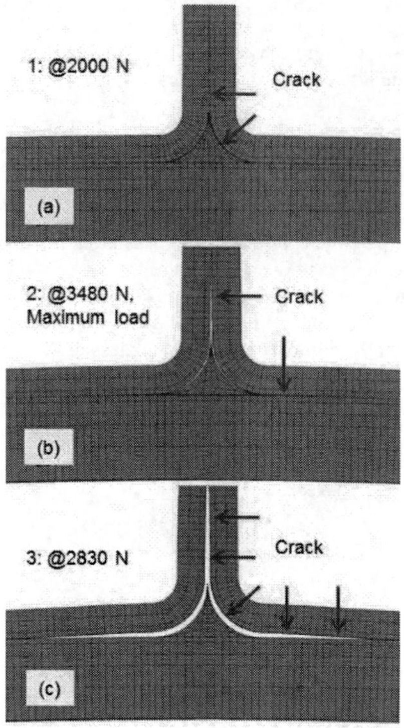

Fig. 14. Simulated images showing damage processes of T-joints without through-thickness reinforcements.

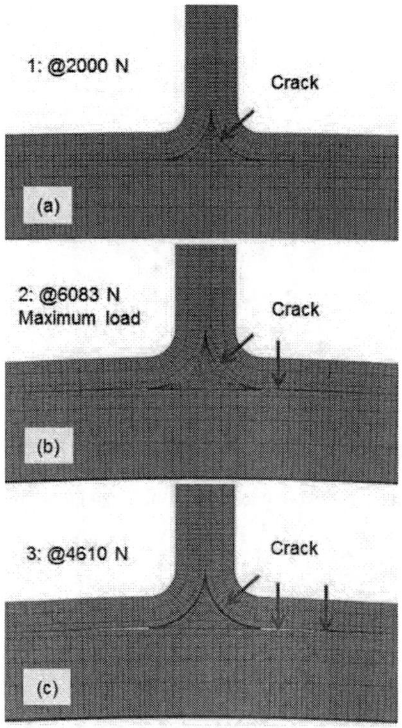

Fig. 15. Simulated images showing damage processes of T-joints with through-thickness reinforcements.

6 Conclusions

A comprehensive study was carried out on fatigue failure mode of fibre-reinforced composite T-joints subjected to a tensile pull-out loading. Fatigue tests were performed on four groups of composite T-joints including baseline, inter-laminar veil, tufting and 3D weaving to investigate its failure mechanisms. A finite element failure model was created using ABAQUS which was used to simulate the mechanical response of the T-joint. The main observations from the present study are as follows:

1. 3D woven type T-joints always show the best performance in both static and fatigue loading. Increasing the static properties increases fatigue performance and a good agreement was seen between the static and fatigue strength.

2. The application of through-thickness reinforcements into the web is very important to improve fatigue strength in that it habits fatigue crack growth in the web and this leads to slow crack growth in the flange-skin region as the driving force for crack growth is significantly reduced if the crack does not exist in the web. This has been confirmed by the FE simulations.

3. In the flange-skin-stiffener region, through-thickness reinforcements (e.g. tufted T-joint) do not improve the fatigue life performance of the T-joint, although the tensile pull-out strength is increased by 27% compared to Baseline type T-joint.

4. Selecting the proper inter-laminar veil (e.g. polyester veil) may increase the fatigue life performance of the T-joints, but not significantly, although the ultimate strength under a tensile pull-out loading is increased significantly. Fatigue life performance is reduced for the T-joint containing the polyamide veil.

5. An FE failure analysis model was developed which can be used to determine delamination behaviour of composite T-joints under a tensile pull-out loading. The FE predictions were validated by experimental measurements and observations.

References

1. E. Sesto and C. Casale, J Wind Eng Ind Aerod. **74-76**:375-387 (1998)
2. J.-S. Chou and W.-T. Tu, Eng Fail Anal **18**:295–313 (2011)
3. R.A. Shenoi and G.L. Hawkins, Composites. **23**(5):335–345 (1992)
4. V.V.S. Rao, K.K. Veni, and P.K. Sinha, Aircr Eng Aerosp Tec. **76**(4):404-413 (2004)
5. A.R. Rispler, G.P. Steven, and L. Tong, J Reinf Plast Comp. **16**(18):1642-1658 (1997)
6. J.I.R. Blake, R.A. Shenoi, J. House, and T. Turton, Composites Part A. **32**:641–653 (2001)
7. D.D.R. Cartie, G. Dell'Anno, E. Poulin, and I.K. Partridge, Eng Fract Mech. **73**:2532–2540 (2006)
8. F. Dharmawan, R.S. Thomson, H. Li, I. Herszberg, and E. Gellert, Compos Struct. **66**:181–187 (2004)
9. L. Burns, A. Mouritz, D. Pook, and S. Feih, Composites Part A. **43**(11):1971–1980. (2012)
10. A.P. Mouritz, K.H. Leong, and I. Herszberg, Composites Part A. **28**(12):979–991 (1997)
11. Y. Wang and C. Soutis, Engineering and Computational Mechanics **169**(EM4):158-170 (2016)
12. K. Dransfield, C. Baillie, and Y.-W. Mai, Compos Sci Technol. **50**(3):305–317 (1994)
13. M. Kuwata and P.J. Hogg, Composites Part A **42**(10):1551–1559 (2011)
14. Y. Wang, C. Soutis, A. Hajdaei, and P.J. Hogg, Plast Rubber Compos. **44**(3):87-97 (2015)
15. K. Leong, S. Ramakrishna, Z. Huang, and G. Bibo, Composites Part A. **31**(3):197–220 (2000)
16. A.P. Mouritz, M.K. Bannister, P.J. Falzon, and K.H. Leong, Composites Part A. **30**(12):1445-1461 (1999)
17. B.N. Cox, M.S. Dadkhah, W.L. Morris, and J.G. Flintoff, Acta Metall Mater. **42**(12):3967-3984 (1994)
18. G. Allegri and J. Zhang, Composites Part A. **38**(4):1107–1115 (2007)
19. F. Bianchi, T.M. Koh, X. Zhang, I.K. Partridge, and A.P. Mouritz, Compos Sci Technol **73**:48-56 (2012)
20. A.P. Mouritz, Composites: Part A. **38**(12):2383–2397 (2007)
21. A. Baldi, A. Airoldi, M. Crespi, P. Iavarone, and P. Bettini, Procedia Engineering **10**:3483–3489 (2011)
22. F. Bianchi and X. Zhang, Compos Sci Technol. **71**:1898–1907 (2011)
23. J. Chen, Fatigue Fract Eng M. **34**:123–130 (2010)
24. J. Chen, E. Ravey, S. Hallett, M. Wisnom, and M. Grassi, Compos Sci Technol. **69**(14):2363–2367 (2009)
25. R.A. Shenoi, P.J.C.L. Read, and G.L. Hawkins, Int J Fatigue. **17**(6):415-426 (1995)
26. P.J.C.L. Read and R.A. Shenoi, Int J Fatigue **21**:281–296 (1999)
27. V. Marcadon, Y. Nadot, A. Roy, and J.L. Gacougnolle, Int J Adhes Adhes. **26**:481–489 (2006)
28. J. Degrieck and W.V. Paepegem, Appl Mech Rev. **54**(4):279-300 (2001)
29. B.L.V. Bak, C. Sarrado, A. Turon, and J. Costa, Appl Mech Rev. **66**(11):060803-1-14 (2014)
30. Y. Wang and C. Soutis, Appl Compos Mater. **24**:461–475 (2017)
31. A. Hajdaei, *Extending the Fatigue Life of a T-joint in a Composite Wind Turbine Blade*, in *School of Materials*. PhD Thesis, The University of Manchester (2014)
32. A.C. Ugural and S.K. Fenster, *Advanced Strength and Applied Elasticity* (1995)
33. ABAQUS Analysis User's Manual (2016)

Fatigue in Fiber-Metal Laminates for Small Wind Turbine Blades Application

Wei Sai[1,*], *Gin Boay* Chai[1]

[1]Energy Research Institute (ERI@N), Nanyang Technological University, 50 Nanyang Drive, Singapore 639798
[2]Mechanical and Aerospace Engineering School, Nanyang Technological University, 50 Nanyang Drive, Singapore 639798

Abstract. A methodology to study the fatigue of a wind turbine blade in a 10KW small wind turbine is proposed in this paper. Two working conditions (namely normal fatigue operation condition and extreme wind condition) are considered based on IEC61400-2. The maximum load calculated from both cases were used as a reference to perform material sample fatigue study. Fiber-metal laminate – GLARE 3/2 with a centre 1mm notch on the external aluminium layers was modelled based on fracture mechanics approach to calculate the stress intensity factor and fatigue crack growth rate at maximum applied stress of 240Mpa. GLARE panel fabrication and tensile tests were included. The fatigue tests were performed on unnotched samples with stress range from 80Mpa to 300Mpa and plotted into S-N curve.

1 Introduction

To estimate the fatigue life of wind blades in the natural environment, local wind load history is needed. This load history is based on the local wind history and the load induced on the wind blade. The load-time history data requires further processing using rain flow to count the peaks and valleys. The processed data is plotted into histogram that displays the stress range versus its frequency. The stress range would be tested through fatigue tests or modeled and plot into S-N curve. WISPER and WISPERX are specially developed spectrum loading sequence based on the actual wind-load data captured in Europe[1, 2] to simulate or do the testing of wind blades. To estimate the fatigue life of wind blades at specific load ranges, either structural level models or sample tests are to be done.

The load calculation performed in this paper is based on a simplified average wind speed of 8.5m/s for a 10KW small wind turbine with blade diameter 6m. The induced moment and stress at blade root section based on standard for small wind turbine IEC61400-2[3]. The load calculated for the wind blade root section was taken as reference for the sample modeling and testing. The fatigue analysis was performed on fiber-metal laminate (GLARE) samples for a stress range from 80Mpa to 300Mpa.

Currently the choices of wind blade materials are glass fiber composite, carbon fiber composite, wood, steel and aluminum alloy[4]. Glass fiber composites and carbon fiber composites are the most commonly used for wind blade materials and lots of fatigue studies have been done on fiber reinforced composites[5-7].

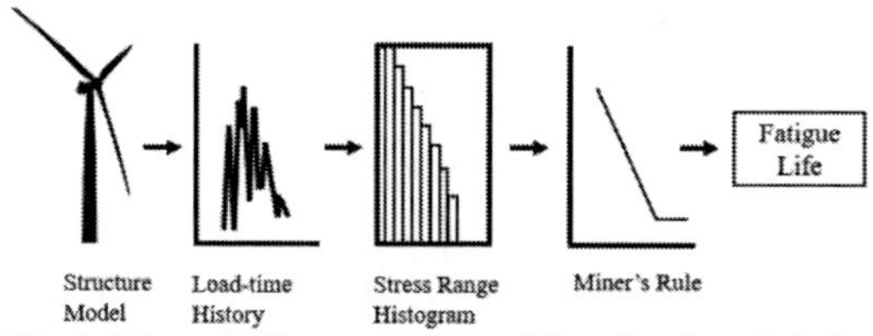

Fig. 1. Schematic diagram for fatigue life estimation of wind blades.

The designed working life for wind turbine is normally 20-25 years. Fatigue tests and fatigue modeling on fiber reinforced composites for wind blades are normally up to a few million cycles. A typical fiber-metal laminate such as GLARE was developed about thirty years ago and successfully applied in the fuselage skin for Airbus 380[8] due to its low specific weight and high fatigue strength compared with pure alloy materials[9]. The main reason of the high fatigue probability for GLARE is the fiber bridging effect to restraint the crack tip opening at the aluminum layers to reduce the stress intensity factor and decrease the crack growth rate[10]. In contrast to commercial aircraft application, wind turbine blades require higher fatigue cycle since the required working life is longer due to the vibration induced by environment (mainly wind). Up until now, GLARE has yet to be applied to the wind turbine blades. Fatigue modeling based on fracture mechanics and tension-tension fatigue tests were performed based on the highest extreme load calculated at the small wind blade root section.

* Corresponding author: mgbchai@ntu.edu.sg

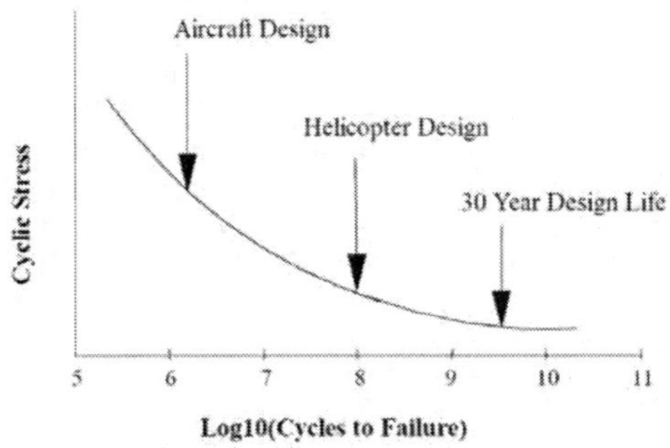

Fig. 2. Schematic S-N plot for various fatigue critical application[11].

2 Fatigue Model

The fatigue model applied is based on Paris' law for monilithic metals. Assumptions have been made that the fatigue data of aluminum 2024 could be applied for GLARE laminate fatigue life[12, 13]. To estimate the number of fatigue cycles, effective stress intensity factor is required which require the stress level in the aluminum layer to be calculated first.

2.1. Wind blade load calculation

A 10KW small wind turbine blade fatigue loading under normal operation and ultimate load under extreme wind loading are calculated based on IEC61400-2[5]. The basic design parameters are taken from Windlite 10KW DC wind turbine. The loads and stresses are calculated at blade root section where the highest bending moment is found. An average wind speed of 8.5m/s with tip speed ratio of 4.68 is applied in the load calculation. The radial distance between the center of gravity of the blade and the rotor center is estimated by assuming trapezoid shape plane cross-section with blade root air foil NREL S822 and tip air foil NREL S823. The formula involved in the calculation in Table 2 could be found in IEC61400-2, and the related design parameters are shown in Table 1.

Table 1. Design parameters for 10KW wind turbine

R	Rotor radius	3m
n_{design}	Rotor speed at rated wind speed	130rpm
n_{max}	Max rotational speed	260rpm
V_{design}	Designed wind speed	8.4m/s
P_{design}	Designed electrical power	10Kw
Q_{design}	Designed shaft torque	918Nm
η	Efficiency (assumed)	0.8
m_B	One blade mass	25Kg
R_{cog}	The radial distance between the center of gravity of a blade and the rotor center	1.5m
B	Number of blades	3
λ_{design}	Designed tip speed ratio	4.86
C_d	Drag coefficient	1.5
$C_{l,max}$	Max lift coefficient	2
V_{e50}	Small wind turbine class IV maximum gust	42m/s
$A_{proj,B}$	Blade section area projected on to a plane perpendicular to wind direction	$1.05m^2$

The loads in the normal operation is considered as fatigue load caused by constant amplitude cyclic loading at blade root section. The loads at the extreme wind loading condition are the ultimate loads when wind blade is still functioning. Loads from both cases are taken as reference for the following crack growth rate modelling and fatigue tests.

Table 2. Wind blade load calculation

Normal operation			
Centrifugal load	13900	N	ΔF_{ZB}
Edgewise bending	1041	Nm	ΔM_{XB}
Flapwise bending	1488	Nm	ΔM_{YB}
Axial loading stress	7.08	Mpa	σ_{ZB}
Bending stress	148	Mpa	σ_{MB}
Combined load	155	Mpa	σ_{eq}
Extreme wind loading			
Edgewise bending	1041	Nm	ΔM_{XB}
Flapwise bending (parked rotor)	2563	Nm	ΔM_{YB}
Flapwise bending (spinning rotor)	2278	Mpa	ΔM_{YB}
Axial loading stress	7.08	Mpa	σ_{ZB}
Bending stress (parked rotor)	225	Mpa	σ_{MB}
Bending stress (spinning rotor)	204	Mpa	σ_{MB}
Combined load (parked rotor)	233	Mpa	σ_{eq}
Combined load (spinning rotor)	211	Nm	σ_{eq}

2.2. Stress in Al layers

The stress level in aluminum layer is calculated based on classical laminate theory. The governing equation is shown in equation (1), and the detail procedure could be found from Ref[10]. The stress in aluminum layer is composed of internal stress due to applied stress to the laminate and the tensile curing stress due to the different thermal expansion coefficient [14].

$$\sigma_{Al} = S_{Al}[S_{lam}^{-1}(\sigma_{lam}) + \Delta T(\alpha_{lam} - \alpha_{Al})] \qquad (1)$$

The maximum load calculated from the wind blade root section is 233Mpa. The stress of 240Mpa was applied in the following stress calculation in aluminum layers as well as the stress intensity factor and fatigue crack growth rate calculation.

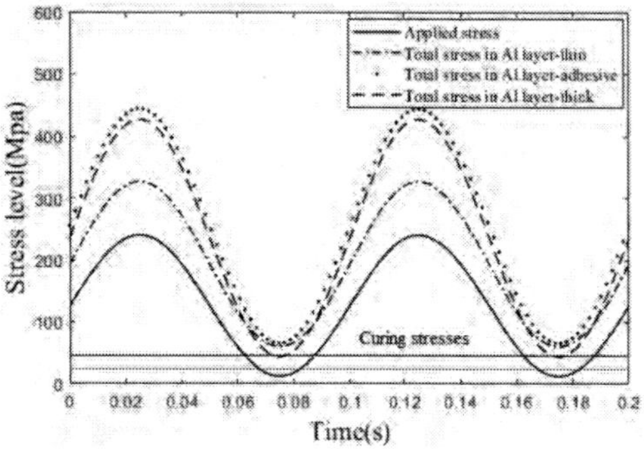

Fig. 3. Stress calculated in an aluminum layer with applied stress 240Mpa and stress ratio R=0.05.

Three different types of modeling are studied regarding the calculation of stress in the aluminum layer:

The total stress in Al layer (adhesive model) – considering the adhesive layers applied at metal and prepreg interface as one separate layer and the adhesive properties are taken from[13].

The total stress in Al layer (thin model)– considering the adhesive layers of negligible thickness and merging it with the prepreg layers which means the thickness of laminate panel less than the actual panel thickness.

The total stress in Al layer (thick model) – assuming the adhesive is cured into prepreg layer and the thickness of the adhesive is averaged and added for each prepreg layer which reflects the actual panel thickness.

As the thickness of prepreg increases, the internal stress due to the externally applied load in the prepreg layers and aluminum layers also increases. As a result, the total stress level in the individual aluminum layer is higher as it accounts for the extra adhesive layers or thicker prepreg layers due to the merger with extra adhesive layers. For the rest of the modelling, the "total stress in Al layer (thin model) is applied.

2.3. Bridging stress

The bridging stress is defined as the stress pulled up the cracked metal layers in fiber-metal laminates by the intact fibers and prevent the cracks further opening. The bridging stress controls the crack growth rate, and it is a prerequisite to calculate the laminate crack growth rate. One of the approaches to calculating the bridging stress is the model proposed in [15] and further improved in [12].

The governing equation is shown below based on the GLARE laminate crack opening displacement caused by applied stress and compensate partially by the fiber bridging stress which equals to the fiber elongation and prepreg deformation along the loading direction. The illustration of the crack opening displacement with elliptical delamination in shown in Figure 4.

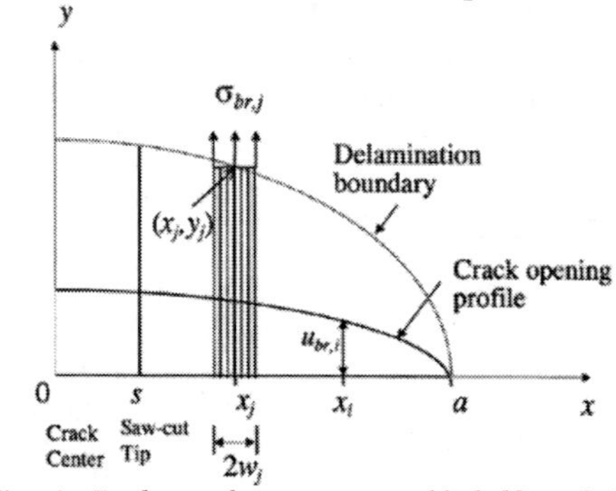

Fig. 4. Crack opening geometry with half crack length a divided into N bar elements[15].

$$u_\infty(x) - u_{br}(x) = \delta_f(x) + \delta_{pp}(x) \qquad (2)$$

The governing equation could be further transformed into the equation below:

$$\sigma_{br} = H^{-1}Q \qquad (3)$$

$$H = \sum_{j=1}^{N} \frac{u(x_i, x_j)}{\sigma_{br}(x_j)} - \frac{b(i)}{E_f}\delta(i, j)$$

$$Q = v_{ap}(x) - \delta_{pp}(i) - \frac{S_f}{E_f}b(i)$$

And σ_{br} represent the bridging stress in the composite; $u(x_i, x_j)$ is the crack opening displacement; $b(i)$ is the delamination shape function; and $\delta(i, j)$ is Kronecker.

To calculate the bridging stress, 60 bar elements are used, and matrix operation is used to solve the equation. The function of Kronecker is like an identity matrix, and it forms a square matrix with the same dimension as the first term in the equation of "H".

Based on the model in [12], the bridging stress in GLARE 3/2 with glass fiber prepreg layup $[0^\circ]_4$ was calculated. The triangular and elliptical shapes were assumed for delamination due to the notch on the external aluminum layers. Different delamination shape functions could be found in [12, 15]. The half crack size is modelled at 12.5mm and half delamination size at 5mm with a center notch size 1mm. The applied stress to the composite sample is 240Mpa with stress ratio R=0.05. The results are shown in Figure 5.

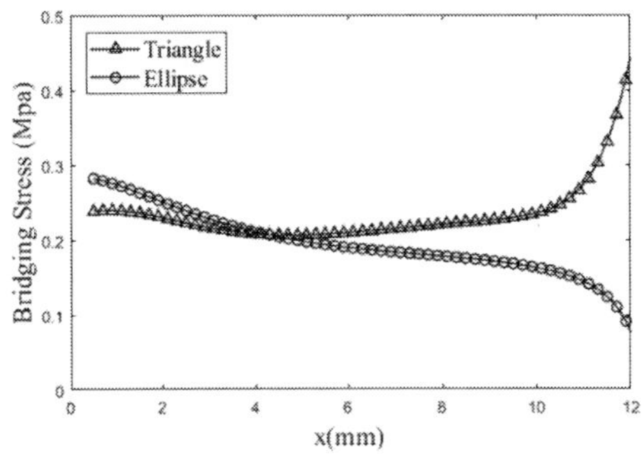

Fig. 5. Bridging stress for GLARE 3/2-$[0°]_4$-0.4 with a=10mm, b=5mm and central notch radius 0.5mm.

2.4. Stress intensity factor

The effective stress intensity factor includes far-field stress intensity factor and the bridging stress intensity factor. The equation for center-cracked specimen is shown in [12].

$$K_{far} = \sigma_{Al}\sqrt{\pi a} \tag{4}$$

$$K_{br} = 2\sum_{i}^{N} \frac{\sigma_{br,Al}(x_i)w}{\sqrt{\pi a}} \frac{a}{\sqrt{a^2 - x_i^2 + b_i^2}} \left(1 + \frac{1}{2}(1+\upsilon)\frac{b_i^2}{a^2 - x_i^2 + b_i^2}\right) \tag{5}$$

Where σ_{Al} is the stress level in aluminum layer which could be calculated based on classical laminate theory[10]; $\sigma_{br,Al}$ is the bridging stress in aluminum layer and it is calculated using the rule of the mixture; and w is the bar element width.

The bridging stress intensity factor could be calculated once the bridging stress is obtained from equation (3).

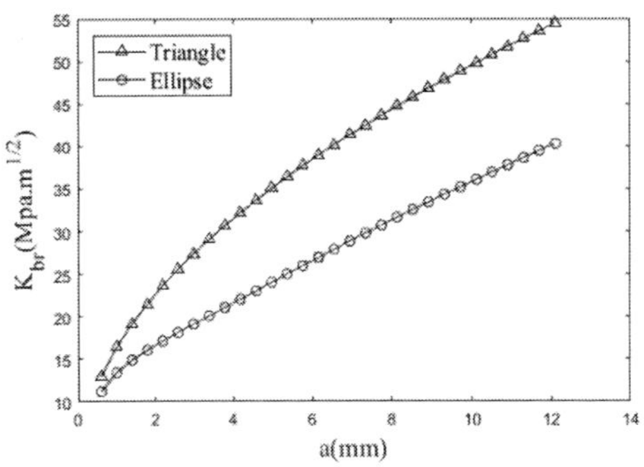

Fig. 6. Bridging stress intensity factor for applied stress 240Mpa and stress ratio R=0.05.

2.4. Fatigue crack growth rate

The fatigue crack growth rate is modeled based on Paris' law. The model does not include the crack initiation for the metal layer since it is notched with a 1mm cut in the center of the specimen. The unit for da/dN is (m/cycle).

$$\frac{da}{dN} = C\Delta K_{eff}^{n} \tag{6}$$

$$\Delta K_{eff} = (1 - R^{1.35})(K_{far} + K_{br}) \tag{7}$$

Where C $=2.17 \times 10^{-12}$ and n=2.94 are aluminum 2024-T3 material constants. Different value of C and n could also be found in [13]. Once the crack growth rate is calculated, the fatigue cycle N could be obtained by doing the integration of the equation (6).

The plot of fatigue crack growth rate is shown in Figure 6. The correction factor for the sample width geometries is shown in equation (8) below.

$$\beta = \frac{1}{\sqrt{1-(\frac{2a}{W})^2}} \tag{8}$$

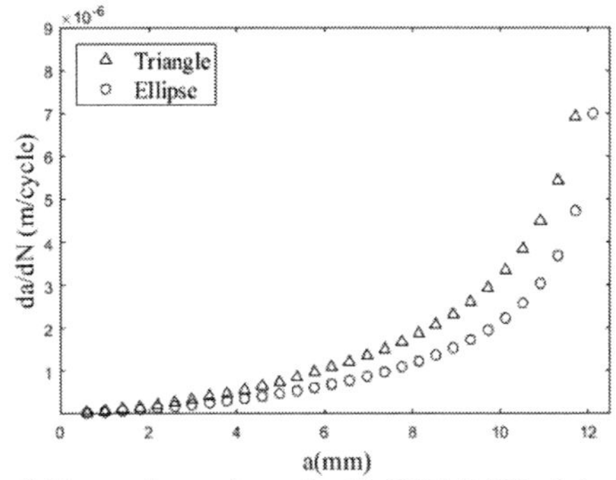

Fig. 7. The crack growth rate for GLARE 3/2-$[0°]_4$-0.4 with triangular and elliptical delamination shapes.

3 Experimental data

GLARE 3/2 panels were fabricated with aluminum 2024-T3 of thickness 0.3mm and 0.4mm, and Hexcel 913 glass fiber prepreg with $[0°]_4$ and $[0°/90°]_s$ layup. Redux film adhesive was applied at the interface of aluminium layer and prepreg layer to promote the adhesion between metal and composite layer.

3.1. Fabrication

GLARE panels were fabricated at curing temperature of 120°C as shown in Figure 7 at pressure of 80psi and vacuum pressure of -20psi. The total autoclave cycle time is around 2-3 hours. The error percentage is controlled within 5% during the heat up and curing stages.

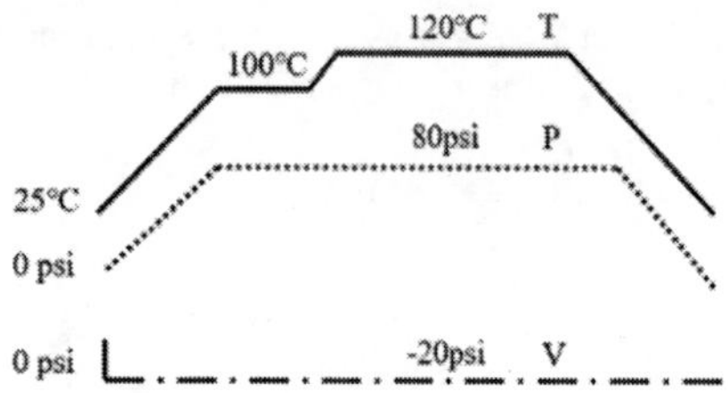

Fig. 8. GLARE panel curing cycle

The fabricated GLARE panel is shown in Figure 8. Part of the excess adhesive applied at the interface between aluminum layers and prepreg layers was cured and absorbed by the bleeder fabric.

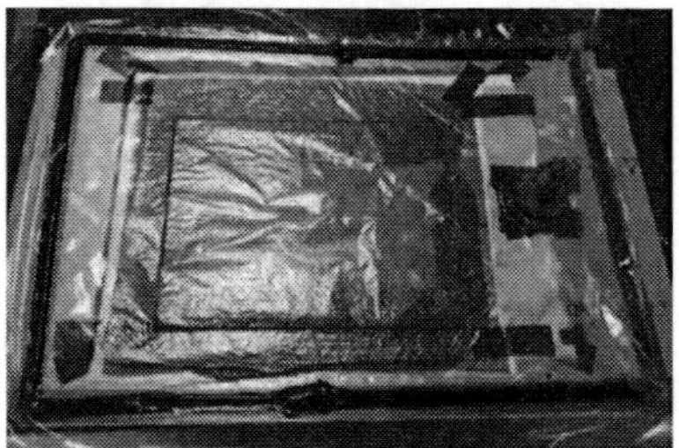

Fig. 9. GLARE 3/2-$[0°/90°]_s$-0.3 was cured at 120°C with autoclave machine.

3.2 Tensile test

Tensile tests were performed based on standard ASTM D3039[16]. The sample size is 25mm in width and 210mm in length. Thickness is around 2-3mm. The test results for GLARE 3/2 with prepreg layup $[0°/90°]_s$ were compared for GLARE 3/2 with prepreg layup $[0°]_4$. The GLARE samples with unidirectional fiber layup show 27.7% higher in ultimate tensile strength and 13.04% lower in yield strength. The stress-strain plot is shown in Figure 9.

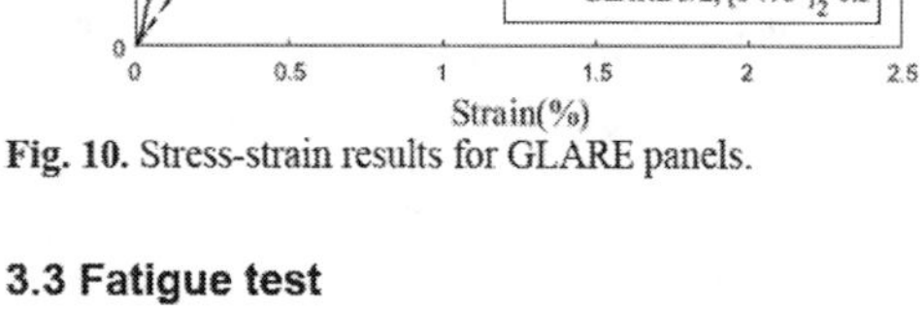

Fig. 10. Stress-strain results for GLARE panels.

3.3 Fatigue test

Tension-tension constant amplitude fatigue tests were performed using MTS 810 servo-hydraulic testing machine and based on ASTM D3479[17]. The maximum stress applied is from 80Mpa to 300Mpa at a stress ratio of R=0.05 and frequency of 5Hz on GLARE 3/2-$[0°]_4$-0.4 samples. Fatigue test stopped when the samples are total fractured to the whole width. The sample tested at a maximum stress of 80Mpa showed a run-out at 10^6 cycle and no cracks were found at the external aluminum layers.

To study the total fatigue life including the fatigue initiation and crack propagation for GLARE panels with unidirectional fiber layup, no predefined notches are made onto the samples.

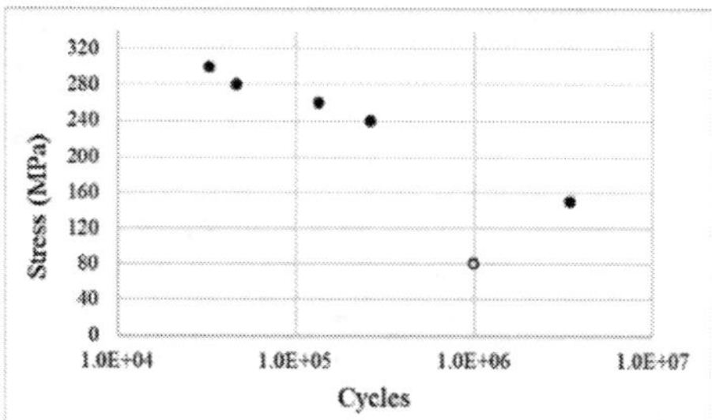

Fig. 11. S-N plot for GLARE 3/2-$[0°]_4$-0.4 samples

4 Summary and conclusion

The model applied for the calculation of the stress level in the aluminum layer was based on classical laminate theory. Three cases were proposed regarding the extra layers of adhesive applied at the metal to prepreg interfaces. The stress value used for the following calculation was the one without considering the extra adhesive layers.

The model for bridging stress was based on the assumption of the triangular and elliptical delamination shapes with a center notch on the specimen external aluminum layers. It's applicable only for center-cracked constant amplitude tension-tension fatigue test for unidirectional fiber layup.

The stress intensity factor and crack growth rate are positively related to the bridging stress, higher the bridging stress higher the crack growth rate. The material constant of C and n are assumed to be accurate although different value was used at other publications.

To study the fatigue life which includes fatigue initiation and crack propagation, the fatigue tests were performed on un-notched test specimens. The maximum stress applied was based on the maximum load calculated from the wind blade root section and the S-N curve showed quite good fatigue life.

Acknowledgements: This work was supported by the scholarship and findings from Interdisciplinary Graduate School and Energy Research Institute(ERI@N)at Nanyang Technological University, Singapore. The laboratory and computation supports of the School of Mechanical and Aerospace Engineering is also much appreciated.

References

[1] A. Tenhave, "WISPER and WISPERX: Final definition of two standardised fatigue loading

sequences for wind turbine blades," *NASA STI/Recon Technical Report N,* vol. **94**, (1992)

[2] A. Ten Have, "WISPER: a standarized fatigue load sequence for HAWT-blades," Nationaal Lucht-en Ruimtevaartlaboratorium(1988)

[3] *Wind Turbines - Part 2: Small wind turbines. IEC 61400-2:2013.*, (2013)

[4] E. Hau, *Wind Turbines Fundamentals, Technologies, Application, Economics*. (2013)

[5] A. Poursartip and P. W. R. Beaumont, "The fatigue damage mechanics of a carbon fibre composite laminate: II-life prediction," Compos. Sci. Technol., Article vol. **25**, no. 4, pp. 283-299, (1986)

[6] J. Degrieck and W. Van Paepegem, "Fatigue damage modeling of fibre-reinforced composite materials: Review," Appl. Mech. Rev., Review vol. **54**, no. 4, pp. 279-299, (2001)

[7] J. Degrieck and W. Van Paepegem, "Fatigue damage modeling of fibre-reinforced composite materials," Appl. Mech. Rev., vol. **54**, no. 4, pp. 279-300, (2001)

[8] G. Wu and J.-M. Yang, "The mechanical behavior of GLARE laminates for aircraft structures," JOM, vol. **57**, no. 1, pp. 72-79, (2005)

[9] G. B. Chai and P. Manikandan, "Low velocity impact response of fibre-metal laminates–A review," Compos. Struct., vol. **107**, pp. 363-381, (2014)

[10] J. J. Homan, "Fatigue initiation in fibre metal laminates," Int. J. Fatigue, Article vol. **28**, no. 4, pp. 366-374, (2006)

[11] H. Sutherland, "On the fatigue analysis of wind turbines, Sandia National Laboratories," *SAND99-0089, Albuquerque, New Mexico,* (1999)

[12] R. Alderliesten, "Analytical prediction model for fatigue crack propagation and delamination growth in Glare," Int. J. Fatigue, vol. **29**, no. 4, pp. 628-646, (2007)

[13] P.-Y. Chang and J.-M. Yang, "Modeling of fatigue crack growth in notched fiber metal laminates," Int. J. Fatigue,, vol. **30**, no. 12, pp. 2165-2174, (2008)

[14] T. Sinmazçelik, E. Avcu, M. T. Bora, and O. Çoban, "A review: Fibre metal laminates, background, bonding types and applied test methods," Mater. Des., Review vol. **32**, no. 7, pp. 3671-3685, (2011)

[15] Y.-J. Guo and X.-R. Wu, "Bridging stress distribution in center-cracked fiber reinforced metal laminates: modeling and experiment," Eng. Fract. Mech., vol. **63**, no. 2, pp. 147-163, (1999)

[16] A. C. D.-o. C. Materials, *Standard test method for tensile properties of polymer matrix composite materials*. ASTM International, (2008)

[17] I. De Baere, W. Van Paepegem, M. Quaresimin, and J. Degrieck, "On the tension–tension fatigue behaviour of a carbon reinforced thermoplastic part I: Limitations of the ASTM D3039/D3479 standard," Polym. Test., vol. **30**, no. 6, pp. 625-632, (2011)

A bandwidth correction to the Allegri-Zhang solution for accelerated random vibration testing

Denis Benasciutti*, *Marta* Carlet, and *Davide* Zanellati

University of Ferrara, Department of Engineering, via Saragat 1, 44122, Ferrara, Italy

Abstract. In 2008, Allegri and Zhang published a study [Int. J. Fatigue. 2008, 30(6):967-977] in which they provided an exact analytical solution to the inverse scaling law for accelerated vibration tests of linear systems submitted to stationary Gaussian excitations By combining finite element analysis with multiaxial spectral methods defined in the frequency-domain, their solution generalised the simple inverse power law model suggested in some standards. The solution adopted the "equivalent von Mises stress" multiaxial criterion combined with the narrow-band damage expression. This work aims to propose a bandwidth correction to the original Allegri-Zhang solution to account for the actual spectral bandwidth of the local multiaxial stress. The corrected Allegri-Zhang solution is also extended to another multiaxial spectral method, namely the "Projection-by-Projection" criterion. A numerical example is finally discussed, in which the corrected solution is applied to an L-shaped beam submitted to random accelerations.

1 Introduction

Mechanical systems are often exposed to vibrations, which induce randomly varying stresses that may be responsible for fatigue damage and even lead to failures. To verify the component service life in the laboratory, it is customary to perform accelerated vibratory tests in which the excitation level is considerably higher than the actual one, which allows the testing time to be greatly shortened compared to the actual life of the system (which may be thousands of hours). Inverse power scaling laws are commonly used to correlate the time to failure in the accelerated laboratory test to the life estimated under real excitations. The following simple power law model is commonly used [1-3]:

$$\frac{T_s}{T_r} = \left(\frac{\bar{g}_r}{\bar{g}_s}\right)^{\alpha} \tag{1}$$

where T_r is the time to failure that a mechanical system is expected to survive in the actual operative environment, T_s is the time to failure in the laboratory test, $\bar{g}_r$ and $\bar{g}_s$ are the root mean square (RMS) values of acceleration in the actual and laboratory environments, respectively. In the original derivation of Eq. (1), the exponent α represents the inverse slope of the S-N fatigue curve. In practical applications, however, α is often treated as an empirical factor (ranging from 5 to 8) that summarises all the uncertainties and approximations in Eq. (1) and its value is often adjusted based on previous experience [3].

Due to its simplicity, the scaling law (1) has been largely used in industry, especially at the design stage, and it is currently included in MIL Standard 810 [1]. Such a simple law, however, only provide a rough approximation of the complex stresses really experienced by a structure under a random input. For example, Eq. (1) neglects the actual shape of the acceleration PSD, which is only synthesised by its RMS value. It also ignores the structure dynamic behaviour, which may directly affect the output stress.

In 2008, Allegri and Zhang published a research study [3] in which they contextualized the scaling law (1) in a more general theory based on finite element analysis and multiaxial spectral methods defined in the frequency-domain through a Power Spectral Density (PSD) matrix – their theory adopted the "equivalent von Mises stress" (EVMS) criterion. The Allegri-Zhang's study derived an exact solution of a general scaling law model, of which the empirical relationship (1) is only a simplified case.

Despite its novelty, their solution seems not to have received so much attention, nor was it further applied in practical case studies. In Allegri-Zhang (A-Z) approach, though, some issues are worth to be investigated, as for example the spectral bandwidth of the multiaxial stress state, the accuracy of the EVMS criterion, or the possibility to implement other multiaxial criteria. With these premises, this work will investigate the A-Z solution with the aim to:

i) propose a modified solution that includes the spectral bandwidth of the local stress state, which was ignored in the original approach;
ii) discuss the role of S-N parameters in EVMS criterion and, accordingly, to suggest possible corrections;
iii) extend the A-Z solution to the "Projection-by-Projection" multiaxial criterion.

* Corresponding author: denis.benasciutti@unife.it

A simple numerical example will be finally discussed, in which the obtained scaling-law solutions are applied to an L-shaped beam submitted to random accelerations imitating the real and accelerated laboratory environments. A comparison with the simple scaling law (1) will be investigated by analysing the results for different acceleration PSD shapes.

2 Spectral description of uniaxial and multiaxial random stresses

Let $x(t)$ be a uniaxial zero-mean stationary and Gaussian random stress with two-sided auto-PSD $S(f)$, which is characterised by the set of spectral moments [4]:

$$\lambda_n = \int_{-\infty}^{\infty} |f|^n S(f)\,df \qquad n = 1,2,\ldots \tag{2}$$

The variance is $\lambda_0 = Var(x(t))$ and the root mean square (RMS) value is $\bar{x} = \sqrt{\lambda_0}$ [4]. The frequency of mean up-crossings and the peaks rate are, respectively [4]:

$$\nu_0 = \sqrt{\lambda_2/\lambda_0} \quad ; \quad \nu_p = \sqrt{\lambda_4/\lambda_2} \tag{3}$$

The following bandwidth parameters will be considered:

$$\alpha_1 = \frac{\lambda_1}{\sqrt{\lambda_0 \lambda_2}} \quad , \quad \alpha_2 = \frac{\lambda_2}{\sqrt{\lambda_0 \lambda_4}} \tag{4}$$

where $0 \le \alpha_m \le 1$ and $\alpha_1 \ge \alpha_2$. Note that $\alpha_2 = \nu_0/\nu_p$ ("regularity index"). Two limiting cases exist: a narrow-band PSD that has $\alpha_1 \to 1$, $\alpha_2 \to 1$, a wide-band PSD that has $\alpha_1 < 1$, $\alpha_2 < 1$ (e.g., a white noise spectrum has $\alpha_2 \cong 0.745$). If $\alpha_2 << 1$, the signal has many small oscillations between two successive zero up-crossings ($\nu_0 << \nu_p$).

The previous frequency-domain description can be extended to a multiaxial stationary random stress. For the sake of simplicity, equations will be presented only for a biaxial stress, although the same formalism applies to a three-dimensional stress. Let $\mathbf{x}(t) = (\sigma_{xx}(t), \sigma_{yy}(t), \tau_{xy}(t))$ be a biaxial stress (σ is normal stress, τ shear stress), characterised by the Hermitian PSD matrix:

$$\mathbf{S}_\sigma(f) = \begin{bmatrix} S_{xx}(f) & S_{xx,yy}(f) & S_{xx,xy}(f) \\ S^*_{xx,yy}(f) & S_{yy}(f) & S_{yy,xy}(f) \\ S^*_{xx,xy}(f) & S^*_{yy,xy}(f) & S_{xy}(f) \end{bmatrix} \tag{5}$$

in which the diagonal terms are auto-PSDs and the out-of-diagonal terms are cross-PSDs (complex functions).

3 Spectral methods and fatigue life

For a narrow-band uniaxial stress of time duration T, the fatigue damage is [5,6]:

$$D_{NB} = \frac{T\,\nu_0}{C}\left(\sqrt{2\lambda_0}\right)^b \Gamma\left(1+\frac{b}{2}\right) \tag{6}$$

where C, b are the parameters of the S-N curve $s^b N = C$. The coefficient $C = s_A^b N_A$ relates to the reference stress amplitude s_A at $N_A = 2\times10^6$ cycles.

The time to failure follows by settling $D_{NB} = D_{cr}$ (where D_{cr} is a critical damage value):

$$T_{f,NB} = \frac{C\,D_{cr}}{\nu_0\left(\sqrt{2\lambda_0}\right)^b \Gamma\left(1+\frac{b}{2}\right)} \tag{7}$$

In the narrow-band case Eq. (7) is exact, while in the wide-band case it is largely conservative and needs to be corrected by accounting for the actual PSD bandwidth [5,6]. A very simple approach is to introduce a correction factor $\rho \le 1$ in Eq. (7), so that for a wide-band stress:

$$T_f = \frac{C D_{cr}}{\rho\,\nu_0\left(\sqrt{2\lambda_0}\right)^b \Gamma\left(1+\frac{b}{2}\right)} \tag{8}$$

Factor ρ depends on the PSD shape and it is a function of spectral moments and bandwidth parameters. In the narrow-band case $\rho \to 1$, while in the wide-band case $\rho \to 0$ (factor ρ then gives an indirect measure of the spectral width of a PSD). Several methods adopt such a correction factor approach, as the "TB method" [5,6]:

$$\rho = \left[q + (1-q)\alpha_2^{b-1}\right] \tag{9}$$

where $q = q(\alpha_1, \alpha_2)$ is a coefficient depending on α_1, α_2 (its analytical expression can be found in [5,6]).

Other methods also exist (e.g. Dirlik) which replace the single factor ρ with more elaborated expressions [6]. However, to keep things simple and without any loss of generality, it will be assumed that bandwidth effects are summarised by factor ρ and that the time to failure T_f for a wide-band uniaxial stress is estimated by Eq. (8).

With a multiaxial stress, a suitable criterion needs to be introduced for estimating the time to failure. A criterion largely used is that based on the "Equivalent Von Mises Stress" (EVMS) $\sigma_{eq}(t)$, which is a zero-mean Gaussian uniaxial stress, described by the following PSD:

$$S_{eq}(f) = S_{xx}(f) + S_{yy}(f) - \mathrm{Re}\left[S_{xx,yy}(f)\right] + 3S_{xy}(f) \tag{10}$$

In matrix form ("Tr" is the trace operator):

$$S_{eq}(f) = \mathrm{Tr}\{\mathbf{Q}\,\mathbf{S}_\sigma(f)\} \quad , \quad \mathbf{Q} = \begin{bmatrix} 1 & -0.5 & 0 \\ -0.5 & 1 & 0 \\ 0 & 0 & 3 \end{bmatrix} \tag{11}$$

The RMS value of $\sigma_{eq}(t)$ is $\bar{\sigma}_{eq} = \sqrt{\lambda_{0,eq}}$, where λ_{0eq} is the zero-order moment of $S_{eq}(f)$. Note that in general $S_{eq}(f)$ is wide-band, which then requires that Eq. (8) is applied to estimate the fatigue life.

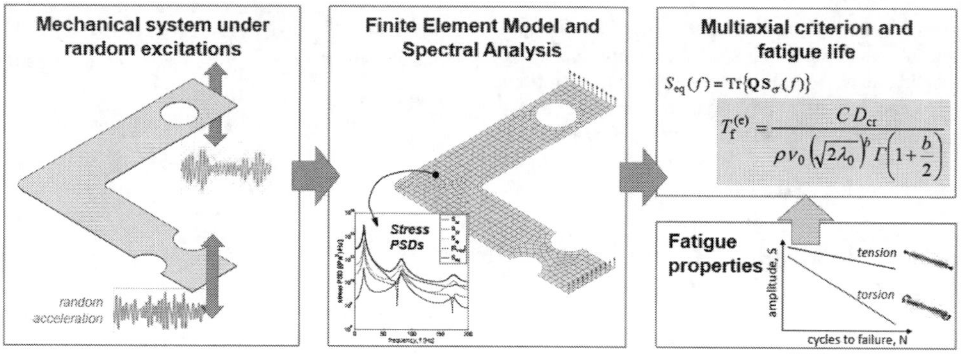

Fig. 1. Analysis steps (for both "real" and "lab" environment): finite element model, spectral analysis to compute the stress PSDs at each element, multiaxial spectral method to estimate the time to failure.

4 Accelerated tests and scaling law

4.1 Allegri-Zhang (A-Z) original solution

Assume that a mechanical system is subjected to a random base acceleration. The finite element method can profitably be exploited to evaluate the structure dynamic response and the stress PSD matrix $\mathbf{S}_\sigma^{(e)}(f)$ in each element (e) of the model (Fig. 1). For a plane stress state, $\mathbf{S}_\sigma^{(e)}(f)$ is a 3×3 matrix, see Eq. (5).

In the A-Z approach [3], the exact scaling law was derived by considering two main assumptions. The first one was to use the EVMS criterion, Eq. (11), to transform $\mathbf{S}_\sigma^{(e)}(f)$ of a multiaxial stress into $S_{eq}^{(e)}(f)$ for the equivalent stress $\sigma_{eq}(t)$.

The second assumption was the choice of the statistical parameter of $\sigma_{eq}(t)$ to construct a stress-life relationship. The stress level was quantified by $\overline{\sigma}_{eq}$, i.e. the RMS value of $\sigma_{eq}(t)$. The number of cycles to failure, instead, was measured by the number of zero up-crossings N_0 observed in $\sigma_{eq}(t)$. For a stationary stress $N_0=\nu_{0eq}\cdot T_f$, where T_f is the fatigue life (in seconds) and ν_{0eq} the rate of zero up-crossings for $\sigma_{eq}(t)$. An S-N type equation $\overline{\sigma}_{eq}^b N_0 = C$ was then introduced and the fatigue life estimated as:

$$T_f = \frac{C}{\nu_{0eq}\,\overline{\sigma}_{eq}^b} \tag{12}$$

Although b, C virtually represent the parameters of the constant amplitude S-N curve, their value may change under a random stress, as also pointed out in [3] (this issue will be further discussed in Section 6).

To derive a scaling law, Eq. (12) has to be applied twice: first for estimating the fatigue life $T_{f,r}$ of the mechanical system exposed to the "real" environment (actual excitation), secondly for assessing the fatigue life $T_{f,s}$ in the laboratory accelerated test. The following ratio of time durations then results:

$$\frac{T_{f,s}}{T_{f,r}} = \frac{\nu_{0eq,r}}{\nu_{0eq,s}}\left(\frac{\overline{\sigma}_{eq,r}}{\overline{\sigma}_{eq,s}}\right)^b \tag{13}$$

where $\nu_{0eq,r}$, $\overline{\sigma}_{eq,r} = \sqrt{\lambda_{0eq,r}}$ are the spectral parameters of the equivalent von Mises stress in the "real" environment, while $\nu_{0eq,s}$, $\overline{\sigma}_{eq,s} = \sqrt{\lambda_{0eq,s}}$ are those for the laboratory experiment. Being a scaling law, Eq. (13) only depends on the slope b, but not on the coefficient C. Expanding Eq. (13) in terms of spectral moments yields:

$$\frac{T_{f,s}}{T_{f,r}} = \sqrt{\frac{\lambda_{2eq,r}}{\lambda_{2eq,s}}}\left(\frac{\lambda_{0eq,r}}{\lambda_{0eq,s}}\right)^{\frac{b-1}{2}} \tag{14}$$

Note that Eq. (14) only depends on two spectral moments λ_{0eq} and λ_{2eq} of $S_{eq}^{(e)}(f)$. By writing explicitly the definition of such spectral moments, see Eq. (2), the previous expression returns the final equation in [3]:

$$\frac{T_{f,s}^{(e)}}{T_{f,r}^{(e)}} = \sqrt{\frac{\mathrm{Tr}\left[\mathbf{Q}\int_0^\infty f^2\,\mathbf{S}_{\sigma,r}^{(e)}(f)\,df\right]}{\mathrm{Tr}\left[\mathbf{Q}\int_0^\infty f^2\,\mathbf{S}_{\sigma,s}^{(e)}(f)\,df\right]}}\left\{\frac{\mathrm{Tr}\left[\mathbf{Q}\int_0^\infty \mathbf{S}_{\sigma,r}^{(e)}(f)\,df\right]}{\mathrm{Tr}\left[\mathbf{Q}\int_0^\infty \mathbf{S}_{\sigma,s}^{(e)}(f)\,df\right]}\right\}^{\frac{b^{(e)}-1}{2}} \tag{15}$$

The superscript (e) is reintroduced to remind that the quantities in Eq. (15) refer to element (e) in the finite element model. Also the slope $b^{(e)}$ depends on the material which the element is made of.

Equation (15) is the original A-Z solution and it represents the exact scaling law for linear systems undergoing Gaussian random accelerations.

Allegri and Zhang also demonstrated that Eq. (15) returns exactly the scaling law (1) in the special case in which the acceleration PSD for the "lab" environment is simply scaled with respect to that characterising the "real" environment [3].

It has to be emphasised that the scaling law (15) is based on the hypothesis to take N_0 as a measure of the number of rainflow cycles, which is exact only for a narrow-band stress (where $\nu_0 \cong \nu_p$). In a wide-band stress, it is $N_0 < N_p$, where N_p is the number of peaks, which in turn is coincident with the number of rainflow cycles (in rainflow counting, a cycle is attached to each peak). In a wide-band stress, N_0 would then underestimate the actual number of rainflow cycles.

This conclusion is also confirmed by noting that the relationship (14) only involves two spectral moments λ_0 and λ_2, which implies that it does not accounts for the actual spectral bandwidth of $\sigma_{eq}(t)$, i.e. it provides the same result for any type of narrow-band and wide-band processes sharing the same values of λ_0 and λ_2. For example, one may construct an infinite set of wide-band spectra $S_{eq}^{(e)}(f)$ that have common values of λ_0 and λ_2, but have different values of other spectral moments (e.g. λ_1 and λ_4). Such spectra would have identical values of ν_{0eq}, $\overline{\sigma}_{eq}$, and also the same value of T_f from Eq. (12).

These arguments motivate the attempt to introduce a bandwidth correction to the A-Z solution, as explained in the next section.

4.2 Bandwidth correction of A-Z solution

The hypothesis in A-Z solution that $\sigma_{eq}(t)$ is considered to be always narrow-band is confirmed by noting that Eq. (13) can also be obtained more directly by applying the narrow-band expression (7) to the "real" and "lab" environments.

To account for the actual bandwidth of $\sigma_{eq}(t)$, the wide-band solution (8) has to be used in place of Eq. (7). In the "real" environment, the time to failure is:

$$T_{f,r} = \frac{C\,D_{cr}}{\rho_{eq,r}\,\nu_{0eq,r}\left(\sqrt{2\lambda_{0eq,r}}\right)^b \Gamma\left(1+\frac{b}{2}\right)} \qquad (16)$$

where subscript r stands for "real". As before, symbols $\nu_{0eq,r}$, $\overline{\sigma}_{eq,r} = \sqrt{\lambda_{0eq,r}}$ identify the zero up-crossings and RMS value of $\sigma_{eq}(t)$, whereas $\rho_{eq,r}$ is the bandwidth correction factor (9) according to the TB method. For the "simulation" ("lab") environment, the time to failure $T_{f,s}$ is given by an expression identical to Eq. (16), in which symbols $\nu_{0eq,s}$, $\overline{\sigma}_{eq,s} = \sqrt{\lambda_{0eq,s}}$, $\rho_{eq,s}$ are used:

$$T_{f,s} = \frac{C\,D_{cr}}{\rho_{eq,s}\,\nu_{0eq,s}\left(\sqrt{2\lambda_{0eq,s}}\right)^b \Gamma\left(1+\frac{b}{2}\right)} \qquad (17)$$

The scaling law is constructed as the ratio:

$$\frac{T_{f,s}}{T_{f,r}} = \frac{\rho_{eq,r}}{\rho_{eq,s}}\frac{\nu_{0eq,r}}{\nu_{0eq,s}}\left(\frac{\overline{\sigma}_{eq,r}}{\overline{\sigma}_{eq,s}}\right)^b \qquad (18)$$

which generalises the original A-Z solution (13) by introducing a spectral bandwidth correction factor ρ for the PSD of the equivalent stress. Note that, in general, $\rho_{eq,r} \neq \rho_{eq,s}$. In fact, if the same structure is excited by "real" and "lab" acceleration PSDs with different shapes, it may be expected that different stress PSDs will also be obtained. For example, if the "real" acceleration is narrow-band, the output stress is close to narrow-band. Instead, if the "lab" acceleration is wide-band, it is more likely that at some node the stress PSD is more wide-band. The condition $\rho_{eq,r} = \rho_{eq,s}$ occurs only if the "real" and "lab" acceleration PSDs are exactly scaled.

Similarly to the A-Z solution (15), the ratio of time durations can be further manipulated to obtain:

$$\frac{T_{f,s}^{(e)}}{T_{f,r}^{(e)}} = \sqrt{\frac{\rho_{eq,r}\,\mathrm{Tr}\left[\mathbf{Q}\int_0^\infty f^2\,\mathbf{S}_{\sigma,r}^{(e)}(f)\,df\right]}{\rho_{eq,s}\,\mathrm{Tr}\left[\mathbf{Q}\int_0^\infty f^2\,\mathbf{S}_{\sigma,s}^{(e)}(f)\,df\right]}}\left\{\frac{\mathrm{Tr}\left[\mathbf{Q}\int_0^\infty \mathbf{S}_{\sigma,r}^{(e)}(f)\,df\right]}{\mathrm{Tr}\left[\mathbf{Q}\int_0^\infty \mathbf{S}_{\sigma,s}^{(e)}(f)\,df\right]}\right\}^{\frac{b^{(e)}-1}{2}} \qquad (19)$$

in which superscript (e) reminds that quantities refer to element (e) in the finite element model.

Equation (19) is the bandwidth correction of the original A-Z solution (15). Compared to Eq. (15), the two factors $\rho_{eq,r}$, $\rho_{eq,s}$ appear. If in each element (e) the equivalent stress $\sigma_{eq}(t)$ is always narrow-band, $\rho_{eq,r} = \rho_{eq,s} = 1$ and Eq. (19) converges to the original A-Z scaling law (15).

5 Numerical example

The purpose of this numerical example is to show the practical application of the analytical scaling laws discussed in the previous sections. The example considers an L-shaped beam excited by a base random acceleration. The beam geometry imitates that one already proposed in [7], although some dimensions have been slightly changed to enhance the stress concentration effect at the hole and two lateral notches.

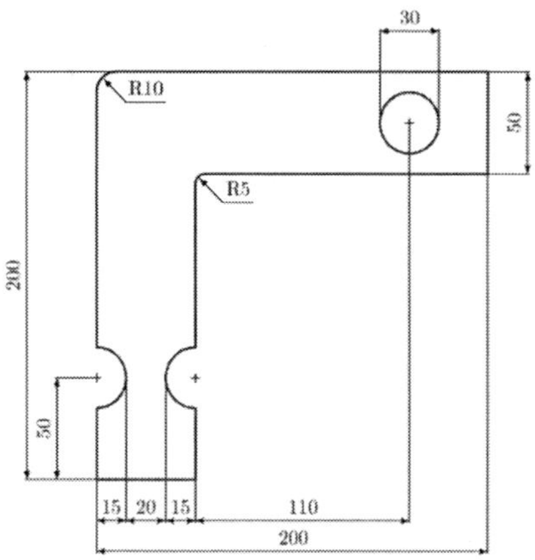

Fig. 2. Beam dimensions (in mm). Thickness is 0.5 mm.

The specimen is clamped at both ends, where a random acceleration is imposed along the direction normal to the specimen plane. The input accelerations applied at the two clamped ends are assumed to be fully correlated (their cross-PSD is different from zero).

The beam is made of steel; its tension S-N curve has parameters (taken from [7]) b=9.82, C=4.0641×10^{88} (the reference strength amplitude is s_A=241 MPa).
A 'shell' finite element model is used to discretise the beam (elements have an average size of 5 mm, resulting in 641 elements and 735 nodes), see Fig. 3.

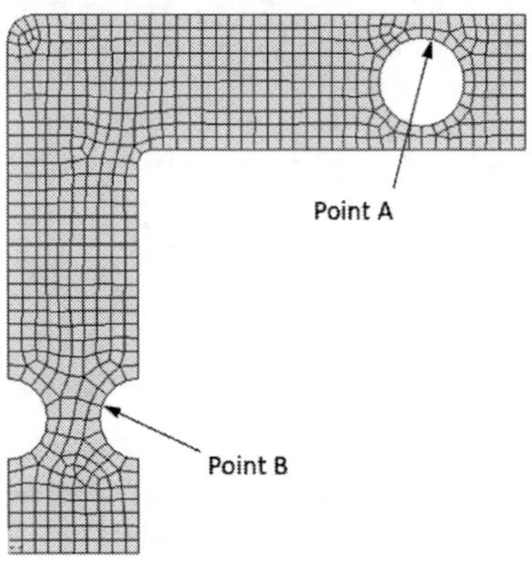

Fig. 3. Finite element model (shell elements).

In the finite element simulation (Fig. 1), a frequency-domain spectrum analysis is carried out to determine the natural frequencies and the stress PSDs at each node. Stress spectra are next processed to determine the time to failure used in the inverse scaling laws discussed in the previous sections. The whole numerical analysis is performed by software ANSYS with APDL language, which is also used to implement the scaling laws.
A modal analysis returns the following first natural frequencies: f_1=15.9 Hz, f_2=67.1 Hz, f_3=81.9 Hz, f_4=175.1 Hz, f_5=178.9 Hz. A harmonic analysis is also carried out to determine the frequency response function and the contribution of each mode, see Fig. 4. The figure clearly shows the peaks at each resonant frequency (two of them are less pronounced).

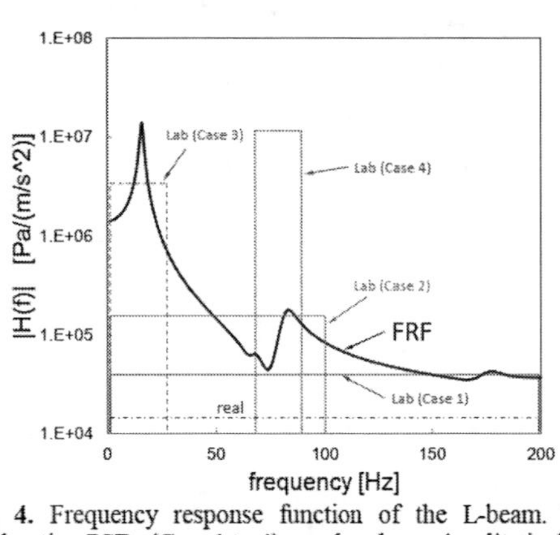

Fig. 4. Frequency response function of the L-beam. The acceleration PSDs (Case 1 to 4) are also shown (qualitatively).

In order to imitate the application of the inverse scaling laws, the PSD of the input acceleration has to be defined for both the "real" and "laboratory" environments.

Since the purpose of this example is only to demonstrate the practical application of scaling laws (without any claim to provide results of direct practical use), the acceleration PSD is idealised as a band-limited rectangular spectrum.
For the "real" environment, a rectangular PSD from 1 to 200 Hz is chosen; its RMS value $\bar{g}_r = 2.8833g$ (g=9.81 m/s^2 is the gravity acceleration) is so adjusted in order that, in numerical simulations, the beam has a "real" time duration T_r=10 years (3.1536·10^8 seconds) in the most stressed point.

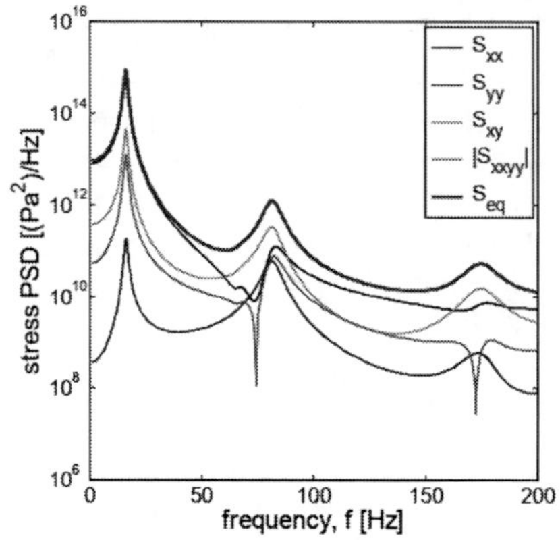

Fig. 5. PSDs of stress components in point A.

Fig. 5 displays the auto- and cross-PSD of the biaxial stress calculated at point A, for the "real" case. The figure highlights how the rectangular PSD of input acceleration is filtered out, resulting in stress PSDs peaked at resonance frequencies. The figure also plots the power spectrum $S_{eq}(f)$ of the equivalent stress, which is not exactly narrow-band, as it has spectral parameters α_1=0.924, α_2=0.447.
For the "lab" environment, a target time duration of $T_{s,0}$=1 day (86400 seconds) is taken as the period to be achieved in an hypothetical accelerated testing. The resulting scaling ratio is $(T_{s,0}/T_r)$=0.274·10^{-3}.
Also the "lab" acceleration PSD is idealised as a rectangular (band-limited) spectrum. Four different cases are distinguished (see Table 1), based on the following frequency ranges:

- Case 1: rectangular PSD from 1 to 200 Hz (same shape as the "real" spectrum);
- Case 2: rectangular PSD from 1 to 100 Hz;
- Case 3: rectangular PSD from 1 to 30 Hz;
- Case 4: rectangular PSD from 70 to 90 Hz.

All the previous PSDs share the same value $\bar{g}_s = 6.6473g$ of acceleration RMS, which is determined through Eq. (1) by targeting the laboratory time duration $T_{s,0}$=1 day and assigning the specifications T_r=10 years, $\bar{g}_r = 2.8833g$ for the "real" environment (the exponent in Eq. (1) is taken equal to the S-N slope α=b=9.82).
Fig. 4 compares (qualitatively) the four PSD types to the frequency response function. Being the RMS value the same for all spectra, the height of each rectangle changes depending on the frequency range of each spectrum.

Simulation results of several test cases are summarised from Fig. 6 to Fig. 9. For the Case 1 acceleration PSD, Fig. 6 displays the RMS value $\bar{\sigma}_{eq,s}$ of the equivalent stress, while Fig. 7 shows the corresponding distribution of the logarithm $\log(T_s)$ of the time duration T_s computed in each finite element by Eq. (17).

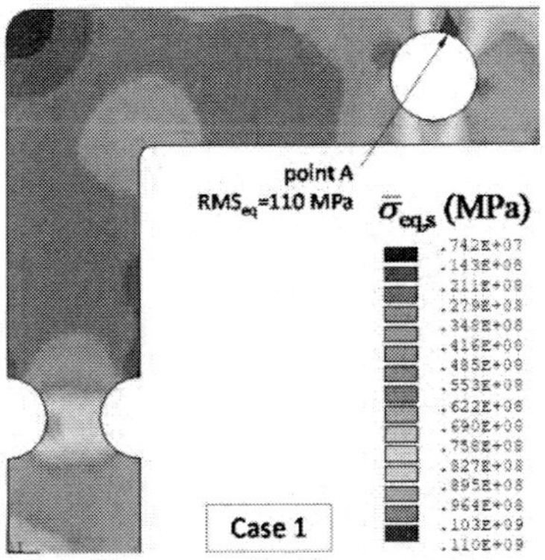

Fig. 6. Distribution of RMS of eq. stress σ_{eq} (Case 1).

The highest RMS value (110 MPa) occurs at point A, where also the shortest time duration T_s=86400 seconds is obtained. This time duration exactly matches the target value $T_{s,0}$ predicted by Eq. (1). This result confirms that if "lab" and "real" PSDs of the input acceleration have the same shape and are simply scaled, both scaling laws (1) and (19) return the same result. This also holds true for the ratio (T_s/T_r) in each node of the model, which is always equal to the ratio $(T_{s,0}/T_r)$=$0.274 \cdot 10^{-3}$.

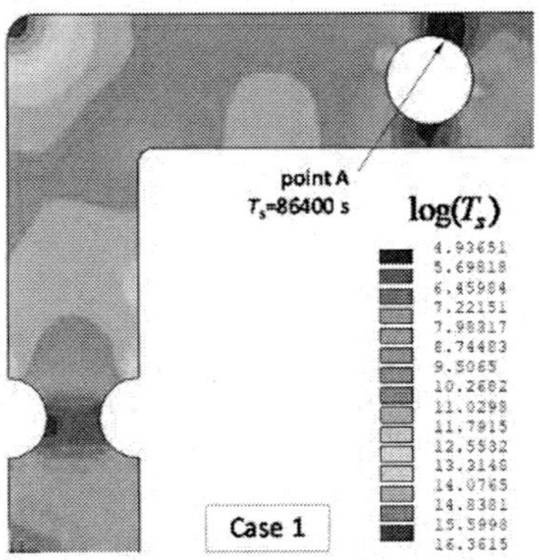

Fig. 7. Distribution of $\log(T_s)$ (Case 1).

For the Case 2, the distribution of $\log(T_s)$ is shown in Fig. 8. The shortest fatigue life $T_{s,A}$=2890 seconds still occurs at point A, where the highest RMS value (156 MPa) is also identified.
Surprisingly, now the value $T_{s,A}$ returned by Eq. (17) through numerical analysis is much shorter than the target $T_{s,0}$ estimated by Eq. (1). Differently from Case 1, in this Case 2 in which "lab" and "real" acceleration PSDs are not simply scaled, a disagreement is observed in the time scaling estimated from Eq. (1) and Eq. (19).

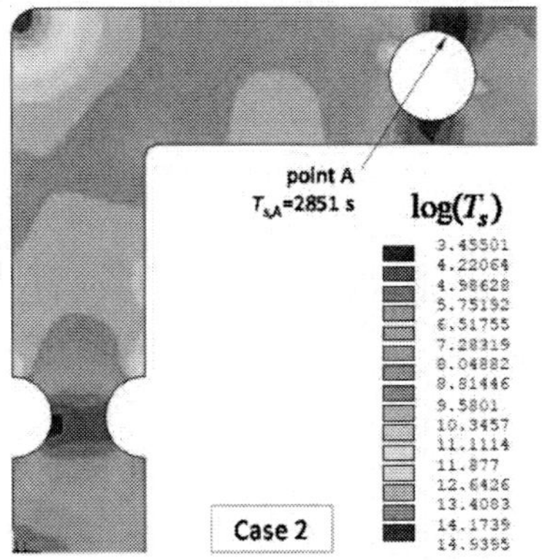

Fig. 8. Distribution of $\log(T_s)$ (Case 2).

This disagreement can be explained by considering that the simple scaling law model (1) provides only a rough approximation of the relationship of the system response in the real and laboratory environments.
First, it takes the acceleration RMS value as the only one parameter synthesising the input acceleration spectrum (the actual spectrum shape and frequency range are not considered). Secondly, the scaling law model (1) also ignores the actual system dynamic response (especially the coupling between modes) and the resulting multiaxial state of stress in each part of the system. On the other hand, these aspects are correctly evaluated by the Allegri-Zhang model in its original as well as modified version.
This behaviour can be explained by observing that the acceleration PSD in Case 2 (ranging in 1÷100 Hz) has double the height of Case 1 (which ranges in 1÷200 Hz) and it is more concentrated towards the first natural frequency. This makes the contribution of the first vibration mode enhance the RMS of the output stress.

Table 1. Comparison of simulation results for the "real" and "lab" environments.

Type of PSD	Acceleration RMS	$T_{s,0}$ from Eq. (1)	RMS of eq. stress, $\bar{\sigma}_{eq}$ (MPa)	T_s from Eq. (19) - numerical analysis	New exponent α
"real"	2.8833g	–	47.7 (in A)	315360013 s	–
Case 1	6.6473g	1 day (86400 s)	110 (in A)	86400 s	9.82
Case 2	6.6473g	1 day (86400 s)	156 (in A)	2851 s	13.9
Case 3	6.6473g	1 day (86400 s)	287 (in A)	8.6 s	20.9
Case 4	6.6473g	1 day (86400 s)	43 (in B)	17419281 s	3.47

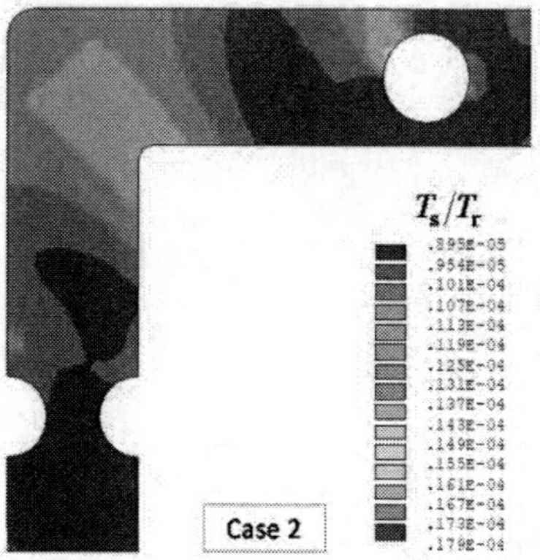

Fig. 9. Distribution of T_s/T_r (Case 2).

This effect is further confirmed by considering the Case 3, in which the input acceleration PSD is almost entirely centred on the first natural frequency. In this case, as can be seen in Table 1, the difference between Eq. (1) and (19) is largely increased.

On the other hand, one may presume that the observed discrepancies between the scaling law models are caused from taking the exponent α in Eq. (1) equal to the material inverse slope, instead of an "empirical" value around 3-5, as suggested in [3]. This conclusion is, however, not confirmed by the results in the last column of Table 1, which lists the values of α which would make Eq. (1) fit the time duration T_s resulting from the A-Z model (19), as implemented in numerical modelling. These values of α are largely dependent on the type of input acceleration PSD and it would then rather difficult to determine in advance a proper value of α.

To overcome the above limitations of the simplified scaling law (1), the use of the Allegri-Zhang model – preferably in its bandwidth corrected form (19) – is thus recommended. An example of output (T_s/T_r) as given by Eq. (19) is shown in Fig. 9 for the Case 2.

6 The role of S-N parameters

According to the Palmgren-Miner rule, the fatigue parameters b, C in the expressions presented so far should be those from the constant amplitude S-N curve. Experimental results, however, seem to highlight a change in fatigue parameters in random loading tests. In this case the S-N curve is generally located on the left of the corresponding S-N curve for constant amplitude loading, if the RMS value is used for comparison (limitations in the use of RMS are discussed in [8]). A higher b is also observed in the random loading case. The shape of the stress PSD seems to have a limited effect [3,9].

A more important aspect is the role of tension and torsion S-N parameters in scaling laws, and especially with regard to the EVMS criterion adopted by A-Z method. It has been recognised (see [10]) that the EVMS criterion neglects the S-N parameters for torsion loading (in particular, it implicitly assumes that the torsion S-N line is parallel to the tension one and exactly scaled by √3). These assumptions mean, for example, that Eq. (15) and (19) would return the same ratio of time durations, regardless of the torsion S-N curve parameters. Also note how the simple power law (1) does not even consider the torsion properties at all. Materials, on the other hand, are characterised by a wide range of torsion fatigue properties [10].

To account for both tension and torsion S-N parameters in the A-Z scaling law, other multiaxial spectral methods, among those currently available in the literature [11-13], should be used in place of the EVMS criterion.

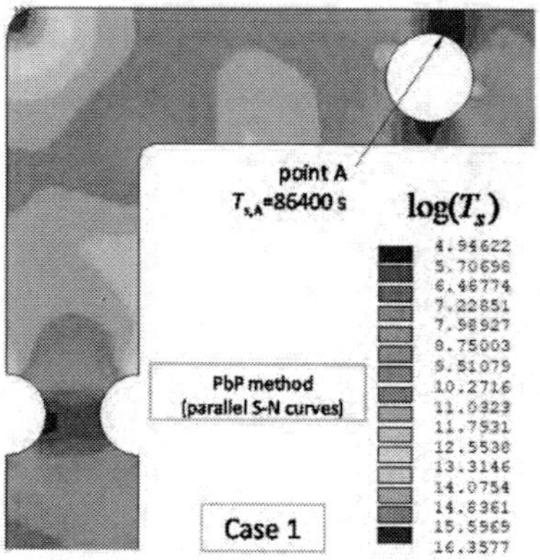

Fig. 10. Distribution of log(T_s) (Case 1) by PbP criterion (material with parallel S-N lines).

A proposal is made here to reformulate the A-Z scaling law by using the so-called "Projection-by-Projection" (PbP) criterion, which is a stress invariant-based multiaxial criterion that can account for both torsion and tension S-N parameters (the main theoretical aspects are outside the goal of this work and they can be found elsewhere [14,15]). The PbP method computes the fatigue damage in time T as:

$$D_{\mathrm{PbP}} = T\left[\sum_{i=1}^{5} d_i^{\frac{2}{k_{\mathrm{ref}}}}\right]^{\frac{k_{\mathrm{ref}}}{2}} \tag{20}$$

where d_i is the damage intensity (damage/s) of the i-th stress projection in the deviatoric space, which can be estimated by uniaxial spectral methods, see Section 3.1.

If the time to failure (i.e. the inverse of fatigue damage) for each stress projection is estimated by the TB method for wide-band stress (see Eq. (8)), the following scaling law is obtained:

$$\frac{T_{\mathrm{f,s}}^{(\mathrm{e})}}{T_{\mathrm{f,r}}^{(\mathrm{e})}} = \frac{C_{\mathrm{ref,s}}^{(\mathrm{e})}\,\Gamma\left(1+\frac{b_{\mathrm{ref,r}}^{(\mathrm{e})}}{2}\right)\left[\sum_{i=1}^{5}\left(\lambda_{0i,\mathrm{r}}^{(\mathrm{e})}\left(\rho_{i,\mathrm{r}}^{(\mathrm{e})}\,\nu_{0i,\mathrm{r}}^{(\mathrm{e})}\right)^{\frac{2}{b_{\mathrm{ref,r}}^{(\mathrm{e})}}}\right)\right]^{\frac{b_{\mathrm{ref,r}}^{(\mathrm{e})}}{2}}}{C_{\mathrm{ref,r}}^{(\mathrm{e})}\,\Gamma\left(1+\frac{b_{\mathrm{ref,s}}^{(\mathrm{e})}}{2}\right)\left[\sum_{i=1}^{5}\left(\lambda_{0i,\mathrm{s}}^{(\mathrm{e})}\left(\rho_{i,\mathrm{s}}^{(\mathrm{e})}\,\nu_{0i,\mathrm{s}}^{(\mathrm{e})}\right)^{\frac{2}{b_{\mathrm{ref,s}}^{(\mathrm{e})}}}\right)\right]^{\frac{b_{\mathrm{ref,s}}^{(\mathrm{e})}}{2}}} \tag{21}$$

https://doi.org/10.1051/matecconf/201816507006

in which λ_{0i}, ν_{0i}, ρ_i are the variance, zero up-crossing rate and bandwidth correction factor of the i-th stress projection. In Eq. (21), C_{ref} and b_{ref} are the parameters of a reference S-N curve in a Modified Wöhler Diagram, defined in the deviatoric space. They are linearly interpolated from the tension and torsion S-N parameters (see [14,15]).

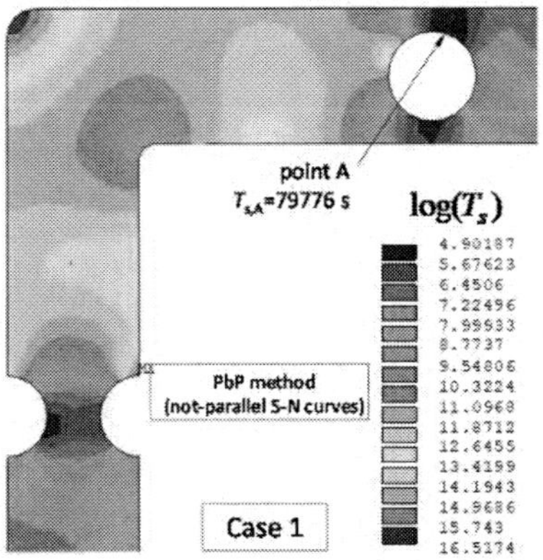

Fig. 11. Distribution of $\log(T_s)$ (Case 1) by PbP criterion (material with non-parallel S-N lines).

To appreciate the effect of S-N parameters, Fig. 10 and Fig. 11 compare the distribution of $\log(T_s)$ estimated by the PbP criterion, when considering two types of materials with: i) S-N curves parallel and scaled by $\sqrt{3}$ (Fig. 10), ii) none of the previous (Fig. 11). The figures refer to the Case 1 acceleration PSD. The comparison demonstrates that the PbP is sensitive to the torsion S-N parameters and the time duration changes from material i) to material ii). The difference would be more pronounced if the input acceleration PSD were centred around a single vibration mode characterised by a torsional deformation. By contrast, no difference would be observed in the time duration for material i) and ii), if the EVMS were used in the A-Z solution, see Eq. (15) or (19). In both cases, the distribution of $\log(T_s)$ would be that shown in Fig. 7.

7 Conclusions

This work dealt with inverse scaling laws used to calibrate accelerated vibratory tests for linear systems undergoing stationary Gaussian excitations. A comparison was made between a simple inverse power law model and the Allegri-Zhang (A-Z) solution, which was first reviewed and next modified by introducing a bandwidth correction factor to account for the actual shape of the frequency spectrum for the "equivalent von Mises stress" (EVMS).

A numerical example (L-beam submitted to random base accelerations) solved by the finite element method was used as a test case for the comparison. Several types of idealised rectangular acceleration PSDs for the laboratory environment were considered, to investigate the sensitivity of the scaling law to the type of input.

The results confirmed that both scaling laws agree only for "real" and "lab" acceleration PSDs which are exactly scaled. In the other cases, large differences occur and the power law model estimates a much longer time duration than the A-Z solution. This difference was explained in terms of the influence of the system dynamics on the time duration estimated by numerical analysis, which instead is ignored by the simple power law model. Numerical results also showed that the values of the exponent α in the power law model largely dependent on the type of acceleration PSD; choosing *a priori* a suitable value of α would then be rather difficult.

An attempt was finally made to reformulate the A-Z scaling law in terms of the "Projection-by-Projection" multiaxial spectral method, in place of the EVMS criterion. This modification permitted the A-Z solution to be sensitive also to the torsion S-N properties.

In summary, the results presented in this work emphasised the limitations in the use of the simple power law model and they suggested that the A-Z solution should be preferred.

References

1. MIL-STD-810 CF, *Military standard environmental test methods for aerospace and ground equipment, Method 514: Vibration* (1962)
2. A. Angeli, B. Cornelis, M. Troncossi, Mech. Systems and Signal Proc. **103**, 340 (2018)
3. G. Allegri, X. Zhang, Int. J. Fatigue **30(6)**, 967 (2008)
4. L.D. Lutes, S. Sarkani, *Random vibrations: analysis of structural and mechanical systems* (Elsevier, USA, 2004)
5. D. Benasciutti, R. Tovo, Int. J. Fatigue **27(8)**, 867 (2005)
6. D. Benasciutti, R. Tovo, Probab. Eng. Mech. **21(4)**, 287 (2006)
7. X. Pitoiset, A. Preumont, Int. J. Fatigue **22**, 541 (2000)
8. C.M. Sonsino, Int. J. Fatigue **11(3)**, 142 (1989)
9. C. Lalanne, *Mechanical Vibration and Shock Analysis - vol. 4: Fatigue Damage* (ISTE Ltd, UK, 2009)
10. D. Benasciutti, J. Sound Vib. **333(18)**, 4326 (2014)
11. D. Benasciutti, F. Sherratt, A. Cristofori, Procedia Engineering **101(C)**, 34 (2015)
12. D. Benasciutti, F. Sherratt, A. Cristofori, Int. J. Fatigue **91(2)**, 397 (2016)
13. K. Bel Knani, D. Benasciutti, A. Signorini, R. Tovo, Int. J. Mater. Prod. Technol. **30(1–3)**, 172 (2007)
14. A. Cristofori, D. Benasciutti, R. Tovo, Int. J. Fatigue **33(7)**, 887 (2011)
15. A. Cristofori, D. Benasciutti, SAE Technical Paper 2014-01-0924 (2014)

Fatigue behaviour of open-hole samples and automotive mini-structures made of woven glass-fibre-reinforced polyamide 6,6

Amélie Malpot[1,*], *Fabienne* Touchard[2], *Sébastien* Bergamo[1], *Catherine* Peyrac[3], *Richard* Montaudon[1], and *Jean-Baptiste* Blumenfeld[1]

[1]Renault SA, 78084 Guyancourt, France
[2]Département Physique et Mécanique des Matériaux, Institut Pprime, 86961 Futuroscope Chasseneuil, France
[3]Pôle Fatigue des Composants Mécanique, CETIM, Senlis, France

Abstract. In the automotive industry, the integration of thermoplastic composite components represents a high-potential solution to the mass reduction challenge. In this study, a woven glass-fibre-reinforced composite with a polyamide 6,6 matrix is considered for the purpose of being integrated into automotive parts. Tension-tension fatigue tests were conducted on [$(0/90)_3$] open-hole samples. These tests were instrumented with non-destructive techniques, namely acoustic emission and infrared thermography. Acoustic emission results showed fibre-matrix debonding and fibre breakages in open-hole samples, located around the hole. Furthermore, 3-point bending fatigue tests were performed on "omega" mini-structures. A semi-empirical model was used in order to predict the fatigue lives of both open-hole coupons and automotive mini-structures. Predictions of the model for open-holes samples underestimate experimental fatigue lives. Nevertheless, the semi-empirical model showed good results for the fatigue life prediction of composite mini-structures.

1 Introduction

Glass-fibre-reinforced composites with thermoplastic matrix are strategic for automotive manufacturers because of their strength-to-weight ratio, their recyclability, their production rate and their cost. Many vehicle parts are subjected to fatigue loading. The origin of this loading may be the road itself, the users or the engine. Thus, it is important for car manufacturer to be able to predict the fatigue behaviour of composite automotive parts.

According to Degrieck & Van Paepegem [1], fatigue models can be classified in three groups: fatigue life models, phenomenological models to predict residual stiffness/strength and progressive damage models. In the context of this study, fatigue life models seem to be particularly appropriate since they allow the direct determination of the number of cycles at failure for a given set of experimental conditions [2, 3, 4].

Fatigue life model proposed by Epaarachchi & Clausen [4] has been used by several authors in order to model the fatigue behaviour of composite materials. Mortazavian et al. [5] have developed a model based on the work of Epaarachchi & Clausen [4] in order to predict the fatigue life of short-fibre-reinforced polymer. The authors showed that this model is able to predict accurately the fatigue life of short glass fibre/PA66 composite for several temperatures, stress ratios and fibres orientations. Other authors have used the model presented in [4] in order to predict the fatigue life of composite reinforced with continuous fibres [6, 7, 8, 9].

This study is focused on a woven glass-fibre-reinforced polyamide 6,6 (referred as GFRPA66). In a previous study, the influence of fabric orientation and conditioning on fatigue damage of this composite was studied [10]. Moreover, the fatigue life model proposed by [4] was investigated [11]. This model showed a good ability to predict the fatigue lives of the composite material studied for different fabric orientations and conditionings. Thus, the present study is a straight continuation of these studies and tends to evaluate the capacity of this model to predict the fatigue life of open-hole samples and automotive mini-structures.

2 Material and Methods

2.1 Tested material

The composite material studied is made of three plies of a 2/2 twill woven glass fabric impregnated with polyamide 6,6 resin. The glass fibre fabric has a weight of 600 g/m^2 and a warp to weft ratio of 50/ 50. The fibre mass fraction (m_f) is equal to 0.63 and the void content is below 1%. The resulting composite plates are characterised by a density of 1.78 g/cm^3. The material is provided as plates of 1.53 mm thick and coupons are cut using water jet cutting technique. It has been checked

* Corresponding author: amelie.malpot@renault.com

that this technique has no significant influence on the material moisture content. The specimen edges were polished to remove mechanical damage caused by the cutting.

The glass fibre fabric, referred as [(0/90)$_3$], has the warp direction of each ply oriented at 0° from the tensile axis (x axis). Polyamide 6,6 being known to be very sensitive to moisture, all samples used in this study were conditioned at RH50.

Mechanical properties of the material are detailed in Table 1.

Table 1. Mechanical properties of [(0/90)$_3$] GFRPA66 conditionned at RH50

σ_u [MPa]	ε^R [%]	E [GPa]	ν
343.5 ± 1.0	2.05 ± 0.04	17.5 ± 0.5	0.07 ± 0.01

2.1.1 Open-hole samples

Rectangular [(0/90)$_3$] samples were drilled in the center of the coupon in order to create open-hole samples. The general dimensions of the coupon were 200 x 20 x 1.57 mm. Three diameters were used for the hole, namely 6, 7 and 8 mm.

2.1.2 Mini-structures

Mini-structures studied are shaped by thermoforming of a [(0/90)$_3$] woven glass-fibre-reinforced polyamide 6,6 plate. The general dimensions of the mini-structure are 730 mm length and 140 mm width. The central part has an omega section (Figure 1).

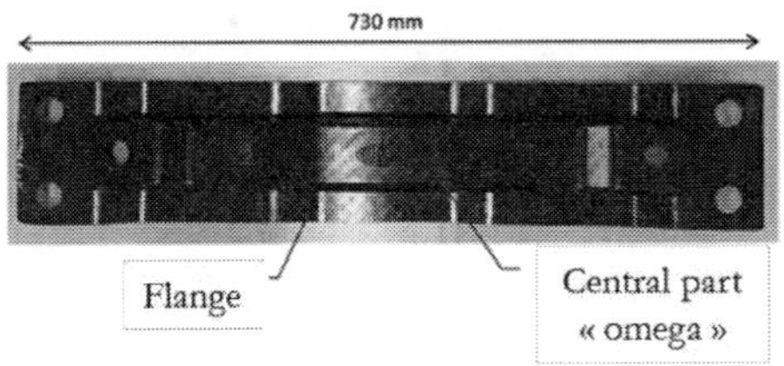

Fig. 1. GFRPA66 mini-structure

2.2 Mechanical testing

Tension-tension fatigue tests on open-hole samples were performed by using an INSTRON 8501 servo-hydraulic machine. The jaws of test machine clamp 40 mm of each specimen extremity and 80 grit sand papers were used in the jaws to improve clamping. Constant amplitude loads were applied in a sinusoidal waveform at the frequency of 1 Hz in order to limit self-generated heating of the specimen. The stress ratio (R), i.e. ratio between minimum (σ_{min}) and maximum (σ_{max}) stresses, was equal to 0.1 for all tests.

3-point bending fatigue tests on mini-structures were performed by using a servo-hydraulic machine, equipped with a 100 kN load cell. Stress ratio is set equal to 0.1 and constant amplitude loads were applied in a sinusoidal waveform at the frequency of 2.5 Hz. The space between the two lower support spans is 480 mm and the span diameter is 30 mm. The load is applied in the middle of the mini-structure using also a 30 mm diameter span.

2.3 Non-destructive techniques and observations

2.3.1 Acoustic Emission (AE)

Acoustic emission monitoring was performed on open-hole samples by using the AE system from Mistras Group. Two sensors Micro-80 with a resonant frequency of 300 kHz and an active surface diameter of 10mm were used. They were placed at the gauge extremities of the specimen using silicon grease as the coupling agent. Sensors are separated with a distance of 100 mm between their centers. The amplitude threshold has been chosen equal to 35 dB. Table 2 shows the settings of the AE system used.

Table 2. Acoustic Emission settings

Preamplifier gain	40 dB
Threshold of detection	35 dB
Type of sensors	Micro-80
Couplant	Silicon grease
PDT (Peak Definition Time)	30 μs
HDT (Hit Definition Time)	100 μs
HLT (Hit Lockout Time)	1000 μs
Resonant Frequency	300 kHz
Bandwidth	1 kHz – 3 MHz

Each test was preceded by a data acquisition calibration step. Using a pencil lead break procedure, the acoustic wave speed as well as the attenuation phenomenon were measured. For the latter, the lead breakage operation was repeated several times between the two sensors, at regular intervals (Hsu-Nielsen method). This procedure has shown that the attenuation phenomenon is negligible in the present work.

Post-processing was done using a multi-parametric identification, based on the k-means algorithm. The k-means algorithm aims to partition observations into k clusters by minimizing the Euclidian distance between each observation and the nearest center ($C_k = c_1, c_2, \ldots, c_k$). This algorithm is unsupervised, which means that the number of clusters k has to be known a priori.

Observations are assimilated to a n-dimensional vector ($X = x_1, x_2, \ldots, x_n$). The k-means algorithm procedure can be detailed as follow:

1. Random initialization of the cluster center for all k-classes ($C_k = c_1, c_2, \ldots, c_k$).
2. Euclidian distance calculation between each observation and cluster centers.

3. Assignment of each observation to the cluster which minimize the Euclidian distance between the observation and the cluster center.
4. Calculation of the new cluster centers for the new k-classes created.
5. Go to step 2 while there is change in the coordinates of the cluster centers.

In this study, observations are acoustic events and n is chosen equal to five among all AE descriptors: amplitude, duration, rise time, energy and number of counts. Each cluster is then associated to one damage mechanism. Studies dealing with woven composite damage process have highlighted three major damage types: matrix cracking, interface damage and fibre breakage [12, 13, 14, 15]. Thus, it was chosen to create three clusters. Attribution of each cluster to one particular damage mechanism was done by using previous results for clustering based on the amplitude only. Several authors [16, 17, 18, 19] have shown that acoustic events with lower amplitudes are associated to matrix cracking whereas those with higher amplitudes are associated to fibre breakage. The intermediate range corresponds to interface damages. Based on these results, each cluster was associated to one damage mechanism depending on its amplitude center value.

2.3.2 Infrared Thermography

An infrared camera from Cedip Infrared Systems with a detector resolution of 90 mm/pixel was used. The energy radiated by the specimen can be converted into temperature levels assuming that the specimen emissivity is known. In this study, this parameter could not be determined experimentally. Thus, instead of the absolute temperature, the temperature variation at the surface of the specimen has been considered.

2.4 Fatigue life model

The fatigue life model used in this study is proposed by Epaarachchi and Clausen [4]. This model allows the prediction of the fatigue life using a very limited amount of experimental data. The model is based on the hypothesis that the material strength undergoes a continuous decay, following a power law as proposed by Caprino and D'Amore [3] (Eq. (1))

$$\frac{d\sigma_N}{dN} = -a \cdot N^{-b} \quad (1)$$

where σ_N is the residual strength after N cycles, b is a positive definite constant, dependent on the material and the mode of loading, a is assumed to increase linearly with the stress amplitude. Finally, the model is presented in Eq. (2) [4].

$$N_f = \left(1 + \left(\frac{\sigma_u}{\sigma_{max}}\right) \frac{f^{\beta}}{\alpha(1-R)^{\lambda - R|sin\theta|}} \left(\frac{\sigma_u}{\sigma_{max}}\right)^{\lambda - 1 - R|sin\theta|}\right)^{1/\beta} \quad (2)$$

where N_f is the fatigue life, σ_{max} is the maximum fatigue stress, σ_u is the ultimate strength, f is the frequency, R is the stress ratio and θ is the smallest angle between the loading axis and the fibres. The parameter λ is assumed to be equal to 1.6 according to Epaarachchi et al. [4].

Hence, α and β are the only two material parameters (dependent on the mode of loading) that need to be determined using experimental data. Only one S-N curve for a given stress ratio, frequency and lay-up is necessary to determine these parameters. In this study, α and β were determined by using the Wöhler curve obtained at RH50 on $[(0/90)_3]$ plain coupons.

3 Results and Discussion

In a previous study [11], the ability of the fatigue model presented in part 2.4 to predict the fatigue life of the GFRPA66 for plain samples with different fibre orientations and conditioning was evaluated. Based exclusively on the Wöhler curve obtained at RH50 on $[(0/90)_3]$ fabric, the model is able to predict correctly the fatigue life of GFRPA66 for any other fabric orientation and conditioning.

For example, Figure 2 shows the fatigue lives determined with this model on $[(\pm45)_3]$ RH100 GFRPA66 plain samples.

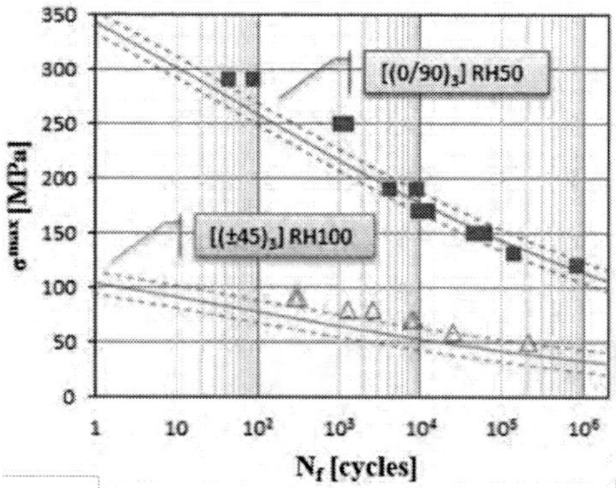

Fig. 2. Experimental (symbols) and estimated (line) fatigue lives of $[(0/90)_3]$ RH50 and $[(\pm45)_3]$ RH100 plain samples (f = 1 Hz, R = 0.1)

This study aims to evaluate the capability of this fatigue life model to predict the fatigue life of open-hole samples and mini-structure.

3.1 Open-hole samples

Monotonic and fatigue tests were conducted on $[(0/90)_3]$ rectangular samples, with a 7 mm diameter hole. Fatigue tests were instrumented with infrared thermography and acoustic emission.

3.1.1 Fatigue life

In first stage, monotonic tensile tests were conducted on open-hole samples at a crosshead speed of 1 mm/min in order to determine the tensile strength. The stress value is calculated far from the hole and is noted σ_∞.

Fatigue tests were performed in order to determine the fatigue lives of open-hole samples. Results are shown on Figure 3.

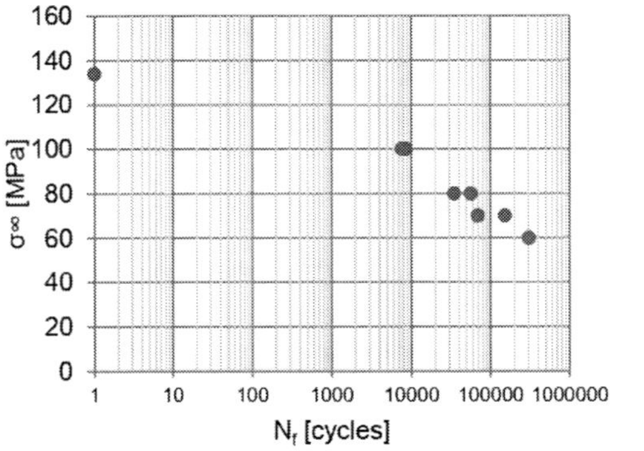

Fig. 3. Experimental fatigue life of [(0/90)$_3$] open-hole samples (f = 1 Hz, R = 0.1)

In order to apply the fatigue life model (equation 2), it is necessary to determine the local ultimate strength (σ_u) and the maximum local fatigue stress (σ_{max}) around the hole. For that purpose, finite elements simulation was used.

3.1.2 Finite Element Modelling

A simplified FE model (Figure 4) was used by considering that one fabric ply, noted (0/90), is equivalent to the stacking of a 0° UD ply and a 90° UD ply. In order to ensure the symmetry of the stacking, the central woven ply is modelled as the stacking of 4 equivalent UD plies. Finally, the 3 plies [(0/90)$_3$] woven composite is modelled with the following equivalent UD layup : [0°/90/0°/90°/90°/0°/90°/0°].

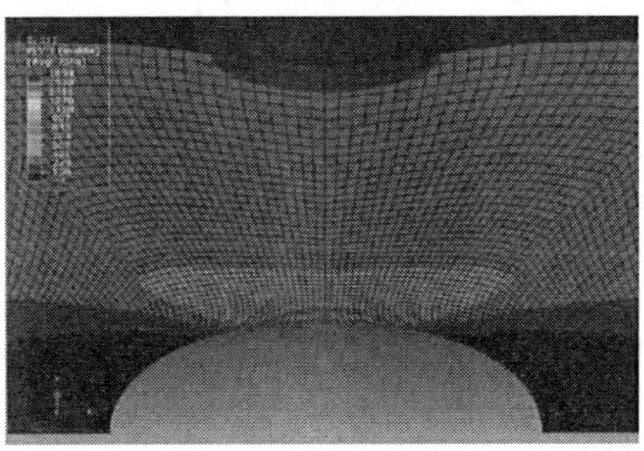

Fig. 4. FE model around the hole (Ø7 mm) of drilled GFRPA66 [(0/90)$_3$] RH50 specimen.

The elastic coefficients of the equivalent UD plies were determined in order to ensure that the stacking of 0° and 90° plies is equivalent to the woven composite (Table 3).

Table 3. Elastic coefficient of the equivalent UD ply.

E_1 [MPa]	E_2 [MPa]	G_{12} [GPa]	ν_{12}
32800	2200	690	0.33

In order to determine the ultimate local strength around the hole, point stress criterion (PSC) were used. This semi-empirical method states that the material failure occurs when the local stress at a characteristic length (d_0 = 0.42 mm) along the ligament reach the material strength. This characteristic length was determined on 7 mm diameter open-hole samples and then check on 6 mm and 8 mm diameter samples (Table 4).

Table 4. Ultimate local strength for plain material and open-hole samples calculated with PSC method.

	σ_u [MPa]
Plain	344
Ø6	341
Ø7	338
Ø8	339

3.1.3 Fatigue Life Estimation

The model parameters, α and β, are taken equal to those obtained on plain samples. Fatigue life estimation is shown on Figure 5. Results show that the model tends to underestimate the fatigue life of open-hole samples. It is worth noting that the higher the fatigue stress level, the higher the model underestimation. This observation is consistent with the assumption made by Epaarachchi & Clausen [4], i.e. the ultimate strength has to be obtained at the same strain rate as the specimen subjected to fatigue loading. So far, this hypothesis was not taken into account since monotonic tensile tests were performed at 1 mm/min.

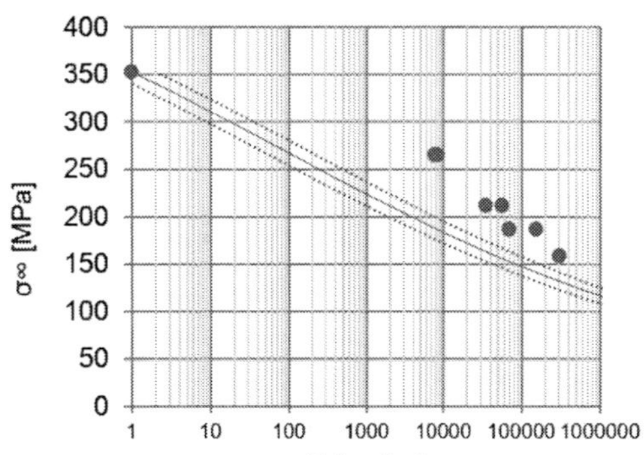

Fig. 5. Fatigue life estimation of open-hole samples (f = 1 Hz, R = 0.1)

3.1.4 Damage evolution

Fatigue tests on open-hole samples were instrumented with infrared thermography and acoustic emission. Figure 6 shows the surface temperature variation observed at the end of a fatigue test.

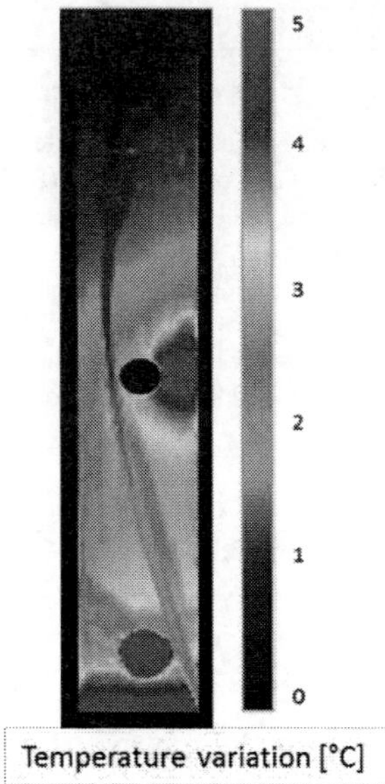

Fig. 6. Temperature variation at the end of a fatigue test.

Figure 6 shows a localized heating of the coupon on the right side of the hole indicating a damage initiation on the edge of the hole.

This observation was confirmed by acoustic emission results (Figure 7). Results show an accumulation of events related to interface and fibre damage around the hole (position = 100 mm). Meanwhile, matrix damage is recorded in the entire coupon.

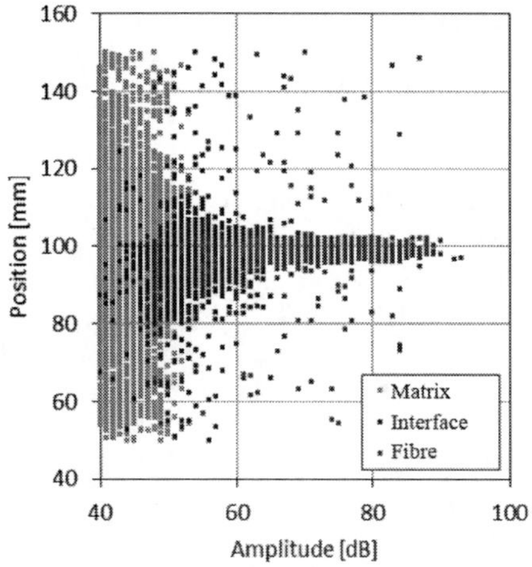

Fig. 7. Acoustic events location recorded during a fatigue test conducted on open-hole sample.

3.2 Automotive mini-structures

3.2.1 Monotonic 3-point bending

GFRPA66 mini-structures were studied in 3-point bending. First, monotonic tests were conducted at a speed of 3 mm/min. Force vs. span displacement curve is shown on Figure 8.

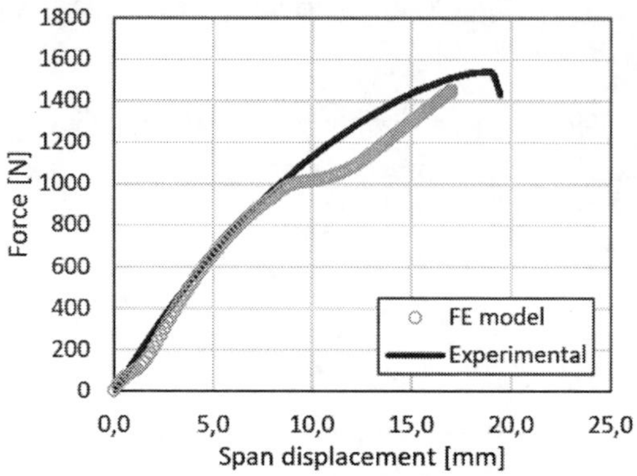

Fig. 8. Numerical and experimental force vs. span displacement curves.

A FE model was developed in order to simulate the behavior of the mini-structure (Figure 9). The woven GFRPA66 material was modelled by the equivalent UD layup [0°/90/0°/90°/90°/0°/90°/0°].

Figure 8 shows the force vs. span displacement curve obtained with FE model. The numerical model shows a good fitting to experimental data until 1000 N. Beyond, some geometrical effects appear in the FE modelling leading to a deviation from the experimental data.

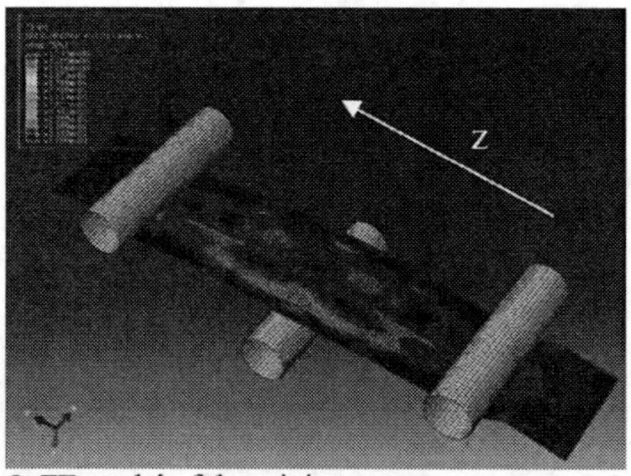

Fig. 9. FE model of the mini-structure.

3.2.2 3-point bending fatigue

Fatigue tests were performed on automotive mini-structures at three force levels. The experimental fatigue lives are shown on Figure 10.

A preliminary study was done by applying the fatigue life model with force instead of stress. For the force-based version of the model, the tensile strength (σ_u) and the maximum fatigue stress (σ_{max}) were respectively replaced by the force at failure and the maximum fatigue force. Moreover, the model parameters α and β being dependent of the loading mode, 3-point bending fatigue tests were conducted on rectangular coupons in order to determine a new set of parameter. The estimated fatigue life of the composite mini-structure are shown on Figure 10 together with experimental data.

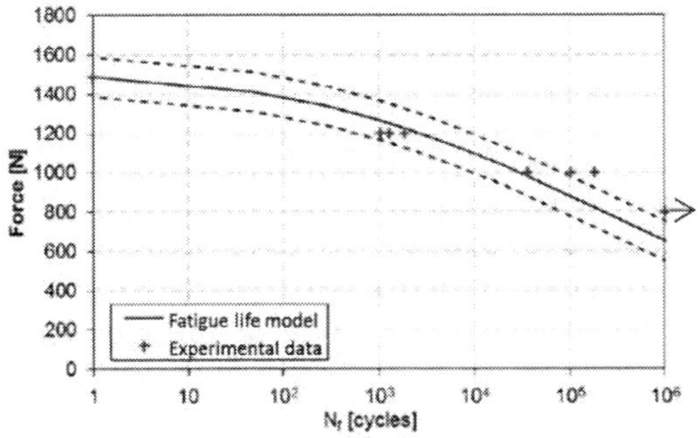

Fig. 10. Experimental and estimated fatigue life of the GFRPA66 mini-structure with the force-based model.

This preliminary version of the model gives satisfactory results in terms of fatigue life estimation. However, it can be improved in order to be applied in an industrial context.

In order to use the semi-empirical model as it is given in equation 2, a representative local stress has to be chosen. In a first stage, the maximum local stress in the z direction (as given in Figure 9) has been chosen. The FE model was used in order to determine this local stress value (noted σ_{zz}) for the different maximum force levels used in fatigue tests and for the force at failure.

Then, the model presented in equation 2 was applied with the set of parameter determined for 3-point bending. Results are shown on Figure 11. The model gives a very good fitting to experimental data, proving its ability to be used in order to evaluate the fatigue life of a composite mini-structure.

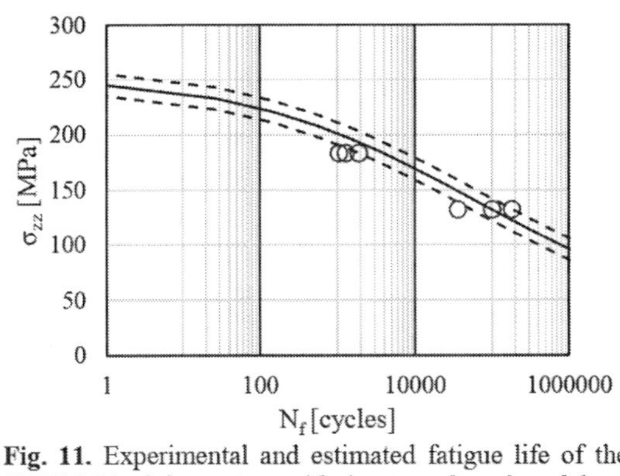

Fig. 11. Experimental and estimated fatigue life of the GFRPA66 mini-structure with the stress-based model.

Additional fatigue tests on mini-structures with a different fabric orientation, $[(\pm45)_3]$ for instance, could be used in order to validate the semi-empirical model for a different fibre orientation. This could also validate the initial choice of using the local stress in the z direction.

4 Conclusions

This study deals with the fatigue behavior of a woven glass-fibre-reinforced composite with PA66 matrix. Tension-tension fatigue tests were performed on open-hole samples. These tests were instrumented with acoustic emission and infrared thermography. Monitored data have highlighted a damage accumulation around the hole, identified as fibre-matrix debonding and fibre breakages. A fatigue life model was applied on open-hole sample, with material parameters determined on plain samples. Results show a slight underestimation of the fatigue life.

In addition, 3-point bending fatigue tests were performed on automotive mini-structures. The fatigue life model tested on open-hole samples were also applied on mini-structure. As a reminder, the inputs needed for the use of this model are:

- A Wöhler curve obtained on rectangular samples tested in 3-point bending ;
- A numerical model of the mini-structure ;
- A monotonic 3-point bending test for the validation of the numerical model ;

With these data, it is then possible to estimate the fatigue life of the mini-structure in 3-point bending mode. Estimated fatigue lives of the mini-structure were very close to those obtained experimentally, indicating a good capacity of this model to predict the fatigue life of GFRPA66 mini-structure.

References

1. J. Degrieck, W. Van Paepegem. Appl. Mech. Rev., **54**(4), 279-300 (2001)
2. R. Sarfaraz, A.P. Vassilopoulos, T. Keller. Compos. Part A Appl. Sci. Manuf. **43**(3), 445-453 (2012)
3. G. Caprino, A. D'Amore. Compos. Sci Technol, 58(97), 957-965 (1998)
4. J.A. Epaarachchi, P.D. Clausen, Compos. Part A Appl. Sci. Manuf. **34**, 313–326 (2003)
5. S. Mortazavian, A. Fatemi, Int. J. Fatigue, **77**, 12–27 (2015)
6. C. Colombo, F. Libonati, A. Salerno, L. Vergani. Procedia Eng. **10**, 3518–3527 (2011)
7. M.R. Satapathy, B.G. Vinayak, K. Jayaprakash, N.K. Naik, Mater. Des. **51**, 347–356 (2013)
8. R. Ben Toumi, J. Renard,M. Monin, P. Nimdum. Procedia Eng. **66,** 723–736 (2013)
9. D.S. de Vasconcellos, F. Touchard, L. Chocinski-Arnault. Int. J. Fatigue **59,** 159–169 (2014)
10. A. Malpot, F. Touchard, S. Bergamo. Compos. Part B **130,** 11-20 (2017)
11. A. Malpot, F. Touchard, S. Bergamo. Mater. Des. **98**, 12-19 (2016)
12. T. Fujii , S. Amijima, K. Okubo. Compos Sci Technol **49**, 327-333 (1993)
13. S.D. Pandita, G. Huysmans, M. Wevers, I. Verpoest. Compos Part A Appl Sci Manuf **32**, 1533 – 1539 (2001)
14. S. Daggumati, I. De Baere, W. Van Paepegem, J. Degrieck,J. Xu, S.V. Lomov, et al. Compos Sci Technol, **70,** 1926 - 1933 (2010)

15. T. Osada, M. Mizoguchi, M. Kotaki. ICCM13, Beijing (China) (2001)
16. S. Barré, M.L. Benzeggagh. Compos Sci Technol, **3538**, 369 – 376 (1994)
17. D. Scida, Z. Aboura, M.L. Benzeggagh. Compos Sci Technol, **62**, 551 – 557, (2002)
18. D. Silva de Vasconcellos. Thèse, ENSMA, 2013.
19. G. Kotsikos, J.T. Evans, A.G. Gibson, J. Hale. Polym Compos, **20**, 689 - 696 (1999)

Rapid evaluation of S-N curve for composite laminates on thermographic approach

Jia Huang[1], *Christian* Garnier[2,*], *Marie-Laetitia* Pastor[1], and *Xiaojing* Gong[1]

[1]Univ. Toulouse, IUT Tarbes, ICA, 1 rue Lautréamont, F-65016 Tarbes, France
[2]Univ. Toulouse, INP-ENIT, LGP, 47 avenue d'Azereix, F-65016 Tarbes, France

Abstract. A new fatigue life prediction model is developed for composite laminates based on both infrared thermographic data and stiffness degradation process. The fatigue limit is firstly determined by the improved thermographic method and then a two-parameter function is proposed to describe the normalized stiffness degradation of specimens subjected to various maximum loading stresses. These two parameters can be calibrated by the stiffness degradation data and stabilized temperature data. Thereafter, the calibrated model can be used to predict S-N curve. The proposed model is applied the experimental data from literature and the results show that the predicted S-N curve has a well agreement with the experimental one.

1 Introduction

Fatigue property of composite laminates receives more and more concerns since they are usually used in aerospace, automotive and marine load-bearing components. Investigating the fatigue performance f composite laminates by traditional methods is a costly and time-consuming process. Moreover, due to the rich diversity (different orientations, fiber ratio, nature of fibers and matrix, etc.) in composite laminates, rapid evaluation of their fatigue property is of great significance for choosing proper structure of laminates in component designing.

It is a common phenomenon that the plastic deformation of the material is accompanied by the generation of a large amount of heat. This inspires researchers to measure temperature change as an effective approach of studying fatigue behavior. With appearance of high-speed, high-resolution infrared thermography (IRT), the slight change of temperature during fatigue, which is due to the thermal-elastic–plastic effect, can be detected. Based on IRT data analysis, many criteria have been developed to rapidly determine fatigue limit and obtain S-N curves for metallic alloys [1-7]. However, those criteria may not be accurate any more for composite laminates because damage and failure mechanisms are different and more complicated (compared to metallic materials).

Unlike metals, the damage mechanism in composite laminates under cyclic loading usually includes matrix cracking, delamination and fiber breakage etc. In general, there are three distinctive damage stages during the damage accumulation process in composite laminates under fatigue loading [8]: matrix cracking and interface debonding (I), delamination (II), delamination growth and localized fiber breakage (III). Since the stage III corresponding to final fracture stage is short-period where unstable fracture occurs, the first two stages are usually used to estimate the residual life of laminates. It is well known that the stiffness degradation in fatigue loaded specimen has also three stages due to material damage [9], thus some of the damage accumulation model was developed based on stiffness degradation in the reference [10].

In the present paper, a new fatigue life prediction model by combining infrared thermographic data and damage accumulation process is developed. By applying the improved thermographic method which is recently proposed, the fatigue limit can be easily and rapidly obtained. This work tries to give an answer to the question: how to quickly determine fatigue resistance in term of S-N curve by using thermographic method? Herein, curve fitting method is used to estimate the value of threshold of stiffness under the load corresponding to fatigue limit. After that, damage evolution curves under different maximum loading stresses can be calculated. Then, a two-parameter damage accumulation model is proposed to characterize the damage evolution curves. One parameter characterizes the influence of temperature rising on damage accumulation speed and the other one is used to regulate amplitude of function. After parameter calibration with the experimental stiffness degradation and IRT data, the model can be used to predict S-N curve.

2 IRT technique foundation

Generally, fatigue behavior can be considered as an irreversible process of mechanical property degradation

* Corresponding author: christian.garnier@enit.fr

under cyclic loading. There are two main approaches in this irreversible process [11]: (i) a chemical-physical process, such as dislocation, chemical bond rupture, creep deformation, etc. (ii) a physical separation of the material, such as cracks, cavitations, etc. Both of those two approaches will cause heat release which is called intrinsic dissipation. By using infrared thermographic camera, the variation of temperature can be measured with high precision. A local heat conduction equation [12-14] can be employed here to explain the mechanism above:

$$\rho C\dot{T} - \mathrm{div}(h\mathrm{grad}T) = d_i + s_{\mathrm{the}} + s_{\mathrm{ic}} + r_e \quad (1)$$

where, ρ, C, T, h are mass density, heat capacity, temperature and heat conduction coefficient, respectively. The first term $\rho C\dot{T}$ on the left side is the heat storage rate characterized by temperature change, and the second left-hand term $-\mathrm{div}(k\mathrm{grad}T)$ characterizes heat loss rate induced by conduction. The term group on the right side represents the different heat sources: d_i denotes the intrinsic dissipation source; s_{the} is the thermoelastic source; sic describes the heat source induced by the coupling effect between internal variables and temperature; and r_e denotes the external heat supply. When components or specimens are under homogeneous uniaxial loading, we can have the following hypothesis:

(1) Parameters ρ, C, h are material constants and independent of the internal states.
(2) Thermoelastic source s_{the} only causes small fluctuations of temperature but does not contribute to mean temperature rising [15].
(3) Internal coupling source between internal variables and temperature, sic, can be neglected [16].
(4) External heat supply re is time-independent and can be removed by using reference specimen placed nearby.

Based on the hypothesis above, the intrinsic energy dissipation d_i can be identified as the main contributor to the total heat generated. Particularly, as can be seen from Eq. (1), if d_i remains unchanged, with the increase of temperature, the heat loss rate, $-\mathrm{div}(h\mathrm{grad}T)$, will also increase. It can be deduced that when temperature T reaches a certain value, there will be a balance between heat loss and intrinsic energy dissipation. In this situation, the heat storage rate $\rho C\dot{T}$ equals to zero, so that T will remain stable. This conclusion has been proved by a number of experiments (with constant stress ratios) in the previous works [14, 17-20]. As shown in Fig. 1, there are three phases from the beginning of test to the final breakage of specimen. The temperature increases from initial value (room temperature) in phase I, then remains stable for the longest period in phase II and finally experiences a sharp increase just prior to the final breakage in phase III. Thus, it can be derived that d_i keeps unchanged in phase II, which indicates that the damage accumulation rate is stable in this phase. If different loading stresses are applied, different stabilized temperature rising ΔT_{stab} will be obtained. Thus, the relationship of ΔT_{stab} and maximum applied loading stress σ can be plotted, as shown in Fig. 1. Based on this, two straight lines are used to characterize the relationship and the intersection can be considered as fatigue limit [11, 21].

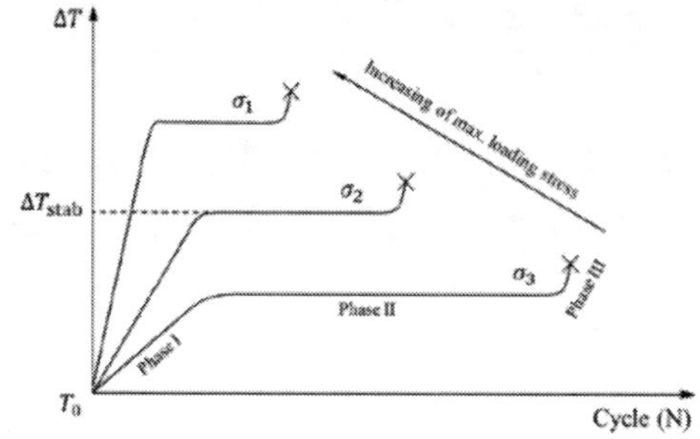

Fig. 1. Typical temperature evolution during fatigue test.

3 Stiffness degradation model

As known from literature, the main stiffness degradation process observed in composites laminates under fatigue loading can also be divided into three main stages [8, 10], as shown in Fig. 2. In stage 1, the normalized stiffness decreases rapidly with micro cracks initiation in the first few cycles and some fibers with low strength may break during this stage. Then the normalized stiffness experiences a slow and steady reduction mainly due to delamination and matrix crack propagation in stage 2. Prior to final breakage, the normalized stiffness drops down dramatically with a mass of fibers failure in stage 3. It is commonly known that stage 3 are short-period and unstable so that the first two stages are usually used to estimate the residual life of laminates. Similar as the temperature evolution curves, the stage 2 with stable damage growth accounts for the most parts of the total cycle numbers. Therefore, we speculate that there is a certain relationship between ΔT_{stab} and the normalized stiffness degradation rate.

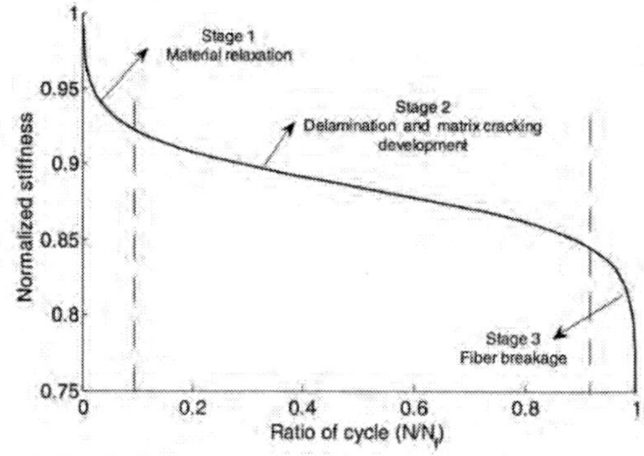

Fig. 2. Typical temperature evolution during fatigue test [9].

In order to propose a practical stiffness degradation model, two reasonable simplifications are made here. Firstly, only stage 1 and stage 2 are taken into account because stage 3 is short-period, unstable and difficult to capture [22, 23]. Secondly, for given composite laminates with same stacking sequence, ply orientation, geometry and fabrication process, if they are subjected to cyclic loading with fixed loading frequency and loading stress ratio, the critical values of normalized failure threshold stiffness k_f can be considered to be independence of the maximum loading stress level. This has been proved by previous experimental work [9]. Thereafter, the normalized stiffness degradation model is proposed as:

$$k(N) = 1 - p\Delta T_{\text{stab}} \ln\left(\frac{N}{q} + 1\right) \quad (2)$$

where p and q are two constant parameters which need to be calibrated by experimental data. It will be introduced in details in Section 4. The term $N^{1/q}$ characterizes the shape of normalized stiffness with the increase cyclic number during stage 1 and stage 2. ΔT_{stab} is used to describe the general degradation speed, which is related to damage accumulation rate. In the proposed model, the normalized stiffness $k(N)$ starts from 1, decreases sharply and then becomes stable when cyclic number gets larger, which is similar as experimental normalized stiffness evolution. For Eq. (2), if cycle number N is replaced by the final failure cycle number N_f, we can obtain:

$$N_f = q\left(e^{\frac{1-k_f}{p\Delta T_{\text{stab}}}} - 1\right) \quad (3)$$

As can be observed from Eq. (3), for different stabilized temperatures we can obtain different failure cycle number N_f. And the different stabilized temperatures is depended on different maximum loading stress, as shown in Fig. 1. Thus, the new proposed model can predict the whole S-N curve.

4 Results and discussion

The experimental data used in the present paper is taken from [18], where both IRT and stiffness degradation data are available, as shown in Fig. 3 and Fig. 4.

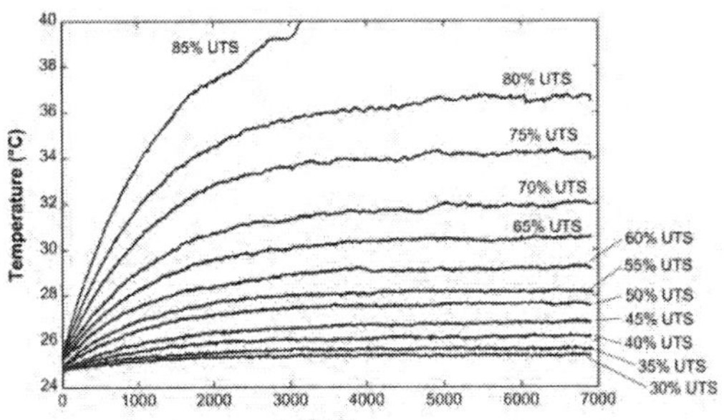

Fig. 3. Experimental IRT data taken from [18]. (UTS: Ultimate Tension Stress)

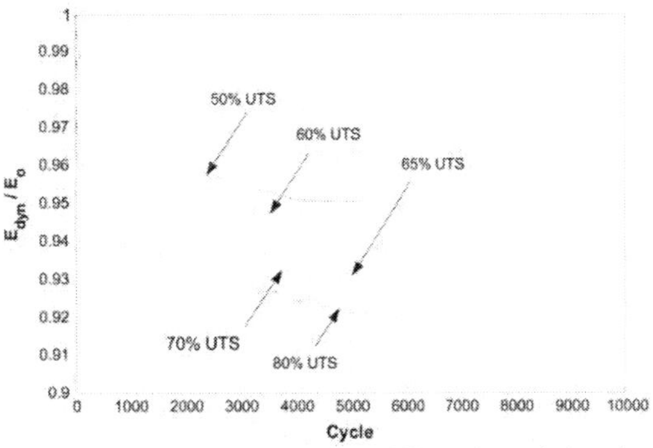

Fig. 4. Experimental normalized stiffness degradation data taken from [18].

The material tested in [18] is carbon fiber reinforced composite plate which is made from triaxially braided carbon fiber (T650/35-6 K) fabric with 0°/±θ braid orientation and thermosetting polyamide resin. The parameter calibration procedure and results are listed as below:

(1) Determine fatigue limit.

The improved thermographic methods in article [15] were firstly applied. There are three different improved graphic methods and we chose Method 2 and Method 3 here because for the IRT data in [18], the values of fatigue limit determined by Method 1 and Method 3 are the same, while Method 3 has a relative larger application scope. The fatigue limit was determined as 58% UTS by Method 2 and 67% UTS by Method 3, as shown in Fig. 5 and Fig. 6, respectively. The stabilized temperature rising corresponding to fatigue limit ($\Delta T_{\text{stab\_fl}}$) is also determined as 4.7 ℃ and 6.3 ℃, which will be used to determine normalized failure threshold stiffness k_f.

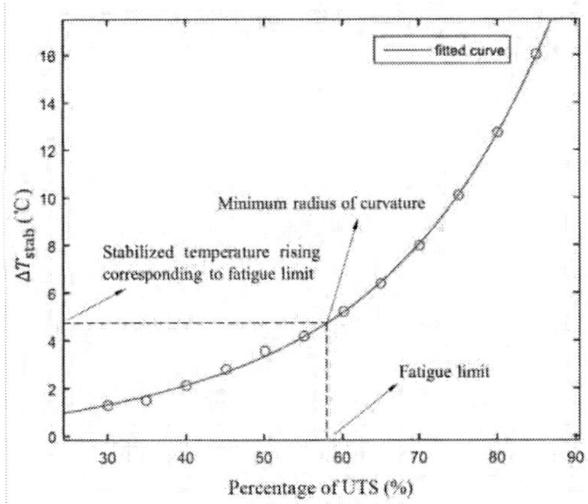

Fig. 5. Fatigue limit determined by Method 2.

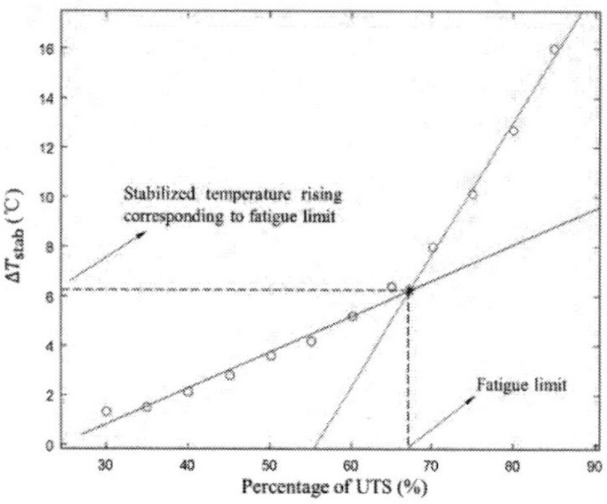

Fig. 6. Fatigue limit determined by Method 3.

(2) Determine the values of parameters p and q.
According to experimental normalized stiffness degradation data in Fig. 4, there are five different stiffness degradation curves. For each curve, 20 points were sampled, so a total of 100 points are used for fitting. Similarly, Eq. (2) is used to fit the data (3D surface fitting in MATLAB Toolbox) and the values of parameters p and q can be calibrated. The fitting results are shown in Fig. 7, and the values of p and q are determined as 0.001042 and 2.035, respectively.

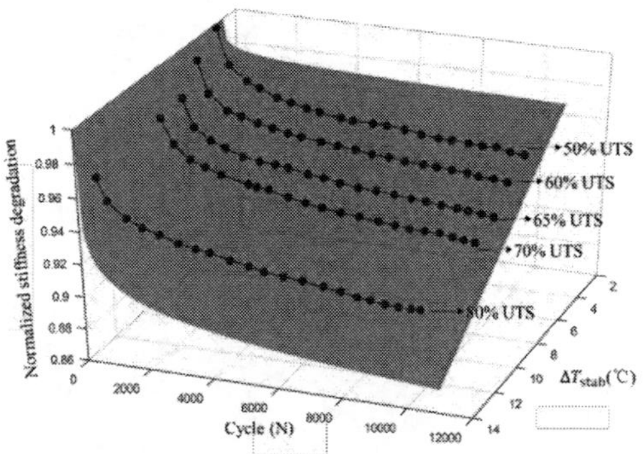

Fig. 7. Surface fitting of Eq. (3) by MATLAB.

(3) Calculate failure threshold stiffness.
Similar as previous case, The failure cycle number corresponding to fatigue limit, N_{fl}, is usually taken between $10^6 - 10^7$. Herein, both the $k(N_{fl} = 10^6)$ and $k(N_{fl} = 10^7)$ are considered to calculate the failure threshold stiffness k_f. According to Eq. (2), we can obtain:

$$k_f = 1 - p\Delta T_{stab\_fl}\ln\left(\frac{N_{fl}}{q}+1\right) \quad (4)$$

Thus, k_f corresponding to different values of N_{fl} and ΔT_{stab_fl} can be calculated.

(4) Calculate $S - N_f$ curve.
After knowing the values of p, q and k_f, we can use Eq. (3) to calculate $S - N_f$ curve. So the relationship between ΔT_{stab} and maximum applied stress (σ_a) should be established. For Method 2, the expression is given here:

$$\Delta T_{stab} = a \cdot \exp\left(b\sigma - \frac{1}{b\sigma}\right) + c\sigma \quad (5)$$

where the first term on right side is used to describe temperature response under high loading stress amplitude, and the second term on the right side is used to represent temperature rising under low loading stress amplitude. a, b, c are three parameters of interpolation to be determined. a is used to regulate the amplitude of exponential function, b is used to adjust σ_a axis range, and c is used to describe the amplitude of temperature rising in low stress cases. By combining Eq. (3) and Eq. (5), we have:

$$N_f = qe^{\frac{1-k_f}{pa\cdot\exp\left(b\sigma-\frac{1}{b\sigma}\right)+c\sigma}} - q \quad (6)$$

By applying Eq. (6), the S-N curve can be predicted, as shown in Fig. 8.

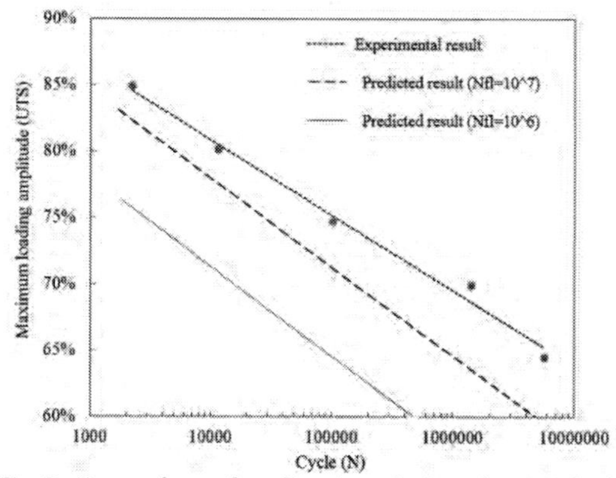

Fig. 8. Comparison of predicted results based on Method 2 and traditional test result.

As can been seen in Fig. 8, for triaxially braided CFRP laminates, the predicted $S - N_f$ curves corresponding to $N_{fl} = 10^7$ and $N_{fl} = 10^6$ are both lower than the experimental curve. The results indicate that although the S-N curve predicted by model relative to $N_{fl} = 10^6$ is very conservative and safe, the error between predicted result and experimental result is too large. So for model based on Method 2, only $N_{fl} = 10^7$ is recommended for engineering applications.

For Method 3, the relationship can be considered as two two-straight-line section function and the expression is given here:

$$\Delta T_{stab} = \begin{cases} a_1\sigma + b_1\ (-b_1/a_1 \le \sigma < \sigma_{fl}) \\ a_2\sigma + b_2\ (\sigma \ge \sigma_{fl}) \end{cases} \quad (7)$$

where a_1, b_1, a_2 and b_2 are four coefficients depending on those two straight lines. σ_{fl} is the value of maximum applied loading stress corresponding to fatigue limit. For engineering applications, we usually just need to know the fatigue life when the maximum loading stress is higher than fatigue limit, so we have

$$N_f = qe^{\frac{1-k_f}{p(a_2\sigma+b_2)}} - q\ (\sigma \ge \sigma_{fl}) \quad (8)$$

By applying Eq. (8), the S-N curve can also be calculated, as shown in Fig. 9.

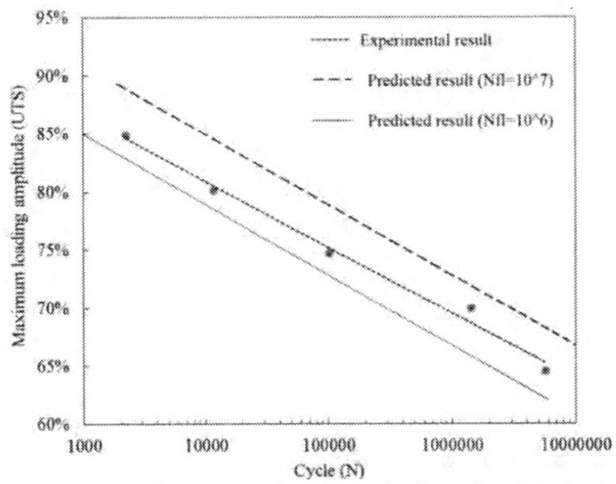

Fig. 9. Comparison of predicted results based on Method 3 and traditional test result.

As can be observed in Fig. 9, the predicted $S - N_f$ curve corresponding to $N_{fl} = 10^7$ is overall higher than the experimental curve and the error between predict value and experimental is relatively larger for low fatigue life (less than 10000 cycles). On the contrary, the predicted $S - N_f$ curve corresponding to $N_{fl} = 10^6$ is lower than the traditional test result. Therefore, due to safety reasons, for model based on Method 3, $N_{fl} = 10^6$ is recommended here for engineering applications because the predicted result is more conservative.

5 Conclusions

The proposed model combined temperature rising with the stiffness degradation process. It was speculated that the stabilized temperature could reflect the rate of damage accumulation. To the authors' knowledge it is the first time both IRT and stiffness degradation data are taken into consideration to predicted fatigue behaviour of composites. A two–parameter model was developed to characterize stiffness degradation process. These two parameters were then calibrated by 3-D surface fitting method. Experimental validation shows that the new proposed model can predicted curves with good accuracy. The whole determination process takes about 10-12 hours, which is much faster than traditional fatigue test method. To sum up, this paper provides a quick and simple way to evaluate fatigue limit and S-N of composite laminates for engineering applications.

The author Jia Huang would like to acknowledge China Scholarship Council for the financial support of 3 years study at the University of Toulouse.

References

1. Huang, Y., S. Li, S. Lin, and C. Shih, Using the method of infrared sensing for monitoring fatigue process of metals. Mater. Eval. (United States), **42**(8) (1984)
2. Jiang, L., H. Wang, P. Liaw, C. Brooks, and D. Klarstrom, Characterization of the temperature evolution during high-cycle fatigue of the ULTIMET superalloy: experiment and theoretical modeling. Metallurgical and Materials Transactions A, **32**(9): 2279-2296 (2001)
3. Jiang, L., H. Wang, P. Liaw, C. Brooks, L. Chen, and D. Klarstrom, Temperature evolution and life prediction in fatigue of superalloys. Metallurgical and Materials Transactions A, **35**(3): 839-848 (2004)
4. Yang, B., P. Liaw, G. Wang, M. Morrison, C. Liu, R. Buchanan, and Y. Yokoyama, In-situ thermographic observation of mechanical damage in bulk-metallic glasses during fatigue and tensile experiments. Intermetallics, **12**(10): 1265-1274 (2004)
5. Curà, F., G. Curti, and R. Sesana, A new iteration method for the thermographic determination of fatigue limit in steels. International Journal of Fatigue, **27**(4): 453-459 (2005)
6. Amiri, M. and M.M. Khonsari, Life prediction of metals undergoing fatigue load based on temperature evolution. Materials Science & Engineering A, **527**(6): 1555–1559 (2010)
7. Amiri, M. and M.M. Khonsari, Rapid determination of fatigue failure based on temperature evolution: Fully reversed bending load. International Journal of Fatigue, **32**(2): 382-389 (2010)
8. Vasiukov, D., S. Panier, and A. Hachemi, Direct method for life prediction of fibre reinforced polymer composites based on kinematic of damage potential. International Journal of Fatigue, **70**: 289-296 (2015)
9. Peng, T., Y. Liu, A. Saxena, and K. Goebel, In-situ fatigue life prognosis for composite laminates based on stiffness degradation. Composite Structures, **132**: 155-165 (2015)
10. Shiri, S., M. Yazdani, and M. Pourgol-Mohammad, A fatigue damage accumulation model based on stiffness degradation of composite materials. Materials & Design, **88**: 1290-1295 (2015)
11. Luong, M.P., Fatigue limit evaluation of metals using an infrared thermographic technique. Mechanics of materials, **28**(1): 155-163 (1998)
12. Boulanger, T., A. Chrysochoos, C. Mabru, and A. Galtier, Calorimetric analysis of dissipative and thermoelastic effects associated with the fatigue behavior of steels. International Journal of Fatigue, **26**(3): 221-229 (2004)
13. Chrysochoos, A., V. Dattoma, and B. Wattrisse, Deformation and dissipated energies for high cycle fatigue of 2024-T3 aluminium alloy. Theoretical & Applied Fracture Mechanics, **52**(2): 117–121 (2009)
14. Guo, Q., X. Guo, J. Fan, R. Syed, and C. Wu, An energy method for rapid evaluation of high-cycle fatigue parameters based on intrinsic dissipation. International Journal of Fatigue, **80**: 136-144 (2015)
15. Huang, J., M.-L. Pastor, C. Garnier, and X. Gong, Rapid evaluation of fatigue limit on thermographic data analysis. International Journal of Fatigue, **104**(Supplement C): 293-301 (2017)
16. Guo, Q. and X. Guo, Research on high-cycle fatigue behavior of FV520B stainless steel based on

intrinsic dissipation. Materials & Design, **90**(Supplement C): 248-255 (2016)
17. La Rosa, G. and A. Risitano, Thermographic methodology for rapid determination of the fatigue limit of materials and mechanical components. International journal of fatigue, **22**(1): 65-73 (2000)
18. Montesano, J., Z. Fawaz, and H. Bougherara, Use of infrared thermography to investigate the fatigue behavior of a carbon fiber reinforced polymer composite. Composite structures, **97**: 76-83 (2013)
19. Wang, X.G., V. Crupi, X.L. Guo, and Y.G. Zhao, Quantitative thermographic methodology for fatigue assessment and stress measurement. International Journal of Fatigue, **32**(12): 1970-1976 (2010)
20. Fargione, G., A. Geraci, G. La Rosa, and A. Risitano, Rapid determination of the fatigue curve by the thermographic method. International Journal of Fatigue, **24**(1): 11-19 (2002)
21. Luong, M.P., Infrared thermographic scanning of fatigue in metals. Nuclear Engineering and Design, **158**(2): 363-376 (1995)
22. Tao, C., H. Ji, J. Qiu, C. Zhang, Z. Wang, and W. Yao, Characterization of fatigue damages in composite laminates using Lamb wave velocity and prediction of residual life. Composite Structures, **166**: 219-228 (2017)
23. Lee, L.J., K.E. Fu, and J.N. Yang, Prediction of fatigue damage and life for composite laminates under service loading spectra. Composites Science and Technology, **56**(6): 635-648 (1996)

A two-parameter analysis of surface fatigue crack growth behavior of fine-grained WC-Co cemented carbide at different stress ratios

Hiroko Mikado[1,*], *Sotomi* Ishihara[2,3], *Noriyasu* Oguma[3] and *Shingo* Kawamura[1]

[1] Department of Machinery and Engineering Group, YKK Corporation, Kurobe, Toyama, 938-8601, Japan
[2] National Institute of Technology, Toyama College, Hongo, Toyama, 939-8630, Japan
[3] Department of Mechanical Engineering, University of Toyama, Gofuku, Toyama, 930-8555, Japan

Abstract. In this study, the fatigue crack growth (FCG) characteristics of the short surface crack was studied at various stress ratios, *R*, using the fine grained cemented carbide. The FCG law for the material was studied using two parameters, the maximum stress intensity factor K_{max} and the stress intensity factor range ΔK. It was found that the relationship between the rate of FCG, d*a*/d*N* and ΔK depends on the *R* value. In the d*a*/d*N* vs. K_{max} relation, at the region of high crack growth rate, the FCG rate was nearly constant by the value of K_{max} without depending on *R*. However, in the low FCG rate region, the FCG rate tended to accelerate as *R* became lower. Therefore, it was difficult to specify d*a*/d*N* as a single value at different *R* ratios using either ΔK or K_{max} from the low speed region to the high-speed region. For that reason, the FCG behaviour of the material was studied based on the crack growth law, $da/dN = AK_{max}{}^{m}\Delta K^{n}$. The values of *m* and *n* were obtained as 6.6 and 1.4, respectively. Therefore, it was found that the FCG rate of the cemented carbide was strongly dependent on K_{max} rather than ΔK, like the FCG characteristics of ceramics.

1 Introduction

Cemented carbide is hard material and has excellent wear resistance, so it has been widely used in cutting tools and so on. In recent years, cemented carbide has been increasingly used for the mold material. In the forging process, compressive stress and shear stress are repeatedly applied to the mold material at the boundary between the work and the mold.

In the above situation, short surface fatigue cracks initiate and grow on the mold, causing chipping and fracture of the mold. It is expected that the initiation and growth behavior of the short surface crack are concerned with wear behavior of the mold [1]. Currently, how to extend the mold life has been a big issue for manufacturing companies.

Researches on the fatigue crack growth (FCG) behavior of the cemented carbide have been mainly conducted targeting the long through crack.

Moreover, these studies were mainly conducted in the high K_{max} area where the FCG speed is faster than 10^{-9} m/cycle or more. For example, Fry et al. [2] studied the FCG behavior of the long through crack, using various cemented carbides. They reported the effect of the WC grain size and Co content on the FCG characteristics of these cemented carbides. In addition, Llanes et al. [3, 4] reported the FCG characteristics of the long through crack of the various cemented carbides. Both of the above studies focused on the high-speed FCG zone, larger than 10^{-9} m/cycle.

On the other hand, there have been few researches on the FCG behavior for the short surface crack, due to difficulty in doing experiments and observations. Ishihara et al. [5] reported the FCG characteristics of the short surface crack at various stress ratios using the cemented carbide with comparatively large WC grain size. Mikado et al. [6] studied the FCG behavior of the short surface crack at stress ratio, $R = -1$ using the fine-grained cemented carbide.

In this study, the FCG behavior of the short surface crack was investigated at various stress ratios using the fine grain cemented carbide. In particular, the FCG behavior was studied in the low FCG zone below 10^{-9} m/cycle, which is important for practical use and has not been clarified so far.

The FCG data obtained were evaluated using two parameters, the maximum stress intensity factor K_{max} and the stress intensity factor range ΔK.

In the two parameters analysis of FCG, the FCG behavior of the short surface were compared with those of the long penetrating crack. Then, the characteristics of FCG behavior of the short surface crack were clarified. Consideration was also given how the prediction accuracy in the crack growth amount was changed depending on whether FCG threshold is taken into consideration or not.

* Corresponding author: h-mikado@ykk.co.jp

2 Two-parameter analysis

FCG behavior of the short surface crack for the fine grain cemented carbide were analysed using two parameters of K_{max} and ΔK [7]. In the FCG law using two parameters, the following equation [7] will be used. Here, m and n are constants determined from experiments. The decision method for the values of m and n will be described later.

$$da/dN = CK_{max}{}^{m}\Delta K^{n} \quad (1)$$

The following equation holds between K_{max} and ΔK, where R is the stress ratio, K_{min} and K_{max} are the minimum stress intensity factor and the maximum stress intensity factor, respectively.

$$K_{max} = \Delta K / (1 - R) \quad (2)$$

$$R = K_{min} / K_{max}$$

Substituting equation (2) into equation (1) and transform it, we obtain the following relation.

$$da/dN / (1 - R)^{n} = CK_{max}{}^{m+n} \quad (3)$$

The constants m and n in equation (3) were determined as follows.

First, FCG data obtained from experiments at various stress ratios R was plotted on the log-log graph. In that case, the vertical axis was $da/dN/(1-R)^{n}$ and the horizontal axis was taken as K_{max}. The first n value used was arbitrary assumed. After that, the n value was varied with a step width of 0.1. Multiple log $(da/dN/(1-R)^{n})$ - log K_{max} relationships were created for different n values.

Thereafter, on the logarithmic graph, the above relation was approximated by a straight line having a slope ($m + n$). The approximation degree of these lines to the experimental data was judged from the value of correlation coefficient.

The relation of log $(da/dN/(1 - R)^{n})$ - log K_{max} where the correlation coefficient becomes maximum was determined, and m and n values at that time were obtained.

3 Material and experimental procedures

3.1 Material

In the present study, a commercially available fine-grained WC-Co cemented carbide was used as the testing material. Tables 1 and 2 show the chemical composition and mechanical properties of the material, respectively. The material contains 13.0 wt% Co and its bending strength is 4100 MPa. Figure 1 shows the microstructure of the material, which is a composite microstructure consisting of both WC and Co phase. A detailed microstructural analysis of Fig. 1 revealed that the material has a log-normal distributed WC grains with a modal value of about 0.35 μm, and has an exponential distributed binder mean free paths, ranging from 0 to 1 μm.

Table 1. Chemical composition of the WC-Co cemented carbide [wt%].

Co	Cr	W and C
13.0	0.51	Bal.

Table 2. Mechanical properties of the WC-Co cemented carbide.

Young's modulus [GPa]	Fracture toughness, K_{IC} [MPam$^{1/2}$]	Bending strength [MPa]	Vickers hardness [HV]
550	12.1	4100	1480

Fig. 1. Microstructure of the WC-Co cemented carbide used in the study.

3.2 Specimens

The specimen used in the present study was a rectangular shaped specimen having a width of 8 mm, a height of 4 mm, and a length of 50 mm, as shown in Fig. 2. A Vickers indentation (Indentation load: 400 N and indentation time 20 sec) was introduced on the centre of specimens to make a pre-crack.

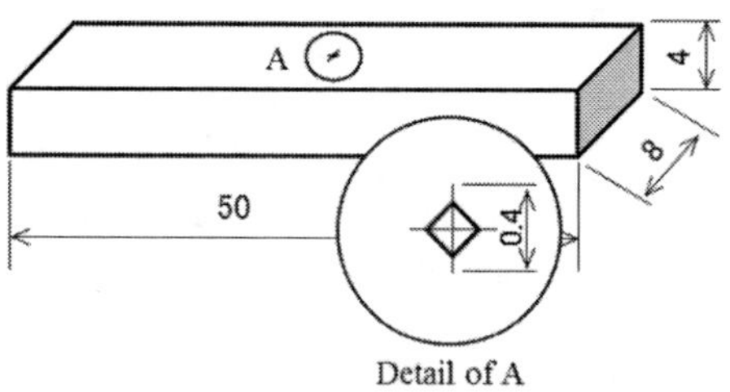

Fig. 2. Specimen used for investigating the FCG rate of short surface cracks at R = 0.1, 0.2 and 0.5.

3.3 Experimental Procedures

Four-point bending fatigue tests (inner span: 10 mm, outer span: 30 mm) were conducted at three different stress ratios, R = 0.1, 0.2 and 0.5, using the rectangular specimens. To eliminate the effects of the indentation, such as those of machining and residual stress, on the FCG behaviour of short surface cracks, the pre-crack was extended to a length of approximately 200 μm by applying a cyclic load. After that, crack growth measurement was started.

A servo-hydraulic fatigue testing machine was used to perform the four-point bending fatigue test. Tests were conducted in laboratory air at room temperature with a stress frequency of 10 Hz. Crack length during the fatigue process was measured using the replication technique [8].

For calculation of the maximum stress intensity factor K_{max}, equation (4) was used assuming that the crack shape was a semicircular.

$$K_{max} = Y\sigma_{max}\sqrt{\pi a} \quad (4)$$

Here, Y is a correction coefficient (= 0.73), σ_{max} is the maximum stress, and a is a half crack length. Equation (5) was used for calculation of the stress intensity factor range ΔK.

$$\Delta K = K_{max} - K_{min}\ (K_{min} > 0),$$

$$\Delta K = K_{max}\ (K_{min} \leq 0) \quad (5)$$

4 Experimental results

4.1 FCG curve at various stress ratios

In this study, FCG behavior of the fine grain cemented carbide was investigated under various stress ratios. To examine the FCG curve, two types of experiments, σ_{max} constant test and K_{max} constant test, were performed. Some of the experimental results are shown in Fig. 3. Fig. 3(a) and Figs. 3(b) – 3(d) show the test results at the σ_{max}-constant condition and the K_{max}-constant condition, respectively. In the former, the approximate curve to the experimental data was drawn, and the rate of FCG, da/dN, was determined from the approximate curve. In the constant K_{max} test, the $2a$-N curve obtained under the constant K_{max} condition was approximated by the straight line and the rate of FCG, da/dN, was measured from its slope. The maximum stress intensity factor, K_{max} and the stress intensity factor range, ΔK were evaluated using the equations (4) and (5), respectively.

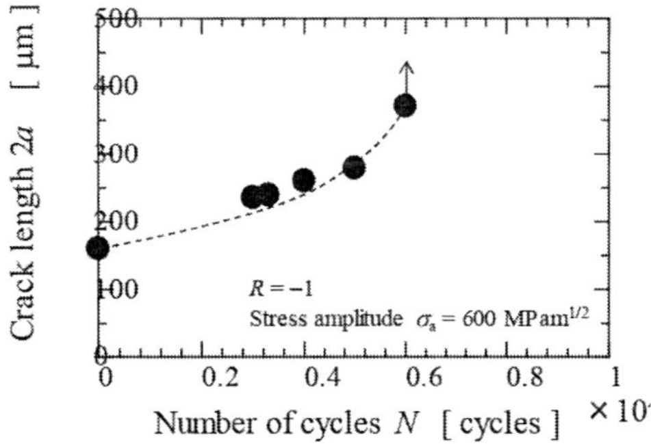

(a) Stress amplitude = constant, R = −1 [5]

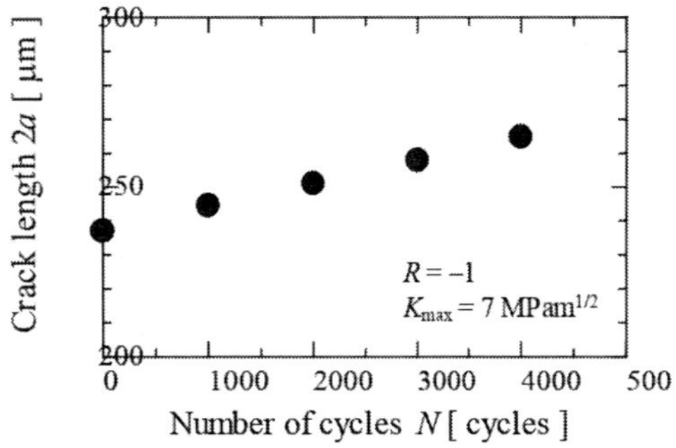

(b) K_{max} = constant, R = −1

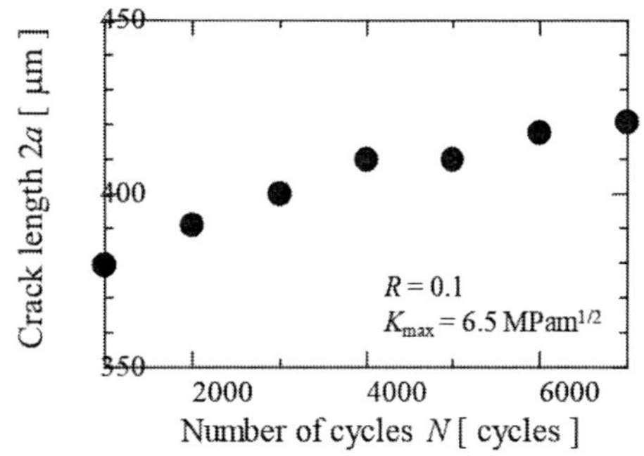

(c) K_{max} = constant, R = 0.1

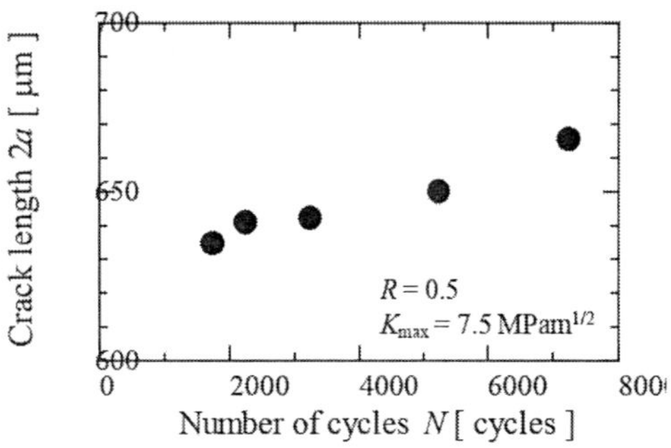

(d) K_{max} = constant, R = 0.5

Fig. 3. Crack growth curve at different stress ratios.

https://doi.org/10.1051/matecconf/201816507009

4.2 FCG law using one parameter and two parameter

4.2.1 One parameter analysis

From the FCG curve described in section 4.1, the rate of FCG, da/dN was obtained. The relations, da/dN vs. K_{max} and da/dN vs. ΔK, are shown in Figs. 4 (a) and 4 (b), respectively.

As can be seen from Fig. 4 (a), at the high K_{max} region where da/dN is larger than 5×10^{-9} m/cycle, da/dN can be determined by the K_{max} value regardless of the R ratio. On the other hand, looking at the da/dN - ΔK relationship in Fig. 4 (b), da/dN is not uniquely determined by ΔK and varies depending on the R ratio. More specifically, the necessary ΔK for a constant FCG rate, da/dN, decreases as the R ratio increases. This tendency is particularly conspicuous in the low FCG region less than 10^{-9} m/cycle.

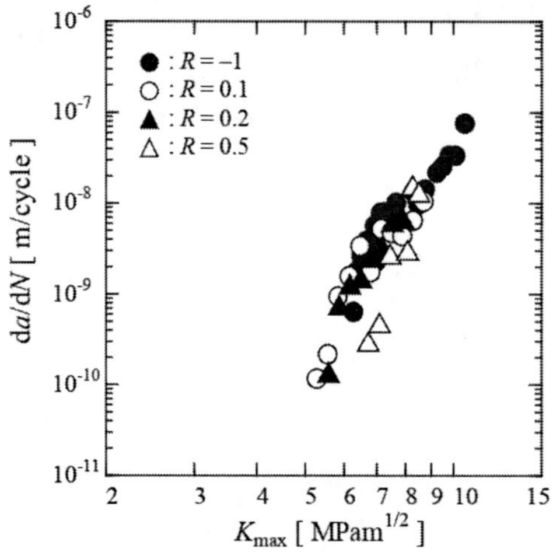

(a) Relationship between da/dN and maximum stress intensity K_{max} for the four different stress ratios.

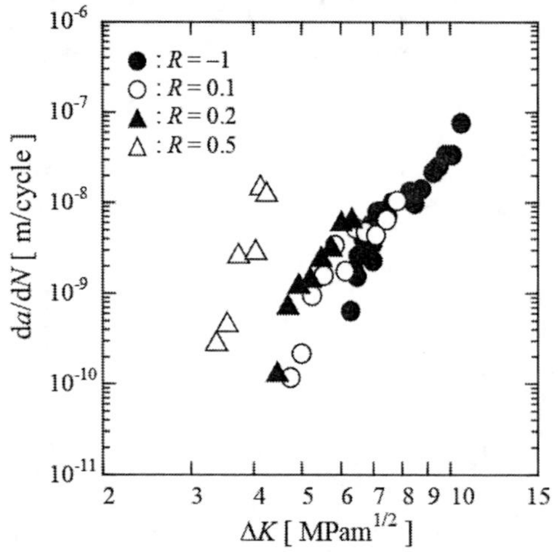

(b) Relationship between da/dN and stress intensity factor range ΔK for the four different stress ratios.

Fig. 4. Crack growth curve at different stress ratios.

The relationships da/dN - K_{max}, and da/dN - ΔK of the above fine grained cemented carbide show the FCG characteristics similar to ceramics [7] rather than metal. Therefore, the FCG law using both parameters of K_{max} and ΔK will be examined next.

4.2.2 Two parameter analysis

FCG law expressed by equation (3) using both parameters of K_{max} and ΔK will be applied to the present experimental data. Figure 5 shows the log-log plot of the FCG data for the fine grain cemented carbide at various stress ratios R. The vertical and horizontal axises of the figure represent $da/dN/(1-R)^n$ and K_{max}, respectively.

For the experimental data of $R = -1$, it was assumed that stress fluctuation on the compressive stress side do not affect the FCG behavior and was handled as experimental result of $R = 0$.

Figure 5 shows the relationship between $da/dN/(1-R)^n$ and K_{max} for the fine cemented carbide. In Fig. 5, the values of $m = 6.6$ and $n = 1.4$ were used. These values were determined using the maximum correlation coefficient method described in Chapter 2. As can be seen from the figure, experimental data of various stress ratios can be approximated by the straight line in the figure.

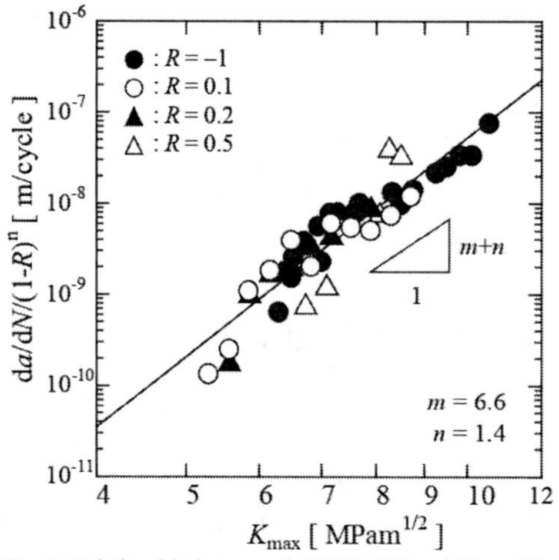

Fig. 5. Relationship between $da/dN/(1-R)^n$ and K_{max} of the fine cemented carbide.

5 Discussion

5.1 Comparison with the FCG behaviour of long through cracks

Llanes et al. [3] investigated the FCG behavior of through cracks using the cemented carbide (WC grain size: 0.5 μm, Co: 16.3 %, K_{IC}: 9.2 $MPam^{1/2}$) of the same grade as the one used in this study. In the following, the FCG characteristics of the short-surface cracks observed in this study and the one for the long through cracks obtained by Llanes et al. [3] were compared with each other.

However, the experimental result of Llanes et al. is targeting to a high-speed region where the FCG rate is 10^{-9} to 10^{-6} m/cycle. On the other hand, in the short-surface crack of this study, the low velocity region of 10^{-10} to 10^{-7} m/cycle, including the threshold level of the FCG, is targeted. So, there is a difference in the measurement area.

Figure 6 shows a log-log plot of the relationship between $da/dN/(1-R)^n$ and K_{max} for the long through crack obtained by Llanes et al. Values of m and n for the long through crack obtained were 22 and 5, respectively. These values are obviously high compared with m value 6.6 and n value 1.4 for the short-surface crack. On the other hand, the ratio n/m of both shows the same value about 0.21.

Therefore, as compared with the short-surface cracks, the long through cracks are highly susceptible to K_{max} and ΔK [6], and shows FCG characteristics of brittle materials such as ceramics.

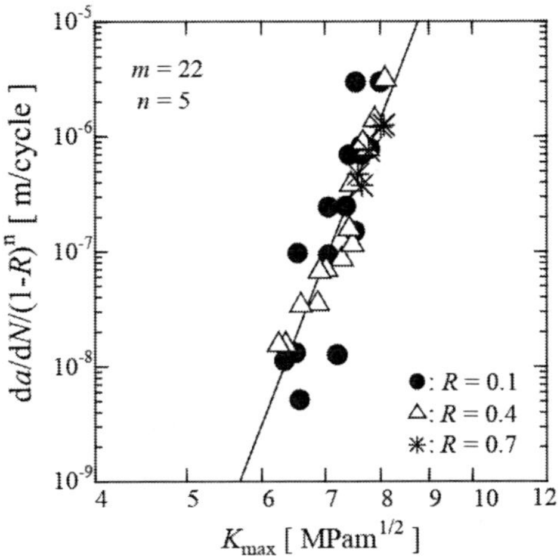

$$da/dN = 2 \times 10^{-30} K_{max}^{22} \Delta K^{5} \quad (6)$$

Fig. 6. Relationship between $da/dN/(1-R)^n$ and K_{max} obtained by Llanes et al. [3]. The data was obtained for the long through crack using the fine grained cemented carbide.

5.2 FCG law considering the threshold level of crack growth

To accurately evaluate the rate of FCG in the low speed region, it is considered necessary to introduce the threshold level of crack growth, K_{th}. Therefore, in this study, the threshold levels of crack growth, $K_{max,th}$ were determined from the experimental data, by defining $K_{max,th}$ as the values for K_{max} at the speed, $da/dN = 10^{-10}$ m/cycle. The value of ΔK_{th} can be obtained from the value of $K_{max,th}$ using the relation, $\Delta K_{th} = (1 - R)\, K_{max,th}$.

Table 3. Threshold level of crack growth at each stress ratio.

MPam$^{1/2}$

Stress ratio, R	-1 (0)	0.1	0.2	0.5
$K_{max,th}$	5.3	5.3	5.6	6.6
ΔK_{th}	5.3	4.8	4.5	3.3

Table 3 shows the values of $K_{max,th}$ and ΔK_{th} obtained for each stress ratio R. As can be seen from the table, as R increases, $K_{max,th}$ increases and conversely ΔK_{th} decreases. With cemented carbide, it is considered that WC grains form crack bridging portions [9]. This crack bridging locally reduces the driving force of fatigue crack [10]. As the R value decreases, the stress variation range increases. In the case, it is considered that elimination of the bridging portion will be promoted. Conversely, as the R value increases, the elimination of the bridging portion does not proceed, and as a result, it is considered that $K_{max,th}$ increases.

By substituting these values into the following equation, K^*_{max} and ΔK^* can be obtained. The FCG law using K^*_{max} and ΔK^* instead of K_{max} and ΔK was derived.

$$\Delta K^*_{max} = K_{max} - K_{max,th} \quad (7)$$

$$\Delta K^* = \Delta K - \Delta K_{th} = \Delta K - (1 - R)\, K_{max,th} \quad (8)$$

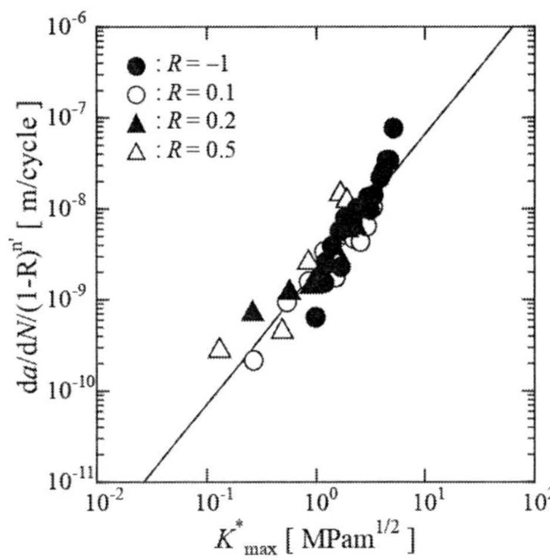

Fig. 7. Relationship between $da/dN/(1-R)^{n'}$ and K^*_{max} of the cemented carbide used in the present study.

Figure 7 shows the log-log plot for the relationship between $da/dN/(1 - R)^{n'}$ and K^*_{max} of the short surface crack. Compared with the $da/dN/(1 - R)^n$ - K_{max} relation in Fig. 5, in the low FCG speed range of Fig. 7, it is found that the fitness with experimental data and FCG law is clearly improved. The m' and n' values obtained were 1.5 and 0, respectively. These values are lower than the m and n values in Table 3.

In addition, by introducing the threshold levels of crack growth into the FCG law, da/dN - $K^{*}_{max}{}^{m'} \Delta K^{*n'}$, the value of n' = 0 was obtained. So, the FCG law with one parameter K^{*}_{max} was obtained. This result is different from the research result of Llanes et al. As one of the reasons for this, their research may be directed to the high-speed regions that do not need to consider FCG threshold levels.

5.3 Prediction accuracy of FCG amount

Next, the FCG speed under various stress ratios was estimated using the obtained FCG law. A comparison between the estimation result and the experimental one is shown in Fig. 8. As can be seen from the figure, the estimation result from the FCG law using two parameters corresponds well with the experimental result.

Moreover, it can be seen that the estimation accuracy of the crack growth amount is higher when the $K_{max,th}$ is introduced into FCG law than the case when FCG law does not take $K_{max,th}$ into consideration. Also, when dealing with FCG behavior of cemented carbide at various stress ratios, it is necessary to know whether only K_{max} parameter is sufficient, or, two parameters of K_{max} and ΔK are necessary. To make above issue clear in future, a systematic study including other cemented carbides is necessary. In addition, it is necessary to consider the physical meaning of the m', n' values.

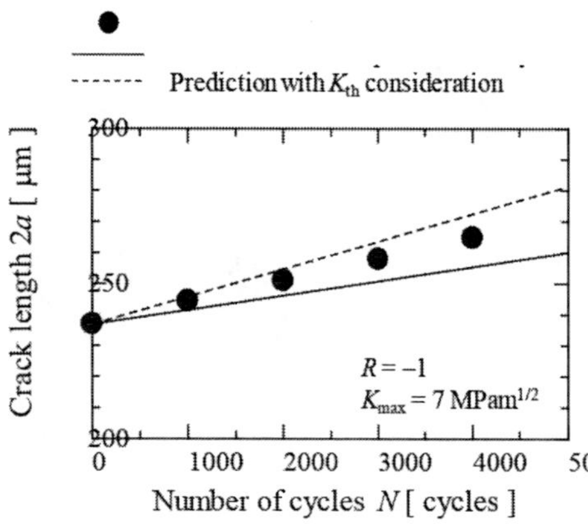

(a) K_{max} = constant, R = −1

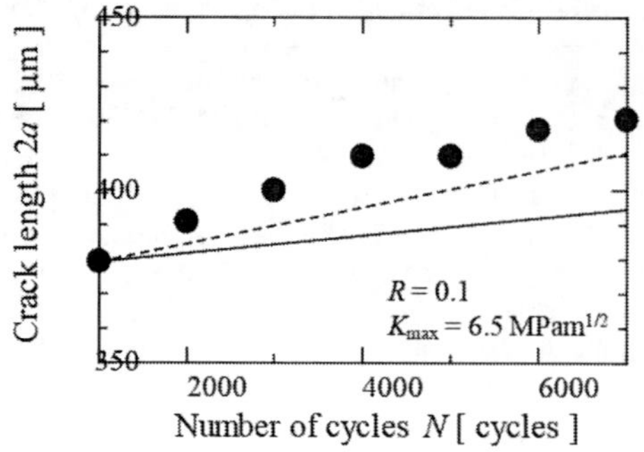

(b) K_{max} = constant, R = 0.1

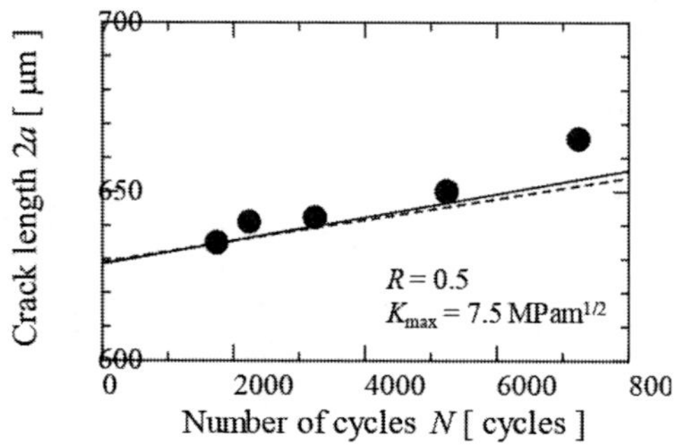

(c) K_{max} = constant, R = 0.5

Fig. 8. Comparison between estimated results and experimental results for the crack growth rate.

6 Concluion

The short surface FCG behavior for the fine grained cemented carbide was studied under different stress ratios. The obtained experimental data was analysed using two parameters, K_{max} and ΔK, and the following results were obtained.

(1) In the relationship da/dN - K_{max}, in the high K_{max} region, da/dN was determined by the K_{max} value regardless of the R ratio, but in the low K_{max} region, the influence of the R ratio appeared. On the other hand, in the da/dN - ΔK relationship, da/dN was not uniquely determined by ΔK but depends on the R ratio.
(2) The short-surface FCG law using two parameters K_{max} and ΔK of the cemented carbide was expressed by $da/dN = A\ K_{max}{}^{m}\ K^{n}$. Here m and n are material constants, and their values were 6.6 and 1.4, respectively. Therefore, FCG rate of cemented carbide strongly depends on K_{max} rather than ΔK, indicating the brittle FCG characteristics like ceramics.
(3) In comparison with the FCG law of the long through crack obtained by Llanes et al., the m and n values for the short surface crack were found to be lower than those of the long through crack. However, the n/m ratio was comparable. Therefore, the FCG law obtained for the long through crack cannot be applied to that of the short surface crack.

(4) By introducing the threshold levels, $K_{max,th}$ and ΔK_{th} into the relation da/dN - $K_{max}{}^{m}$ ΔK^{n}, the value of $n' = 0$ was obtained. So, the FCG law with one parameter K_{max} was obtained.
(5) In the FCG law of the short surface crack using two parameters, when the $K_{max,th}$ was introduced into FCG law, the estimation accuracy of the crack growth amount was higher than the case when the FCG law did not take $K_{max,th}$ into consideration.

Acknowledgments

Authors would like to express their sincere appreciation to the students of University of Toyama who helped us when carrying out the experiment.

References

1. A. Mizoguchi, T. Goshima, M. Shimizu, S. Ishihara, J. Sol. Mech. Mater. Eng. **2**, 1080 (2008)
2. P.R. Fry, G.G. Garret, J. Mat. Sci. **23**, 7, 2325 (1988)
3. L. Llanes, Y. Torres, M. Anglada, Acta Mater. **50**, 2381 (2002)
4. Y. Torres, J. M. Tarrago, D. Coureaux, E. Tarrés, B. Roebuck, P. Chan, M. James, B. Liang, M. Tillman, R.K. Viswanadham, K.P. Mingard, A. Mestra, L. Llanes, Int. J. Refract. Met. Hard Mater. **45**, 179 (2014)
5. S. Ishihara, H. Shibata, T. Goshima, M. Shimizu, J. Thermal Stresses, **31,** 1025 (2008)
6. H. Mikado, S. Ishihara, N. Oguma, S. Kawamura, Metals, **7**, 7 (2017)
7. R.O. Ritchie, Int. J. Fract. **100**, 55 (1999)
8. S. Ishihara, A.J. McEvily, *Proceeding of 12th International Conference on Fracture,* CD-ROM (Ottawa, 2009)
9. H. Mikado, S. Ishihara, N. Oguma, K. Masuda, S. Kawamura, *Fatigue of Materials III Advances and Emergences in Understanding: Proceedings of the Third Biennial Symposium*, (Pittsburgh, 2014)
10. A.J. McEvily, R.O. Ritchie, Fatig. Fract. Eng. Mater. Struct. **21**, 847 (1998)

Investigation of the fatigue behavior of notched reinforced thermoplastics

Ilan Raphael[1,2,3*], *Nicolas* Saintier[1], *Lucien* Laiarinandrasana[2] and *Gilles* Robert[3]

[1]I2M, Arts et Métiers Paris Tech, Esplanade des Arts et Métiers, 33405 Talence, France
[2]MINES Paristech, Centre des Matériaux, BP 87, 91003 Evry Cedex, France
[3]Solvay Engineering Plastics, Avenue Ramboz, BP 64, 69192 Saint-Fons, France

Abstract. Understanding and modeling of fatigue damage and behavior in short glass fiber reinforced thermoplastics is an important issue in order to improve material processing and the design of composite parts for the industry. The present work aims at a more precise description of the fatigue damage mechanisms and the establishment of a simple, yet efficient, durability criterion. Observations bring new insights on the fatigue damage mechanisms: fiber failure triggered by fretting fatigue and crack propagation by intra-spherulitic failure in their equatorial plane. Finally, a cyclic strain rate criterion is introduced for notched specimens and the identification of a behavior model is performed for later implementation of the durability criterion.

1 Introduction

1.1 Context of the study

In order to comply with new legislation regarding the environment, manufacturers in the car industry need to reduce the weight of parts and structures. One way to do so is to use composite materials rather than metals. Short glass fibers reinforced (SGFR) thermoplastics constitute a promising category of material thanks to their good specific mechanical properties and ease of manufacturing as injection molding allows for high production rates as needed in the car industry.
Short glass fiber reinforced polyamide 6,6 (PA66) is increasingly used, including for crucial components of the vehicle. However in order to engineer more efficient parts, a better knowledge of the mechanical properties of the material is required, especially regarding its durability under fatigue loading. Advances are needed both in our understanding of the damage mechanisms and the modeling of the behavior at the macro and micro scales.
Previous studies have established damage scenarios under quasi-static and fatigue loading using fractography analysis and in-situ tomographic observations [1,2]. The aim is now to refine our knowledge of these mechanisms, especially at the onset of fatigue damage.

1.2 Material

1.2.1 Fiber orientation

In injection molded parts, the short glass fibers are oriented following a well described *core-shell* microstructure [3]. In the shell fibers are aligned with the mold flow direction (MFD). Conversely, in the core at the center of the plate, the fibers are aligned perpendicularly to the MFD.
Specimens are usually described as 0°/45°/90° following the direction of their axis with respect to the MFD.

1.2.2 Spherulitic structure

Semi-crystalline thermoplastics often crystallize as spherulites [4]. These are spherical structures nucleated from a single lamella and composed of a succession of crystalline lamellae and amorphous polymer chains. The width of a lamella and the distance between two lamellae is of about 10nm.
Spherulite diameter depends on the cooling conditions and number of potential nucleating sites. In our material spherulite diameter is generally comprised between 15 and 30μm.

1.3 Notched specimens

Specimens are waterjet-cutted from injection-molded rectangular plates 3.24mm thick. Two notch geometries were considered: symmetric "U" notches 0.5mm in radius and symmetric "U" notches 2mm in radius. For pure PA66, specimens were all cut at 90°. For reinforced PA66, specimens are designated by the direction of the fibers in the shell with respect to the specimen main axis.
The use of notched specimens allows for the localization of the damage. This is useful in order to facilitate the

* Corresponding author: ilan.raphael@ensam.eu

observations of fatigue damage. Furthermore, the different notch geometries result in different local mechanical fields and different gradients of those mechanical fields.

Specimens were conditioned at RH50 prior to testing and experiments were conducted at 23°C. Fatigue loading was performed at 0.5Hz and at a loading ratio R=0.1.

2 Fatigue damage mechanisms

2.1 Usual description

Fatigue damage scenarios traditionally include cavitation and crazing in the polymer matrix, damage at fiber tip, debonding at the fiber-matrix interface and fiber failure.

These mechanisms can be observed on the fracture surface using a scanning electron microscope (SEM) or in the bulk of the material by performing X-ray tomography (fig.1).

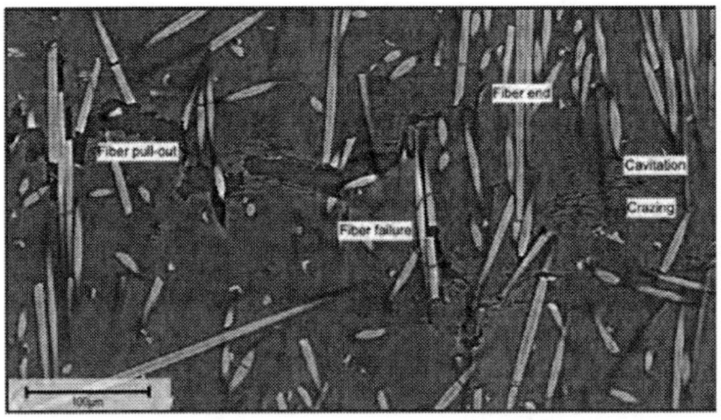

Fig. 1. Fatigue damage mechanisms observed by tomography.

2.2 On the importance of cavitation

Cavitation occurs both in poorly reinforced areas where local strains can be high and close to fibers where confinent is important. Based on in-situ observations and fractography analysis, cavitation in the thermoplastic matrix appears as a precursor of other fatigue damage mechanisms.

Cavitation in the matrix is an early sign of damage that subsequently can lead to the formation of craze and short cracks is the material.

Debonding at fiber-matrix interface is generally described as starting from the fiber tips, where adhesion is poor. However cavitation along fibers can also in turn induce a weakening of the interface (fig.2). This might lead to a local stress redistribution triggering a failure of the fiber.

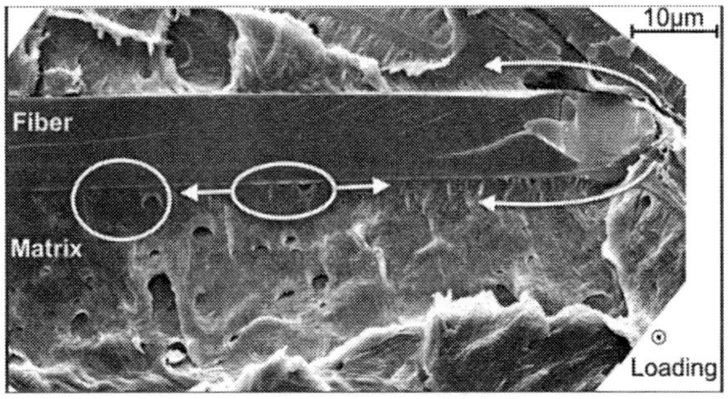

Fig. 2. Damage at the interface induced both by the poor bonding at the fiber tip and cavitation close to the interface.

2.3 Fiber failure by fretting fatigue

Failure of the short glass fibers is one of the best known damage mechanisms in SGFR thermoplastics. In quasi-static tensile tests, fibers break because they are overloaded. Yet in fatigue multiple scenarios can lead to the breakage of glass fibers.

Earlier works have shown that fibers tend to break at the crossing with an other fiber. This crossing of fibers induces a confinement of the matrix in between and a stress concentration in the adjacent fibers, thus triggering the failure (fig.3).

However, in some cases, fibers are in contact. This can lead to indentation and/or fretting fatigue between the fibers. The reinforcement being weakened, its failure become much more likely. This fiber failure triggered by contact between reinforcements in the composite material has, to our knowledge, not been described in previous damage scenarios, yet is can be unambiguously observed by fractography analysis (fig.4).

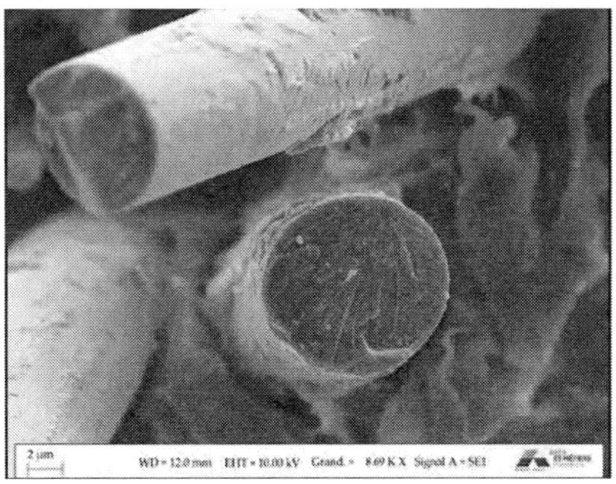

Fig. 3. Failure induced by a crossing with a neighboring fiber.

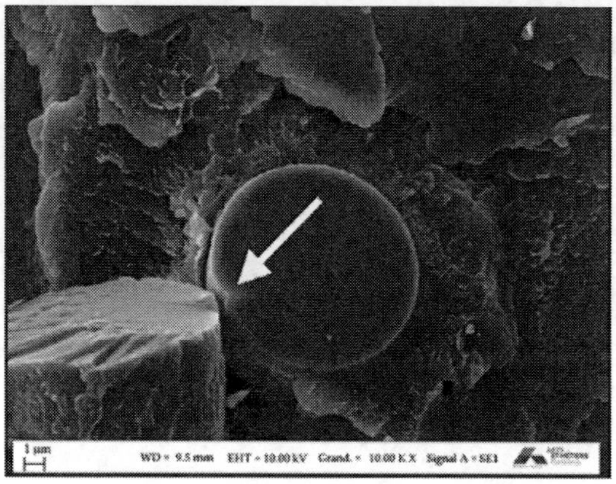

Fig. 4. Failure induced by the contact with a neighboring fiber.

2.4 Spherulitic fatigue damage

2.4.1 Tomographic observations

In-situ tomographic observations have revealed that damage progresses in the matrix with a characteristic length of 20-25μm, close to the spherulite diameter.
Further analysis reveal that these structures are flat penny-shaped disks, that could correspond to the equatorial planes of spherulites.
Subsequently a damage scenario was advanced for damage in the semi-crystalline polymer matrix: fatigue damage progresses by opening the spherulites in their equatorial plane.
In order to confirm such a hypothesis, fractography observations have been performed.

2.4.2 Fractography observations

SEM observations on the fracture surface of notched unreinforced PA66 specimens reveal a ductile layer in the fatigue crack path.
In this area, it is possible to clearly observe spherulites. The radial organisation of the lamellae is visible, as well as the nucleus of the spherulite (fig.5). This confirms the hypothesis of intra-spherulitic failure in the equatorial plane.

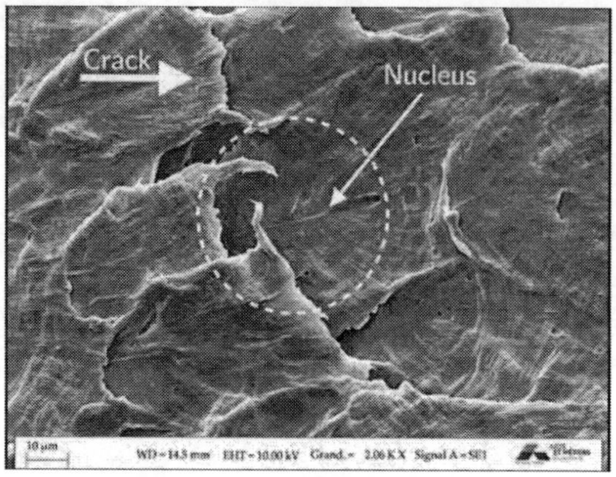

Fig. 5. Spherulites as observed on the fracture surface of PA66 after fatigue loading.

3 Modeling of fatigue behavior

3.1 Towards a pertinent fatigue indicator

3.1.1 Cyclic strain rate in secondary state

In order to unify fatigue results for SGFR PA66, one should determine a pertinent indicator that yields consistent results for the multiple conditions that can be encountered. These conditions can vary greatly with the imposed loading, the structural heterogeneities of the part or the environment. Fatigue results for various fiber orientations, mean stresses, loading ratios, frequencies, notches and relative humidity should all be unified by a pertinent indicator.
Previous results on smooth specimens have shown that the cyclic strain rate during the secondary state of loading unifies fatigue results. This cyclic strain rate can be determined by measuring the slope of the *Mean strain per cycle vs. Number of cycles* curve at half life of the specimen (fig. 6).

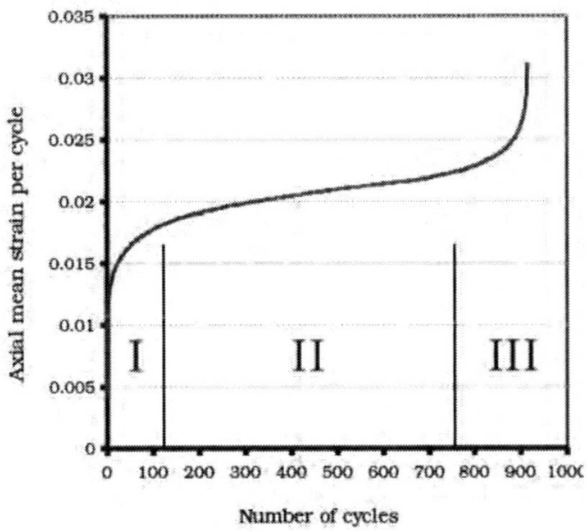

Fig. 6. Mean strain per cycle during the fatigue loading of a pure PA66 specimen.

3.1.2 Experimental results on notched specimens

Fatigue testing on notched SGFR PA66 specimens have shown that the cyclic strain rate at mid-life is also a good indicator. Is can be used to identify a durability criterion:

$$N_r = A\left(\frac{\dot{\epsilon}_{11}}{\dot{\epsilon}_0}\right)^n \quad (1)$$

with the following parameters: A=0.220 and n=-0.81.

It yields a good estimation of Nr, the number of cycles until failure, regardless of fiber orientation and notch geometry (fig.7).

3.1.3 Measuring a local strain rate

In order to obtain a criterion that can be implemented in a Finite-Element (FE) model, more data is needed regarding the local cyclic strain rates close to the notch.
An image acquisition setup, timed with the loading cycles, has been put in place. Images of the specimen are taken every 20 cycles at peak loading.

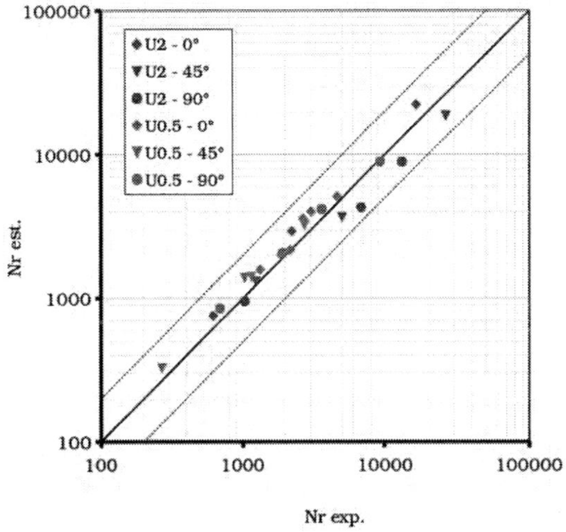

Fig. 7. Durability criterion based on the macroscopic cyclic strain rate.

Digital image correlation (DIC) can then be performed using VIC-2D in order to determine local cyclic strains and strain rates. These values can be compared to those obtained by modeling.
The local data can also be used in order re-identify the parameters of the equation (1) based on the local axial strain rate at notch tip giving A=19.55 and n=-0.421. These can thus be used as a new local durability criterion (fig.8).

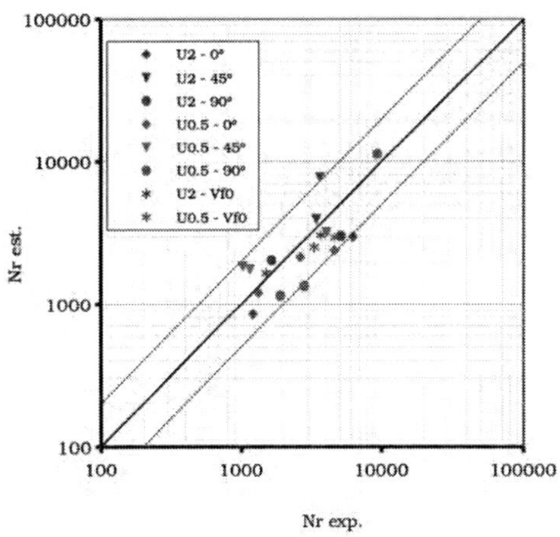

Fig. 8. Durability criterion based on the macroscopic cyclic strain rate.

The criterion unifies fatigue results for different notch geometries, fiber orientation as well as for unreinforced PA66 notched specimens.

https://doi.org/10.1051/matecconf/201816507010

3.2 Modeling

3.2.1 Constitutive equations

Total strain is parted between a linear elastic contribution and a visco-plastic contribution:

$$\epsilon = \epsilon^e + \epsilon^{vp} \tag{2}$$

Plastic strain rate is defined using Norton law:

$$\dot{p} = \left(\frac{\phi}{K}\right)^n \tag{3}$$

with:

$$\phi = \sigma_* - R(p) \tag{4}$$

and the isotropic hardening given by:

$$R(p) = R_0 + Q(1 - exp(-bp)) \tag{5}$$

The GTN model can be used to calculate the effective stress σ_* by taking into account the growth of cavities in the material. This volume fraction of cavities f_* is used as the damage variable.

Thus σ_* is calculated using:

$$\Phi(\Sigma, f_*, \sigma_*) = \frac{\Sigma_{eq}^2}{\sigma_*^2} + 2q_1 f_* cosh\left(\frac{q_2 \Sigma_{kk}}{2\sigma_*}\right) - 1 - q_1^2 f_*^2 = 0 \tag{6}$$

3.2.2 Parameter identification

Parameter identification on notched specimens have first been performed using a genetic algorithm on quasi-static tests at $\dot{\epsilon} = 10^{-2}s^{-1}$ and $\dot{\epsilon} = 10^{-3}s^{-1}$ in order to match the stress-strain curves and the notch opening displacement (NOD).
Parameters have then been adjusted for the cyclic loading. Focus was primarily set on obtaining a consistent value for the cyclic axial strain rate in order to later be able to apply the durability criterion (see Eq.1).

3.2.3 Results for quasi-static tensile loading on PA66

Quasi-static tensile tests were performed at two strain rates $\dot{\epsilon} = 10^{-2}s^{-1}$ and $\dot{\epsilon} = 10^{-3}s^{-1}$. Load and NOD are plotted and compared to the experimental data (fig.9 and fig.10).

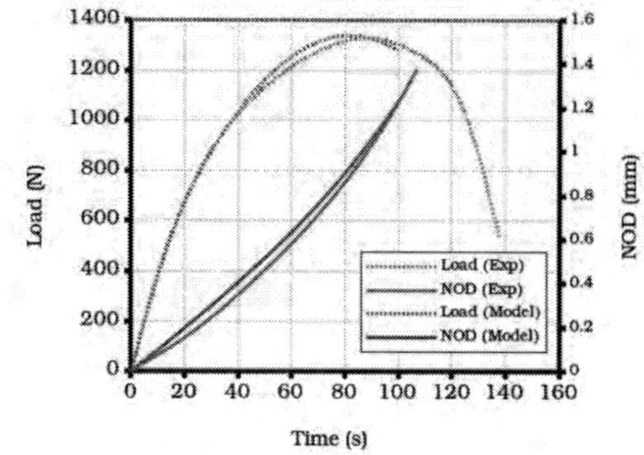

Fig. 9. Load and NOD for a U2 notched PA66 specimen during tensile testing performed at $\dot{\epsilon} = 10^{-3}s^{-1}$.

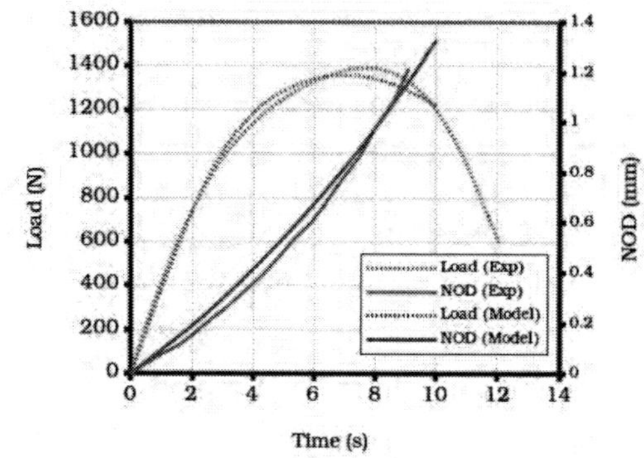

Fig. 10. Load and NOD for a U2 notched PA66 specimen during tensile testing performed at $\dot{\epsilon} = 10^{-2}s^{-1}$.

3.2.4 Results for cyclic loading on PA66

First cyclic tests were performed at 0.5Hz and R=0.1. Axial strain is averaged on each cycle to be plotted and compared to the experimental data (fig.11 and fig.12).

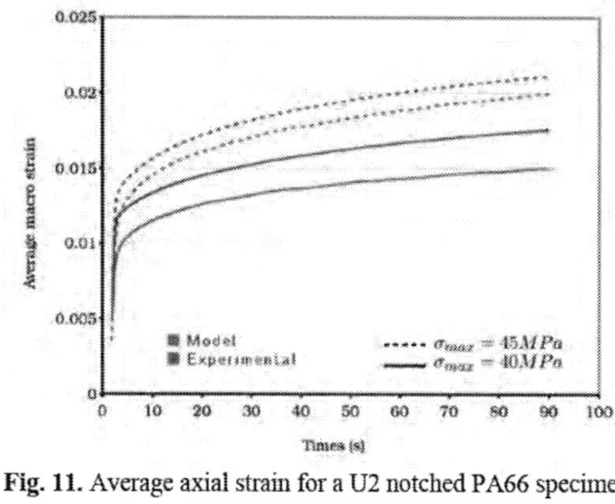

Fig. 11. Average axial strain for a U2 notched PA66 specimen.

 https://doi.org/10.1051/matecconf/201816507010

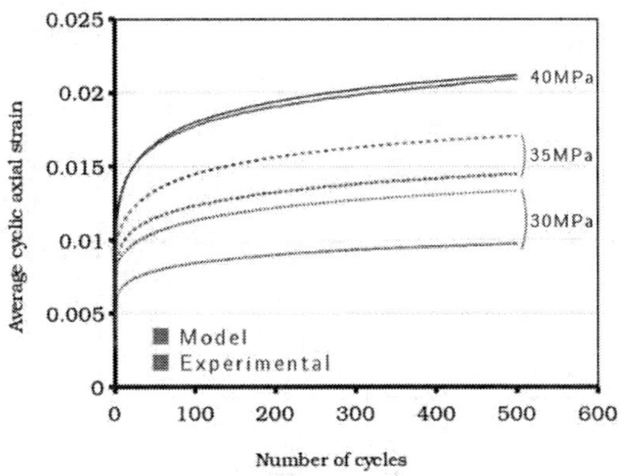

Fig. 12. Average axial strain for a U0.5 notched PA66 specimen.

It can be noted that the average axial strain differs from the experimental data, especially at lower stress levels. However, in all cases, the strain rate in accurately determine as the slopes of the experimental and model curves are alike.

The computed NOD can also be compared to the experimental data (fig.13), as well as the cyclic strain rate between notches at half the fatigue life of the specimen (fig.14), giving a good agreement between the model and the experimental results.

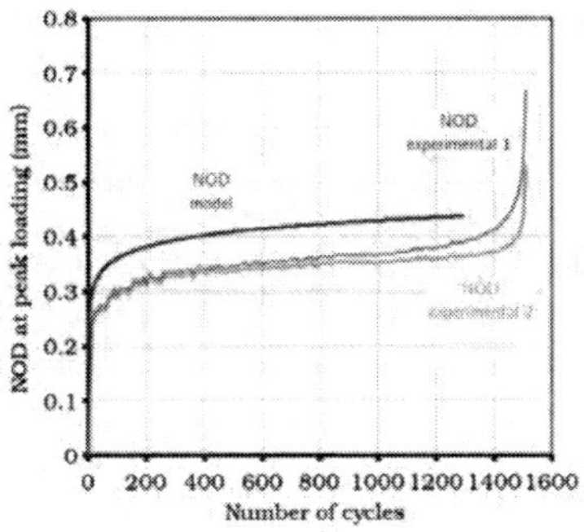

Fig. 13. NOD for a U2 notched PA66 specimen during fatigue loading.

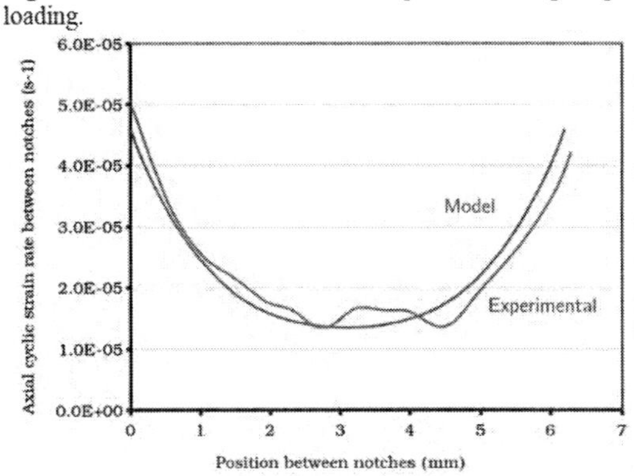

Fig. 14. Axial cyclic strain rate between the notches for a U2 notched PA66 specimen.

Conclusion

Our fractography analysis and tomographic observations have shed light on the importance of better understanding fatigue damage mechanisms. Observations have thus shown that cavitation is a precursor of other fatigue damage mechanisms, and in particular that it can lead to a damage of the fiber-matrix interface resulting in the breakage of the fiber. Fractography analysis also showed that fiber failure can be triggered by the contact between crossing fibers, resulting in a weakening of the reinforcement by fretting fatigue. Moreover, analysis of the fracture surface of pure PA66 specimens showed that fatigue damage progress is closely linked to the semi-crystalline microstructure of the polymer. Further work should evaluate the impact of the spherulitic structure, size and anisotropy at a finer scale on the fatigue damage progress and localization.

Experimental results on notched specimens have shown that a simple two-parameters cyclic strain rate criterion can efficiently unify fatigue results for different fiber orientations, notch geometries and loading intensities. The criterion can be tested and later improved, for example by considering an equivalent strain and/or a non-local measurement around the notch tip.

A model has been identified that can match the experimental cyclic strain rate. The model should now be tested on various conditions of loading and notch geometries.

References

1. Horst JJ. PhD. 1997.
2. Rolland H, Saintier N, Raphael I, Lenoir N, King A, Robert G. *Compos Part B Eng*. (2018);143:217-229. doi:10.1016/j.compositesb.2017.12.051.
3. Arif MF. PhD. 2014.
4. Crist B, Schultz JM. *Prog Polym Sci*. (2016);56:1-63. doi:10.1016/j.progpolymsci.2015.11.006.

Fast screening of the fatigue properties of thermoplastics reinforced with short carbon fibers based on a heat build-up protocol

Louis Leveuf[1,2,*], *Yann* Marco[1], *Vincent* Le Saux[1], *Libor* Navratil[1], *Sylvain* Leclercq[3], and *Jérôme* Olhagaray[2]

[1] ENSTA Bretagne - iRDL, UMR CNRS 6027, 29200 Brest, France
[2] Safran Composites, 91760 Itteville, France
[3] Safran Landing Systems, 78140 Vélizy-Villacoublay, France

Abstract. This study deals with the characterization of the fatigue lifetime of a short carbon fiber reinforced PEEK matrix thermoplastic composite using a heat build-up protocol. Several commercial grades and fillers ratios are considered in order to challenge the ability of the technique to capture the influence of these variations. First, the materials investigated and the experimental protocols are described. The way to build the heat build-up curves is outlined. Then the results obtained from the heat build-up and the fatigue experiments are presented. Finally these results are used to discuss two main points. The first one is the validity of the hypotheses needed to apply the approach, ranging from the evaluation of the dissipation to the response during the fatigue tests. The second one is the capability to predict the fatigue curves accurately throughout an energetic criterion and to catch the influence of the variation of material on the fatigue properties. The results presented here are published with more details in [47].

1 Introduction

Short-fiber-reinforced thermoplastics (SFRT) provide a major opportunity today to obtain lightweight parts at a reasonable cost and have been widely investigated [1–6]. Over the last few years, this kind of materials has been used for structural components in the automotive industry to replace die casting aluminum or stamped steel parts [7–11]. The possible use of these materials for aeronautic applications is also under consideration. The main reasons are the reasonable properties of these materials, especially when using a thermostable matrix, and the use of the injection molding process which allows both a cost reduction and the definition of complex parts with stiffeners and ribs, to circumvent the loss in rigidity compared to metallic solutions. Fatigue campaigns are usually very time consuming , especially for polymeric materials as the test frequency cannot be very high to avoid any bias brought about by the heat build-up induced under cyclic solicitations. As these materials also provide a high versatility in fiber content, grades and suppliers, a screening tool, able to discriminate good candidates for fatigue applications quickly, prior to a full fatigue campaign, would be a crucial asset. One of the experimental tools to do so could be based on the analysis of the heat build-up response. Since the pioneer studies [12, 13], this technique has become of considerable interest over the last few years, on metallic materials for empirical approaches [14–17] or throughout constitutive modeling [18–20]. This is explained mainly by the considerable development of thermography devices and analysis [16, 21, and 22]. The extension of the technique to other materials has also been tested [17] and is now applied, following different approaches, to rubbers [23– 25], continuous fiber composites [26, 27] and short fiber reinforced plastics [11, 28, 29]. All these approaches use the evolution of the sample temperature but can be quite different. Some of them rely on the temperature only, which is quite limited as temperature is dependent on the frequency, on the samples' geometry and on the thermal boundary conditions. These analyses therefore give no access to an intrinsic parameter. Most of the protocols now seek the dissipated energy associated with the temperature variations. This evaluation is not trivial and requires solving the heat equation according to a wide range of hypotheses both on the temporal aspects (adiabatic or stabilized configurations) and on the spatial distribution of the dissipation sources. This evolution of the dissipated energy versus the amplitude of solicitation (strain or stress, depending on the materials tested) is then generated and is called the heat build-up curve. The following step is to relate it to the "fatigue properties", which could cover a pretty wide range of ambitions. The most basic is to evaluate an unlimited fatigue lifetime or at least a lifetime high enough compared to the range of cycles aimed at by the application. This analysis uses the change of shape of the heat build-up curve to identify this threshold. A more interesting objective is to predict the fatigue mean curve. This approach relies on the use of an energy based fatigue criterion and could be determined either empirically [23, 25, 29], with the help

* Corresponding author: louis.leveuf@ensta-bretagne.org

of a constitutive model [18, 30], or using additional microstructural measurements [31, 32]. In all cases, it is necessary to check if the method can be applied, because a main difference of this approach compared to classic fatigue tests is that it relates the response over a few initial cycles to the fatigue failure occurring after several days or weeks of testing. It is therefore necessary to check that the constitutive response under fatigue conditions reaches a stationary state within this limited number of cycles [33, 34], before relating the cyclic thermomechanical response under several loading amplitudes to the fatigue lifetime. An even more desirable objective is to predict the full fatigue curve including the fatigue scattering from the heat build-up curve. This has been proven to be possible on several metallic materials [18–20], by relating the dissipated energy to the damage mechanisms throughout a constitutive modeling including a probabilistic failure criterion. This last approach is clearly the most appealing but it faces several difficulties for short fiber reinforced composites. The main difficulty is the extremely numerous dissipation phenomena possibly involved in the material response under fatigue loading, ranging from visco-elasticity to damage and plasticity [35–38]. Despite numerous studies [6, 39, 40], relating the macroscopic dissipation to the basic fatigue mechanisms remains an open issue. A second issue is that predicting a relevant local dissipation and thus reliable fatigue predictions requires a very accurate evaluation of the hysteretic loop, which remains clearly difficult for these materials [41, 42, 46]. A last point is that a rich modeling approach usually also requires a rich testing campaign, involving a long preliminary study for each compound tested. This could not be integrated easily in a fast testing approach.

This study focuses on a thermostable matrix reinforced with short carbon fibers and aims at three main objectives. The first is to check if the hypotheses required to apply the approach is valid for these materials, both from a measurement point of view (adiabaticity, spatial distribution of the dissipation sources) and from a criterion point of view (evaluation possible over a limited amount of cycles). The second is to check the accuracy of the technique to predict the mean fatigue curve of these materials. The last is to challenge its capacity to be used as a fast screening tool to discriminate quickly several possible materials, for different grades, suppliers and carbon fiber amount. This study provides two main original aspects compared to the team's previous studies. The first one is the application of the heat build-up approach to these materials that have not been investigated yet. The second is the use of a specific identification of the temperature variation giving access to both dissipation and thermoelastic couplings, for each pixel of the measured area.

A first section presents the material investigated and the samples' geometry. In the second section, the experimental protocols used for the fatigue tests, the thermal characterization and the determination of the heat build-up curve are presented. The third section presents the results obtained both for the fatigue and the heat build-up tests. The fourth section provides an analysis of the results obtained and discusses the objectives of the paper, i.e. the consistency of the hypotheses and the capacity of the approach to predict the fatigue properties accurately and to discriminate quickly the fatigue properties of several close materials.

2 Experimental settings

2.1 Materials and samples

Four different materials were tested in this study, with different kinds of polymer matrix and amounts of carbon fibers (referenced as A, B, C, D). All the polymer matrices were thermostable materials sourced from several suppliers and are either PEEK or PAEK matrix reinforced with 30% to 40% (in weight) of short carbon fibers. It is important to underline that all these materials are close and are potential candidates for industrial applications.

Two different geometries were tested in this study. Figure 1 presents the shape and the geometrical parameters and Table 1 provides the parameters for the two different geometries. These samples were milled from a rectangular plate (nominal dimensions 250mm, 100 mm), injection molded through a film gate on the shorter edge. The longitudinal axis of the samples was the main flow direction. Table 2 presents an overview of the materials tested and of the samples considered.

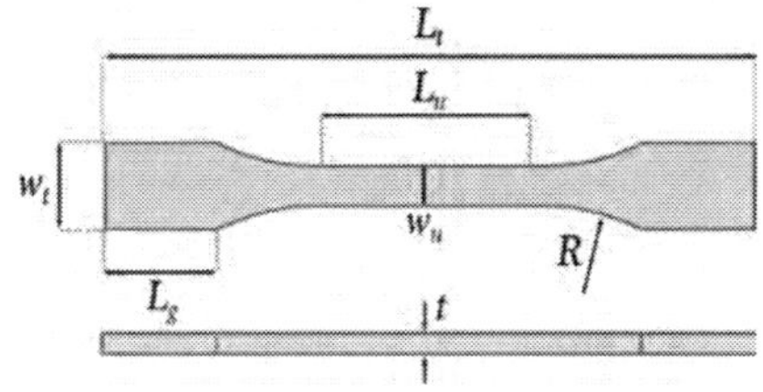

Fig. 1. Sample geometry. The values of the parameters are given in Table 1.

Table 1. Dimensions of the specimens (in mm).

Parameter	Type 1	Type 2
L_t	170	99
L_u	50	29
L_g	30	18.75
t	4	3.25
w_t	20	18
w_u	10	11.9
R	60	45

2.1 Fatigue tests

The tests were performed on an INSTRON servo-hydraulic testing machine (capacity of 100kN). The tests were force controlled, the load ratio was set to R=0 and the frequency was 2 Hz. All the tests were performed at room temperature (25°C). The number of cycles between the initiation and the final failure was less than 3% of the total lifetime. To simplify the analysis, the number of cycles to initiation was therefore considered equal to that leading to the final failure. For each test, an evaluation of the dissipated energy according to the protocol described in Section 2.4.2 was achieved. For confidentiality issues, all the energetic and mechanical data were divided by arbitrary values, kept the same for all materials to allow comparison.

Table 2. References of the materials tested

Reference	Polymer Matrix	Carbon fibers (%wt)	Geometry
A_1	Grade 1	30	Type 1
A_2	Grade 1	30	Type 2
B	Grade 2	30	Type 2
C	Grade 3	30	Type 2
D	Grade 2	40	Type 2

2.3 Thermal device

The thermal acquisition was performed using a FLIR SC7600-BB infrared camera. The device is equipped with a Stirling-cycle cooled Indium-Antimonide (InSb) Focal Plane Array (FPA). The FPA is a 512x640 array of detectors digitized on 14 bits and sensitive in the 1.5-5.1µm spectral band with a 15µm pitch (distance between two detectors) . The focal length of the optical lens is 50mm, and its transmissivity is 0.94. For all the experiments, the integration time was set to 1800µs. The chosen thermal acquisition frequency was 30Hz. In order to improve the thermal resolution (compared to classic thermal calibration), a compensated pixelwise calibration was used [22]. After this specific calibration, a NETD (Noise Equivalent Temperature Difference) of 8 mK was obtained for differential measurements.

2.4 Generation of the heat build-up curves

2.4.1 Heat build-up protocol

All the tests were carried out at room temperature (25°C). Special attention was paid to the experimental conditions in order to avoid the influence of the environment on the measurements as much as possible. The sample was painted with high emissivity paint (measured value of 0.95), surrounded by a black box in order to limit any reflections and a dark sheet covered the experimental set-up.

The principle of a heat build-up protocol basically consists in submitting a sample to a succession of cyclic tests of increasing stress amplitude, and to record the evolution of the temperature for each sequence. The analysis of the dissipated energy from the thermal resolution requires the heat equation to be resolved, as presented in Section 2.4.2. This analysis can be based either on a stationary state or on an adiabatic hypothesis [17, 25, 43]. The choice of the mechanical protocol depends of course on the chosen analysis.

In this study, the adiabatic analysis was selected, which led to a mechanical protocol focused on the first loading cycles. Each loading block was composed of two stages. The first was the loading step: the sample was submitted to 30 cycles according to the stress amplitude considered in the block. The tests were force controlled, the load ratio was set to R=0 and the frequency was 2Hz. During this stage, a thermal acquisition was performed, supplying the thermal images that would be analyzed to follow the temperature evolution and to determine the dissipated energy. The second stage was a cooling step lasting between 3 to 15 minutes, in order to let the sample cool down to the ambient temperature and to have comparable initial conditions from one loading block to another. This mechanical protocol was repeated for each block, and allowed the dissipated energy to relate to the force amplitude. The last block ran until the failure of the sample.

2.4.2 Thermal measurements and determination of the cyclic dissipated energy

The experimental determination of the dissipated energy fields relies on the solving of the heat equation [21, 44]. Under some classic assumptions (including adiabaticity) detailed in former publications [11, 25, 43, 44], the heat equation can be written as

$$\rho c_p \left.\frac{\partial \theta}{\partial t}\right|_{t\to 0} = f_r \Delta^* + \mathcal{C}_e$$

With θ the temperature variation, f_r the mechanical frequency, Δ^* the cyclic dissipated energy and $\mathcal{C}_e$ the thermo-elastic coupling contribution, while ρ and c_p stand for the specific mass and specific heat, respectively. Under sinusoidal cyclic loading, the solution of this equation writes [16]

$$\forall t \in \mathbb{R}^+, \theta(t) = \frac{f_r}{\rho c_p}\Delta^* t + c_e \sin(2\pi f_r t + \phi)$$

This equation is local and the rigorous identification of the cyclic dissipation Δ^* and elastic coupling $\mathcal{C}_e$ is not straightforward because the sample is moving whereas the camera is fixed. It is therefore necessary to compensate the displacement of the sample and to transpose the fields measured into the reference configuration before performing the temporal analysis. To do so, the displacement of a high emissivity marker printed on the sample is analyzed using a home developed software Belenos [46]. The thermal fields are then transposed in the reference configuration prior to any other analysis.

Figure 2 presents the typical evolution of the temperature during one block (circular plain markers), obtained for one pixel of the observed zone. Two main classic phenomena are clearly observed here. The first is a global rise in temperature, which is related to dissipation. The second is the oscillation related to the thermo-elastic coupling. This signal is used to identify the dissipation and the amplitude of the thermo-elastic coupling, by fitting the unknowns of Equation 2 (*i.e.* $\Delta^{\star}$and c_e). This identification is illustrated in Figure 2 (continuous line). As highlighted by that figure, the next question is the choice of the temporal window used to perform the identification. In order to make this choice, the respect of two hypotheses has to be considered. The first is that the cyclic dissipation is constant from one cycle to another. For the materials studying the mechanical hysteresis is stabilized within 4 cycles [47]. The temporal window should therefore start ideally after 4 cycles (*i.e.* 2 seconds). The second hypothesis to be checked is the adiabatic condition, i.e. that the time considered to perform the evaluation is far below the characteristic time allowing for thermal exchanges. In the case considered here, this characteristic time (evaluated from the temperature drop during the cooling steps was around 300s). A temporal window within the 5 first seconds from the start of the experiment seemed reasonable. The identification was therefore performed over a temporal window of 3 seconds after the 4th cycle, as illustrated in Figure 2. One can see that the function identified also fitted the experimental signal very well over the full duration of the loading step, which validates the adiabaticity hypothesis.

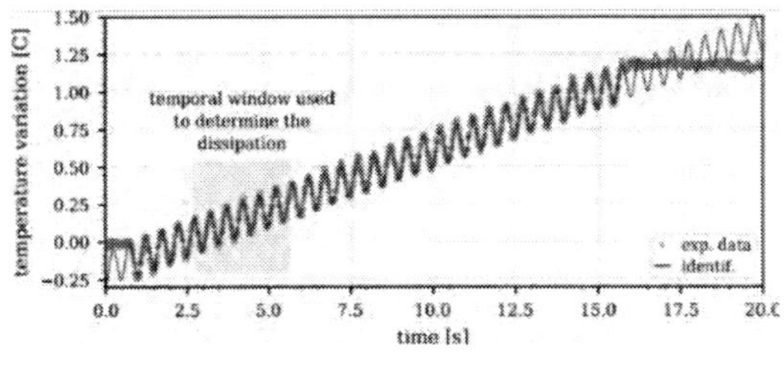

Fig. 2. Typical thermal measurements during one block of the heat build-up protocol.

This identification performed for every pixel of the observed zone, leading to the generation of the fields of dissipation and thermo-elastic coupling, identified over the temporal window. As explained in Section 2.4.1, several loading blocks with increasing force amplitudes were performed. For each loading block, the fields of dissipated energy and thermo-elastic coupling were evaluated. As illustrated Figure 3, the mean dissipation was computed over a given area, which led to the generation of the full heat build-up curve. As the dissipation field is homogenous over the gauge length of the sample, this evaluation is not dependant on the size of the averaging area.

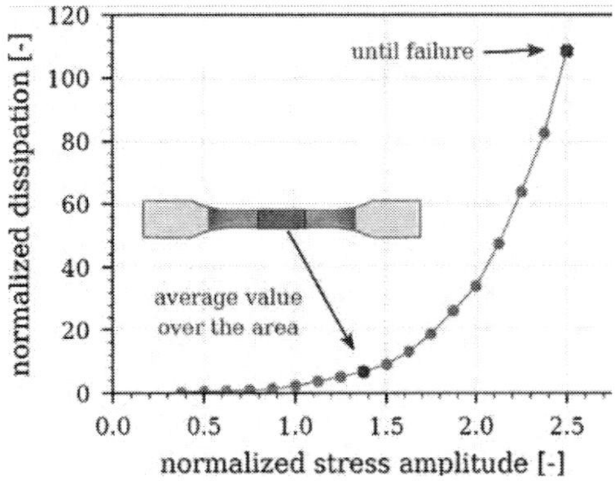

Fig. 3. Heat build-up curve relating the dissipated energy per cycle (normalized) to the maximum nominal principal strain (normalized). Curve associated to the material A2.

3 Results

3.1. Fatigue tests

The evolutions of the cyclic mechanical features were recorded during the fatigue tests. The secant modulus, the hysteretic loop and the residual strain (*i.e.* the strain at the end of each cycle) were computed from the nominal strain and stress derived from the extensometer and the load cell. These evolutions illustrate that the hysteresis remained stable over the full fatigue test and that the value evaluated at the beginning of the fatigue test was relevant up to the fatigue failure [47]. The evolutions of the secant modulus and of the residual strain were limited [47] for these materials compared to other thermoplastics [3, 29]. The initial evolution seems consistent with the amount of cycles needed to reach a stabilized temperature.

3.2 Fatigue curves

In order to keep a reasonable amount of figures, the fatigue results are plotted in the Figures 7 and 8 comparing the criterion prediction to the experimental data. At this stage, only the experimental data are of interest. First of all, it is important to underline that these curves can obviously not be used for fatigue design as the amount of samples is too limited to afford a statistical approach. The goal of the fatigue test was only to provide a few failure data to check the tendencies predicted by the heat build-up predictions. The curve obtained for the material A with the sample geometry of type 1 nevertheless provides a richer database and it seems that the scattering remained limited. This also seems to be confirmed by the reasonable alignment obtained for the other materials and sample types. As mentioned previously, the dissipated energy (cf. Section 2) was measured. This allows both the classic S-N curves and a fatigue curve based on dissipated energy to be plotted.

3.3 Heat build-up curves

The repeatability of the tests was systematically checked. Figure 4 illustrates the very good match between two different tests. It is also important to underline that these two heat build-up curves were generated throughout two different sequences of loading (different amount and amplitudes of loading blocks). This shows that there was no dependency of the heat build-up curve on the loading history. This is furthermore confirmed by the observation that the dissipated energy measured at the beginning of the fatigue tests (*i.e.* for one single block), was consistent with the heat build-up curve. This is also illustrated in Figure 4.

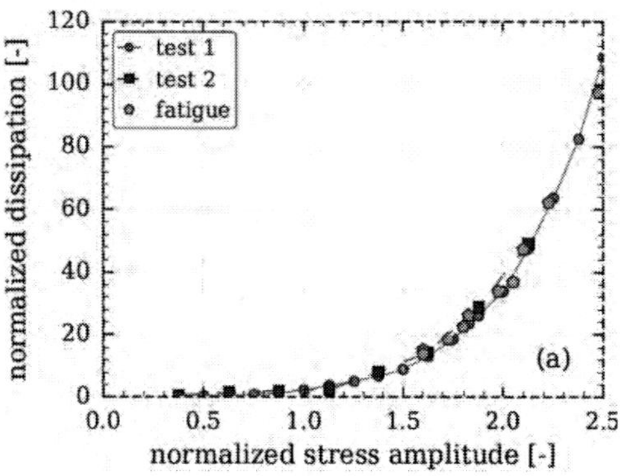

Fig. 4. Effect of the experimental scattering on the heat build-up curves. These data are obtained for the material A_2.

The heat build-up curves obtained for the 4 materials studied are presented in Figure 5. One can observe that the dissipation responses can be markedly different, depending on the grade and the amount of fibers. One can also notice that the heat build-up curves obtained for the same materials, but using two different geometries (references A_1 and A_2), were different. This was due to their different thicknesses, leading to different microstructures.

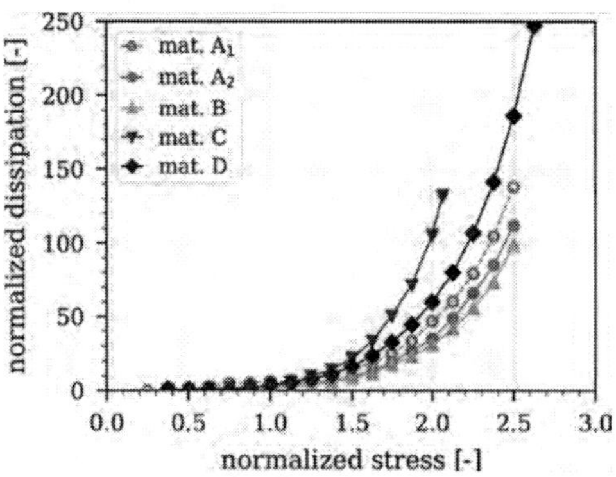

Fig. 5. Heat build-up curves for the studied materials.

4 Analysis and discussion

4.1. Validation of the hypothesis for the heat build-up curves

The objective of this paragraph is to sum-up and check the hypotheses required to deduce the fatigue properties from the heat build-up curve. A first set of hypotheses is related to the evaluation of the dissipated energy. In this case, a protocol valid for adiabatic conditions was applied, on the additional assumption of constant dissipation per cycle. Both conditions were validated in Section 2.4.2. Moreover, the energy evaluated seems representative of the whole section of the sample. Finally, the heat build-up test should be repeatable, which was checked for all the materials tested. A second set of hypotheses is related to the connection to the fatigue tests and the tentative prediction of the fatigue lifetime. First of all, the energy should be evaluated over the failure location, which is consistent here with a failure occurring in the gauge volume. Then, the dissipation evaluated for a fatigue test should be consistent with that obtained for the heat build-up test. That was checked in Section 2.4.2. Finally, the energy evaluated over a limited amount of cycles should be consistent with that dissipated during the fatigue test, which was checked in Section 2.4.2.

4.2. Identification of the dissipated energy criterion parameters

The heat build-up curve provides a relation between the stress amplitude and the dissipated energy. The next step is to use an energy based criterion to relate the dissipated energy to the fatigue lifetime. In former studies [23, 25, 28, 29], protocols using thermal measurements to predict the fatigue life throughout an energy based criterion proved to afford a very efficient prediction of the deterministic Wöhler curve (one sample, within less than one day) and for various materials. This very efficient approach is based on thermal data only and uses a fatigue criterion relating the number of cycles N to initiation to the cyclic dissipated energy $\Delta^\star$ with two parameters b and C that need to be identified

$$\Delta^\star N^b = C \qquad (3)$$

The approach considered in [11, 25] consists in identifying regimes on the heat build-up curves based on a graphical analysis of the curve. This approach is possible only with a large amount of experimental data and thus a high number of blocks. This is possible here, thanks to the evaluation on a few cycles, avoiding any damage accumulation during the heat build-up test. To highlight these regimes (see Figure 8a), we propose here a new approach based on the cyclic dissipation rate computed using finite difference according to

$$\dot{\Delta}^\star\left(\sigma_n + \frac{1}{2}[\sigma_{n+1} - \sigma_n]\right) = \frac{\Delta^\star_{n+1} - \Delta^\star_n}{\sigma_{n+1} - \sigma_n} \qquad (4)$$

These variations are computed using two following points and are associated to the mean stress values related to these points. Figure 6(a) shows the evolution of the cyclic dissipation rate for the experiments presented in Figure 4. The regimes can then be identified easily.

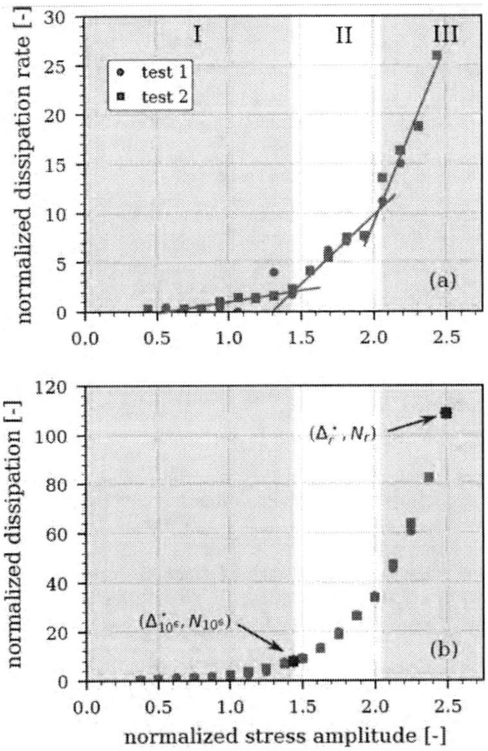

Fig. 6. Determination of the parameters based on a regimes approach for the material A2.

Two sets (Δ^*, N) are needed to identify the parameters analytically. The first set (Δ_R^*, N_R) was obtained from the last block that ran until the sample failure. The second set $(\Delta_{10^6}^*, N_{10^6})$ was obtained from the heat build-up curve at the transition between regimes I and II, as illustrated in Figure 6b. Here, the hypothesis is that the energy evaluated for this transition leads to a lifetime of 10^6cycles.This hypothesis is highly disputable as these materials do not show any clear fatigue limit. This point, initiated in other references [5, 44], deserves further discussion. In particular, the physical nature of the regimes identified is to be clarified as a clear split between the nonlinear effects of viscoelasticity, visco-plasticity and damage remains difficult for these materials [9, 38, and 42]. Modeling, of course, is fundamental in clarifying this point. A very recent paper [45] presents a comparison of the experimental curves to numerical simulations for one of the material investigated here. The model includes elasticity, plasticity and non-linear viscoelasticity coupled to plasticity and strain rate. The identification is achieved on a extensive database and validated throughout the heat buildup curve prediction. The agreement is good and it appears that the first regime can be related to pure visco-elastic effects, that the second one is correlated to the onset of plasticity even if the contribution remains limited to the overall dissipation and that the third regime is induced both by plastic effects and by the coupling of plasticity to non-linear visco-elasticity. In this study, the focus is on the application of the existing protocol and checking its relevancy for the materials tested. The identification procedure was applied to the 4 materials studied and the two samples' geometry (material A).

4.3. Comparaison between fatigue curves and energetic predictions

Figures 7 and 8 present the comparison between the experimental data obtained from the fatigue tests and the prediction of the dissipated energy criterion. Above, the fatigue curves are plotted using the dissipated energy as the fatigue indicator. Below, the classic S-N curve is plotted.

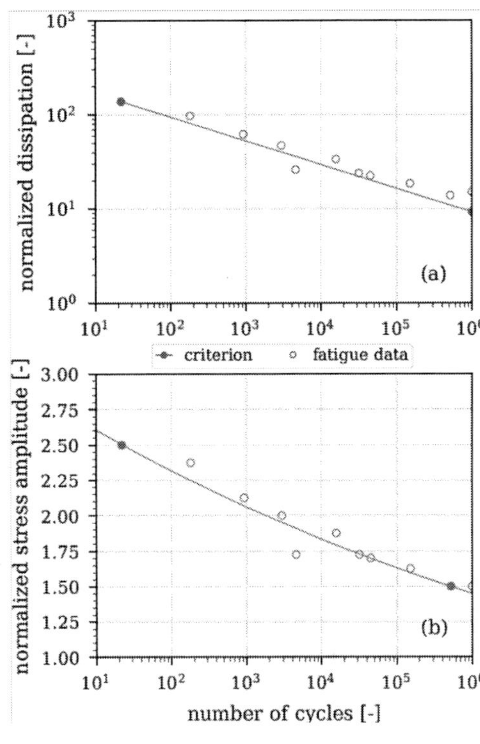

Fig. 7. Comparison between the fatigue data (empty markers) and the energetic criterion (line and filled markers) for the material A1 in terms of dissipation versus number of cycles (a) and nominal stress versus number of cycles (b).

In both cases, the correlation obtained is fairly good for all the materials and samples considered. It is important to underline that different heat build-up curves could nevertheless lead to similar S-N curves. A very interesting result is that, for the same material, different fatigue properties can be found due to different microstructures induced by injection. In any case, the technique manages to describe well the fatigue resistance of the materials investigated. It therefore appears to be a relevant tool to evaluate the best candidates for fatigue applications, prior to generating full fatigue curves using classic fatigue campaigns.

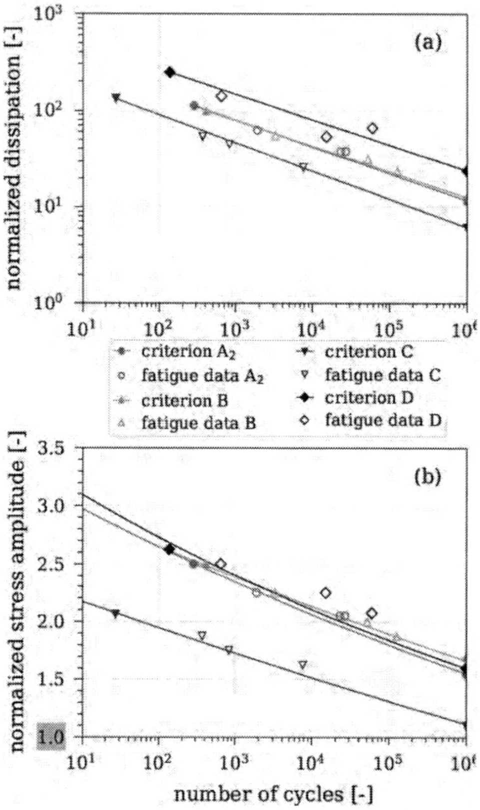

Fig. 8. Comparison between the fatigue data (empty markers) and the energetic criterion (line and filled markers) for the type 2 specimen in terms of dissipation versus number of cycles (a) and nominal stress versus number of cycles (b).

5 Conclusions

This paper focused on thermostable matrices, reinforced with short carbon fibers and aimed at three main objectives. The first was to highlight and investigate if a heat build-up approach can be applied to these materials. This was checked both from a measurement point of view (adiabaticity, spatial distribution of the dissipation sources) and from a criterion point of view (evaluation possible over a limited amount of cycles). Compared to previous papers on other materials published by the authors, the adiabatic analysis was here performed over each pixel, throughout the analysis of the temporal variation. A second objective was to check the ability of the technique to predict the mean fatigue curve of these materials. This was assessed on a wide database for one material and one sample geometry. The last objective was to challenge the capacity of the heat build-up technique to be used as a fast screening tool, i.e. to discriminate several possible materials quickly, for different grades, suppliers and carbon fiber amounts. The results obtained on 4 materials illustrated that the heat build-up protocol, defined and applied in this study, managed to predict the fatigue resistance of the materials investigated well, with one sample and within one day per material tested. It therefore appears to be a relevant tool to select the best candidates for fatigue applications from a given panel, prior to generating full fatigue curves using classic fatigue campaigns.

References

1. A. Zago, G. Springer, Fatigue lives of short _ber reinforced thermoplastics parts, Journal of Reinforced Plastics and Composites 20 (2001) 606-620.
2. C. Sonsino, E. Moosbrugger, Fatigue design of highly loaded short-glass-fibre reinforced polyamide parts in engine compartments, International Journal of Fatigue 30 (2008) 1279-1288.
3. Klimkeit, B., Etude expérimentale et modélisation du comportement en fatigue multiaxiale d'un polymère renforcé pour application automobile, Ph.D. thesis, ENSMA (2014).
4. A. Bernasconi, P. Davoli, C. Armanni, Fatigue strength of a clutch pedal made of reprocessed short glass fibre reinforced polyamide, International Journal of Fatigue 32 (2010) 100-107.
5. Serrano, L., Thermomechanical characterization of the fatigue behaviour of short fibers reinforced thermoplastic, Ph.D. thesis, Université de Bretagne Occidentale (2015).
6. S. Mortazavian, A. Fatemi, Fatigue behavior and modeling of short fiber reinforced polymer composites: A literature review, International Journal of Fatigue 70 (2015) 297-321.
7. M. De Monte, E. Moosbrugger, M. Quaresimin, Inuence of temperature and thickness on the off-axis behaviour of short glass fibre reinforced polyamide 6.6 - quasi-static loading, Composites Part A: Applied Science and Manufacturing 38 (2010) 859-871.
8. B. Klimkeit, Y. Nadot, S. Castagnet, C. Nadot-Martin, C. Dumas, S. Bergamo, C. Sonsino, A. Buter, Multiaxial fatigue life assessment for reinforced polymers, International Journal of Fatigue 33 (2011) 766-780.
9. A. Launay, M. Maitournam, Y. Marco, I. Raoult, Multiaxial fatigue models for short glass fiber reinforced polyamide. Part II: fatigue life estimation, International Journal of Fatigue 47 (2013) 390-406.
10. A. Launay, Y. Marco, M. Maitournam, I. Raoult, Modelling the inuence of temperature and relative humidity on the time-dependent mechanical behaviour of a short glass fibre reinforced polyamide, Mechanics of Materials 56 (2013) 1-10.
11. L. Serrano, Y. Marco, V. Le Saux, G. Robert, P. Charrier, Fast prediction of the fatigue behavior of short-fiber-reinforced thermoplastics based on heat build-up measurements: application to heterogeneous cases, Continuum Mechanics and Thermodynamics 29 (2017) 1113-1133.
12. C. Stromeyer, The determination of fatigue limits under alternating stress conditions, Proceedings of the Royal Society of London. Series A. 90 (1914) 411-425.
13. H. Moore, J. Kommers, Fatigue of metals under repeated stress, Chemical and Metallurgical Engineering 25 (1921) 1141-1144.

14. A. Galtier, Contribution à l'étude de l'endommagement des aciers sous sollicitations uni ou multiaxiales, Ph.D. thesis, ENSAM (1993).
15. M. Luong, Fatigue limit evaluation of metals using an infrared thermographic technique, Mechanics of Materials 28 (1998) 155-163.
16. J. Krapez, D. Pacou, G. Gardette, Lock-in thermography and fatigue limit of metals, in: Proceedings of QIRT'2000, Reims, 2000, pp. 277-282.
17. G. La Rosa, A. Risitano, Thermographic methodology for rapid determination of the fatigue limit of materials and mechanical components, International Journal of Fatigue 22 (2001) 65-73.
18. C. Doudard, S. Calloch, F. Hild, P. Cugy, A. Galtier, Identification of scatter in high cycle fatigue from temperature measurements, Comptes Rendus M_ecanique 10 (2004) 795-801.
19. A. Ezanno, C. Doudard, S. Calloch, J. Heuzé, A new approach to characterizing and modeling the high cycle fatigue properties of cast materials based on self-heating measurements under cyclic loadings, International Journal of Fatigue 47 (2004) 232-243.
20. R. Munier, C. Doudard, S. Calloch, B. Weber, Determination of high cycle fatigue properties of a wide range of steel sheet grades from self-heating measurements, International Journal of Fatigue 63 (2014) 46-61.
21. A. Chrysochoos, H. Louche, An infrared image processing to analyse the calori_c e_ects accompanying strain localisation, International Journal of Engineering Science 38 (2000) 1759-1788.
22. V. Le Saux, C. Doudard, Proposition of a compensated pixelwise calibration for photonic infrared cameras and comparison to classic calibration procedures: case of thermoelastic stress analysis, Infrared Physics and Technology 80 (2017) 83-92.
23. V. Le Saux, Y. Marco, C. Doudard, S. Calloch, P. Charrier, Fast evaluation of the fatigue lifetime of rubber-like materials based on a heat build-up protocol and micro-tomography measurements, International Journal of Fatigue 32 (2010) 1582-1590.
24. C. Cruanes, G. Berton, F. Lacroix, S. Meo, N. Ranganathan, Study of the fatigue behavior of the chloroprene rubber for uniaxial tests with infrared method, Elastomery 18 (2014) 3-9.
25. Y. Marco, I. Masquelier, V. Le Saux, P. Charrier, Fast prediction of the W□ohler curve from thermal measurements for a wide range of NR and SBR compounds, Rubber Chemistry and Technology 90 (2017) 487-507.
26. T. L., M. Karama, B. Lorrain, Damage evolution and infrared thermography in woven composite laminates under fatigue loading, International Journal of Fatigue 28 (2006) 1867-1872.
27. M. Liakat, K. Khonsari, Analysis and life prediction of a composite laminate under cyclic loading, Composites Part B: Engineering 84 (2016) 98-108.
208. G. Meneghetti, M. Quaresimin, Fatigue strength assessment of a short fiber composite based on the specific heat dissipation, Composites Part B: Engineering 42 (2011) 217-225.
29. L. Jégou, Y. Marco, V. Le Saux, S. Calloch, Fast prediction of the wöhler curve from heat build-up measurements on short fiber reinforced thermoplastics, International Journal of Fatigue 47 (2012) 259-267.
30. M. Poncelet, C. Doudard, S. Calloch, B. Weber, F. Hild, Probabilistic multiscale models and measurements of self-heating under multiaxial high cycle fatigue, Journal of the Mechanics and Physics of Solids 58 (2010) 578-593.
31. V. Le Saux, Y. Marco, S. Calloch, P. Charrier, Evaluation of the fatigue defect population in an elastomer using x-ray computed micro-tomography, Polymer Engineering and Science 51 (2011) 1253-1263.
32. Y. Marco, B. Huneau, I. Masquelier, V. Le Saux, P. Charrier, Prediction of fatigue properties of natural rubber based on the descriptions of the cracks population and of the dissipated energy, Polymer Testing 59 (2017) 67-74.
33. A. Constantinescu, K. Dang Van, M. Maitournam, A unified approach for high and low cycle fatigue based on shakedown concepts, Fatigue and Fracture of Engineering Materials and Structures 26 (2003) 1065-1077.
34. A. Benoit, M. Maitournam, L. Rémy, F. Oger, Cyclic behaviour of structures under thermomechanical loadings: Application to exhaust manifolds, International Journal of Fatigue 38 (2012) 65-74.
35. A. Drozdov, A. Al-Mulla, R. Gupta, Finite viscoplasticity of polycarbonate reinforced with short glass fibers, Mechanics of Materials 37 (2005) 473-491.
36. H. Nouri, F. Meraghni, P. Lory, Fatigue damage model for injection-molded short glass fibre reinforced thermoplastics, International Journal of Fatigue 31 (2009) 934-942.
37. A. Andriyana, N. Billon, L. Silva, Mechanical response of a short fiber-reinforced thermoplastic: experimental investigation and continuum mechanical loading, European Journal of Mechanics A 29 (2010) 1065-1077.
38. A. Launay, M. Maitournam, Y. Marco, I. Raoult, F. Szmytka, Cyclic behaviour of short glass fibre reinforced polyamide: experimental study and constitutive equations, International Journal of Plasticity 27 (2011) 1267-1293.
39. M. Arif, F. Meraghni, N. Chemisky, Y. ad Despringre, G. Robert, In situ damage mechanisms investigation of PA66/GF30 composite: effect of relative humidity, Composites Part B: Engineering 58 (2014) 487-495.
40. H. Rolland, N. Saintier, P. Wilson, J. Merzeau, G. Robert, In situ X-ray tomography investigation on damage mechanisms in short glass fibre reinforced thermoplastics: Effects of fibre orientation and relative humidity, Composites Part B: Engineering 109 (2017) 1359-1368.
41. A. Launay, M. Maitournam, Y. Marco, I. Raoult, Multiaxial fatigue models for short glass fiber reinforced polyamide. Part I: nonlinear anisotropic constitutive behavior for cyclic response, International Journal of Fatigue 47 (2013) 382-389.
42. A. Krairi, I. Doghri, A thermodynamically-based constitutive model for thermoplastic polymers coupling viscoelasticity, viscoplasticity and ductile damage, International Journal of Plasticity 60 (2014) 163-181.
43. I. Masquelier, Y. Marco, V. Le Saux, S. Calloch, P. Charrier, Determination of dissipated energy fields from temperature mappings on a rubber-like structural sample: Experiments and comparison to numerical simulations, Mechanics of Materials 80 (2015) 113-123.
44. Y. Marco, V. Le Saux, L. J_egou, A. Launay, L. Serrano, I. Raoult, S. Calloch, Dissipation analysis in SFRP structural samples: thermomechanical analysis and comparison to numerical simulations, International Journal of Fatigue 67 (2014) 142-150.
45. V. Le Saux, Belenos users manual, Internal report. ENSTA Bretagne (2017).
46. L. Leveuf, L. Navratil, V. Le Saux, Y. Marco, J. Olhagaray, S. Leclercq, Constitutive equations for the cyclic behaviour of short carbon fibre-reinforced thermoplastics and identification on a uniaxial database, Continuum Mechanics and Thermodynamics accepted (2018) 1-20.
47. L. Leveuf, Y. Marco, V. Le Saux, J. Olhagaray, S. Leclercq, Fast screening of the fatigue properties of thermoplastics reinforced with short carbon fibers based on thermal measurements, Submitted to Polymer Testing (January, 2018)

Effect of UV Ageing on the fatigue life of bulk polyethylene

Hamza Lamnii[1,2], *Moussa* Nait-Abdelaziz[1,*], *Georges* Ayoub[2,3], *Jean-Michel* Gloaguen[4], *Ulrich* Maschke[4] and *Bilal* Mansoor[3]

[1]Universite de Lille, FRE 3723 - LML - Laboratoire de Mécanique de Lille, 59000 Lille, France
[2]Mechanical Engineering Program, Texas A&M University at Qatar, Education City, Qatar
[3]Industrial and Manufacturing System Engineering department, University of Michigan Dearborn, 48128 Dearborn, USA
[4]Université de Lille, Unité Matériaux Et Transformations (UMET), UMR CNRS 8207, F-59650 Villeneuve d'Ascq, France

Abstract. Polymers operating in various weathering conditions must be assessed for lifetime performance. Particularly, ultraviolet (UV) radiations alters the chemical structure and therefore affect the mechanical and fatigue properties. The UV irradiation alters the polymer chemical structure, which results into a degradation of the mechanical and fatigue behavior of the polymer. The polymer properties degradation due to UV irradiation is the result of a competitive process of chain scission versus post-crosslinking. Although few studied investigated the effect of UV irradiation on the mechanical behaviour of thermoplastics, fewer examined the UV irradiation effect on the fatigue life of polymers. This study focuses on investigating the effect of UV irradiation on the fatigue properties of bulk semi-crystalline polymer; the low density Polyethylene (LDPE). Tensile specimens were exposed to different dose values of UV irradiation then subjected to fatigue loading. The fatigue tests were achieved under constant stress amplitude at a frequency of 1Hz. The results show an important decrease of the fatigue limit with increasing absorbed UV irradiation dose.

1 Introduction

1.1 Context of the study

Because of their low cost, wide range of mechanical properties, and adaptability, polymers are extensively used in a large range of applications, such as aerospace and automotive industries, fluid transport, medical sectors and many others. Polymers have a large variety of molecular and supramolecular structures, separated on two basic macromolecular arrangements: amorphous and semi-crystalline structures [1]. Polyethylene (PE) is a semi-crystalline polymer that is a largely produced polymer, because of its cheap cost, outstanding electrical and mechanical properties, suitable chemical resistance, toughness, flexibility and transparency when used as thin films [2]. Polymers and particularly polyethylene microstructure and mechanical properties degrade at different rates depending on the operation environment. Studies show that polymer degradation can be induced by five processes [3]: photo-degradation by UV exposure [4], oxidation by chemical additives [5], thermal degradation by heat [6], mechanical degradation by mechanical effects [7] and biodegradation by micro-organisms [8]. Depending on the environmental and operating conditions, a degradation mechanism might be preponderant relatively to the others[9]. Hence, it is necessary to identify this preponderant degradation mechanism to control through design the material's durability.

1.2 Ultraviolet effects on polymers

Because most of the materials operate under various weathering conditions, their lifetime must be assessed [10]. Photo-oxidation in polymers initiates a competitive degradation mechanism of chain scission versus crosslinking. Although few studied investigated the effect of UV irradiation on the mechanical behaviour of thermoplastics, fewer examined the UV irradiation effect on the fatigue life of polymers. The photo-degradation consists in the absorption of a photon of light by the polymer resulting into an alteration of the chemical and macromolecular structure of the polymer[11]. During polyolefin photo-oxidation the rates of macromolecular scission and crosslinking changes with the absorbed UV dose[12]. Chain scission mechanism is driven by the free radicals resulting from the decomposition of hydroperoxide in the presence of UV and oxygen [13]. This mechanism takes place in the amorphous phase and in the amorphous-lamellar interface [14], where the diffusion of oxygen in those phases is easier[9]. On the other hand, the crystalline phase remains typically inert [15]. This unwelcome degradation can be accelerated by increasing the absorbed UV radiation by the polymer through longer exposure time[16][17]. Several experimental techniques could be employed to point out changes of the chemical and mechanical properties undergone by polymers during and after degradation. The UV-degradation is a surface mechanism which

* Corresponding author: moussa.nait-abdelaziz@polytech-lille.fr

 https://doi.org/10.1051/matecconf/201816508002

generally affects a thin surface layer, which thickness ranges from a few micrometres up to 1 mm at maximum [18]. At last, growth in crystallinity [19][20], tensile strength [21], modulus of elasticity [22], hardness [23], density [24], etc, is an indicator of the prevalence of crosslinking mechanism.

1.3 Equivalence principle

As polyethylene pipes are widely used in water and natural gas distribution system, they could be subjected to both UV radiations and variable mechanical loading. It is therefore important to examine the effects of UV aging on the ultimate properties and more precisely on the fatigue limit.

The time–temperature equivalence principle is generally used to build a master curve of a given material mechanical property from a set of data obtained at different temperatures or strain rates [25]. In other words, the time–temperature principle consists in making a master curve from the measurements of a mechanical property (fatigue life for example) in a specified range of temperatures (or strain rates). The resulting curves are then superimposed by shifting them via a shift factor noted a_T that is determined from equation (1). This equivalence, suited for viscoelastic materials, allows to estimate the chosen property in the range of temperatures or strain rates experimentally unreachable (very long-time process for example). One of the most known model is the Arrhenius type law, which expresses the shift factor in the following form:

$$a_T = exp\left(\frac{E_a}{RT}\right) \quad (1)$$

where E_a, is the effective activation energy, R is the gas constant, and T is absolute temperature.

Furthermore, when radiation (e.g., UV) is added to temperature in the aging process, this latter becomes more complex. Thermal aging leads to establishing free radicals at a thermally weak region of the polymer chains. With UV radiation, the energy involved produces more varied free radical species.

When specimen are submitted to UV irradiation, several attempted to extract an equivalence principle [25][26][27]. The most suitable in our case is based on dose equivalency which is consistent with the concept of a limited diffusion inducing an oxidative degradation in a thin layer of the specimen [27]. In this case, the shift factor is simply expressed as follows:

$$a_R = 1 + R^x \quad (2)$$

$$t' = \frac{t}{a_R} \quad (3)$$

Where R is the dose in [J/m²] and x is a parameter to be determined through the experimental results. this value being chosen to get the best fitting of the experimental data. The reduced time (or service lifetime under irradiation ageing) is expressed according to equation (3).

2 Materials and Experimental Setups

2.1 Material

The material is a low-density polyethylene (LDPE)which was supplied by QAPCO©. The physical properties of the material as provided by the supplier are as follows: Melting temperature, 109°C; Melt Flow Index 0,30 (g/10min), density (0.92 g/cm^3).

The geometry of the specimen is shown in figure 1.The fatigue samples were extracted from plates of 2 mm thickness. This geometry of the fatigue sample was chosen to localize the deformation in the centre of the specimen allowing fatigue crack location in the smallest cross section. Note that the radius is sufficiently large to ensure a triaxiality ratio close to that of uniaxial tension.

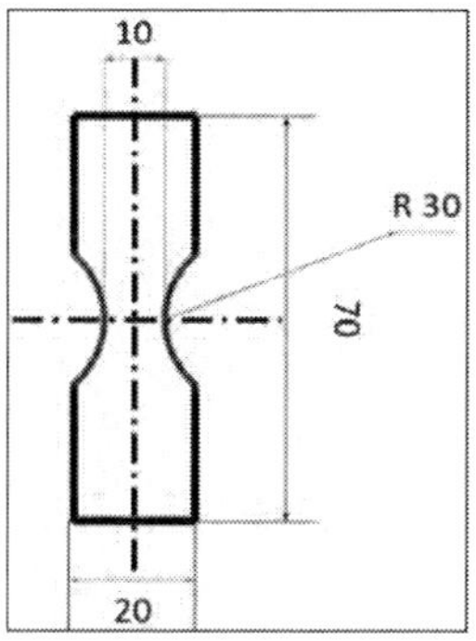

Fig. 1. Specimen geometry (dimensions in mm)

2.2 Mechanical analysis

The fatigue tests were performed on an electromechanical device (INSTRON Electro-pulse E3000) as shown in figure. 2.

Fig. 2. Experimental setup (Instron E 3000)

The fatigue tests were achieved under force control with a R ratio equal to 0. The load control condition was more

https://doi.org/10.1051/matecconf/201816508002

desirable than displacement control condition as creep is induced during fatigue test (the minimum strain increases) which results into compression loading and hence buckling of the sample as shown in figure 3.

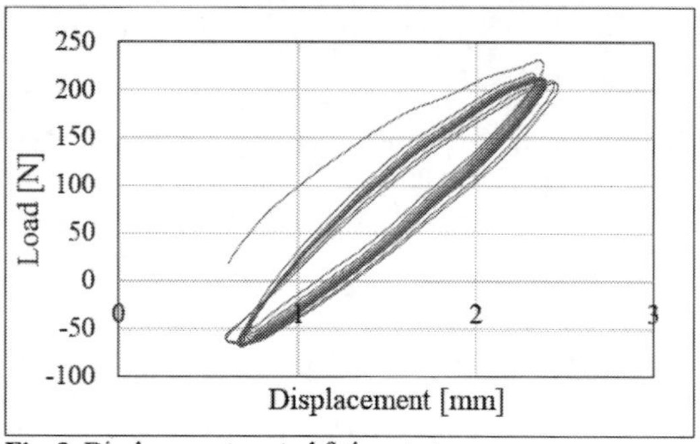

Fig. 3. Displacement control fatigue test

The tensile tests were achieved using a video extensometer (Video-Traction) device which allows keeping the strain rate at a constant value of $10^{-3}s^{-1}$during the test. The method principle is based upon following 4 markers painted on the front surface of the sample, two marks are aligned along the tensile direction and the two other marks are aligned along the transverse direction in the minimum cross section.

2.3 UV aging procedure

The aging procedure consists in subjecting the specimen to UV radiations. During the procedure, exposure time is controlled in order to control the amount of energy absorbed by the specimen. A Dymax UV light-curing flood (2000) lamp was used to irradiate the specimens.

Fig. 4 .Experimental set up for UV radiations

The specimens were placed side by side under the UV lamps and exposed to irradiation (figure 4). The irradiation process is achieved in the absence of controlled temperature and humidity. Temperature measurements at the surface of the specimen raises up to 69°C due to the irradiation lamps. At mid-test, the specimen is turned on the other side in order to obtain the same exposure times on both sides. Table 1gives the corresponding values of the UV dose correlating with a given exposure time.

Table 1. Time exposure and corresponding UV dose

UV duration [hour]	Dose per side [mJ/mm²]
0	0
5	580,03
9	1038,98
13	1497,93

3 Experimental results

3.1Tensile tests

The tensile tests were performed on as received and aged samples. The video extensometer allows to calculate the true strain and consequently the true stress in the minimum cross section. Figure 5 shows the true stress as a function of true strain for the as received material and two samples with different time exposures to UV aging. It is clearly shown that both the stiffness and the yield stress are not affected by the UV aging. The yield stress of the material is approximately 10 MPa. Nevertheless, the ultimate properties are affected by UV aging as indicated by table 2 for the strain at failure.

Table 2. Engineering strain at failure

UV[hours]	Engineering strain at failure %
0	425%
9	239%
13	95%

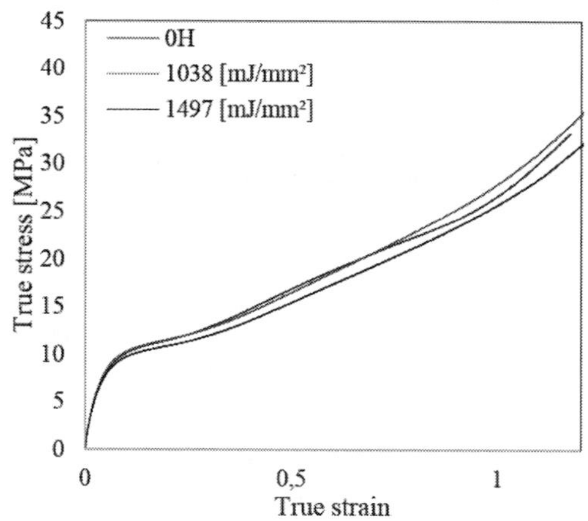

Fig. 5. Tensile tests of as received and UV-aged LDPE

3.2 Fatigue results

The fatigue tests under load control and constant amplitude were carried out up to the total failure of the specimens. The number of cycles to failure was therefore taken (as a first approximate) as the fatigue life. Figure 6 shows the evolution of the applied stress as a function of the fatigue life (Wohler curve). The data, plotted in a semi-logarithmic scale, correspond to the results obtained for 2 aging conditions: as received specimen and aged specimen (in this case, time exposure is 13 hours corresponding to an energy of 1497 mJ/mm²).

As clearly shown, the data are quite scattered. Even scattering is generally observed when dealing with fatigue, one can note that in this case, it is relatively important. This could be the consequence of the self-heating induced by cyclic loading, the material being viscoelastic.

Nevertheless, when fitting the two set of data separately with straight lines (in the logarithmic scale), the effect of aging is clearly highlighted.

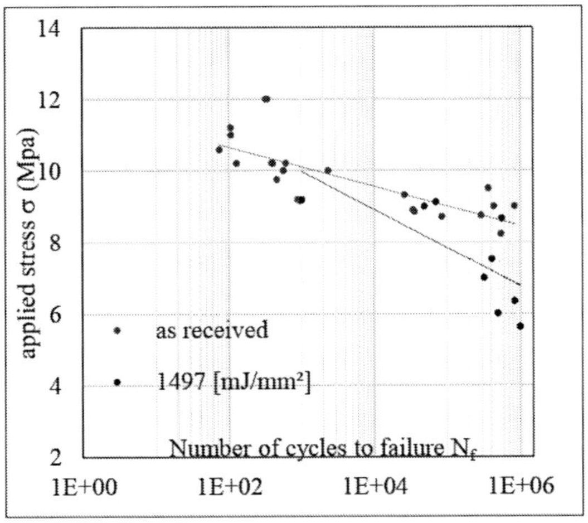

Fig. 6. Wohler curve for as received and UV-aged LDPE

Indeed, the equations of the straight lines are written in the following form:

$$\sigma = a.log_{10}(N_f) + b \qquad (4)$$

Either the slope or the fatigue limit decreases with UV aging, as reported in the following table. This kind of behaviour was already reported in the literature but for varying R ratios [25] . In our case, the load ratio is constant and equal to 0. But, if we assume that UV aging induces mechanical properties change in a thin layer, then the specimen must be seen as a composite structure. Therefore, even the global R ratio is 0, residual stresses acting in the layers could modify locally the value of the load ratio.

Table 3. Wohler curves slopes

UV [mJ/mm²] per side	a (MPa)
0	-0,238
1038	-0,355
1497	-0,464

Figure. 6 shows the evolution of the shift factor (equation 2) as function of the dose undergone by the specimen. In our case, the value of the parameter x is 6.

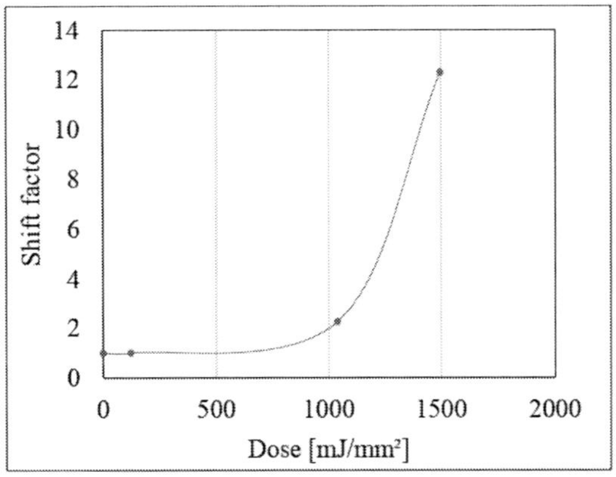

Fig. 7. shift factor function of absorbed dose

According to figure. 7, the power law evolution is an indicator of the damage severity provoked by UV radiations.

Applying the equivalence principle by multiplying the fatigue life data by the corresponding shift factor allows to obtain a master curve as shown in figure. 8.

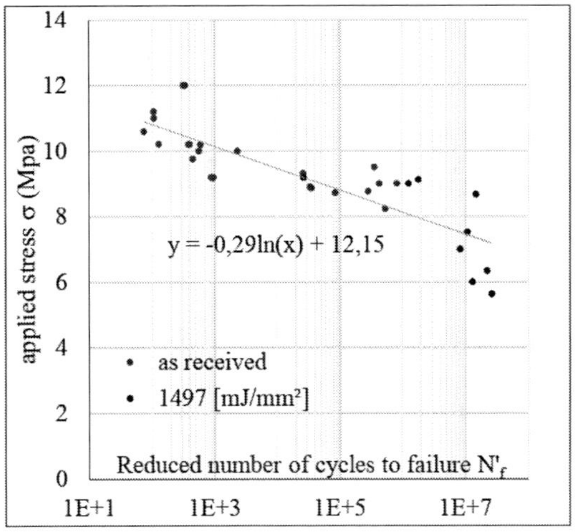

Fig. 8. Fatigue life master curve

4 Conclusion

In this paper, we have investigated the fatigue life behaviour of bulk low-density polyethylene submitted to UV-radiations. UV radiations generally modify the molecular structure inducing chain scission mechanism in a thin layer of the irradiated surface which thickness depends on the absorbed energy.
We have found that UV aging does not affect the mechanical response of the bulk material but the fracture properties. Indeed, stresses and strains at break decreases with increasing UV absorbed dose.
Fatigue tests were performed under load control to avoid compression loading induced by creep during fatigue cycling. We have noted an important scattering of the fatigue life data partly induced to self-heating.

Nevertheless, the whole data can be unified using a master curve which is obtained by applying an equivalence principle based upon the radiation dose undergone by the specimen.
The decrease of the fatigue life due to aging may be explained by the existence of micro cracks initiated at the surface, which density clearly depends on the UV exposure time. Another source of the degradation of the mechanical ultimate properties could be the modification of the microstructure in the thin layer affected by UV radiations. That means the specimen behave globally as a composite structure, the mechanical properties exhibiting a gradient along the thickness.
As an outlook, this work must be completed by a deeper analysis of the fracture surface and by an estimation of the affected thickness as a function of the irradiation dose. An analysis of the microcracks in terms of size and density is also required to estimate the damage induced by UV aging.

Acknowledgement

This work was made possible by NPRP Grant # 7-1562-2-571 from Qatar National Research Fund (A member of Qatar Foundation). The statements made herein are solely the responsibility of the authors.

References

1. Goerg H. Michler. *Electron Microscopy of Polymers*. (Springer Berlin Heidelberg, 2008). doi:10.1007/978-3-540-36352-1
2. Brydson, J. *Plastics Materials (7th Edition) - 10 Polyethylene - Knovel*. (1999).
3. Calmon-Decriaud, A., Bellon-Maurel, V. & Silvestre, F. Standard Methods for Testing the Aerobic Biodegradation of Polymeric Materials. Review and Perspectives. in *Blockcopolymers - Polyelectrolytes - Biodegradation* (eds. Bellon-Maurel, V. et al.) 207–226 (Springer Berlin Heidelberg, 1998). doi:10.1007/3-540-69191-X_3
4. Babaghayou, M. I. *et al.*Photodegradation characterization and heterogeneity evaluation of the exposed and unexposed faces of stabilized and unstabilized LDPE films. *Mater. Des.***111,** 279–290 (2016).
5. Michelot, H. *et al.* Effect of drug precursors and chemicals relevant to clandestine laboratory investigation on plastic bags used for collection and storage. *Forensic Sci. Int.***273,** 106–112 (2017).
6. El-Mazry, C., Ben Hassine, M., Correc, O. & Colin, X. Thermal oxidation kinetics of additive free polyamide 6-6. *Polym. Degrad. Stab.***98,** 22–36 (2013).
7. White, J. R. &Teh, J. W. Fatigue of viscoelastic polymers: 2. Fractography. *Polymer***20,** 764–771 (1979).
8. Kumar Sen, S. & Raut, S. Microbial degradation of low density polyethylene (LDPE): A review. *J. Environ. Chem. Eng.***3,** 462–473 (2015).
9. Kyrikou, I. &Briassoulis, D. Biodegradation of Agricultural Plastic Films: A Critical Review. *J. Polym. Environ.***15,** 125–150 (2007).
10. Feldman, D. Polymer Weathering: Photo-Oxidation. *J. Polym. Environ.***10,** 163–173 (2002).
11. Rabek, J. F. Physical aspects of the photodegradation of polymers. in *Polymer Photodegradation* 1–23 (Springer, Dordrecht, 1995). doi:10.1007/978-94-011-1274-1_1
12. White, J. R. &Shyichuk, A. V. Macromolecular scission and crosslinking rate changes during polyolefin photo-oxidation. *Polym. Degrad. Stab.***92,** 1161–1168 (2007).
13. Hsu, Y.-C. *et al.* A fundamental study on photo-oxidative degradation of linear low density polyethylene films at embrittlement. *Polymer***53,** 2385–2393 (2012).
14. Roy, P. K., Surekha, P., Raman, R. & Rajagopal, C. Investigating the role of metal oxidation state on the degradation behaviour of LDPE. *Polym. Degrad. Stab.***94,** 1033–1039 (2009).
15. Koutny, M., Lemaire, J. &Delort, A.-M. Biodegradation of polyethylene films with prooxidant additives. *Chemosphere***64,** 1243–1252 (2006).
16. Tidjani, A. Comparison of formation of oxidation products during photo-oxidation of linear low density polyethylene under different natural and accelerated weathering conditions. *Polym. Degrad. Stab.***68,** 465–469 (2000).
17. Şener, S. *et al.* World Conference on Technology, Innovation and EntrepreneurshipImpact of Solar Radiation Effects on the Physicochemical Properties of Polyethylene (PE) Plastic Film. *Procedia - Soc. Behav. Sci.***195,** 2210–2217 (2015).
18. Cunliffe, A. V. & Davis, A. Photo-oxidation of thick polymer samples—Part II: The influence of oxygen diffusion on the natural and artificial weathering of polyolefins. *Polym. Degrad. Stab.***4,** 17–37 (1982).
19. Yakimets, I., Lai, D. &Guigon, M. Effect of photo-oxidation cracks on behaviour of thick polypropylene samples. *Polym. Degrad. Stab.***86,** 59–67 (2004).
20. Valadez-Gonzalez, A., Cervantes-Uc, J. M. &Veleva, L. Mineral filler influence on the photo-oxidation of high density polyethylene: I. Accelerated UV chamber exposure test. *Polym. Degrad. Stab.***63,** 253–260 (1999).
21. Kelly, C. T. & White, J. R. Photo-degradation of polyethylene and polypropylene at slow strain-rate. *Polym. Degrad. Stab.***56,** 367–383 (1997).
22. Carrasco, F., Pagès, P., Pascual, S. & Colom, X. Artificial aging of high-density polyethylene by

ultraviolet irradiation. *Eur. Polym. J.***37,** 1457–1464 (2001).

23. Tavares, A. C., Gulmine, J. V., Lepienski, C. M. &Akcelrud, L. The effect of accelerated aging on the surface mechanical properties of polyethylene. *Polym. Degrad. Stab.***81,** 367–373 (2003).
24. Gulmine, J. V., Janissek, P. R., Heise, H. M. &Akcelrud, L. Degradation profile of polyethylene after artificial accelerated weathering. *Polym. Degrad. Stab.***79,** 385–397 (2003).
25. Guedes, R. M. *Creep and Fatigue in Polymer Matrix Composites*. (Elsevier, 2010).
26. Gillen, K. T. & Clough, R. L. Time-temperature-dose rate superposition: A methodology for extrapolating accelerated radiation aging data to low dose rate conditions. *Polym. Degrad. Stab.***24,** 137–168 (1989).
27. Burnay, S. G. &Hitchon, J. W. Prediction of service lifetimes of elastomeric seals during radiation ageing. *J. Nucl. Mater.***131,** 197–207 (1985).

Fatigue of short fibre reinforced polymers: from material process to fatigue life of industrial components

Pablo Wilson[1*] and *Peter* Heyes[2]

[1]Renault S.A.S., 1 avenue du Golf, 78084 Guyancourt, France
[2]HBM Prenscia, Catcliffe, Rotherham S60 5WG, England

Abstract. For many years, SFRPs (short fibre reinforced polyamides) have been used in the automotive industry as a means to reduce vehicle weight. However, their complex anisotropic and heterogeneous microstructure requires sophisticated material characterisation and simulation. This study aims at presenting the simulation strategy adopted by an automotive company address these challenges. The manufacturing process is first simulated and correlated with tomography analysis. Then, based on the numerical microstructure, integrative simulation is used to analyse and predict mechanical behaviour of fatigue coupons and industrial parts. Lastly, two fatigue criteria based on strain energy density are presented and fatigue lives of coupons and industrial parts assessed.

1 Introduction

Faced with the environmental challenges of the 21st century, the automotive industry has had to adapt its material strategy. Indeed, current and future regulations are ever more demanding. To minimize the environmental footprint, vehicle weight has to be kept to a minimum to lower gas consumption, while engine performance has to be maximized to increase vehicle efficiency.

Plastics in general, and specifically short fibre reinforced plastics (SFRP) are often a good choice for more mechanically demanding applications: they combine a good specific mechanical strength, high production rate and low manufacturing costs, while providing the capability to be moulded in complex geometries.

However, designing such mechanical parts has generated new difficulties: the material microstructure caused by the fibre orientation produces material anisotropy. All other things being equal, the average fibre orientation can vastly modify mechanical characteristics [1].

In order to take into account the effect of the microstructure on the mechanical behaviour of SFRP, new lifetime assessment methods based on finite element analysis (FEA) have been developed. Firstly, the injection moulding is simulated and the fibre orientation in the part estimated. Secondly, integrative simulation is performed: this allows capture of the local anisotropy of the material and provides reliable mechanical fields (stress and strain). Lastly, since mechanical behaviour is here considered uncoupled with damage, the results can be post-processed and, using an appropriate fatigue criterion, the fatigue life assessed.

In this study, the different steps of this fatigue life assessment are presented on fatigue coupons and tested on industrial prototyped parts.

2 Material and testing

For confidentiality reasons, the results have been normalized. In order to remain intelligible, stress values have all been normalized by the same value σ_{UTS}(ultimate tensile stress at 23°C, 50% relative humidity of the 0° specimen).

2.1 Coupons material and preparations

The material used in this study is a polyamide reinforced with 50% weight glass fibres (hereafter referred to as PA66GF50). Further details on material are not disclosed for confidentiality reasons.

The fatigue test coupon geometry is presented in Figure 1, accordingly to ISO 527B standard.

These are machined from injected plates (see Figure 2). By using this method, different specimen "orientations" can be used. The "0°" orientation hence corresponds to the specimen machined following the injection direction, while the "90°" one is taken perpendicular to it. The "45°" specimen is cut at a 45 degree angle from the injection direction.

Considering the sensitivity of the material to humidity (see [2]), the material is first conditioned at

* Corresponding author: pablo.wilson@renault.com

50% relative humidity using an accelerated conditioning method described in the ISO 1110 standard.

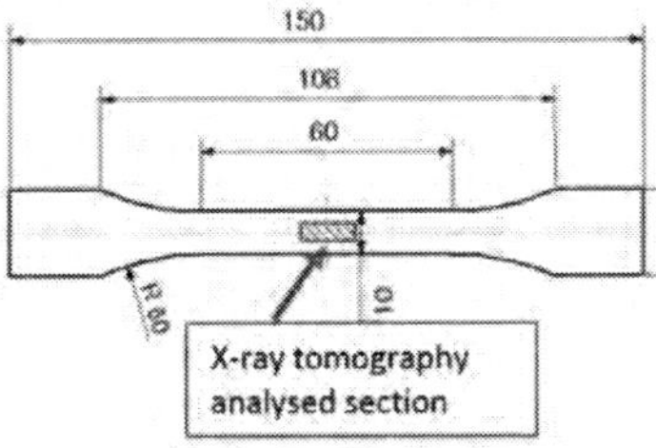

Figure 1 : Fatigue test coupon used (ISO 527B). The plate thickness is 3mm.

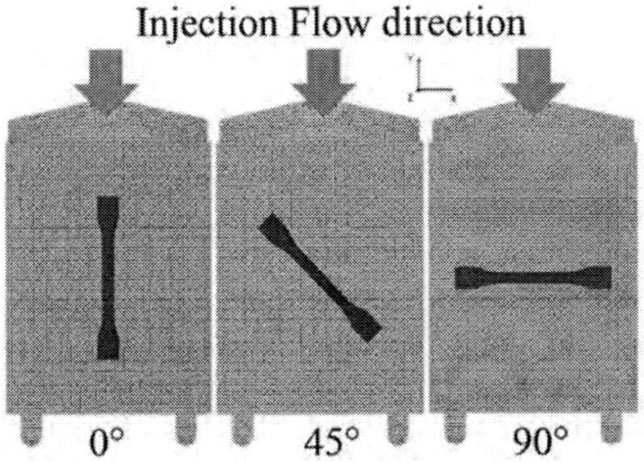

Figure 2: Plate injection and fatigue coupon machining

2.2 Industrial parts

The part used in this study is a prototype engine mount. As for the fatigue test coupons, the parts are conditioned using ISO 1110 standard.

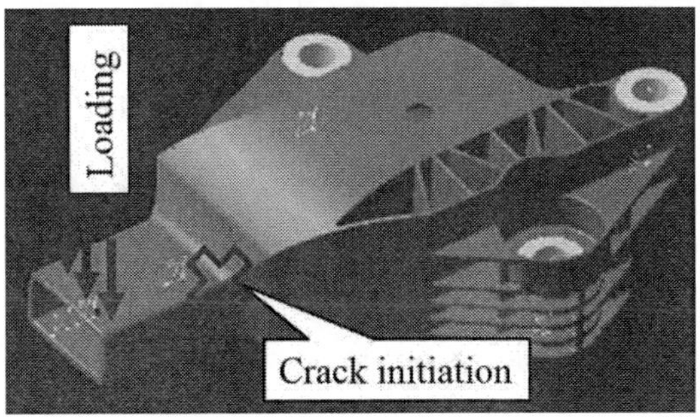

Figure 3: Engine mount studied.

2.3 Testing

Fatigue coupon testing was carried out at HBM Prenscia's testing facilities using a compressive tensile machine. To ensure test repeatability and representativeness, the tests were done in a climatic chamber at 23°C and 50% relative humidity. The tests were all conducted in load control using 1 to 3Hz test frequency (see part 3.1 for more details). For such long fatigue specimens, undesirable buckling in compression can often occur and a R=0.1 load ratio was therefore used. Further tests, using shorter specimens could be used to investigate compressive loadings. For all tests, the fatigue life is considered as the specimen failure.

Figure 4: Climatic chamber and fatigue testing rig.

In order to test the engine mount, a specific testing rig was developed (see Figure 5). It allows testing of the part in different orientations. The tests were performed in load control using a load ratio of R=0.1 and a test frequency of 3Hz. Test were conducted until part failure (severe drop in part rigidity). It must be noted the applied load shown here does not represents loadings occurring in real part situations and was mainly designed to investigate the accuracy of the fatigue criterion. Most of the tests were also monitored using a thermographic camera in order to detect crack initiation (see section 4.2 for more details).

Figure 5: Engine mount testing rig showing loading direction.

3 Injection moulding process

3.1 Tomography Observations

In order to better understand the mechanical tests, tomography observation and simulation of the moulding process is of assistance. Samples of material were taken from the central zone of the fatigue specimen (see Figure 1) and analysed by X-Ray tomography. The resulting scan has a 4µm resolution and covers a volume approximately 3x3x3mm. A slice of the sample analysed can be seen Figure 6. It shows the typical structure of fibre reinforced thermoplastics:

- Surface: while difficult to see on the figure, the surface effect can be seen in 3D, close to the plate surface, the fibres are randomly oriented.
- Skin: this section constitutes the main part of our volume. The fibres are mainly oriented in the injection direction
- Core: this section, of approximately 500µm thick, has fibres perpendicular to the injection direction.

Based on this tomographic data, the fibre tracing algorithm of FEI® Avizo® software and a specific data processing routine was used to study quantitatively the fibre orientation in the specimen. The xx component of the experimentally determined fibre orientation tensor (corresponding to the degree of fibre alignment in the flow direction) is plotted against depth in Figure 7. The 0mm depth corresponds to the plate top surface while 3mm is the bottom surface. The different sections can clearly be seen (surface, skin and core)

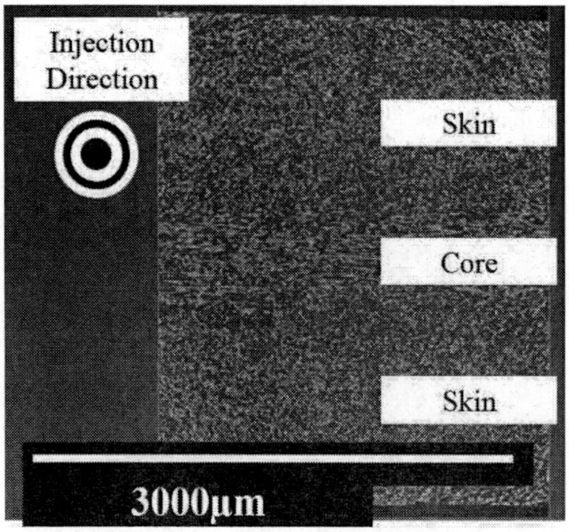

Figure 6: Slice of sample analysed by tomography

3.2 Injection moulding simulation

While tomography analysis helps better understand the material microstructure, it remains technically difficult to analyse large volumes, and even more so for parts with complex geometry such as the engine mount studied here, where machining samples can be cumbersome. Therefore, the injection process of the plates and engine mounts was simulated using Autodesk® Moldflow®. The moulding simulation generates estimated fibre orientation tensors which for the moulded plates may be compared with the experimental data as shown in Figure 7. The agreement is quite good: the injection simulation displays the same sections as the observed ones, and orientation values appear close. However, the simulation appears to slightly underestimate the degree of fibre orientation in the different layers.

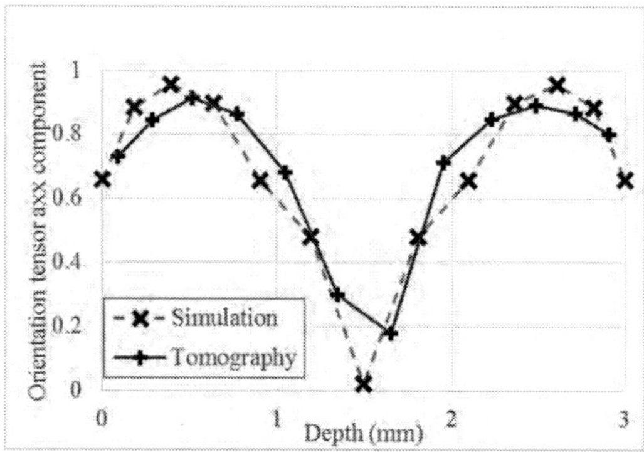

Figure 7: Experimentally observed fibre orientation tensor. The X direction matches the injection direction.

Using the same method, the industrial part injection process was also simulated. The results, shown in Figure 8, can then be used as an input to the integrative mechanical simulation.

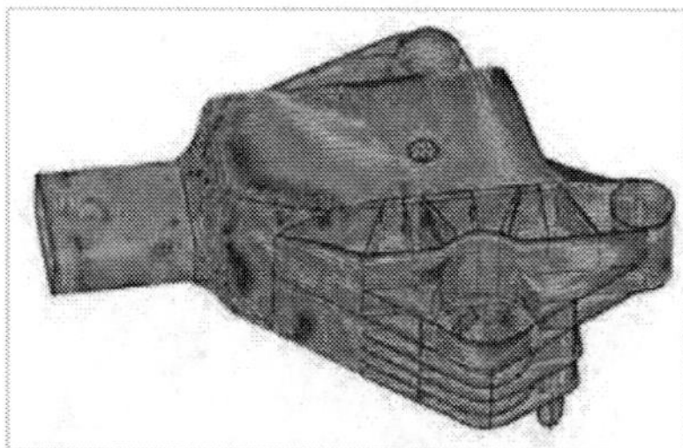

Figure 8: Injection moulding simulation of engine mount showing fibre alignment proportion with local principal direction. A higher value (red) means the fibres are locally aligned, whereas for the lower values (blue), the fibres are locally more randomly oriented.

4. Experimental results

4.1 Specimen heating

Polyamide based materials are highly sensitive to loading frequency effects [3], even for the lower load levels (i.e. the higher fatigue lives) due at least partly to hysteretic heating. In order to maintain the heating level below a certain threshold, a series of incremental tests were first undertaken to define testing frequency: increasing loads were applied to the specimen and, using thermal imaging, the average surface temperature was measured after temperature stabilisation (10000 cycles). These tests were done at 3Hz and provide a design of experiment for the fatigue tests. For a given load, if the temperature rise is higher than 5°C at 3Hz, the test frequency is lowered and set to 1Hz. The results of these tests can be seen in Figure 9 for all three orientations considered.

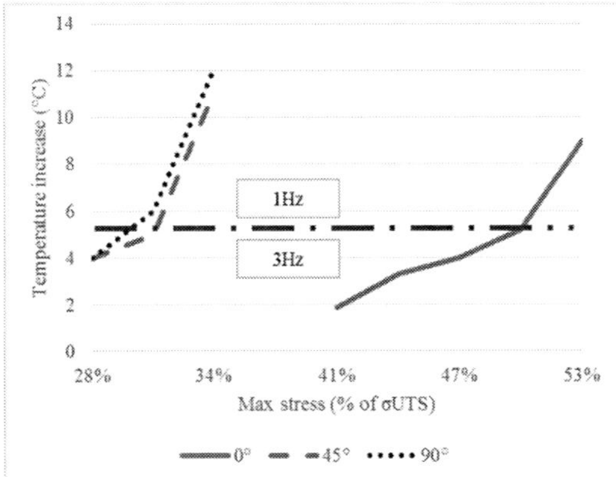

Figure 9 : Surface temperature increase for increasing loadings at 3Hz of different specimen orientations

4.2 Fatigue results

The results of the coupon fatigue tests are presented in Figure 10 for the three different specimen orientations. The 0°specimen, which is aligned with the injection direction has the best fatigue resistance. Using the microstructure analysis presented in section 3, the influence of fibre orientation appears clearly: when the skin fibres are mainly aligned with the loading direction, the material exhibits the best fatigue resistance. For 45° specimens, the fatigue resistance lies between that of the 0° and the 90° specimens, although closer to 90° ones.

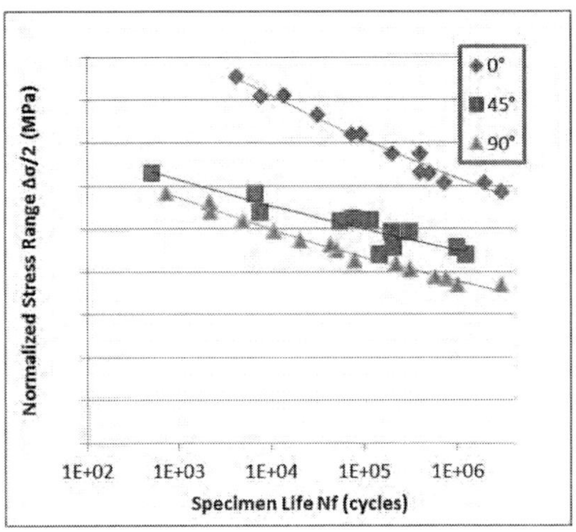

Figure 10: Fatigue test results on PA66GF50.

The results of the fatigue tests on engine mounts are shown in Figure 11, while thermoelasticimetry of the crack initiation location is shown in Figure 12. Rather than standard thermal imaging, thermoelasticimetry allows detection of small variations in stress (see [4] [5]) by synchronizing the measurement signal (here the thermal imaging) with the applied loading signal. Indeed, for a homogenous elastic solid under adiabatic conditions, the temperature variation ΔT can be linked to the stress tensor $\boldsymbol{\sigma}$:

$$\Delta T = -K.T.\mathrm{Tr}(\sigma) \quad (1)$$

Where T is the temperature, Tr the trace operator and K the thermoelastic constant. (This last constant was not measured so the numerical values of the stress change in Figure 12 are not meaningful.) The interest of this measurement is twofold. Firstly, it allows us to detect hot spots in the structure, which may becompared to the simulation results in section 5.3. Secondly, it allows for a precise detection of crack initiation, which is key to fatigue life assessment.

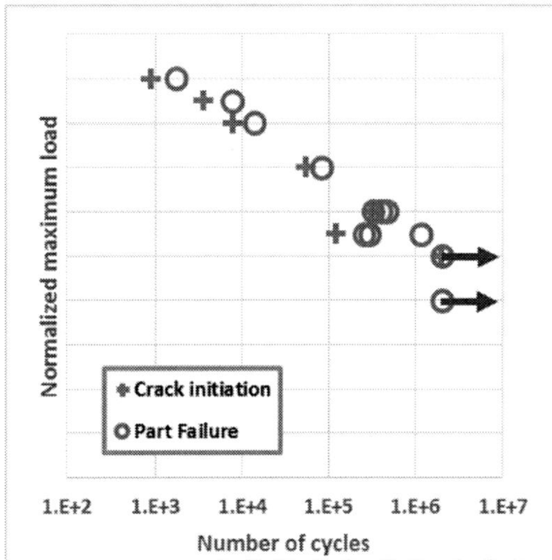

Figure 11: Fatigue test results of engine mount.

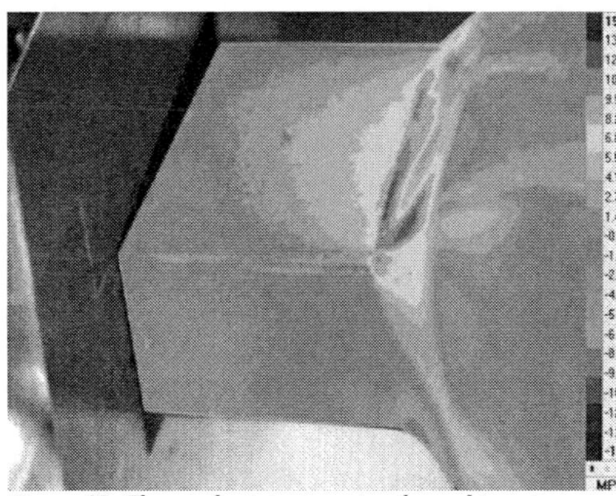

Figure 12: Thermoelasticimetry using thermal imaging camera.

It must be noted Figure 11 is in logarithmic scale. The crack propagation phase, for this specific configuration, represents between 33 and 54% (average 44%) of the total fatigue life, the part failure being considered as a severe loss of part rigidity. This shows for SFRP parts and long fatigue lives, the crack propagation phases cannot be neglected and can represent half of the total fatigue life.

5 Mechanical simulation results

5.1 Simulation model

In order to simulate the mechanical tests of fatigue coupons and parts, an integrative simulation method provided by Digimat® and coupled with Abaqus® was used. Based on Mori-Tanaka homogenization models, it allows us to take into account the local microstructure of the material for each layer. All simulations are done using elastic material hypothesis.

The fatigue coupons and the engine mounts were meshed using 2D shell elements. Each element contains 12 layers (where layer 1 is the top surface layer, 6-7 the central layers of the plate, and 12 the bottom surface layer), each defined using the local fibre orientation tensor. Additional microstructural and material data are required: fibre length and aspect ratio, fibre density, Young modulus and Poisson ratio of fibre. The Young modulus of the matrix was not used directly, but rather as a parameter optimised to provide the best fit between test and simulation for the 0° specimens. Results are shown Table 1 and shows the model is able to correctly simulate the different specimen orientations.

Table 1 : Comparison of experimental and simulated equivalent Young Modulus of tensile tests. The results are normalized by the value of 0° specimen.

Specimen	Experimental ()	Numerical ()
0°	1.000	0.955
45°	0.485	0.510
90°	0.465	0.497

The optimised material data was then used to simulate the engine mount test. The part stiffness for this loading configuration was found to be very close to the experimental results (less than 10% error in the part stiffness for the maximum load used in the fatigue tests).

5.2 Fatigue criteria

As seen in Figure 10, using equivalent stress (i.e. load divided by specimen section) does not yield satisfactory results. Therefore, a local strain energy density approach is used. The elastic strain energy density ΔW_e is defined by:

$$\Delta W_e = \frac{1}{2}\sigma : \varepsilon_e$$

Where $\boldsymbol{\sigma}$ is the stress tensor and $\boldsymbol{\varepsilon_e}$ the elastic strain tensor. The fatigue life N can then be estimated by:

$$N = A \times \Delta \boldsymbol{W_e}^{b}$$

Where A and b are material parameters. Even though the parameter A can be modified to take into account the stress ratio, it is not detailed in this study, since all tests are done using identical stress ratios. Using HBM Prenscia's nCode DesignLife® software, the elastic strain energy quantity was calculated for the 0, 45 and 90° specimens. Two approaches were then used to calculate the maximum value:

1. The "W max" approach: in this case, the critical value for a specimen is taken as the maximum value all over the elements and all the layers.
2. The "W mean" approach: in this non-local approach, the results are averaged over the specimen thickness (i.e. across all layers). Indeed, each element contains 12 different stress/strain tensors values corresponding to the different layers. For this method, the elastic strain energy density of an element is first calculated by averaging the 12 layers. The critical value of the specimen is then chosen as the maximum value over the specimen.

5.3 Experimental vs simulation results

The abilities of the two methods to correlate the coupon test results are shown Figure 13 and Figure 14. These must first be compared to results of stress-life ones (shown in Figure 10): the elastic strain energy density methods minimize differences between the different specimen orientations, making them suitable for complex simulations of industrial parts. Indeed, for these more complex simulations, it is impossible to define an orientation and any fatigue criteria needs unify the test results. For fatigue coupons specimens, no significant differences between the "W max" and the "W mean" approach appear.

The two different approaches were then applied to the engine mount parts; see Figure 15 and Figure 16. While results of the "W max" approach appear over-conservative, the "W mean" show promising results. Especially for the longer fatigue lives, which are of interest, the results using this last criteria are close to the /10 x10 scatter band. An explanation of the different results between coupons and part simulation could be attributed to the different mechanical loadings: while coupons are loaded in uniaxial tensile loadings, the engine mount undergoes multiaxial and bending loads. Other effects that might be considered are the notch effect (local plasticity and size effect), and mean stress relaxation at stress concentrations when $0 < R < 1$ due to accumulating creep strain.

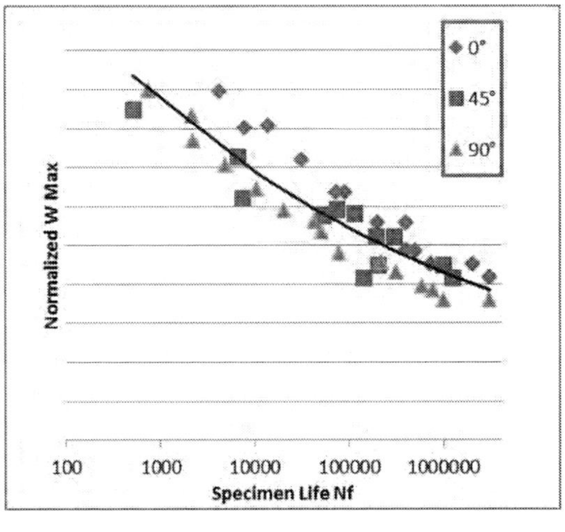

Figure 13: Fatigue results of coupons using the "W max" approach

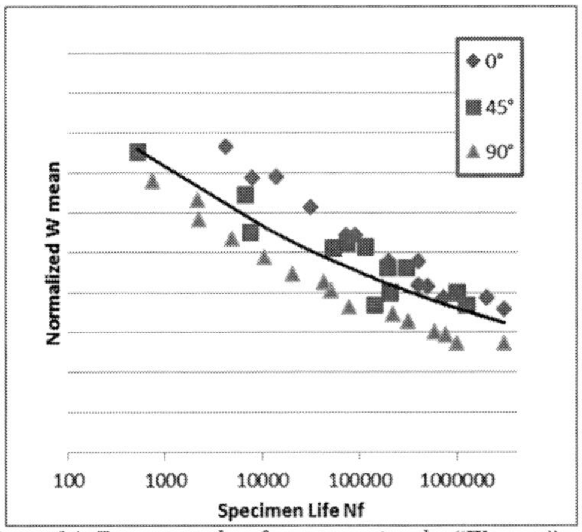

Figure 14: Fatigue results of coupons using the "W mean" approach

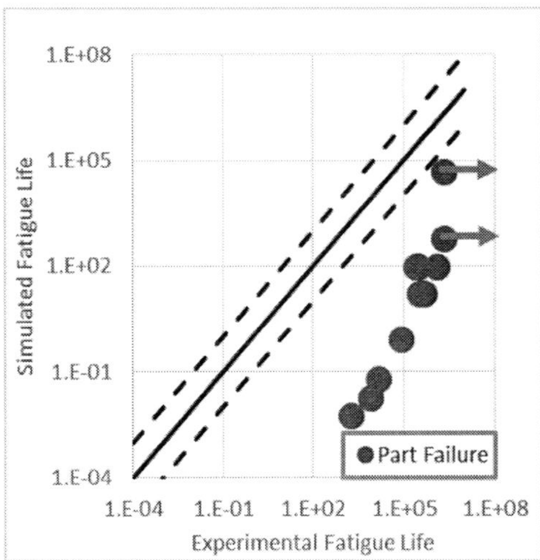

Figure 15: Fatigue results of parts using the "W max" approach

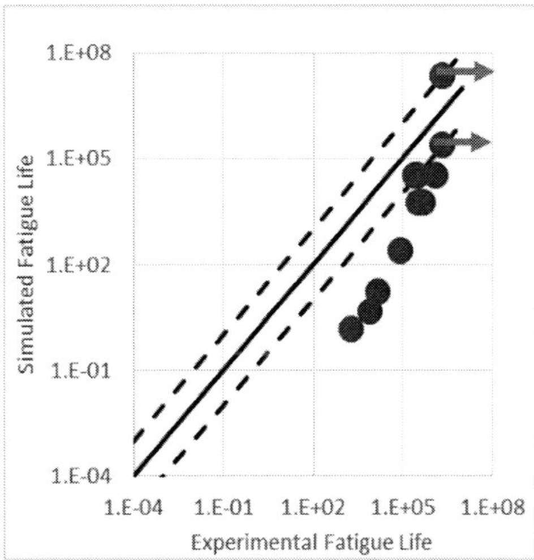

Figure 16: Fatigue results of parts using the "W mean" approach

6 Conclusions

An integrative simulation method is used to simulate the fatigue life of a short fibre reinforced plastic (PA66GF50). First, by using X-Ray laboratory tomography and specific image analysis software, the microstructure of the material is investigated, and compared to the results of the moulding injection simulation. Secondly, based on the simulated microstructure, the anisotropic and heterogeneous mechanical behaviour is identified. It shows integrative simulation methods are able to capture and simulate different fibre orientations of specimens. Lastly, using a strain energy based fatigue criterion, the mechanical simulations are used to calibrate the fatigue model based on coupon tests, and assess the life of structural parts. Both local and non-local criteria show promising results.

Further investigation should focus on detailed description of material mechanical behaviour. Indeed, the tests conducted in this study show the material has a strong creep behaviour, especially for the lesser orientated specimens (45 and 90° specimens), for which the matrix behaviour prevails. Recent studies (see for example [6]) have shown the precise study of this creep-fatigue behaviour could help better understand fatigue life assessment.

References

[1] B. L. Shao-Yun Fu, «Effects of fiber length and fiber orientation distributions on the tensile strength of short-fiber-reinforced polymers,» *Composites Science and Technology*, **56**, Issue 10, 1179-1190, (1996).

[2] H. Rolland, N. Saintier, P. Wilson, J. Merzeau et G. Robert, «In situ X-ray tomography investigation on damage mechanisms in short glass fibre reinforced thermoplastics: Effects of fibre orientation and relative humidity,» *Composites Part B: Engineering*,

Volume **109**, 170-186, (2017).

[3] S. Mortazavian, A. Fatemi, S. R. Mellot et A. Khosrovaneh, «Effect of cycling frequency and self-heating on fatigue behavior of reinforced and unreinforced thermoplastic polymers,» *Polymer engineering and Science*, **55** Issue 10,2355-2367, (2015).

[4] W. J.M.B., «Principles of IR Measurements and Review of Instrumentation Techniques for Therlmoelastic Stress Analysis,» *Stress Analysis by Thermoelastic Techniques*, (1987).

[5] S. Offermann, E. Merienne et J. L. Beaudoin, «*Statistical Method to Evaluate the Amplitude of Cyclical Noisy Signal in IR Thermography*,» pp. 43-48, (1994).

[6] A. Bernasconi et R. M. Kulin, «Effect of frequency upon fatigue strength of a short glass fiber reinforced polyamide 6: A superposition method based on cyclic creep parameters,» Polymer Composites, **30**, issue 2, pp. 154-161, (2009).

Investigation of thermo-oxidative ageing effects on the fatigue design of automotive anti-vibration parts

Morgane Broudin[1,*], *Yann* Marco[1], *Vincent* Le Saux[1], *Pierre* Charrier[2], *Wilfried* Hervouet[2] and *Pierre-Yves* Le Gac[3]

[1]ENSTA Bretagne, Institut de Recherche Dupuy de Lôme (IRDL), UMR CNRS 6027, 2 Rue François Verny, 29200, Brest, France
[2]Vibracoustic, CAE & Durability Prediction, 1 Rue du Tertre, 44474, Carquefou, France
[3]IFREMER (French Ocean Research Institute), Centre de Brest, Materials and Structures Group, BP 70, F-29280, Plouzané, France

Abstract. Elastomeric parts found in automotive anti-vibration systems are usually massive. Ageing therefore leads to heterogeneous properties, usually induced by several mechanisms due to the availability, or not, of oxygen in the part's bulk. To better understand the effects of oxygen in the degradation process and on the fatigue properties, this paper aims at studying the ageing of a rubber material (semi-efficient vulcanization system) in aerobic (with oxygen) and anaerobic (without oxygen) conditions for a wide range of temperatures, relevant for under hood applications. The material studied here is a fully formulated compound NR/IR blend reinforced with carbon black. A specific protocol to perform ageing under anaerobic conditions was set up and validated. Numerous tests have been carried out to evaluate the consequences of ageing on monotonic tension and fatigue properties. A comparison of these consequences and of their kinetics is finally presented for ageing with or without oxygen.

1 Introduction

1.1 Industrial context

Rubber parts for anti-vibration systems applications need to be designed taking into account several parameters: efficient damping on given ranges of frequencies, reliable static response over the service time, resistance to creep and fatigue [1-2]. The automotive applications require, in addition, the assessment of these features for thermal conditions inducing ageing. This ageing is related both to the effect of oxygen but also, for these massive parts, to the effect of temperature only. Several recent features illustrate the need to go beyond the empirical rules. The first one is an ever reducing development time, which leads to assess compounds faster than ever. A second point is the need to design parts for worldwide purposes increasing the severity profiles of users (roads and weather conditions). A last one is connected to the environmental standards (REACH and EURO6) that restrict the use of specific ingredients and lead to higher service temperatures. Integrating efficiently the ageing effects in the design loop therefore becomes a crucial need.

1.2 Scientific background

Thermo-oxidative degradation of rubber components has been widely investigated in the literature [3-10]. Ageing could indeed induce severe modifications of the mechanical properties both in terms of constitutive response [11], failure properties under monotonic tension [12], crack propagation [13] and duration life [14]. Nevertheless, relating the degradation mechanisms to the evolution of the mechanical properties is not an easy task because there are so many mechanisms at stake. A first range of mechanisms is related to oxidation, leading to important modifications of the elastomer network (main-chain scissions, crosslink formation and/or crosslink breakage) [15]. Large-sized automotive parts age heterogeneously due to the availability, or not, of oxygen in the part's bulk. A classic analysis could be done using a DLO approach based on the competition between the kinetics of oxygen diffusion and consumption [16]. However hardness profiles along the thickness of thick parts or specimens also illustrate that ageing leads to mechanical changes even without oxygen [17]. Numerous studies investigated the effect of temperature ageing in the absence of oxygen on the mechanical properties of rubber materials. A first range of studies investigated the evolution of the tensile properties for anaerobic conditions and have found a limited increase of stiffness and little reduction of the ultimate tensile properties [18-19]. Then, a change of fatigue crack growth resistance have been noticed for fully formulated compound and different vulcanization systems under various ageing environments. A higher fatigue crack growth rate is observed under air than under nitrogen whatever the vulcanization system. The oxidative degradation of the rubber causes a high reversion inducing a dominant effect on the mechanical resistance [20]. This effect was confirmed later to a lesser extent [21].
There is limited data available on the effect of thermo-oxidative ageing on fatigue properties. Moreover the

* Corresponding author: morgane.broudin@ensta-bretagne.fr

origins of the evolutions of mechanical properties both for stiffness, failure or fatigue still remain unclear. Several explanations could be suggested: plasticizers diffusion-evaporation [22], sulfur network evolution [10,17], fillers network evolution [23] and/or evolution of the interface rubber/fillers [24].

The present paper aims providing an extensive comparison of the mechanical consequences of ageing under anaerobic and aerobic conditions, based on the results of monotonic tensile tests and fatigue tests. Secondly, the goal is to better understand the common mechanism at stake under anaerobic and aerobic ageing conditions and to investigate the hypothesis of a physical ageing induced by the loss of low molecular species.

2 Experimental settings

2.1 Materials

2.1.1 Fully formulated compound

The material studied here is a carbon black-filled NR/IR blend. All samples were made using the same material batch to limit properties scattering due to mixing. This material will be called "C" in the Figures presented in the following.

2.1.2 Simplified compound

To determine the influence of low molecular weight components including plasticizers on the durability, it would be very convenient to consider the same compound but without these ingredients. Nevertheless, a filled material, as the material C, could not be mixed without plasticizers. An extraction of low molecular weight components was therefore performed thanks to swelling in toluene. Then, the specimen is dried at room temperature to get back to a geometry close to the initial one. This material will be called "B" in the Figures hereafter.

2.2 Accelerated ageing conditions

2.1.1 Thermo-oxidative ageing

Thermo-oxidative ageing was carried out in ovens with air-flow control (Fig. 1). To get reliable tests, several authors noticed the importance to specify the ageing conditions, such as air speed and air exchanges rate [25]. Therefore, to have quite similar ageing conditions, all used ovens have been checked in terms of air speed and air renewal rates thanks to an anemometer [26]. A thermocouple checks the temperature close to the samples and is connected to continuous strip-chart recorders. The table 1 gives the tested temperatures and durations.

Fig. 1. Picture of the specimens in the oven during thermo-oxidative ageing.

Table 1. Thermo-oxidative ageing conditions.

	Ageing temperature [°C]						
	40	50	60	80	100	110	120
Ageing duration [days]							0.25
						0.5	0.5
				1	1	1	1
				2			
				3	3	3	3
				7	7	7	
			14	14	14		
	21	21	21	21	21		
	42	42	42	42	42		
			90				
		161	161				
		362	316				
		771					

2.1.2 Thermal ageing

An experimental protocol was set up to achieve thermal ageing. To package the specimens, a multilayer film is used (SIDEC CA40 NF H 00 310 Class IV) that is impermeable to oxygen and temperature resistant. To avoid bending / torsion when the vacuum is performed, wedges are put on the samples (Fig. 2, on the left). To avoid any preload due to differential thermal expansions, the wedges material is chosen to present a thermal expansion coefficient close to the one of rubber. Finally, to remove oxygen inside the package, a vacuum chamber machine is used (Audionvac VMS 133) (Fig. 2, on the right). The Table 2 sums up the tested temperatures and durations for thermal ageing.

Fig. 2. AE2 specimen with PTFE wedges (left); sealed package of AE2 specimens for ageing without oxygen (right).

Table 2. Thermal ageing conditions.

	Ageing temperature [°C]						
	40	50	60	80	100	110	120
Ageing duration [days]						0.5	0.5
				3	3	3	3
				14	14		
			21	21	21		
	42	42	42				
				90			
		204	204				

2.3 Tensile tests

S2 specimens have been used. The S2 samples are cut from plates and then put in oven to perform the various ageing conditions. The tensile tests were achieved on a testing machine equipped with a 1kN load cell. The tests were displacement controlled (grip speed of 200mm/min). An optical extensometer was used to measure the local elongation. For each ageing condition, at least three samples were tested and the results averaged [26].

2.4 Fatigue campaign

Diabolo shape samples are obtained by injection moulding. The fatigue tests were displacement controlled with adapted frequencies to have a limited heat build-up. All tests were performed at a controlled temperature of 23°C. The fatigue samples geometry and the fatigue protocol have been detailed in a former publication [28]. To keep a reasonable number of samples despite the numerous ageing conditions investigated, for each condition, 5 AE2 specimens are aged and then tested in fatigue for a given enforced displacement. In this campaign, 3 levels of displacement have been studied, 4.6mm, 8.1mm and 14.7mm relating respectively to local principal strains of 80%, 150% and 280%. These strains are evaluated by FE simulations from the geometry of the unaged sample.

3 Results and discussion

3.1 Thermo-oxidative ageing

Figures 3 to 5 illustrate the results obtained for thermo-oxidative ageing.

Figure 3 highlights typical results obtained for tensile tests on samples aged in presence of oxygen (here for 80°C and several ageing durations). These observations are valid whatever the ageing temperature. One can notice a drop of the nominal strain at break and a stiffening of the mechanical response, which is a classical result and could be related to the changes of the vulcanized network. Nevertheless, it should be noticed that the evolutions in stiffness and strain at break observed for short durations is already significant. Figure 4 shows the results of the fatigue campaigns for the same ageing conditions. The drop of the duration life with the ageing durations is clearly visible. It is worth noticing that the curves keep the same slope, except for the last ageing condition, considered here to be extremely severe.

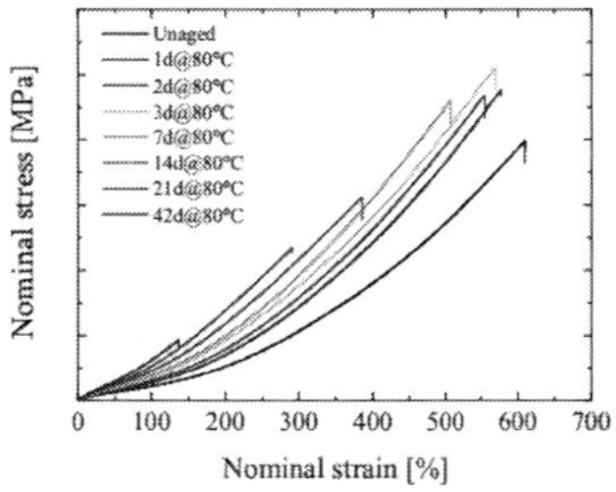

Fig. 3. Material C. Aerobic ageing. Strain – stress curves after various preliminary thermal ageing (from 1 to 42 days at 80°C).

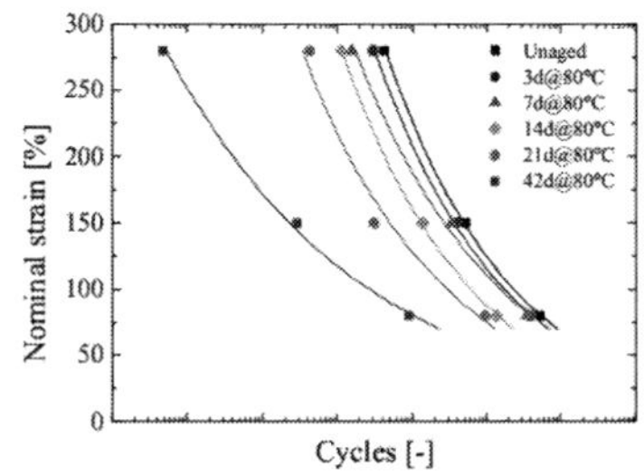

Fig. 4. Material C. Aerobic ageing. Wöhler curves after various preliminary thermal ageing (from 3 to 42 days at 80°C).

A single ageing mechanism is clearly not probable for this complex ageing. Nevertheless, the identification of a time-temperature superposition seems possible, as illustrated by the master curve obtained on the relative evolution of the nominal strain at break (Fig. 5). The deduced activation energy is of 90kJ/mol. A comparable result can be obtained from the fatigue tests with an activation energy of 95kJ/mol [4]. Both values are very close to other values found in the bibliography for various ageing indicators [16], [17], [29], [30].

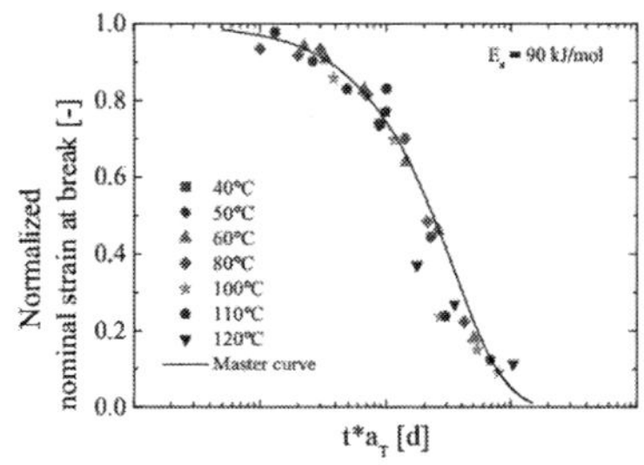

Fig. 5. Material C. Aerobic ageing. Master curve obtained on the relative evolution of the nominal strain at break. Reference temperature = 80°C.

3.2 Thermal ageing

Figures 6 to 8 illustrate the results obtained for thermal ageing conditions.

Figure 6 highlights a noticeable increase of the mechanical response for short ageing durations. This increase appears more gradually with increasing aging time. A progressive drop of failure properties is also observed. Both variations seems to occur early and then to stabilize. The stabilized values reached are less severe than for oxidative ageing. These evolutions are observed whatever the ageing temperature.

Figure 7 presents the results of the fatigue campaigns for the same ageing conditions. Here again, the drop of the duration life with the ageing duration occurs early and then stabilizes to a level much less severe than for long oxidative conditions.

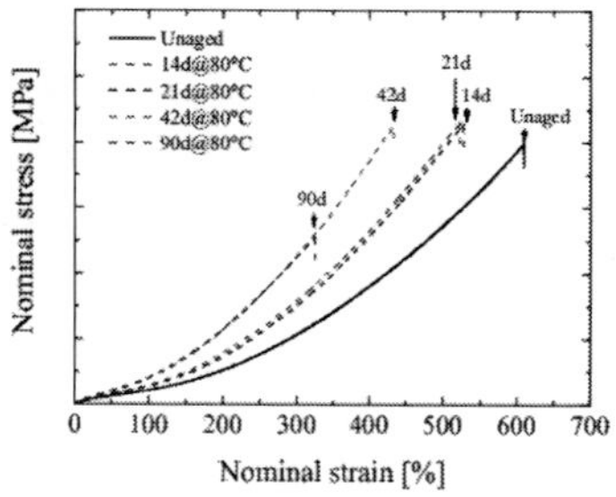

Fig. 6. Material C. Anaerobic ageing. Strain – stress curves after various preliminary thermal ageing (from 14 to 90 days at 80°C).

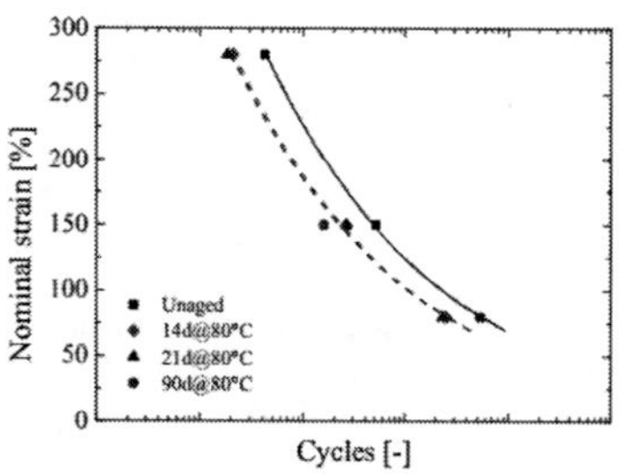

Fig. 7. Material C. Anaerobic ageing. Wöhler curves after various preliminary thermal ageing (from 3 to 90 days at 80°C).

Figure 8 illustrates the master curve obtained on the relative evolution of nominal strain at break. The deduced activation energy would be of 72kJ/mol. This value seems lower than the one obtained for thermo-oxidative aging but quite consistent with those found for anaerobic aging in the literature [31]. However, contrary to thermo-oxidative ageing, an identification of a time-temperature superposition from fatigue tests is not reasonable in this case.

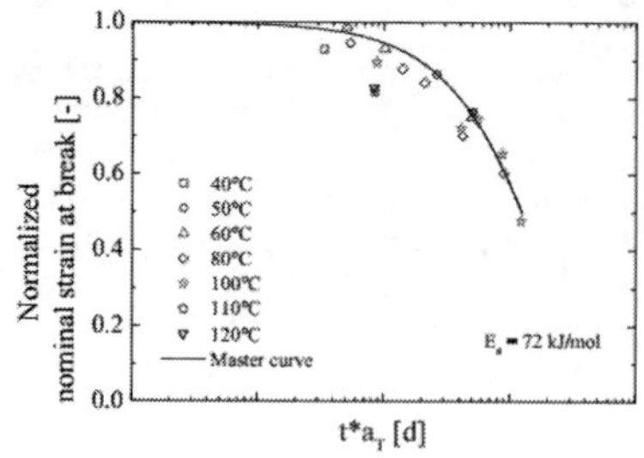

Fig. 8. Material C. Anaerobic ageing. Master curve obtained on the relative evolution of the nominal strain at break. Reference temperature = 80°C.

3.3 Thermal ageing vs thermo-oxidative ageing

Figure 9 summarizes the evolution of the relative nominal strain at break for tensile tests performed on samples after aerobic and anaerobic ageing with several durations at 80°C. Figure 10 provides the same synthesis for the variations of the relative fatigue lifetime (for a given enforced displacement).

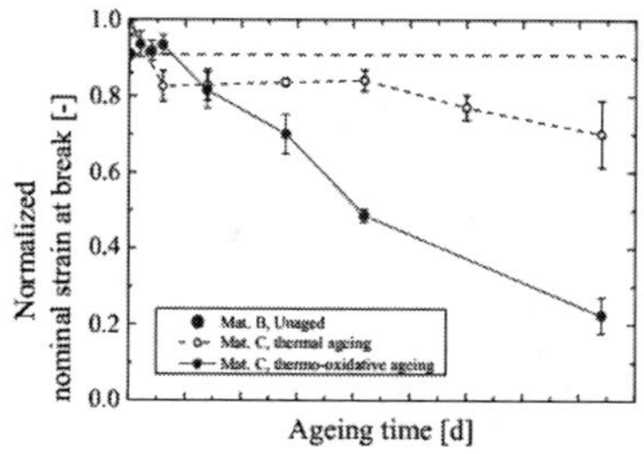

Fig. 9. Evolution of the relative nominal strain at break obtained after anaerobic and aerobic ageing at 80°C and after swelling in toluene (mat. B).

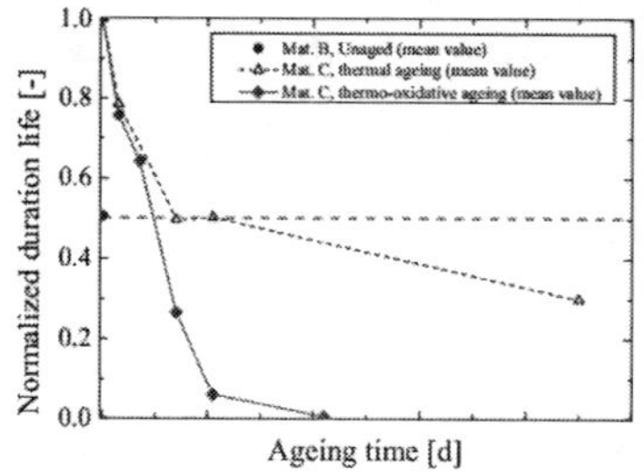

Fig. 10. Evolution of the relative nominal duration life obtained after anaerobic and aerobic ageing at 80°C and after swelling in toluene (mat. B). Displacement level = 8.1mm.

In order to check these first comparisons, the values obtained for the samples tested after swelling in toluene (material B) were compared with the evolutions of anaerobic and aerobic ageing conditions. These comparisons are illustrated in Figures 9 and 10. The dotted line is the value obtained for the samples tested after swelling in toluene (material B).

One can observe on these curves that the drops of tensile and fatigue properties seem to present a common kinetic for the short ageing durations. Then, the loss of the failure properties reaches a stabilized level for the ageing conditions without oxygen, whereas the loss of failure properties continues for the ageing conditions with oxygen.

It should be underlined here that the results obtained are presented for 80°C only due to the restricted format of the paper, but similar conclusions can be done from the results obtained for lower and higher ageing temperatures as 60°C and 100°C.

Figures 9 and 10 illustrate that the drop of strain at break and fatigue lifetime for anaerobic conditions stabilize for values correlated to the ones obtained for the material after swelling in toluene. It therefore seems that the initial drops of strain at break and fatigue properties observed both under aerobic and anaerobic conditions seem well correlated to the loss of low molecular weight components. The nature of these components is still to be determined even if it seems very probable that they are plasticizers. TGA and GC-MS measurements are currently done in order to clarify this point.

This evaluation performed on nominal strain at break and a given displacement for fatigue test are also valid for the global tension curve (see Figure 11) and the full fatigue curve (see Figure 12).

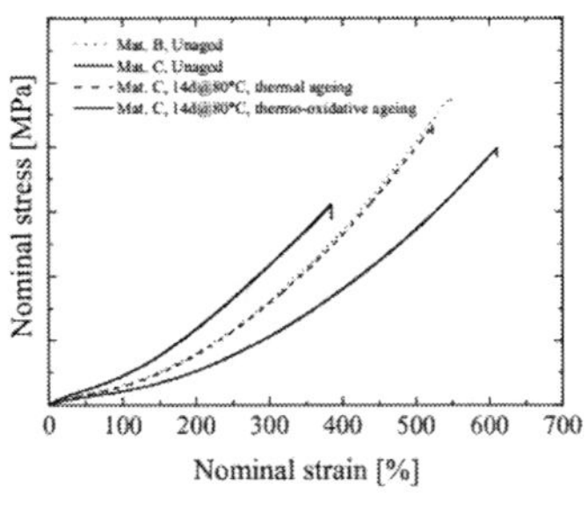

Fig. 11. Comparison of nominal stress – strain curves obtained for material C after anaerobic and aerobic ageing for 14 days at 80°C and for material B.

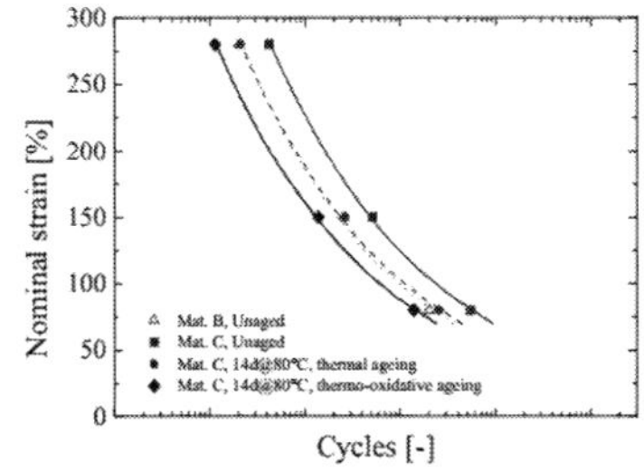

Fig. 12. Comparison of Wöhler curves obtained for material C after anaerobic and aerobic ageing for 14 days at 80°C and for material B.

Figure 13 compares the master curves built from the relative evolutions of the nominal strain at break for anaerobic and aerobic ageing conditions. It allows us to underline that the kinetics of aerobic ageing is substantially faster than anaerobic ageing (about a decade of difference between the two kinetics).

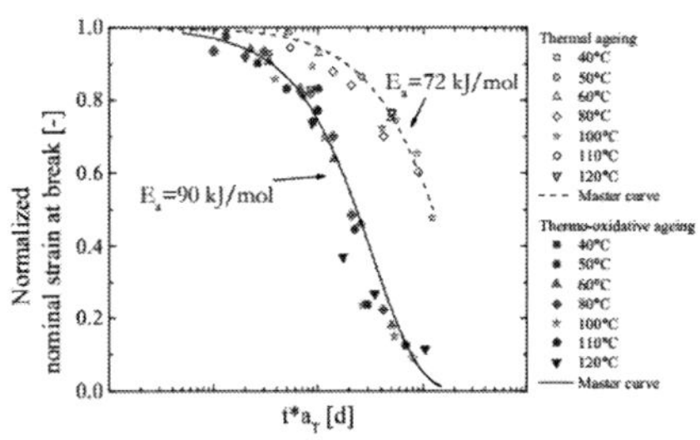

Fig. 13. Comparison of time-temperature superposition for thermal and thermos-oxidative conditions for normalized strain at break. Reference temperature = 80°C.

4 Conclusion and perspectives

In this paper, we focused on the ageing of a rubber material in aerobic (with oxygen) and anaerobic (without oxygen) conditions, for a wide range of temperatures relevant for under the hood applications. Tensile and fatigue tests are analysed. For both ageing cases, progressive drops of the strain at break and the fatigue properties are observed as well as a progressive increase of the stiffness. Under anaerobic and aerobic conditions, these evolutions seem to exhibit the same kinetics for the short durations. Nevertheless, for anaerobic ageing, while ageing duration increases, the evolution of the mechanical indicators stabilizes, whereas the evolutions go on for aerobic ageing.

In order to investigate if these evolutions can be related to a physical ageing induced by the extraction of low molecular weight components (plasticizers, for example),

tensile and fatigue tests have been performed on samples after swelling in toluene (material B). It appears that the values obtained on these samples are well correlated to the stabilization observed for anaerobic ageing. This hypothesis seems therefore reasonable and this mechanism could be common to ageing conditions with or without oxygen.
Further investigations are in progress to identify the nature of the components extracted in toluene and to understand better the other possible ageing mechanisms at stake, both for anaerobic and aerobic ageing.

References

[1] NF EN ISO 5349-1, *Vibrations mécaniques - Mesurage et évaluation de l'exposition des individus aux vibrations transmises par la main - Partie 1 : exigences générales* (2002)
[2] NF EN ISO 5349-2, *Vibrations mécaniques - Mesurage et évaluation de l'exposition des individus aux vibrations transmises par la main - Partie 1 : guide pratique pour le mesurage sur le lieu de travail* (2001)
[3] J. L. Bolland and G. Gee, Trans. Faraday Soc., **42**, 236 - 243 (1946)
[4] M. Broudin, V. Le Saux, Y. Marco, P. Charrier, and W. Hervouet, *Proc. Ninth Eur. Conf. Const. Models Rubber*, 53 (2015)
[5] A. Herzig, M. Johlitz, and A. Lion, *Proc. Ninth Eur. Conf. Const. Models Rubber,* 23 (2015)
[6] S. Kamarudding, P. Y. Le Gac, Y. Marco, and A. Muhr, *Proc. Seventh Eur. Conf. Const. Models Rubber*, 197 (2011)
[7] C. Naumann and J. Ihlemann, *Proc. Eigth Eur. Conf. Const. Models Rubber*, 43 (2013)
[8] C. Schlomka, J. Ihlemann, and C. Naumann, *Proc. Tenth Eur. Conf. Const. Models Rubber*, 77 (2017)
[9] C. Neuhaus, A. Lion, M. Johlitz, P. Heuler, M. Barkhoff, and F. Duisen, Int. J. Fatigue, **104**, 72–80 (2017)
[10] F. Grasland, J. M. Chenal, L. Chazeau, J. Caillard, and R. Schach, *Proc. Tenth Eur. Conf. Const. Models Rubber*, 65 (2017)
[11] M. Johlitz and A. Lion, Contin. Mech. Thermodyn., **25**, 605–624 (2013)
[12] M. Celina, K. T. Gillen, and R. A. Assink, Polym. Degrad. Stab., **90**, 395–404 (2005)
[13] J. T. South, S. W. Case, and K. L. Reifsnider, Rubber Chem. Technol., 76, 785–802 (2003)
[14] P. Charrier, Y. Marco, V. Le Saux, and R. K. P. S. Ranaweera, *Proc. Seventh Eur. Conf. Const. Models Rubber, 381* (2011)
[15] J. Verdu, *Oxidative ageing of polymers (*Wiley-ISTE, 2012)
[16] M. Celina, J. Wise, D. K. Ottesen, K. T. Gillen, and R. L. Clough, Polym. Degrad. Stab., **68**, 171–184 (2000)
[17] V. Le Saux, *Fatigue et vieillissement des élastomères en environnements marin et thermique : de la caractérisation accélérée au calcul de structure*, Université de Bretagne occidentale, France (2010)
[18] G. T. Kohman, J. Phys. Chem., **33**, 226–243 (1929)
[19] J. R. Shelton and H. Winn, Ind. Eng. Chem., **36**, 728–730 (1944)
[20] S. G. Kim and S.-H. Lee, Rubber Chem. Technol., **67**, 649–661 (1994)
[21] P. Soma, N. Tada, M. Uchida, K. Nakahara, and Y. Taga, J. Solid Mech. Mater. Eng., **4**, 727–737 (2010)
[22] P. D. Calvert and N. C. Billingham, J. Appl. Polym. Sci., **24**, 357–370 (1979)
[23] A. Roychoudhury and P. P. De, J. Appl. Polym. Sci., **50**, 181–186 (1993)
[24] S. T. Tchalla, P. Y. Le Gac, R. Maurin, and R. Créac'hcadec, Polym. Degrad. Stab., **139**, 28–37 (2017)
[25] G. Spetz, *Polym. Test.*, 13, 239–270 (1994)
[26] ASTM D 1765, *Standard classification system for carbon blacks used in rubber products* (2016)
[27] NF ISO 37, *Caoutchouc vulcanisé ou thermoplastique. Détermination des caractéristiques de contrainte-déformation en traction* (2012)
[28] E. Ostoja Kuczynski, P. Charrier, E. Verron, G. Marckmann, L. Gornet, and L. Chagnon, *Proc. Third Eur. Conf. Const. Models Rubber* (2003)
[29] M. Celina, J. Wise, D. K. Ottesen, K. T. Gillen, and R. L. Clough, Polym. Degrad. Stab., **60**, 493–504 (1998)
[30] K. T. Gillen, R. Bernstein, and D. K. Derzon, Polym. Degrad. Stab., **87**, 57–67 (2005)
[31] J. T. South, *Mechanical properties and durability of natural rubber compounds and composites,* Faculty of the Virginia Polytechnic Institute, United-States (2001)

Fatigue initiation mechanisms in elastomers: a micro-tomography-based analysis

Thomas Glanowski[1,*], *Bertrand Huneau*[2], *Yann Marco*[3], *Vincent Le Saux*[3], *Clément Champy*[1] and *Pierre Charrier*[1]

[1]Vibracoustic, CAE & Durability Prediction Department, 44474 Carquefou, France
[2]Institut de Recherche en Génie Civil et Mécanique (GeM), UMR CNRS 6183, Ecole Centrale de Nantes, France
[3]Institut de Recherche Dupuy de Lôme (IRDL), UMR CNRS 6027, ENSTA Bretagne, France

Abstract. The fatigue properties of filled elastomers are strongly related to the population of inclusions induced by their complex recipes and mixing/injection processes. The description and the understanding of the basic damage and dissipation mechanisms involved around these inclusions, the influence of their nature, geometry, size, interface and cohesion properties are especially important to optimize the fatigue design of industrial compounds and parts. The objective of the study is to take advantages of tomography observations to characterize the effects of the 3D inclusions on the fatigue resistance of rubber filled with carbon black. An images processing is necessary to dissociate accurately the inclusions of the matrix. The three dimensional observations provide many information on the inclusion's morphology (size, shape) and their spatial distribution. Moreover, the proposed images processing allows detecting cracks inside and on the surface of the specimen, which offers a better insight on the early stages of the fatigue damage scenario. Then, a numerical model and thermo-mechanical measurements are used to understand why some inclusions become initiation sites. Finally, one activated site is fully described as example.

1 Introduction

The fatigue properties of elastomeric parts are strongly related to the microstructure features (inclusions nature, shape, size, size scattering, spatial distribution) because these reinforced materials are heterogeneous. The industrial development of materials resistant to fatigue is therefore very complex because the parameters driving the microstructure are incredibly numerous, including original ingredients and the processing parameters, from mixing to injection and curing. Inclusions above ten micrometres exert an influence on the fatigue properties of elastomeric compounds. They are commonly estimated with the "Dispergrader" device [1]. More sophisticated technics can be used like micro-tomography [2] to describe the inclusions populations. The accurate description of the initiation sites' morphology [3-7] allows the understanding of the early stages of fatigue damage mechanisms but only from a geometrical point of view. Some recent studies tried to relate thermo-mechanical measurements and micro-tomography analysis to characterize the fatigue properties of elastomeric materials [8, 9]. But the link between macroscopic mechanical criteria and local failure mechanisms are still missing because the main failure mechanisms identified in the literature are only described from a geometrical point of view.

The aim of this study is to understand why some inclusions become an initiation site and others don't. First of all, a tool is developed to provide a full description of the microstructure features such as the shape, the size, and the spatial distribution of inclusions. Then, this microstructural information feeds simulations in order to evaluate the criticality of inclusions properties. Finally, the case of one inclusion is presented and a thermo-mechanical experimental protocol is defined to assess the response at the inclusion scale.

2 Tools for the description of the inclusions and cracks population

2.1. Material and specimen

The material studied is a fully formulated natural rubber reinforced with carbon blacks. Classic hourglass samples (Figure 1) are used to characterize fatigue damage scenarios. These samples are injected and representative of the manufacturing process used for automotive anti-vibration parts.

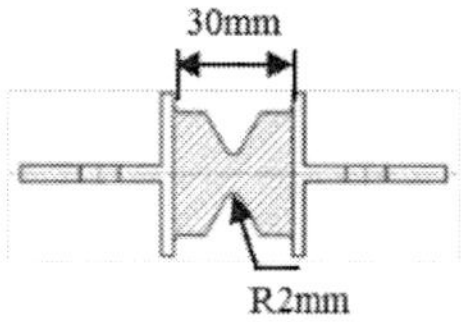

Fig. 1. Geometry of the AE2 hourglass sample.

* Corresponding author: thomas.glanowski@vibracoustic.com

2.2 Motivations for using X-ray computed micro-tomography

Micro-tomography is a remarkable tool to provide a non-destructive internal observation. This method is widely used for the study of materials microstructure [10] including rubber [4]. Compared with usual 2D analysis methods such as the scanning electron microscope, the 3D reconstruction of the microstructure by micro-tomography grants to:

- Quantify statistically the number, size, shape and orientation of objects;
- Study the spatial distribution of objects;
- Avoid artefacts due to cutting and surface preparation for 2D observations.

However, the micro-tomography has some drawbacks:

- The resolution is about a few micrometre (much lower than that SEM);
- The possibly duration of acquisition;
- The huge size of the result files;
- Lack of access to the chemical composition (unlike SEM with EDS probe).

Furthermore, the X-ray micro-tomography is an excellent tool to follow 3D damage evolution [4]. The samples of the study were analysed with the X-Radia XCT micro-tomograph, located at Ecole Centrale de Nantes (GeM). A quite high resolution is reached with a voxel size at 8.32μm. The specimen is slightly stretched to open the cracks.

2.3 Images processing

Many software are available to process and visualize 3D micro-tomographies, such as Avizo3D, myVGL, Paraview and ImageJ. However, our choice was to use the Python programming language which is widely used in the scientific world and can provide account to more specific developments. Indeed, the use of a programming language for tomography processing allows finer control and more complex analysis possibilities. Gouillart et al. [11] show the ability of Python and the scikit-image library to analyse micro-tomography data.

The goal of the images processing is the segmentation of the different types of inclusions in the material. The segmentation consists of partitioning the tomography into several regions, in which the voxels satisfy a certain homogeneity criterion, based for example on the gray level. The segmentation algorithm must minimize the loss of information on the microstructure caused by the simplification of the images, due to the different transformations applied. Thus, segmentation remains a difficult step since the choice of the segmentation algorithm can lead to significant differences in the morphological characterization of the microstructure. The algorithms used to segment regions of interest depend on the properties of the images. They can be classified in different families such as the thresholding algorithms, the super pixel algorithms [12], the region growing algorithms [13] and the active contour algorithms [14]. Thresholding methods are the most common. They use the histogram data of the images. The threshold gray level is chosen manually or automatically to select the objects of interest. These techniques consider voxels independently and don't use consistency with their neighbors. These techniques are effective when the objects and the background have two distinct peaks on the grayscale histogram.

2.3.1 Pre-treatment

An exposure adjustment is done in order to increase the contrast and delete extreme values due to experimental noise. Then, a 3D median filter is applied to make the segmentation easier afterwards. It reduces noise while preserving edges.

2.3.2 Cracks segmentation

A thresholding operation is applied to the sample in order to separate it from the background. Then, it is necessary to reconstruct the specimen without cracks in order to make the segmentation of cracks. Mathematical morphology operations are applied to fill inner cracks. The outer cracks are filled with an active contour method [14]. Finally, a subtraction of the reconstructed sample from the sample obtained after threshold is done to have only cracks. The Figure 2 shows the cracks detected on one slice of the micro-tomography.

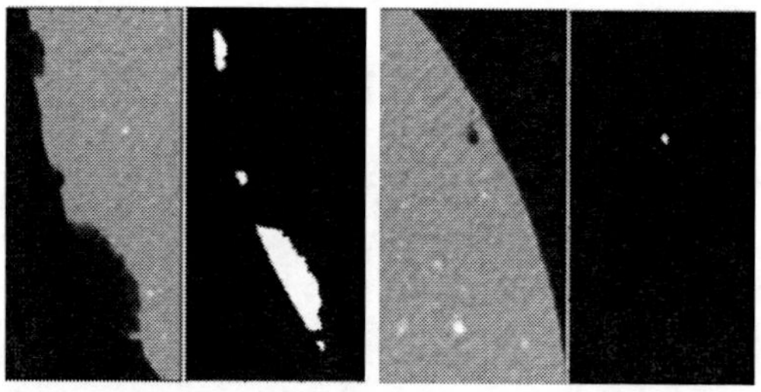

Fig. 2. Cracks segmentation on the sample's surface (left) and inside the specimen (right).

2.3.3 Inclusions segmentation

The thresholding method is not used for the inclusions segmentation because the chosen value has a strong influence on the size of the biggest detected inclusions. Thus, in order to have an accurate inclusions segmentation, a 3D region growth algorithm that combines information on the value of the gray level and the position of the voxels is used [13]. First of all, the user defines markers labeling regions of the image belonging to the inclusion or to the matrix. Some regions remain indeterminate like highlighted in the Figure 3. Then, an anisotropic diffusion equation is solved with tracers initiated at the marker's position. The local diffusivity coefficient is greater if neighboring voxels have similar values, so that diffusion is difficult across high gradients. The label of each unknown voxel is set to the label of the known marker that has the highest probability to be reached first during the diffusion process. This algorithm needs one parameter which controls the split between the two phases. Then, only inclusions above 30μm are kept. However, this size is relevant for the study of the fatigue properties, considering that the most critical inclusions for the enforced fatigue loading are those which size exceeds 20 to 40μm.

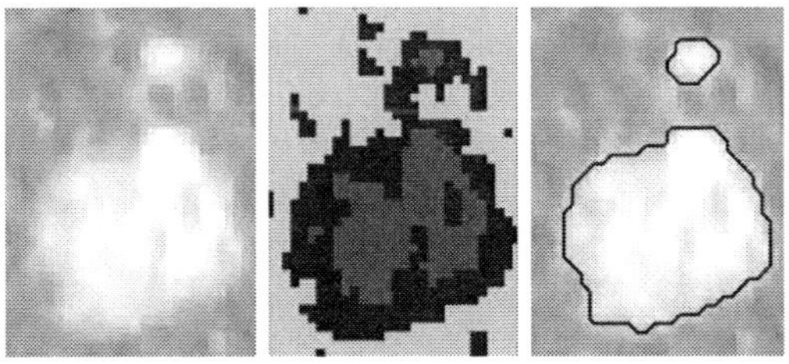

Fig. 3. Inclusions segmentation steps: denoised data (left), markers (center), result of the algorithm superposed to the denoised data (right).

Thanks to this segmentation method, the size of the biggest inclusions is insensitive to the user defined markers (see Figure 3). The Figure 4 shows on one slice the result of the inclusions segmentation. All detected inclusions have their outline drawn.

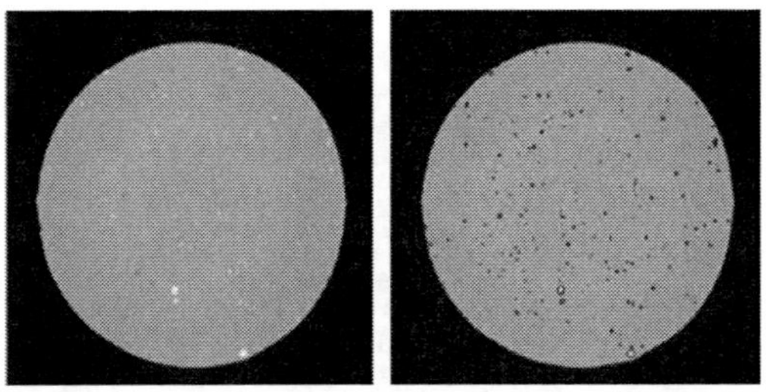

Fig. 4. Result of the inclusions' segmentation: denoised image (left), inclusions detected (right).

2.4 Statistical treatment of inclusions

The segmentation give a binarized matrix of regions of interests. The characteristic of each labelled region can be measured. The boundary between the inclusion and the background is approximated with a triangular representation, obtained from the Lewiner marching cube algorithm [15]. According to Nyström et al. [16], the estimations of the volume and the surface area are more accurate with the triangular representation than the voxelized representation of the inclusions. The statistical post treatment is made on about 5 000 inclusions.

2.4.1 Inclusions' size

A first microstructural indicator is the size of the inclusion. The equivalent diameter of a sphere with the same volume as the labelled region is used.

2.4.2 Inclusions' shape and orientation

The shape of inclusions can be assessed with a compactness factor CF, function of the surface area A and the volume V:

$$CF = \frac{\pi^{1/3}\,(6V)^{2/3}}{A} \tag{1}$$

The shape factor expresses the sphericity of the inclusion: in the case of a perfect sphere, it is equal to 1, otherwise $CF<1$. The shape factor gives a good evaluation of the inclusions' shape, but is strongly dependent of the voxel size. To solve this issue, the optimal ellipsoid with the smallest volume enclosing all voxels is calculated [17] (Figure 5). The surface area and the volume can be calculated on this ellipsoid. Furthermore, the orientation of the inclusions in the sample can be obtained.

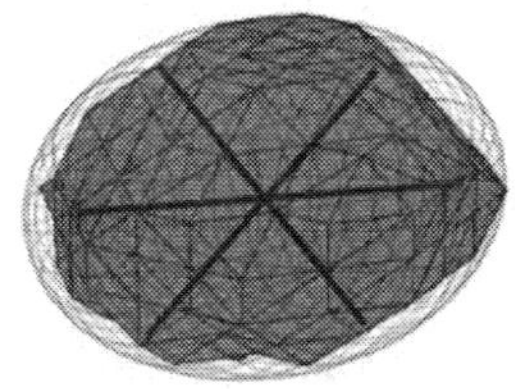

Fig. 5. Best ellipsoid found with its axis for an inclusion meshed with triangles.

2.4.3 Inclusions' spatial distribution

In the following, the inclusions are considered only by their centroid. Several methods are used to study the spatial distribution of inclusions. The first one can be the measurement of the nearest neighbour spacing distribution. This method is able to describe the inclusions distribution as being ordered, random, clustered or composed of clusters superimposed on random inclusions distribution. However, it is not a very sensitive method for clusters evaluation since only the nearest neighbour is detected. No description of the cluster size and spacing can be provided. Another method often used to assess the spatial distribution is the K-function of Ripley [18]. This function is the average number of inclusions in a sphere of radius r, centered on each inclusion. The K-function is reasonably sensitive to local disturbance of inclusions pattern and is able to characterize inhomogeneous population of points as well as mixture of several types of point's pattern [19].

2.5 Observed initiation mechanisms

The segmentation data of the inclusions and of the cracks are superposed in order to observe the damage mechanism. This analysis exhibits two types of damage

https://doi.org/10.1051/matecconf/201816508005

mechanisms. The first one is the cavitation at the poles of the inclusion. The other type of damage mechanism is the failure of inclusions (see Figure 6).

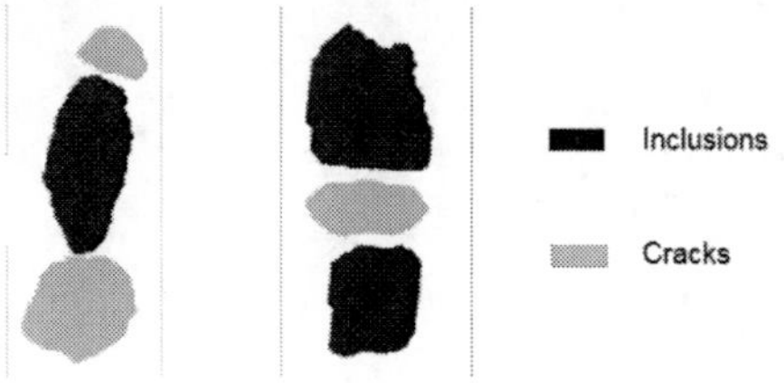

Fig. 6. Damage mechanism observed: cavitation at the pole (left) and failure of the inclusion (right).

3 Tools for the definition of a local criterion

3.1. Criticity of inclusion

A detailed study on the initiation of fatigue cracks in carbon black filled natural rubber was done by Huneau et al. [5]. The criticity of inclusions are assessed thanks to the scanning electron microscope observations of interrupted fatigued samples and fatigue fracture surfaces. The use of the scanning electron microscope, coupled with energy dispersive spectrometry of X-rays, enables the analysis of the chemical nature of inclusions detected as activated sites. The study highlights that fatigue damage generally initiates on carbon black agglomerates or on metallic oxide such as ZnO. However, those two types of inclusions generate different crack initiation mechanism. It is shown that the ones on carbon black agglomerates are almost always followed by a crack propagation leading to the failure of the sample. A possible explanation is that carbon black agglomerates have a stronger cohesion than ZnO and a stronger adhesion to the matrix.

In our study, the inclusions detected by the image processing are composed of metallic oxide and of inclusions called "bumps" (Figure 7). This last type of inclusion is an agglomerate of rubber, carbon black and zinc oxide according to the energy dispersive spectrometry of X-rays analysis.

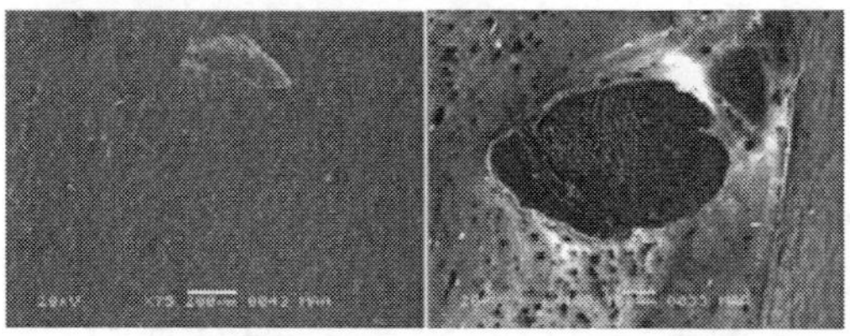

Fig. 7. Scanning electron microscope observation of the inclusions called "bumps".

It was observed with the same method than used in Huneau et al. [5] that "bumps" are possible initiation sites that can lead to failure. More precisely, we observed a cavitation process at the pole of the inclusion in this case as described in the study of Saintier et al. [20]. The cavitation is the process of void nucleation under a given stress state. In our study, it was also observed that ZnO are initiation sites but the created cracks don't propagate. In order to understand, why some inclusions initiate and others don't, we used some simple finite element model and temperature measurements.

3.2 Determination of a local criterion

3.2.1 Simulation tool

A parametric finite element study can be carried out in order to understand the influence of inclusions parameters such as the type, the size, the shape, the inclusions spacing on the fatigue resistance. The simulations are fed with the micro-tomography results found in the previous section. A cylindrical cell is considered with a central ellipsoidal inclusion. Due to symmetry, only a quarter of the cell is considered (Figure 8). The radius of the cell used is set according to the micro-tomography results giving account of the average distance to the nearest inclusion. The size of the inclusion considered is the average of the equivalent diameter of all inclusions detected by the image processing.

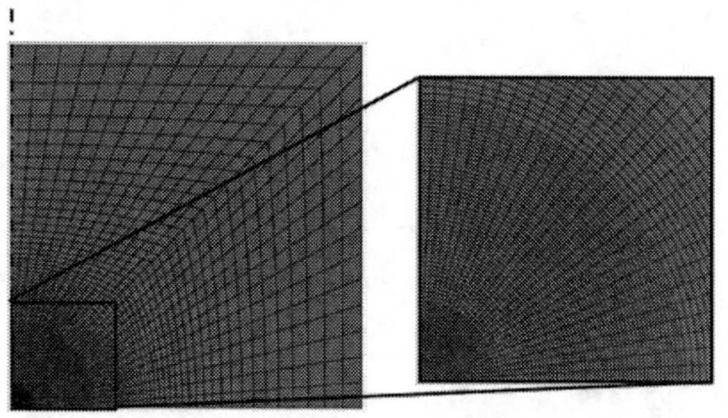

Fig. 8. Axisymmetric finite element model of a cylindrical cell with a central ellipsoidal inclusion.

The nodal displacement of the upper part is imposed and the lateral border of the cell verifies the periodicity hypothesis. A Neo-Hookean behaviour law is used. The hyperelastic potential parameters are identified on the stabilized tensile behaviour of the material using a specific protocol applied on AE2 samples. The transposition of constitutive models and failure criteria from the macroscopic scale to microscopic scale seems possible [21]. The failure criteria used here is related to the cavitation process. Hou and Aberyartne [22] established a cavitation criterion defined for multiaxial loading with a cavitation surface. Considering a Neo-Hookean behaviour law this criterion is written:

$$S: (4\sigma_1 - \sigma_2 - \sigma_3)(4\sigma_1 - \sigma_2 - \sigma_3)(4\sigma_1 - \sigma_2 - \sigma_3) - (5\mu)^3 = 0 \quad (1)$$

with σ_1 , σ_2 , σ_3 the principal stresses and μ the shear modulus. In this study, it is not expected to give an

https://doi.org/10.1051/matecconf/201816508005

accurate prediction of the strain and stress level that leads to cavitation. The use of this criterion aims at assessing the criticity of inclusion parameters. We considered that the interface between the inclusion and the matrix is perfect.

The first studied parameters are the shape and the Young's modulus of the inclusion. According to micro-tomography results, most of inclusions have a compactness factor between 0.9 and 1, so the inclusions are close to spherical object. Therefore, we define three ellipsoidal inclusions with the same volume (see Figure 9). The sphere, the oblate ellipsoid and the prolate ellipsoid have respectively compactness factor of 1, 0.93 and 0.91 The description of these three ellipsoids can also be done thanks to the shape factor f defined by: $f = a/b$. In order to evaluate the influence of the shape and of the Young's modulus on the cavitation, we considered the strain needed to reach the cavitation criterion (see Equation 1).

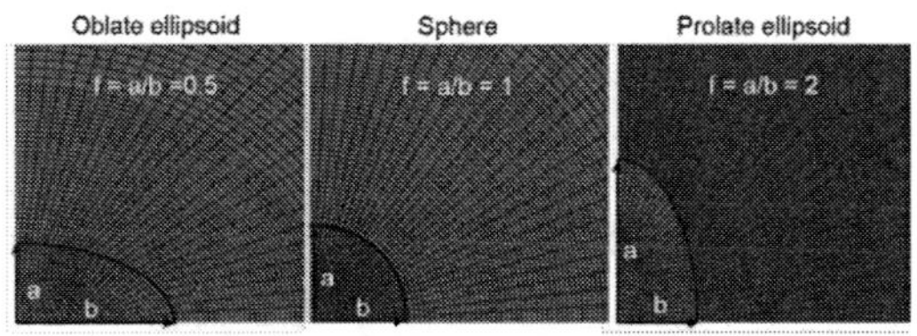

Fig. 9. Ellipsoids considered to study the influence of the shape on the cavitation.

The results show that the cavitation is localized above the inclusion in the spherical and prolate ellipsoid. For the oblate ellipsoid, the cavitation is close to the inclusion's interface. Whatever the shape of the inclusion, an increase of the Young's modulus induces a decrease of the strain leading to cavitation (Figure 10). A low strain value leading to cavitation means a poorer fatigue properties of the compound. In our material, the inclusion can be considered as fully rigid compared to the elastomeric matrix above a Young's modulus of 500MPa. We observe no evolution of the strain at cavitation for a higher value of Young's modulus. This effect of the Young's modulus can be explained easily by the less stiff inclusion and the lower stress values [23]. Regarding the inclusion's shape, the most critical inclusion is the prolate ellipsoid ($f=2$). If we consider a fully rigid inclusion, the strain at cavitation is reduced of 38% in the case of a prolate ellipsoid compared to a sphere. Moreover, the increase of the Young's modulus on prolate ellipsoid has more effect than for the sphere and oblate ellipsoid cases. The oblate ellipsoid is the less critical for the cavitation phenomenon. However, due to the location of the cavitation at the inclusion's interface, this shape of inclusion could be the most critical if a debonding criterion is used.

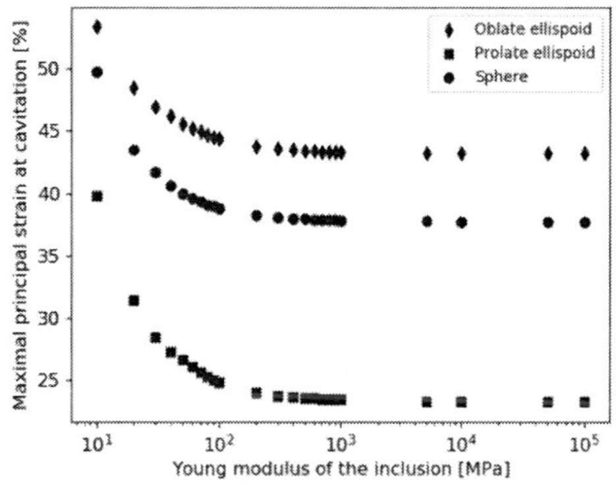

Fig. 10. Effect of the inclusion's size and Young's modulus on the cavitation.

Furthermore, the influence of the size of the three ellipsoidal inclusions is studied. The Young's modulus is fixed at 1GPa. Thus, the inclusion is fully rigid compared to the elastomeric matrix. The Figure 11 shows that the most critical inclusion is always the prolate ellipsoid whatever its size. The increase in size induces a decrease of the strain at cavitation.

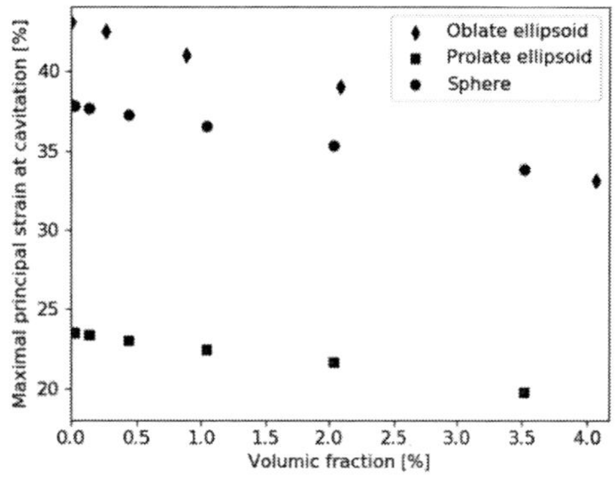

Fig. 11. Effect of the inclusion's size and shape on the cavitation.

Finally, we focus on the effect of the distance between two inclusions. The range of distance between inclusions is determined thanks to the evaluation of distance to the nearest neighbour inclusion. In average, the inclusions' centroids are separated by 245μm. The shortest distance between two inclusions is 32μm. Figure 12 highlights the axisymmetric model used to assess the spherical inclusions inter-distance effect. We considered only the most critical case where inclusions are aligned along the tensile direction. The size of the sphere is determined as the average size of inclusions detected by micro-tomography.

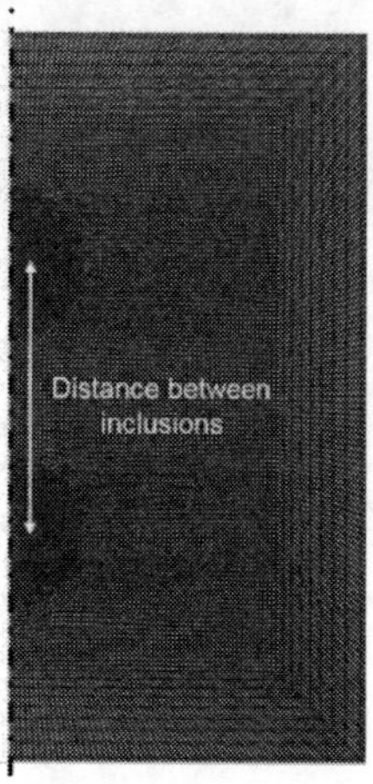

Fig. 12. Finite element model considered for two spheres aligned in the tensile direction.

Figure 13 shows that the proximity of two inclusions in the tensile direction can decrease dramatically the strain at cavitation. For a distance above 225µm, the inclusion has no more effect on the other one for this average size of inclusions. Thus, in the case studied here, few inclusions have influence on each other.

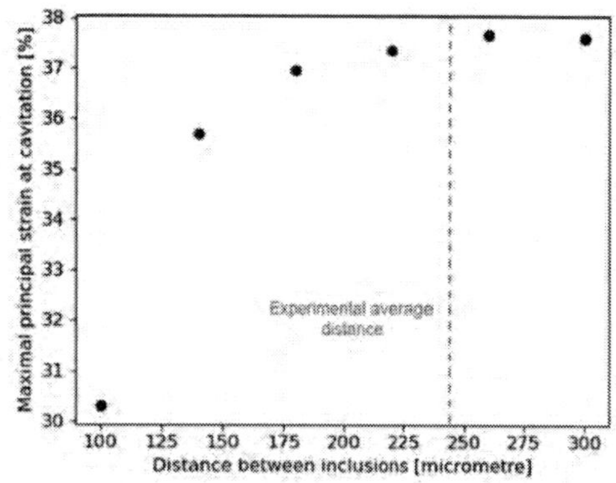

Fig. 13. Effect of the distance between inclusions on the cavitation.

3.2.2 Experimental tool

The thermo-mechanical observations at the scale of inclusions can help to characterize the early stages of the fatigue damage scenario. The experimental approach is fully detailed in a previous communication [21]. Only the main points are recalled here.

The tests are achieved on a Bose electro-dynamical testing machine equipped with a 3.2kN load cell and are displacement controlled. The thermal acquisition is performed thanks to a FLIR SC7600-BB infrared camera with the so-called “G1” lens with an analyzed zone of 7mm*5mm. the thermal resolution is improved with a pixelwise calibration taking into account the housing temperature [24].

The sample is submitted to succession of cyclic sinusoidal loading blocks increasing displacement amplitudes. Each block is composed of an accommodation stage (10 cycles at 30Hz) to stabilize the Mullins effect. Then, the sample is kept to the maximal deformation during cooling to return to thermal equilibrium. Finally, the thermal acquisition is performed using a rate of 200 frames/s, during the loading step (50 cycles at a mechanical frequency of 30Hz).

The evaluation of the thermal sources from the temperature measurements requires solving the heat equation. The way to obtain the dissipated energy from the temperature is explain in another study [25].

4 Example on one activated site

4.1. Infrared response and micro-tomography of one activated site

An initiation site in the AE2 sample was observed with the infrared measurement. The acquisition of the thermal response of the inclusion was obtained during the heat build-up test (Figure 14).

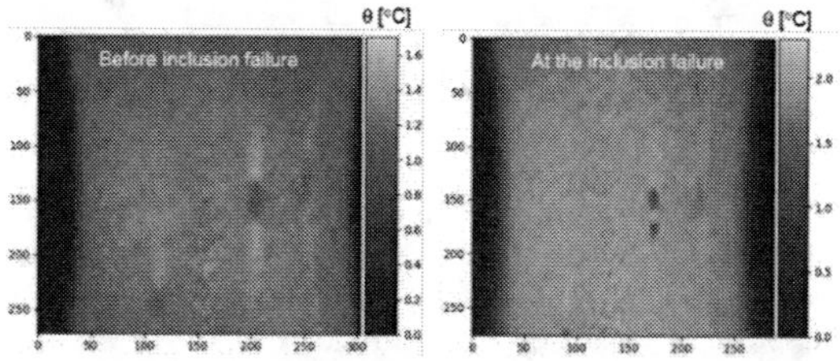

Fig. 14. Infrared measurement before (left) and after (right) the activation of the site.

After the activation of the site, it was observed by micro-tomography and optical microscopy. It was noticed that the inclusion is localized below the surface of the sample (Figure 15).

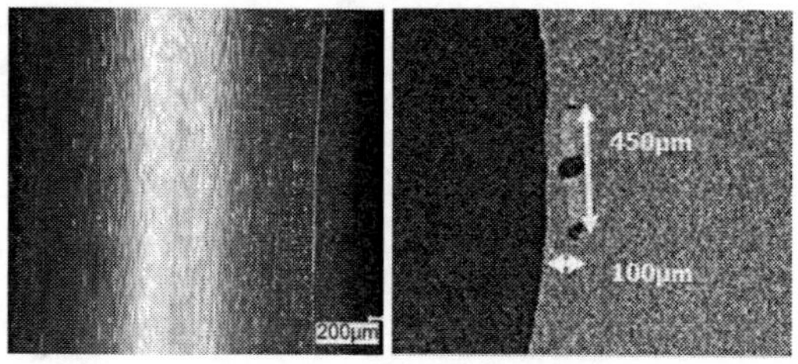

Fig. 15. Observations of the sample with a numerical microscope (left) and with micro-tomography (right).

4.2. Comparison of the infrared measurement with the simulation

To predict the dissipated energy, Le Chenadec et al. [26] model can be used. This approach is based on the idea to relate the cyclic hyperelastic energy to the dissipated energy. Le Saux et al. [27] showed the efficiency of this model to describe and predict the temperature and energy fields for various loading conditions of macroscopic samples. Moreover, Masquelier et al. [25] illustrated the ability of this approach to give a good correlation between the experimental and simulated fields. The parameters of this model are identified on the AE2 sample [28]. We are using this energetic model in order to evaluate the dissipated energy of this inclusions numerically. The comparison between experimental and finite elements simulation enables to challenge the experimental analysis protocols and opens the way to the validation of microscopic failure criteria.

5 Conclusion

This paper presents the approach developed to better understand and describe the early stages of the fatigue damage scenarios. The micro-tomography and the image processing used enable to have quantitative 3D information on the size, shape and spatial distribution of inclusions in the material. This information feeds simulations in order to better understand the impact of some inclusion's parameters on the cavitation phenomenon. As expected, it was shown that the parameters such as the Young's modulus and the shape have a strong influence on the strain leading to cavitation and thus on the fatigue properties. This study has to be pursued by taking into account the quality of the interface between the matrix and the inclusion.

The authors would like to thank the ANRT for its financial support and all the partners of the PROFEM 2 project, Vibracoustic, GeM, IRDL and François Bertrand (GeM) for providing micro-tomography images.

References

1. S. Persson, *Polymer Testing* **4**, 45 (1984)
2. K. Le Gorgu Jago, ECCMR V, 173-177 (2008)
3. A. N. Gent and B. Park, *Journal of Materials Science* **19**, 1947-1956 (1984)
4. K. Le Gorgu Jago, *Rubber Chemistry Technology* **85**, 387-407 (2012)
5. B. Huneau, I. Masquelier, Y. Marco, V. Le Saux, S. Noizet, C. Schiel and P. Charrier, *Rubber Chemistry and Technology* **89**, 126-141 (2016)
6. J. B. Le Cam, B. Huneau and E. Verron, *International Journal of Fatigue* **52**, 82-94 (2013)
7. V. Le Saux, Y. Marco, S. Calloch and P. Charrier, *Polymer Eng. and Science* **51**, 1253-1263 (2011)
8. V. Le Saux, Y. Marco, S. Calloch, C. Doudard and P. Charrier, *Int. J. Fatigue* **32**, 1582-1590 (2010)
9. Y. Marco, I. Masquelier, V. Le Saux, S. Calloch, B. Huneau and P. Charrier, *ECCMR VIII* (2013)
10. L. Salvo, P. Cloetens, E. Maire et al., *Nucl. Intrum. Meth.* **200B**, 273-286 (2003)
11. E. Gouillart, J. Nunez-Iglesias and S. van der Walt, *Adv. Struct. Chem. Imag.* (2018)
12. P.F. Felzenswalb, D. P. Huttenlocher, *International Journal of Computer Vision* **59**, 167-181 (2004)
13. L. Grady, *IEEE Trans. Pattern Analysis and Machine Intelligence* **28**, 1768-1783 (2016)
14. M. Kass, A. Witkin, D. Terzopoulos, *International Journal of Computer Vision* **1**, 321-331 (1988)
15. T. Lewiner, H. Lopes, A. Vieira and G. Tavares, *Journal of Graphic Tools* **8**, 1-15 (2003)
16. I. Nyström, J. K. Udupa, G. J. Grevera and B.E. Hirsch, *Medical Imaging 2002: Visualization, Image-Guided Procedures Display* **4681**, 669-680 (2002)
17. N. Moshtagh, *Convex Optimization.*, 111-112 (2005)
18. B. D. Ripley, *Journal of the Royal Statistical Society* **39**, 172-212 (1977)
19. R. Pyrz, *Mechanics of Microstructural Materials*, 173-233 (2004)
20. N. Saintier, G. Cailletaud, R. Piques, *International Journal of Fatigue* **28**, 61-72 (2006)
21. T. Glanowski, Y. Marco, V. Le Saux, B. Huneau, C. Champy, P. Charrier, ECCMR X, 341-346 (2017)
22. H. Hou and R. Abeyaratne, *J. Mech. Phys. Solids* **92**, 571 (1992)
23. N. Saintier, PhD, 2001
24. V. Le Saux, C. Doudard, Infrared Physics and Technology **80**, 83-92 (2017)
25. I. Masquelier, Y. Marco, V. Le Saux, S. Calloch, P. Charrier, Mechanics of Materials **80**, 113-123 (2015)
26. Y. Le Chenadec, I. Raoult, C. Stolz, M. Nguyen-Tajan, Mech. Material Structure **4**, 309-318 (2009)
27. V. Le Saux, Y. Marco, S. Calloch, P. Charrier and D. Taveau, *Rubber Chem. Techn.* **86**, 38-56 (2013)
28. T. Glanowski, V. Le Saux, C. Doudard, Y. Marco , C. Champy, P. Charrier, Continuum Mechanics and Thermodynamics **29**, 1163-1179 (2017)

Effect of Surface Mechanical Attrition Treatment on the very high cycle fatigue behavior of TC11

Tao Gao[1], *Zhidan* Sun[2], *Hongqian* Xue[1*], and *Delphine* Retraint[2]

[1]School of mechanical engineering, Northwestern Polytechnical University, Xi'an 710072 China
[2] ICD, P2MN, LASMIS, University of Technology of Troyes, CNRS, Troyes, France

Abstract. As an important engine component material, TC11 (Ti-6.5Al-3.5Mo-1.5Zr-0.3Si) titanium alloy is subjected to high frequency cyclic loading and its failure occurs beyond 10^9 cycles. It is thus essential to investigate the very high cycle fatigue (VHCF) behavior of this alloy. Surface Mechanical Attrition Treatment (SMAT) is a promising surface treatment technique to improve fatigue strength by modifying the surface microstructure. Therefore, it is important to understand the fatigue damage and failure process of SMATed titanium alloy in the VHCF regime. In this work, VHCF tests of TC11 before and after SMAT under fully reversed loading were conducted at room temperature by using an ultrasonic fatigue testing machine at a frequency of 20 kHz. The preliminary results seem to indicate that SMAT can reduce fatigue strength and fatigue life of TC11. Fracture surface analysis of the specimens before and after SMAT was performed using scanning electron microscope (SEM) to investigate the mechanisms of crack initiation and propagation. Particular attention was paid to fatigue crack initiation sites. The effect of SMAT on damage mechanism of SMATed TC11 in the VHCF regime was discussed.

1. Introduction

Many titanium components used in the field of aeronautical engineering such as blade, compressor and gear, are subjected to high frequency cyclic loading and failure occurs beyond 10^7 cycles and even up to 10^{11} cycles. This fatigue cycle range is called very high cycle fatigue (VHCF) regime. Beyond the fatigue limit determined using conventional fatigue testing method, many investigations of titanium alloy in the VHCF regime show S-N curves with stepped shape, which continuously decreases with significant slope at cyclic loading cycles greater than 10^7 cycles [1-3]. Moreover, the crack initiation site is transferred from the surface to the subsurface or even the interior of materials in the VHCF regime [1, 4], which implies that the damage and failure mechanism in VHCF could be different from those in low cycle fatigue (LCF) and high cycle fatigue (HCF). Therefore, understanding the fatigue damage and failure process of titanium alloy in the VHCF regime is important for ensuring reliability and safety of aircrafts during the period of service.

It is well known that engineering components are subjected to fatigue loading and failure from flaws or cracks at the surface. Surface modification techniques have been successfully applied to improve the fatigue strength of materials by forming a hardened surface layer and introducing superficial compressive residual stresses. Surface Mechanical Attrition Treatment (SMAT) is a relatively new method of surface severe plastic deformation (SPD), which is attracting increasing attention [5]. SMAT process is similar to conventional shot peening except that the shot present a larger diameter in the case of SMAT and are propelled by a sonotrode vibrating at ultrasonic frequency instead of compressed air. The main beneficial characteristic of the SMAT process is to transform ordinary coarse grains to nano-sized grain in the surface layer of a part without changing the chemical composition of materials [6]. The SMAT process has been proved to be efficient to nanocrystallize surface layer of various materials [7] including aluminum, stainless steels and titanium alloys [8-11]. Extensive research has shown that fine grains can improve the plastic strength and inhibit crack initiation, while coarse grains exhibit better crack propagation resistance due to their good ductility and toughness [12]. Considering the fact that most fatigue cracks of components initiate at the surface and propagate inward,

* Corresponding author: xuedang@nwpu.edu.cn

a nanostructured surface layer and coarse grained interior bulk can be expected to effectively improve fatigue life. Furthermore, the compressive residual stresses and work hardening induced by SMAT can also considerably delay fatigue crack initiation and propagation.

It has been demonstrated that SMAT can improve the fatigue life of different materials [13, 14]. However, most studies focused on low and high cycle fatigue improvement and considered the effect of surface residual stresses, roughness and microstructures induced by surface treatment. Only few investigations focus on the application of surface treatment in the VHCF regime and no study on the VHCF properties of SMATed titanium alloy has been reported in the literature.

This paper is dedicated to investigate the VHCF performance of a TC11 titanium alloy before and after SMAT. Ultrasonic fatigue tests at a frequency of 20 kHz are conducted to examine the VHCF properties of TC11. Two series of dumbbell-shaped ultrasonic fatigue specimens, mechanically polished and SMATed respectively, are prepared to investigate the fatigue behavior and to study the effect of SMAT on fatigue life under VHCF. Fracture surfaces are analyzed by scanning electron microscope (SEM) in order to examine the mechanism of crack initiation and propagation involved during fatigue loading. The influence of SMAT on the fatigue behavior of the untreated and SMATed specimens is then discussed.

2. Material and experimental methods

The material used in this work is an α+β titanium alloy, TC11 (Chinese standard) in a mill-annealed condition. The chemical composition of this alloy is shown in Table 1. This alloy has a duplex microstructure, which contains about 60% volume fraction of primary α phase. The typical values of basic physical/mechanical properties are shown in Table 2.

Table 1. Chemical composition of TC11 (in weight-%)

Al	Mo	Zr	Si	Fe	Ti
6.45	3.45	1.69	0.24	0.01	Bal.

Table 2. Basic physical/mechanical properties of TC11

ρ (kg·m^{-3})	UTS (MPa)	$R_{p0.2}$ (MPa)	A (%)
6.45	1132	971	16.5

The shape and dimension of specimens for ultrasonic fatigue tests are shown in Fig. 1. These specimens have a diameter of 4 mm in the center. The surface of the specimens' central region was polished using abrasive paper with a grit scale of 500, 1000, 1500 and 2000 to remove machining marks that could lead to stress concentration and promote premature fracture. A set of specimens was subjected to SMAT with two levels of intensity. The SMAT process was performed at room temperature with 100Cr6 steel shot of 2 mm diameter at a sonotrode vibration amplitude of 20 μm (denoted as SMATed 20) and 45 μm (denoted as SMATed 45) for 20 minutes. A rotating motor with a speed of 80 rpm was adapted to turn the specimens in order to get a homogenous treatment on the surface [13]. According to previous studies, a nanostructured layer can be generated on the top surface of the specimens, especially under high treatment intensity[11]. More detailed description and principle of SMAT can be found in [13, 15].

Fatigue experiments were conducted using an ultrasonic fatigue system at room temperature with a resonance frequency of 20 kHz, and a stress ratio of R=-1. During ultrasonic fatigue test, the specimens were cooled by compressed air, and ultrasonic interval loading was applied to avoid heating in order to keep specimen temperature below 35°C.

After the fatigue tests, the fracture surfaces of failed specimens were observed using SEM, and the chemical composition of the crack initiation site on fracture surface was analyzed by Energy Dispersive X-ray Analysis (EDX) in the SEM.

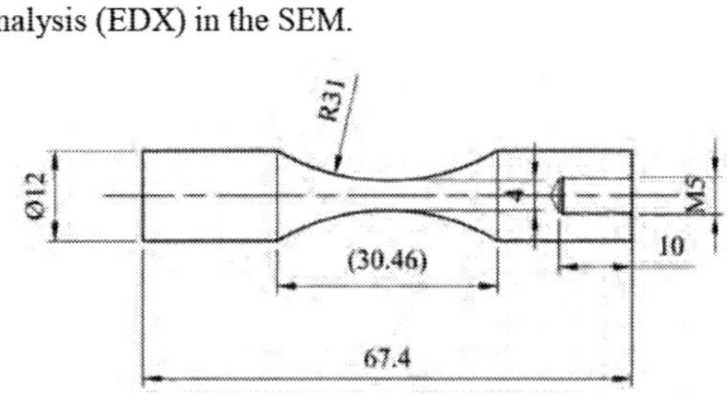

Fig. 1. Shape and dimension of specimens for ultrasonic fatigue tests

3. Results and discussion

3.1 Nanostructured layer generated by SMAT

There are many studies reported in the literature on surface nanocrystallisation of material surface by SMAT process. The deformation and nanocrystallisation of this alloy is essentially related to twinning and dislocation activities. The nanocrystallisation and grain refinement mechanism of titanium was investigated by Zhu et al. [16].

SEM micrographs of fracture surface of TC11 specimens processed with a sonotrode vibration amplitude of 20 μm (SMATed 20) and 45 μm (SMATed 45) are shown in Fig. 2. It is shown that a nanostructured and strongly plastically deformed layer generated by SMAT can be distinguished on the fracture surface, which is similar to Ni_3Al [17] and S45C steel [18] presented in the literature. When SMAT is carried out at a vibration amplitude of 20μm, the nanostructured and strongly plastically deformed layer is generated in the near surface layer, and it seems there is a transition boundary between the strongly plastically deformed layer and the bulk, as shown in Fig. 2a. As the sonotrode vibration amplitude increased from 20μm to 45μm, the thickness of the nanostructured and strongly plastically deformed layer is increased from approximately 6μm to 12μm, which can be seen in Fig. 2b. Besides, there are no obvious boundary between deformable layer and the bulk for SMAT 45.

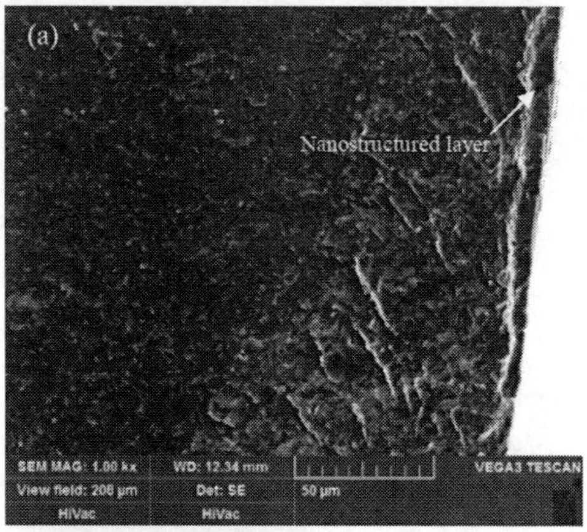

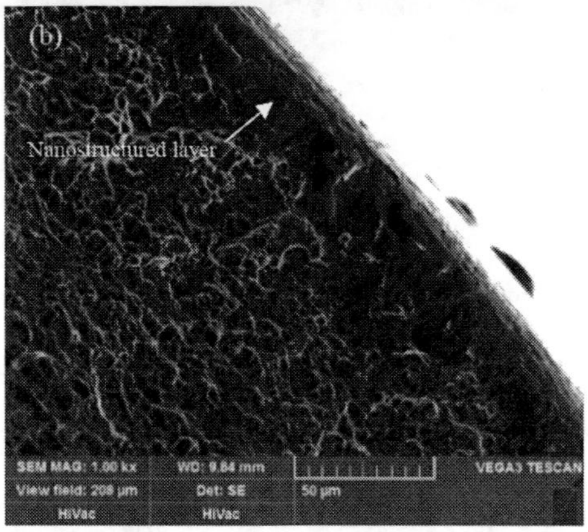

Fig. 2. SEM micrographs of fracture surface of SMATed TC11 specimens; nanostructured and strongly plastically deformed layer produced by SMAT with a sonotrode vibration amplitude of (a) 20 μm and (b) 45 μm

3.2 Fatigue life

The ultrasonic fatigue test results of both untreated and SMATed samples are shown in Fig. 3 in the form of S-N plot. The surface and interior crack initiation of untreated specimens are represented by the hollow circles and solid circles, respectively. Triangles are used to indicate the fatigue test results of specimens SMATed with a sonotrode vibration amplitude of 20 μm and 45 μm, which are respectively referred to red and blue triangles. The fatigue life of the untreated specimens increases with the decrease of the applied stress, varying from 2×10^7 cycles to 1.7×10^9 cycles in a range of applied stress levels from 720 to 660 MPa. It seems that at a same stress level, specimens with surface crack initiation failed earlier than specimens with crack initiation sites located in the bulk. Unexpectedly the fatigue life and fatigue strength of the SMATed specimens significantly decrease compared to the untreated specimens, especially for higher vibration amplitude, (SMATed 45). Qualitatively this could be explained by a higher roughness of SMATed specimens due to shot impacts in comparison to the polished untreated specimens. As a matter of fact, higher SMAT intensity (represented by vibration amplitude) leads to higher surface roughness of the specimens. The higher roughness induced by SMAT could have played the role of a micro notch, and consequently promoted the crack nucleation at surface due to stress concentration. This assumption is consistent with the fact that the crack initiation occurred at surface for all the SMATed specimens, according to fracture surface observation using SEM. Deterioration in fatigue life by mechanical surface treatment was also reported by other authors [19]. In a further investigation, attention will be paid to the effect of SMAT on the crack initiation process of the TC11 alloy.

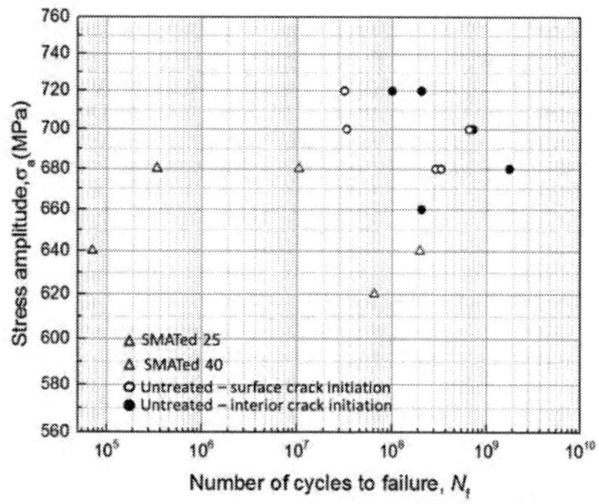

Fig. 3. S-N plot of the preliminary ultrasonic fatigue test results for the studied TC11 alloy.

3.3 Microscopic morphology analysis of fracture surfaces

The fatigue crack initiation site and the fracture surface morphology were observed by SEM. Based on the SEM observation of all the fracture surfaces, the crack initiation of untreated specimens can be classified into two crack initiation modes: surface induced failure and interior induced failure, as shown in Fig. 4a and Fig. 4b, respectively. The fatigue crack initiation of the titanium alloy studied in this work is similar to Ti-6Al-4V investigated by Zuo et al. [2]. The fracture surface morphologies revealed by SEM in the present work and in other investigations show that the location for main crack initiation in the VHCF regime is always determined by a competitive effect between the surface defects and the internal inhomogeneous microstructures such as grain orientation.

There is only one crack initiated and propagated through the cross section for surface induced failure of untreated specimens, as shown in Fig. 5a. The fracture surface shows radial lines and these lines radiate out from the center of the surface crack source position. Two

https://doi.org/10.1051/matecconf/201816509001

types of surface crack initiation of Ti-6Al-4V alloy were observed in [3, 20], namely surface without facets (Type I) and surface with facets (Type II). Similar to the Ti-6Al-4V alloy, both Type I and Type II can be found in the surface crack initiation area by using SEM observation. For type II, Fig. 5b shows that the diameter of the facets is about 6μm. In the area close to the crack initiation center, there are a lot of small facets with approximately the same size. These facets should have been caused by the cleavage fracture of the α grains located in the subsurface region, according to the literature [3].

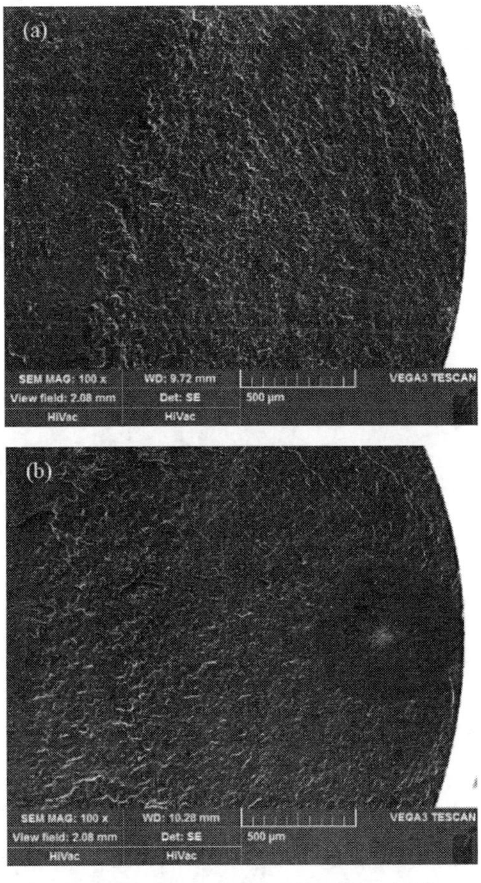

Fig. 4. SEM micrographs of (a) surface induced failure of a specimen without SMAT failed at 3.35×10^7under σ_a=700 MPa, (b) interior induced failure of a specimen without SMAT failed at 1.27×10^8 under σ_a=660 MPa

Interior fatigue failure occurred at more than 10^8 cycles and formed a typical fisheye area at the crack initiation site on the fracture surface. Fig. 6 shows an example of interior crack initiation of a specimen failed in the VHCF regime at 2.27×10^8 cycles under 660 MPa. A fine granular area (FGA) in the center of the fisheye region with similar fractographic characteristics as Ti6Al4V titanium alloy [1] is shown in Fig. 6a and Fig. 6b. The fisheye can be divided into two areas, namely FAG and flat surface around the FGA, which are marked in Fig. 6a. The FGA is approximated to a circle and the diameter is about 420 μm with a very rough surface, which can be considered as the result of crack initiation process.

Fig. 6b shows that there are no inclusions and pores in FGA. The chemical composition of the FGA was measured using electron dispersive spectroscopy (EDS). Fig. 6c shows that the chemical composition of this area is similar to that of the bulk material without inclusion. Therefore, the interior crack initiation of the present TC11 might be associated with microstructural inhomogeneities. This point will be checked in our future work.

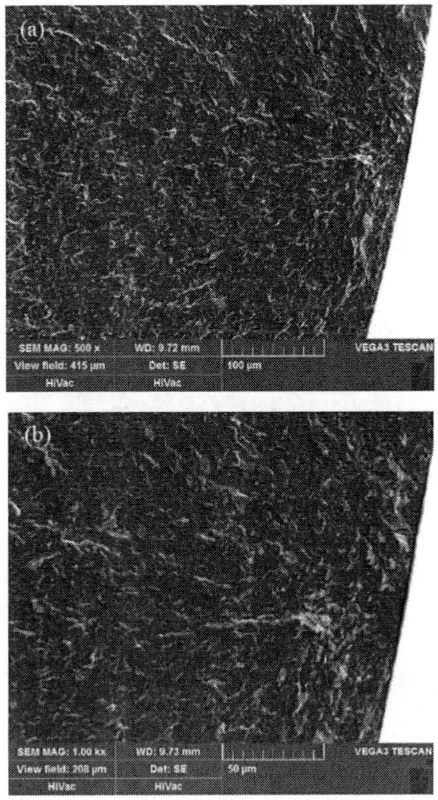

Fig. 5. (a) Crack initiation at the surface and (b) type II "surface with facets" of a specimen without SMAT failed at 2.90×10^8under σ_a=680 MPa

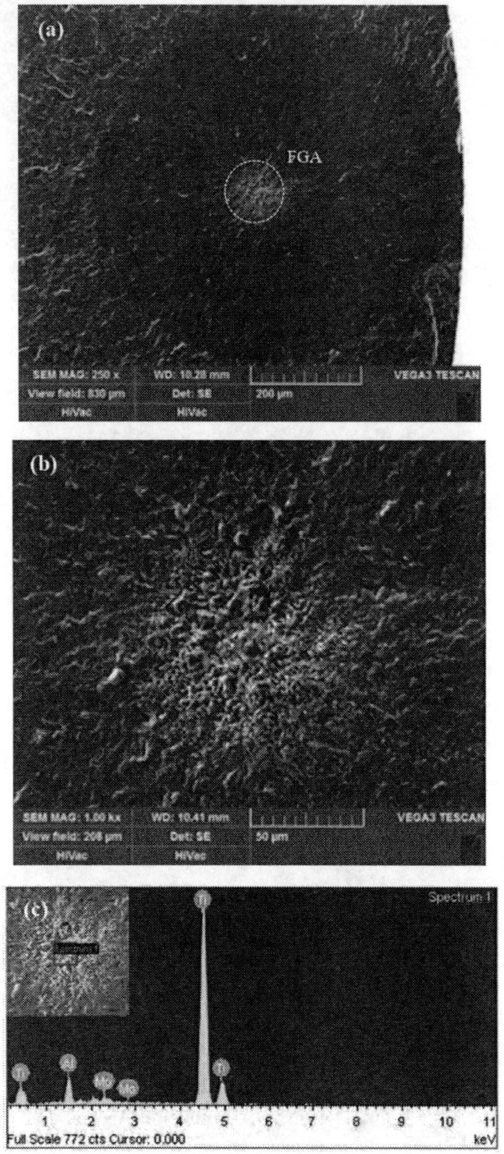

Fig. 6. (a) Fracture surface of interior induced failure with fisheye of a specimen failed at 2.05×10^{8} under σ_a=720 MPa, (b) detailed observation of FGA (fine granular area), (c) chemical composition of the FGA analyzed using EDS

Furthermore, the surface roughness of the fisheye, as shown in Fig. 6, was analyzed by a surface profilometer (VeecoWyko NT1100) to derive quantitative geometric morphology information of the FGA and the flat surface in the fisheye region. Fig. 7 shows the 2D graph and the corresponding 3D scan from the surface profilometer of a 1.20 mm×0.93 mm area. The mean surface roughness Ra of the FGA and the other region in the fisheye is approximately 32 μm and 3.4 μm, which was measured and calculated automatically by Vecco vision software.

In the case of specimens treated by SMAT, the fatigue failures are all initiated at the surface, as illustrated in Fig. 8a and Fig. 8b. The fatigue crack initiated at the surface of SMATed specimens without the presence of obvious facets. Compared to the surface initiation region of untreated specimens (Fig. 5), the surface crack initiation source of SMATed specimens is a larger area around the surface as shown in Fig. 8b. It is assumed that a micro notch caused by very rough surface induced by SMAT could be the main factor affecting the very high cycle fatigue life of the SMATed specimens.

Due to the fact that SMAT only affects surface and subsurface layers of specimens, the aspect of fatigue crack propagation should be similar to the case of the untreated specimens with surface-induced fatigue fracture. Fig. 8c is an example of fatigue crack propagation for failed specimens in the VHCF regime. In the crack propagation zone, many facets and second cracks can be observed with the presence of obvious typical characteristics of inter-crystalline and trans-crystalline fatigue fracture in the VHCF regime.

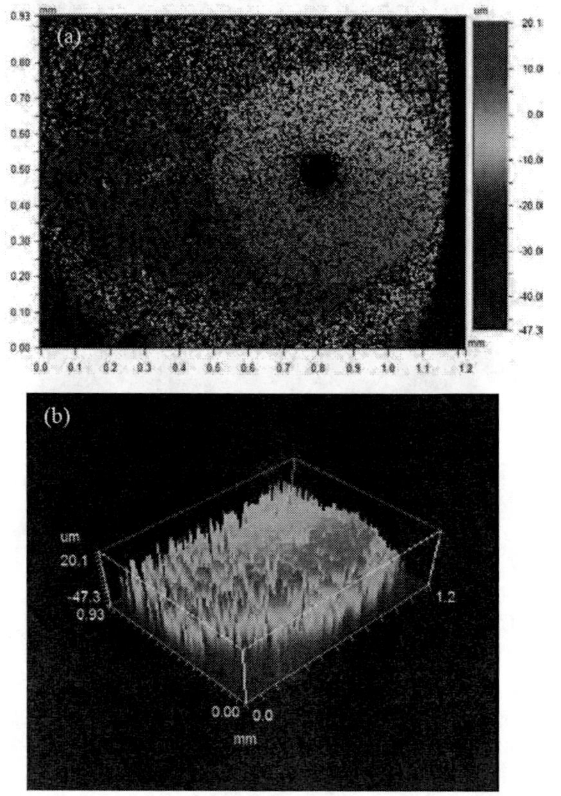

Fig. 7. (a) 2D and (b) 3D profiles of the fisheye of a specimen failed at 2.05×10^{8}under σ_a=720 MPa

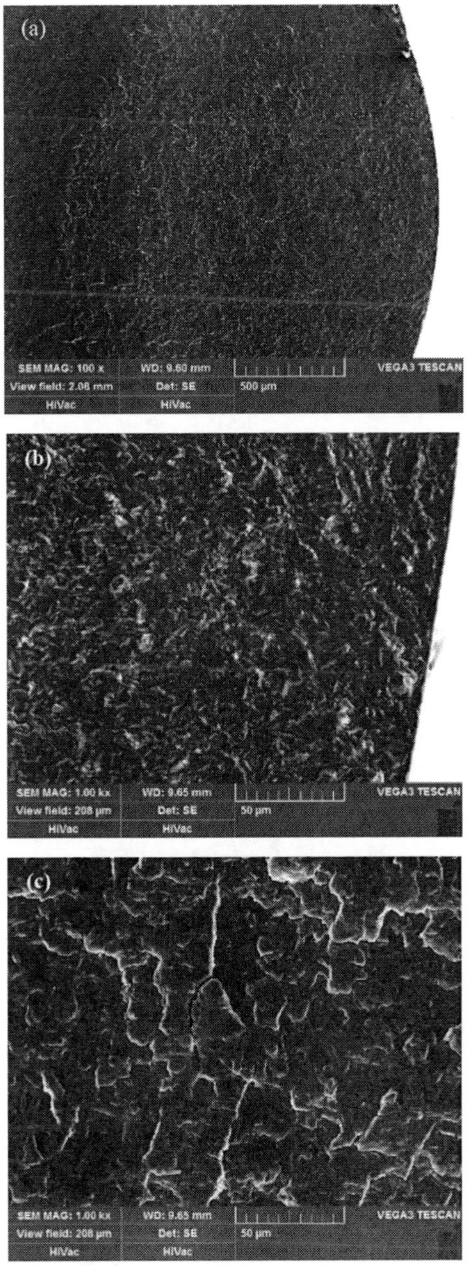

Fig. 8. (a) Fracture surface of a SMATed specimen (SMATed 45) failed at 1.05×10^7 under σ_a=680 MPa, (b) detail of the crack initiation source, (c) enlargement of the crack propagation zone

4. Conclusion

In this work, the fatigue behavior of a TC11 titanium alloy before and after SMAT has been studied in the very high cycle regime. A nanostructured and strongly plastically deformed layer has been generated in the surface region by SMAT. Contrary to the beneficial effect of SMAT in LCF and HCF regimes, the preliminary results presented in this paper show that SMAT has a negative effect on the fatigue properties under VHCF, It reduces the fatigue life compared to specimens without SMAT.

Two types of crack initiation modes, surface crack initiation and interior crack initiation in VHCF regime are observed for TC11 without SMAT. Interior fatigue failures occur at more than 10^8 cycles and are characterized by the presence of fisheye. There are no inclusions and pores in FGA, and the interior crack initiation of TC11 in the VHCF regime could be related to microstructural inhomogeneities for example inhomogeneous grains orientation. The fatigue failures of SMATed specimens all initiate at the surface, and the fatigue crack propagation is the same as the surface-induced fatigue fracture of untreated specimens. The crack propagation zone shows obvious characteristics of inter-crystalline and trans-crystalline fatigue fracture.

References

1. S. Heinz, F. Balle, G. Wagner, D. Eifler, Ultrasonics **53**(8) 1433-1440 (2013).
2. J. Zuo, Z. Wang, E. Han, Materials Science and Engineering: A **473**(1) 147-152 (2008).
3. X. Liu, C. Sun, Y. Hong, Materials Science and Engineering: A **622** 228-235 (2015).
4. Z.Y. Huang, H.Q. Liu, H.M. Wang, D. Wagner, M.K. Khan, Q.Y. Wang, International Journal of Fatigue **93** 232-237 (2016).
5. Z. Sun, D. Retraint, T. Baudin, A. Helbert, F. Brisset, M. Chemkhi, J. Zhou, P. Kanouté, Materials Characterization **124** 117-121 (2017).
6. J. Zhou, Z. Sun, P. Kanouté, D. Retraint, International Journal of Fatigue **103** 309-317 (2017).
7. K. Lu, J. Lu, Materials Science and Engineering: A **375** 38-45 (2004).
8. Y. Liu, B. Jin, J. Lu, Materials Science and Engineering: A **636** 446-451 (2015).
9. T. Roland, D. Retraint, K. Lu, J. Lu, Materials Science Forum **490-491** 625-630 (2005).
10. G. Proust, P. Trimby, S. Piazolo, D. Retraint, Journal of Visualized Experiments : JoVE (122) 55506 (2017).
11. Y. Pi, G. Agoda-Tandjawa, S. Potiron, C. Demangel, D. Retraint, H. Benhayoune, Journal of Nanoscience and Nanotechnology **12**(6) 4892-4897 (2012).
12. L. Yang, N.R. Tao, K. Lu, L. Lu, Scripta Materialia **68**(10) 801-804 (2013).
13. T. Roland, D. Retraint, K. Lu, J. Lu, Scripta Materialia 54(11) 1949-1954 (2006).
14. S.A. Kumar, S.G.S. Raman, T.S. Narayanan, Transactions of the Indian Institute of Metals **67**(1) 137-141 (2014).

15. Sandá, V.G. Navas, O. Gonzalo, Materials & Design **32**(4) 2213-2220 (2011).
16. K. Zhu, A. Vassel, F. Brisset, K. Lu, J. Lu, Acta Materialia **52**(14) 4101-4110 (2004).
17. J. Ren, A. Shan, J. Zhang, H. Song, J. Liu, Materials Letters **60**(17) 2076-2079 (2006).
18. X. Cao, Y. Pyoun, R. Murakami, Applied Surface Science 256(21) 6297-6303 (2010).
19. M.-S. Suh, C.H. Suh, S.-H. Nahm, C.-M. Suh, Advances in Mechanical Engineering **7**(1) 126848 (2015).
20. V. Crupi, G. Epasto, E. Guglielmino, A. Squillace, International Journal of Fatigue **95** 64-75 (2017).

Short fatigue crack growth in welded joints described by the effective cyclic J-integral

Désiré Tchoffo Ngoula[1,*], *Michael* Vormwald[1]

[1]Materials Mechanics Group, Technische Universität Darmstadt, Franziska-Braun-Str. 3, 64287 Darmstadt, Germany

Abstract. The purpose of the present contribution is to predict the fatigue life of welded joints by using the effective cyclic *J*-integral as crack driving force. The plasticity induced crack closure effects and the effects of welding residual stresses are taken into consideration. Here, the fatigue life is regarded as period of short fatigue crack growth. The node release technique is used to perform finite element based crack growth analyses. For fatigue lives calculations, the effective cyclic *J*-integral is employed in a relation similar to the Paris (crack growth) equation. For this purpose, a specific code was written for the determination of the effective cyclic *J*-integral for various lifetime relevant crack lengths. The effects of welding residual stresses on the crack driving force and the calculated fatigue lives are investigated. Results reveal that the influence of residual stresses can be neglected only for large load amplitudes. Finally, the predicted fatigue lives are compared with experimental data: a good accordance between both results is achieved.

1 Introduction

Welding is one of the most important joining methods in technology, particularly for metallic structures. Thus, improving the design of welded structures is of high technical use. Welded structures are generally exposed to fatigue failure: under cyclic external loading, an immense part of the crack growth takes place in the weld notch area. As a result of the notch effect, the local behaviour of the weld in this area might be elastic-plastic. Thereby, the material is subjected to transient processes like the cyclic hardening/softening, the cyclic relaxation and ratcheting.

During the welding process, residual stresses can arise in the joint due to the heating and cooling cycles, which belongs to this process. In case of an external cyclic loading, these stresses can have an influence on the fatigue life of the welded structure: the presence of a tensile residual stress field at the failure critical notch for example reduces the fatigue life, while a compressive residual stress field extends it [1,2].

Crack closure effects were observed for the first time by Elber [3]. During cyclic loading, crack closure means that the two crack faces contact each other during the unloading phase of a cycle before the minimum load is reached, or that the crack remains closed during the loading phase of a cycle until a certain load level [4]. Therefore, only the load portion, in which the crack is open, is responsible for the propagation of the crack (keyword "*effective load*"). Closure effects lead to a reduction of the crack driving force and may influence the crack propagation rate. Therefore, to avoid significant errors in the fatigue life assessment of welded joints, residual stresses and crack closure effects should be considered.

Different approaches can be used to assess the fatigue life of welded joints: the nominal stress approach, the structural stress approach, the notch stress approach, the notch strain approach and the crack propagation approach [5, 6]. The main different between these approaches is the parameter employed for the description of the fatigue life. Many studies [7] have shown that, the lifetime estimation of welded joints can be done by using the fracture mechanics concept (crack propagation approach) alone, because the crack nucleation stage is very short (due to the presence of defects in the weld) and can be neglected compared to crack propagation stages. Thus, the main part of the fatigue life falls on the crack propagation, precisely on the short crack growth [8]. Accordingly, the application of the elastic-plastic fracture mechanics will be required because the length of the plastic zone at the crack tip (in case of short cracks) is not negligible, compared to the crack length.

For the modelling of fatigue crack growth in elastic-plastic fracture, Dowling and Begley [9] proposed to use the cyclic *J*-integral, ΔJ, as crack driving force. ΔJ is an extension of the monotonic *J*-integral introduced by Rice [10]. This parameter has successfully been applied as an appropriate crack driving force for the description of fatigue crack growth, when the elastic-plastic material exhibits Masing's behaviour [11,12]. ΔJ can be related to the increase in crack length per cycle, da/dN, by the following power-law expression:

$$\frac{da}{dN} = C_J \cdot (\Delta J)^{m_J} \quad (1)$$

* Corresponding author: tchoffo@wm.tu-darmstadt.de

where, C_J and m_J are material dependent parameters which can be determined experimentally.

The purpose of this work is to calculate the fatigue life of welded joints by taking into account plasticity induced crack closure effects as well as the effect of welding residual stresses. Here, the fatigue life is regarded as period of short fatigue crack growth. Analyses are performed with two and three dimensional finite element models, by using the crack propagation approach. Closure effects were considered by defining contact boundary conditions between crack faces and by using the effective cyclic *J*-integral (ΔJ_{eff}) as crack tip parameter for the description of the crack propagation rate. The advantage of this approach (fracture mechanics concept) is that, residual stresses can be treated explicitly. However, the numerical determination of ΔJ_{eff} appeared to be complex. A specific code was written for the determination of ΔJ_{eff} at each crack length configuration.

The research results in this contribution were achieved at the Materials Mechanics Group of Technische Universität Darmstadt in the scope of a project called IBESS ("*Integrale Bruchmechanische Ermittlung der Schwingfestigkeit von Schweißverbindungen*" [13]). The aim of IBESS is the fracture mechanics based simulation of Wöhler curves (S-N curves) in welded joints. IBESS is divided into 8 subprojects and the authors were working on the subproject titled: "*Modelling of fatigue crack growth in welded joints under consideration of the transient plastic deformation behaviour.*"

2 Geometry and material

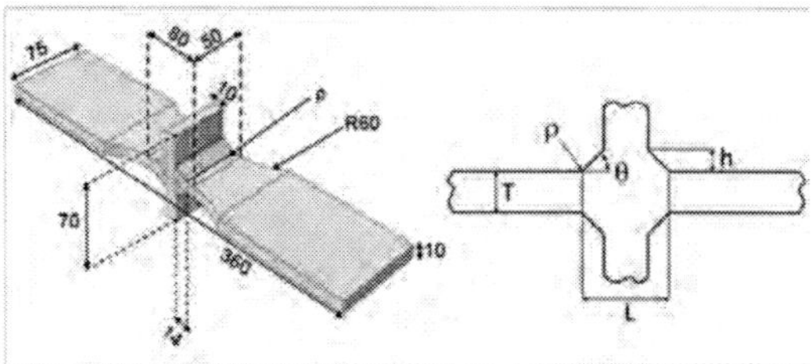

Fig. 1. Geometry and sizes (in mm) of the cruciform welded joint examined [14].

The structure used in this contribution is a cruciform welded joint. In this case, it is a full penetration weld. The geometry and dimensions of the structure are shown in *Fig. 1*. This specimen has a thickness T of 10 mm and a length of about 360 mm. The weld width L is about 38 mm and the investigated cross section is $50\text{ mm} \times 10\text{ mm}$, i.e 500 mm^2. Other sizes can be found in the figure.

The weld toe radius ρ and the weld flank angle θ are parameters which can scatter. They were measured and documented by a project partner [15,16]. According to the project partner, ρ has a mean value of approximately 1mm and varies from 0,2 mm to 3,0 mm. In addition, θ varies between 3 and 40 degrees.

The tested specimens were made from medium strength structural steel S355NL. The cyclic material data for the Ramberg-Osgood equation,

$$\varepsilon = \varepsilon_{el} + \varepsilon_{pl} = \frac{\sigma}{E} + \left(\frac{\sigma}{K'}\right)^{\frac{1}{n'}}, \tag{2}$$

were determined by a project partner (IEHK, RWTH Aachen): $K' = 1115\text{ MPa}$, $n' = 0{,}1961$ and $E = 210000\text{ MPa}$. By using these cyclic material data, the Chaboche's material [17] parameters were also determined:

$$\sigma = \sigma_y + \sum_{i=1}^{5} \left[\frac{C_i}{\gamma_i}\left(1 - exp(-\gamma_i \varepsilon_{pl})\right)\right]. \tag{3}$$

These parameters (the coefficients C_i, γ_i as well as the yield stress σ_y) are summarized in *Table* 1 and were used in FE calculations.

Table 1. Chaboche's material parameters of steel S355NL.

i	1	2	3	4	5
C_i [MPa]	305034	98322	90427	26762,3	7607
γ_i	12318,92	3654,02	1204,24	230,50	21,07
σ_y [MPa]	152,17				

3 Calculation of the cyclic J-integral

Under non-monotonic loading, the ΔJ-integral proposed by Dowling and Begley [9] is generally used in the framework of elastic-plastic fracture mechanics as crack tip parameter and is defined as:

$$\Delta J = \int_{\Gamma} \left(\Delta W dx_2 - \Delta t_i \frac{\partial \Delta u_i}{\partial x_1}\right) ds \tag{4}$$

with

$$\Delta W = \int_0^{\Delta\varepsilon_{ij}} \Delta\sigma_{ij} d\Delta\tilde{\varepsilon}_{ij}\,, \qquad \Delta t_i = \Delta\sigma_{ij} n_j \tag{5}$$

and with

$$\Delta\sigma_{ij} = \sigma_{ij} - \sigma_{ij}^0\,, \Delta\varepsilon_{ij} = \varepsilon_{ij} - \varepsilon_{ij}^0\,, \Delta u_i = u_i - u_i^0. \tag{6}$$

Thereby, Γ is a crack tip enclosing path, ds is a small element on Γ, W is the deformation energy density, t_i is the stress vector on the path Γ, u_i is the displacement vector of the path's points and n_j is a local outward normal vector to Γ. $\Delta\sigma_{ij}$, $\Delta\varepsilon_{ij}$ and Δu_i are respectively the ranges of stresses, strains and displacements. The index 0 in Eq. (6) symbolizes an initial state which is the state of the structure at load reversal. Note that ΔJ and ΔW are not ranges (maximum value minus minimum value in a cycle), but absolute values. By assuming the deformation plasticity, ΔJ is like J also path independent and can therefore be used as crack tip parameter.

The integration of Eq. (4) must be done with a post-processing code after finite element calculations have been performed. This means that, at every step increment, the finite element results for stresses, strains and displacements must be available in the area of interest. First of all, the integration path Γ must be defined and the local outward normal vectors, n_j (Eq. 8), at its points must also be calculated. Secondly, the integration limits of Eq. (4) are to be defined. This integration can be done from the lower load reversal point to the upper one or vice-versa (if closure effects are not considered).

The trapezoidal rule is used for the determination of ΔW (cyclic counterpart of the strain energy density):

$$\Delta W \approx \sum_{L=L_{lower}}^{L_{upper-1}} \left[\frac{\Delta\sigma_{ij}^{L+1} + \Delta\sigma_{ij}^{L}}{2} \cdot \left(\Delta\varepsilon_{ij}^{L+1} - \Delta\varepsilon_{ij}^{L}\right) \right] \quad (7)$$

Thereby, L is the load increment counter, L_{lower} and L_{upper} symbolize the lower and upper load reversal points respectively, which are the integration limits in this case. The components of the local stress vector range, ΔT_i, are calculated at every load increment by using the following relationship:

$$\Delta T_i^{\,L} = \Delta\sigma_{ij}^{L} \cdot n_j = \left(\sigma_{ij,\max}^{L} - \sigma_{ij,\min}^{L}\right) \cdot n_j \quad (8)$$

The partial derivative of the local displacement vector, $\frac{\partial \Delta u_i}{\partial x}$, is approximated with difference quotients. To do this, the path Γ is to be shifted left (negative direction, index 1 in Eq. (10)) and right (positive direction, index 2 in Eq. (10)) along the x-axis (parallel to the crack line) by an amount of $\delta x/2$:

$$\frac{\partial \Delta u_i^L}{\partial x} \approx \frac{\Delta u_i^{L,(2)} - \Delta u_i^{L,(1)}}{\delta x} \quad (9)$$

where

$$\Delta u_i^{L,(1)} = \Delta u_i^L \,\Big|_{\Gamma - \frac{\delta x}{2}} , \Delta u_i^{L,(2)} = \Delta u_i^L \,\Big|_{\Gamma + \delta x/2} \quad (10)$$

Besides, the previously determined auxiliary variable, δx, is to be chosen very small. Eq. (4) can now be evaluated completely.

3.1 Effective cyclic J-integral

If crack closure effects are to be considered, the ΔJ-integral is replaced by the effective ΔJ-integral, ΔJ_{eff}. Equation (4) is utilized for the calculation of ΔJ, as well as ΔJ_{eff}. However, the integration limits are different. The integration limits for ΔJ are the load reversal points whilst ΔJ_{eff} is computed during the unloading stage of a cycle and the integration limits are respectively the upper load reversal point and the crack closure time.

In order to capture crack closure effects [12,18,19] in the FE-model, contact boundary conditions have to be defined between crack faces. The contact opening displacements of all the crack face nodes were monitored during the cyclic loading for the determination of the crack closure time, t_{cl}. The crack is considered to be closed if one of these contact opening displacements becomes zero during the unloading stage of a cycle.

Post-processing codes were written to determine ΔJ and ΔJ_{eff} as function of the crack length/depth for 2D as well as 3D geometry [20]. It was possible to evaluate ΔJ and ΔJ_{eff} numerically at each crack length configuration. Note that, different to J, the estimation of ΔJ and ΔJ_{eff} cannot be performed directly in a finite element program.

3.2 Path-independency of the cyclic J-integral

A 3D model of a cruciform joint was used to study the path-independency of ΔJ. A semi-circular surface crack was placed at one of the weld toes, such that there is symmetry with respect to the mid-plane (perpendicular to the welds length direction). By exploiting this symmetry plane, only half of the investigated structure needs to be modelled (see *Fig. 2*). The commercial finite element program ABAQUS was used for the modelling. The upper part of *Fig. 2* is a detail of the FE-model which shows the prescribed crack geometry. The maximum element length in the predefined crack area is 0,05mm. Elements with linear shape functions were applied. The specimen was loaded with a cyclic nominal stress S in the direction perpendicular to crack faces (x-direction). Closure effects were not considered and the node release technique [12, 19] was used for the gradual extension of the crack. The ratio $a/c = 1$ was kept constant during node release. The increase in crack depth Δa was chosen to be 0,05 mm and only two load cycles were applied for every crack length configuration in order to restrict the FE-calculation time.

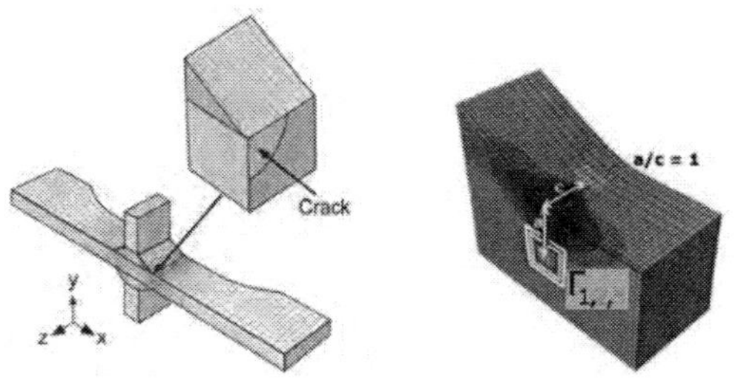

Fig. 2. FE modelling of a cruciform joint: the symmetry with respect to the mid-plane (parallel to the xy-plane) is exploited. A semi-circular crack (a/c=1) is inserted at the weld toe, exactly at the position of maximum stress concentration.

The ΔJ-integral was always calculated at the deepest point of the crack front (crack depth). Examples of calculation paths are presented in Fig. 2 right. For a crack of length $a = 0{,}5$ mm and a stress range of $\Delta S = 240$ MPa at $R = 0$, the ΔJ-values obtained for the cruciform joint (by considering the elastic-plastic material behaviour) on the paths Γ_1, Γ_2 and Γ_3 (see Fig. 3 right) were respectively 0,330 N/mm, 0,331 N/mm and 0,334 N/mm, which means that ΔJ can be considered as path independent (see Fig. 3a).

The influence of the nominal stress ratio on ΔJ was studied, see Fig. 3b. As expected, the results show that the R-ratio has almost no influence on the ΔJ-integral as long as crack closure effects are not considered.

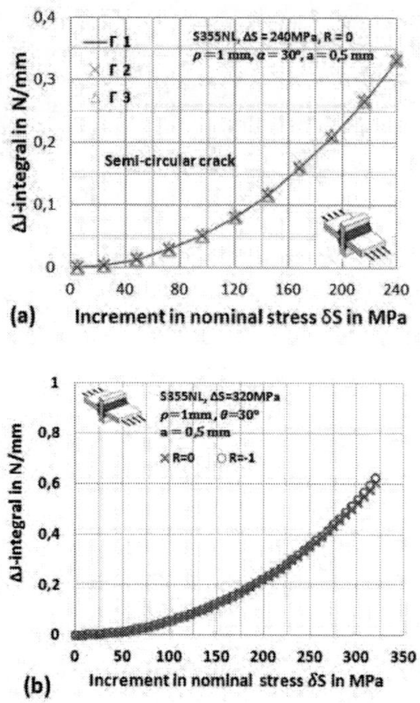

Fig. 3. (a)-Path-independency of the J-integral. (b)-Influence of the load ratio on the J-integral.

3.3 Validation example

For validation purposes, the finite element based ΔJ results have been compared with those of an analytical model known as *IBESS approach* [21]. The approach is named after the IBESS-project, on which the present research took place. Zerbst et al. [21] calculated ΔK via weight functions and determine the ΔJ-integral by using the expression

$$\Delta J = \frac{(\Delta K)^2}{\mathrm{E}'} \cdot [f(\Delta L_\mathrm{r})]^{-2} \;. \tag{11}$$

Thereby, $f(\Delta L_\mathrm{r})$ represents a plasticity correction function and depends on a cyclic ΔL_r given as

$$\Delta L_\mathrm{r} = \frac{\Delta\sigma_\mathrm{app}}{2\cdot\sigma_0} \;. \tag{12}$$

$\Delta\sigma_\mathrm{app}$ corresponds to the applied cyclic stress (referring to the gross cross-section) and σ_0 is a reference yield stress, which was introduced to improve the accuracy of the estimate. For more details, a thorough description, including the parameter equations, is provided in [22].

Fig. 4 shows results obtained for a cruciform joint with semi-circular surface crack and a weld flank angle $\theta = 30°$. Elastic-plastic material behaviour was applied and the calculated ΔJ values were transformed into "plasticity-corrected" ΔK_p-values by using Eq. (13). A very good agreement can be observed between numerical results and the analytical model.

$$\Delta K_\mathrm{p} = \sqrt{\Delta J \cdot E'} \tag{13}$$

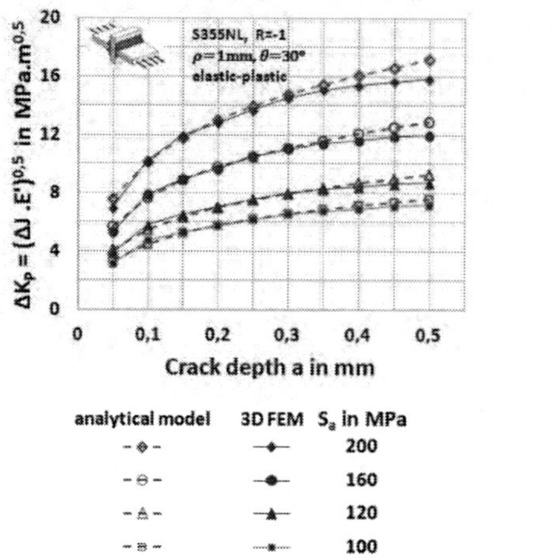

Fig. 4. Comparison between analytical and finite element based $\Delta K_P = \sqrt{\Delta J \cdot E'}$ for cruciform joints with a flank angle of $\theta = 30°$. S355NL, $\rho = 1\,mm$, $a/c = 1$.

4 Fatigue crack growth simulations

4.1 Application of the node release technique to welded joints

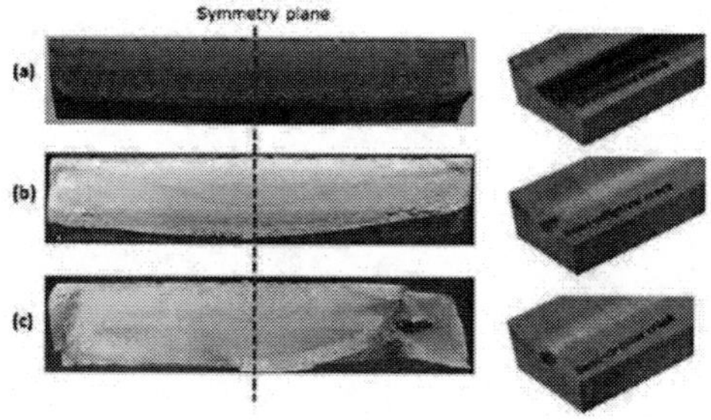

Fig. 5. Crack configurations: continuous crack (a); semi-elliptical crack (b); semi-circular crack (c).

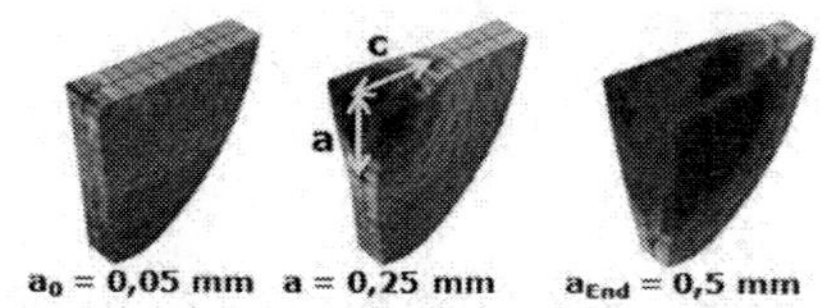

Fig. 6. Evolution of a semi-circular crack.

By assuming that the crack path and crack geometry are known in advance, the node release technique can be applied for the gradual extension of a crack from an initial crack length, a_0, up to a final crack length, a_End. In experiments, different types of crack geometry were observed for the joint: continuous, semi-elliptical as well as semi-circular cracks (see *Fig. 5*). Simulations that are carried out in this contribution only considered continuous (lead to shorter fatigue lives) and semi-circular (lead to longer fatigue lives) cracks, i.e. $a/c \cong 0$ and $a/c = 1$ respectively. Here, a represents the crack depth and c the half of the surface crack length. In *Fig. 6*, the evolution of a semi-circular crack is shown in some configurations. An initial crack depth $a_0 = 0{,}05\ \text{mm}$ was selected, which approximately corresponds to the maximum defect size of 25 to 75 µm that was measured in the specimens. These measurements were carried out by a project partner, see [15,16]. FE calculations were performed up to a final crack depth $a_{\text{End}} = 0{,}5\ \text{mm}$, which corresponds to the chosen numerical fracture criterion at short crack initiation. Only two load cycles were applied for every crack length configuration in order to restrict the FE-calculation time. In addition, two load cycles were applied before the first node release so that the residual stress relaxation/redistribution can be taken into account [18,23]. The crack extension was done at positive nominal stress (in order to satisfy the physical requirement that only an opened crack can propagate) and the increase in crack depth Δa was chosen to be $0{,}05\ \text{mm}$ for semi-circular cracks (3D calculations) and $0{,}01\ \text{mm}$ for continuous cracks (modelled by using 2D models) respectively. The element length in the area of the crack was set equal to Δa. Note that the use of elements with linear shape functions is inevitable when contact constraints are applied.

4.2 Algorithm of crack growth simulation

The complete algorithm of crack growth simulation is presented in *Fig. 7*. This algorithm consists of 5 points:

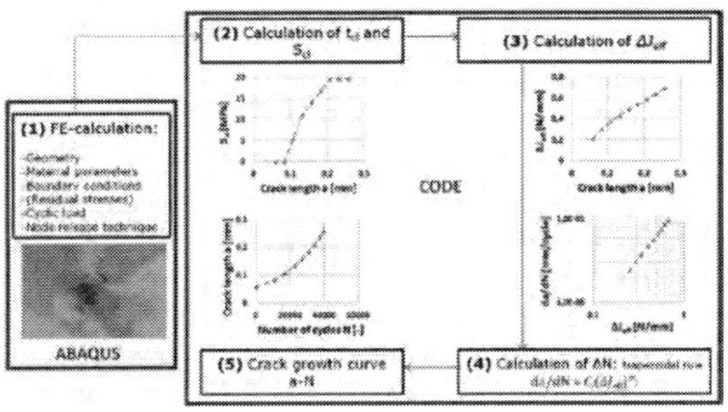

Fig. 7. Algorithm of crack growth simulation.

(1) FE-calculation: the complete structure calculation is carried out in a FE program (here ABAQUS) by using the "node release technique" to insert a crack and to extend it gradually until the final crack length (fracture criterion) is reached. The specimen geometry, material parameters, boundary conditions (including the gradual definition of contacts between the crack faces) and applied loads are needed here. If residual stresses are to be considered, they are inserted by using a procedure described by the authors in [18,23].

(2) Identification of crack closure times (t_{cl}) as well as crack closure stresses (S_{cl}): after the FE calculation is done, the result file is accessed with a post-processing code in order to identify t_{cl} and S_{cl} for every crack length configuration.

(3) Calculation of ΔJ_{eff}: if the crack closure times are known, the effective cyclic *J*-integral ΔJ_{eff} can be determined as function of the crack length.

(4) Calculation of ΔN (number of load cycles which are needed to extend the crack from a length a_i to a length a_{i+1}): the crack growth equation,

$$\frac{\mathrm{d}a}{\mathrm{d}N} = C_{\mathrm{J}} \cdot (\Delta J_{\mathrm{eff}})^{m_{\mathrm{J}}} \tag{14}$$

is integrated by using the trapezoidal rule in order to determine ΔN step by step from a_0 to a_{End}. The material parameters C and m of steel S355NL at $R = -1$ of the Paris equation,

$$\frac{\mathrm{d}a}{\mathrm{d}N} = C \cdot (\Delta K_{\mathrm{eff}})^{m} \tag{15}$$

were determined by a project partner [13]: $C = 8{,}37 \cdot 10^{-9}\ \text{mm/cycl.} \cdot \left(\text{MPa} \cdot \text{m}^{1/2}\right)^{m}$ and $m = 3$. These parameters were used for all calculations with steel S355NL at $R = -1$ as well as $R = 0$.

By setting the crack propagation rate $\left(\frac{\mathrm{d}a}{\mathrm{d}N}\right)$ equal in Eq. (14) and Eq. (15), and by using $K = \sqrt{J \cdot E'}$ in case of linear elastic fracture mechanics, the following relationships can be identified for the determination of C_{J} and m_{J}:

$$m_{\mathrm{J}} = \frac{m}{2}, \quad C_{\mathrm{J}} = C \cdot (E')^{m_{\mathrm{J}}}\ . \tag{16}$$

Note that the calculated fatigue lives strongly depend on C_{J} and m_{J}.

(5) Crack growth curve (*a*–*N* curve): the crack length is plotted together with the calculated number of cycles.

5 Numerical results and discussions

5.1 Example of a cruciform joint made from steel S355NL

Results obtained by using the procedure of crack growth simulation (see Fig. 7) are presented in this section. The selected crack geometry was semi-circular. The Chaboche's material parameters of steel S355NL were employed (see Table 1). Two different stress ratios ($R = 0$, $R = -1$) as well as three different stress amplitudes ($S_{\mathrm{a}} = 120\ \text{MPa}$, $S_{\mathrm{a}} = 160\ \text{MPa}$ and $S_{\mathrm{a}} = 200\ \text{MPa}$) were used during computations.

The normalized crack closure stress, $S_{\mathrm{cl}}/S_{\max}$, and the crack closure function U_{eff} (Eq. 17) are presented in *Fig. 8*. The function U_{eff} determines the proportion of the load cycle in which the crack is opened and it is equal 1 for a fully opened crack or zero in case the crack is completely closed ($0 \leq U_{\mathrm{eff}} \leq 1$). In both functions ($S_{\mathrm{cl}}/S_{\max}$ and U_{eff}), a gradual build-up of crack closure effects can be observed. Two effects are to be distinguished: *a transient crack closure behaviour* that changes at the beginning of crack propagation with increasing crack length – this phenomenon only occurs at the short crack stage – and a *stabilized crack closure behaviour* that no longer changes with increasing crack length. This is typical for long cracks. It is also seen that the crack closure behaviour depends on the R-ratio: by maintaining the stress amplitude fixed, the functions S_{cl}/S_{max} and U_{eff} increase with increasing *R*.

$$U_{eff} = (S_{max} - S_{cl})/(S_{max} - S_{min}) \tag{17}$$

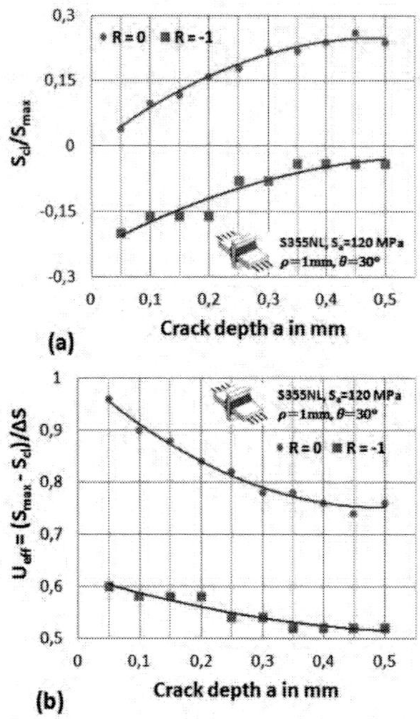

Fig. 8. Normalized crack closure stress (a) and crack closure function depending on the R-ratio (b), 3D-cruciform joint.

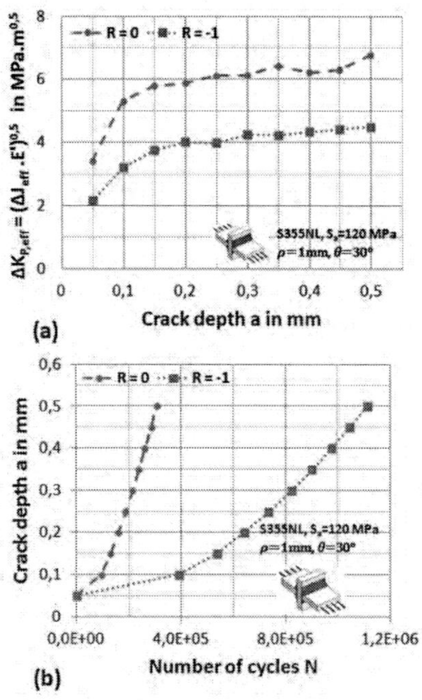

Fig. 9. Effective crack driving force $\Delta K_{p,eff}$ (a) and crack growth curve depending on the R-ratio (b).

In Fig. 9, the $\Delta K_{p,eff}$-curve (effective value of the plasticity corrected range of the stress intensity factor, Fig. 9a) and the crack growth curve (Fig. 9b) are shown as function of the R-ratio. As expected, the effective crack driving force, $\Delta K_{p,eff}$, increases with increasing crack depth as well as with increasing R-ratio. This also leads to the fact the fatigue life decreases with increasing R when the stress amplitude is kept constant. The sudden discontinuities in the numerical solutions (for example in Fig. 9a at $R = 0$ and $a = 0{,}35$ mm) may be due to plasticity in combination with variations of crack closure along the crack front or discontinuous crack closure. It should be mentioned that continuous functions were not fitted from this figure onward like in Fig. 8.

In Fig. 10, the $\Delta K_{p,eff}$*-curve* as well as the crack growth curves are plotted as function of the nominal stress amplitude. As expected, $\Delta K_{p,eff}$ increases with increasing stress amplitude. This leads to a decrease of the fatigue life when the stress amplitude increases.

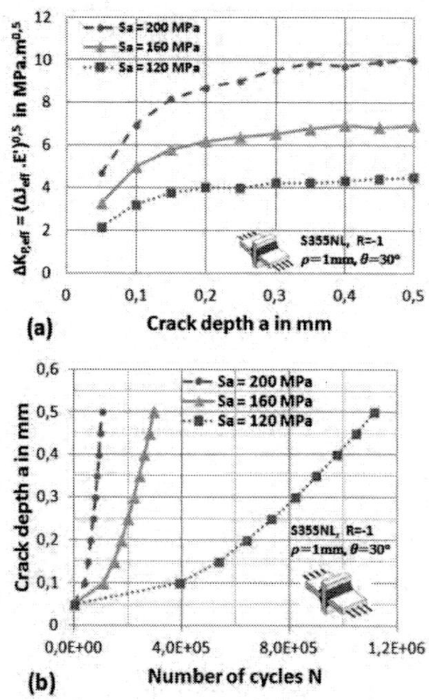

Fig. 10. Effective crack driving force $\Delta K_{p,eff}$ (a) and crack growth curve depending on the stress amplitude (b).

5.2 Influence of residual stresses on the calculated fatigue lives

The residual stress field that arises from the welding process can have a significant influence on the fatigue life [1, 18]. Tensile and compressive residual stresses were generated in the joint by means of structural calculations (see [24] for more details). The value of this stress at the critical weld notch was +100 MPa in case of a tensile residual stress field or -100 MPa in case of a compressive field. These stresses were then used for the numerical investigation of the residual stress influence on the fatigue crack growth.

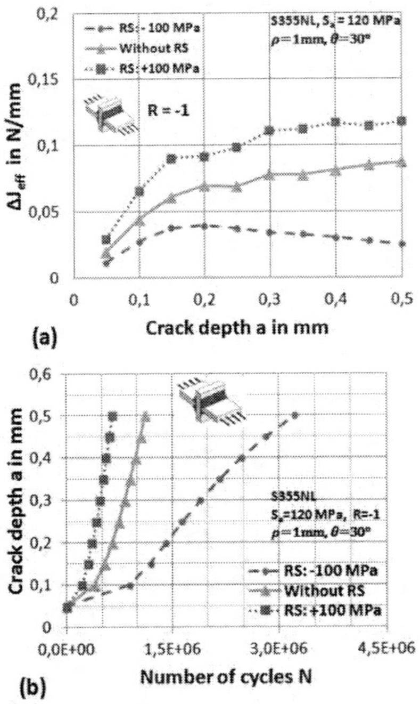

Fig. 11. Influence of residual stresses on ΔJ_{eff} (a) and on the crack growth curve (b).

The effects of residual stresses on the effective cyclic J-integral, ΔJ_{eff}, are illustrated in Fig. 11a. As expected, tensile and compressive residual stresses lead to larger and smaller ΔJ_{eff} values respectively, in comparison to ΔJ_{eff} values that were calculated without residual stresses. The effects of residual stresses on the calculated fatigue life are illustrated in Fig. 11b. As expected, the consideration of tensile or compressive residual stresses leads to shorter or longer fatigue lives respectively. The reason for this is the fact that a compressive residual stress field increases crack closure effects, while a tensile residual stress field prevents these effects to occur.

6 Comparison between numerical and experimental results

In this Section, the fatigue lives at short crack initiation that were calculated for cruciform joints are compared with experimental results (see Fig. 12). Calculations were performed with and without residual stresses and the numerical fracture criterion at crack initiation was chosen to be $a = 0{,}5$ mm. Experimental data were determined in various research groups, so that the experimental short crack initiation criterion depends on the method for short crack detection. The experimental fracture criterion at short crack initiation for S355NL was the appearance of small changes in the thermographic image [15]. This corresponds to a surface crack length of approximately $2c = 1$ mm.

Numerical and experimental results for cruciform joints S355NL in a stress relieved state (SR) are depicted in Fig. 12a. With regard to the different crack geometries (see Fig. 5), calculations were performed with 2D-models (corresponds to a continuous crack in 3D-model, which leads to shorter fatigue lives) as well as with 3D-models (for the case of a semi-circular crack, which leads to longer fatigue lives). Experimental fatigue life results almost lie between the 2D and 3D numerical fatigue life results, which represent the calculated boundaries. Thus, a good accordance between both results is achieved.

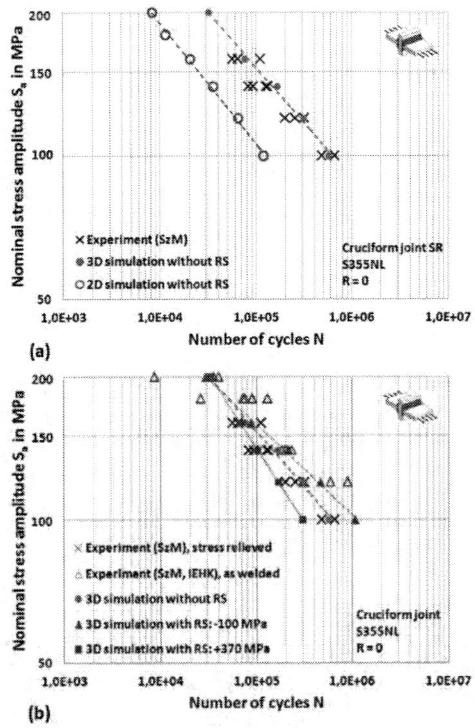

Fig. 12. Fatigue life at short crack initiation – comparison between numerical and experimental results. (a)-stress relieved (SR) specimens, (b)- stress relieved vs. as welded, R = 0.

The influence of compressive residual stresses and of high tensile residual stresses (equal to the yield stress) on the fatigue life was studied. Results are displayed in Fig.12b. Numerical results of computations with compressive residual stresses fit well to test results of as-welded specimens. A good accordance between experiments and calculations is achieved again. At lower stress amplitude, the stress relieved specimens lead experimentally to shorter fatigue lives compared to the as-welded specimens. This can be explained by the fact that according to residual stress measurements [23], the as-welded specimens exhibit compressive residual stresses, which lead to longer fatigue lives. At higher stress amplitude, both experimental and numerical results show almost no influence of welding residual stresses on the fatigue life.

Calculations with high tensile residual stresses (equal to the yield stress) lead to conservative lifetimes at lower amplitudes. Generally, in a case where the welding residual stress field is unknown, it is often recommended in the literature [25] to use high tensile residual stresses for the fatigue life prediction. Therefore, the conservative effect of such an approach for the consideration of welding residual stresses is confirmed.

7 Summary and conclusion

Fatigue crack growth simulations have been performed by taking into consideration plasticity induced crack closure effects as well as the effects of welding residual stresses. For this purpose, an algorithm of fatigue life prediction was presented. Thereby, the effective cyclic *J*-integral (ΔJ_{eff}) was used as crack tip parameter in a relation similar to the Paris equation for the calculation of the fatigue life. Analyses were performed with two and three dimensional FE models by using the node release technique together with the FE program ABAQUS.

Crack closure functions and the effective crack driving force (ΔJ_{eff} or $\Delta K_{p,eff}$) were determined numerically for surface cracks in butt and cruciform joints. Numerical results show that the closure functions and the effective crack driving force strongly depend on the load ratio as well as the load amplitude. It was possible to describe the transient crack closure behaviour that occurs at the short crack stage as well as the stabilized crack closure behaviour for long cracks.
The numerical fatigue life results were compared with experimental data: a good accordance between numerical and experimental results was achieved for both stress relieved and as-welded specimens. Thus, ΔJ_{eff} is obviously an appropriate parameter for calculations concerning fatigue life prediction in welded joints.

The influence of welding residual stresses on the fatigue life was investigated numerically as well as experimentally. As expected, numerical results show that at lower stress amplitude, a compressive residual stress field is favourable to the fatigue life, while a tensile residual stress field is unfavourable. The effect of residual stresses can be neglected only for large load amplitudes. Experimental observations show that the fatigue lives of the stress relieved specimens are smaller than those of the as-welded specimens. This was due to the fact that the as-welded specimens exhibit compressive residual stresses (according to residual stress measurements), which are favourable to the fatigue life.

The German Research Foundation (Deutsche Forschungsgemeinschaft) is greatly acknowledged by the authors for financing this project under grant no. V0729/14-1.

References

1. C.M. Sonsino, Int. J. Fatigue *31*, pp. 88-101, (2009)
2. D. Tchoffo Ngoula, H. Th. Beier, M. Vormwald, DVS-Bericht *315*, pp. 294-299, (2015)
3. W. Elber, Engng Fract Mech *2*, pp. 37-45 (1970)
4. H.A. Richard, M. Sander, 3. erweiterte Auflage, Springer Vieweg, Wiesbaden (2012)
5. W. Fricke, Mar. Struct. *16*, pp. 185-200, (2003)
6. D. Radaj, C.M. Sonsino, W. Fricke, Int. J. Fatigue *31*, pp. 2-11 (2009)
7. S.J. Maddox, Metal Const. 2, pp. 327-331, (1970)
8. J. Polak, Comprehensive Structural Integrity, *4*, 1-39 (2003)
9. N.E. Dowling, J.A. Begley, ASTM STP *590*, 82-103 (1976)
10. J.R. Rice, ASME J. Appl. Mech., 35, 376–386, (1968)
11. McClung, R.C., Chell, G.G., Lee, Y.D., Russel, D.A. and Orient, G.E, Report NASA/CR-1999-209428, Marshall Space Flight Center, 1999.
12. Vormwald, M., In Radaj, D. and Vormwald, M., Springer, Heidelberg, p. 391-481, 2013.
13. U. Zerbst: Schussbericht zu den IGF-Vorhaben 17518 N/1 , IBESS-A3, 2016.
14. Tchoffo Ngoula, D., Beier, H. Th., Vormwald, M.: Abschlussbericht zu den DFG-Vorhaben V0729/14-1, IBESS-G3, 2016.
15. H. Th. Beier, B. Schork, J. Bernhard, D. Tchoffo Ngoula, T. Melz, M. Oechsner, M. Vormwald, Mat.-wiss. U. Werkstofftech. *47*, No. 10, (2016)
16. B. Schork, U. Zerbst, P. Kucharczyk, J. Bernhard, M. Kaffenberger, J. Baumgartner, M. Oechsner, DVS-Bericht 315, pp. 271-276, 2015.
17. J.L. Chaboche, Int. J. Plast. 24, 1642-1693, 2008.
18. Tchoffo Ngoula, D., Beier, H. Th., Vormwald, M., Int. J. Fatigue 101(2), pp. 253-262, 2017.
19. de Matos, P.F.P., Nowell, D., Engineering Fracture Mechanics, 75, pp. 2087-2114, 2008.
20. Tchoffo Ngoula, D., Madia, M., Beier, H. Th., Vormwald, M., Zerbst, U., Engng Fract Mech , 2017. http://dx.doi.org/10.1016/j.engfracmech.2017.06.023.
21. Zerbst, U., Madia, M., Hellmann, D., Engng. Fracture Mech. 82, 115-134, 2011.
22. Madia, M., Zerbst, U., Beier, H.Th. and Schork, B., Engng Fract Mech , 2017. http://dx.doi.org/10.1016/j.engfracmech.2017.08.033.
23. Hensel, J., Nitschke-Pagel, T., Tchoffo Ngoula, D., Beier, H.-Th., Tchuindjang, D. and Zerbst, U., Engng Fract Mech , 2017. http://dx.doi.org/10.1016/j.engfracmech.2017.10.024.
24. Tchoffo Ngoula, D., Rep 117, Inst Stahlbau Werkstoffmech, TU Darmstadt, 2018.
25. Hobbacher, F. (2016), Second Edition, IIW Document IIW-2259-15, Springer.

Fatigue life assessment of a railway wheel material under HCF and VHCF conditions

H. Soares[1], *P. Costa*[1], *M. Freitas*[2], *L. Reis*[2]*

[1]Instituto Superior Técnico, Universidade de Lisboa, Av. Rovisco Pais, 1, 1049-001 Lisbon, Portugal
[2]IDMEC, Instituto Superior Técnico, Universidade de Lisboa, Av. Rovisco Pais, 1, 1049-001 Lisbon, Portugal

Abstract. Fatigue damage and life assessment is still an issue and a challenge nowadays. Many different tests can be performed for the assessment of fatigue properties of any given material. In the present study a worn out railway wheel goes under uniaxial fatigue analyses for the high cycle and very high cycle fatigue regimes through the use of a conventional hydraulic machine and an ultrasonic fatigue test, both with a stress ratio of R=-1. For every used specimen, a life cycle is obtained for the corresponding induced specimen and the data is afterwards plotted. All the fracture surfaces of all failed specimens were analysed. The study and initial application of multiaxial ultrasonic fatigue condition and the design of the corresponding specimen for evaluating the wheel material is also presented.

1 Introduction

The fatigue study is of the most importance for any system subjected to dynamic loadings being the most common and unpredictable type of failure. It has been shown that fatigue in metals may occur in any range of cycles, even beyond what was once considered the fatigue life limit [1].

To study the fatigue of a certain material several tests can be performed to relate the fatigue life for low cycle fatigue (LCF), high cycle fatigue (HCF) and very high cycle fatigue (VHCF). In any given range different specimens and displacements can be induced depending on the regime and type of load of interest, uniaxial or multiaxial. For LCF and HCF regimes, conventional hydraulic machines are employed which induce stress cycles at a low rate of around 50 Hz, easy to apply and control. In VHCF regime, in the 10^6 and 10^9 cycle range, ultrasonic resonance methods are employed to induce stress cycles with a frequency of around 20 kHz. This type of tests allows to perform a huge number of cycles which it would be unpractical for a conventional machine.

Ultrasonic fatigue tests apply resonance principles for the achievement of high enough stress at a high frequency to an excited specimen. Mason was the first to achieve a working system in 1950 [2] with a frequency of work of 20 kHz, establishing the standard frequency of all methods created afterwards. Since then, many researchers have dedicated to the development of new variations of this type of test capable of applying different types of loadings, stress ratios and also in different ambient conditions, like those seen in [3;4].

The present study applies uniaxial stress, tensile and compression cycles, with a stress ratio of R=-1 to a worn-out railway wheel using a conventional hydraulic machine Instron 8874 for HCF, and an ultrasonic fatigue testing with a device built in the Instituto Superior Técnico at the IDMEC laboratories [5] for the study of fatigue life in the VHCF regime. All material samples were machined from the railway wheel rim itself. Soul components used for ultrasonic fatigue test are showed in figure 1.

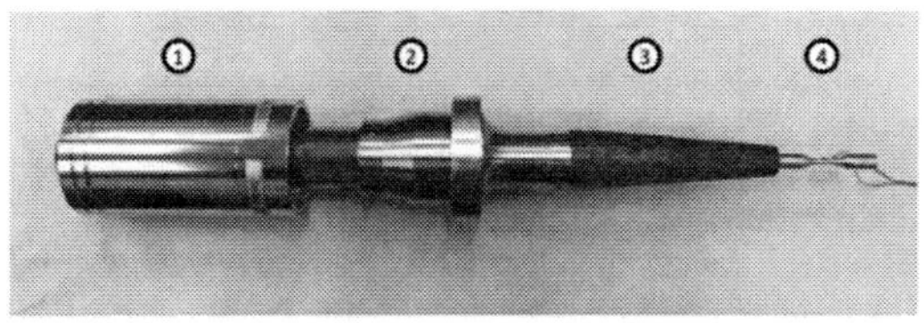

Fig. 1. Components to perform a uniaxial ultrasonic fatigue test: 1-Piezoelectric transducer; 2- Booster; 3-Horn; 4-Test Specimen.

The piezoelectric transducer transforms electric pulses in mechanical pulses and is responsible for the resonance excitement of all the attached components. The main purpose of the Booster and the horn is to amplify the displacement of the transmitted transducer pulses. The specimen has an hourglass shape for the creation of higher stresses. Thereby with the use of the booster, horn and the specimens shape it is possible to achieve high enough stresses with a low power and a high frequency of 20 kHz.

Many different studies of uniaxial ultrasonic fatigue were already made by several different authors, as it can be seen in [6;7].

A multiaxial ultrasonic fatigue test is also mentioned, built in the same facilities using the same piezoelectric machine [8] and some of the preliminary trials preformed.

* corresponding author: luis.g.reis@tecnico.ulisboa.pt

https://doi.org/10.1051/matecconf/201816509003

2 Methodology

The material in study is a worn-out railway wheel that should follow the European norm EN_13262+A1-2009.

For both the HCF and VHCF tests Young Modulus is required for the specimen's creation and applied stress. Material density is also required in the ultrasonic fatigue test for the determination of resonance frequency. These material properties were obtained through a tensile stress test and by measuring the weight and volume of a certain part of the railway wheel rim.

2.1 High cycle fatigue

For any conventional HCF test a standard specimen is need it. The used geometry follows the standard ASTM E606 (2003) for axial fatigue tests. The machined geometry and the respective dimensions are presented in figure 2.

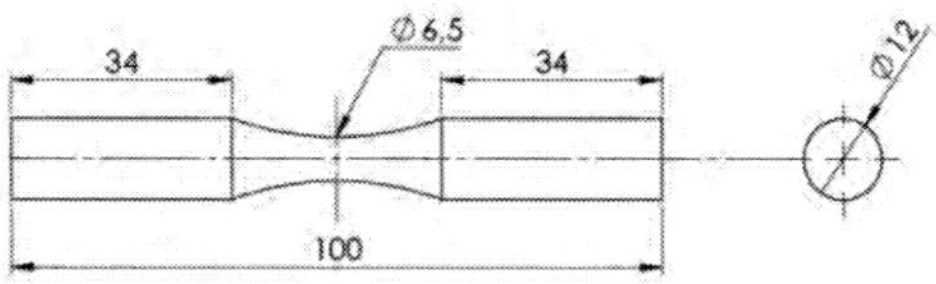

Fig. 2. HCF specimen geometry and dimensions following the standard ASTM E606 (2003) (mm).

As for the ultrasonic test specimen the conventional specimens are machined directly from the railway wheel rim.

All tests are executed through load control. A stress of interest for the study is chosen and with the Young Modulus and the measuring the radius of the smallest cross section the control force to be apply can be calculated. The specimen is then subjected to stress cycles until failure. If after 10E06 cycles the specimen didn't show any sign of failure it is withdrawn from the machine and higher stress is applied. When the specimen fractures the machine automatically stops, the cycles are registered and the specimen is withdrawn for the analysis of its fracture surface.

2.2 Very high cycle fatigue

Using the determined material proprieties, a preliminary geometry for the uniaxial ultrasonic test specimen was created through the base geometry and equations methodology found in [2].

After the initial calculations the geometry is applied in the finite element software Abaqus to be readjusted to a final design. The readjustment is made because the analytical method gives, according to finite element modal analysis, a significant lower frequency than 20 kHz. Thereby a reduction of the cylindrical part of the specimen is made for increase of frequency until the resonance mode of interest is as close as possible to 20kHz, maintaining a reasonable dimension for machining.

With the final geometry presented in figure 3 the specimens were machined delicately from the railway wheel rim.

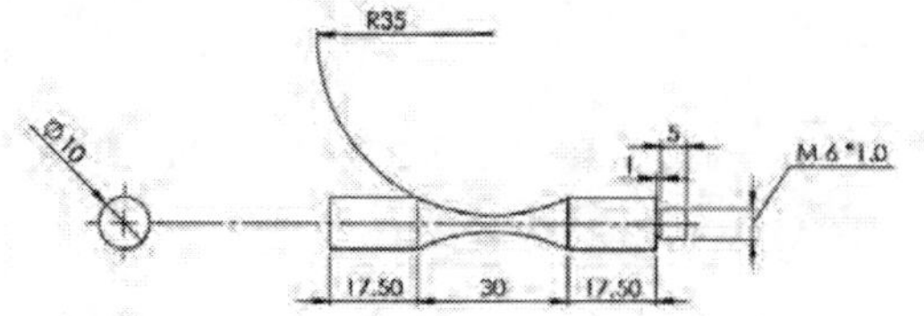

Fig. 3. Uniaxial ultrasonic fatigue test final specimen geometry and dimensions (mm).

Before applying a constant stress cycle to any given specimen, a preliminary research of the created stresses is required. This was made through the usage of a strain gage to one of the machined specimens. For the first specimen in question several tests with different power control settings are applied. In each test the strains are measured, registered and worked through the use of the software MATLAB, thus obtaining the stress peak for each power setting applied. A trendline is then obtained linking the applied power with the resulting stresses.

2.2.1 Strain measurements

In order to measure the specimen's strain an HBM strain gage is applied to the smallest cross section where the strain should be the highest and consequently highest stress. Figure 4 shows a photograph of the strain gage applied. The strain gage is aligned with the induced displacement.

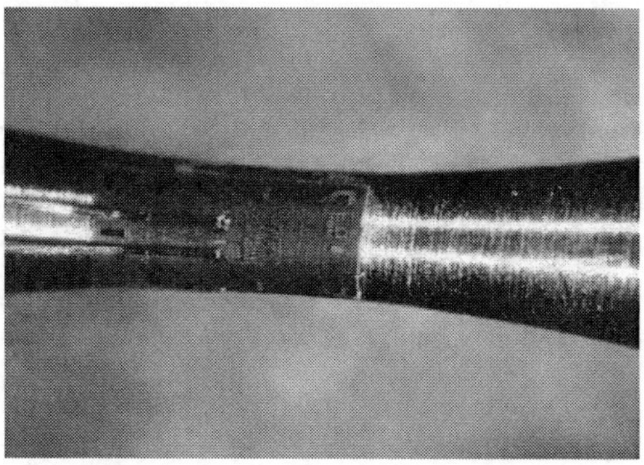

Fig. 4. HBM strain gage used and bounded in the smallest cross section.

The strain gage is connected to a data acquisition system for the recording of the strain data. Four power settings were applied and the corresponding strains were measured. The strains are then transformed to stress using the determined Young Modulus. Figure 5 and 6 show an interval of strain and derivative stress values respectively for a power setting of 10%.

Through the use of the MATLAB software the maximum and minimum stresses are taken from a certain stable span and the average is considered to be the stress equivalent for the related power setting. Figure 7 shows the relation between stress and power setting and the corresponding trendline linear equation.

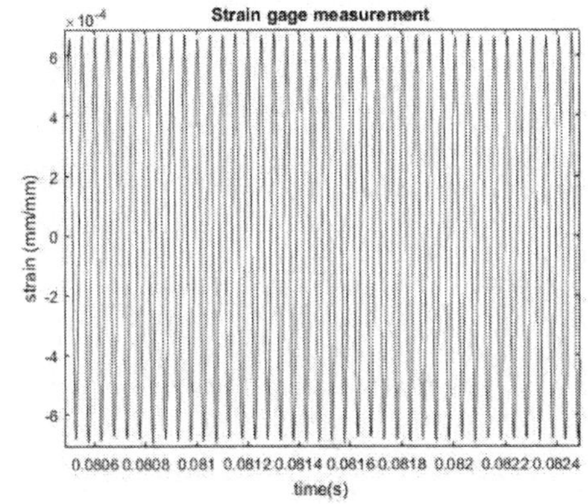

Fig. 5. Span of the measure strains for an applied power setting of 10%.

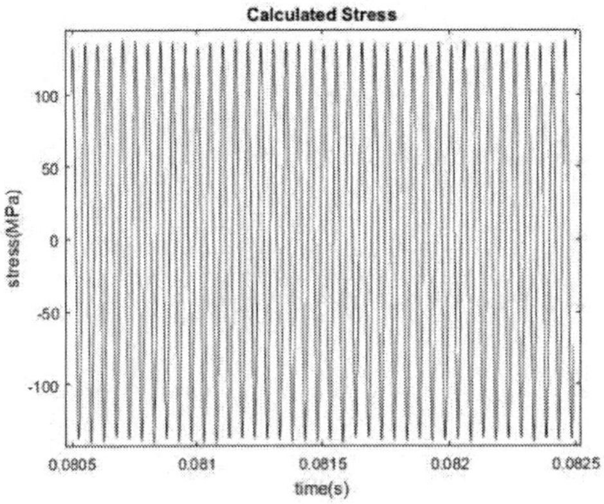

Fig. 6. Resulted stress from a span of measure strain for an applied power setting of 10%.

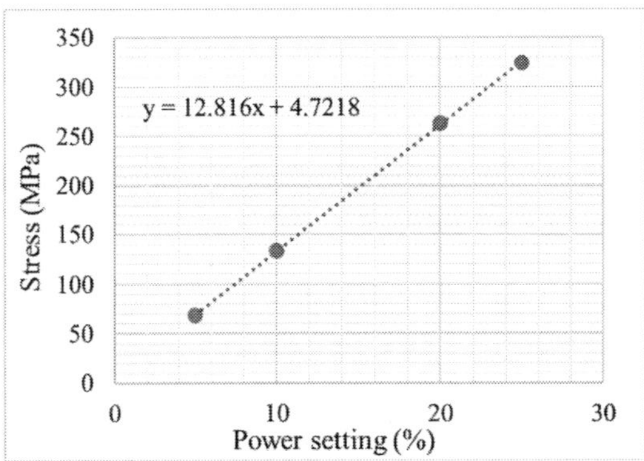

Fig. 7. Obtained relation between stress and power.

Using the obtained linear equation power/stress and knowing the supposed fatigue limit by HCF tests previously preform, ultrasonic fatigue tests were them carried out. The power settings used and the corresponding stresses of interest is presented in table 1.

It is important to mention that the used piezoelectric transducer can only vary the power setting by half percent.

Table 1. Calculated stresses for the piezoelectric transducer power settings of interest

Power (%)	Average Stress (MPa)
23.0	299.49
23.5	305.90
24.0	312.31
24.5	318.71
25.0	325.12

2.2.2 Stress cycles application

Through the previously performed HCF tests, it was determined the stress at which the material won't break after being subjected to 10E06 cycles. This stress level corresponds to what it is called the fatigue limit. This and slightly lower stresses are the range of interest for the VHCF tests to be performed by the ultrasonic fatigue machine.

Several specimens are then tested below the threshold obtained by the HCF tests performed beforehand. Throughout any test a temperature control is applied due to the high heat generation during the test. A temperature limit is established and when reached the transducer is turned off temporarily until the specimen is cooled. For the measurement of the specimen's temperature all are previously painted in black. The higher heat generation comes from the region with higher induced stress where the damping effects are more influent. Figure 8 shows thermographic images of one specimen where it can be seen the heat generation on its smallest cross section.

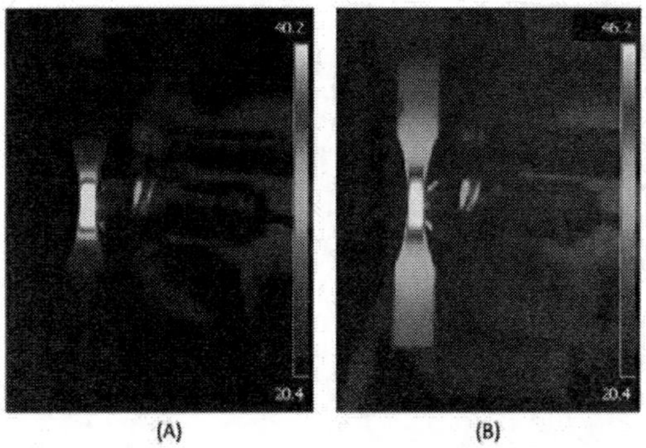

Fig. 8. Thermographic image of two tested ultrasonic uniaxial specimens: (A) beginning of a teste; (B) few minutes later.

For counting the applied cycles, a laser was used. The laser points directly to the free end of the specimen where the displacements are the highest, thus being able to count the cycles. It can also provide the frequency at which the test is currently working.

Opposite to the HCF conventional test the ultrasonic specimen does not completely break while in the ultrasonic fatigue machine. The test is finished when the resonance of the components has deviated from the range of work of the transducer. This happens when the fatigue damage to the specimen is big enough for it to lose integrity altering considerably his resonance frequency. Afterwards the specimen is withdrawn from the ultrasonic fatigue machine and a tensile test is made until the specimen's total failure.

2.3 Multiaxial ultrasonic fatigue test

The multiaxial ultrasonic test has the same methodology as the uniaxial. It only varies the horn and specimen geometry that allows to create the biaxial stress state of tension/compression and torsion.

The horn has a special geometry that partially transforms the longitudinal pulses created by the transducer to rotational pulses. Both types of displacement and rotation are then transmitted to the specimen. The specimen has also a special geometry that is able to be excited in two resonance modes simultaneously, the first longitudinal mode and the third rotational mode. With both of the referenced resonance modes it is possible to have one single area where both types of stresses achieved the highest value.

To create a final specimen geometry several finite element modal analysis are required until a final geometry with both resonance modes of interest have similar frequencies around 20 kHz. The final geometry and its dimensions are shown in figure 9.

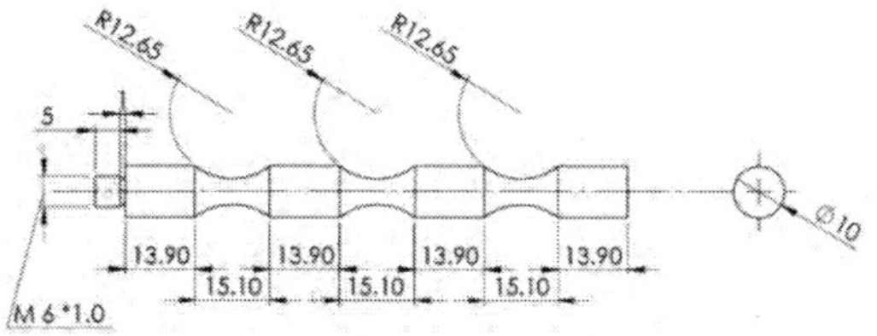

Fig. 9. Multiaxial ultrasonic fatigue test final specimen geometry and dimensions (mm).

After machining the multiaxial specimen, it is first validated through the generated temperature. The specimen is painted in black and excited while the temperature is observed by a thermal camera. The results have proven to be promising since two machined specimens have shown to have the higher heat generation on the intended location, as it can be seen in figure 10.

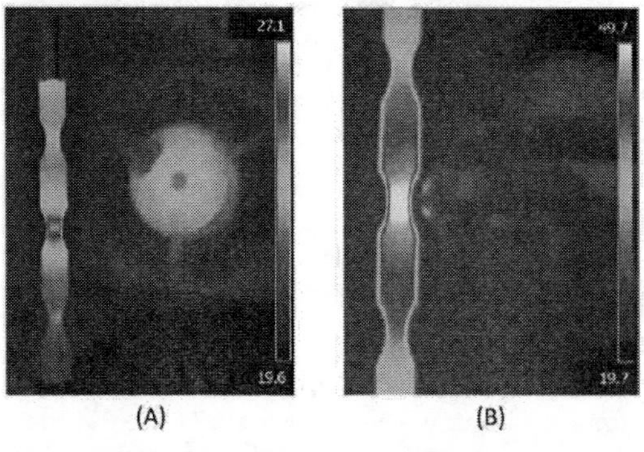

Fig. 10. Thermographic image of two tested ultrasonic multiaxial specimens at: (A) low power setting; (B) higher power setting.

The next step for the study is the introduction of a strain gage with three measuring gages thus providing the ability to obtain the shear stress created by the induced torsion.

3 Results and Discussion

3.1 High cycle fatigue

Studying the results, we can confirm that the material reaches 10E06 cycles around 325 MPa. Thereby it should be applied to the ultrasonic fatigue specimen stresses below this value.

All fracture surfaces of failed specimens showed to have a zero degree as it is expected for a conventional tension/compression fatigue test. Through the analysis of the fracture surface of all failed specimens the fatigue propagation initiated at the surface. Figure 11 shows the fracture angle and the initial site of propagation.

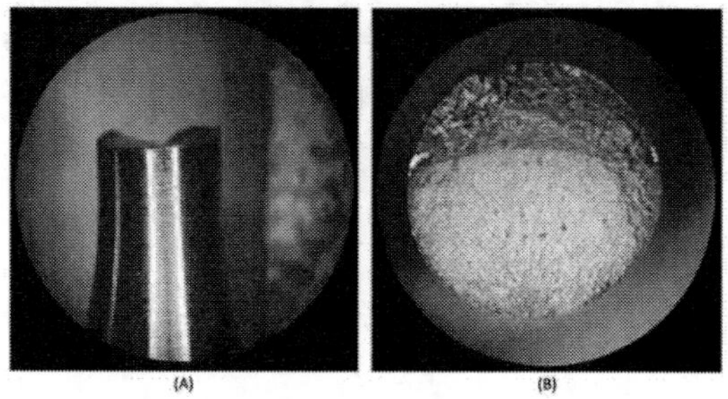

Fig. 11. (A) Fracture angle; (B) fracture surface.

3.2 Very high cycle fatigue

The ultrasonic fatigue specimens showed as with the conventional test results to have the fracture surface with a zero-degree angle of failure. The fracture surface also shows to have initiated at the specimen's surface. Figure 12 shows the fracture angle and the initial site of propagation.

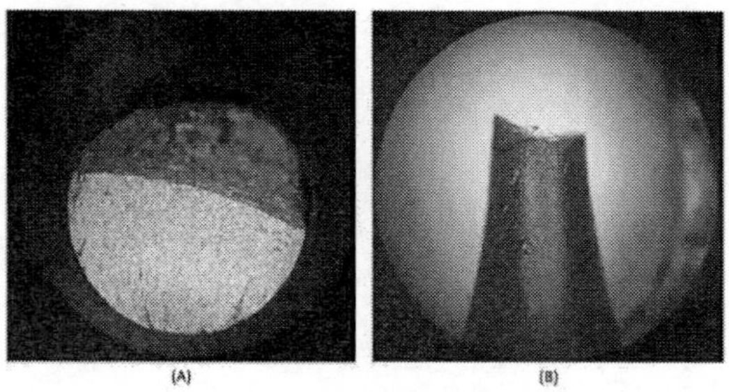

Fig. 12. (A) Fracture surface; (B) fracture angel.

https://doi.org/10.1051/matecconf/201816509003

3.3 S-N curve of HCF and VHCF results

Putting all the results of both preformed tests a stress life cycle graph is plotted, as it can be seen in Figure 13.

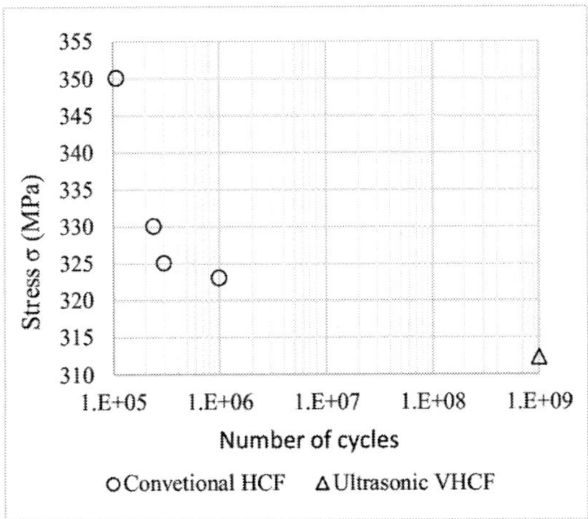

Fig. 13. Stress life results of conventional HCF and ultrasonic VHCF tests.

Observing the results, it is not possible with the points so far obtained to fully characterize the material behaviour in the VHCF regime. The HCF results show with clarity the increase of material's life with stress reduction until the VHCF regime. In the VHCF regime, a stress reduction shows to have a much more significant impact on material life.

4 Final Remarks

Comparing the obtained fracture surfaces between the performed tests there are no obvious differences. For the two studied fatigue regimes fracture initiation site showed to be in the specimen's surface. The fracture surface marks, angle of failure and even the shape seem to be similar at eye sight and through a microscope, as it can be compared between figures 11 and 12. A detailed and more powerful magnification of the surface would be required to completely assess the differences between the fracture morphologies of both tests.

More tests are being done in order to fully characterize the material behaviour in the VHCF regime.

References

1. Bathias, C., Fract Engng Mater Struct, **22**, 559-565 (1999).
2. Bathias, C., & Paris, P. C. (2005). *Gigacycle Fatigue in Mechanical Practice (1st ed.)*. New York: Marcel Dekker.
3. Bathias, C., International Journal of Fatigue, **28**, 1438-1445 (2006).
4. Xue, H. Q., Tao, H., Montembault, F., Wang, Q., & Bathias, C., **29**, 2085-2093 (2007).
5. Freitas, M., Reis, L., Anes, V., Montalvão, D., Ribeiro, A. M., & Fontes, M. Anales de Mecânica de la Fractura, **1**, 335-340 (2011).
6. Mayer, H., International Journal of Fatigue, **28**, 1446-1455 (2006).
7. Marines, I., Dominguez, G., Baudry, G., Vittori, J. F., Rathery, S., Doucet, J. P., & Bathias, C., International Journal of Fatigue, **25**, 1037-1046 (2003).
8. Vieira, M., Freitas, M. d., Reis, L., Ribeiro, A. M., & Fonte, M. d., **37**, 131-137 (2016).

MATEC Web of Conferences **165**, 09004 (2018) https://doi.org/10.1051/matecconf/201816509004
FATIGUE 2018

Development of a technique for the real-time determination of crack geometries in laboratory samples

Thomas M Buss, James P Rouse, Christopher J Hyde and Seamus D Garvey*
University of Nottingham, University Park, Nottingham, NG7 2RD, UK.

Abstract.

Crack size determination using electrical potentials both in service and in the laboratory has been undertaken for many years. In the laboratory this has mainly concentrated on the measurement of crack depth, with either alternating current (AC) or direct current (DC) supplies. Some work to determine the varying depth along the width of cracks as an inspection tool of in service parts using mapping methods has been done. This has used both AC and DC utilising various models to understand the data recorded, in Alternating Current Potential Drop (ACPD) a range of frequencies have been used to give various skin depths.

The resulting analyses have been grouped into two groups 'thin skin' and 'thick skin', in the thin skin case the skin depth is significantly smaller than the depth of the crack 1/10th of the crack depth whereas in the thick skin cases are for cases where skin depth is over this limit. Some work has been carried out to try and unify these two approaches.

The work presented here looks to develop a method using variable frequency ACPD to resolve further information about cracks growing in laboratory specimens. A system has been developed to rapidly sweep a wide frequency band and record voltage drop across a crack or feature. A selection of steel samples with known geometries and features have been used to trial and benchmark the technique. These samples have a range of cross sections as well as machined features or a range of shapes and sizes to simulate a range of crack geometries. This work has been approximated using a 2D computational model. This has been done using a reduced thickness approach.

* Corresponding author: thomas.buss@nottingham.ac.uk

1 Introduction

Direct Current Potential Drop (DCPD) as well as ACPD are often used to monitor crack length in crack initiation and crack growth testing. For both techniques a constant current is injected into the sample. In DCPD the voltage measured across the crack increases with crack length due to the reduction in area of the remaining ligament. High frequency ACPD work utilises the skin effect, this means that the current flows in a thin skin along the surface of the sample, as the crack grows the path that the current flows along increases, this again leads to an increase in the voltage measured across the crack. A schematic representation of this can be seen in Figure 1.

It has been show that the thin skin only works where the crack is greater than ten times the skin depth, at intermediate frequencies neither of these situations are true. In the transitional frequencies there is significant error in both methods, some empirical work has been published bridging these two conditions[1]–[3].

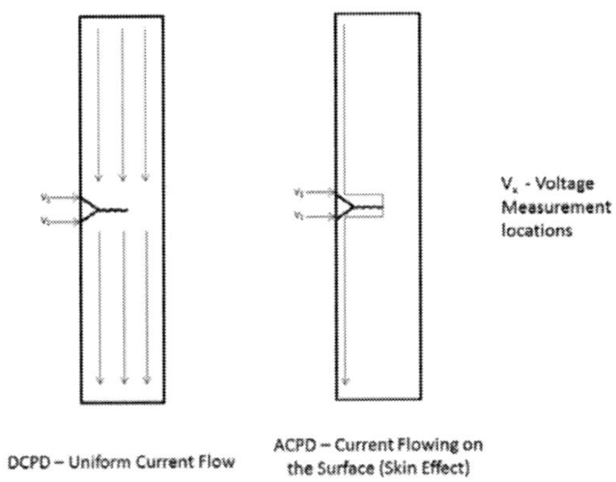

Fig. 1. Current flow in DCPD and ACPD.

There has been a range of work concerning crack measurement on components using movable probes, this is termed Alternating Current Field Measurement (ACFM), but uses the same theories as ACPD but the movable probes can be used to map the response at different points along or across a crack[4]–[7].

This work summaries the current state of work being undertaken at the University of Nottingham, the aim of the work is to be able to monitor the shape of fatigue cracks growing in a range of fatigue conditions. Most importantly in Low Cycle Fatigue (LCF) and Thermo-mechanical Fatigue (TMF) where is has been shown that non-uniform crack growth can be prevalent and the measurement of these can be difficult.[8] Quite often efforts are made to stop tunnelling occurring in the test, by modified specimen geometries[9], interrupting the test to put different loading on the sample which has been shown to return the crack to a uniform shape or post processing to try to understand the stress intensity factor[10]. Current work however is not being undertaken in live crack growth experiments, efforts are concentrating on characterising response using a range of samples with known shape. The aim being to understand sample shape and the shape of features in samples by simple two point measurements, without the need for mapping the potentials over the surfaces. This is being undertaken by utilising a wide range of Alternating Current (AC) frequencies and linking the results to the expected results generated from computational models. The work presented here looks at the first set of very simple samples. The samples chosen have identical cross sectional areas and do not contain any features or cracks, the only difference is the cross sectional shape. This has been done to enable the understanding of shape, the aim is to understand the fundamentals of the technique before embarking on more complex problems.

* Corresponding author: thomas.buss@nottingham.ac.uk

2 Experimental

Four samples have been manufactured with the same length (55mm) and cross sectional area ($100mm^2$), but different cross section shapes. The four chosen are circular, hexagonal, square and triangular. The material used is 316 stainless steel. An image of the experimental setup can be seen in Figure 2.

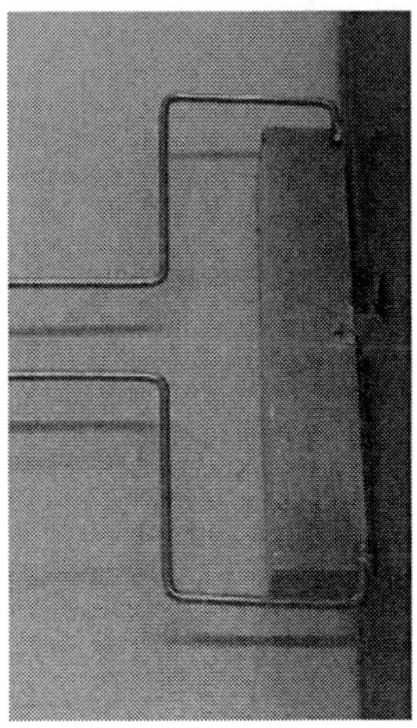

Fig. 2. Square Section Sample

A swept sine wave signal was used to sample frequencies from 15Hz to 220kHz covering skin depths from quasi-DC to thin skin at 220kHz. At 15Hz the skin depth is approximately 28mm which is larger than the width of any of the samples and at 220kHz the skin depth is 0.22mm which falls in to the thin skin range[1], [11]. Throughout all the experiments the amplitude of the current injected was kept at 3 amps. The voltage between two points symmetrically arranged around the midpoint of the length of the samples was measured, for all sample geometries these were placed with a separation of 2mm (1mm either side of midpoint) and on the same apex as the current injection points. Induced voltages due to electromagnetic (EM) pickup were controlled by a repeatable current input path consisting of some custom made steel wires 1.14mm in diameter, the shape and location of these could be easily controlled. These were connected to the current input lead by a screw terminal block and connected to the samples using spot welding. This enabled good and repeatable electrical connection for the input current.

The measurement wires were also shielded by the use of a die-cast box which can be seen on the right hand side of Figure 2 this was electrically insulated from the sample by means of a thin polymer tape. The box shielded the connection wires and the connector block from EM interference, once leaving the die cast box shielded instrument wire was used to continue the shielding to the amplifier. This enabled nearly the whole length of the measurement wires to be shielded from EM pickup. Repeat experiments were carried out to understand the level of repeatability of this setup, the data from two experiments on the square cross section sample can be seen in Figure 3, in these experiments all the wires were removed and new ones installed so this replicates a full 'new' test.

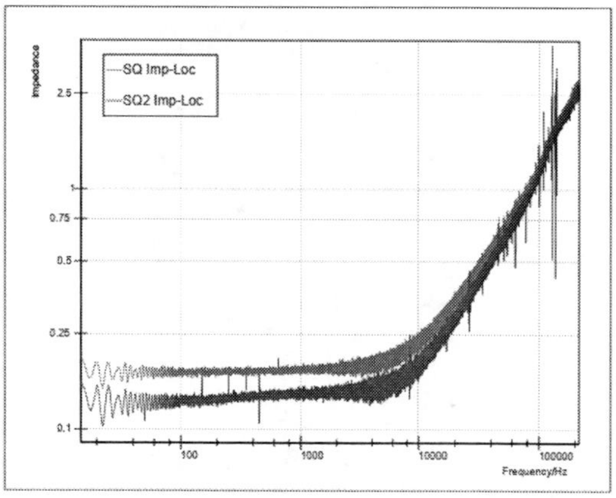

Fig. 3. Square Section Data

It can be seen in Figure 3 that there is an offset between the two curves across the frequency range. Due to the large skin depth at low frequencies the impedance at low frequencies if only governed by the cross sectional area and the path length. As the cross sectional area in all the samples is the same (100mm^2) the differences at low frequencies are due to variation in the experimental setup, the most significant variation is likely to be due to small variation in the spacing of the voltage measurement wires. Figure 3 only shows the square cross section result but this is also true for the other geometries, in the data processing this has been accounted for and the curves have all been shifted onto the same impedance as the circular cross section sample based on the readings at 100Hz as the response at this frequency can be considered quasi-DC.

The input was generated using a piece of LabVIEW code that generated a swept sine signal this was output through an NI-PXIe-6124 input/output module. This signal was fed into a bespoke amplifier supplied by Fylde Electronics which amplifies the voltage supplied and outputs a current in real time that is proportional to the voltage input. The measured signals were fed back in to the same PXI unit. Due to the small magnitude of the voltages being measured the signals were amplified 1000 times using a Fylde FE-H793-TA before the PXI unit. The output data rate was 2.5x10^6 samples per second and the signals were read back at 4x10^6 samples per second. The data was processed using the National Instruments DIAdem software. The fast furrier transform was run on the signals and the root mean squared (RMS) amplitude of this was used to represent the magnitude of the signals in all these analyses. As the input sweep was logarithmic the RMS amplitude of the current diminished with time therefore it was necessary to calculate the impedance between the measurement points, V_1 and V_2 in Figure 1. To do this the FFT RMS amplitude of the voltage between these points was divided by the FFT RMS amplitude of the input current. The resulting data for all the samples can be seen in Figure 4.

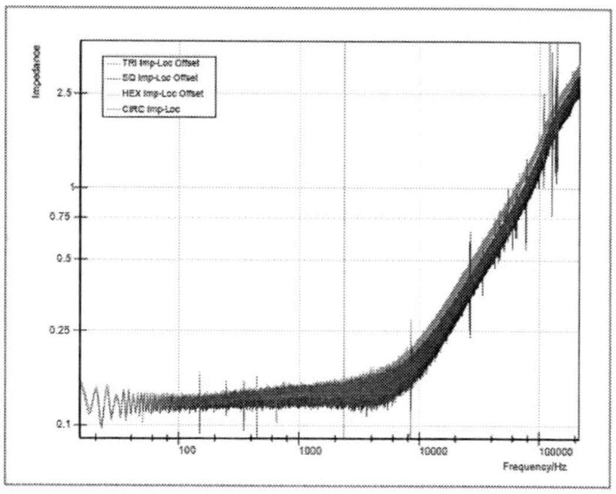

Fig. 4. Circular, Hexagonal, Square and Triangular Section Data

 https://doi.org/10.1051/matecconf/201816509004

3 Modelling

In this paper the theoretical analysis has been limited to a reduced thickness model. Given knowledge of the electrical and magnetic properties the skin depth at any given frequency can be calculated. If this is less than half the effective diameter of the samples this can be taken as the thickness of the conductor, this is termed a reduced thickness approach. This has been done in two ways, calculations of the resistance of the element between the measurement points and a computational model to simulate the voltages over the surface of the sample given the voltage input and a symmetry plain at the midpoint of the sample.

The calculations of the resistance between the two measurement points have been done for a number of frequencies. In this the conductor is considered to have a width equal to the perimeter of the sample and a thickness equal to the skin depth. Given this and the knowledge of the resistivity of the material, the resistance can be calculated by Equation 1, x being the distance between the measurement points and A the effective cross sectional area based on the perimeter of the sample and the skin depth at that frequency. This has been done for four frequencies across the range.

$$R = \rho \frac{x}{A} \qquad (1)$$

The computational model was undertaken using a finite element technique in Abacus, images from this can be seen in Figure 5. As a DC problem is analogous to a heat transfer problem, the heat transfer functions have been used. This was set up as a shell model using DS4 elements the thickness of which were altered to mimic the skin depth at a range of frequencies. As mentioned above, only half of the sample has been modelled as the sample is symmetric about the centre line of the measurement points. The boundary conditions were set such that a point heat was positioned at the end of the sample on the same apex as the measurement points this is to simulate the input voltage and the face at the other end of the model, which is effectively the midline of the real sample, was set to zero temperature. The problem was solved for the steady state condition using the Abacus standard method.

The temperature at the point 1mm from the 'zero end', on the same apex as this input heat, was calculated this can be considered as half the voltage between the measurement points and therefore can be compared to other results.

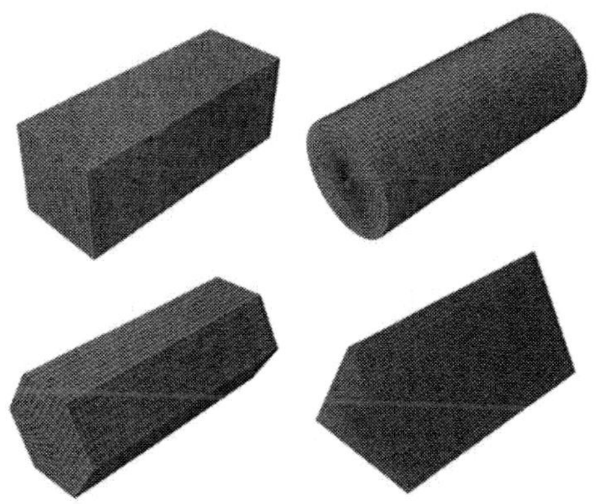

Fig. 5. FE Model

All the results both experimental and calculated have been normalised with respect to the low frequency quasi-DC response to aid comparison. Plots of the results from the resistance calculations and heat flow model can be seen in Figure 6 and 7.

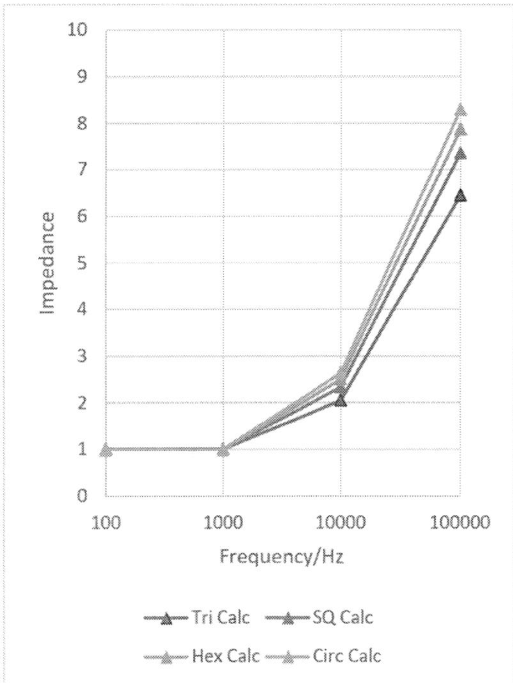

Fig. 6. Calculation Predictions

https://doi.org/10.1051/matecconf/201816509004

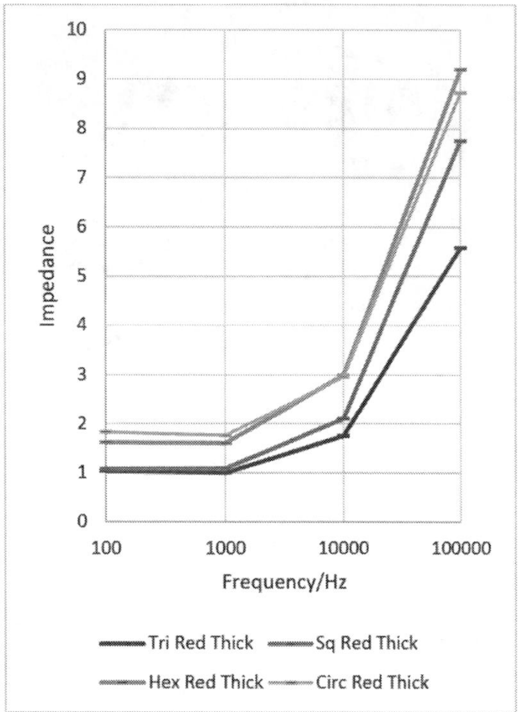

Fig. 7. Model Predictions

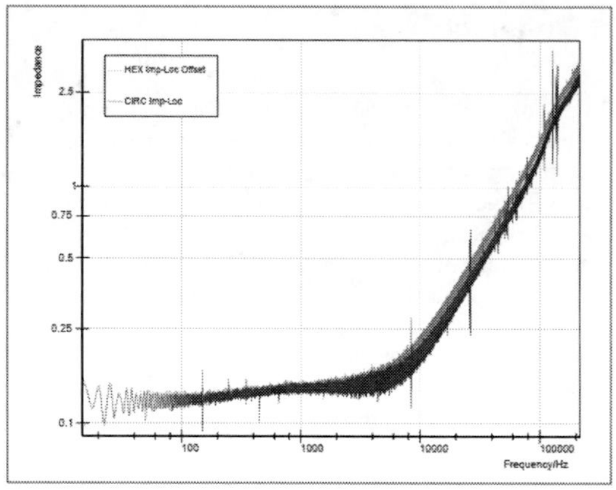

Fig. 8. Circular and Hexagonal Cross Sections Data

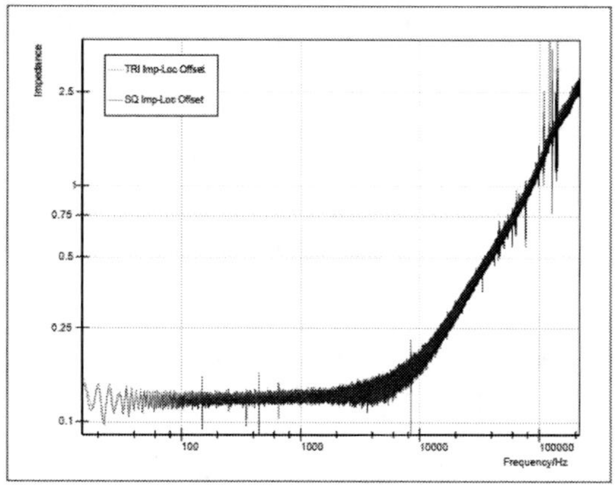

Fig. 9. Square and Triangular Cross Sections Data

4 Discussion

It can be seen in the experimental data that the broad response for all the specimens follows the expected trends, at low frequency where the skin depth is greater than half the effective diameter the impedance is independent of frequency as the resistance is governed by the path length and real cross sectional area of the samples, therefore is constant. As the frequency rises the skin becomes less than the effective half the effective diameter and the impedance increases.

In this 316 stainless steel the skin depth reaches half the effective diameter at about 500Hz. It can be seen in the circular and hexagonal plots (Figure 8) that the impedance starts to rise at about this point, although does not immediately move onto the expected line for the thin skin assumption. This is less clear on the square and triangular cross section samples (Figure 9) and the impedance seems to remain flat until higher in the frequency range. It is likely that this is due to the hexagonal and circular samples being most similar to an idealised conductor, there are no sharp features to effect the distribution of the current in the intermediate skin thickness range. Whereas in the triangular and square samples the sharp corners will affect the current flow and this may be affecting the voltage measurement points in relation to the thinning skin.

In both the calculated impedance results and the Abacus model it can be seen that there is some reasonable correlation between the circular and hexagonal cross section samples, these giving the higher impedances and in the experimental results these two samples also give the highest impedances. In all the calculations and measured data the square and triangular section samples exhibit lower impedances although they match less well. This highlights some of the limitations of the current analytical approach.

It can also be seen that these theoretical methods are not sufficiently sensitive to pick up the differences highlighted above, at around 500Hz, between the circular and hexagonal when compared to the square and triangular samples.

5 Conclusions

In samples of constant cross sectional area but with differing cross sectional shapes it can be seen that there are differences in impedances measured between two points centrally positioned on the test piece, as is customary in fatigue crack growth experiments. This has been considered theoretically using two methods. Calculations of resistance between the two measurement points and the results form a computational model based on voltages on a surface have been shown to correlate with the experimental results at lower frequencies, however deviations are significant at higher frequencies.

These theoretical techniques show some correlation to the experimental but also deviate in certain areas and lack some of the necessary sensitivity. There have been significant simplifications made in some of the calculations in relation to the current flow at the corners of the specimens. Also this approach, approximating the AC skin to a think DC conductor[7], does not have any capacity to account for phase shifting which may give further information on samples geometry and eventually crack shape.

It is therefore suggested that further simulation work is required to probe further the added information available in the AC phase data and a better understanding of current flows will also feed into the impedance analysis.

References

[1] H. Saguy and D. Rittel, "Bridging thin and thick skin solutions for alternating currents in cracked conductors," *Appl. Phys. Lett.*, vol. 87, no. 8, pp. 68–70, 2005.

[2] H. Saguy and D. Rittel, "Alternating current flow in internally flawed conductors: A tomographic analysis," *Appl. Phys. Lett.*, vol. 89, no. 9, pp. 214–216, 2006.

[3] H. Saguy and D. Rittel, "Flaw detection in metals by the ACPD technique: Theory and experiments," *NDT E Int.*, vol. 40, no. 7, pp. 505–509, Oct. 2007.

[4] M. Akama and M. Saka, "Nondestructive sizing of a 3D surface crack generated in a railway component using closely coupled probes for direct-current potential drop technique," *Eng. Fract. Mech.*, vol. 72, no. 2, pp. 319–334, Jan. 2005.

[5] Y. Hashimoto, Y. Urabe, S. Masamori, Y. Kamiwaki, and K. Baba, "Procedure of crack shape determination by reversing DC potential method," *Nucl. Eng. Des.*, vol. 138, no. 3, pp. 259–268, Dec. 1992.

[6] G. Sposito, "Advances in Potential Drop Techniques for Non-distructive Testing," Imperial College London, 2009.

[7] G. Sposito, P. Cawley, and P. B. Nagy, "An approximate model for three-dimensional alternating current potential drop analyses using a commercial finite element code," *NDT E Int.*, vol. 43, no. 2, pp. 134–140, Mar. 2010.

[8] J. D. Landes, J. A. Begley, and G. A. Clarke, "Elastic-Plastic Fracture," in *Elastic-Plastic Fracture ASTM STP 668*, 1977.

[9] S. B. Narasimhachary and A. Saxena, "Crack growth behavior of 9Cr 1Mo (P91) steel under creep fatigue conditions," *Int. J. Fatigue*, vol. 56, pp. 106–113, Nov. 2013.

[10] ASTM, *ASTM E647-13a - Standard Test Method for Measurement of Fatigue Crack Growth Rates*. 2014, pp. 1–50.

[11] K. Chen, F. P. Brennan, and W. D. Dover, "Thin-skin AC field in anisotropic rectangular bar and ACPD stress measurement," *NDT E Int.*, vol. 33, no. 5, pp. 317–323, Jul. 2000.

Simulation of Structure Dynamic Stress under Multi-load Condition in Resonance State

Fansong Li*, *Pingbo* Wu, and *Chaotao* Liu

Traction Power State Key Laboratory, Southwest Jiaotong University, 610031 Chengdu, China

Abstract. Some metro vehicle lifeguard structure cracking emerged during operation, and the reason is due to resonance. In order to simulate the dynamic stress accurately in resonance state, the formula for calculating the dynamic stress under multi-load condition was deduced based on structural dynamics mechanics. During the dynamic stress simulation, firstly, the frequency response function (FRF) of each external load was calculated, which including the phase information. Then, the method was based on Fast Fourier Transform (FFT) to get the frequency domain with real and imaginary part of each external load rather than power spectrum density (PSD) calculation, avoiding loss of the phase information due to PSD method. Finally, the dynamic stress frequency domain data could be calculated by FFT results and FRFs. Based on above, the dynamic stress time history could be acquired by Inverse FFT (IFFT) of stress frequency domain data, and the stress simulation method was verified by the line test result. Compared with the conventional stress PSD calculation method, the method in this paper can retain the phase information of the each external load. Meanwhile, the conventional stress simulation method based on PSD could not get satisfactory result.

1 Introduce

In dynamic theory, when the frequency of external load on structure is close to the structural mode frequency, and the external load direction is the same with the mode direction at the external load point, it will cause structural resonance easily, which may lead to fracture. Due to the fatigue test cost is high, and it could not be carried out before the structure prototype finished, as a result, the numerical simulation of the dynamic stress level under working condition becomes a necessary tool during the design stage. In general, the structure resonance cracking is mainly due to that the main frequency of external load in the working environment is unknown during structure design stage, but how to accurately simulate the dynamic stress is also a problem when we get the external excitation, only the accurate dynamic stress is obtained can we predict the fatigue life. Therefore, it is very important to accurately and quickly simulate the dynamic stress of structure under the resonance state, which could be used to guide the structure optimization.

There are two common methods in structural dynamic stress simulation [1], one is quasi-static superposition method, and the other is the modal superposition method. Quasi-static superposition method [2] is suitable when the external load frequency is much lower than the structural first elastic modal frequency, but this method is not suitable for dynamic stress prediction when structure is under resonance condition [1]. In modal superposition method, the external load frequency and amplitude is the system input, and the dynamic stress is predicted according to structure inherent properties. It is a suitable method for vibration and stress prediction of structure under resonance state. However, in present study [3, 4], the input loads are often processed as PSD, which ignoring the phase information of the load in the frequency domain. When the input loadings are multiple, the phase between the loadings could not be separated. So, it could not simulate the structure dynamic stress factually under multi-load input condition.

In this paper, the simulation method of structural dynamic stress under multi-load input condition was studied, and the phase relationship between the input loadings was considered.

2 Background

2.1 Research background

As shown in **Fig. 1**, lifeguard structure cracking emerged in metro vehicles. And the cracking reason was found by line test, which is that the first two directional bending mode of lifeguard structure was aroused by the wheel excitation from ninth-order wheel polygon. And the cracking event is a typical vibration fatigue problem.

During the line test, the different directional acceleration of lifeguard fixed end and the strain (**Fig. 2**) at the crack were recorded. In order to redesign a new lifeguard structure, we should know how to calculate the dynamic stress accurately according to external load in resonance

* Corresponding author: lifansong2013@163.com

state. Then the new designed structure could be verified whether satisfy the design life by the test environment load.

(a) The position of lifeguard of metro vehicle

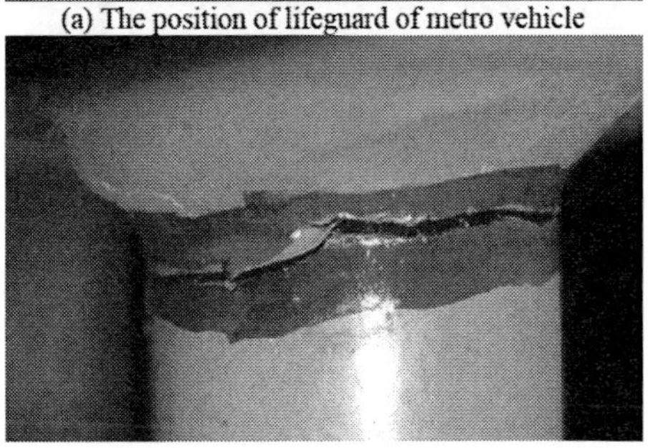

(b) Crack of lifeguard structure

Fig. 1. Lifeguard structure cracking

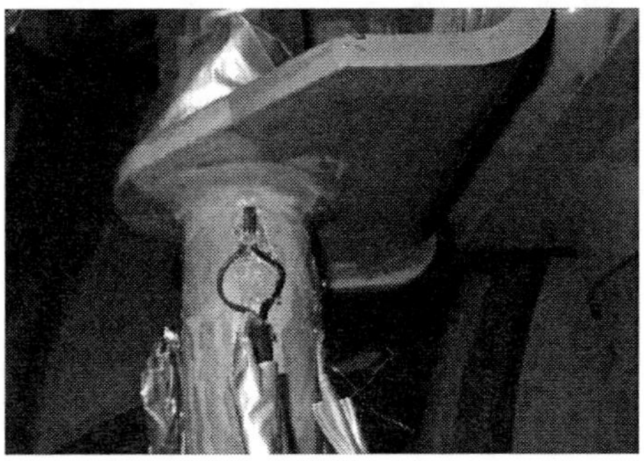

Fig. 2. Line test of strain of lifeguard structure

2.2 Theoretical background

The motion equation of structure could be given in Equation (1) [4].

$$[\mathbf{M}]\{x\} + [\mathbf{C}]\{x\} + [\mathbf{K}]\{x\} = \{\mathbf{f}\} \qquad (1)$$

Where [M] is the mass matrix; [C] is the damping matrix; [K] is the stiffness matrix; {x} is the vector of degrees of freedom and {f} is the excitation force vector. Furthermore, through modal analysis the FRFs from one point on the structure to the other can be deduced.
According to the dynamic theory, the load-stress FRFs could be written in Equation (2) [5].

$$\mathrm{H}(\omega) = \mathrm{E} \cdot \sum_{r=1}^{R} \frac{\{\psi_r^{\varepsilon}\}\{\varphi_r\}^T}{k_r - \omega^2 m_r + j\omega c_r} \qquad (2)$$

Where m_r is the rth modal mass, k_r is the rth modal stiffness, c_r is the rth modal damping, ω is the frequency of excitation. $\{\varphi_r\}$ is the rth real value mode vector, R is the total number of modes considered, $\{\psi_r^{\varepsilon}\}$ is called the rth strain mode corresponding to $\{\varphi_r\}$, j is $\sqrt{-1}$, E is material elastic module, $\omega=2\pi f$, f is frequency.
According to [6], when the external load is single, the stress PSD G(f) can be found very rapidly by multiplying the load PSD W(f) and FRF H(f), which is shown in Equation (3).

$$\mathrm{G}(f) = |\mathrm{H}(f)|^2 \cdot \mathrm{W}(f) \qquad (3)$$

When the external load is not single but multiple, for a general circumstances with n simultaneous loadings, the stress PSD G(f) can be determined by Equation (4).

$$G(f) = \sum_{a=1}^{n}\sum_{b=1}^{n} \mathrm{H}_a(f) \cdot \mathrm{H}_b(f)^* \cdot \mathrm{W}_{ab}(f) \qquad (4)$$

Where $\mathrm{H}_a(f)$ and $\mathrm{H}_b(f)$ are FRFs relating to different two loadings, and a and b could be equal. $\mathrm{W}_{ab}(f)$ is the cross power spectrum density (CPSD) of load a and load b.

3 The FFT stress simulation method

However, the PSDs alone do not provide information on the phase relations between diferent loadings. So, the method above of calculating stress will cause something wrong under multi-load condition. In order to overcome the problem, this paper suggest to use FFT method rather than PSD method. From Equation (2) we know the FRF was complex, which including phase relations between input and output (load and stress), so if FFT of external loadings is adopted, the phase information of loading itself will be retained in frequency domain, which could known in Equation (5). In the case of multi-load input, that is, when the number of external loadings is more than one, assuming that the number is k. The ith external loading FFT could be described as:

$$\mathrm{F}_i(\omega) = \sum_{m=0}^{N-1} f_i(n)\cos m\omega - j\sum_{m=0}^{N-1} f_i(n)\sin m\omega \qquad (5)$$

Where N is the sampling length of load.
Then, the frequency domain of stress could be calculated by Equation (6) based on dynamic mechanics.

$$\sigma(\omega) = \sum_{i=1}^{k} F_i(\omega) \cdot H_i(\omega) \qquad (6)$$

In Equation (6), the FRFs and FFTs are all complex, not real. And the time domain of stress can be obtained by IFFT in Equation (7).

$$\sigma(n) = \mathrm{real}\left(\frac{1}{N}\sum_{\mathrm{m}=0}^{N-1} \sigma(\omega)e^{j\omega m}\right) \qquad (7)$$

Then, the stress time history can be used to evaluate the fatigue strength or life.

4 Stress simulation

Firstly, the simulation finite element model of lifeguard was established by ABAQUS and modal analysis was conducted, the modal results are shown in **Fig. 3**, and three different directions (X, Y, Z) are defined in **Fig. 3.**

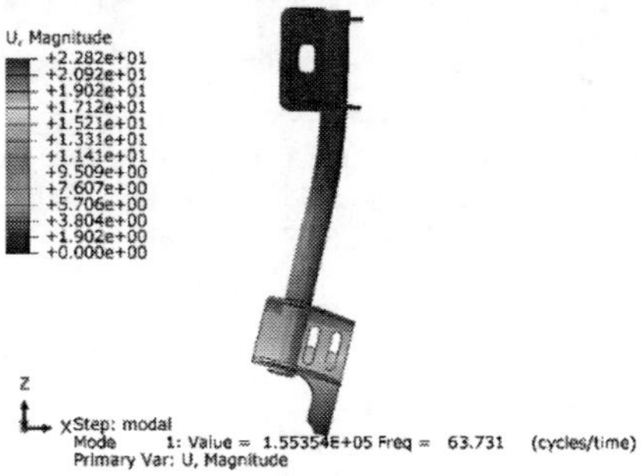

(a) First bending mode

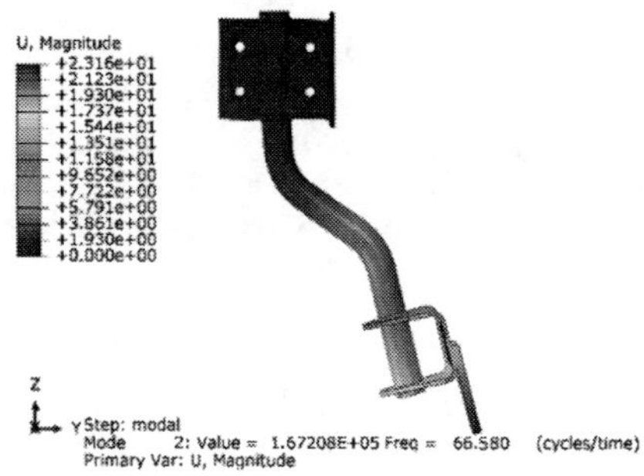

(b) Second bending mode

Fig. 3. The first two modes of lifeguard

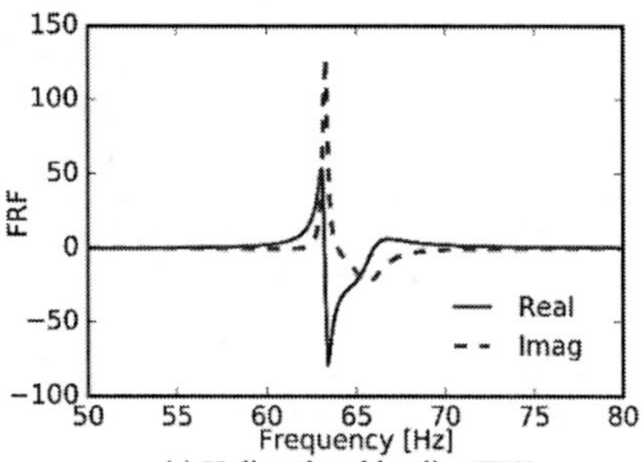

(a) X directional loading FRF

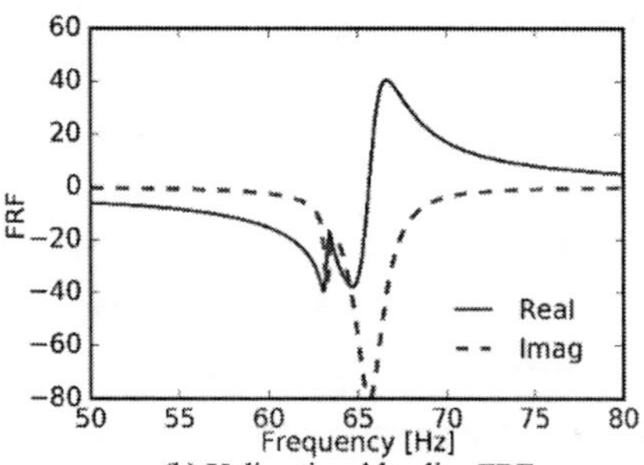

(b) Y directional loading FRF

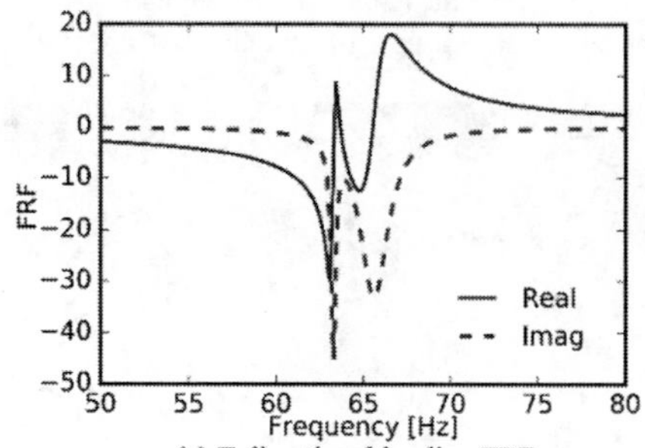

(c) Z directional loading FRF

Fig. 4. FRFS of three external loadings

After the modal analysis, the cracking position acceleration-stress FRFs of three different directional loadings were calculated, and the modal damper ratio of first two modes is 0.23% and 0.61% respectively. And the three FRFs are shown in **Fig. 4**.

4.1 PSD stress simulation method

Firstly, we just use the single load condition stress simulation method, the stress PSD could be calculated by Equation (8). We just call it "PSD method1".

$$G(f) = |H_x(f)|^2 \cdot A_x(f) + |H_y(f)|^2 \cdot A_y(f) + |H_z(f)|^2 \cdot A_z(f) \quad (8)$$

Secondly, the Equation (4) was used to calculate the stress PSD, which is "PSD method2". The simulation results are shown in **Fig. 5**, which are compared to the test stress result.

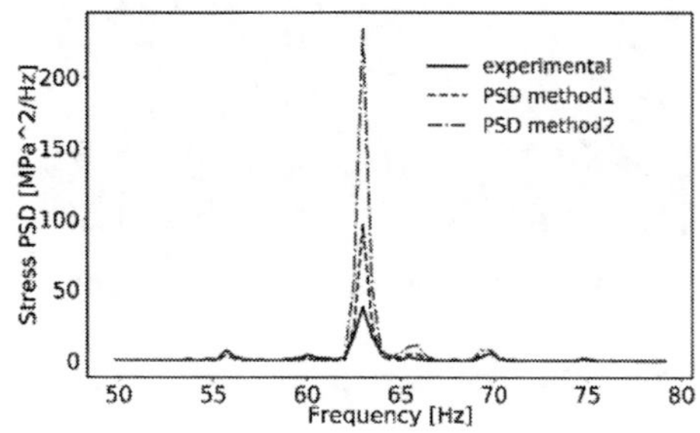

Fig. 5. Stress comparison of test and simulation by PSD

From the comparison result, we can know that the PSD stress simulation method is conservative. The reason maybe is that the PSD method just consider the different external loadings reach the maximum amplitude simultaneously. Actually, every two external loadings have the phase difference.

4.2 FFT stress simulation method

In this section, the FFT stress simulation method will be illustrated. First, the FFTs of external loadings are calculated, and the phase should be retained. Then, the Equation (6) will be adopted. We could get the FFT of dynamic stress calculated, which is used to compare with the FFT result of experimental stress in **Fig. 6**.

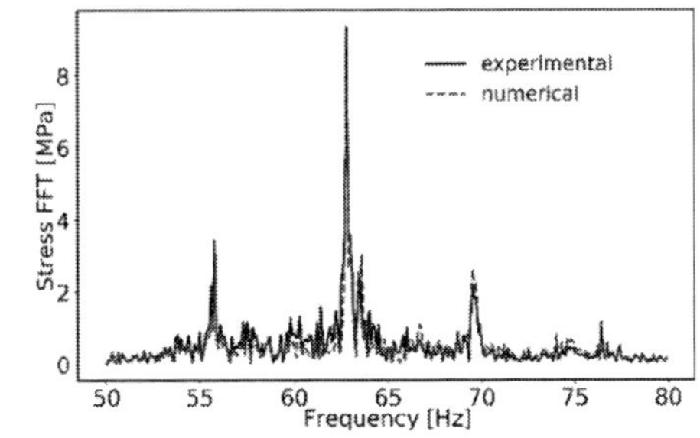

Fig. 6. Stress comparison of test and simulation by FFT

It can be seen from the **Fig. 6** that the numerical results are almost same with the experimental result. The time history of stress could be calculated by IFFT according to Equation (7). Which could be seen in **Fig. 7**.

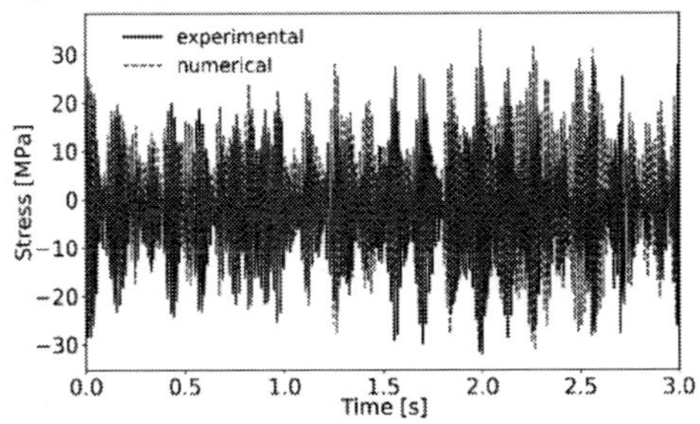

Fig. 7. Stress comparison of test and simulation by FFT

The stress PSD from three simulation methods are shown in **Fig. 8**. It can be seen from **Fig. 8** that under multi-load condition, the dynamic stress simulation method in this paper is accurate and acceptable, however, the PSD method is conservative, especially PSD method 2.

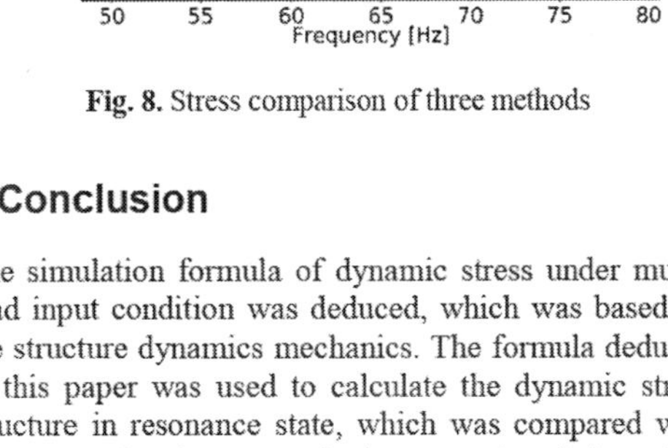

Fig. 8. Stress comparison of three methods

5 Conclusion

The simulation formula of dynamic stress under multi-load input condition was deduced, which was based on the structure dynamics mechanics. The formula deduced in this paper was used to calculate the dynamic stress structure in resonance state, which was compared with the conventional PSD method. It was found that when use the PSD method to simulate the stress under multi-load condition, the simulated result is conservative, however, the FFT method has the advantages in solving the phase information under this condition. In conventional method, the phase information is neglected when calculating the PSD of data, so, it will not predict the dynamic stress accurately. Therefore, the method in this paper is recommended, and which was verified by experimental result.

The authors gratefully acknowledge the financial support of Traction Power State Key Laboratory Research Project (Grant No. 2015TPL_Z03), Science and technology support program of China (Grant No. 2015BAG13B01-03) and Shandong Science and Technology Special Project (Grant No. 2015ZDXX0102A01)

References

1. M. Haiba, D.C. Barton, P.C. Brooks, M.C. The development of an optimisation algorithm based on fatigue life. Int J Fatigue 25, 4 (2003)
2. P. Heyes, J. Dakin, St.John C, The Assessment and Use of Linear Static FE Stress Analyses for Durability Calculations, SAE Technical Paper 951101 (1995).
3. Haiba M, Barton DC, Brooks PC, Levesley MC, Review of life assessment techniques applied to dynamically loaded automotive components, Comput Struct 80,5 (2002).
4. Moon SI, Cho IJ, Yoon D, Fatigue life evaluation of mechanical components using vibration fatigue analysis technique, J Mech Sci Technol 25, 3 (2011).
5. Yam LY, Leung TP, Li DB, Xue KZ, Theoretical and experimental study of modal strain analysis, J Sound Vib 191, 2 (1996).
6. A. Halfpenny. A frequency domain approach for fatigue life estimation from finite element analysis. Key Engineering Materials 167 (1999).

 https://doi.org/10.1051/matecconf/201816510002

Fatigue of polyamide mooring ropes for floating wind turbines

Yoan Chevillotte[1,2*], *Yann* Marco[1], *Peter* Davies[2,3], *Guilhem* Bles[1], and *Maël* Arhant[2]

[1]ENSTA Bretagne, Institut de Recherche Dupuy de Lômes (iRDL), FRE CNRS 3744, 29200 Brest, France
[2]IFREMER, Marine Structures Laboratory, 29200 Brest, France
[3] France Energies Marines, 29280 Plouzané, France

Abstract. This paper describes a study of the fatigue characterization of polyamide mooring ropes for floating wind turbines. Under some conditions polyester ropes, which are favoured for offshore platform station-keeping, are too stiff for wind turbine moorings, and polyamide may be a suitable alternative. While early studies on fatigue of braided nylon ropes showed very short lifetimes some recent results have indicated that it is possible to significantly enhance lifetime by modifying rope construction and improving fibre coatings [1]. The fatigue results presented here for ropes from a different supplier, confirm this result. In order to develop an accelerated evaluation of the fatigue performance, heat build-up tests have been performed, and promising first results are shown. Finally, the influence of coating is examined by microscopy and yarn-on-yarn tests, in order to improve understanding of the fatigue mechanisms leading to failure.

1 Introduction

As the development of Marine Energy Converter systems (such as wave energy generators, tidal current turbines and floating wind turbines) starts to flourish, there is a need for mooring lines in shallow water (between 50 and 100 m). Since these applications aim at harnessing marine energies, the systems are located on exposed area where dynamic loading are significant. In addition, the inherent moderate to small water depth induce a global stiff behaviour of the mooring system in using standard components as per done for deepwater floating oil platforms [1] (steel chains or polyester rope). The system integrity will be maintained by damping the dynamic loadings, so requiring the use of more stretchable fibre ropes than polyester.

Mainly because of their use in other marine applications, their low price and high breaking strain (up to 20%) polyamide (Nylon) is a good candidate for these applications. However, studies on braided Nylon ropes in the 1980's [2] have shown relatively poor fatigue performance which provided little incentive for studies on Nylon mooring ropes. Indeed, Nylon single point moorings offshore are typically changed every one or two years. More recently however, Ridge et al. [3] have shown significantly improved fatigue results for twisted nylon ropes, and those authors qualified these ropes as ideal for wave energy convertor moorings. Weller et al [4] also examined a nylon rope which had been tested at sea for 18 months on a wave energy buoy, and showed the importance of loading history and operating conditions on nylon performance.

The idea of using long term mooring lines in this material is relatively new, and there is little knowledge on the durability behaviour of such structures. It is therefore crucial to further investigate the behaviour of polyamide mooring ropes, in order to optimise their utilisation.

The database is still very limited, only the paper by Ridge et al provides long lay length stranded PA rope fatigue results [3]. Thus the deployment of these fibres for permanent mooring is currently impossible due to the lack of data on their long term mechanical properties.

The goals of the present study are therefore to provide a better understanding of the mechanical behaviour and the fatigue damage of these polyamide ropes, in order to help designing floating wind turbine mooring lines. However, fatigue studies are quite long and costly, this is why an evaluation of rapid characterization by heat build-up measurements [5] [6] has been included in the test programme. This kind of test has proven to be efficient on a wide range of materials [7]. These techniques have been developed and successfully applied in recent years to various polymeric materials included polyamide, from single specimens to structural specimens [8]. However, heat build-up characterization has not been used to date on materials such as ropes or cables. Moreover, a specific challenge is to perform the tests in realistic testing conditions, i.e. with over 90% humidity. One of the objectives of the study presented here is to investigate if the analysis of heat build-up measurements, which works well on dry solid materials under cyclic loading, can be applied to ropes which damage due to internal friction, and which are tested wet.

* Corresponding author: yoan.chevillotte@ensta-bretagne.org

Finally, the influence of coating will be assessed by some yarn-on-yarn abrasion tests and microscopy.

2 Materials and Test methods

2.1 Fibres and ropes

Polyamide 6 (PA6), often described as Nylon 6 yarns and ropes provided by BEXCO have been investigated in this study. The yarns used are made by Nexis fibers with a tex of 188, with and without coating. The nylon rope samples used in this series of tests were specially manufactured for the research project. The rope is a six-meter-long (pin to pin) three stranded rope of outer diameter around 15 millimeters for a nominal break load of 75 kN. The lay length is similar to polyester subrope constructions for deepwater mooring ropes and imposes splices of more than 2 meters at each end of the rope, leaving a 1.5-meter central area.

2.2 Yarn on Yarn tests

The yarn-on-yarn abrasion tests were performed as described in CI (Cordage Institute) 2009-02P "Performance requirements for marine grade polyester yarn for fiber rope" [9] and ASTM D 6611, "Standard Test Method for Wet and Dry Yarn-on-Yarn Abrasion Resistance" [10]. This test is the standard method for testing inter-yarn abrasion of synthetic fibers and provides a qualitative value of the abrasion resistance. The only difference from the standard test procedure was that tests were performed in natural sea water here.

These tests were carried out at IFREMER with the configuration presented on Figure 1. A yarn is inter-wrapped with itself (by turning it by 3.5 turns) between three pulleys. At one end of the yarn a weight is attached to apply a constant tension, and at the other end a motor drives the cable back and forth with a 1 Hz frequency until the yarn breaks due to abrasion in the wrapped region.

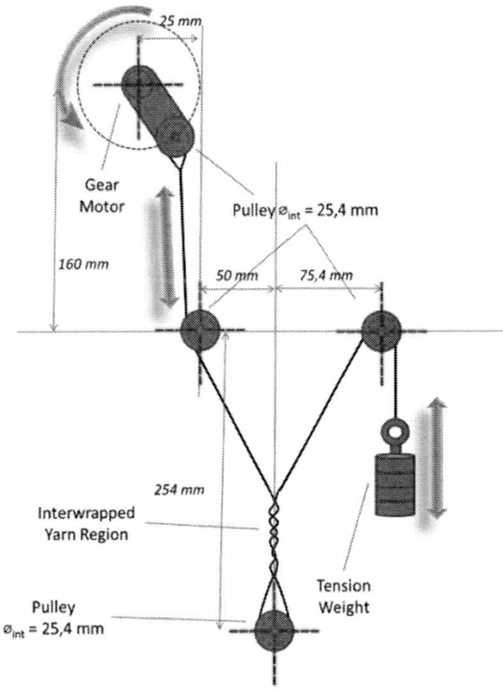

Fig. 1. Yarn on yarn test machine

The inter-wrapped region is kept in sea water throughout the test until failure.

Two types of specimens have been tested with this test: yarn without coating and yarn with a special marine finish.

2.3 Fatigue tests

Fatigue testing has been performed on a servo-hydraulic test machine at the IFREMER Marine Structures Laboratory in Brest with a load capacity of 300 kN and a piston stroke of 3 meters.

Before each test the rope sample was immersed in water for 4 hours on the testing machine, and it was then continuously sprinkled with tap water during the test (Figure 2).

Fig. 2. Rope during a fatigue test

The fatigue test results presented here were performed with a mean-load of 40% of nominal break load, at a frequency of 0.1 Hz. Before each test, a pre-cycling protocol called "bedding in" was applied. This consists of 5 load-unload cycles from 2 to 50% of nominal break load.

2.4 Heat build-up tests

The heat build-up tests were performed on the same testing device as the fatigue ones. A specific plastic box is used in order to control the reflexion conditions and to simplify the thermal measurements, Figure 3. The temperature was recorded using a Photon™ camera (FLIR SC7600-BB). The thermal resolution was evaluated around 50mK, which was enough for the protocol applied. The mechanical protocol applied is close to the one applied during the fatigue tests, except that no bedding-in was applied. As the analysis is based on the stabilized response (see below), the transposition of the results obtained from heat build-up tests to the fatigue prediction remains reasonable. Several loading steps with increasing load levels were applied. Each loading step consists of 150 cycles between 2 kN and XX kN (XX ranging from 10 to 52, see Figure 5).

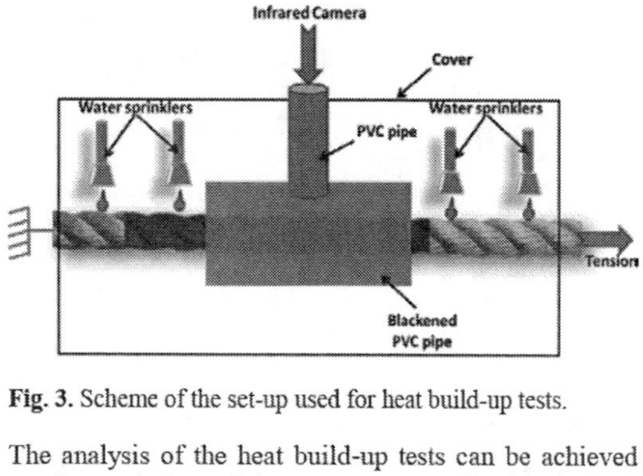

Fig. 3. Scheme of the set-up used for heat build-up tests.

The analysis of the heat build-up tests can be achieved on the evolution of the temperature only, but it is of course much better to investigate the evolution of the dissipated energy, which is an intrinsic parameter. To do so, one can consider either an adiabatic analysis [5] [11] or a stationary analysis [12].

In our study, the stationary state analysis was chosen, in order to catch the thermal response over the full diameter of the ropes. To do so, the thermal evolution is recorded over the full loading steps and averaged over the cable and over the surrounding box (Figure 4). Then these temperatures are combined to compensate the evolution of the ambient temperature during the test. Finally, the relative evolution with time is identified as a saturating exponential function, using the in-house developed Belenos® software [13]. This analytical fit provides the stabilized temperature $\bar{\theta}$ and the characteristic time τ_{eq} . These two parameters are then used to deduce the mean cyclic dissipated energy Δ^*, according to Equation (1):

$$\Delta^* = \left(\frac{\rho c \bar{\theta}}{f_r \tau_{eq}}\right) \quad (1)$$

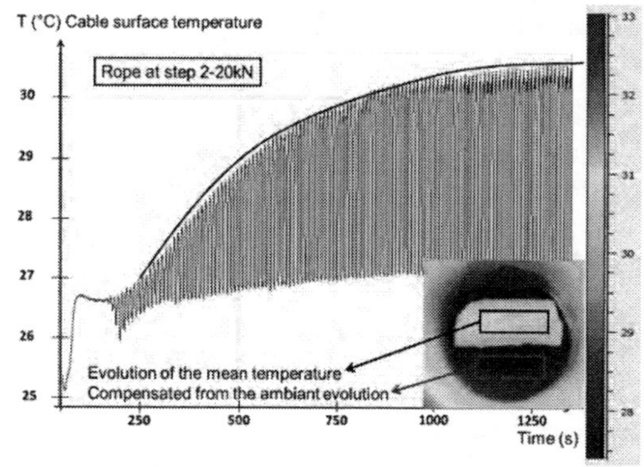

Fig. 4. Typical evolution of the temperature and of the analytical fit.

For each loading step, the mean cyclic dissipated energy Δ^* is evaluated, leading to the construction of the heat build-up curve, relating Δ^* to the maximal force applied (see Figure 5).

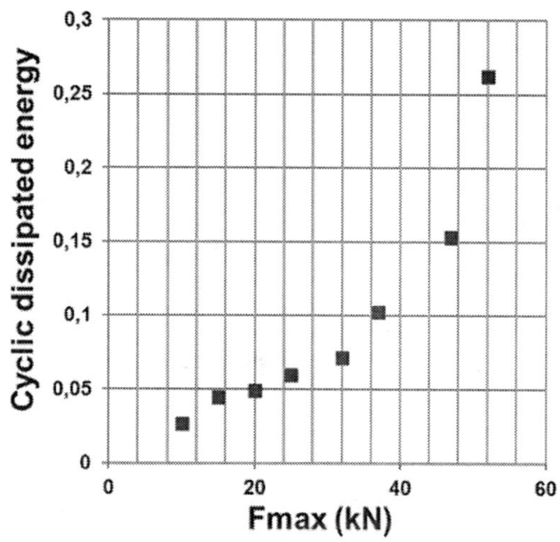

Fig. 5. Heat build-up curve on rope.

3 Results and discussion

3.1 Yarn on yarn

Yarn on yarn tests were performed up to failure, which occurred in the inter-wrapped section of the yarn in all cases reported on Figure 6. On the curve, each point is made from 3 to 8 tests depending on the dispersion. The error bars represent the minimum and maximum number of cycles found at this load.

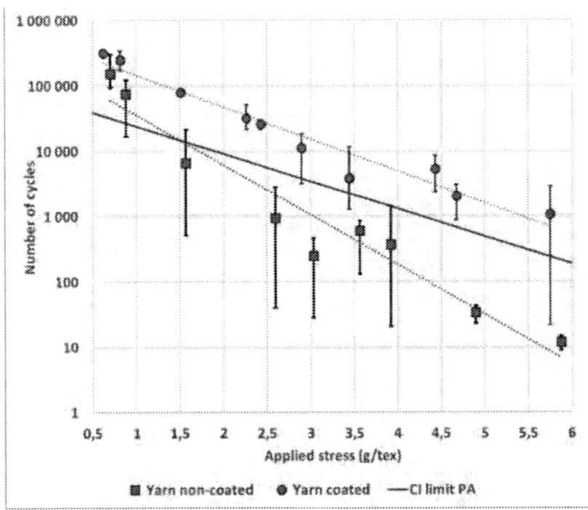

Fig. 6. Yarn on yarn results for uncoated and coated yarns.

Polyamide yarn has a minimum performance criterion (number of cycles versus applied tension) given by the CI document, which is represented as a red line on Figure 6.

The main point of these results is to demonstrate the strong effect of the coating on the abrasion resistance, upgrading the number of cycles to failure by one or two decades, which allow our product to be above the CI limit.

3.2 Fatigue results

Fatigue tests were performed up to failure, which occurred in the central section of the rope in all cases. The fatigue data points are shown in Figure 7.

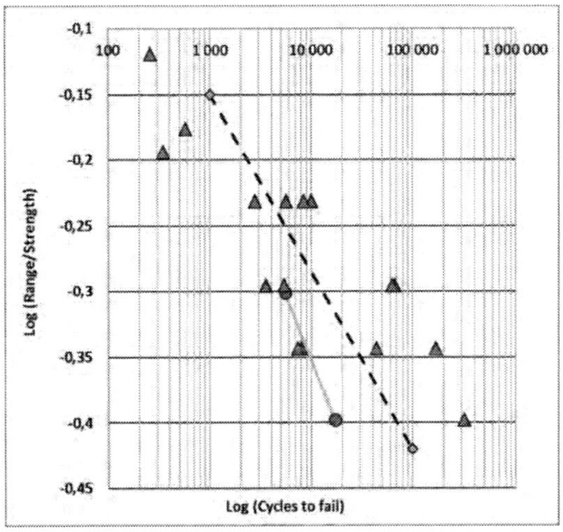

Fig. 7. Example of fatigue test results for rope samples.

Dotted line between the two hexagons shows results from Ridge et al; on their long term mooring polyamide rope for Wave Energy Converters [3]. Our results are fairly close to the one found by Tension Technology International with a different rope manufacturer, and reasonably above the reference curve (grey line between the two red dots) in ISO 18692 (fibre ropes for station keeping) [14] which reinforces the idea that provided appropriate rope constructions and coatings are adopted, nylon fatigue may no longer be an issue for mooring line applications.

3.3 First analysis of the heat build-up tests and comparison to the fatigue tests

The analysis of the heat build curve in order to predict the fatigue properties can be more or less complex and accurate, depending on the maturity of the constitutive description [15] or on the quality of the microstructural description [16]. In the case investigated here, these are the very first tests performed on ropes and the analysis will remain very simple. Two features of the heat build-up curve can tentatively be related to the fatigue properties. The first one is the fatigue lifetime obtained for the last loading block that leads to failure (see the top blue arrow on Figure 8). It seems that the lifetime obtained is fully consistent with the fatigue results obtained (see Figure 9). The second one, that is the key point of the heat build-up based analysis, is related to the analysis of the shape of the curve. As explained earlier, the dissipative mechanisms are very numerous (visco-elasticity of the material, friction at several scales in the rope) and an accurate understanding and therefore analysis of the heat build-up curve requires a much deeper investigation. This investigation is underway and should provide a better insight within the next months.

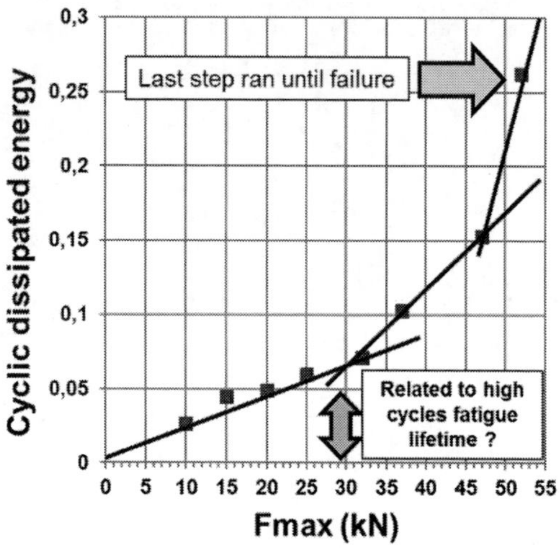

Fig. 8. Illustration of the analysis of the heat build-up curve on rope.

Nevertheless, a rough analysis can be provided by considering that the change of shape of the heat build-up curve could be related to a change of the intensity of the dissipation mechanisms. For several materials [5] [11], this change can be related to the acceleration of the fatigue mechanisms, leading to a very reasonable evaluation of a given fatigue lifetime (usually 10^6 cycles). This evaluation is applied to the curve obtained, as illustrated on Figure 8 (bottom yellow arrow). The force evaluated is also used to plot the predicted lifetime on the same graph as the fatigue tests (see Figure 9). One can conclude that this evaluation, seems also very consistent and that the application of the heat build-up approach to these ropes deserves a closer look.

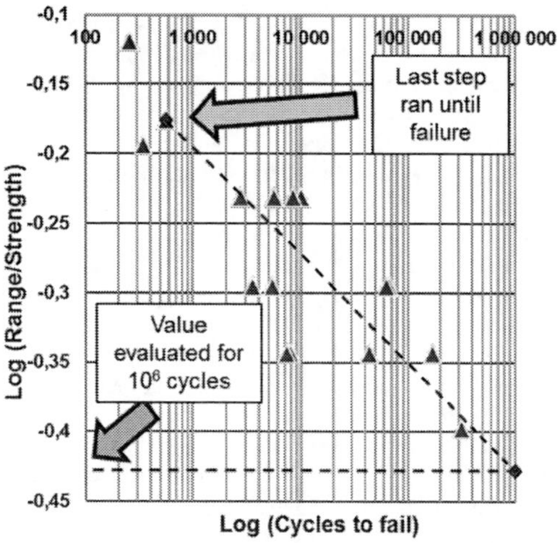

Fig. 9. Comparisons between the fatigue data and the prediction coming from the heat build-up tests.

3.4 Scanning Electron Microscope SEM

The aim of this study with SEM (Quanta 200, FEI) analysis was to assess the main phenomena behind fatigue failure and validate if the yarn-on-yarn test samples show the same damage mechanisms as those in rope fatigue testing.

3.4.1 Rope after fatigue testing

Samples were taken from broken ropes at 200mm from the final failure zone. A significant feature we have found in our observations of broken ropes in fatigue is the presence of melted fibres (Figure 10), but it is not clear if the melting happened during the test or during the breakage of the rope when all the energy was freed. Further observation on samples from tests interrupted before final failure should allow us to determine when the melted zone appears.

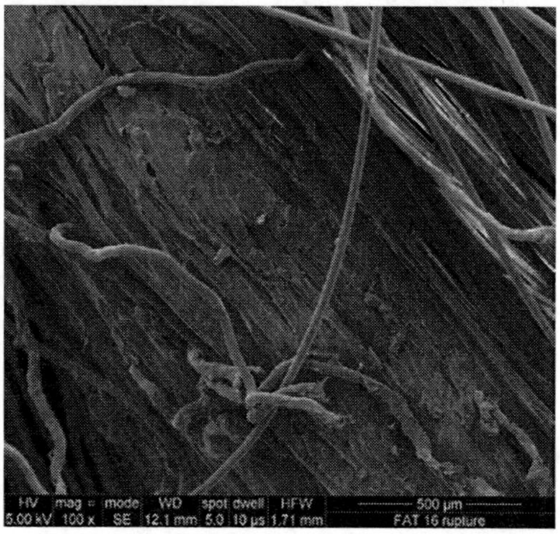

Fig. 10. Example of melted zone in a fatigue rope.

Another feature which seems to be characteristic of the damage developing in this material, is the peel observed on some fibers (Figure 11 and Figure 12), similar behaviour was documented by Hearle in the Atlas of fibre fracture and damage to textiles [17].

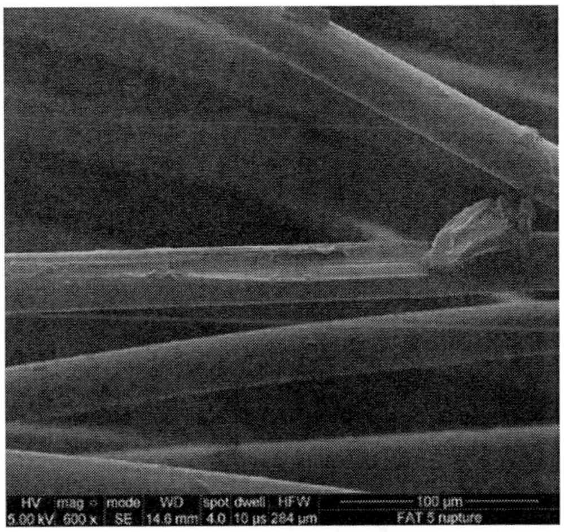

Fig. 11. Example of peeled zone on a fiber after rope fatigue test.

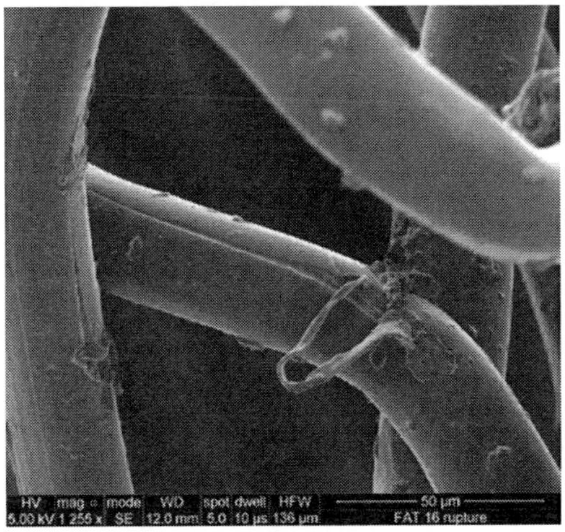

Fig. 12. Example of peeled zone on a fiber after rope fatigue test.

3.4.2 Yarn samples after Yarn on Yarn abrasion tests

On specimens after yarn-on-yarn tests, we haven't found melted fibers on specimens observed at low and mid-range load. However even at quite low load, we see some marks on broken fibers, which are similar to the peeling seen in the rope fatigue specimens. Mainly, on broken fibers we see either a skin shape where the center of the fiber is gone, Figure 13 or part of the fiber surface is removed like in Figure 14 or damaged like in Figure 15.

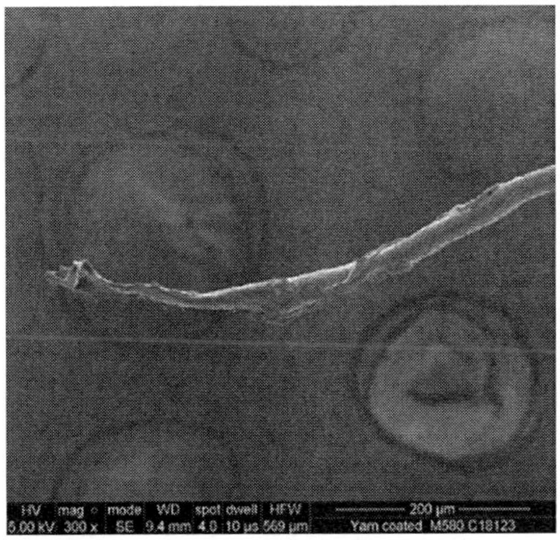

Fig. 13. Example of peeled zone on a fiber after yarn on yarn.

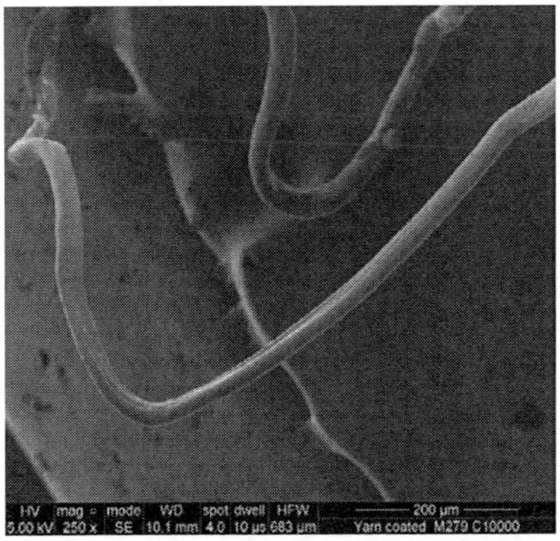

Fig. 14. Example of peeled zone on a fiber after yarn on yarn.

Fig. 15. Example of peeled zone on a fiber after yarn on yarn.

More tests are underway to examine in more detail these mechanisms.

4 Conclusion

The results from abrasion tests by yarn-on-yarn loading show a significant increase in abrasion resistance by the addition of a specially developed coating. That and the long lay length rope construction enables these polyamide ropes to exhibit very good fatigue performance, similar to those demonstrated by Ridge et al.

A first assessment by heat build-up analysis, demonstrates that measurement is possible on a wet rope and shows a promising result which seems consistent with fatigue results even if the loading was not identical. However, an analysis based on temperature alone is not satisfactory because of variations of the heat exchange conditions, which oblige us to adapt the characteristic time. Another campaign of testing will be performed shortly on a different scale of rope to further investigate this aspect.

SEM observations shows two particular failure mechanisms, the first and most visible is melted fibres usually on a large area of the size of a yarn. The second mechanism is the peeling of fibres, this is probably indicative of abrasion in our rope as we found similar behaviour on broken fibres from the yarn on yarn test samples.

5 Acknowledgements

This work was performed within the FEM/ANR POLYAMOOR project (ANR-10-IEED-0006-16). This is led by France Energies Marines with partners Naval Energies, Bureau Veritas, Bexco Ropes, Ensta Bretagne and IFREMER.

Thanks also to Nicolas GAYET of IFREMER for SEM photos.

6 References

1. Weller, S. D., Johanning, L., Davies, P., & Banfield, S. J. Synthetic mooring ropes for marine renewable energy applications. *Renewable Energy, 83*, 1268-1278. (2015).
2. Kenney, M. C., Mandell, J. F., & McGarry, F. J. Fatigue behaviour of synthetic fibres, yarns, and ropes. *Journal of materials science, 20*(6), 2045-2059. (1985).
3. Ridge, I. M. L., Banfield, S. J., & Mackay, J. Nylon fibre rope moorings for wave energy converters. In *OCEANS 2010* (pp. 1-10). IEEE. (2010, September).
4. Weller, S. D., Davies, P., Vickers, A. W., & Johanning, L. Synthetic rope responses in the context of load history: The influence of aging. *Ocean Engineering, 96*, 192-204. (2015).
5. La Rosa, G., & Risitano, A. Thermographic methodology for rapid determination of the fatigue limit of materials and mechanical components. *International journal of fatigue, 22*(1), 65-73. (2000).
6. Le Saux, V., Marco, Y., Calloch, S., Doudard, C., & Charrier, P. Fast evaluation of the fatigue lifetime of rubber-like materials based on a heat build-up protocol and micro-tomography measurements. *International Journal of Fatigue, 32*(10), 1582-1590. (2010).
7. La Rosa, G., and A. Risitano. "Thermographic methodology for rapid determination of the fatigue limit of materials and mechanical components." *International journal of fatigue* 22.1 (2000): 65-73.
8. Marco, Yann, Vincent Le Saux, L. Jégou, A. Launay, Leonell Serrano, Ida Raoult, and Sylvain Calloch. "Dissipation analysis in SFRP structural samples: Thermomechanical analysis and comparison to numerical simulations." *International Journal of Fatigue* 67 (2014): 142-150.
9. Flory, J. F. Cordage institute guidelines for marine grade nylon and polyester rope-making yarns. In *Oceans-San Diego, 2013* (pp. 1-6). IEEE. (2013, September).
10. ASTM D 6611, "Standard Test Method for Wet and Dry Yarn-on-Yarn Abrasion Resistance, ASTM, West Conshohocken, PA, 2000, re-approved 2007.
11. Marco, Y., Masquelier, I., Le Saux, V., & Charrier, P. Fast prediction of the wöhler curve from thermal measurements for a wide range of nr and sbr compounds. *Rubber Chemistry and Technology, 90*(3), 487-507. (2017).
12. Jegou, L., Marco, Y., Le Saux, V., & Calloch, S. Fast prediction of the Wöhler curve from heat build-up measurements on Short Fiber Reinforced Plastic. *International Journal of fatigue, 47*, 259-267. (2013).
13. V. Le Saux, Belenos users manual, Internal report. ENSTA Bretagne (2017).
14. International Organization for Standardization, "Fibre ropes for offshore station keeping – Polyester", Norme ISO 18692, 2007.
15. Ezanno, Anthony, Cédric Doudard, Sylvain Calloch, and Jean-Loup Heuzé. "A new approach to characterizing and modeling the high cycle fatigue properties of cast materials based on self-heating measurements under cyclic loadings." *International Journal of Fatigue* 47 (2013): 232-243.
16. Marco, Yann, Bertrand Huneau, Isaure Masquelier, Vincent Le Saux, and Pierre Charrier. "Prediction of fatigue properties of natural rubber based on the descriptions of the cracks population and of the dissipated energy." *Polymer Testing* 59 (2017): 67-74.
17. Hearle, J. W., Lomas, B., & Cooke, W. D. *Atlas of fibre fracture and damage to textiles*. Elsevier. (1998).

Methods for Structural Stress Determination according to EN 13445-3 Annex NA – Comparison with other Codes for Unfired Pressure Vessels

Ralf Trieglaff[1,*], *Jürgen* Rudolph[2], *Martin* Beckert[1], and *Daniel* Friers[3]

[1]TÜV NORD EnSys GmbH & Co. KG, 22525 Hamburg, Germany

[2]Framatome GmbH, 91052 Erlangen, Germany

[3]Krones AG, 93073 Neutraubling, Germany

The European Pressure Vessel Standard EN 13445 provides in its part 3 (Design) a simplified method (Clause 17) and a detailed method for fatigue assessment (Clause 18). Clause 18 "Detailed Assessment of Fatigue Life" is under revision within the framework of the European working group CEN/TC 54/WG 53 – Design methods with the aim of reaching a significant increase in user-friendliness and a clear guideline for the application.

This paper is focused on the new informative annex NA "Instructions for structural stress oriented finite elements analyses using brick and shell elements". As an essential amendment for the practical user, the determination of structural stress ranges for fatigue assessment of welds is further specified in this new annex. Different application methods for the determination of structural stresses are explained in connection with the requirements for finite element models and analyses.

This paper will give a short overview of the proposed approaches of structural stress determination in annex NA of the revised EN 13445-3. It will present the status of the approaches based on the results of fatigue analyses according to EN 13445-3 Clause 18 for different application examples.

For verification purposes, the results of the approaches proposed in EN 13445-3 are compared with the results of other pressure vessel design codes for nuclear and non-nuclear application.

1 Introduction

The "Recommendations for fatigue design of welded joints and components" by Professor A. Hobbacher (see [1] to [3]) opened new horizons in the determination of the hot-spot structural stress and the fatigue assessment of welded joints in different engineering disciplines in general. The adoption of the concept of structural stress in the standard EN 13445-3 started as early as in the 1990ies by the work of Professor Stephen Maddox from TWI.

Generally speaking, the European Pressure Vessel Standard EN 13445 provides in its part 3 (Design) a simplified method for fatigue assessment (Clause 17) and a detailed method of fatigue assessment (Clause 18) [4]. Clause 18 "Detailed Assessment of Fatigue Life" is currently under revision within the framework of the European working group CEN/TC 54/WG 53 – Design methods. An extensive overview of the baseline of revision was given in [5]. All amendments primarily aim at a significant increase in user friendliness and clear guidelines for application.

Three items of the revised clause 18 and related annexes are outstanding:

- Detailed guidelines for the fatigue assessment of welded pressure vessel components based on the structural hot-spot stress approach
- Detailed guidelines for the determination of the relevant stresses and stress ranges based on shell type and brick type finite element models
- Detailed guidelines for the cycle counting issue.

The basic rules for unwelded components essentially remain unchanged.

As an essential amendment for the practical user the determination of structural stress ranges for the fatigue assessment of welds is further detailed in the new informative annex NA "Instructions for determining structural hot-spot stress by finite element analysis using shell and brick elements". Here, different applicable methods for the determination of structural stresses are explained in connection with the requirements of the finite element models and analyses.

* Corresponding author: rtrieglaff@tuev-nord.de

As mentioned before the concept of structural stress was introduced into Clause 18 "Detailed Assessment of Fatigue Life" of the standard EN 13445-3 as early as 1990 by Professor Stephen Maddox from TWI. This concept was quite new for pressure vessel designers who usually performed fatigue analysis using fatigue curves based on the tests of smooth, unnotched and polished standard specimens in connection with some weld factor. In 2002, the standard EN 13445-3 was published with rules transposing the results of (component) fatigue tests on welded specimens performed in laboratory. Nevertheless, it paved the way to catch up with overall technical developments of the fatigue design of welds in other engineering fields. Besides fracture mechanics based crack propagation approaches the widely applied methods for the fatigue design of welded structures are (see [1] to [3]):

- Nominal stress approach
- Structural stress approach
- Notch stress approach.

2 Overview on the revised EN 13445-3, Clause 18 [4]

The nominal stress approach (e.g. design S-N curves of Eurocode 3 [6] and IIW Guidelines [1]-[3]) constituted the basis of the EN 13445 S-N curves for weld details. The design S-N curves are described by the equivalent stress range Δσ, and the fatigue life in cycles N. C is a constant dependent on the fatigue class and a slope of m = 3 was imposed when deriving component design curves up to $5 \cdot 10^6$ cycles (m = 5 up to $1 \cdot 10^8$ cycles for variable amplitude loading). The design curves refer to particular weld details, because the stress concentration effect of the weld detail is already included in the S-N curve. The assignment of the welding details to the design curves (fatigue classes) is based on (the equally revised) table 18-4 in [4].

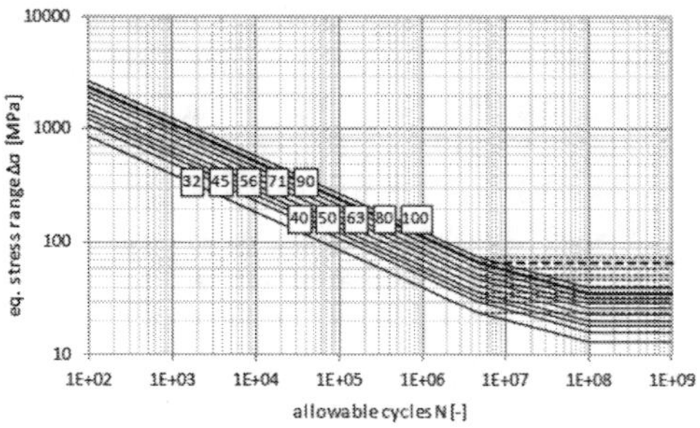

Figure 1: Fatigue design curves for welded components according to [4]

The application of the structural hot spot stress concept in EN 13445-3, Clause 18 [4] was discussed in detail in [5]. Detailed in Annex NA of EN 13445-3 [4] the determination of the structural hot-spot stress can be based on (see Fig. 2):

- Linear extrapolation of the stress on the surface (usually at distances l_1=0.4e and l_1+l_2=1.0e) in shell type or brick type finite element models
- Quadratic extrapolation of the stress on the surface usually at distances l_1=0.5e and l_1+l_2=1.5e and $l_1+l_2+l_3$=2.5e) in shell type or brick type finite element models
- Linearizing the stresses through the section thickness (path linearization in brick type finite element models)
- Geometric removal of the stress singularity by rounding the weld with an appropriate radius (CAB concept according to [9])
- Use of the stress at a position of 2.5 mm in front of the weld toe (Haibach concept [10]) usually in brick type finite element models.

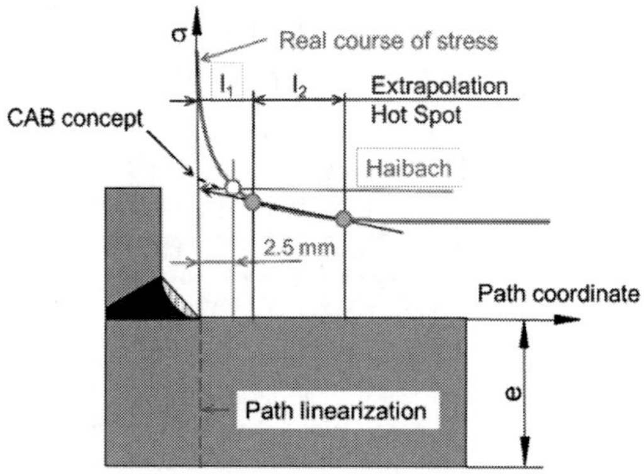

Figure 2: Methods for the determination of the structural hot-spot stress according to [4]

In the case, that the calculated linearized stress range $\Delta\sigma_{eq,mech}$ exceeds twice the material's yield strength, a correction factor should be applied to account for effects of material plasticity [4]. The following defines this factor k_e:

$$k_e = 1 + A_0\left(\frac{\Delta\sigma_{eq,mech}}{2R_{p0.2T}} - 1\right) \quad (1)$$

The factor A_0 depends on the materials tensile strength. For ferritic material with $500\ MPa \le R_m \le 800\ MPa$, A_0 is calculated by the following formula:

$$A_0 = 0.4 + \frac{R_m - 500MPa}{3000MPa} \quad (2)$$

The corrected stress range is given by the following formula:

$$\Delta\sigma_{eq} = k_e \cdot \Delta\sigma_{eq,mech} \quad (3)$$

k_e is the plasticity correction factor for stress due to mechanical loading

$\Delta\sigma_{eq,mech}$ is the equivalent stress range due to mechanical loading

$\Delta\sigma_{eq}$ is the equivalent stress range with plasticity correction (mechanical and thermal)

2.1 General comparison of the fatigue assessment with different codes

The conceptual basis for the fatigue evaluation of welds is:

- The factorial consideration of the weld seam
- The fatigue assessment based on strain-life fatigue curves (of unnotched and unwelded polished round or hourglass standard specimens)

In fatigue analyses, the fatigue strength reducing influences of weld seams are to be considered as a function of the weld seam treatment. Notice: Stress indices with regard of fatigue are part of the different codes. The background of these Fatigue Strength Reduction Factors (FSRF) are given in WRC 432 [15] and is a base for the codes ASME III/ VIII-2 (US), RCC-M (French) and KTA (German).

Table 1 shows a section of table 8.4-1 of the German Nuclear Code KTA 3201 [16], where several FSRF are given.

Table 1: Examples of FSRF in German Code KTA [16]

Piping products and joints	FSRF for pressure loading (K_1)
Straight pipe, remote from welds or other discontinuities	1.0
Butt girth welds between straight pipes or pipe and butt-welded item a) flush b) as-welded	 1.1 1.2
Longitudinal butt welds in straight pipe a) flush b) as-welded $s_c > 5$ mm c) as-welded $s_c \leq 5$ mm	 1.1 1.2 2.5
Curved pipe or elbows	1.0
Branch connections	2.0
Butt welding tees	4.0

In the European Code EN 12952-3 [18] for Water-tube boilers and auxiliary installations Fatigue Strength Reduction Factors are given for the categories "mild", "middle" and "strong" depending on the ultimate strength of the material and calculated allowable numbers of cycles.

Table 2: EN 12952-3 – Fatigue strength reduction factors for welds [18]

Tensile Strength *Rm*	Welding Class Middle fatigue reduction K2
MPa	C_{k2}
400	$1.6 \leq 0.21 \lg N_a + 0.79 \leq 2.1$
600	$1.9 \leq 0.40 \lg N_a + 0.60 \leq 3.1$
800	$2.1 \leq 0.56 \lg N_a + 0.44 \leq 4.0$
1000	$2.5 \leq 0.75 \lg N_a + 0.25 \leq 5.0$

2.2 Comparison of the Design Fatigue Curves between the Codes – KTA, ASME III, RCC-M, EN 12952-3 and EN 13445-3

In Figure 3 the design fatigue curves based on unnotched and unwelded polished round or hourglass standard specimens of the codes KTA [16], ASME III [17], RCC-M [19] and EN 12952-3 [18] reduced by the FSRF as defined in the codes are shown and compared with the component fatigue design curves as given in EN 13445-3 [4].

Note that this comparison is done under the retention of the different methodological bases as explained above. In the case of the component fatigue curves as given in EN 13445-3 [4] the stress range is proportional to the load of the underlying load controlled tests while in the case of the other mentioned design codes the stress range is plotted in terms of pseudoelastic stress resulting from a simple multiplication of the results of the underlying strain-controlled fatigue tests with Young's modulus and applying a constant FSRF. Comparison results are almost necessarily different, particularly in the high cycle fatigue range (HCF) where the component fatigue curve represents the decrease of the endurance limit due to the weld. Obviously, the application of a constant FSRF is not sufficient to provide the correction for the entire fatigue lifetime range from the low cycle fatigue range (LCF) to the high cycle fatigue range (HCF) and the endurance limit.

Figure 3 displays the comparison of the fatigue curves with middle fatigue strength reduction, in this case the FAT40 of EN 13445 with the factors K3 of EN12952 and FSRF=3.0 of ASME, RCC-M, KTA.

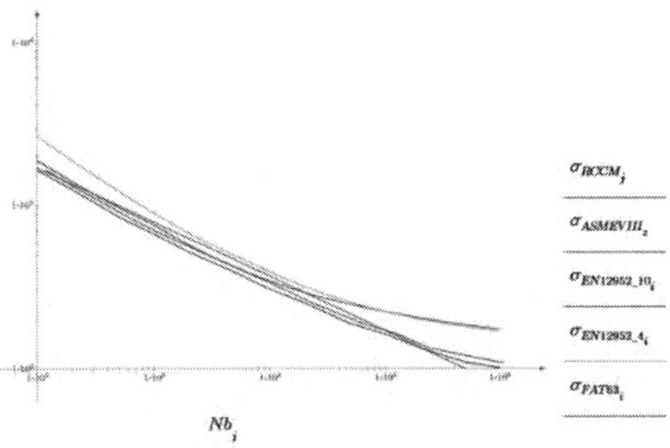

Figure 3: Fatigue curves for middle weld reduction

The comparison shows, that the FAT63-curve is conservative in the high cycle fatigue regime. Below $2 \cdot 10^5$ cycles the FAT63-curve should give results above the other curves.

2.3 Comparison of mechanical correction factors between the Codes – KTA, AD 2000 and EN 13445-3

The plasticity correction factor for stress due to mechanical loading k_e differs between EN 13445-3, AD 2000 [20] and KTA. For a limited material range (given tensile stress and steel alloy type), it is possible to compare the correction factors of the different design codes with each other.

The mechanical correction factors k_e for the French and American nuclear Code (RCC-M and ASME III) are calculated in a similar way as in the German KTA Code.

Figure 4 displays the correction factor depending on the ratio of mechanical stress range $\Delta\sigma_{eq,mech}$ and yield stress $R_{p0.2}$. The curves are only valid for ferritic material with a tensile strength between 500 MPa and 800 MPa.

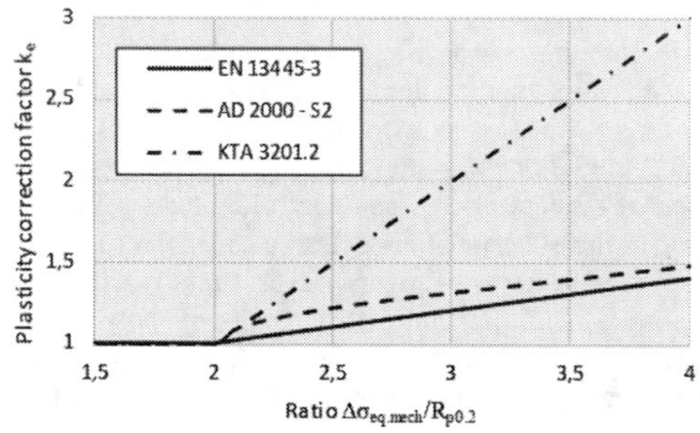

Figure 4: Plasticity correction factor for stress due to mechanical loading

As shown in Figure 4, the correction factor k_e need to be taken into consideration only when the mechanical stress range $\Delta\sigma_{eq,mech}$ exceeds twice the material's yield strength.

The correction factor k_e in the nuclear KTA code produces the most conservative values, especially for high stress/strength ratios.

3 Calculation examples for consolidation of EN 13445-3, clause 18 and annex NA [4]

3.1 Specification of Round Robin

Different calculation examples were analyzed with the target to compare the different proposed approaches of structural stress determination in annex NA of the revised EN 13445-3 based on the results of the fatigue analysis according to Clause 18 of EN 13445-3. It should be noted that two methods for linear and quadratic extrapolation were used for the hot-spot stress approach and the results of volume models with and without modeled welding overlap were compared. Originally, four examples were analyzed for structural stress determination. [14]

The following three calculation examples were analyzed:

(1) Flat head with cylindrical shell

(2) Central nozzle in elliptical head

(3) Nozzle in a cylindrical shell

The calculation example 1, flat head with cylindrical shell, refers to the fatigue tests described in [11]. Figure 5 shows the essential dimensions and cyclic loads for this calculation example 1.

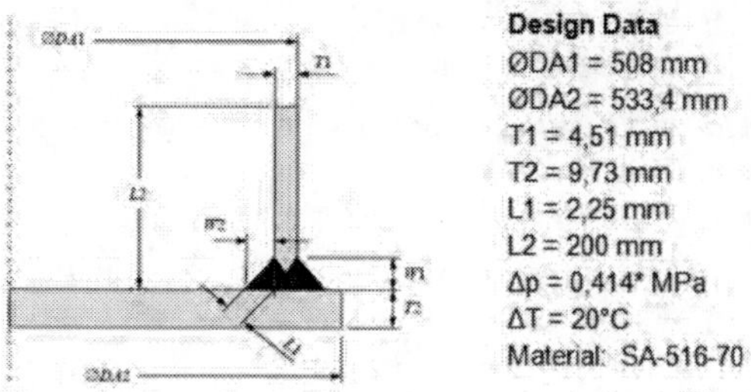

Figure 5: Example 1 – Flat head with cylindrical shell

The calculation example 2, central nozzle in elliptical head, and the calculation example 3, nozzle in a cylindrical shell, based on the document CEN/TC 54 N 1715 [12], where an application example of EN 13445-3, a vertical vessel with bracket supports, is given. Figure 6 shows the essential dimensions and cyclic loads for example 2 and Figure 7 shows the essential dimensions and cyclic loads for example 3.

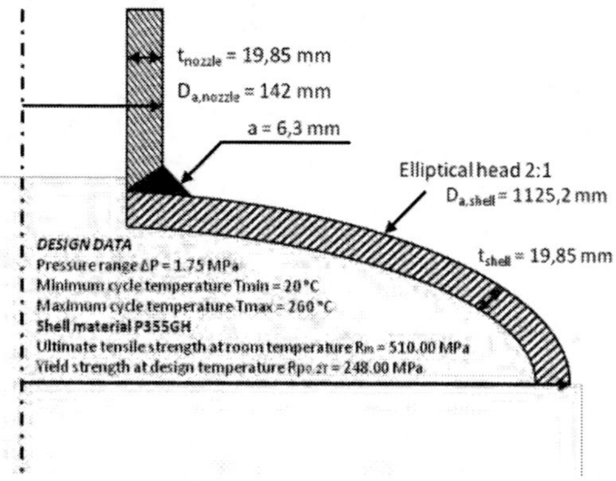

Figure 6: Example 2 – Central nozzle in elliptical head

The calculation examples 1 and 2 could be analyzed based on rotationally symmetrical FEA-models. Therefore, a calculation model has been added so that a 3D model is also investigated.

The example 3, a nozzle in a cylindrical shell, is based on the proceedings paper no. PVP2013-97568 [13]. Before, the example was used to demonstrate the simplified method for fatigue assessment of the European Pressure Vessel Standard EN 13445. Figure 7 shows the essential dimensions and cyclic loads for this example 3.

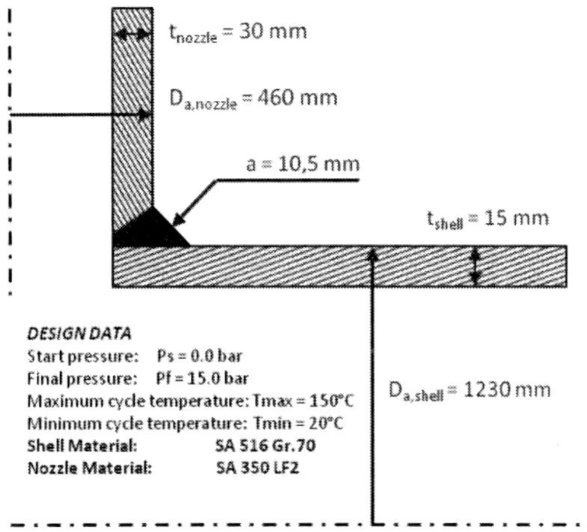

Figure 7: Example 3 – Nozzle in a cylindrical shell

3.2 Results of Round Robin

The different methods described in part 2 of the paper were applied by the round robin participants. The stress results are presented on a tabular basis for each example. Further, the resulting allowable load cycles were also analyzed according to EN 13445-3 Chapter 18 and presented in bar graphs.

The extrapolation method (linear and quadratic) uses the stress at the surface of the adjacent structure at different locations. It was not explicitly regulated which locations (distance from hot-spot point) have to be used for the stress extrapolation.

When using shell element formulation, the hot-spot point was defined for all models at the intersection of the shell geometry in the simulation model (equivalent to the mid-surface of the components). The fillet weld, which acts as an additional reinforcement of the butt weld, was only considered, when a volumetric (brick) element formulation was used.

The through-wall linearization is presented using different options of calculating the resulting structural stress. The first option (here named equivalent stress method) uses all structural stress components for calculating the equivalent structural stress. The second option uses only the structural stress normal to the hypothetical crack plane, as proposed by ASME VIII-2 [21].

Example 1: Weld at flat end of a cylindrical shell

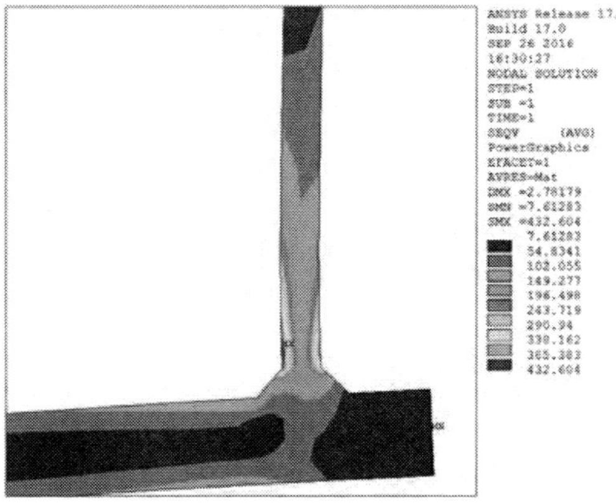

Figure 8: Local stress results (von Mises stress) at the weld of the flat head (with weld reinforcement)

In the Figure 8 above, the equivalent stress results for example 1 are presented. The highest stress occurs at the weld toe inside the vessel, close to the cylindrical shell.

Table 3: Results for different structural stress evaluation concepts

	Min. stress [MPa]	Max. Stress [MPa]	Median stress [MPa]	Max. scatter [MPa]
FE-model with shell element formulation				
Direct Method	434	436	435	2
Linear Extrapolation	515	519	517	4
Quadratic Extrapolation	520	521	521	1
Haibach concept (2.5 mm)	-	-	477	-
FE-model with volumetric element formulation				
Linear Extrapolation	450	452	451	2
Quadratic Extrapolation	446	453	450	7
Linearization *equivalent stress method*	311	353	332	42
Linearization *normal stress method*	488	496	492	8
Haibach concept (2.5 mm)	409	410	410	1
CAB concept	495	496	496	1

Table 3 shows that there is less scatter between the participants of the working group for all concepts except for the linearization using the equivalent stress method. The structural stress for this method is also very low compared to the other concepts.

Furthermore, there are only small differences between linear and quadratic extrapolation methods in the case of using shell or volumetric elements. In addition to that, various extrapolation schemes have been applied. The different distances 0.5e/1.5e or 0.5e/1.5e/2.5e and 0.4e/1.0e or 0.4e/0.9e/1.4e to the hotspot for deriving stresses have minor influence on the stress results.

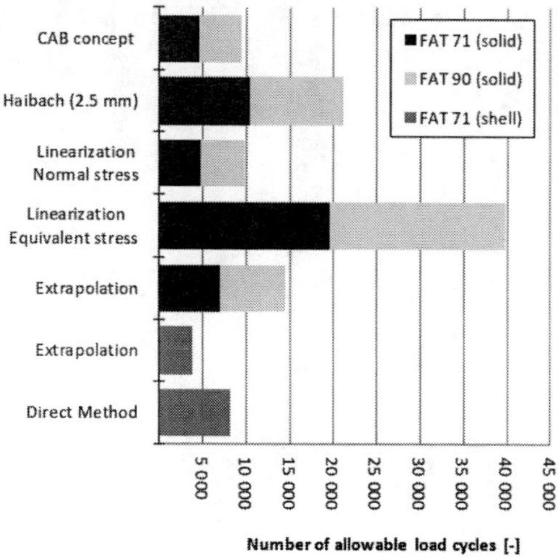

Figure 9: Comparison of allowable load cycles

It should be noted that in the bar graphs the average stress value of the linear and quadratic extrapolation using the same hotspot has been considered in the calculation of the allowable load cycles. The analogous procedure has been executed for volumetric elements.

It can be seen that the equivalent stress method estimates a significantly higher service life than the other methods because of the low stress result. The Haibach method produces a slightly higher number of allowable cycles than for example the linearization (equivalent stress method and normal stress method) or the CAB-concept. Using FAT 71 for shell and solid elements produces comparable results.

Example 2: Nozzle weld in an elliptical head

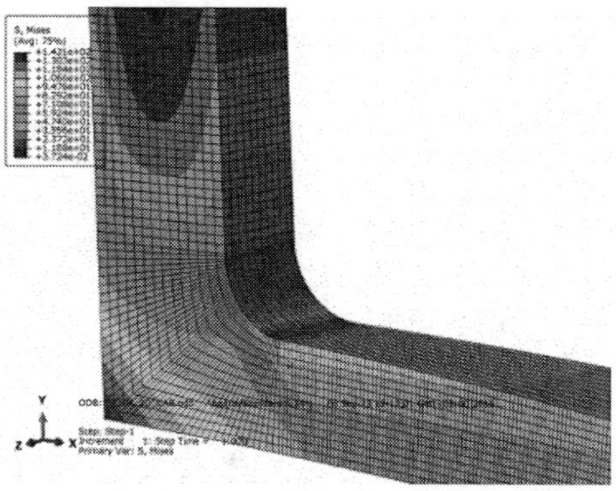

Figure 10: Local equivalent stress results (von Mises) at the nozzle weld (model for CAB-concept)

Table 4: Results for different structural stress evaluation concepts

	Min. stress [MPa]	Max. Stress [MPa]	Median stress [MPa]	Max. scatter [MPa]
FE-model with shell element formulation				
Direct Method	100	101	101	1
Linear Extrapolation	101	108	105	7
Quadratic Extrapolation	102	109	106	7
Haibach concept (2.5 mm)	-	-	100	-
FE-model with volumetric element formulation				
Linear Extrapolation	105	113	109	8
Quadratic Extrapolation	105	113	109	8
Linearization *equivalent stress method*	66	99	83	33
Linearization *normal stress method*	108	110	109	2
Haibach concept (2.5 mm)	107	108	108	1
CAB concept	115	118	117	3

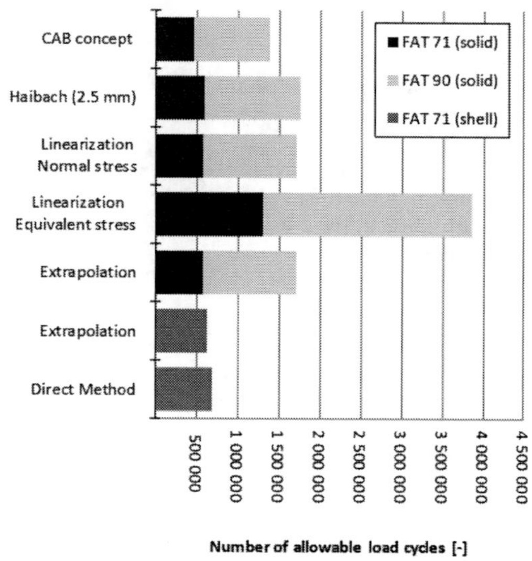

Figure 11: Comparison of allowable load cycles

Regarding the scatter of the methods, the same behavior of the methods as in example 1 can be observed. In this example the observed discrepancies in stress results by choosing different hotspots for shell models is less pronounced than in example 1. Again, the equivalent stress method of the linearization estimates more than double the number of load cycles than all other structural stress methods. The CAB concept is more conservative compared to all other methods using volumetric elements. The Haibach method without the modeled weld reinforcement delivers nearly the same number of allowable load cycles compared to a model with weld reinforcement.

It should be noted that in the bar graphs the average stress value of the linear and quadratic extrapolation using the same hotspot has been considered in the calculation of the service lives. The analogous procedure has been executed for solid elements.

The use of FAT 71 in connection with volumetric elements would result in a comparable number of load cycles for shell models and contribute to the expectation as stated in example 1 (see Fig. 11).

Example 3: Nozzle weld in a cylindrical shell

The FE-model for the nozzle weld in a cylindrical shell is the only example using a 3D modeling space instead of an axisymmetric model.

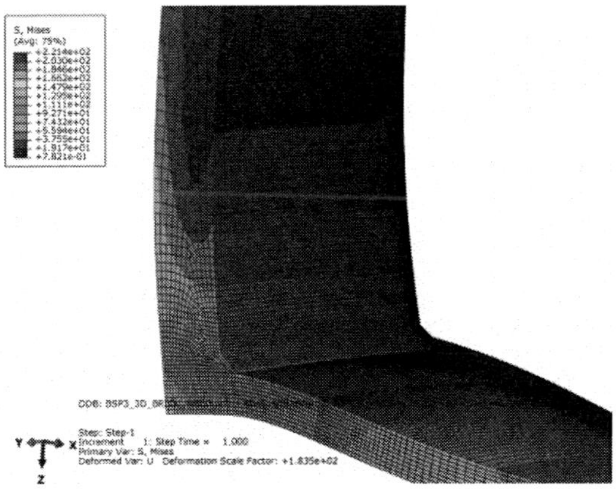

Figure 12: Local equivalent stress results (von Mises) and deformation around the nozzle weld (with modeled weld)

In Fig. 12 the deformation, mesh density and equivalent stress (von Mises) are illustrated. The highest stress level occurs as expected at the weld toe.

Table 5: Structural stress results for the nozzle weld

	Min. stress [MPa]	Max. stress [MPa]	Median stress [MPa]	Max. scatter [MPa]
FE-model with shell element formulation				
Direct Method	-	-	145	-
Linear Extrapolation	176	177	177	1
Quadratic Extrapolation	178	178	178	0
FE-model with volumetric element formulation				
Linear Extrapolation	137	138	138	1
Quadratic Extrapolation	137	139	138	2
Linearization *equivalent stress method*	126	128	127	2
Linearization *normal stress method*	131	134	131	3
Haibach concept (2.5 mm)	143	144	144	1
CAB concept	158	159	159	1

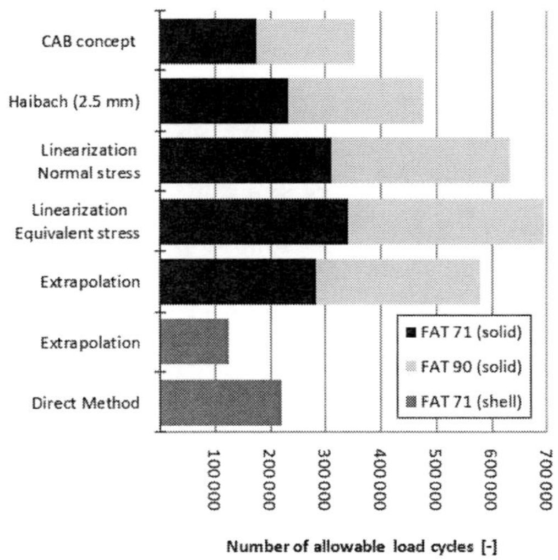

Figure 13: Comparison of the number of allowable load cycles for FAT-71 and FAT-90

The results in Table 5 show very little scatter for all methods used. Even for the through-wall linearization (equivalent stress method), the results are in-line with the other methods.

Remarkable is also the significant difference when using shell or volumetric element formulation.

As the level of stress is lower, when using a volumetric element formulation, the number of allowable load cycles differs a lot between shell and volumetric element formulation.

3.3 Comparison of the allowable cycles with different design codes

For comparison of the calculated allowable number of cycles with different design codes, the results of the volumetric FE-models were used. The structural stresses were estimated based on the quadratic extrapolation method. For the assessment procedure of the codes ASME III, ASME VIII-2 [21], RCC-M and KTA, the results of the CAB concept were used as notch stresses.

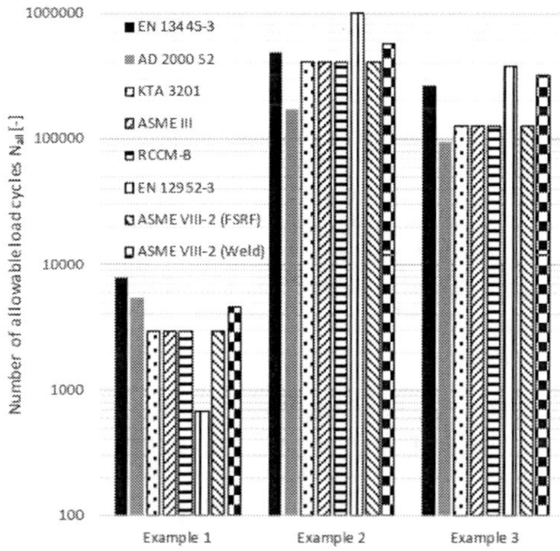

Figure 14: Allowable load cycles for the analyzed examples as a comparison between different codes

The comparison of the results show no fundamental deviations to the other assessment procedures. The results of the weld assessment concept in ASME VIII-2 [21] are close the results in EN 13445-3. The results of EN 12952-3 seems to be conservative in the LCF-region and non-conservative in the HCF-region.

The results according to AD 2000 are comparatively low for example 2 and 3, because of the classification of the weld geometry in class K2. For example 1, the weld detail was classified in class K1. For the analysis according to EN 13445, all weld details were classified in FAT71.

4 Summary and Conclusions

In the following, the results for the calculation examples with regard to the concepts for the determination of the

https://doi.org/10.1051/matecconf/201816510003

structural stresses are evaluated and recommendations are derived.

Hot-spot approach

The quadratic extrapolation procedure delivers only slightly higher stress values compared to the linear extrapolation. The different methods based on different distances between the surface stress extrapolation points have only negligible influence on the stress result. The use of shell elements instead of brick elements leads in general to higher stress values.

If the fatigue evaluation is based on the weld design fatigue curve FAT90 in general instead of the recommended FAT Class in table 18.4 of EN 13445-3 [4] for brick elements, larger differences for the number of allowable cycles between shell and brick elements will result. As a results, the only use of the FAT Class acc. to 18.4 of EN 13445-3 [4] is proposed.

Through-wall linearization

The results for all calculation examples show the lowest stress level compared to all other concepts when using the equivalent linearized stresses. In addition, the deviation between the results of the participants of the Round robin is the biggest. This appears to be due to the mesh dependency of the results. Therefore, a modification of this method was investigated. The modification uses the largest stress perpendicular to the expected crack plane, as proposed by ASME VIII-2 [21]. For the three calculation examples, the modification shows different success. Further investigations are necessary.

Haibach-concept

The calculated structural stresses of the Haibach-concept for the calculation examples 2 and 3 are in the same level as the Hot-spot-approach. For the example 1, this concept underestimated the structural stress. The reason is the thin shell thickness of 4.51 mm compared with the evaluation distance of 2.5 mm from the weld toe, where the stress has to be taken. For thin-walled structures, the result may be non-conservative. Therefore, a different method is proposed in the preliminary annex NA of EN 13445 [4]. The Haibach stress is located at 2.0 mm from the weld toe and for a smaller wall thickness than 8 mm, the stress is taken from the point 0,25*wall thickness.

CAB-concept

The calculated structural stresses of the CAB-concept have a higher stress level than all other methods for all calculation examples. This concept shows for all applications stable results regardless of different grid sizes in the FE-models of the participants of the round robin test. For mechanical loads, full penetration welds and testing group 1 or 2, the stresses estimated by the CA-concept can be reduced by a correction factor of $f_{CAB} = 0.95$ [4].

The comparison with other codes in section 3.3 of the paper shows, that the proposed calculation procedure in EN 13445-3 lead to results closed to the results of other well-established codes.

References

[1] Hobbacher, A.: *Recommendations for fatigue design of welded joints and components.* IIW document IIW-1823-07 ex XII-2151r4-07/XV-1254r4-07, 2008.

[2] Hobbacher, A.: *Recommendations for Fatigue Design of welded joints and components.* International Institute of Welding document IIW document IIW-1823-07. Welding Research Council Bulletin 520, New York (2009)

[3] Hobbacher, A.: *Recommendations for Fatigue Design of Welded Joints and Components.* Second Edition. IIW document IIW-2259-15 ex XIII-2460-13/XV-1440-13, revision of XIII-2151r4-07/XV-1254r4-07, Springer Cham Heidelberg New York Dordrecht London, 2016

[4] CEN/TC 54; EN 13445-3:2014/prA3:2015; 2015-10.
Clause 18: „Detailed assessment of fatigue life"
Annex NA: „Instructions for determining structural hot-spot stress by finite element analysis using shell and brick elements"
Annex NB: „Cycle counting and determination of equivalent stress range"

[5] Rudolph, J.; Baylac, G.; Wilhelm, P.; Wintle, J.; Buennagel, E.: *Recent Amendments of EN 13445-3, Clause 18 and Related Annexes: Detailed Assessment of Fatigue Life.* Proceedings of PVP2014. Paper No. PVP2014-28129. ASME Pressure Vessels & Piping Conference. July 20-24, 2014, Anaheim, California, USA

[6] Eurocode 3 - Design of steel structures - prEN 1993.
European Committee for Standardization, Brussels, 1992

[7] Maddox, S. J.: *Key developments in the fatigue design of welded constructions.* 2003 IIW Portvin Lecture, Proc. IIW Int. Conf. on „Welded Construction for Urban Infrastructure", ISIM, Timisoara, Romania, 2003.

[8] Niemi, E.; Fricke, W.; Maddox, S. J.: Fatigue Analysis of welded Components – Designer´s guide to the structural hot-spot stress approach. IIW-1430-00, first published 2006, reprinted 2011

[9] J. Rudolph, M. Rauth, E. Weiß: *Modellgebundener Strukturspannungsnachweis für zyklisch beanspruchte Zylinder-Stutzen-Verbindungen ohne Schweißnahtnachbearbeitung.* TU-Zeitschrift 43 (2002), No. 6, pp. 19/26

[10] Haibach, E.: *Die Schwingfestigkeit von Schweißverbindungen aus der Sicht einer örtlichen Beanspruchungsmessung. (The fatigue strength of welded joints considered based on local stress measurement).* LBF Report FB77, Fraunhofer-Inst. für Betriebsfestigkeit Darmstadt Germany (1968)

[11] Hinnant, C.: *Fatigue Testing and Life Estimates of Welded Flat Head Pressure Vessel Joints.* Proceedings of PVP2006. Paper No. PVP2006-93967. ASME Pressure Vessels & Piping Conference July 23- 27, 2006, Vancouver, BC, Canada

[12] Document CEN/TC 54 N 1715: Example of Application of EN 13445 – Vertical Vessel with Bracket Supports (Cen/Tr 13445-102), 10 February 2014

[13] Lidonnici, F.: *The Simplified Method for Fatigue Assessment of the European Pressure Vessel* Standard EN 13445, Proceedings of PVP2013. Paper No. PVP2013-97568, ASME Pressure Vessels & Piping Conference July 14-18, 2013, Paris, France

[14] Trieglaff, R.; Rudolph, J.; Beckert, M.; Hauser, F.: *Comparison of Methods for Structural Stress Determination according to EN13445-3 Annex NA* Proceedings of PVP2017. Paper No. PVP2017-66112. ASME Pressure Vessels & Piping Conference. July 16-20, 2017, Hawaii, USA

[15] Welding Research Council (WRC) bulletin 432 Fatigue strength reduction and stress concentration factors
for welds in pressure vessels and piping, June 1998

[16] KTA Program of Standards, Standard No. 3201.2, *Components of the Reactor Coolant Pressure Boundary of Light Water Reactors, Part 2: Design and Analysis,*
Issue November 2013

[17] ASME Boiler and Pressure Vessel Code Section III, *Rules for Construction of Nuclear Facility Components*, Div. 1
The American Society of Mechanical Engineers, New York, 2017 edition

[18] EN 12952-3*Water-tube boilers and auxiliary installations – Part 3: Design and calculation for pressure parts of the boiler*; German version EN 12952-3:2011

[19] RCC-M, *Design and Construction Rules for Mechanical Components of PWR Nuclear Islands*, afcen, 2012 edition

[20] AD 2000 Merkblatt S2 *Analysis for cyclic loading* German version TÜVIS Prüfgrundlagen, Edition 07.2012

[21] ASME Boiler and Pressure Vessel Code Section VIII-2, *Rules for Construction of Pressure Vessels Division 2 - Alternative Rules*
The American Society of Mechanical Engineers, New York, 2017 edition

Fatigue Life Prediction of Self-Piercing Rivet Joints Between Magnesium and Aluminum Alloys

Hong-Tae Kang[1,*] and *Sai* Boorgu[1]

[1]The University of Michigan-Dearborn, 4901 Evergreen Road, Dearborn, MI, USA

Abstract. Various light materials including aluminum alloys and magnesium alloys are being used to reduce the weight of vehicle structures. Joining of dissimilar materials is always a challenging task to construct a solid structure. Self-piercing rivet (SPR) joint is one of various joining methods for dissimilar materials. Front shock tower structures were constructed with magnesium alloy (AM60) joined to aluminum alloy (Al6082) by SPR joints. To evaluate the durability performance of the SPR joints in the structures, fatigue tests of the front shock tower structures were conducted with constant amplitude loadings. Furthermore, this study investigated fatigue life prediction method of SPR joints and compared the fatigue life prediction results with that of experimental results. For fatigue life prediction of the SPR joints in the front shock tower structures, lap-shear and cross-tension specimens of SPR joint were constructed and tested to characterize the fatigue properties of the SPR joint. Then, the SPR joint was represented with area contact method (ACM) in finite element (FE) models. The load-life curves of the lap-shear and cross-tension specimens were converted to a structural stress-life (S-N) curve of the SPR joints. The S-N curve was used to predict fatigue life of SPR joints in the front shock tower structures. The test results and the prediction results were well correlated.

1 Introduction

Automotive industries continue to develop fuel-efficient vehicles by reducing their weight. Magnesium (Mg) and Aluminum (Al) alloys are a viable option for the weight reduction of vehicle structures due to their low density. Currently automotive companies are producing various vehicle structures with aluminum alloys. Many efforts are also focusing on developing vehicle structures and components with magnesium alloys. Furthermore, some researchers demonstrated for constructing vehicle structures and components with dissimilar lightweight materials to reduce the weight of vehicles.

Joining technology plays key role to construct vehicle structures with dissimilar light materials including aluminum alloys and magnesium alloys. The self-piercing rivet (SPR) is a widely employed mechanical joining technology ductile materials [7]. This joining method is similar to the traditional riveting but without a pre-drilling procedure. Thus, SPR technology can achieve rapid assembly time and high joint integrity. Many vehicle manufacturers replaced the traditional electrical resistance spot welds to SPR joints to join Al/Al and Al/steel. Some researchers also demonstrated the employment of SPR for joining magnesium alloys with aluminum alloys.

In this research front shock tower structures of a passenger vehicle were constructed with magnesium alloy (AM60), aluminum alloys (Al6082) and sheet steel (HSLA350). The main part of the shock tower is made of AM60 and the upper rail and the lower rail of the structures are made of HSLA350 and Al6082, respectively. AM60 and HSLA350 are joined with adaptable insert weld (AIW) that is a hybrid joining technique of electrical resistance spot welding and riveting. AM60 and Al6082 are joined with SPR.

Durability tests of the constructed structures were conducted to evaluate the structural performance of the joints and the base material of the shock tower components. The loading direction applied for the durability tests was determined to evaluate a specific joint or base material. X-direction loading as shown in Figure 1 was intended to assess durability of SPR joints between AM60 and Al6082. To evaluate AIW joints, Y-direction loading was applied to the loading fixture as shown in Figure 1. The test results and prediction activities were reported in [1]. The fatigue performance for the base material of AM60 was assessed by applying load in Z-direction as shown in Figure 1. The test results and fatigue life prediction activities were reported in [2].

This study reports a fatigue life prediction method for SPR joints between AM60 and Al6082 and validate the method using front shock tower structures. To achieve the goal, lap-shear (LS) and cross-tension (CT) specimens were fabricated and tested with quasi-static loading and constant amplitude cyclic loading. The fatigue test results

* Corresponding author: htkang@umich.edu

of LS and CT specimens were correlated with structural stress amplitude of SPR joints. This relationship between the structural stress and fatigue life becomes the joint fatigue property of SPR joints for AM60 and Al6082. Then, this joint fatigue property is used to predicted fatigue life of SPR joints in the front shock tower structures. Finally, the prediction results and test results were compared to ensure the effectiveness of the method used in this study.

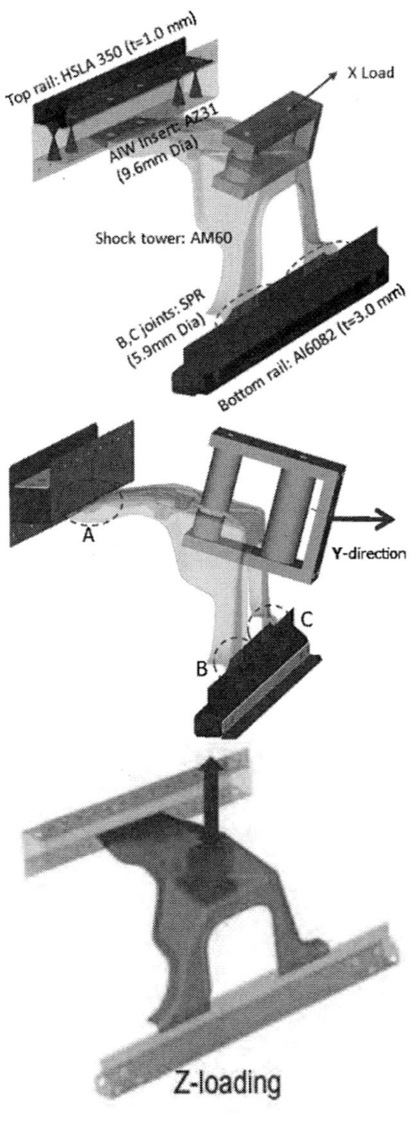

Fig. 1. Fatigue loading directions for the front shock tower structures.

2 Specimen Geometry and Dimensions

Specimens of lap-shear and cross-tension were used to characterize the performance of SPR joints in this study. The upper sheet was made of AM60 with thickness of 2.0 mm and lower sheet was Al6082 of 3.0 mm thick. The dimensions of the specimens are shown in Figure 2. The cross section of a typical SPR joint between AM60 and Al6082 is shown in Figure 3.

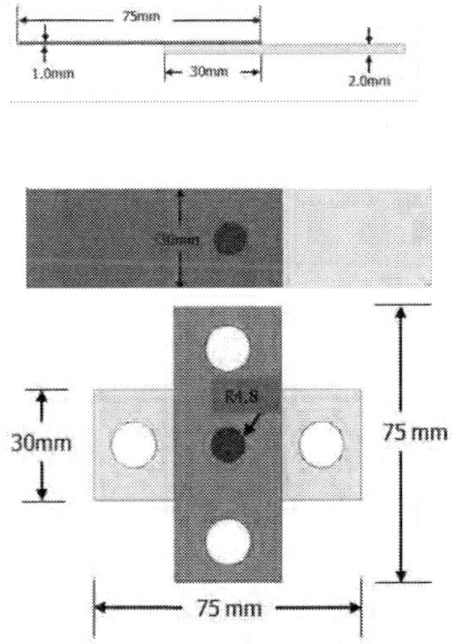

Fig. 2. Specimen geometry and dimension for the lap-shear and cross-tension specimen of SPR joint.

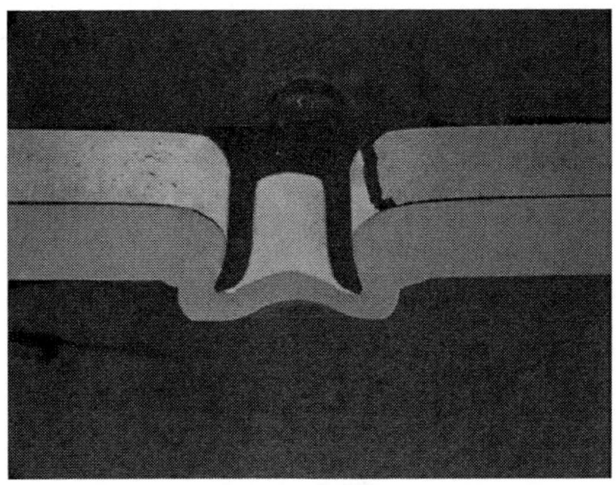

Fig. 3. A cross section of SPR joint between AM60 and Al6082.

3 Fatigue Test Results of Coupons

Fatigue tests for the specimens were conducted with constant amplitude loads at a load ratio ($R = (P_{min}/P_{max})$ of 0.1. These tests were performed at six different load levels for lap-shear specimens with two repetitions at each load level. For cross tension specimens fatigue tests were conducted at four different load levels with three repetitions at each load level. The load ranges for fatigue testing of specimens were determined from quasi-static tests of the specimens so that the fatigue life ranges between 1×10^3 and 5×10^6 cycles. The definition of fatigue failure was a complete separation of the specimen in two pieces. The typical failure mode observed in fatigue tests was cracked at the cast AM60 in both the specimen geometries as shown in Figure 6. For all these tests the runout was defined at 5×10^6 cycles. The applied maximum load versus the cycle to failure of the SPR joint specimens are plotted in Figure 4. The runouts are presented as hollow circle and rectangle in the Figure 4.

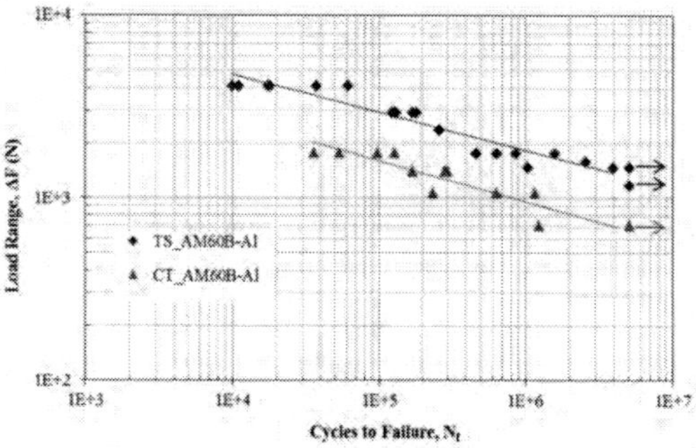

Fig. 4. Load range versus cycles to failure for the SPR specimens.

4 Finite Element Representation of LS and CP Specimens

The bottom sheet (Al6082) was represented with shell elements along the mid plane of the sheet. The element size of the bottom sheet was 5 mm, which is generally accepted element size for vehicle structures in automotive industries. The lower sheet (AM60) was modeled using three dimensional 2nd-order tetrahedral elements since AM60 in the shock tower structures is a casted component. Three dimensional 2nd-order tetrahedral elements are an easy and accurate way of representing the casting features in the shock tower structures. One strip of the LS and CP specimens was constrained in all-degrees of freedom and the other strip was constrained all-degrees of freedom except the loading direction of the specimen as constrain conditions were applied during the specimen testing. Figure 5 shows meshed finite element (FE) models of LS and CP specimens used in this study.

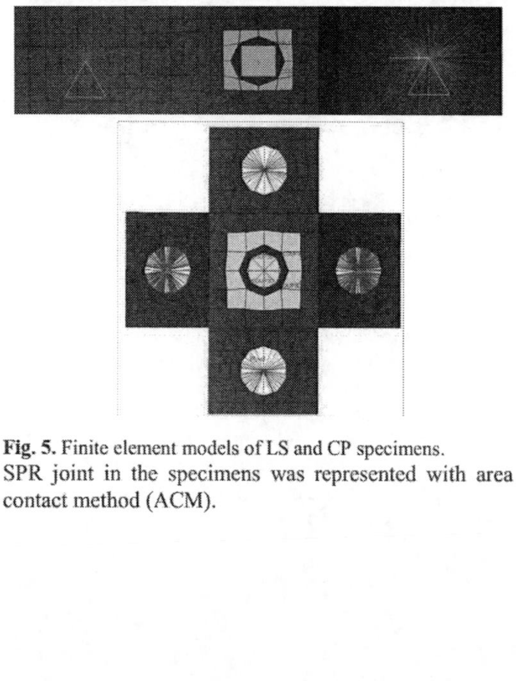

Fig. 5. Finite element models of LS and CP specimens.

SPR joint in the specimens was represented with area contact method (ACM).

In this method the joint is represented using a solid hexahedral element with distributed coupling elements. The distributed coupling elements connect the nodes of the solid hexahedral element and the nodes of the top and bottom sheets as shown in Figure 5. The material properties of AM60 are E=44 GPa, υ=0.3, and ρ=1.6E-9 tonne/mm³. The material properties of Al6082 are E= 70 GPa, υ=0.33, and ρ=1.6E-9 tonne/mm³.

5 Structural Stress Calculation

Structural stress is calculated from the forces and moments acting on the joint using Rupp and co-workers' approach [3]. The mechanics around SPR joint is not exactly the same with that of spot welded joint. However, the failure modes of SPR and spot welded joint are very similar. Thus, this study used the same structural stress equations developed for spot welded joints. This approach enables to avoid using fine meshed finite element model of SPR joints in vehicle body structures. This capability is so important in FE modeling of vehicle body structures that have many joints. Fine meshed representation for an SPR joint is only possible for a specimen level not for a structural level.

The stresses/strains obtained from FE analysis at the joints are not representing realistic ones because the joints are coarsely meshed. Thus, structural stresses are calculated at the joints using force and moment information less sensitive to the mesh density variations. The structural stress is, then used for constructing the Stress-Life (S-N) curve that defines fatigue property of SPR joints. The structural stress equations used in this study is briefly summarized below in Figure 6 [4].

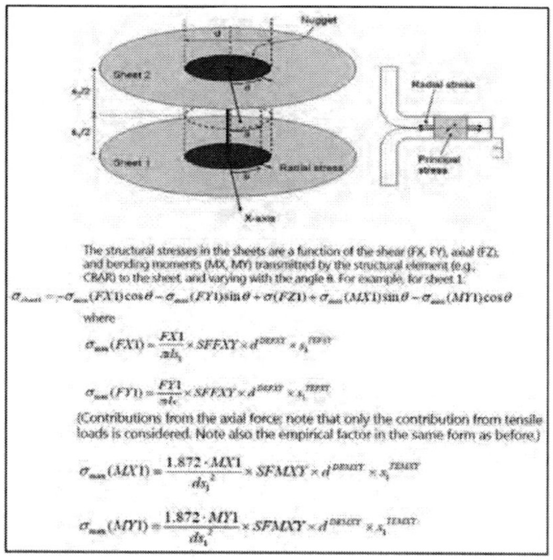

Fig. 6. Schematic drawing of spot weld presentation and structural stress equations [4].

To calculate structural stress, it is necessary to obtain the shear force components (FX1 and FY1), axial component

(FZ1), and bending moment components (MX1 and MY1). It also requires empirical factors such as SFFXY, DEFXY, TEFXY, SFMXY, DEMXY, and TEMXY. These factors play an important role in collapsing structural stresses for different dimensions of sheet thickness, joint diameters and different joint geometries. These parameters can be determined by nonlinear optimization described in [5]. To construct S-N curve of SPR joints in this study, LS and CP specimens were fabricated and tested as described previous section. Also FE representations for those specimens were described. At each fatigue test load level for LS and CP specimens, forces and moments were obtained at the joint from FE analysis. The structural stress equations shown above were used to calculate the structural stress corresponding to the specific load level applied for fatigue testing. These structural stresses were then used to plot the structural stress range versus cycles to failure curve on a log-log scale as shown in Figure 7.

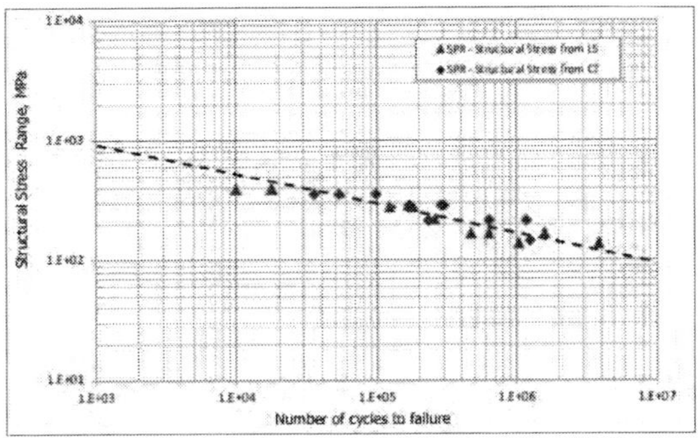

Fig. 7. S-N curve using structural stress ranges for LS and CP specimens.

6 Testing of Front Shock Tower Structure

Three quasi-static tests were conducted for the front shock tower structure subjected to X-direction loading as shown in Figure 1. The shock tower was constrained at the top and bottom rails by using fixtures and bolts. The load is applied using a jig that pulls the shock tower in X-direction until failure is observed. During this ultimate strength testing the failure occurred at SPR and the average ultimate strength was 9,700 N. The failure was occurred at AM60 near a SPR joint location. The failure location of the ultimate strength testing is shown in Figure 8. This ultimate strength was then used to estimate the fatigue test load such that the result of fatigue life ranges between 1E3 to 1E6 cycles. Three load amplitude ranges were applied to investigate durability performance of the structures. A series of nine structures were tested in this study. As shown in Figure 8, all structures tested were failed at the SPR joint on left leg of the shock tower structure. Failure of the structure was defined as any visible crack is observed in the structure.

Fig. 8. Failure locations ofSPR joints subjected to quasi-static and fatigue loading.

7 Fatigue Life Prediction of Joints in Shock Tower

Before testing of the front chock tower structures, FE model of the structure was developed and analyzed under the boundary and loading conditions to identify the critical locations. The upper rail (HSLA350) and the lower rail (Al6082) were represented with shell elements and the shock tower (AM60) was represented with 2nd-order tetrahedral elements. The joints connecting the upper rail and the lower rail were presented with ACM as described in previous section.

The fatigue life prediction of SPR joints in the front shock tower structures was performed and compared with that of experimental results. The S-N curve obtained from LS and CP specimens of the SPR joint was used to predict fatigue life of SPR joints in the front shock tower structures. The FE prediction identified correctly the failure location of the SPR joint and fatigue life prediction results are within factor of 3 lines sown in Figure 9. The solid line represents the perfect correlation between the test results and prediction results. The upper and lower dotted lines represent the band for the factor of 3.

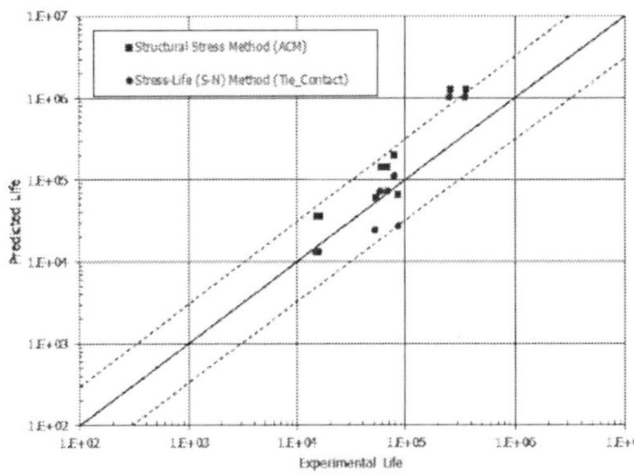

Fig. 9. Comparison between experimental results and prediction results for fatigue life of SPR joints in the front shock tower structures.

8 Summary and Conclusions

Lap shear and cross tension specimens were used to characterize the fatigue performance of SPR joints between AM60 and Al6082. Based on the observation of the failure modes of the joints, it was assumed that the mechanics at the SPR joint are similar to those of the

 https://doi.org/10.1051/matecconf/201816510004

electrical resistance spot weld joint. Therefore, the same structural stress equations for spot welded joints were used for SPR joints between AM60 and Al6082. Fatigue test results of LS and CP specimen configurations were correlated well with a log linear function after the applied load ranges were converted to structural stress ranges. The S-N curve based on the structural stress ranges was then used to predict the fatigue life of SPR joints in the front shock tower structures. Fatigue test results of the front shock tower structures and their prediction results were compared. From this study, the following conclusions are made:

1. While the fatigue test results in load range versus life of LS and CT specimens were quite different, these differences are minimized after the test results were presented with structural stress range versus life for the LS and CP specimens.
2. Structural stress equations and methodology developed for electrical resistance spot welded joints are still applicable for SPR joints between AM60 and Al6082.

The authors would like to express thanks to A.K. Khosrovaneh (GM), X. Su (Ford), M. Guo (FCA), C. Jiang (AET), and Y. Lee (FCA) for their valuable helps for the data and discussions.

References

1. H. Kang, Z. Li, A.K. Khosrovaneh, X. Su, M. Guo, Y. Lee, and C. Jiang, *SAE International Journal of Materials and Manufacturing*, 8(3): 789-794, 2015.
2. H. Kang, A.K. Khosrovaneh, X. Su, M. Guo, Y. Lee, and C. Jiang, 73rd IMA Conference, May 15-17, 2016, Rome, Italy.
3. Rupp, A., Storzel, K., and Grubisic, V., *SAE technical report No. 950711*, Detroit, Michigan, 1995.
4. nCode, DesighLife User Manual, 2017.
5. Kang, H., Wu, X., Khosrovaneh, A.K., Li, Z., *Fatigue and Fracture Test Planning, Test Data Acquisitions and Analysis*, 2016, STP1598, pp. 198-211.

https://doi.org/10.1051/matecconf/201816510005

Significance of Fatigue Damage from Overload Trucks on Bridge Load Rating

Bora Jang[1,*], and *Jamshid* Mohammadi[2]

[1]PhD Candidate, Department of Civil, Architectural and Environmental Engineering, Illinois Institute of Technology, Chicago, Illinois USA
[2]Professor, Department of Civil, Architectural and Environmental Engineering, Illinois Institute of Technology, Chicago, Illinois USA

Abstract. Overload trucks constitute a major portion of truck load populations on highways. Specific to bridges, the increase in the frequency of overloads may trigger shortening of service life of critical components because of fatigue damage, considering the fact that fatigue damage accelerates at higher loads. This is especially important to older bridges that have been designed for loads lower than the standard for current highway bridge design in the United States. As the number of requests for special permits to use bridges by overload trucks increases, the accumulated damage may substantially reduce the service life of bridges resulting in costly repair and replacements. In this paper, the significance of overloads in causing fatigue damage to bridges is discussed. Samples of truck load data have been compiled and used to determine the contribution of overloads to fatigue damage for steel girder bridges made up of welded cover plates. The paper discusses using the expected fatigue life expended (FLE) as an index to determine the impact of overloads on bridge fatigue damage. The FLE is then determined to be an important factor that can be used to modify bridge rating equations as a way of controlling the number of overload permits for a given bridge.

1 Introduction

1.1 Truck load population and overloads

The intensity and frequency of load occurrences in the truck load population vary to a great extent for a given vehicle class. Load populations mostly exhibit an inconsistent pattern often with two or more distinct peaks. Figure 1 shows the distribution of truck load data, designated as the gross vehicle weight (GVW), for one data sample compiled by weigh-in-motion (WIM) system in Illinois (by the Illinois Department of Transportation (IDOT) from June, 2014 to May, 2015). As seen in Figure 1, the distribution appears with two peaks in the entire truck load population. That is because of the variety of loads in the population. The appearance of two peaks may be due to a combination of loaded and empty trucks in most cases. In addition, the population also contains overload frequencies that may be significant in some cases resulting in multiple peaks in the population. It is noted that the term overload, as used here, refers to truck weights in excess of the 356-kN (80-kip) limit. Studies on the effect of truck loads on highway systems have shown that truckload spectra often contain occurrences of overloads [1-3]. In certain cases, overloads are rather frequent; and as such, they may significantly affect the service life of bridges [1,4]. Using the weigh-in-motion (WIM) system, truck load data gathered in several stations in Illinois and Michigan indicate that the percentage of overloads can be as high as 10% of the total truck load population. It is rather large and can be very critical for certain bridges with critical components prone to fatigue damage. A traffic volume for vehicle class 9 in 5-axle truck category is considered as a fatigue loading, since this type of truck appears to make up a significant portion of the entire truck load population.

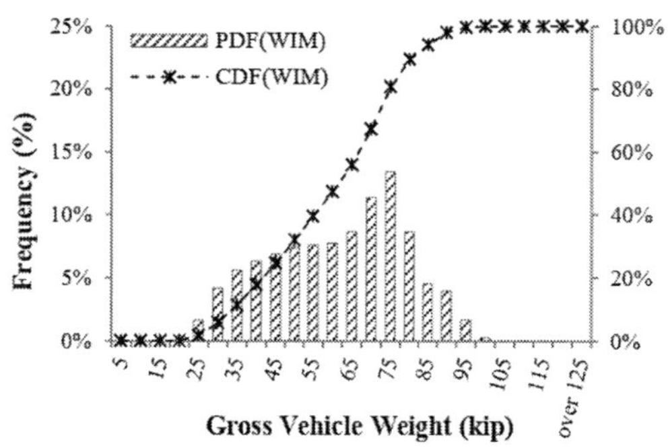

Fig. 1. Frequency distribution of gross vehicle weight for vehicle class 9 in 5-axle truck from WIM data.

1.2 Fatigue damage potentials

* Corresponding author: bjang5@hawk.iit.edu

As it is known, fatigue damage initiates cracks within structural components of bridges. These cracks grow in length with the repeated application of loads leading to a potential failure. Fatigue cracks grow more rapidly as the intensity of truck weights and the frequency of truck load applications increase. As the requests for special permits to use bridges by heavy trucks increases, there is a concern over the potential for rapid damage to bridges if the number of such permits is issued frequently and without any control. Furthermore, some bridges have been built for low truck weights (some for as little as 240-kN (54-kip), especially along feeder ramps) than current design load, which is 356-kN (80-kip) in the United States. Hence, some of these bridges may have a shorter life than expected and thus may become subject to premature failure, if the loading spectra contains many occurrences of overloads.

Mohammadi et al. report on a study of fatigue damage in steel girders in bridges with welded cover plates to determine their potential fatigue damage from overloads [5]. The study indicated that the rate of damage accumulation because of overloads can be several magnitudes larger than those because of regular loads. In the long run, the situation may become critical, if the damage from overloads is not controlled. These bridges may then be subject to premature costly repair and replacement.

In this paper, (1) the significance of overloads in causing fatigue damage to bridges is discussed; (2) truck load data have been compiled; and (3) the contribution of overloads to fatigue damage for steel girder bridges made up of welded cover plates is determined using the compiled truck load data. The paper discusses using the expected fatigue life expended (FLE) as an index to determine the impact of overloads on bridge fatigue damage. The FLE is then determined to be an important factor that can be used to modify bridge rating equations as a way of controlling the number of overload permits for a given bridge. It is demonstrated that by limiting the FLE and modifying the bridge rating equations, a reduction in the number of permits is achieved, which in turn will prevent bridges from rapidly approaching the end of their service life. Illustrative examples are provided in the paper to demonstrate the applicability of the modified rating equations along with discussions and conclusions.

2 Distributions of stress ranges at fatigue critical locations because of overloads

In this study, the stress range is determined by conducting a series of structural analyses using various truck load data sets obtained from the WIM system. The structural analysis was conducted considering designs conforming to the AASHTO Specifications in terms of load distribution factors, the number of lanes, design requirements, etc [6].

As mentioned earlier, a major portion of the overloaded trucks in the entire truck load population is made up of the vehicle class 9 in the 5-axle truck category. The stress range is computed from the gross vehicle weight (GVW) with the critical axle load configuration shown in Figure 2. The corresponding frequencies for stress ranges are considered and counted based on the fact that usually the passage of a truck will result in one major stress cycle at a given critical location. Figure 3 presents the stress range history at a fatigue-critical location in four bridges in which the span lengths are 21.3 meters (for Bridge #1), 18.3 meters (for #2), 15.2 meters (for #3) and 12.5 meters (for #4), respectively. The coefficient of variation in stress range is computed to be between 4% to 12%, indicating the variability in the GVW data. Figure 4 presents the stress range data for Bridge #1; and as seen, the stress range distribution also has two distinct peaks, as expected.

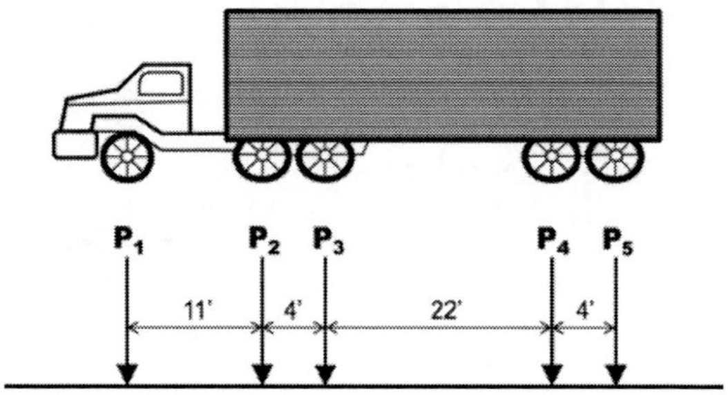

Fig. 2. Type 3-S2 truck configuration (in the United States)

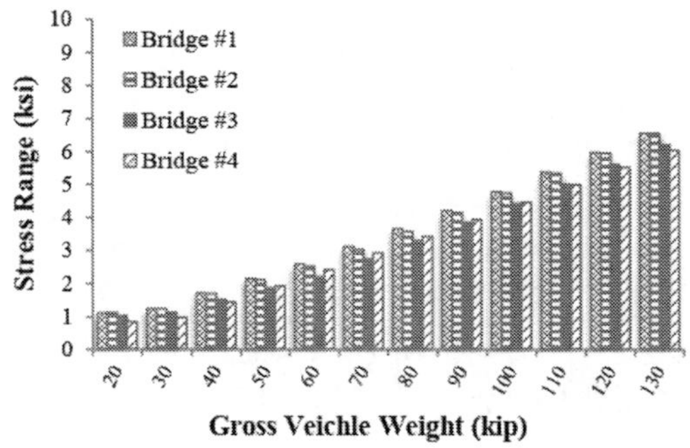

Fig. 3. Stress range history at the critical point on the given bridges

https://doi.org/10.1051/matecconf/201816510005

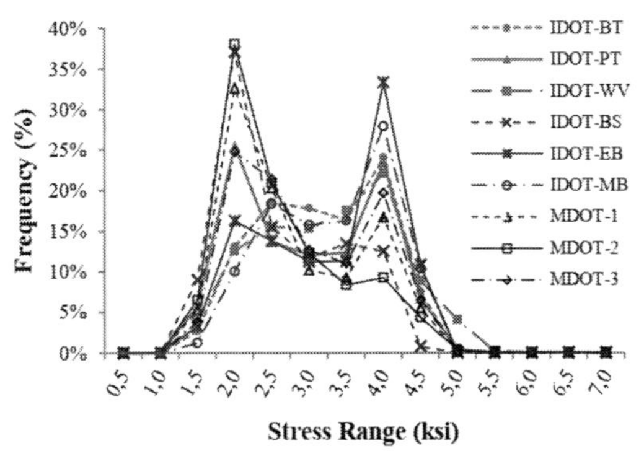

Fig. 4. Stress range distribution for all WIM stations for Bridge #1

3 Fatigue damage expected from overloads

In this paper, the focus is on damage that can occur to bridges in cases where overloads become a frequent occurrence. Fatigue damage from overloads is estimated using the Miner linear-damage rule at a critical location and a detail with a known S-N relationship. Bridges investigated are considered functionally failed when damage at the critical detail equals unity. Using this notion, the bridge fatigue life is then estimated.

To estimate the fatigue damage and the remaining fatigue life, a deterministic or probabilistic method may be used [7]. The probabilistic method allows incorporating uncertainties inherent in Miner damage rule, S-N relation curve, and stress range data in the fatigue damage estimation. Because of its formulation process, the probabilistic process provides fatigue life estimates that are based on specific values of accepted failure probabilities (for example, 2.3% for non-redundant structures). The deterministic method, on the other hand, offers life estimations in simple forms that can readily be used for practical applications without specific reference to any probability of failure. In this paper, the deterministic method is used.

The damage equation is as follows:

$$D = \sum n_i / N_i \quad (i = 1 \text{ to } k) \tag{1}$$

In which D is the total damage accumulated, k is the number of stress ranges developed from the load population on a given bridge, n_i is the number of applied cycles for i^{th} stress range, and N_i is the total number of cycles to failure of the ith stress range. The number of applied cycles is from the truck load population as indicated earlier; and the total number of cycles to failure is defined from S-N curve that is component-dependent. The following equation is used to represent the S-N curve:

$$N = C / S^m \tag{2}$$

In which N is the number of cycles to failure corresponding to stress range S, and C and m are constants specific to the structural detail (see for example, Munse et al [8]). For steel beams with welded cover plates corresponding to the fatigue detail categories E and E' in AASHTO manual, C and m are 4.218E09 and 3.256, respectively [9]. This type of detail is used in this paper.

4 Determination of fatigue damage potentials from overloads, FLE

The computation of the stress range at a fatigue-critical location requires a structural analysis of the bridge when it is subjected to truck axle loads. It is noted that the Miner damage rule assumes constant load amplitude and a linear formulation for the computation of fatigue damage. Since the stress data has many ranges each with its own number of cycles, using the damage equation incrementally, damage for each stress range is computed individually for the specific number of cycles for that range. The total damage is the summation of these incremental damage values.

The fatigue damage is estimated for a one-year period and then extrapolated considering traffic growth in future years. The accumulation of damage over 25, 50, 75, and 100 years can be obtained using the annual damage with and without traffic growth over this period. In this paper, specifically, the stress ranges of Bridge #1 are used as a demonstration of the procedure in computing the typical trend in fatigue life expended (FLE). Tables 1 and 2 summarize the FLE results for Bridge #1 with and without traffic growth. Figure 5 represents the annual fatigue damage due to overloads as a proportion of total fatigue damage when the entire load population is considered.

As seen, the fatigue life expended (FLE) varies depending on the volume of traffic and the percentage of overloads in the entire truck load population. Considering the data from all WIM stations, it is observed that overloads will use up about 0.2% to 7.8% and 0.6% to 23% of the fatigue life of the bridge component in 25 and 75-year periods, respectively. These are if no annual traffic increase is considered. According to a NCHRP 2016 report, from 2005 to 2012, a trend in oversize and overweight transportation showed approximately 3% annual increase in the truck traffic [10]. Thus, it is imperative to consider some level of steady traffic growth over a long period of time. For example, considering a 2% annual increase in truck traffic, the damage over a 25-year period for Bridge #1 can amount to about 0.26% to 10% of the fatigue life of the component. In addition, for 50 and 75-year periods, the cumulated fatigue damage potentials significantly increase with a 2% annual traffic growth. The percentage reduction in fatigue life may be especially important for older bridges that are designed for a lower truck load than current design requirements. For these bridges, any additional reduction in fatigue life may be critical resulting an early replacement or costly repairs.

Table 1. Cumulative fatigue damage over years without a traffic growth– Bridge #1.

Location of Station	FLE_{1yr} [a]	Without Traffic Growth			
		FLE_{25yrs}	FLE_{50yrs}	FLE_{75yrs}	FLE_{100yrs}
IDOT-BT	0.00099	0.0248	0.0496	0.0743	0.0991
IDOT-PT	0.00094	0.0235	0.0471	0.0706	0.0941
IDOT-WV	0.00312	0.0781	0.1562	0.2342	0.3123
IDOT-BS	0.00008	0.0020	0.0040	0.0060	0.0080
IDOT-EB	0.00076	0.0191	0.0381	0.0572	0.0762
IDOT-MB	0.00189	0.0472	0.0945	0.1417	0.1889
MDOT-1	0.00016	0.0040	0.0079	0.0119	0.0159
MDOT-2	0.00074	0.0185	0.0370	0.0554	0.0739
MDOT-3	0.00107	0.0269	0.0537	0.0806	0.1075

[a] *FLE* is the fatigue life expended.

Table 2. Cumulative fatigue damage over years with a traffic growth (2%) – Bridge #1.

Location of Station	FLE_{1yr} [a]	With Traffic Growth (2%)			
		FLE_{25yrs}	FLE_{50yrs}	FLE_{75yrs}	FLE_{100yrs}
IDOT-BT	0.00099	0.0317	0.0838	0.1693	0.3094
IDOT-PT	0.00094	0.0301	0.0796	0.1607	0.2939
IDOT-WV	0.00312	0.1000	0.2641	0.5334	0.9751
IDOT-BS	0.00008	0.0026	0.0068	0.0137	0.0250
IDOT-EB	0.00076	0.0244	0.0645	0.1302	0.2380
IDOT-MB	0.00189	0.0605	0.1598	0.3227	0.5899
MDOT-1	0.00016	0.0051	0.0135	0.0272	0.0496
MDOT-2	0.00074	0.0237	0.0625	0.1263	0.2308
MDOT-3	0.00107	0.0344	0.0909	0.1835	0.3355

[a] *FLE* is the fatigue life expended.

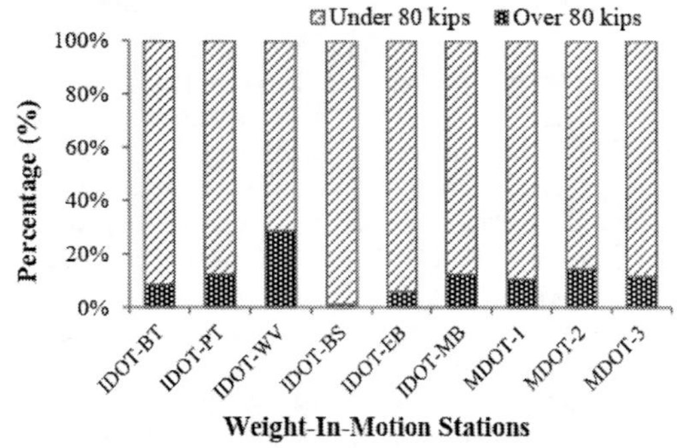

Fig. 5. Proportions of the annual fatigue damage due to overloads for Bridge #1

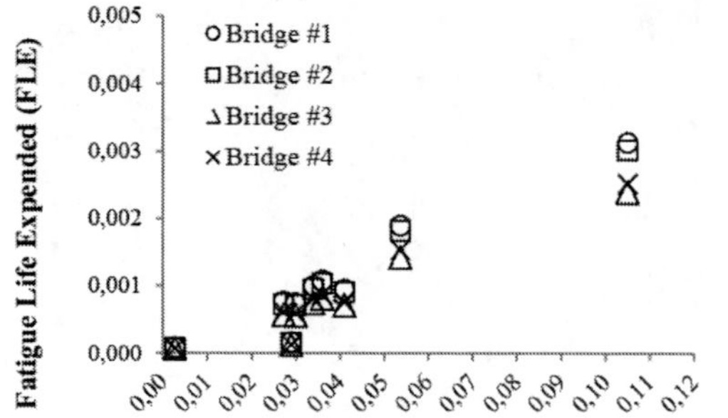

Fig. 6. Annual fatigue damage from overloads in terms of the percentage of overload in the entire truck load population

Figure 6 shows the significance of percentage of overload trucks in the truck population. It shows that the relationship between the fatigue life expended (FLE) and the percentage of overloads (in the entire truck load population) is almost linear (at least for the percentages considered in this study, i.e., up to 12%). This type of linear relationship may be helpful in offering a straightforward approach in incorporating the FLE as an index in bridge rating equations to account for the significance of fatigue damage from overloads.

5 Illustrative Examples and Discussions

The fatigue life expended (FLE) is determined to be an important factor that can be used to modify bridge rating equations to account for the significance of fatigue damage from overloads. It is demonstrated that by limiting the FLE and modifying the bridge rating equations, a reduction in the number of permits is achieved, which in turn will prevent bridges from rapidly approaching the end of their service life. It is noted that the rating factor represents the safety and serviceability of the bridge by evaluating the effect of passage of the overload truck for the purpose of the permit-load rating. In general, a rating of greater than 1 is considered satisfactory unless otherwise specified. A modification factor, α_{FLE}, is proposed in the bridge rating equation to account for the significance of the fatigue damage potentials for the overloads [4]. This modification factor is determined based on the cumulative fatigue damage and is formulated as a bilinear function in terms of a current age of bridge (a) and the percentage of overloads in the truck load population (OL). The modification factor is used in the Load and Resistance Factor Rating (LRFR) method as given in the AASHTO's Manual for Bridge Evaluation [9].

5.1. Examples

For illustrative purposes, a simply supported steel girder bridge with a welded cover plate (Category E in fatigue details) is considered. The span length of the bridge is 21.3-m (70-ft). The parameters used in this example are: ADTT = 2623 trucks/day and OL = 0.105 (10.5 %), φ

(capacity reduction factor) = 0.95, γ_{DC} = 1.25, γ_{DW} = 1.5, γ_L = 1.441 (these are the load factors), IM = 1.2 (for minor surface deviations), traffic growth = 2%, fatigue strength coefficient = 11E08 (for fatigue detail category E). These parameters were used for estimating (1) the remaining fatigue life of the example bridge using the equation provided by the Manual for Bridge Evaluation [9]; and (2) the rating with and without the modification factor. Several different ages for the bridge are considered (25, 50 and 65 years) to estimate and compare the remaining fatigue life and the rating factor. The remaining fatigue life is presented for the 25 and 50-year bridges; while the modified rating values are shown for the 50-year bridge as well as for a case where the age is as long as 65 years.

The example bridge was analyzed using the 5-axle truck configuration as mentioned earlier. Structural analyses were conducted to obtain the bending moment and the corresponding stress ranges at the critical component in the steel girders. Figures 7 and 8 present the results for the remaining fatigue life and the rating factor, respectively for the two different bridge ages.

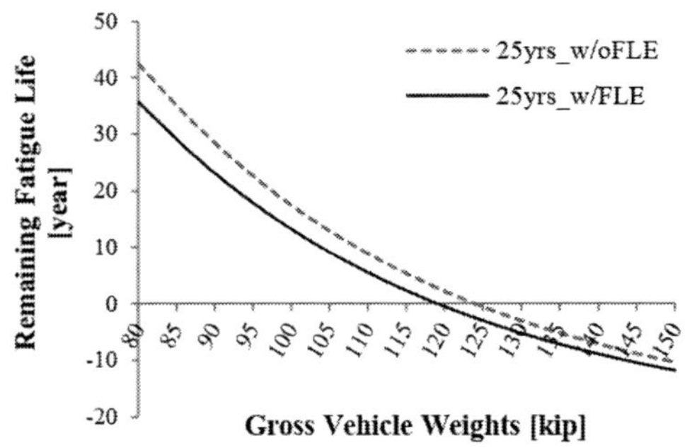

(a) Bridge with 25 years of life

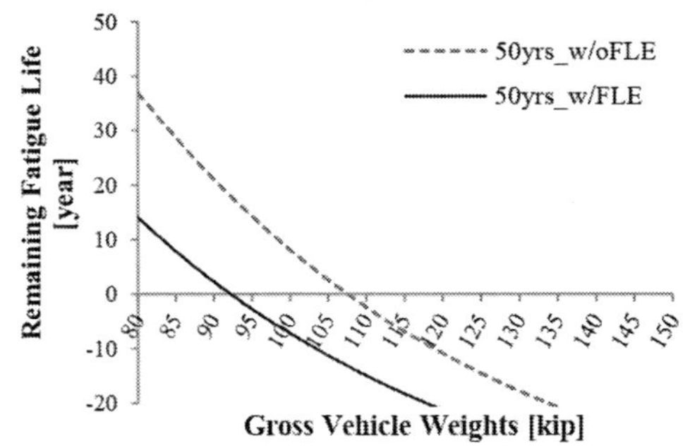

(b) Bridge with 50 years of life

Fig. 7. Remaining fatigue life with and without FLE for two different ages for a given bridge

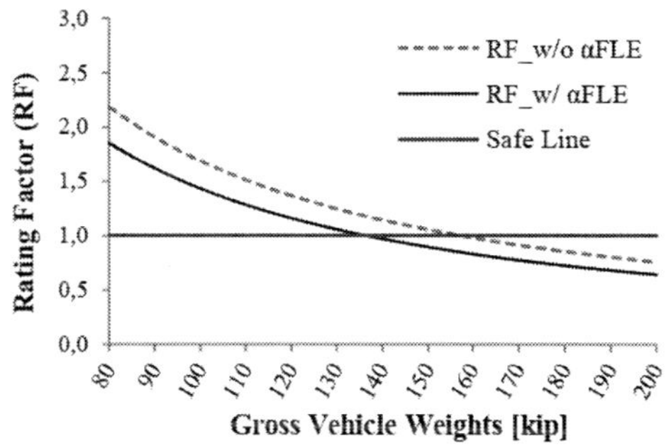

(a) Bridge with 50 years of life

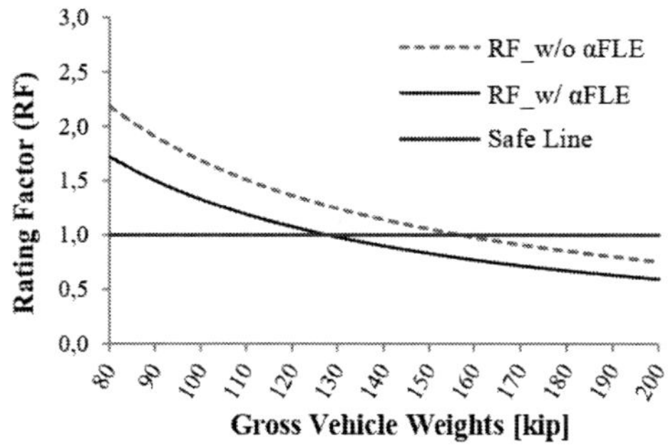

(b) Bridge with 65 years of life

Fig. 8. Rating factor with and without the modification factor, α_{FLE}, for two different ages for a given bridge

5.2 Discussions

As expected, the remaining fatigue life for the 25-year and 50-year bridges decreases with the growth vehicle weight as shown in Figure 7. It is noted that a negative fatigue life is only a mathematical estimate and corresponds to a practical case when the bridge age exceeds the predicted cyclic life. This situation happens when the gross vehicle weight is greater than 556-kN (125-kip) for the 25-year bridge and 480-kN (108-kip) for the 50-year bridge without consideration for the fatigue life expended (FLE). However, if FLE with traffic growth for overloads is considered, the remaining fatigue life decreases even more dramatically compared with the cases where FLE is not considered. The 50-year bridge has a significantly more reduction in its service life. This indicates that when the bridge ages, the remaining fatigue life dramatically decreases. This is mainly attributed to the traffic growth and to a lesser extent to the percentage of the overloads in the truck load population. As shown in Figure 7, if the overloaded truck with only 414-kN (93-kip) applies on the 50-year bridge, the negative remaining fatigue life is resulted and field inspection has to be initiated to ascertain the condition of the critical details for any significant cracks.

The values of the modification factor, α_{FLE}, were determined to be 0.1790 for the 50-year bridge and 0.2687 for the 65-year bridge. In this example bridge, the rating values are satisfactory without the modification factor if the gross vehicle weight is less than 689-kN

(155-kip) for both cases. However, the ratings with considerations for the fatigue damage result in values less than 1 for overloads at 601-kN (135-kip) and 556-kN (125-kip), for the 50-year and 65-year bridges, respectively as shown in Figure 8. As is evident from Figure 8, using the modification factor has resulted in certain number of overloads to be disallowed due to the fatigue damage potentials. This in turn indirectly imposes a limit on the number of permits.

6 Conclusions

The following presents the main conclusions from this study.

1) Fatigue damage potentials from overloads may become significant, if the number of special permits for overloads is issued frequently and without any control.

2) Fatigue life expended (FLE) varies depending on the volume of traffic (ADTT) and the percentage of overloads in the entire truck load population (*OL*). FLE shows a linear trend with the percentage of overloads (*OL*).

3) This linear relationship offers a straightforward equation for incorporating FLE in the bridge condition assessment.

4) Incorporating the FLE in the load rating equations, may be an indirect method in limiting the number of permits for overloads.

The assistance provided by David C. Thompson of the Illinois and Thomas Foltz of the Michigan Department of Transportation, respectively, in making the WIM data available to the authors is appreciated.

References

1. J. Mohammadi, N. Shah, J. Trans. Eng. ASCE, *Statistical evaluation of truck overloads*, **118** (5), 651-665 (1992)
2. R.E. Snyder, G.E. Likins, F. Moses, *Loading spectrum experienced by bridge structures in the United States*, Report RD-851012, (FHWA, Washington DC. 1985)
3. D.H. Timm, S.M. Tisdale, R.E. Turochy, J. Trans. Eng. ASCE, *Axle load spectra characterization by mixed distribution modeling*, **132** (2), 83-88 (2005)
4. B. Jang, J. Mohammadi, Brdg. Strc., *Bridge rating modification to incorporate fatigue damage from truck overloads*, **13** (2-3), 101-107 (2017)
5. J. Mohammadi, R. Polepeddi, J. Brdg. Eng. ASCE, *Bridge rating with consideration for fatigue damage from overloads*, **5** (3), 259-265 (2000)
6. AASHTO, *AASHTO LRFD Bridge Design Specifications, 8th edition*, (AASHTO, Washington DC. 2017)
7. J. Mohammadi, S.A. Guralnick, R. Polepeddi, FFRC of SEI of ASCE, *Use of Stress Range Data in Fatigue Reliability Assessment of Highway Bridges*, in *NDT Methods Applied to Fatigue Reliability Assessment of Structures* (J. Mohammadi, ed), 56-71 (2004)
8. W. Munse, T. Wilbur, M. Tefralian, K. Nicoll, K. Wilson, *Fatigue characterization of fabricated ship details for design*, Report SSC-318, (Ship Strc. Cmt. Washington D.C. 1983)
9. AASHTO, *Manual for bridge evaluation, 2nd edition*, (AASHTO, Washington DC. 2011)
10. CPCS, Perkins Motor Transport, Inc., Portscape, Inc., *Multi-State, Multimodal, Oversize/Overweight Transportation*, NCHRP Report 830, (TRB, Washington DC. 2016)

Failure analysis of a 40 ton crane hook at a Hot Strip Mill

Souvik Das*, Goutam Mukhopadhyay, Sandip Bhattacharyya

R&D and Scientific Services, Tata Steel Limited, Jamshedpur-831 001, India

*Corresponding author: Address: Metallurgical Laboratories and QA Group, R&D and Scientific Services, Tata steel Ltd., Jamshedpur 831 001, India Tel.: +91 809284714; fax: +91 657 234395.

E-mail address souvik.das@tatasteel.com

Abstract:

There have been accounts of repeated failure of crane hooks at the coil yard of a Hot Strip Mill which pose a serious threat to safety in the area. More than 4 hooks failed in less than 5 years. The crane hook (rated for 36000 kg) failed from the threaded shank while lifting a load of 18143 kg. The metal in the hook was revealed by chemical analysis to be killed IS: 4367 20C15 steel. The hook rod failed from a step where there was a cross sectional change and the locations were associated with machining and chatter marks. Such cross-sectional changes are the potential sites of stress concentrations leading to crack initiations. Fracture surfaces of broken pieces of hook reveal initiation of beach marks from both sides with granular rough surface at the middle of fracture zone. Beach marks initiated from both sides indicate origin of reverse bending fatigue. Distinct granular rough zone at the middle is due to final brittle fracture. Microstructure of the polished sample revealed numerous inclusions which indicate that the steel was not clean .Such a huge number of inclusions are not desirable as they can act as stress concentration sites and lead to fatigue crack initiation. Etched microstructure of failed hook reveals coarse cast structure having inhomogeneous microstructure with a mixture of ferrite and pearlite (which meant lower fatigue strength). This inhomogeneous coarser cast structure is outcome of lower reduction ratio during rolling followed by improper heat treatment process. The fracture was concluded to have occurred due to stress concentration from the step region due to inferior material (Inclusion and Improper Heat treatment process). Preventive maintenance and condition monitoring procedures should be applied to identify and minimize the risk(s) Establishment of an NDT procedure for regular basis inspection of the structural members (welded joints and hook-shaped steel rods during incoming inspection and in-service).

Key Words: **Crane Hook; Hot Strip Mill; Beach marks; cast structure; Fatigue**

Introduction

Crane hooks are highly liable components and are always subjected to premature failure due to accumulation of unwanted stresses which can eventually lead to its failure. Crane hooks are the primary components used in industries and constructional sites to elevate the heavy load. A crane is a machine, equipped with a hoist, wire ropes or chains and sheaves used to lift and move heavy material. Cranes are mostly employed in transport, construction and manufacturing industry. Overhead crane, mobile crane, tower crane, telescopic crane, gantry crane, deck crane, jib crane, loader crane are some of the commonly used cranes. A crane hook is basically a hoisting fixture designed to engage a ring or link of a lifting chain or the pin of a shackle or cable socket. Catastrophic failure of carne hook is one of the main reasons for industrial disaster [1-4]. In this study a 40 ton crane hook has been analysed which failed during lifting of 24 ton coil at a Hot Strip Mill (HSM). There was no human loss due to this incident but it had huge potential for fatal accident. There have been accounts of repeated failure of crane hooks at the coil yard of HSM which pose a serious threat to safety in the area [5]. More than four hooks failed in less than 5 years. The objective of this study is to find out the root cause of the failure so that proper actions can be taken to prevent the reoccurrence of such failures.

Investigation Methods and Results

Visual Observation

Overall view of the sample referred for analysis is shown in Fig. 1(a). The hook failed from a step where there was a cross sectional change (Fig. 1b); rough-machining marks and chatter marks were observed near the location of failure (Fig.1c). Such locations involving cross-sectional change are the potential sites of stress concentrations leading to crack initiations. Fracture surfaces of broken pieces of hook reveal presence of beach marks from both sides with granular rough surface at the middle of fractured zone (Fig. 1d). Beach marks initiated from both sides indicate origin of reverse bending fatigue. Distinct granular rough zone at the middle is due to final brittle fracture.

 https://doi.org/10.1051/matecconf/201816510006

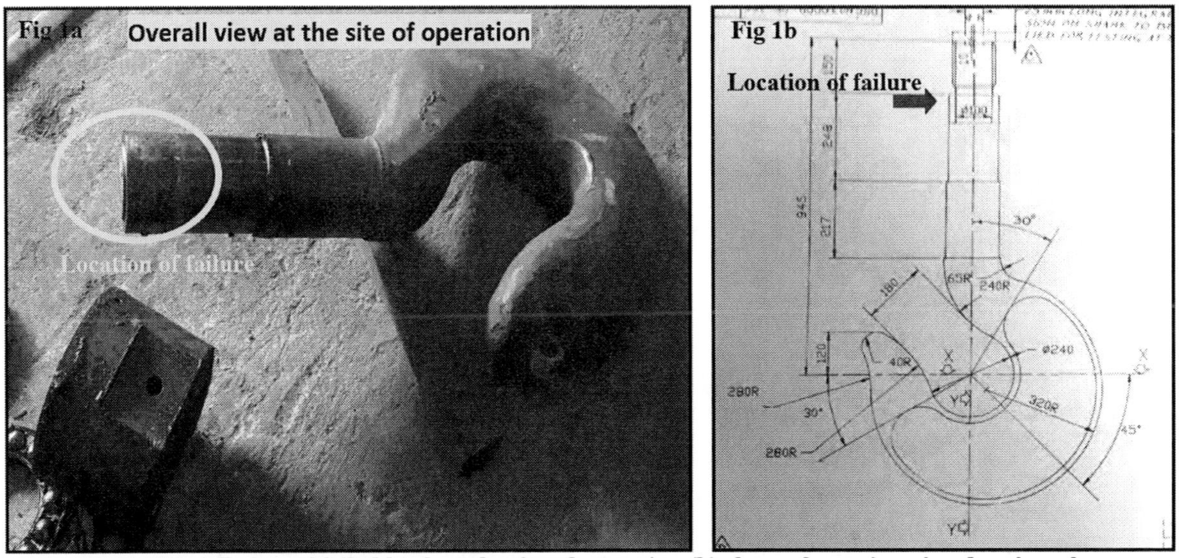

Fig. 1(a) overall view of failed hook at the site of operation (b) shows the engineering drawing of the hook with the location of failure.

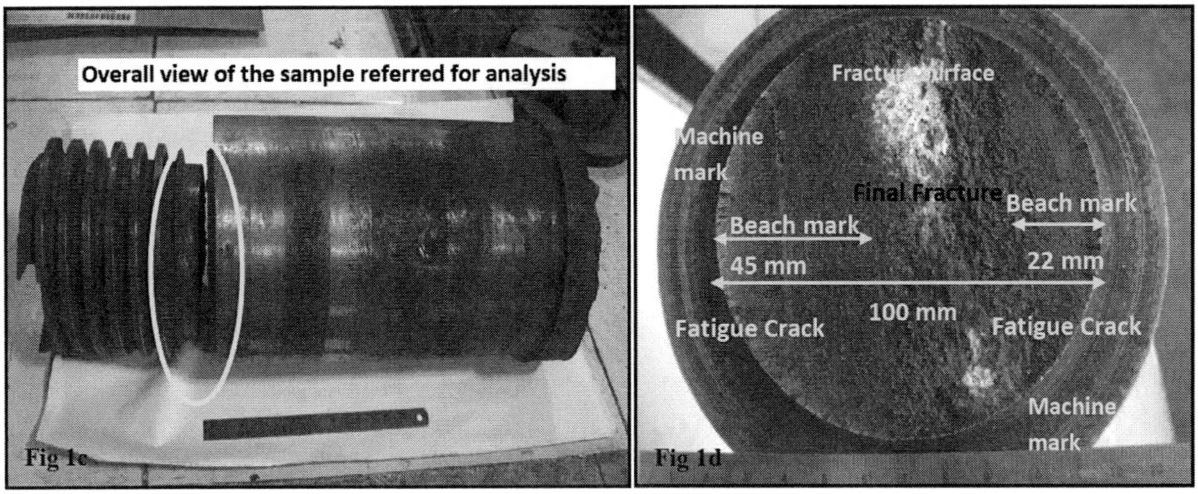

Fig. 1(c) Overall view of the sample referred for analysis, (d) Closer view of fracture surface of hook

Chemical Analysis

The chemical analysis of the failed hook sample was carried out by ICP spectroscopy. The chemistry of the failed hook matches with IS: 4367 20C15 grade of steel which was specified in the drawing. The chemical compositions of the failed sample are compiled in Table 1.

Table 1: Chemical Analysis (Wt. %):

Sample	C	Mn	S	P	Si	V	Al	Ti	Cr	Mo	Cu	Ni
Hook	0.16	1.31	0.040	0.053	0.11	0.0003	0.001	0.0007	0.022	0.001	0.009	0.012
IS:4367 20C15 grade	0.16-0.24	1.30-1.70	--	--	0.10-0.35	--	--	--	--	--	--	--

Experimental Procedure

The failed hook was collected from the plant for investigations. The samples were cleaned with acetone to remove dirt for visual examination prior to metallographic sample preparation. Transverse and longitudinal specimens were made from the fractured end of the failed samples for conducting optical microscopic examination. These samples were individually mounted in conductive mounting and polished by conventional metallographic techniques for scratch free surface. The polished samples were etched in 3% nital solution (3 mL HNO_3 in 97 mL ethyl alcohol), and both un-etched and etched samples were examined under an optical microscope. The hardness of different location was observed in the failed samples was determined in a pneumatically controlled automatic micro hardness tester (Leco-$LM247_{AT}$). An applied load of 100gf was used during testing, and several indentations were made to determine the hardness of the failed component. Field Emission Gun Scanning Electron Microscopy (FEG-SEM) of the samples was also carried out to identity fracture surface at higher magnifications. The analyses were performed at 15 keV accelerating voltage and 5^{10-8} A probe current.

Fractography

Field Emission Gun Scanning Electron Microscopy (FEG-SEM) of the samples was also carried out for fractography analysis to find out the mode of failure. The analyses were performed at 15 keV accelerating voltage and 5^{10-8} A probe current. The extent of fatigue zone and striation spacing is consistent to high cycle fatigue (HCF). FEG-SEM revealed cleavage nature of the final fracture surface suggesting brittle fracture due to over load.

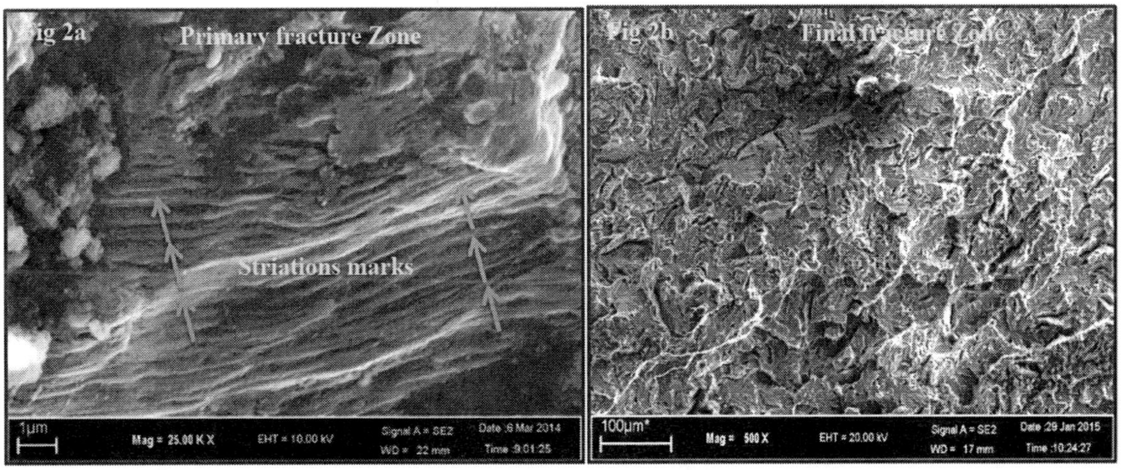

Fig 2a. FEG-SEM analysis shows fatigue striations marks in the primary fracture surface, (b) showing the microstructure of final overload area (fast crack growth zone).

Microstructure Analysis

Microstructure of the polished sample revealed numerous inclusions which indicate that the steel was not clean (Fig. 3a). Such a huge number of inclusions (as shown in Table 2) are not desirable as they can act as stress concentration sites and may lead to fatigue crack initiation.

Etched microstructure of failed hook as shown in Fig. 3(b-c) reveals coarse cast structure having inhomogeneous microstructure with a mixture of ferrite and pearlite. This inhomogeneous coarser cast structure is outcome of lower reduction ratio during rolling followed by improper heat treatment process.

. **Table 2: Inclusion rating:**

Sample	**Type A (T/H)**	**Type B (T/H)**	**Type C (T/H)**	**Type D (T/H)**
Hook	**2.0/1.0**	0.0/0.0	0.0/0.0	0.5/0.0

Inclusion rating of the failed component was carried out as per ASTM E-45.Un-etched microstructure of hook shows that thin and thick sulfide (Type A) inclusions (Conformed by

EDS analysis) are present with a severity of 2.0 and 1.0 respectively. Some inclusions of Type D (oxides) are observed with a severity of 0.5.

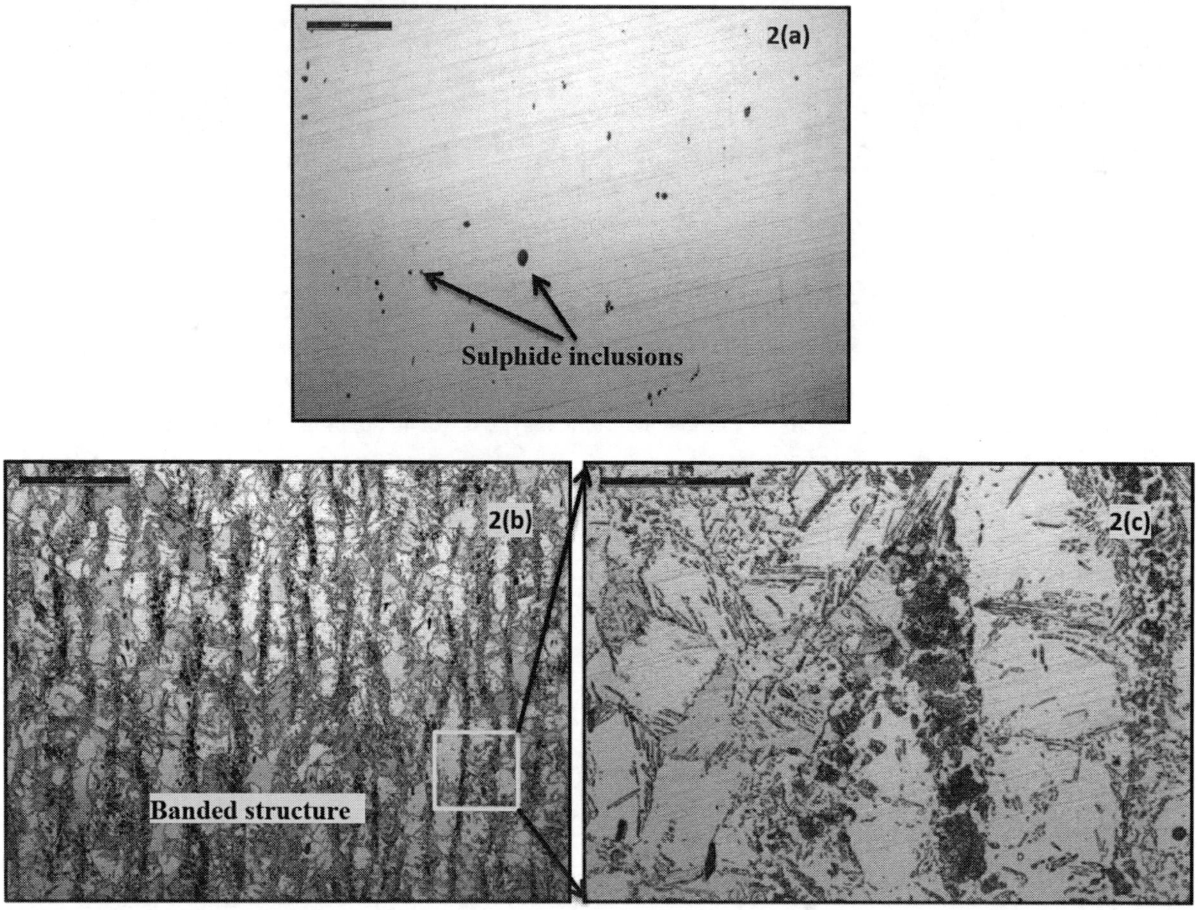

Fig. 3: (a) Un-etched microstructure of hook (b -c) Etched microstructures of hook sample at different magnifications.

X-Ray Mapping of the Different Phases

X-ray mapping was carried out on the failed hook sample to determine the elemental concentration in the inclusion (Fig. 4). Elemental analysis under SEM reveals that the inclusions were observed to be rich in Sulphur (S) and Manganese (Mn) indicating them to be manganese sulphide inclusions.

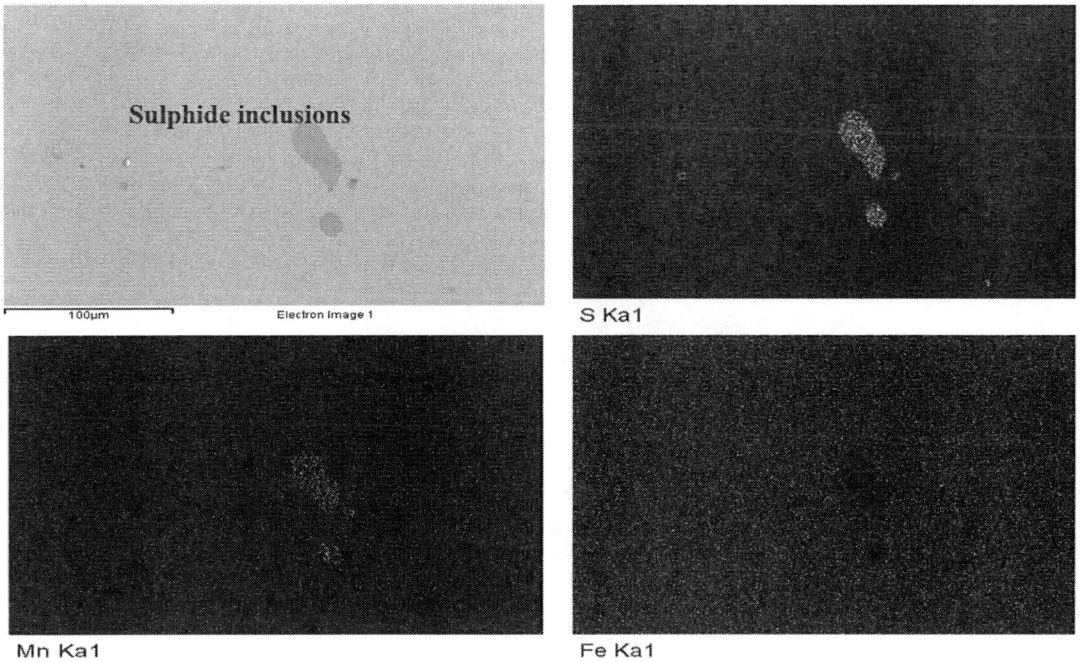

Fig. 4 showing X-ray mapping of failed hook sample.

Mechanical Property Analysis

Vickers hardness tester was used to analysis the hardness of the failed sample. The average hardness of the hook was found to be 163 BHN. But the Yield Strength (Y.S) and the UTS (Ultimate Tensile Strength) of the hook was found significantly lower than the specification due to inhomogeneous coarser cast structure as shown in Table 3.

Table 3: Result of Mechanical properties:-

Sample ID	Hardness (BHN)	YS (Mpa)	U T S (Mpa)	%EL	Avg. Impact Value (Joules)
Hook	163	**214.5**	**443**	32	54.9
IS:4367 20C15 grade	200 Max	400 Min.	600-750	18 Min.	50 Min.

Discussion

There have been accounts of repeated failure of crane hooks at the coil yard of Hot Strip Mill which pose a serious threat to safety in the area. More than 4 hooks failed in less than 5 years. Fracture surfaces of broken pieces of hook reveal initiation of beach marks from both sides with granular rough surface at the middle of fracture zone. Beach marks initiated from both sides (Area 1 and Area 2) indicate origin of reverse bending fatigue.

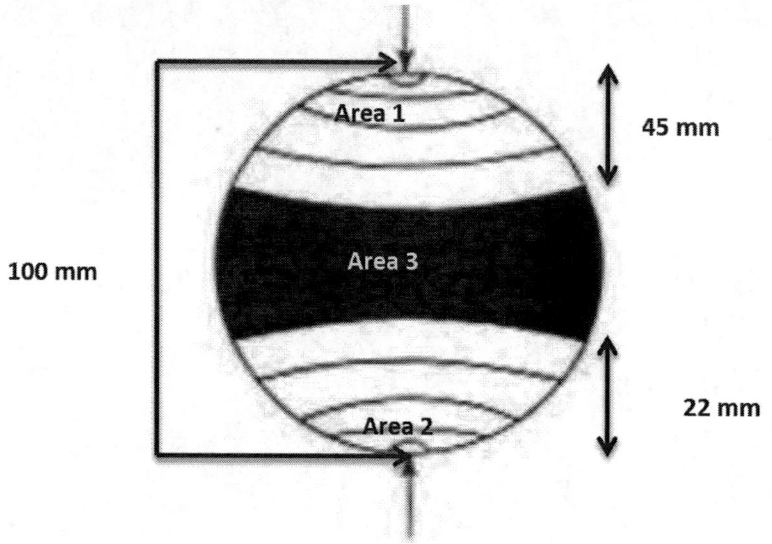

Fig 5: Schematic illustration of the failed hook showing the approximate location of the fracture

Calculation of Area 1 (Fatigue Failure)

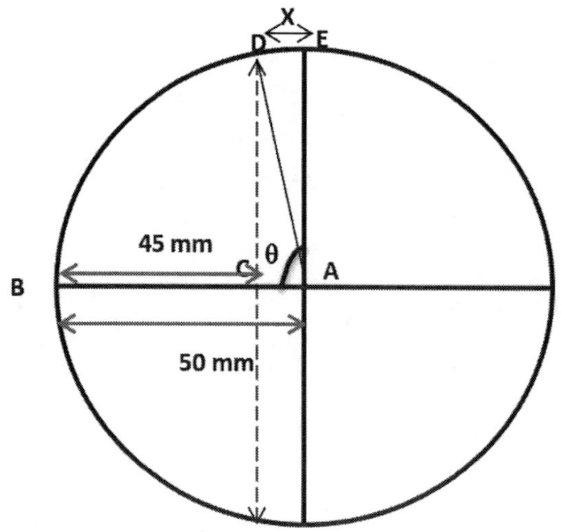

From the drawing we found the radius of the failed sample is 50 mm and the length of the first fatigue 1 is 45 mm (BC).Lets take distance of x to be zero. Thus D is equal to E

Now Length of CD = $\sqrt{(50^2\text{-}5^2)}$ = 49.74 mm^2

And Cos θ = AB-BC/50 = 5/50.

$\therefore \theta = 84$ degree.

Now Total Area 1 = 2 $[(\theta/2\Pi)*\Pi*R^2) - (1/2*49.7*5)]$ mm2

Thus total area of Fatigue **Area 1** is 3414 mm^2.

Form this way we have calculated the total area of Fatigue **Area 2** is 1456 mm^2.

Now Brittle fracture Area is (Area 3) = Π * 502 – (Area 1+ Area 2)

= 2980 mm^2.

Thus the area of final fracture is small, approximately 30% of total area, indicating that the material was in-adequate for the applications.

This indicates the hook underwent cyclic bending stresses [6] at the fracture site. Distinct granular rough zone at the middle is due to final brittle fracture (Area 3). The hook rod failed from a step where there was a cross sectional change. Such locations involving cross-sectional change are the sites of stress concentrations leading to crack initiations which thereby lead to final failure by the propagation of crack under fatigue mode. Microstructure of the polished sample revealed numerous inclusions which indicate that the steel is not clean. Such a huge number of inclusions are not desirable as they can act as stress concentration sites and might lead to fatigue crack initiation. Etched microstructure of failed hook reveals coarse cast structure having inhomogeneous microstructure with a mixture of ferrite and pearlite. This inhomogeneous coarser cast structure is outcome of lower reduction ratio during rolling followed by improper heat treatment process. Elemental analysis under SEM reveals that the inclusions were found to be rich in Sulfur (S) and Manganese (Mn) indicating them to be manganese sulphide inclusions. The hardness of the hook was found satisfactory as per drawing. But the hardness seems to be lower for the application, thus the specification should be reviewed. The Y.S and the UTS of the hook found significantly lower than the specification due to inhomogeneous coarser cast structure.

Conclusion

The 40 ton hook of crane failed in fatigue mode from the step due to inferior material. Life of the hook was shortened considerably because the material was in inhomogeneous coarser cast condition, which resulted in low fatigue strength.

Recommendations

- **Heat Treatment:** The crane hook should normalize properly after forging to produce the most desirable structure.
- **Health monitoring of plant equipment**: Preventive maintenance and condition monitoring procedures should be applied to identify and minimize the risk(s) Establishment of an NDT procedure for regular basis inspection of the structural members (welded joints and hook-shaped steel rods during incoming inspection and in-service).

References:

[1] F.L. Jamieson: Fatigue Fracture of a 1020 Steel Crane Hook Failures of Lifting Equipment, *Failure Analysis and Prevention*, Vol 11, *ASM Handbook*, ASM International, 1986, p 514–528

[2] Roe, G.J., Bramfitt, B.L.: ASM Handbook, Properties and Selections; Irons, Steels, and High Performance Alloys, vol. I, Notch Toughness of Steels. ASM, Metals Park, OH (1990)

[3] Lu, S., Su, J., Liao, S., Wang, B., Yu, L., Jiang, Y., Wen, S.:Finite element analysis on fatigue failure prediction of a rear axle housing of vehicle based on cosmos. Appl. Mech. Mater. 121(126), 843–848 (2012)

[4] Qinghua, M., Huifeng, Z., Fengiun, L.: Fatigue failure fault prediction of rear axle housing excited by random road roughness. Int. J. Phys. Sci. 6, 1563–1568 (2011)

[5] Asi, O.: Fatigue failure of a rear axle shaft of an automobile. Eng. Fail. Anal. 13, 1293–1302 (2006).

[6] Souvik Das, GoutamMukhopadhyay, Sandip Bhattacharyya.: Failure analysis of axle shaft of a fork lift: Case Studies in Engineering Failure Analysis 3 (2015) 46–51

Development of a test bench for vibratory fatigue experiments of a cantilever beam with an electrodynamic shaker

Aymeric Appert[1,*], *Christophe Gautrelet*[1], *Leila Khalij*[1], and *Renata Troian*[1]

[1]Normandie Université, LMN/INSA Rouen Normandie, 76000 Rouen, France

Abstract. When fatigue tests are carried out using a shaker, dynamic phenomena often appear and lead to a non-linear behaviour in the experiment. In this paper, the dynamic response of the specimen is studied in order to evaluate the impact on the results. The experiment consists in a notched beam fixed on the armature of an electro-dynamic shaker. The transfer function between the center of the armature and the tip of the beam is obtained with a swept sine near the first resonance frequency of the specimen. The drive is a constant acceleration amplitude. An evolution in the transfer function of the beam is observed when the drive is modified: the resonance frequency decreases and the damping increases. This non-linearity is investigated by studying the movement of the shaker, the fixture and the beam. The results show that the beam close to its resonance disrupt the imposed movement. An experimental setup correcting those defects is proposed.

1 Introduction

Conventional fatigue tests performed with servo-hydraulic testing machines can be costly and time-consuming for high-cycle fatigue studies. One way to accelerate the tests is to carry out vibration testing with an electro-dynamic shaker. Different configurations of this experimental setup can be found in the literature.

The simplest method is to fix the specimen directly on the shaker armature by means of a clamp. Different shapes can be used for the specimen such as plate, beam or other complex designs to study multiaxial loading. In all cases, the fixture consists in constraining the sample between two parts tightened with multiple screws. The screws are placed either close to the embedding or going through the specimen. Generally, the choices made for the fixture are not explained in the literature.

Abdullah [1] fixed multiple beams on the shaker's table with 4 screws through the upper blocks and the specimen. He observed that damping increases with the augmentation of specimen tip amplitude and explained it by air damping. A similar fixture is used by Angeli et al. [2], Ge et al. [3] and Zhou et al. [3]. The specimens are fixed with 2 through-the-thickness screws. A groove is added in the clamp by Ge et al. and Zhou et al. to prevent the specimen rotation. Hooreweder et al. [5] used this specificity to ensure accurate contact surfaces. In [6], George et al. maintained the plate-shaped specimen between two blocks with two screws positioned close to the embedded part. One objective was to reduce the potential damping in the clamp, that's why the maximum torque was applied to the mounting bolts. Habtour et al. [7] precise that non-linearity may appear due to the clamping but assume that it is negligible. No more precisions are given on the used fixture as in Hammami et al. [8], Paulus et al. [9] or Li et al. [10]. Only pictures of the experiments show the fixture used. Ellyson et al. [11] don't even show the clamping but precise that their results exhibit localized slipping at the clamped edge. The importance of the clamping strategy is highlighted. Rognon [12] and Zanellati et al. [13] used rod-shaped specimens, but the clamping was still made of multiple parts tightening the beam. Rognon added a flat part to the beam to ensure a good positioning of the specimen and control the tightening torque.

The literature points out the importance of the clamping. In vibration testing, this part has a huge impact on the dynamic response and modal parameters are directly used in fatigue tests to predict the damage [14, 15, 16]. However, we have seen that phenomena induced by the fixture are often neglected or taken into account as a structural damping. In this work, we have highlighted the importance of the fixture design for accurately assessing the modal parameters.

The first part of this paper presents an experimental setup to carry out fatigue tests on a beam with an electrodynamic shaker. The behaviour near the first resonance frequency is studied by doing swept sines at different excitation levels. The results reveal unexpected non-linearities in the structure and the causes are investigated. The second part shows the study achieved in order to eliminate non-linearities and suggests improvements in the experimental setup. Those improvements are tested in the third part and the results are discussed.

* Corresponding author: aymeric.appert@insa-rouen.fr

2 Experimental setup

2.1. Equipment

The test rig in **Fig. 1** consists in an air-cooled electrodynamic shaker driven by an acquisition control peripheral (ACP) with the help of a power amplifier. The ACP is used to control the shaker in a closed loop. Thus, an acceleration, velocity or displacement profile is applied on a localised point of the structure instead of an excitation force.

The specimen is fixed directly on the armature of the shaker, clamped between two rectangular plates by means of four screws at the corners. The screws are tightened with the same torque. A wedge is added between the two plates to force them to remain parallel and ensure a good contact surface on the beam.

The driving point is an accelerometer placed in the center of the clamp, on the axis of the shaker. The transmissibility function is obtained thanks to a vibrometer measuring the velocity at the tip of the beam. In this paper, only swept sines are used, thus the acceleration is easily obtained from this velocity. A vibrometer was preferred to an accelerometer on this location, because the mass and rigidity of the accelerometer wire have a significant effect on the beam dynamic behaviour [17].

2.2 Beam design

The shape of the beam was determined by carrying out a modal analysis with Code_Aster. The objective is to find the resonance frequencies and the associated modal shapes of the sample. The modal shape gives the type of loading the area of interest is subjected to. In our case, we use the first flexural mode of vibration of a steel beam. Another point to consider about the geometry is that the maximum stress will be located in the fixture. The crack cannot be studied in this area. The most common way to avoid this problem is to create a stress concentration by adding a section reduction far from the clamp [18].

The mesh visible in **Fig. 2** is made of quadrangular plate elements based on the Kirchhoff theory, to avoid shear locking. The properties used for the simulation are given in **Table 1**.

Table 1. Mechanical properties of the steel specimen

E (GPa)	ν	$\rho\ (kg/m^3)$	$h\ (mm)$
205	0.3	7696	1

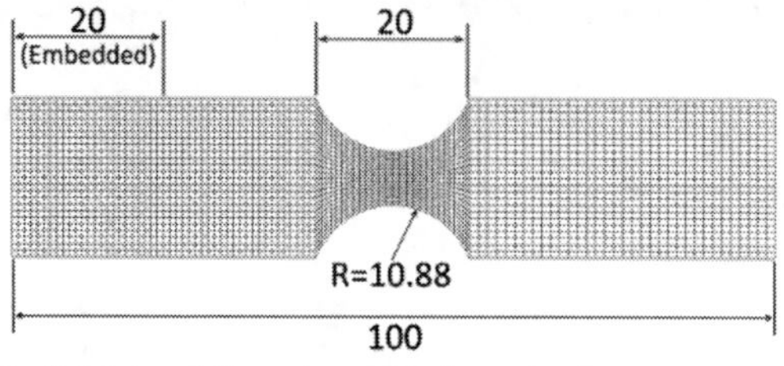

Fig. 2. Mesh and dimensions of the beam (mm)

The first natural frequency at 114.12 Hz corresponds to the first flexural mode of the beam. This mode is far from the second one which is also a flexion at 810.63 Hz. The maximum stress is not located close to the fixture thanks to the section reduction (**Fig. 3**). Thus, the fracture will occur in this stress concentration zone and will be easy to detect. In comparison, the second mode has relatively high stresses close to the fixture. This configuration can't be used for fatigue tests because the fracture location is not well controlled.

Fig. 3. Modal shapes of mode at 114.12 Hz (above) and 810.63 Hz (below) with stress repartition.

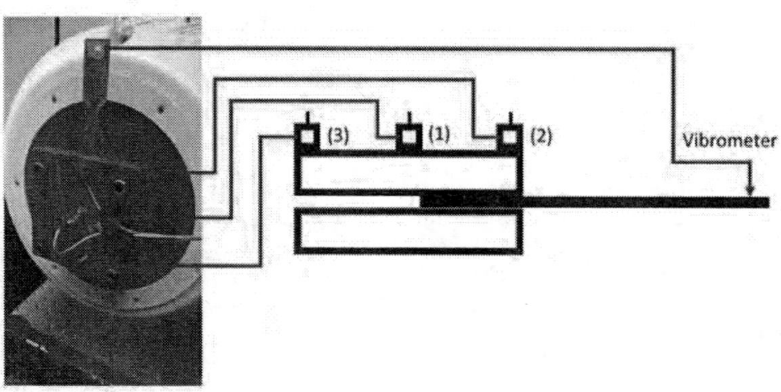

Fig. 1. The experimental setup

2.3 Experimental test

The tests performed are swept sines at different levels of acceleration, from 0.5 g to 1 g every 0.1 g, in a frequency range surrounding the first resonance (100 Hz-120 Hz). The sweep speed is 0.1667 Hz/sec. The control is carried out by a triaxial accelerometer (noted (1) on **Fig. 1**), which is placed in the center of the fixture. Obviously, only the acceleration in the axis of the shaker is used for the drive, but we also measure the transversal accelerations. Two other accelerometers are fixed on the clamp, one where the specimen is fixed (noted (2)) and one at the opposite (noted (3)), symmetrically to the shaker axis.

2.4 Results and discussion

The transmissibilities plotted in **Fig. 4** reveal non-linearities in the experiment and a few hypotheses can be made about their sources. The resonance frequency is shifting to the left and the damping increases with the excitation level. The tests were made at very low acceleration levels to ensure that there is no geometrical non-linearity nor plastification of the beam. Moreover, after all the swept sines realized at growing accelerations, a last sweep was carried out at the smallest level. The transmissibility between the first and the last test at the same excitation were identical which reject any plastic strain hypothesis.

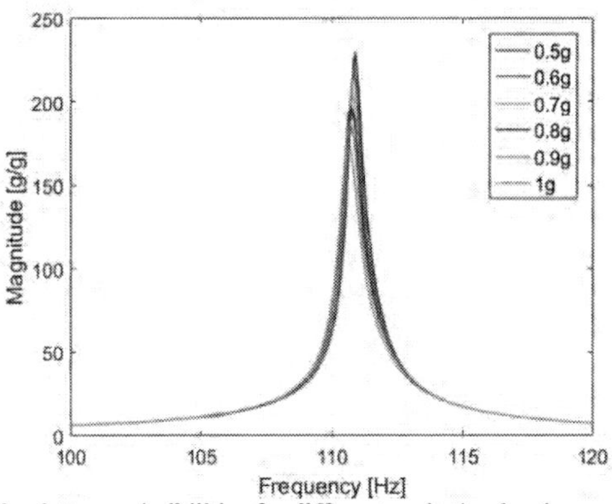

Fig. 4. Transmissibilities for different excitation levels

The accelerometers (2) and (3) placed on the fixture reveals a spurious displacement of the shaker when the swept is close to the resonance of the beam. **Fig. 5** represents the accelerations at those points for an excitation of 0.5g and 1g. The difference between the acceleration given by the control accelerometer and the others reaches 30% of the driving acceleration.

The pattern of the curve tends to show that the specimen drives the armature in a rotation movement (see **Fig. 6**). Before the resonance frequency, the beam is amplifying the movement of the shaker on its side. Then the phase is shifting, and the rotation is in opposition to the shaker displacement. This shift is clearly visible with the accelerations measured by (2) and (3).

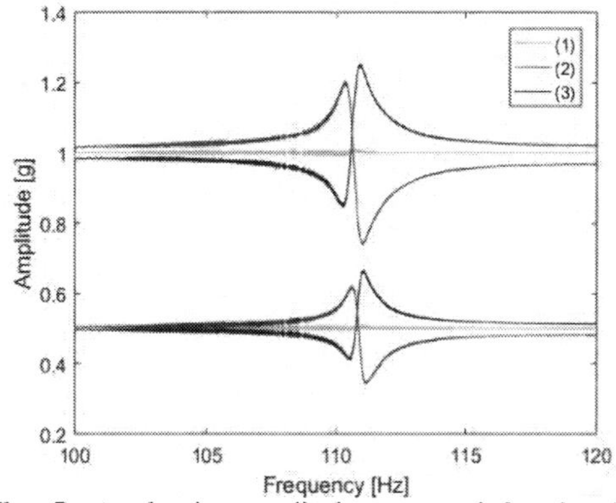

Fig. 5. Acceleration amplitudes measured by the three accelerometers for excitation levels of 0.5 g and 1 g

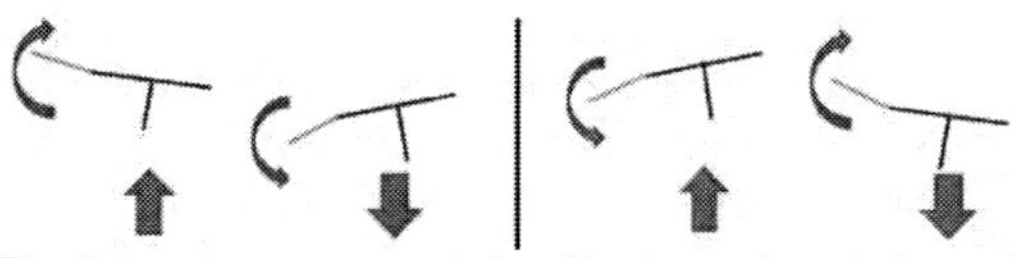
Fig. 6. Rotation movements induced by the specimen before and after the phase shifting.

With the triaxial accelerometer, a high acceleration amplitude is measured in the direction of the beam when the frequency is close to the resonance of the specimen (see **Fig. 7**). In the perpendicular direction, only noise can be measured. This observation suits with the hypothesis described above.

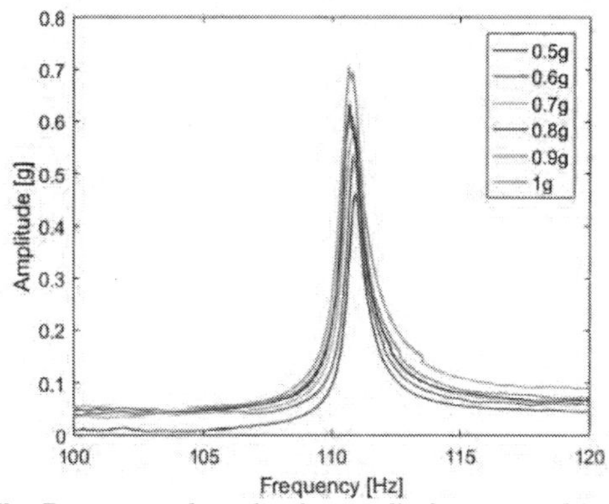

Fig. 7. Transversal acceleration amplitudes measured by the triaxial accelerometer (1) in the direction of the beam

One major issue with this phenomenon is that you can plot different transmissibilities with the same test, giving different modal parameters. This is illustrated in **Fig. 8**, where the accelerometers (1), (2) and (3) are used successively as the input for an excitation of 1g. Here we can find three resonance frequencies (respectively 110.6 Hz, 110.9Hz and 110.5Hz) with three associated amplification (175.4, 196.6 and 175.7), which is not logical. There shouldn't be a choice for the input acceleration.

 https://doi.org/10.1051/matecconf/201816510007

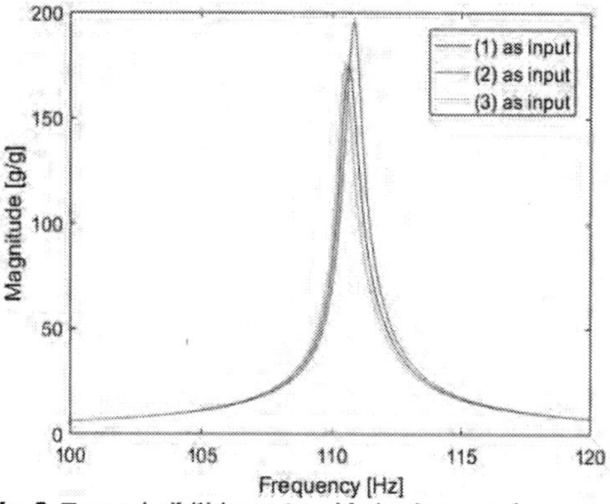

Fig. 8. Transmissibilities at 1g with the three accelerometers as input acceleration

Another cause of this non-linearity could be the fixture. Without the wedge to ensure that the two plates remain parallel, the resonance frequency is lowered by more than 10Hz. In our case, the upper plate will inevitably bend under the pressure of the screws, such that the contact surface is located on the edges closest to them and not on the edge of the fixture. The free part of the specimen is not clearly dissociated and the resonance mode used depends strongly on the length of the beam. This deformation can be seen in the **Fig. 9**. The plates on the fixture have been used for high-cycle fatigue tests [19]. The successive mounting and unmounting of the specimens resulted in a permanent deformation of the upper plate. This picture shows clearly how the contact is located in the clamp, even with a thicker plate that will not be irreversibly deformed.

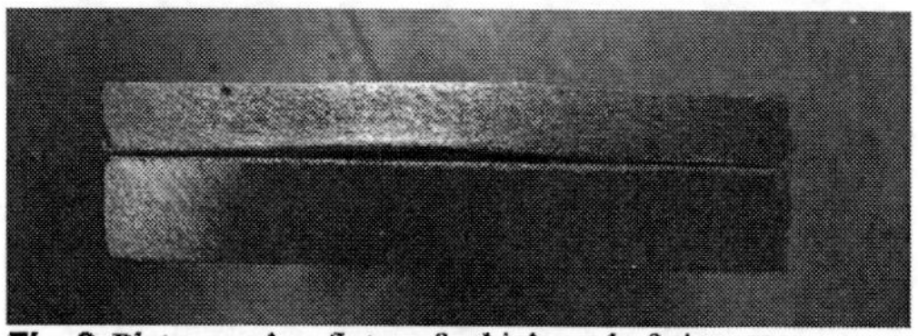

Fig. 9. Plates used as fixture for high-cycle fatigue tests

2.5 Verification of the shaker

The shaker was verified by carrying out the same sweeps as described in the previous part, but without the beam. The two plates were stilled fixed on the armature with the same tightening torque. The accelerations of the three accelerometers (1), (2) and (3) are plotted in **Fig. 10** for each level of excitation. The transversal accelerations are negligible. The spurious displacements observed in the experiments of the previous section are completely caused by the beam resonance.

The mass of the fixture and the specimen is only 1.8 kg. This is way below the shaker's capacity according to its documentation. We have observed the same phenomenon on other shakers.

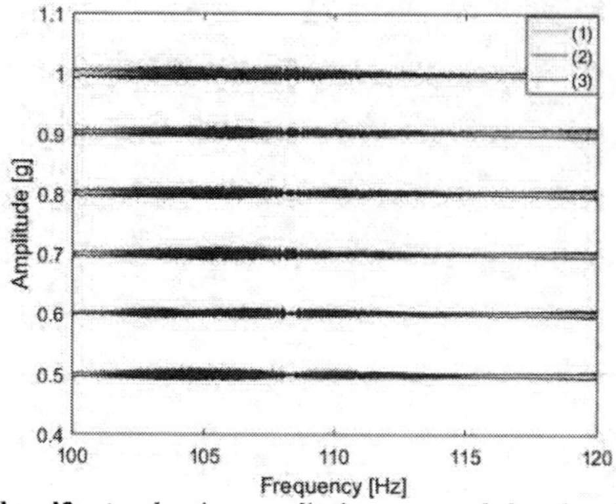

Fig. 10. Acceleration amplitudes measured by the three accelerometers without the beam

3 Improvement of the test bench

3.1. Rotation of the armature

Firstly, we tried to equilibrate the axis by placing two beams symmetrically on the armature, but we didn't observe any improvement. Indeed, the movement is disturbed because of the specimen resonance. Regarding the variabilities of the geometrical dimensions and of the positioning in the clamp, two beams cannot have exactly the same resonance frequency, and then the same dynamic response. As a consequence, the resonance of one specimen disrupted the resonance of the other and reciprocally.

A similar observation was noted by Rognon [12]. He carried out fatigue tests on an electrodynamic shaker with three specimens placed at 120°. This configuration was sufficient to reduce the transversal movements, but when a crack occurred on one of the beams, the modal parameters changed. It was then impossible to continue the test for the two other beams because of the change in the excitation.

Our proposal is to ensure a high rigidity in the direction of the beam. For that, we realised a mounting using roller bearings, separated from the shaker. A sketch presenting the setup is visible on **Fig. 11**.

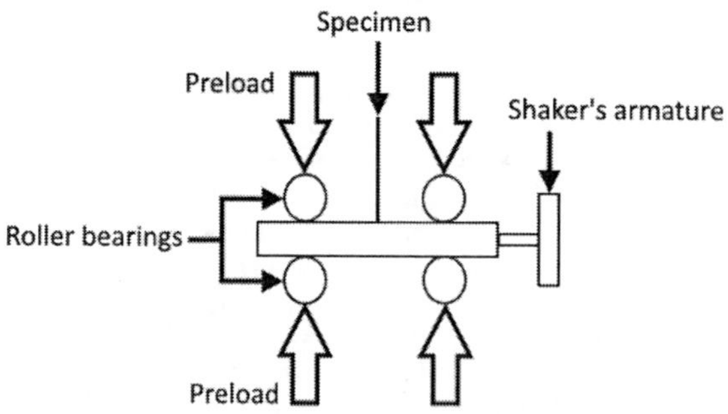

Fig. 11. New setup designed to avoid rotation of the fixture

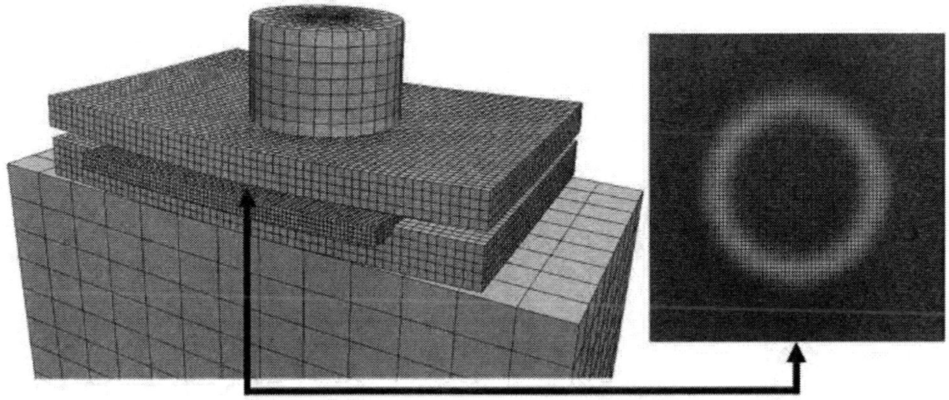

Fig. 12. Deformed shape and contact pressure between the upper plate and the beam. Red color (external area) corresponds to no pressure and blue color (internal area) to maximal pressure. The arrows indicate the same point on both views.

The preload and the high rigidity of the chosen roller bearings guarantee that not a rotational movement will appear. The shaker axis is not perturbed by the resonance of the sample. Another advantage is that it is possible to mount and unmount the specimen without submitting the shaker to tightening forces. Realizing such a setup is also less expensive than using an oversized electrodynamic shaker.

However, the preload will increase the required force to drive the experiment. The maximal attainable acceleration will be reduced because not only the mass of the moving part has to be considered. There is a dilemma between this maximal capacity of the shaker and the transverse rigidity.

3.2. Contact surface in the clamp

A non-linear quasi-static finite element analysis has been carried out in Code_Aster to study the contact in the fixture. The two objectives are:

- have a continuous contact along the edge of the fixture, where the beam is located.
- use only one screw for the tightening. Indeed, using multiple screw yields to a complicated mounting and affects repeatability. It's easier to control only one tightening torque.

A first study has been realized to see if it's possible to find a particular shape for the clamping parts. **Fig. 13** presents a sketch of the simulation. The load is applied thanks to a cylinder which represents the screw. A displacement is imposed on its upper face. The specimen is sandwiched between two parts and the whole is pressed against a last part embedded at the opposite. This is useful to let the part under the sample deform freely. Contacts are considered without friction on each surface because displacements are very small. The surface of the cylinder where the load is applied is blocked in the transverse directions x and y. The two plates and the specimens can't only be maintained by contact in Code_Aster, so springs with negligible stiffness have been added in the x direction and they are clamped in the z direction on the faces where the beam is not emerging.

Fig. 12 shows the deformed shape with a scale factor of 100 and the contact pressure on the upper surface of the specimen. Almost the same contact pressure can be observed on the lower surface. It is clear that the specimen is clamped only close to the screw. The free length of the beam with such a fixture is not well defined. The conclusion was the same with other shapes for the two plates, with different thickness, angles, …

Some setups can be found in the literature where the fixture is realized with cylinders. This kind of embedding can be used when the specimen is symmetric. If not, the vibration of the beam from one side of the cylinder will have an impact on the other side. This phenomenon is problematic in fatigue tests. Even if we consider that the specimen is perfectly symmetric at the beginning of the test, cracking will not appear symmetrically.

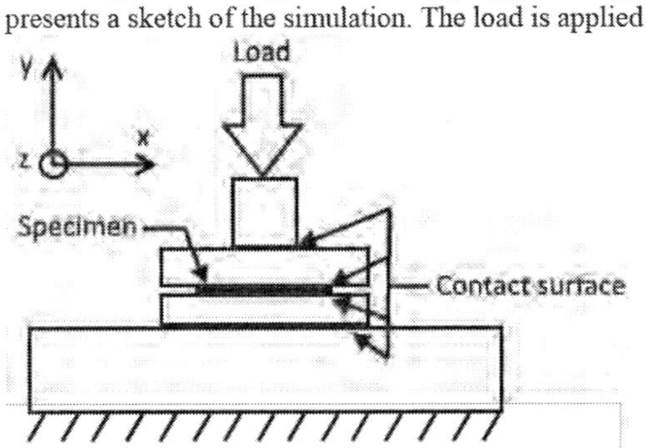

Fig. 13. Boundary conditions for the simulation in Code_Aster

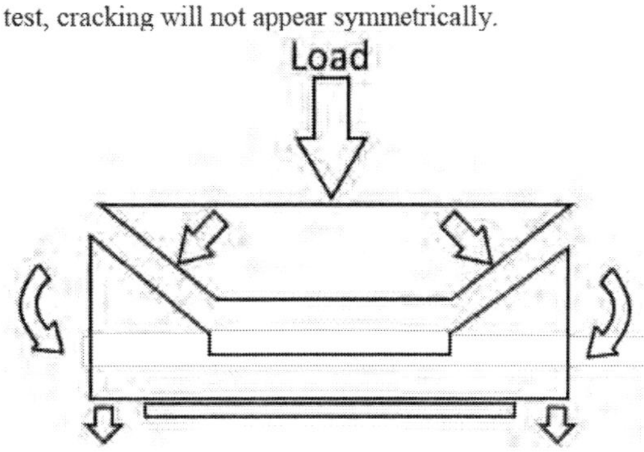

Fig. 14. Representation of the forces in the upper parts to ensure a good fixture

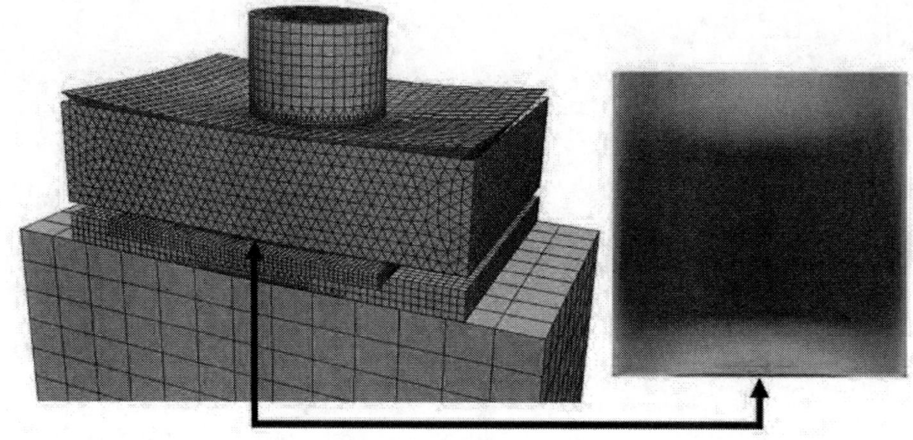

Fig. 15. Deformed shape and contact pressure between the upper piece and the beam. Red color (internal area) corresponds to no pressure and blue color (external area) to maximal pressure. The arrows indicate the same point on both views.

Finally, the solution found to this problem is to use two pieces instead of one for the upper plate. **Fig. 14** explains the principle used. In this case, the contact pressure will be maximal on all the fixture edges of the specimen, determining clearly the limits of the embedding. The results of a new simulation with this geometry are shown on **Fig. 15**.

The pressure distribution is at its expected maximal where the beam goes out the fixture. On this side, the contact force is not constant along the edge. This is due to the shape of the embedding pieces: their width is bigger than their length. Using a square instead of this rectangle would give a more regular contact pressure, but we want to keep the possibility of using beam with variable width.

4 New test bench

4.1. Experimental setup

The test bench shown on **Fig. 16** was created using the improvements detailed previously. Some attention was paid to the alignment and the preloading of the roller bearings.

The tests performed with the new setup are exactly the same as in part 2.3. The same specimen is used and accelerometers are placed such that equivalences can be found with the experiment in **Fig. 1**. The drive is controlled with accelerometer (1). The same acceleration is measured by the triaxial accelerometer (2) and compared. The rotational displacement observed with the previous experiment is measured by the accelerometers (2) and (3). Velocity is measured with the vibrometer at the same point.

4.2. Results and discussion

We can see on **Fig. 17** that the fixture is not rotating because of the resonance of the beam. The curves correspond to the acceleration given by (2) in the direction of the specimen and (3) for an excitation of 1 g. They can be compared to the previous setup with **Fig. 7**. There is still a perturbation around the resonance frequency, but it is highly reduced. With the first test rig, this transversal acceleration represented more than 70% of the driving acceleration. Now, the signal is relatively noisy. A better alignment should reduce the mean value observed but the impact of the beam resonance represents only a change in the amplitude of approximately 0.04 g.

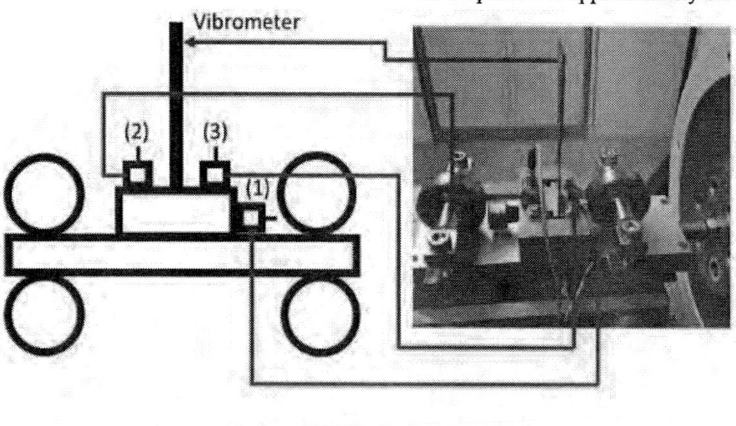

Fig. 16. Illustrative and real view of the setup

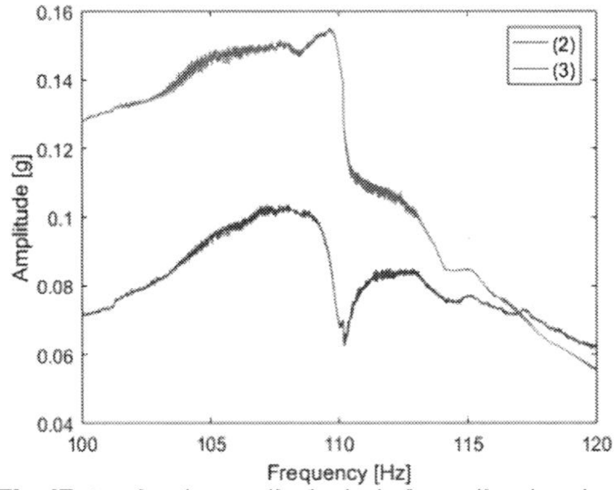

Fig. 17. Acceleration amplitudes in the beam direction given by accelerometer (2) and (3) for and excitation level of 1g.

This observation is confirmed if we compare the results of the driving accelerometer (1) and accelerometer (2) (see **Fig. 18**). They are placed on two opposite points on the fixture and not at the same height. If there is a rotation, we should see the same pattern as in **Fig. 5**. The two signals are identical, only a little peak of 0.01 g can be observed at the resonance, which is negligible. The transmissibility is not dependent from which accelerometer is chosen for the input.

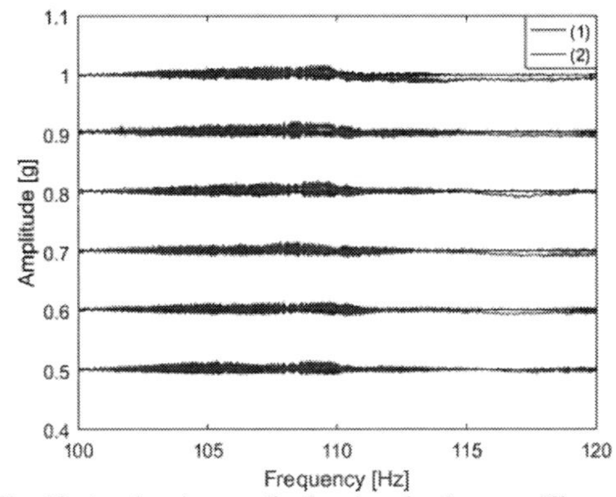

Fig. 18. Acceleration amplitudes given by the controlling accelerometer (1) and accelerometer (2) in the direction of the excitation.

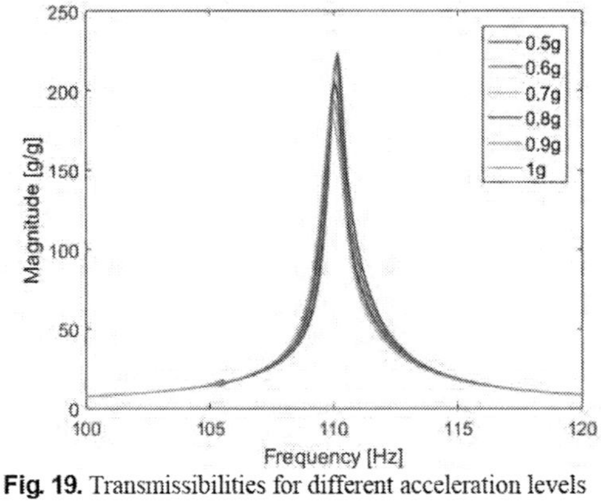

Fig. 19. Transmissibilities for different acceleration levels

Transmissibilities are plotted on **Fig. 19**. There is still a modification in the response when the excitation is changing. However, this effect is reduced. The increase in damping (**Fig. 20**) and the softening (**Fig. 21**) are less important than with the first setup.

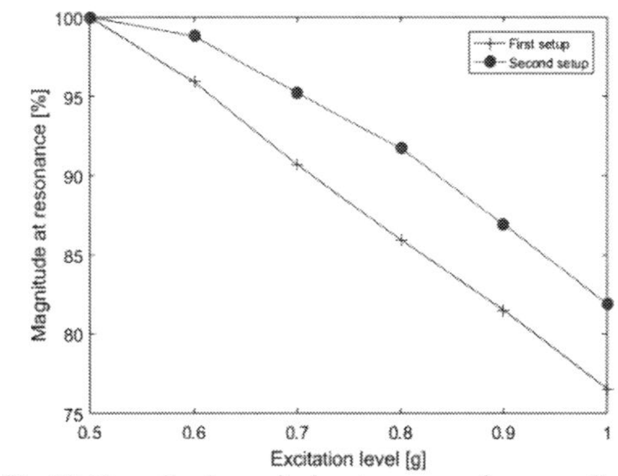

Fig. 20. Normalized magnitudes at resonance frequency for both setups according to the 0.5 g excitation level.

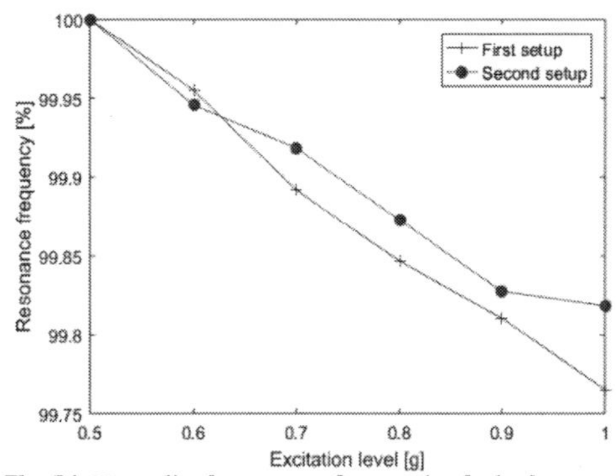

Fig. 21. Normalized resonance frequencies for both setups according to the 0.5 g excitation level.

The improvement in the results show that there are multiple causes for this non-linearity. Other hypotheses are still under study and will be presented in future works.

Conclusion

A test bench for vibratory fatigue experiments was developed. Non-linearities were observed with a classical setup for this kind of test. The fixture has a huge impact on the results and the specimen can distort the excitation movement. Solutions were investigated and taken into account for the design of a new test rig. The proposed fixture enables a better reproducibility in the experiment by ensuring a well-defined contact surface and an easy control of the tightening with only one screw. The transversal rigidity is also controlled and ensures a good excitation. However, the non-linearity was not completely eliminated and the study is still under consideration.

Acknowledgements

The authors gratefully acknowledge the financial support provided by "La Région Normandie" and the European Union through the large-scale research network TERA (MRT MADNESS) – "Grands Réseaux de Recherche Territoire, Environnement, Risques, Agronomie" (Technological Risks Control Network).

References

1. A.B.M Abdullah, Development of a closed-loop resonant fatigue testing methodology and experimental life test of aluminum alloy, University of Akron (2010)
2. A. Angeli, B. Cornelis, M. Troncossi, Synthesis of Sine-on-Random vibration profiles for accelerated life tests based on fatigue damage spectrum equivalence, *Mechanical Systems and Signal Processing* **103**, 340-351 (2018)
3. J. Ge, Y. Sun, S. Zhou, L. Zhang, Y. Zhang, Q. Zhang, A hybrid frequency-time domain method for predicting multiaxial fatigue life of 7075-T6 aluminium alloy under random loading, *Fatigue Fract. Eng. Mater. Struct.* **38**, 247-256 (2015)
4. S. Zhou, Y. Sun, L. Guo, Random fatigue life prediction of carbon fiber-reinforced composite laminate based on hybrid time-frequency domain method, *Random Advanced Composite Materials* (2016)
5. B. Van Hooreweder, D. Moens, R. Boonen, P. Sas, Fatigue strength analysis of notched aluminium specimens using the highly stressed volume method, *Fatigue Fract. Eng. Mater. Struct.* **35**, 154-159 (2011)
6. T. J. George, J. Seidt, M.-H. Herman Shen, T. Nicholas, C. J. Cross, Development of a novel vibration-based fatigue testing methodology, *International Journal of Fatigue* **26**, 477-486 (2004)
7. E. Habtour, D. P. Cole, J. C. Riddick, V. Weiss, M. Robeson, R. Sridharan, A. Dasgupta, Detection of fatigue damage precursor using a nonlinear vibration approach, *Struct. Control Health Monit.* **23**, 1442-1463 (2016)
8. M. Hammami, A. El Mahi, C. Karra, M. Haddar, Vibration non linéaire des composites stratifiés sains et endommagés par délaminage, *CFA 2016 / VISHNO*, 2435-2440 (2016)
9. M. Paulus, A. Dasgupta, E. Habtour, Life estimation model of a cantilevered beam subjected to complex random vibration, *Fatigue Fract. Eng. Mater. Struct.* **35**, 1058-1070 (2012)
10. Y. Li, B. Sun, J. Fang, X.-F. Liu, T. Liang, G.-B. Cai, *Fatigue Fract. Eng. Mater. Struct.*, 1-8 (2017)
11. B. Ellyson, M. Brochu, M. Brochu, Characterization of bending vibration fatigue of SLM fabricated Ti-6Al-4V, *International Journal of Fatigue* **99**, 25-34 (2017)
12. H. Rognon, Comportement en fatigue sous environnement vibratoire : Prise en compte de la plasticité au sein des méthodes spectrales, Ecole Centrale Paris (2013)
13. D. Zanellati, D. Benasciutti, R. Tovo, Vibration fatigue tests by tri-axis shaker: design of an innovative system for uncoupled bending/torsion loading, *Procedia Structural Integrity* **8**, 92-101 (2018)
14. H. Sohn, C. R. Farrar, F. M. Hemez, D. D. Shunk, D. W. Stinemates, B. R. Nadler, J. J. Czarnecki, *A Review of Structural Health Monitoring Literature: 1996-2001* (2004)
15. O. S. Salawu, Detection of structural damage through changes in frequency: a review, *Engineering Structures*, Vol. **19**, No. 9, 718-723 (1997)
16. A. C. Altunisik, F. Y. Okur, V. Kahya, Modal parameter identification and vibration based damage detection of a multiple cracked cantilever beam, *Engineering Failure Analysis* **79**, 154-170 (2017)
17. J.-S. Dupuy, Identification des propriétés mécaniques de matériaux composites par analyse vibratoire, Université Montpellier II – Sciences et Techniques du Languedoc (2008)
18. T. Straub, *Experimental Investigation of Crack Initiation* (2016)
19. L. Khalij, C. Gautrelet, A. Guillet, Fatigue curves of a low carbon steel obtained from vibration experiments with an electrodynamic shaker, *Materials and Design* **86**, 640-648 (2015)

Reliability-based fatigue life estimation of shear riveted connections considering dependency of rivet hole failures

Davide Leonetti[1,*], *Johan* Maljaars[1, 2], and *H.H. (Bert)* Snijder[1]

[1]Eindhoven University of Technology, Department of the Built Environment, Eindhoven, the Netherlands
[2]TNO, Delft, the Netherlands

Abstract. Standards and guidelines for the fatigue design of riveted connections make use of a stress range-endurance (S-N) curve based on the net section stress range regardless of the number and the position of the rivets. Almost all tests on which S-N curves are based, are performed with a minimum number of rivets. However, the number of rivets in a row is expected to increase the fail-safe behaviour of the connection, whereas the number of rows is supposed to decrease the theoretical stress concentration at the critical locations, and hence these aspects are not considered in the S-N curves. This paper presents a numerical model predicting the fatigue life of riveted connections by performing a system reliability analysis on a double cover plated riveted butt joint. The connection is considered in three geometries, with different number of rivets in a row and different number of rows. The stress state in the connection is evaluated using a finite element model in which the friction coefficient and the clamping force in the rivets are considered in a deterministic manner. The probability of failure is evaluated for the main plate, and fatigue failure is assumed to be originating at the sides of the rivet holes, the critical locations, or hot-spots. The notch stress approach is applied to assess the fatigue life, considered to be a stochastic quantity. Unlike other system reliability models available in the literature, the evaluation of the probability of failure takes into account the stochastic dependence between the failures at each critical location modelled as a parallel system, which means considering the change of the state of stress in the connection when a ligament between two rivets fails. A sensitivity study is performed to evaluate the effect of the pretension in the rivet and the friction coefficient on the fatigue life.

1 Introduction

Most of the steel bridge structures designed in the 19th century until about the second half of the 20th century were built by employing built-up structural members assembled by hot riveting, i.e. the manufacturing process of installing rivets by hot-forming one of the heads, usually done in-situ. To date, a number of riveted bridges is still in service. Making available models and methods able to estimate the safety of the riveted connection with respect to fatigue is a topic of high relevance for these bridges [1].
Several variables have been shown to affect the fatigue performance of riveted structures. Extensive reviews are reported in [2, 3], among others. In case of a shear connection, there is experimental evidence that fatigue failure does not originate in the rivets but in the plate. This means that the fatigue life depends on features such as the relative size of the hole with respect to the plate, the method of hole forming, the bearing condition of the rivet with respect to the hole and the clamping force [4]. The clamping force provided by the rivet is due to the circumstance that when the rivet cools down it tends to reduce in length, but it is constrained by the plates, determining the grip length. The scatter shown in fatigue test data of riveted joints has been attributed not only to the variability of the fatigue properties within the material, but also to the clamping force that varies from rivet to rivet [5]. It has been observed that a higher clamping force induces a beneficial effect on the fatigue life [5]–[7]. Several investigations have been performed to evaluate the clamping force. It has been found to be dependent on the type of steel [8] and to increase with the grip length [8] [9]. In addition, it has been shown to be independent of the hot forming process [8], which can include manual or hydraulic hammering, but dependent on the hammering time and temperature [10]. Based on the few results reported in [8], the clamping force has been related to yield stress of the rivet material by evaluating the ratio between the nominal tensile stress in the rivet shank cross-section due to the clamping force, namely the clamping stress, and the measured yield stress of the rivet [11]. The analyses conducted report an average value of 0.70. However, due to the dependence on the grip length, it has to be indicated that average values of 0.61 and 0.77 result

for grip lengths of 75 mm and 125 mm, respectively. In [12], based on a comparison between static load displacement diagrams and numerical simulations, a value of 0.5 is suggested as an optimal value.

The failure of a shear joint with multiple rivets in a row involves a complex response to fatigue loading. Because of the stochastic nature of fatigue, cracks nucleate at the critical locations not at the same time, even when the local stresses are equal. In addition, during crack propagation, both the local stiffness of the plate and the distribution of the load among the fasteners change. This process is known as multiple- site damage [3].

Recently, research has been done in order to assess the reliability of riveted structural details and bridge structures [13]. Based on the test data available in the literature, also new fatigue classes were proposed according to their proneness to fatigue failure [14]. Reliability models have been formulated to quantify the safety of both structural details [15] and bridges [16]. An extensive review of the method employed is reported in [17]. In [18] a probabilistic analysis was performed to estimate the probabilistic stress range vs endurance field (P-S-N curves) of a riveted shear splice. In order to do so, the strain-life data related to the plain material have been employed in combination with the results of a finite element (FE) model by employing linear elastic material properties. In [19] a probabilistic approach has been presented to estimate the fatigue resistance and its scatter for a double shear riveted butt joint with one rivet per side. The material used was tested to determine the monotonic and cyclic static properties as well as the fatigue crack propagation data. The stress range vs endurance relation was simulated in a Monte Carlo analysis estimating the number of cycles to crack initiation and the subsequent propagation. The results were compared with fatigue test data showing a good agreement. In addition, the results confirm that a large portion of the fatigue life of the connection analysed is determined by the crack nucleation period. A system reliability model has been proposed in [15] in order to estimate the reliability of a cross-girder-to-stringer connection under variable amplitude fatigue loading. A FE model and the theory of critical distances have been used to estimate the fatigue life in each Monte Carlo sample. The model is based on the assumption that collapse due to fatigue may occur in several potential failure modes, resulting from the failure at specific hot-spot. The probability underlying each identified mode of failure can be estimated from the probabilities of failure evaluated at the hot-spots that contribute to it, which are assumed to be statistically independent. Although this assumption may be a good approximation for a particular type of connection, its use might determine a wrong estimation of the survival probability of the considered mode of failure for shear riveted connections. Indeed, after that a failure occurs at a hot-spot, the remaining hot-spots determining the same mode of failure will experience a more severe state of stress with respect to fatigue.

Fig. 1. Definition of the three geometries: 1) riveted double cover butt joint, 2) riveted double cover butt joint with two rivet rows, 3) riveted double cover butt joint with two rivets.

In this paper, the authors present a method to estimate the fatigue life distribution of a double-covered butt riveted connection with multiple rivets

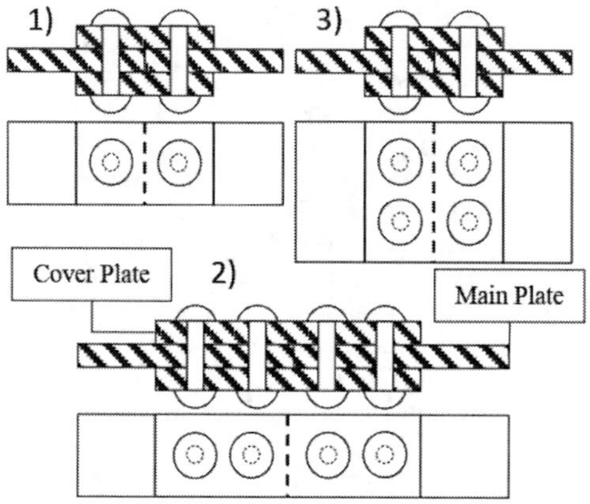

based on a system reliability model. The proposed formulation is able to consider statistical dependency between failures at different hot-spots, using a limited number of FE simulations, which results in modelling the multiple-site damage phenomenon. For the purpose of this study, three geometries have been defined as shown in Figure 1.

2 Models and methods

2.1 Finite element model

The FE model has been generated in ANSYS 17.2 for the purpose of evaluating the state of stress in the considered connection under the external loads. For each geometry, the FE model consists of the following bodies:

- The main plates.
- The cover plates.
- The rivets.

The diameter of the holes and the rivet shank are set equal to 20 mm, the diameter of the rivet head is 30 mm, the ratio between the gross and net section areas of the plates is 1.4 and the edge distance in the loading direction is 70 mm. The thickness of the main plate is set to be equal to the thickness of each cover plate. For this reason, failure is expected to occur in either one of the main plates that are assumed to be independent with respect to their failure behaviour. If a crack is not present, the

https://doi.org/10.1051/matecconf/201816510008

geometry is three times symmetric. When a ligament fails, only one plane of symmetry exists. Thus, half of the connection geometry is modelled to reduce the computational effort, see Figure 2. The element type
employed to discretize the geometrical model is a 20-node isoparametric element (SOLID186).

Contacts and interaction between the plates and the rivets are simulated by applying contact and target elements, (CONTA174) and (TARGE170) respectively, further meshing the surfaces that might penetrate each other. The Augmented Lagrange formulation is used in order to simulate contact interaction. The Coulomb friction model is used to simulate the friction between plates and between the rivet and plate. The displacement vectors in the tangential direction at the two contact surfaces equal each other when the following

Fig. 2 FE model of half the connection.

condition occurs:

$$\|\boldsymbol{t}_T\| < \mu_s \boldsymbol{t}_N \tag{1}$$

where $\|\boldsymbol{t}_T\|$ is the vector of the tangential stress evaluated at the contact nodes, μ_s is the static friction coefficient and $\boldsymbol{t}_N$ is the vector of the normal stresses evaluated at the contact nodes. Based on the values reported in [20], [21], the friction coefficient has been set equal to 0.33. The rivet pretension has been implemented by placing a negative thermal load to the rivet shank. The value of the thermal load applied to the rivet is such that the average tensile stress in the rivet is 188 MPa, which is estimated by dividing the mean yield stress for a St44 steel grade, evaluated according to [22], and equal to 312MPa by 0.6, according to the findings of [8, 11]. The connection is then loaded by applying a far field pressure to the main plate. The thermal and pressure loads have been implemented in three "Steps":

1. The thermal load is applied (and hereafter kept constant);
2. The far field load is applied to the maximum value;
3. The far field load is reduced to zero.

The third step is performed to evaluate the effects of the load-sequence, which is expected since the model is non-linear.

A convergence study has been performed to find the mesh size, non-linear solution controls and the contact settings, i.e. the normal contact stiffness and allowable penetration, to be used in the analysis.

2.2 Notch stress estimation

The stress range to be used for the fatigue life estimation has been estimated by employing the FE method in conjunction with the gradient method in the FKM Guidelines [23], in order to consider both the influence of the stress concentration and the stress gradient on the fatigue life. In this method, the fatigue notch stress is defined as

$$\sigma = \sigma_{nom} K_f \tag{2}$$

$$K_f = K_t / n_\sigma \tag{3}$$

where K_f is the fatigue notch factor, K_t is the theoretical stress concentration factor and n_σ accounts for the stress gradient effect on the fatigue life. It depends on both the material, by the constants a_G and b_G and the ultimate tensile strength σ_u [MPa], and the stress gradient G_σ [MPa/mm] at the notch root, by the following:

$$n_\sigma = \begin{cases} 1 + G_\sigma \cdot mm \cdot 10^{-\left(a_G - 0.5 + \frac{\sigma_u}{b_G \cdot MPa}\right)} & G_\sigma \leq 0.1 \\ 1 + \sqrt{G_\sigma \cdot mm} \cdot 10^{-\left(a_G + \frac{\sigma_u}{b_G \cdot MPa}\right)} & 0.1 < G_\sigma \leq 1 \\ 1 + \sqrt[4]{G_\sigma \cdot mm} \cdot 10^{-\left(a_G + \frac{\sigma_u}{b_G \cdot MPa}\right)} & 1 < G_\sigma \leq 100 \end{cases} \tag{4}$$

where a_G and b_G are dimensionless material constants respectively equal to 0.50 and 2700 for non stainless steels. The maximum fatigue notch stress, σ_{max}, and the minimum fatigue notch stress, σ_{min}, are evaluated at the maximum applied load ($P = P_{max}$) and the minimum applied load ($P = P_{min}$), respectively. The mean stress (σ_m^*), and the stress amplitude (σ_a) at the notch can be evaluated. The mean stress is marked with an asterisk because it might be corrected in case elastic shakedown would occur. Indeed, when dealing with an elastic ideally plastic material, elastic shakedown occurs if the maximum notch stress exceeds the yield stress (σ_y) and the related stress amplitude is smaller than the yield stress. This assumption is also realistic for steels, which exhibit a yield plateau. The corrected mean stress (σ_m) is:

$$\begin{cases} \sigma_m = \sigma_y - \sigma_a & \text{if } \sigma_{max}^* > \sigma_y \text{ and } \sigma_a < \sigma_y \\ \sigma_m = \sigma_m^* & \text{if } \sigma_{max}^* \leq \sigma_y \end{cases} \tag{5}$$

The mean stress and the stress amplitude characterizing the stress fluctuation at each hotspot location are then further modified by employing the Goodman mean stress correction model in order to evaluate the equivalent stress amplitude ($S_a = \Delta S/2$) at a stress ratio (R) equal to -1:

$$S_a = \frac{\sigma_a}{1-\frac{\sigma_m}{\sigma_u}} \qquad (6)$$

2.2 Fatigue resistance estimation

In [22] a method to estimate the tensile strength properties of structural steels from the nominal yield stress is proposed based on an extensive analysis of test data available in literature. In particular, the ultimate tensile strength and the yield stress of low strength structural steels are found to be distributed following a Bivariate Normal Probability density function (Pdf). The mean values of the distribution, the correlation matrix and the coefficient of variation of the Pdf are reported in [22]. According to [22], the mean yield stress and the mean ultimate tensile strength for St52 steel grade are 409 MPa and 613 MPa, respectively. The stress range vs. endurance relation is formalized by employing the Basquin relation

$$log_{10} N = log_{10} C_0 + m\, log_{10}(\Delta S) \qquad (7)$$

where, $\log_{10} C_0$ and m can be derived based on an estimation of the fatigue limit (ΔS_0), the fatigue resistance at two million cycles, and the fatigue strength coefficient (σ_f'), the intercept of the elastic S-N curve with the stress range axis, as follows:

$$m = \frac{log_{10} N_0}{log_{10}(\Delta S_0) - log_{10}(2\sigma_f')} \qquad (8)$$

$$log_{10} C_0 = log_{10} N_0 - m\; log_{10}(\Delta S_0) \qquad (9)$$

The fatigue limit is determined by making use of the relation prosed in [24], where the stress amplitude S_0 determining the fatigue limit at $N_0 = 2$ million cycles and R=-1is related to the ultimate tensile strength by:

$$S_0 = 0.5\; N(1,0.15)\; \sigma_u k_a \qquad (10)$$

Where N denotes a Normal distribution with average and standard deviation equal to 1 and 0.15, respectively. The surface factor (k_a) is here employed in order to take surface finishing into account. The following relation, valid for machined surfaces, is considered [24]:

$$k_a = 4.45\; \sigma_u^{-0.2653}\; LN(1,0.060) \qquad (11)$$

Where the second term of the equation is a lognormal distribution with average and standard deviation equal to 1 and 0.06, respectively. The evaluated fatigue surface factor, k_a, also takes into account the variability of the surface finishing. The mean and the standard deviation of the fatigue strength coefficient are estimated by employing the relations reported in [25], [26] for carbon steels:

$$E(\sigma_f') = 0.95\; E(\sigma_u) + 400 \qquad (12)$$

$$std(\sigma_f') = 0.95 \frac{std(\sigma_u)}{\rho_{\sigma_u,\sigma_f'}} \qquad (13)$$

where $\rho_{\sigma_u,\sigma_f'}$ is the correlation coefficient, which is reported to be equal to 0.8, and E(·) is the expectation operator. In the framework of this paper, the mean value of the ultimate tensile strength and the yield stress are assigned to each hotspot. Then, to estimate the fatigue variability at each critical location, Equations (8)-(13) have been employed.

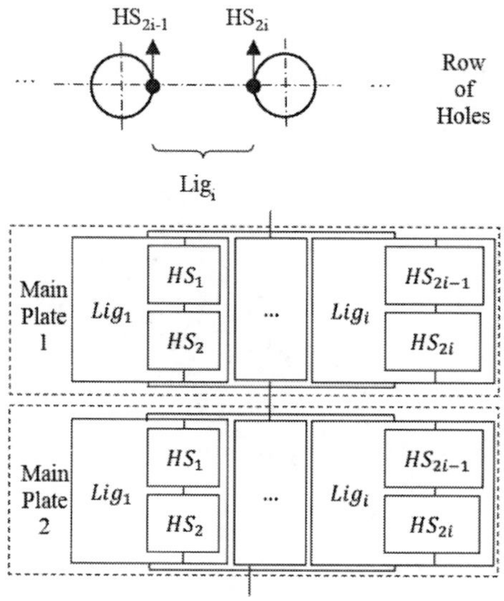

Fig. 3 Relation between ligament and hot spot location for a row of holes and box diagram showing the system reliability model employed to describe the failure in each main plate.

2.3 System reliability modelling

To evaluate the fatigue life of the selected geometries a system reliability model should be set. Given the considered geometry, the fatigue life is determined by the failure of either one of the main plates.

In general, the connection consists of a number of components and thus different failure modes might be expected, each one of these determined by failure at one or more hot-spots. Thus, since the occurrence of any of the failure modes determines the failure of the connection, the connection is considered as a series system of the failure modes. In this paper, since only one failure mode is considered, the probability that the system fails equals the probability that failure occurs in the considered failure mode. However, it is recognized that other modes of failure might exist.

According to Figure 3, the considered mode of failure is assumed to take place when all the ligaments in one of the two main plates have failed.

Whit respect to one main plate, its failure can be assumed as a parallel system:

$$Pr\{Mode\} = Pr\{Lig_1 \cap ... \cap Lig_i\} \qquad (1)$$

The failure of a ligament is directly associated to the failure at either one of the hot-spots related to it, see Figure 3. This means that the probability that a ligament fails is equal to:

$$Pr\{Lig_i\} = max[Pr\{HS_{2i-1}\};\ Pr\{HS_{2i}\}] \qquad (2)$$

since the failure events at the hotspots are fully correlated. The full correlation is due to the fact that if one of the two hotspot fails also the ligament fails, which automatically means failure at the other hot-spot, see Figure 3.

2.4 Solution of the reliability model

The Monte Carlo method is employed to evaluate the probability of failure of the considered mode at a certain stress range value. The FE model described in section 2.1 is used to determine the stress state in the connection in any possible combination of ligament failure that does not cause static failure of the main plate at $P=P_{max}$, or plastic shakedown under cyclic loading. At each stress range the simulated dataset of $k=10000$ samples has been produced by employing the following algorithm:

a) Based on the selected geometry a FE model is created to evaluate the state of stress in the connection for the identified failure scenarios of interest.
b) The value of the stress range to be applied to the far field cross-section is chose together with the stress ratio R. The FE model is solved.
c) At each hot-spot, the maximum principal stresses are extracted from the FE model at $P=P_{max}$ and $P=P_{min}$.
d) The stress values are corrected using Equations (4)-(6) to estimate the equivalent stress amplitude for a fully reversed cyclic loading.
e) At each hot-spot, the following are assigned:
   - The values of the monotonic strength properties, which are the same at each hot-spot;
   - The parameters of the Basquin relation, by randomly sampling a value of the fatigue limit and the fatigue strength coefficient;
   - The value of the cumulative, the critical and the residual Palmgren-Miner damage;
   - The hot-spot status, which is "not failed" at the first iteration, and can be "failed" in a later stage.
f) A state parameter storing the calculated number of cycles to failure for the entire connection is initialized for the current k-th sample.
g) For each hot-spot, the predicted number of cycles to failure is determined. The minimum non-zero value obtained represents the number of cycles to the next failure at a certain hot-spot. Based on this value, the partial damage due to the current iteration is calculated for each hot-spot location. The cumulated damage, the cumulated number of cycles and the failure status are updated at each critical location.
h) The damage at a certain hot-spot exceeds the critical damage, and the corresponding ligament is assumed to be cracked, determining the failure scenario of interest in the next iteration. Thus, based on the results of the current iteration, the failed critical location is determined as well as the failed ligament. Thus, the stress state belonging to this new configuration is evaluated and employed in the next iteration.
i) The steps from g) to i) are repeated until either plastic shakedown occurs ($\sigma_a > \sigma_y$, see Equation (5)), or the remaining net-section is no longer able to carry the maximum applied load.
j) The steps from e) to j) are repeated until the total number of samples is reached.

The steps from a) to d) have been performed in ANSYS Workbench environment. Then, the results have been exported to be processed in MatLab for the following steps from e) to k).

3 Results and discussion

Geometries 1 and 2 are compared in terms of S-N curves in which the nominal net section stress range is plot against the number of cycles to failure, assumed to occur when a hot-spot fails.

In order to compare geometries 1 and 3, the results are expressed in terms of cumulative distribution functions (Cdf) of the fatigue life at a given net stress range. For the analysis, two arbitrary net section stress ranges have been selected, 140 MPa and 161 MPa, considering a zero pulsating stress cycle, i.e. stress ratio $R = 0$.

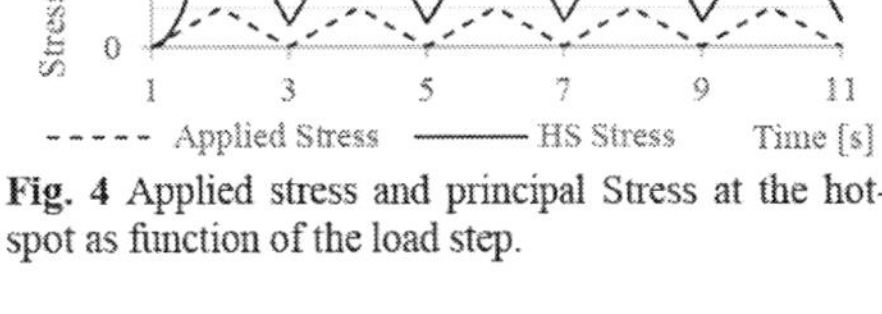

Fig. 4 Applied stress and principal Stress at the hot-spot as function of the load step.

3.1 Geometry 1

When the external load drops to zero after reaching the maximum load in the first cycle, a residual state of stress remains in the main plate. This phenomenon is found to stabilize after the first cycle, as shown in Figure 4 where five constant amplitude stress cycles ($\Delta\sigma = 100$ MPa, R = 0) have been applied at the gross section and the corresponding stress at the hot-spot has been plot. This can be addressed to the nonlinearities introduced by slipping at the contacting surface. Despite at macro level the friction condition ($P < \mu F_c$) may be satisfied and slipping should not occur, in the vicinity of highly stressed locations, the friction condition imposed by Equation (1) might not be satisfied. This results into local slipping that is not recovered during the unloading phase. Thus, a load sequence effect occurs. Each critical location is subjected to a fatigue loading characterized by both stress concentration and a stress ratio higher than the nominal one. In addition, when the contact is not able to carry the load by friction, the rivet carries part of the total load by shear, so the higher the maximum net section stress applied, the higher is the theoretical stress concentration factor (K_t) and thus, the fatigue notch factor (K_f). At a relatively low far field stress range level, the load is transferred mainly by friction having smaller stress concentration at the hotspot. The results, in terms of nominal S-N curve, in which the stress range refers to the net section, are plot in Figure 5 where a comparison is made with the S-N curve of the parent material for R=-1. In the finite life region, the median fatigue life is directly obtained by the Monte Carlo samples resulting from the reliability model for each selected stress ranges. The median fatigue limit has been estimated from the same Monte Carlo samples by employing a Probit analysis: at each stress range level, the ratio between failed and non-failed simulations has been interpolated by a normal, lognormal and generalized extreme value Cdf using the least square method in order to estimate the parameters of the distribution. Based on the coefficient of determination, the generalized

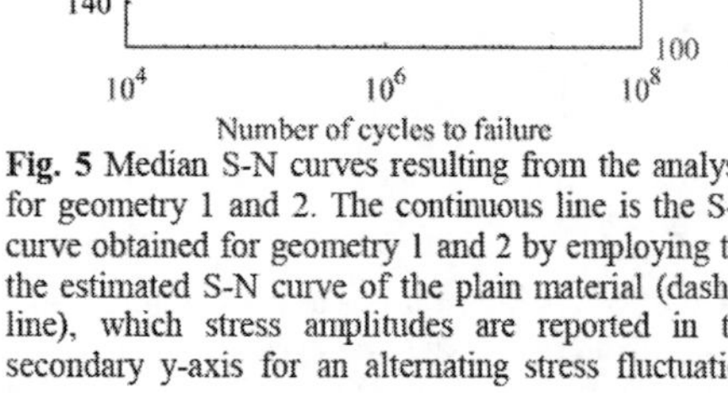

Fig. 5 Median S-N curves resulting from the analysis for geometry 1 and 2. The continuous line is the S-N curve obtained for geometry 1 and 2 by employing the the estimated S-N curve of the plain material (dashed line), which stress amplitudes are reported in the secondary y-axis for an alternating stress fluctuation (R=-1).

extreme value distribution has been selected and employed to estimate the median value of the fatigue limit (154 MPa).

At high far field stress ranges, both the stress concentration and the stress ratio at the hotspot increase and consequently, the resulting S-N curve tends to be steeper than the S-N curve of the plain material. At low stress range levels, the S-N curve becomes shallower and then approaches the fatigue limit. This is in line with the findings of [14] in terms of having a stepwise S-N curve describing the fatigue life of shear riveted connections.

3.2 Geometry 2

For geometry 2, the generalized extreme value distribution has been selected to estimate the fatigue limit. The estimated median value of the fatigue limit is 172 MPa. As it is shown in Figure 5, at the same net section stress range level the fatigue life of geometry 2 is higher compared to geometry 1. This because the maximum principal stress resulting from the FE analysis at the most stressed critical location is lower for geometry 2 as compared to geometry 1. Moreover, a higher percentage of the total load is transferred by friction and, eventually, transferred by bearing of the second rivet. Hence, it is advantageous to have multiple rows of rivets.

3.3 Geometry 3

At the hot-spot locations, the maximum principal stresses obtained by the FE model for this case are not significantly different from those obtained for geometry 1. Despite this, the first ligament failure occurs on average at a lower number of cycles. This is caused by the fact that first failure occurs at the weakest hot-spot. Thus with respect to the first failure, the higher the number of rivets in a row, the higher is the number of the hot-spots, the lower is the average life to first failure and the more this distribution is skewed towards lower life. On the other hand, the fail-safety of the connection is higher than for geometry 1, because the connection has some residual fatigue life after failure of the first ligament. This aspect determines a longer detection period with respect to the total life of the joint, meaning higher probability of detecting the damage. The higher is the applied stress range, the lower is the contribution of the residual life after failure of the first ligament to the total fatigue life, see Figure 6. According to these findings, when considering the entire joint, failure of the first ligament occurs earlier than in the geometry 1 but, since there is more redundancy, the additional life associated with failure of the entire joint after failure of the first ligament depends on the stress range level.

3.4 Sensitivity study

For geometry 1, a variation of the clamping force and the friction coefficient is considered and the maximum and minimum principal stresses at the hot-spots are calculated for a far-field stress range of 100 MPa at R=0. The friction coefficient is assumed to
vary between 0.1 and 0.6 whereas the nominal stress in the rivet due to the clamping force is assumed to vary between zero and the nominal yield stress of the rivet material, which is 275 MPa. The results are expressed in terms of maximum principal stress, fatigue notch stress range and stress ratio and are shown in Figure 7. In any case, the theoretical stress concentration factor (K_t) ranges between 2 and 5. It results that at low clamping, K_t is almost independent of the friction coefficient and is close to 5. It is clear that the lower the friction coefficient, the less K_t depends on the clamping. At relatively low clamping and low friction values, no residual stress is present at the hot-spots. At relatively high clamping and high friction values, when the load is mainly transferred by friction, the stress after unloading also tends to approach a minimum. In the region of interest, when $0.2 < \mu < 0.4$ and the clamping stress is about 0.5-0.7 times the yield stress, the higher stress ratios occur (>0.3) and the stress ranges are in the order of 250-400 MPa. The obtained variability suggests that the estimated fatigue life is strongly dependent on the friction coefficient and the clamping stress, confirming the findings in [4]. Moreover, the analysis shows that the phenomena involved determine a complex response to loading and unloading, making it difficult for the double cover riveted butt joint to formulate any "rule of thumb".

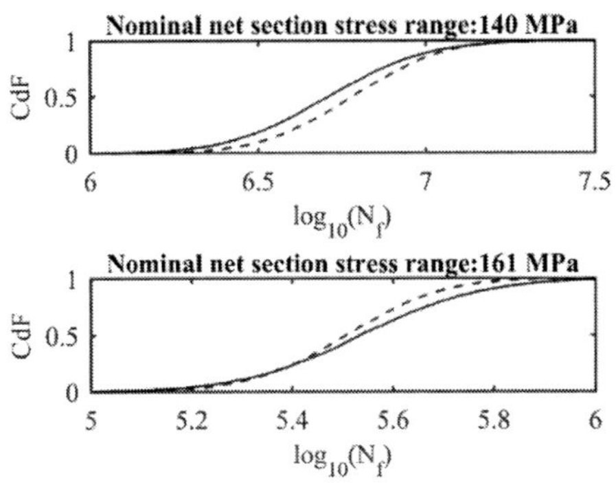

Fig. 6 Cumulative distribution of the number of cycles to failure for geometries 1 (continuous) and 3 (dashed).

4 Conclusions

The position and the number of the rivets have been shown to affect the state of stress in the connection under external loading. In order to evaluate the fatigue life of shear riveted connections, a reliability model and an algorithm to solve it have been proposed in the present paper.

The results highlight that the fatigue life increases with increasing number of rivet rows. By increasing number of rivets in a row, the fail-safety of the connection increases. However, this does not necessarily mean a higher fatigue life since the number of cycles to first ligament failure is found to be decreasing with the increasing number of rivets in a row.

It is recognized that the S-N curve employed in guidelines for fatigue design of riveted connections, as [27], does not address the effect of the rivet number and position on the fatigue life. This because the number and the position of the rivets employed in the specimens used to carry out fatigue tests significantly differs from the details often employed in practice.

The formulated FE model estimates that high stress levels cause higher stress ratios and stress concentration factors than low stress levels. This justifies the findings reported in [14] of a two-slope S-N curve to represent the fatigue life of shear riveted connections.

In accordance with the findings reported in [4]–[7], the fatigue life and stress state at the hotspot location strongly depend on the friction coefficient of the contacting surfaces and on the clamping force exhibited by the rivet. Further investigation is needed to quantify and predict these aspects. Further development is also needed in order to increase the accuracy of the estimation of the probabilistic stress range vs. endurance field. In particular, the presented algorithm does not consider the uncertainties related to the friction coefficient and the clamping force. In addition, since the use of the stress approach is limited to elastic shakedown, employing the notch strain approach would increase the accuracy and the domain of application of the fatigue life estimation.

The authors would like to acknowledge the Dutch infrastructural asset owners Rijkswaterstaat and ProRail and the research organization TNO for their support.

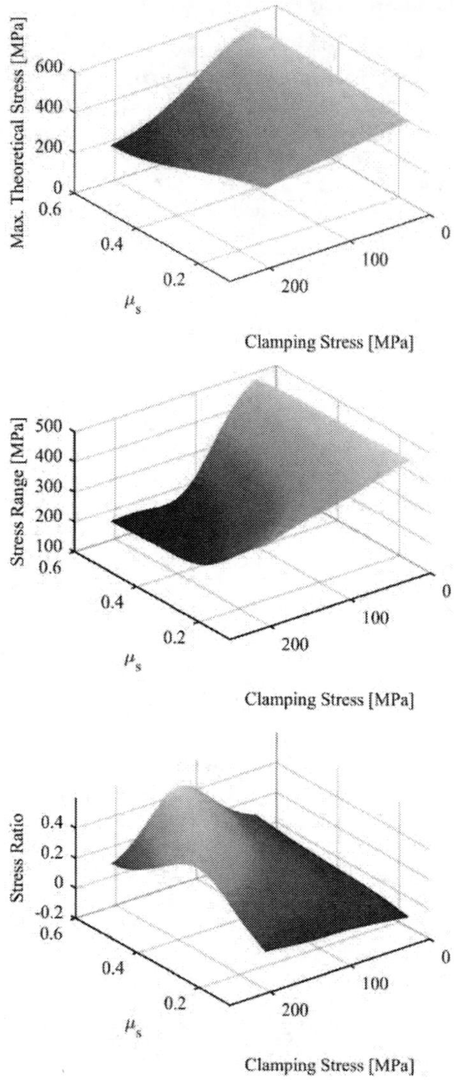

Fig. 7. Maximum theoretical stress, stress range and stress ratio evaluated at a critical location as function of the friction coefficient and the clamping for a stress range of 100 MPa applied to the gross section.

References

[1] K. A. Baker and G. L. Kulak, "Fatigue of riveted connections," *Can. J. Civ. Eng.*, vol. 12, no. 1, pp. 184–191, 1984.

[2] G. L. Kulak, J. W. Fisher, and J. H. A. Struik, *Guide to Design Criteria for Bolted and Riverted Joints*. 2001.

[3] G. T. Hahn and C. A. Rubin, *Structural shear joints analyses, properties and design for repeat loading*. ASME Press, 2005.

[4] J. D. DiBattista, D. E. J. Adamson, and G. L. Kulak, "Fatigue Strength of Riveted Connections," *J. Struct. Eng.*, vol. 124, no. 7, pp. 792–797, Jul. 1998.

[5] J. W. Carter, "Fatigue in Riveted and Bolted Single Lap JointsTitle," *Proc. Am. Soc. Civ. Eng.*, vol. 80, no. 8, pp. 1–35, 1954.

[6] J. F. Parola, E. Chesson, Jr., and W. H. Munse, "Effect of bearing pressure on fatigue strength of riveted connections," 1965.

[7] W. H. Munse and E. Chesson Jr, "Riveted and bolted joints: net section design," *J. Struct. Div.*, vol. 89, no. 1, pp. 107–126, 1963.

[8] W. M. Wilson and F. P. Thomas, "Fatigue Tests of Riveted Joints," no. 79, pp. 173–195, 1938.

[9] F. Baron and E. W. Larson, "High-Strength Bolts in Structural Joints: A Symposium: Comparison of Bolted and Riveted Joints," *Trans. Am. Soc. Civ. Eng.*, vol. 120, no. 1, pp. 1322–1334, 1955.

[10] O. Graf, *Versuche mit nietverbindungen*. J. Springer, 1941.

[11] B. Åkesson, "Fatigue Life of Riveted Railway Bridges," p. 169, 1994.

[12] R. Marmo, "Numerical and experimental investigation on the shear behaviour of riveted connections," 2011.

[13] C. Cremona, B. Eichler, B. Johansson, and T. Larsson, "Improved Assessment Methods for Static and Fatigue Resistance of Old Metallic Railway Bridges," *J. Bridg. Eng.*, vol. 18, no. 11, pp. 1164–1173, 2013.

[14] A. Taras and R. Greiner, "Development and

Application of a Fatigue Class Catalogue for Riveted Bridge Components," *Struct. Eng. Int.*, vol. 20, no. 1, pp. 91–103, Feb. 2010.

[15] B. M. Imam, M. K. Chryssanthopoulos, and D. M. Frangopol, "Fatigue system reliability analysis of riveted railway bridge connections," *Struct. Infrastruct. Eng.*, vol. 8, no. 10, pp. 967–984, 2012.

[16] C. S. Wang, Y. Xu, A. R. Chen, and W. Z. Chen, "System Fatigue Damage Reliability Assessment of Railway Riveted Bridges," *Key Eng. Mater.*, vol. 347, pp. 173–178, 2007.

[17] X. W. Ye, Y. H. Su, and J. P. Han, "A state-of-the-art review on fatigue life assessment of steel bridges," *Mathematical Problems in Engineering*, vol. 2014. Hindawi, pp. 1–13, 10-Feb-2014.

[18] A. M. P. De Jesus, H. Pinto, A. Fernández-Canteli, E. Castillo, and J. A. F. O. Correia, "Fatigue assessment of a riveted shear splice based on a probabilistic model," *Int. J. Fatigue*, vol. 32, no. 2, pp. 453–462, Feb. 2010.

[19] R. F. Sanches, A. M. P. De Jesus, J. A. F. O. Correia, A. L. L. Da Silva, and A. A. Fernandes, "A probabilistic fatigue approach for riveted joints using Monte Carlo simulation," *J. Constr. Steel Res.*, vol. 110, pp. 149–162, 2015.

[20] G. L. Kulak, J. W. Fisher, and J. H. A. Struik, *Guide to design criteria for bolted and riveted joints*, vol. 15, no. 1. 1988.

[21] M. Veljkovic *et al.*, "High-strength tower in steel for wind turbines (Histwin)," 2009.

[22] JCSS, "Probabilistic Model Code - Part 3," *Jt. Comm. Struct. Saf.*, 2000.

[23] B. Hänel, E. Haibach, T. Seeger, G. Wirthgen, and H. Zenner, "Analytical strength assessment of components in mechanical engineering." Forschungskuratoirium Maschinenbau - VDMA Verlag, Frankfurt am Main, 2012.

[24] C. R. Mischke, "Prediction of Stochastic Endurance Strength," *J. Vib. Acoust. Stress Reliab. Des.*, vol. 109, no. 1, p. 113, Jan. 1987.

[25] S. K. Park and F. V Lawrence, "Monte Carlo Simulation of Weldment Fatigue Strength," *J. Constr. Steel Res.*, vol. 12, pp. 279–299, 1989.

[26] J. C. McMahon and F. V. Lawrence, "Predicting fatigue properties through hardness measurements," 1984.

[27] B. Kühn, M. Lukic, A. Nussbaumer, H. P. Günther, and R. Helmerich, "Assessment of Existing Steel Structures: Recommendations for Estimation of Remaining Fatigue Life," Aachen, Delft and Ispra, 2008.

Fatigue life assessment of bolted connections

Aliou Badara Camara[1,*], *Fabienne* Pennec[1], *Sébastien* Durif [1], *Jean-Louis* Robert [1], and *Abdelhamid* Bouchaïr [1]

[1] Université Clermont Auvergne, CNRS, Institut Pascal, F-63000 Clermont-Ferrand, France
aliou_badara.camara@uca.fr, fabienne.pennec@uca.fr, sebastien.durif@uca.fr, j-louis.robert@uca.fr, abdelhamid.bouchair@uca.fr

Abstract.
The work presented in this paper deals with the fatigue damage assessment of bolted joints. The influence of the preload is particularly underlined as it is shown that it strongly improves the fatigue resistance of the bolt whatever the geometrical parameters are. A 3D tee-stub model with the bolt submitted to preload force is simulated by using the Salome-Meca FEM software. A parametric study is also carried out to analyse the influence of the bolt location m and the column flange thickness t_f onto the bolt loading. In order to analyse the effect of the stress concentration generated by the head-shank transition of the screw, a second 3D finite bolt element model is developed. This allows to exhibit multiaxial stress states in the vicinity of the fillet radius. Finally, a multiaxial fatigue post-treatment tool has been implemented on Matlab software for damage assessment purpose. Two multiaxial fatigue criteria approaches contribute to this tool and may be used for fatigue behaviour prediction. The so-called critical plane approach (Dang Van criterion) and the integral approach (Zenner criterion) may consequently be compared for that analysis.

1 Introduction

Bolted joints are used in many industries as steel construction or transportation engineering (for railways and bridges for instance). Indeed bolted assemblies are the most widely used mechanical connections and are often subjected to rough design methods, which generally lead to oversize the assemblies for some aspects without however assuming their safety [1]. These structural elements may have complex behaviour due on the one hand to the load distribution in the threaded connection and on the other hand to the phenomenon of stress concentration. Nowadays when the reduction of the material volume becomes an economical concern, it is generally necessary to take into account the fatigue phenomenon in the design step in order to optimize and to ensure the reliability and safety of the bolted assemblies. The damaging fatigue effect is one process which may modify the local material properties and can lead to cracks occurrence and possibly to the structure failure under the influence of repeated or alternating solicitation. Fatigue failures are often more dangerous than other kind of failures because they generally occur suddenly, without significant prior deformations. It is thus important to correctly design structures submitted to dynamic load especially bolted joints in order to avoid fatigue failures.

For a right understanding of the fatigue behaviour of bolted connections, a tee-stub in which the bolt is submitted to a pretension force is studied. The improvement of the precision of the tightening is really one of the fundamental problems of the modern bolted joints technology [2]. We analyze first the influence of the preload on the bolt behaviour, especially from the stress cycle it experiences. Then a parametric analysis is made on the tee-stub. Finally a fatigue post-treatment tool is implemented on MATLAB software and uses two multiaxial fatigue criteria approaches. These are used to assess the fatigue damage under multiaxial constant amplitude cyclic stress states. The two multiaxial fatigue criteria approaches we implemented are the so called critical plane approach (Dang Van criterion) and the integral approach (Zenner criterion).

2 Presentation of a tee-stub

For metallic construction, a bolted beam-to-column assembly usually ensures the connection between the column flange and the end plate. This connection is most often subjected to a bending moment (Fig.1). The component method consists in decomposing the bolted

aliou_badara.camara@uca.fr

beam-to-column connection into its elementary components among which the most important one is the equivalent tee-stub in the tension zone.

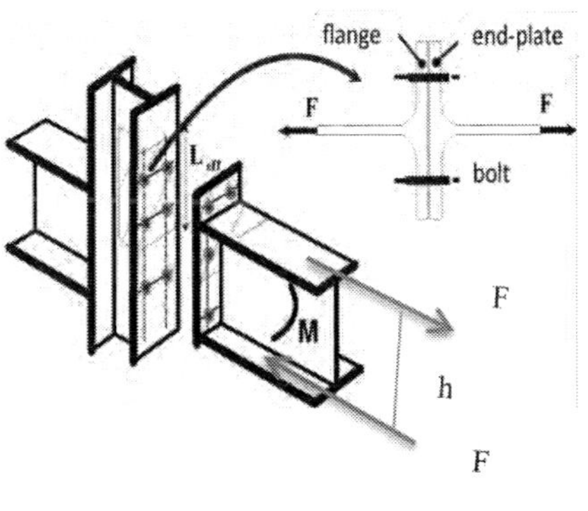

Fig. 1. : Beam-to-column connection and definition of the tee-stub connection

As a matter of fact, the elementary tee-stub is used in the actual standard EN-1993-1-8 to predict the mechanical behaviour of steel bolted connections under static loading [3].

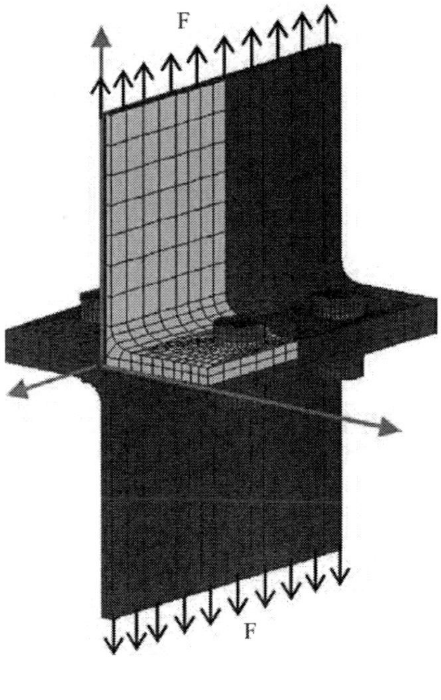

Fig. 2. : Numerical model of Tee-stub [4]

3 Finite element modelling of the tee-stub

3D finite element models have been developed using SALOME-MECA software to quantify the axial force F_B and the bending moment M_B in the bolt throughout the loading of the tee-stub. These models allow thus the calculation of the axial stresses distribution on cross-section of the bolt shank.

3.1. Geometry

One eighth of the tee-stub is modelled by taking into account all the symmetry conditions and the contact zones (Fig.2 and Fig.3).

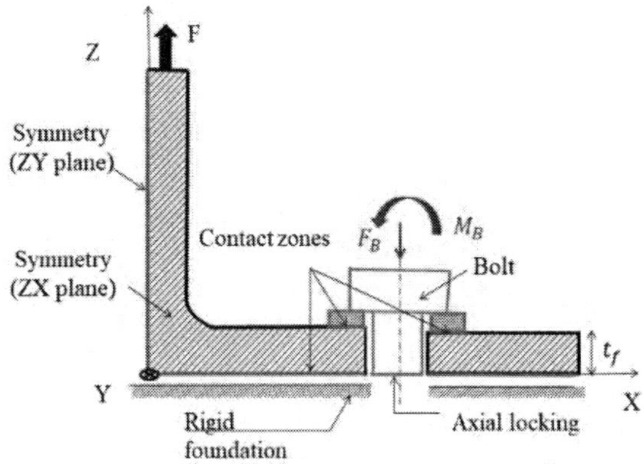

Fig. 3. : Description of symmetry planes and contact zones of the tee-stub model

The two geometrical parameters of the tee-stub are the bolt location m and the column flange thickness tf (fig.4.). Constant amplitude cyclic loading F is applied upon the tee-stub.

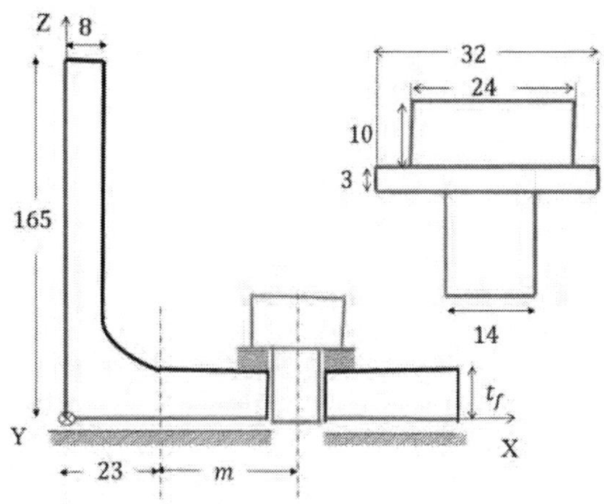

Fig. 4. : Geometrical parameters of the tee-stub

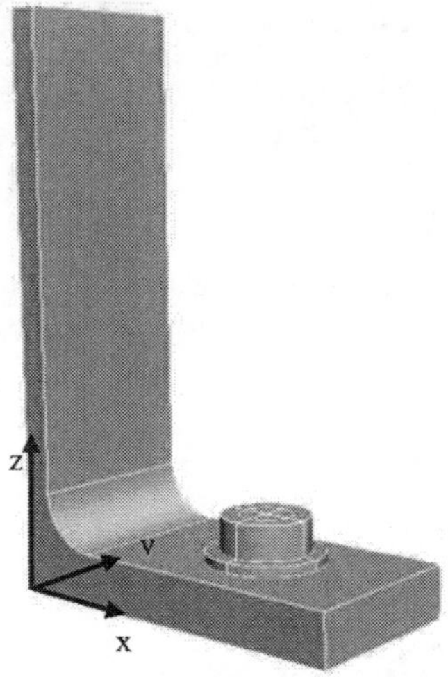

Fig. 5. : 3D tee-stub model using Salome-Meca (m = 40 mm and $tf = 15$ mm)

Once the assessment of axial load (F_B) and bending moment (M_B) is made, half the bolt is modelled using SALOME MECA software. We take into account the fillet radius r or the first thread (Fig.6) as previous studies showed that they support a large part of the load [5]. Then the axial load (F_B) and the bending moment (M_B) are applied to the bolt as in the tee-stub. The goal is to analyze the impact of the multiaxial stresses distribution upon the bolt fatigue behaviour in the head-shank transition and then in the first fillet of the shank where it engages in the nut.

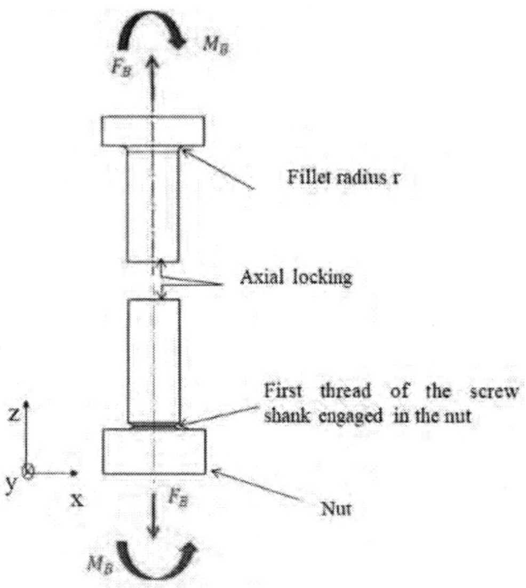

Fig. 6. : Bolt model with fillet radius and first thread submitted to F_B and M_B

3.2. Material properties and model mesh

Materials have a bi-linear elastic-plastic behaviour and the bolt is referred to 8.8 class.

Table 1. : Materials properties

	E (GPa)	E_T (GPa)	υ	σ_e (MPa)
Bolt	210	10	0.3	640
Tee-stub	210	10	0.3	355

Tetrahedral solid elements are used for the numerical analysis.

Fig. 7. : Tee-stub mesh (m =40 mm and t_f= 15 mm)

3.3. Preload and boundary conditions

The preload of the bolt is modelled by imposing a relative displacement between the nodes of the inner surface of the nut and those of the outer surface of the screw in contact with the nut.

A friction free unilateral contact condition is applied on the one hand between the lower surface of the tee-stub and the upper surface of the rigid foundation and on the other hand between the lower surface of the washer and the upper surface of the tee-stub (Fig.3).

4. Fatigue analysis tools

A fatigue post-treatment tool is implemented on MATLAB software in order to analyse the fatigue behaviour of the bolt. In the case of constant amplitude multiaxial stress states, multiaxial fatigue criteria are useful tools for assessing the fatigue resistance of the material submitted to any periodical stress states cycle [6]. As multiaxial stress states are observed in stress concentration zones, we have implemented Dang Van and Zenner criteria. The purpose is also to investigate the validity of the so-called critical plane approach versus the integral approach. Dang Van criterion is related to the critical plane approach and Zenner criterion deals with the integral approach.

4.1. Dang Van criterion

The Dang Van criterion in its first version (1973) is the first multiaxial fatigue criterion that was introduced in the French industry (PSA manufacturer). It was originally presented through considerations estalished at the microscopic scale, even if the criterion uses macroscopic stresses which are the alternating part of the shear stress τ_{ha} and the hydrostatic pressure p [7]. The damage fatigue function E of the criterion is a maximization of the damage indicator by plane E_h, defined on the material plane which unit normal vector is h:

$$E_h = \underset{t}{Max}\left\{\frac{\tau_{ha}(t) + \alpha.p(t)}{\theta}\right\} \quad (1)$$

Where $p(t) = \frac{I_1(t)}{3} = \frac{\sigma_{11}(t) + \sigma_{22}(t) + \sigma_{33}(t)}{3}$,

$I_1(t)$ is the first invariant of the stress tensor at time t,

$\tau_{ha}(t)$ is the alternating shear stress acting at time t on the material plane; it is obtained by determining the smallest circle surrounding to the load trajectory [6].

The plane where E_h is maximal is called critical plane. It allows the criterion to express the fatigue severity of the multiaxial loading cycle.

The fatigue function of the criterion is then written as:

$$E_{DV} = \underset{h}{Max}(E_h) \quad (2)$$

The fatigue function of the criterion is equal to unity when the fatigue limit of the material is reached by the analysed multiaxial stress cycle. The fatigue function of the criterion is also equal to unity for particular material fatigue limits as fully reversed tension fatigue limit (σ_{-1}, R = -1), fully reversed torsion fatigue limit (τ_{-1}, R = -1). This allows to calibrate the criterion, i.e to determine its parameters α and θ by:

$$\begin{cases} \alpha = 3\left(\frac{\tau_{-1}}{\sigma_{-1}} - \frac{1}{2}\right) \\ \theta = \tau_{-1} \end{cases}$$

Dang Van's criterion is valid when $\frac{\tau_{-1}}{\sigma_{-1}} > \frac{1}{2}$

4.2. Zenner criterion

This criterion defines its damage indicator per plane by:

$$E_h = a_{zn}\tau_{ha}^2\left(1 + m_{zn}\tau_{hm}^2\right) + b_{zn}\sigma_{hha}^2\left(1 + n_{zn}\sigma_{hhm}\right) \quad (3)$$

The fatigue function of the criterion is established as a quadratic mean value of the indicator E_h all over the possible material planes :

$$E_{zn} = \frac{1}{\sigma_{-1}}\sqrt{\frac{1}{S}\int_{\gamma=0}^{\pi}\int_{\varphi=0}^{\pi} E_h dS} \quad (4)$$

Where S = 2π (area of half the sphere of unit radius) and $dS = \sin\gamma d\varphi\, d\gamma$

The criterion includes four constants denoted a_{zn}, b_{zn}, m_{zn}, n_{zn}, which are established from four fatigue limits (fully reversed and repeated tension and torsion fatigue limits: σ_{-1} and τ_{-1}, σ_0 and τ_0). The repeated torsion fatigue limit τ_0 is assessed from the three other fatigue limits according to the following relation [8].

$$\tau_0 = \frac{4\tau_{-1}}{\frac{2\sigma_{-1}}{\sigma_0} + 1} \quad (5)$$

The criterion involves two terms related to the shear stress (τ_{ha} et τ_{hm}) and two others related to the normal stress (σ_{hha} et σ_{hhm}) [6] acting on the plane which unit normal vector is h :

- τ_{ha} et τ_{hm} : amplitude and mean value respectively of the shear stress acting on the h-plane,
- σ_{hha} et σ_{hhm} : amplitude and mean value of the normal stress acting on the h-plane.

The parameters obtained by the calibration of the criterion are thus:

$$a_{zn} = \frac{9}{2}\left(\frac{\sigma_{-1}}{\tau_{-1}}\right)^2 - 6$$

$$b_{zn} = 9 - 3\left(\frac{\sigma_{-1}}{\tau_{-1}}\right)^2$$

$$a_{zn} * m_{zn} = \frac{35}{8}\frac{(\sigma_{-1})^2 - \left(\frac{\tau_0}{2}\right)^2\left(\frac{\sigma_{-1}}{\tau_{-1}}\right)^2}{\left(\frac{\tau_0}{2}\right)^4}$$

$$b_{zn} * n_{zn} = \frac{(\sigma_{-1})^2\left[63 - 7\left(\frac{\sigma_0}{\tau_0}\right)^4\right] - 63\left(\frac{\sigma_0}{2}\right)^2 + 7\left(\frac{\sigma_0}{\tau_0}\right)^4\left(\frac{\sigma_{-1}}{\tau_{-1}}\right)^2\left(\frac{\tau_0}{2}\right)^2}{9\left(\frac{\sigma_0}{2}\right)^3}$$

5. Results analysis

5.1. Influence of the bolted joint preload

One external force F is applied to the tee-stub. It generates inside the bolt one tensile force F_B and one bending moment M_B due to the prying effect in the tee-stub column flange.

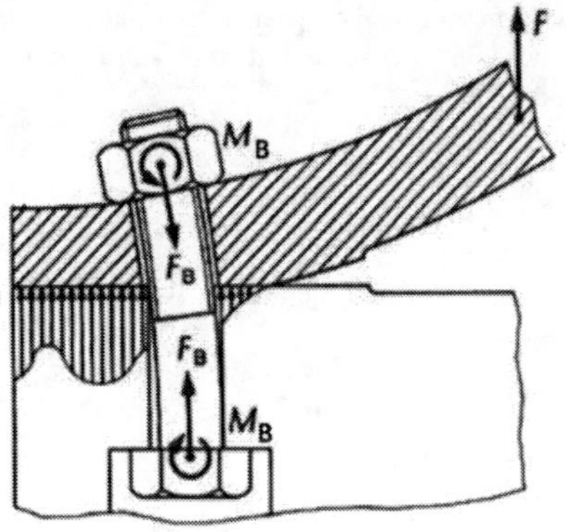

Fig. 8. : Bending of the bolt in the tee-stub [2]

To highlight the bending stresses that are generated within the bolt, the stress cycle is plotted for three points (A, I and B) in the middle of the bolt shank as shown in Fig. 9.

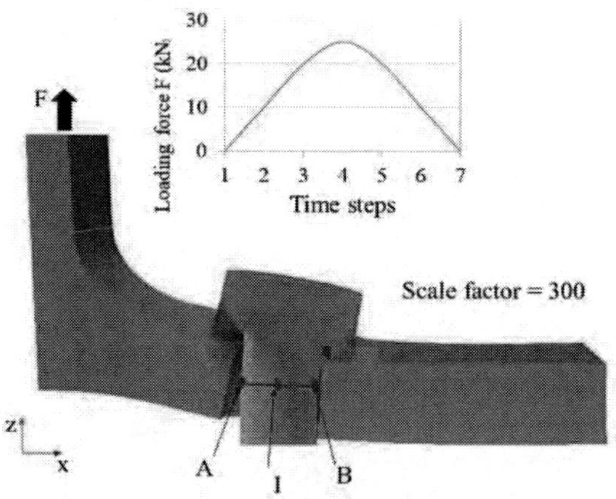

Fig. 9. : Bending deformation of the assembly (m = 20 mm and t_f = 15 mm)

The external force F applied to the tee-stub varies from 0 kN to 25 kN and then back to 0 kN. Point A is on the most stressed side of the bolt as it can be seen from the evolution of the stresses (Fig.10) during the loading cycle at points A, B and I.

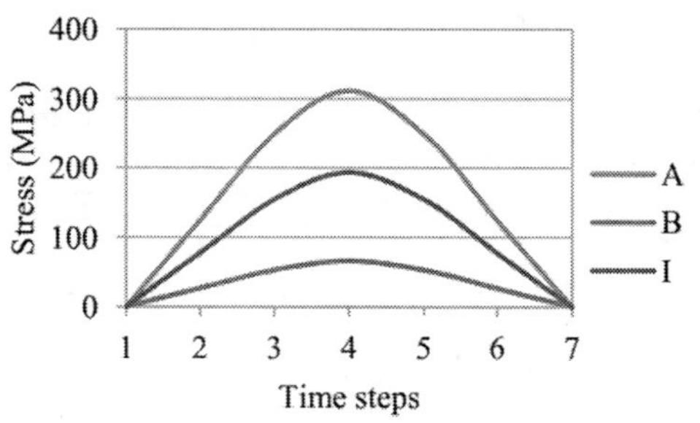

Fig. 10. : Stress cycle in points A, I and B

For the maximum value of the external force, Table 2 shows that the higher the preload, the lower the bending stress in the bolt.

Table 2. : Axial stresses calculated at point A due to tensile force and bending moment as a function of the preload

F	25 kN	
Preload (kN)	$\sigma_{tensile}$ (MPa)	$\sigma_{bending}$ (MPa)
0	188.87	122.77
20	224.15	90.81
40	300.47	51.58

The beneficial effect of the preload on the fatigue strength of the bolt is also underlined. The increase of the preload value makes greatly reduce the alternating stress in the bolt and makes increase its mean stress. As the stress amplitude is the predominant fatigue damaging parameter, its reduction makes it possible to lower the fatigue damage and thus to grow longer the fatigue life of the bolt.

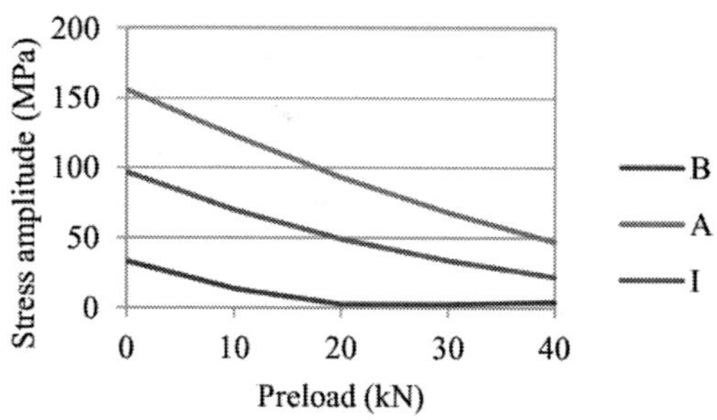

Fig. 11. : Decrease of the stress amplitude as a function of the preload

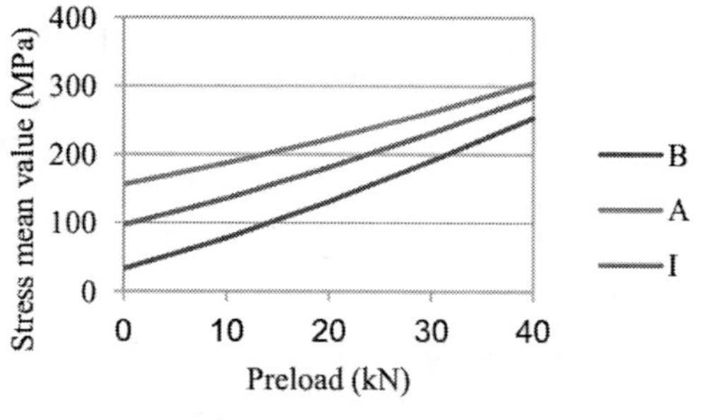

Fig. 12. : Increase of the mean stress value as a function of the preload

The figure 13 summarizes the bolt stress cycle transformation as a function of the preload. It can be seen that the stress amplitude gradually decreases when the preload of the bolt increases.

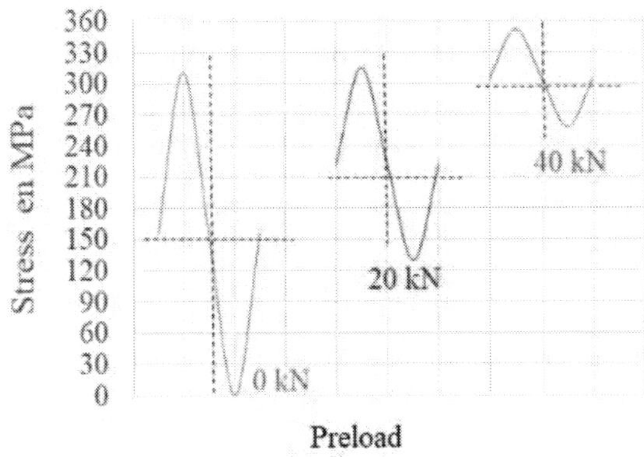

Fig. 13. : Bolt stress cycle versus the applied preload force

5.2. Parametric study

The parametric study is carried out in order to analyze on the one hand the influence of the bolt location m and on the other hand the column flange thickness tf on the stress experienced by the bolt in the tee-stub. Two values of the bolt location (20 mm and 40 mm) and two values of the column flange thickness (7.5 mm and 15 mm) are used for that purpose. The calculations are carried out for the external force F equal to 10 kN and the 20 kN preload in the bolt. The results obtained are the normal force F_B and the bending moment M_B in the bolt. They are summarized in tables 3 and 4.

The increase of the bolt location is harmful for the bolt resistance against fatigue. In fact, the increase of the bolt location significantly makes enlarge the resulting normal force and bending moment in the bolt because of the leverage effect. For a given flange thickness (7,5 mm for example in table 3), the 20 mm increase of the bolt location makes increase twofold the moment in the bolt and at the same time makes increase the normal force up to 25 percents.

Table 3. : Influence of bolt location m

tf (mm)	m (mm)	F_B (kN)	M_B(N.m)
7.5	20	24.61	15.12
	40	30.29	33.41
15	20	21.68	4.15
	40	24.11	9.71

The increase of the flange thickness results in the increase of the tee-stub rigidity. This leads to the reduction of the normal force and the bending moment encountered by the bolt when the external load is applied. Table 4 shows the decrease of the normal force and the bending moment in the bolt versus the tf thickness increase.

Table 4. : Influence of the column flange thickness

m (mm)	tf (mm)	F_B (kN)	M_B(N.m)
20	7.5	24.61	15.12
	15	21.68	4.15
40	7.5	30.29	33.41
	15	24.11	9.71

The study of the two parameters indicates their respective influence on the solicitations applied to the bolt. It can be deduced from this study that the increase of the column flange thickness is beneficial for the bolt whereas the bolt location m is detrimental to it both from static and dynamic loadings point of view.

5.3. Influence of the fillet radius on the stress distribution within the screw

Accounting for the fillet radius considerably modifies the stress distribution in the stress concentration zone, i.e. in the connection zone of the screw head and its shank. Two cases without fillet radius and with a 2 mm fillet radius have been studied. The fillet radius makes decrease the maximum value of the stress up to 40 MPa (Fig 15. and Fig.16).

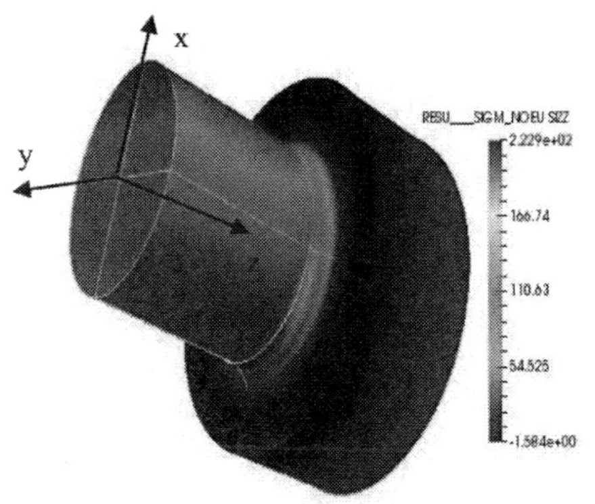

Fig. 14. : Mapping of the dominant stress σ_{zz} on the screw with fillet radius

Stress states remain uniaxial in the bolt shank but they are indeed multiaxial at the stress concentration zone. That justifies the use of multiaxial fatigue criteria to assess the fatigue damage in the screw.

https://doi.org/10.1051/matecconf/201816510009

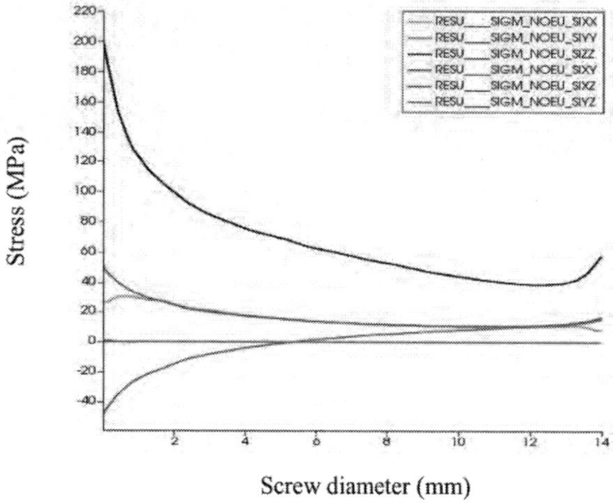

Fig. 15. : Multiaxial stress states at the stress concentration zone with the 2 mm fillet radius

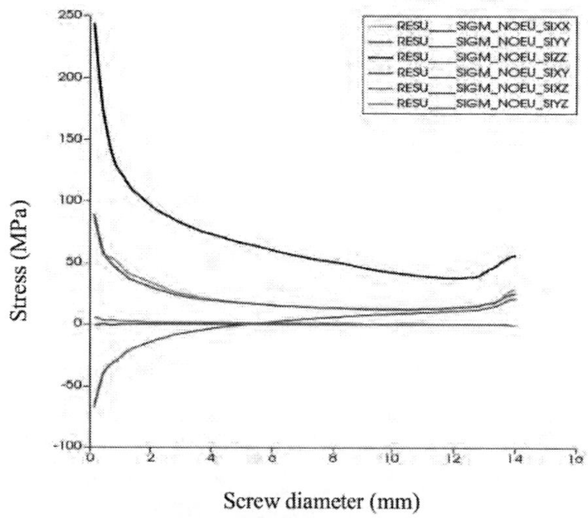

Fig. 16. : Multiaxial stress states at the stress concentration zone without fillet radius

In the following section the tee-stub is subjected to the dynamic load, for the purpose of the bolt fatigue analysis. We use here the post-treatment tool that is implemented on MATLAB software with Dang Van and Zenner multiaxial fatigue criteria.

5.4. Influence of the preload on the bolt fatigue damage assessment

The external loading cycle considered for the study of the bolt fatigue behaviour varies from 0 kN to 10 kN and back to 0 kN.

The fatigue damage function E in the screw is assessed with the following configuration of the tee-stub:

- Bolt location m : 20 mm
- column flange thickness tf : 15 mm

As the S-N curves of the bolt material are not precisely known at the present time, some fatigue data are taken from the literature to analyze the influence of the preload on the fatigue behaviour of the bolt in the tee-stub.

The considered S-N curve derives from the Wöhler model and is expressed for fully reversed tensile load (R=-1) as:

$$\sigma_{-1}(N) = 404 - 31 \log(N) \qquad (6)$$

Where N is the fatigue lifetime.

The damage function E is calculated at the fatigue strength of the material at 2.10^6 cycles.

The analysis of the damage evaluated with both criteria (Dang Van and Zenner) provides the same critical zone.

It can be noted that there is a beneficial effect of the preload on the fatigue strength of the screw. In fact, the maximum fatigue damage in the case where the fillet radius is equal to 2mm, decreases from 0.668 (Fig.14) in the case where no preload is applied to 0.345 (Fig.15) when a 20 kN preload is applied according to Dang Van criterion. For Zenner criterion the fatigue damage decreases from 0.560 (Fig.16) in the case where no preload is applied to 0.130 (Fig.17) and (Table 5) when a 20 kN preload is applied.

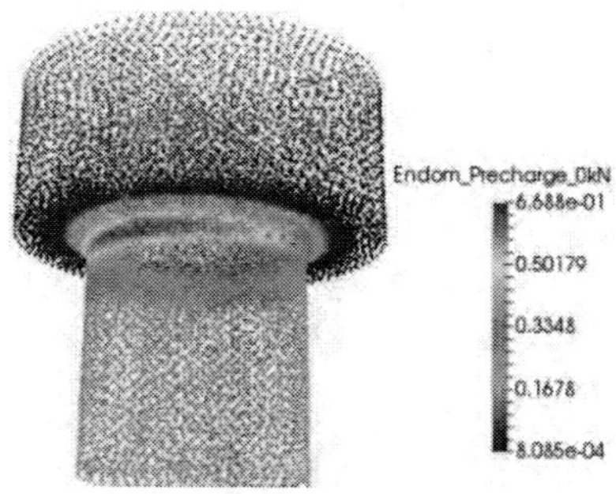

Fig. 14. : Fatigue damage mapping with Dang Van criterion in the case where no preload is applied

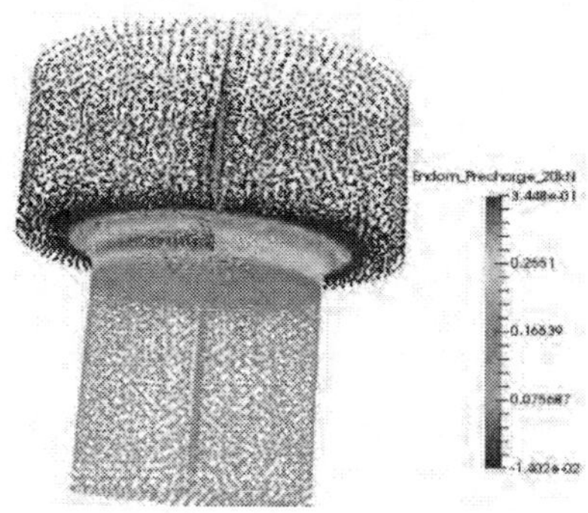

Fig. 15. : Fatigue damage mapping with Dang Van criterion in the case where 20 kN preload is applied

https://doi.org/10.1051/matecconf/201816510009

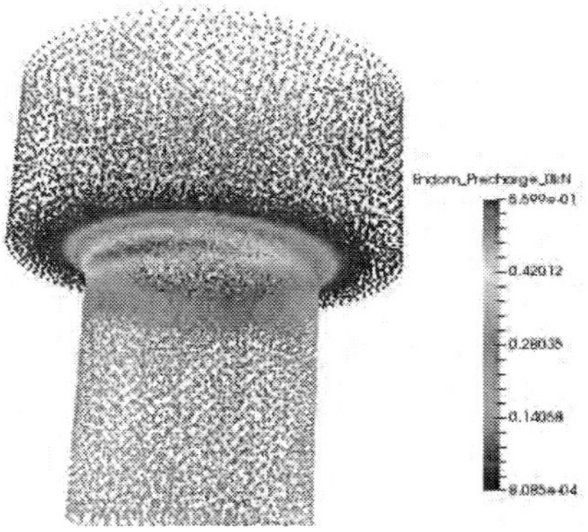

Fig. 16. : Fatigue damage mapping with Zenner criterion in the case where no preload is applied

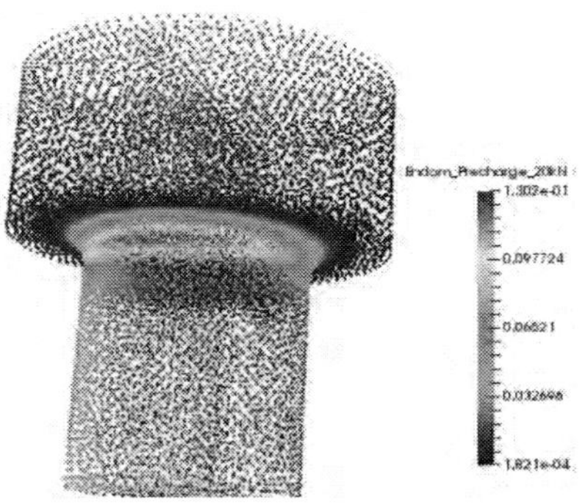

Fig. 17. : Fatigue damage mapping with Zenner criterion in the case where a 20 kN preload is applied

Both criteria predict no failure of the screw up to 2.10^6 cycles with the loading cycle applied to the tee-stub and according to the fatigue properties that were assumed for the screw.

Table 5. : Maximum fatigue damage for r =2 mm

	Maximum fatigue damage function	
Preload (kN)	Dang Van	Zenner
0	0.669	0.560
20	0.345	0.130

Table 6. : Maximum fatigue damage for r =0 mm

	Maximum fatigue damage function	
Preload (kN)	Dang Van	Zenner
0	0.911	0.780

It is also noted that the fatigue damage of the bolt with no fillet radius is greater than for the case where a 2 mm fillet radius is modelled (Tables 4 and 5). The reason is related to higher stress concentration in the first case.

Conclusion

The main result of the present work is that the preload improves the fatigue behaviour of the bolted joints by making decrease the stress amplitude even if the stress mean value increases. Additionally, the parametric study of the tee-stub shows that the increase of the column flange thickness is beneficial for the fatigue behaviour of the bolt whereas the increase of the bolt location has a detrimental influence. Finally, taking into account the fillet radius under the bolt head greatly improves the fatigue resistance of the bolt by reducing the stress concentration factor. In this area stress states induced by the loading of the tee-stub are multiaxial and require thus to be analysed by multiaxial fatigue criteria.

The future work aims to analyse also the fatigue behaviour of the first thread of the shank engaged in the nut and to completely determine the fatigue properties of the bolt assembly materials.

References

[1] J. Guillot, T.I, *Assemblages par éléments filetés*, 62 (1989)

[2] M. Paredes, A. Daidie, *Modélisation des systèmes mécaniques* (2010)

[3] Eurocode 3, *Partie 1-8 : Calcul des assemblages*, **33**, 130 (2005)

[4] A. Abidelah, A. Bouchaïr, D. Elddine, E.S, *Influence of the flexural rigidity of the bolt on the behavior of the T-stub steel connection*, **81**, 181–194 (2014)

[5] M. Tanaka, K. Hongo, JSME, *Stress Analysis of Threaded connections by Finite Element Method*, **24,** 8 (1981)

[6] B. Weber, B. Kenmeugne, J.C. Clement, J. L. Robert, CMS, *Improvements of multiaxial fatigue criteria computation for a strong reduction of calculation duration,* **15**, 381–399 (1999)

[7] B. Weber, *Fatigue multiaxiale des structures industrielles sous chargement quelconque*, 243 (1999)

[8] H. Zenner, A. Smbürger, J. Liu, *On the fatigue limit of ductile metals under complex multiaxial loading* ,137-145 (2000)

Fatigue design of cast iron components with surface discontinuities

Márton Gróza[1,*], *Yves* Nadot[2], and *Károly* Váradi[1]

[1] TU Budapest, Department of Machine and Product Design, Műegyetem rkp. 3, Budapest 1111, Hungary

[2] Institut Pprime, CNRS, ISAE-ENSMA, Université de Poitiers, Téléport 2, 1 Avenue Clément Ader, 86961 Futuroscope Chasseneuil Cedex, France

Abstract. Surface discontinuities are unavoidable properties of production castings, and have fundamental effect on in-service reliability. In the current study a multiaxial probabilistic fatigue analysis model has been developed for the description of local (e.g.: notches or surface defects) and structural gradient (e.g: bending) effects on crack initiation life. Identification of the model parameters and the validation has been done on the ferrite-pearlitic NCI grade EN GJS 500-7. The survival probability type simulation results are proven to be highly accurate with 5% average, and 11% maximum prediction error. In theory, the model might be able to assess the effect of arbitrarily complex defects on the fatigue strength, which is to be investigated in the future.

1 Introduction

Nodular cast irons (NCI) are widely used due to their favorable mechanical properties coupled with economical manufacturability of components with complex geometrical features. One of the most important design goals of NCI components in the transportation industry is to ensure safe in-service life. Fatigue is strongly linked with surface quality, which is a well-known weakness of NCI. Shrinkages, surface pores and gas blisters are frequent on cast surfaces, and they are usually responsible for the initiation of the fatigue process.

From a mechanical standpoint the different types of surface defects are usually modelled as notches or equivalent cracks. A large number of deterministic approaches based on the continuum and fracture mechanics have been developed over the years, from which probably the semi-empirical approach from Murakami [1] is the best known in the scientific and industrial community.

Due to the large number of parameters involved (material properties and component geometry, load history, environmental effects) result scatter is inseparable from the cyclic loading governed material degradation process.

One way to describe the experienced scatter in the fatigue strength and life is to account for the uncertainties of the input parameters, and use the established deterministic models for the computation of the expected value and the deviation. Beretta and Regazzi [2] developed a probabilistic extension to the standard railway axle calculation method which computes failure probability utilizing Monte Carlo simulations considering the position of the S-N curve and the load spectrum as random variables. Kim et al. [3] analyzed multiple crack growth with the NASGRO model and XFEM simulations considering the variability of initial crack length, material parameters and crack growth rates.

Probabilistic fatigue models are usually built with the introduction of different damage parameters into specific distribution functions. Pessard et al. [4] have built a probabilistic modelling framework for the description of the Kitagawa diagram where different damage parameters are employed for crack initiation and threshold crack propagation. The weakest link hypothesis is then applied to compute the probability of failure from the two damage mechanisms. The two parameter Weibull distribution is used to describe the threshold of the damage parameter corresponding to the failure of one grain.

Correia et al. [5] introduced several damage parameters into a general probabilistic model using the three parameter Weibull distribution. Their model with energy and critical plane based fatigue damage parameters led to promising results for multiaxial loading conditions as well.

In the current paper the probabilistic approach introduced in Simon's PhD Thesis [6] for the modelling of the notch effect under proportional high-cycle fatigue conditions is developed further. The updated formulation leads to more accurate description of the notch effect on the high-cycle fatigue strength considering both the expected value and the scatter of results. The subject of the study is a ferritic-pearlitic nodular cast iron, which classifies as grade EN GJS 500-7 according to EN 1563:2012 [7].

* Corresponding author: groza.marton@gt3.bme.hu

2 Experimental results

To identify the parameters of the probabilistic model introduced in Section 3. experimental data is needed related to stress multiaxiality and the notch effect, and information related to the scatter of the fatigue strength.

The experimental results concerning the notch, load ratio and load type effects are derived from the work of Deguchi et al. [8]–[10]. Since this data cannot be used to draw conclusions related to result scatter, a set of experimental data is introduced by the authors for the same material grade to analyze of the estimated result scatter.

2.1 Notch, load ratio and load type effects

Deguchi et al. [8]–[10] introduced a set of experimental data related to a ferrite-pearlitic nodular cast iron material with 10.5% graphite, 45.3% ferrite, and 44.2% perlite content. The investigated material classifies as grade EN GJS 500-7 according to EN 1563:2012 [7] with a yield strength of $R_{p0.2} = 335$ [MPa], an ultimate tensile strength of $R_m = 552$ [MPa].

Rotating bending and tension compression fatigue tests were conducted on circumferentially notched and unnotched polished specimens, and on specimens with a drilled hole. Fig. 1 shows the geometry of the test specimens. The fatigue limit was determined as the stress amplitude which the specimen has endured for 10^7 load cycles. Table 1. summarizes the test results, where D is the diameter of the proof section, $\sigma_a^{10^7}$is the stress amplitude leading to the survival of 10^7 load cycles (fatigue limit), RB designates rotating-bending, T stands for pulsating tension, TC is for tension-compression, R is the load ratio, t is the notch depth, D is the diameter of the drilled hole and h is the depth of the drilled hole.

2.2 Scatter of fatigue strength

Fatigue tests were conducted at three different load ratios (R-1, R0.05 and R0.5) on cylindrical polished EN GJS 500-7 ferrite-pearlitic nodular cast iron test specimens for the investigation of the mean stress effect. The investigated material contained 12.8% graphite, 51.5% ferrite, 35.7% pearlite, and the basic mechanical properties were: $R_{p0.2} = 325$ [MPa], $R_m = 504$ [MPa] and the elongation at fracture $A = 7.1$ [%]. The chemical composition is detailed in Table 2., whereas Fig. 2 shows the microstructure of the material on LOM and SEM images. Fatigue tests were conducted at 22°C room temperature with a force controlled 30 Hz sinusoidal load curve on an Instron 8874 servohydraulic fatigue testing machine according to ISO 1099:2017 [11]. The fatigue limit ($\sigma_a^{10^6}$) has been determined with the step-by-step method [12] for 10^6 load cycles. The fracture surfaces were analyzed by means of optical and electron microscopy. The crack initiation sites were small subsurface defects not exceeding the 100 μm size measured in the $\sqrt{area}$ parameter from Murakami [1]. Specimen geometry is shown in Fig. 3, results of the fatigue tests are summarized in Table 3.

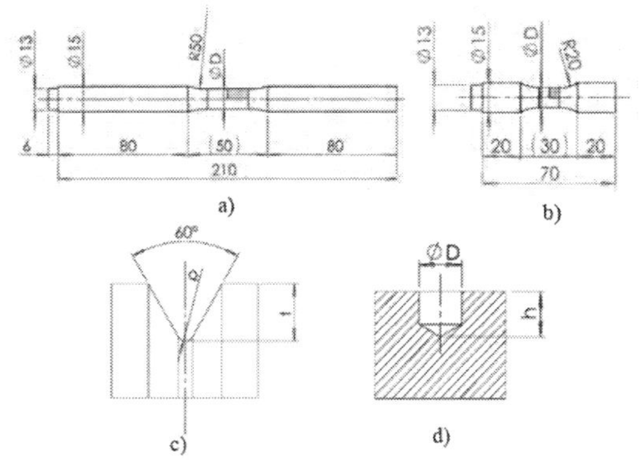

Fig. 1. Geometry of test specimens [8]–[10], a.) rotating bending specimen, b.) tension compression specimen, c.) circumferential notch geometry, d.) drilled hole geometry.

Table 1. Summary of test results [8]–[10].

Load type	D [mm]	Specimen type	$\sigma_a^{10^7}$[MPa]
RB, R=-1	10	smooth	260
T, R=0.1	10	smooth	170
RB, R=-1	12	notched, t=100 μm	195
		notched, t=200 μm	165
		notched, t=1000 μm	92.5
TC, R=-1	15	notched,=2500 μm	52.5
T, R=0.1	15	notched, t=100 μm	125
		notched,=2500 μm	45
RB, R=-1	12	hole, d=h=200 μm	220
		hole, d=h=500 μm	175
T, R=0.1	15	hole, d=h=200 μm	150
		hole, d=h=500 μm	120

Table 2. Average chemical composition of the test specimens.

C	Si	Mn	P	S	Cr	Cu	Mg	Ni
3.57	2.1	0.18	0.023	0.006	0.02	0.32	0.031	0.03

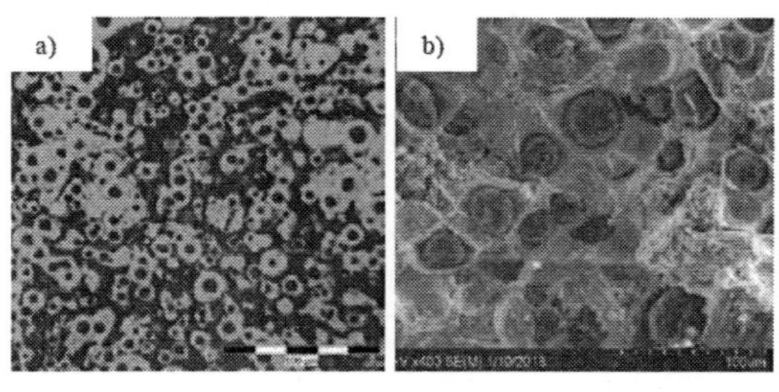

Fig. 2. a) LOM cross-sectional micrograph 50x, b) SEM image from the fracture surface of a fatigue test specimen 400x.

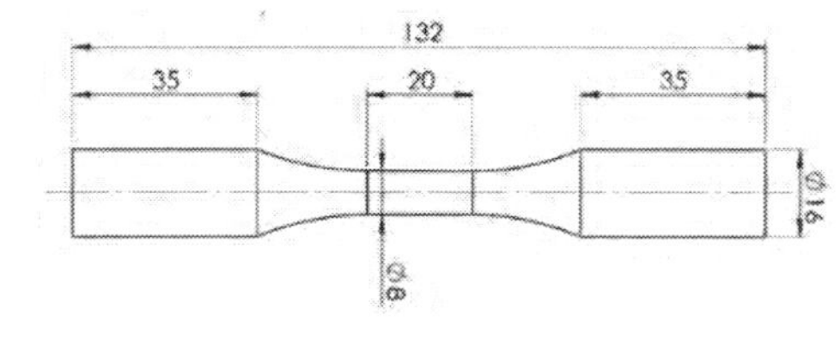

Fig. 3. Geometry of cylindrical polished test specimens.

Table 3. Summary of test results, estimated stress amplitudes at 10^6 load cycles and the corresponding Crossland equivalent stress.

R=-1		R=0.05		R=0.5	
$\sigma_a^{10^6}$ [MPa]	$\sigma_{Cr}^{10^6}$ [MPa]	$\sigma_a^{10^6}$ [MPa]	$\sigma_{Cr}^{10^6}$ [MPa]	$\sigma_a^{10^6}$ [MPa]	$\sigma_{Cr}^{10^6}$ [MPa]
258	308	146	274	99	301
252	300	152	284	97	294
244	291	154	288	99	300
247	295	151	273	98	297
263	314	-	-	99	301

In order to obtain comparable data from the tests conducted at three different load ratios, the corresponding Crossland equivalent stress ($\sigma_{Cr}^{10^6}$) has been calculated for each data point. The choice of the Crossland equivalent definition is justified by the nature of the analysis; the aim is to analyze high cycle fatigue behavior under proportional loading conditions. The Crossland equivalent stress (σ_{Cr}) is defined:

$$\sigma_{Cr} = \sqrt{J_{2,a}} + \alpha_{Cr}\sigma_{h,max}\,, \tag{1}$$

where $\sqrt{J_{2,a}}$ is the square root of the amplitude of the second invariant of the deviatoric stress tensor, $\sigma_{h,max}$ is the maximum of the hydrostatic stress in each load cycle and α_{Cr} is a model parameter. The identification process is illustrated in Fig. 4, and the value of α_{Cr} is 1.85.

In the probabilistic model introduced in Section 3, the two parameter Weibull distribution is utilized to describe the scatter of the fatigue limit. The probability density function of the Weibull distribution is given by:

$$f(x;\lambda,k) = \begin{cases} 1 - e^{-(x/\lambda)^k} & x \geq 0 \\ 0 & x \leq 0 \end{cases}, \tag{2}$$

where x is the variable ($\sigma_{Cr}^{10^6}$ in the current case), λ is the scale and k is the shape parameter. The maximum likelihood estimators describing both result scatter and inaccuracy of the Crossland criterion are $\lambda = 299.54$ MPa and $k = 30.49$. The dispersion of the corresponding Crossland equivalent stress corresponding to the fatigue limit $\sigma_{Cr}^{10^6}$ is displayed in Fig. 5 alongside the fitted Weibull probability distribution function.

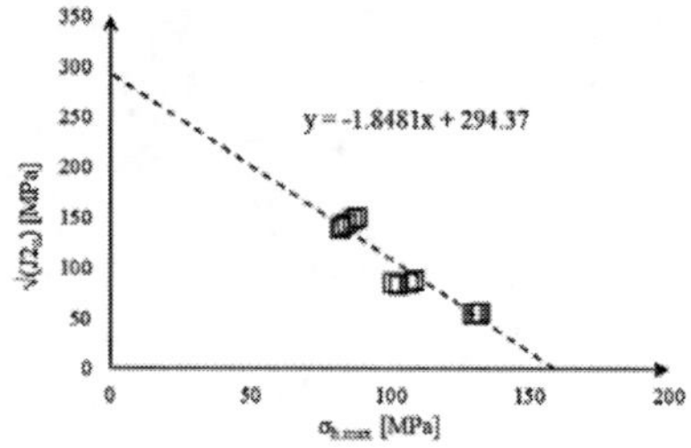

Fig. 4. Identification of the Crossland parameters on the $\sigma_{h,max}$-$\sqrt{J_{2,a}}$ plane based in R=-1, R=0.05 and R0.5 tensile fatigue test results at 10^6 cycles.

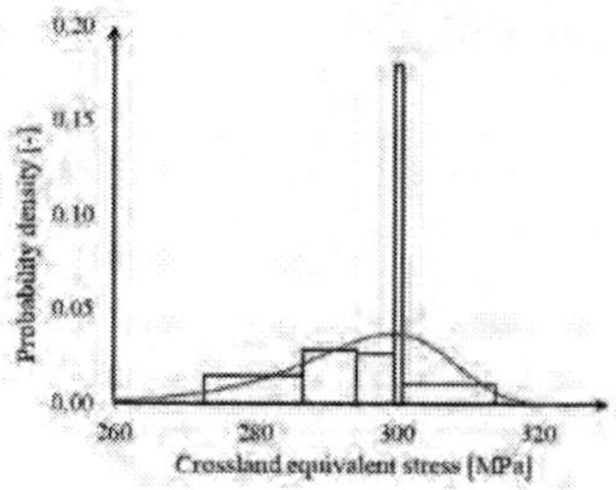

Fig. 5 Histogram and fitted Weibull distribution for the fatigue strength at 10^6 cycles.

3 Probabilistic approach

The first iteration of the current probabilistic model was introduced in by Simon [6]. The current probabilistic criterion is based on the spatial distribution of the damage parameter and a selected distribution function. Due to the current application (proportional high cycle fatigue loading) the chosen damage parameter is the Crossland criterion. The choice of the two parameter Weibull distribution is justified by its convenient application with the weakest link theory and the low amount of parameters to identify.

In general, the probability of survival according to the two parameter Weibull distribution function:

$$S(x;\lambda,k) = \begin{cases} e^{-(x/\lambda)^k} & x \geq 0 \\ 1 & x \leq 0 \end{cases}, \tag{3}$$

where x is the variable, λ and k are the scale the shape parameters, respectively. Let this function define the probability of survival for one infinitesimal volume element in the component.

According to the weakest link theory, the failure of the weakest link leads to the failure of the whole structure. Similarly, the survival probability of the component is the product of the survival probability of the infinitesimal volume elements:

$$S_{comp} = \prod_{comp} S(x;\lambda,k)\,. \tag{4}$$

The following substitutions introduce the mechanical variables into the distribution function: $x := \sigma_{eq}, \lambda := \sigma_u \sqrt[m]{V_{ref}/\ln 2}$ and $k = m$. The equivalent stress σ_{eq} is the Crossland equivalent stress according to Equation 1. The σ_u is the ultimate value of σ_{eq}, which is denoted as β_{Cr} in the Crossland criterion. The reference volume V_{ref} is linked with σ_u, the ultimate equivalent stress leads to failure in the reference volume. In this context the failure event is crack initiation.

The survival probability of the ith infinitesimal volume element:

$$S_{E_i} = \begin{cases} 1 & \sigma_{eq} < 0 \\ e^{-ln2\left(\frac{\sigma_{eq}(E_i)}{\sigma_u}\right)^m \frac{1}{V_{ref}}} & 0 \leq \sigma_{eq} \leq \sigma_c\,. \\ e^{-ln2\left(\frac{\sigma_c}{\sigma_u}\right)^m \frac{1}{V_{ref}}} & \sigma_c < \sigma_{eq} \end{cases} \tag{5}$$

For equivalent stresses below 0 MPa, the survival probability of the volume element is 1. Between 0 and the critical value σ_c the survival probability is computed according to the distribution function. The critical equivalent stress σ_c is a model parameter, which has been introduced to achieve more accurate results, which are less effected from localized disturbances in the stress field. Equivalent stresses above σ_c lead to the same survival probability for an element of a fixed size.

The survival probability of the component is the product of the elemental survival probabilities:

$$S_{comp} = \prod_{E_i} S_{E_i} \,. \tag{6}$$

In the current evaluation the volume discretization into infinitesimal volume elements will be linked with the FE (Finite Element) mesh used for stress computation.

3.1 Parameter identification

The probabilistic approach requires parameter identification for the damage parameter, and for the distribution function.

3.1.2 Identification of the Crossland parameters

The parameters of the Crossland criterion can be determined with a least squares linear fit on the test data displayed on a $\sqrt{J_{2,a}} - \sigma_{h,max}$ plot. The parameters have been identified on the unnotched results at the load ratios R-1 and R0.1 listed in Table 1. The identified parameters are $\alpha_{Cr} = 1.32$ and $\beta_{Cr} = 265\ MPa$.

3.1.2 Identification of σ_u and V_{ref}

The ultimate value of the equivalent stress is $\sigma_u = \beta_{Cr}$, which is the consequence of the applied Crossland criterion. The reference volume is the quarter of the proof volume of the unnotched specimens, in order to be coherent with the FE models built for the stress computation in the notched specimens: $V_{ref} = 504\ mm^3$.

3.1.3 Identification of the m parameter

In the first formulation from Simon [6] the m parameter was used in the identification process on notched fatigue results. This approach led to a good description of the notch effect, but the ability to describe result scatter was impaired. In the current formulation m is identified on distribution of the fatigue limit expressed in the used equivalent stress parameter. Through the identification of m on a different set of experimental data, the aim was to gain a realistic description of result scatter in the model. With the process detailed in Section 2.2 $m = k = 30.49$.

3.1.4 Identification of the σ_c parameter

The critical equivalent stress σ_c was introduced as parameter in order to readjust the model, since m is not used as a fitting parameter for the accurate description of the notch effect any more. Using a limiting value for the linear stresses has the beneficial effect of leading to more robust results, which are less dependent on localized stress peaks and mesh quality. The identification has been done through minimization of errors, the value of σ_c is 420 MPa in the current application. Using one notched result as identification point would lead to different model prediction results.

4 Calculations

Linear elastic FE simulation models were built in ANSYS Mechanical to compute elemental stress results for each specimen and load scenario, which are necessary for the probabilistic model. The post processing of results to obtain elemental survival probabilities has also been coded in ANSYS.

The symmetry of the geometry and load conditions made it possible to utilize quarter symmetry of the mechanical problem. Analysis setup is displayed in Fig. 6, the modelled portion of the test specimen is shown as a dashed area in Fig. 1. Tension compression load scenarios are modelled with a distributed load at the end surface of the proof section, rotating bending is simulated with a moment load. In accordance with the method from Deguchi et al. [8]–[10], loads are considered to lead to the reported stress amplitudes on a cross-section corresponding the nominal proof section diameter for the specimens with drilled holes and small notches. For larger notch depths ($t = 1000$ and $2500\ \mu m$) the actual cross section is considered when computing tensile and bending stresses.

At the bottom of the drilled holes a fictitious rounding of 50 μm has been introduced, in order to obtain a more realistic representation of the drilled hole geometry and a non-singular stress field.

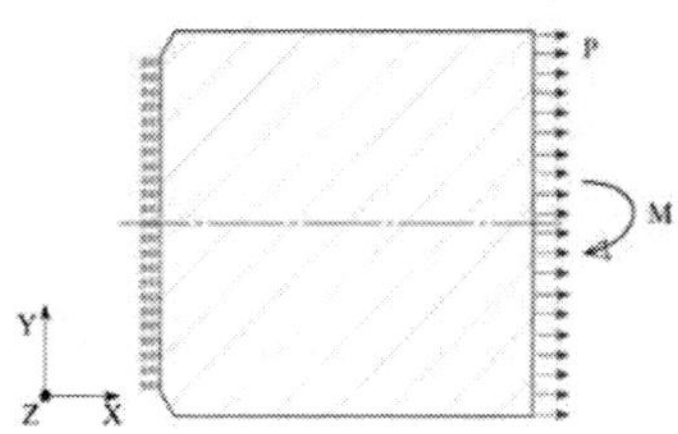

Fig. 6. Analysis setup for the FE computations.

Mesh convergence studies have been conducted for the circumferential notches and the drilled holes. Considering ensured mesh convergence and solution stability notched specimens were meshed with 20 node hexahedron elements (SOLID186), with mesh size defined by a 7.5 deg angle between nodes (midside nodes considered) along the 50 μm radius at the bottom of the notches of various depth. Specimens with drilled holes were meshed with 10 node tetrahedrons (SOLID 187) with a mesh size of $R/25\ \mu m$. The ferrite-pearlitic NCI material is considered with the modulus of elasticity $E = 169\ GPa$ and Poisson's ratio $\upsilon = 0.275$ material parameters in the calculation models. For each analysis model a half load cycle is simulated for the computation of the stresses corresponding to the load maximum and amplitude.

Fig. 7.a shows the mean elemental Crossland equivalent stress results (calculated according to Equation

https://doi.org/10.1051/matecconf/201816510010

1.) for the rotating bending test specimen with 1000 μm notch depth. Fig. 7.b) introduces the cut-off corresponding to the σ_c critical equivalent stress model parameter. Fig. 8 shows the same results for the rotating bending specimen. Fig. 9 shows the computed elemental survival probability (calculated according to Equation 5.) for the rotating bending specimen with a notch depth of 1000 μm.

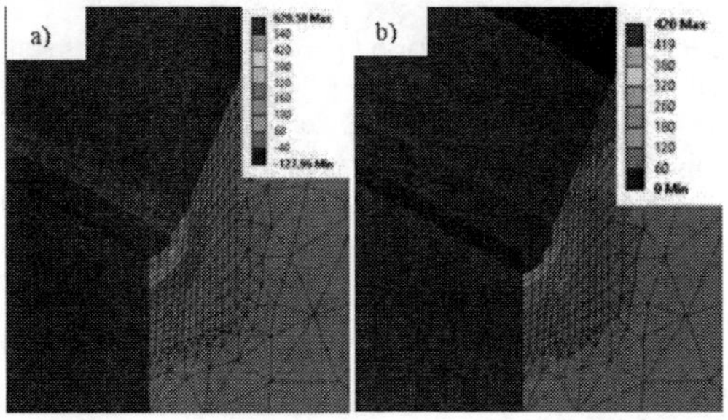

Fig. 7.a) Mean elemental Crossland equivalent stress results, b) Mean elemental Crossland equivalent stress results with the lower (0) and upper (σ_c) cut-off for the RB t=1000 μm specimens.

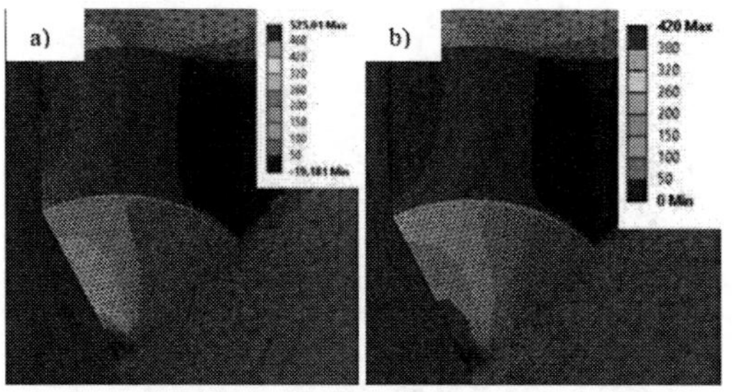

Fig. 8.a.) Mean elemental Crossland equivalent stress results, b.) Mean elemental Crossland equivalent stress results with the lower (0) and upper (σ_c) cut-off for the RB d=h=500 μm specimens.

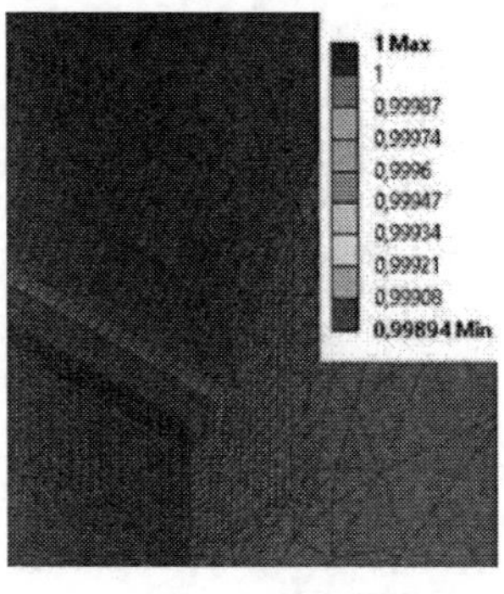

Fig. 9. Elemental survival probability for the RB t=1000 μm specimen.

5 Results and discussion

Comparison of computational and experimental results is shown in Fig. 10. Each simulation leads to cumulative distribution function type survival probability curves as function of the applied stress amplitude during testing. The curves corresponding to the unnotched specimens are in perfect fit with the experiments since they have been used in the identification process. The computational results are in excellent agreement with the experiments for different load types (rotating bending, alternating tension-compression and pulsating tension), notch types (circumferential and drilled hole) and sizes (100-2500 μm). Fig. 11. illustrates the prediction accuracy of the model, the average and maximum prediction error is 5 and 11%, respectively.

In addition to the accurate description of local and structural gradient effects, the scatter of the fatigue limit is predicted in a correct manner. For a quantitative evaluation of the prediction accuracy related to result scatter a large experimental data set would be required with more than 10 specimens tested for every test case.

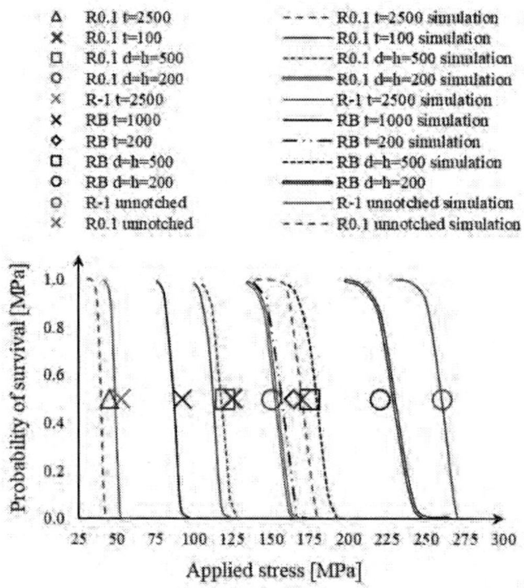

Fig. 10. Comparison of experimental and simulated survival probabilities.

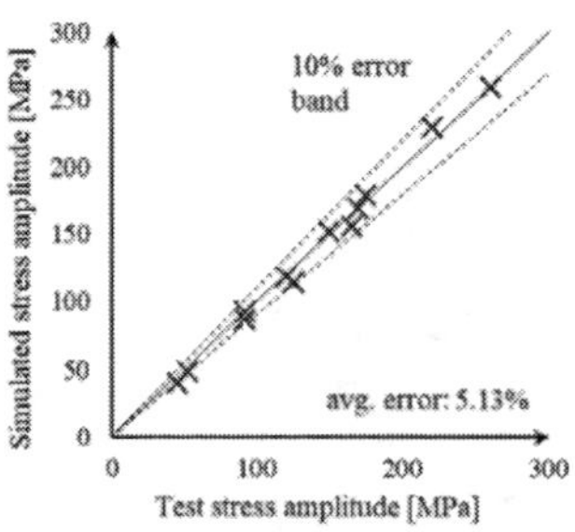

Fig. 11. Model prediction accuracy.

6 Conclusion

A multiaxial probabilistic fatigue analysis model has been developed for the description of local and structural gradient effects on crack initiation. Identification of the model parameters and the validation has been done on the ferrite-pearlitic NCI grade EN GJS 500-7 [7].

- The probabilistic model describes load ratio, structural and local stress gradient effects induced by notches and surface holes of different size without the need for elastoplastic computation. The average and maximum estimated prediction error was 5 and 11%, respectively.
- The model gives not only deterministic answers; it predicts the result scatter to be expected for a certain test case. The results scatter description accuracy has not been quantitatively evaluated.
- Theoretically, the model is able to describe the effect of arbitrarily complex defects and notches, and is not highly dependent on mesh quality. Further analyses are to be conducted to investigate these possibilities.
- Due to the application of the Crossland criterion as damage parameter, the application is limited to high cycle fatigue and proportional loading conditions in the current formulation.

This recent study was realized within the Knorr-Bremse Scholarship Program supported by the Knorr-Bremse Rail Systems Budapest.

References

[1] Y. Murakami, Metal Fatigue*: Effects of Small Defects and Nonmetallic Inclusions*, Oxford: Elsevier Science Ltd, 2002.

[2] S. Beretta and D. Regazzi, Int. J. Fatigue, **86**, pp. 13–23, (2016)

[3] J. H. Kim, T. Chau-Dinh, G. Zi, W. W. Lee, and J. S. Kong, Eng. Fract. Mech., **156**, pp. 182–196, (2016)

[4] E. Pessard, D. Bellett, F. Morel, and I. Koutiri, Eng. Fract. Mech., **109**, pp. 89–104, (2013)

[5] J. Correia, N. Apetre, A. Arcari, A. De Jesus, M. Muñiz-Calvente, R. Calçada, F. Berto, and A. Fernández-Canteli, Int. J. Fatigue, **100**, pp. 187–194, (2017)

[6] S. Julien, PhD Thesis: *Influence de micro-entailles sur le comportement en fatigue à grand nombre de cycles d'un alliage de TA6V; comparaison avec le fretting-fatigue* ISAE-ENSMA, (2017)

[7] EN 1563:2012 Founding – Spheroidal graphite cast irons. (2012)

[8] T. Deguchi, Adv. Exp. Mech., **2**, pp. 87–91, (2017)

[9] T. Deguchi, H. J. Kim, T. Ikeda, and K. Yanase, *Influence of mean stress on fatigue strength of ferritic-pearlite ductile cast iron with small defects*, J. Phys. Conf. Ser., **843**, no. 1, (2017)

[10] T. Deguchi, H. J. Kim, and T. Ikeda, *Fatigue limit prediction of ferritic-pearlitic ductile cast iron considering stress ratio and notch size,* J. Phys. Conf. Ser., **842**, no. 1, 2017.

[11] ISO 1099:2017 Metallic materials - Fatigue testing - Axial force-controlled method. (2017)

[12] R. S. Bellows, S. Muju, and T. Nicholas, Int. J. Fatigue, **21**, no. 8, pp. 687–697, (1999)

The relevance of non-stationarities and non-Gaussianities in vibration fatigue

Martin Česnik[1], *Janko* Slavič[1,*], *Lorenzo* Capponi[2], *Massimiliano* Palmieri[2], *Filippo* Cianetti[2], and *Miha* Boltežar[1]

[1]University of Ljubljana, Faculty of Mechanical Engineering, 1000 Ljubljana, Slovenia
[2]University of Perugia, Department of Engineering, 06125 Perugia, Italy

Abstract. In classical fatigue of materials, the frequency contents of dynamic loading are well below the natural frequencies of the observed structure or test specimen. However, when dealing with vibration fatigue the frequency contents of dynamic loading and structure's dynamic response overlap, resulting in amplified stress loads of the structure. For such cases, frequency counting methods are especially convenient. Gaussianity and stationarity assumptions are applied in frequency-domain methods for obtaining dynamic structure's response and frequency-domain methods for calculating damage accumulation rate. Since it is common in real environments for the structure to be excited with non-Gaussian and non-stationary loads, this study addresses the effects of such dynamic excitation to experimental time-to-failure of a structure.
Initially, the influence of non-Gaussian stationary excitation is experimentally studied via excitation signals with equal power density spectrum and different values of kurtosis. Since no relevant changes of structure's time-to-failure were observed, the study focused on non-stationary excitation signals that are also inherently non-Gaussian. The non-stationarity of excitation was achieved by amplitude modulation and significantly shorter times-to-failure were observed when compared to experiments with stationary non-Gaussian excitation.
Additionally, the structure's time-to-failure varied with the rate of the amplitude modulation. To oversee this phenomenon the presented study proposes a non-stationarity index which can be obtained from the excitation time history. The non-stationarity index was experimentally confirmed as a reliable estimator for severity of non-stationary excitation. The non-stationarity index is used to determine if the frequency-domain methods can safely be applied for time-to-failure calculation.

1 Introduction

Before it's exploitation a product (structure) must undergo a number of tests to prove its durability when exposed to harsh environmental conditions (vibrations, temperature, humidity, etc.). When discussing product's resistance to vibrations an automotive, aerospace and other testing standards [1] specify vibration loads that should be applied to a product in a controlled laboratory conditions via electro-magnetic shaker. Potential failure of product during vibration testing mainly occurs due to wear or fatigue fracture. In such cases of fatigue failure, the damage accumulation is not directly correlated to vibration load, but is rather coupled in frequency domain with structure's dynamic properties [2].

Thus, when dealing with vibration fatigue of structure under random vibration loads the counting methods in frequency domain (Dirlik [3], Tovo-Benasciutti [4]) are significantly more applicable, computationally effective and faster [5] compared to counting methods in time domain (rainflow, range count), especially when long time-histories need to be evaluated. Fatigue life estimation with frequency counting method is based on power spectral density (PSD) function of structure's stress response and properties of multiaxial stress criterion [6, 7]. Inherently to this, a frequency counting methods assume signal's stationarity [8]. Furthermore, due to reasons of analytical deduction, an additional assumption of Gaussian nature of stress signal should be sufficed. When both assumptions are respected, the frequency methods give reliable fatigue life estimation [9].

Considering standard vibration testing with random signal with given frequency profile both assumptions for frequency domain methods are satisfied. However, when dealing with real service loads or environment conditions, the non-Gaussian and non-stationary signals are commonly observed. To what extent does the invalidity of those assumptions result on fatigue estimation has been studied from different aspects in classical [10, 11] and vibration fatigue. In latter a non-Gaussian and non-stationary excitation signal is mechanically filtered by dynamic structure. An influence of non-Gaussian excitation to response signal and fatigue life was numerically investigated by Braccesi *et al.* [12], Rizzi *et al.* [13] and Kihm *et al.* [14], who used kurtosis and skewness as estimators of non-Gaussianity. Non-Gaussianity was also studied for vibration fatigue of composite structures [15] and MIMO test systems [16].

This manuscript presents an experimental and numerical analysis of non-Gaussianity and non-stationarity on vibration fatigue. Firstly, a dynamic structure of Y-shaped specimen [17] was experimentally excited with random stationary Gaussian signal to obtain fatigue parameters of used aluminium alloy. Secondly, a number of Y-specimens was excited with different sets of random signals, obtained by combining different kurtosis values and non-stationarities. As the non-

* Corresponding author: janko.slavic@fs.uni-lj.si

stationary non-Gaussian excitation was experimentally shown to be significantly more damaging compared to stationary non-Gaussian [18] the research later focused on experimental and numerical evaluation of non-stationarity rate of excitation signal. In this manuscript a run-test method [19] is proposed to obtain a non-stationarity index γ [20]. Its applicability will be tested on a large number of fatigue tests and the relation between the amplitude-modulated non-stationarity and time-to-failure will be researched.

2 Theoretical background

In this chapter a condensed review of fundamental theory will be presented that is later applied to actual experimental data. Firstly, methods for identifying non-Gaussianity and non-stationarity of the random signal will be shown. Secondly, general equations of motion for dynamic system are presented, which result in stress response PSD. Finally, a Tovo-Benasciutti frequency counting method is presented, which is used to obtain damage accumulation from stress response PSD.

2.1. Random signal properties

Given a discrete time-series of random process as $\mathbf{x} = [x_1, \dots, x_i, \dots, x_n]$, its j-th central moment M_j and mean value μ are given as:

$$M_j = \frac{1}{n}\sum_{i=1}^{n}(x_i - \mu)^j, \tag{1}$$

$$\mu = \frac{1}{n}\sum_{i=1}^{n} x_i \tag{2}$$

and are based on Probability Density Function $P(x)$ (PDF). The PDF of Gaussian process is defined by mean value and second central moment (variance) of the process. For PDF definition of non-Gaussian processes a higher number of parameters is needed, introducing kurtosis k_u as a measure of non-Gaussianity and defined as:

$$k_u = M_3/M_2^{3/2}, \tag{3}$$

which characterizes the sharpness of the PDF peak and the width of the PDF tails. By definition, Gaussian process has kurtosis of 3. Non-Gaussian processes with high peak excursions have kurtosis higher than 3.

On the other hand, quantification of non-stationarity in signal is not that straightforward, due to nature of non-stationarity itself. However, certain methods exist that monitor the changes in signal's statistical properties, which are for stationary process constant. Run-test [8] is a non-parametric method that divides a signal in time windows and calculates a variation of a chosen statistical variable over the whole signal. For each time window a chosen variable is met with a criterion related to whole signal; in presented study a condition was defined as:

$$V(n) = \begin{cases} 1; & |R_W(n) - R_T| > \sigma_R \\ 0; & |R_W(n) - R_T| \le \sigma_R \end{cases}, \tag{4}$$

where $R_W(n)$ is a RMS value of a n-th window, R_T a RMS value of the whole signal and σ_R is a standard deviation of all window RMSs, as illustrated on signal time history on Fig. 1. A *run* is a sequence of identical observations V, followed and preceeded by different observation. Too many or too little runs in a whole signal is a proof of signal's non-stationarity. If probability of observation 1 or 0 is equal for each window, N_i is a number of i observations and N is a total number of windows, then a mean value μ_r and variance σ_r^2 of runs is:

$$\mu_r = \frac{2\,N_0 N_1}{N} + 1, \quad \sigma_r^2 = \frac{2N_0N_1(2N_0N_1 - N)}{N^2(N-1)}. \tag{5}$$

Any difference of actual number of runs r from expected mean indicates a non-stationarity in signal. If a number of runs lays outside $\mu_r \pm \sigma_r$, a signal is considered non-stationary. Additionally, a non-stationarity index γ is defined as:

$$\gamma = r/\mu_r \quad [\%]. \tag{6}$$

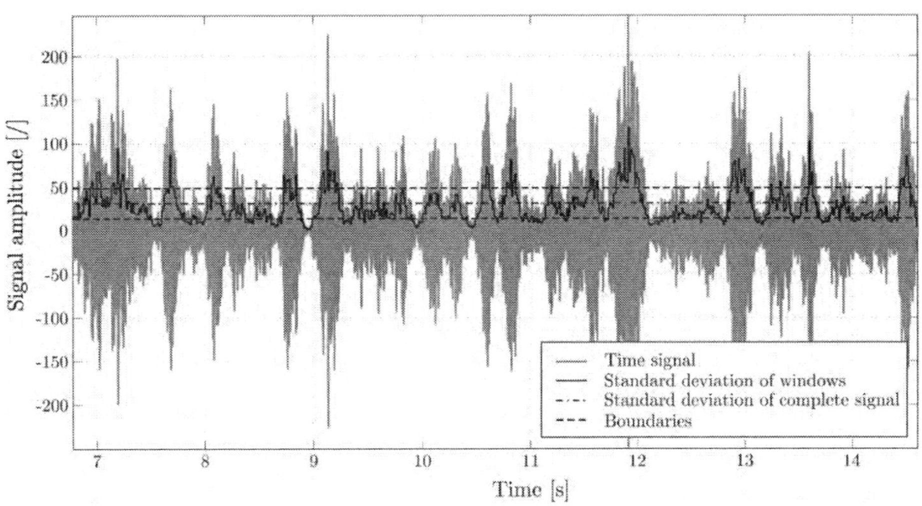

Fig. 1. Run-test evaluation of a non-stationary time signal.

2.2. Structural dynamics

Flexible multi-degree-of-freedom structures respond to force excitation by respecting the equation of motion [21]:

$$\mathbf{M}\,\ddot{\mathbf{x}}(t) + \mathbf{D}\,\dot{\mathbf{x}}(t) + \mathbf{K}\,\mathbf{x}(\mathrm{t}) = \mathbf{F}(\mathrm{t}). \quad (7)$$

In Eq. (7) **M**, **D** and **K** present mass, damping and stiffness matrices, respectively; **x** and **F** are displacement and excitation force vectors. This coupled system of differential equations can be decoupled into

$$\mathbf{I}\,\ddot{\mathbf{q}}(t) + \mathbf{I}\,[2\boldsymbol{\xi}\boldsymbol{\omega}_0]\,\dot{\mathbf{q}}(t) + \mathbf{I}\,[\boldsymbol{\omega}_0^2]\,\mathbf{q}(\mathrm{t}) = \boldsymbol{\Phi}^{\mathrm{T}}\mathbf{F}(\mathrm{t}), \quad (8)$$

by introducing modal coordinates $\mathbf{x} = \boldsymbol{\Phi}\mathbf{q}$, where $\boldsymbol{\Phi}$ is a modal matrix, consisting of mode shapes. $\boldsymbol{\omega}_0$ and $\boldsymbol{\xi}$ are natural frequency and damping ratio vectors, respectively. By solving eigenvalue problem to obtain r natural frequencies and modeshapes, uncoupled system of equations yields frequency response function on j^{th} position for excitation on k^{th} position:

$$H_{jk}(\omega) = \sum_{r=1}^{n} \frac{\phi_{jr}\phi_{kr}}{\omega_r^2 - \omega^2 + 2\mathrm{i}\xi_r\omega_r\omega}. \quad (9)$$

Combining response functions into response function matrix and knowing stress states of mode shapes, one can obtain stress response power spectral density (PSD) for given excitation PSD $\mathbf{S}_{\mathbf{ff}}(\omega)$ as:

$$\mathbf{S}_{\boldsymbol{\sigma\sigma}}(\omega) = \mathbf{H}_{\boldsymbol{\sigma}\mathbf{f}}^{*}(\omega)\,\mathbf{S}_{\mathbf{ff}}(\omega)\,\mathbf{H}_{\boldsymbol{\sigma}\mathbf{f}}^{\mathrm{T}}(\omega). \quad (10)$$

2.3. Damage accumulation

Continuing from stress response PSD $\mathbf{S}_{\boldsymbol{\sigma\sigma}}(\omega)$ an equivalent stress PSD $\mathbf{S}_{\mathbf{eq}}(\omega)$ can be quickly obtained [22]. To employ frequency methods for fatigue calculation, m-th spectral moment should be defined [7]:

$$\lambda_m = \int_0^\infty \omega^m\, S_{eq}(\omega)\mathrm{d}\omega. \quad (11)$$

Depending on the frequency method, a damage accumulation is a function of a list of spectral moments and material fatigue parameters C and b, as included in Basquin's equation $\sigma = C\,N^{-1/b}$ [23]. In this research, a Tovo-Benasciutti method [4] is proposed, that obtains damage accumulation D using following equation

$$D = \left(B_{TB} + (1 - B_{TB})\alpha_2^{B_{TB}-1}\right)\alpha_2\, D_{NB}, \quad (12)$$

where B_{TB} is a factor obtained from spectral moments λ_0, λ_1, λ_2 and λ_4. D_{NB} is a damage for narrow band stress signal and depends on spectral moments (λ_0, λ_1, λ_2, λ_4) and material fatigue parameters (b, C) [3, 4].

3 Non-Gaussianity vs. non-stationarity

The first part of the study focuses on non-Gaussianity and non-stationarity of random excitation signal and its influence on actual vibration fatigue life. For this purpose, a Y-shaped specimen is used, Fig. 2. Its geometry consists of three beams that are arranged at 120° angles around the main axis and have a rectangular cross-section of 10 × 10 mm. The Y-shaped specimens were made from the aluminium alloy A-S8U3; additional steel dead-weights with a mass of 52.5 g were added for natural-frequency tuning. The specimen was attached to electro-dynamic shaker and excited with flat-shaped force PSD profile (Fig. 3) with frequency range 600-850 Hz so that specimen's 4th natural frequency of 775 Hz lied within excited frequency band.

Fig. 2. Fixed Y-specimen with excited modeshape.

The procedure for evaluating non-Gaussianity and non-stationarity is following: first a set of specimens is excited with Gaussian stationary signals with flat PSD profiles at different amplitude levels (Fig. 3). By minimizing error between experimental and numerical fatigue lives material fatigue parameters C and b are obtained. Secondly, random signals with similar PSD profiles but different kurtosis and non-stationarities (Tab. 1) are generated and applied to the Y-specimens. Influence of non-Gaussianity and non-stationarity is then studied by comparing actual and predicted fatigue lives for given excitation signal type.

Tab. 1. Excitation signal types

Nr.	Signal type	k_u
1.	Gaussian stationary	2.96
2.	Non-Gaussian stationary	7.36
3.	Non-Gaussian stationary	5.43
4.	Non-Gaussian non-stationary	7.08

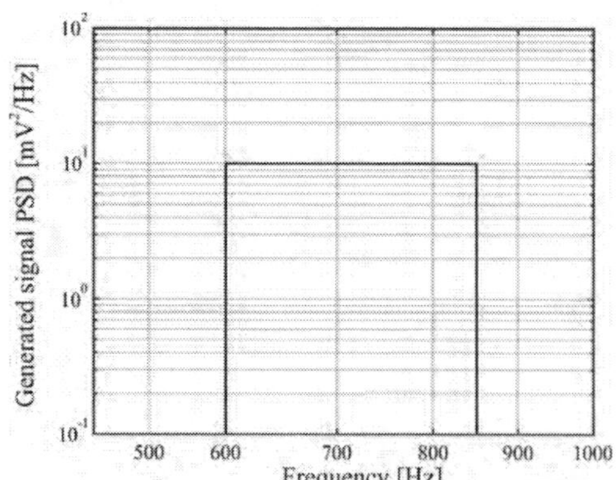

Fig. 3. Flat PSD profile with amplitude level 10 mV²/Hz.

Four Y-shaped specimens were excited with random stationary Gaussian signal until fatigue failure occurred. The obtained fatigue lives are presented in Fig. 4. In order to obtain fatigue material parameters a numeric model was built and validated (Fig. 5). A function for numerical minimization of numerical error was proposed as:

$$\Delta T\ (b, C) = \sum_i \left(\log_{10} T_{\mathrm{act},i} - \log_{10} T_{\mathrm{est},i}(b, C)\right)^2, \tag{13}$$

where i denotes a single test specimen, T_{act} is the measured fatigue life and T_{est} is the estimated fatigue life of the specimen based on the numerical model. The minimization method showed strong convergence. Using the identified fatigue parameters, the Basquin's equation can be written as:

$$\sigma = 987.5 \cdot N^{-0.169}. \tag{14}$$

Once fatigue parameters were known, a number of samples were excited with signals, where non-Gaussianity and non-stationarity was combined according to Tab. 1. Time signals for signal types 3. and 4. with identical PSD profiles are presented in Figs. 6 and 7. The excitation force PSD level was precedently determined by measuring acceleration response of shaker's armature without Y-specimen [18]. Non-stationary signal was generated using amplitude modulation of stationary signal with beta-distribution signal [18, 20].

A total of nine Y-shaped specimens were tested under non-Gaussian excitation, three for each signal type 2-4. In order to evaluate the influence of the kurtosis on the fatigue life of tested specimens, the fatigue parameters obtained under the Gaussian condition are also used for a numerical estimation of the fatigue life for the case of non-Gaussian excitation. In order to investigate when it is justified to apply Gaussian-based counting methods to non-stationary excitation, the Tovo-Benasciutti counting method was also used in the case of the non-Gaussian excitations.

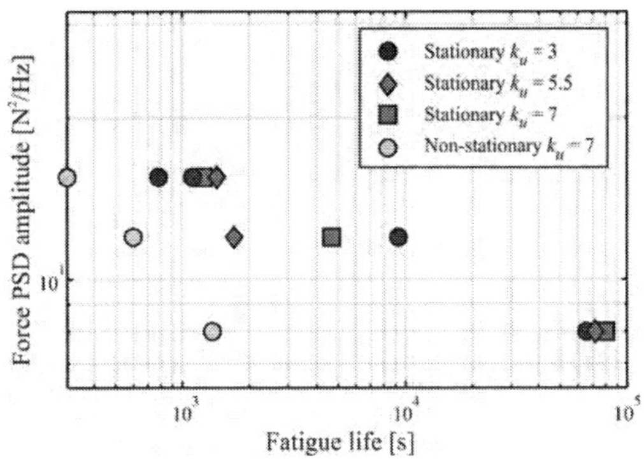

Fig. 4. Experimental fatigue lives.

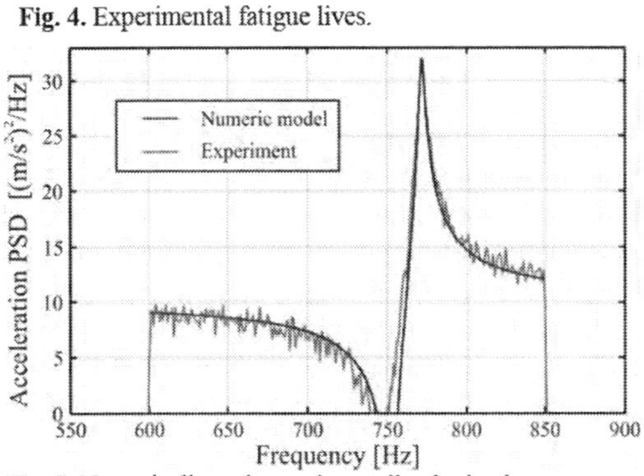

Fig. 5. Numerically and experimentally obtained response acceleration PSD.

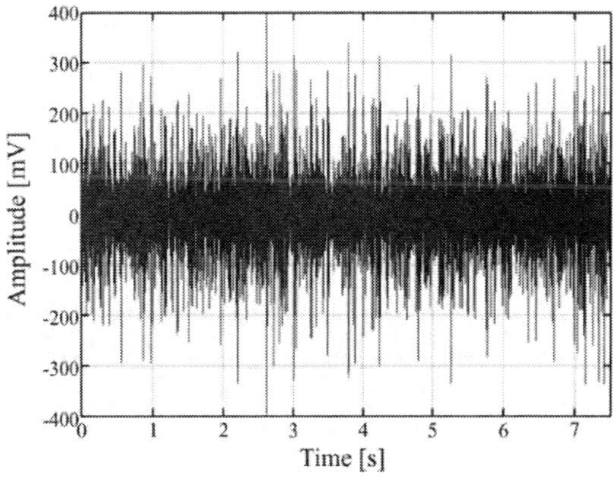

Fig. 6. Stationary non-Gaussian signal with k_u = 7.

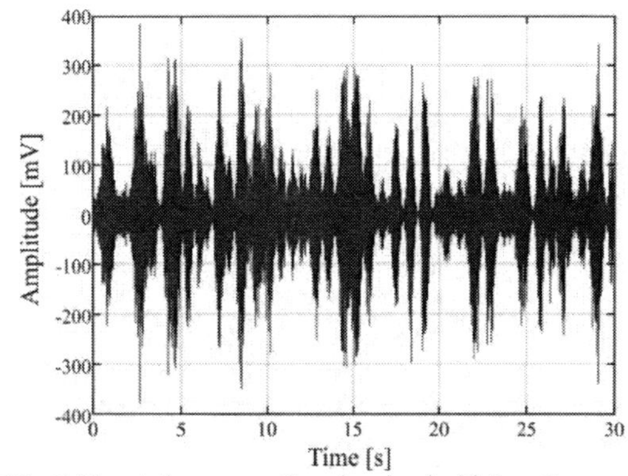

Fig. 7. Non-stationary non-Gaussian signal with k_u = 7.

A comparison between the experimental and calculated fatigue lives is shown in Fig. 8. From Fig. 8 it is noticeable that the stationary, non-Gaussian excitation differs only slightly if compared to the calculated fatigue life. In contrast to this case, for the case of non-stationary non-Gaussian excitation, the comparison shows how the difference between the calculated fatigue lives and the actual fatigue lives is significant. From this comparison one can conclude that for the case of stationary excitation the use of standard frequency-counting methods [9] available in the literature gives reliable results comparable to reality, while in the case of strongly non-stationary signals the use of the same counting methods supplies uncorrected results. Also, Fig. 4 shows the relation between the excitation force PSD amplitude and the experimental fatigue life; again, the non-stationary experiments clearly differentiate from the stationary experiments.

The latter difference between the calculated and measured fatigue lives arises due to the non-Gaussianity of the stress response. For this reason, a piezo-electric strain gauge was applied on the specimen's beam, as shown in Fig. 2, to monitor the stress response. For the case of stationary non-Gaussian loadings the kurtosis of the stress is always around the value 3, confirming that due to the stationarity of the excitation, the response of the structure is Gaussian. However, for the case of non-stationary non-Gaussian loadings the output kurtosis is significantly higher, although it is still lower than the kurtosis of the input signal. In any case, we can conclude that if the input signal is quasi-stationary, the output kurtosis always tends to the Gaussianity; however, if the input signal is non-stationary, the stress response remains strongly non-Gaussian and leads to shorter fatigue lives than expected.

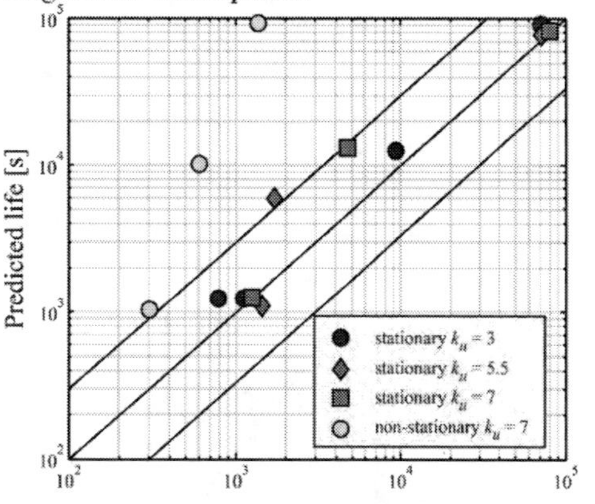

Fig. 8. Experimental and calculated fatigue lives.

4 Non-stationarity index

The following section focuses on quantification of non-stationarities in excitation signal and on identification of non-stationary signals that result in reduced vibration fatigue life.

4.1. Signal generation

In order to investigate this first a set of time signals with different non-stationarity rates was obtained. For the bulk of physical phenomena the non-stationarity is exhibited as the time variance of a signal's power; the fluctuations in signal's frequency content are less commonly observed.

Therefore, beta-distribution amplitude modulation of a stationary signal with flat-shaped PSD profile was initially performed with a carrier wave, as proposed in Sec. 3 (Fig. 7) and described in detail in [14] and [18].

In the next step the initial carrier wave with time duration of 30 minutes was then squeezed by a factor of 2 and again multiplied with flat-profiled stationary signal . This signal was denoted as SQ-2, whereas the amplitude modulated signal with original (non-squeezed carrier wave) was denoted as SQ-1. In a similar manner, additional 5 nonstationary signals were generated for different squeezing factors 4, 10, 50, 500 and 10000. After squeezing, each signal was repeated to reach the

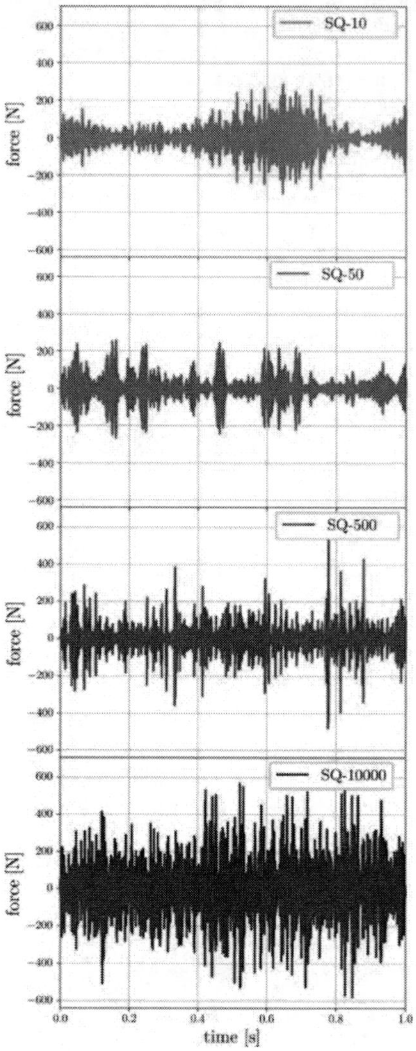

Fig. 9. Time history of measured excitation signals SQ-10, SQ-50, SQ-500 and SQ-10000 with PSD level of 13.5 N^2/Hz.

time length of 30 minutes. All time signals had the same PSD, the same time length, the same kurtosis, but different levels of non-stationarity. Squeezed signals SQ-10, SQ-50, SQ-500 and SQ-10000, that were measured on electro-dynamic shaker are presented in Fig. 9.

4.2. Signal evaluation with run-test method

Here, the signals later used for Y-specimen excitation and with PSD level of 13.5 N^2/Hz are evaluated with two-sided run-test method, presented in Sec. 2.1 and Fig. 1. As known [3] run-test method is sensitive to the width of the windows, to which the signal is divided. Consequently, all 8 signals (7 non-stationary and 1 *de facto* stationary) were evaluated for different time lengths of dividing windows. In total seven time-window widths were applied to the run-test, ranging from 0.005 to 1 s. In Tab. 2 the resulting non-stationarity indexes γ are given. Moreover, the influence of the window's width is graphically presented in Fig. 10. According to Rizzi *et al.* [13] study the optimal window width is expected to be related to the period of [14]; in the case of the Y-specimen the period of the impulse response was experimentally determined to be in the range up to 0.2 s. By inspecting Fig. 10 a significant difference between low squeezing and high squeezing is observed; at a window width of 0.04 s the signals SQ-1 to SQ-4 were identified as non-stationary, while the SQ-500, SQ-10000 and also *de facto* stationary signal were identified as stationary. Similar results were found for other RMS values of the excitation signal.

Tab. 2. Non-stationarity index γ for two-sided method: results in bold identify signal as stationary.

Signal type	Window width						
	0.005 s	0.01 s	0.0125 s	0.02 s	0.04 s	0.1 s	1 s
	Non-stationarity index γ [%]						
Stationary	96	**99**	**100**	**101**	**102**	**99**	72
SQ-10000	96	**101**	**101**	**101**	**102**	**100**	**101**
SQ-500	95	97	**99**	**99**	**102**	105	**90**
SQ-50	73	86	92	98	95	96	114
SQ-10	65	62	61	61	72	**96**	**109**
SQ-4	64	60	57	50	46	67	**105**
SQ-2	65	60	58	50	40	39	**104**
SQ-1	66	62	60	53	40	31	**96**

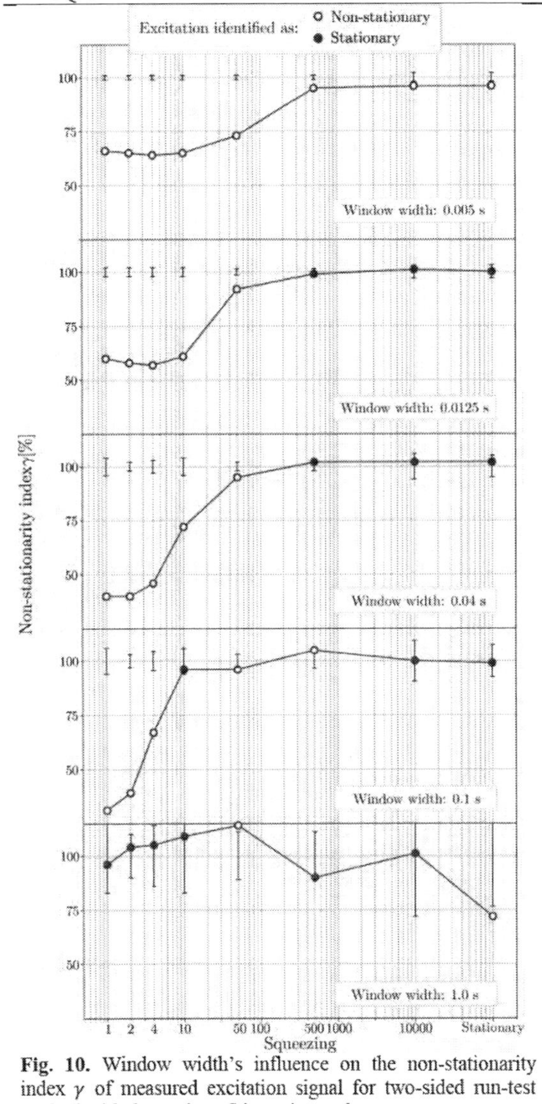

Fig. 10. Window width's influence on the non-stationarity index γ of measured excitation signal for two-sided run-test method with denoted confidence intervals.

the system's impulse response, *i.e.*, the time in which the response amplitude reduces to 10% of the initial value

4.2. Vibration fatigue testing

Here, the *de facto* stationary and squeezed signals SQ-1 - SQ-10000 were applied to the Y-specimen. According to a preliminary analysis of the excitation signals the non-stationarity index γ shows that the signals SQ-500, SQ-10000 can be considered as stationary and the signals SQ-1, SQ-2 and SQ-4 can be considered as non-stationary. This finding will now be tested against the actual vibration fatigue life.

For the sake of experimental comprehensiveness the time signals were applied to the Y-specimen at four different force PSD levels, as shown in Fig. 11. For each of the 19 combination pairs of excitation PSD level and signal type, two samples were teste. In total, 41 samples were broken. Certain load level and signal type combinations were left untested either due to instantaneous failure or due to the absence of failure after 2×10^7 load cycles.

The fatigue life for all tested specimens is given in Tab. 3. It is clear that squeezing the carrier wave and thus

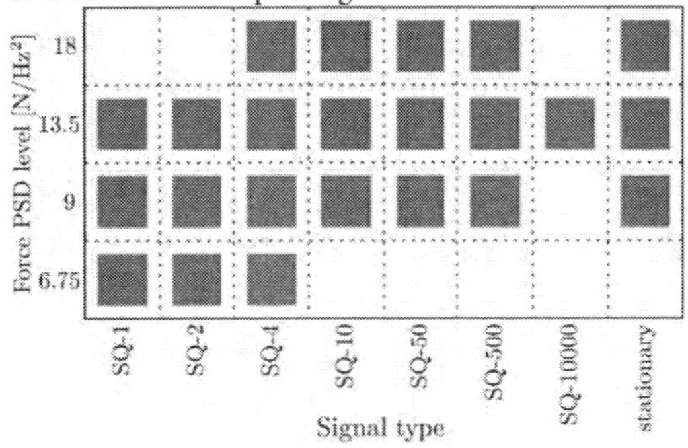

Fig. 11. Experimentally tested combinations of signal types and excitation levels.

changing the non-stationarity rate significantly changes the fatigue life, see Fig. 12. From Fig. 12 it is clear that for the SQ-4 the fatigue life is significantly shorter than

Tab. 3. Experimental fatigue lives [s] of tested Y-specimens.

Force PSD level [N^2/Hz]	Signal type							
	SQ-1	SQ-2	SQ-4	SQ-10	SQ-50	SQ-500	SQ-10000	Stationary
18	/	/	643	598	999	1624	/	3157
			341	497	1069	2793		
13.5	961	871	737	1175	1120	4069	6299	4428
	857	907	899	1333	5234	6411	6309	
9	2364	2853	3279	3177	3722	>12600	/	>12600
	2994	2391	2899	3824	5894			
6.75	5112	2737	4429	/	/	/	/	/
	4865	4405	5435					

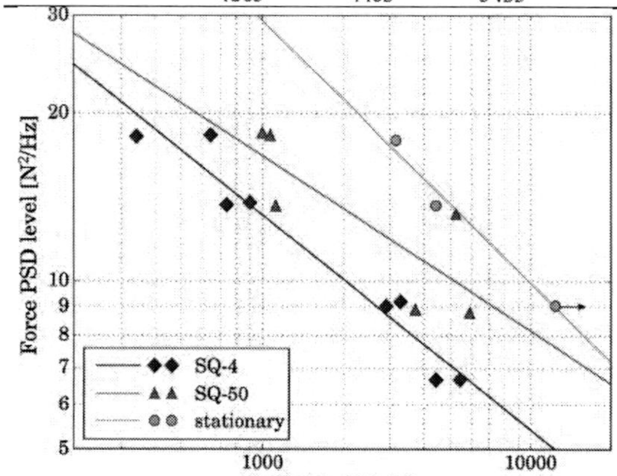

Fig. 12. Experimental fatigue life for the representative excitation signal types: SQ-4, SQ-50 and *de facto* stationary.

for the *de facto* stationary signal, while SQ-50 is in between.

Due to the considerable influence of the non-stationarity rate on the fatigue life only tests with an excitation force PSD level of 13.5 N^2/Hz were possible on the complete set of generated non-stationary signals SQ-1 - SQ-10000 and on the *de facto* stationary signal, see Tab. 3. To better understand Fig. 12, a detailed analysis of the PSD level of 13.5 N^2/Hz is shown in Fig. 13: one axis shows the fatigue life, the other axis shows the non-stationarity index. It was previously shown that the excitation signals from SQ-1 - SQ-4 can be considered as non-stationary, while signals SQ-500 and above can be considered as stationary. As all the time signals had the same PSD, the same time length, the same kurtosis, but different rates of non-stationarity one can conclude that of non-stationarity results in significant differences in fatigue life. The difference between non-stationary and stationary conditions is approximately 5-fold.

Fig. 14 shows the fatigue life normed to the fatigue life during stationary excitation. From the results of all 41 tested samples it is clear that the excitation identified as non-stationary resulted in a reduced fatigue life to approximately 20%. Signals that are not clearly stationary, nor are they non-stationary, have a fatigue life between the two groups. Based on presented results it is reasonable to conclude that non-stationary index γ provides reliable results to distinguish between non-stationary excitation signals that do or do not reduce vibration fatigue life when compared to stationary excitation.

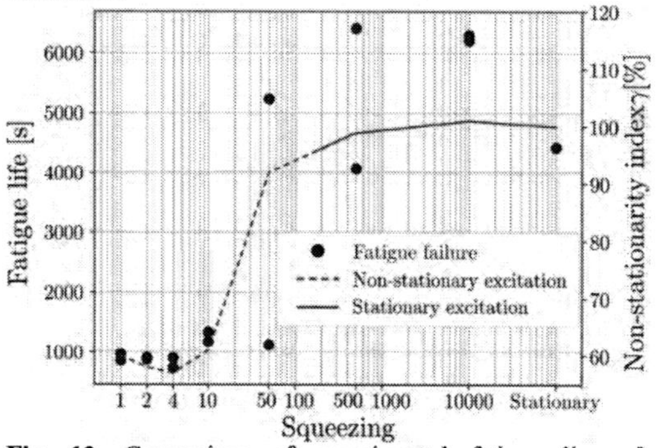

Fig. 13. Comparison of experimental fatigue lives for excitation PSD level of 13.5 N^2/Hz and excitation signal's non-stationarity index γ for window width of 0.0125 seconds using two-sided run-test method.

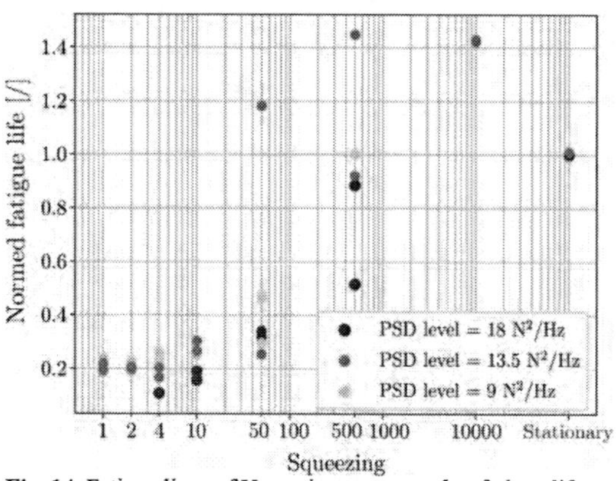

Fig. 14. Fatigue lives of Y-specimens, normed to fatigue life under *de facto* stationary excitation signal.

5 Conclusions

In the presented research the fatigue life of a simple Y-shaped specimen was investigated in order to determine how a change in the excitation kurtosis and stationarity affects the fatigue life of a real structure. Firstly, several experimental verifications of random excitation signals with the same PSDs, but with different kurtosis were performed. In the case of stationary random excitations, it was found that non-Gaussian signals with a stationary rate produced a Gaussian response, while in the case of non-stationary random signals, the response of the structure is non-Gaussian. The obtained results show that the fatigue life due to amplitude-modulated non-stationary excitations is significantly shorter when compared to the fatigue lives obtained under the valid condition of stationarity. Moreover, it was found that if the non-Gaussian excitation is stationary, the calculated fatigue lives with classical frequency-counting methods are comparable to the fatigue lives under Gaussian excitation; therefore, justifying the use of frequency-counting methods, even if validated only in the case of Gaussian excitations. In contrast, for the case of burst, non-Gaussian excitation, the obtained fatigue life exhibits a significantly higher damage accumulation compared to the fatigue lives attained under Gaussian signals. For this reason, considering a non-stationary non-Gaussian excitation as Gaussian and, consequently adopting the classic frequency-counting methods may result in a wrong fatigue-life estimation.

To this end, a study later focused on identifying critical non-stationary amplitude-modulated in terms of actual fatigue life. Excitation is frequently non-stationary in terms of time-varying power and the question is what rate of non-stationarity can still be considered as stationary and how does the rate of non-stationarity effect the fatigue life? To answer this, additional experimental tests of Y-specimens were performed using squeezed signals with the same PSD and kurtosis, but different rates of non-stationarity. An enhanced method to identify the non-stationarity was proposed and resulted in a clear differentiation of the non-stationarity in the signal.

The signals that were identified as non-stationary resulted in a significantly shorter fatigue life of the sample than the ones that were identified as (or were de facto) stationary. Finally, the non-stationarity identification relates on the natural frequencies of the researched structure. However, if the non-stationarity is identified during the excitation, the resulting fatigue life was shown to significantly decrease (in this research to 1/5th).

1. United States Department of Defense, *MIL-STD-810F, Department of defense test method standard for environmental engineering consideration and laboratory tests* (2000)
2. N. W. M. Bishop, *The use of frequency domain parameters to predict structural fatigue*, University of Warwick (1988)
3. T. Dirlik, *Application of computers in fatigue analysis*, University of Warwick (1985)
4. D. Benasciutti, R. Tovo, *Spectral methods for lifetime prediction under wide-band stationary random process*, Int. J. Fatigue, **27**(8): pp. 867-877 (2005)
5. C. Braccesi, F. Cianetti, L. Tomassini, *Fast evaluation of stress state spectral moments,* Int. J. of Mech. Sci., **127**: pp.4-9 (2017)
6. M. Mršnik, J. Slavič, M. Boltežar, *Multiaxial vibration fatigue - a theoretical and experimental comparison*, Mech. Sys. Signal Proc., **76-77**: pp. 409-423 (2016)
7. M. Mršnik, J. Slavič, M. Boltežar, *Vibration fatigue using modal decomposition*, Mech. Sys. Signal Proc., **98**: pp. 548-556 (2018)
8. S. Bendat, G. Piersol, *Random data: analysis and measurement procedures*, 4th edition, Wiley&Sons (2010)
9. M. Mršnik, J. Slavič, M. Boltežar, *Frequency-domain methods for a vibration-fatigue-life estimation – application to real data,* Int. J. Fatigue, **47**: pp. 8-17 (2013)
10. D. Benasciutti, *Fatigue analysis of random loadings*, University of Ferrara (2004)
11. A. Niesłony, M. Bohm, T. Łagoda, F. Cianetti, *The use of spectral method for fatigue life assessment for non-Gaussian random loads*, Acta Mech. Automat., **10**(2): pp. 100-103 (2016)
12. C. Bracessi, F. Cianetti, G. Lori, D. Pioli, *The frequency domain approach in virtual fatigue estimation of non-linear system: the problem of non-Gaussianity state of stress*, Int. J. Fatigue, **31**(4): pp. 766-775 (2009)
13. S. A. Rizzi, A. Prezekop, T. Turner, *On the response of a non-linear structure to high-kurtosis non-Gaussian random loadings*, in: 8th Eurodyn, Leuven (2011)
14. F. Kihm, S. A. Rizzi, N. S. Fergusson, A. Halfpenny, *Understanding how kurtosis is transferred from input acceleration to stress response and its influence on fatigue life*, in: 11th RASD, Pisa (2013)
15. Z. Fan, Y. Jiang, S. Zhang, J. Tao, X. Chen, *Research on vibration fatigue of carbon fibre-reinforced composites under non-Gaussian random load,* In: PHM-2017, Harbin (2017)
16. R. Zheng, H. Chen, X. He, W. Zheng, *Probability distributions control for multi-input multi-output stationary non-Gaussian random vibration test*, J. Vib. Control, DOI: 10.1177/1077546317747503 (2017)
17. M. Česnik, J. Slavič, M. Boltežar, *Uninterrupted and accelerated vibrational fatigue testing with simultaneous monitoring of the natural frequency and damping*, **331**(24): pp. 5370-5382 (2012)
18. M. Palmieri, M. Česnik, J. Slavič, F. Cianetti, M. Boltežar, *Non-Gaussianity and non-stationarity in vibration fatigue*, Int. J. Fatigue, **97**: pp. 9-19 (2017)
19. V. Rouillard, *Quantifying the non-stationarity of vehicle vibrations with the run test*, Packag. Technol. Sci., **27**(3): pp. 203-219 (2014)
20. L. Capponi, M. Česnik, J. Slavič, F. Cianetti, M. Boltežar, *Non-stationarity index in vibration fatigue: Theoretical and experimental research*, Int. J. Fatigue, **104**: pp. 221-230 (2017)
21. N. M. M. Maia, J. M. M. Silva, *Theoretical and experimental modal analysis*, 1st edition, Research Studies Press (1997)

22. X. Pitoiset, A. Preumont, *Spectral methods for multi-axial random fatigue analysis of metallic structure*, Int. J. Fatigue, **22**(7): pp. 541-550 (2000)
23. Y.-L. Lee, J. Pan, R. Hathaway, M. Barkey, *Fatigue testing and analysis: Theory and practice*, 1st edition, Elsevier (2004)

Reliability prediction, scatter-band analysis and fatigue limit assessment of high-cycle fatigue properties in EN-GJS700-2 ductile cast iron

Mohammad Jamalkhani Khameneh[1] and *Mohammad* Azadi[2,*]

[1]Faculty of Mechanical Engineering, Semnan University, Semnan, Iran
[2]Faculty of Mechanical Engineering, Semnan University, Semnan, Iran

Abstract. In this article, high-cycle fatigue properties of the EN-GJS700-2 ductile cast iron were experimentally examined, using the four-point rotary-bending fatigue machine. Such material has been widely utilized for manufacturing crankshafts in automotive industries. Therefore, standard specimens were extracted from the crankshaft to consider the manufacturing effect. The other aim of this work is to develop a model, anticipating the probabilistic behavior of the EN-GJS700-2 ductile cast iron. For such objective, the parametric analysis of the most commonly used statistical distributions were performed and compared. Based on the best fitted distribution function on experimental data, fatigue lifetime reliability specifications of the material were analyzed and represented through probabilistic modeling. In addition, the S-N diagram with its scatter-band were conducted at different values of the confidence level. As another investigation, the criterion for designing the crankshaft is the infinite lifetime, or in more practical terms, is the design with allowable stress levels, below the fatigue limit. Based on the staircase test methodology, fatigue endurance limit tests were conducted. Then, statistical trends of the fatigue limit were analyzed by the Dixon-Mood method, based on the maximum likelihood estimation. Obtained results represents that the fatigue limit was 211 MPa for the ductile cast iron.

1 Introduction

Ductile cast irons have been extensively utilized in crankshafts, as a main part of automotive engines [1]. Such engine components are subjected to the fatigue phenomenon, due to constant amplitude loading, during their usage. On the other hand, fatigue test results (S-N data) usually expose remarkable scatters, even when tests are intently performed to diminish the experimental error [2-4]. As a result of this inherent variation in fatigue data, median S-N fatigue lifetime curves are not adequate for the fatigue analysis in order to design components [4]. Therefore, the statistical analysis through testing activities must be considered to evaluate the fatigue property of such material, for the design of the engine crankshaft, which are subjected to the high-cycle fatigue (HCF) damage.

In this field of study, several researches have been implemented by researchers. Caiza and Ummenhofer [5] introduced a procedure to model S-N curves, subject to the Weibull model. They presented two applications, considering experimental data of girders and welded specimens, made of the S690QL steel. They obtained that estimated S-N curves, through the Weibull model onto HCF or very HCF regions, were more precise than those obtained by applying the Basquin model. Chandrappa and Biligiri [6] investigated the effect of stress levels and frequencies on the fatigue behavior of pervious concrete paving mixtures. Non-parametric and parametric methods showed that the fatigue lifetime were significantly affected by the stress level and the stiffness, compared to the loading frequency in the range of 2-10 Hz. Kaplan-Meier survival curves demonstrated a sudden drop in the survival proportion at higher stress levels. They also fitted three-parameter Weibull and log-normal distributions to fatigue data, where the log-normal distribution performed a better fitness. Zhu et al. [7] represented a probabilistic approach for modeling the damage accumulation to investigate the fatigue reliability of railway axle steels. They performed the fatigue reliability analysis under variable amplitude loadings. They acquired an agreement between predicted results and experimental measurements under different loading conditions. Feng et al. [8] evaluated mechanical properties, fatigue properties and damage characteristics of the rubber-modified recycled aggregate concrete. They conducted the statistical analysis on test results, using the Weibull distribution and also suggested double logarithm fatigue equations for the material at different failure probabilities. Chen et al. [9] investigated the crankshaft fatigue phenomenon, using the experimental method. In addition, they performed the statistical regression analysis of eight hypothesis distributions, based on experiment data and evaluated the degree of the fitting effect, between experimental results with chosen statistical models.

* Corresponding author: m_azadi@semnan.ac.ir

Finally, they discovered that the three-parameter Weibull distribution model fitted data, in a superior condition.
Borgia et al. [10] investigated a test methodology based on accelerated lifetime tests, capable to evaluate the reliability of a domestic washing machine. They utilized two-parameter Weibull, as a probability function, to assess the reliability of the machine, at each overstress level. Cevik and Gurbuz [11] evaluated the effect of fillet rolling on the fatigue behavior of the EN-GJS800-2 ductile cast iron crankshaft, with the application of diesel engines. They used the staircase test methodology to demonstrate S-N curves and to measure fatigue endurance limits. They also revealed that induced compressive residual stresses increased the surface hardness and increased the dislocation density, which provided an improvement in the fatigue strength. Zhai and Li [3] performed a conditional probability density distribution surface, by two functions such as a function of the mean value of the population and the stress level and the other one was a function of the standard deviation and the stress level. They also proposed a theoretical fatigue limit, according to the small probability event.
Klemenc and Fajdiga [12] presented an estimation of the S-N curve and its scatter, by applying the two-parameter Weibull distribution to demonstrate the scatter of the lifetime, at a stress level. They described the S-N curve and its scatter, by three parameters, including the constant term, the scale coefficient and the shape parameter of the Weibull distribution. Mohd et al. [13] evaluated scatter characteristics of the fatigue lifetime, by a three-parameter Weibull distribution in a die-cast magnesium alloy. They also investigated the correlation between the scatter behavior of the fatigue lifetime and the pore size, observed on the cross section, as well as on the fracture surface. They concluded that the three-parameter Weibull distribution was suitable to describe the scatter behavior of both the fatigue lifetime and the pore size. Rathod et al. [14] presented a methodology for modeling the probabilistic distribution of the fatigue damage accumulation for single- and multi-stress level loading. The methodology used the linear damage accumulation model of Palmgren-Miner, as a probabilistic S-N curve, and an approach for a one-to-one transformation of probability density functions. Then, they predicted the reliability by the proposed methodology, using the dynamic statistical model of the cumulative fatigue damage.
Bouraoui et al. [15] developed a probabilistic approach to investigate the HCF behavior reliability of shot peened metallic components. They evaluated the reliability in different values, using the strength-load method and computed by the numerical Monte-Carlo simulation. Then, shot peening HCF reliability predictions and obtained iso-probabilistic Crossland diagrams showed a good agreement with experimental results. Sghaier et al. [16] presented a methodology to predict the reliability of metallic parts, according to their HCF behavior. They utilized the multi-axial Crossland criterion, in the case of normally distributed in-phase fully reversed torsion and bending loadings and HCF material characteristic parameters. Their results showed that the proposed method was qualitatively consistent with physical observations. Spiteri et al. [17] evaluated the fatigue limit of rolled ductile cast iron crankshafts, under bending. They used the staircase test methodology to find the fatigue strength of the material.
Schijve [18] compared three statistical distribution functions, containing the normal distribution, the three-parameter Weibull distribution and the three-parameter normal distribution for the logarithmic lifetime. The second and the third function showed a good agreement of fitted data of 30 similar test results. Khashaba [19] performed rotating bending fatigue tests on glass fiber reinforced polyester composites. He conducted the statistical analysis of experimental fatigue lifetime results, using the two-parameter Weibull distribution and calculated the safe design of the lifetime, based on the time to the first failure concept for 99% of the confidence level and at two values of the reliability, 0.368 and 0.990. He concluded that these values of the reliability and the confidence level were appropriate and could be utilized to predict the flexural fatigue lifetime of such composites. Shen [20] developed a probability-based methodology to predict the reliability of gas turbine engine blades, exposed to HCF loadings. Doremus [21] compared three statistical distribution functions such as normal, Weibull and extreme value distributions, to fit glass fracture strengths. He concluded that the accuracy of the extreme value distribution was much better than those of others.
A review of literatures has indicated that the assessment of statistical fatigue problems of structures and materials has been mentioned in numerous publications. In addition, it is concluded that in most cases, Weibull and normal distributions have been used for describing the reliability. However, investigating the reliability and the fatigue limit of crankshafts is still rare. To bridge this gap, in this article, the reliability of the EN-GJS700-2 ductile cast iron has been predicted. For such objective, several statistical distributions have been investigated and compared. The article continues with evaluating the endurance limit, according to the staircase test methodology. Then, obtained results have been presented in figures, including probability, reliability and the S-N curve with its scatter-band.

2 Materials, Fatigue Testing and Statistical Analysis

2.1. Studied materials and experiments

In this research, the studied material was a ductile cast iron, entitled EN-GJS700-2, which has been utilized widely for manufacturing engine crankshafts in automotive industries [1]. The chemical composition of such material (obtained by crankshaft webs) is shown in Table 1. The microstructure of the studied material has been comprehensively investigated in the previous work [1].
Tests were carried out, under fully reversed bending loading, by a four-point rotary-bending high-cycle fatigue (HCF) testing machine. Related equipments including the picture of the fatigue testing machine and its components, besides the real specimen and the geometry of standard

https://doi.org/10.1051/matecconf/201816510012

specimens were demonstrated in the literature [1]. Sinusoidal waves with a frequency of 100 Hz and a stress ratio (R_σ) of −1 (the fully reversed condition), was utilized for fatigue tests, at the room temperature of 25°C. It should be noted that the loading frequency has been chosen base on the rated power condition (equal to highest mechanical loads), about 6000 rpm of the engine speed. The applied force for stress amplitude levels was considered as 110, 100, 90, 80, 70, 60, 55 and 50 N. It should be added that the roughness of machined specimens was measured and compared with standards in the literature [1]. As a recommendation in references [4, 22], the sample number for the fatigue test to generate the S-N curve, should be in the range of 12-24 specimens. Given this, the S-N curve was characterized by testing 24 specimens, 4 specimens at each stress level (6 different levels of stress amplitude) and the fatigue endurance limit was denoted by the staircase test methodology, with 7 specimens. The total number of specimens was 31 samples, which were extracted from engine crankshafts.

Table 1. The chemical composition (wt.%) of the ductile cast iron, entitled EN-GJS700-2.

C	Si	Mn	Cu	Ni	P	Sn
3.50	2.18	0.45	0.48	0.02	0.014	0.050
Cr	Al	Ti	V	W	Co	Fe
0.030	0.016	0.035	0.003	0.002	0.004	based

2.2. Statistical distribution functions and fatigue reliability

Good understanding of the statistical behavior for fatigue properties of the ductile cast iron is necessary for the design of engine crankshafts, due to the importance of the component performance. On the other hand, to predict the reliability (the probability of survivals) of such material, it is essential to discover the suitable probability function that fitted well on data, collected at each stress level. Therefore, the statistical analysis of experimental fatigue lifetime results was investigated, utilizing six common distribution functions such as Weibull (two- and three-parameter), normal, log-normal, extreme maximum value and smallest extreme value distributions. These distribution functions were characterized by a probability density function (PDF), $f(t)$, and the associated cumulative distribution function (CDF), $F(t)$ for failures and $R(t)$ for non-failures. In this case study, t (random variable) refers to the fatigue lifetime. Probability functions of such distributions are shown in Table 2.

The PDF, $f(t)$, specifies the distribution of a failure through the fatigue lifetime and indicates the absolute failure speed [23]. In addition, the probability that a test sample will fail during an interval, $[0, t]$, is defined by the CDF, $F(t)$, which could be found as follows, [23],

$$F(t) = \int_0^t f(t)dt \tag{1}$$

In addition, the probability that a test sample will survive during an interval $[0, t]$ (the probability of non-failure) could be derived by the CDF, $R(t)$, called as "reliability" and is given by the following relation, [23],

$$R(t) = 1 - F(t) \tag{2}$$

The probability plot and the maximum likelihood estimator (MLE) was employed in the MINITAB statistical software program to perform the distribution fitting process. The MLE method is a statistical procedure for the parameter estimation by maximizing the likelihood function. This method is recommended, since maximum likelihood estimations often have desirable statistical properties [4].

Table 2. Probability formulations for six distribution functions.

Distributions	$f(t)$	$F(t)$
Two-parameter Weibull	$\frac{\beta}{\theta}\left(\frac{t}{\theta}\right)^{\beta-1} e^{-\left(\frac{t}{\theta}\right)^{\beta}}$	$1 - e^{-\left(\frac{t}{\theta}\right)^{\beta}}$
Three-parameter Weibull	$\frac{\beta}{\theta}\left(\frac{t-\gamma}{\theta}\right)^{\beta-1} e^{-\left(\frac{t-\gamma}{\theta}\right)^{\beta}}$	$1 - e^{-\left(\frac{t-\gamma}{\theta}\right)^{\beta}}$
Normal	$\frac{1}{\sqrt{2\pi}\sigma} e^{-\left(\frac{(t-\mu)^2}{2\sigma^2}\right)}$	$\emptyset\left(\frac{t-\mu}{\sigma}\right)$
Log-normal	$\frac{1}{\sqrt{2\pi}\sigma t} e^{\left(-\frac{(\ln(t)-\mu)^2}{2\sigma^2}\right)}$	$\emptyset\left(\frac{\ln(t)-\mu}{\sigma}\right)$
Extreme maximum value	$\frac{1}{\sigma} e^{\left(-\frac{t-\mu}{\sigma}\right)} e^{\left(-exp\left(-\frac{t-\mu}{\sigma}\right)\right)}$	$e^{\left(-exp\left(-\frac{t-\mu}{\sigma}\right)\right)}$
Smallest extreme value	$\frac{1}{\sigma} e^{\left(\frac{t-\mu}{\sigma}\right)} e^{\left(-exp\left(-\frac{t-\mu}{\sigma}\right)\right)}$	$1 - e^{\left(-exp\left(\frac{t-\mu}{\sigma}\right)\right)}$

For two- and three-parameter Weibull functions, mentioned in Table 2, β is the shape parameter. Then, θ is the characteristic lifetime, which is also called the scale parameter. The three-parameter Weibull function has an additional parameter, γ, which is called the location parameter [23]. In addition, in four other functions, μ is the location parameter (the mean value of experimental data at each stress level) and σ is the scale parameter (the variance of experimental data at each stress level). The function $\emptyset(t, \mu, \sigma)$ indicates the standard normal CDF [23]. The value of above parameters in six distribution functions is denoted by rewriting PDFs in the form of the standard linear equations through a mathematical transformation: $Y = A + BT$ [9,19]. These linear equations of six distribution functions with their variables, Y and T, are illustrated in Table 3. Values of A and B in all linear equations of distributions have been specified by following equations, using the MLE [22]:

$$\hat{A} = \bar{Y} - B\bar{T} \tag{3}$$

$$\hat{B} = \frac{\sum_{i=1}^{k}(T_i - \bar{T})(Y_i - \bar{Y})}{\sum_{i=1}^{k}(T_i - \bar{T})^2} \tag{4}$$

The symbol "over-bar" denotes the average value and k is the total number of specimens in each stress level [22].

Table 3. Standard linear equations of six PDFs and variables of $Y = A + BT$.

Distributions	Y	T
Two-parameter Weibull	$\ln\left(\ln\left(\frac{1}{1-F}\right)\right)$	$\ln(t)$
Three-parameter Weibull	$\ln\left(\ln\left(\frac{1}{1-F}\right)\right)$	$\ln(t-\gamma)$
Normal	$\emptyset^{-1}(F)$	t
Log-normal	$\emptyset^{-1}(F)$	$\ln(t)$
Extreme maximum value	$\ln\left(\ln\left(\frac{1}{F}\right)\right)$	t
Smallest extreme value	$\ln\left(\ln\left(\frac{1}{1-F}\right)\right)$	t

In the MINITAB statistical software program, the Anderson-Darling (A-D) statistic was utilized to test the goodness of fitting, by measuring of how far plot points fall from the fitted line, in a probability plot. This process helps to determine the best fitted distribution function to the fatigue lifetime data at each stress level. Substantially, for comparing fitting of several distributions to data, the one with the lowest value of A-D statistic has best fitting to experimental data [24]. However, the A-D statistics, calculated for different distributions, may not be directly comparable or it is not enough for comparing when the A-D values are close, since A-D values are distributed differently for different distribution functions. To have a better comparison for the fit evaluation of different distribution functions, an additional criterion such as the P-value should be utilized [24]. The P-value is a probability that measures the evidence against the null hypothesis that data follow the distribution. For an A-D test, the null hypothesis is that data follow the distribution. Therefore, higher values of the P-value are associated with a better fitting distribution to the data [24]. The P-value should be compared for each distribution function to the significance level. Generally, a significance level (denoted as α) of 0.05 works well. A significance level of 0.05 shows a 5% risk of concluding data do not follow the distribution, when they actually do follow the distribution [24]. If the P-value is less than or equal to the significance level ($P \leq \alpha$), the decision is to reject the null hypothesis and conclude that the data do not follow the distribution. If the P-value is greater than the significance level ($P > \alpha$), the decision is to fail to reject the null hypothesis [24]. Hence, to assess the appropriateness of a distribution fit, the A-D value and P-value must be considered. For comparing the fit of several distributions, the one with the largest P-value and the lowest A-D value generally has best fitting to experimental data [24].

2.3. Fatigue limit testing

The aim of fatigue limit testing (also called the fatigue strength or endurance test) is to evaluate a statistical distribution of the fatigue limit at a specific high-cycle fatigue lifetime. The staircase test methodology is the most recommended one among many fatigue limit test methods that has been adopted by many standards to investigate the statistical behavior of a fatigue limit. Hence, staircase fatigue tests for the fatigue limit of the EN-GJS700-2 ductile cast iron were conducted by cyclic loading, considering $R_\sigma = -1$.
Fatigue endurance tests are aborted by the specimen fail, which represents a complete fracture of specimens or when specimens have sustained at 2×10^6 cycles; then, they are considered as run-outs. It should be noted that the specific fatigue lifetime, mentioned above, was also considered in the literature [11] and by the data bank in the FEMFAT software [25].
According to the staircase method [4], the first test was conducted at a predetermined stress level. Then, if the specimen fails prior to the lifetime of interest, the stress level is decreased one step for the next test. Otherwise, if the specimen does not fail within the lifetime of interest, one step higher stress level should be utilized for the following test. Therefore, each test is dependent on previous test results. Tests continue until a valuable number of data are obtained to predict the fatigue endurance limit. Totally, 7 fatigue limit tests with standard specimens, which were extracted from engine crankshaft, were implemented. In addition, stress levels were considered at 60, 55 and 50 N. Therefore, the stress increment of 5 N (18.88 MPa) was utilized.
The Dixon-Mood method [4], based on the MLE was implemented as a typical data reduction technique to obtain statistical parameters of test results, according to a normal distribution. This method evaluates the mean value (μ_s) and the standard deviation (σ_s) of a fatigue limit (S_e), through approximate formulations. The equally spaced stress levels (S_i) are numbered by the index, i, with a selected increment, d. The lowest stress level (S_0) is denoted by $i = 0$. Two values, C and D could be specified by determining the number of the less frequent event (n_i) at the numbered stress level i [4], as follows,

$$C = \sum i \times n_i \tag{5}$$

$$D = \sum i^2 \times n_i \tag{6}$$

The mean fatigue limit is then calculated by the following equation [4],

$$\mu_s = S_0 + d \times \left(\frac{C}{\sum n_i} \pm 0.5\right) \tag{7}$$

The plus sign (+) is utilized, if the survival event is less repeated. If the failure event is less repeated, the minus sign (−) is used.
The standard deviation could be evaluated by following formulations [4],

$$\sigma_s = 1.62 \times d \times \left[\frac{D \times \sum n_i - C^2}{(\sum n_i)^2} + 0.029\right] \tag{8}$$

if $\frac{D\times\sum n_i - C^2}{(\sum n_i)^2} \geq 0.3$
or

$$\sigma_s = 0.53 \times d \tag{9}$$

if $\frac{D\times\sum n_i - C^2}{(\sum n_i)^2} < 0.3$.
It should also be emphasized that the stress increment value should be in the range of $0.5\sigma_s - 2.0\sigma_s$ [4].
The confidence interval value could be determined as follows [4],

$$S_e = \mu_s \pm Z \times \sigma_s \tag{10}$$

It should be mentioned that the Z value, in the above equation, is the pth percentile of the standard normal distribution and associated with the confidence level. When the confidence level was considered as 99, 95 and 90%, the Z value was 2.576, 1.960 and 1.645, respectively [4]. Besides, it should be again noted that the mean value of the fatigue limit (μ_s) and the standard deviation (σ_s) were obtained for a normal distribution.

3 Results and Discussions

3.1. Distribution functions analysis

Selecting an appropriate distribution function is an essential first step in performing the reliability analysis

[24]. If the selected distribution does not fit the data well, then the reliability estimation will be inaccurate. Hence, according to experimental data, the goodness of fitting through most commonly used distribution functions, such as Weibull (two- and three-parameter), normal, log-normal, smallest extreme value and extreme maximum value functions, were checked. Based on the A-D statistical method, six distribution functions were compared and obtained results are illustrated in Table 4, for corresponding stress levels, including 110 and 60 N, for samples of obtained results. In addition, the rank (1 to 6) of each distribution function at each stress level, due to evaluate the goodness of fitting, is shown in Table 5. In this figure, the averaged rank of distribution functions was also determined to find the best function for estimating the reliability performance. As can be seen in Tables 4 and 5, the three-parameter Weibull distribution function was generally the best fitted probability function to the experimental fatigue lifetime at different stress levels. Besides, the extreme maximum value function had the worst fitting effect, in this case study. Therefore, the three-parameter Weibull distribution was selected for EN-GJS700-2 fatigue strength data fitting, due to its adaptability to experimental data. Such function was also recommended in the literature [9], for the crankshaft. As mentioned before, all calculations were performed by the MINITAB software. Table 6 depicts parameters of the three-parameter Weibull distribution function with standard error and 95% confidence level at different stress levels.

The cumulative distribution curve, contains the relation between the failure probability and fatigue lifetime of the studied material, is predicted by the three-parameter Weibull previously chosen from six kinds of distributions. This curve including the cumulative probability versus the fatigue lifetime at different stress levels is shown in Figure 1. As can be seen from Figure 1, the percentage of failures was greater at higher values of the fatigue lifetime. As a sample plot, the cumulative probability versus the fatigue lifetime at a constant stress level (at 60 N) with scatter-bands can be seen in Figure 2, for the confidence level of 99, 95 and 90%.

Table 4. The A-D value and the P-value for fitting of distribution functions.

Distributions	110 N		60 N	
	A-D	P-value	A-D	P-value
Two-parameter Weibull	0.33	>0.25	0.41	>0.25
Three-parameter Weibull	0.28	>0.50	0.28	>0.50
Normal	0.24	0.54	0.44	0.20
Log-normal	0.23	0.58	0.28	0.52
Extreme maximum value	0.25	>0.25	0.36	>0.25
Smallest extreme value	0.34	>0.25	0.60	0.10

Table 5. The rank of distribution functions.

Distributions	Stress levels (N)						Averaged rank
	110	100	90	80	70	60	
Two-parameter Weibull	5	3	6	5	3	4	4.33
Three-parameter Weibull	4	1	4	2	1	1	2.16
Normal	2	4	1	3	4	5	3.16
Log-normal	1	5	2	6	5	2	3.50
Extreme maximum value	3	6	5	4	6	3	4.50
Smallest extreme value	6	2	3	1	2	6	3.33

Table 6. Parameters of the three-parameter Weibull distribution function for the experimental data for 95% confidence level, as a sample for obtained results.

Stress Level	Parameter	Estimate	95% Confidence Level	
			Lower	Upper
110 N	Shape	3.81	1.79	8.09
	Scale	1.66E+3	1260.97	2176.59
	Location	5.13E+3	5125.20	5125.20
100 N	Shape	8285.00	749.96	91527.10
	Scale	6.02E+6	632710.00	57328757.00
	Location	-6.01E+6	-19580380.00	7561027.00
90 N	Shape	1557.21	672.70	3604.77
	Scale	1.07E+7	10709754.00	10723905.00
	Location	-1.07E+7	-10677269.00	-10677269.00
80 N	Shape	886.62	377.57	2081.98
	Scale	2.83E+7	28279349.00	28344572.00
	Location	-2.82E+7	-28161329.00	-28161329.00
70 N	Shape	2856.59	16.37	493444.00
	Scale	1.22E+8	745140.00	1.97E+10
	Location	-1.21E+8	-737672475.00	496053350.00
60 N	Shape	1.54	0.87	2.71
	Scale	5.35E+5	321029.00	892119.00
	Location	3.81E+5	380746.00	380746.00

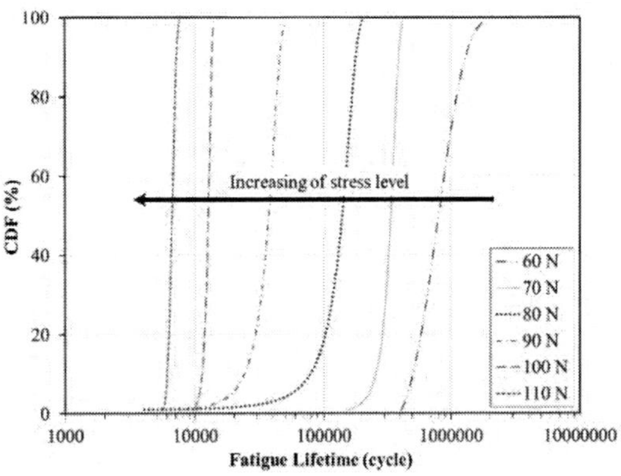

Fig. 1. The cumulative probability versus the fatigue lifetime at different stress levels.

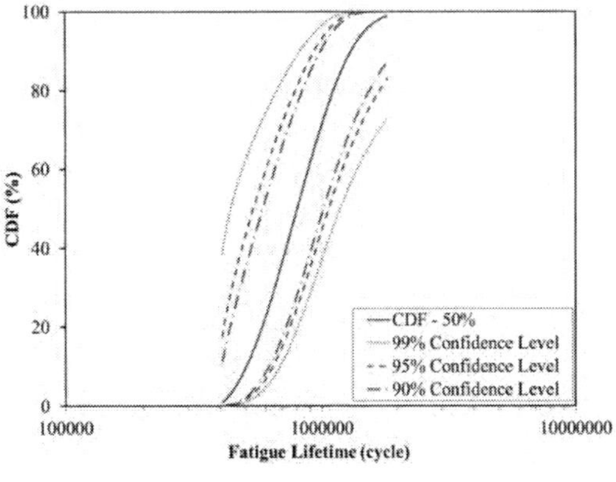

Fig. 2. The cumulative probability versus the fatigue lifetime with scatter-band at 60 N of the stress level for the confidence level of 99, 95 and 90%.

3.2. Fatigue reliability analysis

The term "reliability" in engineering refers to the probability that a product, or a system, will perform its designed functions under a given set of operating conditions, for a specific period of the time. The reliability is also known as the "probability of survival" [19].
Using the MLE method and considering the three-parameter Weibull distribution function, Figure 3 illustrates a reliability plot for different stress levels for the EN-GJS700-2 ductile cast iron. In Figure 4, as a sample plot, the reliability versus the fatigue lifetime can be seen at a constant stress level (at 60 N) with scatter-bands, for the confidence level of 99, 95 and 90%.
These figures could be a considerable value for designers, where the fatigue lifetime could be easily determined at any degree of the reliability for such material. For example, if a value of 90%R95%C is used for the component design, this particular value indicates that there is a 90% possibility of the survival (or the reliability) with a 95% confidence level for a fatigue lifetime at a specified stress level. Besides, Figure 3 shows that the trend of the reliability loss by increasing the fatigue lifetime, under constant amplitude loading.

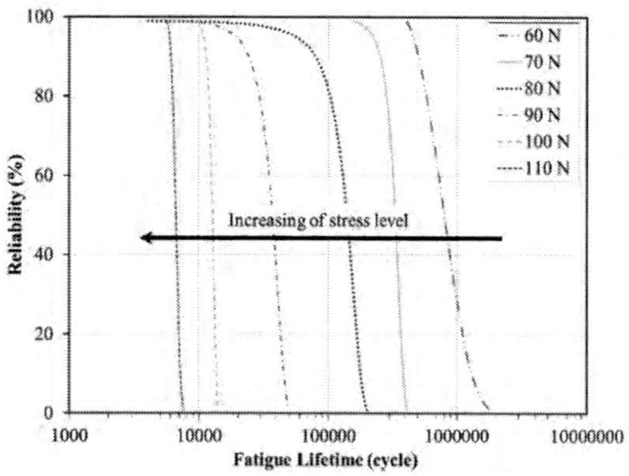

Fig. 3. The reliability versus the fatigue lifetime at different stress levels.

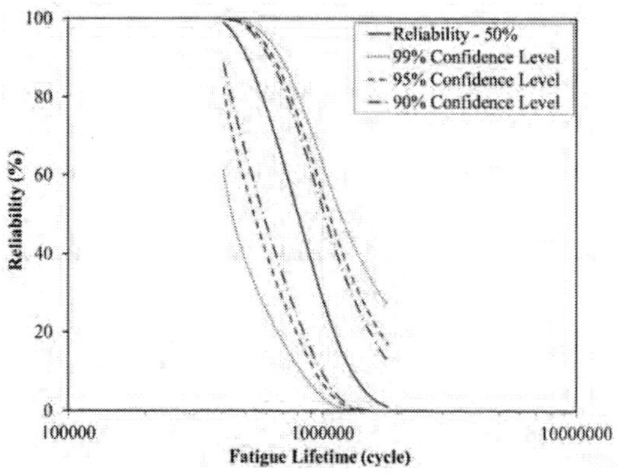

Fig. 4. The reliability versus the fatigue lifetime with scatter-band at 60 N of the stress level, with the scatter-band for the confidence level of 99, 95 and 90%.

In addition, these results are fundamental for the estimation of the optimal warranty period and for sizing of the supply system involved with the service lifetime. Once that the reliability period is known, the manufacturer could plan and manage all activities, necessary for the service lifetime. In this way, costs would be reduced, while ensuring a high quality of the service offered [10].
A careful analysis of these reliability plots indicated that the reliability remained higher (almost constant) for the initial period and later on, started declining with the usage cycle. This phenomenon explained existing understanding of periods of the crack initiation and the crack propagation. It should be noted that he reliability varied by changing the stress level. It is also clear from these plots that the crack initiation period was smaller for higher stress levels and the reliability loss was faster, during the crack propagation, indicating a faster degradation or a higher rate of the damage accumulation. However, when a product is subjected to the high stress, the rate of the reliability loss is higher and starts at the early stage of the lifetime. This happens due to a shorter crack initiation period under high stress levels and relatively a longer crack propagation period [14].

3.3. S-N curve and scatter-band analysis

The Analysis of the S-N curve was performed for the purpose of determining constant values (or properties) of the material during the high-cycle fatigue regime. Figure 5 shows series of S-N diagrams for the EN-GJS700-2 ductile cast iron, with various values of the confidence level at 90% of the reliability. Obtained results in this figure indicate that the scatter-band in the fatigue lifetime became wider at 10^5 to 10^6 cycles. A similar behavior was also observed by the literature [19], for composite materials under rotating bending fatigue testing.
In general, the variance of the logarithmic lifetime increased, as the stress level decreased. It has been observed that once cracks nucleated in the material at high stress levels, these cracks had a better chance of overcoming the surrounding microstructure. Therefore, there was a long period to propagate. Most cracks could successfully nucleate at low stress levels, but only few cracks could overcome surrounding obstacles to grow. Thus, the crack propagation period was so small [4].
For the S-N curve, a relation between the fatigue lifetime (N_f) and the stress amplitude (σ_a) could be written [1], as follows,

$$\sigma_a = \sigma_f'(2N_f)^b \quad (11)$$

Where σ_f' and b are the fatigue strength coefficient and the fatigue strength exponent, respectively. Obtained results based on these material constants for the EN-GJS700-2 ductile cast iron can be observed in Table 7, based on 90% of the reliability and various confidence levels. It should be noted that in Figure 5 and Table 7, there was no data for the lower-band of 99% confidence level; therefore, there was no line for the scatter-band.

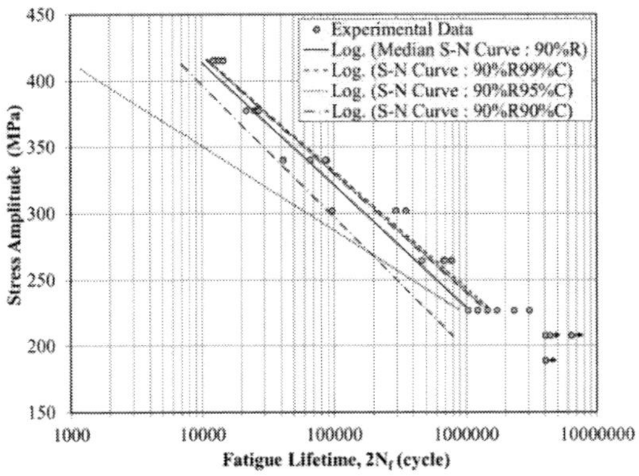

Fig. 5. The S-N curve and scatter-bands for different confidence levels in the EN-GJS700-2 ductile cast iron.

Table 7. The characterization of the high-cycle fatigue behavior in the EN-GJS700-2 ductile cast iron.

Reliability (R) and confidence level (C)	R^2	Fatigue strength coefficient (MPa)	Fatigue strength exponent (-)
Median S-N Curve	0.9885	1378.1611	-0.1283
90%R99%C (Upper)	0.9919	1322.5131	-0.1218
90%R99%C (Lower)	-	-	-
90%R95%C (Upper)	0.9916	1334.7502	-0.1230
90%R95%C (Lower)	0.4547	809.4685	-0.0929
90%R90%C (Upper)	0.9917	1340.9111	-0.1236
90%R90%C (Lower)	0.8924	1457.1357	-0.1408

* R^2: Coefficient of determination

As it can be observed from obtained results in Table 7, high-cycle fatigue properties (including the fatigue strength coefficient and the fatigue strength exponent) of the material varied, when the confidence level changed, both at the lower and upper scatter-band. It should be mentioned that the coefficient of determination (R^2) was more than 89%, unless one data point for the confidence level of 95%, which generally showed proper modeling to obtain fatigue properties.

3.4. Fatigue limit assessment

In accordance with mentioned steps in Part 2.3, the fatigue endurance limit of the EN-GJS700-2 ductile cast iron was experimentally examined. Figure 6 illustrates results of the fatigue endurance limit, including failure/run-out records of the studied material, at different confidence levels, including values of 90, 95 and 99%. It should be noted that in Figure 6, black dots and open circles represented test data, failed prior to 2×10^6 cycles and suspended at 2×10^6 cycles, respectively. Besides, red lines represents the mean and the interval (or the scatter-band) of the endurance limit at different confidence levels. In addition, all determined values are also listed in Table 8, including the mean endurance limit, the standard deviation and lower and upper scatter-band of the endurance limit at different confidence levels.

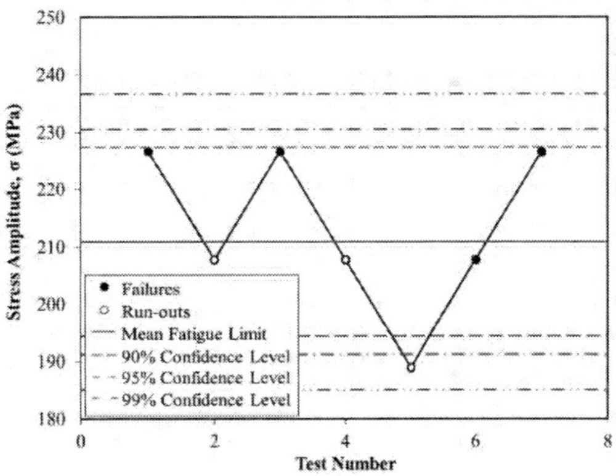

Fig. 6. Results of the Dixon-Mood analysis for the EN-GJS700-2 ductile cast iron.

Table 8. Results of fatigue limit testing for the EN-GJS700-2 ductile cast iron.

Fatigue limit properties (MPa)	Confidence level	Experimental data
Mean fatigue limit (μ_s)	-	210.899
Fatigue limit scatter-band	90%	194.433-227.365
Fatigue limit scatter-band	95%	191.280-230.519
Fatigue limit scatter-band	99%	185.114-236.685
Standard deviation (σ_s)	-	10.010
Stress increment (d)	-	18.887
d/σ_s	-	1.880

As it can be seen in Table 8, the stress increment value was equal to $1.88\sigma_s$, which was in the recommended range in the literature [4]. In addition, the mean fatigue endurance limit measured for the studied material was about 211 MPa. The fatigue limit of the studied material in the FEMFAT data bank was 240 MPa [25]. As the first discussion in this section, the mean fatigue limit value was 12% lower than the endurance limit from the FEMFAT data bank. For the reason, it could be stated that the endurance limit in the FEMFAT data bank was determined by standard specimens, made from the material, and not from the component. However, presented experimental data were conducted by standard specimens, extracted from the crankshaft web, due to consider the manufacturing effect. This emphasizes that the manufacturing effect could decrease the fatigue strength of the raw material. Additionally, as seen in Figure 6 and Table 8, when a higher value for the confidence level was utilized, a wider scatter-band was obtained. It is obvious that to have higher reliable confidence intervals, higher degrees of the confidence level should be considered.

4 Conclusions

In the present research, the reliability prediction, the S-N curve, the scatter-band analysis and the fatigue limit assessment of high-cycle fatigue properties in the EN-GJS700-2 ductile cast iron were performed. For this objective, experimental data, acquired from four-point rotary bending fatigue testing on standard specimens (extracted from engine crankshafts) were analyzed.

Obtained results could be listed as follows,

- Based on the distribution functions analysis, the three-parameter Weibull distribution was optimum for the experimental data adaptation of the studied material.
- The reliability of the material was calculated at each stress level, considering the three-parameter Weibull distribution.
- The scatter-band for the S-N curve at different confidence levels was also obtained to find high-cycle fatigue properties of the material.
- The fatigue limit of the studied material was obtained as 211 MPa, considering the effect of manufacturing approaches in the component, based on the staircase test method.

As the acknowledgement, Authors would like to thank Irankhodro Powertrain Company (IPCo) for financially supporting of the presented work.

References

1. M.J. Khameneh, M. Azadi, Eng. Fail. Anal., **85** (2018) 189–200
2. *Metallic materials - Fatigue testing Statistical planning and analysis of data*, Standard No. ISO 12107, ISO International Standard, (2012)
3. J.M. Zhai, X.Y. Li, Int. J. Fatigue, **44** (2012) 107–115
4. Y.L. Lee, J. Pan, R. Hathaway, M. Barkey, *Fatigue Testing and Analysis*, John Wiley and Sons, (2004)
5. P.D. Toasa Caiza, T. Ummenhofer, Int. J. Fatigue, **106** (2018) 70–80
6. A.K. Chandrappa, K.P. Biligiri, Const. Build. Mater., **153** (2017) 1–15
7. S.P. Zhu, Q. Liu, H.Z. Hong, Struct. Integ. Proc., **4** (2017) 3–10
8. L. Feng, M.L. Yu, N.G. Fang, L.L. Juan, Const. Build. Mater., **95** (2015) 207–217
9. X. Chen, X. Yu, R. Hu, J. Li, Eng. Fail. Anal., **42** (2014) 210–220
10. O. Borgia, F.D. Carlo, N. Fanciullacci, M. Tucci, *Accelerated life tests for new product qualification: a case study in the household appliance*, 11th IFAC Workshop on Intelligent Manufacturing Systems, Brazil, (2013)
11. G. Cevik, R. Gurbuz, Eng. Fail. Anal., **27** (2013) 250–261
12. J. Klemenc, M. Fajdiga, Int. J. Fatigue, **43** (2012) 90–97
13. S. Mohd, Y. Mutoh, Y. Otsuka, Y. Miyashita, T. Koike, T. Suzuki, Eng. Fail. Anal., **22** (2012) 64–72
14. V. Rathod, O.P. Yadav, A. Rathore, R. Jain, Int. J. Qual., Statis. Reliability, (2011) doi:10.1155/2011/718901 (to be published)
15. C. Bouraoui, R. Ben Sghaier, R. Fathallah, Mater. Des., **30** (2009) 475–486
16. R.B. Sghaier, C. Bouraoui, R. Fathallah, T. Hassine, A. Dogui, Int. J. Fatigue, **29** (2007) 209–221
17. P. Spiteri, S. Ho, Y.L. Lee, Int. J. Fatigue, **29** (2007) 318–329
18. J. Schijve, Int. J. Fatigue, **27** (2005) 1031–1039
19. U.A. Khashaba, J. Compos. Mater., **37** (4) (2003) 317–331
20. M.H.H. Shen, Int. J. Fatigue, **21** (1999) 699–708
21. R.H. Doremus, J. Appl. Phys., **54** (1983) 193–198
22. *Standard practice for statistical analysis of linear or linearized stress–life (S–N) and strain-life (ε-N) Fatigue Data*, Standard No. ASTM E739–10, ASTM International, (2015)
23. G. Yang, *Life cycle reliability engineering*, John Wiley and Sons, USA, (2007)
24. MINITAB software, Help of the software
25. FEMFAT software, Help of the software

Fatigue of automotive engine cylinder heads – A new model based on crack propagation

Mathieu Béranger[1,*], *Guillaume* Morin[1], *Sarah* Saanouni[2], *Alain* Koster[2], and *Vincent* Maurel[2]

[1]Direction de la Mécanique, RENAULT, 78084 Guyancourt cedex, France
[2]MINES ParisTech, PSL Research University, MAT - Centre des Matériaux, CNRS UMR 7633, BP 87 91003 Evry, France

Abstract. The main reliability issue encountered during cylinder head development is the appearance of fatigue cracks during severe endurance tests. Some of these cracks can initiate in the coolant water jacket and lead to the complete part failure. They are mainly associated with the combination of combustion pressure alternate stresses and high cycle fatigue load in infinite life domain. However, Haigh or Dang Van analysis issued from 3D Finite Element Analysis reveals several issues: crack initiations are not correctly located and scatter of crack sizes are not predicted. To overcome these difficulties, a new fatigue model dedicated to cylinder heads in Aluminum - Silicon alloy has been developed in Renault powertrain division. The model is based on fatigue crack growth modelling in order to take into account various and combined load ratio in one hand, and typical characteristics of the microstructure of casting alloy in the other hand. This paper presents the theoretical basis of this model and the associated identification methodology for two different aluminum alloys.

1 Cylinder Head background

Design of modern automotive internal combustion engines has become an increasing challenging task in the past years due to increasing constraints in terms of performance, fuel economy and emissions.
Mechanical and thermal loads increase on high specific power output engines, mainly turbocharged direct injection Diesel and gasoline engines, has made more and more difficult to reach reliability targets. In parallel, development costs reduction calls for an increase in simulation model accuracy in order to get directly the optimized design with reduced physical validation tests.

1.1 Fatigue failure mode

Cylinder head is a critical and complex part in an internal combustion engine. It has a great impact on performance via intake and exhaust aerodynamic and it undergoes high mechanical and thermal loads directly linked to combustion process and design of the part.

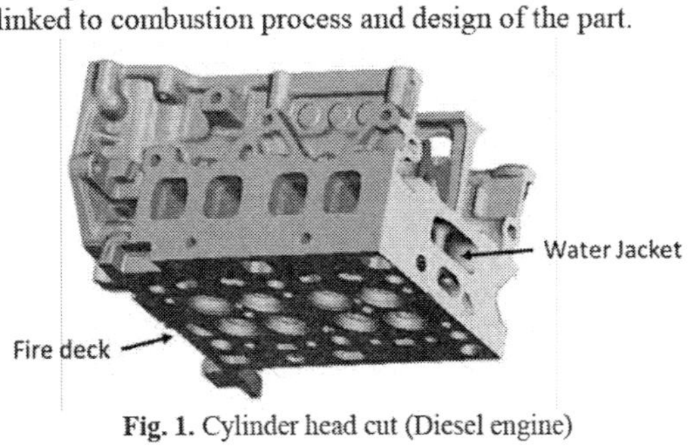

Fig. 1. Cylinder head cut (Diesel engine)

The main reliability issue encountered during cylinder head development is the appearance of fatigue cracks during severe endurance tests.
These cracks can initiate in the coolant water jacket and can lead to the complete part failure.

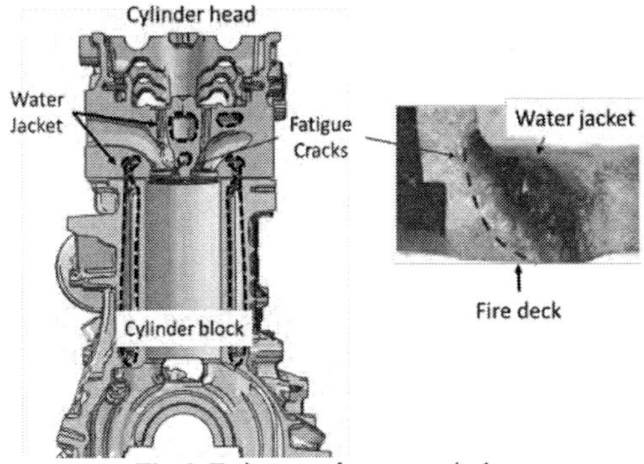

Fig. 2. Fatigue crack on water jacket

1.2 Water jacket loading

Cracks initiated in the coolant water jacket are associated mainly with combustion pressure alternate stresses and high cycle fatigue (HCF).
However, in the next discussion we will consider several kinds of loadings:

- Static loading due to tightening on cylinder block, interference fitting of seats;
- Thermal loading due to gas combustion flow;

* Corresponding author: mathieu.beranger@renault.com

Because of the engine power variation, that loading is cyclic and therefore creates thermo-mechanical stresses.

- Pressure loading at each combustion in the cylinder

The range and frequency of both thermal and pressure cyclic loadings are not the same. Actually, thermal loading will induce about 10^4 cycles of low cycle fatigue (LCF) range (typically with a load ratio R=0.1) while pressure loading induces about 10^7 cycles of high cycle fatigue (HCF) range (typically with a load ratio R=0.7). Figure 3 illustrates the decomposition of the different loading during an engine elementary test bench.

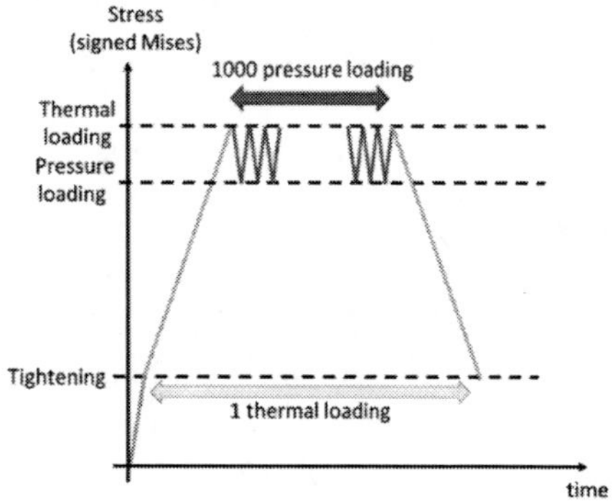

Fig. 3. Decomposition of loading on water jacket

The quantity of interest is the ratio of the number of pressure loadings added on water jacket for each thermal loading, e.g. 1000 in Fig. 3. This ratio depends on the type of engine test bench, in particular the engine speed applied during the test.

1.3 Material characteristic

Aluminum – Silicon alloys are currently used for cylinder head for decades to reduce weight and therefore increase fuel efficiency. Material casting (in our case gravity casting) are known to generate coarse microstructure and shrinkage cavities associated to the dendritic structure of these alloys. In our case, the size of typical shrinkage cavities is ranging from 0.1 mm to 0.5 mm.

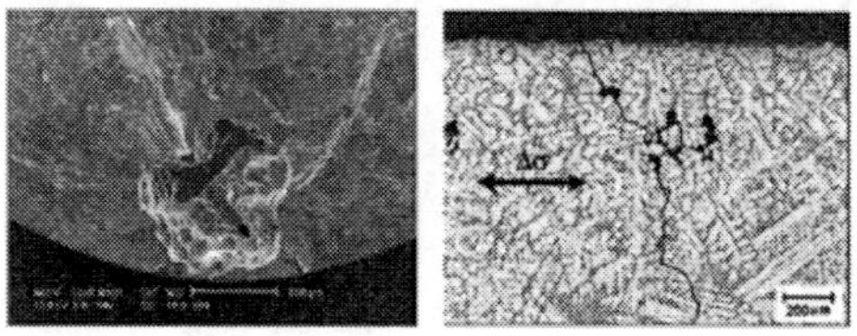

Fig. 4. Crack initiation on shrinkage cavity and subsequent crack propagation [1, 2]

For both HCF [3, 4] and LCF loadings [5], this kind of defects can be considered as the main sites of crack initiation. After initiation, cracks firstly develop in the so-called short crack range and can propagate to reach long crack domain [6].

This study will cover the analysis of two kinds of aluminum-silicon alloy: the first one is a 3% copper alloy (historical choice of Renault cylinder head) and the second one is a new 0.5% copper alloy developed for high performance engines.

2 Crack propagation model for water jacket

2.1. Historical context

Tests bench performed on 2L Diesel engine have shown the role of both thermal and combustion loading on water jacket damage.
Measurements of crack lengths have been achieved for different test benches associated to different thermal-mechanical fatigue loadings, Fig. 5:

- P_{max} tests: engine is maintained at maximum power. This test induces only pressure loading stresses.
- *Hot-Cold* tests: both thermal and pressure loadings are applied on cylinder head.

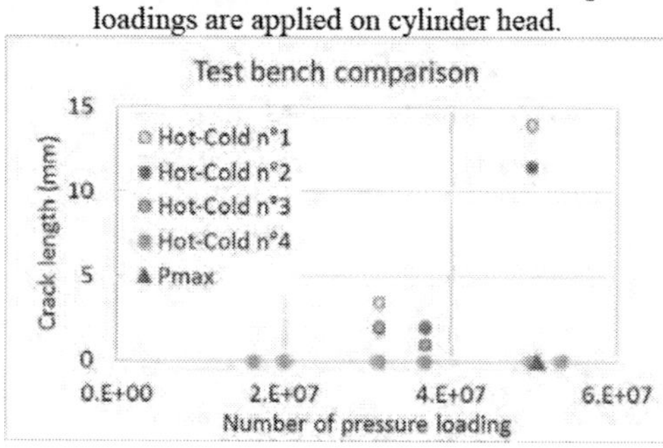

Fig. 5. Thermal-Mechanical loading effect on crack length for bench P_{max} and *Hot Cold* tests

While many cracks are observed at the end of *Hot-Cold* tests, for higher number of pressure loading, P_{max} test has not revealed any crack on water jacket.

To prevent appearance of cracks, classical methodologies use stresses estimated from 3D Finite Element Analysis to determine the level of a fatigue safety factor based on Dang Van criterion [7] as illustrated in Fig. 6. In this case, the stress path is represented in a mesoscopic shear τ(t) versus mesoscopic hydrostatic stress $\sigma_H(t)$ diagram.

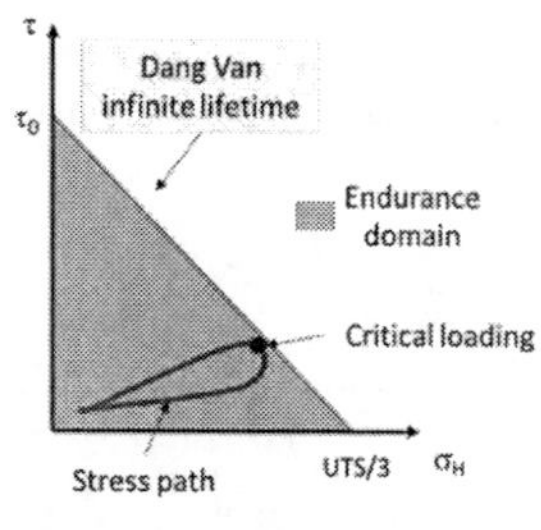

Fig. 6. Illustration of the Dang Van criterion

The criterion is defined by the boundary between infinite life (endurance domain) and fatigue. Component life for a load path contained completely below the Dang Van

Criteria line will be infinite. However, this approach does not allow to consider simultaneously low and high cycles fatigue.

Another classical methodology to take into consideration variable amplitude stresses, is to consider damage accumulation.

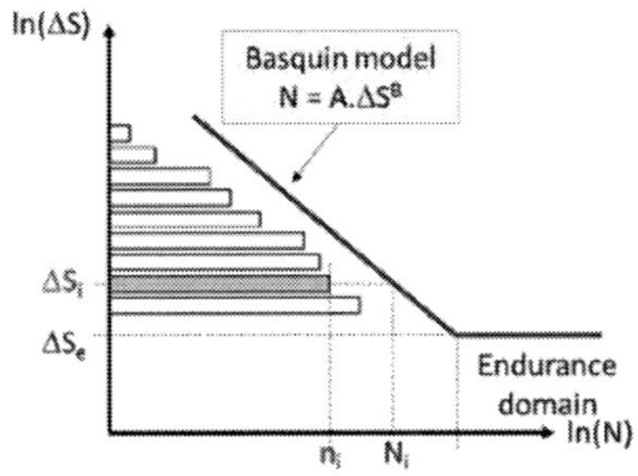

Fig. 7. Damage calculation on S-N curve

Using the linear damage rule of Miner – Palmgreen, it is possible to estimate, for various loading the total damage D:

$$D = \sum_i \frac{n_i}{N_i} \quad (1)$$

Classically, High Cycle Fatigue and Low Cycle Fatigue are implemented separately.

2.2 Crack propagation model

To overcome these difficulties, a new fatigue model, based on fatigue crack growth modelling, has been developed in Renault powertrain division.

This approach has two main advantages:

- It takes into account typical characteristics of microstructure of casting alloy. Distribution of size and localisation of porosities allow performing probabilistic analysis of fatigue lifetime, and can therefore explain observed experimental scatter.
- Effect of various and combined loading is used to evaluate fatigue lifetime.

The model formulation is defined by:

$$\frac{da}{dN} = \frac{da}{dN}(\Delta K_{LC}) + \frac{N_{SC}}{N_{LC}} \times \frac{da}{dN}(\Delta K_{SC}) \quad (2)$$

With the following notations:

- ΔK_{LC} is the amplitude of stress intensity factor for cycles of large amplitude, such as thermal loading ;
- $\frac{da}{dN}(\Delta K_{LC})$ is the crack growth rate associated to large cycles;
- ΔK_{SC} is the amplitude of stress intensity factor for cycles of small amplitude, such as pressure loading;
- $\frac{da}{dN}(\Delta K_{SC})$ is the crack growth rate associated to large cycles;
- $\frac{N_{SC}}{N_{LC}}$ is the ratio of numbers of small cycles and large cycles

Initial crack growth rate of water jacket is then estimated based on a typical flaw size of 300 µm.

Next paragraph details this model, and describes the global methodology applied to identify those parameters: crack propagation tests, mathematical formulation and correlation with variable stress loadings.

3 Variable amplitude crack growth methodology

The design of the crack propagation model, developed in equation (2), is divided into four steps, described hereafter:

- Estimation of Paris' law with threshold using Compact Tension tests,
- Validation of long crack propagation model on small cracks using notched specimen,
- Crack propagation tests with variable amplitude loadings,
- Correlation with engine tests.

The following nomenclature will be used in the next sections

a crack length (mm)
σ stress (uniaxial formulation)
K stress intensity factor, SIF, $K = f(\sigma, a)$
R load ratio $R = {K_{min}}/{K_{max}}$
CT Compact Tension tests

3.1 Long crack propagation model

The determination of propagation model is based on the determination of the coefficients of the Paris law:

$$\frac{da}{dN} = C.\Delta K^m \quad (3)$$

Where

- m and C are two material dependent parameters,
- ΔK the amplitude of stress intensity factor
- $K = \sigma.FI(a).\sqrt{\pi a}$ (4)

  with FI the shape function associated to the specimen shape

CT tests, according to ASTM standard [8] have been achieved for different load ratios R ranging from 0.1 to 0.7, at 150 °C, which is the average temperature of in use water jacket. Those tests are currently used to characterize FCGR for long cracks. Crack length measurements is achieved using continuous potential drop technique calibrated on optical inspection of crack length.

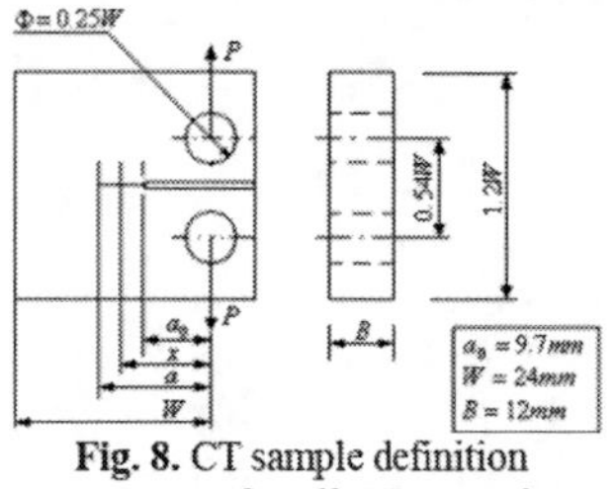

Fig. 8. CT sample definition

The next figure presents for all CT tests the evolution of fatigue crack growth rate (FCGR) vs the amplitude of stress intensity factor, these tests corresponds to the 0.5% copper aluminum alloy.

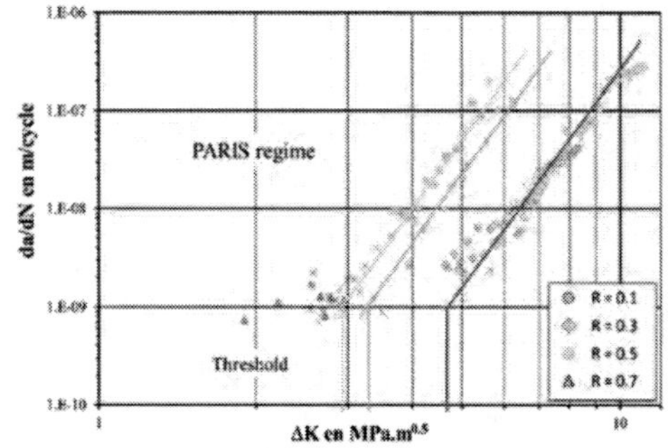

Fig. 9. CT tests results for 0.5% copper aluminum alloy

CT tests show an obvious dependency of the threshold ΔK_t as a function of the load ratio R, Fig. 9. According to Elber [9], such a dependency can be corrected by taking into account only a part of the SIF amplitude, considering the SIF value needed to open the crack K_{open}. Thus, the effective stress intensity factor ΔK_{eff} is defined as:

$$\Delta K_{eff} = K_{max} - K_{open} \tag{5}$$

and the crack propagation model by:

$$\frac{da}{dN} = C.\Delta K_{eff}{}^{m} \tag{6}$$

with C and m material parameters being independent of R load ratio.

To determine the K_{open} value as a function of R, we first assume that for $R \geq 0.7$, the crack remains opened during the whole fatigue cycle. For other load ratio R, K_{open} is obtained as the gap between the asymptotic threshold value and the linear interpolation of experimental ΔK_t, Fig. 10. It is thus assumed that for SIF values lower than K_{open}, cracks will not propagate.

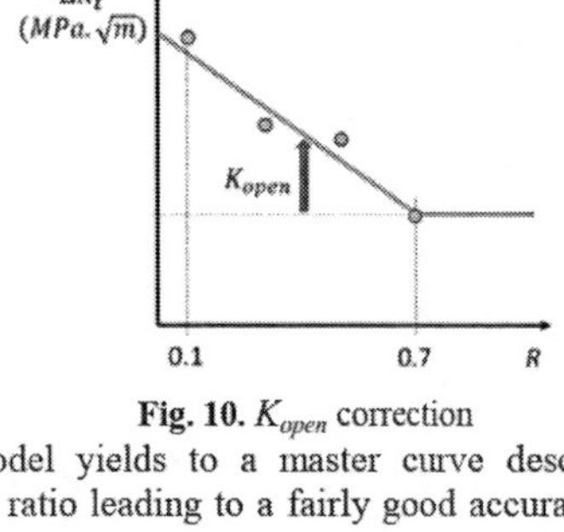

Fig. 10. K_{open} correction

This model yields to a master curve describing any tested R ratio leading to a fairly good accuracy for both aluminum alloys with 3% and 0.5% copper, respectively Fig11 and 12.

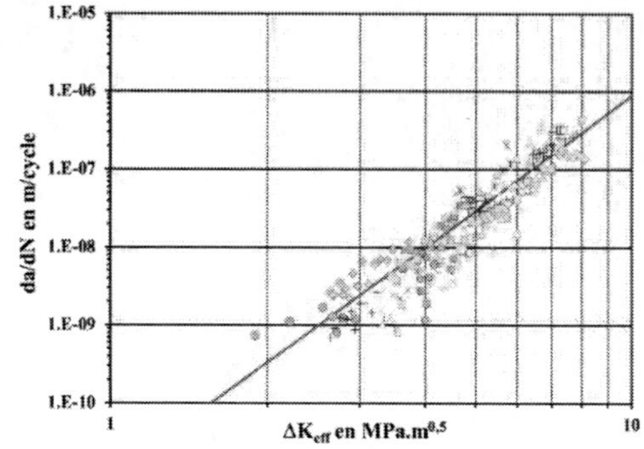

Fig. 11. Long crack growth rate model (3% copper alloy)

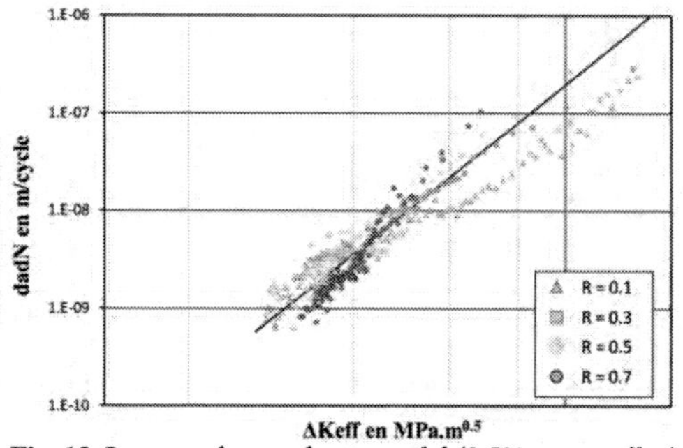

Fig. 12. Long crack growth rate model (0.5% copper alloy)

It is worth noting that observed scatter, is the direct consequence of CT tests results for highest applied ΔK values. However, this study aims at focusing on intermediate ΔK values where the model is very consistent.

3.2 Small crack propagation validation

Crack propagation mechanisms are not similar for short and long crack [6]. As our objective is to take into account initiation of small crack on water jacket, it is important to test the proposed crack propagation model issued from long cracks tests onto small cracks propagation ones.

Special geometry sample has been developed for that purpose [4] using notched sample with rectangular section. The initial notch is a semi-elliptical cut of about 500 μm in width achieved by electro-discharge machining (EDM) on the largest flat face of the gage area.

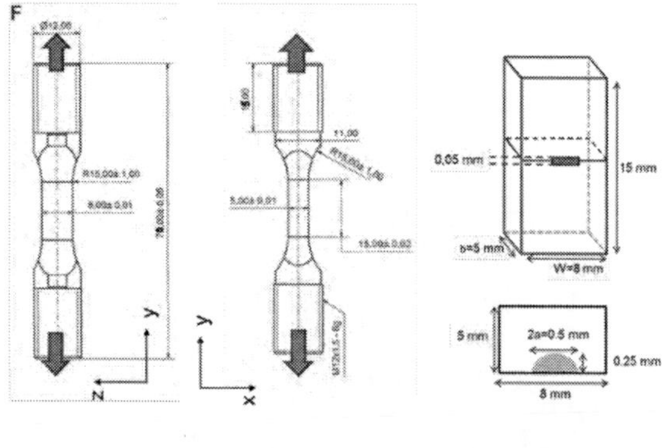

Fig. 13. Sample definition for short cracks study

Unfortunately, due to both the small size of the gage area and of the notch, this geometry prevent from the use of potential drop technique for crack length tracking.
Therefore, crack length evolution is derived from *in situ* image analysis using an optical microscope.

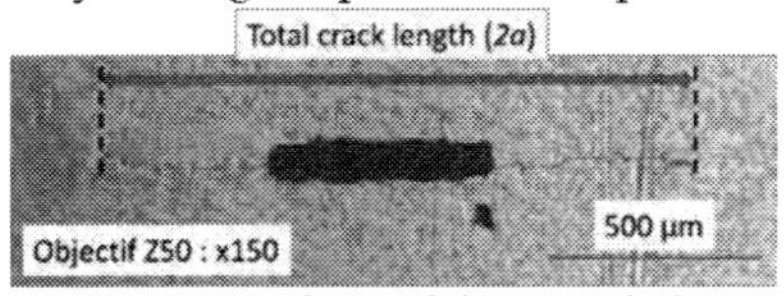

Fig. 14. Crack growth image analysis

Tests have been driven at 150 °C, for different load ratio ranging from -1. to 0.7.
The following picture represents the crack growth rate $\frac{da}{dN}$ as a function of the effective stress intensity factor ΔK_{eff}. It is worth noting that ΔK_{eff} takes into consideration the K_{open} correction yielded by equation (5).

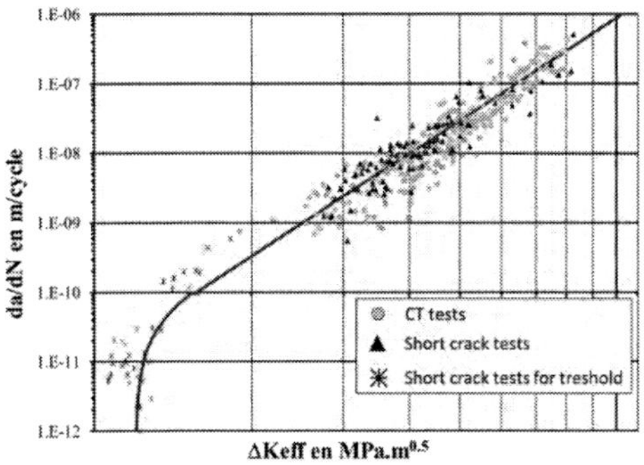

Fig. 15. Short crack growth rate model (3% copper alloy)

First analysis of those results confirms the consistency of the long crack model for cracks initiated on a small notch.
However, for low ΔK_{eff} values, significant increase of FCGR is observed as compared to the unified Paris model. This locally high FCGR is consistent with the influence of short crack on FCGR [6].

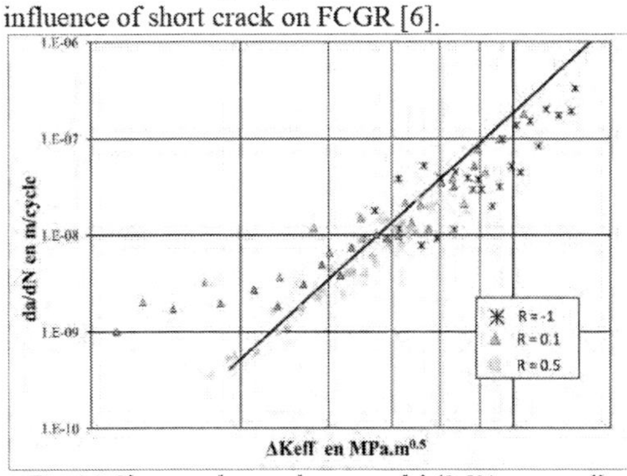

Fig. 16. Short crack growth rate model (0.5% copper alloy)

Those results and their impact on global methodology are still under analysis for the new aluminum alloy.

3.3 Crack propagation with variable amplitude loading

As discussed in the section devoted to the context of this study, water jacket of cylinder head bears complex loadings: first, the tightening of the cylinder head on peripheric parts, which is kind of a permanent static load, and then 2 dynamic loadings: thermal one and cylinder pressures.
To get as close as possible to real loads conditions, specific propagation tests have been carried out. Based on the same geometry of small cracks tests, we proposed to combine low amplitude and high frequency cycles referred to "small cycles" (corresponding to cylinder pressure cycles), on high amplitude and low frequency cycles, designated by "large cycles" (corresponding to the starting and stopping of the engine).
Those tests are defined with 3 parameters: the ratio between the respective number of small and large cycles (from 500 to 1000 in our case), the load ratio for small and large cycles (0.7 and 0.1 in our application).
The next figure shows the diagram load program test.

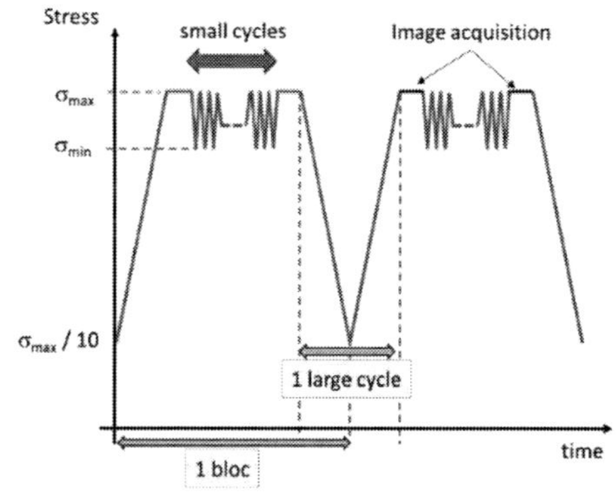

Fig. 17. Crack tests with combined loading

The intent of the study is to understand the effect of combination of short cycles and long cycles on FCGR, thus we proposed a first simple analysis of tests results.
Only 3% copper alloy results will be discussed in the next paragraph as recent tests for 0.5% copper alloy are still in progress and need additional analysis.
Firstly, we consider only the effect of small cycles to evaluate the FCGR, which assumption yields to:

$$da/dN \approx \Delta a/\Delta N_{SC}$$

where ΔN_{SC} is the number of small cycles measured during the increment of crack Δa.

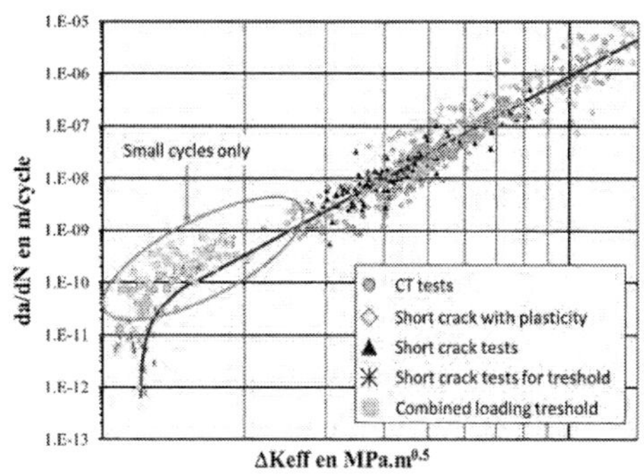

Fig. 18. Comparison of small cycles effect with Paris model for long cracks (3% copper aluminium alloy)

This assumption, leads to an overestimation of the FCGR with respect to the master curve, Fig. 18. Moreover, whereas these tests have been achieved using notched specimens, no evidence of short crack behavior is observed.

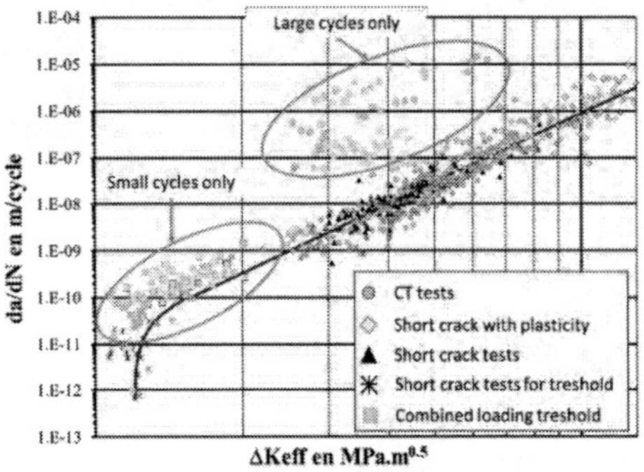

Fig. 19. Comparison of large cycles effect with Paris model for long cracks (3% copper aluminium alloy)

Secondly, we consider only the effect of large cycles to evaluate the FCGR, which assumption yields to:

$$da/dN \approx \Delta a/\Delta N_{LC}$$

where ΔN_{LC} is the number of large cycles measured during the increment of crack Δa. In this case, FCGR is ten times higher than the master curve.

As a partial conclusion, variable amplitude loading shows a dependency of crack growth on both large and small cycles.

3.4 First interpretation and limits

The first interpretation is to separate, with a cumulative damage model, the effect of large and small cycles.

For large cycles, an incremental approach is applied to estimate, for a constant step of crack length (da = 50 µm):

- on one hand, the number of large cycles dN_k performed during the tests with respect to the following picture where one block is equalled to one large cycle.

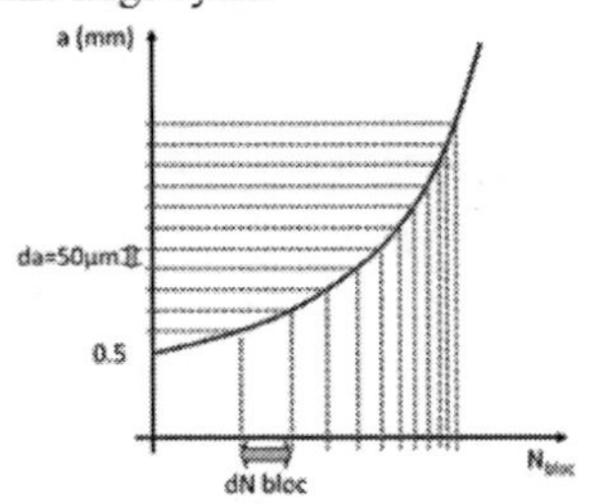

Fig. 20. Estimation of realised dN_k large cycles

- on the other hand, the level of ΔK_{eff} given by equations (4) and (5). The considered SIF values are limited to large cycles (here $R=0.1$) and are associated to a given dN_P_k according to the Paris' model (6).

Finally, the increment of damage associated to crack growth due to large cycles is given by:

$$d_i = \frac{\sum_{k=1}^{i} dN_k}{\sum_{k=1}^{i} dN\_P_k} \qquad (7)$$

For small cycles, the described methodology is slightly adapted. Considering only small cycles, the associated SIF values (here for $R=0.7$) are observed to be very close to SIF threshold derived from long crack analysis and the master curve. This is the reason why, we choose to extend the Paris law to low values of ΔK_{eff}.

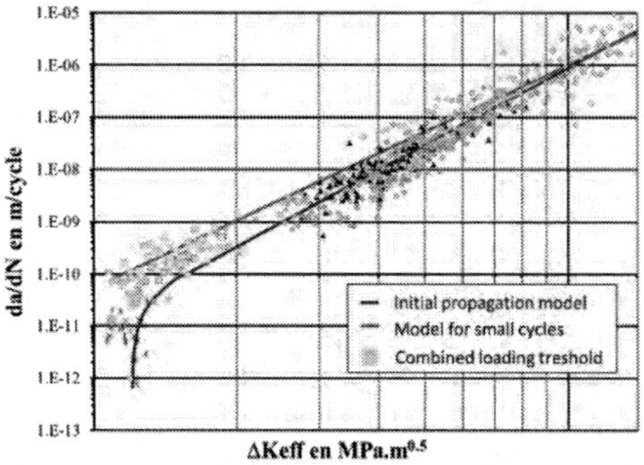

Fig. 21. Paris model extension for small cycles (3% copper aluminium alloy)

This methodology is finally applied to combined loading testing analysing both the FCGR, Fig. 21, and, after integration of the Paris law, the crack length plot as a function of the number of cycles, Fig. 22.

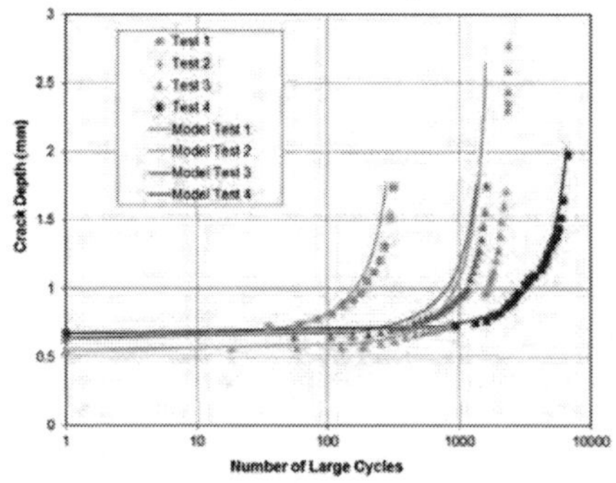

Fig. 22. Validation of crack growth model on combined loading (3% copper aluminium alloy)

Finally, the chosen assumptions have led to a fairly good evaluation of the measurements, validating the interest of such a methodology, at least for that kind of aluminium alloy.

3.5 Initial size of water jacket's flaw

Once the proposed model (2) has been fitted on different kinds of tests, it is necessary to define the initial size of critical flaw leading to crack initiation. This should be determined for the considered water jacket's.

First analysis is based on small cracks propagation tests. We have noticed that for machined elliptical notch of less than 300 µm, cracks started on other flaw induced by the process, initially present in the specimens. However, for an elliptical notch size of above 500 µm, cracks always iniate from the notch [4].

Another way to validate that initial size of the water jacket's flaw is based on S-N curve dispersion. Wöhler analysis allows us to define the dispersion of S-N curve.

On the other hand, lifetime curve can be estimated with the crack propagation model (6) for different given size of initial flaw.

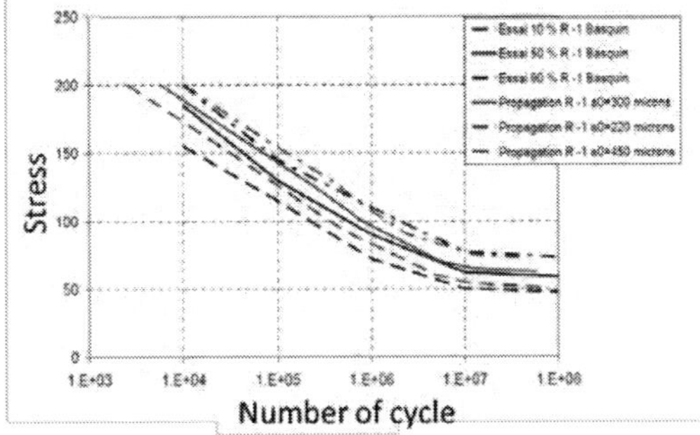

Fig. 23. Estimation of flaw size on S-N curve (3% copper aluminium alloy)

As we can see in Fig. 23, a flaw size of 300 µm is consistent with the average lifetime of S-N curve. This value of initial flaw will be used to estimate the initial FCGR for any location in the water jacket.

As a validation step, evaluation of cracks propagation has been achieved on the 2L Diesel engine for different kind of loadings.

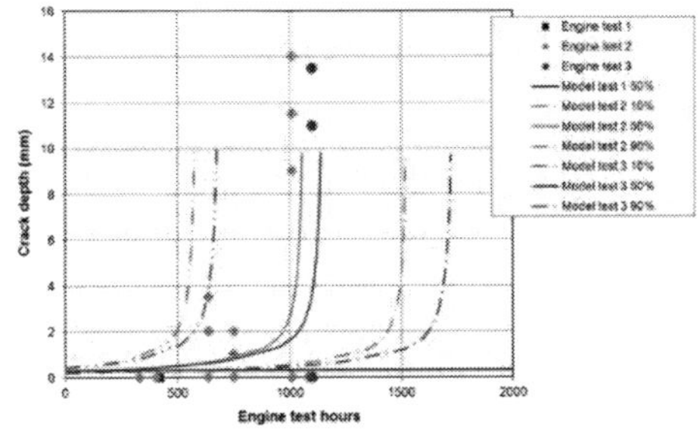

Fig. 24. Correlation of crack length on engine tests (3% copper aluminium alloy)

For the *Pmax* test, the model confirms that the crack cannot propagate without thermal loading while a good correlation of crack length is noticed for *Hot-Cold* tests.

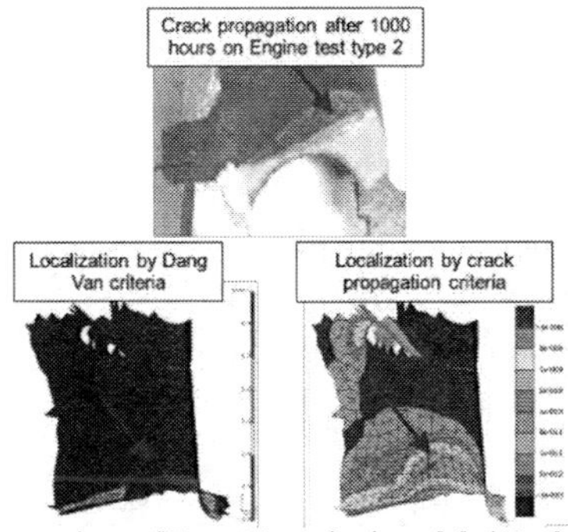

Fig. 25. Comparison of Dang Van criteria and dadN criteria on water jacket

Moreover, the proposed methodology is able to obtain a ranking of critical areas consistent with observed in-service damage, instead of classical initiation method leading to artefacts in damage localization and ranking. This kind of ranking is of first importance to focus on the main issues to solve for a reliable design of the cylinder head.

4 Conclusions and perspectives

The proposed methodology and criterion is used as a simulation standard for the cylinder head design for about 10 years.

The first interest is to improve the localization of critical area as compared to classical models based on crack initiation analysis. Actually, this criterion allows us to identify cracks due to thermal and pressure loadings where classical multi-axial criteria failed.

The same approach is currently under construction for a new aluminium silicon alloy with low copper content, for which first results have been presented in this study. Additional tests to understand deeper the interaction between large and small cycles will be discussed in a future work.

References

1. W.W. Bose-Filho, *Al-Si cast alloys under isothermal and thermomechanical fatigue conditions*, IJF (2007)
2. H. Mayer, *Influence of cyclic loads below endurance limit or threshold stress intensity on fatigue damage in cast aluminium alloy 319-T7*, IJF 27, p 129-141 (2005)
3. Shyam, et al. Scripta Materialia, 50(8) (2004), pp 1109-1114.
4. R. Salapete, PhD Thesis in French, *Propagation de fissures dans les culasses Diesel* (2007), Mines ParisTech
5. Dezecot, S., et al, Acta Materialia, 123 (2017), pp 24-34.
6. S. Suresh, R.O Richie, *Propagation of short fatigue cracks*, Int. Met. Rev. 29 (1984), pp 445-476
7. K. Dang Van, *Advances in multiaxial fatigue*, ASTM STP 1191 (1993), pp 120-130
8. *Standard Test Method for Measurement of Fatigue Crack Growth Rates*, ASTM E647 (2015)
9. W. Elber, *The significance of fatigue cracks closure*, ASTM STP, 486 (1971), pp 230–242

Introduction to the new FKM guideline which considers nonlinear material behaviour

Melanie Fiedler[*1], *Michael Vormwald*[1]

[1]Materials Mechanics Group, Technische Universität Darmstadt, D-64285 Darmstadt, Germany

Abstract. In 2018 the German Research Association of Mechanical Engineering (FKM) will present a new guideline for calculating the fatigue crack initiation life of structures and components.
The user of this guideline
... is able to calculate the fatigue life for crack initiation for structures of steel, cast steel and aluminium alloy
... has to perform an elastic FE-calculation and gets the quality of an elastic-plastic calculation
... is able to determine the cyclic material parameters and the damage parameter Wöhler curve by the ultimate tensile strength R_m
... is able to calculate the fatigue life in **HCF** and **LCF** for constant and variable amplitude loading
This paper presents the algorithms of the new guideline, gives an overview about the mechanical background and discusses the results for specimens of steel and aluminium alloys. It also shows results comparing the new to the current FKM guideline for a given steel structure.

1 Introduction

The Research Association of Mechanical Engineering (FKM) is a cooperating network of German industry and research to identify and solve current mechanical problems in industry. During the last years various guidelines have been published by the FKM in order to improve the competitiveness of the small and medium-sized industry. The FKM guideline for calculating the fatigue life of structures of steel or aluminium for the failure criteria fracture is available for a couple of years [1]. Currently a new guideline which accounts for the nonlinear material behaviour in the fatigue life calculation for failure criteria crack initiation is going to be released.

1.1 The FKM guideline for fracture

The fatigue part of the FKM guideline [1] lead the user through a calculation of the fatigue live for fracture of a structure of steel or aluminium alloy based on stresses according to the Theory of Elasticity. For steel, the FKM guideline proposes a Wöhler curve with an exponent of $k = 5$ and a fatigue limit for a fatigue life of $N_D = 10^6$.

$$\sigma_{BK} = \sigma_{AK} \cdot \left(\frac{N_D}{N_{BK}}\right)^{1/k} \tag{1}$$

The amplitude for the fatigue limit σ_{AK} depends on the material, the geometry and the stress ratio of a constant amplitude loading, see [1].

1.2 The FKM guideline for crack initiation

The fatigue part of the new guideline is based on the Local Strain Approach (LSA) [2]. Using standard algorithms of the LSA linear elastically calculated stresses and strains in a notched area are transformed to elastic-plastic stresses and strains using approximation formulas. Based on these elastic-plastic stresses and strains, the stress-strain-curves for a given loading sequence can be simulated and the damage effect for the structure can be predicted using damage parameters and a damage accumulation rule. The basic material parameters used in describing the cyclic stress-strain-behaviour and the damage parameter Wöhler curves can be estimated applying simple approximation methods which only require input of the ultimate tensile strength R_m and the material group.

2 The algorithms of the new guideline

2.1 Cyclic stabilized stress strain curves

The material behaviour under cyclic loading can be described by the cyclic stabilized stress-strain curve by Ramberg-Osgood [3]

$$\varepsilon = \frac{\sigma}{E} + \left(\frac{\sigma}{K'}\right)^{\frac{1}{n'}} \tag{2}$$

Hysteresis loop branches are described using Masing's law [4]:

$$\varepsilon = \frac{\sigma}{E} + 2 \cdot \left(\frac{\sigma}{2K'}\right)^{\frac{1}{n'}} \tag{3}$$

with material parameters E, K' and n'. Using an approach by Wächter [5] Young's modulus E can be set to 206 000 MPa for steel and 70 000 MPa for aluminium alloy. Wächter also proposes the FKM method to estimate the cyclic material parameters K' and n' based on the ultimate tensile strength.

$$K' = \frac{a_\sigma \cdot \left(\frac{R_m}{\text{MPa}}\right)^{b_\sigma}}{\left(\min\left(\varepsilon_{limit}\,, a_\varepsilon \cdot \left(\frac{R_m}{\text{MPa}}\right)^{b_\varepsilon}\right)\right)^{n'}} \tag{4}$$

For $n', a_\varepsilon, b_\varepsilon, a_\sigma, b_\sigma, \varepsilon_{limit}$ see Table 1.

Table 1 : Constants for cyclic stabilized stress strain curve [5]

Material group	Steel	Cast iron	Aluminium alloy
n'	0.187	0.176	0.128
a_σ	3.1148	1.732	9.12
b_σ	0.897	0.982	0.742
a_ε	1033	0.847	895.9
b_ε	−1.235	−0.181	−1.183
ε_{limit}	0.338	∞	∞

* Corresponding author: fiedler@wm.tu-darmstadt.de

2.2 Flow curves

Flow curves connect the loading of the structure or the elastically calculated stresses with the elastic plastic strains in the notched area, see [6, 7]. Two different approximation formulas for flow curves are presented in the new guideline.

2.2.1 Flow curve by Neuber

The approach by Neuber [8] is a simple approach for a flow curve. It connects the elastic stresses σ_e with the elastic-plastic stresses and strains σ, ε using the limit load factor K_p.

$$\varepsilon = \frac{\sigma_e}{\sigma} \cdot K_p \cdot e^* \tag{5}$$

The strains e^* can be calculated with

$$e^* = \frac{\sigma_e/K_p}{E} + \left(\frac{\sigma_e/K_p}{K'}\right)^{\frac{1}{n'}} \tag{6}$$

2.2.2 Flow curve by Seeger and Beste

The approach by Seeger and Beste [6] is another approach for a flow curve with a better fitting for the high strain area where the strains are mainly plastic. It also connects the elastic stresses σ_e with the elastic-plastic stresses and strains σ, ε using the limit load factor K_p.

$$\varepsilon = \frac{\sigma}{E} \cdot \left[\left(\frac{\sigma_e}{\sigma}\right)^2 \cdot \frac{2}{u^2} \cdot \ln\left(\frac{1}{\cos(u)}\right) - \frac{\sigma_e}{\sigma} + 1\right] \cdot \left(\frac{e^* \cdot E \cdot K_p}{\sigma_e}\right) \tag{7}$$

With

$$u = \frac{\pi}{2} \cdot \left(\frac{\sigma_e/\sigma - 1}{K_p - 1}\right) \tag{8}$$

And using (6) for e^*.

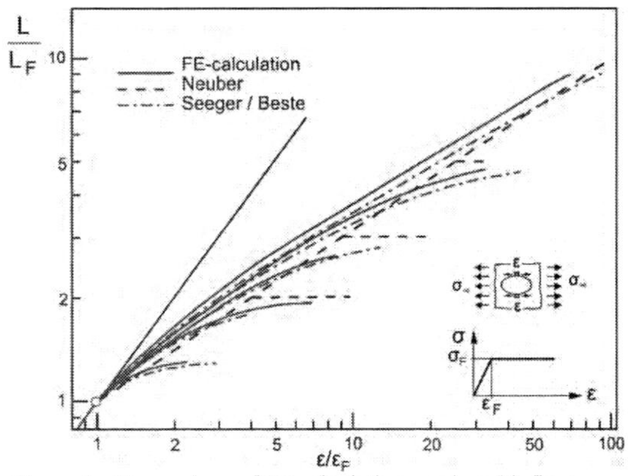

Figure 1 : Comparison of FE-calculation results with flow curves [9]

Fig. 1 shows the difference between the approximation formulas by Neuber and by Seeger and Beste compared with FE-results based on linear elastic ideal plastic material behaviour for a sample of notched plates.

2.3 Hysteresis Counting Method (HCM)

The stress-strain-path in the notched area has to be simulated for the whole loading sequence to identify the damage effects. The HCM algorithm as described by Clormann [10] is used in the new guideline to simulate the elastic-plastic stress-strain-path, identify closed hysteresis loops and calculate the required parameters for the damage accumulation. This algorithm is able to distinguish between the initial and reloading stabilized stress-strain-curve and considers the 3 memory effects.

2.4 Damage Parameter Wöhler curve

2.4.1 Damage parameter P_{RAM}

The damage parameter Wöhler curve is based on the material group and the ultimate tensile strength R_m, see [5, 11]. The grid point $P_{RAM,Z}$ for the damage parameter Wöhler curve can be calculated using

$$P_{RAM,Z} = \frac{a_{P,Z}}{f_{RAM}} \cdot \left(\frac{R_m}{\text{MPa}}\right)^{b_{P,Z}} \tag{9}$$

for $a_{P,Z}$ and $b_{P,Z}$ see Table 2, for factor f_{RAM} which includes size effects and a safety concept see [11].

Table 2 : Constants for the P_{RAM} Wöhler curve [5]

Material group	Steel	Cast iron	Aluminium alloy
d_1	−0.302	−0.289	−0.238
d_2	−0.197	−0.189	−0.167
$a_{P,Z}$	20	25.56	16.71
$b_{P,Z}$	0.587	0.519	0.537

The damage parameter Wöhler curve is described by:

$$N_{CI} = 10^3 \cdot \begin{cases} \left(\frac{P_{RAM}}{P_{RAM,Z}}\right)^{\frac{1}{d_1}}, & P_{RAM} \geq P_{RAM,Z} \\ \left(\frac{P_{RAM}}{P_{RAM,Z}}\right)^{\frac{1}{d_2}}, & P_{RAM} < P_{RAM,Z} \end{cases} \tag{10}$$

2.4.2 Damage parameter P_{RAJ}

The damage parameter Wöhler curve is based on the material group and the ultimate tensile strength R_m. The reference point $P_{RAJ,Z}$ for the damage parameter Wöhler curve can be calculated using

$$P_{RAJ,Z} = \frac{a_{P,Z}}{f_{RAJ}} \cdot \left(\frac{R_m}{\text{MPa}}\right)^{b_{P,Z}} \quad (11)$$

for $a_{P,Z}$ and $b_{P,Z}$ see Table 3.

Table 3 : Constants for the P_{RAJ} Wöhler curve [5]

Material group	Steel	Cast iron	Aluminium alloy
d	-0.56	-0.66	-0.61
$a_{P,Z}$	1.173	10.03	101.7
$b_{P,Z}$	1	0.695	0.26
$a_{P,D}$	$3.33 \cdot 10^{-5}$	$5.25 \cdot 10^{-6}$	$5.18 \cdot 10^{-7}$
$b_{P,D}$	1.55	1.63	2.04

Considering the actual fatigue limit $P_{RAJ,D}$ (25) the damage parameter Wöhler curve is described by:

$$N_{CI} = \begin{cases} \left(\frac{P_{RAJ}}{P_{RAJ,Z}}\right)^{\frac{1}{d}}, & P_{RAJ} \geq P_{RAJ,D} \\ \infty, & P_{RAJ} < P_{RAJ,D} \end{cases} \quad (12)$$

The fatigue limit for the mechanical short crack $P_{RAJ,D,0}$ can be described by

$$P_{RAJ,D,0} = \frac{a_{P,D}}{f_{RAJ}} \cdot \left(\frac{R_m}{\text{MPa}}\right)^{b_{P,D}} \quad (13)$$

for $a_{P,D}$ and $b_{P,D}$ see Table 3 for f_{RAJ} which includes size effects and a safety concept see [11].

2.5 Damage Parameter P_{RAM} and P_{RAJ}

In the new guideline two different damage parameters P_{RAM} and P_{RAJ} can be used for calculating the fatigue life for crack initiation.

2.5.1 Damage parameter P_{RAM}

The damage parameter P_{RAM} is based on Smith, Watson and Topper's [12] approach, modified by an approach to respect the material dependent mean stress sensitivity by Bergmann [13].

$$P_{RAM} = \left\{ \sqrt{(\sigma_a + k \cdot \sigma_m) \cdot E \cdot \varepsilon_a} \right. \quad (14)$$

In Bergmann [13] k is material depending but not specified for the material groups, in the new guideline the parameter k is described by

$$k = \begin{cases} M_\sigma \cdot (M_\sigma - 2), & \sigma_m \geq 0 \\ \frac{M_\sigma}{3} \cdot \left(\frac{M_\sigma}{3} + 2\right), & \sigma_m < 0 \end{cases} \quad (15)$$

with mean stress sensitivity M_σ as defined in [1]

$$M_\sigma = a_M \cdot 10^{-3} \cdot \frac{R_m}{\text{MPa}} + b_M \quad (16)$$

with material group specific parameters a_M and b_M.

2.5.2 Damage parameter P_{RAJ}

The damage parameter P_{RAJ} has a fracture mechanics background and considers the influence of crack opening and closing and the loading sequence, see [14]. These effects [15, 16] can be considered by using effective stresses and strains, adapting the crack opening strain and reducing the fatigue limit with increasing damage during variable amplitude loading.

Using effective parameters

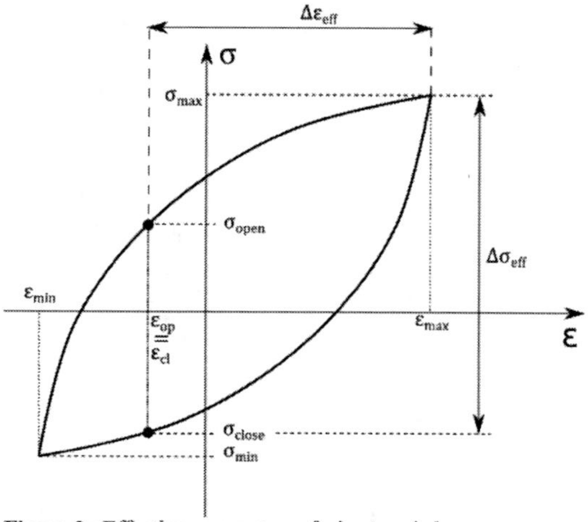

Figure 2 : Effective parameters of a hysteresis loop

The P_{RAJ} uses effective stresses and strains when calculating the damage effect of a hysteresis loop. The effective parameters are illustrated in Figure 2, where σ_{open} and ε_{op} are the crack opening stress and strain and σ_{close} and ε_{cl} the crack closure stress and strain. The effective stresses and strains, $\Delta\sigma_{\text{eff}}$ and $\Delta\varepsilon_{\text{eff}}$, are calculated for an open crack, the influence of the hysteresis parameters for a closed crack is neglected.
The damage parameter is formulated as

$$P_{RAJ} = \left[1.24 \cdot \frac{(\Delta\sigma_{\text{eff}})^2}{E} + \frac{1.02}{\sqrt{n'}} \cdot (\Delta\sigma_{\text{eff}}) \cdot \left((\Delta\varepsilon_{\text{eff}}) - \frac{(\Delta\sigma_{\text{eff}})}{E} \right) \right] \quad (17)$$

with E, K' and n' see section 2.1. The crack opening stress is determined by an approach of Newman [17] with

$$\sigma_{open} = \begin{cases} \sigma_{max} \cdot (A_0 + A_1 \cdot R + A_2 \cdot R^2 + A_3 \cdot R^3), & R \geq 0 \\ \sigma_{max} \cdot (A_0 + A_1 \cdot R), & R < 0 \end{cases} \quad (18)$$

For constants A_i see [11].

Adapting the crack opening strain

For constant amplitude loadings the crack opening strain can be determined using the ascending hysteresis arm by

$$\varepsilon_{op,const} = \varepsilon_{cl,const} = \varepsilon_{min} + \frac{(\sigma_{open} - \sigma_{min})}{E} + 2 \cdot \left(\frac{\sigma_{open} - \sigma_{mi}}{2 \cdot K'}\right. \quad (19)$$

For variable amplitude loadings the sequence effects of the loading history has to be taken into account when calculating the crack opening strain. Therefore, four different strains have to be defined: $\varepsilon_{op,const}$, for a fictitious constant amplitude loading using (19), $\varepsilon_{op,hist}$ for the influence of history of the former hysteresis loops, $\varepsilon_{max,old,hy}$ and $\varepsilon_{min,old,hy}$ for the maximum and minimum strain of all former hysteresis loops and $\varepsilon_{max,old,sq}$ and $\varepsilon_{min,old,sq}$ for the maximum and minimum strain of all former loading points.

Four different effects has to be considered:

3 For $\varepsilon_{max} \leq \varepsilon_{op,hist}$ is $\varepsilon_{open} = \varepsilon_{op,hist}$ and $P_{RAJ} = 0$

4 a) For $\varepsilon_{max,old,hy} < \varepsilon_{max,old,sq}$ is $\varepsilon_{open} = \varepsilon_{op,const}$
b) For $\varepsilon_{min,old,hy} > \varepsilon_{min,old,sq}$ is $\varepsilon_{open} = \varepsilon_{op,const}$

3. For $\varepsilon_{op,const} > \varepsilon_{op,hist}$ is $\varepsilon_{open} = \varepsilon_{op,hist}$

4. For $\varepsilon_{op,const} \leq \varepsilon_{op,hist}$
a) For $\sigma_a \geq 0.4 \cdot \sigma_F$ is $\varepsilon_{open} = \varepsilon_{op,const}$
b) For $\sigma_a < 0.4 \cdot \sigma_F$ is $\varepsilon_{open} = \varepsilon_{op,hist}$

For specific case $\varepsilon_{min} \geq \varepsilon_{open}$ the crack opening stress is $\sigma_{open} = \sigma_{min}$ and $\Delta\sigma_{eff} = \sigma_{max} - \sigma_{min}$, $\Delta\varepsilon_{eff} = \varepsilon_{max} - \varepsilon_{min}$ when using (17).
Else for $\varepsilon_{min} < \varepsilon_{open}$ the crack opening stress is determined by

$$\varepsilon_{max} - \varepsilon_{open} = \frac{\Delta\sigma_{eff}}{E} + 2 \cdot \left(\frac{\Delta\sigma_{eff}}{2K'}\right)^{1/n'} \quad (20)$$

with $\Delta\sigma_{eff} = \sigma_{max} - \sigma_{open}$ and $\Delta\varepsilon_{eff} = \varepsilon_{max} - \varepsilon_{open}$ when using (17).

For case 3 with $\varepsilon_{op,const} > \varepsilon_{op,hist}$ the crack opening strain history needs to be updated by

$$\varepsilon_{op,hist} = \varepsilon_{op,const} - (\varepsilon_{op,const} - \varepsilon_{open}) \cdot \exp(15 \cdot N_{CI}^{-1}) \quad (21)$$

with N_{CI} the fatigue life for the current hysteresis loop calculated by (12). For any other case, the new history crack opening strain is

$$\varepsilon_{op,hist} = \varepsilon_{open} \quad (22)$$

Decreasing the fatigue limit

A mechanical short crack length a_0 has to be determined using material constants m and C, the crack length for crack initiation $a_{end} = 0.5$ mm and the grid point of the damage parameter Woehler curve $P_{RAJ,Z}$.

$$a_0 = \left(a_{end}^{1-m} - (1-m) \cdot C \cdot P_{RAJ,Z}^{m}\right)^{\frac{1}{1-m}} \quad (23)$$

C can be determined using an approach by [14] with

$$C = 10^{-5} \cdot (5 \cdot 10^5)^m \cdot (E)^{-m} \quad (24)$$

and $m = -\frac{1}{d}$ while d is the exponent of the damage parameter Woehler curve. During the loading sequence, the mechanical short crack opens and closes and growths until the crack length a_{end} is achieved. This effect is taken into account in the calculation of the decreasing fatigue limit $P_{RAJ,D}$ with

$$P_{RAJ,D} = \Delta J_{eff,th} \cdot \left(\left((a_{end}^{1-m} - a_0^{1-m}) \cdot D + a_0^{1-m}\right)^{\frac{1}{1-m}} + \frac{\Delta J_{eff,th}}{P_{RAJ,D,0}} - a_0\right)^{-1} \quad (25)$$

which is equivalent to the consequent Miner rule with actual damage sum D. The threshold for the effective ΔJ can be approximated by

$$\Delta J_{eff,th} = \frac{E}{(5 \cdot 10^6 \text{ MPa})} \quad (26)$$

2.6 Damage Accumulation Rule

2.6.1 Damage parameter P_{RAM}

The damage accumulation rule for variable amplitude loading is equivalent to the Miner rule.

$$D = \sum_i \frac{n_i}{N_{CI,i}} \quad (27)$$

Each loading sequence has to be calculated for a second time to close all possible hysteresis loops. Therefore the damage sums of first D_1 and second run D_2 can differ. A factor x has to be calculated by

$$x = \frac{1 - D_1}{D_2} \quad (28)$$

To calculate the fatigue life for crack initiation with

$$N_{CI} = (1 + x) \cdot H_0 \quad (29)$$

2.6.2 Damage parameter P_{RAJ}

The damage accumulation rule for variable amplitude loading is equivalent to a consequent Miner rule. First of all, the maximal possible P_{RAJ} and the fatigue limit for crack initiation $P_{RAJ,D,e}$ has to be determined and 200 classes have to be created.

$$\frac{P_{RAJ,max,class}}{P_{RAJ,1}} = \frac{P_{RAJ,1}}{P_{RAJ,2}} = \dots = \frac{P_{RAJ,199}}{P_{RAJ,D,e}} \quad (30)$$

Second, the actual fatigue limit $P_{RAJ,D}$ has to be classified within this 200 classes.

$$P_{RAJ,q-1} \leq P_{RAJ,D} < P_{RAJ,q} \quad (31)$$

Third, with the class index q the amount of cycles to crack initiation after the second run through the loading sequence N_{-2} can be determined using

$$N_{-2} = H_0 \cdot \sum_{j=q}^{199} \frac{f(j+1) - f(j)}{\sum_{i=1}^{j} \frac{h_i}{N_{CI,i}}} \quad (32)$$

With

$$f(j) = \frac{a_0^{1-m} - \left[\left(\frac{\Delta J_{eff,th}}{P_{RAJ,j}} + \frac{\Delta J_{eff,th}}{P_{RAJ,D,0}}\right) - a_0\right]^{1-m}}{a_0^{1-m} - a_{end}^{1-m}} \quad (33)$$

Last, the fatigue life for crack initiation can be calculated by

$$N_{CI} = 2 \cdot H_0 + N_{-2} \quad (34)$$

with the scope of the hysteresis collective H_0 from the second run through the loading sequence.

2.7 Extension of the new guideline to fracture

The new guideline in the current version is only approved to calculate the fatigue life for crack initiation. A first and simple approach to calculate the fatigue life for fracture N_{Fr} based on the fatigue life for crack initiation N_{CI} is the U-concept.

The U-concept was determined on a database with more than 1000 datasets for steel. It simply shifts the fatigue life for crack initiation based on damage parameter P_{RAJ} using

$$N_{Fr} = U \cdot N_{CI}^{1,15} \quad (35)$$

With

$$U = \begin{cases} 37.4 \cdot \left(\frac{\sigma_a - \sigma_m}{R_m}\right)^5 \cdot |R| \geq 0.85, \ R \leq -1 \\ 0.93 \cdot e^{-3.7R} \cdot \left(\frac{\sigma_a + \sigma_m}{R_m}\right)^4 \geq 0.42, R > -1 \end{cases} \quad (36)$$

3 Results

During a former FKM project a large structural database with results and datasets for crack initiation and fracture fatigue life experiments was build. The database called DaBef [18] was the basis for the following calculation results for constant (CA) and variable amplitude (VA) loadings.

3.1 Samples with constant amplitude loading

480 datasets are available for notched geometries of steel under constant amplitude loading in DaBef. Table 4 gives an overview about the range of the datasets. Calculation results with damage parameter P_{RAJ} in contrast to the experimental results are illustrated in Figure 3. For the calculation results the crack initiation was set to $a_{end} = 0.5$ mm, for experimental results the criteria for crack initiation differs between the different references stored in DaBef. In addition there are results for stress- and strain-controlled experiments and also for tension and bending experiments combined in this figure, which explain the range between experimental and calculated fatigue lives, see [19] for comparable results. No safety concept was applied on the calculation results.

Table 4 : Experimental datasets for CA for crack initiation

Material group	Ultimate tensile strength R_m [MPa]	Elastic stress concentration factor K_t [-]	Stress ratio R
Steel	476 - 1136	1.49 -22.2	-1, ..., 0.33

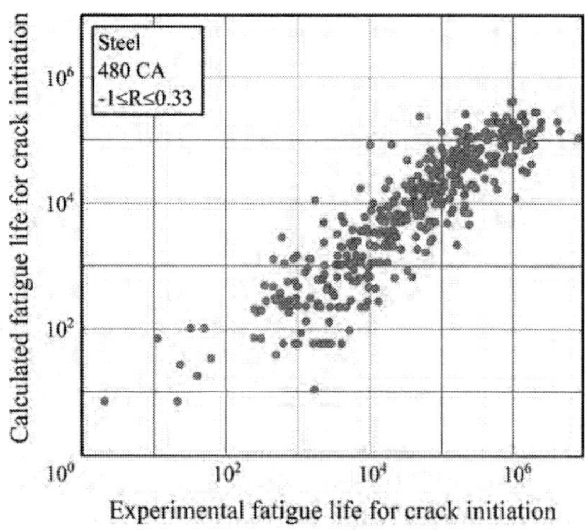

Figure 3 : Experimental vs. calculation results for steel calculated with P_{RAJ} for crack initiation and constant amplitude loading

Extending the calculation with the U-concept, the calculation results show a good accordance with the experimental results from DaBef, see Figure 4. For fracture the failure criteria hardly differs between the different references of DaBef in contrast to the failure criteria for crack initiation. The range between calculated and experimental results is smaller although 1097 experimental datasets for notched structures of steel were included. Based on this results one could argue that the calculated fatigue life for crack initiation also have a good accuracy compared with experimental results if all experimental results would have been obtained accurately applying the same failure criterion as used in the numerical approach. Figure 5 compares the experimental results with calculation results using the FKM guideline for fracture.

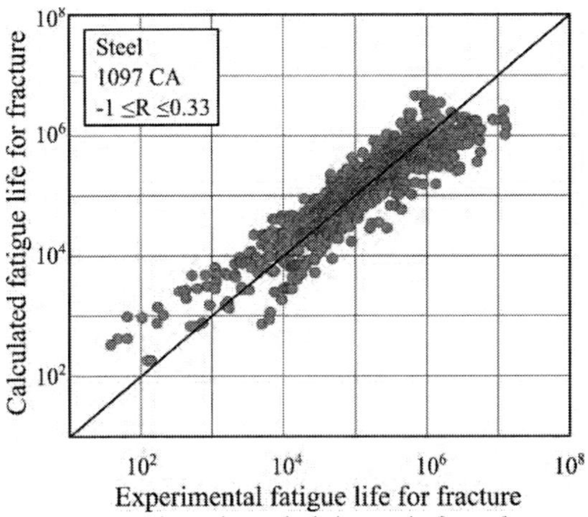

Figure 4 : Experimental vs. calculation results for steel calculated with P_{RAJ} for fracture and constant amplitude loading

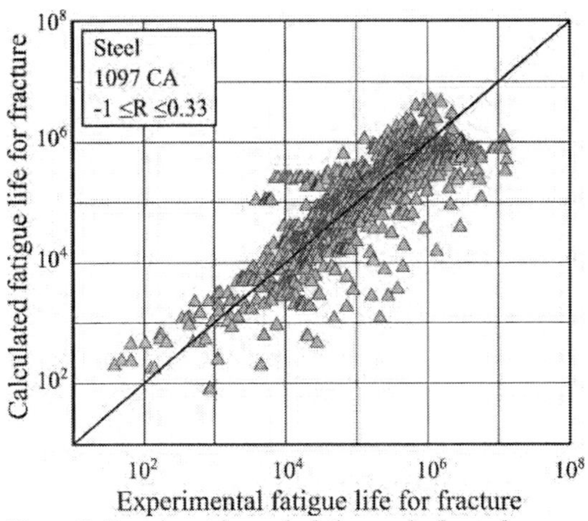

Figure 5 : Experimental vs. calculation results for steel calculated with FKM guideline for fracture

Table 5 : Experimental datasets for CA for fracture

Material group	Ultimate tensile strength R_m [MPa]	Elastic stress concentration factor K_t [-]	Stress ratio R
Steel	476 - 1136	1.01 -22.2	-1, …, 0.33

3.2 Samples with variable amplitude loading

360 structural datasets are available in DaBef [18] with structures under variable amplitude loading for failure criteria crack initiation, see Table 6.

Table 6 : Experimental datasets for VA for crack initiation

Material group	Ultimate tensile strength R_m [MPa]	Elastic stress concentration factor K_t [-]	Stress ratio R
Steel	544 - 935	1.49 -3.48	-1, …, 0

Figure 6 shows calculation versus experimental results. Again the range between experimental and calculation results is higher than expected because of the large difference in the experimental conditions. Figure 7 shows calculation versus experimental results for the failure criteria fracture with 657 experimental datasets, see Table 7. Again there is a better accordance between experimental and calculated fatigue lives for fracture compared with crack initiation. With a life variance T_N defined as

$$T_N = \frac{\left(\frac{N_{exp}}{N_{calc}}\right)_{90\%}}{\left(\frac{N_{exp}}{N_{calc}}\right)_{10\%}} \qquad (37)$$

The variance for the results for calculations with variable amplitude loadings is $T_N = 6.49$ and the mean value is $m = 3.79$. The results for variable amplitude loadings for fracture were calculated with $P_{RAJ,D,0} = 0.001$ so the fatigue limit were neglected.

Table 7 : Experimental datasets for VA for fracture

Material group	Ultimate tensile strength R_m [MPa]	Elastic stress concentration factor K_t [-]	Stress ratio R
Steel	544 - 1097	1.01 -3.48	-1, …, 0

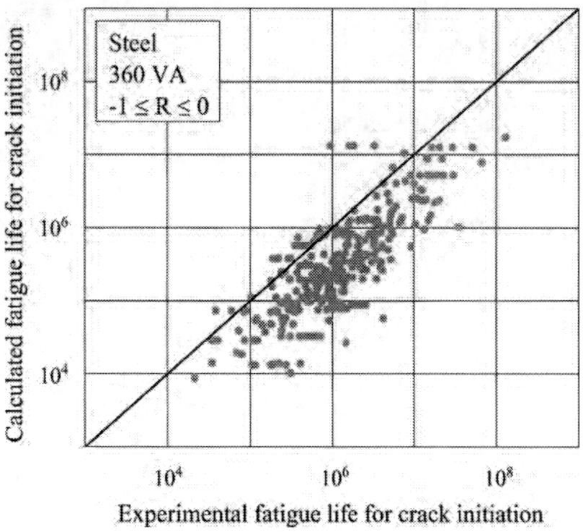

Figure 6 : Experimental vs. calculation results for steel calculated with P_{RAJ} for crack initiation and variable amplitude loading

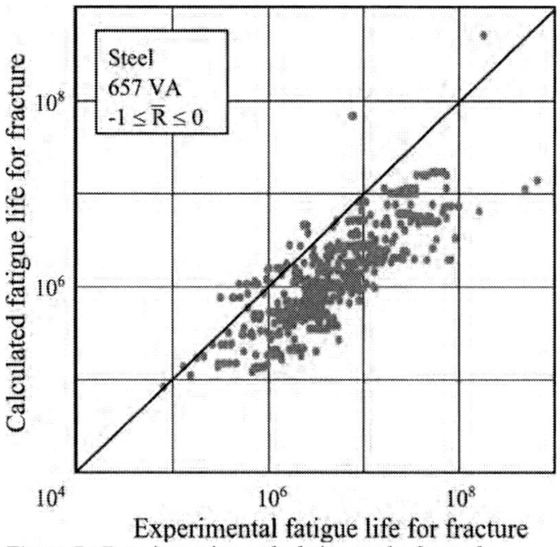

Figure 7 : Experimental vs. calculation results for steel calculated with P_{RAJ} for fracture and variable amplitude loading

4 Summary and Conclusions

This paper gives an overview about the new FKM guideline. The guideline provides a recipe for calculations of fatigue lives for notched structures of steel and aluminium alloy under constant or variable amplitude loadings. The guideline presumes a crack initiation criteria $a_{end} = 0.5$ mm for all structures independent of experimental conditions. Because of this independence the accuracy between the experimental and calculated fatigue lives differs, which is similar to observations in the literature. A simple concept to calculate the fatigue live for fracture based on the calculations for crack initiation was shown. This concept shows a good accordance with experimental results even without a crack growth simulation.

Acknowledgement

The authors expresses their sincere gratitudes to the German Federal Ministry of Economics and Energy (Bundesministerium für Wirtschaft und Energie, BMWi) for financial support via AiF (Arbeitsgemeinschaft industrieller Forschungsvereinigungen „Otto von Guericke“ e.V.) of the research project „Rechnerischer Bauteilfestigkeitsnachweis unter expliziter Erfassung nichtlinearen Werkstoff-Verformungsverhaltens“ under the IGF project no. 17.612 N/1.

References

1. R. Rennert, E. Kullig, M. Vormwald, A. Esderts, D. Siegele, *Analytical strength assessment of components made of steel, cast iron and aluminum materials in mechanical engineering*, **6** (VDMA-Verlag, Frankfurt, 2012)
2. N. Dowling, Comprehensive structural integrity, **4**, 77 (2003)
3. W. Ramberg, W. Osgood, Technical Note No. 942 (1943)
4. G. Masing, Proc. of the 2nd Int. Congress of Applied Mechanics, 332 (1926)
5. M. Wächter, *Zur Ermittlung von zyklischen Werkstoffkennwerten und Schädigungsparameterwöhlerlinien* (TU Clausthal, 2016)
6. T. Seeger, A. Beste, *Zur Weiterentwicklung von Näherungsformeln für die Berechnung von Kerbbeanspruchungen im elastisch-plastischen Bereich* (Düsseldorf, 1977)
7. T. Seeger, A. Beste, H. Amstutz, *ICF4*, (Waterloo (Canada) 1977)
8. H. Neuber, *Trans. ASME J. Appl. Mech.*, **28**, 544 (1961)
9. T. Seeger, *Betriebsfestigkeit auf Grundlage örtlicher Beanspruchungen*, **3** (2016)
10. U. H. Clormann, T. Seeger, *Der Stahlbau*, **55 (3)**, 65 (1986)
11. M. Fiedler, M. Wächter, I. Varfolomeev, M. Vormwald, A. Esderts, *Rechnerischer Bauteilfestigkeitsnachweis unter expliziter Erfassung nichtlinearen Werkstoff-Verformungsverhaltens*, **AiF-Projekt Nr. 17612** (Frankfurt, 2015)
12. K. Smith, P. Watson, T. Topper, *Journal of Materials*, **5 (4)**, 767 (1970)
13. J. Bergmann, *Zur Betriebsfestigkeitsbemessung gekerbter Bauteile auf der Grundlage der örtlichen Beanspruchungen* (TU Darmstadt, 1983)
14. M. Vormwald, *Anrisslebensdauervorhersage auf Basis der Schwingbruchmechanik für kurze Risse* (TU Darmstadt, 1989)

15. M. Vormwald, *International Journal of Fatigue*, **82**, 80 (2016)
16. M. Vormwald, *Procedia Engineering*, **101 (C)**, 534 (2015)
17. J. C. Newman, *ASTM STP 748,* 53 (1981)
18. F. Ellmer, A. Esderts, K.-G. Eulitz, K. Hinkelmann, *Datenbank und Auswertsystem Betriebsfestigkeit, FKM-Vorhaben Nr.288*, **313** (2011)
19. J. Tölle, H. Schmidt, *FAT-Schriftenreihe*, **221** (2009)

Numerical fatigue life estimation of aluminium 6201-T81 wires containing geometric discontinuities.

Jesús Martínez, *Vitor* Adriano, *José* Araújo, *Jorge* Ferreira, and *Cosme* da Silva.

University of Brasilia, Department of Mechanical Engineering, Brasilia CEP 70910-900, Brazil

Abstract. Previous works have shown that fretting fatigue can be modelled as a notch problem due to the mild superficial damage produced in the partial slip regime during fretting. In this context, the aim of this numerical and experimental study is to estimate the fatigue life of aluminium 6201-T81 wires, used in overhead conductors, containing geometric discontinuities. These discontinuities induce a stress field analogous to the fretting problem. To perform the fatigue life assessments the L_M-N_f relationship, which is an extension of the Theory of Critical Distance (TCD), was applied. The L_M-N_f relationship was calibrated using two *S-N* curves, one of a plain wire and other of a notched wire. The validation of the numerical approach estimations was performed using a wire containing a through hole. The estimations were fallen almost entirely among factor 3 bands when compared with the experimental results. This factor 3 represents the biggest scatter in the *S-N* curves used to calibrate the L_M-N_f relation. The TCD approach applied, in terms of point method, provide accurate life estimations in the range 10^5 and 10^6 cycles.

1 Introduction

Failure in overhead conductor, used in power transmission lines, is mainly associated to the fretting fatigue. Such cables are normally composed of aluminium wires twisted around a core and it is perhaps the most important component in the transmission lines [1–3]. The fatigue failure in these cables are induced by Aeolian vibration and frequently occur inside the "rigid" device where the conductor's motion is constrained, such as suspension clamps, spacers, spacer dampers or Stockbridge dampers. The suspension clamp is one of the most critical devices in terms of conductor fretting fatigue [4]. The Aeolian vibration is caused by the vortices and these are formed alternately from the top and bottom of the conductor. Hence, this vibration causes the cable to bend in the vicinity of a "rigid" device [5, 6]. The combination of the bending stress, the mean stretching load of the conductor and the clamping force generate complex mechanical interactions due to the relative movements between cable wires and between the cable external wires and the suspension clamp. Some of this interactions causes slip and high multiaxial stress state in the contact zones [4, 7]. For this reason, the failure in the conductor is recognized as a fretting fatigue problem [8].

The fatigue phenomenon appears to be a well-understood problem for some structures however this is not the case for overhead conductors. Currently, most of the publications [1–4, 7] in fatigue failure of overhead conductor are oriented to assessment the nominal stress, related to the bending movement of the conductor, and the mean stretching loads using the *Poffenberger-Swart* (P-S) formula [9]. However, the P-S formula is semi-empirical and provides nominal idealized stress which is used to establish and compare the fatigue strength of different conductors based on bending stress. Nevertheless, this approach cannot allow describing the fretting fatigue phenomenon in the failure regions. The numerical simulation tools are solutions to obtain the state of the stress in this specific region where the accessibility is quite difficult. However, the numerical simulation could be quite complex and very expensive in some cases [10].

On the other hand, some works [11–13] have shown that the fretting fatigue can be modelled as a notch problem, due to the fact that in both cases the fatigue exist, a multiaxial stress state and high stress gradients. In this setting, this approach allows the use of different methods to estimate the fatigue life, normally used to assessment the notched components, in some fretting fatigue cases.

The study of the notched components is quite important in mechanical engineer studies. Early work to estimate the fatigue strength were based on the concentration factor (k_t). This estimation can be correct for high-cycle fatigue strength of blunt notches, but not for sharp notches, for which the estimations are too conservative [14]. For a better representation of fatigue strength reduction at long lives, the fatigue notch factor

* Corresponding author: author@e-mail.org

(k_f) was introduced to the analyses. This factor is the relation between plain and notched specimen fatigue strength and is formally defined only for completely reversed stresses [15]. The discrepancy between k_f and k_t can be quantified by the "notch sensitivity" (q). This factor is used to describe the effect of a notch on the fatigue strength of the specimen in linear elastic analyses. In this setting, Neuber [16] and Peterson [17] proposed the notch-sensitivity formulas in terms of a "empirical material constant". They argue that the stresses around the notch should be averaged to obtain an effective stress (σ_{eff}) that could be linked to the fatigue damage process zone. Latter, the "material constant" was formalized in a unifying theoretical model by Taylor [18] in four methods to evaluate the equivalent stress. The group of this methods is known as the Theory of Critical Distance (TCD) that have shown high accurate results to predict the fatigue limit of notched components [19, 20].

The TCD is a nonlocal approach to predict the fatigue limit of notched component by using the stress field around the notch. This theory was extended by Taylor [18] who applied the same theory to the components with other geometrical discontinuities. In addition, Susmel and Taylor [14] reviewed the critical distance concepts and extended the TCD to perform fatigue life assessments in the medium cycle regime based on the assumption that the critical distance (L_M) changes with the number of cycles to failure (N_f): the L_M -N_f relationship. This idea, allow to estimate the fatigue strength/life, in the same load ratio, and not just the fatigue limit like an early TCD approach.

The L_M -N_f relationship was used in this study to perform the fatigue life estimation of aluminium wires containing geometric discontinuities. A sharp notch and a through hole. The aim of this experimental and numerical work is to evaluate this approach in the aluminium 6201-T81 wire. For this reason, it was necessary to establish the methodology for an experimental campaign to obtain the *S-N* curves and calibrate L_M -N_f relationship. The *point method* (PM) of the TDC was used for the calibration stage and for the fatigue life estimation. This is a preliminary study in single wires, which could contribute to analyses of fretting fatigue behaviour and life fatigue estimation on overhead conductor cables.

2 L_M-N_f Relationship

The Theory of Critical Distance (TCD) is a group of methods to assess the effective stress in a specific distance. Experimental results shown [18, 21] that the critical distance values, for any method, depends of the material characteristic length (L) [19]:

$$L = \frac{1}{\pi}\left(\frac{\Delta K_{th}}{\Delta\sigma_0}\right)^2 \qquad (1)$$

where $\Delta\sigma_0$ is the plain fatigue limit and K_{th} is the threshold value of the stress intensity factor. The two parameters, $\Delta\sigma_0$ and K_{th}, are determined at the same load ratio, R. The main objective of the critical distance theory is to estimate the effective stress which exceeds the plain fatigue limit in the linear-elastic stress distribution around the notch (Fig. 1). Then, this approach assess a region and not the hot spot of notch. This region is known like fatigue process zone. Therefore, the effective stress can be estimated from analyses of the maximum principal stress (σ_1) distribution around the geometric discontinuity by four methods: (*i*) Point Method (PM) at certain distance from the hot-spot of the stress concentrator; (*ii*) Line Method (LM) from the averaged over a line; (*iii*) Area Method (AM) from averaged over an area, and (*iv*) Volume Method (MV) from the averaged over a volume [14, 18, 19].

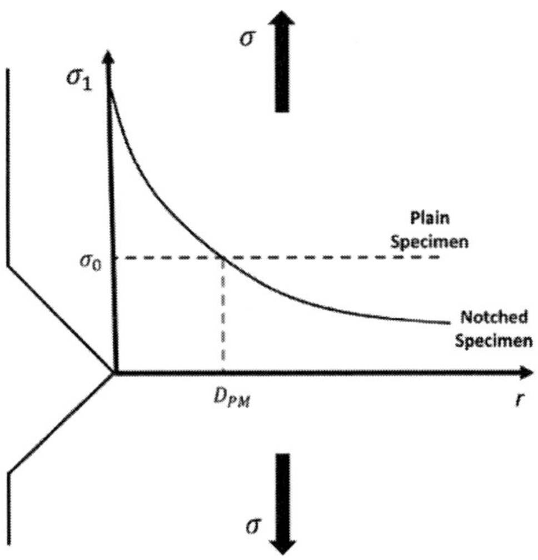

Fig. 1. Linear-elastic stress distribution and effective stress calculated according to the Point Method.

The methods of the TDC are based in the linear elastic mechanics concepts and experimental results which define the critical distance [18, 19]. According to the above definitions, the PM (2), LM (3), AM (4) and MV (5) can be formalized in the fallowing expressions:

$$\Delta\sigma_{eff} = \Delta\sigma_1(r) = \Delta\sigma_0 \qquad (2)$$

$$\Delta\sigma_{eff} = \frac{1}{l}\int \Delta\sigma_1(r)dr = \Delta\sigma_0 \qquad (3)$$

$$\Delta\sigma_{eff} = \frac{1}{A}\int \Delta\sigma_1(\theta, r)\, dA = \Delta\sigma_0 \qquad (4)$$

$$\Delta\sigma_{eff} = \frac{1}{V}\int \Delta\sigma_1(\emptyset, \theta, r)\, dV = \Delta\sigma_0 \qquad (5)$$

These expressions are used to predict the fatigue limits in notched specimens [19, 20]. One of the applications of this theory is to estimate lifetime of notch components in the medium-cycle fatigue regime proposed by Susmel and Taylor [14]. They proposed that the material characteristic length changes with the number of the cycles to failure (N_f). Then, the material characteristic determined in the medium-cycle fatigue regime (L_M) is defined as:

$$L_M(N_f) = A(N_f)^B \tag{6}$$

where A and B are material constants that result of the calibration process. These material constants are expected to be different for dissimilar materials and varying load ratios, R. Also, Susmel and Taylor [14] show that the better strategy to calibrate the L_M-N_f relation is using two fatigue curves (S-N curves), one for a plain specimens and the other for the notch specimens, in the same load ratio. The same strategy used in this work and the calibration process are explained in the following sections.

3 Methodology

This section describes the experimental details and the numerical simulations used to estimate the fatigue life of the wires containing geometrical discontinuities.

3.1. Experimental procedure

The test specimens are wires taken from the aluminium conductor cable used in the overhead transmission lines. These wires were machined to transform in test specimens of the three different geometries. The dimensions of these specimens are shown in Fig. 2.

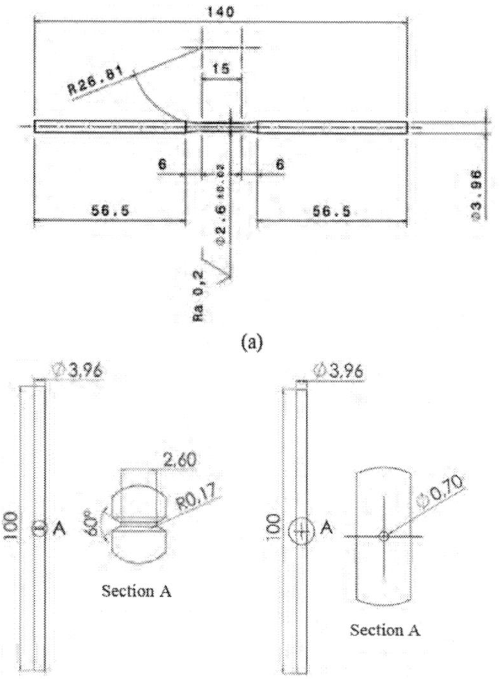

Fig. 2. Geometries of the test specimens (a) Plain wire, (b) Notch wire and (c) Wire containing a through hole.

The plain specimen was machined with special careful until a superficial roughness ≤ 0.2 Ra was reached. The roughness measurements were performed by a Confocal Laser Microscopy Olympus LEXT OLS4100 in different regions of the reduced cross-section. Also, the geometrical characteristics for the sharp notched wires ρ=0.17 were verified.

All fatigue tests were load controlled and conducted in a stress ratio R=-1 with frequencies varying from 10 to 40 Hz in the servohydraulic machine MTS. The fatigue test were conducted until the failure or the specimens were considered "run-out" if they survived 5 x 10^6 cycles.

3.2 Finite element analysis (FEA)

The TDC is based on the determination of the stress field into the material in the stress concentration region or process zone. To determine the stress field the use of numerical analyses like a finite element method is a powerful tool. However, this method has some limitations with regard to the size and complexity of components that can be modelled [19]. Furthermore, the computational cost is another topic to be considered. In this study, this stress field around the notched wire and the wire containing a through hole were obtained by using the commercial finite element software *Abaqus*. Therefore, these geometries were designed in the simplified three dimensional model with 1/4 of the specimen (Fig 3). In the initial model were established symmetry conditions and a low mesh density with ten-node tetrahedral elements (C3D10). Afterward, it was created a submodel around the geometrical discontinuities with a very fine mesh with eight-node hexahedral elements (C3D8R). Converge analyses helped to establish the mesh refinement. These strategies allowed to obtain good results of the stress field around the geometrical discontinuity in a reasonable computes times.

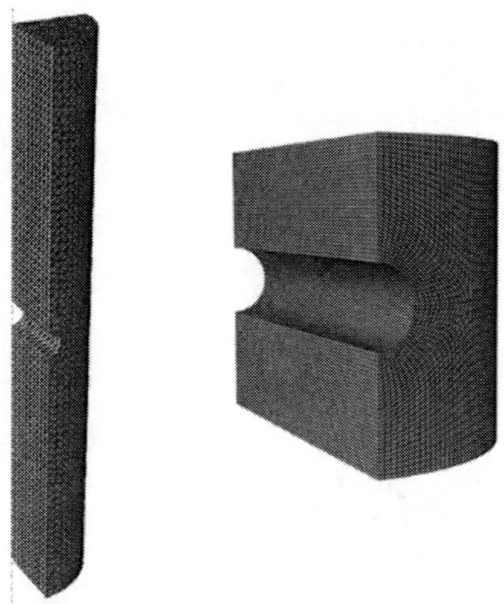

(a) (b)

Fig. 3. The simplified three dimensional model of the wire containing a through hole, (a) initial model and (b) submodel.

3.2 Calibration of the L_M-N_f relationship

The strategy used to calibrate the L_M-N_f relationship in this work was presented by Susmel and Taylor [14]. This calibration was performed by using two fatigue curves, one for a plain wire and the other for the notch wire, for the same load ratio. In the other hand, to determine the stress field, it was necessary to create a three-dimensional model of the sharp notch and thereafter to perform the FEA. This analyse gives the stress distribution in the region of the stress concentrator, as shown in Fig. 1. At the end, it was necessary to establish the critical distance for the both specimen, plain and notched, which present approximately the same fatigue life. This process is repeated several times to find the enough points to fit a L_M-N_f relationship. The Fig. 4 presents the flowchart which summarizes the calibration process described above.

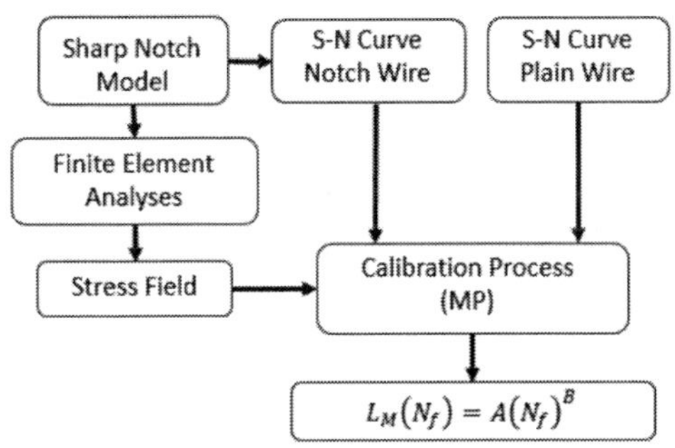

Fig. 4. Flowchart summarizing the calibration process.

3.3 Procedure to estimate the fatigue Life

The life estimation was performed by using an iterative process in the numerical algorithm. This algorithm was developed in a *Matlab* code to compare two fatigue lives: an initial arbitrary value and a material fatigue life obtained from the *S-N* curve of the plain wire. The initial arbitrary fatigue life was used to obtain the critical distance. From this critical distance, the maximum principal stress (σ_1) is determinated from the linear-elastic distribution around the geometric discontinuity. The calculated value of σ_1 allows to estimate the "first failure life" from the *S-N* relation of the plain wire. Now, the two fatigue lives are compared. The iterative process was stopped when the both fatigue lives are approximately the same (tolerance < 5%). If not, the arbitrary fatigue life is incremented and the procedure is repeated until the code has reached its convergence [14].

4 Results

This section presents the results of fatigue test for the plain wire, notch wire and wire containing a through hole with diameter of 0.7 mm. The L_M-N_f relationship in terms of the PM and the fatigue life estimation are also presented.

4.1 Fatigue tests

The fatigue tests results are show in log-log graph (Fig. 5) and the amplitude stress levels correspond to the nominal gross stress. The *S-N* curves for the plain wire, notch wire and wire containing a through hole are in the same graph aiming the evaluation of notch effect at fatigue lives of aluminium wires. Also, in the graph are presented the 95% confidence bands. Each point in the graph (Fig. 5) represents the fatigue life and the points with arrow represent the "run-out" tests.

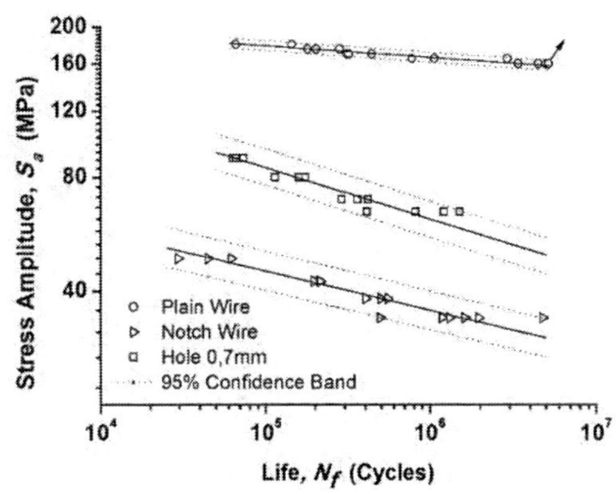

Fig. 5. *S-N* Curves of the plain wire, notch wire and wire containing a through hole 0.7 mm of diameter.

4.2 L_M-N_f Relationship by point method

From results of the calibration procedure made using the two S-N curves, the L_M-N_f relationship was established in terms of the PM:

$$L_M(N_f) = 0{,}6388(N_f)^{-0{,}1963} \quad (7)$$

where the constants from the calibration procedure can only be used for this specific material and for the fully reverse uniaxial loadings. The above equation was determined using L_M values calculated between 3×10^4 and 5×10^6 cycles to failure, approximately. This relationship was used in the algorithm to determine the critical distance in terms of the PM and in consequence, the fatigue life for the medium-cycle regime.

4.3 Determination of the fatigue life estimation

The results of the fatigue life estimation for the medium-cycle regime are show in Fig. 6. This graph compares the experimental life (N_f) and the estimated life (N_e). The straight line represents the perfect fatigue life estimation. Region above this line is considered a conservative life estimation and below of the line is a non conservative region. Furthermore, the dashed lines represent the factor 3 bands as the admissible limit error for the estimated life. The factor 3 was select because it represents the biggest scatter of the experimental fatigue data in the *S-N* curves.

5 Discussion

The fatigue test results shown that aluminium alloy 6201-T81 is extremely sensitive to the fatigue phenomenon. In the *S-N* curves, little changes (5 MPa) in the stress amplitude causes change of one order of magnitude in the fatigue life. This proves that the estimation of the fatigue life is a better approach to describe the aluminuim 6201-T81 fatigue behavior rather than use the fatigue strength.

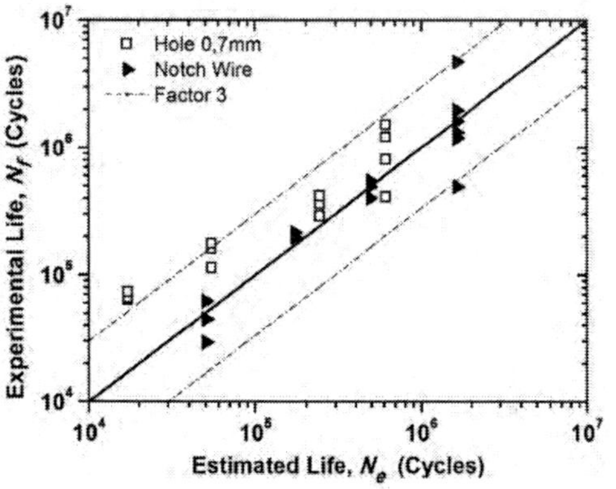

Fig. 6. Fatigue life estimation for the notch wire and wire containing a through hole 0.7 mm.

The chosen parameter to estimate the fatigue damage is the maximum principal stress. This choice is in accordance with the most modern methods to predict the uniaxial high-cycle fatigue strength of notched components [16–18, 22, 23]. Moreover, this idea is supported by the fact that the stage II of the propagation failure is assumed to be predominant over other failure mechanisms. This means that the fatigue damage occurs at the plane perpendicular to the loading direction (maximum loading), so that, the fracture is characterized to be a Mode I from the linear-elastic mechanics. This fact was observed in some test specimens. All wires containing the through hole of 0.7 mm failed in the plane perpendicular to the loading direction.

The fatigue life estimation for the wire containing a through hole present a good correlation with the experimental results. The majority of data points fall within the scatter band. However, it was observed that the fatigue lives below 10^5 cycles had poor estimations, falling in the conservative region. This result is probably associated with the fact that the experimental data of the through hole have smaller fatigue lives than the fatigue lives of the plain wire in the calibration procedure. It is possible to appreciate a trend in the fatigue life estimation that is conservative for the lives below 10^5 cycles and not conservatives for the lives above 10^6 cycles. In general, the accuracy for the life estimation is between 10^5 and 10^6 cycles.

The accuracy of L_M-N_f relationship approach to estimate the fatigue life was proved by Susmel and Taylor [14]. However, their results are not enough to establish that this approach always works in different materials, geometric discontinuities and specimen's types. The study of 6201-T81, plain wire and those containing geometric discontinuities, as depicted in the present work offers new challenges which are the geometric and dimensional features. According to this observation, it is important to validate this life estimation approach and to establish some strategies that will be useful to study geometric discontinuities with complex shape. This is the same reason that motivates the use a three-dimensional model for the sharp notch and the through hole in the FEA.

The three-dimensional model is indispensable for the geometric discontinuities with complex shape because of the absence of the symmetry conditions. In this case, the trajectory to describe the stress field in the geometries shapes, such as a notch, is an intuitive process. However, to establish this trajectory for the through hole in the cylindrical specimen is not a simple task. Therefore, the use of the FEA is necessary. To get experience, knowledge and create news strategies are steps to study geometric discontinuities with complex shape. The intermediate steps are necessary because when the complexity of the phenomenon increases, the numerical simulations can be a high demand of computer resources, with expensive costs and unsatisfactory results.

The authors would like to acknowledge the financial support of TAESA by means of the R&D program of ANEEL. The support of Finatec, CAPES and CNPq are also acknowledged.

6 Conclusions

The present work is a numerical and experimental study to estimate the fatigue life of aluminium alloy 6201-T81 wires containing geometric discontinuities. The present work is part of the studies that try to find news methodologies to the fatigue life estimation in the system cable-suspension clamp electric conductors. The main conclusions are summarized as follows:

- The L_M-N_f relationship, when calibrated with fatigue data from the plain wire and the notched wires, generated good life estimations for the wires containing through hole of 0.7 mm of diameter. In terms of the PM, the majority of the estimations were located between factor 3 bands and these results were considered satisfactory for the aim of this work.
- The PM need a path to define the stress distribution. In this sense, the strategy to chose the trajectory, for a three-dimensional model of the through hole, was considered successful and gave accurate life estimations between 10^5 and 10^6 cycles.
- The L_M-N_f relationship and some strategies developed in this study for the AA6201-T81 wires can be used to perform futures wires fretting fatigue analysis based on the TCD. Therefore, this study not only validates the use of the TCD to predict fatigue life in the above mentioned aluminium alloy but also contributes to further fretting fatigue analyses that would improve the life estimation of overhead conductors.

References

1. Z. R. Zhou, A. Cardou, M. Fiset, and S. Goudreau, Wear, **173**, 179–188 (1994).
2. R.B. Kalombo, J.M.G. Martínez, J.L.A. Ferreira, C.R.M. Da Silva, and J.A. Araújo, Procedia Eng., **133**, 223–232 (2015).
3. C.R.F. Azevedo, A.M.D. Henriques, A.R. Pulino Filho, J.L.A. Ferreira, and J.A. Araújo, Eng. Fail. Anal., **16**, 136–151 (2009).
4. R. B. Kalombo, M. S. Pestana, J. L. A. Ferreira, C. R. M. da Silva, and J. A. Araújo, Tribol. Int., **108**, 141–149 (2017).
5. CIGRÉ-SCB2-WG11-TF7-06-01 (2006).
6. D. Committee, I. Power, and E. Society, IEEE Guide for Aeolian Vibration Field Measurements of Overhead (2007).
7. A.A. Fadel, D. Rosa, L.B. Murça, J.L.A. Fereira, and J.A. Araújo, Int. J. Fatigue, **42**, 24–34 (2012).
8. Z.R. Zhou, A. Cardou, S. Goudreau, and M. Fiset, "Fundamental investigations of electrical conductor fretting fatigue," Tribol. Int., vol. 29, no. 3, pp. 221–232, (1996).
9. J.C. Poffenberger and R.L. Swart, IEEE Trans. Power Appar. Syst., **84**, 508–513 (1965).
10. F. Lévesque, S. Goudreau, L. Cloutier, and A. Cardou, Tribol. Int., **44**, 1014–1023 (2011).
11. A.E. Giannakopoulos, T.C. Lindley, and S. Suresh, Acta Mater., **46**, 2955–2968 (1998).
12. J.A. Araújo, L. Susmel, D. Taylor, J.C.T. Ferro, and E.N. Mamiya, Int. J. Fatigue, **29**, 95–107 (2007).
13. J.A. Araújo, F.C. Castro, S. Pommier, J. Bellacave, and J. Mériaux, **33**, 427–433 (2015).
14. L. Susmel and D. Taylor, Fatigue Fract. Eng. Mater. Struct., **30**, 567–581 (2007).
15. N.E. Dowling, *Mechanical Behavior of Materials: Engineering Methods for Deformation, Fracture, and Fatigue*, 4th ed. Pearson Education (2013).
16. H. Neuber, *Theory of notch stresses: principles for exact calculation of strength with reference to structural form and material*, 2nd ed. U.S. Atomic Energy Commission, Office of Tech. Inf. (1958).
17. R. Peterson, *Notch Sensitivity*, McGraw Hill, 293–306 (1959).
18. D. Taylor, Int. J. Fatigue, **21**, 413–420 (1999).
19. D. Taylor, *The Theory of Critical Distances: A New Perspective in Fracture Mechanics*, Elseiver Science Ltda, (2007).
20. L. Susmel, Eng. Fract. Mech., **75**, 1706–1724 (2008).
21. K. Tanaka, Y. Nakai, and M. Yamashita, Int. J. Fract., **17**, 519–533 (1981).
22. E. I. Haddad, K. N. Smith, and T. H. Topper, J. Eng. Mater. Technol., **101**, 42–46 (1979).
23. D. Taylor and G. Wang, Fatigue Fract. Eng. Mater. Struct., **23**, 387–394 (2000).

Uncertainty analysis of constant amplitude fatigue test data employing the six parameters random fatigue limit model

Davide Leonetti[1,*], *Johan* Maljaars[1,2], and *H.H. (Bert)* Snijder[1]

[1]Eindhoven University of Technology, Department of the Built Environment, Eindhoven, the Netherlands
[2]TNO, Delft, the Netherlands

Abstract. Estimating and reducing uncertainty in fatigue test data analysis is a relevant task in order to assess the reliability of a structural connection with respect to fatigue. Several statistical models have been proposed in the literature with the aim of representing the stress range vs. endurance trend of fatigue test data under constant amplitude loading and the scatter in the finite and infinite life regions. In order to estimate the safety level of the connection also the uncertainty related to the amount of information available need to be estimated using the methods provided by the theory of statistic. The Bayesian analysis is employed to reduce the uncertainty due to the often small amount of test data by introducing prior information related to the parameters of the statistical model. In this work, the inference of fatigue test data belonging to cover plated steel beams is presented. The uncertainty is estimated by making use of Bayesian and frequentist methods. The 5% quantile of the fatigue life is estimated by taking into account the uncertainty related to the sample size for both a dataset containing few samples and one containing more data. The S-N curves resulting from the application of the employed methods are compared and the effect of the reduction of uncertainty in the infinite life region is quantified.

1 Introduction

Performing constant amplitude (CA) fatigue tests on structural details, plain material or assemblies is a necessary step to either perform or validate the design of these components against fatigue failure. When undamaged specimens are tested, each fatigue test results in two types of data: failure and right censored (runout), which consists of a test terminated without failure. According to common practice [1] the fatigue test data related to the finite life region and those related to fatigue limit, the stress range at which no failure is assumed to occur, should be analyzed separately employing two different statistical techniques: the linear regression and the staircase method. Recently, several models have been proposed in order to infer CA fatigue test data irrespectively on the type of data. In this case, the inferential process results in the estimation of the parameters related to the location and scatter in the finite life region and in the fatigue limit. The estimation of the uncertainty underlying the analysis of fatigue test data is of primary importance in order to assess the safety of structures subjected to fatigue loads by employing probabilistic methods. According to the categorization made in [2], reference is made to the aleatory uncertainty to represent the inherent variability of a stochastic quantity and to the epistemic uncertainty for the lack of knowledge [3]. In practice, the aleatory uncertainty is quantified through the evaluation of the scatter produced by the experimental data. Instead, the epistemic uncertainty is quantified by evaluating the degree of belief attributed to the estimators of the model parameters, which increases with increasing the number of experimental observations. Estimating both the aleatory and the epistemic uncertainty allows to correctly estimate the safety margin to be applied in order to reach a predetermined safety level.

Table 1. Description of the dataset related to cover-plated steel beam with welded and unwelded ends, from [4] determining the "large sample" data.

Reference	Steel grade	Failures	Right Censored
[4], Table F1	A36	34	0
[4], Table F2	A441	33	0
[4], Table F9	A36	30	0
[4], Table F7	A36	29	0
[4], Table F8	A36	30	0
[5], Table 2	A36	12	14
[4], Table F10	A36	30	0

* Corresponding author: d.leonetti@tue.nl

The present paper is concerned with the estimation and reduction of aleatory and epistemic uncertainties of constant amplitude fatigue test data of the cover-plated steel beam detail, which are summarized in Table 1. A "large sample" dataset containing all the test data, reported in Table 1, and a "small sample" containing a random subset of the data, see Table 2, are inferred using the linear regression model (LRM) for comparison with common practice and to explore the affinity of the employed methods in estimating the epistemic uncertainty. Then, a regression model for fatigue is employed to infer the "large sample" data, to estimate, to reduce the epistemic uncertainty, and to evaluate their effect on the predicted fatigue life distribution.

Table 2. "Small sample" dataset, R=0

ΔS [MPa]	N [kcycle]	Type
165.00	47,5	Failure
165.00	77,4	Failure
138.00	231,4	Failure
138.00	186,3	Failure
110.00	208,9	Failure
110.00	344,1	Failure
83.00	907,5	Failure
83.00	1113,3	Failure
41.40	100000,0	Runout
34.50	8451,0	Failure
34.50	89314,0	Runout
32.40	37714,0	Failure
32.40	34930,0	Failure
31.70	100000,0	Runout

2 Models and Methods

In order to model the trend of the fatigue test data, the Basquin relation [6] is often employed in its logarithmic form:

$$\log_{10}(N) = \beta_0 + \beta_1 \log_{10}(\Delta S) \qquad (1)$$

Where ΔS is the stress range and β_0 and β_1 are parameters that respectively control the location and the slope of the curve. The aleatory uncertainty are often [1] quantified assuming that the logarithm of the fatigue life at a given stress range is normally distributed and the data are assumed to be heteroscedastic, i.e. the scatter is assumed to be independent on the stress range. On the other hand, the epistemic uncertainty are considered by determining the sampling distribution of the model parameters using ancillary functions. Indeed, this method is often employed to estimate prediction intervals when the Basquin relation and a normal log-fatigue life distribution are used to infer the fatigue test data, reference is made to this model as the LRM. Despite this, it is well established that the fatigue life shows an increasing variability at lower stress ranges, i.e. approaching the fatigue limit, and that the most suitable distribution function to model the fatigue life is actually unknown [7]. In order to better model the trend of the test data, their variability and the fatigue limit, several models have been formulated in the last decades. Pascual and Meeker [8] proposed the Random Fatigue Limit model in order to mimic the curvature at low stress ranges and to better describe the variability of the fatigue life at stress ranges approaching the fatigue limit, ΔS_0, which is defined as a random variable. To model the aleatory uncertainty, they formulated the distribution of the fatigue life (f_W) based on the distributions of the fatigue life given the fatigue limit ($f_{W|V}$) and the fatigue limit (f_V). Moreover, by using the maximum likelihood method (MLM), runouts and failure data can be analysed together. D'Angelo et alii [9-10] proposed the Bilinear Random Fatigue Limit model, which is based on the formulation of the Random Fatigue Limit model, in order to obtain linear S-N curves, similar those employed in the standard and to define a strategy to consider the epistemic uncertainty without assuming a predetermined confidence level. Moreover, their model implies a very weak correlation between the estimators of location the fatigue limit and the fatigue life. In [11] the Six-Parameter Random Fatigue Limit model (6PRFLM) has been introduced in order to explicitly model the curvature from the finite life region towards the fatigue limit and keep the added values of the two previous model: the increasing variability of the log-fatigue life at lower stress ranges, the description of the logarithm fatigue limit as a random variable f_V having location parameter μ_V and scale parameter σ_V, and the low correlation between model parameters related to the location $f_{W|V}$ and f_V. The equation proposed is:

$$\log_{10}(N) = \beta_0 + \beta_1 \log_{10}(\Delta S) - p \log_{10}\left(1 - \frac{\Delta S_0}{\Delta S}\right) \qquad (2)$$

where p is the model parameter which controls the curvature from the finite life region to the fatigue limit. For the cover-plated steel beam datasets inferred in [11], the equation employed in the 6PRFLM revealed to better mimic the trend of the test data with respect to the Random Fatigue Limit model and the Bilinear Random Fatigue Limit model. The MLM has been employed to evaluate the estimator of the model parameters, that is the maximum likelihood estimator (MLE), $\hat{\theta}_{MLE}$. Pascual and Meeker alternatively employed the lognormal distribution and the smallest extreme value distribution to describe the variability of the fatigue limit and the fatigue life. Thus, the assumption made about the shape of these two random variables affects the description of the uncertainty, more details can be found in [5, 7, 8]. However, in the literature, less attention is paid to the estimation of the epistemic uncertainty. In [9-10] the authors proposed a Monte Carlo based approach to evaluate confidence bounds on quantiles by employing a Nested sampling of the epistemic and the aleatory uncertainty. As described later, the epistemic uncertainty was assumed to follow a normal distribution and was estimated using the Likelihood function.

2.1 Epistemic uncertainty in Frequentist Statistic

2.1.1 The linear regression theory

The linear regression model is based on the assumption that the observations are normally distributed. The estimators are evaluated by minimizing the sum of the residuals squared and assuming the stress range to be the independent variable. Thus, by employing a linear relation, the estimators of the slope and the intercept are a linear combination of the observation. Then, it is possible to construct in a closed form their sampling distribution. Because of the normality of the residuals, $\varepsilon \sim N(0,\sigma)$, given an n-dimensional dataset, the Student-T distribution is employed as ancillary function and the following relations apply:

$$t_{n-2} \sim \frac{\hat{\beta}_1 - \beta_1}{\sqrt{\frac{\hat{\sigma}^2}{Sxx}}} \quad (3)$$

$$t_{n-2} \sim \frac{\hat{\beta}_0 - \beta_0}{\sqrt{\hat{\sigma}^2 \left(\frac{1}{n} + \frac{\overline{\log_{10} \Delta S_i}}{Sxx}\right)}} \quad (4)$$

where n is the number of tests, t_{n-2} is the Student-T distribution with n-2 degrees of freedom, S_{xx} is the corrected sum of squares of the $\log_{10}(\Delta S_i)$, the hat denotes the estimator of the parameters of Equation (1) and the upper-score denotes the mean value. Equations (3) and (4) denote the sampling distributions for the model parameters

In the same way, the sampling distribution for the standard deviation, σ, is denoted by the following:

$$\chi_{n-2} \sim \frac{(n-2)\hat{\sigma}^2}{\sigma} \quad (5)$$

where χ_{n-2} is the chi-square distribution (ancillary function) with n-2 degrees of freedom. By combining these formulations it is possible to determine confidence bounds on any quantile of the model.

2.1.1 The Maximum Likelihood Method

When the distribution of the fatigue life is not normal, the Student-T distribution is no longer suitable to describe the epistemic uncertainty. Thus, the distribution of the model parameter has to be known. The likelihood function is a function of the model parameter $\boldsymbol{\theta}$ proportional to the probability of the data that were observed. The maximum likelihood method (MLM) involves the maximization of the likelihood function in order to estimate the model parameter. Moreover, the MLM can be employed to evaluate the uncertainty affecting the estimation of the model parameter. For this purpose, the Wald statistic can be employed. It assumes that the (long-run) distribution of the model parameters is a multivariate normal distribution. For each model parameter $\boldsymbol{\theta}_i$, the following relation apply:

$$N \sim \frac{\hat{\boldsymbol{\theta}}_{MLE} - \boldsymbol{\theta}_i}{\sqrt{\left[I\left(\hat{\boldsymbol{\theta}}_{MLE}\right)^{-1}\right]_{ii}}} \quad (6)$$

Where $I(\hat{\boldsymbol{\theta}}_{MLE})$ is the observed Fisher information matrix and $\mathcal{N}$ is the standard normal distribution. The multivariate normal distribution has mean equal to the MLE and the covariance matrix can be obtained from the Fisher information matrix, which is calculated from the second order derivatives of the log-likelihood function at the MLE. More details can be found in [7, 8] where the epistemic uncertainty have been estimated by employing the Wald statistic. The main drawback of this approach consist in the fact that the resulting distribution might be non-meaningful in some cases [12], especially when a parameter is close to the boundary of its domain. This has been found for the 6PRFLM fitting cover plate data [11]. More details about these methods can be found in textbooks dedicated to the MLM, such as [12], among others.

2.1.1 The Bootstrap Method

Another way to estimate the epistemic uncertainty is the Bootstrap [14]. It is a resampling method, which is able to provide non-parametric statistical inferences, without any assumption about the distribution describing the quantity under consideration. In order to estimate a certain statistic and its uncertainty, the method involves the use of the sampling distribution, which is a probability distribution that shows how the statistic of interest would vary if the sample was collected several times from the population. The bootstrap distribution, an approximation of the sampling distribution, is obtained by sampling with replacement a certain number of bootstrap samples from the original dataset and estimate the statistic of interest for each one of them. It is necessary that the dimension of each bootstrap sample is the same as the dataset. Thus, a certain number of estimators of the statistic of interest are obtained, determining the bootstrap distribution, which can be intended as a frequentist estimation of the epistemic uncertainty. The main drawback of the bootstrap is the intense computational effort with respect to the other methods, which increases with increasing the dimension of the original dataset. Indeed, since the resampling is performed with replacement, the number of possible bootstrap samples would be equal to $(2n-1)!/[n!(n-1)!]$, which makes the analysis time consuming if the original dataset consists of more than 25-30 data. For this reason, the Monte Carlo method is employed to randomly draw bootstrap samples in order to approximate the bootstrap distribution.

2.2 Posterior estimation using Bayesian Analysis. The Random-Walk Metropolis-Hastings algorithm.

The Bayesian statistic explicitly threats the model parameters as probability density functions. For this reason, the output of an inference performed using Bayesian statistics already includes the information about the uncertainty underlying the model parameters. The Bayesian statistics involves the use of the Bayes theorem and gives a method for updating the probabilities of unobserved events, given that another related event has occurred. The Bayes theorem allows the evaluation of the conditional probability of observing the event A given that the event B has occurred:

$$P(A|B) = \frac{P(B|A)P(A)}{P(B)} \tag{7}$$

where P(A), the probability of observing A, is the prior information about the event A; P(B) is the probability of observing B and P(B|A) is the conditional probability of observing B given that A has occurred. P(A|B) in Equation (7) is usually referred to as the ''posterior". Recently, it has been applied to infer fatigue test data in [8, 11]. It is usually referred as non-informative (NBI) when no prior information is employed. Instead, the informative (IBI) involves the use of prior information on the model parameters. The random walk Metropolis-Hastings algorithm [12, 13] has been employed to perform the analysis. It is a Markov Chain Monte Carlo (MCMC) method giving as output a Markov Chain which allows estimating the posterior distribution. The algorithm is based on the following steps:

1) Start from an initial arbitrary set of values for the model parameters;
2) Using a symmetric candidate distribution (e.g. normal), whose mean is the current state of the Markov Chain and with standard deviation to be tuned, sample a new candidate set for the model parameters;
3) Based on the criterion described in [16–18], accept or reject the candidate value based on the likelihood function and the prior knowledge. In case of acceptance, the candidate value becomes the state of the Markov Chain at the current time, otherwise, the previous state is kept at the current time.
4) Repeat the steps 2 and 3 until the Markov Chain converges to its limiting distribution.

In order to reach convergence, the candidate distribution should not be too narrow or too wide. The limiting distribution obtained through his procedure is a non-parametric estimation of the posterior. More details about the random walk Metropolis-Hasting algorithm and the tuning process can be found in [12–16], among others. As result of this procedure, the posterior distribution is approximated and its mean value identifies the Bayesian mean estimator of the parameter.

3. Results and Discussion

In order to show the degree of affinity of the proposed methods, the two fatigue datasets are inferred by employing the LRM, which is based on the Basquin relation, Equation (1), assuming that the base 10 logarithm of the fatigue life follows a normal distribution. The analysis involves the estimation of the parameters and both the aleatory and epistemic uncertainty according to Table 3. The effect of the epistemic uncertainty on the quantiles of the fatigue life is estimated using the approach proposed in [6, 7]. Then, the 6PRFLM is employed in order to infer the "large sample" data. The uncertainty is estimated and reduced following the scheme in Table 3. The different methods are discussed and the effect of the reduction of the uncertainty is also quantified.

Table 3. Simulation scenario

Fit and epistemic uncertainty		LRM	6PRFLM
Least Square Method	Student T-distribution	Y	N
Maximum Likelihood Method	Wald Statistic	Y	Y
	Bootstrap	Y	Y
Non-Informative Bayesian Inference	MCMC	Y	Y
Informative Bayesian Inference	MCMC	N	Y

3.1 The Linear Regression Model

The methods summarized in Table 3 can be employed to infer the data, estimate the model parameters β_0, β_1, σ and the associated uncertainty. In order to perform the fit using the Bootstrap and the Bayesian inference, the number of samples generated is equal to one million.

3.1.1 The large sample dataset

For the failure data considered in the large sample

Table 4. Results of the LRM fitting the "large sample" data. The estimators of the model parameters and the COV are reported for each method employed.

	Model Parameters		
	β_0	β_1	σ
	Estimators		
LSM	1.22E+01	-3.29E+00	1.47E-01
MLM	1.22E+01	-3.28E+00	1.46E-01
NBI	1.22E+01	-3.29E+00	1.47E-01
	Coefficient of Variation [%]		
LSM	0.99	1.83	5.05
Wald	0.98	1.81	4.98
Bootstrap	1.04	2.53	6.11
NBI	0.97	1.80	5.09

https://doi.org/10.1051/matecconf/201816510016

dataset, the estimators of the LRM parameters and the coefficient of variation are reported in Table 4. The value of the log-likelihood function at the MLE is 102.18. However, such value is related to the failure data only. By inferring the whole dataset using the LRM and

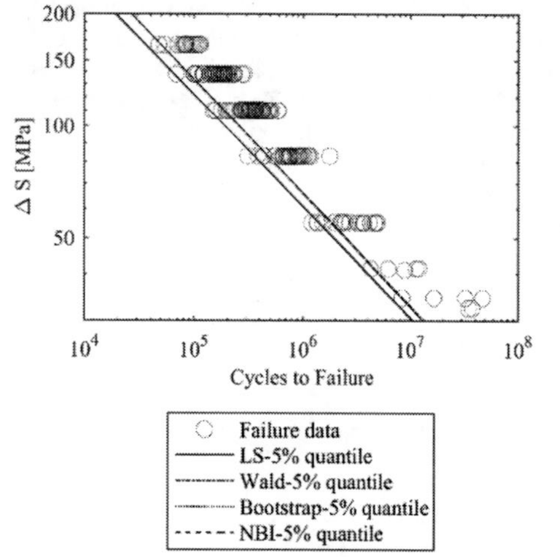

Figure 1. Fatigue test data belonging to the large sample dataset and 5% quantile of the fatigue life obtained for the linear regression model considering the epistemic uncertainty.

the MLM thus, including the contribution of runouts, the value of the log-likelihood function drops to 3.43. The estimators obtained using the different methods are equal. It is known that the mean Bayesian posterior estimator tends to be equal to the MLE with increasing the sample size. The small difference observed is addressed to the use of the numerical procedure employed to determine the MLM, the Bootstrap and the NBI based estimators.

The normality of the distributions describing the epistemic uncertainty has been verified in each case by using the Kolmogorov-Smirnoff test, leading in any case

to "p-values" close to 1. In this case, the LSM, the MLM and the NBI are equivalent and would lead to a close estimation of each quantile of the fatigue life,

Table 5. Results of the LRM fitting the "small sample" data. The estimators of the model parameters and the COV are reported for each method employed.

	Model Parameters		
	β_0	β_1	σ
	Estimators		
LSM	1.28E+01	-3.56E+00	2.05E-01
MLM	1.28E+01	-3.56E+00	1.84E-01
	Coefficient of Variation [%]		
LSM	3.77	6.96	29.74
Wald	3.01	5.56	21.32
Bootstrap	4.32	7.64	26.75

when epistemic uncertainty is not considered. In order to consider both the epistemic and the aleatory uncertainty, a nested sampling has been performed in a Monte Carlo framework as described in [9-10].

The 5% quantile of the linear regression model has been estimated considering the effect of the epistemic uncertainty. From Figure 1 it results that, despite the estimators of the model parameters and the coefficient of variation of the marginal distribution are comparable for each model parameter, the 5% quantile predicted by the least square method predicts a lower fatigue life (-21.93% at Δσ=100 MPa) than the other methods, that at maximum differ by 3%, in absolute value. This large difference obtained by employing the linear regression theory is mostly associated with the fact that in the linear regression theory the correlation between the model parameters is not taken into account. Instead, the correlation is estimated when the Wald statistic and the Bootstrap method are employed. However, in the former case the correlation coefficient is estimated at the mode of the Likelihood function through the Fisher Information Matrix, and, in the latter case, the correlation is evaluated through the resampling procedure.

3.1.1 The small sample dataset

For the 'small sample' analysis, the estimators of the model parameters are evaluated only using the LSM and the MLM because the NBI using the Random Walk Metropolis-Hastings algorithm results in a Markov-Chain that hardly converges to its limiting distribution. In other words, the amount of test data is so small that the resulting Markov-Chain was very sensitive to the values of the tuning parameter, the standard deviation of the candidate distribution. For this reason, it was not possible to obtain a converging Markov Chain by using a "trial and fail" procedure.
As shown in Table 5, the estimators of the parameters β_0 and β_1 are in good agreement. Because of the small amount of test data, the LS estimator of the parameter σ is 1.1 times higher than that one obtained by the MLM. Moreover, as it can be seen in Figure 2, also the estimation of the epistemic uncertainty led to different results. In general, the coefficient of variation is in any case higher than for the "large sample" dataset, meaning that due to the smaller amount of data the epistemic uncertainty increased. When the Wald statistic is employed, the epistemic uncertainty are normally distributed. Instead, the distribution resulting by employing the Bootstrap is of a different type. For example, in Figure 2 it results that the bootstrap distribution associated to the parameter σ is bi-modal; this can be seen because the trend of the cumulate distribution function shows two inflection points, the

Table 6. Prior distributions employed in the informative Bayesian inference, from [11].

6PRFLM Parameter	Estimators	
$-\beta_1$	$\mu_{\log(-\beta_1)}$	1.15E+00
	$\sigma_{\log(-\beta_1)}$	2.00E-01
μ_v	$m'=\mu_{\mu_v}$	1.55E+00
	n'	20
σ_v	$s'=\mu_{\sigma_v}$	7.24E-02
	v'	20

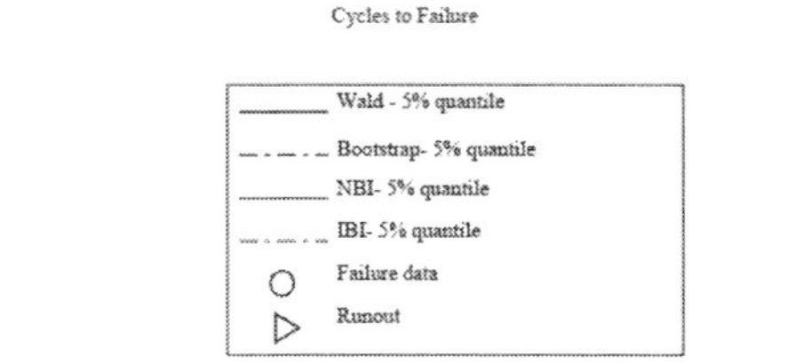

Figure 4. Large sample data inferred using the 6PRFLM. Only data at ΔS<60MPa are shown. The median curve is obtained using the MLE. Moreover, the 5% quantile of the fatigue life is plot considering the epistemic uncertainty.

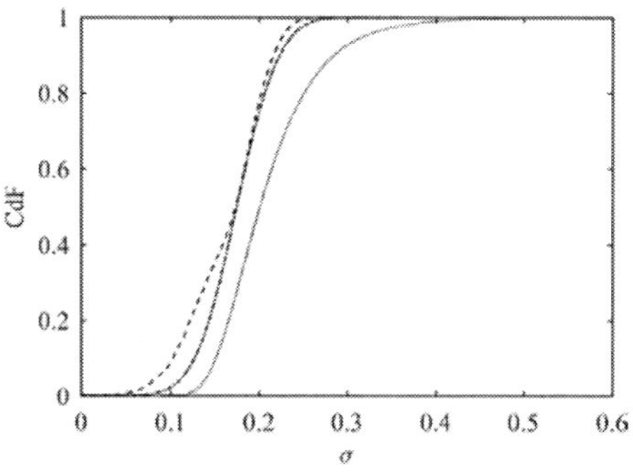

Figure 2. Epistemic uncertainty related to the parameter σ of the LRM fitting the small sample data.

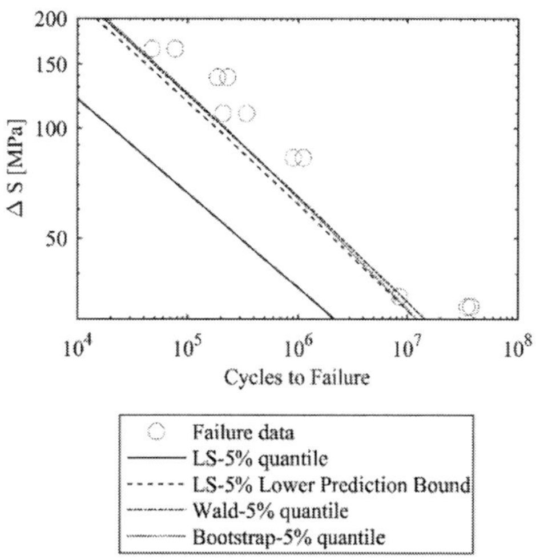

Figure 3. Fatigue test data belonging to the small sample dataset and 5% quantile of the fatigue life obtained for the linear regression model considering the epistemic uncertainty. For comparison, also the 5% lower prediction bound is plotted.

first at a cumulate probability value around 0.2 and a second one which is around 0.6.
The estimated quantiles of the fatigue life are shown in Figure 3. As for the "large sample" dataset, the results are obtained considering the epistemic uncertainty. LSM predicts a 5% quantile of the fatigue life that is 0.11 times the life predicted by the lower prediction bound as defined in the linear regression theory. The 5% quantile of the fatigue life estimated considering the evaluation of the uncertainty given by the Bootstrap and the Wald statistic are closer to each other and predict a fatigue life respectively 1.18 and 1.19 times higher than the 5% lower prediction bound as defined in the linear regression theory.

3.2 The Six-Parameters Random Fatigue Limit Model

The 6PRFLM has been employed to infer the large sample data according to the scheme reported in Table

3. In the IBI, two prior distributions have been formulated for the parameters related to the slope of the finite life region and the parameters associated with the location and the scatter of the fatigue limit.
In the first case, the prior information is obtained from the fatigue test data of transverse loaded welded connections reported in [19], where the LRM has been employed in order to estimate the slope parameter for each sample. To formalize the prior distribution of $-\beta_1$, in [11] its estimators have been collected for each data sample and a lognormal distribution have been fitted to them, its parameters are summarized in Table 6.
In order to formalize the prior information on the location and scatter of the fatigue limit, the probability distribution introduced in [20] has been employed:

$$f=k\sigma_v^{-[v'+\delta(n')+1]}\exp\left\{-\frac{1}{2\sigma_v^2}\left[v's'^2+n'\left(\mu_v-m'\right)^2\right]\right\} \quad (8)$$

where s' is the prior expectation of the standard deviation, ν' is the number of degrees of freedom for s',.m' is the prior expectation of the mean value and n' is the number of degrees of freedom for m'. If $n'=0$, the function $\delta(n')=0$, otherwise $\delta(n')=1$. In case of vague information, the degree of freedom should be zero
whereas a value between 20 and 40 can be assigned if experience leads to almost deterministic knowledge.
The parameters have been estimated in [11] and are reported in Table 6. The estimators of the model parameters and the coefficient of variation and the skewness related to their distributions are reported in Table 7. The value of the log-likelihood function at the MLE is 100.26. This is lower compared to the value obtained by the LRM, when runouts were not included. Instead, it is significantly higher than the value of the log-likelihood function obtained by inferring the complete dataset using the LRM and the MLM. The MLE and the non-informative Bayesian estimator are almost equal to each other and also the estimated coefficients of variation are close. The differences are addressed to the small skewness of the distributions that never exceeded the interval ±2, except for the parameter p, which is also the parameter more affected by uncertainty. As result of the IBI, both the skewness and the uncertainty affecting the parameter p are significantly reduced.
It should also be noted that with respect to the determination of quantiles of the fatigue life considering the epistemic uncertainty, it is possible that in the Monte Carlo procedure negative values of the standard deviation of the fatigue limit and the parameter p might be sampled if the results of the Wald statistic are employed. For this reason, when the quantiles of the fatigue life are computed by nested sampling the aleatory and the epistemic uncertainty estimated by the Wald statistic, the parameters σ, σ_V and p cannot be negative. Thus, a truncation scheme should be applied in order to limit the sampling to the domain of the model parameter.
The highest difference in the fatigue life predicted in the different frequentist scenarios is expected to be in the transition from the finite to the infinite life region. By comparing the predicted 5% quantiles of the fatigue limit distribution, the maximum difference obtained is less than 1% for the frequentist methods and using the Wald statistic the predicted 5% quantile is 25.27MPa. IBI predicts 26.17MPa, which is 1.16 times higher than the NBI. Despite the Bootstrap allows a better estimation of the epistemic uncertainty, the fatigue life predicted by the 5% quantile does not significantly differ by that one estimated using the MLM and the Wald statistic, even at relatively low stress ranges, see Figure 4.

4. Conclusions

By the analyses performed in the present work, the following conclusions can be drawn:

(1) In order to include epistemic uncertainty in the estimation of the fatigue life distribution by performing nested sampling of epistemic and aleatory uncertainty in a Monte-Carlo simulation, the correlation between the model parameters should be estimated. Thus, the linear regression theory cannot be employed to estimate the uncertainty underlying the model parameters.

(2) If the number of test data is relatively small, the Bayesian Inference employing the

Table 7. Results of the LRM fitting the "large sample" data. The estimators of the model parameters and the COV are reported for each method employed.

	Model Parameters					
	β_0	β_1	σ	μ_v	σ_v	p
	Estimators					
MLM	1.11E+01	-2.78E+00	1.32E-01	1.54E+00	7.27E-02	4.74E-01
NBI	1.09E+01	-2.70E+00	1.29E-01	1.52E+00	8.98E-02	6.70E-01
IBI	1.12E+01	-2.82E+00	1.34E-01	1.54E+00	7.36E-02	4.31E-01
	Coefficient of Variation [%] / Skewness					
Wald	2.88 / -0.05	5.17 / 0.05	5.87 / -0.00	1.47 / -0.01	28.20 / 0.02	38.21 / 0.08
Bootstrap	3.42 / 0.30	6.11 / 0.25	7.29 / 0.26	1.88 / -0.83	31.40 / -0.34	46.45 / 1.62
NBI	3.15 / -0.33	5.61 / 0.26	6.59 / -0.09	2.19 / -1.16	24.14 / 0.48	44.05 / 2.19
IBI	2.08 / -0.11	3.72 / 0.07	5.58 / 0.28	0.83 / -0.18	14.19 / 0.61	27.93 / 0.58

Random-Walk Metropolis-Hastings algorithm determines a Markov-Chain appears to be sensitive to the scale parameter of the candidate distribution. This resulted in difficulties in estimating the posterior distribution of the model parameters.

(3) The inference performed using the 6PRFLM determines a higher value of the log-likelihood function than the LRM when the "large sample" dataset is employed.

(4) For the cover plate steel detail, the fatigue life predicted in the transition from finite to infinite life region is affected by large uncertainty. This is due to the small amount of test data executed at relatively small stress range levels. It has been found that by employing a prior distribution for the parameters of the fatigue limit distribution this uncertainty decreases but remains still high.

(5) Despite the Bootstrap, different from the Wald statistic, allows any shape of the distribution of the model parameters, the difference of the 5% quantile of the fatigue life obtained considering the epistemic uncertainty for the 6PRFLM fitting the "large sample" data is negligible. Thus, for relatively large samples inferred with complex models such as the 6PRFLM, the computational effort spent to perform the Bootstrap estimation of the epistemic uncertainty does not seem to lead to a better prediction.

The authors would like to acknowledge the Dutch infrastructural asset owners Rijkswaterstaat and ProRail and the research organization TNO for their support.

References

[1] C. Schneider and S. Maddox, "Best practice guide on statistical analysis of fatigue data," 2003.

[2] M. B. A. Van Asselt, *Perspectives on Uncertainty and Risk - the PRIMA Approach to Decision Support*. 2000.

[3] M. H. Faber, "On the Treatment of Uncertainties and Probabilities in Engineering Decision Analysis," *J. Offshore Mech. Arct. Eng.*, vol. 127, no. 3, p. 243, Aug. 2005.

[4] J. Fisher, P. Albrecht, B. Yen, and D. Klingerman, "Effects of weldments on the fatigue strength of steel beams-transverse stiffness and attachments, September 1972," *Fritz Lab. Reports*, Jan. 1972.

[5] R. Slockbower, "Fatigue resistance of full scale cover-plated beams.," *Theses Diss.*, Jan. 1976.

[6] O. H. Basquin, "The exponential law of endurance tests," *Proc ASTM*, vol. 10, pp. 625–630, 1910.

[7] J. Schijve, "Statistical distribution functions and fatigue of structures," *Int. J. Fatigue*, vol. 27, no. 9, pp. 1031–1039, 2005.

[8] F. G. Pascual and W. Q. Meeker, "Estimating Fatigue Curves with the Random Fatigue-Limit Model," *Technometrics*, vol. 41, no. 4, pp. 277–290, 1999.

[9] Luca D'Angelo and Alain Nussbaumer, "Estimation of fatigue S-N curves of welded joints using advanced probabilistic approach. Under review: IJFATIGUE-D-16-00685," *Int. J. Fatigue*, vol. 97, pp. 98–113, 2017.

[10] L. D'Angelo, M. Rocha, A. Nussbaumer, and E. Brühwiler, "S-N-P Fatigue Curves Using Maximum Likelihood: Method for fatigue resistance curves with application to straight and welded rebars," *Eurosteel 2014*, no. 1, pp. 1–8, 2014.

[11] D. Leonetti, J. Maljaars, and H. H. (Bert. Snijder, "Fitting fatigue test data with a novel S-N curve using frequentist and Bayesian inference," *Int. J. Fatigue*, 2017.

[12] Y. Pawitan, *In all likelihood: statistical modelling and inference using likelihood*. Clarendon Press, 2001.

[13] S. S. Wilks, "The Large-Sample Distribution of the Likelihood Ratio for Testing Composite Hypotheses," *Ann. Math. Stat.*, vol. 9, no. 1, pp. 60–62, 1938.

[14] P. Diaconis and B. Efron, "Computer-intensive methods in statistics," *Sci. Am.*, vol. 248, no. 5, pp. 116–130, 1983.

[15] I. Babuska, Z. Sawlan, M. Scavino, B. Szabó, and R. Tempone, "Bayesian inference and model comparison for metallic fatigue data," Dec. 2015.

[16] W. K. Hastings, "Monte carlo sampling methods using Markov chains and their applications," *Biometrika*, vol. 57, no. 1, pp. 97–109, 1970.

[17] N. Metropolis, A. W. Rosenbluth, M. N. Rosenbluth, A. H. Teller, and E. Teller, "Equation of State Calculations by Fast Computing Machines," *J. Chem. Phys.*, vol. 21, no. 6, pp. 1087–1092, Jun. 1953.

[18] W. M. Bolstad, *Understanding Computational Bayesian Statistics*. 2012.

[19] G. Sedlacek, A. Hobbacher, J. B. Schleich, A. Nussbaumer, S. J. Maddox, and J. Brozzetti, "First draft of the background document prEN 1993-1-9."

[20] JCSS, "Background documentation Eurocode 1 (ENV 1991) - Part 1: Basis of design.," 1996.

MATEC Web of Conferences **165**, 10017 (2018) https://doi.org/10.1051/matecconf/201816510017
FATIGUE 2018

Evaluating the Impact of Conservatism in Industrial Fatigue Analysis of Life-Limited Components

Joshua Hoole[1*], *Pia* Sartor[1], *Julian* Booker[2], *Jonathan* Cooper[1], *Xenofon V.* Gogouvitis[3] and *R. Kyle* Schmidt[4]

[1]Department of Aerospace Engineering, University of Bristol, Bristol, BS8 1TR, United Kingdom
[2]Department of Mechanical Engineering, University of Bristol, Bristol, BS8 1TR, United Kingdom
[3]Safran Landing Systems, Gloucester, GL2 9QH, United Kingdom
[4]Safran Landing Systems, Ajax, ON, L1S 2G8, Canada

Abstract. This paper presents a review of the conservatism approaches applied by different industrial sectors to the stress-life (S-N) analysis of 'life-limited' or 'safe-life' components. A comparison of the fatigue design standards for 6 industrial sectors identified that the conservatism approaches are highly inconsistent when comparing the areas of variability and uncertainty accounted for along with the conservatism magnitude and method of application. Through the use of a case-study based on the SAE keyhole benchmark and 4340 steel S-N data, the industrial sector which introduces the greatest reduction of a component life-limit was identified as the nuclear sector. The results of the case-study also highlighted that conservatism applied to account for scatter in S-N data currently provides the greatest contribution to the reduction of component life-limits.

1 Introduction

Across the different engineering sectors (e.g. aerospace, nuclear, offshore, etc.), many life-limited components are designed against fatigue failure using analysis methods comprising of stress-life (S-N) curves and Miner's Rule [1]. This approach to fatigue design and analysis is also known as 'safe-life' [1]. The 'safe-life' or 'life-limit' of a component represents the number of duty cycles after which the component must be removed from service [1].

Fatigue design of life-limited components contains significant areas of variability and uncertainty. 'Variability' is defined as the 'randomness' or scatter that can be observed within fatigue design parameters, resulting from the probabilistic nature of fatigue [2]. 'Uncertainty' is defined as the lack of knowledge or assumptions required about a specific element of the S-N analysis process [2]. This variability and uncertainty propagates through the analysis process, resulting in significant variability and uncertainty in a component's life-limit [1]. In order to account for this variability and uncertainty, and to ensure components retain their structural integrity in-service, conservatism is currently applied within the analysis process [1]. This conservatism effectively reduces the life-limit of a component and often takes the form of statistical reductions on material properties or as 'safety factors' [1]. The conservatism required during design is usually prescribed in the relevant design standards for each industrial sector.

This paper will compare the different approaches to conservatism used by various industrial sectors when analysing life-limited components. In addition, the impact of each conservatism approach on the life-limit of a component will be evaluated using a case-study.

1.1 Justification for Investigating the Impact of Conservatism

Investigating the impact of conservatism on component life-limits during fatigue design provides the opportunity to compare the conservatism approaches used in various industrial sectors. This includes a comparison of: the areas of variability and uncertainty accounted for, the method of conservatism application and the magnitude of the conservatism currently required. The introduction of conservatism into the analysis process can lead to the potential life of a component not being fully exploited due to retiring a component from service at an earlier life-limit. Conservatism can also result in components being 'over-sized' and heavier, increasing the total life-cycle cost of a component and reducing the overall performance of the structure [1]. By comparing the conservatism approaches applied by various industrial sectors to the S-N fatigue analysis of life-limited components, it will be possible to identify which industrial sectors introduce a significantly greater magnitude of conservatism. For sectors that are shown to be more conservative, an investigation into areas of conservatism that could be reduced can be initiated, potentially resulting in more efficient components with longer life-limits, which remain safe and reliable in-service.

* Corresponding author: jh12317@bristol.ac.uk

1.1.1 Previous Literature Comparing Industrial Conservatism Approaches

Previous studies that compare the conservatism approaches within different industrial sectors have been identified within the literature. Work performed by Micone & De Waele compared the safety factors and statistical reductions on material properties across various design standards relating to the fatigue design of steel structures and offshore structures [3]. However, they concluded [3] that the conservatism approaches were not comparable and therefore, it was not possible to identify whether one fatigue design standard was more conservative than other. This paper aims to further their work by quantifying the impact on a component life-limit of the various conservatism approaches through the use of a case-study. Fatigue design standards for the offshore industry and nuclear industry have also been qualitatively compared in [4] and [5] respectively. However, to the authors' knowledge there is yet to be a comparison of the conservatism approaches from fatigue design standards across various industry sectors. This paper aims to contribute to this gap in the literature by comparing the aerospace, nuclear, wind turbine, offshore and steel structures sectors.

1.2 Included Industrial Sectors

Table 1 shows the different industrial sectors reviewed, along with the supporting design standards, advisory material/industrial practice and typical examples of life-limited components. References from the literature for each sector are also shown.

Other industrial sectors that could have been included within this review were the automotive sector [1] and the rotorcraft sector (e.g. helicopters) [6]. All of these sectors feature fatigue critical components due to high cyclic loading. Unfortunately, insufficient literature in the public domain was available in the time-frame of this work to identify the conservatism used by these industries.

1.3 Scope of Paper

In order to enable a comparison across the different industrial sectors, which often have different fatigue design drivers based upon the loading and consequence of failure of the component, the scope of this paper has been focused in a number of ways. This sub-section defines the scope of this paper.

1.3.1 Component Characteristics

The design standards shown in Table 1 represent a wide range of potential component and material types. Therefore, a 'common' component must be defined that is applicable to each of the design standards. This component has been identified to be a 'life-limited', single-load path and monolithic (i.e. no weldments) component manufactured from high-tensile strength steel, that is not inspected in-service. Whilst this is a simple component compared to the welded components that feature heavily in the offshore and nuclear sectors [11, 13], it does represent the typical characteristics of an aircraft landing gear component [6].

1.3.2 Stress-Life (S-N) Fatigue Analysis

The scope of this paper has also limited the analysis methods to be considered. The fatigue analysis method for life-limited components that is consistently represented across the various design standards is a stress-life (S-N) approach to fatigue analysis. Whilst many of the design standards provide alternative routes to fatigue certification, such as a fracture mechanics approach, this is beyond the scope of this paper and therefore, only an S-N approach will be considered. In addition, there are subtle differences in the S-N analysis methods prescribed in each design standard (e.g. the mean-stress correction required). However, as this paper is concerned with the impact of the conservatism approach on a component's life-limit, a 'common' S-N fatigue analysis process has been defined. As S-N fatigue analysis is a 'traditional' and well-known analysis approach, a detailed description of the method can be found in many textbooks (e.g. [1]). The common S-N fatigue analysis process is defined as follows:

1) The load spectrum is converted into load cycles 'i', comprised of a load amplitude and mean load using rainflow counting.
2) Analytical stress analysis is used to convert the loads into a stress amplitude (σ_{a_i}) and mean stress (σ_{m_i}).
3) The Goodman mean stress correction is used to covert the stress cycles into fully-reversed (i.e. $\sigma_m = 0$) stress amplitudes (σ_{s_i}).
4) From a mean S-N curve, which indicates how the number of cycles to failure varies with the applied fully-reversed stress amplitude, the number of cycles to failure (N_{f_i}) for each applied stress cycle

Table 1. The industrial sectors included within this paper and supporting references.

Industrial Sector	Design Standard	Advisory Material	Typical Components
Large Aircraft	EASA CS25 [7]	EASA AMC25 [7]	Aircraft Landing Gear [6, 7]
Light Aircraft	FAA Part 23 [8]	FAA AC23 [8]	Fuselage and Wing Structures [8]
Nuclear	ASME III NB [9]	ASME III NB [9]	Piping and Boilers [9]
Wind Turbines	BS61400-1 [10]	-	Turbine Blades, Hubs, Towers [10]
Offshore Structures	DNVGL-RP-C203 [11]	DNVGL-OS-C101 [12]	Offshore Platforms, Jackets and Piles [11]
Steel Structures	Eurocode 3 [13]	-	Steel Bridges [13]

 https://doi.org/10.1051/matecconf/201816510017

is identified.

5) Using Miner's Rule, the damage accumulated for each applied stress cycle (d_i) is computed.
6) The total damage accumulated for the loading spectrum (D_T) is computed as the sum of the individual damages 'd_i'.
7) The life-limit (i.e. the number of loading spectrum applications to failure) of the component is computed (assuming failure occurs when $D_T = 1$):

$$Life\ \ Limit = \frac{1}{D_T} \quad (1)$$

When using a mean S-N curve, this infers that half of the components manufactured will have failed at a damage value of 1.

2 Methods of Introducing Conservatism into S-N Fatigue Analysis

There are a number of distinct methods that enable conservatism to be introduced into the S-N fatigue analysis of life-limited components. These methods can be differentiated into statistically-derived reductions and single-value (i.e. 'deterministic') reductions. Deterministic reductions will be referred to as 'safety factors' for the remainder of the paper. The first of the conservatism application methods is to apply a safety factor directly to the computed component life-limit. The application of this conservatism results in the component only being permitted to be in-service for a design life equal to the life-limit as computed using S-N analysis divided by the safety factor. A similar approach is to divide the Miner's Rule failure criterion (which is typically taken as $D_{fail} = 1$) by a safety factor. Another method of applying conservatism is to use a safety factor on the applied stress amplitudes. The final method of application is to modify the S-N curve from the mean S-N curve that would be fitted to S-N data. This reduced curve is typically known as a 'design', 'working' or 'characteristic' curve. This reduction can either be achieved by dividing the stress amplitude or the number of cycles to failure (i.e. a factor on 'life') for the mean curve by the selected safety factor. These factors shift the S-N curve 'down' and 'left' respectively. Figure 1 presents a flowchart that demonstrates how conservatism is applied at the different stages of S-N fatigue analysis.

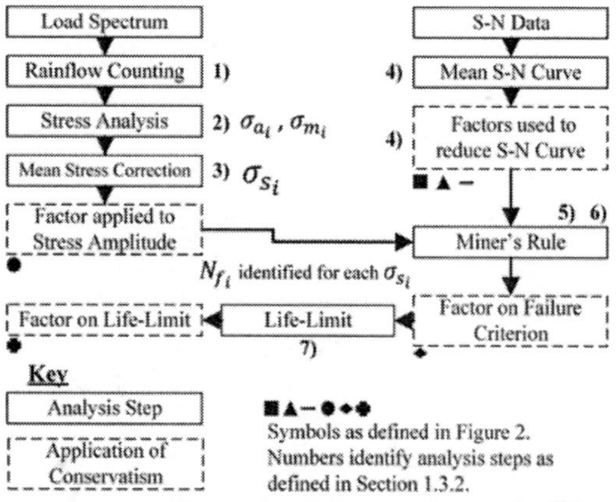

Figure 1. A flowchart showing the S-N fatigue analysis stage at which each conservatism method is applied.

2.1 Statistical Reduction of S-N Curves

The final approach for introducing conservatism into S-N fatigue analysis is through applying statistically-derived reductions to the S-N data to produce design curves. These curves are typically defined using a Probability of Survival (PoS) and a Confidence Level (CL). For a set of S-N data the PoS value represents the proportion of specimens that would be expected to survive at a given stress level.

If the number of cycles to failure (or the logarithm of cycles to failure) are Normally distributed, tables exist [14] for identifying the statistical reduction factor '$k_{PoS/CL}$' (for the desired PoS/CL level) that computes the reduced number of cycles to failure '$N_{PoS/CL}$' using:

$$N_{PoS/CL} = \overline{N_f} - \left(k_{PoS/CL} \times S\right) \quad (2)$$

where '$\overline{N_f}$' is the mean number of cycles to failure from the S-N dataset at a given σ_{s_i} and 'S' is the sample standard deviation of the number of cycles to failure at the same σ_{s_i}. As '$\overline{N_f}$' and 'S' are estimates of the population based on the sample data, they must be stated with a level of confidence and this is captured by the CL value. It should be noted that a PoS/CL value of 50/50 represents the mean S-N curve.

3 Qualitative Comparison of Industrial Conservatism Approaches

A comprehensive review of the fatigue design standards for each industrial sector from Table 1 enabled the population of the matrix shown in Figure 2. Each populated cell in Figure 2 permits the identification of the area of variability and uncertainty accounted for by each conservatism approach, along with the method of application and the magnitude of the conservatism approach for each industrial sector. From reviewing the matrix in Figure 2, qualitative comparisons regarding the three elements of the conservatism approaches listed above can be drawn across the different industrial sectors.

Firstly, the areas of variability and uncertainty accounted for vary significantly across the industrial sectors. The only area consistently accounted for is scatter within S-N data, as shown by the fully populated row in Figure 2. Most sectors, excluding nuclear and steel structures, also account for deviations in the loading applied to a component. The large aircraft, wind turbine, offshore and steel structures sectors all account for the consequence of failure of a component using conservatism. It is interesting to note that despite the widely discussed uncertainty present within the assumptions of an S-N approach based upon Miner's Rule [1, 2], only light aircraft structures and offshore structures apply conservatism that directly accounts for uncertainty within analysis methods. The conservatism approach used by the wind turbine sector appears to be the most comprehensive approach by being the only sector to use conservatism to account for: dimensional

Symbol	Meaning
■	"Stress" factor applied to stress amplitude on S-N curve
▲	"Life" factor applied to number of cycles to failure on S-N curve
▬	PoS/CL statistical reduction on mean S-N Curve
●	Factor applied to stress cycle amplitude
◆	Factor applied to Miner's Rule failure criterion
✚	Factor applied to component life-limit
*	Conservatism applied to multiple areas of variability / uncertainty

	Area of Variability or Uncertainty	Industrial Sector: Large Aircraft (EASA CS25) [7]	Light Aircraft (FAA Part 23) [8]	Nuclear (ASME III NB)^{A} [9]	Wind Turbines (BS61400-1)^{B} [10]	Offshore Steel Structures (DNVGL-RP-C203) [11, 12]	Steel Structures (Eurocode 3) [13]
Materials Data	S-N curve material scatter	▬ 99/95	✚ >8*	■ 2* ▲ 2	▬ 97.7/95	▬ 97.7/75	▬ 95/75
Materials Data	Deviation of properties from design data				● 1.1*		
Materials Data	Difference between coupon/specimen used to generate S-N curve and component	✚ 3*	✚ >8*		● 1.1*		
Materials Data	Effect of surface finish and material treatment			■ 2* ▲ 4*			
Materials Data	Size effect			▲ 2.5			
Materials Data	Environmental degradation / no environmental protection applied to component	✚ 3*		■ 2* ▲ 4*			
Loading	Deviations in applied loads from those assumed during analysis / uncertainty in loading	✚ 3*	✚ >8*		● 1^{C}	◆ 10*^{E}	
Loading	Accuracy of stress concentration factors		✚ >8*				
Geometry	Variability or uncertainty in geometrical properties (e.g. manufacturing tolerances)				● 1.1*		
Geometry	Inaccurate assessment of component resistance (e.g. structural sections)				● 1.1*		
Component	Consequence of failure for component / level of required safety	✚ 3*			● 1.15^{D}	◆ 10*^{E}	■ 1.35^{F}
Component	Inspectability of components					◆ 10*^{E}	
Analysis	Uncertainty in analysis methods		✚ >8*			◆ 10*^{E}	
Analysis	Uncertainty in conversion factors				● 1.1*		

A: Only the most conservative of either the "stress" factor or "life" factor on the S-N curve is applied.
B: Assuming Steel material (other conservatism values are provided for other materials).
C: This conservatism is used to account for loading uncertainty and deviation during static design.
D: Class 2 "non fail-safe" components whose failure may lead to the failure of a major wind turbine part.
E: Either a multiplication of the accumulated damage or as a divisor on the Miner's rule failure criterion.
F: Required for a safe-life component (i.e. non inspected) with a high consequence of failure.

Figure 2. A matrix showing the different conservatism approaches used by various industrial sectors, including the areas of variability and uncertainty accounted for, the method of conservatism application and the magnitude of the conservatism applied.

variability (e.g. manufacturing tolerances), deviations from anticipated material properties and differences between the tested specimens and the final component. The ASME III nuclear conservatism approach accounts for material properties beyond other sectors, such as surface finish effects and size effect. Figure 2 shows that Eurocode 3 steel structures account for the fewest areas of variability and uncertainty, only using conservatism to account for S-N data scatter and the consequence of failure.

As most of the areas of variability and uncertainty shown in Figure 2 are common to fatigue design, regardless of application/industrial sector, these findings suggest that 'lessons learned' could be shared between the different sectors, to ensure that each industrial sector is accounting for all potential areas of variability and uncertainty within fatigue design. For example, most of the industrial sectors could introduce conservatism to account for uncertainty in analysis methods.

Concerning the method of application of conservatism, there is little consistency across the industrial sectors. As can be seen in Figure 2, large aircraft employ both a PoS/CL curve with a safety factor on the life-limit, whereas light aircraft structures only use a safety factor on life-limit to introduce conservatism. The offshore sector uses a PoS/CL curve and a safety factor on the value of Miner's Rule failure criterion (known as a Design Fatigue Factor – DFF [11]). The wind turbine sector introduces yet another approach for introducing conservatism, using a PoS/CL curve and a safety factor on the applied stress amplitude. It is interesting to note that the ASME III approach for nuclear components applies the most conservative out of a deterministic stress or life factor on the mean S-N curve. Finally, steel structures under Eurocode 3 use a stress factor on the statistically reduced PoS/CL S-N curve. It can be seen that there is significant inconsistency in the method used to apply conservatism across the industrial sectors.

The methods of conservatism application also vary considerably for each individual area of variability or uncertainty. For example, Figure 2 shows that S-N data scatter is accounted for using statistically reduced PoS/CL S-N curves, deterministically reduced S-N curves as well as a safety factor on the component life-limit. Likewise, deviations in applied loading is accounted for using: a safety factor applied to the stress amplitude, a safety factor applied to Miner's Rule failure criterion or using a safety factor on the component life-limit. Figure 2 also shows that often a conservatism method is used that does not directly address the origin of the area of variability and uncertainty being accounted for. For example, the wind turbine sector uses a safety factor on the stress amplitude to account for deviations in the materials data, which one would expect to be accounted for using conservatism applied to material properties directly. This discussion has shown that the methods of applying conservatism are highly inconsistent across the industrial sectors.

On the other hand, one area of consistency regarding the conservatism methods is the use of PoS/CL S-N curves, as these are used by the majority of industrial sectors. However, as Figure 2 shows, different sectors use different PoS/CL levels. In order to demonstrate the impact of applying conservatism to S-N curves, a rich S-N dataset for 4340 Steel (typical of the high-tensile strength steel used in landing gear components [6]) was sourced from the Engineering Sciences Data Unit (ESDU) [15]. Two significant challenges exist when working with S-N data. Firstly, steels exhibit a fatigue limit 'σ_{FL}' - an applied cyclic stress amplitude that could theoretically be applied indefinitely without fatigue failure. As Miner's Rule assumes $d_i = 0$ for such stress amplitudes, the σ_{FL} must also be statistically reduced to the desired PoS/CL level using the same method as for $N_{PoS/CL}$ (shown previously in Equation 2). The mean and sample standard deviation of the fatigue limit was calculated using the 'Probit' method as described by Schijve [2]. Secondly, S-N data typically contains 'run-outs' where the fatigue test is stopped at a pre-defined number of cycles, even if the specimen has not failed. As run-outs require complex statistical treatment, which is beyond the scope of this paper, these data points were omitted, and the sample Coefficient of Variation (CoV) was assumed to be constant for all stress levels that demonstrated run-outs. The CoV for S-N data is shown in Equation 3 [1].

$$CoV = \frac{S}{\overline{N_f}} = 0.0323 \quad (3)$$

Following the computation of the $N_{PoS/CL}$ values (using Equation 2) for each required PoS/CL level, the S-N curve was fitted to the S-N data using the ESDU curve-fitting program, which is based upon maximum likelihood estimates [16]. The ESDU program identified the curve shape shown in Equation 4, where 'P_2' is equivalent to 'σ_{FL}'. To account for the statistically reduced 'σ_{FL}' values, the appropriate σ_{FL} PoS/CL value was used for P_2.

$$\sigma_a = P_1 N_{PoS/CL}{}^q + P_2 \quad (4)$$

In instances where $N_{PoS/CL}$ values fell below the resulting curve, the curve-fit was adjusted to enclose all values of $N_{PoS/CL}$. The resulting PoS/CL curves are shown in Table 2. It is appreciated that a more statistically rigorous approach for producing the PoS/CL curves would be desirable, and this will be considered in Section 4. This section has shown that challenges exist when working with highly populated S-N datasets that contain run-outs.

Table 2. Statistically reduced 4340 Steel S-N curves.

PoS/CL	P1	q	P2 (σ_{FL}) MPa
50/50 Mean	15020	-0.4245	457.0
99/95	4523	-0.3281	421.6
97.7/95	5073	-0.3377	426.5
97.7/75	6206	-0.3516	429.3
95/75	6993	-0.3619	434.1

Figure 3 shows the PoS/CL curves and the S-N dataset for 4340 steel [15]. The S-N curves produced when applying the safety factors used in ASME III for nuclear components and Eurocode 3 for steel structures are also shown. Figure 3 shows that the S-N curves that

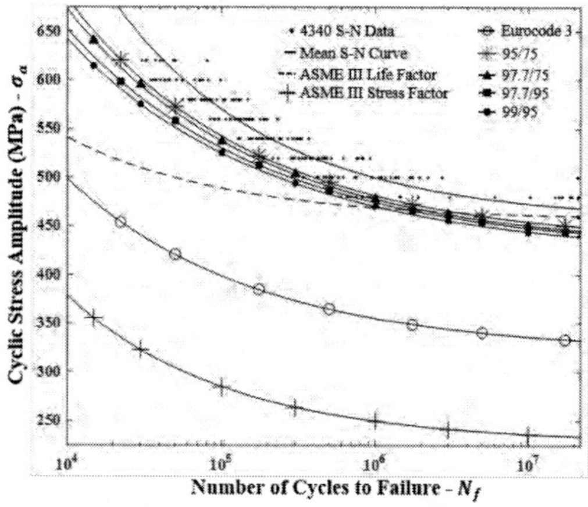

Figure 3. S-N curves from applying each conservatism approach to ESDU 4340 Steel S-N data

are based on safety factors produce significantly more conservative S-N curves than the statistically reduced PoS/CL curves. This could suggest that increased conservatism is introduced when using deterministic safety factors on S-N curves compared to statistically reduced curves.

Finally, it can be seen from the wide variety of PoS/CL curves (and their different impacts on the S-N curve shown in Figure 3) and safety factor values (shown by the numbers on Figure 2), that the magnitude of the conservatism applied is also inconsistent across the industrial sectors. As each industrial sector uses different methods of applying conservatism, coupled with the fact that some sectors use multiple methods of applying conservatism to S-N fatigue analysis (e.g. using a PoS/CL S-N curve and a life-limit safety factor), it is not possible to directly compare the total magnitude of the conservatism introduced by each sector [3]. The reduction in a component life-limit (i.e. 'impact') could however, be used as a measure of the total magnitude of selected a conservatism approach. Therefore, a case-study can be used to quantify the impact on a component life-limit when using each of the industrial conservatism approaches. This is presented within the remainder of the paper.

This section has presented a comprehensive review of the conservatism approaches adopted across different industrial sectors. As has been discussed and shown in Figure 2, the approaches to conservatism are inconsistent across the different sectors. This inconsistency could lead to either under- or over-conservatism, as industrial sectors may not be appropriately accounting for the areas of variability and uncertainty present within the S-N fatigue analysis and design of life-limited components.

4 Case Study: Evaluating the Impact of Industrial Conservatism

In order to quantify the reduction of a component life-limit resulting from the different conservatism approaches, a case-study was developed. The case-study retained the component characteristics as defined within Sections 1.3.1 and 1.3.2: a single-load path, monolithic component manufactured from high-tensile steel and designed using a 'life-limit' and S-N approach. This section of the paper defines the case-study and presents the resulting life-limits of the component when applying the different conservatism approaches.

4.1 The SAE Keyhole Benchmark

A case-study familiar to the fatigue design community was selected in the form of the SAE keyhole benchmark shown in Figure 4 [17, 18]. Whilst the geometry of the benchmark has been unaltered, due to the richness of the ESDU S-N data [15], it was assumed that the case-study component was manufactured from 4340 steel. Various potential loading spectra are supplied with the SAE Keyhole benchmark and the 'transmission' spectrum (also shown in Figure 4) was selected due to the presence of tensile mean-stress, which would increase the number of damaging stress cycles extracted from the

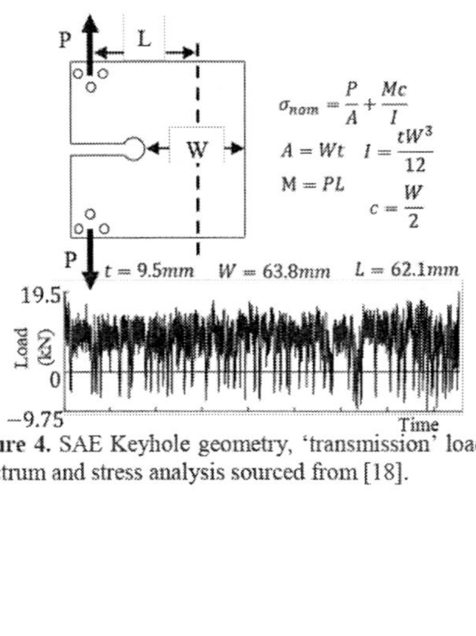

Figure 4. SAE Keyhole geometry, 'transmission' loading spectrum and stress analysis sourced from [18].

spectrum.

4.1.1 Computing the Component Life-Limit

The life-limit of the case-study component was computed using the S-N analysis process described in Section 1.3.2 for each conservatism approach shown in Figure 2. The conservatism was applied at the appropriate analysis stage, as shown in Figure 1. Where PoS/CL S-N curves were required, the appropriate curve equation from Table 2 replaced the mean S-N curve during analysis.

Following rainflow counting of the transmission loading spectrum, the cyclic loads were converted to cyclic stresses using the equations in Figure 4 [18]. These equations compute the nominal stress 'σ_{nom}' in the keyhole section. An elastic stress concentration factor of $K_T = 3$ was assumed to compute the stresses present at the keyhole as suggested in [18].

4.2 Impact of Conservatism Approach on Component Life-Limit

The case study component life-limits (e.g. the number of times that the 'transmission' load spectrum can be applied) resulting from each conservatism approach is shown in Table 3. The total accumulated damage from Miner's Rule 'D_T' and the percentage reduction in the life-limit from the 'baseline' case (i.e. the mean S-N curve and no consevatism applied) is also shown.

Table 3. Life-limit reduction for the case-study component.

Sector	D_T	Life-Limit	% Reduction in Life-Limit
Baseline	$3.02x10^{-5}$	33157	-
Large Aircraft	$1.02x10^{-4}$	3273	90.13 %
Light Aircraft	$2.41\ x10^{-4}$	4144	87.50 %
Nuclear (Stress)	$1.06x10^{-2}$	94	99.72 %
Wind Turbines	$2.47x10^{-3}$	404	98.78 %
Offshore	$7.00x10^{-4}$	1428	95.69 %
Steel Structures	$1.63x10^{-3}$	615	98.15 %

4.2.1 Discussion

As can be seen from Table 3, the baseline component life-limit of 33157 cycles is significantly reduced following the application of the conservatism approaches, with a maximum life-limit of 4144 cycles for light aircraft structures, whilst the minimum life-limit was 94 cycles for nuclear components under the ASME III conservatism approach. Therefore, the conservatism approach that has the largest impact on the component life-limit is that used by the nuclear sector, reducing the life-limit by 99.7%. It is interesting to note that the conservatism impact for both large aircraft and light aircraft structures are similar in magnitude, producing a reduction in the component life-limit of 90.1% and 87.5% respectively. These results from the aerospace sectors contrast with the results from the remaining 'land' based sectors, which all show consistency in the percentage reduction in the life-limit. These range from 95.7% for offshore steel structures, 98.1% for Eurocode

3 steel structures, 98.8% for wind turbine components and 99.7% for nuclear components. This trend suggests that the conservatism approaches of the aerospace sectors introduce a smaller magnitude of conservatism (and hence impact on life-limit) compared to all other sectors. This may result from the requirement to minimise weight and the stringent characterisation of materials and loading data within aerospace design [19]. In addition, the aerospace sectors are the only sectors that require a full-scale test of the complete structure, whilst due to size and economic considerations in the other sectors, only full-scale testing of components (e.g. welded joints in the offshore sector) is performed [20]. As the full-scale test provides a "safety-net" for the fatigue analysis, this could also permit the aerospace sectors to introduce reduced conservatism. Previous work by Sutherland [21] also highlights that the different industrial sectors have different fatigue design cases. For example, wind turbine components have long target design lives (30+ years) and are exposed to many fatigue cycles due to the rotational nature of components (e.g. blade hubs) [21]. Offshore structures, with a 20-year target design life, are exposed to significant cyclic loading from wave and wind loads [11]. Nuclear components can have long design lives (40 years) and are also safety-critical [9]. The most severe fatigue design case would be a component that is highly cycled over a long target design life. On the other hand, aerospace structural components, whilst having typical design lives of 20 years [19] (which is lower than wind turbines, nuclear and steel structures) will also be exposed to lower level of cycling, due to the reduced rotational nature of components. These comparisons are shown in Figure 5, after [21]. Therefore, a trend exists between the more severe fatigue design cases (i.e. long design lives with highly cycled components) and the conservatism introduced by each sector. Super-imposed

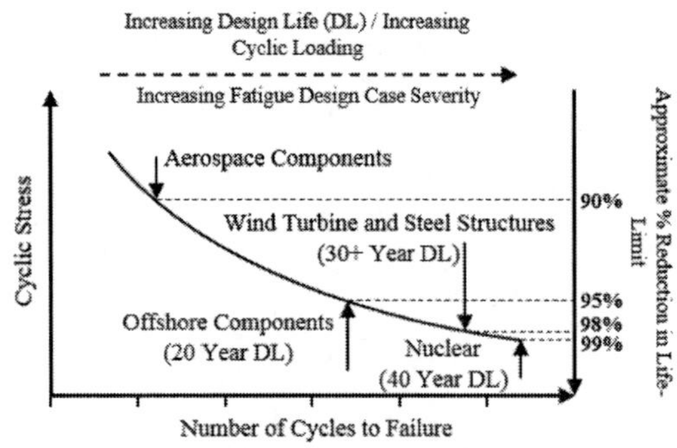

Figure 5. A comparison of the severity of the fatigue design case and percentage reduction of component life-limit for each industrial sector, modified from [21].

on Figure 5 are the percentage reductions in life-limit, showing that as the target design life or the cyclic nature of loading increases, the percentage reduction in the component life-limit also increases. It can therefore be suggested that sectors which anticipate more severe fatigue design cases introduce increased conservatism into S-N analysis. A final trend that the authors wish to explore in future work is the trend between the conservatism impact and the production volumes of each sector, which vary from potentially thousands in the aerospace sector to one-off components found in steel structures and the nuclear sector.

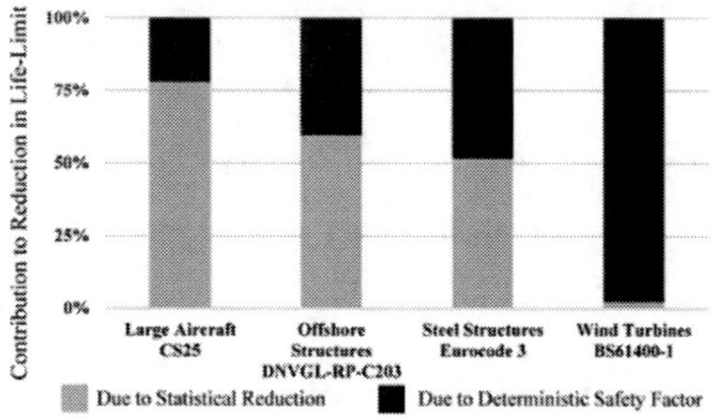

Figure 6. Breakdown of life-limit reduction into contribution from statistical reductions and safety factors.

4.3 Deterministic Safety Factors vs. Statistical Reduction Factors

As discussed in Section 3, a number of the industrial sectors reviewed use a mixture of both safety factors and statistical reduction factors within their conservatism approaches. Figure 6 shows the breakdown of the percentage reduction in the life-limit between the safety factors and statistical reduction factors.

Figure 6 shows that for large aircraft, offshore structures and steel structures, the majority of the life-limit reduction originates from the use of PoS/CL curves. Wind turbines on the other hand receive 97.7% of life-limit reduction from deterministic safety factors. This shows the significant impact that a safety factor applied to the cyclic stress magnitude can have. This is as a result of increasing the number of cyclic stress amplitudes that will lie above the fatigue limit. A similar effect is introduced by reducing the value of the fatigue limit, either through a deterministic stress safety factor on the S-N curve or through using PoS/CL S-N curves. This effect is shown in Table 4, whereby the different conservatism approaches are compared to the number of damaging cycles above σ_{FL} extracted by rainflow counting. Table 4 also shows the sensitivity of the reduction in component life-limit to the reduction in the

Table 4. Demonstration of how factors on stress can dramatically increase the percentage reduction in life-limit.

Sector	Factor on Stress	Number of Damaging Cycles Extracted	% Reduction in Life-Limit
Baseline	-	10	-
Large Aircraft	Statistically Reduced Fatigue Limit	21	90.13%
Nuclear	Stress Factor on S-N Curve	112	99.72%
Wind Turbines	Factor on Applied Stress, Statistically Reduced Fatigue Limit	66	98.78%
Offshore	Statistically Reduced Fatigue Limit	19	95.69%
Steel Structures	Statistically Reduced Fatigue Limit, Stress Factor on S-N Curve	57	98.15%

fatigue limit of the material.

Regarding safety factors and statistical reductions in general, statistical reductions are a more 'flexible' approach to applying conservatism than safety factors. This is because they are based upon the datasets of the specific design case, rather than the 'general' and deterministic nature of safety factors. It is interesting to note from Table 3 and Figure 2 that the two conservatism approaches that produced the minimum and maximum life-limit reductions (light aircraft and nuclear respectively) are both sectors that use only deterministic safety factors. This suggests that the introduction of conservatism through statistical reductions will be more representative of the specific design case at hand, potentially reducing the risk of producing an under- or over-conservative design. Therefore, the increased use of statistical reductions within fatigue design is suggested, to mitigate against un-safe or inefficient designs.

4.3.1 Identifying the Most Influential Area of Variability and Uncertainty.

From the split between the safety factors and statistical reductions shown in Figure 6, it can be seen that in most instances that when statistical reductions are applied, they account for at least 50% of the reduction in the component life-limit. Referring back to the matrix in Figure 2, it can be seen that statistical reductions are only applied to account for scatter within the S-N data. It can also be seen that when PoS/CL curves are used in conjunction with safety factors, that the safety factors are applied to account for various areas of variability and uncertainty. This shows that for most sectors, most of the conservatism applied accounts for scatter in S-N data. This suggests that the various industrial sectors consider scatter in S-N data to be the most influential area of variability and uncertainty within fatigue design. As Figure 6 suggests that S-N data scatter provides the largest contribution to the conservatism magnitude for each sector, increased statistical characterisation of S-N data could support a potential reduction in the conservatism currently required in the fatigue design of life-limited components.

5 Conclusions and Future Work

This paper has presented a comprehensive review and comparison of the conservatism approaches used during S-N fatigue design of life-limited components across a number of industrial sectors. A case-study, based upon the SAE keyhole benchmark has also been used to identify the impact on the component life-limit of each conservatism approach. The conclusions of this paper are:

- Conservatism approaches are inconsistent regarding the areas of variability and uncertainty accounted for, method of conservatism application and magnitude of conservatism.
- The conservatism approaches used by the large aircraft and light aircraft structures sectors result in a similar reduction in a component's life-limit.
- A trend exists between the reduction in a component's life-limit due to a sector's conservatism approach and the severity of the fatigue design case for that sector.
- The conservatism approaches identify scatter in S-N data as the most influential area of variability.
- Increased use of statistical reduction factors and increased characterisation of the variability in S-N data could support a reduction in conservatism currently required in S-N fatigue analysis.

It should be noted that the findings of this paper are case-study specific and therefore, future work should apply further case-studies to observe whether the same conclusions can be drawn. In addition, future work should account for the variations in S-N fatigue analysis methods used within each industrial sector and should also use a more statistically rigorous approach to handling the variability in the fatigue limit and 'run-outs' in S-N data ([2] provides information on such an approach).

This paper represents work performed as part of the Aerospace Technology Institute (ATI) funded "Large Landing Gear of the Future" Project in collaboration with Safran Landing Systems.

References

1. S. Suresh, *Fatigue of Materials*, 2nd Edition, Cambridge University Press, 1998.
2. J. Schijve, *Fatigue of Structures and Materials*, Springer, 2009.
3. N. Micone, W. De Waele, *Proc. of the ASME 2015 34th Int. Conference on Ocean, Offshore and Artic Engineering*, Newfoundland, Canada, 2015.
4. Health & Safety Executive, *Comparison of fatigue provisions in codes and standards*, 2001.
5. J. Shi, et al, *Proc. of the ASME 2016 Pressure Vessels and Piping Conference*, Vancouver, Canada, 2016.
6. D. F. O., Braga, et al, Prog. Aerosp. Sci., **69**, 29-39, 2014.
7. European Aviation Safety Agency, *Certification Specifications and Acceptable Means of Compliance for Large Aeroplanes CS-25 Amendment 18*, 2016.
8. Federal Aviation Administration, *Fatigue, Fail-Safe and Damage Tolerance Evaluation of Metallic Structure for Normal, Utility, Acrobatic, and Commuter Category Airplanes AC23-13A*, 2005.
9. Sandia National Laboratories, *Evaluation of Conservatisms and Environmental Effects in ASME Code, Section III, Class 1 Fatigue Analysis*, 1994.
10. British Standards Institute, *BS EN 61400-1:2005+A1:2010, Wind Turbines, Part 1: Design Requirements*, 2016.
11. DNV GL, *RP-C203: Fatigue design of offshore steel structures*, 2014.
12. DNV GL, *OS-C101: Design of offshore steel structures, general LRFD method*, 2015.

13. British Standards Institute, *Eurocode 3: Design of steel structures – Part 1-9: Fatigue*, 2010.
14. Sandia Corporation, *Factors for One-Sided Tolerance Limits and for Variables Sampling Plans*, 1963.
15. Engineering Sciences Data Unit, *Endurance of high-strength steels*, Data Item 04019a, 2006.
16. Engineering Sciences Data Unit, *Derivation of endurance curves from fatigue test data, including run-outs,* Data Item 97024, 2003.
17. Society of Automotive Engineers, *Fatigue Under Complex Loading: Analyses and Experiments,* 1977.
18. D. Socie, *eFatigue*, www.efatigue.com [Accessed 09/02/18].
19. M. C. Y., Niu, *Airframe Stress Analysis and Sizing*, 2nd Edition, Hong Kong Conmilt Press Ltd., 2001.
20. K. J. Marsh, *Full-scale fatigue testing of components and structures,* Butterworth & Co. 1988.
21. H. J. Sutherland, Wind Energ., **3**, 1-34, 2000.

Mean stress effect in stress-life fatigue prediction re-evaluated

Jan Papuga[1,*], *Ivona* Vízková[1], *Maxim* Lutovinov[1], and *Martin* Nesládek[1]

[1] Department of Mechanics, Biomechanics and Mechatronics, Faculty of Mechanical Engineering, Czech Technical University in Prague, Technická 4, 166 07 Prague 6, Czech Republic

Abstract. The paper compares various methods for computing the equivalent stress amplitude for stress cycles of non-zero mean value in stress-life fatigue prediction. A set of 11 calculation methods is evaluated. In addition to formulations based on common static or fatigue properties, the Walker formula and the generalized Linear formula are included in the investigation. These two methods use an optimization routine to find the material parameters. The final response of the methods is compared and discussed. The Walker method provides a better solution. The generalized Linear method produces inferior results, i.e. the linear fit of the segment of the Haigh diagram is not an optimal solution.

1 Introduction

This paper focuses on the domain in fatigue life estimation that deals with evaluations of the mean stress effect (MSE) in the stress-life approach. The problem is solved without hesitation in industrial practice, because the traditional approaches (the Goodman formula and the Gerber formula) are assumed to be valid universally and in the case of the Goodman method safely. A considerable number of relatively recent works in fatigue journals have dealt with this topic, e.g. Chu [1], Wang et al [2], Kwofie [3], Dowling et al. [4], Sekercioglu et al [5] and Niesłony et al [6], if we consider only stress-based methods.

These papers reveal a clear mismatch between general practice and the results presented by researchers. The Goodman solution is found to be unreliable and conservative. The Smith-Watson-Topper (SWT) method is shown to work better, and its generalization into the Walker formula provides even better results. New, superior MSE models are often proposed, and are proved to be valid. However, the test sets used for the validation are small, consisting of less than 10 materials (Kwofie [3], Nieslony et al [6], Sekercioglu et al [5], Wang et al [2]).

Recent work presented by Dowling et al [4] seems to be well substantiated by the number of data sets that are evaluated. However, Dowling at al. do not cover the newer MSE models. They claim that the Walker method offers the best solution. A closer look reveals, however, that the analysis of the prediction quality is mostly based on strain-controlled experimental data, or on fatigue curves established after the specimens were initially pre-strained by a larger cycle.

Our paper re-analyses some of Dowling's conclusions on the basis of our own validation data set of experiments, which should better conform to the requirements of a stress-based analysis. Multiple methods are evaluated, and conclusions are drawn. In [4], the Walker method provided the most promising features, but this was.thanks to the way in which it was optimized. Here, we optimize the generalized Linear method in a similar way, and we compare it with the Walker method. The results of this comparison allow some interesting conclusions to be drawn.

2 Equivalent completely reversed stress amplitude

There are two concepts for including the mean stress in a stress-based analysis. The first approach is to reduce the stress amplitude for a given mean stress. The second approach refers to the equivalent completely reversed stress amplitude as the stress amplitude, which would cause the same damage as would be caused by the pair of stress amplitude and mean stress. This second approach was selected for our investigation.

A list of the methods chosen for the analysis can be found in Table 1. Their graphic interpretation for one particular material, Ck35, is shown in Fig. 1. Many of the methods can be ranked as simple linear methods (Goodman, Half-slope, Morrow, Soderberg), while others are exponential (ASME, Bagci, Gerber). The SWT method is of another type. The Walker method is a generalization of the SWT method. The Walker method in Table 1 is an important exception in comparison with most other methods – it includes the additional □ Walker exponent. If the value is 0.5, the method degenerates to the SWT parameter.

In the analyses in [4], the Walker method was optimized in the multiple linear regression over all experimental data to obtain the optimum value of the Walker exponent. It is no wonder that the final application of this optimized method led to better results than those provided by the other methods presented here,

* Corresponding author: jan.papuga@fs.cvut.cz, papuga@pragtic.com

where the material parameters remained fixed the way they were set by their authors.

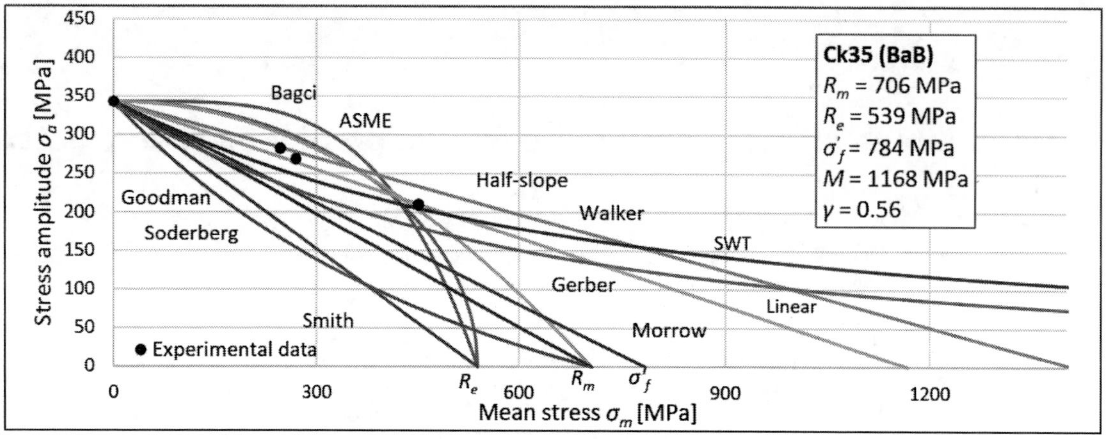

Fig. 1. Overview of various mean stress effect methods treated in this paper for the case of the BaB data set (100,000 cycles).

The complete family of linear methods differs only in one parameter, which is the stress value at which the method intersects the horizontal (mean stress) axis. These methods are among the most widely used solutions in engineering practice. The Linear method Eq. (6) was therefore prepared as a generalization of all these methods. The optimum value of its M stress parameter was obtained by non-linear regression on an available experimental data in individual data sets.

Because the Walker method and the Linear method now include one optimized parameter, it is obvious that they should outperform the prediction quality of the other methods evaluated here. A more important question is, Which of these two methods will result in lower prediction errors.

3 Composition of the data set

As was noted in the Introduction, Dowling et al. [4] presented an extensive validation data set of experiments in their work, but many of their experimental data refer to strain-controlled experiments, and the results of some of the experiments are likely to be affected by initial overload cycles. The data set presented in Table 2 is therefore built as a new validation data set, and the following rules were applied to each partial data set:

- Only the axial loading was analysed, and a distinction was made between plane bending cases and push-pull load cases.
- Only S-N curves with more than 4 data points were admitted into the analyses.
- Each data set had to include the fully reversed axial load case.
- All S-N curves of each data set had to be prepared for a material manufactured by the same processes from the same semi-product.
- The same dimensions of the specimens (at least in the critical cross-section) were valid for all tests within each data set.
- The same environment and testing conditions had to be kept for all curves in the same data set.
- Stress-controlled tests accepted with a minimum applicable lifetime of 10 000 cycles.
- Only experiments on smooth unnotched specimens were admitted.
- Only fatigue strengths within the interpolation regions of all regression curves evaluated by the MSE method were used.

Table 1. List of methods for setting the equivalent stress amplitude.

ASME	$\sigma_{a,eq}=\dfrac{\sigma_a}{\left[1-\left(\frac{\sigma_m}{R_e}\right)^2\right]^{1/2}}$	(1)
Bagci	$\sigma_{a,eq}=\dfrac{\sigma_a}{1-\left(\frac{\sigma_m}{R_e}\right)^4}$	(2)
Gerber	$\sigma_{a,eq}=\dfrac{\sigma_a}{1-\left(\frac{\sigma_m}{R_m}\right)^2}$	(3)
Goodman	$\sigma_{a,eq}=\dfrac{\sigma_a}{1-\frac{\sigma_m}{R_m}}$	(4)
Half-slope	$\sigma_{a,eq}=\dfrac{\sigma_a}{1-\frac{\sigma_m}{2\cdot R_m}}$	(5)
Linear	$\sigma_{a,eq}=\dfrac{\sigma_a}{1-\frac{\sigma_m}{M}}$	(6)
Morrow	$\sigma_{a,eq}=\dfrac{\sigma_a}{1-\frac{\sigma_m}{\sigma'_f}}$	(7)
Soderberg	$\sigma_{a,eq}=\dfrac{\sigma_a}{1-\frac{\sigma_m}{R_e}}$	(8)
Smith	$\sigma_{a,eq}=\dfrac{\sigma_a\cdot\left(1+\frac{\sigma_m}{R_m}\right)}{1-\frac{\sigma_m}{R_m}}$	(9)
SWT	$\sigma_{a,eq}=\sqrt{(\sigma_a+\sigma_m)\cdot\sigma_a}$	(10)
Walker	$\sigma_{a,eq}=(\sigma_a+\sigma_m)^{1-\gamma}\cdot\sigma_a^{\gamma}$	(11)

Table 2. Composition of the data set. Explanation of notes: *1Finish turned, *2Ground, *3Normalized, *4Hardened, *5Longitudinal direction, *6Transverse direction, *7Not stated in the reference

Data set mark	Material	Material type	R_m [MPa]	R_e [MPa]	σ'_f [MPa]	Axial load mode	Applicable curves with MSE	N. of datapoints		Source type	Ref.
								All curves	$R = -1$		
BaB	Ck35	Steel	706	539	784	Ten	3	28	8	graph	[7]
Bai	34CrMo4	Steel	902	706	1022	Ten	3	35	12	graph	[7]
FAD30	2124-T851*1	Al alloy	477	440	1828	Ten	2	39	13	raw	[8]
FAD40	2124-T851*2	Al alloy	477	440	1146	Ten	3	36	11	raw	[8]
Fin	76S-T61	Al alloy	508	470	901	PB	4	39	7	graph	[9]
GBA	24S-T3	Al alloy	503	372	1963	Ten	4	100	43	raw	[10], [11]
GBB	75S-T6	Al alloy	569	524	2286	Ten	4	53	10	raw	[10]
GBC	SAE 4130*3	Steel	807	679	1142	Ten	3	31	5	raw	[10]
GBD	SAE 4130*4	Steel	1241	1200	3917	Ten	1	18	13	raw	[11]
GrN	GGG-60	Cast iron	635	425	1109	Ten	1	28	19	graph	[12]
Han	SAE 4340	Steel	844	732	3611	PB	4	71	20	graph	[13]
KLN	2017A-T4	Al alloy	545	395	1215	PB	3	57	19	raw	[14], [15]
No40	JIS SPV 50	Steel	628	-*7	4519	Ten	1	14	7	raw	[16]
No53	SUS304-HP	Steel	618	-*7	481	Ten	2	30	10	raw	[17]
Ral	EN-GJV-450	Cast iron	498	368	1158	Ten	5	147	29	graph	[18]
SaL	Alcoa 14S-T	Al alloy	498	441	995	PB	3	25	7	raw	[19]
SHM	Ck45	Steel	690	384	995	Ten	1	12	6	raw	[20]
SiB*5	Ck45	Steel	850	812	927	Ten	1	34	17	graph	[21]
SiB*6	Ck45	Steel	856	807	1493	Ten	1	19	10	graph	[21]

Table 2 explicitly notes cases where the referred source failed to provide the necessary information in raw values (typically in a table), and the values had to be retrieved by a graph-reading procedure.

4 Methods of comparison

All the methods, with the exception of the Walker method and the Linear method, are analysed in the same way. Each fatigue curve is analysed by a linear regression, using the Basquin formula. The interpolation region of each curve is then set, and it is compared with the interpolation region of the fully reversed S-N curve. Their final intersection corresponds the applicable validation range to prevent extrapolations.

The material parameters necessary for running the methods, Eqs. (1-5, 8, 9), were available for individual materials in the referred sources. The deduction of material parameters γ and M is described below. The only remaining parameter is the σ'_f fatigue strength coefficient. This value was obtained from the Basquin regression analysis of the S-N curve in fully reversed axial loading.

The fatigue strengths corresponding to the 1, 2 and 5-multiple of each decade are established, starting at 10 000. These fatigue strengths relate to a combination of the stress amplitude and the mean stress for which the equivalent stress amplitude $\sigma_{a,eq}$ is set by the methods presented in Table 1. This equivalent stress amplitude is then compared with the fatigue strength of the fully reversed cycle $\sigma_{a,-1}$ for the same lifetime. On this basis, the fatigue index error ΔFI is computed as the relative deviation from the perfect estimate:

$$\Delta FI = \frac{\sigma_{a,eq} - \sigma_{a,-1}}{\sigma_{a,-1}} \quad (12)$$

The negative values mean that an excessively low equivalent stress cycle was set, leading to a non-conservative response. A perfect estimate results in $\Delta FI = 0$.

Fig. 1 shows at a glance that some of the methods must inevitably fail to result in any reasonable error. There are many cases where the mean stress value exceeds the yield stress. In these cases, none of the models using the yield stress as the material parameter could be evaluated, because the only intersection with the vertical axis would be in negative values for the Soderberg method, or it would be non-existent for the Bagci method. These cases are therefore not included in the final summary. In bending load cases, the mean stress even sometimes exceeded the tensile strength of the material. This situation can occur when there is an uneven stress distribution over the cross-section, and the material can switch to plastic behaviour. Here, too, models using tensile strength will obviously be useless.

In order to obtain some comparable results, the following limitations were set:

- Only load cases with mean stress lower than 75% of R_e were analysed for the methods presented by ASME, Bagci and Soderberg
- Only load cases with mean stress lower than 75% of R_m were analysed for the methods presented by Goodman, Gerber and Smith.
- Only load cases with mean stress lower than 75% of σ'_f were analysed for the Morrow method.

The output of these methods should be evaluated accordingly, because these limitations are obviously too strict for commonly observed material behaviour.

In addition to the overall statistics for the ΔFI parameter, five sub-groups were formed on the basis of material (steel, aluminium alloy, cast iron) and of the type of axial loading (plane bending, push-pull).

In the case of the Walker method and the Linear method, linear and non-linear regression analyses were run directly on experimental data for all applicable fatigue curves in the partial data set, i.e. without involving regression analysis in the process. The regression analysis included the fatigue curve in fully reversed loading.

Using the Basquin formulation for the S-N curve:

$$\sigma_{a,eq} = AN^b \tag{13}$$

the Walker formula Eq. (11) was transformed by logarithms into a multiple linear formulation, as Dowling et al. suggested in [4]:

$$\log N = \frac{1}{b}\left[\log\sigma_{max} + \gamma \cdot \log\left(\frac{1-R}{2}\right) - \log A\right] \tag{14}$$

This process was also implemented on the Linear method Eq. (6), and leads to the non-linear formulation:

$$\log N = \frac{1}{b}\left[\log\sigma_a + \log\frac{M}{A} - \log(M - \sigma_m)\right] \tag{15}$$

The main purpose of this process was to obtain parameters γ_3 and M_3 (the subscript refers to the method by which the values are obtained – here by the optimization routine). The minimized parameter was the sum of squares of the logarithms of the cycles.

5 Results, and a discussion

The type of analysis favours more sophisticated optimized methods. The final results depicted in Table 3 are therefore discussed separately for optimized methods and non-optimized methods.

5.1 Non-optimized methods

Unlike the two optimized methods, the intersection of the functions with the horizontal axis is fixed by these non-optimized methods (with the exception of the SWT method, where there is no intersection). This property enforced the removal of some data items from the evaluation, when the mean stress was too close to the intersection point (see Sec. 4). All these methods would otherwise result in extensive errors, which would disallow any reasonable analysis.

Apart from limitation by R_e, R_m or σ'_f and the exception of the SWT method, as mentioned above, only one other method is available: the Half-slope method. Here, the Half-slope method has the best score. Sec. 5.4 will show the reason for its best performance, mostly for steels loaded in push-pull, and its closest similarity with the generalized Linear method.

The exponential rules (ASME, Bagci, Gerber) use a similar intersection with the mean stress axis, but they lie higher in the area of modest mean stresses than most linear methods. Their general response is therefore non-conservative. However, the trend toward high mean stresses again cannot be correct, because the material response quite frequently exceeds R_e in the mean stresses.

The Smith method is not suitable for any evaluated sub-group, and it can be removed from any further attempts at an improvement. However, the SWT parameter, which does not contain any material parameter at all, acts almost miraculously in comparison with other non-optimized parameters, and above all its good performance for aluminium alloys should be highlighted.

5.2 A comparison between optimized methods

In addition to the statistical summary presented in Table 3, histograms of the ΔFI errors are depicted in Fig. 2 (left) for the Linear method and in Fig. 2 (right) for the Walker method. In both cases, the M_3 and γ_3 parameters were retrieved by the optimization routine on all available experimental data, as described above.

Table 3. Overall comparison of statistical evaluations of individual methods. Note that only the SWT, Walker, Linear and Half-slope methods also present results for high mean stresses, as described in Sec. 4. Al ~ aluminium alloys, St ~ steels, C ~ cast irons. The coloured backgrounds highlight the best results (ΔFI from -5 to 5% in the average, and less than 10% in the standard deviation).

	Average fatigue index error ΔFI [%]								Standard deviation ΔFI [%]							
	Plane Bending			Tension				Total	Plane Bending			Tension				Total
	Al	St	Total	Al	C	St	Total		Al	St	Total	Al	C	St	Total	
ASME	-13.4	23.7	-6.9	-23.4	-7.1	-4.8	-13.0	-11.3	8.6	15.3	17.3	13.2	33.0	7.6	18.8	18.6
Bagci	-17.1	16.0	-11.3	-26.7	-11.5	-10.8	-17.6	-15.9	7.6	15.5	15.7	14.1	32.8	8.0	18.6	18.1
Gerber	-8.0	35.3	-0.4	-19.1	-10.7	-2.8	-10.6	-8.1	12.6	22.1	22.0	12.1	32.0	8.0	17.6	19.2
Goodman	23.2	103.9	37.3	9.4	1.8	33.6	18.6	23.1	33.7	49.9	47.7	21.1	12.2	16.7	22.3	31.3
Half-slope	-5.8	42.0	5.2	-18.9	-22.2	1.3	-11.4	-7.1	9.6	16.6	23.3	12.4	25.3	9.0	18.3	20.9
Linear M_3	-1.4	1.7	-0.6	-4.3	1.8	3.1	0.0	-0.2	6.0	2.9	5.5	11.6	16.3	5.9	11.5	10.2
Morrow	-5.1	15.0	0.4	-26.0	-23.8	19.1	-7.4	-5.3	9.7	5.5	12.6	14.6	27.6	17.7	29.0	25.8
Soderberg	36.6	151.2	56.6	18.9	18.1	62.7	36.7	42.0	46.5	86.0	69.7	27.9	24.0	33.7	36.7	48.3
Smith	68.5	209.2	93.1	51.1	24.7	85.0	61.2	68.8	70.8	100.5	92.7	48.7	33.8	37.6	47.3	62.5
SWT	6.6	36.6	14.8	-1.5	-6.6	14.8	4.1	7.0	10.3	9.2	16.8	7.7	10.3	9.8	12.9	14.8
Walker γ_3	1.9	3.0	2.2	2.6	0.5	4.5	2.9	2.7	4.9	2.4	4.4	5.4	9.6	5.8	6.8	6.2
Walker γ_4	8.6	17.3	11.0	0.8	-	-1.5	-0.4	3.3	11.1	3.3	10.4	7.9	-	8.6	8.3	10.4

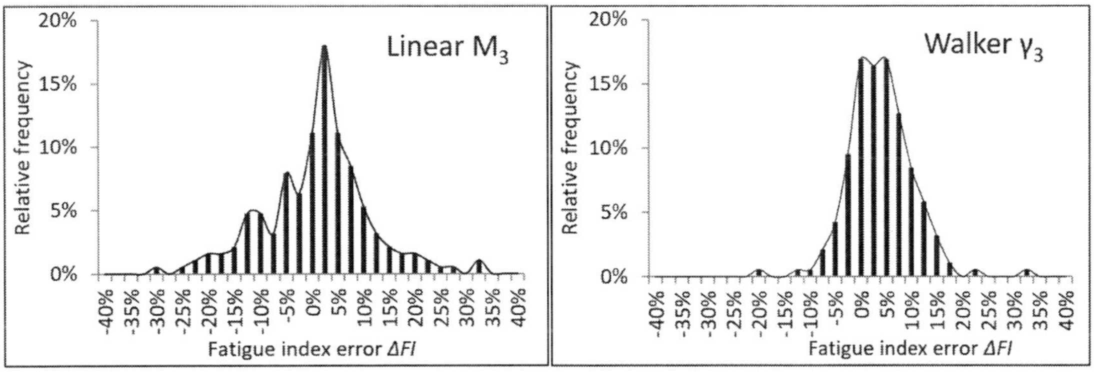

Fig. 2. (left) Histogram of the results of the Linear method, **(right)** Histogram of the results of the Walker method.

Though both methods contain one optimized parameter, the prediction quality of the Walker method exceeds the prediction quality that can be retrieved from the Linear method. The bottom of the histogram is not so broad when the Walker method is used. Only 71.4% of all data items lie within the range of ±10% for the Linear method, in comparison with 87.3% when the Walker solution is applied. These levels would change to 93.1% for the Linear method and 98.4% for the Walker method, if the allowed error range is raised to ±20%.

An additional histogram presented in Fig. 3 was prepared for a very special case designed as a "worst case scenario". While the optimum Walker parameter is obtained by best-fit analysis over a number of fatigue curves, most engineers are not so well equipped with data. A typical best-case situation in practice is if there is some curve including the mean stress effect in addition to the fully reversed case.

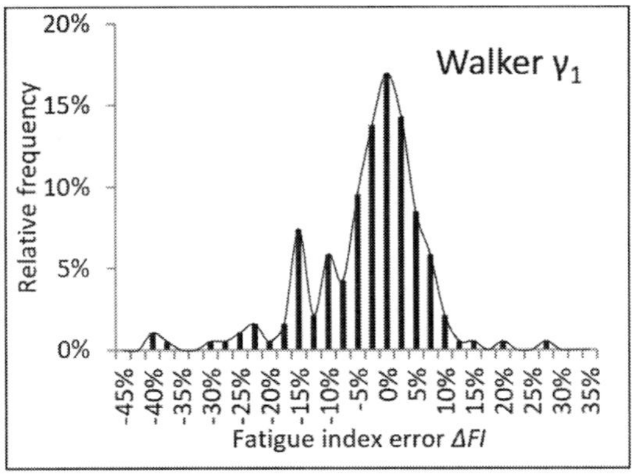

Fig. 3. Histogram of all results for the worst-case scenario for the Walker method.

The worst-case analysis is derived from a regression analysis of just two curves, one of which corresponds to fully reversed loading, while the other corresponds to any other curve with non-zero mean stress. From all such combinations within each partial data set, the combination leading to the maximum γ_1 value is selected (subscript 1 is added to distinguish it from other cases). A search is made for the maximum value, because the Walker formulation then leads to the smallest equivalent stress value and to the most non-conservative prediction.

Due to this intentional selection of the unsafest cases, the output is no longer either symmetrical or Gaussian, and the median value is shifted more to the non-conservative side. However, if the content of the ±10% and ±20% intervals is checked, the resulting values are 75.1% for ±10% and 93.7% for ±20%, i.e. the values obtained by the best-fit optimization procedures of the Linear method. We can conclude that the use of the Walker method with the exponent tuned to any additional experimental data, including the mean stress effect, leads to results that are at worst similar in quality to those produced by any of the linear methods.

5.3 Estimates of Walker parameter γ

Dowling et al. in [4] proposed estimates of γ exponents for steels and aluminium alloys. Constant values were set for aluminium alloys: 0.651 for low-strength alloys, and 0.473 for high-strength alloys. For steels, the function

$$\gamma = -0.0002 \cdot R_m + 0.8818 \qquad (12)$$

was set. These proposals are confronted in Figs. 4-5; all values are presented in Fig. 6.

Regression models would lead to slightly different values and formulas, but the differences are not excessive. Scarce evidence from a limited number of tested data sets suggests that it does not make sense to provide an improved formula. This is particularly evident for steels presented in Fig. 4, where only a single high-strength steel serves as a clue to the material behaviour.

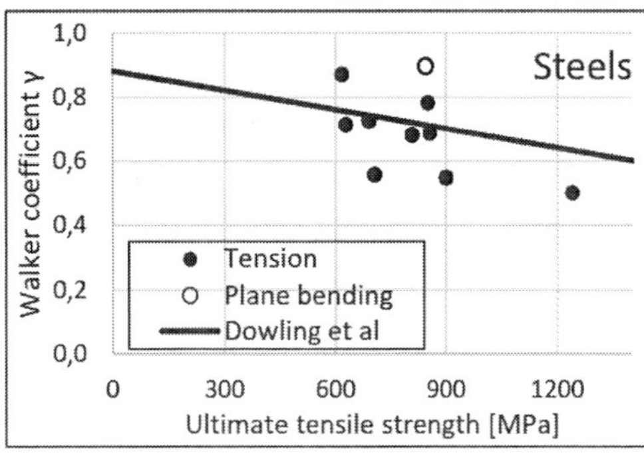

Fig. 4. A comparison between the proposal of Dowling et al. [4] for an estimate of the Walker exponent with the best-fit parameters from the experiments on steels.

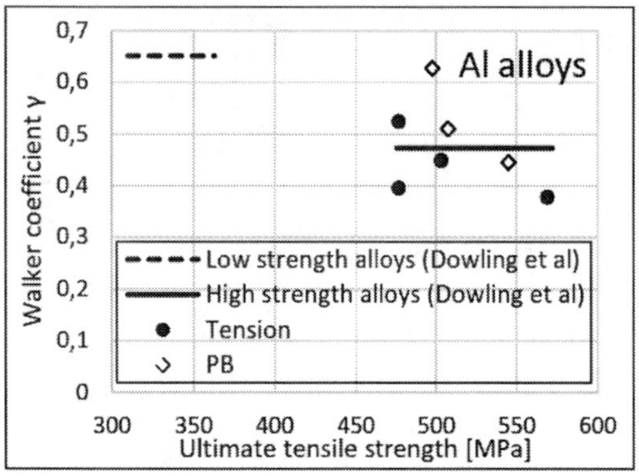

Fig. 5. A comparison of the proposals of Dowling et al. [4] for an estimate of the Walker exponent with the best-fit parameters from the experiments on aluminium alloys.

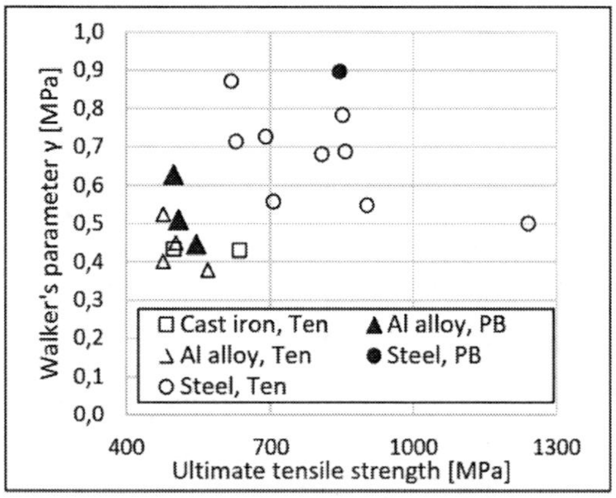

Fig. 6. Dependency of the γ exponent of the Walker method on the tensile strength. Ten ~ push-pull, PB ~ plane bending.

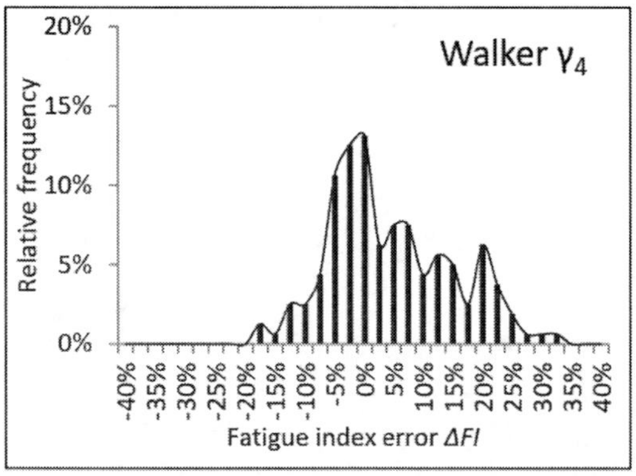

Fig. 7 Histogram of fatigue index errors for the Walker method and the g exponents retrieved by the proposal by Dowling et al. [4]. Note that only steels and aluminium alloys are evaluated.

The use of the coefficients proposed by Dowling et al. on the validation set resulted in the histogram presented in Fig. 7. Subscript 4 is added to the Walker exponent to distinguish it from other solutions aimed at deriving the exponent. The histogram is much lower than the histograms presented above, and it is more spread to conservative values of the prediction. If the content of the ±10% and ±20% intervals is checked, values of 66.3% for ±10% and 92.5% for ±20% are obtained. This makes the prediction more dispersed than when the generalized Linear method is applied.

The range of the γ exponent from 0.5 to 0.9 for steels does not seem to be significantly affected by the value of R_m – there is only a weak relation (see Fig. 6). Nevertheless, even the use of the proposed relation is better than the straightforward use of $\gamma = 0.5$, as in the SWT method (see Table 3 - compare the results for the optimized Walker method γ_3 with the proposal of Dowling et al γ_4 and SWT). Fig. 4 shows that the SWT method has to be generally safe for steels, because $\gamma = 0.5$ intrinsic to the SWT method forms the lower envelope of the optimized parameters, which results in the maximum mean stress effect.

For the high-strength aluminium alloys depicted in Fig. 5, the best-fit results are gathered close to a value of 0.5. This can lead to the conclusion that even the SWT method is a good fit for the general MSE behaviour for this material group.

5.4 *M* parameter of the Linear method

A similar comparison between the optimized *M* parameter and the tensile strength is presented in Fig. 9. The quadrant axis depicted in the figure shows that the expectation $M = R_m$ used by the Goodman method is generally too conservative. It also shows that the Half-slope formula crosses the cluster of points much better.

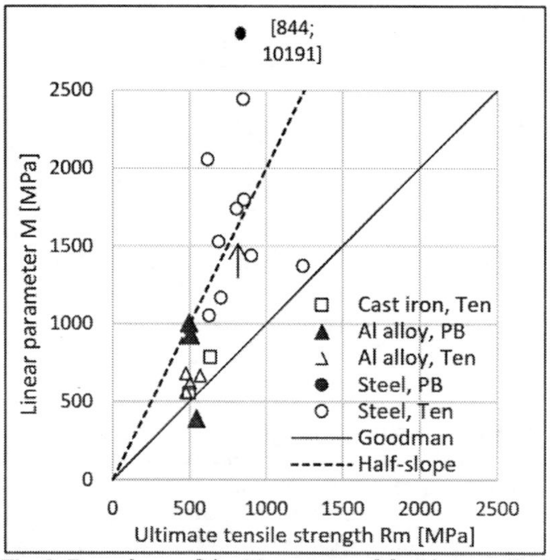

Fig. 9. Dependency of the *M* parameter of the Linear method on tensile strength. Ten ~ push-pull loading, PB ~ plane bending.

Fig. 10 was prepared in order to provide a better overview of how well each of the linear methods corresponds with the optimum Linear method. The ratios shown in the figure refer to various methods presented here:

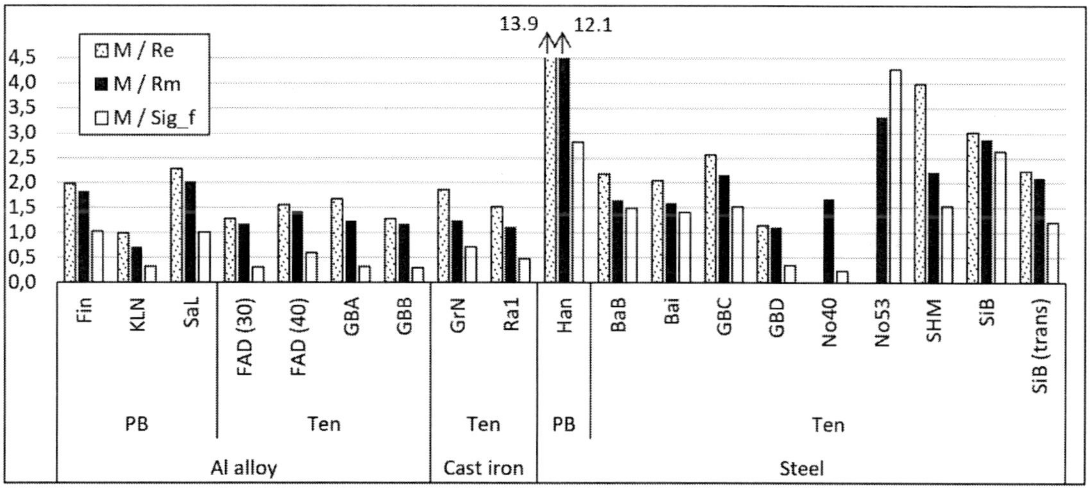

Fig. 10. A comparison between the optimized *M* parameter and other constants of linear methods. Ten ~ push-pull, PB ~ plane bending.

Table 4. The sum of squares of the relative errors from ideal values in the four items in the list above, as evaluated on the individual data sets depicted in Fig. 10. Al~aluminium alloys, St~steels, C~cast irons.

	Plane Bending			Push-pull				Total
	Al	St	Total	Al	C	St	Total	
Goodman	1.3	11.1	11.2	0.5	0.3	3.8	3.8	11.8
Half-slope	1.3	10.1	10.2	1.5	1.2	1.9	2.7	10.5
Morrow	0.7	1.8	1.9	1.3	0.6	3.9	4.2	4.6
Soderberg	1.6	12.9	13.0	0.9	1.0	4.4	4.6	13.8

- The Goodman method would fit with $M/R_m = 1$
- The Half-slope method corresponds to $M/R_m = 2$
- Morrow method ~ $M/\sigma'_f = 1$
- Soderberg method ~ $M/R_e = 1$

It should be noted that the Soderberg formula using the yield stress is not optimal, even for the brittle materials (cast irons here) for which it is usually proposed. The use of R_m is generally conservative, and the M/R_m ratio of 2 is obviously a closer estimate.

Table 4 was prepared to improve the comprehensibility of Fig. 10, and to quantify the differences between these four methods and a perfect prediction. For each data set, Table 4 presents the sum of squares of the absolute deviations from the equalities defined in the list above. The values presented in Table 4 are obviously extremely affected by the plane bending case of the Han data set. Because the Han set is what produces the enormous differences, Table 5 was prepared without this test set. Table 5 shows that, when the Han data set is removed, the Half-slope method presents the smallest deviation from the optimum value.

There is a simple answer to the question, Why is the Han data set well modelled only by the Morrow method? It is caused by the way in which the σ'_f parameter was defined. It was not obtained from the hysteresis loop in the push-pull test, but as the best fit to the fully reversed plane bending loading. The stress distribution effect, and the induced plastic redistribution, are similar for all S-N curves in this test set. However, parameters R_m and R_e, which were used in the analyses by other methods, were not impacted at all by the stress distribution effect. This case illustrates very clearly how dubious it is to include plane bending cases in these analyses, because the stress gradient effect is involved.

Table 5. The sum of squares of the relative errors from the perfect values as in Table 4, but without the Han data set.

	PB	Push-pull				Total
	Al	Al	C	St	Total	
Goodman	1.3	0.5	0.3	3.8	3.8	4.0
Half-slope	1.3	1.5	1.2	1.9	2.7	3.0
Morrow	0.7	1.3	0.6	3.9	4.2	4.2
Soderberg	1.6	0.9	1.0	4.4	4.6	4.9

Table 5 shows that the Goodman method is not far from the optimum values for aluminium alloys and cast irons. For steels, the Half-slope formula achieves the best results. Nevertheless, these conclusions should be taken only with care, because there is a substantially bigger number of items in the steel group than in the aluminium alloy and cast iron groups.

6 Conclusion

This paper has presented a comparison of various methods for computing the equivalent stress amplitude for a given mean stress value. In comparison with most methods of fixed common material parameters, the Walker method and the generalized Linear formula were treated separately, because values of their additional parameters were optimized.

The following major conclusions can be drawn:

1. The Half-slope formula provides the best results for simple non-optimized solutions based on static material parameters.
2. The SWT method is even better, and its performance for aluminium alloys is close to the optimum values provided by the optimized Walker method.
3. A comparison of the optimized formulas indicated that the Walker method leads to a better solution than the generalized Linear formula.
4. The dependency of the Walker exponent on the tensile strength proposed by Dowling et al. in [4] is reasonable, though the dependency is quite weak. Direct use of this parameter results in a scatter of the results similar to that provided by the optimized Linear method. These estimates provide better prediction results than widely-used SWT.
5. If there is any additional S-N curve with non-zero mean stress to accompany the S-N curve with fully reversed loading, the use of non-linear regression to obtain the optimum Walker's exponent is likely to result in a better prediction than the use of the fully optimized Linear method, or at least an equally good prediction.
6. All methods, with the exception of the optimized methods, SWT method an Half-slope method, had to be analysed only for cases where the mean stress was lower than 75% of their intersection with the mean stress axis (be it tensile strength, yield stress or fatigue strength coefficient). For higher mean stresses, these methods are meaningless.
7. The plane bending load case is affected by the non-uniform stress distribution. If it is necessary to use one of the simpler methods, on which the 75% limit noted above must be imposed, the plane bending load effect should be treated separately.

Acknowledgements: The authors thank the Ministry of Education, Youth and Sport and the Czech Technical University in Prague for its financial support for the research presented here within the framework of grant No. SGS17/175/OH K2/3T/12.

References

1. C.-C. Chu, In: *SAE 2000 World Congress*. SAE International, Detroit (2000)
2. S. Wang, M. Dixon, C.O. Huey, S. Chen, Journal of Mechanical Design. **122**, 143-146 (2000)
3. S. Kwofie, Int Jnl of Fatigue. **23**, 829-836 (2001)
4. N. Dowling, C. Calhoun, A. Arcari, Fat & Fract of Engng Maters & Structs. **32**, 163-179 (2009)
5. T. Sekercioglu, Materialwissenschaft und Werkstofftechnik. **40**, 713-717 (2009)
6. A. Niesłony, M. Böhm, Int Jnl of Fatigue. **52**, 49-56 (2013).
7. F. Baier, *Zeit- und Dauerfestigkeit bei überlagerter statischer und schwingender Zug-Druck- und Torsionbeanspruchung*. (Universität Stuttgart, Stuttgart 1970)
8. J. Papuga, F. Fojtík, M. Vargas, A. Hodr, A. Karolczuk, M. Fusek, R. Halama, *Summary of experiments on 2124-T851 realized within FADOFF project. [FAD/14/001]*. (CTU in Prague, Prague 2014)
9. W.N. Findley, *Combined-stress fatigue strength of 76S-T61 aluminum alloy with superimposed mean stresses and corrections for yielding. [NACA TN-2924]*. (NACA, Washington, 1953)
10. H.J. Grover, S.M. Bishop, L.R. Jackson, *Fatigue strengths of aircraft materials: Axial-load fatigue tests on unnotched sheet specimens of 24S-T3 and 75S-T6 aluminum alloys and of SAE 4130 steel. [NACA TN 2324]*. (NACA, Washington, 1951)
11. W. Illg, *Fatigue tests on notched and unnotched sheet specimens of 2024-T3 and 7075-T6 aluminum alloys and of SAE 4130 steel with special consideration of the life range from 2 to 10,000 cycles. [NACA-TN-3866]*. (NACA, Washington, 1956)
12. V. Grubisic, J. Neugebauer, Giessereiforschung **31**, 123-128 (1979)
13. B.C. Hanley, *The effect of range of stress and state of stress on the fatigue strength of SAE 4340 steel.* (University of Illinois Urbana, Illinois, 1951)
14. K. Kluger, T. Łagoda, Int Jnl of Fatigue **66**, 229-245 (2014).
15. A. Niesłony, T. Łagoda, K. Walat, M. Kurek, Materialwissenschaft und Werkstofftechnik **45**, 947-952 (2014)
16. *Data sheets on fatigue properties for butt welded joints of SPV50 steel plate for pressure vessels, Effect of stress ratio. [NRIM Fatigue Data Sheet No. 40]*. (National Research Institute for Metals, Tokyo, 1984)
17. *Data sheets on fatigue properties for butt welded joints of SUS304-HP (18Cr-8Ni) hot rolled stainless steel plate, Effect of stress ratio. [NRIM Fatigue Data Sheet No. 53]*. (National Research Institute for Metals, Tokyo, 1986)
18. T. Rausch, *Zum Schwingfestigkeitsverhalten von Gusseisenwerkstoffen unter einachsiger und mehrachsiger Beanspruchung am Beispiel von EN-GJV-450*. (Shaker Verlag, Aachen, 2011)
19. M. Shariati, H. Mehrabi, Fat & Fract of Engng Maters & Structs **38**, 489-502 (2015)
20. A. Simbürger, *Festigkeitsverhalten zäher Werkstoffe bei einer mehrachsigen phaseverschobenen Schwingbeanspruchung mit körperfesten und veränderlichen Hauptspannungsrichtungen*. (TH Darmstadt, Darmstadt 1975)
21. J.A. Sauer, D.C. Lemmon, Transactions of the American Society for Metals **42**, 559-576 (1950).

The effect of mean stress on the fatigue behaviour of overhead conductor function of the *H/w* parameter

Remy Badibanga[1,*], *Thiago* Miranda[1], *Pedro* Rocha[1] , *Jorge* Ferreira[1], *Cosme* da Silva[1], *José* Araújo[1,*].

[1]UnB, University of Brasília, Department of mechanical Engineering, CEP 70910-900

Abstract. The objective of this work is to evaluate the effects of mean stress on the fatigue behaviour of an All Aluminium Conductor (AAC Orchid), Aluminium Conductor Steel Reinforced (ACSR Tern), and an Aluminium Conductor Alloy Reinforced (ACAR 750 MCM). In this sense, 72 fatigue tests on overhead conductors were performed using different values of *H/w* parameter. Based on the experimental results, the parameters which describe the fatigue behaviour of the conductors were determined after generating theirs S-N curves. In the assessment of the mean stress effects on the fatigue life, Goodman and Gerber's relations were fitted to evaluate the use of such models for the conductors. It was observed that the evaluation of the mean stress effect on the overhead conductor could be made by using the fatigue relations.

1 Introduction

The overhead conductor is the only part carrying electricity, so its contribution has been estimated as carrying up to 40% of all power line transmission costs. However, this transmission line is susceptible to wind, snow, earthquakes and other weather conditions [1]. Aeolian vibration is the main cause of conductor fatigue failure, especially at devices which restrain its movements [2-4]. Frequently, the observed fatigue failures on overhead conductors occurs near or inside the suspension clamp devices, where there are many loads acting on the assembly conductor/clamp [5,6]. Among the loads, there are the stretching load of the conductor, the bending displacement due to the Aeolian vibration which generates the bending stress and the clamping torque.

The control of the stretching tension of conductors at the design stage has been a concern for many years [1]. One of the reasons is that the tension of conductors during the most severe climate period should not exceed the allowable tension of the conductor. Also, the stretching tension must be controlled in order to restrict the violation of the line clearance and also to protect the conductor against the harmful Aeolian vibration. It is well-known that an overhead conductor becomes more vulnerable to Aeolian vibration when its tensile tension increases. Therefore, the *Conseil International des Grands Électriques* (International Council on Large Electric Systems), abbreviated CIGRÉ, has found necessary to establish an upper limit for conductor tension that can prevail for a significant period of time. The Every Day Stress (EDS) panel was created by CIGRÉ to investigate the safe parameter design of power line conductors [7,8]. The EDS is the maximum tension load to which the conductor can be subjected, at the temperature which will occur for the longest period of time in one year, without any risk of damage from Aeolian vibrations. It is expressed in percentage of the ultimate tensile tension of the conductor, and CIGRÉ has fixed its values according to the type of the conductor. However, failure on overhead conductors due to Aeolian vibration has been observed in situ even when the transmission line has been designed with the recommended EDS value. For instance, a survey conducted by CIGRÉ revealed that 78% of lines failed by fatigue before 20 years after their launching [8]. Thus, such a kind of study makes it clear the need of another parameter to design the lines against fatigue

The *H/w* parameter has recently been suggested by CIGRÉ to better explain and describe the fatigue damage occurring on power line conductors due to the Aeolian vibration. The tenants of this parameter suggest that all overhead conductors stretched with the same value of *H/w* will have a similar fatigue life [9]. Although, no experimental laboratory on overhead conductors was performed to corroborate this idea. After a thorough literature investigation, some publications have been identified regarding fatigue of power lines related to the EDS parameter [1,5,6]; however, there are few publications related to the *H/w* parameter. This present work, then, has the objective of conducting an experimental campaign to access the effect of the mean stress on the fatigue behaviour of conductor tested under different values of *H/w* parameter. Seventy two (72) fatigue tests were carried out on three types of conductors, namely an All Aluminium Conductor (AAC Orchid), Aluminium Conductor Steel Reinforced (ACSR Tern), and an Aluminium Conductor Alloy Reinforced (ACAR 750 MCM). Additionally, the mean stress,

* Corresponding author: badibanga12@gmail.com jaaunb@gmail.com

https://doi.org/10.1051/matecconf/201816511001

which is function of *H/w*, and the bending stress (Poffenberger Swart stress) were fitted through different prediction models of the mean stress effect.

2 Bending stress and *H/w* parameter

2.1 Bending stress

The main cause of fatigue failure of strands in overhead conductors is Aeolian vibration, a fatigue which generally occurs at points where conductor motion is constrained against transverse vibration, such as at suspension clamps. During Aeolian vibration, the conductor undergoes cyclic movement which induces bending stress in power line conductor wires.

Most fatigue failures in strands of conductors occurs in the suspension clamp at the contact points of wires-to-wires and wires-to-suspension clamp [1,10]. Because of the difficulty on accurate measurement of mechanical strain or stress at these points, the adoption of some assumptions is vital. A mathematical model was developed by Poffenberger and Swart [11] to calculate the bending strain or stress in the outer layer of the conductor's aluminium wire. This mathematical model considers the cable, near the suspension clamp, as an Euler beam (Fig.1).

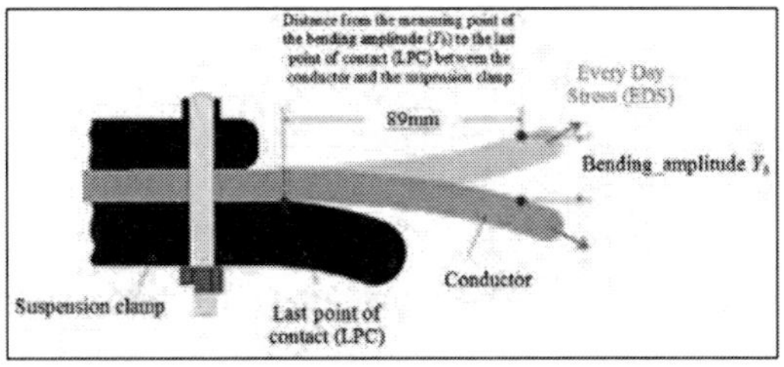

Fig. 1. Schematic montage of the conductor and the suspension clamp showing the standard position to measure the bending amplitude

Thus, the vertical amplitude, measured peak-to-peak at 89 mm (3.5 in) from the last point of contact (LPC) between cable and suspension clamp, can be converted to the bending strain or stress in the outer layer of the aluminium wire conductor using Eq. 1. The Poffenberger-Swart Equation has been developed as follows:

$$\sigma_a = KY_b \tag{1}$$

where σa is the dynamic bending stress amplitude (0-to-peak); Y_b is the conductor's vertical amplitude range (peak-to-peak) measured at 89 mm from LPC; and:

$$K = \frac{E_a d p^2}{4(e^{-px} - 1 + px)} \tag{2}$$

where E_a (MPa) is the aluminium Young's modulus; d (mm) is the diameter of wire in the outer layer; x is the distance on the conductor from the LPC (between the conductor and the suspension clamp) and the vertical displacement measuring point, (usually x = 89 mm), as one can see on Fig 1; and:

$$p = \sqrt{\frac{T}{EI}} \tag{3}$$

where T (N) is the static conductor tension at average ambient temperature during test period; and EI (N.mm2) is the flexural stiffness of the conductor, whose minimum value is as follows:

$$EI_{min} = n_a E_a \frac{\pi d_a^4}{64} + n_s E_s \frac{\pi d_s^4}{64} \tag{4}$$

where n_a, E_a, d_a are the number, individual diameter and Young's modulus of the aluminium wires; and n_s, E_s, d_s are the respective values for the steel wires. In this approach, the conductor is considered as a bundle of individual wires free to move relative to each other; the flexural stiffness takes its minimum value, EImin. For smaller bending amplitudes, the individual wires would stick together; thus the conductor would behave as a solid rod, increasing the flexural stiffness to its maximum. Formulae that consider the stick-slip theory to compute EI and hence the dynamic bending stress were proposed by Papailiou [12,13].

2.2 *H/w* parameter

The design of overheard power lines has been guided by the control of the conductor tension. Many reasons have justified this option: among them is the objective of ensuring that the maximum tension of the conductor corresponding to the assumed most severe climatic loading does not exceed a predefined load. Another reason is related to the maximum operation temperature of the conductor, which could allow it to work with respect to the conductor safe clearance. It is well-known that the conductor becomes increasingly vulnerable to Aeolian vibration when its static tension increases. Therefore, it is necessary to establish an upper limit for conductor tension that can prevail for a significant period of its lifetime. Consequently, the Every Day Stress (EDS) panel was created by CIGRÉ to investigate a limiting parameter capable to safely design line conductors against fatigue due to Aeolian vibration. Many parameters have been proposed for the safe design of overhead conductors, but two – Every Day Stress and the catenary parameter *H/w* – are the most prevalent in the literature.

The Every Day Stress (EDS) parameter was initially proposed by CIGRÉ in 1960 [9]. The EDS, expressed as a percentage of the conductor rated tensile stress (RTS), is defined as the maximum tensile load to which the conductor can be subjected, at the temperature which will occur for the longest period of the time without any risk of damage due to Aeolian vibrations. CIGRÉ recommends different values of EDS for overhead conductors and for conductors with dampers only, armor rods only, as well as for conductors with both dampers and armor rods. However, field observations have reported fatigue of power lines after the application of the recommended EDS values by CIGRÉ [1,7,9].

Consequently, the EDS parameter appears to be insufficient for explaining the recent damage found on power lines. Clearly, the requirement for a new parameter is evident.

Another parameter adopted by CIGRÉ, is defined as the ratio between the initial horizontal tensile load (*H*) and the conductor weight (*w*) per unit length, *H/w*; this parameter is also called the catenary parameter. The tensile load (*H*) is the initial horizontal tension before any significant wind and ice loading and before creep at the average temperature of the coldest month at the site of the power line [1,7]. Compared to the EDS, the *H/w* presents several advantages, as it affects a number of parameters involved in the fatigue characteristic of conductors. One of them is that the *H/w* parameter takes into account the diameter of the power line conductor, which influences the energy induced by the wind and the frequency of vortex formation [7,9].

3 Materials and methods

3.1 Materials

Three conductors named AAC Orchid, ACSR Tern and ACAR 750 MCM were chosen for this work, thereafter the fatigue behaviour of conductors was investigated. The AAC Orchid is a conductor composed only of aluminum wires Al 1350 H19, the ACSR Tern cable has steel core (7 wires) and two layers of Al 1350 H19 aluminum wires, and cable ACAR 750 has aluminum alloy core (Al 6201 T81) and the wires from external layers are made of Al 1350 H19 aluminum. The basic characteristics of these conductors are presented in Table 1.

Table 1. Mechanical and geometrical properties of the three conductors

		AAC Orchid	ACSR Tern	ACAR 750 MCM
Conductor diameter (mm)		23.3	27.03	25.32
Number of wires on each layer	*Aluminium*	18-12-6-1	21-15-9	18-12-6-1
	Steel	-	6-1	-
Diameter of wires (mm)	*Aluminium*	3.33	3.38	3.617
	Steel	-	2.25	-
Linear mass (kg/m)		0.889	1.339	1.046
Rated tensile strength (kgf)		5143	10010	8635

3.2 Strategy to evaluate the adherence of the mean stress models.

The strategy used to evaluate the adherence of the fatigue models consist of the use of three parameters that characterize a mechanical fatigue behavior: the mean stress, S_m, alternating stress, S_a, and the resulting life, *N*. As shown in Fig. 2, the application of the data that characterize the mean and alternating stress in one mean stress model allows to evaluate theirs effect on the fatigue behavior. The evaluation of the mean and the alternating stress is made by extrapolating the S-N equation when $S_m = 0$ and the equivalent fatigue strength, S_{ar} Model, to the specific fatigue model. In a similar way, the application of the fatigue life, *N*, in the Basquin's equation allows the estimation of the new value for the equivalent endurance limit, called S_{ar} Basquin. If the prediction model correlate well to the experimental results then the values of S_{ar} Model and S_{ar} Basquin should be identical statistically. The equations used to estimate the equivalent alternating stress, S_{ar} Model are shown in Eq. 5-6, which correspond respectively to Goodman and Gerber´s models.

$$S_{ar} = \frac{\sigma_a}{1-\left(\frac{\sigma_m}{S_{rt}}\right)} \quad (5)$$

$$S_{ar} = \frac{\sigma_a}{1-\left(\frac{\sigma_m}{S_{rt}}\right)^2} \quad (6)$$

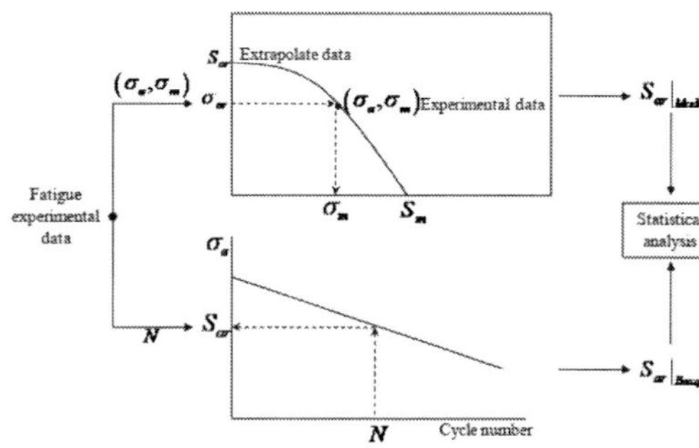

Fig. 2. Strategy used to evaluate the adherence of the mean stress models.

4 Experimental procedure

At the GFFM (Fatigue, Fracture and Materials Research Group) laboratory (University of Brasília), three similar resonance fatigue test benches of conductors were used. A brief description of the bench is provided below as it was previously described in other publications [2,5,14]. Each bench has a length of 46 m divided in two spans – the active and the passive span – which have, respectively, 40 and 6.8 m. The scheme of the three benches is shown in Fig. 3(a), while Fig. 3(b) represents the benches side view. The test starts by anchoring the conductor on the two fixed blocks (fixed blocks 1 and 3) and thereafter mounting the suspension clamp on the support which is rigidly fixed on the adjustable block 3. The conductor is stretched at the recommended value of *H/w* through the strain clamp by using the hand traction winch and weight. To simulate the vibration, an electrodynamic shaker is connected to the conductor via device which allows a good alignment between the shaker and the conductor axes. A wire break detector is attached to the conductor at the first node of the conductor from the suspension clamp toward the shaker.

All fatigue tests were performed according to IEEE (Institute of Electrical and Electronics Engineers) and CIGRÉ standards which establish the criterion to stop the conductor's fatigue test when the number of strands broken is 10% of the total number of conductor aluminium wires [15,16]. To measure the bending stress of the conductor, three strain gauges were glued on the most upper conductor's wires (being one strain gauge by wire) diametrically opposed to the LPC between the suspension clamp and the conductor. The bending displacement peak-to-peak (Y_b) measured at 89 mm from the LPC between the conductor and the suspension clamp was controlled, along with the vibration frequency, during all fatigue tests. For each conductor, the S-N graph was generated by performing nine fatigue tests (three bending stress and three tests by bending stress). The three levels of bending stress considered during the experimental program and the corresponding bending displacement peak-to-peak generated by applying the Poffenberger-Swart constant (K) are presented in Table 2 for the three values of H/w parameter.

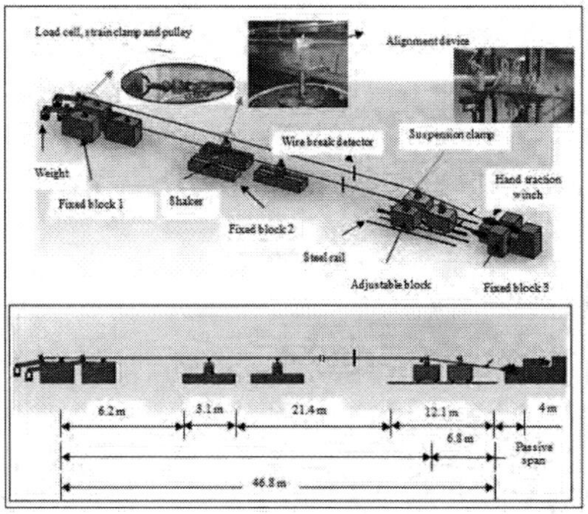

Fig. 3. Three fatigue test benches for overhead conductors at the University of Brasilia: (a) overall three dimensional view and (b) side view

Table 2. The bending amplitude (Y_b) at 89 mm from the LPC between the suspension clamp and the conductor and the Poffenberger-Swart constant (K) calculated using Eq. 1 and Eq. 2 for different conductors at a specific mean stress.

Conductor	H/w [m]	K [MPa/mm]	Poffenberger-Swart stress [MPa] 26.8	28.22	31.35
ACSR Tern	1820	32	0.84	0.88	0.98
	2144	33.66	0.80	0.84	0.93
	2725	36.38	0.74	0.78	0.86
AAC Orchid	1820	30.91	0.87	0.91	1.01
	2144	32.49	0.82	0.87	0.96
	2725	35.07	0.76	0.80	0.89
ACAR 750 MCM	1820	31.98	0.84	0.88	0.98
	2144	33.53	0.80	0.84	0.93
	2725	36.09	0.74	0.78	0.87

5. Results and discussions

5.1 Fatigue tests

A total of seventy-two tests were performed to generate nine S-N curves under three values of H/w parameter. Each H/w value is related to the mean stress of the conductor as shown experimentally by Kalombo et al [17]. For each conductor, 24 fatigue tests were performed, nine tests with H/w value of 1820 m; nine tests with H/w = 2144 m; and six tests with H/w = 2725 m. Figures 4-6 present the S-N curves obtained by means of the experimental procedure presented in the preview section. In these graphs, the results are grouped by H/w value and the dependent variable, the fatigue life N, Number of cycles, is plotted in abscissa on a logarithmic scale and the independent variable, Poffenberger-Swart stress (Expressed in MPa), is plotted on the ordinate, also on logarithmic scale. The fatigue test was stopped when the number of wires broken is equal to either three or 10% of the total number of aluminium wires, whichever is greater [16].The synthesis of the results obtained from the above experimental procedures is presented in Table 3. The S-N curves of the three conductors tested at the same value of H/w = 1820 m are shown in Fig. 4. Meanwhile, Figs. 5-6 present respectively the graphs correspond to the fatigue curves behaviour of the conductors tested with the H/w value equal to 2144 and 2725 m.

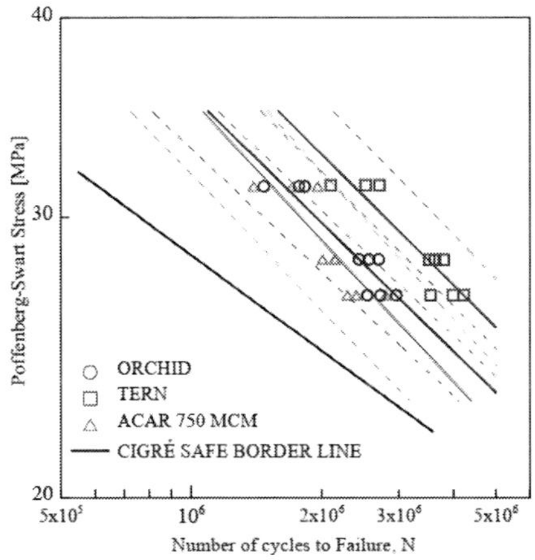

Fig. 4. S-N graphs of different conductors tested with the same value of *H/w* =1820 m.

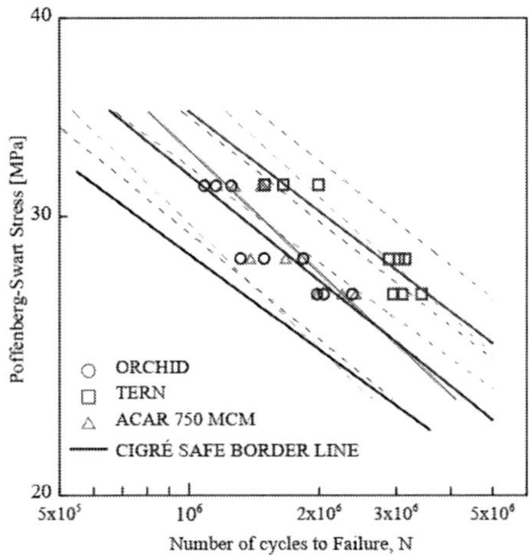

Fig. 5. S-N graphs of different conductors tested with the same value of *H/w* = 2144 m.

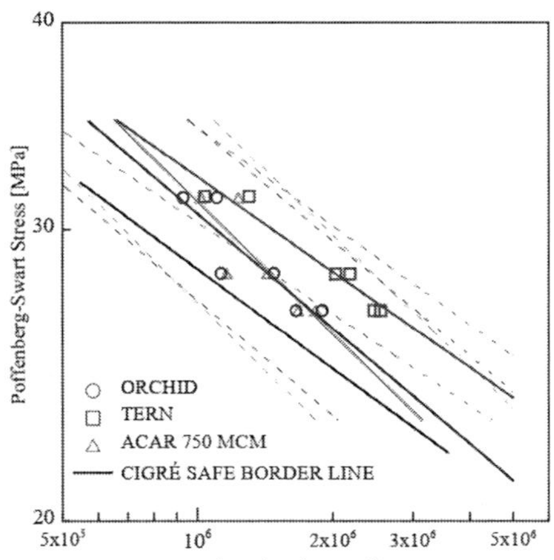

Fig. 6. S-N graphs of different conductors tested with the same value of *H/w* =2725 m.

In order to evaluate the accuracy level of the process of obtaining the S-N curves, whose parameters are listed in the Tables 3, the dispersion diagram was constructed (Fig. 7). This figure relates the experimental fatigue life observed for each conductor tested, *N*, with the estimated fatigue life calculated from the S-N relationships, N_{test}. In the graph of Fig. 7, the dashed lines and dot-dash lines parallel to the one that passes through the origin (perfect correlation) represent respectively a range of life variations with factors of 1.5 and 3. The graph presented in Fig. 7 shows that the prediction based on the CSBL (CIGRÉ Safe Border Line) is the most conservative one, as presented by CIGRÉ [1,16]. Additionally, the predictions based on the S-N curves represented in Table 3 showed a strong correlation between the observed and the respective expected lives. In general, the predicted values are distributed symmetrically around the perfect correlation line.

Table 3. The synthesis of the experimental results conducted on different conductors

Conductor	*H/w* (m)	S_m (MPa)	$S_a = AN^b$	
			A	b
AAC Orchid	1820	48.3	1.88E+04	-0.443
	2144	56.9	5.80E+03	-0.372
	2725	72.3	1.45E+03	-0.278
ACSR Tern	1820	48.3	4.70E+03	-0.340
	2144	56.9	1.72E+03	-0.277
	2725	72.3	9.74E+02	-0.244
ACAR 750 MCM	1820	48.3	9.75E+03	-0.401
	2144	56.9	3.71E+03	-0.339
	2725	72.3	3.87E+03	-0.347

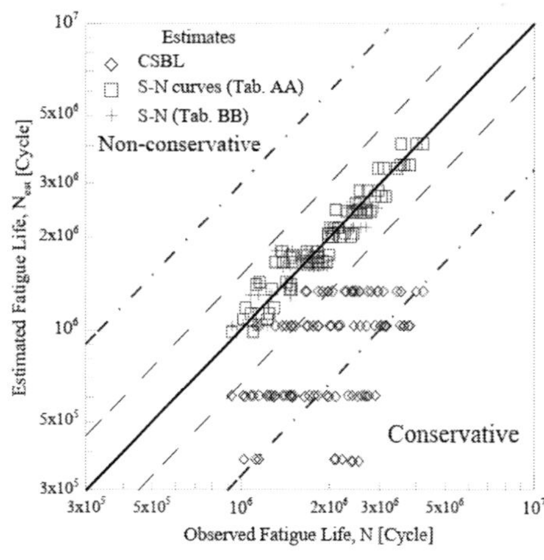

Fig. 7. Accuracy of the fatigue life predictions obtained from the experimental S-N curves of conductors with respect to the CSBL (CIGRÉ safe border line).

5.2 Adherence of the models to the experimental results

The comparisons are made between the predicted models of the mean stress effect by considering the experimental lives obtained in section 3.1 (Fig. 8). These graphs show the dispersion diagrams which correlate the observed fatigue lives and the respective fully reversed equivalent stress estimated by Goodman and Gerber's model, respectively. In addition, the S-N curves that best represent each model is also plotted. Thus, replacing the terms representing the fatigue strength limit for the reverse loading condition in Eq. 7, the function describing the Goodman and Gerber's models will take the form of the expression presented in Eq. 8 when $k = 1$, the function represents the Goodman's model and when $k = 2$, the function represents the Gerber's model.

$$S_{ar} = AN^{b} \quad (7)$$

Where A and b are respectively the constant and the exponent of the Sar-N curves.
From Eq. (7) and the fatigue models Eq. 5-6, the fatigue life was estimated using the following equation:

$$N = \left[\frac{1}{A}\left(\frac{\sigma_a}{1-\left(\frac{\sigma_m}{S_{ut}}\right)^{k}}\right)\right]^{\frac{1}{b}} \quad (8)$$

The analysis of the results from Table 4, which summarizes the results of the non-linear regression process of the experimental data, shows that the models to estimate the mean stress effect of the fatigue resistance of overhead conductors had the same capacity to correlate a fully reversed equivalent stress, S_{ar}. Despite having very different adjustment parameters in terms of the fit quality of the analyzed models (quantified by r^2), the experimental fatigue life shows that the correlation coefficient has a value of 0.5, with small marginal variations between the two models.

A significant part of the experimental points related to the AAC Orchid and ACAR 750 MCM conductors are positioned to the left of the trend curves, indicating that the fatigue failure occurred with the number of cycles which is smaller than the one expected. This position of the experimental points indicates that the predictions are non-conservative. However for the ACSR Tern conductor, the experimental points were positioned to the right of the trend curves (in this case the predictions are conservative). This characteristic in the distribution of the results also helps to explain the deviation from the trend curve of the lines representing the boundaries of the mean response prediction interval. The analysis of the graphs presented in this figure indicates that, regardless of the model used, all prediction were within the region bounded by the bands of variation of two lives.

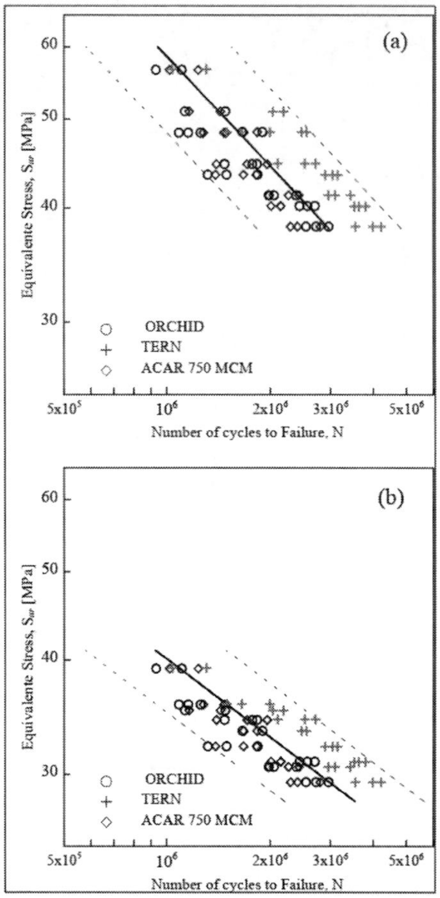

Fig. 8. Comparison between the experimental fatigue lives, *N*, and the fully reverse equivalent stress amplitudes, S_{ar}, predicted by various fatigue models: (a) Goodman, (b) Gerber

Table 4. Stress-life fitting constants with the correlation coefficient r^2

Model	S-N Curve Parameter		r^2
	A[MPa]	b	
Goodman	1.13E+04	-0.381	0.55
Gerber	2.80E+04	-0.435	0.49

6. Conclusions

In the present publication, the fatigue behavior of the three conductors named AAC Orchid, ACSR Tern and ACAR 750 MCM are presented using the *H/w* value of 1820, 2144 and 2725 m. The S-N curves were generated using three values of *H/w* which correspond to the following values of mean stress (48.30, 56.89 and 72.31 MPa). Thereafter the comparison was done between the Basquin alternate stress and different fatigue models. The mean stress effect was predicted using the Goodman and Gerber's models. Based on the results presented, the following conclusions could drawn:

- The fatigue life of conductors is strongly influenced by the mean stress which is a function of the *H/w* value;
- Using Goodman and Geber's fatigue models, most of the estimates for our experimental data fell within a factor two bandwidth.
- The Goodman and Geber's fatigue models were more conservative to estimate the mean stress on the AAC Orchid and ACAR 750 MCM conductors.

Acknowledgements

The authors would like to acknowledge the financial support of TAESA, ATE II, ATE III, EATE, TME, AETE, TBE and Brasnorte for this research (Grant no. 4600001511) by means of the R & D program of ANEEL. The support of Finatec is also acknowledged.

References

1. J. Chan, *Transmission Line Reference Book: Wind-Induced conductor motion* (EPRI, Palo Alto, California, 2006)
2. A.A. Fadel, D. Rosa, L.B. Murça, J.L.A. Ferreira, and J.A. Araújo, Int. J. Fatigue **42** 24-34 (2012)
3. IEEE, *Guide for Aeolian Vibration Field Measurements of Overhead Conductors* (2007)
4. A. Cardou, L. Cloutier, M. St-Louis, A. Leblond, Standardisation of Fretting Fatigue Test Methods and Equipment. 231-242. (1992)
5. V.F. Volker, R.B. Kalombo, R.M.S. Cosme, M.N. Nogueira, J.A. Araújo, Electr. Pow. Syst. Res **116** 346–358 (2014)
6. G.E. Ramey, R. R., Duncan, R.M. Brunair, J. Energ **112** 138-151 (1986)
7. CIGRÉ Report 273, *Overhead conductor safe design tension with respect to aeolian vibrations* (Task Force B2.11.04, CIGRÉ, 2005)
8. CIGRÉ, *Engineering Guidelines Relating to Fatigue Endurance Capability of Conductor-clamp Systems* (SC B2, CIGRÉ, 2010)
9. J.S. Barrett, Y. Motlis, IEEE Proc. Gener. Transm. Distrib **148** 54-59 (2001)
10. W.G. Fricke, C.B. Rawlins, IEEE Trans. Power Appar. Syst **87** (1968)
11. J.C. Poffenberger, R.L. Swart, IEEE Trans. paper **84** 281-289 (1965)
12. K.O. Papailiou, IEEE Trans. Power Del. **12** 1576-1588 (1997)
13. K.O. Papailiou, CIGRÉ, SC 22 WG11, (1995).
14. C.R.F. Azevedo, A.M.D. Henriques, A.R. Pulino Filho, J.L.A. Ferreira, J.A. Araújo, Eng. Fail. Ana **16**, 136-151 (2009)
15. IEEE Std 1368 (2007)
16. CIGRÉ WG 04 SC 22 - 02, **63** (1979)
17. R.B. Kalombo, M.S. Pestana, J.L.A. Ferreira, C.R. M. da Silva, J.A. Araújo, Tribol. Int **108** 141–149 (2017)

Observation of Flaking Process in Rolling Contact Fatigue by Laminography Using Ultra-bright Synchrotron Radiation

Yoshikazu Nakai[1,*], *Daiki* Shiozawa[1], *Shoichi* Kikuch[2], *Hitoshi* Saito[1], *Takashi* Nishina[1], *Taizo* Makino[3], and *Yutaka* Neishi[3]

[1]Kobe University, Department of Mechanical Engineering, 1-1, Rokkodai, Nada, Kobe 657-8501, Japan
[2]Shizuoka University, Department of Mechanical Engineering, 3-5-1, Johoku, Naka, Hamamatsu 432-8561, Japan
[3]Nippon Steel & Sumitomo Metal Corporation, Research & Development, 1-8, Fuso, Amagasaki 660-0891, Japan

Abstract. The flaking failure in rolling contact fatigue (RCF) results from crack initiation and propagation has been believed to originate from non-metallic inclusions located beneath the surface. With conventional microscopies, however, damage process in the internal region of materials could not be observed, then RCF crack initiation and propagation behaviours were observed by using synchrotron radiation computed laminography (SRCL) in the brightest synchrotron facility in Japan, and the effect of the inclusion orientation on the RCF property was examined. In our previous studies, crack initiation and propagation behaviours caused by extended MnS inclusions distributed in depth or transverse (width) direction was observed by the SRCL. In the present study, the fracture mechanism under RCF was discussed on specimens with MnS inclusions distributed in the rolling direction. As a result, vertical cracks were initiated on the surface, parallel to the ball-rolling direction in specimens. The crack propagation direction was then changed perpendicular to the rolling direction. Thereafter, similar with our previous studies, vertical cracks caused the horizontal cracks beneath the surface, when the vertical cracks reached to a critical length. The ratio of the vertical crack initiation life to the flaking life was higher than specimens with other inclusion orientation.

1 Introduction

Roller bearings are used in various machinery such as cars, air planes and electrical appliances. In order to improve bearing life, it is necessary to investigate the rolling contact fatigue (RCF) mechanism. Bearings are generally applied under a pure rolling contact condition with oil lubrication without macroscopic slip, and RCF is a crucial factor for bearings. Since non-metallic inclusions are known to have a detrimental effect on the fatigue strength of high-strength steels, the effects of the composition (chemical/mechanical properties), shape, and orientation of inclusions on RCF should be clarified and controlled.

It has been recognized that flaking starts from inclusions beneath the surface where the ball track is located and the shear stress is greatest, and the cracks grow from them to the surface [1, 2]. Martin et al. dealt with microstructural alterations, which develop with cyclic stressing under rolling contact, and attempted to define their nature and formation mechanisms [3]. In many cases, 'winglike' structural alterations, commonly called 'butterflies', were associated with the inclusions, as were microcracks [4-7]. Nagao et al., however, reported that the mechanism depended on the shape of the inclusions [8]. From spherical inclusions, shear-type cracks parallel to the rolling direction and surface were formed, while vertical cracks perpendicular to the rolling direction were first formed from stringer-shaped inclusions. In this case, shear-type cracks were formed after the formation of the vertical cracks. The formation and propagation of vertical cracks were considered to be controlled by the normal stress, while the propagation of horizontal cracks were controlled by the shear stress. Tsuchida and Tamura observed that horizontal cracks were formed from spherical inclusions, however, they considered that the crack initiation was controlled by the normal stress [9].

Since these phenomena occur beneath the surface, observations have been conducted destructively by cutting samples. Thus, successive observations of RCF have not been conducted. Synchrotron radiation micro computed tomography imaging (SRCT) has been applied for the non-destructive observation of the RCF process. Stiénon et al. calculated the stress field around non-metallic inclusions in bearing steels in RCF tests using three dimensional (3D) shapes obtained by SRCT and at the European Synchrotron Radiation Facility (ESRF) [10,11]. Shiozawa et al. [12] and Makino et al. [13] used SRCT imaging to observe samples exhibiting flaking

* Corresponding author: nakai@mech.kobe-u.ac.jp

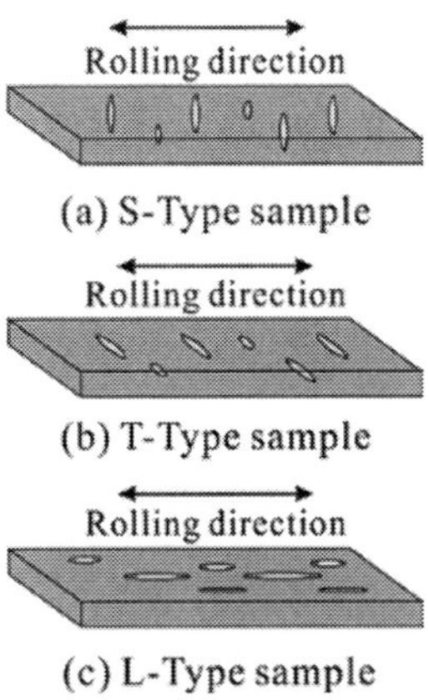

Figure 1. Orientation of inclusion.

damage and RCF-induced cracks. In their studies, samples were cut from normal-size RCF specimens so that they included damaged areas, and the 3D imaging of the damage before flaking provided useful information on the RCF crack initiation and propagation processes. To investigate the effect of the shape of inclusions on the crack initiation, artificial defects that simulate stringer-shaped inclusions were introduced into the specimens, and the crack initiation and propagation from the artificial defects were observed. For the successive SRCT imaging of the RCF process, samples must be sufficiently small to allow the transmission of X-rays, and the cross section must be smaller than 500 μm × 500 μm for steels. Shiozawa et al. showed that the mechanism of RCF in a thin sample is different from that in a bulk sample [14]. Then, Nakai et al. [15] used synchrotron radiation computed laminography (SRCL) [16], which allows the high-resolution, non-destructive imaging of thin plates, to perform successive observations of the flaking process in RCF.

Nakai and others examined the effects of size and orientation of S-Type and T-Type samples those are designated in Figure 1 [17]. In most of previous RCF

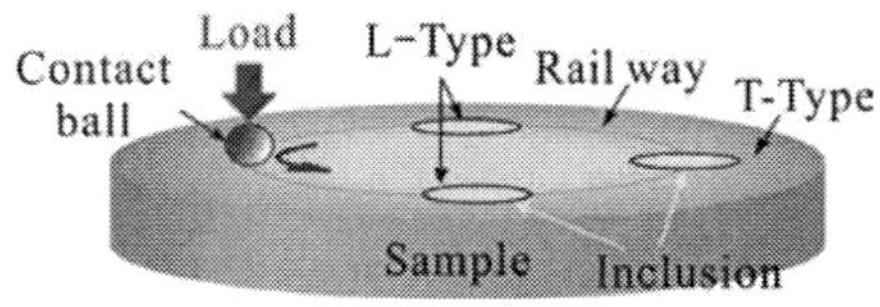

Figure 2. Thrust type rolling contact fatigue tester.

studies, thrust type machines with rotation of samples, shown in Figure 2, have been employed. Using this type machine, pure S-Type RCF test can be conducted. Pure T-Type or L-Type tests, however, cannot be performed because the rolling direction of each inclusion changes during the rotation of the sample. To accomplish pure T-Type RCF test, RCF tester with reciplocal motion has developed in our previous study [14,15].

The flaking process in S-Type and T-Type samples is summarized in Figure 3 [17]. First a crack perpendicular to the rolling surface and the rolling direction forms from an inclusion that is adjacent to the rolling surface. Then the crack propagates vertically in the depth (thickness) direction. After the vertical crack propagates to a critical depth, a horizontal crack forms, which is parallel to the rolling surface. The horizontal crack propagates to form flaking.

In the present study, the RCF crack initiation and propagation in L-Type samples are examined on the basis of the observations obtained using SRCL.

2 Experimental procedure

2.1. Material

The material for the present study is a bearing steel (modified JIS SUJ2), whose chemical composition (in mass %) is as follows: 1.00C, 0.35Si, 0.47Mn, 0.006P, 0.017S, 1.50Cr, and balance Fe. The material has intentionally contained high concentration of sulfur for the observation of the crack initiation from MnS inclusion. It was forged from an ingot with diameter of

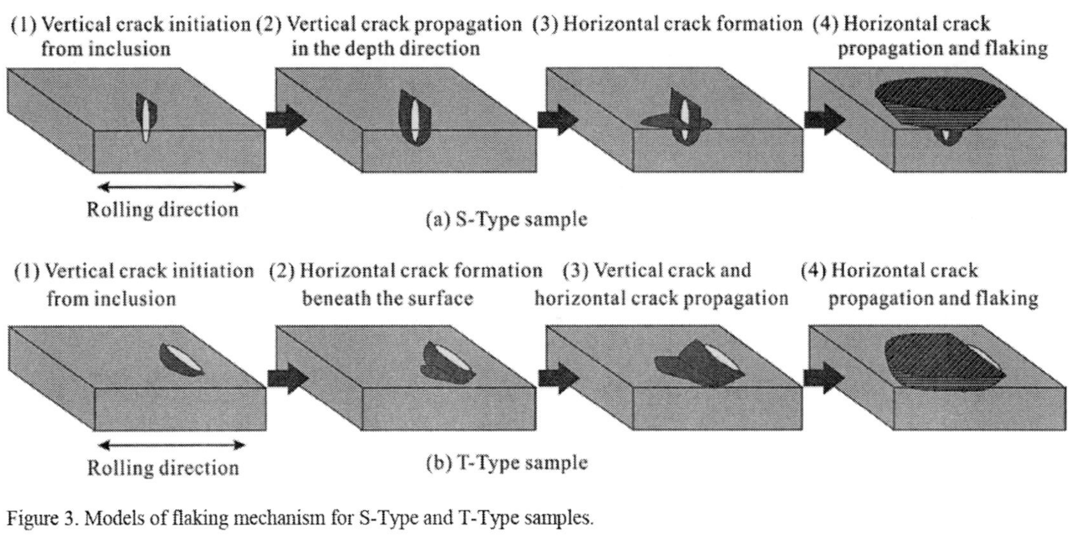

Figure 3. Models of flaking mechanism for S-Type and T-Type samples.

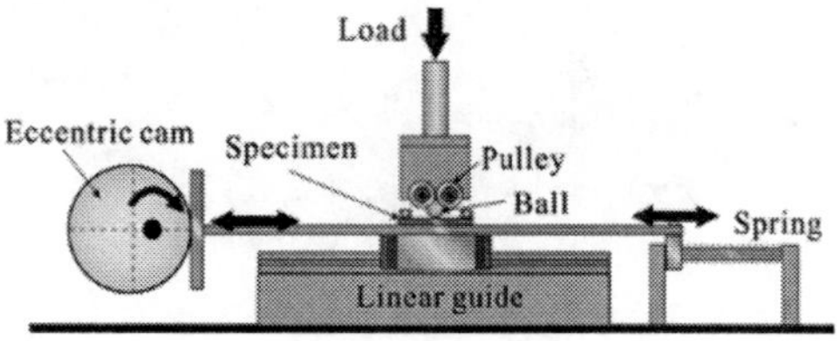

Figure 4. Schematic illustration of compact rolling fatigue testing machine.

65 mm, and inclusions are intergranular with a preferential alignment along the forging direction. After the spheroidizing annealing of cementite particles, specimens were cut from the forged bar, where the longitudinal direction of the bar corresponds to the rolling direction. The specimen was quenching at 1103 K for 0.5 h and tempered at 453 K for 2 h. The dimension of the specimen for SRCL imaging was 10 mm in width, 24 mm in length and 1 mm in thickness. The thickness of the specimen was determined to allow transmitted X-ray with enough intensity for imaging.

2.2. Rolling contact fatigue test

To conduct RCF test for thin plate nearby the experimental hatch of beam lines of the synchrotron radiation facility, SPring-8, a special testing machine was developed. The schematic illustration of the new RCF testing machine is shown in Figure 4. The developed testing machine is boll-on-disk type contact tester. In this testing machine, reciprocal sliding motion is generated by a liner guide and an eccentric cam, and then the ceramic ball with diameter of 6.0 mm and Young's modulus of 300 GPa rolls on the specimen linearly and reciprocally in napthenic lubricant oil with kinetic viscosity of 8.46 mm^2/s, unlike the thrust type rolling contact fatigue testing machine in which rolling is in a single direction, and the direction relative to the inclusions is changed during the rotation of specimen.

All RCF tests were conducted at the maximum Hertz stress of 5.39 GPa, where the sliding distance of this testing machine was 3.0 mm. For observations of the crack initiation and propagation behaviours, fatigue tests were interrupted to conduct SRCL imaging. In this developed testing machine, a sample can be attached and removed easily.

2.3. Laminography setup

The measurement of SRCL imaging was carried out at BL46XU beam lines of Spring-8, which is the brightest synchrotron radiation facility in Japan. The method and apparatus of SRCL is shown in Figure 5. The axis inclination angle for the laminography, ϕ, was 30°, and a monochromatic X-ray beam with 37 keV was employed which was generated by undulator source. In the present study, the effective voxel size in the reconstructed 3D image was 0.74 μm. For 3D reconstruction, a set of 720 radiographs of a specimen were recorded over 360° rotations, where each rotation

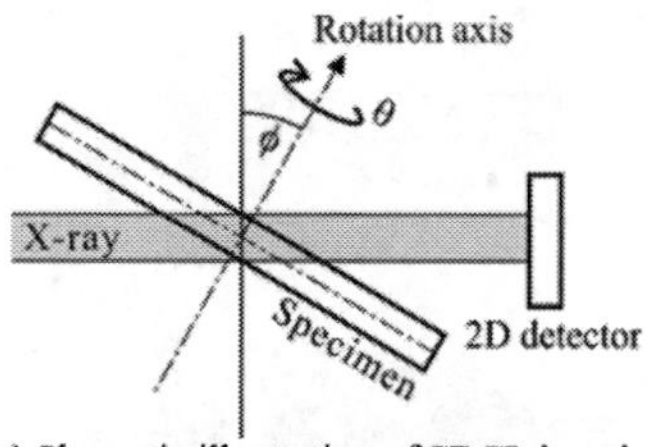

(a) Shematic illustration of SRCL imaging

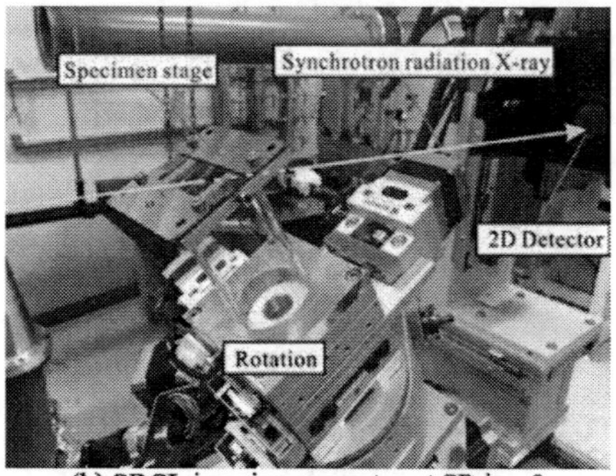

(b) SRCL imaging apparatus at SPring-8

Figure 5. SRCL imaging apparatus.

angle was 0.5°. The exposure time for each angle was 4 s, and 1.5 h for one rotation including rotation of stage. To utilize the refraction contrast effect [18], the X-ray area detector was set by 0.35 m behind the sample. Once reconstructed, the 3D images have been visualized using software, Image-J and Amira. For 3D representation of inclusions and cracks, region growing method was employed to segment them [19], where gray value threshold between matrix and air was employed to produce binarized 3D images.

3. Experimental results

3.1. Crack morphology at surface

Figure 6 is representative micrographs showing the crack profiles at the surface those are observed by SEM. It is clear that cracks were formed from the tip of inclusions because of the stress concentration. The crack propagation direction depended on the orientation of inclusion. For S-Type sample and T-Type sample, the cracks propagate perpendicular to the rolling direction, and zigzag. It means that their propagations are controlled by the in-plane shear stress near the surface. For L-Type sample, however, cracks first propagated to the rolling direction, where the crack is almost straight. It means that the crack propagated by the tensile mode. After the crack length reached a length of 122 μm, the propagation direction was changes perpendicular to the rolling direction. The increase of stress intensity by crack extension may affect this change. After the change of the direction, the crack zigzag similar to S-type and T-type inclusions.

For L-Type inclusion shown in Figure 6 (c), the flection of crack already occured at the first observation

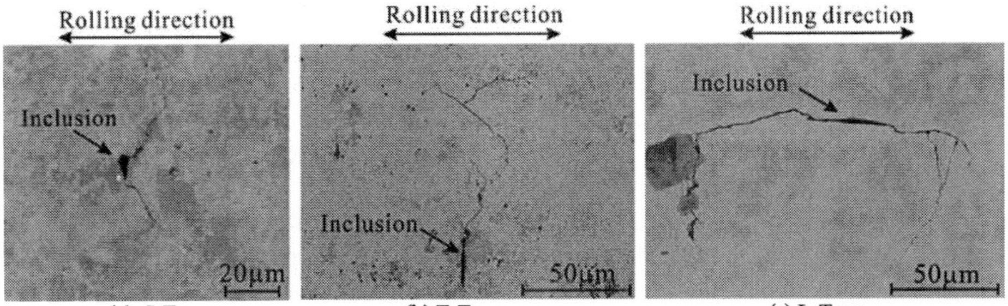

(a) S-Type (b) T-Type (c) L-Type

Figure 6. Crack initiation from inclusion and propagation at surface determined by replicas.

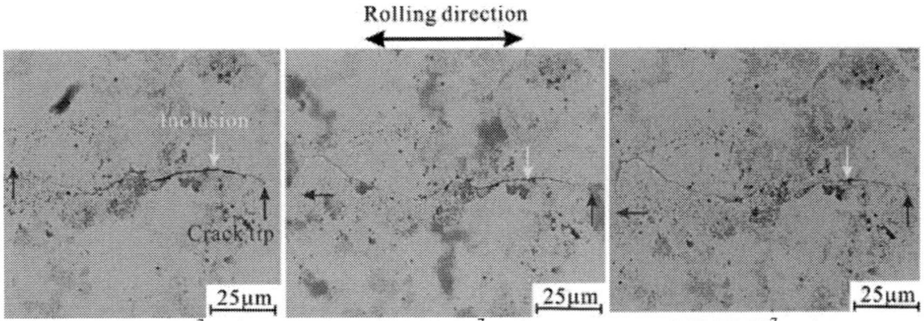

(a) 2.18×10^7 cycles (b) 2.32×10^7 cycles (c) 2.37×10^7 cycles

Figure 7. Crack initiation from inclusion and propagation at surface in L-Type sample determined by replicas..

after the start of RCF test, another observation was conducted, and the resuls are presented in Figure 7. In this observation, it is clear that the crack was initiated at the longitudinal tip of an inclusion, and propagated pararell to the rolling direction. After the crack reached a length of 113 μm, only the left side crack propagation changes perpendicular to the rolling direction between 2.18×10^7 and 2.32×10^7 cycles. The critical length of the crack propagation direction from longitudinal to the transverse direction is almost identical to that in Figure 5 (c).

3.2. Laminography

3D images of inclusions and cracks presented in Figure 6 (c) observed by SRCL at N = 1.50×10^7, 1.53×10^7, and 1.58×10^7 cycles, and shown in Figures 8 and 9, where (A) shows top views, (B) shows views from the rolling direction, and (C) shows side views of the specimen, and inclusions are indicated in pale yellow. Other color codes indicate the depth of the crack from the surface. As shown in the figures, the surface crack in Figure 6 (c) appears to form from a stringer-shaped inclusion with a length of about 30 μm that adjacents to the surface, and the crack face is parallel to the rolling direction and vertical to the surface. At the first ofservation (a), however, both vertical and horizontal cracks were alrady formed, then it is not clear from these figure whether the formation of vertical crack or horizontal crack was first.

Then, the laminography of the crack shown in Figure 7 was conducted. The results are presented in Figure 10, where inclusions are indicated in pale yellow. Note that the other color code is different from that in Figures 8 and 9. Those indicate the number of cycles at each observation in this figure. In this case also a crack was first initiated from a tip of a longitudinal inclusion that located adjacent to the surface, and propagate to the rolling direction and vertical to the surface. After the longitudinal vertical crack propagated 113 μm, it changed perpendicular to the rolling direction between 2.18×10^7 and 2.32×10^7 cycles without forming a horizontal crack, and still vertical to the surface.

The flaking process of L-Type sample is summarized in Figure 11. (1) First, vertical cracks parallel to the rolling direction are formed and propagated until it reached a critical length. (2) Then, the vertical crack propagates perpendicular to the rolling direction. (3) When the vertical crack reached a critical depth, horizontal crack formed at the bottum of the vertical crack. (4) As a result of horizontal crack propagation, flaking is formed. Processes shown from (2) to (3) are similar to T-Type sample, while process (1) is an addition for L-Type sample.

4. Discussion

4.1. Crack paths

Figure 12 shows the potential mechanics for the difference of crack path between T-Type and L-Type samples. The stresses under contact with spherical ball are always axisymmetric. Because of the axisymmetry,crack can propagate to any directions, but considering the movement of rolling ball, only the

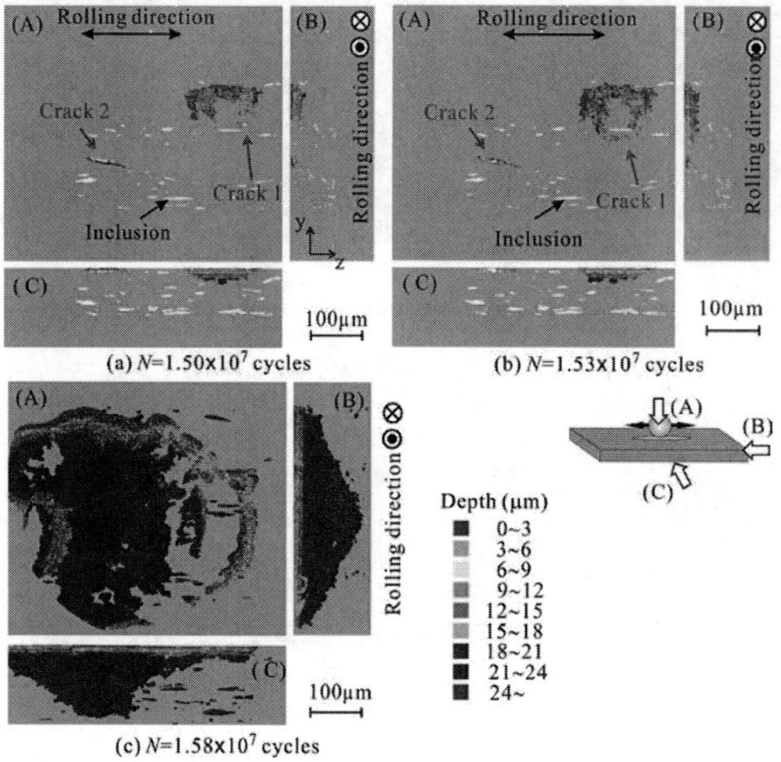

Figure 8. Laminographies showing flaking process for L-Type sample under p_{max}=5.39 GPa, where color code indicates depth of crack from the surface.

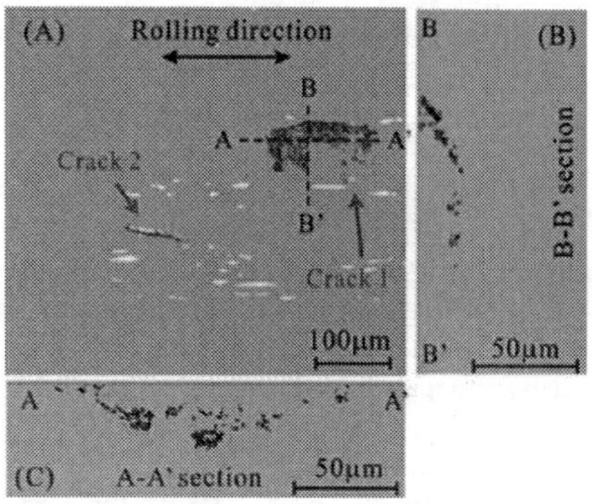

Figure 9. Crosssection of crack in Figure 7 (a) where color code is identical to that for Figure 8.

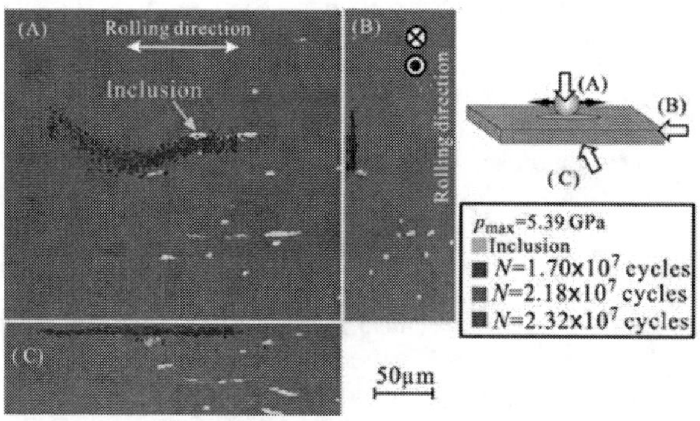

Figure 10. Laminography showing flaking process for L-Type sample, where color code indicates the number of cycles at each observation..

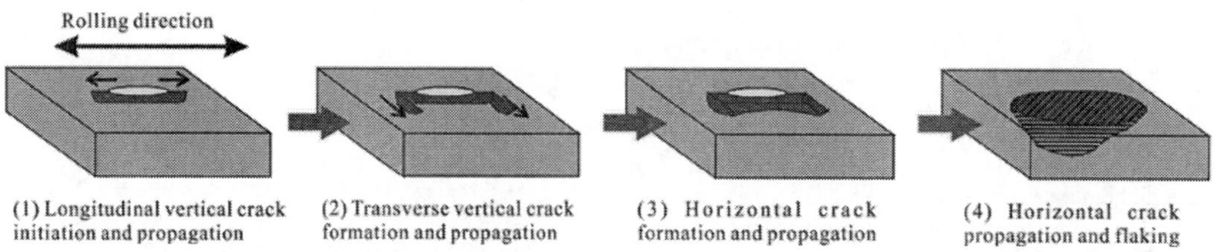

Figure 11. A model of flaking mechanism for L-Type sample.

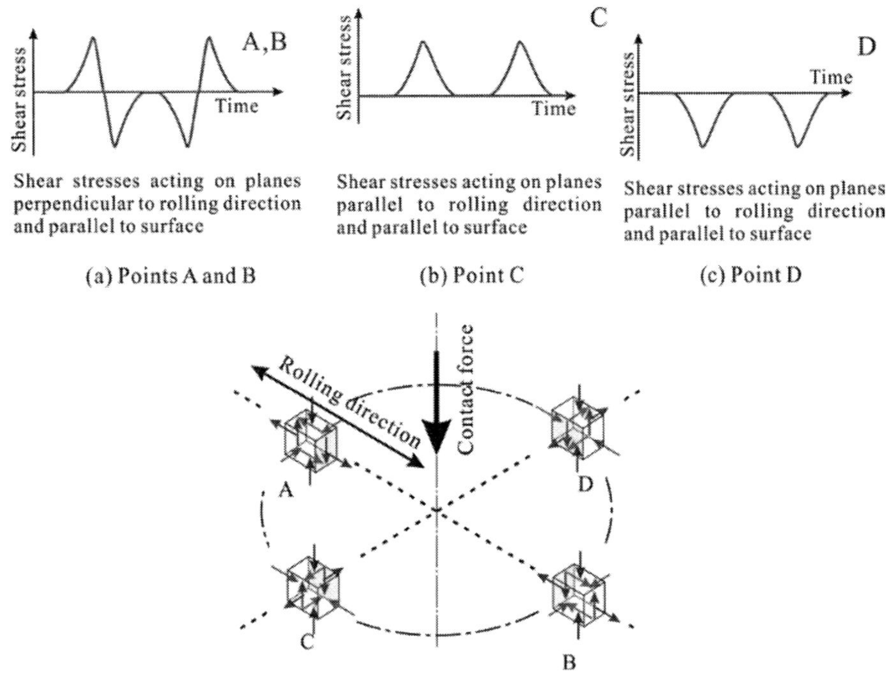

Figure 12. Axisymmetric stress field due to contact force.

change of the anti-plane shear stress can explain the difference of crack growth direction between S-Type, T-Type and L-Type samples because only the antiplane shear stress has different values in Cartesian coordinate system fixed to the sample. For S-Type and T-Type samples, the dominant stress component of the vertical crack propagation is acting on the plane perpendicular to the rolling direction. This shear stress changes alternately as shown in Figure 12 (a). Contrtary to them, the dominant stress component of the vertical crack propagation in L-Type sample is acting on the plane parallel to the rolling direction, and it is pulsating as shown in Figure 12 (b) and (c), where the stress

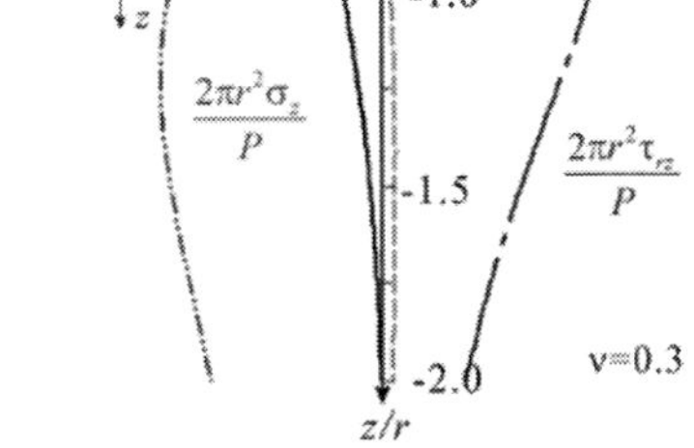

Figure 13. Stress distributio due to contact stress.

amplitude is half for L-Type sample. Although L-Type sample has the same antiplane shear stress component at the transverse side as S-Type and T-Type sample, the stress concentration by inclusion [20] for the direction may not be enough for the crack initiation.

While the above discussion is valid for overall vertical crack path, the difference of crack morphology observed at the specimen surface cannot be explained. For more precise discussion, the stress components by the contact stress should be considered, those are given by the following equations and shown in Figure 13 [21].

$$
\begin{aligned}
\frac{2\pi r^2\sigma_r}{P(1-2\nu)} &= 1 - \frac{1}{\sqrt{1+(r/z)^2}} - \frac{3(r/z)^4}{\sqrt{\left[1+(r/z)^2\right]^5}} \\
\frac{2\pi r^2\sigma_\theta}{P(1-2\nu)} &= -1 + \frac{1}{\sqrt{1+(r/z)^2}} + \frac{(r/z)^2}{\sqrt{\left[1+(r/z)^2\right]^3}} \\
\frac{2\pi r^2\sigma_z}{3P} &= -\frac{(r/z)^2}{\sqrt{\left[1+(r/z)^2\right]^5}} \\
\frac{2\pi r^2\tau_{rz}}{3P} &= \frac{(r/z)^3}{\sqrt{\left[1+(r/z)^2\right]^5}}
\end{aligned}
\tag{1}
$$

This figure shows the stress distribution by concentrated contact point force, and the actual stress distribution by the distributed pressure induced by the ball contact should be obtained by integrating them, then the following discussion must be valid for the actual contact stress.

Since the anti-plane shear stress indicated in Figure 13 is vanished at the specimen surface, the stress component which derive the crack initiation should be the radial stress, σ_r, and/or the circumferential stress, σ_θ. Since $\sigma_r = -\sigma_\theta$, pure shear stress state is achieved at the specimen surface. Then the jigzag crack path induced by the shear stress is appeared for cracks whose face is perpendicular to the rolling direction. For L-Type sample, however, the path whose face is parallel to the rolling direction, is streight indicates that the crack path is controlled by the positive radial stress, σ_r.

Beneath the surface, the anti-plane shear stress componet controlls the crack path as shown in Figure 12, which has a conjugate components acting on the horizontal plane parallel to the specimen surface, while those two stress component have the same amplitude. A question arising here is which shear stress controlls the crack path. Just below the surface, this may be the stress actiong on the vertical plane because the normal stress on the plane is tension and the normal stress on the horizontal plane is compression. The compression stress leads the friction and reduces the driving force of the crack extension.

In deep site, however, this is the horizontal plane because the amplitude of the anti-plane shear stress reduces with the crack extension to the depth direction. On the other hand, the amplitude of the shear stress along a plane parallel to the surface keeps constant with crack extension to the longitudinal direction of the sample because of the horizontal movement of the contact point to the rolling direction.

The effect of the stress concentration induced by inclusions was not consided in the above discussion. Makino et. al. conducted the finite element analysis of the stress distribution for S-Type inclusion under the contact surface. Their analysis, however, the change of the stress component due to the movement of contact ball was not discussed.

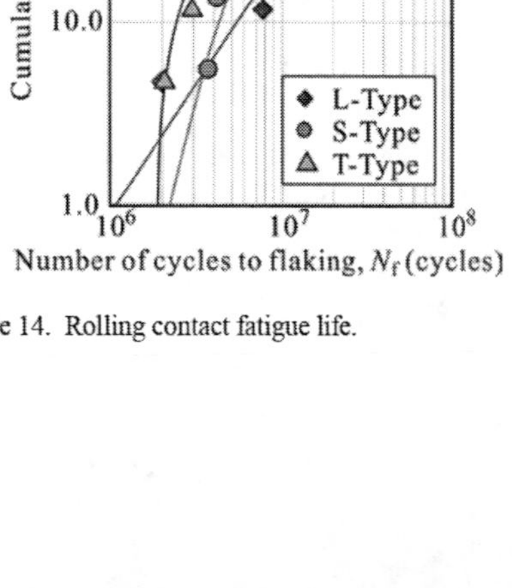

Figure 14. Rolling contact fatigue life.

4.2. Fatigue life

Weibull plot of the flaking life is shown in Figure 14. The flaking life at the probability of 50 % is almost the same for S-Type and T-Type samples, but L-Type sample has much longer life.

To clarify which process defined in Figures 3 and 11 is responsible for the difference, the crack propagation life for each process is examined. Figure 15 shows the comparison of the crack propagation life for each type of sample where the abscissa is the number of cycles up to the flaking. The plots show the number of cycles when each type of crack was observed. Except Inclusion F in T-Type sample which was not cause the first flaking, the propagation life of the horizontal crack is almost similar for all types of sample because this process occured far away from the inclusion at their initiation site and the macroscopic crack propagation behavior may not be affected by the orientation of inclusions, then the difference of fatigue life between the orientation of inclusions should be attributed to the difference in the vertical crack initiation and the propagation lives. Then, the main difference in the flaking life for L-Type sample between other samples must originates from the crack initiation and the vertical crack propagation in the rolling direction.

5. Conclusions

In the present study, 4D observations of the formation and propagation of cracks in the rolling contact fatigue (RCF) tests were performed on a high-strength steel by combining a newly developed compact RCF testing machine with synchrotron radiation computed laminography (SRCL), and the crack propagation behavior and the fracture mechanism under RCF was discussed on specimens with MnS inclusions elongated in the rolling direction (L-Type). The following results were obtained.

(1) A vertical crack was initiated on the surface, parallel to the ball-rolling direction in specimens. The crack propagation direction was then changed perpendicular to the rolling direction. Thereafter, similar with S-Type or T-Type samples, vertical cracks then cause the horizontal cracks, parallel to the sample surface, when the vertical cracks reached a critical depth.

(2) The main difference for L-Type sample between other samples must originated from the crack initiation

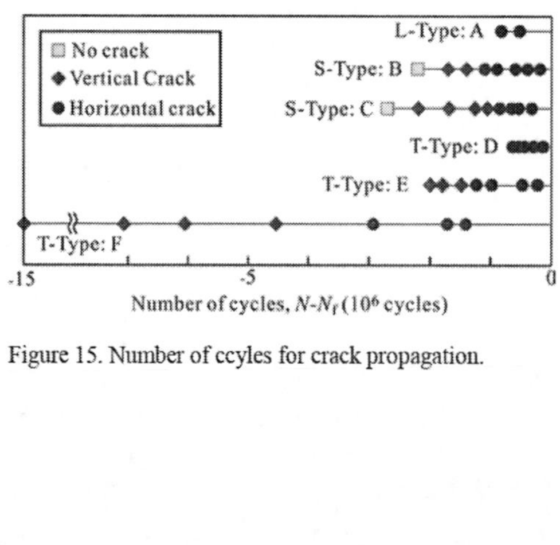

Figure 15. Number of ccyles for crack propagation.

and propagation in the rolling direction. The fatigue life from the horizontal crack formation to the flaking is almost identical independent of the inclusion orientation.

Acknowledgment

The synchrotron radiation experiments were performed at beam line BL46XU of SPring-8 with the approval of the Japan Synchrotron Radiation Research Institute (JASRI) under proposal numbers 2017A1581, 2017B1610, and 2017B1908. The authors are grateful for the technical support of Dr. K. Kajiwara (JASRI).

References

1. L. Gustave, and A. Palmgren, J. App. Mech., Trans. ASME, **16**, 165 (1949).
2. H. Styri, Proc.ASTM, **51**, 682-700 (1951).
3. J.A. Martin, S.F. Borgese, and A.D. Eberhardt, J. Basic Eng., Trans. ASME, **88**, 555-565 (1966).
4. K. Sugino, K. Miyamoto, M. Nagumo, and K. Aoko, Trans. ISIJ, **10**, 98-111 (1970).
5. P.C. Becker, *Metals Technology*, 234-243 (1981).
6. A. Grabulov, U. Ziese, and H.W. Zandbergen, Scripta Materialia, **57**, 635-638 (2007).
7. A. Grabulov, R. Petrov, and H.W. Zandbergen, Int. J. Fatigue, **32**, 576-583 (2010).
8. M. Nagao, K. Hiraoka, and Y. Unigame, Sanyo Technical Report, **12**: 38-45 (2005).
9. T. Tsuchida, and E. Tamura, Kobe Steel Engineering Reports, **61**, 62-65 (2011).
10. A. Stiénon, A. Fazekas, and J.-Y. Buffière, A. Vincent, P. Daguier, F. Merchi, Materials Sci. Eng. A, **513-514**: 376-383 (2009).
11. A. Stienon, A. Fazekas, J.-Y. Buffiere, P. Daguier, F. Merchi,and A. Vincent, J. ASTM Int., **7**, 1-14 (2010).
12. D. Shiozawa., Y. Nakai., Y. Fukuda, Y. Neishi., and T. Makino., Proc. 15th Int. Conf. Exp. Mech., CD-ROM, No. 2635 (2012).
13. T. Makino., Y. Neishi, D. Shiozawa, Y. Fukuda, K. Kajiwara., and Y. Nakai, Int. J. Fatigue, **68**, 168-177 (2014).
14. D. Shiozawa., T. Makino., Y. Neishi., and Y. Nakai., Procedia Mat. Sci., **3**, 159-164 (2014).
15. Y. Nakai, D. Shiozawa., S. Kikuchi., K. Sato, K., T. Obama, T. Makino., and Y. Neishi, Frattura ed Integrità Strutturale, **34**, 267-275 (2014).
16. S. Gondrom, J. Zhou, M. Maisl, H. Reiter, M. Kroning, and W. Arnold, Nuclear Eng. Design, **190**, 141-147 (1999).
17. Y. Nakai, D. Shiozawa, S. Kikuchi, T. Obama, H. Saito, T. Makino, and Y. Neishi, Eng. Fract. Mech., **183**, 108-189 (2017).
18. Y. Nakai, D. Shiozawa, T. Kurimura, and K. Kajiwara, Proc. SPIE, **7522, 75224B**: 1-7 (2010).
19. R. Adams, and L. Bischof, Analysis and Machine Intelligence, 16: 641 – 647 (1994).
20. T. Makino, Y. Neishi, D. Shiozawa., S. Kikuchi., T. Okada., K. Kajiwara., and Y. Nakai, Int. J. Fatigue, **92**, 507-516 (2016).
21. S.P. Timoshenko and J.N. Goodier, *Theory of Elasticity*, Third Ed., 398 (McGraw-Hill Kogakusha, Ltd., Tokyo, 1970).
22. T. Makino, Y. Neishi, D. Shiozawa, S. Kikuchi, H. Saito, K. Kajiwara and Y. Nakai, Procedia Struct. Integ., **7**, 468-475, (2017).

MATEC Web of Conferences **165**, 11003 (2018) https://doi.org/10.1051/matecconf/201816511003
FATIGUE 2018

A novel approach to predict fretting fatigue crack initiation

Guillaume **Rousseau**[1,2,*], *Claudio* **Montebello**[2], *Yoann* **Guilhem**[1], **and** *Sylvie* **Pommier**[1]

[1]Ecole normale supérieure Paris-Saclay, 94230 Cachan, France
[2]Safran Aircraft Engines, 77550 Moissy-Cramayel, France

Abstract. This paper proposes a simplified approximation of the velocity fields close to the contact edge using non-local intensity factors and reference fields. This non-local description is not affected by size or gradient effects and leads to unique crack initiation boundaries. Linear intensity factor ***I*** represents the contribution of the elastic field on the total one and can be determined thanks to the macroscopic load even with coarse mesh. While the complementary intensity factor I^c describes the friction effect and is proportional to the size of the partial slip zone. The prediction of I^c using ***I*** through an incremental approach allows to predict the velocity fields for complex loadings and can be used to set up criteria to predict crack initiation.

1 INTRODUCTION

Fretting-fatigue occurs in engineering assemblies subjected to vibration loads. Micro-slip caused by oscillating forces, between the contacting bodies, leads to surface damage, and then crack initiation. Combined with cyclic bulk fatigue loading, crack propagation happens. Consequently, fretting-fatigue is a frequent source of failure constituting a major concern in safety-critical industries such as aerospace. Experimental test campaigns have shown that this phenomenon has an important negative effect on the material fatigue limit [1] due to the rise of embryo cracks very early in life.

One of the major difficulties in dealing with the fretting fatigue problem is the severe stress gradient close to the surface. Local approaches used to predict the fatigue life based on stress fields at the hot spots are over-conservative for fretting-fatigue because they do not account for the possibility of crack arrest due to a strong stress gradient from the surface to the bulk. Moreover, the crack initiation threshold is not unique in terms of local quantities (Fig. 1) and shows a strong dependency to the geometry [2-3]. However, using the theory of critical distance [4], it is possible to predict crack arrest and consequently to estimate fretting fatigue life and the gradient effect with a good degree of accuracy [3,5].

In spite of the good predictions of fretting-fatigue threshold obtained using the theory of critical distance, there are criticisms with respect to the use of critical length as a material property [6]. Thus, the transferability of results obtained on laboratory samples (cylinder-plane contact) to industrial cases (disk-blade interface), which are complex and 3D geometry, is not guaranteed.

The goal of this paper is to propose a solution to predict the velocity field close to the contact edge by means of non-local intensity factors, independent from geometry, and to use these factors to set up a crack initiation criterion.

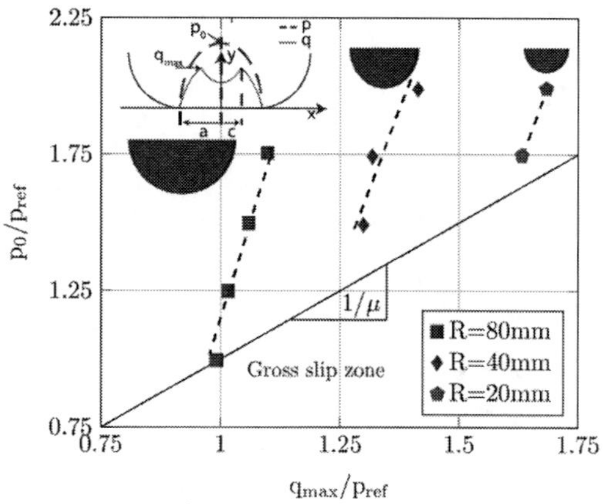

Fig. 1. Experimental crack initiation for Inconel 718 in plain fretting [2].

2 MODELING

2.1 Background

The interest of this approach, developed by Montebello [7] is shown in Fig. 2 where the description of the crack initiation boundaries through non-local intensity factors shows a unique threshold. Hence the

* Corresponding author: guillaume.rousseau@ens-paris-saclay.fr

geometry and the gradient effect have no effect on these variables.

In this part, the computation of the reference fields and the related intensity factors will be explained and an extension to fretting-fatigue problem will be made.

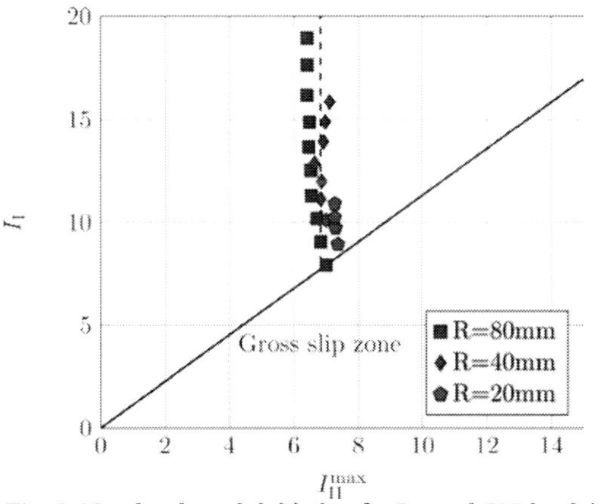

Fig. 2. Non-local crack initiation for Inconel 718 in plain fretting [8].

2.2 Mechanical field partitioning

Analogous to facture mechanics [9], a local self-similar geometry and a strong stress gradient induced by the contact, makes the spatial distribution of the mechanical fields imposed by the geometry while its intensity is determined by the macroscopic loads. Consequently, in the coordinate system R' attached to the contact edge (see Fig. 3) where $\boldsymbol{P}$ is the position coordinates vector, I_{I} and I_{II} the elastic intensity factors, I_{II}^c the complementary intensity factor and $\boldsymbol{\phi}_I$, $\boldsymbol{\phi}_{II}$ and $\boldsymbol{\phi}_{II}^c$ the associated reference fields. The velocity field can be expressed through a sum of products between an intensity factor (time dependent) and a reference field (geometry dependent):

$$\boldsymbol{v}_e(\boldsymbol{P},t)_{R'} \simeq I_{\text{I}}(t)\boldsymbol{\phi}_I(\boldsymbol{P}) + I_{\text{II}}(t)\boldsymbol{\phi}_{\text{II}}(\boldsymbol{P}) \quad (1)$$

$$\boldsymbol{v}_c(\boldsymbol{P},t)_{R'} \simeq I_{\text{II}}^c(t)\boldsymbol{\phi}_{\text{II}}^c(\boldsymbol{P}) \quad (2)$$

$$\boldsymbol{v}(\boldsymbol{P},t)_{R'} \simeq \boldsymbol{v}_e(\boldsymbol{P},t)_{R'} + \boldsymbol{v}_c(\boldsymbol{P},t)_{R'} \quad (3)$$

This field is partitioned into two terms: the elastic one $\boldsymbol{v}_e$ (Eq. 1) represents the elastic response of the structure; the complementary one $\boldsymbol{v}_c$ (Eq. 2) describes the non-linear contribution of the friction in the slip zone. In turn $\boldsymbol{v}_e$ is separated into symmetric and antisymmetric parts with respect to the contact plane, analogous to mode I and mode II in LEFM. These fields are respectively generated by a variation of normal and tangential loads. Then the intensity factors are computed by projecting the velocity field $\boldsymbol{v}$ to the reference field $\boldsymbol{\phi}_*$:

$$I_*(t) = \frac{\int_\Omega \boldsymbol{v}.\boldsymbol{\phi}_*}{\int_\Omega \boldsymbol{\phi}_*.\boldsymbol{\phi}_*} \quad (4)$$

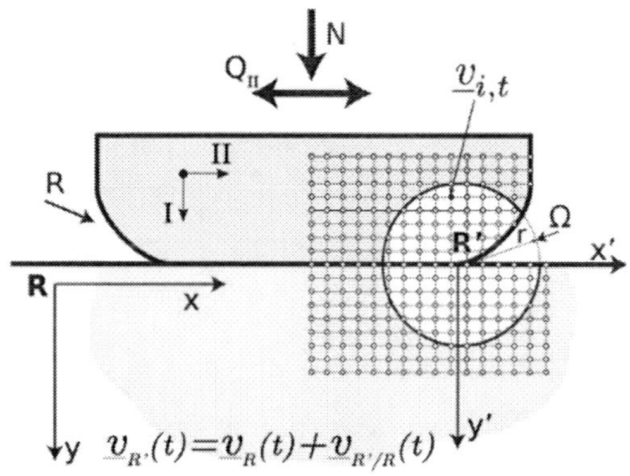

Fig. 3. FE parameters and extraction area.

2.3 Extension to fretting fatigue

The previous modelling is only usable for plain fretting because the fatigue loading σ_F cannot be taken into account. Fig 4 (F-el and F-tot) shows the fretting modelling is unable to describe the velocity field. To solve this, the elastic velocity field must be enriched (Eq. 5) by an additional term. This termn takes into account the residual velocity field from the fatigue loading which cannot be described by the other intensity factors. The Karhunen-Loeve decomposition [10] is used to partition the residual field between two functions depending separately on time I_F and space $\boldsymbol{\phi}_F$.

$$\boldsymbol{v}_e(\boldsymbol{P},t)_{R'} = \boldsymbol{v}_e^{Fretting}(\boldsymbol{P},t)_{R'} + I_F(t)\boldsymbol{\phi}_F(P) \quad (5)$$

2.4 Quality of the assumption

The final approximation of the velocity field for fretting-fatigue is obtained (Eq. 3) and the elastic (Eq. 6) and total (Eq. 7) error of the approximation are defined.

$$\xi_{el} = \frac{\sqrt{\int_\Omega (\boldsymbol{v}-\boldsymbol{v}_e)^2}}{\sqrt{\int_\Omega \boldsymbol{v}^2}} \quad (6)$$

$$\xi_{tot} = \frac{\sqrt{\int_\Omega (\boldsymbol{v}-\boldsymbol{v}_e-\boldsymbol{v}_c)^2}}{\sqrt{\int_\Omega \boldsymbol{v}^2}} \quad (7)$$

The error evolution during a proportional fretting-fatigue cycle is shown in Fig. 4: the errors defined with the fretting modelling (F-el and F-tot) remains high. However, with the introduction of the fatigue field (Eq. 5), the total error remains low. It is worth noting that the elastic and total errors are equal when the whole contact is stuck.

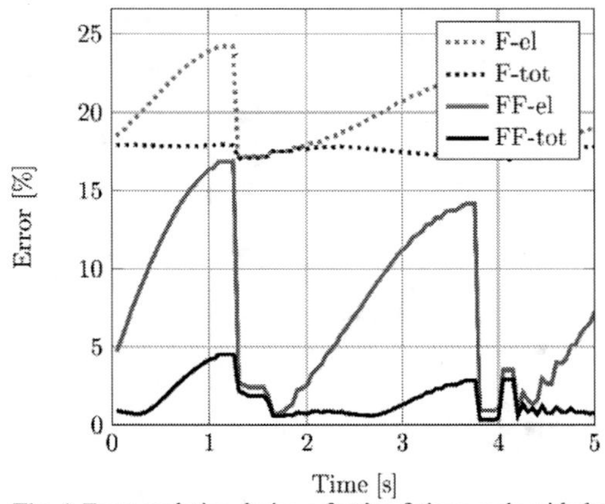

Fig. 4. Error evolution during a fretting-fatigue cycle with the plain fretting (F) and the fretting-fatigue (FF) modeling.

The additional fatigue intensity factors provide a good approximation of the velocity field in all the extraction area. However to characterize crack initiation, using the intensity factors from fretting is sufficient: different fatigue loadings imply different Wöhler curves (see Fig. 5), but when the intensity factor I_{II} is used to plot the Wöhler curve, a unique crack boundary is highlighted (see Fig 6).

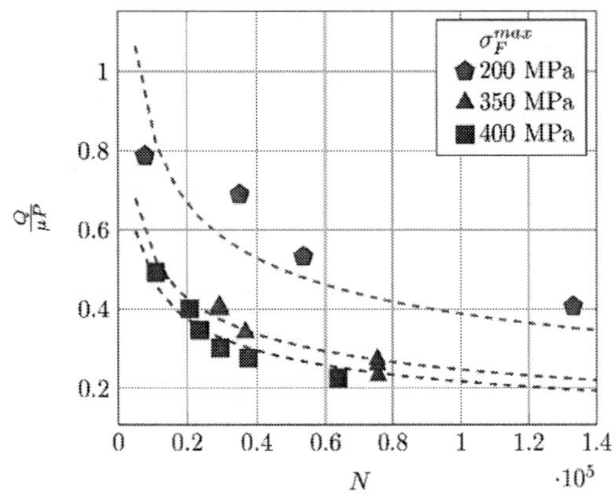

Fig. 5. Experimental Wöhler curve for fretting-fatigue crack initiation for Ta6V [11]

Fig. 2 and Fig. 6 show that linear intensity factors can be used to describe a unique crack initiation boundary for fretting and fretting-fatigue problem. These factors are directly linked to the macroscopic loading, so they can be computed with a coarse mesh FE simulation. The prediction of the non-linear intensity factor (which needs a fine mesh) thanks to the linear ones can lead to a good approximation of the velocity field close to the contact edge even with a coarse mesh.

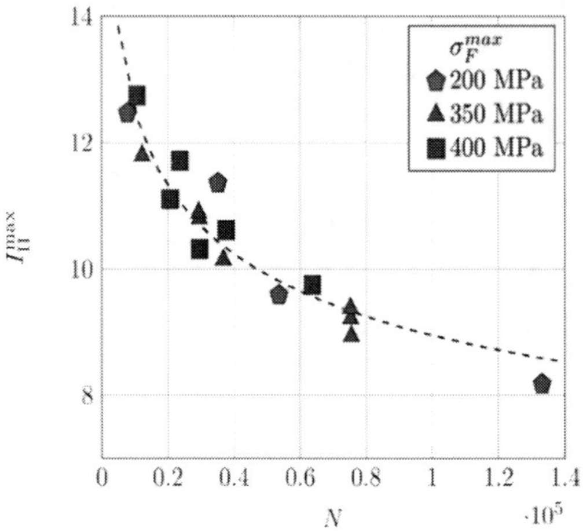

Fig. 6. Non-local Wöhler curve for fretting-fatigue crack initiation for Ta6V.

3 INCREMENTAL MODEL

The prediction of the non-linear intensity factor I_{II}^c will be made through an incremental approach which will be easy to use for complex loadings.

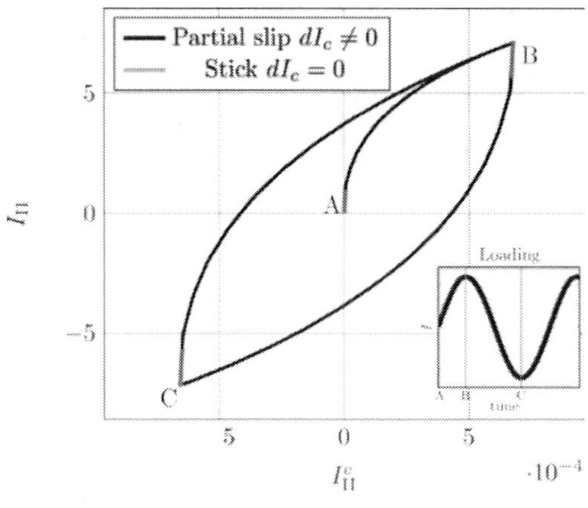

Fig. 7. Evolution of the intensity factors during a fretting cycle.

To model the non-linear behaviour (see Fig 7), a bilinear kinematic hardening law is chosen to provide a simple model along with a good precision. This law allows to predict the evolution of I_{II}^c during a fretting cycle with a complex tangential amplitude (see Fig. 8).

To estimate more precisely the error, the complementary velocity field predicted by the model (Eq. 8) and the error associated (Eq. 9) are defined.

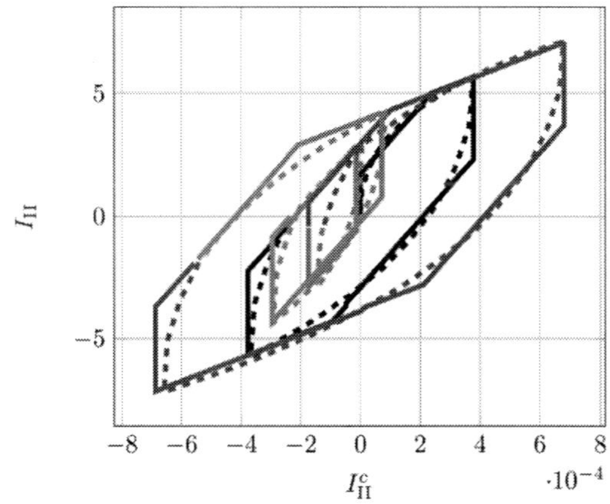

Fig. 8. Sliding prediction during a proportional fretting-fatigue cycle with variable amplitude.

$$v_c^{model} = \dot{I}_{II}^{c\,model}(t)\boldsymbol{\phi}_{II}^c(\boldsymbol{P}) \tag{8}$$

$$\xi_{model} = \frac{\sqrt{\int_\Omega (\boldsymbol{v} - \boldsymbol{v}_e - \boldsymbol{v}_c^{model})^2}}{\sqrt{\int_\Omega \boldsymbol{v}^2}} \tag{9}$$

With regard to the error on the velocity fields (Fig. 9), it appears that the complementary field predicted by the model permit to reduce the error, not as much as the complementary field issue from the simulation but to a reasonable level.

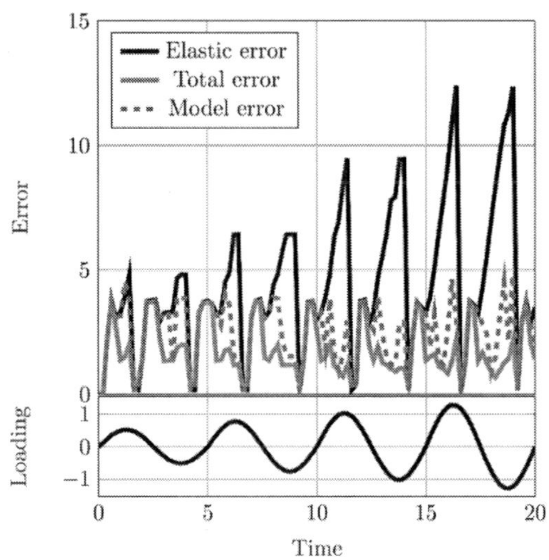

Fig. 9. Error evolution introduced by the approximation during a proportional fretting- cycle.

4 CRACK INITIATION CRITERIA

The previous parts highlighted a modelling allowing to plot unique crack initiation boundary for fretting but also for fretting-fatigue whereas the incremental model can predict the sliding I^c_{II} with I_{II}. The aim of this part is to predict the crack boundaries using the intensity factors and to determine a material property characterizing the crack initiation threshold.

4.1 Criteria

To set up a criterion, non-local intensity factors will be used: (i) dI^c_{II} which is directly linked to the size of the partial slip zone; (ii) I_{II} which is proportional to the shear stress. This can be demonstrated using the analytical solution from LEFM [12] for mode I and II. Only three criteria are taken into account here but others can be formulated with these intensity factors.

Experiments are made with constant loading amplitude, so only one cycle is considered. Moreover, the spatial distribution of the damage is not taken into account because the contact front remains motionless and the maximum damage appears in the same area.

4.1.1 Total sliding

The first criterion is based on the accumulation of the micro sliding dI^c_{II} (the non-linear variable).

$$C_1 \sim \int_t |dI^c_{II}| \tag{10}$$

4.1.2 Energy dissipated by friction

The second criterion is based on the energy dissipated by friction on the contact surface.

$$C_2 \sim \int_t |I_{II}\, dI^c_{II}| \tag{11}$$

4.1.3 Energy release rate

The last criterion is analogous to the energy release rate in LEFM and express the hypothesis that failure occurs when the mechanical power reaches a critical value.

$$C_3 \sim \max({I_{II}}^2) \tag{12}$$

4.2 Methodology

1. The experimental works highlight some crack initiation frontiers expressed in (q_{max}-p_0) or ($Q/\mu P$-N) coordinates. For each point on these frontiers the relative macroscopic loads are computed.
2. A FE computation, using the macroscopic loads calculated previously, is performed to obtain the relative velocity field close to the contact edges.
3. The non-local intensity factors are computed.
4. An average value of each criterion is computed through all the data.
5. Non-local (and consequently local) quantities are predicted thanks to the criteria and a crack initiation map is built.

It is worth mentioning that experimental data are enriched; a linear interpolation of the experimental data point is done. This permits to apply the steps listed above on a more complete set of data.

4.3 Results

The three criteria have been processed with the previous experiments to evaluate their ability to predict crack initiation boundary for constant and variable lifetimes.

4.3.1 Constant lifetime

The non-local boundary is predicted with the methodology describes in section 4.2 and the inverse post-treatment is applied to plot the local map (Fig. 10)

The three criteria are able to predict boundaries for the highest radii (40 and 80 mm). The results are less satisfactory for the smallest radius, where criteria 2 and 3 respectively overestimate and underestimate the boundary. Concerning the criterion 1, it seems unable to predict correctly the boundary.

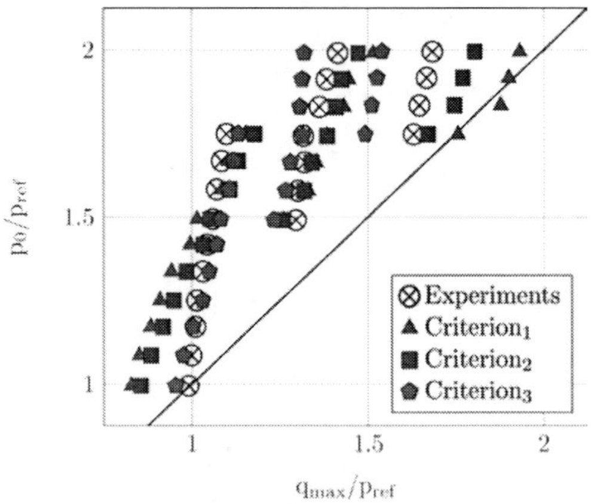

Fig. 10. Local crack initiation boundary predicted by the criteria for Inconel 718.

4.3.2 Variable lifetime

The three criteria are able to predict the boundaries when the lifetime is higher than 20000 cycles. Until this limit, only the criteria based on the dissipated energy by friction gives good results. However, for short lifetime, stress is high and plasticity may not be neglected.

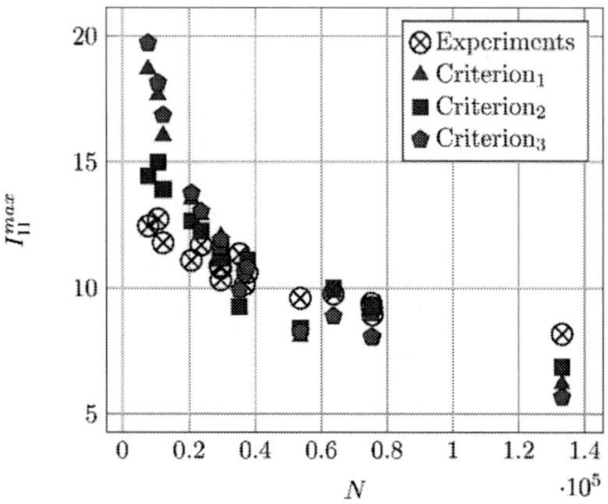

Fig. 11. Non-local crack initiation boundary predicted by the criteria for Ta6V.

5 3D MODELING

The contact between disc and root blade corresponds to a complex geometry and is submitted to non-proportional loading. Consequently, a 3D modelling is required to study such industrial cases. The fretting modelling (Eq. 2) can be extended in 3D with the addition of two new intensity factors and associated reference fields: (i) a linear one I_{III} to take into account the out of plane displacement (Eq. 13); (ii) a complementary one I_{III}^c (Eq. 14) to describe the friction effect in this new direction.

$$v_e(\boldsymbol{P},t)_{R'} \simeq I_{\mathrm{I}}(t)\boldsymbol{\phi}_{\mathrm{I}}(\boldsymbol{P}) + I_{\mathrm{II}}(t)\boldsymbol{\phi}_{\mathrm{II}}(\boldsymbol{P}) + I_{\mathrm{III}}(t)\boldsymbol{\phi}_{\mathrm{III}}(\boldsymbol{P}) \quad (13)$$

$$v_c(\boldsymbol{P},t)_{R'} \simeq + I_{\mathrm{II}}^c(t)\boldsymbol{\phi}_{\mathrm{II}}^c(\boldsymbol{P}) + I_{\mathrm{III}}^c(t)\boldsymbol{\phi}_{\mathrm{III}}^c(\boldsymbol{P}) \quad (14)$$

The incremental model introduced previously is still able to predict (I_{II}^c, I_{III}^c) using (I_{II}, I_{III}) for complex non-proportional tangential loading conditions. Fig. 12 shows the different errors under non-proportional tangential loading which remains low for the model.

However, the lack of experiments with complex and non-proportional loadings makes the identification of criterion difficult.

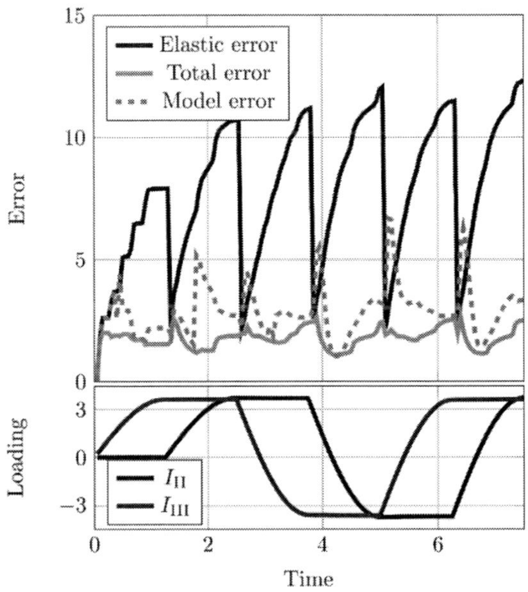

Fig. 12. Errors evolution introduced by the approximation during a non-proportional fretting-fatigue cycle.

5 CONLUSION

A novel approach to predict the local velocity field of a fretting-fatigue case, based on a fracture mechanics analogy, has been presented. Once the reference fields are identified for a specific geometry, it is possible to predict the velocity field close to the contact edge thanks to the non-local intensity factors directly from the macroscopic loadings. Finally, using a criterion based on the energy dissipated by friction on the contact surface, the crack initiation boundary can be predicted. Moreover, as the size and gradient effects are included in the reference fields, the use of criteria and laws based on non-local intensity factors are not geometry dependent. Consequently, they can be identified on laboratory tests and easily transferred to industrial structure.

The displacement of the contact edge and the damage accumulation should be taken into account to study normal load variations and all kinds of complex loadings. To complete this work, fretting tests in non-proportional and variable amplitude could be performed to verify the results obtained by simulation.

The authors are grateful to Safran Aircraft Engines (Safran group) for its financial support within the IRG COGNAC project framework.

REFERENCES

1 TC Lindley. *Fretting fatigue in engineering alloys*. Int J Fatigue **19**:39–49 (1997).

2 R Ferre, S Fouvry, B Berthel, JA Ruiz-Sabariego. *Stress gradient effect on the crack nucleation process of a Ti-6Al-4V titanium alloy under fretting loading: Comparison between non-local fatigue approaches*. Int J Fatigue **54**:56–67 (2013).

3 R Amargier, S Fouvry, L Chambon, C Schwob, C Poupon. *Stress gradient effect on crack initiation in fretting using a multiaxial fatigue framework*. Int J Fatigue **32**:1904–12 (2010).

4 D Taylor. *The Theory of Critical Distances: Basics*. Eng Fract Mech **75**:1696–705 (2008).

5 JA Araújo, L Susmel, D Taylor, JCT Ferro, EN Mamiya. *On the use of the Theory of Critical Distances and the Modified Wöhler Curve Method to estimate fretting fatigue strength of cylindrical contacts*. Int J Fatigue **29**:95–107 (2007).

6 FC Castro, JA Araújo, N Zouain. *On the application of multiaxial high-cycle fatigue criteria using the theory of critical distances*. Eng Fract Mech **76**:512–24 (2009).

7 C Montebello. *Analysis of the stress gradient effect in Fretting-Fatigue through a description based on nonlocal intensity factors*. Université Paris-Saclay, 2015.

8 C Montebello, S Pommier, K Demmou, J Leroux, J Meriaux. *Analysis of the stress gradient effect in fretting-fatigue through nonlocal intensity factors*. Int. J. Fatigue, vol. 82, Elsevier; , p. 188–98.

9 A. E Giannakopoulos, TC Lindley, S Suresh. *Aspects of equivalence between contact mechanics and fracture mechanics: theoretical connections and a life-prediction methodology for fretting-fatigue*. Acta Mater **46**:2955–68 (1998).

10 M Loève. *Probability Theory I* **45**:1–10 (1977).

11 J Meriaux. *Etude expérimentale de modélisation de l'endommagement d'un contact aube disque de soufflante soumis à des chargements de fretting fatigue*. 2010.

12 HM Westergaard. *Bearing pressures and cracks*. J Appl Mech **61**:A49–53 (1939).

Phase transformation in white etching area in rolling contact fatigue

Yun-Shuai Su, *Shu-Xin* Li*, *Si-Yuan* Lu, and *Li-Biao* Wan

School of Mechanical Engineerying and Mechanics, Ningbo University, Ningbo 315211, PR China

Abstract. Rolling contact fatigue (RCF) involves microstructural change in the subsurface of contact. The changed microstructure is generally termed as white etching area (WEA) as it appears white under optical microscope when etching in nital solution. WEA has been acknowledged as one of the primary failure modes in RCF since it causes severe local inhomogeneity of microstructure. It was reported that WEA consists of nano ferrites as martensite grains and carbides are significantly refined in the WEA. Some carbides are dissolved. In some cases, an amorphous-like structure was occasionally observed in the WEA, indicating that phase transformation may possibly occur. The WEAs were studied by using scanning electron microscopy (SEM), transmission electron microscopy (TEM) and Electron back scattered diffraction (EBSD). The result showed that WEA is dominated with an amorphous phase with martensite, austenite and carbides embedded interior. A distinct interface between the matrix and the WEA was present. In addition to grain refinement down to nanometers, phase transformation including amorphization and austenitization happened in WEAs. The content of austenite was increased from 2% in the matrix to 20% in the WEA. The analysis showed that phase transformation is controlled by plastic deformation mechanism.

1 Introduction

The rolling bearings are always failed under rolling contact fatigue with much shorter lives than that designed, especially in wind turbine gearbox bearings. Plastic deformation is highly localized into a narrow region at both contact surface and subsurface until the number of cycles reaches a certain level. When inspecting cross-sections of failed bearings or samples after RCF using optical microscope, white etching matters such as white etching bands [1,2], white etching areas (WEAs) [3-7] and white etching layers (WELs) [8] are observed in different depths below the contact surface. These white microstructural changes are different from base materials in terms of compositions and hardness.

White bands and WEAs are commonly observed in bearing steels and have been extensively studied, especially WEAs [1-7]. As a result of microstructural alteration, WEA has been accepted as one of the primary flaking failure mode in RCF since it causes severe local inhomogeneity of microstructure and cracks associated. Extensive investigations have been carried out on WEAs regarding microstructural evolution and formation mechanisms. The results showed that WEAs are composed of nano-sized ferrites [1,2,5]. The hardness of WEAs varies with materials. They are harder than the base materials of martensitic steels [9-11], but softer than the base materials of carbide-free nano-sized bainitic steels [12]. RCF is a cyclic dynamic loading mode involves repeated rolling and sliding. Shear morphology present in WEA such as elongated carbides [5] once plastic deformation accumulates to an extent level during RCF. WEA composed of nano-sized ferrites with grain size of tens nanometers. Some authors proposed that WEAs may possible form as a result of grain refinement under mechanism of dynamic recovery or low temperature recrystallization [5,9]. An amorphous-like region was also presented in WEA except the nanocrystallites [9,10]. We have provided the evidence for an amorphous phase exist in the WEA [3,4].

Although WEAs have been studied extensively, but an in-depth comparison of this deformed matter has rarely been given. This paper presents a better understanding of the composition in the WEA and analyse the formation mechanism in detail. There may have relations since they are both the result of microstructural changes in bearing steel under RCF.

2 Materials and methods

* Corresponding author: li_shuxin@163.com

The failed sample with WEAs in this paper is tempered martensite steel of AISI52100 with the main chemical composition of (wt.%): C, 0.09; Cr, 1.87; Si, 0.31; Mn, 0.32; P, ≤0.027; S, ≤0.020; Fe, balance [3]. The materials were austenitized at 860°C for 2 hours and quenched in oil to room temperature, subsequently followed by tempering at 160 °C for 1 h. This material consists of matensite, bainite and retained austenite (less than 2 vol%). The tests were carried out on a RCF machine with various SRRs of 5%-20% and contact pressure ranging from 0.9 GPa to 2.0 GPa.

The RCF sample bearing were sectioned along circumferential and axial direction. The inspection was conducted using scanning electron microscopy (SEM), transmission electron microscopy (TEM), and electron backscattered diffraction (EBSD) and hardness test. Especially, focused ion beam (FIB) installed in SEM was used to prepare a sample for TEM analysis from the deformed location.

3 Results

3.1 SEM morphology of WEA

Fig. 1(a) displays a typical WEA at the depth of 810 μm below the contact surface [4]. It has inclined to 45 degrees with the contact surface. According to the researches of others, WEA can be formed at any angle with the contact surface [13-16]. WEA was distributed along the crack and split into two parts with the widest part of 2.6 μm on the left side. The microstructure next to the WEA does not show an obvious difference with the matrix except slight sheared morphology. Fig. 1(b) shows another WEA initiated at non-metallic inclusion, some researchers call it butterfly [7,9]. The cracks initiated around inclusion and propagated parallel to contact surface. Besides, there is another WEA distribution along the cracks which propagated parallel to roller surface, as shown in Fig. 1(c).

3.2 Stress-Number of cycles curve

Fig.2 presents the maximum contact pressure versus the number of cycles of RCF. It can be seen that the RCF lives decrease as the maximum contact pressure decreases with the exception of the point at 1.5 GPa with the SRR of 5%. WEA was observed at various contact pressure levels as marked in Fig.2. The effect of the SRR on the formation of WEA exhibits much scatter. WEA was formed at the contact pressure of 0.8 GPa with the SRR of 10%, but was absent at 1.0 GPa and 15%. The same phenomenon occurred at 1.2 GPa and 10% and 1.4 GPa and 15%, where WEA was observed under the former condition but not under the latter. This indicates that a higher contact pressure and a larger SRR do not necessarily generate WEA as the formation of WEA depends not only on stress and deformation state but on local material microstructure as well. Coincidently, WEA was observed at all SRRs of 10% regardless of the contact pressure level.

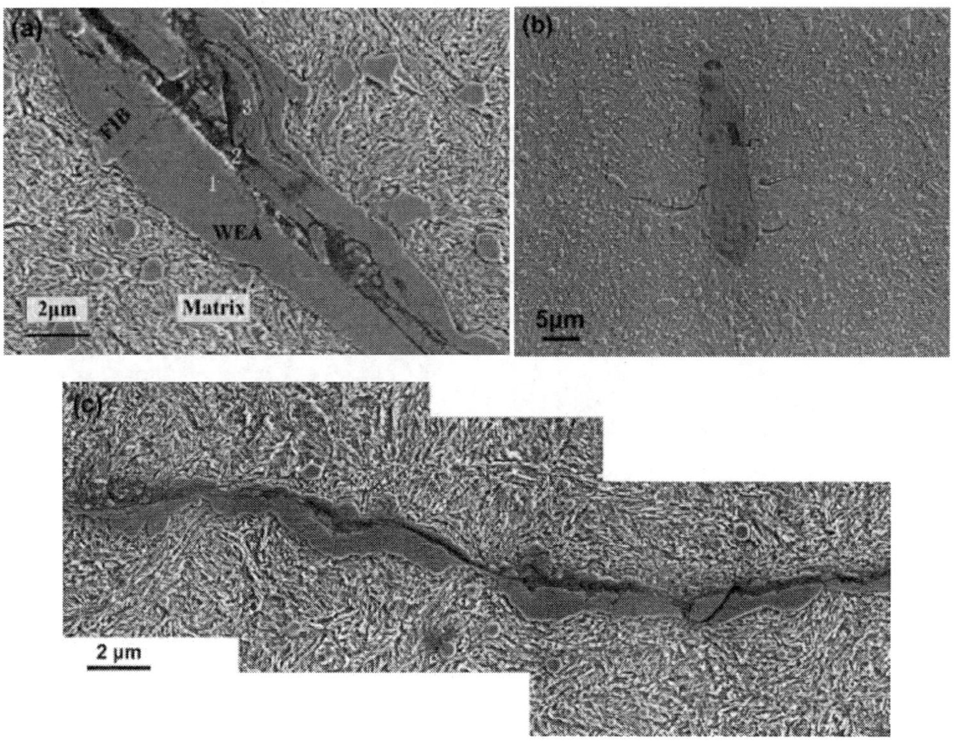

Fig. 1. SEM images (a)WEA along the crack [4]; (b) WEA initiated at non-metallic inclusion; (c) WEA propagate along the crack which parallel to contact surface

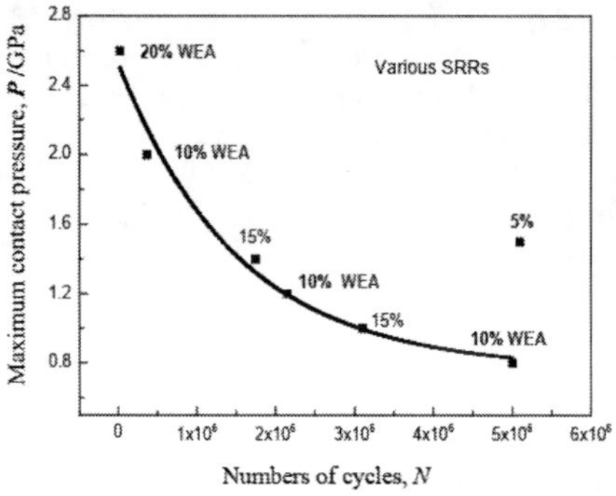

Fig. 2. Contact pressure versus the number of cycles of RCF at various SRR

3.3 TEM microstructure

TEM sample of WEA were taken from designated location as showed in Fig. 1(a) and the bright field TEM images in Fig. 3. The deformed matrix consists of refined grains and deformed matrix, such as Fig. 3(a). Grains were refined compared to the base material. We can hardly identify the original tempered martensite and the microstructure filled with dislocation cluster as circled. The size of the grains in WEA refined down to nano-sized ranging from 3 nm to 50 nm, as shown in Fig. 3(b). The WEA and matrix have different microstructure and the carbides were disappeared. The sheared morphology is no longer exist in the WEA and it contains amorphous phase based on deep inspection, such as Fig. 3(b).

Fig. 3(a) also shows the selected area diffraction (SAD) patterns in order to further understand the composition of the WEA. The thick halo ring that occurred in area "B" indicated an amorphous structure, which is different from "A". Fig. 3(b) is the HRTEM image of amorphous structure found in the WEA around place "B" in Fig. 3(a). Besides, martensite α', austenite γ, and carbide θ were indexed in the WEA [4], which is also different with the SAD pattern in the matrix.

3.4 EBSD inspection

The size and the shape of grains can be measured by EBSD. Besides, phase identification and percentage calculation can also be performed. Fig. 4 showing the EBSD-orientation maps of the WEA and matrix, grains size distribution was also presented. The grains were seriously refined in the WEA compared with the matrix, as shown in Fig. 4(a). The WEA present an amorphous feature. The sizes of the matrix near the WEA were found from approximately 100 nm to 600 nm and the WEA filled with the size of 30 nm particles, as described in Ref [4]. Fig. 5 gives the EBSD phase maps in WEA. The red is austenite and the green is martensite. The content of austenite in WEA is 20% and in the matrix is 2%-3%. It shows that the content of austenite in the WEA is much higher than matrix. Austenite is uniformly distributed in WEA.

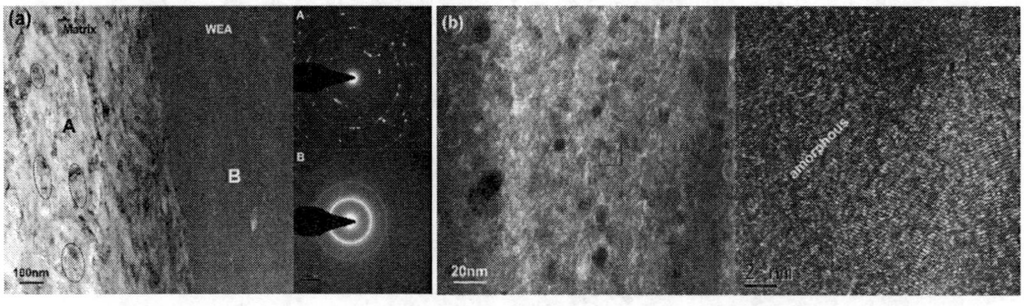

Fig. 3. TEM images (a) Interface between the matrix and the WEA; (b) HRTEM image of morphous structure in the WEA

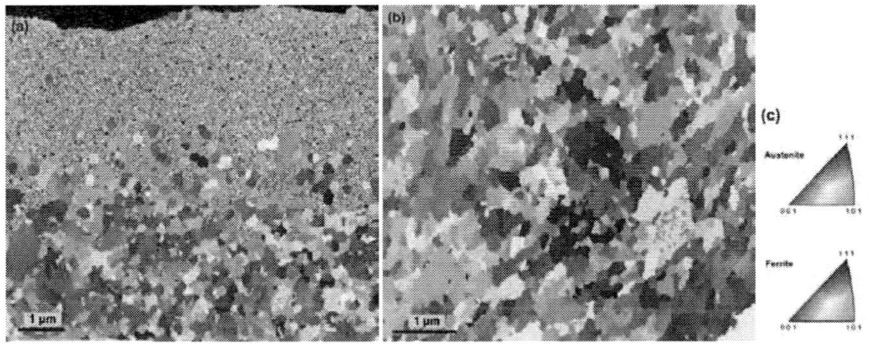

Fig. 4. EBSD-orientation maps [4] of (a) the interface; (b) the matrix; (c) standard inverse pole figure maps

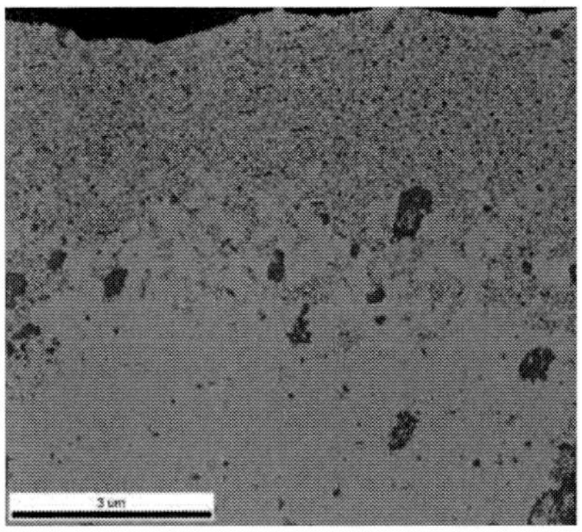

Fig. 5. EBSD phase maps with martensite bcc colored green and austenite fcc colored red

4 Discussion

4.1 Contact pressure threshold for WEA formation

The P-N diagram (Contact Pressure-Number of cycles) proposed by Warhadpande et al [12] suggested that there exists a threshold contact pressure of 2.5 GPa, below which no microstructural change would occur in RCF. WEA was largely observed under the contact pressure of 2.5-5.6 GPa [17,18]. However, in some cases where WEA is formed at a low contact pressure of 1-2 GPa, sliding has to be introduced. A typical example was given by Evan's et al [7] that WEA was generated at 1.2 GPa with the SRR 14.8%. A large number of WEAs were found at 1.9 GPa with the slip ratio of -30% [13]. But no WEA was formed in the positive sliding even when the SRR reached 30% and only cracks were produced, implying that microstructure alternation does not happen in positive sliding. However, this observation conflicts with Evan's [7] and our current result that WEA forms in the positive sliding direction. The current result shows that WEA can be generated at the contact stress of 0.8 GPa with the SRR of 10%. In view of the above discussion, it is rational to suggest that it is the combined co-action of the contact pressure and the sliding that determines the formation of WEA. The threshold of the contact pressure may not exist under the current RCF condition. As mentioned above, WEA is the result of microstructural alternation. Its generation is significantly dependent on local material microstructure. This can be proved that a higher contact pressure with a larger SRR (for example, comparing the points of 1.2 GPa and 10% (WEA) and 1.4 GPa and 15% (without WEA) in Fig.1) did not generate WEA.

4.2 Grain refinement in the WEA

Grains in the WEA were refined to nano-sized and grains next to the WEA were refined to micro or nano scaled based on TEM inspection. It is reported that the materials would became unstable under certain loading condition in Ref. [19]. Under the large deformation, grains are refined through disintegrating and breaking down the elongated grains by entangled dislocation clusters. The process is also accompanied with splitting and twisting, which can further accelerate the refinement, as indicated in Fig. 3(a). Dougherty et al [20] stated that shear localization is a mode of stress accommodation, plastic flow is confined to zones of intense shear surrounded by regions of lower, more homogeneous plastic deformation.

Multiple mechanisms were proposed to explain this phenomenon, one of which stated that the white etching matter formed around a subsurface crack is a symptom of crack faces rubbing which leads to local changes in the material microstructure [21]. There was a strong evidence that deformed microstructure dominate the WEA, supporting the theory that it is a deformation-induced but not a thermal-induced process.

4.3 Phase transformation and amorphization in the WEA

The amount of austenite increased a lot and consists of an amorphous phase indicated that phase transformation occurred in the WEA. This essential microstructural change appeared during the formation of the WEA compared to previous study, which consist of nanocrystalline ferrites [5,9,22]. It is generally accepted that the existence of austenite provided an evidence of phase transformation [23,24].

Baumann [25] stated that temperature can induce ferrite transformed to austenite. Pyzalla [26] reported that the transformation from ferrite and pearlite to austenite occurred due to austenitization, and austenite transformed to martensite during cooling. The heat that raise up the temperature comes from the friction during the sliding of two contact surfaces. The WEA were not generated at the surface even though the highest temperature appeared at top surface [4]. And the temperature cannot reach the austenitization temperature of 720 °C because lubrication play a role in cooling during RCF. Another phenomenon that WEA formed with various depth below the surface support that the friction of contact surface cannot became the formation condition of WEAs.

Phase transformation has been proved under mechanically induced mechanism. Ivanisenko [27] proposed that shear deformation drove the transformation from bcc to fcc in Fe-C steel after grains are refined into nano scale. Molecular dynamic simulation confirmed that the formation of nanocrystalline structure and dissolved carbides increased Gibbs free energy, providing driving force from bcc to fcc. Laptaoie [28] stated that the transformation from bcc to fcc was controlled by the stress. The fcc structure is not stable because it has a high energy make the transformation from bcc to fcc is reversible. It will transforme back to bcc in the condition of room temperature and unloading. Phase transformation is under formation induced mechanism according to above discussion.

A narrow transition zone of 3-5 nm happened in the interface of the WEA and matrix showed in Ref. [4] explained that the ordered crystals transformed into a less ordered and then a disordered amorphous phase. An average size of nanocrystallites observed in the amorphous is 5 nm implies that may exist a transformation size. But this critical size in microstructural changes cannot be captured during RCF. When the size of nanocrystallite reached a critical value, it would preferentially become amorphous. Partial amorphization lead to the amorphous phase and nanocrystallites coexist in the WEA, as shown in Fig. 3(b). A full amorphization occurs when the energy of the system is sufficient for all nanocrystallites to reach a critical size.

5 Conclusion

The microstructural changes under RCF is a complex process. We can draw the following conclusions based on the above analysis:

1) The hardness of WEA increased significantly due to the refinement of grains.
2) WEA involves not only nanocrystallization, but amorphization and phase transformation as well. It consists of an amorphous phase with nano martensite, austenite and carbides embedded interior. The content of austenite in WEA is 20% and in the matrix is 2%-3%. The content of austenite in the WEA was increased 10 times compared to the matrix.
3) The nanocrystallization was initiated under large shear plastic deformation. When the grain size were refined into 3-5 nm, the amorphization occurred. The amorphization is partial, leading to the coexistence of nanocrystallites and an amorphous phase. During this process, a phase transformation from martensite to austenite were stimulated under the large deformation.

References

[1] J.H. Kang, B. Hosseinkhani, P.E.J. Rivera-Diaz-del-Castillo, Mater. Sci. Technol. **28**, 44-49 (2012)
[2] W. Holweger, M. Wolf, D. Merk, et al, Tribol. Trans. **58**, 59-69 (2015)
[3] S.X. Li, Y.S. Su, X.D. Shu, J.J. Chen, Wear **380-381**, 146-153 (2017)
[4] Y.S. Su, S.X. Li, S.Y. Lu, et al, Int. J. Fatigue **105**, 160-168 (2017)
[5] M.H. Evans, Mater. Sci. Technol. **11**, 1-37 (2016)
[6] J. Lai, K. Stadler, Wear **364-365**, 244-256 (2016)
[7] M.H. Evans, A.D. Richardson, L. Wang, et al, Tribol. Int. **75**, 87-97 (2014)
[8] L.B. Wan, S.X. Li, S.Y. Lu, et al, Wear **396-397**, 126-134 (2018)
[9] M.H. Evans, J.C. Walker, C. Ma, et al, Mater. Sci. Eng. A **570**, 127-134 (2013)
[10] H. Harada, T. Mikami, M. Shibata, et al, ISIJ. Int. **45**, 1897-1902 (2005)
[11] G. Baumann, K. Knothe, H.J. Fecht, Nanostruct. Mater. **9**, 751-754 (1997)
[12] W. Solano-Alvarez, E.J. Pickering, M.J. Peet, et al, Acta Mater. **121**, 215-226 (2016)
[13] B. Gould, A. Greco, Tribol Lett. **62**, 1-14 (2016)
[14] A. Warhadpande, F. Sadeghi, R.D. Evans, Tribol. Trans. **56**, 349-358 (2013)
[15] J. Lai, K. Stadler, Wear **s364–365**, 244-256 (2016)
[16] Š. Viktorija, S. Alexander, L. Wang, et al, Int. J. Fatigue **100**, 148-158 (2017)
[17] J.H. Kang, B. Hosseinkhani, C.A. Williams, et al, Scripta Mater. **69**, 630-633 (2013)
[18] J.H. Kang, R.H. Vegter, Mater. Sci. Eng. A **607**,

328-333 (2014)
[19] Q. Wei, D. Jia, E. Ma, K.T. Ramesh, Appl. Phys. Lett. **81**, 1240-1242 (2012)
[20] L.M. Dougherty, E.K. Cerreta, E.A. Pfeif, et al, Acta. Mater. **55**, 6356-6364 (2007)
[21] Y. Kadin, M.Y. Sherif, Int. J. Fatigue **96**, 114-126 (2017)
[22] A. Grabulov, U. Ziese, H.W. Zandbergen, Scripta Mater. **57**, 635-638 (2007)
[23] Y. Zhou, J.F. Peng, Z.P. Luo, et al, Wear **362-363**, 8-17 (2016)
[24] M. Steenbergen, R. Dollevoet, **47**, 361-372 (2013)
[25] G. Baumann, H.J. Fecht, S. Liebelt, Wear **191**, 133-140 (1996)
[26] A. Pyzalla, L. Wang, E. Wild, T. Wroblewski, Wear **251**, 901-907 (2001)
[27] Y. Ivanisenko, I. MacLaren, X. Sauvage, et al, Acta. Mater. **554**, 1659-1669 (2006)
[28] A. Latapie, D. Farkas, Model. Simul. Mater. Sci. Eng. **11**, 745-753 (2003)

MATEC Web of Conferences **165**, 11005 (2018) https://doi.org/10.1051/matecconf/201816511005
FATIGUE 2018

Finite Element Analysis of Fretting Fatigue Fracture in Lug Joints Made of High Strength Steel

Reza Hojjati Talemi[1,2,*], *Jie* Zhang[3,4], *Stijn* Hertelé[3], *Wim* De Waele[3]

[1]ArcelorMittal Global R&D Gent-OCAS N.V., Pres. J.F. Kennedylaan 3, 9060 Zelzate, Belgium
[2]Present address: Department of Materials Engineering, KU Leuven, Kasteelpark Arenberg 44, 3001 Leuven, Belgium
[3]Soete Laboratory, Department of Electrical Energy, Metals, Mechanical Constructions and Systems, Ghent University, Belgium
[4]SIM vzw, Technologiepark 935, BE-9052 Zwijnaarde, Belgium

Abstract. Many structural applications are aiming for weight reduction by using high strength steel. In a lug joint the load is transmitted by a pin, which leads to a pressure distribution on the hole in the lug. When a lug joint is subjected to axial cyclic loading conditions, the stress distribution becomes multiaxial, i.e. a combination of normal and tangential stresses. In such loading case, a fretting crack initiates at the contact interface between the pin/lug connection which is followed by a fatigue crack propagation up to the final rupture of the lug. In this study, the fretting fatigue crack initiation and propagation in a pin/lug joint are simulated using multiaxial fatigue criterion and fracture mechanics, respectively. To do so, first a 2D finite element model is developed for obtaining stresses and strains at the contact interface in a pin/lug joint. Using the extracted data, fretting fatigue failure parameters are analysed. Next, the obtained stresses and strains are used to estimate the crack initiation lifetime using a fatigue multiaxial critical plane model. A 3D model is set-up to simulate the crack propagation using eXtended Finite Element Method (XFEM). Eventually, the predicted total fatigue lifetimes are compared against experimental observations taken from literature.

1 Introduction

The production of lightweight and reliable structures is primarily achieved by using high-strength materials. This special place is occupied by micro-alloyed High Strength Steels (HSS), due to the wide availability of steel as a material, and their cost being significantly more acceptable in comparison to other alternative materials. The use of HSS may reduce the weight, and thus lower the costs of procurement, manufacturing, installation and testing of a steel structure. Nowadays, it is impossible to imagine the construction of mobile and fixed cranes, passenger and freight cars (road or rail), or equipment for mining, without the use of HSS. Given its high strength, and considering the type of structures to be used for, these steels should meet essential mechanical, technological and design requirements.

Namely, the higher the yield stress and tensile strength, design codes are considered to require greater "reserve" in case of overloading, and thus significant plastic strain that would precede the eventual fracture or damage of a structure. Thus, in order to ensure higher safety in the application of HSS, standard tensile and impact tests must be accompanied by appropriate tests in terms of fatigue and fracture mechanics. This is particularly important considering the fact that the highly efficient methods and procedures for the assessment and design of various types of structures, are based on the knowledge of these parameters (Eurocode and so on).

Figure 1 shows a crawler crane in which the counter weight is supported by a steel chain, which consists of a series of high loaded bars made of HSS e.g. S1100Q as reported in [1].

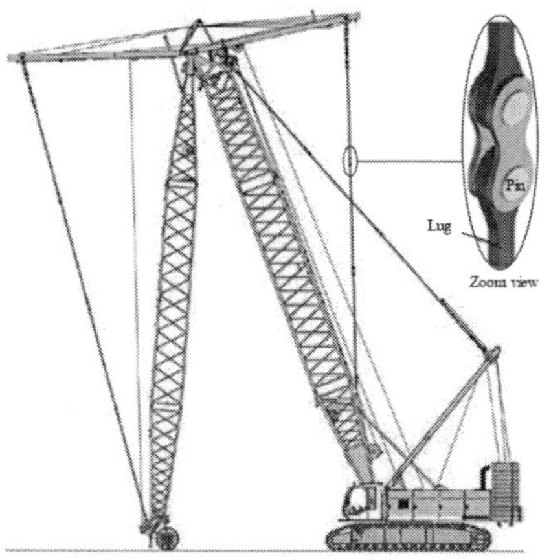

Fig. 1. Schematic view of a crawler crane and pin/lug joint made of HSS.

* Corresponding author: reza.hojjatitalemi@kuleuven.be

The steel chain is made of different pin-lug connections. In a lug joint the load is transmitted by a pin, which leads to a pressure distribution on the hole in the lug. When a lug joint is subjected to axial cyclic loading conditions, the cyclic stress distribution becomes multiaxial, i.e. a combination of normal and tangential stresses. In such loading case, a fretting crack may initiate at the contact interface between the pin/lug connection which is followed by fatigue crack propagation up to the final rupture of the lug. This phenomenon is known as fretting fatigue failure, and is well documented in literature for different applications such as automotive industry [2, 3], bolted and riveted connections [4, 5], steel cables, bearing shafts and steam/gas turbines. Fretting fatigue may reduce the lifetime of a component by half or even more, in comparison to plain fatigue [6]. This reduction is due to a complex multiaxial and non-proportional stress state which occurs at a contact interface under partial slip loading conditions [7-10]. Therefore, it is very important to understand the underlying failure mechanism. This goal could be achieved by separating the failure process into different stages, and subsequently estimating the fretting fatigue crack initiation and fatigue crack propagation lifetimes from the final lifetime.

In this study, the fretting fatigue crack initiation and propagation in a pin/lug joint are simulated using a multiaxial fatigue criterion and fracture mechanics, respectively. To do so, first a 2D plane strain finite element model is developed for obtaining stresses and strains at the contact interface in a pin/lug joint. Using the calculated stress distributions, fretting fatigue failure parameters are analysed. Next, the extracted data are used to estimate the fatigue crack initiation lifetime using the Smith, Watson, and Topper (SWT) damage model. To analyse the fretting fatigue crack propagation part of the total lifetime, a 3D model is set-up to simulate the fatigue crack propagation using eXtended Finite Element Method (XFEM). A cubic spline description for crack geometry is used to characterize the evolving crack front, whose rate and three-dimensional growing shape is controlled by a crack propagation law. Eventually, the predicted total fatigue lifetimes are compared against experimental observations taken from literature [1].

2 Theoretical background

2.1 Fatigue crack initiation

To numerically estimate the number of cycles spent in the initiation phase, the SWT multiaxial fatigue criterion was used based on local stress and strain distributions. SWT is a local approach, meaning that it recognizes the localized nature of fatigue damage, and correlates a local damage parameter (e.g. strain, energy) with the number of cycles required to initiate a macroscopic crack.

In the lug joint which is subjected to cyclic external loads, the behaviour of material at the contact interface between the pin and the lug is best considered in terms of strain. Since fatigue damage is assessed directly in terms of local strain, this approach is also called the "local strain approach". A reasonable expected fatigue life, based on the initiation or formation of small macro-cracks, can then be determined. The most well-known relations in this area are the proposals by Basquin [11], Coffin [13], Manson [12] and Morrow [13]:

$$\Delta\sigma/2\mathrm{E} = \Delta\varepsilon_e/2 = \frac{\sigma_f'}{E}(2\mathrm{N_i})^{b} \quad (1)$$

$$\Delta\varepsilon_p/2 = \varepsilon_f'(2\mathrm{N_i})^{c} \quad (2)$$

$$\varepsilon_a = \Delta\varepsilon_e/2 + \Delta\varepsilon_p/2 = \frac{\sigma_f'}{E}(2\mathrm{N_i})^{b} + \varepsilon_f'(2\mathrm{N_i})^{c} \quad (3)$$

where $2N_i$ is the number of reversals to local failure; $\Delta\varepsilon_t$, $\Delta\varepsilon_e$ and $\Delta\varepsilon_p$ are, respectively, the total, elastic and plastic strain ranges; $\Delta\sigma$ is the stress range and E is the Young's modulus. The constants in these relations may be determined from fatigue tests on smooth specimens under strain-controlled conditions. These constants can be used to evaluate fatigue performance of different materials and can be defined as:

- σ_f'= Fatigue strength coefficient,
- b= Fatigue strength exponent,
- ε_f'= Fatigue ductility coefficient,
- c= Fatigue ductility exponent.

2.1.1. SWT fatigue damage model

In this criterion, the fatigue damage parameter SWT is described as a stress and strain product and the critical plane is the plane where the normal strain range is maximum [14]. The SWT parameter proposes a suitable relationship that includes both the cyclic strain range and the maximum stress. The SWT parameter is used in the analysis of both proportionally and non-proportionally loaded components for materials that fail primarily due to mode I tensile cracking. The SWT parameter for multiaxial loading is based on the principal strain range, $\Delta\varepsilon_1$ and maximum stress on the principal strain range plane, $\sigma_{n,max}$. The SWT fatigue damage parameter is defined as:

$$\sigma_{n,max}(\Delta\varepsilon_1/2) = \frac{\sigma_f'^2}{E}(2\mathrm{N_i})^{2b} + \varepsilon_f'\sigma_f'(2\mathrm{N_i})^{c+b} \quad (4)$$

2.2 Fatigue crack propagation

In terms of fatigue crack propagation analysis, the framework [15] based on extended finite element is used to calculate fatigue crack propagation lifetime under complex loading automatically.

2.2.1 Extended finite element method

XFEM was developed to relieve a local discontinuity by introducing enriched functions to the element, as shown in Figure 2. It solves the singularity problem by adding more integration points within elements instead of having to remesh and relax nodes. Due to this method, accuracy and computational time consumption are drastically improved.

A key development that facilitates the treatment of cracks in the XFEM analysis is the description of the crack geometry, because the mesh is not required to conform to the crack geometry. The level set method, which is a powerful numerical technique for analyzing and computing interface motion, fits naturally with the XFEM and makes it possible to model arbitrary crack growth without re-meshing. Figure 2 shows the schematic representation of an XFEM crack in a single mesh as reported by [16].

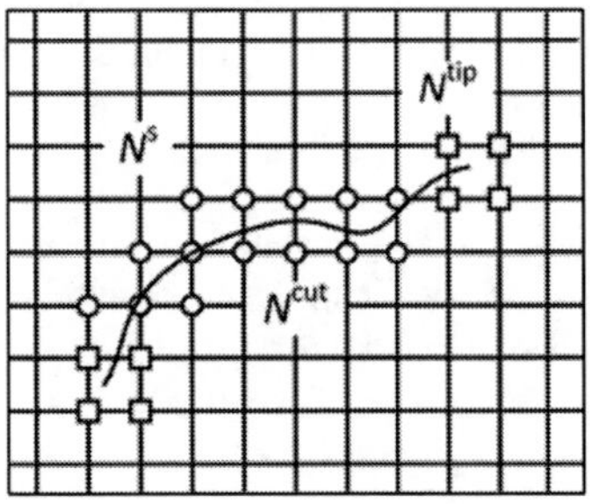

Fig. 2. Schematic of extended finite element method [16].

For the purpose of fracture analysis, the enrichment functions typically consist of near-tip asymptotic functions that capture the linear-elastic singularity around the crack tip and a discontinuous function that represents the jump in displacement across the crack surfaces. The approximation for a displacement vector function u with the partition of unity enrichment is

$$\boldsymbol{u} = \sum_{i \in N^s} N_i(x) u_i + \sum_{i \in N^{cut}} N_i(x) a_i H(x) + \sum_j b_j^i B_j(x) \quad (5)$$

where N^s, $N_i(x)$, N^{cut}, $H(x)$, N^{tip} and $B_j(x)$ are usual nodal elements, normal node functions, through crack elements, discontinuous jump functions, partial crack elements and crack tip functions, respectively.

2.2.2 Automated framework for three-dimensional crack propagation

The basic idea of the framework is using information of discrete stress intensity factors along the fatigue crack front to determine how it propagates (comprising both the amount and trajectory of crack growth). The framework consists of pre-processing, post-processing and analysis modules, communicating with ABAQUS, in Python programming language.

The pre-processing module establishes the model by constructing crack geometry, initializing material properties and defining boundary conditions and external load. The post-processing module extracts and processes discrete necessary data. The analysis module creates the simulation iteratively depending on the results generated in the previous increment. The extended finite element feature within ABAQUS is hereto utilized. Increment and direction of propagating crack front are calculated based on the crack propagation rate curve of the material and an out-plane direction criterion.

2.2.3 Life prediction of fatigue crack propagation

The stress intensity factor range can be expressed as a function of crack length. In this study a cubic spline interpolation function is used, as shown in in Figure 3. The total life of crack propagation can be calculated by summing up all amounts of cycles corresponding to the crack increments using Equations 6 and 7.

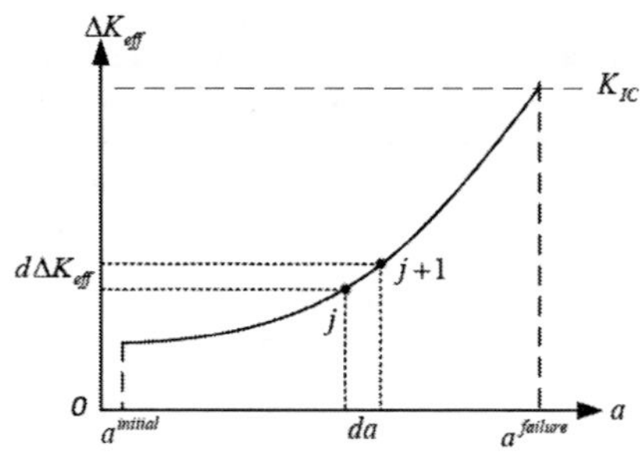

Fig. 3. Schematic of stress intensity factor versus the crack length growth rate.

$$da/dN = C(\Delta K_{eff})^m \quad (6)$$

$$N = \int_{a^{initial}}^{a^{failure}} \left(da/C\left(\Delta K_{eff}\right)^m\right) \quad (7)$$

3 Numerical model and material

3.1 Finite Element Model

2D and 3D models were developed to simulate the fretting fatigue crack initiation and propagation lifetimes using ABAQUS commercial software, respectively, as shown in Figures 4 and 5.

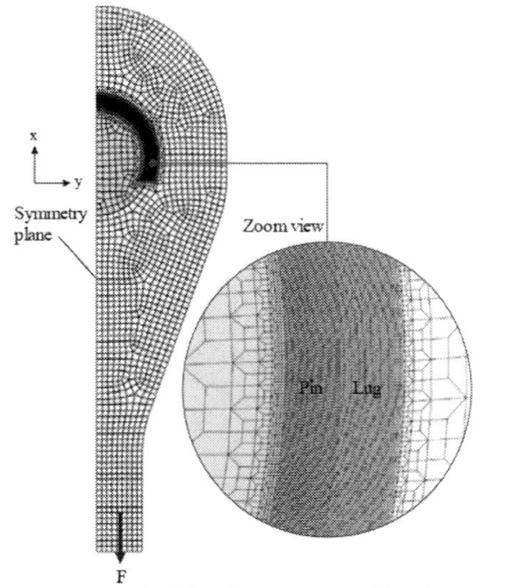

Fig. 4. 2D plane strain finite element model with a fine mesh at the contact interface between the pin and the lug.

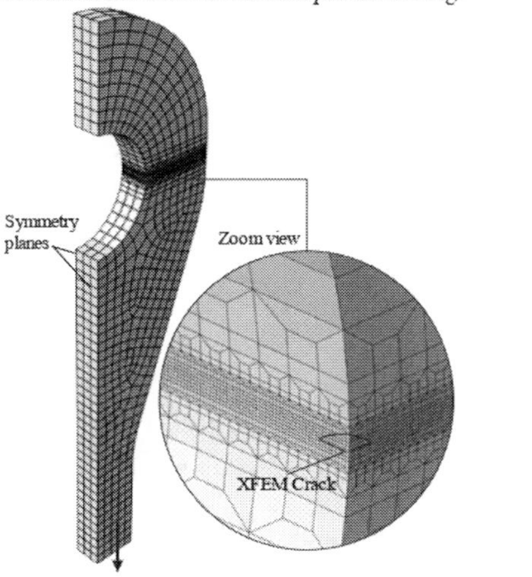

Fig. 5. 3D Finite element model with relatively coarser mesh for inserting the XFEM crack to study the fatigue crack propagation lifetime in the pin/lug joint.

For 2D and 3D models, as shown in Figure 4 and 5, half and quarter of the lug geometry were simulated, respectively. The loading and boundary conditions for both models were selected based on the study published by S.Glodez et al. [1]. Five fatigue axial loads were applied ranging from 500kN to 880kN with loading ratio of 0.1. The applied loads result in axial stresses in the cross section of the lug which are 333.33MPa, 400MPa, 466.67MPa, 533.33MPa and 586.67MPa.

For the 2D model, a two-dimensional, plane strain, 4-node (bilinear), quadrilateral, reduced integration element (CPE4R) was used. A mesh size of 0.1m × 0.1mm was considered at the contact interface between the pin and the lug and decreased gradually far from the contact region, as depicted in Figure 4. The contact between the pin and the lug was defined using a master-slave algorithm. The outer surface of the pin was defined as a slave surface and the inner surface of the lug was defined as a master surface. A penalty formulation of friction was included in the contact pair to define the frictional behaviour of the connected parts. The contact was considered as lubricated, having a low friction coefficient of $\mu = 0.05$.

For the 3D model, a three-dimensional, eight-node linear hexahedral continuum element (C3D8R) was used in order to mesh the lug component. A mesh size of 0.2 × 0.2 × 0.2 mm^3 was considered at the potential crack propagation regions and increased gradually away from the area of interest. Moreover, a rigid part was used to represent the pin.

3.2 Material parameters

In this study S1100Q HSS grade was used to analyse the fretting fatigue failure mechanism of the lug joint. Table 1 shows all material parameters that were used for simulating both 2D and 3D models, along with the parameters needed for estimating the fretting fatigue crack initiation and propagation lifetimes of the lug joint taken from the literature [1]

Table 1. Material properties of S1100Q HSS grade [1].

Elastic modulus	E [GPa]	194889
Poisson's ratio	ν [-]	0.3
Yield stress	σ_y [MPa]	1148
Ultimate stress	σ_{ult} [MPa]	1450
Strength coefficient	K [MPa]	2272
Strain-Hardening exponent	n [MPa]	0.109
Fatigue crack initiation parameters		
Fatigue strength coefficient	σ_f' [MPa]	2076
Fatigue ductility coefficient	ε_f' [-]	9.93
Fatigue strength exponent	b [-]	-0.0997
Fatigue ductility exponent	c [-]	-0.978
Fatigue crack propagation parameters		
Threshold stress intensity factor	ΔK_{th} $[MPa\sqrt{mm}]$	315
Critical stress intensity factor	K_{IC} $[MPa\sqrt{mm}]$	2100
Paris's law coefficient	C $[mm/(cycle \cdot MPa\sqrt{mm})]$	2.02×10^{-11}
Paris 's Law exponent	m	2.761

4 Result and discussion

4.1 Fretting Fatigue Stress Distribution

Figure 6 shows the von Mises stress distribution of the pin/lug joint. Figure 7(a) to (d) show different stress distributions, which are normalized using the material's yield stress, plotted versus the normalized contact distance. Figures 7(e) and (f) plot the slip amplitude and the normalized fretting fatigue damage parameter (FFD) at the contact interface for all applied axial loads.

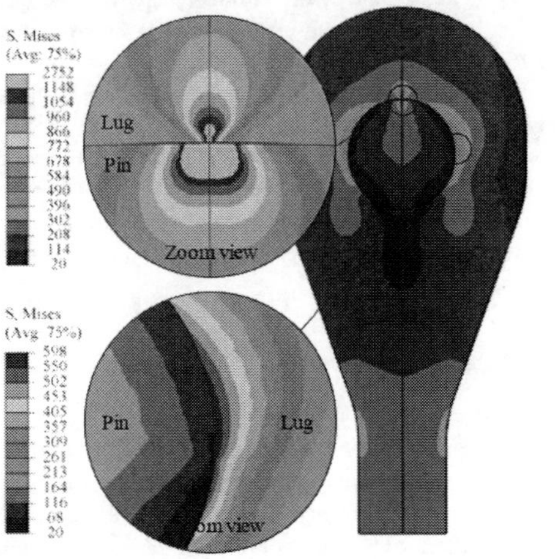

Fig. 6. Von Mises stress distribution at maximum applied load of 500kN ($\sigma_{axial} \approx$ 333MPa).

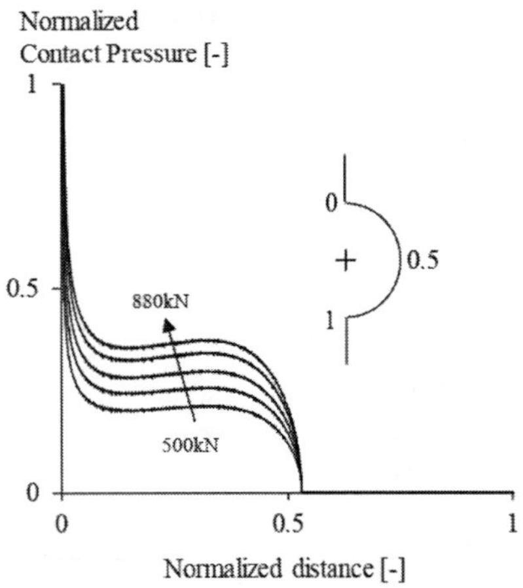

(a)

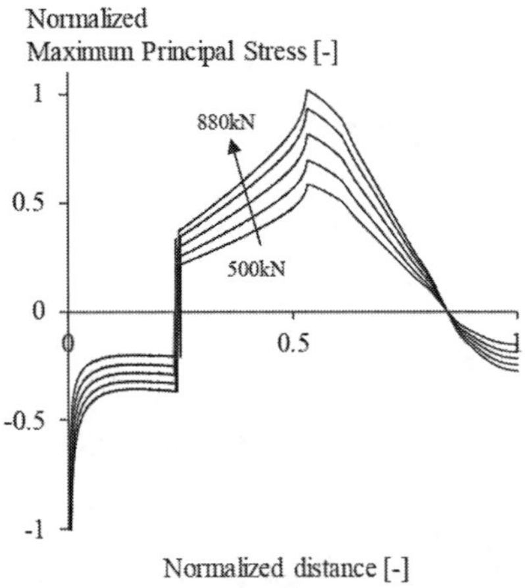

(b)

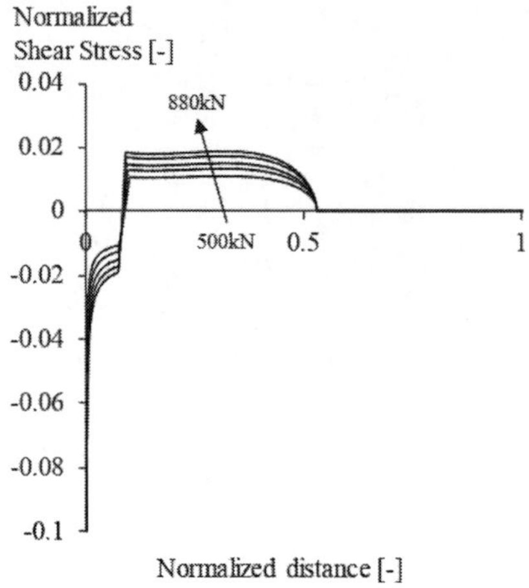

(c)

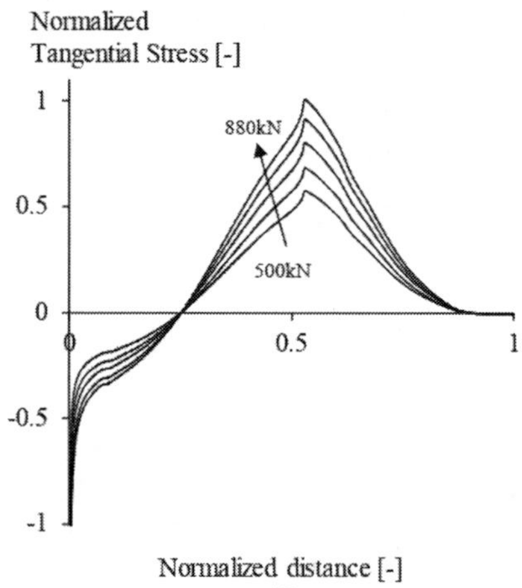

(d)

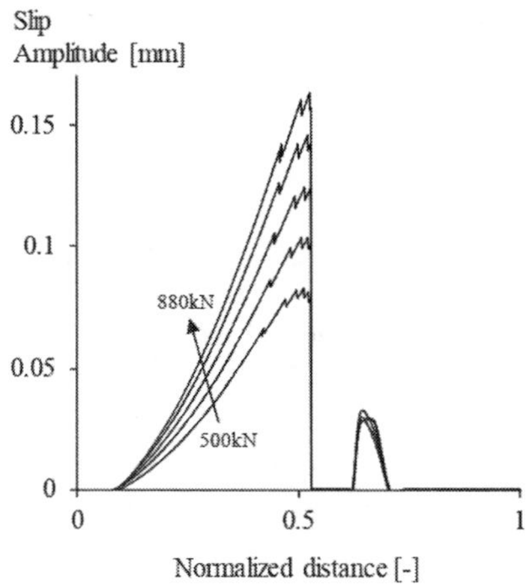

(e)

https://doi.org/10.1051/matecconf/201816511005

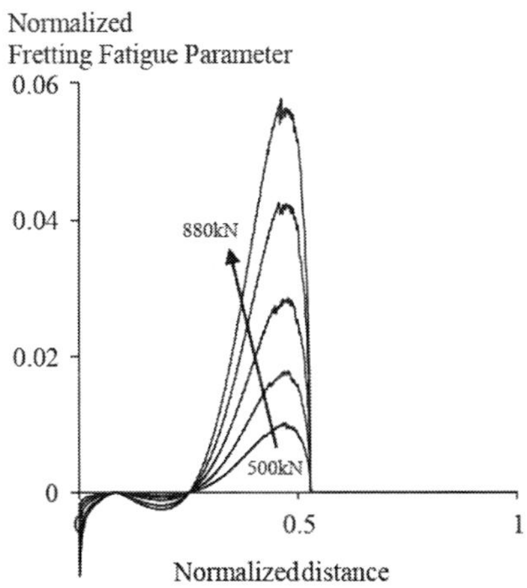

(f)

Fig. 7. Different normalized stress distributions plotted versus the normalized distance. (a) contact pressure, (b) maximum principal stress, (c) shear and (c) tangential stresses along with (e) slip amplitude. (f) the normalized fretting fatigue parameter (FFD) at the contact interface.

As can be seen from Figures 6 and 7(a), the maximum contact pressure is located at the 0 position (see Figure 7). However, the fatigue crack initiates almost at the centre of the contact between the pin and the lug, as illustrated in Figures 6 and 7(b). Figure 7(c) shows that the shear stress at the contact interface increases by raising the applied axial fatigue load from 500kN to 880kN. Nonetheless, as expected because of the low friction coefficient, the magnitude of the shear stress is low compared to other stress values such as contact pressure, maximum principal and tangential stress (σ_{xx}) plotted in Figures 7(a), (b) and (d). This low shear stress results in higher slip amplitude, as shown in Figure 7(e), which is more damaging concerning the fatigue damage accumulation according to the fretting map reported by Vingsbo and Söderberg [17].

Figure 7(f) plots the fretting fatigue damage parameter (FFD) at the contact interface for all applied stress levels. The FFD parameter is a relationship between fretting damage and stress distribution at the contact interface, as first introduced by Ruiz et al. [18]. This parameter is a measure of the frictional work at contact interface which can be written as; $\tau.\delta$, where τ is local value of frictional shear stress and δ is a local slip amplitude within the contact region. To take into consideration the effect of applied axial fatigue load, the axial stress at the contact interface along the loading direction, i.e. tangential stress (σ_{xx}) is multiplied by the frictional work, which results in the FFD parameter. In this study the FFD parameter was normalized against the product of yield stress (σ_y) and lug hole radius (R). The location of maximum FFD is near the centre of the hole for all applied stress levels, where the initial fatigue crack is expected to be initiated, as indicated in Figure 7(f).

4.1 Lifetime Estimation

To calculate the fretting fatigue lifetime, the stress and strain were extracted at the maximum location of FFD parameter along the contact interface for all applied stress levels. Then the SWT parameter in Equation 4 was calculated by means of a Python script using the principal strain range, $\Delta\varepsilon_1$ and maximum stress on the principal strain range plane, $\sigma_{n,max}$. Eventually, the fretting fatigue lifetimes of the pin/lug joint were estimated for all applied axial loads.

Concerning the fatigue crack propagation, an initial fatigue elliptical crack with two axes equal to 0.5mm were imported in the three dimensional FE model, as shown schematically in Figure 5. A pre-defined crack propagation increment was set to 0.5mm. Figure 8 shows the initial XFEM crack inserted in the 3D model and the in-plane fatigue crack propagation lines at a low stress level. Using the approach elaborated in section 2.2.3, the fatigue crack propagation lifetimes were calculated for all applied load levels.

Eventually, the predicted fatigue crack initiation and propagation lifetimes were added for each applied stress level to estimate the total lifetime of the pin/lug connection subjected to the fretting fatigue loading conditions. Figure 9 shows the axial stress versus the number of cycles to final rupture for all applied fatigue load levels. The predicted fretting fatigue lifetime at high stress level, i.e. 880kN axial fatigue load, was compared against the observed experimental data reported in [1]. The figure shows that the estimated results at high stress level are in good agreement with the measured experimental lifetime.

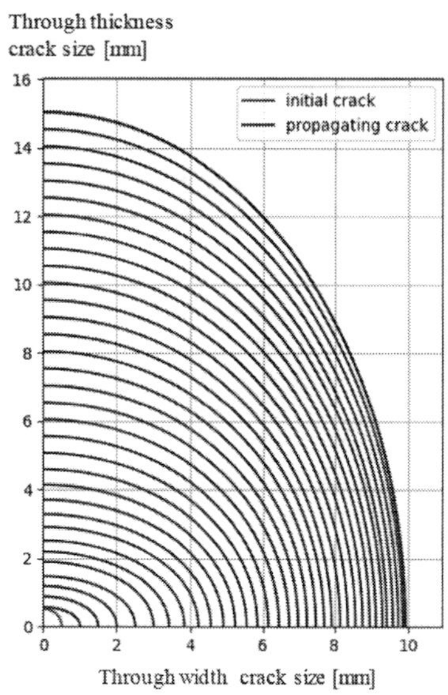

Fig. 8. In-plane fatigue crack propagation line at a low stress level (σ_{axial}= 200MPa).

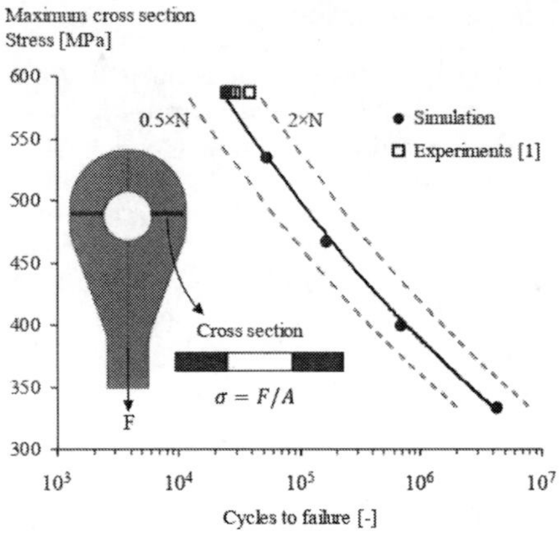

Fig. 9. Cross section stresses versus estimated lifetime of a pin/lug joint subjected to fretting fatigue lifetime. The predicted results at high stress level were compared against the observed data from literature [1].

5 Conclusion

In this study, the fretting fatigue crack initiation and propagation in a pin/lug joint are simulated using a multiaxial fatigue damage criterion and linear-elastic fracture mechanics, respectively. First a 2D finite element model was developed for obtaining stress and strain distributions at the contact interface in the pin/lug joint. Using the extracted data, the fretting fatigue failure parameters were analysed. Next, the obtained stress and strain were used to estimate the fatigue crack initiation lifetime using the multiaxial SWT damage model. Next, a 3D model was set-up to model the fatigue crack propagation using eXtended Finite Element Method (XFEM). A cubic spline description for crack geometry was used to characterize the evolving crack front, whose rate and three-dimensional shape of growth was controlled by the crack propagation law.

Eventually, the predicted total fatigue lifetimes were compared against experimental observations taken from literature. Form the predicted results it could be concluded that the fretting fatigue mechanism is the dominated failure mechanism in a pin/lug joint subjected to cyclic loading conditions. A good correlation was found between the estimated results and the observed experimental results from the literature.

References

1. S. Glodež, M. Knez, N. Jezernik, and J. S. Kramberger, Eng. Failure Anal. **16**, p. 2348-2356 (2009).
2. R. Hojjati Talemi, *Numerical modelling techniques for fretting fatigue crack initiation and propagation*. (Ghent University, 2014).
3. R. Hojjati Talemi, A. Zahedi, and P. De Baets, Int. J. Fatigue **73** p. 58-65 (2015).
4. A. Ferjaoui, T. Yue, M.A. Wahab, R. Hojjati-Talemi, Int. J. Fatigue **73** p.66-67 (2015)
5. C. Jiménez-Peña, R. Hojjati Talemi, B. Rossi, D. Debruyne, Tribol. Int. **108** p. 128-140 (2017).
6. D.A. Hills, Wear, **175** p.107-113 (1994).
7. R.H. Talemi, M.A. Wahab, and P. De Baets, J. Phys.: Conf. Ser. **305** p. 012061 (2011).
8. R. Hojjati-Talemi, M.A. Wahab, and P. De Baets, Proc. Inst. Mech. Eng., Part J 228 p. 470-479 (2014).
9. J. De Pauw, W. De Waele, R. Hojjati-Talemi, and P. De Baets, Int. J. Solids Struct. **51** p. 3058-3066 (2014).
10. R.H Talemi, Numerical measures of non-proportionality degree in incomplete contact subjected to fretting fatigue loading. Theor. Appl. Fract. Mech. 90 p. 33-42 (2017).
11. O. Basquin, *The exponential law of endurance tests in Proc. ASTM* (1910).
12. Jr. Coffin, *A study of the effects of cyclic thermal stresses on a ductile metal, trans. ASME* **76** p. 931-950 (1954).
13. J. Morrow, *Cyclic plastic strain energy and fatigue of metals, in Internal friction, damping, and cyclic plasticity* (ASTM International 1965).
14. K.N. Smith, P. Watson, and T.H. Topper, Journal of Materials **4** p. 767-778 (1970).
15. J. Zhang, S. Hertelé, N. Micone, and W. De Waele, Modelling framework for 3D fatigue crack propagation in welds of offshore steel structures. *in International Conference on Integrity: Reliability: Failure INEGI/FEUP* (2016).
16. N. Moës, J. Dolbow, and T. Belytschko, Int. J. Numer. Meth. Eng. **46** p. 131-150 1999.
17. O. Vingsbo, and S. Söderberg, Wear **126** p. 131-147 (1988).
18. C. Ruiz, P.H.B. Boddington, and K.C. Chen, Exp. Mech. **24** p. 208-217 (1984).

Probabilistic fatigue analysis of ultra-high-performance fibre-reinforced concrete under thermal effects

José D. Ríos, *Héctor* Cifuentes*

ETS de Ingeniería, Universidad de SevilPrrla, Camino de los Descubrimientos s/n, 41092, Sevilla, Spain

Abstract. This paper describes the influence of the temperature and the fibre reinforcement on the flexural fatigue behaviour of an ultra-high-performance fibre-reinforced concrete. Three-point bending fatigue tests were carried out for an ultra-high-performance concrete subjected to different temperatures ranging from room temperature up to 300 °C and considering three different types of reinforcement: a) steel fibres, b) hybrid steel and polypropylene fibres and c) non-reinforced (reference matrix). The fatigue behaviour was assessed from the S-N fields obtained through a probabilistic fatigue model developed by Castillo and Fernández-Canteli. The influence of the type of reinforcement on the fatigue behaviour was analysed by S-N curves. An analysis of the thermal effects in the fatigue life of the ultra-high-performance concrete has also been carried out. The results showed the most suitable fibre reinforcement among the analysed options to get the best fatigue behaviour in accordance to the exposure temperature.

1 Introduction

Ultra-high-performance fibre-reinforced concrete (UHPFRC) is characterized by a significant enhancement of its mechanical properties, especially the tensile strength, in comparison to conventional or high strength concrete [1]. To achieve this ultra-high strength, a low water-to-binder ratio (usually $w/b \leq 0.2$) is used with the addition of ultra-fine admixtures [2], [1]. A denser material is then achieved but with a strongly brittle behaviour. In that way, the addition of steel fibres (in a volume ratio between 2 and 2.5 %) is essential to obtain a more ductile material and improve the tensile strength [3].

The excellent mechanical properties of this type of concrete lead to a significant reduction in the weight of structures [4]. As result, the application of this type of concrete in situations where traditionally, only metallic materials were used, has been extended. Consequently, the interest in the fatigue analysis of fibre-reinforced concrete has grown significantly, as can be seen from the number of recent works related to the fatigue failure under compressive [5]–[8], tensile [9] or flexural loading [10]–[12]. The most common approach to assess the fatigue life of a material is the stress-life (S-N) curves [13]. In this way, it is possible to design structures under fatigue loads (wind, wave or traffic) by more efficiently designs for each application.

Due to the thermal storage capability of concrete [14], its workability and availability, an emerging technology is the use of concrete as thermal energy storage system in solid media. This technology has attracted great attention by researchers. The most of current studies are mainly focused on the energetic performance of concrete storage systems [15]–[17]. Nevertheless, the main concern of the application of this technology is the durability of concrete due to the degradation of cementitious matrix at moderately high temperatures and the bearing of thermal-mechanical fatigue for long-periods. With respect to the influence of temperature, there are many studies focused on the residual [18]–[23] or hot [20], [24]–[26] mechanical properties of concrete subjected to high temperatures. Other authors have studied the thermal effect on concrete reinforced with polymeric fibres [21], [22], [24], [27] and demonstrated as the addition of polymeric fibres prevent the premature cracking due to pore pressure (spalling) and improve the mechanical properties of concrete at high temperatures [27]–[29]. However, no information is available on the fatigue behaviour of ultra-high-performance fibre-reinforced concrete subjected to moderately high temperatures.

In this paper the flexural fatigue behaviour of an ultra-high-performance fibre-reinforced concrete and its corresponding matrix without any reinforcement are reported in terms of its S-N curves at different temperatures (room temperature, 100 °C, 200 °C and 300 °C). It was obtained through a probabilistic analysis that the thermal damage in the cement paste strongly decrease the endurance limit and increase the scattering in the S-N curves. On the contrary, the thermal damage leads to a more ductile behaviour of the concrete matrix and consequently to a softer trend of the S-N curves. Spalling effects were observed at 200 °C for concrete without reinforcement and 300 °C for concrete with fibre

* Corresponding author: bulte@us.es

reinforcement. The additional reinforcement of concrete with PP micro-fibres was thought as an alternative to improve the fatigue behaviour at higher temperatures and satisfactory results were obtained.

The paper is organized as follows. A description of materials, mix preparation and testing procedure is presented in Section 2. The probabilistic fatigue model is summarized in Section 3. The results and discussion of the influence of the addition of fibres and the temperature on the fatigue life is exposed in Section 4. Finally, conclusions are given in Section 5.

2 Experimental Program

2.1 Materials and Mix Preparation

The materials consisted of an ultra-high-performance fibre-reinforced concrete (UHC), the same fibre-reinforced concrete with additional polypropylene (PP) micro-fibres (UHC_PP) and its corresponding cementitious matrix without any type of reinforcement, used as reference concrete (RC). The mix proportions are given in Table 1. The UHC is reinforced with 2.5% in volume ratio of steel fibres (1.25% of microfibres with 13 mm in length, 0.2 mm in diameter and straight end and 1.25% of macrofibres of 30 mm in length, 0.38 mm in diameter and with hooked end).

Table 1. Mix Proportions and Nomenclature

Constituent	RC (kg/m^3)	UHC (kg/m^3)	UHC_PP (kg/m^3)
Cement	544	544	544
Silica fume	214	214	214
GGBS	312	312	312
Quartz sand < 0.315 mm	470	470	470
Quartz sand < 0.800 mm	470	470	470
Superplasticizer	42	42	42
w/binder	0.17	0.17	0.17
Steel microfibre (13 mm in length)	-	98	98
Steel macrofibre (30 mm in length)	-	98	98
PP microfibre (24 mm in length)	-	-	1.2

The PP micro-fibres were of 24 mm in length. Four batching's of UHC and RC type of concrete were prepared to be tested under fatigue loading at four different temperature conditions (room temperature, 100 °C, 200 °C and 300 °C). Two batching's of UHC_PP were casted to be tested at room temperature and 300 °C. The nomenclature of each experimental test is shown in Table 2. All specimens were demoulding after 24 hours of casting, cured in water for 28 days and dried at air after curing process for 24 hours.

Table 2. Nomenclature of Test Conditions

	RC	UHC	UHC_PP
Room temperature	RC_RT	UHC_RT	UHC_PP_RT
100 °C	RC_100	UHC_100	-
200 °C	RC_200	UHC_200	-
300 °C	RC_300	UHC_300	UHC_PP_300

2.2 Three-Point Bending Tests on Prismatic Notched Specimens

Previously to the fatigue life tests, is required to obtain the monotonic three-point flexural strength. The test was carried out in accordance to the EN-14651 standard [30] for fibre-reinforced concrete. Three-point bending tests were performed on four 440x100x100-mm notched specimens (1/6 of height), for each temperature and type of concrete tested, simply supported over a span of 400 mm.

2.3 Flexural Fatigue Tests on Prismatic Notched Specimens

The fatigue tests were performed on fourteen prismatic notched specimens subjected to three-point bending tests, as monotonic tests. A servo-hydraulic machine with a 200-kN dynamic-static actuator was used. The frequency of loading and stress ratio were the same for each test, with values of 4 Hz and 0.2, respectively. The tests were carried out at different stress levels in the range between 0.5 and 0.9 of the monotonic strength. The tests were stopped after specimen failure or after 2 million cycles, whichever occurred first.

In the cases of the tests at high temperature, the testing equipment was specifically designed to perform the temperature tests, and to insert the hydraulic actuator, the specimen and the supporting structure into the furnace. The heating rate of the specimens was 10 °C/min and once the target temperature was reached, it was kept up to 24 hours before starting the fatigue test, to ensure the highest thermal homogeneity in the specimen.

3 Probabilistic Fatigue Model Applied

In this study has been applied a probabilistic fatigue model for the determination of S-N curves proposed by Castillo and Fernandez-Canteli [31]. The probabilistic model allows the assessment of experimental fatigue data to obtain S-N fields. The probabilistic approach of the model, rather than mechanical, makes it suitable for any type of material (ceramic, metallic or concrete) independently of the mechanism of failure [11]. In most of models used, the high scatter of the lifetime at different stress is not considered and that has consequences in the fatigue design of structures. The model used in this study consider both physical (weakest link principle and limited range) and statistical conditions related to extreme value analysis (stability, limit behaviour). The model approach establishes a necessary compatibility between the cumulative distribution functions for fatigue lifetime at given stress range and stress range at given lifetime (Fig .1).

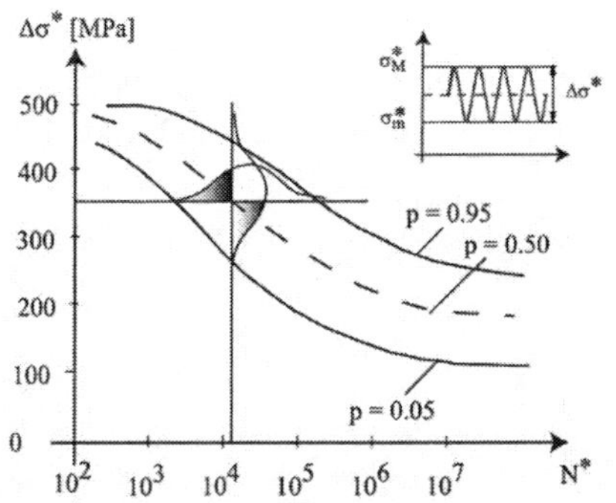

Fig. 1. Compatibility condition approach for the cumulative distribution functions [31].

Such condition derives into a functional equation which only can be solve by a Weibull or Gumbell models:

$$p = 1 - \exp\left[-\left(\frac{N-\lambda}{\delta}\right)^{\beta}\right]; N \geq \lambda \quad (1)$$

$$p = 1 - \exp\left[-\exp\left(\frac{N-\lambda}{\delta}\right)\right]; N \in \Re$$

where B is a threshold value of lifetime, C is the endurance limit or fatigue limit for N→∞, and λ, δ, β are location, scale and shape Weibull parameters, respectively.

For more information of the probabilistic model is recommended consult Castillo and Fernández-Canteli work [31]. To facilitate the application of the model from experimental fatigue data, Castillo and Fernández-Canteli developed a software so-called ProFatigue [32] which has been used.

4 Results and Discussion

In this study has been shown the effect of fibre reinforcement (for steel and polypropylene fibres) and temperature separately to remark more clearly their influences. With the aim of greater clarity, only the S-N curves for 20 (lower curve) and 80 (upper curve) percentile have been plotted for each type of test. The logarithm of fatigue life, N, on x-axis and the stress normalised level, S, on y-axis, are plotted.

4.1. Influence of Fibre Reinforcement

Fig. 2 shows the S-N curves of UHC and UHC_PP at room temperature. The addition of PP fibres in the concrete matrix might decrease the flexural monotonic strength and increase the energy absorption capacity in accordance to the fibre content [33]. In this case the monotonic flexural strength decreased from 26 MPa to 19 MPa which, in the fatigue behaviour derived into a slight reduction of the endurance limit as can be also seen in Table 3 and a softer trend of the S-N curves respectively.

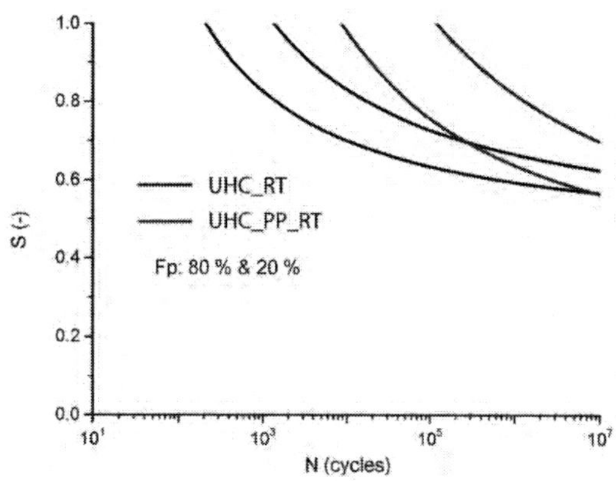

Fig. 2. Influence of PP fibre reinforcement on the S-N curves at room temperature.

Table 3. Endurance Limit

	UHC	UHC_PP	RC
Room temperature	0.488	0.375	0.210
100 °C	0.263	no tested	0.060
200 °C	0.083	no tested	spalling
300 °C	spalling	0.370	spalling

It was well established that the contribution of steel fibres increases the ductile behaviour and the tensile strength [2], [34]. Both effects are shown in the fatigue curves in Fig. 3 by means of a more progressive tendency of the S-N curves and a noticeably higher endurance limit respectively, in comparison with the reference concrete (RC_RT). Nevertheless, the addition of steel fibres might lead to results with a greater scatter, as was postulated by [11], [35] and observed in the higher distance between the S-N curves for UHC_RT.

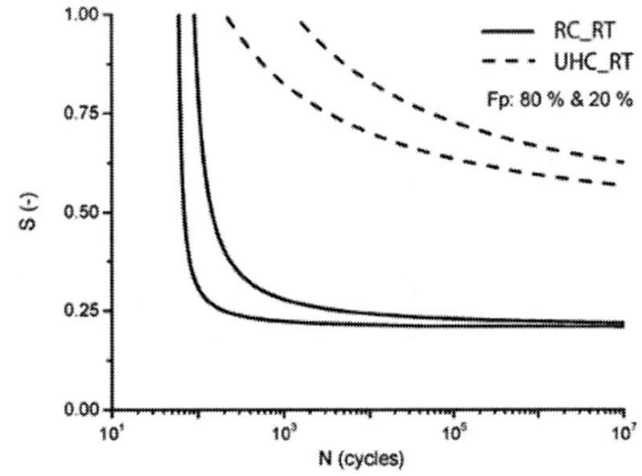

Fig. 3. Influence of steel fibre reinforcement on the S-N curves at room temperature.

4.2. Influence of Temperature

The effect of temperature in concrete matrix has been studied in previous works by several authors [36], [37]. Some conclusions extracted from these studies are that between 20 and 110 °C take place the evaporation of free water and between 160 and 350 °C an important dehydration derived by C-S-H gel is produced [37]. Both effects lead to a high pressure in the matrix, which generate cracks in concrete (spalling phenomenon). The intensity of thermal damage depends on the concrete matrix, especially its pore network, which allows evacuate the internal high pressure [38].

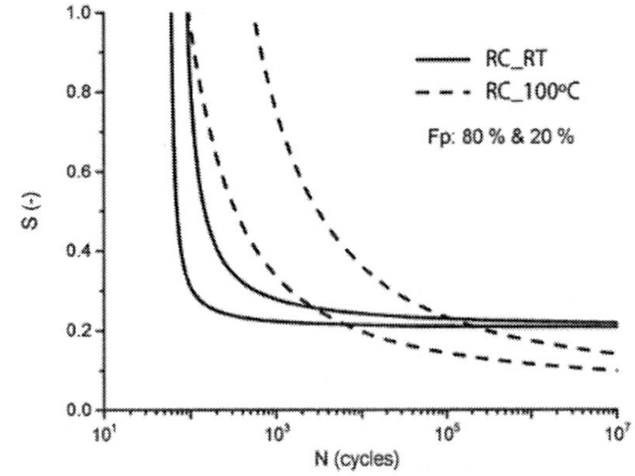

Fig. 4. Effect of temperature on the S-N curves for the RC.

The ultra-high-performance concrete used in this study has a high density (low pore network) which confers excellent mechanical properties but sooner thermal damage. Fig 4. shows the effect of temperature on the reference concrete but only reliable results at room temperature and 100 °C could be obtained because the spalling appeared from 200 °C and above. The fatigue life is significantly low at RT because of the brittle behaviour and the absence of barriers (fibre reinforcement) so that when the cracking appears is quickly propagated [39]. The thermal damage, at 100 °C, leads to a lower endurance limit and close to zero which shows as in ultra-high-performance concrete is essential the fibre reinforcement. Additionally, it is observed an increment of scattering by the thermal damage, reflected through the distance between the 80 and 20 percentile S-N curves.

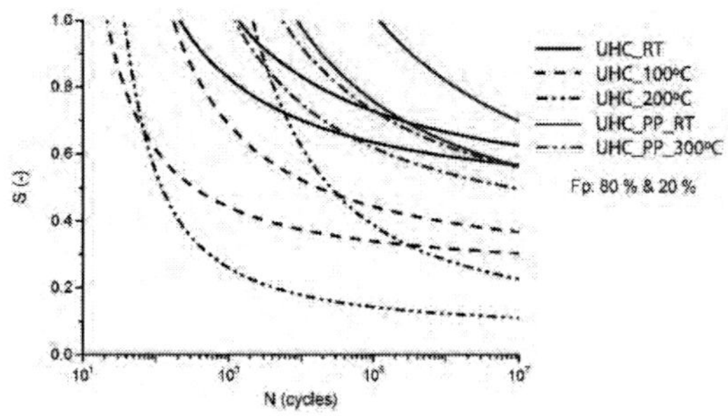

Fig. 5. Effect of temperature on the S-N curves on the UHC and UHC_PP.

Regarding to the effect of temperature on the fibre reinforced concrete (UHC), the observations are qualitatively the same than in RC. Nonetheless, it is worth noting as on the UHC, that significant spalling effects not appear up to 300 °C because the steel fibres avoid the free propagation of cracks and increase the ductility of the matrix. The scattering inferred by the thermal damage is more relevant than RC due to the higher temperature reached in the test and the influence of the fibre distribution [35].

The addition of PP micro-fibres in the UHC was needed to reach 300 °C without spalling failure. The PP fibres are melted at 160 °C and create a widespread micro-channel network which allows the pressure evacuation more easily and delay the spalling [38]. The influence of the addition of PP micro-fibres at RT were discussed in the section 4.1. At 300 °C, the spalling clearly does not affect and the endurance limit reaches a value very close to the RT. Thus, the addition of PP micro-fibres might be recommended to enhance the fatigue life behaviour at 300°C.

5 Conclusions

From results, the following conclusions can be drawn:

- The addition of steel fibres was essential for the enhancement of the endurance limit and the fatigue life because it avoids the free crack propagation.
- The steel fibre reinforcement slightly increases the scattering of S-N curves due to the fibre distribution.
- The use of PP micro-fibres does not improve the endurance limit at RT but is crucial to reduce spalling effect and to achieve a higher endurance limit at 300 °C.
- The effect of high temperatures is reflected in a damage that infer a significant decrease of the endurance limit and a higher scattering of the S-N curves.

Acknowledges

Authors would like to acknowledge the financial support from the Ministerio de Economía y Competitividad de España through the projects BIA2016-75431-R.

References

[1] D. Y. Yoo, N. Banthia, "Mechanical properties of ultra-high-performance fiber-reinforced concrete: A review," *Cem. Concr. Compos.*, **73**, (2016).

[2] D. Y. Yoo, S. Kim, G. J. Park, J. J. Park, and S. W. Kim, "Effects of fiber shape, aspect ratio, and volume fraction on flexural behavior of ultra-high-performance fiber-reinforced cement composites," *Compos. Struct.*, **174**, (2017).

[3] B. S. Al-Azzawi, B. L. Karihaloo, "Flexural Fatigue Behavior of a Self-Compacting Ultrahigh Performance Fiber-Reinforced Concrete," *J. Mater. Civ. Eng.*, **29**, 11, (2017).

[4] C. M. Tam, V. W. Y. Tam, and K. M. Ng, "Assessing drying shrinkage and water permeability of reactive powder concrete produced in Hong Kong," *Constr. Build. Mater.*, **26,** 1, (2012).

[5] G. Ruiz, A. Medeiros, X. X. Zhang, "Experimental Study of Loading Frequency Effect on Compressive Fatigue.", *An. Mec. Fract.* **29**, 9, (2011)

[6] Y. Chen, J. Ni, P. Zheng, R. Azzam, Y. Zhou, and W. Shao, "Experimental research on the behaviour of high frequency fatigue in concrete," *Eng. Fail. Anal.*, **18**, 7, (2011)..

[7] P. B. Cachim, J. A. Figueiras, P. A. A. Pereira, "Fatigue behavior of fiber-reinforced concrete in compression," *Cem. Concr. Compos.*, **24**, 2, (2002).

[8] M. Á. Pindado, A. Aguado, A. Josa, "Fatigue behavior of polymer-modified porous concretes," *Cem. Concr. Res.*, **29**, 7, (1999).

[9] R. Tepfers, "Tensile Fatigue Strength of Plain Concrete," *ACI J. Proc.*, **76**, 8, (1979).

[10] Y. Mohammadi, S. K. Kaushik, "Flexural Fatigue-Life Distributions of Plain and Fibrous Concrete at Various Stress Levels," *J. Mater. Civ. Eng.*, **17**, 6, (2005).

[11] J. D. Ríos, H. Cifuentes, R. Yu, G. Ruiz, "Probabilistic Flexural Fatigue in Plain and Fiber-Reinforced Concrete," *Materials*, **10**, 7, (2017).

[12] X. P. Shi, T. F. Fwa, S. A. Tan, "Flexural Fatigue Strength of Plain Concrete,", *Materials Journal*, **90**, (1994).

[13] B. S. Al-Azzawi, B. L. Karihaloo, "Flexural Fatigue Behavior of a Self-Compacting Ultrahigh Performance Fiber-Reinforced Concrete," *J. Mater. Civ. Eng.*, **29**, 11, (2017).

[14] M. C. Alonso, J. Vera-Agullo, L. Guerreiro, V. Flor-Laguna, M. Sanchez, M. Collares-Pereira, "Calcium aluminate based cement for concrete to be used as thermal energy storage in solar thermal electricity plants," *Cem. Concr. Res.*, **82,** (2016).

[15] E. Özrahat, S. Ünalan, "Thermal performance of a concrete column as a sensible thermal energy storage medium and a heater," *Renew. Energy*, **111,** (2017).

[16] F. Bai, C. Xu, "Performance analysis of a two-stage thermal energy storage system using concrete and steam accumulator," *Appl. Therm. Eng.*, **31,** 14–15, (2011).

[17] M. Hazami, S. Kooli, M. Lazâar, A. Farhat, A. Belghith, "Energetic and exergetic performances of an economical and available integrated solar storage collector based on concrete matrix," *Energy Convers. Manag.*, **51**, 6, (2010).

[18] F. B. Varona, F. J. Baeza, D. Bru, S. Ivorra, "Influence of high temperature on the mechanical properties of hybrid fibre reinforced normal and high strength concrete," *Constr. Build. Mater.*, **159**, (2018).

[19] A. H. Akca, N. Özyurt, "Effects of re-curing on residual mechanical properties of concrete after high temperature exposure," *Constr. Build. Mater.*, **159**, (2018).

[20] I. Hager, "Behaviour of cement concrete at high temperature," *Bulletin of the Polish Academy of Sciences*, **61**, (2013).

[21] H. Cifuentes, C. Leiva, F. Medina, C. Fernández-Pereira, "Effects of fibers and rice husk ash on properties of heated high-strength concrete," *Mag. Concr. Res.*, **64**, 5, (2012).

[22] T. Drzymała, W. Jackiewicz-Rek, M. Tomaszewski, A. Kuś, J. Gałaj, R. Šukys, "Effects of High Temperature on the Properties of High Performance Concrete (HPC)," *Procedia Eng.*, **172**, (2017).

[23] G. F. Peng, W. W. Yang, J. Zhao, Y. F. Liu, S. H. Bian, L. H. Zhao, "Explosive spalling and residual mechanical properties of fiber-toughened high-performance concrete subjected to high temperatures," *Cem. Concr. Res.*, **36**, 4, (2006).

[24] N. Zhang, B; Bicanic, "Fracture energy of high-performance concrete at high temperatures up to 4508C: the effects of heating temperatures and testing conditions (hot and cold)," *Mag. Concr. Res.*, **58**, 5, (2006).

[25] W. Zheng, B. Luo, Y. Wang, "Microstructure and mechanical properties of RPC containing PP fibres at elevated temperatures," *Mag. Concr. Res.*, **66**, 8, (2014).

[26] W. Khaliq, V. Kodur, "Thermal and mechanical properties of fiber reinforced high performance self-consolidating concrete at elevated temperatures," *Cem. Concr. Res.*, **41**, 11, (2011).

[27] W. Zheng, B. Luo, Y. Wang, "Microstructure and mechanical properties of RPC containing PP fibres at elevated temperatures," *Mag. Concr. Res.*, **66**, 8, (2014).

[28] A. H. Akca, N. Ö. Zihnioğlu, "High performance concrete under elevated temperatures," *Constr. Build. Mater.*, **44**, (2013).

[29] G-F. Peng, Y-C. Jiang, B-H. Li, J. Zhang, Y-X. Shi, "Effect of high temperature on normal-strength high-performance concrete," *Mater. Res. Innov.*, **18**, 2, (2014).

[30] "EN 14651: 2007+A1 Test method for metallic fibered concrete- Measuring the flexural tensile strength (limit of proportionality (LOP), residual), AENOR .," (2007).

[31] E. Castillo, A. Fernandez-Canteli, *A unified statistical methodology for modeling fatigue damage*. (Springer, 2009) ISBN: 978-1-4020-9182-7.

[32] A. Fernández-Canteli, C. Przybilla, M. Nogal, M. L. Aenlle, E. Castillo, "Profatigue: A software program for probabilistic assessment of experimental fatigue data sets," *Procedia Eng.*, **74**, (2014).

[33] A. A. Ramezanianpour, M. Esmaeili, S. A. Ghahari, M. H. Najafi, "Laboratory study on the effect of polypropylene fiber on durability, and physical and mechanical characteristic of concrete for application in sleepers," *Constr. Build. Mater.*, **44**, (2013).

[34] A. Abrishambaf, M. Pimentel, S. Nunes, "Influence of fibre orientation on the tensile behaviour of ultra-high performance fibre reinforced cementitious composites," *Cem. Concr. Res.*, **97**, (2017).

[35] E. S. Lappa, (2007) *High Strength Fibre Reinforced Concrete: Static and fatigue behaviour in bending*. (Thesis dissertation).

[36] A. Lau, M. Anson, "Effect of high temperatures on high performance steel fibre reinforced concrete," *Cem. Concr. Res.*, **36**, 9, (2006).

[37] C. Alonso, L. Fernandez, "Dehydration and rehydration processes of cement paste exposed to high temperature environments," *J. Mater. Sci.*, **39**, 9, (2004).

[38] H. Cifuentes, F. García, O. Maeso, F. Medina, "Influence of the properties of polypropylene fibres on the fracture behaviour of low-, normal- and high-strength FRC," *Constr. Build. Mater.*, **45**, (2013).

[39] R. L. Norton, *Machine design. A integrated approach*, 5th Editi. (Pearson, 2013) ISBN-13: 978-0133356717.

MATEC Web of Conferences **165**, 13001 (2018) https://doi.org/10.1051/matecconf/201816513001
FATIGUE 2018

On the transition of short cracks into long fatigue cracks in reactor pressure vessel steels

Rajwinder Singh[1,*], *Amanjot* Singh[1], *Aman* Arora[1], *Pawan* K. Singh[2] and *Dhiraj* K. Mahajan[1]

[1]Ropar Mechanics of Materials Laboratory, Department of Mechanical Engineering, Indian Institute of Technology Ropar, Rupnagar, Punjab, India, 140001
[2]Reactor Safety Division, Bhabha Atomic Research Centre (BARC), Mumbai, India, 400085

Abstract. Short fatigue cracks, having dimension less than 1 mm, propagate at much faster rates (*da/dN*) even at lower stress intensity factor range (*da/dN*) as compared to the threshold stress intensity factor range obtained from long fatigue crack growth studies. These short cracks originate at the sub-grain level and some of them ultimately transit into critical long cracks over time. Therefore, designing the components subjected to fatigue loading merely on the long crack growth data and neglecting the short crack growth behavior can overestimate the component's life. This aspect of short fatigue cracks become even more critical for materials used for safety critical applications such as reactor pressure vessel (RPV) steel in nuclear plants. In this work, the transition behaviour of short fatigue crack gowth into long fatigue crack is studied in SA508 Grade 3 Class I low alloy steel used in RPVs. In-situ characterization of initiation, propagation and transition of short fatigue cracks is performed using fatigue stage for Scanning Electron Microscope (SEM) in addition to digital microscopes fitted over a servo-hydraulic fatigue machine and correlated with the microtructural information obtained using electron backscatter diffraction (EBSD). SA508 steel having an upper bainitic microstructure have several microstructural interfaces such as phase and grain boundaries that play a significant role in controlling the short fatigue crack propagation. Specially designed and prepared short fatigue specimens (eletro-polished) with varying initial crack lengths of the order of tens of microns are used in this study. The transition of such short initial cracks into long cracks is then tracked to give detailed insight into the role of each phase and phase/grain boundary with an objective of establishing Kitagawa-Takahashi diagram for the given RPV steel. The behavior of the transited long cracks is then compared with the crack propagation behavior obtained using conventional CT specimens. The outcome of this research will enhance information on the integrity of the components made from RPV steel used in Indian nuclear power plants.

1 Introduction

A significant proportion of component's fatigue life is spent during short crack propagation [1]. Short cracks are mainly classified as mechanically short cracks (plastic zone at crack tip is less than the crack length), microstructurally short cracks (crack length is less than grain size of the material) and physically short cracks (crack length less than 1 mm and up to which crack closure is not eminent) [2]. The initiation of short crack propagation takes place at lower value of threshold stress intensity factor range as defined for the conventional long cracks. After initiation short cracks propagate at higher rate as compared to the long cracks for the same value of applied stress intensity factor range. Linear elastic fracture mechanics (LEFM) is not applicable in short cracks regime and the propagation behaviour of short fatigue cracks cannot be predicted by the famous Paris-Erdogan law [3]. Unlike long cracks, the short cracks during propagation highly interact with the inherent microstructural features of the material such as grain boundaries, phase boundaries and inclusions [4–6]. These interactions of short cracks with the microstructural features in the material will result in fluctuations in the propagation rate of the short cracks. Under the action of external loading, slip bands are formed at the crack tip due to the emission of dislocations along the active slip planes at the crack front. The short cracks tend to propagate along these slip bands by shearing due to combination of mode I and mode II loading [2]. Since the orientation of the active slip bands change with the change in the grain orientation, therefore these short fatigue cracks change their direction as they travel from one grain to the other by crossing the grain boundary [5]. The natural occurring defects in a material are of the order of short cracks, therefore it is more logical to study short fatigue crack propagations in a material.

SA508 Gr.3 Cl.I low alloy steel (LAS) is used in the manufacturing of shell of nuclear power plants [7]. The safe operation of this shell is very important for the safety of the nuclear power plant as well as the surrounding living beings and environment. As surveyed through SCOPUS database, lot of experimental investigations are being conducted by the researchers on long fatigue crack

* Corresponding author: rajwinder.singh@iitrpr.ac.in

propagation in this steel. These investigations are related to the effect of cyclic loading ratio [8], testing environment [9–12] and heat treatment [13] on the long fatigue crack propagation in this LAS. Apart from long fatigue crack propagation studies only one research paper regarding the experimental study related to the difference in short and long fatigue crack propagation rates in the LAS is available in literature [14]. Due to the abnormal propagation behaviour of short cracks and inability of the LEFM theory to model the propagation rate of short cracks, a detailed understanding regarding the interaction of short cracks with the microstructural feature in this SA508 LAS is necessary.

In the present work, the experimental investigations of mechanically short fatigue crack propagation in SA508 Gr.3 Cl.I LAS is carried out. The short fatigue crack propagation is studied at applied stress levels near to the yield strength of the material to investigate the short fatigue crack propagation rate and interaction of short crack with the microstructural features in the material. Comparison between the propagation rate of short and long cracks under cyclic loading is also done. The outcome of these investigations will enhance information on the integrity of the components made from RPV steel, which are used in Indian nuclear power plants.

2 Experimental procedure

2.1. Material characterization

SA508 Gr.3 Cl.I LAS test material in the form of pipe is received from BARC (Mumbai). The chemical composition of the material is as shown in Table 1.

Table 1. Chemical composition of SA508 Gr.3 Cl.I LAS in weight %

C	S	P	Mn	Si	Cr
0.19	0.002	0.018	1.3	0.23	0.17
Ni	**Mo**	**Cu**	**Al**	**Ti**	**Fe**
0.7	0.44	0.13	0.02	--	96.8

For microstructural study of the material, the cube sample is polished with emery papers and alumina paste to obtain a mirror like finish. This mirror finished samples are then etched with 10% Nital solution for a period of 8 seconds to reveal the microstructure of the material under the SEM.

2.2. Long fatigue cracks propagation test

The compact tension (CT) specimens as shown in Figure 1(a) are used to study the long fatigue crack propagation in the material. Both constant load and load shredding experiments as mentioned in ASTM E647-11 are performed at a cyclic loading ratio ($R = P_{min}/P_{max}$) of 0.1 and frequency of 35 Hz to explore the both stage I and stage II long fatigue crack propagation behaviours of the material. The crack length during fatigue loading is measured with crack opening displacement (COD) gauge. Incremental polynomial method as mentioned in ASTM E647-11 is used to compute the long crack growth rate.

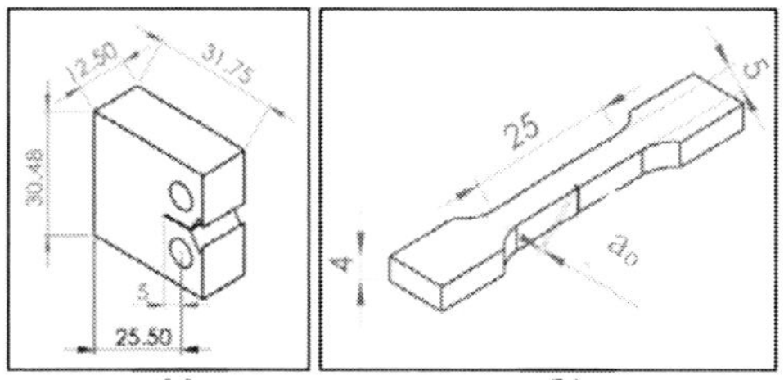

(a) (b)

Fig. 1. (a) Compact tension (CT) specimen to study long crack propagation (b) single edge notched (SEN) specimen for short crack propagation study.

2.3. Short fatigue crack propagation test

The single edge notched (SEN) specimens with an initial notch size (a_o) varying from 50 µm to 100 µm as shown in Figure 1(b) are used to study the short fatigue crack propagation in the material. The short fatigue crack propagation tests are carried out at applied stress range $\Delta\sigma$ of 475 MPa at a cyclic loading ratio of 0.1 and frequency of 35 Hz. The value of $\Delta\sigma$ is chosen such that the value of σ_{max} during cyclic loading is near to the yield strength of the material. The yield strength and ultimate tensile strength of the material under investigation are determined experimentally by Singh et al. [15]. A new technique of short crack measurement by optical microscope is used [15]. The length of short crack during cyclic loading is measured with moving digital microscope by interrupting the cyclic loading at regular intervals. The specimen surfaces are polished with emery papers and alumina paste to obtain mirror finish, so that the short cracks propagation at the surface can be easily tracked by digital microscope. The polished and cracked specimen is then etched to reveal the inherent microstructural features of the material intercepted by the short crack path. EBSD analysis of the cracked surface is also carried out to understand the relative orientation of the grains through which the short fatigue crack passed during propagation. Secant method as mentioned in ASTM E647-11 is used to compute the short crack growth rate as.

$$da/dN = (a_{i+1} - a_i)/(N_{i+1} - N_i) \quad (1)$$

where, $(a_{i+1} - a_i)$ is the increment in crack length after $(N_{i+1} - N_i)$ loading cycles.

3 Results and discussion

3.1. Microstructure

The inherent microstructure of SA508 Gr.3 Cl.I LAS as revealed through SEM is shown in Figure 2. The microstructure of this material is tempered upper bainitic consisting of prior austenite grain in which the carbides are distributed inside and outside the ferrite (α-Fe) laths. M_3C type coarse carbides are precipitated along the α-Fe lath boundaries whereas, M_2C type fine carbides are distributed inside the α-Fe laths. Rod shaped carbides are also distributed along the prior austenite grain boundaries

(PγGBs) [13]. The average prior austenite grain size of the material is 30 μm.

Fig. 2. Microstructure of SA508 Gr.3 Cl.I LAS as obtained through SEM

3.2. Short and long fatigue crack propagation

The graph of short crack propagation length with the number of cycles at applied $\Delta\sigma$=475 MPa is as shown in Figure 3(a). Three numbers of samples are tested at the same loading conditions to get the short crack data. The crack growth measurement is done up to crack length of approximately 1 mm to predict the short crack growth behaviour. The crack initiated after 10000, 12000 and 16000 numbers of loading cycles in sample 1, sample 2 and sample 3 respectively. The number of cycles required to initiate the short crack in all the three samples is almost the same. After the initiation of the crack, the propagation of short crack with the number of cycles is different in the three samples as per graph shown in Figure 3(a). Since the short crack propagation under cyclic loading is affected by the microstructural features in the material, therefore the difference in graph shown in Figure 3(a) for the three tested samples is because of the difference in the microstructural features and orientation of the grains being intercepted by the short crack during propagation in all the three samples. The graph of short crack propagation rate (da/dN) with the crack length (a) is shown in Figure 3(b). The acceleration and deceleration in short crack propagation rate is observed due to the interaction of short cracks with the microstructural features such as grain boundaries in the material. During propagation, the interaction of short crack with the microstructural features in the material will be discussed in Section 3.3.

The variation in plastic zone size r_p (μm) ahead of the crack tip with crack length for different applied stress levels $\Delta\sigma$ is as shown in Figure 4. The plastic zone size r_p ahead of the crack tip under cycling loading can be evaluated as [16]

$$r_p = (1/24\pi)(\Delta K/\sigma_Y)^2 \quad (2)$$

The plastic zone size ahead of the crack tip is exceeding the value of average grain size i.e. 30 μm of the material when the crack length is approximately 600 μm for applied $\Delta\sigma$. Therefore, the interaction of short crack with the microstructural features is reduced after crack length of 600 μm as shown in Figure 3(b).

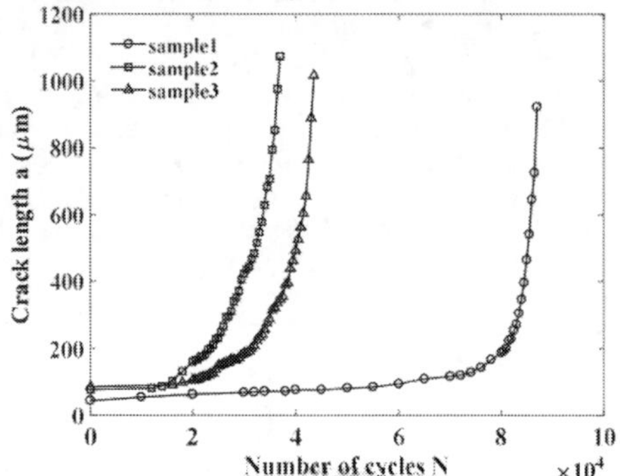

(a)

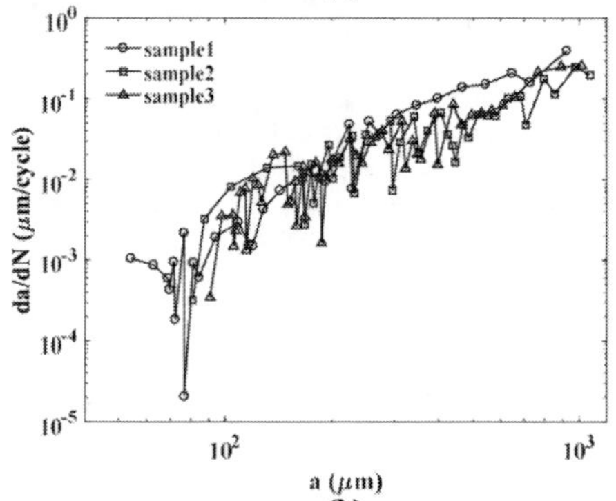

(b)

Fig. 3. (a) Short crack growth 'a' with numbers of cycles 'N' and (b) short crack growth rate 'da/dN' with crack length 'a' at applied $\Delta\sigma$=475 MPa in SA508 Gr.3 Cl.I LAS

The concept of linear elastic fracture mechanics (LEFM) and small-scale yielding (SSY) is applicable if

$$a/r_p \geq 50 \quad (3)$$

The value of this ratio is less than 50 in short crack regime because of small value of crack length and high value of plastic zone size. Thus, LEFM and SSY are not applicable in short crack regime.

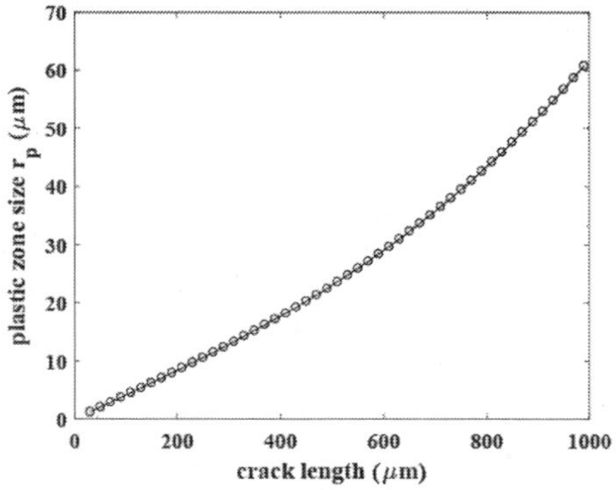

Fig. 4. Variation of plastic zone size r_p with crack length at applied stress range $\Delta\sigma$=475 MPa in SA508 Gr.3 Cl.I LAS.

https://doi.org/10.1051/matecconf/201816513001

Even though the LEFM theory is not valid in short crack regime but a graph between crack growth rate (da/dN) and stress intensity factor range ΔK is plotted for short cracks to compare the propagation rate of short and long cracks under cyclic loading as shown in Figure 5. The details regarding the long crack growth under cyclic loading and Paris law for SA508 Gr.3 Cl.I LAS can be found elsewhere [15]. The propagation rate of short cracks is higher as compared to the long cracks at higher applied $\Delta\sigma$ and thus the propagation rate of short cracks cannot be modelled by conventional Paris law as shown in Figure 5. Similar higher rate of crack propagation at higher applied stress levels in aluminium alloy is also reported in literature [17]. Therefore, the design of the pressure vessels made from this LAS merely on long crack growth data is not recommended.

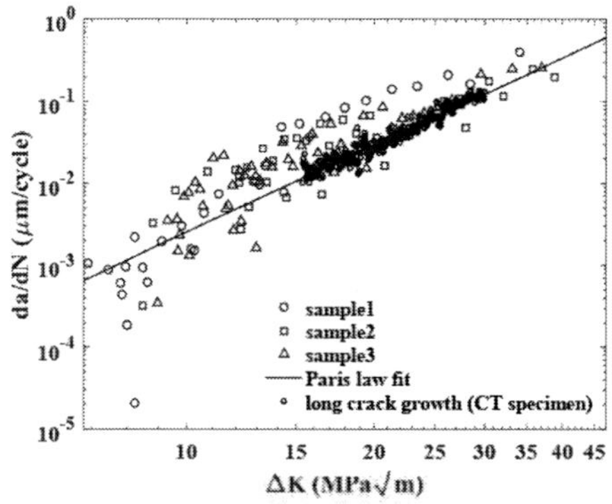

Fig. 5. Comparison of short and long crack propagation in SA508 Gr.3 Cl.I LAS under cyclic loading

3.3. Short crack interaction with microstructure

For studying the short crack interaction with the microstructural features in the material, the crack is made to propagate to the length of approximately 150 μm under cyclic loading. In this case, the graph of the increase in short crack length with the number of applied loading cycles is shown in Figure 6(a). The corresponding graph of short crack propagation rate (da/dN) with the crack length (a) is shown in Figure 6(b). The retardation in short crack propagation rate is observed at the points 1, 2 and 3 as marked in the graph shown in Figure 6(b). The surface of the cracked sample is etched to observe under SEM the microstructural features in the material corresponding to the marked points 1, 2 and 3. The path of short crack as observed under the SEM is shown in Figure 7(a). The zoomed view of the short crack path is shown in the inset of Figure 7(a). The points 1 to 3 corresponding to the retardation of short crack growth shown in Figure 6(b) are marked by white circles on the short crack path observed under the SEM as shown in Figures 7(b)-7(e).

Figure 7(b) shows that the retardation at point 1 is due to the PAγBs. The crack has changed its path as it crossed the PAγBs due to the change in the grain orientation. Another short crack parallel to the main crack can be seen in Figure 7(c). Away from the short crack tip another crack initiated from the triple point of the PAγBs as marked by point 2 in Figure 7(d). Tripe point is referred to as a point where three grain boundaries meet. These triple points act as stress raisers as reported in the literature [18]. The coarse rod type carbides distributed along the PAγBs at the triple point as shown in Figure 7(d) also favoured the initiation of the crack. This crack initiated from the triple point propagated in the backward directions and met the main crack at the position marked by white square in Figure 7(c). The two short cracks then met at the triple point and then propagated together as a single short crack. The retardation in crack propagation rate at point 2 as shown in Figure 6(b) is due to the arrest of the short crack at the triple point shown in Figure 7(d) for some duration. The deceleration in crack growth rate at point 3 is due to the presence of another PAγBs as shown in Figure 7(e). At higher applied cyclic stress levels, the only retardation to short crack growth is offered by PAγBs in SA508 Gr.3 Cl.I LAS.

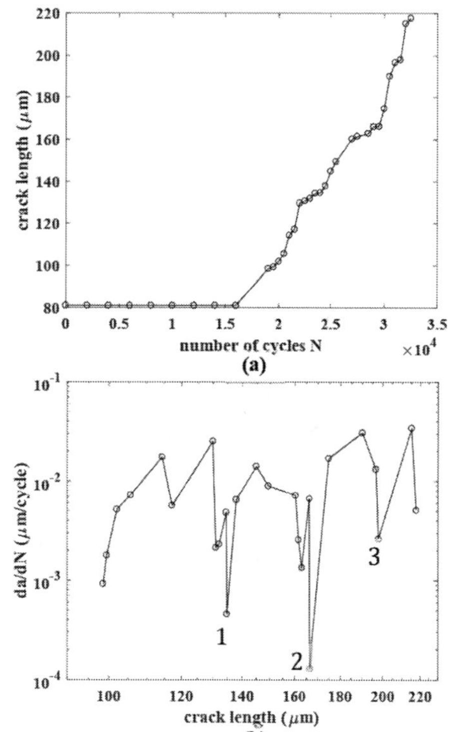

Fig. 6. (a) Short crack growth 'a' with numbers of cycles 'N' and (b) short crack growth rate 'da/dN' with crack length 'a' at applied $\Delta\sigma$=475 MPa in SA508 Gr.3 Cl.I LAS (the crack is propagated to approximately 150 μm length)

https://doi.org/10.1051/matecconf/201816513001

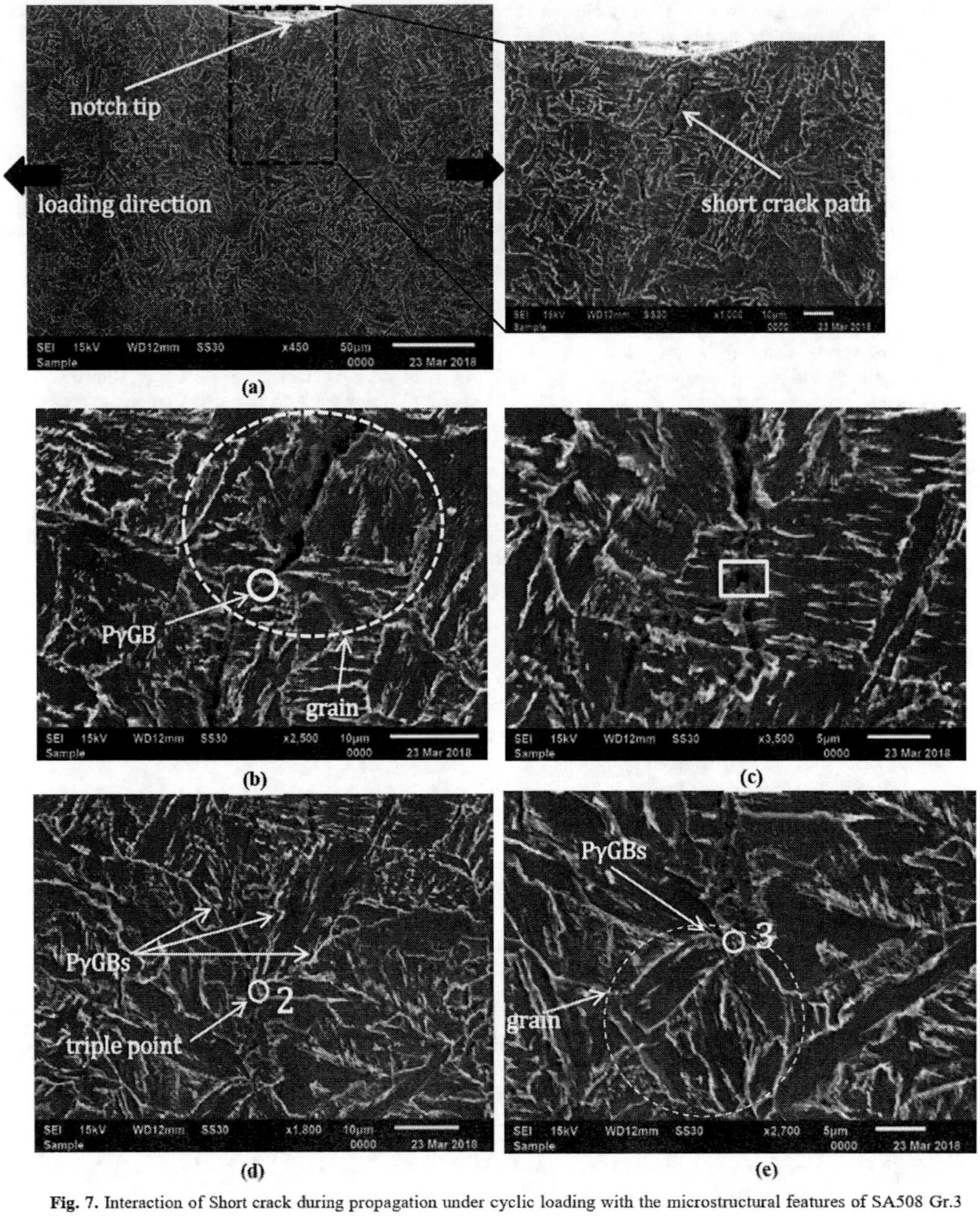

Fig. 7. Interaction of Short crack during propagation under cyclic loading with the microstructural features of SA508 Gr.3 Cl.I LAS as observed through SEM

The details regarding the interaction of short crack propagation with the microstructural features in SA508 Gr.3 Cl.I LAS through EBSD and EDS will be presented during the 12th international fatigue congress and a full-length paper of the detailed work will be submitted in a reputed journal.

4 Conclusion

The following conclusions are drawn from the comparative study of short and long crack propagation under cyclic loading in SA508 Gr.3 Cl.I LAS:

1. The short crack propagation rate at higher stress levels i.e. above the yield strength of the material cannot be predicted by the Paris-Erdogan law. Therefore, the design

of pressure vessels installed in nuclear power plants merely on long crack growth data is not recommended.

2. The short crack propagation rate is highly influenced by the PγGBs in SA508 Gr.3 Cl.I LAS.

3. The triple point in the material acted as a source for the initiation of the short cracks under cyclic loading.

References

1. K. J. Miller, Fatigue of engineering materials and structures 5, 223 (1982).
2. A. J. McEvily, Transactions on Engineering Sciences 13, 93 (1996).
3. C. Bjerken and S. Melin, International Journal of Solids and Structures 46, 1196 (2009).
4. J. C. Newman, E. P. Phillips, and M. H. Swain, International journal of fatigue 21, 109 (1999).
5. R. Strubbia, S. Hereñú, L. Alvarez-Armas, and U. Krupp, Materials Science & Engineering A 615, 169 (2014).
6. C. Kaynak, A. Ankara, and T. J. Baker, International journal of fatigue 18, 17 (1996).
7. M. Kim, S. Park, K. Lee, and B. Lee, International Journal of Pressure Vessels and Piping 131, 60 (2015).
8. W. A. Logsdon and P. K. Liaw, Engineering Fracture Mechanics 22, 509 (1985).
9. J. Y. Huang, J. J. Yeh, R. C. Kuo, S. L. Jeng, and M. C. Young, International Journal of Pressure Vessels and Piping 85, 772 (2008).
10. S. G. Lee and I. S. Kim, Journal of pressure vessel technology 123, 173 (2001).
11. I. S. Kim and S. S. Kang, International Journal of Pressure Vessels and Piping 62, 123 (1995).
12. L. Dong, Q. Peng, E. Han, W. Ke, and L. Wang, Journal of Materials Science & Technology (2017).
13. J.-T. Kim, B.-I. Yang, and H.-K. Kwon, ASME 2002 Pressure Vessels and Piping Conference, 181.
14. J. L. Breat, F. Mudry, and A. Pineau, Fatigue of engineering materials and structures 6, 349 (1983).
15. R. Singh, A. Singh, P. K. Singh, and D. K. Mahajan, International journal of fatigue (under review).
16. R. I. Stephens, A. Fatemi, R. R. Stephens, and H. O. Fuchs, John Wiley & Sons, Inc. 2nd edition (2001).
17. R. D. Brown and J. Weertman, Engineering Fracture Mechanics 10, 757 (1978).
18. A. Manonukul and F. P. E. Dunne, Proceedings of the Royal Society A: Mathematical, Physical and Engineering Sciences 460, 1881 (2004).

The stress ratio effect on plastic dissipation during fatigue crack growth

H. Quan[1,*], *R.C.* Alderliesten[1], and *R.* Benedictus[1].
[1]Structural Integrity & Composites, Faculty of Aerospace Engineering, TU Delft, Netherlands

Abstract. Plastic energy dissipation is inevitable during fatigue crack growth. There have been previous attempts reported in literature to correlate the plastic dissipated energy (*dW/dN*) to fatigue crack growth rate (*da/dN*). However, at a given *dW/dN*, the *da/dN* changes with the ratio of minimum and maximum loads, known as the stress ratio. This paper describes an experimental study carried out on 2024-T3 central crack tension specimens to quantify the relation between *dW/dN* and *da/dN*. By selecting different stress ratios in the individual tests, the experiments reveal the influence of the stress ratio on this relationship. It is evident that *dW/dN* has no unique relationship with *da/dN* valid for the tested stress ratios. Instead, the relationship for each stress ratio is different. This is illustrated with the value of plastic dissipation per unit of fatigue crack growth (*dW/da*), representing the effective resistance to the crack increment. This value is not a constant, but changes with the stress ratios and *da/dN* values. Hence the plastic energy dissipation cannot be used directly for predicting crack growth.

1. Introduction

Fatigue failure is a major failure type in aerospace engineering, so it is important to have a deep understanding of this failure type, in order to mitigate it. Although Linear Elastic Fracture Mechanics (LEFM) has been applied successfully to practical engineering fatigue problems through the use of the Paris relation, LEFM still fails in explaining the physical nature of fatigue phenomena in metallic materials. This is because metallic materials are ductile, and the plasticity phenomena strongly influences the fatigue crack growth rate (*da/dN*). Therefore, if a deeper understanding of the nature of fatigue is needed, it has to be answered how plasticity influences fatigue crack propagation in metallic materials. In this paper, the fatigue crack phenomena will be discussed from a physical perspective. An energy approach is chosen because the energy approach shows the universality among various materials and the plasticity itself represents a typical type of energy dissipation.

There is literature [1-17] on the topic of the plastic energy dissipation during the fatigue crack propagation. Most of the papers aim at finding the relation between the plastic energy dissipation per cycle (*dW/dN*) and *da/dN*, and their work could be summarized as:

$$\frac{dW}{dN}=\frac{da}{dN}\frac{dW}{da} \qquad (1)$$

The *dW/da* represents the plastic energy dissipated per unit fatigue crack extension. There are some differences among the studies mentioned above, mainly in the definitions and in the method of obtaining those values of *dW/dN* and *dW/da*. Klingbeil et al. [1-6] relate the fatigue crack growth to the static crack propagation and claim that *dW/da* is a material property and its value should be equal to the static fracture toughness G_c. Then, *dW/dN* is the total plastic dissipation energy per cycle per unit width. Karlsson et al.[7-11] use the finite element method to predict *da/dN*. They believe that the fatigue crack propagation is the result of the degradation in a process zone ahead of the crack tip where the plastic deformation occurs. So Karlsson et al.[7-11] link *da/dN* with the plastic dissipation in the process zone, which is represented by the reverse cyclic plastic zone ahead of the crack tip. Karlsson et al. [7-11] believe *dW/da* should be a constant that is not necessarily equal to the value of G_c. Ranganathan et al. [12-13] perform an experimental investigation on the relation between the total energy dissipated per cycle and *da/dN*. Their experimental results show that the total energy dissipated per unit fatigue crack extension (*dW/da*) is not a constant and could depend on the different mechanism of crack growth. Zheng [14] simulates the fatigue crack propagation based on the plastic dissipation value of critical distances. Smith [15] assumes that the energy dissipation to create a unit of new crack surface could be considered as a material property and applies this assumption to simulate fatigue crack growth with 3D finite element model. Besel et al. [16-17] study AA2024-T3 plastic dissipation with digital image correlation (DIC) method and coupled FE simulation. From the work mentioned above, although with some slight differences, there seems to be a contradiction between the work of [1-6,15] and [12-13]: some researchers seem to believe that the *dW/da* value is a material property,

* Corresponding author: h.quan@tudelft.nl

but the others disagree with this. Therefore the research questions in this paper are as follows:

1. Is the value dW/da (it is defined as the total plastic dissipation per unit fatigue crack growth) a material property that is approximately constant?
2. Does the same value of dW/dN correspond to the same da/dN among various stress ratios?

2. EXPERIMENT METHODOLOGY

2.1 Fatigue experiments description

In order to obtain the plastic energy dissipation during fatigue crack growth, Mode I fatigue crack propagation experiments of the 2024-T3 aluminium alloy CCT specimens were carried out. The geometry of the specimen is shown in **Fig. 1**. The thickness of 2024-T3 specimens was 4mm. The initial crack was made in 3 steps: first a small hole was drilled at the centre of the plate; then on each side of the hole a small notch was created by fret saw; finally the fatigue cracks were produced after fatigue precracking.

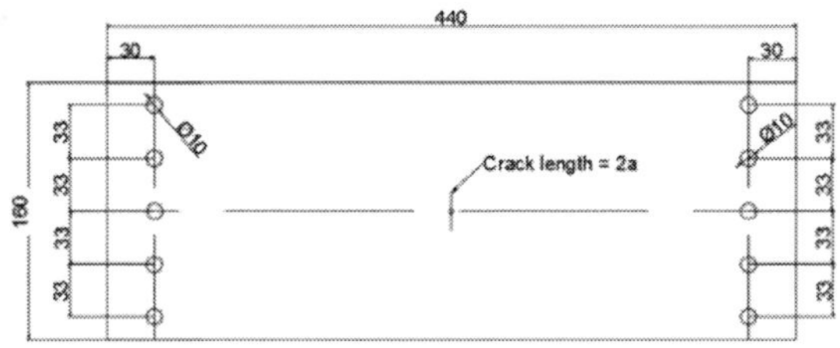

Fig. 1. The CCT specimens(The unit is mm)

Fig. 2. The set-up of the experiments

During the fatigue experiments, there is one single 4 Mpix camera on one side of the specimens in order to measure the fatigue crack length visually after the experiment. The other side of the specimens were painted with a speckle pattern for the purpose of measuring the strain and displacement field with DIC method. For the DIC measurement, a pair of 80mm lenses cameras were used to observe the deformation of the test specimens. The set-up is shown in **Fig. 2**.

The fatigue tests were carried out under a combination of baseline cycles and slow cycles. The frequency of baseline cycles is 10Hz. The slow cycles were used to make measurements and take photos, because the camera system needed some time to capture images. In the fatigue experiments, the slow cycles were used after every 2000-3000 baseline cycles. The specimens were tested at a load controlled condition with MTS 250kN fatigue machine. For the stress ratio of R=0.05 the maximum load is 45kN, while for R=0.5 the maximum load is 60kN.

2.2 Data acquisition and post-processing

The photos taken by 4 Mpix camera during the fatigue experiments were used to measure the crack length at the number of cycles the image was captured, which was evaluated to obtain da/dN. The photos taken for DIC were processed with DIC post-processing software.

The energy dissipation per cycle during the fatigue experiments was determined by the integration of the hysteresis loop of the load-displacement curve measured throughout an entire fatigue load cycle. The load values were obtained by the load cell of the fatigue machine, while the displacement was measured from DIC results at a location defined hereafter. Theoretically the energy dissipation could be calculated with:

$$\frac{dW}{dN} = \int F du \qquad (2)$$

in which F represents the total force on the specimen and u represents the displacement values in the direction parallel to the load direction in the far field where the displacement distribution is nearly uniform.

However, it is not possible to measure the dissipated energy with the displacement directly from the far field away from the crack area. For the reason that the load-displacement curves for loading and unloading during one fatigue cycle could be approximately coincide, such that the hysteresis loop could not be observed. This is because the amount of plastic dissipation in one cycle is very small if it is compared with the total strain energy change in single loading or unloading. Moreover, the plastic dissipation only has strong effects on a small local area around crack tip, but in the far field the effects could be negligible. Therefore, in this study the displacement measurements were made around crack tip area.

Some "virtual extensometers" were used, which could be applied in the DIC post-processing software, in the post-processing procedure to obtain local displacement values. The "virtual extensometers" were located at crack tip area, perpendicular to the crack plane and with a length of 1mm. Meanwhile, the central points of the "virtual extensometers" should approximately coincide with the crack tip point. The displacement values used in the load-displacement hystereses are defined as the elongation of the "virtual extensometers" at the given locations mentioned above on the specimens, and as shown in **Fig. 3**.

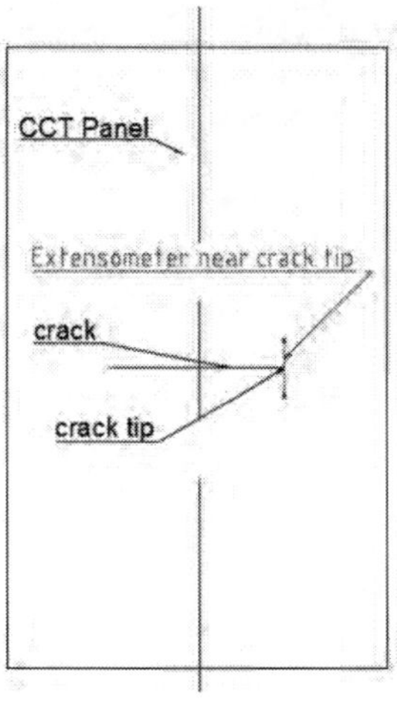

Fig 3. The extensometer at crack tip area to obtain the displacement values in hysteresis loops for measuring the dissipated energy

However, the energy dissipation measured with the displacement values at crack tip area is not the total energy dissipation during fatigue crack growth. In this case, a correction factor α is induced to obtain the total energy dissipation value. The correction factor α is defined as the ratio of energy dissipation calculated with far field displacement, which represents the total energy dissipation during one fatigue cycle, to the energy dissipation calculated from the displacement measurement by extensometers at crack tip area. The value of the correction factor α changes with different crack lengths and various load cases. The value of the correction factor α could be determined by either analytical method or numerical method. In this paper, the value of α was obtained with finite element method. It is assumed that the value of α in the fatigue experiment is approximately equal to the value of α obtained in the corresponding finite element model.

The finite element analysis (FEA) is performed with the commercial finite element software ABAQUS. In this paper, the 2D plane stress model is chosen to simulate the fatigue experiments. The FEA mesh of this model is shown in **Fig. 4**.

The FEA model is chosen to be ¼ model because the symmetricity of Mode I CCT specimen. Then the symmetric boundary conditions are applied on the symmetric edges, except for the crack surface which is considered as a free surface. In order to take the crack closure effect into account, a discrete rigid line is positioned at the position of the crack surface, and the vertical hard contact and tangential frictionless contact properties are chosen to simulate the possible contact of the crack surface caused by crack closure, following the same procedures in [9]. The material properties are simulated with the nonlinear kinematic hardening model in ABAQUS, using the full range stress-strain relation for 2024-T3 obtained from [18]. In this paper the simulation method follows the similar way as the simulation work presented in [1-6,12]. In the FEA model, a stationary crack with no crack propagation is used to model the fatigue crack because the crack growth in one fatigue cycle is very small, and its influence is quite limited and could be neglected. Two fatigue cycles are simulated. The first cycle is used for generating an initial residual plastic field as the plastic energy dissipation becomes stable afterwards. The second cycle is used to obtain both the value of plastic dissipation and the value of the correction factor α. Then the value of α in various load cases was used as a correction factor to obtain the total energy dissipation value during the fatigue experiment.

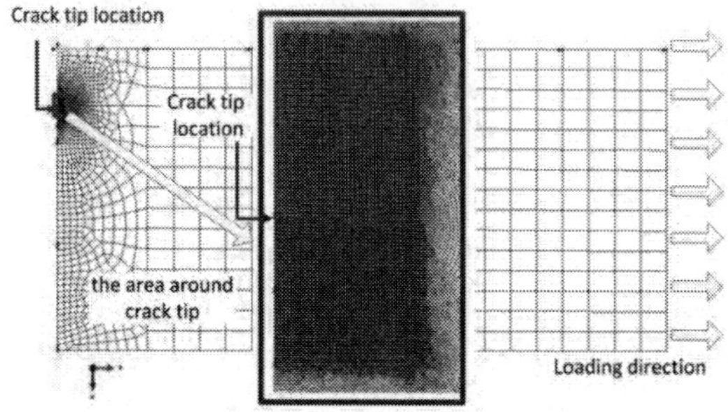

Fig. 4. The FEA model of simulation

As is shown in **Fig. 4**, a very fine mesh was applied around the crack tip area where the plastic deformation during one cycle is significant. The element size of the crack tip area is around one-tenth of the reverse plastic zone size during the simulated fatigue cycles. And the size of the fine mesh part should be large enough to make sure that all the elements in the plastic zone fall within the fine mesh part. In the field which is far away from the crack tip, a coarser mesh is used to reduce calculation time. The element type used are 2D four-nodes bilinear quadrilateral elements with reduced integration. The simulation results of the shapes of the forward plastic zone and the reverse plastic zone in the first fatigue cycle of 2024-T3 specimen at R=0.05 and ΔK=15MPa$\sqrt{m}$ are shown in the **Fig. 5**.

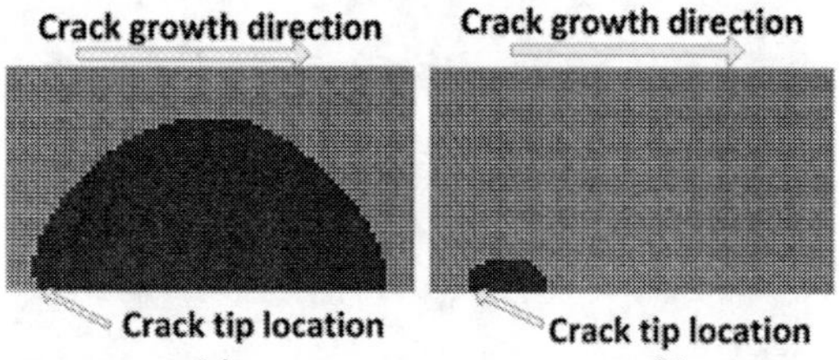

Fig 5. The 2024-T3 specimen at R=0.05. FEA results of (a) the shape of the forward plastic zone at the 1st cycle (b) the shape of the reverse plastic zone at the 1st cycle.

3. EXPERIMENTAL RESULTS AND DISCUSSION

3.1 Experimental results

In this part, the raw experimental results are presented.

The fatigue tests were done with an initial crack length $2a\approx$16mm. The fatigue crack length a at different number of cycles with different stress ratios is shown in **Fig. 6**.

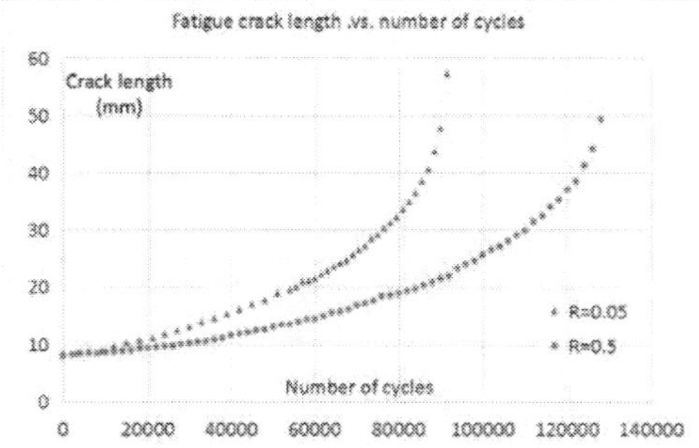

Fig. 6. The crack length versus number of cycles in different stress ratios

The energy dissipation per cycle *dW/dN* was obtained for about every 1/10 of N_f, where N_f is defined as the total fatigue life of the specimen. The trend is shown in **Fig. 7**.

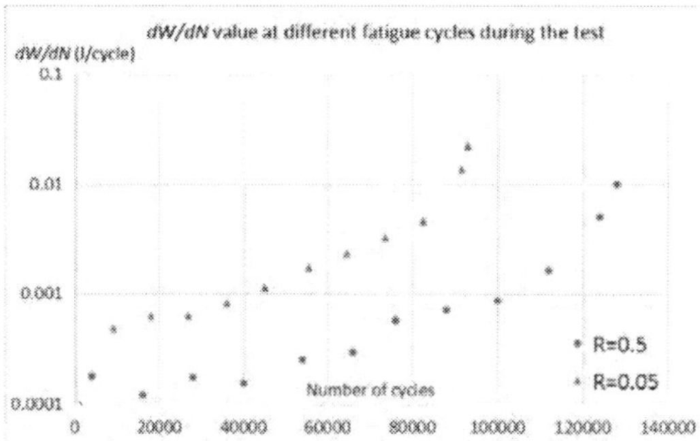

Fig. 7. *dW/dN* versus number of cycles in different stress ratios

3.2 Discussion based on experimental results

3.2.1 da/dN versus ΔK and dW/dN versus ΔK

The fatigue crack growth rate are shown against ΔK in **Fig. 8**. This figure shows that the *da/dN* data could approximately fit the Paris relation, although there is some error. The Paris relations for the curves are

for R=0.05:

$$\frac{da}{dN} = 1.6499 \times 10^{-7} \left(\Delta K\right)^{2.70844} \tag{3}$$

for R=0.5:

$$\frac{da}{dN} = 1.9704 \times 10^{-7} \left(\Delta K\right)^{2.90346} \tag{4}$$

The energy dissipation per cycle for both R=0.5 and R=0.05 is plotted against ΔK in **Fig. 9**.

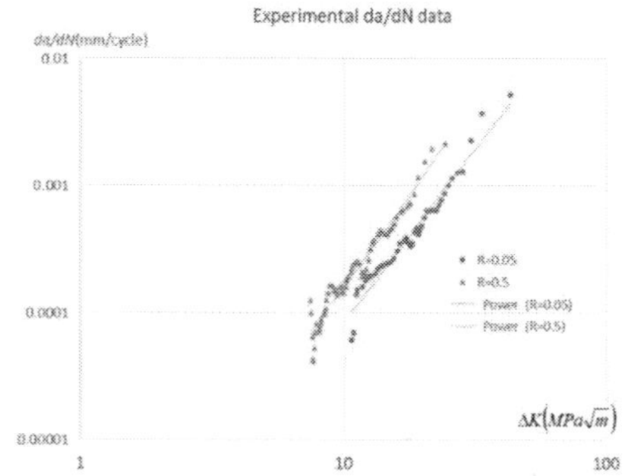

Fig. 8. *da/dN* versus ΔK for 2024-T3

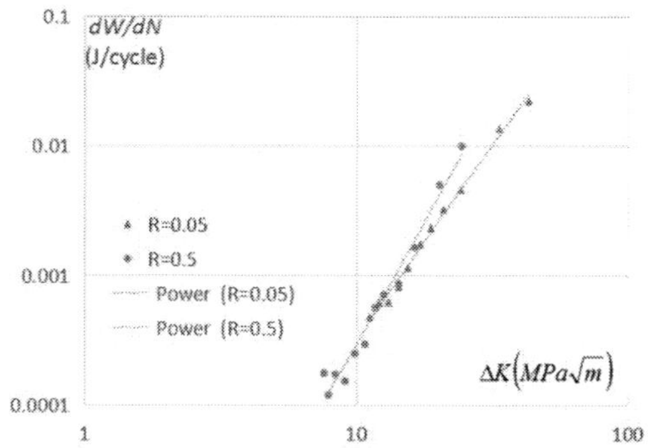

Fig. 9. The experimental energy dissipation per cycle *dW/dN* versus ΔK

From **Fig. 9**, it could been observed that the *dW/dN* versus ΔK curves of both R=0.5 and R=0.05 seems to fall into the same scatter band, which agrees with [12]. The power law fit for the data in **Fig. 9** are

for R=0.05:

$$\frac{dW}{dN} = 3.48547 \times 10^{-7} \left(\Delta K\right)^{2.97284} \tag{5}$$

for R=0.5:

$$\frac{dW}{dN} = 5.30572 \times 10^{-8} \left(\Delta K\right)^{3.74486} \tag{6}$$

3.2.2 dW/da during fatigue crack propagation

From **Eq.(1)** it is known that the value of the *dW/da* is equal to *dW/dN* divided by *da/dN*. So in **Fig. 10**, the energy dissipation per cycle *dW/dN* versus fatigue crack growth rate *da/dN* is plotted. For nearly constant values of *dW/da*, the curves in **Fig. 10** should be nearly straight lines. Thus the curves could be fitted to a power law whose exponent is close to one. The resulting equations after curve fitting in **Fig. 10** are

for R=0.05:

$$\frac{da}{dN} = 0.1563 \left(\frac{dW}{dN}\right)^{0.9277} \tag{7}$$

for R=0.5:

$$\frac{da}{dN} = 0.0773 \left(\frac{dW}{dN}\right)^{0.7573} \tag{8}$$

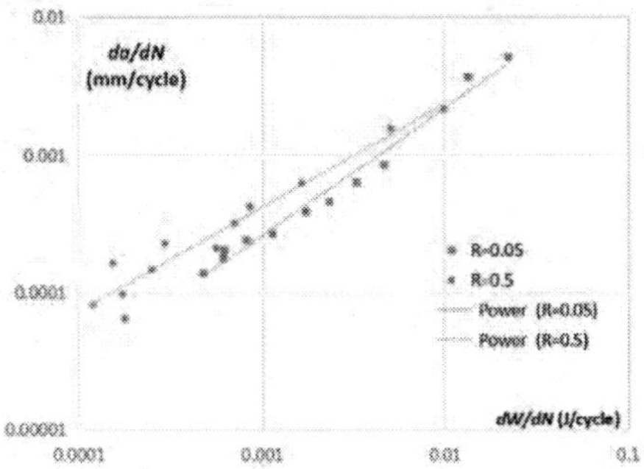

Fig. 10. da/dN versus dW/dN

However, from **Eq.(7)** and **Eq.(8)** it could be concluded that both curves in **Fig. 10** are not linear, which means the dW/da does not remain constant for the tested stress ratios. Moreover, the curve of R=0.5 is clearly nonlinear, while the curve of R=0.05 is much closer to a linear trend. Therefore, the trend of how dW/da changes differs for different stress ratios. These results answer the first research question raised in the introduction: dW/da is not a material property that is approximately constant.

Then the second research question needs to be answered, because it is important for answering whether dW/dN is an effective similitude parameter to evaluate the fatigue crack growth regardless of the influence of stress ratio.

In order to answer the second question raised in the introduction, the dW/dN versus da/dN data in **Fig. 10** for R=0.05 and R=0.5 is compared. It could be observed that for the same da/dN value, the dW/dN for R=0.05 is larger than dW/dN for R=0.5. Therefore, the dW/da value of R=0.05 is larger than the value of R=0.5. This is the answer to the second research question: the same value of dW/dN does not mean the same da/dN.

The same result could also be obtained from both **Fig. 8** and **Fig. 9**. For the same ΔK value, the dW/dN values for both stress ratios seems to be nearly the same, while the fatigue crack growth rate of R=0.5 is much larger than the fatigue crack growth rate of R=0.05. So the dW/da value of R=0.5 should be smaller. Meanwhile from **Fig. 9**, it is possible to conclude that the dW/dN could be considered as an equivalent similitude to ΔK, for the dW/dN stays nearly the same at the same ΔK for both stress ratios. Obviously, the amount of two tests is too small to derive firm conclusions. Hence, more test data at more stress ratios is required to confirm this hypothesis.

3.2.2 The energy balance equation for the fatigue crack growth phenomenon

From the discussion above, it could be concluded that the plastic energy dissipation per unit of fatigue crack growth dW/da is not an effective parameter to evaluate the fatigue crack growth rate, because the same dW/dN could lead to different da/dN values. However, is there any reason to explain this conclusion?

Here the phenomena of crack propagation is considered at a more fundamental level from a physical point of view. The analysis here is started from the example of the static crack propagation. Irwin and Orowan proposed that crack resistance is equal to the sum of the surface energy and the plastic strain work accompanying crack extension [19].

In a similar way, then the static crack propagation could be thought of as energy conservation. The energy balance equation of the crack propagation will then be:

$$\dot{W} = \dot{U}_a + \dot{U}_{pl} + \dot{U}_e \tag{9}$$

The physical meaning of the equation in differential form is that: the external work done to the solid $\dot{W}$ is equal to the sum of the energy cost by the formation of the new crack surface $\dot{U}_a$, the plastic energy dissipation $\dot{U}_{pl}$, and change of the elastic strain energy of the solid $\dot{U}_e$. That also agrees with the opinion of Xiao [20]: the surface-forming energy release is equal to the power of the external force minus the rate of the stored elastic strain energy and the rate of dissipation by plastic deformation. The three variables in the right side of the equation are coupled. But the crack propagation is only directly related to $\dot{U}_a$, although the other two variables would change as $\dot{U}_a$ changes, because $\dot{U}_{pl}$ and $\dot{U}_e$ are only the consequences rather than the reasons or driving force of the crack propagation. For metals, the value of the $\dot{U}_{pl}$ is significantly larger than $\dot{U}_a$ and $\dot{U}_e$.

Then the **Eq.(9)** is extended to the fatigue crack growth situation. When the equation is integrated over one fatigue cycle, it becomes:

$$\frac{dW}{dN} = \frac{dU_a}{dN} + \frac{dU_{pl}}{dN} + \frac{dU_e}{dN} \tag{10}$$

The physical explanation of the equation above is that the external work done (mainly by fatigue loading here) during one cycle dW/dN is equal to the sum of the energy dissipation for the fatigue crack propagation dU_a/dN, the plastic dissipation during one cycle dU_{pl}/dN, and the change of the elastic strain energy after one cycle dU_e/dN. In this concept, the external work represents a driving force for fatigue crack growth, while the energy storage and dissipation over the created crack surface represents effectively the resistance against crack propagation. Both the driving force and resistance depend on the applied fatigue load cycle, the geometry of the problem, crack length and the material properties.

Similar to the static situation, the three variables on the right hand side of the equation are coupled. The fatigue crack propagation is only directly related to dU_a/dN, while dU_e/dN and dU_{pl}/dN are only the consequences of the fatigue crack propagation. The value of the dU_{pl}/dN is much larger than dU_a/dN and dU_e/dN for metallic materials. Therefore, in general it holds approximately that:

$$\frac{dW}{dN} \approx \frac{dU_{pl}}{dN} \tag{11}$$

The term dU_a/dN is directly linked to the fatigue crack growth, which is related to the fatigue crack growth rate and fatigue crack surface characteristics. The fatigue crack surface characteristics are influenced by the mechanics of fatigue crack propagation and could be represented by the roughness of the crack surface. The term dU_{pl}/dN should be established based on the entire load cycle, including the forward plastic zone development during loading and the formation of the reverse plastic zone during unloading.

Now the answer to the question of the varying value of dW/da among different load cases is clear: the correct similitude to describe the fatigue crack propagation is dU_a/dN rather than dU_{pl}/dN, and the plastic energy dissipation is only the consequence accompanying the fatigue crack growth. The change of the ratio of dU_a/dN to dU_{pl}/dN leads to the variance of dW/da.

However, the quantity of the dU_a/dN value is too small to be measured with fatigue experiments. And it is also impractical to obtain the dU_a/dN value for its extremely small value compared with dU_{pl}/dN. So for further research and engineering practices, the value of dU_{pl}/dN seems to be the only choice if there are no extraordinary methods to quantify dU_a/dN for real tests. But extra care has to be taken since dU_{pl}/dN is not directly related to da/dN, and that dW/da cannot be considered as constant or as a material property.

3.2.3 Limitations of the work in this paper

First, there are some shortcomings with the experimental study in this paper. Only 2 stress ratios were tested and only a few dW/dN data points were obtained for each stress ratio. While this is enough for the first step of the study to observe a trend, it is not sufficient to come to firm conclusions. Therefore more stress ratios will be tested and more data points will be obtained in future work. Besides, the dW/dN is obtained by a combination of experimental measurements and FEA calculation, such that errors in numerical simulation might affect the final results.

Second, both the cases in fatigue experiments and FEA simulations are assumed to be pure Mode I. However, in reality it is not strictly true, because the fatigue crack direction is not strictly transverse to the load direction and crack surfaces are not smooth and plain. Therefore, the crack is in mixed Mode I-II-III rather than pure Mode I, which would lead to more energy dissipation than currently considered. Meanwhile, the actual crack surface area is much larger than assumed by taking the of crack length a multiplied by thickness t because of roughness of the crack surface. Hence, that will result in larger dU_a/dN and dU_{pl}/dN values than assumed in both experimental and numerical cases. For the experimental study, the energy dissipation measured is not the theoretical energy dissipation of pure Mode I, and in the FEA modelling those effects cannot be simulated and would result in some error in the simulation.

Moreover, the stress and strain distributions are not uniform through the thickness direction of the specimen, therefore the energy dissipation properties are not the same for the through-the-thickness direction. On the mid-plane of the specimen, it is more like plane strain, but on the surface of the specimen, it is more like plane stress. Nevertheless, in this paper for both experimental and numerical study, it is assumed to be nearly uniform through thickness direction for the thickness is relatively small compared with other geometry parameters: the thickness effect is ignored. In fatigue experiments, the displacement measured on the surface is chosen to represent the whole specimen, and in FEA only the 2D model is applied.

Finally, there must be some error in the FEA modelling. Besides neglecting the rough crack surface, curved crack path and the differences through thickness direction mentioned above, the FEA model still brings in some errors. For one thing, in the simulations, a stationary crack is used, so it fails to illustrate the influence of residual plastic wake behind crack tip and the effect of crack propagation on the energy dissipation. For another, the material model used and the material data from [18] could be a little different from the material properties of the specimens used, based on material production variability.

Therefore, there is still a lot of work to be done in the future. More experimental data at various stress ratios should be gathered, and more care could be taken on the differences between assumption and reality. Meanwhile, a more detailed numerical study which is closer to reality is also needed in the future work.

4. CONCLUSION

In this paper, an experimental study on 2024-T3 fatigue crack growth test is carried out. The plastic energy dissipation at two different stress ratios is measured. The experiment shows that plastic dissipation per unit of fatigue crack growth (dW/da) does not stay constant for different load cases. Therefore the value of dW/da cannot be regarded as a material property, and cannot be used directly to predict da/dN.

The reason is that the dU_{pl}/dN is not an effective similitude parameter. It is the crack surface forming energy that is directly linked to fatigue crack growth, which is also the proper similitude parameter rather than plastic dissipation. The plastic dissipation is only the consequence which accompanies the fatigue crack growth. However, the value of dU_a/dN is too small to be measured accurately in experiments. Then, the plastic dissipation seems to be the only choice whose value could be obtained in reality. But it should be aware that plastic dissipation does not directly link to fatigue crack growth and some extra efforts are needed to bridge this gap.

There are still a lot of limitations in this paper, which result from the differences between assumption and reality. So some future work is still needed to overcome the shortcoming of the current work.

Finally, if we consider the contradiction between the work of [1-6,15] and [12-13], this paper is on the side of [12-13] stating that dW/da is variable. Considering that in the work done by Karlsson et al. [7-11] and Zheng[14] the definitions of dW/dN are slightly different, further numerical details need be given in order to make a comparison between their work and the work in this paper.

References

1. J.S. Daily, N. W. Klingbeil, Int J Fatigue, **32** (2010) 1710-1723.
2. C.M. Baudendistel, N.W. Klingbeil, Int J Fatigue, **51** (2013) 96-104.
3. J.S. Daily, N.W. Klingbeil, Int J Fatigue, **28** (2006) 1725-1738.
4. J.S. Daily, N.W. Klingbeil, Int J Fatigue, **26** (2004) 727-738.
5. N.W. Klingbeil, Int J Fatigue, **25** (2003) 117-128.
6. N. Klingbeill, J. Daily, C. Baudendistel. Key Eng Mater, **378-379** (2008) 385-404.
7. D. Cojocaru, A.M. Karlsson, ADV ENG SOFTW, **39** (2008) 995-1009.
8. D. Cojocaru, A.M. Karlsson, Int J Fatigue, **31** (2009) 1154-1162.
9. P.G. Nittur, A.M. Karlsson, L.A. Carlsson, Eng Fract Mech, **124-125** (2014) 155-166.
10. P.G. Nittur, A.M. Karlsson, L.A. Carlsson, Int J Fatigue, **54** (2013) 47-55.
11. G. Ding, A.M. Karlsson, M.H. Santare, Int J Fatigue, **94** (2017) 89-96.
12. N. Ranganathan, F. Chalon, S. Meo, Int J Fatigue, **30** (2008) 1921-1929.
13. M. Mazari, B. Bouchouicha, M. Zemri, M. Benguediab, N. Ranganathan, Comput Mater Sci, **41** (2008) 344-349.
14. X. Zheng, H. Cui, X. Su, C.C. Engler-Pinto Jr., W. Wen, Eng Fract Mech, **114** (2013) 151-165.
15. K.V. Smith, Eng Fract Mech, **78** (2011) 3183-3195.
16. M. Besel, E. Breitbarth, Int J Fatigue, **93** (2016) 92-108.
17. E. Breitbarth, M. Besel, Int J Fatigue, **100** (2017) 263-273.
18. Military Handbook - MIL-HDBK-5H: Metallic Materials and Elements for Aerospace Vehicle Structures (Knovel Interactive Edition), U.S. Department of Defense.
19. M. Janssen, J. Zuidema, R. Wanhill, Fracture Mechanics, Second Edition: Fundamentals and Applications, vol. 2nd ed, CRC Press, 2004, Page 96.
20. S. Xiao, H. Wang, B. Liu, K. Hwang, J Mech Phys Solids, 84 (2015) 336-357.

Uniaxial equivalent of stress and stress intensity factor in Mode I crack opening for fatigued metals subjected to multi parametric external loading

Andrey Shanyavskiy[1,2,*]

[1]Aviation Register of the Russian Federation, 14142, Sheremetyevo-1 Airport, Moscow Region, Russia
[2]Moscow Aviation Institute (National Research University), 125993, Volokolamskoe street, 4, Moscow, Russia

Abstract. The principles of physical mesomechanics and synergetics, provides for a measure of energy involved in the ability of a material to resist deformation and fatigue cracking used for introducing a concept of equivalent stress and stress intensity factor in Mode I crack opening of fatigued metals. Acquired under the standard test conditions, the curves of unidirectional tension and those of fatigue duration and of crack-growth kinetics show themselves applicable as universal master-diagrams descriptive of the material behavior though under multiparametric loading. Specimen patterns of fracture, revealed by the test pieces of metallic alloys given to asymmetrical biaxial loading or multiparametric loading, passed careful research. In the biaxial case, this research facilitated finding of the view of the correction function for the level of equivalent stress and, in the multi-axial case, employing of the above subject ideas to ascertain the very level of equivalent stress.

1 Introduction

In terms of physical mesomechanics or of synergetics, the evolution behavior of externally affected metallic materials involves a sequence (cascade) of self-organized and self-consisted energy-storing processes of differing scale levels [1–3]. It is a property of metallic materials to meet an external attack by starting a cascade of the evolution processes, each cascade pattern inherently peculiar to a particular metal. Those processes do (or do not) arise depending not on the mode of the external energy supply but on whether this supply does take place at all; in fact, the processes answer *any* mode of external attack and arise on a scale consistent with the growth of the material volume to absorb supplied energy [2]. As the energy intensity of cyclic loading begins to grow, the test metal first absorbs this energy free of fracture effects, i.e., preserves its integrity by switching on still new intrinsic mechanisms of storing energy and thereby avoids creation of new free surface. The greater a scale level the more complex the energy storing process ensuring the metal soundness (integrity): indeed, the volume ΔV, to store in the energy amount ΔW, is greater in the latter case [2, 3].

In real service conditions, external action by cyclical loading on an engineering structure is likely to happen at various combinations of the cycle parameters. Therefore, we may talk about *multiparametric field* of external attack on a metallic material [3]. In addition, it may be that the operating fracture mechanisms follow one another in a sequence, varying with time (*i.e.,* with growing amount of damage) and depending on the sequence in that greater and smaller energy per loading cyclic is supplied to the test structure. Consequently, one can characterize the variety of loading conditions, using quite a number of parameters, including those reflecting the metal reaction to an external attack.

They most often consider in separate, i.e., for each of the single parameters, those effects that those parameters have on the process of building up damage. For instance, they look in separate into the effects of temperature, stress ratio, cycling frequency, etc., on material durability and crack propagation rate. In some instances, either of the two latter characteristics may be a subject examined as depending jointly on temperature and stress ratio [6]. Nevertheless, for *however differing* conditions of external action, they employ *always the same* structural characteristics of the material reaction in order to describe the process of building up material damage at the stage of crack nucleation and propagation. Typically, they consider the ordered cascade of dislocation-structure types, which sequentially replace one another [4], the crack-propagation rate, and/or the elements of the fracture-surface morphology, *e.g.*, fatigue striation patterns relevant to various stages of crack propagation under various loading conditions, *etc.* [3].

Metallic materials respond to multiparametric cyclic loading by absorbing energy in an adequate way. Such a

* Corresponding author: 106otdel@mail.ru

response turns out similar to that under the standard conditions of uniaxial loading provided the mode of final deformation remains the same for the varying multiparametric attacks. One can see the similarity of this kind between the cases of monotonic uniaxial tension and cyclic loading showing the normal (Mode I) crack-opening behavior [3].

We may, therefore, describe a material behavior with the use of a single approach to the idea of the density of deformation energy [5, 6] and of the density of fracture energy [3, 7]; in so doing we take, for an integrated characteristic, the parameters of the material reaction to an external affection for any combination of the loading-cycle parameters. Here, the critical requirement is to ensure that, under the external action of an arbitrary kind, material stretching would dominate at the single-path loading and Mode I normal crack opening, at the cyclic loading.

2 Durability

In the materials under cyclic loading, damage of various scales builds up, to agree with the tensile stress-strain diagram [2, 3, 7]. In metallic materials, the built up damage forms in [8] and beneath [9] the surface layer, and the oncoming of the limiting state, with its concomitant initiation and propagation of a fatigue crack, takes a period depending on the level of the applied stress; characteristic of the limiting state is the number N_f of cycles to failure. Each of the loading cycles brings in an energy density dW/dV. At various scale levels, the latter quantity is characteristic of the limiting state of a material in terms of the number of cycles that the material can resist until the fracture of all volume ΔV.

Let us consider the stress factor in its dimensional form kg/mm^2. Having multiplied this ratio by mm/mm, we come about a dimension, kg×mm/mm^3, peculiar to the deformation work $(dW/dV) = \sigma_e$ of the unite volume of a material, i.e., come about the density of strain energy. With any method of affecting a material, the attained ε_j value relates to the only magnitude $(dW/dV)_{ej}$ of energy density or the only level σ_{ej} of equivalent stress.

The idea that, in metals, energy builds up on the differing scales makes it necessary, when describing the metal behavior, to consider the bifurcation diagram [9] adopted for the specified standard experiment conditions. We use preserved the entire test parameters of an external cyclic influence, and only vary applied uniaxial tensile stress, setting it different for different test specimens.

Based on the concept of the equivalent tensile stress [3, 7], one can reduce the earlier drawn bifurcation diagram of fatigue curve to a general form of fatigue durability versus equivalent stress or strain-energy density, Fig. 1. Equality in the fatigue durability provides for a criterion that the etalon test conditions (relevant to the etalon fatigue diagram) are equivalent to the test conditions in a multiparametric space, *i.e.*, relevant to the master diagram.

A metallic material achieves its limiting state of crack nucleation as soon as the strain energy builds up to the critical amount, which is independent of the applied stress level. Therefore, one can accept for every scale that

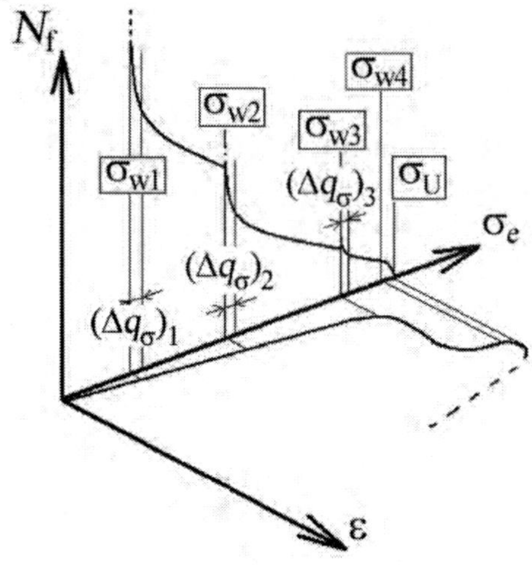

Fig. 1. A bifurcation bearing plot $N_f - \sigma_e$ (schematic) of metal fatigue, drawn in agreement with the diagram σ_e–ε of tension tests and in terms of "equivalent stress level σ_e" or "strain energy density dW/dV". The bifurcation ranges $\Delta q_{\sigma i}$ of the transitions to micro- or nano- $(\sigma_{w1} - \sigma_{w2})$, meso- $(\sigma_{w2} - \sigma_{w3})$, and macro-scale $(\sigma_{w3} - \sigma_{w4})$ levels of fracture [9].

$$N_f\,(\Delta W/\Delta V) = \text{const.} \tag{1}$$

In terms of the equivalent stress, Eq. (1) coincides qualitatively with the well-known equality from Wöhler [10],

$$\sigma_e^{m} N_t = C_f. \tag{2}$$

Equation (2) involves the exponent $m_f = 2$ as long as $\Delta W/\Delta V = \sigma_e^2/(2E)$ [8, 9], *e.g.*, for the macro elastic high-cycle fatigue behavior of metallic materials. Actually, the exponent m_f magnitude differs significantly from two to demonstrate that, in fact, the consumed amount of strain energy is characteristic of different scales of the material behavior. In a metallic material the defects generated by cyclic loading build up predominantly in the surface layer of a test piece; accordingly, the test piece consumes the strain energy supplied by cyclic loading, and crack nucleation occurs in the sites of particularly high concentration of the defects. Therefore, one would be wrong employing a *macroscopic* approach in order to describe an operative *mesoscopic* pattern (building up damage) of material behavior.

The surface layer differs significantly from the interior part of the material body [8]. In the surface layer, damage builds up of both the macroscopic and mesoscopic scales. The $\Delta W/\Delta V$ magnitude is a macroscopic characteristic of metal deformation all over the test-piece body, in accordance with the tensile stress-strain diagram. Hence, to ascertain the whole number of cycles to failure, one has to measure the dW/dV magnitude as related to the scale level of the mechanisms that control effects of building up damage, including those taking place in the surface layer of the test sample.

Let us regard the mesoscopic scale of the strain-energy magnitude, which controls the surface crack-nucleation effects under high-cycle fatigue, and estimate how much this magnitude deviates from that inferred from Eq. (1). In so doing, assume that the limiting state of a material, described by Eqs. (1) and (2), requires the same level of the limiting energy (characteristic of the breaking point) for both macroscopic and mesoscopic scales; then, the local energy density at the loading part of the cycle is $m_f/2$ times as big as the dW/dV value, inferred from the tensile stress-strain graph.

Thus, for each of the scales, equations descriptive of the relation of the value of test-piece durability to the applied stress or to the strain-energy density, expressed in terms of "equivalent stress", get the view as presented in [9].

One should particularly mind that, in case of cyclic loading, another parameter, the stress ratio $R = \sigma_{min}/\sigma_{max}$, appears. Its effect on the material behavior is similar to that of the deviations of equivalent stress or of strain-energy density; likewise, some of the parameters influence the test-piece behavior under tension.

3 Crack Propagation

K.J. Miller [11] first suggested to regard the magnitudes of crack-growth rate and the size of a crack-tip zone as an integrated characteristic of a material reaction to the propagation of fatigue-cracks starting from the test-piece surface and having the crack front of various shape. For two differing test conditions, the stress-intensity factors K_{Ii} and K_{Ij} will be equal to one another as long as the two crack-growth rates or (which is the same) the sizes of the two crack-tip plastic zones are, respectively, also equal to one another, but crack lengths a_i and a_{i+1} will be different for two compared cases. The two crack lengths a in the ratio a_i/a_{i+1} correspond to the above mentioned two different test conditions; and the ratio figure indicates how much the crack length must change, with the test conditions, to thereby keep unchanged the crack-growth rate (or the plastic-zone size).

The above criterion was a subject of investigation as applied to the case of uniaxial or biaxial tension (also to a tension-compression case) of flat models [11]. A significant requirement is that one supports the normal (Mode I) crack-opening behavior and similarity of the fracture mechanisms for the two compared test conditions. We introduce a correction $f(a_i/a_{i+1})$ for the crack length, which, in fact, is a dimensionless correction for the stress ratio $\lambda = \sigma_1/\sigma_2$ in the case of biaxial loading, where σ_1 is the stress of uniaxial tension and σ_2 is the stress applied normal to σ_1.

Another approach, which helps one to universally describe the growth behavior of a fatigue crack in a metallic material, makes use of the concept of *effective stress-intensify factor* ΔK_{eff} [12], its magnitude inferred from the effect of incompletely closed fatigue crack (crack lips). McClung [13] found out that the three kinetic curves (patterns of biaxial-tension or of tension-compression included) plotted as functions of stress ratio, are equidistant with respect to one another and can be reduced to a single curve, as applied to the specified standard (adopted) conditions of the experiment.

As a matter of fact, the effect of incompletely closed crack lips can vary in its nature, *i.e.*, in the mechanisms that control the crack-tip behavior [14]. Owing to such an uncertainty, the similarity might be lacking between the growth patterns of the fatigue cracks, despite still maintained equality between the magnitudes of the above mentioned effective stress-intensity factor; hence, the prospects become uncertain of employing the latter as a universal characteristic of a multiparametric affection.

Now we are going to demonstrate greater universality of a criterion based on the idea of the *density of fracture energy*.

Imagine a fatigue-crack front line as a non-stretchable string moved along two parallel strings by the action of a force F (Fig. 2). In proportion to the string displacement (crack increment), the work $dW = Fda$ is being done.

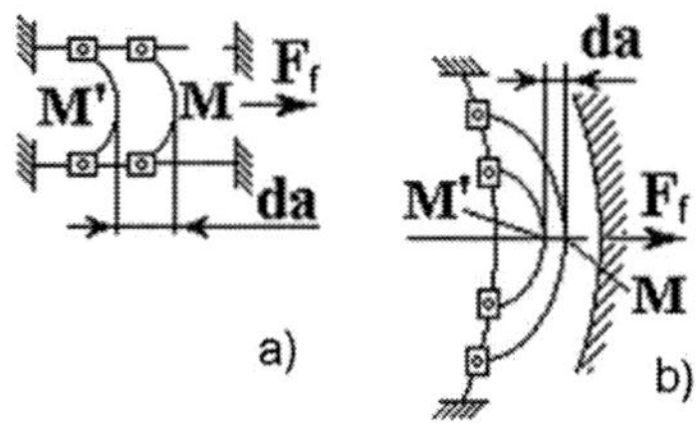

Fig. 2. The crack frontline (schematic) in the form of a stretchable string for the cases of (a) through fatigue crack or (b) surface fatigue crack; the frontline shifts, moved by a force applied to a material point M, in the crack-growth direction.

Returning to the pattern of crack, we mind that the crack lengthens owing to the event of breaking down some volume dV of the material within the limits of the plastic zone. Having considered some area ds, normal to the crack-growth direction, we multiply and divide by ds the right part of the above-derived relationship to have a result as follows,

$$dW = (F/dS)dads = \sigma_f dV, \text{ or } dW/dV = \sigma_f. \quad (3)$$

Thus obtained stress magnitude is proportional to the total effect of the combined attacks terminated at the crack increment, *i.e.*, proportional to the equivalent stress σ_e; one should also mind that, for the same stress, the crack-growth rate is the higher the less the elasticity modulus of the material. One can take account of this interrelation using the value of strain revealed by the loaded test piece. Then we finally have

$$(dW/dV)_f = (1 - \nu^2)\sigma_e^2/E. \quad (4)$$

In Eq. (4), we take into account the Poisson's effect of the transition between the two mutually normal stress vectors. This equation is descriptive of the crack-increment in a fatigue process; here, equivalent stress σ_e represents the density of fracture energy $(dW/dV)_f$

peculiar to the crack-increment event in a material during a loading cycle and can be estimated by the next equation:

$$\sigma_e = \sigma_1 F[f(X_1), f(X_2), \ldots, f(X_i)] \quad (5)$$

In Eq. (5), σ_e is the value of equivalent stress characteristic of an integrated reaction of a material on the external multiparametric attack, $F[f(X_1), f(X_2), \ldots, f(X_i)]$ is the correction function for the effect of the parameters X_i on the deformation behavior of a material, and σ_1 is the stress of standard uniaxial tension representing one of the constituents of the multiparametric attack for the case of $[f(X_1), f(X_2), \ldots, f(X_i)] = 1$.

Bearing in mind Eq. (3), we may suggest the way of the universal description of fatigue cracks propagation; in so doing, we employ the entities of crack-growth rate and of fatigue-striation spacing δ_i in the relationships for different types of Al, Fe, Ti based alloys [3], [7], [15-17].

Thus, the concept of *equivalent stress* provides for a single approach to the problem of having described the behavior of metallic materials in terms of a universal characteristic, the *energy density*, applicable to whether the fracture or the deformation events.

Let us now ground on the principles of physical mesomechanics and synergetics to see, from examples, how efficient is to employ the above concept when producing a universal description of the behavior of metallic materials under multiparametric cyclic loading.

3.1 Example 1

Shanyavsky *et al.* [15] reported on the tests of cruciform test specimens (imitational model) by cyclic biaxial tension, with the concurrently changed cycle parameters, as well as under compression-tension conditions with a diverse stress ratio. The 5-mm thick specimens, made of D16 aluminum-based alloy, passed the tests in the range $-1 < \lambda < 1$ of principle-stress ratio ($\lambda = \sigma_1/\sigma_2$) and in the range 0.1 to 0.8 of the stress ratio R.

The variations of the cycle parameters involved correction functions found from a relationship

$$C_{Is0}/C_{Isi} = F(R, \lambda). \quad (6)$$

The authors [15] acquired the $C_{Is0} = (1 - v^2)/(8\pi\sigma_{0.2}E)$ magnitude of relation (6) as applied to the standard test conditions when crack-growth rate has proportion to the $(K_e)^2$ of stress equivalent factor. They inferred the C_{Isi} magnitude from the test results, including the measured crack-growth rates and striation spacing for various combinations of the value R and the ratio λ of principle stresses when kinetic curves have equidistant position of one to another one.

They found out that, throughout the said range of the combined loading-cycle parameters, the curves of fatigue-crack growth remain equidistant with respect to the one and another for the crack-growth rate or for striation spacing [16]. The same (equivalent) crack-growth kinetics persisted all over the examined range of variously combined parameters R and λ of the loading cycle. For the aluminum-based D16T alloy, the authors [16] determined the magnitudes of equivalent stress (with respect to the stress used in the uniaxial cyclic loading). They found out that, in the mentioned range of the parameters R and λ, the correction functions for the mentioned magnitudes of the equivalent stress take the view of polynomials as follows for the striation spacing δ:

$$\begin{aligned} F^2(\lambda_\sigma, R)_\delta &= 0.953 - 0.547\lambda_\sigma - 0.099\lambda_\sigma^2 + 0.588R + \\ &+ 0.839R\lambda_\sigma - 0.201R\lambda_\sigma^2 - 5.184R^2 - 0.329R^2\lambda_\sigma + \\ &+ 2.091R^2\lambda_\sigma^2 + 3.855R^3 + 0.08R^3\lambda_\sigma - 2.093\lambda_\sigma^2 R^3 \end{aligned} \quad (7)$$

It follows from Eq. (7) that, in the case of multiparametric cycling, nonlinear functions describe the effect of the cycling parameters on a material behavior and indicate to the maintained law of similarity between the fracture mechanisms or between the crack-growth kinetics.

3.2 Example 2

The author [17] examined the cases of in-service failure of the longerons of helicopter blades (Mi-family helicopters) manufactured from Al-alloy AVT. In service, those longerons experience the complicated multiparametric effects of tension, torsion, and bending, as well as of one-atmosphere excessive internal pressure. A pressure gage, built in the longeron, responds to depressurization, thus warning about an occurred discontinuity.

Fatigue cracking, ordinarily starting from various material defects, might be a result of loading over a limit preset by a designer. Therefore, the author's aim had to answer if the in-service stresses did not exceed a designed limit.

On the strength of that, the author performed a set of fractographic examinations, the special bench tests of the longeron fragments, as well as the standard tensile tests of the specimens cut out of the longerons. The author answered the question about the in-service loading pattern of the longerons by having performed a two-stage research. At the beginning, the author examined the specimens that passed the tests by bi-parametric loading: by tension and torsion. Fractographically, the specimens showed fatigue striations formed all along the crack-growth direction; this finding confirmed the maintenance of the similarity law for the fracture mechanism in a wide range of the loading-cycle parameters as concerns the stress ratio R of the loading cycle and the angle ω of torsion. Of the correction functions $F(R,\omega)$ involved in the relationship

$$C_{Is0}/C_{Isi} = F(R,\omega), \quad (8)$$

a magnitude C_{Isi} answers the test data, including the measured values of crack-growth rate and of striation spacing for the variously combined magnitudes of stress ratio R and torsion angle ω.

The test data, including those on the crack-growth rates, show that, plotted as the functions of the stress-intensity factor K_I, all the curves of the cracking kinetics appear equidistant with respect to one another. A correction function had a view [17],

$$F_i^2(\omega) = A_1 + A_2\omega + A_3\omega^2 + A_4\omega^3 + A_5\omega^4 \quad (9)$$

for each of the stress ratio (with R = const) with different A_i constants.

Using the values of coefficients for unified fatigue crack growth description [17] made it possible to reduce all the curves to a single one, related to an equivalent stress-intensity factor K_e (Fig. 3).

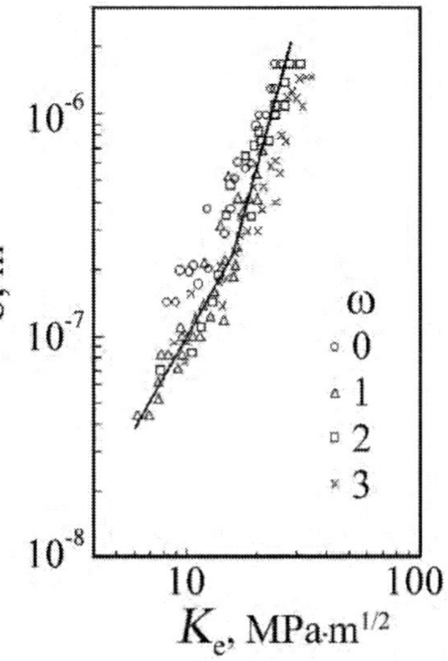

Fig. 3. The test data, on the rectangular specimens of AVT alloy, obtained at various torsion angles ω and a magnitude $R = 0.1$ of stress ratio. A single unified curve of crack-growth kinetics (solid line) as compared with the superimposed data points of striation spacing δ, plotted as functions of the stress-intensity factor $K_e = K_{Imax}F(R,\omega)$ [17].

At the second stage of these investigations, the author employed the formulas of fracture mechanics to have calculated the levels of equivalent stress. The calculations concerned various crack lengths, various cross sections of rotor blades, and various helicopter units, *i.e.*, covered different loading conditions typical of the longeron in service.

Thus acquired, the data turned out to agree with those obtained, by designers, in the framework of traditional calculation techniques with 15 to 20% calculation accuracy.

3.3 Example 3

The helicopter V-3 passed through a severe flying accident after 90-hour operation in service. The accident circumstances and all structure elements of the helicopter were examined to find out that the braking events mostly began from the pitch levers of the rotor. And one of the levers (No. 4) appeared broken by fatigue Fig. 4.

The fracture began from a bottom ledge of the I-shaped part of the lever manufactured from Fe-based alloy with 1800 MPa of ultimate tensile stress. In its starting zone, the fracture looked quite smooth and free of any distinct plastic-shear lips. The fracture path is illustrated by the sequence of fracture micrographs in Fig. 5. From the micrographs one can see the fracture surface cracked and showing (locally) fatigue striations, whose spacing achieved 5×10^{-7} m quite close to the fracture-origin zone. Such a fracture pattern persisted to the crack lengths of 5–7 mm. As the crack length grew further the cracking effects of the fracture surface became more extensive.

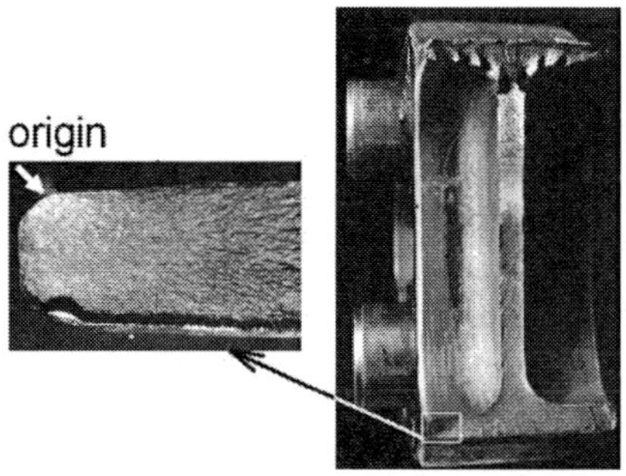

Fig. 4. View of the lever fracture surface with indication area of crack origination.

The most important here is the problem of crack-growth period and equivalent-stress level; both were estimated based on the single curve of fatigue kinetics based on discussed concept of equivalent value of stress intensity factor.

a)

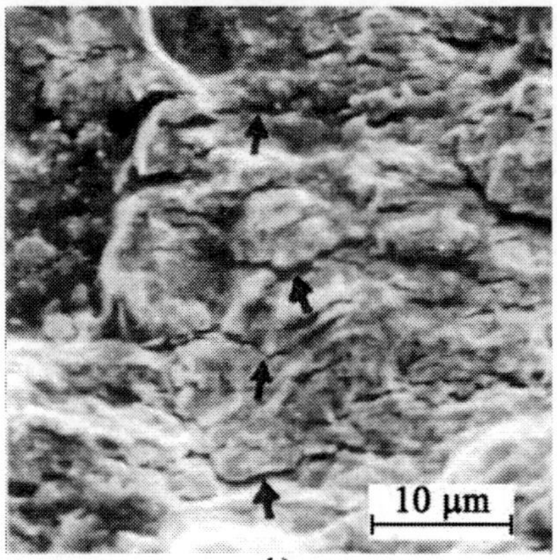

b)

Fig. 5. Areas with cracks (indicated by arrows) and fatigue striations (a), (b) on the different distance from the origin.

The average crack-growth rate approximates 1.2×10^{-6} m per loading cycle at the fracture stage where this rate is proportional to the value of equivalent stress-intensity factor to the power of four in accordance with unified kinetic curve for Fe-based alloy [3], [7]. In the subject case the distance of stable crack growth measured about 30 mm. So we can estimate that the crack-growth period took at least 30000 loading cycles. If, instead, we had used the smallest crack-growth rate of 5×10^{-7} m per loading cycle, the greatest value of crack-growth period would achieve 70000 cycles. As follows from the data of in-flight and bench tests, a basic revolution frequency (4 Hz) of the rotor makes a basic loading frequency of the propeller-pitch levers. This means that the crack propagation took 2 to 5 hours and could come to the critical crack length during the last flight.

Judging from the designed stress level – 50 MPa – and the portion – 80% – of fatigue fracture in the total cross-section area, the starting (crack-nucleation) and final (end-rupture) levels of equivalent stress differed appreciably. The threshold stress-intensity factor is at least 10 MPa $m^{1/2}$ for the high-strength subject steel that was used for manufactured structure. This value predicts that an initial defect (or inherited crack) not shorter than 10 mm is to exist to make possible initiation of a fatigue crack at the stress level of 50 MPa. Yet no initial crack was revealed at all in the discussed lever. Therefore, the nucleation and growth of the fatigue crack should have occurred at a substantially greater stress level, which we estimated based on the following concepts.

As follows from the single fatigue-kinetics curve, an equivalent value, about 50 MPa $m^{1/2}$, of stress intensity factor is required to maintain the growth rate of fatigue crack near 5×10^{-7} m per loading cycle. Such a growth rate was measured in the region about 3 mm ahead of the crack-origin site. At this stage, the steady-state growth of the crack occurred and we could analyze this growth like in the case of steady-state growth of a through crack in a specimen of rectangular cross-section. The crack here was assumed to reach far enough from its origin site, though its length remained rather small as compared with the general dimension of the lever cross-section. Therefore, we estimated the equivalent stress level using the following simple relationship,

$$\sigma_e = K_e/\sqrt{\pi a} = 50/\sqrt{3.14\times3\cdot10^{-3}} \geq 500MPa \quad (10)$$

This stress value is somewhat underestimated because we neglected here certain corrections, including those for the crack-front geometry. Moreover, we only considered the smallest values of crack-growth rate and equivalent stress-intensity factor.

In parallel, flat specimens, specially cut out of similar levers, were tested bend-loaded. Judging from the test results, the fracture morphology revealed by the broken lever was characteristic of cracking at a high nominal tensile (uniaxial) stress and a high stress-intensity factor. This finding, which showed again to the efficiency of the unified fatigue-kinetics curve as used in calculations of an equivalent stress, made it clear that one still should reveal the way in that the loads applied to the lever in service became greater than predicted by design.

Based on the discovered stress equivalent value more than 500 MPa, special (short-term) in-flight tests were conducted with strain-gages installed in the crack-initiation zone. The test conditions were applied with due account of the fact that the subject helicopter operated being 2-ton overloaded, which should have influenced somehow the operation conditions of all its structure elements.

The in-flight tests showed that stresses as high as 700 MPa can really arise in the propeller-pitch levers of a helicopter overloaded to the level mentioned above. This finding conformed very well with the above stress-calculation results, acquired by processing of the qualitative-fractography data in accordance with the concept of single fatigue-kinetics curve. (Besides, thus estimated figure of the crack-growth period appeared quite consistent with the documented in-flight time-to-failure of the lever.) Done based on those results, the in-flight tests appeared well programmed, properly performed, and informative. The test results definitely confirmed that the total overload of the helicopter was what actually initiated and controlled fatigue failure of the lever and, consequently, caused this severe flight accident of the vehicle.

The three presented examples show true the above-discussed concept of equivalent stress, the latter acquired from the material reaction to an external multiparametric attack; the concept conforms to the well-known alternatives of the fracture-mechanical approach to the problem of estimating the stress level of metallic constructions.

4 Conclusions

In terms of physical mesomechanics and synergetics, a concept of equivalent stress, put forward in the present effort, considers such a stress as a physical characteristic of the material behavior and of the energy stored in a unite material volume under the dominating effects of single-acting and/or cyclic tension.

The present research snows a cyclic lifetime (durability under cyclic loading) as informing on the material reaction to a multiparametric attack; at whichever scale level, the durability figure strongly depends on the portions of strain energy supplied with each loading cycle. The magnitude of energy density per loading cycle, as available from the tensile stress-strain diagram, *i.e.*, relating to a macro-scale level, cannot be characteristic of the material durability if in the range of high-cycle fatigue. Indeed, in this range, one should regard the said energy density as related to the surface layer, in that fatigue cracks nucleate, *i.e.*, to the meso-scale level of deformation.

At the stage of fatigue-crack propagation, an ordered sequence (cascade) of the fracture mechanisms shows itself; each of those mechanisms is uniquely relative to one of the respective scale levels of each material reaction against the multiparametric attack and, therefore, furnishes the way for having identified a level of equivalent stress.

Judging from data on experimental bi-parametric attacks on a material as well as from the findings brought in by the analysis of the broken parts of real structures subjected to multiparametric attacks, the estimations of the level of equivalent stress are proving reliable.

This study received a Supporting Grant from the Russian Scientific Foundation (Project No. 16-19-10376).

References

1. V.S. Ivanova, *Synergetics, Strength and Fracture of Metallic Materials* (Cambridge Intern. Science Publ., 1998)
2. V.E. Panin, V.E. Egorushkin, A.V. Panin, A.G. Chernyavskiy, Fiz. Mesomech., **19**, 1 (2016)
3. A.A. Shanyavskiy, *Modeling of Fatigue Failure of Metals. Synergetics in Aviation* (Ufa, Monograph, 2007)
4. N.A. Koneva, D.V. Lychagin, L.A. Teplyakova, E.V. Kozlov, *Theory and Experiment in the Study of Disclinations*, 116-126 (Leningrad, Physical-Technical Institute, 1986)
5. G.C. Sih, X.S. Tang, Theor. Appl. Fract. Mech., **43**, 2 (2005)
6. P. Lazzarin, F. Berto, F.J. Gomez, M. Zappalorto, Int. J. Fatigue, **30**, 8 (2008)
7. A.A. Shanyavskii (1998) *PROBAMAT-21st Century: Probabilities and Materials*, **46**, 11-44 (Dordrecht, Springer, 1998)
8. V.E. Panin, Fiz. Mesomech., **2**, 6 (1999)
9. A.A. Shanyavskiy, Fiz. Mesomech., **17**, 6 (2014)
10. A. Wöhler, Zeitschr. für Bauwesen, **13**, pp. 233-258 (1863)
11. K.J. Miller, Met. Sci., **11**, 8-9 (1977)
12. W. Elber, *Damage Tolerance in Aircraft Structures*, pp. 230-242 (ASTM STP 486, Philadelphia, 1971)
13. R.C. McClung, Fatigue Fract. Eng. Mater. Struct., **12**, 5 (1989)
14. R.O. Ritchie, Mater. Sci. Eng.: A, **103**, 1 (1988)
15. A.A. Shanyavsky, E.F. Orlov, M.Z. Koronov, Fatigue Fract. Eng. Mater. Struct., **18**, 11 (1995)
16. A. Shanyavskiy, Engng Fract. Mech, **78**, 8 (2011)
17. A.A. Shaniavski, Fatigue Fract. Eng. Mater. Struct., **19**, 9 (1996)

Microstructurally small fatigue crack growth rates in aluminium alloys for developing improved predictive models

Madeleine Burchill[1,*], *Simon* Barter[1], *Lok Hin* Chan[2], *Michael* Jones[3]

[1]Defence Science and Technology Group, Fishermans Bend, VIC, Australia
[2]Hong Kong University of Science and Technology, Hong Kong
[3]Fortburn PTY, LTD, Wendoree, VIC, Australia

Abstract. The fatigue or durability life of a few critical structural metallic components often sets the safe and/or economic useful life of a military airframe. In the case of aluminium airframe components, growth rates, at or soon after fatigue crack nucleation are being driven by near threshold local cyclic stress intensities and thus are very low. Standard crack growth rate data is usually generated from large cracks, and therefore do not represent the growth of small cracks (typically <1mm). Discussed here is an innovative test and analysis technique to measure the growth rates of small cracks growing as the result of stress intensities just above the cyclic growth threshold. Using post-test quantitative fractographic examination of fatigue crack surfaces from a series of 7XXX test coupons, crack growth rates and observations of related growth phenomenon in the threshold region have been made. To better predict small crack growth rates under a range of aircraft loading spectra a method by which standard material data models could be adapted is proposed. Early results suggest that for small cracks this method could be useful in informing engineers on the relative severity of various spectra and leading to more accurate predictions of small crack growth rates which can dominate the fatigue life of airframe components..

1 Introduction

The life of type (LOT) of a military airframe in service maybe based on the fatigue or durability life of a few critical structural metallic components, due to safety and/or maintainability concerns [1]. In the case of aluminium alloy aircraft structures, it has been found that the fatigue life is dominated by the initial growth rate of the *lead crack* in any component, i.e. the crack that leads to the component failure [2]. The majority of the component's fatigue life is therefore consumed while such cracks are small [3, 4], and of the dimensions of the microstructure; typically nucleating from small naturally occurring material discontinuities, with an equivalent initial crack size, a_i of ~20μm [5]. For many airframe aluminium alloys, *microstructurally small* fatigue cracks initially have very low growth rates; when loaded with representative load spectra. These low growth rates are often being driven by near threshold local cyclic stress intensities (ΔK), and are not well represented by standard fatigue lifing methodologies, further such methods often under estimate growth rates in this regime [6-8], resulting in un-conservative fatigue life predictions.

One reason for this discrepancy is that standard crack growth rate data is usually generated from much larger cracks by methods such as those set out in the ASTM E647 "Standard test method for measurement of fatigue crack growth rates" [9] and as a result, they do not represent the growth of small cracks (typically <1mm).

A range of different strategies to develop accurate models of small crack growth, in particular for critical aircraft structural components, have been investigated. These include:

- assuming plasticity induced crack closure on the measured long crack data and adjusting the predicted growth for this [i.e. adjusted compliance ratio] [7];
- correlating crack growth to the crack tip opening displacement parameter [10];
- predicting formation of dislocations ahead of the crack front to correlate with crack growth [11];
- adjusting the effective threshold down [12];
- assuming exponential growth and a very low (or zero) growth threshold [13];
- determining a critical strain/distance threshold based on local notch/crack geometry [14, 15]; and
- using the average crack growth for spectra that contain repeating blocks of variable loading, [16].

The Threshold and Small Crack (TASC) research program at DST has used another approach: an innovative test and analysis technique to *directly measure* the crack growth rates of small cracks growing as the result of stress intensities just above the cyclic threshold region (ΔK_{th}). Here, the method applies a loading spectrum that includes alternating constant amplitude (CA) load sub-blocks, to small aluminium alloy test specimens. Post-test microscopy of the fracture

* Corresponding author: Madeleine.Burchill@dst.defence.gov.au

surfaces, using both optical and electro-optical inspection allows the crack growth rates to be measured from the CA growth bands. Such measurements include growth rates well below the standard threshold values typically published using other measurement techniques.

This paper outlines the TASC test and analysis process, and other interesting results, particularly the *subtle* crack growth retardation associated with CA only loading as well as general retardation and acceleration phenomena. A new data model of the crack growth rate was also developed from the data generated in this test program. Finally, a validation of the crack growth data model is presented to predict a fatigue crack growth in similar coupons that were tested with a variable amplitude (VA) spectrum.

2 Measuring Threshold Crack Growth

The TASC methodology has been previously reported, including in [17] where a brief synopsis of one principle methodology employed was given to measure near threshold and small crack growth. This is reproduced below, along with a material crack growth data model informed by this TASC data for AA7050-T7451. The TASC program at DST aims to:
- improve DST's fatigue crack growth predictive capacity and tools to support the Australian Defence Force capability, and
- improve general understanding of the retardation/acceleration mechanisms present in the fatigue cracking process and their various sources.

2.1 General Test and Analysis Methodology

The TASC coupon testing program uses specially engineered novel test spectra that combine small sequences or *sub-blocks* of CA and VA loading. An example of a normalised TASC spectrum is shown as an insert in Fig. 1, where this load spectrum was used to generate crack growth data for two load (stress, σ) ratios, R ($\sigma_{min}/\sigma_{max}$): R=0.5 and R=-0.5 and produced the fracture surface shown in Fig 1. Here, the normalised spectrum is scaled to achieve 180MPa net peak stress in the test region. The coupons were tested to failure after which the fracture surfaces were inspected using the following quantitative fractography methods: optical microscopes were used with magnifications up to X1000 to measure crack growth increments as small as 1 µm in width. Field emission electro-optical scanning electron microscopes (FEGSEM) were used for higher levels of magnification, where crack growth increments as small as 20 nm could be measured. A typical FEGSEM image of the fracture surface at a small crack depth (a~0.4mm) is also given in Fig. 1, showing the differences crack growth topography that can be observed at this scale from testing with the spectrum shown in the figure.

The content of the CA and VA sub-blocks in each TASC spectra was selected to ensure the growth bands could be identified. Depending upon the material to be tested, some trial-and-error is required to ensure good visibility. In particular it is worth noting, and can be seen in Fig. 1 that the incorporation of the VA[a] sub-block greatly aided identification of CA growth bands on the fracture surface by producing topography and growth plane differences [18].

Test coupons are (typically) of a *hour-glass* shape with a large radius and thus a low net stress concentration, (Kt) with the dimensions given in Table 1. To ensure fatigue cracks nucleated in the test region, the coupon surfaces were prepared in one of two ways, either: (i) *etched* entirely and shot-peened outside of the test area or (ii) polished all over and laser slotted [19] within the test area.

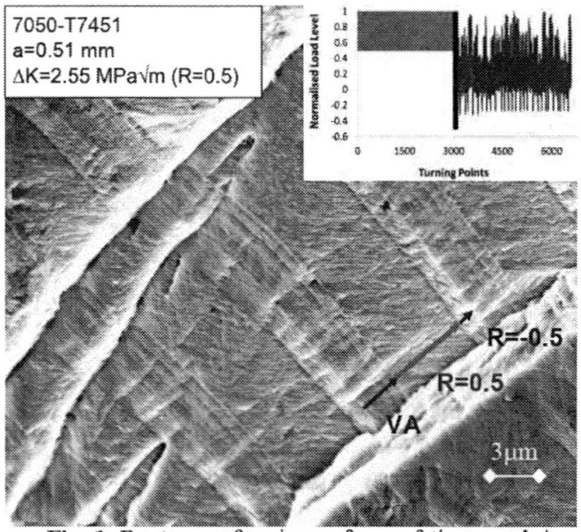

Fig. 1. Fracture surface image from a fatigue crack in a AA7050-T7451 coupon tested with the spectrum shown in the insert; topography differences are used to identify growth from CA (R=0.5 and R=-0.5) and VA load sub-blocks.

Table 1. Dimension details of hour-glass coupons (mm)

overall dimensions			test (net) section			
length	width	thick-ness	radius	width	Kt net	Kt gross
160	40	6.35	100	25	1.08	1.73

2.2 Crack growth rate data

Crack growth rate data (da/dN) for small fatigue cracks in AA7050-T7451 has been collected from 20 coupons, tested with TASC spectra, shown in Fig. 1 for R=0.5 only, or from similar spectra for R=0.1 and R=0.8 as reported in [17]. This data is presented in Fig. 2 along with the standard material curve taken from the a linear elastic fracture mechanics crack growth data included with AFGROW [20]. Here it can be seen that the TASC method enables the measurement of crack growth well below the standard threshold ΔK and at lower rates of growth than can be measured using those *long crack* standard techniques.

Table 2 provides a summary of a material data set for three R values developed by DST (referred to as DST16)

[a] Segments of a sequence from an aircraft wing root bending moment loading spectrum are used, such as the industry standard *TWIST* or *FALSTAFF*

that was designed to allow prediction of the growth of small cracks in AA7050-T7451, informed by the TASC data. Validation studies have shown that these curves when used with a crack growth algorithm, such as that in AFGROW, with no closure (retardation) model activated, predicts well coupons tested with VA spectra based on combat aircraft wing root bending moment loading [21]. However for VA spectra, that contained significant number of cycles with high mean and small amplitude[b] (e.g. R=0.8), the DST16 model did not correlate as well. The following sections detail research undertaken to investigate how crack growth is influenced by prior load cycles, such as long sequences of loading that approximates CA.

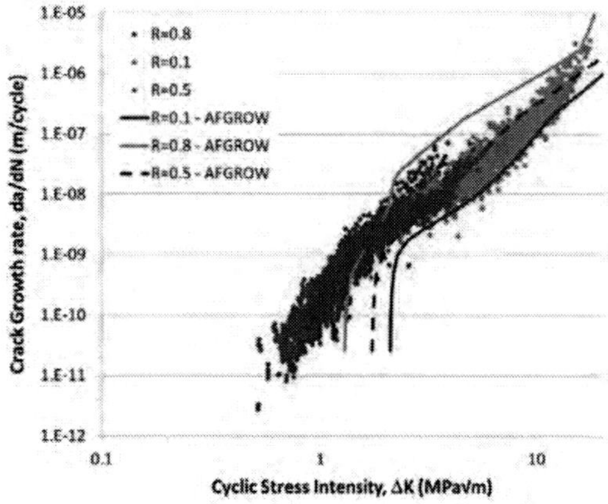

Fig. 2. Crack growth data generated during the TASC program for AA7050-T7451 coupons and standard (AFGROW) growth models.

Table 2. Crack growth rate data set (DST16) for aluminium alloy AA7050-T7451 at R=0.1, R=0.5 and R=0.8.

da/dN (m/cycle)	Cyclic Stress Intensity, ΔK (MPa√m)		
	R=0.1	R=0.5	R=0.8
1.00E-12	0.5720	0.5165	0.4745
2.00E-12	0.5723	0.5168	0.4749
5.00E-12	0.5738	0.5182	0.4762
1.00E-11	0.5747	0.5190	0.4770
2.00E-11	0.5759	0.5202	0.4781
5.00E-11	0.5772	0.5215	0.4794
1.00E-10	0.5779	0.5221	0.4799
2.00E-10	0.8034	0.6864	0.5914
5.00E-10	1.2171	1.0153	0.8598
1.00E-09	1.6988	1.4011	1.1692
2.00E-09	2.3705	1.8951	1.5768
5.00E-09	3.6720	2.7516	2.3019
1.00E-08	5.1016	3.7316	3.1078
2.00E-08	6.9502	5.0564	4.2937
5.00E-08	10.4032	6.9159	5.7607
1.00E-07	13.7856	8.7249	6.7573
2.00E-07	17.9561	10.8514	7.8065
5.00E-07	23.7728	13.7536	9.0629

[b] for example blade-pass loading in rotor aircraft or highly dynamic (buffet) loading on combat aircraft structures

3 Growth Retardation and Acceleration

3.1 Retardation from repeated R=0.8 cycles

Previously DST [22] has reported on small crack growth retardation with increasing number of CA cycles. Evidence of this phenomenon was observed in the measured crack growth rates from a series of coupons tested with the TASC methodology. In this research, identical AA7050-T7451 coupons to those used in the above work, were tested with a spectrum that included CA R=0.8 sub-blocks with varying numbers of cycles (N). Fig. 3 is an example of a typical fatigue crack fracture surface, with the TASC spectrum shown in the insert, here the peak load was scaled to achieve 200MPa at the net test section. From measurements of nine coupons tested in this study, the changes in average crack growth with increasing R=0.8 cycles can be determined. A summary of such data is presented in Table 3, where such a trend can be seen, for example at a ΔK=1.00 MPa√m, the growth of the 5000th R=0.5 load cycle will be approximately half that of the 500th load cycle.

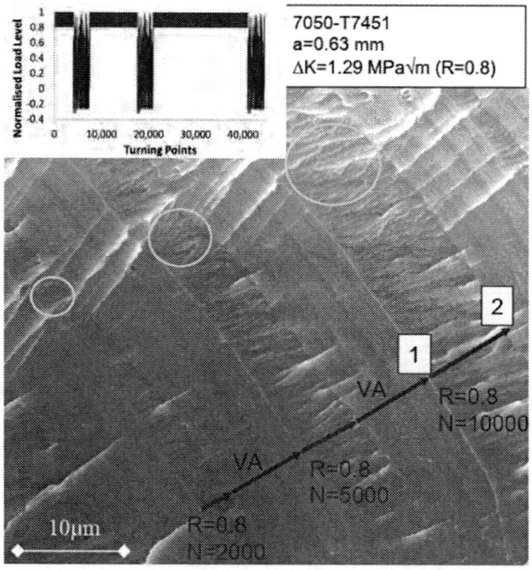

Fig. 3. Fracture surface image from a fatigue crack in a AA7050-T7451 coupon, showing micro-river patterns (circled) the crack front produced under VA loading (labelled 1) and R=0.8 CA loading (labelled 2).

Table 3. Average crack growth rates (nanometres/cycle) for increasing R=0.8 CA cycles (N)

ΔK MPa√m	Average Crack Growth Rates (da/dN), nm/cycle				
	N < 500	500 < N < 1,000	1,000 < N < 2,000	2,000 < N < 5,000	5,000 < N < 10,000
0.60	0.105	0.032	0.057	0.040	0.031
0.80	0.253	0.105	0.144	0.112	0.080
1.00	0.499	0.254	0.296	0.247	0.170
1.50	1.721	1.192	1.090	1.029	0.659
2.00	4.139	3.464	2.747	2.831	1.726
3.00	14.261	15.131	10.103	11.738	6.696

It is hypothesised that, in the case of small cracks, repeating high R (e.g. R=0.8) load cycles are insufficient to maintain a continuous consistent crack front (labelled 2 in Fig. 3). On the local scale, the crack follows varying paths though the material influenced by the local microstructure, crack orientation and the damage state ahead of the crack tip. As a result, this may be one *subtle* mechanism, of several, that reduces growth efficiency and therefore growth rates. Other possible evidence of this is highlighted (circled) in Fig.3; in the variation in the fracture surface from crack growth under CA R=0.8 load cycles where *micro-river patterns* form as the crack front separates and re-joins. Compare this to the smoother fracture surface that results from crack growth under CA R=0.1 and VA loading.

3.2 Retardation from repeated R=0.5 cycles

The retardation results were even more striking from a similar series of coupons tested with spectrum containing sub-blocks of CA R=0.5 cycles with varying load cycles. This test spectrum is shown insert in Fig.4, where it should be noted that the σ_{max} of each CA sub-block is less than σ_{max} of the VA sub-block. In this case, as shown in the fracture surface image in Fig.4, crack growth where the crack growth is observed to essentially stop for increasing N (i.e. the CA growth bands appear to be of the same width regardless of N).

The average crack growth rate data collected from two coupons tested (with the spectrum shown insert in Fig. 4), is shown in Fig. 5. In this figure the variation in average growth rate between the different length R=0.5 CA sub-blocks; of 2000, 5000 and 10000 cycles can be seen. Observed in this data, are the significant differences between the average crack growth rates during each CA sub-block of loading up to ΔK~2.2MPa√m. Since no difference was observed on the fracture surface in the wake of these CA sub-block growth bands, plastic-wake induced closure is not considered to be the cause of the apparent retardation. Rather the growth within the CA bands appears to *hit-a-wall* or slow to a very slow rate. Noting that grain boundaries may often create barriers for crack growth, observations of the fracture surfaces ruled out this potential cause.

For crack lengths greater than 1.5 mm (ΔK ~ 2.2MPa√m), no retardation effect could be identified in this data from repeated R=0.5 cycles. This could be the result of larger cracks being less influenced by local microstructural variations (e.g. crystalline orientation) compared to smaller cracks. Also shown in Fig.5 is the R=0.5 curve from the DST16[c] material model (Table 2), as this model has been optimised for combat aircraft wing root bending (VA) load spectra, thus this curve could be considered the rate for a single R=0.5 load cycle of a lead crack (i.e. fastest predicted).

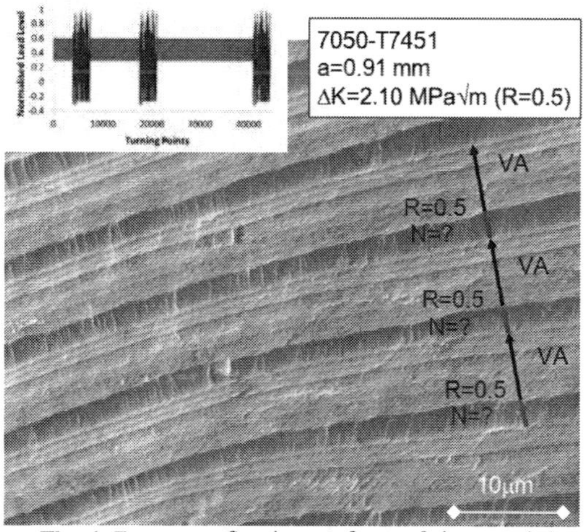

Fig. 4. Fracture surface images from a fatigue crack in a AA7050-T7451 coupon, a=0.91mm, showing evidence of crack growth retardation; individual CA R=0.5 sub-block cannot be identified (labelled N=?).

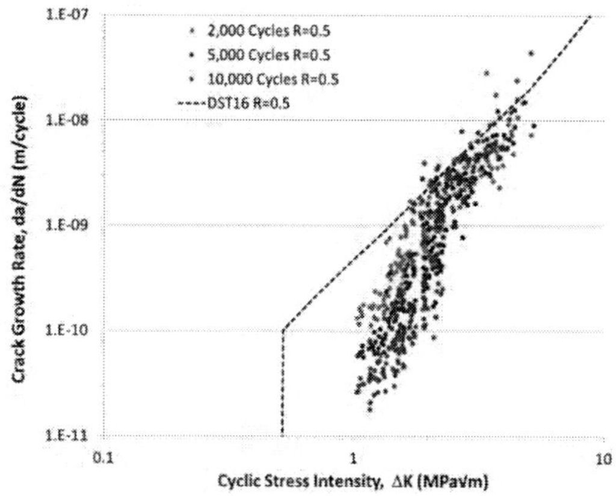

Fig. 5. Variation in average crack growth from R=0.5 CA sub-blocks (insert Fig.4) for N=2000, 5000 and 10000 cycles. Compared to DST16 crack growth model (Table 2 for R=0.5).

3.3 Acceleration from underloads R=-1.0 cycles

Further coupon testing was undertaken to investigate the extent to which the preceding load cycles influence the subsequent crack growth rates. In the case of the

[c] the DST16 material model (Table 2) was optimised to predict small crack growth for a sub-set of aircraft wing root bending moment loading. Validation of this model, including the threshold limit ΔK_{th}, was made against a different VA data set [21].

spectrum shown in Fig.4, it was noted that the VA sequence, with a higher peak load, σ_{max} could influence the retardation effect measured in the following CA sub-blocks. To investigate this, another series of coupons were tested with *under-loads*, i.e. $\sigma_{min}<0$ applied to the specimen prior to the application of each CA sub-block. Either; one, four or ten under-loads of R=-1.0 were applied.

As expected [23] the presence of underloads had a significant effect on the crack growth; possibly through the creation of alternate crack paths that promote crack front cohesion [19], or other mechanisms such as a reduction in the *effective stress intensity* [24]. Regardless, the effect on the crack growth rate can be directly measured over a range of crack lengths. An example of such measurements is shown below in Fig. 6, on the average growth rate from the 10,000 cycle R=0.5 CA sub-block without under-loads (No UL) compared to that with ten preceding under-loads (10 UL). This data is typical of that observed in the range of coupons tested in this program, with the underload effect being dependent upon the crack length (therefore K_{max}), and most pronounced in the threshold ΔK region. Of note in Fig. 6 is the effect of underloads on the longer crack growth rates, which are greater than would be predicted by the DST16 model.

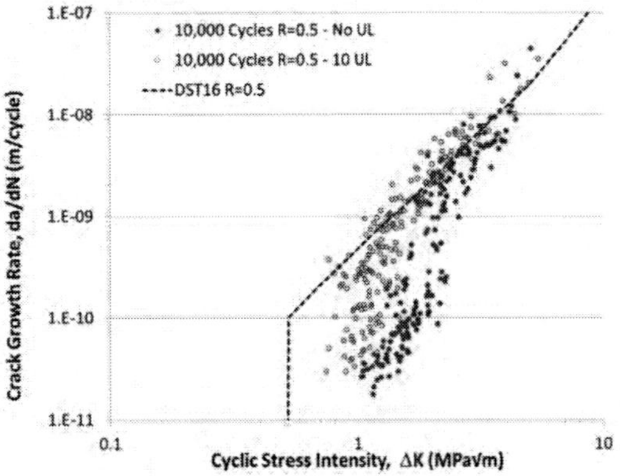

Fig. 6. Variation in average crack growth from 10,000 R=0.5 load cycles – both with and without 10 under-loads (insert Fig.4) applied prior. Compared to DST16 crack growth model (Table 2 for R=0.5)

4 Predicting Small Cycle Retardation

As discussed above while, the DST16 material data model was found to be good at predicting small crack growth from a range of VA[d] loading; the model did not predict well those spectra that contained load sequences that contained load sequences approximating high R CA loading. A small study was undertaken to explore a method by which crack growth model data (e.g. Table 2) could be used to predict small crack growth for a larger range of VA spectra types. It also may provide a method by which a comparison of spectrum truncation or filtering processes can be made to support full scale test activities at DST.

For this work a modified material model was proposed, based on DST16, whereby the growth rate was varied as a function of the number of the repeated *small* cycles applied prior to the current cycle. An example of the output from such a function for R=0.8 is shown below in Fig. 7, along with the DST16 material model curve and the R=0.8 TASC test data. In this figure are the proposed crack growth rates after 1, 500 and 10000 R=0.8 cycles.

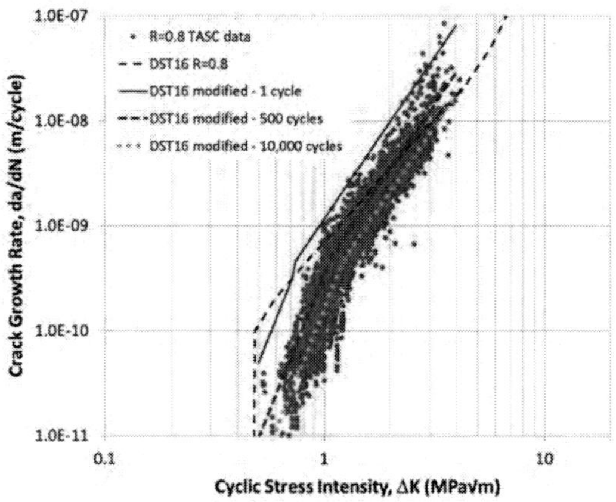

Fig. 7. Modification to R=0.8 DST16 model material model (Table 2) for repeated cycles.

To assess the validity of this method, crack growth data results from an associated small fatigue test coupon were used [19]. In this program, five coupons of identical material and design as those discussed above were tested with a spectrum, shown as an insert in Fig.8. This spectrum was designed to allow the direct measurement of how sequence order within a VA spectrum influences crack growth. Shown in Fig. 8, the spectrum includes: a VA sequence based of a combat aircraft wing root bending moment with 1000 R=0.8 cycles intertwined (blue), the same VA sequence and 1000 R=0.8 cycles *blocked* (red), and an additional sequence of R=0.8 cycles (green) used to aid post-test inspection – a marker band.

Small crack growth results from the coupon tests (Fig. 8) showed that the intertwined spectrum (blue points) resulted in faster crack growth than the block spectrum (red points). Over the range of the small crack depths measured here (0.3-5.0mm) the intertwined spectrum is determined to be approximately 24% *more severe* than the blocked spectrum. Predictions based on the DST16 material data that included a retardation function based on cycle count is shown in Fig.8. Here the modification resulted in the intertwined spectrum being predicted to be 19% (in terms of crack growth rates) more severe than the blocked spectrum. This compares to the un-modified DST16 model that only predicted a difference of 3% between the spectra for

[d] it should be noted that typical aircraft spectra used for validation testing based on recorded flight data will often require spectrum manipulation such as rainflow counting to create a sequence of minimum-maximum load turning points.

small crack growth rates. (Only the un-modified DST16 model for the intertwined crack growth is shown in Fig.8 since the blocked result is almost identical and hard to differentiate at this scale).

This result suggests that accounting for retardation effects associated with repeated small load cycles may improve the accuracy of prediction of small crack growth rates under certain VA load spectra. It should be noted that a growth model should also include other load history effects (e.g. under- and over-loads) to enable robust crack growth predictions for VA spectra.

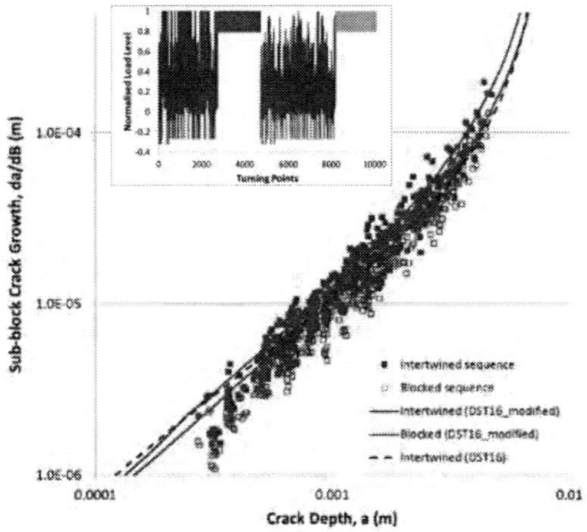

Fig. 5. Improved correlation from the modified DST16 model.

5 Discussion

Shown in the results presented here are the significant influences that load sequence has on crack growth rates of small cracks in AA7050-T7451. Of particular interest to DST is the effect that repeated small amplitude cycles, with a high mean, has in retarding crack growth in the threshold region. Due to the inherently exponential crack growth rates [2, 25] observed in common airframe alloys under aircraft loading, retardation of growth at the very early stages of fatigue nucleation/growth can have a significant impact on the total fatigue life of a component. A modification to a material model for AA7050-T7451 has been developed that accounts for such a subtle crack growth retardation produced by the use of CA cycles when blocked.

Further, in the threshold region, the use of crack growth data models to predict variable loading such as that occurring in aircraft platforms should not be based on data derived from CA testing only. Noting that the effect of overloads to crack growth has long been investigated since there is significant retardation from the assumed large compressive yield zones that result from overloads. Less well recognised is the acceleration in growth that can occur, and observed here in the threshold region, when significant underloads are applied. This may be another reason CA crack growth data is a poor predictor of fatigue growth for small cracks under VA spectra, such as aircraft wing root bending moment loading. Such VA spectra often contain cycles with compressive loads which, suggested by the results presented here, may result in the increased effectiveness of subsequent small load cycles in growing the crack.

Finally, when truncation or *spectrum blocking* are required (often to ensure efficient testing) for validation activities, the results suggest that the application of relatively large underloads when applied prior to a sequence of CA load cycles, can result in a significant reduction in the retardation.

6 Conclusion

Undertaking quantitative fractographic examination of fatigue crack surfaces that have been tested with specially engineered load sequences has enabled DST to accurately measure threshold crack growth rates and correlate growth retardation and acceleration measurements with load events. One phenomenon explored was, where loading spectra contain sequences of many ('000's) small amplitude/high mean cycles, crack growth retardation for small cracks that was identified as a possible reason for poor correlation with traditional predictive algorithms. (The influence on crack growth rates from repeated small cycles by the prior application of an underload was also discussed.) To better predict this retardation a method has been proposed by which standard material data models for crack growth could be adapted. Early results suggest that for small cracks this method could be useful in informing engineers on the relative severity of various loading spectra. This could have direct consequences for validation testing through full scale fatigue tests, to reduce the likelihood of un-conservative findings through the removal of too many small load cycles from spectra or the use of blocked spectra which may result in un-representative loading.

References

1. MIL-HDBK-1530 (2016) *Standard Practice Aircraft Structural Integrity Program (ASIP)*. US, Department of Defense
2. L. Molent, S. A. Barter and R. J. H. Wanhill (2011) The lead crack fatigue lifing framework. *International Journal of Fatigue* **33** 323–331
3. C. Wallbrink, P. Jackson and W. Hu (2011) Crack Growth Rate Curves: Which Part Dominates Life Prediction And When? In: *International Committee on Aeronautical Fatigue,* Montreal, Canada: May 29th - June 3rd
4. K. F. Walker, Barter, S. A. (2011) The Critical Importance of Correctly Characterising Fatigue Crack Growth Rates in the Threshold Regime. *26th ICAF Symposium* 1-3 June pp 249-263
5. S. A. Barter, L. Molent and R. J. H. Wanhill (2012) Typical fatigue-initiating discontinuities in metallic aircraft structures. *International Journal of Fatigue* (41) 2012 11-22

6. S. Pearson (1975) Initiation of fatigue cracks in commercial aluminium alloys and the subsequent propgation of very short cracks. *Eng. Fract. Mech.* **7** 235-247
7. J. C. Newman Jr, E. P. Phillips and M. H. Swain (1999) Fatigue-life prediction methodology using small-crack theory. *International Journal of Fatigue* **21** 109-119
8. J. Schijve (2003) Fatigue of Structures and Materials in the 20th Century and the state of the art. *International Journal of Fatigue* **25** (8) 2003 679-702
9. (2014) West Conshohocken, PA 19428-2959, USA, *Standard Test Method for Measurement of Fatigue Crack Growth Rates.*
10. M. Liao (2010) Dislocation theroy based short crack model and it's application for aircraft aluminium alloys. *Engineering Fracture Mechanics* **77** 22-36
11. A. Shyam, et al. (2007) Small fatigue crack growth in metallic materials: A model and it's application to engineering alloys. *Acta Materialia* **55** 6606-6616
12. A. K. Vasudevan, K. Sadananda and G. Glinka (2001) Critical parameters for fatigue damage. *International Journal of Fatigue* **23** 2001 39-53
13. R. Jones, L. Molent and K. Walker (2012) Fatigue crack growth in a diverse range of materials. *Journal of Fatigue* **40** 43-50
14. M. H. El Haddad, K. N. Smith and T. H. Topper (1979) Fatigue crack propagation of short cracks. *Journal of Engineering Materials and Technology, Trans. ASME* **101** 42-46
15. D. Taylor (2008) *The theory of critical distances*. UK, Elsevier
16. M. McDonald, L. Molent and A. J. Green (2006) *Assessment of Fatigue Crack Growth Prediction Models for F/A-18 Representative Spectra and Material.* DSTO-RR-0312, Melbourne, Australia, Defence Science and Technology Organisation
17. S. Barter, M. Burchill and M. Jones (2015) Improving the prediction of small crack growth in 7XXX aluminium alloys In: *28th ICAF Symposium,* Helsinki: 3–5 June 2015
18. P. White, S. Barter and L. Molent (2007) Observations of crack path changes caused by periodic underlaods in AA7050-T7451. *International Journal of Fatigue* **30** (7) 1267-1278
19. S. Barter, P. White and M. Burchill (2016) Fatigue crack path manipulation for crack growth rate measurement. *Journal of Engineering Fracture Mechanics* **167** 224-238
20. J. A. Harter (2014) *AFGROW Users Guide and Technical Manual*. Lex Tech,
21. M. Burchill, S. Barter and L. H. Chan (2017) Improving fatigue life predictions with a crack growth rate material model based on small crack growth & legacy data. In: *AIAC,* Melbourne, Australia: February, 2017
22. M. Burchill, S. A. Barter and M. Jones (2014) The effect of crack growth retardation when comparing constant amplitude to variable amplitude loading in an aluminium alloy. *Advanced Materials Research* **891-892** 948-54
23. P. White, S. A. Barter and C. Wright (2009) Small crack growth rates from simple sequences containing underloads in AA7050-T7451. *International Journal of Fatigue* **31** 2009 1865-1874
24. J. Newman and Y. Yamada (2009) Crack-Closure Behavior of 7050 Aluminum Alloy near Threshold Conditions for Wide Range in Load Ratios and Constant Kmax Tests. In: *ASTM/ESIS Fatigue and Fracture Mechanics Symposium,* Vancouver, CANADA,
25. R. J. H. Wanhill (1994) Flight simulation fatigue crack growth testing of aluminium alloys. *Fatigue* **16** 1994 99-110

Predicting creep-fatigue crack growth rates in Alloy 709 using finite element simulations of plasticity and creep-induced crack closure

Jose Ramirez[1], *Gabriel* P. Potirniche[1*], *Henry* Pugesek[1], *Nicholas* Shaber[1], *Martin* Taylor[2], *Robert* Stephens[1], *Indrajit* Charit[2].

[1]Mechanical Engineering Department, University of Idaho, 875 Perimeter Drive MS 0902, Moscow, ID 83844-0902, USA
[2]Chemical and Materials Engineering Department, University of Idaho, 875 Perimeter Drive MS 3024, Moscow, ID 83844-0902, USA

Abstract. This paper reports on a computational study and experimental validation of creep-fatigue crack growth rates at high temperature in two structural materials. The objectives are to develop a methodology to predict creep-fatigue crack growth rates using plasticity-induced crack closure under creep-fatigue loading conditions by characterizing the effect of hold time on crack growth rates during cyclic loading. In this study, the computation of fatigue crack growth rates is based on the crack closure phenomenon. The total crack growth rate during creep-fatigue loading is based on the addition of fatigue crack growth rate during cyclic loading and creep crack growth rate during hold time. The study identifies the effects of frequency and shape of loading cycle on crack-tip opening stresses induced by the combined action of the plasticity-induced crack closure and creep relaxation at the crack tip. Two-dimensional finite element analyses of compact tension specimens are performed to simulate crack growth under cyclic and time-dependent loading conditions. Elastic-plastic-creep material behavior is considered in these simulations. Closure levels are computed for high temperature structural materials such as 9Cr-1Mo steel and Alloy 709. The numerical predictions provide satisfactory agreement with experimental data of creep-fatigue crack growth rates in modified 9Cr-1Mo and Alloy 709 steels at high temperatures.

1 Introduction

Innovative structural materials are a keystone in improving the design and development of advanced energy systems. The performance of advanced energy systems, hereby advanced nuclear reactors, require that structural materials to be employed in advanced reactor components go through critical stages of development and qualification. In order to ensure adequate performance of these components in service and under extreme conditions, these materials must push the current boundaries of in-service conditions, namely, elevated temperatures and long service life expectations. Furthermore, selected materials must provide greater safety margins and facilitate the implementation of challenging designs while simultaneously leading to the reduction of capital costs and ensuring structural integrity [1].

Sodium Fast Reactor (SFR) is a leading concept and promising candidate in the U.S. power sector to reduce CO_2 emissions and to make nuclear energy production more efficient and sustainable. The design, construction, and operational costs associated with these advanced reactors are highly dependent on the selected high temperature structural materials. Alloy 709 is an advanced heat and corrosion resistant austenitic stainless steel derived from the original Nb/Ti stabilized Fe-20Cr-25Ni-1.5Mo-Nb (NF 709) steel. Due to its superior creep resistance compared to conventional 304 and 316 stainless steels, and due to the cost to produce it, which is estimated to be 2 to 4 times that for 304 SS, and far below the cost of Ni-based superalloys with comparable strength [2], alloy 709 was down-selected as a candidate material for structural applications, namely, for reactor vessels, core supports, and primary and secondary piping components in SFRs.

Creep-fatigue (CF) is a dominant loading mode expected in fast reactor structural materials. Fatigue and CF characterization are necessary to construct the creep-fatigue interaction diagram required to build the future code case submittal for alloy 709. In this regard, structural integrity assessment of these power generation components requires rigorous creep-fatigue crack growth characterization and predictive computational models for the creep-fatigue crack growth behavior that takes into account creep-fatigue interactions in Alloy 709 at temperatures between 600°C and 700°C.

2. Plasticity-induced crack closure

The relevance of plasticity-induced crack closure (PICC) on fatigue crack growth was first identified experimentally by Elber [3]. Since then, numerous developments on fatigue crack growth modeling have been performed. In this work, it was revealed that the

* Corresponding author: gabrielp@uidaho.edu

fatigue crack closure phenomenon was observed along the crack faces due to permanent plastic deformation in the wake of a propagating crack. Moreover, it was observed that a propagating crack in a fatigue specimen is completely open only for a portion of the loading cycle. The main implication of this observation is that the crack faces of a growing crack contact each other even before reaching the minimum load during the unloading portion of a cycle. This premature contact generates stresses that reduce the effective stress intensity factor at the crack tip, decreasing the crack advance rate. Several crack closure mechanisms were revised in detail in [4], and the possibility of a synergistic operation of these mechanisms was also considered. However, it has been established that in engineering problems, the main contribution to fatigue crack closure is material plasticity due to the existence of compressive zones at the crack tip, and the plastic wake generated by permanently deformed material along the crack faces [5].

Two prior studies have been identified in the area of finite element computation of crack closure levels with hold periods in high temperature fatigue crack growth [6, 7], and only Pommier et al. [7] computed crack growth rates based on measured crack-tip opening stresses. Therefore, in this study, the proposed model aims at predicting fatigue crack propagation rates based on crack closure computation under cyclic and time-dependent loading conditions.

3 Finite element modelling

3.1 Specimen geometry

In this study, a two-dimensional compact tension (CT) specimen was used to simulate creep-fatigue crack growth behavior. The study was conducted on a standard CT specimen of dimensions shown in Fig. 1. The selection of this specimen was done in accordance with the standard ASTM E2760 on testing methods for creep-fatigue crack growth [8]. CT specimens were machined and tested in the laboratory air under fatigue and creep-fatigue loading.

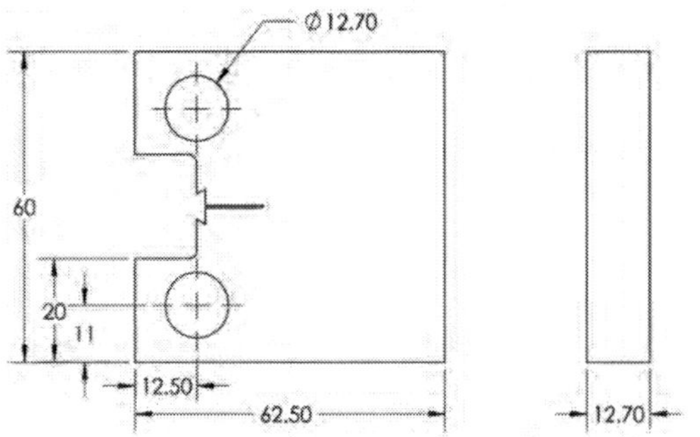

Fig. 1. C (T) geometry. Dimensions are in millimeters.

Initial crack length to width ratio was between 0.35 and 0.4. The starting ΔK evaluated ranged between 20 and 23 MPa$\sqrt{m}$. The total crack growth simulated in this study corresponded to 2.0 mm. Due to symmetry, only one half of the specimen was modelled in the finite element method. Loading of the CT specimen was imposed in the numerical model from a pin to the internal nodes of the hole through a multipoint constraint.

The numerical validation of this process entailed the computation of elastic load line displacements, stress intensity factors, J integrals, and computation of the reaction forces at each node in the crack plane. Moreover, a common parameter used as a crack-tip driving force in creep and creep-fatigue crack growth experiments, the $C(t)$ contour integral parameter, was computed numerically with ABAQUS for stationary cracks. The prediction of the load line displacements, stress intensity factors and $C(t)$ contour integral were in good agreement with analytical computations from the theory of fracture mechanics at elevated temperatures.

3.2. Meshing

Two-dimensional plane-stress and plane-strain models of CT specimens were built in ABAQUS in order to simulate creep-fatigue crack growth behaviour. Four-node quadrilateral elements of type CPS4R for plane stress, and CPE4R for plane strain were employed for meshing. Reduced integration with hourglass control was employed to prevent shear locking. In order to capture the forward and reversed plastic zones around the crack tip, refined areas near the crack tip were required. The mesh of the two-dimensional CT specimen is shown in Fig.2.

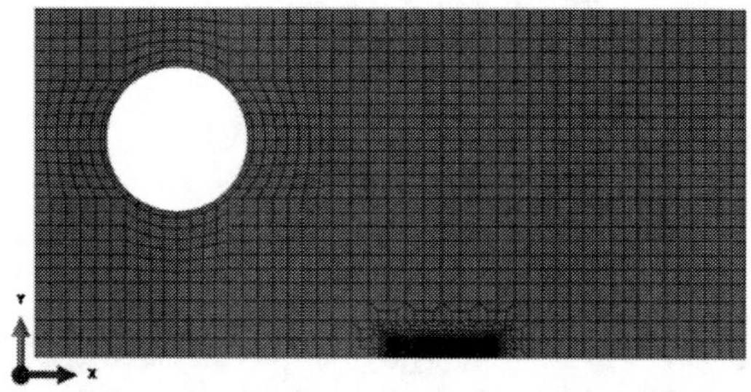

Fig. 2. 2D finite element mesh for CFCG and FCG predictions.

The magnified area near the crack tip with the largest mesh refinement is shown in Fig. 3. For this simulated specimen, the element size in the most refined area near the crack tip is 0.0125 mm.

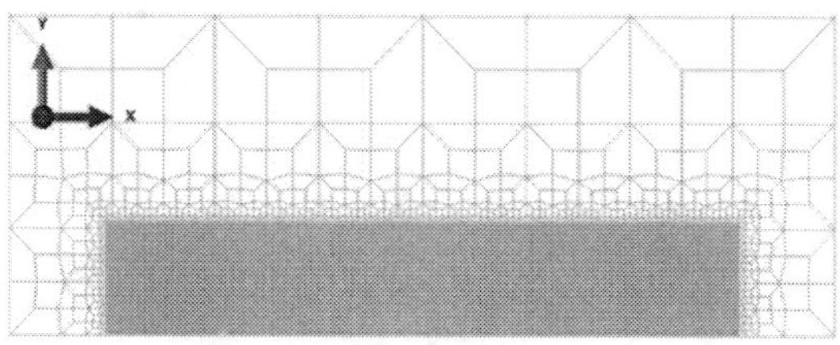

Fig. 3. Refined area around the crack tip.

3.3 Boundary conditions

Symmetry boundary conditions were applied to the finite element mesh. All nodes along the crack plane (from the crack tip to the right-hand side of the specimen) were constrained in the vertical direction (Y=0). In addition, the load point displacement was constrained in the horizontal direction (X=0). These boundary conditions are shown in Fig. 4.

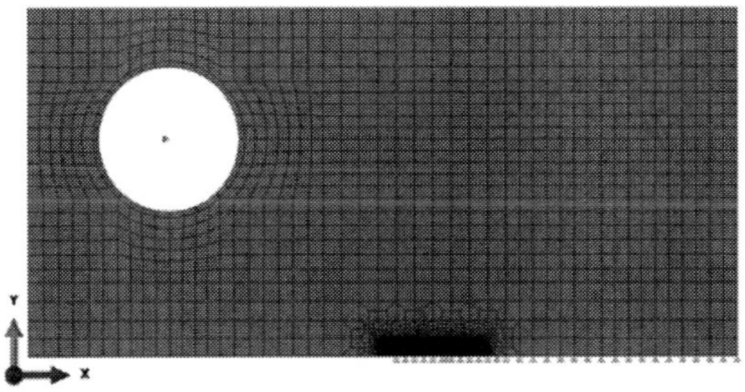

Fig. 4. Boundary conditions applied to the FE model.

3.4 Loading waveform

Trapezoidal waveforms simulated include a cyclic dependent and a time dependent step. In the first cyclic step, a tensile loading is applied during a specific time of loading, and a similar time is applied to the unloading step (L_T, U_T). For the experiments on alloy 709 at 600°C and 700°C, $L_T = U_T = 0.25$s. For the P91 625°C steel study, $L_T = U_T = 2$s [9]. In the second step, a time-dependent step with a hold time t_H ranging from 60s to 600s according to experiment conditions was applied. A schematic of the waveform is illustrated in Fig.5. In the finite element simulations, crack advance was performed by a node release scheme. In the case of two-dimensional numerical simulations, two loading cycles were applied before node release. The procedure can be observed in Fig. 5.

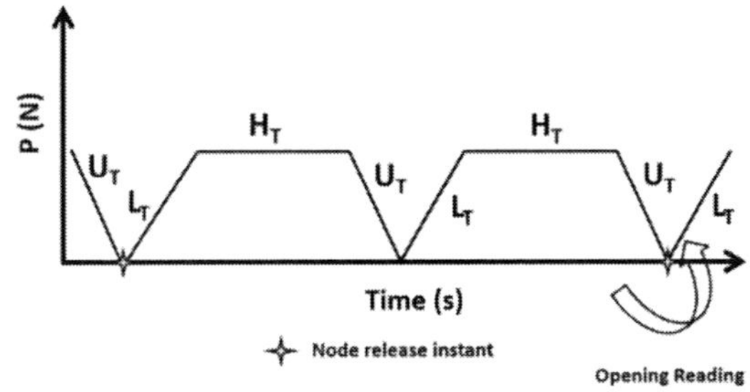

Fig. 5. Loading waveform and cycles applied before node release.

Crack propagation schemes to simulate crack closure are generally approached by the nodal release technique. The nodal release technique lacks physically realistic mechanism of crack growth; however, it is a very effective approach to simulate the plastic wake development with a steady state closure behaviour. In this study, the nodal release technique is adopted by node release at minimum load. Crack-tip opening stresses were computed with two methods, namely, the node displacement method and the stress method. The node displacement method defines as opening stress the remotely applied stress value that produces the first increment in vertical opening displacement for the last node to open behind the crack tip (most commonly, this is the first node behind the crack tip). The stress methods defines as opening stress the value of the remotely applied stress that causes the first increment in which all compressive stresses vanish for the nodes on the crack surface behind the crack tip. It was observed that both methods provide similar results. Therefore, the node displacement method was retained for its practicality.

3.5 Material model

Isotropic hardening of the material is considered by using an elastically-perfectly plastic material with the yield stress equal to the flow stress, which is calculated as the average between the yield and flow strengths. The choice of this material model is due to the early characterization stage of the material and the fact that low-cycle fatigue data is currently not available for elevated temperatures in Alloy 709. Moreover, in order to measure the impact of hardening and softening behavior of the material, the basic material models need to be considered first. The creep behavior was modeled with the strain-hardening version of Norton power law for secondary or steady state creep

$$\dot{\bar{\varepsilon}}^{cr} = A\sigma_e^{\,n} \tag{1}$$

3.6 Creep-fatigue crack growth equations

Paris et al. [10] proposed a relationship between fatigue crack growth rate da/dN and stress intensity factor range ΔK, of the form

$$\frac{da}{dN} = C \cdot \Delta K^m \tag{2}$$

where C and m are material parameters. Stress intensity factor range ΔK is defined as:

$$\Delta K = F(S_{\max} - S_{\min})\sqrt{\pi a} \tag{3}$$

where, a is crack length, and S_{max}- S_{min} is applied stress range. The term F is a geometrical factor which depends on crack length and specimen geometry. In real cyclic loading conditions, the presence of repeated compressive loads are expected. The compressive residual stresses generated in front of the crack tip and between the crack surfaces decrease the crack-tip driving force to an effective value ΔK_{eff}.

Fatigue crack propagation based on the crack closure concept states that crack growth rates are computed by the equation

$$\frac{da}{dN} = A \cdot \left(\Delta K_{eff}\right)^B \tag{4}$$

In a loading cycle, the effective part of the cycle is defined by $K_{eff} = K_{max} - K_{op}$, where crack opening levels are defined by either the corresponding stress intensity factors or stresses at which the crack tip begins to open during the loading cycle. K_{op} is defined as the stress intensity factor at which the crack starts to open. The effective stress intensity factor range in terms of the maximum stress intensity factor is given by

$$\Delta K_{eff} = K_{\max} - K_{op} = K_{\max} \cdot \left(1 - \left(\frac{K_{op}}{K_{\max}}\right)\right) \quad (5)$$

The effective stress range ratio or the closure factor can be written in terms of the normalizing opening loads P_{open}/P_{max}

$$U = \frac{\Delta K_{eff}}{\Delta K} = \frac{\left(1 - \frac{P_{open}}{P_{\max}}\right)}{(1-R)} \quad (6)$$

The total amount of crack growth in a creep-fatigue cycle is generally modelled as the added contribution of the cyclic dependent portion (loading and unloading parts of the trapezoidal waveform) and the tensile hold period part of the waveform.

$$\left(\frac{da}{dN}\right) = \left(\frac{da}{dN}\right)_L + \left(\frac{da}{dN}\right)_H \quad (7)$$

Crack closure effect is considered by introducing the effective fatigue crack-tip driving force in the crack growth rate equation (4). The second contribution is the creep crack growth rate, which is given by:

$$\left(\frac{da}{dN}\right)_H = \left(\frac{da}{dt}\right)_H \cdot t_H \quad (8)$$

where the first term on the right hand side is computed based on the assumption that K is a dominant parameter during crack growth

$$\left(\frac{da}{dt}\right)_H = A' \cdot K_{MAX}^{B'} \quad (9)$$

3.7 Model validation

The numerical model has been validated first with respect to crack closure. Crack-closure measurements have been corroborated with two different numerical methods, namely strip yield modelling (SYM) and finite element (FE) model. The first validation corresponds to predictions performed for two-dimensional specimens in plane stress and plane strain conditions. Several studies [11-14] presented normalized opening stresses with respect to load ratio obtained from SYM. The comparison of results from the current finite element simulations with those from other studies [11-14] is given in Fig. 6. The higher dotted line represents plane stress, whereas the lower dotted line represents plane strain computations. These predictions have generally shown very good agreement with the values predicted for plane stress and plane strain when compared to the published data. Particularly, the normalized opening stress values for the load ratio ranging from 0 to 0.8 are satisfactory. It is important to point out that the future predictions of the CFCG of the present model are performed in the loading range between 0 and 0.7. Therefore, good confidence of the predictions can be given to the proposed model.

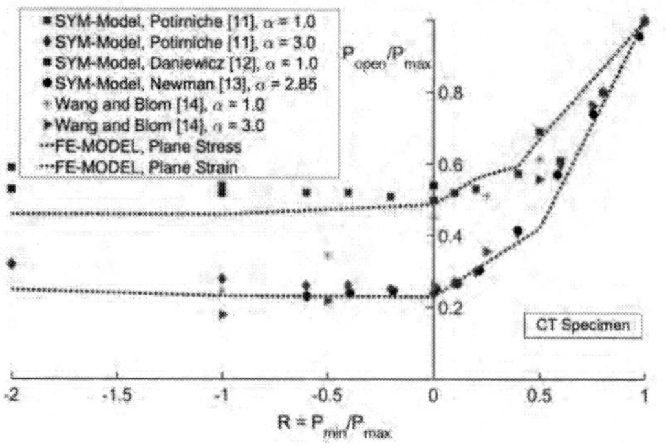

Fig. 6. Predicted normalized opening loads plotted as a function of the load ratio in compact tension specimens.

The second validation of predicted crack opening stresses was performed by comparing the current model results with finite element results of Solanki et al. [15]. Normalized opening stresses are presented in Fig. 7 as a function of the normalized element length with respect to the size of forward plastic zone. Normalized opening stress levels predicted are in good agreement when the same normalized stress intensity factor is introduced. Moreover, different conditions are evaluated to assess the model predictions at higher normalized stress intensity factors with different *a/W* ratios.

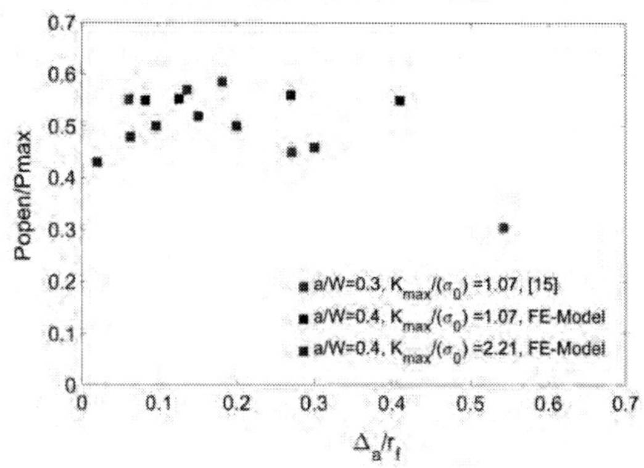

Fig. 7. Comparison of predicted opening stresses by this study and the finite element results of Solanki et al. [15] for the plane stress case.

3.8 Experimental testing

Fatigue and CFCG tests were conducted on Alloy 709 at 600°C and 700°C in laboratory air conditions. Applied hold times in the CFCG tests were 60s and 600s, in a trapezoidal loading waveform. Specimens were pre-cracked to the initial crack length using fatigue crack growth at room temperature. CT specimens were side-grooved, with 4% of cross-section reduction on each side in order to keep crack growth planar. Experimental data of FCG and CFCG rates of Alloy 709 as a function of stress intensity factor range are presented in Fig 8.

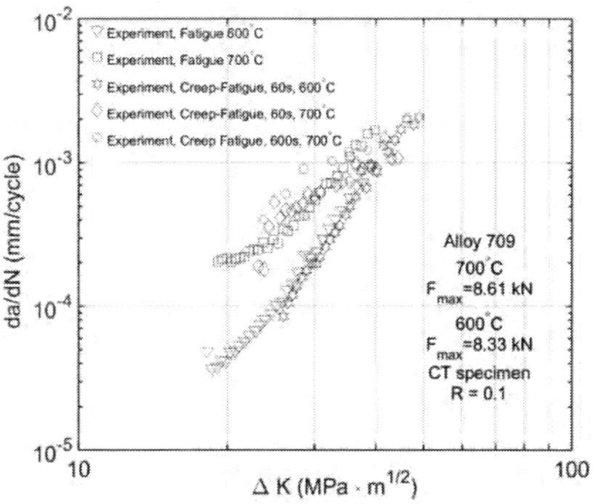

Fig. 8. Crack growth rate behavior of Alloy 709.

Fatigue crack growth rates of modified 9Cr-1Mo steel at 625°C have been retrieved from a published study by Narasimhachary and A. Saxena [9] and they are shown in Fig. 9. In the next section, the modeling predictions of these results are predicted.

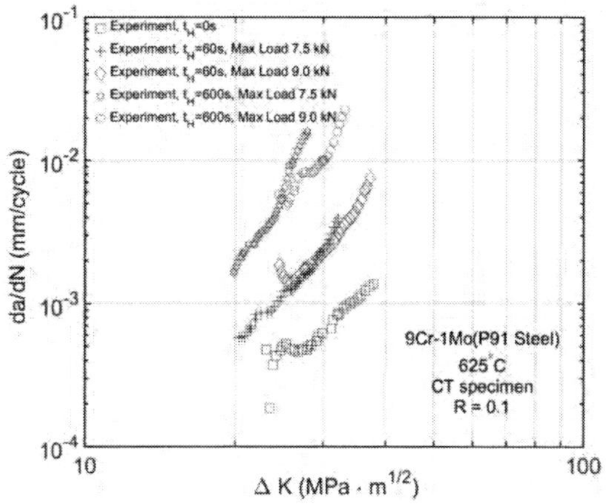

Fig. 9. Crack growth rate behavior of modified 9Cr-1Mo at 625°C according to Narasimhachary, and A. Saxena [9].

4 Results

In this section, the finite element predictions for the modified 9Cr-1Mo steel at 625°C, and Alloy 709 at 600°C and 700°C are presented. First, the fatigue crack growth rates, as a function of ΔK with different hold times, namely, 60s and 600s, for 9Cr-1Mo steel are discussed. Narasimhachary and Saxena [9] reported significant contribution of time dependent crack growth to the total crack growth rate for creep-fatigue loading with various hold times and maximum applied loads. For this reason, the first set of CFCG rates for hold times of 60s and 600s, in CT specimens with initial crack length (a/W) =0.4 and maximum load F_{max} = 7.5 kN are presented. The finite element predictions are carried out by incorporating the normalized opening stress values computed with ABAQUS and the corresponding constants that allow the computations of total crack growth rate due to the time dependent deformation. The variation of the opening loads with hold time predicted by the finite element simulations is shown in Fig. 10. The material properties used in this study were: (a) Norton creep constants, A = $9.53 \cdot 10^{-21}$ MPa^{-n} h^{-1} and n = 8.24, (b) mechanical properties, E = 125 GPa, σ_0 = 325.1 MPa, σ_{UTS} = 343.7 MPa at 625 °C [9]. The material constants used in the fatigue and fatigue crack growth equations (4) and (9) were $A = 8.50 \cdot 10^{-8}$, $B = 3.0$, and $A' = 1.30 \cdot 10^{-7}$, $B' = 3.5$, respectively.

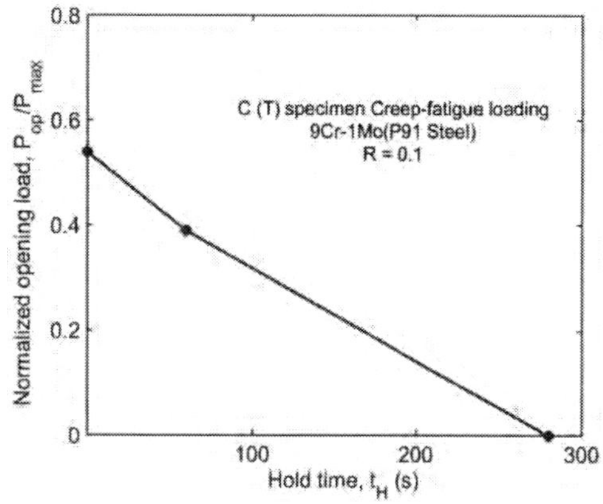

Fig. 10. Normalized opening loads as a function of hold time for F_{max} = 7.5 KN.

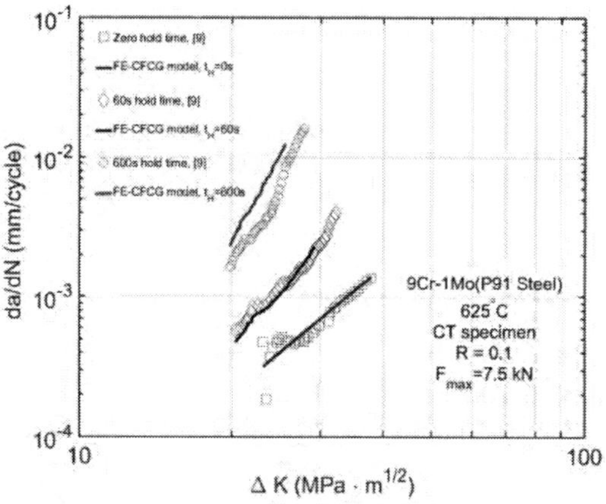

Fig. 11. Comparison between the current finite element predictions and experimental data ([9]) of creep-fatigue crack growth rate for different hold times in modified 9Cr-1Mo steel at 625°C at F_{max} = 7.5 KN.

Fig. 11 shows the model prediction for hold times of 60s and 600s. Fatigue crack growth rates per cycle predicted for 600s show a slight over-prediction. This behavior is due to the fact that the fatigue crack growth rate per cycle is predicting using ΔK values and coefficients A and B that satisfy the fatigue crack growth rates corresponding to a higher load. More precisely, the fatigue crack growth rate employed to make the predictions for 7.5 kN correspond to a growth rate generated by a load of 9 kN. In contrast, when the adequate fatigue crack growth rate data is employed to make the computations, more accurate predictions can be made by the model.

These predictions are shown in Fig.12 for 60s and 600s when a maximum load of 9kN load is applied. It can be noticed that the predictions are in good agreement with the experimental data. The corresponding normalized opening loads employed in these computations are given in Figure 13. The material properties used in this study were: (a) Norton creep constants, $A = 9.53{\cdot}10^{-21}$ MPa^{-n} h^{-1} and $n = 8.24$, (b) Material properties, $E = 125$ GPa, $\sigma_0 = 325.1$ MPa, $\sigma_{UTS} = 343.7$ MPa at 625 °C [9]. The material constants used in the fatigue and fatigue crack growth equations (4 and 9), were $A = 1.95{\cdot}10^{-7}$, $B = 3.0$, and $A' = 2.50{\cdot}10^{-7}$, $B' = 3.4$, respectively.

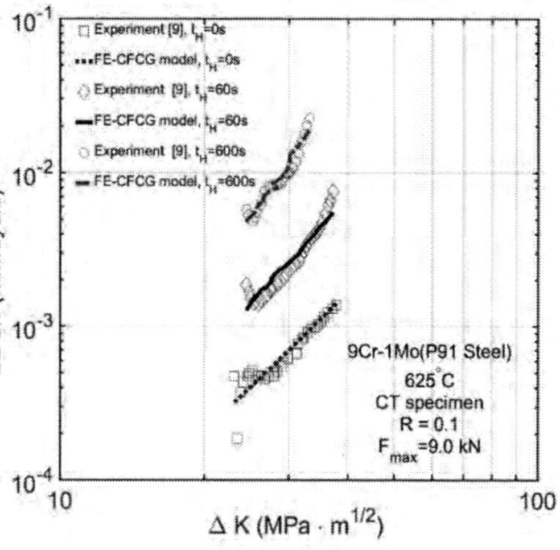

Fig. 12. Comparison between the current finite element predictions and experimental data ([9]) of creep-fatigue crack growth rate for different hold times in modified 9Cr-1Mo steel at 625°C at $F_{max} = 9.0$ KN.

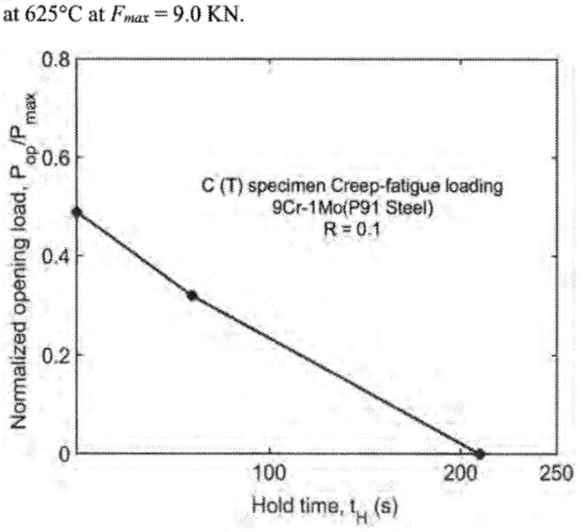

Fig. 13. Normalized opening loads as a function of hold time for $F_{max} = 9.0$ KN.

Next, predictions for alloy 709 are presented. These predictions are shown in Fig. 14. From the performed finite element simulations, it was observed that opening loads for Alloy 709 do not change with increasing hold times. Therefore, it can be concluded that there is no influence of the normalized opening load in the predictions of the CFCG rate in Alloy 709. In Fig. 14, crack growth rates per cycle predicted for 600s shows a slight over-prediction for lower values of ΔK. However, between ΔK values of 27 and 35 MPa √m, the computational predictions are in good agreement with experimental data. The material properties used in this simulations for Alloy 709 were: (a) Norton creep constants, $A = 4.96{\cdot}10^{-18}$ MPa^{-n} h^{-1} and $n = 5.9$, (b) Material properties, $E = 148.9$ GPa, $\sigma_0 = 178$ MPa, $\sigma_{UTS} = 449$ MPa at 700 °C. The normalized opening stress predicted value is 0.62. The material constants used in the fatigue and fatigue crack growth equations (4) and (9) were $A = 2.0{\cdot}10^{-7}$, $B = 2.98$, and $A' = 1.50{\cdot}10^{-7}$, $B' = 3.2$, respectively.

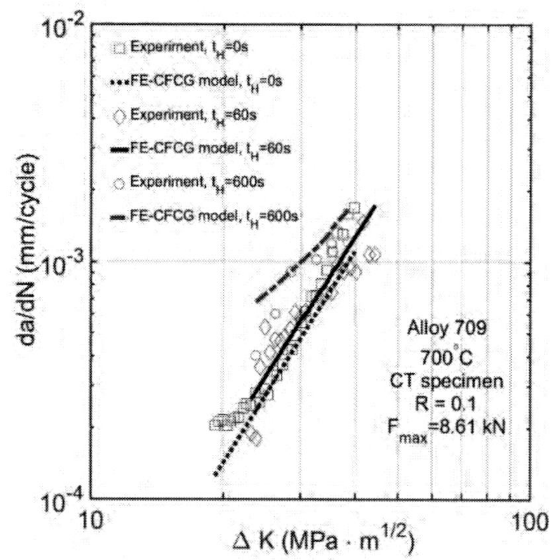

Fig. 14. Comparison between computational prediction and experimental measurements of creep-fatigue crack growth rates at various hold times in Alloy 709 at 700°C.

Fatigue crack growth rates in Alloy 709 CT specimens were experimentally measured at 600°C and compared with finite element modeling predictions. This comparison is illustrated in Fig. 15. Only predictions for 60s hold time were carried out, as experimental data for longer hold times was not available given the excessively long testing time that would have been needed for creep testing at 600°C.

Overall, model predictions compare satisfactorily with the experimental data at both temperatures of 600°C and 700°C. Material properties used in this prediction were: (a) Norton creep constants, $A = 2.83{\cdot}10^{-22}$ MPa^{-n} h^{-1} and $n = 6.7$, (b) The material properties, E = 155.8 GPa, $\sigma_0 = 191.4$ MPa, $\sigma_{UTS} = 583$ MPa at 600 °C. The predicted normalized opening stress value was found to be 0.65. The material constants used in the fatigue and fatigue crack growth equations (4) and (9) were $A = 9.0{\cdot}10^{-9}$, $B = 4.26$, and $A' = 1.50{\cdot}10^{-7}$, $B' = 3.2$, respectively.

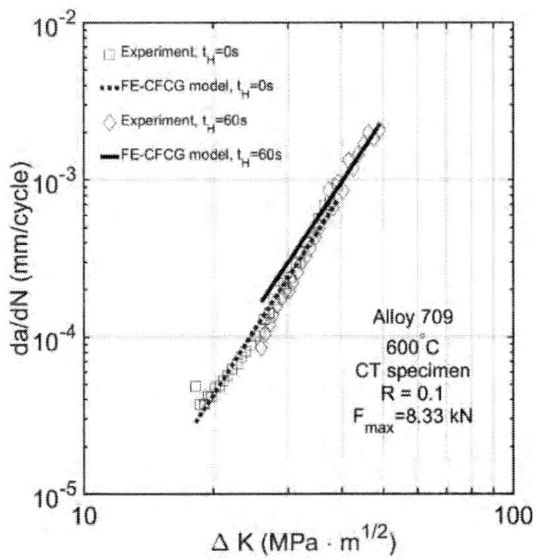

Fig. 15. Comparison between computational prediction and experimental measurements of creep-fatigue crack growth rates at various hold times in Alloy 709 at 600°C.

5 Conclusions

A two-dimensional finite element model was proposed to predict experimental creep-fatigue crack growth rates in modified 9Cr-1Mo and Alloy 709 steels. Plasticity-induced crack closure simulations in modified 9Cr-1Mo have shown a decrease of the opening stress with increasing hold time, whereas in Alloy 709 this influence does not exist. While creep-fatigue interaction in modified 9-Cr-1Mo are significant, in Alloy 709 they are almost inexistent. This can be attributed to high resistance to creep deformation of Alloy 709. The proposed model provides good agreement with experimental data, independently of the computational method used to measure the opening levels, i.e. the nodal displacement or contact stress method. In these simulations, the number of cycles between releases have an effect on the computed opening stresses but this is due primarily to the applied hold times. The proposed finite element modeling can account for creep-fatigue interactions on creep-fatigue crack growth in alloys at high temperatures, and it can be reliably used to make life prediction estimates in structural components.

The funding support of this work by the U.S. Department of Energy under the Nuclear Energy University Program (Grant DE-NE0008443) work is gratefully acknowledged.

References

1. T.L. Sham, K. Natesan, Code Qualification Plan for an Advanced Austenitic Stainless Steel, Alloy 709, for Sodium Fast Reactors Structural Applications, International *Conference FR17* (2017)
2. J.T. Busby, Candidate Developmental Alloys for Improved Structural Materials for Advanced Fast Reactors. ORNL/GNEP/LTR-2008-023. 2008.
3. W. Elber, Eng Fract Mech, **2**, 37-45 (1970)
4. S. Suresh, Fatigue of Materials (Cambridge University Press, 1998)
5. P.F.P. de Matos, D. Nowell, Eng Fract Mech, **75**, 2087-2114 (2008)
6. H. Sehitoglu, W. Sun, Eng Fract Mech, **33**, 371-388 (1989)
7. S. Pommier, Fatigue Fract. Eng. Mater. Struct, **19**, 1117-1128 (1996)
8. ASTM Standard E2760-10, (ASTM International, West Conshohocken, PA, 2010)
9. S.B. Narasimhachary, A. Saxena, Int. J. Fatigue, **56**, 106-113 (2013).
10. PC. Paris, F. Erdogan, J Basic Eng, **85**, 528-534 (1960)
11. G.P. Potirniche, *12th International Fatigue Congress* (2018) (to be published)
12. S.R. Daniewicz, J.A. Colins, D.R. Houser, Fatigue, 16 (1994)
13. J.C. Newman, Jr., *Design of Fatigue and Fracture resistant Structures, ASTM STP 761* (1982)
14. G.S.Wang, F. Blom, Eng Fract Mech, 40, 507-533 (1991)
15. K. Solanki, S.R. Daniewicz, J.C. Newman Jr, Eng Fract Mech, **70**, 1475-1489 (2003)

Energy Dispersive X-ray technique applied to fatigue fracture surfaces oxidation to provide information on growth rate

Iñigo Hernandez[1,*], *Edward A.* Saunders[2], and *Iñaki* Madariaga[1]

[1], Industria de Turbopropulsores, SAU, Materials & Processes Department, Parque tecnológico nº300, 48170 Zamudio, Spain
[2], Rolls-Royce plc, Materials – Failure Investigation, Bristol BS34 7QE, United Kingdom

Abstract. Gas turbine components are strongly affected by cyclic loading during operation, which makes fatigue development the most significant cause of failure or cracking within them. The nature and growth rate of that fatigue is dependent on the stress experienced at a particular area of a part, the temperature, the material properties and the applied loading frequency. Various techniques, such as striation and/or band counting can, in some cases, provide information on fatigue life. However, often such techniques are rendered useless by lack of formation or obliteration by damage and/or oxidation of such features. Once a crack has initiated and is exposed to the engine atmosphere, the level of oxidation and other contaminant elements observed on the crack surface will evolve with time, and, correspondingly, crack depth. Element distribution analysis by an Energy Dispersive x-ray Spectroscopy (EDS) technique applied along a fatigue fracture surface could help to understand the evolution of the crack with time.

Using EDS, an experimental procedure was carried out to ascertain a perceived measured percentage of oxygen plus other additional chemical species at a large number of stations along the length of a fatigue crack. The analysis was performed from the initiation site to the end of the propagation area, and also within the final rupture zone which had less time exposed to air and gasses compared to the fatigue system. An exercise of comparison between that progression in oxidation across cracked in-service components versus perceived measured oxidation level analyses on trial fracture surfaces exposed in atmospheric oven conditions at high temperatures for a number of different durations was performed. The obtained results provided information on specific questions concerning the fatigue propagation life of the in-service components; thereby amplifying the use of the EDS technique in this aspect of the materials forensic field.

1 Introduction

An explanation of a gas turbine engine broken blade through fatigue propagation is described herein. Some examples of blade release investigations may be found in [1-6]. Evaluation of the root cause is assisted through usage of the EDS technique, helping understanding of the evolution of the crack with time.The EDS technique and fundamentals is described in [7-10].

Strip inspection of an engine that suffered an event during running, revealed damage to different components, including a broken blade as one of the most significant finding. Therefore, the understanding of the blade fracture was considered essential in terms of knowing what the primary failure of the turbine was.

Fracture surface evaluation of the blade partially assembled in the disc showed evidence of a fatigue mechanism, but in spite of that, certain doubts about when the crack was produced arose. In particular, there was uncertainty about whether that blade was the primary root cause and produced the rest of the damage in the turbine, or it was a consequence of the failure of another component.

2 Fracture surface assessment

2.1. Visual inspection

The fracture surface of the blade was produced near the inner shroud of the component, about 10-15% of the aerofoil span (see figure 1).

The fracture of the blade started in the suction side of the aerofoil profile close to the trailing edge (see Fig.2). An area approximately 11mm long with thin thickness (about 0,45mm) in that particular region of the surface was detected. That related to a thickness reduction in that area from nominal of around 65-70%, whilst the remainder of the section in the blade was close to nominal (see figure 2).

* Corresponding author: inigo.hernandez@itpaero.com;

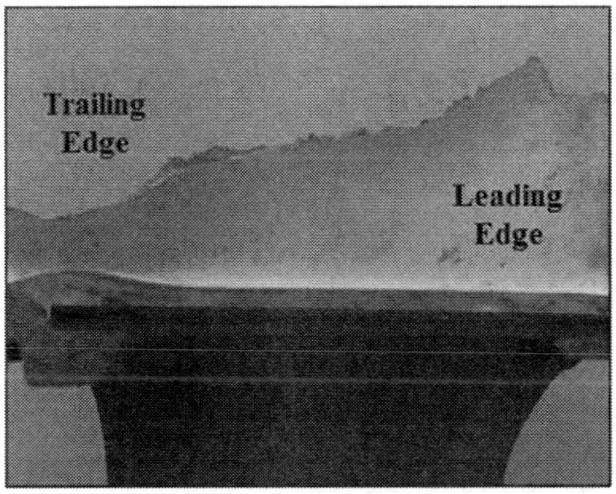

Fig. 1.Pressure side view of the broken blade.

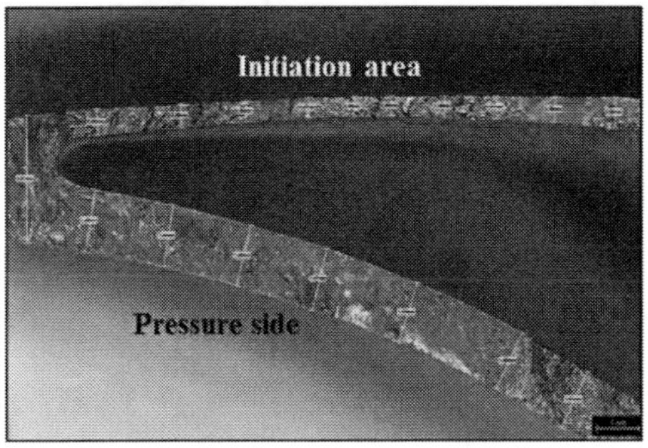

Fig. 2.Initiation in a thickness reduction area at suction side.

2.2. Fractographic assessment

The appearance of the blade fracture surface was mainly flat, except in the leading edge area and the pressure side adjacent to the LE that was more irregular and rough.

Beach and river marks typical of fatigue mechanisms were observed on the fracture surface in both directions from the initiation area of the crack (see Fig.3). Additionally, fatigue striations were observed only at the end of the fracture surface (see Fig.4). The features observed were commensurate with fatigue propagation in predominantly high cycle modes.

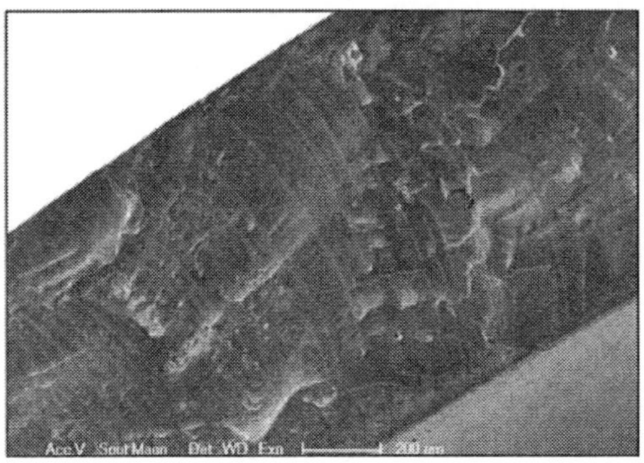

Fig. 3. Beach and river marks observed along the fatigue propagation area.

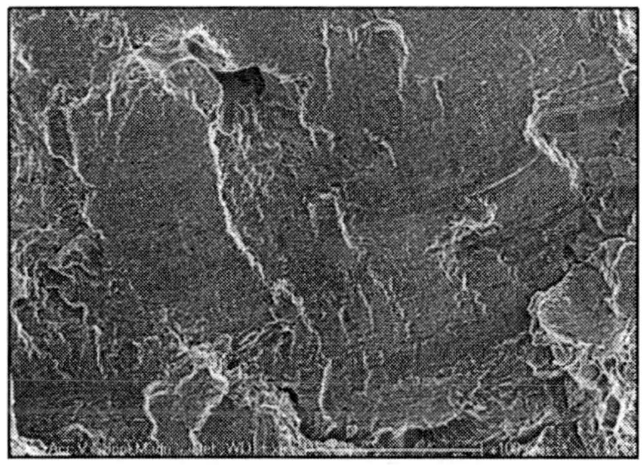

Fig. 4. Fatigue striations at the end of the propagation area, close to leading edge static fracture surface.

Summarizing, the assessment of the broken blade fracture surface showed an initiation area at the suction side with a thin thickness comparing to the rest of the aerofoil walls. Fatigue propagated from that position in both directions, toward leading and trailing edges, and going forward through a partial area of the pressure side. Accordingly, the fracture surface of the leading edge area and part of the pressure face showed ductile fracture features typical from static and sudden breakage (see Fig. 5).

3 Energy Dispersive Spectroscopy on fracture surface

3.1. EDS on blade fracture surface

The crack surface is exposed to the engine gas environment and subjected to oxidation and other contaminant elements (amongst other elements, deposits of silicon and phosphorous are often detected on the surface of turbine components). The level of that contamination evolves with the exposure time and with crack depth. Therefore, an Energy Dispersive x-ray Spectroscopy (EDS) technique applied along a fatigue fracture surface could help to understand the evolution of the crack with time.

The purpose of the analysis was to obtain the percentage of oxygen and silicon at plenty of stations along the length of the fatigue fracture surface. The analysis was carried out from the end of the fracture surface to the initiation area, including chemical analysis at the final rupture (the static one) which would have been exposed to the engine-running environment for less time than the fatigue propagation area.

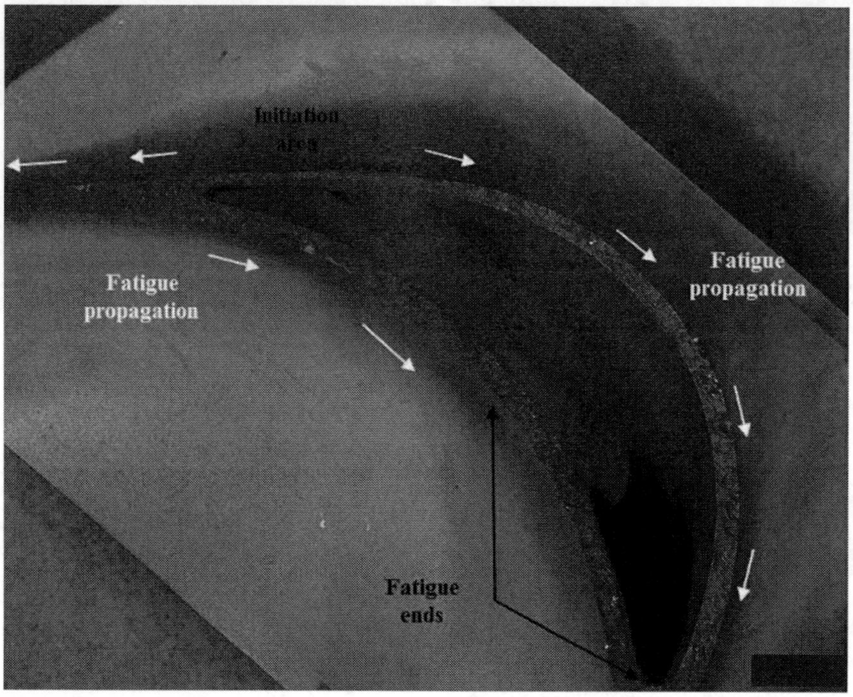

Fig. 5. General view of blade fracture surface.

Measurements started in a static fracture region in the area close to the leading edge, with "point zero" being the transition point from the end of the fatigue system to static fracture at the suction side of the aerofoil.

Those analyses were carried out under low pressure/ high vacuum conditions, using a low accelerating potential (voltage). Spot size and working distance were slightly modified to achieve the same test conditions along the fracture surface (keeping Counts and Dead Time parameters within the same ranges along the extension of the analysed area).

A clear increasing oxygen percentage was observed starting from the end of the fatigue propagation limit to the origin of the crack (see Fig. 6). Silicon, from surface contamination, showed a similar behaviour, although the quantity of this element was lower. Oxidation, therefore, had been evolving with time and in this particular way was "measurable".

It is recognised that the oxygen percentage measured is not absolute, as it depends on the x-ray escape depth/volume ('bowl') with respect to the thickness of the oxide layer and how much underlying parent material that encompasses [7-10]. That escape depth/volume is governed in part of the applied accelerating voltage. By use of a relatively low voltage (eg. 10kV) this sensitises the surface (an oxide layer) to a greater extent as the escape volume is smaller, so encompasses a greater proportion of the oxide layer.

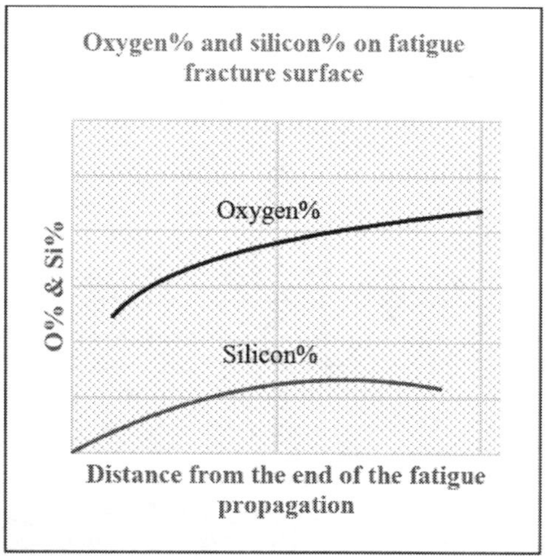

Fig. 6.Oxygen and silicon percentage along fatigue fracture surface.

Accordingly, in the current paper, the absolute oxygen percentage has not been sought, nor has it been the intent to measure an oxide thickness by this methodology. However, the work is intended to demonstrate a changing/increasing oxygen percentage that relates to a thickness as a comparative exercise.

* Corresponding author: inigo.hernandez@itpaero.com;

3.2. EDS on treated test samples

Fresh fracture faces (in tensile overload) were generated by breaking open blade material of the same alloy as the blade (IN713LC), as initially unaffected by temperature. These fracture faces were then over-exposed to the typical temperature as seen in service in the engine – over nearly 20 different exposure times from 1 hour to 800 hours (at 1 atmosphere).

EDS analysis of those samples after each heat treatment was performed to generate an oxygen percentage value under the same analysis conditions as used on the blade fatigue fracture surface. It is appreciated that there will be variance in oxidation on directly-exposed tensile overload samples compared to that seen on a growing fatigue crack, but it was found that the oxygen percentage measured was similar enough to provide an estimate of the growth time of the fatigue crack – as a comparative measure (as described below).

The results of the tests again showed a clear increase in O% content as the sample is exposure for longer times, although the level of O% content after long exposed times was found to be lower than the O% content calculated in areas near the fatigue initiation in the broken blade (see Fig. 7).

It was observed that, the tendency of increasing O% content as the fracture is exposed for longer times at relatively high temperature is similar to the change in O% content observed on the fracture surface of the broken blade. The difference of the O% content could be explained by several reasons:

- The difference of the surface being exposed under a clean atmosphere in a furnace during the tests compared to under gas path conditions in the production engine.
- There been other contaminants in the analysis of the broken blade surface (%Si and %P content) not observed in the tests carried out in the laboratory.
- The differences between the flatness of the surface of the broken blade associated with fatigue and the fracture caused in the lab (static fracture) that was rougher.
- The dispersion of the results observed during the different analyses conducted in samples treated under the same condition.
- It was found in general for both studies that the EDS O% results exhibited a logarithmic trend, as it is observed in general oxidation theory – in that oxide thickness development under temperature exposure shows a logarithmic trend with time [11, 12].

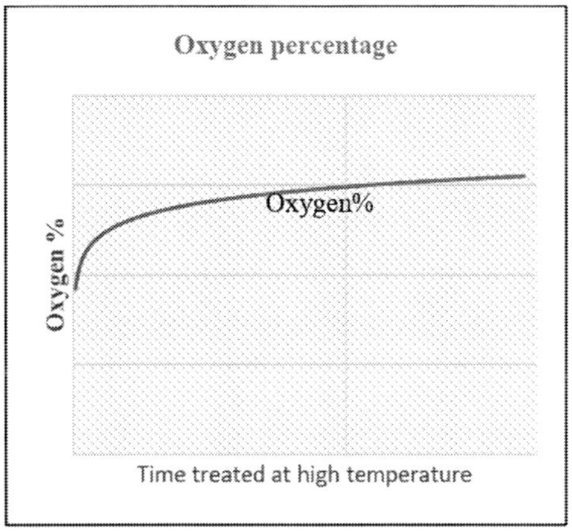

Fig.7.Oxygen percentage on test samples treated long exposure time.

4 Summary

As soon as a crack initiates it was exposed to the turbine atmosphere, and as a consequence of that the conditions of the surface change with air exposure with time, and, correspondingly, crack depth.

Energy Dispersive x-ray Spectroscopy (EDS) technique applied along a fatigue fracture surface has helped understand the evolution of the oxide layer on the crack surface with time for the oxygen percentage increases as the analysed surface is more exposed to the air providing the propagation took a significantly large number of hours. The level of contaminant silicon is another indicator of the time that a surface is exposed to the air.

A similar (logarithmic) oxygen percentage increase was also observed doing the same EDS exercise using laboratory-broken fracture surface samples treated from 1 hour to more than 800 hours at high temperature.

The sequence of component part failure in this engine/turbine was thereby established – in that the blade fracture was primary. This is because the blade fatigue surface had taken hours to develop (as recorded by the oxidation curve), and no rapidly if it had been instigated through failure of another part in the turbine.

The EDS results provided information concerning the fatigue propagation life of the in-service components, amplifying the use of this technique in the failure analysis field.

References

1. Y.-S. Choi and K.-H.Lee., *Investigation of blade failure in gas turbine*, 1969-1974, Journal of mechanical Science and Technology **24** (10) (2010)
2. X. Xu and Z. Yu, *An investigation on the failed blades in a locomotive turbine*, Engineering Failure Analysis, **14** (7) (2007) 322-1328.
3. W. Wang, F. Xuan, K. Zhu and S. Tu, *Failure analysis of the final stage blade in steam turbine*, Engineering Failure Analysis, **14** (4) (2007) 632-641.

4. Z. Mazur, A. Hernandez-Rossette and R. Garcıa-Illescas, *Investigation of the failure of the L-0 blades*, Engineering Failure Analysis, **13** (8) (2006) 1338-1350.
5. Z. Mazur, A. Luna-Ramırez, J. A. Juarez-Islasb and A.Campos-Amezcua, *Failure analysis of a gas turbine blade made of Inconel 738LC alloy*, Engineering Failure Analysis, **12** (13) (2005) 474-486.
6. T. Carter, *Common failures in gas turbine blades*, Engineering Failure Analysis, **12** (2) (2005) 237-247.
7. JI Goldstein, DE Newbury, JR Michael, NWM Ritchie, JHJ Scott, DC Joy. *Scanning Electron Microscopy and X-ray Microanalysis*. Springer, US, 2003
8. JC Russ, M Ashby Frs, R Kiessling, J Charles. *Fundamentals of Energy Dispersive X-ray Analysis*. Elsevier, 1984. https://doi.org/10.1016/B978-0-408-11031-0.50002-X
9. *Oxford Instruments Energy Dispersive Spectroscopy (EDS) Operating Manual*. http://research.engineering.ucdavis.edu/cnm2/wp-content/uploads/sites/11/2017/04/EDXS-SOP_033017.pdf
10. B. Hafner. *Energy Dispersive Spectroscopy on the SEM: A Primer*. http://www.charfac.umn.edu/instruments/eds_on_sem_primer.pdf
11. IM Ritchie. *A logarithmic law of metal oxidation which is controlled by a surface reaction*. The Philosophical Magazine: A Journal of Theoretical Experimental and Applied Physics, **Volume 19**, (1969) - Issue 158
12. HH Uhlig, *Initial oxidation rate of metals and the logarithmic equation*, Acta Metallurgica, **Volume 4**, Issue 5, (1956), Pages 541-554

Acknowledgement to Richard Hanlon from the Rolls-Royce Derby Failure Investigation team as he came up with the idea originally.

Understanding the near-threshold crack growth behavior in an aluminum alloy by x-ray tomography

Marcel Wicke[1,*], *Angelika* Brueckner-Foit[1], *Tina* Kirsten[2], *Martina* Zimmermann[2], *Fatih* Buelbuel[3], and *Hans-Juergen* Christ[3]

[1]Institute for Materials Engineering, University of Kassel, D-34125 Kassel, Germany
[1]Institute for Materials Engineering, Technical University of Dresden, D-01069 Dresden, Germany
[1]Institute for Materials Engineering, University of Siegen, D-57068 Siegen, Germany

Abstract. The near-threshold behavior of long cracks is evaluated using pre-cracked flat dogbone specimens of a commercial aluminum alloy in two heat treatment states. Once the threshold was known, the crack initially introduced by cyclic compression was propagated under load control with approximately constant range of the stress intensity factor at values close to the threshold values. The 3D morphology of the crack and the spatial distribution of the primary precipitates obtained by μ-CT were reconstructed with the aim of analyzing the role of the microstructure and primary precipitates on crack growth. Results pointed out that there were two major mechanism keeping the crack from continuous extension. First, the crack front was pinned by primary precipitates leading to kinking of the crack front. The second mechanism was pronounced shear-controlled crack growth of long cracks seen in terms of hilly structures on the crack faces. The first mechanism was found to be dominant at R=-1, whereas tests performed with R=0.1 showed an increased propensity towards shear controlled crack growth.

1 Introduction

Unusual crack extension mechanisms may occur if components containing pre-existing flaws are subjected to cyclic loads at very small amplitudes. This is related to the fact that the energy available for cyclic crack extension is very low in this case, and that the cyclic plastic zone ahead of the crack tip may be somewhat incomplete. Understanding the phenomenon implies that crack extension has to be studied in the near-threshold regime in a very well-controlled way.

The method proposed in ASTM E 647 standardizing the determination of the long-crack threshold is based on the so-called load-shedding procedure. However, it has been reported (e.g. [1-5]) that crack closure effects originating from the stepwise decrease of the stress intensity range can shift the threshold towards higher values causing non-conservative estimates. With the aim of circumventing the problems related with ASTM E 647, Pippan et al. [6, 7], among others, proposed an alternative procedure in which the threshold is approached from below. After pre-cracking specimens containing a sharp notch by cyclic compression, the load amplitude at the desired stress ratio R is increased in steps until the crack propagates stably. Cracks generated by compression pre-cracking and propagating at a constant stress amplitude have been shown to be well suited for analyzing crack extension at very low cyclic loads.

In a previous study [8] on the near-threshold behavior of long cracks in a commercial aluminum alloy at R = -1, it was found that local pinning of the crack front caused by primary precipitates and shear-dominated crack extension very similar to the stage-I extension of small cracks keep the crack from extending continuously. The last effect was most surprising as the cracks analyzed in [8] were definitely long cracks, which extended over several grains. However, shear-dominated effects may have been enhanced since the tests in [8] were performed at R=-1 and hence not in a typical fracture mechanics testing scenario. Some minor crack closure effects may even have developed in the compression phase of the loading cycle when the crack faces were in contact making it worthwhile to verify the unexpected crack extension behavior found in [8] under tension-tension loading.

Therefore, fatigue crack growth tests at a stress ratio of R = 0.1 were performed in the present study using the aluminum alloy investigated in [8]. Two heat treatment states were examined with the aim of getting in-depth information on the influence of the microstructure and precipitates on crack growth.

The paper is structured as follows: The description of the material and the experimental methods in section 2 is followed by an overview of the fracture mechanics tests. In section 4, crack growth of long cracks in the near threshold region and crack advance of long cracks are discussed with some concluding remarks given in section 5.

2. Material and experimental methods

The material examined in this study is the aluminum alloy EN-AW 6082, which was peak-aged (6PA) and overaged (6OA). $R_{p0.2}$ and R_m were determined in a

* Corresponding author: marcel.wicke@uni-kassel.de

tensile test as 336 MPa and 344 MPa for material 6PA, and 262 MPa and 294 MPa for material 6OA, respectively. A rolled sheet with a thickness of 20 mm and the microstructure shown in Fig. 1 taken in rolling direction served as base material. The grains are elongated with a maximum extension of up to 2000 – 3000 µm in rolling direction and 100 - 200 µm in transverse direction. Two types of primary precipitates can be detected in the microstructure: Mg-based ones (dark spots in Fig. 1) and Fe-based ones (bright spots in Fig. 1), which are arranged in a line-like fashion in rolling direction.

Fig. 1. EN-AW 6082 SEM micrograph of grain structure taken in rolling direction

Flat dogbone specimens with a total length of 40 mm and the geometry shown in Fig. 2a were machined out of the sheet material both in rolling (-LS) and transverse direction (-TS). After mechanical and electrolytic polishing, a part-through notch with a depth of 50 - 150 µm depending on test conditions and a notch radius smaller than 20 µm was cut in the specimen radius using a razor blade polishing technique similar to that proposed by Nishida et al. [9], see Fig. 2b. This technique is common practice in ceramics, but can also be used for metals, if the load applied to the razor blade is small enough. The main advantage of this technique is that extremely sharp notches can be made exhibiting only minor plastic deformation.

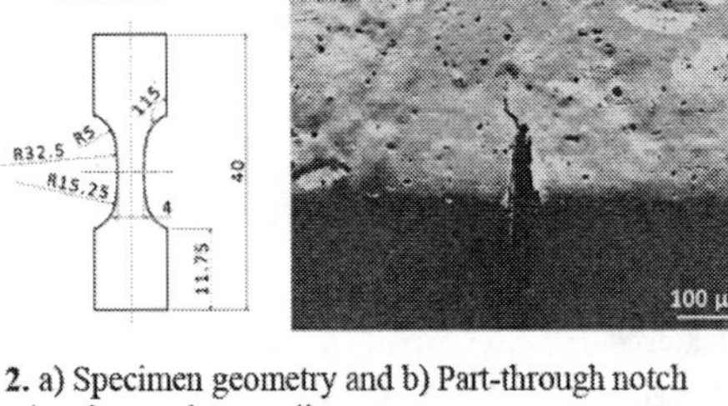

Fig. 2. a) Specimen geometry and b) Part-through notch located at the specimen radius

Fatigue crack growth tests were performed at a stress ratio of R = 0.1 on a Rumul Mikrotron resonance machine equipped with a 20 kN load cell and a sinusoidal force with a frequency of ca. 150 Hz in laboratory atmosphere. The test procedure was quite similar to the one used in [8] and is described again for the sake of completeness.

A long-distance microscope was used to monitor the crack growth on the specimen surface, enabling the documentation of the crack propagation when the cyclic load is interrupted by stopping the testing machine. The crack growth rate was calculated using the measured crack length and the number of test cycles between the scans. The specimens were pre-fatigued in compression to introduce a pre-crack which is open when unloaded. The crack growth test was started after applying up to 500.000 cycles at a stress ratio of R = 20 and a stress level of $\sigma_{min} = -290\ MPa$ for material 6PA and $\sigma_{min} = -208\ MPa$ for material 6OA, respectively.

The threshold of the stress intensity range ΔK was determined using the stepwise increasing load amplitude crack growth test described in [7]. After pre-cracking in compression, testing was changed to a pull-pull load (R = 0.1) at a low stress intensity range of about $\Delta K = 0.6\ MPa\sqrt{m}$ in order to let the crack propagate. If no crack growth was detected after 500.000 cycles, the stress amplitude was increased by 5 - 10 %. This procedure was repeated until the crack propagated in a stable manner. The threshold was then defined using the stress amplitude which first led to continuous crack growth.

Once the threshold was known, specific values of the stress amplitude were selected corresponding to ranges to the stress intensity factor (SIF) close to the threshold value for the starting crack length of 250 µm. The crack was then propagated at a constant stress amplitude according to the procedure proposed in [8] until the crack advance, visible on the surface, amounted to 50 µm. The stress amplitude was then decreased such that the initial value of the range of the SIF was recovered.

This procedure, allowing a crack propagation at a nearly constant stress intensity range with a maximum variability of ΔK of 10 %, was repeated until the crack reached a length of 1000 µm or the crack growth rate fell below 10^{-11} m/cycle, which was defined as the criterion for a crack stop. The fractured specimens were analyzed both in a scanning electron microscope (Zeiss Ultra 55) and in high-resolution X-ray computed tomography (µ-CT). The three-dimensional measurements were carried out with a ZEISS Xradia 520 Versa X-ray microscope at a voltage of 40 kV and a power of 3 W. Up to 2501 radiographs were captured with a resolution of about 3.4 µm/voxel and the 4x objective. Application of image correction filters was followed by reconstruction of the acquired radiographs with TXM Reconstructor (Zeiss) software. The image analysis was carried out with the rendering software Avizo (VSG), in which the crack morphology and spatial distribution of the primary precipitates was segmented manually after applying the non-local means filter.

3. Experimental results

3.1 Long-crack threshold

Three tests on specimens of both materials in LS- and TS-orientation were performed to determine the long-crack threshold according to the procedures described in section 2. The resulting da/dN vs. ΔK-curve of one experiment is shown in Fig. 3, where the cutting direction corresponds to the rolling direction (6PA-LS). Left to the dashed line representing the threshold value of ΔK_{th}, the crack stopped repeatedly as schematically illustrated by the circles on the x-axis, and the stress amplitude had to be increased by 5 – 10 %. After reaching the threshold value, the crack grew in a stable manner at a constant stress amplitude and showed long crack behavior.

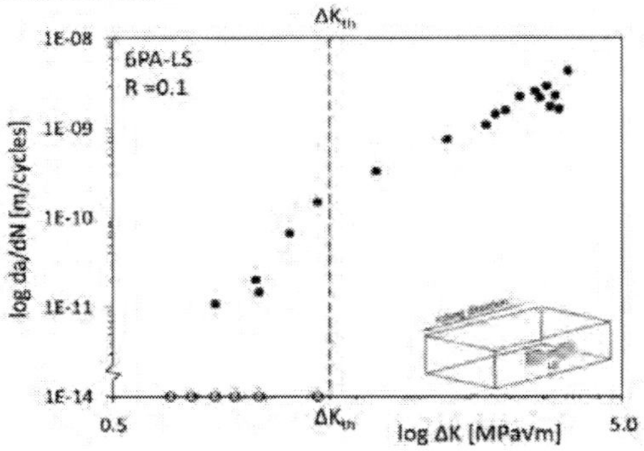

Fig. 3. da/dN vs. ΔK-curve with threshold determination (ΔK_{th})

An overview of the determined threshold values, which are discussed in detail in [10], is given in table 1.

Table 1. Threshold data for material 6PA and 6OA.

	6PA-LS	6PA-TS	6OA-LS	6OA-TS
$\Delta K_{th,1}$	1.34	1.30	1.10	1.13
$\Delta K_{th,2}$	1.13	1.51	1.17	1.33
$\Delta K_{th,3}$	1.38	1.32	1.20	1.14
$\overline{\Delta K}_{th}$	1.28	1.38	1.16	1.20

Peak-aged material has been found to exhibit a higher threshold compared to the overaged material, which is in agreement with results presented in the literature for various aluminum alloys [11-13]. Specimens taken in transverse direction (-TS) have a slightly better near-threshold crack growth resistance as the samples machined in rolling direction (-LS) independent of the heat treatment condition.

3.2 Crack growth at ΔK = const.

The influence of the microstructure and precipitates on crack growth in dependency of the heat treatment condition and specimen orientation was investigated in experiments at quasi-constant ΔK. On the basis of the determined thresholds, SIF values between 1.2 and 1.5 MPa√m were defined.

3.2.1 Crack paths

A comparison of the crack paths of the investigated materials 6PA and 6OA tested at a ΔK-value of 1.3 MPa√m is shown in Fig. 3. No results are depicted for material 6OA taken in rolling direction, as the crack repeatedly stopped at this load before reaching the starting crack length of 250 µm. The crack in the TS-oriented specimen of material 6PA stopped after ca. 310 µm, while it grew to a length of 410 µm (6PA) and 435 µm (6OA), respectively, in the LS-oriented samples. Crack arrest before reaching a crack length of 1000 µm was observed even in the experiments at higher load amplitudes.

No significant differences can be found between the crack paths of material 6PA in both LS- and TS-orientations and material 6OA in LS-orientation. The predominant crack propagation direction is mode-I with a certain amount of crack kinking (black arrows) or even crack branching when the crack path encounters primary precipitates. It can be seen that these precipitates are aligned in rolling direction (-LS) and form irregular clusters in transverse direction (-TS).

Taking a closer look at the crack paths reveals that there are occasionally sections in which the crack seems to propagate in a shear-dominated mode (white arrow in Fig. 4a). This effect is independent of the microstructure and rolling texture, as it was observed in both the LS- and TS-oriented specimens.

3.2.2 3D crack morphology

The 3D morphology of the cracks and spatial distribution of the primary precipitates obtained by µ-CT was reconstructed to further investigate the crack growth behavior at $\Delta K = const$. Fig. 5 shows the results for the LS-oriented specimen of material 6OA tested at a load of 1.2 $MPa\sqrt{m}$. As it can be seen, the crack already stopped after ca. 400 µm on the observed specimen front shown on the left side of Fig. 5, while it grew to a length of about 1000 µm on the specimen backside

The reconstructed crack front is of irregular form exhibiting some kinks, which are related to discontinuities in the crack morphology (yellow ellipses in Fig. 5). A closer inspection reveals that these discontinuities are not restricted to kinks but showed up as isolated features on the crack faces too. In the near-surface region, the crack seems to propagate in a shear-dominated mode as indicated by the "hilly" features highlighted by the blue square.

3.2.3 Fracture surfaces

The fracture surfaces were analyzed to gain additional information on the crack propagation behavior in dependency of the heat treatment condition and specimen orientation. For this purpose, the specimens were cooled down in liquid nitrogen and cracked in tension in order to

https://doi.org/10.1051/matecconf/201816513007

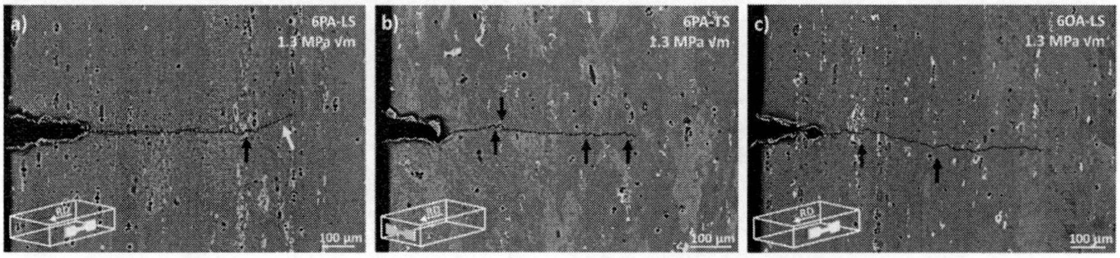

Fig. 4. SEM-micrograph showing the crack paths at 1.3 MPa√m: a) 6PA-LS; b) 6PA-TS; c) 6OA-LS

Fig. 5. 3D crack morphology of material 6OA tested in LS-orientation at 1.2 MPa√m obtained by µ-CT

introduce as little plastic deformation as possible. In Fig. 6-9, the fracture surfaces of the samples in peak-aged and overaged state tested at a 1.5 $MPa\sqrt{m}$ are presented.

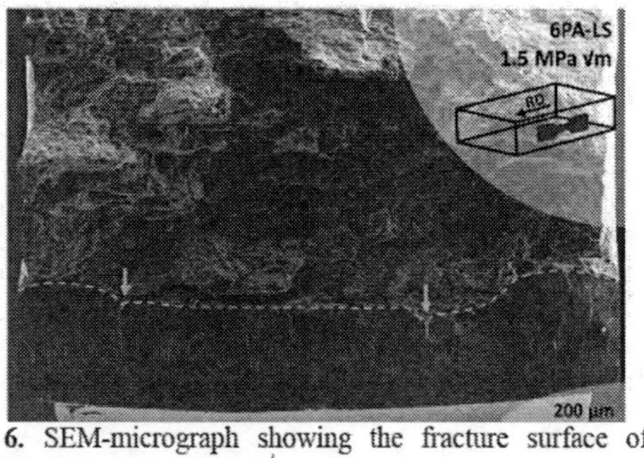

Fig. 6. SEM-micrograph showing the fracture surface of material 6PA-LS at 1.5 MPa√m

The fracture surface of material 6PA in LS-orientation is shown in Fig. 6. Only small irregularities (marked with orange arrows) can be detected along the crack front, which are related to fine white lines extending into the fracture surface. The final crack length on the observation side and specimen backside differs by about 40 µm. All in all, the fracture surface is smooth.

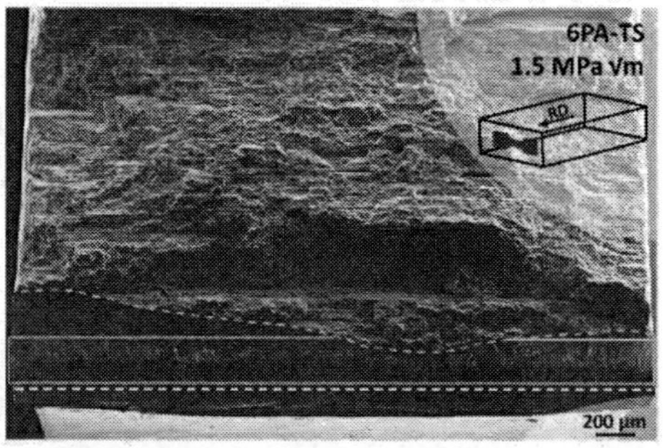

Fig. 7. SEM-micrograph showing the fracture surface of material 6PA-TS at 1.5 MPa√m

Different characteristics were detected on the fracture surface of the specimen taken from material 6PA in TS-orientation. As it can be seen in Fig. 7, a pronounced ridge parallel to the crack front is visible (blue square in Fig. 7) indicating that the crack extension is shear-dominated across the complete specimen. The difference in crack length between the observed specimen front and backside is about 180 µm.

* Corresponding author: marcel.wicke@uni-kassel.de

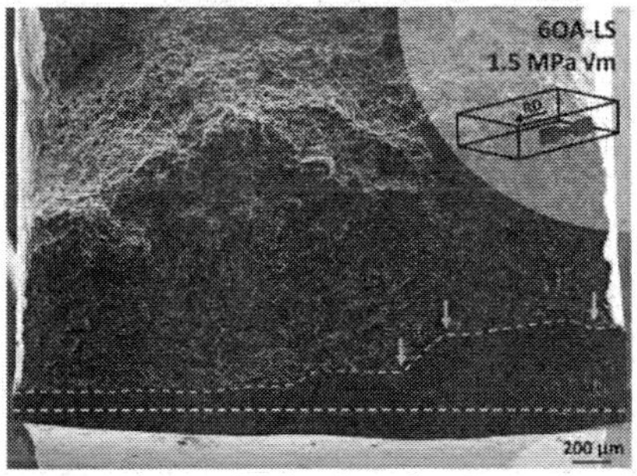

Fig. 8. SEM-micrograph showing the fracture surface of material 6OA-LS at 1.5 MPa√m

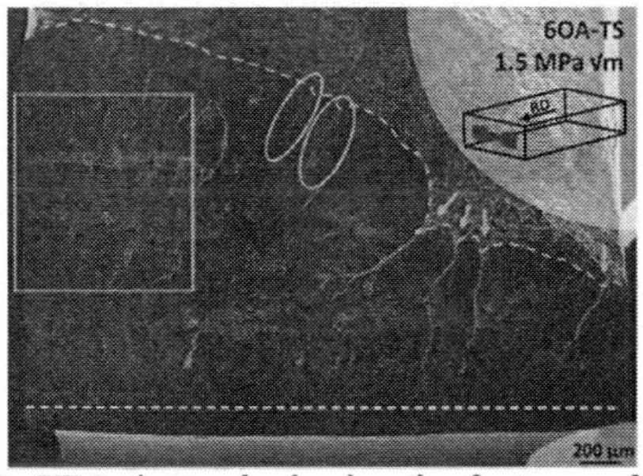

Fig. 9. SEM-micrograph showing the fracture surface of material 6OA-TS at 1.5 MPa√m

Fig. 8 shows the fracture surface of the LS-oriented specimen of material 6OA. While the crack grew to a length of ca. 435 µm on the observed specimen front, shown on the right side of Fig. 8, it already stopped after a length of ca.120 µm on the specimen backside. Similar to specimen 6PA-LS, only small irregularities in the form of kinks extending as fine lines into the fracture surface can be detected.

The fracture surface of material 6OA in TS-orientation depicted in Fig. 9 shows significantly different characteristics. As it can be seen, the crack front lagged behind on the specimen backside causing a difference in crack length of about 310 µm between the observed specimen front (right side in Fig. 9) and its backside. The distinct kinks in the irregular-shaped crack front are related to both lines extending into the fracture surface (orange arrows in Fig. 9) and ridges (green ellipses in Fig. 9). These ridges were not only restricted to kinks in the crack front, but showed up as isolated features on the fracture surface. A certain amount of shear-dominated crack growth can be observed in the near-surface region (blue square in Fig. 9).

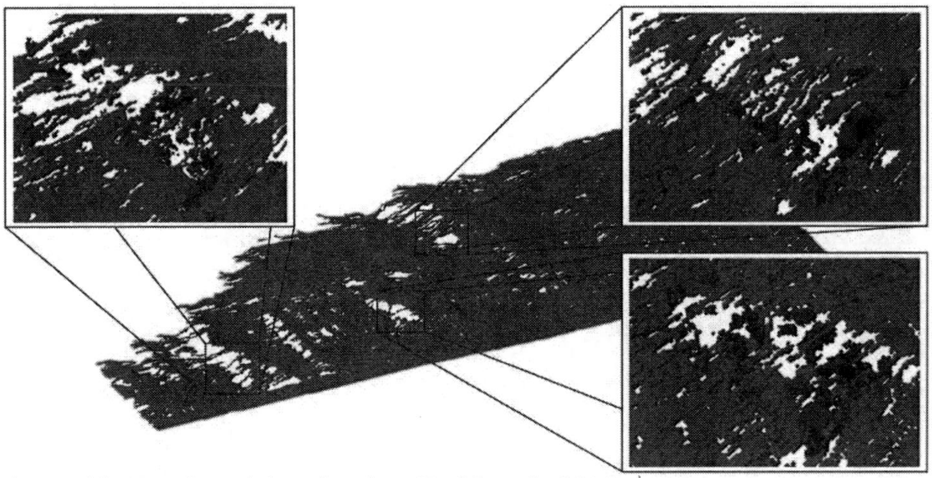

Fig. 10. Close-up of the 3D crack morphology of specimen 6OA-LS tested at 1.2 MPa√m showing primary precipitates located at the root of ridges

4. Discussion

From the crack paths shown in Fig. 4, it can be concluded that the primary precipitates act as microstructural barriers causing deviations of the crack. These barriers apparently pinned the crack front at several locations as indicated by the 3D crack morphology (Fig. 5) and fracture surfaces (Fig. 6-9). A closer inspection of the discontinuities in the 3D crack morphology of specimen 6OA-LS tested at $\Delta K = 1.2\, MPa\sqrt{m}$ reveals - after reconstruction of the 3D morphology of the primary precipitates - that large precipitates are located at the root of the ridges (see Fig. 10). These precipitates cause a pinning of the crack front eventually leading to a crack arrest.

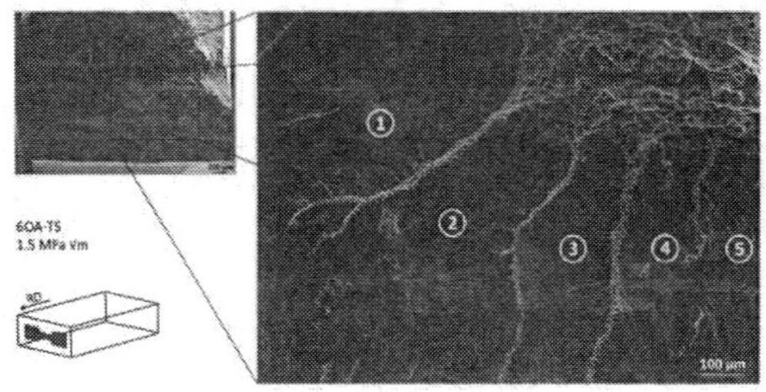

Fig. 11. Close-up of the fracture surface of specimen 6OA-TS tested at 1.5 MPa√m showing lines of ductile fracture

Detailed analysis (see Fig. 11) of the white lines depicted in Fig. 9 indicates that these lines are ridges of ductile fracture decorated by dimples. The crack front was thus locally pinned and the crack faces were bridged by material ridges, which failed by ductile fracture.

The primary precipitates seem to have a stronger influence in the TS-direction as the crack front is held up several times. This effect may be correlated with the spatial distribution of the primary precipitates. As it can be seen in Fig. 12, the crack front passes through well-defined lines of primary precipitates decorating the grain boundaries in the LS-specimen, while local precipitate clusters are formed in the TS-oriented samples acting as pinning sites. Local propagation of the crack front is thus possible in precipitate free areas leading to kinked crack fronts (see Fig. 9). Overaging seems to increase the pinning potential of the primary precipitates as ridges of ductile fracture and pronounced crack front kinking were observed even in the LS-oriented specimen (see Fig. 5).

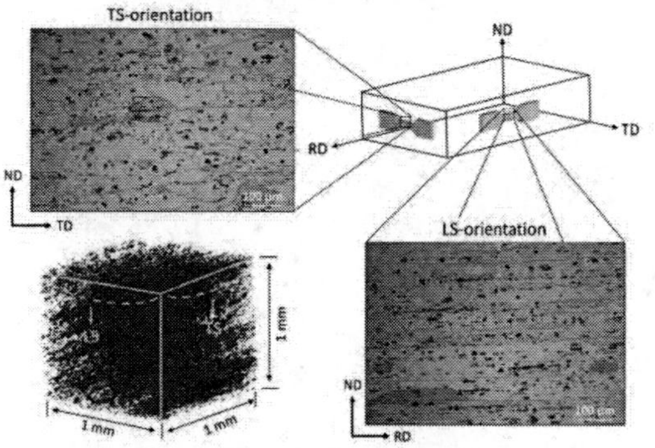

Fig. 12. CT scan and micrographs illustrating the position of the crack front according to the primary precipitates.

Even though a local pinning of the crack front was observed at R = -1, there are pronounced differences as to the pinning potential of the precipitates which occur in a less pronounced form at R = 0.1. This is attributed to the increased propensity of the primary precipitates to cracking rather than to pinning, as the crack front advance at R = 0.1 requires higher K_{max}-values at the same ΔK-value than a test at R=-1. Higher K_{max}-values also seem to enhance crack propagation in a shear-dominated mode according to the results presented above.

5. Conclusion

Results of near-threshold fatigue crack growth experiments at R = 0.1 on peak-aged and overaged aluminum specimens pre-cracked by cyclic compression were presented. Based on the long-crack thresholds initially determined on samples taken in both rolling and transverse direction using the load-increasing procedure proposed in [7], crack growth experiments were performed with a nearly constant ΔK-value resulting in a constant propagation rate. Detailed analysis of the fractured samples in SEM and µ-CT points out that crack advance in a shear-dominated mode is enhanced at R = 0.1 compared to R = -1 [8] whereas the pinning potential of the primary precipitates is significantly decreased. This is related to the fact that the primary precipitates tend to crack with increasing R-ratio and are consequently less able to form substantial barriers for crack advance.

The authors would like to thank the German Research Foundation (DFG) for financial support.

6. References

1. J.C. Newman Jr., AGARD Conf. Proc. **328**, 6.1-6.26 (1983)
2. J.C. Newman Jr., ASTM STP **1372**, 227-251 (2000)
3. S.C. Forth, J.C. Newman, Jr., R.G. Forman, Int. J. Fatigue **25**, 9-15 (2003).
4. J.C. Newman, Jr., J. Schneider, A. Daniel, D. McKnight, Int. J. Fatigue **27**, 1432-1440 (2005).
5. J.C. Newman, Jr., Y. Yamada, Int. J. Fatigue **32**, 879-885 (2010)
6. R. Pippan, Fatigue Fract. Eng. Mater. Struct. **9**, 319-328 (1987)
7. R. Pippan, L. Plöchl, F. Klanner, H.P. Stüwe, ASTM J. Test. Eval. **22**, 22 (1994).
8. T. Stein., M. Wicke, A. Brueckner-Foit, T. Kirsten, M. Zimmermann, F. Buelbuel, H.-J. Christ, J. Mat. Research **32**, 4354-4361 (2017)
9. T. Nishida, G. Pezzotti, T. Mangialardi, A.E. Paolini, Fract. Mech. Ceram. **11**, 107 (1996)
10. M. Wicke, A. Brueckner-Foit, T. Kirsten, M. Zimmermann, F. Buelbuel, H.-J. Christ, Proc. Struc. Int. **7**, 235-241 (2017)
11. A.K. Vasudevan, S. Suresh, Met. Trans. **13A**, 2271-2280 (1982)
12. M.C. Lafarie-Frenot, C. Gasc, Fat. Fract. Eng. Mat. Struct. **6**, 329-344 (1983)
13. S. Suresh, A.K. Vasudevan, P.E. Bretz, Met. Trans. **15A**, 369-379 (1984)

The coupled criterion for description of fatigue fracture. Material embrittlement in pre-fracture zone

Vladimir M. Kornev*

Lavrentyev Insti subareas tute of Hydrodynamics SB RAS, Novosibirsk, 630090, Russia

Abstract. Step-wise extension of a crack in quasi-brittle materials under low-cycle loading conditions is considered. Both steady and unsteady loadings in pulsating loading mode are studied. It is proposed to use quasi-brittle fracture diagrams for bodies under cyclic loading conditions. When diagrams are plotted, both necessary and sufficient fracture criteria by Neuber-Novozhilov are used. A specific implementation is made on the base of the Leonov-Panasyuk-Dugdale model for the mode I cracks when the pre-fracture zone width coincides with the plasticity zone width near the crack tip. The condition of a step-wise crack tip extension has been derived. A crack extends only in the embrittled material of the pre-fracture zone. The number of cycles between jumps of the crack tip is calculated by the Coffin equation, when damage accumulation in material in the pre-fracture zone is taken into account. Critical fracture parameters under low-cyclic loading conditions have been obtained in a closed form. Estimates of the average rate of crack tip advance for a loading cycle at step-wise crack extension and $S-N$ curves have been obtained.

1 Introduction

In order to obtain estimates for the number of cycles, the diagram of quasi-brittle fracture under low-cycle fatigue is used [1, 2].

The problems to be settled in this work are as follows: *i)* to derive relations for critical stresses above which the growth of this macrocrack takes place in specimens having a short macro-crack; *ii)* in the plane stresses vesus crack length plotting a subarea where damage accumulation occurs after single loading; *iii)* description of the process of step-wise propagation of short macro-cracks under fatigue when material embrittlement in the pre-fracture zone takes place ; *iv)* selection of parametrs for the proposed model acording to the resuls of laboratory tests.

2 Diagrams of quasi-brittle fracture under single loading

Apply the simplest approximation of a $\sigma-\varepsilon$ diagram for elastic-plastic material when the diagram is approximated by a double-link straight line. The parameters of this approximation are as follows: E is the modulus of elasticity, σ_Y is the yield strength of material, and σ_Y is constant stress acting in accordance with the modified Leonov-Panasyuk-Dugdale model, ε_0 is the maximum elastic material strain ($\sigma_Y = E\varepsilon_0$), ε_1 is the maximum material strain. Let r be the effective diameter of fracture structures. The Neuber-Novozhilov approach [3, 4] makes it possible to use solutions having a singular function for structured media.

Now consider an internal I mode crack. Assume that an internal plane I mode crack extends rectilinearly. In addition to the internal rectilinear crack-cut of length $2l_0$, model crack-cuts of length $2l = 2l_0 + 2\Delta$, each of pre-fracture zones Δ being located on the continuation of a real crack ($2l$ and Δ are lengths of model crack and pre-fracture zones) are introduced into consideration. The problem of fatigue fracture has two linear scales: if a grain diameter is defined by a material structure, then the second linear size is governed by the system itself. Under low-cycle fatigue conditions, the second linear scales serve as pre-fracture zone lengths Δ, which change in accordance with changing *i)* the length of a real step-wise extending crack, and *ii)* the intensity of loading conditions. Emphasize that under single loading conditions, the critical pre-fracture zone length Δ^* is a completely definite parameter [1, 2] and $2l^* = 2l_0 + 2\Delta^*$ is the critical macro-crack length, that is $\Delta^* / l_0 = 1$ for quasi-brittle materials.

When diagrams of quasi-brittle fracture under conditions of low-cycle loading are plotted in [1, 2, 5], sufficient fracture criteria are used when the mode I cracks are considered

$$\frac{1}{r}\int_0^r \sigma_y(x,0)dx = \sigma_Y \, , \; 2v(x,0) = \delta^* . \qquad (1)$$

Here $\sigma_y(x,0)$ are normal stresses on the crack continuation; Oxy is the rectangular coordinate system oriented about right crack sides (the coordinate origin coincides with the model crack tip in the modified Leonov-Panasyuk-Dugdale model [1, 2, 5]); $2v = 2v(x,0)$ is CTOD (crack tip opening displacement, $x < 0$), δ^* is the critical CTOD.

The field of normal stresses $\sigma_y(x,0)$ on the model crack continuation $x > 0$ can be represented as a sum of two summands

* Corresponding author: kornev@hydro.nsc.ru

$$\sigma_y(x,0) \cong K_{\rm I} / (2\pi x)^{1/2} + O(1), \quad (2)$$

$$K_{\rm I} = K_{\rm I\infty} + K_{\rm I\Delta}, \quad K_{\rm I\infty} > 0, \quad K_{\rm I\Delta} < 0,$$

where $K_{\rm I} = K_{\rm I}(l,\Delta)$ is the total stress intensity factor (SIF) at the tips of a model crack, $K_{\rm I\infty}$ is SIF generated by stresses σ_∞ specified at infinity, $K_{\rm I\Delta}$ is SIF generated by constant stresses σ_Y. The first and second summands in relation (2) are singular and smooth parts of solution, respectively. The first equality in criterion (1) controls stresses on the model crack continuation after averaging, these stresses being coincident with the yield strength σ_Y, and the second equality in criterion (1) describes real blunting of a crack at its tip. The sufficient fracture criterion (1) simultaneously takes into consideration both strength and deformation fracture criteria at specific points of a pre-fracture zone. Selected specific points of the pre-fracture zone are well-adapted to description of step-wise crack tip advance and accumulation of damages in the pre-fracture zone material under fatigue conditions [1, 2].

The simplest analytical representation of the field of stresses $\sigma_y(x,0)$ on the model crack continuation has the following form ($2L$ is the plain specimen width)

$$\sigma_y(x,0) = \left(K_{\rm I\infty} + K_{\rm I\Delta}\right) / \sqrt{2\pi x} + \sigma_{nom} = \quad (3)$$

$$= \sigma_\infty Y(l/L)\sqrt{l/2x} + 2\sigma_Y\sqrt{\Delta/x}/\pi + \sigma_\infty/(1-l/L)$$

Introduce into consideration critical stresses calculated via the necessary fracture criterion σ_∞^0 and sufficient one σ_∞^* and Δ^*. System (1) is simplified when fracture of quasi-brittle materials is studied under conditions of small-scale yielding $\Delta^*/l_0 = 1$ [2]. In what follows, dimensionless variables are used (μ is Poisson's ratio)

$$\bar\sigma_\infty^* = \sigma_\infty^*/\sigma_Y, \quad \bar\sigma_\infty^0 = \sigma_\infty^0/\sigma_Y, \quad \bar\Delta^* = \Delta^*/r, \quad \bar l^* = l^*/r,$$

$$r = (2/\pi)\left(K_{\rm Ic}/\sigma_Y\right)^2\left(1 - A\bar\varepsilon_p\right)^2, \qquad \bar\varepsilon_p = (\varepsilon_1 - \varepsilon_0)/\varepsilon_0,$$

$\bar L = L/r, \quad A = \left[3 + 2(1-2\mu)^2\right] / \left[8\pi\left(1-\mu^2\right)\right]$. The approximate system of equations is valid for short macrocracks if $2\bar l^*/r \ge 0.1 \div 0.2$.

Fracture diagrams for quasi-brittle materials are pairs of curves $\bar\sigma_\infty^0$ and $\bar\sigma_\infty^*$ Analytical representations of critical fracture parameters for quasi-brittle materials has the form for a plain specimen of the finite width [1,5]

$$\bar\sigma_\infty^* \approx \left[1/\left(1 - \bar l^*/\bar L\right) + \left(1 - A\bar\varepsilon_p\right)Y\left(\bar l^*/\bar L\right)\sqrt{2\bar l^*}\right]^{-1},$$

$$\bar\sigma_\infty^0 \approx \left[1/\left(1 - \bar l_0/\bar L\right) + Y\left(\bar l_0/\bar L\right)\sqrt{2\bar l_0}\right]^{-1}, \quad (4)$$

$$\bar\Delta^*/\bar l^* \approx \left(\pi^2/8\right)\left[A\bar\varepsilon_p Y\left(\bar l^*/\bar L\right)\bar\sigma_\infty^*\right]^2.$$

Expressions (4) make sense if $A\bar\varepsilon_p < 1$.

The finite specimen width effects upon critical fracture curves on the plane "crack length versus internal load".

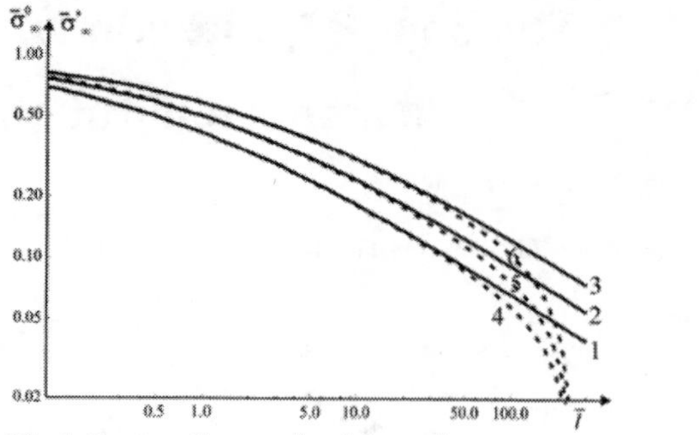

Fig. 1. Fracture diagrams for plain specimens.

The calculation results are given in Fig. 1 in the log-log coordinates, when $\bar L = 250$ ($Y(l/L) = \sqrt{\sec(\pi l/2L)}$). In this Figure, solid curve 1 and dashed curve 4 are plotted via relation (4) for $\bar\sigma_\infty^0$ at $\bar L = \infty$ and $\bar L = 250$, respectively; solid curves 2 and 3 ($\bar L = \infty$), and dashed lines 5 and 6 ($\bar L = 250$) are plotted via relation (4) for $\bar\sigma_\infty^*$ at $\bar\varepsilon_p = 1.5$ and $\bar\varepsilon_p = 2.5$. Pairs of curves 1-2, 1-3, and 4-5, 4-6 in Fig. 1 are diagrams of quasi-brittle fracture for plain specimens of infinite and finite width, respectively. Under cyclic loading conditions the subareas between curves $\bar\sigma_\infty^0$ and $\bar\sigma_\infty^*$ are areas, in which damage accumulation in pre-fracture zone material takes place.

3 Step-wise crack propagation

When the proposed diagram under low-cycle fatigue [1, 2, 5] is plotted, there is no need to use SIFs. It can be plotted depending on both elastic-plastic material properties and a crack length.

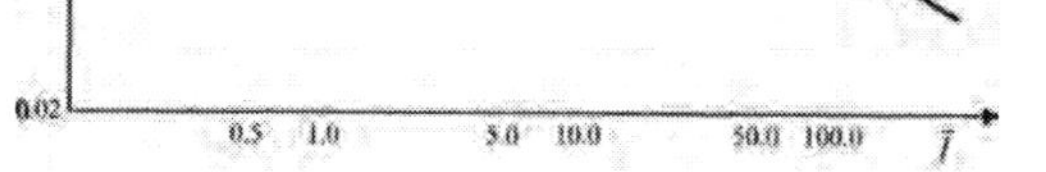

Fig. 2. Diagram of fatigue fracture.

This diagram in Fig. 2 consisting of a pair of curves divides the plane of "crack length versus. external load amplitude" $(\bar l, \bar\sigma_\infty)$ into three subareas: *1)* the subarea where fatigue fracture is not observed is located to the left and lower the curve $\bar\sigma_\infty^0$; *2)* the subarea of fracture at single loading is located to the right and above the curve $\bar\sigma_\infty^*$; *3)* the subarea of step-wise crack propagation due to embrittlement of material under cyclic loading conditions is located between curves $\bar{\sigma}_{\infty}^{0}$ and $\bar{\sigma}_{\infty}^{*}$. When repeated loads are applied, the curve $\bar{\sigma}_{\infty}^{*}$ is transformed, embrittlement of the pre-fracture zone material being considered. The proposed diagram has two threshold values $\bar{\sigma}_{\infty}^{0}$ and $\bar{\sigma}_{\infty}^{*}$ in relation (4).

Consider limitations typical for one-frequency loading

$$\bar{\sigma}_{\infty}^{0} < \bar{\sigma}_{\infty} < \bar{\sigma}_{\infty}^{*}. \quad (5)$$

Here $\bar{\sigma}_{\infty}$ is amplitude. When limitations (5) are obeyed, the low-cycle fatigue may be implemented. Changes in the subarea of the diagram of quasi-brittle fracture with account for damage accumulation are described for the normal loading regime

$$\bar{\sigma}_{\infty}^{0} < \bar{\sigma}_{\infty} < \bar{\sigma}_{\infty}^{*(s)}, 1 \le s_j \le N_j - 1, j = 1,2,...,j^* - 1,$$

$$\bar{\sigma}_{\infty}^{0} < \bar{\sigma}_{\infty}^{*(s)} < ... < \bar{\sigma}_{\infty}^{*(2)} < \bar{\sigma}_{\infty}^{*(1)} = \bar{\sigma}_{\infty}^{*};$$

$$\bar{\sigma}_{\infty} \ge \bar{\sigma}_{\infty}^{*(s)}, s_j = N_j, j = 1,2,...,j^*, \bar{\sigma}_{\infty}^{0} < \bar{\sigma}_{\infty}^{*(s)},$$

when the critical number of loading cycles N^* is calculated in such a way

$$N^* = 1 + \sum_{1}^{j^*-1} N_j, N_j \ge 2, j = 1,2,...,j^* - 1, N_{j^*} = 1. \quad (6)$$

Here $\sigma_{\infty}^{*(s)}$ is the critical load obtained via the sufficient fracture criterion in s-th loading cycle up to $j = 1,2,...,j^* - 1$ crack tip advance, s_j is number of cycle between $j-1$ and j crack tip advances, $s_j = 1$ corresponds to the state of original material after each crack tip advance, N_j is the number (group) of cycles between $(j-1)$ and j crack tip advances. For the loading level $\bar{\sigma}_{\infty}$ of a real crack of length $2l_{j-1}$, equations for specimens of finite width (4) are as follows

$$\bar{\sigma}_{\infty} \approx \left[1/\left(1 - \bar{l}_{j-1}/\bar{L}\right) + \left(1 - A\bar{\varepsilon}_{\infty,j-1}\right) Y\left(\bar{l}_{j-1}/\bar{L}\right)\sqrt{2\bar{l}_{j-1}}\right]^{-1},$$

$$\bar{\Delta}_j / \bar{l}_{j-1} \approx \left(\pi^2/8\right)\left[A\bar{\varepsilon}_{\infty,j-1} Y\left(\bar{l}_{j-1}/\bar{L}\right)\bar{\sigma}_{\infty}\right]^2,$$

$$2\bar{l}_j = 2\bar{l}_{j-1} + 2\bar{\Delta}_j, j = 1,2,...,j^*. \quad (7)$$

The number of cycles N_j between $j-1$ and j crack tip advances at the loading level $\bar{\sigma}_{\infty}$ is calculated as follows

$$N_j = \left(\bar{\varepsilon}_p / \bar{\varepsilon}_{\infty,j-1}\right)^{1/C}, j = 1,2,...,j^* - 1. \quad (8)$$

Here $0.2 \le C \le 1$ are Coffin's constants [6], numerical values of which depend on material properties. In fatigue failure, it is reasonable to obtain estimates of average dimensionless rate $\bar{V}_j = 2\bar{\Delta}_j / N_j$ of crack advance per one loading cycle for plain specimens of finite width at the appropriate loading mode (7) and (8)

$$\bar{V}_j = \frac{\pi^2 A^2}{4} \frac{\left(\bar{\varepsilon}_{\infty,j-1}\right)^{2+1/C}}{\left(\bar{\varepsilon}_p\right)^{1/C}} \left[Y\left(\bar{l}_{j-1}/\bar{L}\right)\bar{\sigma}_{\infty}\right]^2 \bar{l}_{j-1},$$

$$j = 1,2,...,j^* - 1. \quad (9)$$

4 Averaged rates of macro-crack tip advance

Intsead of the discontinous function $\bar{V}_j$ in relation (9), we consider continous function $\bar{V} = \bar{V}\left(\bar{l}\right)$, argument of which is continuously changing. The continuous function $\bar{V} = \bar{V}\left(\bar{l}\right)$ and the discontinuous function $\bar{V}_j$ describe all three process stages: initiation, evolution, and completion of the process of macro-crack tip advance (load level $\bar{\sigma}_{\infty} = const$, characteristic of material $\bar{\varepsilon}_p$, and specimen dimensions $\bar{L}$ are assumed to be given). Two-sided estimates of crack lengths, for which the Paris law is appropriate, have been obtained (ζ is small positive number)

$$\left[\left(1/\bar{\sigma}_{\infty} - 1\right)/\zeta\right]^2 \le 2\bar{l} \le 2\bar{L}\left(1 - \bar{\sigma}_{\infty}\right). \quad (10)$$

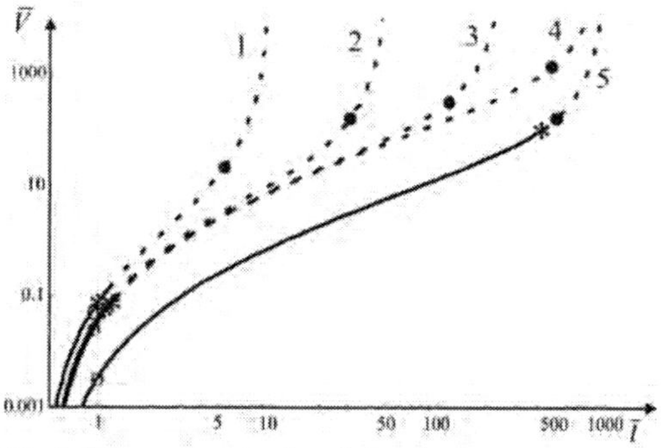

Fig. 3. The average rate of crack tip advance.

In Fig. 3, plots of averaged dimensionless rate of crack tip advance $\bar{V} = f\left(\bar{l}, \bar{\sigma}_{\infty}, \bar{\varepsilon}_p, \bar{L}, C\right)$ are given in log-log coordinates. Curves in Fig. 3 have been constructed via relation (9) for linear damage summation $C = 1$: when curves 1 – 4 were plotted, the parameters $\bar{\sigma}_{\infty} = 0.5$, $\bar{\varepsilon}_p = 2$, and $\bar{L} = 10, 50, 250$, and 1000, respectively, were used; calculations for curve 5 were performed with parameters $\bar{\sigma}_{\infty} = 0.5$, $\bar{\varepsilon}_p = 20$, and $\bar{L} = 1000$. In Fig. 3, light and dark dots on curves correspond to the left- and right-side inequalities in relation (10). Middle sections of curves are lengths of cracks for which the Paris law "works" well. To the right of asterisks depicted on curves 1 – 5, failure of the specimen takes place in accordance with relation (4), therefore, the conditional sections of curves 1 – 5 are continued as dotted lines. Curve 1 almost completely falls into the region of short macro-cracks.

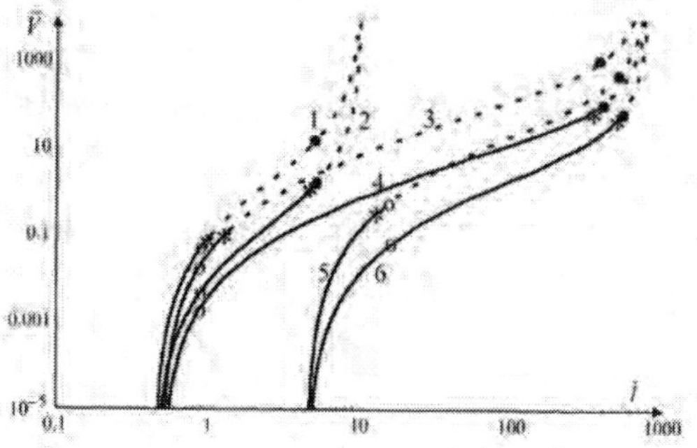

Fig. 4. The average rate of crack tip advance for linear damage summation.

Plots of averaged dimensionless rate of crack tip advance $\bar{V}=\bar{V}\left(\bar{l},\bar{\sigma}_{\infty},\bar{\varepsilon}_p,\bar{L},C\right)$ in Fig. 4 are given in double logarithmic coordinates. Curves in Fig.4 have been plotted via relation (9) taking into account the linear damage summation $C=1$: the following parameters were used for construction of the curves: $\bar{\sigma}_{\infty}=0.5$ and $\bar{L}=10$ for curves 1 and 2, $\bar{\sigma}_{\infty}=0.5$ and $\bar{L}=800$ for curves 3 and 4, and $\bar{\sigma}_{\infty}=0.25$ and $\bar{L}=800$ for curves 5 and 6 ($\bar{\varepsilon}_p=2$ and $\bar{\varepsilon}_p=20$ are relatuve margin of plastic strain of material for uneven and even curves, respectively, and $\mu=0$). Light and dark dots on curves in Fig.4 correspond to the left- and right-side inequalities in relation (10). To the right of asterisks depicted on these curves, failure of the specimen takes place in accordance with relation (4), therefore, the conditional sections of the curves are continued as dotted lines. Each of these curves descrides initiation, evolution, and completion of the process of macro-crak tip advance. Middle sections of all the curves correspond to Paris' curves with some degree of certainty.

The exellent representation of all stages of the process is shown in monograph [6, p.27, Fig. 1.18]. In this mongraph and handbook [7], all three stages of the process and transition modes are discussed in detail. Howeever, the main attention of these authors [6, 7] is concentrated on the representation of these diagrams in the form of dependencies on SIF that does not agree with the results obtained for small-scale yielding in quasi-brittle materials (9).

Given in Fig. 5 are curves characterizing the influence of relative plastic strain margin of material $\bar{\varepsilon}_p$ on the rate of crack tip advance $\bar{V}$. Curves 1 – 6 have been plotted with account of linear damage summation $C=1$ for the parameters $\bar{\varepsilon}_p=1,2,5,10,20$, and 40, respectively ($\bar{\sigma}_{\infty}=0.25$, and $\bar{L}=800$). Light and dark dots on the curves in Fig, 5 correspond to the left- and right-side inequalities from relation (10). To the rigth of asteriks depicted on the curves, failure of the specimen takes place in accordance with relation (4), therefore, the conditional sections of these curves are continued as dotted lines.

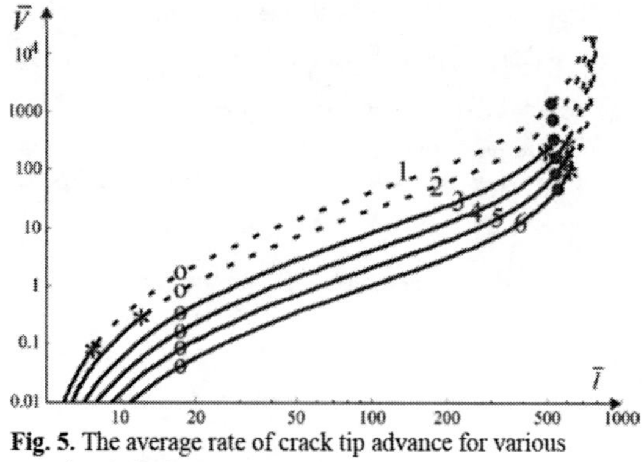

Fig. 5. The average rate of crack tip advance for various margin of plastic material strains $\bar{\varepsilon}_p$.

Plastic strain margin of material $\bar{\varepsilon}_p$ significantly affects the rates of crack tip advances $\bar{V}$. Let us consider two materials which differ only in margin of plastic strains of material $\bar{\varepsilon}_{p0}$ and $\bar{\varepsilon}_{p1}$. Then, using relation (9), the estimate can be written as

$$\frac{\bar{V}_0\left(\bar{l},\bar{\sigma}_{\infty},\bar{\varepsilon}_{p0},\bar{L},C\right)}{\bar{V}_1\left(\bar{l},\bar{\sigma}_{\infty},\bar{\varepsilon}_{p1},\bar{L},C\right)}\approx\left(\frac{\bar{\varepsilon}_{p1}}{\bar{\varepsilon}_{p0}}\right)^{1/C}. \tag{11}$$

The margin of plastic strain after preliminary plastic deformation of material $\bar{\varepsilon}_{p1}$ may be a small part of the plastic strain margin of basic material $\bar{\varepsilon}_{p0}$, i.e., $\bar{\varepsilon}_{p0}$? $\bar{\varepsilon}_{p1}$. Therefore, taking into account relation (11), the ratio of rates of crak tip advance $\bar{V}_0/\bar{V}_1$ for basic material and material after significant plastic deformation can reach from one to two orders of magnitude.

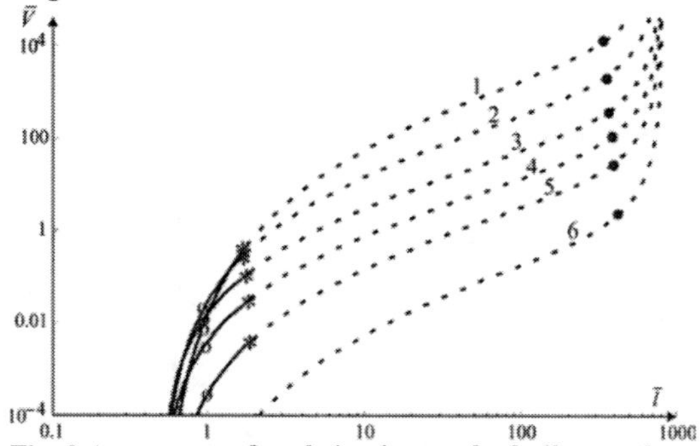

Fig. 6. Average rate of crack tip advances for for linear and nonlinear damage summation.

Given in Fig, 6 are curves reflecting effect of linear and nonlinear summation of materal damages on the averaged rate of crak tip advance $\bar{V}$. These curves have been plotted via relation (9) taking into account linear ($C=1$ for curves 3 and 6) and nonlinear ($C=0.25$ for curves 1 and 4 and $C=0.5$ for curves 2 and 5) damage summation. The parameters $\bar{\sigma}_{\infty}=0.5$, $\bar{L}=800$, and $\bar{\varepsilon}_p=2$ were used for construction the curves. The remaining notations are the same as those in Figs. 3 – 5. If the conditional sections of these curves are excluded, then there are no sections on these curves, which can be treated as Paris' curves.

5 The interrelation between the Coffin constant and characteristic rate of crack tip advance

The proposed approximate relation (9) permits relate the Coffin constant with the charcteristic rate of crak tip advance drawing on their association at the second stage of crack tip advance when the Paris law "works" well. Let the characteristic rate $\tilde{V}_0=\tilde{V}_0\left(\bar{l}_0,\bar{\sigma}_{\infty 0},\bar{\varepsilon}_{p0},\bar{L}_0,C\right)$ of crack tip advance be determined for some parametrs $\bar{l}_0,\bar{\sigma}_{\infty 0},\bar{\varepsilon}_{p0},\bar{L}_0$ at the second stage of a process during laboratory experiment. The procedure for measurung this rate under laboratory conditins is well developed [8]. After simple transformations of relation (9), we obtain

$$\begin{gathered}(1/C)\left(\ln F_0-\ln A\bar{\varepsilon}_{p0}\right)=\ln\tilde{V}_0-\ln\left(\pi^2/4\right)-\\ -2\ln F_0-2\ln\left[Y\left(\bar{l}_0/\bar{L}_0\right)\bar{\sigma}_{\infty}\right]-\ln\bar{l}_0\end{gathered}, \quad (12)$$

$$F_0=Y^{-1}\left(\frac{\bar{l}_0}{\bar{L}_0}\right)\left[\frac{1}{1-\bar{l}_0/\bar{L}_0}-\frac{1}{\bar{\sigma}_{\infty 0}}+Y\left(\frac{\bar{l}_0}{\bar{L}_0}\right)\sqrt{2\bar{l}_0}\right]\sqrt{\frac{1}{2\bar{l}_0}}.$$

Relation (12) associates the constant C and characteristic rate $\tilde{V}_0$ based on the Paris law, i.e. $C=C\left(\tilde{V}_0\right)$.

It is proposed to use in relation (9) the experimentaly determined C constant from relation (12) for description of all three stages of crack tip advances including closely adjacent value of parameters $\bar{\sigma}_{\infty}$ and $\bar{L}$. It is to be recalled that derived relation (9) permits one to study behavior of cracks, lengths of which can be assignd just as to short, so to long cracks. Three stages of the process of crack tip advances are roughly estaimated by relations (10). Implementation of some or other stages of the process should be agreed with constrains imposed by the presence of two certain threshold values of crack lengths with the given loading intensity $\bar{\sigma}_{\infty}$. Relation (9) can be used for variable loading condititions. With a gradual decrease of loading intensity, the system can fall several times into the region, which corresponds to the region of short crack tip advance (10) (as compared with the behavior of curves 3 and 4 with that of curves 5 and 6 in Fig. 4).

6 $S-N$ fracure diagrams

The analysis of the crack tip advance does not fully satisfy the requirements of practice since, from a practical point of view, one has to know how many cycles are left before complete failure when the stress intensity of loading $\bar{\sigma}_{\infty}$ and the length of initial crack $2\bar{l}_0$ are given. Suppose that this crack length $2\bar{l}_0$ has been obtained at the latest inspection of the structure.

Let us study relation (9) in more detail. Intsead of the discontinous function $\bar{V}_j$ in relation (9), we consider continous function $\bar{V}=\bar{V}\left(\bar{l}\right)$, argument of which is continuously changing. Assume that specimens of the infinite widh, i,e., $\bar{L}=\infty$, are under consideration. Let the right-hand side of relation (9) be a continuos function. The left-side of relation (9) can be represented as $\bar{V}\approx d2\bar{l}/dN$, if we assume that N is the continous function. Then relation (9) is approximately rewritten in thr form of a differential equation

$$d2\bar{l}/dN\approx f\left(2\bar{l}\right), \quad (13)$$

$$f\left(2\bar{l}\right)\approx\frac{\pi^2}{8\left(A\bar{\varepsilon}_p\right)^{1/C}}\left[1+\left(1-\frac{1}{\bar{\sigma}_{\infty}}\right)\sqrt{\frac{1}{2\bar{l}}}\right]^{2+1/C}\bar{\sigma}_{\infty}{}^2 2\bar{l}.$$

This ordinary differential equation (13) may be integrated

$$N^*\approx\int_{2\bar{l}_0}^{2\bar{l}^{(2)}}f^{-1}\left(2\bar{l}\right)d2\bar{l}+b, \quad (14)$$

$$\bar{\sigma}_{\infty}\approx\left(1+\sqrt{2\bar{l}^{(1)}}\right)^{-1},\ \bar{\sigma}_{\infty}\approx\left[1+\left(1-A\bar{\varepsilon}_p\right)\sqrt{2\bar{l}^{(2)}}\right]^{-1},$$

where N^* is the critical number of cycles before failure, b is the constant of integration, $2\bar{l}_0$ is the initial crack length satisfying the constrains $2\bar{l}^{(1)}<2\bar{l}_0<2\bar{l}^{(2)}$, and $2\bar{l}^{(2)}$ is the critical crack length. This critical crack length $2\bar{l}^{(2)}$ is the second threshold value of a crack length at the given loading level $\bar{\sigma}_{\infty}<1$ of quasibrittle material. Thus, both threshold values of crack lengths $\bar{l}^{(1)}$ and $\bar{l}^{(2)}$ in relation (14) are used at the given loading level $\bar{\sigma}_{\infty}$ with account of constrains,. The constant b is chosen from the condition: for $2\bar{l}_0=2\bar{l}^{(2)}$, the critical number of cycles before failure $N^*=1$, i.e., $b=1$, that corresponds to monotonic loading. Approximate relation (14) is one again approximate alternative of the damage tolerant criterion.

Integral (14) is an improper one since $2\bar{\Delta}\to 0$ for $2\bar{l}_0\to 2\bar{l}^{(1)}$. This integral permits representation in the closed form when the parameter $2+1/C$ is the the integer value, i.e., e. $2+1/C=3,4,5,6$, and $2+1/C=7$.

Given in Figs. 7 and 8 are the $N-l_0$ fracture diagrams, which are estimates of lifetime for specimens with an initial crack at the given loading level $\bar{\sigma}_{\infty}$. Curves 1, 2, and 3 shown in Fig. 7 are plotted via relation (14) for the loading intensities $\bar{\sigma}_{\infty}=0.55;0.5$, and $\bar{\sigma}_{\infty}=0.4$ on linear damage summation $C=1$ ($\bar{\varepsilon}_p=5$, $\mu=0.33$). Points marked with asterisks indicate the places where a specimen is finally broken down. Thick segements on the basic curves reflect the selected complex three-level loading mode (load range shiftings are labeled by thin dotted lines). Due to change of the loading level, such modes of fatigue fracture are practible when the system can fall several times into areas of short crack lengths.

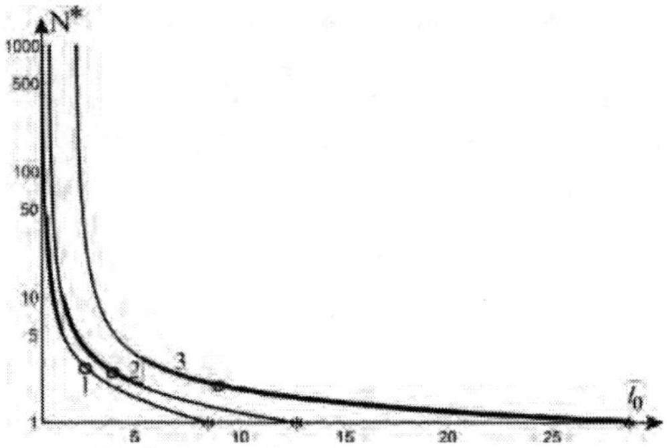

Fig. 7. $N - l_0$ fracture diagrams.

Curves 1, 2, and 3 in Fig. 8 have been plotted via relation (14) for low loading intensities $\bar{\sigma}_\infty = 0.2; 0.1$, and $\bar{\sigma}_\infty = 0.05$ on linear damage summation $C = 1$ ($\bar{\varepsilon}_p = 5$, $\mu = 0.33$). Points marked with asterisks indicate the places where a specimen is finally broken down. At low loading intensities $\bar{\sigma}_\infty$, such modes are specific in which an intial crack does not grow.

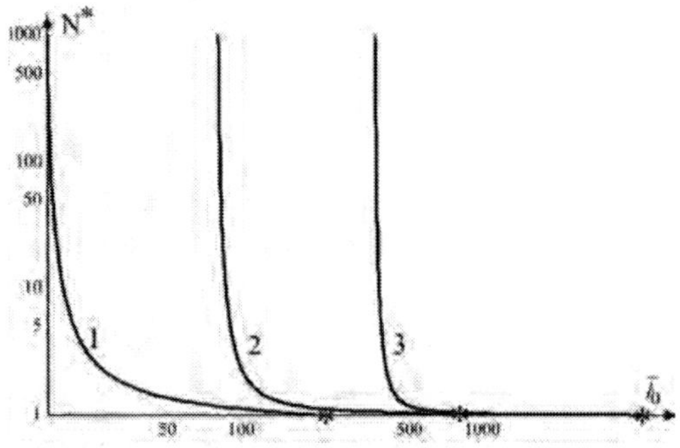

Fig. 8. $N - l_0$ fracture diagrams for different loading intensities.

If $S - N$ or $N - l_0$ fracture diagrams and the crack growth rate $\bar{V}$ are taken into account, then it will be possible to estimate transition from relatively secure operating conditions of a structure with a crack to dangerous loading mode.

7 Conclusion

Step-wise crack propagation in quasi-brittle materials under cyclic loading conditions has been considered. For a model of deformable body, the model of elastic ideally-plastic material having the ultimate tensile strain is taken. When basic material undergoes nonlinear deformation, the material is embrittled.

For analysis of the process mentioned above, it is proposed to use modified diagrams of quasi-brittle fracture of deformable bodies. Reference diagrams have been plotted within the context of quasi-linear fracture mecanics when the small-scale material yelding is realized in the vicinity of a crack tip and the singularity of the stress field at the tip of a crak is partally retained.

Two theshold values of critical parameters are proposed for description of the process of step-wise crack tip advance in responce to fatigue loading modes. Both theshold values are given in the explicit form including that for short macro-cracks. The second theshold value is a variable and it is directly associated with material damage in a pre-fracture zone. Under nonlinear deformation of material in the pre-fracture zone, the plasticity margin is spent due to material embrittlement during each loading cycle.

Constants in the described model are proposed to be selected for individual material from the data obtained during three laboratory experiments based on *i)* approximation of a real diagram stress versus strain, *ii)* searching critical SIF, and *iii)* Coffin's constants or the characteristic rate of crack tip advance for the second process stage.

The analysis of all three process stages of crack tip afvances and their dependence on geomentric parameters of cracks and specimens, characterisics of material, and intensity of fatigue loading under pulstaing loading appliance has been performed in detail. For the averaged process, crude rwo-sided estimates for transition from the first to the second stage and from the second to the third stage have been pointed out, fairly simple analitical expressions for describing the whole process, including the case when some stages are absent have been derived. The plotted $N - S - \bar{l}_0$ diagrams charcterize lifetime of specimens having initial defects. All the constructions have peformed in terms of cracks. There is the fracture mechanics approach, which specifically treats growing cracks by the methods of fracture mechanics.

This work was supported by the Russian Foundation for Basic Research (Grant No. 16-08-00483).

References

1. V.M. Kornev, Physical. Mesomech., **14,** 5, pp. 31-45 (2011) (in Russian)
2. V.M. Kornev, Physical. Mesomech., **19,** 2, pp. 80-99 (2016) (in Russian)
3. H. Neuber, Kerbspannungslehre: Grunglagen fur Genaue Spannungsrechnung (1937)
4. V.V. Novozhilov, Prikl. Mat. Mekh., **33,** 2, pp. 212-222 (1969) (in Russian)
5. V.M. Kornev, Engin. failure analysis, **35** (2013), pp. 533-544
6. V.V. Bolotin, Mechanics of fatigue (1998)
7. O.N. Romaniv, S.Ya. Yarema, G.N. Nikiforchin, N.A. Makhutov, M.M. Stadnik, Fracture Mechanics and strength of materials, Vol. 4 (1990)
8. R.O.Ritchie, Intern. J. Fracture. **100**, pp 55-83 (1999)

Fatigue crack surface area and crack front length: new ways to look at fatigue crack growth

Jesse van Kuijk[1,*], *René* Alderliesten[1], *Rinze* Benedictus[1]

[1]Structural Integrity & Composites, Faculty of Aerospace Engineering, TU Delft, Netherlands

Abstract. This paper discusses the appropriateness of crack length as a reference dimension for fatigue damage. Current discussion on short crack versus long crack data is still divided between various approaches to model small crack growth. A proper physical explanation of the probable cause of the apparent differences between short crack and long crack data is not yet provided. Long crack data often comprises crack growth in constant thickness specimens, with a through crack of near constant crack front geometry. This is not true for corner cracks or elliptical surface crack geometries in the small crack regime where the crack front geometry is not symmetric or through-thickness. This affects similitude parameters that are based on the crack length. The hypothesis in this paper is that a comparison between long crack data and short crack data should be made using similar increments in crack surface area. The work applied to the specimen is dissipated in generation of fracture surface, whereas fracture length is a result. The crack surface area approach includes the two-dimensional effect of crack growth geometry in the small crack regime. A corner crack and a through crack are shown to follow the same power law relationship when using the crack area as base parameter. The crack front length is not constant, and its power law behaviour for a corner crack is shown.

List of Symbols

ΔK_a	: Stress intensity factor, f(a)
ΔK_{th}	: Threshold stress intensity factor, f(a)
ΔK_A	: Stress intensity factor, f(A)
β	: Elliptical aspect ratio modifier factor
γ	: Aspect ratio of ellipse; $= b/a$
a	: Crack length
b	: Crack length through thickness
da/dN	: Crack length growth rate per cycle
dA/dN	: Crack area growth rate per cycle
f	: Crack front length
A	: Crack area
N	: Number of cycles or iterations
R	: Stress ratio
S_{max}	: Maximum applied stress
S_{ol}	: Maximum applied overload stress
T	: Specimen thickness
W	: Specimen width

1 Introduction

Most fatigue crack growth data is published as da/dN versus ΔK.[a] The data often appears as one or more straight lines in a graph with log-log scales, and is known as the 'Paris law'. In industry, this Paris relation is widely used for its simplicity. However, small crack data is not so well-defined and a mismatch with Paris relations is often seen in this region.

Alderliesten [1,2] questions the general idea that the Paris relation in all its forms should be taken as a 'law', as there is no physical basis for the power law. He suggests that a better understanding of fatigue crack growth could be obtained by looking at the energy balance throughout fatigue cycles. Amsterdam et al. [3] agree with the notion that the power law approach is flawed from a physical point of view, as the equation is not dimensionally correct.

The relationship of da/dN versus ΔK_a on the 'base parameter' crack length a seems to work fine for through cracks, where the crack front is ideally straight, and perpendicular to side of the specimen. For any other crack front shape, this does not necessarily hold. The mean a along the crack front is different from the

[a] As the stress intensity factor is a function of the crack length, the notation ΔK_a will be used to distinguish it from another stress intensity factor introduced in this paper being a function of the crack area: ΔK_A.

* Corresponding author: J.J.A.vanKuijk-1@tudelft.nl

https://doi.org/10.1051/matecconf/201816513009

observed a at the specimen edge, and different local da/dN values are present along the crack front. Another choice of base parameter might improve the understanding of fatigue crack growth, and could help in understanding the energy balance cycle during crack growth.

2 Hypothesis

Given the large spread in small crack da/dN data, and the dependency on a which is questionable for several crack front geometries, further search for a more suitable base parameter seems warranted.

It is hypothesized that the crack area A is a better choice than the crack length a, because the area can be related to the energy in the cross-section, providing a path to relate the energy input to the crack growth. A schematic view of a specimen cross-section with cracked surface and geometric parameters is shown in **Fig. 1**.

During crack growth the cross-sectional area decreases, whereas the crack length is a result of this area decrease along one dimension. Only for an ideal through crack with straight crack front are a and A directly related by the specimen thickness. With literature mostly dealing with through cracks and a being readily measurable from the specimen surface, it is understandable why a is such a popular parameter to describe crack growth (rate).

Note that potential drop measurements are essentially crack area measurements, as the area is related to the electrical resistance of the current. When the area measurement is normalised by the specimen thickness T it directly transforms into a mean a measurement.

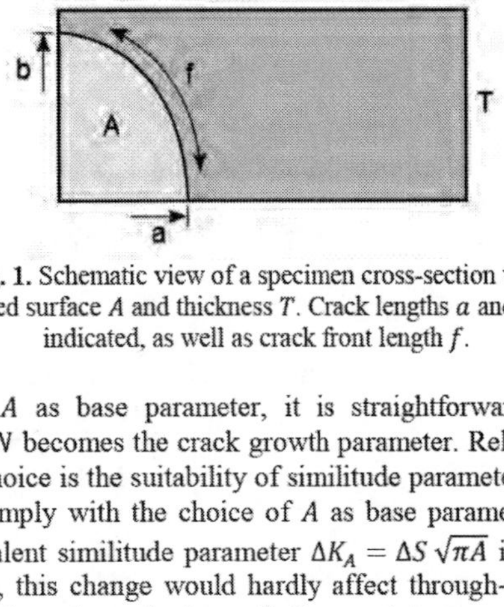

Fig. 1. Schematic view of a specimen cross-section with cracked surface A and thickness T. Crack lengths a and b are indicated, as well as crack front length f.

With A as base parameter, it is straightforward that dA/dN becomes the crack growth parameter. Related to this choice is the suitability of similitude parameter ΔK_a. To comply with the choice of A as base parameter, an equivalent similitude parameter $\Delta K_A = \Delta S\sqrt{\pi A}$ is used. Again, this change would hardly affect through-cracks, or the through-crack phase of other crack types.

A corner crack grows along two dimensions (a and b) instead of one (a), and as such the crack front length f, measured from free surface to free surface along a crack front, is also growing per cycle. It is hypothesized that this lengthening of f can be correlated to da/dN or dA/dN as well, as the available energy for crack growth per cycle is divided over da and f.

3 Examples from literature

Two examples from literature are shown. The first contains data from fatigue tests, while the second presents a common analytical corner crack model.

3.1 Corner cracks in PMMA

There is ample literature on fatigue crack growth rate da/dN versus ΔK_a data. Very few include crack front geometry data. Grandt et al. [4] discuss various corner crack and surface crack tests by Snow [5] where the crack front geometry was recorded during the test. An example is given in **Fig. 2**. From this data both a and A can be obtained. The number of analysed crack fronts and data points per front are rather limited, but it provides insight into the change of a and A during the growth of a corner crack.

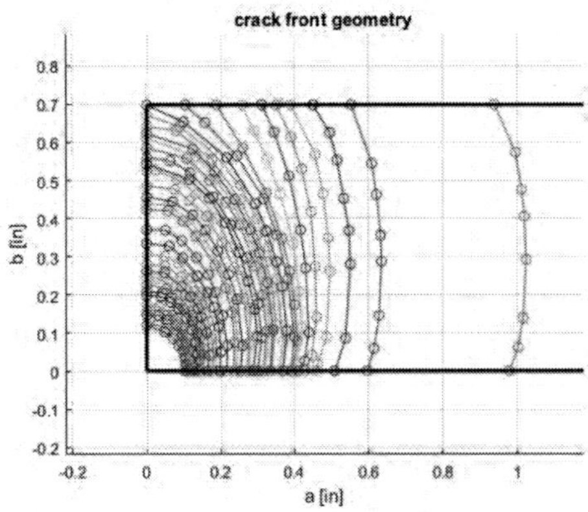

Fig. 2. Corner crack geometry of several crack fronts of Test 6 by Snow [5]. Measurement points belonging to identical crack fronts have been joined by lines for clarity.

The corresponding da/dN versus ΔK_a plot is given in **Fig. 3**. The scatter in the corner phase is evident, and when the crack becomes a through crack, the curve becomes a smooth power law. From textbook crack growth curves, one would expect a through crack to have a steeper curve during small crack growth, starting at a ΔK_{th} (e.g. Schijve [6], Fig. 8.6). For the corner crack shown in **Fig. 3**, the opposite is seen.

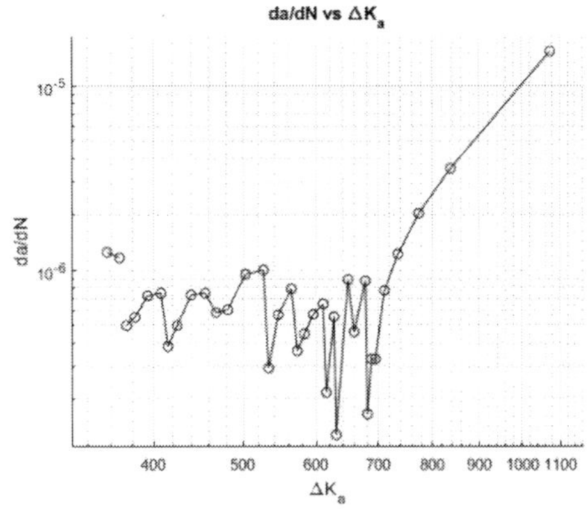

Fig. 3. da/dN versus ΔK_a belonging to Test 6 by Snow [5].

3.2 Corner crack model of Newman and Raju

A simple model of a corner crack is presented by Newman and Raju [7]. A quarter ellipse corner crack is modelled, after which it becomes instantly a through crack. There is no transition phase between the corner- and through crack phases, and the model is based on the crack length throughout all phases; which gives the false impression that the da/dN curve is smooth. **Fig. 4** shows this transition at $a/T = 0.675$. This model is of interest, however, because a good comparison example is given with experimental data of Hsu et al. [8].

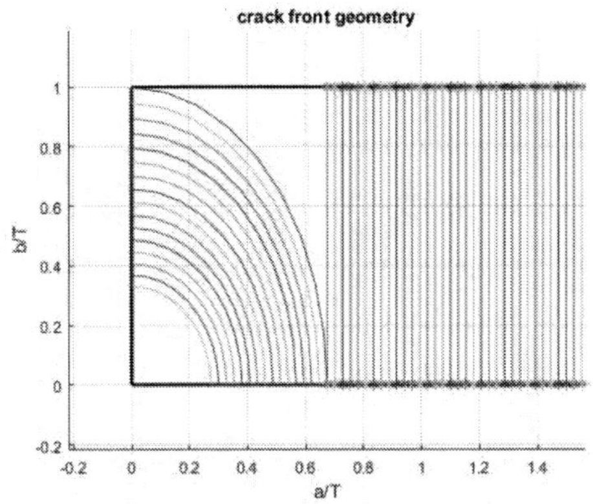

Fig. 4. Model of a corner crack. Redrawn from Newman and Raju [7]. With a as base parameter, da/dN appears smooth while there is a large, unrealistic step in A at $a/T = 0.675$.

4 Numerical modelling

Two numerical models are presented here that are used to predict crack front geometries during crack growth.

4.1 Corner crack and through crack comparison model

The premise here is that dA/dN versus ΔK_A is a power law. A could be a more suitable parameter for comparison of several crack types, and it would still compare with da/dN versus ΔK_a for the through crack.

A model was created which simulates a corner crack, together with an equivalent through crack. Although the appearance of the corner crack might be similar to the previously discussed Newman-Raju model, it is partially more detailed as the three phases are modelled separately. By modelling the transition region with a smooth dA/dN, the unrealistic behaviour of the Newman-Raju model (a significant dA between two cycles, both at identical a) is largely mitigated. Furthermore the crack length a or crack growth ratio da/dN is not used as base parameter, but dA/dN is.

The corner crack is modelled as a quarter ellipse crack front, whose aspect ratio (minor axis over major axis) γ is varied exponentially by a factor β during growth to mimic crack behaviour seen in literature and in reality. Per iteration N, γ is multiplied by β^N. The area of a quarter ellipse area is $A = \pi ab/4$. With $b = a\gamma$, reworking gives $a = \sqrt{4A/(\pi\gamma)}$, and a and b are obtained. Then b is calculated using a and γ. Input consists of specimen width, thickness, and the stress range[b] ΔS. The values of dA/dN and β are given, and the model grows a corner crack until it becomes through-thickness; at $b \geq T$.

For the transition phase, a given number of cycles is chosen such to represent real crack front development when transforming a corner crack into a through crack. The dA/dN power law to drive the area growth is still used, while graphically the crack is deformed from a quarter ellipse to a straight line, by flattening the elliptical curve gradually into a straight line, i.e. it takes account of the growth of dA and A but does only approximate the crack front development graphically.

The model then also simulates a true through crack for a given number of cycles, matching up with the through crack phase from the corner crack model.

For two cases, each containing a corner and a through crack, the da/dN versus ΔK_a and dA/dN versus ΔK_A are obtained, and discussed below. An example of the crack geometry of the first case is given in **Fig. 5**.

[b] Strictly speaking, ΔS is a constant here, and can be set to unity. As the model does not take into account the applied stress or load cycles, an R-effect is not present.

https://doi.org/10.1051/matecconf/201816513009

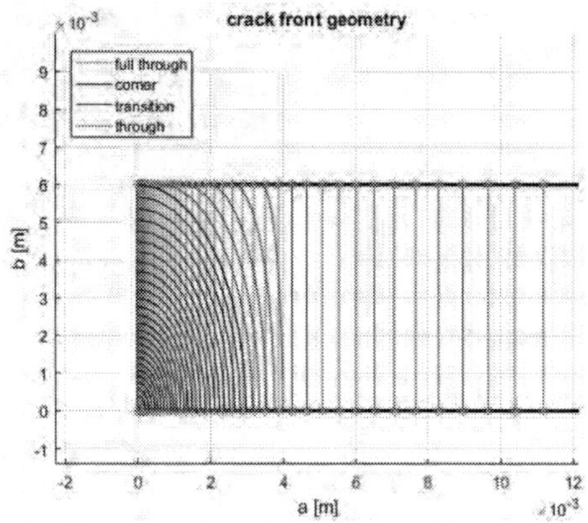

Fig. 5. A corner crack in three phases, with an overlaid true through crack, both following the same dA/dN power law.

The da/dN data from this model is presented in **Fig. 6**. It is not unreasonable as it correlates to real results, e.g., **Fig. 2**. When looking at the dA/dN data in **Fig. 7**, the values are found to be continuous and increasing all the time. There is some slight deviation from the linear power law behaviour in this log-log plot, in the very beginning of the crack growth. This is because the power law[c] $f(x) = ax^b + c$, has $c \neq 0$ because of a finite starting crack area present. Nevertheless it is evident that various crack geometries can be compared much better when based on A instead of a.

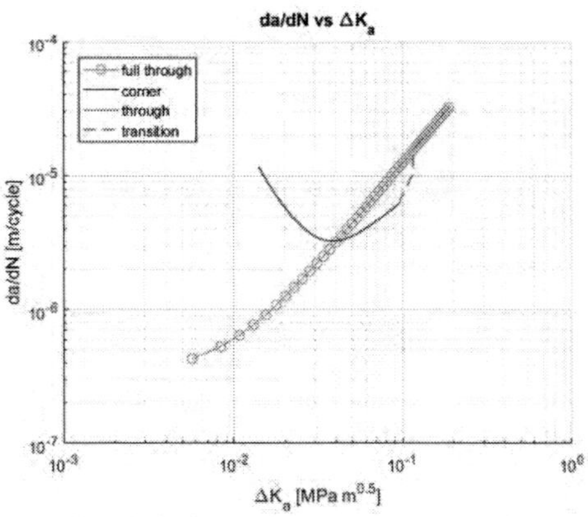

Fig. 6. The standard da/dN versus ΔK_a curve of this model. Note how irregular the corner and transition phases behave, while the true through crack is faithful to the power law.

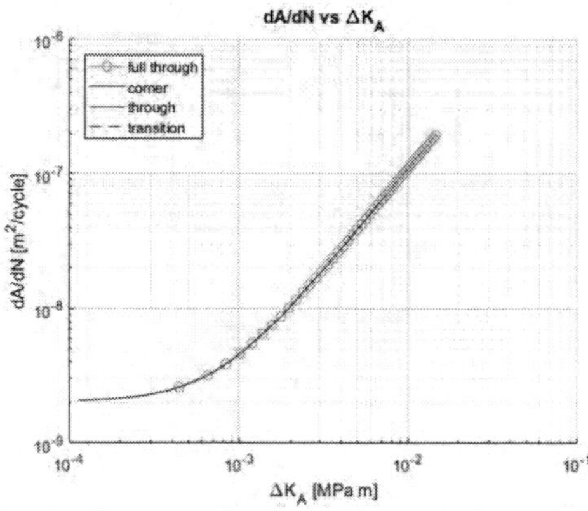

Fig. 7. When presenting the crack growth data as dA/dN versus ΔK_A, both crack types predictably collapse onto the same power law (since it was programmed to follow this relation), even though they are geometrically different.

Slight changes in starting parameters (mainly constant β, which acts as an exponential function on γ), can alter the ellipse aspect ratio γ during crack growth such that the da/dN actually decreases during the corner crack phase. It increases again through the transition and through crack phases. This is the second case. An example of that is given in **Fig. 9**, together with the crack front geometry in **Fig. 8**. The dA/dN data of this case is equal to the data shown in **Fig. 7**.

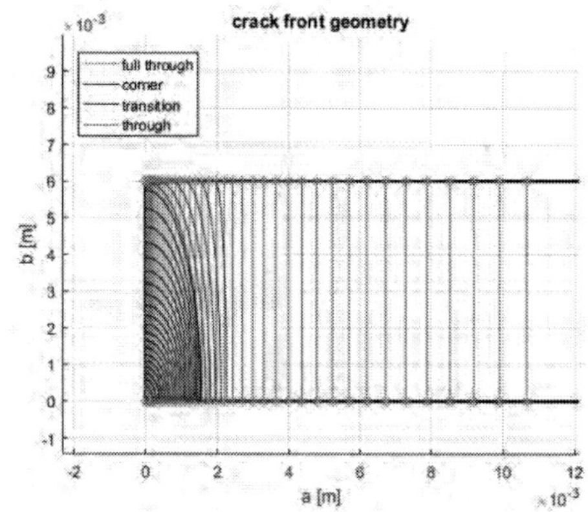

Fig. 8. A corner crack in three phases, with an overlaid true through crack, both following the same dA/dN power law.

---

[c] In this example equation, a, b, and c, are generic constants, with x a generic variable.

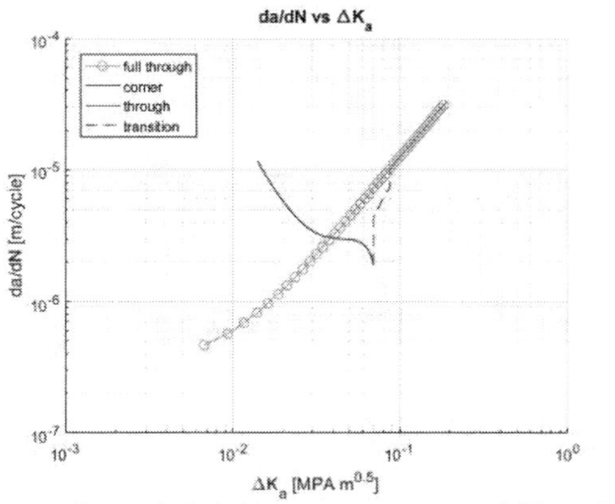

Fig. 9. The standard da/dN versus ΔK_a curve of this model. Note how irregular the corner and transition phases behave, while the true through crack is faithful to the power law.

4.2 Cellular Automaton

A cellular automaton procedure was programmed to predict crack front growth, based on unpublished work by Conen [9]. The specimen cross-section is modelled as a matrix. Every iteration N, all indices of a probability matrix are updated based on certain surrounding indices, and a binary version of this matrix is stored, indicating which part of the cross-sectional area is still solid and which part has disappeared (cracked). The crack fronts are generated every N, and as such do not follow any growth law. A crack growth relation is introduced separately to read out the correct crack fronts. A power law relation on dA is chosen to obtain the growth intervals. From the data also the crack length a is determined, to calculate ΔK_a. **Fig. 10** shows an example of a grown corner crack. (Note the similarity with **Fig. 9**). **Fig. 11** gives the corresponding da/dN versus ΔK_a.

Given the nature of the simulation, da/dN here appears also realistic, but given the model resolution, less smooth.

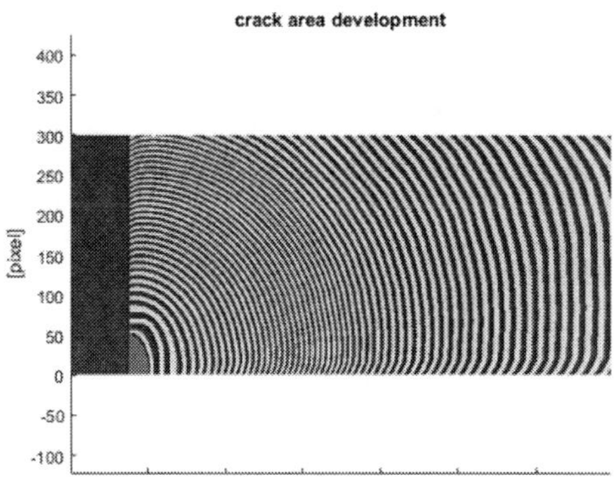

Fig. 10. Cellular Automaton has grown a corner crack, from left to right. Half of the centre hole is visible as the dark rectangle on the left, and the initial crack in turquoise. Dimensions in pixels.

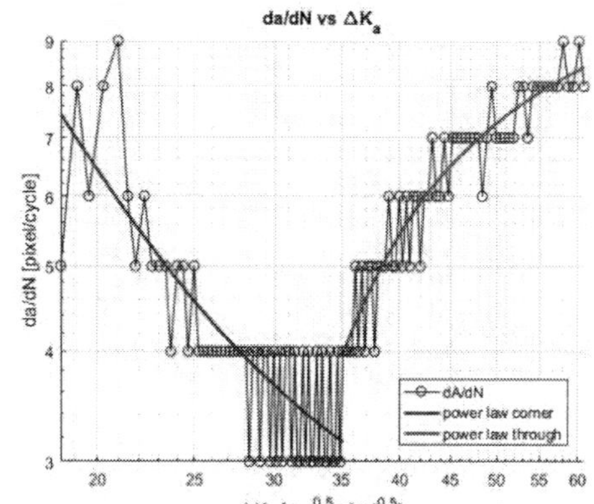

Fig. 11. The standard da/dN versus ΔK_a curve of this model. The corner phase clearly has a decreasing growth rate, an only after becoming through-thickness it assumes a normal growth behaviour.

And again, when plotting the growth with A as base parameter, a smooth power law shows up: **Fig. 12**.

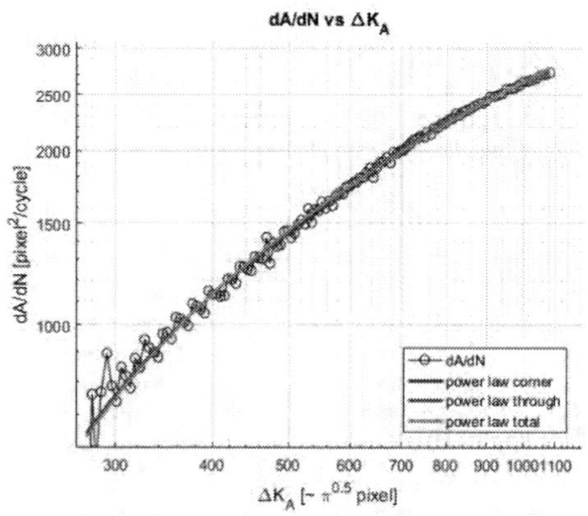

Fig. 12. When showing the crack growth data as dA/dN versus ΔK_A, both crack phases connect with nearly equal slopes The complete crack follows a power law relationship neatly.

To show the power of the cellular automaton, consider the example crack growth in **Fig. 13**. A corner crack is present, as well as a single slit radiating perpendicular from the centre hole. The zebra-striped pattern shows clearly how the cracks start to grow independently, and then link up to form one crack front.

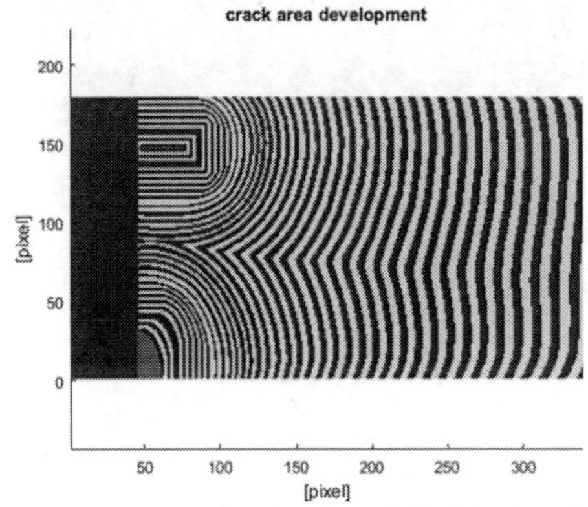

Fig. 13. A corner crack and a slit crack (both in turquoise) are grown. Note how the crack fronts join and become one through crack.

5 Crack front length

Another parameter often overlooked in (small) crack literature is the crack front length f. There is a marked increase and decrease of f in respectively the corner and transition phase. Energy is used to increase the crack surface area along two dimensions here, which also modifies f. For an ideal through-crack, only the surface area is increased, and f is constant.

When plotting the crack front length development for the literature case of Grandt et al. [4] in **Fig. 14** again a power law relationship is found for the corner phase. The decrease in the transition phase is small, and not enough data points are available to make reliable curve fit estimations. The power law is not a surprise, as it follows from the power law dependency of dA/dN.

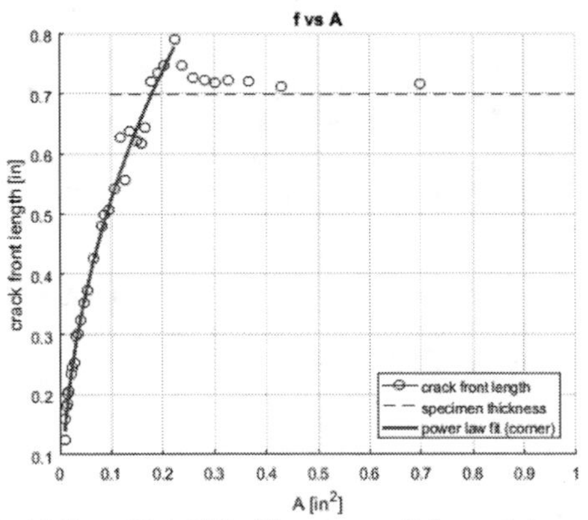

Fig. 14. Snow Test 6 [5]. f increases until the crack becomes a through crack, after which it decreases to a constant value. This end value is slightly larger than the specimen thickness since the crack front is slightly curved. Some data scatter is present.

Evaluating the crack front lengths of the cases presented in **Section 4.1**, very good power law fits are found for both the corner and transition phases, as shown in **Fig. 15** and **Fig. 16**.

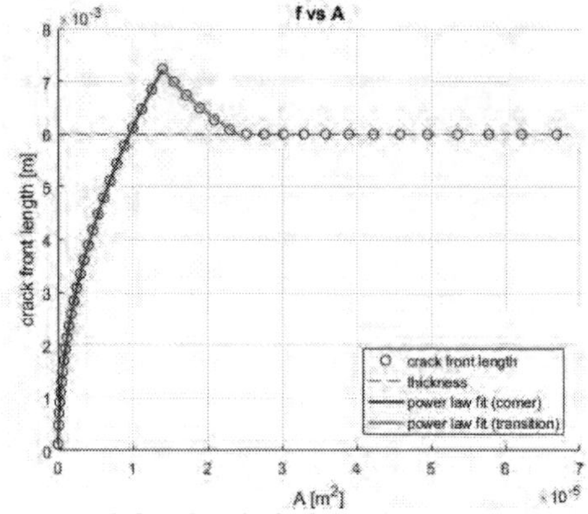

Fig. 15. Crack front length plot belonging to **Fig. 5**. f increases with a near perfect power law fit until becoming a through crack. Beyond that, f decreases to the specimen thickness (straight through crack), again following a power law.

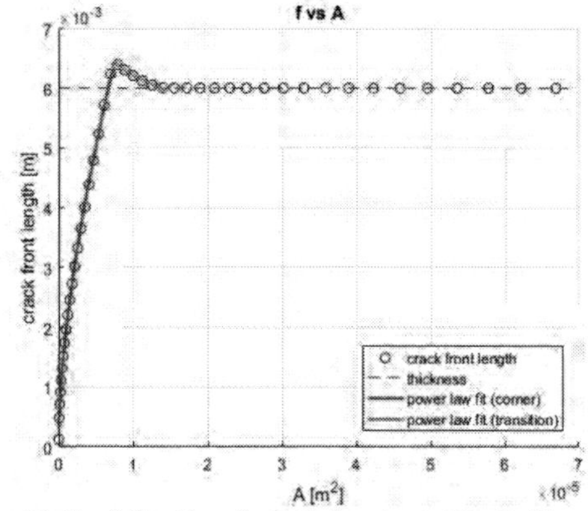

Fig. 16. Crack front length plot belonging to **Fig. 8**. f increases with a near perfect power law fit until becoming a through crack. Beyond that, f decreases to the specimen thickness (straight through crack), again following a power law.

This power law relation for f versus A also holds in the cellular automaton simulation, see **Fig. 17**. Only here a small step in absolute value is seen when the crack goes 'around the corner'. The crack is not exactly a quarter-ellipse, so some crack length is lost in the very corner at $b = T$. This is visible from the yellow markings in the top left corner in **Fig. 10**. Beyond the corner phase, the transition phase asymptotically nears a real through crack phase, but again with a good power law fit.

https://doi.org/10.1051/matecconf/201816513009

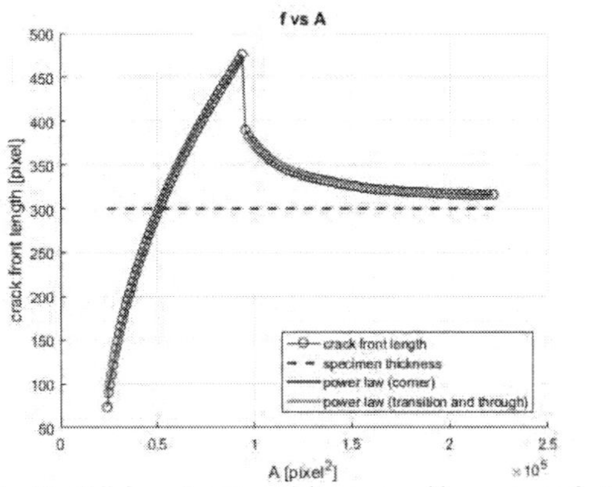

Fig. 17. Cellular automaton: f increases with a near perfect power law fit until becoming a through crack. Beyond that, f decreases asymptotically to a value slightly larger than the specimen thickness (curved through crack), again following a power law. See also **Fig. 10**.

6 Discussion

The figures from Grandt et al. [4] are very illustrative. However, these cracks were grown in PMMA material, a polymer. Most fatigue crack growth data in literature is gathered from metals (aluminium/steel/titanium), which tend to be more isotropic than a polymer. Nevertheless these results are very similar to the results found in metals. The crack growth mechanism for cracks at this scale might not be very sensitive to anisotropic material structures.

Given the behaviour of da/dN versus ΔK_a as seen in **Fig. 6** and **Fig. 9**, the use of a Paris relationship for a corner crack is incorrect. The model of the author (**Section 4.1**) does not show such a change in slope when using dA/dN versus ΔK_A, see **Fig. 7**. However, introducing a different crack growth rate power law for the small crack regime in this model does not make it exactly match up with corner crack growth rate curves either, although close. The geometry effect can significantly affect the crack growth rate in the small crack growth regime.

Furthermore, consider a surface crack inside a hole in a plate. Such a crack also grows through the three phases. In the first phase, by definition, a is constant as the crack first has to grow to a through-thickness crack. In the standard da/dN versus ΔK_a plot, $da/dN = 0$ here, such that these points cannot be shown on the log-log plot. The ΔK_a parameter is constant during this phase, while the crack is growing. This shows up as a ΔK_{th}, while the second dimension is not present in this similitude parameter.

The cellular automaton is an interesting method, in the sense that it has absolutely no physical connection with fatigue crack growth, yet it produces eerily similar crack growth geometries. Two major arguments can be made against using this method:

- The growth dA/dN is artificially and a-priori introduced using a power law. The automaton grows crack fronts, but lacks a coupling with the number of cycles. The relevant crack fronts are found using an area-related relationship.
- The growth rate along a single crack front is not constant, but the variation in local stress intensity factors, linked to local growth rate, seems to be smaller than observed in reality.

Regarding the latter argument, the Newman-Raju model and the author's model correct slightly for this by changing γ of the ellipse during growth, but do limit the shape to the ellipse. The cellular automaton tends to make crack fronts slightly too curved or too much alike a circular arc compared to reality, but captures the transition region wonderfully realistic.

While these arguments are valid concerns, the results do appear similar to geometries observed in practice and reported in literature.

3 Conclusions and recommendations

The choice of crack length a as base parameter for fatigue crack growth measurements is questionable for crack types other than through cracks. It is shown that corner cracks behave differently in the small crack growth region. The crack surface area A is shown to be a more suitable parameter, as the results for both through cracks and corner cracks are now similar in magnitude and slope on a graph with log-log scales, making comparisons easier. This change in base parameter affects the similitude parameter too. The common da/dN versus ΔK_a plot can then be transformed into an equivalent dA/dN versus ΔK_A.

It is shown that in a quarter-elliptical corner crack, the crack front length during the corner phase grows along a similar power law behaviour as introduced for dA. The energy used for crack growth is basically split into crack area increase and crack front length extension.

It is shown that a corner crack has three distinct phases; corner, transition, and through crack. The crack length growth rate behaves markedly different from a power law during the corner phase, and the transition phase links the corner and through phases together.

The cellular automaton approach can create great insight in the behaviour of the crack growth, especially when unique and/or multiple initial cracks are present. Although the automaton routine has no physical link to fatigue crack growth, it can be used as a prediction tool for fatigue crack growth geometry, especially when multiple starter cracks are present.

As there is scarcely crack growth data available which also includes the crack front geometry development, more fatigue tests are needed with high accuracy measurements of the crack front geometry. This has proven to be difficult but not impossible, and warrants further study to create data sets to compare the models with.

Another recommendation is to include more fatigue crack types into the models, to see if the results for the corner crack versus through crack also hold for these types.

This research was carried out under project number S21.5.15581 in the framework of the Partnership Program of the Materials innovation institute M2i (www.m2i.nl) and the Technology Foundation STW (www.stw.nl), which is part of the Netherlands Organisation for Scientific Research (www.nwo.nl).

References

1. R.C. Alderliesten, *How proper similitude can improve our understanding of crack closure and plasticity in fatigue*, Int. J. of Fatigue, **82** 263-273 (2016)
2. R.C. Alderliesten, *How proper similitude could have improved our understanding about fatigue damage growth,* Proc. 28th ICAF Sym, 47-57 (2015)
3. E. Amsterdam, F. Grooteman, *The influence of stress state on the exponent in the power law equation of fatigue crack growth,* Int. J. of Fatigue, **82** 572-578 (2016)
4. A.F. Grandt, J.A. Harter, D.E. Tritsch, *Semielliptical cracks along holes in plates and lugs,* AFWAL-TR-83-3043 (1983)
5. J.R. Snow, *A stress intensity factor calibration for corner flaws at an open hole,* AFML-TR-74-282 (1975)
6. J. Schijve, *Fatigue of Structures and Materials* (2009)
7. J.C. Newman, I.S. Raju, *Prediction of fatigue crack-growth patterns and lives in three-dimensional cracked bodies,* Proc. 6th Int. Conf. on Fracture (ICF6), 1597-1608 (1984)
8. T.M. Hsu, W.M. McGee, J.A. Aberson, *Extended study of flaw growth at fastener holes*, AFFDL-TR-70-149 (1978)
9. M. Conen, *Appendix B : Cellular Automator,* unpublished work (2008)

Influence of static strength on the fatigue resistance of welds

Ceferino Steimbreger[1,*], *Nenad* Gubeljak[2], *Norbert* Enzinger[3], *Wolfgang* Ernst[4], and *Mirco* Chapetti[1]

[1] Laboratory of Experimental Mechanics (LABMEX), INTEMA (Institute for Material Science and Technology), CONICET - University of Mar del Plata, J.B. Justo 4302 (B7608FDQ), Mar del Plata, Argentina
[2] Faculty of Mechanical Engineering, University of Maribor, Slovenia
[3] Institute for Materials Science and Welding, Graz University of Technology, Kopernikusg. 24 (8010), Graz, Austria
[4] Voestalpine Stahl GmbH, Linz, Austria

Abstract. The present paper deals with a fracture mechanic approach that employs the Resistance-Curve concept, in order to predict fatigue endurances of welded components, with different tensile strengths of the base metal. The Resistance-Curve method compares the total driving force applied to a crack with its threshold for propagation, both defined as a function of crack length. The former depends on load scheme and weld geometry and it can be obtained from finite element analyses, while the second is inherently related to weld resistance. Results obtained herein showed that threshold curve shape is changed when static strength of the base material is modified. Consequently, its interaction with the driving force differed, giving raise to different fatigue endurances for various values of the tensile strength. However, this effect is only likely to be leveraged, provided that the initial crack length is small enough. In real welded structures, the presence of defects demands longer initial crack lengths to be used in calculations, at which the benefit of enhanced strength is minimised or even inverted. Moreover, at these lengths, the growing process is mainly controlled by weld geometry and long crack propagation threshold, whereas local properties become less important in fatigue limit prediction.

1 Introduction

In the last decades, industries have invested much effort and time in developing safe structures with minimum costs. Advances in materials and inspecting techniques have led to lighter structures that safely withstand design loads, including either static or cyclic stresses. Among these materials, high strength steels (HSS) with yield strength over 600 MPa have particularly caught the attentions of companies and researchers.

It is normally believed, that tensile strength of base material does not have an influence on final fatigue resistance of welded components [1-5]. On the other hand, short crack behaviour is known to determine the fatigue limit of metals, which is therefore affected by microstructure [6-8]. Furthermore, fusion welding processes alter base material (BM) microstructures and may lead to harder regions, presenting local properties with enhanced plain fatigue limit in the heat affected zone (HAZ). Considering these facts, it can be thought that an increase in the tensile strength of the BM would lead to improved fatigue resistance. However, in this analysis, initial crack length and overall weld geometry is not being considered. As-welded joints are likely to present defects that range in size from some μm to 150 μm or even more [1-3]. The effect of residual stresses, inhomogeneity and stress concentrators that usually appear in welds (slag inclusions, undercuts, overlaps, among others) essentially overwhelm the influence of other factors, such as the static strength of the main plate. Therefore, enhancing this mechanical parameter may not improve fatigue strength of the joint. This statement can be questioned by observing limitation of material yield strength in current fatigue design documents, to stress up to ca. 960 MPa [4, 9, 10]. Although this is mainly related to a lack of experimental data for ultra high-strength steels, subject is still at open debate, and much work is needed for better understanding of the damaging mechanisms and controlling variables in this kind of materials.

2 Methodology

Methodology employed in the present research to predict fatigue strength of welded components is thoroughly described in previous publications from the authors [11-13]. It is based on the resistance curve method that compares the total driving force applied to a crack with its threshold for propagation, resulting in the effective driving force. The latter is the energy necessary for crack growth.

The applied driving force depends on load scheme, geometry and crack length. It can be properly described by the applied stress intensity factor range, ΔK. Likewise, crack propagation threshold, ΔK_{th}, is a function of crack size and in the present approach it includes the short crack regime. Relationship between

* Corresponding author: cefesteim@gmail.com

these quantities can be expressed as a modified Paris Law, like Eq. (1).

$$da/dN = C\,(\Delta K - \Delta K_{th})^m \tag{1}$$

where C and m are constants that depend on material and environment. Note that in the case of fatigue limit assessment the value adopted by constants in Eq. (1) is irrelevant, since for da/dN approaching to zero only equality between ΔK and ΔK_{th} defines the critical stress range.

Chapetti proposed a method to calculate crack propagation threshold [14] based on location d of the strongest microstructural barrier, which defines a microstructural threshold for short crack propagation. This variable can be expressed as follows:

$$\Delta K_{dR} = Y\,\Delta\sigma_{eR}\,(\pi d)^{1/2} \tag{2}$$

where Y is a geometrical factor and $\Delta\sigma_{eR}$ is the plain fatigue limit, described as the nominal stress range below which cracks do not propagate in a smooth sample. Since it depends on the stress ratio, R, the microstructural fatigue threshold also does. The value of microstructural barrier, d, is usually determined by ferrite grain size and bainite or martensite lath length [8, 14].

In the long crack range, threshold adopts a constant value, represented by ΔK_{thR}, for a given stress ratio R. Difference between this mechanical threshold for long cracks and microstructural fatigue threshold, ΔK_{dR}, gives ΔK_{CR}, according to Eq. (3).

$$\Delta K_{CR} = \Delta K_{thR} - \Delta K_{dR} \tag{3}$$

where ΔK_{CR} is constant and depends on the stress ratio R. In order to connect ΔK_{dR} to ΔK_{thR} a transition region should be defined. Chapetti suggested [14] that the development of the extrinsic component ΔK_C can be calculated with Eq. (4).

$$\Delta K_C = \Delta K_{CR}\,(1 - e^{-k(a-d)}) \tag{4}$$

where k is a material constant that controls the shape of the transition zone for each stress ratio, and a is the crack length in mm, measured from the free surface.

Finally, the shape of a single threshold curve that describes the resistance of a material to crack growth can be described by Eq. (5).

$$\Delta K_{th} = \Delta K_{dR} + \Delta K_C = Y\,\Delta\sigma_{th}\,(\pi a)^{1/2} \tag{5}$$

Replacing Eq. (3) and (4) into (5), the following form of the threshold curve is obtained:

$$\Delta K_{th} = \Delta K_{dR} + (\Delta K_{thR} - \Delta K_{dR})(1 - e^{-k(a-d)}) \tag{6}$$

which is valid for $a \geq d$.

It was found that that the value of k in Eq. (7) gives a threshold for fatigue crack propagation that correlates well with experimental data [14].

$$k = \Delta K_{dR} / [4d\,(\Delta K_{thR} - \Delta K_{dR})] \tag{7}$$

3 Finite element model

A finite element software [15] was employed to simulate an 8 mm butt-welded joint, having a reinforcement angle, α, of 147°. Two-dimensional linear elastic model was assumed and symmetry was considered in order to improve simulation performance. Figure 1 shows symmetric model and it schematically displays load configuration, boundary conditions and overall geometry. Four point bending was considered. Minor and major span are represented by s and L, respectively, t is the plate thickness and P is the applied load. All these variables define the maximum nominal stress on the surface that is used in the following calculations.

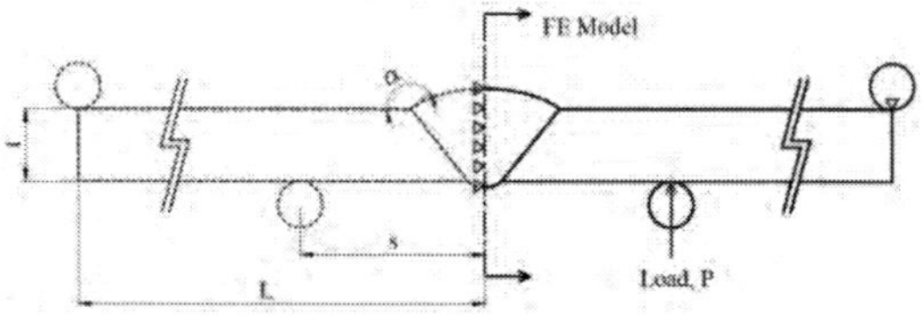

Fig. 1. Geometry, load configuration and boundary conditions of the finite element model. t is the plate thickness, α is the reinforcement angle, s is half the minor span, L is half the major span and P is the applied load.

Stress intensity factors were obtained following software procedure for fracture mechanic simulations [15]. Cracks were introduced as "seam cracks" growing from the weld toe and the maximum energy release criterion was selected to determine the crack propagation direction. Concentric circular partitions were done from the crack tip, and the area defined by the first circle was considered to be the crack front, which will later be computed as the first contour integral. Mesh was constructed with 6-node quadratic plane strain triangles in the crack front, which use a modified second-order interpolation, whereas 8-node biquadratic plane strain quadrilateral elements were assigned to the rest of the mesh. The software converts the elements in the crack front to collapsed quadrilateral elements. In order to improve accuracy, a square root singularity is recommended to be assigned to the crack tip, which constrains the collapsed nodes to move together.

It must be mentioned, that the analysis was purely geometric and residual stresses resulting from the welding process were not considered at this first instance.

Stress intensity factor depends on weld geometry, crack length and applied remote stress. Having defined the weld detail and load in the bending test, their values can be obtained for different crack lengths by means of the finite element analysis described previously. This procedure was repeated for 17 different crack lengths ranging from 20 μm to 2.5 mm, following a simplified path perpendicular to the plate surface, i.e. normal to the maximum nominal stress. Step size was varied from 25 μm to 100 μm in the first millimetre, in order to obtain an accurate profile in the short crack range. For larger crack lengths, 0.5 mm was selected.

Having calculated ΔK vs. a for a predefined $\Delta\sigma$ and a propagation threshold curve, ΔK_{th}, fatigue limit can be

Table 1. Mechanical properties and important parameters for the model. $R = 0.1$.

	$\sigma_{ys0.2}$ [MPa]	σ_{UTS} [MPa]	ΔK_{thR} [MPa√m]		$\Delta\sigma_{eR}$ [MPa]		d [μm]	H_{V10}	
Zone	BM	BM	BM	HAZ	BM	HAZ	HAZ	BM	HAZ
A36 [12, 13]	286	472	7.03	-	-	360	28	180	360
S960MC	990	1030	6.25*	5.8*	550	650*	20	320	380

determined by equalising Eq. (1) to zero. There is a value of $\Delta\sigma$ for which both curves touches at a single point. This is the critical stress or fatigue limit of the configuration, and the contact point between the two curves is the non-propagating crack length.

4 Materials

Materials used in the present analysis are a structural steel, A36, and a high strength steel S960MC, in accordance with standards ASTM A36 [16] and EN 10149-2 [17], respectively. Welds were manufactured with gas metal arc welding (GMAW) in the structural steel and laser hybrid welding (LHW) in the HSS.

Mechanical properties and important variables for the model are displayed in Table 1. In the HSS, some parameters had to be estimated as described below. They are indicated with an asterisk in Table 1. It is also shown the relevant zone associated to each parameter. Hardness values were measured at the HAZ, close to the weld toe, where the cracks are likely to grow. Similarly, value of d was determined in the same region as the average size of martensite laths, in accordance with previous studies [8, 14].

The plain fatigue limit was estimated with the properties of the HAZ. Assuming that fatigue limit is proportional to hardness, ratio of HV between HAZ and BM approximates the increase in the fatigue limit in the former. Since the average HV in the HAZ is around 380 HV_{10}, and 320 HV_{10} in the BM, and the experimental fatigue limit at $R = 0.1$ in the BM is $\Delta\sigma_{eR} = 550$ MPa (see Table 1), then $\Delta\sigma_{eR} \approx 650$ MPa in the HAZ.

Long crack propagation threshold of S960MC at $R = 0.1$ was obtained from Figure 2, by means of the ultimate tensile strength of the BM, giving 6.25 MPa√m.

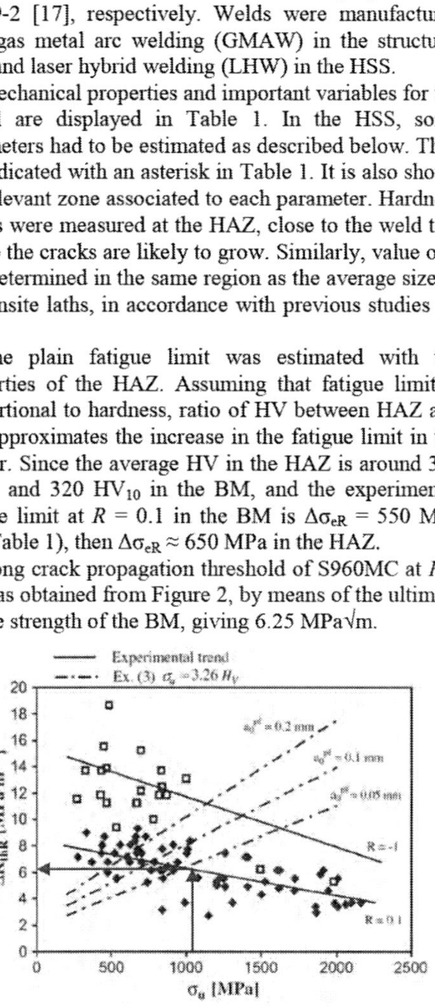

Fig. 2. Variation of long crack propagation threshold with the static strength of material [18, 19].

Due to the geometry of the weld pool and the final bead appearance, it can happen in LHW that the crack initiates in the HAZ and propagates mainly into that region. Therefore, the long crack propagation threshold should also be corrected accordingly with HAZ mechanical properties. By considering the well-known relation $\sigma_{UTS} = 3.26\ H_V$, the ultimate tensile strength in the HAZ of S960MC can be estimated. After this, Figure 2 can be used to estimate the long crack propagation threshold in the HAZ, giving a value of 5.8 MPa√m.

5 Results

At this first stage of the assessment, the weld geometry was simplified in accordance with former studies [13]. A butt-weld with toe radius equal to zero and reinforcement angle equal to 147° was used in the FE model. This procedure allows comparing previous results on structural steel [13] with the present analysis on high strength steel.

In order to compare results when variables are changed, a reference case must be assumed. Due to the reasons given before, the HAZ properties best describe the crack growing process in the S960MC weld. Then, $\Delta\sigma_{eR} = 650$ MPa, $d = 20$ μm and $\Delta K_{thR} = 5.8$ MPa√m are the reference values. Under these circumstances, prediction for two different initial crack lengths is shown in Figure 3. It is important to mention, that values adopted for $\Delta\sigma_{eR}$, d and ΔK_{thR} determine a flat threshold curve with almost no noticeable transition at low crack lengths. This implies that the fatigue limit depends mainly on the long crack propagation threshold, which is constant, and the geometry of the weld, given by the applied stress intensity factor curve.

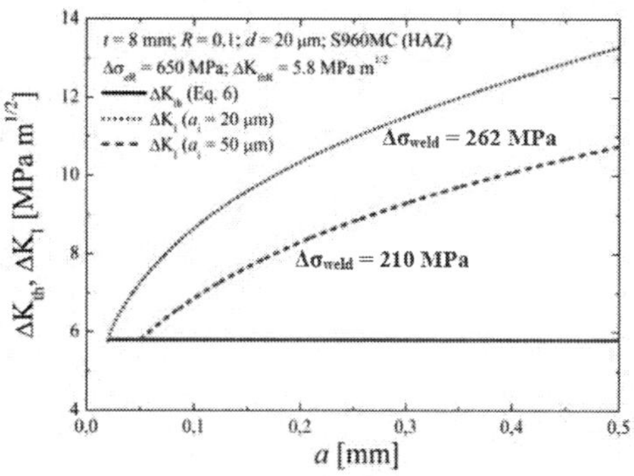

Fig. 3. Fatigue limit prediction in S960MC (HAZ), for $\Delta\sigma_{eR}$ = 650 MPa, d = 20 μm and ΔK_{thR} = 5.8 MPa√m, at two different initial crack lengths: 20 and 50 μm. R = 0.1 and t = 8 mm.

It can be argued that determination of the fatigue limit in the HAZ from hardness values is very empirical. Likewise, long crack propagation threshold from Figure 2 obeys an empirical relation. In order to assess the effect of each variable that affects the threshold curve and estimate an error when using empirical deductions, sensitivity analyses were conducted. Value of d was allowed to vary from 10 μm to 20 μm, plain fatigue limit from 550 (BM) to 650 MPa (HAZ), and long crack propagation threshold from 5.8 (HAZ) to 6.25 MPa√m (BM).

5.1 Effect of the strongest microstructural barrier, *d*.

Figure 4 illustrates variations the threshold curve for three values of *d*. It can be seen that shorter microstructural barriers give lower fatigue limits in the welds, when initial crack length is small enough to be affected by these changes. Note that plateau is practically developed in all cases for crack lengths above 0.1 mm, where the long crack propagation threshold controls the growing process.

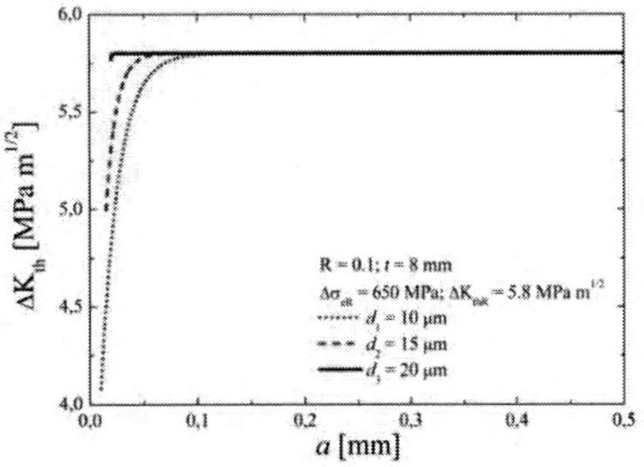

Fig. 4. Change in threshold curve with variation of the strongest microstructural barrier.

5.2 Effect of the plain fatigue limit, $\Delta\sigma_{eR}$.

The effect of the plain fatigue limit of the material at the toe region can be analysed, keeping *d* and ΔK_{thR} constant. This behaviour can be seen in Figure 5, in which a minimum plain fatigue limit corresponding to the BM (experimentally determined) was employed. The highest value of 650 MPa was calculated by relating hardness with fatigue strength values.

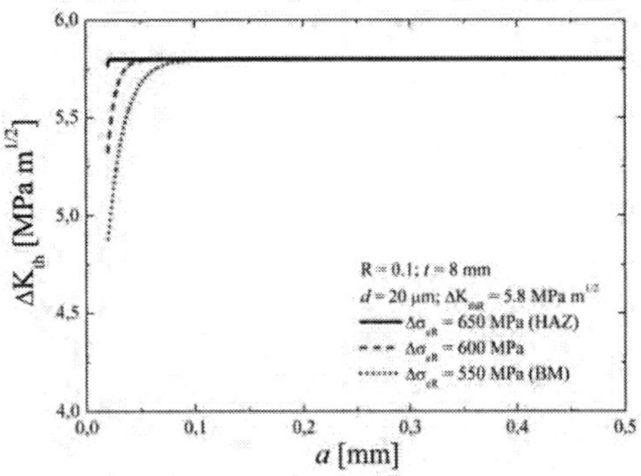

Fig. 5. Change in threshold curve with variation of the plain fatigue limit at the toe region.

5.3 Effect of the long crack propagation threshold, ΔK_{thR}.

Finally, the change in the threshold curve with variation in ΔK_{thR} can be assessed. This is represented in Figure 6, where an increase in the plateau ordinate can be related to a higher ΔK_{thR}, as it was expected.

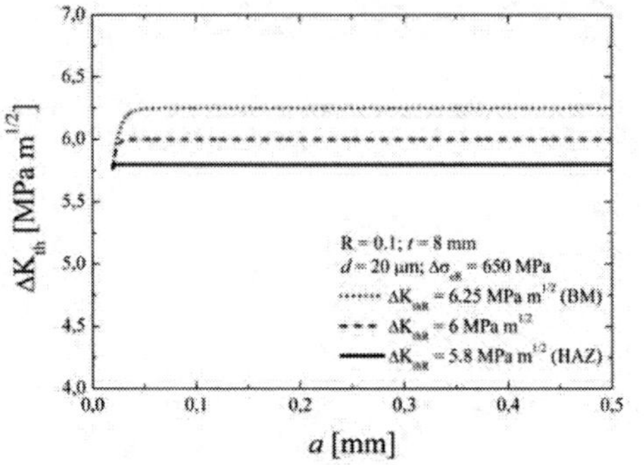

Fig. 6. Change in threshold curve with variation of the long crack propagation threshold.

5.4 Prediction of the fatigue limit of welded joints.

Results showed in Figures 4 to 6 can be used to predict the fatigue limit in each case, for the weld under consideration, and for different initial crack lengths. This is achieved by comparing threshold curves with the applied stress intensity factor curve, as shown in Figure 3. Figure 7 displays a Kitagawa-Takahashi-like chart, showing the effect of the strongest microstructural barrier, *d*, on the predicted fatigue strength of the weld. It is evident that *d* only influences results when $a_i < 50$ μm, a value that is very short to be used in real weld calculations. Additionally, it can be seen that long crack propagation threshold defines the fatigue limit for longer initial crack length.

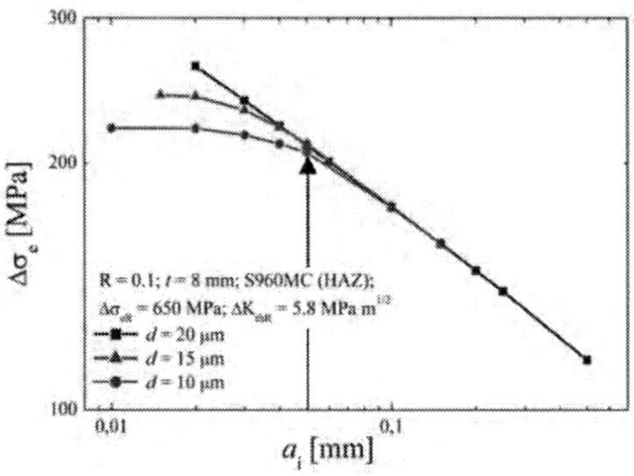

Fig. 7. K-T plot, showing the effect of the strongest microstructural barrier, *d*, on the fatigue strength of the HSS joint.

Figure 8 displays the K-T diagram when plain fatigue limit is varied. Results are similar as before, and the influence of this variable is only noticeable when $a_i < 50$ μm.

 https://doi.org/10.1051/matecconf/201816513010

A similar procedure can be done with ΔK_{thR}, when $\Delta\sigma_{eR}$ = 650 MPa and d = 20 μm are held constant. This situation measures the influence of variation in the long crack propagation threshold, on fatigue strength prediction. Figure 9 illustrates the case. It is important to note that difference in results is little, and it presents a maximum discrepancy close to a_i = 50 μm, above which it starts to decrease with crack length. It must be highlighted, that in the log-log scale, the three curves tend to be parallel with increasing initial crack lengths.

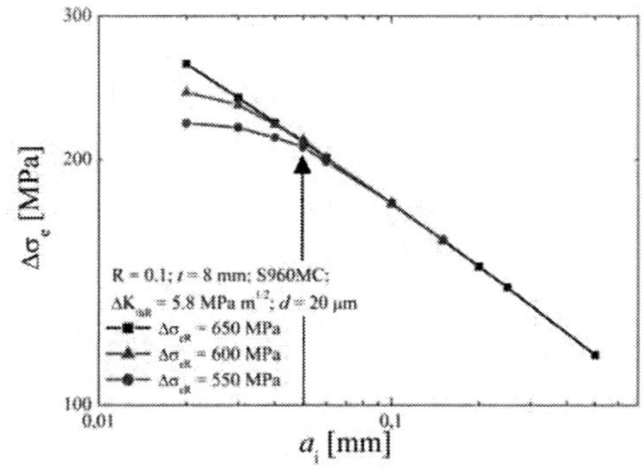

Fig. 8. K-T plot, showing the effect of the plain fatigue limit at the toe region, $\Delta\sigma_{eR}$, on the fatigue strength of the HSS joint.

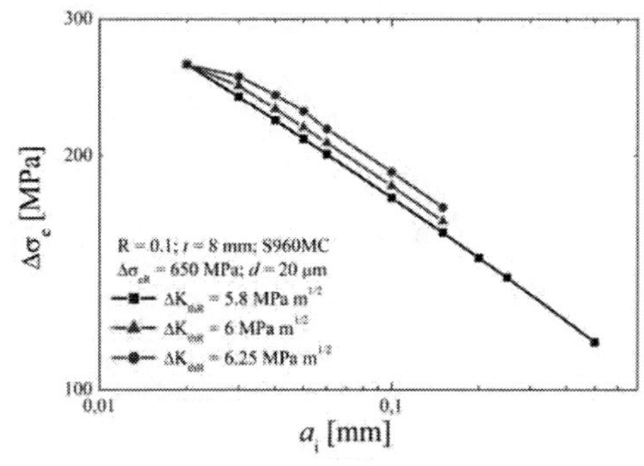

Fig. 9. K-T plot, showing the effect of long crack propagation threshold, ΔK_{thR}, on the fatigue strength of the HSS joint.

6 Analyses and discussions

An increase in the BM strength reduces the long crack propagation threshold, as it is shown in Figure 2. At the same time, a reduction in ΔK_{thR} results in slightly lower fatigue limits, depending on the initial crack length, as it is depicted in Figure 6 and 9. Regarding initial crack length, it is important to consider that codes and standards usually recommend at least 50 μm for integration. Common values are around 100 and 150 μm [4, 9, 20-22]. Only in the case of high quality welds, smaller lengths can be allowed.

On the other hand, an increase in tensile strength is usually accompanied by an enhanced plain fatigue limit, which consequently leads to a higher ΔK_{dR}, in accordance with Eq. (2). This is reflected in curves from Figures 4 to 6, where stronger materials present a sharp transition in the threshold curve, with a squared appearance. Consequently, fatigue limit of the welded component is less affected by plain fatigue limit and is mainly defined by weld geometry (applied stress intensity factor curve) and long crack propagation threshold. This is evident in Figures 7 to 9 for S960MC joints containing normal flaws ($a_i \geq 50$ μm).

This result is very important for assessing fatigue behaviour of welded components, since the variables that control the damaging mechanisms can be isolated and improved (if possible).

In order to clarify these concepts, three representative welds are considered. The first corresponds to the reference case previously described ($\Delta\sigma_{eR}$ = 650 MPa, d = 20 μm and ΔK_{thR} = 5.8 MPa√m), associated to S960MC HAZ. The second option is the same weld modelled with the properties of the BM ($\Delta\sigma_{eR}$ = 550 MPa, d = 20 μm and ΔK_{thR} = 6.25 MPa√m). Finally, a weld made of A36 was considered, with local properties corresponding to the HAZ ($\Delta\sigma_{eR}$ = 360 MPa and d = 28 μm) and ΔK_{thR} = 7.03 MPa√m of the BM [12, 13]. Predicted fatigue strengths of the joints are displayed in Figure 10, as a function of the initial crack length.

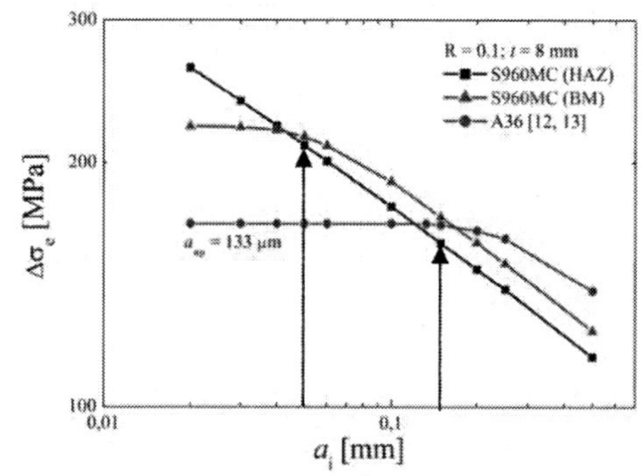

Fig. 10. K-T plot showing the fatigue strength of welds as a function of initial crack length, for three distinct cases.

If welds are performed under high quality control procedures, it is reasonable to assume an initial crack length equal to 50 μm. Under this assumption, increasing the strength of the BM leads to a higher fatigue strength of the weld.

Since the non-propagating crack length that determines the fatigue limit in the A36 steel joint is around 130 μm, then comparison between materials should be done at a longer initial crack length. At the same time, joints in the as-welded condition with scarce quality controls are usually related to longer initial crack length. By assuming an average value of 150 μm, based on recommendations in design documents [20-22], it can be seen in Figure 10 that trend is inverted and HSS welds start to become weaker than their structural steel counterparts. Particularly, for a_i = 150 μm HSS weld fatigue limit is around 160 MPa, whereas that of A36 weld is approximately 168 MPa.

In spite of this, differences between curves are not significant above 100 μm, as is depicted in Figure 11. Below this value, the effect of static strength of BM may be considered in design.

 https://doi.org/10.1051/matecconf/201816513010

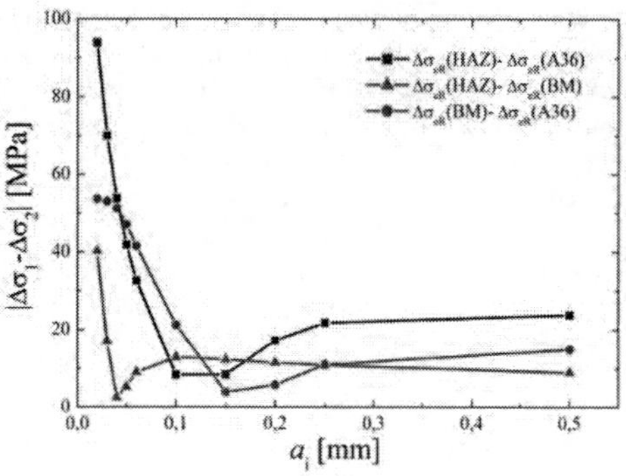

Fig. 11. Absolute differences between curves in Figure 10.

Technically, these results imply that low or medium quality joints, where flaws are likely to be found at the toe, might not experience any benefit from stronger materials, whereas high quality welds (for example, FAT values above 100) may be more susceptible to changes in BM strength. It can be further deduced that joints presenting improved weld profiles by grinding or TIG-dressing, may also experience the benefit from enhanced strength of the BM, as it was observed in different studies [2, 3, 23, 24]. These improvement techniques not only re-shape weld profile, but also reduce the amount of defects at the weld toe [25]. Therefore, smaller initial flaw sizes can be assumed.

A final comment can be done in relation to slope in the log-log scale at long crack lengths. Figures 7 to 10 show that at long initial crack length, slope in the K-T plot is constant. This implies that the fatigue limit of the weld depends mainly on the long crack propagation threshold (which is constant) and the geometry of the joint. Therefore, relation between fatigue strength of the weld and initial crack length can be deduced and it is generally described by Eq. (8).

$$\Delta\sigma_{eR} = \Delta K_{thR} \,/\, [Y\,(\pi a_i)^n] = A\,.\,a_i^{-n} \qquad (8)$$

where A is a constant that depends on geometry and long crack propagation threshold, and n is the exponent that also depends on the geometry of the weld and loading scheme (i.e. bending). Figure 12 illustrates results for the reference case (S960MC HAZ), together with the fitting curve. It was found that a power relation best fit results, in accordance with Eq. (8), presenting an exponent n around 0.25. This value can be analysed by calculating the derivative of the applied stress intensity factor curve. Figure 13 depicts the latter in the log-log scale, and the corresponding slope for the whole range of crack lengths. It can be seen in this figure, that the slope is approximately 0.25 for crack lengths up to 0.3 mm. Beyond this value, the slope starts to increase markedly.

7 Conclusions

Resistance curve concept was employed to determine the influence of the ultimate tensile strength on the fatigue behaviour of welded components. It was shown that an increase in the static strength of the BM enhances the value of ΔK_{dR} and reduces ΔK_{thR}, generating a threshold curve that presents a sharp transition above the value of d. Under these circumstances, fatigue threshold of the weld becomes less dependent on the plain fatigue limit, $\Delta\sigma_{eR}$, and it is increasingly defined by the long crack propagation threshold, ΔK_{thR}. Since the applied stress intensity factor curve depends only on geometry and loading conditions and it is not altered by changes in BM properties, fatigue strength of the welded joint is principally defined by the overall geometry and ΔK_{thR}.

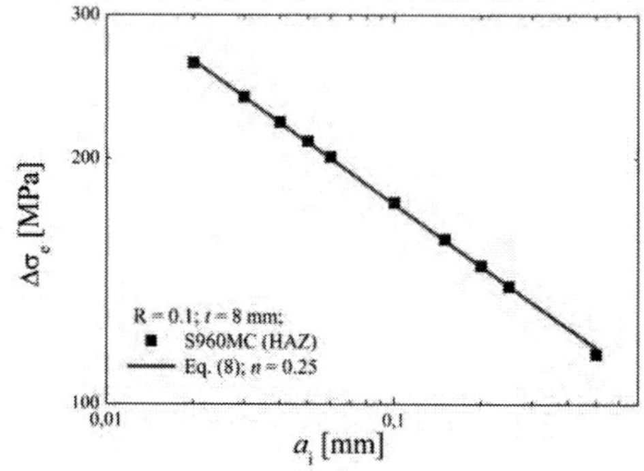

Fig. 12. Fatigue strength of S960MC (HAZ), and power fitting.

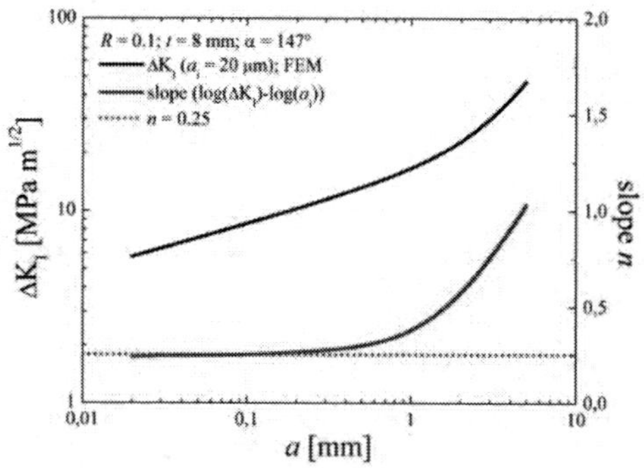

Fig. 13. Slope n of the log ΔK_I vs. log a plot.

Methodology additionally revealed that either a decrease in $\Delta\sigma_{eR}$ or a shorter microstructural barrier, d, reduced the fatigue strength of the weld, provided that the initial crack length is sufficiently small (a_i < 50 µm). However, in the presence of larger flaws, their influence becomes less important. Since the natural presence of defects in normal welds demands a larger crack length to be assumed as the initial defect, then, the possible improvement in fatigue behaviour by employing stronger materials vanishes accordingly. Furthermore, the benefit of altering ΔK_{thR} by modifying the static strength is small.

The historical believe that fatigue limit of welded components do not change with yield strength is based on experimental observations generally made on as-welded joints. Defects normally found in these samples greatly affect fatigue behaviour as was demonstrated in previous publications. Then, the influence of static properties is overwhelmed by other parameters, such as undercut depth or underfill depth. This implies that weld profile improvement methods (TIG-dressing and grinding) that remove undesired defects from the toe, can experience the benefits of employing HSS in fatigue loaded welds, as it is mentioned in literature.

For the particular case presented in this work, initial crack lengths as short as 0.1 mm are enough to hide any benefit from elevated static strength. For shorter crack lengths, the presence of non-propagating crack length in the former material cause a strong influence of the static strength. However, no standard, code or recommendation allows integration from such small flaws in normal situations.

In summary, methodology presented in the present work, although applied to particular steels, showed its usefulness for analysing fatigue limit of welded components and its dependence on certain variables. The influence of the most important parameters can be assessed in order to have a better insight of the damaging phenomenon. Control over these variables may contribute to a better performance of welded components under cyclic loading.

Authors wish to express their gratitude to the funding provided by CONICET (Consejo Nacional de Investigaciones Científicas y Técnicas), and by Agencia Nacional de Promoción Científica Tecnológica, Argentina (PICT 2013 Nro.0110). Support was also provided by the K-Project Network of Excellence for Metal JOINing which is fostered in the frame of COMET - Competence Centers for Excellent Technologies by BMWFW, BMVIT, FFG, Land Oberösterreich, Land Steiermark, Land Tirol and SFG. The programme COMET is handled by FFG.

References

1. T.R. Gurney, *Fatigue of welded structures*, 2nd ed., London: Cambridge UP (1979)
2. S.J. Maddox, *Fatigue strength of welded structures*, 2nd ed., Abington, Cambridge (1991)
3. D. Radaj, *Design and analysis of fatigue resistant welded structures*, Abington Publishing (1990)
4. A.F. Hobbacher, *Recommendations for Fatigue Design of Welded Joints and Components*, 2nd ed., Springer International Publishing (2016)
5. EN 1993-1-9:2010-12, *Eurocode 3: Design and construction of steel components – Part 1-9: Fatigue*.
6. D. Taylor, J.F. Knott, Fat. Eng. Mat. Struct. **4**(2), 147-155 (1981)
7. J. Lankford, Fat. Eng. Mat. Struct. **8**(2), 161-175 (1985)
8. K.J. Miller, Fat. Eng. Mat. Struct. **16**(9), 931-939 (1993)
9. British Standard, *BS 7608-2014: Guide to fatigue design and assessment of steel products*, BSI Standards Limited (2014)
10. Det Norske Veritas As, *DNVGL-RP-C203: Recommended Practice-Fatigue Design of Offshore Steel Structures* (2016)
11. M.D. Chapetti, J. Belmonte, T. Tagawa, T. Miyata, Sci. Tech. Weld. Join. **9**(5), 430-438 (2004)
12. M.D. Chapetti, L.F. Jaureguizahar, Int. J. Fatigue, **43**, 43-53 (2012)
13. C. Steimbreger, M.D. Chapetti, Int. J. Fatigue, **105**, 296-304 (2017)
14. M.D. Chapetti, Int. J. Fatigue, **25**(12), 1319–26 (2003)
15. Dassault Systèmes, Simulia. *Abaqus 6.13 User manual*, Providence, RI, USA (2013)
16. ASTM A36 / A36M-14, *Standard Specification for Carbon Structural Steel*, ASTM International, West Conshohocken, PA (2014)
17. British Standard, *BS EN 10149-2:2013, Hot rolled flat products made of high yield strength steels for cold forming. Technical delivery conditions for thermomechanically rolled steels*, BSI Standards Limited (2013)
18. M. Jono (Ed.), *Data book on the fatigue crack growth rates of metallic materials*, Society of Materials Science, Japan (JSMS) (1983)
19. M.D. Chapetti, Int. J. Fatigue, **33**, 833-841 (2011)
20. British Standard, *BS 7910:2013+A1:2015: Guide to methods for assessing the acceptability of flaws in metallic structures*, BSI Standards Limited (2015)
21. Det Norske Veritas As, *DNVGL-RP-C210: Recommended Practice-Probabilistic methods for planning of inspection for fatigue cracks in offshore structures* (2016)
22. B. Jonsson, G. Dobmann, A.F. Hobbacher, M. Kassner, G. Marquis, *IIW Guidelines on weld quality in relationship to fatigue strength*, IIW Collection, Springer International Publishing, Switzerland (2016)
23. T. Dahle, Int. J. Fatigue, **20**(9), 677–681 (1998)
24. H.C. Yildirim, Int. J. Fatigue, **79**, 36-45 (2015)
25. P.J. Haagensen, S.J. Maddox, *IIW Recommendatons on methods for improving the fatigue strength of welded joints (IIW-2142-10)*, Woodhead Publishing Limited, Cambridge, UK (2013)

Fatigue and fracture resistance of ferritic ductile cast iron: the effect of Sb and solidification time

Matteo Benedetti[1,*], *Vigilio* Fontanari[1], and *Danilo* Lusuardi[2]

[1]Department of Industrial Engineering, University of Trento, 38123 Trento, Italy
[2]Fonderie Ariotti S.p.A., 25030 Adro BS, Italy

Abstract. In this paper, we explore the effect of inoculants and solidification time on the mechanical properties of an EN-GJS-400-type ferritic ductile cast iron (DCI). For this purpose, static tensile, rotating bending fatigue, fatigue crack growth and fracture toughness tests are carried out on three different material conditions. They are produced under fast cooling (solidification time 2h45min), representative of thin walled castings, and very slow cooling (solidification time 10—13h), representative of thick walled castings, this latter with and without the addition of Sb. It has been found that the long solidification time leads to an overgrowth and degeneration of the spheroidal graphite nodules. The addition of Sb avoids the formation of chunky graphite observed in the slowly cooled condition but results in large exploded graphite nodules. These effects impact negatively on tensile strength, total elongation and fatigue strength. Conversely, the resistance to fatigue crack growth is even superior and the fracture toughness comparable to that of the fast cooled condition. Metallurgical and fractographic analyses are carried out to explain this behaviour.

1 Introduction

Low production cost and excellent castability make ductile cast iron (DCI) the preferred material choice when low-to-moderately stressed mechanical components of complicated shape and large dimensions must be manufactured. The microstructural control of DCI is of paramount importance as it greatly influences a wide range of mechanical properties. The typical microstructure consists of graphite nodules dispersed in a matrix that can be ferritic, ferritic and/or pearlitic depending on the alloy formulation, the casting control and the final heat treatment. The ferritic matrix displays usually the lower tensile and fatigue strength but also a noticeable ductility, which have promoted the DCI in the production of thick walled parts undergoing in service low/medium levels of mechanical stresses.

Casting large volume parts of weight on the order of tens of tons is particularly critical as the control of the final microstructure is rather difficult [1-3]. In fact the resulting long solidification time introduces into the material intrinsic defects, such as solidification cavities or poor graphite nodularity, dross and degenerated (chunky and spiky) graphite, which can be tolerated up to a certain level. It was found that the increase in non-nodular graphite content decreases the mechanical strength; indeed, the ultimate tensile strength and mostly the elongation to fracture are severely lowered, despite hardness and yield strength remain nearly unaffected. In particular, the increase of chunky content up to 20% and more induces a significant decrease in fatigue lifetime, with a drop of fatigue limit on the order of 15%–20% [4]. Sometimes, small amounts of rare earth elements such as Ce, Nd, Pr, Sm and Gd are inoculated into the casting to induce graphite spheroidization but at the cost of the formation of chunky graphite [5-7].

While the effect of chunky graphite on static and cyclic properties has been largely investigated in the literature, its effect on fatigue crack growth rate and fracture toughness is not yet clear. This is however of paramount importance when designing high-added value components of large dimensions, such as those employed in MW-series wind turbines, which cannot rely on the knowledge of the fatigue strength only. In fact, infinite life design approaches usually take large safety margins thus resulting in very heavy constructions, while safe life approaches suffer from large uncertainty on the actual lifetime of the component prior to its withdrawal from service. Damage tolerant design and structural health monitoring approaches have been proposed to overcome these limitations. According to these approaches, the fatigue life is exclusively thought of as the propagation of cracks up to a critical size leading to structural collapse or functionality loss. Therefore, they require a detailed characterization of fracture toughness and fatigue crack growth resistance of the castings, in order to assess the critical crack size and the time to propagation to this critical value. In the present paper, we address this latter issue by investigating the crack growth resistance of a DCI under three different microstructural conditions. They are produced under fast, representative of thin walled castings, and very slow

* Corresponding author: matteo.benedetti@unitn.it

cooling, representative of thick walled castings, this latter with and without the addition of inoculants able to promote graphite spheroidization. Static tensile, fatigue and fracture mechanics tests are performed to quantify the resistance to fatigue crack initiation and propagation as well as the fracture toughness. The discussion is complemented with metallurgical and fractographic analyses.

2 Material and Experimental Procedures

The experimentation is performed on an EN-GJS-400-type ferritic DCI. The condition representative of thin-walled castings, hereinafter denoted as "fast cooling" (FC), is obtained from the solidification of a cylinder with 300 mm diameter and 520 mm height exposed to natural air convection (solidification time of 2h 45min). Long solidification times representative of thin-walled castings are obtained from an innovative production line described in [8]. It consists of an electric furnace where the solidification of a cylinder with 240 mm diameter and 260 mm height (about 80 kg mass) takes place under controlled cooling conditions to achieve solidification time of 10-13h. In this way, two material variants are obtained, without and with the addition of Sb and denoted as "slow-cooling" (SC) and "slow-cooling+Sb" (SC+Sb), respectively. Specifically, 50 ppm of Sb are introduced into the casting to suppress the formation of chunky graphite observed in the SC condition.

Monotonic tensile tests are performed according to the standard UNI EN ISO 6892-1 on dog-bone coupons (14 mm gauge diameter, 84 mm parallel length), using a servo-hydraulic universal testing machine, equipped with hydraulic grips, a load cell of 200 kN. The yield strength is determined as the 0.2% offset yield stress.

Fatigue tests are performed on hourglass specimens according to the standard ASTM E466. Rotating bending fatigue tests (load ratio R = -1) are carried out under load-control at a nominal frequency of 100 Hz in laboratory environment (25 °C, 60% R.H.). The fatigue strength corresponding to a fatigue life of 5×10^6 cycles is obtained by a staircase procedure, employing 15 samples and 10 MPa stress increments.

The fracture mechanics tests are carried out on compact C(T) specimens whose geometry, compliant with the standard ASTM E1820-09, is illustrated in Figure 1. Specifically, K_{Ic} fracture toughness and fatigue crack growth rate tests are performed on 32 and 10 mm thick specimens, respectively. The initial notch of all C(T) specimens is Electro Discharge Machined (EDM) with a wire of 0.5 mm diameter to facilitate the follow-up pre-cracking process.

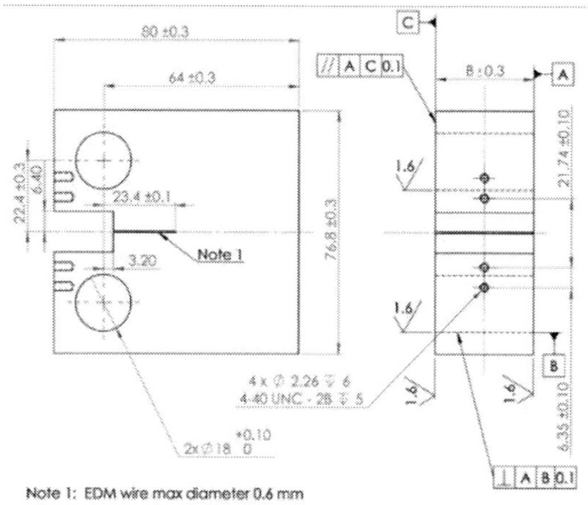

Fig. 1. Compact tension C(T) specimens used in fracture mechanics tests. Dimensions are given in mm. The thickness B of the C(T) specimens is equal to 10 mm and to 32 mm for fatigue crack growth rate and fracture toughness tests, respectively.

The fatigue crack growth rate testing is performed according to the standard ASTM E647-08. The experiments are conducted in the laboratory environment on a resonant testing machine Rumul Testronic 50 kN. A sinusoidal pulsating load waveform is applied at a frequency of ~100 Hz. Tests are performed at three load ratios R, namely 0.1, 0.5 and 0.75. A Fractomat® apparatus based on the indirect potential drop method is used to continuously measure the crack length. Periodically, the crack length on the back face of the sample is inspected using a travelling microscope to make sure that the two crack lengths do not differ by more than 0.25 times the sample thickness, as prescribed by the standard ASTM E647-08. The crack growth rates are calculated from discrete crack length increments of 0.3 mm, necessary for sampling the fatigue crack growth resistance of the fairly coarse material microstructure. The tests are performed in two distinct phases. In the initial force-shedding phase, the applied ΔK is exponentially reduced by setting the decay constant C=0.08mm^{-1} until reaching near-threshold fatigue crack growth conditions. A specific testing campaign was carried out in [9] to determine the minimum crack length sufficient for the material to fully deploy extrinsic crack growth resistance mechanisms. Only after reaching this crack size, the final part of the experiments is performed under constant force amplitude and the crack growth rate data are used to build the da/dN vs ΔK curves. After the fatigue tests, the specimens are sectioned across the thickness and metallographic samples are extracted in order to investigate the crack front profile. Fracture surfaces are investigated using a JEOL JSM-IT300LV SEM.

Specimens for fracture toughness tests are fatigue precracked to a crack-length-to-specimen-width of about 0.5 using the same resonant testing machine in laboratory environment under K-controlled conditions (load ratio R = 0.1) suitable to maintain the crack growth rate below 10^{-8} m/cycle. After precracking, to enforce the plane strain condition, 20% V-notch side grooves are machined on the samples, resulting in a net thickness (BN) of 25.6 mm. Subsequently, the samples are monotonically loaded up to fracture to determine the Kc fracture toughness (dK/dt = 1 MPa $\sqrt{m}\cdot s^{-1}$, room temperature air) using a 100 kN servo-hydraulic testing machine equipped with a clip-gauge mounted on knife edges attached to the specimen to measure the crack mouth opening displacement. All data of load versus clip gauge displacement are acquired with a sampling rate of 50 Hz. After fracture, the crack length prior to final monotonic loading is measured using an optical microscope following the ASTM E1820-09 standard procedure.

3 Results and discussion

3.1 Microstructure

The optical micrograph shown in Figure 2a-c illustrates a representative example of the FC, SC, and SC+Sb material microstructure, respectively. It can be noted that the microstructure is highly inhomogeneous, being composed of a predominantly ferritic matrix in which graphite domains of different size are unevenly distributed. In some regions, the matrix is even pearlitic. The effect of the cooling rate on the graphite nodules is evident, as their size is much smaller in the FC than in the SC and SC+Sb conditions. To better characterize type, shape and distribution of the graphite particles, ten cross-sections randomly extracted from the samples are metallographically prepared and statistically analysed. Specifically, the deviation from the spherical shape is estimated by the shape factor expressed as:

$$f_0 = \frac{4\pi A}{U^2} \qquad (1)$$

where A and U are the area and perimeter of the particles, respectively. Graphite nodules are identified as spheroidal if their shape factor f_0 is larger than 0.35. Table 1 summarizes the results of the statistical analysis of the graphite morphology.

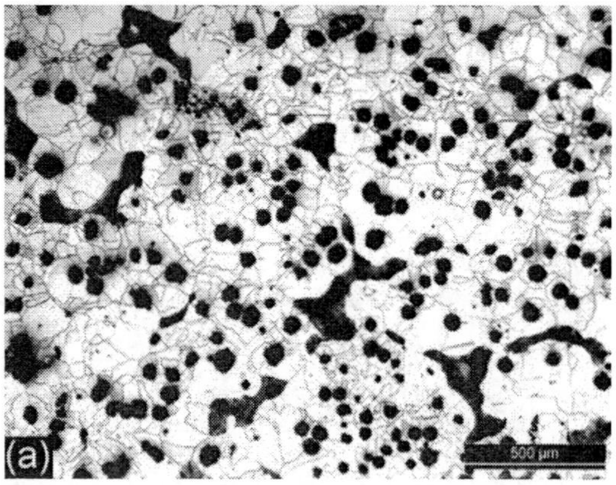

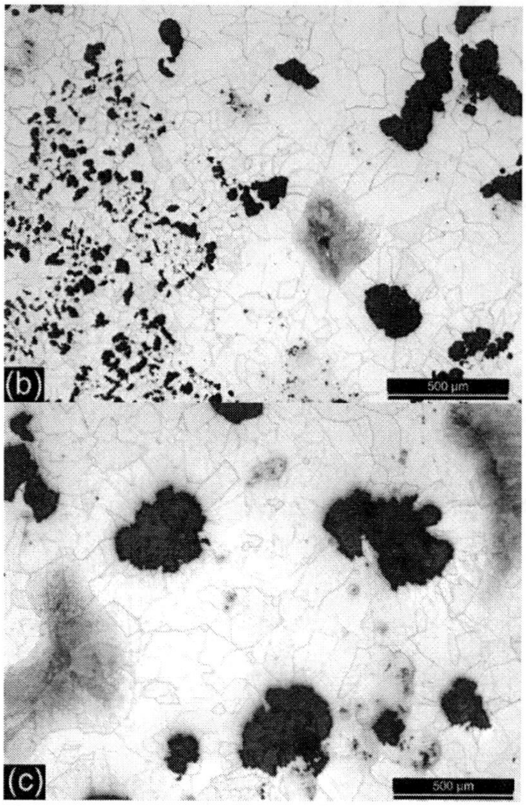

Fig. 2. Optical micrograph of the microstructure (etching with 2% Nital reagent). (a) FC, (b) SC, and (c) SC+Sb condition.

Table 1. Results of quantitative analysis done on 10 metallographic samples. Standard error corresponds to 1σ uncertainty band.

Condition	Total Graphite Content (%)	Fraction of Spheroidal Graphite (%)	Diameter of Spheroidal Graphite (µm)	Shape Factor Spheroidal Graphite f_0
FC	9.9±1.7	79±7	41±23	0.67±0.15
SC	9.3±2.1	71±15	310±110	0.49±0.19
SC+Sb	12±5	72±18	104±95	0.63±0.16

It can be noted that the fraction of spheroidal graphite increases from 70 to 80% with increasing cooling rate. The size of the nodules of spheroidal graphite is on the order of 40 micron in the FC cooling and increases in the slow cooled conditions, especially in the absence of inoculants. The remaining fraction of graphite has very different morphology in SC and SC+Sb conditions. In the former, it is mainly in the form of a fine distribution of chunky graphite, while in the latter in the form of large exploded nodules, as shown in Fig. 1b and c, respectively.

3.2 Monotonic and high-cycle fatigue properties

Mean values of materials parameters obtained from the analysis of tensile, hardness and fatigue data are summarized in Table 2. For comparison purposes, Table 2 reports also the requirements prescribed by the standard EN-1563 [10] on the tensile properties of samples extracted from bars and guidance values of mechanical properties measured on samples cut from moderately-thick walled castings (60 mm < $t \leq$ 200 mm) of EN-GJS-400-15.

Table 2. Monotonic tensile (based on four replicated tests) and rotating bending fatigue properties. Standard error corresponds to 1σ uncertainty band.

Condition	σ_{YS} (MPa)	σ_U (MPa)	T.E. (%)	HB	$\sigma_{lim,-1}$ (MPa)
FC	290±2	405±4	11±1	152±3	199±13
SC	275±4	295±8	2.9±0.2	144±2	138±4
SC+Sb	290±2	310±10	1.9±1.1	156±6	156±10
EN-GJS-400-15 bar [10]	240	370	11	135-180	-
EN-GJS-400-15 casting [10]	230	350	8	-	-

σ_{YS}: 2% yield stress; σ_U: ultimate tensile strength; T.E.: total elongation; HB: Brinell hardness; $\sigma_{lim,-1}$: rotating bending fatigue strength at five million cycles.

As expected, the FC condition displays the highest tensile and fatigue strength. The former are higher than that suggested in [10] for moderately thick-walled castings. Conversely, the very long solidification time and the resulting microstructural alterations lead to low values of tensile strength, total elongation, and fatigue strength; the addition of Sb partially limits the decrement in tensile and fatigue strength. SEM analyses are carried out to shed light on the damage mechanisms acting in the high-cycle fatigue regime.

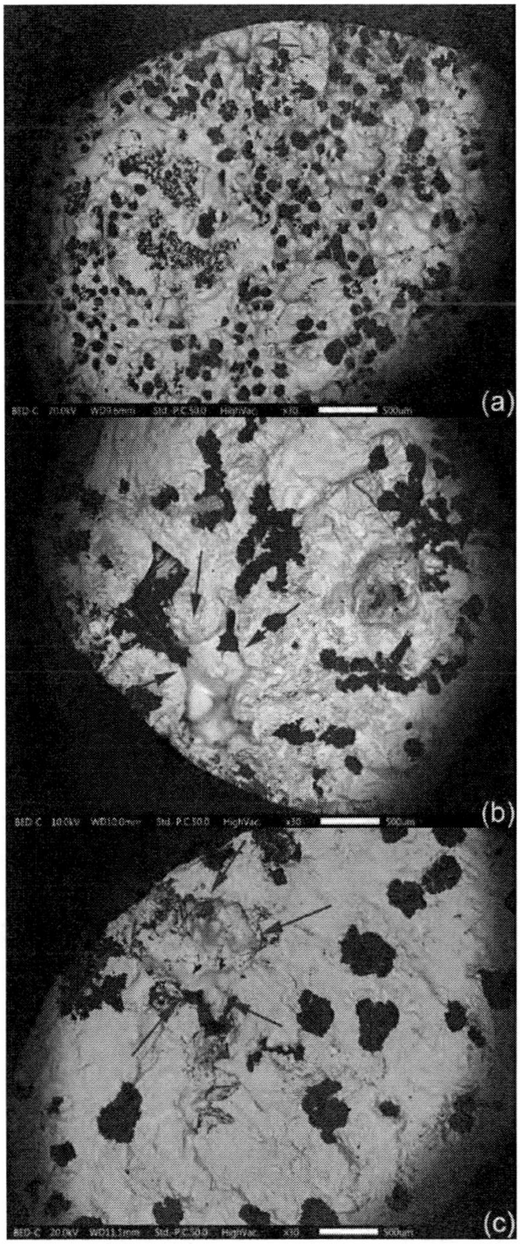

Fig. 3. SEM micrographs of the fracture surfaces of rotating bending fatigue samples around the largest shrinkage pore marked by red arrows: (a) FC, (b) SC, and (c) SC+Sb condition.

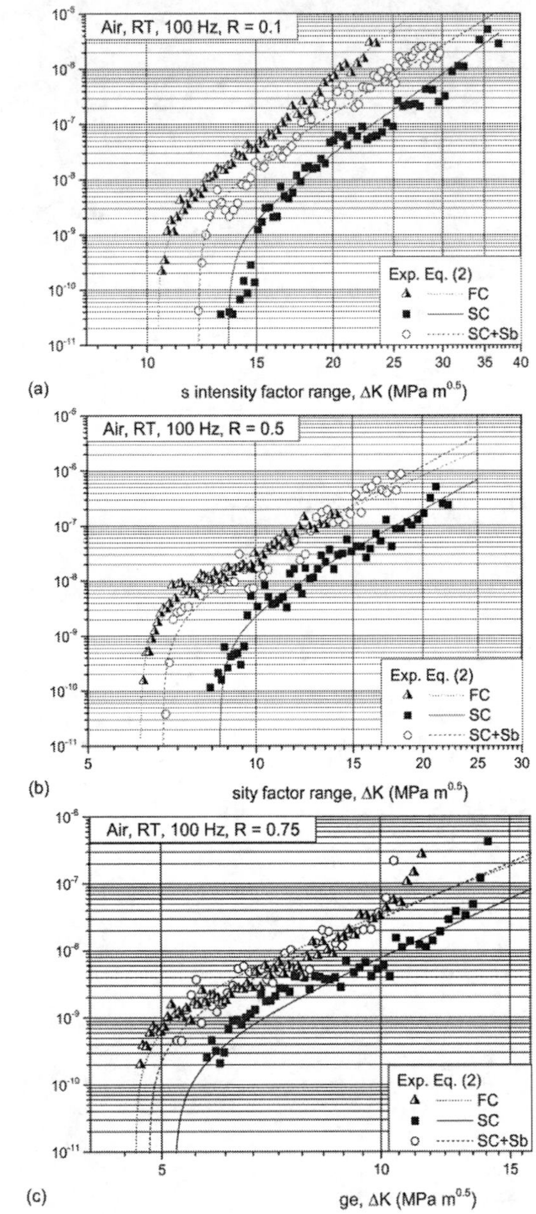

Fig. 4. Fatigue crack growth curves plotted as a function of the nominal stress intensity factor range. Load ratio (a) R=0.1, (b) R=0.5, and (c) R=0.75.

Figure 3a-c shows the fracture surface of fatigue samples in FC, SC, and SC+SB condition, respectively, tested at stress levels close to the fatigue limit. It can be noted that shrinkage defects (marked by red arrows) are preferential site of crack initiation. Their size is significantly lower in the FC than in the SC and SC+Sb condition. The graphite nodules are fairly uniformly distributed in the FC condition, while in the SC and SC+Sb conditions the crack initiation site is located in regions of low graphite content. According to Murakami and Endo model [11], the fatigue limit of DCI is related to the size of the critical defect and to the hardness of the ferritic matrix. By the light of this model, the highest fatigue limit of the FC condition is essentially due to the smaller defect size. The superior fatigue strength of the SC+Sb with respect to the SC condition is supposed to be related to the absence of chunky graphite.

3.3 Fatigue crack growth curves

The resulting da/dN-ΔK curves obtained at load ratios R=0.1, 0.5 and 0.75 are compared in Figure 4a-c, respectively. At all the explored load ratios, the highest fatigue crack growth resistance is exhibited by the SC condition, while the SC+Sb material shows an intermediate behaviour. The experimental data are fitted by the exponential law proposed by Klesnil and Lukáš [12]:

$$\frac{da}{dN} = C\left(\Delta K^m - \Delta K_{th}^m\right) \tag{2}$$

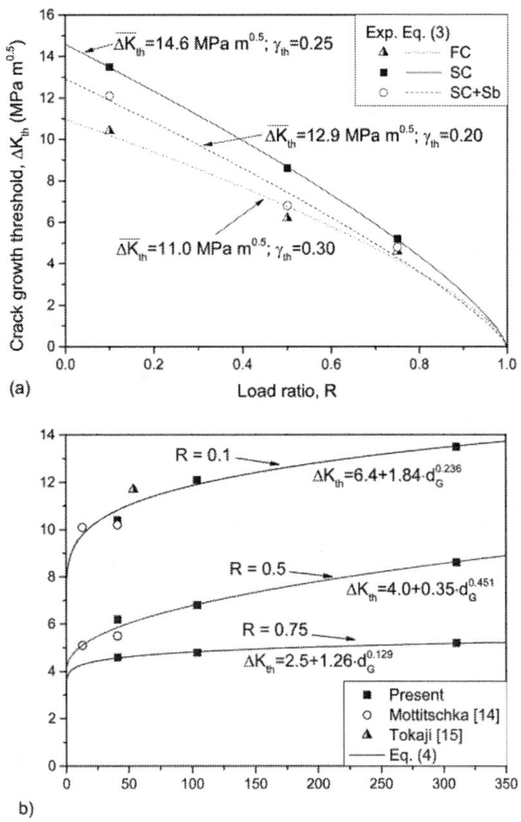

Fig. 5. Threshold dependency upon (a) load ratio R and (b) mean diameter of spheroidal graphite.

The best-fit parameters are given in Table 3 for the different load ratios used in the tests.

The R-dependency of the threshold ΔK_{th} is well represented by the Walker equation [13]:

$$\Delta K_{th} = \overline{\Delta K_{th}}\left(1-R\right)^{1-\gamma_{th}} \tag{3}$$

https://doi.org/10.1051/matecconf/201816513011

as shown in Figure 5a, where the least-square fit of Equation (4) is compared for all the material types. Figure 5b depicts the threshold ΔK_{th} as a function of the mean graphite diameter d_G and compares the outcomes of the present paper with the results published in the technical literature [14,15] for EN-GJS-400 DCI with similar chemical composition.

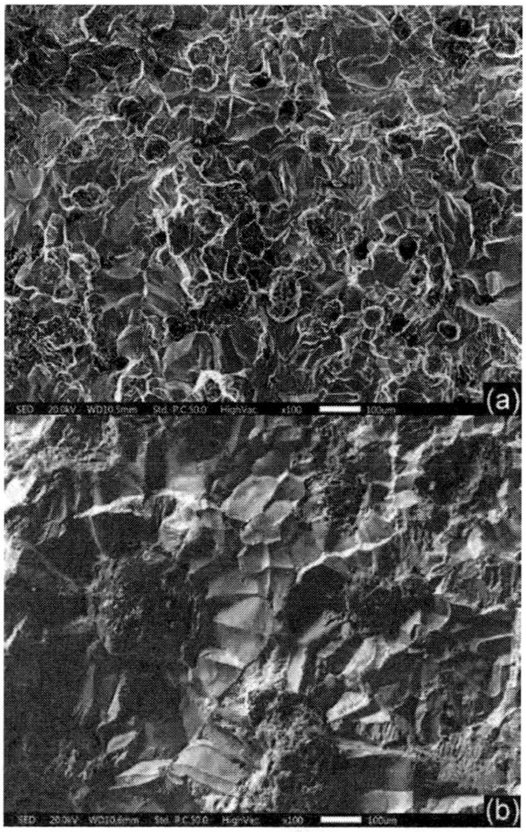

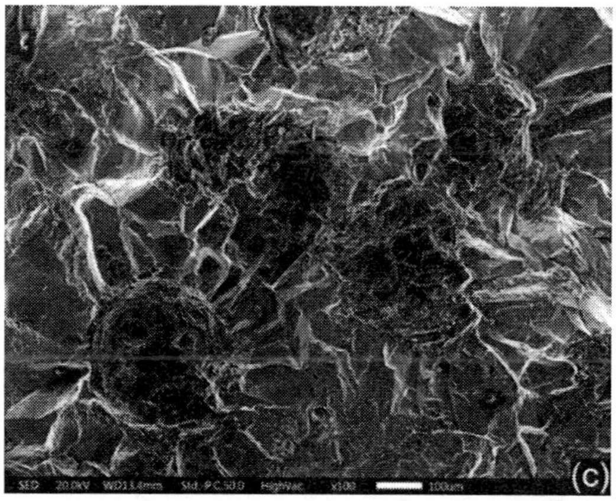

Fig. 6. SEM) micrographs of the fatigue fracture surfaces of the C(T) sample tested at R = 0.1: (a) FC, (b) SC, and (c) SC+Sb condition.

Table 3. Best-fit parameters of the crack propagation law expressed by Equation (2).

Condition	Load ratio *R*	ΔK_{th} (MPa·m$^{0.5}$)	*C*	*m*
FC	0.1	10.4	8.33×10^{-19}	9.1
	0.5	6.2	1.03×10^{-12}	4.5
	0.75	4.6	3.13×10^{-12}	4.1
SC	0.1	13.5	7.72×10^{-19}	8.1
	0.5	8.6	9.48×10^{-15}	5.6
	0.75	5.2	9.59×10^{-14}	4.9
SC+Sb	0.1	12.1	1.56×10^{-16}	6.9
	0.5	6.8	6.32×10^{-14}	5.6
	0.75	4.8	7.49×10^{-13}	4.6
Units in Eq. (2) are m/cycle and MPa m$^{0.5}$				

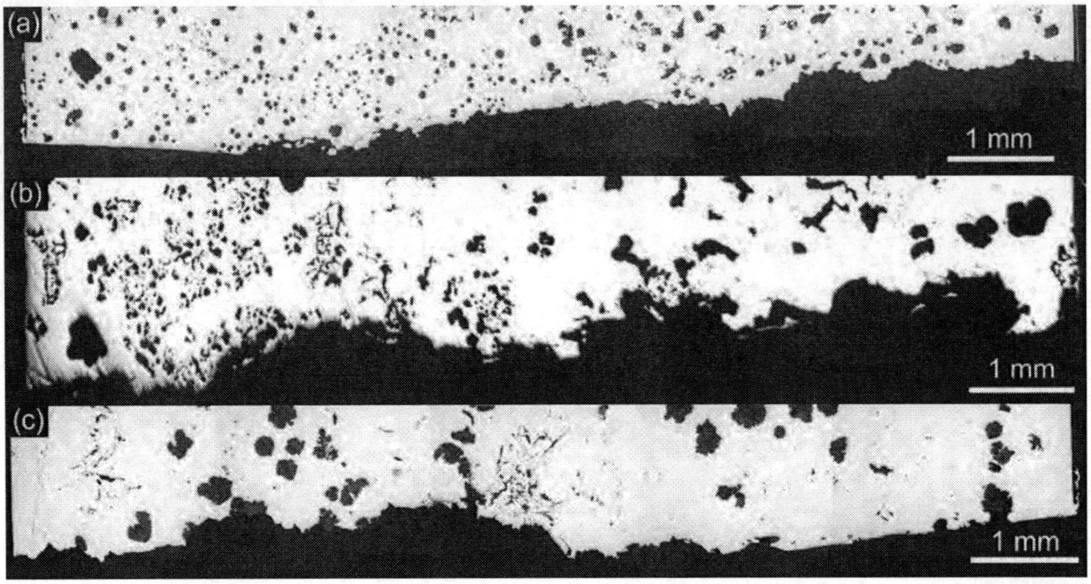

Fig. 7. Light microscope images of crack front profiles taken in the near-threshold crack propagation region of the C(T) samples tested at R = 0.1: (a) FC, (b) SC, and (c) SC+Sb condition.

Figure 5b shows that there exists a clear tendency of ΔK_{th} to increase with increasing d_G. As shown in Fig. 5b, this trend is well represented by the following power equation:

$$\Delta K_{th} = \Delta K_{th}^0 + a \cdot d_G^b \quad (4)$$

where the intercept ΔK_{th}^0, representing the growth threshold in the absence of graphite, is expressed by the lower bound of ΔK_{th} proposed in [16] for various steels:

$$\Delta K_{th}^0 = 7.0(1 - 0.85R) \quad MPa\sqrt{m} \quad (5)$$

It can be noted that the effect of the graphite nodule size on the growth threshold is more pronounced at R=0.1, while it becomes almost negligible at R=0.75.

Figure 6a-c illustrates SEM overviews of the fatigue facture surface in the vicinity of the near-threshold propagation region for the FC, SC, and SC+Sb condition, respectively. All the fracture surfaces are characterized by the presence of different graphite morphologies, ranging from nearly spherical nodules to fine dispersion of chunky graphite. Clearly, the size of the nodules is much smaller in the fast than in the slow cooled conditions. It is worth noticing that the regions containing no graphite at all or surrounding graphite spheroids show large evidence of brittle cleavage and intergranular fracture. On the contrary, the presence of chunky graphite is associated to a less brittle fracture mode resulting in rougher fracture surface.

Similar indications are given by crack front profiles taken from regions under the same crack propagation regime and shown in Fig. 7a-c for the FC, SC, and SC+Sb condition, respectively. Indeed, it can be noticed that where the crack front encounters regions characterized by degenerated graphite and large graphite nodules the crack profile is significantly rougher than where only small spherical graphite nodules are present or where graphite is even absent. The former scenario, typical of the slow cooled conditions, especially of the SC material, will result in more effective crack retardation mechanisms, such as roughness-induced crack closure and crack front geometry effects [17], with respect to the latter one prevailingly occurring in the FC condition.

3.4 Fracture toughness

K_{Ic} fracture toughness tests are performed on C(T) specimens depicted in Figure 1, whose geometry is taken as large as possible with the aim of satisfying the linear elastic fracture mechanics requirement expressed by [18]:

$$B, a, W - a > 2.5\left(\frac{K_{Ic}}{\sigma_{YS}}\right)^2 \quad (6)$$

where a, B and W are the initial crack length, specimen thickness and width, respectively. Considering the value of the material yield strength σ_{YS} listed in Table 2, the maximum theoretical fracture toughness that can be estimated using this C(T) sample geometry is about 32 MPa $m^{0.5}$.

Table 4. Results of the fracture toughness tests and comparison with literature data.

Condition	Specimen	K_Q (MPa·m$^{0.5}$)	K_{max} (MPa·m$^{0.5}$)

FC	1	23.5	50.9
	2	23.1	52.5
SC	1	18.0	57.3
	2	16.6	50.2
SC+Sb	1	27.9	51.3
	2	26.5	44.8
Ref. [19]	C(T)	20	39

Table 4 summarizes the results of the two fracture toughness tests and compares them with data found in [19] regarding similar EN-GJS-400 DCI subjected to normal solidification conditions. K_Q is the apparent fracture toughness determined from load P_Q, viz. the intersection of the force versus crack opening displacement record with the 95% secant. K_{max} is obtained from the maximum value of the load record. In order for K_Q to be qualified as linear elastic, plane-strain fracture toughness, the ratio K_{max}/K_Q must be ≤ 1.10. It can be noted that, despite the use of thick-walled and side-grooved specimens, no test listed in Table 3 met this requirement, thus resulting in thickness-dependent fracture toughness values. The FC condition displays an apparent fracture toughness slightly higher than that reported in [19] for a similar material, even though a much thicker specimen was used (32 mm vs. 10 mm thickness). The effect of slow cooling conditions is not evident, in the sense that the marked overgrowth of the nodules occurred in SC seems to be slightly detrimental effect on the fracture toughness. On the contrary, the addition of inoculants, which limits the graphite degeneration, seems to have a beneficial effect, resulting in a small increment in fracture toughness.

4 Conclusions

In this paper, the effect of Sb and solidification time on the mechanical properties of an EN-GJS-400-type ferritic ductile cast iron (DCI) was explored. The following conclusions can be drawn:

1) The microstructure consists of a prevailing ferritic matrix, in which 3/4 of graphite is interspersed in form of spheroidal nodules. Their size is smaller in the fast cooled condition and largest in the slow cooled condition without the addition of Sb. The rest of graphite is present in the form of chunky (SC) or exploded (SC+Sb) graphite.
2) The slow cooled conditions exhibit a tensile strength and a total elongation 20% and 75% lower than the fast cooled material.
3) The fatigue limit of the slow cooled condition is about 30% lower than that of the fast cooled material. The addition of Sb limits this decrement down to 20%. The fatigue limit is strongly correlated to the dimension of the largest shrinkage pore found on the fracture surface.
4) The slow cooled conditions, especially that without the addition of Sb, display a significantly higher fatigue crack growth resistance, especially in the near-threshold regime and at low load ratio, with respect to DCI subject to fast solidification conditions. The fatigue crack growth threshold seems to be correlated to the mean size of the graphite nodules. Their presence in conjunction with chunky graphite makes the crack front profile much rougher, resulting in more effective crack retardation mechanisms. Further investigations are needed to reveal a possible correlation of the crack threshold with the size of the chunky graphite domains.
5) The effect of slow cooling condition on the fracture toughness is not evident, in the sense that the marked overgrowth of the nodules occurred in SC seems to be slightly detrimental effect on the fracture toughness. On the contrary, the addition of inoculants, which limits the graphite degeneration, seems to have a beneficial effect, resulting in a small increment in fracture toughness.

References

1. H. Kaufmann, D.B. Wolters, Giess **27**, 4–27 (2002).
2. P. Minnebo, K.F. Nilsson, D. Blagoeva, J. Mater. Eng. Perform. **16**, 35–45 (2007).
3. M. Kobayashi, J. Yamabe, Fatigue Fract. Eng. Mater. Struct. **29**, 403–415 (2006).
4. H. Kaufmann, D.B. Wolters, Konstr. Giess. **27**, 4–27 (2002).
5. I. Riposan, M. Chisamera, V. Uta, S. Stan, Int. J. Metalcast. **8**, 65–80 (2014)
6. M.I. Onsøien, Ø. Grong, T. Skaland, K. Jørgensen, Mater. Sci. Technol. **15**, 253–259 (1999).
7. J.O. Choi, J.Y. Kim, C.O. Choi, J.K. Kim, P.K. Rohatgi, Mater. Sci. Eng. A **383**, 323–333 (2004).
8. E. Foglio, M. Gelfi, A. Pola, S. Goffelli, D. Lusuardi, Int. J. Metalcast. **11**, 33–43 (2017).
9. M. Benedetti, E. Torresani, V. Fontanari, D. Lusuardi, Metals **7**, 88 (2017).
10. European Standard EN 1563:2011. Founding—Spheroidal Graphite Cast Iron; CEN: Brussels, Belgium, 2011.
11. Murakami, Y.; Endo, M. In *The Behaviour of Short Fatigue Cracks* (Institution of Mechanical Engineers: London, UK, 1986).
12. M. Klesnil, P. Lukáš, Eng. Fract. Mech. **4**, 77–92 (1972).
13. E.K. Walker, In *Effect of Environment and Complex Load History on Fatigue Life* (American Society for Testing and Materials, West Conshohocken, PA, USA, 1970).

14. T. Mottitschka, G. Pusch, H. Biermann, L. Zybell, M. Kuna, Procedia Eng. **2**, 1557–1567 (2010).
15. K. Tokaji, T. Ogawa, K. Shamoto, Int. J. Fatigue **16**, 344-350 (1994).
16. N. E. Dowling, *Mechanical Behavior of Materials* (International Edition Pearson, 2007).
17. M. Benedetti, J. Heidemann, J.O. Peters, G. Lütjering, Fatigue Fract. Eng. Mater. Struct. **28**, 909–922 (2005).
18. T.L Anderson, *Fracture Mechanics: Fundamentals and Applications* (CRC Press, Boca Raton, FL, USA, 2005).
19. H.R. Zambrano, G. Härkegård, K.F. Stärk, Fatigue Fract. Eng. Mater. Struct. **35**, 374–388 (2011).

AUTHOR INDEX

AUTHOR INDEX

AUTHOR INDEX

AUTHOR INDEX

AUTHOR INDEX

AUTHOR INDEX

AUTHOR INDEX

EDP Sciences – Web of Conferences
17, Avenue du Hoggar
Parc d'Activité de Courtabouf
BP 112
F-91944 Les Ulis Cedex A
France

ISBN 978-1-5108-6386-6

9 781510 863866

CURRAN ASSOCIATES INC.
proceedings.com

12th International Fatigue Congress (FATIGUE 2018)

MATEC Web of Conferences Volume 165 (2018)

Poitiers Futuroscope, France
27 May – 1 June 2018

Part 2 of 2

edp sciences

12th International Fatigue Congress (FATIGUE 2018)

MATEC Web of Conferences Volume 165 (2018)

Poitiers Futuroscope, France
27 May – 1 June 2018

Part 2 of 2

Editor:

Gilbert Henaff

ISBN: 978-1-5108-6386-6

Printed from e-media with permission by:

Curran Associates, Inc.
57 Morehouse Lane
Red Hook, NY 12571

Printed by Curran Associates, Inc. (2018)

For additional information, please contact EDP Sciences – Web of Conferences at the address below.

EDP Sciences – Web of Conferences
17, Avenue du Hoggar
Parc d'Activité de Courtabœuf
BP 112
F-91944 Les Ulis Cedex A
France

Phone: +33 (0) 1 69 18 75 75
Fax: +33 (0) 1 69 28 84 91

contact-edps@webofconferences.org

Additional copies of this publication are available from:

Curran Associates, Inc.
57 Morehouse Lane
Red Hook, NY 12571 USA
Phone: 845-758-0400
Fax: 845-758-2633
Email: curran@proceedings.com
Web: www.proceedings.com

TABLE OF CONTENTS

PART 1

PART 2

Crystallographic crack propagation rate in single-crystal nickel-base superalloys

Christian Busse[1,*], *Frans* Palmert[2,3], *Paul* Wawrzynek[4], *Björn* Sjödin[3], *David* Gustafsson[3], and *Daniel* Leidermark[1]

[1]Division of Solid Mechanics, Linköping University, SE-58183 Linköping, Sweden
[2]Division of Engineering Materials, Linköping University, SE-58183 Linköping, Sweden
[3]Siemens Industrial Turbomachinery AB, SE-61283 Finspång, Sweden
[4]Fracture Analysis Consultants, Inc., Ithaca, New York, USA

Abstract. Single-crystal nickel-base superalloys are often used in the hot sections of gas turbines due to their good mechanical properties at high temperatures such as enhanced creep resistance. However, the anisotropic material properties of these materials bring many difficulties in terms of modelling and crack growth prediction. Cracks tend to switch cracking mode from Mode I cracking to crystallographic cracking. Crystallographic crack growth is often associated with a decrease in crack propagation life compared to Mode I cracking and this must be taken into account for reliable component lifing. In this paper a method to evaluate the crystallographic crack propagation rate related to a crystallographic crack driving force parameter is presented. The crystallographic crack growth rate is determined by an evaluation of heat tints on the fracture surface of a specimen subjected to fatigue loading. The complicated crack geometry including two crystallographic crack fronts is modelled in a three dimensional finite element context. The crack driving force parameter is determined by calculating anisotropic stress intensity factors along the two crystallographic crack fronts by finite-element simulations and post-processing the data in a fracture mechanics tool that resolves the stress intensity factors on the crystallographic slip planes in the slip directions. The evaluated crack propagation rate shows a good correlation for both considered crystallographic cracks fronts.

1 Introduction

There is always a strive towards higher combustion temperatures when designing industrial gas turbines used for power generation, due to the resulting increase in efficiency and reduction in terms of pollution [1]. These ever increasing temperatures put high requirements on many components in the hot sections of the gas turbines. One crucial component that is exposed to this harsh environment is the blade in the first turbine stage. These blades are therefore often manufactured from nickel-base superalloys cast as single-crystals, due to their great material properties under these conditions. Nevertheless, the complex nature of the anisotropy of these materials is often a challenge in terms of testing, modelling and understanding the fracture behaviour [2-8]. Cracks have been observed to change cracking mode from Mode I to cracking on crystallographic slip planes [9-10], *cf.* Fig.1. Since single-crystal nickel-base superalloys contain an inherent internal structure of distinct crystallographic slip planes and slip directions, it is physically motivated to assume that these crystallographic planes are potential critical planes for crack initiation and crack growth [10]. Crystallographic cracking differs from Mode I cracking in its mechanisms, resulting in a different resistance

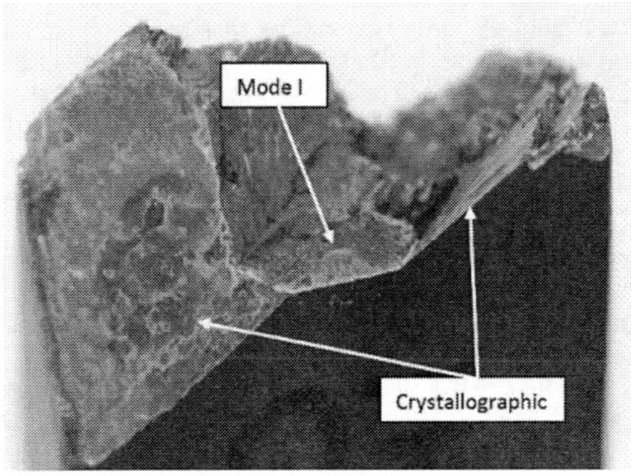

Fig. 1: Fracture view of the specimen Kb1 with visible crystallographic and Mode I cracking

towards crack growth. The Mode I crack growth behaviour is well understood and easier to model compared to the crystallographic cracking mode, where this knowledge is lacking. One major aspect complicating the modelling of the behaviour of single-crystal materials is the misalignments from the nominal crystal orientation influencing the stress state and thus the fracture behaviour [11]. Especially three-dimensional

* Corresponding author: christian.busse@liu.se

crack growth modelling, incorporating crystallographic cracking, has not been attempted due to its complexity. The aim of this paper is to evaluate and model the crack growth rate for crystallographic cracking. This has been done by evaluating a Resolved Stress Intensity Factor (RSIF) on the crystallographic crack fronts and correlating it to a crack growth law. The parameters in the crack growth law have been calibrated by evaluating heat-tinted fracture surfaces of performed experiments.

2 Material and experiments

The investigated material is a single-crystal nickel-base superalloy of face-centred cubic structure (FCC), similar to the alloy described by Reed *et al.*[12]. Its main alloying elements, in order of decreasing wt %, are as follows: Ni-Cr-Ta-Co-Al-W-Mo-Si-Hf-C-Ce.

As the specimen is cast as a single grain they possess anisotropic properties, where the elastic stiffness constants for 500 °C are presented in Table 1. The temperature of 500 °C has been chosen since it corresponds to the temperature during normal operation at the blade fir tree root, which is a critical location in the blade design.

Table 1. Elastic stiffness constants, units in GPa.

Temp. [°C]	C_{11}	C_{12}	C_{44}
500	201.9	127.0	112.8

To study the crack growth behaviour and its effects on the studied single-crystal nickel-base superalloy material, an isothermal fatigue crack growth experiment at 500 °C was performed. Isothermal conditions were chosen due to their simpler nature compared to thermo-mechanical fatigue conditions. The test specimen geometry was a surface flawed fatigue crack growth specimen of Kb-type [13], henceforth denoted as Kb specimen. The geometry of the Kb specimen can be seen in Fig. 2. The specimen was manufactured from a cast bar with the casting direction parallel to the nominal [001] crystallographic direction. The secondary orientation [010] was normal to the wide surface on which the flaw was introduced. The surface flaw, of semi-elliptical shape, was introduced by electro

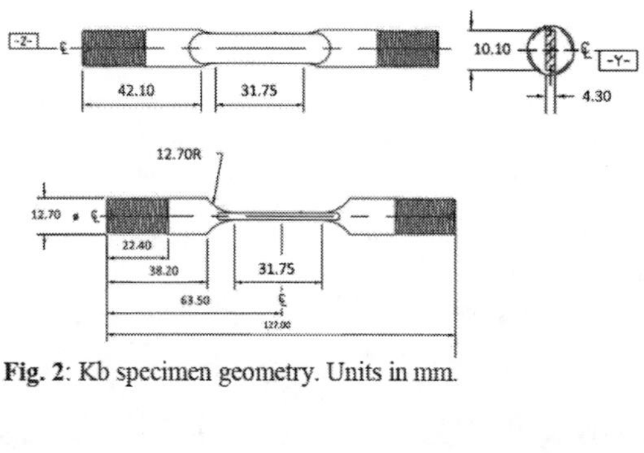

Fig. 2: Kb specimen geometry. Units in mm.

discharge machining to a depth of 0.127 mm

and width of 0.245 mm centred on the gauge section of the specimens. The misalignments from the nominal crystallographic orientations in terms of sequential rotations about the nominal [010], [100] and [001] directions are 1.7°, 0.1° and -2.8°, respectively. The misalignments are determined by measuring the visible dendrites on three polished and etched orthogonal surfaces on the specimen. For a detailed descrition see [14].

The fatigue crack growth testing of the Kb specimen was done according to the guidelines provided in the ASTM E647 [15] test specification. The test was run in an MTS servo-hydraulic testing machine with a test capacity of 89 kN. The specimen was precracked at room temperature, with a frequency of 10 Hz to initiate a crack. After this, the fatigue crack growth test was performed according to the test conditions shown in Table 1. Three heat tints were created in each test by maintaining an open crack and increasing the temperature to 704°C under a hold time of eight hours. The crack length was monitored using direct current potential drop measurement (DCPD) according to [16].

Table 2. Test conditions.

Temperature [°C]	Stress ratio R_σ [-]	Frequency [Hz]
500	0.05	0.5
Max. Force [N]	Nominal loading Direction	
19857	[001]	

2.1 Fracture surfaces

After the experiments were finished, the fracture surfaces were examined, and as can be seen in Fig. 1, the cracking mode transitions from Mode I into crystallographic cracking on each side. Further examination in a Scanning Electron Microscope (SEM) at the instances where the two cracking modes intersect was performed. An example is depicted in Fig. 3, where the cracking modes transition on the right side of the Mode I crack is shown. It can also be observed that the progressing crystallographic crack front advanced on a crystallographic slip plane beneath the Mode I crack front. The higher magnification on the right image suggests that the crystallographic crack grew beneath the Mode I crack all the way from the free surface.

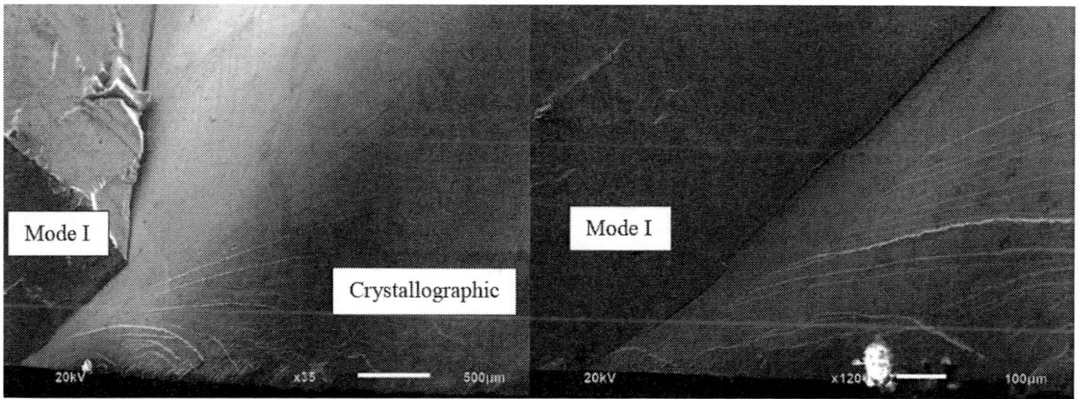

Fig. 3: SEM image of the transition of cracking modes, showing (left) the crystallographic crack front advancing beneath the Mode I crack front and (right) the mutual crack front at the surface.

Furthermore, the fracture surfaces have been studied in an optical microscope in order to evaluate the crack front shapes for the different heat tints. The heat tints are shown for the visible parts of the crystallographic and the Mode I crack in Fig. 4, where they are highlighted in blue, green and red. It is to be pointed out that only the last two heat tints provide information about the crack front shapes of the crystallographic cracks fronts. The crack shapes of the hidden portions of the two crystallographic crack fronts (on the left and on the right side of the Mode I crack front) have been approximated and are represented by the dashed contours in Fig. 4. In the following, the crack instances outlined by the green and red contours and associated to the second and third heat tints are referred to as crack instance 1 and 2, respectively. From the analysed fracture surfaces it can be concluded that each crack instance includes three intersection crack fronts, *i.e.* two crystallographic fronts and one Mode I front.

Each heat tint is associated to a number of loading cycles until that specific crack shape is formed. The heat tints associated with the number of cycles are presented in Table 3, where N_0, N_1 and N_2 represent the number of cycles until crack instance 0, 1 and 2 is reached. Crack instance 0 is disregarded in the following since no crystallographic cracking is associated to it. Furthermore, the fracture surfaces were scanned in an Atos Triple Scan machine to acquire a 3D representation of the fracture surfaces in terms of mesh data. This data, in conjunction with the optical images is used to create a realistic mapping of the crack shape into a mesh representation. This procedure is elaborated in Section 4.1.

Table 3. Experimental results of the number of cycles for the different heat tints.

Heat tints	N_0	N_1	N_2
Cycles	203678	217426	218827

The 3D representations of the fracture surfaces have also been used to determine the crystallographic cracking

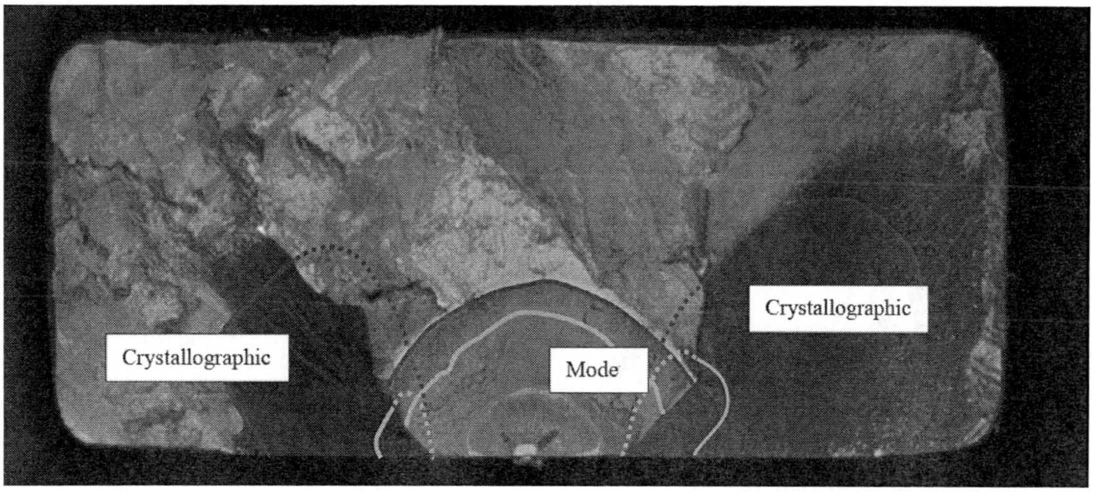

Fig. 4: Heat tints on fracture surface of specimen Kb1. Cracks instances 0, 1 and 2 are highlighted in blue, green and red, respectively.

planes by determining its normal vector. The active crystallographic cracking planes are $\bar{1}1\bar{1}$ and $\bar{1}\bar{1}1$ in terms of Miller notation for the left and right side of the Mode I crack front, respectively.

3 Crystallographic crack driving force and crack growth rate

In order to evaluate the crystallographic crack growth rate an appropriate crack driving force parameter has to be adopted. A RSIF k_{EQ} for the crystallographic portions of the crack was used. This parameter is employed in a crack growth model as proposed by Paris *et al.* [17] as:

$$\frac{dL}{dN} = C_c\left(k_{EQ}\right)^{n_c}, \qquad (1)$$

where Eq. (1) represents the crack growth model for crystallographic cracking with the parameters C_c and n_c, and dL/dN is the crystallographic crack growth rate. The entity L is defined as the orthogonal distance between the crystallographic crack fronts of the two crack instances, see the dotted line Fig. 5, where the discretized crack fronts are schematically illustrated. Based on the performed experiments, only the shape of the crystallographic fronts of the crack instances 1 and 2 are known. Under the assumption, that crack growth occurs orthogonal to the crack front, the orthogonal distance between two points along the crack fronts of the two crack instances corresponds to $L(i)$, where i describes the positions along the crack front. For each position i, two k_{EQ} values and associated number of cycles are determined. These relations are then used to calibrate the parameters C_c and n_c by fitting the crack growth data in a log-log diagram of the crack growth rate over the crack driving force parameter. This is schematically shown in Fig. 6, where the subscript of N and the superscript of k_{EQ} indicate the crack instance number. From two corresponding $k_{EQ}(i)$ values the mean value was used for the crack propagation rate evaluations below. The difference in loading cycles ΔN_{2-1} between crack instance 2 and 1 is constant for all points along the crack fronts and has the value 1401 cycles. Thus, the following approximation is used: $dL/dN \approx L/\Delta N_{2-1}$, where $k_{EQ} = (k_{EQ}^1 + K_{EQ}^2)/2$. When considering the discretised crack front in Fig. 5, it can be seen that the crack fronts of both crack instances can have a different number of nodes. Thus not all nodes on both crack fronts are accounted for.

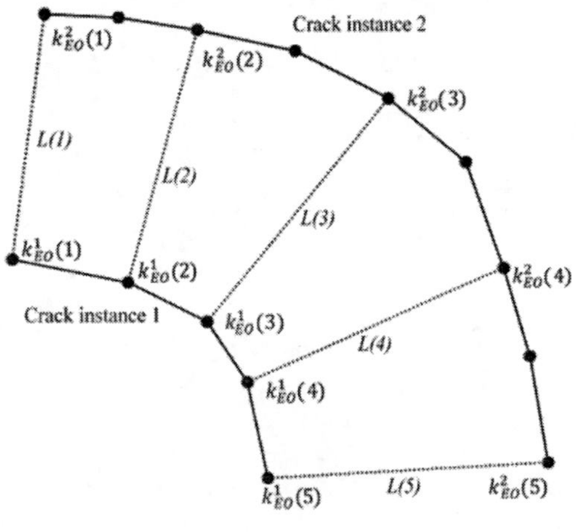

Fig. 5: Representation of the nodes pairs of crack instance 1 and 2 and corresponding RSIF and L.

The correct choice of crack driving force parameter is vital for a meaningful prediction of the crack growth rate. The equivalent RSIF k_{EQ} used in this work describes the singular stress field resolved onto the crystallographic slip planes in the Burgers' directions. This is due to the fact that the deformations in these materials are localised to the crystallographic slip planes in the Burgers' directions, and are associated with the dislocation motion [18]. As there are three Burgers' directions per crystallographic slip plane, the direction resulting in the highest k_{EQ} is chosen for each point along the crack fronts in the crack growth model evaluation shown below. Thus, the RSIF parameter k_{EQ} is evaluated for each point along the two crack fronts in the three Burgers' directions on the corresponding crystallographic slip planes.

The equivalent RSIF k_{EQ} is based on anisotropic Stress Intensity Factors (SIFs) (K_I, K_{II} and K_{III}), which can be evaluated in different ways. The approach in this work has been by using the Displacement Correlation Method (DCM) with incorporation of anisotropic material properties and accounting for the material misalignments. The DCM matches crack opening displacements computed using the finite element method with theoretical expressions for the near crack-front displacement fields developed by Hoenig [19] as follows:

$$u_i = \frac{\sqrt{2r}}{\pi} Re\left(\textstyle\sum_{j=1}^{3} m_{ij} N_{jl}^{-1} K_l Q_j\right), \qquad (2)$$

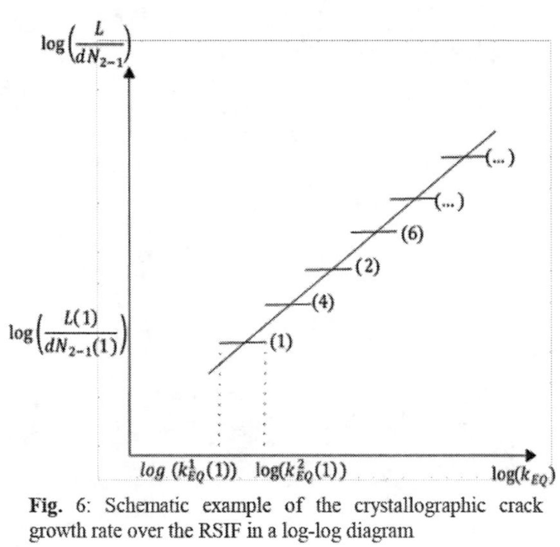

Fig. 6: Schematic example of the crystallographic crack growth rate over the RSIF in a log-log diagram

 https://doi.org/10.1051/matecconf/201816513012

$$\begin{pmatrix} u_1 - u_1' \\ u_2 - u_2' \\ u_3 - u_3' \end{pmatrix} = -\sqrt{\frac{8l}{\pi}} \begin{bmatrix} Re\left(m_{1j}N_{j1}^{-1}\sqrt{p_j}\right) & Re\left(m_{1j}N_{j2}^{-1}\sqrt{p_j}\right) & Re\left(m_{1j}N_{j3}^{-1}\sqrt{p_j}\right) \\ Re\left(m_{2j}N_{j1}^{-1}\sqrt{p_j}\right) & Re\left(m_{2j}N_{j2}^{-1}\sqrt{p_j}\right) & Re\left(m_{2j}N_{j3}^{-1}\sqrt{p_j}\right) \\ Re\left(m_{3j}N_{j1}^{-1}\sqrt{p_j}\right) & Re\left(m_{3j}N_{j2}^{-1}\sqrt{p_j}\right) & Re\left(m_{3j}N_{j3}^{-1}\sqrt{p_j}\right) \end{bmatrix} \begin{Bmatrix} K_I \\ K_{II} \\ K_{III} \end{Bmatrix}, \quad (3)$$

where Re represents the real part of the quantities in parenthesis, two repeated indices obey the summation convention from 1 to 3, the directions 1, 2 and 3, and coordinates r and θ refer to the crack-front coordinate system illustrated in Fig 7., and K_j are the anisotropic SIFs K_I, K_{II}, and K_{III}. It is assumed that the stresses and displacements are functions only of the in-plane coordinates r and θ. The quantities, m_{ij}, N_{ij}, and Q_i are defined in [20]. Consider the near crack-front configuration shown in Fig. 8, with two points A and A', which are on opposing crack faces a distance l from the crack front. Substituting $Q_j = \sqrt{\cos\theta + p_j \sin\theta}$, $\theta = \pi$,

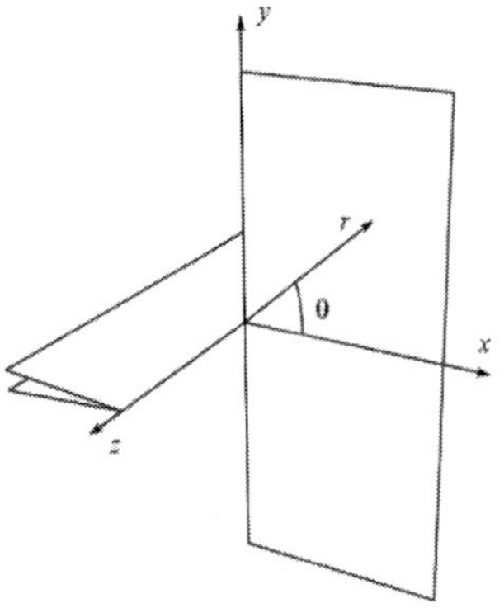

Fig. 7: Crack front Cartesian and cylindrical coordinate system.

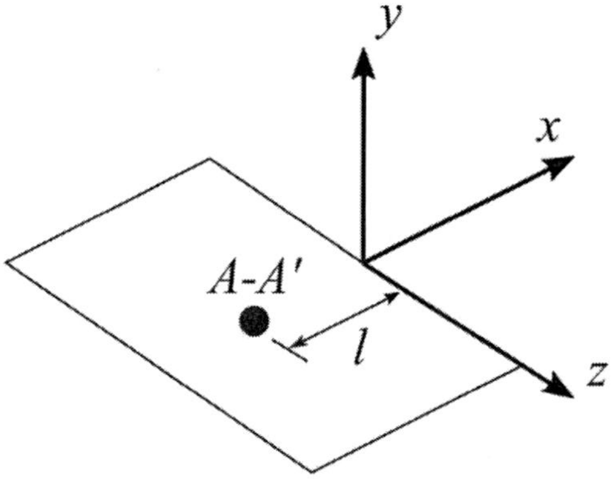

Fig. 8: Local coordinate system and configuration used to compute stress intensity factors with DCM.

and $r = l$ into Eq. (2) results in Eq. (3), which can then be solved for the stress intensity factors with a summation over $j = 1, 2, 3$.

Based on the anisotropic SIFs, the stress field around the crack tip is calculated using the descriptions of the stresses near the crack tip in an anisotropic material developed by Hoenig [19]. The use of a 3D-analysis enables the evaluation the RSIFs for every point along the crack fronts allowing for stress evaluations throughout the thickness of the model. The local stress tensor is given by:

$$\boldsymbol{\sigma}' = \frac{1}{\sqrt{2\pi r}} \boldsymbol{f}'(K_I, K_{II}, K_{III}, \theta), \quad (4)$$

where the entities r and θ are the distance from the crack front and the angle from which the crack front is approached. The superscript denotes that the stress tensor is scaled to account for Burgers' direction from which the crack front is approached, which does not necessarily lie in the plane orthogonal to the crack front. Further information and the definition of the tensor $\boldsymbol{f}'$ is given in the submitted manuscript by Busse *et al.* [14].

The stress tensor in Eq. (4) is then projected on the crystallographic slip planes with the normal **n** in the directions **n**, **s** and **t** and multiplied by $\sqrt{2\pi r}$, yielding the definitions of the RSIFs in the three modes of fracture on a crystallographic plane:

$$k_I = \boldsymbol{n} \cdot \boldsymbol{f}'(K_I, K_{II}, K_{III}, \gamma, \beta) \cdot \boldsymbol{n}, \quad (5)$$

$$k_{II} = \boldsymbol{s} \cdot \boldsymbol{f}'(K_I, K_{II}, K_{III}, \gamma, \beta) \cdot \boldsymbol{n}, \quad (6)$$

$$k_{III} = \boldsymbol{t} \cdot \boldsymbol{f}'(K_I, K_{II}, K_{III}, \gamma, \beta) \cdot \boldsymbol{n}, \quad (7)$$

This is illustrated in Fig. 9. As all modes of fracture are assumed to influence the crack growth behaviour, an equivalent crystallographic crack driving force parameter is defined by:

$$k_{EQ} = \sqrt{\psi k_I^2 + k_{II}^2 + k_{III}^2}, \quad (8)$$

The Mode I RSIF is scaled with ψ, which is a calibration parameter based on experiments in order predict correct

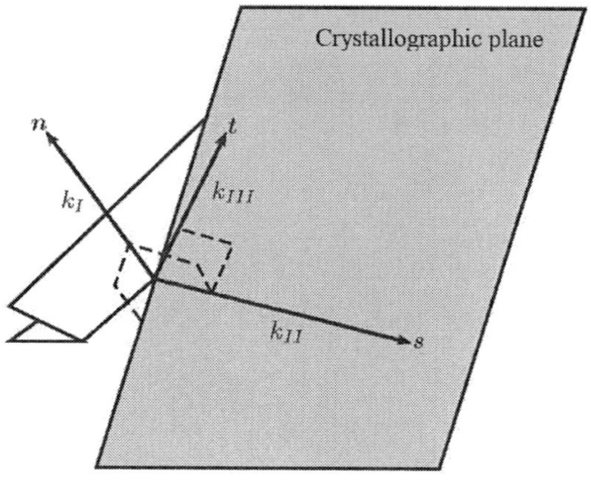

Fig. 9: Illustration of the directions of the RSIFs.

crystallographic cracking planes along a Mode I crack front. Further details can be found in [14].

4 Simulations and evaluations

In order determine the crystallographic crack driving forces for both crack instances an FE-model of the Kb specimen has been built. The following boundary conditions were applied; a fixed support at one of the circular surfaces and a concentrated force to one master node on the other circular surface. This force corresponds to the one in the experiment. The master node was coupled to all other nodes of the surface by a constraint that restrained motion in the plane and gave an equal deformation in the axial direction. The equivalent RSIF was then calculated along the crack fronts where crystallographic cracking is present and the conventional anisotropic SIFs are calculated for the parts of Mode I cracking. This is done for the cracks corresponding to the heat tints, since the crack geometry can be approximated best. The anisotropic SIFs are calculated by the crack evaluation tool FRANC3D [21]. The FE-solver of ABAQUS [22] is used to calculate the displacements. It should be pointed out that no singular elements were used at the crack fronts. This may lead to a less accurate capturing of the $\sqrt{r}$ singularity and thus affecting the quality of the calculated SIF values.

4.1 Fracture surfaces

In order to represent the experiments in meaningful manner the crack geometries corresponding to the above described heat tints are modelled. For this, the outlined crack front contours, *cf.* Fig. 4, are projected from the 2D top view on the 3D representation of the scanned fracture surface. The created 3D crack geometry corresponding to the heat tints of crack instance 1 and 2 are then transformed into a FE-mesh for the further analysis. The 3D mesh consisting of triangle shell elements was then converted into a volumetric void with a distance of 0.1 mm between the crack flanks allowing the computation of the SIFs for the complex crack geometry with three intersecting crack fronts. The corresponding 3D mesh for crack instance 1 is shown in Fig. 10, where the three different sections of the crack can be seen. On each crack geometry three different crack fronts are defined, corresponding to both crystallographic and the Mode I crack fronts as explained above.

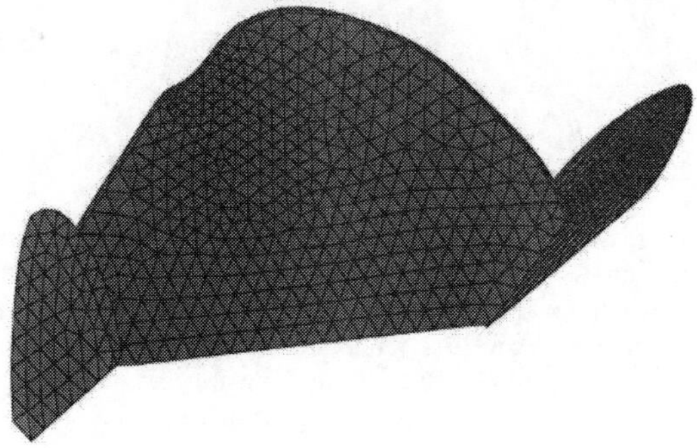

Fig. 10: 3D FE-mesh representation of the geometry of crack instance 1.

5 Results and Discussion

The equivalent RSIFs for the crystallographic crack fronts for the two considered crack instances are evaluated along the parts of the crack fronts, which are visible in the optical microscope. Furthermore, the nodes close to the free surface were disregarded due to the inaccuracies in the determination of the anisotropic SIFs. This is mainly due to poorly shaped elements causing local errors and a different order of the singularity. Even though not all points along the crack fronts were used to evaluate the crystallographic crack growth rate, it was important to model the whole geometry of the crack to capture the state more accurately.

The calibration parameter ψ necessary to predict the correct crystallographic cracking plane after a transition from Mode I cracking, as described in [14], was initially set to a value of 0.1 based on experimental results. In the referenced study, the RSIF was calculated along a Mode I crack and not along crystallographic crack fronts as in the current study. In this work, the RSIFs along the crystallographic crack front were evaluated using different ψ values, where the best correlation to the considered crack growth law was achieved by a ψ value of zero, meaning that once the crack grows on a crystallographic plane, the resolved shear stresses are the main driving forces. The resulting Resolved Shear Stress Intensity Factor (RSSIF) is then

$$k_{RSS} = \sqrt{k_{II}^2 + k_{III}^2}. \qquad (9)$$

The evaluated entities are then plotted in log-log diagram of the crystallographic crack growth rate of the crack driving force as described above. The results are shown Fig. 11, where it can be seen that all data points for both crystallographic cracks fall into a narrow band. The blue squares correspond to the left and the orange squares to the right crystallographic crack front. A power fitting was used to determine the parameters C_c and n_c to the values of $C_c = 3.155 \cdot 10^{-4}$ mm/cycle and $n_c = 0.6594$. The corresponding fitting curve is also shown in

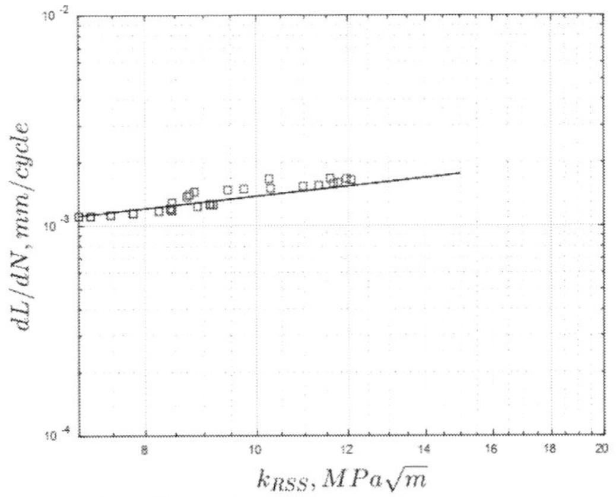

Fig. 11: Crystallographic crack propagation rate

5.1 Uncertainties and future work

It should be noted that the actual crack geometries for the crack instances 1 and 2 are only partly known and, as described above, the sections hidden beneath the Mode I portion of the crack are approximated. Thus, the stress state along the crack fronts in the experiment deviates from the one in the simulation. The error depends on the quality of the approximation. Furthermore, uncertainties in the determination of the misalignments can alter the stress state and thus influence the modelling of crack driving forces. As described above, the quality of the evaluated SIFs should be verified as future work by a comparison to a different method for calculating SIFs. Work to compute SIFs by a conservative crack front integral is ongoing. Furthermore, comparisons to different experiments are needed to verify the determined parameters.

6 Conclusions

- The data points of the crystallographic crack propagation rate collapse on a small band for both crystallographic cracks indicating a correlation between crystallographic crack propagation rate and the determined crystallographic crack driving force.
- A comparison to different experiments is needed to verify the determined parameters.
- 3D crack growth modelling of complex crack geometries with multiple intersecting crack fronts could be achieved, but the quality of the determined SIFs needs to be verified

Acknowledgments

This research has been funded by the Swedish Energy Agency and Siemens Industrial Turbomachinery AB through the Research Consortium of Materials Technology for Thermal Energy Processes, Grant no. KME-702, the support of which is gratefully acknowledged.

References

[1] A. Pineau and S. D. Antolovich, "High temperature fatigue of nickel-base superalloys - A review with special emphasis on deformation modes and oxidation," *Eng. Fail. Anal.*, vol. **16**, no. 8, pp. 2668–2697, (2009).

[2] D. Leidermark, J. J. Moverare, K. Simonsson, S. Sjöström, and S. Johansson, "Room temperature yield behaviour of a single-crystal nickel-base superalloy with tension/compression asymmetry," *Comput. Mater. Sci.*, vol. **47**, no. 2, pp. 366–372, (2009).

[3] D. P. Pope and S. S. Ezz, "Mechanical properties of Ni3Al and nickel-base alloys with high volume fraction of gamma prime," *Int. Met. Rev.*, vol. **29**, no. 3, pp. 136–167, (1984).

[4] M. Segersäll, J. J. Moverare, D. Leidermark, and K. Simonsson, *Low-cycle fatigue behaviour of a Ni-based single-crystal superalloy*, vol. **891–892**. (2014).

[5] D. Leidermark and M. Segersäll, "Modelling of thermomechanical fatigue stress relaxation in a single-crystal nickel-base superalloy," *Comput. Mater. Sci.*, vol. **90**, pp. 61–70, (2014).

[6] R. L. Amaro, S. D. Antolovich, R. W. Neu, and A. Staroselsky, "Physics-Based Modeling of Thermo-Mechanical Fatigue in PWA 1484," in *Superalloys 2012*, (2012), pp. 481–490.

[7] B. F. Antolovich, A. Saxena, and S. D. Antolovich, "Fatigue crack propagation in single-crystal CMSX- 2 at elevated temperature," *J. Mater. Eng. Perform.*, vol. **2**, no. 4, pp. 489–495, (1993).

[8] J. Telesman and L. J. Ghosn, "The unusual near-threshold FCG behavior of a single crystal superalloy and the resolved shear stress as the crack driving force," *Eng. Fract. Mech.*, vol. **34**, no. 5–6, pp. 1183–1196, (1989).

[9] J. Telesman and L. J. Ghosn, "Fatigue crack growth behavior of PWA 1484 single crystal superalloy at elevated temperatures," *J. Eng. Gas Turbines Power*, vol. **118**, no. 2, pp. 399–405, (1996).

[10] D. Leidermark, J. Moverare, K. Simonsson, and S. Sjöström, "A combined critical plane and

critical distance approach for predicting fatigue crack initiation in notched single-crystal superalloy components," *Int. J. Fatigue*, vol. **33**, no. 10, pp. 1351–1359, (2011).

[11] C. Busse *et al.*, "A finite element study of the effect of crystal orientation and misalignment on the crack driving force in a single-crystal superalloy," in *Proceedings of the ASME Turbo Expo*, (2016), vol. **7A–2016**.

[12] R. C. Reed, J. Moverare, A. Sato, F. Karlsson, and M. Hasselqvist, "A New Single Crystal Superalloy for Power Generation Applications," *Superalloys 2012*, pp. 197–204, (2012).

[13] A. Coles, R. E. Johnson, and H. G. Popp, "Utility of surface-flawed tensile bars in cyclic life studies," *J. Eng. Mater. Technol. Trans. ASME*, vol. **98**, no. 4, pp. 305–315, (1976).

[14] C. Busse *et al.*, "Prediction of crystallographic cracking planes in single-crystal nickel-base superalloys," *Manuscr. Submitt. Publ.*, (2018).

[15] ASTM, "Test Method for Measurement of Fatigue Crack Growth Rates," (2012).

[16] K. Schwalbe and D. Hellmann, "Application of the Electrical Potential Method to Crack Length Measurements using Johnson's Formula," *J. Test. Eval.*, vol. **9**, no. 3, pp. 218–221, (1981).

[17] P. C. Paris and F. Erdogan, "A Critical Analysis of Crack Propagation Laws," *Journal of Basic Engineering*, vol. **85**, no. 4. p. 528, 1963.

[18] D. R. Askeland and P. P. Phule, *The Science and Engineering of Materials*, Fifth Edit. Toronto: Thomson, Canada, (2006).

[19] A. Hoenig, "Near-tip behavior of a crack in a plane anisotropic elastic body," *Eng. Fract. Mech.*, vol. **16**, no. 3, pp. 393–403, (1982).

[20] L. Banks-Sills, P. A. Wawrzynek, B. Carter, A. R. Ingraffea, and I. Hershkovitz, "Methods for calculating stress intensity factors in anisotropic materials: Part II-Arbitrary geometry," *Eng. Fract. Mech.*, vol. **74**, no. 8, pp. 1293–1307, (2007).

[21] FRANC3D, *FRANC3D Reference Manual*. Ithaca, USA: Fracture Analysis Consultants Inc., (2016).

[22] ABAQUS, *ABAQUS 6.12 Documentation*. Providence, USA: Dassault Systèmes, (2014).

In-situ SEM and optical microscopy testing for investigation of fatigue crack growth mechanism under overload

Wei Zhang[1,*], *Liang* Cai[2]

[1]Reliability and System Engineering Department, Beihang University, Beijing, China

Abstract. In this paper, the in-situ scanning electron microscope (SEM) and optical microscopy experiments are performed to investigate the crack growth behavior under the single tensile overload. The objectives are to (i) examine the overload-induced crack growth micromechanisms, including the initial crack growth acceleration and the subsequent retardation period ; (ii) investigate the effective region of single overload on crack growth rate. The specimen is a small thin Al2024-T3 plate with an edge-crack, which is loaded and observed in the SEM chamber. The very high resolution images of the crack tip are taken under the simple variable amplitude loading. Imaging analysis is performed to quantify the crack tip deformation at any time instant. Moreover, an identical specimen subjected to the same load condition is observed under optical microscope. In this testing, fine speckling is performed to promote the accuracy of digital imaging correlation (DIC). The images around the crack tip are taken at the peak loads before, during and after the single overload. After that, the evolution of local strain distribution is obtained through DIC technique. The results show that the rapid connection between the main crack and microcracks accounts for the initial crack growth acceleration. The crack closure level can be responsible for the crack growth rate during the steady growth period. Besides that, the size of retardation area is larger than the classical solution.

1 Introduction

Many structural components are usually subjected to the moderate amplitude fatigue loading with occasional large spike loads, which can be simplified as constant amplitude loading with overloads. Many studies claim that the applied overload can lead to the typical nonlinear crack propagation, which includes instant crack growth rate acceleration, a long-time crack growth retardation and sometimes even crack arrest[1-4]. It is essential to understand the exact fatigue crack growth mechanism under the constant amplitude load with overload. A great amount of theoretical and experimental explorations proposed in the past to investigate mechanism of the transient acceleration and the following retardation period [5-9]. Among them, the crack closure concept proposed by Elber [10] is widely recognized in which the effective stress intensity factor is considered as the unique driving force of the crack growth. Crack closure is successfully used to explain the retardation phenomenon [11]. Some researchers use the crack closure to explain the post-overload transient acceleration [12,13]. They claim that the overload-induced enlarged plastic zone leads to crack tip blunting, which causes the concentration of high stress. However, the increase of effective stress intensity factor cannot fully accounts for the acceleration phenomenon, which makes the crack closure theory controversial. Many studies use the plasticity and residual stress to investigate the overload effect on fatigue crack growth. For instance, Willenborg et al. described the retardation phenomenon by proposing a relationship between the effective stress intensity factor and the plastic zone size [14]. However, the transient acceleration period is not taken into account in their model. Besides, their models are based on the assumption of monotonic plastic zone without consideration of the effect of residual stress by the previous loading sequence and the shape of the large plastic zone. In addition, the overload ratio for crack arrest equals to 2 according to Willenborg's model. However, many experimental results show that this value is greater than 2 in the actual situation [15]. Although, many studies are proposed to modify Willenborg's model [16-18], there is still a lack of convincing experimental observation. Therefore, a reliable approach to determine the crack closure level and the failure phenomenon is important to understand the crack tip driving force under variable amplitude load.

In this paper, an in-situ SEM fatigue testing approach is proposed to directly observe the crack closure within one loading cycle and the transient acceleration under single overload case. Meanwhile, an investigation of the plastic zone evolution near the fatigue crack tip is carried out to assist the examination of fatigue crack growth mechanism.

2 Experimental setup and specimen design

* Corresponding author: zhangwei.dse@buaa.edu.cn

Two different experimental approaches are applied in the current studies: an in-situ scanning electron microscope (SEM) testing and an in-situ optical microscopy testing. The in-situ SEM experiment is conducted to investigate the overload-induced crack growth micromechanisms by analyzing the high resolution images taken around the fatigue crack tip. The detailed illustration of the experimental setup is shown in Fig.1. It mainly includes a palm-sized load stage installed in a field emission SEM (FEG 650), inside of which a secondary electron beam is applied. The load stage has a maximum load capacity of 5kN. The specimen is clamped on the load stage and cyclic load is applied during the testing. The vacuum level remains lower than 3×10^{-2} MPa. The working distance is approximately between 15 to 20 mm. A desktop is used to monitor the deformation process near the crack tip and collect those images taken under the SEM.

Fig.1. Experimental setup of in-situ SEM testing (a) and the load stage installed in SEM (b)

The specimen is a single edged notch plate made of aluminum alloy sheet (Al2024-T3), as shown in Fig.2. Its width W is 10 mm, length L is 32 mm and thickness T is 1.02 mm. The edged notch of 1mm is machined by low speed wire-cutting electronical discharge machining (LSWEDM) for efficient avoidance of the residual stress in this cutting process. After that, the specimen is pre-cracked under a low stress level using an electromagnetic fatigue loading machine M-3000. Following this, the specimen with an initial crack is polished by using a metallographic polishing machine (1 um polishing suspension in the final step), which facilitates the observations of the crack tip deformation. The mechanical properties of Al2024-T3 are given in table 1.

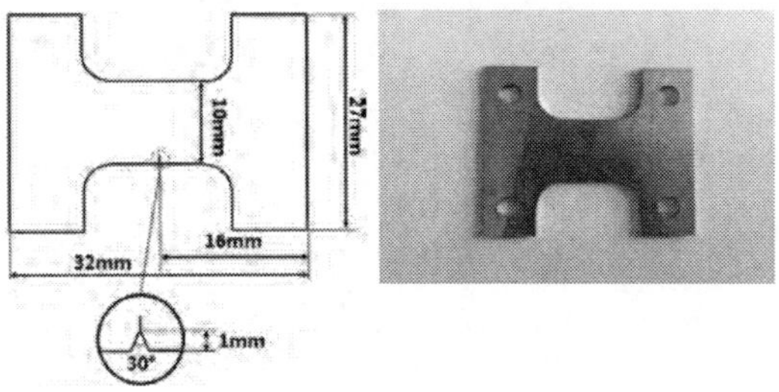

Fig. 2. The illustration the specimen geometry and specimen configuration

Table.1. mechanical properties of Al2024-T3

Ultimate tensile strength	453-459 MPa
Tensile yield stress	306-310 MPa
Young's Module	71.7 GPa

3 Experimental procedure and analysis

3.1 In-situ SEM testing

3.1.1 Experimental procedures

Tensile-tensile cyclic load is applied on the specimen and the crack growth behavior is monitored in-situ under the SEM. The overloading spectrum includes a constant amplitude loading part (the maximum loading 1.5kN and the minimum loading 0kN), and a single spike load with an overload ratio of 1.2. Firstly, the 30-50 cycles of loading is applied on the specimen to ensure the steady growth of the crack tip. Then, the crack tip deformation is traced before, during and after the single overload. Two major characteristics are measured: crack tip opening displacement (CTOD) and the crack growth. The CTOD under the constant amplitude load is defined as the opening displacement at the previous crack tip location. During the CTOD measurement, the load cycle is divided into many steps. When the stress level remains stable at each step, the images in the vicinity of the crack tip are taken under very high magnification. Reference points on the crack surface are chosen to ensure that the images are collected in-situ. The images taken during the loading process within one load cycle before overload are shown in Fig.3. The yellow lines represent the current crack tip positions and the horizontal brackets indicate the CTOD.

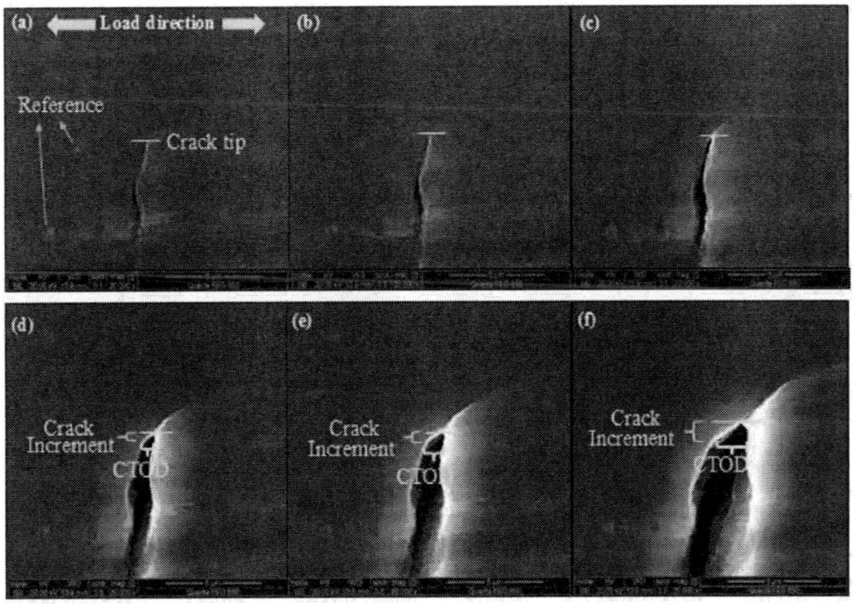

Fig.3. Illustration of CTOD measurements under different load levels during the retardation period (from a to f: K=0, 2.4, 4.8, 7.3, 9.7, 12.1Mpa•m$^{0.5}$, magnification 20000x)

The crack growth rate is defined as the crack increments between two maximum loads within one load cycle, where the crack increments are quantified by measuring the distance of the new crack tip to the original crack tip along the crack direction. The micrographs of crack tip at maximum load are taken cycle by cycle before the overload, shown in Fig.4. It is clearly seen that the fatigue crack tip produce a steady crack growth rate under constant amplitude load.

Fig. 5 shows the SEM micrographs near the crack tip at maximum load before and during the overload (Fig.5b and Fig.5c), and in the unloaded condition before overload (Fig.5a) and immediately after overload (Fig.5d). It is found that the CTOD significantly increases and plenty of shear bands appear in front of the crack tip during the overload. The crack tip is completely closed at unloaded condition before overload, whereas the crack blunting is found in the same loading condition right after the overload.

Fig.4. Crack increments at three consecutive peak loads before overload (magnification 20000x)

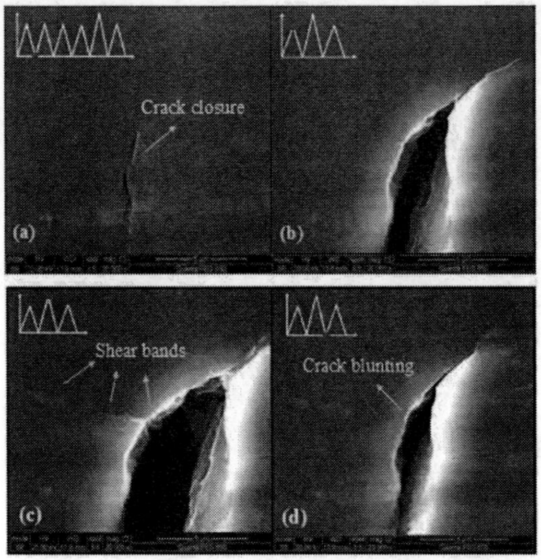

Fig.5. SEM micrograph: (a) minimum load before overload; (b) maximum load before overload; (c) maximum load during overload;(d) minimum load right after overload (magnification 20000x)

Fig.6 gives the crack tip behavior at maximum load immediately after the single overload. The magnification is adjusted to 5000x. It can be seen that the shear bands in Fig.6(c) develop into microcracks rapidly right after the overload. Then the microcracks connect with the main crack and produce a fast crack growth rate within several load cycles, as shown in Fig.6 (a)-(c). It is also found the crack tip usually tends to bifurcates after the overload. The main crack propagates along one of these two directions. Subsequently, a long-time crack growth retardation is observed. Fig.6 (d) illustrates the crack tip that just enters the retardation period. The crack growth rate during the acceleration period is calculated by the ratio of the entire crack increments during this whole period to its corresponding load cycles.

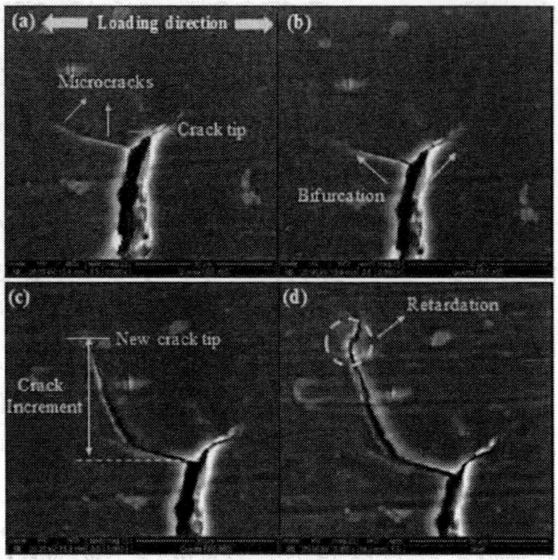

Fig.6. The crack tip behavior at maximum loading within several load cycles immediately after overload (magnification 5000x)

The illustrations of the crack tip deformation from the minimum load to the maximum loading within one load cycle during the retadation period are shown in Fig.7(a)–7(f). The measuring position corresponds to the yellow circle in Fig.6 (d). It is observed that no evident crack growth is found in one loading cycle during this period. Therefore, the CTOD is defined as the opening displacement with a measurement distance identical to that under the same load value before overload. Take Fig.7 (f) as an example, the measurement distance a is equal to the crack increment indicated in Fig.3 (f).

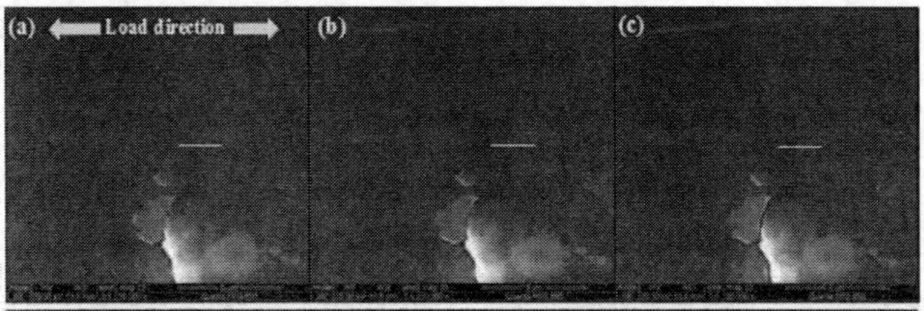

Fig.7. Illustration of CTOD measurements under different load levels during the retardation period (from a to f: K=0, 2.4, 4.8, 7.3, 9.7, 12.1, magnification 20000x)

https://doi.org/10.1051/matecconf/201816513013

Due to the overload-induced crack growth retardation, the crack increments are measured every ten load cycles in order to reduce the error. Fig.8 shows the two images of crack tip between two adjacent measuring position with an interval of ten load cycles. It is clearly seen that the crack growth rate decrease to approximately 0.2um/cycle.

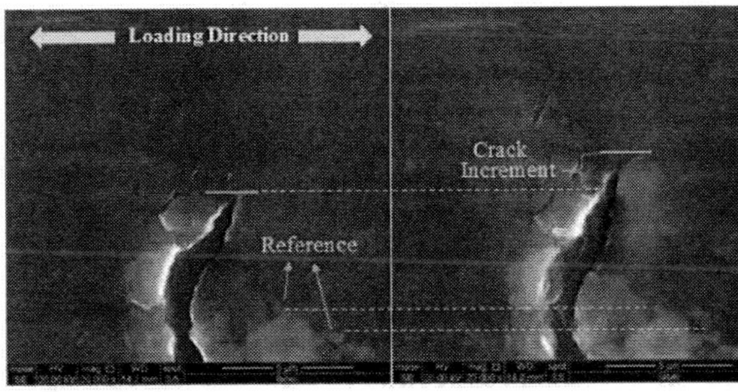

Fig.8. The crack increments between two adjacent measuring position during the retardation period (magnification 20000)

3.1.2 Analysis of experimental results

For the constant amplitude load part, it is observed the CTOD at peak load of one cycle before overload is approximately 2.5um and this value decrease to 1.2um at the beginning of the retardation period. The CTOD is correlated with the effective stress intensity factor(SIF) range modified by the crack closure level[19], given in equation (1). The equation (2) gives the relationship between effective SIF and crack growth rate[20]. Equation (3) can be drawn from equation(1) and (2).

$$CTOD = \begin{cases} \frac{1}{2\alpha}\left(\frac{K-K_{op}}{E\sigma_y}\right)^2 & K > K_{open} \\ 0 & K \le K_{open} \end{cases} \quad (1)$$

$$\frac{da}{dN} = C(K - K_{op})^m \quad (2)$$

$$\frac{da}{dN} = C'(CTOD)^{\frac{m}{2}} \quad (3)$$

In the current experiment, the measured crack growth rate before overload and during the retardation period is plotted against its corresponding CTOD in logarithmic coordinate, shown in Fig.9. It is seen that they exhibit line relationship under the logarithmic coordinate, where m approximately equals to 3 according to the slope. This indicates that CTOD can be used to estimate the crack growth rate. In other words, the crack growth rate can be reflected by the crack closure level.

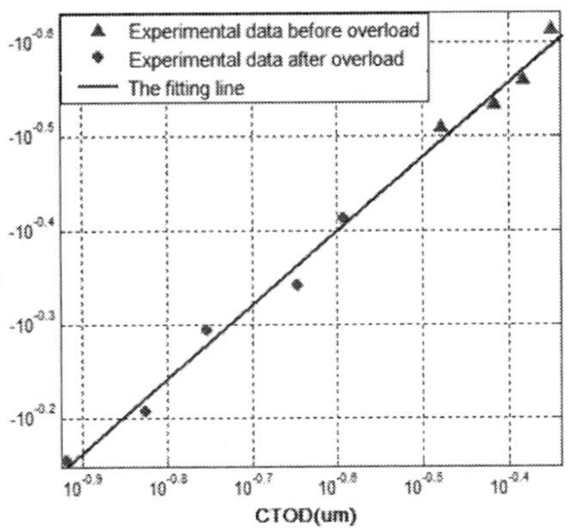

Fig.9. CTOD vs. crack growth rate in the single tensile overload case

The crack growth rate in the current overload case is plotted against the load cycles in Fig.10. The steady crack growth rate before overload is approximately 0.6 um/cycle. The single overload is applied on the specimen at 290th load cycles. The crack growth rate during the post-overload acceleration period is reaches 3.1 um/cycle. After that, the crack growth rate decreases sharply and experiences a long-time retardation. The trend of crack growth rate without overload are visualized as the green line by using linear fit of the experimental data acquired before overload. As the crack continues to advance, the crack growth rate in the sample cycles shows a gradual restoration to the trend before the spike load at approximately 1220th load cycles. The crack increment from the recovery position to the overload site is approximately 430 um according to the experimental measurement.

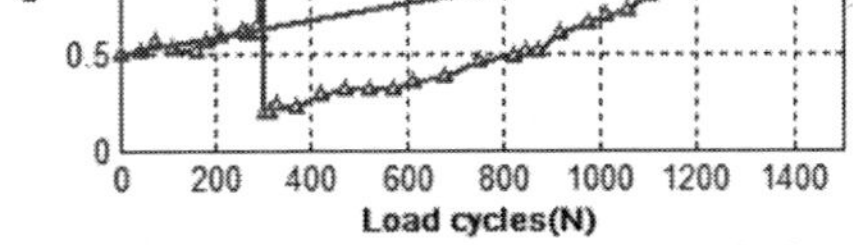

Fig.10. Load cycles vs. crack growth rate in the single tensile overload case

The theoretical predictions for the diameter of the forward plastic zone considering the crack closure effect can be expressed as

$$\rho_f = \begin{cases} \frac{1}{\pi}\left(\frac{K-K_{op}}{\sigma_y}\right)^2 & K_{max} \ge K > K_{op} \\ 0 & 0 \le K \le K_{op} \end{cases} \quad (2)$$

* Corresponding author: zhangwei.dse@buaa.edu.cn

$$K = F \times \sigma\sqrt{\pi a} \tag{3}$$

$$F = 1.12 - 0.23 + 10.6\left(\frac{a}{W}\right) - 21.7\left(\frac{a}{W}\right)^3 + 30.4\left(\frac{a}{W}\right)^4 \tag{4}$$

The diameter of the overload-induced enlarged forward plastic zone in current study equals to 403 um by using equation (2), which is slightly less than the experimental results of the distance from the crack growth rate recovery position to the overload site.

3.2 In-situ optical microscopy testing

In our previous study, an equivalent specimen subjected to the same load condition is observed under optical microscope. The experimental set-up for the in-situ optical microscopy testing is shown in Fig.11, which mainly includes: an in-situ fatigue testing stage, an optical measuring microscope, and an imaging processing system. The loading stage is installed on the platform of measuring microscope, whose magnification is adjusted to 50x. The tested specimen is fixed on the stage and cyclic loading is applied on the specimen. A microscopy camera is used to acquire the images in high resolution during the loading process. Afterwards, these images are collected and processed by digital imaging correlation (DIC) system.

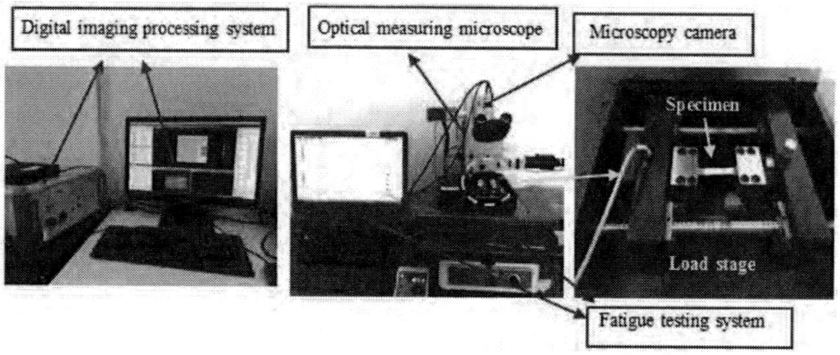

Fig.11 The experimental set-up of in-situ optical microscopy testing

In this testing, the specimen surface is speckled with tiny dense spots to improve the accuracy of DIC. For the measurement of strain distribution, the images of the crack tip taken at the previous minimum loads are set as reference and the image taken at current maximum load are compared with it. The distance from the crack tip to the plastic zone boundary along the crack direction is defined as the diameter of the plastic zone. The strain at boundary of the plastic zone is approximately equal to yield strain. Once the strain distribution along the diameter direction is averaged measured, the diameter can be easily determined.

The strain evolution in the vicinity of the crack tip under the single overload are acquired and measured. The strain contours near the fatigue crack tip at maximum loads before, during and right after the overload are given in Fig.12 (a)-(c), where the white line indicates the crack tip position. It is can be seen that plastic zone near the crack tip dramatically declines right after overload, compared with that before overload under the same loading condition. The variations of plastic zone at the maximum loads and the corresponding crack increments are measured every one hundred cycles after the single overload. It is observed that, as the fatigue crack advanced, the plastic zone size gradually recovers to the level before the overload. Fig.12 (d) shows the restoration of plastic zone at the 2000th load cycle after the spike load.

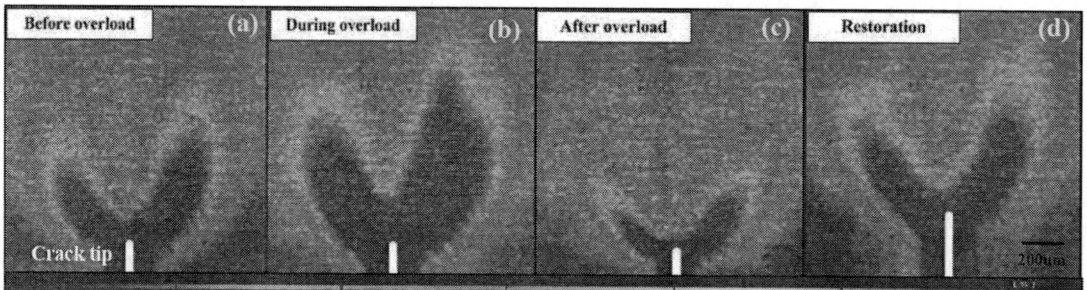

Fig.12 The strain evolution near the crack tip at maximum loading in overload case (from left to right: before, during, right after the overload and 2000th after overload)

The evolution of plastic zone size during the whole single overload case is plotted against the crack growth in Fig.13. It is observed that after the single overload, the plastic zone size returns to the original level before overload until the crack growth reaches 280 um. This distance is almost equal to the diameter of large plastic zone caused by the single overload.

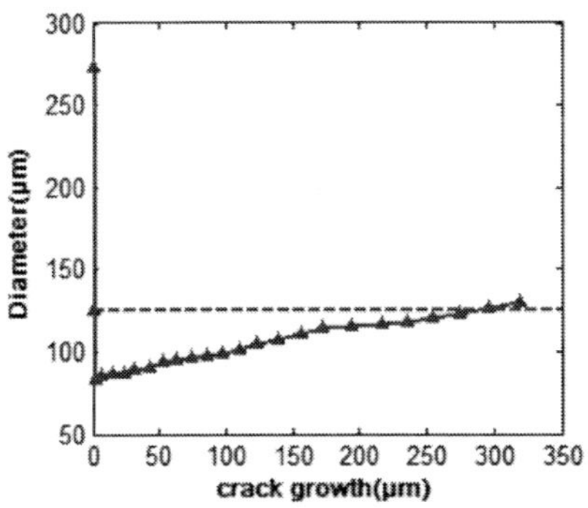

Fig.13 Crack growth vs. measured plastic zone diameter

4 Discussions

Based on the experimental results and analysis, it is found that the CTOD can be used to estimate the crack growth rate during the steady growth stage, which suggests that the crack closure level can be a dominant parameter of the crack growth rate.

There is no crack closure in the unloaded condition immediately after the overload, which might contribute to the initial acceleration due to the high stress concentration in the crack blunting region. However, the crack growth rate is 5 times than that of this value in the steady growth period before overload. The increase of effective stress intensity factor cannot fully accounts for the post-overload transient acceleration. Another micromechanism might be due to the overload-induced enlarged plastic deformation ahead of the fatigue crack tip, shown in Fig.12 (b), which produces a great number of shear bands, microcracks and potential damage. Then, the rapid connection of the microcracks and the primary crack generates a fast crack growth rate. The crack bifurcation caused by the single tensile overload may also reduce the stress intensity factor of each crack tip, which contributes to the crack growth retardation.

As for most of the plasticity-based models describing the crack growth retardation, it is supposed that once the current plastic zone reaches the boundary of large plastic zone, the retardation phenomenon vanishes [11,21,22,23]. However, according to the experimental observations in the SEM and optical microscopy testing, both of the crack growth rate recovery location and the plastic zone restoration position are further away from the overload site than the classical solution. This phenomenon is probably related with the shape effect of the large plastic zone, which is not taken into account in those classical plasticity-based models.

5 Conclusions and future work

In this paper, in-situ SEM and optical microscopy testing are performed to investigate the fatigue crack growth mechanism under the single overload. The SEM experiment successfully investigated the micromechanism of the transient crack growth acceleration and the following retardation period. Both the in-situ SEM and optical testing results suggest a revision of the classical plasticity-based retardation model. Several conclusions are drawn based on current study.

- The crack closure is considered as a dominant parameter of the crack growth rate in the steady growth stage.
- The rapid coalescence of the microcracks with the primary crack is a significant mechanism for transient acceleration right after the overload. The decrease or vanishing of the crack closure caused by large plastic deformation of the crack tip cannot fully explain this transient acceleration.
- The crack bifurcation caused by the single tensile overload contributes to the crack growth retardation.
- The actual region of retardation is larger than the assumption of classical solution. The shape effect of the large plastic zone probably contributes to this phenomenon, which is ignored in the classical models

In the future, the overload-induced crack growth micromechanisms of different metallic materials will be investigated. Besides, multiple overload ratio cases, even the crack growth arrest, need further exploration.

Acknowledgements

This research reported in this paper was financially supported by National Natural Science Foundation of China (No. 51405009).

References

1. Damri. D, Knott, J.F. Fatigue & Fracture of Engineering Materials & Structures, **14(7)**, 709-719 (1991)
2. Seifi R, Bahrami R. Engineering Failure Analysis, **17(6)**,1475-1482(2010).
3. Corbly D M, Packman P F. Engineering Fracture Mechanics, **5(2)**, 479-497(1973).
4. Rao K T V, Ritchie R O. Acta Metallurgica, **36(10)**, 2849-2862(1988)
5. Newman, J. C. Astm Stp, 1981.
6. Bichler C, Pippan R. Engineering Fracture Mechanics, **74(8)**, 1344-1359(2007).
7. Borrego L P, Ferreira J M, Cruz J M P D, et al. Engineering Fracture Mechanics, **70(11)**, 1379-1397. (2003)
8. Robin C, Pelloux R M. Materials Science & Engineering, **44(1)**, 115-120(1980)
9. Elber W. Astm Stp, 486(1971).

10. Yang J, Zhang W, Liu Y. International Journal of Fatigue, **62(2)**, 144-153(2014).
11. Lee S Y, Liaw P K, Choo H, et al. Acta Materialia, **59(2)**: 485-494(2011).
12. Lee S Y, Choo H, Liaw P K, et al. Acta Materialia, **59(2)**, 495-502(2011).
13. J. Willenborg, R.M. Engle, H.A. Wood, 1971.
14. Manjunatha M, Parida B K., **42(8)**:1536-1542. (2004)
15. Ziegler B, Yamada Y, Jr J C N., **78(14)**, 2609-2619. (2011)
16. Sun G, Chen J. Journal of Mechanical Strength, **10(2)**, 161-173(1992).
17. Ray A, Patankar R. Applied Mathematical Modelling, **25(11)**, 979-994(2001).
18. Ray A, Patankar R. Applied Mathematical Modelling, **25(11)**, 995-1013(2001).
19. Zhang W, Liu Y. International Journal of Fatigue, **42(4)**,14-23(2012).
20. P.C. Paris, F. Erdogan, Journal of Basic Engineering, **85(4)**,528-533(963).
21. O.E. Wheeler, **94(1)**, 181–186.2(1972)
22. B.K.C. Yuen, F. Taheri, International Journal of Fatigue, **28(12)**, 1803-1819(2006).
23. M.V.S. Pereira, F.A.I. Darwish, A.F. Camarão, et al. Materials Research, **10(2)**, 101-107(2007).

Influence of orientation and temperature on the fatigue crack growth of directionally solidification superalloys DZ125

Yi Shi[1,*], *Xiaoguang* Yang[1], *Didi* Yang[1], *Guolei* Miao[2] and *Duoqi* Shi[1]

[1]School of Energy and Power Engineering, Beihang University, 100191 Beijing, China
[2]Chengdu Holy Industry & Commerce Corp.LTD, 611936, Sichuan Province, China

Abstract. The influence of orientation and temperature on the fatigue crack growth of a directionally solidification superalloys named DZ125 were investigated in this study. The specimens were designed based on ASTM E647 standard and the compliance method was employed to monitor the crack growth. Tests were conducted at room temperature, 760℃ and 850℃ respectively. The longitudinal and transverse orientation were both studied in this tests. Results show that the orientation dependence is weak at elevated temperature compared with that at room temperature. However, the temperature influences the fatigue crack growth rate significantly and the largest difference can be an order of magnitude on the da/dN-ΔK graph. But the reduction in higher ΔK area of elevated temperature can still be observed. Potential reasons and hypothesis were explained in this paper. In addition, the crack growth mode appears to transfer from type I with the increase of temperature which also reflects the influence of temperature.

1 Introduction

DZ125 is a Chinese made nickel based directionally solidification superalloys, which is similar to the Rene'125 in USA. Due to its good resistance to creep and excellent strength at elevated temperature, it is applied for blades and vanes of advanced aero-engine [1]. However, the working condition of turbine blades is so harsh that it is close to the materials' capability limit. To ensure the safety of operation, understanding the behavior of fatigue crack growth at elevated temperature is important for design and maintenance. Many previous studies focused on DZ125's behavior of low cycle fatigue life at elevated temperature [2-3], TMF [4], and environment and coating influence [5-6] by conducting experiments on smooth cylinder specimen, while less efforts are put on the fatigue crack growth behavior of DZ125. Other directionally solidification material like IN738, GTD-111's fatigue crack growth behavior were studied.

This study aims at investigating the influence of orientation and temperature on the fatigue crack growth behaviour of DZ125. The fatigue crack growth rates were obtained by conducting experiments and the crack growth path was also compared to analyze the temperature and orientation effects.

2 Experiment

2.1 Material and specimen

The material was subjected to the following heat treatment: 1180℃/2 h→1230 °C/3 h, AC + 1100 °C/4 h, AC + 870 °C/20 h, and AC (AC: air cooling) before machined. The nominal chemical composition is presented in **Table. 1** and mechanical properties at different temperature are listed in **Table. 2**.

All the test specimens were machined from three pads whose dimension was 2000×60×20mm. The directional solidification direction was shown in **Fig. 1**. The compact tension (CT) specimens, shown in **Fig. 2** were designed based on the ASTM E647 standard [7], with 50mm wide and 10mm thick. The notch length of the CT specimen is a_0=8mm. They specimens are divided into two groups, T and L. For the loading direction is parallel to the directionally solidification direction, these specimens are denoted as L-CT. While for the loading direction is perpendicular to the directionally solidification direction, such specimens are denoted as T-CT, as shown in **Fig. 1**.In addition, to accommodate the extensometer, the knife edge at the side of CT specimen for elevated temperature is different from that of a room temperature specimen.

2.2 Test method

In this study, the compliance method was applied to monitor the length of crack according to ASTM E647.

* Corresponding author: shiyi_sepe@buaa.edu.cn

The definition of compliance, $c = \delta/P$, is the instantaneous ratio between displacement and load [8], which is also the reciprocal of the stiffness D. Also, the compliance would remain constant until the crack has new advance. The compliance value increases with the crack growth, which also indicates the degraded resistance to the external load. At the premise of linear condition, such relationship can be analytically expressed in a dimensionless value, (EvB/P, E is the elastic modulus, v is the displacement between measurement points, B is the specimen thickness and P is force) and later is correlated to the crack length, which is similar to the potential drop method [9]. The theoretical compliance expression for CT specimen is given in the E647 standard, see equation (1)

$$a/W = N_0 + N_1U + N_2U^2 + N_3U^3 + N_4U^4 + N_5U^5 \quad (1)$$

Where: $U = 1/\left[A^{\frac{1}{2}} + 1\right]$; $A = \varepsilon EBW/P$; $N_0 = 1.0033$; $N_1 = -2.35$; $N_2 = 1.3694$; $N_3 = -15.294$; $N_4 = 63.182$; $N_5 = -74.42, for\ 0.2 < a/W < 0.95$

Table. 1. Nominal chemical compositions of DZ125 (wt. %)

Ni	C	Cr	Co	W	Mo	Al	Ti	Ta	B	Hf
Bal	0.096	8.77	9.88	7.29	2.00	5.13	1.03	3.70	0.015	1.42

Table. 2. Material tensile properties

Orientation	Longitude				Transverse			
Temperature(℃)	20	760	850	900	20	760	850	900
Young's modulus (GPa)	125.5	99.5	91.0	82.0	152.5	115.5	117.0	110.0
Tensile strength (MPa)	1220	1200	1040	860	990	940	915	810
0.2% offset yield strength (MPa)	915	960	825	635	815	730	760	625
Elongation (%)	11.0	13.0	16.0	19.0	13.0	9.7	9.0	8.7
Reduction in area (%)	12.5	20.0	21.0	26.0	20.0	26.0	24.0	23.0

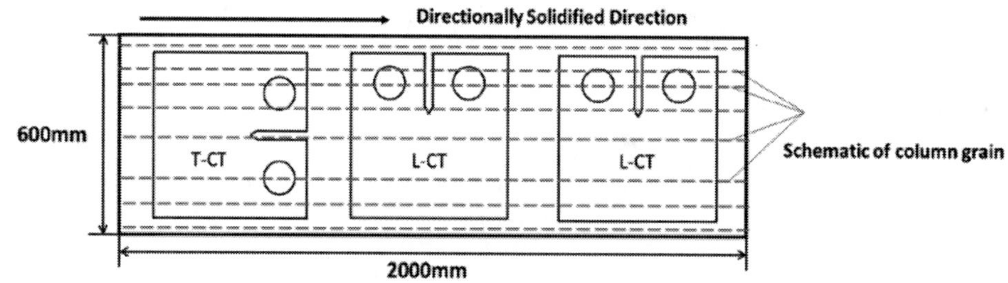

Fig. 1. The schematic of spemiens cut from a tab

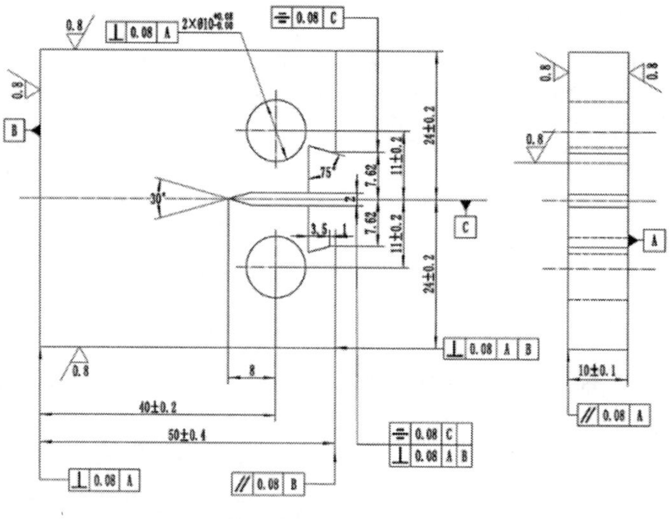

Fig. 2. The geometry of the compact tension specimen

The optical method was not utilized in this study due to the extremely high temperature. According to the physical law: anything will emit various radiation at any temperature. When the surface temperature is below 500 ℃, it mainly emits infrared light with longer wave length. Generally, the sensor of camera is not sensitive to the long wave length. Therefore, there will be no influence on the brightness of the captured images. However, with the temperature increases, the corresponding wavelength shifts to shorter area which falls into the sensitive area of the sensor. Thus, the image will be too bright to observe and measure the crack length [10].

2.3 Experimental Procedure

All the fatigue crack growth tests were conducted on the MTS 370 servo-hydraulic system. The load was monitored by the system load cell. Prior to the fatigue crack growth test, all specimens were pre-cracked for 1mm at room temperature to ensure the shape of crack inside meets the corresponding requirement. Then the knife edge was machined to accommodate the extensometer. The gage whose length was 15.2mm and the diameter was 5mm was assembly at the knife edge of the specimen. This is designed to extend to the furnace, shown in **Fig. 3**. To ensure the accuracy of measurement, moderate cyclic loading was applied and the gage was adjusted until the curve of COD and force was approximately two close line during the loading and unloading stage, shown in **Fig. 4**. Some space is left for the tip of the rod and the knife edge considering the expansion of specimen after heating.

Later, the specimens were heated in a resistance-type furnace and center- temperature of the notch was measured by the thermal-couples. The specimens were set at minimum loading during the heating process and they were soaked at the expected temperature for 1h before the fatigue crack growth test. The thermal gradient was strictly controlled below 5 ℃ and the response of COD was also checked under cyclic loading.

The condition of fatigue tests were shown in **Table. 3**. The maximum loading ranges from 8.6kN to 9.4 kN and the reason for this difference will be explained later. The test frequency was fixed at 10Hz and the stress ratio was set constant at R=0.1. The specimens were conducted at room temperature, 760 ℃ ,850 ℃ ,respectively. The stress intensity factor, K is calculated according to the formula from the E647 standard , see eqution (2), where $\alpha = a/W$, a is crack length, W and B is 40 and 10mm for this study respectively. ΔP means the external loading range. When the crack length reached 25mm, the test terminated.

$$\Delta K = \frac{\Delta P}{B\sqrt{W}} \frac{(2+\alpha)}{(1-\alpha)^{3/2}} (0.886 + 4.64\alpha - 13.32\alpha^2 + 14.72\alpha^3 - 5.6\alpha^4) \quad (2)$$

Table. 3. Test conditions for specimens

Test Specimens	Temperature ℃	Maximum Loading kN	Loading Ratio	Frequency	Ending cycles
L1	RT	8.6	0.1	10Hz	163017 cycles
T1	RT	8.6			341389 cycles
L4	850	8.6			53517 cycles
L5	850	8.6			Fail, 618600 cycles, No Propagation
L6	850	9.4			Fail, 20503 cycles, No Propagation
L9	760	9.4			32919 cycles
T3	760	9.4			31460 cycles
T8	850	10.0			19433 cycles

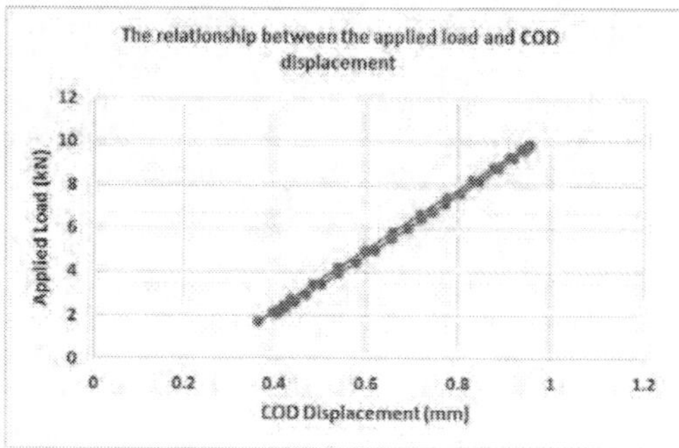

Fig. 3. The relationship between the applied load and COD displacement

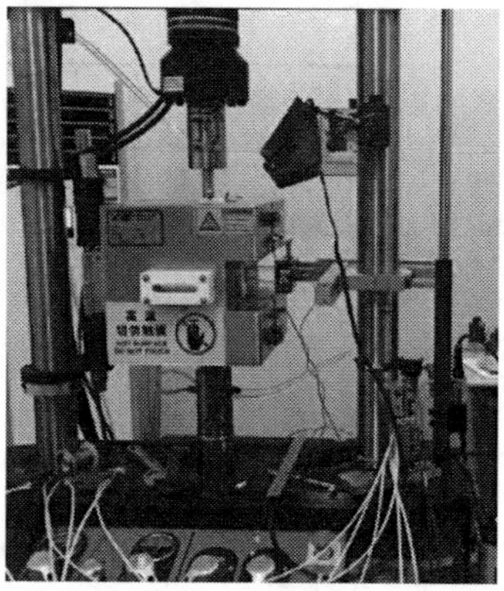

Fig. 4. The setup of the elevated temperature

3 Result and discussion

3.1 The fatigue crack growth rates

The overall test results are plotted in the Paris regime, shown in **Fig. 5**. The influence of temperature is obvious, especially for the difference between elevated temperature and room temperature. Meanwhile, the fatigue crack growth (FCG) rate curves were also correlated according to the Paris Law, the result are shown in **Table. 4**. The correlation coefficients have an average of 0.97, which means the curve fits the Paris Law well.

$$\frac{da}{dN} = C(\Delta K)^m \quad (3)$$

Table. 4 Material constant and correlation coefficient for various temperature and orientation

Temperature	Orientation	C	m	R
Room temperature	L	1.63×10^{-11}	4.74	0.99
	T	5.65×10^{-11}	4.34	0.99
760°C	L	3.86×10^{-10}	4.09	0.97
	T	2.17×10^{-8}	2.90	0.97
850℃	L	3.35×10^{-11}	5.05	0.98
	T	3.79×10^{-9}	3.57	0.99

The change of coefficient C and m with the temperature is shown in **Fig. 6**. Both orientation exhibits the same trend in coefficient C and m respectively, but the trends between two coefficients are converse. Coefficient C for orientation T first increased with the growth of temperature. After peaking at the 760°C, it dropped. Whereas, coefficient m for orientation T first decreased and then reached its bottom at 760°C before increased at 850℃. In addition, the coefficient C for orientation T is always higher than that of L, while the coefficient m for orientation T is much lower than L. It seems to be that the coefficient m and C are independent of the temperature for both directions.

 https://doi.org/10.1051/matecconf/201816513014

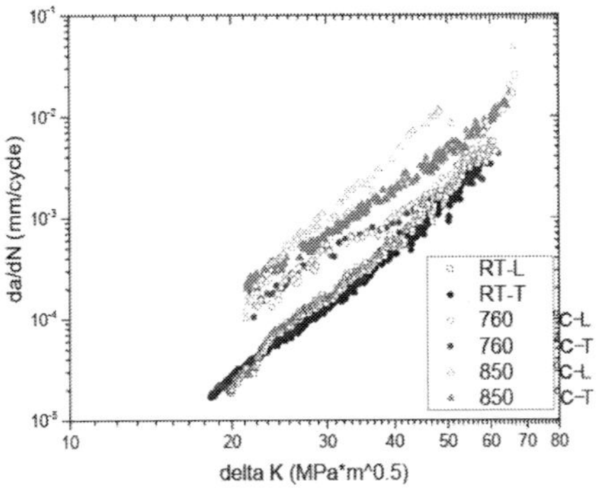

Fig. 5. The da/dN-ΔK curve of six specimens

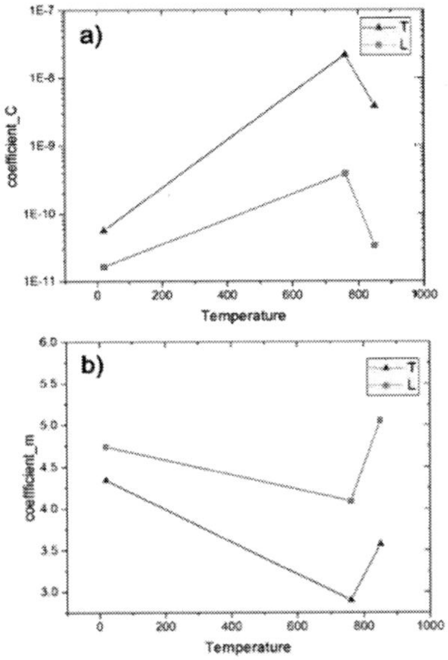

Fig. 6. The relationship between temperature and coefficients, a) C, b) m

Fig. 7 shows the comparison between parts of the test results and handbook data[11]. The hollow points represent the data collected from the handbook. The test points are almost consistent with the presented data in the medium and high ΔK area, which also means that the result obtained by compliance method is also confident.

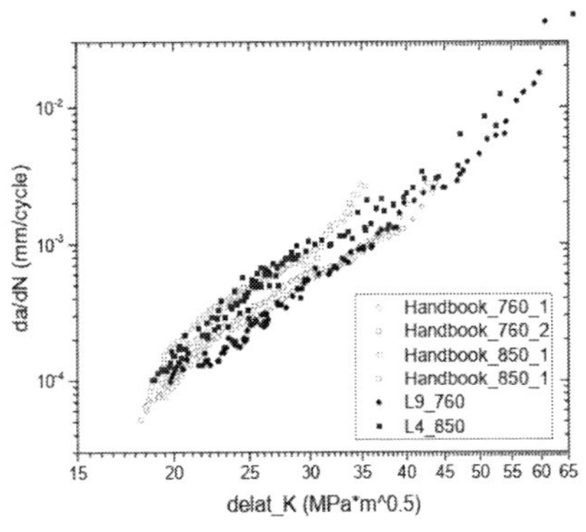

Fig. 7. The comparsion between the curve from test and handbook

3.2 Influence of orientation on FCG rates

Fig. 8 shows the crack growth behavior for T & L orientation at three different temperatures. At the room temperature, the specimen of L exhibited higher FCG rate than that of T through all stages of ΔK, and more specifically the difference seems to be 3-5times at the medium ΔK level. When it comes to the elevated temperature, there is no such significant difference at lower and medium ΔK level between the L and T orientation. However, for 760℃,in higher ΔK level, the FCG rates between orientation L and T separate obviously. The FCG rate of L is much higher than that of T at the same stress intensity factor, which is not observed in other temperature. The orientation seems to less influence the FCG rates with the increase of temperature. For 850℃, the gap between red curve and blue curve in higher ΔK level is reduced compared with 760℃. In general, the FCG rates of L orientation is higher than T at the three test temperature. However the difference becomes slighter with the increase of temperature, except the obvious difference in higher ΔK level at 760℃. In other words, the influence of orientation of FCG rates is limited at elevated temperature. This is also consistent with other studies: Shi [12] studied the directionally solidification nickel based superalloys PW1484 at high temperature and pointed that the effect of orientation on the crack propagation is very small. Yoon [13] also founded that for the GTD-111 the orientation dependence was marginal at the temperature of 649, 761 or 871℃.Although Yoon did not point out the orientation influence at room temperature, an obvious difference can be found in the corresponding graphs.

https://doi.org/10.1051/matecconf/201816513014

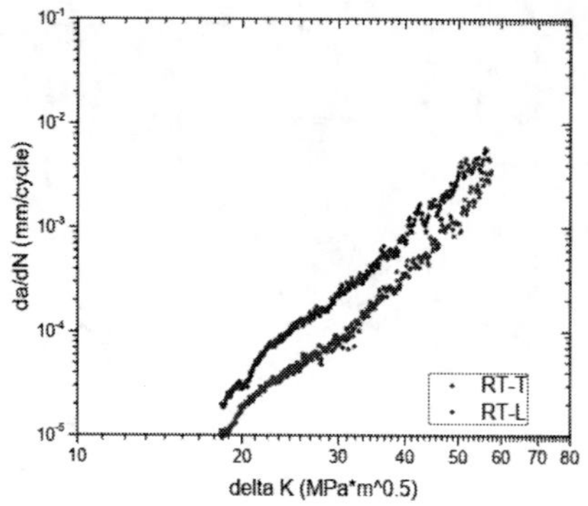

(a) room temperature

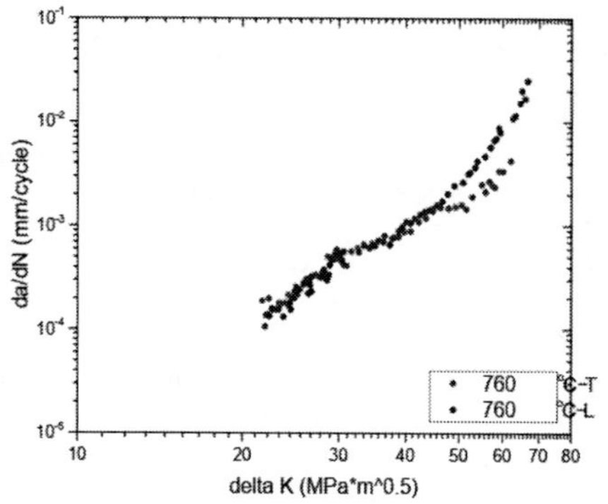

(b) 760℃

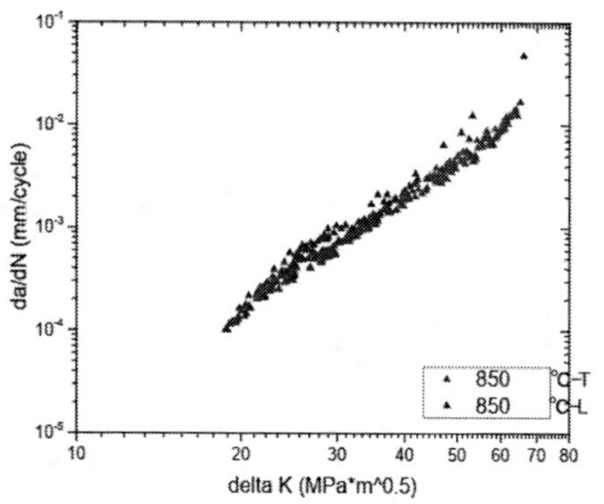

(c) 850℃

Fig. 8. Inlfluence of orientation on FCG

3.3 Influence of temperature on FCG rates

Fig. 9 shows the FCG rates of cracks for orientation L. The temperature dependence is strong at low and medium ΔK level. The elevated temperature seems to accelerate the FCG rates 10 times larger than that at room temperature. However, in the higher stress intensity factor, the FCG rate at the room temperature even slightly exceeds that at 760 ℃ .For high temperature, 90℃'s increase can also exhibit obvious difference. As shown in **Fig. 9**, the curve for 850℃ is always above the curve of 760℃ and room temperature. Even in the higher stress intensity level, there is no trend of FCG rates reduction.

For the orientation T, similar temperature independence can be found. This also indicates the weak orientation dependence. The FCG rates of 850℃ is still the highest than any other temperature through all the ΔK level. Meanwhile, at higher ΔK level, the FCG rates of room temperature 'caught up' that of 760℃.

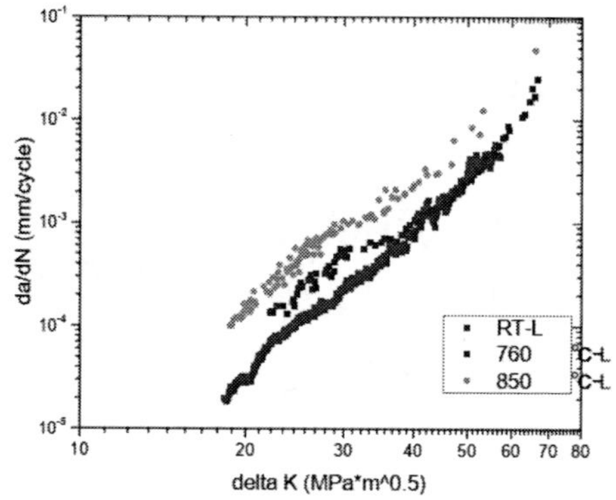

(a) Longitundial orientation

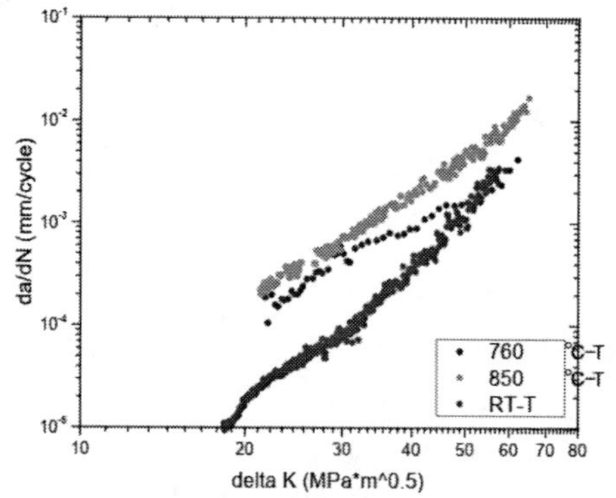

(b) Transeverse orientation

Fig. 9. Influence of temperature on FCG

In conclusion, the increase of temperature can exert strong influence on FCG rates, which speeds up the crack propagation. However, even the room temperature can still have higher FCG rate than 760℃ in higher ΔK level. Generally speaking, high temperature is likely to activate the dislocation motion. And it is also postulated that higher temperature will cause more oxidation, which degraded the material resistance and thus increasing the FCG rates. The reason for the ' collapse ' for 760℃ in higher stress level is likely to be the crack closure effect induced by the oxidation. However, it seems to be paradox for the trend observed at 850℃. Perhaps the reason for accelerating the FCG rates is determined by other factors. For 760℃, the crack closure plays a more leading role and thus decreasing the FCG rates in higher stress intensity factor level. While for 850℃, the crack closure effect seems to be weak compared with others, thus the curve of lower temperature fails to exceed 850 ℃ in higher ΔK level.

In the test process, high temperature is also likely to influence the fatigue crack growth threshold. As is shown in **Table. 3** two tests of specimens L5 and L6 could last 20503 and 617600 cycles without any signal of propagation. However, under the same test condition, specimens L4 only took 53517 cycles to the length of 25mm which was set as the terminal condition. For L5 and L6 specimens, we also examined the CT specimens under optical microscope and no sign of propagation can be found. Therefore, we inferred that the threshold is might larger than that of room temperature. The maximum load for L6 was 0.8kN larger than that of L4, L5 and other specimens conducted at room temperature before. In the subsequent tests, the temperature was adjusted to 760℃ with maximum 9.4kN. For tests of L9 and T3, approximately 35000 cycles were consumed before reaching the terminal condition and no 'not propagation' phenomenon were observed with higher maximum loading. It is hypothesized that the crack closure induced by oxidation is more serve than that at room temperature, and thus increase the threshold level. Telesman[14] reported that the fatigue crack growth threshold for a PM nickel based superalloy was influence by temperature and frequency. The increase of temperature will lead to the increase of threshold value. Bouvard [15] and K.S Chan [16] also mentioned that the thickening of oxide layer on the crack lips tends to increase the opening threshold. However, at present we still lack some evidence to confirm that the oxidation in the main cause for this. In addition, it is also hard to explain the test for L4 whose maximum force was 8.6kN. Further efforts should be made to investigate by studying the fractography under SEM.

4 FCG mode

All the test specimens were separated into two parts shown in **Fig. 10**. For specimens at room temperature, all the crack paths were perpendicular to the loading direction, which satisfies the type I crack growth. However, at the elevated temperature, deflections can be observed in all directions at different high temperature. For L specimen under 760 ℃, the crack path grew upward with a certain angle, while the counterpart of T exhibited a 'strange' and downward path. At 850℃, L and T specimens travelled in a similar downward path. From the result, it seems that both orientation and temperature could change the fatigue crack growth path. The crack path belongs to the type I-II combined mode. Therefore, the fatigue crack growth path can also to some extent reflect the influence of temperature. The reason for this deflection is likely to be that thermal activated some mechanism and leads to the crack did not travel perpendicularly to the loading direction. And it is also likely to be caused by the barrier of grain in the crack advancing. In other words, the effect of column grain boundary in the FCG process in unneglectable at elevated temperature. More detailed study should be conducted to confirm it.

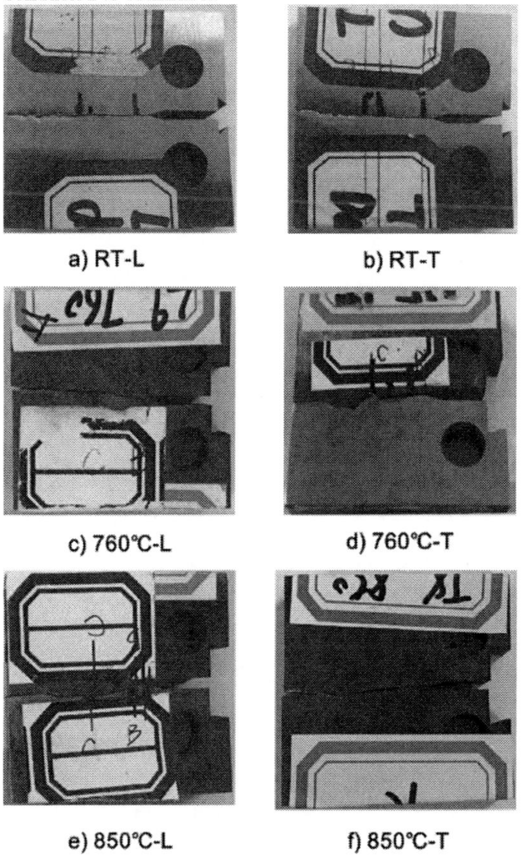

Fig. 10. The FCG path of CT

5 Conclusion

The influence of temperature and orientation on the fatigue crack growth behavior of directionally solidification superalloys named DZ125 have been investigated in this study. Six CT specimens fall into two orientation (longitudinal and transverse) or three temperature (room temperature, 760℃, 850℃). Relevant conclusions are given below:

1) The orientation influence seems to be most evident at room temperature, and the temperature effect becomes less significant in low and medium ΔK level with the increase in temperature. For room temperature, 760℃ and 850℃, orientation L exhibits slightly higher FCG rates than that of T.
2) The increase of temperature accelerated the FCG rates and the difference between the room temperature and elevated temperature is almost an order of magnitude. However, at higher ΔK level, the curve of room temperature could exceed that of 760 ℃ for both orientations, which is mainly contributed

by the crack closure effect.

3) The fatigue crack growth threshold is also likely to be influenced by the increasing temperature, and it is also hypothesized that oxidation is responsible for this.
4) The fatigue crack growth path is different from room temperature and elevated temperature. Both L and T orientation travelled perpendicularly to the loading direction which belongs to the type I crack at room temperature while they both exhibited the combined type I-II crack growth mode at elevated temperature.

References

1. M.Qiao,C.Zhou, Corros. Sci.,**63(10)**, 239-245 (2012)
2. D.Shi,J.Liu,X.Yang,H.Qi,J.Wang, Mater. Sci. Eng. A, **528(1)**,233-238 (2010)
3. J.Liang,A.Xing,Z.Wen,Z.Yue, Eng. Failure Anal., **59**,326-333 (2015)
4. X.Hu,D.Shi,X.Yang, Mater. Sci. Eng. A, **674**,451-458 (2016)
5. S.Li,X.Yang,G.Xu,H.Qi,D.Shi, Int. J. Fatigue., **84**,91-96 (2016)
6. T.Liang,H.Guo,H.Peng,S.Gong,Chin.J.Aeronaut., **25(5)**,796-803 (2012)
7. ASTM International. *Standard test method for measurement of fatigue crack growth rates E647-15* (2015)
8. J.Renart,J.Vicens,S.Budhe,A.Rodriguez-Bellido,J.comas, Int. J. Fatigue., **82**, 634-642 (2016)
9. D.Ewest,P.Almroth,B.Sjodin,K.Simonsson,D.Leidermark, Int. J. Fatigue., **92**, 61-70 (2016)
10. B.Pan, Acta Opt. Sin., **31(2)**, (2011)
11. China aviation material manual editorial board , *China aeronautical material manual second volume superalloy casting high temperature alloy* (2002)
12. X.He,Y.Zhang,H.Shi,J.Gu,C.Li, Mater. Sci. Eng. A, **618**,153-160 (2014)
13. K.Yoon,T.Park,A.Saxena, Stren. Fract. & Complex., **4(1)** (2005)
14. J.Telesman,T.Smith,T.Gabb,A.Ring, Mater. Sci. Eng. A,**708** (2017)
15. J.Bouvard,F.Gallerneau,P.Paulmier,J.Chaboche, Int. J. Fatigue., **38(5)**,130-143 (2012)
16. K.Chan,High.Temp.Technol.,**33(4-5)**,425-438 (2016)

Fatigue behavior of Ti-6Al-4V alloy modified by plasma immersion ion implantation: temperature effect.

Verônica Velloso[1,*], *Leonardo* Nozaki[1], *Diego* Tapia[2], *Maria Odila* Cioffi[1], *Rogério* Oliveira[3], *Miguel* Barboza[4], and *Herman* Voorwald[1]

[1]Faculdade de Engenharia de Guaratinguetá – FEG/UNESP, Materials and Technology Department, Fatigue and Aeronautical Materials Research Group, Av. Ariberto Pereira da Cunha 333, Brazil
[2]Universidad Nacional de Ingeniería – UNI, Faculty of Mechanical Engineering, Av. Túpac Amaru 210, Peru
[3]Instituto Nacional de Perquisas Especiais – INPE, Associated Laboratory of Plasmas, Av. dos Astronautas 1758, Brazil
[4]Escola de Engenharia de Lorena – EEL/USP, Materials Department, Pólo-Urbo Industrial Gleba AI-6, Brazil

Abstract. This research studied Ti-6Al-4V alloy behavior with two (2) different microstructure subjected to nitrogen addition by PIII treatment, with and without sample heating, under cyclic load. PIII conditions, at 390 °C, were DC voltage of 9.5 kV, frequency of 1.5 kHz and pulse of 40 μs. PIII conditions, with sample heating at 800 °C, were 7 kV, 0.4 kHz and 30 μs. Axial fatigue tests were performed on untreated and treated samples for resistance to fatigue comparison. The untreated Ti-6Al-4V had an annealed microstructure, PIII treatment at 390 °C resulted in a microstructure that has no nitride layer or diffusion zone. In the PIII treatment at 800 °C, the microstructure presented nitride layer and diffusion zone. Resistance to fatigue decreased with PIII treatments in both temperatures. At 390 °C, the treatment created deformation regions and cracks on surface due to nitrogen implantation that formed solid solution with titanium and imposed lattice strains on the crystal lattice. At 800 °C, bulk ductility decrease, increasing of αTi proportion in microstructure due to α case formation and the presence of a ceramic layer dropped fatigue resistance of Ti-6A-4V alloy.

1 Introduction

Industry applies titanium alloys in the biomedical, petrochemical, automotive and aeronautical field. The aeronautical industry accounts for about 80% of all worldwide titanium demand and is, therefore, the main application field, being Ti-6Al-4V alloy the most used of the titanium alloy family. Its main properties are mechanical resistance at room and high temperatures (up to 350 °C), corrosion resistance, biocompatibility, toughness, formability and, most important for the aircraft industry, high specific strength and good damage tolerance. Titanium alloys have the highest specific resistance among structural materials, which means that they are as light as aluminum alloys and mechanical resistant as microalloyed steel [1-4].

Most of structural components, during their useful life, are subjected to cyclic loading. Even though the applied stress is below the yield strength (σ_{ys}), these components may fail due to fatigue. This failure mode involves nucleation, crack propagation stages and subsequent catastrophic failure (including materials that exhibit ductile behavior). Among the aspects that change fatigue life, surface characteristics are critical because, in most cases, crack nucleation starts at the surface [5, 6].

Aircraft turbine engines are made of titanium and nickel superalloys. Aeronautical applications search for lightweight materials. In the case of turbines, however, the material must be able to withstand high loads, during service, at high temperatures. Therefore, the material needs to ensure safe and continuous operation of the engine [7].

Surface treatments intend to improve the surface quality of the metallic component increasing its useful life. The literature reported several studies that investigated the mechanical behavior of metallic alloys subjected to surface treatments [8-13]. Plasma immersion ion implantation (PIII) is a technique that showed good results related to the increase in surface hardness, corrosion resistance and decrease friction coefficient, producing no residues [14-17]. However, there is a lack of studies relating PIII treatment with the fatigue behavior of titanium alloys. This research aims to study Ti-6Al-4V alloy behavior with two (2) different microstructure subjected to nitrogen addition by PIII treatment, with and without sample heating, submitted to axial fatigue.

2 Materials and methods

The Ti–6Al–4V alloy was used in rod form with a diameter of 14 mm and a length of 1 m. Specimen preparation followed metallographic patterns: hot mounting, polishing and etching with Kroll's solution

* Corresponding author: veronicamcaoliveira@gmail.com

(1% HF, 4% HNO3 and distillate water). Optical microscopy was used to image the samples.

The PIII system consists of a vacuum chamber with a specimen holder, plasma source and a high voltage pulses modulator. The treatment conditions were T = 390 °C, DC voltage of 9.5 kV, frequency of 1.5 kHz and pulse of 40 μs, t = 2 h; and with an auxiliary system of sample heating at 800 °C, were 7 kV, 0.4 kHz and 30 μs, t = 2 h.

Ti-6Al-4V alloy was characterized by X-ray diffraction. The following conditions were used: 40 kV; 30 mA; angle (2θ) ranging from 10 to 90°; angular step of 0.05° and 1s count time per point. The analysis was performed at room temperature with Cu-Kα radiation and a graphite monochromator.

An optical profilometer, Leica DCM3D model, was used to calculate the surface roughness parameters. The measured area was 1.27 x 0.95 mm². The surface roughness was measured before and after PIII treatment. The parameters were calculated using Gwyddion software and used in notch sensitivity analysis.

Axial fatigue tests were conducted until fracture or 10^6 cycles, for R = 0.1 and 10 Hz. S x N curves were raised for untreated and PIII treated conditions. Data were analyzed statistically by Log-Normal. Figure 1 shows the specimen made according ASTM E466.

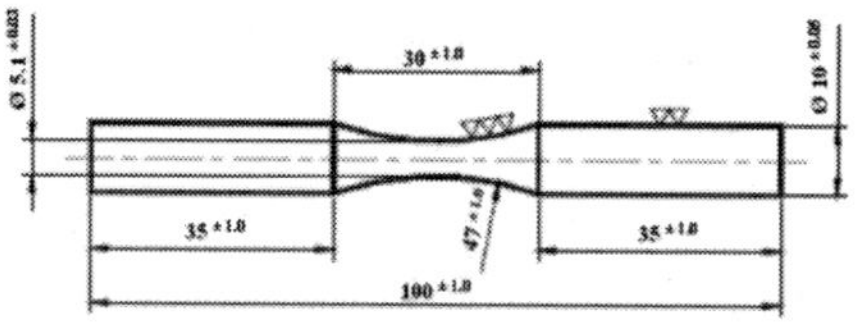

Fig. 1. Axial fatigue testing specimen.

3 Results and discussion

Figures 2 and 3 show Ti6Al4V alloy microstructure at annealed and lamellae conditions. The latter is also known as Widmanstätten morphology, characterized by colonies arrangement and the coarse grains.

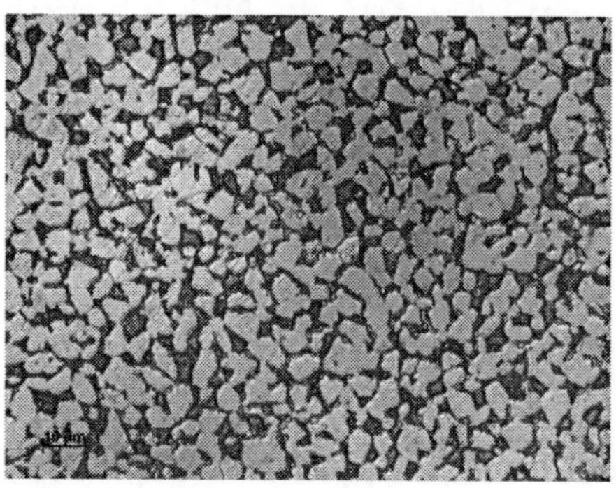

Fig. 2. Optical microscopy image of Ti-6Al-4V alloy: annealed condition.

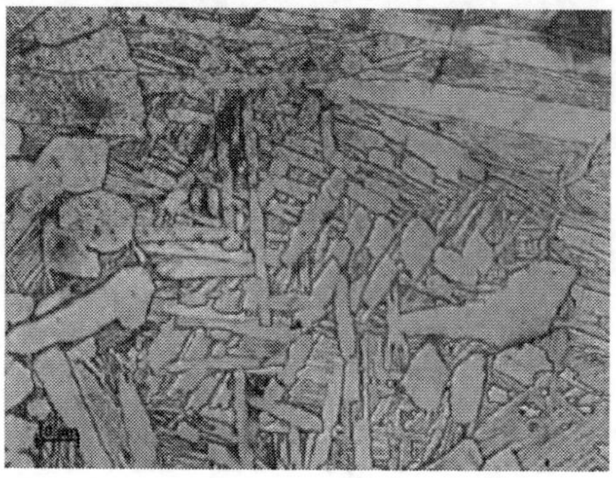

Fig. 3. Optical microscopy image of Ti-6Al-4V alloy: Widmanstätten condition.

Annealed Ti6Al4V alloy was subjected to PIII treatment at 800 °C and had the microstructure changed during the treatment. This transformation was due to the treatment temperature which is close to β *transus* and, therefore, allowed the diffusion process in the titanium alloy microstructure [18].

The PIII treatment carried out at 390 and 800 °C modified the Ti6Al4V alloy in the following ways:

- 390 °C (PIII - 390), the treatment temperature resulted from nitrogen bombardment on surface through the combination of voltage (power) and frequency parameters. The power and N^+ collisions amount on the alloy surface release energy as heat, increasing the environment temperature. In this case, it is said that the treatment was cold, because the sample was not heated and the work temperature remained 40% lower than Ti6Al4V alloy melting temperature. Thus, the nitrogen implantation on surface was ballistics and allowed nitrogen diffusion in some planar distances [19,20]. As a result, there was not observed a nitride layer or a stabilized α phase zone (since nitrogen is an α stabilizer) on surface. This treatment allowed the nitrogen solubilization on titanium, straining the titanium crystalline lattice solid-solution;
- 800 °C (PIII - 800), the treatment temperature was reached through an auxiliary sample heating system. The combination of the heat generated by the auxiliary heating system + nitrogen bombardment resulted in local temperature increase. Therefore, the treatment temperature achieved levels higher than β *transus*, ie exceeded the transformation limit $\alpha \rightarrow \beta$ and provided atomic diffusion even in the bulk microstructure [21]. As the annealed type microstructure is unstable at high temperatures, the bulk morphology changed to a lamellae type, also known as Widmanstätten morphology. In addition, nitrogen implantation modified the surface, creating a layer of 13.4 μm thickness (nitride layer, 1.4 μm + diffusion region, α case). Figure 4 shows the modified surface of Ti6Al4V through nitrogen implantation in PIII-800 condition.

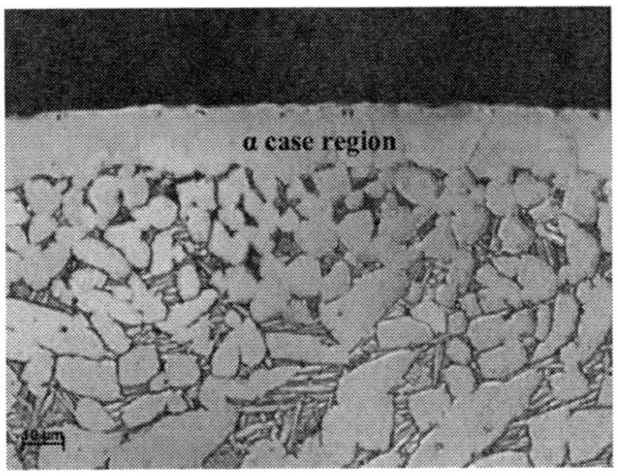

Fig. 4. Optical microscopy image of Ti-6Al-4V alloy: nitride layer + α case.

Figure 5 shows XRD analysis of Ti6Al4V alloy at untreated, PIII - 390 and PIII – 800 conditions.

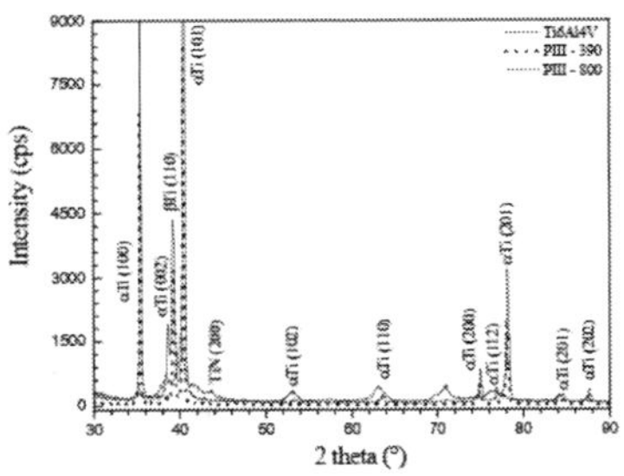

Fig. 5. DRX analysis of Ti6Al4V alloy for untreated, PIII - 390 and PIII - 800 treated conditions.

The highest intensity peaks were related to the α phase, which was found at 2θ = 35.4°, 38.4°, 40.1° and 77.7°. A small amount of β phase was observed, characterized by the peak related to (110) plane and found at 2θ = 39.6°. The DRX analysis at PIII – 390 condition did not present nitrites. Since, at this condition, there was no formation of a nitrided layer and only nitrogen solubilization in the titanium crystalline lattice. While at PIII – 800 condition, the presence of the peak related to the plane (200) that corresponds to TiN compound was observed.

Figure 6 shows the S *vs* N curves of Ti6Al4V alloy at the untreated, PIII – 390 and PIII – 800 conditions.

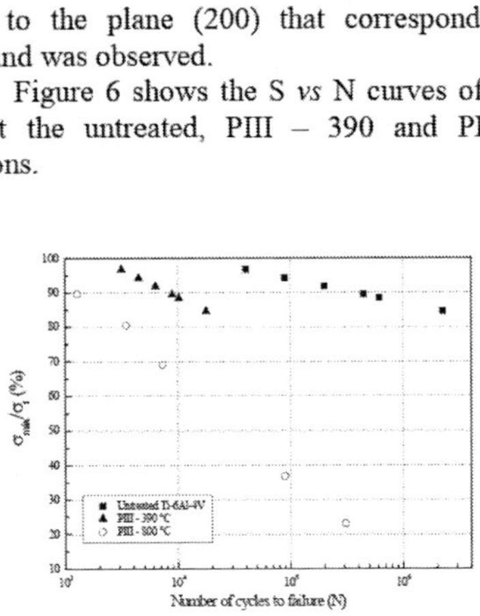

Fig. 6. S x N curves of Ti-6Al-4V alloy for untreated, PIII - 390 and PIII - 800 treated conditions.

Equations 1 to 3 represent the curves generated by the Log-Normal model for untreated, PIII – 390 and PIII – 800 conditions, respectively:

$$\log N = 18.6 - 0.014\sigma \tag{1}$$

$$\log N = 9.5 - 0.006\sigma \tag{2}$$

$$\log N = 6.1 - 0.0029\sigma \tag{3}$$

The tensile testing results of the sample treated at PIII - 800 condition showed that the tensile strength limit (σ_t) increased by 5% and that the ductility decreased by half. Thus, Figure 6 shows on the ordinate axis a relationship $^{\sigma_{max}}/_{\sigma_t}$.

Figure 6 analysis showed that, for a stress level of 89% σ_t, the fatigue life time was equal to 4.5×10^5 cycles on untreated condition, 8.9×10^3 cycles on PIII – 390 condition and 1.3×10^3 cycles on PIII – 800 condition. Meaning a decreasing in fatigue life by the PIII treatment of 50 times for PIII – 390 condition and 346 times for PIII – 800 condition.

Based on the equations 1 to 3 generated by the statistical analysis Log-Normal, fatigue resistance was calculated for N = 10^3 and for N = 10^7. The values were 1058 MPa, when N = 10^3, and 840 MPa, when N = 10^7 for untreated condition. At PIII – 390 condition, fatigue resistance was equal to 1080 MPa, when N = 10^3, and 417 MPa, N = 10^7. Finally, at PIII – 900 condition, fatigue resistance was equal to 1120 MPa, when N = 10^3, and 117 MPa, N = 10^7. The results showed that the surface treatments influenced the fatigue behavior of Ti6Al4V alloy in the high cycle fatigue region.

One of the goals of the notch sensitivity tests to fatigue behavior studies is to evaluate the crack initiation process. Simulating, thus, a real situation where a structural component contains in its design and structure defects and irregularities that act as stress concentrators points [22]. Arola and Ramulu [23] proposed a mathematical model, based on the notch sensitivity calculations, to define the surface features effect on stress concentration factor ($\overline{K_t}$) of ceramic-polymer composites. Later, Arola and Williams [24] applied this mathematical model together with the Neuber rule for HSLA steels to evaluate the fatigue behavior dependence of average roughness (Ra). The equations involved in this study are [24]:

$$\overline{K_t} = 1 + n\left(\frac{R_a}{\bar{\rho}}\right)\left(\frac{R_y}{R_z}\right) \tag{4}$$

$$\overline{K_f} = 1 + q(\overline{K_t} - 1) \tag{5}$$

$$q = \frac{1}{\left(1+\frac{\gamma}{\bar{\rho}}\right)} \tag{6}$$

where n is a constant that relates the load type applied to the material. According to Neuber rule, $n = 1$ to shear loads, and $n = 2$ to tensile loads. Ra, Ry and Rz is the average roughness, peak-to-valley height and 10-point roughness, respectively. The parameter ρ corresponds to the effective radius profile valley. $\overline{K_f}$ and q correspond to effective fatigue stress concentration factor and notch sensitivity, respectively. Finally, γ is a constant related to the material and to the tensile strength limit [24]. In this work, the values of n and γ are equal to 2 and 0087, respectively.

Table 1 shows the roughness parameters calculated for Ti6Al4V alloy at the untreated, PIII – 390 and PIII – 800 conditions.

Table 1. Main surface roughness parameters estimated for untreated, PIII-390 and PIII-800 treated conditions.

Sample	Ra (μm)	Rz (μm)	Ry (μm)	ρ (μm)
Ti6Al4V	1.5±0.2	10.4±2.2	12.6±3.4	25.8±8.9
3IP – 390	1.5±0.0	11.1±1.9	14.9±7.1	19.8±1.7
3IP – 800	1.4±0.3	8.9±2.5	10.2±3.7	18.0±2.3

Table 2 presents the notch sensitivity parameters based on the mathematical model of Arola and Ramulu and Neuber rule [24].

Table 2. Notch sensitivity parameters.

Sample	$\overline{K_t}$	Q	$\overline{K_f}$
Ti6Al4V	1.14	0.23	1.03
3IP – 390	1.20	0.19	1.04
3IP – 800	1.17	0.17	1.03

Figures 7 shows the relationship between notch sensitivity q and average roughness Ra.

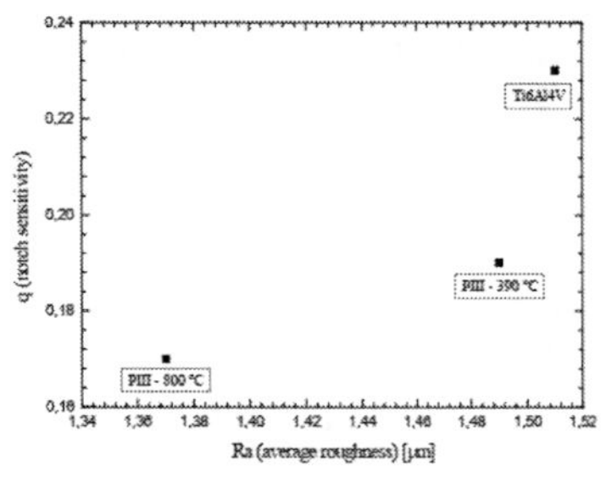

Fig. 7. Notch sensitivity analysis: q vs Ra.

Figure 8 shows the relationship between the stress concentration factors, $\overline{K_t}$ and $\overline{K_f}$, and average roughness Ra.

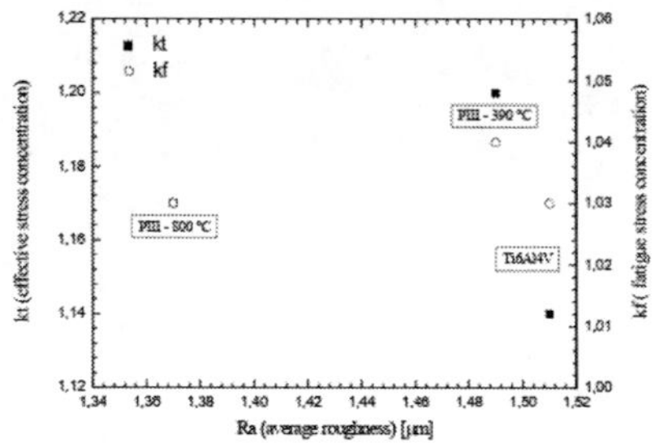

Fig. 8. Notch sensitivity analysis: $\overline{k_t}$ vs Ra and $\overline{k_f}$ vs Ra .

Tables 1 and 2 analysis together with Figures 7 and 8 show that the roughness increase raises the notch sensitivity. This is an usual observation, since the surface roughness acts as a defect, increasing the stress concentration at the surface and accelerating the crack nucleation process. About the stress concentration factors, it was observed that the highest values of $\overline{K_t}$ and $\overline{K_f}$ are related with the PIII - 390 condition.

Correlating the fatigue results with the notch sensitivity analyzes and with the fracture mechanisms, they can be explained as follows:

- the first point to be raise is, for this set of studied samples, Ra does not influence the fatigue behavior of Ti6Al4V alloy in all the experimental conditions. If the average roughness contributes to the increase of stress concentrators points, accelerating the crack nucleation, it should be observed in terms of fatigue strength, an increase in fatigue life with the PIII treatments following the scheme: $N_{Ti6Al4V} < N_{PIII-390} < N_{PIII-800}$;
- in the PIII – 390 condition, the parameter that controls the fatigue fracture mechanisms is the strain generated in the titanium crystalline lattice by solid solution. This strain has reached a limit, due to the high frequency adopted, f = 1.5 kHz, resulting in microdeformations and microcracks to relieve strain in the crystal lattice [25]. Interestingly, the parameter $\overline{K_f}$, that is related to a stress concentration factor is the highest for this condition. Mathematically corroborate the hypothesis that the titanium crystalline lattice strained in such a level that, to relieve, generated on the sample surface pre-microcrack. If the fatigue failure mechanism is the nucleation and growth of a crack, the surface treatment carried out in this condition eliminated the nucleation process. These microcracks generated in the order of a few interplanar distances, when subjected to cyclic loading and combined with the favorable crystallographic orientations of dislocations slipping, coalesce and contribute to crack propagation [26, 27], reducing the fatigue life;
- in the PIII – 800 condition, the morphology transformation from annealed microstructure to Widmanstätten was responsible for the decrease in the

Ti6Al4V alloy mechanical strength. This microstructure did not offer resistance to crack propagation. Ra and $\overline{K_f}$ values show that the surface parameters are suitable for fatigue application. The microstructure presented here is coarse grained, 350 μm, 90 times larger than the size of the grains in the annealed condition. The coarse-grained microstructure contributes to a decreased in fatigue toughness, favoring crack propagation and reducing the component lifetime. Moreover, the α + β colonies present on Ti6Al4V alloy with lamellar morphology do not offer barrier to the growth of the crack since the effective slip distance is not between lamellae and colonies but between the coarse grains of prior β phase. Regarding to nitrogen surface implantation, the nitride layer composed of nitrides + α case further contributes to embrittlement of the Ti6Al4V alloy. The nitrogen diffusion region, known as α case is characterized by a surface layer composed only of α phase, hexagonal crystal structure known to have only 5 slip systems. The thin nitride layer, 1.4 μm, could act as a hard compound capable of slowing the crack nucleation process [21, 26, 27]. Its effect, however, can be considered negligible due to the contributions of the coarse grained and α case that embrittled the Ti6Al4V alloy and dropped the fatigue life of Ti6Al4V alloy.

Recent papers showed that PIII treatment increases the fatigue life of steels and aluminum alloys, both applied as landing gear components [28, 29]. And when this treatment had a deleterious effect on the fatigue behavior compared to the base material, this drop in performance was lower than that caused by other more conventional surface treatments [25, 29, 30, 31].

4 Conclusions

This study reported the results involving untreated and PIII treated Ti-6AL-4V alloy, at different work temperatures, and submitted to axial fatigue loading.

Resistance to fatigue decreased with PIII treatments in both temperatures. The fatigue resistance decreasing was of 50 times for PIII – 390 condition and 346 times for PIII – 800 condition. At 390 °C, the treatment created deformation regions and microcracks on surface due to nitrogen implantation that formed solid solution with titanium and imposed lattice strains on the crystal lattice. At 800 °C, bulk ductility decrease, increasing of αTi proportion in microstructure due to α case formation and the presence of a ceramic layer dropped fatigue resistance of Ti-6A-4V alloy.

More studies should be done about PIII and fatigue behavior, varying temperature and frequency, in order to improve the fatigue life of metallic alloys applied in structural components. It is also worth considering a route of heat treatment to provide the best combination of grain size and fracture toughness in a Widmanstätten type morphology. Finally, the effect of the treatment temperature can only be evaluated in a situation in which the bulk microstructure is maintained during the treatment.

Acknowledgments

The authors would like to acknowledge FAPESP (Fundação de Amparo à Pesquisa do Estado de São Paulo) and CNPq (Conselho Nacional de Desenvolvimento Científico e Tecnológico) for their continued financial support (Process FAPESP 2015/00331-2, CNPQ 303832/2014-2).

References

1. R.R. Boyer, Adv. Perform. Mater., **2** (1995).
2. C. Cui, B. Hu, L. Zhao, S. Liu, Mater. Des., **32** (2011).
3. V.M.C.A. Oliveira, C. Aguiar, A.M. Vazquez, A. Robin, M.J.R. Barboza, Corros. Sci., **88** (2014).
4. O. Schauerte, Adv. Eng. Mater., **5** (2003).
5. R.G. Bonora, H.J.C. Voorwald, M.O.H. Cioffi, G.S. Junior, L.F. V Santos, Procedia Eng. **2** (2010).
6. W.J. Evans, **243** (1998).
7. J.C. Williams, E.A. Starke, Acta Mater. **51** (2003).
8. L. Chen, Y. Du, X. Xiong, K.K. Chang, M.J. Wu, Int. J. Refract. Met. Hard Mater. **29** (2011).
9. M.Y.P. Costa, M.O.H. Cioffi, M.L.R. Venditti, H.J.C. Voorwald, Procedia Eng. **2** (2010).
10. C. Fei, Z. Hai, C. Chen, X. Yangjian, Prog. Org. Coatings. **64** (2009).
11. A. Höling, L. Hultman, M. Odén, J. Sjölén, L. Karlsson, Surf. Coatings Technol. **191** (2005).
12. P. Pérez, Surf. Coatings Technol. **191** (2005).
13. H.J.C. Voorwald, L.F.S. Vieira, M.O.H. Cioffi, Procedia Eng. **2** (2010).
14. G.F. Gomes, M. Ueda, H. Reuther, E. Richter, A.F. Beloto, Surf. Coatings Technol. **196** (2005).
15. K. Kostov, M. Ueda, M. Lepiensky, P. Soares, G. Gomes, M. Silva, H. Reuther, Surf. Coatings Technol. **186** (2004).
16. C.B. Mello, M. Ueda, M.M. Silva, H. Reuther, L. Pichon, C.M. Lepienski, Wear. **267** (2009).
17. G. Silva, M. Ueda, C. Otani, C.B. Mello, C.M. Lepienski, Surf. Coatings Technol. **204** (2010).
18. L. Kuncická, R. Kocich, T.C. Lowe, P. Mater Sci. **88** (2017).
19. X. Tian, P.K. Chu, Phys. Letter. **277** (2000).
20. A.V. Byeli, S.K. Shikh, V.V. Khatko, Wear **159** (1992).
21. L. Wanying, L. Yuanhua, C. Yuhai, S. Taihe, A. Singh, Rare Metal Mater Eng. **46** (2017).
22. L. Jianhui, Z. Ruilong, W. Yaobing, L. Shanshan, Theor. App. Fract. Mech. **93** (2018)

23. D. Arola, M. Ramulu, J. Compos. Mater. **33**, 02 (1999)
24. D. Arola, C.L. Williams, Inter. J. Fatigue **24** (2002).
25. V.M.C.A. Oliveira, M.O.H. Cioffi, M.J.R. Barboza, R. Landers, B. Schmitt, D.C.A.R. Tapia, H.J.C. Voorwald, Inter. J. Fatigue **109** (2018)
26. K.L. Biavant, S. Pommier, C. Prioul, Fatigue Fract. Eng. Mater. Struct. **25** (2002).
27. K.S. Chan, Inter. J. Fatigue, **32** (2010).
28. R. Bonora, M.O.H. Cioffi, H.J.C. Voorwald, J. Physic.: Conf. Series **843** (2017).
29. T. A. Minto, V.M.C.A. Oliveira, H.J.C. Voorwald, Inter. J. Fatigue **103** (2017).
30. H.J.C. Voorwald, M.P. Silva, M.Y.P. Costa, M.O.H. Cioffi, Fatigue Fract Eng Mater Struct **32** (2009).
31. R.C. Souza, H.J.C. Voorwald, M.O.H. Cioffi, Surf Coat Technol **203** (2008).

Fatigue strength characterization of Al-Si cast material incorporating statistical size effect

Roman Aigner[1*], *Martin* Leitner[1], and *Michael* Stoschka[1]

[1]Christian Doppler Laboratory for Manufacturing Process based Component Design, Montanuniversität Leoben, Chair of Mechanical Engineering, 8700 Leoben, Austria

Abstract. Cast aluminium components may exhibit material imperfections such as shrinkage and gas pores, or oxide inclusions. Therefore, the fatigue resistance is significantly influenced by the size and location of these inhomogenities. In this work, two different specimen geometries are manufactured from varying positions of an Al-Si-Cu alloy casting. The specimen geometries are designed by means of shape optimization based on a finite element analysis and exhibit different highly-stressed volumes. The numerically optimized specimen curvature enforces a notch factor of only two percent. To enable the evaluation of a statistical size effect, the length of the constant testing region and hence, the size of the highly-stressed volume varies by a ratio of one to ten between the two specimen geometries. Furthermore, the location of the crack initiation is dominated by the comparably greatest defects in this highly-stressed volume, which is also known as Weibull's weakest link model. The crack initiating defect sizes are evaluated by means of light microscopy and modern scanning electron microscope methods. Finally, the statistical size effect is analysed based on the extreme value distribution of the occurring defects, whereby the size and location of the pores is non-destructively obtained by computed tomography (CT) scanning. This elaborated procedure facilitates a size-effect based methodology to study the defect distribution and the associated local fatigue life of CPS casted Al-Si lightweight components.

1 Introduction

Aluminium cast parts enable quite complex geometries thus featuring lightweight goals. On the other hand, the service strength must be well known already in early stages of the design process. For such applications, Al-Si-Cu cast alloys are commonly used materials. One of the most important limiting factors regarding fatigue strength is the size of the intrinsic defects in the casting. Microstructural parameters like the grain size or the dendrite arm spacing (DAS) show a secondary influence [1]. Numerous preliminary studies showed an influence of the solidification time of the melted aluminium on the dendrite arm spacing, and hence, additionally on the resulting fatigue strength of the bulk material [2]. The dominant material flaws, who scale as well with the local solidification time, cause this.

As exemplified in other preliminary studies [3, 4], Tiryakioglu showed that in the HCF region a statistic model can be applied to link the fatigue life with the area of the fracture-initiating micropores [5]. The aim of this work is to investigate the fatigue strength of Al-Si-Cu cast alloy at three sampling positions, whereby each series exhibits two varying highly-stressed volumes.

In addition, μCT scans of the differently sized specimens are compared by means of the location and size of the micropores. Furthermore, the fracture surfaces of the tested specimens are subsequently analysed to gain information about the crack initiating pores, their position as well as their spatial extent. Subsequently, the extreme value distribution of the fracture-initiating defects are evaluated on the basis of a most likelihood type of function.

2 Experimental

2.1. Investigated Material

The examined material is an Al-Si-Cu cast alloy in T6 heat-treated condition. The specimens are manufactured from six different locations, which possess by simulation only minor differences in cooling rate. They are arranged into two regions featuring specimens of different sizes for examination of size effect. The two regions exhibit a similar local cooling rate and therefore, a similar DAS is expected. Measurements on metallographic sections revealed that the DAS in these two testing regions is nearly identical; see fig. [1].

This supports the assumption that the subsequent experimental and analytical investigations focus on the size effect and are not affected by local microstructural parameters.

* Corresponding author: roman.aigner@unileoben.ac.at

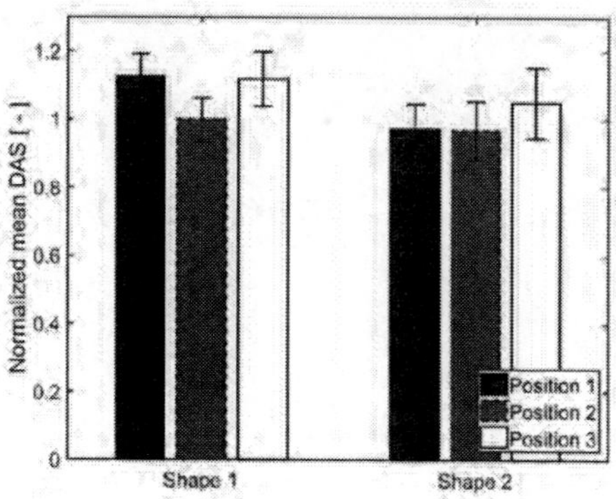

Fig. 1. Microstructural properties in the testing region

As aforementioned, the two different specimen shapes vary only by means of the highly-stressed volume. The larger specimens are labelled as shape 2. The constant testing diameter ensures a specimen failure at the most critical defect, thus depending on its size, spatial location and orientation, in agreement with Weibull's Weakest-Link concept. As seen in fig. [2], due to variation of the testing length between the two specimen shapes, the ratio of the highly-stressed volumes exhibits a value of about ten if evaluating stresses down to ninety-five percent of the peak values. Therefore, the statistical size effect for this cast aluminium alloy is properly determinable with these two specimen designs.

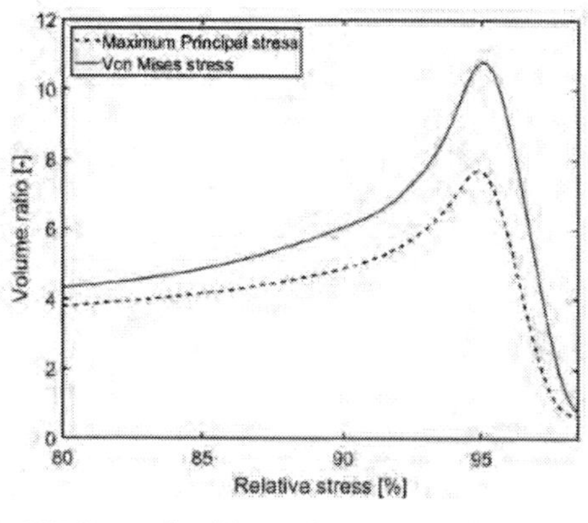

Fig. 2. Volume ratio of the varying specimen shapes (shape 1 to shape 2)

To assess the fatigue strength of the investigated material, quasi-static tensile tests are performed. These tensile test specimens are taken out of shape 2 position thus ensuring identical microstructural parameters.

Table 1. Normalized mechanical and microstructural properties of the investigated positions.

Position	UTS [-]	YS [-]	A_{565} [%]	DAS [-]
1	3.35	2.84	1.62	0.97
2	3.30	2.78	1.62	0.96
3	3.20	2.80	1.08	1.05

The results, listed in tbl. [1], show that all three positions inherit similar quasi-static characteristics by means of the ultimate tensile strength and yield strength. On the other hand, the ductility of position 3, represented by the elongation after fracture, is lower than on the other two positions, due to the higher porosity in position 3. But yield strength and tensile strength are almost identical, only the DAS value at this position is slightly increased, though it differs only in the scale of standard deviation. Summing up, the evaluated DAS scatter index is about ten percent, but possible effects in extremal pore distribution are further on discussed.

2.2. Fatigue results

High-Cycle Fatigue tests are performed with each of the two specimen shape geometries. The tests are executed at a load ratio of R=-1 under compression-tension loading with a testing frequency of about 100 Hz. The run-out cycle number is set at ten million load-cycles because preliminary studies confirmed that defect-based specimen failures occur also in the endurance region between one and five million cycles. The slope of the high cycle fatigue region k_2 of the S/N curve is assumed to be five times as high as the slope k_1 in the high cycle fatigue region as proposed in [6]. The evaluated fatigue data is statistically assessed in the finite life area invoking ASTM E739 [7]. In addition, the fatigue strength in the transition region is evaluated by means of the arcsin$\sqrt{p}$ methodology [8].
The results of the fatigue tests regarding specimen shape 1 show, that the fatigue strength differs only slightly between the three sampling positions, as seen in fig. [3].

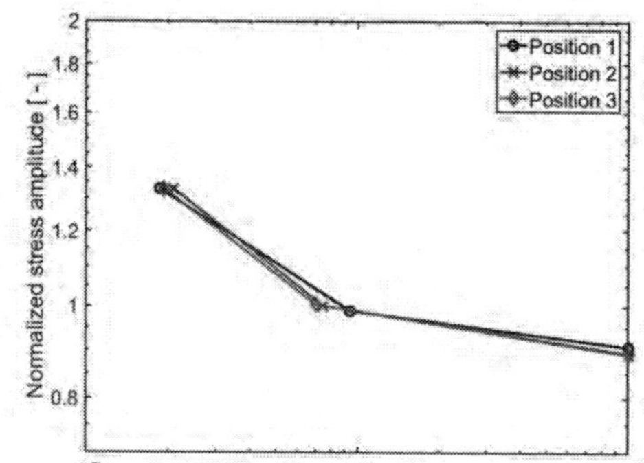

Fig. 3. Fatigue data from specimen shape 1 at different positions

As depicted in fig. [4], the specimens with shape 2 exhibit a distinctive lower fatigue strength than those with shape 1. This reduction is expected as the highly-stressed volume is ten times larger in case of specimen with shape 2 compared to shape 1.

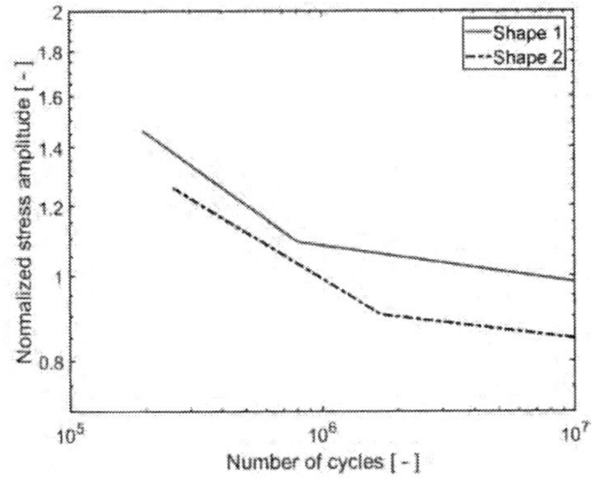

Fig. 4. Fatigue data commonly evaluated from all positions for two different specimen shapes to study size effect

This goes along with the statistical size effect, implying the elevated probability of shape 2 specimens to include greater defects than in shape 1, as shown in [9]. It must be pointed out that the results shown in fig. [4] incorporate all specimens of the three different positions, separated only by their geometric shape. The assessment of the shape 2 specimen S/N curves shows a slightly higher scatter of the fatigue life compared to shape 1. Furthermore, the slope and transition knee point N_K of the shape 2 Wöhler curve is somewhat increased. In tbl. [2], the fatigue data of the six test series is listed in detail.

Table 2. Fatigue data from the varying sample positions.

Position	Shape	k_1	N_K	$S_{a, P50, LLF}$
1	1	5.45	0.938E6	0.99
2	1	4.46	0.747E6	1.00
3	1	4.57	1.084E6	0.98
1	2	6.65	1.993E6	0.87
2	2	5.54	1.615E6	0.86
3	2	5.07	1.210E6	0.83

As exemplified in [2], the microstructure, by means of the DAS, has a significant influence on spatial extent of the immanent defects.

In case of increased dendrite arm spacings, an augmented sealed-off volume between two adjacent solidified dendrite arms is obtained thus increasing the likelihood of shrinkage pores. Numerous metallographic analyses are performed, which confirm the statistical relationship between the DAS and microporosity. In this context, the pore size is denoted as estimated mean equivalent circle diameter (ECD). The ECD is evaluated by the projected area of the defect in fractographic analysis and can be mathematically described by Eq (1).

$$ECD = \sqrt{\frac{4\,Area}{\pi}} \qquad (1)$$

2 Evaluation of defects

2.1. Fractography

In order to locate the fracture-initiating defect and to evaluate its geometrical characteristics, each tested specimen is subsequently analysed by either digital microscope or scanning-electron-microscopy (SEM). The geometrical shape of each defect is manually determined by applying commercial image analysis software and further on statistically evaluated.

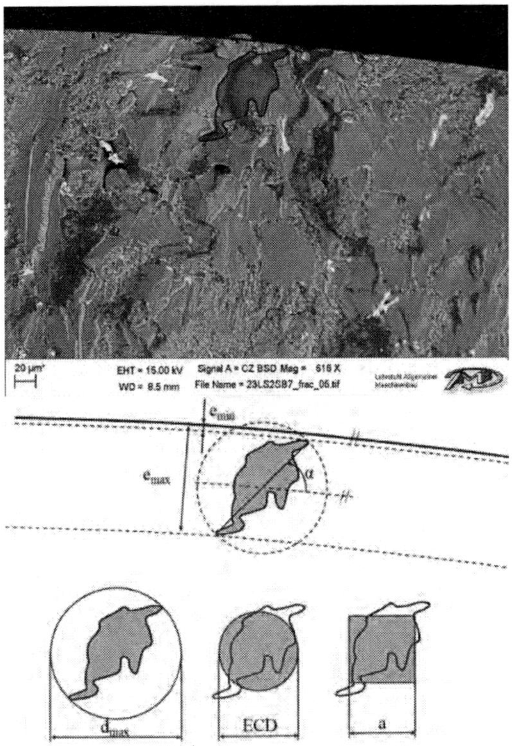

Fig. 5. Representative crack-initiating micropore and its geometrical characterization

Fig. [5] shows the crack initiation at a representative micropore. In addition, geometrical parameters such as the square root of the effective defect area a, its largest extent d_{max}, the equivalent circle diameter (ECD) and the minimal respectively maximum rectangular distance to the surface (e_{min}, e_{max}) are evaluated accordingly. The orientation of the flaws is determined by means of the angle α, which is measured between the largest geometric extent of the project defect area and the specimen surface. The analysis of the initiating cracks of shape 1 specimen deduces that in most cases the technical crack initiates at surface near defects. In detail, fig. [6] displays that the crack initiating defects are mostly surface-intercepting as the normalized minimum distance gets close to zero. This implies that these pores are truncated during the final machining process of the round specimens.

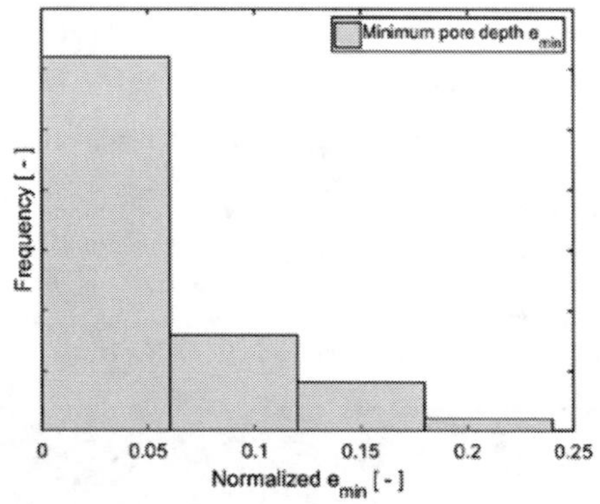

Fig. 6. Frequency histogram of the minimum pore depth e_{min}

Subsequently, these intrinsic defects can be treated like predominant surface cracks, whereat the crack initiation phase is lowered compared to closed defects within the volume. A study to evaluate the lifetime within the crack-initiating phase on microstructural defects is scheduled as further work.

In case of higher loaded specimens, load level at about twenty percent enhanced of the normalized stress amplitude, a second failure mechanism is recognizable. The crack starts from a surface-near microdefect until it progresses through large slip plane areas.

To characterize the local fatigue strength of near surface cracks, the determined geometric values can be used as initial approach. As the micropores are often cut-open, the maximum spatial extent e_{max} of the defect can be considered as the intrinsic crack length. Fig. [7] shows the frequency distribution of the maximum spatial extent e_{max}. It also illustrates a fit of the probability density function of type one generalized extreme value distribution. This type one distribution, also known as Gumbel distribution [10, 11], possesses the following probability density function (pdf), see Eq (2).

$$f(x)= \frac{1}{\sigma} \exp\left(-\frac{x-\mu}{\sigma}\right) \exp\left[-\exp\left(-\frac{x-\mu}{\sigma}\right)\right] \tag{2}$$

Herein, x represents the pore size, by means of e_{max}, μ is the location parameter and σ is the scale parameter of the distribution. Tiryakioglu showed [5] that the relationship between fatigue life and the corresponding defect size in Al-Si cast alloys can be well described by means of a Gumbel distribution. Furthermore, the area and hence the ECD of the fracture initiating micropores is evaluated. Therefore, the scale and the location parameters of the Gumbel distribution can be investigated by means of several different methodologies, including the method of moments, the maximum likelihood method and the least square method. According to [12], the maximum likelihood method is suited to be the best matching method to meet both parameters μ and σ.

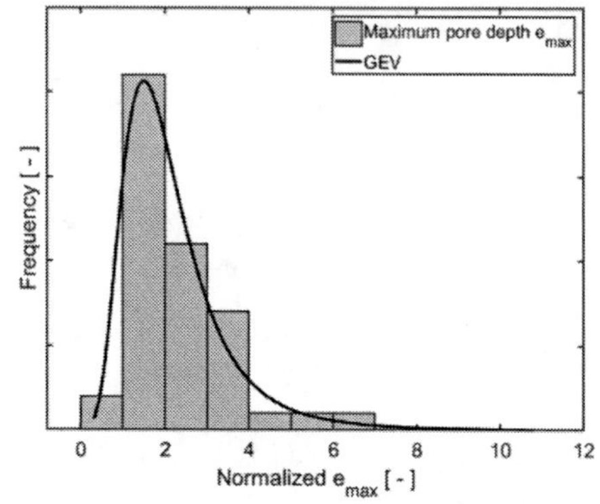

Fig. 7. Frequency histogram of the maximum pore depth e_{max}

In table tbl. [3], the different normalized parameters as well as their relating ninety-five percent confidence intervals for both shape 1 and shape 2 specimens are given. All the distribution parameters are evaluated on the basis of the maximum likelihood method, as recommended in [13].

Table 3. Estimated distribution parameters incorporating the 95% confidence intervals.

Shape	μ [-]	σ [-]	CI (μ) [-]		CI (σ) [-]	
			5%	95%	5%	95%
1	2.09	0.30	1.18	1.44	0.22	0.41
2	1.31	0.33	1.69	2.47	0.13	0.84

The cumulative distribution function (cdf) for the Gumbel distribution of both specimen shapes is subsequently calculated as denoted in Eq. (3). It is displayed along with the probability density function in fig. [8]. The maximum occurring defect size elevates the density function, which is in agreement to Weibull's Weakest-Link concept [14] as specimens of shape 2 possess an increased highly-stressed volume.

$$F(x)= \exp\left[-\exp\left(-\frac{x-\mu}{\sigma}\right)\right] \tag{3}$$

The probability density function of both tested sample volumes also show a similar shape parameter σ. In detail, specimen geometries of shape 1 lead to a value of $\sigma = 0.30$ and a value of 0.32 for shape 2 samples, which confirms the assumption that the cast manufacturing process is equal for both shape geometries.

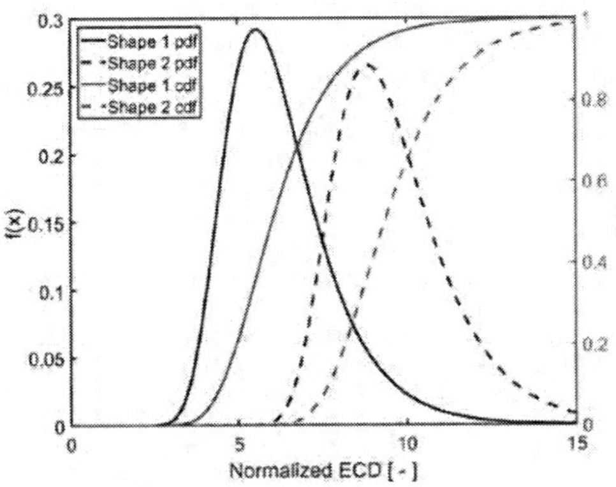

Fig. 8. Probability distributions of both tested specimen shapes possessing two different highly-stressed volumes

2.1. Computed tomography

In order to characterize not only the fracture initiating defect but also the completely tested volumina in terms of intrinsic inhomogenities, non-destructive high resolution μCT scans at selected samples are performed. Due to the low density of the tested material, the transmission by means of radiography is comparably high. Therefore, defects of about two to three times the voxel size can be properly evaluated. One voxel in CT-based image analysis refers to one pixel in the three-dimensional space. This defines the resolution of the μCT scan. Fig. [9] demonstrates an X-ray tomography scan result of a representative testing volume.

Fig. 9. μCT scan of a representative specimen

In numerous preceding studies e.g. [15, 16], the standards for computed tomography investigations as well as the description of experimental procedures and the subsequent image analysis are investigated. It is important to scan the tested volume with a reasonably high resolution to evaluate the micropore shape accordingly.

Thus a voxel size of about 5.5 μm is applicable. The assessed spatial extent of the defects by μCT reveals that the evaluated size of the micropores is somewhat larger than by means of fractographic analysis. This results from the location of the defects. As previously mentioned, the crack-initiating pores are mostly in the surface near region though not necessarily being the largest defect in the tested volume.

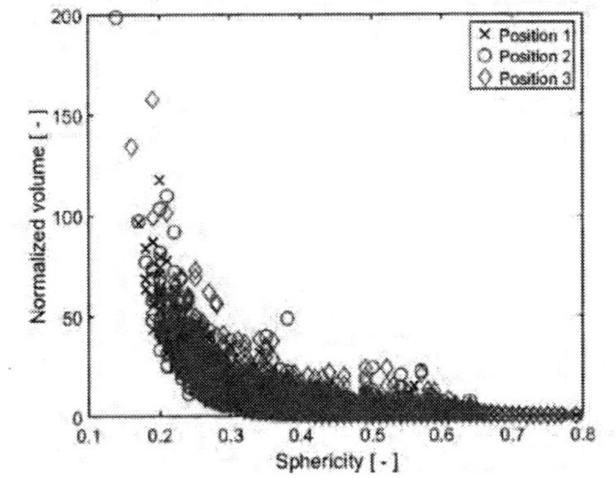

Fig. 10. Sphericity of the flaws in specimen shape 1

On the one hand, the fatigue limit of the bulk material slightly decreases in the surface layers due to a lower dislocation density. Furthermore, flaws with an orientation α close to ninety degrees inherit a faster crack propagation due to its dominant growth direction towards the interior of the specimen. Subsequently, the cut-open pores act like a predominant crack and inherit a shorter crack-initiating phase.

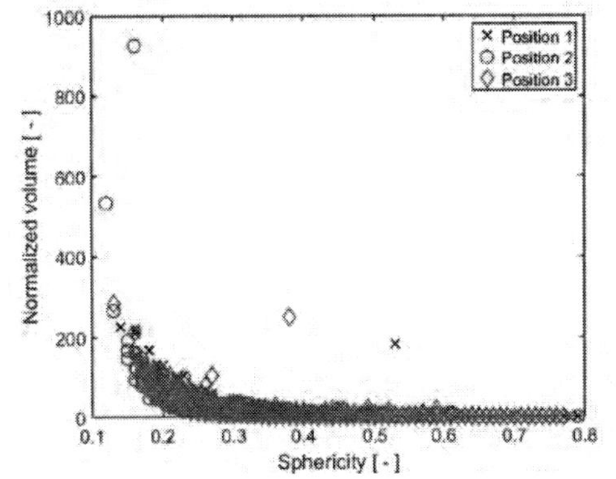

Fig. 11. Sphericity of the flaws in specimen shape 2

Finally, the size effect on pore sphericity can be evaluated. The investigated volumes show that the sphericity of all μCT-detected micropores decreases with lowered volume. As one can see in fig. [10] and fig. [11], larger defects possess a rather sharp needle-shaped structure, while the smaller defects are more spherical.

These figures additionally highlight that differences in extremal imperfection size exist for the two evaluated

https://doi.org/10.1051/matecconf/201816514002

specimen shapes. In detail, the extreme volume is about four times if the highly-stressed volume ratio is set ten to one.

2.2. Statistical size effect

To study the statistical size effect, the ratio between the highly-stressed volume V_2 of shape 2 is ten times the highly-stressed volume V_1 of shape 1. Therefore, the probability to contain larger defects of V_2 increases [17-19]. The relationship between the volume size and the maximum intrinsic defect size is given by the Gumbel distribution; see Eq. (2). The cumulate probability distribution function F(x) defines the probability of occurrence for a certain imperfection size P(x) within the examined volume. The ratio between the highly-stressed volume and the reference control volume is denoted as return period T. The probability that the variable x exceeds the return variable X_t is given as P(x). If the variables X_1, X_i...X_n are a sequence of random variables representing the maximal size of a defect within a given volume V, then the maximal size Y in an increased volume V_n is given by Eq. (4).

$$Y = \max\{X_1, X_i, \dots X_n\} \tag{4}$$

The probability of a n- times greater volume is given by Eq. (5), postulating a Gumbel type extreme value distribution leads to Eq. (6). Therefore, the fatigue life of a corresponding specimen with a larger tested volume is proportionally reduced.

$$P_n(x) = P(x)^n \tag{5}$$

$$P_n(x) = \exp[-\exp\left(-\frac{x-\mu}{\sigma}\right)]^n \tag{6}$$

Another method to describe the statistical size effect, as suggested in [20, 21], is based only on the highly-stressed volume, as it features an elevated probability of crack initiations. The two presented models differ only in the relative stress limit of the highly-stressed volume; in detail Sonsino suggests to use the ninety percent limit whereas Kuguel recommends to take the volume with relative stresses above ninety-five percent into account. Both models are mathematically described by the following Eq. (6), in which V_n represents the limit value of the highly-stressed volume.

$$\frac{\sigma_{a,V1}}{\sigma_{a,V2}} = \left(\frac{V_{n,1}}{V_{n,2}}\right)^{\kappa} \tag{7}$$

The material dependent parameter κ, acts as Weibull parameter and can be calculated from Eq. (8), depending on the probability distribution of the fatigue data [22]. According to Eq. (8), the calculated Weibull parameter κ has a value of about 11.2, based on the evaluated fatigue scatter index T_S. It should be noted that other guidelines [23] defines the parameter as a constant κ=10 for aluminium castings.

$$\kappa = \frac{1.3151}{\log(T_S)} \tag{8}$$

A more thorough comparison of the different models to describe a statistical size effect in the high cycle fatigue region is given in [24].

3 Results and analysis

As listed in table tbl. 3, the investigated fatigue strength shows a significant distinction between both specimen shapes. In order to assess the statistical size effect, the fatigue strength is evaluated at a survival probability of ninety percent instead of using the proposed mean value. The evaluation takes all sample positions commonly into account as microstructural differences are negligible. The influence of the highly-stressed volume as statistical size effect is calculated by means of Eq. (7). Furthermore, the Weibull exponent is conservatively defined as κ=10, as recommended.

Fig. [12] demonstrates the adapted model, based on the volumetric approach from Sonsino [18]. The model calculates the expected fatigue strength levels for the two investigated specimen shapes quite exactly. In detail, the deviation for specimen shape 2 is only about 0.7 % lower compared to the experiments. Therefore, the approach according to Sonsino in combination with an increased survival probability can be further used to assess the statistical size effect in Al-Si-Cu cast alloys.

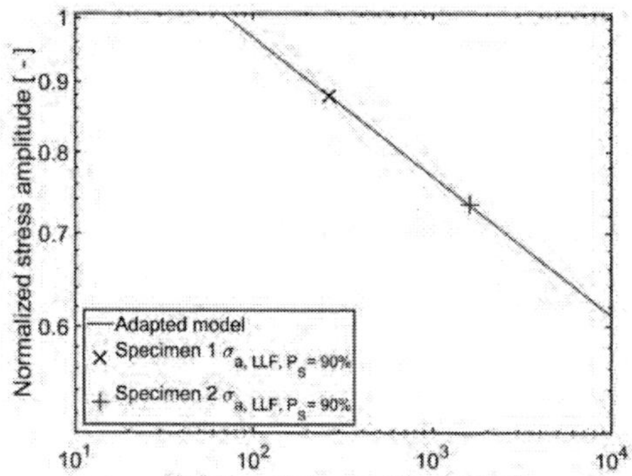

Fig. 12. Calculated fatigue strength in dependence of highly-stressed volume

4 Conclusions

This paper assesses the defect-correlated fatigue strength of Al-Si-Cu cast alloy incorporating the effect of the sampling position as well as the highly-stressed volume as statistical size effect. The fatigue tests are performed under alternating tension-compression. The different sampling positions and specimen shapes inherit similar microstructural properties to ensure only minor influences of varying DAS as microstructural parameter.

Thus, the technological size effect is minimized but the presented test data can be used as reference in future work.

The fatigue test data reveals that the fatigue limit decreases by increasing the highly-stressed volume. Subsequent fractographical analysis is comparative conducted to characterize the crack initiating defects. Moreover, μCT scans are performed to investigate the defect population within the tested volumina, whereat the shape and spatial extent of the pores are assessed. Based on the evaluated defect data, the different probability distributions of both specimen shapes are derived by maximum likelihood regression. The obtained location and scale parameters of the extreme value distribution function are nearly identical for both specimen shapes, which confirms the assumption that the technological effect is negligible and the subsequent evaluation focusses only on the statistical size effect. Finally, a fatigue life model to assess the statistical size effect based on Weibull's approach is adopted towards higher probability of survival, which matches quite well and reveals only a minor difference with an error of about 0.7% compared to the experiments.

The financial support by the Austrian Federal Ministry for Digital and Economic Affairs and the National Foundation for Research, Technology and Development is gratefully acknowledged. Furthermore, I would like to thank the industrial partners BMW München Group and Nemak Dillingen GmbH for the excellent mutual scientific cooperation within the CD-laboratory framework.

References

1. A.M.A. Mohamed, F.H.Samuel, Saleh Al kahtani, Mat. Sci. & Eng. A **577**, (2013)
2. R. Minichmayr, W. Eichlseder, Giessereirundschau, **51**, (2004)
3. C. Garb, M. Leitner, F. Grün, Procedia Eng., **160**, (2016)
4. M. Leitner, C. Garb, H. Remens, M. Stoschka, Mat. Sci. & Eng., A **707**, (2017)
5. M. Tiryakioglu, The Minerals, Met. And Mat. Society and ASM International, **40**, (2009)
6. H. Leitner, PhD Thesis, (2001)
7. ASTM International ASTM E739, (1998)
8. D. Dengel, Journal of Materials Technology, **8**, (1975)
9. H. Friederich, B. Kaiser, K. H. Kloos, Mat.-wiss. U. Werkstofftech., **29**, (1998)
10. E.J. Gumbel, Statistics of Extremes (1958)
11. B.V. Gnedenko, Ann. Math., **44**, (1943)
12. S. Beretta, Y. Murakami, Fatigue & Fract. of Mat. & Struct., **21**, (1998)
13. S. Mahdi, M. Cenac, Revista de Matematica, **12**, (2005)
14. W. Weibull, Ingenieur Archiv, **2**, (1959)
15. DIN EN 16016, (2011)
16. VDI/VDE-Richtlinie 2630, (2009)
17. K.D. Tocher, E.J. Gumbel, Journal of the Royal Statistical Society, (1955)
18. K. Heckel, J. Köhler, Materialwissenschaft und Werkstofftechnik, **6**, (1975)
19. M. Tiryakioglu, Mat. Sci. and Eng. A **497**, (2008)
20. C.M. Sonsino, Materialwissenschaft und Werkstofftechnik, **36**, (2005)
21. R. Kuguel, ASTM Proceedings, **61**, (1961)
22. J. W. Bergmann, A. Diemar, K. Linne, M. Vormwald, FVV, **784**, (2004)
23. FKM Richtlinie, **6**, (2012)
24. T. Tomaszewski, J. Sempruch, T. Piatowski, Journal of applied mechanics, **52**, (2014)

Influence of inclusion type and size on the fatigue strength of high strength steels

Jens Schumacher[1,*], *Brigitte* Clausen[1], and *Hans-Werner* Zoch[1]

[1]Leibniz Institute for Materials Engineering IWT, Badgasteiner Str. 3, 28359 Bremen, Germany

Abstract. In this investigation the influence of the inclusion type and size on the fatigue strength of different steels is analysed. As database case hardened, quenched and tempered as well as bearing steels in different heat treatment conditions, which have been investigated in several research projects over the last two decades, are used. Based on the approach of Murakami the local stress intensity at inclusions was determined to evaluate the influence of fracture causing inclusions on the fatigue strength. Different fatigue criteria have been used to calculate the local equivalent stress amplitudes considering residual stresses in the specimens, mean stresses during loading and multiaxial stresses in notched specimens. Since many run out specimens were subsequently fatigued at a higher stress amplitude, the critical inclusion type and size which have survived the initial stress amplitude could be determined. As a result the local stress intensity at inclusions which leads to no fatigue failure could be calculated as well. With the knowledge of the stress intensity factor range of the fractured and run out specimens a threshold stress intensity factor range could be derived for the different steels and inclusion sizes.

1 Introduction

In fatigue tests steel specimens fail at stresses below the static strength [1]. Cracks are formed either at the surface or in the interior of the specimen which propagate and lead finally to the fracture of the specimen. In case of internal failures inclusions frequently act as crack initiating sites, where oxides, nitrides and sulphides are the most frequently mentioned inclusion types [2-5]. It is well known that with increasing inclusion size the fatigue strength of steel specimens is reduced [6]. Different statements are published about the influence of the inclusion type on the fatigue strength. According to Murakami the inclusion type has nearly no influence on the fatigue strength. On the other hand, depending on the inclusion type, different harmful effects on fatigue strength are observed by other authors [7]. Hence, in this investigation the influence of the inclusion type and size on the fatigue strength of different steels is analysed. Different high strength steels in different heat treatment conditions, which have been investigated in several research projects over the last two decades, are used as database. A fracture mechanical approach was used to evaluate the influence of fracture causing inclusions on the fatigue strength. Different fatigue criteria have been used to calculate the local stress amplitude to consider residual stresses in the specimens, mean stresses during loading and multiaxial stresses in notched specimens. The threshold stress intensity factor range was determined separately for different inclusion types. For that reason the influence of the inclusion types oxide, sulphide, carbide and nitride on the fatigue strength of steels with varying hardness could be evaluated.

2 Calculation approach

For the calculation of the influence of the inclusion size and type on the fatigue strength of steel specimens a fracture mechanical approach is used. The inclusions are regarded as a crack with the same size as the inclusion in a plane perpendicular to the loading direction as described by Murakami [6].

According to Murakami the stress intensity factor range ΔK depends on the equivalent stress amplitude σ_{ae} and the size of the inclusion $area^{0.5}$ which equals the square root of the projected area of the inclusion in the fracture plane:

$$\Delta K = \sigma_{\text{ae}} \cdot \sqrt{\pi \sqrt{area}} \quad (1)$$

The stress intensity factor range which leads to a fatigue failure of a specimen can be calculated with Equation 1 after the size of the crack initiating inclusion was measured in the fracture surface. In case of run out specimens, which did not fail after a fixed number of load cycles the stress intensity factor range cannot be calculated since the inclusion size is unknown. For that reason most run out specimens were fatigue tested again at a higher load amplitude. At higher loads the specimens fracture and the critical inclusion type and size can be observed in the fracture surface. With this information the stress intensity factor range, which led

* Corresponding author: schumacher@iwt-bremen.de

previously to no fatigue failure, can be calculated. With this procedure the stress intensity factor ranges for fractured and run out specimens can be determined. However, the exact threshold stress intensity failure range ΔK_{th}, which leads to a fatigue failure if this value is exceeded, cannot be measured directly at one single specimen. Generally, even specimens of the same batch have scattering mechanical properties so that a scattering threshold stress intensity failure range is expected as well. For a mathematical description of the scattering threshold value a two parametric Weibull distribution is used:

$$P_f = 1 - 2^{-\left(\frac{\Delta K}{\Delta K_{th}}\right)^m} \quad (2)$$

With Equation 2 the fracture probability P_f of a specimen which is loaded with a stress intensity factor range ΔK can be calculated. The threshold stress intensity failure range ΔK_{th} describes the stress intensity factor range which leads to a fracture probability of 50 % and the parameter m is a measure for the scatter of the fatigue strength. Based on the experimental results the parameter ΔK_{th} and m are determined by maximum likelihood estimation. In case of specimen failure a parameter set which maximizes the product of all fracture probabilities has to be determined. At the same time the product of survival probabilities P_s (Equation 3) of all run out specimens have to be maximized with the same parameter set.

$$P_s = 1 - P_f \quad (3)$$

Hence, the parameter combination of ΔK_{th} and m which leads to maximum L value is the best description of the experimental fatigue results:

$$L(\Delta K_{th}, m) = \prod_{\text{Fracture}} P_f \cdot \prod_{\text{Run out}} P_s \quad (4)$$

According to Murakami and Yamashita [8] the threshold stress intensity failure range ΔK_{th} depends for an inclusion in the interior of the specimen on the Vickers hardness H_V of the steel and the inclusion size $area^{0.5}$ which causes the failure and can be calculated as follows:

$$\Delta K_{th} = 0.00277 \cdot (H_V + 120) \cdot \left(\sqrt{area}\right)^{1/3} \quad (5)$$

In this empiric formula ΔK_{th} is expressed in terms of MPa $m^{0.5}$ and the $area^{0.5}$ in µm. To take the effect of both the hardness and the inclusion size into account a similar approach was used in this investigation. As displayed in Equation 6 the parameters C_1, C_2 and C_3 have to be determined based on the experimental results.

$$\Delta K_{th} = C_1 \cdot (H_V + C_2) \cdot \left(\sqrt{area}\right)^{C_3} \quad (6)$$

Depending on the loading condition, mean stresses have to be considered at the fatigue strength evaluation. Multiaxial loading condition are possible or can occur in notches at uniaxial loading. Furthermore, multiaxial residual stresses caused by the specimen manufacturing or heat treatment exist in the highly stressed volume and have to be considered at the evaluation. Four different fatigue criteria were used in this investigation to take multiaxial stresses, mean stresses and residual stresses into account. The equivalent stress amplitude σ_{ae} has been calculated according to Bomas [9] (Equation 7), Crossland [10] (Equation 8), Dang Van [11] (Equation 9) and Sines [12] (Equation 10).

Bomas [9]: $$\sigma_{ae} = 2 \cdot (\tau_{a,max} + \alpha \cdot p_m) \quad (7)$$

Crossland [10]: $$\sigma_{ae} = \frac{3}{\sqrt{2} + \alpha}(\tau_{a,oct} + \alpha \cdot p_{max}) \quad (8)$$

Dang Van [11]: $$\sigma_{ae} = \frac{(\tau_{a,max} + \alpha \cdot p_{max})}{\frac{1}{2} + \frac{\alpha}{3}} \quad (9)$$

Sines [12]: $$\sigma_{ae} = \frac{3}{\sqrt{2}}(\tau_{a,oct} + \alpha \cdot p_m) \quad (10)$$

In the fatigue criteria of Bomas and Dang Van the maximum shear stress amplitude $\tau_{a,max}$ is required. Crossland and Sines are using the octahedral shear stress amplitude $\tau_{a,oct}$ instead. Besides the stress amplitudes the hydrostatic stress state of the fatigue specimens are considered in all fatigue criteria. Bomas and Sines use the mean hydrostatic pressure p_m. Crossland and Dang Van consider the maximum hydrostatic pressure p_{max}. All fatigue criteria have in common that the parameter α has to be known in advance to calculate the equivalent stress amplitudes. This parameter depends on the material and is usually determined experimentally.

Schleicher et al. determined the parameter α for Dang Van's and Bomas' fatigue criterion in dependence of the Vickers hardness H_V of the steel [13]:

$$\alpha = \frac{H_V}{H_\alpha} \quad (11)$$

In case of Dang Van's fatigue criterion the parameter H_α is 1263 HV, while in case of Bomas' fatigue criterion H_α is 1500 HV [13]. In this approach the parameter α for the different fatigue criteria is correlated with the mean stress sensitivity M. Furthermore, a linear dependence of the mean stress sensitivity on the hardness is assumed, which can be described by the parameters A and B:

$$M = A \cdot H_V + B \quad (12)$$

At a uniaxial loading condition, the mean stress sensitivity M allows to calculate the fatigue limit $\sigma_D(\sigma_m)$ in dependence of the mean stress σ_m with knowledge of the mean stress free fatigue limit σ_W.

$$\sigma_D(\sigma_m) = \sigma_W - M \cdot \sigma_m \tag{13}$$

Under the assumption that this mean stress sensitivity approach works, this correlation can also be obtained at a uniaxial load case with the different fatigue criteria. For example at a uniaxial stress amplitude σ_a and a mean stress σ_m Bomas' fatigue criterion (Equation 7) results in the following expression:

$$\sigma_a = \sigma_{ae} - \frac{2}{3} \cdot \alpha \cdot \sigma_m \tag{14}$$

Since the equivalent stress amplitude σ_{ae} corresponds to a mean stress free uniaxial loading condition it is comparable with the mean stress free fatigue limit σ_W. In the event that Equation 13 and 14 are both valid, the following relationship between the mean stress sensitivity M and the parameter α of Bomas' fatigue criterion exists:

Bomas: $$\alpha = \frac{3}{2} \cdot M = \frac{3}{2} \cdot A \cdot H_V + \frac{3}{2} \cdot B \tag{15}$$

By combining Equation 15 and 12 the parameter α can be expressed with the parameters A and B in dependence of the hardness of the steel specimens:

$$\alpha = \frac{3}{2} \cdot A \cdot H_V + \frac{3}{2} \cdot B \tag{16}$$

The same approach was used to find a correlation between the mean stress sensitivity M and the parameter α for the three other fatigue criteria used in this investigation. The correlation for Crossland's fatigue criterion is displayed in Equation 17, for the Dang Van's fatigue criterion in Equation 18 and for Sines' fatigue criterion in Equation 19.

Crossland: $$\alpha = \frac{\sqrt{2} \cdot A \cdot H_V + \sqrt{2} \cdot B}{1 - A \cdot H_V - B} \tag{17}$$

Dang Van: $$\alpha = \frac{\frac{3}{2} \cdot A \cdot H_V + \frac{3}{2} \cdot B}{1 - A \cdot H_V - B} \tag{18}$$

Sines: $$\alpha = \sqrt{2} \cdot A \cdot H_V + \sqrt{2} \cdot B \tag{19}$$

3 Results and discussion

As described before, the parameters C_1, C_2 C_3, A, B and m are required to describe the fatigue behaviour in case of an inclusion failure. The fatigue test results of case hardened, quenched and tempered as well as bearing steels in different heat treatment conditions, which have been investigated in several research projects over the last two decades, are used as database. Only fatigue specimens which failed at an inclusion were suitable for this investigation. A total of 603 specimens were taken into account in the parameter determination. Oxides, sulphides, carbides or nitrides were the failure causing inclusions. The local stress amplitudes and mean stresses at the position of the fracture causing inclusion have to be calculated. In case of notched specimens or complex loading conditions the entire multiaxial stress state has been regarded. Furthermore, the residual stresses and the hardness at the crack initiation site have been determined. The entire database is displayed exemplarily in Figure 1.

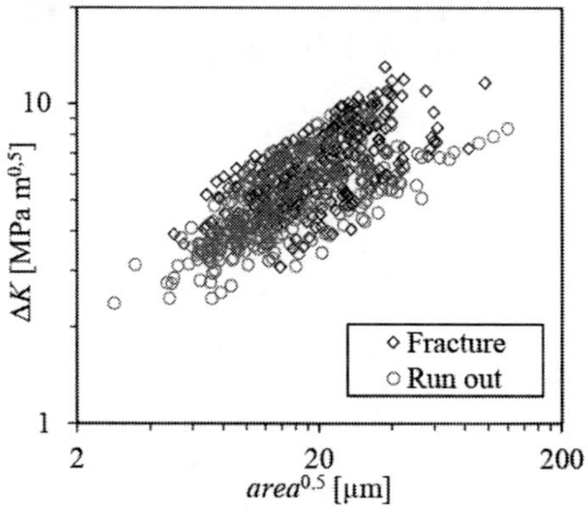

Fig. 1. Specimen used as database for the adaption of the parameters. Dang Van's fatigue criterion was used for the calculation of ΔK.

The diagram shows the stress intensity factor range ΔK according to Murakami (Equation 1) of the fractured and run out specimens in dependence of the size $area^{0,5}$ of the failure causing inclusion. For the calculation of the equivalent stress amplitude σ_{ae} the fatigue criterion of Dang Van (Equation 9) was used and the parameter α was calculated according to Schleicher et al. (Equation 11) with a H_α of 1263 HV [13]. The inclusion size ranges between 2.8 and 119 µm. Since the hardness of the investigated specimens ranged between 357 and 776 HV the stress intensity factor range ΔK, which leads to a fracture or a run out specimen, is scattering as well. Nevertheless, an increase of the stress intensity factor range ΔK with increasing inclusion size is visible. However, it must be considered, that the stress intensity factor range ΔK is influenced by the chosen fatigue criterion.

3.1. Influence of the inclusion type on the fatigue strength

For a detailed investigation of the influence of the inclusion type on the fatigue strength, steel batches with a constant hardness and different crack initiating inclusions have been analysed. In that case the influence of the hardness on the fatigue strength did not superimpose the result. Exemplarily, Figure 2 shows the fatigue test results of two bearing steel batches. Figure 2 a shows batch A with a hardness of 765 HV and Figure 2 b shows batch B with a hardness of 756 HV. Both batches have been tested at push-pull-loading at the stress ratio R = -1. Additionally, batch A was tested at alternating tensile loading at the stress ratio of R = 0.1

and batch B was tested at alternating tensile loading at the stress ratio $R = 0$.

In batch A oxides and carbides are the dominant failure types. One failure at a sulphide and some fractures at nitrides occur as well. However, the number of crack initiations at nitrides is significantly higher at $R = 0.1$ than at $R = -1$. On average, the oxide inclusion are larger than the nitrides, carbides or the sulphides. For the same inclusion size the stress intensity factor range ΔK is higher at a stress ratio of $R = -1$ compared to $R = 0.1$. This gives an indication that the the different loading conditions are described inaccurately by the fatigue criterion. At both stress ratios the stress intensity factor range increases with increasing inclusion size. However, an influence of the inclusion type on the fatigue strength is not visible.

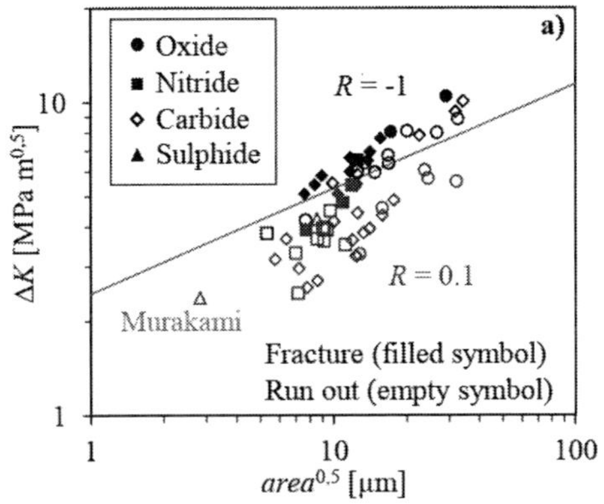

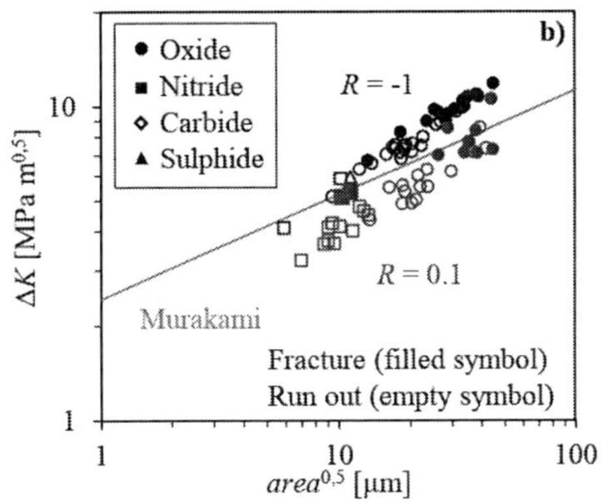

Fig. 2. Stress intensity factor range at different inclusion types in fractured and run out specimens of **a)** batch A and **b)** batch B.

At batch B oxides are the dominant failure type. Nitrides occur as failure causes as well, especially at $R = 0$. The average size of the oxides is larger than that of the nitrides. At the same inclusion size the stress intensity factor range ΔK is higher at $R = 1$ compared to $R = 0$. This gives again an indication that the fatigue criterion describes the different loading conditions inaccurately. At both stress ratios the stress intensity factor range increases with increasing inclusion size. Again, an influence of the inclusion type on the fatigue strength is not visible.

The prediction of the threshold stress intensity factor range ΔK_{th} according to Murakami and Yamashita [8] (Equation 5) is displayed for both steel batches in Figure 2 a and b as well. For both batches the prediction separates the values at push pull and alternating tensile loading. Since the predicted threshold stress intensity factor range ΔK_{th} describes the 50 % fracture probability of the fatigue specimens the line should be between the fractured and run out specimens. Both the chosen fatigue criterion as well as the predicted threshold stress intensity failure range can be responsible for this deviation. Furthermore, the experimental stress intensity factor range values increase more strongly with increasing inclusion size than the predicted threshold stress intensity factor range.

In summary different failure types lead to fatigue failure of the investigated specimens. Depending on the inclusion type different average inclusion sizes are observed. Taking into account the influence of the inclusion size on the fatigue strength, no difference between the inclusion types is recognizable. For the calculation of the fatigue strength follows, that no distinction has to be made between the inclusion types.

3.2. Influence of the inclusion size, hardness and loading condition on the fatigue strength

The entire database displayed in Figure 1 is used to determine the most likely parameters C_1, C_2 C_3, A, B. The parameter m depends on the individual scatter of each batch and cannot be used for a prediction of the scatter of other batches. For that reason the parameter m was set to 20 for the fitting procedure. The parameter sets for the four fatigue criteria which lead to the maximum likelihood values according Equation 4 are displayed in Table 1.

Table 1. Parameter C_1, C_2, C_3, A, and B determined at the entire database for the four investigated fatigue criteria.

Fatigue criterion	C_1	C_2	C_3	A	B
Bomas	0.00290	95	0.384	0.00128	-0.383
Crossland	0.00287	100	0.382	0.00120	-0.343
Dang Van	0.00285	104	0.383	0.00122	-0.359
Sines	0.00291	94	0.382	0.00125	-0.358

There are no major differences between the parameter sets for the different fatigue criteria. The resulting mean stress sensitivity M and the weighting factor α, which are both calculated based on the parameters A and B, are displayed as a function of the hardness in Figure 3 and Figure 4, respectively.

For all fatigue criteria, an almost identical trend of the mean stress sensitivity depending on the hardness can be observed. The curves show an increase of the mean stress sensitivity with increasing hardness. The weighting factor α shows, depending on the chosen fatigue criterion, a different curve shape. For Bomas' and Sines' fatigue criterion the curves increase linear with increasing hardness. For both fatigue criteria the weighting factor is nearly identical. For Crossland's and Dang Van's criterion a progressive increase of the weighting factor α with increasing hardness was determined. For these two fatigue criteria the weighting factors are nearly identical as well.

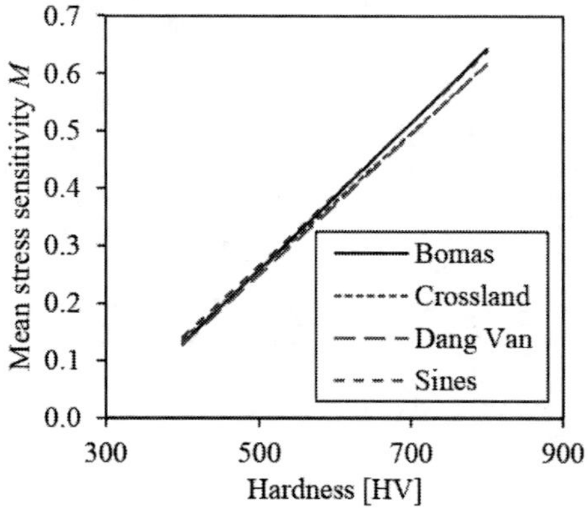

Fig. 3. Mean stress sensitivity M determined with different fatigue criteria as a function of the specimen hardness.

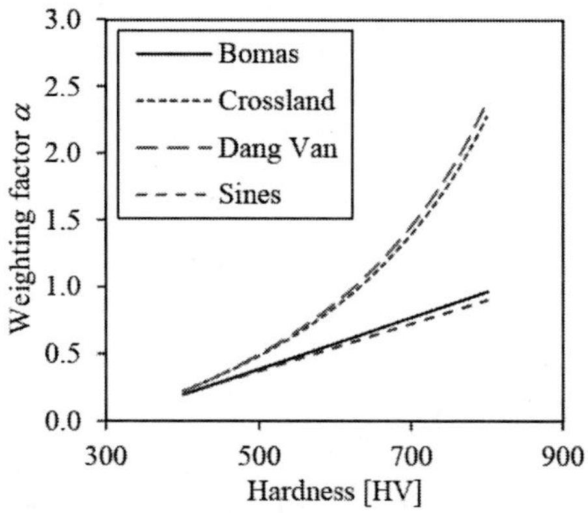

Fig. 4. Weighting factor α determined for the different fatigue criteria as a function of the specimen hardness.

Figure 5 shows the influence of the hardness on the threshold stress intensity factor range ΔK_{th} exemplarily for an inclusion size of 50 µm. All investigated fatigue criteria lead to nearly identical threshold values. Compared to the threshold value according to Murakami (Equation 5) the calculated threshold values are located above the Murakami curve and the increase with hardness is somewhat steeper.

Both resulting parameters, the mean stress sensitivity M and the threshold stress intensity factor range ΔK_{th} are nearly identical for the different fatigue criteria investigated. However, it should be noted that a large number of the specimens examined were subjected to uniaxial tensile compression or alternating tensile loads. In combination with the chosen calculation approach for weighting parameter α, the four investigated fatigue criteria lead to the same equivalent stress amplitude in case of uniaxial tensile compression or alternating tensile loads. Due to the small number of multi-axially loaded samples, the determined parameters for the different fatigue criteria are close to each other.

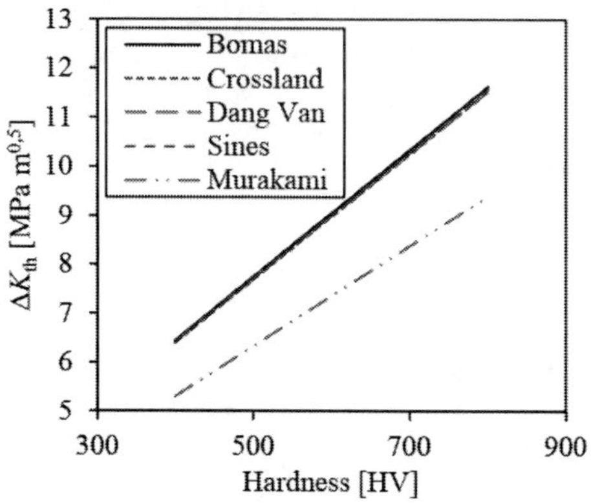

Fig. 5. Threshold stress intensity factor range determined for different fatigue criteria and threshold stress intensity factor range according to Murakami as a function of hardness displayed for an inclusion size of 50 µm.

To show the influence of the new parameters the data of the batches A and B, which are displayed in Figure 2, are used. As an example, the parameters of Bomas' fatigue criterion are used and displayed in Figure 6.

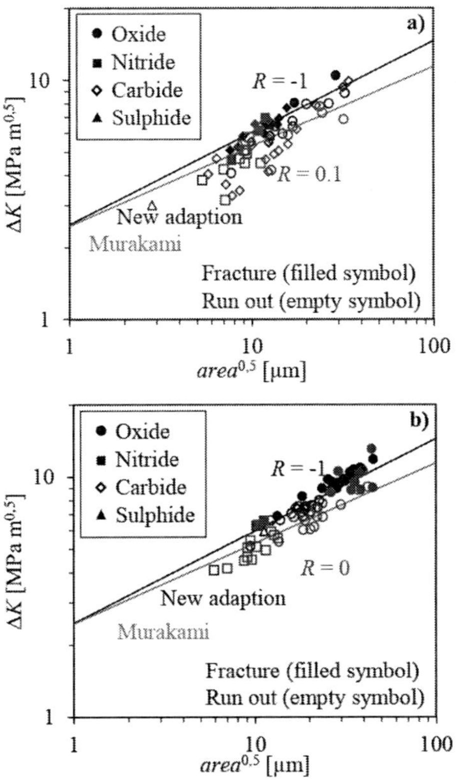

Fig. 6. Stress intensity factor range calculated with Bomas' fatigue criterion with the parameters, *A* and *B* of Table 1 in fractured and run out specimens of **a)** batch A and **b)** batch B. The threshold stress intensity range values have been calculated according to Murakami and the new adaption for Bomas' fatigue criterion.

Compared with the results displayed in Figure 2 the gap between the specimens tested with and without mean stress disappears with the new calculation approach.

The determined threshold stress intensity factor range and the threshold value according to Murakami are shown in Figure 6 as well. In case of small inclusion sizes, both approaches supply almost identical values for the threshold stress intensity factor range. In the new approach, the threshold stress intensity factor range increases more strongly with increasing inclusion size, so that a higher threshold stress intensity factor range is calculated with this approach for large inclusions. Since both threshold stress intensity factor range calculation approaches only depend on the size of the inclusion and the hardness of the specimen not all fatigue strength determining factors are considered. Therefore, the threshold stress intensity factor range of different steel batches is described with varying degrees of accuracy. For example, the threshold value according to Murakami separates the fractured and run out specimens of batch A slightly more accurate than the threshold value calculated in this investigation, as can be seen in Figure 6 a. In comparison, the newly calculated threshold value describes the data of batch B much more precisely than the threshold value of Murakami, as can be seen in Figure 6 b.

4 Summary

An empiric fracture mechanical approach is presented describing the fatigue strength of high strength steel specimens. This calculation method combines the approach of Murakami and different fatigue criteria assuming that inclusions are responsible for the fatigue failure. The hardness of the specimens and the crack initiating inclusion sizes are the fatigue strength determining material parameters. The required model parameter have been determined at 603 fatigue specimens which have been tested in several research projects over the last two decades and failed at inclusions.

Different inclusion types like oxides, carbides, nitrides or sulphides showed the same effect on the fatigue strength. Hence the calculation approach does not distinguish between different inclusion types.

Based on the experimental data the threshold stress intensity factor range for a fracture probability of 50 % was determined. Compared with the threshold value according to Murakami the new approach results in a slightly higher threshold value and a steeper increase of the threshold stress intensity factor range with increasing inclusion size.

Additionally, a calculation approach for the material depending weighting factor α of the different fatigue criteria is described. Individually, for each fatigue criterion the weighting factor can be determined on the basis of the mean stress sensitivity.

In summary, with the knowledge of the hardness and the expected failure-critical inclusion size, the introduced calculation approach enables the prediction of the fatigue strength for arbitrary steel batches and loading conditions.

The IGF Project (18973 N) of the Arbeitsgemeinschaft Wärmebehandlung und Werkstofftechnik e. V. (AWT) was funded by the Arbeitsgemeinschaft industrieller Forschungsvereinigungen „Otto von Guericke" e. V. (AiF) through the IGF program of the Bundesministerium für Wirtschaft und Energie (BMWi) due to a resolution of the Deutscher Bundestag. The authors are grateful for the financial support. We further wish to thank the Expert Committee 21 "Microstructure and mechanical properties of heat treated materials" of AWT for scientific support and supervision of the project. Our special thank is given to the members of the Board of Experts for their collaboration and interest in the project: Deutsche Edelstahlwerke GmbH, Georgsmarienhütte GmbH, Robert Bosch GmbH, SchaefflerTechnologies AG & Co.KG, Steinhoff GmbH & Cie. OHG, Thermo Fisher Scientific GmbH and ZF-Friedrichshafen AG.

References

1. J. Rösler, H. Harders, M. Bäker, *Mechanisches Verhalten der Werkstoffe* (Vieweg + Teubner Verlag, Wiesbaden, 2008)

2. A. Melander, M. Rolfsson, A. Nordgren, B. Jansson, H. Hedberg, T. Lund, Scand. J. Metall. **20** (1991)
3. H. Bomas, M. Schleicher, Fatigue Fract. Eng. Mater. Struct. **28**, 11 (2005)
4. H. Bomas, M. Bacher-Hoechst, R. Kienzler, S. Kunow, G. Loewisch, F. Muehleder, R. Schroeder,. Fatigue Fract. Eng. Mater. Struct. **33**, 2 (2010)
5. J. Ma, B. Zhang, D. Xu, E.-H. Han, W. Ke, Int. J. Fatigue **32** (2010)
6. Y. Murakami, *Metal Fatigue: Effects of Small Defects and Nonmetallic Inclusions* (Elsevier, Oxford 2002)
7. J. Monnot, R. Heritier, J. Y. Cogne, *Effect of Steel Manufacturing Processes on the Quality of Bearing Steels* (ASTM STP 987, 1988)
8. Y. Murakami, Y. Yamashita, Procedia Eng. **74** (2014)
9. H. Bomas, T. Linkewitz, P. Mayr, Fatigue Fract. Eng. Mater. Struct. **22** (1999)
10. B. Crossland, *Proceedings of the International Conference on the Fatigue of Metals* (Institute of Mechanical Engineers, London, 1956)
11. K. Dang Van, B. Griveau, O. Message, *Biaxial and Multiaxial Fatigue (EGF 3)* (Mechanical Engineering Publications, London, 1989)
12. G. Sines, *Metal Fatigue* (McGraw Hill, New York, 1959)
13. M. Schleicher, H. Bomas, P. Mayr, HTM Haerterei-Techn. Mitt., **56** (2001)

Interactions between geometrical defects and microstructure during high cycle fatigue of polycrystalline aluminium with different grain sizes

Benoît Bracquart[1], *Charles* Mareau[1], *Nicolas* Saintier[2], and *Franck* Morel[1,*]

[1]Arts et Métiers ParisTech, Campus d'Angers, LAMPA, 2 bd du Ronceray, 49035 Angers, Cedex 1, France
[2]Arts et Métiers ParisTech, Campus de Bordeaux-Talence, I2M, Esplanade des Arts et Métiers, 33405 Talence, France

Abstract. In this work, the influence of the geometrical defect size on the high cycle fatigue behavior of polycrystalline aluminium with different grain sizes is investigated, to better understand the role of internal length scales. Two sizes of grains and defect are used: 100 µm and 1000 µm, the grain size being controlled with thermomechanical treatments. Fully reversed stress-controlled fatigue tests are then carried out. According to fatigue test results, surface crack initiation is delayed when the grain size is reduced, while an approximation of the fatigue limit shows that it is not much influenced by the average grain size. The relative defect diameter (compared to the grain size) seems to be the leading parameter influencing fatigue crack initiation from a defect. Finally, Electron BackScattered Diffraction (EBSD) maps are collected for specimens with large grains and small defects. Fatigue crack initiation from a defect is found to be strongly impacted by the crystallographic orientation of the surrounding grain, crack initiation preferably occurring in crystals being favorably oriented for plastic slip.

1 Introduction

Defects in metallic alloys can originate from various factors [1, 2, 3, 4], have either a geometrical (e.g. notches) or metallurgical (e.g. inclusions) nature and can be of various shapes and sizes. Because they act as stress concentrators, such defects usually have a detrimental influence on the high cycle fatigue (HCF) resistance of metallic alloys [5]. The fatigue resistance reduction associated with the presence of defects is however strongly dependent on their size: the reduction is higher for a larger defect, and there exists a critical size below which the fatigue limit is no longer affected [6, 7]. Although the knowledge of the critical size is important to design fatigue-resistant structures, the dependence of this critical size regarding microstructural features remains unclear.

The role of microstructural features on the fatigue behavior in the absence of significant geometrical defects has been widely studied. For instance, the influence of grain size on the fatigue behavior of smooth specimens is largely documented [8, 9, 10]. The fatigue limit is usually found to increase when the grain size is reduced and the influence of grain size on the fatigue limit is often correctly described with an Hall-Petch [11,12] type of relation [13]. According to Chan [14], the influence of grain size is explained by the fact that polycrystals with large grains are more prone to initiate cracks in slip bands than their smaller grained counterparts, leading to earlier crack initiation in the former. Regarding the influence of grain size on fatigue crack propagation in aluminium polycrystals, several authors give contradictory results: Saga et al. [15, 16] conclude that the larger the grains, the higher the crack growth rate and the lower the number of cycles to failure, whereas Thompson et al. [17] found no significant impact on the fatigue resistance at 10^6 cycles, in the range 20-200 µm. The interpretation of the latter is that cross slip can easily be activated in pure aluminium, and that the resulting cellular dislocation structures tend to hide grain boundaries in the crack propagation stage.

Few attempts have been made to investigate the impact of microstructure on the fatigue behavior of metallic specimens with geometrical defects. Guerchais et al. [18] studied the influence of local microstructure around the defect for different loading conditions, but only the defect size and local microstructure were varied, not the grain size. Sweeney et al. [19] and Wan et al. [20] relied on a comprehensive experimental procedure, including HCF tests and systematic Electron BackScattered Diffraction (EBSD) maps, to set up numerical models in order to predict fatigue crack initiation. In their case, the microstructure is varied around the defect, but grain and defect sizes are held constant. To explore the connection between grain size and defect size, Karry et al. [21] carried out an experimental investigation on brass specimens with defects of different sizes and shapes, as well as on

* Corresponding author: franck.morel@ensam.eu

smooth specimens. They observed that when a defect is introduced, the decrease in fatigue limit compared to the corresponding smooth specimen was lower for a higher grain size. The defect sensitivity is thus higher for small grained microstructures. Lorenzino *et al.* [22] carried out fatigue tests on pure aluminium specimens with different grain and defect sizes. Besides the fact that a larger grain size and/or notch size reduces the fatigue limit, they showed that, as far as crack propagation is concerned, the governing parameter is the relative defect size (the defect size divided by the mean grain size) rather than the absolute physical one. Indeed, by normalizing the fatigue limit of specimens with defects by the fatigue limit of smooth specimens with the same grain size, and the defect size by the grain size, all data points fall onto the same line in a fatigue limit/notch size diagram. Vincent *et al.* [23], using various defect sizes and grain sizes, came to the same conclusion for commercially pure iron. Those two studies focus on the number of cycles to failure rather than to crack initiation, but they demonstrate the interest of varying both characteristic sizes.

In the present work, an experimental campaign is carried out to investigate the joint influence of geometrical surface defect size and grain size on the HCF behavior of pure aluminium polycrystals. In contrast with the work of Lorenzino *et al.* [22], who focused on crack propagation, this study concentrates on crack initiation in the presence of defects. Indeed, the influence of the defect is most important in the crack initiation stage, as it acts as a stress concentrator and a preferential crack initiation site; its influence becomes lower once a crack is initiated and starts to propagate. Also, to better understand the factors governing the initiation of a fatigue crack, Electron Backscattered Diffraction (EBSD) techniques are used to evaluate the influence of crystallographic orientation.

1 Experimental procedure

1.1 Material description

The material used in this study is commercial purity aluminium (AA1050), for which the aluminium weight concentration exceeds 99.5 %. The alloy is supplied in the form of 6 mm-thick cold-rolled sheets. The initial microstructure is composed of elongated grains along the rolling direction.

Two different thermomechanical sequences have been applied to control the grain size Φ, which is evaluated by the average grain intercept method. For the first sequence, in addition to the strain-hardening associated to the rolling process, aluminium samples are held at 550°C for 45 minutes and then air cooled. The resulting microstructure is referred to as "small" grain microstructure (denoted by subscript *s*) with an average grain size Φ_s of about 100 µm along the rolling direction. For the second sequence, samples are firstly maintained at 550° C for 2 hours and then air-cooled. A tensile strain of 12 % is then applied at room temperature in the rolling direction, followed by a second heat treatment at 600° C for 2 hours, before a final air cooling. The corresponding grain size Φ_l is approximately of 1000 µm along the rolling direction. This second microstructure is referred to as "large" grain microstructure (denoted by subscript *l*). Both sequences ending with a recrystallization treatment, they provide fully annealed microstructures with different grain sizes.

Grain sizes have been evaluated in the three principal sheet directions from several micrographs. For both microstructures, grain sizes along the rolling and transverse directions are identical. The average grain size is however smaller, by a factor of 2, along the normal direction. It should be noticed that, for a given microstructure, all samples have been extracted from the same aluminium sheet. Also, during the preparation stage, some samples have been randomly selected to check the repeatability of the preparation procedure.

Tensile tests have been performed on both microstructures, in the rolling direction, using a strain rate $\dot{\varepsilon} = 10^{-3}\,s^{-1}$. The corresponding mechanical properties are presented in Table 1. One can note that the values of $R_{p0.2}$ and R_m are lower for a higher grain size, in accordance with the classical Hall-Petch law.

Table 1. Tensile mechanical properties of polycrystalline aluminium for both grain sizes.

	$R_{p0.2}$ (MPa)	R_m (MPa)	A (%)
Φ_s = 100 µm	17.8	61.7	50.3
Φ_l = 1000 µm	16.1	60.7	48.1

1.2 Specimen preparation

Fatigue specimens have been machined from previously prepared aluminium samples. The geometry of fatigue specimens is presented in Figure 1. The gauge section is 15 mm-long and 30 mm-wide, the thickness is 5 mm and the loading axis coincides with the rolling direction. For the large grain microstructure, particular care has been taken to extract the gauge part of fatigue specimens from the area where the applied tensile strain, hence the grain size, had been controlled.

Geometrical defects have been introduced in the gauge area of fatigue specimens with a hemispherical drill, the defect size *d* being thus given by the drill diameter. A high cutting speed and a low translation speed (1 µm.s^{-1}) have been used to ensure that the local microstructure is not affected by drilling operations. In the present work, two defect sizes have been considered: d_s = 100 µm and d_l = 1000 µm, where subscripts *s* and *l* denote respectively the small and the large defects. Defect depths and shapes have been systematically inspected with a 3D profilometer. For small defects, the relative difference between the nominal depth and the actual depth is under 5 % in most cases, and under 10 % in all cases. For large defects, this difference is always inferior to 1 %.

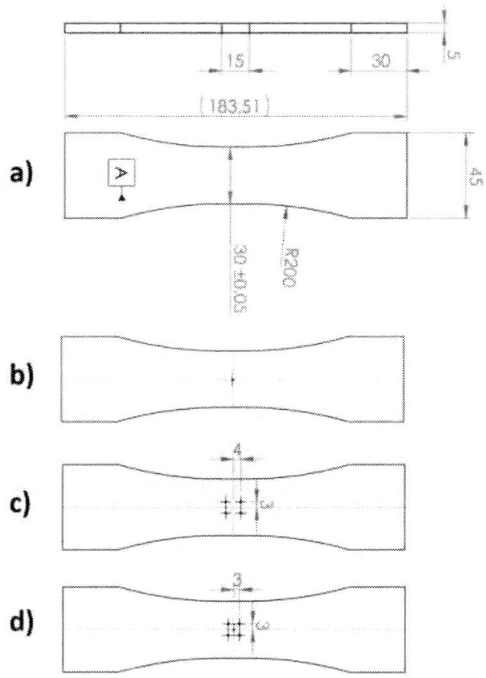

Fig. 1. Geometry of fatigue specimens. a) smooth, b) single large defect (SLD), c) multiple large defect (MLD) and d) multiple small defects (MSD) specimens

Four different types of specimens are used: smooth specimens, specimens with a single large defect (denoted SLD specimens), specimens with multiple large defects (MLD specimens) and specimens with multiple small defects (MSD specimens). Respectively 4 and 5 defects are machined in one specimen, for MLD and MSD specimens. Figure 1 displays the location(s) of defect(s) relative to the center of the gauge section.

Fatigue specimens are then submitted to mechanical polishing (FEPA abrasives 600, 1200 and 2000), followed by a final heat treatment (260° C for 2 hours and air-cooling) to relieve the residual stresses resulting from machining and polishing operations. The temperature for this heat treatment is chosen low enough not to engage recrystallization and modify the grain size.

1.3 Experimental set-up

Fully reversed load-controlled push-pull fatigue tests (i.e. $R_\sigma = -1$) are carried out on an Instron 8802 servohydraulic testing machine. The tests are thus nominal-stress-controlled; for the sake of simplicity, they are referred to as "stress-controlled" in the following. Different tests frequencies, namely 1 Hz and 12 Hz, are used, depending on the kind of test performed.

To prevent buckling of the specimen while allowing elongation measurement by an extensometer, a specific anti-buckling fixture has been developed. As shown in Figure 2, it consists of two opposing asymmetrical steel plates which are assembled to exert a normal effort on the specimen during the compression stage. A teflon film is placed between the specimen and the fixture to reduce friction.

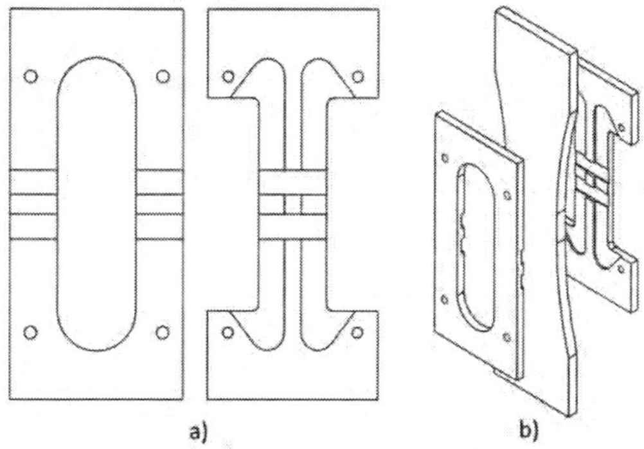

Fig. 2. a) Geometries of the two opposing plates of the anti-buckling fixture, and b) exploded view of the assembly on fatigue specimen.

Preliminary tests have been conducted on a specimen equipped with strain gauges to verify that, despite the additional effort due to the fixture, the stress state in the center of the gauge area is not altered. However, when no defect is introduced, cracks initiate and propagate from the interface between the specimen and the rectangular plate in a region where the stress field might be perturbed. As a result, smooth specimens cannot be used to characterize the fatigue behavior since the stress state might not be correctly controlled in the crack initiation area. Smooth specimens are thus solely used for the evaluation of the cyclic behavior.

For specimens with defects, the final failure of fatigue specimens does not always result from a defect-induced crack. As a consequence, when evaluating the influence of geometrical defects on the number of cycles to failure, only tests for which fatigue failure is caused by defects have been considered. This aspect is not problematic for large defects (d_l = 1000 μm), as final failure is caused by defects in most cases. However, for small defects (d_s = 100 μm), final failure often results from another crack than the one(s) initiated from the defect. As a consequence of the propagation of non defect-induced cracks, the stress field in the vicinity of defects can be strongly perturbed. Therefore, for small defects, only crack initiation, for which perturbations caused by cracks initiated away from defects are negligible, is investigated.

Different experimental configurations are used to investigate the influence of grain size and/or defect size. In the first one, the cyclic elastic-plastic behavior of pure aluminium is evaluated from smooth specimens of both grain sizes equipped with a clip-on Instron extensometer with a 12.5 mm spacing. The second configuration uses SLD specimens (d_l = 1000 μm). This configuration aims at measuring the length of cracks having been initiated from the defect. For this application, the area surrounding the defect is observed with an optical camera and pictures are recorded every 1000 loading cycles. While an *in-situ* monitoring of the crack length is possible with this second configuration, the detection is limited to surface cracks with a minimum length of about 100 µm. Due to the potential influence of non defect-induced cracks, small defects are not studied in this configuration. The third configuration uses MLD and MSD specimens (d_l = 1000 µm and d_s =100 µm). For a given low stress amplitude, preventing macroscopic crack propagation, it allows determining the number of cracked defects from *a posteriori* Scanning Electron Microscopy (SEM) observations. Some of the MSD specimens are also selected to carry out EBSD observations of the microstructure surrounding the defect.

2 Results

2.1. Smooth specimens

Fully reversed stress-controlled cyclic tests have been performed with a loading frequency of 1 Hz and a constant stress amplitude σ_a, for both grain sizes, to investigate the influence of grain size on the cyclic elastic-plastic behavior of pure aluminium. Most tests have been stopped after 10^4 cycles, but some have been continued up to 10^5 or 2.10^5 cycles to determine whether elastic shakedown is observed or not. The evolution of the axial plastic strain range $\Delta\varepsilon^p$ as a function of the number of cycles N is plotted for different stress amplitudes in Figures 3 and 4, respectively for small and large grains.

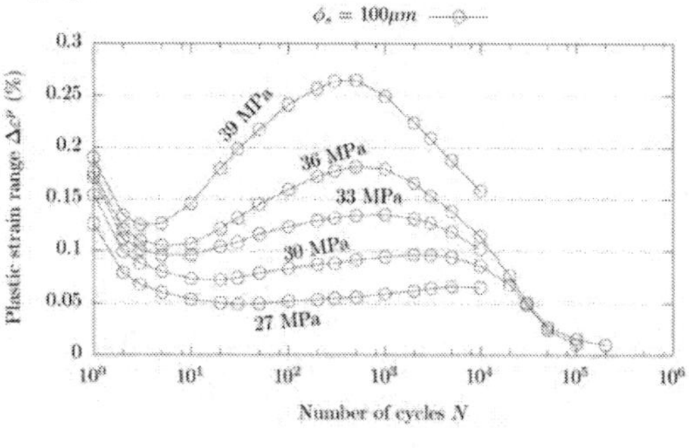

Fig. 3. Evolution of Δε^p during cycling for Φ_s, for different stress amplitudes.

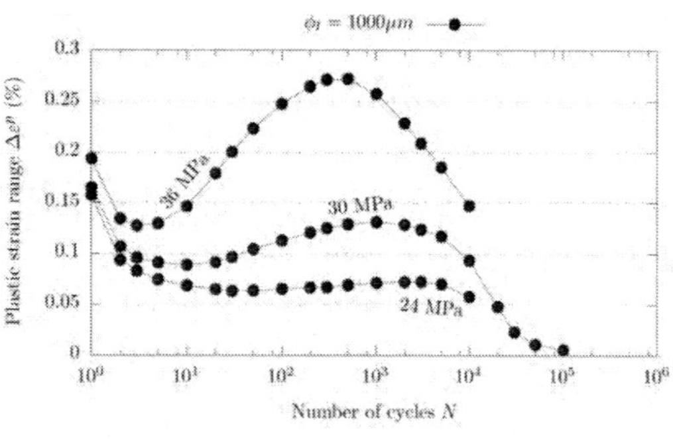

Fig. 4. Evolution of Δε^p during cycling for Φ_l, for different stress amplitudes.

2.1. Specimens with a single large defect (SLD)

Stress-controlled fatigue tests have been carried out on SLD specimens (d_l = 1000 µm) to evaluate the impact of grain size on fatigue crack initiation and specimen fracture. For these tests, the loading frequency is set to 12 Hz. In the absence of failure, fatigue tests have been interrupted after 10^6 cycles. The observation of the area surrounding the defect with the optical camera allows monitoring early crack propagation from the defect, on the specimen surface, as illustrated in Figure 5 for the (d_l-Φ_s, σ_a = 30 MPa) configuration, for which failure occurred at 292,000 cycles.

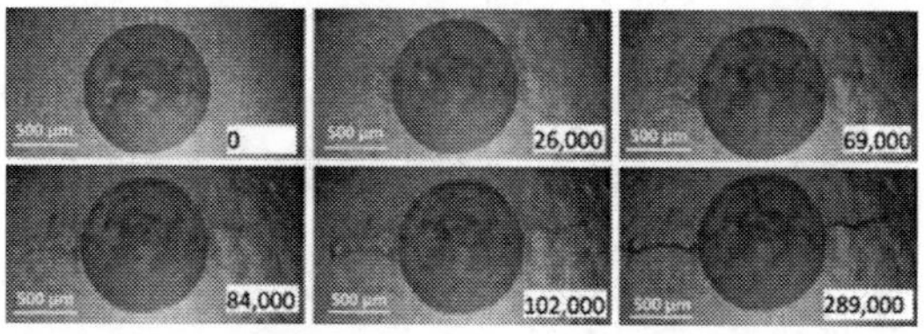

Fig. 5. Example of crack monitoring during fatigue test of a SLD specimen. The number of cycles is indicated in the bottom right corner of each picture. Loading axis is vertical.

Based on fatigue test results, two different S-N diagrams have been plotted. The first S-N diagram, Figure 6, represents the evolution of the stress amplitude σ_a as a function of the number of cycles to failure N_f, the failure being defined as the separation of the specimen into two distinct parts. Note that for large grains (Φ_l = 1000 µm), for respectively σ_a = 30 MPa and σ_a = 33 MPa, only 2 and 1 point(s) were representative (failure from the crack initiating from the defect). The evolution of the stress amplitude as a function of the number of cycles to crack initiation N_i is displayed in a second S-N diagram, Figure 7. In the present work, the crack initiation criterion, defining Ni, is arbitrarily set to a total surface crack length of 100 µm, which is the shortest detectable crack length. One should note that, as a result, especially for the small grain microstructure, the number of cycles to crack initiation may include a portion of early crack propagation.

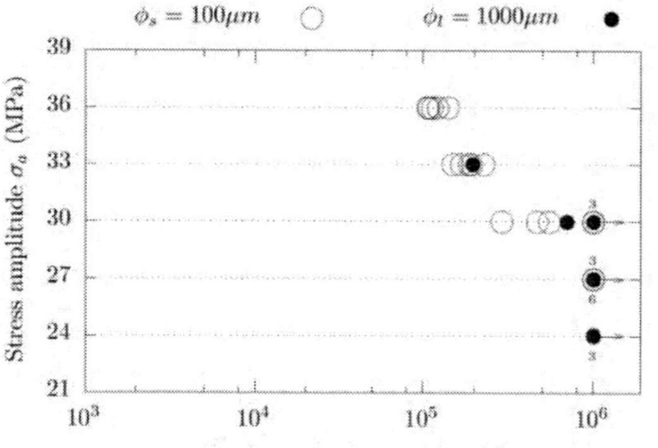

Fig. 6. S-N curves for failure, obtained for SLD specimens (d_l = 1000 µm) and two sizes of microstructure. Arrows represent run-out specimens. Number indicates the superimposed points (above for Φ_s, below for Φ_l).

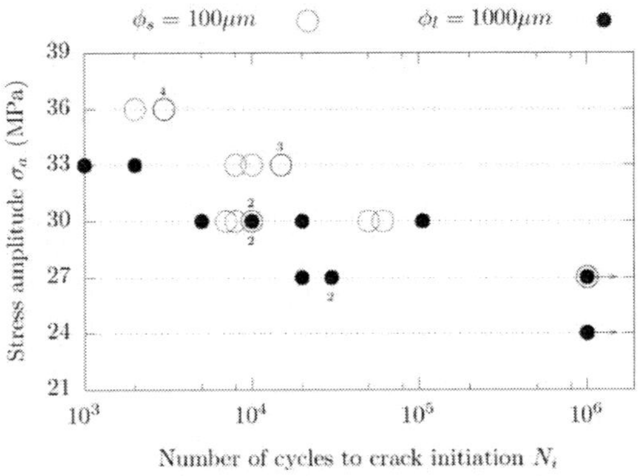

Fig. 7. S-N curves for initiation, obtained for SLD specimens (d_l = 1000 μm) and two sizes of microstructure. Crack initiation criterion: 100 μm-long surface crack. Arrows represent run-out specimens for which this criterion was not reached. Number indicates the superimposed points (above for Φ_s, below for Φ_l).

2.1. Specimens with multiple defects (MLD and MSD)

To investigate the influence of defect size on fatigue crack initiation, some specimens with multiple defects have been submitted to stress-controlled fatigue tests. The fatigue tests are conducted with a frequency of 12 Hz and up to 10^6 loading cycles. In this case, both the grain size Φ and the defect size d are varied. Stress amplitudes are selected low enough to prevent macroscopic crack propagation (hence failure) and exclude interactions between defects.

After fatigue testing, SEM observations of each defect are carried out to determine whether a crack initiated or not. In comparison with the *in-situ* monitoring used earlier, short cracks are more accurately detected with SEM because it allows observing the bottom pole of hemispherical defects. The results obtained for different grain sizes, defect sizes and stress amplitudes are presented in Table 2, which shows the proportion of cracked defects in the different configurations.

Table 2. Proportion of cracked defects for the four d-Φ configurations (MLD and SLD specimens), for different stress amplitudes σ_a. n is the total number of defects studied in the configuration and n_c the number of cracked defects after cycling (* denotes a high stress level leading to a macroscopic crack propagation and possibly failure of the specimen).

	d_s = 100 μm		d_l = 1000 μm	
	σ_a (MPa)	n_c/n	σ_a (MPa)	n_c/n
Φ_s = 100 μm	27	1/15	24	0/8
	28.5	8/15	27	10/12
	30	11/15	28.5	11/11
Φ_l = 1000 μm	24	1/15	24	0/17
	25.5	4/15	25.5	7/16
	27	9/15	27*	7/8

Based on these results, the stress amplitude corresponding to a 50 % probability of crack initiation to a defect, $\sigma_a(P_{50})$, has been estimated for each d-Φ configuration, by linear interpolation (see Figure 8).

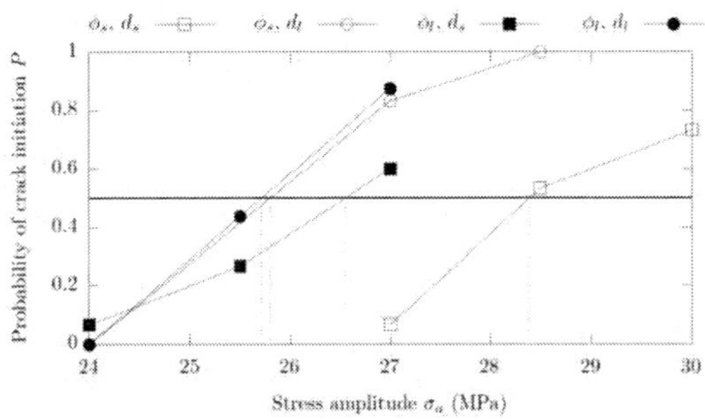

Fig. 8. Probability P of crack initiation to the defect, for the different d-Φ configurations (MLD and MSD specimens), as a function of the applied stress amplitude σ_a.

3 Discussion

3.1. Influence of grain size and defect size

Based on the experimental results, the influence of grain size and/or defect size on the HCF behavior of pure aluminium is firstly discussed.

According to Figures 3 and 4, for both grains sizes, the cyclic behavior is composed of three successive stages: primary hardening, softening and secondary hardening. This specific behavior, which is explained by the formation and evolution of dislocation structures [24, 25, 26, 27], has already been highlighted [28, 29].

Figure 9 shows, for all SLD specimens that failed from the defect before 10^6 cycles, the ratio between the crack initiation life and the total fatigue life. Most of the fatigue life is spent in the propagation stage, the ratio N_i/N_f being under 0.1 in all cases. Since, as shown in Figure 6, the fatigue limit (estimated as the higher stress level for which no failure occurs) is not much influenced by grain size, this gives credit to the results of [17].

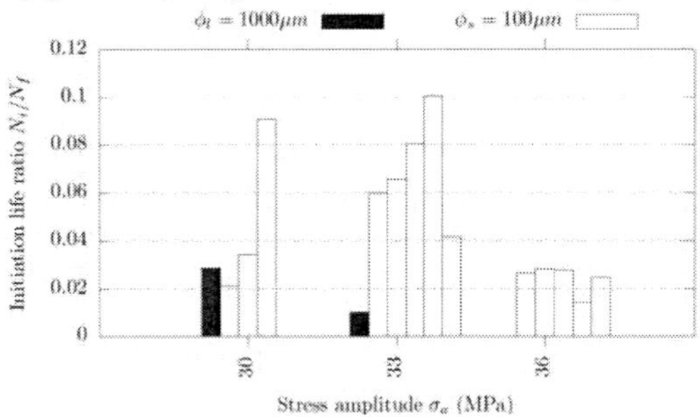

Fig. 9. Ratio between the number of cycles to surface crack initiation N_i and the number of cycles to failure N_f, as a function of the stress amplitude σ_a, for all SLD specimens that failed from the defect before 10^6 cycles.

On the contrary, the crack initiation stage may be impacted by the average grain size. Indeed, as illustrated by Figure 7, the mean number of cycles to crack initiation N_i is reduced when the grain size is increased. However, this last conclusion must be taken with caution, as it depends on the crack initiation criterion used. Indeed, as can be seen in Figure 8, there are no differences in the stress amplitudes necessary to initiate a crack from a large defect, between the two grain sizes. The different conclusions can be explained by the fact that a small portion of early crack propagation is necessary to reach the crack initiation criterion in the results of Figure 7, whereas the criterion used in the results of Figure 8 is more precise.

This influence of grain size on fatigue crack initiation is also observed for specimens with multiple defects. Indeed, as shown in Table 3, the stress amplitude corresponding to a 50 % crack initiation probability, $\sigma_a(P50)$, is lower for the large grain microstructure, when small defects are introduced (reduction of 1.8 MPa, *i.e.* 6.3 %, between both grain sizes). For large defects, however, the difference between both grain sizes is negligible (this difference with the *in-situ* crack monitoring approach on SLD specimens presented above being attributable to the different crack initiation criteria used, as discussed earlier). The small grain microstructure is thus more sensitive to defect size.

Table 3. For each of the four d-Φ configurations (MLD and MSD specimens): macroscopic stress amplitude for 50 % probability of crack initiation at the defect, $\sigma_a(P50)$, and corresponding values normalized by the yield strength.

	d_s = 100 µm		d_l = 1000 µm	
	σ_a(P50) (MPa)	σ_a(P50) /$R_{p0.2}(\Phi)$	σ_a(P50) (MPa)	σ_a(P50) /$R_{p0.2}(\Phi)$
Φ_s = 100 µm	28.4	1.60	25.8	1.45
Φ_l = 1000 µm	26.6	1.65	25.7	1.60

To compare the results from both microstructures, a normalized stress level is used, in a similar way to the works of Lorenzino *et al.* [22] and Vincent *et al.* [23]. As the fatigue limit of smooth specimens could not be determined due to the anti-buckling fixture, the yield strength of the smooth specimen is used. As can be seen on Table 3, when increasing the defect size, the change in normalized stress level is more significant for the small grain microstructure (relative reduction of 9 % and 3 %, from a small to a large defect, respectively for the small grain and the large grain microstructures).

Figure 10 shows that, for a constant relative defect size d/Φ but different absolute sizes, the normalized stress level corresponding to a 50 % crack initiation probability is the same for both grain sizes. This is consistent with the results of Lorenzino *et al.* [22] and Vincent *et al.* [23] who observed that the relative defect size d/Φ is more relevant than the actual physical defect size d. The equivalence of the ratio between the stress amplitude necessary to initiate a crack and the yield strength suggests that, on a macroscopic point of view, regardless of the local microstructural features (e.g. crystallographic orientation, grain shape) which can influence variability of crack initiation behavior for a given d-Φ-σ_a configuration, crack initiation is essentially governed by plasticity.

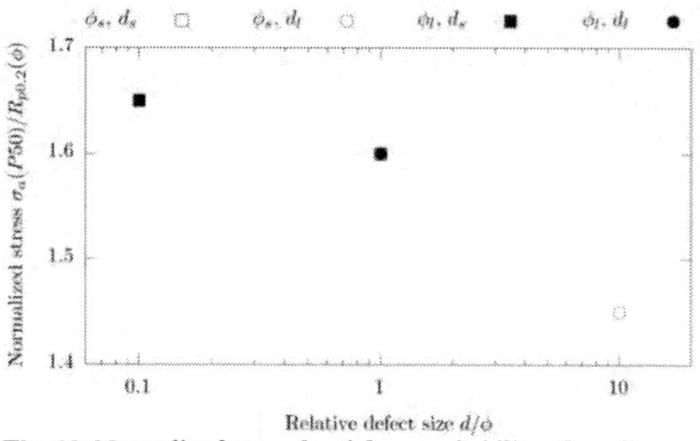

Fig. 10. Normalized stress level for a probability of crack initiation to a defect of 50 %, $\sigma_a(P50)/R_{p0.2}(\Phi)$, as a function of the relative defect size d/Φ, for the four d-Φ configurations (MLD and MSD specimens).

3.2. Influence of local microstructure

To better understand the role of local microstructure on fatigue crack initiation, EBSD analyses have been carried out on specimens which have been submitted to a reversed stress-controlled fatigue test up to 10^6 cycles. The analyses have been performed *a posteriori* on MSD specimens with large grains (d_s-Φ_l). In this specific configuration, defects are usually embedded within a single grain. It is therefore possible to associate a unique crystallographic orientation to each defect (among the 33 defects studied, only one lies on a grain boundary, and was therefore excluded from the analysis). The same stress amplitude (σ_a = 27 MPa) is used for all observed specimens.

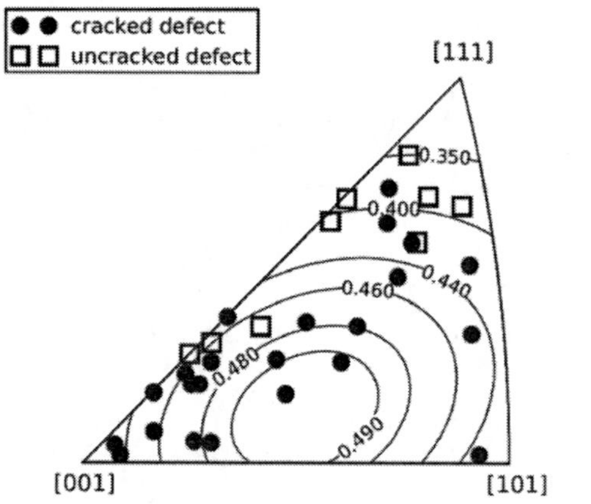

Fig. 11. Orientation of the grains embedding defects, regarding the loading axis, with values of the Schmid factor on the primary slip system.

To evaluate the influence of crystallographic orientation on fatigue crack initiation, the maximum Schmid factor has been determined for each defect. The maximum Schmid factor corresponds to the ratio between the shear stress acting on the primary slip system and the applied uniaxial stress. Grain orientations of cracked and uncracked defects are plotted in Figure 11 in the stereographic unit triangle. Crack initiation is found to be highly dependent on crystallographic orientation, as grains being favorably oriented for plastic slip are more prone to crack initiation.

To evaluate the possible influence of normal stresses, the ratio between the maximum normal stress acting on the slip systems and the applied uniaxial stress has been determined for each defect, with a similar approach. The results are presented in Figure 12. One can note that normal stresses have no obvious influence on fatigue crack initiation.

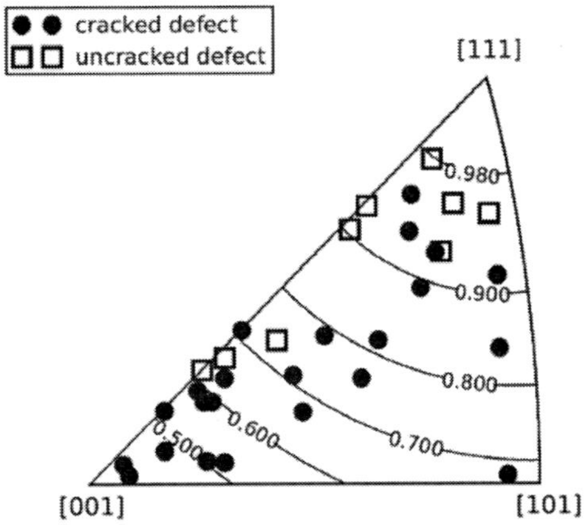

Fig. 12. Orientation of the grains embedding defects, regarding the loading axis, with values of the ratio between the maximum normal stress on the slip planes and the applied uniaxial stress.

A posteriori SEM observations of cracked defects often do not allow for a clear identification of the crack initiation plane. However, for one cracked defect, this plane can be identified. From EBSD data, the maximum Schmid factor has been calculated for each of the four {111} slip planes of the surrounding grain. Figure 13 shows this cracked defect, the slip planes traces (*i.e.* the directions of the intersections of slip planes with a surface parallel to the free surface, as is approximately the defect root), and the corresponding values of maximum Schmid factors. The crack at the defect root (circled one) is parallel to one of the two planes of maximum Schmid factors: the crack initiation plane does not correspond to the primary slip plane (with Schmid factor of 0.48), but to the secondary slip plane (with a slightly lower Schmid factor of 0.47). This observation, although limited to one specimen among all specimens tested, is in agreement with the previous remark regarding the influence of shear stresses on fatigue crack initiation.

The present analysis, which requires the determination of Schmid factors, assumes the stress state to be uniaxial. However, while this simple approach seems sufficient for interpreting the influence of crystallographic orientation in many situations, there are some uncracked defects for which the maximum Schmid factor is significant (up to 0.47, see Figure 11). Such results could be explained by the fact that the stress field is not strictly uniaxial because of the defect and microstructural heterogeneities. Further investigations will therefore be carried out using appropriate computational methods to obtain more accurate estimations of the stress field.

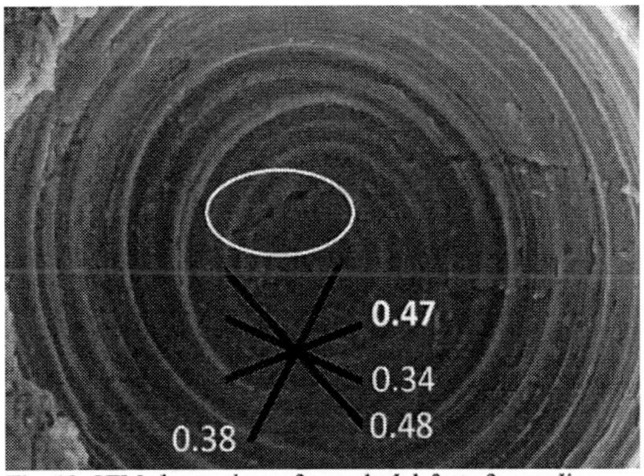

Fig. 13. SEM observations of a cracked defect after cycling, from a MSD specimen, with traces of the slip planes (black lines) and values of the associated maximum Schmid factor.

4 Conclusion

In this work, the impact of geometrical defects on the HCF behavior of pure aluminium (AA1050) has been investigated. More specifically, experimental efforts have been made to better understand the role of local microstructure on fatigue crack initiation in the presence of defects. Two sizes of microstructure (Φ_s = 100 µm and Φ_l = 1000 µm) and defect (d_s = 100 µm and d_l = 1000 µm) are used.

Stress-controlled fully reversed uniaxial HCF tests have then been carried out on specimens with defects and two grain sizes. According to fatigue test results, for a large defect, the number of cycles to obtain a 100 µm-long surface crack is higher when the grain size is reduced. However, fatigue limit, estimated as the higher stress level for which no failure occurs, is not much influenced by grain size, and the fatigue life is spent mostly in the propagation stage in all cases.

The small grain microstructure is observed to be more sensitive to the defect size. Also, when the ratio between the defect size d and the grain size Φ is held constant, the normalized stress amplitude associated with a 50 % crack initiation probability to the defect (normalization by $R_{p0.2}(\Phi)$) is equivalent for both grain sizes. This suggests a strong connection between fatigue crack initiation and plasticity (as usually observed for smooth specimens) even in the presence of defect.

Finally, EBSD maps of the defect surroundings microstructure have been collected to evaluate the influence of crystallographic orientation on fatigue crack initiation. The results show that, when the defect size is small compared to the grain size, crack initiation is strongly impacted by crystallographic orientation. More specifically, crack initiation preferably occurs in crystals being favorably oriented for plastic slip. Crack initiation thus seems to be governed mostly by shear stresses, a conclusion that is consistent with the observation of an actual crack initiation plane.

To better estimate the role of microstructural heterogeneities, future work will focus on the construction of a polycrystalline plasticity model. Strain-gradient plasticity [30] will be used to introduce an internal length scale in constitutive relations. This model will provide quantitative information regarding the stress field perturbations caused by geometrical defects.

This work was financially supported by Institut Carnot ARTS and Région Pays de la Loire.

References

[1] : R.S. Qi, M. Jin, X.G. Liu and B.F. Guo, Formation Mechanism of Inclusion Defects in Large Forged Pieces, J Iron Steel Res Int 23 (6), 531-538, 2016.
[2] : H. Dehmani, C. Brugger, T. Palin-Luc, C. Mareau and S. Koechlin, Experimental study of the impact of punching operations on the high cycle fatigue strength of Fe–Si thin sheets, Int J Fatigue 82, 721-729, 2016.
[3] : E. Pessard, B. Abrivard, F. Morel, F. Abroug, P. Delhaye, The effect of quenching and defects size on the HCF behaviour of Boron steel, Int J Fatigue 68, 80-89, 2014.
[4] : J.O. Peters, B.L. Boyce, A.W. Thompson, R.O. Ritchie, O. Roder, Role of foreign-object damage on thresholds for high cycle fatigue in Ti-6Al-4V, Metall Mater Trans A 31 (6), 1571-1583, 2000.
[5] : Y. Murakami and M. Endo, Effects of defects, inclusions and inhomogeneities on fatigue strength, Int J Fatigue 16 (3), 163-182, 1994.
[6] : M. Endo and Y. Murakami, Effects of an Artificial Small Defect on Torsional Fatigue Strength of Steels, J Eng Mater Tech 109 (2), 124-129, 1987.
[7] : P. Lukas, L. Kunz, B. Weiss, R. Stickler, Notch size effect in fatigue, Fatigue Fract Eng Mater Struct 12 (3), 175-186, 1989.
[8] : H.W. Höppel, L. May, M. Prell and M. Göken, Influence of grain size and precipitation state on the fatigue lives and deformation mechanisms of CP aluminium and AA6082 in the VHCF-regime, Int J Fatigue 33 (1), 10-18, 2011.
[9] : A. Järvenpää, L.P. Karjalainen and M. Jaskari, Effect of grain size on fatigue behavior of Type 301LN stainless steel, Int J Fatigue 65, 93-98, 2014.
[10] : G.J. Deng, S.T. Tu, X.C. Zhang, Q.Q. Wang and C.H. Qin, Grain size effect on the small fatigue crack initiation and growth mechanisms of nickel-based superalloy GH4169, Eng Fract Mech 134, 433-450, 2015.
[11] : E.O. Hall, The deformation and ageing of mild steel, Proc Phys Soc London, Sec B 64 (9), 742, 1951.
[12] : N.J. Petch, The cleavage strength of polycrystals, J Iron Steel Inst 174, 25-28, 1953.
[13] : A. Turnbull and E.R. De Los Rios, The effect of grain size on the fatigue of commercially pure aluminium, Fatigue Fract Eng Mater Struct 18 (12), 1460-2695, 1995.
[14] : K.S. Chan, Roles of microstructure in fatigue crack initiation, Int J Fatigue 32 (9), 1428-1447, 2010.
[15] : J. Saga, M. Hayashi and Y. Nishio, Effect of Grain Size on Fatigue Damage in Pure Aluminium, J Soc Mater Sci Japan 26 (282), 289-295, 1977.
[16] : J. Saga, M. Hayashi and Y. Nishio, Effect of Grain Size on Fatigue Crack Propagation in Aluminium, J Soc Mater Sci Japan 26 (291), 1202-1207, 1977.
[17] : A.W. Thompson and W.A. Backofen, The effect of grain size on fatigue, Acta Metall 19 (7), 597-606, 1971.
[18] : R. Guerchais, N. Saintier, F. Morel and C. Robert, Micromechanical investigation of the influence of defects in high cycle fatigue, Int J Fatigue 67, 159-172, 2014.
[19] : C.A. Sweeney, W. Vorster, S.B. Leen, E. Sakurada, P.E. McHugh and F.P.E. Dunne, The role of elastic anisotropy, length scale and crystallographic slip in fatigue crack nucleation, J Mech Phys Solids 61 (5), 1224-1240, 2013.
[20] : V.V.C. Wan, D.W. MacLachlan, F.P.E. Dunne, A stored energy criterion for fatigue crack nucleation in polycrystals, Int J Fatigue 68, 90-102, 2014.
[21] : R.W. Karry and T.J. Dolan, Influence of grain size on fatigue notch-sensitivity, ASTM Proceeding (53), 789-804, 1953.
[22] : P. Lorenzino and A. Navarro, Grain size effects on notch sensitivity, Int J Fatigue 70, 205-215, 2015.
[23] : M. Vincent, Y. Nadot, C. Nadot-Martin and A. Dragon, Interaction between a surface defect and grain size under high cycle fatigue loading: Experimental approach for Armco iron, Int J Fatigue 87, 81-90, 2016.
[24] : K.U Snowden, Dislocation arrangements during cyclic hardening and softening in Al crystals, Acta Metall 11 (7), 675-684, 1963.
[25] : Y.B. Xia, Z.G Wang and R.H. Wang, Secondary Hardening and Dislocation Evolution in Low Cycle Fatigue of Polycrystalline Aluminium, Phys Status Solidi A 120 (1), 125-132, 1990.
[26] : R. Fougères, Early stages of fatigue damage in aluminium and aluminium alloys, J Phys IV (C7), 669-678, 1993.
[27] : T. Fujii, N. Sawatari, S. Onaka and M. Kato, Cyclic deformation of pure aluminum single crystals with double-slip orientations, Mater Sci Eng A 387, 486-490, 2004.
[28] : A. Giese, A. Styczynski and Y. Estrin, Cyclic hardening behaviour of polycrystalline aluminium under tension-compression, Mater Sci Eng A 124 (2), L11-L13, 1990.
[29] : M. Videm and N. Ryum, Cyclic deformation and fracture of pure aluminium polycrystals, Mater Sci Eng A 219 (1), 11-20, 1996.
[30] : A. Acharya, J.L. Bassani, Lattice incompatibility and a gradient theory of crystal plasticity, J Mech Phys Solids 48 (8), 1565-1595, 2000.

The significance of microstructure heterogeneities on the fatigue thresholds of aluminum castings

Sascha Gerbe[1,*], *Stephan* Knorre[1], *Ulrich* Krupp[1], and *Wilhelm* Michels[1]

[1]Institute of Materials Design and Structural Integrity, University of Applied Sciences Osnabrück, 49076 Osnabrück, Germany

Abstract. Industrial in-series aluminum castings contain a wide range of microstructural heterogeneities like differences in secondary dendrite arm spacing (SDAS), eutectic silicon and intermetallic precipitates of varying morphologies and diverse-shaped and -sized porosity. Regarding to technical and economic limitations, the complete elimination of them is hard to achieve, which requires conservative design, i.e., increased wall thicknesses to accommodate the failure tolerance. To improve the performance of cast aluminum products concerning safety and fatigue properties, the present work deals with the significance of such structures with respect to the threshold for crack propagation $\Delta K_{I,th}$ under pure bending and the fatigue behaviour in the high- and very-high-cycle-fatigue regime (HCF and VHCF). Therefore, two automotive cast alloys taken from engine blocks (AlSi8Cu3) and cylinder heads (AlSi7Cu0.5Mg) and a gravity die cast set (AlSi7Mg0.3), either T6 conditioned or additionally hot isostatic pressed (HIP), were used. For in-series castings, two positions of maximal difference in cooling rate and respective microstructure were extracted. With this set of specimens, the significance of SDAS in interaction with (i) eutectic silicon regions, (ii) intermetallic precipitates in varying occurrence, (iii) the crystallographic orientation, and (iv) the porosity in correlation with the fatigue threshold is shown and compared with first results of fatigue damaging mechanisms in quasi pore-free material.

1 Introduction

Cast aluminum alloys are frequently used materials in the automotive sector if a good strength to weight ratio is required. Especially, if lightweight solutions at large quantities are demanded in highly-stressed parts (i.e., in automobile drive trains) such alloys representing a good choice. Alloyed with silicon, aluminum provides excellent casting conditions due to better filling properties and a reduced shrinkage behaviour. When considering the extreme complex casting geometries of, e.g., the highly-loaded drive train, engine blocks or cylinder heads, with wide ranges in wall thickness, defect-free castings are technically hard to realize and hence, a compensation by respective conservative design guidelines is required. Since these design guidelines are correlated with high safety factors the full potential of aluminum for lightweight and downsizing concepts cannot be used. Furthermore, complex geometrical circumstances are leading to a strong variation in the local cooling rate of cast products, which acts as a major driving force in determining the value of the secondary dendrite arm spacing (SDAS), the shape and size of the eutectic silicon particles as well as the occurrence of intermetallic precipitates and porosity. In previous studies it was shown that fatigue properties are strongly influenced by heterogeneity in SDAS [1-4], morphology of eutectic silicon [5-6] and porosity [7-10]. The mentioned complexity of typical industrial castings related to respective differences in local cooling rates leads to a more or less wide-ranged appearance of such heterogeneities in microstructure [11]. Furthermore, interactions between some of these characteristics exist, since for example the porosity is influenced by the presence of plate-shaped intermetallic precipitates (i.e., α- and β-AlFeSi) as they are able to hinder the solidification front [11]. The occurrence of porosity also leads to a strong scattering in results of fatigue investigations [9] according to their random distribution and potential of crack initiation. In a further work by [12] it was shown for quasi pore-free specimens of AlSi7Mg0.3, realized by hot isostatic pressing procedure (HIP) after casting, that crack initiation takes place in the α solid solution, when porosity is absent. This was realized by in-situ high-resolution thermography observation of a specimen equipped with a shallow-notch during fatigue tests.

Takin into account the current knowledge about the occurrence of microstructural heterogeneities, like pores, various shaped and sized intermetallic precipitates, or eutectic silicon particles and different heights in the SDAS in in-series cast products, the mechanism-based understanding of the correlation between the cooling-rate-dependent microstructure heterogeneities and the fatigue behaviour of cast aluminum alloys are of an essential meaning. Thus, the aim of the present work is

* Corresponding author: sascha.gerbe@hs-osnabrueck.de

to describe the different influences of typical microstructure heterogeneities on fatigue damaging mechanisms in cast aluminum alloys, with focus on the threshold stress intensity range for technical crack propagation, microstructure-controlled crack propagation and crack initiation. Specimens were extracted from industrial in-series castings of engine blocks (AlSi8Cu3) and cylinder heads (AlSi7Cu0.5Mg); and by means of laboratory gravity-die-cast specimens (AlSi7Mg0.3). Cyclic loading tests were performed under pure bending to observe crack propagation and under uniaxial loading in the high-cycle-fatigue (HCF, up to 107 cycles) and in the very-high-cycle-fatigue (VHCF, from 10^7 cycles up to 10^9 cycles) regime. The long-term objective is to use the resulting findings to develop and to adapt a computer-based procedure to create synthetic microstructures of cast aluminum alloys, which serves as input data to extend an existing and well described short crack model (cf. [13-14]). This model is based on the boundary element model, which follows the approach of microstructural barrier effects on fatigue crack propagation as described by Navarro and de los Rios [15].

2 Experimental

Experiments in this work were performed using three conventional hypoeutectic cast aluminum alloys. For two of them, specimens were extracted from in-series castings of engine blocks (secondary metallurgy alloy AlSi8Cu3) and cylinder heads (primary metallurgy alloy AlSi7Cu0.5Mg), both T6 heat treated. For the engine block as well as for the cylinder head the alloy extraction position were chosen with respect to a maximum difference in cooling rates and the related variation in SDAS values. Furthermore, another set of specimens of the primary metallurgy alloy AlSi7Mg0.3 was laboratory-produced by means of gravity die casting. For some of the experiments, additional hot isostatic pressed (HIP) specimens were studied to eliminate and to work out the influence of porosity on the fatigue properties. Details of specimen extraction and production can be found in Fig. 1; important microstructure properties are summarized in Table 1.

The crack propagation and ΔK_I threshold tests were carried out under pure bending using a resonance testing machine Rumul Cracktronic (f = 100 Hz). For the uniaxial cyclic loading experiments in the HCF and VHCF regime an ultrasonic resonance testing machine from BOKU Vienna (f = 20 kHz) was used. Close-ups of the experimental setups linked to the used specimen geometries are shown in Fig. 2.

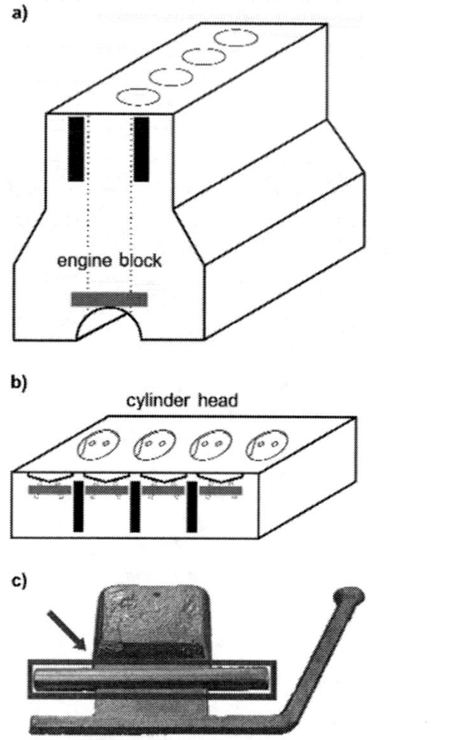

Fig. 1. Specimens were extracted from a) engine blocks (AlSi8Cu3) at stud bolt (black) and bearing seat position (grey); b) cylinder heads (AlSi7Cu0.5Mg) at stud bolt (black) and combustion chamber position (grey); and c) gravity die casting (AlSi7Mg0.3) with specimen extraction position (framed).

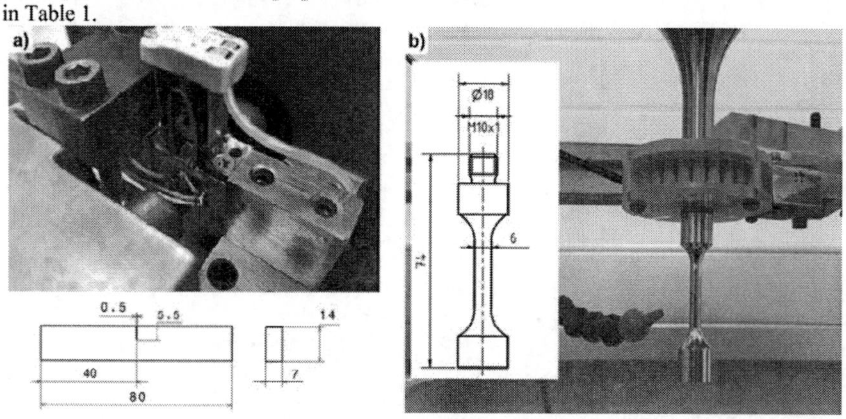

Fig. 2. Close-up of the experimental setups and the respective specimen geometries; a) SENB specimen in the resonance bending system Cracktronic with applied crack/ opening gages; b) uniaxial fatigue testing in the ultrasonic resonance testing machine.

Table 1. Extraction/ treatment details for the investigated alloys linked to important microstructure characteristics.

alloy	position/ state	SDAS [µm]	porosity fraction [%]	av. pore diameter d_p [µm]
AlSi8Cu3 engine block	stud bolt	65	1.81	47.7
	bearing seat	18	0.12	9.3
AlSi7Cu0.5Mg cylinder head	stud bolt	26	0.17	8.5
	combustion chamber	20	0.11	5.5
AlSi7Mg0.3 gravity die cast	cast condition	45	1.96	-
	HIP	41	<0.1	-

The measurement of the crack length during the crack-propagation test was realized by means of crack gages. These gages operate on the principle of alternating-current-potential-drop and were attached to the surface of the SENB specimen directly below the notch (maximal measurable crack length a_{max} = 5 mm). To determine the threshold for technical crack propagation of the stress intensity factor range (SIF) $\Delta K_{I,th}$ the tests were performed in the load shedding procedure under a stress ratio R = -1 (R = 0.1 for AlSi7Mg0.3). For this purpose, an initial crack with a length of a_{pre} = 1.3 mm is applied to the SENB specimen by a stepwise reduction of the bending moment. After this pretreatment, with the objective to guarantee a small plastic zone at the crack tip, the threshold tests were started immediately with 10 % load increase in relation to the last precrack step. The operating SIF range ΔK_I during the crack-propagation tests is calculated according to Equation 1 with the bending stress range $\Delta\sigma_b$, the total crack length a_{tot} and the geometry factor $Y(a_{tot})$ for SENB specimens (cf. [16]). Further crack-propagation tests were performed to obtain the constant C and the exponent m of the Paris-Erdogan law (Equation 2) [17-18], which states the correlation between the crack-propagation rate da/dN and the SIF range ΔK_I as shown in Equation 2.

$$\Delta K_I = \Delta\sigma_b \cdot \sqrt{\pi \cdot a_{tot}} \cdot Y(a_{tot}) \qquad (1)$$

$$\frac{da}{dN}(\Delta K) = C \cdot \Delta K^m \qquad (2)$$

Following the specification of ASTM E 647 [19] the threshold $\Delta K_{I,th}$ is reached if a minimum of five measurement points with a sufficient linearity were detected for 10^{-10} m/cycle < da/dN < 10^{-9} m/cycle. The resulting linear formula has to be extrapolated to $da/dN = 10^{-10}$ and thus returns the respective threshold value for the SIF range $\Delta K_{I,th}$. In case of the examination of the Paris law, experiments were carried out with constant bending moment (R = -1 for in-series casting specimens; R = 0.1 for laboratory AlSi7Mg0.3). Due to the propagating crack, the geometry factor $Y(a_{tot})$ and according to this the SIF range ΔK_I rises until $a_{tot} = a_{max}$. Selected SENB specimens were analyzed post-testing by means of light microscopy and high-resolution scanning electron microscopy (SEM), Zeiss Auriga FEG, equipped with an electron-back-scatter-diffraction detector (EBSD), which allows to correlate the crack growth to crystallographic orientations of the grains.

All uniaxial cyclic loading tests in this work were performed with a stress ratio of R = -1 and limited by the defined abort criterions of a maximum decrease in resonance frequency (Δf = -2 %, fail) or a maximum number of cycles ($N = 10^9$ cycles, run-out). To avoid specimen heating during the high-frequency fatigue test, a pulse-pause mode was applied (200 ms loaded state followed by 800 ms unloaded state). Additionally, an air cooling system was attached. To obtain statistically confirmed values for the fatigue limit σ_f, the test series were performed following the step method by Hück [20], accounting for a fatigue limit criterion of $N = 10^7$ cycles.

3 Results and Discussion

3.1 Crack-propagation experiments

During the crack-propagation tests under pure bending, different microstructural heterogeneities were observed as being of high influence on the threshold value $\Delta K_{I,th}$. The significance of the SDAS value is shown for the specimens extracted from engine blocks. The difference in SDAS is the highest in comparison between the stud bolt (low cooling rate) and the bearing seat (high cooling rate). Significantly lower values for the threshold $\Delta K_{I,th}$ were obtained for the high cooling rate (low value of SDAS) as shown in Fig. 3a. Also, the crystallographic orientation in interaction with the morphology and size of the eutectic was found to have an appreciable influence on the crack-propagation rate for low SIF ranges near the threshold $\Delta K_{I,th}$ (cf. Fig. 3b) in the test series of the cylinder head. Another strong effect is provoked by the occurrence of porosity as it is obvious form the crack-propagation rate vs. SIF range plot in Fig. 3c. In this example, a laboratory AlSi7Mg0.3 specimen was tested in the as-cast T6 heat-treated condition and compared to an analogues produced but additionally HIP treated specimen to determine the influence of porosity. Due to the high fraction of pores in the material, locally increased SIF ranges are leading to substantially lower global threshold values $K_{I,th}$.

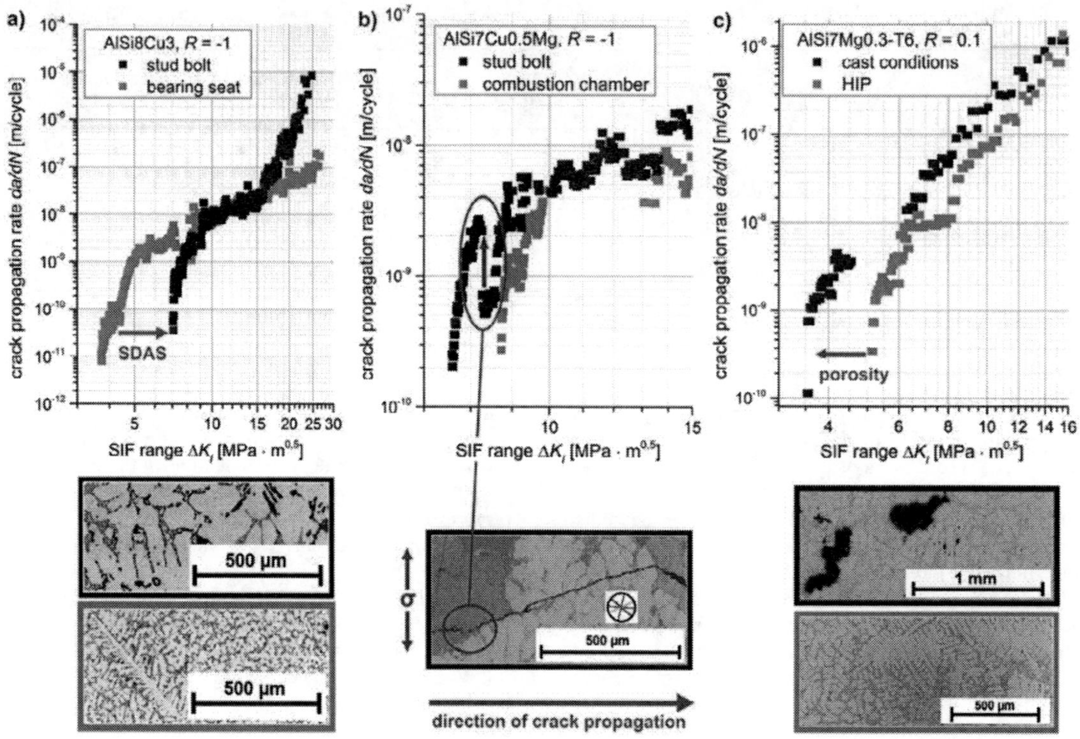

Fig. 3. The threshold value for technical crack propagation $\Delta K_{I,th}$ is influenced by different microstructure heterogeneities as shown via crack propagation graphs and light microscopy/ EBSD scans for a) the cooling-rate-dependent SDAS value represented by micrographs of stud bolt (black frame) and bearing seat microstructure (grey frame); b) the crystallographic orientation regarding to the direction of principal stress with highlighted position of entering the next grain via (111) slip plane (circled); c) the occurrence of shrinkage or gas porosity in AlSi7Mg0.3-T6 specimens in conditions as cast with T6 treatment (black frame) and additionally HIP treated (grey frame).

Since the occurrence of interdendritic eutectic silicon is directly related to the SDAS value and acting as a barrier against crack growth, the SDAS is a most significant parameter in increasing fatigue resistance. However, a reduction in SDAS leads to lower threshold values because of the linear characteristic of the eutectic regions between two secondary dendrite arms. If a crack for example propagates shear-stress controlled on a (111) slip plane (cf. Fig. 3b, EBSD scan) such line-like barriers are showing just a small reduction in crack propagation rate. In a coarser microstructure (cf. Fig. 3a, stud bolt, framed black), the eutectic regions are present in a bimodal distribution. Between the secondary dendrite arms they occur in analogy to microstructures with lower SDAS but in bigger and rather round-shaped cells accompanied by more plate-shaped silicon particles inside the grains as well as on grain boundaries. If the crack path enters such a eutectic cell loaded with low SIF ranges, the crack-tip will be deflected multiple times and propagation takes place at the phase boundary of α solid solution and eutectic silicon as shown in Fig. 4. This leads to an effective drop in the crack propagation rate (*da/dN* decreases from 10^{-9} m/cycle to $3 \cdot 10^{-11}$ m/cycle) giving rise to the higher threshold value of the material extracted from engine blocks stud bolt (cf. Fig. 3a). Also the crack path was deflected along with a reduction in crack propagation rate *da/dN* by intermetallic precipitates of species Al_2Cu or α- and β-AlFeSi.

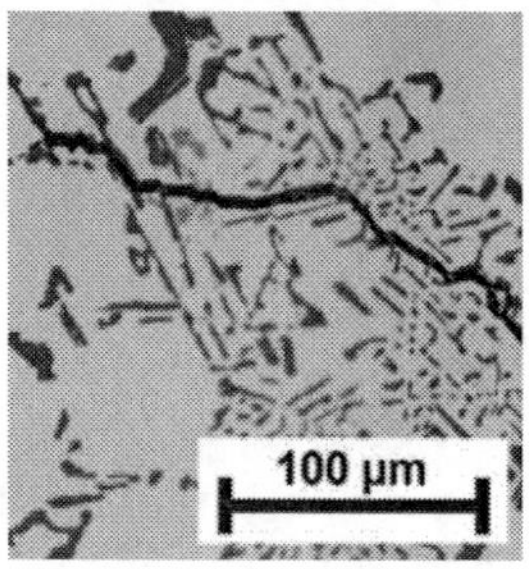

Fig. 4. Crack propagation rate drops due to passing a eutectic cell under low SIF range. At this point, the crack-propagation test was stopped, followed by further investigations regarding to observe the Paris-law behaviour.

The crystallographic orientation of the grains is another important influence factor to the threshold value $\Delta K_{I,th}$. As shown in Fig. 3b, the crack propagation rate was already low enough to stop the experiment (according to [19]; cf. Fig. 3b for $\Delta K_I \approx 8.5$ MPa $\cdot$ m$^{0.5}$). However, *da/dN* suddenly increases by a factor of 6. As determined by post-testing EBSD analysis, this can be attributed to the crack tip penetrating the next grain along a highly loaded (111) slip plane (Schmid factor $M_S = 0.46$). Otherwise, the crack propagation curves for both extraction positions of the cylinder heads (stud bolt and combustion chamber) would be almost identically. This would be expectable due to the low variation in microstructural characteristics (cf. Table 1).

Furthermore the significance of porosity to the threshold regime is shown in Fig. 3c. The occurrence of pores leads to multiple regions with an increased SIF range during cyclic loadings. Due to this, the crack path is influenced by pores since the direction of crack growth is attracted by these local heterogeneities of the stress intensity field. Such a behaviour of a crack in the near of a pore is shown exemplarily for a specimen from engine blocks stud bolt in Fig. 5.

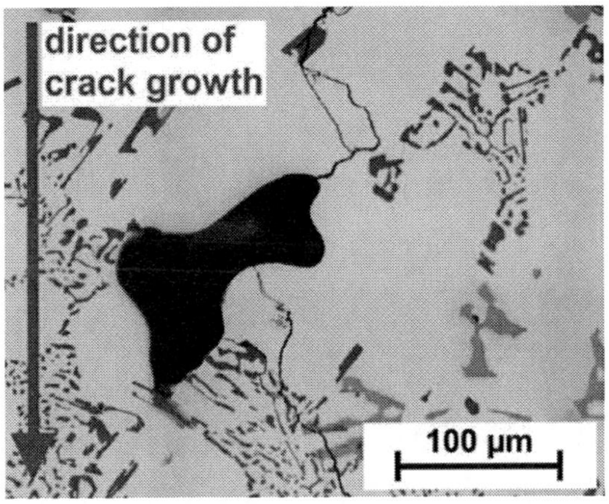

Fig. 5. Crack path through a SENB specimen of AlSi8Cu3 (engine block, stud bolt) influenced by the presence of a pore.

To determine the local influence of events like this during crack-propagation test a closer look to the *da/dN* measurements was made. In the experiments regarding to the results in Fig. 3c the crack propagation rate is calculated with respect to an incremental crack growth of $\Delta a_{inc} = 0.1$ mm. In this procedure the necessary number of cycles will be count and divides the incremental grown distance to generate the average crack propagation rate *da/dN* which overlaps such small microstructural influences. The closer examination of *da/dN* regarding to a smaller increment of $\Delta a_{inc} = 10$µm enables to detect high accelerations if the crack passes pores. This local effect in multiple occurrence leads to a global effect that results in the shown reduction of the threshold value $\Delta K_{I,th}$ (cf. Fig. 3c)

3.2 Fatigue experiments

3.2.1 Tests on porosity-containing material

During uniaxial cyclic loading tests carried out using the ultrasonic resonance testing machine, porosity was identified as major crack-initiating defect. In the case of the material with the more coarse microstructure (AlSi8Cu3, engine block, stud bolt, low cooling rate) having a higher fraction of lager porosity, the crack starts at surface pores. Compared to this, crack initiation for the material with lower SDAS value (AlSi8Cu3, engine block, bearing seat) was found at surface-near pore accumulations. Exemplarily, both cases are shown in Fig. 6. Nevertheless, the fatigue limit σ_f (criterion $N = 10^7$ cycles) is significantly higher for the specimens with lower SDAS. Results of the fatigue tests linked to them from crack-propagation for this alloy set can be found in Table 2.

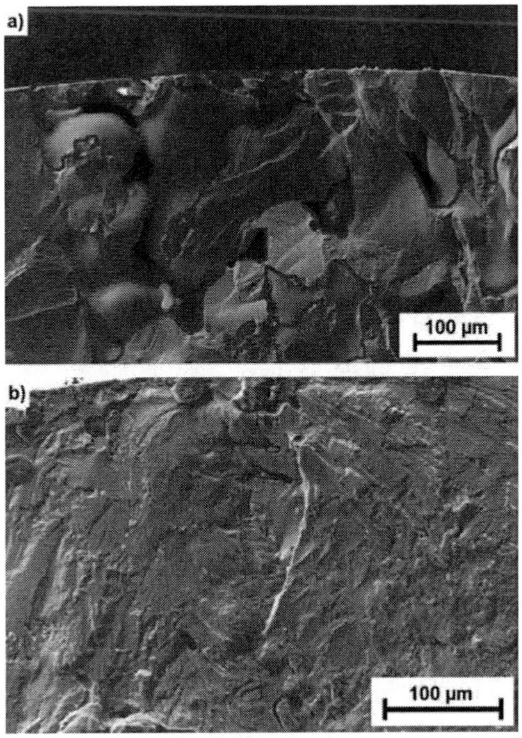

Fig. 6. Micrographs of fracture surfaces from fatigue tests for engine block specimens from a) stud bolt and b) bearing seat with different crack provoking porosities.

Table 2. Summarized results of fatigue and crack-propagation tests for the cast aluminum alloy AlSi8Cu3.

engine block position	**fatigue limit σ_f [MPa]**	**threshold $\Delta K_{I,th}$ [MPa · m$^{0.5}$]**
stud bolt	68 ±1	7.0
bearing seat	114 ±23	4.4

Using the results given in Table 2, a threshold diagram according to Kitagawa and Takahashi [21] including the modification by El Haddad [22] can be established using Equation 3 and 4. The Kitagawa-Takahashi diagram represents the correlation between a present defect with the size a (e.g. a critical shaped intermetallic precipitate or the projected diameter of a pore) and the sustainable cyclic stress amplitude $\Delta\sigma_{th}$ which is limited by the

https://doi.org/10.1051/matecconf/201816514005

fatigue limit σ_f until the threshold crack length a_0 is exceeded. For higher defect sizes, the threshold $\Delta\sigma_{th}$ decreases (cf. Equation 3).

$$\Delta\sigma_{th} = \frac{\Delta K_{I,th}}{\sqrt{\pi\cdot(a+a_0)}} \qquad (3)$$

$$a_0 = \frac{1}{\pi}\cdot\left(\frac{\Delta K_{I,th}}{\Delta\sigma_f}\right)^2 \qquad (4)$$

The resulting Kitagawa-Takahashi diagrams for both microstructure conditions from engine block are shown in Fig. 7, where the influence of SDAS on the progression of the threshold value $\Delta\sigma_{th}$ is highlighted by arrows.

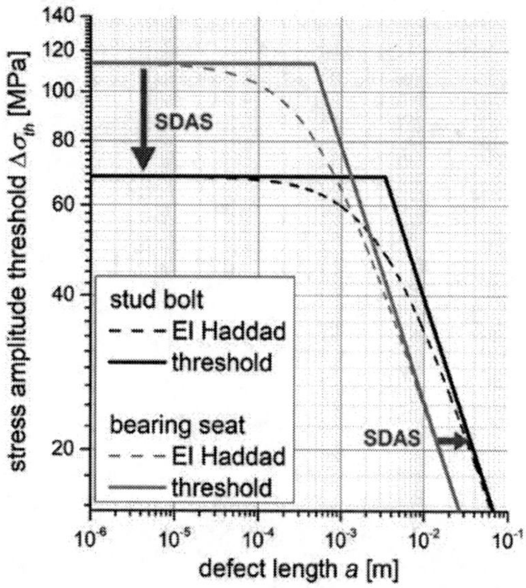

Fig. 7. Kitagawa-Takahashi diagrams for the varying microstructure characteristics in AlSi8Cu3 (engine block) resulting from different cooling-rate-dependent SDAS values. The arrows represent the gradient of SDAS values.

As it is shown above, the SDAS value has a significant influence on the fatigue and crack-propagation properties. The Kitagawa-Takahashi diagram indicates the better resistance against fatigue for lower SDAS values in the first horizontal part of the graph but at the same time the reduced defect tolerance (shift of the decreasing part of the graph to lower defect lengths a).

3.2.2 Tests on quasi pore-free material (HIP treated)

If the major crack initiating defect in form of pores is eliminated, a closer examination of the early VHCF damage mechanisms in correlation with intrinsic metallurgical parameters is possible. Former results on this aspect were shown for the alloy AlSi7Mg0.3 and published in [12]. By means of high-resolution thermography, crack initiation was found to take place within the dendritic α solid solution. Via the formation of a well visible hotspot, the initiation process was detectable at approximately 40 % of the number of cycles to failure (here $N_f \approx 3.8\cdot 10^8$ cycles). Additionally, first experiments on HIP treated cylinder head specimens (AlSi7Cu0.5Mg) were performed on stress amplitudes σ where for non-HIP treated specimens failures in the region of $N_f \approx 10^8$ cycles occur. On this stress level after $N = 8\cdot 10^9$ cycles (run out) no crack was detected for the quasi pore-free cylinder head (combustion chamber) set. The post-testing observation carried out using the SEM showed some traces of microstructural fatigue damage (slip markings) distributed over the complete surface of the specimen's gage length (cf. Fig. 2a). Results of the SEM investigation are shown in Fig. 8.

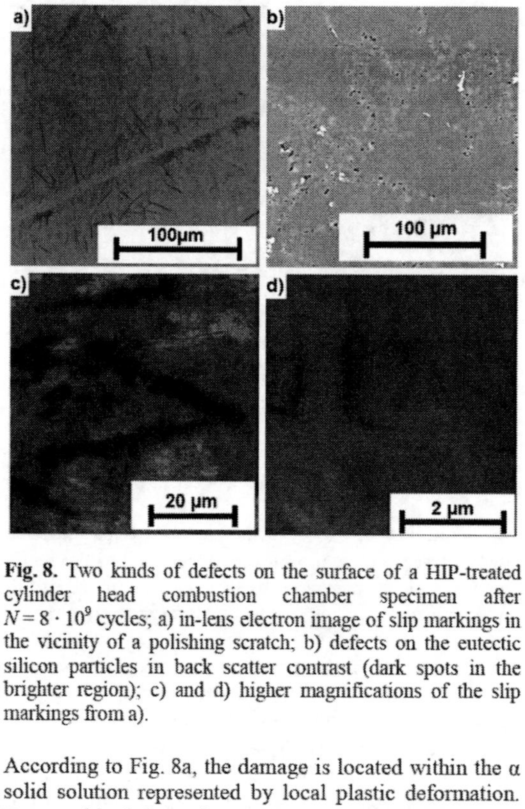

Fig. 8. Two kinds of defects on the surface of a HIP-treated cylinder head combustion chamber specimen after $N = 8\cdot 10^9$ cycles; a) in-lens electron image of slip markings in the vicinity of a polishing scratch; b) defects on the eutectic silicon particles in back scatter contrast (dark spots in the brighter region); c) and d) higher magnifications of the slip markings from a).

According to Fig. 8a, the damage is located within the α solid solution represented by local plastic deformation. Due to this deformation, EBSD measurements did not lead to useful results and hence, a correlation of the slip markings to the potential traces of the slip traces was not possible yet. Furthermore, it was shown that the eutectic silicon particles at the surface revealed debonding from the surrounding α solid solution (cf. Fig. 8b). This supports the assumption that at stress levels close to the fatigue limit the crack initiation mechanism takes place in both, the interdendritic eutectic and the dendrites. However, this represents the present state of preliminary experimental work on the HIP treated specimens, where further investigations are in progress.

4 Conclusions

By means of crack propagation and fatigue experiments on different cast aluminum alloys with varying microstructure conditions, the significance of heterogeneities like SDAS, intermetallic precipitates, eutectic regions and porosity was quantified. Lower SDAS, fine and only line-shaped eutectic regions between secondary dendrite arms and a higher amount of porosity fraction lead to lower threshold values $\Delta K_{I,th}$ for technical fatigue crack propagation. Furthermore, the crystallographic orientation of the grains affects the resistance against crack growth at near-threshold SIF ranges due to preferred propagation on (111) slip planes. If in such a case the material contains a low SDAS value and accordingly fine defined eutectic regions, the barrier effect is decreased dramatically. Near crack tip porosity influences the crack path by a locally increased stress intensity field, which is able to force the crack to propagate directly through them. Furthermore, crack initiation for porosity-containing material was shown to occur either at large surface pores (high fraction of porosity) or at pore accumulations (low fraction of porosity). If porosity is eliminated due to a HIP treatment, the fatigue limit will be increased significantly.

It is the objective of current work to systematically expand the range of microstructure properties (with regard to cooling rate and the respective SDAS values) to obtain a more detailed and quantitative definition of microstructural barriers to fatigue crack propagation. The results will serve to define reference microstructures of cast aluminum alloys to be implemented in a short crack modelling environment. For this purpose, wedge-shaped castings have been produced. Due to their step-wise decreasing thickness, a systematic variation in cooling rate and SDAS is possible.

The authors acknowledge the financial and technical support by the German Federal Ministry of Education and Research (BMBF), Germany, by Nemak Dillingen GmbH, RWP GmbH and deutsche Forschungsgemeinschaft (grant DFG KR1999/19).

References

1. R. Gonzáles et al., Int. J. Fatigue **54**, 118-126 (2013)
2. D.A. Lados, D. Apelian, Mater. Sci. Eng. A **385**, 200-211 (2004)
3. B. Zhang, D.R. Poirier, W. Chen, Metall. Mater. Trans. A **30**, 2659-2666 (1999)
4. M.I. Houria, Y. Nadot, R. Fathallah, M. Roy, D.M. Maijer, Int. J. Fatigue **80**, 90-102 (2015)
5. K. Gall, N. Yang, M. Horstemeyer, D.L. McDowell, J.H. Fan, Metall. Mater. Trans. A **30**, 3079-3088 (1999)
6. S. Siegfanz, A. Giertler, W. Michels, U. Krupp, Mater. Sci. Eng. A **565**, 21-26 (2013)
7. D.L. McDowell, K. Gall, M.F. Horstemeyer, J. Fan, Eng. Fract. Mech. **70**, 49-80 (2003)
8. L. Ceschini, A. Morri, G. Sambogna, J. Mater. Process. Technol. **204**, 231-238 (2008)
9. J.Z. Yi, Y.X. Gao, P.D. Lee, H.M. Flower, T.C. Lindley, Metall. Mater. Trans. A **34**, 1879-1890 (2003)
10. A. Ben Ahmed, A. Nasr, A. Bahloul, R. Fathallah, Int. J. Adv. Manuf. Technol. **92**, 1113-1125 (2017)
11. L. Lu, A.K. Dahle, Metall. Mater. Trans. A **36**, 819-835 (2005)
12. S. Knorre, J. Tenkamp, U. Krupp, W. Michels, F. Walther, *7th International Conference on Very High Cycle Fatigue*, 222-227 (DVM, Dresden, 2017)
13. B. Künkler, *Mechanismenorientierte Lebensdauer-vorhersage unter Berücksichtigung der Mikro-struktur,* Diss. (VDI-Verlag, Siegen, 2006)
14. U. Krupp, *Fatigue Crack Propagation in Metals and Alloys* (Wiley VCH, Weinsheim, 2007)
15. A. Navarro, E.R. de los Rios, Philos. Mag. A **57**, 15-36 (1988)
16. D. Gross, T. Seelig, *Bruchmechanik*, 5th ed. (Springer, Berlin, 2011)
17. P.C. Paris, M.P. Gomez, W.E.A. Anderson, Trends Eng. **13**, 9-14 (1961)
18. P.C. Paris, F.A. Erdogan, J. Basic Eng. **85**, 528-534 (1963)
19. ASTM E 647, *Standard Test Method for Measurement of Fatigue Crack Growth Rates* (ASTM Inter-national, West Conshohocken, 2000)
20. M. Hück, Matterwiss. Werksttech. **14**, 15-36 (1983)
21. H. Kitagawa, S. Takahashi, *ASM Proceedings of 2nd International Conference on Mechanical Behavior of Materials*, 62731 (Metalspark, Ohio, 1976)
22. M.H. El Haddad, T.H. Topper, K.N. Smith, Eng. Fract. Mech. **11**, 573-584 (1979)

The effect of microstructural heterogeneities on the High Cycle Fatigue scatter of cast aluminium alloys: from an elementary volume to the structure

Driss EL KHOUKHI[1, 2, 3,*], *Franck* MOREL[1], *Nicolas* SAINTIER[2], *Daniel* BELLETT[1] and *Pierre* OSMOND[3]

[1]LAMPA, Arts et métiers Paris Tech, 49100 Angers, Cedex, France
[2]I2M, Arts et métiers Paris Tech, 33170 Bordeaux, Cedex, France
[3]PSA Groupe, 92250 La Garenne-Colombes, Cedex, France

Abstract. Cast Al-Si alloys have been widely used in automobile applications thanks to their low density and excellent thermal conductivity. A lot of components made of these alloys are subjected to cyclic loads which can lead to fatigue failure. Furthermore, the well know size effect in fatigue, whereby the fatigue strength is reduced in proportion to an increase in size, can be important. This is caused by a higher probability of initiating a crack in larger specimens (i.e. statistical size effect). This paper analyses the role of casting defects on the statistical size effect. For that, a uniaxial fatigue testing campaign (R=0.1) has been conducted using two cast aluminium alloys, fabricated by different casting processes (gravity die casting and lost foam casting), associated with the T7 heat treatment, and with different degrees of porosity. Different specimens (smooth and notched) with different stressed volumes have been investigated. The first part of this article is dedicated to the experimental characterization of the statistical size effect in both alloys via the concept of the Highly Stressed Volume. The second part investigates the effect of the Highly Stressed Volume on the critical defect size via diagram of Kitagawa-Takahashi. The results show that the presence of statistical size effect is strongly linked to the characteristics of the pore population present in the alloy. A numerical approach, linking the observed pore distribution to the volume of loaded material, is proposed and discussed.

1 Introduction

1.1 Context and objectives

This work focuses on cast Al-Si alloys, widely used in the automotive industry. In order to manufacture engine components, the PSA Groupe uses two foundry processes that result in components containing Microstructural Heterogeneities (MH) with different characteristics. These Microstructural Heterogeneities control the fatigue behaviour of these Al-Si alloys. These components are subjected to cyclic mechanical loads (in High Cycle Fatigue) that can result in the appearance of cracks and thus lead to the failure of the structure. The effect of the MH on the fatigue behaviour has been well documented in references [1-3].

A typical engineering design practice [4] involves the identification of the fatigue criteria, used to size and validate real components, based on tests carried out on laboratory specimens, which have a specific volume, under simple loading modes (tension, bending, and torsion). However, real component loading conditions often prove to be different from those applied to laboratory test specimens, particularly in terms of the size of the critical zone (size effect) and the homogeneity of the stress state in this zone (stress gradient effect). Hence, the industrial objective of this work is to develop predictive tools to ensure the passage from an elementary volume to the structure taking in account the size and stress gradient effects. In particular, the effect of microstructural heterogeneities, principally shrinkage porosity, on the statistical size effect in cast Al-Si alloys is investigated.

1.2 State of the art

It is well known that the fatigue strength of certain materials decreases with an increase in the volume [6-7]. This is referred to as either the "volume effect", the "size effect" or the "scale effect" and is often simply explained by an increase in the probability of encountering a large material defect. The effect is influenced by factors related to the random distribution of defects in the material and the impact of the manufacturing process on the homogeneity of the material.

Kloos et al. [8] has classified the size effect phenomenon into the following categories: (i) statistical size effect induced by the high probability of defects in larger specimens, (ii) geometrical size effect attributed to stress inhomogeneity from different notch types, (iii) production size effect generated by production technology and (iv) surface size effect caused by the surface characteristics such as roughness.

* Corresponding author: driss.el-khoukhi@ensam.eu

https://doi.org/10.1051/matecconf/201816514006

For Makkonen [9] the size effect can be explained by the statistical size effect using the Weibull [10] approach and the weakest link theory. Makkonen [9] experimentally showed the influence of the statistical size effect on the high cycle fatigue behaviour of both smooth and notched cylindrical specimens.

The Highly Stressed Volume (HSV) approach was first introduced by Kuguel [11] in 1961, and then employed by Sonsino and Fischer [12], and others [13]. The terminology Vn% is used to define the volume of material that is subjected to at least n% of maximum stress (σn%=n%×σmax). This volume of material is assumed to have an increased probability of fatigue crack initiation. The percent is assumed to be 95% by Kuguel [11] and 90% by Sonsino and Fisher [12]. The concept has been verified for metallic components with good success using the V90% volume. For short-fiber reinforced specimens and components, a value of V80% proved to be more appropriate than V90%.

In the present work the influence of the statistical size effect on the fatigue strength of cast aluminium alloys is investigated using the HSV concept. A value of V80% is used for the reason explained in section 2.4.

Due to its significance in structural reliability, certain design standards require that the statistical size effect be taken into consideration [14-15]. Even so, at the present time, it remains an open topic of discussion in the scientific literature, in particular with respect to the HCF strength cast Al-Si alloys. Therefore, in this work an in-depth assessment of the statistical size effect on the HCF strength of two cast Al-Si alloys, is underway with the aim of developing a predictive model, based on the theory of Monte-Carlo, to evaluate the size effect using knowledge of the defect population.

Two Al-Si alloys with different porosity distributions were chosen in order to understand how the characteristics of the defect populations affect the statistical size effect. Different specimen shapes, including both smooth and notched specimens have been machined from these alloys, in order to obtain geometries with different Highly Stressed Volumes. High cycle fatigue tests were conducted to investigate the size effect on the fatigue strength using the HSV concept. Fractography analyses were conducted to investigate the relationship between the critical defects and the fatigue strength via the Katagawa-Takahashi diagram.

2. Experimental investigation

2.1 Materials

To study the effect of casting defects on the statistical size effect, two cast aluminium alloys, referred to as alloys A and B, were used [Fig. 1]. These alloys were fabricated by different casting processes (gravity die casting and lost foam casting), associated with the T7 heat treatment [Table1]. Each process results in different porosity populations (i.e. volume fraction, defect size…) and different mechanical properties [Table2].

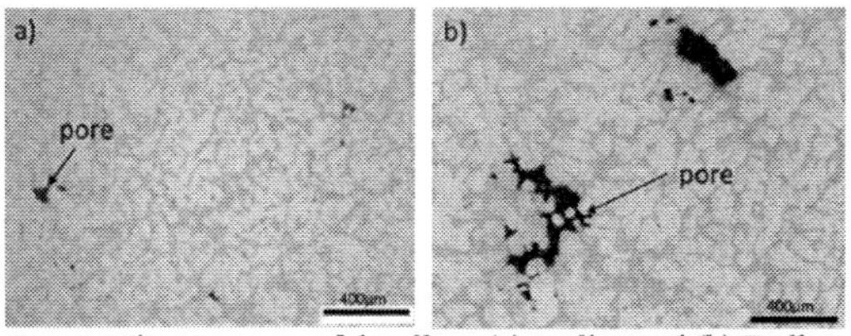

Fig. 1. Microstructure of the alloys (a) A alloy and (b) B alloy

Table 1. Chemical composition, casting process and heat treatment of the two alloys

Alloy	Chemical composition (wt)	Process	Heat treatment
A	7%Si, 0.3%Mg, 0.5%Cu, Al Rem.	Gravity Die Cast	T7
B	7%Si, 0.3%Mg, Al Rem.	Lost Foam Cast	T7

Table 2. Mechanical properties of the two alloys

Alloy	SDAS (µm)	Micro-hardness (Hv0.025)	$\sigma_{Y0.2\%}$ (MPa)
A	42.3±9.7	113.8±2.8	260.4±1.6
B	77.3±18.9	99.0±9.0	240.1±4.5

In the work of LE [1] these alloys were characterised in terms of their casting defect size distributions using the Murakami method with a standard inspection area of S_0=2.55mm^2.

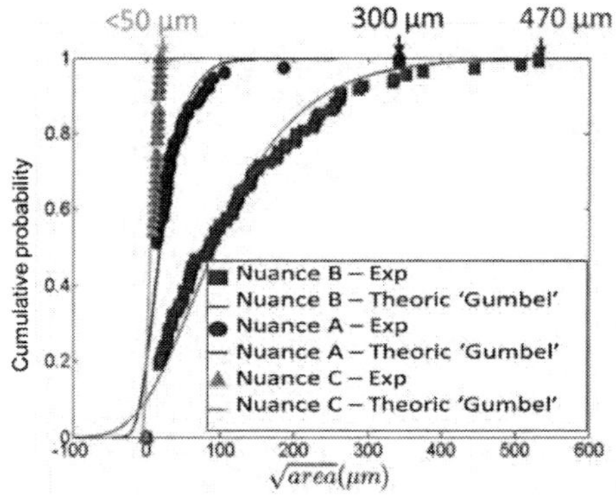

Fig. 2. Defect size distributions of the three alloys using the Murakami Method

Fig 2 shows the defect size distributions for three alloys, A, B and C. Note that alloy C used by LE corresponds to alloy B subjected to High Isostatic Pressing (HIP). The maximum defect sizes observed on the polished surfaces of alloys A and B are 300µm and 470µm. 80% of the defects in alloy A have an √Area lower than 50µm and 20% of the defects in alloy B have a √Area greater than 200µm.

2.2 Experimental conditions

All fatigue tests were performed in uniaxial tension with stress ratio R = 0.1 using a Rumul Testronic resonant testing machine at approximately 100 Hz. The stopping criterion used is a drop in the resonance frequency of 1 Hz, corresponding to the presence of a fatigue crack of approximately 2 mm. All tests were performed at room temperature in a laboratory air environment. The tests were conducted following the staircase technique with a maximum life of $2x10^6$ cycles. The not-failed samples were retested until failure in order to identify the critical defect that initiates a crack. A step of 5 MPa stress amplitude is used in the staircase procedure.

2.3 Specimens geometry

To study the effect of volume on the fatigue strength of alloys A and B, different fatigue specimen geometries were defined, corresponding to different loaded volumes. They are referred as: “V1-Small Volume”, “V2-Reference Volume”, “V3-Large Volume” and “VEE-Notch specimen” (Fig. 3).

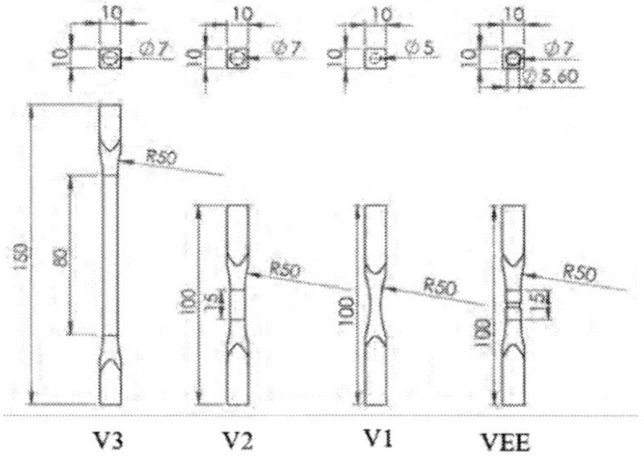

Fig. 3. Specimens for tensile fatigue tests

2.4 Estimation of the Critical Volume

In the following the estimated critical volume is chosen taking into account the combination of two criteria:
- The damage mechanism: observations of the fatigue failure surfaces of the specimens show that 90% of the critical defects are located in a sub-surface layer with a thickness of about 500 micrometers.
- Stress heterogeneity: A Highly Stress Volume of V80% is chosen. The justification for choosing this value is that some of the V1 specimen which has an hourglass shape broke in zones where the stress level is approximately 80% less than the maximum principal stress σ_{Imax} at the centre of the specimen. Figure 4 shows the case of a specimen in which the fatigue crack is located approximately 5.2mm from the middle of the specimen.

Fig. 4. V1 sample showing a crack after failure

The combination of the two criteria above gives a highly stressed volume of $0.8\times\sigma_{Imax}$ located in a subsurface layer of thickness 500µm. In the following section, this definition is referred to as the Critical Volume (Vc). For each specimen geometry, the estimation of the Highly Stressed Volume was performed using a general purpose finite element code. For a static load, the highly stressed zones in each specimen are identified by linear elastic finite element analysis. Finite element models of the specimens are shown in Fig. 5.

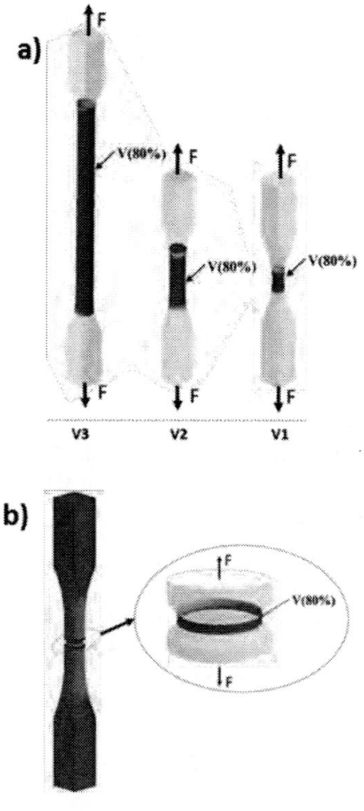

Fig. 5. Highly stressed volume for (a) smooth specimens V1, V2 and V3 and (b) notched specimens

After determination of the highly stressed volume, and in order to take in account the damage mechanism, an analytical calculation was conducted to estimate the critical volume (Vc) under a layer of 500µm. The critical volume corresponding to each sample and other characteristics are shown in Table 3.

Table 3. Characteristics of the samples

Sample	Number of specimens		Kt	Vc [mm³]
	Alloy A	Alloy B		
VEE	5	5	1.78	4.80
V1	13	3	1	90
V2	13	4	1	320
V3	12	0	1	912

3. Results

3.1 Mechanical results

The fatigue strength of specimens V1, V2 and V3 of alloy A were estimated using the Dixon-Mood [16] method. For the notched specimens (VEE) of alloy A and all of specimens of alloy B the fatigue strengths were estimated using the method proposed by Houria et al. [17] because of the small number of tests.
The product of the nominal stress amplitude σ_a and stress concentration factor K_t gives the local stress amplitude at the notch-tip. In [18-20] it was shown, that there is a continuous relationship between the fatigue strength expressed as $K_t\sigma_a$ and the highly stressed volume. In this work, it will be correlated to the critical volume Vc.

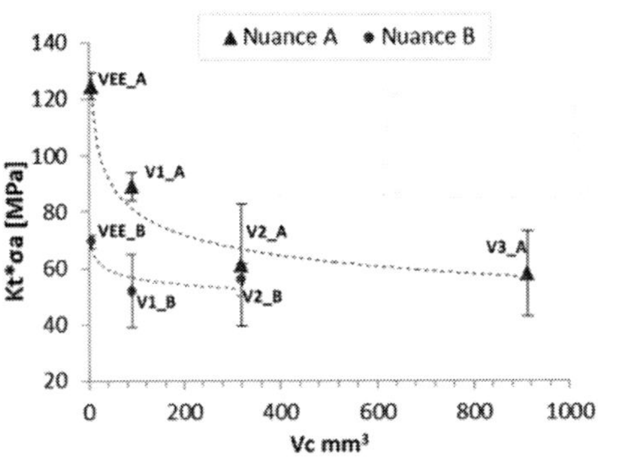

Fig. 6. Relationship between highly stressed volume and local stress amplitude for A and B alloys

Fig. 6 shows the effect of highly loaded volume (Vc) on the local fatigue strength amplitude Kt*σa for a stress ratio of R=0.1 at 2×10^6 cycles for 50% probability of failure. For both alloys, the local fatigue strength amplitude decreases with increasing critical volume Vc. The notched specimens have the highest strength in terms of the local stress amplitude. This can be explained by the fact that the probability of failure decreases with a smaller material volume, the local stress amplitude is higher than for bigger material volumes such as the axially loaded smooth specimens with homogeneous stress distributions (V1, V2 and V3).
It is also noted, that for alloy A, the dependence on the critical volume is more pronounced compared to alloy B. For alloy A, saturation of the volume effect occurs at a specific volume of 320mm³. Following these observations, it can be concluded that:
- the alloy A has a more pronounced volume effect than alloy B. This is probably due to the characteristics of the defect population (i.e. pore size, distance between large pores, etc….).
- for alloy A, the fatigue strength becomes stabilized from a specific volume of 320mm³. This could be considered as the Fatigue Representative Elementary Volume of the alloy A.
- for alloy B, the fatigue strength stabilizes rapidly at a specific volume of 90mm³, which can be considered as the REV of the B alloy.

An analysis of the critical defects on the fatigue failure surfaces can confirm this tendency.

3.2 Kitagawa-Takahashi diagram

SEM microscopy was used to measure the √Area parameter of the critical defects on the fatigue failure surfaces of tested specimens. It was observed that for the four notched specimens of the B alloy, the fracture surfaces show two initiation sites (Site 1 and Site 2). This result reflects the higher density of large pores in alloy B even in a small critical volume (Fig. 7).

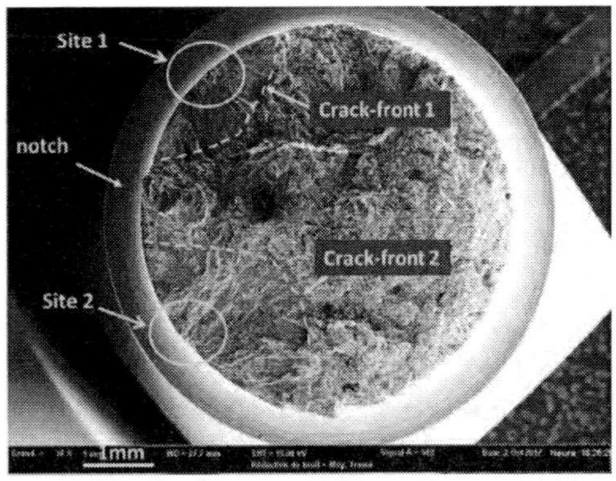

Fig. 7. Fatigue fracture surface of a notched specimen from the B alloy

The summary of the size of critical defects present on the fatigue fracture surface of notched samples from the B alloy are shown in the Table 4.

Table 4. Critical defects for notched specimens of the B alloy

Specimen	√Area : Site 1 [μm]	√Area : Site 2 [μm]
VEE_B_01	176	248
VEE_B_02	243	333
VEE_B_03	204	320
VEE_B_04	176	330

To link the fatigue strength of the samples to the critical defects size, Kitagawa-Takahashi diagrams were established. The estimation of fatigue strength of each specimen was carried out using the Houria *et al.* Method. The results are represented on Fig.8.

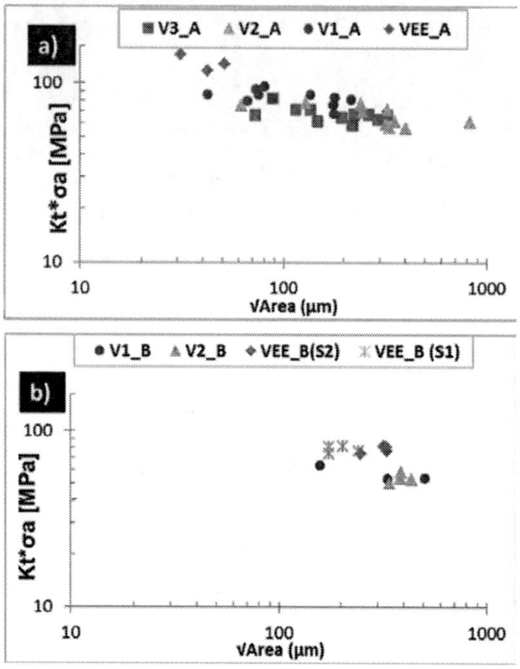

Fig. 7. Kitagawa-Takahashi diagram with R=0.1 (a) for the A alloy (b) for the B alloy.

From Fig. 7, the following conclusions can be made:

i) The fatigue strength of alloy A [Fig.7, a] has a tendency to decrease when the size of the critical defects increases. Note also that the size of the critical defects increases when the critical volume increases. Indeed, the notched specimens (VEE) with the lowest critical volume (4.80mm^3) have the smallest defect sizes. By changing to a critical volume of 90mm^3 for the V1 specimens then to 320mm^3 for the V2 specimens, the size of the critical defects has a tendency to increase and the fatigue strength drops.

ii) Between specimens of size V2 and V3, there is almost no difference in the average value of the fatigue strength. It can be concluded that the fatigue strength is stabilised, most probably due to the saturation in the maximum size of the critical defects [Fig.7, a].

iii) The size of the critical defects in alloy B is relatively large for all types of tested specimens. Their √Area are greater than 160µm. For this range of defect size, the fatigue strength of the specimens does not change significantly compared to the variation in alloy A [Fig.7, b].

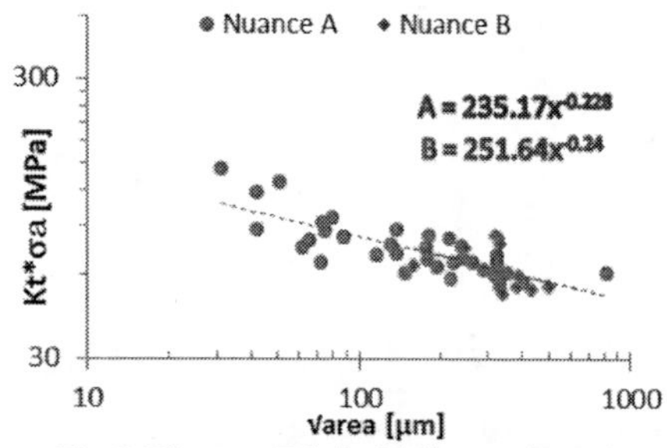

Fig. 8. Kitagawa-Takahashi diagram (R=0.1)

Fig. 8 shows the Kitagawa-Takahashi diagram in terms of the local stress amplitude (Kt*σa) as a function of the square root of the critical defect size. The points on the graph are the critical defects observed on the fatigue failure surfaces of the tested specimens of the two alloys. Note that, for the notched specimens of alloy B (with multi-site initiation), only the largest defect is shown in Fig 8.

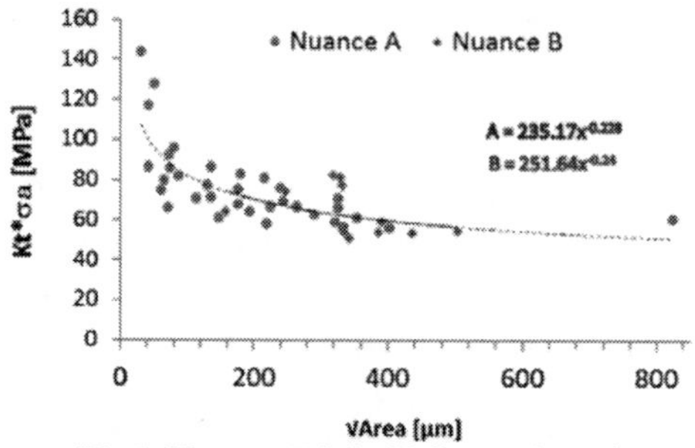

Fig. 9. Kitagawa-Takahashi diagram (R=0.1)

From the Kitagawa-Takahashi diagram represented in linear axes on (Fig. 9), a better understanding of the weak presence of the size effect in alloy B can be seen. The critical defects are generally at the level where the fatigue strength begins to stabilize. Therefore, and since the size of the critical defects remain relatively large (√Area greater than 160µm), the difference in the fatigue strength remains low. Indeed, when the size goes from 160µm to 825µm the local stress amplitude goes from approximately 67MPa to 60MPa (Fig. 8). So, even for the different specimen sizes, the strength does not significantly change, hence the weak size effect. Also from Fig. 8 two parallel trend lines are obtained for the two materials. The parallelism indicates that the failure in both alloys is controlled by the same mechanisms.

Normally, the fatigue strength of the B alloy is lower than that for the A alloy, due to the traces of Copper present in the A alloy [Table 1]. But in the Fig. 8 and Fig. 9 the curves are almost superimposed. This is can be explained by the lower number of specimens tested in the B alloy.

4. Modelling approach

A modelling approach, linking the observed pore distribution to the volume of loaded material is in the process of development. The basic methodology is to use the Monte-Carlo theory to understand the relationship between the size effect and the parent defect distribution (size and spatial distribution) for a cast material. It is hoped that this approach will also yield insight in the experimentally observed scatter in the fatigue data.

The principal steps in the modelling approach are:

- Identification of the parent defect population based on experimental observations (tomography or optical microscopy).

- Generation of synthetic material with the parent defect distribution.
- Sampling of different sized volumes from the synthetic material.
- Statistical extreme value analyses to determine the extreme value distribution of the critical defects for different volumes.
- Use of an appropriate fatigue criterion to define the relationship between the defect size and the fatigue strength for different volumes.

The parent distributions for both alloys are derived from micro-tomography analysis. Synthetic materials, of volume [30x30x3mm] for alloy A and [70x70x3mm] for alloy B, were formed with these parent distributions. These volumes are equivalent to the volumes scanned via micro-tomography [Fig. 10]. The pore positions are generated randomly within the synthetic material. Sampling volumes (SVi) are then taken randomly from the parent synthetic material [Fig. 11]. Then, extreme values analyses were conducted on these volumes. Note that, for the volumes SV1 to SV5, only a sub-surface volume is considered in the analysis (i.e. a hollow cylindrical volume including the exterior surface and a thickness of 0.5mm). The volumes SV6 to SV8 are used to determine the stabilized extreme value distribution of each material. The dimensions of these volumes are summarized in [Table 5]. This process is repeated 10 000 times according to the Monte Carlo principle. The distributions of the extreme values corresponding to the critical defect size are shown in Fig. 12.

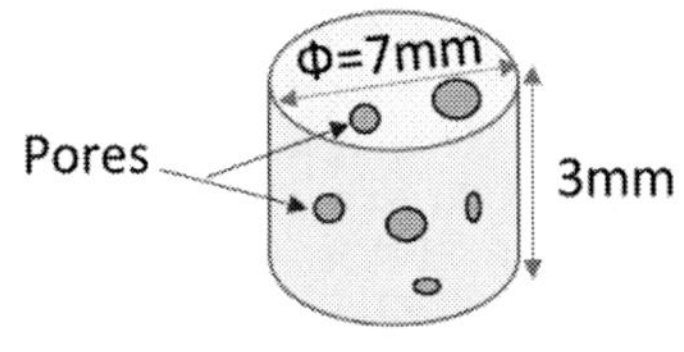

Fig. 10: Schematic representation of Micro-tomographic scanned volumes

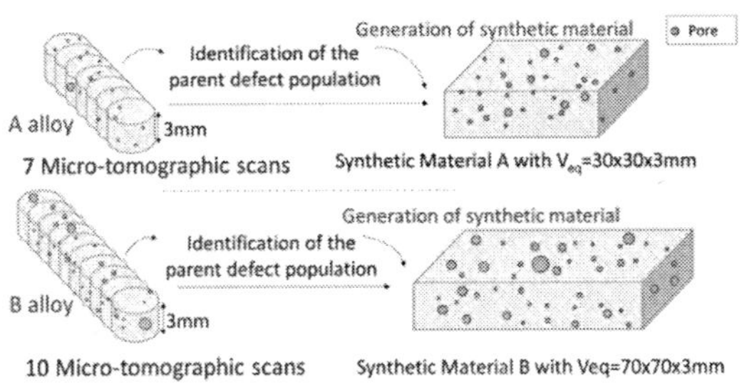

Fig. 11: Principle of identification of parent defect population and generation of synthetic materials for both alloys A and B.

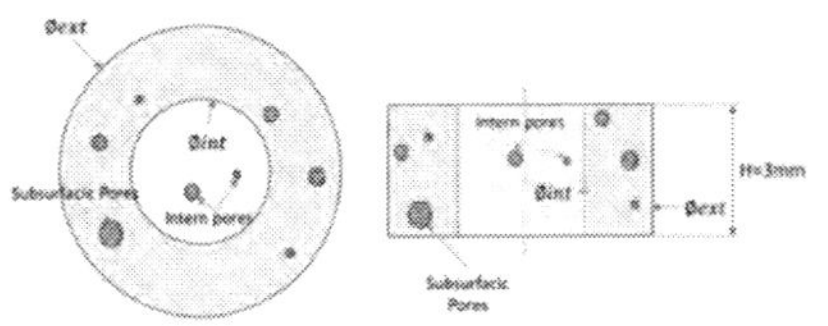

$sub\ surface\ thickenss = 0.5\text{mm}$

Fig. 12: Definition of the sampling volume

Table 5: Sampling volumes

Sampling Volume	Rint [mm]	Rext [mm]	Volume [mm3]
SV1	2	2.5	21
SV2	3	3.5	31
SV3	3.5	4	92
SV4	9.5	10	139
SV5	14.5	15	2121
SV6	0	15	4882
SV7	0	70	46181
SV8	0	80	60319

4.1 Numerical results

In the following, the preliminary results of this approach are presented. Specifically, the results of the extreme value analyses of the critical defect size as a function of the sampling volume are discussed in order to explain the observed experimental tendencies.

Fig. 13 presents the results of the extremes value analysis of the distribution of the critical defects for different sampled volumes (SVi) in alloys A and B. It can be seen that the extreme value distributions are associated with the larger defects in the tail of the parent distributions. For alloy B, the extreme value distribution is flatten out when the sample volume is increased and tends towards a stabilized distribution (Fig. 13b). For alloy A, the opposite trend is observed. That is, the distribution becomes narrower when increasing the sample volume from SV1 to SV8 and it also tends towards a stabilized distribution (Fig. 13a). It can be seen that the mean critical defect size, as a function of volume, for alloy A remains less than 100 μm. But for alloy B, it goes from 167μm to 712μm.

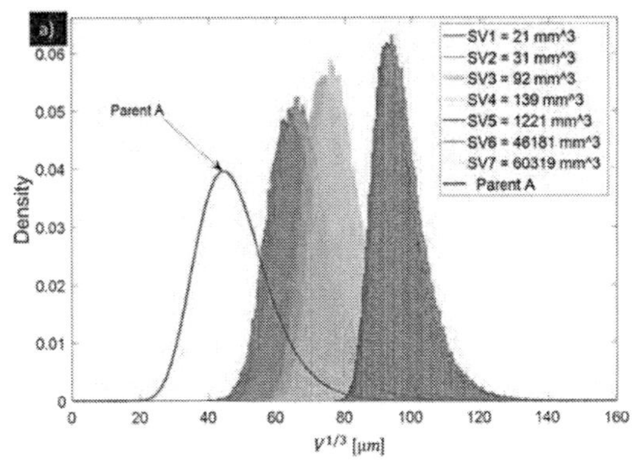

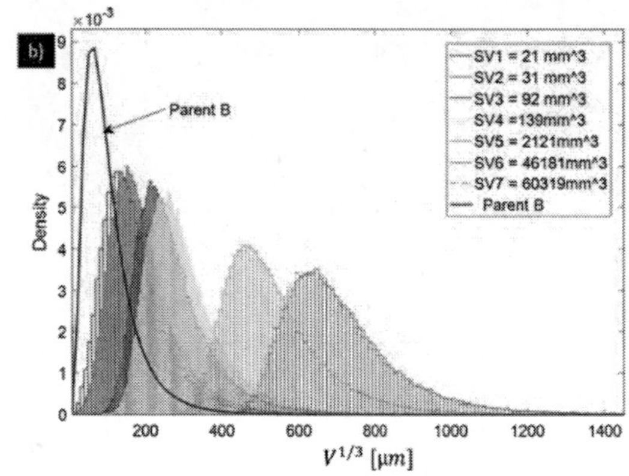

Fig. 13: Extreme value distributions for (a) A alloy and (b) B alloy as a function of sampling volume

In the following section the influence of the sampling volume on the Arithmetic Mean and Coefficient of Variation of critical defects size is discussed.

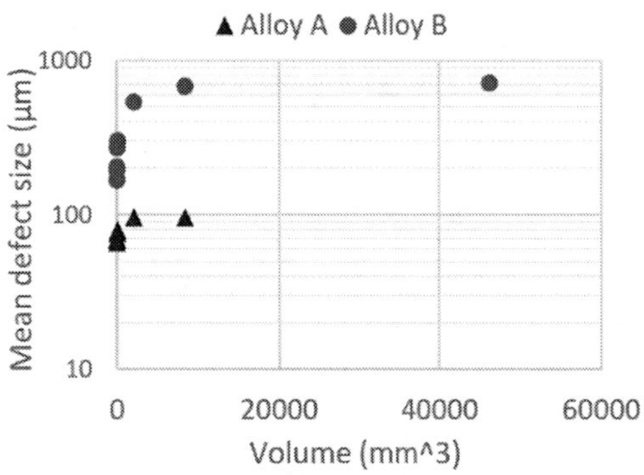

Fig. 14. Influence of the sampling volume on the arithmetic mean value of the critical defect size distributions.

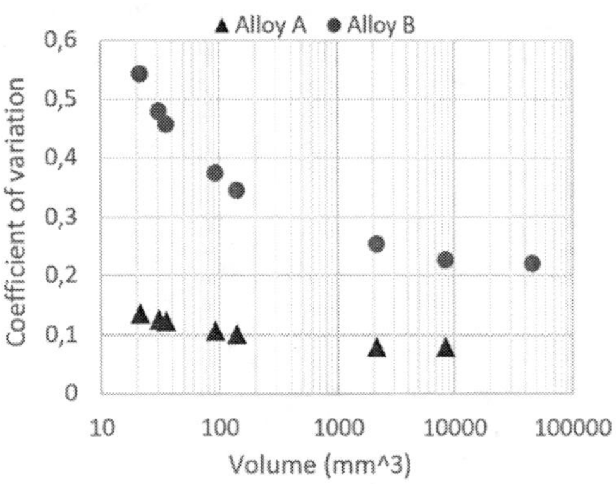

Fig. 15. Influence of the sampling volume on the coefficient of variation of critical defect size distributions.

Fig. 14 shows the evolution of the arithmetic mean value of the critical defect size distribution as a function of the sampling volume for both alloys.

In order to analyse the influence of the sampling volume on the scatter, Fig. 15 shows the evolution of the coefficient of variation as a function of the sampling volume. The coefficient of variation is defined as:

$$C_v = \frac{standard\ deviation,\ s}{mean\ value,\ \mu} \quad (1)$$

Both Fig. 14 and 15 highlight the fact that the distributions tend to a stabilized state as the sampling volume increases.

- For alloy A, it starts to stabilize from a volume of 2120mm^3. This can therefore be considered as the minimum value, above which the volume has no effect on the critical defect size for alloy A.
- For alloy B, it starts to stabilize from a volume of 8482mm^3. Hence, the minimum value for alloy B, above which the volume has no effect on the critical defect size, is much greater compared to alloy A.

It should be noted that these volumes are relatively large compared to the critical volumes determined for the fatigue specimens tested experimentally (see Table 3). Hence, a volume effect should be observed experimentally in terms of the critical defect size.

Also, it is important not to confuse these volumes (which are related to the defect size only) with the Fatigue ERV discussed in section 3.1. There is undoubtedly a direct relationship between the influence of the volume on the critical defect size and the influence of the volume on the fatigue strength (via the Kitagawa- Takahashi diagram), however this work is in-progress and will be discussed in future publications.

5. Conclusions

In this study, an experimental investigation has been conducted to evaluate the statistical size effect on the fatigue strength of cast Al-Si alloys. The main results can be summarized as follows:

- A variation in the highly stressed volume is accompanied by a change in the fatigue strength of the material. This result confirms the existence of a size effect in cast Al-Si alloys. Furthermore, alloy A has a more pronounced volume effect than alloy B. This must be due to the characteristics of the defect population of each alloy (pore size, distance between large pores, etc…).
- Good correlation between the local fatigue strength amplitude (Kt*σa) and the critical volume was found. Hence, it can be concluded that the highly stress volume approach is an effective way to describe the statistical size effect in cast Al-Si alloys.

- The fatigue strength for alloy A stabilizes from a critical volume of 320mm^3, which can be considered as the Fatigue Representative Elementary Volume.
- The fatigue strength of alloy B stabilizes at a smaller critical volume of 90mm^3 which can be considered as the Fatigue REV of this alloy.
- Both alloys are degraded by the same fatigue damage mechanism. In fact, on the Kitagawa-Takahashi diagram, both materials show the same tendency.
- The preliminary results of the numerical modelling approach highlight the influence of the volume on the critical defect size. It is demonstrated that the mean value of the critical defect size decreases with decreasing volume, but stabilises when the volume is greater than 2100mm^3 for Alloy A and greater than 8500 mm^3 for Alloy B.
- The inverse trend is predicted for the dispersion. That is, the coefficient of variation of the critical defect distributions increases with decreasing volume.

PSA Groupe and the French National Agency for Research and Technology (ANRT) financially supported this work.

References

1. V. D. Le, Etude de l'influence des hétérogénéités microstructurales sur la tenue en fatigue à grand nombre de cycles des alliages d'aluminium de fonderie. (Doctoral dissertation, Angers, ENSAM). 2016.
2. I. Koutiri, Effet des fortes contraintes hydrostatiques sur la tenue en fatigue des matériaux métalliques. (Doctoral dissertation, Angers, ENSAM), 2011
3. C. E. Phillips, R. B. Heywood, "The size effect in fatigue of plain and notched steel specimens loaded under reversed direct stress.," p. 113–124., 1951.
4. J. E. Shigley, C. R. Mischke, Jr. T. H. Brown, "Standard Handbook of machine design. Third Edition". McGraw-Hill, New York 2004.
5. H. F. Moore, D. Morkovin, "Third progress report on the effect of size of specimen on fatigue strength of three types of steel." 1944.
6. N. Saintier, T. Palin-luc, J. Bénabes, F., "Cocheteux, Non-local energy based fatigue life calculation method under multiaxial variable amplitude loadings, International Journal of Fatigue," Sep-2013.
7. A. Carpinteri, A. Spagnoli, Size effect in S-N curves: A fractal approach to finite-life fatigue strength. International Journal of Fatigue, 31, 927-933. 2009
8. K. H. Kloos, A. Buch, D. Zankov, Pure geometrical size effect in fatigue tests with constant stress amplitude and in program tests. Materialwissenschaft und Werkstofftechnik. 12, 40-50. 1981
9. M. Makkonen. Size effect and notch size effect in metal fatigue. Thesis for the degree of Doctor of Science (Technology), Lappeenranta University of Technology, Acta Universitatis Lappeenrantaensis 83; 1999
10. W. Weibull, 1939, the phenomenon of rupture in solids, In: Proc., Royal Swedish Institute of Engineering Research (Ingenioersvetenskaps Akad. Handl.), vol. 153, pp. 1-55 (Stockholm)
11. R. Kuguel, A relation between theoretical stress concentration factor and fatigue notch factor deducted from the concept of highly stressed volume, ASTM STP Proc; 61:732–48. 1961
12. C. M. Sonsino, G. Fischer, Local assessment concepts for the structural durability of complex loaded components. Materialwissenschaftund Werkstofftechnik. 36, 632-641. 2005.
13. C. K. Lin, W. J. Lee, Effects of highly stressed Volume on fatigue strength of austempered ductile irons. International Journal of Fatigue. 20. 301-307. 10.1016/S0142 1123(97)00134-5. 1998
14. DNV-RP-C203. Fatigue design of offshore steel structures; 2012.
15. API RP 2A-WSD. Recommended practice for planning, designing and constructing fixed offshore platforms – working stress design. 21st ed.; December 2000.
16. W. J. Dixon, A. M. Mood A Method for Obtaining and Analyzing Sensitivity Data, Journal of the American Statistical Association, p 43-241.1948
17. M. I. Houria, Y. Nadot, R. Fathallah, M. Roy D. M. Maijer, "Influence of casting defect and SDAS on the multiaxial fatigue behaviour of A356-T6 alloy including mean stress effect." 2015
18. P. Beiss, A. Zafari, K. Lipp, J. Baumgartner: Fatigue Behavior of a Sintered Steel Containing 4 % Ni, 1.5 % Cu, 0.5 % Mo and 0.6 % C; Int. J. Powder Metal 48, p. 19-34. 2012
19. A. Zafari, P. Beiss, C. Broeckmann, K. Lipp: Assessing the Fatigue Strength of a Sintered Steel as affected by the Highly Stressed Volume; Proc. Euro PM, Barcelona, Vol.1, p.15-20; EPMA, Shrewsbury, 2011
20. A. Moesenbacher, C. Guster, G. Pinter, W. Eichlseder. Investigation of concepts describing the influence of stress concentration on the fatigue behaviour of short glass fibre reinforced polyamide. ECCM 2012 - Composites at Venice, Proceedings of the 15th European Conference on Composite Materials. 2012

Influence of Surface Finish and Porosity on the Fatigue behaviour of A356 Aluminium Casting Alloy

Raquel Hidalgo[1]*, *Jon Ander* Esnaola[1], *Miren* Larrañaga[1], *Iñigo* Llavori[1], *Nuria* Herrero-Dorca[1], *Iñaki* Hurtado[1], *Edurne* Otxoa[2], *Patxi* Rodríguez[2], *Aitor* Kortabarria[2]

1 Mondragon University, Loramendi 4, 20500, Arrasate-Mondragón, Guipúzcoa, Spain.
2 Edertek Research Center, Uribarri Auz., 43, 20500 Arrasate-Mondragón, Guipúzcoa, Spain.

Abstract. In casting parts, due to the manufacturing process, the presence of defects such as porosity, inclusions and oxide films is unavoidable. All these irregularities have a negative effect on the component performance. Several works have demonstrated that, among them, porosity is especially detrimental to the fatigue properties. As most fatigue failures nucleate at the surface of a material, casting defects at or near the surface and surface roughness become an extremely important factor in determining the fatigue strength of cast components. Very little research has been conducted into the influence of both surface quality and porosity on the fatigue behaviour of aluminium castings parts. In the present work, the effects of two different surface qualities (machined and as-cast) on fatigue behaviour of an A356 casting alloy were studied. The S-N curves obtained showed that the cast surface had higher fatigue strength than the machined one. The failure in cast specimens initiated predominantly from valleys of the rough surface near pores or inclusions. On the other hand, in machined surfaces, the cracks initiated directly from surface pores. Thus, the improvement in fatigue life was attributed to a longer crack "initiation" period.

1 Introduction

Steel components are being replaced with aluminium equivalents due to the need for lighter and more fuel efficient automobiles. Between them, A356 alloy is employed in automotive and aeronautical industry because of its low strength/weight ratio, good castability and treatability. The problem of casting aluminium alloys is the presence of defects, such as inclusions and pores. These defects are detrimental to the fatigue properties(1). On the other hand, surface roughness can also significantly reduce the fatigue strength.

Fatigue cracks are initiated, as a rule, at the surface or subsurface of structural components, and, therefore, fatigue life is sensitive to surface conditions. In some casting components, the highly stressed surface areas where fatigue cracks initiate are machined. Nevertheless, in others, the as-cast surface skin is left intact. Most of the laboratory fatigue tests are performed on specimens with machined surfaces, which do not represent some of the real component surfaces. Thus, the data obtained from test specimens may not reflect the fatigue behaviour of the actual structural components(2). To accurately predict the fatigue life of an actual component, the effects of surface conditions on the fatigue life must be taken into account.

As most fatigue failures nucleate at the surface of a material, casting defects at or near the surface become an extremely important factor in determining the fatigue strength of cast components.

Very little research has been conducted into the influence of both surface quality and porosity on the fatigue behaviour of aluminium castings parts. All authors agree that fatigue strength of specimens with as-cast surfaces is lower than that with polished surfaces (2–6). As for the different behaviour of as-cast and machined surfaces on the fatigue behaviour of casting aluminium alloys, the studies are even fewer. S. Jiang (3) showed that specimens with as-cast surfaces presented higher fatigue strength than those with machined surfaces. Nevertheless, this may be questionable because the microstructural aspects were not constant for all the specimens.

Therefore, in the present work, the effects of two different surface qualities (machined and as-cast) on fatigue behaviour of an A356 casting alloy were studied. All specimens were cast in permanent a mould under the same conditions to avoid the influence of microstructural variables (SDAS, grain size, composition…). Then, some of them were machined before testing.

* Corresponding author: rhidalgo@mondragon.edu

The S-N curves for each condition were obtained and the fatigue cracks initiation sites were identified using Scanning Electron Microscope (SEM).

2 Experimental procedure

Samples were obtained from A356 alloy ingots (Al-7Si-0.3 Mg). The molten alloy was poured into a preheating permanent mould from which two dog bone specimens were obtained each time. Next, all cast specimens were heat treated in the T6 condition: solution treatment at 540°C for 6 hours followed by a warm-water quench and age hardening at 160 °C for 5 hours.

The dimensions of the cast samples were: total length 190 mm, diameter of the grip parts 19 mm, gauge length and diameter of 50 and 12 mm, respectively. As two different surface conditions were needed, some samples were machined to a final diameter of 11 mm. The dimensions of the samples can be seen in Fig. 1

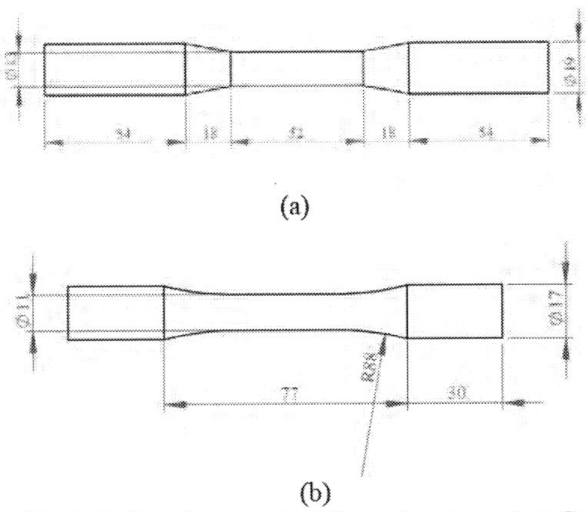

Fig. 1. Fatigue test samples (dimensions in mm) a) Cast sample b) Machined sample.

The roughness of the samples was measured with a Mitutoyo Portable Surface Roughness Tester. To measure the shape and analyse the surface, an Alicona InfiniteFocusSL was employed.

The high-cycle fatigue tests were conducted according to the ASTM-E466 test procedure specifications with sinusoidal loading using an MTS810 Materials Testing System at room temperature. The fatigue test was performed using a stress ratio of R=-1 with a frequency of 10 Hz.

The fatigue strength coefficient and exponent in Basquin's equation were obtained via regression analysis of the experimental data using the least-squares method.

To identify the places where the fatigue cracks initiated for each condition, a scanning electron microscope FEI Nova Nano SEM 450 was used.

3 Results and Discussion

3.1 Surface roughness

The surface roughness was measured on as-cast and machined surfaces with a Mitutoyo Portable Surface Roughness Tester. Two parameters were studied: the Ra and Rz. The Ra is a widely used parameter for assessment of surface roughness, its value reflects the average length between the peaks and valleys of the profile. However, for fatigue analysis. The Rz parameter, which corresponds to the maximum peak-to-valley height, is a more suitable measure. Fig. 2 compares these parameters for both samples, each value being the average of three measurements.

What stands out in the graph is the significant difference in Ra and Rz parameters. Based on these results it can be concluded that cast specimens presented greater roughness than machined ones.

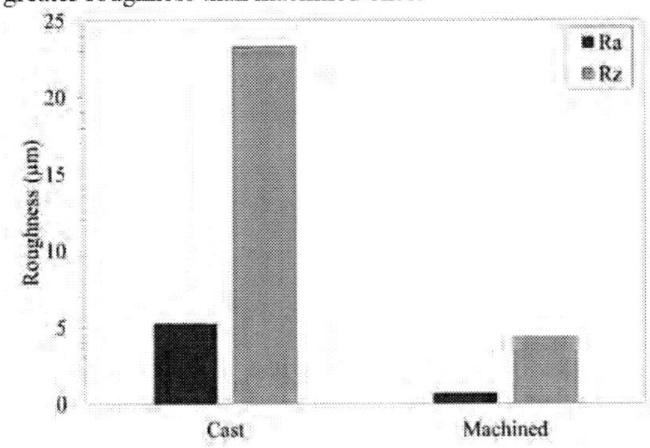

Fig. 2. Surface Roughness (Ra and Rz) of as-cast and machined samples.

The considerable difference between the two surfaces are also clearly seen in the 3D image in Fig. 3 obtained with Alicona InfiniteFocusSL.

The as-cast surface was very irregular as a result of deep hollows and steep peaks which came from the paint used in the mould. On the other hand, the machined surface was smoother presenting only superficial machined grooves.

Similarly, Fig. 4 compares the topographic profile of the two samples. The maximum peak-to-valley height was around 30 µm for the cast sample and 5 µm for the machined one. This supported the results measured in Fig. 2.

The big difference of parameters in Fig. 2 is clearly explained by the two different surfaces observed in Fig. 3 and the profiles in Fig. 4.

Fig. 3. Image showing the 3D surface roughness of a) as-cast and b) machined samples.

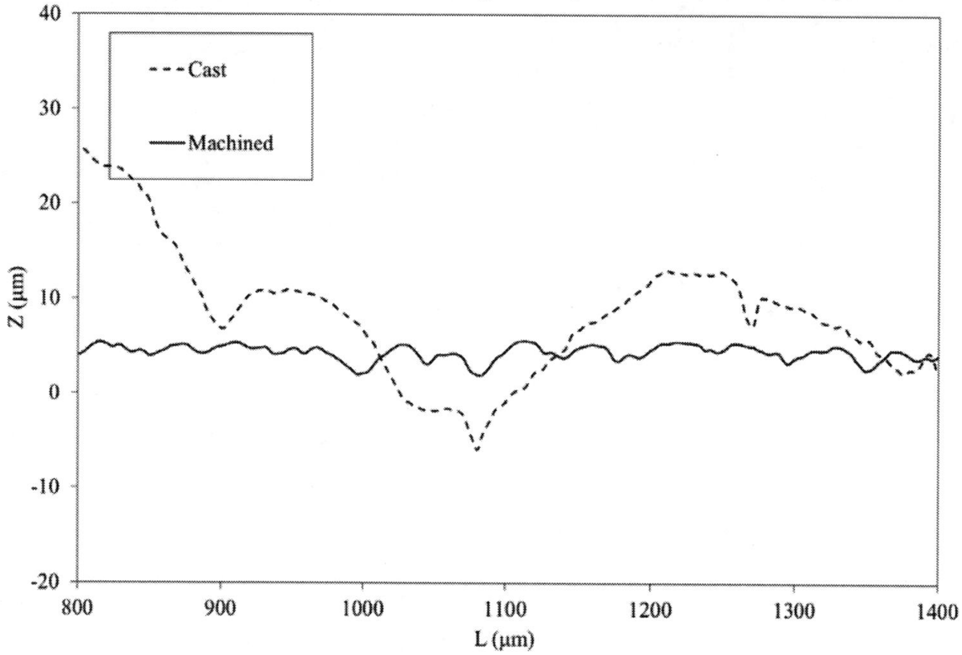

Fig. 4. Topographic profile of the samples.

3.2 Porosity

The microstructure of A356 alloy consisted of Al dendrites surrounded by eutectic silicon phase.

There were two types of porosity found in the samples as shown in Fig. 5. Shrinkage pores were the predominant porosity found in the samples. This kind of porosity results from insufficient metal flow into the space between connected dendrites during the solidification process. Only a few defects were gas pores. The shrinkage pores had very irregular three-dimensional shapes and varied sizes, whereas the gas pores were usually roughly spherical (circular in cross-section). Typical examples of a shrinkage pore and a gas pore are shown in Fig. 5

3.3 Fatigue Tests

To compare the difference between the fatigue behaviour of cast and machined specimens, fatigue tests were carried out in samples with both surface conditions. Fig. 6 shows S-N curves obtained where the number of cycles to failure (N_f) is plotted against stress amplitude applied (S_a). As can be seen in the graph, for a given stress amplitude, the machined surface present a higher fatigue life than the cast one.

Previous investigations indicated that almost all fatigue fractures start from the sites of stress concentrations at structural discontinuities such as holes, notches, grooves, cracks, defects, and scratches(3). Thus, to understand the difference in the fatigue behaviour among the two types of surface conditions, the places where the fatigue failure starts must be studied.

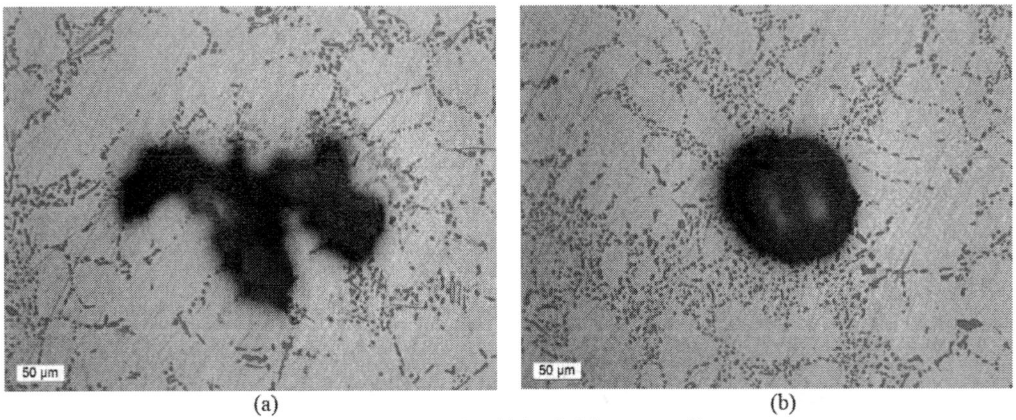

Fig. 5. Optical micrographs of (a) a shrinkage pore.(b) a gas pore.

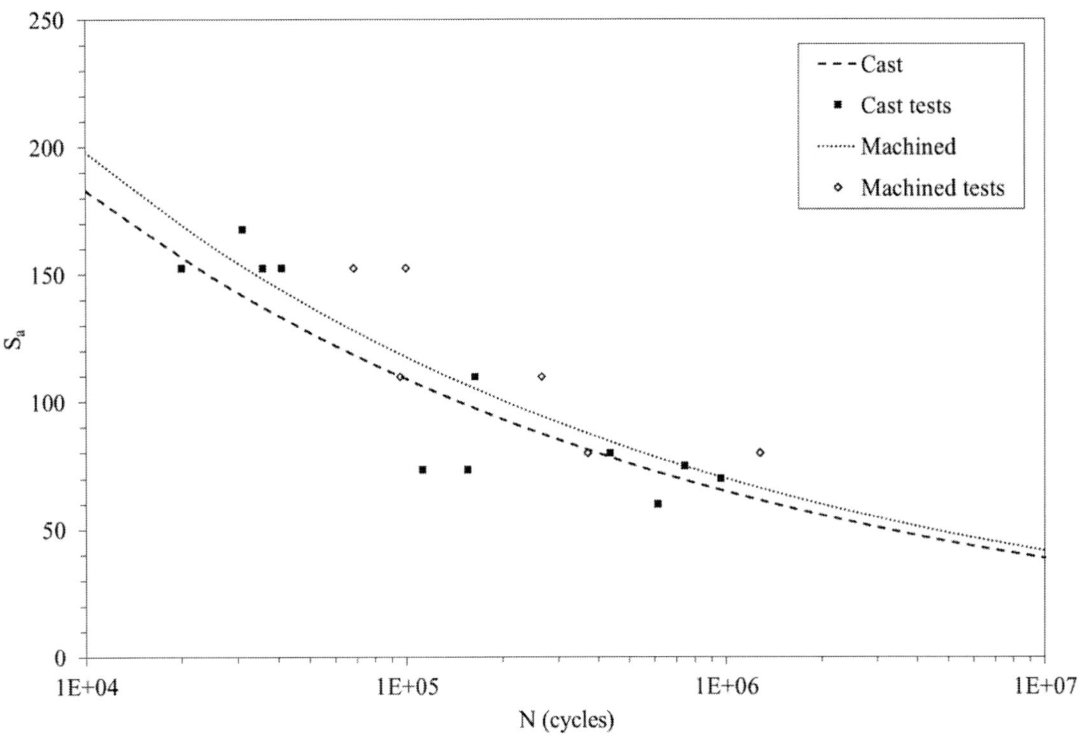

Fig. 6. Fatigue life curves for cast and machined samples.

SEM was used to carry out the fractographic study. For cast specimens, failure initiated predominantly from valleys of the rough surface near pores or inclusions as shown in Fig. 7. On the other hand, in machined surfaces, the cracks were initiated directly from surface pores.

Taking into account that all the samples were cast under the same conditions, no microstructure variations existed and the average porosity can be considered constant in all the samples. For this reason, the only difference was the roughness. The fatigue results in Fig. 6 can be related to roughness in Fig. 2, Fig. 3 and Fig. 4. It can be concluded that a decrease in surface roughness in the machined materials increased the difficulty of crack initiation and, hence, increased the fatigue life. Therefore, in this case, the improvement in fatigue life was attributed to a longer crack "initiation" period. This explained the results in Fig. 6 and it correlated well with previous studies (2,5)

However, this outcome is contrary to that of Jiang S. et al. (3) who found that grooves on specimens with machined surface roughness of Ra=1.6 and 3.2 µm were deep and sharp, which reduced the fatigue strength considerably compared with as-cast specimens. In this study, the grooves were not deep and sharp as shown in

* Corresponding author: rhidalgo@mondragon.edu

 https://doi.org/10.1051/matecconf/201816514007

Fig. 4 which could explain the different results obtained.

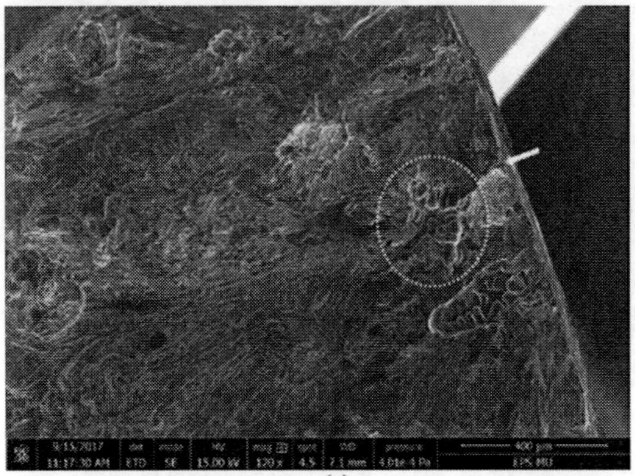

(a)

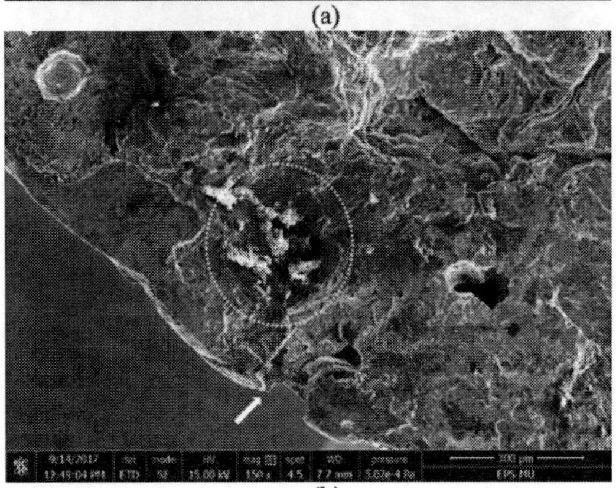

(b)

Fig. 7. Fatigue fracture surface of a cast specimen showing initiation from a valley near (a) a shrinkage pore (b) an inclusion.

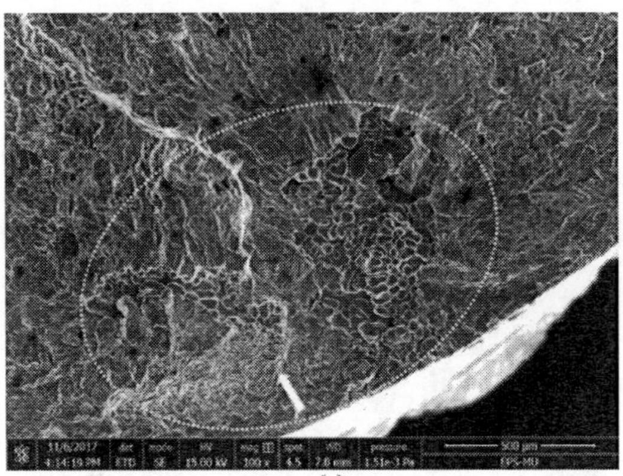

(a)

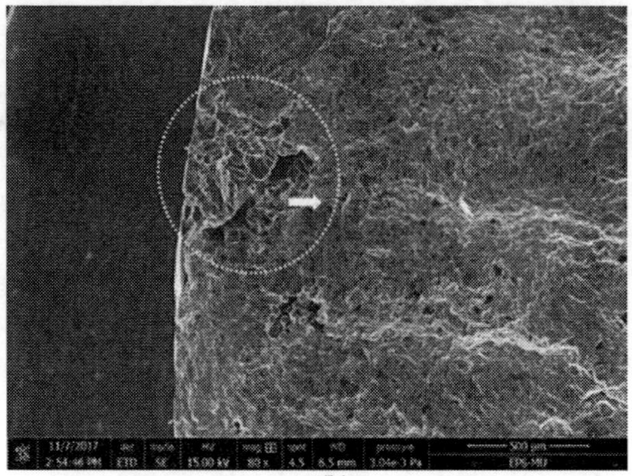

(b)

Fig. 8. Fatigue fracture surface of machined samples showing crack initiation from surfaces pores.

4 Conclusions

The aim of this investigation was to study the difference in the fatigue behaviour of A356 alloys considering two different surface conditions; machined and cast.

- The measurements of roughness parameters, the 3D images and the topographic profiles indicated that the as-cast specimens had greater surface roughness than machined ones.
- The S-N curve showed that fatigue life fatigue life of machined specimens was higher than that of cast ones.
- The cracks in as-cast samples initiated from valleys of the rough surfaces near pores or inclusions. In the machined samples, the surface roughness was reduced and the failure was observed to start in pores near the surface.
- A decrease in the roughness of machined samples changed the mechanisms of crack initiation which explained the results of fatigue life curves:

The authors acknowledge the financial support provided by the Ministry of Industry Economics and Competitiveness of the Spanish Government by means of ALPRE project (RTC-2015-3822-4) and the FEDER funds from the European Union.

References

1. Linder J, Axelsson M, Nilsson H. The influence of porosity on the fatigue life for sand and permanent mould cast aluminium. 2006;
2. Jiang H, Bowen P, Knott JF. Fatigue performance of a cast aluminium alloy Al-7Si-Mg with surface defects. J Mater Sci. 1999;34(4):719–25.
3. Jiang SN, Xu SY. Influence of Surface Conditions on the Fatigue Behavior of A357 Aluminum Alloy. Mater Sci Forum [Internet].

2016;850:113–9. Available from: http://www.scientific.net/MSF.850.113
4. Serrano-Munoz I, Buffiere J-Y, Verdu C, Gaillard Y, Mu P, Nadot Y. Influence of surface and internal casting defects on the fatigue behaviour of A357-T6 cast aluminium alloy. Int J Fatigue. 2016;
5. Yamabe J, Kobashi M. Influence of casting surfaces on fatigue strength of ductile cast iron. Fatigue Fract Eng Mater Struct. 2006;29(6):403–15.
6. Svenningsson R, Svensson H, Borgström H, Gotte A. Influence of the surface roughness on the fatigue properties in ausferritic ductile irons (ADI). MATEC Web Conf [Internet]. 2014;12:6006. Available from: http://www.scopus.com/inward/record.url?eid=2-s2.0-84903288711&partnerID=tZOtx3y1

FES of the effect of free vibration treatment on fatigue damage recovery for notched copper film

Hai-Meng Zhang, De-Guang Shang, Shuai Lv*

College of Mechanical Engineering and Applied Electronics Technology, Beijing University of Technology, Beijing 100124, China

Abstract. In this paper, finite element simulation (FES) of the effect of free vibration treatment on fatigue damage recovery for notched copper film with the thickness of 33 um was investigated. The damage process analyzed by static analysis of original specimen and vibration process analyzed by transient dynamic analysis of damaged specimen were simulated respectively. The transient dynamic analysis was conducted with different frequencies and acceleration amplitudes. It is found that the plastic strain was produced as a result of cyclic loading and reduced after the vibration treatment. The values of plastic strain are different after treatment with different vibration frequencies and acceleration amplitudes. Furthermore, the plastic strain in notch reduces more than that of other areas of the specimen. According to the analysis, the vibration treatment can reduce the plastic strain of notched specimen significantly by which the micro-properties are improved so that fatigue damage can be recovered.

1 Introduction

The VSR (vibratory stress relief) process is a green and emerging technology that refers to the relaxation of residual stress via cyclic loading treatments. VSR methods have been successfully used to relieve or redistribute stress and therefore guard against service failures, such as stress corrosion cracking. Heat treatment is a popular stress relief method in the metal forming industry. While this treatment yields stress-free materials, it has numerous disadvantages, such as along process time, significant energy consumption, oxide scale formation on the surface, and substantial effects on the mechanical properties. VSR does not suffer from these disadvantages. Although the stress relief performance of the VSR process remains inferior to that of heat treatment, this method is widely used in various industrial fields as a result of its time-saving, small equipment volume requirements, low cost, minimal energy consumption [1-5].

The relationship between VSR and fatigue damage recovery have been studied by many researchers. A. S. M. Y. Munsi et al. [6] investigated the effect of VSR treatment on fatigue life improvement for weld components. The results indicated that the fatigue life of welds increased 17%-30% after VSR. D.L. Rao et al. [7] found that the residual stress measurement results show that the average principal stresses decrease by about 56% for the drum of hoist machine and decrease by about 31% for the stainless steel 304L plate after VSR. Y.P. Lu et al. [8] compared the mechanisms of VSR and TSR(thermal stress relief). They indicated that both methods are similar in reducing macroscopic residual through plasticity deformation to reduce residual elasticity strain. Y.D. Zhang et al. [9-11] investigated the effect of VSR. That low amplitude VSR saves more energy than the high amplitude VSR with similar stress relieving effect. They believed that low amplitude VSR has positive influences on the fatigue whereas high amplitude VSR has negative influences. M. Jafari et al. [12] proposed a mathematical model based on plasticity theorem with linear kinematic hardening to explain the mechanism of reduction of residual stresses of materials experiencing cyclic loading. The numerical results from the mathematical model and the analytical results with existing experimental data for rectangular cross-section beams showed a similar trend. However, D. Djuric et al. [13] found that no clear tendencies could be seen. Nevertheless, fatigue damage due to the VSR treatment could be found. The changes in residual stresses did not cause macroscopic deformations of the specimens.

Since one end of the specimen was clamped to a vibration stage with fixture, and the other end is free to move in the method mentioned above, it may induce additional damage which similar with vibration fatigue during treatment.

Finite element method has been widely used to investigate mechanical properties of materials and structures. It can be used to simulate various test schemes and optimal design to reduce test time and expense [14-16].

The aim of this paper is to investigate the effect of unconstraint vibration treatment on fatigue damage recovery by means of finite element simulation which is a preliminary study for improving fatigue life. Both two ends of the specimen are free to move during vibration treatment and the direction of vibration is perpendicular

* Corresponding author: author@e-mail.org

 https://doi.org/10.1051/matecconf/201816514008

to the longitudinal direction. The effects of vibration parameters including different vibration frequency and vibration acceleration amplitude on plastic strain reduction are also analysed.

2 Simulated cases

A notched copper film model is simulated in this investigation. The shape and dimensions of the copper film specimen are shown in Fig.1.

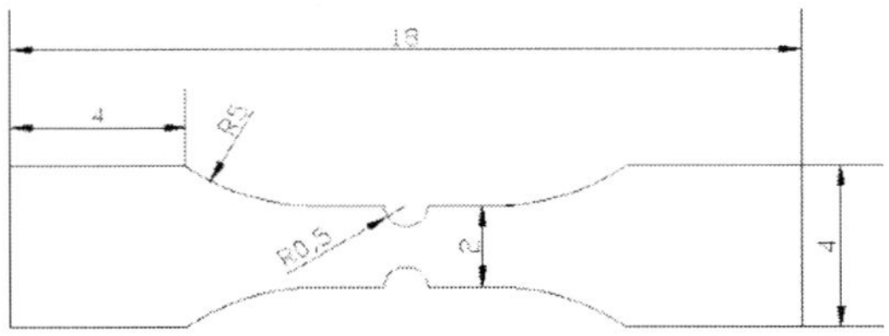

Fig. 1. Shape and dimensions of copper film specimen.

The element type selected for the specimen is shell 181. Meanwhile, a combination element named combin14 is also selected to apply constrains since the analysis process cannot conduct without any constrains on the model in transient dynamic analysis. The parameters of the real constants of the combination element are shown in table 1.

Table 1. Parameters of the combination element

Spring constant	10
Damping coefficient	0.02

Combined nonlinear isotropic-kinematic hardening is used to describe the elastic-plastic material behaviour. The material parameters of the specimen are listed in Table 2.

Table 2. Material parameters of the specimen

Young's modulus(MPa)	42000
Poisson's ratio	0.3
yield strength(MPa)	252
tensile strength(MPa)	418
elongation(%)	6

2.1 Meshing

The element edge length of the model is 0.0001 m and division arc of the notch of the model is 3° . The model is convergent by choosing these parameters. Meshing of the model after free mesh is shown in Fig.2.

Fig. 2. Meshing of copper film specimen.

2.2 Loading

The damage process analyzed by static analysis of original specimen and vibration process analyzed by transient dynamic analysis of damaged specimen are simulated respectively. A transverse non-load-carrying attachment subjected to axial loading is modeled using elastic-plastic finite elements.The force appiled to the axial direction in fatigue process is 9N. Two cycles including four load steps are loaded and each load step including one substep. The load-time history is shown in Fig.3. The axial direction is the X direction and width direction is the Y direction during analysis.

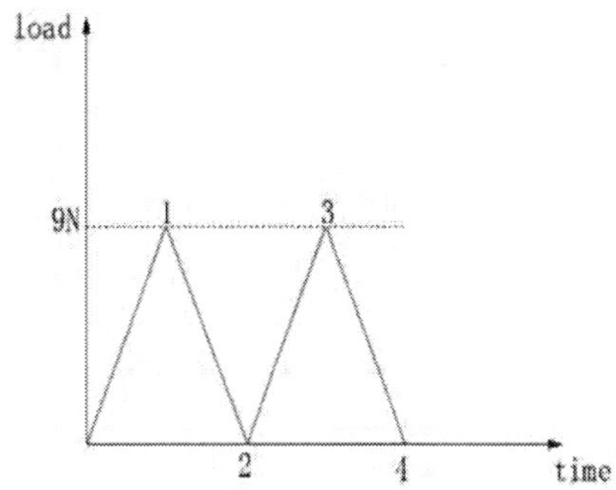

Fig. 3. Load-time history of fatigue process.

A new node perpendicular to the specimen center is created, then it is a combined with the node in the center of the specimen as a combination element. Displacement constraints were applied on the new node. A prestress which produced in fatigue damage process is guided into the model before transient dynamic analysis is conducted. The specimen is excited under a sinusoidal vibration waveform which is perpendicular to the XY plane and action time is one cycle. A vibration data file which consists of vibration time and vibration acceleration is guided into the model as the loading during transient dynamic analysis. The load-time history of vibration process is shown in Fig.4. Different time represents different vibration frequencies in the data file.

https://doi.org/10.1051/matecconf/201816514008

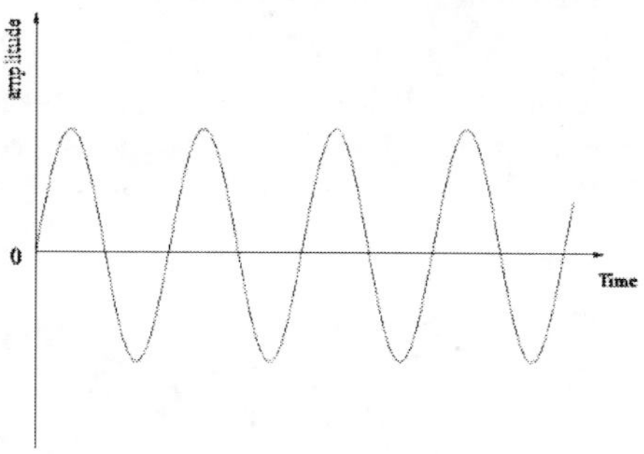

Fig. 4. Load-time history of vibration process.

3 Results and discussion

The X directional plastic strain distribution of the notch area after fatigue loading is shown in Fig 5.

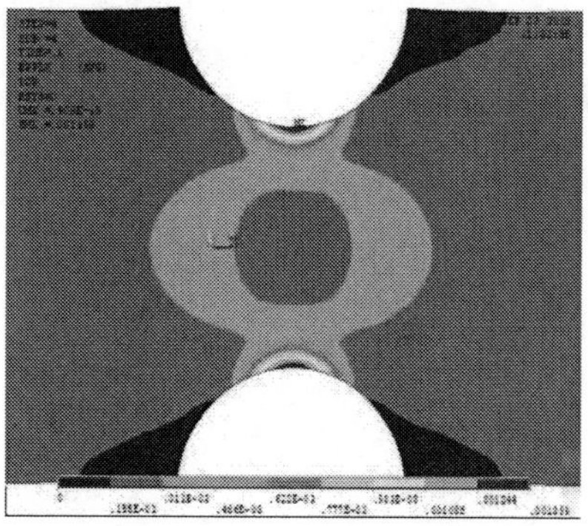

Fig. 5. X direction plastic strain distribution of the notch area after fatigue loading.

The results show that the maximum plastic strain is located at the root of specimen notch which was the stress concentration district and the value is 1.399×10^{-2}. In general, it is superior to initiate fatigue crack in this district.

3.1 Effect of vibration frequency

Vibration process is simulated under different vibration frequencies including 40Hz, 60Hz, 80Hz, 100Hz, 150Hz respectively. All of the vibration acceleration amplitude applied in these simulations is 4g. The X directional plastic strain distribution of the notch is shown in Fig 6. Different colours represent different strain values, and the value of the strain in red region is maximum.

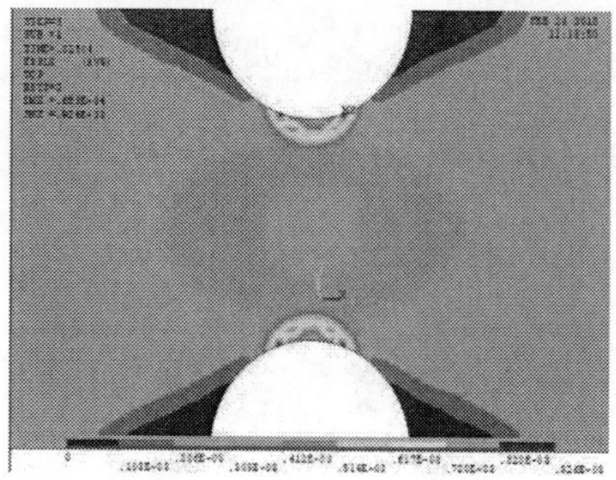

(a)

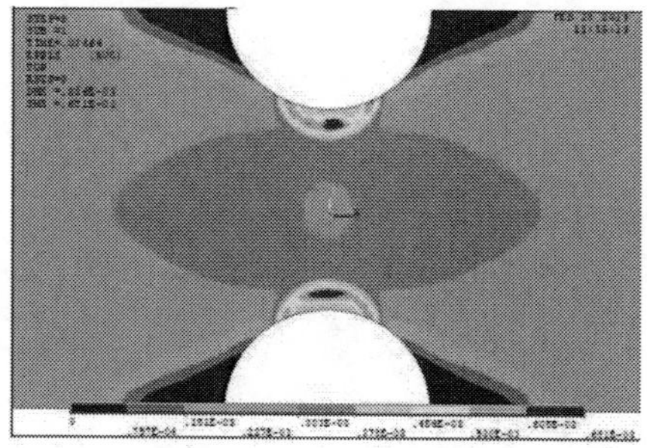

(b)

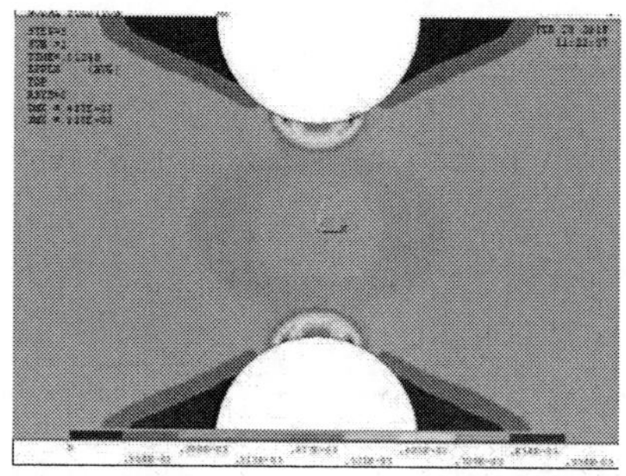

(c)

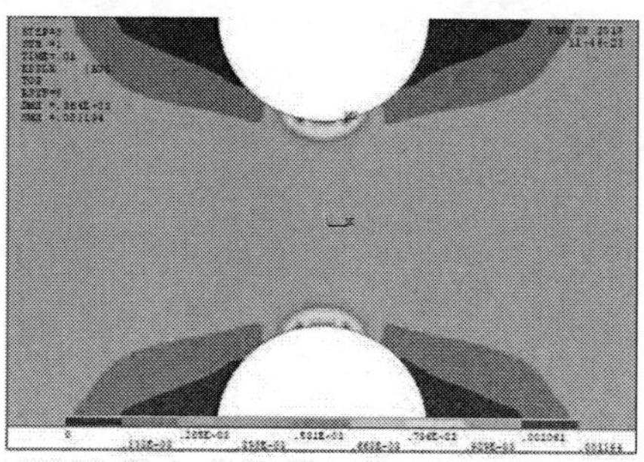

(d)

https://doi.org/10.1051/matecconf/201816514008

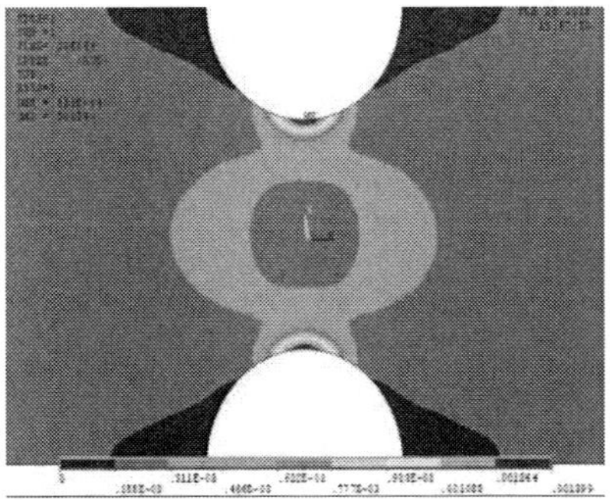

(e)

Fig.6. X directional plastic strain distribution of the notch area (a) 40Hz, (b) 60Hz, (c) 80Hz, (b) 100Hz, (e) 150Hz.

The maximum value of the X directional plastic strain under different vibration frequencies in the notch is shown in Fig 7.

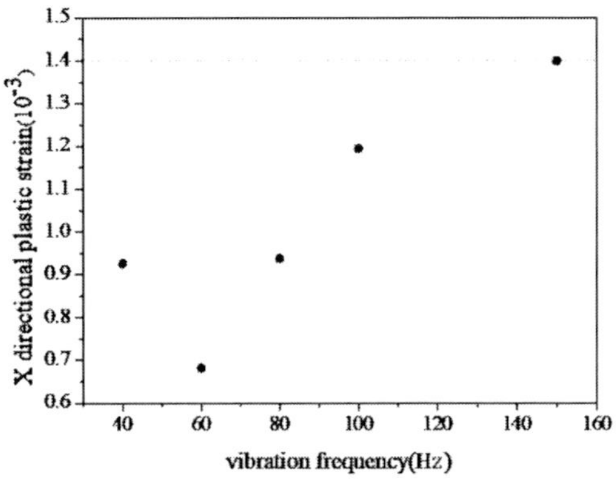

Fig.7. Maximum value of the X directional plastic strain under different vibration frequencies.

As can be seen in the figure 7, all of the X directional plastic strain are decreased after vibration treatment expect 150Hz compared to that without vibration treatment and the value of the plastic strain is 0.926×10^{-3}, 0.681×10^{-3}, 0.938×10^{-3}, 1.194×10^{-2}, 1.399×10^{-2}, respectively. The results show that X directional plastic strain decreases firstly and then increases with the increase of vibration frequency and the value is minimum when the vibration frequency is 60Hz. The plastic strain decreases only if with proper vibration frequency, the plastic strain cannot decrease even increase with other vibration frequencies. Frequency is related to the mode of vibration. When the vibration frequency is low, vibration is irregular according to the results of simulation which may affect the effect of vibration treatment on fatigue damage recovery, so that the reduction of plastic strain is little. When the vibration frequency is too high, the mode of vibration cannot produce effective stress to recover damage defect, so that the plastic strain does not decrease.

3.2 Effect of vibration acceleration amplitude

Vibration process under different vibration acceleration amplitudes is simulated including 3g, 4g, 5g, 6g, 7g, respectively. All of the vibration frequencies applied is 60Hz. The X directional plastic strain distribution of the notch is shown in Fig 8. Different colours represent different strain values, and the strain of red region is maximum.

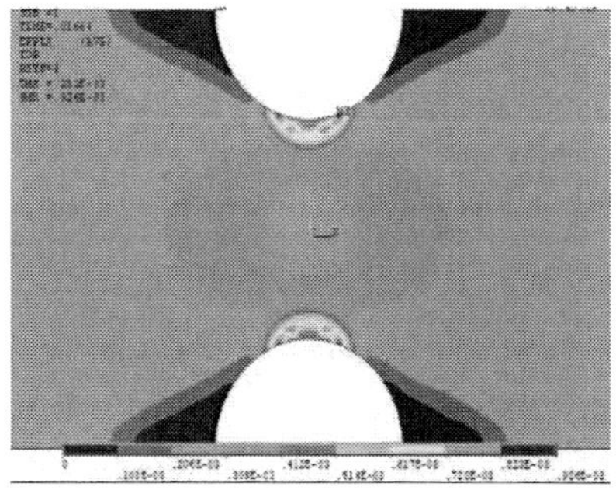

(a)

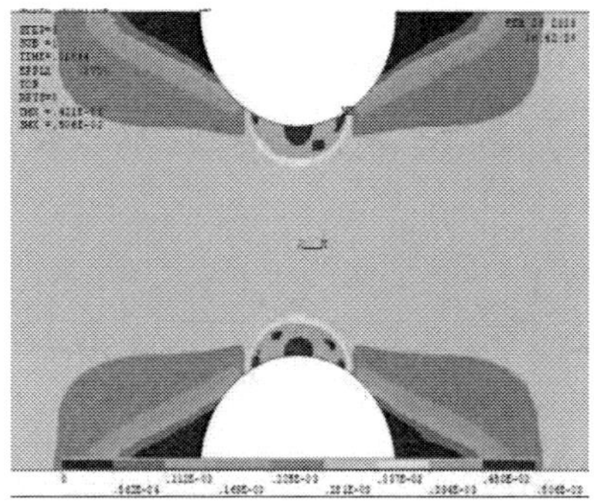

(b)

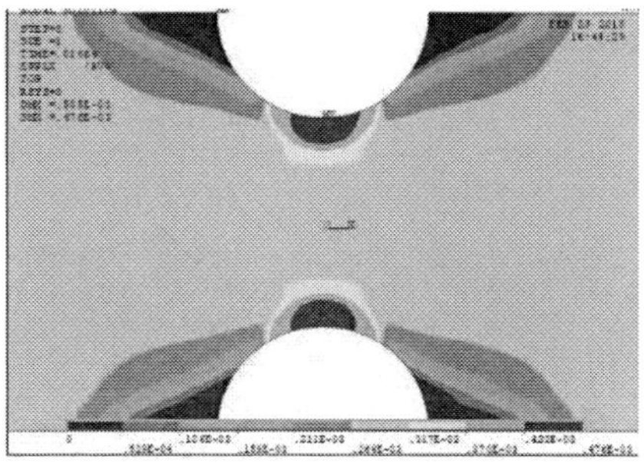

(c)

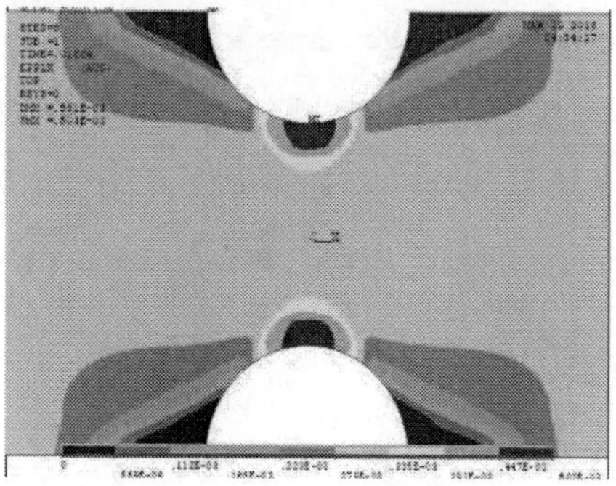

Fig. 8. X directional plastic strain distribution of the notch area (a) 3g, (b) 5g, (c) 6g, (d) 7g.

It can be found that the maximum plastic strain point is located at the verge of the specimen notch but not in the stress concentration district when the vibration acceleration amplitude is 3g. The maximum plastic strain point is located inside the verge of the specimen notch and away from the stress concentration district when the vibration acceleration amplitude is 4g as listed in figure 6. However, the maximum plastic strain point is located at the verge of the specimen notch and also in stress concentration district when the vibration acceleration amplitudes are from 5g to 7g.

Values of the X directional plastic strain of the notch root under different vibration acceleration amplitudes are shown in Fig 9.

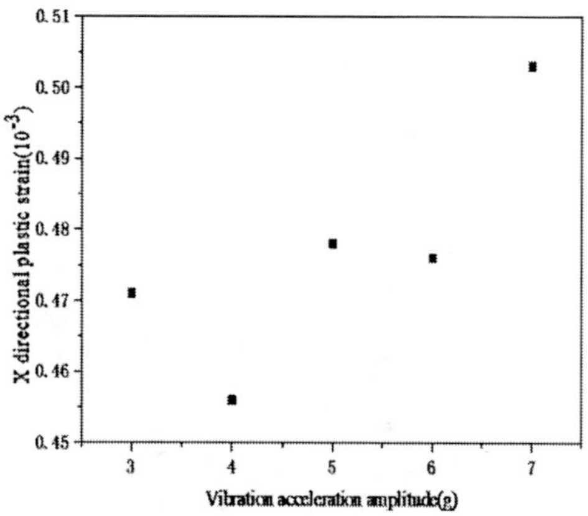

Fig. 9. Values of the X directional plastic strain of the notch root under different acceleration amplitudes.

As can be seen in the figure 9, all of the X directional plastic strain of the notch root are decreased after vibration treatment compared to that without vibration treatment and the value of the X directional plastic strain is 0.471×10^{-3}, 0.456×10^{-3}, 0.478×10^{-3}, 0.476×10^{-3}, 0.503×10^{-3}, respectively. The results show that X directional plastic strain decreases firstly and then increases with the increase of vibration acceleration amplitude and the value is minimum when the vibration frequency is 4g. Vibration acceleration amplitude is related to the height of specimen vibration and the height is related to the exciting force during vibration. The plastic strain decreases only if with proper vibration acceleration amplitude, the plastic strain cannot decrease even increase with other vibration acceleration amplitudes. Too little exciting force is not enough for the driving force to eliminate the dislocation and too much force will produce additional damage.

Compared to conventional VSR which one end of the specimen is champed to a vibration stage with fixture and the other end is free to move, both ends are free in the approach proposed in the present study and could avoid additional damage which is induced during conventional VSR.

Simulation results indicate that the proposed method is effective on fatigue damage recovery. Vibration treatment with proper parameters can produce exciting force which acts as a driving force that return the internal structure to a stable state. It is known that the energy of plastic strain is stored in dislocation structures during fatigue. Plastic strain decrease even disappear means dislocation decrease and microstructure improvement so that fatigue crack initiation is delayed and fatigue life is improved.

4 Conclusions

In this work, finite element method is utilized to derive the change of the notched copper film before and after unconstraint vibration treatment with different frequencies and different acceleration amplitudes. The conclusions are as follows:

1. The plastic strain of the specimen decreases effectively and the reduction is different after vibration treatment with different frequencies and different acceleration amplitudes. The optimal parameters are 60Hz and 4g respectively.

2. Vibration frequency determines the value of the axial direction plastic strain and vibration acceleration amplitude determines the site of maximum value of the axial direction plastic strain.

3. With the decrease of the plastic strain, the microstructure of the specimen is improved and the specimens are in a stable state which is good for delaying fatigue crack initiation and improving fatigue life.

The authors would like to acknowledge the support of the financial support of the National Natural Science Foundation of China(No.51535001,11572008)

References

1. J.S. Wang, C.C. Hsieh, C.M. Lin, E.C. Chen, C.W. Kuo, W.T. Wu. Materials Science&Engineering A. **605** (2014)
2. J.S. Wang, C.C. Hsieh, C.M. Lin, E.C. Chen, C.W. Kuo, W.T. Wu. metallurgical and Materials Transactions **44A** (2013)
3. J.S. Wang, C.C. Hsieh, H.H. Lai, C.W. Kuo, Y. Wu P. Ti, W.T. Wu. Materials Characterization **99** (2015)

4. D. Rao, D. Wang, L. Chen, C. Ni. International Journal of Fatigue **29** (2007)
5. C.M. Sonsino, F. Muller, J De Back, A. M. Gresn. Fatigue Fract Eng M **19**, 6 (1996)
6. A. S. M. Y. Munsi, A. J. Waddell, C. A. Walker. Strain **37**, 4 (2001)
7. D.L. Rao, Z.Q. Zhu, Chen L.G, C.Z. Ni. Materials Science Forum **490-491** (2005)
8. Y.P. Lu, Z.Y. Ma, Q.S. Jia, R.R. Gu. Mechanical Science and Technology **20** (2001)
9. J. Song, Y.D. Zhang. Advances in Mechanical Engineering **8**, 6 (2016)
10. H.J. Gao, Y.D. Zhang, Q. Wu, J. Song. Metals **7** (2017)
11. H.J. Gao, Y.D. Zhang, Q. Wu, J. Song, K. Wen. International Journal of Fatigue **108** (2018)
12. M. Jafari Vardanjani, M. Ghayour, R. Mokhtari Homami. Experimental Techniques **40** (2016)
13. D. Djuric, R. Vallant, K. Kerschbaumer, N. Enzinger. Welding in the World **55** (2011)
14. J. Lee, J. Kim, S. Kang, M. Hirohata, K. Chang. International Journal of Steel Structures **17** 2 (2017)
15. S.Q. Li, H.Y. Fang, X.S. Liu, W. Wang, Q. Wang, W. Cui. Journal of vibroengineering **18** 3 (2016)
16. M.M. El-Sayed, A.Y. Shash, M. Abd-Rabou. Journal of Materials Processing Tech **252** (2018)

STRAIN ENERGY-AND STRESS-BASED APPROACHES REVISITED IN NOTCH FATIGUE OF DUCTILE STEELS

Bruno Atzori*, *Mauro* Ricotta, and *Giovanni* Meneghetti

[1]Department of Industrial Engineering, University of Padova, 35131 Padova, Italy

Abstract. The constant amplitude, zero-mean stress, axial-fatigue behaviour of plain and bluntly notched AISI 304 L stainless steel specimens is investigated in terms of strain energy density. Concerning plain material, it was found that at the fatigue knee the plastic strain energy density is 1.49 times higher than the elastic strain energy density. In the authors' opinion, the presence of plasticity at the fatigue knee is responsible for the unsuitableness of classical stress - based approaches to synthesise the fatigue behaviour of this material. On the contrary, the elastic-plastic strain energy density was found an efficient parameter to rationalise in a single scatter band fatigue data of plain and bluntly notched specimens. Based on this result, the classic stress- and the point stress- based approaches were revisited taking into account the presence of plasticity at the fatigue knee, by introducing an equivalent fully elastic material having a linear elastic strain energy density at the fatigue knee equal to that of the actual material. Accordingly, a coefficient of plasticity K_p was successfully introduced to modify the classical definition of fatigue strength reduction factor, K_f.

1 Introduction

Design against fatigue is a key aspect for structural designers, in particular when notch effects are present. Usually, the fatigue strength of notched components is assessed starting from the fatigue strength of plain specimens, which is expressed by means of equations relating a damage variable with the number of cycles to failure.

Stress-, strain- or energy-based approaches are available in literature to correlate the fatigue strength of plain materials. Concerning the energy-based methods, some of them has been developed under the hypothesis of linear elastic material [1-4], others require the knowledge of cyclic elastic-plastic material behaviour (see 5-11, as examples).

Referring to AISI 304L stainless steel, the authors showed that neither net-section nor linear elastic peak stresses are able to rationalise in a single scatter curve the fatigue life of bluntly notched specimens. On the contrary, it was successfully synthesised in a single scatter band, by considering the dissipated heat energy density per cycle (the Q parameter) as fatigue damage indicator [12].

In this paper, first the elastic-plastic SED was successfully confirmed as an effective index able to rationalise fatigue data, obtained by carrying out fully reversed fatigue test on plain and bluntly notched specimens made from hot rolled AISI 304L stainless steel sheet. After that, with the aim to provide designers with a relative simple tool to make linear elastic calculations also in the presence of plasticity at the fatigue knee, the classical stress- and the point stress-based approaches were revisited, in terms of energy-based concept. As a result, the use of a modified fatigue notch factor allowed evaluating the fatigue strength of notched specimens in very good agreement with the experimental data.

2 Theoretical background

In this paper the elastic-plastic Strain Energy Density is assumed the damage variable able to correlate the fatigue strength of plain and notched components. The underlying concept is that two components have the same fatigue life when they have the same level of elastic-plastic SED. Usually at the fatigue knee, the plain material behaviour is supposed to be elastic and consequently the relevant strain energy is equal to the elastic strain energy density, W_{LE}

$$W_{LE}(\sigma_0)=\frac{\sigma_0^2}{2\cdot E} \quad (1)$$

being σ_0 the material fatigue limit and E its elastic modulus.

Nevertheless, when at the fatigue knee plasticity can not be neglected, the relevant elastic-plastic SED can be calculated according to Eq. 2:

$$W_{CC}(\sigma_0)=W_{CC}^E(\sigma_0)+W_{CC}^P(\sigma_0) \quad (2)$$

* Corresponding author: bruno.atzori@unipd.it

where $W^E_{CC}(\sigma_0)$ and $W^P_{CC}(\sigma_0)$ is the elastic and the plastic component of $W_{CC}(\sigma_0)$, respectively. The material cyclic behaviour can be described by the Ramberg-Osgood law, according to which the strain is the sum of its elastic, ε^E, and its plastic component, ε^P, as follows:

$$\varepsilon = \varepsilon^E + \varepsilon^P = \frac{\sigma}{E} + \left(\frac{\sigma}{K'}\right)^{\frac{1}{n'}} \quad (3)$$

where K' is the cyclic strength coefficient and n' the cyclic hardening exponent. Therefore, it can be derived that:

$$W^E_{CC}(\sigma_0) = \frac{1}{2}\sigma_0 \cdot \varepsilon^E_0 = \frac{\sigma_0^2}{2E} \quad (4a)$$

$$W^P_{CC}(\sigma_0) = \frac{1}{1+n'} \cdot \sigma_0 \cdot \varepsilon^P_0 = \frac{1}{1+n'} \cdot \sigma_0 \cdot \left(\frac{\sigma_0}{K'}\right)^{\frac{1}{n'}} \quad (4b)$$

ε^E_0 and ε^P_0 being the elastic and the plastic component of the strain at the fatigue knee, respectively. Therefore, the high cycle fatigue strength of bluntly notched components can be performed by comparing the elastic-plastic SED, evaluated at the notch root, to the elastic-plastic SED of plain material. Since elastic-plastic SED evaluation on components requires non-linear finite element analyses and often structural designers are more familiar with stress- than energy-based approaches, in this paper the nominal stress- and the point stress- based approaches are revised to take into account the presence of plasticity at the fatigue knee, on the basis of energy considerations.

The nominal stress approach is based on the assumption that a notched component reaches the fatigue knee when an effective stress acting at the notch tip, σ_{eff} is equal to the fatigue limit of plain material. Usually, σ_{eff} is evaluated by starting from the evaluation of a nominal stress σ_{nom} properly increased by using the fatigue notch factor K_f [13,14]. In general, K_f is an experimental parameter defined as:

$$K_f = \frac{\text{fatigue limit of unnotched specimen}}{\text{fatigue limit of notched specimen}} \quad (5)$$

It is well known that when the K_f value is not available, it can be estimated by starting from the theoretical stress concentration factor K_t and the notch sensitivity index q, according to Peterson [13]:

$$q = \frac{K_f - 1}{K_t - 1} \quad (6)$$

The values of q vary from q=0 for no notch effect (K_f=0) to q= 1 for the full theoretical effect (K_f=K_t).

In the classical approach, the behaviour of plain material is supposed to be fully elastic, at least at the fatigue limit. In order to overcome this limitation, in this paper an equivalent fatigue limit $\sigma_{0,eq}$ is introduced such that the SED experimentally measured at the fatigue knee for plain material is equal to that of an equivalent fully elastic plain material, following the original idea proposed by Glinka [15,16], regarding notched components:

$$W_{LE}(\sigma_{0,eq}) = W_{CC}(\sigma_0) = W^E_{CC}(\sigma_0) + W^P_{CC}(\sigma_0) \quad (7)$$

where $W_{LE}(\sigma_{0,eq}) = \dfrac{\sigma_{0,eq}^2}{2 \cdot E}$.

Let us to define a coefficient of plasticity K_p, as follows:

$$K_p = \frac{\sigma_{0,eq}}{\sigma_0} = \sqrt{\frac{W_{LE}(\sigma_{0,eq})}{W^E_{SCC}(\sigma_0)}} = \sqrt{1 + \frac{2}{1+n'} \cdot R_p} \quad (8a)$$

being $R_p = \varepsilon^P_0 / \varepsilon^E_0$ the plasticity ratio evaluated at the fatigue limit. The K_p coefficient versus R_p of plain material is plotted in Fig. 1 for different values of n'. Other approaches are available in literature to correlate elastic-plastic to linear elastic analyses. Of these, Neuber's rule is the most widely used and the relevant K_p versus R_p trend is plotted in the same figure, according to Eq. 8b

$$K_p^N = \sqrt{1 + R_p} \quad (8b)$$

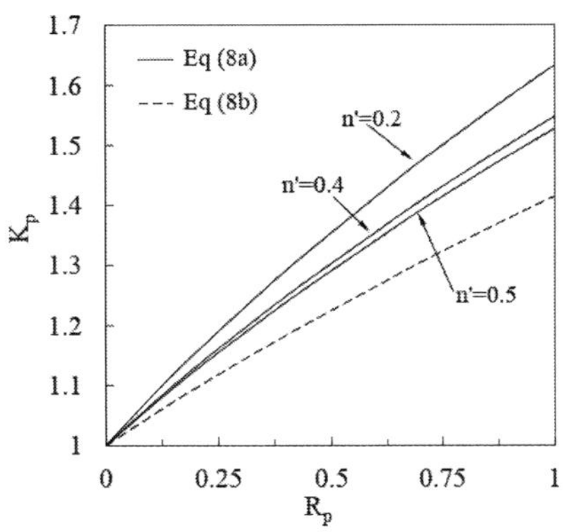

Fig. 1. K_p versus R_p trend for different values of cyclic hardening exponent.

Therefore, the classical approach can be extended to materials characterised by plasticity at the fatigue knee, by imposing that a notched component reaches the fatigue knee when the effective stress σ_{eff} is equal to $\sigma_{0,eq}$. In the case of full theoretical effect q=1 (i.e. K_f=K_t) and by using the net-section stress as nominal stress, it follows:

$$\sigma_{eff} = \sigma_{0,eq} \Rightarrow K_{tn} \cdot \sigma_{0,net} = K_p \cdot \sigma_0 \quad (9)$$

where $\sigma_{0,net}$ is the nominal stress to be applied to the notched component to reach the fatigue knee. Therefore, a new strength reduction factor can be defined which takes into account for the plasticity of plain material at the fatigue knee:

$$K_{fn,p} = \frac{K_{tn}}{K_p} \quad (10)$$

Eq. 10 being valid when $K_{tn} > K_p$. On the contrary, $K_{fn,p}=1$ when $K_p < K_{tn}$, since in this case one can suppose that the fatigue knee of notched component is equal to that of plain material. When a partial theoretical effect is considered (i.e $0 < q < 1$):

$$K_{fn,p} = \frac{1+q\cdot(K_{tn}-1)}{K_p} \quad (11)$$

Once defined the fatigue knee of notched component, the S-N curve can be completely defined, assuming that in the low cycle fatigue regime the notch effect vanishes, thanks to the diffuse plasticity, that can completely involve the net-section. Usually, in the classical nominal stress approach the stress-life curve is defined for a number of cycles to failure starting from 1000 cycles (e. g [17]). In view of this, it is assumed that the fatigue strength of notched component is equal to that of the plain material evaluated at N=1000 cycles.

In the '90s, different approaches were proposed to evaluate the fatigue strength of cracked or severely notched components, based on Linear Elastic Fracture Mechanics concepts. The underlying idea was the possibility to evaluate the fatigue limit of components weakened by sharp notches, by means of a linear elastic finite element analysis and comparing the stress value acting at a given distance (or averaged on a suitable distance, area or volume) to the fatigue limit of plain material [18,19]. Since this approach is based on the linear elastic analysis of the stress state of notched component, it can be adopted as long as the plain material behaviour is elastic. In order to overcome this limitation, in this paper the same procedure adopted in the case of nominal stress-based method will be adopted to revise the point stress approach. Therefore, at the fatigue knee we have:

$$\sigma_{point} = \sigma_{0,eq} \Rightarrow K_{tn,point}\cdot\sigma_{0,net} = K_p\cdot\sigma_0 \quad (12)$$

or alternatively stated:

$$\frac{\sigma_0}{\sigma_{0,net}} = \frac{K_{tn,point}}{K_p} \quad (13)$$

where σ_{point} is the stress acting at a given distance x_0 from the notch tip, calculated according to [19], as follows:

$$x_0 = \frac{a_0}{2} = \frac{1}{2\pi}\left(\frac{\Delta K_{th}}{\Delta\sigma_0}\right)^2 \quad (14)$$

where a_0 is the El Haddad et al. [20] material parameter length and ΔK_{th} the threshold range value of the mode I stress intensity factor.

The stress- [21,22] as well as the strain- [23] point approach has been extended to estimate the S-N curve from low to high cycle fatigue regime. The more simple and diffused approach is to assume x_0 as a constant and to define fatigue design curves on the basis of a statistical analysis of a significant number of experimental results. In the authors' knowledge, in literature are not available point stress-based approaches developed to evaluate the point-stress design fatigue curve by starting from the fatigue life curve of plain material. In this paper, it is proposed to extend to local approaches, the methodology usually adopted in the case of ductile materials. Therefore, it is assumed that for a fatigue life equal to 1000 cycles, the fatigue strength of notched specimens is equal to that of a plain specimen at the same number of cycle. In fact, the fatigue strength of notched component is driven by the net-stress amplitude, since the notch-effect can be assumed ineffective due to the presence of diffused plasticity. Accordingly, the relevant point stress value can be evaluated simply multiplying the fatigue strength of plain specimen evaluated for N=1000 cycles by $K_{tn,point}$.

3 Material and specimen's geometry

Constant amplitude, fully-reversed, stress-and strain controlled fatigue tests were carried out on plain specimens, machined from 6-mm-thick, hot rolled AISI 304L stainless steel plate. The mechanical properties of tested material are listed in Table 1, where R_s and R_m is the static and the proof strength, respectively. Details of specimens' geometry and test conditions are reported in [12]. The notched specimens were characterised by a 3D stress concentration factor referred to the net section, K_{tn}, equal to 2.25, 3.00 and 3.70, for r_n=8 mm, 5 mm and 3 mm, respectively.

Fig. 2a reports all fatigue data referred to plain specimens along with the mean stress-life curve, the 10-90% survival probability curves, the inverse slope k and the scatter index T_σ defined as $\sigma_{a,10\%}/\sigma_{a,90\%}$. Finally, the constant amplitude material fatigue limit σ_0=225 MPa, evaluated by means of a stair case procedure at 10 million cycles, is reported.

The Ramberg-Osgood law was adopted to define the material cyclic stress-strain curve. Accordingly, K' and n' were calculated by fitting the experimental data shown in Fig. 2b, obtained by carrying out the strain controlled fatigue tests [24].

Having E, K' and n', according to Eq. (3), the elastic ε^E and plastic ε^P component of strain at the fatigue knee resulted equal to $1.16\cdot10^{-3}$ and $1.24\cdot10^{-3}$, respectively. Therefore from Eq. 4a and 4b one can calculate $W_{CC}^E(\sigma_0)$= 0.130 MJ/m^3 and $W_{CC}^P(\sigma_0)$ =0.194 MJ/m^3. Finally, Eq. 8a gives K_p=1.58. It is worth nothing that, at the fatigue knee, $W_{CC}^{P}(\sigma_0)$ is 1.49 times higher than $W_{CC}^{E}(\sigma_0)$, showing that the plasticity plays a significant role on the fatigue behaviour of AISI 304L stainless steel. The fatigue data relevant to the notched specimens are shown in Fig. 3a, 3b and 3c in terms of net stress amplitude $\sigma_{a,net}$, together with the reference fatigue strength evaluated at N_0=160000 cycles and the experimental fatigue notch factor $K_{fn,exp}$. It was found σ_0= 186.4, 114.5 and 121.4 MPa, for r_n=8 mm, 5 mm and 3 mm, respectively, i.e $K_{fn,sper}$ = 1.21, 1.97 and 1.85, that are significantly lower than the relevant K_{tn} value. Values of $K_{fn,sper}$ are listed in Table 2.

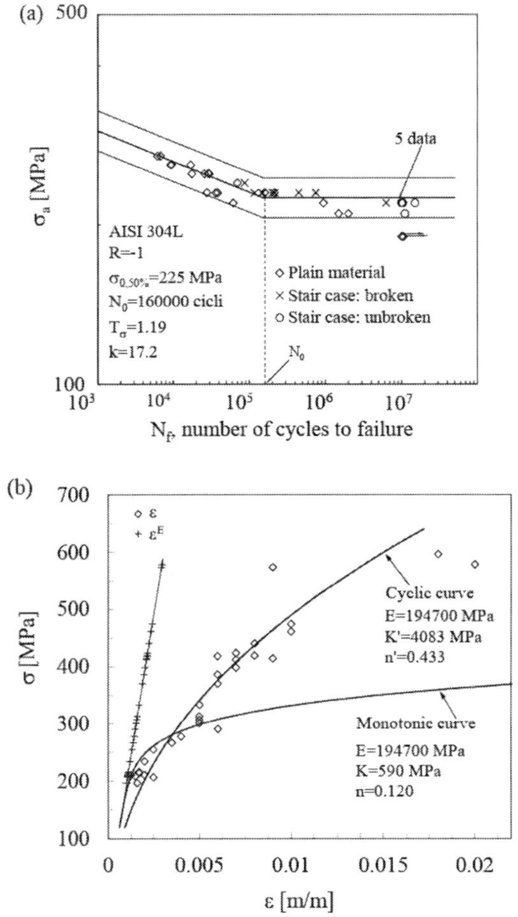

Fig. 2. Stress-life (a) and cyclic stress-strain curve of AISI 304L (b).

Table 1. Mechanical properties of AISI 304L stainless steel.

E [MPa]	R_s [MPa]	R_m [MPa]	K' [MPa]	n'	σ_0 [MPa]
194700	327	690	4083	0.433	225

Table 2. Comparison between the experimental and the calculated values of K_{fn}, taking into account plasticity of plain material.

r_n [mm]	$K_{fn,exp}$	q	K_{tn}	$K_{fn,p}$*	$K_{fn,p}$**
3	1.85	0.94	3.70	2.34	2.25
5	1.97	0.97	3.00	1.90	1.86
8	1.21	0.98	2.25	1.42	1.41

*Eq.10; **Eq. 11

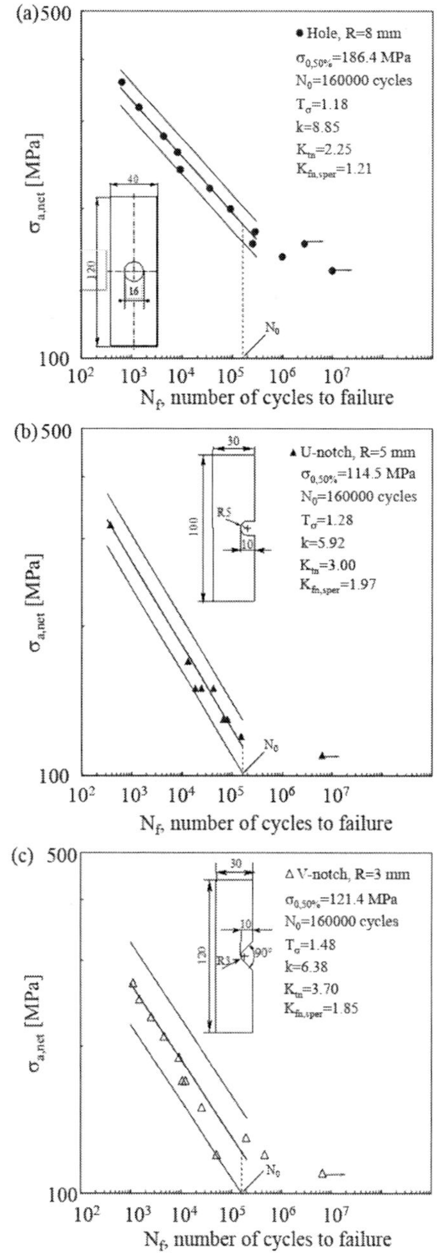

Fig. 3. Stress-life curves of r_n=8mm (a), r_n=5 mm (b) and r_n=3 mm (c) notched specimens.

4 Fatigue data synthesized in terms of elastic-plastic SED and modified

nominal stress- and point stress-approaches

Concerning the plain material, the elastic-plastic SED can be easily calculated from Eqs 4, by substituting σ_0 with the experimentally applied stress amplitude σ_a. On the contrary, for notched specimens, 2D plane-strain finite element analyses were performed by using the 8-node PLANE183 element of Ansys® commercial software, by using an element size at the notch tip equal to $2 \cdot 10^{-2}$ mm. The cyclic curve shown in Fig. 2b was implemented, along with Von Mises plasticity rule and isotropic hardening behaviour. Fig. 4 shows the fatigue data reanalysed in terms of W_{cc}, which was evaluated at the notch tip in the case of notched specimens. One can see that at the fatigue knee data relevant to plain and notched specimens can be rationalised with a higher level of accuracy by using W_{cc}, rather than $\sigma_{a,net}$.

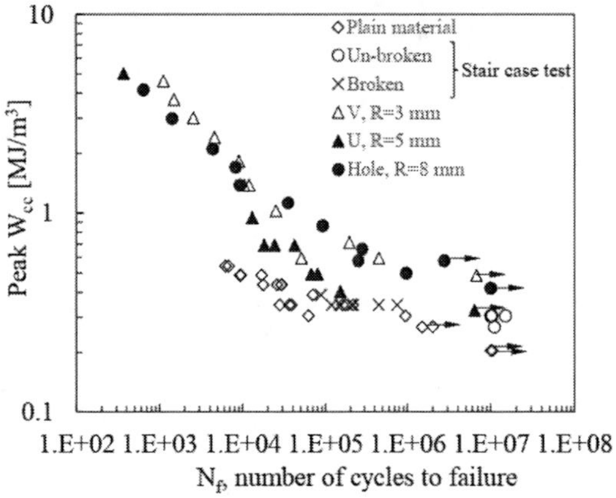

Fig. 4. Fatigue data reanalysed in terms of peak elastic-plastic Strain Energy Density.

Concerning the material analysed in this paper, the values of K_{fn} calculated according to [13] were found equal to 2.17, 2.90 and 3.55, for r_n=8, 5 and 3 mm, respectively, and they are summarised in Table 2. The $K_{fn,p}$ values evaluated according to Eq. 10 and 11 are listed in the two last columns of this table. Moreover, as it can be expected considering the notch radii analysed in this paper, negligible differences were found comparing the $K_{fn,p}$ values calculated considering partial- and full-theoretical notch effect. Figure 5a-c shows the fatigue curves relevant to r_n=3, 5 and 8 mm, respectively, obtained by using the revised stress approach, compared to that calculated according to Peterson [13], here named as "classical approach". Once evaluated the fatigue strength at N_0=160000 cycles, the fatigue strength at 1000 cycles was assumed equal to that of a plain specimen at the same number of cycles. Therefore, considering the S-N curve shown in Fig. 1a, it can be calculated equal to 301.4 MPa. Fig. 5 shows that a satisfactory agreement with the experimental data can be achieved by using the proposed modified stress-based approach.

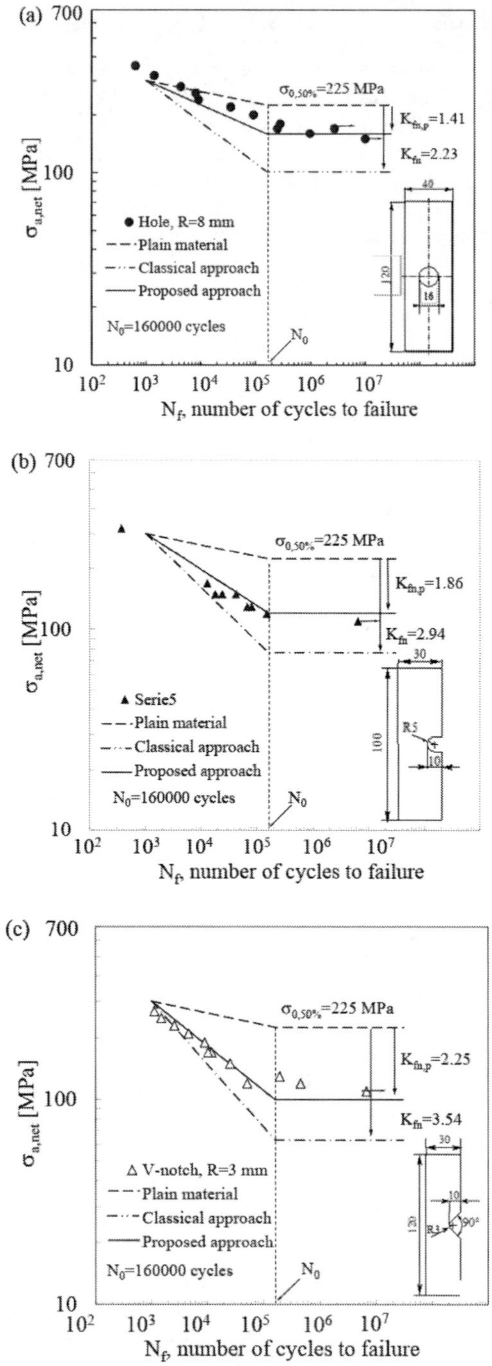

Fig. 5. Fatigue data reanalysed in terms of net-stress amplitude compared to the classical approach (Eq. 5) and the proposed approach (Eq. 10), for r_n=8 mm (a), r_n=5 mm (b) and r_n=3 mm (c).

The application of point stress method requires the evaluation of the distance x_0 [19]. Having ΔK_{th} = 8.69 $MPa \cdot \sqrt{m}$ [25], on can calculate x_0=0.06 mm, from Eq. 14. Figure 6 shows the experimental results reanalysed

 https://doi.org/10.1051/matecconf/201816514009

in terms of point stress σpoint. It is worth noting that the fatigue knee of notched specimens is significantly different from σ_0, while it is in very good agreement with $\sigma_{0,eq}$ value.

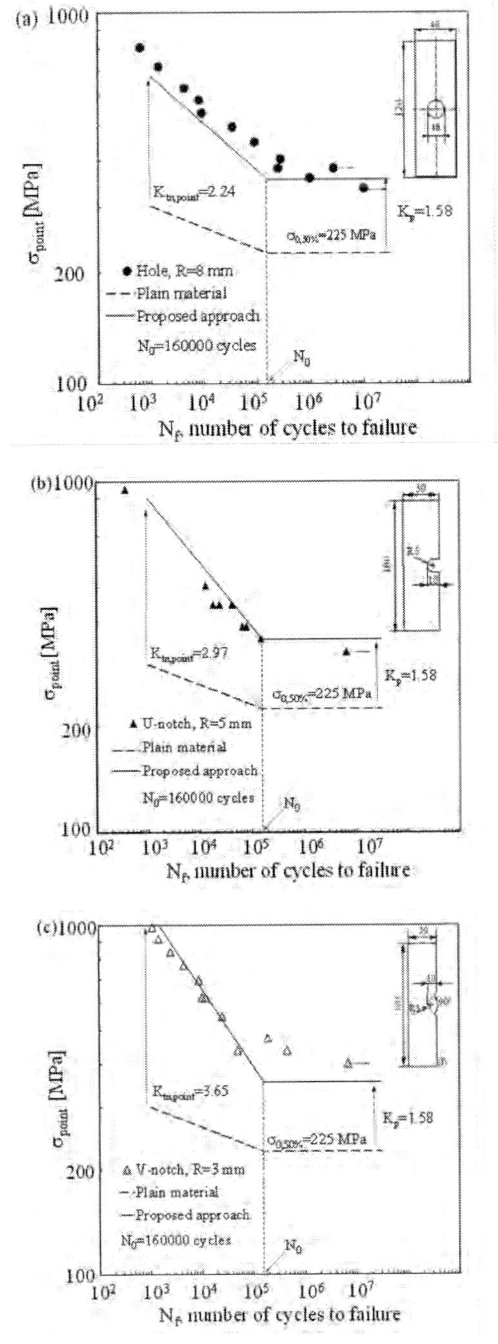

Fig. 6. Fatigue curves evaluated according to the proposed approach (Eq . 13) compared to the experimental data, reanalysed in terms of point-stress, for r_n=8 mm (a), r_n=5 mm (b) and r_n=3 mm (c).

Finally, according to the theoretical background proposed above, the stress point-life curves were completely defined by evaluating the fatigue strength of notched component at N=1000 cycles, multiplying the fatigue strength of plain material evaluated at 1000 cycles (σ_a=301.4 MPa) by $K_{tn,point}$. The results are shown in Figure 6 and it can be noticed a very good agreement with the experimental data.

The proposed methodology was compared to the usual approach of defining a fatigue design scatter band, obtained by a statistical analysis of the experimental results reanalysed in terms of point stress, σ_{point}. Figure 7 shows the experimental results of Figure 3 reanalysed in terms of σ_{point}, the relevant mean curve, evaluated by assuming a log-normal distribution of the number of cycles to failure and the 10%-90% survival probability scatter band. It was found a reference fatigue strength evaluated at N_0=160000 cycles equal to $\sigma_{0,point}$=337.8 MPa, that is close to that evaluated by starting from the fatigue limit of plain material, corrected by means of the K_p coefficient ($\sigma_{0,point}$ =225·1.58=355.5 MPa). Moreover, one can see that the fatigue curves relevant the notched geometries fall in the experimental scatter band.

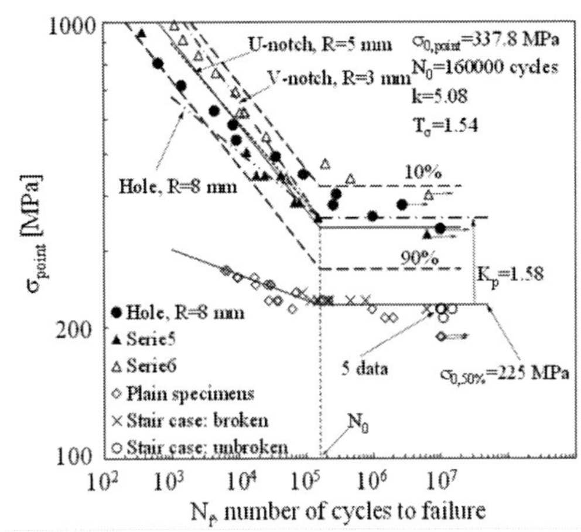

Fig. 7. Fatigue curves evaluated according to the proposed approach (Eq. 13) compared to the experimental scatter band defined by the statistical analysis of fatigue data reanalysed in terms of stress-point.

5 Conclusions

In this paper the elastic-plastic Strain Energy Density (SED) was adopted as fatigue index to successfully correlate fatigue data relevant to bluntly notched specimens and plain material made from ductile steels, characterised by the presence of plasticity at the fatigue knee. In general, in the case of notched components, the elastic-plastic SED requires non-linear finite element analyses. On the contrary, the elastic-plastic SED of plain material can be easily evaluated analytically. With the aim to provide structural engineers with a relative simple tool to make linear elastic calculations in the presence of notches, the classical nominal stress approach and the point stress method were revisited, on the basis of elastic-plastic SED considerations, saving the advantage of carrying out linear elastic calculations in the presence of notches. In view of this, a K_p parameter to account for plasticity at the fatigue knee was defined, on the basis of an equivalent fatigue limit, which assures that the elastic-plastic SED evaluated at the fatigue knee is maintained. Once defined the K_p parameter, the classical stress approach, based on the nominal stress concept, and the point stress method were modified. Finally, the proposed methodology was successfully applied to evaluate the fatigue curve of 6-mm-thick, hot-rolled AISI 304 L stainless specimens, weakened by blunt notches, having radius equal to 8, 5 and 3 mm, starting from the fatigue curve of plain material.

References

1. G.C. Sih GC, Int J Fracture, **10**, 305-321, (1974).
2. P. Lazzarin, R. Zambardi, Int. J. Fracture, **112**, 275-298, (2001).
3. Z. Yosibash, O. Adan, R. Schneck, H. Atlas H, Int J Fracture **122**, 47-64 (2003).
4. P. Lazzarin, F. Berto, Int J Fracture, **135**, 161-185 (2005).
5. C.E. Feltner, J.D. Morrow. Trans ASME Ser D J Basic Eng, **83**, 15-22, (1961).
6. J.D. Morrow, *International Friction, Damping and Cyclic plasticity*, ASTM STP 378, 45-84, American Society for Testing and Materials, Philadelphia, PA (1965).
7. G.R. Halford, J Mater, **19**, 3-18, (1966).
8. J. H. Kuang, Y:C. Chen, Eng Fract Mech, **53**, 571-80 (1996).
9. F. Ellyin, *Fatigue damage, crack growth and life prediction*. Chapman&Hall, (1997).
10. R.P. Skelton, T. Vilhelmsen , G:A.Webster, Int J Fatigue, **20**, 641-49, (1998)
11. N.W. Klingbeil. Int J Fatigue, **25**, 117-128, (2003).
12. G. Meneghetti, M. Ricotta, B. Atzori, Fatigue Fract Eng Mater Struct, **36**,1306-1322, (2013).
13. R.E. Peterson, Notch sensitivity. In: Sines G, Waisman JL, editors. *Metal fatigue*. New York: MacGraw-Hill;. p. 293–306, (1959).
14. M. Ciavarella, G. Meneghetti, Int J Fatigue, **26**, 289-298, (2004).
15. K. Molski, G. Glinka, Mater Sci Eng, **50**, 93-100 (1981)
16. G. Glinka, Eng Fract Mech, **22**, 485-508 (1985).
17. N.E. Dowling, Mechanical behaviour of materials, Pearson Prentice Hall, (2013).
18. K. Tanaka, Int J of Fracture, **22**, R39-R46, (1983).
19. D. Taylor, Int J Fatigue, **21**, 413-420 (1999)
20. M.H. El Haddad, T.H. Topper, K. N Smith, J of Eng Mater Tech (ASME Trans.) **101**, 42-45 (1979).
21. M. Ciavarella, F. Monno, Int J Fatigue, **25**, 1826-1837 (2006).
22. L. Susmel, D. Taylor D, Fatigue Fract Eng Mater Struct, **30**, 567-581 (2007).
23. S. Bentachfine, G. Pluvinage, J. Gilgert, Z. Azari, D. Bouam, Int J Fatigue, **21**, 421-430 (1999).
24. G. Meneghetti, M. Ricotta, Eng Fract Mech, **81**, 2-17, (2012).
25. G. Meneghetti, M. Ricotta, D. Rigon, *Proceedings of International Fatigue Conference*, Cambridge, UK, 3-5 July 2017.

Development of a generalized Kitagawa diagram for cast G21Mn5 steel

Manuel Schuscha[*], *Martin Leitner*, and *Michael Stoschka*

Christian Doppler Laboratory for Manufacturing Process based Component Design, Montanuniversität Leoben, Chair of Mechanical Engineering, Leoben, Austria

Abstract. An accurate assessment of cast steel components including cast imperfections minimizes production rejects and moreover facilitates lightweight design. In order to improve the fatigue assessment of cast steel structures exhibiting macroscopic imperfections, a generalized Kitagawa diagram based on the linear elastic fracture mechanical theory is presented. Therefore, single-edge-notch-bending crack propagation tests as well as extensive fatigue tests utilizing un-notched base material specimens are conducted. Both the stress intensity factor threshold range and the plain material fatigue limit act as basis to set-up the generalized Kitagawa diagram. In addition, fatigue tests under rotating bending as well as axial loading using round V-notched specimens with varying opening angles are additionally conducted for validation. It was observed that the notch-stress-intensity-factor (NSIF) calculation based on El-Haddad's material length leads to slightly non-conservative designs for this cast material. But if the intrinsic material length is calculated by Neuber's stress averaging approach or Peterson's microstructural length instead, the presented NSIF-based fatigue assessment reveals a sound comparability to the experimentally determined fatigue reference strength. Summing up, the presented generalized Kitagawa diagram provides a feasible engineering-applicable fatigue assessment tool incorporating varying imperfection opening angles and load conditions for G21Mn5 cast steel.

1 Introduction

The aspect of a lightweight as well as economical design affects the dimensioning of modern high-performance components. To overcome this issue, it is required to enhance manufacturing process dependent fatigue assessment tools for the deployed materials. Despite quite high standards in casting simulation and manufacturing process technologies, the occurrence of imperfections, such as shrinkage-based porosities as well as sand, slag, or non-metallic inclusion, is unavoidable in cast steel.

Several fatigue approaches like Murakami's √area concept [1] are suggested to properly assess such local imperfections, whereby recent studies [2, 3, 4] showed sound accordance with experimentally evaluated fatigue data. It should be noted that thereby, the fatigue dominating imperfection is characterized by the projected cross-section of the crack initiating defect, which is limited to comparably smaller defects with √area<1000 µm. Supplementary to such stress-based methods, the linear elastic fracture mechanics (LEFM) considers the local stress field around the crack tip. In case of cast steel components, shrinkage porosities tend to reach larger dimensions due to the higher contraction during cooling, which corresponds to the multiple of the material intrinsic length, generally known as the El-Haddad length [5]. Due to this significantly enhanced dimensions of the imperfections, the application of Murakami's √area concept is basically not valid anymore. Considering the size and spatial arbitrary shape of pores, one can treat flaws as cracks or internal notches, which enables the assessment of crack-like defects by LEFM and notch-like defects in a uniform way utilizing linear elastic notch fracture mechanics (LENFM).

Early studies [6, 7] observed that a linear elastic stress field in front of notch tip can be expressed in terms of plane problems, the so-called notch-stress-intensity-factor (NSIF), labelled as K_I^V for mode I and K_{II}^V for mode II: They generalize the stress intensity factor of a sharp crack toward the stress singularity at the notch tip for varying opening angles.

Considering a sharp V-notched component under tension mode I loading, the definition of the notch stress intensity factor K_I^V corresponding to Gross and Mendelson [7] equals to:

$$K_I^V = \sqrt{2\pi} \lim_{r \to 0} \left[\sigma_{\theta\theta}(r,\theta)\, r^{1-\lambda_I} \right] \quad (1)$$

Herein, r and θ equal the polar coordinates around the notch tip, and λ_I is the Eigenvalue for mode I loading [6]. Furthermore, one can express the mode I NSIF K_I^V more generalized as follows:

[*] Corresponding author: manuel.schuscha@unileoben.ac.at

https://doi.org/10.1051/matecconf/201816514010

$$K_I^V = \alpha_\gamma \sqrt{\pi a^\gamma} \sigma_g = \sqrt{\pi}\left(\alpha_\gamma^{1/\gamma} a\right)^\gamma \sigma_g \qquad (2)$$

Herein,

- α_γ is a geometry shape factor, which depends on the overall geometry of the component, the opening angle, and the loading type
- $\gamma = 1 - \lambda_I$ is the degree of the stress singularity for mode I loading
- a is the characteristic dimension, i.e. the notch or crack depth.

Originating from the applicability on sharp notches or re-entrant corners, Atzori et al. [8] extended the Kitagawa diagram to blunt notches with a notch root radius ρ and established a bridging between the defect sensitivity, referred to LENFM, and the notch sensitivity correlated to linear notch mechanics. Several studies [9-11] applied this approach mainly focusing on carbon steels and aluminum alloys. A sound conformity between the evaluated fatigue test results and the NSIF-concept regarding different notch acuities was achieved.

For sharply notched components, the fatigue strength is affected by their stress singularity exponent γ. Evaluating parts with varying opening angles φ leads to different stress singularities and therefore disables the possibility to compare the NSIF's directly as fatigue strength qualifier. Former studies [12, 13] proposed an advanced energy-based approach to estimate static and cyclic behavior of cracks and notches.

2 Numerical Analysis

Due to the spatial arbitrary shape of porosities, which are found in steel castings, it is intended to cover the fatigue assessment of different types of macroscopic defects in a uniform way. Therefore, three round specimens with circumferential notches were defined representing cast steel imperfections with varying opening angles. To correlate the fatigue behavior of crack-like defects as well as more globular shrinkage porosity, the following opening angles φ were defined: 45°, 90° and 135°. To obtain fatigue test results similar to steel castings with macroscopic imperfections, the notch depth was consistently set to 3 mm. It should be noted that the fatigue effective notch depth a_{eff} increases based on the overall specimen geometry and hence, additionally the applicable shape factor α_γ enhances as exemplified in Eq.(3). To facilitate the utilization of the NSIF concept, the notch root radius was set to ρ = 0.1 mm to ensure the condition ρ << a.

Numerical analyses of the aforementioned three specimen geometries were conducted as linear-elastic studies utilizing the software package Abaqus to evaluate the associated NSIF's. The material and simulation parameters, which were used to set-up the numerical model, are listed in Table 1.

Table 1. Material and simulation parameters

Parameter	Value
Young's Modulus [GPa]	210
Poisson's ratio [-]	0.3
Element type	CAX8 / CAXA84

At this point it should be pointed out that all stress values within this paper, except the plain fatigue limits and the notch stresses, are referred to the gross section of the specimens. Referencing to the gross stress enables a fatigue assessment without defining an actual net cross-section. This procedure is feasible especially in case of complexly-shaped components, at which a net cross-section is generally not simply definable. For example in case of a shrinkage porosity-afflicted component under axial loading a definition of the size of the supporting cross-section may not be simple. The reference to the gross section and an accordingly defined imperfection based geometry factor allows a more engineering-feasible straightforward assessment of local fatigue strength.

The numerical analyses of the specimen geometries were done as linear-elastic simulation runs with unified nominal gross stress amplitude of one Mega-Pascal. The notches were sketched with a notch root radius of zero to obtain the stress distribution and its singularity quite close to the notch root, which is shown in Fig.1. Due to the double logarithmic scale, the slope of the resulting curve represents the degree of singularity γ given by the specific notch opening angle. In case of the 45° notched specimen, the slope equals to 0.495. This is similar to the well-known singularity of a crack, which possesses a value of one-half. The calculation of the notch stress intensity factor K_I^V was conducted on the basis of Eq.(1). Hence, the related NSIF keeps constant over the radial distance r from the notch tip.

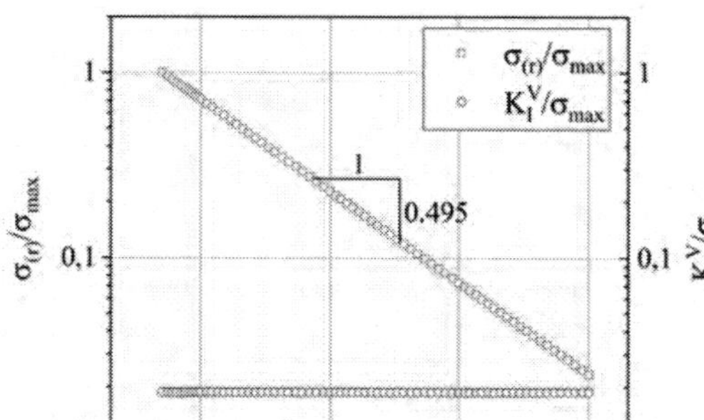

Fig.1 Normalized Stress and KIV relative to the notch root

Furthermore, the mode I notch stress intensity factors for the remaining opening angles under bending as well as for axial loading were calculated in the same manner. The resulting NSIF factors are listed in Table 2.

Table 2. Mode I NSIF's for bending and axial loading related to a unified nominal gross stress

Opening Angle	NSIF K_I^V/σ_g [mm$^\gamma$]	
	Bending	Axial Loading
45°	6.02	4.80

90°	6.61	5.22
135°	7.80	5.89

Moreover, the specimen's notch stress intensity factors can be utilized to calculate the effective notch depth a_{eff} by means of Eq.(2), which can be understood as the equivalent size of a crack in an infinite plate, based on the derived NSIF value. In regard to the calculation of a_{eff} for different opening angles φ, it is necessary to apply the corresponding degree of stress singularity γ for each notch shape. In addition, the shape factor α_γ, which depends on the overall sample geometry, the load case, and the notch shape, can be directly deduced from a_{eff}, as exemplified in Eq.(3).

$$a_{eff} = \alpha_\gamma^{1/\gamma} a \quad (3)$$

Finally, the corresponding effective notch sizes a_{eff} are determined for each load case and opening angle based on the acting NSIF-values. The results are summarized in Table 3.

Table 3. Mode I NSIF's for bending and axial loading related to a uniform nominal gross stress

Opening Angle	a_{eff} [mm]	
	Bending	**Axial Loading**
45°	11.82	7.48
90°	17.95	10.68
135°	94.18	39.79

The obtained values of a_{eff} significantly increase with the notch opening angle. Hence, both the notch depth as well the notch opening angle affect the endurable fatigue strength.

Based on these results, the fatigue strength for the three samples with varying notch acuity can be assessed in terms of the generalized Kitagawa diagram. The location on the abscissa is defined by the effective notch depth of each specimen geometry. In order to estimate the fatigue limit of the specimens, experimental crack growth tests ought to supply the materials threshold value, which can be further used to determine the intrinsic El-Haddad value for a specific load ratio. The ratio of the NSIF values compared to the $K^V_{I,th}$ supports a proper estimation of the specimen's fatigue limit.

Concluding the numerical analyses, the following procedure has to be maintained to obtain proper NSIF-values for fatigue assessment purposes.

- A very fine mesh, down to 10^{-4}-10^{-5} mm, should be utilized to properly consider the notch stress singularity.
- To obtain the NSIF value, the notch has to be modelled without a notch tip radius.
- The evaluation of the NSIF uses the principal stress component, which is in agreement with the axial specimen axis.

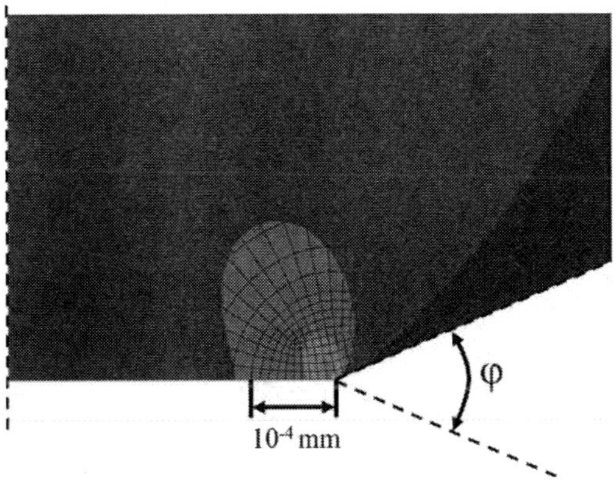

Fig.2 Mesh size at the notch root of the axisymmetric model

3 Development of the generalized Kitagawa Diagram

In contrast to the Kitagawa-Takahashi diagram for crack-like defects, the generalized form enables the application for open notches. The generalized Kitagawa diagram (GKD) can be divided in three basic sections, which feature different fatigue approaches dependent on the effective imperfection size.

The first section, located on the left side of the diagram below the intrinsic material length a_0 [5, 14], generally known as the El-Haddad length, can be defined as the short crack region. In case of a GKD, the intrinsic material length of V-notches is defined similarly to El-Haddad-Smith-Topper length parameter as the intersection of $\Delta\sigma_{g,th}$ and $\Delta\sigma_0$ [15].

$$a_0^V = \left(\frac{1}{\sqrt{\pi}} \frac{\Delta K^V_{I,th}}{\Delta\sigma_0} \right)^{1/\gamma} \quad (4)$$

A defect of this length has supposedly no characteristic impact on the component's fatigue resistance. Therefore, the short defect region, strictly assuming no further short crack growth during cyclic loading, exhibits an endurance limit as characteristic fatigue strength. If the effective defect increases, the GKD is controlled by the LENFM and is therefore represented by a straight line in a double-logarithmic scale. The area on the right hand side of the diagram illustrates the third section, which is dominated by the linear elastic notch mechanics in case of extremely huge imperfections. If a defect size is greater than the intersection point a^*, the LENFM is no longer applicable to assess the fatigue strength properly. This intersection point is defined by the geometric correlation:

$$a^* = K_t^2 \cdot a_0^V \quad (5)$$

https://doi.org/10.1051/matecconf/201816514010

Figure 3 provides an overview about the three basic regions of the generalized Kitagawa diagram.

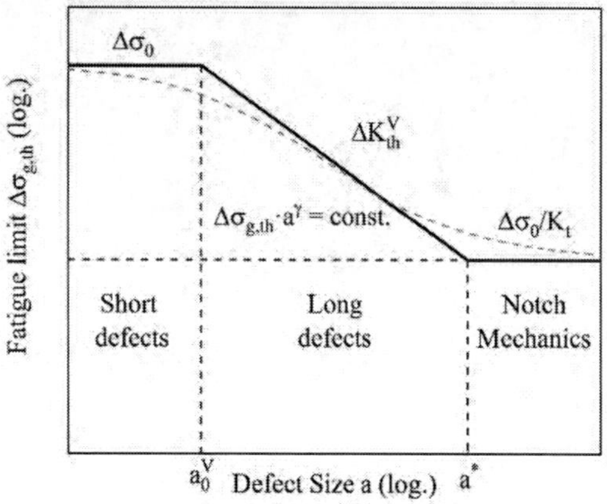

Fig.3 Structure of the generalized Kitagawa diagram (GKD)

Several parameters, e.g. notch opening angle, stress concentration factor and load ratio, affect the shape of the GKD. The material's plain fatigue limit influences the short defect region in case of $\Delta\sigma_0$ and the intrinsic material length as stated in Eq.(4). Based on the opening angle φ and Williams' equation, the Eigenvalue λ and the degree of singularity γ can be determined, which affect both the position of the intersection point on the left side as well as the slope of the LENFM section. An increase of the opening angle φ leads to a lowered stress singularity as well as to larger values of a_0^V. Furthermore, the slope of the LENFM region decreases with rising φ. In the notch mechanics applicable design region on the right hand side of the diagram, it is worth mentioning that the stress concentration factor K_t affects the position of the transition point a^*. A comparably high notch acuity leads to a high stress concentration factor and therefore, the value of a^* is shifted towards larger defect sizes.

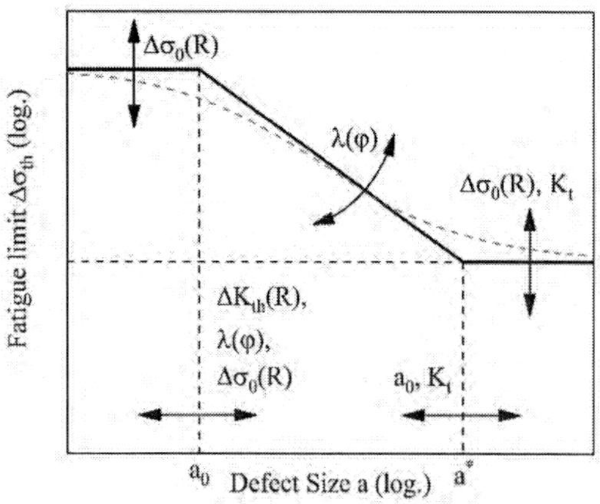

Fig.4 Influencing parameters on the GKD

To draw the generalized Kitagawa diagram, two base material parameters need to be determined first, namely the plain fatigue limit $\Delta\sigma_0$ and the threshold stress intensity factor range ΔK_{th}. The plain fatigue resistance can be obtained by testing un-notched specimens under different load cases and load ratios. Concerning the determination of the long crack threshold, the crack growth behavior of single-edge notched bending (SENB) specimens can be assessed for same load ratios as in case of cyclic testing un-notched specimens. Based on these two material parameters, the El-Haddad's intrinsic material length a_0 can be estimated.

4. Fatigue testing

The alternating cyclic tests were performed on a rotating bending test rig as well as on an axial fatigue test machine. The corresponding load stress ratio is set to a value of $R = -1$. The abort criterion of the fatigue tests is total burst failure in the finite life region or defined as a run-out if ten million load cycles are reached. It should be noted that no endurance limit is feasible within this limited load cycle range [16]. The assessment of the test results was conducted by the ASTM E739 [17] and the arcsin√P [18] approach for the finite-life and high-cycle fatigue region, respectively.

 https://doi.org/10.1051/matecconf/201816514010

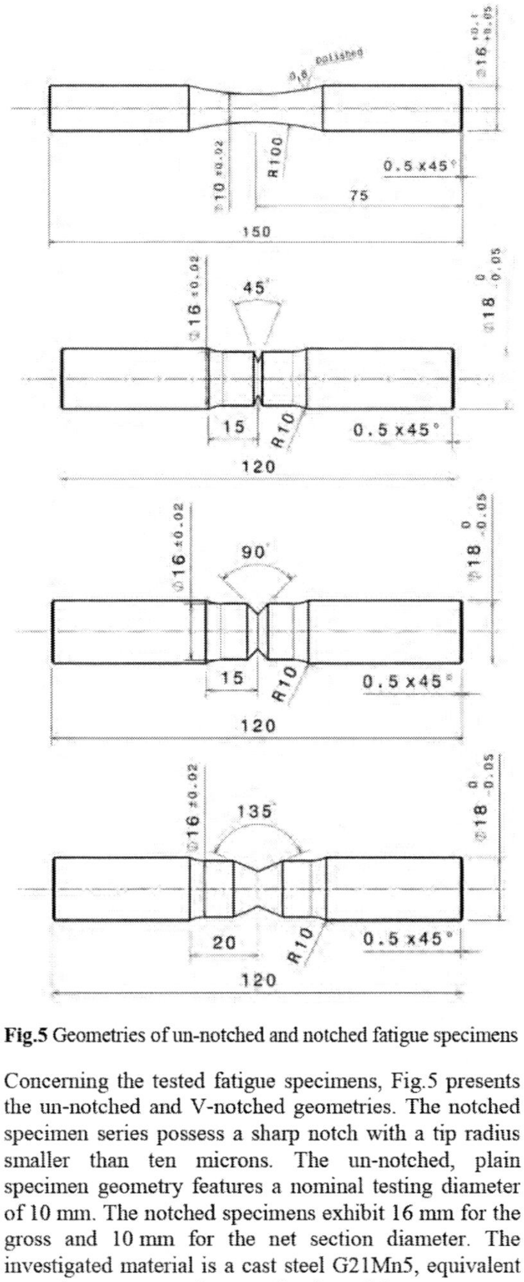

Fig.5 Geometries of un-notched and notched fatigue specimens

Concerning the tested fatigue specimens, Fig.5 presents the un-notched and V-notched geometries. The notched specimen series possess a sharp notch with a tip radius smaller than ten microns. The un-notched, plain specimen geometry features a nominal testing diameter of 10 mm. The notched specimens exhibit 16 mm for the gross and 10 mm for the net section diameter. The investigated material is a cast steel G21Mn5, equivalent to grade ASTM 352, in normalized condition.

Table 4. Chemical composition of the evaluated cast steel G21Mn5 in weight percent

C	Si	Mn	P	S
0.21%	0.47%	1.1%	0.01%	0.004%

In addition, the microstructure of the heat-treated cast steel is investigated. A three per cent alcoholic nitric acid is used as etchant.

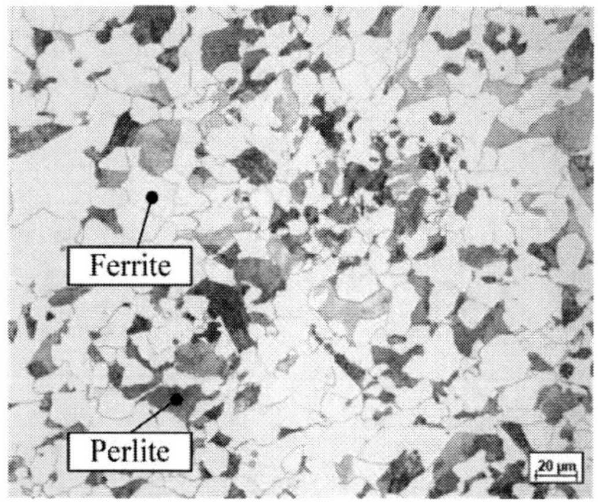

Fig.6 Microstructure of normalized G21Mn5

The perlitic islands in Fig.6 occur quite uniformly distributed within the ferritic matrix in metallographic sections taken out of fatigue test samples.

Regarding the rotating bending fatigue test results, the evaluated S/N-curves reveal an increase in fatigue limit with an enhanced notch opening angle. This behavior is expected as an increasing notch stress intensity factor reduces the nominal gross section based fatigue strength. The statistically evaluated high-cycle fatigue strength values of the 45° and 90° notched V-specimen exhibit only a fraction of about 75% to 80% compared to the 135° notched samples. Furthermore it can be stated that a rising stress concentration factor leads to a narrower scatter band of the results as well as an increase of the slopes in the finite-life region. As the investigated material is a cast steel thus featuring imperfections such as shrinkage based micro-pores in the base material, the second slope of the S/N curves in the high-cycle fatigue region was set to a decrease of ten percent stress level per decade load cycles [19]. Moreover, a direct comparison of the results indicates that the fatigue strength relation between the different specimens is non-linear, despite an equidistant net section. The difference among the 45° and 90° geometries is smaller than those two compared to the 135° samples. This tendency is valid both for the fatigue limits as well as the slope and the scatter band values. It deduces that the notch opening angle has a severe effect on the gross section based fatigue strength. Fig.7 depicts an overview of the test results of the rotating bending test series for all three specimen types. All fatigue test values are normalized to the statistically evaluated high-cycle fatigue strength at ten million load-cycles for the 135° specimen at a survival probability of 50 %.

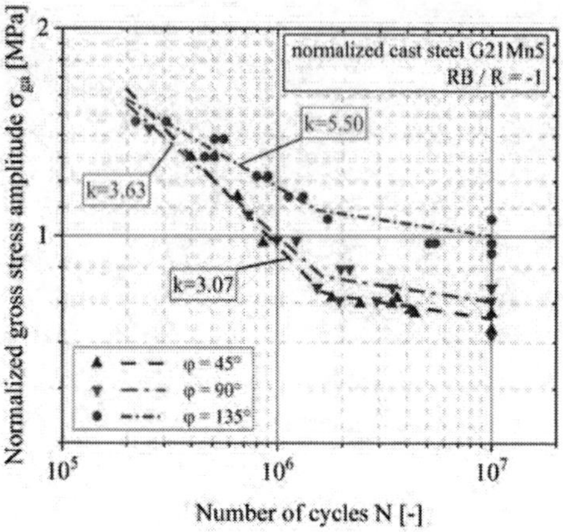

Fig.7 S/N curves of the rotating bending test

Subsequently, light optical microscopy (LOM) and scanning electron microscopy (SEM) were conducted on specimens under bending as well as under axial loading to analyze the location crack initiation. Fracture surface analysis of the V-notched specimens showed an equally distributed circumferential crack initiation for the entire test series. Fig.8 and Fig.9 illustrate the fracture surfaces of two rotating bending samples with 90° opening angle, tested at different load-levels.

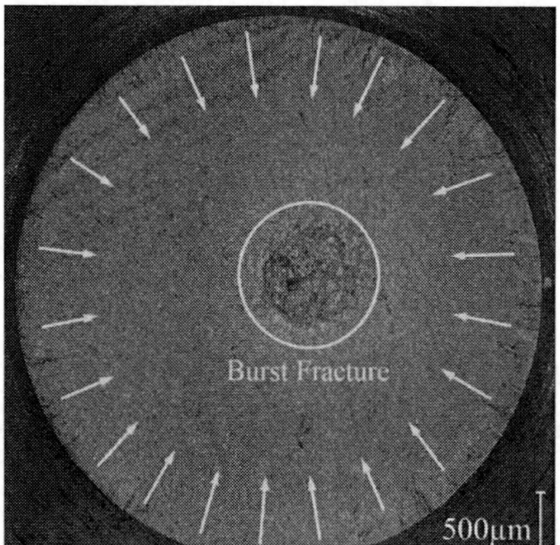

Fig.8 Rotating bending specimens: σ_{norm}=0.81; N=3·10^6

Due to the quite high notch acuity both specimen show similar behavior regarding crack initiation. Several microscopic crack initiation planes can be found by scanning electron microscopy (SEM), which coalesce during further cycling loading. In addition, some minor defects are recognizable on the fracture surfaces. Despite the occurrence of such local shrinkage porosities, the rotating bending and axial loading test results show no significant influence of these small defects on the high-cycle fatigue endurance limit of notched specimen.

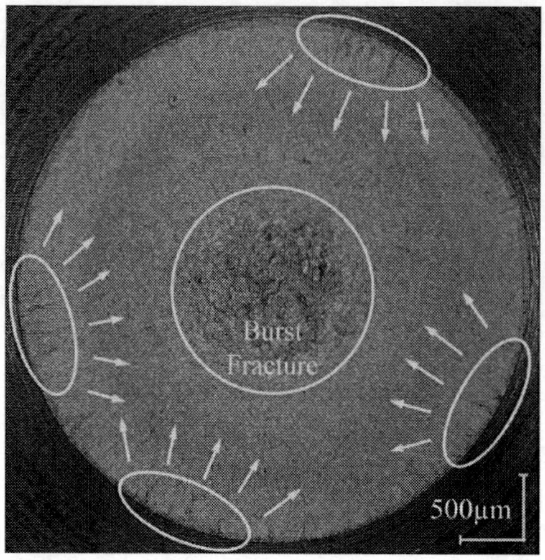

Fig.9 Rotating bending specimen: σ_{norm}=1.43; N=2.5·10^5

A scanning-electron-microscopy image of a small defect found at an un-notched rotating bending specimen is shown in Fig.10. However, it is recognizable by fracture surface analysis that these pores do not majorly affect the short crack propagation phase and hence, no significant impact on the lifetime is given.

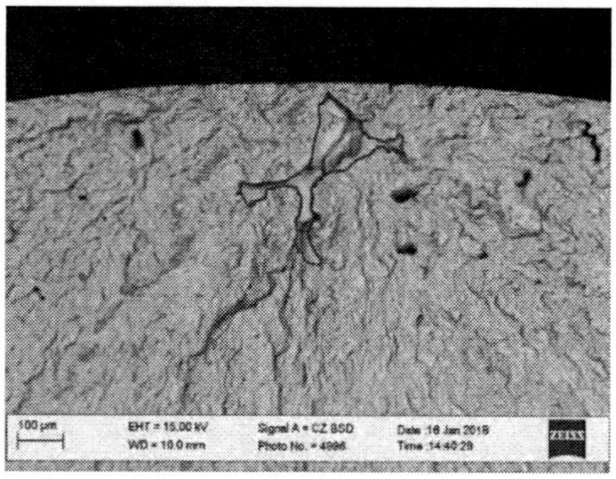

Fig.10 SEM image of an defect (un-notched RB): σ_{norm}=1.08; N=1.9·10^6

On the contrary to notched specimen, the test results as well as the fracture surfaces of the un-notched, axial loaded specimens exhibit a pronounced effect on bulk fatigue strength by shrinkage porosity. The uniform stress distribution under axial loading in case of un-notched specimens features the crack growth of defects. For example, the specimen displayed in Fig.11 failed at N=1.7·10^6 load cycles even though the load level is below the evaluated endurance limit of the cast steel. As the goal of the un-notched test series is to determine the plain high-cycle fatigue strength, test results with such huge casting imperfections are excluded from the statistical evaluation.

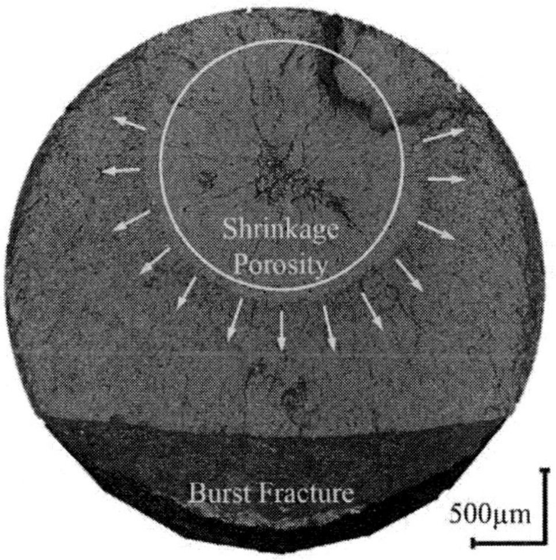

Fig.11 Critical defect within un-notched specimens

Summing up, the occurrence of shrinkage porosity is unavoidable in cast materials, especially in cast steel due to the high casting temperature and subsequent contraction of the material during cool-down. As long as the equivalent square defect length is below one millimeter, the application of Murakami's √area approach can be easily utilized, but for larger imperfections or even networks of shrinkage porosity the NSIF-values may be used to gain insight into the local fatigue strength of arbitrary shaped defects.

5. Discussion

The assessment of the fatigue tests including plain and V-notched specimens as well as the crack growth tests were utilized to set-up the generalized Kitagawa diagram for the investigated G21Mn5 cast steel. It was found that the calculation of the material's intrinsic length a_0 based on the El-Haddad-Smith-Topper equation leads to a value of about one-hundred and eighty microns. This would lead to a non-conservative deviation of the NSIF-calculation based fatigue limits compared to the experimentally obtained endurance fatigue strength at ten million load cycles. This discrepancy persists approximately constant for all three evaluated notch geometries and both load cases and ranges up to about thirty percent in fatigue strength. It should be mentioned that this deviation can be traced back to a change in stress intensity threshold value of about one to a square-root of two.

Due to the alternating testing implying a stress ratio of R=-1, crack closure effects may occur during the crack initiation and crack propagation phase [20]. This effects builds up during the growth of the crack; hence, the effect enhances from short to long crack growth [21]. Therefore, the effective threshold stress intensity factor for short cracks is lower than the threshold value for long cracks. This behavior is known as the crack resistance (R)-curve. For the determination of the effective SIF, the 45° results were taken to recalculate the long crack threshold value and further on the El-Haddad length for sharp cracks is very close to the singularity of the V-notch. Therefore, the V-notch can be treated as a crack of the same length [22]. The results show an accurate, even slightly conservative estimation of the NSIF approach, which is illustrated in Fig.12. It should be mentioned that this recalculated intrinsic length matches exactly if El-Haddad's length is divided by a factor of two.

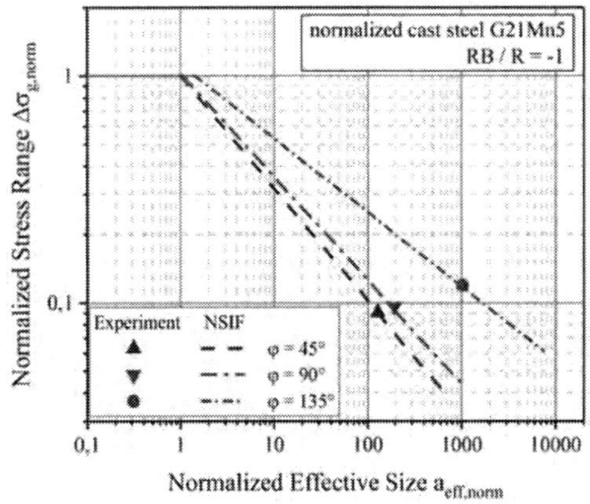

Fig.12 Generalized Kitagawa diagram in comparison with RB fatigue test results

A comparison of the NSIF-based calculated and experimentally evaluated fatigue strength values up to ten million load cycles reveals a sound agreement with a maximum scatter of only three percent.

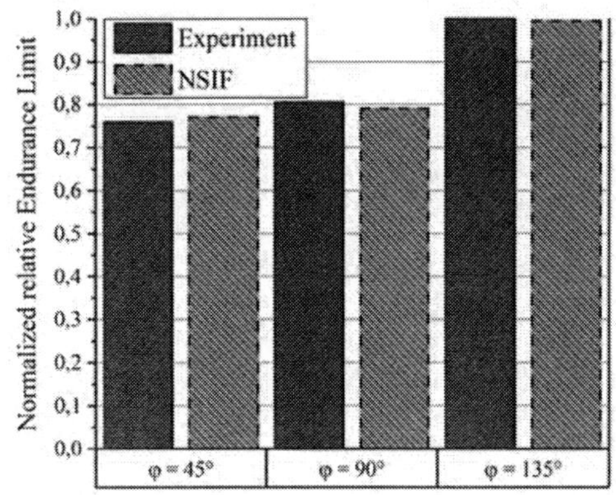

Fig.13 Normalized estimated vs. experimental results

Finally, it should be noted the critical distance theory [14] provides an alternative fatigue test sample based approach to determine the microstructural support length if S/N-curves with different notch factors are available. Applicable methods are the point and the line method which are both applied for the un-notched and notched fatigue results. Both approaches lead to a reduction of the proposed intrinsic length value to about ninety microns. This leads also to a sound match between the experimental fatigue results and the proposed mean fatigue limit deduced by the generalized Kitagawa-diagram.

6. Summary and Conclusions

Firstly, a numerical simulation of V-notched specimens with opening angles of 45°, 90° and 135° under bending and axial loading was conducted. Further on, the essential values for the application of the linear elastic notch fracture mechanics in terms of the notch stress intensity factor and an effective notch depth, were calculated. Fatigue test results under rotating bending and axial loading illustrate primarily a relation between the notch acuity, the opening angle and the resulting S/N-curve slope as well as the individual fatigue limits.

Secondly, the NSIF approach was utilized to calculate the essential values of the generalized Kitagawa-diagram for varying specimen geometries and notch acuities. It enables the study of notch designs on gross fatigue strength in a unified manner.

Thereby, the calculated values for G21Mn5 cast steel showed a non-conservative deviation of up to thirty percent compared to validating fatigue test results if El-Haddad's material length is used. This intrinsic material length leads to about one-hundred and eighty microns in case of calculation with the long crack growth threshold.

To overcome this discrepancy, Taylor's theory of critical distance, applied both for line- and point method, lead to a value of exactly half of El-Haddad's length. This heads to a reduced value of ninety microns as microstructural support length for G21Mn5, which matches the experimental data well. Since the degree of stress singularity of a crack and the V-notch with an opening angle of 45° are very close, a final recalculation of the stress intensity factor based on the notched specimen is supported. The results show a sound accordance of the estimated and experimentally determined fatigue limits for varying notch acuities. They prove the basic applicability of the NSIF-concept both for rotating bending and axial loading as fatigue design strength parameter for G21Mn5 cast steel.

Further work will focus on the effect of crack closure effects on the effective stress intensity factor range in terms of R-curves and especially its subsequent impact on generalized Kitagawa diagram based fatigue strength.

Summing up, the presented approach facilitating the generalized Kitagawa diagram for cast steel G21Mn5 demonstrates an unrestricted applicability of the NSIF concept and will be further investigated for porosity afflicted cast steel components.

The financial support by the Austrian Federal Ministry for Digital and Economic Affairs and the National Foundation for Research, Technology and Development is gratefully acknowledged.

References

1. Y. Murakami, *Metal Fatigue : Effects of Small Defects and Nonmetallic Inclusions* (2002)
2. Y. Murakami, Int. J. Fatigue, **41**, 2-10 (2012)
3. Y. Yamashita, Y. Murakami, Int. J. Fatigue, **93**, 406-414 (2016)
4. D. Krewerth, T.Lippmann, A. Weidner, H. Biermann, Int. J. Fatigue, **84**, 40-52 (2016)
5. M.H. El Haddad, T.H. Topper, K.N. Smith, Eng. Fract. Mech., **11**, 573-584 (1979)
6. M. L.Williams, J. Appl. Mech., **19**, 526-528 (1952)
7. B. Gross, A. Mendelson, Int. J. Fract., **8**, 267-276 (1972)
8. B. Atzori, P. Lazzarin, Int. J. Fatigue, **107**, 1-8 (2001)
9. B. Atzori, P. Lazzarin, G. Meneghetti, Fatigue Fract. Eng. Mater. Struct., **26**, 257-267 (2003)
10. B. Atzori, P. Lazzarin, Int. J. Fract., **118**, 271-284 (2002)
11. P. Lazzarin, F. Berto, Int. J. Fract., **135**, 33-38 (2005)
12. G. Glinka, Eng. Fract. Mech., **22**, 485-508 (1985)
13. P. Lazzarin, R. Zambardi, Int. J. Fracture, **112**, 275-298 (2001)
14. D. Taylor, Int. J. Fatigue, **21**, 413-420, (1999)
15. B. Atzori, P. Lazzarin, G. Meneghetti, Int. J. Fract., **133**, 61-87 (2005)
16. B. Pyttel, D. Schwerdt, c. Berger, Int. J. Fatigue, **33**, 49-58 (2011)
17. ASTM E739
18. D. Dengel, J. Materialwiss. Werkstofftech, **8**, 253-288 (1975)
19. C.M. Sonsino, Int. J. Fatigue, **29**, 2246-2258 (2007)
20. R. Pippan, A. Hohenwarter, FFEMS, **40**, 471-495 (2017)
21. B. Tabernig, R. Pippan, Eng. Fract. Mech., **69**, 899-907 (2002)
22. G. Meneghetti, S. Masaggia, World Foundry Congress, (2012)

Continuum damage mechanics based approach to the fatigue life prediction of cast aluminium alloy with considering the effect of porosity

Xiaojia Wang[1], *Weiping* Hu[2,*], and *Qingchun* Meng[3]

[1]Institute of Solid Mechanics, School of Aeronautic Science and Engineering, Beihang University, China
[2]Institute of Solid Mechanics, School of Aeronautic Science and Engineering, Beihang University, China
[3]Institute of Solid Mechanics, School of Aeronautic Science and Engineering, Beihang University, China

Abstract. A damage mechanics based approach is applied for the study of fatigue behaviour of high pressure die cast ADC12 aluminium alloy. A damage coupled elasto-plastic constitutive model is presented according to the concept of effective stress and the hypothesis of strain equivalence. An elastic fatigue damage model taking into account the pore-induced stress concentration is developed to investigate fatigue damage evolution of the specimens subjected to cyclic loading. The predicted lives for the specimens with different sizes of pores are consistent with the experimental data. The pore-induced fatigue damage and the variation of fatigue life along with the size of pores are also investigated.

1 Introduction

In current industrial practice, cast aluminium alloy is widely applied in the automotive and aerospace industries due to their advantage of lightweight, relatively high efficiency, and low cost. However, various kinds of microscopic defects, such as porosities, silicon particles, and cold flakes, are known to be an obstacle concerning fatigue resistance [1]. Among these defects, porosity is one of the most typical forms and inevitably formed during the casting process. Many experimental studies have investigated that the stress concentrations induced by pores become the primary cause of initiation of fatigue cracks and have a considerable influence on fatigue failure [2-3]. Therefore, the effect of porosities is of significant concern in assessing the fatigue life of cast aluminium alloy.

Presently, numerous methods have been adopted to assess the fatigue life of cast aluminium alloy with considering the effect of defects. The most commonly used methods are Murakami's square-root area parameter model [4] and some crack growth models [5-6]. These models have been applied to predict the fatigue life based on the configuration characters such as the maximum area or diameter of defects. However, there are few damage mechanics models are developed to consider the effect of porosity on the fatigue damage of the cast aluminium alloy. An approach based on continuum damage mechanics (CDM) can not only describe mechanical behaviour of a deteriorated material at the macroscopic scale by introducing damage variables, but also evaluate the progressive damage based on the evolution laws derived from thermodynamics.

This study is aimed at predicting the high cycle fatigue life of specimens obtained from high pressure die-cast ADC12 aluminium alloy by using a continuum damage mechanics based approach with considering the effect of porosity. First, from the framework of CDM, a macroscopic scalar damage variable D is introduced to describe the mechanical deterioration of material under cyclic loading. Second, a fatigue damage model, which includes an elasto-plastic constitutive model and an elastic damage evolution model, is developed to evaluate the fatigue damage of cast aluminium alloy with different sizes of pores under cyclic loading. The damage coupled elasto-plastic constitutive equations are adopted as the material model according to the concept of effective stress and the hypothesis of strain equivalence. Then, a dimensionless quantity related to the diameter of pore is introduced to the damage evolution equations to reflect the effect of stress concentration induced by a pore. It is found that the calculated fatigue lives are in accordance with the experimental results [7].

2 Fatigue damage model

2.1 Damage-coupled elastic-plastic constitutive model

In the framework of CDM, the deterioration of material can be represented by the damage variable. The damage variable D for the case of isotropic damage of isotropic

* Corresponding author: huweiping@buaa.edu.cn

https://doi.org/10.1051/matecconf/201816514011

materials, can be represented by the deterioration ratio of the stiffness of representative volume element (RVE):

$$D=\frac{E-E_D}{E}, \qquad (1)$$

where E is the Young's Modulus of the material and E_D is the effective Young's Modulus of the RVE with damage. The value of E_D ranges from E to 0, and the value of D varies between 0 and 1.

For the case of a small deformation, the total strain ε_{ij} can be decomposed into:

$$\varepsilon_{ij}=\varepsilon_{ij}^{e}+\varepsilon_{ij}^{p}, \qquad (2)$$

where ε_{ij}^{e} and ε_{ij}^{p} are the elastic and plastic strains, respectively.

Based on the hypothesis of strain equivalence, the elastic strain of damaged material takes the form of:

$$\varepsilon^{e}=\frac{1+v}{E}\left(\frac{\sigma_{ij}}{1-D}\right)-\frac{v}{E}\left(\frac{\sigma_{ij}\delta_{ij}}{1-D}\right), \qquad (3)$$

where E, v are the Young's Modulus, Poisson's ratio respectively, The variable σ_{ij} is the Cauchy stress of the damaged material. δ_{ij} is the Kronecker delta: $\delta_{ij}=1, i=j$; $\delta_{ij}=0, i\neq j$ and the subscripts i and j have values of 1, 2, or 3.

The evolution of plastic strain is defined as:

$$\dot{\varepsilon}_{ij}^{p}=\dot{\lambda}\frac{\partial f}{\partial\sigma_{ij}}, \qquad (4)$$

where λ is the plastic multiplier. f is the von Mises yield function with damage and the nonlinear kinematic hardening model proposed by Chaboche [8] is expressed as:

$$f=\sqrt{\frac{3}{2}\left(\frac{S_{ij}}{1-D}-\alpha_{ij}\right)\left(\frac{S_{ij}}{1-D}-\alpha_{ij}\right)}-Q, \qquad (5)$$

where S_{ij} is the deviatoric part of the stress. Q is the radius of the yield surface. α_{ij} is the deviatoric part of the back stress.

The evolution of plastic strain components [9] can be obtained as:

$$\dot{\varepsilon}^{p}=\frac{3}{2}\frac{\dot{\lambda}}{1-D}\frac{S_{ij}/(1-D)-\alpha_{ij}}{(S_{ij}/(1-D)-\alpha_{ij})_{eq}}, \qquad (6)$$

$$\dot{p}=\sqrt{\frac{2}{3}\dot{\varepsilon}_{ij}^{p}\dot{\varepsilon}_{ij}^{p}}=\left(\frac{\dot{\lambda}}{1-D}\right), \qquad (7)$$

the kinematic hardening law is expressed as:

$$\alpha_{ij}=\sum_{k=1}^{K}\alpha_{ij}^{(k)}, \qquad (8)$$

$$\dot{\alpha}_{ij}^{(k)}=(1-D)\left(\frac{2}{3}C_{k}\dot{\varepsilon}_{ij}^{p}-\gamma_{k}\alpha_{ij}^{(k)}\dot{p}\right), \qquad (9)$$

where K is the number of the back stress components. Parameters C_k and γ_k are material constants obtained from experiments.

2.2 Fatigue damage evolution model

In the regime of high cycle fatigue, the cyclic stress determines the fatigue lifetime of the material. Without considering the effect of defect contained in the material, for the multiaxial stress state, the elastic damage evolution equation is given by [10]:

$$\frac{dD}{dN}=[1-(1-D)^{\beta+1}]^{\alpha}\cdot\left[\frac{A_{II}}{M_0(1-3b_2\sigma_{H,m})(1-D)}\right]^{\beta}, \qquad (10)$$

where N is the number of cycles until failure. A_{II} is the amplitude of the octahedral shear stress, which is defined by:

$$A_{II}=\frac{1}{2}\left[\frac{3}{2}(S_{ij,\max}-S_{ij,\min})\cdot(S_{ij,\max}-S_{ij,\min})\right]^{1/2}, \qquad (11)$$

where $S_{ij,\max}$ and $S_{ij,\min}$ are the maximum and the minimum values of the deviatoric stress tensor i, j components during one loading cycle. The variable $\sigma_{H,m}$ is the mean hydrostatic stress and is defined by:

$$\sigma_{H,m}=\frac{1}{6}[\max(tr(\sigma))+\min(tr(\sigma))], \qquad (12)$$

where, $tr(\sigma)=\frac{1}{3}(\sigma_1+\sigma_2+\sigma_3)$.

The parameter α is defined by:

$$\alpha=1-a\left\langle\frac{A_{II}-A_{II}^{*}}{\sigma_u-\sigma_{eq,\max}}\right\rangle, \qquad (13)$$

The Sines fatigue limit criterion in this model is formulated by:

$$A_{II}^{*}=\sigma_{l0}(1-3b_1\sigma_{H,mean}). \qquad (14)$$

where $\sigma_{eq,\max}$ is the maximum equivalent stress over a loading cycle. σ_u is the ultimate tensile stress and σ_{l0} is the fatigue limit at the fully reversed loading condition. Five parameters, a, β, M_0, b_1 and b_2 are determined by using plain fatigue tests of standard specimens.

In the above evolution equation, the term $A_{II}/(1-D)$ can be regarded as the effective stress amplitude, representing the average effect of damage on the reduction of load-resisting area. However, the stress concentration effect of pores is not considered in this model. It was reported that the fatigue crack initiated around the pore with maximum diameter in cast aluminium alloy [11], which indicates that the stress concentration induced by pore relates to the fatigue life of specimen. Accordingly, in our study, a dimensionless quantity related to the maximum pore diameter of porosity is introduced to the above evolution equation to describe the stress concentration effect on fatigue life. This method is much simpler than using the stress field obtained by building a meso-scale model of RVE involving a pore. Moreover, considering that the stress concentration factor increases as the pore diameter increases, correspondingly, the fatigue life decreases [12], the damage evolution model for cast aluminium alloy is proposed as:

$$\frac{dD}{dN}=[1-(1-D)^{\beta+1}]^{\alpha}\cdot[\frac{A_{u}[1+(\frac{d_{max}}{l})^{c}]}{M_{0}(1-3b_{2}\sigma_{H,m})(1-D)}]^{\beta}\ , \quad (15)$$

where d_{max} is the diameter of the largest pore, which can be obtained through 3D X-ray tomography inspection. It should be noted that all the pores are assumed to be spheric for the simplicity. The term $[1+(\frac{d_{max}}{l})^{c}]$ is used to reflect the effect of stress concentration induced by porosities. l is a characteristic dimension for describing the effect of a pore on the stress field around the pore. The length of l is taken to be about twice as long as the diameter of the largest pore. c is a material parameter that can be determined through the fatigue testing of materials.

3 Fatigue life prediction of material

A recent experimental study on die cast aluminium alloy indicated that porosities have a significantly effect on fatigue behaviour. The specimens for the X-ray tomography scan and fatigue test were machined from a practical high pressure die casting (HPDC) ADC12 aluminium alloy [7]. During the die casting process, the melt pouring temperature was 680°C, and the shot speeds were 0.1 to 0.8 and 4.0 m/s during the slow and high injection periods, respectively. The intensification pressure was set to 65 MPa. The compositions of the material are presented in Table 1. The fatigue specimens were in accordance with the ASTM standard. The geometry of specimens is shown in Fig. 1. The thickness of specimens is 4mm. The pores in the gauge segment of all specimens were detected with high resolution X-ray tomography nondestructively before the fatigue test.

Table 1. Chemical composition of the material (wt. %).

Si	Cu	Fe	Zn	Mn	Mg	Al
10.77	1.87	0.78	0.78	0.2	0.17	Bal.

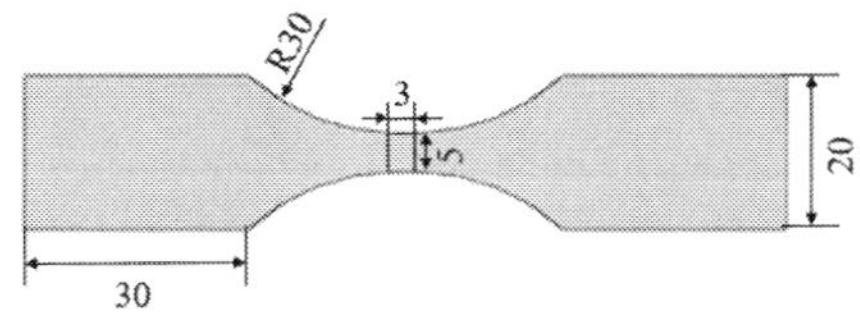

Fig.1. The geometry of the fatigue specimen (all dimensions are in mm).

The specimens were classified according to the average porosity of them. The high cycle fatigue tests were carried out at five stress levels of 110, 130, 145, 160, and 190 MPa for each group. The stress ratio of fatigue loading was R=0.1 and the test frequency was f=30Hz under the controlled temperature of 20°C. The fractured surfaces of all specimens were observed with a scanning electron microscope (SEM). Part of the results [7] are shown in Table 2.

Table 2. Experimental results under different pore conditions.

Maximum Stress	Volume fractions	Maximum diameter(mm)	fatigue life(cycles)
190	0.63%	0.452	21856
190	0.41%	0.301	36594
190	0.17%	0.137	134966
160	1.02%	0.315	84886
160	0.57%	0.171	120615
160	0.17%	0.105	260788
145	0.65%	0.482	182796
145	0.25%	0.136	1044904
145	0.15%	0.201	807453
130	0.94%	0.448	372192
130	0.45%	0.131	387274
130	0.15%	—	$>10^7$
110	0.91%	0.367	2360642
110	0.56%	—	$>10^7$
110	0.15%	—	$>10^7$

3.1 Determination of material parameters for damage analysis

The stress-strain curve obtained from the uniaxial tensile test [13] shown in Fig. 2 is used to determine the parameters in the elasto-plastic constitutive equations. The integration of Eq. (9) under the uniaxial loading leads to the stress-strain curve expressed as:

$$\sigma = \sum_{k=1}^{K} \frac{C_k}{\gamma_k}(1 - e^{-\gamma_k \varepsilon_p}) + \sigma_y. \tag{16}$$

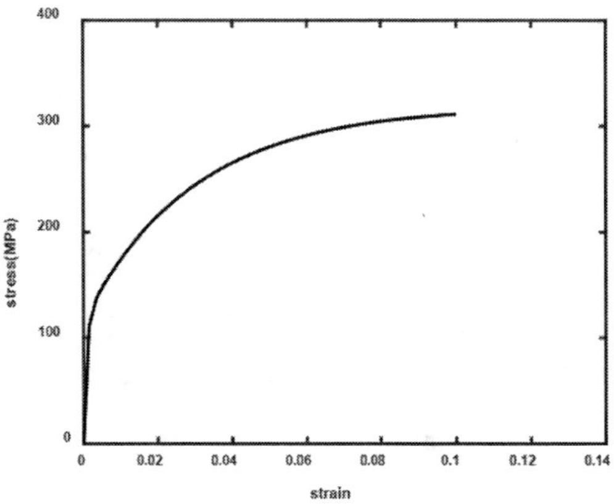

Fig.2. Stress–strain curve for ADC12 cast aluminium alloy.

In this study, three components of back stress are adopted to represent the nonlinear kinematic hardening behaviour. The static mechanical and calculated material parameters of the constitutive model are presented in Table 3.

Table 3. Static mechanical and material parameters for ADC12 cast aluminium alloy.

E(GPa)	ν	σ_y (MPa)
76	0.3	110
C_1 (MPa)	C_2 (MPa)	C_3 (MPa)
85610.7	2017.5	6342.5
γ_1	γ_2	γ_3
7053.87	373.2	33.2

For the elastic damage evolution model, there are five parameters a , β , M_0 , b_1 and b_2 needed to be determined. The parameters are determined through fatigue testing of the smooth specimens under uniaxial fatigue loading with different stress ratios. The fatigue test data for ADC12 cast aluminium alloy is taken from the literature [14]. The calculated material parameters in the fatigue damage model are listed in Table 4.

Table 4. Material parameters in the fatigue damage model.

a	β	M_0	b_1	b_2	c
0.6	7.09	1129.3	0.0176	0.00075	0.5757

The integration of Eq. (15) gives:

$$N_F = \frac{1}{1+\beta} \cdot \frac{1}{aM_0^{-\beta}} \cdot \left\langle \frac{\sigma_u - \sigma_{eq,\max}}{A_{II} - A_{II}^*} \right\rangle \left(\frac{A_{II}[1+(\frac{d_{\max}}{l})^c]}{1-b_2\sigma_{H,m}} \right)^{-\beta} . \tag{17}$$

The parameter c is needed to determine through the experimental data of fatigue specimens. In this study, the testing data with fatigue life in terms of loading cycles under different pore conditions in Table 3 are used to determine this parameter. The calculated material parameter c is listed in Table 4. The value of l is taken as 1mm, which is about twice as long as the diameter of the largest pore, 0.482 mm.

3.3 Calculation of stress and algorithm of fatigue damage calculation

In the ABAQUS platform, only 1/2 of the specimen is built with the symmetry boundary conditions at the plane of symmetry. A 3-D 8-node solid element is employed to accomplish the calculation. The FE model after meshing is shown in Fig. 3. A mesh discretization of high quality in the middle region of specimen was performed to accurately acquire the stress field under cyclic loadings. A minimum element size of 0.3 mm×0.5 mm×1.0 mm, was adopted to compromise between the accuracy of numerical simulation and the computational time.

The entire numerical computation of the fatigue life of ADC12 cast aluminium alloy includes the calculation of stresses and strains and the elastic-plastic fatigue damage. The flowchart of calculation is showed in Fig. 4.The user-defined subroutines called at all material integration points at the beginning of each time increment are used to implement the damage-coupled elasto-plastic constitutive model and damage evolution model. The detailed computational procedures can be referred in literature [15].

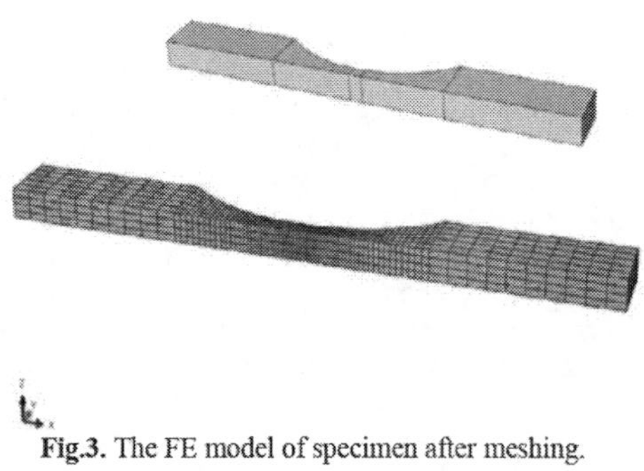

Fig.3. The FE model of specimen after meshing.

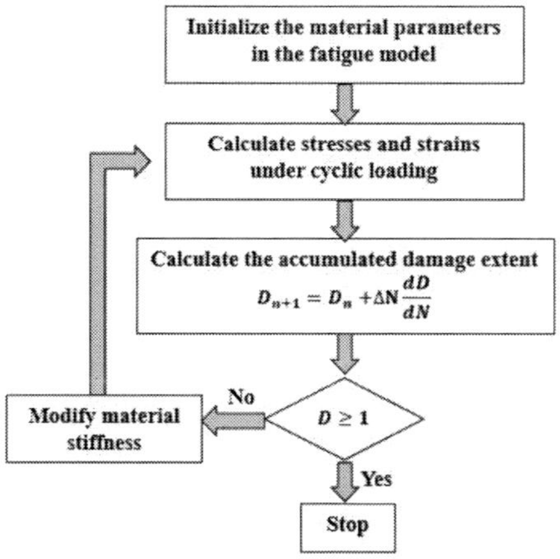

Fig.4. The flowchart for the FE computational methodology.

4 Results and discussion

4.1 fatigue damage analysis

Fig.5 depicts the damage distribution zone of the specimen right before the initiation of fatigue crack under the conditions σ_a=190 MPa and d_{max}=0.301. The maximum fatigue damage occurs at the middle of the cross section of specimen.

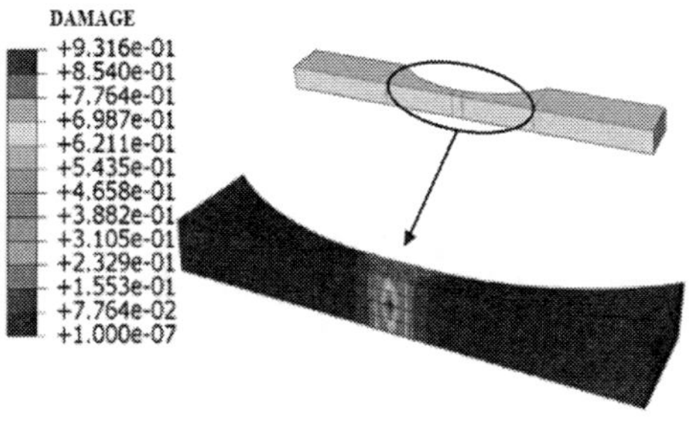

Fig.5. Damage contour on the middle cross section for σ_a=190 MPa, d_{max}=0.301.

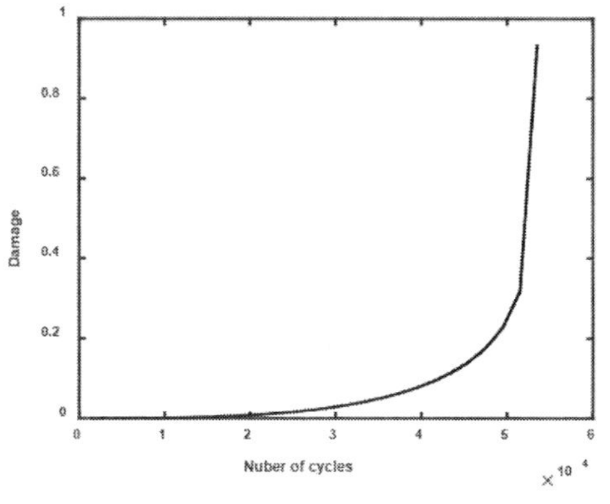

Fig.6. Evolution of the fatigue damage at the critical point for σ_a=190 MPa, d_{max}=0.301.

The extent of damage versus the number of cycles at the critical point is shown in Fig. 6. During a long period of the entire lifetime, the elastic damage increases slowly as the number of cycles increases. After the fatigue damage reaches around 0.2, it increases significantly, the specimen becomes failure after experiencing a small amount of cycles.

4.2 Fatigue life results

Based on the aforementioned computational method, the fatigue lives of specimens with porosities are calculated. Fig.7 illustrates the predicted fatigue lives under different pore conditions and their comparation with the experiment data. It can be seen that the calculated results agree well with the experimental data and all of the data fall within the factor of twice error bands.

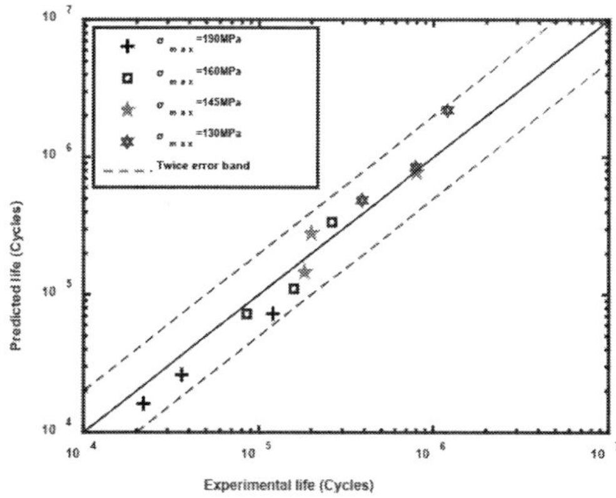

Fig.7. Comparison between the calculated results and experimental data

This fatigue model is then applied to predict the S-N curves of material with different pore conditions. Table 5 illustrates the assumed diameter of porosities of ADC12 cast aluminium alloy and the calculated stress concentration factors induced by pores. Fig.8 depicts the predicted S-N curves under these conditions. The vertical axis is the applied maximum fatigue stress and the horizontal axis is the logarithm scale of number of failure cycles. It is shown that the fatigue strength decreases as the diameter of the largest pore increases.

Table 5. Stress concentration factor under varies pore conditions.

d_{max} (mm)	0.1	0.3	0.5
$1+(\frac{d_{max}}{l})^c$	1.267	1.500	1.671

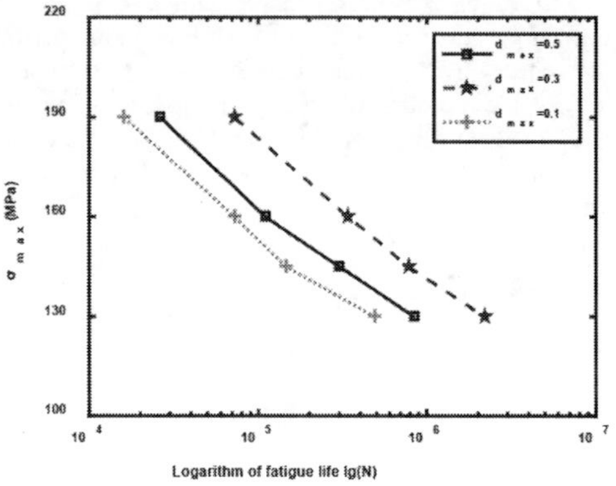

Fig.8. Predicted S-N curves with various pore conditions.

5 Conclutions

In this study, a new approach is proposed to calculate the fatigue life of ADC12 cast aluminium alloy considering pore-induced stress concentration based on the continuum damage mechanics theory. The following conclusions can be made:

(1) A dimensionless quantity related to the diameter of the largest pore is employed to the elastic fatigue damage evolution model. The proposed model can be used to well predict the fatigue life of ADC12 cast aluminum alloy considering the effect of porosity.

(2) During a long period of the entire lifetime, the elastic damage increases slowly as the number of cycles increases. After the fatigue damage reaches around 0.2, it increases significantly, the specimen becomes failure after experiencing a small amount of cycles.

(3) The calculated fatigue lives of ADC12 cast aluminum alloy under different sizes of pores are in accordance with the experimental data available in the literature. Moreover, the predicted S-N curves provide an important reference to investigate the fatigue property of ADC12 cast aluminum alloy.

References

1. Y.X. Gao, et al., Acta Mater, **52**(19):5435-5449 (2004).
2. Q. G. Wang, D. Apelian and D. A. Lados, J. Light Met,**1**:73–84(2001).
3. P.Wu,Y.Nadot, et al. Int. J Fatigue,**63**:97–109(2014).
4. Y. Murakami, M. Endo, Int. J Fatigue,**16**(1994).
5. J. P. Weiler, J. T. Wood, R. J. Klassen, E. Maire, R. Berkmortel and G. Wang, Mater. Sci. Eng. A, **395**, 315(2005).
6. H. Hu, M. Zhou, Z. Sun and N. Li. J. Mater. Process. Technol., **201**, 364(2008).
7. Q. Wan, H.D. Zhao and C. Zou, ISIJ Int., **54**,511–515 (2014).
8. J. L. Chaboche, Int. J Plasticity,**7**: 661–678 (1991).
9. J. Lemaitre, J. L. Chaboche, *Mechanics of solid materials*. (Cambridge University Press, 1990).
10. J. L. Chaboche, P.M. Lesne, Fatigue Fract Eng Mater Struct,**11**(1):1–17(1988).
11. K.S. Chan, Int. J Fatigue. **32**(9):1428–1447(2010).
12. F. Bouafia, S. Boualem, MM. El Amin, B. Benali. Comput Mater Sci,**50**(4):1450–1459(2011).
13. C. Dørum, H. I. Laukli, O. S. Hopperstad, M. Langseth, Eur J Mech A-Solid, **28**:1–13(2009).
14. O. Kuwazuru , Y. Murata, Y. Hangai, et al., J. Solid Mech.& Mater. Eng., **2**(9):1220-1231(2008).
15. Z.X. Zhan, W.P. Hu, M. Zhang, et al., Int. J Fatigue,**74**:173–182(2015).

Prediction of fatigue crack initiation life of aluminium alloy joints using cyclic elasto-plasticity FEM analysis

Tsutsumi Seiichiro [1,*], *Sano* Moe [1], *Fincato* Riccardo [1]

[1] Joining and Welding Research Institute, Osaka University, 11–1 Mihogaoka, Ibaraki, Osaka, JAPAN

Abstract. Finite element analyses (FEA) are particularly useful for investigating fatigue problems since it is possible to carry out elasto-plastic simulations for any configuration and to predict the material behaviour for a large number of loading cycles. This study aims to investigate the fatigue life for Al-Mg alloy A5083-O joints by means of numerical simulations. The Proposed method needs to give a precise description of the elasto-plastic behaviour of the alloy together with an appropriate definition of the criteria for the fatigue crack initiation. In this paper, the elasto-plastic behaviour of the A5083-O alloy was investigated by FE analyses. On the other hand, the fatigue crack initiation criterion is provided based on strain ranges observations. In detail, the finite element analyses focused the attention on the study of the service life of a butt-weld join.

1 Introduction

Fatigue is one of the main causes of failure of welded structures. Finite element analyses recently became an essential tool for fatigue life estimations, especially when utilizing the advanced fatigue assessment methods for welded structures. The reason for this choice is mainly due to the versatility of the FEM which makes possible to carry out elasto-plastic simulations for any configuration and to predict the material behaviour for a large number of loading cycles.

Fatigue life is usually considered to be divided into two phases: the crack initiation and the crack propagation. Among fatigue failure processes, the prediction of fatigue initiation as a series of damage accumulation and crack initiation is one of the most challenging issues. For the purpose of describing and evaluating the mechanical fatigue processes of materials, a phenomenological plasticity model is proposed [1-4]. This model is based on a cyclic plasticity model [5-9] and has been extended to describe the behaviour of materials, especially steel, and applied to several case studies in low and high cycle fatigue life regions. In this study, we extend the field of investigations to aluminium alloys.

This study aims to investigate the fatigue life for Al-Mg A5083-O aluminium alloy joints by means of numerical simulations. This specific material is very common in the automotive, construction and shipbuilding industries due to the moderately high strength, excellent corrosion resistance and very high toughness, and therefore, the characterization of its fatigue life is crucial to the technological development of the alloy.

In order to characterize the crack initiation life, it is important to give a precise description of the elasto-plastic behaviour, which was here investigated by [1-4, 10-15]. In addition, the criteria for the fatigue crack initiation was defined based on the strain ranges.

The numerical examples reported in the next sections 3 and 4 show the evaluation of the service life of a butt-weld join, modelled with a 2D geometry under the assumption of a plane strain condition. The model geometry, boundary conditions and material properties were assumed referring to the low-cycle fatigue test carried out by Iida [16]. Three different analyses were carried out in order to investigate the influence of the angular distortion on the service life performances. The fatigue life was judged by observing the local behaviour in terms of plastic strain accumulation and by the definition of a fatigue crack initiation criteria. An investigation in the field of high cycle fatigue ($>10^5$) was also conducted in the same manner. In this last case, the model geometry was assumed referring to the experimental results carried out by Hirukawa [17].

2 elasto-plastic model

The elasto-plastic behaviour was investigated with a plasticity model based on the subloading surface model, and extended to include the cyclic damage concepts [1-4]. A detailed explanation of the model features is beyond the scope of this paper, and the reader is referred to Ref. [1] for a complete discussion. The extended subloading surface model [5-9] describes the generation of plastic strain within the yield surface (subloading surface), which can be obtained through a similarity transformation from the conventional yield surface (referred to as the normal-yield surface in Fig. 1).

* Corresponding author: tsutsumi@jwri.osaka-u.ac.jp

Classical theories distinguish elastic and plastic regions, allowing an irreversible stretch only in the plastic region. In contrast, the subloading surface model abolishes the separation into domains, stating that a plastic response can be realized for every change in the stress state that satisfies the loading criterion. Furthermore, the use of a mobile similarity center, which is a function of the plastic strain, makes this theory particularly suitable for studying cyclic mobility problems.

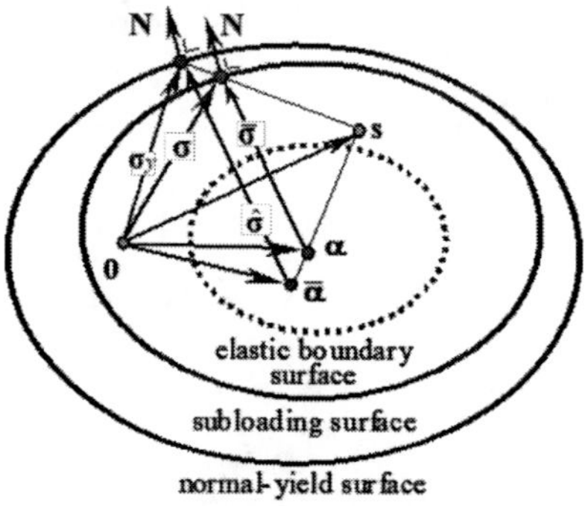

Fig. 1. : Sketch of the Subloading, elastic boundary and normal yield surfaces in the stress space.

Starting from the original subloading surface model a scalar damage variable was introduced to simulate the loss of stiffness induced by the material mechanical degradation. The damage is coupled with a special variable of the cumulative plastic strain H_d according to the following formula:

$$D(H_d)=(1-d_2)\left[1+\left(\frac{d_1}{H_d}\right)^{d_3}\right]^{-1} \tag{1}$$

where d_1, d_2, and d_3 are material parameters that regulate the damage rate evolution, as shown in Fig. 2, and they must be calibrated through an experimental campaign.

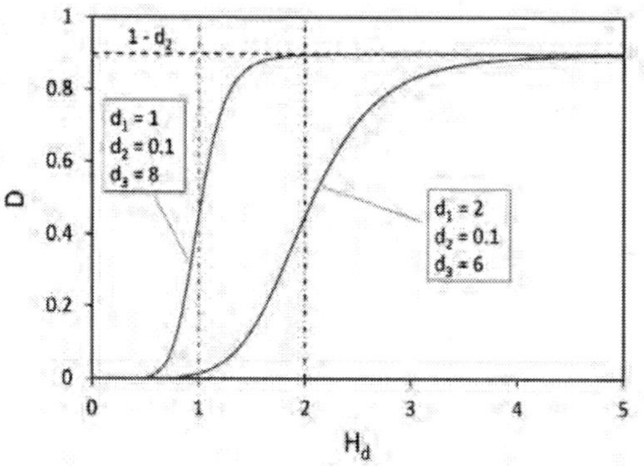

Fig. 1. Damage evolution for two sets of material parameters.

The material hardening behaviour was described by the combination of isotropic and kinematic hardening laws, in Eq.(2) and Eq.(3) , respectively.

$$F=F_0\left[1+h_1\left\{1-e^{(-h_2H)}\right\}\right],\quad F'=F_0h_1h_2e^{(h_2H)} \tag{2}$$

$$\boldsymbol{\alpha}=a_1\left(a_2F\mathbf{N}-\sqrt{\frac{2}{3}}\boldsymbol{\alpha}\right)\left\|\varepsilon^p\right\| \tag{3}$$

where h_1, h_2, a_1 and a_2 are four material constants defined by the user.

Previous works carried out by the authors [1-4, 10-15], focused the attention on steel components; in this study, we extend the field of investigations to the aluminium alloy A5083-O.

The identification of the plastic parameters was carried out through inverse analyses by minimizing the difference between the experimental data found in Horikawa and Cho [18] and the numerical results obtained by the FE analyses. Fig. 3 reports the comparison of the model response against the experimental results, showing a good agreement between the curves. The material parameters used in the computation are reported in the following in Table. 1.

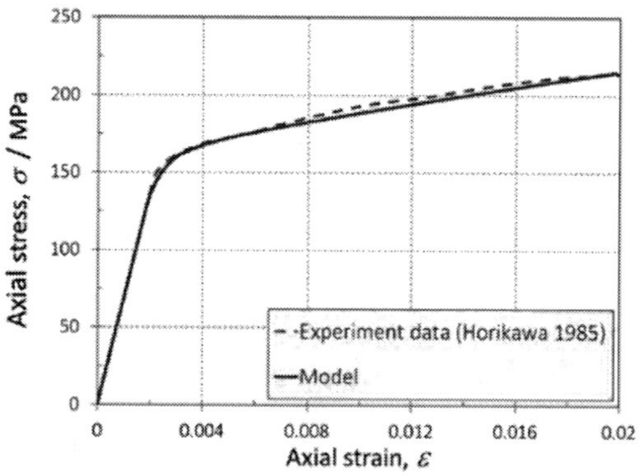

a) Tensile stress-strain curve

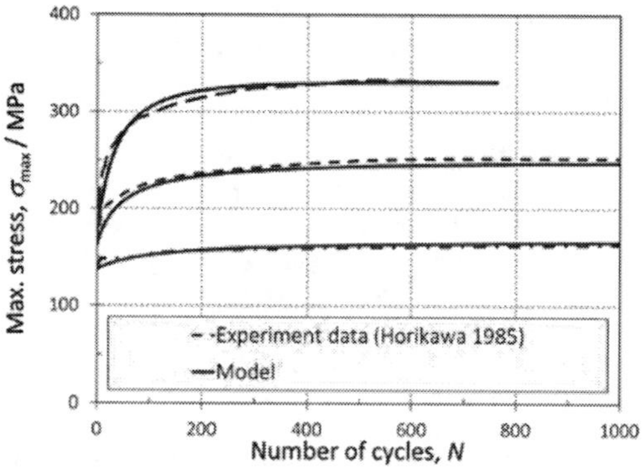

b) Max stress versus number of cycles in the fatigue test (displacement control boundary condition).

Fig. 3. Comparison of the numerical model response against the experimental results.

Table 1. Material Parameter used in this paper

Material parameter	value
E	68,600 [MPa]
v	0.33
u	200
F_0	168
R_e	0.4
h_1; h_2	1.05; 1.8
, a_a; γ_a	20; 0.6

3 Fatigue crack initiation criteria

In order to characterize the fatigue life, it is also important to give an appropriate definition of the criteria for the fatigue crack initiation.

The relation between the strain ranges and fatigue life is defined based on previous experimental data in [16], [18, 19]. The following Eq. (4) was formulated by means of the least square method on all the experimental results, obtaining the solid line in Fig.4:

$$\frac{\Delta\varepsilon}{2} = 14.5N_f^{-0.488} + 0.168 \quad (4)$$

where $\Delta\varepsilon$ and N_f are the total strain range and the fatigue life, respectively.

In the same manner, the relation between the crack initiation life and fatigue life can be analytically given by the following expression:

$$N_c = 0.654N_f \quad (5)$$

N_c is the crack initiation life. In this paper, the surface crack initiation size is defined as a crack of 0.5 mm length. Finally, combining Eq. (4) and (5), the relation between the total strain ranges and crack initiation life is obtained as follows:

$$\frac{\Delta\varepsilon}{2} = 11.9N_f^{-0.475} + 0.168 \quad (6)$$

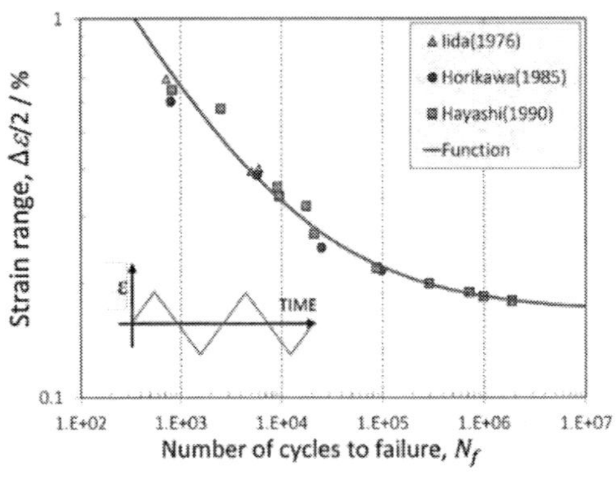

Fig. 4. Strain range vs number of cycles to failure.

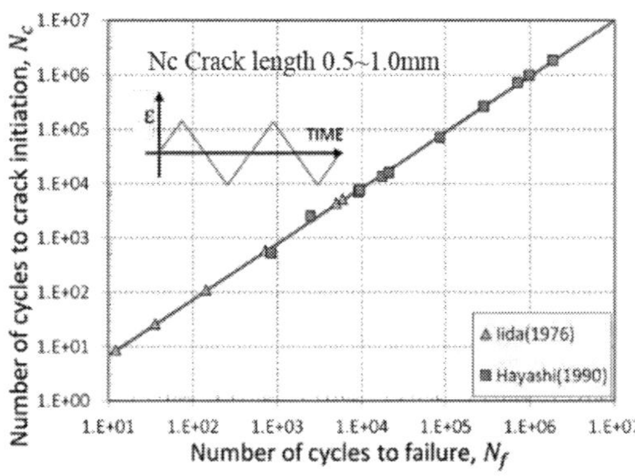

Fig. 5. Crack initiation life vs number of cycles to failure

4 Analysis of Low cycle fatigue

4.1. FE model

The constitutive equations of the phenomenological model in [1-4] were implemented via user subroutine for the commercial code *Abaqus* (ver. 6.14-4).

The numerical models were created referring to the experimental results carried out by Iida [16]. The finite element analyses focused the attention on the study of the service life of a butt-weld join, modelled with a 2D geometry under the assumption of a plane strain condition. A mesh refinement (0.05 mm minimum element dimension) was performed around the welding toe, where the largest stress concentration tends to appear. A total of 4840 plate elements were used in the discretization, amounting to a total of 5874 nodes. The numerical simulations were conducted for cyclic loading with a constant amplitude (i.e. unidirectional cycling loading R = 0).

As a result of the welding process, aluminum alloys show higher angle distortions compared to common carbon steels, which impacts significantly to the fatigue life of the components. In this study, three cases were reproduced simulating three different angular distortions; 0°, 0.5° and 1.0° (see Fig.7 as reference).

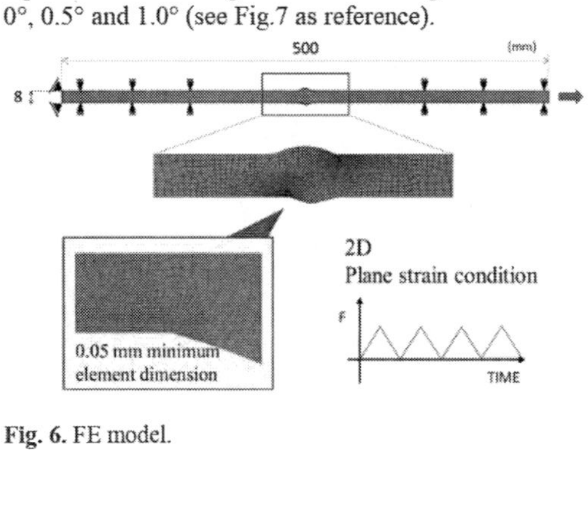

Fig. 6. FE model.

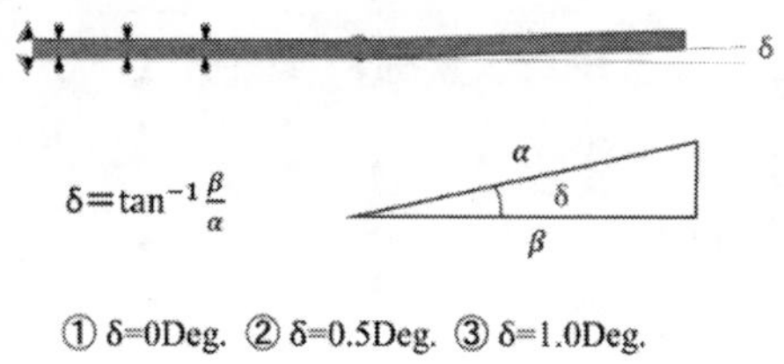

Fig. 7. Sketch of the angular distortion

4.2. Result

Fig. 9 shows the local S-N curves obtained from FE analyses. The local stress-strain response, taken in correspondence of the element which shows a higher plastic accumulation, is used for the life prediction. The result is particularly relevant for the expected fatigue life of the components, as shown in Fig. 10.

The black dashed line indicates the experimental results carried out by Iida [16]. The blue solid line indicates the numerical results of the 0° angular distortion case. Analogously, the green and red lines represent the 0.5° and 1.0° angular distortion cases, respectively. In the low cycle region, the simulations showed good agreement with the experimental results [16], confirming the predicting capability of the numerical model. Whereas, in the high cycle region, the estimated crack initiation life is longer than the experimental results.

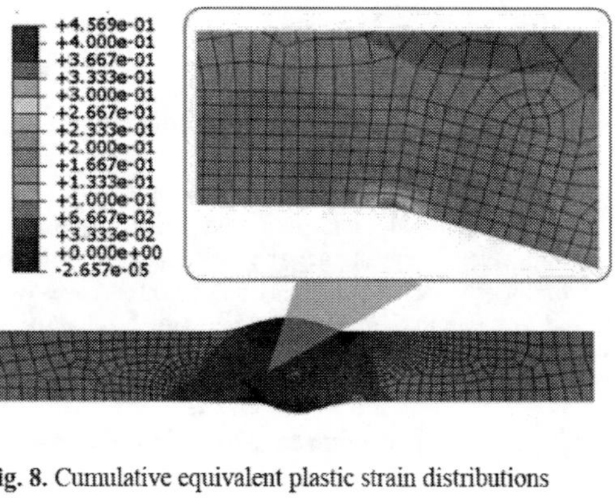

Fig. 8. Cumulative equivalent plastic strain distributions

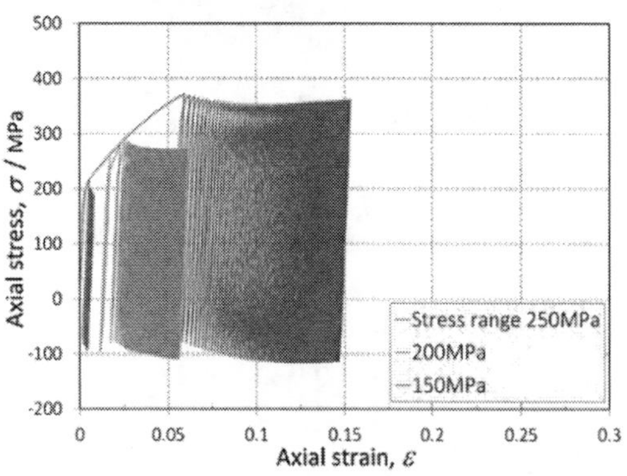

Fig. 9. Predicted local stress-strain curves.

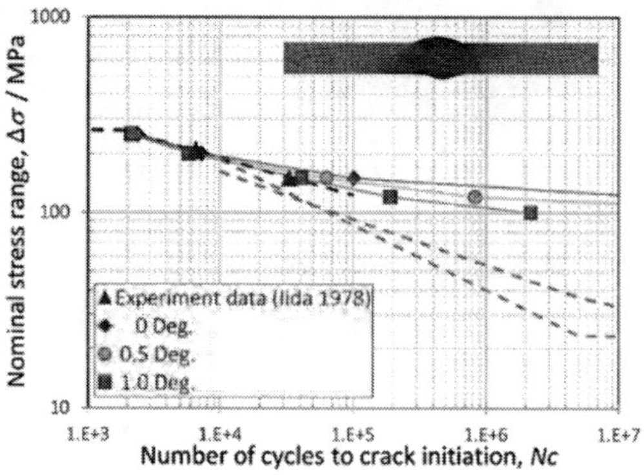

Fig. 10. Predicted S-N curves for crack initiation.

5 Analysis of High cycle fatigue

5.1. FE model

In this case, the numerical models were created referring to the experimental results carried out by Hirukawa [17]. A butt-weld join was modelled with a 2D geometry under the assumption of a plane strain condition. A mesh refinement (0.05 mm minimum element dimension) was performed around the welding toe. A total of 6400 plate elements were used in the discretization, amounting to a total of 6615 nodes.

The numerical simulations were conducted for cyclic loading with a constant amplitude (R = 0). Here, two cases were reproduced simulating two different angular distortions; 0°, and 1.0°.

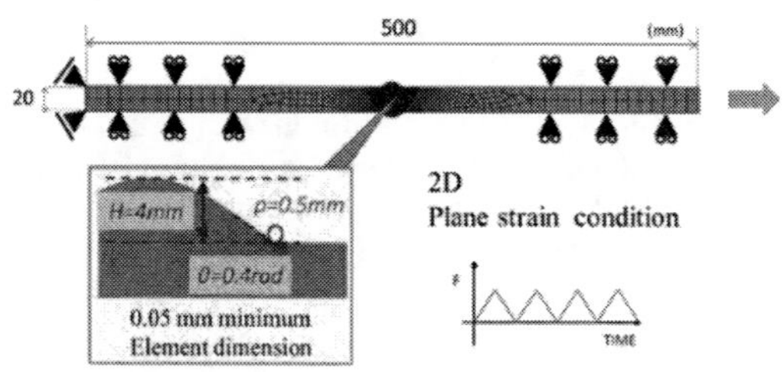

Fig. 12. FE model

5.2. Fatigue crack propagation analysis

In this case, fatigue crack propagation analysis was carried out using X-FEM, after predicting crack initiation life. The analysis was carried out using the open-source Salome-Meca software. A linear static analysis is performed based on the assumption of a small scale yielding condition of the fatigue crack. Paris' law is employed as crack propagation law.

$$\frac{da}{dN} = C\Delta K^{m} \tag{5}$$

where da/dN is crack velocity and ΔK is the range of stress intensity factor. The material parameters for the Paris' law are presented in Table 2. These parameters were experimental data obtained by Takaai [20].

Table 2. Material parameters of the Paris' law.

m	2.7
C	3.3×10^{-10}

The model geometry and the predicted crack location were taken from the crack initiation analysis.

4.3. Results

Fig. 13 shows the local S-N curves obtained from FE analysis. The local stress-strain response, taken in correspondence of the element which shows a higher plastic accumulation, is used for the crack initiation life prediction. Fig. 14 shows the schematic representation of fatigue crack path that was obtained from the fatigue crack propagation analysis. The fatigue life was obtained by adding the predicted crack initiation life to the propagation life. The result is particularly relevant for the expected fatigue life of the components, as shown in Fig. 15. The black dots indicate the experimental results carried out by Hirukawa [17]. The green solid line indicates the numerical results of the 0° angular distortion case. Analogously, the blue line represents the 1.0° angular distortion cases. If the angular distortion is taken into account, the simulations come closer to the experimental results. However, in the low cycle region, the estimated fatigue life is shorter than experimental results. Whereas, in the high cycle region, the estimated one is longer than the experimental results.

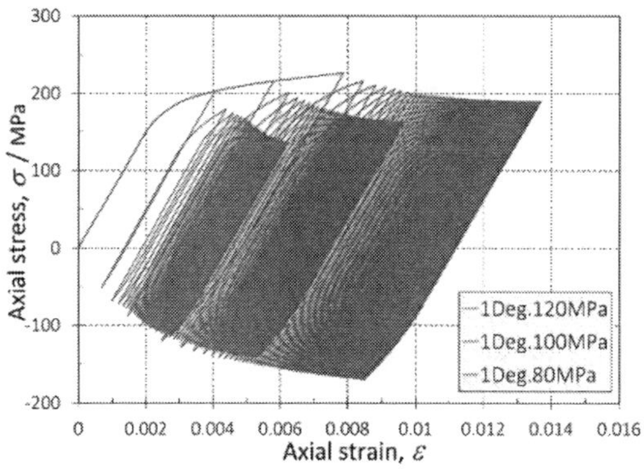

Fig. 13. Predicted local stress-strain responses.

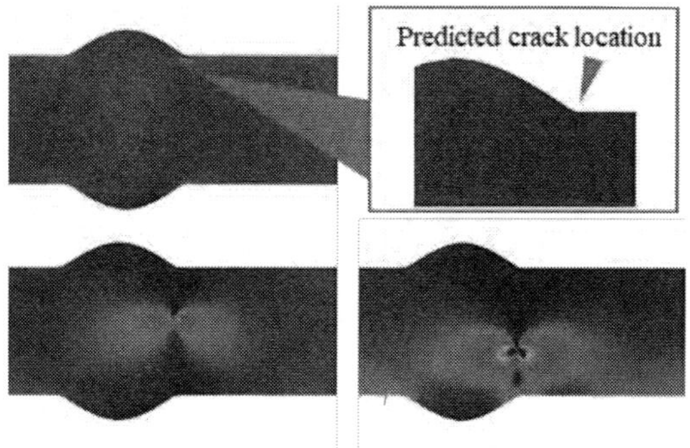

Fig. 14. Schematic representation of fatigue crack path.

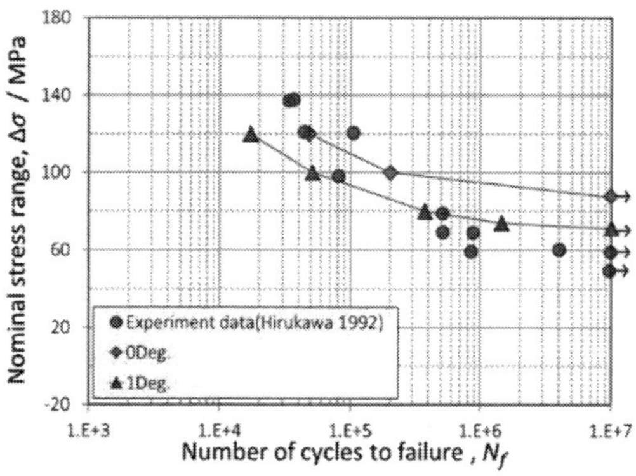

Fig. 15. Predicted S-N curves for total life

5 Conclusion

The paper presented a series of investigations on the fatigue crack initiation life for aluminum alloy A5083-O joints, by means of numerical simulations. The work can be summarized in the following points:

- the elasto-plastic behavior of the material was investigated using an unconventional plasticity model;
- a crack initiation criterion was formulated;
- The numerical simulations under uniaxial condition were conducted with constant strain rate in two cases. The first case in low cycle fatigue field, and other in the high cycle fatigue region. The results were compared against the experimental data.

The simulations in the low cycle fatigue region showed good agreement with the experimental results, confirming the predicting capability of the numerical model. However, the estimated fatigue life was longer than the experimental one in the high-cycle fatigue regime.

The simulations in the high cycle fatigue field were conducted using FEM and X-FEM. The fatigue crack propagation life was estimated together with the crack initiation life. The numerical results indicate the same tendency as experimental results. However, in the low cycle region, the estimated fatigue life is shorter than experimental results. Whereas, in high cycle region, the estimated one is longer than experimental results.

Further studies are needed in order to widen the application range of the fatigue assessment method.

References

1. S. Tsutsumi, K. Murakami, K. Goto, M. Toyosada, Journal of the Japan Society of Naval Architects and Ocean Engineers, **7** (2008)
2. S. Tsutsumi, M. Toyosada, K. Hashiguchi, J. of. App. Mech. JSCE, **9**, 455 (2006)
3. S. Tsutsumi, M. Toyosada, K. Murakami, Transactions of the Japan Society of Mechanical Engineers Series A, JSME, **73** (2007)
4. S. Tsutsumi, M. Toyosada, F. Dunne, Procedia Engineer, **2(1)** (2010)
5. K. Hashiguchi, Int. J. Solids Structures, **25** (1989)
6. K. Hashiguchi, T. Yoshimaru, Int. J. Plast. **11**, 347 (1995)
7. K. Hashiguchi, S. Tsutsumi, Int. J Plasticity, **17** (2001)
8. K. Hashiguchi, *Elastoplasticity Theory* (Springer, Berlin, 2009)
9. R. Fincato, S. Tsutsumi, Fracture and Structural Integrity **38**, 231 (2016)
10. S. Tsutsumi, K. Morita, R. Fincato, H. Momii, Fracture and Structural Integrity **38**, 240 (2016)
11. R. Fincato, S. Tsutsumi, Marine Structures, **55**, (2017)
12. S. Tsutsumi, K. Morita, R. Fincato, Journal of Structural Engineering A, JSCE, **63**, (2017)
13. S. Tsutsumi, R. Nagao, R. Fincato, T. Ishikawa, R. Matsumoto, Q. J. Jpn. Weld. Soc., 35 (2017)
14. R. Fincato, S. Tsutsumi, Q. J. Jpn. Weld. Soc. **35**, 185s (2017)
15. H. Momii, S. Tsutsumi, R. Fincato, J. of. App. Mech. JSCE, **73**, 2 (2017)
16. K. Iida, K. Minida, Y. Kho, Journal of the Society of Naval Architects of Japan, **144** (1978)
17. H. Hirukawa, S. Matsuoka, E. Takeuchi, S. Nishijima, Transactions of the Japan Society of Mechanical Engineers Series A, **58** (1992)
18. K. Horikawa, S. Cho, Transactions of JWRI, **14**(2), (1985)
19. M. Hayashi, T. Kuriyama, Proceedings of the Faculty of Engineering of Tokai University **30.1** (1990)
20. T. Takaai, Y. Nakayama, Japan Inst. Light Metal, **42** (1992)

The use of areal surface topography characterisation in relation to fatigue performance

Alaitz Zabala[1,*], *Liam* Blunt[2], *Wilson* Tato[1], *Andrea* Aginagalde[1], *Xabier* Gomez[1], and *Iñigo* Llavori[1]

[1]Mondragon Unibertsitatea, Loramendi 4, 20500 Arrasate-Mondragón, Spain
[2]EPSRC Advanced Metrology Hub, University of Huddersfield, Queensgate, Huddersfield UK

Abstract. Although the effect of surface topography on fatigue life is widely accepted, the underlying role of surface roughness from first principles is still poorly understood. Currently approaches which consider the influence of surface roughness on fatigue life prediction can be broadly classified into surface corrector factors (C_s) and stress concentration factors (K_t). Those approaches describe the surface according to the manufacturing process (machined, grounded…), or using 2D height descriptor parameters (R_a, R_z…). However, these approaches are not able to correctly describe the effect of roughness on the fatigue performance where it is anticipated a richer set of surface descriptors would show correlation. The present work aims to highlight the inherent limitations of the most commonly employed 2D surface measurement and characterization techniques, and provides an insight into the application of 3D areal surface characterization processes including the use of the latest areal surface topography parameters quantifying amplitude, spatial, and hybrid topographical information which is considered to be important for fatigue performance correlation.

1 Introduction

The influence of surface topography on fatigue strength is widely accepted. The micro-geometrical irregularities that constitute surface roughness can be treated as microscopic notches, which promotes crack initiation through local stress concentrations, reducing life-to-first-crack and hence total fatigue life [1-2]. Although other surface quality parameters can also affect fatigue behaviour (*i.e.* microstructure and residual stresses) [3-4], the impact of surface topography becomes more significant when crack initiation life is notable (*i.e.* in high cycle fatigue, HCF) [5-6].

One of the earliest published work on the effect of workpiece surface topography on fatigue life was conducted by Thomas in 1923 [7], which identified the detrimental effects of microgeometrical irregularities on fatigue behaviour. Since then, several studies on the role of roughness on fatigue life have been published [8-9].

Technical literature provides an empirical correction factor, generally known as surface factor C_s, [10] that can be used to adjust the endurance limit if surface roughness is different from standard specimen conditions. This surface finish corrector factor categorizes finish in qualitative terms according to the manufacturing process (machined, forged...), or presents the surface finish correction factor in a more quantitative way by using quantitative measure of surface roughness according to the average roughness amplitude (R_a). The asperities that constitute surface roughness can also be treated as microscopic notches by introducing a stress concentration factor K_t. This factor can be calculated either from averaged geometrical parameters of the surface [11], and or by finite element analysis (FEA) [12].

The average roughness amplitude, R_a, remains the most widely used parameter to describe fatigue specimen roughness. However, as the critical functional role played by the surface roughness on fatigue performance gains recognition, there is an increasing need for more function related surface characterization procedures in order to describe and identify the features affecting fatigue performance.

The present work aims at highlighting the inherent limitations of the most commonly employed 2D surface measurement and characterization techniques, and provide an insight into the 3D/areal characterization process and latest areal surface topography parameters important in fatigue performance correlation.

2 Critical Analysis of R_a

Current approach to characterize surface roughness of fatigue specimens typically rely on a single amplitude parameter, R_a, which describes the arithmetic average value of the deviation of the trace above and below the centre line (see equation 1) [13]. Higher values of R_a are usually related to lower fatigue strength.

$$R_a = \frac{1}{N}\sum_{i=1}^{N}|y_i| \tag{1}$$

* Corresponding author: azabalae@mondragon.edu

https://doi.org/10.1051/matecconf/201816514013

While R_a remains useful for quality control once a well defined manufacturing process is shown to be sufficiently stable, it typically proves too general to describe the surface's functional nature. Topography characteristics that have high impact on fatigue strength, such as tip radius of micro-notches and spacing between notches are not taken into account, since R_a only quantifies vertical values of the surface.

In order to show the influence of the tip radius and micro-notch spacing on fatigue, 2 simulated sine wave surfaces were generated with the same R_a value (see Figure 1). By changing the frequency of the sine, different tip radii and notch spacings were obtained. It is noteworthy that these profiles also present the same R_t and R_z values, defined as the total height of the profile evaluation length, and the average maximum height of the profile calculated at each sampling length, respectively [13]. Those parameters describes the extreme height characteristics of the profile and has been suggested as good fatigue performance predictors due to their capability to reflect the 'worst' defects present in the workpiece [14].

Two dimensional (2D) uniaxially loaded FEA has been performed (Figure 2) to obtain the K_t generated by these profiles, computed as:

$$K_t = \sigma_{11}/\sigma_\infty \,, \qquad (2)$$

where, σ_{11} is the normal stress and σ_∞ is the applied remote stress. Since $\sigma_\infty = 1$ MPa, K_t is equal to σ_{11}.

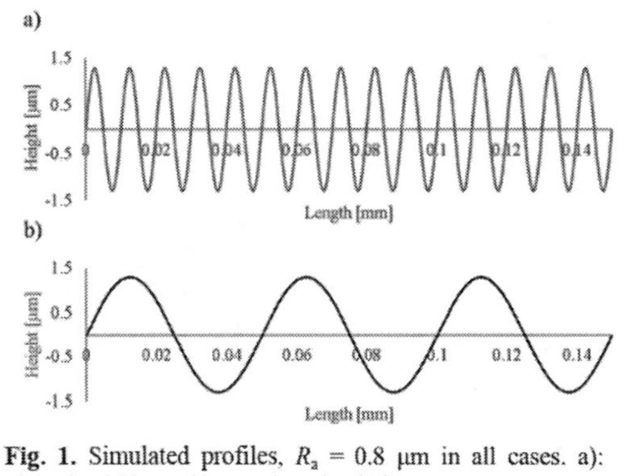

Fig. 1. Simulated profiles, R_a = 0.8 µm in all cases. a): wavelength of 50 µm. b) wavelength of 10 µm.

Studies of crack nucleation under uniaxial and multiaxial fatigue loading reveal that fatigue damage starts in an individual crystal, usually at the surface [15]. From this standpoint, higher values of K_t will lead to lower cycles to crack initiation and therefore to shorter lives.

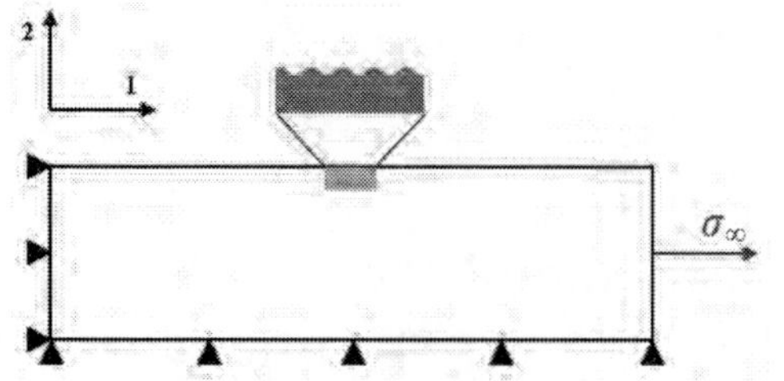

Fig. 2. Applied boundary condition and forces of the numerical models.

Figure 3 illustrates the results of the 2 monotonic tensile simulations. As expected, the maximum K_t is associated with the deepest topographic location in all simulations. Additionally, the minimum K_t are linked with the highest points due to the stress relaxation produced by the local loss of stiffness.

It can be observed that the first profile, with the smallest tip radius and shortest notch spacing, presents the maximum stress concentration. From a fracture mechanics point of view, small tip radii are linked with shorter fatigue lives [16]. Regarding the notch spacing effect on fatigue life, literature shows that shorter notch spacing suggests longer lives due to the so called 'notch interference effect'[6]. It should be highlighted that such statements correspond to the spacing effect for equal valley tip radius, which is not the case of the present profiles. However, the notch interference effect is visible in the present profiles. It can be observed that the stress relaxation produced at the highest topographic location are more pronounced in the shortest spacing profile. Therefore, lower K_t could be expected on shorter spacing profiles for the same theoretical tip radius, which is consistent with the literature.

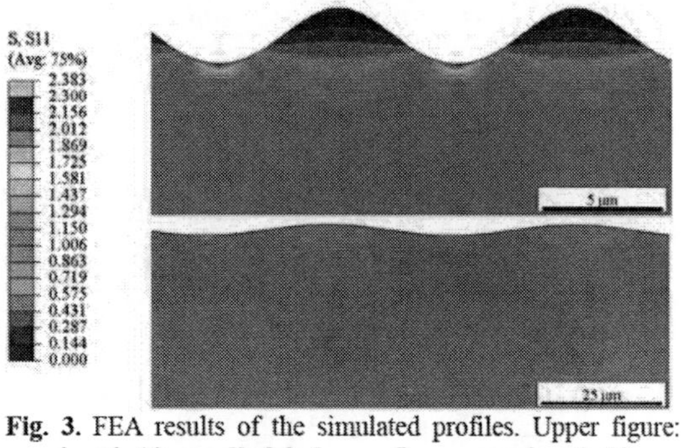

Fig. 3. FEA results of the simulated profiles. Upper figure: wavelength 10 µm, K_t=2.3; Lower figure: wavelength 50 µm, K_t=1.4.

As mentioned previously, higher K_t values are associated with shorter crack nucleation cycles. However, another important role on fatigue is the stress gradient effect. High stress gradient indicates that the stress decays rapidly near the hot spot. Thus, for the same K_t value, a crack of certain length will grow faster if a low stress gradient exists, thus leading to shorter fatigue life.

Figure 4 shows the K_t evolution for each profile below the surface. On the one hand, it is observed that the highest K_t profile shares also the highest stress gradient. On the other hand, the lowest K_t profile also has the lowest stress gradient. Therefore, from a certain depth onwards a crack will growth faster in the simulation with the lowest K_t [17].

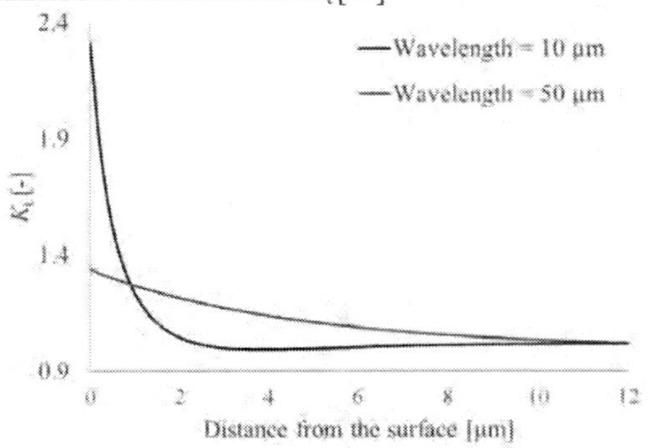

Fig. 4. Stress concentration factor (K_t) at different depths below the surface for the 2 simulated surfaces.

This analysis highlights the inherent limitations of the R_a parameter. Different profiles with the same R_a could show different characteristics in terms of stresses leading to different fatigue lives. Therefore, the average roughness amplitude parameter may not be best suited to describe the role of surface roughness on fatigue strength.

3 2D (profile) Vs 3D (areal) Characterization

2D profile measurement and its analysis is still playing an important role in the assessment of surface topography. This traditional technique is based on a contact stylus measurement system, where the stylus is moved across the surface to be measured for a pre-determined trace length while recording the vertical deviations of the stylus and thus the surface. The positive features of the technique that makes it currently preferred in the industry are mainly based on the short time needed to measure the profile, the lower cost of the instrument, and the fact that the process is well-established [18]. The surface topography is however three dimensional in nature, and there are inherent limitations to the use of these 2D techniques.

Profile measurement and its characterisation makes it difficult to determine the exact nature of the areal surface topography. Figure 5 presents a profile (2D) and areal (3D) representation of the same component covering the same measurement area. While the valleys present on the surface are clearly identified in the areal representation, it is impossible to identify whether they arise from pits or valleys in the 2D profile.

Areal measurement is able to provide a more complete information regarding the surface topography, not only in the qualitative identification of the surface features (pits, troughs, lay…), but the sizes, shapes and volumes can also be quantitatively calculated. Since 2D profiles coincides with an intersection of a vertical plane with a measured surface, it may not cross surface summits or valleys contained within a real surface.

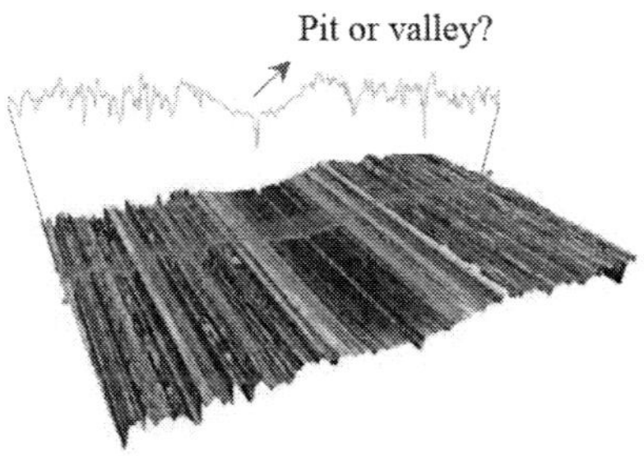

Fig. 5. A profile taken from an areal measurement showing the possible ambiguity of 2D measurement and characterization

It should be also considered that areal measurement presents more statistical significance than the equivalent profile measurement. The larger volume of data obtained increases the independence of the data, reducing the variance of parameters and resulting in a closer representation of the 'real surface' [19].

Finally, it should be noted that the stylus geometry influences the amount of information that can be extracted from a given surface. The stylus tip, which is typically 4 micron diameter 60o subtended cone, produces smoothing of the true profile, due to a mechanical filtering effect (Figure 6).

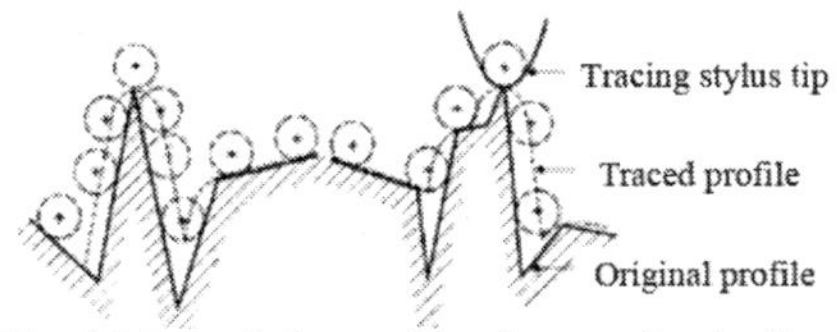

Fig. 6. Mechanical smoothing due to stylus tip diameter [20].

Areal (3D) characterization should therefore provide a more precise representation of the surface topography, and may allow a more precise correlation with fatigue behaviour.

4 3D Topographical parameters for fatigue

As stated by X. Jiang *et al.* the surface metrology discipline is undergoing a huge paradigm shift over the last decades, from profile to areal (3D) characterization [21]. The recently published ISO 25178:2 [22] standard encompass over 30 topographical parameters for surface topography characterization. Regarding the naming rules of the areal parameters, the upper letter '*S*' (for 'surface') is used in 3D instead of the '*R*' (for 'roughness') in 2D. It should also be noted that unlike 2D parameters, which are evaluated along several sampling length, the 3D parameters are determined within one sampling area and there is no distinction in nomenclature as to whether they derive from primary, roughness or waviness information.

Historically adopted taxonomies such as roughness and waviness has been replaced by the scale-limited surface concept, surface from which the topographical parameters are computed. The scale-limited surface represents the resulting surface after the post-processing, which is based on: (i) the removal of the irrelevant form and translation errors, and (ii) filtering (if applicable) to remove non desired frequency components from the study. Unlike 2D characterization, where filtering is systematically applied, the idea of functional filtering is gaining recognition. As stated by Stout et al. [23], filtering should be applied only when the bandwidth of the frequency contents affecting the functional properties is understood. Therefore, it is suggested that surfaces should not be further filtered after eliminating the form components for fatigue performance analysis.

Areal parameters are classified into field and feature parameters. The field parameters are based on statistics of the whole surface whereas the feature parameters are defined over data from previously identified segments of the surface texture. Several areal parameters have been identified as potential indicators of fatigue performance. A summary and description of the most significant areal parameters is presented in this section, where:

• The discrete approximations of the formulas are given since surfaces are always sampled and digitised (for continuous descriptions of the formulas the reader is referred to [24]).

• Figure 7 shows the coordinate system used to define 3D topographic parameters, where η is the mean value surface (the reference datum).

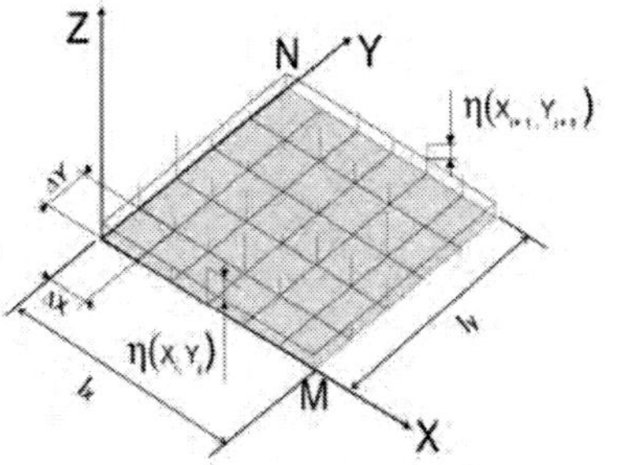

Fig. 7. Coordinate system used to define 3D topographic parameters.

4.1 Height parameters

Fatigue life is strongly dependant on surface roughness when the average asperity heights exceed the intrinsic defects in material [16]. The height descriptor parameters are therefore necessary for surface characterization of fatigue specimens. Among height descriptors, the average roughness (S_a), analogous to the 2D R_a is defined as:

$$S_a = \frac{1}{MN}\sum_{j=1}^{N}\sum_{i=1}^{M}\left|\eta\left(x_i, y_j\right)\right|, \quad (3)$$

where M and N are the number of samples in each axis. As mentioned in the section 2, the characterization should include however, other parameters in order to characterize all attributes that affects fatigue life. As demonstrated by Andrews *et al.*, the average fatigue life decreases with increasing the standard deviation in asperity heights [16]. Therefore, the root mean square height parameter, which corresponds to the standard deviation of heights, should also be included in the analysis:

$$S_q = \sqrt{\frac{1}{MN}\sum_{j=1}^{N}\sum_{i=1}^{M}\eta^2\left(x_i, y_j\right)}. \quad (4)$$

Among parameters describing the extreme characteristics of surface height, the S_v parameter, which describes the maximum valley depth has been highlighted. As expected, high values of S_v has been correlated to lower fatigue strength [24].

Due to the predominant functional role of the valleys over the peaks on fatigue performance, the Skewness parameter has the potential to deliver functionally useful results regarding surface valleys. The skewness quantifies the symmetry of the height distribution and mathematically is evaluated as follows:

$$S_{sk} = \frac{1}{MNS_q^3}\sum_{j=1}^{N}\sum_{i=1}^{M}\eta^3\left(x_i, y_j\right). \quad (5)$$

A negative S_{sk} indicates a predominace of valleys, whereas a positive S_{sk} indicates a predominance of peaks.

Additionally, the Kurtosis of the surface describes the spread of the height distribution, and is described as:

$$S_{ku} = \frac{1}{MNS_q^4}\sum_{j=1}^{N}\sum_{i=1}^{M}\eta^4\left(x_i, y_j\right). \quad (6)$$

A Kurtosis value higher than 3 (S_{ku}>3) indicates a narrowing of the height distribution function and presents physically as a more 'sharply peaked' surface. A purely Gaussian surface would have an S_{sk} value of 0 and an S_{ku} value of 3

Consequently, a negative S_{sk} and high value of S_{ku} could indicate a reduced fatigue life. However, it should be noted that the S_{ku} parameter in particular is very sensitive to surface outliers and as a result can present a high dispersion.

4.2. Spatial parameters

Spatial parameters are determined and quantified by tools based upon the Fourier transform and autocorrelation, detailed descriptions can be found in [19].

As demonstrated in section 2, the spatial properties of the texture has a strong impact on the stress concentration, and by extension, in fatigue life. Accordingly, the fastest decay autocorrelation length parameter, S_{al}, has been suggested as good fatigue performance indicator [24]. This parameter is described as the horizontal distance of the autocorrelation function having the fastest decay to 0.2, and is a quantitative measure as to the distance along the surface by which one would find a texture that is statistically different from the original location. Therefore, a large value of S_{al} denotes that the surface is dominated by long wavelength components, *i.e.* that the texture presents big spacing and vice versa. Suraratchai *et al.* [12] analysed the effect of the texture orientation relative to the axial loading direction on the fatigue life, demonstrating that when the groove direction is parallel to the loading axis, the stress concentration factor generated is very low. The characterization of the directionality of surface texture is therefore highly interesting. The texture aspect ratio parameter, S_{tr}, allows to identify texture pattern, *i.e.* to evaluate if the texture is directional or not. The value of this unit-less parameter ranges from 0 to 1. If S_{tr} value is near 1 this indicates that the surface is isotropic (*i.e.* it has the same characteristics in all directions, *e.g.* sand blasted surface), and in case of being near 0 corresponds to anisotropic surfaces (*i.e.* has an oriented and/or periodic structure, *e.g.* machined surface). For cases where the directionality is identified (S_{tr}~0) the texture direction parameter, S_{td}, calculates the texture direction of the surface expressed in degrees, anticlockwise.

4.3. Functional parameters

Functional parameters are calculated from the Abbott-Firestone curve (also called the areal material ratio curve) obtained by the integration of the height distribution of the whole surface. It represents the percentage of material traversed (in relation to the area covered) for a given depth, and allows analysing the surface properties at different regions (peak, core, valley). The ISO 25178-2 standard defines two arbitrary thresholds (10% and 80%) to define these regions, and the functional parameters are calculated following this definition. The void volume parameter, V_{vv}, has been identified as good fatigue performance indicator. This parameter is defined as the void volume in the valley zone from 80% to 100% surface material ratio, as shown in Figure 8).

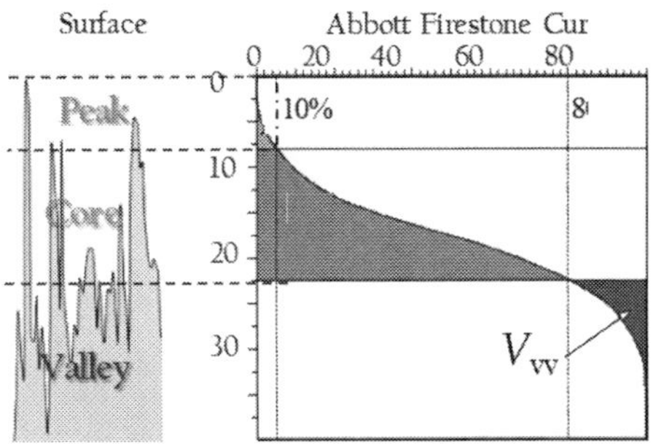

Fig. 8. Abbot Firestone Curve (in red). Peak (pink), core (green) and valley (blue) regions identified, according to the default threshold defined in ISO 25178-2 values (10%, 80%).

Smaller void volume values could correspond to larger tip radius at the root of valleys, which turns to better fatigue performance [23].

5. Conclusions

In this work, the inherent limitations of the most commonly employed 2D surface measurement and characterization techniques has been presented. It has been shown that defining topography based on the average roughness (R_a) is inadequate for correlating with fatigue performance.

Areal (3D) topographic characterization allows a better representation of the surface. An insight into the 3D areal characterization process and latest areal surface topography parameters critical for fatigue performance prediction has been given. Based on the current knowledge, the following areal parameters are suggested to use for topographic characterization for correlation with fatigue performance: S_a, S_q, S_{al}, S_{sk} and V_{vv}.

References

1. MR. Bayoumi, A.K. Abdellatif. Effect of surface finish on fatigue strength. Eng. Fract. Mech., 51, 861-70, (1995).
2. S.K. Ås, B. Skallerud, B.W. Tveiten, B. Holme. Fatigue life prediction of machined components using finite element analysis of surface topography. Int. J. Fatigue , 27, 1590-6, (2005).
3. D. Novovic, R.C. Dewes, D.K. Aspinwall, W. Voice, P. Bowen. The effect of machined topography and integrity on fatigue life. Int. J. Mach. Tool Manu., 44, 125-134, (2004).
4. H. Itoga, K. Tokaji, M. Nakajima, H.N. Ko. Effect of surface roughness on step-wise S–N characteristics in high strength steel. Int. J. Fatigue, 25, 379-385, (2003).
5. Y. Murakami, K. Takahashi, T. Yamashita. Quantitative evaluation of the effect of surface roughness on fatigue strength (Effect of depth and pitch of roughness). Nippon Kikai Gakkai Robunshu, A Hen/Transactions of the Japan Society of Mechanical Engineers Part A, 63, 1612-1619, (1997).
6. M. Yukitaka. Effect of Surface Roughness on Fatigue Strength. Metal Fatigue, Elsevier Science Ltd, Oxford, pp. 305-320, (2002).
7. TR. Thomas. Effect of scratches and of various workshop finishes upon the fatigue strength of steel. Eng., **16**, 449-85, (1923).
8. P.G. Fluck. The influence of surface roughness on fatigue life and scatter of test results of two steels. Proccedings ASTM, 51, 584-92, (1951).
9. R. Konečná, M. Kokavec, G. Nicoletto. Surface conditions and the fatigue behavior of nodular cast iron. Procedia Eng., 10, 2538-43, 2011.
10. Y.L. Lee, ME Barkey, HT Kang. Metal Fatigue Analysis, Elsevier, (2012).
11. D. Arola, C.L. Williams. Estimating the fatigue stress concentration factor of machined surfaces. Int. J. Fatigue , **24**, 923-930, (2002).
12. M. Suraratchai, J. Lmido, C. Mabru, R. Chieragatti. Modelling the influence of machined Surface roughness on the fatigue life of aluminium alloy. Int. J. Fatigue , **30**, 2119-2126, (2008).
13. ISO 4287: 1997. Geometrical Product Specifications (GPS)—Surface texture: Profile methoed—Terms, definitions and surface texture parameters.

14. D. Novovic, R.C. Dewes, D.K. Aspinwall, W. Voice, P. Bowen. Effect of machined topography and integrity on fatigue life. Int. J. Mach. Tool Manu., , **44**, 125-34, (2004).
15. D. Socie, G. Marquis. Multiaxial Fatigue, SAE, Portland, OR, (2000).
16. S. Andrews, S. Huseyin. A computer model for fatigue crack growth from rough surfaces. Int. J. Fatigue , 22, 619-630, (2000).
17. D. Taylor, The theory of critical distances: a new perspective in fracture mechanics, Elsevier, (2010).
18. K. Stout, L. Blunt. Three Dimensional Surface Topography. Butterworth-Heinemann, London, (2000).
19. R.K. Leach. Characterisation of areal surface texture. Springer, (2013).
20. B. Bhushan. Introduction to tribology. John Wiley & Sons, (2013).
21. X. Jiang, P.J. Scott, D.J. Whitehouse, L. Blunt. Paradigm shifts in surface metrology. part i. historical philosophy. Proc. Royal Soc. A: Math. Phys. Eng. Sci., 463 (2085), 2071.2099, (2007).
22. ISO 25178-2 2012 Geometrical Product Specification (GPS)—Surface Texture: Areal (International Organisation for Standardization).
23. K. Stout. The Development of Methods for the Characterisation of Roughness in Three Dimensions. Publisher, (1993).
24. D.T. Ardi, YG. Li, L. Blunt, MR. Bache. Surface topography and the impact on fatigue performance. Surface Topography: Metrology and Properties, 3, 015007, (2015).

Influence of graphite morphology on static and cyclic strength of ferritic nodular cast iron

Christian Gebhardt[1*], *Geng* Chen[1], *Alexander* Bezold[1] and *Christoph* Broeckmann[1]

[1]Institute for Materials Applications in Mechanical Engineering, RWTH Aachen University, 52062 Aachen, Germany

Abstract. High silicon alloyed nodular cast iron consists of a purely ferritic matrix and graphite nodules, mainly. Varying wall thicknesses and manufacturing conditions result in different graphite morphologies throughout a structural component. From an experimental point of view, axial fatigue and tensile tests were carried out on specimens with differently degraded graphite. From a numerical point of view, the microstructure has been modelled using a finite element (FE) approach with representative volume elements (RVE). The RVE models were built according to micrographs of fatigue specimens. The generated RVEs determine effective material properties through elasto-plastic homogenization and were subsequently analysed using a shakedown approach. In shakedown theory, the material re-enters the elastic regime after a few cycles of initial plastic deformation. This work uses the shakedown theorem to derive a lower bound estimation of the endurance limit from a non-incremental simulation. Here, the material has to be modelled elastic-perfectly plastic. The major challenge in modelling nodular cast iron is to determine suitable material parameters for the graphite and ferrite phase, revealed by parameter studies on the static and cyclic model. By using reasonable material parameters, fundamental effects, observed in the fatigue tests, were reproduced on the model level.

1 Introduction

Ductile cast iron is a material that meets a wide range of engineering applications due to its ductility, fatigue and corrosion properties. Since cast irons are near eutectic, they are well suited for complex and thin walled castings. As a consequence, the demand of ductile iron castings has significantly increased in the last decades, especially in sectors with extreme demands for mechanical properties such as the automotive and wind energy industry [1]. Next to classic ferritic and ferritic-pearlitic ductile cast iron materials (DCI), the innovative high silicon alloyed materials GJS-500-14 and GJS-600-10 gain new popularity. The materials consist of a pure ferritic matrix and mostly nodular graphite inclusions. Fatigue data is, however, rare for these materials, which restricts its use in structural applications and leads to conservative design. Some of these applications feature a varying wall thickness, which yields different graphite morphologies throughout the component. The graphite morphology influences the endurance limit of nodular cast irons [2, 3]. In many industrial applications, during the design process elastic stresses are calculated using FE-analysis. Most often, fatigue design is performed by applying a single S-N-curve to the entire part using commercial software. Hence, using local endurance limits in the fatigue design process saves resources and leads to an optimised part geometry.

In general, conventional DCI lacks fatigue data in torsional and axial testing. Sofu [4] noticed that an increased number of nodules leads to higher endurance limits. Siefer [5] argues that varying microstructure does not influence the fatigue behaviour significantly. Sobota and Stets [6] stated similar theses. In circumferential bending and axial fatigue testing, the nodularity did not influence GJS-400-15 fatigue behaviour significantly. Small deviations are attributed to varying graphite shapes of the non-nodular graphite. These results contradict the findings of many other works, in which a decreasing nodularity evokes a decreasing endurance limit.

Circumferential axial bending tests of high silicon alloys GJS-450-18, GJS-500-14 and GJS-600-10 were performed by Löblich [7]. Löblich obtained small improvements of the endurance limit of silicon alloyed DCI compared to conventional GJS-500-7. However, this data is statistically secured poorly and large deviations occurred during testing. La Torre [8] detected high disparity between silicon alloyed and conventional DCI. In tensile fatigue testing, a 3.78 % Si alloy reached 358 MPa fatigue limit and a low silicon alloyed DCI 170 MPa.

Describing the fatigue lifetime with models considering the microstructure is a challenging task. Three groups of empirical models stand out in the literature considering the microstructure explicitly. In fracture mechanics based models, the inclusions are considered as a crack [9, 10]. Secondly, the graphite can be modelled as a notch in an otherwise defect free material [11, 12]. Lastly, when the graphite is considered as a defect the

* Corresponding author: c.gebhardt@iwm.rwth-aachen.de

fatigue lifetime is predicted according to Murakami and De Kazinczy [13, 14] using the area parameter or a surrounding circle respectively. These empirical models resemble two main disadvantages. First, the assumptions may lead to errors in the fatigue lifetime predictions due to the experimental character of the model parameters. Secondly and more importantly, they fail to respond to changing boundaries, such as new micro constituents and their various shapes. Additionally, fatigue lifetime predictions depend highly on the applied loading pattern. With respect to the costs of fatigue tests and the graphite morphologies possible, a numerical methodology provides useful insights into the graphite morphology's influence on the static and cyclic strength on a model level. The approach required should account for the varying microstructure. On account of this, a suitable approach based on the representative volume element (RVE) was introduced by Hill in 1963 [15]. Over the past decades, virtual material testing using RVE models has gained wide technological and commercial significance [16]. Here, computational micro-macro mechanics determine relationships between the microstructure and the macroscopic response using models on the microscale as simple as possible, commonly referred to as RVE [17]. Appropriate boundary conditions and the size of the RVE are required to perform an adequate homogenization. The size of the RVE depends on the mechanisms investigated. An RVE can contain a single inclusion surrounded by a matrix material, often referred to as unit cell model (UCM). Apart from that, material parameters for both matrix and inclusion phase are required.

The concept of RVE modelling may be combined with cyclic models. Performing incremental FE simulations is one method to solve this task. However, the method is costly with respect to the computational power and time. Moreover, introducing and simulating the loading patterns requires great efforts. This problem is evaded using the shakedown theorem based on Melan's theorem [18], where the loading history is considered through the bounding envelope of the loading domain. Thus, the specific loading history is not required and complex loading conditions can be considered with a reasonable effort [19], which is why shakedown approaches are often referred to as direct methods (DM). Based on the condition applied, DM approaches historically ensued the static theorem by Melan or the kinematic theorem introduced by Koiter in 1956 [20]. In shakedown theory, the structure deforms plastically in the first few cycles. Then, the material returns to the elastic regime and behaves pure elastically in subsequent cycles. Mathematically, all conditions of the theorem form an optimisation problem whose result is a so-called load factor, which leads to a lower bound estimation of the endurance limit of the structure. In the original theorem, the application is restricted to elastic-perfectly plastic material.

In modelling nodular cast iron, the choice of appropriate material parameters is the major challenge, especially when the phases are modelled elastic or elastic-perfectly plastic. Material parameters for the matrix phase are found easily compared to the graphite phase. Here, comparing grain size and chemical composition to single phase materials is one possible way. A matrix equivalent material was casted in [21]. Uniaxial tension tests at different temperatures and nominal strain rates provide material parameters subsequently used in micromechanical simulations. The graphite phase, however, shows strong anisotropic behaviour , indicated by nanoindentation tests performed in [22]. There, indentations on varying graphite shapes yielded a high diversity of material responses. The issues when measuring elastic properties of particles in composites were investigated in [23]. Resultant, the Oliver-Pharr method cannot be applied to accurately determine the elastic modulus, especially when the indentation depth is not within a particle-dominated depth. Therefore, a high scatter of Young's moduli and yield strength for the graphite are reported in the literature, resulting likewise partly from nanoindentation tests [24, 25]. Other values for the Young's modulus of the graphite nodules result from assumptions made and results obtained in micromechanical simulations. Early studies [26] modelled the graphite as voids because a single phased graphite material has a negligible stiffness. An early debonding observed in uniaxial tension tests supports this assumption. Several studies investigate DCI with unit cell models [27–33]. These models are suitable to investigate material parameters and models. They are oversimplified when graphite morphologies should be investigated. RVE models obtained directly according to micrographs are shown in [24, 34–36]. Lastly, a 3D UCM of a single inclusion has been developed in [37] by performing metallographic investigation of up to 100 slices of the same cross section. Assembling the micrographs yielded a model of a single nodule. Summarizing the literature, accurate material parameters for the graphite phase are hard to find. The Young's modulus varies considerably from 0 – 375 GPa, pointed out by [28].

2 Experimental Procedure

2.1 Fatigue Testing

The setup for the casting of the raw material consist of 12 cuboids (100 x 200 mm). A single sprue is used in the middle of the setup. Each cuboid has a single riser. Each cuboid yields four specimens, manufactured in vertical direction from the raw material. Varying the magnesium and cerium content spawned four different graphite morphologies. The first alloy *R* is taken as reference due to its high nodularity. A graphite morphology with reduced nodularity is called *RN*. The abbreviation *BN* indicates bigger nodules. Finally, *V* refers to a vermicular-like graphite morphology. Thermal analysis determined the carbon content. Spectroscopy provided silicon-, magnesium and cerium content, illustrated in Tab. 1.

Table 1. Chemical composition of investigated alloys in wt %.

	R	RN	BN	V
C	3,02	3,07	3,00	3,07

Si	3,78	3,85	3,95	3,95
Ce	<0,0033	<0,0033	*0,0470*	<0,0033
Mg	0,0327	0,0280	*0,0416*	0,0218

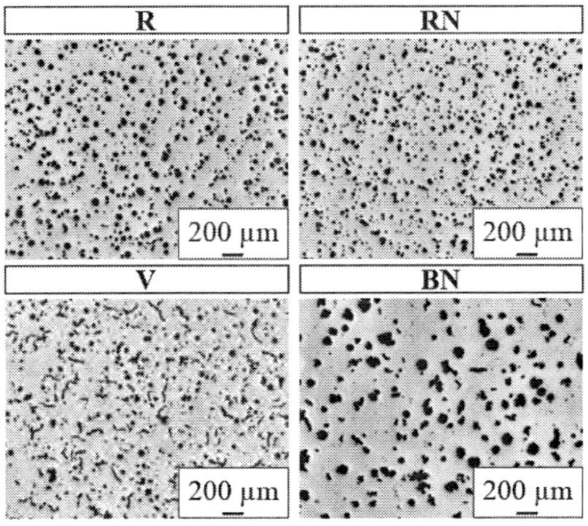

Fig. 1. Representative graphite morphologies: reference (R), reduced nodularity (RN), vermicular (V) and bigger nodules (BN).

The carbon- and silicon contents do not vary significantly, whereas the cerium and magnesium contents are high for the material with bigger nodules *BN*. The magnesium content is slightly elevated for R leading to bigger nodules compared to *RN*. Micrographs ensured the specimens taken from the blocks have a constant graphite morphology. The fatigue specimens were drilled and axial polished to a diameter of 12 mm. The length of the specimen is 200 mm. Axial fatigue testing was performed on electronically modernized 100 kN Amsler HFP 422 pulsator fatigue testing machines using a stress ratio $R = 0$. A crack is detected on a frequency drop of 5 Hz. The testing procedure applied is described in [38]. The endurance limit is obtained with a stair-case procedure using at least 15 specimens. An additional amount of 5 specimens is then tested on the highest stress level to statistically secure the transition region to long life fatigue. At most, this method saves number of specimens. No additional information is adversely obtained when few fractures occur on the highest stress level. The fatigue tests stop at $N = 10^7$ loading cycles.

2.2 Uniaxial tension tests

Tensile tests were carried out at room temperature using cylindrical specimens. The specimens were machined from the cast samples with a diameter of 10 mm and 58 mm of gauge length according to the DIN 50125-B. An amount of eight specimens was tested for each alloy *R, RN, V*, and *BN*. The specimens of *R, RN* and *V* were manufactured using material from the same batch as the fatigue specimens, whereas the specimens for *BN* were machined from a comparable alloy with comparable graphite morphology. Metallographic samples of an elastically deformed region (A), a mostly elastically deformed region (B) and a fully plastic region of a tensile specimen are shown in Fig 2. The frequency of debonded nodules increases with increasing plastic deformation. It has to be noted, that decohesion takes place already in the almost elastically deformed region (A). Besides the tensile tests, this effect could also be attributed to grinding during preparation of the metallographic specimens. Nevertheless, the micrographs reveal a weak connection between the two phases.

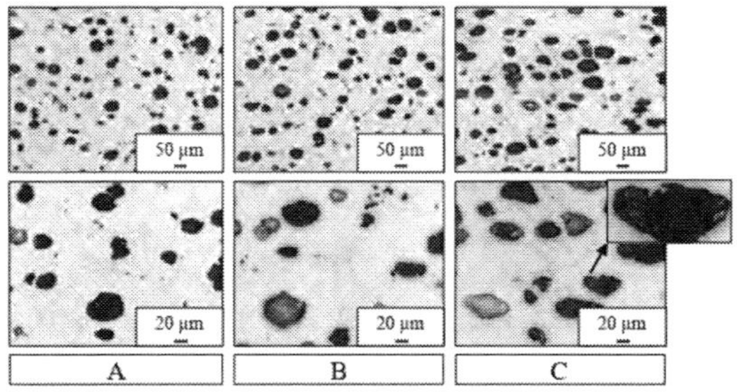

Fig. 2. Elastic (A) and plastic (B, C) deformation GJS-500-14 in a uniaxial tension test.

3 Numerical Modelling

3.1 Model

The RVE-Models are built directly according to binarized micrographs, where the pixels are classified according to Otsu's method. To determine a magnification that represents all geometrical features of the microstructure, the same area is analysed using 50x, 100x and 200x magnification. A single pixel of a micrograph taken with 50x magnification covers an area of 4,19 µm². Thus over- or underestimates the graphite area depending on the threshold used to classify the pixels. The graphite area of the 100x- and 200x magnification is approximately equal. However, large scatter is obtained comparing single micrographs of the 200x magnification. Therefore, the size of a 200x magnification is not representative. Consequently, the RVE is built directly from a micrograph of 100x magnification. Beyond the graphite area, shape parameters such as circularity, aspect ratio etc. are analysed. The metallographic specimens were ground and polished. The microscope used is *AxioImager M2m Zeiss* with the camera *ProgRes SpeedXTcore5 Jenoptik*. A 2.5D RVE-Model is built according to [39], as the complexity of the microstructure prohibits the use of 3D RVE. The 2.5D approach is a compromise between plane stress and strain approach and is not mesh sensitive compared to a conventional 2D approach. The algorithm compares positions of the binarized micrograph to an initial hexahedral mesh. The graphite inclusions are identified, sketched and remeshed using wedge elements. This methodology allows more DOF in mesh generation compared to substituting techniques. A convergence analysis shows that a hexahedral mesh converges slightly faster while a relative error of 0.99 % compared to the tetrahedral mesh occurs in the converged state. The edge length of elements is 3 µm leading to ~250.000 elements per RVE-Model. After the 2D-Meshing, the model is extruded to 2.5D in *Hypermesh*. Multiple-point constraints introduce the boundary conditions respectively. Using the shakedown model presented in [40], unfavourable simplifications are required. Both phases are elastic-perfectly plastic. Additionally, the graphite and ferrite phase are modelled perfectly bonded.

Table 2. Summary of the RVE-model.

Micrograph	100x
Type	2.5D
Size	1 mm x 1 mm x 1µm
Element type	C3D6
Elements quantity	~250.000
Ferrite / graphite	Elastic-perfectly plastic

3.2.1 Elasto-plastic homogenisation

The relations

$$\boldsymbol{\Sigma} = \frac{1}{\Omega}\int \boldsymbol{\sigma}\, d\Omega \tag{1}$$

$$\boldsymbol{E} = \frac{1}{\Omega}\int \boldsymbol{\varepsilon}\, d\Omega \tag{2}$$

hold for small deformations in absence of body forces and perfectly bonded phases relating the microscopic stresses $\boldsymbol{\sigma}$ and strains $\boldsymbol{\varepsilon}$ to the macroscopic stress $\boldsymbol{\Sigma}$ and strain $\mathbf{E}$ tensors on the RVE-domain Ω. The equations are referred to as average stress- and strain theorems respectively. When all phases are elastic and perfectly bonded

$$\boldsymbol{\Sigma} = \bar{C} : \boldsymbol{E} \tag{3}$$

holds and the effective Young's modulus is taken directly from the effective elastic tensor $\bar{\mathrm{C}}$. In plasticity, the micro strains are additively decomposed into an elastic and plastic part:

$$\varepsilon(\mathbf{u}) = \varepsilon^e + \varepsilon^p \tag{4}$$

Suquet [41] proposed a macroscopic strain measure

$$\boldsymbol{E}^p = \boldsymbol{E} - \bar{\mathrm{C}}^{-1} : \boldsymbol{\Sigma} \tag{5}$$

which is used to get the macroscopic yield strength defined as follows :

$$\Sigma_Y = \arg(E^p_{eq}(\Sigma_{eq}) = 0.1\ \%) \tag{6}$$

Kinematic uniform boundary conditions in two independent directions are imposed. Pierard [42] showed for a sufficient size of the RVE, the macroscopic response is independent from its chosen boundary condition which was ensured for the model.

3.2.2 Shakedown model

The static shakedown theorem by Melan is used to determine the shakedown factor α. The materials are assumed to be Mises-type, therefore the existence of a convex yield function $f(\boldsymbol{\sigma}(\boldsymbol{x},t),\sigma_Y)$ is assumed. The deformation is restricted to elastic-perfectly plastic. According to [43] when shakedown happens the total plastic dissipation is bounded, therefore the plastic strain $\dot{\varepsilon}^p$ would stop developing. Consequently, a time-independent residual stress field $\bar{\boldsymbol{\rho}}$ exists and the total stress can be expressed as follows:

$$\boldsymbol{\sigma}(\boldsymbol{x},t) = \alpha\boldsymbol{\sigma}^E(\boldsymbol{x},t) + \bar{\boldsymbol{\rho}}(\boldsymbol{x},t) \tag{7}$$

The index E denotes the stress field in a purely elastic reference body. Furthermore, König showed that for materials with a convex yield function the material will shake down over any given load path within a load domain. Then Melan's shakedown theorem is given as

$$\begin{aligned} &\max_{\alpha,\bar{\boldsymbol{\rho}}} && \alpha \\ &\text{subject to} && \boldsymbol{\nabla}\cdot\bar{\boldsymbol{\rho}} = \boldsymbol{0} \text{ in } \Omega \\ & && \bar{\boldsymbol{\rho}}\cdot\boldsymbol{n} = \boldsymbol{0} \text{ on } \Gamma_1 \\ & && F(\alpha\boldsymbol{\sigma}^E(\boldsymbol{P}_k) + \bar{\boldsymbol{\rho}}) \le \sigma_Y^2 \end{aligned} \tag{8}$$

The static shakedown theorem is expressed in finite element formulation as:

$$\begin{aligned} &\min_{\alpha,\bar{\boldsymbol{\rho}}} && -\alpha \\ &\text{subject to} && \textstyle\sum_{i=1}^{NG} \boldsymbol{C}_i\boldsymbol{\rho}_i = \boldsymbol{0} \\ & && F(\alpha\boldsymbol{\sigma}^e_{ik} + \bar{\boldsymbol{\rho}}_i) - \sigma_{Y,i}^2 \le 0 \end{aligned} \tag{9}$$

The matrix $\boldsymbol{C}$ takes the role of the gradient-operator for k load vertices and i Gaussian points. The problem is reformulated according to [44] for numerical reasons. The RVEs are modelled as previously explained. The loadings are introduced through static variables, meaning forces or stresses. Two independent uniaxial loadings are considered. The elastic stress fields are calculated using the commercial FE-solver Abaqus. The lower vertex $\boldsymbol{P}$ is zero respectively, corresponding to a pulsating cyclic loading. The optimisation problem is then formulated using an in-house Matlab based code. The problem is solved by the commercial optimisation toolbox Gurobi.

4.2 Parametric studies

4.2.1 Static model

The procedure previously mentioned is used to model a micrograph from a uniaxial tension test. The goal of this study is to understand the influence of the material parameters on a model level to clarify the uncertainty about the material parameters, pointed out in the literature review. Thus, the material parameters for the elastic-perfectly plastic phases vary systematically, as illustrated in Tab. 3. The Young's modulus for the matrix phase is estimated based on its silicon content and an empirical relation given in [45].

Table 3. Parametric study of static model.

Parameter	Ferrite	Graphite
Young's modulus *E* [MPa]	206 660	1 – 300 000
Yield strength [MPa]	350 – 1000	1 - 1000

The phases are perfectly bonded. Consequently, the model is highly idealistic with respect to the results of the uniaxial tension tests. Regardless, the frequency of debonded nodules is not high and as the macroscopic Young's modulus and the macroscopic yield strength are investigated, the effect of decohesion is neglected. The admissible range of the macroscopic Young's modulus is set to 160.000 ±5.000 MPa. Furthermore, the stress-strain curve obtained from the simulation is compared to an experimental curve until 1 % macroscopic plastic strain using squared errors. Based on these criteria, the red markers in Fig 3 indicate admissible sets of material parameters for an elasto-plastic model of idealised nodular cast iron.

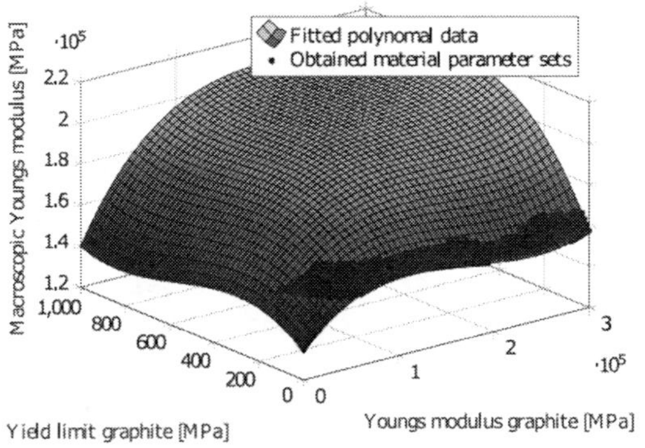

Fig. 3. Exemplary parameter study for a matrix yielding of 600 MPa.

None of the values from the literature for the graphite Young's modulus can be excluded. Moreover, multiple combinations of graphite Young's modulus (1 MPa – 300 GPa) and yield strength (1 – 172.5 MPa) are possible. Here, high values of one quantity correlate with low values of the other and vice versa. Applying the balance of mechanical energy, the external applied energy density splits into an averaged strain energy density and an averaged dissipation density. By further splitting the microstress into a self-equilibrated residual stress field and an elastic stress occurring in a purely elastic reference body, the average strain energy density separates in three parts. The applied external energy density is partly globally reversible, partly stored in residual strains and partly dissipated. Manipulating the material parameters of the elasto-plastic phases will influence the macroscopic response enabling multiple possible sets of material parameters.

4.2.2 Shakedown model

Two parametric study were carried out on the shakedown model to understand the influence of the graphite's material parameters under cyclic loads (Tab. 4). First, a study with a small step size was taken out in the low graphite Young's modulus range. Secondly, the influence of the graphite material parameters was studied in the high graphite Young's modulus range.

Table 4. Parametric study of cyclic model.

Parameter	Ferrite	Graphite
Young's modulus *E* [MPa]	206 660	1 – 300 000
Yield strength [MPa]	600	1 - 170

The material parameters of the matrix phase remain constant. In Fig. 4, the data points are linearly interconnected for constant values of the graphite yield strength $\sigma_{Y,G}$. Close to a graphite Young's modulus of 0 MPa, all curves tend towards a unique value. Afterwards, the shakedown load converges. At roughly 20,000 MPa no significant change on the shakedown load occurs, suggesting it became independent from the graphite Young's modulus.

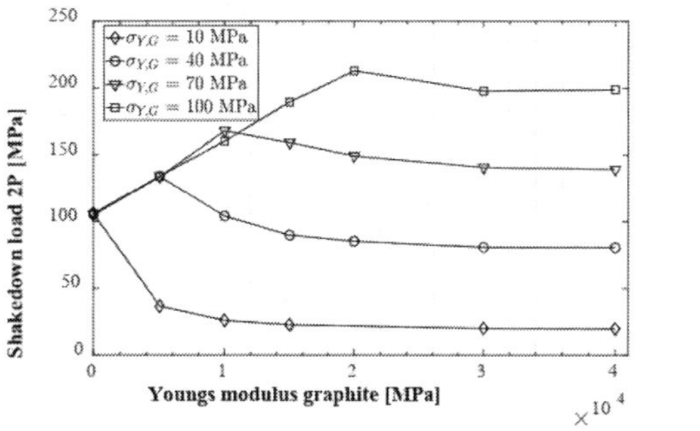

Fig. 4. Influence of low graphite Young's modulus on macroscopic shakedown load.

Fig. 5 reveals this assumption is hardly true for graphite Young's moduli higher than approximately 150,000 MPa. Why the shakedown load is independent from the Young's modulus in this range requires further investigation. A conceivable explanation reads: The location of failure changes from the nodules to the matrix when the curves become stable at roughly 20,000 MPa in Fig 4. Consequently, the shakedown limit drops due to an increase of the Young's modulus and vice versa. When the stiffness becomes sufficiently high, it induces high local stresses in the matrix. The deviatoric parts of the stress intuitively favour the failure of the matrix.

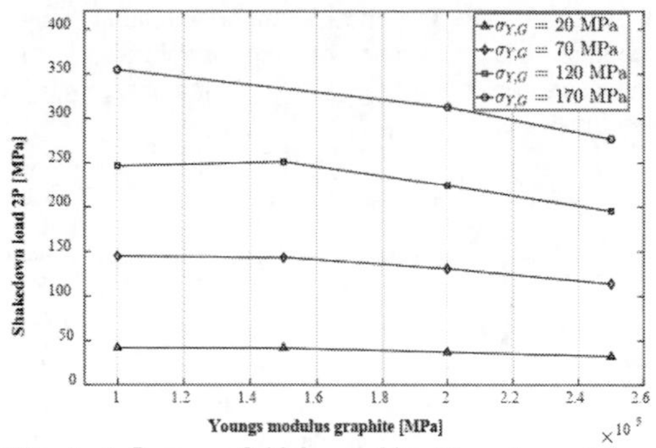

Fig. 5. Influence of high graphite Young's modulus on macroscopic shakedown load.

5 Results and Discussion

5.1 Experimental investigation– Endurance limit and tensile strength

The testing is statistically analysed with SAFD software [46]. A combined procedure is applied, where the stair step procedure yields a data point for 50 % failure probability. All stress levels with more than five specimens provide additional data points in the regression of the stress amplitude over the failure probability. The data is weighted with respect to the number of specimen. Tab. 5 shows the pulsating fatigue strength $\sigma_{A,R=0}$ for the investigated alloys with a failure probability of 50 %.

Table 5. Results of fatigue and tensile testing and microstructural analysis.

	R	RN	BN	V
$\sigma_{A,R=0,P=50\,\%}$ [MPa]	148.9	156.3	153.2	137.8
Ultimate Strength R_m [MPa]	536	528	*535*	487
Nodularity DIN 459-4 [%]	82.7	70.5	76.7	46.9
Nodularity ASTM E2567 [%]	72.0	54.3	51.8	33.0
Nodule Area (Avg / StD) [µm²]	738/ 859	546/ 659	1860 /264 9	680/ 951
Nodule Density [1/mm²]	163	176	63	143
Max Feret (Avg / StD) [µm]	33.0/ 20.7	30.5/ 18.7	50.6/ 43.6	38.3/ 31.4
Min Feret (Avg / StD) [µm]	23.9/ 15.7	21.0/ 12.8	33.5/ 29.9	22.3/ 16.0

The transition region was obtained for *R*, *RN* and *V*. Too many runouts prohibited an analysis of the transition region in the *BN*-series. Comparing the endurable stress amplitudes, vermicular graphite decreases the endurance limit, while there is no significant difference between the reference graphite, the graphite with reduced nodularity and the bigger nodules. Exemplary, fig. 6 shows a statistical evaluation of the fatigue testing for the vermicular-like graphite morphology *V*. The combined procedure yields three points for the stress levels and an additional point resulting from the stair-case procedure. The good correlation of the stress amplitude with the failure probability has a square error of $R = 0.9681$. The ultimate strength of *R*, *RN*, and *BN* shows no significant differences, whereas a reduction is obtained for the vermicular-like graphite morphology. Additional metallographic samples were taken from fatigue testing specimens of each series. The microstructure was analysed using an in-house MatLab code. Nodules with an area greater than 30 µm² were considered. The nodularity was analysed according to the standard ASTM E2567. Additionally, the bounds from Velichko [47] were used to classify the nodules into graphite types and determine the nodularity according to standards DIN 459-4. The nodularity is the highest for the reference *R*. *RN* and *BN* have values of 70.6 % and 76.7 % respectively. The nodularity is low for the vermicular like graphite *V* with a value of 46.9 %. The nodularity calculated according to the ASTM standard deviates significantly from those calculated with the DIN standard. The nodule density is high for the *RN*-morphology where the endurance limit is also the highest. As the *BN*-series has mostly bigger nodules compared to the others, the nodule density is low with 63 1/mm².

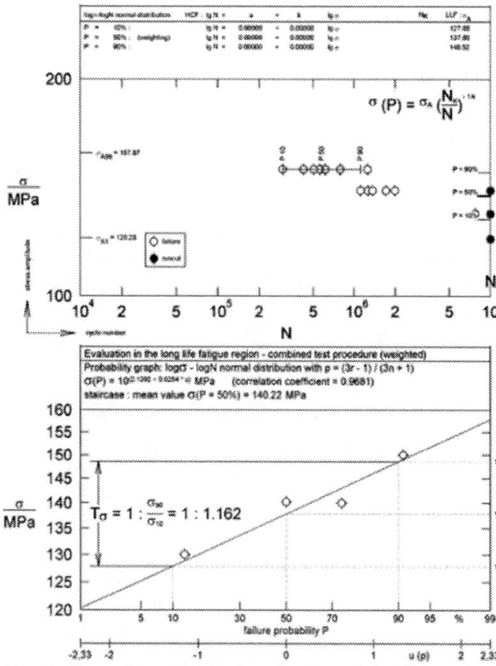

Fig. 6. Exemplary statistical evaluation of the fatigue testing for the vermicular-like graphite morphology *V*.

5.2 Shakedown

An amount of eleven micrographs per graphite morphology *R*, *RN*, *BN* and *V* was taken from specimens of the fatigue testing to model the RVEs. The matrix yield strength was set to 600 MPa and the Young's modulus to 206,660 MPa. A graphite Young's modulus of 15 GPa and a yield strength of 40 MPa are chosen because here, failure is expected to occur in the matrix and the macroscopic response is reasonable with respect to the parametric study. The highest relative error between two loading directions is 1.1 %. Therefore, the data points in the boxplots in Fig. 7 are mean values respectively.

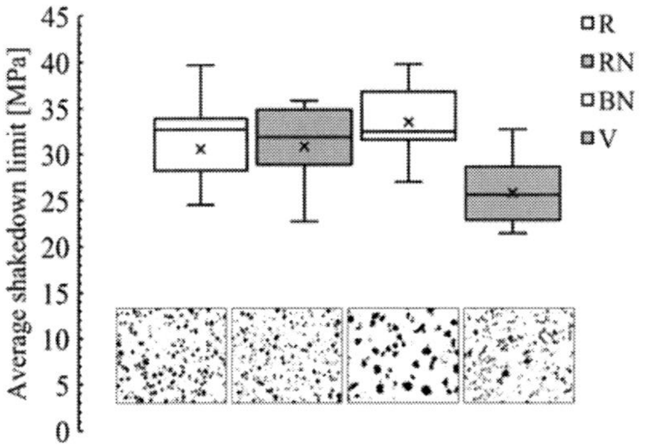

Fig. 7. Boxplots of averaged shakedown limits. Eleven micrographs were analysed per sample group.

The vertical lines mark the maximum and minimum values of a sample group. A single RVE in the R-series was excluded as the value was outside of the 1.5 interquartile range. Here, a single extremely degraded nodule resulted in a low shakedown limit of 15.4 MPa. The average shakedown limit for the sample group *V* is the lowest, corresponding to the experiment. High deviatoric stress parts occur at the graphite phase. Intuitively, high deviatoric parts decrease the shakedown limit as they contribute to the Mises stress. The sample groups *R, RN* and *BN* are higher with small differences between the groups. The shakedown limits of the bigger nodules seem surprising with respect to defect based approaches. However, as the nodule density is low the equality constraints are fulfilled easier. Moreover, deviatoric stress merely occurs on the microscale as the shape of the nodules is mostly round. Comparing the grain sizes of the investigated alloys, assuming a constant yield strength for the matrix does not seem reasonable. The scatter within a sample group is considerably high and is attributed to high geometrical variance of the microstructure within the groups. Consequently, the microstructure should be evaluated on a statistical basis with a large number of samples.

6 Conclusion

This work compared four different graphite morphologies on an experimental and numerical basis. The experiments reveal that defect-based approaches might be unsuitable for estimating the fatigue limit of high silicon alloyed nodular cast iron, as bigger nodules had a higher fatigue limit. As expected, vermicular graphite decreases the fatigue limit. However, the fatigue limit does not correlate with the nodularity. Broadly speaking, DCIs with round nodules favour a higher lifetime independent from the nodule's size. When the graphite is degraded, the size controls the fatigue limit. However, it is difficult to draw conclusions about the graphite morphology because, in fact, four different alloys were investigated.

The core issue in modelling nodular cast iron is to define suitable material parameters for the graphite and ferrite phase. A set of material parameters was presumed where the failure seems to take place in the matrix according to the experimental observations. Fundamental effects of the graphite morphology were reproduced using the shakedown theorem on simplified RVE models obtained from micrographs of the fatigue specimens. The values are not comparable to the experimental endurance limit due to various reasons, primary because the shakedown limit does not correspond to a failure probability of 50 %. The high values for the *BN* graphite morphology qualitatively agree with the results of the experiments. This is attributed to the high hydrostatic parts, which do not contribute to the Mises stress. However, effects of the matrix were completely neglected and hardening was not considered. The matrix yield strength depends on the grain size and therefore on the graphite morphology.

Acknowledgements

The project has been funded through the German ministry for economy and technology (BMWi) by a resolution of the German Bundestag. IGF 18524N

References

1. N.S. Tiedje, MST **26**, 505-514 (2013). doi: 10.1179/026708310X12668415533649
2. N. Costa, N. Machado, F.S. Silva, CTMAT **20**, 120-128 (2008)
3. M. Endo, JSMS **38**, 1139-1144 (1989). doi: 10.2472/jsms.38.1139
4. M. Sofu, S. Okada, A. Sobota, AFS Transaction, 173-182 (1978)
5. W. Siefer, K. Orths, Gießerei-Forschung **4**, 3–7 (1969)
6. W: Stets, A. Sobota, (AiF Nr. 13507 N/1)
7. H. Löblich, (AiF Nr. 41 EN, 2012)
8. U. de La Torre, A. Loizaga, J. Lacaze, J. Sertucha, MST **30**, 1425-1431 (2013). doi: 10.1179/1743284713Y.0000000483
9. M.H. El Haddad, T.H. Topper, K.N. Smith, EFM **11**, 573-584 (1979). doi: 10.1016/0013-7944(79)90081-X
10. N. Costa, N. Machado, F.S. Silva, Int. J. Fatigue **32**, 988–995 (2010)
11. P. Lukas, L. Kunz, B. Weiss, R. Stickler, FFEMS **3**, 195–204 (1980)
12. M.R. Mitchell, J. Eng. Mater. Technol. **99**, 329 (1977). doi: 10.1115/1.3443549
13. F. de Kazinscky, J. Iron Steel Inst. **20**, 851-855 (1970)
14. Y. Murakami, S. Nemat-Nasser, EFM **16**, 373-386 (1982). doi: 10.1016/0013-7944(82)90115-1
15. R. Hill, J. Mech. Phys. Solids **11** 357-372 (1963). doi: 10.1016/0022-5096(63)90036-X

16. J. Llorca, C. González, J.M. Molina-Aldareguía, J. Segurado, R. Seltzer, F. Sket, M. Rodríguez, S. Sádaba, R. Muñoz, L.P. Canal, *Advanced materials (Deerfield Beach, Fla.)* **23**, 130-147 (2011). doi: 10.1002/adma.201101683
17. S.J. Hollister, N. Kikuchi. Comput. Mech. **10**, 73–95 (1992). doi: 10.1007/BF00369853
18. E. Melan, *Der Spannungszustand eines Mises Henckyschen Kontinuums bei veränderlicher Belastung*. (Hölder-Pichler-Tempsky; Akad. d. Wiss, Wien, Leipzig, 1938)
19. J.W. Simon, D. Weichert, Comput. Mech. **49**, 477–485 (2012). doi: 10.1007/s00466-011-0656-8
20. W.T. Koiter: KNAW 14-34 (1956)
21. T. Andriollo, K. Hellström, M.R. Sonne, J. Thorborg, N. Tiedje, J. Hattel, J. Mech. Phys. Solids **111** 333-357 (2018). doi: 10.1016/j.jmps.2017.11.005
22. S.K. Pradhan, B.B. Nayak, S.S. Sahay, B.K. Mishra, Carbon **47**, 2290–2292 (2009). doi: 10.1016/j.carbon.2009.04.024
23. W. Yan, C.L. Pun, Z. Wu, G.P. Simon, Composites Part B: Engineering **42**, 2093-2097 (2011). doi: 10.1016/j.compositesb.2011.05.002
24. D.O. Fernandino, A.P. Cisilino, S. Toro, P.J. Sanchez, Int. J. Fract **207**, 1-26 (2017). doi: 10.1007/s10704-017-0215-1
25. P. Dierickx, C. Verdu, A. Reynaud, R. Fougeres, Scripta Materialia **34**, 261-268 (1996). doi: 10.1016/1359-6462(95)00496-3
26. S.H. Pundale, R.J. Rogers, G.R. Nadkarni, AFS Transactions **106**, 99-106 (1998)
27. T. Andriollo, J. Thorborg, N.S. Tiedje, J. Hattel, Mater. Sci. Eng. **84**, 12-27 (2015). doi: 10.1088/1757-899X/84/1/012027
28. T. Andriollo, J. Hattel, Mech. Mat. **96**, 138–150 (2016). doi: 10.1016/j.mechmat.2016.02.007
29. T. Andriollo, J. Thorborg, J. Hattel, Int. J. Solid. Struct. **100-101**, 523–535 (2016). doi: 10.1016/j.ijsolstr.2016.09.023
30. N. Bonora, A. Ruggiero, Int. J. Solid. Struct. **42**, 1401–1424 (2005). doi: 10.1016/j.ijsolstr.2004.07.025
31. W. Brocks, S. Hao, D. Steglich, J. Phys. IV France **6**, 43- 52 (1996). doi: 10.1051/jp4:1996605
32. M.J. Dong, G.K. Hu, A. Diboine, D. Moulin, C. Prioul, J. Phys. IV France **3**, 643- 648 (1993). doi: 10.1051/jp4:19937103
33. F. Rabold, M. Kuna, Comput. Mat. Sci **32**, 489–497 (2005). doi: 10.1016/j.commatsci.2004.09.016
34. D.O. Fernandino, A.P. Cisilino, R.E. Boeri, Mech. Mat. **83**, 110–121 (2015). doi: 10.1016/j.mechmat.2015.01.002
35. F.D. Carazo, S.M. Giusti, A.D. Boccardo, L.A. Godoy, Comput. Mat. Sci. **82**, 378–390 (2014). doi: 10.1016/j.commatsci.2013.09.044
36. F. Carazo, S. Giusti, A. Boccardo, P. Dardati, L Godoy, Mec. Comput. **30**, 611–629 (2011)
37. V. Di Cocco, D. Iacoviello, F. Iacoviello, A. Rossi, Pro. Eng. **109**, 135–143 (2015). doi: 10.1016/j.proeng.2015.06.223
38. F. Klubberg, P. Beiss, *Effizienzsteig. d. innov. Werkstofft: Wkstfftag'95*, 777–780 (1995)
39. Chen, G, *Strength Prediction of Particulate Reinforced Metal Matrix Composites* (Shaker, Aachen, 2016)
40. G. Chen, A. Bezold, C. Broeckmann, Comp. Struct. **189**, 330-339 (2018). doi: 10.1016/j.compstruct.2018.01.005
41. P. Suquet, *Elements of homogenization for inelastic solid mechanics* (Universite des Sciences et Technique du Languedoc, 1985)
42. O. Pierard, *Micromechanics of inclusion-reinforced composites in elastoplasticity and elasto-viscoplasticity: modeling and computation.* (Universite Catholique de Louvain, 2006)
43. J.A. König, *Shakedown of elastic-plastic structures. Fundamental studies in engineering*, **7** (Elsevier, Amsterdam, 1987)
44. A. François, H. Abdelkader, T. Le An, M. Said, P. Tao, J Glob Optim **37**, 609–630 (2007). doi: 10.1007/s10898-006-9069-1
45. W. Bleck, *Werkstoffkunde Stahl für Studium und Praxis* (Mainz, Aachen, 2010)
46. F. Klubberg, M. Hempen, H. Schäfer, P. Beiss, *7th Ams. Symp. Wrld. Dyn. Tes.*, 318-342 (2000)
47. A. Velichko, F. Mücklich, PM **43**, 192-207 (2006) doi: 10.3139/147.100297

MATEC Web of Conferences 165, 14015 (2018) https://doi.org/10.1051/matecconf/201816514015
FATIGUE 2018

Fatigue size effect due to defects in an AA7050 alloy

Foued Abroug[1-2], *Etienne* Pessard[1,*], *Guénaël* Germain[1] and *Franck* Morel[1]
[1]LAMPA, Arts et Métiers ParisTech Angers, 2 Bd du Ronceray, 49035 Angers, France
[2]LaMCoS, INSA -Lyon, 20 Avenue Albert Einstein, 69 621 Villeurbanne, France

Abstract. One objective of this project is to propose a fatigue design approach that is able to account for a large range of machining surface defects and different component sizes and geometries. Due to the huge size difference between a typical fatigue specimen and large aircraft components it was first necessary to confirm if a size effect can indeed be observed. This was done by introducing different numbers of artificial surface defects on smooth specimens. The material investigated is a 7050 Aluminium alloy (Al Zn6CuMgZr). Plane bending specimens both without and with artificial hemispherical surface defects were tested. The number of defects was varied from 1 to 44 defects per specimen and the defect size ranged from 60 μm to 800 μm in diameter. The test results allow the characterization of both the defect effect and scale effect on the fatigue response of the material. A probabilistic approach based on the weakest link concept together with a proper fatigue crack initiation criterion is used to account for the stress distribution and the size of the highly stressed volume. Predictions using FE simulations show a good agreement with experimental results and illustrate the importance of taking the scale effect into account in HCF.

1 Introduction

The effect of component size on fatigue strength is a well know problem in fatigue and has been investigated by many researchers since many years. Pogoretskii and Karpenko in 1965 [1] observed for example a high fatigue strength decreasing with the specimen length in rotating pure bending on a 37Cr4 steel (figure 1).

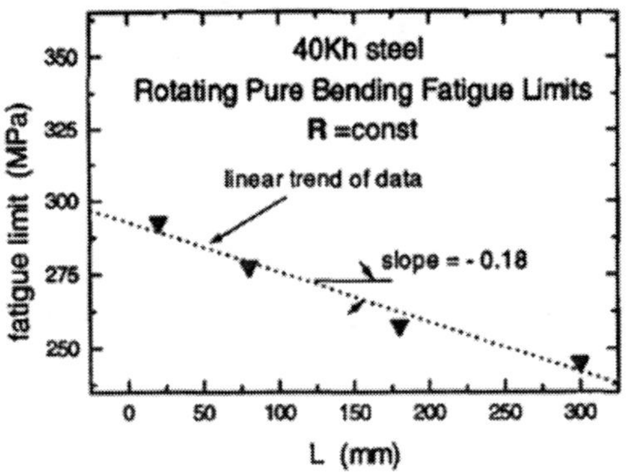

Fig. 1. Evolution of the fatigue strength in function of the specimen length [1]

The drop in the fatigue strength that is observed with increasing volume is very often explained by an increase in the probability of finding a critical defect or a weak zone as the volume of highly stressed material increases.

The impact of defect size on the fatigue strength is also well known and the Kitagawa-Takahashi diagram [2] is a useful tool to design material containing defects (see figure 2). The Kitagawa-Takahashi diagram in this case differed from his initial definition and the defect are considered as initial cracks.

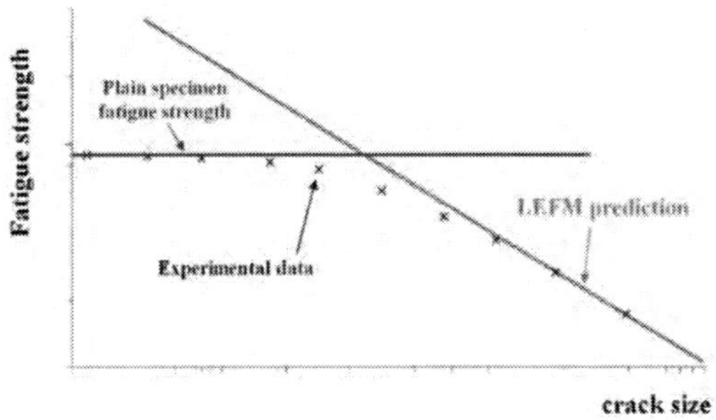

Fig. 2. Schematic representation of the Kitagawa–Takahashi diagram

The premise of this work is that these two phenomena are linked. That is, in a uniform stress field, without the presence of notches, it is really the defect size and the number of defects that control the stressed volume. Therefore, the Kitagawa-Takahashi diagram can be justified as an illustration of the scale effect.

In the present paper, the material and the experimental procedure will be firstly presented. Plane bending results with specimens both without and with artificial hemispherical surface defects to characterize both the defect effect and scale effect will be discussed. Then the observed specific size effect will be modeled by using Finite Element simulations together with the weakest link concept.

* Corresponding author: etienne.pessard@ensam.eu

2 AA7050-T7451 alloy, experimental procedure and results

2.1 Material

The material under investigation is the Al Zn6CuMgZr Aluminum alloy, also called AA 7050. The composition is presented in Table 1. Specimens are machined out of a 30 mm thick sheet in the rolling direction. The laminate operation creates a particular microstructure usually called "pan cake" microstructure with elongated grains in the rolling direction (from 5 to 300 µm) and thinner in the transverse direction (and 5 to 50 µm in depth) (See Fig.3).

Table1. Chemical composition of the AA7050-T7451 alloy, in weight percentage.

Element	Zn	Cu	Mg	Zr	Ti	Si	Fe	Mn
Weight%	6.03	2.22	1.85	0.10	0.039	0.038	0.098	0.01

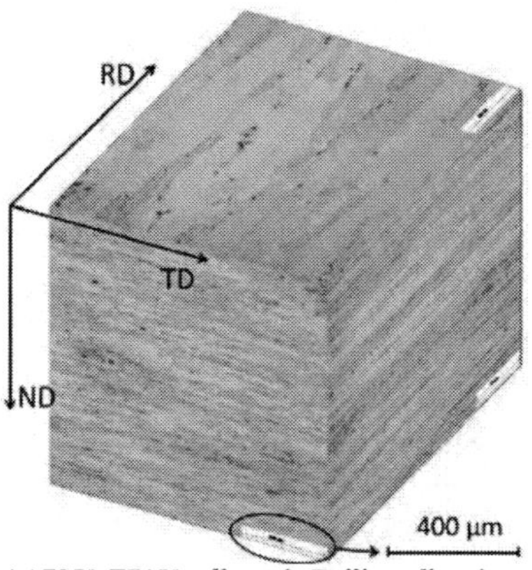

Fig. 3. AA7050-T7451 alloy a) Rolling direction, b) long transverse direction, c) short transverse direction

Different types of intermetallic particles are present at the grain boundaries and in the recrystallized grains (Mg2S i, Al7Cu2Fe, MgZn2, AlCuZn, etc...). These particles are brittle and can be at the origin of crack initiation under fatigue loading [3,4]. Monotonic tests show that the monotonic properties are slighty marked. The average Yield Strength and Ultimate Tensile Strength are respectively: σ_y=460MPa, σ_{UTS}=556MPa. The microhardness is also constant in the sheet depth with a value around 157Hv/0.5

2.2 Fatigue tests conditions

All the fatigue tests are carried out under fully reversed plane bending (R=−1) using a resonant testing machine at room temperature and ambient air at a frequency around 75Hz. The details of the machine and the geometry used are presented in Fig.4.

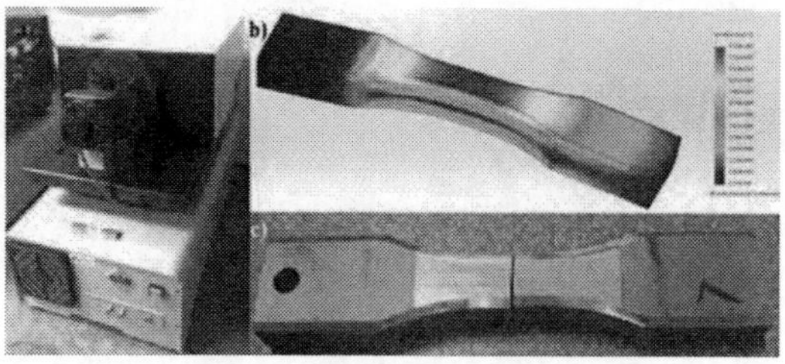

Fig. 4. AA7050-T7451 alloy a) Rolling direction, b) long transverse direction, c) short transverse direction

The endurance limit is determined at 2×10^6 cycles using Staircase or Locati methods. Stress value is decreased and increased by 10 MPa step for Staircase and Locati method.

Due to the high number of cycles applied on each steps, fatigue limit is estimated as the mean stress value between the final loading step leading to failure and the step before. Specimens are polished and artificial defects of 6 different sizes are introduced by micro-drilling technique (Ø60, 100, 200, 400, 600, and Ø800 µm). After machining, defects are scanned with a 3D BRUKER type profilometer (see Fig. 6). As it can be seen in Fig. 5, defect depths and diameters are carefully adjusted and controlled.

In order to characterize the scale effect, 2–3 specimens per defect size are tested with 1 and 22 defects per specimen. In addition, 3 specimens are tested in presence of 44 defects of Ø400 µm. To complete this large experimental campaign also presented in [5], a specimen containing 30 defects Ø200µm and one defect Ø400 µm is tested.

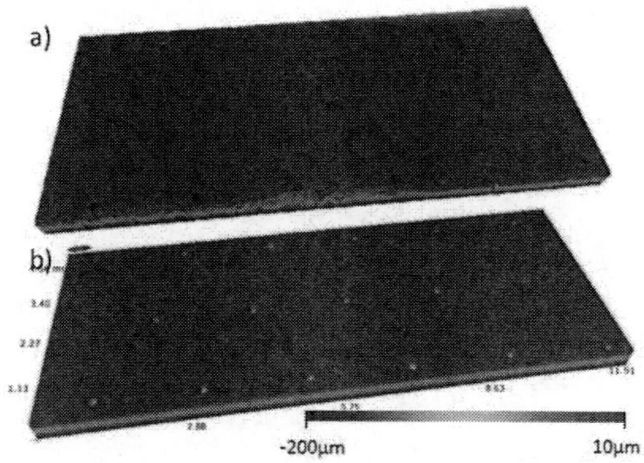

Fig. 5. Surface topography of the surface specimen a) face I containing 13 defects Ø200µm and one defect Ø400µm, b) face II containing 17 defects Ø200µm

2.3 Fatigue results

All fatigue tests are summarizing on Kitagawa Takahashi diagram in Fig. 6 and on the Table 1. On the Table 2, the standard deviation is estimated only for the 3 fatigue limits characterized via the Staircase approach.

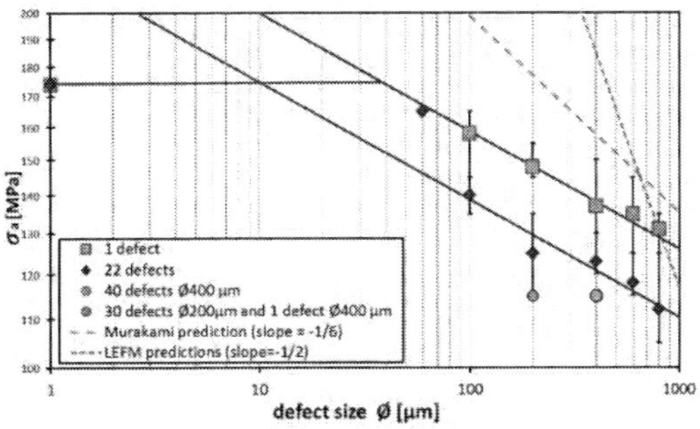

Fig. 6. Kitagawa-Takahashi diagram, AA7050alloy: experimental results for different defect size and number.

Table 2 : Fatigue test results

Defect size (µm)	Number of defect	Number of specimens	Fatigue limit at $2x10^6$ cyles (MPa)
0-Polished	0	15	174 ± 5.3
Ø 60	22	2	165
Ø 100	1	3	158
	22	3	140
Ø 200	1	3	148
	22	3	125
Ø 400	1	10	137 ± 9.5
	22	10	123 ± 5.3
	44	3	115
Ø 600	1	3	135
	22	3	118
Ø 800	1	3	131
	22	3	112
30 defects Ø200µm and 1defect Ø400µm		1	115

In presence of a single defect, above a critical defect size, the fatigue decreases progressively and follows a straight line in the log (σ_a) - log (defect-size) diagram (see Fig. 6). This behaviour is commonly known as the Kitagawa effect [2]. As shown in Fig. 6, the slope of decrease is the same in case of 1 and 22 defects and around (−1/10). Note that the defect sensitivity is lower compared to the Murakami criterion [6] and Linear Elastic Fracture Mechanics with slopes (−1/6) and (−1/2) respectively.

As shown on the Kitagawa-Takahashi diagram (Fig. 6) the fatigue limit decreases with the number of introduced defects. A drop of 20 MPa is observed when the number of defects is multiplied by 22. For specimens containing Ø400µm defects the fatigue limit decreasing seems to be less pronounced when the number of defects goes from 22 to 40 compare to 1 to 22.

The effect of the defect number on the fatigue limit can be explained by the scale effect. When increasing the number of defects, the highly stressed volume increases and the probability of finding a heterogeneity (i.e particle) or an unfavorably oriented grain increases. Hence, crack initiation is more likely to occur at a lower stress value than in the case with a single defect.

The fatigue test result on the specimen containing 30 defects Ø200µm and one defect Ø400µm illustrates that the size effect controls the impact of the defect size and the number on fatigue (see Fig.7 and Fig.8). For this particular specimen, initiation doesn't start from the largest defect (Ø400µm) but from one of the thirty smallest defect (Ø200µm). Even if the Ø200µm defects are smallest, the thirty Ø200µm defects create a largest highly stressed volume compared to the single defect Ø400µm. The probability to find a heterogeneity around a Ø200µm defect is therefore higher and initiation does not start from the largest defect.

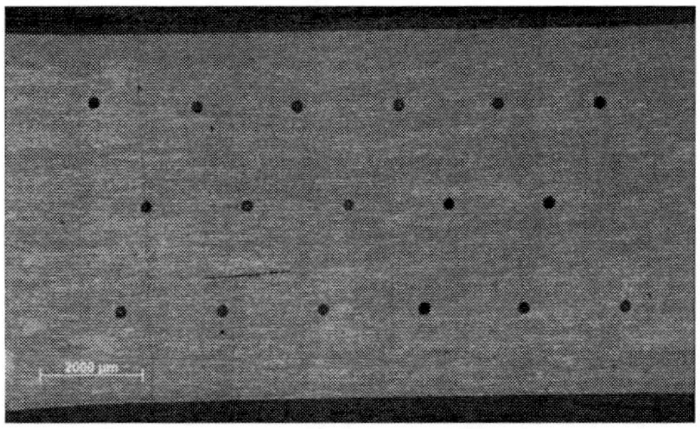

Fig. 7. Surface specimen after fatigue test showing a crack starting from a Ø200µm defect on the face II on the specimen containing 30 defects Ø200µm and one defect Ø400µm (see figure 5).

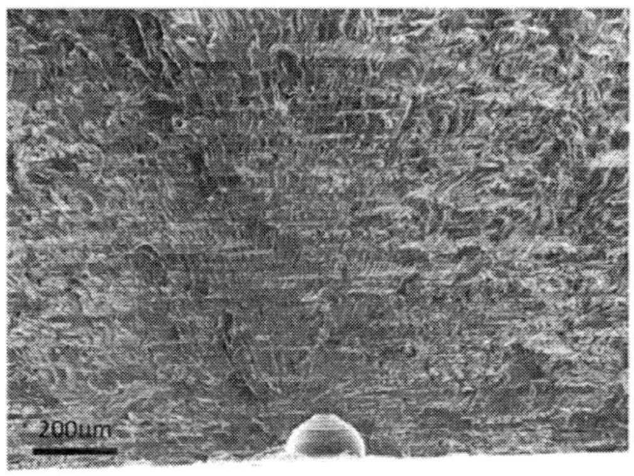

Fig. 8. Failure surface of the specimen containing 30 defects Ø200µm and one defect Ø400µm: initiation start from a Ø200µm defect.

4 Probabilistic fatigue criterion

4.1 Probabilistic framework

This section is about the modelling of the crack initiation damage mechanism in a polycrystalline aggregate containing geometric defects.

As a general rule most fatigue criteria can be defined by an equivalent mechanical quantity (i.e. stress, strain, energy, etc.), which is compared to a threshold quantity, for a given number of cycles. Crack initiation is subsequently predicted if the equivalent quantity is greater than or equal to the threshold quantity. In the following, for convenience, these will be referred to stress quantities. Hence, the condition for crack initiation is defined by:

$$\sigma_{eq} \geq \sigma_{th} \quad (1)$$

In order to take into account the inherent stochastic nature of the fatigue phenomenon it is proposed, as part of the modelling framework, to use a two parameter Weibull distribution [7, 8] to describe the threshold stress σ_{th}

$$f_0(\sigma_{th}) = \frac{\mathrm{m}}{\sigma_0}\left(\frac{\sigma_{th}}{\sigma_0}\right)^{m-1} exp\left\{-\left(\frac{\sigma_{th}}{\sigma_0}\right)^m\right\} \quad (2)$$

where σ_0 is the scale parameter and m is the shape parameter (or the Weibull exponent) used to reflect the scatter associated with the threshold stress and hence the fatigue limit.
The probability of microcrack initiation corresponds to the probability of finding a threshold stress σ_{th} that is less than the applied equivalent stress σ_{eq}. This can be expressed as:

$$P_{F_0} = P(\sigma_{th} < \sigma_{eq}) = \int_0^{\sigma_{eq}} f_0(\sigma_{th})\, d\sigma_{th} \quad (3)$$

Using the simple equation of Weibull's distribution function, the integration allows to express the failure probability as follows:

$$P_{F_0} = 1 - \exp\left[-\left(\frac{\sigma_{eq}}{\sigma_0}\right)^m\right] \quad (4)$$

To obtain the probability of crack initiation for the complete structure P_F , the weakest link theory is used [9]. The probability of survival of the structure $(1-P_F)$ is defined as the product of the survival probabilities of each elementary volume. The quantity V_0 corresponds to a reference volume and V_Ω is the total volume considered.

$$1 - P_F = \prod_{V_\Omega}(1 - P_{F_0}) \quad (5)$$

Hence the probability of survival of the complete structure is given by:

$$P_F = 1 - \exp\left[-\frac{1}{V_0}\iiint_{V_\Omega}\left(\frac{\sigma_{eq}}{\sigma_0}\right)^m dV\right] \quad (6)$$

Introducing the surface stress heterogeneity factor on the component surface H_m [10]:

$$H_m = \frac{1}{V_\Omega}\left[\iiint_{V_\Omega}\left(\frac{\sigma_{eq}}{\sigma_{max}}\right)^m dV\right] \quad (7)$$

with σ_{max} the maximum equivalent stress on the volume considered. The expression for the failure probability of the total structure becomes:

$$P_F = 1 - \exp\left[-\frac{V_\Omega}{V_0}H_m\left(\frac{\sigma_{max}}{\sigma_0}\right)^m\right] \quad (8)$$

In order to simplify the expression and to make a link with the high cycle fatigue limit, the macroscopic fatigue limit σ_a can be expressed via the following expression:

$$\sigma_a = \frac{\sigma_0}{K_{tmax}}\left[\frac{V_0}{V_\Omega}\frac{ln\left(\frac{1}{1-P_F}\right)}{H_m}\right]^{1/m} \quad (9)$$

Where K_{tmax} is the maximal equivalent stress concentration factor which corresponds to the maximal equivalent stress divided by the nominal equivalent applied stress.
In the case of polished specimen and a simple uniaxial loading, the surface stress heterogeneity factor H_m could be easily obtained by an analytical calculation [10]. In our study, the surface stress field due to hemispherical defect is more complex and numerical finite element simulation must be conducted to assess this quantity.

4.2 Finite element simulation

To simplify the calculation, the specimen is divided in different cells. Using the weakest link theory (Eq 5) allows calculating the failure probability of the entire specimen from the products of the failure probability of cells containing or not a defect.

The finite element simulation is performed on a cube containing a hemispherical defect on its upper surface (see Fig. 9). Calculation is conducted in order to assess the stress in each element. Elements are tetrahedral with reduced integration and convergence of calculation results is verified with the chosen size of elements. The material is supposed to be homogeneous and an elastic behaviour is used. The potential residual stresses are also neglected. The boundary conditions are as follows:

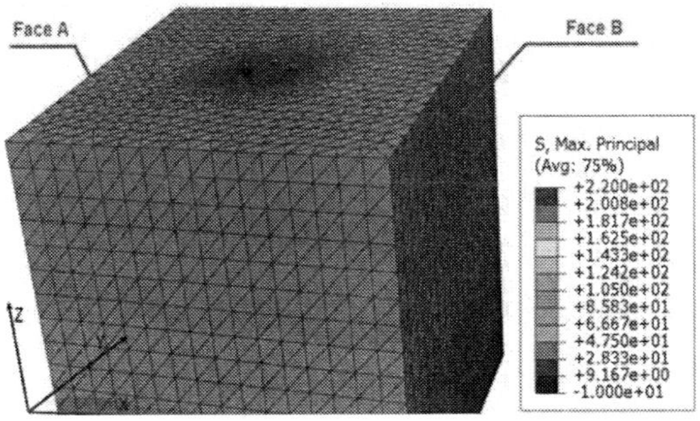

Fig. 9. FE simulation of a cube with hemispherical defect under bending loading

- Displacement Ux=0 along the plane X = 0 (surface A).
- Displacement Uy=0 along the line X = 0, Y = 0.
- Displacement Uz=0 at the point X = 0, Y = 0, Z = 0.

A bending pressure function representative of experimental bending loading is applied along the surface B (see Fig. 9).

It is well known in fatigue that initiation in metallic alloys mostly occurs in the surface or subsurface. Failure surface observations show for specimen with artificial defects that intermetallic particles are mostly localized in the first 50 µm in depth around the defect. In our failure probability calculation only the first 50 µm in depth will therefore be considered (see Fig. 10).

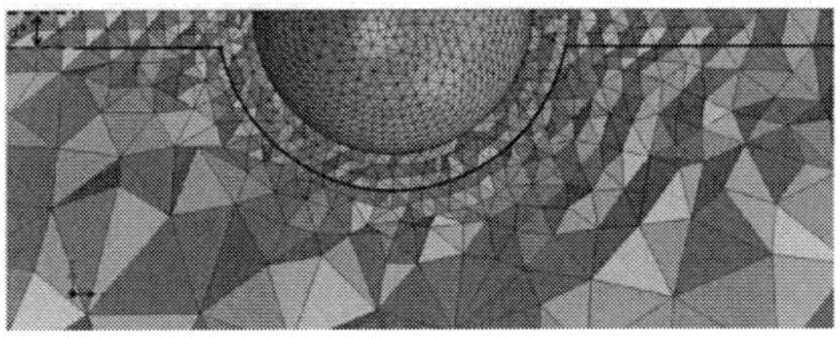

Fig. 10. Illustration of the d*= 50µm parameter in the Finite Element Model.

To identify the two model parameters m and σ_0 that defines the Weibull distribution of the equivalent stress threshold, the staircase results obtained on the specimens containing Ø400µm defect are used. Finite Element Simulation of the bending test on a specimen containing a hemispherical defect of Ø400µm has been conducted with different value of m to calculate the stress heterogeneity factor. In the simple case of a bending loading study in this works the maximal principal stress is considered. The maximal stress concentration factor K_{tmax} is the same for the different defects size and is equal to 2.2.

The scale parameter σ_0 is then calculated using Eq. (9). The couple of value (m, σ_0) that provides the lower error value between predictions and all experimental results is the couple (m = 22; σ_0 = 226 MPa).

4.3 Predictions

The predictions using the developed approach are in rather good agreement with the experimental results. The criterion correctly predicts the fatigue strength of the polished samples and the effect of the size and number of defect. The maximal error is observed for the specimen containing 22 defects of Ø60µm. The approach for the smallest defect size is conservative and overestimates the material sensitivity to the defects.

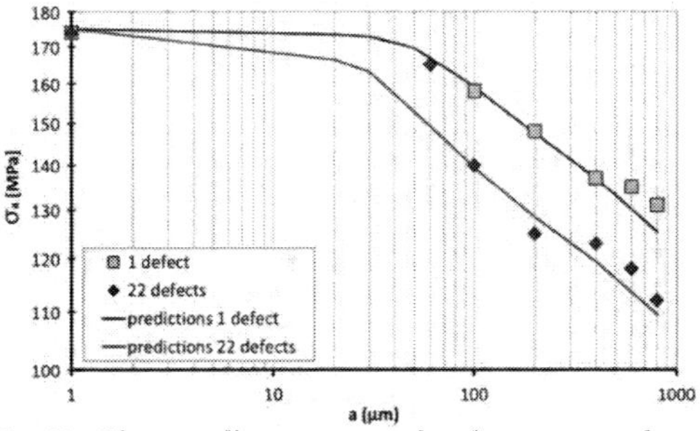

Fig. 11. Kitagawa diagram: comparison between experimental results and fatigue criterion predictions

The Fig. 12 shows the evolution of the fatigue strength with the number of defects.
Predictions show a good agreement with experimental results. Furthermore, it is shown that the decrease in fatigue limit slows down with the increase of defect number. The probabilistic model predicts that after a number of defects around 30, fatigue limit decrease is no longer significant. This tendency is usual for statistical size effect [11,12].

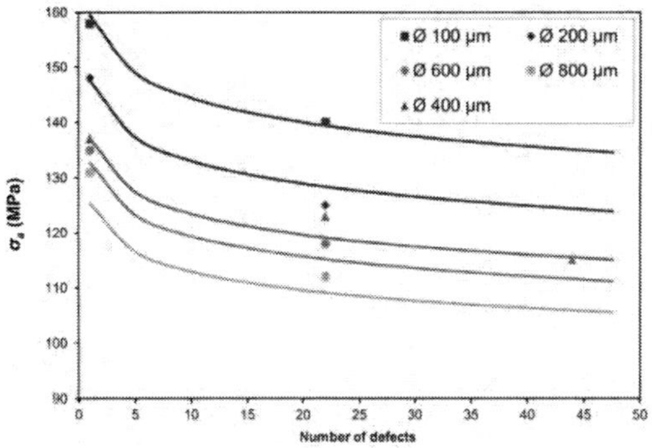

Fig. 12. Evolution of the fatigue limit with the number of defects comparison between experimental and predictions

It is not possible to compare in the Kitagawa diagram Fig.11 and on the diagram Fig. 12 the result obtained on the last specimen containing 30 defects Ø200µm and one defect Ø400µm. Based on the expression (Eq.9) of the fatigue strength using the criterion developed, the evolution of the fatigue strength in function of the product of $K_{tmax}\ H_m^{1/m}$ is plotted in Fig.13. In our case with a Kt_{max} equal to 2.2 for all specimens, this figure illustrates that the number and the size of defects control the stress heterogeneity factor H_m. It is clear in this figure that a single defect Ø400μm is less critical in fatigue that 20 defects Ø200μm and so that 30 defects Ø200μm. The same conclusion is observed between one defect Ø200μm and 20 defects Ø100μm.

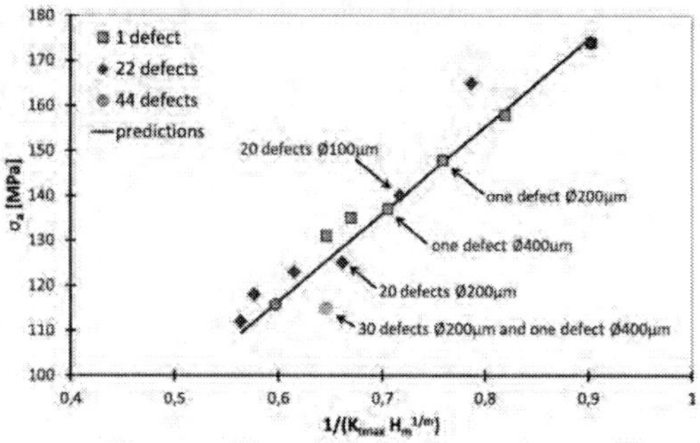

Fig. 13. Evolution of the fatigue limit in function of the $(K_{tmax} H_m^{1/m})^{-1}$: comparison between experimental results and predictions

The quantity K_{tmax} $H_m^{1/m}$ seems to be a relevant parameter to quantify the impact of defects or surface topography on the fatigue limit. The Figure 13 also illustrates that, knowing the K_{tmax} $H_m^{1/m}$ generated by any surface defect population allows predicting the high cycles fatigue strength of the component using the proposed approach

4 Conclusions

The main conclusions of this work are:

- A vast experimental plane bending fatigue campaign on specimens containing different defect size and number have been conducted.
- The Kitagawa diagram obtained shows a low sensitivity of the AA7050 alloy to the defect size compared to usual fatigue criterion approach (LEFM and Murakami).
- The studied material shows a high sensitivity to the number of defects.
- An approach based on Weibull's distribution law and the weakest link concept is proposed to take into account the high stressed volume size of each specimen. It allows to accurately predict the fatigue limit for all specimens (containing or not a single or several defects).

Beyond the results obtained on the AA7050 alloys, these works illustrate the importance of taking the scale effect into account when designing in fatigue. Usually only the impact of the largest defect is considered. However for some materials and manufacturing processes, the defect density on the surface or in the core is of great importance in fatigue.

Acknowledgements

This work was carried out within the FUI QUAUSI project with the support of industrial (Dassault Aviation, Europe Technologie, Figeac Aéro, Mecachrome, Spring Technologie, Precise, GEBE2, CETIM) and academic partners (LSN2-IUT Carquefou).

References

1. R. G. Pogoretskii and G. V Karpenko, Fiziko-Khimicheskaya Mekhanika Materialov, pp. 1, 90–94, (1965).
2. H. Kitagawa, and S. Takahashi, Metalspark, Ohio, p. 627-731, (1976).
3. M. Shahzad et al., Procedia Engineering, 2, pp 1015–1024 (2010).
4. V. K. Gupta and S. R. Agnew, Int J Fatigue, Vol 33 (2011).
5. F. Abroug et al. , Int J Fatigue 110 81-94 (2018)
6. Y. Murakami Metal fatigue: effects of small defects and nonmetallic inclusions. Elsevier; (2002).
7. W. Weibull A statistical theory of the strength of materials. Roy. Swed. Inst. Eng. Res. Report 151; (1939).
8. Weibull W. ASME J Appl Mech 1951;18:293–7 (1951).
9. A.M. Freudenthal and H. Liebowitz, editor. Fracture, vol. 2. New York: Academic Press;. p. 591–619 (1968).
10. L. Flaceliere and F. Morel. Fatigue Fract Eng Mater Struct 27(12):1123–35, (2004).
11. I.V. Papadopoulos. G, Lindley TC, editors. Multiaxial fatigue and design, ESIS 21. London: Mechanical Engineering Publications; p. 349–64. (1996).
12. T. Delahay and T. Palin-Luc Int J Fatigue 28:474–84, (2006).

Study of load history effects on the high cycle fatigue properties of high-strength low-alloy steel from self-heating measurements

Julien Louge[1,2,*], *Cédric* Doudard[1], *Sylvain* Calloch[1] and *Bastien* Weber[2]

[1]Institut de Recherche Dupuy de Lôme (FRE 3744), UBS/UBBO/ENSTA-Bretagne/ENIB – 2, rue François Verny 29806 Brest cedex 9, France
[2]ArcelorMittal Maizières Research, Voie Romaine 57283 Maizières-lès-Metz cedex, France

Abstract. In the context of high cycle fatigue (HCF), the experimental characterization of the fatigue properties is often performed by using specimens in a virgin state (i.e., without preliminary loading), and with a constant stress amplitude for each specimen. However, the load history applied to a real structure is more complex and the fatigue life prediction remains a difficult task because of the time dedicated to the classical fatigue tests (i.e., the specimen is loaded until failure) and the dispersion of fatigue lives. The load history effects on the HCF properties is characterized using an alternative method: self-heating measurements under cyclic loadings. This method is based on the observation of the mean steady state temperature evolution of a specimen under a successive series of cyclic loadings with increasing stress amplitude for each loading series. A probabilistic two-scale model was developed from the self-heating method able to predict HCF properties. Some self-heating tests are performed to study the influence of a load history effects. It seems that the plasticity is the most influential factor. So, the evolution of the plasticity is observed at the surface of the material under cyclic loading. There is a significant evolution in function of the plastic pre-strain.

1 Introduction

The standard method to characterize high cycle fatigue properties of steel grade requires one month of tests. To reduce the time of characterization, some fast methods have been developed, such as the self-heating method. This method, associated with a two scales probabilistic model, allows to obtain the fatigue limit with only one specimen in less than one day. This kind of model assumes that the damage is mainly due to microplasticity activation. Many steel grades have been studied with the self-heating method [1,2]. To go deeper in this way, and to expand the two scales probabilistic model, two kind of load history are studied with the self-heating method: a plastic pre-straining and a cyclic-pre-loading.

First, the self-heating method and the two scales probabilistic model associated are presented. The heat conduction equation is reminded, the test procedure and heat acquisition are introduced. Then, the two kind of load history are studied to show their influence on the self-heating curves. Finally, the mechanisms strain are analysed by observing the emergence of plasticity at the surface of the material under cyclic loading. The material studied is a High Strength Low Alloyed steel HSLA320 2mm thick (bare product) which was hot rolled. The mechanical properties the rolling direction (direction of interest) are: Yield Stress 340 MPa, Tensile Strength 445 MPa and a total elongation 31%.

2 Self-heating method

2.1. Self-heating test procedure

The self-heating method consist on apply a successive series of cyclic loadings, with increasing the stress amplitude, Σ_0 (the amplitude is constant per series) (Fig.1). A 0D approach is chosen and the corresponding heat conduction equation is given by

$$\dot{\theta}^{0D} + \frac{\theta^{0D}}{\tau^{0D}} = S_t^{0D}, \quad (1)$$

Fig.1. Successive series of cyclic loadings while increasing stress amplitudes used for self-heating tests.

* Corresponding author: julien.louge@ensta-bretagne.org

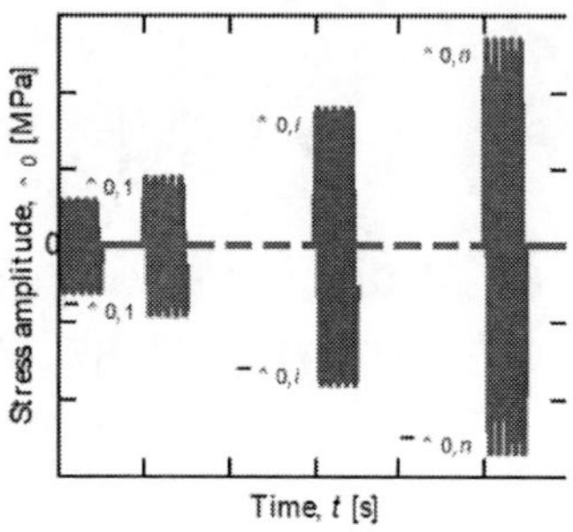

The temperature is measured by three thermocouples, one fixed on the center of the specimen and the two other are fixed on the grips of the hydraulic machine. The evolution of the mean temperature elevation during a series of cyclic loading is given by

$$\theta^{0D} = T_{\text{specimen}} - \frac{T_{\text{uppergrip}}+T_{\text{lowergrip}}}{2}, \quad (2)$$

and is shown in (Fig.2). This evolution reached a steady-state, called $\bar{\theta}^{0D}$.

Fig.2. Evolution of the mean temperature elevation during one series of cyclic loadings.

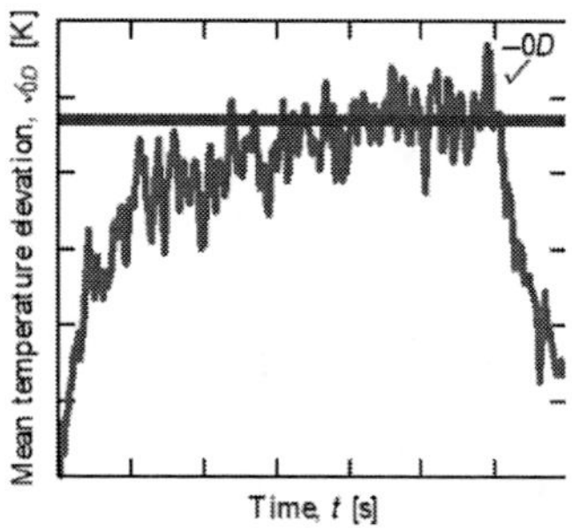

After all the series of cyclic loadings are performed on a specimen, the self-heating curve is obtained, representing the mean steady-state temperature evolution $\bar{\theta}^{0D}$ relating to the stress amplitude Σ_0 (Fig.3). Two self-heating regimes are observed. The first one for low amplitudes of load, called primary regime, and the second one for the highest amplitudes, called the secondary regime. It has been shown empirically that the intersection between the two self-heating regimes, in a logarithmic scale, provides a good approximation of the mean fatigue limit of the studied material $\bar{\Sigma}_\infty$.

Fig.3. Self-heating curves in a logarithmic *scale*.

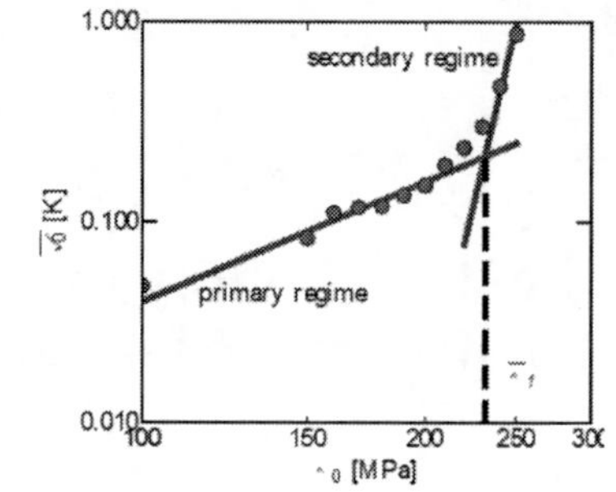

2.2. Two scales probabilistic model

To describe the self-heating curves and predict the fatigue properties, a two scales probabilistic model has been proposed [1-3]. A REV (Representative Elementary Volume) is defined as an elasto-platic matrix containing a set of inclusions with an additional plastic hardening (Fig.4). Under low amplitude of cyclic loadings, the dissipation is due to the elasto-plastic behaviour of the matrix. Then, above a given stress amplitude, the dissipation is due to the behaviour of the matrix combined with the activated inclusions.

Fig.4. Principle of the two scales probabilistic model.

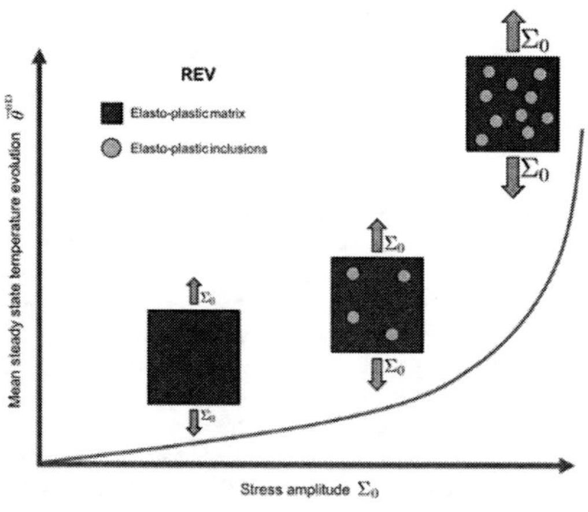

The dissipation associated with the matrix corresponds to the primary regime. The secondary regime is due to the dissipation of the matrix associated to the set of inclusions. Thus, the mean steady-state temperature elevation $\bar{\theta}^{0D}$ can be written as

$$\bar{\theta}^{0D} = \alpha\left(\frac{\Sigma_0}{\Sigma_{max}}\right)^2 + \beta\left(\frac{\Sigma_0}{\Sigma_{max}}\right)^{m+2}, \quad (3)$$

with the parameters α and β, which correspond respectively to the intensity of the primary and the secondary regime. The parameter m corresponds to the

https://doi.org/10.1051/matecconf/201816514016

slope of the secondary regime, related to the gradual appearance of active sites.

In the model, it is assumed that the site activation governs the fatigue damage initiation. Within the weakest link theory, the probability of failure, called P_F, is the probability of finding at least one active site. To determine the number of cycles to failure $N(P_F)$, an energetic criterion is used. The cyclic dissipation energy is constant for each cycle, so the critical dissipated energy can be written as a function of $N(P_F)$. Then, the number of cycle to failure is given by

$$N = \frac{A\Sigma_{\max}^2}{\overline{\Sigma}_\infty(P_F)(\Sigma_0 - \overline{\Sigma}_\infty(P_F))}, \quad (4)$$

with $\overline{\Sigma}_\infty(P_F)$ the fatigue limit associated with a given probability of failure. Finally, the ability of the model to predict S-N-P curves is shown Fig.5.

Fig.5. Validation of the method with an S-N curve.

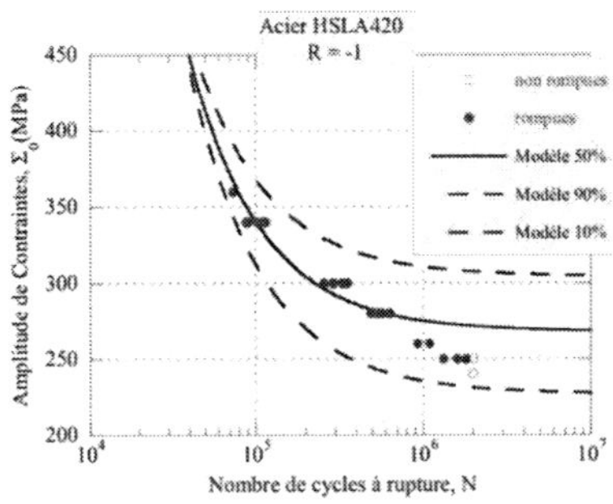

3 Load history effects influence on the self-heating measurements

The self-heating (SH) method has been developed [1-3] to characterize the fatigue properties using specimen in a as-delivered state (*i.e.*, without preliminary loading). The first aim of this study is to analyse the influence of load history effects on HCF properties. For that, two loads history are studied: a plastic pre-strain range and an initial series of cyclic loading.

In order to observe the evolution of the self-heating curves after performed a load history, two indicators are introduced. The first one represents the evolution of the intensity of the primary regime, called α_{norm}. It corresponds to the ratio between the amplitude value for the as-delivered state, α_{virgin}, and the value for the specimen where a history loading was applied, α_{history}, given by

$$\alpha_{\text{norm}} = \frac{\alpha_{\text{virgin}}}{\alpha_{\text{virgin}}}. \quad (5)$$

The parameter β and the exponent $m+2$ (Eq.3) represent respectively the intensity and the slope of the secondary self-heating regime. The slope is determined for the as-delivered specimen, $m=23$, and it is considered to be identical for the other self-heating curves. It has been shown that the secondary regime can be related to the fatigue life properties. So, the evolution of the secondary regime intensity corresponds directly to the evolution of the fatigue limit $\overline{\Sigma}_\infty$, estimated by the self-heating method. So, the second indicator, called β_{norm} given by

$$\beta_{\text{norm}} = \frac{\overline{\Sigma}_\infty^{\text{history}}}{\overline{\Sigma}_\infty^{\text{virgin}}} = \left(\frac{\beta_{\text{virgin}}}{\beta_{\text{history}}}\right)^{1/m}, \quad (6)$$

represents the ratio between the fatigue limit for the as-delivered state, $\overline{\Sigma}_\infty^{\text{virgin}}$, and the fatigue limit for the specimen where an history loading was applied, $\overline{\Sigma}_\infty^{\text{history}}$, determined by the self-heating method. For example, a value 0.95 for this indicator corresponds to a 5% decrease of the fatigue limit.

3.1. Plastic pre-strain effect

In order to study a plastic pre-strain effect on the self-heating measurements, a tensile test is performed on a specimen until a target strain level is reached. Then, a self-heating test is carried out (Fig.6).

The result of the self-heating test for a plastic pre-strain level of 1% is shown in Fig.7. The amplitude of the primary regime increased. It is not so easy to appreciate the evolution of the secondary regime in this figure, but the evolution of the parameters is given in Tab.1. The parameter β_{norm} is equal to 0.94, so the fatigue limit for the pre-strain material is around 6% lower compared to the as-delivered state.

Fig. 6. Principle of the test performed to study the influence of a plastic pre-strain effect.

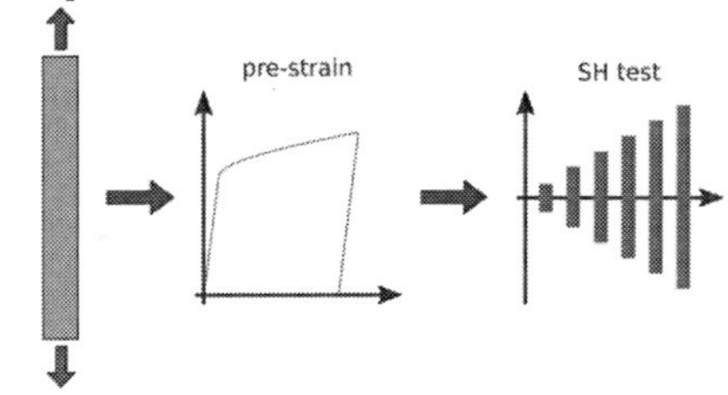

Fig. 7. Comparison of the self-heating curves for the virgin state and the 1% level of plastic pre-strain state.

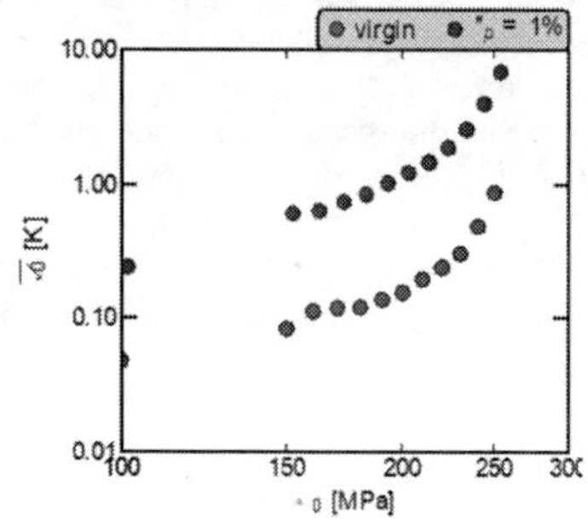

Tab. 1. Values of the model parameters for the as-delivered state and the 1% level of plastic pre-strain state.

State	α [K]	β [K]	α_{norm} [-]	β_{norm} [-]
as-delivered	0.251	0.696	1.00	1.00
Pre-strain 1%	1.558	5.048	6.200	0.924

3.2 Initial series of cyclic loading effect

For this study, an initial series of cyclic loadings is applied to the specimen, then a self-heating test is performed (Fig.8). First, the influence of the stress amplitude loading is investigated for a constant number of cycles (N=50 000 cycles). Then, the influence of the cycle number is analysed for a constant stress amplitude $\Sigma_0 = 210$ MPa (transition between the primary and the secondary regime).

Fig. 8. Principle of the test performed to study the influence of an initial series of cyclic loading.

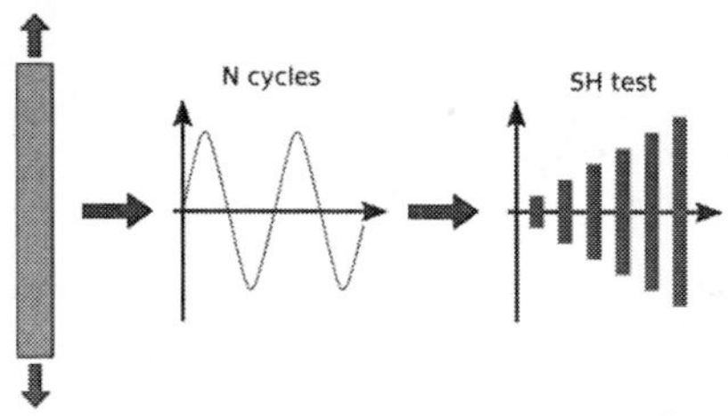

3.2.1 Influence of the stress amplitude

A substantial evolution of the self-heating curves can be observed for the different stress amplitudes (Fig.9). However, the maximum temperature elevation is lower than the elevation obtained previously with 1% of plastic pre-strain.

Fig. 9. Influence of the stress amplitude of a cyclic pre-loading on self-heating curves, with N=50 000 cycles.

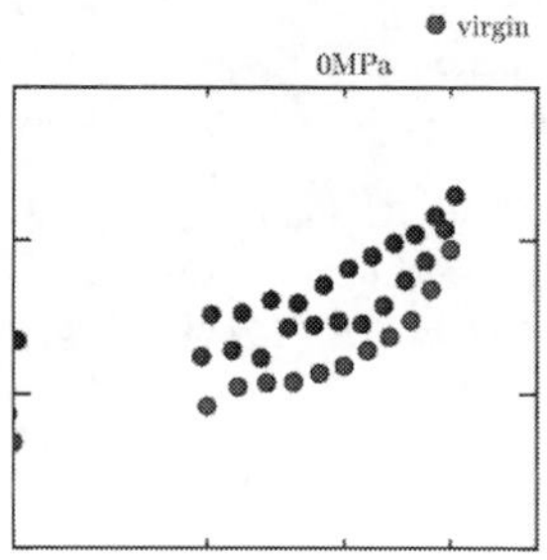

The evolution of the two parameters α and β, as well as the normalized parameters α_{norm} and β_{norm} are shown in Tab.2. This history loading seems to have less impact than a plastic pre-strain. Indeed, with only 1% of plastic pre-straining, the primary regime intensity is multiplied by around 5 whereas a cyclic pre-loading with $\Sigma_0 = 240$ MPa leads to a factor 1.5. The fatigue limit varies very slightly.

Tab. 2. Influence of the stress amplitude of a cyclic pre-loading on the model's parameters, with N=50 000 cycles.

State	α [K]	β [K]	α_{norm} [-]	β_{norm} [-]
as-delivered	0.251	0.696	1	1
50 000cycles 210 MPa	0.482	1.168	1.918	0.980
50 000cycles 240 MPa	0.856	1.568	3.405	0.968

3.2.2 Influence of the number of cycles

Fig. 10. Influence of the number of cycles on self-heating curves after a cyclic pre-loading, with $\Sigma_0 = 210$ MPa.

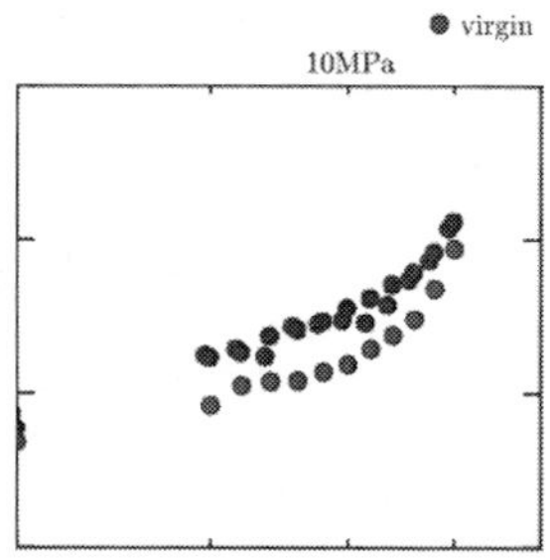

The chosen values to study the influence of the number of cycles are 50 000 and 500 000 cycles, for the same stress amplitude $\Sigma_0 = 210$ MPa. The results show that the self-heating curves are of the same order of magnitude as previously (Fig.10). The indicators show that the both self-heating regimes are almost no affected by the number of cycles (Tab.3).

Tab. 3. Influence of the number of cycles on the model's parameters after a cyclic pre-loading, with $\Sigma_0 = 210$ MPa.

State	α [K]	β [K]	α_{norm} [-]	β_{norm} [-]
as-delivered	0.251	0.696	1	1
50 000cycles 210MPa	0.482	1.168	1.918	0.980
500 000cycles 210MPa	0.531	1.058	2.114	0.984

3.3 Combined effects of a plastic pre-straining and an initial cyclic loading

To show which history loading (pre-straining or pre-cycling) has the most influence, a 1% of plastic pre-straining is performed on a specimen and then a cyclic pre-loading with N=50 000 cycles and $\Sigma_0 = 210$ MPa is applied. Finally, a self-heating test is carried out (Fig.11). The self-heating curves are provided in Fig.12.

Fig. 11. Principle of the test performed to study the combined effects of a plastic pre-straining and an initial cyclic loading.

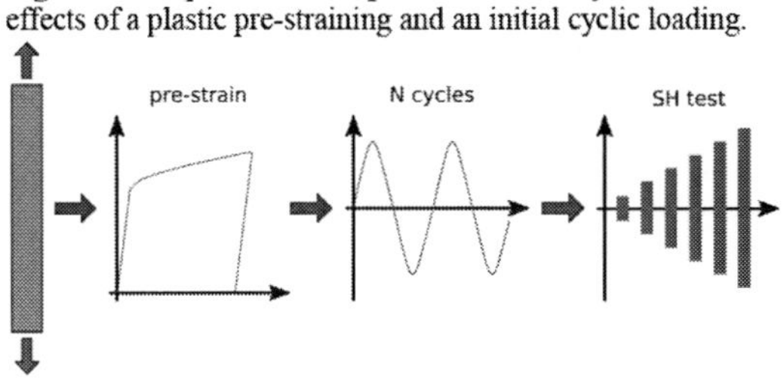

Fig. 12. Self-heating curves obtained for 1% of plastic pre-strain range and with or without a cyclic pre-loading (N=50 000 cycles at $\Sigma_0 = 210$ MPa).

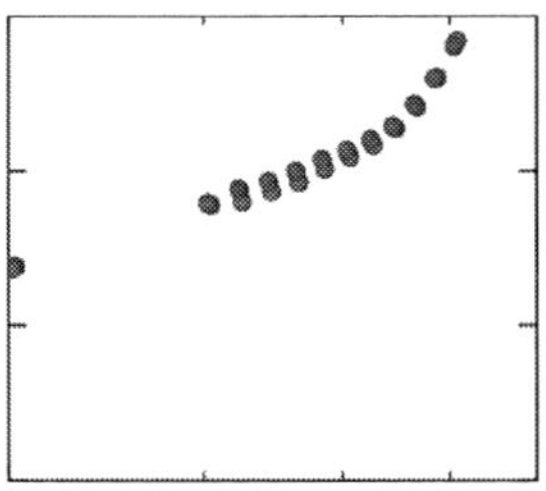

Tab. 4. Model's parameters obtained for 1% of plastic pre-strain range and with or without a cyclic pre-loading (N=50 000 cycles at $\Sigma_0 = 210$ MPa).

State	α [K]	β [K]	α_{norm} [-]	β_{norm} [-]
Pre-strain 1%	1.558	5.048	1.000	1.000
Pre-strain 1% 50 000 cycles 210 MPa	1.990	4.738	1.227	1.002

Here, to determine the parameters α_{norm} and β_{norm}, the reference state is no longer the as-delivered one, but the material pre-strained at 1% of plastic strain (Tab.4).

The main conclusion is that after a plastic pre-straining, the self-heating curve is almost the same with performing an initial series of cyclic loading.

3.4 Conclusions

In this study, it has been shown that a plastic pre-strain range has an important impact on the self-heating curves. However, an initial series of cyclic loading has a slightly influence and this influence can explain the difference between the fatigue limit determined from a two blocks fatigue test. Indeed, if a such as fatigue test is performed with the first block amplitude higher than the second one, the fatigue limit determined is lower than if the two blocks are inverted. For the modelling, the microplasticity would be govern the fatigue properties.

4 Observation of the deformation mechanism

For a self-heating test, the temperature elevation is due to dissipative mechanisms which occur in the volume of the material. It has been shown in the previous part that these mechanisms are caused by the micro-plastic strain. This micro-plasticity is represented by activated inclusions in the two scales probabilistic model and the activation evolution of these inclusions is given by

$$\lambda = \frac{1}{V_0}\left(\frac{\Sigma_0}{S_0}\right)^m, \quad (7)$$

with λ the activated inclusions density, V_0 and ${S_0}^m$ material parameters.

The evolution of this micro-plasticity can be observed at the surface of the specimen by means of optical microscope. The procedure is explained hereafter and in [1]. These observations allow to justify the form of the equation (Eq.7).

4.1 Experimental protocol

4.1.1 Specimens

https://doi.org/10.1051/matecconf/201816514016

The shape of used specimens has a varying cross-section so that the studied mechanisms are localized in its center (Fig.13).

Fig. 13. Shape of the used specimens to observe deformation mechanisms.

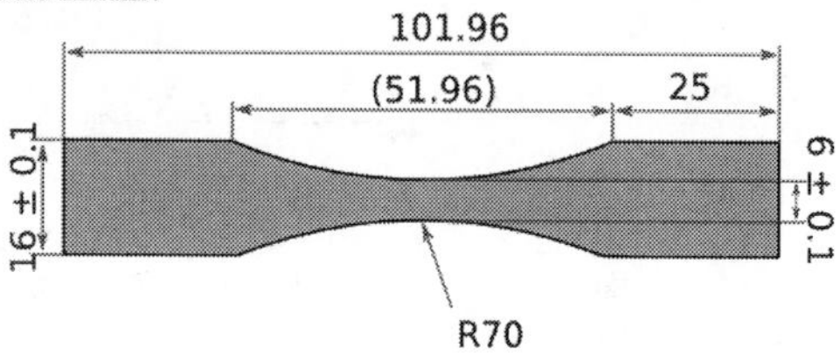

Fig. 14. Principle to apply uniform plastic pre-strain on the specimen.

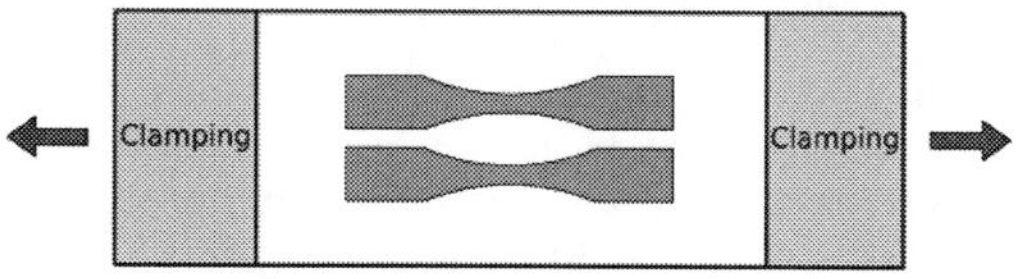

The material surface has its initial roughness. To observe the deformation mechanism at the face of the specimen, a mechanic follows by an electrolytic polishing were applied. This last polishing is achieved at the center of the specimen and has many advantages. Indeed, the repeatability of the surface state is guaranteed, the residual stress induced by the mechanical polishing is vanished and the microstructure is revealed.

Plastic straining was applied to a prismatic coupon and the specimens extracted by electro-discharge machining (Fig14).

4.1.2 Loading

Interrupted fatigue test was performed by block of N_{block} cycles at a constant stress amplitude, Σ_O, until 80%-90% of the number of cycles to failure, N_r (Fig.15). The initial picture is taken before any load is applied (*i.e.* N=0 cycle). Then, after each block of N_{block} cycles another picture is taken and so on. This particular protocol is applied for several stress amplitude and different initials states of material (as-delivered or with an initially plastic pre-strained).

Fig. 15. Principle of the interrupted fatigue test to observe the deformation mechanisms at the surface of the material.

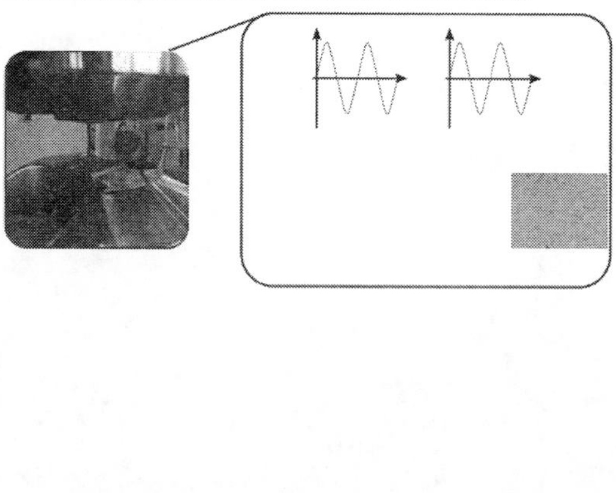

4.1.3 Observations

The pictures are taken by means of an optical microscope at magnification factor x1000. After cyclic loading with stress amplitude below the macroscopic elastic limit, some "dark stains" are observed (Fig.16). It has shown that these "dark stains" corresponding to PSB (Persistent Slip Bands) [1,4] (Fig.17). These PSB demonstrate the occurence of micro-plasticity in the material. From these pictures, the number and the spatial distribution of PSB can be analysed.

Fig. 16. Evolution of the PSB occurence under cyclic loading at $\Sigma_O = 260$ MPa.

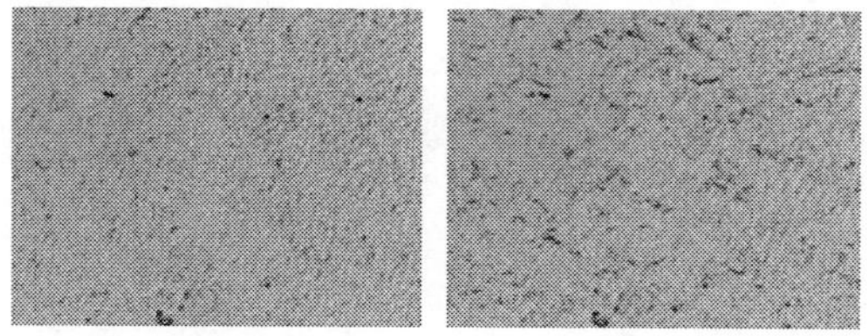

(a) Initial picture: 0 cycle (b) Final picture: 100 000 cycles

Fig. 17. Observation of the PSB at the surface of the material and representation of the mechanisms.

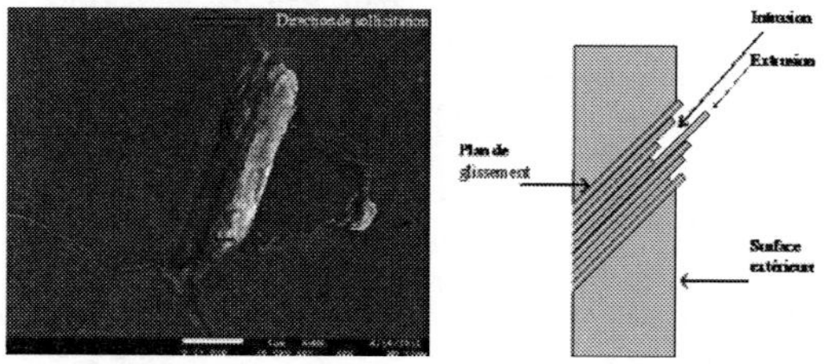

(a) SEM observation (b) Representation of the PSB

4.2 Results

This experimental protocol is applied for several stress amplitudes. Two states of the material are studied: the as-delivered state and the 1% plastic pre-strain (Fig.18). These results show that a plastic pre-straining has an influence on the occurence of the PSB, so the evolution of the plasticity into the material. With a 1% plastic pre-strain, the emergence of PSB appears for lower stress amplitude than for the as-delivered state. So, for the same stress amplitude, the dissipation is most important for the pre-strained state, which is consistent with the self-heating results.

Fig. 18. Evolution of the number of the PSB for different stress amplitude under cyclic loading.

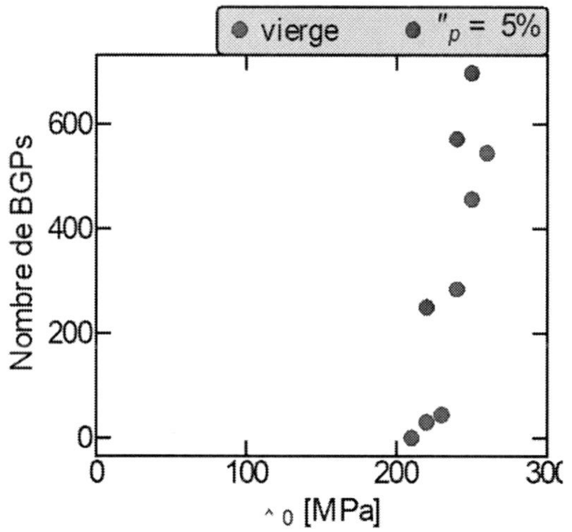

5 Conclusions

The goal of this study was to study the influence of load history effects on the self-heating measurements under cyclic loading. On the one hand, it has shown that the load history effects impacted the material response (without question the presence of the two self-heating regimes). On the other hand, an analyse combining two load history effects was proposed if the initial state is defined be the plasticity level reached during the pre-load (either a plastic pre-strain or an initial loading).

Then, to validate the interpretation of the previous result (*e.g.*, presence of both self-heating regimes) an experimental protocol was developed to study the emergence of microplasticity under cyclic loadings at the surface of a specimen. These observations are in accordance with the evolution of the secondary regime of the self-heating curves.

The next step consists on developing a particular model to describe these results: the probabilistic two-scale model developed by [1,2] have to be extended to take into account this load history effects on fatigue properties.

References

1. R. Munier, PhD thesis, (2012)
2. R. Munier, C. Doudard, S. Calloch, B. Weber, Int J Fatigue, **63**, 46-61, (2014)
3. C. Doudard, PhD thesis, (2004)
4. S. Suresh, Fatigue of Materials, (1991)

Evaluation of notch effects in low cycle fatigue of alloy 718 using critical distances

Robert Eriksson[1,*], *Kjell* Simonsson[1], *Daniel* Leidermark[1], and *Johan* Moverare[2]

[1]Division of Solid Mechanics, Department of Management and Engineering, Linköping University, 58183 Linköping, Sweden
[2]Division of Engineering Materials, Department of Management and Engineering, Linköping University, 58183 Linköping, Sweden

Abstract. Gas turbine disks contain many notch-like features acting as stress raisers. The fatigue life based on the notch root stress may be overly conservative as the steep stress gradient in front of the notch may give rise to so-called notch support. In the current work, the theory of critical distances was applied to the prediction of the total fatigue life of low cycle fatigued, notched specimens made from alloy 718. The fatigue tests were performed at 450 °C and 550 °C. It was found that, for lives shorter than 5000–10000 cycles, the notched specimens had longer lives than would have been expected based on the notch root strain. For lives longer than 5000–10000 cycles, there were no notch support. The life prediction for notched specimens could be significantly improved by basing the prediction on the strain chosen some distance from the notch (the critical distance). An expression for calculating the critical distance based on the notch root strain was suggested.

1 Introduction

Materials used for gas turbine disks should have properties such as high yield and tensile strength, reasonable ductility and fracture toughness, good resistance to fatigue crack initiation and low crack propagation rates [1]. One alloy commonly used as disk material in gas turbines is alloy 718 [1].

Turbine disks may have complex shapes where several geometric features act as notches (e.g. the blade attachments). Such notches act as stress raisers and influence the load at fracture as well as the fatigue life. Fatigue life predictions of turbine disks thus involve taking notch effects into account.

Predictions of failure based on the notch root stress may be overly conservative. A steep stress gradient at the notch can cause the stress to drop significantly with distance from the notch resulting in the observed fatigue life to exceed the predicted life based on the notch root stress, so called notch support [2, 3].

One way of reducing conservatism is through the use of critical distances [4]. Rather than basing the failure criterion on the notch root stress (or strain), the stress/strain at some distance from the notch is thought to be more representative for the severity of the load at the notch. For example, in the case of monotonic failure, the failure is assumed to occur when the stress at a critical distance, r, reaches a critical value, σ_0,

$$\sigma(r) = \sigma_0 \tag{1}$$

where

$$r = \frac{1}{2\pi}\left(\frac{K_c}{\sigma_0}\right)^2 \tag{2}$$

with K_c and σ_0 being (at least similar to) the fracture toughness and the ultimate tensile strength. The theory of critical distances has been shown to be applicable also for high cycle fatigue (HCF) in which case [4]

$$r = \frac{1}{2\pi}\left(\frac{K_{th}}{\sigma_0}\right)^2 \tag{3}$$

with K_{th} being (at least similar to) the threshold stress intensity factor for crack growth; σ_0 may, in this case, be more suitably chosen as the fatigue limit.

Leidermark et al. [5] have successfully used critical distances for predicting crack initiation from the strain range at low cycle fatigue (LCF) of notched specimens made from a Ni-based single-crystal alloy at 500 °C. This was achieved by using a critical distance that depended on the fatigue life.

The purpose of the current work is to test the applicability of critical distances to predict the total life (i.e. initiation and propagation) of low cycle fatigued notched specimens made from alloy 718. Fatigue data for a reference case (smooth specimens) were taken from literature. The theory of critical distances was applied to experimentally determined fatigue lives for notched specimens to see if the curves could be collapsed, thus enabling the LCF life of notched specimens to be predicted from data for smooth specimens.

Fatigue data for 450 °C, from notched specimens, were used for calibration of needed equations and fatigue

* Corresponding author: robert.eriksson@liu.se

data for 550 °C, from notched specimens, were used for validation.

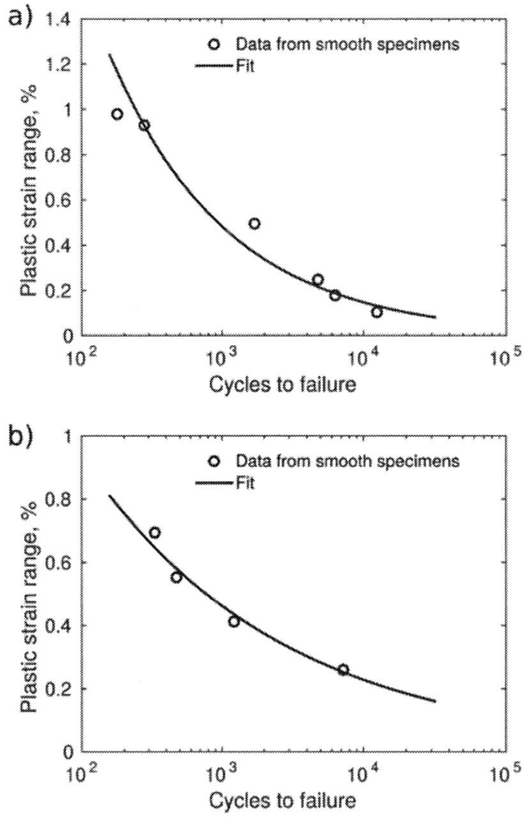

Fig. 1. Plastic strain range versus cycles to failure for smooth specimens at: a) 450 °C and b) 550 °C. Experimental data from Moverare et al. [6] and curve fit according to Equation 4.

2 Fatigue data

2.1. Literature data for smooth specimens

Fatigue data (mainly in the LCF region) for smooth alloy 718 specimens are available from the work by Moverare et al. [6]. The tests were performed in strain control with $R_\varepsilon = 0$ at 450 °C and 550 °C. The fatigue lives at 450 °C and 550 °C are shown in Figure 1 versus the plastic strain range.

The data in Figure 1 lie mainly in the LCF region and is described reasonably well by a Manson-Coffin type equation [7]

$$\frac{\Delta\varepsilon_p}{2} = \varepsilon_f'(2N_f)^c \qquad (4)$$

where $\Delta\varepsilon_p$ is the plastic strain range, N_f is cycles to failure and ε_f' and c are fitting parameters given in Table 1 for 450 °C and 550 °C (based on a fit to the data by Moverare et al. [6]).

Table 1. Material data for alloy 718 for two different temperatures, T.

T	E	K'	n'	ε_f'	c
°C	GPa	MPa			
450	179.0	1328.8	0.056	0.119	-0.513
550	172.3	2406.1	0.184	0.023	-0.305

The $\Delta\varepsilon_p$–N_f curves shown in Figure 1 (and described by Equation 4) were used as a reference for comparison with the fatigue life of the notched specimens.

2.2 Experiments: notched specimens

The number of cycles to failure was established at 450 °C and 550 °C for notched specimens made from alloy 718. The specimen geometry was similar to that of a compact tension (CT) specimen but had a notch root radius of 2 mm (here referred to as a blunt notch CT specimen), see Fig. 2. The outer dimensions were 32.4 mm × 31.2 mm × 12 mm.

The fatigue tests were conducted in load control with $R_\sigma = 0.05$. Several load levels were chosen in the interval 4–10 kN. The tests were run until final rupture of the specimens.

Fig. 2. The blunt notch CT specimen.

3 An equation for the critical distance

Equations 2 and 3 give the critical distance for the limiting cases of monotonic failure and infinite life. For crack initiation, the critical distance must vary between these extremes. It has been suggested that this can be accounted for by a cycle dependent critical distance [4, 5]

$$r = AN_f^B \qquad (5)$$

where r is the critical distance and A and B are fitting parameters. However, Equation 5 is somewhat inconvenient to use for life predictions where N_f is unknown.

Instead, inserting Equation 4 in 5, the expression for the critical distance takes the form

$$r = A\left(\frac{1}{2}\left(\frac{\Delta\varepsilon_p}{2\varepsilon_f'}\right)^{1/c}\right)^B = p'\Delta\varepsilon_p^{q'} \qquad (6)$$

where p' and q' are fitting parameters. Hence, the critical distance becomes a function of the plastic strain range. Furthermore, it seems intuitive that a higher load (and thus a higher strain gradient at the notch) would motivate that the life prediction is based on the state farther from the notch in order for it not to be overly conservative.

The strain range in Equation 6, $\Delta\varepsilon_p$, should really be taken at distance r. Here, however, the plastic strain range at the notch root, $\Delta\varepsilon_p^{root}$, will be used. The rationale being that a high value of $\Delta\varepsilon_p^{root}$ implies a high value of $\Delta\varepsilon_p$. Hence, it would not be too far-fetched to assume that

$$r = p\left(\Delta\varepsilon_p^{root}\right)^q \qquad (7)$$

where p and q are fitting parameters.

The number of cycles to final rupture, N_f, is the sum of the number of cycles to crack initiation, N_i and the number of cycles of crack propagation, N_p, i.e. $N_f = N_i + N_p$. An objection to the use of critical distances for predicting the total fatigue life might be that, after initiation, the crack will grow and extend outside the region influenced by the notch. However, it is argued here that, for low cycle fatigue, the final rupture happens at crack lengths short enough to still be located in the region where the stresses are largely determined by the notch (the average crack length at final rupture was a little over 2 mm with the longest crack, i.e. lowest load, being 3.5 mm). Furthermore, a significant part of the total life is still made up by the crack initiation phase and, consequently, the total life is largely influenced by the stresses at the notch (the number of cycles to initiation, on average, constituted ~60 % of the total life in the current fatigue tests). Since the theory of critical distances could be expected to account for the notch influence on the crack initiation life for low cycle fatigue [5], it is believed that it could also be used for explaining differences in the total life.

4 Results for notched specimens

4.1 Calculation of strain ranges

The stresses in the notched specimen were calculated by a linear-elastic finite element analysis (FEA) in the commercial finite element code Abaqus. The stress around the notch scales linearly with the applied load, which enables it to be derived for all load levels from a single elastic analysis. The stress component parallel to the direction of the applied load, here referred to as the vertical component (the y-direction in Figure 3), was used for all calculations; it coincides with the maximum principal stress for this specimen geometry.

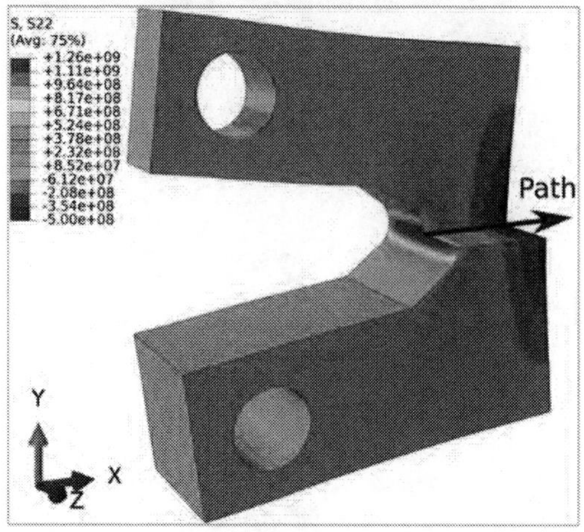

Fig. 3. A linear-elastic finite element analysis of the loaded blunt notch CT specimen. The path at which the vertical stress component (y-direction) is taken is marked by an arrow.

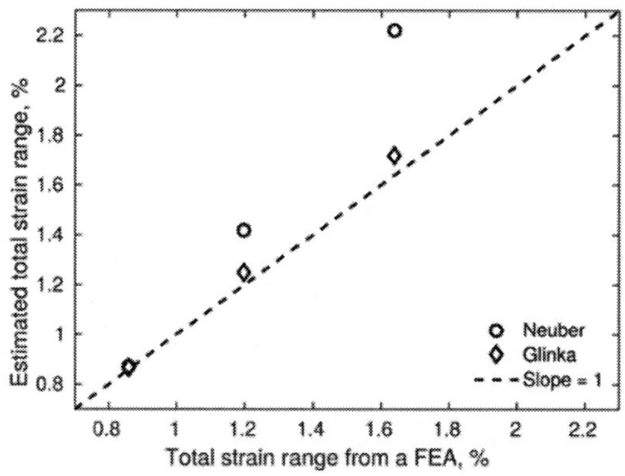

Fig. 4. Comparison of strain ranges (vertical component) at the notch root established by Neuber's rule, Glinka's rule and an elasto-plastic finite element analysis (FEA) at 450 °C.

The total and plastic strain ranges can be established from the linear-elastic FEA stress using the methods of either Neuber or Glinka, together with a Ramberg-Osgood type equation. The total strain range is obtained by solving

$$\begin{cases} \Delta\varepsilon\Delta\sigma = \dfrac{K_f^2\Delta\sigma_{nom}^2}{E} \\ \Delta\varepsilon = \dfrac{\Delta\sigma}{E} + 2\left(\dfrac{\Delta\sigma}{2K'}\right)^{1/n'} \end{cases} \qquad (8)$$

based on Neuber's rule or

$$\begin{cases} \int_0^{\Delta\varepsilon} \Delta\sigma d\Delta\varepsilon = \frac{K_f^2 \Delta\sigma_{nom}^2}{2E} \\ \Delta\varepsilon = \frac{\Delta\sigma}{E} + 2\left(\frac{\Delta\sigma}{2K'}\right)^{1/n'} \end{cases} \tag{9}$$

based on Glinka's rule. The fatigue notch factor is set to $K_f = 1$ since the nominal stress, $\Delta\sigma_{nom}$, is taken locally along the path shown in Figure 3. The needed material parameters (the Young's modulus, E, and the cyclic Ramberg-Osgood parameters K' and n') are establish from References [6, 8, 9] and are given in Table 1. Equations 8 and 9 are solved for $\Delta\varepsilon$ and $\Delta\sigma$.

The strain ranges estimated from Equation 8 and 9 were compared to an elasto-plastic finite element analysis at a few selected load levels using an Armstrong-Frederick type material model calibrated using data from Gustafsson et al. [10, 11]. The comparison is shown in Figure 4 where it can be seen that Glinka's rule is closer to the finite element results. Therefore, Equation 9 was used to calculate the strain ranges. The reason for using Glinka's rule instead of finite element modelling for establishing the strain ranges is the considerable gain in computation time thus achieved.

Figure 5 gives an example of the stress and strain ranges as function of distance from the notch (see Figure 3) for linear-elastic results from a finite element analysis and the same data recalculated using Glinka's rule.

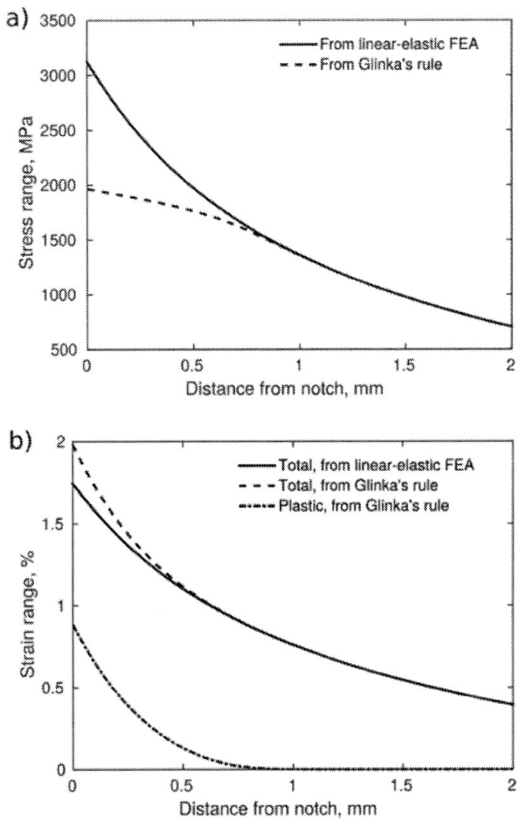

Fig. 5. Comparison of stress and strain ranges (vertical component) from a linear-elastic finite element analysis (FEA) and Glinka's rule at 450 °C. a) Stress range as function of the distance from the notch at 9.9 kN load. b) Total and plastic strain ranges as function of the distance from the notch at 9.9 kN load.

4.2 Calibration of critical distance at 450 °C

Fatigue data for smooth and notched specimens at 450 °C were used for calibrating the expression for the critical distance (see Equation 7).

Figure 6 shows the plastic strain range as function of number of cycles to failure; for the blunt notch CT specimens, the strain has been taken at the notch root. It can be seen that the $\Delta\varepsilon_p$–N_f curve does not coincide with the curve for the smooth specimens from Moverare et al. [6]. The predicted life to failure based on data for smooth specimens and the notch root strain range from the notched specimens would obviously be lower than the observed life.

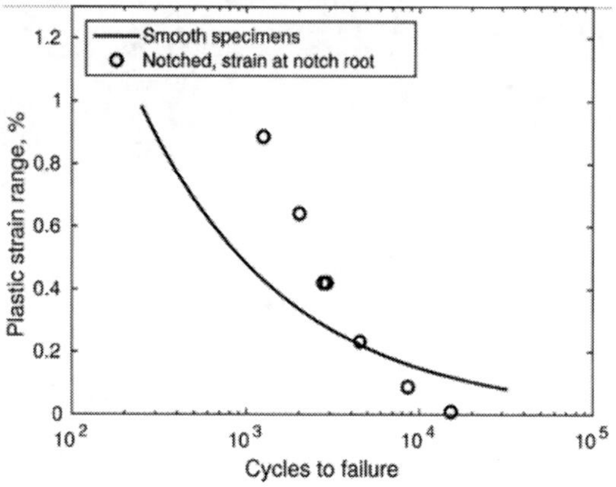

Fig. 6. Plastic strain range versus number of cycles to failure at 450 °C. Comparison between data for smooth specimens (from Moverare et al. [6]) and notched specimens. For the latter, the plastic strain range has been taken at the notch root.

It should be noted, however, that the notch support seems to diminish around 5000–10000 cycles. As seen in Figure 6, this occurs as the plastic strain range becomes low and thus coincides with the transition from LCF to HCF behaviour. Consequently, notch support should only be expected in the LCF regime and, in the current work, data points with lives longer than ~5000 cycles will be ignored.

To find the fitting parameters p and q in Equation 7, the following procedure was used:

1. Using Equation 4 fitted to data for smooth specimens, the expected plastic strain range, $\Delta\varepsilon_p^*$, was established for each experimentally observed life for the notched specimens, N_f^{notch}, such that $\Delta\varepsilon_p^* = 2\varepsilon_f'(2N_f^{notch})^c$.
2. For each load level, the distance from the notch, r^*, corresponding to $\Delta\varepsilon_p^*$ was established from curves as that shown in Figure 5 b).
3. The distances, r^*, were plotted versus the plastic strain range from the blunt notch CT specimens taken at the notch root and Equation 7 was fitted.

Figure 7 shows the obtained critical distances for data at 450 °C plotted versus the plastic strain range at the notch root from the blunt notch CT specimens; as seen, Equation 7 provides a reasonable fit to the data (albeit less accurate for lower loads). The fitting parameters in Equation 7 are $p = 0.0741$ m and $q = 1.2184$.

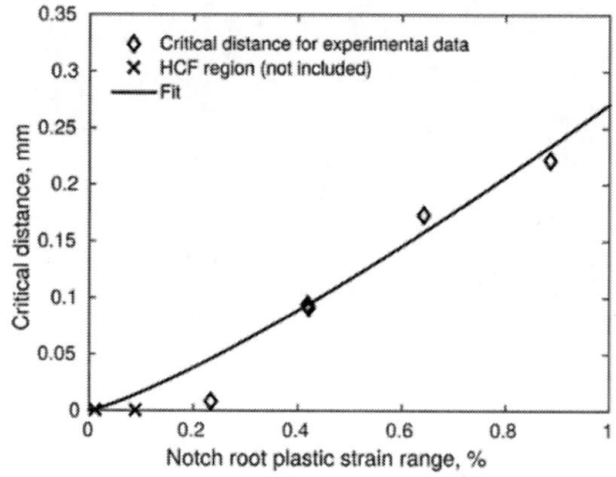

Fig. 7. The distance from the notch (the critical distance) at which the plastic strain range should be extracted from the blunt notch CT specimens in order for the fatigue life to be predicted by $\Delta\varepsilon_p$–N_f curves for smooth specimens; plotted versus the plastic strain range at the notch root.

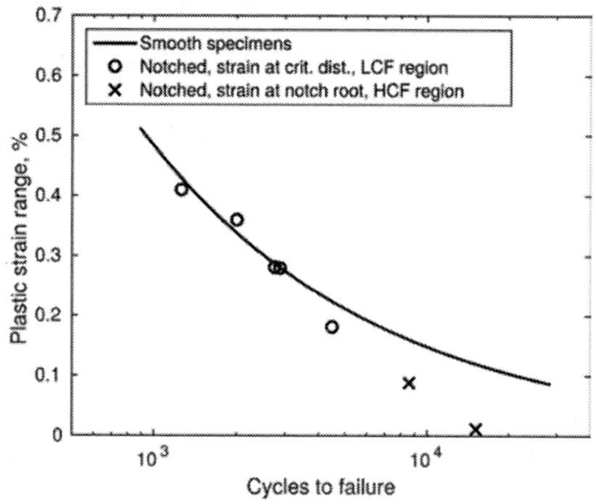

Fig. 8. Plastic strain range versus cycles to failure for smooth and notched specimens. For the notched specimens, the strain has been taken at the critical distance. Above ~5000 cycles, no notch support is expected and those data points should be ignored.

Figure 8 shows a comparison of $\Delta\varepsilon_p$–N_f curves for smooth specimens and blunt notch CT specimens with the strain taken at the critical distance as determined by Equation 7. As seen, the curves collapse reasonably well. Fatigue data for smooth specimens can thus be used for the total life prediction of notched specimens if the prediction is based on the strain range taken at the distance r from the notch where r is given by Equation 7.

It is important to note, though, that notch support only seems to have an effect for shorter lives. Specimens that lived longer than ~5000 cycles should hence be ignored here. Since the region around 5000–10000 cycles coincides with very low plastic strain ranges, it can be concluded that notch support only takes effect in the LCF region.

4.3 Validation at 550 °C

To justify the chosen method for calculating the critical distance, a validation was performed for fatigue data from tests at 550 °C.

Figure 9 shows the plastic strain range versus cycles to failure for fatigue at 550 °C; for the blunt notch CT specimens, the strain has been taken at the notch root. Just as for 450 °C (see Figure 6), the blunt notch CT specimens had longer lives than would be predicted from data from smooth specimens. Again, notch support only seems to exist for lives shorter than about 5000 cycles.

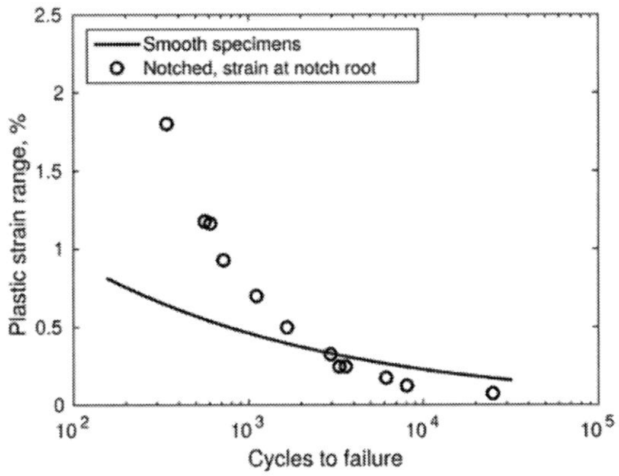

Fig. 9. Plastic strain range versus number of cycles to failure at 550 °C. Comparison between data for smooth specimens (from Moverare et al. [6]) and notched specimens. For the blunt notch CT specimen, the plastic strain range has been taken at the notch root.

A life prediction was made for the blunt notch CT specimens using data for smooth specimens at 550 °C by the following procedure:

1. Stress versus distance from the notch was obtained from an elastic FE analysis for each load level.
2. Equation 9 (Glinka) was used to establish the plastic strain range versus distance from the notch.
3. The critical distance, for each load level, was calculated from the notch root plastic strain range and Equation 7 calibrated with data for 450 °C.
4. The plastic strain range, $\Delta\varepsilon_p^r$, was taken at the critical distance, r, for each load level.
5. The fatigue life of the notched specimens, N_f^{notch}, was predicted as $N_f^{notch} = \frac{1}{2}\left(\frac{\Delta\varepsilon_p^r}{2\varepsilon_f'}\right)^{1/c}$ with ε_f' and c calibrated to data for smooth specimens at 550 °C.

The thus predicted life is shown in Figure 10; as seen, the use of critical distances reduces the conservative prediction (would it have been made based on the notch root strain) and thus accounts for the notch support in the LCF region.

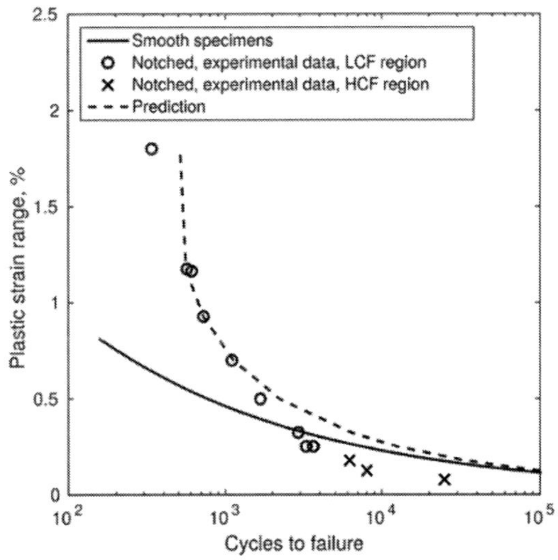

Fig. 10. Plastic strain range versus number of cycles to failure at 550 °C. For the notched specimens, the prediction has been made using the plastic strain range at the critical distances and a $\Delta\varepsilon_p$–N_f curve calibrated for smooth specimens.

5 Conclusions

The theory of critical distances was applied to low cycle fatigue of blunt notch CT specimens made from alloy 718 in an attempt to allow for total life prediction of notched specimens to be made from data for smooth specimens. It was argued that, in the low cycle fatigue regime, the final rupture occurs at crack lengths short enough to still be in the stress field influenced by the notch; additionally, the initiation part of the fatigue life (which is heavily influenced by the notch stress) constitutes a significant part (~60 %) of the total life. A critical distance based life prediction approach should therefore be applicable also for the total life in the low cycle regime.

It was found that the critical distance, with reasonable accuracy, could be calculated from the notch root plastic strain range via a simple power law. It was also found, however, that the notch support diminishes at around 5000–10000 cycles with the transition into the high cycle regime. The critical distance based life prediction approach should hence be applied only in the low cycle regime.

The use of critical distances collapsed the $\Delta\varepsilon_p$–N_f curves for smooth and blunt notch CT specimens at 450 °C and was justified by validation at 550 °C. The method of critical distances could arguably be used for accounting for the notch support in low cycle fatigue total life predictions of notched components.

This project has received funding from the European Union's Horizon 2020 research and innovation programme under grant agreement No. 653941.

References

1. R. Reed, *The superalloys — fundamentals and applications* (Cambride University Press, The Edinburgh Building, Cambride CB2 8RU, UK, 2008)
2. C. Kontermann, H. Almstedt, A. Scholz, M. Oechsner, Procedia Structural Integrity **2**, 3125 (2016)
3. L. Mäde, H. Gottschalk, S. Schmitz, T. Beck, G. Rollmann, *Probabilistic LCF Risk Evaluation of a Turbine Vane by Combined Size Effect and Notch Support Modeling*, in *ASME. Turbo Expo: Power for Land, Sea, and Air, Volume 7A: Structures and Dynamics* (2017)
4. D. Taylor, *The theory of critical distances* (Elsevier, The Boulevard, Langford, Kidlington, Oxford, OX5 1GB, UK, 2007)
5. D. Leidermark, J. Moverare, K. Simonsson, S. Sjöström, Int. J. Fatigue **33**, 1351 (2011)
6. J. Moverare, G. Leijon, H. Brodin, F. Palmert, Mater. Sci. Eng. A **564**, 107 (2013)
7. S. Manson, G. Halford, *Fatigue and Durability of Structural Materials* (ASM International, 2006)
8. M. Fukuhara, A. Sanpei, J. Mater. Sci. Letters **12**, 1122 (1993)
9. H. Ledbetter, Cryogenics **December**, 653 (1982)
10. D. Gustafsson, Ph.D. thesis, Linköpings universitet (2012)
11. D. Gustafsson, J.J. Moverare, K. Simonsson, S. Sjöström, J. Eng. Gas Turbines Power **133**, 094501-1 (2011)

Low cycle fatigue of 316L stainless steel processed by surface mechanical attrition treatment (SMAT)

Zhidan Sun[1], *Jianqiang* Zhou[1], *Delphine* Retraint[1,*], *Thierry* Baudin[2], *Anne-Laure* Helbert[2], *François* Brisset[2], and *Pascale* Kanouté[1,3]

[1]ICD, P2MN, LASMIS, University of Technology of Troyes, CNRS, Troyes, France
[2]ICMMO - UMR 8182, Université Paris-Sud, 91405 Orsay, France
[3]ONERA, The French Aerospace Lab, 29 avenue de la Division Leclerc, 92322 Chatillon Cedex, France

Abstract. In this work, the effect of surface mechanical attrition treatment (SMAT) on the cyclic behaviour of a 316L stainless steel under low cycle fatigue (LCF) is investigated. The LCF results are presented in the form of cyclic stress amplitude evolution for both untreated and SMATed samples. In order to better understand the microstructure change due to cyclic loading, electron backscatter diffraction (EBSD) is used to characterize the microstructure of the SMATed samples before and after fatigue tests. A microstructure gradient is highlighted for samples after SMAT from the top surface layer in nanocrystalline grains to the interior region non-affected by impacts. Under LCF loading, new slip systems are activated in the work hardened region, whereas no plastic slip is activated in the nanostructured layer. The residual stresses generated by SMAT are measured using X-ray diffraction (XRD), and their relaxations under cyclic loading are studied by taking into account the microstructure change. The cyclic behaviour of the samples in different material states is interpreted based on these investigations.

1 Introduction

Fatigue properties of metallic materials are important for design and life prediction of mechanical components subjected to cyclic loading in service. It is well known that the fatigue properties of materials are sensitive to grain size as well as residual stresses. Small grains can effectively increase the yield strength of a material and consequently enhance its fatigue crack initiation resistance [1, 2], whereas coarse grains exhibit better crack propagation resistance due to the good ductility and toughness of materials. Since most fatigue cracks initiate from the surface and propagate to the interior of a structure, a component with a nanostructured surface layer and an inner coarse grained matrix can have improved fatigue resistance [3]. Furthermore, the presence of compressive residual stress is beneficial for fatigue life improvement through increasing the resistance to both crack initiation and propagation.

Surface mechanical attrition treatment (SMAT) is a surface treatment technique which can transform the coarse grained surface layer of a material into nanosized grains by means of severe plastic deformation (SPD) [4-6]. SMAT is based on mechanical impacts on the surface of a structure by metallic balls with high kinetic energy. This technique is able to generate a large quantity of crystallographic defects such as dislocations, deformation twins, which can lead to the formation of refined grains down to the nanometer scale at the surface. The obtained nanostructured layer coupled with compressive residual stresses generated by SMAT can effectively improve the fatigue strength of materials [7]. To promote SMAT for example in engineering applications, it is important to have a good understanding of all the properties of materials processed by this technique including fatigue properties. It has been experimentally demonstrated that SMAT can significantly improve the fatigue resistance of metallic materials [5, 8-10]. On the one hand, the near surface compressive residual stresses can improve fatigue life by increasing the resistance to crack initiation and propagation. On the other hand, a work hardened layer generated by mechanical impacts can enhance the material strength by increasing for example the elastic limit [11, 12]. In addition, in the case of SMAT, the presence of a superficial nanostructured layer can effectively increase the yield strength and consequently enhance the materials' fatigue crack initiation resistance [8]. However, most of the studies performed in the literature concerning the effect of SMAT on fatigue properties were performed in high cycle fatigue (HCF) regime [5, 13-16]. It is well documented that the residual stress relaxation is strongly dependent on load amplitude. When shot peened samples are subjected to high strain/stress amplitudes, the generated compressive residual stresses can be quickly relaxed, i.e. during the first several cycles [16-18].

In this work, low cycle fatigue (LCF) behaviour of a SMATed 316L stainless steel is investigated compared to its untreated state. Electron backscatter diffraction (EBSD) observation and microhardness test are conducted to characterize the SMAT affected region before and after fatigue test. The effect of SMAT on the mechanical behaviour of this material under cyclic loading is analysed. A comparison between SMATed and untreated samples is given based on the results including cyclic stress amplitude evolution, cyclic hardening/softening behaviour, and hysteresis loop. The results are discussed through analysing the eventual

* Corresponding author: delphine.retraint@utt.fr

microstructure change and residual stress relaxation due to cyclic loading.

2 Material and experimental procedures

The material used in this work is a biomedical grade 316L ASTM F138 austenitic stainless steel. Its chemical composition is given as follows (wt.%): 17.37Cr, 14.52Ni, 1.7Mn, 2.8Mo, 0.26Si, 0.013C, 0.017P, 0.08Cu, 0.003S, 0.07V, 0.088N, <0.005Ti, and balance with Fe. The as-received material has an initial average grain size of about 15 μm. For fatigue tests, dumbbell shape samples with an overall length of 110 mm were used. The parallel length of the reduced section is 12 mm with a diameter of 6 mm. SMAT was carried out so as to cover the entire parallel length of the reduced section. SMAT is based on the projection of spherical shots (3 mm diameter) boosted by a high frequency (20 kHz) ultrasonic generator. More detailed description of SMAT can be found in the literature [4]. In this work, the SMAT conditions used corresponds to a treatment of 15 minutes with a generator power of 30%, followed by a treatment of 5 minutes with a generator power of 50%.

LCF tests were conducted with untreated (UNT) and SMATed samples at room temperature using a servo-hydraulic machine. Uniaxial tension-compression fatigue tests were performed under total strain control with a load ratio of $R_\varepsilon = -1$. An extensometer with a gauge length of 10 mm was used to control the total strain, and a strain rate of $4 \times 10^{-3}s^{-1}$ was applied for all the fatigue tests. When the stress amplitude drop reached 30%, the samples were considered failed and the fatigue tests were stopped.

Residual stress was measured on the SMATed samples before and after fatigue tests by X-ray diffraction (XRD) with a Cr-Kα radiation using classical $\sin^2\psi$ method. To determine the in-depth residual stress variation, iterative electrolytic polishing was used to remove material from the surface of samples. Subsequent XRD measurements were performed in order to plot in-depth residual stress variation curve for the SMATed samples before and after fatigue tests in order to study the residual stress relaxation.

3 Microstructure and hardness

3.1 Microstructure observation

EBSD observation on transverse cross-section reveals a microstructure gradient in the SMAT affected region, as illustrated in Fig. 1. It can be seen that the severe plastic deformation induced by the impingement of shots flying at high speed generated a refinement of coarse grains down to nanometer scale in the top surface layer (Fig. 1b). The depth of the nanostructured layer is about 5 μm, followed by a transition region in which the grains are plastically deformed by impacts during SMAT. In this work hardened region, the impact intensity is actually lower than in the top surface nanostructured layer, and thus only many plastic slips have been activated without the formation of small grains. This is because the grain refinement requires not only a high accumulated plastic strain but also a high strain rate. In the region far from the top surface, both the accumulated plastic strain and the strain rate are significantly reduced. In the work hardened region, the grains are subdivided by transgranular traces which correspond to plastic slips activated by multi-directional impacts of shots during SMAT. Notice that there are few deformation-induced twins in the SMAT affected region including the nanostructured layer and the work hardened region [19].

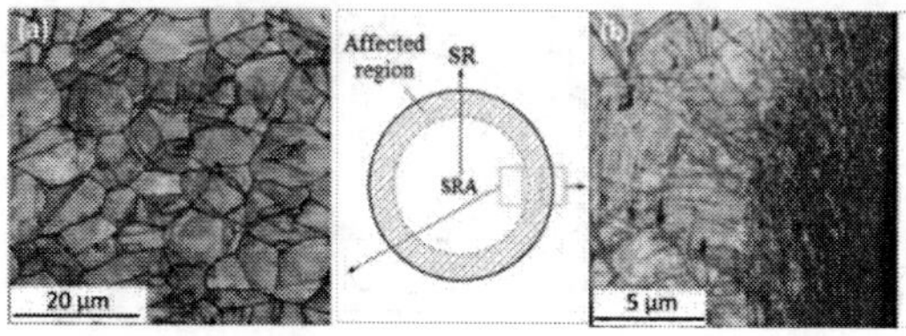

Figure 1. Typical transverse cross-sectional EBSD micrographs showing typical regions of different microstructures for a sample treated by SMAT: (a) non-affected interior region, and (b) SMAT affected region including the nanostructured layer and a part of the work hardened region.

3.2 Hardness measurement

Microhardness was measured to assess the change of mechanical strength in the SMAT affected region, as frequently done in the literature [4, 20, 21]. Fig. 2 presents the in-depth hardness variation from the treated surface for a SMATed sample. The microhardness was also measured for an untreated sample and its variation was plotted in Fig. 2 for comparison. The hardness was determined by using a Vickers microhardness tester, with a load of 0.025 kg. From the hardness variation curves shown in Fig. 2, it can be seen that there is a significant hardness increase in the SMAT affected region with respect to the untreated sample. Close to the treated surface, the hardness reaches a value as high as 400 HV0.025 which is about twice that in the central non-affected region of the sample. The hardness remains almost constant between 0.5 mm and 2 mm below the treated surface. This signifies that the depth within which the material is affected by SMAT is at least 500 μm. Fig. 2 also shows that in the interior region, the microhardness is obviously higher for the untreated sample than for the SMATed sample. It means that there is a decrease of hardness in the interior region caused by SMAT. However, as observed by EBSD, beyond the depth of about 500 μm, there is almost no plastic slip. Therefore, this decrease of hardness could be due to the presence of tensile residual stress in the interior region. Notice that the existence of tensile residual stress in the interior region is needed to compensate for the compressive residual stress present in the near surface region. The residual stress in-depth variation will be presented after in this paper.

4 Cyclic behaviour

4.1 Cyclic stress amplitude

Cyclic stress amplitude evolution during cyclic loading gives a good description of the cyclic behaviour of materials in terms of hardening/softening as function of the number of cycles. The hardening/softening behaviour of a material can be altered by mechanical surface treatment techniques such as SMAT.

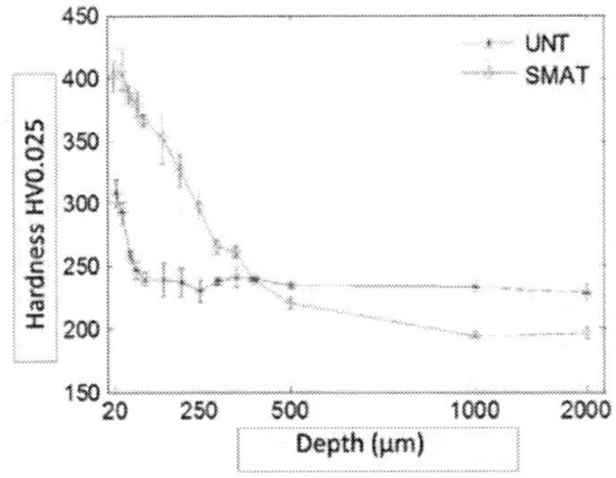

Figure 2. In-depth hardness variations from the top surface for untreated (UNT) and SMATed samples.

Fig. 3 shows the cyclic stress amplitude evolution with the increase of the number of cycles for untreated and SMATed samples obtained under different total strain amplitudes from ±0.3% to ±1.25%. Different strain amplitude levels were used in order to discern the eventual load amplitude dependent behaviour of the material. Globally it can be observed, by comparing Fig. 3a and Fig. 3b, that SMAT significantly enhanced the mechanical strength and changed the cyclic hardening/softening behaviour of the samples. From Fig. 3, it can also be seen that higher imposed strain amplitude gives higher cyclic stress amplitude. Furthermore, for a given strain amplitude, the cyclic stress amplitudes of the SMATed samples are obviously higher than that of the untreated ones.

The hardening/softening behaviour (characterized by the stress increase/decrease with respect to the initial state) of the material throughout fatigue tests can also be observed from the curves shown in Fig. 3. At the beginning of cyclic loading, there is an increase of stress amplitude during the first several cycles, which corresponds to the initial hardening of the material. It can be distinguished that under lower strain amplitude (±0.3% for example), the maximum stress amplitude is reached earlier, and there is an evident longer softening phase which follows the short period of initial hardening. Under strain amplitude of ±1.25%, untreated samples undergo more obvious initial hardening than the SMATed samples. In addition, the phenomenon of secondary hardening which follows the cyclic softening period can be observed for untreated samples under high strain amplitude (±1.25% and ±0.8%). In the case of low strain amplitude (±0.5% and ±0.3%), the secondary hardening is not present, and the samples undergo long term period softening. For the SMATed samples, the stress amplitude is nearly stabilized by the end of the fatigue tests under high stress amplitudes (±0.8% and ±1.25%), while under low strain amplitudes (±0.3% and ±0.5%), only the long term softening period is present (Fig. 3b).

The change of cyclic behaviour of the SMATed samples with respect to the untreated ones is due to the SMAT affected region which generally enhances the global mechanical strength of the SMATed samples under monotonic loading [4, 20, 21]. In this paper, the strengthening effect of shot impacts is manifested by the increase of the global stress level induced by SMAT, as shown in Fig. 3. On the one hand, strain hardening induced by shot impacts can improve the mechanical strength due to dislocation entanglement which can impede permanent deformation of the material during subsequent cyclic loading. On the other hand, the generated nanostructured layer has much higher yield strength than its initial non-treated state according to Hall-Petch relationship. However, the nanostructured layer might have a small contribution to the global mechanical behaviour of the samples because of its very thin thickness (5 μm) compared to the depth of the work hardened region (more than 500 μm according to the microhardness curves) and the diameter of the sample (6 mm).

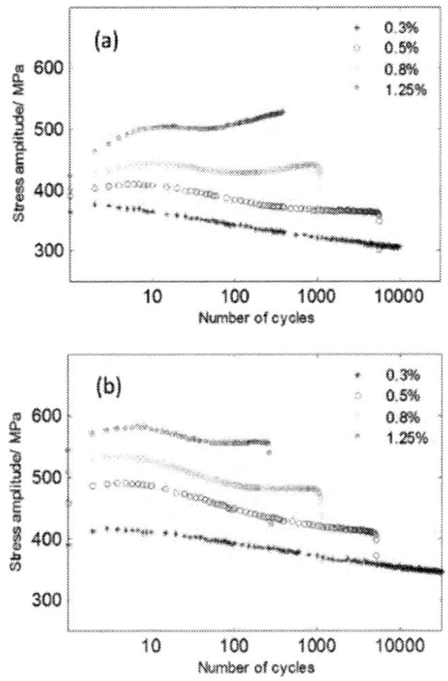

Figure 3. Cyclic stress amplitude evolution curves obtained under different strain amplitudes for: (a) untreated samples, and (b) SMATed samples.

In addition, as presented above, the interior region of the SMATed samples has no plastic deformation. The interior region should also undergo the cyclic secondary hardening as undergone by untreated samples (Fig. 3a). This implies that the disappearance of secondary hardening for SMATed samples under high strain amplitude is related to the SMAT affected region. Therefore, this SMAT affected region should exhibit cyclic softening behaviour, which allows to reduce the secondary hardening effect of the interior region on the whole SMATed sample.

4.2. Hysteresis loops

LCF hysteresis loop describes the relationship between strain and stress during one cycle, and characterizes more specifically the plastic cyclic deformation behaviour of materials. Fig. 4 illustrates the comparison of hysteresis loops for the untreated and SMATed samples under different strain amplitudes at different number of cycles. It can be seen that at the 2nd cycle the hysteresis loops of SMATed samples have larger height and smaller width (which represents plastic strain range) than untreated samples, under both high (Fig. 4a) and low (Fig. 4b) strain amplitudes. This means that the SMATed samples exhibit higher stress amplitude and lower plastic strain at the beginning of cyclic loading. As cyclic loading continues, the difference between these hysteresis loops obtained with different material states gradually decreases, and they tend to become superposed by the end of fatigue tests, especially under high strain amplitude (see the two dotted curved presented in Fig. 4a obtained at 260th cycle). This result implies that SMAT mainly affects the cyclic behaviour during the early stage of cyclic loading, and its effect is gradually reduced with the increase of the number of cycles.

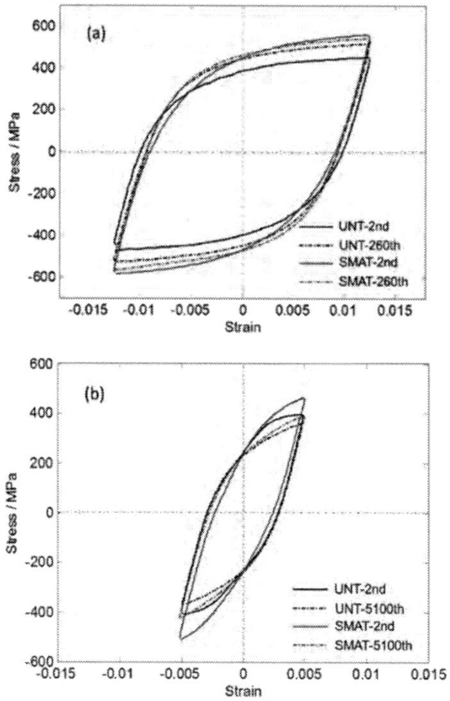

Figure 4. Hysteresis loops obtained with the untreated and SMATed samples at the beginning of fatigue tests (2nd cycle) and by the end of fatigue tests (260th cycle) under strain amplitude of (a) ±1.25% and (b) ±0.5%.

Other information which can be obtained from Fig. 4 concerns the hardening/softening. Under high strain amplitude (±1.25%), comparing the hysteresis curves obtained at 2nd and 260th cycles, it can be seen that the untreated sample undergoes cyclic hardening, whereas the SMATed sample undergoes a slight cyclic softening (Fig. 4a). In contrast, under low strain amplitude (±0.5%), both the untreated and the SMATed samples undergo softening, through comparing the hysteresis curves obtained at 2nd and 5100th cycles (Fig. 4b). These observed hardening/softening phenomena are consistent with the analysis given in Section 4.1 regarding the stress amplitude evolution.

5 Analysis and discussion

The in-depth hardness variation of SMATed samples after fatigue tests was measured in order to reveal the change in material properties. Fig. 5 shows an example of results obtained with strain amplitude of ±1.25%. It can be seen that after fatigue loading, compared to the state without fatigue, the hardness in the work hardened region of the SMATed sample decreases about 50 HV0.025, whereas the hardness in the interior region increases from 200 HV0.025 to about 230 HV0.025. This means that the work hardened region of the SMATed sample is softened, and the interior region (which also represents the untreated state of material) is hardened due to fatigue loading, with respect to their initial states. This investigation is consistent with the result presented in Fig. 3 in which it is shown that untreated samples underwent cyclic hardening under high strain amplitudes (Fig. 3a). By contrast, the cyclic hardening under high strain amplitude illustrated in Fig. 3b is less pronounced for the SMATed samples. This must be due to the cyclic softening of the work hardened region, as revealed by the hardness measurement shown in Fig. 5.

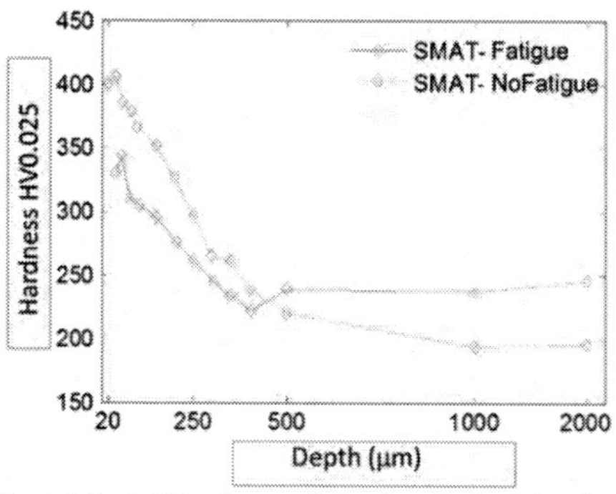

Figure 5. In-depth hardness variation from the treated surface for a SMATed sample after fatigue. The curve for a sample after SMAT (without fatigue) is also given for comparison.

It is worth mentioning that the residual stress could also have a contribution to the hardness change, as indicated above. If the tensile residual stress in the interior region has been (partially) relaxed during cyclic loading due to plastic deformation, it could also lead to an increase of hardness. According to the measurement of residual stress with DRX, it seems that this assumption could be confirmed. Fig. 6 illustrates the in-depth longitudinal residual stress variation for SMATed samples before and after fatigue tests with different strain amplitudes (±0.5% and ±1.25%). It can be seen that under stain amplitude of ±1.25%, nearly most residual stress is relaxed by the end of fatigue test, whereas under ±0.5%, the residual stress is partially relaxed and remains equal to about 30% of the initial residual stress generated by SMAT.

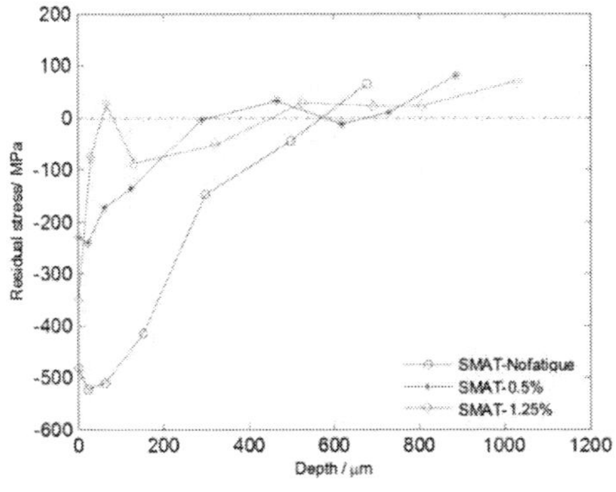

Figure 6. In-depth longitudinal residual stress variation for SMATed samples showing residual stress relaxation due to cyclic loading with different strain amplitudes. The curve obtained with the SMATed sample without fatigue is also given for comparison.

For fatigue tests, particular attention was paid to the eventual microstructure evolution of the SMAT affected region including the nanostructured layer and the work hardened (or plastically deformed) region. The results show that no obvious difference in grain size distribution can be noticed in the interior region for the samples before and after fatigue tests. This means that no grain size modification occurred in this region due to fatigue loading. However, according to Fig. 7, in all the regions except the nanostructured layer (Fig. 7a), grain orientation spread (GOS) has been increased under the effect of cyclic fatigue loading (Figs. 7b and 7c), with respect to the material state just obtained by SMAT. Notice that GOS characterizes the mean intragranular misorientation, i.e. the misorientation within a grain. Thus the GOS changes shown in Figs. 7b and 7c must be due to the plastic deformation activated by fatigue loading. This point of view is consistent with the literature. As indicated in [22], the plastic deformation in polycrystalline materials is able to generate intragranular disorientation, and accordingly an increase of GOS.

Fig. 8 shows the plastic slip traces present in the work hardened region of the SMATed samples before and after fatigue tests observed by EBSD. As already described above, SMAT induced plastic slips due to shot impacts (Fig. 1b and Fig. 8a). During the subsequent fatigue loading, the slip plane trace density is increased (Fig. 8b). Note that the hardening/softening behaviour as well as the residual stress relaxation is strongly related to the plastic slip activities due to fatigue loading. The red crosses shown in Fig. 8 indicate the planes {111} of the activated systems which are confirmed to be {111} <110> in this study. As a matter of fact, plastic slips newly activated by fatigue loading can further split the grains and consequently increase the intragranular disorientation inside each grain. This investigation is in good agreement with the GOS distribution presented in Fig. 7.

However, in the nanostructured layer, no change of GOS was caused by fatigue loading (Fig. 7a), which means that there is almost no plastic slip activated by cyclic loading in this area. This could be due to the fact that the yield stress of the nanostructured layer (with nanoscale grain size) is much higher than that of the work hardened layer or the interior non-affected region of the sample where the grains have an ordinary size. It is well known that the grain interior slip resistance increases with the decrease of grain size. The activation of plastic slips is thus more difficult in the nanostructured layer, which is consistent with the previous results on the mechanical behaviour of nanocrystalline or ultra-fine grained materials reported in the literature [23-25].

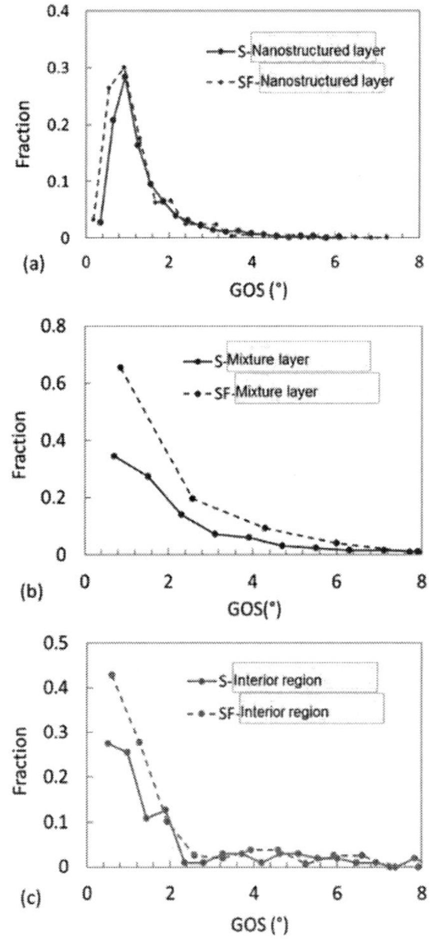

Figure 7 – GOS distribution calculated in different areas: (a) nanostructured layer, (b) work hardened region, and (c) interior region, for SMATed state (S) as well as SMATed and fatigued state (SF).

Furthermore, under cyclic loading with high strain amplitude, as indicated above, new plastic slips can be activated, which could lead to the restructuring of the dislocations introduced by SMAT in the work hardened region. As cyclic loading continues, the dislocation structure in the work hardened region induced by SMAT could tend to become similar to that in non-affected region of the sample. This constitutes the reason that the two curves corresponding to the untreated and SMATed samples tend to be superposed by the end of fatigue tests (Fig. 4a). In addition, accompanied by these dislocation activities, most of the compressive residual stresses introduced by SMAT are relaxed, especially during the first several cycles under high strain amplitude [16], and the remained compressive residual stresses may not be high enough to significantly influence the fatigue properties at later stage of fatigue tests. This investigation concerning residual stress relaxation has been confirmed by the residual stress measurement using XRD, as shown in Fig. 6.

Figure 8 – Image quality (IQ) maps illustrating (a) the presence of plastic slip traces in the work hardened area without fatigue, and (b) the state of plastic slips in the work hardened area of a SMATed sample after fatigue. {111} plane traces are indicated by red crosses [19].

6 Conclusions

In this work, the effect of SMAT on low cycle fatigue behaviour of a 316L austenitic stainless steel was investigated, and the following conclusions can be drawn:

(1) After SMAT, a microstructure gradient is generated in the SMAT affected region, from the nanostructured layer to the non-affected bulk of the sample, through the transition region where many plastic slips are present.

(2) Cyclic stress amplitude of SMATed samples is enhanced due to higher strength of the SMAT affected region. During cyclic loading, the SMAT affected region undergoes cyclic softening, whereas the non-affected interior region is hardened under high strain amplitude (±1.25%), and softened under low strain amplitude (±0.5%).

(3) Comparison of hysteresis loops between different cycle numbers (at the beginning and by the end of fatigue tests) indicates that SMAT mainly affects the cyclic behaviour of the studied material in the early stage of fatigue test, and this effect is gradually attenuated as cyclic loading goes on.

(4) Under the effect of fatigue loading, new plastic slips are activated in the work hardened region, which accordingly leads to an increase of GOS. However, in the nanostructured layer, no obvious change of GOS can be observed. This could be due to the fact that the slip resistance is higher in the nanocrystalline or ultra-fine materials.

Acknowledgements

This work was supported by the Champagne-Ardenne Research Council through NANOSURF and NANOTRIBO projects with grateful acknowledgement.. It is also a pleasure to acknowledge the University of Technology of Troyes, Grand Troyes and Conseil Départemental de l'Aube for financial support through CompNano project. J. Zhou acknowledges the MESR for the financial support of his PhD.

References

1 T. Hanlon, Y.N. Kwon, S. Suresh, Scripta Mater. **49**, 675 (2003).
2 L. Yang, N.R. Tao, K. Lu, L. Lu, Scripta Mater. **68**, 801 (2013).
3 Y. Li, L. Zhou, W. He, G. He, X. Wang, X. Nie, B. Wang, S. Luo, Y. Li, Sci. Technol. Adv. Mater. **14**, 055010 (2013).
4 K. Lu, J. Lu, Mater. Sci. Eng. A 375-377, 38 (2004).
5 T. Roland, D. Retraint, K. Lu, J. Lu, Scripta Mater. **54**, 1949 (2006).
6 Y. Pi, G. Agoda, S. Potiron, C. Demangel, D. Retraint, H. Benhayoune, J. Nanosci. Nanotechnol. **12**, 4892 (2012).
7 T. Roland, D. Retraint. K. Lu, J. Lu, Mater. Sci. Forum **490-491**, 625 (2005).
8 H.A. Padilla, B.L. Boyce, Exp. Mech. **50**, 5 (2010).
9 T. Hanlon, E.D. Tabachnikova, S. Suresh, Fatigue Damage Struct. Mater. V **27**, 1147 (2005).
10 K.A. Gujba, M. Medraj, Materials 7, 7925 (2014).
11 J. Zhou, Z. Sun, P. Kanouté, D. Retraint, Int. J. Fatigue **103**, 309 (2017).
12 K. Dalaei, B. Karlsson, L-E. Svensson, Mater. Sci. Eng. A **528**, 1008 (2011).
13 J.C. Villegas, L.L. Shaw, K. Dai, W. Yuan, J. Tian, P.K. Liaw, D.L. Klarstrom, Philos. Mag. Lett. **85**, 427, (2005).
14 D. Li, H.N. Chen, H. Xu, Appl. Surf. Sci. **255**, 3811 (2009).
15 L.L. Shaw, J.W. Tian, A.L. Ortiz, K. Dai, J.C. Villegas, P.K. Liaw, R. Ren, D.L. Klarstrom, Mater. Sci. Eng. A **527**, 986 (2010).
16 K. Dalaei, B. Karlsson, Int. J. Fatigue 38, 75 (2012).
17 R. John, D.J. Buchanan, M.J. Caton, S.K. Jha, Procedia Eng. **2**, 1887 (2010).
18 J.C. Kim, S.K. Cheong, H. Noguchi, Int. J. Fatigue **56**, 114 (2013).
19 Z. Sun, D. Retraint, T. Baudin, A.L. Helbert, F. Brisset, M. Chemkhi, J. Zhou, P. Kanoute, Mater. Charact. **124**, 117 (2017).
20 X.H. Chen, J. Lu, L. Lu, K. Lu, Scripta Mater. **52**, 1039 (2005).

21 J. Petit, L. Waltz, G. Montay, D. Retraint, A. Roos, M. François, Mater. Sci. Eng. A **536**, 124 (2012).
22 J. Oddershede, J.P. Wright, A. Beaudoin, G. Winther, Acta Mater. **85** 301 (2015).
23 Y.J. Wei, L. Anand, J. Mech. Phys. Solids **52** 2587 (2004).
24 Y.J. Wei, A.F. Bower, H.J. Gao, Acta Mater. **56** 1741 (2008).
25 M.A. Meyers, A. Mishra, D.J. Benson, Prog. Mater. Sci. **51** 427 (2006).

Effects of mean strain and tensile pre-strain on torsional fatigue behaviours of duplex stainless steel SAF2205

Bingbing Li[1], *Dunji* Yu[1], and *Xu* Chen[1,*]

[1]School of Chemical Engineering and Technology, Tianjin University, Tianjin 300350, China

Abstract. Effects of mean strain and tensile pre-strain were investigated on the torsional fatigue behaviours of duplex stainless steel SAF2205. Two equivalent strain amplitudes (0.5%,0.7 %), three strain ratios (-1, -0.5, -0.25) and 5% tensile pre-strain were chosen. Results indicated that the mean strain had no distinct influences on the torsional fatigue behaviours in terms of cyclic stress reponse and fatigue life while tensile pre-straining made a significant increase in cyclic stress response which was mainly attributed to the cross hardening derived from the loading sequence of monotonic tension preceding to cyclic torsion and led to a reduction in fatigue life. The failure mechanisms were revealed by scanning electron microscope characterized by microcracks initiation at the extrusions in ferrite and phase boundary inhibited further propagation. Additionally, the fractography of all fatigued specimens revealed a quasi-cleavage brittle mode with features of distinct tearing ridges and cleavage facets.

1 Introduction

Duplex stainless steels (DSS) beneficially combining the mechanical properties of both ferritic and austenitic phases in terms of strength, ductility and corrosion resistance are increasingly employed in highly aggressive environments such as nuclear power plants, automobile, chemical and petrochemical industries as a potential alternative to austenitic stainless steels (ASS) [1-6]. The structural components usually suffer alternative loadings resulting from the variable temperature fields and corresponding high stress fields. Hence, all the parameters which can influence the cyclic stress response and microstructural evolution should be evaluated comprehensively.

Investigations about the low cycle fatigue behaviours of DSS have been published mainly focused on the cyclic stress response, mechanisms of cracks nucleation and fatigue life [7-9,10-12]. PK Chiu et al. [7, 8] founded that a higher strain amplitude and tensile mean strain would promote the occurrence of strain-induced martensite (SIM) formation which could result in secondary hardening (SD). Additionally, V Aubin et al. [9] explored the fatigue properties for DSS X2CrNiMoN25-07 under complex biaxial loadings. The results showed that this material was not susceptible to the loading path as ASS did and presented a lower extra-hardening on cyclic stress response, which should be attributed to the performance of each single phase under non-proportional fatigue where the austenite exhibited a strong over-hardening and contrarily little for the ferrite. However, information about the DSS LCF behaviours are too limited. The microcracks usually initiated along the extrusions in ferrite [10,11] and ferrite grain boundaries [12]. Moreover, the phase boundaries usually worked as an effective barrier for microcracks to further propagate.

Pre-straining (cold work) recognized as an important fabrication processing variable has been universally employed in engineering construction during the fabrication of vessels, tanks, piping, heat exchangers, etc. Substantial references have been published focusing on the mechanisms of cold work/pre-straining on the LCF behaviours of ASS, aluminium and magnesium alloy [13-17], among which CL Zhou et al. [18, 19] investigate the effects of prior tensile plastic deformation on the torsional fatigue properties of a structural steel. As for the DSS, however, the researchers are mainly concerned with the effects of pre-straining/cold-work on the basic mechanical properties, microstructural evolution during monotonic loadings [20-24]. Much more efforts should be made to explore how it works on the subsequent LCF behaviours and reveals the corresponding mechanisms.

The present work aims at characterizing the torsional fatigue behaviours of DSS SAF2205 under various mean strain. In addition, much attention has been paid to the effects of monotonically tensile pre-straining on the subsequent torsional fatigue behaviours in order to get a better knowledge of the LCF behaviours of DSS sufficiently under complex loading conditions and supplement experimental data.

2 Experimental procedure

The material studied in the current research was DSS SAF 2205 containing a volume fraction (%) of ferrite (α) and austenite (γ) of 53:47 which was manufactured in the

* Corresponding author: xchen@tju.edu.cn

process of cold-rolled and a thermal treatment at 1050°C followed by a rapid water quench. The chemical composition of alloying elements in weight percent (%) is: C 0.019, N 0.192, Ni 4.74, Cr 22.3, Mn 1.64, Mo 3.13, Si 0.41, S 0.001, P 0.018, Fe Bal. For the metallography microstructures observation of both phases, the material of as-received were ground and polished with sequentially finer grits down to 0.25 μm diamond paste and then electrolytically etched in a solution of 10% oxalic acid at 20V for 5s. It can be found that the austenitic islands were embedded in ferrite matrix with both phases distributed homogeneously in the transversal section meanwhile a highly elongated shape along the rolling direction was revealed (Fig.1). The surface and fractographic morphology of specimens after failure were observed using a HITACHI S-4800 scanning electron microscope (SEM).

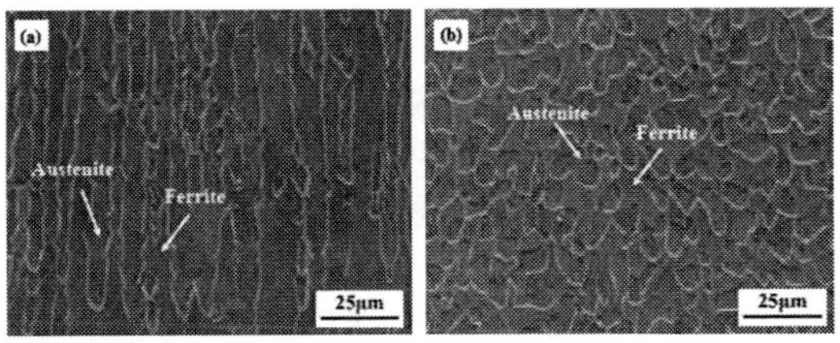

Fig. 1. Microstructure of DSS SAF2205 in longitudinal (a) and transversal (b) sections of as-received specimen.

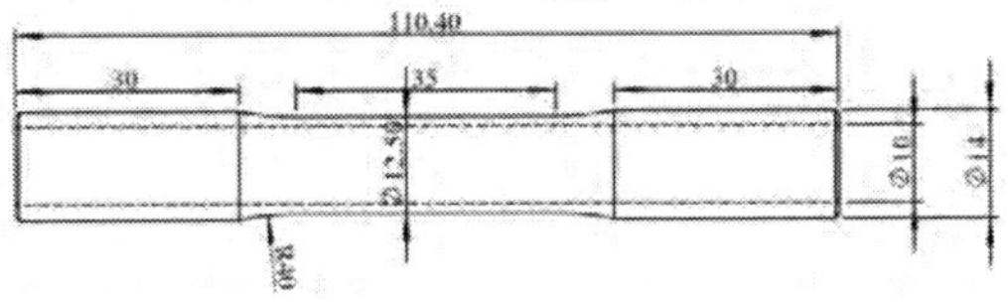

Fig. 2. Dimension of specimens used in current work.

Two equivalent strain amplitudes of 0.5% and 0.7% were chosen. Three strain ratios (-1, -0.5 and -0.25) corresponding to various mean strain and 5% tensile pre-straining were employed to investigate the effects of mean strain and monotonically tensile pre-straining on subsequent torsional fatigue behaviours. A servo-hydraulic MTS axial-torsional instrument and a tension-torsion extensometer for measuring and controlling the shear strain were used. The thin wall-hollow tubular specimen was chosen in order to obtain the uniform distribution of shear stress over the transverse section and the dimension was schematically presented in Fig.2, as per ASTM E2207-08.

3 Results and discussion

3.1 Effects of mean strain

Table 1 and Fig. 3 exhibited the relevant parameters configuration and evolution of cyclic stress response of tests under various mean strain, respectively. Tests with strain range of 1.0% presented a consistent initial hardening followed by a continuous softening and finally a drastic reduction of stress amplitude resulting from the formation of macrocracks whatever the strain ratio was. Namely, the mean strain had almost no influence on the torsional fatigue behaviours in terms of cyclic stress response as shown in Fig. 3(a). From the perspective of fatigue life, the same conclusion seemed to be true yet in which specimens with different strain ratios gave a nearly equal life except tests with strain ratio of -0.25 which given a relatively longer life owing to the lower stress amplitude in the course of cycling. As for tests with strain range of 1.4%, an extra stage of cyclic stabilization after a prolonged period of softening (Fig. 3(b)) developed comparing to that of test with strain range of 1.0%. Fig.4(a) and (b) showed the two groups of hysteresis loops of tests with strain range 1.0% and 1.4% at half life time, respectively. It could be found that the hysteresis loops corresponding to different mean strain almost coincided at constant strain range, which indicated that the mean strain didn't bring about extra influence of beneficial or detrimental to the torsional fatigue behaviours viewed from the plastic strain energy point. Besides, the value of mean stress caused by asymmetrical strain was small relative to the cyclic stress amplitude and relaxed quickly with cycling (Fig. 5). Furthermore, the phenomenon of secondary hardening before final failure developed at tests with strain ratio of -0.25. The stage of secondary hardening, even though the degree of which was quite slight lasted more than half of the fatigue life. The local amplification at the lower-left of Fig. 3(b) was made for clearer observation.

Table 1. Test parameters with mean strain.

R	$\Delta\gamma$ (%)	$\gamma_{min}/\gamma_{max}$	γ_{mean}	Life
-1	1.0	-0.50/0.50	0	7126
	1.4	-0.70/0.70	0	3304
-0.5	1.0	-0.33/0.67	0.17	6260
	1.4	-0.47/0.93	0.23	3174
-0.25	1.0	-0.20/0.80	0.3	10366
	1.4	-0.28/1.12	0.42	5438

The fatigue behaviours resulting from cyclic plasticity is linked to the formation, movement and annihilation of dislocations. Therefore, the evolution of cyclic stress response is strongly dependent on the dislocation configuration and evolution of microstructures. GVP Reddy et al. [25] determined that the cyclic hardening was the result of the individual or combined of 1) dislocation reproducible and mutual interactions, 2) dislocation-precipitations interactions acting as local obstacles on the glide plane, 3) dislocation-mobile dislocations, namely DSA. [26] measured the Vickers hardness and fraction of martensite of specimens at cycles corresponding to the initial hardening region and found a large increase in hardness with almost no increase in the amount of martensite, which implied the initial hardening should be attributed to the increase in dislocation density during fatigue. Additionally, the progressive build-up of dislocation pile-ups during cycling which will induce an increase in the resistance of grain boundaries or obstacles to the further pile-ups might be another source of initial hardening in addition to the increase in

https://doi.org/10.1051/matecconf/201816515003

dislocation density. With regard to the softening which is quitely sensitive to the microstructural changes mainly to the rearrangement of dislocation configuration, factors which can influence the movement and evolution of dislocation, such as chemical composition, temperature, strain amplitude and strain rate will make a distinct contribution to the cyclic softening. The addition of N into the matrix will promote the trend of slip planarity and slip reversibility. The nitrogen addition to 316LN stainless steel has been reported to induce cyclic softening at room temperature [27, 28]. Similarly, the same effect was also observed in DSS Fe-Cr-Mn-N. The higher strain amplitude will bring out a larger plastic component which can result in a high density of mobile dislocations to make further efforts to the softening. In addition, the stabilization was closely related to the formation of well-defined low energy configuration (such as dislocation cell).

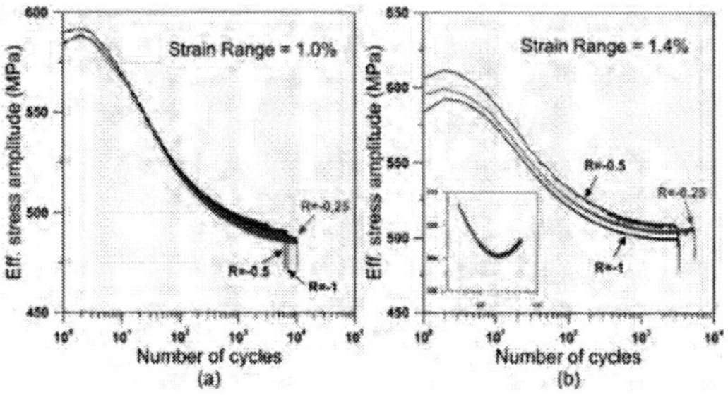

Fig. 3. Cyclic stress response under different strain ranges and strain ratios.

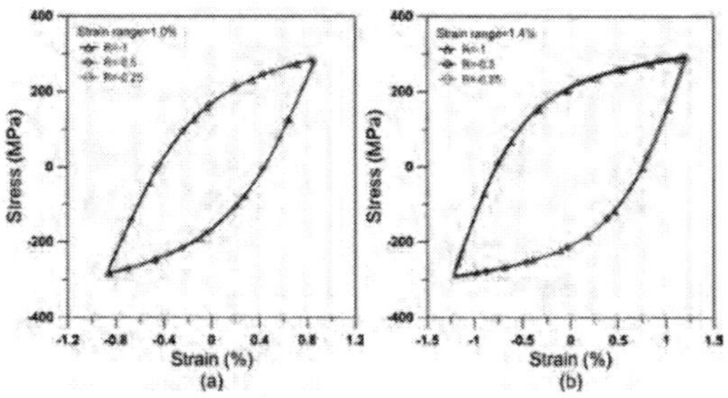

Fig. 4. Hysteresis loops at half life time.

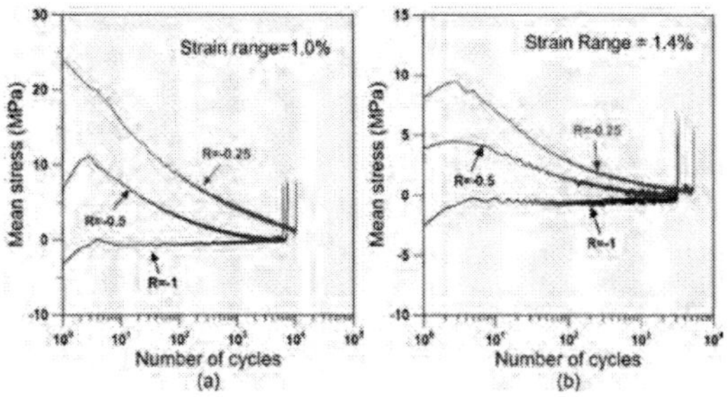

Fig. 5. Mean stress versus number of cycles at various loading conditions.

Plenty of investigations have attributed the occurrence of secondary hardening to the strain induced martensite (SIM) transformation pertaining to stainless steel especially ASS or DSS at room temperature [29, 30]. Despite the low stacking fault energy (SFE) which can prompt the SIM formation of DSS, the steel was found to be more stable than other stainless-steel grades, such as AISI 304 or 316 and its variants [23] which was agreement with the results in the present work in which the degree of secondary hardening was quite slight. There were usually two types of martensite formed during the course of SIM, hexagonal ε-martensite phase and BCC-α'-martensite occurred directly from austenite or via the hexagonal ε-martensite phase [23,31]. Additionally, P.K. Chiu [7] proposed that the presence of strain-induced intersected ε-martensite bands is more beneficial and preferential than the α'-martensite to the fatigue properties of SAF2205. In conclusion, there was no distinct influence of mean strain on the torsional fatigue behaviours at shear strain range of 1.0% and 1.4% in the current work.

3.2 Effects of tensile pre-straining

The consistent trend of continuous cyclic softening until final failure characterized was detected irrespective of strain amplitudes illustrated in Fig. 6. Additionally, the higher the shear strain amplitude, the higher softening rate obtained. Reference [32] reported that the fatigue behaviours showed strong propensity to the loading history for materials with a low SFE. The cyclic softening which can reduce dislocation density and promote microstructural rearrangement leading to finally the formation of low energy dislocation structures was a dynamic recovery process induced by cyclic loading [13, 33, 34]. It has been extensively investigated that the material subjected to a prior plastic deformation will present continuously cyclic softening in the subsequent pull-push fatigue tests [14, 31-33, 35, 36]. As mentioned in section 3.1, the high dislocation density was responsible to the initial hardening, thus, the pre-straining stage producing plenty of dislocations could be regarded as an equivalent process to the initial cycling. What's more, the stored strain energy ascribed to the high tensile pre-deformation will work as the driving force to accelerate the rearrangement process, both of which worked together to cause the continuous softening during the whole life process. C. larid et.al [33] reported that the softening would occur if a specimen initially cycled at a high strain amplitude and subsequently cycled at a low strain amplitude, or if severe monotonic deformation was followed by fatigue. Results showed that the dislocation structures were strongly strain amplitude dependent in which loop patches at low strain amplitudes, mixed with dipolar wall structures at intermediate amplitudes and increasingly cellular at high strain amplitudes was revealed. In addition, the dislocation cell became increasingly equiaxed and small size with increase of strain amplitude. The initial cyclic hardening stage preceding to the softening can be found for 316 ASS in the annealed state owing to the thermal activation recovery process which can reduce the original dislocation density before the followed fatigue tests and provide enough capacity for new dislocation

https://doi.org/10.1051/matecconf/201816515003

nucleation while a substantial cyclic softening was observed, as for the cold worked condition [36]. The strain-controlled fatigue behaviour of cold-drawn type 316 austenitic stainless steel underwent various levels of pre-straining has been investigated at strain amplitudes varying from 0.22% to 0.75% and at room temperature. The results showed that continuous cyclic softening occurred over the whole lifetime at strain amplitude higher than 0.35% in the 20% cold-drawn steel, and 0.31% in the 30% cold-drawn steel. The speed for dislocation formation was still higher than annihilation at small strain amplitude leading to the occurrence of initial hardening, which implied that there was a threshold strain required for cyclic softening during fatigue. Only the applied strain amplitude was high enough, can the material obtain sufficient driving force for the rearrangement [13, 26].

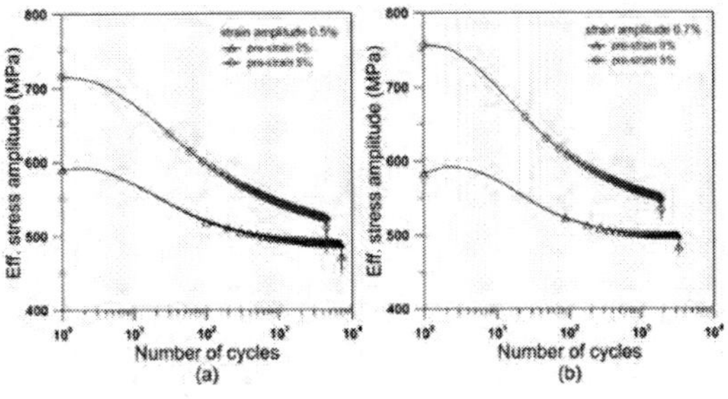

Fig. 6. Cyclic stress response under different strain amplitudes and pre-straining levels.

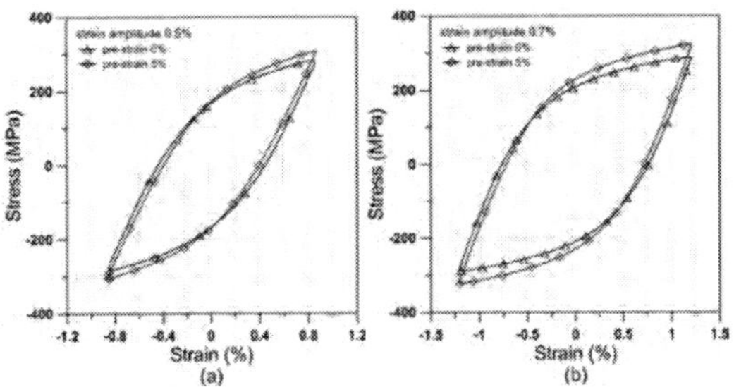

Fig. 7. Hysteresis loops at half life time.

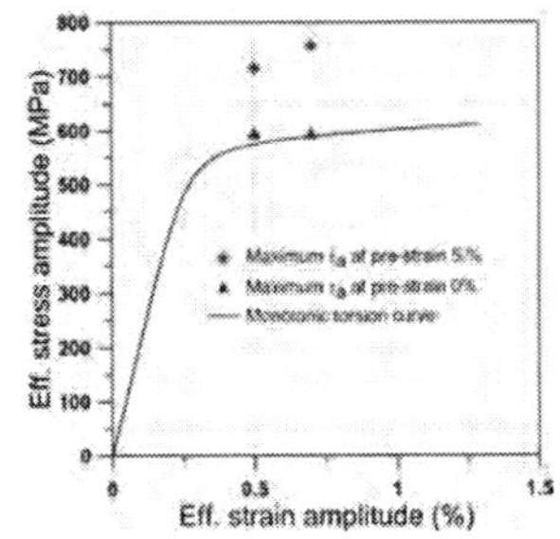

Fig. 8. Schematic illustration to significantly cyclic hardening for tests with pre-straining.

Fig. 6 depicted the relation of stress amplitude to number of cycles at strain amplitude 0.5% and 0.7% after 5% tensile prior plastic deformation where a significantly enhanced cyclic hardening can be observed. The pre-straining could improve strength and decrease ductility which can be verified by Fig. 7 in which a gradual reduction in the plastic strain amplitude with the increase of pre-straining levels can be observed. Tensile pre-straining indicates loading the material to one certain value monotonically and then unloading, which has been recognized as an important processing variable for industrial application to improve the material strength [13, 14, 26]. Huang [37] gave the explanation for the considerable cyclic hardening from perspective of dislocation glide, deformation induced twinning and martensite transformation. The increase of dislocation density enhancing the interaction in a short-range order which further inhibited the dislocation movement and formation of twinning particularly for material with low SFE was the predominant mechanism at the early stage. The strain-induced martensite transformation began to control the deformation process after the amount of twinning reaching saturation. The loading sequence of tensile pre-straining followed by torsional fatigue which was quite similar to the non-proportional cross loading path which will introduce a considerable cross-hardening effect. However, the loading profile in the present work just equaled to only one cycle of cross loading path for the tensile loading applied in the pre-straining process was just monotonic. It can be observed that the maximum stress amplitude of tests without pre-straining at different strain amplitudes were larger than that determined by monotonic curve at the identical strain amplitude which can be ascribed to the effect of cyclic hardening (Fig.8). Therefore, the difference between maximum stress amplitudes at tests without pre-straining and that at tests with pre-straining can be taken as simply the effect of cross-hardening. Moreover, the effect of cross-hardening degraded gradually with cycling which indicated that the effect of cross-hardening would make an influence to the stress response in the subsequent cycling which was consistent with the progressively decreasing difference between stress amplitudes. The effect of cross-hardening derived from non-proportional cross-shaped loading path has been published [38, 39]

3.3 Microstructural analysis

The failure model of all the tests in the present work irrespective of strain amplitudes, mean strain and pre-straining can be consolidated into the shear-type, which was characterized by a primary macrocrack propagate along the maximum shear stress plane, namely parallel to the specimen axis (Fig. 9(a)). The Fig. 9(b) demonstrated the manner of cracks propagation of one representative fractography morphologies in detail. The outer surface usually became the preferential sites for cracks initiation and then propagating along radial direction toward inner surface by a manner of circular-arc-shaped. The propagation toward the two ends of specimen began simultaneously after approaching a large depth, which finally induced the primary macrocracks. The well-defined tearing ridges at the lower left of Fig. 9(c) presented the typical features of brittle quasi-cleavage fracture. Fig. 10 compared the morphology of stable propagation regions of tests with 5% pre-straining and without pre-straining to investigated the effects of tensile prior deformation on the cracks behaviours. The well-defined cleavage facets could be found at tests without pre-straining. Further, the number of secondary cracks decreased meanwhile the size increased considerably attributing to the coalescences of small cracks owing to the pre-straining, which finally evolved into the macrocracks parallel to the specimen axis. In addition, the characteristics of brittle fracture mainly referring to the cleavage facets became less and less obvious.

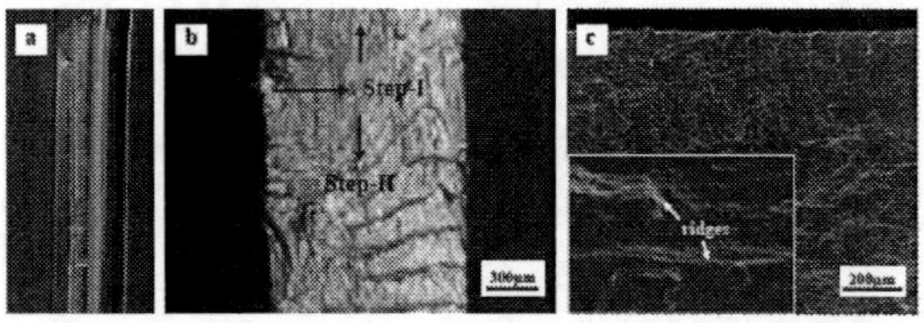

Fig. 9. (a)Morphology of primary cracks by camera, (b) Illustration of cracks propagation behaviours, (c)Different regions of fracture surface over radial direction

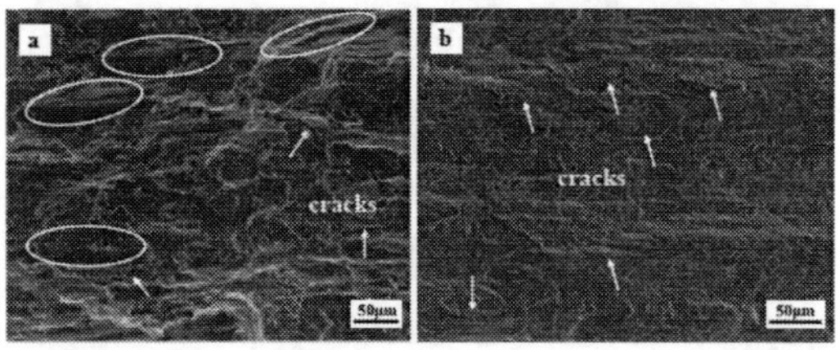

Fig. 10. Morphology of propagation regions on longitudinal surfaces at strain amplitude of 0.5% under loading conditions of (a)without pre-straining, (b)5.0 pre-straining.

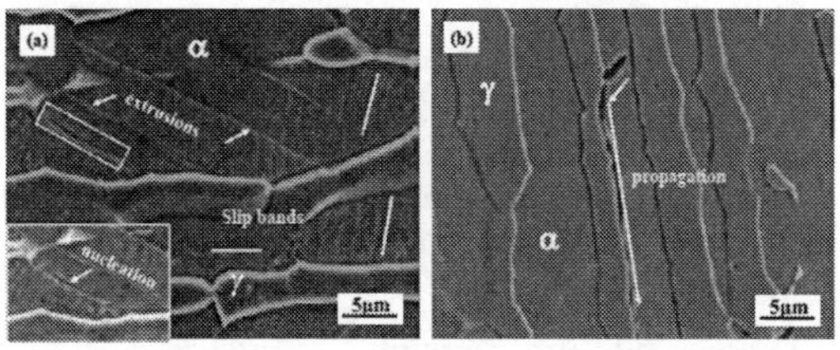

Fig. 11. Microcracks nucleation and propagation at extrusions in ferrite.

The majority of the plastic deformation should occur mainly in the ferritic phase which was consistent with the SEM observations of fatigued specimens as shown in Fig. 11(a). The substructure of intense slip marks formed in ferrite firstly and then microcracks nucleated along the extrusions. Besides, the flow of plasticity was nearly inhibited in the ferrite and the slip marks cannot pass through the phase boundary when the crystallographic Kurjumov–Sachs (K–S) relationships was not fulfilled. Thus, after nucleation the microcracks continued propagating along the phase boundary which was parallel to the specimen axis (Fig. 11(b)). Similar conclusions have also been found that the preferential sites for microcracks nucleation are the extrusions in ferritic phase and the phase boundary will hinder further propagation as an effective barrier [40].

4 Conclusions

A series of strain-controlled fatigue tests were carried out to investigate the effects of mean strain and prior tensile plastic deformation on torsional fatigue behaviours of DSS SAF2205 primarily in terms of cyclic stress response, fractographic morphology analysis and mechanisms of microcracks nucleation. After careful observation and discussion, conclusions can be drawn:

(1) A mixed model including initial cyclic hardening followed by a long period of softening and/or stabilization was observed in the current research. The slight secondary hardening developed at strain amplitude 0.7% jointed with a relative higher mean strain (R=-0.25) attributed to the occurrence of SIM transformation.

(2) The mean stress resulted from asymmetrical loading profile was small and relaxed quickly, which made little influence on the torsional fatigue behaviours in terms of cyclic stress response and fatigue life.

(3) Specimens subjected to the tensile pre-straining exhibited a continuous cyclic softening over the whole lifetime. Moreover, cross-hardening derived from the loading sequence of large tensile plastic deformation preceding to the torsional cyclic loading made a significantly extra cyclic hardening.

(4) A shear-type failure with features of a primary long crack propagating along the maximum shear stress plane was illustrated. More specifically, the microcracks nucleated at the outer surface and then propagated along the radial direction and further propagation along the axial direction. The fractography morphology of all fatigued specimens revealed a quasi-cleavage brittle mode characterized by distinct cleavage steps, tearing ridges and small cleavage facets.

References

1. R. Lillbacka, G. Chai, M. Ekh, P. Liu, E. Johnson, K.
Runesson, Acta Mater. **55**, 5359 (2007)
2. R. Strubbia, S. Hereñú, M.C. Marinelli, I. Alvarez-Armas, Mater. Sci. Eng., A **659** 47 (2016)
3. S. Emami, T. Saeid, R.A. Khosroshahi, J Alloys Compd **739** 678 (2018)
4. H. Luo, X.G. Li, C.F. Dong, K. Xiao, Arabian J. Chem. **10** S90 (2017)
5. L. Jinlong, L. Tongxiang, W. Chen, D. Limin, Mater.
Sci. Eng., C **62** 558 (2016)
6. A.G. Terricabras, C.G. Schön, Mater. Sci. Eng., A **717** 20 (2018)
7. P.K. Chiu, K.L. Weng, S.H. Wang, J.R. Yang, Y.S. Huang, J. Fang, Mater. Sci. Eng., A **398** 349 (2005)
8. P.K. Chiu, S.H. Wang, J.R. Yang, K.L. Weng, J. Fang,

Mater. Chem. Phys. **98** 103 (2006)

9. P.Q. Ve´ronique Aubin, S. Degallaix, Mater. Sci. Eng., A, **346** 208 (2003)
10. R. Strubbia, S. Hereñú, M.C. Marinelli, I. Alvarez-Armas, Int. J. Fatigue **41** 90 (2012)
11. R. Strubbia, S. Hereñú, A. Giertler, I. Alvarez-Armas, U. Krupp, Int. J. Fatigue **65** 58 (2014)
12. M.-C. Marinelli, A. El Bartali, J.W. Signorelli, P. Evrard, V. Aubin, I. Alvarez-Armas, S. Degallaix-Moreuil, Mater. Sci. Eng., A, **509** 81 (2009)
13. S.G.S. Raman, K.A. Padmanabhan, Int. J. Fatigue **18** 71 (1996)
14. K.B.S. Rao, M. Valsan, R. Sandhya, S.L. Mannan, P. Rodriguez, Metall. Trans. A **24** 913 (1993)
15. B. Wang, Z.J. Zhang, C.W. Shao, Q.Q. Duan, J.C. Pang, H.J. Yang, X.W. Li, Z.-F. Zhang, Metall. Mater. Trans. A **46** 3317 (2015)
16. K.S. Al-Rubaie, M.A. Del Grande, D.N. Travessa, K.R. Cardoso, Mater. Sci. Eng., A **464** 141 (2007)
17. T. Hama, H. Nagao, Y. Kuchinomachi, H. Takuda, Mater. Sci. Eng., A **591** 69 (2014)
18. C.L. Zhou, S.I. Nishida, N. Hattori, Key Eng. Mater. **353-358** 94 (2007)
19. C.L. Zhou, S.I. Nishida, N. Hattori, Key Eng. Mater. **340-341** 507 (2007)
20. J.L. KH Lo, CH Shek, D.J. Li, Mater. Sci. Eng., A **s452–453** 78 (2007)
21. S.S.M. Tavares, J.M. Pardal, M.R.D. Silva, C.A.S.D. Oliveira, Mater. Res. **17** 381 (2014)
22. M. Breda, K. Brunelli, F. Grazzi, A. Scherillo, I. Calliari, Metall. Mater. Trans. A **46** 577 (2014)
23. S.S.M. Tavares, M.R. da Silva, J.M. Pardal, H.F.G. Abreu, A.M. Gomes, J. Mater. Process. Technol. **180** 318 (2006)
24. S. Pramanik, S. Bera, S.K. Ghosh, Steel. Res. Int. **85** 776 (2014)
25. G.V.P. Reddy, R. Sandhya, S. Sankaran, P. Parameswaran, K. Laha, Mater. Des. **88** 972 (2015)
26. Y.C. Jeon, C.S. Kim, H. Ki, S.I. Kwun, J.W. Byeon, Mater. Sci. Forum **580-582** 597 (2008)
27. B.K. Choudhary, Metall. Mater. Trans. A **45** 302 (2013)
28. G.V.P. Reddy, R. Sandhya, S. Sankaran, M.D. Mathew, Mater. Trans. A **45** 5044 (2014)
29. M. Okayasu, H. Fukui, H. Ohfuji, T. Shiraishi, J. Mater. Sci. **48** 6157 (2013)
30. Z.S. XUE Zong-yu, WEI Xi-cheng, J. Iron. & Steel Res. (Int.) **17** 51 (2010)
31. J. SINGH, J. Mater. Sci. **20** 3157 (1985)
32. A. Belattar, L. Taleb, A. Hauet, S. Taheri, Mater. Sci. Eng., A **536** 170 (2012)
33. Z.W. C Laird, B.T. Ma, H.F. Chai, Mater. Sci. Eng., A **113** 245 (1989)
34. C.L. HF Chai, Mater. Sci. Eng. **93** 159 (1987)
35. J.T. LF Coffin, Trans. Met. Soc. AIME 215 794 (1959)
36. M.D. WJ Plumbridge, PJ Castle, Fatigue Fract. Eng. Mater. Struct. **3** 177 (2010)
37. K. HUANG Wenke, Acta Metall. Sinica (English Letters) **45** 275 (2009)
38. E.T.H. Okuchi, Trans. Japan. Soc. Mech. Eng. A **54** 1588 (1988)
39. L. Taleb, A. Hauet, Int. J. Plast., **25** 1359 (2009)
40. R. Strubbia, S.Hereñú, I.Alvarez-Armas, U.Krupp, Mater. Sci. Eng., A **615** 169 (2014)

Influence of the microstructure on the cyclic stress-strain behaviour and fatigue life in hypo-eutectic Al-Si-Mg cast alloys

Jochen Tenkamp[1,*], *Alexander* Koch[1], *Stephan* Knorre[2], *Ulrich* Krupp[2], *Wilhelm* Michels[2], and *Frank* Walther[1]

[1]TU Dortmund University, Department of Materials Test Engineering (WPT), D-44227 Dortmund, Germany
[2]University of Applied Sciences Osnabrueck, Institute of Materials Design and Structural Integrity, D-49009 Osnabrueck, Germany

Abstract. Aluminium alloys are promising candidates for energy- and cost-efficient components in automotive and aerospace industries, due to their excellent strength-to-weight ratio and relatively low cost compared to titanium alloys. As modern cast processing and post-processing, e.g. hot isostatic pressing, result in decreased frequency and size of defects, the weakest link depends on microstructural characteristics, e.g. secondary dendrite arm spacing (SDAS), Si eutectic morphology and α-Al solid solution hardness. Hereby, fatigue investigations of the effect of the microstructure characteristics on the cyclic stress-strain behaviour as well as fatigue mechanisms in the low cycle and high cycle fatigue regime are performed. For this purpose, samples of the aluminium cast alloy EN AC-AlSi7Mg0.3 with different Si eutectic morphology and α-Al solid solution hardness were investigated. To compare the monotonic and cyclic stress-strain curves, quasistatic tensile tests and incremental step tests were performed on two microstructure conditions. The results show that the cyclic loading leads to a hardening of the material compared to monotonic loading. Based on damage parameter Woehler curves, it is possible to predict the damage progression and fatigue life for monotonic and cyclic loading in hypo-eutectic Al-Si-Mg cast alloys by one power law.

1 Introduction

Aluminium alloys are promising candidates for energy- and cost-efficient components in automotive and aerospace industries, due to their excellent strength-to-weight ratio and relatively low cost compared to titanium alloys. Hereby, Al-Si alloys belong to the most common alloys in industry, as they achieve a market share of 70 to 80%. As modern cast processing and post-processing, e.g. hot isostatic pressing, result in decreased frequency and size of defects (oxides, porosity) [1-5], the weakest link depends on microstructural characteristics, e.g. secondary dendrite arm spacing (SDAS) [6-8], Si eutectic morpho-logy [6, 9-12] and Al solid solution hardness. They are relevant to ensure a safe design for safety-relevant components made of aluminium cast alloys. While the impact of porosity and SDAS on S-N curve has already been investigated, the effect of microstructure characteris-tics as Si eutectic morphology and α-Al solid solution hardness on the cyclic stress-strain behaviour as well as fatigue mechanisms in the low (LCF), high (HCF) and very high cycle fatigue (VHCF) regime is not well understood.

For this purpose, samples of the aluminium cast alloy EN AC-AlSi7Mg0.3 were casted in a steel mold with grain refinement and Sr modification. To avoid cast porosity, all samples were hot isostatically pressed (HIP). To investigate the influence of Si eutectic morphology and Al solid solution hardness, T6 heat treatment was variated. To compare the monotonic and cyclic stress-strain curves, quasistatic tensile tests and incremental step tests (IST) were performed on two microstructure conditions, which vary in solution annealing temperature resulting in different Si eutectic morphologies and α-Al solid solution hardness. Based on incremental step tests, cyclic stress-strain curves were determined. To validate the determined cyclic stress strain curves based on IST, strain- and stress-controlled fatigue tests at constant amplitude in the LCF and HCF regime from 10^2 to 10^7 cycles were performed. The fatigue progress was monitored by using correlated online hysteresis analysis and fractographic investi-gations. Finally, micro hardness mappings were carried out on longitudinal cross sections of fractured specimens to determine the effect of cyclic loading related to the specific Si eutectic morphology and the α-Al solid solution hardness.

2 Material and experimental setup

2.1 Material and microstructure

* Corresponding author: jochen.tenkamp@tu-dortmund.de

 https://doi.org/10.1051/matecconf/201816515004

For the characterisation of the effect of microstructural features on the fatigue behaviour, samples of the aluminium cast alloy EN AC-AlSi7Mg0.3 were casted in a steel mold with grain refinement ($AlTi_5B_1$) and Sr modification. The alloys were molten in an electric resistance furnace, including the alloying procedure. The melt was heated up to 730 °C and poured into a steel mold, which was pre-heated to 450 °C. From the cast samples (Ø 19 mm×185 mm) the sprue and riser were cut off. To avoid the influence of cast porosity, all samples were hot isostatically pressed by Densal^{TM}. To investigate the influence of Si eutectic morphology and α-Al solid solution hardness, the T6 heat treatment was variated in form of different solution annealing temperatures of (A) 530 °C and (B) 545 °C for one hour in an air-circulated furnace, quenched in water at room temperature and afterwards artificially aged at 160 °C for five hours. The two T6 heat treatment conditions are further called T6/A (530 °C/1 hrs) and T6/B (545 °C/1 hrs). The chemical composition was analysed by means of spark emission spectroscopy and is shown in Table 1.

Table 1. Chemical composition of the aluminium cast alloy EN AC-AlSi7Mg0.3, all data in wt-%.

Si	Mg	Fe	Mn	Ti	Al
7.5	0.288	0.094	0.0018	0.146	Bal.

Samples for basic characterisation were cold-embedded, ground and polished to a grit size of 0.3 µm using oxide polishing suspension (colloidal SiO_2). The grain size was determined in micrographs by means of Barker etching technique and quantified by linear intercept method (Fig. 1a) under polarised light using a stereoscopic light microscope. The secondary dendrite arm spacing (SDAS) and the formation of silicon particles were analysed on cross sections polished finally by oxide dispersion suspension (OPS) (Fig. 1b). Hereby, the SDAS was determined by linear intercept method based on a minimum of ten dendrites at 50× magnification. The grain size was determined to 620 ± 330 µm and the SDAS to 35 ± 5 µm.

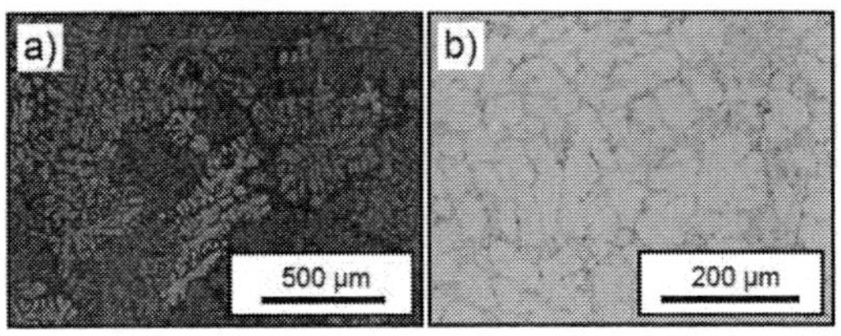

Fig. 1. Light micrographs: a) Barker etched (grains) and b) ODS polished (SDAS and Al-Si interdendritic eutectic).

To evaluate the characteristics of the Si particles within the interdendritic Al-Si eutectic, the image processing program ImageJ has been used to quantify size and shape by a binary greyscale analysis on micrographs with 500× magnification on a minimum of five micrographs.

The effects of casting and T6 heat treatment on the morphology of Si particles are illustrated in Fig. 2 by representative micrographs of as-cast material without T6 heat treatment (Fig. 2a), heat treatment T6/A (Fig. 2b) and T6/B (Fig. 2c). The morphology of Si particles was quantified by means of average value of area, perimeter, major and minor dimension as well as sphericity for the Si particles. The sphericity S was determined by following equation

$$S = 2(\pi \cdot A_{real})^{0.5} / P_{real} \quad (1)$$

using real area A_{real} and perimeter P_{real}. The sphericity takes values between 0 and 1, whereby 1 is attributed to an ideal spherical particle. For each batch, around 1,300 Si particles were analysed to determine representative values. In Table 2, the values of Si eutectic morphology are compared for botch batches.

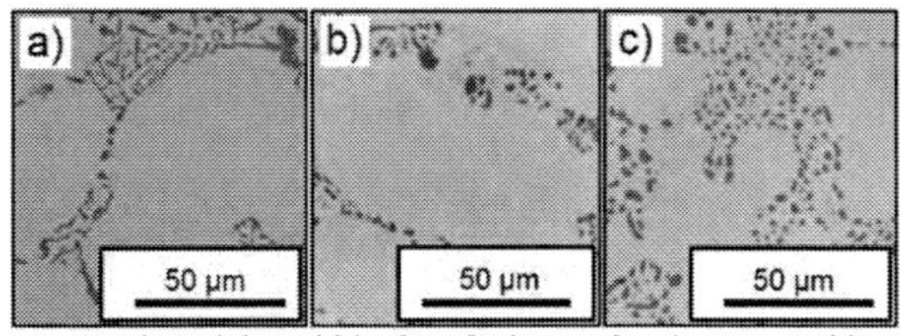

Fig. 2. Si particles within the Al-Si eutectic: a) as-cast without T6 heat treatment as well as after b) T6/A and c) T6/B heat treatment.

Table 2. Morphology characteristics of Si particles in interdendritic Al-Si eutectic for batches T6/A and T6/B

Batch	A_{real} [µm^2]	P_{real} [µm]	Dimension [µm]		S [-]
			Major	Minor	
T6/A	3.82	7.63	2.64	1.56	0.86
T6/B	5.69	9.69	3.25	1.79	0.84

The microstructure analyses show that the T6 heat treatment leads to a coarsening of Si particles at higher solution annealing temperatures. Botch batches show a high and comparable sphericity of Si particles due to the heat impact of HIP treatment and T6 heat treatment, respectively.

The effect of age hardening on the hardness of microstructure was measured by a micro hardness testing system using a force of 0.09807 N (HV0.1). The micro hardness was determined to 103 ± 4 HV0.1 and 110 ± 4 HV0.1 for batch T6/A and T6/B, respectively. The increased hardness of α-Al solid solution in batch T6/B is attributed to the increased solubility of magnesium at solution temperatures of 545 °C compared to 530 °C. Due to this, an increased age hardening of the α-Al solid solution by the formation of Mg_2Si precipitates can be realised.

 https://doi.org/10.1051/matecconf/201816515004

2.2 Quasistatic testing

To determine the monotonic stress-strain behaviour of batches T6/A and T6/B, quasistatic tensile tests were performed using a universal testing machine (Instron 3369) with a load cell of 50 kN. The strain was determined by means of tactile extensometer with a gage length of 10 mm. The test was performed strain-controlled at a strain rate of $0.07\cdot10^{-3}$/s in elastic region and $0.25\cdot10^{-3}$/s in elastic-plastic region of stress-strain curve related to DIN 6892-1. The specimen geometry used for both quasistatic and low cycle fatigue (LCF) tests is shown in Fig. 3a. The specimen geometry used for high cycle fatigue (HCF) tests is shown in Fig. 3b. The specimens were machined in accordance with DIN EN 6072:2011-06. Hereby, the gage sections were ground and polished up to 1 µm diamond paste.

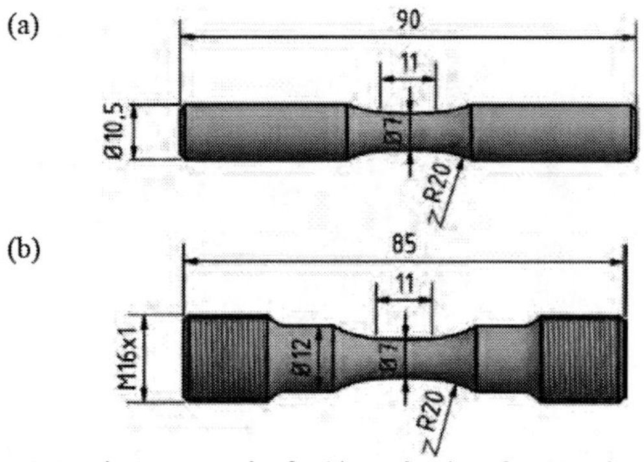

Fig. 3. Specimen geometries for (a) quasistatic and LCF testing as well as (b) HCF testing.

2.3 Fatigue testing

To characterise the fatigue behaviour in the low cycle fatigue (LCF) and high cycle fatigue (HCF) regime, tests were performed at servo-hydraulic as well as resonant testing systems, all at room temperature.

For LCF tests a servo-hydraulic testing system Schenck PC63M equipped with a load cell of 63 kN at a frequency of 0.1 Hz was used. Hereby, the LCF tests were carried out total strain-controlled with a strain ratio of R_ε = -1. Incremental step tests (IST) and constant amplitude tests (CAT) were performed to determine the LCF behaviour. Within the IST, the total strain amplitude $\varepsilon_{a,t}$ was increased each cycle by $\Delta\varepsilon_{a,t}$ = 0.01%, starting from $\varepsilon_{a,t,min}$ = 0.05% until reaching $\varepsilon_{a,t,max}$ = 0.7%, following a decrease of total strain amplitude by $\Delta\varepsilon_{a,t}$ until $\varepsilon_{a,t,min}$. These blocks of increasing and decreasing total strain amplitude were repeated until failure. CAT were carried out at a total strain amplitude of $\varepsilon_{a,t}$ = 0.7%.

HCF tests were conducted on resonant fatigue testing system Rumul Testronic equipped with a load cell of 20 kN at a frequency of around 65 Hz. The HCF behaviour was characterised in stress-controlled CAT with a load ratio of R = -1 at various stress amplitudes σ_a between 100 and 260 MPa.

3 Results and discussion

3.1 Monotonic stress-strain behaviour

To determine the monotonic properties of batches T6/A and T6/B quasistatic tensile tests were performed. The determined stress-strain curves are plotted in Fig. 4. Batch T6/B shows an increased yield strength compared to T6/A, while both batches show comparable Young's modulus, tensile strength and fracture strain. The quasistatic properties are quantified in Table 3.

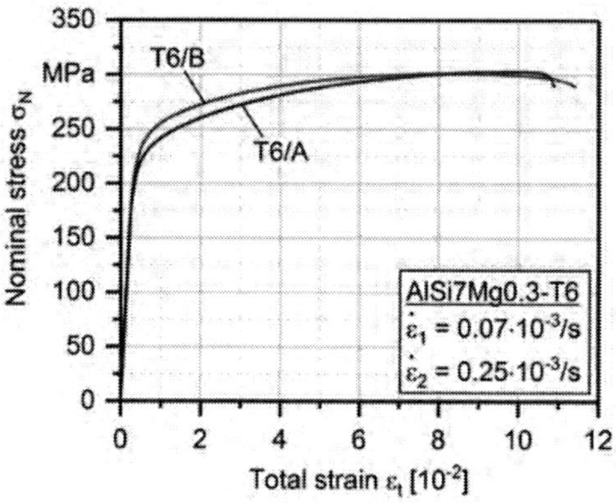

Fig. 4. Monotonic stress-strain curves for batches T6/A and T6/B based on quasistatic tensile tests.

The differences in yield strength can be attributed to the increased age hardening of α-Al solid solution in T6/B compared to T6/A due to higher solution temperature during T6/B heat treatment, which could be determined by micro hardness measurements.

Table 3. Monotonic properties based on quasistatic tensile tests.

Batch	E [GPa]	$\sigma_{y0.02}$ [MPa]	$\sigma_{y0.2}$ [MPa]	σ_{UTS} [MPa]	ε_t [10^{-2}]
T6/A	72.5	183.7	220.6	303.2	10.9
T6/B	72.8	187.4	233.4	301.3	11.4

3.2 Cyclic stress-strain behaviour

The cyclic stress-strain behaviour was determined by incremental step tests (IST). The correlation between hysteresis curves in IST and the cyclic stress-strain behaviour or curve, respectively, is shown in Fig. 5 and detailed explained in [13].

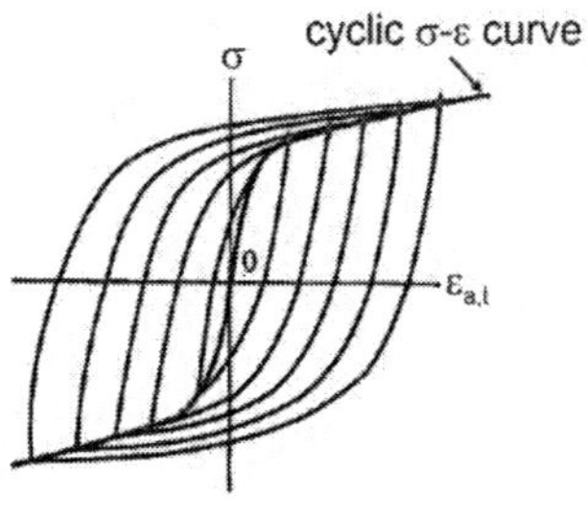

Fig. 5. Correlation of hysteresis curves of IST and the cyclic stress-strain curve [13].

The cyclic stress-strain curves are shown in Fig. 6 for batch T6/A and in Fig. 7 for batch T6/B. Hereby for each batch, the progression of cyclic stress-strain curve related to each block of IST is shown for three characteristic blocks 1st, 2nd and cyclic saturation (4th for T6/A and T6/B) compared to last block before failure (7th for T6/A and 6th for T6/B).

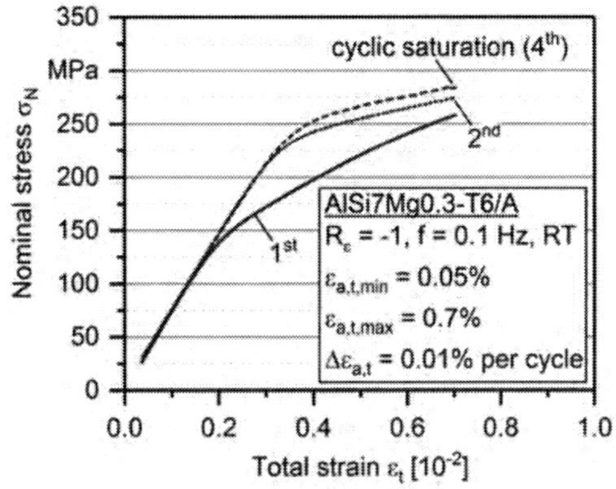

Fig. 6. Cyclic stress-strain curves for batch T6/A based on IST for three characteristic IST blocks (1st, 2nd and 4th).

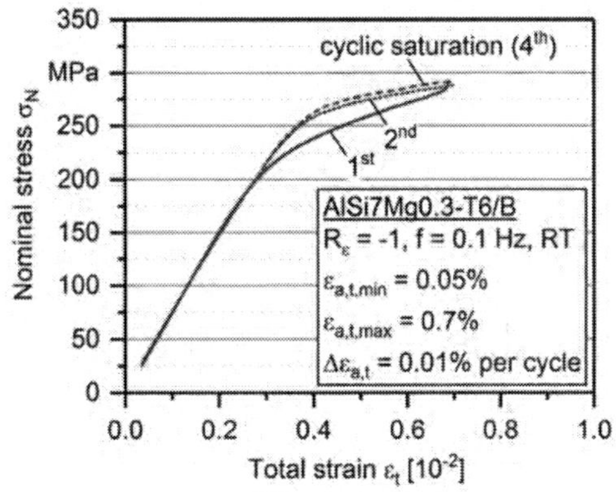

Fig. 7. Cyclic stress-strain curves for batch T6/B based on IST for three characteristic IST blocks (1st, 2nd and 4th).

For each batch, the material shows a significant cyclic hardening after 1st IST block, whereby batch T6/A shows a significantly decreased cyclic yield strength in the first block compared to batch T6/B. A cyclic saturation of the stress-strain behaviour could be determined for botch batches. Hereby, batch T6/B shows a nearby cyclic saturation for the 2nd block of IST, as no significant change could be determined compared to the cyclic saturated curve of 4th block. For batch T6/A, a cyclic saturated stress-strain curve could be determined for 4th block of IST.

In Fig. 8, the monotonic and cyclic stress-strain curves are compared for batches T6/A and T6/B. Related to IST, the last full block before failure (6th for T6/A and 4th for T6/B) was used for the cyclic saturated stress-strain curve for each batch. Botch batches show a significant cyclic hardening of the material. Hereby, the tendency of an increased strength or yield strength, respectively, can be determined for batch T6/B for both monotonic and cyclic stress-strain behaviour. Additionally, the CAT are plotted in Fig. 7. The stress and total strain values were deter-mined for strain- and stress-controlled CAT at half number of cycles to crack initiation. The procedure to determine the fatigue crack initiation during fatigue testing will be explained in 3.3. The cyclic stress-strain curves determined by IST and CAT fit very well (Fig. 8). Related to [14], this indicates that the investigated alu-minium alloys feature a planar slipping character.

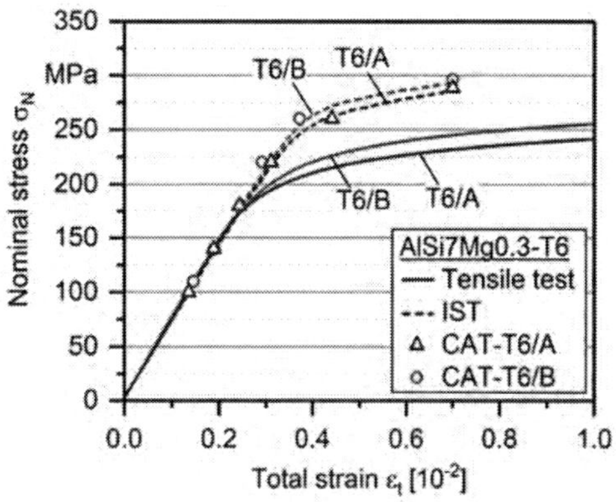

Fig. 8. Comparison of monotonic and cyclic stress-strain curves for batches T6/A and T6/B based on tensile tests and IST (6th block for T6/A and 4th block for T6/B) and validated by CAT ($σ_a$ and $ε_{a,t}$ at $N_i/2$).

The stress-strain behaviour of quasistatic tensile tests and IST was quantified by the following equations of Morrow:

Tensile test: $$\sigma_{a1/4} = K \cdot (\varepsilon_{a,p1/4})^{n} \quad (2)$$

IST : $$\sigma_a = K' \cdot (\varepsilon_{a,p})^{n'} \quad (3)$$

In Table 4, the monotonic and cyclic strain hardening exponents n and n' and the monotonic and cyclic strength coefficients K and K' as well as the cyclic yield strength $\sigma_{y0.02'}$ are compared for batches T6/A and T6/B.

Table 4. Comparison of monotonic and cyclic stress-strain properties for both batches based on tensile tests and IST.

Batch	n [-]	K [MPa]	n' [-]	K' [MPa]	$\sigma_{y0.02'}$ [MPa]

T6/A	0.0824	368.9	0.0534	388.4	243.7
T6/B	0.0758	371.6	0.0461	381.8	246.6

3.3 Fatigue behaviour

The fatigue behaviour was determined by total strain- and stress-controlled CAT from 10^2 to 10^7 cycles. The total strain-controlled CAT performed with an amplitude of $\varepsilon_{a,t} = 0.7\%$ ($R_\varepsilon = -1$, $f = 0.1$ Hz) and stress-controlled CAT at stress amplitudes between 100 and 260 MPa ($R = -1$, $f = 65$ Hz). The crack initiation could be determined for each test and was used as failure criteria for strain- and stress-controlled CAT. Hereby, the crack initiation under strain control was determined by a stress drop of 1% related to steady state stress amplitude σ_{a^*} before failure: T6/A $\sigma_{a^*} = 293.0$ MPa, T6/B $\sigma_{a^*} = 296.7$ MPa. For stress-controlled CAT, the crack initiation was determined in resonant testing system by a resonant frequency drop of 0.01% for specimens related to the steady state resonant frequency $f^* \approx 65$ Hz.

In Fig. 9, the S-N diagrams are plotted for both batches. Hereby, for strain-controlled CAT the stress amplitude σ_a was determined at half of number of cycles to crack initiation $N_i/2$. Moreover, the ultimate tensile strength determined by quasistatic tensile test is plotted in the S-N diagram, using $N_i = 0.25$.

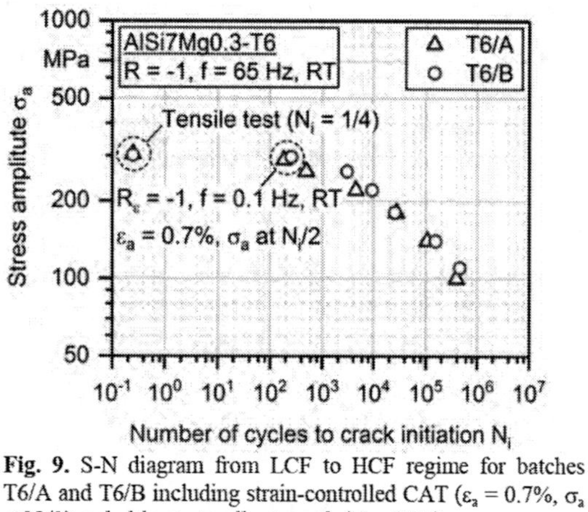

Fig. 9. S-N diagram from LCF to HCF regime for batches T6/A and T6/B including strain-controlled CAT ($\varepsilon_a = 0.7\%$, σ_a at $N_i/2$) and ultimate tensile strength ($N_i = 0.25$).

As batches T6/A and T6/B show significant differences in hardness, quasistatic and cyclic stress-strain curves, batch T6/A shows a decreased fatigue lifetime in the LCF regime, but only a slightly decreased lifetime in the HCF regime. Hereby, the most significant differences for both batches could be determined in the fatigue regime between 10^2 and 10^4 cycles to crack initiation. While the fatigue strength for each batch shows a linearly decrease in log-log diagram at higher lifetime $N_i > 10^4$, the fatigue strength approximates to the ultimate tensile strength in the LCF regime.

Based on the monotonic and cyclic tests, the damage parameter P_{SWT} related to Smith, Watson and Topper was determined by following equation

$$P_{SWT} = (\sigma_{max} \cdot \varepsilon_{a,t} \cdot E)^{0.5} \tag{4}$$

containing maximum stress σ_{max}, total strain amplitude $\varepsilon_{a,t}$ and Young's modulus E. The damage parameters P_{SWT} for each test (tensile test, IST, CAT) and botch batches are plotted versus number of cycles to crack initiation in Fig. 10 as P-N diagram.

The P-N diagram underlines the results of S-N diagram, in which an influence of microstructure could be determined for LCF regime, but no significant changes in lifetime related to HCF regime. In contrast to S-N diagram, the correlation between damage parameter and lifetime can be described by a power law for each batch over the complete lifetime from tensile test to CAT in the HCF regime, as shown by a straight line in Log(P)-Log(N) diagram.

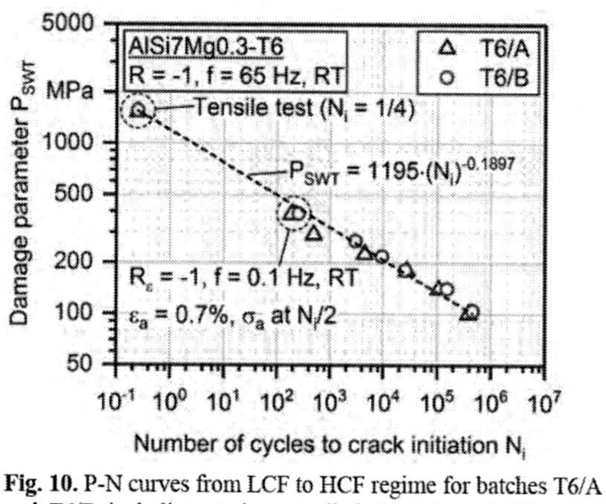

Fig. 10. P-N curves from LCF to HCF regime for batches T6/A and T6/B including strain-controlled CAT ($\varepsilon_a = 0.7\%$, σ_a at $N_i/2$) and quasistatic tensile tests ($N_i = 0.25$).

3.4 Fractographic investigations

For batches T6/A and T6/B the fracture surfaces were analysed to determine the role of microstructural features like α-Al solid solution hardness and interdendritic Si particles on the fatigue behaviour.

As the most significant differences could be determined in the LCF regime, the focus is set on the total strain-controlled CAT at an amplitude of 0.7%. In Fig. 11, the fracture surface is shown for batch T6/A. The fatigue crack initiates at small remnant pores (Fig. 11a) and interacts with the dendritic and interdendritic micro-structure during fatigue crack propagation phase. Hereby as exemplarily shown in Fig. 11b, a high amount of cracked Si particles or particles, respectively, in the interdendritic Al-Si eutectic could be determined along the fatigue crack on the fracture surface.

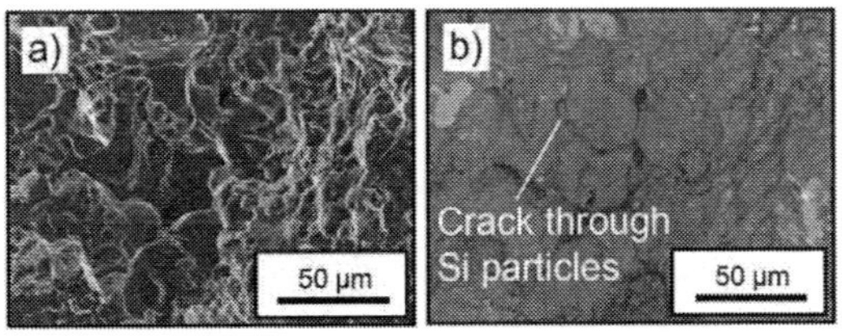

Fig. 11. Characteristic fracture surface in LCF regime for batch T6/A: a) Crack initiation at remnant porosity, b) cracking through Si particles.

In Fig. 12, the fracture surface of T6/B specimens in the LCF regime is shown. Hereby, the crack initiation also starts from remnant pores (Fig. 12a). In contrast to T6/A, the crack propagates between silicon particles and α-Al solid solution, leading to a debonding of Si particles.

Further investigations were performed to understand the interaction of fatigue loading on the cyclic hardening and softening of the microstructure. Therefore, micro hardness mappings using a force of 0.09807 N (HV0.1) were carried out. Hereby, longitudinal cross sections of the specimen of the strain-controlled CAT ($\varepsilon_{a,t}$ = 0.7%) were cut perpendicular to the crack surface including the clamped region in order to enable a direct comparison of loaded and unloaded regions as schematically shown in Fig. 13.

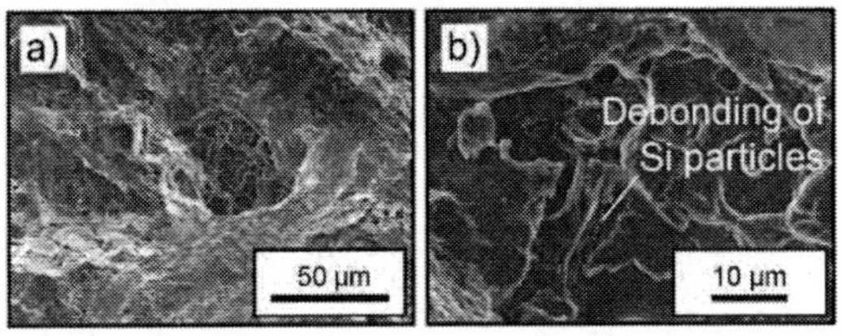

Fig. 12. Characteristic fracture surface in LCF regime for batch T6/B: a) Crack initiation at remnant porosity, b) debonding of Si particles within α-Al solid solution.

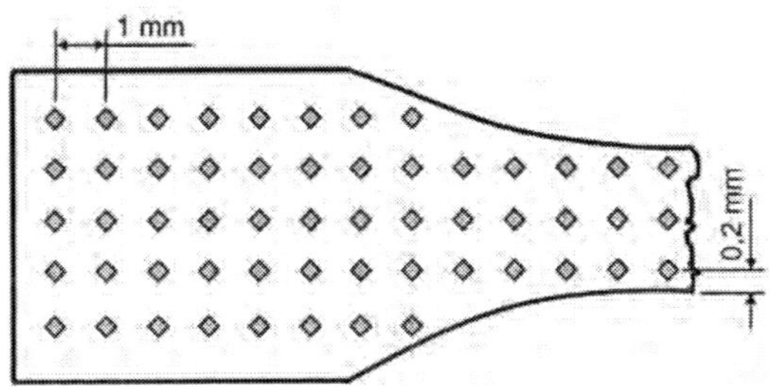

Fig. 13. Schematic illustration of micro hardness mappings on longitudinal cross sections for fractured CAT specimens.

In Fig. 14, the micro hardness mappings for the longitudinal cross sections are compared for batches T6/A (a) and T6/B (b). While the fatigue loading in the LCF regime leads to an increased hardness in loading region of specimen T6/B due to cyclic hardening as determined in Fig. 6, a decreased hardness in the loaded region compared to unloaded region could be determined for batch T6/A.

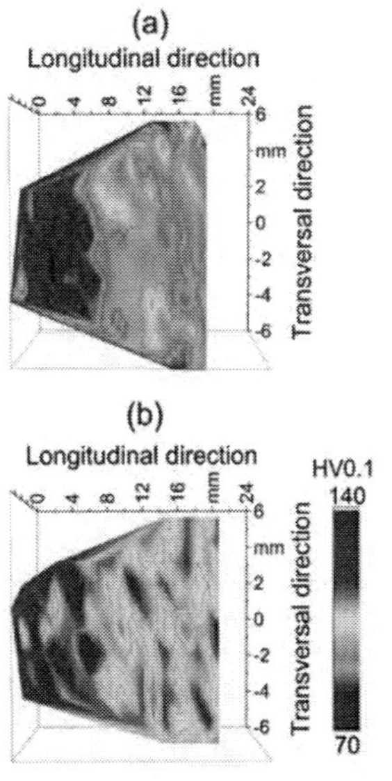

Fig. 14. Hardness mappings on fatigue loaded longitudinal cross sections of a) T6/A and b) T6/B.

For a better comparison of hardness distribution due to fatigue loading, the average micro hardness over longitudinal distance from fracture surface is plotted for both batches in Fig. 15. The initial hardness of each batch could be determined in the unloaded regions to around 95 HV0.1 for batch T6/A and 108 HV0.1 for batch T6/B. As already mentioned, the fatigue loading leads to an increased hardness of 131 HV0.1 (closest to fracture surface) for batch T6/B. For batch T6/A, a decrease of hardness could be determined in the loaded region with a hardness minimum of 76 HV0.1.

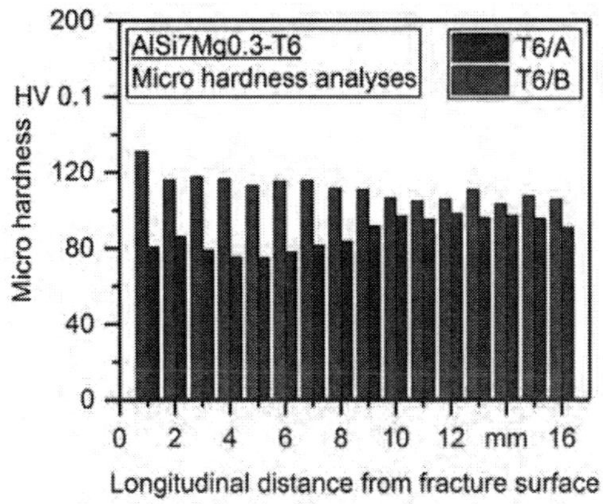

Fig. 15. Distribution of hardness related to distance from fracture surface due to fatigue loading for batch a) T6/A and b) T6/B

As the fracture surfaces in Fig. 11b showed a dominating cracking of Si particles in the interdendritic Al-Si eutectic, it is assumed that this phenomenon can be attributed to the decreased hardness in T6/A. The reason for cracked Si particles in T6/A and debonded Si particles in T6/B is related to the coarsening of the eutectic Si particles in batch T6/B due to higher solution temperatures as already mentioned and shown in Table 2. The interaction of particles with dislocations dependent on the applied shear stress can be schematically described by Fig. 16 [15].

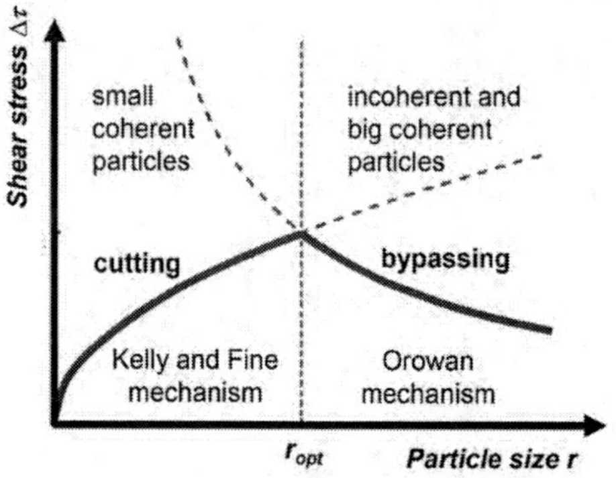

Fig. 16. Interaction of particle size r and applied shear stress $\Delta\tau$ related to Kelly and Fine (cutting of dislocations) and Orowan mechanism (bypassing of dislocations) [15].

The dominating interaction mechanisms of particle size and applied stress (here: shear stress) can be explained by mechanisms by the Kelly and Fine (cutting) and Orowan (bypassing). In regions of dominating Kelly and Fine mechanism, the particles are cut by the applied stresses. This happens for small particles with a size smaller than r_{opt} related to Fig. 16. For bigger particles, the Orowan mechanism dominates and the stress induced dislocations do not cut the particles, as bypassing the particles efforts less energy. The enveloped dislocation rings around the particles during bypassing are the reason for material hardening during monotonic and fatigue loading. As a dominating cutting of Si particles could be determined for batch T6/A and not for T6/B, it is assumed that the critical size of Si particles is within these two batches.

4 Conclusions and outlook

The effect of T6 heat treatment on the microstructure of aluminium cast alloy AlSi7Mg0.3 was investigated by two different solution annealing conditions of 530 °C/ 1 hrs for T6/A and 545 °C/1 hrs for T6/B, while using the same age hardening condition of 160 °C/5 hrs. Due to the difference in solution annealing temperature a significant effect on the morphology of the silicon particles in interdentritic Al-Si eutectic as well as the age hardening due to increased solubility of magnesium could be determined leading to an increased micro hardness of α-Al solid solution by formation of Mg_2Si precipitates. Therefore, T6/B features coarser Si particles compared to T6/A and an increased micro hardness of α-Al solid solution.

The effect of coarsening of Si particles and hardness of α-Al solid solution on the monotonic and cyclic mechanical behaviour was determined by quasistatic tensile tests, incremental step tests as well as strain- and stress-controlled constant amplitude test in the LCF and HCF regime. Hereby, an increased monotonic and cyclic yield strength could be determined for T6/B, which correlates to the increased micro hardness. Based on IST, it was possible to determine the cyclic stress-strain curve of botch batches, which underlines the planar slipping character of the age hardened aluminium cast alloy. The cyclic stress-strain curve was validated by comparison with strain- and stress-controlled CAT. While a significant increase in LCF lifetime could be determined for T6/A, which is attributed to the increased micro hardness and cyclic yield strength, both batches show a comparable lifetime in the HCF regime.

Fractographic investigations illustrate that there is a high amount of cracked Si particles within fracture surface of batch T6/A, whereby batch T6/B shows a debonding phenomenon instead of cracking of Si particles along the fatigue crack path. By a micro hardness mapping on longitudinal cross sections for fractured specimens of T6/A and T6/B in the LCF regime, a decreased micro hardness of loaded section compared to unloaded section could be determined for batch T6/A. Batch T6/B shows an increased micro hardness in the loaded section compared to unloaded section, which could be assumed due to cyclic hardening of the material as also could be determined in the IST investigations. It is assumed, that the decreased micro hardness for LCF specimens of T6/A is attributed to the cracking of Si particles.

To validate these results and assumptions TEM investigations should be performed to determine the influence of Mg_2Si precipitates on the monotonic and cyclic stress-strain behaviour and on the decreased micro hardness of batch T6/A in the LCF regime. Moreover, further fractographic investigations are planned for the HCF regime to determine, if a cracking of Si particles can be detected as well.

Acknowledgements: The authors acknowledge the German Research Foundation (DFG) for the partial funding of the investigations presented within the framework of the research project WA 1672/15-1 and KR 1999/19-1 "Identification and modeling of damage mechanisms in Al-Si-Mg cast alloys during HCF and VHCF". Additionally, the authors express their special thanks to Ohm & Haener Metallwerk GmbH for supplying A356 base alloy for casting.

References

1. J.Z. Yi, Y.X. Gao, P.D. Lee, T.C. Lindley, Metall. Mater. Trans. **37B** (2006) 301-311
2. J. Tenkamp, A. Koch, S. Knorre, U. Krupp, W. Michels, F. Walther, Int. J. Fatigue **108** (2018) 25-34
3. Q.G. Wang, D. Apelian, D.A. Lados, J. Light Metals **1** (2001) 73-84
4. H.R. Ammar, A.M. Samuel, F.H. Samuel, Mater. Sci. Eng. **473A** (2008) 65-75
5. J. Campbell, Metall. Mater. Trans. **37B** (2006) 857-863
6. Q.G. Wang, D. Apelian, D.A. Lados, J. Light Metals **1** (2001) 85-97

7. B. Zhang, D. Poirier, W. Chen, Metall. Mater. Trans. **30A** (1999) 2659-2666
8. M. Roy, Y. Nadot, C. Nadot-Martin, P. Bardin, D. Maijer, Int. J. Fatigue **33**, 6 (2011) 823-832
9. L. Ceschini, A. Morri, G. Sambogna, J. Mater. Process Technol. **204** (2008) 231-238
10. C. Bowles, J. Schijve, Int. J. Fracture **9**, 2 (1973) 171-179
11. M. Joyce, C. Styles, P. Reed, Int. J. Fatigue **25** (2003) 863-869
12. K. Shiozawa, Y. Thoda, S. Sun, Fatigue Fract. Eng. Mater. Struct. **20**, 2 (1997) 237-247
13. C. Hollmann, TU Dresden (2004), PhD. Dissertation
14. H. Richard, B. Schramm, T. Zipsner, Springer (2017), ISBN 978-3-658-17779-9
15. F. Ostermann, Springer (2007), ISBN 978-3-540-71196-4

Strain characteristics of non-ferrous metals obtained on the basic of different loads

*Andrzej Kurek, Justyna Koziarska, and Tadeusz Łagoda**

Opole University of Technology, Faculty of Mechanical Engineering, Ul. Mikołajczyka 5, 45-271 Opole, Poland

Abstract. The paper presents the comparison of the results of experimental fatigue tests for the 6082-T6 and 2017-T4 aluminium alloys. Tests performed under two types of load conditions. That is tensile-compressive and oscillatory bending. The results are obtained by applying the Basquin's stress-life equation and strain-life equations: Kandil's equation, Langer's equation and the authors' own equation. On the basis of own experimental studies and the relevant physical relations, the stress and strain amplitudes occurring in bent rods under smooth specimens, were calculated according to the elastoplastic model. The results were then used to compare both types of loads with different types of control..

1 Introduction

The subject of material fatigue is an important issue in our economy every day. The effects of tension and bending are known virtually in each branch of industry [1], therefore it is not a surprise that these two loading states are also considered with reference to material fatigue [2, 3]. The majority of current fatigue characteristics are developed in tension-compression conditions. Unfortunately, this state of loading is very rare in real mechanical structures subject to fatigue loads [4]. Variable bending occurs more often [5]. As a result of this, the relation between fatigue characteristics for tension-compression and oscillatory bending constitutes an interesting and up-to-date subject for considerations [6]. It should be emphasised here that in the case of bending, these characteristics are most often developed using the model of a perfectly elastic body.

This paper compares the characteristics of tension-compression and oscillatory bending for the model of a perfectly elastic and elasto-plastic body, using the models of strain characteristics [7-9]. The analysis has been performed on the basis of fatigue tests completed for the 6082-T6 and 2017(A)-T4 aluminium alloys in the two considered loading states. It has been proven that in the case of the elasto-plastic body model, fatigue characteristics for oscillatory bending do not lie under the fatigue characteristic for tension-compression. This shows that we are safe using conventional tension-compression characteristic to calculate fatigue life of structures subject to oscillatory bending [10].

When analysing the issue of tension-compression we can't forget to mention the Manson-Coffin-Basquin model (**MCB**) [11-13]:

$$\varepsilon_{a,t} = \varepsilon_{a,e} + \varepsilon_{a,p} = \frac{\sigma'_f}{E}\left(2N_f\right)^b + \varepsilon'_f\left(2N_f\right)^c \tag{1}$$

where: $\varepsilon_{a,t}$ – total strain amplitude expressed as the sum of the amplitudes of elastic strain $\varepsilon_{a,e}$ and plastic strain $\varepsilon_{a,p}$, $2N_f$ – the number of loading reversals (semi-cycles), E – the Young's modulus,

σ'_f, b – coefficient and exponent of the fatigue strength, respectively, ε'_f, c – coefficient and exponent of the plastic strain, respectively.

The original **MCB** characteristic has been developed for tension-compression while analysing the strain, stress and the number of cycles until destruction.

Model (1) is used only in the case, when it is possible to determine separately both elastic ε_{ae} and plastic ε_{ap} component of total strain ε_{at} [14].

Then, for cyclic loading we obtain:

$$\varepsilon_{ae} = \frac{\sigma_a}{E} \tag{2}$$

and

$$\varepsilon_{ap} = \varepsilon_{at} - \varepsilon_{ae} \tag{3}$$

This relation is defined by the Ramberg-Osgood equation [15]:

$$\varepsilon_{a,t} = \varepsilon_{a,e} + \varepsilon_{a,p} = \frac{\sigma_a}{E} + \left(\frac{\sigma_a}{K'}\right)^{\frac{1}{n'}}, \tag{4}$$

where: σ_a – stress amplitude, K' – cyclic strength coefficient, n' – cyclic strengthening exponent.

In 1910 Basquin [11] proposed the fatigue Table showing the relation between the number of cycles until destruction and stress amplitude in a double-logarithmic

* Corresponding author: t.lagoda@po.opole.pl

system log(σ_a)-log(N_f), and the approximating formula, which can be expressed as below in the exponential form for tension-compression:

$$\sigma_a = \sigma_f'(2N_f)^b \quad (5)$$

or

$$\log N_f = A + m \log \sigma_a \quad (6)$$

where: N_f – fatigue life in cycles, σ_a – stress amplitude for tension-compression or bending, A, m – regression model constants.

Another issue has been shown in the study [16], where it is pointed out that the sense of plastic strain amplitude in the expression (1) depends on fatigue life, and thus c is not a constant value.
Moreover, various authors proposed other empirical model making total strain amplitude dependent on the number of cycles. Among these models there is the Langer [17] proposal, which is used in numerous studies and promoted e.g. by Manson [18], [19] and Chrop [20].

$$logN_f = A - B\, log(\varepsilon_{ac} - C), \quad (7)$$

where: A, B, C – constants to select special form of the characteristic for a given material.
Another characteristic is proposed by Kandil [21] and Gorash [22], in the following form:

$$log\varepsilon_{ac} = A - B\, log(N_f) + C\, log^2(N_f), \quad (8)$$

where: A, B, C – constants to select special form of the characteristic for a given material.
Since in case of bending it is not possible to separate elastic and plastic component, then characteristic (1) cannot be used but it is possible to use characteristics (7) or (8), or other empirical form of a strain characteristic. For example, this may be a combination of characteristics (7) and (8) in the following form:

$$log(\varepsilon_{ac} - D) = A - B\, log(N_f) + C\, log^2(N_f), \quad (9)$$

where: A, B, C, D - constants to select special form of the characteristic for a given material.
An extensive review of fatigue characteristics can be found e.g. in the study [23]. The new form, which is proposed there, requires 4 material constants to be determined, same as for the popular characteristic MCB (1).

In the literature there is no simple model allowing determination of elasto-plastic strains and stresses for smooth rods at bending. However, it has been confirmed empirically many times that normal strain distribution for bending is linear in cross-section:

$$\varepsilon_a(x) = \varepsilon_{a\max} \frac{x}{R} \quad (10)$$

where: x – the distance from bending plane, R – maximum height (radius in case of round component /rod/).

The basic relationship that has to be satisfied is that normal stresses, which appear both in an elastic and elasto-plastic model, must compensate the preset bending moment, that is:

$$M_b = \int_S \sigma(x,y)x dS \quad (11)$$

while, the relationship between stress and strain amplitude may be expressed using the Ramberg-Osgood equation (4). Fig. 1 shows an example distribution of stresses and strains at bending.

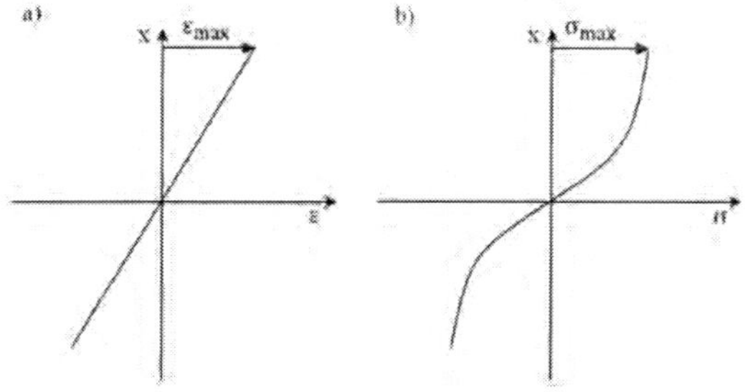

Fig.1. Distribution of strain (a) and stress (b) in a test piece subject to bending

Whereas, Fig. 2 shows values in the test piece cross-section, required to calculate distributions of stresses and bending moment according to the elasto-plastic body model.

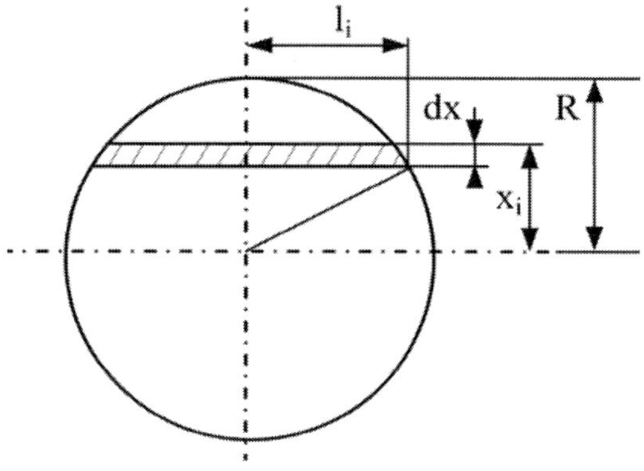

Fig. 2. Cross-section of test piece subject to bending

2 Experimental studies

The studies have been carried out for test pieces made of the 6082-T6 and 2017-T4 aluminium alloys.

Chemical constitution of the tested material is shown in Table 1. The basic mechanical properties of the discussed material are specified in Table 2.

Table 1. Chemical constitution of the aluminium alloys (in %) – Al – the rest

6082

Cu	Mg	Mn	Si	Fe	Zr+Ti	Zn	Cr
<0.1	0.6÷1.2	0.4÷1	0.7÷1.3	<0.5	<0.1	<0.2	<0.25

2017A

Cu	Mg	Mn	Si	Fe	Zr+Ti	Zn	Cr
3.5 ÷4.5	0.4 ÷1	0.4 ÷1	0.6	<0.7	<0.25	<0.25	<0.25

Table 2. The basic mechanical parameters of aluminium alloys

Aluminium	E, GPa	$R_{p0.2}$, MPa	R_m, MPa	A_5 %	ν
6082	77	365	385	27.2	0.32
2017A	72	395	545	21	0.32

3 Test for tension-compression

The low-cycle tests for tension-compression were performed in cooperation with the Institute Laboratory for Materials and Structures Testing at UTP - University of Science and Technology in Bydgoszcz [24] for 6082-T6 and at Opole University of Technology [25] for 2017(A)-T4.

The purpose of tests was to find basic fatigue characteristics for test pieces made of aluminium alloy at ambient temperature.

16 test pieces were used in the tests for 6082-T6 and 10 test pieces for 2017(A)-T4. The tests were performed according to the standard PN-84/H-04334.

The results of fatigue tests at single-axis tension-compression are compared in Table 3. Material constants appearing in the Manson-Coffin-Basquin (1) and Ramberg-Osgood (4) characteristics have been determined on the basis of these results. These values are compared in Table 4.

Table 3. Cyclic parameters of aluminium alloys

	K', MPa	n'	σ'_f, MPa	ε'_f	b	c
6082-T6	616	0.099	533	0.185	-0.066	-0.634
2017A-T4	617	0.066	643	1.879	-0.065	-0.988

The third tests type for 2017(A)-T4 [26] was tension-compression at a constant amplitude of energy parameter (W_a =const) for 19 test pieces [27], where:

$$W(t)=0.25\{\,|\sigma(t)|\,\varepsilon(t)+\sigma(t)\,|\varepsilon(t)|\,\}, \qquad (12)$$

which in load amplitude-based form eventually gives

$$W_a=0.5\sigma_a\,\varepsilon_a, \qquad (13)$$

where W_a is amplitude of strain energy parameter.

4 Test for oscillatory bending

Fatigue tests for oscillatory bending were carried out using fatigue-testing machines belonging to the laboratory of the Department of Mechanics and Machine Design at Opole University of Technology.

The "diabolo" type cylindrical test pieces without geometrical notch were used in fatigue tests. The geometry of test pieces used results from simplified localisation of spot characterised by highest stresses, and shown in Fig. 3. The starting material was round rod made of the 6082 and 2017(A) aluminium alloys, ϕ16mm in diameter. The tests in cyclic conditions under controlled moment involved using 25 test pieces, and under controlled strain - 25 test pieces as well for 6082-T6 and 21 test pieces for 2017(A)-T4).

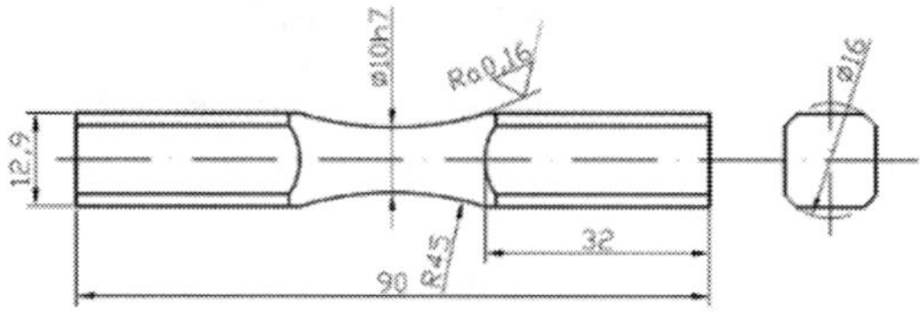

Fig. 3. The shape and dimensions of a test piece for fatigue tests

The cyclic (constant-amplitude) tests were carried out on the MZGS-100 test bench designed by Achtelik. Fig. 4 shows the view of a fatigue-testing machine. The MZGS-100 test bench consists of a propulsion system, head, loading system, and control and measurement setup. The trajectory of bending moment M loading the test piece was the parameter monitored during the tests.

Fig. 4. The setup for fatigue tests under controlled moment

Fatigue tests under controlled strain were performed using a new setup shown in Fig. 5. In this case, lever excursion amplitude has been controlled, then ensuring control of test piece strain.

Fig.5. The setup for fatigue tests under controlled strain

In case of fatigue tests carried out both under controlled moment and strain amplitude, the stiffness or moment amplitude drop by 15% respectively was taken as fatigue life, this corresponds with a moment when crack became visible with bare eyes (ca. 1 mm).

Fig. 6 presents typical test pieces cross-section after oscillatory bending. In this photo we can clearly see the initiation point and neutral plane, in relation to which bending has been occurring. On the figure you can also see the long furrows which start on the surface of the material in the origin of the crack and they extend deeper into the material perpendicular to the bending surface.

a)

b)

Fig.6. An example test piece photo taken after oscillatory bending test a) 6082-T6 b) 2017(A)-T4

5 Comparison of test results

Completed tests and respective calculations provided grounds for determining fatigue characteristics, cumulatively shown in the illustrations. Constants appearing in the formulae are compared in successive Tables.

Successive analyses concern strain characteristics. The Manson-Coffin-Basquin strain-life curve (1) has been the first one analysed. The Fig. 7 and Table 4 - 5 allows observing that in practice all three characteristics overlap and it is possible to find one cumulative MCB curve. However, it should be noted that in the case of bending the MCB characteristic can be determined only owing to previously defined Ramberg-Osgood model (4), this allows separating elastic and plastic component of strain. Considering this, it seems necessary to seek other characteristics, in which there is no need to divide strain into an elastic and plastic component. Finding such a model will allow determining fatigue characteristics under controlled strain at bending on relatively simple and inexpensive machines.

The Langer characteristic (7) has been employed first. It is the simplest one, and originally it assumes linear character in a double-logarithmic system. Fig. 8 compares these characteristics for individual tests. Parameters of these characteristics are confronted in Table 6 and 7. Completed data analysis indicates that the characteristic for tension-compression and the cumulative characteristic approximate the empirical data incorrectly. As a result of this, it cannot be used for correct description of the empirical data. The Kandil characteristic (8) has been taken as the second one. It is more complex than the Langer characteristic, and as it may be concluded from Fig. 9 and data collected in Table 8 and 9, it describes experimental test results by far better. This applies both to the description of these results for individual tests, and the cumulative characteristic determined for all tests. The last characteristic used is the one proposed by the Authors of this study in form of (9). Another, fourth parameter to be determined appears in this characteristic. Analysis of Fig. 10 and data contained in Table 10 and 11 proves that, same as in case of the Langer characteristic, this characteristic describes all test results very well, both individually and for all tests jointly. Analysis of correlation coefficients for all three analysed strain characteristics (Tables 4 - 11) indicates that the best match of combined characteristics is ensured in case of the model proposed by the Authors of this study.

Table 4. Comparison of parameters of the analysed characteristics according to the Manson-Coffin-Basquin model for 6082-T6

	b	c	σ'_f, MPa	ε'_f
Bending (M_a=const)	-0.101	-0.657	733	0.188
Bending (ε_a =const)	-0.113	-0.636	823	0.275
Tension-compression (ε_a =const)	-0.066	-0.634	533	0.185
All (combined)	-0.091	-0.582	649	0.148

Table 5. Comparison of parameters of the analysed characteristics according to the Manson-Coffin-Basquin model for 2017(A) – T4

	b	c	σ'_f, MPa	ε'_f
Tension-compression (ε_a=const)	-0.063	-0.962	627	1.305
Tension-compression (W_a=const)	-	-	-	-
Bending (M_a=const)	-	-	-	-
All (combined)	-	-	-	-

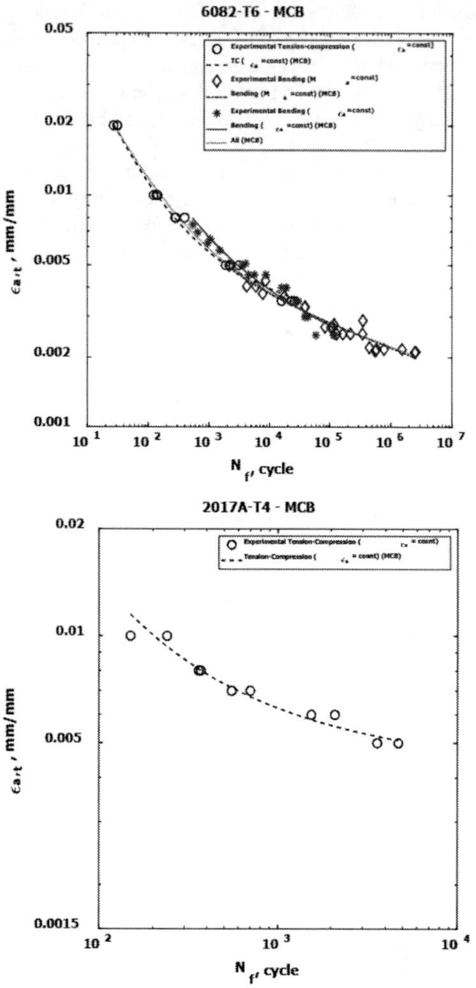

Fig. 7. Fatigue characteristics according to the Manson-Coffin-Basquin model a) 6082-T6, b) 2017(A)-T4

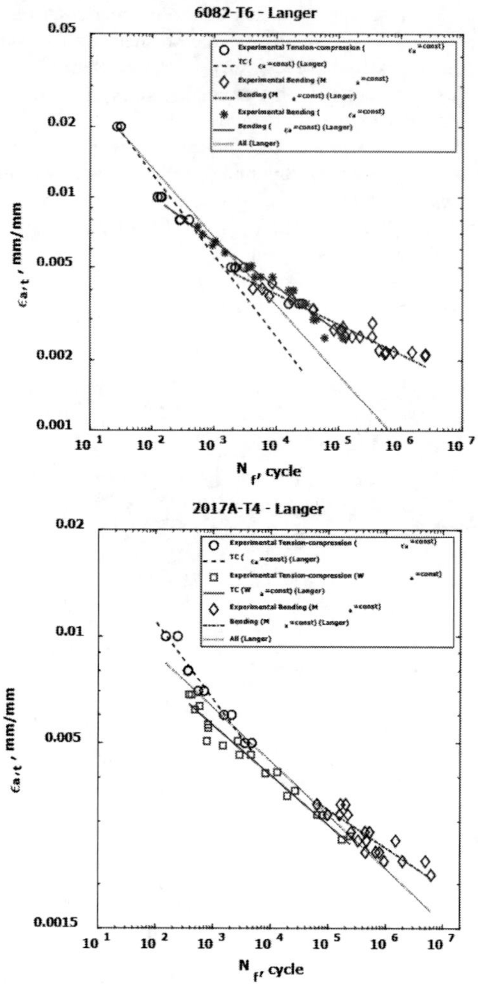

Fig.8. Fatigue characteristics according to the Langer model a) 6082-T6, b) 2017(A)-T4

Table 6. Comparison of parameters of the analysed characteristics according to the Langer model for 6082-T6

	A	B	C	R^2
Bending (M_a=const)	-1.898	0.129	0.002	0.94
Bending (ε_a =const)	-1.611	0.193	1.068	0.98
Tension-compression (ε_a =const)	-1.200	0.350	0.002	0.96
All (combined)	-1.284	0.295	0.200	0.93

Table 7. Summary of the parameters of analysed fatigue life values according to Langer's model for 2017(A)-T4

	A	B	C	R^2
Tension-compression (ε_a=const)	-1.523	0.218	1.247	0.95
Tension-compression (W_a=const)	-1.829	0.1407	0	0.95
Bending (M_a=const)	-1.949	0.1079	0	0.78
All (combined)	-1.746	0.1516	0.0016	0.91

Table 8. Comparison of parameters of the analysed characteristics according to the Kandil model for 6082-T6

	A	B	C	R^2
Bending (M_a=const)	-1.571	0.274	0.0154	0.95
Bending (ε_a =const)	-1.72	0.132	-0.0082	0.98
Tension-compression (ε_a =const)	-0.835	0.703	0.7033	0.99
All (combined)	-1.035	0.524	0.0436	0.99

Table 9. Summary of the parameters of analysed fatigue life curves according to Kandil's model for 2017(A)-T4

 https://doi.org/10.1051/matecconf/201816515005

	A	B	C	R^2
Tension-compression (ε_a=const)	-1.139	0.4945	0.04851	0.96
Tension-compression (W_a=const)	-1.1611	0.2636	0.01642	0.96
Bending (M_a=const)	-1.283	0.3432	0.02059	0.79
All (combined)	-1.305	0.3944	0.03007	0.94

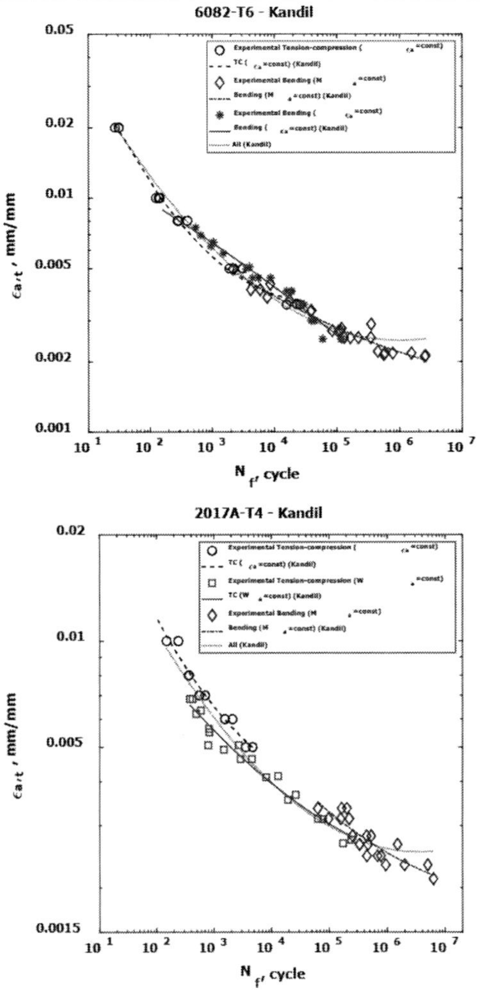

Fig.9. Fatigue characteristics according to the Kandil model a) 6082-T6, b) 2017(A)-T4

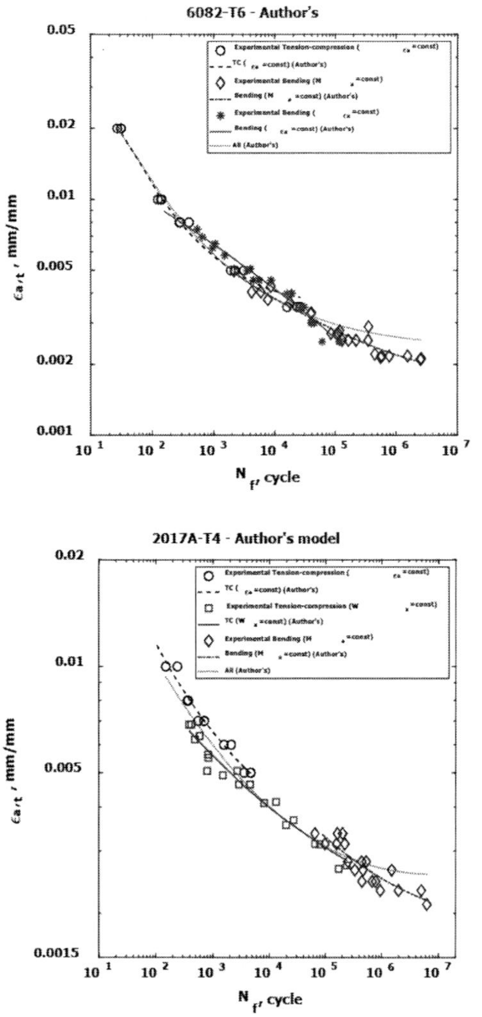

Fig.10. Fatigue characteristics according to the model proposed by the Authors a) 6082-T6, b) 2017(A)-T4

Table 10. Comparison of parameters of the analysed characteristics according to the model proposed by the Authors for 6082-T6

	a	b	c	d	R^2
Bending (M_a=const)	-1.572	0.270	0.016	-0.00018	0.95
Bending (ε_a =const)	-1.709	0.136	-0.007	-0.00023	0.98
Tension-compression (ε_a =const)	-0.788	0.761	0.079	0.00141	0.99
All (combined)	-0.965	0.595	0.038	0.00192	0.99

Table 11. Summary of the parameters of analysed fatigue life values according to the authors' for 2017(A)-T4

	A	B	C	D	R^2
Tension-compression (ε_a=const)	-1.15	0.485	0.048	-0.00022	0.96
Tension-compression (W_a=const)	-1.62	0.267	0.016	0.00036	0.96
Bending (M_a=const)	-1.28	0.343	0.021	-0.00010	0.79
All (combined)	-1.29	0.425	0.031	0.00083	0.94

6 Conclusions

1. Experimental tests carried out for test pieces made of the 6082-T6 and 2017(A)-T4 aluminium alloys under strain control in conditions of single-axis tension-compression, and both under strain control and moment control for oscillatory bending are much the same and independent on loading method.
2. The tests under controlled strain may be replaced by oscillatory bending carried out on a simple, modern test bench instead of the basic tension-compression test performed using large fatigue-testing machines.
3. The best known Mason-Coffin-Basquin strain characteristic describes the results of all experimental tests very well; however, it can be used only when there is a possibility to divide strains into the elastic and plastic components. Analysing the results of specific tests, it becomes clear that fatigue life is almost identical across all tests; therefore, a common fatigue life can be determined for each of the analysed models.
4. The Langer strain characteristic describes fatigue test results incorrectly, and it is not recommended for use to characterise fatigue test results.
5. The best match of fatigue characteristic to all experimental data is ensured by the four-parameter characteristic proposed by the Authors, which is a fragment of a parabola in a double-logarithmic system, and it is slightly better than the three-parameter Kandil characteristic.

References

1. A. Kurek, J. Koziarska, T. Łagoda, Energetyka, **11**, 673-675 (2016)
2. M. Kurek, T. Łagoda, D. Katzy, Material Testing, **56**, 92-95 (2014)
3. A. Kulesa, A. Kurek, T. Łagoda, H. Achtelik, K. Kluger, Acta Mechanica et Automatica, **10**, 62-65 (2016)
4. K. Walat, T. Łagoda, M. Kurek, Material Testing, **57**, 160-164 (2015)
5. S. S. Manson, U. Muralidharan, Fatigue & Fracture Engineering Materials & Structures, **9**, 357-372 (1987)
6. A. Kulesa, A. Kurek, T. Łagoda, H. Achtelik, K. Kluger, Solid State Phenomena, **250**, 85-93 (2016)
7. V. Troschenko, Multiaxial and Fatigue Design, ESIS 21, (Edited by A. Pinueau, G. Cailletaud and T. C .Lindley), *Mechanical Engineering Publications, London*, 335-348 (1996)
8. M. M. Megahed, Fatigue & Fracture of Engineering Materials & Structures, **13**, 361-374 (1990)
9. D. Krzyżak, M. Kurek, T. Łagoda, D. Sówka Materialwissenschaft und Werkstofftechnik, **45**, 1018-1029 (2014)
10. A. Niesłony, A. Kurek, Applied Mechanics and Materials, Trans Tech Publications, **104**, 197-205 (2012)
11. O.H. Basquin, Am. Soc. Test. Mater. Proc., **10**, 625-630 (1910)
12. L. F. Coffin, Trans ASME, **76**, 931-950 (1954)
13. S. S. Manson, Experimental Mechanics, **5**, 193-226 (1965)
14. V. M. Radhakrishnan, NASA Technical Memorandum 105840, *NASA-TM-105840, E-7283, NAS 1.15:105840*, 11 (1992)
15. W. Ramberg, W. R. Osgood, Technical Note No. 902, *National Advisory Committee for Aeronautics, Washington DC* (1943)
16. V. M. Radhakrishnan, *NASA Technical Memorandum 105840*, p.11, (1992)
17. B. F. Langer, ASME J. Basic Eng, **84**, 389-402 (1962)
18. S. S. Manson, Fatigue of Engineering Materials and Structures, **1**, 37-57 (1979)
19. S. S. Manson, Fatigue: Experimental Mechanics, 193-226 (1965)
20. O. K. Chopra, *U.S. Nuclear Regulatory Commission* (1999)
21. F. A. Kandil Standards Measurement & Testing Project No. SMT4-CT97-2165, **1**, 1-26 (2000)
22. Y. Gorash, H. Chen, Engineering Failure Analysis, **34**, 308-323 (2013)
23. A. Niesłony, A. Kurek, Ch. El Dsoki, H. Kaufmann, International Journal of Fatigue, **39**, 88-94 (2012)
24. St. Mroziński St. Raport z badań, UTP w Bydgoszczy, Bydgoszcz (2012)
25. W. Będkowski, T. Łagoda, J. Słowik, Materials Science, **43**, 492-498 (2007)
26. W. Będkowski, J. Słowik, Raport Nr 1/WM/2004, Wydział Mechaniczny, Politechnika Opolska (2004)
27. T. Łagoda, Int. J. Fatigue, **23**, 481-489 (2001)

This work has been carried out under the grant of National Science Centre (Poland) no. 2015/19/B/ST8/01115

Effect of size of alpha phases on cyclic deformation and fatigue crack initiation during fatigue of an alpha-beta titanium alloy

Qiaoyan Sun[1*], *Changsheng Tan*[1], Lin Xiao[1], Jun Sun[1]
1. State Key Laboratory for Mechanical Behavior of Materials, Xi'an Jiaotong University, Xi'an, Shaanxi, 710049, P.R. China
**Corresponding author. Tel.: +86 29 82668614; fax: +86 29 82663453.*
E-mail address: qysun@mail.xjtu.edu.cn (Qiaoyan Sun)

Abstract. Alpha phase exhibits equiaxed or lamellar morphologies with size from submicron to microns in an alpha-beta titanium alloy. Cyclic deformation, slip characteristics and crack nucleation during fatigue in different microstructures of TC21 alloy (Ti-6Al-2Sn-2Zr-3Mo-1Cr-2Nb-0.1Si) were systematically investigated and analyzed. During low-cycle fatigue, equiaxed microstructure (EM) in TC21 alloy exhibits higher strength, ductility and longer low-cycle fatigue life than those of the lamellar microstructure (LM). There are more voids in the single lamellar alpha than the equiaxed alpha grains. As a result, voids more easily link up to form crack in the lamellar alpha phase than the equiaxed alpha phase. However, during high-cycle fatigue, the fine lamellar microstructure (FLM) shows higher fatigue limit than bimodal microstructure (BM). The localized plastic deformation can be induced during high-cycle fatigue. The slip bands or twins are observed in the equiaxed and lamellar alpha phases(>1micron), which tends to form strain concentration and initiate fatigue crack. The localized slip within nanoscale alpha plates is seldom observed and extrusion/intrusion dispersedly distributed on the sample surface in FLM. This indicates that FLM show super resistance to fatigue crack which bring about higher fatigue limit than BM.

1 Introduction

Due to their high specific ratio of strength to density, high temperature properties and fatigue properties, titanium alloys have been attracting increasing emphasis on the application in aerospace and aircraft industries, such as landing gear and springs [1-3]. The majority of engineering titanium alloys is consisted of alpha or beta phase or a combination of the two phases with different volume fractions and morphologies. TC21 titanium alloy is a typical two phase alloy which possesses high strength and toughness, and has been successfully applied in the aerospace as a structural material [4-6]. During these applications, the structural member is mainly subjected to cyclic fatigue loading. Therefore, it cannot avoid referring to the fatigue failures in engineering materials.

The significance of fatigue in engineering applications is evident and it is reported that at least of half of all mechanical failures are caused by fatigue [7-8]. It has been reported that a fatigue crack is formed due to the localized plastic deformation during fatigue loading [9-14]. These strain concentration and gradient are primarily caused through the dislocation motion within the phase or grains of materials. During the fatigue, because of the cyclic forward and reverse loading, dislocation slip moves in some specific paths within each cycle, however, some of the dislocation is not returned completely in the reverse loading, which results in accumulated strain in some weak sites [13-14, 15-19]. In many materials, the strain is localized in the form of slip bands and grain boundary. Especially, plastic deformation is induced in each cycle in low cycle fatigue [3, 6, 15]. Thus, the microstructure has a significant influence on fatigue crack initiation and propagation behaviour in titanium alloys [2-3, 6, 9-14]. Some researchers found that decreases the size of β grain and volume fraction of primary α phase would improve the low cycle fatigue life [3]. It has been realized that the microstructural parameters, including morphology, size, distribution, volume fraction and orientation et al, have important effect on the cyclic deformation and damage mechanism. For these reason, it is essential to investigate the effect of microstructure parameters on the cyclic deformation and fatigue crack initiation and coalescence behavior.

Although the exact numbers are not available, it has reported that over half of fatigue failures in the manufactured products are caused by high cycle fatigue (HCF) failure, particularly in aviation and aerospace industries [8, 11, 20]. Due to the unpredictable and overwhelming failure, the investigation on the fatigue crack initiation has attracted a great amount of attention [9-13, 21-23]. Wu et al [13] found that high cycle fatigue strength of Ti-6Al-4V is dependent on the microstructures and increases in the order of equiaxed microstructure, lamellar microstructure and bimodal microstructure. However, opposite results was obtained

[*] Corresponding author: qysun@mail.xjtu.edu.cn

in the Ti-6Al-2Zr-1Mo-1V and TC21 titanium alloy [24]. These results indicate that microstructure has a domain influence of high cycle fatigue properties.

In the last decades, the thermos-mechanical processing [4-5, 25], phase transformation [26-27] and relationships between microstructures and properties [25, 28-29] of TC21 titanium alloy can be found in many researches. However, the cyclic deformation and fatigue crack initiation behaviour of TC21 titanium alloy almost have not been reported [6, 9]. Due to the significant of microstructure in initiation of crack, the effect of size of α phases on cyclic deformation and fatigue crack initiation during fatigue of TC21 titanium alloy was investigated in present work.

2 Materials

The as-received material of TC21 titanium alloy was produced by the Northwest Institute for Nonferrous Metal Research in China. Then, different heat treatments were used to obtain the equiaxed microstructure (EM), lamellar (widmanstatten) microstructure (LM), bimodal microstructure (BM) and fine lamellar microstructure (FLM), as described in details in Ref. [6, 9].

The tensile tests were carried out on tested on INSTRON 1195 Testing Machine. Symmetrical push–pull low-cycle fatigue tests controlled by total strain were carried out on INSTRON 1342 with triangular waveforms at room temperature. The selected total strain amplitudes were 0.6%, 0.8%, 1.0%, 1.2% and 1.5%. The symmetrical push-pull stress controlled high-cycle fatigue tests were tested on ZwickAmsler HFP 5100 with triangular waveforms at room temperature under a load ratio of R= -1.

Fractographic surface and deformation behaviour after fatigue were observed using HITACHI S-2700 and a field emission scanning electron microscope (FESEM) of HITACHI SU6600. The observation of the crack and voids was preformed near the site of crack initiation, as described in Ref. [6, 9]. Thin foils for TEM observation were prepared in the fracture surface to explore the dislocation structures during fatigue. The dislocation structures were observed with transmission electron microscope (JEM-200CX) operating at 120 KV. The corresponding microstructure parameters for the four microstructures, such as size, volume fraction, aspect, were quantitatively analysed by the Image Pro-Plus 6.0 software.

3 Results and discussion

3.1 Microstructures before fatigue

Various microstructures in two phase titanium alloys can be obtained through different heat treatments. Fig. 1 shows four typical microstructures in TC21 titanium alloy. EM consist of 45% equiaxed α phase with 3 μm in diameter, as shown in Fig. 1a. The average width of α phase in LM is about 520 nm (Fig. 1b). BM contains primary equiaxed α phase and primary α lath with size of 10 μm and 2 μm, respectively, as seen in Fig. 1c. Fig. 1d displays the morphology of FLM whose width of α phase is about 180 nm.

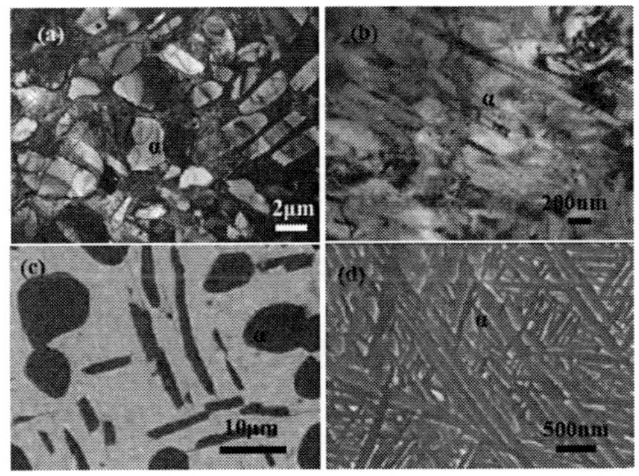

Fig. 1. The microstructures before fatigue: (a) equiaxed microstructure, (b) lamellar (widmanstatten) microstructure, (c) bimodal microstructure, (d) fine lamellar microstructure.

The quantitative analysis of microstructural parameters is listed in table 1.

Table 1. Quantitatively analysis of microstructural parameters

Structures	Morphology of α phase	Size
EM	Equiaxed	3 μm
	Lamellar	700 nm
LM	Lamellar	520 nm
BM	Equiaxed	10μm
	Lamellar	2.0μm
FLM	Lamellar	180 nm

3.2 Mechanical properties

Fig. 2 shows the tensile stress strain relationship for the four microstructures. It can be found that the strength increases in the order of LM, EM, BM and FLM. The ductility increases in the order of FLM, LM, EM and EM.

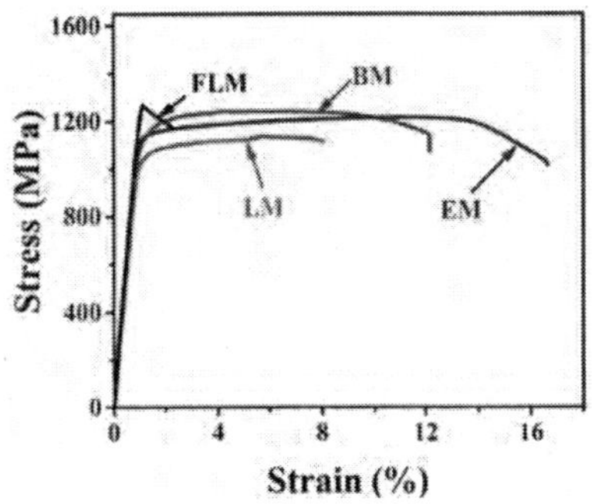

Fig. 2. The engineering stress-strain curves for different microstructures

The corresponding tensile results are listed in table 2. An excellent comprehensive mechanical property was obtained in the BM. Based on the fatigue tests, equiaxed microstructure (EM) in TC21 alloy exhibits higher strength, ductility and longer low-cycle fatigue life than those of the lamellar microstructure (LM) during low-cycle fatigue. However, during high-cycle fatigue, the fine lamellar microstructure (FLM) shows higher fatigue limit than bimodal microstructure (BM).

Table 2. Tensile results for the four microstructures

Structures	σ_s (MPa)	σ_b (MPa)	ε(%)	A (%)
EM	1120	1210	16.6	35
LM	1030	1130	7.5	11
BM	1110	1240	12	20
FLM	1210	1280	2	3

3.3 The crack initiation and coalescence behavior

To further investigate the effect of microstructure parameter on the crack initiation and propagation behavior, the crack initiation and coalescence behaviour at different strain amplitude or stress amplitude are explored in detail in the following section

The detail of crack nucleation and propagation in the EM and LM at strain amplitude of 0.6% are shown in Fig. 3. Fatigue cracks nucleate at the equiaxed α phase or the α/β interface in EM, as shown in Fig. 3a. In some cases, it initiates at the α lath interface in EM (Fig. 3b). Whereas, some microcracks were observed at the α/β interface within the α colony in LM, as shown in Fig. 3c. Additionally, the crack propagates along the large α colony and deviates from original path (Fig. 3d). Due to orientation difference and short slip length, poor coordination during cyclic deformation is obtained in the LM [30], resulting in stress concentration in the lamellar α.

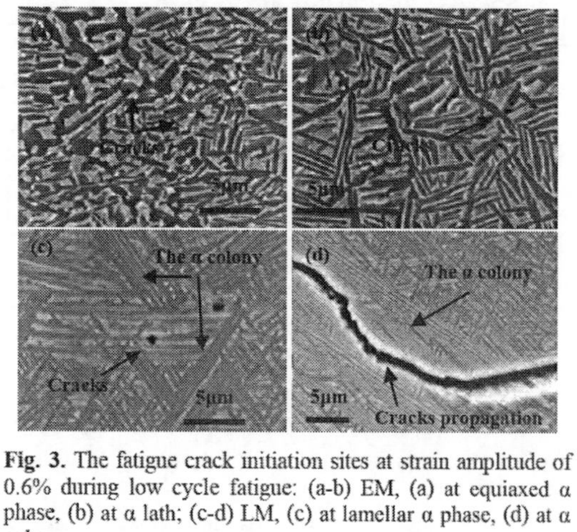

Fig. 3. The fatigue crack initiation sites at strain amplitude of 0.6% during low cycle fatigue: (a-b) EM, (a) at equiaxed α phase, (b) at α lath; (c-d) LM, (c) at lamellar α phase, (d) at α colony

Fig. 4 and Fig. 5 show the crack initiation behaviour for the BM and FLM, respectively. Fatigue cracks nucleate at the primary α lath and primary equiaxed α phase and coalesce along the interface , as shown in Fig. 4a and b. Once the crack initiated, it will grow along the interface but lie within an individual or occupy several equiaxed α phases, as shown in Fig. 4c. Whereas, it easily concatenate with each other to form a long crack when encounters the in primary α lath (Fig. 4d).

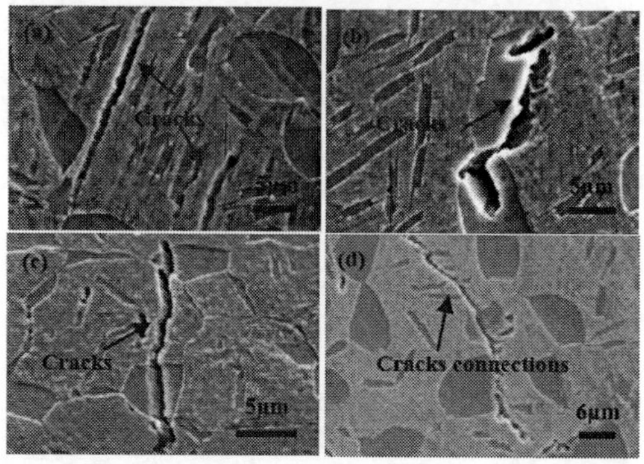

Fig. 4. The fatigue crack initiation and coalescence at stress amplitude of 600MPa in BM during high cycle fatigue: (a) at primary α lath, (b) at equiaxed α phase; (c) connects along α interface, (d) coalescence of crack through transferring α lath

While in the FLM, some parallel intrusions are induced on the sample surface and the fatigue cracks nucleate at the intrusions (Fig. 5a). Fig. 5b shows some voids initiate at the concave edge of the intrusions. The voids then coalesce with each other along the intrusions to form large cracks, as shown in Fig. 5c. The width of the β grain boundary is in the same magnitude of the fine α laths in FLM. Majority of the cracks nucleate at the intrusions and extrusions which are observed in Fig. 5a, b and c. These parallel cracks dispersedly distributed within the large β grain may improve the resistance for crack initiation and coalescence, as shown in Fig. 5d.

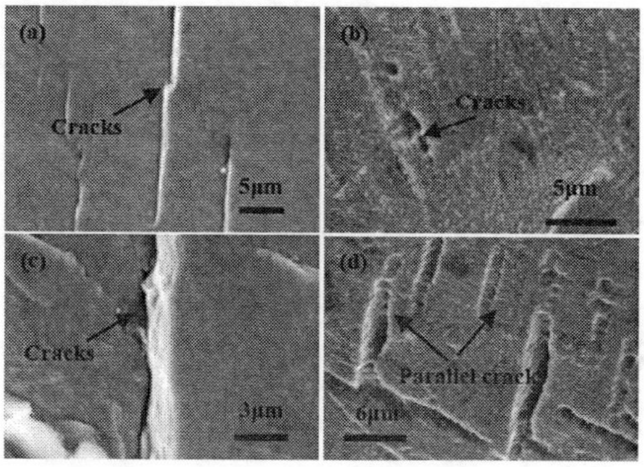

Fig. 5. The fatigue crack initiation and coalescence at stress amplitude of 660MPa in FLM during high cycle fatigue: (a) intrusions at surface, (b) crack initiation at intrusions; (c) connects along intrusions, (d) parallel cracks dispersedly distribute at the surface

3.4 The fatigue dislocation structures

To reveal the internal cause of fatigue crack initiation behaviour, typical dislocation structure during fatigue

 https://doi.org/10.1051/matecconf/201816515006

should be studied. Fig. 6 and Fig. 7 show the dislocation structure at the strain amplitude of 0.6% and 1.5% for the EM and FLM during the low cycle fatigue. The dislocation structures during the high cycle fatigue with different stress amplitude are shown in Fig. 8, Fig. 9 and Fig. 10.

Fig. 6a shows a large dislocation pileups and tangles at the α/β interface at strain amplitude of 0.6%, which indicates that interaction between dislocation and interface takes place [6]. Additionally, dislocation pileups are observed at the α lath interface, which can induce stress concentration at the interface [7, 15]. With increasing the strain amplitude to 1.5%, much more dislocation pileups and tangles at equiaxed phase boundary generate to accommodate plastic strain, as shown in Fig. 6c. In the α lath, a high density of dislocation tangles and debris result from the huge plastic deformation during one cycle, and some slip bands are produced within the α lath (Fig. 6d).

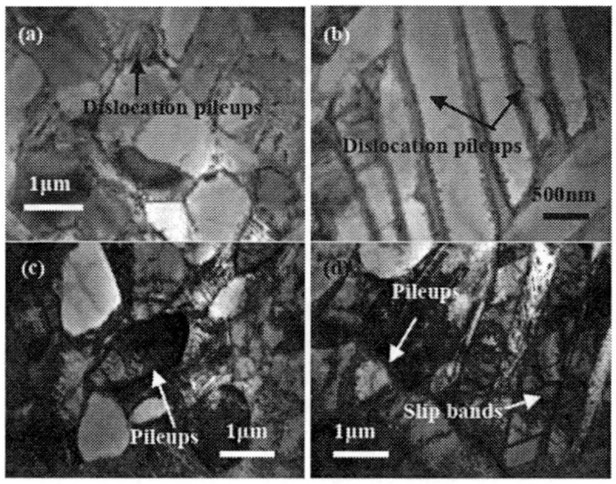

Fig. 6. The fatigue dislocation structure at low strain amplitude of 0.6% (a-b) and high strain amplitude of 1.5% (c-d) in EM during low cycle fatigue: (a) at equiaxed α phase, (b) at α lath; LM, (c) at lamellar α phase, (d) at α colony

At low strain amplitude of 0.6%, dislocation tends to pile up at the interface (Fig. 7a). Some parallel slip bands are observed in the α lath at low strain amplitude, as shown in Fig. 7b. More dislocation pileups and tangles appear in the α phase at the high strain amplitude of 1.5% (Fig. 7c). The deformation is mainly induced in the α lath. Fig. 7d shows that much more short dislocations are observed, together with a high density of small dislocation tangles and fragments. Due to the small size of α lath, the strain localization occurs in the formation of slip bands in the α lath, which are the precursors to crack initiation [7, 9, 15].

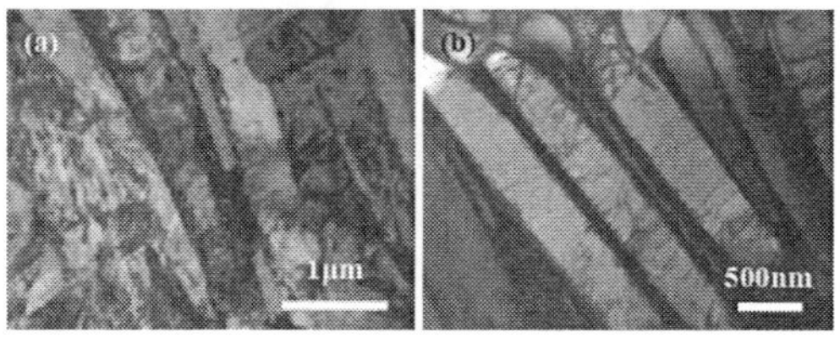

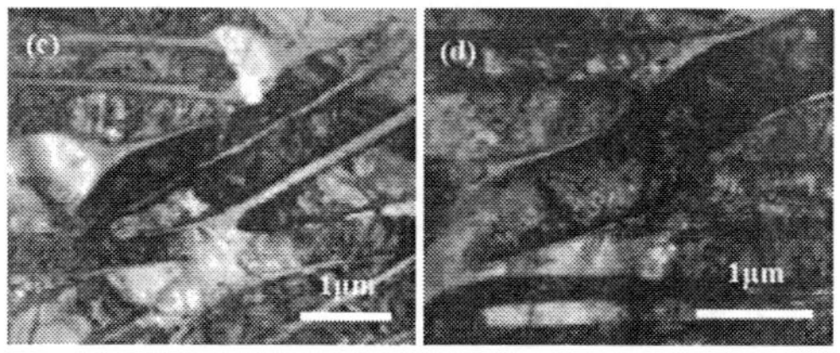

Fig. 7. The fatigue dislocation structure at low strain amplitude of 0.6% (a-b) and high strain amplitude of 1.5% (c-d) in LM during low cycle fatigue: (a) dislocation pileups at interface, (b) dislocation slip bands in α lath; (c) large dislocation pileups and tangles, (d) multi-slip bands

There are some dislocation debris and tangles at the primary α lath in BM at the stress amplitude to 530 MPa, as shown in Fig. 8a, which indicates that a drastic heterogeneous plastic deformation is induced within the large α phase. Besides the dislocation pileups, some prismatic slip bands in the primary α lath and steps at the interface are observed, as shown in Fig. 8b. These steps are induced by the cyclic interaction between dislocation and interface [9]. Fatigue cracks nucleate at the steps preferentially with local strain incompatibility and stress concentration [9-10, 22, 31-32]. Fig. 8c and Fig. 8d show that multi slip bands are prevalent in the equiaxed primary α phase. These results indicates that the cyclic deformation is easy to localize within the large primary α phase which results in more heterogeneous deformation.

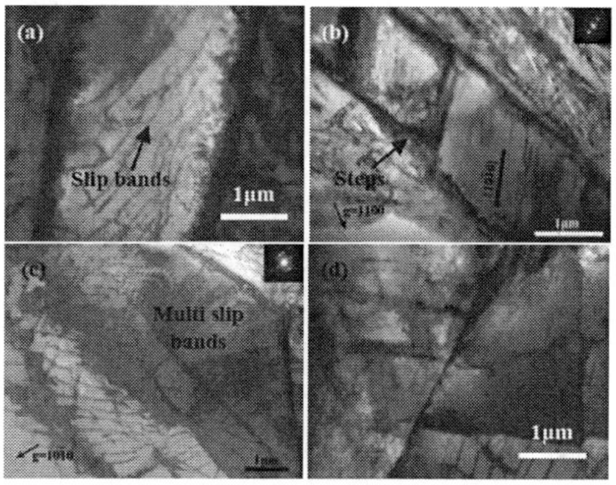

Fig. 8. The fatigue dislocation structure at low stress amplitude of 530 MPa in BM during high cycle fatigue: (a) dislocation pileups at interface, (b) prismatic slip bands in α lath; (c) multi-slip bands and dislocation network, (d) multi-slip bands

At high stress amplitude of 600 MPa, some deformation twins are produced in the primary α laths, as shown in Fig. 9a, which has been reported in reference [6-7, 9]. Fig. 9b is the dark field image for the deformation twins in Fig. 9a. In addition to twins, lots of parallel dislocation slip bands are observed in the primary α lath, as shown in Fig. 9c. Much more dislocation tangles generate in the β_{trans} matrix (Fig. 9d).

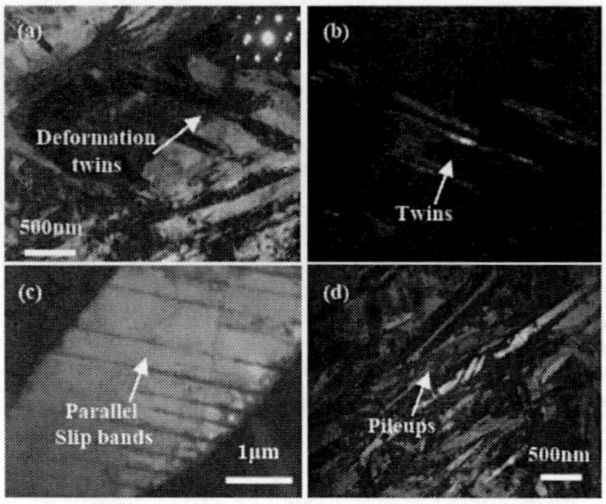

Fig. 9. The fatigue dislocation structure at high stress amplitude of 600 MPa in BM during high cycle fatigue: (a) deformation twins were induced in primary α lath, (b) the corresponding dark field image for the twin in α lath; (c) some parallel slip bands were produced in primary α lath, (d) lots of dislocation pileups and tangles in β_{trans} matrix

To compare with BM, the dislocation structure in FLM at different stress amplitudes is shown in Fig. 10. At stress amplitude of 690MPa, some dislocation slip bands and dislocation arrays are observed in the α phase, as shown in Fig. 10a and b. Cyclic deformation both exists in the α and β phase, which indicates that a relatively uniform deformation is induced in FLM (Fig. 10b). With increasing the stress amplitude to 705MPa, a large of dislocation pileups and tangles are observed in the α phase, which indicates a serious heterogeneous cyclic deformation generates. However, the distribution of dislocation is relatively homogeneous both in the two phases, as Fig. 10c shows. Fig. 10d displays that lots of deformation twins with size of about 50nm in widths are produced in α phase. It has been reported that deformation twins will be induced in many metals when the dislocation slip is not sufficient to accommodate the cyclic plastic deformation [6, 9-10, 33-34].

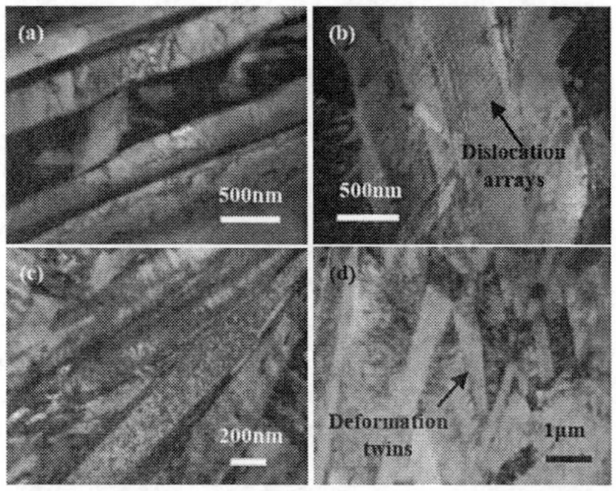

Fig. 10. The fatigue dislocation structure at different stress amplitudes in FLM during high cycle fatigue: (a) dislocation slip within α lath at 690MPa, (b) dislocation arrays in α lath at 690MPa; (c) much more dislocation pileups and tangles at 705MPa, (d) abundant of deformation twins in primary α lath at 705MPa

In general, assuming everything else is equal, fatigue crack tends to nucleate in coarse grained material than the fine grained materials [6]. The grain size is expected to affect the resistance against fatigue crack initiation and propagation is in opposite manners [35]. It is well known that decreases the grain size would increase the crack initiation life and decrease the crack propagation life [6]. Thus, due to the similar size and uniform distribution of α phase, FLM with nanoscale α phase possesses super resistance against fatigue crack initiation than BM. Moreover, Pan et al [36] has found that the dislocations move collectively back and forth along the nanotwins during cyclic loading, which indicated that nanotwins promotes the reversible dislocation slip and increases the fatigue life. Whereas, cyclic plastic deformation is produced at each cycle during low cycle fatigue. Abuzaid and Sangid found that the materials deforms heterogeneously and the strain varies across the microstructure, especially in the presence of interface and grain boundary [15, 37]. The plastic deformation easily localizes at the interface in the LM. As indicates by the SEM and TEM results, voids tend to initiate and link up to a large crack more easily in the LM than in the EM at the same strain amplitude. It is reasonable to conclude that fatigue crack nucleates more easily in LM than in EM, which results in shorter low-cycle fatigue life in LM.

4 Conclusions

The dislocation structure and crack nucleation during fatigue in different microstructures of TC21 alloy were systematically investigated and analyzed.

(1) It can be found that the strength increases in the order of LM, EM, BM and FLM. The ductility increases in the order of FLM, LM, EM and EM.

(2) During low-cycle fatigue, EM exhibits higher strength, ductility and longer low-cycle fatigue life than those of the LM. The voids tend to initiate and link up to a large crack more easily in the single lamellar alpha than the equiaxed alpha grains.

(3) However, during high-cycle fatigue, the fine lamellar microstructure (FLM) shows higher fatigue limit than bimodal microstructure (BM).

(4) The localized plastic deformation at the primary α phase can be induced during high-cycle fatigue. The slip bands or twins are observed in the equiaxed and lamellar alpha phases(>1micron), which tends to form strain concentration and initiate fatigue crack. The localized slip within nanoscale alpha plates is seldom observed and extrusion/intrusion dispersedly distributed on the sample surface in FLM. This indicates that FLM show super resistance to fatigue crack which bring about higher fatigue limit than BM.

Acknowledgments

This project was financially supported by the 973 Program of China (2014CB644003), the National Natural Science Foundation of China (51671158, 51471129, 51621063), and the 111 Project of China (B06025).

References

1. C. Leyens, M. Peters, *Titanium and titanium alloys: fundamentals and applications, John Wiley & Sons.* (2003)
2. D. Banerjee, J.C. Williams. Acta Mater. **61**(3) (2013)
3. G. Lütjering, J.C. Williams. *Titanium, Berlin: Springer*. (2003)
4. Y.Q. Zhao, G.J. Yang. Titanium Industry Progress. **23**(5) (2006)
5. X.N. Wang, Z.S. Zhu, L. Tong, Y. Zhou, X.H. Zhou, H.Q. Yu. Rare Metals Letters.**27** (2008)
6. C.S. Tan, X.L. Li, Q.Y. Sun, L. Xiao, Y.Q. Zhao, J. Sun. Int. J. Fatigue. **75** (2015) 1-9
7. K.S. Chan. Int. J. Fatigue. **32**(9) (2010)
8. R.K. Nalla, B.L. Boyce, J.P. Campbell, J.O. Peters. Metall. Mater. Trans. A **33** (13)(2002)
9. C.S. Tan, Q.Y. Sun, L. Xiao, Y.Q. Zhao, J. Sun. Mater. Sci. Eng. A. **711C** (2018)
10. C.W. Huang, Y.Q. Zhao, S.W. Xin, C.S. Tan, W. Zhou, Q. Li, W.D. Zeng. Int. J. Fatigue. **94** (2017)
11. C.W. Huang, Y.Q. Zhao, S.W. Xin, C.S. Tan, W. Zhou, Q. Li, W.D. Zeng. Mater. Sci. Eng. A. **682** (2017)
12. C.W. Huang, Y.Q. Zhao, S.W. Xin, W. Zhou, Q. Li, W.D. Zeng, C.S. Tan. J. Alloys Compd. **695** (2017)
13. G.Q. Wu, C.L. Shi, W. Sha, A.X. Sha, H.R. Jiang. Mater. Des. **46** (2013)
14. S.K. Jha, C.J. Szczepanski, R. John, J.M. Larsen. Acta Mater. **82** (2015)
15. M.D Sangid. Int. J. Fatigue. **57** (2013)
16. H. Mughrabi. Acta Mater. **61**(4) (2013)
17. H. Mughrabi. Metall. Mater. Trans. B. **40**(4) (2009)
18. H. Mughrabi. Fatigue. Fract. Eng. Mater. Struct. **22**(7) (1999)
19. A. Berg, J. Kiese, L. Wagner. TMS; (1995)
20. A. Pineau, D.L. McDowell, E.P. Busso, S.D. Antolovich. Acta Mater. **107** (2016)
21. J.R. Mayeur, D.L. McDowell. Int. J. Plasticity. **23** (2007)
22. F. Bridier, D.L. McDowell, P. Villechaise, J. Mendez. Int. J. Plasticity. **25** (2009)
23. F. Bridier, P. Villechaise, J. Mendez. Acta Mater. **56** (2008)
24. S.K. Li, B.Q. Xiong, S.X. Hui, W.J. Ye, Y. Yu. Mater. Sci. Eng. A **460** (2007)
25. C.S. Tan, Q.Y. Sun, L. Xiao, Y.Q. Zhao, J. Sun. J. Alloys. Compd. **724** (2017)
26. H. Fei, L. Zhou, H.L. Qu, Y.Q. Zhao, C.Z. Huang. Mater. Sci. Eng. A. 494(1) (2008) .
27. Y.H. Wang, H.C. Kou, H. Chang, Z.S. Zhu, X.F. Su, J.S. Li, L. Zhou. J. Alloys. Compd. 472(1) (2009)
28. H. Shao, Y.Q. Zhao, P. Ge, W.D. Zeng. Mater. Sci. Eng. A. 559 (2013)
29. H. Shao, Y.Q. Zhao, P. Ge, W.D. Zeng. Mater. Sci. Eng. A. 586 (2013)
30. R Liu, S Hui, W Ye. Rare Metals. **31**(2012).
31. K.S. Chan, M.P. Enright. Metall. Trans. A. **36**A (2005)
32. K.S. Chan, Y.D. Lee. Metall. Trans. A. **39**A (2008)
33. J. Huang, Z.R. Wang, K.M. Xue. Mater. Sci. Eng. A. **528**(29-30) (2011).
34. J. Huang, Z.R. Wang, J. Zhou. Metall. Mater. Trans. A **42**(9) (2011)
35. K.S. Chan. Scripta Metall. Mater. **32** (1995)
36. Q.S. Pan, H.F. Zhou, Q.H. Lu, H.J. Gao, L. Lu. Nat. **551** (2017)
37. W. Abuzaid, M.D. Sangid, J.D. Carroll, H. Sehitoglu, J. Lambros. J. Mech. Phy. Solids. **60**(6) (2012)

Multiaxial life prediction of hydrodynamic plain bearings

Henrik Wünsch[1,*], *Christopher* Sous[2] and *Christoph* Broeckmann[1]

[1]Institute for Materials Applications in Mechanical Engineering, RWTH Aachen University, 52062 Aachen, Germany
[2]Institute for Machine Elements and Systems Engineering, RWTH Aachen University, 52062 Aachen, Germany

Abstract. Journal bearings are installed in turbomachinery and in all kind of components related to drive technology. All of these applications expect a high degree of reliability. The functional interaction of shaft, lubricant and bearing shell result in a multiaxial stress state within the bearing lining. However, the life assessment of the bearing material exposed to these multiaxial loads is mostly neglected in the design process. The presented research work formulates a complete methodology for a long-term fatigue assessment, based upon the Quadratic Failure Hypothesis (QFH) This approach satisfies the invariant requirements of coordinate transformation, takes into account positive and negative stresses and is applicable to non-proportional multiaxial loads in 3-dimensional geometric configurations.

1 Introduction

Journal plain bearings are widely used machine elements. Machine and plant constructors appreciate among other advantages, e.g. towards roller bearings, their efficiency in high-speed applications, their favourable damping characteristics and reliability. As long as journal bearing operate in a hydrodynamic state they are insusceptible to wear phenomena. A wide used parameter for journal bearing design is the specific load p_{spec}, which is the applied load divided by the projected bearing area. This parameter was refused by other authors in the past upon good reason [1,2]. In a dynamically loaded bearing, not the specific load is a parameter for assessing fatigue but rather the multiaxial and often non-proportional stress quantities, induced by the oil film pressure into the lining material. Other influences as mounting stresses and thermal induced stresses in bimetallic designs enhance the multiaxiality of the stress state. In order to overcome the uncertainty of the specific load parameter the design strategy is often to enlarge the projected bearing area to a maximum in order to lowering the specific load. This best practise approach contradicts the ideas of lightweight design and resource efficiency.

To overcome this dilemma other researchers spend great efforts into calculating the oil film pressure in the gap between shaft and bearing under both static und dynamic loads. The solution of the Reynolds equation for lubricating films in which viscosity varies three-dimensionally were focus of numerous researches in the last decades and are published in several review papers [3-5]. The output of these calculations are the hydrodynamic pressure (HD) in the lubricating clearance. Extended analysis consider the elasticity of the bearing lining. This elastohydrodynamic (EHD) analysis is refined when temperature terms are taken into account (T-EHD). At this point one can conclude that the hydrodynamic contact with its pressure distribution, clearance height and temperature distribution in the lubrication gap is mostly understood.

However, from this point on, the design chain stops abruptly when it comes to the assessment of fatigue in the lining material. The computed maximum oil pressure is compared with the static compressive strength of the lining material. The German VDI-guideline, e.g., proposes a "thumb rule" for transient operating bearings whereupon the maximum dynamic pressure shall not exceed one sixth of the lining material's compressive strength [6]. For further analysis, the guideline proposes the von Mises criterion. It is well known that for non-proportional loading this approach is often not successful [7]. Until now, there is no reliable study on fatigue in plain bearings known, that considers multiaxial non-proportional stresses. Therefor the authors present here a comprehensive methodology for the fatigue assessment in plain bearings.

2 Failure Criterion – Quadratic Failure Hypothesis

Over the past decades, numerous fatigue failure criteria were proposed. The overarching classification is made according to the damage parameter and can be divided up into stress, strain and energy based criteria. [8] The local stress state in a structure can be proportional and non-proportional, which is decisive for the fatigue design process. Based on the common perception, a non-proportional stress state is present when the principal stress axes are rotating during a load cycle. For this case, three groups of different approaches are established and can be subdivided as follows: empirical approach, critical plane approach and integral approach.

* Corresponding author: h.wuensch@iwm.rwth-aachen.de

The presented fatigue failure criterion in this paper dates back to the work of Troost and El-Magd [9-11]. They translated the plastic yield criterion for anisotropic materials [12, 13] to a fatigue approach. The approach evaluates the fatigue life in a critical oriented volume element. Because of the quadratic term, they named it "Quadratic Failure Hypotheses" (QFH). The general formula can be written as the quadratic potential:

$$Q_{\mathrm{ijkl}}\,\sigma_{ij}\,\sigma_{kl} = 2F \tag{1}$$

Where σ_{ij} is the stress tensor and Q_{ijkl} is the anisotropy tensor. They interpreted the anisotropy coefficients as the direction-dependent fatigue strength. The rewritten formulation of Eq. (1) by Troost and El-Magd [10] reads as follows:

$$F = \sum_{\substack{i,j=1\\ i\leq j}}^{3} \frac{\left(\sigma_{\mathrm{a}}(t)\right)_{\mathrm{ij}}^{2}}{A_{\mathrm{ij}}^{2}} - \sum_{\substack{i,j=1\\ i\leq j}}^{3} \frac{\left(\sigma_{\mathrm{a}}(t)\right)_{\mathrm{ii}} \cdot \left(\sigma_{\mathrm{a}}(t)\right)_{\mathrm{jj}}}{A_{\mathrm{ii}} \cdot A_{\mathrm{jj}}} = 1 \tag{2}$$

In this orientation-dependent equation $\boldsymbol{\sigma}_{\mathrm{a}}(t)$ are the direction- and time-dependent stress amplitudes of the stress tensor and $\boldsymbol{A}$ is the fatigue strength in the respective direction depending on the mean stress $\boldsymbol{\sigma}_{\mathrm{m}}$ acting in this orientation. Based on the basic formulation the criterion was made more convenient and user-friendly by breaking it down to plain stress conditions and by the consideration that a failure under torsional loading accurses always under an angle of 45°. Furthermore the sustainable torsional fatigue strength A_{ij} for $i \neq j$ was formulated as a function of the uniaxial fatigue strength by introducing the ratio $T = \left(\sigma_f/\tau_f\right)^2$ with the fully reversed uniaxial and torsional fatigue strength, which made it applicable for brittle and ductile materials. For the particular case of synchronous and reversed alternating tension-compression load, the criterion was written in its last form as [14]:

$$\left(\frac{\sigma_{xa}}{\sigma_f}\right)^2 + \left(\frac{\sigma_{ya}}{\sigma_f}\right)^2 - (T-2)\frac{\sigma_{xa}\sigma_{ya}}{{\sigma_f}^2} + T\left(\frac{\tau_{xya}}{\sigma_f}\right)^2 = 1 \tag{3}$$

Eq. 3 is for a ratio of $\tau_f/\sigma_f = 1/\sqrt{3}$ equivalent to the von Mises equivalent stress criterion for alternating stresses. For out-of-phase loading from **Eq. 2** a failure value F_{V} can be derived which is a linear combination of the maximum peak value $\tilde{F}_{\max}$ of the quadratic potential of all orientations and the maximum integral value $\bar{F}_{\max}$ of all orientations.

$$F_{\mathrm{V}} = \tilde{F}_{\max}(\cos\psi_c)^{v} + \bar{F}_{\max}(\sin\psi_c)^{w} \tag{4}$$

By means of the "characteristic principal direction" ψ_c and the "coupling exponents" v and w peak value and integral value are weighted. Thereby it is taken account of the experimental proven fact that loading histories containing shear stresses can be better assessed with the integral value and loading histories in principal stress direction better with the peak value [15]. The characteristic principal direction is calculated by means of the loading stresses:

$$\psi_c = \tan^{-1}\frac{2\left[\left|\tau_{xym}\right| + \left(1 + 3\frac{\sigma_{ya}}{\sigma_{xa}}\right)\tau_{xya}\right]}{\left|\sigma_{xm} - \sigma_{ym}\right| + \sigma_{xa} + \sigma_{ya}} \tag{5}$$

Default values for the coupling exponents are $v, w = 2$, which ensure a quasilinear coupling of peak value and integral value [15].

2.1 Calculation proposal

The present study is based on the basic considerations of the QFH. Today's computing power allow a numerical solution of the quadratic potential in **Eq. 2** for all directions in three-dimensional space. For that reason, complex loading path can be calculated by means of QFH without reducing it to plane stress conditions or analytical solutions in general. Because of the not invariant formulation of ψ_c in **Eq. 5**, the definition was revised. According to the consideration that the peak value is similar to the maximum principal stress criterion the concept of Dietmann [16] and later other authors [17,18] was adapted after which there is beside a critical plane also a critical point of time during a load cycle that is damage relevant. Therefore, the angle ψ_c, which is of somewhat empirical nature is replaced by the angle between the orientation of $\tilde{F}_{\max}$ and the principal direction at the time $\tilde{t}_{max}$. The fundamental steps of the fatigue calculation according to QFH are shown in **Fig. 1**.

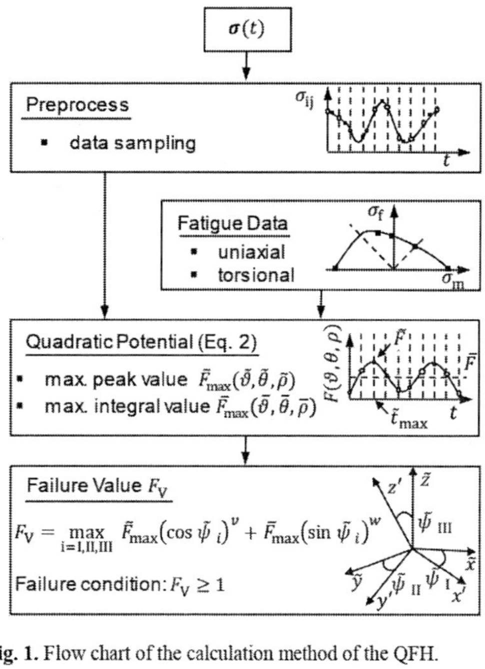

Fig. 1. Flow chart of the calculation method of the QFH.

Input data for the calculation is the time-varying stress tensors for one load cycle. The tensor, which can be the result of a finite element analysis is pre-processed in a initial step. The entries are ordered and sampled in equidistant time intervals. In the next step, the quadratic potential is calculated. There for the stress tensor has to be rotated in all spatial directions. The rotation from $\boldsymbol{\sigma}$ to $\boldsymbol{\sigma}^*$ through the angles ϑ, ϕ and ρ about the coordinate axis is done using a rotation matrix $\boldsymbol{R}$.

$$\boldsymbol{\sigma}^*(t) = \boldsymbol{R}^{\mathrm{T}}\,\boldsymbol{\sigma}(t)\,\boldsymbol{R} \tag{6}$$

In all spatial directions the entries of the stress tensor can be subdivided in a static part $\boldsymbol{\sigma}^*_{\mathrm{m}}$ and a dynamic part $\boldsymbol{\sigma}^*_{a}$.

$$\boldsymbol{\sigma}^*(t) = \boldsymbol{\sigma}^*_{\mathrm{m}} + \boldsymbol{\sigma}^*_{\mathrm{a}}(t) \tag{7}$$

From the mean stress entries, the sustainable stress amplitudes $A(\boldsymbol{\sigma}^*_{\mathrm{m}})$ are derived from experimentally determined fatigue strength diagrams for uniaxial and torsional loading. The calculation of the quadratic potential F is performed according to **Eq. 2** initially in a rough screening with angle increments of 20°. Via bisection both maxima $\tilde{F}_{\mathrm{max}}$ and $\bar{F}_{\mathrm{max}}$ can be determined. Both maxima are not necessarily located in the same orientation. In order to weight both maxima, the alignment of the axis system of the $\tilde{F}_{\mathrm{max}}$ orientation towards the principal stress directions is decisive. As far as both orientations are identic at the time point $\tilde{t}_{max}$ when $\tilde{F}_{\mathrm{max}}$ occurs, only the maximum peak value is considered as damage relevant. The greater the misalignment of both orientations the more influence gains the integral value $\bar{F}_{\mathrm{max}}$. **Fig. 2** shows the maximum peak value $\tilde{F}$ as a spatial representation.

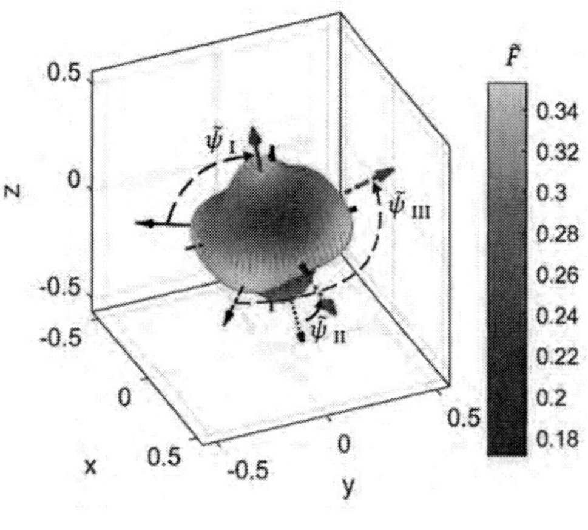

Fig. 2. Maximum peak value $\tilde{F}$ for all spatial directions.

The angles $\tilde{\psi}_{I,II,III}$ represent the misalignment between the orientation of $\tilde{F}_{\mathrm{max}}$ and the principal stress orientation with the case differentiation:

$$\tilde{\psi}_i = \begin{cases} \tilde{\psi}_i & 0° \leq \tilde{\psi}_i \leq 90° \\ 180 - \tilde{\psi}_i & 90° < \tilde{\psi}_i \leq 180° \end{cases} \tag{8}$$

The weighting is carried out according to the linear combination:

$$F_{\mathrm{V}} = \max_{i=\mathrm{I,II,III}} \tilde{F}_{\mathrm{max}}\left(\cos\tilde{\psi}_i\right)^{v} + \bar{F}_{\mathrm{max}}\left(\sin\tilde{\psi}_i\right)^{w} \tag{9}$$

The angle $\tilde{\psi}_i$ for which the linear combination reaches its maximum is used to calculate the failure value F_{V}. The failure condition is $F_{\mathrm{V}} \geq 1$.

3 Object of study and material

The plain bearing under investigation is a fully enclosed journal bearing. This bearing was designed for comparative studies regarding fatigue life on a bearing test-rig. The tested bearings had a width to diameter ratio of 0.4 and an inner diameter of 70 mm. The bearing lining of variable thickness was produced by centrifugal casting. The bearings were subjected to a high sinusoidal load F_{R}, resulting in a pulsating pure compressive load. A sketch of the test-rig and the concept of load application is shown in **Fig. 3**.

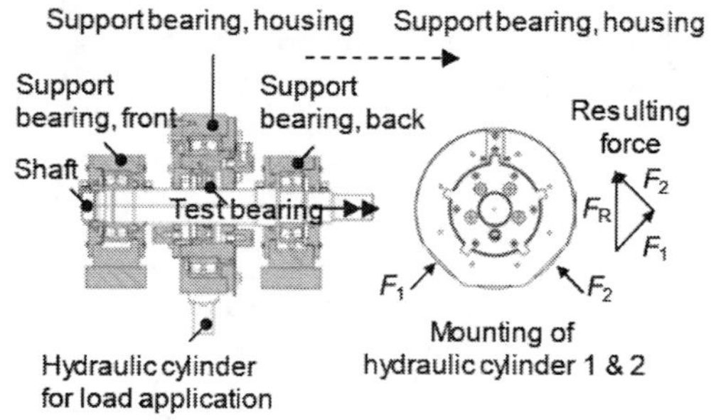

Fig. 3. Journal plain bearing test-rig.

The lining material is a tin-based white metal (SnSb12Cu6ZnAg). **Fig. 4** shows a microsection of SnSb12Cu6ZnAg etched with Iron trichloride. Characteristic is its heterogeneous microstructure with its precipitates SnSb and Cu_6Sn_5, which are embedded in a matrix of saturated tin solid solution. The intermetallic phases, Cu_6Sn_5 and SnSb, lead to an increase in static material strength because of precipitation hardening. Antimony causes solid solution hardening, which ensures a reasonable fatigue strength. Traces of silver have a grain refining effect.

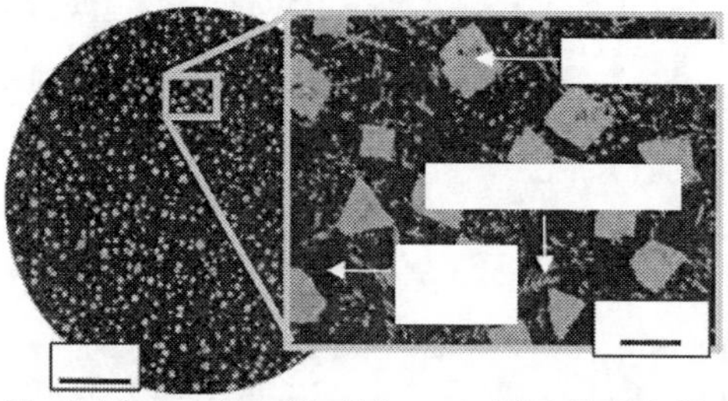

Fig. 4. Microsection of the lining material SnSb12Cu6ZnAg.

Fig. 5 shows the fatigue strength diagram according to Haigh of SnSb12cu6ZnAg for uniaxial loading with a probability of failure of 50 %. The maximum numbers of load cycles was chosen to 10^7 cycles. The load was

https://doi.org/10.1051/matecconf/201816516001

applied in a sinusoidal oscillation at an ambient temperature of $T = 100$ °C.

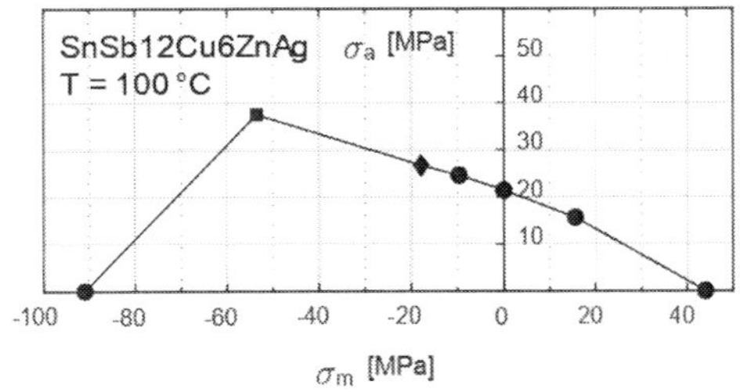

Fig. 5. Fatigue strength diagram according to Haigh of SnSb12Cu6ZnAg for uniaxial loading. ■ interpolated, ♦ Nesic [19].

Accordingly, **Fig. 6** shows under otherwise equal conditions the fatigue strength diagram for torsional loading. In both cases, the solid lines represent the allowable stress amplitude depending on the static mean stress. These are input data for the fatigue calculation below.

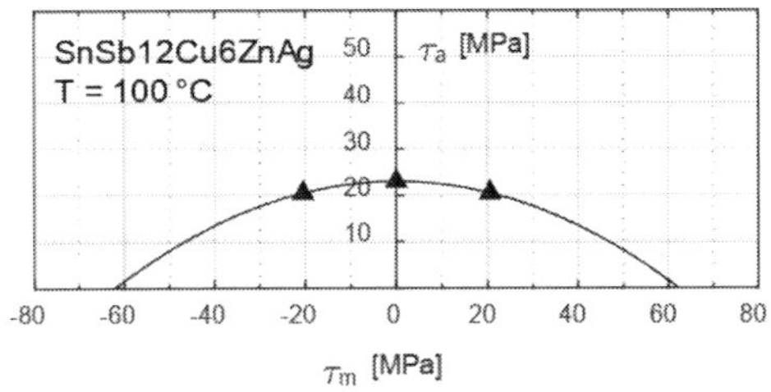

Fig. 6. Fatigue strength diagram according to Haigh of SnSb12Cu6ZnAg for torsional loading.

4 Stress analysis

The geometry and a high number of boundary conditions of this bearing were transferred to a Finite Element Model. The loading of the bearing was calculated by means of the before mentioned T-EHD approach and was imprinted to the FE-Model. A more detailed description of this procedure was published before by Sous et al. in [1]. **Fig. 7** shows the bearing under investigation with the definition of the circumferential angle θ, which rotates analogous to the shaft clockwise.

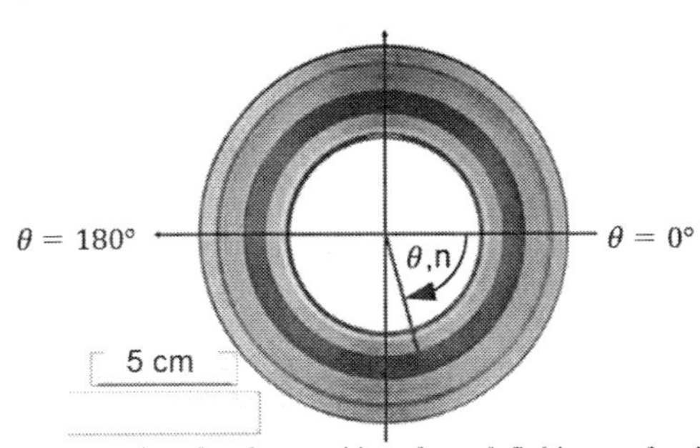

Fig. 7. Test-rig bearing with the definition of the circumferential angle θ.

4.1 Principal stress in bearing lining

In order to gain an idea of the stress distribution within the bearing lining **Fig. 8** shows the mean principal stress distribution $\sigma_{I,II,III,m}$ 125 µm below the sliding surface of a bearing with a layer thickness of 1 mm loaded and applied with a specific load of 20 MPa. Except in the case of the first mean principal stresses, which act in axial direction, the static mean stresses are compressive. The highest compressive stresses act in radial direction. There are considerable compressive mean stresses in circumferential direction caused by mounting stresses and stresses due to thermal expansion. These are responsible for the asymmetric stress distribution.

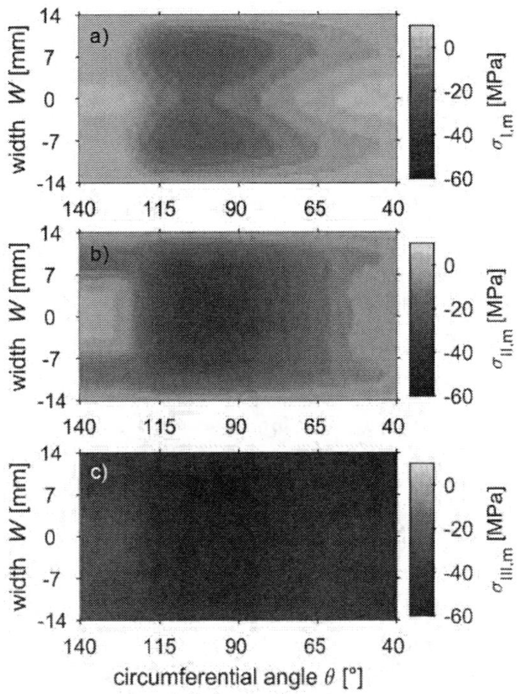

Fig. 8. Distribution of principal mean stresses $\sigma_{I,II,III,m}$ 125 µm below the projected sliding surface for $p_{spec} = 20$ MPa and layer thickness of 1 mm.

The amplitudes of the principal stresses $\sigma_{I,II,III,a}$ are shown in **Fig. 9**. The highest amplitudes act in circumferential direction. This is an indication that a possible crack initiation occurs in direction of the $\sigma_{II,a}$ direction because of the orientation perpendicular to the free surface.

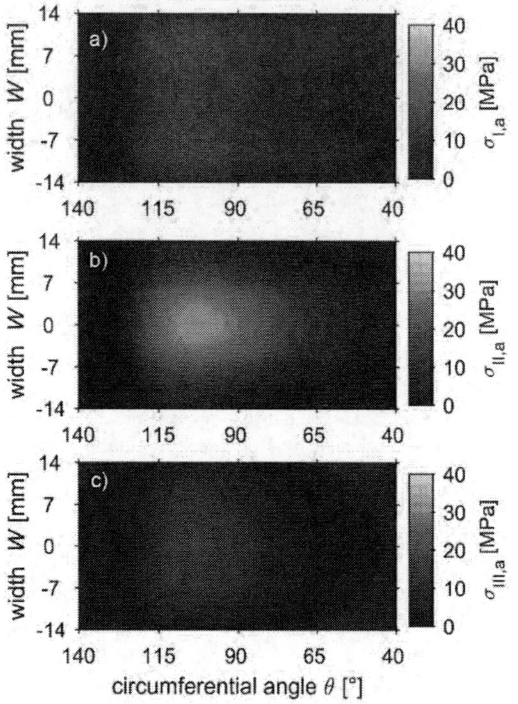

Fig. 9. Distribution of principal stress amplitudes $\sigma_{I,II,III,a}$ 125 µm below the projected sliding surface for p_{spec}= 20 MPa and layer thickness of 1 mm.

Fig. 10 shows the distribution of the maximum shear stresses. The static mean shear stresses increase in rotating direction of the shaft. The time-dependent shear stress amplitudes are of low magnitude and act in areas of low oil-film pressure.

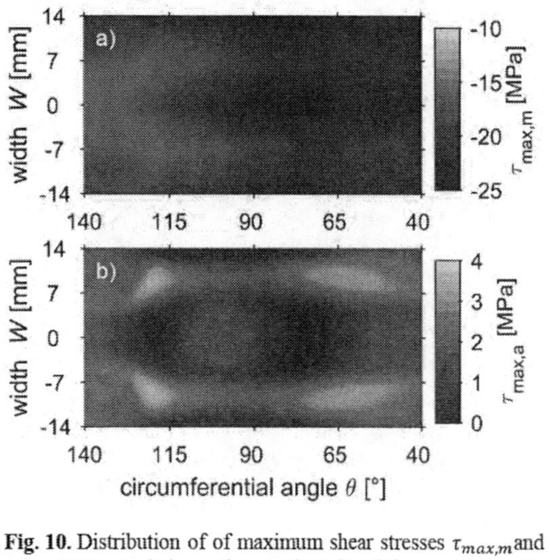

Fig. 10. Distribution of of maximum shear stresses $\tau_{max,m}$and $\tau_{max,a}$ 125 µm below the projected sliding surface for p_{spec}= 20 MPa and layer thickness of 1 mm.

3.1 Evaluation of the stress proportionality

An often-used criterion for proportionality of stress states is the rotation of principal directions. **Fig. 11** shows the respective change of the principal direction angles $\Delta\varphi_{I,II,III}$ during one load cycle.

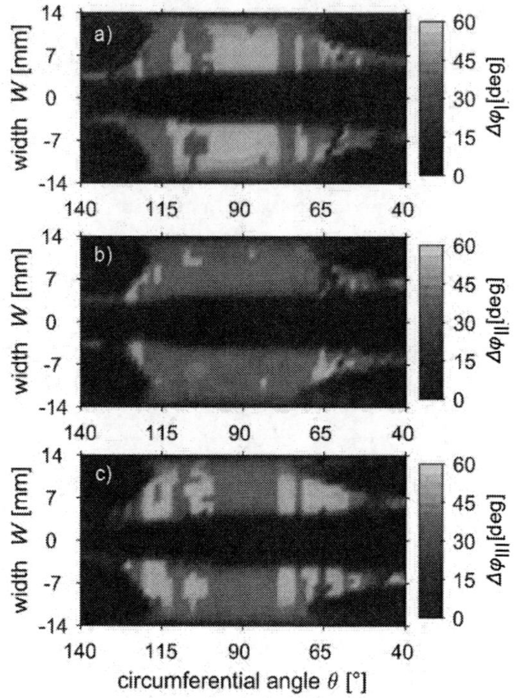

Fig. 11 Change of the principal direction angles $\Delta\varphi_{I,II,III}$ during one load cycle.

According to that, the stress state in the axial middle of the sliding surface where normal stresses are high is proportional. The stresses are non-proportional in areas shifted from the axial middle where normal stresses act in axial direction.

5 Some calculation examples

For the same bearing configuration fatigue calculations according to the proposed methodology were carried out. From FE analysis stress tensors for specific loads of 20, 30 and 36 MPa were post-processed via QFH. **Fig. 12** shows from a) to c) the distribution of the failure value F_V for increasing specific loads on the projected sliding surface.

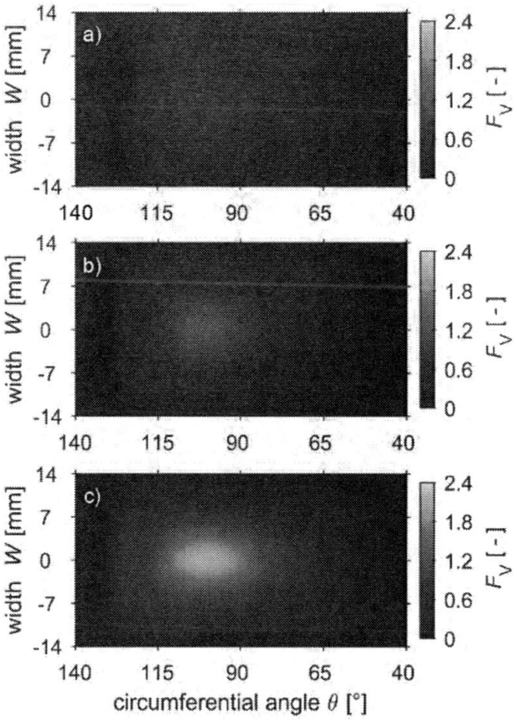

Fig. 12. Distribution and amount of failure value F_V for a) p_{spec}= 20 MPa, b) p_{spec}= 30 MPa and c) p_{spec}= 36 MPa and layer thickness of 1 mm.

For all loading cases the general appearance and distribution of F_V is similar. Logically, the intensity of the failure value increases with increasing specific load. At a circumferential angle of 115°, a discontinuity is apparent. This is the location of the narrowest gap of the oil-film where a transition from compressive stress to tension stress in tangential direction takes place.

In **Fig. 13**, both summands of the linear combination in **Eq. 4** are compared. The exponentials are set to $v = w = 1.7$. The peak part clearly predominates the linear combination. The influence of the integral part is marginal.

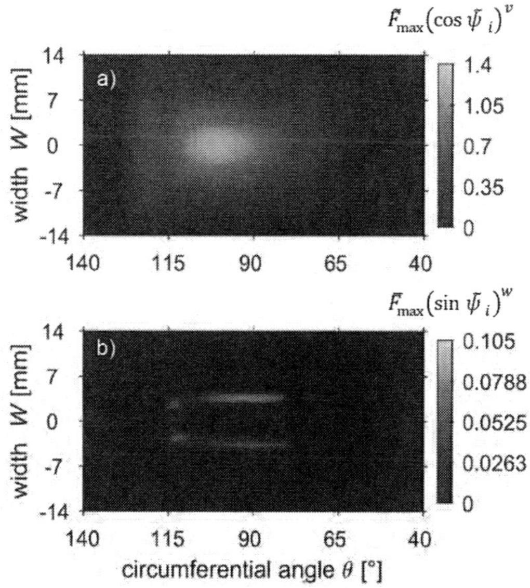

Fig. 13. a) Peak value $\tilde{F}_{max}$ under consideration of thee weighting factor, b) integral value $\bar{F}_{max}$ under consideration of thee weighting factor.

The integral part has a minor influence because the angle $\tilde{\psi}_i$ according to **Eq. 9** is in most areas small. This means automatically that the stress state in these regions is in at least one direction close to the principal stress state.

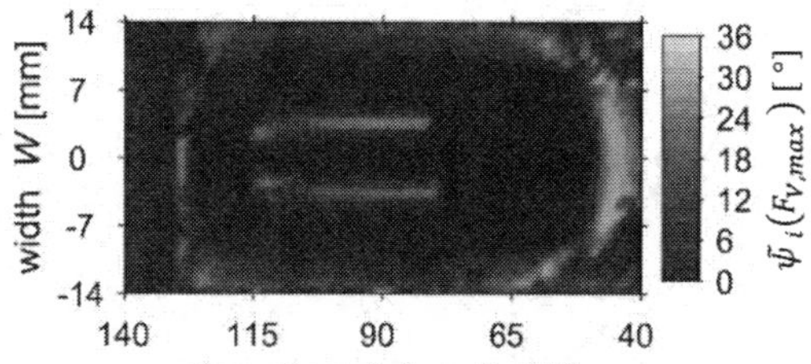

Fig. 13. Weighting angle $\tilde{\psi}_i$ for which F_V reaches its maximum.

In accordance to this finding **Fig. 14** shows the weighted composition of the failure value F_V at the spot where F_V reaches its absolute maximum at a circumferential angle of $\theta = 101.5°$.

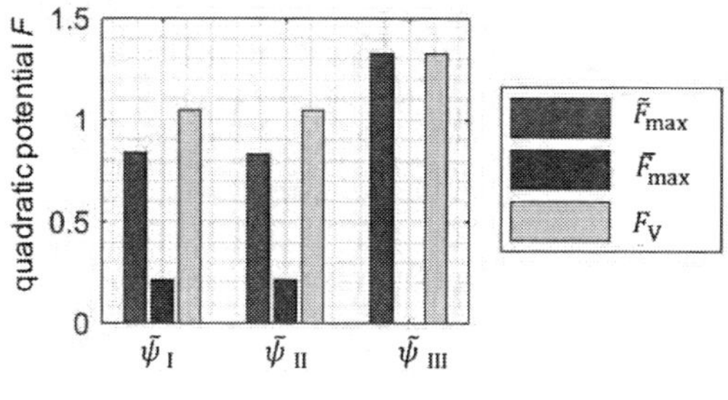

Fig. 14. Peak value $\tilde{F}_{\max}$, integral value $\bar{F}_{\max}$ and failure value F_V depending on the angles $\tilde{\psi}_{I,II,III}$.

Fig. 15 shows the failure value in the axial middle of the sliding surface (at width $W = 0$). The load cases for a specific load of 36 MPa and 30 MPa exceed the failure condition of $F_V \geq 1$. The area in which the circumferential angle exceeds the failure condition extends with increasing specific load.

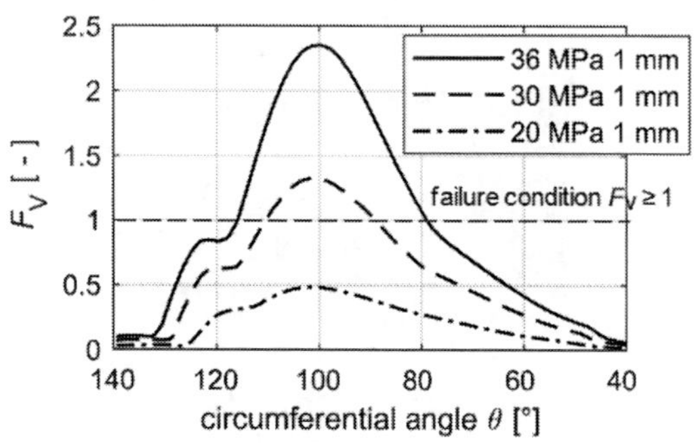

Fig. 15. Failure value F_V for different specific loads.

Therefore, the possibility that a crack initiation takes place at a circumferential angle, which is lower than the one where the maximum is calculated, increases with increasing specific load. **Fig. 16** compares the influence of different lining thicknesses on the failure value. It can be shown that a higher layer thickness is more prone to a fatigue damage than a thinner layer. The reason are probably lower compressive mean stresses according to amount and as a consequence thereof, lower sustainable stress amplitudes.

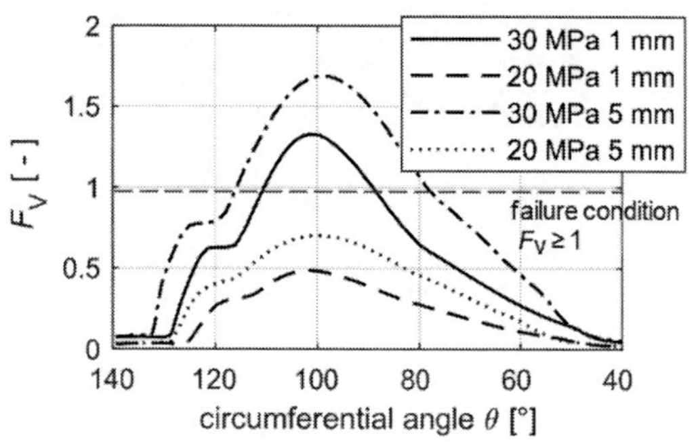

Fig. 16. Failure value F_V for different specific loads and layer thicknesses.

Fig. 17 summaries some results of the described calculation method. The intersection points with the failure condition at $F_V = 1$ deliver the maximum allowable specific loads p_{spec} fort the respective bearing configuration. A fatigue calculation according to Sines [20] is compared to the results of the QFH:

$$\frac{\sqrt{J_{2,a}} + (\sqrt{3}\sigma_f/\sigma_{UTS})\sigma_{H,m}}{\tau_f} = a \qquad (10)$$

The results of the calculations show a good agreement with the experimentally determined fatigue strength. The Sines criterion, in this case, gives a conservative fatigue life prediction.

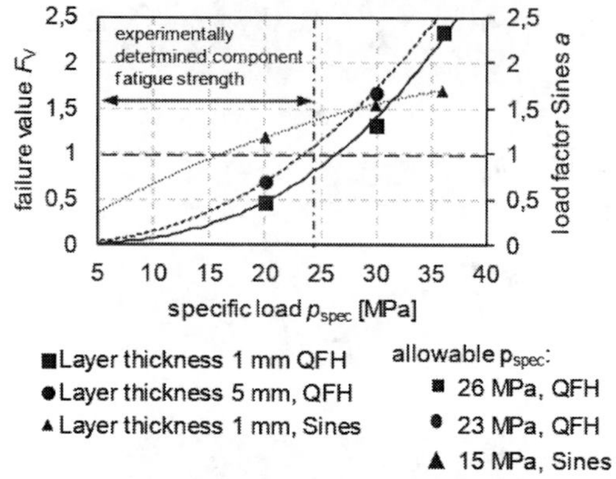

Fig. 17. QFH calculation for different specific loads and layer thicknesses compared to the Sines criterion and the experimentally determined component fatigue strength

6 Conclusion and outlook

The results of the stress analysis and the fatigue assessment based on the QFH provide a unified picture of the fatigue behaviour of journal plain bearings. The regarded boundary conditions including the temperature dependent EHD contact, mounting stresses and thermal expansion within the lining material result in a multiaxial stress state within the sliding layer. However, contrary to the original assumption, the stress state is less non-proportional than expected. This is because shear stresses induced by the viscous oil-film in the lubrication gap are small compared to the induced normal stresses. Superimposed static stresses lead to rotating principal axes in areas of the sliding layer, which are only moderately loaded. According to these observations, the QFH weights the failure relevant mechanisms correctly. The calculated critical orientations considering the anisotropic fatigue strength of the material are consistent with experimental observations, according to which, crack initiation occurs under an angle of 45° in sliding direction of the shaft. The revised weighting factor between peak and integral part of the failure value is feasible and improves the approach particular in terms of a fully invariant formulation. The calculations have shown that especially normal stresses represented by the peak value $\tilde{F}_{\max}$ are the fatigue promoting stress quantities.

This is consistent with observations made in SEM-insitu four point bending tests on the lining material, **Fig. 18**. According to this, the fatigue damage mechanism is as follows: Normal stresses lead to a brittle fracture of Cu_6Sn_5 and SnSb. These rod-shaped and cube-shaped phases are cross-linked in the lining volume. Once a fracture in these phases occurs by exceeding a certain threshold, the initiated crack jumps to neighbouring precipitations and propagates unstable through the volume. Consequently, in the next stage of the study the presented methodology will be translated from a macroscopic view to a microscopic. With the help of Representative Volume Elements (RVE), stress distributions will be understood on microscopic scale. Because of their general formulation, the QFH seems capable when parameterized with phase specific strength properties to assess fatigue also on microscopic scale.

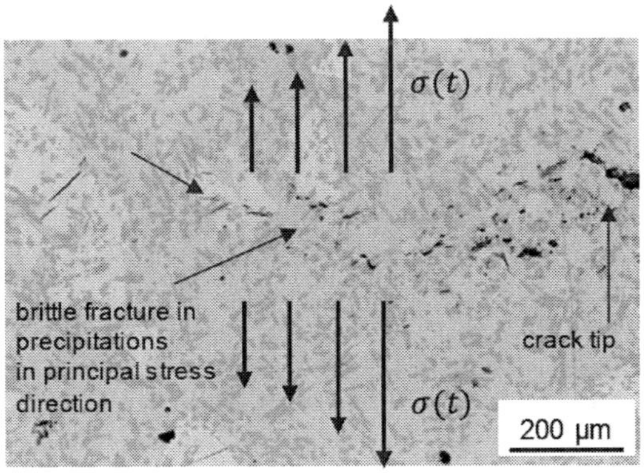

Fig. 18. Mechanism of fatigue crack propagation in front of the crack tip during an insitu four point bending test.

The research for this paper was financially supported by Deutsche Forschungsgemeinschaft (DFG) and Forschungs-vereinigung Verbrennungskraftmaschinen e.V. (FVV) in the research project BR 1844/10-2 / JA 1940/4-2 (FVV 1252).

References

1. C. Sous, H. Wünsch, G. Jacobs, C. Broeckmann, Ind Lubrication and Tribology, **68**(3) (2016)
2. F.A. Martin, D.R. Garner, D.R Adams, J. Tribol. **103**(1), p. 150 (1981)
3. E. Kutnetsov, S. Glavatskih, Tribology International, **94**, pp. 288-305 (2016)
4. M.M. Khonsari, ASLE Transactions, **30**(1), pp. 26-33 (1987)
5. M. Tanaka, Proceedings of the Institution of Mechanical Engineers, Part J, Journal of Engineering Tribology, **214**(1), pp. 107-122 (2005)
6. Verein Deutscher Ingenieure, VDI-Handbuch Produktwentwicklung und Konstruktion, VDI2204 (2007)
7. A. Fatemie, D.F. Socie, Fat Frac Eng Mat Struct, **11**(3), pp. 107-122 (1988)
8. T. Łagoda, P. Ogonowski, Mat.-wiss. u. Werkstofftech. **36** , No. 9 (2005)
9. A. Troost, O. Akin, F. Klubberg, (1987), Konstruktion, **39** No. 12, pp. 479–488.
10. A. Troost, O. Akin, F. Klubberg, F., Materialwissenschaft und Werkstofftechnik, Vol. **19** No. 12, pp. 406–407 (1988)
11. A. Troost, E. El-Magd, Metall, Vol. **31,** pp. 759-764 (1977)
12. R. Hill, *The mathematical theory of plasticity,* Oxford Univ. Press, (1971)
13. Z. Sobotka, ZAMM **49** (1/2), pp. 25-32 (1969)
14. F. Klubberg, C. Broeckmann, P. Beiss, Materials Testing **54**, 9, pp. 569-577 (2012)
15. C. Kaniut, Dissertation, RWTH-Aachen Univerity (1983)
16. H. Dietmann, Z. f. Werkstofffechnik II. of Materials Technology, **6** (1973)
17. K. Dang-Van, Advances in Multiaxial Fatigue, ASTM STP 1191, pp.120-130 (1993)
18. A. Carpinteri, A. Spagnoli, S. Vantadori, International Journal of Fatigue **33,** pp. 969–976 (2011)
19. D. Nešić, Dissertation, RWTH Aachen University (2013)
20. G. Sines, Metal Fatigue, pp. 145-169 (1959)

Multiaxial fatigue life estimation in low-cycle fatigue regime including the mean stress effect

Daniela Scorza*, *Andrea* Carpinteri, *Giovanni* Fortese, *Camilla* Ronchei, *Sabrina* Vantadori, and *Andrea* Zanichelli

University of Parma, Department of Engineering and Architecture, Parco Area delle Scienze 181/A, 43124 Parma, Italy

Abstract. The goal of the present paper is to discuss the reliability of a strain-based multiaxial Low-Cycle Fatigue (LCF) criterion in estimating the fatigue lifetime of metallic structural components subjected to multiaxial sinusoidal loading with zero and non-zero mean value. Since it is well-known that a tensile mean normal stress reduces the fatigue life of structural components, three different models available in the literature are implemented in the present criterion in order to take into account the above mean stress effect. In particular, such a criterion is formulated in terms of strains by employing the displacement components acting on the critical plane and, then, by defining an equivalent strain related to such a plane. The Morrow model, the Smith-Watson-Topper model and the Manson-Halford model are applied to define such an equivalent strain. The effectiveness of the new formulations is evaluated through comparison with some experimental data reported in the literature, related to biaxial fatigue tests performed on metallic specimens under in- and out-of-phase loadings characterised by non-zero mean stress values.

1 Introduction

Engineering metallic components and structures are frequently subjected to complex cyclic loading during their service life, which produces a multiaxial stress field generally characterised by tensor components with mean values different from zero.

In order to investigate the effect of mean stress and mean strain on fatigue life, two different tests are usually performed: stress-controlled cyclic test with constant mean stresses, or strain-controlled cyclic test with constant mean strains [1,2].

In High-Cycle Fatigue (HCF) regime, since cyclic behaviour of material can be considered as linear-elastic, the above tests are essentially equivalent, and one or the other can be indifferently used to examine the effect of mean stress or mean strain on the fatigue life.

As is well-known, a tensile mean normal stress/strain superimposed upon an alternating normal stress/strain strongly reduces the fatigue resistance of structural components, while a compressive mean normal stress/strain has a beneficial effect [1,3]. This can be explained by noting that a tensile mean normal stress increases the crack opening and accelerates the fatigue damage accumulation, whereas a compressive normal stress reduces the crack propagation through the friction between the two surfaces of the crack [1,4].

Stress-based methods are usually employed in HCF regime, since the material behaviour can be considered as linear-elastic and, consequently, stresses can be easily estimated using elastic analysis. In the literature, several models based on the S-N approach have been developed to properly evaluate the mean stress effect: for instance, the Soderberg relationship, the Goodman relationship, and the Gerber parabola [5-8].

In Low-Cycle Fatigue (LCF) regime, instead, the cyclic response of material is elasto-plastic. Consequently, stress-controlled cyclic test and strain-controlled cyclic test may produce quite different results.

In particular, in the case of *stress-controlled cyclic tests* with constant mean stress, the occurrence of a plastic deformation causes a ratcheting strain (also named cyclic creep strain) [1,2]. The ratcheting strain progressively increases cycle by cycle, and the ratcheting strain accumulation intensifies the fatigue damage, resulting in shorter fatigue life [9,10].

It follows that, in stress-controlled condition, a fatigue failure model should be able to take into account both mean stress and ratcheting strain effects (an interesting review concerning the constitutive models for the ratcheting behaviour of metals is reported in Ref. [9]). However, the analysis of the ratcheting strain phenomenon is complex, since the ratcheting behaviour depends on many factors including the value of mean stress and stress amplitude, the loading history, the loading rate and the microstructural characteristics. Due to such variables, it is very difficult to find fatigue damage models concerning with the mean stress effect and containing a unique parameter able to take into account the ratcheting strain. An interesting energy-based model has been proposed by Xia and co-workers in order to properly evaluate the mean stress and the ratcheting strain effect on fatigue life [1,2].

* Corresponding author: daniela.scorza@unipr.it

In LCF regime under *strain-controlled cyclic tests*, the occurrence of plastic deformations usually results in a mean stress which may relax fully or partially during the loading cycle (i.e. mean stress relaxation). The mean stress usually quickly decreases in the early fatigue life, and then reaches a constant value.

Fatigue damage models may be developed only considering the mean stress effect, since the effect of the applied mean strain is negligible. In particular, it has experimentally been observed that the mean strain does not contribute to the crack opening mechanism, when the mean stress relaxes to a stable value (about one thousand cycles) [11]. Mean stress relaxation depends on the magnitude of the applied strain amplitude [11-13], that is, the rate of mean stress relaxation is greater for higher values of strain amplitude.

The evaluation of the mean stress influence on fatigue life is a complex problem. Different models are available in the literature in order to take into account the above mean stress: the most popular ones are those by Smith [14], Morrow [15,16], Smith-Watson-Topper (SWT) [17], Walker [18] and Manson-Halford (MH) [19]. Note that the above models are currently used, but nowadays none of them is considered better than the others.

In the present paper, three different models are implemented in the critical plane-based multiaxial fatigue criterion for LCF regime recently proposed by Carpinteri et al. [20, 21], in order to take into account the above mean stress effect. Such a criterion is formulated in terms of strains by employing the displacement components acting on the critical plane and defining an equivalent strain as the parameter to quantify the fatigue damage. The Morrow model, the SWT model and the MH model are applied to define such a strain.

The effectiveness of the new formulations is evaluated through comparison with experimental data [22] related to biaxial fatigue tests performed on metallic specimens under both proportional and non-proportional loadings characterised by non-zero mean values.

2 Mean stress effect on multiaxial fatigue life estimation

The present critical plane-based multiaxial fatigue criterion, recently proposed by some of the present authors [20,21], is formulated in terms of strains in order to estimate the fatigue lifetime under LCF regime. Such a criterion aims to reduce the multiaxial strain state to an equivalent uniaxial one.

Figure 1 summarises how to use the above-mentioned criterion to estimate fatigue lifetime of structural components in LCF regime. In more detail, by starting from the strain state at a material point P, the averaged directions of the principal strain axes may be computed on the basis of their instantaneous directions by means of the averaged values of the principal Euler angles. The normal **w** of the critical plane (see Figure 1) is linked to the above directions through the off-angle δ, defined as follows:

$$\delta = \frac{3}{2}\left[1-\left(\frac{1}{2\left(1+\nu_{eff}\right)}\cdot\frac{\gamma_a}{\varepsilon_a}\right)^2\right]45° \tag{1}$$

being ν_{eff} the effective Poisson ratio. Moreover, ε_a and γ_a are determined through the tensile and torsional Manson-Coffin equations, respectively:

$$\varepsilon_a = \frac{\sigma'_f}{E}\left(2N_f\right)^b + \varepsilon'_f\left(2N_f\right)^c \tag{2a}$$

$$\gamma_a = \frac{\tau'_f}{G}\left(2N_f\right)^{b_0} + \gamma'_f\left(2N_f\right)^{c_0} \tag{2b}$$

where N_f is the number of loading cycles to failure, E is the Young modulus, G is the shear modulus and σ'_f, ε'_f, b, c, τ'_f, γ'_f, b_0 and c_0 are material constants to be determined by running appropriate axial fatigue tests.

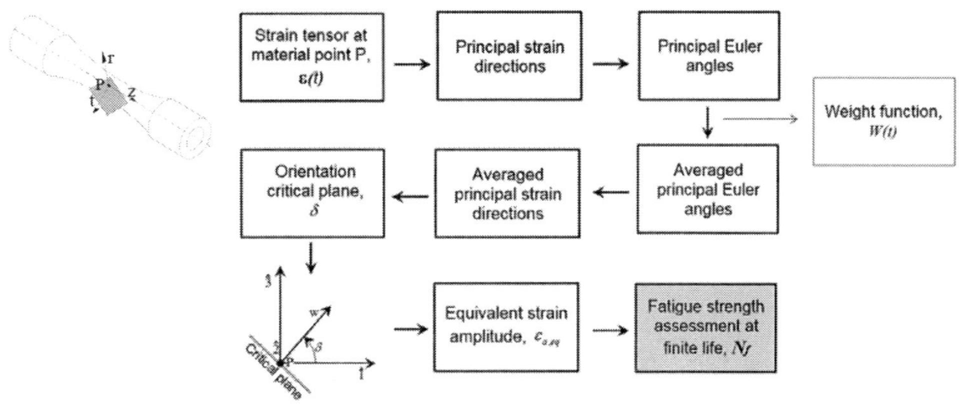

Fig 1. Graphical representation of the low-cycle multiaxial fatigue criterion.

Let us consider a local frame $Puvw$, attached to the critical plane, and the corresponding strain tensor. Details about the definition of such a frame may be found in Ref. [20].

An equivalent strain amplitude, $\varepsilon_{eq,a}$, is computed through a quadratic combination of the normal displacement amplitude, $\eta_{N,a}$ (being $\eta_{N,a} = \varepsilon_{w,a}$), and the shear displacement amplitude, $\eta_{C,a}$ (being $\eta_{C,a} = \sqrt{(\gamma_{uw,a}/2)^2 + (\gamma_{vw,a}/2)^2}$), acting on the critical plane:

$$\varepsilon_{eq,a} = \sqrt{(\eta_{N,a})^2 + \left(2(1+\nu_{eff})\cdot\frac{\varepsilon_a}{\gamma_a}\right)^2 (\eta_{C,a})^2} \quad (3)$$

All terms in Eq. (3) depend on the number of loading cycles to failure, N_f. Therefore, by equaling Eq. (3) to Eq. (2a), the value of N_f can be worked out by means of an iterative procedure.

The original formulation of the criterion can be used to predict the fatigue life at zero mean stress. Now three different models are implemented in such a formulation in order to include the effect of non-zero mean stress:

(i) The Morrow model [15-16]. Such a model introduces the mean stress correction into the elastic term of the tensile Manson-Coffin equation (Eq.(2a)):

$$\varepsilon_{a,M} = \frac{\sigma'_f}{E}\left(1 - \frac{\sigma_{z,m}}{\sigma'_f}\right)(2N_f)^b + \varepsilon'_f (2N_f)^c \quad (4)$$

where $\sigma_{z,m}$ is the value of the mean axial stress at point P, measured during the strain-controlled cyclic test.

Consequently, Eqs (1) and (3) are rewritten by assuming $\varepsilon_a = \varepsilon_{a,M}$;

(ii) The Smith-Watson-Topper (SWT) model [17]. Such a model assumes that the fatigue life depends on the product of the maximum normal stress and the normal strain amplitude and, consequently, the tensile Manson-Coffin equation (Eq. (2a)) turns out to be:

$$\varepsilon_{a,SWT} = \frac{1}{\sigma_{z,max}}\left[\frac{(\sigma'_f)^2}{E}(2N_f)^{2b} + \sigma'_f \varepsilon'_f (2N_f)^{c+b}\right] \quad (5)$$

where $\sigma_{z,max}$ is the maximum value of the axial stress at point P, measured during the strain-controlled cyclic test.

Consequently, Eqs (1) and (3) are rewritten by assuming $\varepsilon_a = \varepsilon_{a,SWT}$.

(iii) The Manson-Halford (MH) model [19]. Such a model is an extension of the Morrow model by assuming that also the plastic term of the tensile Manson-Coffin equation (Eq. (2a)) is affected by the presence of the mean stress:

$$\varepsilon_{a,MH} = \frac{\sigma'_f}{E}\left(1 - \frac{\sigma_{z,m}}{\sigma'_f}\right)(2N_f)^b + \varepsilon'_f \left(1 - \frac{\sigma_{z,m}}{\sigma'_f}\right)^{\frac{c}{b}} (2N_f)^c \quad (6)$$

Consequently, Eqs (1) and (3) are rewritten by assuming $\varepsilon_a = \varepsilon_{a,MH}$.

3 Criterion validation and discussion

In the present paper, the different relationships reported in Eqs (4)-(6) are implemented in the original formulation of the criterion in order to verify whether they are able to estimate the lifetime of some experimental tests available in the literature [22].

The examined experimental data are related to 27 specimens made of Inconel 718 [22, 23], whose static and fatigue properties are listed in Table 1. The above specimens are subjected to synchronous sinusoidal in- and out-of-phase loading with zero and non-zero mean value. In particular, 19 specimens are subjected to in-phase loading (i.e. phase angle, Φ, equal to 0°), 2 specimens to out-of-phase loading with $\Phi = 45°$, and the remaining 6 specimens are characterised by a phase angle equal to 90°.

The value of the effective Poisson ratio ν_{eff} is here obtained from a best fitting procedure by considering only the experimental data related to loading histories with zero-mean strain value [22]. The following error index, I, has been optimised by varying the value of ν_{eff} from 0.1 to 0.5:

$$I = \frac{\varepsilon_{eq,a} - \varepsilon_a}{\varepsilon_a} \quad (7)$$

being ε_a defined by Eq. (2a). At the end of such a best fitting procedure, the value of ν_{eff} is taken equal to 0.34.

The accuracy of the proposed fatigue lifetime estimation has been evaluated by means of the root mean square error method [24]. In more detail, the value of the root mean square logarithmic error is computed as follows:

$$E_{RMS} = \sqrt{\frac{\sum_{i=1}^{n} \log^2(N_{f,\exp}/N_{f,cal})}{n}} \quad (8)$$

Table 1. Static and fatigue properties of Inconel 718 [22, 23].

E (MPa)	σ'_f (MPa)	b (-)	ε'_f (-)	c (-)	G (MPa)	τ'_f (MPa)	b_0 (-)	γ'_f (-)	c_0 (-)
208500	3950	-0.151	1.5	-0.761	77800	2146	-0.148	18.0	-0.922

* Corresponding author: daniela.scorza@unipr.it

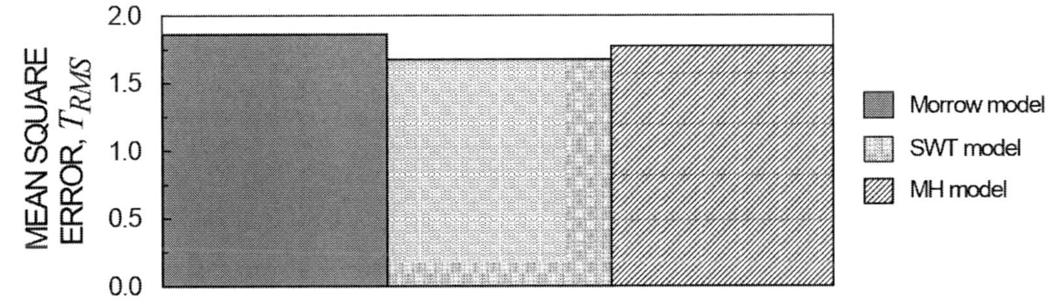

Fig. 2. Mean square error related to Inconel 718, determined by applying the present criterion together with the Morrow model or the SWT model or the MH model.

where n is the total number of data, $N_{f,exp}$ is the experimental multiaxial fatigue life, and $N_{f,cal}$ is the theoretical multiaxial fatigue life determined by considering the different approaches presented in Eqs (4)-(6).

Finally, the mean square error T_{RMS} is given by

$$T_{RMS} = 10^{E_{RMS}} \tag{9}$$

Figure 2 shows the mean square error computed for all the values of fatigue life, in accordance with the three different models presented in Section 2 in order to take into account the effect of the mean stress. The analysis of the results clearly proves that the present criterion together with the three models produces evaluations within the scatter band 2, being the value of the T_{RMS} lower than 2. Moreover, it can be noticed that the lifetime estimated through the SWT model (Eq. (5)) is slightly more accurate than the other two ones, being T_{RMS} equal to 1.68 whereas the mean square error values related to the MH model and the Morrow model are equal to 1.77 and 1.86, respectively.

As a consequence of such a remark, the comparison between experimental data and theoretical estimations in terms of fatigue life is limited to the results determined by implementing Eq. (5) in the present criterion (see Figure 3).

In Figure 3, the solid line indicates $N_{f,cal} = N_{f,exp}$, the dashed lines correspond to $N_{f,cal} / N_{f,exp}$ equal to 0.50 and 2 (scatter band 2), and the dot-dashed lines correspond to $N_{f,cal} / N_{f,exp}$ equal to 0.33 and 3 (scatter band 3).

Figure 3 demonstrates that 78% of the estimated results fall within the scatter band of 2, whereas all of them fall within the scatter band of 3, and this holds true independent of both the value of the mean stress and the degree of non-proportionality.

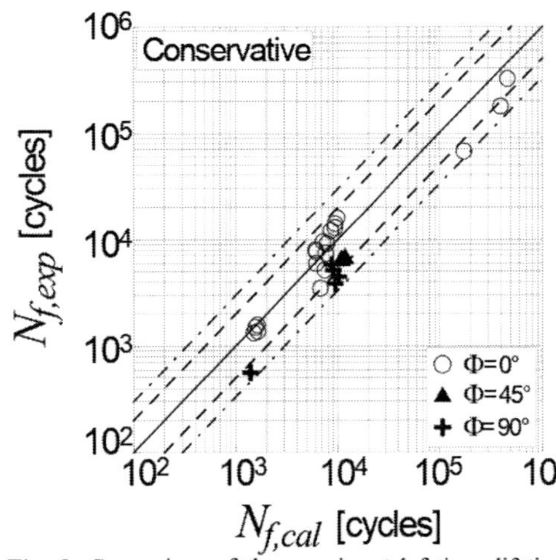

Fig. 3. Comparison of the experimental fatigue lifetime, $N_{f,exp}$, [22] and the theoretical one, $N_{f,cal}$, estimated through the present criterion together with the SWT model (Eq. (5)).

4 Conclusions

In the present paper, the reliability of a strain-based multiaxial LCF criterion in estimating the fatigue lifetime of metallic structural components subjected to sinusoidal loading with zero and non-zero mean stress value has been discussed.

The effect of the non-zero mean stress has been taken into account by implementing three models available in the literature: the Morrow model, the Smith-Watson-Topper model, and the Manson-Halford model. Since such a criterion is formulated in terms of strains, the above-mentioned models are implemented in the definition of both the critical plane orientation and the equivalent strain amplitude, which is the parameter quantifying the fatigue lifetime.

The effectiveness of the new formulations is evaluated through some experimental data reported in the literature, related to biaxial fatigue tests performed on metallic specimens under in- and out-of-phase loadings characterised by non-zero mean values. The accuracy of the proposed formulations has been evaluated by means of the root mean square error method, which proves that all the implemented models provide values of the mean square error lower than 2. According to the mean square error value, the SWT model results to be slightly more accurate than the other two models.

By comparing the experimental data with the theoretical estimations determined by means of the SWT model in terms of fatigue life, it has been observed that the agreement is satisfactory, and this holds true for both compressive and tensile non-zero mean stress values.

Therefore, the implementation of the SWT model in the present criterion seems to be a promising tool to assess the fatigue lifetime of metallic components in LCF regime with non-zero mean stress, although different materials need to be processed in order to validate the robustness of the criterion for practical applications.

The authors gratefully acknowledge the financial support provided by the Italian Ministry for University and Technological and Scientific Research (MIUR), Research Grant PRIN 2015 No. 2015JW9NJT on "Advanced mechanical modeling of new materials and structures for the solution of 2020 Horizon challenges".

References

1. Z. Xia, D. Kujawski, F. Ellyin, Int J Fatigue, **18**(5), 335-341 (1996).
2. F. Ellyin, *Fatigue damage, crack growth and life prediction*, (Chapman & Hall, London, 1997).
3. I.V. Papadopoulos, Int. J. Fatigue, **16**, 377-384 (1994).
4. H.J. Gough, H.V. Pollard, Proc. Instn. Mech. Engrs., **131**, 3-54 (1953).
5. J. Goodman, *Mechanics Applied to Engineering,* (Longmans Green, London, 1899).
6. J. Marin, In: *Proceedings of international conference fatigue metals*, London, 184-192 (1956).
7. G.P. Sedenckyj, Int. J. Fatigue, **23**, 347-353 (2000).
8. L. Susmel, R. Tovo, P. Lazzarin, Int. J. Fatigue, **27**, 928-943 (2005).
9. G. Kang, Int. J. Fatigue, **30**, 1448-1472 (2008).
10. S.K. Paul, N. Stanford, A. Taylor, T. Hilditch, Int. J. Fatigue, **80**, 341-348 (2015).
11. F. Ellyin, J. Eng. Mater. Technol., **107**, 119-125 (1985).
12. S.K. Koh, Int. J. Fatigue, **21**, 1019-1032 (1999).
13. H. Hao, D. Ye, Y. Chen, F. Mi, J. Liu, Mater. Des., **67**, 272-279 (2015).
14. J.O. Smith, Bulletin No. 334, University of Illinois, Engineering Experiment Station, Urbana, IL, Feb. 1942.
15. J. Morrow, *Fatigue Properties of Metals*, Section 3.2 of *Fatigue Design Handbook*, (Pub. No. AE-4, Society of Automotive Engineers, Warrendale, PA 1968).
16. J.D. Morrow, in: *Internal Friction, Damping and Cyclic Plasticity, ASTM STP 853* (American Society for Testing and Materials, Philadelphia, 45-87, 1965).
17. R.N. Smith, P. Watson, T.H. Topper, J. Mater., **5**(4), 767-778 (1970).
18. K. Walker, in: *Effects of Environment and Complex Load History on Fatigue Life, ASTM STP 462* (Am. Soc. for Testing and Materials, West Conshohocken, PA, 1-14, 1970).
19. S.S. Manson, G.R. Halford, Int. J. Fract. **17**, 169-172 (1981).
20. A. Carpinteri, C. Ronchei, A. Spagnoli, S. Vantadori, Theor. Appl. Fract. Mech., **73**, 120-127 (2014).
21. A. Carpinteri, C. Ronchei, D. Scorza, S. Vantadori, Theor. Appl. Fract. Mech., **79**, 77-83 (2015).
22. D.F. Socie, P. Kurath, J. Koch, In: *Biaxial and Multiaxial Fatigue* (EGF 3, Mechanical Engineering Publications, London, 1989).
23. D.L. Morrow, Report No 147, UILU-ENG 89-3610, University of Illinois at Urbana-Champaign, 1989.
24. T. Łagoda, K. Walat, Int. J. Fatigue, **67**, 73-77 (2014).

Thermomechanical fatigue behaviour of ferritic stainless steel grades for high temperatures applications

Cloé Prudhomme[1], *Pierre-Olivier* Santacreu[1*], *Isabelle* Evenepoel[1], and *Benoit* Proult[1]

[1] APERAM, Centre de Recherche, 62330 Isbergues, France

Abstract. Nowadays high temperatures resistant materials are needed to resist to high temperature applications (up to 1000°C), such as automotive exhaust gas manifolds. Some developed stainless steel grades, including ferritic grades or austenitic refractory grades, can be used in this temperature range and both in continuous or cyclic thermal conditions. In order to predict the thermomechanical fatigue damage of stainless steel parts submitted to cyclic thermal loading and constrained bonding conditions, the elasto-viscoplastic model by Chaboche is determined for a wide range of temperatures, of strain amplitudes and strain rate levels thanks to isothermal traction-compression tests. The validation procedure is performed afterward by comparison with stabilized behavior under non isothermal conditions on a dedicated thermal fatigue test performed on V-shape specimens. Results of simulation show very good fitting with the experimental curves which would lead to a more accurate fatigue life prediction. A damage model was derived from Taira's thermal low-cycle fatigue model to include dwell-time period at high temperature and creep-oxidation effect. In this paper the example of K44X, a dedicated grade for high temperatures applications, is presented.

1 INTRODUCTION

Although electric vehicles start to be highly developed nowadays, thermal engine cars but also hybrid vehicles still need appropriate materials for the exhaust line. As a consequence of the decrease of the pollutants emissions of vehicles (high reduction of NOx, HC, PM from Euro 3 to Euro 6), the temperature of the exhaust gas tends to increase. Therefore, corrosion and thermomechanical fatigue resistance of exhaust line need to be improved by the use of stainless steel rather than cast iron [1].

Today some austenitic and ferritic grades are proposed as high temperatures resistant stainless steels. Austenitics stainless steels grades present higher mechanical properties at high temperatures but the advantage of ferritic grades is a lower coefficient of thermal expansion. K44X (AIISI 444, EN 1.4521), a high temperature ferritic grade, offers an important improvement in terms of thermal fatigue and creep resistance in comparison with AIISI441 (EN 1.4509) grade, can overtake temperature peak of about 1 000°C. It has to be pointed out that this grade was especially developed for high temperature applications and that it could not be substituted by a regular 1.4521/444 grade. The other advantage of ferritic stainless steel grades is their free Nickel content which involved higher price stability.

An example of high temperature application is exhaust manifolds which are now commonly made of stainless steels. However, increasing their durability is still a challenge. Indeed, these components are the hottest parts of the exhaust system, bolted on the motor block to collect the hot burned gas and to feed them into the catalytic converter, those parts are subjected to creep, oxidation and thermomechanical fatigue. As motor bench tests are expensive, modelling the manifold under the thermal loading by finite element analysis is a cost effective and a time saving solution.

In order to predict lifetime of such components, a V-shape fatigue test was developed to reproduce in laboratory typical thermomechanical loading of the hot end of the exhaust line. High temperature and thermomechanical properties of our grades are investigated through this test and through isothermal tests since few years. In addition, damage models are identified at high temperatures and used to predict lifetime of components. To get those damage predictive models, a behavior model need first to be identified thank to isothermal cyclic tests [2-4].This paper presents the identification of the behavior and damage models for K44X at high temperatures.

2 MATERIAL AND EXPERIMENTAL PROCEDURE

2.1. Materials

* Corresponding author: pierre-olivier.santacreu@aperam.com

The K44X is a monostabilized niobium 1.4521 ferritic material dedicated to high temperature applications; its typical chemical composition is given Table 1. The 2% Molybdenum content increases the high temperature tensile strength compared to an AIISI441 (EN 1.4509). The Niobium prevents against chromium carbides precipitation during heat treatment and welding and therefore provides a good corrosion resistance [5]. In parallel, the excess of Niobium leads to the precipitation of niobium at grain boundaries which improve the creep resistance of the material ("pinning effect" of the grain boundaries avoiding grain boundaries sliding)[6]. The material keeps its complete ferritic structure up to 1025°C and presents a very good oxidation behavior because its low coefficient of thermal expansion is also close to the one of its oxide layer: unlike austenitic grades, spallation of the oxide at high temperatures is unlikely. Even if the mechanical properties of ferritic stainless steels are intrinsically lower than austenitic grades, the advantages of cyclic oxidation, creep and thermomechanical fatigue behavior make the K44X a suitable material for high temperature components.

EN grade designation	C	Cr	Si	Mn	Mo	N	Nb	Fe
1.4521	0.02	19	0.6	0.3	1.9	0.015	0.6	Bal.

Table 1. Chemical composition of K44X – EN1.4521

The considered thickness for exhaust application is often below 2mm and the surface finished of the tested specimens for this study is 2D (non skinpassed).

2.2. High temperature tensile test

High temperature tensile tests are carried out according to EN ISO 6892-2 standard on a 100KN INSTRON tensile machine with a 1000°C-resistance furnace. Flat specimens are cut along the rolling direction. In general two or three samples are tested per condition. After a 2h heating procedure, tensile tests are performed at a constant displacement rate of 0.54mm/min.

2.3. Isothermal cyclic test

In order to identify the elastoviscoplastic behavior model, isothermal test are performed on LCF flat specimens cut along the rolling direction. The cyclic tests are carried out from 20°C to 1000°C with an INSTRON 8802 tensile machine with a resistance furnace, according to a specific strain signal. The tests are performed with different strain levels (maximal strain value: 0.2, 0.4 and 0.6%) and with two different strain rates (0.1 and 0.02%/s) at a fixed temperature. The first part of the signal is composed of 20 triangular waves (for material accommodation) with a maximal strain value of 0.2% and a strain rate of 0.1%/s. The next parts of the signal are composed of three triangle waves piloted by deformation and three square waves (relaxation phenomena, piloted by displacement) with a strain ratio of R=-1 for each strain level/strain rate condition (i.e: 6 conditions). An example of signal for one strain level/strain rate condition is given Figure 1.

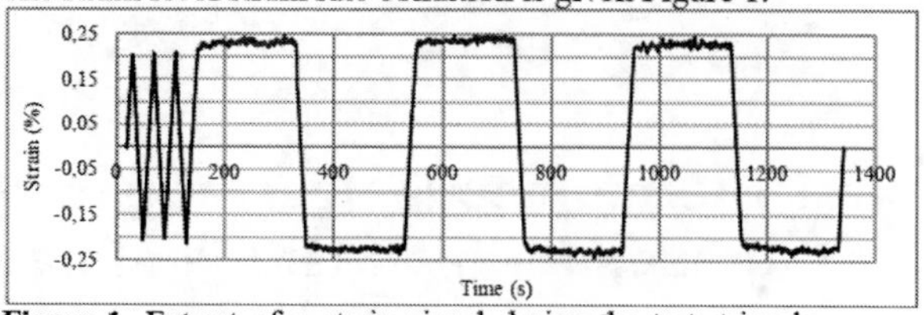

Figure 1. Extract of a strain signal during the test, triangle parts are piloted by deformation and the relaxation parts by displacement.

For this test, at high temperatures, only one sample is used for time saving due to 3 hours of heating procedure. This experimental procedure is validated through additional intermediate signal parts which have to be reproducible, the microstructure and grain size are also checked. To prevent buckling of the flat sample, an anti-buckling system is used (see next figure).

Figure 2. Anti buckling equipments developed for light-gauge flat samples

2.4. V-shape thermal fatigue test

This V-shape test was initially developed to reproduce typical thermomechanical fatigue (TMF) conditions that hot end line undergoes. Today, this test also enables to classify the different stainless steel grades in term of thermomechanical fatigue lifetime but also in terms of damage mechanisms. From the results, a damage behavior model is identified to predict, through finite element simulation analysis, the fatigue lifetime of hot application components and their risky zones.

The specimens are taken from stainless steel sheets of thickness between 1 and 2 mm along the rolling direction and then bended to obtained a V-shape. During the TMF test, the V-shape sample is held by its ends in fixed jaws and is heated up by Joule effect which causes a stress strain response. The thermal stresses in the jaws and the temperature are monitored during the test by means of a force cell and a thermocouple welded on the sample surface respectively. The thermal cycle is composed of a heating, a natural convection air cooling phase and can also include a dwell time (up to 180s).

Interactions between creep and cyclic oxidation phenomena are less or more pronounced depending on the maximal temperature and the holding time. The intrados of the sample is solicited out of phase (compression during heating) and the extrados is solicited in phase.

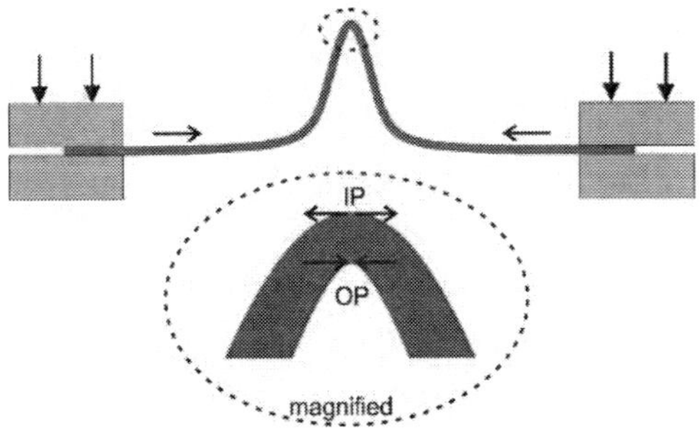

Figure 3. Stress loading of V-shape sample during heating

The lifetime is determined at 50% decrease of the maximal stress compared to the stabilized value in order to avoid an increase of current density. Depending on the maximal and minimal temperature, the dwell time but also the specimen thickness, the test duration is around 3 weeks or more.

3 CONSTITUTIVE MODELS

3.1. Behaviour model

No strain measurement can be carried out during V-shape tests. The cumulated plastic strain for each test condition is calculated thanks to a behavior model and to finite element analysis with a post processor of Abaqus. The elastoplastic model by Chaboche is an accurate model to describe stainless steel behavior at high temperature, especially when a material is subjected to cyclic loading, because it takes in account the viscous effects present at high temperatures unlike an elastoplastic model (especially when T>500°C)[7].
In this model, a non linear kinematic hardening is combined with an isotropic hardening and the Norton viscoplastic law. Strain partition is assumed [8].

$$\underline{\underline{\varepsilon_{tot}}} = \underline{\underline{\varepsilon_{th}}} + \underline{\underline{\varepsilon_{el}}} + \underline{\underline{\varepsilon_{vp}}} \quad (1)$$

The elastic part and the thermal part of the strain are described as:

$$\underline{\underline{\varepsilon_{el}}} = \frac{(1+\nu)}{E}.\underline{\underline{\sigma}} - \frac{\nu}{E}.Tr(\underline{\underline{\sigma}})\underline{\underline{I}} \quad (2)$$

$$\underline{\underline{\varepsilon_{th}}} = \alpha.\Delta T.\underline{\underline{I}} \quad (3)$$

The yield criterion includes isotropic and kinematic hardening and is defined by:

$$J_2(\underline{\underline{\sigma}} - \underline{\underline{X}}) - R \geq 0 \quad (4)$$

From the yield criterion, the viscoplastic strain tensor is deduced:

$$\dot{\underline{\underline{\varepsilon}}}_{vp} = \frac{3}{2}\cdot\dot{p}\frac{\underline{\underline{\sigma}}'-\underline{\underline{X}}'}{J_2(\underline{\underline{\sigma}}-\underline{\underline{X}})} \quad \text{where} \quad \dot{p} = \left\langle \frac{J_2(\underline{\underline{\sigma}}-\underline{\underline{X}})-R}{K} \right\rangle^n \quad (5)$$

The subscript (') indicated the deviatoric part of the tensor and the dot (·) stands for the time derivative. The non linear back stress X (kinematic hardening) and the isotropic hardening R are defined by:

$$\dot{\underline{\underline{X}}} = \frac{2}{3}C\cdot\dot{\underline{\underline{\varepsilon}}}_{vp} - D\cdot\underline{\underline{X}}.\dot{p} \quad (6)$$

$$\dot{R} = b(Q-R).\dot{p} \quad (7)$$

As the behavior material is determined on stabilized material, the isotropic hardening is constant since no change of the viscoplastic radius is assumed.

$$R = R_0 \quad (8)$$

3.2. Taira model

The thermomechanical fatigue damage model is based on Taira's model which is an adaptation of the Manson-Coffin formula to non isothermal low cycle fatigue [9]. The number of cycle up to failure is described as a function of the cumulated plastic strain and of the temperature. The predicted thermal fatigue lifetime is expressed in number of cycle as followed:

$$N_f = \lambda(T_{eq}) \,.\, \Delta\varepsilon_{pcum}{}^{n} \quad (9)$$

$$\text{where } \Delta\varepsilon_{pcum} = \int_{cycle}\sqrt{\frac{2}{3}(\varepsilon_{ij}^{p}:\varepsilon_{ij}^{p})dt} \quad (10)$$

$\Delta\varepsilon_{pcum}$ stands for the cumulated plastic strain during one stabilized cycle. The exponent n may depend on an effective temperature for a given grade. λ is a function of the equivalent temperature and of the dwell time defined first by [4] in order to take into account time-dependent effect like oxidation. The parameters a and b are parameters evaluated from experimental data.

$$T_{eq} = \frac{T_{max}}{1 - aT_{max}\ln(1+b\tau)} \quad (11)$$

4 RESULTS

4.1. Isothermal tests and Chaboche model identification

The Chaboche model, already determined up to 850°C, is described in [10]; until now for higher temperatures, the materials parameters were carefully extrapolated. Materials data have been updated and determined at 920 and 1000°C. The results are mainly focused on higher temperatures such as 1000°C. The isothermal results for 0.1%/s strain rate at the different strain levels are given Figure 4.

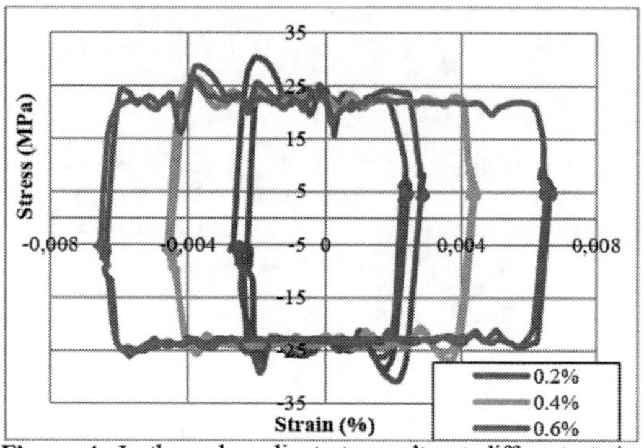

Figure 4. Isothermal cyclic tests results at different strain levels for 0.1%/s strain rate at 1000°C

The results above show that stress strain signal amplitude of the signal does not exceed 25MPa which highlights a very low mechanical behavior at 1000°C. At this temperature, the material is highly viscous and a difference of 6MPa can be observed between the maximal stress reached with the low and the high strain rate tests. Given the flat plastic part, kinematik hardening part is almost negligible (C/D ratio →0) such as the elastic part.

A microstructure study enables to check eventual grain growth during the test. For the "as received sample", very few thin intragranular Nb-precipitates are observed. For the LCF tested specimen, the head of the specimen, where the grains are the less deformed, was studied. The grain size measurement shows no grain growth during the cyclic test. Few niobium precipitates are observed in the grains but a lot can be observed at the grain boundaries and where they have coarsened. This precipitation of Niobium reduces the grain boundary sliding by pinning effect providing a high creep resistance for K44X, the precipitation does not modify the behavior model along the cyclic test and the amplitude of the intermediate additional stress-strain signal is reproducible from a test part to another.

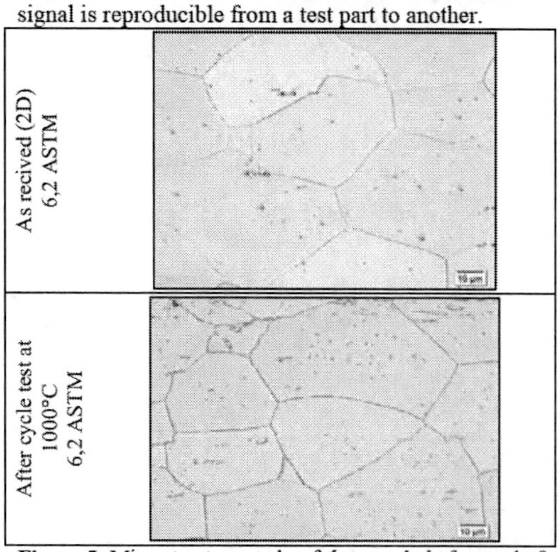

Figure 5. Microstructure study of the sample before and after isothermal cycle test at 1000°C

The 6 tested conditions enable to determine the Chaboche model at 1000°C, the parameters are validated in isothermal conditions but also an isothermal conditions.

Isothermal validation

To validate the obtained model in isothermal conditions, an additional test with different strain rate (0.08%/s) and strain amplitude (1.0%) was carried out. This test, not taking into account during the identification, was simulated with the determined Chaboche parameters. At 1000°C, a very good correlation is obtained between the two signals.

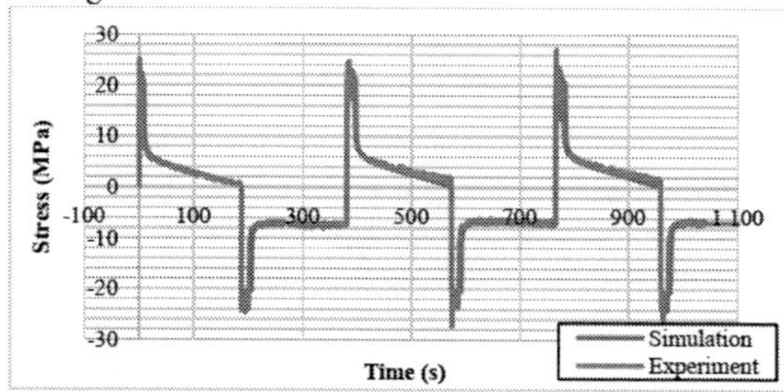

Figure 6. Validation of the behavior model in isothermal conditions at 1000°C

Anisothermal validation

Although the Chaboche model is calibrated with isothermal tests, the model is used in non isothermal conditions. In order to validate the Chaboche model in anisothermal loading, comparison of the experimental data of V-tests (recorded loading and temperature) and a simulated V-shape specimen with the obtained behavior model and finite element simulation with Abaqus is carried out.

The simulation of a 2mm thick V-shape sample between 950 and 250°C without dwell time was performed with Abaqus. The results are compared with the first and the 1000th cycles of the experimental test. A good correlation is obtained in terms of load amplitude but also in terms of area under the curve.

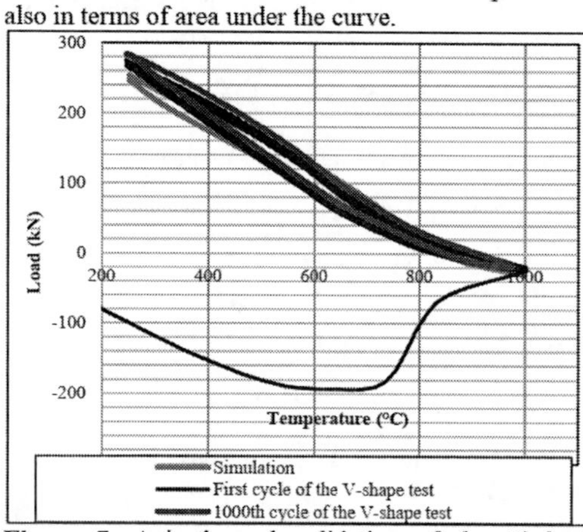

Figure 7. Anisothermal validation of the Chaboche model between 250 and 1000°C

Comparison with tensile test results

The elastoviscoplastic behavior of the material can be related with general mechanical properties, especially the ultimate tensile stress (UTS). An elastoplastic stress value (EVP stress), close to the maximal stress of a strain-stress curve with strain rate of 0.1%/s, was calculated for each temperature (see Figure 8). A linear relation between the UTS and the EVP stress is obtained highlighting the close relation between the identified parameters and typical mechanical properties. Indeed, with the increase in temperature, there is a general drop in mechanical properties with an increase in viscosity: the proportion of the viscosity part of the material behavior increases. With the increases in temperature there are also drops in elastic modulus but also in the kinematic and isotropic hardening. The proportions of those parts with respect to the EVP stress at each temperature are given Figure 9.

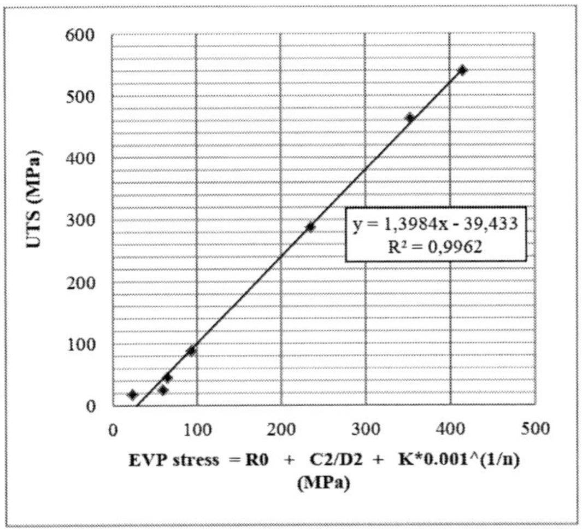

Figure 8. Ultimate tensile stress and behavior model correlation for K44X

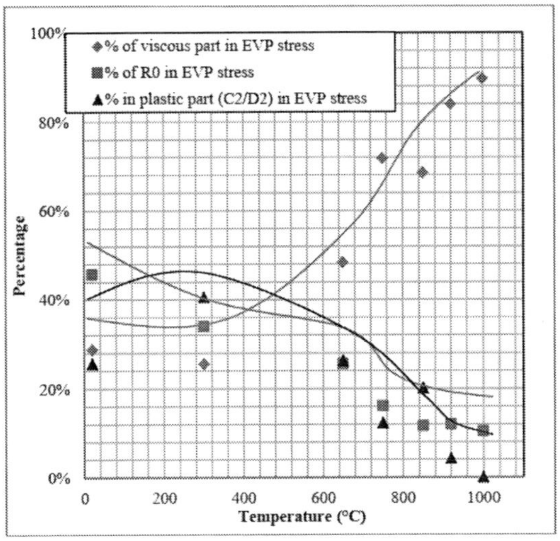

Figure 9. Material behavior evolution with the temperature

4.2. Thermomechanical fatigue tests and Taira model identification

The elastoviscoplastic model but also a large database of thermomechanical fatigue tests are needed to identify the damage model. TMF tests are carried out up to 1050°C and post mortem samples analysis enables to better understand damage mechanisms. For ferritic grades, failure is driven by out of phase mechanisms in the intrados of the V-shape specimen due to the low cyclic oxidation sensitivity and the low thermal expansion. Indeed, for austenitic grades, at high temperature, the cracks tip tends to open in extrados (tensile loading) increasing the oxidation risk. The K44X high creep resistance leads to a conservation of the V-shape of the specimen and to a high lifetime in thermomechanical loading up to 1025°C. Above 1050°C, the damage is more pronounced and grain growth can occur.

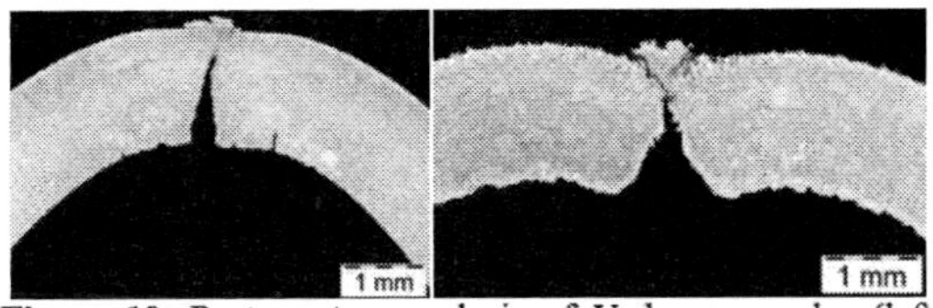

Figure 10. Post mortem analysis of V-shape samples (left maximal temperature of 950°C/ right maximal temperature of 1050°C)

Depending on the thickness of the V-shape specimen, the length between the jaws, the height of the specimen, the maximal temperature reached during the test, the lifetime can be less or more important. The increase of thickness leads to higher thermal constraints and so to a decrease of lifetime. Figure 11 shows, for tests with a minimal temperature of 250°C, the effect of the thickness on the lifetime of the V-shape sample.

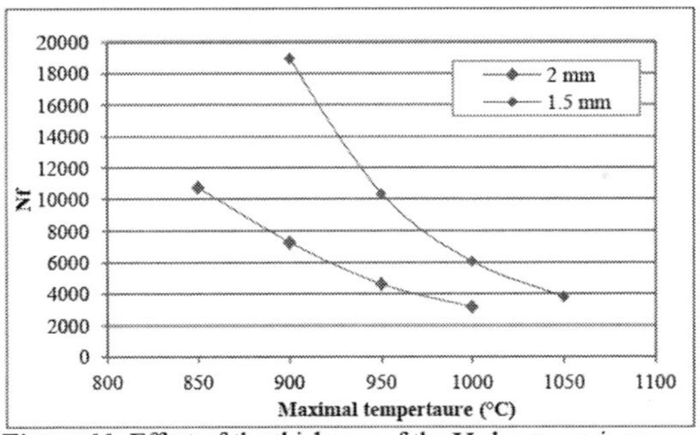

Figure 11. Effect of the thickness of the V-shape specimen on fatigue lifetime (Minimal temperature of 250°C, no dwell time)

From the thermomechanical V-test results and the simulations of this test through Abaqus (elastoviscoplastic model), determination of $\Delta\varepsilon pcum$ values on the top of the V are carried out enabling the identification of the Taira model. The thermomechanical fatigue lifetime of component can be predicted thank to a post processor of Abaqus, Xhaust_Life®, critical areas are highlighted and different design of component can be tested, compared and enhanced.

5 DISCUSSION

The identification of the behavior model (Chaboche) shows good agreement with the experimental data. In the case of thermal fatigue damage model, no measurement can be carried out during the tests. That is why a finite element analysis is required to get the cumulated plastic strain for each test conditions before identifying the damage model. The lifetime prediction demonstrates good correlation with experimental data. Comparison between lifetime calculations on manifold and motor bench test confirms those assessments.

However, some critical areas observed after the simulation and by OEM are often close to welded area. Studied were performed on welded V-shape specimens nonetheless, testing a welded sample is challenging because the configuration of the specimen has to be well chosen. From the different tested welded specimen geometries, the “edge join” geometry seems the most relevant (Figure 11). The lap join specimens were highly deformed during the tests leading to a high scattering in the results and the butt join are less representative of real manifold conditions. A large amount of tests with different grades, thickness, temperatures need to be first performed to determine the damage criteria.

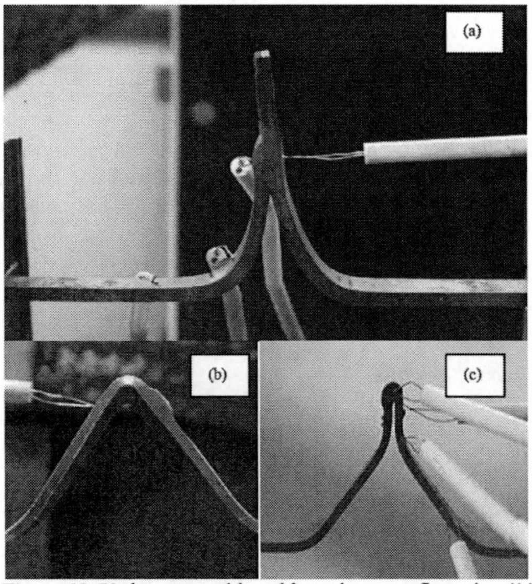

Figure 12. V-shape test with weld specimen configuration ((a) lap join; (b) butt join; (c) edge join)

6 CONCLUSION

K44X is a cost effective solution for high temperature applications because of its really good thermal fatigue behavior due to its lower thermal expansion coefficient and its oxidation resistance at high temperatures. Its chemical composition with high niobium content enables a high creep resistance; the molybdenum increases the tensile strength making this grade suitable for exhaust manifold.

The elastoviscoplastic mechanical behaviour of K44X was determined for a wide range of conditions and up to 1000°C thank to isothermal cyclic traction-compression tests. In parallel, thermomechanical fatigue tests were also performed to classify TMF resistance of the stainless steel grades and studying the damage mechanisms involved. The TMF tests also enable to determine Taira fatigue damage model proposed to predict lifespan of components. With the identified damage model the simulation of thermal fatigue loading in exhaust parts is carried out and critical area in terms of fatigue lifetime can be detected. The comparisons with OEM observation show really good agreement with the simulation. However, critical areas subjected to thermomechanical fatigue are sometimes observed in welded zones. Thus, the resistance of thermal fatigue of weld joins need also to be investigated through new welded V-shape samples.

7 REFERENCES

1. Gocmez, Deuster : “An integral engineering solution for design of exhaust manifolds, *SAE Tech Paper 2009-01-1229* (2009)
2. Chinnouilh, P.O.Santacreu, J.M.Herbelin, Thermal Fatigue Design of Stainless Steel Exhaust Manifolds, SAE Tech. Paper, 2007 – 01 – 0564, (Detroit 2007)
3. C.Simon, P.O. Santacreu, Life time prediction of exhaust manifold in TMF’, Proc. CAMP2002-High-Temperature Fatigue, Eds. G.Ballias et al., pp.257-267, (Bonifatius GmbH Druck 2002)
4. P.O. Santacreu, L. Faivre, A. Acher, damage mechanisms of stainless steels under thermal fatigue, SAE Tech. Paper, 2014-01-0917, (Detroit 2014)
5. Renaudot, N., Santacreu, P.O., Ragot, J., and Molron, J.L., “430LNb: A New Ferritic Welding Wire for Automotive Exhaust Applications, SAE Technical Paper 2000-01-0314, doi:10.4271/2000-01-0314, (2000)
6. Erneman J.and Al. Acta Mater.,35 705-710 (2006)
7. L.Bucher, PO Santacreu, A Koster, L.Remy: Elasto-viscoplastic behavior of the ferritic stainless steel AISI441-EN1.4509 from Room temperature to 850°C, Journal of ASTM International, Vol.3, No.7, Paper ID JAI1397, (2006)
8. J.Lemaitre, J.-L Chaboche : Mécanique des Matériaux Solides, Dunod Eds., (1985)
9. Taira, S.: “Relationship between thermal fatigue and low-cycle fatigue at elevated temperature, Fatigue at elevated Temperatures”, ASTMSTP 520, ASTM, (1973)
10. PO Santacreu, L.Faivre, A Acher, Life prediction for stainless steel exhaust manifolds, SAE int. J. Passeng. Cars-Mech syst. 5(2):2012, DOI 10.4271/2012-01-0732, (2012)

8 NOMENCLATURE

σ	Stress tensor
X	Tensor variable of Kinematic hardening
R	Isotropic hardening variable
R0,	Isotropic hardening constant
Q, b	Material parameters for isotropic hardening
C	Initial hardening variable of Frederick-Armstrong formula
D	Recovery parameter of Frederick-Armstrong formula
ν	Poisson's ratio
K, n	Norton law parameters
T	Temperature
εtot	Strain tensor
εvp	Viscoplastic strain tensor
εth	Thermal strain
εel	Elastic strain tensor
p	Accumulated plastic strain
E	Young Modulus
Δεpcum	Equivalent viscoplastic strain amplitude during a stabilized cycle

Comparing in-service multi-input loads applied on non-stiff components submitted to vibration fatigue to provide specifications for robust design

Gwenaëlle Le Corre (Genet), PhD

[1]Volvo Group Truck Technology, Durability & Reliability department, 99 route de Lyon, 69800 Saint Priest cedex, France

Abstract. This study focuses on applications from the automotive industry, on mechanical components submitted to vibration loads. On one hand, the characterization of loading for dimensioning new structures in fatigue is enriched and updated by customer data analysis. On the other hand, the loads characterization also aims to provide robust specifications for simulation or physical tests. These specifications are needed early in the project, in order to perform the first durability verification activities. At this time, detailed information about the geometry and the material is rare. Vibration specifications need to be adapted to a calculation time or physical test durations in accordance with the pace imposed by the projects timeframe. In the trucks industry, the dynamic behaviour can vary significantly from one configuration of truck to another, as the trucks architecture impacts the load environment of the components. The vibration specifications need to be robust by taking care of the diversity of vehicles and markets considered in the scope of the projects. For non-stiff structures, the lifetime depends, among other things, on the frequency content of the loads, as well as the interactions between the components of the multi-input loads. In this context, this paper proposes an approach to compare sets of variable amplitude multi-input loads applied on non-stiff structures. The comparison is done in terms of damage, with limited information on the structure where the loads sets are applied on. The methodology is presented, as well as an application. Activities planned to validate the methodology are also exposed.

1 Introduction

Early in development projects, loads are needed to enrich the first simulations on the first concepts. At that time, little information on the new concepts, regarding geometry and material, is known (Johannesson & Speckert, 2014). For static or quasi static structures, equivalent loads are usually helpful to describe uni-variate or multivariate loads (Bignonnet & Thomas, 2001) (Le Corre (Genet), 2006). For non-stiff components, input loads can be described with power spectral densities. Each direction is applied on the specimen sequentially. Limitations regarding the representation of singularities like peaks are met in some applications (Colin, 2016).

This paper deals with loads coming from the roads, applied at different points of a structure placed on trucks, attached to the chassis. This structure is considered as non-stiff, i.e the damage induced by these loads is dependent on the frequency content of these loads.

In order to assure the robustness of the loads, they have to be measured on the most damaging configuration of trucks, among the full scope of trucks configurations in the project. The loads can be measured on a physical truck, when driving on the obstacles of the proving ground. They can also come from simulated trucks, virtually driving on numerical obstacles of the proving ground.

In this context, a tool is needed to identify the most severe trucks configurations, in terms of damage applied on component of interest. This needs to be done without detailed information about the geometry and the material of the structure itself. The theory is detailed in the following paragraphs, under several assumptions on the structure. The application chosen here is the cooling package, placed at the front of the truck, in the chassis.

2 Expression of the stress from the multi-input accelerations

This paragraph is dedicated to the expression of the stress tensor from acceleration components, under several working hypotheses.

2.1 Hypotheses and target

The aim is to express the stress from a multi-input acceleration applied on a non-stiff structure. The following hypotheses are taken into account.

Let's consider a structure called (S),

- **H1**: The behavior of (S) is linear,
- **H2**: (S) is considered as non-stiff, i.e. the damage depends on the loads frequency.
- **H3**: A frequency f0 emerges from the other eigen modes, and is the most damaging at the critical point. The frequency f0 highly depends on the dynamic characteristics of (S). The stiffness of the chassis, where the cooling package is mounted on, has a limited impact on f0.
- **H4**: At the critical point, the stress tensor is mainly unidirectional. The direction of the

* Corresponding author: gwenaelle.genet@volvo.com

unidirectional stress is the same, do not vary with time.

2.2 Expression of the uniaxial stress

Let's call A(t), the acceleration set,

$$A(t) = (a_u(t))$$

The integer value u is described as $1 \leq u \leq U$, where U is the number of load components applied on the structure. In the principal stress axis system, the stress tensor $\mathbf{\Sigma}$ at the critical point X_c of a structure S is expressed as follows,

$$\mathbf{\Sigma}(X_c, A(t)) = \sigma(X_c, A(t)) \begin{pmatrix} 1 \\ 0 \\ 0 \end{pmatrix}$$

where $\sigma(X_c, A(t))$ in a scalar. Considering the modal superposition method, the unidirectional stress time history at the critical point can be written as follows,

$$\sigma(X_c, A(t)) = \sum_{i=1}^{m} \sigma_i (X_c). q_i(A(t))$$

The magnitudes σ_i are the modal stresses. They are dependent on the structure, its material and its geometry at the location of the critical point. The magnitudes q_i are the modal coordinates. They are dependent on the structure as well as on the acceleration inputs. The modal coordinates are time series. The magnitude m is the total number of modes taken into account to calculate the stress with the modal superposition method.

By the hypothesis called H3, the resonance frequency is at the origin of most of the damage. Let's consider,

$$\tilde{\sigma}(X_c, A(t)) = \sigma_{iR}(X_c). q_{iR}(A(t)) \qquad (1)$$

The magnitude σ_{iR} is the modal stress and q_{iR} is the modal coordinate at the resonance frequency. The following approximation is considered, under the hypothesis described in the paragraphs above,

$$\sigma(X_c, A(t)) \sim \tilde{\sigma}(X_c, A(t)) \qquad (2)$$

From the hypothesis previously mentioned (H1), the modal coordinates can be expressed in terms of accelerations, from (2)**Error! Reference source not found.**, as follows,

$$\tilde{\sigma}(X_c, A(t)) = \sigma_{iR}(X_c). \sum_{u=1}^{U} b_u(\mathrm{S})\, a_u(t)$$

In order to make the expression of $\tilde{\sigma}$ more tractable mathematically, the following normalization has been chosen,

$$\tilde{\sigma}(X_c, A(t)) = \sigma_{iR}(X_c). B(\mathrm{S}) \sum_{u=1}^{U} \alpha_u(\mathrm{S})\, a_u(t)$$

where,

$$\textstyle\sum_{u=1}^{U}(\alpha_u(\mathrm{S}))^2 = 1 \text{ and } B(S) > 0 \qquad (3)$$

Let's consider the magnitude a^*,

$$a^*(A(t)) = \sum_{u=1}^{U} \alpha_u(\mathrm{S})\, a_u(t)$$

Thus,

$$\tilde{\sigma}(X_c, A(t)) = \sigma_{iR}(X_c). B(X_c). a^*(A(t)) \qquad (3)$$

The magnitude B depends on the structure and the critical point. The magnitude a* can be described as a linear combination of the acceleration components. The unit vector α_u depends on the structure. The product $\sigma_{iR}(X_c). B(X_c)$ is an amplification factor, dependent on the structure and the location of the critical point.

3 Applications

The aim of this paragraph is to use the theoretical background explained above to rank different trucks in terms of damage. The main inputs are the accelerations measured at the attachment points of (S) to its holder. Let's recall that the geometry of (S) is unknown. However, the four hypotheses are valid for (S).

3.1 The cooling package

The cooling package is placed at the front of the chassis, between the two side rails, as shown in Figure 1. The cooling package is characterized by its mass, stiffness and damping. It is placed on rubber bushings, illustrated by springs in the Figure 2 . The cooling package is called (S).

Figure 1 : The cooling package on the truck.

The structure is usually tested on a shake table, with input vibrations like accelerations measured on obstacles from the proving ground. A test procedure describes the number and the type of obstacles of the proving ground to run on the rig, in order to reproduce the complete life, regarding vibrations. The test procedure depends on the targeted customer's usage level. The types of obstacles can be potholes, Belgian pavés, bumps.

Early in projects, the loads can be computed by simulation on complete vehicles, driving on the obstacles

 https://doi.org/10.1051/matecconf/201816516004

of the proving ground. These obstacles are described numerically in the simulation tool. In this paper, accelerations time signals from complete vehicle models are considered only.

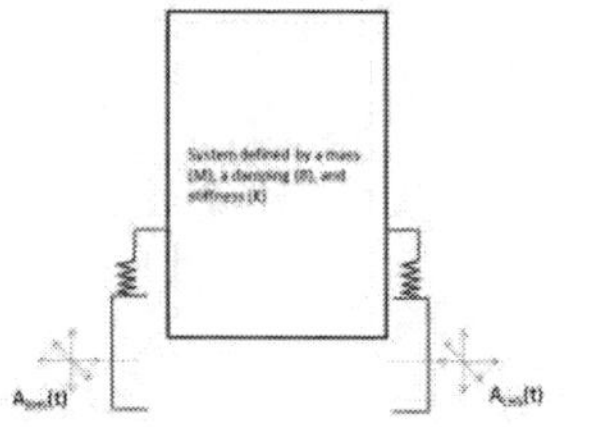

Figure 2 : Model of the cooling package

3.2 Input data

The aim of this paragraph is to present the loads directions and locations. The choice of the linear combinations a* is also described.

3.2.1 The loads

The loads measurements are described as follows.

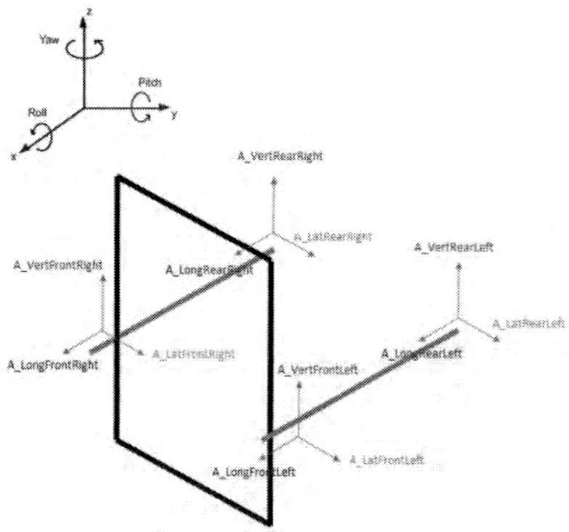

Figure 3: The loads

Accelerations at the attachment points (front) and at the rear of the cooling package are taken into account. Thus, (S) is submitted to 12 acceleration input, as illustrated in Figure 3. Let's consider that the longitudinal loads on the same side, are the same, but they can differ from left to right sides. Lateral accelerations at the left and right side are considered similar as well, but they can differ from front to rear. In this case, the front and rear lateral accelerations are very close: the yaw is mainly due to differences in longitudinal accelerations from left and right sides.

3.2.2 The trucks

Three different trucks will be considered in this study. They are called T27 (6x4 tractor, air suspensions at the front and rear axles), PAVT011 (6x4 tractor, leaf suspensions at the front and rear axles), FH1771 (6x4 tractor, leaf suspensions at the front, air suspensions at the rear). They are modelled in finite elements. The same cooling package is mounted on these three trucks.

3.3 6-DOF approach

For the sake of simplicity, a restricted number of multi-input loads and linear combinations a^* are chosen. The choice is done thanks to the experience of the test engineers and design experts of the cooling package. The mode shapes of the cooling package can be summarized by six movements, 3 translations in X, Y and Z, and 3 rotations, roll, pitch, yaw, as illustrated in the Figure 3. The following linear combinations a* will be considered.

<u>Nomenclature of load names</u>
- A_VFR Vertical acceleration, front right
- A_VFL Vertical acceleration, front left
- A_VRR Vertical acceleration, rear right
- A_VRL Vertical acceleration, rear left
- A_LaFL Lateral acceleration, front left
- A_LoFR Longitudinal acceleration, front right
- A_LoFL Longitudinal acceleration, front left

	PITCH	LAT	LONG
A_VFR	1/2	0	0
A_VFL	1/2	0	0
A_VRR	-1/2	0	0
A_VRL	-1/2	0	0
A_LaFL	0	1	0
A_LoFR	0	0	$(1/2)^{1/2}$
A_LoFL	0	0	$(1/2)^{1/2}$

	YAW	VERT	ROLL.
A_VFR	0	1/2	$(1/2)^{1/2}$
A_VFL	0	1/2	$-(1/2)^{1/2}$
A_VRR	0	1/2	0
A_VRL	0	1/2	0
A_LaFL	0	0	0
A_LoFR	$(1/2)^{1/2}$	0	0
A_LoFL	$-(1/2)^{1/2}$	0	0

In order to compare the trucks regarding their level of severity on (S), these six linear combinations will be analysed. The fatigue damage spectra, also called FDS, (Halfpenny, Investigation of the durability transfer concept for vehicle prognostic applications, 2010) (Halfpenny, Methds for accelerating dynamic durability tests, 2006) are used in this study. This allows comparing them in terms of damage as well as frequencies.

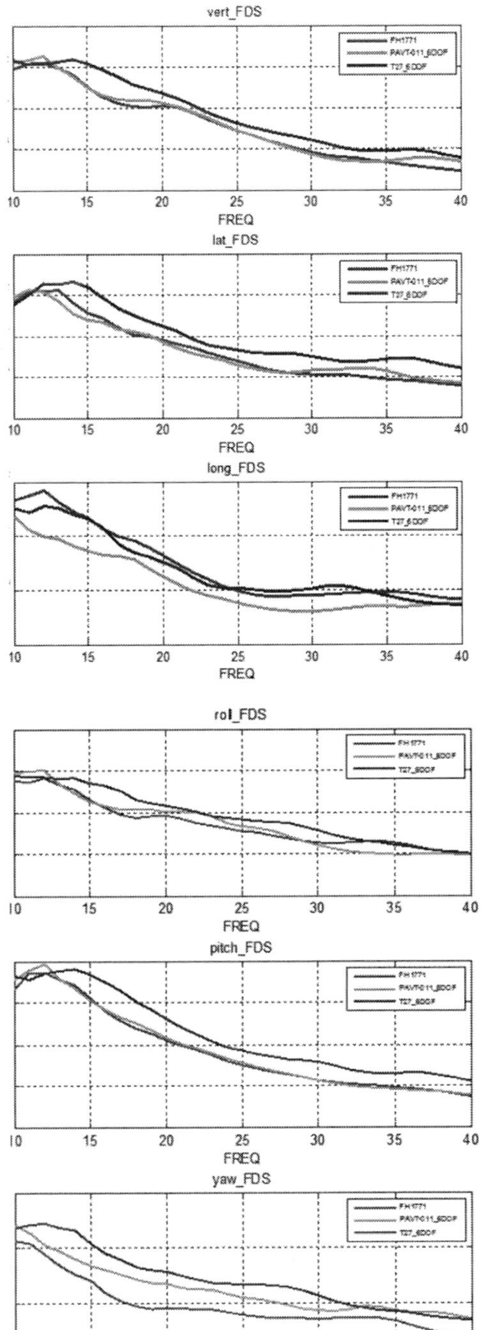

Figure 4: Fatigue damage spectra of the 6 linear combinations

Pitch, Vertical and longitudinal loads combinations are illustrated as examples in

Figure 4. For most frequencies, the truck T27 seems to be more severe, despite for the longitudinal direction. If the damage at the critical point is more influenced by modes with longitudinal mode shape, the choice of the T27 is questioned. In general, this is not the case for the cooling package application.

The advantage of this approach is to refer to mode shapes of the structure, easily identified in a modal analysis of the virtual model. However, the method can be too restrictive as the mode shapes can also be combinations of these 6 movements.

3.4 Unit vectors uniformly spread on the unit hyper sphere,

In this part, the unit vectors are generated from the uniform distribution, according to the Equation (3). As a test, a set of 200 linear combinations have been generated and uniformly spread over the unit hyper sphere. The comparison of the trucks is based on these combinations. Results are expressed in terms of fatigue damage spectra. In Figure 5, only FH1771 and PAVT011 are illustrated, as an example. For the same frequency, a factor of almost 100 in damage can be observed between the 10th and 90th percentile.

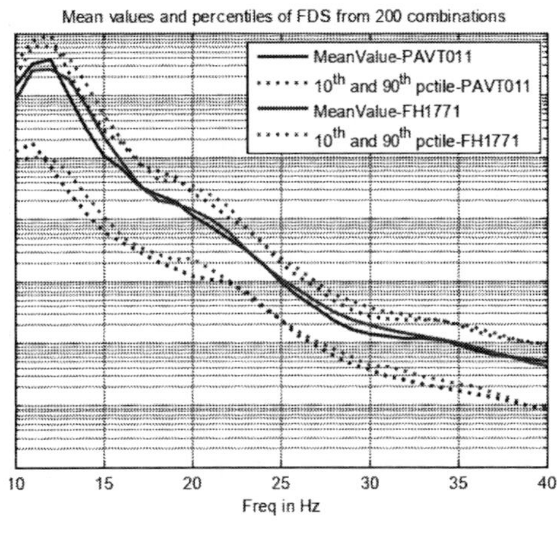

Figure 5 : Scatter of linear combinations of sets of loads from 2 trucks, regarding their fatigue damage spectra

In order to rank the trucks, the relative position of the FDS of each truck, for each combination and frequency, is of interest. For each frequency, we have counted the number of combinations each truck is the most severe.

In the Figure 6, the trucks PAVT011 and T27 are compared. The dark grey part of the histogram represents the number of linear combinations a^* for which the truck T27 is the most severe one. The light grey part represents the number of linear combinations a^* for which the truck PAVT011 is the most severe one. The figures are given in percentage of the total number of linear combinations. In overall, T27 covers 90.2% of the combinations and frequencies when compared to PAVT011. The truck T27 covers 87.9% of the combinations and frequencies when compared with FH1771, as shown in Figure 7.

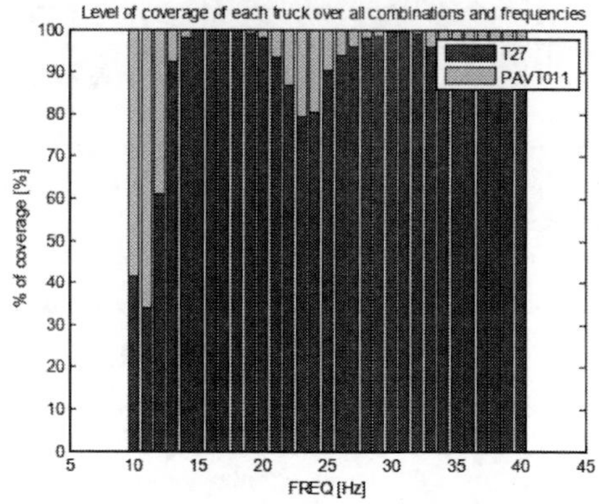

Figure 6 : Damage comparison between T27 and PAVT011

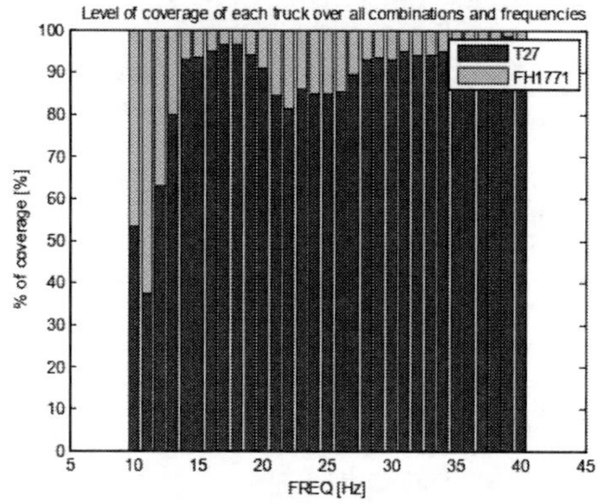

Figure 7 : Damage comparison between T27 and FH1771

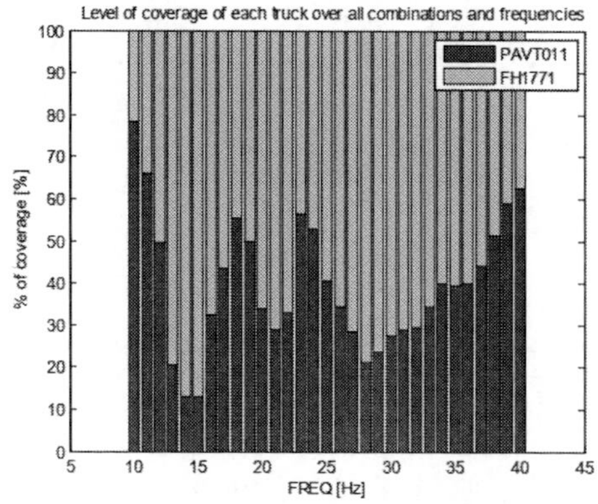

Figure 8 : Damage comparison between PAVT011 and FH1771

In the Figure 8, the same graph is presented for the vehicle FH1771 and PAVT011. The truck FH1771 covers 60.2% of the combinations and frequencies.

By experience, eigen-modes of most of the cooling package installations are higher than 13Hz. Thus, the following conclusion can be drawn. The truck T27 seems to be the most severe truck, followed by both FH1771 and PAVT011.

A special comment can be added regarding the ranking on the FH1771 and PAVT011. In this case, it's more likely to find a cooling package (S) for which the truck PAVT011 would be more severe than FH1771. More information on the cooling package, e.g the impact of each acceleration component on the stress, would help to rank. Another possibility could be to adjust the severity of the loads of the chosen trucks, e.g FH1771, in order to cover more combinations. As these loads are to be used as input of a calculation or a rig, a possibility is to either extend the duration of FH1771 loads application, or increase the amplitude of FH1771 loads (in this last case a calculation is needed to check if the structure doesn't encounter plasticity). If these loads are used as input of durability calculations, another proposal could be to lower the damage criterion when computing damage calculation.

These proposals aim to mitigate the risk that the chosen truck is not the most severe one, among the trucks we have data from.

An extra comment regarding the comparison between FH1771 and PAVT011 can be drawn: it has been observed that, for 73% of linear combinations and frequencies, the damage ratio between the 2 trucks are between 0.5 and 2. It confirms that these 2 trucks are close regarding the damage. It can be hard to distinct them with this method only.

The same exercise has been done with 50 linear combinations, instead of 200. It shows that the truck FH1771 covers 59.8% of the combinations and frequencies, when compared with PAVT011. The truck T27 covers 73.5% of frequencies and combinations when compared with FH1771 and covers 74.5% when compared with PAVT011.

4 Conclusion

A method for storing trucks configurations regarding damage on a common structure has been detailed. Accelerations on same locations in the trucks have been considered. Several hypotheses have been done, in order to express local stresses from accelerations, with limited detailed information on the geometry of the structure and the material.

The main result is a severity indicator, representing the percentage of linear combinations of input loads and frequencies, where a truck is more severe than others. Fatigue damage spectra have been used to compare the loads. This severity indicator is global. In order to reduce the probability that the chosen truck might not be the best choice, a possibility is to increase the severity of the loads of the chosen truck, by increasing the duration of loads, or its amplitude.

The robustness of the severity indicator can be increased with the contribution of FEA analysis, or with the knowledge of the experienced designers and analysts. They are able to give an approximation of the eigenmodes, and, by that, orientate the severity indicator to more precised ranges of frequencies. If the severity indicators of 2 trucks are too close, as it's the case for PAVT011 and FH1771, FEA analysis are eventually helpful to distinct them.

A special care is needed to check if the working hypotheses are valid in the context of the study. This can be done with the help of virtual simulation again. Moreover, the influence of number of linear combinations on the severity indicator can also be investigated.

A perspective is to check the validity of the choice of the most severe truck by measurements. Stress measurements at critical points of a chosen cooling package, on trucks or on rigs, can be helpful to confirm or contradict the choice. It the choice is refuted by measurements, an investigation of the appropriate a* is necessary.

References

Bignonnet, A., & Thomas, J.-J. (2001). Fatigue assessment and reliability in automotive design. *SAE Brazil-International Conference on Fatigue.*

Colin, B. (2016). Caractérisation des processus aléatoires de contraintes Gaussiens et non Gaussiens, en terme de fatigue vibratoire. *Journées de Printemps SF2M*, (p. 15). Paris.

Halfpenny, A. (2006). Methds for accelerating dynamic durability tests. *9th international conference on recent advances in structural dynamics (UK).*

Halfpenny, A. (2010). Investigation of the durability transfer concept for vehicle prognostic applications. *NDIA ground vehicle systems engineering and technology symposium.* US.

Johannesson, P., & Speckert, M. (2014). *Guide to load analysis for durability in vehicle engineering.* Wiley.

Le Corre (Genet), G. (2006). *A statistical approach to multi-input equivalent fatigue loads for the durability of automotive structure.* Göteborg: Chalmers University.

PWR Effect On Crack Initiation Under Equi-biaxial Loading

Cédric Gourdin[1-2,*], Hager Dhahri[1-2], Grégory Perez[1], Stéphan Courtin[3], Jean-Christophe Le Roux[4] and Habibou Matirouman[2]

[1]DEN-Services d'Etudes Mécaniques et Thermiques (SEMT), CEA, University of Paris-Salay, F-91191 Gif-sur-Yvette, France
[2]IMSIA, UMR 9219, CNRS, CEA, EDF, UNIVERSITY OF PARIS-SACLAY, 91762, Palaiseau, Cedex France
[3]FRAMATOME SAS, Tour AREVA F-92084 Paris La Défense, France
[4]EDF, R&D, Site des Renardières, F-77818 Moret sur Loing Cedex, France

Abstract. The lifetime extension of the nuclear power stations is considered as an energy challenge worldwide. That is why, the risk analysis and the study of various effects of different factors that could potentially prevent safe long term operation are necessary. These structures, often of great dimensions, are subjected during their life to complex loading combining varying mechanical loads, multiaxial, with non-zero mean values associated with temperature fluctuations and also PWR environment. Based on more recent fatigue data (including tests at 300°C in air and PWR environment, etc...), some international codes (RCC-M, ASME and others) have proposed and suggested a modification of the austenitic stainless steels fatigue curve combined with a calculation of an environmental penalty factor, namely Fen, which has to be multiplied by the usual fatigue usage factor. The aim of this paper is to present a new device "FABIME2E" developed in the LISN in collaboration with EDF and AREVA. These new tests allow quantifying the effect of PWR environment on disk specimen. This new device combines the structural effect like equi-biaxiality and mean strain and the environmental penalty effect with the use of PWR environment during the fatigue tests.

1 INTRODUCTION

The question of assessing the margins and safety factors in the fatigue analyses which are widely used today (ASME BPV III, RCC-M, JSME, EN-13445-3, etc... [1-4]) is a very challenging one.
The fatigue rules used today in the nuclear industry were initially built and integrated into the ASME code in the 1960's. Establishing fatigue rules is a challenge in itself since fatigue degradation depends on the wear of components which undergo repeated cycling: fatigue tests can therefore be very long and costly, if led on full-size components. As a result, the testing is in practice conducted on small laboratory specimens, which then triggers the question of how to extrapolate results to a full size component. Another difficulty is that the rules need to remain easy to apply in order to be applied for industrial engineering calculations. Since 2007, the USA with the NUREG/CR-6909 [1], have now included the evaluation of environmental effects in their official regulation. Indeed, on the curves presented in Figure 1 and Figure 2, the PWR water environment effect on the fatigue lifetime of material used in the manufacture of reactor components are illustrated.
The 304L and the 316L stainless steels are used for the manufacturing of the pressurized water reactors (PWR). Many components of this type of reactors are subjected to a multiaxial thermo-mechanical cycling [5] and [12]. Therefore, the multiaxial fatigue assisted by environment is considered as one of the possible degradation mechanisms affecting the life of the PWR components.
Based on more recent fatigue data (including tests at 300°C in air and PWR environment, etc...), some international codes (RCC-M, ASME and others) have proposed and suggested a modification of the austenitic stainless steels fatigue curve combined with a calculation of an environmental penalty factor, namely Fen, which has to be multiplied by the usual fatigue usage factor.
Unfortunately, testing on structures representative of real plant components is expensive but should be today increased to help contribute to the general understanding of the various aggravating effects [6-10]. In order to obtain fatigue strength data under structural loading, biaxial test means with and without PWR environment were developed at LISN [11] [13-14].
Two kinds of fatigue devices have been developed. Within the same specimen geometry, structural loads can be applied in varying only the PWR environment. The first device (FABIME2) is devoted to studying the effect of biaxiality and mean strain/stress on the fatigue life. A second and new device named FABIME2e and based on the first device, is for the study of the impact of the environmental effects. With these new experimental results, we will be able to study the interaction between PWR environment and multiaxial loading

* Corresponding author: cedric.gourdin@cea.fr

 https://doi.org/10.1051/matecconf/201816516005

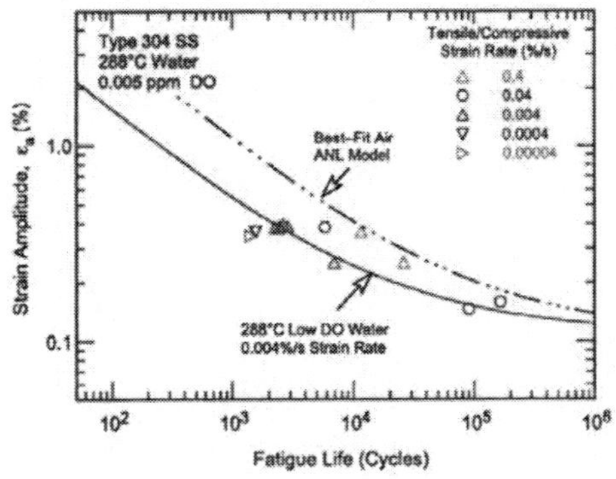

Fig. 1: Fatigue life of 304L steels in PWR water compared with the ANL model Air curve.

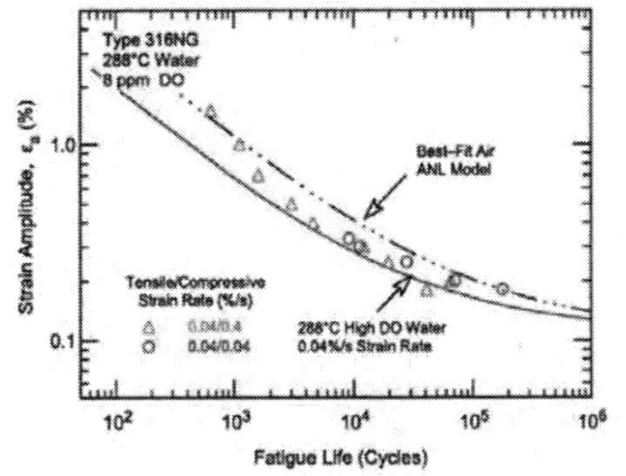

Fig. 2: Fatigue life of 316LN steels in PWR water compared with the ANL model Air curve.

2 THE FIRST EXPERIMENTAL DEVICE

Apply an equibiaxial state loading without mean stress or strain is the objective of this testing. Indeed, we tried to obtain a negative load ratio in order to get the same results as the uniaxial fatigue data and eliminate the residual strain. In this study, equibiaxial state loading generated from spherical bending has been considered. It was used to optimize the geometry of a disk specimen refined in its center. It was used as a circumferentially embedded diaphragm with an applied pressure on both sides in order to obtain an equivalent strain in each loading direction in the plane (Figure 3).

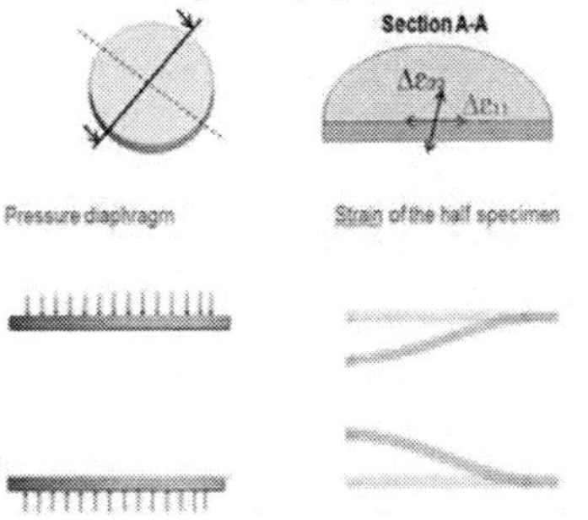

Fig. 3: Principle of the equibiaxial fatigue test.

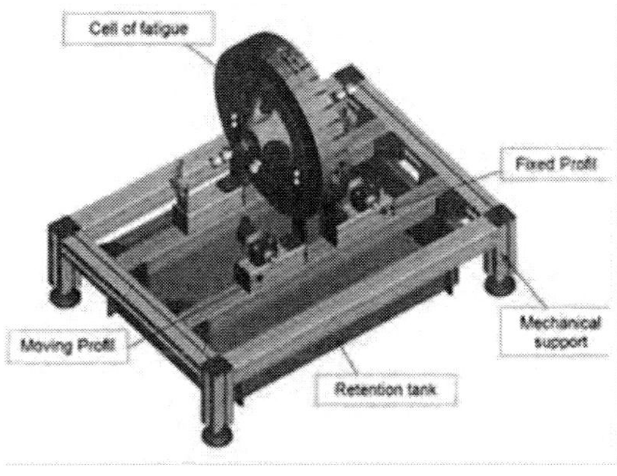

Fig. 4: View of the spherical bending device fatigue cell.

The experimental device called « FABIME2» is divided into four parts:

- Fatigue cell which contains the spherical bending specimen (Figure 4),
- Pressure generating system until 100 bars,
- Electrical enclosure,
- Dedicated CEA in house software developed under LABVIEW that provides control and acquisition data during the tests.

Two half-shells allow the positioning of the spherical bending specimen. Seal and embedment are realized by bolting these two parts.

Maximum experimental conditions are 100 bars for the pressure and 90°C for the temperature. An alternative differential pressure between the two sides of the disk specimen is applied during the fatigue test.

To ensure well-defined experimental conditions, various measuring means are located symmetrically at the two half-shells:

- Pressure sensor with a measuring range between 0 to 100 bars
- Type K thermocouple to measure the temperature of the fluid inside the fatigue cell
- Displacement sensor (LVDT) to measure the deflection at the center of the spherical bending specimen. This sensor has a 5mm range. Realizations of surface observations after the fatigue test show that the contact between LVDT and specimen is negligible (no fretting). No crack initiation is also observed directly under the LVDT.
- Two visualization windows on each half-shell, oriented at 45° with a diameter of 20 mm. The constitutive material is borosilicate glass with a permissible operating pressure of 100 bars.

The fatigue cell was built under European Security directives (Machines 2006/42/CE, Pression 97/23/CE).

3 THE EXPERIMENTAL RESULTS AND INTERPRETATION

Biaxial fatigue tests were carried out on two austenitic stainless steels: "316L THY", and "304L CLI". The first

https://doi.org/10.1051/matecconf/201816516005

material has been provided by Thyssen Krupp Materials France as a 15 mm thickness rolled sheet. The second material supplied by EDF is characterized by a thickness of 30 mm rolled sheet.

An interpretation of the equibiaxial fatigue tests with the definition of the equivalent strain used in the nuclear industry are made. This is an important step to evaluate the impact of an equibiaxial loading on the fatigue life. All tests performed in this study are carried out with imposed displacement with alternating load (displacement load is symmetrical), means value of the displacement is null corresponding to a load ratio R=-1.

To compare the experimental data obtained from uniaxial and equibiaxial tests, it is necessary to define a total equivalent strain. Two definitions of equivalent strain are proposed: the first is based on the definition of von Mises (used in the RCC-MRx) and the second on the definition of Tresca (used in the RCC-M, RSE-M).

Thus, the first equivalent strain used is the von Mises equivalent strain defined by the following equation:

$$\Delta\varepsilon_{eq}^{VM} = \sqrt{\frac{1}{1+\nu'}(\Delta\varepsilon_d : \Delta\varepsilon_d)} = \frac{2}{3}\frac{(1+\nu')}{(1-\nu')}\Delta\varepsilon_1 \qquad (1)$$

with the strain deviatoric component:

$$\Delta\varepsilon_d = \Delta\varepsilon - \frac{1}{3}tr(\Delta\varepsilon) \qquad (2)$$

with ε1 the principal strain and v' the "real" Poisson's ratio (elastic v=0.3 and plastic part v=0.5).

The second equivalent strain is the Tresca equivalent strain defined by the following equation:

$$\Delta\varepsilon_{eq}^{T} = \frac{1}{1+\nu'}Max|\varepsilon_i - \varepsilon_j| \qquad (3)$$

The proposal approach to determine the level of the equivalent strain for each FABIME2 test is as follows:

- Determination of the value of the radial strain corresponding to the imposed deflection from the strain-deflection calibration curve obtained in the paper [14]. With a similar mechanical behavior, the calibration curve can be used for the two materials.
- Determination of the von Mises or Tresca equivalent strain from the relation between the radial strain and the equivalent strain (von Mises or Tresca). This relation has been determined by elasto-plastic calculation of the fatigue test.

This method has been applied to the equi-biaxial fatigue tests presented earlier. The corresponding fatigue life curves are compared to that under uniaxial loading in Figure 5. It appears that there is no impact of equi-biaxial fatigue for the two types of materials, considering von Mises equivalent strains.

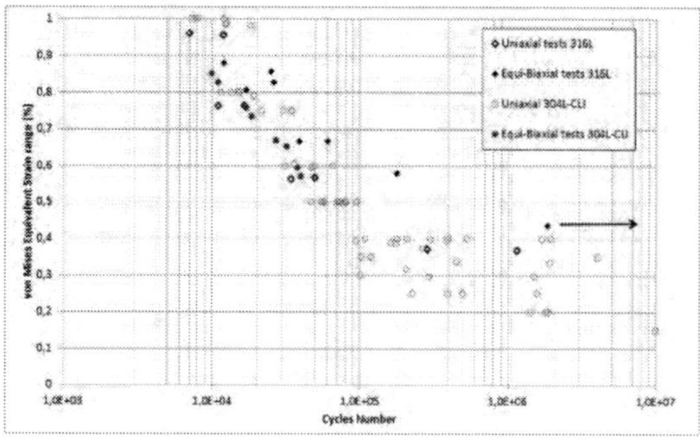

Fig. 5: Fatigue data obtained on the two austenitic stainless steels (316L and 304L-CLI).

4 SPECIFICATION OF THE NEW DEVICE FABIME2E

The second fatigue testing device (FABIME2e, e for environment) has been developed to apply on the same specimen geometry the same structural loads and to only change the testing environment to a PWR environment.

Compared to FABIME2, specification changes for FABIME2e device mainly focused on the following points:

- Specimen is in contact with a PWR environment,
- An operating temperature of 340 °C,
- A maximum pressure of up to 350 bar,
- Monitoring and adjustment of dissolved hydrogen level during testing,
- A perfectly flat and reproductive clamping of the specimen.

With these severe experimental conditions, four major technical difficulties had to be taken into account:

- The cohabitation of the PWR environment with the hydraulic oil at room temperature and 100 bar maximum,
- The PWR environment temperature stability: variations less than 1 °C up to several weeks should be allowed in order to detect the initiation of cracks,
- Monitor and adjust if necessary dissolved hydrogen level,
- The perfect sealing of the device during the tests.

A double cylinder system has been proposed to separate PWR and hydraulic fluids to apply a mechanical loading to the specimen (Figure 6). A double acting cylinder would be moved by the hydraulic unit. Its movement would be mechanically transmitted (by the water incompressibility) to a primary cylinder to modify the volume of the PWR environment contained in each half-shell. Similarly to FABIME2 this system applies a differential pressure, of up to 100 bars, to the specimen. The difference here is that the pressure variation around the specimen is between 150 and 350 bars, respectively the biphasic threshold of the PWR environment and the maximum pressure allowed by FABIME2e.

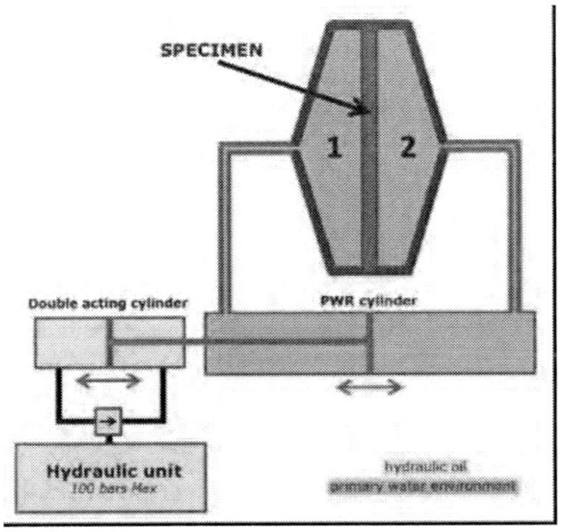

Fig. 6: Double cylinder system for separation of PWR and Hydraulic fluid.

If the required pressure in the PWR environment is obtained by its constraint thermal dilation, this phenomenon must be avoided during the test. A variation of 1 °C would cause a variation of several bars around the specimen. These pressure fluctuations may compromise the detection of the initiation of cracks by compliance. A stronger fluctuation could even lead the PWR environment under its biphasic threshold. This would require a test stop. To avoid this, a temperature regulation system with a great stability was required.
The evolution of the chemical composition of the environment was one of the concerns for this new bench. Hydrogen is the most volatile part of this, so the ability to measure and if necessary adjust its level was needed.
In the end, the requested instrumentation would enable the monitoring of the evolution of the following data, for each half-shell: temperature, pressure, displacement and dissolved hydrogen level.

5 THE NEW EXPERIMENTAL DEVICE: FABIME2E

The manufacturing and commissioning of FABIME2e was ensured by a French company: TOP INDUSTRY. The maximum experimental conditions of the new device are 350 bars and 340 ° C.
The main features of this new device are:

- A cell consisting of two half shells for holding the specimen,
- A clamping system for the cell,
- A sealing system compatible with the PWR environment,
- A closed PWR environment circuit,
- A « double cylinder » system to apply the mechanical solicitations on the specimen,
- An accurate and reliable heating system,
- A system for measuring and adjusting the level of dissolved hydrogen in the PWR environment,

Instrumentation for temperature, pressure, displacement.

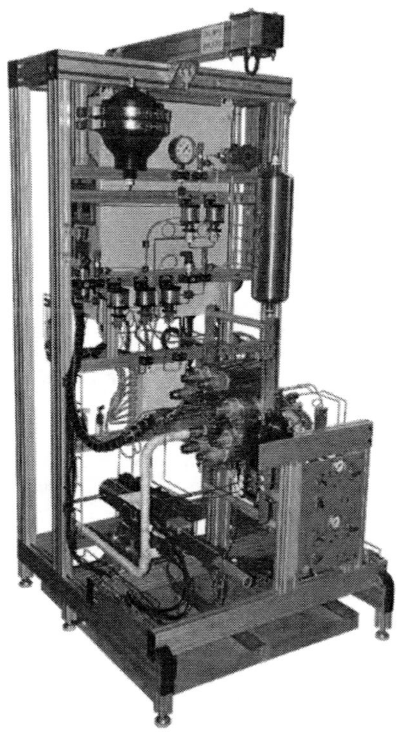

Fig. 7: View of the fatigue new bending device FABIME2e.

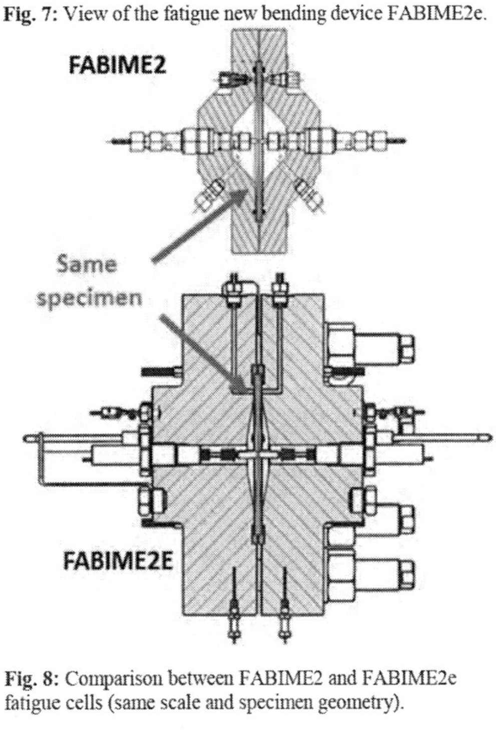

Fig. 8: Comparison between FABIME2 and FABIME2e fatigue cells (same scale and specimen geometry).

Because of much higher pressures and temperatures, the FABIME2e cell has more imposing dimensions than its predecessor FABIME2 (Figure 8). However, the specimen geometry remains absolutely identical to be usable indifferently on the both test benches.

Each half- shell has the following instrumentation:

- Two type K thermocouples, positioned at the top and bottom,
- A pressure sensor with a 0-400 bar range,
- A LVDT compatible with the PWR environment, with a ± 5 mm range to measure the deflection of the specimen,
- Two hydrogen sensors Pd -Ag from AREVA: one for measuring and one for adjusting the dissolved hydrogen level if necessary.

Two metal rings are disposed on each side of the specimen to ensure the sealing of the cell during the test period. Clamping is achieved by means of a hydraulic clamp machine to ensure flatness, sealing and repeatability. Eight heavy section attachment studs ensure the two half shells clamping around the specimen. FABIME2e cell has a 100 ml volume. After filling the cell and high-pressure pipes the heating achieves the desired pressure because of the thwarted thermal dilation of the primary water.

Eight cartridge heaters with an output of 250W each are located on each half-shell to reach the nominal temperature of 340 ° C. The temperature rise is carried out at a maximum rate of 1 ° C / min. The maximum allowable temperature is 400 °C. Although the primary cylinder is at room temperature, a part of the pipes which connects it to each half-shell is also temperature controlled. Two EUROTHERM controllers (NANODAC model) ensure the regulation of the 4 heating zones.

The same hydraulic group provides oil to both FABIME2 and FABIME2e test benches. As shown in Figure 6, a hydraulic cylinder allows transmitting loadings to the primary cylinder to deform the specimen.

As it is possible to do with the bench FABIME2, pressure, displacement or strain control is allowed. The development of the control software in the CEA laboratory allows great flexibility: cycling shape, holds, control mode modifications, mean pressure or strain. The low-level tasks such as security management, hydraulic control and data reading require determinism and speed of processing. That is the reason why they are devolved to real-time autonomous software running on a COMPACT RIO device (NATIONAL INSTRUMENT). The tests management, acquisition and data analysis are performed by software running on a conventional PC. This second software controls each test sequence: from the filling of the PWR fluid untill the crack initiation estimation through sending orders to the CRIO software and the EUROTHERM controllers.

6 FIRST EQUIBIAXIAL FATIGUE TESTS UNDER PWR CONDITIONS

A first series of test was carried out with the FABIME2e device. One of the objectives of these tests was the validation of the concept of the stages separation (oil / primary water) via the intermediate cylinder. The second objective was to demonstrate the criterion for detecting a crack initiation in the centre of the disk specimen.

This series consists of tests with different conditions (pressure and temperature). The environment was composed of distilled water.

Before fatigue tests, a LVDT calibration phase was realized at ambient temperature and at 300°C with a specific specimen. It consists of hollowed specimen out in the centre and contains a zipper with different calibrated height steps (Figure 9).

Fig. 9: Dedicated calibration specimen used to calibrate the LVDT at ambient temperature and at 300°C.

An important software development work has been undertaken to obtain LVDT evolutions without jerk or jump (Figure 10).

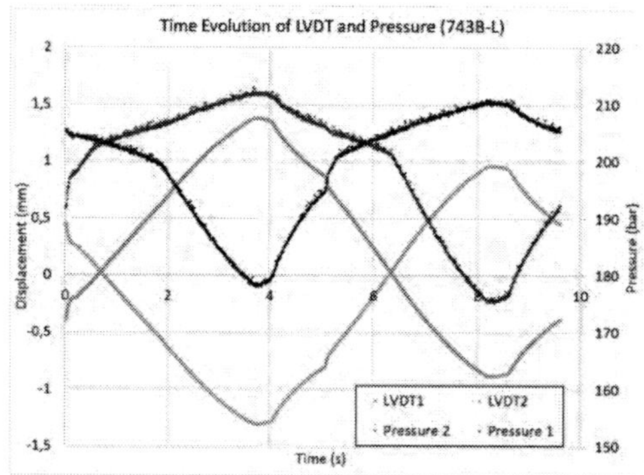

Fig. 10: Time evolution of the displacement (LVDT) and the applied pressure for a spherical fatigue test.

The experimental conditions are: an initial pressure of 50, and 250 barss, an ambient temperature (30 °C.) or 300 °C and an imposed deflection of 1.4 mm, 1.3 mm and 1.2 mm.

The principle of detection of fatigue crack initiation is achieved by studying the evolution of the pressure applied to the fluid via the intermediate cylinder to obtain the imposed deflection. When a "significant" change in this pressure is observed (Figure 11), the test is stopped, the test piece is take off and observed. The crack initiation criterion is defined as having a crack length of 5 mm at the surface on the side 2 (Figure 12), which corresponds to a (downward) variation in the pressure of the order of 1 to 2 bars.

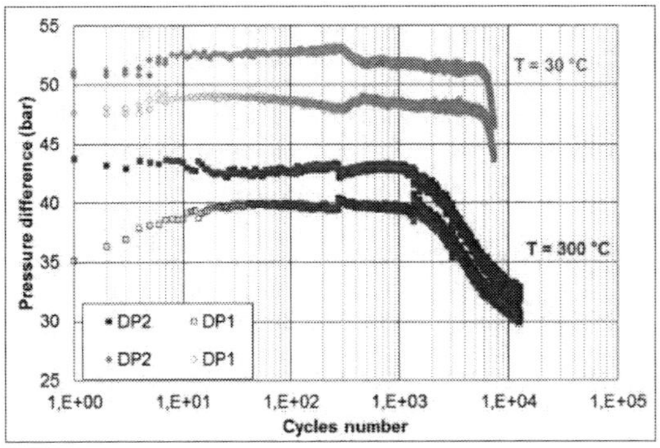

Fig. 11: Evolution of pressure during tests with two temperature conditions (T = 30°C and 300 °C) in distilled water environment.

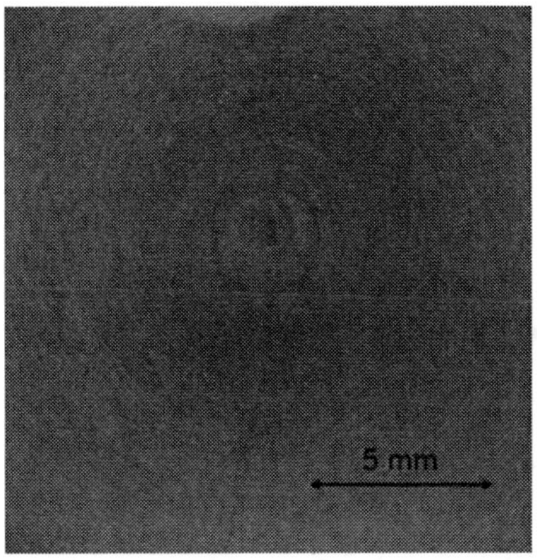

Fig. 12: View of the fatigue crack initiation under PWR conditions on spherical bending specimen.

On the figure 13, the experimental results of spherical bending fatigue tests are represented for different conditions and in distilled water environment. On these first results, the environment effect is highlighted, so for a 5 mm length surface crack, the number of cycles is reduced to a factor 3 (7500 cycles à 30°C to 2500 cycles at 300°C).

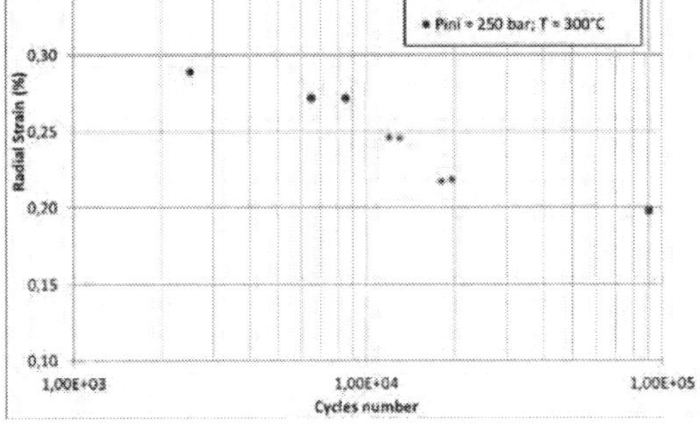

Fig. 13: Experimental data spherical fatigue tests obtained with FABIME2e device for different conditions (temperature, pressure).

7 CONCLUSIONS

This paper is focusing on the description of two kinds of experimental devices to perform fatigue tests on specimen with or without the effect of PWR environment.
The first device (FABIME2) is devoted to study the effect of biaxiality and mean strain/stress on the fatigue life. Biaxial fatigue tests are carried out on two austenitic stainless steels: 316L THY and 304L CLI. The results obtained show that crack initiation under an equibiaxial loads have a low impact on the fatigue life, which remains in the field covered by the design curve defined and used in the codification.
A second and new device named FABIME2e is under development for the study of the impact of the environmental effect. This device will study the impact of the equibiaxial loadings with a primary water environment PWR (300°C with a permanent pressure of 140 bars). The first experimental tests consolidate the conception of this device (FABIME2e).

F. Datcharry and M. Rousseau from CEA is very greatly acknowledged for his participation in the FABIME2E conception.

References

1. Chopra O.K. and Shack W.J. (2007), NUREG/CR-6909, Rev. 0, “Effect of LWR Coolant Environments on the Fatigue Life of Reactor Materials,”.
2. RCC-M (2007) – Design and Construction Rules for mechanical components of nuclear PWR islands – 2007 edition with addenda in 2008, 2009 and 2010.
3. EN-13445-3 V1 standard (2014), “Unfired pressure vessels – Part 3: Design,”.
4. RCC-MRx (2012) « Règles de Conception et de Construction des Matériels Mécaniques des Installations Nucléaires applicables aux structures à haute température et à l'enceinte à vide ITER », AFCEN Code, Association Française pour les Règles de Conception et de Construction des chaudières Électronucléaires. www.afcen.com.
5. Fissolo A. et al. (2009), "Crack Initiation under thermal fatigue: an overview of CEA experience, Part 1: thermal fatigue appears to be more damaging than uniaxial isothermal fatigue", Int. Journal of Fatigue, vol.31(3), p.587-600.
6. Lefebvre D.F. (1989), “Hydrostatic Pressure effect on Life Prediction in Biaxial Low-cycle fatigue”, Biaxial and Multiaxial Fatigue, EGF 3.
7. Itoh T., Sakane M. and al. (2006)," A design procedure for assessing low cycle fatigue life under proportional and non-proportional loading", Int. Journal of Fatigue, 28.
8. Parsons, M. W. Et Pascoe, K. J. (1975), "Development of a biaxial fatigue testing rig", Journal of Strain Analysis, 10, N° 1, p1-3.

9. Shimada H., Shimizu K., Obata M., Chikugo K. & Chiba, M. (1976), "A new biaxial testing machine for the flat specimen and a fundamental study on the shape of the specimen", Technology Reports, Tohoku Univ., 42. N° 2, p351-369.
10. Poncelet M. et al. (2010), "Biaxial High Cycle Fatigue of a type 304L stainless steel: Cyclic strains and crack initiation detection by digital image correlation", European Journal of Mechanics A/Solids.
11. Bradai S. et al. (2013), PVP2013-97200, "Crack Initiation under Equibiaxial Fatigue, Development of a particular Equibiaxial Fatigue Device."
12. De Baglion L. (2006), « Comportement et endommagement en fatigue oligocyclique d'un acier inoxydable austénitique 304L en fonction de l'environnement (vide, air, eau primaire REP) à 300°C », Thèse de l'Ecole Nationale Supérieure de Mécanique et d'Aérotechnique.
13. Bradaï S., Gourdin C. and Gardin C. (2014), "Study of crack propagation under fatigue equibiaxial loading", PVP2014-28417, ASME PVP.
14. Bradaï S., Gourdin C. and al. (2015), "Equi-Biaxial Loading Effect on Austenitic Stainless Steel Fatigue Life", PVP2015-45293, ASME PVP.

An innovative system for uncoupled bending/torsion tests by tri-axis shaker: numerical simulations and experimental results

Davide Zanellati[1*], *Denis* Benasciutti[1], and *Roberto* Tovo[1]

[1]Department of Engineering, University of Ferrara, via Saragat 1, 44122, Ferrara, Italy

Abstract. This work describes a new testing system for applying a coupled/uncoupled bending-torsion loading in vibratory tests by a tri-axis shaker. The system is composed of a cylindrical specimen with eccentric tip masses, excited by horizontal and/or vertical base accelerations. The specimen tip is constrained by a lateral thin and flexible plate which impedes any bending when the specimen is excited horizontally, but which permits the specimen torsional rotation. This layout then allows torsional and bending deformations to be produced and controlled independently, when vertical and horizontal base accelerations are applied simultaneously. A finite element model is first used to estimate the system dynamic response and the stresses in the notched specimen section. The model is then validated through experimental tests under harmonic base accelerations. The strains at clamping system are also monitored to indirectly estimate the bending and torsion moment in the specimen. Comparison of numerical and experimental results showed a close correlation and proved that bending-torsion loading are truly uncoupled. Preliminary fatigue tests with harmonic bending loading (vertical base excitation) are finally compared to the constant amplitude S-N curve, showing a quite satisfactory agreement.

1 Introduction

Mechanical systems and structures are often subjected to vibratory excitations, which generally induce a local multiaxial random stress. In the structural durability assessment, besides the use of theoretical and numerical methods implementing uniaxial/multiaxial criteria (often defined in the frequency-domain [1-5]), it is customary to use experimental tests to directly estimate the material strength against multiaxial loading. The interest is often restricted to bending-torsion random loading, as they are very common in most engineering applications.

Laboratory experiments with multiaxial loading are normally performed in two different ways: by servo-hydraulic machines that impose a force and/or torque (testing frequencies are relatively low), or by shaker tables that apply a base vibratory excitation (thus allowing for higher testing frequencies).

Over the years, a number of testing methodologies that differ in terms of testing machines, specimen geometry and type of excitation, have been proposed [6-10]. While servo-hydraulic machines normally adopt cylindrical specimens, vibratory tests by shakers tables make use of a wider set of specimen configurations, ranging from simple (plate, cylindrical) to more elaborated layouts (Y-geometry), as already commented in [11]. In vibratory testing, the specimen is fixed to the shaker table and excited by a base acceleration.

For what concerns, in particular, tests with a biaxial bending-torsion loading, both testing methodologies offer advantages and disadvantages. While servo-hydraulic machines permit a great flexibility in controlling both the intensity and phase shift of bending and/or torsion loading, shaker tables seem to have greater limitations. For example, in most of vibratory test configurations that use a uniaxial shaker, only a uniaxial stress can be obtained (bending loading). A bending-torsion loading can only be obtained (at a greater cost) by coupling two uniaxial shakers controlled independently [8].

A simple way to obtain a bending-torsion loading by a uniaxial shaker is the use of a cantilever specimen with eccentric mass at the free extremity, as suggested in [10]. In fact, the eccentricity of the tip masses induces a torsion loading (combined to bending), even when the specimen is excited by uniaxial vertical acceleration. The intensity of bending loading is controlled by the mass weight, the intensity of torsion loading is determined by the degree of eccentricity. While a zero eccentricity would give only bending loading (without torsion), the opposite would not be possible (torsion loading is always coupled to bending). Therefore, this configuration cannot produce a torsion loading alone. The use of a tri-axis shaker, which can apply three independent excitations

* Corresponding author: davide.zanellati@unife.it

along three orthogonal directions, seems to open the possibility to overcome this limitation. Inspired by the layout proposed in [10], a new testing system has been proposed, which allowed torsion and bending loading to be controlled independently in a vibratory test with a tri-axial shaker [11]. The purpose of this work is to describe this innovative testing system, to present results from a finite element model and to validate them against experiments, and finally to show preliminary fatigue test results under a harmonic bending loading.

2 Layout of innovative testing system:

The innovative testing system layout (Fig.1) is composed of a T-clamping system, a lateral thin plate and a U-notched cylindrical specimen in S355JR steel. The diameters of gross and net cross sections are 20 and 8 mm, respectively. The notch radius of 2 mm gives a stress concentration factor $K_{t,b}$=1.45 for bending and $K_{t,t}$=1.27 for torsion [12]. A cantilever beam with two equal tip masses is mounted at the specimen free extremity. The overall weight of the system is 4.65 kg.

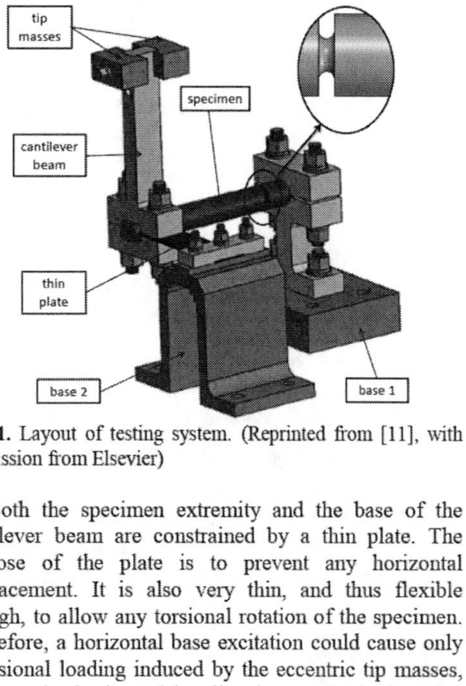

Fig. 1. Layout of testing system. (Reprinted from [11], with permission from Elsevier)

Both the specimen extremity and the base of the cantilever beam are constrained by a thin plate. The purpose of the plate is to prevent any horizontal displacement. It is also very thin, and thus flexible enough, to allow any torsional rotation of the specimen. Therefore, a horizontal base excitation could cause only a torsional loading induced by the eccentric tip masses, whereas the horizontal bending is completely impeded by the thin plate. A vertical excitation, instead, could induce only a bending loading on the vertical plane, without any torsional rotation.

In order to verify this system behaviour, a finite element model was first developed.

3 Finite element model

Through a modal and harmonic analysis, the finite element model aims to prove that the innovative testing layout in Fig. 1 allows bending and torsion loading to be fully uncoupled under a horizontal and vertical base acceleration.

The numerical model assumes a isotropic linear elastic material, with density 8027 kg/m^3, Young's modulus 200 GPa and Poisson's ratio 0.3, typical values of a S355JR steel.

A structured mesh was used everywhere, except on the base 1 due to its complex geometry (Fig. 2). The average element size in the notch is less than 0.5 mm, instead it is 1 mm elsewhere. A 8-node solid finite element, with three degrees of freedom for each node, was adopted. The nodes lying on the bottom surfaces of the clamped base (base 1 and base 2 in Fig. 1) were fully constrained.

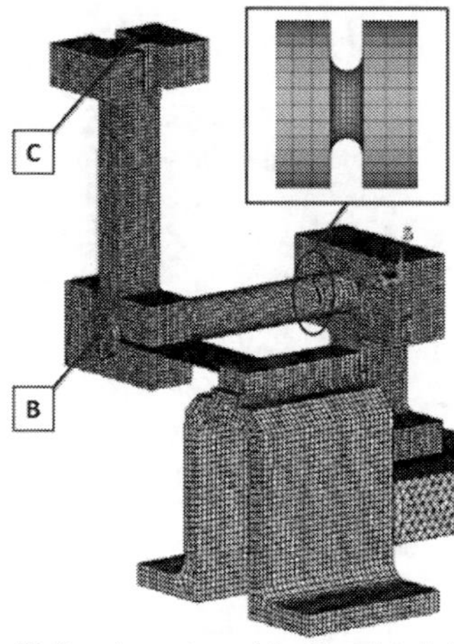

Fig. 2. Finite element model. (Reprinted from [11], with permission from Elsevier).

A modal analysis was performed to identify the first two resonance frequencies of the system; the first bending resonance is at 78.7 Hz, the first torsion resonance occurs at 110 Hz. The corresponding modal shapes are displayed in Fig. 3. The other resonances at higher frequencies were not considered, as they refer to modal shapes of the whole system (not only the specimen).

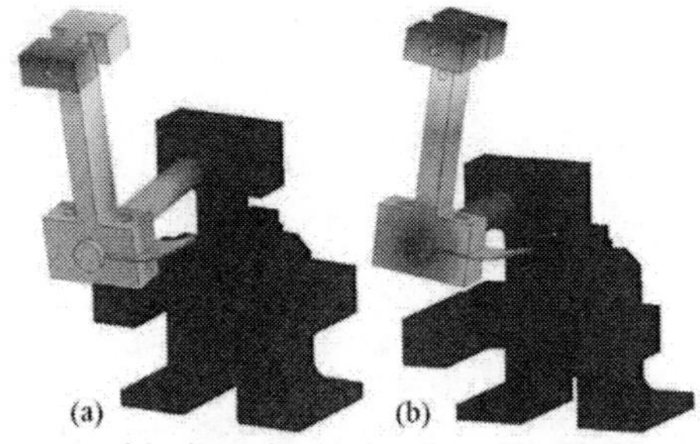

Fig. 3. Modal shape in (a) bending and (b) in torsion. (Reprinted from [11], with permission from Elsevier).

Two harmonic analyses were also performed to determine the frequency response function (FRF) of the system under a vertical or horizontal harmonic acceleration, respectively, which induce a pure bending or torsion loading in the specimen. A harmonic acceleration with 1g of amplitude and frequency range from 20 to 150 Hz was applied as an enforced motion at the base. The acceleration response (vertical and horizontal) was determined at points B and C (Fig. 2).

In the first harmonic analysis, a vertical acceleration was applied to produce a pure bending loading and a corresponding normal stress in the notched section of the specimen. Fig. 4 shows, with a solid line, the FRF in terms of vertical acceleration at the specimen extremity (point B in Fig. 2).

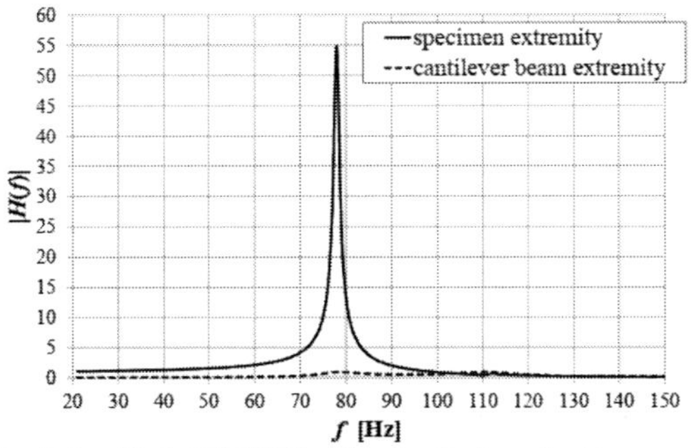

Fig. 4. Amplitude FRF |H(f)| under a vertical acceleration.

The typical resonant behaviour is observed. At low frequencies, the system response is close to one (quasi-static behaviour), whereas when approaching the first resonance frequency, as expected, the system response is greatly magnified. At frequencies higher than resonance, the FRF decreases. Fig. 4 also displays, with a dashed line, the FRF in terms of horizontal acceleration of the cantilever extremity (point C in Fig. 2), which is an indirect measure of the torsional deformation. In this case, the FRF is close to zero at low frequencies and exhibits a little amplification at the two resonances. At frequencies higher than resonance, the system response returns to zero. The two small peaks in Fig. 4 can be explained by the action of the thin plate attached to the specimen: when the specimen bends vertically, the plate cannot elongate and therefore it restraints the specimen, which is then forced to slightly rotate and to produce a corresponding horizontal displacement of point C.

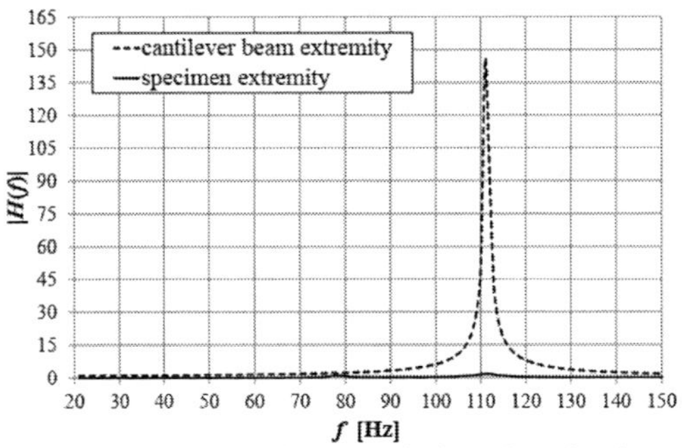

Fig. 5. Amplitude FRF |H(f)| under a horizontal acceleration.

In the second harmonic analysis, a horizontal base acceleration was applied to induce a torsional loading and a corresponding shear stress in the notched section. Fig. 5 shows (dashed line) the amplitude FRF in terms of horizontal acceleration of the cantilever extremity (point C). The trend is the same as that shown previously: it is close to unity at low frequencies and it exhibits a peak around the first torsion resonance. The solid line is the FRF in terms of vertical acceleration of specimen extremity (point B). Here too the FRF is close to zero at low frequencies and displays two peaks around the first two resonances (hardly detectable in Fig. 5), then it becomes almost zero at higher frequencies.

The trends shown in Fig. 4 and 5 point out that the bending and torsional response of the system, undergoing a vertical and horizontal base acceleration, are fully uncoupled (or at least only weakly coupled), in the frequency range far from the resonance.

The finite element analysis also allows two types of information to be obtained: i) an evaluation of the acceleration at shaker table, necessary to break the specimen; ii) an estimate of the strains at the T-clamping system, which are correlated to bending/torsion moments and to the notch stresses. It is very important to know this information, as it ensures that laboratory tests are properly executed.

Since the purpose of the testing system here presented is to carry out fatigue tests in the high-cycle regime, it is required to estimate the acceleration values that lead to specimen failure at $5{\cdot}10^4$ and $2{\cdot}10^6$ cycles (taken as lower and upper limits of life fatigue).

Constant amplitude fatigue tests were preliminary carried out to assess the S-N curve of plain material. Cylindrical smooth specimens were submitted to a tension-compression fully-reversed (R=-1) loading applied by a MTS machine. Regression analysis of experimental results returned the S-N line (Fig. 6), with a reference amplitude strength $\sigma_{A0,-1}$=324.1 MPa (at $2{\cdot}10^6$ cycles) and an inverse slope of S-N curve k=14.78.

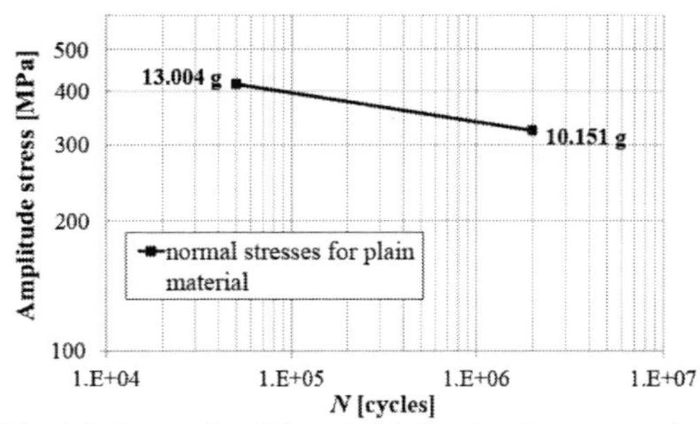

Fig. 6. S-N curve for plain material, from tension-compression tests (MTS).

A reference excitation frequency of 5 Hz was chosen to calculate the input acceleration that needs to be applied by the shaker in fatigue tests. Since the excitation is far from the bending resonance, the intensity of the response acceleration in every point of the testing system (also including the cantilever beam extremity) will be almost equal to the input acceleration applied by the shaker table. In Fig. 6, the numbers close to the S-N line represent the required vertical base acceleration to have for failure at $5{\cdot}10^4$ and $2{\cdot}10^6$ cycles under bending loading.

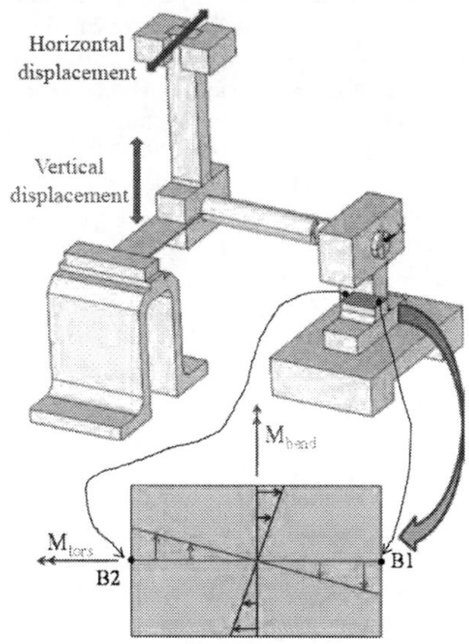

Fig. 7. Strain distribution in the T-clamp, when the specimen is subjected to bending and torsion loading.

While the graph in Fig. 6 gives information about the necessary accelerations to be applied by the shaker, before performing the laboratory tests it is also important to know the strain values in the T-clamping system. In experiments, such strains are measured and used to provide an indirect estimate of the stresses in the notched section, which being so narrow prevents any direct measure by strain gauges. In fact, the strains at the T-clamping system are directly correlated to the bending and torsion moments acting at the specimen fixed end. When the specimen bends vertically, the T-clamp will bend vertically as well. Instead, when the specimen rotates under a torsion loading, the T-clamp will bend in the plane orthogonal to the specimen axis, as sketched in Fig. 7.

Table 1. Results from finite element analysis.

No. cycles	Input acceleration [*g*]	Strain [με]	peak stress σ_{peak} [MPa]
$5 \cdot 10^4$	13.004	34.730	415.2
$2 \cdot 10^6$	10.151	27.115	324.1

When the finite element model is subjected to a vertical acceleration (specimen under bending loading), the simulation returns the values of strains listed in Table 1, calculated as the average values between points B1 and B2.

Table 1 summarises the estimated values of vertical base acceleration, and the corresponding strain and peak stress calculated by simulation, which are estimated to have the specimen failure at $5 \cdot 10^4$ and $2 \cdot 10^6$ cycles.

4 Experimental fatigue tests

Experimental measures are used for validating the finite element model and for proving that the system, when excited by a tri-axis shaker, is really capable to apply (and to control independently) a bending and/or torsion loading .

A Dongling tri-axis electrodynamic shaker (3ES-10-HF-500 model) was used to perform vibration tests (Fig. 8(a)). This brand new machine is capable to apply simultaneous accelerations along 3 orthogonal directions in space, in a frequency range from 5 to 2000 Hz, with a maximum rated acceleration of 9.0*g* and a maximum rated velocity of 1.2 m/s. The controller and acquisition unit is a LMS SCADAS Mobile, driven by LMS Test.Lab software.

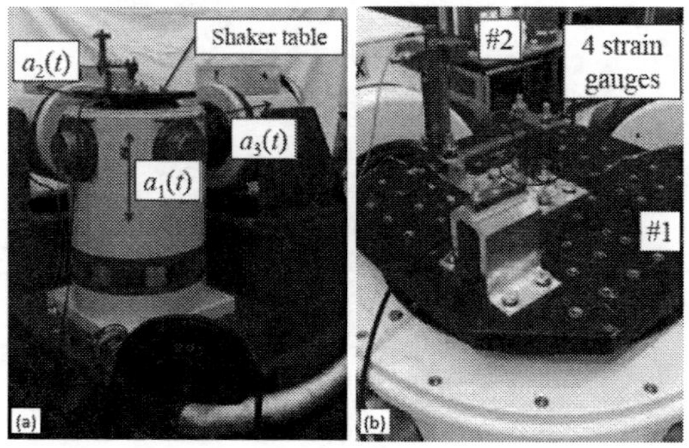

Fig. 8. (a) Dongling tri-axis electrodynamic shaker; (b) testing system prototype mounted to the shaker table.

The shaker is not capable to excite only one single or two axes (channels) at a time, keeping the others at rest. All three axes (channels) must always be activated simultaneously, which could be a limitation if tests with two, or even only one single acceleration, have to be performed. Nevertheless, this restriction can easily be overcome if the excitation level of the "secondary" channel/channels is set to a much lower level than the "primary" one. For example, to apply a vertical acceleration of 1g, the other two channels may be set at 0.1*g*.

A prototype of the new system layout, described in Section 2, is mounted to the shaker head with screws and nuts, as shown in Fig. 8(b). The measurement instruments consist of a tri-axis accelerometer (#1) placed on shaker table in order to measure the input accelerations in closed loop with the control system, and another accelerometer (#2) for monitoring the system response at the extremity of cantilever beam (point C in Fig. 2).

Four strain gauges (with half-bridge connection) have been positioned on the faces of the T-clamp (Fig. 8(b)), because their application on the notch (2 mm of radius) was clearly not feasible.

Preliminary harmonic tests with sinusoidal acceleration were carried out with the aim to perform bending fatigue tests with constant amplitude stress. Before starting the tests, it was noticed that the acceleration values calculated by the numerical model, and required to produce failures while exciting at a frequency far below the bending resonance (78.7 Hz), were actually too higher than the threshold (9.0g) imposed by the shaker available in the laboratory.

Some amplification of the input acceleration then becomes necessary. This can be achieved by exploiting the amplification in the FRF (see Fig. 4), which occurs also at frequency below the resonance. To this purpose, a harmonic analysis was performed to evaluate with greater precision the amplitude FRF of the system under bending. In this case, three harmonic (sinusoidal) accelerations with frequency varying from 20 to 150 Hz (in steps of 0.05 Hz) were applied to the shaker table. The acceleration $a_1(t)$ is along the vertical direction and has amplitude 1g, whereas the other two horizontal accelerations $a_2(t)$, $a_3(t)$ have a amplitude significantly smaller, equal to 0.1g.

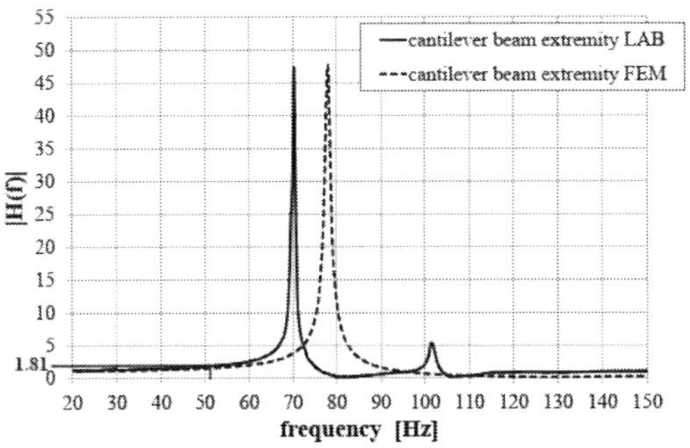

Fig. 9. Numerical (FEM) and experimental (LAB) amplitude FRF |H(f)| under vertical acceleration.

Fig. 9 shows the comparison between numerical simulation and experiments for the FRF relative to base acceleration for the output vertical acceleration at the cantilever beam extremity (accelerometer #2, see Fig. 8(b)). The observed difference can be explained by considering that in the finite element model (Fig. 2) the assembled parts are considered to be perfectly bonded (with no joints).

Considering that the acceleration threshold allowed by the shaker is 9.0g and that, by contrast, the finite element analysis estimated a maximum acceleration of 13.004g to have failures around $5 \cdot 10^4$ cycles (see Fig. 6), the required amplification has to be about 1.44. To keep a safety margin, this requirement can be obtained by choosing a working frequency around 51 Hz, at which the amplification is 1.81 (see Fig. 9). At this frequency, the FRF is reasonably flat and the system acceleration response, being not so sensitive to small changes in the excitation frequency, is easier to be controlled. By contrast, a more precise Input/Output control would be necessary to keep the vibrating system close to the resonance.

With these premises, fatigue tests have been carried out with a vertical harmonic base excitation of varying amplitude, whose intensity is greater than the other two horizontal accelerations (0.1g amplitude).

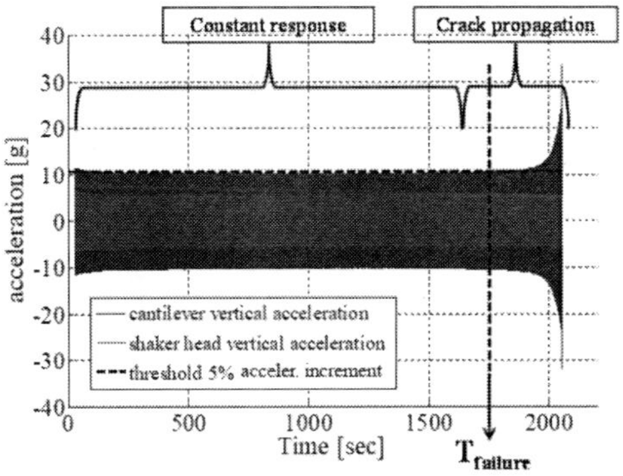

Fig. 10. Fatigue behavior under harmonic excitations.

A typical behaviour observed in this type of test is shown in Fig. 10. A first long part at constant acceleration response is followed by an increase of acceleration due to crack propagation, after which failure occurs very quickly. In the very last part of the test, significantly smaller than the first one, the crack propagation reduces the specimen stiffness, so that the system resonant frequency decreases until it reaches the frequency of the excitation. As a failure criterion, it was established to assume a 5% increment of the measured acceleration, compared to the sector in which the acceleration remains almost constant, see Fig. 10.

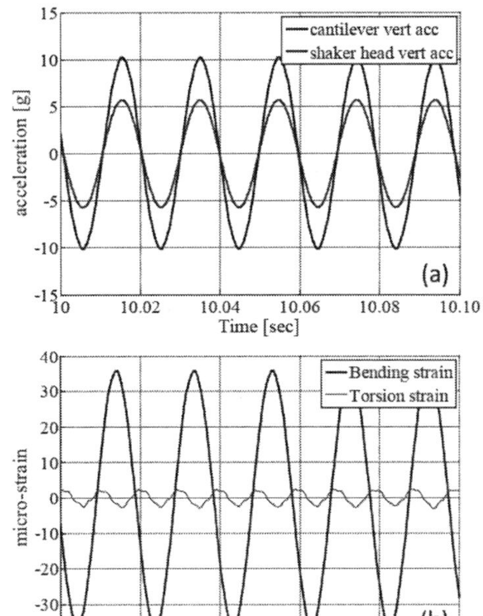

Fig. 11. (a) Acceleration measures and (b) strain measures under harmonic excitations.

Fig. 11(a) also confirms that, if the input acceleration is harmonic, the response acceleration is harmonic too. Fig. 11(b) displays, instead, the time-histories of strain measured at the T-clamp. As can be seen, the strain caused by the bending deformation is largely predominant to that caused by torsion loading (which yet is not zero, as the "secondary" channel for horizontal acceleration cannot be deactivated at all).

Four fatigue tests under harmonic bending have been carried out at different excitation (stress) levels. Results are summarised in Table 2, which also list information about the vertical acceleration measured with the accelerometers #1 and #2, and the corresponding amplitude FRF. Also the averaged strain measures at the T-clamp and the number of cycles to failure are reported. The last column lists the values of peak normal stress in the notched specimen section, which are calculated numerically using as input the measured values of strain.

Finally, Fig. 12 collects fatigue test results obtained with harmonic excitation and with tension-compression loading (MTS testing machine). A close agreement can be clearly appreciated.

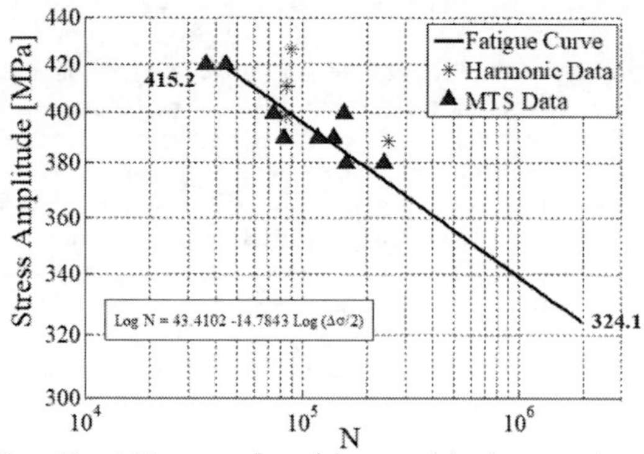

Fig. 12. S-N curve for plain material, from tension-compression tests (MTS) and shaker tests (harmonic data)

Table 2. Results from fatigue tests under harmonic bending.

Test No.	Meas. input acceler. (point #1) [g]	|H(f)|	Meas. vertical acceler. (point #2) [g]	Average strain at B1/B2 [με]	Cycles to failure	Peak stress σ_{peak} [MPa]
1	5.623	1.819	10.230	35.670	89295	426
2	5.459	1.805	9.852	34.353	85238	411
3	5.248	1.824	9.571	33.374	83622	399
4	5.101	1.827	9.318	32.492	249432	388

5 Conclusions

This work presented a new testing system to carry out bending-torsion vibratory tests by a tri-axis shaker. The testing system is composed of a cylindrical specimen that can be excited by three orthogonal base accelerations. An eccentric tip mass is used to produce a torsion loading under a horizontal acceleration, while thin plate attached to the specimen extremity prevents a horizontal bending when torsion is activated. A vertical acceleration, instead, produces a pure bending loading, without torsion. The phase shift between bending and torsion is thus directly proportional to the correlation among base accelerations.

A finite element model was first used to investigate the system dynamic response and to assess the local stress in the notched section of the specimen. A modal analysis allowed the first resonant frequencies in bending and torsion to be identified. A harmonic analysis was next performed by imposing a vertical and horizontal base accelerations in order to quantify the output accelerations in the specimen and the corresponding frequency response function of the system. The observed dynamic behaviour confirmed that bending and torsion deformation are truly uncoupled. The numerical results were also used to establish the excitation frequency to be used in experimental fatigue tests.

Preliminary experimental measures under a harmonic base acceleration, used to validate the results from the finite element model, confirmed the expected behaviour of the system (uncoupled bending-torsion). Four strain gauges attached to the specimen fixture were also used to monitor the strains at the clamping system, which were used to get an indirect measure of bending and torsion moments, and of the corresponding stress, in the specimen. The comparison between numerical and experimental results showed an overall fairly good correlation, thus confirming that bending and torsion loading can be controlled independently.

The works concluded by illustrating few results of constant amplitude fatigue tests obtained by a harmonic (sinusoidal) vertical base acceleration (pure bending loading). These results were in close agreement to the constant amplitude S-N curve for plain material, previously estimated by a tension-compression constant amplitude-reversed loading.

References

1. D. Benasciutti, R. Tovo, Int. J. Fatigue **27(8)**, 867 (2005)
2. D. Benasciutti, R. Tovo, Probab. Eng. Mech. **21(4)**, 287 (2006)
3. D. Benasciutti, J. Sound Vib. **333(18)**, 4326 (2014)

4. D. Benasciutti, F. Sherratt, A. Cristofori, Int. J. Fatigue **91(2)**, 397 (2016)
5. D. Benasciutti, R. Tovo, Mat.-wiss. u. Werkstofftech. **49(3)**, 345 (2018)
6. T. George, Int. Journal of Fatigue **26(5)**, 477 (2004)
7. C. Ghielmetti, R. Ghelichi, M. Guagliano, F. Ripamonti, S. Vezzù, Procedia Engineering 10, 2892 (2011)
8. T. Łagoda, E. Macha, A. Niesłony, Fatigue Fract. Eng. Mater. Struct. 28(4), 409–420 (2005)
9. M. Česnik, J. Slavić, M. Boltežar, J. Sound Vib. 331(24), 5370 (2012)
10. N. Nguyen, M. Bacher-Höchst, C.M. Sonsino, Mat.-wiss. u. Werkstofftech. **42(10)**, 904 (2011)
11. D. Zanellati, D. Benasciutti, R. Tovo, Proc. Struct. Integrity **8**, 92 (2018)
12. W. Pilkey, D. Pilkey, R. Peterson, *Peterson's stress concentration factors* (John Wiley & Sons, Hoboken, NJ, 2008)

Multiaxial fatigue study on steel transversal attachments under constant amplitude proportional and non-proportional loadings

Martin Garcia[1,*], *Claudio A.* Pereira Baptista[2], and *Alain* Nussbaumer[1]

[1]EPFL - Swiss Federal Institute of Technology, Resilient Steel Structure Laboratory (RESSLab), 1015 Lausanne, Switzerland
[2]GRID International - Consulting Engineers, av. João Crisóstomo, 1050-125 Lisboa, Portugal

Abstract. In this study, the multiaxial fatigue strength of full-scale transversal attachment is assessed and compared to original experimental results and others found in the literature. Mild strength S235JR steel is used and an exploratory investigation on the use of high strength S690QL steel and the effect of non-proportional loading is presented. The study focuses on non-load carrying fillet welds as commonly used in bridge design and more generally between main girders and struts. The experimental program includes 33 uniaxial and multiaxial fatigue tests and was partially carried out on a new multiaxial setup that allows proportional and non-proportional tests in a typical welded detail. The fatigue life is then compared with estimations obtained from local approaches with the help of 3D finite element models. The multiaxial fatigue life assessment with some of the well-known local approaches is shown to be suited to the analysis under multiaxial stress states. The accuracy of each models and approaches is compared to the experimental values considering all the previously cited parameters.

Introduction

In bridge design, welded joints submitted to multiaxial stresses interaction are generally fillet welds that can be classified in two types, i.e. load carrying and non-load carrying fillet welds as shown in Fig. 1.

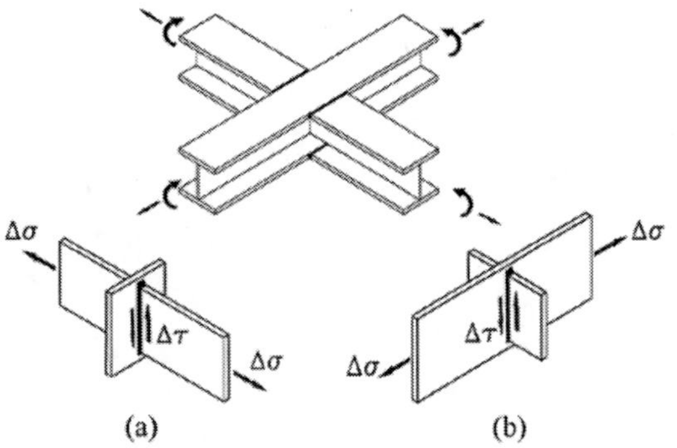

Fig. 1. Multiaxial normal-shear stress interaction in a typical bridge detail according to Baptista [1]

Detail (a) is relatively well reported in literature. It is a cruciform joint (CJ) characterized by the fact that the main tensile or bending efforts are carried by the weld, leading to a higher variability in the critical location under multiaxial stresses, i.e. the weld throat or toe, depending mainly on the size and penetration of the fillet weld. For standard fillet weld, failure usually initiates at the weld root with cracks developing into the throat section [2, 3]. On the contrary, detail (b) represents the so-called transversal attachment (TA), generally seen between for example main girders and secondary elements. It consists in a continuous plate subjected to a main tensile or bending stress, with transversal shear carrying fillet welds on its surface. Indeed, transversal attachments are usually loaded in shear, meaning that the weld toe on the main plate will experience an additional shear stress interacting with the main normal stress. No fatigue tests on this type of detail under multiaxial proportional or non-proportional loads were found in the literature, for any steel grade. Both details are considered in this study.

1 Experimental campaign

The experimental program was mainly carried out on a new multiaxial setup that allows for proportional and non-proportional tests. It includes 33 fatigue tests under uniaxial and multiaxial stress on details CJ (a) and TA (b) according to the matrix of experiments in Table 1. The two details are considered, depending on whether the tests are carried out under uniaxial normal stresses or multiaxial stresses. Effects related to the use of high strength steel are explored on the multiaxial TA detail. Six specimens were tested under uniaxial normal stress on detail CJ and combined to data acquired in a literature review previously done by one of the authors (Baptista [1] in the Appendix A4) to allow for the FAT classification of the detail. The classification in shear, due to some limitation of the setup (see *1.2.2*), is only based on a literature data review (Baptista [1] in the Appendix A6). A total of 27 tests were done on detail (b) with 17 specimens tested proportionally and 10 non-proportionally with a 180° out-of-phase shift (the influence of the phase shift will be studied in a subsequent task). Specimens failed at the weld toe, both under

* Corresponding author: martin.garcia@epfl.ch

Table 1. Matrix of experiments

Steel grade	ΔF_{hor} [kN]	ΔF_{vert} [kN]	$\Delta\sigma$ [MPa]	$\Delta\tau$ [MPa]	R_σ	$\lambda_{\tau/\sigma}$	Phase [°]	# of tests	Detail
S235JR	180	-	180	-	0.1	-	-	1	CP 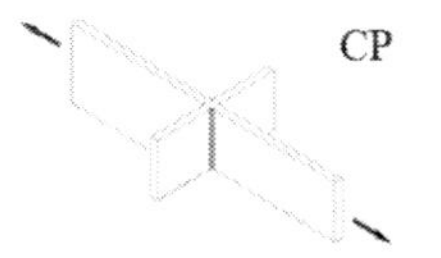
	150	-	150	-	0.1			3	
	120	-	120	-	0.1	-	-	1	
	100	-	100	-	0.1	-	-	1	
S235JR	0	100	54	50	0.1	0.93	0	3	TA 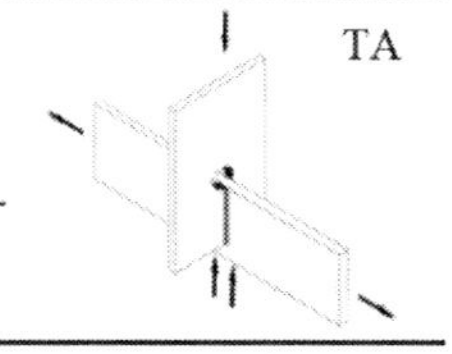
	100	100	154	50	0.1	0.32	0	8	
	100	100	46	50	0.6	1.10	180	10	
S690QL	100	100	154	50	0.1	0.32	0	4	
	38	117	104	60	0.1	0.58	0	2	

uniaxial, multiaxial, proportional and non-proportional stress state.

In order to be consistent with usual bridge state of stress in specific details and because of setup limitations, a unique stress ratio $R = S_{min}/S_{\max}$ of 0.1 is used except in the case of multiaxial non-proportional loading, where an initially unaccounted additional tensile stress (see *2.1*) increased the R ratio to 0.6 . In order to emphasize the detrimental influence of the shear stress on the normal uniaxial fatigue life, the shear to normal stress ratio λ is comprised between 0.32, which correspond to a more realistic bridge loading situation, and 1.1.

1.1 Experimental setup

All the S235JR mild steel specimens were produced by C. Baptista, who is a former welder, at the EPFL-ICOM laboratory with commercial flat bars, whereas all S690QL high strength quenched and tempered steel specimens where fabricated by a steel fabricator. The specimen geometry consists of two orthogonally welded plates loaded in their planes as shown in Fig. 2.

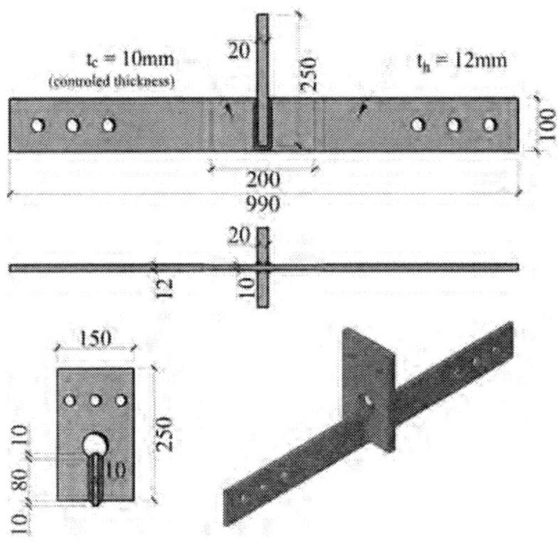

Fig. 2. Transversal attachment (TA) specimen geometry

The longitudinal plate is continuous and has a section of 100*12mm while the second plate is a rigid and not continuous plate with a section of 150*20mm. To ensure crack to initiate at the weld toe on the main longitudinal plate and nor at the weld root neither on the longitudinal material, full penetration welds and a smooth reduction of the thickness to 10mm over 200mm of the plate at the location of interest are applied. To avoid weld defects at the longitudinal plate lower edge, vertical welds length is 80mm, leaving a 10mm distance to the plate edges (see Fig. 2 and Fig. 7). The welding was made using either a MAG 136 (for S235) or a MAG 138 (for S690) with well-defined parameters available upon request to the authors.

The TA setup is based on two orthogonal jacks allowing the application of forces in two directions. The TA specimen is fixed horizontally to a column on one side and to the horizontal jack on the other side. Due to this support condition, a small horizontal displacement is occurring at the middle point and a larger one on the loaded side. To ensure the application of the defined normal stress range at the weld toe and to avoid any interaction between the two orthogonal jacks, the TA specimen is simply supported vertically on its bottom side by two linear roller bearing sliders while the vertical load is applied by the vertical jack above the specimen. Due to the support conditions and plate slenderness, only tensile loads can be applied horizontally to avoid any effects of elastic buckling or slipping at bolted connections. Vertically, in order to maintain contact with the roller bearing sliders, only compressive loads can be applied.

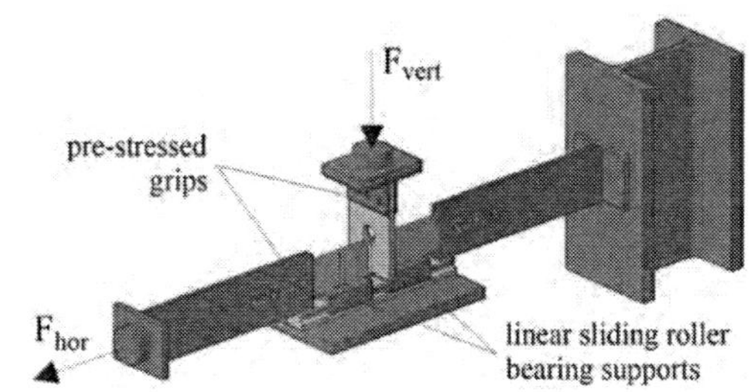

Fig. 3. TA setup operation scheme

1.2 Experiments under uniaxial stresses

1.2.1 Experiments under normal stresses

Six specimens were tested under uniaxial normal stresses. A summary of the results is shown in Table 2. Different stress levels were applied during the tests to allow the definition of the S-N slope and therefore the FAT classification of the detail. The observed cracks are typical of mode I. Their pattern is similar among all the tested specimens, showing multiple semi-elliptical cracks that have initiated at the weld toe, with no preferential initiation point, and that have coalesced and propagated in the longitudinal plate perpendicularly. To ensure this behaviour and that initiation do not start specifically at the weld toe on any of the four longitudinal plate edges, these specific areas were smoothly grinded to avoid any notch or unwanted stress concentration effect. These results under uniaxial normal stress, together with the existing database, were already presented [1] and are shown in Fig. 4 along with their mean S-N curve. As the total number of tests is limited, a fixed slope m=3, is considered, leading to a detail mean strength of $\Delta\sigma_{2\cdot10^6} = 120\,MPa$. The database contains results for specimens with toe-to-toe length smaller than $l = 80mm$ that failed at the weld toe on the main plate and with $R \geq 0$. Procedure for the statistical analysis follows the IIW recommendations [4] as well as Schneider and Maddox [5], see [1] Appendix A for details. Finally, whether or not data from the literature are considered, the detail is classified as FAT80 with a confidence level $p = 95\%$. Results are in good agreement with values given in the IIW [4] and EC3 [6]

1.2.2 Data under shear stresses

One limitation of the setup in its actual design is that it is impossible to create a pure nominal shear stress because of the space between the two roller bearing supports inherent to their geometry (see Fig. 7). Roller bearings are installed as close as possible from the centre axis of the vertical plate, and therefore of the vertical jack, but a bending effect is still present due to the 75mm distance imposed by the bearings design, which induces an additional normal stress due to the vertical load as shown in Fig. 7. However, a minimum span of the toe-to-toe distance is necessary in order to avoid a direct load transfer from the vertical plate to the roller bearing without having full transfer of the shear stress passing thru the weld toe.

As for the uniaxial test data, existing shear fatigue results were collected from the literature and presented previously in [1] (Appendix A.6). Relevant shear fatigue tests found in the literature were mainly done on tube-to-plate specimens and a few on welded plates. An applied stress ratio of generally $R = -1$ was applied and only a few tests were done with $R \geq 0$. The re-analysis is given in Fig. 5 and shows that the mean strength is defined by $\Delta\tau_{2\cdot10^6} = 132\,MPa$ with a fixed slope m=5, the derived category being FAT98.

Table 2. Experimental results for CJ under uniaxial normal stresses (steel S235JR)

Specimen	ΔF_{hor} [kN]	$\Delta\sigma_{hor}$ [MPa]	R_σ	$N_{failure}$
N1	180	180	0.1	261'740
N2	150	150	0.1	657'360
N4	150	150	0.1	1'390'890
N5	150	150	0.1	626'460
N3	120	120	0.1	2'721'530
N6	100	100	0.1	9'000'000*

* run-out

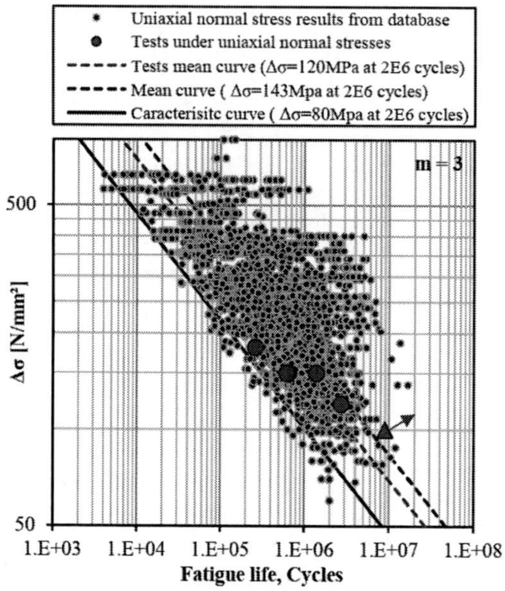

Fig. 4. Uniaxial CJ test results under normal stresses and comparison with results from database

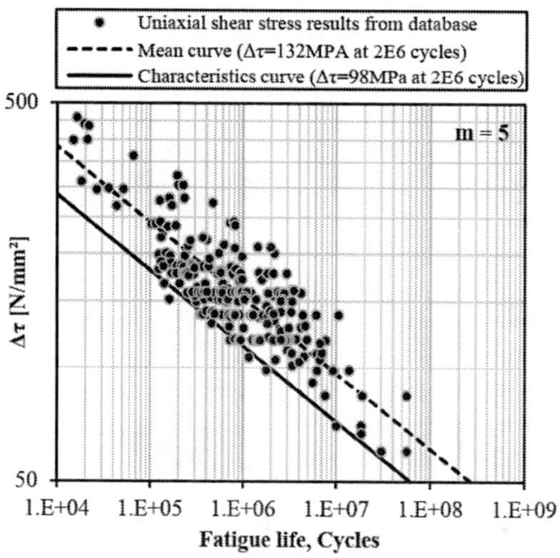

Fig. 5. Uniaxial shear stress database results and analysis (mainly on tube-to-plate specimens)

1.3 Experiments under multiaxial stresses

A total of 27 specimens were tested under various multiaxial states of stress. Twenty-one tests were done on S235JR steel, including 11 tests under proportional and 10 tests under non-proportional loadings. Six tests were done on S690QL steel specimens under proportional loadings. In order to assess the detrimental or beneficial effects of these various situations, results are first shown in the usual nominal normal stress range to number of cycles domain and compared to the uniaxial normal stress results. As a first approach, nominal stresses are defined in a two-dimensional plane at the expected and observed location of crack initiation, i.e. at the bottom of the weld toe, as shown in Fig. 7. Applied load signals and nominal stresses output are presented in Fig. 9. Tests were carried at frequencies between 3 and 4Hz depending on the loading situation.

Table 3. Experimental results on TA under multiaxial proportional (P) and non-proportional (NP) stresses (σ and τ)

	Test	$\Delta\sigma_{hor}$ [MPa]	$\Delta\tau_{vert}$ [MPa]	R_σ	$\lambda_{\tau/\sigma}$	$N_{failure}$
S235 JR	P1	54	50	0.1	0.93	3'780'000
	P2	54	50	0.1	0.93	10'000'000*
	P3	54	50	0.1	0.93	10'000'000*
	P4	154	50	0.1	0.32	174'000
	P5	154	50	0.1	0.32	240'500
	P6	154	50	0.1	0.32	192'500
	P7	154	50	0.1	0.32	306'000
	P8	154	50	0.1	0.32	330'000
	P9	154	50	0.1	0.32	253'000
	P10	154	50	0.1	0.32	291'000
	P11	154	50	0.1	0.32	486'900
	NP1	46	50	0.6	1.10	1'550'000
	NP2	46	50	0.6	1.10	4'052'500
	NP3	46	50	0.6	1.10	3'200'000
	NP4	46	50	0.6	1.10	2'000'000
	NP5	46	50	0.6	1.10	5'000'000
	NP6	46	50	0.6	1.10	5'500'000
	NP7	46	50	0.6	1.10	1'560'000
	NP8	46	50	0.6	1.10	3'282'500
	NP9	46	50	0.6	1.10	10'000'000*
	NP10	46	50	0.6	1.10	10'000'000*
S690 QL	P12	154	50	0.1	0.32	532'500
	P13	154	50	0.1	0.32	503'600
	P14	154	50	0.1	0.32	410'000
	P15	154	50	0.1	0.32	236'000
	P16	143	60	0.1	0.58	151'500°
	P17	143	60	0.1	0.58	208'000°

* run-out
° initiation of the crack on the main plate lower edge

Among all specimens, cracks initiated and propagated in the lower part of the weld toe section into the longitudinal plate, i.e. the same section and crack plane as under uniaxial normal stresses. Multiple semi-elliptical cracks initiated in this area and coalesced into a single one with a high a/c ratio. The bigger the shear ratio, the more the crack pattern tends to a semi-circular shape due mainly to the Mode II. Exceptions to mention are the specimens P16 and P17 where the crack initiated from a small notch on the lower edge of the main plate at the symmetry axis of the vertical plate. This area being subjected to the maximum principal stress when under multiaxial loads, any welding notch or imperfection needs to be carefully eliminated. Unlike the uniaxial case, crack surfaces are rough and presents small branches that are typical of Mode II/III cracks see Fig. 6. The bigger the ratio between shear and normal stress, the more the observable Mode II and III. These tests also show the anisotropic nature of the multiaxial fatigue propagation phenomenon in welded plate details. Most of the crack propagation driving force is lost in friction so these modes have higher lives than their corresponding Mode I cracks. The shear friction may however be severely reduced under the presence of a normal stress. The idea of factorising the shear damage with the normal stress influence can be found in most of the important strain-based criteria, e.g. Findley [7], which however are initiation criteria. In propagation and at macroscopic level, we believe this phenomenon to be the main reason for the lowest fatigue lives of the proportional load case. The normal stress influence, acting in phase with the shear, greatly reduces friction between the crack surfaces, increasing the damaging character of the shear forces.

Fig. 6. Typical crack pattern under multiaxial stresses with a high ratio $\tau/\sigma = 0.93$ (specimen P1)

In bridge and other structures design, fatigue under shear and its relative failure modes are generally neglected in plate details because of the relative lack of tests on these types of details. It is effectively easier to obtain mode II and III in tubular specimens from pure torsion tests. In this regard, the new multiaxial setup presented in this paper showed to be well suited.

The first conclusion that can be drawn is that multiaxial loadings are definitely more damaging than uniaxial loading situations when looking only at the uniaxial stresses. Results in Fig. 8 shows the detrimental effect on the fatigue life of an additional cyclic shear stress acting in combination to the cyclic normal stress. As for uniaxial tests, linear regressions are done with a fixed slope of 3 due to the limited number of specimens tested at only a few different levels of stresses. The linear regression of the proportional results gives a mean strength of $\Delta\sigma_{2\cdot10^6,mean} = 82\,MPa$. The non-proportional load case significantly reduces uniaxial fatigue life with regard to the obtained results expressed in terms of nominal stresses. The corresponding mean fatigue strength is $\Delta\sigma_{2\cdot10^6,mean} = 57\,MPa$. This conclusion is in accordance to what has been observed in multiaxial tests on tubular specimens where the non-proportional load case has been reported as more damaging in ductile material because of grain dislocations

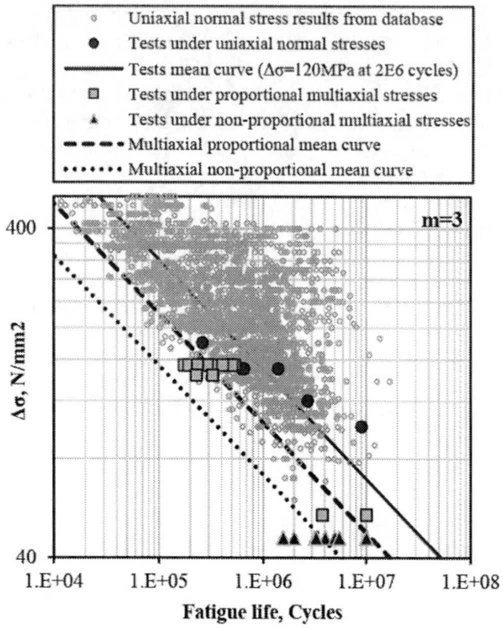

Fig. 2. Multiaxial fatigue tests plotted on the uniaxial normal stress to fatigue life space

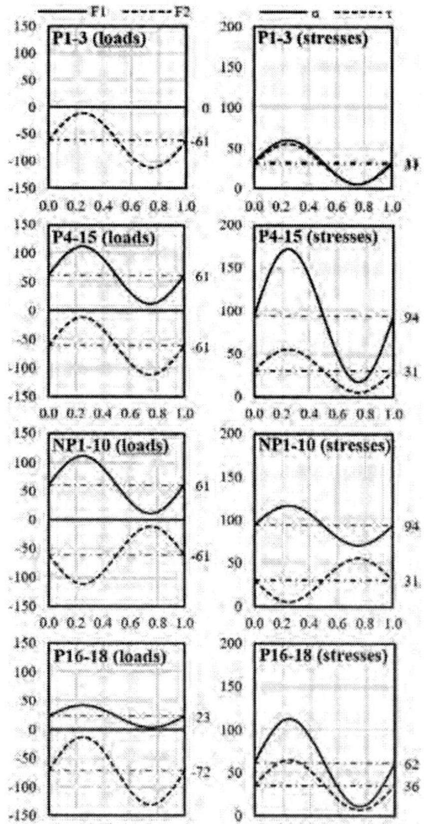

Fig. 3. Loads and nominal stresses signal during one cycle for the four different multiaxial combinations tested

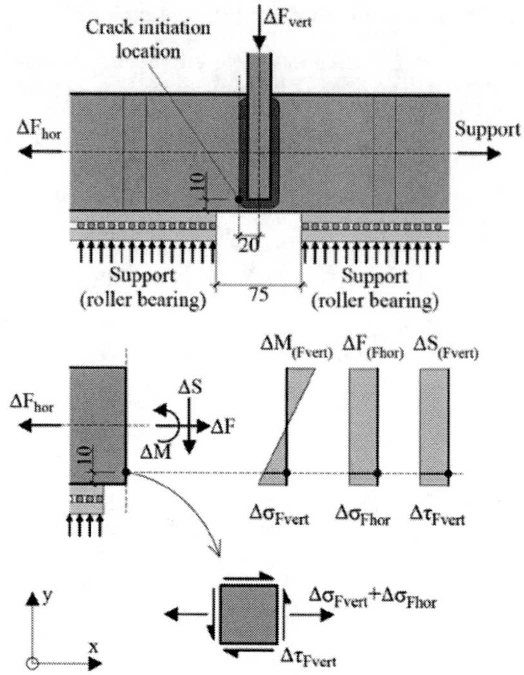

Fig. 1. Detail of the vertical support system and weld geometry. Definition of the crack initiation location and simplified determination of the nominal stresses.

that occur when the principal stress plane angle varies during the loading cycle. The difference may also come from the use of a very high λ ratio of 1.10, but on the other hand, a very low stress amplitude puts the tests very close to the theoretical fatigue limit and thus the fatigue life should increase. This is effectively seen with both specimens NP9 and NP10 whom have reached the fatigue life of 10^7 cycles before being stopped and considered as run-outs. An important remark is that these tests were carried out with a shift angle of 180°. A 90° phase shift should be more damaging but this influence will only be studied in a subsequent task.

It is not possible to draw a general conclusion on the use of high strength steel due to the very low number of specimens tested to date. However, if specimens P16 and P17 are neglected due to their different initiation location, the mean fatigue strength is equal to $\Delta\sigma_{2\cdot 10^6,mean} = 80\,MPa$ for S235JR specimens and $\Delta\sigma_{2\cdot 10^6,mean} = 90\,MPa$ for S690QL specimens indicating a possible improvement in the fatigue strength for the high strength steel.

2 Multiaxial fatigue approaches for fatigue life estimation

Models and criteria presented in this paper can be separated in three groups, i.e. equivalent stresses, equivalent stresses on the critical plane and interaction equations approaches. Most of them were initially used or defined to assess the fatigue life up to crack initiation, to the so-called technical crack size i.e. from 0.01 to 1mm in general. To study the total fatigue life, these models should be combined with propagation approaches based on fracture mechanics. Several authors [8, 9] concluded that even for welded joints an average factor of 0.5 is observed between the time to initiate a 1 mm deep crack and through-thickness cracking, for both as-welded and weld machined and ground [13].

2.1 Modelling and stress definition with the effective notch stress approach

The effective notch stress is part of the various notch stress approaches as explained by Radaj et al. [10]. The fatigue notch factor defining the effective stress concentration is determined with the help of a microstructural notch support hypothesis. The maximum notch stress is determined by an averaging over a material-characteristic small-length, area or volume at the notch, see for example Peterson [11], Neuber [12–14], Taylor [15]. Olivier et al. [16, 17] confirmed the validity of the notch stress approach according to Radaj, for several weld configurations, including the scatter due to different notch values and the effect of stress ratio. They also concluded that even if a higher strength steel offers a better local fatigue strength on smooth specimens, it is also more sensitive to notch sharpness. Thus, in its application, even if the approach was proposed for any notch radius, one generally use the theoretical worst case, leading to the popular fictitious notch radius $\rho_f = 1mm$.

This approach is used to model any notch at the expected crack location, weld toe or root, with a fictitious transition radius of 1mm. The numerical FE model is developed in Abaqus 6.16 (Dassault Systèmes) and a sub-modelling approach is used to model the transition radius. Hexahedral quadratic elements with reduced integration (C3D20R) are used in both global and sub-models. The IIW [4] recommends to discretize the radius by minimum 4 quadratic elements or 6 linear elements. Six elements are used in the model.

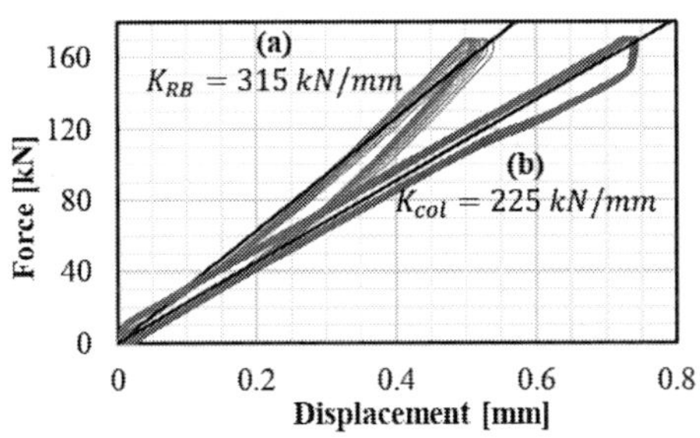

Fig. 10. (a) linear roller bearings vertical stiffness and (b) column horizontal stiffness measured with LVDTs

One important feature of the setup is that it is not purely symmetrical, implying that the FE model cannot be simplified and flexible supports have to be taken into account. The linear roller bearings and the column are modelled with spring elements supporting the edges of the main horizontal plate. Calibration of the spring elements in the FE model is made with the help of measured displacements on the specimens during several quasi-static uniaxial and multiaxial loading situations. As shown in Fig. 10, the behaviour of the spring elements proved to be relatively linear.

The allowable value for structural steels used in bridge and ship welded construction according to the IIW is FAT225/3 (i.e. 225 MPa at 2 million cycles and slope m = 3, with 97,7% survival probability, R ≤ 0.5). Furthermore, the FAT160/5 proposed by Sonsino [18] for the assessment of welded joints loaded in pure torsion/shear is chosen as the shear fatigue strength.

2.2 Equivalent stress approaches

If, in general, most of the multiaxial fatigue situations can be simplified into a two-dimensional system, the studied geometry and its relative stress state intricacy at the location of interest implies to consider a three-dimensional stress tensor when using local approaches. The principal stresses (PS) is used first in the analysis.

Secondly, the von Mises criterion is used. Indeed, Radaj et al. [10] stated that one should use the Von Mises equivalent stress combined with the effective notch stress in case of ductile material instead of the maximum principal normal stress that is more suited to brittle material as cast steel.

2.3 Findley critical plane approach

Critical plane approaches are all based on the research of the one or multiple planes that maximizes a specific damage parameter. The critical plane search method (CPS) used in this paper is similar to the minimum circumscribed circle concept proposed by Dang Van [19] and Papadopoulos [20, 21]. This new CPS method proposed by Karpanan [22] can be used for all stress based proportional and non-proportional loading fatigue analysis and all strain-based fatigue analysis methods. The stress tensor in each material point is needed but in this paper two points (nodes) on the FE models are analysed for sake of simplification, i.e. the points subjected to the maximum principal stress and to the maximum von Mises equivalent stress. Pedersen [23] has recently shown that, when using the effective notch stress approach, for CA proportional loadings, Findley's [7] simple critical plane criterion leads to the safest predictions. The Findley criterion suggest that the maximum normal stress σ_n, acting in combination with the shear amplitude $\Delta\tau/2$ on a specific plane has a detrimental effect on the allowable alternating shear stress. The criterion is defined on the plane subjected to the maximum value of this combination,

$$f = \left(\frac{\Delta\tau}{2} + k * \sigma_n\right)_{max} \quad (1)$$

where k is a material constant related to materials' sensitivity to normal stresses and f is directly related to the materials' shear fatigue strength. For ductile materials, k usually varies between 0.2 and 0.3. A value $k = 0.3$ is used in the present paper. Findley's criterion f is here directly compared to the shear fatigue strength FAT160/5.

2.4 EN1993-1-9 and IIW interaction equations

In the IIW [4], the interaction criterion is a modified Gough-Pollard criterion eq.(2).

$$\left(\frac{\Delta\sigma_S}{\Delta\sigma_R}\right)^2 + \left(\frac{\Delta\tau_S}{\Delta\tau_R}\right)^2 \leq CV \quad (2)$$

The multiaxial comparison value CV may have the value of 1,0 or 0,5 depending on the loading state (proportional or non-proportional) and the ductility of the material.

In the EC3 [6], the criterion is based on a linear damage sum D according to Miner's rule eq.(3).

$$\left(\frac{\Delta\sigma_S}{\Delta\sigma_R}\right)^3 + \left(\frac{\Delta\tau_S}{\Delta\tau_R}\right)^5 \leq D \quad (3)$$

Unlike other models presented, the EC3 and IIW equivalent stresses are dependent of previously defined shear and normal fatigue strengths and comparison value or damage sum.

3 Multiaxial fatigue criteria results

The equivalent stress is calculated for each test and then plotted in SN diagrams for each criterion presented previously. However, in order to analyse the results and present them in a clear and brief manner, Table 4 gives the most important outcomes of the tests data expressed with the help of the five different criteria. A graphical example is given in Fig. 12, showing data plotted in terms of Findley's equivalent stresses defined with the effective notch stress approach. The notch stress has the ability to regroup the different load cases in a single scatter band but all criteria used in this paper lead to a large scatter when considering non-proportional tests except for the Findley critical plane approach, which showed good results. It is interesting to note that the characteristic curve for the uniaxial case with principal stresses is FAT227/3 which is in good agreement with the FAT225/3 from the IIW. In general, the IIW and EC3 criteria give the larger scatter. Table 4 also gives the probability of non-conservative prediction p_f calculated with the help of a life ratio lr as employed by Bruun & Härkegard [24] and Pedersen [23],

$$lr = \frac{\Delta S_{eq,N,i}}{\Delta S_{R,N}} \quad (4)$$

where $\Delta S_{eq,N,i}$ is the equivalent strength at a specific number of cycles for each tests and $\Delta S_{R,N}$ is the design shear or normal fatigue strength at the same number of cycles. A Weibull distribution is then fitted to the life ratios for each situation. The resulting scale α and shape β parameters are then used to compute the non-conservative prediction p_f, i.e. the probability that tests results have life ratios $lr < 1.0$.

In the case of uniaxial and multiaxial proportional load cases, both the maximum principal stress and von Mises give results in the range of 12 to 20% but the principal stress gives overall better results. Again, the combination of Findley's equivalent stress with the effective notch stress method is the safest with a p_f of 1.56%. In the case of non-proportional loadings, all criteria overestimate the fatigue lives, giving a p_f which tends to 0 this is mainly due to the very low strange range level resulting from the out-of-phase angle of 180°. In general, the IIW and EC3 gives surprisingly non-conservative values, with an advantage for the IIW that has better results even in the non-proportional case certainly due to the use of CV = 0.5 in this case. It is important to mention that all these conclusions are mostly indicative because of the very limited number of tests that are analyzed.

Table 2. Evaluation of the different multiaxial fatigue criteria.

	Uniaxial normal			Uniaxial normal + Multiaxial proportional			Uniaxial normal + Multiaxial proportional + non-proportional		
	std. dev. σ	Mean [Mpa]	FAT (p=95%) [Mpa]	std. dev. σ	Mean [Mpa]	FAT (p=95%) [Mpa]	std. dev. σ	Mean [Mpa]	FAT (p=95%) [Mpa]
PS	0.288	304	227	0.235	275	212	0.571	355	286
v. Mises	0.288	272	203	0.231	261	202	0.538	331	180
Findley	0.288	192	144	0.219	178	140	0.328	200	138
IIW				0.531	254	141	1.08	360	106
EC3				0.461	232	139	0.823	302	119

	Uniaxial normal		Multiaxial P		Multiaxial NP		Uniaxial + Multiaxial P		Uniaxial normal + Multiaxial P + NP	
	lr_{mean}	pf	lr_{mean}	pf	lr_{mean}	pf	lr_{mean}	pf	lr_{mean}	pf
PS	1.377	12.09%	1.192	14.54%	2.908	0.31%	1.241	15.06%	1.746	21.43%
v. Mises	1.234	21.10%	1.156	20.50%	2.604	0.53%	1.177	20.23%	1.609	23.04%
Findley	1.581	0.02%	1.61	1.56%	1.841	0.30%	1.603	0.97%	1.675	1.03%
IIW			1.078	28.15%	3.656	0.10%	1.148	25.06%	1.908	27.00%
EC3			0.996	48.16%	2.542	0.60%	1.055	39.13%	1.506	29.18%

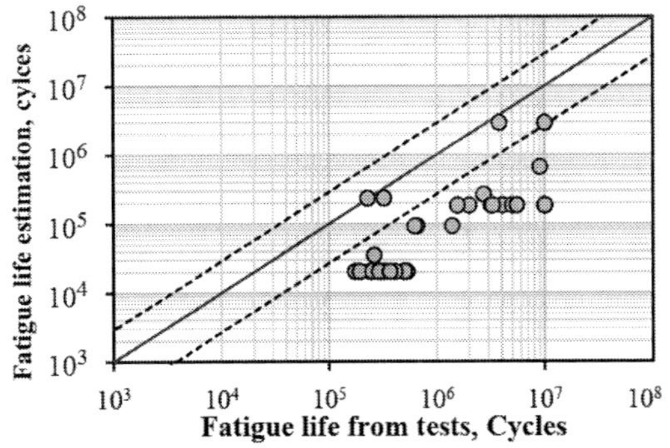

Fig. 11. Fatigue life estimation of all tests using the effective notch stress and Findley equivalent stress

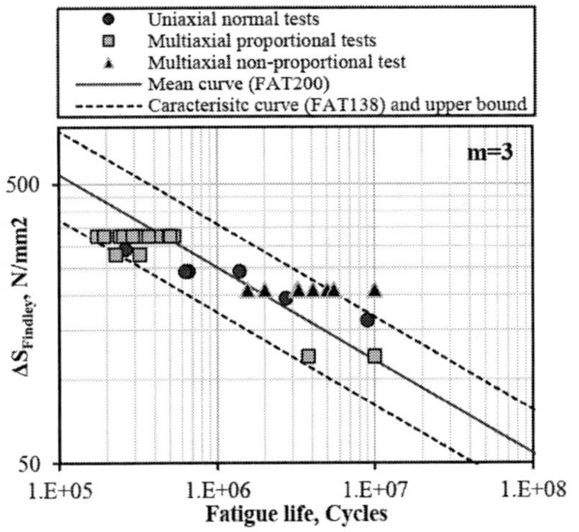

Fig. 12. Example of tests data represented using the effective notch stress and Findley equivalent stress

4 Conclusions

A new multiaxial setup was presented that allowed proportional and non-proportional multiaxial tests in a typical welded plate detail. Mode II/III cracks were obtained under multiaxial loadings tests. The results on a transversal attachments loaded both in normal stress $\Delta\sigma$ and shear stress $\Delta\tau$ clearly showed the detrimental effect of the multiaxial loadings. The local 1mm effective notch stress approach was shown suitable to the analysis under multiaxial stress states. Combined to Findley's criterion and compared to the shear fatigue strength curve FAT160/5 or combined to the maximum principal stress range and compared to the normal fatigue strength curve FAT225/3, it proved to be both safe and the most accurate. Effects of the use of high strength steel was explored and even if results are scarce, one may expect an improvement in the total fatigue life under multiaxial loadings.

Among parameters to be considered in the future to improve the models:

- The influence of transverse residual stresses and their relaxation. Better knowledge would improve initiation life prediction with for example the Manson-Coffin relation using Neuber's notch rule and multiaxial strain based models e.g. Fatemie-Socie [25].
- Propagation life. Propagation measurements would allow calibration of a fracture mechanics model. This would be mostly valuable for the shear propagation under mixed Mode I, II and III. Indeed, mode II/III stress intensity factors are calibrated case by case after experiments as shown for example by Amstutz et al. [27].
- Phase shift angle.

References

1. C. Baptista, thesis, EPFL, Lausanne (2016).
2. I. Lotsberg, *J. Offshore Mech. Arct. Eng.* **131**, 041302-041302–9 (2009).
3. M. Bokesjö, M. Al-Emrani, T. Svensson, *Int. J. Fatigue.* **44**, 21–31 (2012).
4. A. F. Hobbacher, *Recommendations for Fatigue Design of Welded Joints and Components* (Springer International Publishing, Cham, 2016)
5. C. R. A. Schneider, S. J. Maddox, 30 (2003).
6. *Eurocode 3: Design of steel structures - Part 1-9: Fatigue* (Comité européen de normalisation, Bruxelles, 2005)
7. W. N. Findley, *A theory for the effect of mean stress on fatigue of metals under combined torsion and axial load or bending* (1958)
8. T. Bruder, K. Störzel, J. Baumgartner, H. Hanselka, *Int. J. Fatigue.* **34**, 86–102 (2012).
9. C. M. Sonsino, *Int. J. Fatigue.* **17**, 55–70 (1995).
10. D. Radaj, C. M. Sonsino, W. Fricke, *Fatigue assessment of welded joints by local approaches* (2nd edition., 2006).
11. R. E. Peterson, *Proc SESA.* **11**, 199–206 (1950).
12. H. Neuber, *Theory of notch stresses: Principles for exact stress calculation* (JW Edwards, 1946), vol. 74.
13. H. Neuber, *Kerbspannungslehre: Theorie der Spannungskonzentration Genaue Berechnung der Festigkeit* (Springer-Verlag, 2013).
14. H. Neuber, *Konstruktion.* **20**, 245–251 (1968).
15. D. Taylor, *The theory of critical distances: a new perspective in fracture mechanics* (Elsevier, Amsterdam ; Boston, 2007).
16. R. Olivier, V. B. Köttgen, T. Seeger, *Forschungskuratorium Maschinenbau FKM Frankf. Forschungsheft.* **143** (1989).
17. R. Olivier, V. B. Köttgen, T. Seeger, *Forschungskuratorium Maschinenbau FKM FrankfurtMain Forschungsheft.* **180** (1994).
18. C. M. Sonsino, T. Bruder, J. Baumgartner, *Weld. World.* **54**, R375–R392 (2010).
19. K. Dang-Van, in *Advances in multiaxial fatigue* (ASTM International, 1993).
20. I. V. Papadopoulos, *Fatigue Fract. Eng. Mater. Struct.* **21**, 269–285 (1998).
21. Papadopoulos IV, *Fatigue Fract. Eng. Mater. Struct.* **21**, 269–285 (2002).
22. K. Karpanan, (ASME, 2016; http://proceedings.asmedigitalcollection.asme.org/proceeding.aspx?doi=10.1115/PVP2016-63705)
23. M. M. Pedersen, *Int. J. Fatigue.* **83**, 269–279 (2016).
24. A. Bruun, G. Härkegård, *Int. J. Fatigue.* **73**, 1–16 (2015).
25. A. Fatemi, D. F. Socie, *Fatigue Fract. Eng. Mater. Struct.* **11**, 149–165 (1988).
26. H. Amstutz, K. Störzel, T. Seeger, *Fatigue Fract. Eng. Mater. Struct.* **24**, 357–368 (2001).

MATEC Web of Conferences **165**, 16008 (2018) https://doi.org/10.1051/matecconf/201816516008
FATIGUE 2018

Performance of chosen multiaxial cycle counting method under non-proportional multiaxial variable loading.

Matus Margetin[1,*], and *Dominik* Biro[1]

[1]Slovak University of Technology, Faculty of Mechanical Engineering, Namestie Slobody 17, 812 31 Bratislava, Slovakia

Abstract. One of the most crucial tasks in fatigue life-time estimation of components loaded with multiaxial variable amplitude loading is to correctly identify loading cycles that can be used with multiaxial damage criterions. During past years, several cycle counting methods have been proposed for multiaxial loading conditions. The aim of this text is the comparison of selected multiaxial cycle counting methods, namely Wang-Brown's method, Modified Wang-Brown's method, Bannantine-Socie's method and then a critical analysis of the obtained results. For the comparison of chosen methods, a new data set, consisting of experimentally obtained results from multiaxial non-proportional variable amplitude loading tests carried on by authors, has been used. The tested specimens were made from S355J0 structural steel and the testing procedure has been carried out on the MTS Axial/Torsion servo hydraulic testing machine. Findley and McDiarmid multiaxial criterion with Palgren-Miner summation rule have been used for fatigue life-time estimation of the tested specimens.

1 Introduction

Fatigue life time estimation of mechanical components working under real multiaxial variable amplitude operational loadings could be considered as an most complex task in a field of fatigue life time analysis. This complexity of fatigue life time estimation is caused by the fact that the estimation process is composed from multiple independent but crucial tasks which influence final accuracy of fatigue life time estimation.

While topics such as multiaxial damage criterions or damage accumulation rules are research subjects of many scientist and large amount of works in this fields have been published in recent years, there are not so many articles aiming on problem of loading cycle identification in variable multiaxial non-proportional loading signals. And yet, the correct identification (and followed extraction) of loading cycles from variable amplitude multiaxial loading signal is necessary step finite fatigue life time estimation. Nowadays two methods (and their variations) of cycle identification and extraction are most commonly used. They are Bannantine-Socie method and Wang-Brown method.

In this paper authors critically review two most common cycle counting method. Then modelled experiment is presented where result from experimental measurement is compared to estimated fatigue life time using mentioned cycle counting method combined with two multiaxial criterion (Findley and McDiarmid) and Palgren-Miner damage accumulation rule.

2 Cycle extraction technique

The stress (strain) loading signal obtained by measurement or numerical calculation needs to be transform into a form which can be used on finit fatigue life time estimation. Standardized rain flow algorithm can by used for identifying and extracting cycles from loading signal when uniaxial loading conditions occur. In this case general loading signal is transform to set of harmonic cycles with given amplitudes and mean value which has same equivalent effect on component as original signal. However, this technique can not be used for multiaxil loading cases, as opposed to uniaxial loading two independent loading signal acts on loaded component simultaneously (Fig. 1).

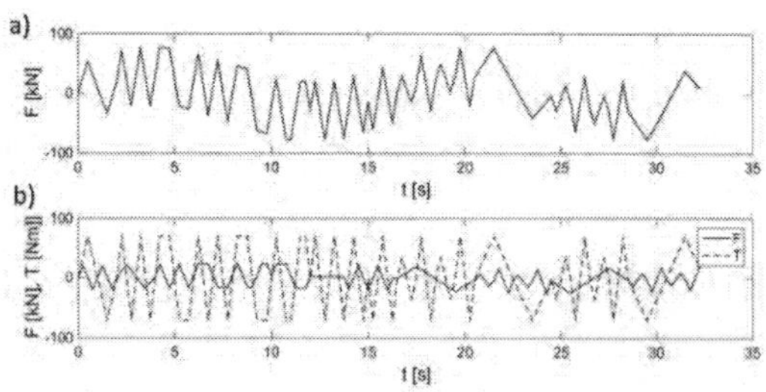

Fig. 1. a) uniaxial history; b) multiaxial history.

As a matter of fact, there is no universally used cycle counting method for fatigue life time analysis of

*Corresponding author: matus.margetin@stuba.sk

component loaded with multiaxial non-proportional variable amplitude loading. Two most common used approaches to cycle identification are Bannantien-Socie method [1] and Wang-Brown method [2,3] (and theirs modifications).

2.1 Wang-Brown method

Wang and Brown developed the method, which can be used for any proportional or non-proportional history of strain or stress. This method identifies cycle based on an equivalent strain or stress. The von Mises equivalent stress (strain) is used to transforms multiaxial state of stress (strains) to uniaxial. The problem with using von Mises is the loss of the sign since von Mises values are always positive. For example in case 90° out-of-phase loading von Mises remains constant, so it leads to the underestimating of loading condition. Because of that, the relative von Mises stress (strain) is introduced, calculated from shifted stress (strain) signals. To obtain the relative von Mises stress (strain) the loading signals have to be rearranged in the way that it starts with the points corresponding to maximal von Mises stress. The signals are then shifted by offset equal to negative value of first point (so the shifted signals starts in zero Fig. 2).

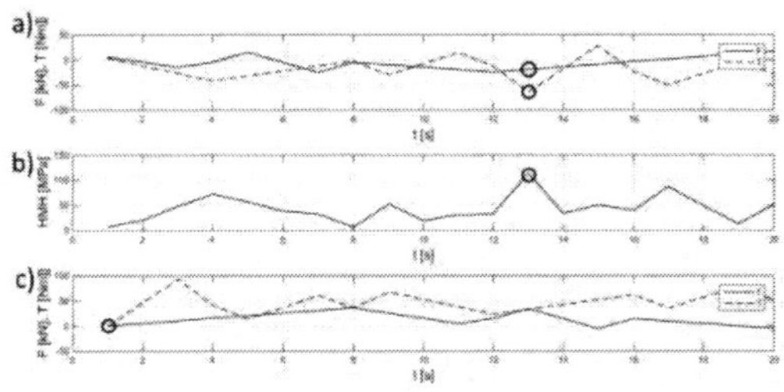

Fig. 2. a) original signal; b) von Mises stress; c) rearranged and shifted signal.

One half-cycle is generated from initial point to point with maximum value of relative von Mises strain/stress in each counting. The remained parts of the signals are then processed in the loop using the same procedure (Fig. 3)

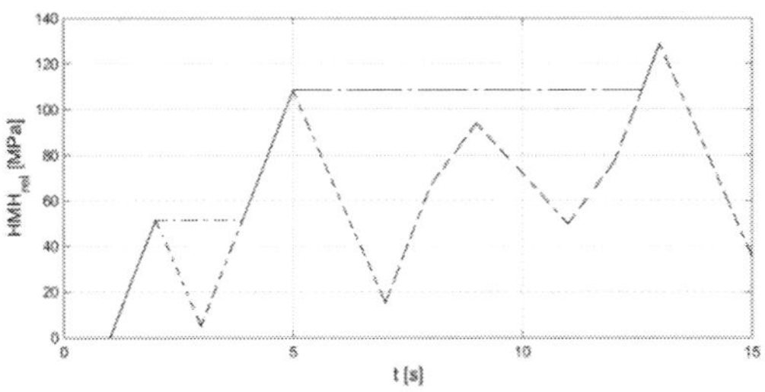

Fig. 3. Extracted half cycle – bold line; remained parts – dashed line.

The main advantage of this method is that whole counting runs only once and the extracted cycles can be used with different multiaxial criterions. Even if we want to used extracted cycles with damage models based on critical plane theory, there is no needed of transformation of loading signals into the different material planes before extraction. The biggest disadvantage of this method is, that it extracts only half-cycles.

2.2 Bannantine-Socie method

Bannantine and Socie proposed methodology that is based on recalculating the loading signals into de plane with different orientation that it provides the normal and shear stress (strain) histories in each plane configuration. Then the modification of standard rain flow method is applied on either shear or normal stress (strain) signal, depending on chosen damage model. Te rain flow method has to keep track of events with occur in the second (non extracted) signal to capture the parameters that are needed in multiaxial fatigue criterion (Fig. 4)

Fig. 4. Tracking events on non extracted signal history.

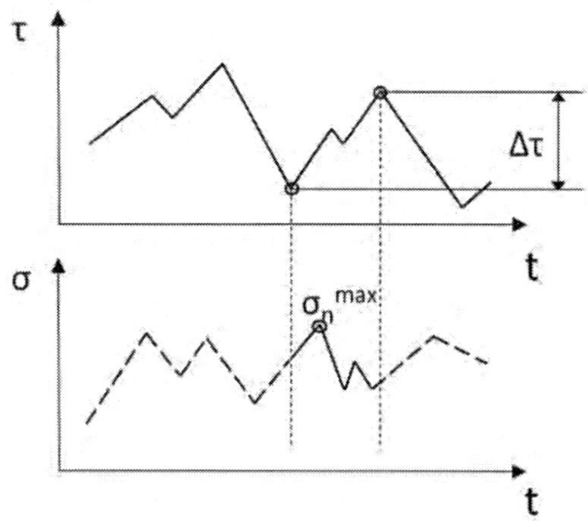

Result from this method can be directly used with critical plane damage models as the recalculation of loadings signals into the different planes is done before extraction. The main disadvantage is that this method is very time consuming as it requires run of rain flow extracting procedure in multiple planes.

3 EXPERIMENTAL PROCEDURE

To verify methods mentioned in previous chapter, an experiment was conducted in which experimental specimens were loaded using various multiaxial loading signals. Experiment was carried out in the Strength and Elasticity Laboratory of the Faculty of Mechanical Engineering STU. Two type of multiaxial variable amplitude loading signal have been applied on round specimens.

3.1 The loading signals

Two types of loading histories have been applied on testing specimens.

The first block was composed of non-proportional combination of force and torque variable amplitude loading signals with zero mean value. The loading histories as well as loading path of the signal can be seen in Fig 5.

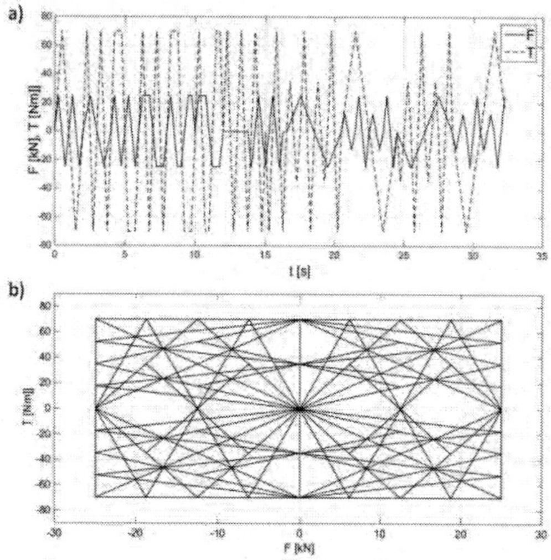

Fig. 5. a) Loading history 1 (LH1); b) Loading path.

The second block was composed from non-proportional combination of force and torque variable amplitude loading signals with variable mean value. The loading histories as well as loading path of the signal can be seen in Fig 6.

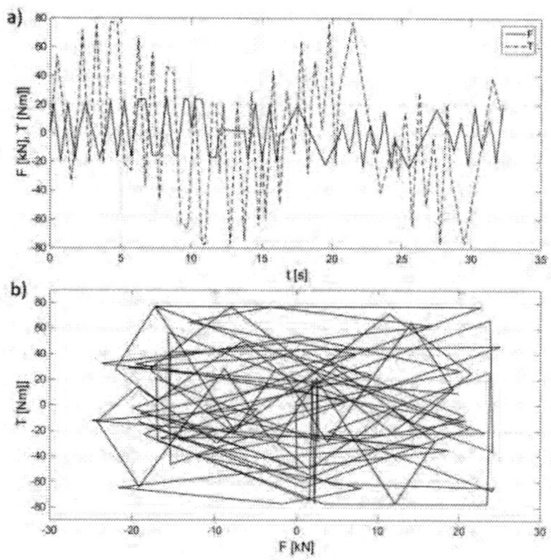

Fig. 6. a) Loading history 2 (LH2); b) Loading path.

3.2 Testing specimens and experimental setup

Testing specimens has been loaded with multiaxial variable amplitude loading signal on MTS Bionix 370.02 Axial/Torsional testing system. The experimental specimens were manufactured from S355J2+C structural steel. The geometry of testing specimen can be seen in Fig. 7.

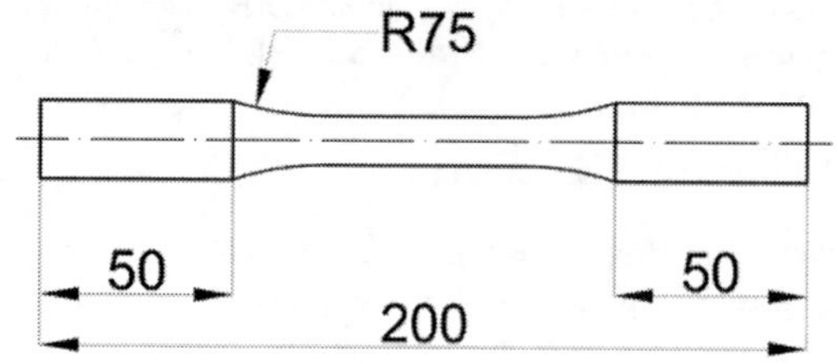

Fig. 7. Testing specimen geometry.

Chemical composition as well as static material parameters obtained from the material a-test can be seen in Table 1.

Table 1. Chemical composition and material parameters of testing specimen.

S355J2+C			
σ_{y02} **[MPa]**	σ_u **[Mpa]**	**C [%]**	**P [%]**
655	680	0,16	0,014
S [%]	**Mn [%]**	**Si [%]**	**Cu [%]**
0,025	1,31	0,18	0,12
Al [%]	**Mo [%]**	**Ni [%]**	**Cr [%]**
0,018	0,01	0,06	0,07

Te tests have been carried on in force and torque controlled mode. The number of loading block repetition prior to the initiation of the fatigue crack was determined on the basis of a continuous measurement of the deformation, and angle of the distortion, response to the loading regime of the test specimen. Failure condition of the experimental specimen was defined by the moment of the so-called "technical initiation of fatigue crack" (0,5–1 mm). The tests have been terminated when either maximal deformation or angle of the distortion of loading block increased by 3 % in reference to previous maximums (Fig. 8).

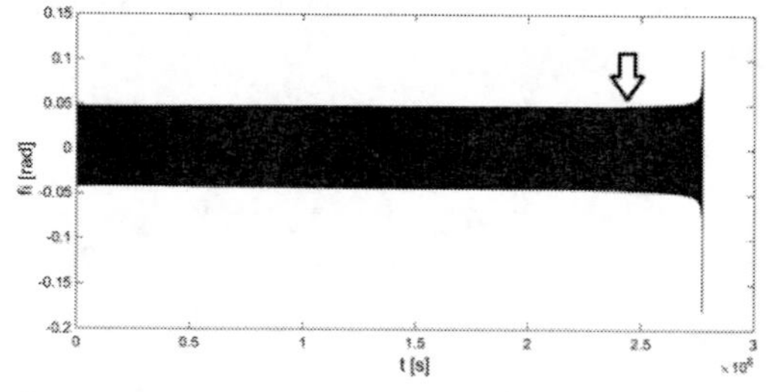

Fig. 8. "Technical" initiation of fatigue crack.

3.3 Experimental results processing

Fatigue life time estimations of measured specimens have been calculated from real measured specimen loading, as the real loading condition slightly differ from modelled loading signal histories (due to regulation imperfection).

The multiaxial criterion has to be used for final fatigue life time estimation based on extracted cycle. Two kind of multiaxial fatigue damage model have been used. Authors chose to use stress based critical plane criterion.

The first one is Findley criterion [4]. Findely propose criterion based on linear combination of shear steress amplitude and maximal normal stress acting on calculated plane. When used for finite fatigue life time estimation, the equivalent shear stress amplitude is compared to shear Basquin equation. Final form is then described by following formulation:

$$\tau_{eq} = \left(\tau_a + k_{fin}\sigma_n\right)_{max} = \tau_f^*\left(2N_f\right)^{b_\tau} \quad (1)$$

The critical plane is defined as a plane with maximal damage parameter. Similarly in the case of variable amplitude loading the critical plane is the plane with maximal accumulated damage.

The second used criterion is criterion proposed by McDiarmid [5]. Similarly to Findely, McDiarmid propose criterion in form of linear combination of shear stress amplitude and maximal normal stress acting on same plane. The main difference, and reason why authors chose this criterion, is the definition of critical plane. Unlike Findley, McDiarmid propose that critical plane is plane with maximal shear stress amplitude. Then again for finite fatigue life time estimation the McDiarmid equivalent shear stress amplitude is compared to shear Basquin equation:

$$\tau_{eq} = \tau_a + \left(\frac{k_{mcd}}{2\sigma_u}\right)\sigma_{n,max} = \tau_f'\left(2N_f\right)^{b_\tau} \quad (2)$$

However when using this criterion for fatigue life time estimation under variable non-proportional loading. The damage in every plane has to be calculated and the final fatigue life time is determined by the plane with maximal accumulated damage.

The material parameters used in multiaxial criteria are summarized in Table 2. and are taken from [6].

Table 2. Fatigue properties.

S355J2+C				
τ_f [MPa]	b_τ [-]	k_{fin} [-]	τ_f^* [MPa]	k_{mcd} [MPa]
550	-0,0736	0,131	550	244

As both criterion take into account only shear stress amplitude the methodology that take into account mean shear stress of extracted cycles has to be implemented in estimation procedure. The mean shear stress correction model proposed by Wang and Miller [7] has been used for life time estimation. The model is described by following formulation:

$$\tau_a = \tau_{am} e^{\frac{|\tau_m|}{\tau_f}} \quad (3)$$

Fatigue damage in each plane has been calculated using well known Palgren-Miner linear damage accumulation rule described by formula (4)

$$D = \sum \frac{n_i}{N_{f_i}} \quad (4)$$

The increment of the angle defining the evaluated plane has been set to 2°.

4 Experimental results

The methodology described in previous chapter have been used for fatigue life time estimation of testing specimens loaded with designed loading histories LH1 and LH2. Estimated results with experimental measurements are compared in following chapter. Three specimens have been tested overall. The specimen nuber 16 and specimen number C2 have been loaded with same loading history with different loading frequencies (loading signal has been applied two time faster on specimen C2).

4.1 Wang-Brown method

The comparison between measured fatigue life time and fatigue life time estimation calculated from cycles extracted using Wang-Brown method can be seen in table 4.

Table 3. Result comparison – Wang-Brown.

Wang-Brown				
Spec. nr.	Load. Hist.	Exp.	Findley	McDia.
16	LH1	6520	11781	6569
C2	LH1	6715	16148	8905
C4	LH2	14480	17054	9584

Fatigue life time estimation as a function of plane orientation for specimen n.16 (loaded with loading history 1) can be seen on Fig. 9.

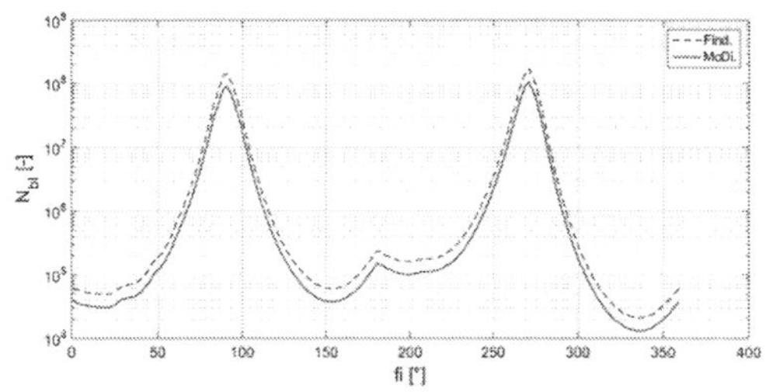

Fig.9. Fatigue life time as a function of plane orientation – Wang-Brown method loading history LH1.

Fatigue life time estimation as a function of plane orientation for specimen n.C4 (loaded with loading history 2) can be seen on Fig. 10.

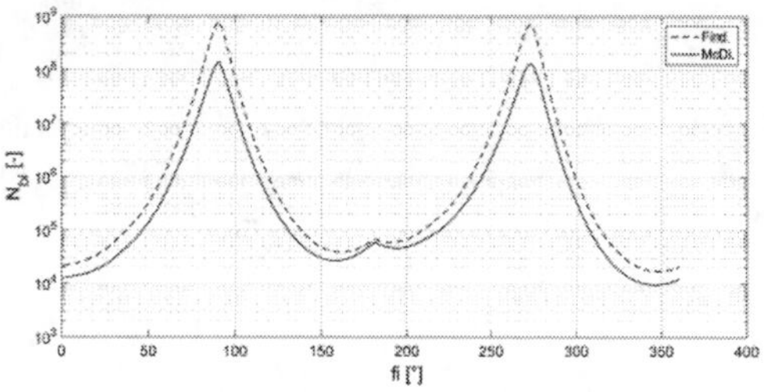

Fig.10. Fatigue life time as a function of plane orientation – Wang-Brown method loading history LH2.

4.2 Banantine-Socie method

The comparison between measured fatigue life time and fatigue life time estimation calculated from cycles extracted using Bannantine-Socie method can be seen in table 4.

Table 4. Result comparison – Bannantine-Socie.

Bannantine-Socie				
Spec. nr.	**Load. Hist.**	**Exp.**	**Findley**	**McDia.**
16	LH1	6520	15647	9146
C2	LH1	6715	18152	10664
C4	LH2	14480	21003	13084

Fatigue life time estimation as a function of plane orientation for specimen n.16 (loaded with loading history 1) can be seen on Fig. 11.

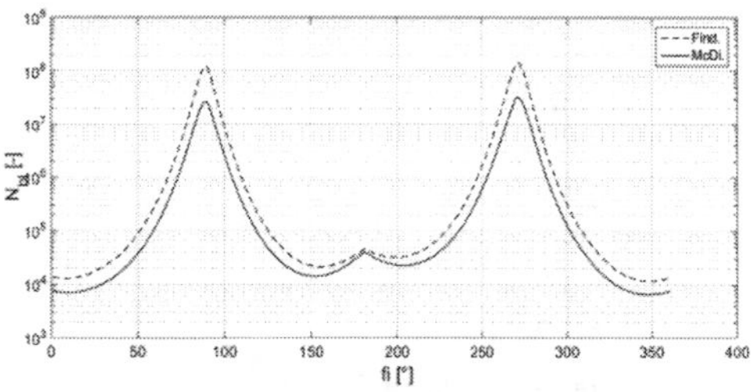

Fig.11. Fatigue life time as a function of plane orientation – Bannantine-Socie method loading history LH1.

Fatigue life time estimation as a function of plane orientation for specimen n.C4 (loaded with loading history 2) can be seen on Fig. 12.

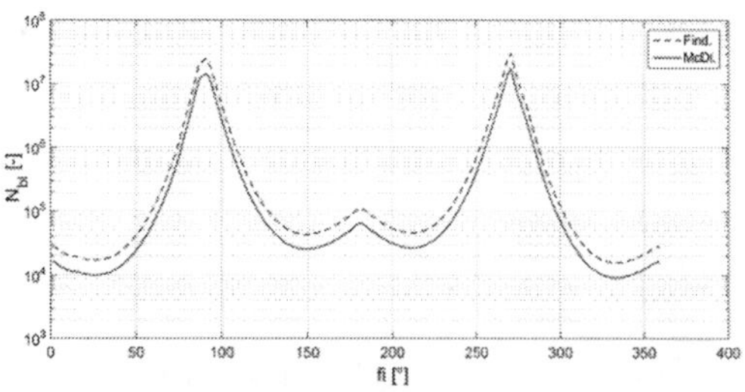

Fig.12. Fatigue life time as a function of plane orientation – Bannantine-Socie method loading history LH2.

5 Conclusion

The fatigue life time estimation of testing specimens loaded with multiaxial non-proportional variable amplitude loading have was carried out in this paper. Two method of loading cycle identification and two different multiaxial criteria was used in the estimating process. The experimental verification of calculated results was realized.

The comparison between measurement and calculated prediction can be found in tables 3 and 4. An accuracy in percentage for all combination of chosen method can be found in table 5.

Table 5. Performance of chosen method

spec. n.	Findley [%]	McDiarmid [%]	method
16	180.7	100.8	WB
C2	240.5	132.6	WB
C4	117.8	66.2	WB
16	240.0	140.3	BS
C2	270.3	158.8	BS
C4	145.0	90.4	BS

Looking at the table 5, following conclusion can be stated:

- Both cycle counting method has similar percentage accuracy for chosen multiaxial damage models.
- The overall results show that Findley criterion underestimates loading conditions for both cycle counting method.
- McDiarmid criterion leads to the more accurate results for both cycle counting methods for the given loading condition.
- When using McDiarmid criterion Wang-Brown cycle counting method seems to be more accurate than Bannantine-Socie method for loading with zero mean stresses while BS method seems to be more accurate in case of non-proportional loadings with variable mean stresses.
- There should be noted the results from both counting method in combination with Mcdiarmid criterion fall within boundaries caused by scatter of material parameters.

This work was supported by the Research & Development Operational Programme funded by the ERDF ITMS: 26240220084 Science city Bratislava

References

1. J.A. Bannantine, D.F. Socie, A variable amplitude multiaxial fatigue life prediction method, in: Fatigue under biaxial and multiaxial loading, European Structural Integrity Society ESIS Publication **10**, Mechanical engineering publications, London, 1991, pp. 367-370
2. C.H. Wang, M.W. Brown, Life prediction techniques for variable amplitude multiaxial fatigue – Part 1: Theories, J Eng Meter Technol **14** (1996) 367-370
3. C.H. Wang, M.W. Brown, Life prediction techniques for variable amplitude multiaxial fatigue – Part 2: Comparison with experimental results, J Eng Meter Technol **118** (1996) 371-374
4. W.N. Findley, Fatigue of metals under combinations of stresses, Trans ASME **79** (1957) 1337-1338
5. D.L. McDiarmid, A general criterion for high cycle multiaxial fatigue failure, Fatigue Fract Eng M **14**, 4 (1991) 429-453
6. M. Margetin, R. Durka, V. Chmelko, Multiaxial fatigue criterion based on parameters from torsion and axial S-N curve, Fattura ed Integrita Strutturale **37**, 10 (2016) 146-152
7. C.H. Wang, K.J. Miller, The effect of mean shear stress on torsion fatigue behavior, Fract Engng Mater Struct **14** (1991) 293-307

https://doi.org/10.1051/matecconf/201816516009

Multiaxial Fatigue Spectrum Editing by Using Combined Wavelet Analysis and Stress Invariant Approach

Anahita Imanian[1,*], *Kelvin Leung*[1], and *Nagaraja Iyyer*[1]

[1]Technical Data Analysis, Inc., 3190 Fairview Park Dr, Suite 650, Falls Church, Virginia, 22042, USA

Abstract. The practicalities of structural fatigue testing limit the fidelity of the cyclic load history that can be applied to a test structure. Testing is, therefore, a compromise between fatigue damage fidelity and test economy. A new methodology is proposed for multiaxial loading spectrum editing to extract cycles that contribute negligible damage during fatigue crack initiation. The method is based on projection by projection (PbP) technique and wavelet transform analysis (WTA) procedure. In this approach, the cycles with negligible contribution to damage in every decoupled projected loading path (i.e. obtained from PbP approach) are extracted using the WTA procedure. Each extracted segment is then replaced with an equivalent cycle that produces the same amount of damage. The effectiveness of the edited spectrums is evaluated by the degree of fatigue damage retention as the original damage and preservation of statistical parameter values.
As a case study, the proposed approach has been applied to the numerically produced random bending-torsion fatigue spectrum in plane-stress condition. The result shows an average of 75% reduction of the original spectrums with retention of 90% of the original spectrums' damage values.

1 Introduction

The service load histories corresponding to fatigue-critical locations on engineering structures can contain extremely large numbers of time points in a single record. Employing these large histories in either analysis or testing can be very time-consuming and therefore very expensive. These challenges could become even more critical when dealing with multi-axial loading spectrums. The accurate development of a representative load spectrum is vital when testing the fatigue performance of a structure. The challenge is, therefore, how to reduce the density of load cycles for test purposes while maintaining fatigue performance accuracy and fidelity, i.e. spectrum editing.

Spectrum editing methods identify and remove the sections of the load history that are relatively insignificant for damage summation, in order to reduce subsequent analysis and testing costs. Various spectrum-editing methods have been developed and studied by fatigue researchers especially in aircraft and ground vehicle industries. Among these techniques, omission based editing methods use a gating or filtering algorithm to maintain sequence effects by directly removing cycles from the time sequence [1, 2]. Further, cycles can be removed based on crack opening [3]; load banding (removing loads within a range, often around zero) [4]; or loading time-frequencies [5, 6]. However, many of these methods challenge a loss of damage content from truncation as a compromise when selecting their truncation level. As a consequence other attempts formed to return damage to the spectrum, e.g. load scaling [7] or, using the principle of equivalent damage [8].

Many structural component or parts are subject to multiaxial loading conditions such as rotorcraft structure where the complex unsteady combination of the dynamic load with vibratory response causes a significant degree of multiaxiality. Current spectrum editing methods were widely examined for scaler valued load spectrums, and to the understating of the author, there is no method available to address multiaxial fatigue spectrum editing problem. This paper proposes a new method to edit the multiaxial fatigue spectrum for high-cycle fatigue loading. The proposed approach is based on the combination of the time-frequency signal analysis procedure, and stress invariant based multiaxial fatigue criterion for fatigue crack initiation damage. The time-frequency signal analysis method is based on WTA which enables to characterize the nonstationary and transitory behavior of random signals and to identify cycles with negligible damage contributions while keeping the sequence of a spectrum. Stress invariant which has been defined as a PbP approach decouples the loading path along fixed directions of a principal reference frame in the deviatory space [2], and define nonlinear cumulative damage criteria based on the damage evaluated at each projection. In the proposed approach, the wavelet transform analysis procedure applies on every projected loading path (i.e. obtained from PbP approach) to identify the segments with negligible damage contributions. Each extracted segment then is replaced with a scaled cycle that produces the equivalent amount of damage.

The remainder of this paper is organized as follows: in Section 2, we provide a detailed description of WTA. Section 3 presents the PbP damage criterion approach. Section 4 explains the wavelet-based editing algorithm and damage equivalent method for replacing extracted segments which have negligible damage contributions to the total damage. In Section 5, the proposed approach is applied to a 2-D fatigue spectrum, followed by conclusions.

2 Wavelet transform analysis

Most fatigue histories exhibit nonstationary characteristics and need time-frequency analysis. Time-frequency methods such as short time Fourier transform (STFT) and Winger-Ville distribution use the same window for the analysis of the entire spectrum, and are unable to analyse spectrums with rapid and slow transient fatigue spectra. The wavelet transform on the other hand uses a different window function for analyzing different frequency band over time, and therefore enables to capture non-stationary features of the spectrums by decomposing the signals into their frequency bands.

The wavelet transform method [9] provides varying time and frequency resolutions by using windows with different lengths. To solve the resolution problem, this technique allows the use of long time window (more precise low-frequency information) and shorter windows (high-frequency information). The wavelet transform calculates a similarity index between the signal being analyzed and the wavelet, called the coefficient. It is the result of the regression of an original signal produced at different scales and different sections on the wavelet. If the index is large, the resemblance is strong otherwise, it is slight. Generally, the wavelet coefficient, C, is expressed with the following integral:

$$C(a,b) = \int_{-\infty}^{\infty} x(t)\psi_{a,b}(t)dt \quad (1)$$

where $x(t)$ is the spectrum/signal to be analysed, $\psi_{a,b}$ is the mother wavelet, parameter a represents the scale factor which is a reciprocal of frequency, the parameter b indicates the time shifting or translation factor, and t is time. The wavelet transform first decompose the spectrum into frequency bands, and then analysing them in time as:

$$CWT(a,b) = \int_{-\infty}^{\infty} x(t)\frac{1}{\sqrt{(a)}}\psi\left(\frac{t-b}{a}\right)dt \quad (2)$$

and

$$x(t) = \frac{1}{C_\psi}\int_{0}^{\infty}\int_{-\infty}^{\infty} CWT(a,b)\psi\left(\frac{t-b}{a}\right)\frac{dadb}{a^2} \quad (3)$$

where C_ψ is the constant that depends on ψ, and $CWT(a,b)$ is the continuous wavelet transform of $x(t)$. The wavelet coefficient indicates how the energy in the signal is distributed in the time-frequency plane.

In this paper, we choose digital wavelet transform (DWT) technique due to its computational efficiency. DWT was first introduced by Daubachies [10]. He realized that by choosing special orthogonal wavelet bases the infinitely redundant CWT can be critically sampled on a dyadic grid. DWT can be computed very efficiently by using techniques from the theory of filter banks. In this method, the DWT of a signal, $x(t)$, is calculated by passing it through a series of filters as shown in Fig. 1. First, the samples are passed through a low pass filter g. The signal is also decomposed simultaneously using a high pass filter h. The outputs include the detail coefficients (from the high-pass filter) and approximation coefficients (from the low-pass). This decomposition is repeated to further increase the frequency resolution and the approximation coefficients decomposed with high and low pass filters and then down-sampled. Fig. 1 shows an example of a three- level filter bank.

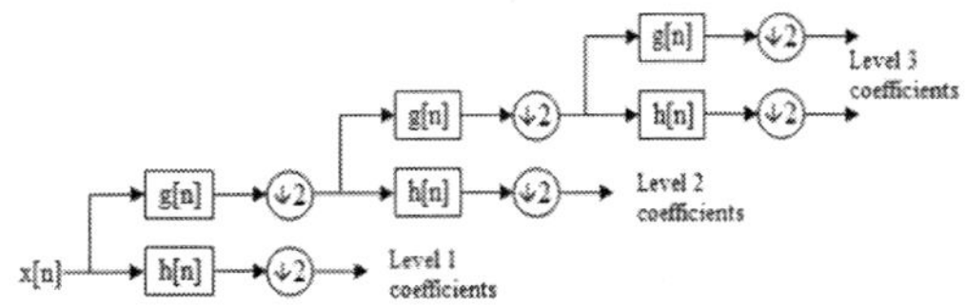

Fig. 1. Three level filter bank.

3 Projection by projection damage criteria

Projection by projection (PbP) method is a type of stress invariant based approaches usually employed in the medium/high-cycle fatigue regimes [11]. The method has shown to provide accurate fatigue damage estimates for variable amplitude multiaxial loading spectrums [12]. This method computes fatigue damage by decoupling the loading path, Ω, along the axes of a principle reference frame in the deviatory space, which are located by the eigenvectors, $\mathbf{U}$, of the covariance matrix, C', of the deviatory stress vector, $\boldsymbol{s}(t)$. The elements of the matrix $C'$are:

$$c_{hk} = cov\big(s_h(t), s_k(t)\big)$$
$$= E[(s_h(t) - s_{m,h}(t))(s_k(t) - s_{m,k}(t))] \quad (4)$$

where $s_{m,h}$ and $s_{m,k}$ are, respectively, the coordinates h and k of the centroid of the path Ω. This matrix has five eigenvalues and five orthogonal eigenvectors for six-dimensional space stress process, $\boldsymbol{\sigma}(t)$. The eigenvalues are the principal moments of inertia, whereas the eigenvectors are the principal directions of the stress tensor path calculated with respect to the centroid itself. For a multiaxial variable amplitude loading spectrum, each projected loading path along the principal direction, $\Omega_{p,i}(t)$, is a uniaxial variable amplitude process. Fatigue cycles in the i_{th} projected load path, $\Omega_{p,i}(t)$, are identified by rainflow method while accounts for both the amplitude, $\left(\sqrt{J_{2a}}\right)_{ij}$, of the deviatoric stress, in each counted cycle (identified by j), and the maximum of hydrostatic stress, $\sigma_{Hmax\,ij}(t)$. Here, $\sqrt{J_{2a}}$ is the square root of the amplitude of second invariant of stress deviator.

Fatigue damage for every projected loading path is evaluated based on a reference fatigue strength curve in a modified Wohler diagram, see Fig. 2.

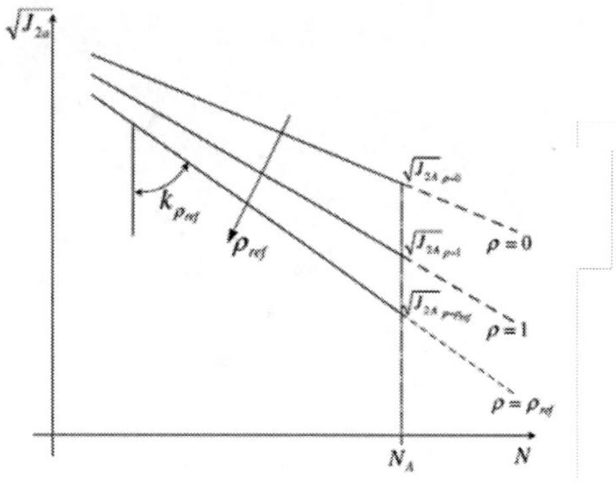

Fig. 2. Modified log–log Wohler diagram for the stress invariant based approach which shows the SN curves for axial ($\rho = 1$), torsion ($\rho = 0$) and general multiaxial stress ($\rho = \rho_{ref}$). (Reprinted from [11]).

The reference curve for an arbitrary multiaxial loading history is located by a reference stress ratio as:

$$\rho_{ref} = \sqrt{3}\frac{\overline{\sigma_{H,max}}}{\sqrt{\sum_i J_{2a_{iref}}}} \tag{5}$$

Reference stress ratio, ρ_{ref}, quantifies the proportionality degree of multiaxial stress in terms of the mean value of the maximum hydrostatic pressure $\overline{\sigma_{H,max}} = mean_{ij}(\sigma_{H\,\max ij})$ and the mean amplitude $J_{2a_{iref}} = \left(mean_j\left(\sqrt{J_{2a}}\right)_{ij}\right)^2$ of all cycle counted in the i_{th} projected loading path, $\Omega_{p,i}(t)$. A value $\rho = 1$ represents pure axial loading, while $\rho = 0$ indicates pure torsion loading.
The calculated ρ_{ref} value allows to estimate the reference fatigue amplitude strength, $\sqrt{J_{2A}}_{\rho_{ref}}$, and the inverse slope k_{ref} of the reference SN curve by a linear interpolation relations [11] as:

$$\sqrt{J_{2A}}_{\rho_{ref}} = \sqrt{J_{2A}}_{\rho=0} + \rho_{ref}\left(\sqrt{J_{2A}}_{\rho=1} - \sqrt{J_{2A}}_{\rho=0}\right) \tag{6}$$

$$k_{ref} = k_{\rho=0} + \rho_{ref}\left(k_{\rho=1} - k_{\rho=0}\right)$$

where $\sqrt{J_{2A}}_{\rho=1}$ and $\sqrt{J_{2A}}_{\rho=0}$ are the fatigue limits, expressed in terms of amplitude of the square root of the second invariant of the stress deviator, under fully-reversed uniaxial and torsional loading, respectively, and $k_{\rho=1}$ and $k_{\rho=0}$ are the corresponding inverse slopes of the Wohler curves. Typically, N_A is considered as 2×10^6 cycles.
Using the above formulations, fatigue damage, D_{ij}, due to each identified cycle, j, can be estimated through the following expression:

$$D_{ij} = \left(\frac{\sqrt{J_{2a_{ij}}}}{\sqrt{J_{2A}}_{\rho_{ref}}}\right)^{k_{ref}} \cdot \frac{1}{N_A} \tag{7}$$

The total damage relative to the i_{th} projection can then be obtained by Palmgren–Miner linear accumulation rule [12] as follows:

$$D(\Omega_{p,i}) = \sum_j D_{ij} \tag{8}$$

A nonlinear accumulation rule finally can be applied to the damage of every projected loading path to obtain the damage for the complete loading path Ω as shown by Eq. 9. This rule was acknowledged by many authors [11, 13].

$$D(\Omega) = \left[\sum_i \left(D(\Omega_{p,i})\right)^{\frac{2}{k_{ref}}}\right]^{\frac{k_{ref}}{2}} \tag{9}$$

4 Wavelet transform and equivalent damage methods for fatigue data editing

The fatigue spectrum editing algorithm uses the wavelet coefficient amplitude as a parameter to determine the threshold value for the low damage cycles identification process. The wavelet coefficient amplitudes are obtained from DWT of each projected loading history, $\Omega_{p,i}$, i=1,3. The extraction processes is performed by searching segments with maximum amplitude less than the specific threshold value over both time history and decomposed levels. After all segments with low contribution to damage are identified, the process of replacing each segment with a cycle with an equivalent damage value is performed in order to compensate for each segment's damage. In this process the equivalent stress amplitude is obtained by the following expression for i=1, 3:

$$\sqrt{J_{2a}}_{eq,z} = \sqrt{J_{2A}}_{\rho_{ref}}\left(N_A \cdot D_z(\Omega_{p,i})\right)^{1/k_{ref}} \tag{10}$$

where $D_z(\Omega_{p,i})$ is the damage produced at z_{th} extracted segment that can be obtained by Eq. 7.
Once the extracted segments are identified and replaced with an equivalent cycle, the statistical features (e.g. root mean square (rms) and kurtosis) of the edited signal will be compared with the original signal. Statistically, the rms is used to quantify the overall energy content of the oscillatory signal and the kurtosis is used as a measure of non-Gaussianity. Mathematically, the rms and kurtosis are defined by the following equations:

$$rms = \left(\frac{1}{n}\sum_{i=1}^{n} x_i^2\right)^{1/2} \tag{11}$$

$$Kurtosis = \frac{1}{n(rms)^4}\sum_{i=1}^{n}(x_i - \bar{x})^4 \tag{12}$$

where x_i is the instantaneous value, n is the number of points, and $\bar{x}$ is the mean of the time history. In each projected axis, the edited spectrum replicates the spectrum statistical parameter and fatigue damaging characteristics of the original time history. The optimum threshold values are accordingly determined and it is based on the effectiveness of retaining the characteristics of the original spectrums.

5 Case study for bending-torsion fatigue history

The discussed wavelet base editing algorithm was applied on an out-of-phase random amplitude bending-torsion fatigue history and for the plane-stress condition. Under bending-torsion condition, the time histories of the random axial stress, $\sigma_{xx}(t)$, and the random shear stress, $\sigma_{xy}(t)$, are numerically produced using following expressions:

$$\sigma_{xx}(t) = \sigma_{xx,m} + \sum_{i=1}^{400} \sigma_{xx,a} \sin(2\pi\omega_i t) \quad (13)$$

$$\sigma_{xy}(t) = \sigma_{xy,m} + \sum_{i=1}^{400} \sigma_{xy,a} \sin(2\pi\omega_i t - \tfrac{\pi}{3}) \quad (14)$$

where the suffix m identifies the mean value of the spectrums, $\sigma_{xx,m} = 300MPa$, the frequency is $\omega_i = 300 \times Normal(0,1)$, $i = 1,2,\ldots,400$, $\sigma_{xy,m} = 200MPa$, the suffix a is the amplitudes, $\sigma_{xx,a} = 3MPa$, $\sigma_{xy,a} = 2MP$, and $\frac{\pi}{3}$ is the phase shift between the applied stress components. It should be noted that the random nature of the frequency in the sinusoidal terms lead to the randomness of amplitudes as well (Please see Fig. 3).

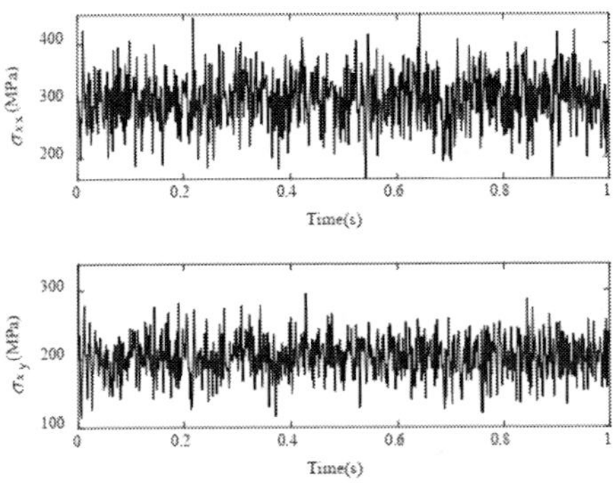

Fig. 3. Random bending-torsion spectrum

The non-null components of the vector associated with the deviatory stresses are:

$$s_1(t) = \frac{1}{\sqrt{3}} \sigma_{xx}(t) \quad (15)$$

$$s_3(t) = \sigma_{xy}(t) \quad (16)$$

Fig. 4 shows the random bending-torsion loading path in the two-dimensional deviatory space, the principal reference frame $(s_{p,1}, s_{p,3})$, and the projected loading paths, $\Omega_{p,i}(t)$, i=1,3.

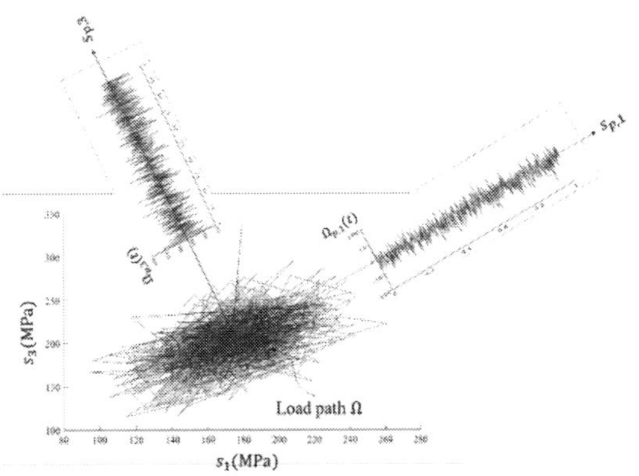

Fig. 4. Bending-torsion random loading path Ω in the two-dimensional deviatory space. The principal reference frame, located by unit vectors $(s_{p,1}, s_{p,3})$, is used to obtain the stress projections, $\Omega_{p,i}(t)$, i=1, 3.

The projected loading elements along the axes of principle reference frame are obtained by projecting the deviatory stresses along the direction of the principal directions. Along these directions, fatigue damage can be expressed as:

$$D_{ij} = \left(\frac{\sqrt{J_{2a_{ij}}}}{\sqrt{J_{2A\ \rho=\rho_{ref}}}} \right)^{k_{ref}} \cdot \frac{1}{N_A} \qquad i = 1,3. \quad (17)$$

The material which is considered for this study is Al 6082-T6, where the parameters of the reference SN curve are adapted from [14], and listed in Table 1:

Table 1. Fatigue properties for the considered material.

Material	Number of available data	ρ	k_ρ	$\sqrt{J_{2A\,\rho}}$	T_σ
Al 6082-T6	45	1	6.90	75	1.07
		0	7.10	77	1.40

After projecting the load history along the principle directions the wavelet-base fatigue editing algorithm is applied on every projected loading path, $\Omega_{p,i}(t)$, i=1, 3. The wavelet transform algorithm uses 12th order (Daubachies) wavelets as the basis functions. Five decomposition levels are selected for DWT of each projected loading history, $\Omega_{p,i}$, i=1, 3. For example, Fig. 5 and 6 depict the decomposition levels for $\Omega_{p,1}$ and $\Omega_{p,2}$, respectively. The wavelet toolbox in MATLAB software was used for DWT.

The low damage contributing cycles identification processes is performed by searching segments with the maximum amplitude less than the threshold values 35 and 60 for $\Omega_{p,1}(t)$ and $\Omega_{p,3}(t)$, respectively, over both time history and decomposed levels. The extracted segments are replaced with an equivalent cycle that produces the same amount of damage in order to retain the damage contributed to each extracted segment as shown in Fig. 7 (a) and Fig. 7(b). Furthermore, the effectiveness of the extracted and replaced segments is evaluated by the preservation of statistical parameter values (kurtosis and rms) for every projected loading, $\Omega_{p,1}(t)$ and $\Omega_{p,3}(t)$, where the difference of $\pm15\%$ in the rms and kurtosis values was used.

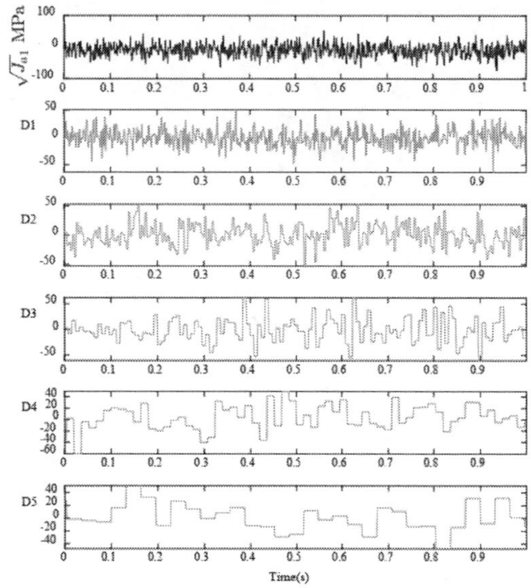

Fig. 5. Wavelet decomposition levels for projected loading path $\Omega_{p,1}$.

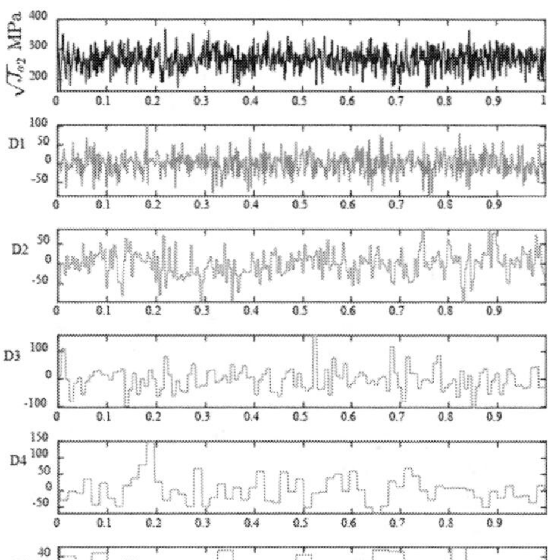

Fig. 6. Wavelet decomposition levels for projected loading path $\Omega_{p,3}$

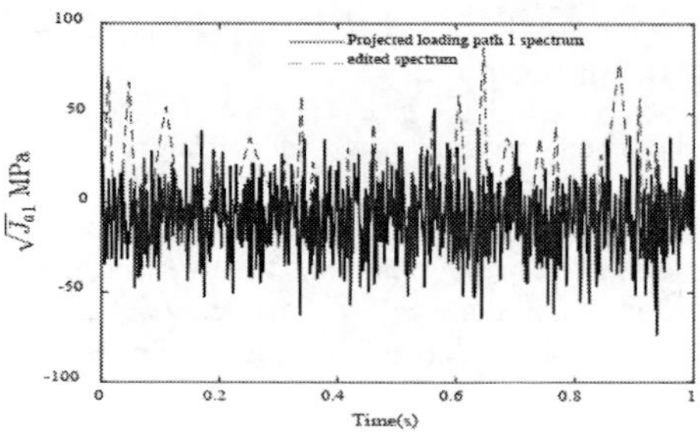

(a)

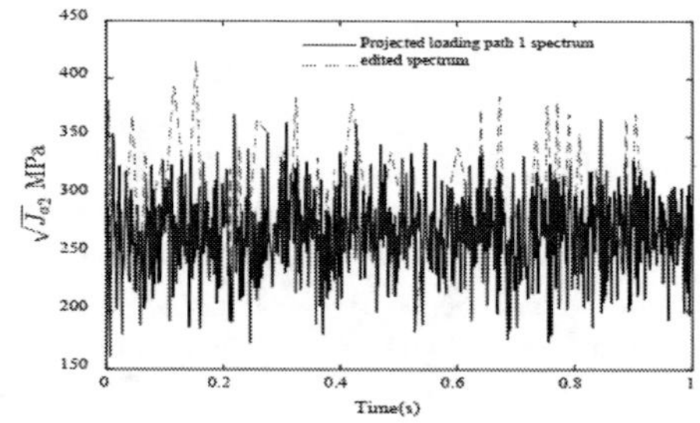

(b)

Fig. 7. Replacement of segments with low contribution to damage with a cycle producing an equivalent damage, (a) for projected loading path, $\Omega_{p,1}$, and (b) for projected loading path, $\Omega_{p,3}$. Blue line depicts the original projected loading path spectrum and red dash line shows the edited spectrum.

The results show that the spectrums in the projected loading paths, $\Omega_{p,1}(t)$ and $\Omega_{p,2}(t)$ are shorted by 73% and 77%, respectively. The error of damage between original projected loadings spectrums $\Omega_{p,1}(t)$ and $\Omega_{p,2}(t)$, and edited spectrums are obtained as -8% and -7.7%, respectively. The total damage error between the original and edited loading spectrum in the principle frame is -7%. The edited load spectrum in each projected axis can finally back-projected to the original frame by using Eq. (15) and (16).

Conclusions

This paper discussed the study of multiaxial fatigue spectrum editing by using the time-frequency wavelet analysis approach and PbP damage criteria method. The newly proposed method allowed for identification of cycles with low contribution to damage in the decoupled loading axes obtained by PbP method. An equivalent damage method was then used to compensate for the damage associated with extracted segments. The numerically produced bending-torsion fatigue history was used to apply the proposed approach. The results of the spectrum editing showed that, on average, 75% of the spectrum lengths in the first and second axes are reduced with the retention of 90% of the original fatigue damage.

Spectrum editing with the proposed approach can accelerate the fatigue test by at least a factor of four to achieve the same or similar fatigue crack initiation damage. The proposed method for multiaxial loading spectrum is entirely new, and the level of acceleration it provides is anticipated to have a significant impact on reducing the costs of testing and provide beneficial solutions not otherwise possible. Further work is planned to systematically investigate the application of the technique for six-dimensional load spectra, evaluation of the proposed approach for real fatigue history spectrums with different material and geometries, and accounting for crack growth damage.

References

[1] A. Lanciotti and L. Lazzeri, Int. J. Fatigue, **14** (1992)
[2] P. R. Stephens, P. M. Dindinger, and J. E. Gunger, Int. J. Fatigue,**19** (1997)
[3] R. Sunder, Int. J. Fatigue, **7** (1985)
[4] R. L. Hewitt, J. P. Weiss, and P. K. Nor, ASTM STP 1439, (2005)
[5] C. S. Oh, Int. J. Fatigue, **23** (2000)
[6] Abdullah. S, J. Eng. Appl. Sci. **2** (2007)
[7] M. McCoy and R. Ogden, Defence Sci. Tech. Org., (2009)
[8] H. J. Grover, Fatigue of Aircraft Structures, (1966)
[9] S. Berry, IEEE, pp. 653-659, (1999)
[10] I. Daubechies, Comm. Pure. Appl. Math, **41**(1988)
[11] A. Cristofori, L. Susmel, R. Tovo, Int. J. Fatigue, **30** (2008)
[12] A. Fatemi, L. Yang, Int. J. Fatigue, **20** (1998)
[13] A. Cristofori, A New Perspective in Multiaxial Fatigue Damage Estimation, Ph.D. thesis, (2007)
[14] L. Susmel L, N. Petrone, Biaxial/ Multiaxial Fatigue Fract. pp. 83–104, (2003)

Study and modelling of anodized 2618 aluminum behavior subjected to multiaxial fatigue

Benaissa Malek*, *Catherine* Mabru and *Michel* Chaussumier

Université de Toulouse, Institut Clément Ader (ICA), UMR CNRS 5312, UPS/INSA/ISAE/ Mines Albi, 3 rue Caroline Aigle, 31400 Toulouse, France.

Abstract: Anodized Aluminum alloys are widely used in aeronautic construction due to their specific mechanical properties. However, anodization process often leads to a decrease of the fatigue resistance of the alloys. In order to identify and characterize the different mechanisms involved in the detrimental effect of anodization of 2618-T851 alloy on its fatigue life and to determine the impact of loading nature, several tests have been performed on specimens with different surface states at various stress ratios. It was found that roughness of machining has no effect unlike the stress ratio or mean stress in tensile tests. The tests on the pickled, anodized, impregnated and sealed specimens showed it was the anodic oxidation step which was the more detrimental for fatigue resistance under tensile loading comparing to the other steps. It has been also observed that no such detrimental effect occurred under torsion loading. Concerning the prediction of fatigue life, two critical plane-based analysis approaches have been used (Morel and Fatemi-Socie criteria) to make fatigue life prediction for uniaxial and multiaxial fatigue test. Comparisons showed that both criteria gives overestimated fatigue life for uniaxial tensile loading under compression mean stress and underestimated fatigue life for tensile-torsion in phase loading.

1 Introduction

In many situations (especially in aircraft construction), it is necessary to modify the surface of previously machined parts by anodizing treatment to improve wear resistance and/or corrosion resistance of aluminum alloys [1]. Anodizing treatment is a successive multi-steps process including: degreasing, pickling, anodic oxidation (growth of anodic film), potentially impregnation and sealing.

However, anodizing treatment may reduce the fatigue resistance [2, 3] and the effect of this surface treatment on the fatigue behavior of aluminum alloys has been extensively studied [2, 3, 4, 5]. According to numerous researchers, the anodic oxidation process can modify the metallurgical composition, the micro-geometry and also internal stresses of the surface. This usually leads to the degradation of the fatigue resistance for aluminum alloys [4, 6, 7].

The available literature in this area has provided evidence that the fatigue life decrease can be related to the brittle properties of the aluminum oxide [3], to the film thickness [8, 9], the type of anodizing process [2, 6, 8], surface pre-treatment [10, 11], sealing step after anodizing [12] and/or substrate microstructure [13].

For 2014 alloy, it has been shown that fatigue cracks of hard anodized samples initiated in the coating in high stress regions and in the interface between coating and substrate in low stress regions [3].

Depending on aluminum alloy microstructure and pickling pre-treatment conditions, several pits could be produced due to galvanic coupling between intermetallic particles and aluminum matrix promoting the dissolution of intermetallic particles or surrounding aluminum matrix. This creates surface defects acting as stress concentrators that promote the initiation of many small fatigue cracks and leads to fatigue resistance decrease [13, 14, 15, 16]. In the case of 2017 alloy, the decrease in fatigue life can also be associated to the brittle nature of the oxide particles containing copper and accumulated at the interface which enhances crack initiation and propagation [13].

In the case of 2A12 and 2024 aluminum alloys, it

*b.malek@isae.fr

 https://doi.org/10.1051/matecconf/201816516010

has been shown that the reduction of its fatigue resistance was principally influenced by anodization treatment creating internal stresses at the interface between the substrate and the anodic layer which resulted in a crack initiation from the interface [2, 6].

Fatigue tests on 2214 aluminum alloy [4], performed on degreased, anodized and sealed specimens have shown that the reduction of the fatigue resistance of 2214 alloy is due to the anodic oxidation-sealing phase. It was observed that anodic layer crazed under fatigue loading and failure occurred from multi-crack initiation sites.

In addition, the nature of loading seems also to have an influence of the effect of anodizing treatment on the fatigue resistance. A reduction of fatigue resistance have been seen on anodized 2014 aluminum alloy subjected to tension fatigue tests, even though the anodized film did not affect the fatigue behavior under the rotating-bending fatigue tests. Under tensile fatigue test, the fatigue resistance of anodized specimen is controlled by the crack initiation behavior in the substrate induced by the rupture of anodized film related to the deformation of substrate during fatigue process [17].

From these various works on 2000 alloys series, it appears that the decrease of the fatigue resistance depends on the anodizing treatment, the loading nature and on the alloy itself. Moreover, all the previous studies deal with the influence of anodization process with respect to uniaxial fatigue. However, industrial parts are often submitted to multiaxial loading in service.

The purpose of the present research project is therefore to quantify the impact of loading nature, surface roughness and anodizing process on fatigue behavior of 2618 alloy and finally to analyze the prediction life models for multiaxial loading and extend them to anodized specimens. In this aim, the fatigue resistance of the bare alloy and then the fatigue behavior of the anodized alloy, considering successively uniaxial and multiaxial loadings is investigated.

The first part of this paper is then devoted to study the fatigue resistance of machined specimens especially regarding surface roughness which can be affected by the surface treatment and has been proved to be able to lead to a decrease of the fatigue resistance [14]. The effect of stress ratio is also investigated in order to prepare multiaxial fatigue modeling. The influence of the different steps of the anodizing process is presented in a second part. At last, these preliminary results are used in two multiaxial fatigue models in order to estimate the fatigue life in the case of tensile-torsion tests.

2 Material and experimental procedures

The material investigated is 2618-T851 aluminum-copper alloy characterized by high mechanical characteristics and low density. The chemical composition is presented on table 1 and the main mechanical properties in table 2.

Table 1: Chemical composition (m%) of 2618 alloy heated at 530°C during 24 hours [18].

Cu	Fe	Mg	Si	Ni	Ti	Zn
1,9-2,7	0,9-1,3	1,3-1,8	0,1-0,25	0,9-1,2	0,04-0,1	0,1

Table 2: Mechanical properties of 2618-T851 alloy.

Young modulus E (GPa)	R_m (MPa)	$R_{p0,2}$ (MPa)	Shear modulus G (GPa)
72	432	372	27

The different steps of heat treatment of the alloy are shown in figure 1. It has also a good resistance in temperature [19] but the corrosion resistance properties of this alloy are bad [20].

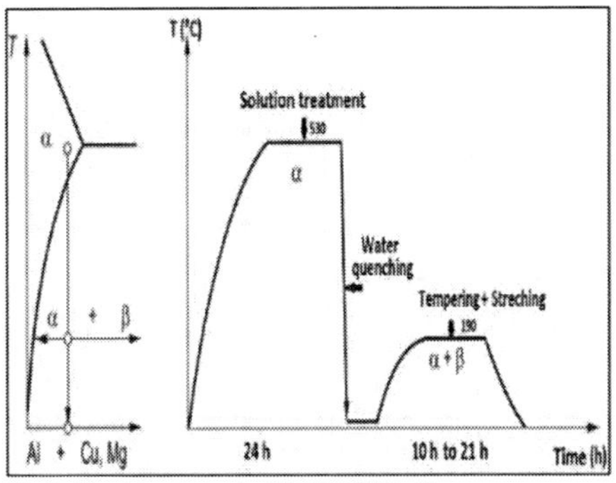

Fig.1. Heat treatment of 2618-T851 alloy.

In terms of the microstructure, 2618-T851 alloy is characterized by the presence of many various intermetallic precipitates localized in grains and grain boundaries as shown in figure 2.

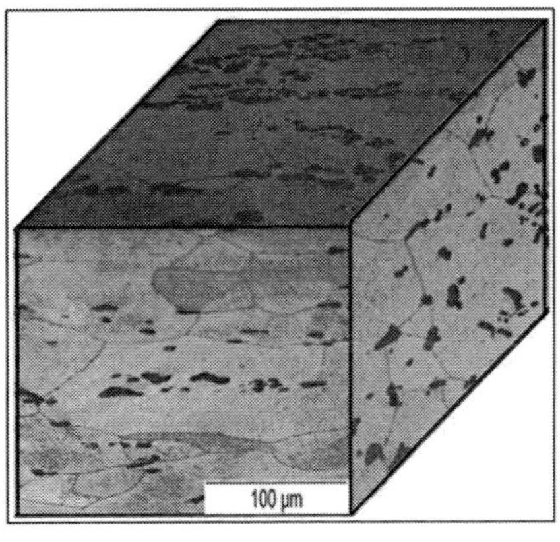

Fig.2. Microstructure of 2618-T851 alloy.

The density of those intermetallic particles is very important and they are rich on aluminum, iron, nickel and copper which is consistent with other works on this alloy [21]. The intermetallic particles are rod-shaped with a size between 1 and 10 µm.

Surface treatments have been performed by IRT M2P (Institut de Recherches Technologiques : Matériaux Métallurgie et Procédés). They are divided into five steps (table 3) in order to get rid of superficial layer, to obtain good corrosion and wear resistance and to strengthen weak areas of the layer.

Table 3: Parameters of different surface treatments processes of 2618-T851 alloy.

Process	Bath nature	Temp (°C)	Time (min)
Degreasing	Sococlean A3431 10%	45	6
Pickling	Socosurf A1858/A1806	50	10
Anodization	Sulfuric acid 200g/L (15V)	18	40
Impregnation	Lanthane 613.3	40	20
Sealing	H_2O	98-100	30

All fatigue tests were performed using cylindrical specimens on a MTS hydraulic multiaxial fatigue machine. Fatigue behavior under uniaxial and multiaxial with constant amplitude was investigated in a preliminary study in tensile and/or torsion at room temperature. The different arithmetic roughness values of the specimens were 0.8, 3.2 and 6.3 µm. Tests have been carried out with different stress ratios (0.1, -1 and -2) and the frequency of all the tests is 10 Hertz.
The specimen geometry and dimensions are given in Fig.3.

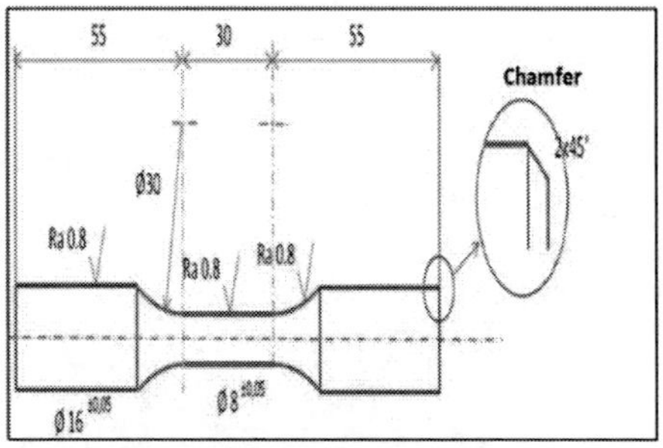

Fig.3. Cylindrical specimen geometry. All dimensions are given in mm.

Same fatigue tests have been conducted on treated specimens with a roughness level of 0.8 µm in order to analyze the effect of each step of the process on the fatigue resistance decrease. Some specimens have been only degreased and pickled, some others have been degreased, pickled and anodized; a third group of specimens has been degreased, pickled, anodized and then impregnated; a last group of specimens have been completely treated (i.e. degreased, pickled, anodized, impregnated and sealed).

3 Results and discussion

3.1 Machined specimens

Fatigue tests have been performed for lives between 10^4 and 10^7 cycles in order to characterize the material parameters needed in the chosen multiaxial fatigue criteria.

3.1.1 Effect of surface roughness

The two following figures (Fig. 4 and 5) show the fatigue test results for the two fatigue loading: uniaxial tension with a stress ratio of 0.1 and torsion with a stress ratio of -1.

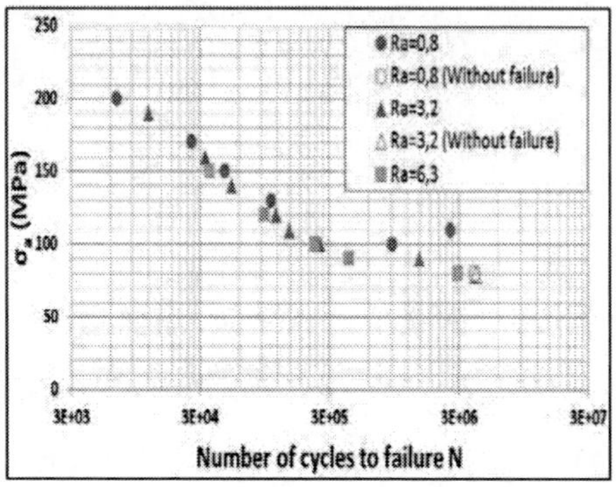

Fig.4. Fatigue test results in tension with different roughness levels, R=0.1.

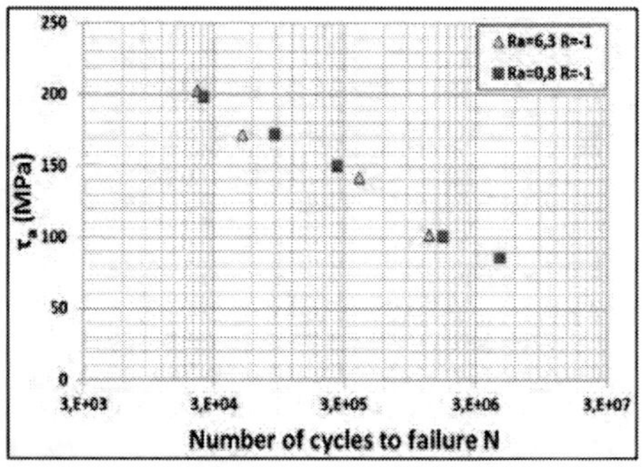

Fig.5. Fatigue test results in torsion with different roughness levels, R=-1.

It appears that roughness level of machined specimens has no effect on the behavior of 2618 alloy whatever the stress level is. It can be concluded that fatigue resistance in not sensitive to surface roughness unlike 2214 aluminum-copper alloy which showed an important sensitivity to surface roughness [14].
These first results suggest that one or both of the two other surface characteristics (metallurgy and residual stresses) is/are the governing factor(s) in fatigue failure. The influence of metallurgy can be intuitively deduced from metallurgical observation which indicated the presence of numerous rod-shaped intermetallic particles inhomogeneously distributed in the material [21, 22].

3.1.2 Effect of load ratio

Uniaxial tensile fatigue tests with various stress ratios have been carried out on machined specimens with arithmetic roughness of 0.8 µm. Fig. 6 illustrates the obtained results for stress ratio of 0.1 and -2.

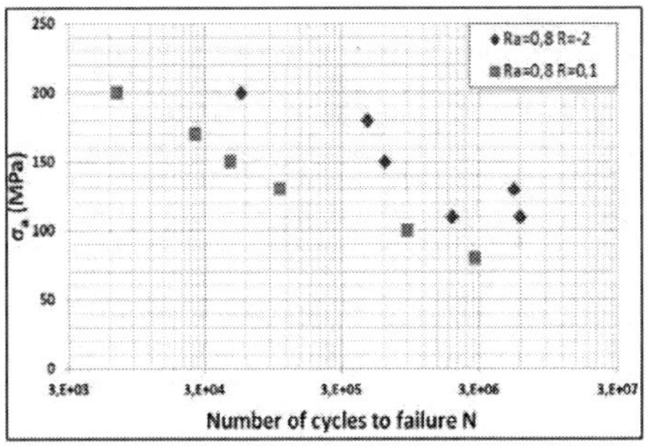

Fig.6. Fatigue test results in tension with different load ratios, Ra=0.8 μm.

The mean stress has an effect on the number of cycles to failure as for numerous metallic alloys [16, 17].

3.2 Surface treated specimens

As mentioned above, fatigue tests were carried out after each step of the whole anodizing process (table 3) on machined specimens with an initial roughness of 0.8 μm.

The objective is to characterize the effect of each step of the anodization process on fatigue life and to define which one is the detrimental step of the whole process and leads to a decrease of the fatigue resistance.

Fig. 7 and 8 show the obtained results for tensile and torsion fatigue tests respectively.

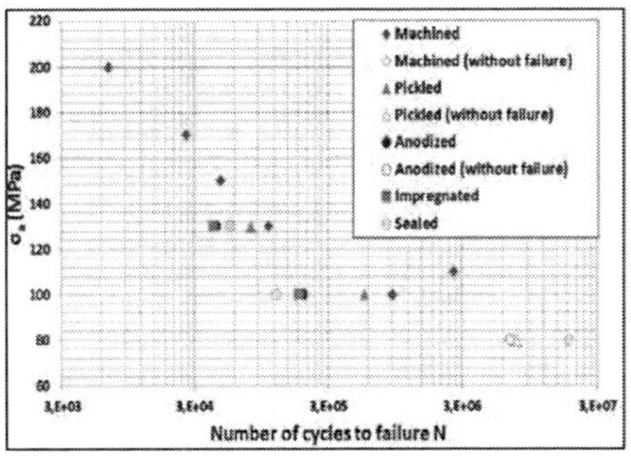

Fig.7. Fatigue tests in tension with different surface treatments, R=0.1 and Ra=0.8 μm.

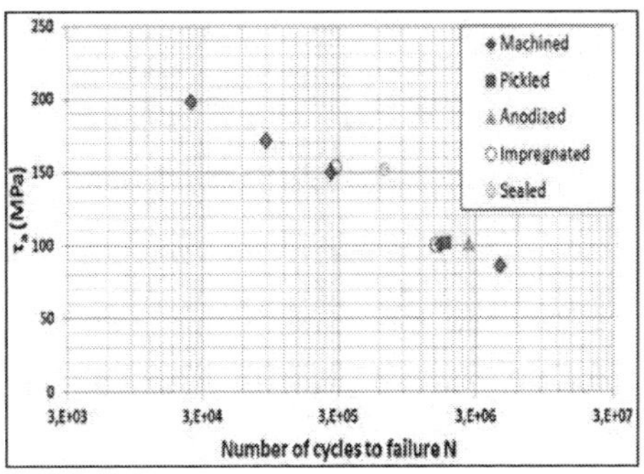

Fig.8. Fatigue test results in torsion with different surface treatments, R=-1and Ra=0.8 μm.

Under tensile fatigue tests, it appears that the main part of the fatigue resistance decrease is mainly due to anodic oxidation: degreasing and pickling have slight influence as it was also observed for 2214 alloy treated with similar conditions [14].

In addition, it seems that impregnation or sealing have quite negligible influence comparing to anodic oxidation.

Preliminary results for torsion fatigue tests show that there is no detrimental effect of the successive surface treatment steps on the fatigue life. It can be concluded that fatigue resistance decrease is mainly due to the anodic oxidation step combined with nature of the loading.

3.3 Fracture surfaces

Some initial observations were made to identify the different modes of fracture of machined and impregnated specimens. The main microscopic fracture features are a combination of ductile intergranular and transgranular fracture. Fracture surfaces of machined specimens exhibit multiple initiation sites. On the contrary, fracture surfaces of impregnated specimens show the existence of a unique crack initiation site associated with cracks in the anodic layer, as shown in Fig. 9. As no crack is observed in other parts of the anodic layer, it could suggest that the fatigue crack initiated from this impregnated layer. This has been observed for tensile and torsion tests indicating an effect of anodization treatment on the mechanisms of crack initiation for both loadings. This is quite surprising concerning torsion as no effect of surface treatment on fatigue life has been detected (Fig. 8).

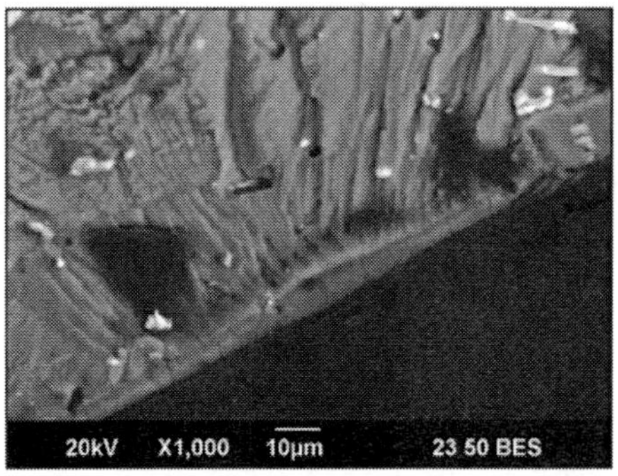

Fig.9. Fracture surfaces of an impregnated specimen tested in torsion, τ_a=150 MPa, Ra=0.8 μm and R=-1.

3.4 Fatigue life predictions

Critical plane-based fatigue life analyses were performed using Morel [28] and Fatemi-Socie [26] models to predict fatigue life of machined specimens in medium cycle fatigue regime for multiaxial loadings in phase with sinusoidal signal and constant amplitude.

3.4.1 Morel's model

In the high cycle fatigue regime, as crack initiation phenomenon takes place at the scale of a few grains, a damage variable at this scale is introduced.

Morel considers accumulated plastic mesostrain on a critical plane as the damage variable. In this model, a crack is considered to be initiated when the damage variable reaches a critical value [25, 27].

The criterion is written as:

$$max_{\theta,\varphi}\,(T_\sigma\,(\theta,\varphi) + \alpha.P_{max}) \leq \beta \qquad (1)$$

where:

P_{max} is the maximum mesoscopic (equal to the macroscopic) hydrostatic stress,

T_σ (θ,φ) is the damage parameter proportional to an upper bound of the plastic mesostrain accumulated on an elementary material shear plane Δ (also average value of T_a), it is a function of the orientation of this plane through the angles θ and φ.

This parameter is estimated by an integration carried out through the whole area of the plane Δ [30]:

$$T_\sigma\,(\theta,\varphi) = \sqrt{\int_{\psi=0}^{2\pi} T_a^2\,(\theta,\varphi,\psi)d\psi} \qquad (2)$$

Where T_a is the amplitude of the macroscopic resolved shear stress acting on a line of the plane Δ directed by **m** as showed in Fig. 10. This line is located by the angle ψ that makes with respect to an arbitrary but fixed axis in Δ [27].

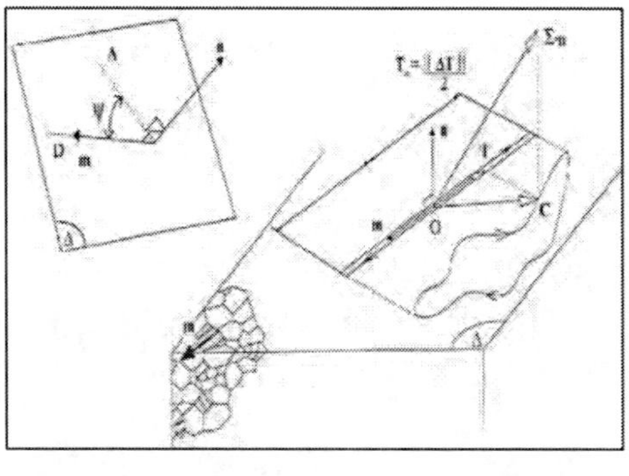

Fig.10. Path of the macroscopic shear stress C acting on Δ and corresponding path of T acting on an easy glide direction **m** [25].

The parameters α and β can be related to the fatigue limits of two standard fatigue tests, e.g. tensile-compression with a stress ratio R, s_R, and fully reversed torsion, t:

$$\beta = \sqrt{\pi}\,t \qquad \alpha = \frac{\sqrt{\pi}(t - \frac{s_R}{(1-R)})}{\frac{2s_R}{3(1-R)}} \qquad (3)$$

where 's_R' and 't' represent respectively the fatigue resistance in tensile test under stress ratio R and in fully reversed torsion.

Fig. 11 shows the result of identification of Morel model parameters from experimental from fitted S-N curves obtained from uniaxial loading in tension (R=0.1) and torsion (R=-1).

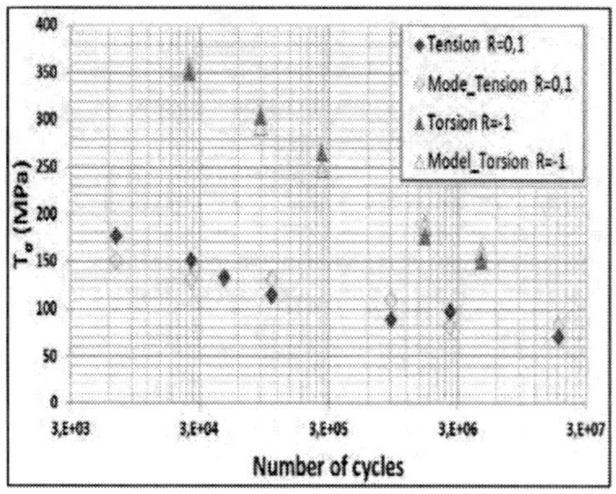

Fig.11. T_σ-N curve corresponding to Morel model and the experimental data, Ra=0.8 µm and f=10Hz.

3.4.2 Fatemi-Socie model

It is usually argued and accepted that fatigue crack initiation involves localized plastic deformation in persistent slip bands even in the high cycle fatigue region [26].
Fatemi-Socie critical plane approach is based on the physical interpretation of fatigue damage. This latter is supposed to be governed by the maximum shear stress amplitude ($\Delta\gamma_{max}/2$) and the maximum normal stress ($\sigma_{n,max}$) that are expressed on the model by the equation 4 below:

$$\frac{\Delta\gamma_{max}}{2}(1 + k\frac{\sigma_{n,max}}{S_y}) = constant \qquad (4)$$

where:

S_y is the material monotonic yield strength; it serves to normalize the maximum normal stress in order to preserve a dimensionless term,

k is a material constant depending on number of cycles to failure.

Equation 4 can be written in terms of shear strain-life properties obtained from fully reversed torsion tests [28, 29] as:

$$\frac{\Delta\gamma_{max}}{2}(1 + k\frac{\sigma_{n,max}}{S_y}) = \frac{\tau_f'}{G}(2N_f)^{b_o} + \gamma_f'(2N_f)^{c_o} \qquad (5)$$

where:
G is the shear modulus, τ_f' is the shear fatigue strength coefficient, γ_f' is the shear fatigue ductility coefficient, b_o and c_o are shear fatigue strength and shear fatigue ductility exponents, respectively.

Fig. 12 shows the result of identification of Fatemi-Socie model parameters from experimental tests obtained from uniaxial loading in tension (R=0.1) and torsion (R=-1) using the least square on the whole set of fatigue data.

 https://doi.org/10.1051/matecconf/201816516010

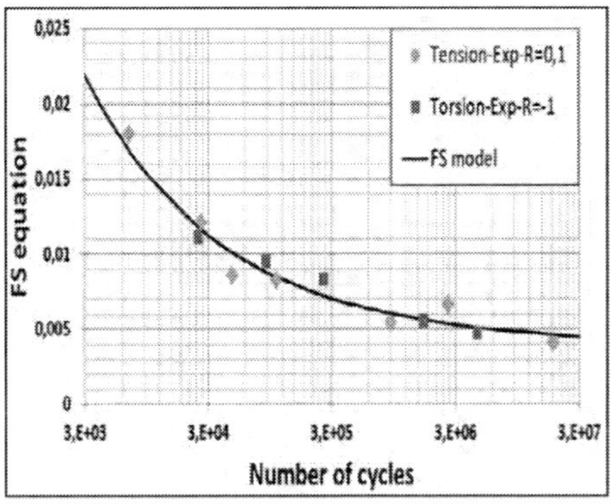

Fig.12. Fatemi-Socie model compared to experimental data, Ra=0.8 μm.

3.4.3 Fatigue life predictions

Uniaxial fatigue test results for R stress ratio equal to -2 have been compared to fatigue life predictions given by both criteria. Some multiaxial fatigue tests have been carried out in tension and torsion in phase and with sinusoidal constant amplitude and results have been also compared to the given predictions. Results of these comparisons are illustrated in Fig. 13 and 14.

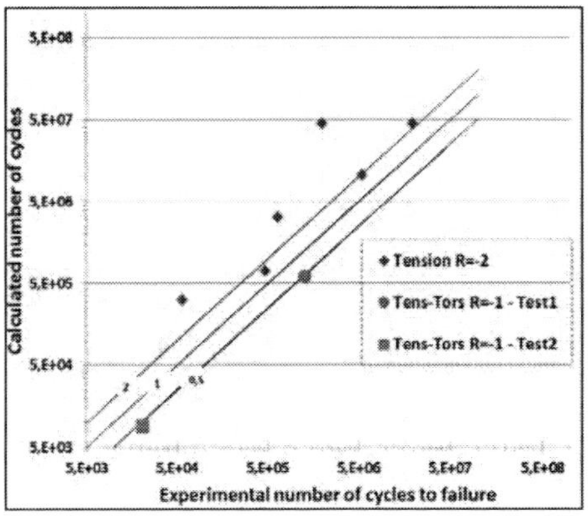

Fig.13. Morel model predicted fatigue life vs experimental fatigue life for multiaxial and uniaxial fatigue tests, Ra=0.8 μm.

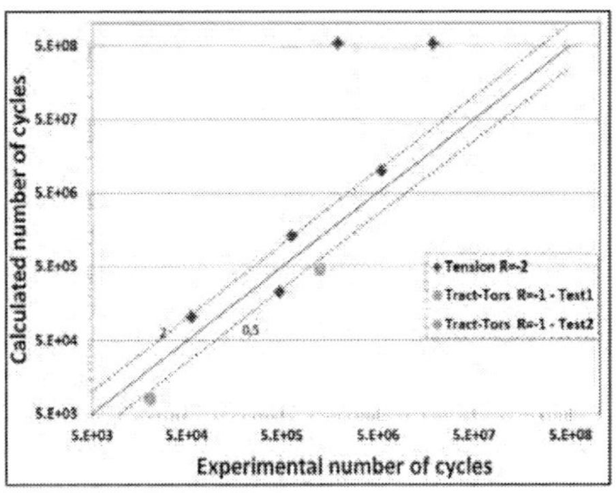

Fig.14. Fatemi-Socie model predicted fatigue life vs experimental fatigue life for multiaxial fatigue tests, Ra=0.8 μm.

Both criteria overestimate the fatigue life for uniaxial tensile test under R=-2. One explanation is that these cases are not in the load ratio range used for material characterization; it would be better to use also these results to perform the fitting of the model. On the contrary, for tension-torsion tests, both criteria give underestimated fatigue life but these predictions are closer to the line 2N which can be considered as quite good. More fatigue test will be performed to ensure these first results.

4 Conclusions and perspectives

Fatigue resistance of bare and anodized 2618 alloy has been investigated. Tests on machined specimens show that the surface roughness has no effect on fatigue lives. As it was expected, the stress ratio has a high influence in tensile loading.

Fatigue tests performed after each step of the anodizing process show that for fatigue tensile tests, the anodic oxidation step in itself is the main detrimental step of the whole process, leading to a decrease of the fatigue life. The other steps such as degreasing and pickling or impregnation or sealing have no or very slight influence on the fatigue resistance.

However, fatigue resistance of anodized 2618 alloy depends also on the loading nature; no effect of the surface treatment on fatigue life is noticed for torsion tests.

Fracture surfaces analysis has shown the existence of one initiation site with the cracking of the anodic layer while bare specimen exhibited multisite initiation.

Considering multi-axial fatigue life prediction, both Morel's and Fatemi-Socie's criteria seem to give underestimated fatigue life under proportional tensile-torsion loading and overestimated fatigue life for tensile tests under a stress ratio of -2. More experimental data and analysis are still needed to improve fatigue life prediction.

Further works are currently performed to investigate existing prediction life models for machined specimens by integrating the internal pressure with tension and torsion.

Next work will be devoted to the extension of predicting models to anodized specimens.

Special thanks are given to Université Paul Sabatier (Toulouse) for financial support of this research study and to IRT M2P, in particular Joffrey Tardelli (IRT M2P), Benjamin Mouls (IRT M2P), Cédric Marchetto (IRT M2P) and Aimé Ramakistin (INEOSURF).

5 References

[1] S. D, Cramer and B.S. Covino, ASM Handbook, **13**, (2003)
[2] B. Nie, Z. Zhang, Z. Zhao and Q. Zhong, Mat Des, **50**, 1005-1010 (2013)
[3] R. Sadeler, J Mat Sc, **208**, 5803-5812 (2006)

[4] M. Shahzad, M. Chaussumier, R. Chieragatti, C. Mabru and F. Rezai-Aria, Surf Coat Tech, **206**, 2733-2739 (2012)
[5] N. Chahboun, E. Rocca, D. Veys-Renaux, M. Augros, M. Boutoba and N. Caldeira, J Elect Soc, **163**, 69-75 (2016)
[6] J. Camargo, H. Voorwald, M. Cioffi and M.Y.P. Costa, Surf Coat Tech, **201,** 9448-9455 (2007)
[7] D. Zavodska, E. Tillova, M. Guagliano, L. Kucharikova and M. Chalupova, Mat Tod, **4**, 6001-6006 (2017)
[8] B. Lonyuk, I. Apachitei and J. Duszczyk, Surf Coat Tech, **201**, 8688–8694 (2007)
[9] E. Cirik and K. Genel, Surf Coat Tech, **202**, 5190–5201 (2008)
[10] M. Chaussumier, C. Mabru, M. Shahzad, R. Chieragatti and F. Rezai-Aria, I J Fatigue, **48**, 205–213 (2013)
[11] M. Chaussumier, C. Mabru, R. Chieragatti and M. Shahzad, Pro Eng, **66**, 300–312 (2013)
[12] X. Zhao, G.Wei, Y. Yu, Y. Guo and A. Zhang, Surf Rev Lett, **22**, 1–7 (2015)
[13] C. Fares, L. Hemmouche, M.A. Belouchrani, A. Amrouche, D. Chicot and E.S. Puchi-Cabrera, Mat Des, **86**, 723-734 (2015)
[14] M. Shahzad, M. Chaussumier, R. Chieragatti, C. Mabru and F. Rezai-Aria, Surf Coat Tech, 2733-2739 (2012)
[15] G. E. Thompson, L.C. Zhang, J. E. Smith and P. Skeldon, Corros Sci, **55**, 1052-1060 (1999)
[16] B. Priet, G. Odemer, C. Blanc, K. Giffard and L. Arurault, Surf Coat Tech, **307**, 206-2019 (2016)
[17] K. Shiozawa, H. Kobayashi, M. Terada and A. Matsui, Jap Soci Mech Eng, **66**, 2170-2175 (2000)
[18] I. Özbek, Mat Char, **58**, 312-317 (2007)
[19] W. Luo, M. Yan and W. Z. Zhu, J Mat Sci, **37**, 2685–2691 (2002)
[20] A. Djakovici, B.F. Dyson and M. Mclean, *2nd International Conference on deformation processing and structure of materials*, (2005)
[21] C. Rockenhäuser, S. Schriever, P. Von Hartrott, B. Piesker and B. Skrotzki, Mat Sci Eng, **716**, 78-86 (2018)
[22] F. Nový, M. Janeček, and R. Král, J All Comp, **487**, 146-151 (2009)
[23] M. Morgantini, D. MacKenzie, T. Comlekci and R-V. Rijswick, Proc Eng, **213**, 581-588 (2018)
[24] J. Janousek, M. Baldab and J. Svobodab, App Comp Mech, **7**, 31-38 (2013)
[25] F. Morel, Fat Frac Eng Mat Str, **21**, 241-256 (1998)
[26] A. Fatemi and D. F.Socie, Fat Frac Eng Mat Str, **11**, 149-165 (1998)
[27] F. Morel, N. Ranganathan, J. Petit and A. Bignonnet, Euro Str Integ Soc, **25**, 87-100 (1999)
[28] N-R. Gates and A. Fatemi, I J Fat, 100, 322-336 (2017)
[29] A. Fatemi, Multixial Stresses, Chapter 10, Multiaxial Fatigue, University of Toledo, *Advances in Fatigue and Fracture mechanics*, (2014)

Mechanical and thermo-mechanical response of a lead-core bearing device subjected to different loading conditions

Todor Zhelyazov[1,*], *Rajesh* Rupakhety[2], and *Simon* Olafsson[2]

[1]Technical University of Sofia, 8 Kliment Ohridski blvd, Sofia 1000, Bulgaria
[2]University of Iceland, Earthquake Engineering Research Centre Austurvegur 2a, 800 Selfoss,Iceland

Abstract. The contribution is focused on the numerical modelling, simulation and analysis of a lead-core bearing device for passive seismic isolation. An accurate finite element model of a lead-core bearing device is presented. The model is designed to analyse both mechanical and thermo-mechanical responses of the seismic isolator to different loading conditions. Specifically, the mechanical behaviour in a typical identification test is simulated. The response of the lead-core bearing device to circular sinusoidal paths is analysed. The obtained shear displacement – shear force relationship is compared to experimental data found in literature sources. The hypothesis that heating of the lead-core during cyclic loading affects the degrading phenomena in the bearing device is taken into account. Constitutive laws are defined for each material: lead, rubber and steel. Both predefined constitutive laws (in the used general-purpose finite element code) and semi-analytical procedures aimed at a more accurate modelling of the constitutive relations are tested. The results obtained by finite element analysis are to be further used to calibrate a macroscopic model of the lead-core bearing device seen as a single-degree-of-freedom mechanical system.

1 Introduction

The lead-rubber bearing is one of the most promising developments in the field of base isolation [5, 6].

Base isolation devices have been increasingly used around the world, specifically in New Zealand, Japan, USA and Italy, in the construction of bridges, large structures and residential buildings.

The lead-core rubber bearing is inherently an elastomeric bearing that is modified by placing a lead plug down its centre.

The lead-plug yields relatively quickly under shear deformation and LCR bearing is then seen to have well-pronounced hysteresis response under cyclic loading paths.

When the lead-core bearing is subjected to repeated loading, the effects of heating of the lead plug should be taken into account because it has reportedly degrading effects on the mechanical properties of the isolator [7, 8, 9]

The lead-core bearing device is considered as a multiple component system. Material models are defined for all the compounding materials – rubber, lead and steel.

In the numerical simulations, the lead-core bearing device is subjected to a time-dependant kinematic excitation.

2 Material models

2.1 Rubber

A Neo-Hookean material model is employed to simulate the hyperelastic response of the rubber. The constitutive relations for rubber are derived from a strain energy potential of the following form:

$$W = \frac{G}{2}\left(\overline{I_1} - 3\right) + \frac{1}{d}(J-1)^2 \tag{1}$$

In equation (1) G stands for the initial shear modulus of the material and $\overline{I_1}$ is the first deviatoric strain invariant [5]:

$$\overline{I_1} = J^{-2/3}\sum_{i=1}^{3}\lambda_i^2 \tag{2}$$

where λ_i are principal stretches and $d=2/K$ with K -the initial bulk modulus of the material which for an isotropic material has the form:

$$K = \frac{E}{3(1-2\nu)} \tag{3}$$

In equation (3) ν is the Poisson's ratio.

J that is defined as the determinant of the deformation gradient:

* Corresponding author: todor.zhelyazov@tu-sofia.bg

$$J = \det\left(F_{ij}\right) \tag{4}$$

2.2 Lead-core

The lead-core is seen to have elastic-plastic behaviour with hardening after yielding [6, 7 8]. More precisely, bilinear isotropic hardening is used (i.e., the yield surface expands uniformly in all directions with plastic flow).

The yield criterion is generally defined as a scalar function of the stress and internal variables (ξ):

$$f(\sigma, \xi) = 0 \tag{5}$$

For isotropic hardening the yield criterion takes the form:

$$f(\sigma) - \sigma_y(\xi) = 0 \tag{6}$$

Alternatively, a fictitious 1-D model of the lead plug is defined. The boundary conditions for the fictitious model are taken from the transient numerical analysis. The fictitious configuration is used to implement a damage-based model for the lead core. Thus, the constitutive relations can be derived from the following state potential [9]:

$$\psi = \frac{1}{\rho}\begin{bmatrix} \frac{1}{2} a_{ijkl}\varepsilon_{ij}^{e}\varepsilon_{kl}^{e}(1-D) + \\ + R_{\infty}\left(r + \frac{1}{b}e^{(-br)}\right) + \frac{X_{\infty}\gamma}{3}\alpha_{ij}\alpha_{ij} \end{bmatrix} \tag{7}$$

In equation (6) ρ stands for the material density, $\mathbf{a}_{ijkl}$ is the elasticity tensor, $\boldsymbol{\varepsilon}_{ij}^{(e)}$- elastic part of the strain tensor, D denotes the damage variable, R_{∞} and b are material parameters characterizing isotropic hardening, X_{∞} and γ which characterize kinematic hardening, $\boldsymbol{\alpha}_{ij}$ is the stress tensor associated with the back stress $\mathbf{X}_{ij}^{D}$.

The law of elasticity coupled with damage reads:

$$\sigma_{ij} = \rho\frac{\partial\psi}{\partial\varepsilon_{ij}} = a_{ijkl}\varepsilon_{kl}^{e}(1-D) \tag{8}$$

Upon further customization of the finite element software, this procedure can be further optimized in order to reduce the computational resources and to minimize the computational time.

2.3 Steel

Generally speaking, the mechanical response of steel can be simulated by assuming isotropic or kinematic hardening after yielding. However, yield stress is not usually reached in the steel components of a lead-core bearing device. Taking into account these considerations, steel is assumed to have linear elastic response.

3 Finite element model and mesh generation

Solid 95 is used to mesh the steel shims, as well as for the upper and for the lower thick steel plates. This finite element is defined by 20 nodes having three degrees of freedom per node: translations in the nodal x, y, and z directions.

Solid186 – a 3-D 20-node solid element that exhibits quadratic displacement behavior is used to mesh the volumes created for the rubber layers. The element is defined by 20 nodes having three degrees of freedom per node: translations in the nodal x, y, and z directions. The finite element is chosen because of the supported mixed formulation capability for simulating deformations of fully incompressible hyperelastic materials.

In order to study the thermos-mechanical response of the lead core, the coupled field finite element: Solid226 a 3-D 20 node coupled-field solid has twenty nodes with up to five degrees of freedom per node is used.

In order to avoid spurious rigidity in the simulated mechanical response, the full integration option is suggested for the finite element analysis.

The employed finite element types are summarized in Table 1.

Table 2. Finite element types.

Finite element type:	Used for:
Solid 95	Steel elements
Solid 186	Rubber
Solid 226	Lead - core

Two test cases are studied. For each case an accurate and explicit 3-D model is built.

3.1 Test case 1

The generated finite element mesh is displayed in Figure 1.

The existing symmetries are taken into account and a half-model spaced is studied. Appropriate boundary conditions are defined on the plane of symmetry (*y=0*).

Nodes on the bottom surface (*z=-25*) are restrained, i.e., displacement along the *x*-, *y*- and *z*-directions of the nodal coordinate system are set to zero.

Figure 1. Finite element mesh. A half-model space: top and bottom thick steel plates, as well as steel shims – in grey, rubber layers – in red and lead plug – in blue.

The half-model space contains a total of 24624 finite elements: 14160 Solid95 finite elements, 6720 Solid186 finite elements and 3744 Solid226 finite elements.

3.2 Test case 2

Simulation of the mechanical response f a 356 x356 mm lead-core rubber bearing that contains seven steel plates 3 mm in thickness each and six 16 mm rubber layers [11].

A total of 39866 finite elements are employed to mesh the quarter-space finite element model: 7658 Solid95 finite elements are used for the steel plates, 26256 Solid186 finite elements – for the rubber layers and 5952 Solid226 finite elements for the lead core.

Nodes on the bottom surface (*z=0*) are restrained, i.e., displacement along the *x*-, *y*- and *z*- directions of the nodal coordinate system are set to zero.

The finite element mesh is displayed in Figure 2.

Figure 2. Finite element mesh. A quarter-model space: steel plates are plotted in grey, rubber layers – in red and lead plug – in blue.

4 Finite element analysis

Time dependant displacements are applied to nodes at the top surface (nodes for which z=142mm for test case 1 and nodes for which z=117 for test case 2).

The time-dependant kinematic excitation is defined as an array that contains the discrete values of the displacements applied to node on the top surface of the bearing device.

To recover the experimental data in [11], the lead-core bearing device (Test case 2) is loaded by using sinusoidal paths of amplitude 120 mm and of frequency of 0.9Hz (Figure 3).

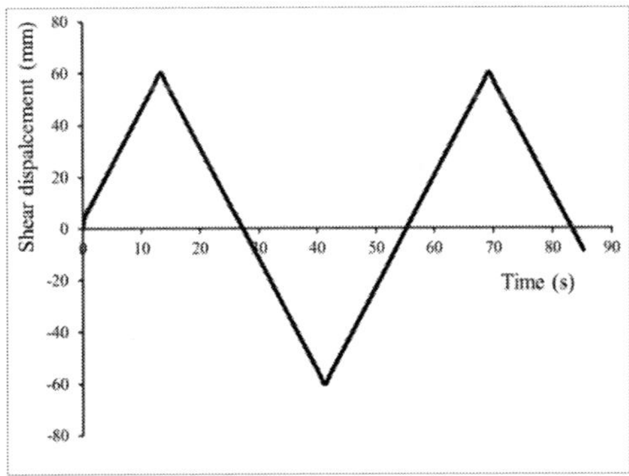

Figure 3. Displacements applied to nodes on the top surface to simulate a typical identification test.

A transient analysis is then performed.

4 Concluding remarks

The mechanical and the thermo-mechanical responses of a lead-core bearing device subjected to different loading conditions have been studied.

Explicit 3-D finite element models have been created for bearing devices of different geometries (test case 1 and test case 2).

Numerical simulations of the mechanical behaviour of the bearing devices to sinusoidal paths of the shear displacement accompanied by vertical load have been carried out.

A coupled-field analysis has been performed in order to study the reportedly degrading effect of the temperature on the mechanical performances of the bearing device subjected to cyclic loading.

References

1. W. H. Robinson, and A.G.Tucker, Bulletin of the New Zealand National Society for Earthquake Engineering, pp.151-153 10, 3 (1977)
2. R. Park and R. W. G. Blakeley, Structures Committee Summary 3, Road Research Unit, National Roads Board, New Zealand (1978)
3. M. C. Constantinou, A. Whittaker, *Technical Report MCEER-07-0012*; Multidisciplinary Center for Earthquake Engineering Research, State University of New York at Buffalo: Buffalo, NY, USA (2008)
4. I. V. Kalpakidis, M. C. Constantinou, J. Struct. Eng., pp. 1450–1461 135 (2009)

5. I. V. Kalpakidis, M. C Constantinou, A. S.Whittaker, Eng. Struct. Dyn., pp. 1533–1549 39 (2010)
6. Crisfield M. A., *Non-Linear Finite Element Analysis of Solids and Structures. Volume 2: Advanced Topics* (John Wiley & Sons, Chichester UK, 1979)
7. Hill R., *The Mathematical Theory of Plasticity* (New York: Oxford University Press, 1983)
8. Rice, J. R., Continuum mechanics and thermodynamics of plasticity in relation to microscale deformation mechanisms. *Constitutive Equations in Plasticity* (Ed. A. Argon. Cambridge, MA :MIT Press, 1975. 23-79).
9. J. L. Chaboche, International Journal of Plasticity **5**, 3 247-302 (1989)
10. Lemaitre J., *A Course on Damage Mechanics* (Springer, 1996)
11. R. I. Skinner, R. G. Tyler, A.J. Heine and W.H. Robinson, Bulletin of the New Zealand National Society for Earthquake Engineering, pp. 22-36 **13**, 1 (1980)

Evaluation of Estimation Methods for Shear Fatigue Properties and Correlations with Uniaxial Fatigue Properties for Steels and Titanium Alloys

Shahriar Sharifimehr[1,*], *Ali* Fatemi[2]

[1]Mechanical, Industrial and Manufacturing Engineering, University of Toledo, Toledo, Ohio, USA
[2]Mechanical Engineering, University of Memphis, Memphis, Tennessee, USA

Abstract. The goal of this study was to evaluate the accuracy of different methods in correlating uniaxial fatigue properties to shear fatigue properties, as well as finding a reliable estimation method which is able to predict the shear fatigue behavior of steels and titanium alloys from their monotonic properties. In order to do so, axial monotonic as well as axial and torsion fatigue tests were performed on two types of steel and a Ti-6Al-4V alloy. The results of these tests along with test results of 23 types of carbon steel, Inconel 718, and three types of titanium alloys commonly used in the industry were analyzed. It was found that von Mises and maximum principal strain criteria were able to effectively correlate uniaxial fatigue properties to shear fatigue properties for ductile and brittle behaving materials, respectively. Also, it was observed that for steels and Inconel 718 obtaining shear fatigue properties from uniaxial fatigue properties which are in turn calculated from Roessle-Fatemi estimation method resulted in reasonable estimations when compared to experimentally obtained uniaxial fatigue properties. Furthermore, a modification was made to the Roessle-Fatemi hardness method in order to adjust it to fatigue behavior of titanium alloys. The modified method, which was derived from uniaxial fatigue properties of titanium alloys with Brinell hardness between 240 and 353 proved to be accurate in predicting the shear fatigue behaviors.

1 Introduction

Mechanical properties, both monotonic and fatigue, are key information in efficient design of components. Monotonic properties are usually easy to obtain and available for many metallic materials. However, characterizing the fatigue behavior of materials requires conducting time consuming fatigue tests. Therefore, fatigue properties of metallic materials are not as commonly available as their monotonic properties. This is in spite of the fact that fatigue is a major consideration in design and up to 90 percent of all mechanical failures are fatigue failures [1].

In many cases engineers tend to estimate the fatigue properties of materials throughout the design process. The estimation methods often correlate the uniaxial monotonic properties of metallic materials, and especially steels, with their uniaxial strain-life fatigue properties. Methods such as four-point correlation proposed by Manson [2], modified four-point correlation proposed by Ong [3], universal slopes proposed by Manson [2], modified universal slopes proposed by Muralidharan and Manson [4], and uniform material law proposed by Bäumel and Seeger [5] have shown to result in close predictions of fatigue behavior for many steels. While the aforementioned studies used monotonic properties in estimation of fatigue properties, Roessle and Fatemi [6] proposed and evaluated a correlation method which only used elastic modulus and Brinell hardness of steels. The Roessle-Fatemi method was shown to result in close predictions for uniaxial fatigue tests on 69 commonly used steels. This method has been widely used by many researchers and many studies such as [7-10] have shown good prediction results using this method.

Uniaxial fatigue properties estimated from the aforementioned methods are able to model the fatigue behavior of components which are subjected to uniaxial cyclic loadings. However, in many cases the load histories subjected to components are either shear loads or a combination of normal and shear loads resulting in a multiaxial stress state. Even under uniaxial loading the stress state may still be multiaxial due to many factors such as complex geometry of components and presence of residual stresses. Shear fatigue properties are used to model the fatigue behavior of materials which are subjected to cyclic shear loadings. Equivalent stress and strain methods such as von Mises are able to correlate the fatigue behavior of many metallic materials under different proportional load paths with their uniaxial fatigue behavior. However, the equivalent stress and strain methods often cannot model the fatigue behavior under non-proportional load paths.

In recent years critical plane approaches have become popular due to their ability to establish a correlation between the fatigue behavior of metallic materials under

* Corresponding author: shahriar.sharifimehr@rockets.utoledo.edu

complex non-proportional load paths and those under uniaxial and proportional load paths. The most commonly used critical plane approaches such as Fatemi-Socie model [11] use shear fatigue properties as the base of their correlations. Therefore, in categorizing the fatigue behavior of materials shear fatigue properties are as important and useful as uniaxial fatigue properties. However, due to complexities of torsional fatigue tests shear fatigue properties are available for a small number of materials, as compared to uniaxial fatigue properties. This highlights the importance of reliable methods which can reasonably estimate the shear fatigue properties of materials based on their uniaxial mechanical properties.

There are very small number of studies in the open literature on the estimation of shear fatigue properties of a small number of metallic materials. For instance, Shamsaei and Fatemi [12] used different correlation methods to estimate the shear fatigue properties of SAE 1050 case-hardened specimens from their uniaxial fatigue behavior. Since the specimens were case-hardened and due to the fact the stress gradient in torsion fatigue tests result in higher shear stress on the surface, they analyzed their test results based on the fatigue properties of the hardened case. The criteria used in this study were Tresca (maximum shear stress), von Mises, and maximum principal strain. It was concluded that correlations based on the maximum principal strain criterion resulted in a close prediction of the shear strain-fatigue life behavior of the specimens while von Mises and Tresca criteria significantly underestimated the fatigue life.

Leese and Morrow [13] also used maximum shear stress and von Mises criteria in order to estimate shear cyclic and fatigue properties of 1045 steel from their axial counterparts. They observed that while maximum shear stress was able to estimate shear cyclic properties with better accuracy compared to von Mises, both equivalent stress criteria resulted in a reasonably close estimation of shear fatigue behavior of the material.

The most notable study on the estimation of shear fatigue properties of steels was conducted by Kim et al. [7]. In their study uniaxial and torsional monotonic and fatigue tests were conducted on eight ductile steels and several methods for estimating uniaxial fatigue properties were evaluated. It was found that modified universal slopes and Roessle-Fatemi hardness method resulted in good life predictions for uniaxial tests. Furthermore, based on von Mises equivalent stress and strain uniaxial fatigue properties estimated from the aforementioned methods were converted to shear fatigue properties. Torsional fatigue life predictions from these estimated properties were reported to be reasonably close to the experimental lives with 85% of life predictions falling within a factor of 3 of the experimental fatigue life data.

The goal of this study was to evaluate the performance of different methods in predicting the shear fatigue behavior of several classes of metallic materials from estimated uniaxial fatigue properties. In order to achieve this goal results of monotonic and fatigue tests on several metallic materials found in the literature were analyzed. In addition, uniaxial and torsional monotonic and fatigue tests were conducted on two types of steel alloy and one wrought titanium alloy, the results of which were added to the data pool. Two general methods were used in the estimation of shear fatigue properties. In the first method the experimentally obtained uniaxial fatigue properties were converted to shear fatigue properties. In the second method the uniaxial fatigue properties were estimated from the Roessle-Fatemi hardness-based method and the estimated uniaxial fatigue properties were then converted to shear fatigue properties. In both methods, depending on the level of hardness of the materials, different criteria were used in correlating uniaxial fatigue properties to shear fatigue properties. In the following sections the experimental program, the materials used in the data pool, the estimation methods, and the performed analysis are presented in detail and the obtained results are discussed.

2 Experimental Program and Results

Two types of steel alloy as well as wrought titanium alloy Ti-6Al-4V were subjected to monotonic and cyclic axial and torsion tests. The two types of steel tested in this study are 44MnSiVS6 and SUPM125 which are used in the crankshafts and coil springs of vehicles in the auto industry, respectively. The 44MnSiVS6 steel specimens were hot forged and controlled cooled, while the SUPM125 steel specimens were hot formed and quenched and tempered. The wrought titanium specimens were made from machining pre-annealed solid bars of grade 5 Ti-6Al-4V with nominal diameter of 28.6 mm. All of the steel and titanium specimens were thin-wall tubular with a wall thickness of 1.25 mm which were designed according to ASTM Standard E2207 [14].

An Instron closed-loop servo-controlled hydraulic axial-torsion load frame was used to conduct the monotonic and fatigue tests. The alignment of the load frame was verified to produce no more than 5% bending which is recommended by ASTM Standard E1012-12 [15]. All of the fatigue tests were performed in fully reversed mode (R = -1) using sinusoidal waveform. The obtained mechanical properties of these materials as well as their Brinell hardness are shown in Table 1. In this table E and G are elastic moduli under axial and shear loadings, respectively. Furthermore, S_u is the ultimate tensile strength, HB is the Brinell hardness, σ_f' is fatigue strength coefficient, ε_f' is fatigue ductility coefficient, b is fatigue strength exponent, c is fatigue ductility exponent, τ_f' is shear fatigue strength coefficient, γ_f' is shear fatigue ductility coefficient, b_0 is shear fatigue strength exponent, and c_0 is shear fatigue ductility exponent.

3 Data Pool from Literature

In addition to the tested materials, shear fatigue data for 23 carbon steels were found in the literature. According to the classification of American Iron and Steel Institute (AISI) among the steels in the data pool 8 were categorized as low-carbon steels with less than

0.3% carbon content, 13 were categorized as medium-carbon steels with 0.3-0.6% carbon content, and 4 were categorized as high-carbon steels with 0.6-1.0% carbon content. 605H32 steel which is one of the medium carbon steels used in the data pool had been tested at elevated temperatures of 300 °C and 600 °C [16]. The test results in the two conditions were analyzed separately. In addition, the 4 types of high-carbon steel used in the data pool were variants of 52100 steel and, thus, shared the same chemical composition [17]. However, since they have different monotonic and fatigue properties they were also analyzed separately.

The ultimate tensile strength of the steels ranged from as low as 316 MPa to as high as 2636 MPa which covers most of the grades of carbon steel used in the industry. Furthermore, test data of commercial pure titanium, titanium alloy BT9, and titanium alloy TC4 were added to the data pool. The data pool also consisted of test data for Inconel 718, an austenitic nickel-chromium-based superalloy commonly used in gas turbine blades. All of the fatigue tests were performed on tubular specimens with machined and smooth surface. The fatigue crack orientation of the materials in each study was also considered. It was found that fatigue cracks in steels with ultimate tensile strength of 1100 MPa or less (HB ≤ 327) initiated on the maximum shear planes, while fatigue cracks in most of the steels with ultimate tensile strength of 1696 MPa or higher (HB ≥ 443) initiated on the plane of maximum principal stress. Spring steel with ultimate tensile strength of 2208 MPa (HB = 615) was the only exception with cracks initiating on the plane of maximum shear strain.

It should be noted that in the process of selecting materials and test data for the data pool the case hardened and case carburized materials were avoided. The reason is that specimens which go through the process of case hardening and case carburizing have a hard case and a softer core. The inhomogeneity of hardness in these specimens in addition to the stress gradient caused by torsional loading result in inconsistencies in correlations between uniaxial and shear fatigue properties. The details of the experiments on the materials used in the data pool can be found in [7, 13, 16-29] and monotonic and fatigue properties are presented in Table 1.

Table 1. Obtained mechanical properties and Brinell hardness of the materials used in the data pool.

Material	E/G (GPa)	S_u (MPa)	HB	σ'_f/τ'_f (MPa)	ε'_f/γ'_f	b/b_0	c/c_0
Steel							
Mild Steel [18]	210.0/74.8	316	-	1185/497	3.341/0.530	-0.167/-0.142	-0.748/-0.478
1045 [13]	202.4/79.1	484	153	948/505	0.260/0.413	-0.092/-0.097	-0.445/-0.445
S25C [17]	209/80	508	153	821/426	0.216/0.249	-0.096/-0.074	-0.458/-0.376
1045 HR [19]	206.0/82.3	565	-	1035/444	0.334/0.326	-0.108/-0.078	-0.491/-0.416
St52-3N [20]	206/78	639	-	564/333	0.156/0.081	-0.058/-0.043	-0.466/-0.323
S460N [21]	208.5/80.2	643	-	1002/477	0.147/0.316	-0.097/-0.075	-0.487/-0.469
AISI 304 SS [22]	183.0/82.8	650	-	1151/781	0.171/0.380	-0.132/-0.131	-0.400/-0.369
Ni-Cr-Mo-V [23]	200.0/80.7	724	-	680/394	1.14/1.12	-0.043/-0.039	-0.690/-0.570
304 SS [24]	171/66	754	-	647/1009	0.105/0.231	-0.056/-0.126	-0.391/-0.386
S45C [25]	186.0/72.7	770	-	923/685	0.359/0.198	-0.099/-0.120	-0.519/-0.360
S45C [7]	206/79	798	234	1400/630	0.449/1.22	-0.107/-0.080	-0.564/-0.564
SF60 [7]	208/79	820	167	978/504	0.187/0.286	-0.082/-0.067	-0.439/-0.417
SFNCM85S [7]	201/80	825	241	1040/533	0.316/0.251	-0.092/-0.071	-0.522/-0.406
SCM435 [7]	210/81	951	300	1100/512	0.996/0.360	-0.067/-0.045	-0.708/-0.519
44MnSiVS6 (Crankshaft Steel)	209.2/80.3	960	280	1188/611	2.01/0.486	-0.070/-0.053	-0.737/-0.539
605H32 at 600 °C [16]	208*/80	972	-	- /455	- /1.02	- /-0.030	- /-0.557
SCM440 [7]	204/80	1000	319	1400/662	0.675/1.31	-0.088/-0.071	-0.650/-0.656
30CrNiMo8HH [26]	208*/80	1025	314	973/1002	1.74/0.148	-0.040/-0.055	-0.732/-0.430
SNCM439 [7]	208/80	1050	323	1380/969	1.89/3.68	-0.072/-0.086	-0.801/-0.765
SNCM630 [7]	196/77	1100	327	1270/858	1.54/1.51	-0.073/-0.061	-0.823/-0.706
605H32 at 300 °C [16]	202.8*/78	1696	-	- /1277	- /0.382	- /-0.078	- /-0.0561
SUPM125 (Spring Steel)	206.8/79.5*	2208	615	6332/1571	- /0.100	-0.185/-0.065	- /-0.487
52100 Type E [17]	210*/80*	2391	-	- /2757	- / -	- /-0.107	- / -
52100 Type C [17]	210*/80*	2527	-	- /2716	- / -	- /-0.104	- / -
52100 Type B [17]	210*/80*	2588	-	- /2776	- / -	- /-0.101	- / -
52100 Type A [17]	210*/80*	2636	-	- /2829	- / -	- /-0.102	- / -
Titanium							
Pure Titanium [27]	112/40	558	240	647/485	0.548/0.417	-0.033/-0.069	-0.646/-0.523
TC4 [28]	108.4/43.2	1044	353	1146/584	0.614/2.39	-0.052/-0.043	-0.690/-0.790
Wrought Ti-6Al-4V	108.2/42.0	1045	353	987/647	0.569/0.352	-0.034/-0.044	-0.636/-0.502
BT9 [27]	118/43	1080	349	1180/881	0.278/0.180	-0.025/-0.082	-0.665/-0.470
Inconel							
Inconel 718 [29]	208.5/77.8	1420	371	1897/1672	5.815/9.500	-0.07/-0.121	-0.906/-0.852

The materials in bold are tested in this study.

* The Poisson's ratios and the elastic moduli were assumed by the authors.

4 Estimation Methods and Analysis

In the presence of uniaxial fatigue properties the shear fatigue properties are usually estimated from these properties using different criteria. The suitable criterion should be chosen based on the behavior of the material. Criteria such as von Mises equivalent stress and strain are used for ductile behaving materials, while criteria such as maximum principal strain are used for brittle behaving materials. The estimation equations based on von Mises and maximum principal strain are expressed as:

$$\tau_f' = \frac{\sigma_f'}{\sqrt{3}}, \gamma_f' = \sqrt{3}\varepsilon_f', b_0 = b, c_0 = c \qquad (1)$$

$$\tau_f' = \frac{\sigma_f'}{1+\nu}, \gamma_f' = 2\varepsilon_f', b_0 = b, c_0 = c \qquad (2)$$

respectively, where ν is the Poisson's ratio. However, when the uniaxial fatigue properties are not available for a material, correlation techniques should be employed to estimate these properties from monotonic tension properties. Roessle-Fatemi equation is one of the most commonly used estimation methods which estimates the uniaxial fatigue properties of steel alloys from their elastic modulus and Brinell hardness (HB):

$$\sigma_f' = 4.25(\text{HB}) + 225 \qquad (3)$$

$$\varepsilon_f' = \frac{0.32(\text{HB})^2 - 478(\text{HB}) + 191000}{E} \qquad (4)$$

In this method, fatigue strength and ductility exponents (b and c) are the same as the exponents used in the Modified Universal Slopes Method and are equal to -0.09 and -0.56, respectively. For materials the hardness of which is not available Brinell hardness can be estimated from the following equation [6]:

$$S_u = 0.0012(\text{HB})^2 + 3.3(\text{HB}) \qquad (5)$$

Depending on the fatigue behavior of the material one of the criteria expressed in Eq. 1 or 2 is used to obtain the shear fatigue properties from the estimated uniaxial fatigue properties.

4.1 Prediction Results for Steels and Inconel 718

Using von Mises criterion for ductile behaving materials and maximum principal strain criterion for brittle behaving materials shear fatigue properties were calculated from both experimental uniaxial properties and uniaxial properties estimated from Roessle-Fatemi hardness method. Then, results of torsional fatigue tests were predicted using the estimated shear fatigue properties and presented in log-log plots of predicted vs. experimental total shear strain amplitudes, as well as predicted vs. experimental fatigue lives. These plots are shown in Figures 1 and 2 for predictions from experimental uniaxial fatigue properties and uniaxial properties estimated from Roessle-Fatemi hardness method, respectively. It should be noted that since experimentally obtained uniaxial fatigue properties of 605H32 and 52100 steels were not available they are not included in Figure 1.

By comparing Figures 1(a) and 2(a) it can be observed that predictions from Roessle-Fatemi method were closer to the experimental values than those from experimentally obtained uniaxial properties as the percentage of data within each scatter band for Roessle-Fatemi method is higher. For example, 74% of data fell within the strain scatter band of 1.2 for Roessle-Fatemi method, while this percentage for predictions from experimentally obtained uniaxial properties was only 38%. Furthermore, the percentage of data falling within a life factor of 3 for uniaxial fatigue properties method and Roessle-Fatemi method were 70% and 83%, respectively. Moreover, both methods showed reasonable predictions for Inconel 718 as well.

4.2 Prediction Results for Titanium

Figure 3 shows the prediction results of shear fatigue tests on titanium alloys based on experimentally obtained uniaxial fatigue properties. All of the titanium alloys were classified as ductile behaving materials and only von Mises criterion was used in the correlation of properties. Good results were observed in predictions of shear strain amplitudes shown in Figure 3(a) in both HCF and LCF regions. However, predictions of fatigue lives shown in Figure 3(b) show scatter in HCF region. This is due to the very shallow slopes of shear strain amplitude-fatigue life curves of titanium alloys in the elastic region.

Since the Roessle-Fatemi prediction method is developed for steels it cannot be used for titanium alloys unless it is modified based on the fatigue behavior of these alloys. However, since only a few number of titanium alloys were found in the literature and used in the data pool, the applicability of the proposed modifications to other types of titanium alloys which are not present in the data pool needs to be evaluated. In this study the estimation of fatigue ductility coefficient for steels, shown in Eq. 4, is modified for titanium alloys.

In order to do so, a procedure similar to the procedure described by Roessle and Fatemi [6] was followed for the titanium materials in the data pool. The following linear relationship was derived for the titanium alloys in the data pool:

$$\varepsilon_f' = \frac{-44(\text{HB}) + 43000}{E} \qquad (6)$$

Fatigue strength and fatigue ductility coefficients calculated from Eqs. 3 and 6 along with fatigue strength and ductility exponents of -0.09 and -0.56, respectively, were used as the Roessle-Fatemi estimation method modified for titanium alloys. The final equation is shown below.

$$\frac{\Delta\varepsilon}{2} = \frac{4.25(HB)+225}{E}\left(2N_f\right)^{-0.09} + \frac{-44(HB)+43000}{E}\left(2N_f\right)^{-0.56} \quad (7)$$

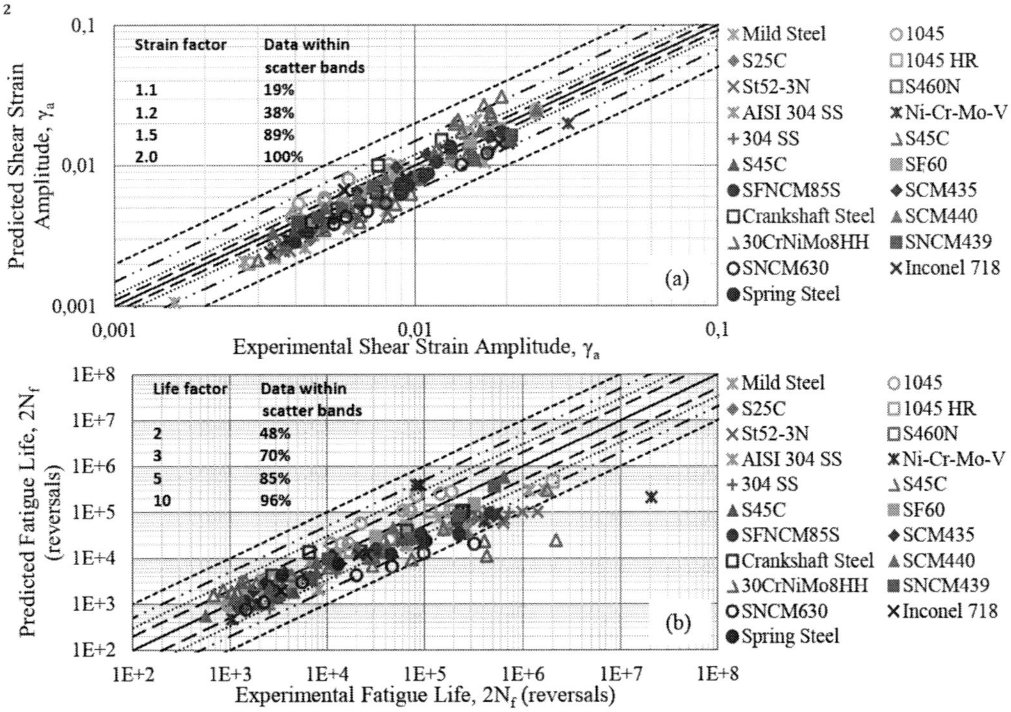

Fig. 1. Results of predictions of (a) shear strain amplitudes and, (b) fatigue lives of torsion fatigue tests on steels and Inconel 718 from shear fatigue properties estimated from experimental uniaxial fatigue properties. Maximum principal strain criterion is used for correlation of spring steel while the rest of the materials are correlated using von Mises effective strain criterion.

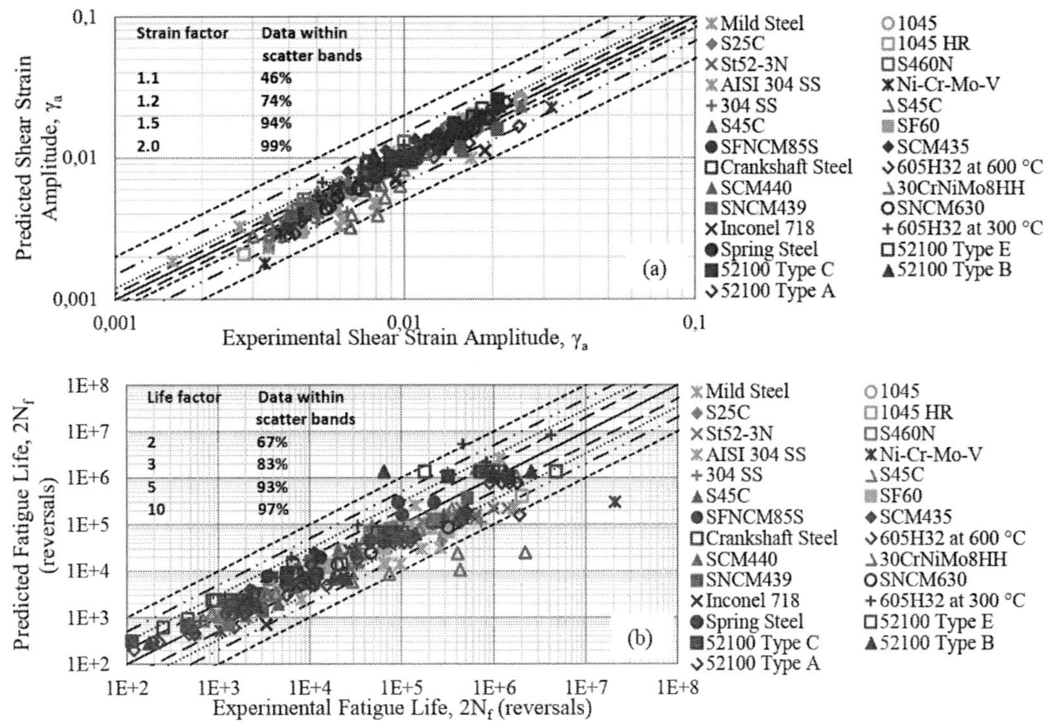

 https://doi.org/10.1051/matecconf/201816516012

Fig. 2. Results of predictions of (a) shear strain amplitudes and, (b) fatigue lives of torsion fatigue tests on steels and Inconel 718 from shear fatigue properties calculated from uniaxial fatigue properties estimated from Roessle-Fatemi hardness method. Maximum principal strain criterion is used for correlation of spring steel, 605H32 at 300°C, and 52100 steels while the rest of the materials are correlated using von Mises effective strain criterion.

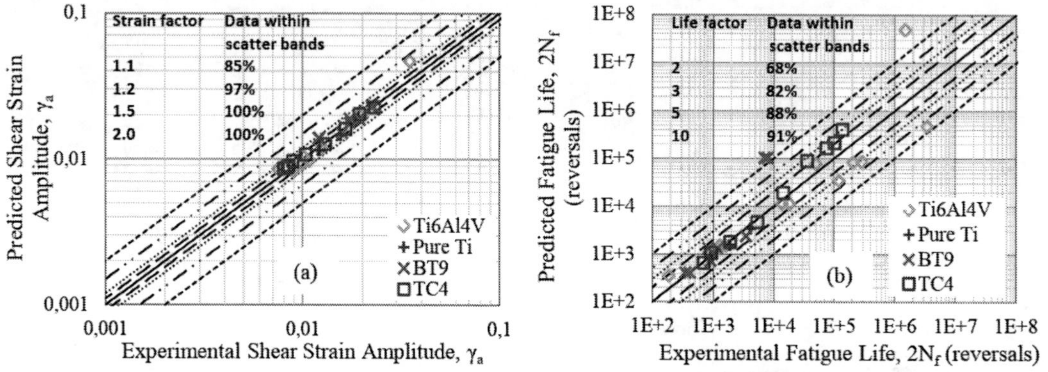

Fig. 3. Results of predictions of (a) shear strain amplitudes and, (b) fatigue lives of torsion fatigue tests on titanium alloys from shear fatigue properties estimated from experimentally obtained uniaxial fatigue properties and based on von Mises effective strain.

The prediction results based on the modified Roessle-Fatemi method for uniaxial and torsion tests are shown in Figures 4 and 5, respectively. It can be observed that using the modified Roessle-Fatemi method results in reasonable predictions of strain amplitude and fatigue life in both uniaxial and torsion fatigue tests. Moreover, a comparison between Figures 3 and 5 shows that the prediction results from the modified Roessle-Fatemi method are as accurate as those from experimentally obtained uniaxial fatigue properties.

It should be noted that the relationship shown in Eq. 7 was calculated based on the uniaxial fatigue properties of the 4 titanium alloys used in the data pool of this study. Therefore, due to the limited number of the materials this equation may not result in reasonable predictions for titanium alloys with Brinell hardness of lower than 240 such as titanium alloy grades 1, 2, and 7, or higher than 353 such as Ti-6Al-4V-2Sn alloy.

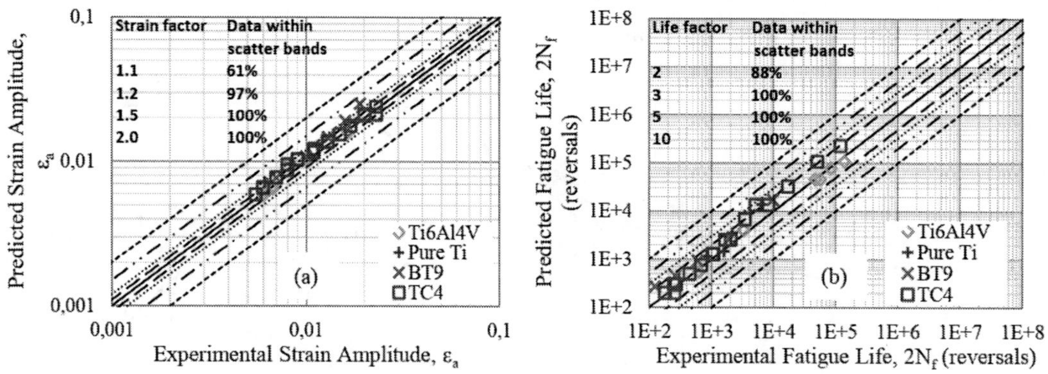

Fig. 4. Results of predictions of (a) uniaxial strain amplitudes and, (b) fatigue lives of uniaxial fatigue tests on titanium alloys estimated from modified Roessle-Fatemi hardness method and based on von Mises effective strain.

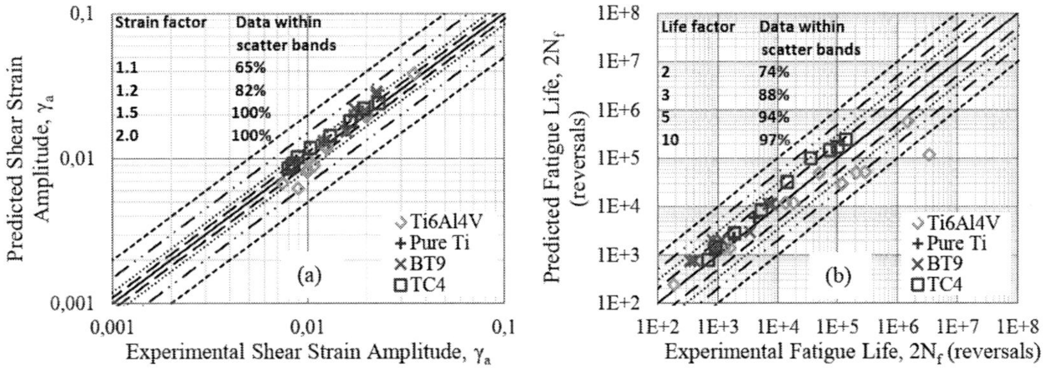

Fig. 5. Results of predictions of (a) shear strain amplitudes and, (b) fatigue lives of torsion fatigue tests on titanium alloys from shear fatigue properties estimated from modified Roessle-Fatemi hardness method and based on von Mises effective strain.

5 Summary and Conclusions

Axial monotonic as well as axial and torsion fatigue tests were performed on two types of steel and one titanium alloy. The test data of these tests along with the test data from 23 types of carbon steel, 4 types of titanium alloy, and Inconel 718 were used to evaluate different criteria in estimating shear fatigue properties from experimentally obtained or estimated uniaxial fatigue properties. The steels used in the data pool of this study consisted of both ductile and brittle behaving steels. Steels with ultimate tensile strength of 1100 MPa or less (HB $\leq$ 327) were classified as ductile behaving and von Mises effective strain criterion was used to correlate their data. Moreover, steels with ultimate tensile strength of 1696 MPa or more (HB $\geq$ 443) were classified as brittle behaving and maximum principal strain criterion was used to correlate their data. For titanium alloys von Mises effective strain criterion was used for the correlations. The main conclusions drawn from the data analysis conducted are as follows:

1. von Mises and maximum principal strain criteria for ductile and brittle behaving steels, respectively, resulted in strong correlations between the uniaxial fatigue properties and shear fatigue data. Similar results were observed for titanium alloys and Inconel 718 using von Mises effective strain.
2. Using the Brinell hardness and elastic modulus uniaxial fatigue properties of steels and Inconel 718 were estimated by Roessle-Fatemi method. Then, using the suitable correlation criterion shear fatigue properties were calculated from estimated uniaxial fatigue properties. This method resulted in predictions which in some cases were better than the predictions from the experimentally obtained uniaxial fatigue properties.
3. A modified Roessle-Fatemi method in which the equation for calculation of fatigue ductility coefficient was modified based on the uniaxial fatigue properties of titanium alloys was developed. The prediction results based on this method were as reasonable as those from the experimentally obtained uniaxial fatigue properties. The modified equation for calculation of fatigue ductility coefficient was derived for titanium alloys with hardness in the range between 240 and 353.

References

1. R.I. Stephens, A. Fatemi, R.R. Stephens, H.O. Fuchs, *Metal fatigue in engineering* (2000)
2. S. Manson, Exp. Mech., **5**, 193-226, (1965)
3. J. Ong, Int J Fatigue, **15**, 213-219, (1993)
4. U. Muralidharan, S. Manson, J Eng Mater Technol, **110**, 55-58, (1988)
5. A. Bäumel, T. Seeger, C. Boller, *Amsterdam: Elsevier*, (1990)
6. M. Roessle, A. Fatemi, Int J Fatigue, **22**, 495-511, (2000)
7. K. Kim, X. Chen, C. Han, H. Lee, Int J Fatigue, **24**, 783-793, (2002)
8. M. Meggiolaro, J. Castro, Int J Fatigue, **26**, 463-476, (2004)
9. A. Fatemi, N. Shamsaei, Int J Fatigue, **33**, 948-958, (2011)
10. K.-S. Lee, J.-H. Song, Int J Fatigue, **28**, 386-400, (2006)
11. A. Fatemi, D.F. Socie, FFEMS, **11**, 149-165, (1988)
12. N. Shamsaei, A. Fatemi, Int J Fatigue, **31**, 1386-1396, (2009)
13. G.E. Leese, J. Morrow, *Multiaxial Fatigue, ASTM International*, (1985)
14. ASTM-E2207-08, *Annual Book of ASTM Standards*, **03.01**, (2012)
15. ASTM-E1012-12, *Annual Book of ASTM Standards*, **03.01**, (2012)
16. N.J. Hurd, P.E. Irving, *Multiaxial Fatigue, ASTM International*, (1985)
17. S.M. Moghaddam, J.A. Bomidi, F. Sadeghi, N. Weinzapfel, A. Liebel, Tribology International, **77**,196-210, (2014)
18. V. Doquet, ICBMFF3, (2013)

19. A. Fatemi, R.I. Stephens, *Multiaxial Fatigue: Analysis and Experiment*, 121-137, (1989)
20. P. Lopez-Crespo, B. Moreno, A. Lopez-Moreno, J. Zapatero, Engineering Fracture Mechanics, **136**, 115-130, (2015)
21. Y. Jiang, O. Hertel, M. Vormwald, Int J Fatigue, **29**, 1490-1502, (2007)
22. D. Socie, J Eng Mater Technol, **109**, 293-298, (1987)
23. R.A. Williams, R.J. Placek, O. Klufas, S.L. Adams, D.C. Gonyea, *Multiaxial Fatigue, ASTM International*, (1985)
24. K. Kim, B. Lee, J. Park, *Fatigue and Fracture Mechanics, ASTM International*, **31**, (2000)
25. K. Kim, J. Park,J. Lee, J Eng Mater Technol, **121**, 286-293, (1999)
26. M. Noban, H. Jahed, S. Winkler, A. Ince, Materials Science and Engineering: A, **528**, 2484-2494, (2011)
27. N. Shamsaei, M. Gladskyi, K. Panasovskyi, S. Shukaev, A. Fatemi, Int J Fatigue, **32**, 1862-1874, (2010)
28. Z.-R. Wu, X.-T. Hu, Y.-D. Song, Int J Fatigue, **59**, 170-175, (2014)
29. J.L. Koch, *M.S. Thesis, Mechanical Engineering, University of Illinios at Urbana-Champaign, Champaign, IL*, (1985)

On the applicability of multiaxial high cycle fatigue criteria to metallic materials

Roberta A. Gonçalves[1], *Marcos* V. Pereira[1,*] and *Fathi* A. Darwish[2]

[1] Catholic University of Rio de Janeiro, Department of Chemical and Materials Engineering, Rua Marques de São Vicente 225, 22453-901 Rio de Janeiro – RJ, Brazil.
[2]Fluminense Federal University, Department of Civil Engineering, Rua Passo da Patria 156, 24210-240 Niteroi – RJ, Brazil.

Abstract. A comparative study is made of the applicability of critical plane based multiaxial high cycle fatigue models to predicting the fatigue behavior of metallic materials. A number of models, namely Matake, McDiarmid, Carpinteri and Spagnoli, Liu and Mahadevan and Papadopoulos, were applied to fatigue limit states, involving synchronous fully reversed in-phase sinusoidal bend and torsion loading. The results obtained indicated a good predictive capability of the models with an average error index situated approximately between -5,5% and 4,5%. However, this average was limited to less than 3% for the latter three models. Finally, the critical plane orientation, which, for a given material, is characteristic of the proper model, is compared with that of the fracture plane, exclusively determined by the ratio between the shear stress and normal stress amplitudes.

1 Introduction

High cycle fatigue under uniaxial loading has shown that many metallic materials possess a fatigue limit, which means that they can sustain a very high (theoretically infinite) number of cycles without fatigue failure. However, many mechanical components, such as railroad wheels, crankshafts, axles and turbine blades are expected to experience multiaxial loading during their in-service lifetime. Accordinglly, the fatigue problem becomes more complex due to the complexity of the stress states, loading histories and different orientations of the initial crack in the components.

Generalization of the fatigue limit concept for multiaxial loading conditions is considered compatible with the idea of dividing the whole stress space in two parts, namely safe and unsafe. The safe part, which contains the origin, is bounded by a closed surface and the fatigue criterion can thus be expressed in terms of an inequality whose satisfaction signifies that the stress state induced by the external cyclic loading remains within the safe part of the stress space [].

Over many decades of research, a large number of models have been proposed to predict fatigue failure under multiaxial loading conditions. As the stress levels involved in high cycle fatigue are kept below the elastic limit, only stress based models, namely Matake (Ma) [1], McDiarmid (Mc) [2,3], Findley (F) [4], Carpinteri and Spagnole (C & S) [5-7], Liu and Mahadevan (L & M) [8] and Papadopoulos (P) [9,10] models, are to be considered in the present work. The underlying purpose is to test the applicability of these six models to some experimental loading conditions, available in the literature [11,12], involving synchronous fully reversed sinusoidal in-phase bend and torsion loading applied to a variety of metallic materials with different fatigue behaviors. At this point, it should be emphasized that all of the chosen loading conditions correspond to the fatigue limit state above which fatigue occurs and below which fatigue life extends over a very high number of cycles, in analogy with the fatigue limit state for uniaxial loading.

Whereas the use of Papadopoulos criterion requires only knowing the applied stress amplitudes, the other models depend for their application on the prior identification of the critical plane, where fatigue damage can occur leading to crack nucleation. Assuming that the critical plane is already known, the normal and shear stress amplitudes can be determined and fatigue failure assessment can thus be presented in the form of inequality. The relative difference between the two sides of the inequality is referred to as the error index and, for a given fatigue limit state, it can be null, positive or negative. As the fatigue criteria in question are to be applied simultaneously to a given loading condition, a comparison of the error index involved is expected to provide a good assessment of their predictive capabilities in defining the fatigue behavior.

Finally, the critical plane orientation, determined for each model, is presented in comparison with that of the fracture plane, for the loading conditions involved.

2 High cycle multiaxial fatigue criteria

The inequalities representative of the Matake, McDiamird, Findley, Carpinteri and Spangnoli, Liu and

* Corresponding author: marcospe@puc-rio.br

Mahadevan and Papadopoulos are given, respectively, by expressions (1) to (6):

$$C_a + \mu N_{max} \leq t_{-1} \quad (1)$$

$$C_a + \frac{t_{-1}}{2\sigma_u} N_{max} \leq t_{-1} \quad (2)$$

$$C_a + k N_{max} \leq f \quad (3)$$

$$\sqrt{N_{max}^2 + (\frac{f_{-1}}{t_{-1}})^2 C_a^2} \leq f_{-1} \quad (4)$$

$$\sqrt{[\frac{N_a + \left(1 + \eta \frac{N_m}{f_{-1}}\right)}{f_{-1}}]^2 + (\frac{C_a}{t_{-1}})^2} \leq \lambda \quad (5)$$

$$\sqrt{\frac{\sigma_a^2}{3} + \tau_a^2} + \alpha \frac{\sigma_a + \sigma_m}{3} \leq t_{-1} \quad (6)$$

C_a and N_{max}, in the expressions above, are, respectively, the shear stress amplitude and the maximum normal stress acting on the critical plane. N_{max} is given by

$$N_{max} = N_a + N_m \quad (7)$$

where N_a is the amplitude and N_m the mean value.

The constants μ, k, f, η, λ and α are material parameters, which depend exclusively, as shown in Table 1, on the fatigue limits for fully reversed bending f_{-1} and fully reversed torsion t_{-1}. Applying the McDiamird criterion, one needs to know, in the addition to t_{-1}, the ultimate tensile strength σ_u.

Different from the critical plane approach, the Papadopoulos criterion is applied by simply substituting the applied normal stress and shear stress amplitudes σ_a and τ_a, together with the mean stress σ_m in expression (6). However, one should note that this type of criterion, which is based on the mesoscopic scale approach, is valid for the category of hard metals where the ratio t_{-1}/f_{-1} lies between $1/\sqrt{3}$ and 0,8 [9].

For fully reserved loading, which is the type of loading considered in the present work, σ_m and N_m are taken to be null and consequently N_{max} is to be replaced by N_a and hence inequalities (5) and (6) simplify to

$$\sqrt{(\frac{N_a}{f_{-1}})^2 + (\frac{C_a}{t_{-1}})^2} \leq \lambda \quad (8)$$

$$\sqrt{\frac{\sigma_a^2}{3} + \tau_a^2} + \alpha \frac{\sigma_a}{3} \leq t_{-1} \quad (9)$$

Table 1. Definition of the pertinent material constants.

$\mu = 2\left(\frac{t_{-1}}{f_{-1}}\right) - 1$
$k = \frac{2 - (\frac{f_{-1}}{t_{-1}})}{2\sqrt{\frac{f_{-1}}{t_{-1}} - 1}}$
$f = \sqrt{\frac{f_{-1}^2}{4(\frac{f_{-1}}{t_{-1}} - 1)}}$
$\eta = \frac{3}{4} + \frac{1}{4}(\frac{\sqrt{3} - \frac{f_{-1}}{t_{-1}}}{\sqrt{3} - 1})$
$\lambda = [cos^2(2\delta)s^2 + sin^2(2\delta)]^{1/2}$
$\alpha = \frac{t_{-1} - (\frac{f_{-1}}{\sqrt{3}})}{\frac{f_{-1}}{3}}$

where $s = t_{-1}/f_{-1}$.

3 Critical plane identification

This can be achieved by first considering a general material plane oriented at angle ψ (Fig.1), where the stress amplitudes N_a and C_a acting on such a plane due to applied synchronous in-phase sinusoidal bending and torsion are given by

$$N_a = \frac{1}{2}\sigma_a + \frac{1}{2}\sigma_a cos2\psi + \tau_a sin2\psi \quad (10)$$

$$C_a = |\frac{1}{2}\sigma_a sin2\psi - \tau_a cos2\psi| \quad (11)$$

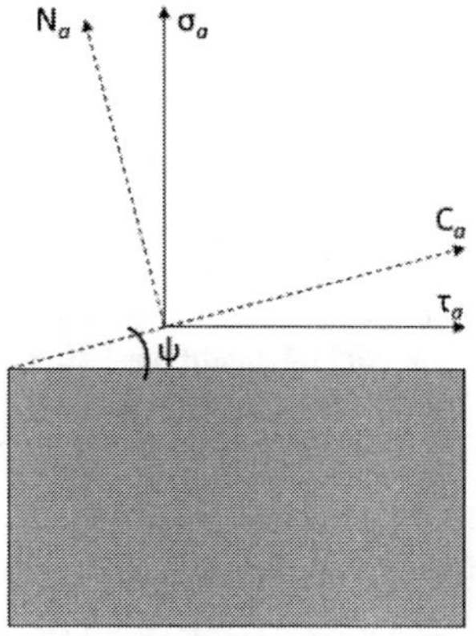

Figure 1: Schematic representation of normal and shear stress amplitudes acting on an arbitrary plane defined by the angle ψ.

Both the Matake and McDiarmid models refer to the critical plane as the plane on which, the shear stress amplitude C_a reaches its maximum. Accordingly the angle $\psi_c$that defines the critical plane orientation is given by

$$tan\psi_c = -\frac{\sigma_a}{2\tau_a} \quad (12)$$

and the corresponding N_a and C_a values can be calculated and then substituted in the left hand side (LHS) of inequalities (1) and (2).

In regard to the Findley model, the critical plane is defined by maximizing the linear combination $(C_a + kN_a)$ and ψ_c in this case will be given by

$$tan2\psi_c = \frac{\sigma_a + 2k\tau_a}{k\sigma_a - 2\tau_a} \quad (13)$$

Again with ψ_c already known, N_a and C_a can be calculated and the LHS of inequality (3) can be determined.

Identification of the critical plane for both C&S and L&M models depend, in the first place, on determining the fracture plane orientation. For a given loading history, the fatigue fracture plane is identified as the material plane normal to the maximum principal stress [5,8]. For a fully reversed loading, the fracture plane is thus oriented at an angle ψ_f given by

$$tan2\psi_f = \frac{-2\tau_a}{\sigma_a} \quad (14)$$

Knowing ψ_f, ψ_c is expressed as [5,8]

$$\psi_c = \psi_f + \delta \quad (15)$$

where δ is given by [5]

$$\delta = \left[1 - \left(\frac{t_{-1}}{f_{-1}}\right)^2\right]\frac{3\pi}{8} \quad (16)$$

for C&S model.

$$\delta = \frac{1}{2}cos^{-1}(A) \quad (17)$$

for L&M model

where $A = \left[\frac{-2+\sqrt{4-4(\frac{1}{s^2}-1)(5-\frac{1}{s^2}-4s^2)}}{2(5-\frac{1}{s^2}-4s^2)}\right]$

Now, with the angle ψ_c already known, one can proceed to apply the C&S and L&M model, substituting N_a and C_a in the LHS of their respective inequalities.

4 Results and discussion

The fracture plane and critical plane orientations, defined by the angle ψ_f, and ψ_c are listed in Tables 2 and 3, in terms of σ_a and τ_a, for the variety of materials indicated in the tables.

Table 2. Fracture and critical plane orientations for different materials, under in phase bend and torsion stresses for Ma, Mc and F models.

			ψ_c (°)		
σ_a(MPa)	τ_a(MPa)	ψ_f (°)	Ma	Mc	F
Material: Hard steel: f_{-1} = 313.9 MPa; t_{-1} = 196.2 MPa; σ_u = 704.1 MPa					
327.7	**0.0**	0.0	45.0	45.0	37.8
308.0	**63.9**	11.3	56.3	56.3	49.0
255.1	**127.5**	22.5	67.5	67.5	60.3
141.9	**171.3**	33.8	78.8	78.8	71.5
0.0	**201.1**	45.0	0.0	0.0	7.2
Material: Hard steel: f_{-1} = 313.9 MPa; t_{-1} = 196.2 MPa; σ_u = 680.0 MPa					
138.1	**167.1**	33.8	168.8	168.8	176.0
245.3	**122.65**	22.5	157.5	157.5	164.7
299.1	**62.8**	11.4	146.4	146.4	153.6
Material: 42CrMo4: f_{-1} = 398.0 MPa; t_{-1} = 260.0 MPa; σ_u = 1025.0 MPa					
328.0	**157.0**	21.9	156.9	156.9	165.8
233.0	**224.0**	31.3	166.3	166.3	175.2
Material: 34Cr4: f_{-1} = 410.0 MPa; t_{-1} = 256.0 MPa; σ_u = 795.0 MPa					
314.0	**157.0**	22.5	67.5	67.5	60.3
Material: 30NCD16: f_{-1} = 660.0 MPa; t_{-1} = 410.0 MPa; σ_u = 1880.0 MPa					
485.0	**280.0**	24.6	69.6	69.6	62.5
Material: Mild steel: f_{-1} = 235.4 MPa; t_{-1} = 137.3 MPa; σ_u = 518.8 MPa					
245.3	**0.0**	0.0	45.0	45.0	40.2
235.6	**48.9**	11.3	146.3	146.3	51.5
187.3	**93.6**	22.5	157.5	157.5	62.7
101.3	**122.3**	33.8	168.8	168.8	74.0
0.0	**142.3**	45.0	0.0	0.0	4.8
Material: Cast iron: f_{-1} = 96,1 MPa; t_{-1} = 91,2 MPa; σ_u = 230.0 MPa					
93.2	**0.0**	0.0	45.0	45.0	13.1
95.2	**19.7**	11.2	56.2	56.2	24.3
83.4	**41.6**	22.5	157.5	157.5	9.4
56.3	**68.0**	33.8	168.8	168.8	20.7
0.0	**94.2**	45.0	0.0	0.0	31.9

https://doi.org/10.1051/matecconf/201816516013

Table 3. Fracture and critical plane orientations for different materials, under in phase bend and torsion stresses for C&S and L&M models.

σ_a(MPa)	τ_a(MPa)	ψ_f (°)	ψ_c (°) C&S	ψ_c (°) L&M
Material: Hard steel: f_{-1} = 313.9 MPa; t_{-1} = 196.2 MPa; σ_u = 704.1 MPa				
327.7	**0.0**	0.0	41.1	39.2
308.0	**63.9**	11.3	52.4	50.5
255.1	**127.5**	22.5	63.6	61.7
141.9	**171.3**	33.8	74.9	73.0
0.0	**201.1**	45.0	86.1	84.2
Material: Hard steel: f_{-1} = 313.9 MPa; t_{-1} = 196.2 MPa; σ_u = 680.0 MPa				
138.1	**167.1**	33.8	74.9	73.0
245.3	**122.65**	22.5	63.6	61.7
299.1	**62.8**	11.4	52.5	50.6
Material: 42CrMo4: f_{-1} = 398.0 MPa; t_{-1} = 260.0 MPa; σ_u = 1025.0 MPa				
328.0	**157.0**	21.9	60.6	58.5
233.0	**224.0**	31.3	70.0	67.9
Material: 34Cr4: f_{-1} = 410.0 MPa; t_{-1} = 256.0 MPa; σ_u = 795.0 MPa				
314.0	**157.0**	22.5	63.7	61.7
Material: 30NCD16: f_{-1} = 660.0 MPa; t_{-1} = 410.0 MPa; σ_u = 1880.0 MPa				
485.0	**280.0**	24.6	66.0	64.2
Material: Mild steel: f_{-1} = 235.4 MPa; t_{-1} = 137.3 MPa; σ_u = 518.8 MPa				
245.3	**0.0**	0.0	44.5	44.1
235.6	**48.9**	11.3	55.8	55.4
187.3	**93.6**	22.5	67.0	66.6
101.3	**122.3**	33.8	78.3	77.9
0.0	**142.3**	45.0	89.5	89.1
Material: Cast iron: f_{-1} = 96,1 MPa; t_{-1} = 91,2 MPa; σ_u = 230.0 MPa				
93.2	**0.0**	0.0	6.7	16.3
95.2	**19.7**	11.2	17.9	27.5
83.4	**41.6**	22.5	29.2	38.8
56.3	**68.0**	33.8	40.5	50.1
0.0	**94.2**	45.0	51.7	61.3

The angle ψ_f, which is invariably determined by the stress amplitudes, is in fact unique for all the models. As can be observed from the same table, the higher the ratio τ_a/σ_a, the higher the angle ψ_f, consistent with the fact that ψ_f tends to zero for uniaxial normal stress and to 45° for pure shear loading.

As one may expect, the critical plane orientation, defined by the angle ψ_c in table, varies from one model to another. Except for the Matake and McDiamird criteria, where the critical plane orientation has the same value of ψ_c, considerable differences can be observed among the other models. These differences are also seen to vary appreciably with the stress amplitudes as well as with the fatigue properties of the material. However, in view of their proper formulation, the C&S and L&M criteria result in essentially the same critical plane orientations, except for a few number of loading conditions associated with the experimental testing of cast iron.

The error index I, associated with the application of any of the six models, refers to the relative difference between the two sides of the inequality. I can thus be expressed as

$$I = \frac{LHS - RHS}{RHS} \times 100 \tag{18}$$

The values of I corresponding to the different loading conditions are listed in Tables 4 and 5, for the variety of materials involved.

Except for a few cases, the vast majority of the I values are situated within the range -10% to 10%, indicating a good predictive capability of the criteria in question. This is also demonstrated by Fig.2, where the overall average values of I are also shown. One can thus conclude that, except for the McDiarmid model, the others are moderately conservative, with the C&S, L&M and P models exhibiting the lowest error compared to the other three.

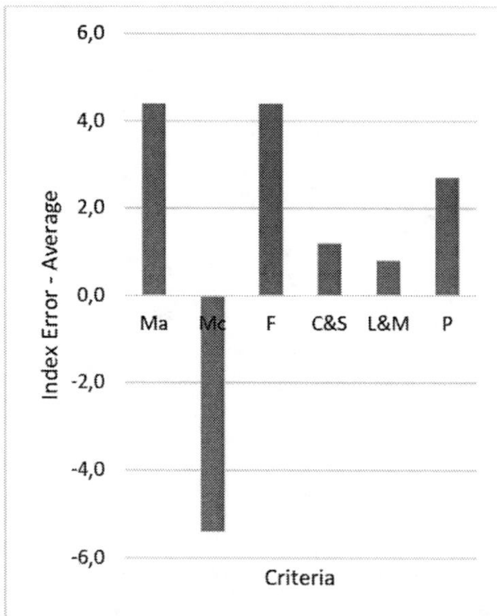

Figure 2: Comparative presentation of the average value of the error index, for the fatigue criteria in question.

https://doi.org/10.1051/matecconf/201816516013

Table 4: Error index, corresponding Ma, Mc and F criteria.

σ_a(MPa)	τ_a(MPa)	Index Error - *I* (%) Ma	Mc	F
Material: Hard steel: f_{-1} = 313.9 MPa; t_{-1} = 196.2 MPa; σ_u = 704.1 MPa				
327.7	**0.0**	4.4	-4.8	4.4
308.0	**63.9**	4.6	-4.1	4.6
255.1	**127.5**	8.2	1.0	8.2
141.9	**171.3**	3.6	-0.4	3.5
0.0	**201.1**	2.5	2.5	2.5
Material: Hard steel: f_{-1} = 313.9 MPa; t_{-1} = 196.2 MPa; σ_u = 680.0 MPa				
138.1	**167.1**	1.0	-2.8	0.9
245.3	**122.65**	4.0	-2.6	4.0
299.1	**62.8**	1.7	-6.3	1.7
Material: 42CrMo4: f_{-1} = 398.0 MPa; t_{-1} = 260.0 MPa; σ_u = 1025.0 MPa				
328.0	**157.0**	6.8	-4.7	6.7
233.0	**224.0**	10.9	2.8	10.8
Material: 34Cr4: f_{-1} = 410.0 MPa; t_{-1} = 256.0 MPa; σ_u = 795.0 MPa				
314.0	**157.0**	2.0	-3.4	2.0
Material: 30NCD16: f_{-1} = 660.0 MPa; t_{-1} = 410.0 MPa; σ_u = 1880.0 MPa				
485.0	**280.0**	4.7	-3.2	4.7
Material: Mild steel: f_{-1} = 235.4 MPa; t_{-1} = 137.3 MPa; σ_u = 518.8 MPa				
245.3	**0.0**	4.2	1.1	4.2
235.6	**48.9**	7.2	4.3	7.2
187.3	**93.6**	7.8	5.5	7.8
101.3	**122.3**	2.6	1.3	2.5
0.0	**142.3**	3.6	3.6	3.6
Material: Cast iron: f_{-1} = 96,1 MPa; t_{-1} = 91,2 MPa; σ_u = 230.0 MPa				
93.2	**0.0**	-3.0	-38.7	-3.0
95.2	**19.7**	3.4	-33.1	3.4
83.4	**41.6**	5.6	-26.3	5.6
56.3	**68.0**	8.5	-13.2	8.4
0.0	**94.2**	3.3	3.3	3.3

Table 5: Error index, corresponding C&S, L&M and P criteria.

σ_a(MPa)	τ_a(MPa)	Index Error - *I* (%) C&S	L&M	P
Material: Hard steel: f_{-1} = 313.9 MPa; t-1 = 196.2 MPa; σu = 704.1 MPa				
327.7	**0.0**	1.8	4.3	4.4
308.0	**63.9**	1.2	3.3	3.8
255.1	**127.5**	3.1	5.3	5.5
141.9	**171.3**	-1.5	0.3	0.2
0.0	**201.1**	1.9	2.3	2.5
Material: Hard steel: f_{-1} = 313.9 MPa; t-1 = 196.2 MPa; σu = 680.0 MPa				
138.1	**167.1**	-3.9	-2.8	-2.3
245.3	**122.65**	-0.8	1.3	1.4
299.1	**62.8**	-1.5	0.3	0.9
Material: 42CrMo4: f_{-1} = 398.0 MPa; t-1 = 260.0 MPa; σu = 1025.0 MPa				
328.0	**157.0**	0.7	4.5	4.2
233.0	**224.0**	4.1	6.6	7.3
Material: 34Cr4: f_{-1} = 410.0 MPa; t-1 = 256.0 MPa; σu = 795.0 MPa				
314.0	**157.0**	-2.8	-0.8	-0.5
Material: 30NCD16: f_{-1} = 660.0 MPa; t-1 = 410.0 MPa; σu = 1880.0 MPa				
485.0	**280.0**	-0.3	1.1	1.8
Material: Mild steel: f_{-1} = 235.4 MPa; t-1 = 137.3 MPa; σu = 518.8 MPa				
245.3	**0.0**	3.9	4.0	4.1
235.6	**48.9**	5.9	6.0	6.3
187.3	**93.6**	4.7	5.0	5.0
101.3	**122.3**	-1.0	-1.0	-0.8
0.0	**142.3**	3.6	4.0	3.6
Material: Cast iron: f_{-1} = 96,1 MPa; t-1 = 91,2 MPa; σu = 230.0 MPa				
93.2	**0.0**	-3.6	-13.9	-3.0
95.2	**19.7**	2.5	-26.3	2.8
83.4	**41.6**	4.0	4.8	3.8
56.3	**68.0**	5.4	6.8	5.9
0.0	**94.2**	-1.7	2.7	3.3

5 Conclusion

Based on what is presented above, the following conclusions can be drawn

- Fracture plane orientation, in fully reversed multiaxial fatigue loading, is exclusively determined by the ratio between the shear and normal stress amplitudes.
- For a given loading condition, the critical plane orientation depends on the adapted multiaxial fatigue criterion. Whereas the Matake and McDiarmid models possess the same critical plane, the C&S and L&M models indicate critical planes with orientations that are close to each other.
- Critical plane orientation predicted by the Findley criterion is generally close to that defined by the Matake model.
- The overall average of the error index I is limited to -5,5% $\leq$ I $\leq$ 4,5%, indicating reasonable predictive capability of the models in question in defining fatigue behavior.
- Except for the McDiarmid criterion, the models are seen to be conservative as they mostly exhibit positive I values.

This work was developed within the scope of the Research and Technological Development of the Brazilian Electric Energy Sector Program regulated by ANEEL, with the support of the Eneva Companies - Pecém II Energy Generation S.A., Itaqui Energy Generation S.A. and Paranaíba I, II and III Energy Generation S.A.

References

1. T. Matake, *An explanation on fatigue limit under combined stress*, Bull JSME **20**, 257 (1977)
2. DL. McDiamid, *Fatigue under out of phase bending and torsion*, Fat. Fract. Engng. Mater. Struct **9**, 457 (1987)
3. DL. Mc Diarmid, *A general criterion for high cycle multiaxial fatigue failure*, Fat. Fract. Engng. Mater. Struct **14**, 429 (1991)
4. WN. Findley, *A theory for the effect of mean stress on fatigue of metals under combined torsion and axial load or bending*, J. Engng. Ind. Trans. ASME **81**, 301 (1959) (Cited in reference 5)
5. A. Carpinteri, A. Spagnoli, *Multiaxial high-cycle fatigue criterion for hard metals*, Int. J. Fat. **23**, 135 (2001)
6. A. Carpinteri, A. Spagnoli, S. Vantadori, *Multiaxial fatigue assessment using simplified critical plane-based criterion,* Int. J. Fat. **33**, 969 (2011)
7. A. Carpinteri, A. Spagnoli, S. Vantadori, C. Bagnic, *Structural integrity assessment of metallic components under multiaxial fatigue: the C-S criterion and its evolution,* Fat. Fract. Engng. Mater. Struct **36**, 870 (2013)
8. Y. Liu, S. Mahadevan, *Multiaxial high-cycle fatigue criterion and life prediction for metals*, Int. J. Fat. **27**, 870 (2005)
9. IV. Papadopoulos, P. Davoli, C. Gorla, M. Fillipini, A. Bernarconi, *A comparative study of multiaxial high-cycle fatigue criteria for metals*, Int. J. Fat. **19**, 219 (1997)
10. IV. Papadopoulos, *Long life fatigue under multiaxial loading*, Int. J. Fat. **23**, 839 (2001)
11. T. Nishihara, M. Kawamoto, *The strength of metals under combined alternating bending and torsion with phase difference*, Mem College Eng., Kyoto Imper. Univ. **11**, 85 (1945) (Reported in reference 5)
12. H. Zenner, R. Heindenreich, IZ. Richter Werkstofftech, **16**, 101 (1985) (Reported in reference 9)

Cost effective strategy using Kriging surrogates to compute fatigue at multiple locations of a structure: Application to offshore wind turbine certification

Quentin Huchet[1,2,*], *Cécile* Mattrand[2], *Pierre* Beaurepaire[2], *Nicolas* Relun[1] and *Nicolas* Gayton[2]

[1]EDF R&D, Dpt ERMES, 7 Boulevard Gaspard Monge, 91120 Palaiseau, France
[2]Université Clermont Auvergne, CNRS, SIGMA Clermont, Institut Pascal, F-63000 Clermont-Ferrand, France

Abstract. Offshore wind energy development has experienced a rapid development over the last few years encouraged by the of carbon reduction policy and the energy mix objectives of several countries worldwide. Nowadays, the offshore projects under investigation are composed of tens to hundreds of units reaching impressive dimensions with total rotor diameters from 150 to 220 meters and production capacity of 5 to 12 MW per turbine. Mechanical analyses have to be performed to validate the design regarding the solicitations it may face during its lifetime (20 years). Because of the high number of solicitation cycles the structure is confronted to, an estimation of the cumulated damage is mandatory and has to be carefully assessed. As presented in standards, this verification requires massive computation investments and is usually a challenging task for project engineers. This paper presents the "MultiSite" extension of the AK-DA numerical strategy ("Adaptive Kriging for Damage Assessment"). After being formalized, an illustration of its behaviour and performances is proposed for the validation of a design regarding its cumulated damages at different locations.

1 Introduction

The reduction of carbon cost is a one the most important challenge in today's energy production industry. Wind energy is one the most developing technology with an increase of more than 10% per year [1]. To ensure the sustainable development of this technology, the reduction of the production cost is required. To that end, the offshore wind production is being increasingly developed because of the greater wind conditions and the shore distance which permits to elevate larger structures with a reduced risk of collapse on people.

To guarantee the project development and secure investments, each of the structural designs has to be certified regarding its mechanical behaviour. In this context, certification bodies propose full-defined analysis frameworks [2] based on standards [3,4] and listing all the numerical assessments engineers have to perform in order to validate the design or not. Because of the high number of cycles the structures undergo during their lifetime, mechanical fatigue analyses have to be carefully assessed. The total lifetime of the structure is considered as a series of elementary "Conceptual Situations" (CS), some of which requiring an estimation of the cumulated damage. An environmental model is then defined and parameterized with *in-situ* observable quantities such as the 10-min wind mean speed or the significant wave-height. The entire solicitation domain is represented as a regular grid of parameter combinations. The size of this grid depends on the number of environmental parameters and the chosen discretization intervals. As an example, the estimation of the cumulated damage during the "Normal production" situation (DLC 1.2 in [3]) involves up to hundreds of thousands combinations of environmental parameters.

For each of the above-defined combinations of parameters, engineers have to perform a fatigue analysis to quantify their potential short-term damage impact on the structure. The computation of this damage, noted d in this paper, is a demanding task. A combination of parameters is firstly used to generate random trajectories of environmental solicitations for the wind and the sea state. Secondly, the structural responses are time computed with multiphysics codes dealing with aerodynamics, hydrodynamics, structural mechanics and turbine servo adaptations. These responses are finally post-treated in order to extract the stress responses and a Rainflow cycle counting is operated for different locations of the analysed design. For each of the location of interest, a short-term damage d is computed by the use of Miner's cumulative law as:

$$d(\boldsymbol{x}, \omega) = \sum_{k=1}^{n_{RF}(\omega)} \frac{1}{N_k(\boldsymbol{x}, \omega)} \quad (1)$$

with $\boldsymbol{x} = [x_1, \dots, x_n]^T$ the vector of the n environmental parameters, n_{RF} the number of cycles extracted by the Rainflow counting and N_k, the number of cycles to failure with a given stress. ω stands for the random

* Corresponding author: quentin.huchet@sigma-clermont.fr

character of the solicitations for a given combination of environmental solicitation. In this paper, the random seed is fixed for the generation of solicitation time series, therefore discarding this variability. Considering this numerical chain, the simulation time is usually larger than the simulated solicitation period, leading to demanding analyses.

A mean damage of the cycle conceptual situation is estimated as a weighted sum of short-term damages as:

$$D = \sum_{k=1}^{n_c} d(\boldsymbol{x}_k)\, p_k \Delta \tag{2}$$

with p_k defined as the probability of occurrence of the combination of environmental parameters $\boldsymbol{x}_k$ ("metocean" data collected *in-situ*) and Δ, the discretization unit provided by the certification guidelines.

Considering n_s sites (structural locations of interest), the design is validated regarding its resistance to mechanical damage if:

$$D_i < D_i^{lim} \quad \text{for all } i \in [1, \ldots, n_s] \tag{3}$$

with D_i^{lim} the damage limit threshold for the location i.

In order to reduce the simulation costs related to the estimation of the global mean damage D, the AK-DA ("Adaptive Kriging for Damage Assessment") algorithm has been proposed in [5]. Based on adaptive Kriging predictions of the short-term model responses d, it aims at iteratively constructing a reduced but informative design of experiments (set of environmental parameter combinations) to accurately approximate the demanding quantity D for a single location of the structure. Although being computationally efficient, only one location at a time could be investigated which reduces its potential industrial applicability. This paper presents an extension of this method named AK-DA MultiSite and developed to tackle this shortcoming, *i.e.* to adress simultaneously several locations. Firstly, the section 2 focuses on a brief explanation of the AK-DA algorithm in its initial formalization (single location approach). Section 3 presents the "MultiSite" formalization, which is illustrated in section 4. Conclusions and future developments are presented in section 5.

2 Estimation of the damage using the AK-DA method (Single location approach)

The AK-DA method was proposed to reduce the simulation costs for the estimation of the global mean damage D, presented in Eq. (2).

1.1 The Kriging Prediction

Kriging is a statistical method aiming at predicting a phenomenon response based on a limited set of realizations (model observations). Its development in the last decades made it widely spread in numerical analysis [6]. Let m be a numerical model mapping the input $\boldsymbol{x} = [x_1, \ldots, x_n]^T \in \mathcal{D}_x$ to a scalar output $y \in \mathcal{D}_y$. This model is assumed to be a "black-box" function from which a set of n_{obs} observations $\{\mathbf{X}_{obs}, \boldsymbol{y}_{obs}\}$ is available (model runs). Kriging assumes the model response y_0 for a non-observed input $\boldsymbol{x}_0 \in \mathcal{D}_x$ is the realization of a Gaussian process presented as:

$$y_0 = \sum_{i}^{p} f_i(\boldsymbol{x}_0)\beta_i + z(\boldsymbol{x}_0) = \boldsymbol{f}_0^{\mathrm{T}}\boldsymbol{\beta} + z_0 \tag{4}$$

This formalization aggregates two contributions. A regressive term is firstly used to capture the global trend of the model response based on the observational set. This part is based on a p-order regression of basis vector $\boldsymbol{f}$. To calibrate the vector of weights $\boldsymbol{\beta}$, statistical analyses are to be performed such as generalized least square estimations. The best estimation of this coefficient vector regarding the observational set is hereafter noted $\widehat{\boldsymbol{\beta}}$. This deterministic part is completed with a stochastic contribution, noted z_0, and assumed to be a realization of a Gaussian process z of zero-mean ($\mathbb{E}[z] = 0$). The covariance kernel of this structure is presented as:

$$COV(z_1, z_2) = \sigma_z^2 r(\boldsymbol{x}_1, \boldsymbol{x}_2, \boldsymbol{\theta}) \tag{5}$$

with σ_z^2 the process variance, r the correlation model of parameter vector $\boldsymbol{\theta}$ and $\boldsymbol{x}_1, \boldsymbol{x}_2 \in \mathcal{D}_x$. This contribution presents the advantage of introducing local adjustments based on a spatial dependence structure between model responses. For accuracy purposes, the calibration of the vector of hyperparameters $\boldsymbol{\theta}$ is crucial. Its best estimator $\widehat{\boldsymbol{\theta}}$ is obtained from a statistical analysis of the observations (*e.g.* maximum likelihood estimations, cross validation methods).

Within this formalization, model responses for non-simulated points can be statistically predicted. The "*Best Linear Unbiased Predictor*" (BLUP) is well suited for numerical science applications. For all $\boldsymbol{x}_0 \in \mathcal{D}_x$, it proposes to compute a predictive density function $\mathcal{N}(\hat{y}_0, \sigma_0)$ of the real response y_0 with parameters defined as:

$$\hat{y}_0 = \boldsymbol{f}_0^{\mathrm{T}}\widehat{\boldsymbol{\beta}} + \boldsymbol{r}_0^{\mathrm{T}}\mathbf{R}^{-1}(y_{obs} - \mathbf{F}\widehat{\boldsymbol{\beta}}) \tag{6}$$

$$\sigma_0^2 = \sigma_z^2[1 - \boldsymbol{r}_0^{\mathrm{T}}\mathbf{R}^{-1}\boldsymbol{r}_0 + \boldsymbol{u}_0^{\mathrm{T}}(\mathbf{F}^{\mathrm{T}}\mathbf{R}^{-1}\mathbf{F})^{-1}\boldsymbol{u}_0] \tag{7}$$

with $\boldsymbol{u}_0 = \mathbf{F}^{\mathrm{T}}\mathbf{R}^{-1}\boldsymbol{r}_0 - \boldsymbol{f}_0$, $\mathbf{F}$ and $\boldsymbol{f}_0$ respectively denoting the matrix of regression evaluated for the observation points and the vector of regression evaluated for the prediction input $\boldsymbol{x}_0$. $\mathbf{R}$ and $\boldsymbol{r}_0$ respectively denote the matrix of correlation evaluated for the observation set and the vector of cross correlation between the input $\boldsymbol{x}_0$ and the observations.

For practical purposes, model predictions are preferred to be deterministic and are thus set to the most likely value, namely the distribution mean: $y_0 \approx \hat{y}_0$. An interesting feature of the Kriging prediction is the interpolating character of the model approximation ($y_0 = \hat{y}_0$ if $\boldsymbol{x}_0$ is contained in the inputs $\mathbf{X}_{obs}$). The prediction variance σ_0^2 gives information on the accuracy of the prediction, evaluating its statistical error.

Additional explanations, demonstrations and proofs are available in [7].

1.2 The AK-DA method

As introduced in Section 1, the estimation of cycle damages presented in Eq. (2) can be demanding during the design stage of the wind turbine structures. The idea of the AK-DA algorithm is to reduce the simulation costs by reducing the number of short-term damage estimations based on the real code. From this reduced set of observations, a Kriging surrogate model $\hat{d}$ of the demanding response d is computed so as the related statistical estimation of the prediction error $\sigma_{\hat{d}}^2$ for the complete grid of environmental combinations from the standards. An estimator $\widehat{D}$ of the cumulative damage D is computable by replacing the real model response d by its computed surrogate model $\hat{d}$:

$$\widehat{D} = \sum_{j=1}^{n_c} \hat{d}(\boldsymbol{x}_j)\, p_j \Delta \qquad (8)$$

The variance $\sigma_{\widehat{D}}^2$ of the estimator therefore reads:

$$\sigma_{\widehat{D}}^2 = \sum_{j=1}^{n_c}\sum_{k=1}^{n_c} p_j p_k COV(\hat{d}(\boldsymbol{x}_j), \hat{d}(\boldsymbol{x}_k), \widehat{\boldsymbol{\theta}})\, \Delta^2 \qquad (9)$$

The quality of this approximation is highly influenced by the model observations. AK-DA proposes an iterative enrichment of an initial set of observation to enhance the accuracy of the estimator $\widehat{D}$.

Figure 1 illustrates the construction of the design of experiments. Considering the $iter^{\text{th}}$ enrichment step, the design of experiments $\{\mathbf{X}_{iter}, \boldsymbol{d}_{iter}\}$ is used to calibrate a Kriging surrogate model $\hat{d}_{iter}$ for the prediction of the damage model responses d for all the non-simulated environmental combinations of parameters. An estimation of the quantity of interest $\widehat{D}_{iter}$ is computed from Eq. 8 as the related coefficient of variation δ_{iter}:

$$\delta_{iter} = \frac{\sigma_{\widehat{D}_{iter}}}{\widehat{D}_{iter}} \qquad (10)$$

While the stopping criterion is not verified ($\delta_{iter} \geq \delta_t$), the design of experiments is enriched. To do so, the variance $\sigma_{\widehat{D}_{iter}}^2$ of the global damage predictor (Eq. 9) is decomposed as:

$$\sigma_{\widehat{D}_{iter}}^2 = \sum_{j=1}^{n_c} \mathcal{C}_j \qquad (11)$$

Each of the contributions $\mathcal{C}_j$ ($j \in [1, \dots, n_c]$) represents the part of a specific environmental combination of parameters $\boldsymbol{x}$ on the uncertainty of the estimator $\widehat{D}_{iter}$. This contribution is presented as:

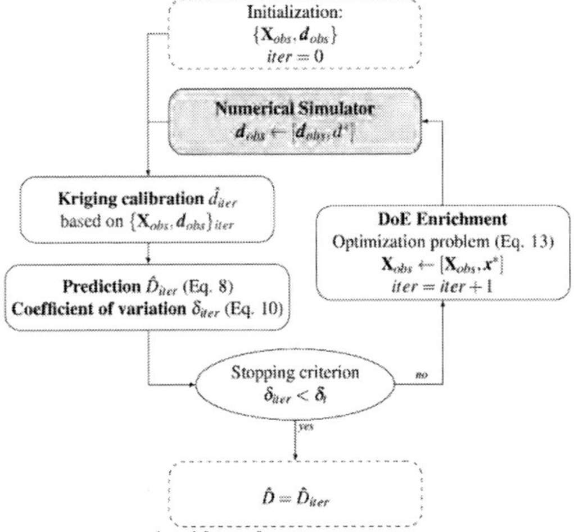

Fig. 1. AK-DA algorithm scheme

$$\mathcal{C}_j = p_i \sum_{k=1}^{n_c} p_k\, COV(\hat{d}(\boldsymbol{x}_i), \hat{d}(\boldsymbol{x}_k), \widehat{\boldsymbol{\theta}})\Delta^2 \qquad (12)$$

The objective is to reduce $\sigma_{\widehat{D}}^2$. Hence, the selection of the environmental combination of parameters $\boldsymbol{x}^* = \boldsymbol{x}_{i^*}$ to enrich the design of experiments with is carried out by solving:

$$j^* = \underset{j \in \bar{S}}{\operatorname{argmax}}\, \mathcal{C}_j \qquad (13)$$

with $\bar{S}$ regrouping the indices of the non-simulated environmental parameters.

A simulator run is then performed to observe the model response d^* for the elected combination of environmental parameters $\boldsymbol{x}^*$ and the observation set is updated. Sequentially, the design of experiments is enriched and the prediction $\widehat{D}_{iter}$ gets more accurate.

Once the convergence criterion is reached ($\delta_{iter} < \delta_t$), an estimator $\widehat{D}$ of the demanding quantity of interest D and based on an observation set $\{\mathbf{X}_{obs}, \boldsymbol{d}_{obs}\}$. The uncertainty of this estimator has to be taken into account in order to accept or to reject the proposed design. For such purpose, the deterministic acceptation criterion presented in Eq. (3) is transformed to consider the uncertainty regarding the estimator $\widehat{D}$ failure probability P_f. The validation criterion is presented as:

$$P_f = P(D > D^{lim} \,|\{\mathbf{X}_{obs}, \boldsymbol{d}_{obs}\}) < P_{admit} \qquad (14)$$

P_{admit} is the probability threshold not to overpass for the design to be accepted.

Although being computationally efficient with accurate estimations and important reductions of the efforts [5], this method is not well-suited for industrial uses in this first formulation. During the design stage of wind turbine structures, several locations of the structures (sites) are to be analysed. For each, an estimation of the long-term damage quantities is to be assessed. The current method does not allow to perform this simultaneous analysis. In the following, an extension of the algorithm is presented to perform damage analyses for multiple locations of a structure in a single process.

3 The AK-DA MultiSite extension

The MultiSite extension of the AK-DA algorithm aims at adapting the approximation method to the analysis of multiple structural sites at the same time. As the single location version, this work proposes to efficiently approximate n_s global damage quantities of interest, namely D_i with $i \epsilon [1, \dots, n_s]$. For each of the analysed sites, the grid of environmental parameters to simulate is unchanged and a single simulator run outputs n_s short term damages, regrouped in an output vector $\boldsymbol{d} = [d_1, \dots, d_{n_s}]^T$. In this sense, the design of experiments is the same for each of the n_s analyses. From the discretization of the environmental parameters, n_s quantities of interest and their predictions are computed as:

$$D_i = \sum_{j=1}^{n_c} d_i(\boldsymbol{x}_j) p_j \Delta \quad \text{for } i \in [1, \dots, n_s] \tag{15}$$

$$\widehat{D}_i = \sum_{j=1}^{n_c} \hat{d}_i(\boldsymbol{x}_j) p_j \Delta \quad \text{for } i \in [1, \dots, n_s] \tag{16}$$

As depicted in Figure 2, the enrichment step is here modified in order to optimize the construction of the design of experiments. The idea is to reduce the simulator calls with respect to a one site to one site approach. The procedure is performed by a two-step strategy. Firstly, a "pilot location" is elected between each of the n_s analysed sites considering its importance regarding the design validation and the uncertainty of the related estimator. This selection is proposed as:

$$i^* = \underset{i \,/\, \delta_i > \delta_t}{\operatorname{argmax}} \; P(D_i > D_i^{lim} | \{\mathbf{X}_{obs}, \boldsymbol{d}_{i,obs}\}) \tag{17}$$

This criterion aims at focusing on risky locations regarding cycle damage response, organizing the enrichments to reduce the related estimation uncertainties. The pilot location is then used to proceed the AK-DA enrichment strategy as defined in Section 2. This iterative enlargement of the observation set occurs until the uncertainty regarding each of the n_s analysed sites is sufficiently reduced. This is controlled by the evolution of their coefficients of variation δ_i and the stopping criterion is presented as:

$$\delta_i > \delta_t \text{ for } i \in [1, \dots, n_s] \tag{18}$$

To accept or to reject the proposed design, engineers could take into account the uncertainties regarding each of the n_s estimators $\widehat{D}_i$. To do so, the method focuses on the related failure events defined as:

$$\mathcal{E}_i = \{D_i > D_i^{lim}\} \text{ for } i \in [1, \dots, n_s] \tag{19}$$

A design is accepted if each of the defined events are verified. This validation criteria is presented as:

$$P_f = P\left(\bigcup_{i=1}^{n_s} \mathcal{E}_i\right) < P_{admit} \tag{20}$$

For practical purpose, the independence hypothesis of the failure events is assumed and the P_f is approximated by the following upper bound:

$$\sum_{i=1}^{n_c} P(\mathcal{E}_i) < P_{admit} \tag{21}$$

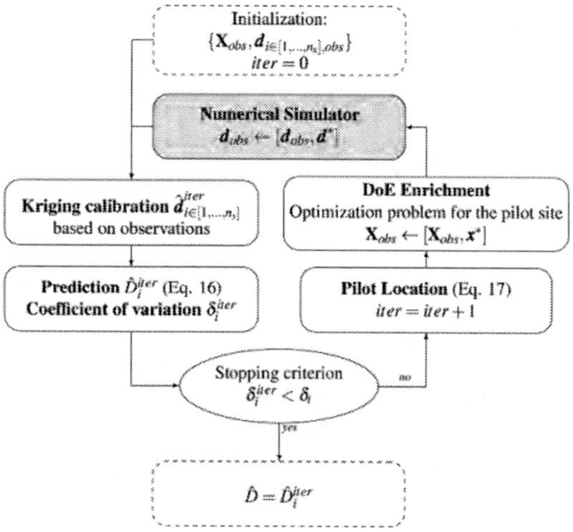

Fig. 2. AK-DA MultiSite algorithm scheme

4 Numerical Illustration

In this section, an industrial case is depicted to illustrate the behaviour of the proposed algorithm. The example, depicted in Figure 3, is a damage estimation of the DLC 1.2 [3] for the NREL OC3 Monopile [8]. Two parameters are set variable: the wind mean speed at hub height (U) and the wind-wave misalignment (θ). The related metocean (probability of occurrence) is extracted from the generic data based proposed in [9].

The discretization of the parameter domain of variation is presented in Table 1. The reference complete design of experiments is composed of $n_c = 276$ combinations of environmental parameters.

The analysis focuses on the four structural locations, presented in Figure 3. Four locations of interest are presented, composed of two points at the mudline and two points at the transition level (azimuts 0 and 90° for each levels). For validation purpose, the complete discretization has been used to obtain the real model responses d_i so as the reference values of the cycle mean damages D_i ($i \in [1, \dots, 4]$). The algorithm is initialised with a reduced grid composed of 20 environmental combinations based on 4 wind speeds (3, 11, 20, 25 m/s) and 5 misalignments (-150, -90, 0, 90, 180 °). The Kriging predictions are all set with a constant basis of regression and the anisotropic 'squared exponential' correlation model. The convergence criterion is set to $\delta_t = 5\,\%$ and the acceptance value is set to $P_{admit} = 10^{-4}$ for the damage limit $D^{lim} = 10^{-5}$. These values are arbitrarily chosen for an illustration purpose

Table 1. Parameters of the study and discretization.

Param.	Domain of definition	Discretization step	Number of Cases
U	3 to 25 m/s	1 m/s	23
θ	[-150,360] °	30°	12
Number of environmental combinations			276

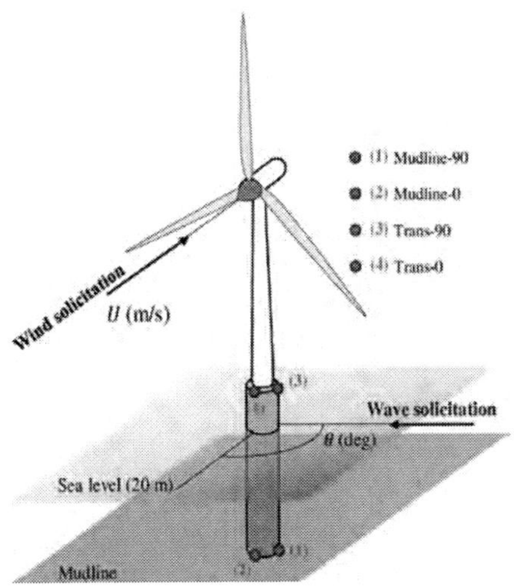

Fig. 3. NREL 5 MW Monopile with 2 environmental parameters and 4 locations of interest.

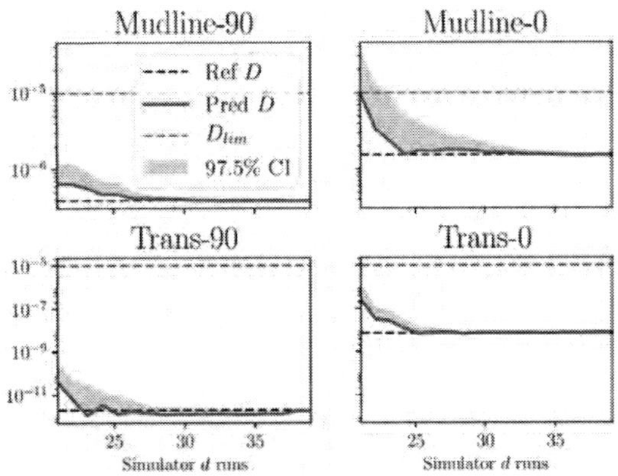

Fig. 4. Convergence of the damage estimators for each of the 4 analysed locations.

Figure 4 shows the evolution of the estimations $\widehat{D}_i$ of the cycle mean damages D_i for the four studied locations of the structure. Firstly, it is to be noted that each of these values converge to the references within a reduced number of simulator calls (39 in total composed of 20 initial calls and 19 enrichments). The statistical errors of the estimations are depicted by the related 97.5% confidence intervals. The uncertainty progressively decreases in function of the iteration because of the information inflow of the model responses (enrichment of the observational set).

Table 2 regroups the results of estimations for the four locations of interest. At convergence, the four estimators perform with different accuracies. The most important locations regarding the damage validation are the mudline points. These two are well approximated with a maximum relative error (regarding the reference value) of -1.4% for the 0°-azimuth location (Mudline-0). The transition locations (Trans-0 and Trans-90) show a lower cumulated damage far below the admitted limit value $D^{lim} = 10^{-5}$. At convergence, the approximations of the damages are not as performant as the Mudline locations, with relative error of -8.9% and 4.3% regarding the reference values.

Table 1. Estimations of the global damage estimated by the AK-DA MultiSite.

	Mudline-90	Mudline-0
Reference D	3.886×10^{-7}	1.548×10^{-7}
AK-DA $\widehat{D}$	3.876×10^{-7}	1.526×10^{-7}
Rel. error (%)	-0.4	-1.4
	Trans-90	Trans-0
Reference D	2.029×10^{-12}	7.362×10^{-12}
AK-DA $\widehat{D}$	1.849×10^{-12}	7.681×10^{-12}
Rel. error (%)	-8.9	4.3

Figure 5 shows the evolution of the "pilot locations" elected by the algorithm at each iterations. This step shows a performant and economical behaviour with enrichments mainly controlled by the sizing locations (Mudline-0 and 90).

In total 39 simulator calls have been performed for the algorithm to converge. These observations are composed of 20 simulations from the initial reduced grid and 19 enrichments control by the AK-DA method. Comparing to the reference case based on a complete grid of size 276, the numerical gain in simulator calls is equal to 7.08.

To validate or to discard the analysed design, the probability of failure P_f is computed (Eq. 20 and 21). Fig. 6 depicts the evolution of this quantity for the 19 enrichments of the algorithm. The uncertainty related to the estimators, especially to the Mudline-0 location, makes the decision quantity high for the initial state ($P_f > 0.1$ for 21 model observations). The AK-DA enrichment strategy aims at reducing this uncertainty adding informative environmental situations to the observational set. The consequence is a rapid decreasing of the probability of failure reaching $P_f < 10^{-11}$ after 5 iterations. From the 6th enrichment, the probability of failure is considered as negligible. For this example, the design is validated regarding the cycle cumulated damages of the four structural locations.

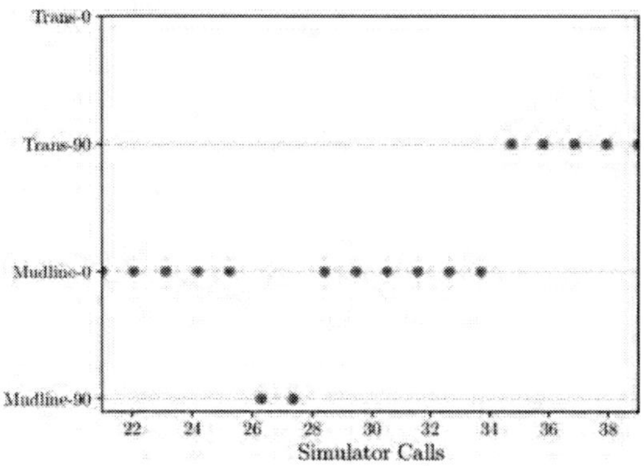

Fig. 5. Evolution of the pilot location for each of the AK-DA enrichments.

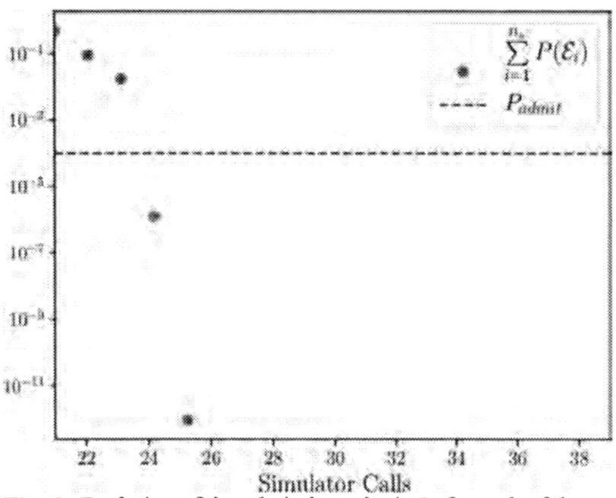

Fig. 6. Evolution of the admission criteria P_f for each of the AK-DA enrichments.

5 Conclusions and perspectives

This paper presents an extension of the AK-DA (Adaptive Kriging for Damage Assessment) algorithm for the complete damage analysis of a design. The proposed MultiSite formalization aims at building a reduced but informative design of experiments to efficiently approximate the demanding cumulated damage quantities of a given structural design.

Based on parallel Kriging estimations, this algorithm is engineered in order to orientate the simulations for the most sizing locations to be accurately approximated with a prioritization of the enrichments regarding the uncertainty of the estimators and the risk of overpassing admitted damage limits. The validation criterion is here modified to take into consideration each of the possible failure events (damage threshold passing) in the decision step of the design validation.

The numerical example, based on an industrial case, illustrates the algorithm behaviour and the prioritization of the enrichments. The results are satisfying with an important reduction of the simulator calls (gain of 7) and effective approximations of the damage quantities.

Future works will focus on the use of this algorithm to perform statistical analyses of the structural damages that require a large number of estimations, *i.e.* uncertainty propagation, sensitivity analysis and reliability analysis.

This PhD project is financially supported by a CIFRE grand industrially supported by EDF (Palaiseau, France) and partially subsidized by the ANRT under grant #2015/0692.

References

1. Global Wind Energy Council, Global wind report 2016. (2016)
2. DNV-GL, DNV-OS-J101 Design Of Offshore Wind Turbine Structures (2014)
3. IEC, 61400-1 Wind Turbines – Part 1; Design requirements (2005)
4. IEC, 61400-3 Wind Turbines – Part 3; Design requirements for offshore wind turbines (2009)
5. Q. Huchet, C. Mattrand, P. Beaurepaire, N. Relun, N. Gayton, A Kriging based procedure for the certification of wind turbine structures: application to large scale models. Procedings of the 54th ESReDA Seminar (2018)
6. T. W. Simpson, J. D. Poplinski, P. N. Koch, J. K. Allen, Metamodels for computer based engineering design : survey and recommendations. Engineering with computers **17**-2 (2001)
7. V. Dubourg, Adaptive surrogate models for reliability analysis and reliability-based design optimization. PhD Thesis, Université Blaise Pascal – Clermont-Ferrand II (2011)
8. J. Jonkman, S. Butterfield, W. Musial, G. Scott, Definition of a 5-MW reference wind turbine for offshore system development. National Renewable Energy Laboratory (NREL) (2009)

9. G. M. Stewart, A. Robertson, J. Jonkman, A. Lackner. The creation of a comprehensive metocean data set for offshore wind turbine simulations. Journal of Wind Energy, **19**-6 (2015)

 https://doi.org/10.1051/matecconf/201816517002

Through-Process Modelling for Accurate Prediction of Long Term Anisotropic Mechanics in Fibre Reinforced Thermoplastics.

Julien Cathelin

SABIC B.V., Application Technology, 6160 AH Geleen, The Netherlands

Abstract. Fibre reinforced thermoplastic (FRTP) materials offer great potential for, among others, weight and cost reduction in a wide range of applications. In this paper, consequences of fiber orientation-induced anisotropy (due to injection molding process) in the development of FRTP parts as well as predictive engineering techniques for part performance evaluation are discussed. A coupled simulation methodology will be used to predict the processing-microstructure-properties relation in FRTP parts, thereby enabling the ability to include fiber orientation-induced anisotropy. In parallel, long term fatigue data was collected on increasingly complex geometries for the purpose of model validation.

1 Introduction

Due to the many advantages of low density, high stiffness and freedom to design geometrically complicated parts, FRTP materials are increasingly replacing metals in a wide range of industrial applications. For instance, metal replacement has accelerated in the automotive industry due to the need for weight reduction to increase fuel efficiency. In water management applications (distribution, measurement, treatment and storage equipment), replacing metals is an approach to meet low-lead regulations, remove system cost and overcome corrosion challenges.

The main process for the production of FRTP parts is injection molding, which gives the engineer freedom to design geometrically complicated multi-functional ribbed parts, needed for robust design, as can be seen in Figure 1.

However, the injection molding process imposes additional complexities to the design and performance of FRTP components by introducing complex glass fiber length and orientation variations in the parts. These variations consequently represent themselves in the form of anisotropic shrinkage during cooling, referred to as warpage, or anisotropic mechanical performance, which might lead to over or under-design in the finished part [1].

One of the important aspects of polymers, either reinforced or non-reinforced, in load-bearing applications is that they display time-dependent response and, ultimately, if not considered in design might lead to failure. It is not the question of whether failure will occur, but rather on what time scale. The assessment of the long-term performance of polymeric products is therefore of the utmost importance. For FRTP components, which are designed for long-term strength, the fiber orientation-induced anisotropy must be taken into account during the development of such parts, since the long-term response and lifetime of the parts can vary depending on the orientation [2].

With the increasing use of FRTP materials in various applications, virtual product development is also growing extensively, as it is an important drive to reduce time-to-market while minimizing (ideally eliminating) physical prototyping. Simulation of the performance of the final product is one of the key components of the virtual product development. Increasing the accuracy of the simulations requires more detailed modeling, such as the incorporation of the fiber orientation-induced anisotropy variable.

Quantitative prediction of fiber orientation-induced anisotropy and utilization of anisotropy in the design process has been a significant challenge for adoption in the water management industry. In this article, a through-process modeling approach for short- and long-term performance evaluation of FRTP parts is discussed. This approach enables establishing a processing-morphology-property relation and taking into account the effect of anisotropy on the performance. Several predictive engineering techniques, including utilization of both isotropic and anisotropic material properties, are also evaluated.

Fig. 1. A complex injection molded thermoplastic flow block (left) and a hydroblock, assembled with a heat exchanger

* Corresponding author: julien.cathelin@sabic.com

https://doi.org/10.1051/matecconf/201816517002

(right), made of NORYL™ resin (short glass fiber reinforced PPO™/PS).

2 Experimental

The material used in this study is a 30% short glass fiber reinforced injection moldable thermoplastic supplied by SABIC (NORYL™ resin, a PPO™/PS material), priamarily used in water management applications (Hydro-block, water meter, faucet, etc) due to its good corrosion properties, high temperature resistance, low water absorption and good mechanical performance [3].

Figure 2 shows the geometries molded and used in this study. A so-called shear plaque was used for mechanical testing (tensile and fatigue), (micro) mechanical model creation, and to fit the models' parameters.

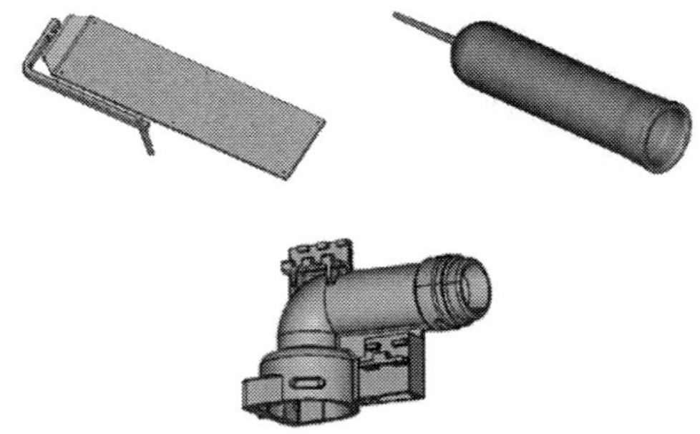

Fig. 2. Various geometries moulded and tested in this study. (Top left): shear plaque. (Top right): Pressure Vessel. (Bottom): Elbow pipe.

Two different geometries with increasing geometrical complexities, i.e. a pressure vessel and an elbow pipe, were also molded for various validation tests (constant internal pressure rate, constant internal pressure and fluctuating internal pressure). The results will be used later to validate the predictive engineering approaches.

2.1 Morphology characterization

Fiber orientation and length distributions were characterized to provide input for micromechanical modelling, including fiber aspect ratio and local fiber orientation distribution. A dynamic image analysis technique, developed within SABIC, was used for fiber length measurement [4]. This technique is a semi-automated approach using Camsizer particle size analyzer for fiber length analysis. It is statistically more robust and faster than the static imaging method.

Local fiber orientation distribution was measured at the location from which tensile specimens were cut. Computed tomography scanning (CT-Scan) was used to directly measure microstructure. Raw data was then analyzed to calculate components of local orientation tensors along the thickness of the test specimen [5]. The process is depicted in Figure 3-4.

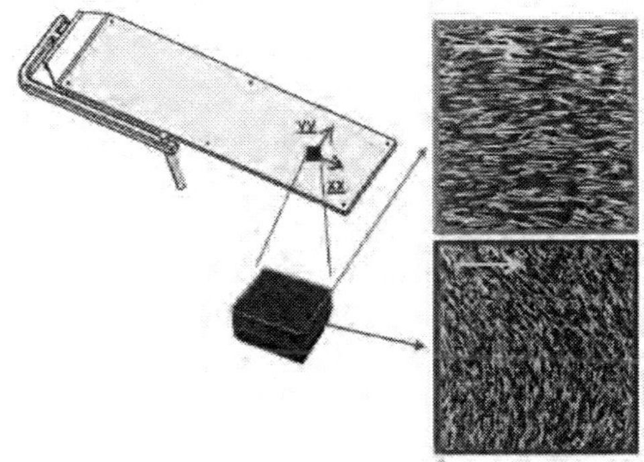

Fig. 3. Left: The shear plaque and the specimen used for CT-Scan. Right: CT-scan images at the region of interest. A distinct difference can be observed in local fiber orientation distributions in the skin (marked as blue) and in the core (marked as red). Arrows indicate flow direction. Right: Diagonal components of local fiber orientation tensor, obtained by post processing of CT-scan images.

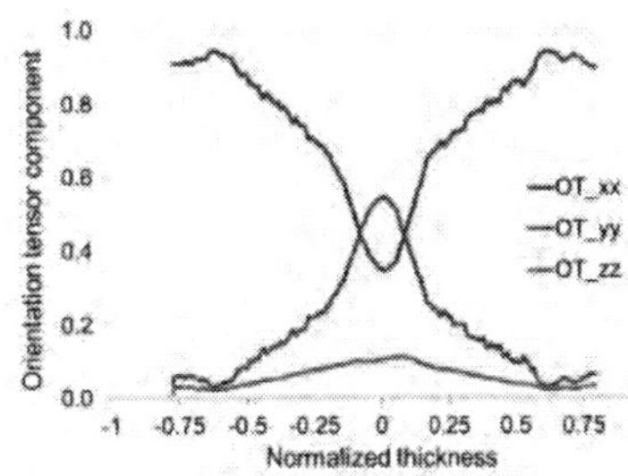

Fig. 4. Diagonal components of local fiber orientation tensor, obtained by post processing of CT-scan images.

2.2 Mechanical testing

In order to characterize the anisotropic performance of NORYL™ resin and create (micro) mechanical models, tensile specimens were cut at 0°, 45° and 90° from a fixed location of the shear plaque. Tensile (constant strain rate) and fatigue (fluctuating stress) tests were performed at various orientations, temperatures and load levels.

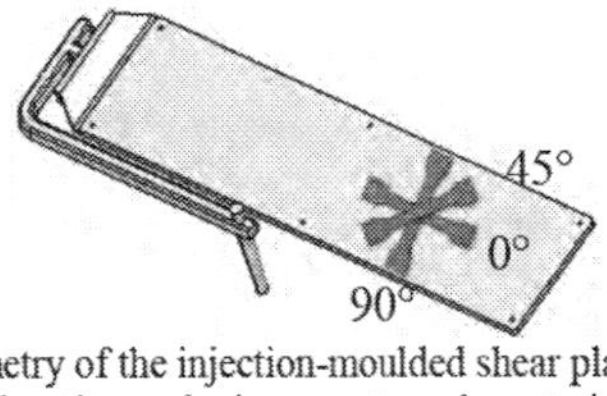

Fig. 5. Geometry of the injection-moulded shear plaque used for mechanical testing and microstructure characterization with location and orientation of tensile specimens.

2.4 Validation test

To evaluate various simulation methodologies, different mechanical tests have been performed on the pressure vessel and the elbow pipe, subject to internal water pressure at 90oC including:

- Constant pressure rate, representing short-term response.
- Fluctuating pressure, representing fatigue performance.

In order to record the evolution of strain under internal pressure, a 12M ARAMIS system was employed. Figure 6 shows the test setup. ARAMIS is a non-contact measuring system, which enables an optical 3D deformation analysis for statically or dynamically loaded test objects.

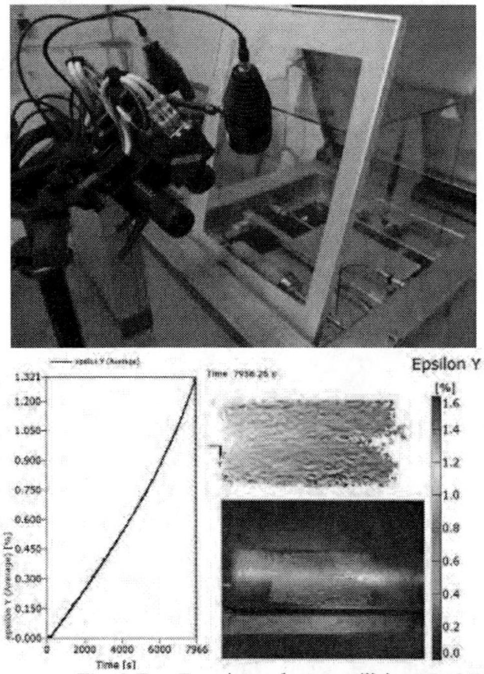

Fig. 6. Top: Experimental setup, utilizing ARAMIS camera system. Bottom: The strain evolution, averaged over the area of interest on the surface of the pressure vessel subject to internal constant pressure rate.

3 Predictive Engineering Approach

In this section, various modelling and simulation methodologies for performance evaluation of FRTP parts are discussed together with their advantages and drawbacks.

3.1 Isotropic approach

The use of isotropic material data for part performance prediction is still common practice in industry, especially for early stages of an application development. There is not yet a standard way on how isotropic material properties are obtained for FRTP materials. Depending on how data is measured (directly injection-molded tensile specimen, tensile samples milled out of injection-molded plaques with various geometries and processing conditions, specimen cut from actual parts, etc.), a wide range of properties are found due to differences in fiber orientation distribution [5].

Measuring material engineering data on a directly injection-molded (DIM) tensile specimen (also referred to as Datasheet values) leads to unrealistically high stiffness and strength values due to an extremely high fiber alignment in a long narrow geometry. This approach could potentially lead to an overestimation of performance and, consequently, under-design of the part in real application [1].

To address this issue, the European Alliance for Thermoplastic Composites (EATC) has developed a methodology to convert the anisotropic mechanical properties, measured on injection molded plaques, into effective (quasi)isotropic properties. This methodology utilizes the classical laminate theory [6], as shown schematically in Figure 7. This method enables having unique isotropic material properties, independent of fiber orientation and processing conditions [5].

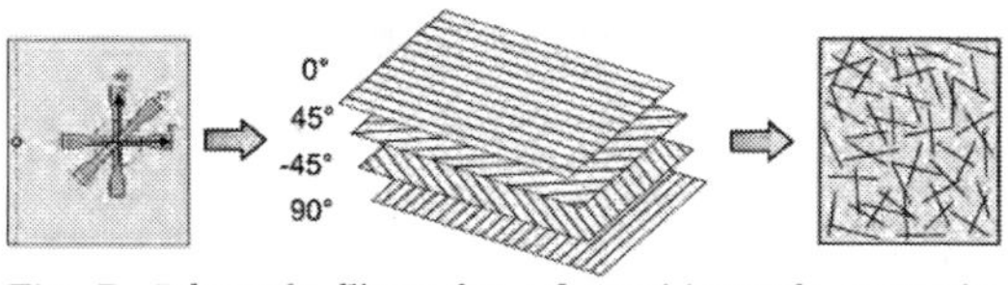

Fig. 7. Schematic illustration of quasi-isotropic properties generation using classical laminate theory.

Engineering datasheets of quasi-isotropic stress-strain and fatigue curves were created based on 0°, 45° and 90° stress-strain curves and fatigue curves.

3.2 Anisotropic approach

However, Using the isotropic modeling approach for FRTP materials does not fully represent actual performance, since fiber orientation-induced anisotropy is not taken into account. Micromechanically based constitutive models [7, 8] are powerful tools, which enable establishing a relation between mechanical properties of FRTP materials and their underlying morphology including being fiber volume fraction, aspect ratio and orientation. Recently, some of these models have been adapted and implemented in several commercial software, which allow their utilization in industrial applications.

A two-step mean-field homogenization procedure [9, 10] is employed here to relate the local mechanical properties of the FRTP part to its processing-dependent morphological features. In this technique, the real representative volume element (RVE) of the reinforced material is replaced with a model RVE of an aggregate of so-called pseudo grains. In each pseudo grain, fibers having similar shape and orientation are grouped together and embedded in the matrix phase. The first homogenization step estimates the properties inside each pseudo-grain using a mean field homogenization scheme, appropriate for basic two-phase composites, e.g. Mori-Tanaka model [11]. Then, a second homogenization step averages these properties, applying fiber orientation distribution, and derives the effective response of the aggregate of homogenized pseudo-grains.

In this study, Digimat software suite is utilized to perform micromechanical modelling. Figure 8 gives the general overview of the micromechanical modelling workflow

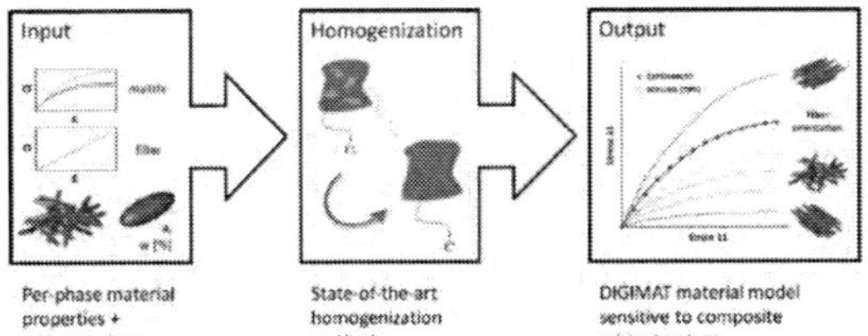

Fig. 8. Schematic overview of Digimat workflow for creating anisotropic material models [12].

Inputs needed are morphology information (local fiber orientation tensor, fiber content, fiber aspect ratio) and per-phase (matrix and fiber) material properties. Morphological information of the RVE required for mean field homogenization are obtained experimentally, as explained in the previous section.

The parameters of the micromechanically based constitutive models of each phase (matrix and fiber) are characterized based on the measurements on the matrix and fiber reinforced material at 0°, 45° and 90° load angles. An elaboration of the material models and reverse engineering procedure for characterization of micromechanical model parameters can be found in [12].

3.2.1 Anisotropic fatigue failure criteria

The fatigue modelling is based on combination of the elastic constitutive model and a macroscopic fatigue failure indicator. The fatigue failure criteria is a modified Tsai-Hill 3D transversely isotropic failure indicator. But in the case of fatigue, strength parameters are function of the number of cycle. This criterion has three input arguments (as it assumes identical strengths in tension and compression) and outputs one indicator. The inputs are:

1. Axial fatigue strength depending on the number of cycle to failure, $S_L(N_c) > 0$.
2. In-plane fatigue strength depending on the number of cycle to failure, $S_T(N_c) > 0$.
3. Transverse shear fatigue strength depending on the number of cycle to failure, $S_{LT}(N_c) > 0$.

and the output is a fatigue failure indicator FC in the form of:

$$FC(N_f) = \frac{\sigma_L^2}{S_L(N_c)^2} - \frac{\sigma_L(\sigma_{T1}+\sigma_{T2})}{S_L(N_c)^2} + \frac{\sigma_{T1}^2+\sigma_{T2}^2}{S_T(N_c)^2} +$$

$$\left(\frac{1}{S_L(N_c)^2} - \frac{2}{S_T(N_c)^2}\right)\sigma_{T1}\sigma_{T2} + \frac{\sigma_{LT1}^2+\sigma_{LT2}^2}{S_{LT}(N_c)^2} + \left(\frac{4}{S_T(N_c)^2} - \frac{1}{S_L(N_c)^2}\right)\sigma_{TT}^2 \quad (1)$$

where σ_L, σ_{T1} and σ_{T2} are the longitudinal and the two transversal stress amplitudes, respectively. Furthermore, σ_{LT1} and σ_{LT2} are the shear stress amplitude between the longitudinal and transverse directions and σ_{TT} is the shear stress amplitude in the plane normal to the longitudinal direction.

Failure occurs when the indicator reaches or exceeds the value of 1.

Like for the short term failure modelling, the problem with these criteria is that you need to enter the fatigue failure stresses in different directions (Digimat notations: X, Y and S). That can be done for a general composite material with known properties. However, for an injection molded material with an arbitrary fiber orientation distribution, this is not known beforehand, as the orientation distribution function will be different for each element. To solve this issue Digimat use a progressive failure mechanism that is called the First Pseudo-Grain Failure (FPGF) model [12].

3.3 Through-process modelling workflow

A through-process modeling methodology (coupled Moldflow-Digimat-Abaqus) is employed here to establish the processing-morphology-property relation in FRTP parts and predict their anisotropic mechanical performance. Moldflow† software is used to simulate the injection molding process and to predict the local fiber orientation distribution in the part. Digimat software suite is utilized to establish a relation between local mechanical properties of a FRTP material and processing-dependent morphology. Structural simulation is then performed in a finite element analysis software, like Abaqus. Workflow of such methodology is depicted in Figure 9.

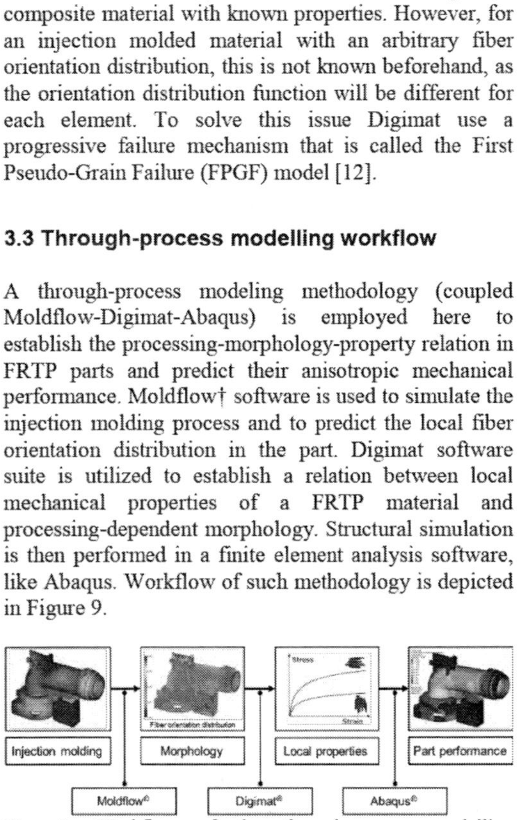

Fig. 9. Workflow of the through-process modelling methodology, employed here for predicting the anisotropic mechanical performance of fibre reinforced thermoplastic parts

For fatigue lifetime prediction, an additional step is required, to relate the stress distribution on the part to the anisotropic time-to-failure.

4 Results and Discussion

In order to evaluate the accuracy of the through-process modeling methodology, validation tests are performed on different geometries, i.e. the pressure vessel and the elbow pipe. Performed tests are burst pressure test (constant applied pressure rate, representing the short-term failure) and fatigue test (fluctuating internal pressure).

4.1 Burst pressure

Burst pressure simulations have been performed on the pressure vessel and the elbow pipe at constant burst pressure rates of 0.01 and 0.1 [bar.s-1], using both isotropic and anisotropic simulation approaches. The results are shown in Figure 10. As can be seen, isotropic prediction using DIM tensile data underestimates the strain evolution, i.e. a much stiffer behavior. It is observed that results can be improved by using quasi-isotropic data for simulation.

However, none of the isotropic approaches can capture the realistic performance of the pressure vessel. Only by taking into account fiber orientation-induced anisotropy via through-process modeling methodology, one can more accurately predict the strain evolution in the part.

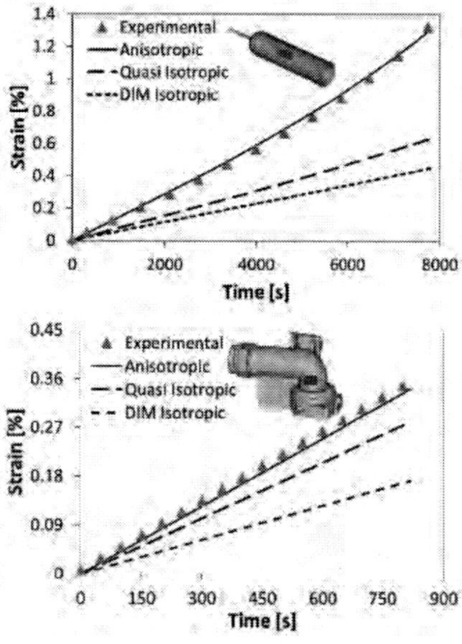

Fig. 10. Burst pressure simulation results on (top): a pressure vessel at a constant pressure rate of 0.01 [bar.s-1] and (bottom): an elbow pipe at a rate of 0.1 [bar.s-1]. The red areas on the surface of the parts, show the location of area of interest, over which strains are measured.

4.2 Fatigue with fluctuating pressure

Fatigue simulations have been done with an internal water pressure oscillating at a constant pressure ratio of 0.1. Figure 11 shows the predicted cycle-to-failure as a function of maximum applied internal pressure, using isotropic and anisotropic approaches.

As can be seen, the isotropic approach strongly overestimate the real cycle-to-failure. For a pressure vessel, with an injection location on top, the first principal direction of stress (Hoop stress) is perpendicular to the fiber alignment at skin. Therefore, fiber reinforcement becomes less effective. This design situation could happen in many of water management applications due to commonality in geometric consideration, in which failure occurs due to an internal water pressure. The anisotropic simulation approach can capture this effect, enables providing a more realistic prediction.

It is worth pointing out that for the case of the pressure vessel, there is a factor of 200 difference in cycle-to-failure prediction of isotropic and anisotropic approaches for a same maximum applied internal pressure. For the elbow pipe, the prediction difference is a factor of 100. As can be seen, fiber orientation-induced anisotropy has a significant influence in the long-term performance and lifetime of FRTP parts and using isotropic approach can lead to a significant overestimation of the durability of the part.

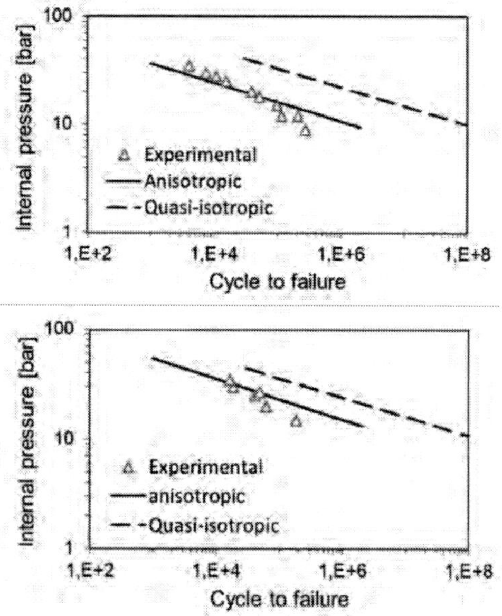

Fig. 11. Fatigue simulation and cycle-to-failure prediction using isotropic and anisotropic approaches at a stress ratio of 0.1. (Top): pressure vessel. (Bottom): elbow pipe.

5 Conclusion

A through-process modeling methodology has been established and evaluated for accurate prediction of short- and long-term performance of short fiber reinforced thermoplastic parts. This methodology establishes a relationship between the processing-dependent morphology of FRTP part and its anisotropic mechanical performance. Simulation results show that anisotropic simulation via the through-process modeling approach can more accurately represent the performance of a FRTP part. An in-depth knowledge and accurate characterization and modeling of material properties is key in this approach.

Quantitative prediction of the performance of FRTP parts via anisotropic simulation and incorporation of that in the application development process is crucial towards sustainable design solutions. Capturing realistic performance of FRTP parts can potentially result in:

- Improved simulation accuracy, leading to accelerated development cycle.
- More weight reduction possibilities via optimum material use, enabling thinner and lighter structures.
- Reliable lifetime prediction, and consequently design for durability.

References

1. A. Sedighiamiri, D. Brands. 16th int. conf. on Deformation, Yield and Fracture of Polymers, Kerkrade, The Netherland, 2015.
2. T. v. Erp, C. Reynolds, T. Peijs, J. v. Dommelen and L. Govaert. Journal of Polymer Science Part B: Polymer Physics, 47:2026 - 2035, 2009.
3. C. Koevoets., SPE Eurotec, Lyon, France, 2013.
4. P. Bajaj. Microscopy and Microanalysis, Indianapolis, USA, 2013.
5. A. Sedighiamiri, T. v. Erp, J. Cathelin and D. Brands. 20th International Conference on Composite Materials, Copenhagen, Danmark, 2015.
6. W. Schijve and A. Rüegg. European alliance for thermoplastic composites, 2008.
7. A. Sedighiamiri, L. Govaert, M. Kanters and J. v. Dommelen. Journal of Polymer Science Part B: Polymer Physics, 50:1664-1679, 2012.
8. I. Doghri and L. Tinel. International Journal of Plasticity, 21:1919-1940, 2005.
9. C. Camacho, C. T. III, S. Yalvac and R. McGee. Polymer Composites, 11:229-239, 1990.
10. S. Kammoun, I. Doghri, L. Adam, G. Robert and L. Delannay. Composites Part A: Applied Science and Manufacturing, 42:1919–1940, 2011.
11. T. Mori and K. Tanaka. Acta Metallurgica, 21:571-574, 1973.
12. Digimat e-Xstream engineering. Available: http://www.e-xstream.com.

Automotive cylinder heads: recent advances on Thermal-Mechanical Fatigue design and upcoming challenges

Fabien Szmytka [1,4], *Pierre Osmond* [2], *Luc Rémy* [3], *Pierre-Damien Masson* [2], *Agathe Forré* [1],

[1] Science and Future Technologies Department, Groupe PSA, Vélizy-Villacoublay, France
[2] Chassis and Engine Material Department, Groupe PSA, La Garenne Colombes, France
[3] Centre des Matériaux, Mines Paristech, UMR CNRS 7633, Evry, France
[4] Institute for Mechanical Sciences and Industrial Applications (IMSIA-UME), ENSTA Paristech, Palaiseau, France

Abstract. Through the peculiar case of an automotive cylinder head, actual and forthcoming challenges for low-cycle and thermal-mechanical fatigue design in an industrial context are presented. A special focus is applied on variable loadings, constitutive models and their interaction with the environment, fatigue criteria and structure validations.

1 Introduction

Evolutions in environmental rules combined with the competitive race to produce cleaner and lighter engines have induced, since about twenty years, a constant increase in thermal fluxes applied to automotive engine parts such as cylinder heads. These loading conditions lead to a well-known serious risk of thermal-mechanical fatigue (TMF) cracks initiation that must be controlled to design reliable structure.

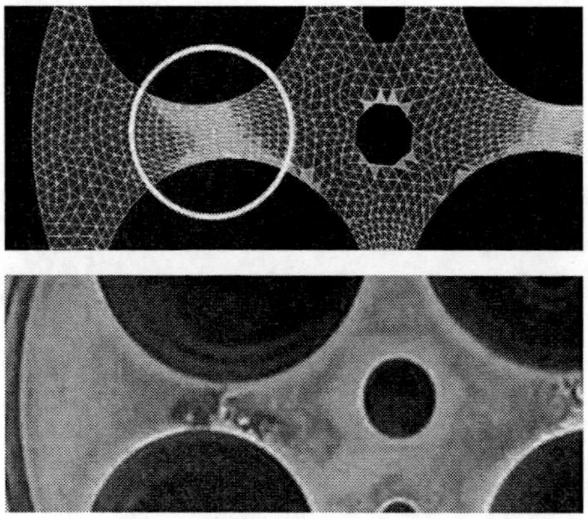

Fig. 1. Localisation of damage for cylinder under deterministic design loading: simulation (top) and experimental test (bottom)

Car manufacturers, and more particularly PSA Group have proposed many evolutions since Charkaluk and co-workers [1] developed the first protocol against TMF for automotive parts (with success in estimating critical zones and fatigue lifetime as seen on Figure 1), beneficiating of a better understanding of the structure behaviour, enhanced computation capacities, and better experimental investigation techniques. This article aims therefore at providing a summary of these improvements and of the upcoming challenges through the example of the cylinder head.

First, the randomness of the thermal loadings is presented as a crucial point for this part fatigue design. For a long time, deterministic design protocols have been used but their high level of conservatism tends to disqualify them today. Huge efforts have been recently made to successfully propose probabilistic design methods, relying on the study of loading and material TMF resistance randomness. Many statistical hypotheses are still questioned and will be briefly discussed. Moreover, aluminium alloys ageing which are of primary importance for the fatigue design, are deeply influenced by these loading sequences. Some reflexions about this phenomenon's impact on TMF resistance evaluation will also be proposed.

In a second time, constitutive models improvements by using inspired-by-physic equations relying on dislocations densities are proposed. Micro-crack growth laws and defects population are introduced in fatigue criterion and the impact on the fatigue design protocol is then evaluated. The reliability of these two elements is presented as crucial for proposing non-linear damage accumulation strategies, more appropriate for probabilistic design protocols. Simplified methods for numerical integration of such models leading to a more flexible use of complex equations are jointly presented.

Finally, a peculiar attention is dedicated to an experimental set-up based on a simplified structure (single cylinder head), half the way from specimens to real industrial structures. They enable to reproduce the complexity of real loadings on structure for which the strain and stress analyses are far easier. Example of

* Corresponding author: luc.remy@mat.ensmp.fr

comparison between models and experiment is detailed to underline the previously discussed improvements.

2 Material

Cylinder heads are currently made from aluminium alloys containing silicon, magnesium and often copper due to their high strength to weight ratio, high thermal conductivity, good casting properties as well as machinability. The A356 alloy is widely used and strengthened by precipitation (T7 heat treatment). Alloys containing copper are also widely used in the automotive industry like the A319 alloy [2]. Many cylinder heads are nowadays produced by die casting. Meanwhile, the lost foam casting process was recently introduced to further reduce the processing cost while opening geometry optimisation new opportunities. However, the slower cooling rates result in coarser microstructures and larger porosity as presented for an A356-T7 alloy on Figure 2.

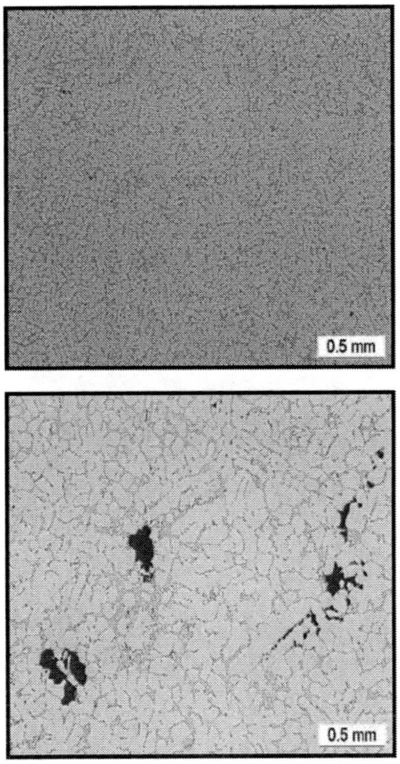

Fig. 2. A356-T7 alloy microstructure for die casting (on top) and lost foam casting (bottom)

The design protocol used today has been established for A356-T7 aluminium alloy that is widely used for PSA cylinder heads in diesel common rail injection engines [1]. Such components can experience operating temperatures that can exceed 250°C in some critical areas. Maximum temperatures thus can exceed the temperature for the precipitation treatment so that overageing occurs during service. That results in a loss of strength and of fatigue resistance. In long term overaged condition, this alloy exhibits a single residual micro-hardness, essentially independent of temperature, as shown on Figure 6- and presented in [3].

3 Variable Loadings and Reliable design

For a long time, deterministic design protocols have been used but their high level of conservatism tends to disqualify them today. Huge efforts have been recently made to successfully propose probabilistic design methods, taking account of the loading randomness, which is a key point for fatigue. If one wishes to estimate the variability in the use of a structure by a given population and the possible dispersion at which may be subject the resistance of this same structure, it is fundamental to experimentally measure this variability and therefore to collect as much information as possible by establishing experimental databases. These latter will then contribute to the improvement of constitutive models and fatigue criterion as mentioned in the following sections.

Measuring the "stress" applied to an automotive structure subjected to thermomechanical fatigue consists in building a distribution of severity of use, taking into account both the usage patterns (road type, vehicle embedded mass, weather conditions, etc.) but also the severity with which a driver will use his vehicle. It is therefore necessary to analyse and measure these two aspects of the "stress" environment. To simplify the problem, one can limit the usage patterns at the travelled types of roads. In order to obtain reliable information on their habits, surveys are conducted on a large population of vehicle drivers. One then classes the vehicle usage according to specific usage criteria such as "Classic Road", "Highway", "City" or "Mountain". It is asked to the vehicle owner to estimate the percentage of time spent or total mileage corresponding to each type of road during a year. This enables to create a customer database in terms of vehicles usage.

To be able to know the vehicles' usage conditions is far from sufficient to estimate a severity of use. Indeed, it is also essential to know the conditions of loading of this structure we want to design on all the specific types of roads, while taking into account the driver behaviour variability. In order to estimate both the drivers' behaviour and structure loads for a specific type of road, the structure one want to design has to be equipped with sensors. This vehicle is then driven on multiple courses taken as references for every possible "real-life" situations by a panel of professional drivers. This produces a matrix of elementary life situation of loading for the vehicle and by the way the studied structure, mixing both road and driver types and helps to understand the loading environment on the structure to be designed and then to adapt the design protocol.

The available data to build the probability density for customers' loads are macroscopic records for overall level of loading (engine speed and torque). However, designing cylinder heads for thermomechanical fatigue is linked to the description of a local phenomenon on the fire deck intervalve bridge and most usually to a finite element analysis in order to quantify the damage per cycle [1]. To take into account the usage variability, one needs to estimate local mechanical answer from global and monitored loadings. Signal processing methods are then developed to exhibit basic loading cycles from global loading signals [4]. In the same time, a global/local transition law is built to estimate the local damage resulting from one global basic loading cycle.

The first step is to choose an observable variable to represent the global loading condition. The relevant boundary condition for a FEM analysis is the thermal flux density applied on the cylinder head fire deck, which can easily be extracted from the engine speed and torque monitoring, coupled with the engine combustion parameters. An engine load is then defined as a random non-zero temporal evolution of thermal flux density applied on the fire deck. The development of a signal processing method, based on the principle of "thermomechanical loading", enables to characterize a material for TMF studies by taking into account the nonlinear parts of its behaviour. This step is in most cases realised with elasto-viscoplastic models where hold times, strain or temperature amplitudes and rates are the most important parameters.

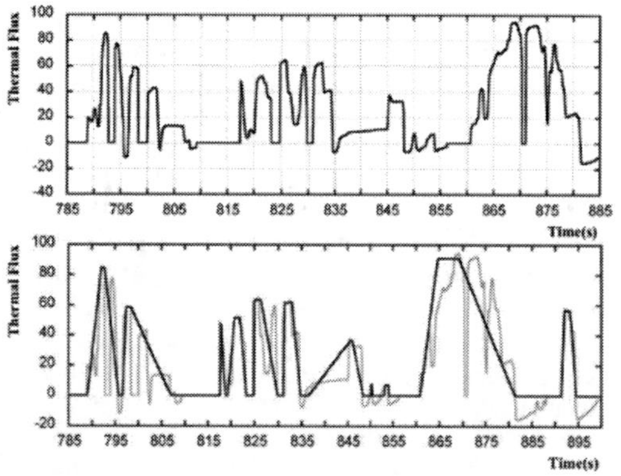

Fig. 3 Real loading sequence for a cylinder head (on top) and filtered load sequence (bottom).

Global temporal signals are thus filtered following the principle described on Figure 3. One aims at detecting cycles which main characteristics are flux variation rates, hold times for high and low flux values and maximal and minimal flux values. The signal processing principles also enable to take into account the thermal inertia of the cylinder head and thus in particular to group basic identified cycles closely spaced together in time and for which the quick flux variations lead only to very limited fire deck temperature variations. It is thus possible to obtain, from a complex and noisy signal, a succession of basic loading cycles as seen once again in Figure 3.

Here, it becomes obvious that deterministic design protocols relying on the succession of standardized severe loading cycles are far from the experimental reality. Mastering the loading conditions is the first step to optimize the fatigue design. Constitutive models and fatigue criteria are the two other ones.

5 Constitutive model and ageing conditions

A common design procedure used to handle the problem of ageing during service is to assume two different constitutive cyclic (and monotonic) stress-strain behaviours. A conventional viscoplastic Chaboche model [5] is identified for initial T7 condition and used for low temperature areas. A second Chaboche model is identified for the long-term overaged condition for areas experiencing high temperatures (Figure 4). However, when casting conditions are changed, when the loading conditions become very variable or when alloy composition is modified, one needs to assess the validity of the present design procedure, or to change it if necessary.

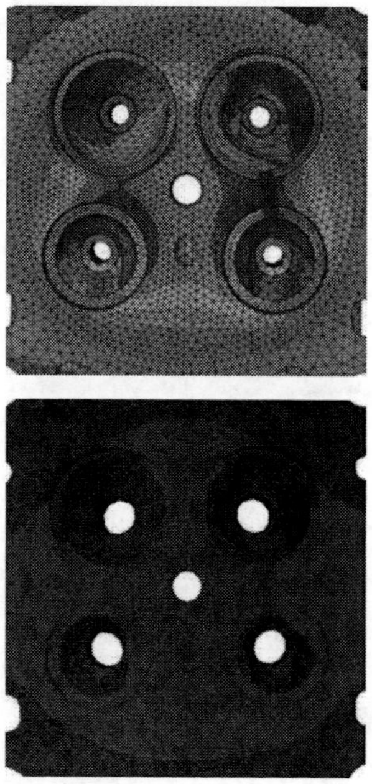

Fig. 4 Example of thermal map for a fire deck and localisation of an "overaged" zone

Figure 5 shows the different scales of microstructure. During the last decade, the influence of variations in compositions and overageing treatments conditions has been investigated. Conventional micro-hardness measurements were combined with transmission electron microscopy observations to quantify the variations in precipitate distribution (nature of precipitates, size and volume fraction). Some specific overageing conditions were selected and cyclic stress-strain behaviour was identified at different temperatures.

https://doi.org/10.1051/matecconf/201816517003

To establish the link between heat treatments and microstructure variation, a microstructure sensitive model was used instead of a Chaboche one. Basically, the model uses dislocation density as internal variable, considering dislocation multiplication and dynamic recovery as proposed by Kocks and Estrin ([6], [7]). Static recovery can be easily introduced and this type of model was shown to be very effective for stainless steels used in exhaust manifolds [8]. Dislocations by-passing precipitates by Orowan mechanism leave loops around precipitates, which is handled by specific contributions in isotropic and kinematic hardenings [9].

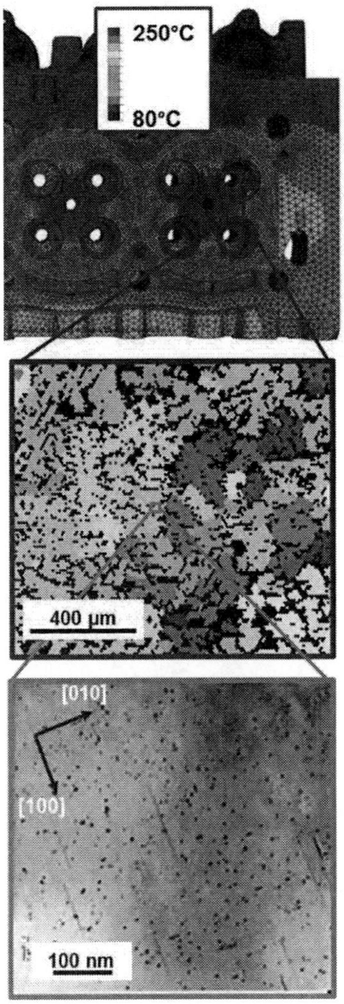

Fig 5: Temperature map on cylinder head (on top), typical grain microstructure of A356-T7 aluminium alloy (middle) with typical strengthening precipitation in T7 condition inside a dendrite (bottom).

This procedure combines physical constants coming from literature and dislocations models, microstructure parameters from TEM observations and optimised parameters from experimental cyclic stress-strain loops. Once these parameters are identified from a series of overageing treatments, they can be used on a range of alloy composition and aging conditions. The model has been validated using out-of-phase thermal mechanical tests [9].

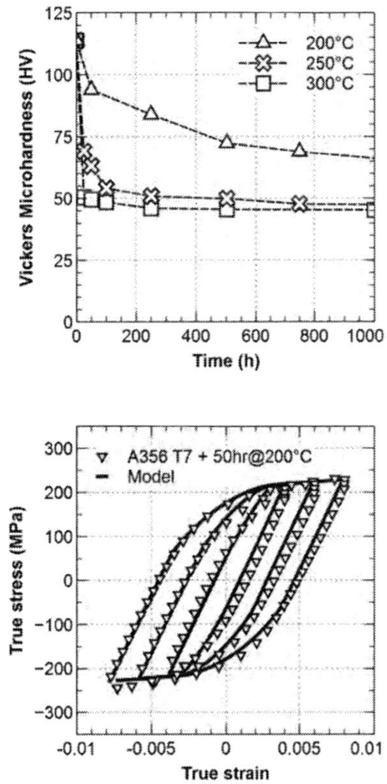

Fig. 6: Change in Vickers microhardness during isothermal ageing (top); Model versus experimental cyclic stress-strain loops for an intermediate overageing treatments (bottom).

Next step is to apply this model to the microstructure gradient that occurs after long-term service operation on a cylinder head. This was tested using an experimental protocol presented above. Depending upon alloy composition, the current design procedure can give rather conservative prediction of fatigue life compared to the improved one [10] but this latter implies longer computation times. Future development must therefore focus on numerical strategies to offer predictive thermal evolution with limited CPU times.

6 Fatigue criteria

Fatigue data are most of the time obtained on over-aged materials to ensure stable mechanical behaviour even if progressive ageing coupled with variable amplitude loading may lead, in the future, in an evolution of standard fatigue tests. A full 3D characterization of the microstructure through X-ray tomography [11] can then be performed in order to measure the size and shape distribution of pores in a Representative Elementary Volume.

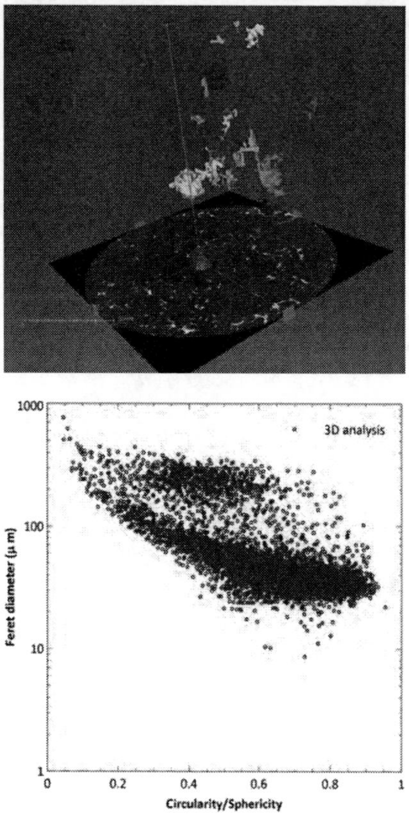

Fig. 7: Example of pore detection by X-ray tomography for a A356 lost foam cast alloy and pore population ranked by size and sphericity.

A quantitative description of the pore population is obtained through the measurement of their Feret diameter and sphericity, which could reveal the presence of gassing porosities for lost foam casting processes which are not observed in standard die casting processes. The largest and sharpest pores have to be accurately observed as they play an important part in the fatigue crack initiation. The pores size distributions are then introduced in the computational approach used to design engine parts against TMF. Under a basic assumption of growth equation of pores from an initial size a_0 to a final size a_f at N_f characterizing a macroscopic crack initiation, a fatigue criterion is developed under a probabilistic form as proposed by [12]. Different damage indicator, as proposed by [13, 14], can enable to take into account micro-cracks propagation and leads to the following improved form of the fatigue criterion:

$$N_f = \int_{a_0}^{a_f} \frac{1}{\left[\left(\frac{w^d}{k_1}\right)^{n_1} a^{n_1} + \left(\frac{w^e}{k_2}\right)^{n_2} a^{n_2}\right]} da \quad (1)$$

where w^d and w^e denotes respectively inelastic and elastic dissipated energy densities per stabilized cycle, and k_1, n_1, k_2 and n_2 are material constants. a consists in the observed statistical distribution of pore size while material parameters are identified from the macroscopic Low-Cycle Fatigue and TMF experiments. N_f and a are thus random variables, which is very important when dealing with variable loading amplitudes.

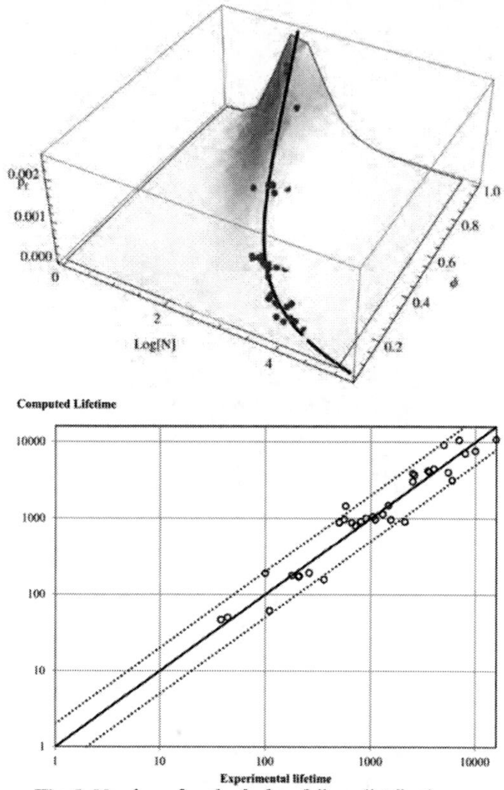

Fig. 8: Number of cycles before failure distribution as computed by [11] and Computed lifetime in terms of maximal probability of failure versus experimental lifetime for à lost foam cast A356 alloy.

When crack propagation is investigated on laboratory specimens, a direct access to the parameters is possible. In other cases, the simple numerical identification of the parameters of such a model is quite complex because no analytical description is possible. While experimental data are missing, treatment is then carried out for class of pore size. For a given pore size and for a given set of parameters $(k_1; n_1; k_2; n_2)$, the estimated lifetime is easily calculated. A histogram and a complete distribution of the number of cycles to failure are then obtained. For a given experimental couple $(w^d; w^e)$, the number of cycles to failure with the highest probability is then finally compared to the experimental lifetime. A least-squares optimization is then performed to obtain the most relevant parameter set. Lifetime prediction results can then be compared with laboratory specimens. One can therefore easily perform non-linear damage accumulation for variable amplitude loading cycles. Following this simple principle, good matching is obtained between estimated and experimental lifetimes as seen on Figure 7. Developments on crack growth law with generalized plasticity and on interactions between microstructural defects and fatigue lifetime are therefore of primary importance to produce robust criteria in the future.

4 Industrial specimens experimental protocols

Thermomechanical fatigue tests on laboratory specimen [15,16], for the sake of analysis simplicity, introduce simplifications with respect to the effective loadings recorded on real structures. Thus, the thermal gradients, which are mainly responsible for damage by thermomechanical fatigue, are suppressed in the experimental protocol for the sake of proposing a direct analysis of the stress-strain curve. When these gradients become very severe, it is most of the time impossible to disregard them in an objective fatigue analysis. In these cases, it then becomes necessary to develop specimens closer to the structure to be designed and therefore often more complex in terms of geometry and loading [16, 17].

Fig. 9: Single cylinder thermal fatigue rig with an acetylene torch heating system

For cylinder heads, Constantinescu and co-authors [1] have proposed using massive specimens for which the microstructure or the surface roughness is not always perfectly representative of the structure to be designed. Thus, the development of a "single-cylinder" specimen as seen on Figure 8 enables loadings as close as possible to the structure ones while respecting both the microstructure and the surface roughness in the critical zone. Development of such rigs is of primary importance in the fatigue design protocols as they enable a precise analysis of the fatigue strength and the validation of constitutive models and fatigue criteria.

7 Conclusions

Taking into account in-use evolutions of thermal and mechanical loadings and thus their variability constantly modifies the way of designing structures against thermomechanical fatigue. The use of simplifying assumptions to take into account aluminium alloys mechanical behaviour and overageing is now clearly inconsistent with the need to produce reliable and robust design protocols. In fact, ensuring reliable and predictive fatigue design has to rely on a careful analysis of the boundary loading conditions, a reliable constitutive model taking into account the physical evolution linked to these loadings and advanced criteria including real damage mechanisms. Finally, development of simplified testing rigs closer to the in-use loadings and to standard production parts process conditions now seems essential to validate models and improve designs.

References

1. A. Constantinescu, E. Charkaluk, G. Lederer, L. Verger, International Journal of Fatigue, **26** (2004) 805-814
2. H. Sehitoglu, X. Qing, T. Smith, H.J. Maier, J. A. Allison, Metallurgical and Materials Transactions A, **31A** (2000) 139-151
3. S. Tabibian, E. Charkaluk, A. Constantinescu, G. Guillemot and F. Szmytka, MSEA, **646** (2015) 190-200
4. F. Szmytka, F., A. Oudin, International Journal of Fatigue, 53 (2013) 82-91
5. J.L. Chaboche, International Journal of Plasticity, **24** (2008) 1642-1696
6. U.F.Kocks, A.S.Argon, M.F.Ashby, Progress in Materials Science, **19** (1975), 110-170
7. Y.Estrin, Journal of Materials Processing Technology, **80** (1998) 33-39
8. L. Rémy, F. Szmytka, L. Bucher, International Journal of Fatigue, **53** (2013) 2-14
9. P. Osmond, Ph. D. Thesis (in French) Mines ParisTech (2010)
10. F. X. Hoche, L. Remy, L. Naze, P. Osmond, A. Koster, This conference
11. N. Limodin, A. El Bartali, L. Wang, J. Lachambre, JY Buffiere, E. Charkaluk,, Nuc. Instru. and Meth. in Phy. Res. B, **324** (2014) 57-62
12. E. Charkaluk, A. Constantinescu, F. Szmytka, S. Tabibian, International Journal of Fatigue, **63** (2014) 127-136
13. L. Rémy, N. Haddar, A. Alam, A. Koster, N. Marchal, MSEA, **468-470** (2007) 40-50
14. V. Maurel, L. Rémy, N. Haddar, F. Damen, International Journal of Fatigue, **31** (2009) 952-961
15. J.L. Malpertu, L. Rémy, Metall. Trans. A, **21A** (1990) 389-399
16. L.Rémy in Comprehensive structural integrity, **5** (2003) 113-200
17. F. Rezai-Aria, B. Dambrine, L. Rémy, Fat. Fract. Eng. Mat. Structures, **11** (1988) 291-302

Rapid estimation of notch stress intensity factors in 3D large-scale welded structures using the peak stress method

Alberto Campagnolo[1] and *Giovanni* Meneghetti[1,*]

[1] University of Padova, Department of Industrial Engineering, 35131 Padova, Italy

Abstract. The Peak Stress Method (PSM) is an engineering, FE-oriented application of the notch stress intensity factor (NSIF) approach to fatigue design of welded joints, which takes advantage of the singular linear elastic peak stresses from FE analyses with coarse meshes. Originally, the PSM was calibrated to rapidly estimate the NSIFs by using 3D, eight-node brick elements, taking advantage of the submodeling technique. 3D modelling of large-scale structures is increasingly adopted in industrial applications, thanks to the growing spread of high-performance computing (HPC). Based on this trend, the application of PSM by means of 3D models should possibly be even more speeded up. To do this, in the present contribution the PSM has been calibrated under mode I, II and III loadings by using ten-node tetra elements, which are able to directly discretize complex 3D geometries without the need for submodels. The calibration of the PSM has been carried out by analysing several 3D mode I, II and III problems. Afterwards, an applicative example has been considered, which is relevant to a large-scale steel welded structure, having overall size on the order of meters. Two 3D FE models, having global size of tetra elements equal to 5 and 1.66 mm, have been solved by taking advantage of HPC, being the global number of degrees of freedom equal to 10 and 140 millions, respectively. The NSIFs values estimated at the toe and root sides according to the PSM have been compared with those calculated by adopting a shell-to-solid technique.

1 Introduction

On the basis of the fundamental contributions of Williams [1], who studied two-dimensional notch problems under mode I (opening) and mode II (sliding) loadings, and Qian and Hasebe [2], who analysed the notch problem under mode III (tearing) loading, the singular linear elastic stress distributions in the neighborhood of a sharp V-notch tip, see the example of a toe side in a welded joint in Fig. 1, can be written as functions of the notch stress intensity factors (NSIFs), which quantify the intensity of the asymptotic stress fields. The following equations define the mode I, II and III NSIFs, respectively, according to Gross and Mendelson [3]:

$$K_1 = \sqrt{2\pi}\cdot\lim_{r\to 0}\left[\left(\sigma_{\theta\theta}\right)_{\theta=0}\cdot r^{1-\lambda_1}\right] \quad (1a)$$

$$K_2 = \sqrt{2\pi}\cdot\lim_{r\to 0}\left[\left(\tau_{r\theta}\right)_{\theta=0}\cdot r^{1-\lambda_2}\right] \quad (1b)$$

$$K_3 = \sqrt{2\pi}\cdot\lim_{r\to 0}\left[\left(\tau_{\theta z}\right)_{\theta=0}\cdot r^{1-\lambda_3}\right] \quad (1c)$$

In Eqs. (1a)-(1c), λ_1, λ_2 and λ_3 are the mode I, II and III eigenvalues, respectively, which are dependent on the opening angle 2α [1,2], while $\sigma_{\theta\theta}$, $\tau_{r\theta}$ and $\tau_{\theta z}$ represent the stress components evaluated along the notch bisector line, which is defined by the angular coordinate $\theta=0$ according to Fig. 1.

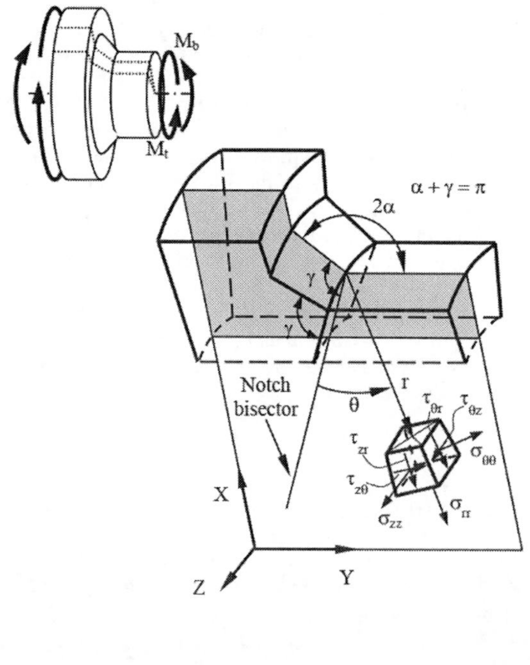

* Corresponding author: giovanni.meneghetti@unipd.it

https://doi.org/10.1051/matecconf/201816517004

Fig. 1. Polar reference system centred at the weld toe of a typical tube-to-flange welded joint geometry subjected to multiaxial bending and torsion loading.

In the literature, the NSIF approach has been used to analyse the medium as well as the high-cycle fatigue strength of specimens weakened by sharp V-notches and made of structural materials [4,5]. Focusing on welded joints, NSIFs allow to correlate the fatigue strength under uniaxial [6–9] as well as multiaxial loadings [10].

Nevertheless, it should be noted that the calculation of NSIFs on the basis of the results of numerical analyses shows a major drawback in engineering applications, since very refined FE meshes (element size on the order of 10^{-5} mm) are required in order to apply definitions (1a)-(1c) to calculate the NSIFs. When dealing with three-dimensional notched components, both the solution of the FE model and the post-processing of numerical results could be even more time-consuming.

In order to overcome this issue, an engineering and rapid technique, the so-called Peak Stress Method (PSM), has been proposed, which allows to speed up the calculation of the NSIFs by adopting coarse FE analyses, the element size being some orders of magnitude larger than that required to apply definitions (1a)-(1c). The second advantage of the PSM is that only a single stress value is necessary to estimate the NSIFs, instead of a number of *stress-distance* results, as usually necessary in order to apply definitions (1a)-(1c). The method takes inspiration from the contribution by Nisitani and Teranishi [11,12], who proposed a technique to readily estimate the mode I Stress Intensity Factor of a crack propagating from an ellipsoidal cavity. The PSM has been theoretically justified and extended to allow the rapid calculation also of the NSIF relevant to sharp and open V-notches under mode I [13,14], the SIF of cracks under mode II [15] and, finally, the NSIF of open V-notches under mode III [16]. It is worth noting that any NSIF-based approach for the structural strength assessment can in principle be reformulated on the basis of the PSM. Recently, the PSM has been applied in combination with the approach based on the averaged strain energy density (SED) to assess the fatigue strength of welded joints under axial [15,17], torsion [16] and multiaxial [18,19] loading conditions.

Practically, the NSIFs K_1, K_2 and K_3 can be readily estimated according to the PSM by adopting the singular, linear elastic, opening (mode I), sliding (mode II) and tearing (mode III) peak stresses $\sigma_{\theta\theta,\theta=0,peak}$, $\tau_{r\theta,\theta=0,peak}$ and $\tau_{\theta z,\theta=0,peak}$, respectively, which are referred to the V-notch bisector line, according to Fig. 2, and calculated at the V-notch tip from FE analyses with coarse meshes. The estimated NSIF values can be obtained from the following expressions [13,15,16]:

$$K_1 \cong K^*_{FE} \cdot \sigma_{\theta\theta,\theta=0,peak} \cdot d^{1-\lambda_1} \quad (2a)$$

$$K_2 \cong K^{**}_{FE} \cdot \tau_{r\theta,\theta=0,peak} \cdot d^{0.5} \quad (2b)$$

$$K_3 \cong K^{***}_{FE} \cdot \tau_{\theta z,\theta=0,peak} \cdot d^{1-\lambda_3} \quad (2c)$$

where a is the characteristic size of the considered sharp V-notch, as an example it is the notch depth in next Fig. 4b, d is the so-called 'global element size', i.e. the average FE size adopted by the free mesh generation algorithm available in the numerical software.

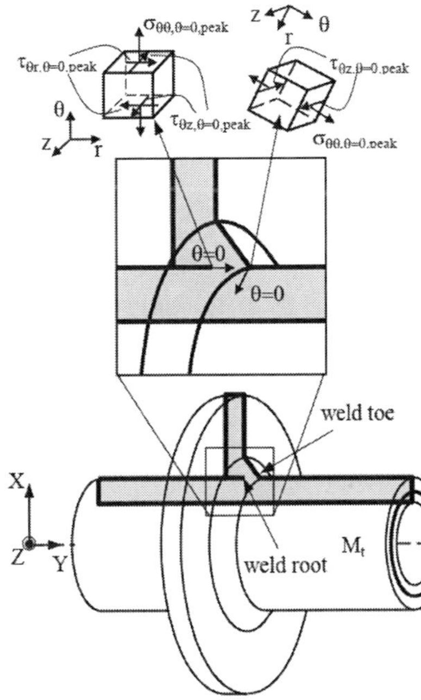

Fig. 2. Sharp V-shaped notches in a welded joint at the weld root (2α= 0°) and at the weld toe (2α typically equal to 135°) sides. Definition of peak stresses $\sigma_{\theta\theta,\theta=0,peak}$, $\tau_{r\theta,\theta=0,peak}$ and $\tau_{\theta z,\theta=0,peak}$.

Parameters K^*_{FE}, K^{**}_{FE} and K^{***}_{FE} depend on the calibration options: (i) element type and formulation; (ii) mesh pattern of finite elements and (iii) procedure for stress extrapolation at FE nodes. See the detailed discussion reported in [20].

Originally, 2D, four-node plane quadrilateral elements of Ansys® element library [13,15,16] were adopted to calibrate parameters K^*_{FE}, K^{**}_{FE} and K^{***}_{FE}, which resulted equal to 1.38, 3.38 and 1.93, respectively, such values being valid under the conditions discussed in the relevant literature [13–16], to which the reader is referred. Recently, K^*_{FE} and K^{**}_{FE} have been also calibrated by adopting six commercial FE packages other than Ansys®, taking advantage of a Round Robin between Italian Universities [20].

Afterwards, the PSM has been extended to be used with 3D, eight-node brick elements [14], taking advantage of the submodeling technique available in Ansys® software. More precisely, when dealing with complex 3D joint geometries a submodel consisting of brick elements was generated after having analysed a main model meshed by employing ten-node tetra elements.

Three-dimensional modelling of large-scale, or even full-scale, structures is increasingly adopted in industrial applications, thanks to the growing spread of high-performance computing (HPC). Based on this trend, the application of PSM by means of three-dimensional models should possibly be even more speeded up. To do this, in the present contribution parameters K^*_{FE}, K^{**}_{FE} and K^{***}_{FE} have been calibrated by using ten-node tetra elements, which are able to discretize complex 3D geometries, making the PSM applicable directly to a single model, without the need for submodels. The calibration of the PSM based on tetrahedral elements has been carried out by analysing several three-dimensional mode I, II and III problems. Afterwards, an applicative example has been considered, which is relevant to a large-scale steel welded structure, having overall size on the order of meters and containing many different welded details.

2 Calibrating the PSM with 10-node tetrahedral elements

When adopting tetrahedral elements to analyse a 3D notch problem, the mesh pattern obtained by the free mesh generation algorithm is intrinsically not regular, so that a node belonging to the notch tip could be shared by a different number of elements having significantly different shape. Therefore, the peak stress could vary along the notch tip profile even in the case of a constant applied NSIF. Figure 3 shows an example of the variability of the peak stress along the notch tip profile of a plate subjected to pure tension loading (see next Fig. 4b) and analysed by adopting a 3D FE model in which a plain strain condition has been simulated, as will be discussed in more detail in the following. For comparison purposes, the reciprocal of the normalised number of elements $(N° FE/max N° FE)^{-1}$, which share the node at which peak stress is evaluated, has been reported in Fig. 3. It can be observed that the higher the number of elements sharing the node, the lower the peak stress.

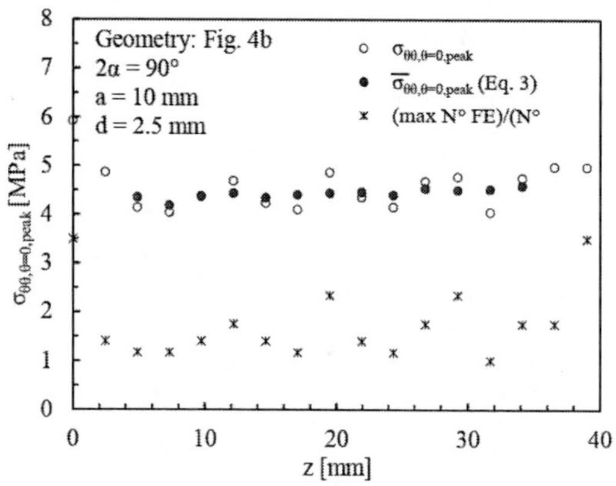

Fig. 3. Example of variability of the peak stress $\sigma_{\theta\theta,\theta=0,peak}$ calculated at vertex nodes belonging to the V-notch tip profile and comparison with the number of elements sharing each node (N° FE). Definition of the average peak stress $\overline{\sigma}_{\theta\theta,\theta=0,peak}$.

Accordingly, to reduce the variability of the peak stress along the notch tip profile, an average peak stress value has been introduced (see red markers in Fig. 3), being defined at the generic node n=k as the moving average on three adjacent vertex nodes, i.e. n= k-1, k and k+1:

$$\overline{\sigma}_{ij,peak,n=k} = \left. \frac{\sigma_{ij,peak,n=k-1} + \sigma_{ij,peak,n=k} + \sigma_{ij,peak,n=k+1}}{3} \right|_{n=node} \quad (3)$$

It is worth noting that only peak stress values calculated at vertex nodes of the quadratic tetrahedral elements have to be input in Eq. (3), i.e. stress values at mid-side nodes located at the notch tip profile, which are provided by Ansys® when "path operations" or "GET commands" are employed in post-processing, must be neglected. On the other hand, when "list nodal results" or "query results" are adopted in the post-processing environment of Ansys® code, stress values at mid-side nodes are automatically excluded.

Given this, the PSM has been calibrated by analysing several 3D notch problems under pure mode I, pure mode II and pure mode III loadings. After having calculated the peak stresses, the parameters K^*_{FE}, K^{**}_{FE} and K^{***}_{FE} have been evaluated from Eqs. (2a)-(2c), re-arranged as follows:

$$K^*_{FE} \cong \frac{K_1}{\overline{\sigma}_{\theta\theta,\theta=0,peak} \cdot d^{1-\lambda_1}} \quad (4a)$$

$$K^{**}_{FE} \cong \frac{K_2}{\overline{\tau}_{r\theta,\theta=0,peak} \cdot d^{0.5}} \quad (4b)$$

$$K^{***}_{FE} \cong \frac{K_3}{\overline{\tau}_{\theta z,\theta=0,peak} \cdot d^{1-\lambda_3}} \quad (4c)$$

The same material properties have been adopted in all FE analyses, consisting in a structural steel, having Young's modulus equal to 206000 MPa and a Poisson's ratio ν of 0.3.

3D, ten-node, quadratic tetrahedral elements (SOLID 187 of Ansys® element library) have been used in all FE analyses. The numerical integration has been carried out by adopting 4 Gauss points, i.e. a reduced integration, being the sole element formulation available in Ansys®. Once selected the proper element type, the average FE size *d* has been the sole parameter used to drive the automatic free mesh generation algorithm.

Dealing with mode I and mode II problems, in order to obtain a uniform distribution of the relevant NSIFs along the notch tip profile, either plane-strain or plane-stress analyses have been performed:

- plane strain state was simulated by constraining the out-of-plane element displacement u_z (and thus the corresponding ε_z strain component, $\varepsilon_z = 0$);
- plane stress state was simulated by defining an ideal orthotropic material having out-of-plane ε_z strain uncoupled from the in-plane components, i.e. ε_x and ε_y. For this aim, ν_{zx} and ν_{zy} Poisson coefficients were set to zero. In absence of coupling of in-plane with out-of-plane strains, being $\varepsilon_z = 0$, then $\sigma_{zz} = 0$, i.e. plane stress conditions are applied.

2.1. 3D problems (plane strain/plane stress), mode I loading, 0°<2α<135°

Several three-dimensional notch and crack problems under pure mode I (see Fig. 4 and Table 1) have been analysed. The considered case studies are a selection of those adopted in the first calibration of the PSM under mode I loading, which had been carried out by considering 2D [13] and 3D [14] problems. Here, those case studies have been extruded to obtain 3D components having thickness equal to *s*.

Table 1. Geometrical and FE parameters considered in the calibration of the PSM with 10-node tetrahedral element under mode I, mode II and mode III loadings.

Figure	Geometrical parameters	Mesh size d [mm]
4a	2a = 3 ÷ 25 mm b = 3 ÷ 24 mm c = 2.2 ÷ 20.5 mm s = 80 mm	0.5 ÷ 1 mm
4b	a = 1 ÷ 20 mm 2α = 0, 90, 135° w = 50 mm s = 80 mm	0.5 ÷ 10 mm
5	2a = 2 ÷ 200 mm w = 400 mm s = 40, 80, 160 mm	0.5 ÷ 5 mm
6	z = 6.3 ÷ 9 mm t = 7 ÷ 10 mm	1 ÷ 10 mm

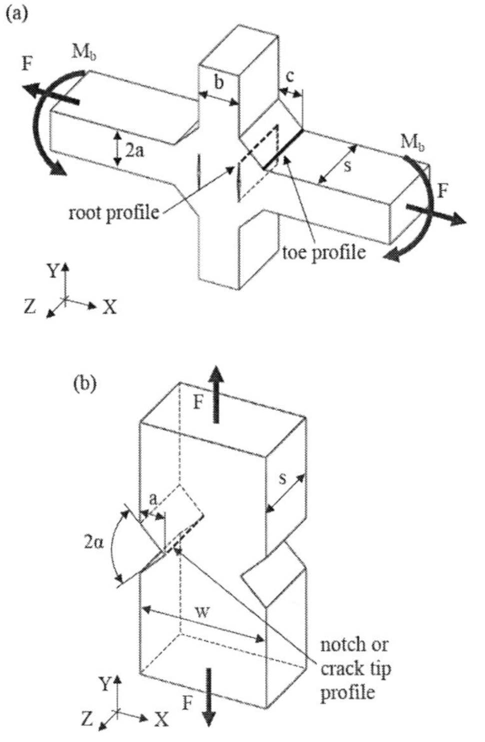

Fig. 4. Geometries of 3D problems under mode I loading: (a) fillet-welded cruciform joint and (b) notched plate.

3D linear elastic analyses have been performed by simulating plane strain or plane stress conditions and by adopting ten-node, quadratic tetrahedral elements, in order to calculate the peak stress values. Only one eighth of each geometry has been modeled by using the triple symmetry condition. The free mesh generation algorithm available in Ansys code has been used after having input the average element size *d*. The mesh density ratio *a/d* has been varied between 1 and 20, by varying either the characteristic size of the notch problem, i.e. *a*, or the element size *d* (see Table 1). Moreover, three values of the notch opening angle, i.e. 2α = 0°, 90°, 135° have been considered, 0° and 135° being typical for weld root and toe side, respectively. A nominal gross-section stress equal to 1 MPa has been applied to each FE model.

After having solved the FE model, the peak value of the opening stress $\sigma_{11,peak}$, being almost equal to $\sigma_{\theta\theta,\theta=0,peak}$ in all cases of Figs. 4a and 4b, has been calculated at vertex nodes belonging to the V-notch tip profile, then Eq. (3) has been applied to derive the average peak stress at each vertex node.

2.2 3D problems (plane strain/plane stress), mode II loading, 2α=0°

A plate having the geometry shown in Fig. 5 (see also Table 1), weakened by a central crack (2α = 0°) and subjected to pure mode II loading has been analysed. The case study has been taken from the first calibration of the PSM under mode II loading for 2D problems [15] and it has been extruded to obtain a 3D plate having thickness equal to *s*.

3D linear elastic analyses have been carried out by simulating plane strain or plane stress conditions and by adopting ten-node, quadratic tetrahedral elements, in order to calculate the peak stress values. The mesh density ratio *a/d* has been varied in a wide range from 1 to 200, being $2<2a<200$ mm. Only one eight of the plate geometry has been modeled by using the double anti-symmetry on planes Y-Z and X-Z and the symmetry on plane X-Y.

The external load has been applied to the model by means of displacements $u_x=u_y=1.262\cdot10^{-3}$ mm at the plate free edges, which correspond to a nominal gross shear stress of 1 MPa when no crack is present. After having solved the numerical model, the peak value of the (mode II) shear stress $\tau_{r\theta,\theta=0,peak} = \tau_{xy,peak}$ has been calculated at the vertex nodes belonging to the crack tip profile, then Eq. (3) has been applied to derive the average peak stress at each vertex node.

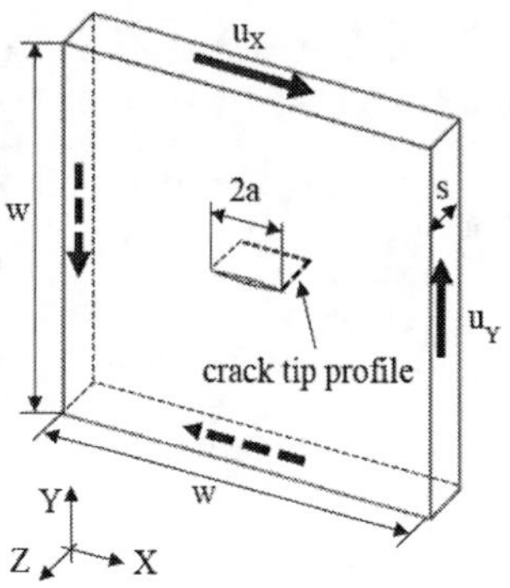

Fig. 5. Geometry of 3D problem under mode II loading: plate weakened by a central crack. $u_x=u_y=1.262\cdot10^{-3}$ mm.

2.3 3D problems, mode III loading, 2α=0°, 135°

The weld root (2α = 0°) as well as the weld toe (2α = 135°) sides of tube-to-flange joints subjected to pure mode III loading, see Table 1 and an example in Fig. 6, have been analysed. The case studies have been taken from the first calibration of the PSM under mode III loading, in which 2D axis-symmetric models were adopted [16]. In the present analysis, the previous 2D geometries have been extruded about the tube axis to obtain the 3D geometries. It should be noted that only one quarter of each geometry has been analysed by taking advantage of the double anti-symmetry boundary conditions.

3D linear elastic analyses have been carried out by adopting ten-node, quadratic tetrahedral elements, in order to calculate the peak stress values. The mesh density ratio *a/d* has been varied in a range from 1 to 10, *a* being the tube thickness *t* which ranged between 7 and 10 mm (see Table 1).

A nominal torsion shear stress equal to 1 MPa has been applied to each FE model at the tube side. After having solved the numerical model, the peak value of the (mode III) shear stress $\tau_{\theta z,\theta=0,peak}$ has been calculated at the vertex nodes belonging to the root or toe profile by adopting a local coordinate system r-θ-z rotated at each node according to Fig. 2. Then Eq. (3) has been applied to derive the average peak stress at each vertex node.

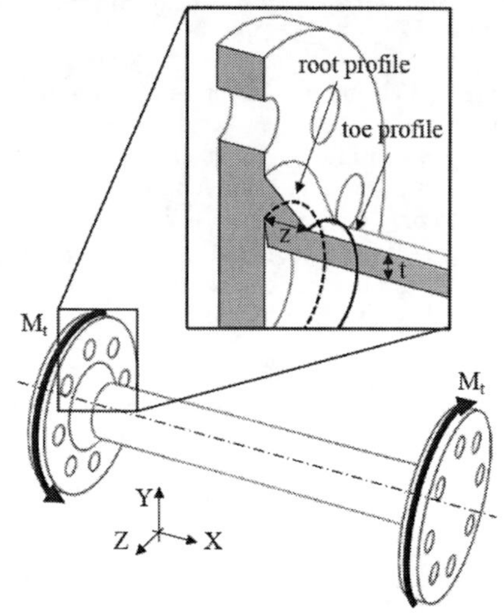

Fig. 6. Geometry of 3D problem under mode III loading: tube-to-flange welded joint.

Finally, the exact values of the mode I NSIF K_1, mode II SIF K_2 and mode III NSIF K_3 to be input in Eq. (4a)-(4c), respectively, have been calculated by applying definitions (1a)-(1c) to the stress-distance numerical results obtained from 2D FE analyses by adopting very refined FE meshes (the size of the smallest element being on the order of 10^{-5} mm). Dealing with mode I and mode II problems, eight-node, quadratic quadrilateral elements (PLANE 183 of Ansys® element library) under plane strain conditions have been adopted, while concerning mode III problems, eight-node, quadratic quadrilateral harmonic elements (PLANE 83 of Ansys® element library) have been employed.

3 Results of FE analyses

The results obtained from the calibration of the PSM under mode I, mode II and mode III loadings by using ten-node tetra elements are shown in Figs. 7, 8 and 9, respectively. The figures report the parameters K^*_{FE}, K^{**}_{FE} and K^{***}_{FE}, calculated from Eqs. (4a)-(4c), respectively, as functions of the mesh density ratio *a/d*. It should be noted that in each FE analysis, parameters K^*_{FE}, K^{**}_{FE} and K^{***}_{FE} exhibited a non-uniform distribution along the notch tip profile, given the variability of the average peak stress shown in Fig. 3. Accordingly, Figs. 7, 8 and 9 report the average value of the K_{FE} parameters evaluated from each FE analysis along with the relevant bar, representing the range between minimum and maximum values calculated along the notch tip profile.

In the case of 3D problems under mode I loading, Fig. 7 shows that $K^*_{FE} \cong 1.01\pm15\%$ for 2α equal to 0° or 90°, while $K^*_{FE} \cong 1.21\pm10\%$ when 2α equals 135°. Convergence is obtained when $a/d \geq 3$ and 1 for 2α equal to 0°, 90° and 135°, respectively.

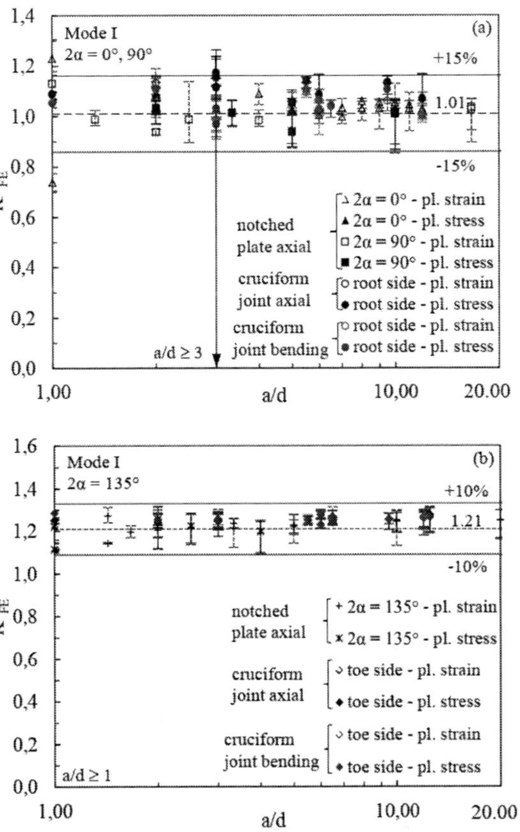

Fig. 7. Calibration of K^*_{FE} (Eq. 4a) for 10-node tetrahedral element under mode I loading: (a) 2α = 0°, 90°, (b) 2α = 135°.

Dealing, then, with mode II loading, the obtained results are reported in Fig. 8, which shows that $K^{**}_{FE} \cong 1.63 \pm 20\%$. Convergence is obtained when the ratio *a/d* ≥ 1.

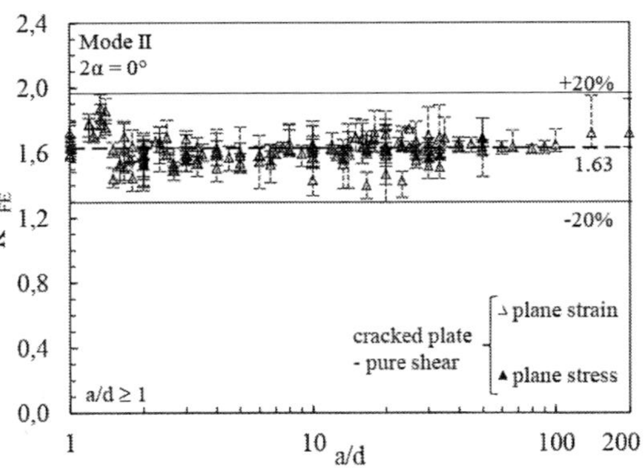

Fig. 8. Calibration of K^{**}_{FE} (Eq. 4b) for 10-node tetrahedral element under mode II loading.

Finally, concerning mode III loading, the obtained results are reported in Fig. 9, which shows that $K^{***}_{FE} \cong 1.37 \pm 10\%$ when root side (2α = 0°) is considered, while $K^{***}_{FE} \cong 1.75 \pm 5\%$ when toe side (2α = 135°) is of interest. Convergence is obtained when *a/d* ≥ 2 both at weld root and toe sides.

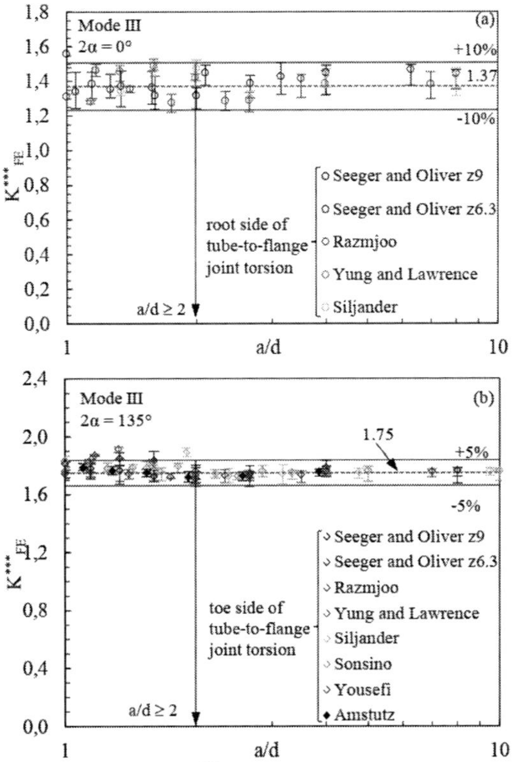

Fig. 9. Calibration of K^{***}_{FE} (Eq. 4c) for 10-node tetrahedral element under mode III loading: (a) 2α = 0°, (b) 2α = 135°.

A summary is reported in Table 2. It can be observed that the scatter bands of K_{FE} parameters are wider at root side (2α = 0°) than at toe side (2α = 135°) for all loading modes. This is due to the presence of elements having significantly different shape and size (the FE size *d* given as input being an average value) at the notch tip profile, especially in the case of cracks or weld root sides as compared to the case of open V-notches or weld toe sides. The non-uniform local mesh pattern has particular effect for cracks under mode II loading, the deviation of K^{**}_{FE} being maximum and equal to ±20% (see Fig. 8).

Table 2. Summary of calibration of K^*_{FE}, K^{**}_{FE} and K^{***}_{FE} for 10-node tetrahedral elements (SOLID 187 of Ansys®).

2α	Mode I		Mode II		Mode III	
[°]	K^*_{FE}	$(a/d)_{min}$	K^{**}_{FE}	$(a/d)_{min}$	K^{***}_{FE}	$(a/d)_{min}$
0	1.01 ± 15%	3	1.63 ± 20%	1	1.37 ± 10%	2
90	1.01 ± 15%	3	/	/	/	/
135	1.21 ± 10%	1	/	/	1.75 ± 5%	2

4 Application to a case study

After presenting the calibration of the three-dimensional PSM based on tetra elements, an applicative example has

 https://doi.org/10.1051/matecconf/201816517004

been considered (see Fig. 10). It is relevant to a large-scale steel welded structure, which represents a detail of a sluice gate having overall size on the order of tens of meters. The considered detail has size on the order of meters and is located at a geodetic height of about 10 meters referred to the free surface.

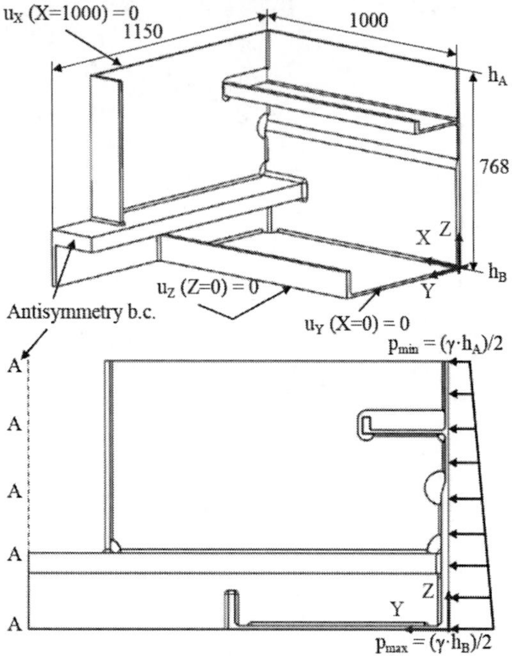

Fig. 10. Geometry (dimensions are in mm) and boundary conditions applied to the detail of the sluice gate. γ is the water specific weight being equal to 10^4 N/m^3, h_A and h_B are the geodetic height referred to the free surface and are equal to 9.232 and 10 meters, respectively.

The detail of Fig. 10 is characterised by many different welded geometries, including T- and cruciform-joints, both fillet and full-penetration welded. The plates thickness ranges between 10 and 58 mm.

The 3D PSM based on tetra elements, previously calibrated, has been applied to estimate the NSIFs at both weld toe and root sides of the considered large-scale steel welded structure. For the sake of brevity, only a selection of toe and root sides undergoing pure mode I loading have been analysed in the following. According to previous Fig. 7, the mesh density ratio a/d must be greater than 1 to analyse weld toe sides and greater than 3 to analyse weld root sides. With the aim of comparing the solution time, two different rather coarse meshes (see Fig. 11) of ten-node tetra elements have been generated by adopting a/d = 1 and 3, respectively. The minimum plate thickness being equal to $2a$ = 10 mm, the mesh density ratio a/d = 1 corresponds to d = 5 mm (see Fig. 11a), while a/d = 3 to a mesh size d = 1.66 mm (see Fig. 11b). The boundary conditions applied to both FE models are reported in Fig. 10. It should be noted that half of the hydrostatic pressure has been applied to the X-Z plane being active an anti-symmetry boundary condition. Both 3D FE models have been solved by taking advantage of Ansys HPC®, the global number of degrees of freedom (dof) being equal to 10 millions, for the case a/d = 1, while it reaches 140 millions, when a/d = 3. The performance of the computer cluster adopted to solve the FE analyses and the solution times are reported in Table 3.

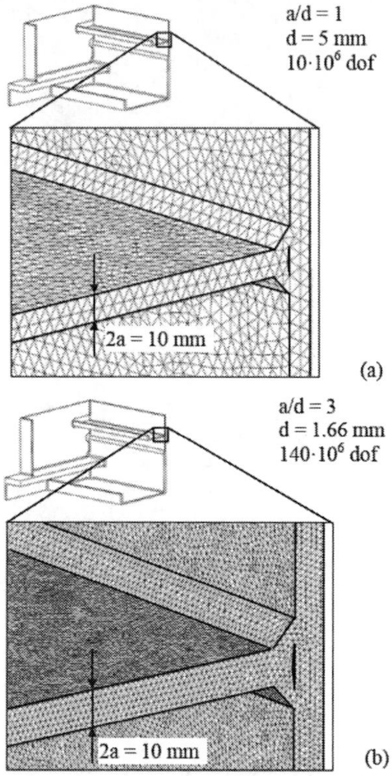

Fig. 11. Coarse meshes of ten-node tetra elements generated in Ansys 18.2 environment and adopted to analyse (a) weld toe sides, being a/d = 1 according to Fig. 7b, and (b) weld root sides, being a/d = 3 according to Fig. 7a. Dof = degree of freedom.

Table 3. Performance of the computer cluster adopted to solve the FE analyses of Fig. 11 by using Ansys 18.2 HPC.

# d.o.f	**CPU model**	**# cores/ RAM [Gb]**	**Solution elapsed time [s]**
$10 \cdot 10^6$ (Fig. 11a)	Intel Xeon CPU E5-2665 2.40GHz	8/300	180
$140 \cdot 10^6$ (Fig. 11b)		64/500	3660

Figures 12 and 13 report the distribution of the mode I NSIF estimated at selected toe and root sides according to the 3D PSM previously calibrated. More in detail, the NSIF K_1 has been estimated at the selected toe side (Fig. 12) by applying Eq. (2a) with K^*_{FE} = 1.21 to the average peak stress values calculated from the FE model of Fig. 11a. On the other hand, the SIF K_1 has been estimated at the selected root side (Fig. 13) by applying Eq. (2a) with K^*_{FE} = 1.01 to the average peak stress values calculated from the FE model of Fig. 11b.

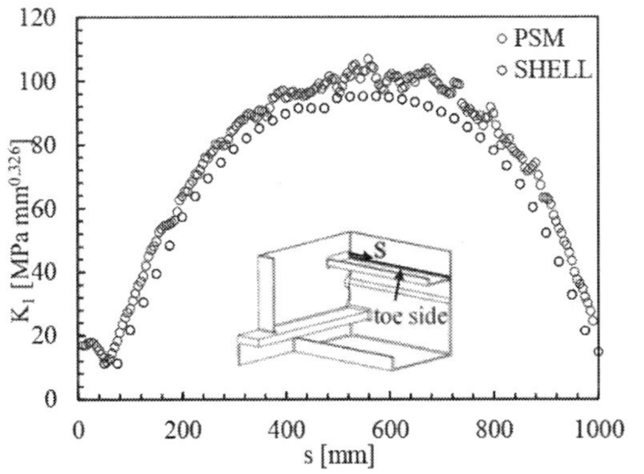

Fig. 12. Mode I NSIF K_1 estimated along a selected toe side of the case study reported in Fig. 10. Comparison between results obtained by applying Eq. (2a) and adopting the FE mesh shown in Fig. 11a and those derived by a shell-to-solid model according to [21].

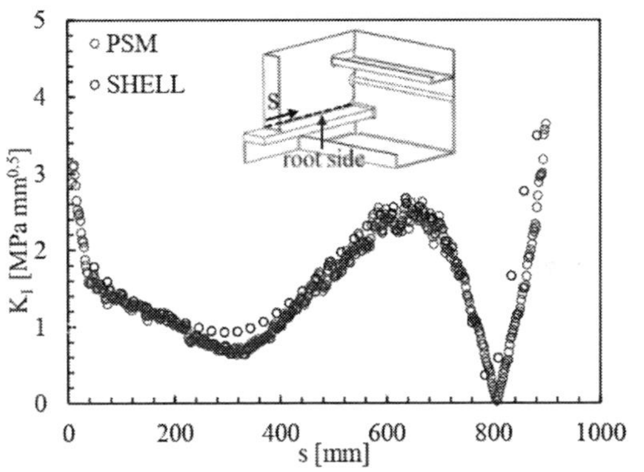

Fig. 13. Mode I SIF K_1 estimated along a selected root side of the case study reported in Fig. 10. Comparison between results obtained by applying Eq. (2a) and adopting the FE mesh shown in Fig. 11b and those derived by a shell-to-solid model according to [21].

For comparison purposes, the mode I NSIF values estimated at the selected toe and root sides according to a shell-to-solid technique [21] have been included in Figs. 12 and 13. A quite good agreement can be observed, the maximum deviation being equal to approximately 10% at the toe side and approximately 15% at the root side.

5 Conclusions

The peak stress method (PSM) employs the singular, linear elastic peak stresses evaluated at the notch tip by means of FE analyses with coarse meshes to rapidly estimate the mode I, mode II and mode III NSIFs. Three calibration constants are needed, namely K^*_{FE} (Eq. (4a)), K^{**}_{FE} (Eq. (4b)) and K^{***}_{FE} (Eq. (4c)), respectively.

Originally, the PSM was calibrated by using 2D plane or 3D brick elements, taking advantage of the submodeling technique. In the present contribution the PSM has been calibrated under mode I, II and III loadings by using ten-node tetra elements, which are able to directly discretize complex 3D geometries without the need for submodels. The following conclusions can be drawn:

- It has been shown that under mode I loading the constant K^*_{FE} to use in Eq. (2a) is 1.01±15% for $2\alpha = 0°$, 90°, while it equals 1.21±10% when $2\alpha = 135°$. Convergence is obtained when the mesh density ratio $a/d \geq 3$ for 2α equal to 0°, 90° and $a/d \geq 1$ for $2\alpha = 135°$.
- Dealing with mode II loading, it has been shown that the constant K^{**}_{FE} to use in Eq. (2b) is 1.63 ± 20% for a mesh density ratio $a/d \geq 1$.
- Concerning mode III loading, K^{***}_{FE} to use in Eq. (2c) results 1.37±10% at root side ($2\alpha = 0°$) and 1.75±5% at the toe side ($2\alpha = 135°$). Convergence occurs when $a/d \geq 2$ both at root and toe sides.
- Finally, an applicative example has been considered, which is relevant to a large-scale steel welded structure, having overall size on the order of several meters. The Ansys High Performance Computing (HPC) has been adopted to solve two 3D FE models, by employing ten-node tetra elements with global element size $d = 5$ and 1.67 mm, respectively. The solution time was on the order of 200 s for the FE model with $d = 5$ mm, i.e. 10 millions degrees of freedom, while about 3500 s were needed to solve the FE model with $d = 1.67$ mm, i.e. 140 millions degrees of freedom. The mode I NSIF values estimated at the toe and root sides according to the PSM have been compared with those calculated by adopting a shell-to-solid technique, showing a quite good agreement.
- Because of the relatively coarse FE analyses required and simplicity of post-processing the calculated peak stresses, the PSM based on three-dimensional models of ten-node tetra elements might be useful in the everyday design practice, even when large-scale structures are considered.

EnginSoft SpA (Padova, Italy) is gratefully acknowledged for making Ansys HPC available for this project.

References

1. M.L. Williams, J Appl Mech. **19** (1952) 526–528.
2. J. Qian, N. Hasebe, Eng. Fract. Mech. **56** (1997) 729–734.
3. B. Gross, A. Mendelson, Int. J. Fract. Mech. **8** (1972) 267–276.
4. S. Kihara, A. Yoshii, JSME Int. Journal. Ser. 1, Solid Mech. Strength Mater. **34** (1991) 70–75.
5. T. Boukharouba, T. Tamine, L. Niu, C. Chehimi, G. Pluvinage, Eng. Fract. Mech. **52** (1995) 503–512.
6. Y. Verreman, B. Nie, Fatigue Fract. Eng. Mater. Struct. **19** (1996) 669–681.
7. P. Lazzarin, R. Tovo, Fatigue Fract. Eng. Mater. Struct. **21** (1998) 1089–1103.
8. P. Lazzarin, P. Livieri, Int. J. Fatigue. **23** (2001) 225–232.
9. B. Atzori, G. Meneghetti, Int. J. Fatigue. **23**

(2001) 713–721.

10. P. Lazzarin, C.M. Sonsino, R. Zambardi, Fatigue Fract. Eng. Mater. Struct. **27** (2004) 127–140.
11. H. Nisitani, T. Teranishi, in: M. Guagliano, M.H. Aliabadi (Eds.), Proc. 2nd Int. Conf. Fract. Damage Mech., 2001: pp. 141–146.
12. H. Nisitani, T. Teranishi, Eng. Fract. Mech. **71** (2004) 579–585.
13. G. Meneghetti, P. Lazzarin, Fatigue Fract. Eng. Mater. Struct. **30** (2007) 95–106.
14. G. Meneghetti, C. Guzzella, Eng. Fract. Mech. **115** (2014) 154–171.
15. G. Meneghetti, Eng. Fract. Mech. **89** (2012) 40–51.
16. G. Meneghetti, Weld. World. **57** (2013) 265–275.
17. G. Meneghetti, P. Lazzarin, Weld. World. **55** (2011) 22–29.
18. G. Meneghetti, A. Campagnolo, D. Rigon, Int. J. Fatigue. **101** (2017) 328–342.
19. G. Meneghetti, A. Campagnolo, D. Rigon, Int. J. Fatigue. **101** (2017) 343–362.
20. G. Meneghetti, A. Campagnolo, M. Avalle, D. Castagnetti, M. Colussi, P. Corigliano, et al., Fatigue Fract. Eng. Mater. Struct. (2018). in press. doi: 10.1111/ffe.12751.
21. M. Colussi, F. Berto, G. Meneghetti, in: P. Bailey, F. Berto, E.R. Cawte, P. Roberts, M.T. Whittaker, J.R. Yates (Eds.), 7th Int. Conf. Durab. Fatigue - Fatigue 2017, Engineering Integrity Society, Cambridge, UK, 2017: pp. 247–258.

A novel approach to simulate the stiffness behaviour of spot welded vehicle structures under multi axial variable amplitude loading

Peter Rösch[1,*], *Halvar* Schmidt[1], and *Thomas* Bruder[1]

[1]BMW AG, Knorrstr. 147, 80807 Munich, Germany

Abstract. Under cyclic loading in combination with high local stress levels a change in stiffness of connection points, e.g. spot welds, self-piercing rivets or flow drill screws can be observed. This paper introduces a numerical approach for the simulation of stiffness changes in vehicle structures under multiaxial cyclic loading as a result of the stiffness degradation of individual spot welds. The basic approach is based on a conventional damage accumulation combined with a power law representing the change in the spot weld's stiffness. Instead of an accumulated damage value D a residual stiffness is set as failure criterion. Extensions to the approach are presented for its application to multiaxial loading with variable amplitudes. A conventional multiaxial fatigue test of a complete vehicle, focusing on its rear end including global stiffness and local strain measurements at the beginning, median and at the end of the test are presented and compared to the simulated results. This computation approach allows a more reliable fatigue assessment in comparison to a fatigue analysis which does not consider stiffness changes. The damage propagation after the crack initiation at an individual spot weld is represented more accurately as subsequent changes in the load paths are considered.

1 Introduction

Strength and stiffness requirements highly influence the dimensions and therefore also the weight of components and structures. In order to meet increasing customer demands and to reduce the overall vehicle emissions, light weight construction is an essential aspect in the vehicle development process. To achieve a lightweight construction a diversity of materials and joining techniques, are applied. Resistance spot welding is one of the most commonly used joining techniques in conventional vehicles. Current research projects, e.g. ALIVE (Advanced High Volume Affordable Lightweighting for Future Electric Vehicles) show a high potential for resistance spot welding or similar techniques (e.g. resistance element welding) to be applied in future vehicle projects [1]. To illustrate the usage of resistance spot welding in the automotive industry, Figure 1 shows the spot welds in a current generation of a sport utility vehicle. In total more than 6500 spot welds are used in the vehicle body in white.

Especially theses joints are known to change their stiffness during fatigue life which can have an impact on the local stress state. For a consequent light weight construction, novel methodologies are needed to predict the material's fatigue behaviour as stiffness changes are not taken into account in currently used fatigue analysis tools yet. Therefore increased safety factors are used in the design process which lead to an unnecessary increase in the overall vehicle weight. As a result of local stiffness changes, fatigue life may be affected positively or negatively. An elongation of fatigue life can occur if, due to changes in the local stiffness, stresses redistribute in areas which have observed lower stresses in the previous fatigue life. Under different conditions a reduction of fatigue life may also be a consequence of degradation. The stiffness behaviour often depends on loading direction and load spectrum, material combination or the used joining technique, respectively.
In this paper the basic approach DegraRP [2], which showed promising results for the stiffness degradation simulation under single channel torsional loading, is extended and applied to a complete vehicle simulation in order to take into account multiaxial loading.

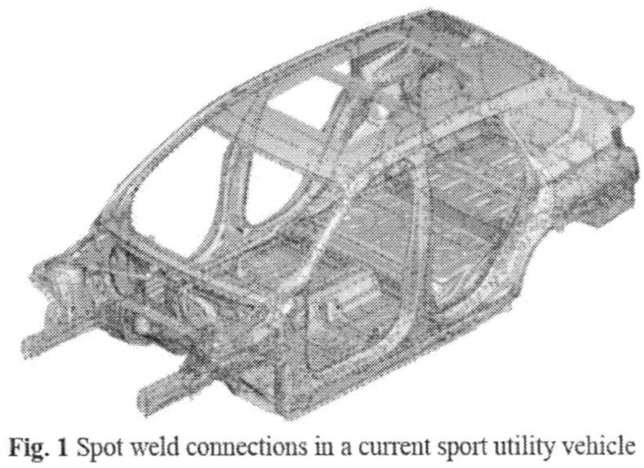

Fig. 1 Spot weld connections in a current sport utility vehicle

* Corresponding author: peter.roesch@bmw.de

 https://doi.org/10.1051/matecconf/201816517005

2 Theoretical Background

Resistance spot welding is widely used in the automotive industry. The joint between two components is produced by heat, generated through the components resistance against an applied current, and pressure [3]. In the following section the theory of different fatigue assessment approaches of spot welded structures and the methodology DegraRP are introduced briefly.

2.1 Fatigue assessment of spot welded structures

The complexity in the assessment of spot welded structures results from various challenges compared to non-welded structures, for example the [4]:

- determination of different material properties and hardnesses within the spot weld (base material, nugget or heat affected zone),
- uncertain residual stresses in the spot weld due to the manufacturing process,
- uncertain micro-geometry on the spot weld's notch,
- unknown crack initiation and propagation areas.

Especially the last bullet point is highly dependent on the local stress state and the load level of the tested specimens and structures and has been focus of a diversity of researchers since the 1970s [5-7].

For the numerical fatigue life assessment of spot welded components the FESPOW concept is often used in the automotive industry [8, 9]. The spot weld nugget with a diameter d connecting two sheets with the according sheet thickness $t_{1,2}$ is represented by beam elements. Six general forces are determined acting on the nugget. Beam cross-section forces F_{xi}, F_{yi}, F_{zi} and moments M_{xi}, M_{yi} are used to derive the equivalent stress state σ_v in the base material which will be evaluated during fatigue life, see equations (1)-(7). Torsion of the spot weld itself M_{zi} is not considered in this model.

$$\boldsymbol{\sigma_{v1}} = (\sigma_{max,Fx1} + \sigma_{max,My1})\cos(\varphi) + (\sigma_{max,Fy1} - \sigma_{max,Mx1})\sin(\varphi) - \sigma_{Fz1} \quad (1)$$

$$\boldsymbol{\sigma_{v2}} = (\sigma_{max,Fx2} - \sigma_{max,My2})\cos(\varphi) + (\sigma_{max,Fy2} + \sigma_{max,Mx2})\sin(\varphi) + \sigma_{Fz2} \quad (2)$$

with:
$$\sigma_{max,Fxi} = F_{xi}/(\pi \cdot d \cdot t) \quad (3)$$
$$\sigma_{max,Fxi} = F_{yi}/(\pi \cdot d \cdot t) \quad (4)$$
$$\sigma_{Fzi} = \begin{cases} c_1 \cdot \sqrt{t} \cdot F_{zi}/t^2 & \text{, if } F_{zi} > 0 \\ 0 & \text{, if } F_{zi} \leq 0 \end{cases} \quad (5)$$
$$\sigma_{max,Mxi} = c_2 \cdot \sqrt{t} \cdot M_{xi}/(t^2 \cdot d) \quad (6)$$
$$\sigma_{max,Mxi} = c_2 \cdot \sqrt{t} \cdot M_{xi}/(t^2 \cdot d) \quad (7)$$

The representative equivalent stress σ_v is the maximum of stress σ_{v1} and σ_{v2} resulting in each sheet. The equivalent stress state is composed of a shear-tension ($\sigma_{max,Fxi}, \sigma_{max,Fyi}$), cross-tension ($\sigma_{Fzi}$) and peel-tension ($\sigma_{max,Mxi}, \sigma_{max,Myi}$) stress portion.

With the FESPOW approach only a single SN curve, derived from a variety of cyclic tests on single spot weld specimens, is necessary compared to an approach in which the stress in the neighboring elements of the nugget is evaluated (FEMFAT Spot) [10]. The associated Finite Element modelling for both approaches is presented in Figure 2. Using the FEMFAT Spot approach different SN curves are applied for damage accumulation, depending on the local stress tensor (tension/compression, shear or a combination of both). The SN curves are derived using a variety of specimens, e.g. H or KS specimens. In addition to the introduced assessment approaches for spot welds notch-stress or strain approaches are available but rarely used in the automotive industry.

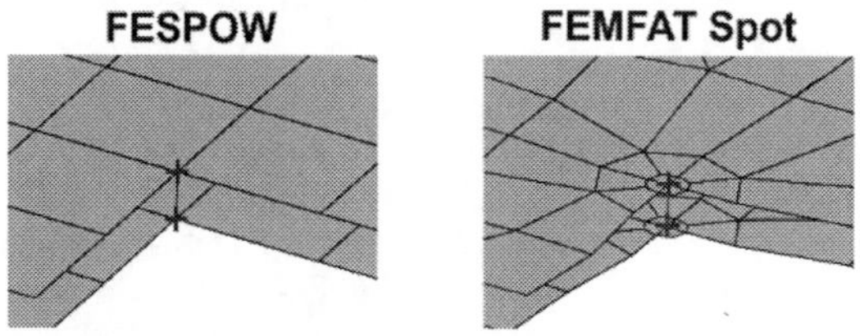

Fig. 2 FESPOW vs. FEMFAT Spot concept

2.2 Degradation simulation of uniaxial loaded structures

To simulate stiffness degradation during fatigue life an incremental computation process – DegraRP – is used. Figure 3 introduces the basic idea of the methodology. A conventional fatigue life simulation (represented with the black outline in Figure 3) is based on the quasi-static FE results, SN curves and damage accumulation using e.g. Miners law and allowable damage sum D. Changes in the local stiffness are not considered at all. To be able to consider the stiffness behavior during fatigue life simulation, the stiffness of the FE model needs to be adapted after certain thresholds have been reached (represented through the additional components with a green, dotted outline in Figure 3). A partial damage accumulation computation is conducted after which an evaluation of the current stiffness degradation state is performed. This evaluation is based on a stiffness degradation curve, describing stiffness as a function of the accumulated damage D derived from experimental data. Two additional parameter sets are necessary to determine boundaries for a recomputation with an adapted FE model. The parameter sets can be divided into material and computation (status) parameters. In a conventional fatigue life computation the result is an accumulated damage D. During the stiffness degradation simulation the main failure mechanism is described using a residual stiffness of each spot weld [2].

In a first step the stiffness degradation curve is assumed to be a simple power law. The material parameters are determined using a method of least squares to fit the functions parameters to the

https://doi.org/10.1051/matecconf/201816517005

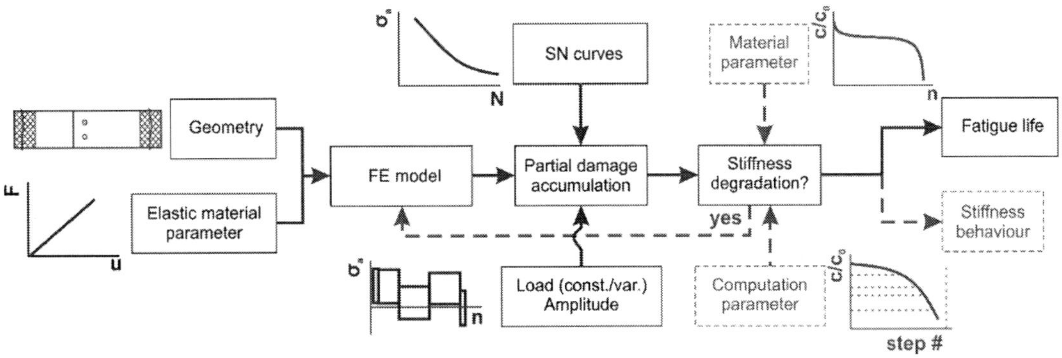

Fig. 3 Numerical stiffness degradation approach DegraRP [2]

experimental data. In a first step the main goal is to show that the principal methodology is capable to simulate stiffness degradation. Deriving a more accurate degradation law will be focus of future research.

The stiffness degradation law is only valid for the loading direction, material combination and weld parameters it has been derived from. The degradation law of another loading direction can be approximated using a combination of known degradation laws. For example for a 45° direction the degradation law may be approximated using 50 % of a 0° and 50 % of a 90° loading direction degradation law. This results in only little testing effort necessary, if applicable. The implemented method uses in total five status parameters for the control of the computation, e.g. when an adaption of the finite element model is necessary or the computation is finished. They have been introduced in detail in previous work not only for joining techniques but also carbon fiber reinforced plastic [2, 11-12].

3 Experimental test setup and results

In order to assess the prediction capabilities of the DegraRP methodology a customized experiment is conducted. The main goal of the experiment is to generate failure of certain, highly loaded spot welds during the test, using variable amplitude loads in different load directions. The failure of a spot weld will induce changes in the connection stiffness which effects the stress state in the base material surrounding the spot weld. In the experiment a sport utility vehicles tow bar is loaded in three independent directions (F_x, F_y and F_z). In the front, the car is mounted in the strut tower. In the back, the rear axle is assembled to the body in white and the fixation is attached to the wheel bearing. An additional support in horizontal direction is attached to one of the front door hinges. A variable amplitude loading is applied which was derived by increasing the loads given by the car loading standard for trailer coupling devices (CARLOS TC) [13]. In order to induce an early failure of certain spot welds the load applied is multiple times higher than it would be under normal testing conditions.

Additionally strain gauges are applied to the vehicles rear end. The strain gauges are used for verification of the complete vehicle's finite element model and the assessment of the local stress state during fatigue testing. The test setup, load directions and the position and orientation of the representative strain gauge rosette #4 are displayed in Figure 4.

A visual inspection of the rear end structure's damage is conducted after n/n_{ref} = 25 %, 50 %, 75 % and 100 %. The strain gauge measurement is conducted prior to the fatigue test, at n/n_{ref} = 50 % and n/n_{ref} = 100 %. The measurement is performed in each load direction independently. The results measured by the strain gauge 4-3 for each load direction are displayed in Figure 5 and compared to the results of a linear Nastran™ simulation of the complete vehicle. Each plot is normalized in x- and y-direction to a representative value (e.g. maximum stress). The experimental data shows an almost linear behavior of the stress state in the base material at strain gauge position #4. The overall fit of the strains in direction 3 between simulation and testing is acceptable. A similar behavior can be observed in direction 1 and 2. Therefore the representation of the stress state at strain gauge #4 by the finite element model can be considered accurate. The experimental results show a high alteration of the σ-F curves slope through different stages of fatigue life indicating a stiffness degradation. Under the application of a load in x-direction the slope $s_\sigma = \Delta\sigma/\Delta F$ increases by 63 % from s_σ = 0.47 at the beginning of the experiment to s_σ = 0.77 at the end of the test duration n/n_{ref} = 100 %. At the test duration only a minor increase of the slope (by approximately 10 %) can be observed.

The marked hot spot positions (HS1-HS3) in Figure 4 indicate spot welds which fail during the fatigue testing with disproportionally high loads. A first crack at the

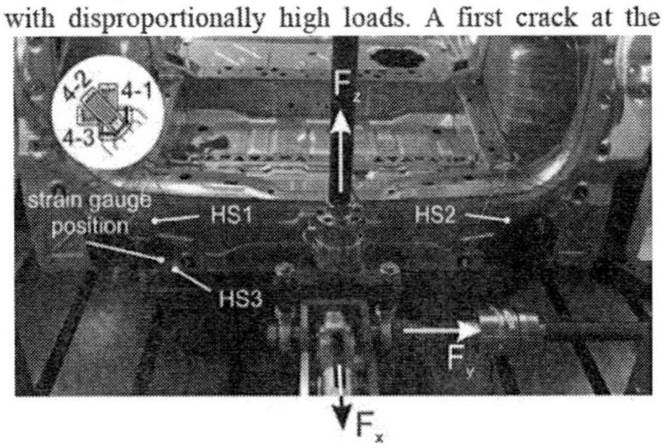

Fig. 4. Test setup, load direction, hot spots, position and orientation of strain gauge #4

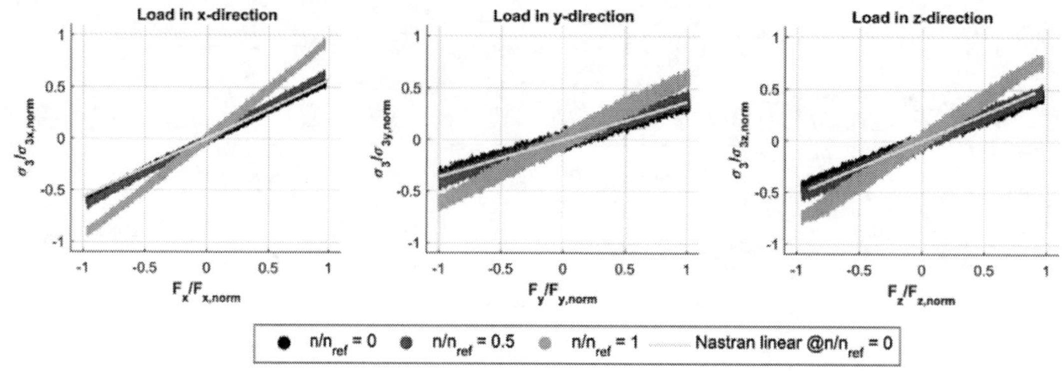

Fig. 5 Strain gauge 4-3 measurement during different stages of fatigue life and comparison to the initial linear Nastran ™ model

spot weld at HS1 can be observed at approximately 25 % of the testing period.

4 Simulation of stiffness degradation under multiaxial loading

In order to simulate stiffness changes under multiaxial, variable amplitude loading several extensions to the existing methodology are necessary which are then tested using a complete vehicle simulation.

4.1 Extension to the methodology

The necessary extension of the methodology mainly refers to the evaluation of the stress state at each spot weld in each computations step and the consideration of the effects of multiaxiality on the degradation behavior.

In the existing DegraRP approach one assumption is the quasi-invariability of the stress state at each spot weld under uniaxial loading through the different degradation stages. The portions of stresses due to a certain local loading type (later referred to as stress type with the three main components shear-, cross- and peel tension) on the stiffness degradation are therefore constant through the computation. The associated portions k_1-k_3, see equations (8)-(10), are determined using the equivalent stress established with the FESPOW approach.

$$\text{Shear:} \quad k_1 = \frac{\sigma_v - \sigma_v(F_{xi},\ F_{yi} = 0)}{\sigma_v} \tag{8}$$

$$\text{Cross:} \quad k_2 = \frac{\sigma_v - \sigma_v(F_{zi} = 0)}{\sigma_v} \tag{9}$$

$$\text{Peel:} \quad k_3 = \frac{\sigma_v - \sigma_v(M_{xi},\ M_{yi} = 0)}{\sigma_v} \tag{10}$$

Example: hat profile (proportional loading)

A first verification of the assumption is performed looking at the cross section forces after each computation step, simulating the stiffness changes of a hat profile under single channel loading. Looking at the shear-tension parameter k_3 the average deviation from its initial value (at $D = 0$) is below 0.1 % for each computation step. The cross-tension and peel-tension portion of the stress type deviate on average by 1.2 % and 0.05 % in each computation step. Considering the small sensitivity of the methodology to changes in the stress-state parameter (k_1-k_3) quasi-invariability of the stress-state through the computation can be assumed for the analyzed example.

Example: tow bar (non-proportional loading)

Under multi-axial loading the stress state at each spot weld can change in each computation step. To illustrate this behavior, Figure 6 shows the portions of each single stress type (shear-, cross- and peel-tension) at three selected hot spots (HS) and for each loading direction acting on the tow bar individually. Looking at HS1 cross-tension is only relevant under loading in x-

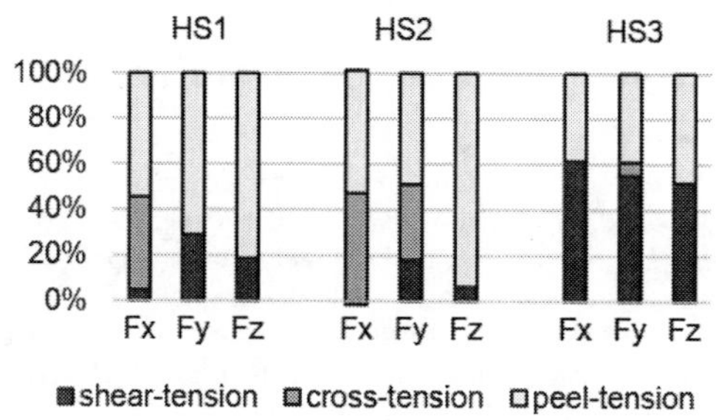

Fig. 6 Stress types at selected hot spots under loads acting in directions (x,y,z)

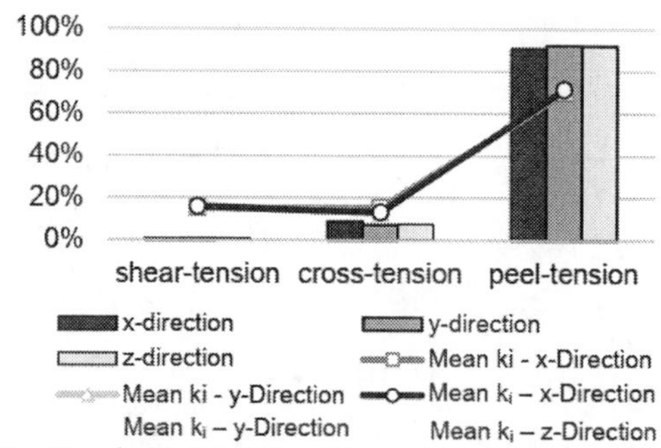

Fig. 7 Dominant stress type and mean stress state on average of all spot welds under loads acting in different directions

 https://doi.org/10.1051/matecconf/201816517005

$$k_i = \sum_{t_j} k_i(t_j) \cdot \frac{\Delta D(t_j)}{\sum_{t_j} \Delta D(t_j)} \qquad \text{for i = 1, 2, 3} \qquad (11)$$

direction. At HS2 the stress state is highly dependent of the cross-tension portion and in HS3 the three portions are almost constant. This diversity of changes in the stress state at each spot weld motivates a stress-state evaluation after each computation step. In some cases it can be observed, that the shear-tension leads to a decrease of the equivalent stress, e.g. in HS2 under load in x-direction. As a result the parameter k_1 is negative and the sum of all parameters $\sum_{i=1}^{3} k_i > 1$. The implications to the computation of a negative k_1 could lead to false conclusions as the stiffness could increase in one computation step, depending on the boundary conditions. In this case the parameter k_1 needs to be set to $k_1 = 0$ and the corresponding k_3 decreased by the previous k_1 in order to keep the sum of all parameters at $\sum_{i=1}^{3} k_i = 1$.

To get an overall impression of the dominating stress type in the complete finite element model, Figure 7 shows the dominating stress type at each spot weld relative to the number of all spot welds under a single constant load at the tow bar F_x, F_y and F_z. The dominating stress type is defined by the maximum of k_1-k_3 at each spot weld (e.g. in HS1 under a load acting in x-direction, peel tension would be the domination stress type). Over 90 % of the spot-welds in the model are dominated by local peel-tension, followed by cross- and shear-tension. This behavior is almost independent of the loading direction actin at the tow bar. On average over all spot welds the stress state is composed of shear-tension $\overline{k_1} \approx 15\%$, cross-tension $\overline{k_2} \approx 12\%$ and peel-tension $\overline{k_3} \approx 73\%$. This behavior is also almost independent of the load direction acting at the tow bar.

The second major extension to the methodology takes into account the combination of multiple load directions. In the tow bar test, three unit load cases are applied on the vehicle's tow-bar and superimposed with the CARLOS load-time series and a constant force representing the tow bar's nose weight. Under multiaxial variable amplitude loading the stress state parameters change from load cycle to load cycle not only in their proportions, but also in their relevance as there are load cycles which add almost no damage and load cycles adding a relative high amount of damage to the individual spot weld. Figure 8 shows the necessary computation results at one specific spot weld node during a short time step ($\Delta t = 0.5\,sec$) of the conventional fatigue life computation, applying CARLOS Module 3. The radial stress state in the critical plane σ_{rad} gives an overview of the minimum and maximum stresses used for the conventional fatigue life computation. At the four points in time t_j (indicated through the dotted line) the damage increment ΔD is accumulated. As mainly the stress state at the minimum and maximum of the reversal are relevant, the existence of a damage increment ΔD can be used as a trigger and the height of the increment as a weighting function for

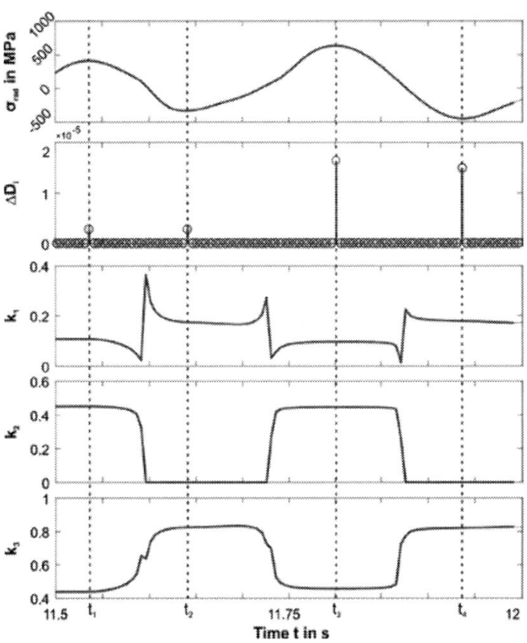

Fig. 8 Clipping of the equivalent stress state, damage increment and stress state parameter k_i in CARLOS Module 3

the computation of the overall modules stress state at one specific spot weld, see equation (11).

The computation needs to be performed for the CARLOS Modules 1-3 and for the critical plane of each spot weld.

4.2 Computation results

In total four equidistant computation steps ($\Delta n/n_{ref} = 25\,\%$) were performed applying the extended DegraRP methodology. Prior to the computation a selection of the relevant spot weld connections is conducted in order to minimize the computation effort of the DegraRP methodology. Therefore a conventional fatigue life computation is performed and the accumulated damage is assessed at each spot weld individually with its accumulated damage D_{sp}. The threshold of relevance of a single spot weld is set relative to the maximum observable damage in the finite element model D_{max} to $D_{sp}/D_{max} \geq 1e^{-5}$. In total 15 % of the spot welds (considering the test setup and loading conditions at hand) in the vehicle's finite element model are left relevant and are evaluated using the introduced methodology.

Figure 9 shows one of the results of the stiffness degradation computation – the accumulated damage for each computation step. A detail of the vehicles tow bar is illustrated and some of the relevant spot welds (SP1-SP4) are highlighted. A redistribution of the accumulated damage after each computation step can be observed. Focusing on SP2 (the spot weld with the highest damage in computation step 1) the damage at the spot weld and the surrounding base material decreases with each

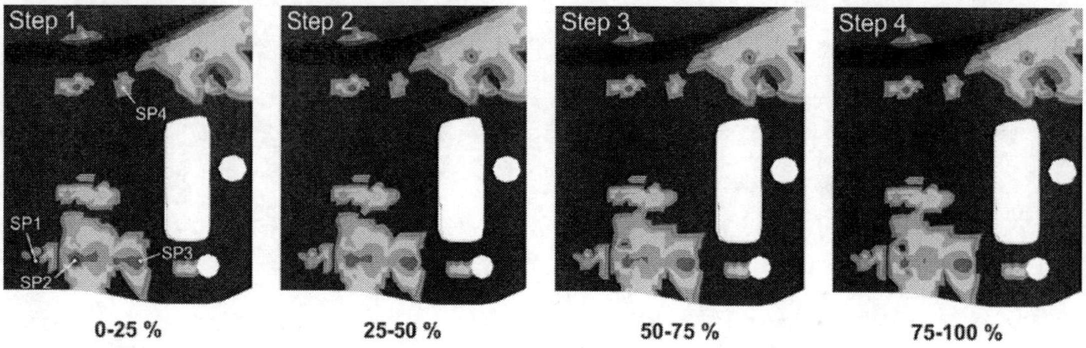

Fig. 9 Fatigue life computation and damage increment in each computation step

computation step due to the simulated stiffness degradation. As a result damage in SP1 and SP3 increases during each computation step. A similar behavior can be observed for SP4. The results show the feasibility and basic simulation capabilities of the extended DegraRP approach.

In order to quantify the simulations result, the stress state at n/n_{ref} = 50 % and 100 % is compared at the strain gauge #4. The degradation of the measured strain is compared to the degradation of the simulated strain at the position of strain gauge 4, see Figure 10.

The assessment of the experimental degradation is conducted based on the increase of the introduced slope s_{σ}. In order to evaluate the simulated degradation at strain gauge #4, a virtual strain gauge (represented by bar elements with a diameter of d_{bar} = 0.1 mm and young modulus E_{bar} = 210 000 MPA) is built in the finite element model. At n/n_{ref} = 50 % the change in the stress-state at strain gauge #4, induced by changes in the spot weld stiffness, are represented quite accurately for a load in x- or z-direction and each measuring direction. In y-direction the experimentally observed degradation differs, especially in measuring direction 4-1 and 4-2. A possible explanation for the deviation is the absolute strain at strain gauge #4 under load in y-direction, which is very small compared to other load and measuring directions. Therefore small changes in the strain results in disproportionally high relative changes. At n/n_{ref} = 100 % the computation highly underestimates the degradation at almost every load- and measuring direction, especially in measuring direction 4-3. The main reason for the deviation is the increasing effect of non-linearity with increasing damage, e.g. evolving cracks in the base material. The consideration of cracks or changes in stiffness of the base material is not possible in the current methodology. To illustrate the limitations of the extended approach a quasi-static computation of the finite element model without any spot welds in the surrounding area of strain gauge #4 is conducted. The results can be interpreted as a computation with 100 % degraded spot welds near the strain gauge. The resulting degradation of the strain is indicated through the boundaries in the numerical section at n/n_{ref} = 100 % for selected load- and measuring directions. It is obvious, that even with a complete degradation of the spot welds, the occurring stresses at n/n_{ref} = 100 % at the location of srain gauge 4-3 cannot be described. For a load applied in x-direction and the measuring direction 4-3 a maximum strain reduction of approximately 42 % can be simulated with the introduced methodology. As a results the measured reduction of over 70 % at strain gauge #4 cannot be reproduced in a simulation which only considers degradation of the spot weld by the reduction of its beam diameter.

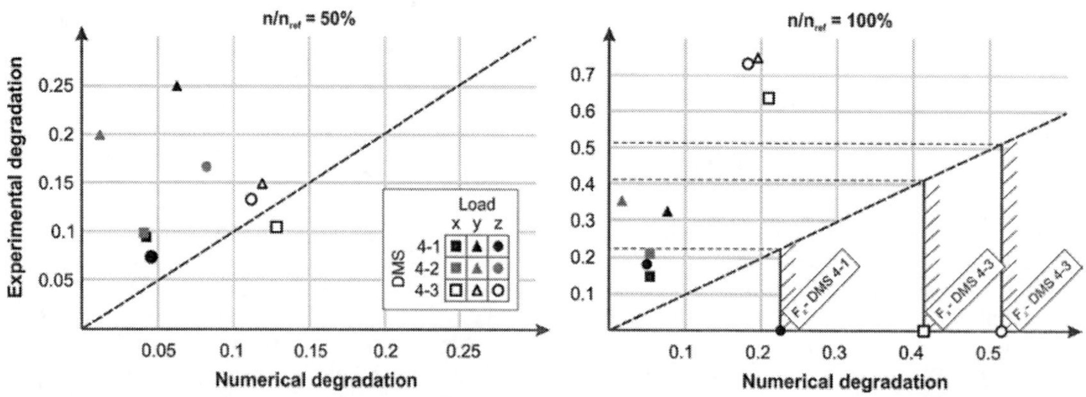

Fig. 10 Comparison of experimental vs. numerical change in strain data - assessed at strain gauge #4

5 Summary and Outlook

A conventional fatigue life analysis does not consider changes in stiffness of connection points due to fatigue loading. An experimental test conducted on a sport utility vehicles tow bar loaded with a CARLOS TC time-series shows a change in stiffness of up to 70 % depending on the boundary conditions, load- and measuring directions. The height of the CARLOS TC load has significantly increased compared to usual testing conditions in order to evoke an early failure of selected spot welds. The already introduced DegraRP approach for uniaxial loaded specimens has been extended considering the altering stress state at each spot weld and the effects of multiaxiality on the degradation behavior. The extended approach has proven feasible and quite accurate in the prediction of local strain redistributions due to stiffness changes of single spot welds.

However there still are certain limitations to the methodology as changes in the stiffness of the base material (induced by cracks) or geometrical non-linearities cannot be considered at the moment. Additionally the underlying material parameter set needs further improvement and refinement. In terms of automatization and performance the implementation still leaves room for improvement. At the moment the computation is conducted using a fixed step size (e.g. n/n_{ref} = 25 %). In future extensions the computation parameters (used for uniaxial loading) need to be adapted to multiaxial, variable amplitude loading.

References

[1] J. Meschke, J. Tölle, L. Berger, ATZ, Automobiltech. Z., **11**, 48-53 (2017)

[2] P. Rösch, T. Bruder, T. Bein, Procedia Eng., **213**, 437-447 (2018)

[3] A. H. Fritz, G. Schulze, *Fertigungstechnik* (Berlin, 2008)

[4] T. Seeger, S. Greuling, J. Brüning, P. Leis, C. M. Sonsino, D. Radaj, FAT, **196** (2015)

[5] J. Draisma, J. L. Overbeeke, Met. Constr. Br. Weld. J., **7**, 213-219 (1974)

[6] R. Yuuki, T. Ohira, H. Nakatsukasa, W. Yi, JSMET, **467**, 1772-1779 (1985)

[7] T. Satoh, H. Abe, K. Nishikikawa, M. Morita, Trans. Jpn. Weld. Soc., **1**, 46-51 (1991)

[8] A. Rupp, K. Störzel, V. Grubisic, SAE Trans, **104**, 635-645 (1995)

[9] H. Hanselka, K. Störzel, T. Bruder, FAT, **239** (2012)

[10] Y. Nakahra, M. Takahashi, A. Kawamoto, M. Fujimotot, N. Tomioka, SAE 2000 World Congress (2000)

[11] P. Rösch, T. Bruder, P. Wagner, Materialwiss. Werkstofftech. (to be published)

[12] M. Hack, S. Straesser, *LMS Virtual.Lab Durability Solver Training* (Kaiserslautern, 2016)

[13] T. Bruder, H. Klätsche, A. Sigwart, J. Riehle, Materialprüfung, **47**, 429-435 (2005)

Multi-axial fatigue life assessment of high speed car body based on PDMR method

Chaotao Liu[1,*], Pingbo Wu[1], and Fansong Li[1]

[1] Traction Power State Key Laboratory, Southwest Jiaotong University, No. 111, North 1st Section of Second Ring Road, Jinniu District, Chengdu City, Sichuan Province, PRC

Abstract. This article mainly introduces a method of converting the acceleration signal of body bolster obtained by the circuit test into the load of the air spring seat of vehicle body. This method mainly decomposes the body's movement posture into the form of ups and downs, roll and nod. Then formulate the test plan according to the performance of the body fatigue test bench. The vertical and horizontal displacements and longitudinal force are used as control commands.Taking advantage of vehicle body fatigue test bench to reproduce these basic types of vibration. Establish the transfer function of the acceleration of the bolster and the displacement excitation of the air spring, and then obtaining the load of the air spring seat. Finally, the multi-axial fatigue life assessment of the vehicle body was performed using the obtained load combined with the Moment of Load Path Method and the Path-Dependent Maximum Range Method.

1 Introduction

Car body is an important bearing structure of high speed train. The key of car body anti-fatigue design is extracting the stress of measuring-point accurately. Simulation analysis and experimental verification is an essential part of the vehicle design stage. Simulation analysis has the advantages of high efficiency, low cost, etc. The test verification not only can verify the accuracy of the simulation calculation model, but also can take into account the incomplete technical considerations in the simulation analysis.

In this paper, firstly obtains the time history of acceleration at the car body bolster through the line test. The time domain signal of the acceleration is decomposed and compressed according to the operating posture of the vehicle body (sinking, rolling, nodding, etc.). Then under laboratory conditions, using the vehicle fatigue test bench to reproduce the actual vibration of the vehicle body. At the same time, the stress at the danger point and the load at loading point are continuously measured. Finally, the dynamic stress response of the vehicle body finite element model is calculated using the load data of loading point, and the fatigue life of the high-speed train body is predicted using PDMR and MLP methods.

2 Determination of the test program

In order to obtain the loading data of the loading point, first of all, the vibration of the vehicle body needs to be reproduced through the test. During the test, the high-speed EMU body is supported on two six-degree-of-freedom platforms. Each platform can realize the translation in three directions along x, y, and z and the rotation in three directions. The connection between the vehicle body and the six-degree-of-freedom platform is set according to the actual force status of the vehicle body. The position of the air spring seat only bears the vertical load, and the position of the center pin bears the lateral and longitudinal loads.

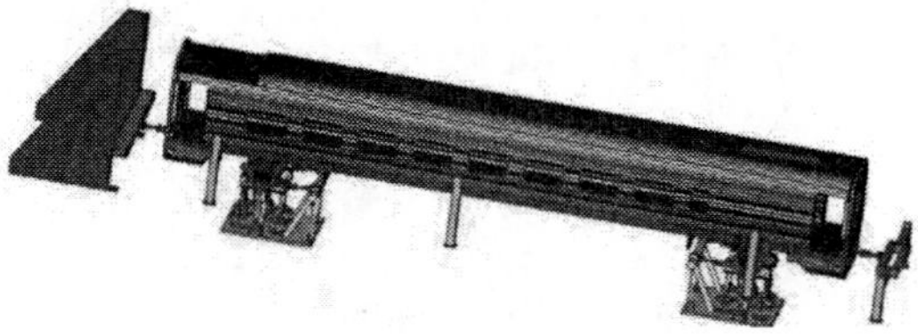

Fig. 1. Overall test program

Secondary vertical loading: Tooling for loading is installed on the six-degree-of-freedom platform. It releases lateral and longitudinal freedom through lateral and longitudinal linear bearings. The force sensor is arranged at the junction of the tooling and the test piece, and the rubber pad with smaller stiffness is arranged in the position of the vehicle-body interface, so as to avoid a rigid impact during the test.

* Corresponding author: liuchaotao@live.cn

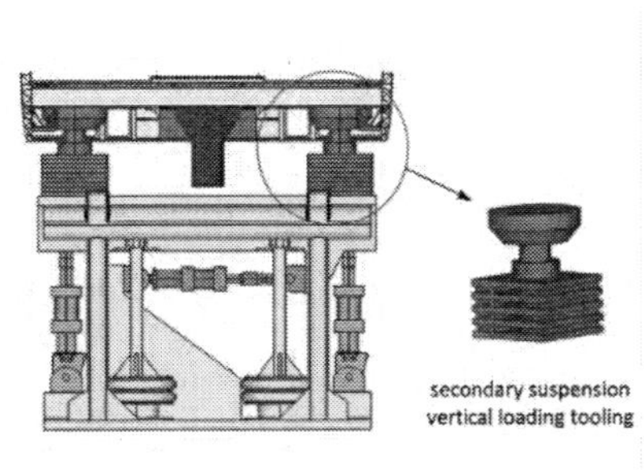

Fig. 2. Secondary suspension vertical loading tooling

The center pin loading tooling is also installed on the six-degree-of-freedom platform, and they are connected through a spherical hinge in order to release the vertical freedom.

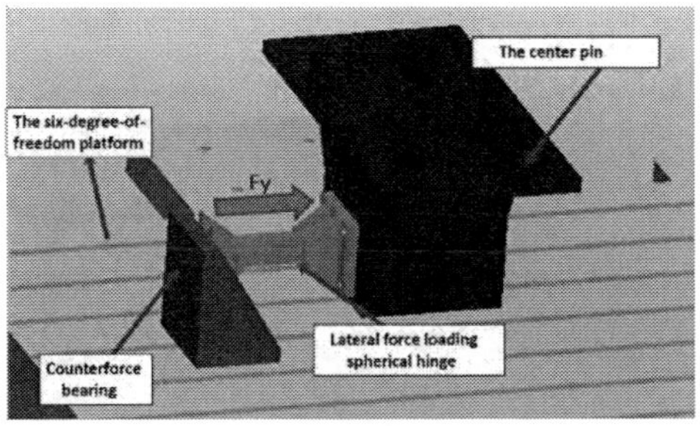

Fig. 3. The center pin loading tooling

Longitudinal actuators are arranged at the two coupler positions, which is used to simulate the coupler force.

Fig. 4. Coupler force loading device

Each six-degree-of-freedom platform is equipped with two longitudinal actuators in the longitudinally and horizontal position for balancing the unbalanced longitudinal forces between the two couplers. The actuator of each platform is orthogonal decoupled in order to realize simultaneous loading of longitudinal, lateral and vertical position at different frequencies.

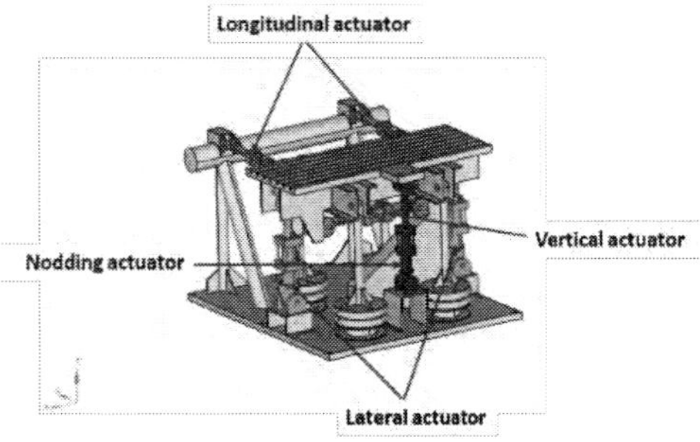

Fig. 5. Six-degree-of-freedom platforms

The above test load is loaded by force on longitudinal load, that is, the actuator is controlled by force, and the lateral and vertical loads are loaded by inertia, which are controlled by displacement.

3 Determination of fatigue test load

Reliable external load is the key of fatigue test. At present there are some international standards specifies the fatigue loads, such as EN 12663-2010 *Railway application - Structural requirements of railway vehicle bodies* [1](EN12663-2010 for short) , JIS E 7106-2006 *Rolling stock-General requirements of car body*[2] (JIS7106 for short), VDV152-2016 *Recommendations on the Design for Strength of Urban Rail Rolling Stock according to BOStrab*[3] (VDV152-2016 for short) and so on. All of these standards stipulate the body's fatigue load through the vibration acceleration of the vehicle body, but these standards are mainly formulated for the operating environment of the relevant countries. So the fatigue loads are not suitable for the track conditions in China. Moreover, these loads are accelerated fatigue tests, and does not appear in the actual operating state under different working conditions of the vehicle body.

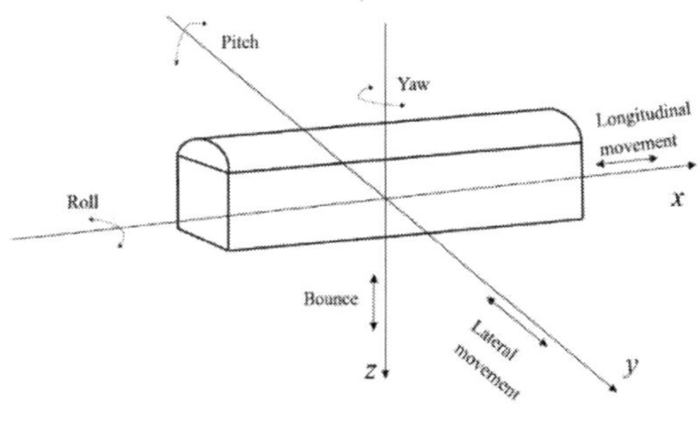

Fig. 6. The coordinate position of the body in space

Through the bench test, we can reproduce the operation posture of the vehicle body under different conditions, which is not only conducive to the research of the fatigue strength of the existing vehicle body, but also to the development of the new vehicle. A vehicle with spring suspension is a multi-degree-of-freedom vibration system. The spatial position of the vehicle body consists entirely of linear motion in three directions along x, y, z and gyration motion in three directions. Therefore, the movement of the vehicle body consists of floating movement, lateral movement, telescopic movement, yaw movement, nod movement and roll movement.

It is assumed that the body's acceleration of bouncing is $a_1(t)$, the acceleration of pitching is $a_2(t)$, the acceleration of rolling is $a_3(t)$, the lateral acceleration is $a_4(t)$, the acceleration of yawing is $a_5(t)$ and the longitudinal acceleration is $a_6(t)$. Four three-way accelerometers are arranged on both sides of the two bolsters, and four three-way accelerometers are arranged on the air spring seats of the two bogie frames. When the contribution of lateral roll to lateral acceleration is not considered, there is:

$$a_1(t) + a_2(t) - a_3(t) = a_{z1}(t) \quad (1)$$

$$a_1(t) + a_2(t) + a_3(t) = a_{z2}(t) \quad (2)$$

$$a_1(t) - a_2(t) - a_3(t) = a_{z3}(t) \quad (3)$$

$$a_4(t) + a_5(t) = a_{y1}(t) \quad (4)$$

$$a_4(t) - a_5(t) = a_{y3}(t) \quad (5)$$

Solved by:

$$a_1(t) = \frac{a_{z2}(t) + a_{z3}(t)}{2} \quad (6)$$

$$a_2(t) = \frac{a_{z1}(t) - a_{z3}(t)}{2} \quad (7)$$

$$a_3(t) = \frac{a_{z2}(t) - a_{z1}(t)}{2} \quad (8)$$

$$a_4(t) = \frac{a_{y1}(t) + a_{y3}(t)}{2} \quad (9)$$

$$a_5(t) = \frac{a_{y1}(t) - a_{y3}(t)}{2} \quad (10)$$

Where $a_{z1}(t)$, $a_{z2}(t)$, $a_{z3}(t)$, $a_{y1}(t)$ and $a_{y3}(t)$ are measured accelerations.

According to the test plan in section 1, the driving document for determining the vertical and lateral test loads is displacement excitation. The four vertical actuators of the two six-degree-of-freedom platform can simulate the three types of movements of the carbody's bouncing, rolling, and pitching, and two lateral actuators can simulate the yawing and rock-rolling of the body. Since the displacement of each actuator is relatively small when reproducing the running posture of the vehicle body, the coupling between the actuators is ignored.

Taking the vertical load as an example, in order to simulate the running conditions of a 350km/h straight-line operation of a high-speed EMU, the acceleration of bouncing, rolling and pitching are calculated by taking 100s test data according to the above formula. In the normal operation, due to the vehicle body is supported on the air spring, high-frequency loads cannot be transmitted to the vehicle body and the rigid body modes of the vehicle body are all within 2 Hz[4], thus, the acceleration signal collected by the position of the bolster beam is low-pass filtered by 20 Hz. The filtered data is subjected to FFT transformation to obtain the basic frequency f_1 of the body bouncing motion, the basic frequency f_2 of the rolling motion and the basic frequency f_3 of the pitching movement.

It is assumed that the acceleration of the body bounce is $1\,m/s^2$ when the vibration amplitudes of the four vertical actuators of the six-degree-of-freedom platform are d_1 and the frequency is f_1. When the two vertical actuators at the front end and two vertical actuators at the end apply an anti-phase displacement excitation d_2 , and frequency is f_2 , the pitching acceleration is $1\,m/s^2$. When the two vertical actuators at the left side end and two vertical actuators at the right side apply an anti-phase displacement excitation d_3, and frequency is f_3, the rolling acceleration is $1\,m/s^2$. The swaying and yawing are also processed in the same way.

It is assumed that the test system is linearly time-domain-invariant in order to get the fatigue test driver file. Based on the above transfer coefficient, the correspondence between the acceleration response of the body's bouncing, rolling and pitching movement and the positional displacement excitation of the air spring can be obtained. When simulating rolling of the vehicle body, the distance l_1 between the two vertical actuators is different from the distance l_2 between the air spring seats of the bogie, therefore, the displacement of the vertical actuator is $l_1 d_2 a_2(t)/l_2$ when the displacement of the air spring seat is excited by $d_2 a_2(t)$.

Table 1. the relationship between motivation and response.

Body movement	**Vibration acceleration**	**Displacement excitation**	**Vibration frequency (Hz)**
bounce	$a_1(t)$	$d_1 a_1(t)$	f_1
Bottom	$a_2(t)$	$l_1 d_2 a_2(t)/l_2$	f_2
Left	$a_3(t)$	$d_3 a_3(t)$	f_3
Right	$a_1(t)$	$d_1 a_1(t)$	f_1

Through the above table, the displacement excitation of the four vertical actuators can be obtained by:

$$d_{z1}(t) = d_1 a_1(t) + l_1 d_2 a_2(t)/l_2 - d_3 a_3(t) \quad (11)$$

$$d_{z2}(t) = d_1 a_1(t) + l_1 d_2 a_2(t)/l_2 + d_3 a_3(t) \quad (12)$$

$$d_{z3}(t) = d_1 a_1(t) - l_1 d_2 a_2(t)/l_2 - d_3 a_3(t) \quad (13)$$

$$d_{z4}(t) = d_1 a_1(t) - l_1 d_2 a_2(t)/l_2 + d_3 a_3(t) \quad (14)$$

For the obtained displacement excitation time history, the data are compressed by deleting the data that contribute less to the damage. The process of compression must maintain the phase relationship between each load to ensure that the medium and high loads are not deleted. The compressed displacement excitation is used as the driver file for body fatigue test. The data obtained by the force sensor feedback of the four air spring seat position $F_{z1}(t)F_{z2}(t)$, $F_{z3}(t)$, $F_{z4}(t)$ can be used as the load spectrum to calculate the body damage.

Table 2. Carbody rigid modal test results

No.	Modal Shape	Modal frequency (Hz)
1	Lower body rolling	0.600
2	Carbody bouncing	0.762
3	Carbody pitching	1.053
4	Carbody yawing	1.138
5	Upper body rolling	1.561

4 Calculation of vehicle body fatigue life

The fatigue life of the vehicle body is evaluated by the load spectrum obtained in Section III. Since the main frequency of the load spectrum is much lower than the lowest frequency of the elastic modal of the vehicle body, the quasi-static method is used to calculate the fatigue life of the vehicle body. Under the effect of the loads of air springs, center pins, couplers, etc., the body structure is subjected to multi-axial cyclic stress during operation, and in particular, the position of the weld seam is often in a multi-axial stress state due to geometrical abrupt changes. Therefore, it is necessary to evaluate the multi-axial fatigue strength of key welds of high-speed train body structure.

The load on the car body is neither in-phase or proportional, and the fatigue damage caused by this non-proportional load is inextricably linked to the load path and material. In order to define the parameters and period of non-proportional variable amplitude multiaxial fatigue damage, Professor Dong has invented a loop counting method based on path-dependency which is called as PDMR (Path-Dependent Maximum Range). Two important parameters can be obtained by PDMR for the stress time courses of a vehicle body measuring point: the maximum effective stress range counted in each half cycle and the resulting path length. The vehicle body fatigue damage can be calculated by combining the two parameters generated by PDMR and a new kind of MLP(Moment of Load Path) also raised by Professor Dong.

5 Test verification

In order to verify the effectiveness of the above method, taking the body of a high-speed EMU as the research object, in order to restore the vibration of the same type of car body running on the Wuhan-Guangzhou line, the vehicle body vibration acceleration data on this line is used as a target for verification. The verification process mainly includes: building body fatigue test device according to test plan, using line test data to compile driver files according to the method presented here , fatigue test , extracting the load spectrum at the point of action and the dynamic stress at key measurement points. based on the load spectrum and using the PDMR method and MLP method for fatigue assessment of key measurement points. Finally, the measured damage values were compared with the simulation values to verify the effectiveness of the method.

Fig. 7. Test devices

Due to the time, only the compilation of test devices and drive files has been completed, and the evaluation of the fatigue life of the vehicle body has not yet been carried out at present.

6 Conclusion and outlook

This article mainly proposes a method of converting the acceleration signal of the bolster beam position into the air spring position force signal. And put forward making use of MLP and PDMR method for the multi-axial fatigue life assessment of the car body. Due to the time, although this method has not been well verified, after the test is completed, the dynamic model of the test bench can be used to restore the test, and the load of the loading point under different working conditions can be restored through the virtual test. And then to achieve the evaluation of the car body fatigue life. Combined with flaw detection results, it is also possible to evaluate the remaining life of existing models.

References

1. EN 12663-2010 Railway application - Structural requirements of railway vehicle bodies
2. JIS E 7106-2006 Rolling stock-General requirements of car body

3. VDV152-2016 Recommendations on the Design for Strength of Urban Rail Rolling Stock according to BOStrab
4. Fansong Li. Research on Vibration and Fatigue of Carbody for EMU under Service Environment. D., 60, (2018)
5. Dong P, Wei Z, Hong J K. A path-dependent cycle counting method for variable-amplitude multi-axial loading. J. International Journal of Fatigue, 720-734, (2010)

MATEC Web of Conferences 165, 17007 (2018) https://doi.org/10.1051/matecconf/201816517007
FATIGUE 2018

Design, service and testing grounds stress spectra and their using to fatigue life assessment of bus bodyworks

Miloslav Kepka[1,*], *Miloslav* Kepka Jr.[1]

[1]University of West Bohemia, Faculty of Mechanical Engineering, Univerzitni 8, 30614 Pilsen, Czech Republic

Abstract. In the development of a road vehicle such as a bus, the input data for the assessment of the fatigue life of the body and other mechanical parts of the vehicle are gradually refined. In the initial phase of the development of a new vehicle, so-called design stress spectra are used. It is necessary to estimate the design spectrum parameters correctly. At the vehicle's prototype testing phase, the so-called service stress spectra can be evaluated. It is essential to determine the characteristic operating modes of the vehicle and to record representative stress-time histories of the critical sections of the construction nodes and vehicle components. Accelerated driving test of fatigue life can be performed on special test routes. The acceleration of such a fatigue test is dependent on the test track composition. The paper generally describes the methodology of such a process and demonstrates it using unique real-time data acquired during the development of a bus.

1 Introduction

In principle, the following input data are required to assess the fatigue life of the construction nodes of vehicles:

- fatigue life curves of the investigated construction nodes (S-N curves),
- information on their loads in operation (representative stress-time histories in critical points of the investigated nodes).

Stress-time histories are converted to stress spectra (histograms of frequency of harmonic stress cycles).

Calculations of fatigue damage and predictions of fatigue life are based on cumulative damage rules proposed by various authors. The input data are refined during the development of a new vehicle. In the design phase, parameters of S-N curves are most often estimated, taken from literature, standards or past solutions. Later, laboratory fatigue tests on structural nodes (which already respect specific structural geometry and manufacturing technology) allow them to determine their corresponding S-N curves.

Unless a mobile vehicle (functional sample, test prototype) is available, it is not possible to perform any strain gauge measurements of service stresses. Therefore, so-called design stress spectra are used in the design phase of vehicles. The design stress spectra are based on empirical experience, sometimes they are part of design standards and regulations, sometimes they represent corporate know-how.

With the vehicle prototype, strain gauge measurements can be made during the vehicle operation on actual lines with various road surface and various unevenness. A differently loaded vehicle (typically empty and fully loaded vehicle) can be measured. By the rain-flow method, the recorded stress-time histories are transformed into so-called service stress spectra.

Measurements should also be carried out on testing grounds. The measurements in this case take place on precisely specified test tracks with precisely given surface roughness. Based on a relative comparison of their damaging effects with fatigue damage generated on real lines, an accelerated running fatigue test of the vehicle can be proposed on testing grounds.

2 Detail of interest

The detail of interest was a severely stressed beam joint in the top corner of the door opening in the bus body shown in Fig. 1.

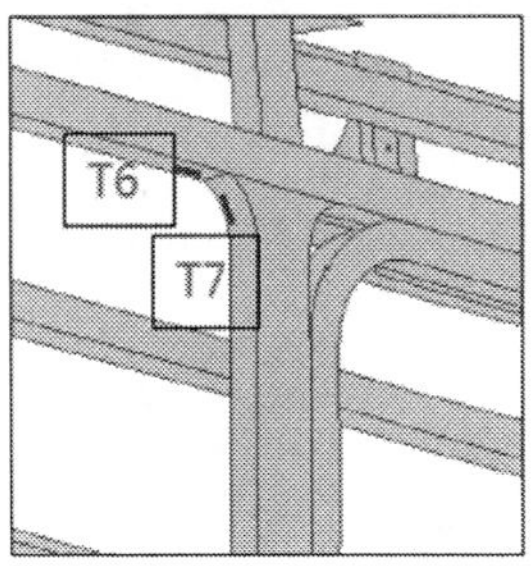

Fig. 1. Detail of interest (T6) - schematic illustration.

The critical cross section was monitored by strain gauge T6. The desired (design) fatigue service life of the vehicle in question was defined L_d = 1,000,000 km-run.

* Corresponding author: kepkam@rti.zcu.cz

This study is a virtual investigation because the specific values used in commercial contracts for various manufacturers are confidential. However, the values and data employed in this study are realistic and can be encountered in real life of a bus or a similar vehicle (trolleybus, battery bus or other vehicle).

2 S-N curve

In order to determine the fatigue strength of the evaluated structural detail, laboratory fatigue testing was carried out. Test specimens were equivalent to the considered beam joint in terms of basic material, shape and manufacturing technology (welding).

The transformation of the nominal loading into a root of notch [1] was not necessary, because the approach of the equivalent structural stress was used.

The test specimens were made from thin-walled welded closed profiles, which had 70×50 mm cross-section and 2 mm wall thickness and were made of S235JR steel. The critical cross-section of the joint was subjected to reverse bending load (the cycle stress ratio was R = -1).

During testing, the stresses acting on the critical cross-section were measured by strain gauges attached approximately 5 mm from the toe of the fillet weld. The measured values by strain gauges T6 can therefore be referred to as the equivalent structural stress.

The limit state was defined by the instant at which a macroscopic fatigue crack forms (1 to 2 mm). In all cases, fatigue cracks initiated in the transition zone of the fillet weld.

Statistical evaluation of the fatigue test data yielded the parameters of the mean S-N curve for the structural detail in the form:

$$log(N_f) = 14.54 - 4.53 \cdot log(\sigma_a)\,; \; \sigma_c = 60\,MPa.$$

Fig. 2 shows a photograph of the test stand. Table 1 summarises test results. The results of laboratory fatigue tests and the evaluated mean S-N curve are shown in Fig. 3.

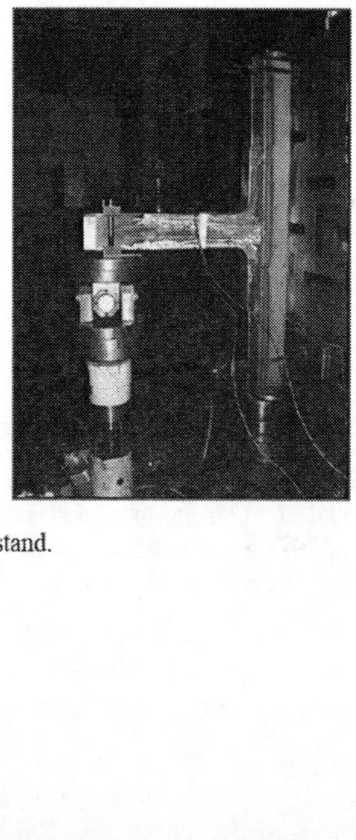

Fig. 2. Test stand.

Table 1. Results of laboratory fatigue tests.

Test specimen	Testing stress amplitude σ_a (Mpa)	Limit fatigue stage N_f (cycles)	Remark
1	140	50,000	
2	120	140,000	
3	110	170,000	
4	100	500,000	
5	80	1,250,000	
6	70	900,000	
7	60	2,000,000	runouts
8	50	2,000,000	runouts

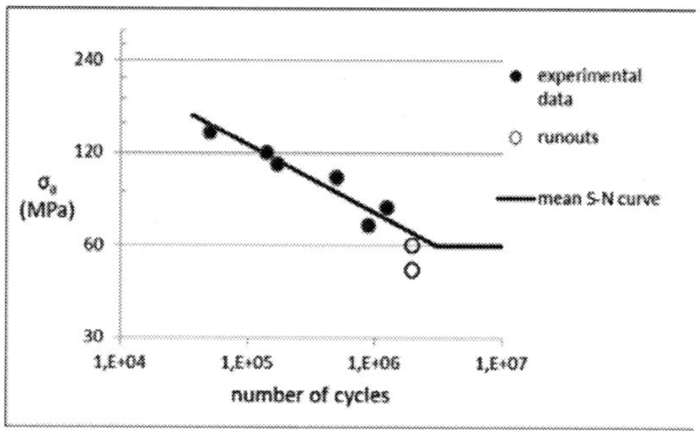

Fig. 3. Results of laboratory fatigue tests and mean S-N curve.

3 Stress spectra

3.1 Design stress spectra

If accurate data from a measurement is not available, the stress spectra should be estimated. The design stress spectra for this parametric study were generated using the relative coordinates $\sigma_{ai}/\sigma_{amax}$ [2].

$$h_i = H_{tot} \cdot \left(\frac{H_{max}}{H_{tot}}\right)^{\left(\frac{\sigma_{ai}}{\sigma_{amax}}\right)^s} \tag{1}$$

σ_{amax} - maximum stress amplitude in the spectrum,
H_{max} - number of cycles with σ_{amax} amplitude,
H_{tot} - total number of cycles in the spectrum,
s - shape parameter of the spectrum,
h_i - cumulative frequency of cycles with σ_{ai} amplitude.

At a constant width of the classification interval $d\sigma_a$, one can derive the discrete stress spectrum σ_{ai} - n_i. Absolute class frequency is calculated: $n_i = h_i - h_{i+1}$, this frequency is assigned to the mid-point of the class (or, safely, to its upper limit) σ_{ai}. With this interpretation of various oscillation processes, the cumulative frequency distribution of cycles h_i is plotted in semi-log graphs using various shape parameters s (Fig. 4):

$s = \infty$ rectangular distribution for a constant-amplitude harmonic process,
$s = 1$ normal distribution for a steady-state random Gaussian process with a standard deviation of σ_{RMS},
$s = 2$ linear distribution for a process consisting of a number of steady-state random Gaussian sub-processes with various values of σ_{RMS}.

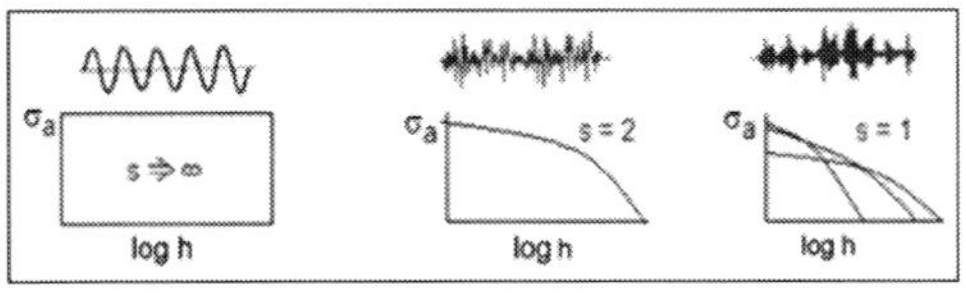

Fig. 4. Characteristic shapes of design stress spectra.

Measurements of service loads in road vehicles were interpreted in literature [3] with the following observations: long-term monitoring of rides on an irregular road surface yields linear distribution stress spectra (s = 1) and manoeuvres (curve riding, braking, etc.) lead to stress spectra with normal distributions (s = 2).

An estimate of the total number of cycles H_{tot} in the design stress spectrum can be obtained from the equation:

$$H_{tot} = \frac{L_d}{v} \cdot 3{,}600 \cdot f \quad (2)$$

L_d - design life of the body (in kilometres),
v - average speed (kph),
f - dominant frequency of vibration (Hz).

The frequency H_{max} of maximum service stress amplitudes σ_{amax} during the design life of the bodywork L_d depends on operating conditions of the vehicle given by the user. For the parametric studies, operating condition indicators P and L_1, were established to indirectly express the severity of the conditions:

$$P = \frac{H_{max}}{H_{tot}} \quad (3)$$

and

$$L_1 = \frac{L_d}{H_{max}} \quad (4)$$

P - probability of occurrence of the maximum service stress amplitude σ_{amax},
L_1 - travelled distance (in kilometres), in which the maximum service stress amplitude σ_{amax} occurs once.

Between both indicators the following relationship is valid:

$$P = \frac{L_d}{L_1 \cdot H_{tot}} \quad (5)$$

Therefore, the design stress spectrum parameters (s, H_{tot}, σ_{amax} and H_{max}) for designing and sizing the beam joints of the body were selected on the basis of the following considerations.

The major cause of the service loads that act on the structural detail is the vibration of the body due to the vehicle's ride on irregular road surfaces of various quality. For this reason, the shape parameter of the design stress spectrum was chosen as $s = 1$.

The design life of the body was set as $L_d = 10^6$ km. The dominant vibration frequency was assumed to be $f = 10$ Hz. The average travel speed was set as $v = 50$ kph. After substituting the values into equation (2), the estimation of total number of cycles in the specified body life was found as $H_{tot} = 7.2 \cdot 10^8$ cycles.

Maximum stress amplitudes σ_{amax} were estimated by strain gauge measurements on model testing track. The model testing track is usually made from artificial obstacles in a form of cylinder segment having the basis of 500 mm and height of 60 mm. The artificial obstacles are laid on an even asphalt road in such composition, that overruns of them occurred gradually by right-hand wheels, by both wheels simultaneously and by left-hand wheels [4].

Two loading states (payloads) were measured (run with empty vehicle and run with fully loaded vehicle). The results of the measurements are collected in Table 2.

Table 2. Measurement on model testing track.

Empty vehicle		Right	Both	Left
σ_{amax}	MPa	45	84	75

Loaded vehicle		Right	Both	Left
σ_{amax}	MPa	44	108	73

As safe values for generation of design stress spectra were finally used the following values: $\sigma_{amax} = 85$ MPa for the empty vehicle and $\sigma_{amax} = 110$ MPa for the loaded vehicle.

Travelled distance L_1, in which the maximum stress amplitude σ_{amax} occurs once, was chosen in three alternative variants, $L_1 = 10$, 100 and 1000 km, it means with three different values $H_{max} = 10^5$, 10^4 and 10^3 cycles. The parameters of all alternative design stress spectra are summarised in Table 3. The obtained design stress spectra are plotted and analysed in chapter 3.4.

Table 3. Parameters of alternative design stress spectra.

Empty vehicle		DSS_10_E	DSS_100_E	DSS_1000_E
σ_{amax}	MPa	85	85	85
H_{max}	cycles	10^5	10^4	10^3
H_{tot}	cycles	$7.2 \cdot 10^8$		
s	-	1		

Loaded vehicle		DSS_10_L	DSS_100_L	DSS_1000_L
σ_{amax}	MPa	110	110	110
H_{max}	cycles	10^5	10^4	10^3
H_{tot}	cycles	$7.2 \cdot 10^8$		
s	-	1		

These alternative design stress spectra were representations of the potential service at various operating condition indicators P and L_1.

In this particular case, the relationship between the operating condition indicators is as follows:

$$P = \frac{L_d}{L_1 \cdot H_{tot}} = \frac{1.3889 \cdot 10^{-3}}{L_1} \qquad (6)$$

3.2 Service stress spectra

If there is a sufficiently representative record of the stress time history, it is more accurate to evaluate the stress spectra from these data. The random processes are converted with the "rain-flow" method into one-parameter or two-parameter histograms of stress cycles depending on the magnitude of their amplitudes and mean values.

The real service stresses are measured by means of strain gauges. In this case, the service stress-time histories were measured for a city bus riding on an irregular surface along a city route whose total length was $L_m \approx 40$ km.

The signals were analysed in the frequency domain, and insignificant high frequencies were eliminated before using the rain-flow technique. As part of this rain-flow step, small cycles with amplitudes below 2MPa were omitted.

In high-cycle fatigue of welded structures, the mean stress of the cycle does not play a major role. Therefore, only the one-parameter stress spectra σ_{ai}-n_i were used for subsequent calculations.

Fig. 5 shows procedure of extrapolation of measured stress-time histories into service stress spectra.

The obtained service stress spectra are plotted and analysed in chapter 3.4.

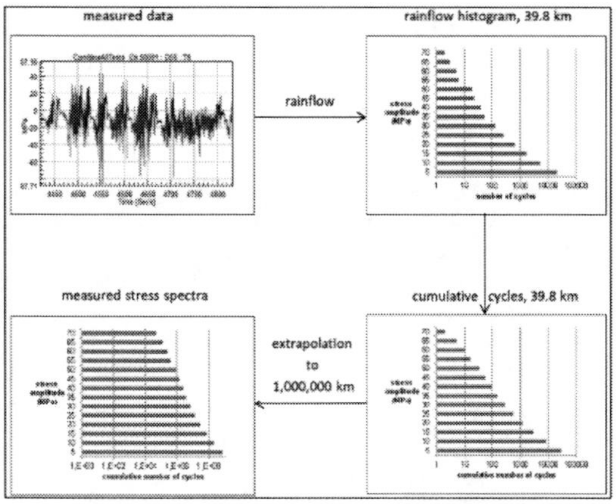

Fig. 5. Extrapolation of measured stress-time history into service stress spectrum (empty vehicle).

3.3 Testing grounds stress spectra

In a similar manner to representative urban traffic, the stress spectra on the test polygon were evaluated. The test polygon offers sections, tracks and roads with various longitudinal road profile and different surface quality. Table 4 specifies the composition of the proposed test route.

The measurement was repeated tree times with empty vehicle and three times with fully loaded vehicle. Total length of the measured route was $L_m \approx 35.5$ km.

The obtained testing grounds stress spectra are plotted and analysed in chapter 3.4.

Table 4. Composition of test route on testing grounds.

Section	km
slope circuit	3.80
speed circuit	2.80
arrival to special roads	0.07
panel road	0.45
exit/arrival	0.13
sinus resonance road	0.40
exit/arrival	0.15
paved road	0.40
exit/arrival	0.15
paved road	0,40
exit/arrival	0.15
paved road	0.40
exit/arrival	0.16
Belgian paving	0.40
exit from special roads	0.08
speed circuit	1.90
Total	11.84

3.4 Analyses of all stress spectra

All theoretically generated design stress spectra and all measured service stress spectra and testing grounds stress spectra are plotted and compared in Fig. 6.

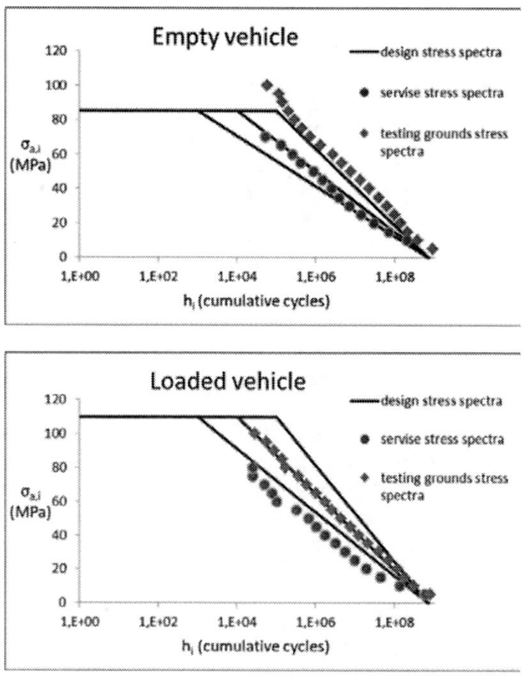

Fig. 6. Stress spectra – empty and fully loaded vehicle.

The difference in the aggressiveness of the individual spectra we can see at first glance. The shape of all measured stress spectra corresponds well to the assumed linear distribution, too.

In the table 5 there are the values of the shape parameter s from equation of design spectra (1). The values were calculated from experimental data by regression analysis. The shape parameter in all cases approaches to $s \approx 1$.

Table 5. Shape parameter of all measured stress spectra.

s – shape parameter	Real operation	Testing grounds
Empty vehicle	1.011	1.092
Loaded vehicle	0.939	1.027

4 Fatigue life calculations

The mostly the fatigue damage D is calculated using the linear cumulative damage rule. According to this rule, the limit state with respect to fatigue is reached (i.e. the fatigue life of the structural part is exhausted) when the following condition is met:

$$D = \sum_i \frac{n_i}{N_i} = D_{lim} \qquad (7)$$

D - fatigue damage caused by the stress spectrum imposed,
n_i - number of cycles applied at the i-th level of stress with the amplitude σ_{ai},
N_i - limit life under identical loading (the number of cycles derived from the S-N curve for the part in question at the amplitude σ_{ai}),
D_{lim} - limit value of fatigue damage.

Various rules apply various boundary conditions to fatigue damage calculation. A schematic representation of these boundary conditions is shown in Fig. 7.

Account is taken of the damage caused by cycles with small amplitudes ($\sigma_{ath} < \sigma_{ai} < \sigma_c$), which occur very frequently. A threshold value σ_{ath} is applied to the conversion of stress to damage, and therefore the damage caused by cycles with amplitudes of ($\sigma_{ai} < \sigma_{ath} < \sigma_c$) is neglected.

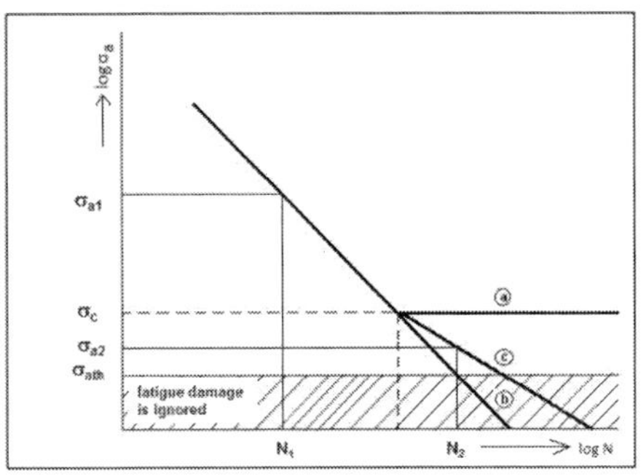

Fig. 7. Boundary conditions for fatigue damage calculations.

A limit value is set for the fatigue damage. According to Miner, $D_{lim} = 1$, but it has been shown experimentally that this value may decline towards hazardous levels with $D_{lim} < 1$.

In the present case, the Haibach-modified version of the Palmgren-Miner rule was chosen for calculating fatigue damage. The limit number of cycles N_i was determined as follows:

$$-\ \sigma_{ai} \geq \sigma_c: \quad N_i = N_c \cdot \left(\frac{\sigma_c}{\sigma_{ai}}\right)^{w} \qquad (8)$$

and

$$-\ \sigma_c > \sigma_{ai} \geq \sigma_{ath}: \quad N_i = N_c \cdot \left(\frac{\sigma_c}{\sigma_{ai}}\right)^{wd} \qquad (9)$$

Haibach recommends the exponent for the lower part of the S-N curve to be set as $w_d = 2w-1$. In this study, the value was $w_d = 8$. The threshold stress amplitude for taking the fatigue damaging into account was given as $\sigma_{ath} = 0.5 \cdot \sigma_c$ in the present case. The limit value of fatigue damage was taken as $D_{lim} = 0.5$.

The computational estimate of service fatigue life (in kilometre run) is obtained from equation:

$$L = \frac{D}{D_{lim}} \cdot L_m \qquad (10)$$

Table 6 provides predicted fatigue life for the service stress spectra and the spectra measured on the test polygon.

The ratio of these predicted fatigue life is an estimation of the acceleration (shortening) of the driving fatigue test that could be achieved on the test polygon compared to normal vehicle operation.

Table 6. Shape parameter of all measured stress spectra.

Empty vehicle	Real operation	Testing ground	Acceleration
L (km)	3,178,000	203,000	15.7

Loaded vehicle	Real operation	Testing ground	Acceleration
L (km)	4,309,000	337,000	12.8

5 Conclusions

Several interesting conclusions and remarks can be formulated on the basis of this case study.

It has been verified that the parameter $s = 1$ can be selected to generate the design stresses spectra in the form (1). This conclusion applies only to the bodywork nodes! For vehicle components, such as axles, this value may be different, because the driving manoeuvres are more important for their stresses.

The estimation of the other parameters of the design stress spectra (H_{tot}, σ_{amax} and H_{max}) is a little more problematic, but the procedure proposed in chapter 3.1 is acceptable. In our case, it proved to be sufficiently conservative, which is essential at the stage of the initial design of a vehicle.

Based on the demonstrated methodology (using fatigue damage calculations), it is possible to determine the acceleration of the driving fatigue tests on the testing grounds. A changing the composition of the test route can result in a change in the acceleration of this test. However, the composition of the test track cannot be any. For example, an intensified driving on special roads could cause an undesirable overheating of shock absorbers.

In the presented case study, the accelerations of the driving fatigue test were calculated separately for a city operation of an empty and fully loaded vehicle. In another practical case it would be possible to determine a more detailed mix of operating conditions.

In the presented case study, the acceleration of the driving fatigue tests were exemplarily determined for the mean fatigue life (because the mean S-N curve with 50 % survival probability was considered). For specific commercial solutions, the required operational reliability should be taken into account. The authors of this paper have published some examples of engineering applications of a probabilistic approach to the fatigue life calculations [5, 6].

The contribution has been prepared under project LO1502 'Development of the Regional Technological Institute' under the auspices of the National Sustainability Programme I of the Ministry of Education of the Czech Republic aimed to support research, experimental development and innovation.

References

1. V. Chmelko, M. Margetin: The methodology of transformation of the nominal loading process into a root of notch. Procedia Structural Integrity, Vol. 5, (2017), 825-831

2. M. Ruzicka, M. Hanke, M. Rost: Dynamic strength and fatigue life (University textbook, Czech Technical University, 1987, in Czech)

3. J. Neugebauer, V. Grubisic, G. Fischer: Procedure for Design Optimization and Durability Life Approval of Truck Axles and Axle Assemblies (SAE Technical Paper Series 892535, 1989)

4. M. Kepka: Durability and Fatigue Life Investigation of Bus and Trolleybus Structures: Review of SKODA VYZKUM Methodology. Proceedings of the 2nd International Conference on Material and Component Performance under Variable Amplitude Loading (VAL2009), DVM, Darmstadt, Germany, March 23 – 26, 2009, pp.1231-1240

5. M. Kepka, M. Kepka Jr.: Identification of Possible Causes of Fatigue Failure in Radius Rod Pins for Trolleybus Rear Axles using Parametric Calculations of Fatigue Life. Proceedings of 7th International Conference on Mechanics and Materials in Design (MCD2017), Albufeira/Portugal, June 11-15, 2017

6. M. Kepka, M. Kepka Jr.: Parametric calculations of fatigue life of critical part of trolleybus rear axle. Procedia Engineering, Vol. 213, (2018), 227–238

https://doi.org/10.1051/matecconf/201816517008

Load spectra and fatigue damage: applications to the automotive industry

Matteo Luca Facchinetti[1,*]

[1]PSA Groupe, Centre Technique de Belchamp, CC SX.BP.06, 25240 Voujeaucourt, France

Abstract. High cycle fatigue behaviour of materials is historically assessed with constant amplitude and variable amplitude loads, respectively. Thus, a long-lasting debate is extremely active in the academic community, trying to link experimental results coming from these different approaches. Overcoming all this, since the 1970s several industrial fields have been choosing to consider representative customer load spectra (in terms of amplitude, not frequency) as the best way to test both materials and structures. In particular, the automotive industry makes use of specific car loading spectra, regularly fed by the customer knowledge and practised on proving grounds. This paper presents a highlight on such spectra, neglecting any sequence effect of the load time history, thus accepting the Palmgren-Miner's rule as an assumption. Whereas a recent communication on this very topic focused on the basic occurrence spectra, which is absolutely independent from the material properties of the car parts, here we deal with the final damage assessment. Obviously, it is worth knowing which part of the spectrum is mainly responsible for the most relevant fatigue damage.

1 Introduction

This paper is devoted to the fatigue phenomenon applied to automotive passenger vehicles. More precisely, we focus on the chassis system highlighted in figure 1, which provides basic functions like braking, steering, vertical suspension, vehicle handling …

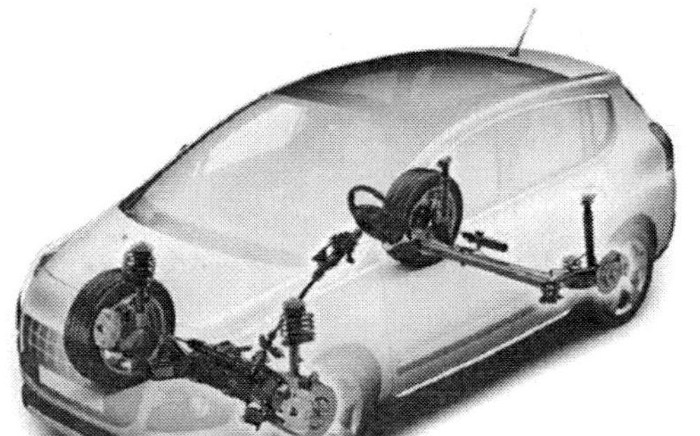

Fig. 1. Passenger car chassis system: front and rear axles.

The chassis system is almost composed of safety parts, whose reliability with respect to fatigue is expected to be proven, by numerical computations and experimental tests, respectively. Fatigue is one of the most relevant design concerns, even if not the only one. In terms of orders of magnitude, considering a life target of about 10^5 km, the customer usage analysis leads to count up to 10^6 elementary braking and cornering events, i.e. 10^6 cyclic load occurrences, thus calling for high cycle fatigue. For the sake of simplicity, within this framework we pay attention on usual driving situations only, coming from everyday life customer usage, thus neglecting any overload related to special driving conditions or misuse.

Road load time series recorded to feed both numerical computations and experimental tests may be long and complex signals. Whether measured data come from proving grounds or customer usage, engineers are more accustomed to reduce them on more simple signals. This may be achieved thanks to a fatigue damage equivalence process, calling for at least three basic problems:

1. identify load cycles (i.e. sinusoidal patterns of defined mean, amplitude) from a time history
2. take into account for any sequence effect, i.e. the relevance of the cycle load order
3. compute a relevant fatigue damage induced by load cycles of different amplitude (and mean)

For the first step, there is a large consensus since the work of Endo & al. [1] and the application of the well-known "rainflow" count method, which has already been published in international standards. We simply accepts this state-of-the-art as an assumption. The second one is still under vigorous discussion: on one hand, there are the advantages and shortcuts of the simplest linear rule proposed by Palmgren and Miner [2,3] nearly 100 years ago; on the other hand, there exists several more recent and complex models for which the material parameters need for complex identifications as well. In this paper,

* Corresponding author: matteoluca.facchinetti@mpsa.com

we just assume and apply the simplest model: please remember that the in-service load history of a passenger car is very long (the order of magnitude is of about 10 years) and somewhat random, from customer to customer and with respect to any possible normal usage operating conditions. At last, the fatigue response may be sensible to the load time history, the gap between variable and constant amplitude points of view resulting in a long-lasting debate among the fatigue community. This paper humbly presents some insights between two frameworks applied by the automotive industry in France and Germany, respectively.

On one hand, the pioneering Wöhler model [4] is the natural reference when dealing with constant amplitude loading and represents the historical point of view of French engineers. The relevant fatigue damage equivalent signal corresponds to a single sinusoidal wave of given amplitude (and frequency), the simplest one to be managed by a testing rig or a FE solver.

On the other hand, standard load spectra are widely applied since the seminal work of Gassner [5] and applied by German engineers since at least 50 years. Here the relevant fatigue damage equivalent signal is composed by blocks of sinusoidal waves of different amplitudes and occurrences, see figure 3.

The key features beyond this apparently nothing but theoretical discussion are practical indeed: it is matter of effective fatigue assessment of chassis parts through numerical computations and experimental tests, quotation of the relevance of a proving ground, acceleration of testing procedures …

In the sequel of a previous recent communication on this very topic [6], this paper is organized as follows. At first, we announce all the assumptions and describe the basic models used (§2). Then, a historical overview on standard load spectra is recalled from the literature, with particular interest on a particular load spectrum family (§3). Thus, special attention is devoted to the fatigue damage assessment (§4). At last, discussions and conclusions end the paper (§5).

2 Assumptions and models

We recall here the main assumptions already presented in [6]. The framework is that of high cycle fatigue phenomenon [7,8], which applies cyclic loads under the yield threshold of the materials used for chassis system parts. Nowadays it is of common practice to represent the load cycle amplitude, S with respect to the life cycle number, N , see figure 2. Following Wöhler [4], constant amplitude fatigue tests provided up to final failure (or any other relevant stop criterion) lead to the so-called Wöhler curve, see figure 2. One of the basic models used to express a relationship between the load cycle amplitude, S with respect to the life cycle number, N is that of Basquin [9], as expressed by the following equation:

$$NS^b = B \tag{1}$$

where the Basquin slope, b ($b > 1$) and the Basquin constant, B ($B >> 1$) are material parameters to be identified. At least for metallic materials, this model is supposed to hold from 10^4 up to 10^6 cycles and leads to a linear relation in the *logS-logN* plane, see figure 2. For some reference values of the Basquin slope and a more comprehensive statistical approach, please refer to [10].

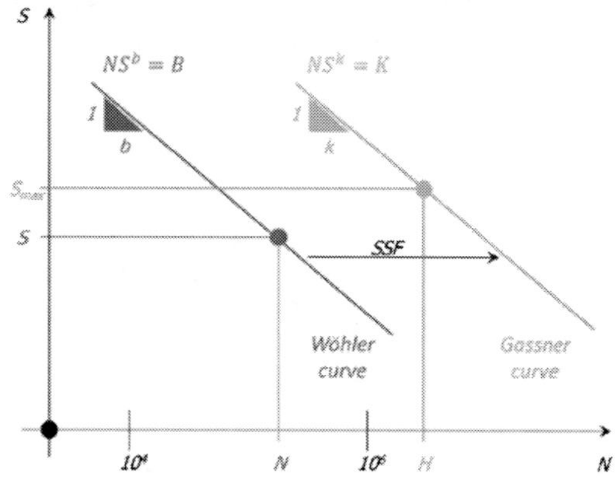

Fig. 2. Wöhler curve (left, dark grey) & Gassner curve (right, light grey).

Moreover, Ernst Gassner seminal work [5] introduces the concept of load spectrum in order to define a variable amplitude lifetime. Please note that the term "spectrum" does not rely here to any frequency analysis, but only deals with the occurrence of load cycles with different amplitudes within the same load sequence. Actually, the load frequency is not expected to have any effect in the range of the load time histories of interest. The original Gassner block schedule, showed in figure 4, is composed of eight load amplitude levels, S_i linearly distributed from the lower, S_{min} up to the maximum, S_{max} and associated to cycle numbers N_i resulting from a Gaussian-like distribution. A cumulative occurrence, H may be introduced by the following equation:

$$H_i = \sum_{j=1}^{i} N_j \tag{2}$$

where the summation is usually started from the maximum amplitude load level (i.e. $S_1 = S_{max}$), which is expected to be associated to the smallest occurrence (i.e. $N_1 = N_{min}$ usually chosen to be equal to 1). Thus, applying the summation over the whole load sequence leads to the spectrum length, H which actually counts the overall cycle number of the spectrum, no matter of the load amplitudes. From the experimental point of view, once the load spectrum pattern is established by the choice of $S_1 = S_{max}$, the load sequence is applied again and again until the specimen failure (or any other

https://doi.org/10.1051/matecconf/201816517008

relevant stop criterion), leading to the overall lifetime H. Within the previous S-N plane (actually $logS$-$logN$), Gassner reports the experimentally achieved overall lifetime, H at the maximum load amplitude, S_{max} of the load spectrum considered, leading to the so-called Gassner curve, see figure 2. Another Basquin-like relation may be used to model Gassner results:

$$HS_{\max}^{k} = K \tag{3}$$

where the slope, k and the constant, K are material parameters to be identified.

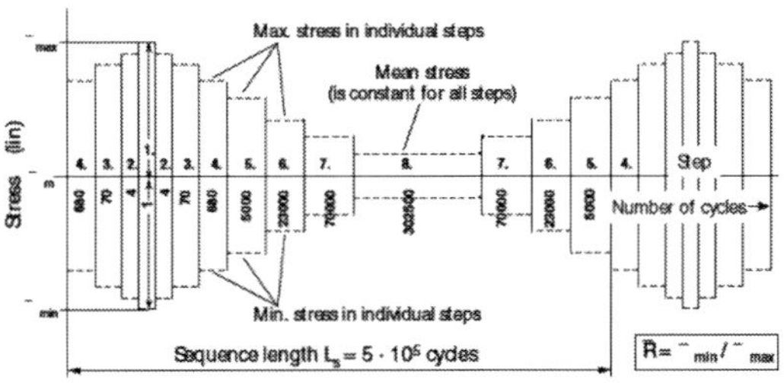

Fig. 3. Gassner load sequence – courtesy from [13].

Several assumptions apply within the Gassner framework. At first, let us recall that the whole spectrum is supposed under the yield threshold of the materials used. Secondly, the load sequence, i.e. the way all the load cycles of different amplitudes are mixed before to be applied to the specimen, is expected to minimize any history effect: please note that since the Gassner original block sequence, see figure 4, both increasing and decreasing load amplitudes are considered within the basic load schedule. Moreover, the sequence length, H should be at least one order of magnitude smaller than the expected lifetime, in order to accept the concept of an elementary spectrum as a relevant metrics. If it were not the case, this would imply that rare maximum load levels are not guaranteed to be applied with the same statistical representation than any other lower load levels, so that fatigue damage would be too sensible to them.

Since the linear Palmgren-Miner's damage rule [2,3] is assumed, it may be proven [11] that the exponents of the Basquin and Gassner laws are expected to be exactly the same:

$$b = k \tag{4}$$

Actually, if no history effects are considered, the only material parameters in use are the pairs (B,b) and (K,k). Whereas the intercepts B and K are associated to the choice of plotting options (S or S_{max} and N or H, respectively), the slopes b and k are expected to describe the same physical behavior: thus they must have the same value.

At last, please note that plotting Gassner lines (S_{max} ; H) in the Wöhler plane (S ; N), see again figure 1, gives the graphical illusion that the Gassner framework offers longer lifetime than Wöhler does. This is nothing but a graphical mirage, due to the choice of S_{max} as the relevant load amplitude for the Gassner plot. According to [12], it may be proven [11] that the horizontal shift of the Gassner line with respect to the Wöhler one is provided by the following formula:

$$SSF = \log\left[\frac{\sum N_i}{\sum N_i \left(\frac{S_i}{S_{\max}}\right)^{1/b}}\right] \tag{5}$$

where SSF stands for spectrum shape factor, which solely depends on the shape of the spectrum and the Basquin slope. Please note that if the spectrum collapses to a constant amplitude load (i.e. $S_i = S_{max}$), the Gassner framework reduces to the original Wöhler one: $SSF = 0$. For any consistent variable amplitude spectrum, SSF is definite positive, because of $b > 1$ and $S_i / S_{max} < 1$. It may be shown that $SSF < 1$ for spectra with many large and few small cycles, whereas SSF > 1 for spectra with larger occurrences for small amplitude cycles. For example, for $b = 5$ and a Gaussian-like load spectrum as the original Gassner one, one has SSF $\sim$ 2.5 , which means an horizontal shift of more than 2 decades on the right with respect to the Wöhler curve.

3 Standard load spectra

Both European and US working groups have been providing reference design load spectra or standard load time-histories since the 1970s, respectively for the aeronautics, the energy industry, the automotive … along with experimental test results and fatigue criteria. Heuler & al. [12,13,14,15] published a comprehensive review where each cited example is collected with its own purpose, the intended structural application, a brief qualification, and some more technical details like the block size.

In order to go further with the comparison and analysis of load spectra, it may be useful to provide a general mathematical formulation. According to [13,14], we make use of the following equation:

$$\frac{\log H_i}{\log H_0} = 1 - \left(\frac{S_i}{S_{\max}}\right)^{\nu} \tag{6}$$

where the pattern of the spectrum is based on an exponential law and ν is the sole spectrum parameter, called the shape exponent. This equation may be plotted in a dimensionless field, see figure 4, considering the amplitude load ratio $r_i = S_i / S_{max}$ on the Y-axis and the occurrences ratio log H_i / log H_0 on the X-axis.

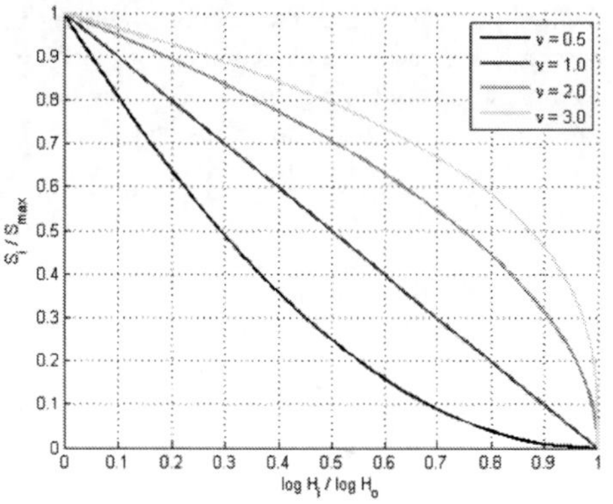

Fig. 4. Standard load spectra according to [13,14] – cumulative occurrences, *H*.

Switching from cumulative occurrences, H_i to usual cycle number, N_i , thus making use of equation (2) on the other way round, leads to the more explicit figure 5. For the sake of precision, whereas figure 4 is full dimensionless, figure 5 needs to define the overall spectrum length, H_o here assumed equal to 10^6.

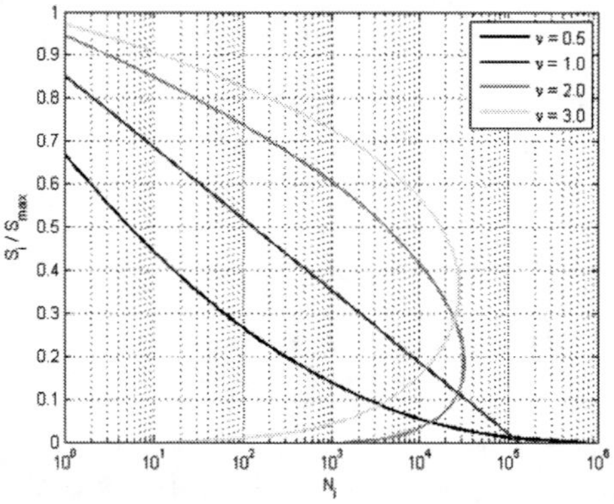

Fig. 5. Standard load spectra according to [13,14] – effective occurrences, *N*.

Figure 5 allows us to give a physical meaning at the shape parameter, ν . A unit value is associated to a linear pattern (please note that abscissa is in log scale), so that $\nu < 1$ corresponds to $\partial^2 N/\partial S^2 < 0$ and a concave spectrum, for which the lower the load, the greater is the occurrence, whereas $\nu > 1$ corresponds to $\partial^2 N/\partial S^2 > 0$ and a convex spectrum, for which the larger occurrence is not associated to the weakest load anymore. It should be noted that figure 5 shows an apparently odd behavior at the upper left corner: the expected maximum load seems to be associated to less than one cycle number, as the curves vanish on the left before attending the maximum load level. Actually, the proposed spectrum model states that the maximum amplitude $S_1 = S_{max}$ always corresponds to a unit cycle indeed $N_{min} = 1$. Then, according to the shape imposed by ν , the following load level expected to have more than 1 cycle is at the intercept between the spectrum curve with the vertical axis at unit abscissa.

Please note that the shape exponent does not depend on the material fatigue parameters like the Basquin slope, because of it results from the cycle count operation prior to any damage ponderation. Thus, since the formulation (6) is accepted, the identification of the shape exponent, ν may be a key to compare load spectra coming from independent sources, histories and practices. According to [13,14], typical values of the shape exponent may be identified and linked to the SSF, equation (5) for a standard Basquin parameter value, b = 5, as reported in table 1.

Table 1. Typical load spectra, according to [13,14].

ν	**SSF (b=5)**	**description**
∞	0	constant amplitude
4	1.5	convex spectrum
2	2.5	Gaussian spectrum
1	3.8	linear spectrum
< 1	4.2	concave spectrum

In a recent paper [6], a comparison between load spectra applied for the automotive has been proposed, based on the identification of the shape exponent, ν and thus limited to the basic occurrence spectra, which is absolutely independent from the material properties. At present, in the following section we focus on the fatigue damage assessment, so that the Basquin slope, *b* is the second relevant parameter to be taken into account.

4 Damage assessment

In the previous section we described a load spectra family already proposed in the literature [13,14], equation (6), and recalled its main characteristics. We now introduce a damage norm, according to the Basquin model, equation (1):

$$D = NS^b = D(\nu, b) \qquad (7)$$

and wish to highlight how the fatigue damage is partitioned through the load spectrum, as a function of both the spectrum shape, ν and the Basquin slope, *b*. As already commented in [16], figure 6 shows the dimensionless damage distribution of the considered spectra family for an usual value of the Basquin slope, *b* = 5.

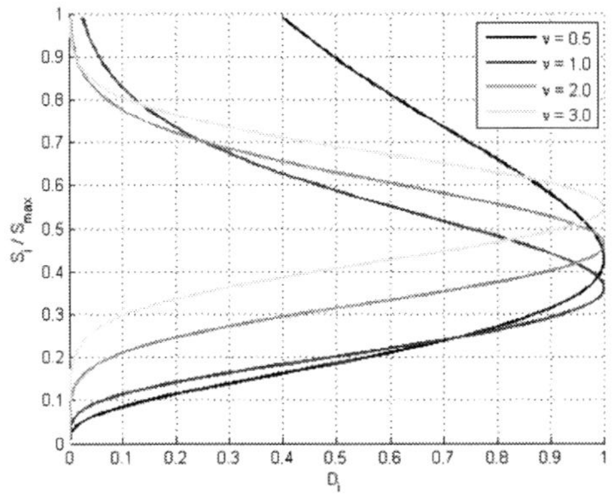

Fig. 6. Standard load spectra according to [13,14] – damage, *D*.

The maximum relative damage is located neither at the maximum load level, nor the minimum one, but somewhere in the middle, not so far from the central kernel of the spectra. On one hand, lower loads than the maximum damage one are candidates for omission, as they may not provide significant damage and are associated to the higher occurrences. On the other hand, higher loads than the maximum damage one may be responsible for lower damage, but are associated to the lower occurrences indeed, so that any omission is more sensible. Depending on the shape exponent, the load level of maximum damage is shown to move upwards as the shape exponent increases. In a more comprehensive view, according to figure 6, the value of the dimensionless load amplitude generating the maximum damage may be defined by:

$$S_{\max D} = \frac{S_i}{S_{\max}}\left(\frac{\partial D}{\partial S} = 0\right) \tag{8}$$

which is a function of the Basquin slope, *b* and the shape exponent, ν. Figure 7 presents a contour plot of S_{maxD}, where *b* spans from 3 to 13 and ν from 0.5 to 3.5 .

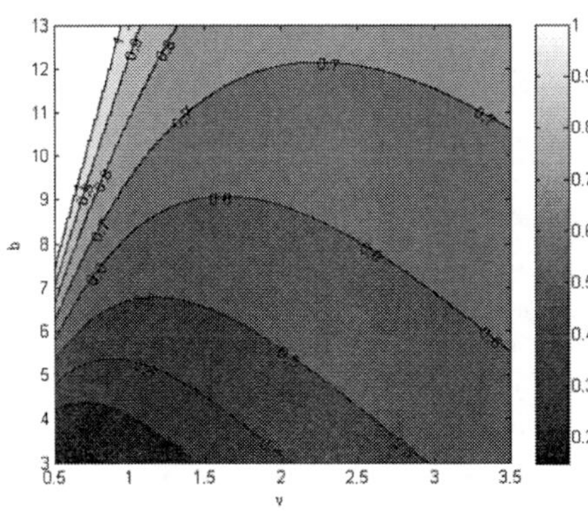

Fig. 7. Load amplitude, S_{maxD} generating the maximum damage as a function of the Basquin slope, *b* and the shape exponent, *ν*.

Most of the contour plot is bounded within 0.3 and 0.7, i.e. the load amplitude associated to the maximum damage lies in the 0.5±0.2 range. Significant deviations appear only for Basquin slopes far from the median value, $b = 8$ or low shape exponent, $\nu < 1$. Please note that the upper left corner may be a critical one indeed, because of the relevant damage of the overall spectrum is provided by its few higher amplitude load cycles only. For a given spectrum shape, i.e. a fixed value of ν, an increasing value of the Basquin slope leads to a higher relevant load amplitude. Whereas for a given Basquin slope, *b* an increasing value of the shape exponent, ν leads to a monotonic decrease of the relevant load amplitude only for lower Basquin slopes.

S_{maxD} provides useful information on the damage content of a spectrum, but does not takes into account what happens just around it. This gap may be closed by introducing the concept of "damage barycentre", i.e. a weighted load with respect to damage [17], defined by:

$$S_{aveD} = \frac{\sum_i \frac{S_i}{S_{\max}} D_i}{\sum_i D_i} \tag{9}$$

From a purely statistical point of view, S_{aveD} represents an overall average, whereas S_{maxD} is a mode of the damage distribution. S_{aveD} is again a function of the Basquin slope, *b* and the shape exponent, ν. Figure 8 presents a contour plot of S_{aveD}, where *b* spans from 3 to 13 and ν from 0.5 to 3.5 .

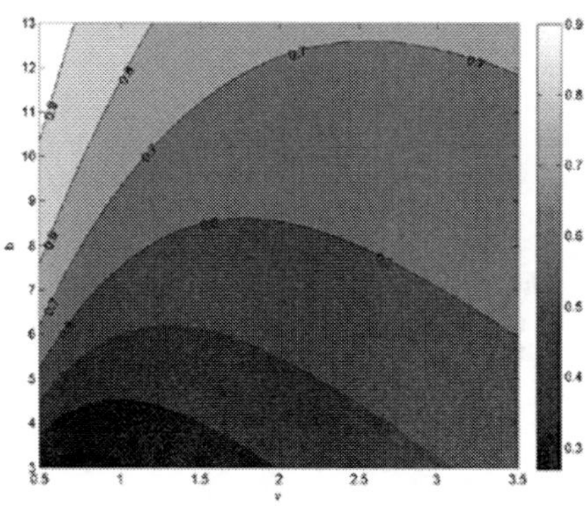

Fig. 8. Load amplitude, S_{aveD} generating the maximum damage as a function of the Basquin slope, *b* and the shape exponent, *ν*.

The comparison of figures 7 and 8 shows that the damage barycentre, S_{aveD} and load generating the maximum damage, S_{maxD} are almost the same. If ν is larger than the unit value, i.e. for convex spectra, the scatter between S_{maxD} and S_{aveD} is lower than 5%. Only for concave spectra the scatter may rise up to 10% or more, especially for higher values of the Basquin slope. Roughly speaking, as far as the damage distribution is bell-like, no significant difference is expected between the average and the mode.

5 Conclusions and outlooks

This paper is devoted to the fatigue phenomenon applied to automotive passenger vehicles. At first (§2), we announced all the assumptions and described the basic models applied in the high cycle fatigue framework. Then (§3), a historical overview on standard load spectra has been recalled from the literature, with particular interest on a particular load spectrum family, equation (6). Thus (§4), a special focus has been provided to the fatigue damage assessment, going further than a previous paper [6].

Let us recall here the main hypothesis of this work.

1. The fatigue damage is based on the Basquin model, applying a unique slope b on the overall load domain. This provides the first relevant parameter, whose significant values may be found in the literature, e.g. in [10]. For fatigue design of chassis parts, a particular interest is usually devoted to lower values, i.e. from 3 to 5, when dealing with welded assemblies and notched parts.
2. A spectrum family is chosen according to [13,14]. This provides the second relevant parameter, the shape exponent, ν for which standard values are already available for several industrial applications [12,13,14]. For fatigue design of chassis parts, some values of interest are reported in [6].

Within this framework, the damage assessment may be highlighted by the analysis herein developed, figures 5 to 8, and the use of the relevant dimensionless loads S_{maxD} and S_{aveD} , equations (8) and (9), respectively.

The present analysis may be easily extended to any other spectra family than (6). Actually, the choice of the underlying mathematical model and its physical meaning should deserve more discussion. Moreover, a full mathematical approach may also be considered. Finding extreme values and locations of continuous functions should help to get more general properties.

This work began from fruitful discussions between German & French automotive researchers within the framework of the DVM – Deutscher Verband für Materialforschung und prüfung & SF2M – Société Française de Métallurgie et des Matériaux. Moreover, the author wishes to acknowledge the IRDL laboratory at ENSTA Bretagne (Brest), whose researchers and students contributed to this work during several internships.

References

1. M. Matsuiski, T. Endo, Fatigue of metals subjected to varying stress, Japan Society of Mechanical Engineering (1969)
2. A. Palmgren, Die Lebensdauer von Kugellagern - The Fatigue Life of Ball-Bearings, Zeitschrift des Vereins Deutscher Ingenieure (1924), 68, 339-341 (in German)
3. M.A. Miner, Cumulative damage in fatigue, Journal of Applied Mechanics (1945), 12, A159-A164
4. A. Wöhler, Versuche zur Ermittlung der auf die Eisenbahnwagenachsen einwirkenden Kräfte und die Widerstandsfähigkeit des Wagen-Achsen, (1860), Zeitschrift für Bauwesen, X, 583-616 (in German)
5. E. Gassner, Festigkeitsversuche mit wiederholter Beanspruchung im Flugzeugbau - Strength tests under repeated loading for aeronautical engineering Luftwissen (1939). 1939;6:61–4 (in German)
6. M. Facchinetti, Fatigue damage of materials and structures assessed by Wöhler and Gassner frameworks: recent insights about load spectra for the automotive, Procedia Engineering (2017) 213, 217-225
7. C. Bathias, J.P. Baïlon, La Fatigue des matériaux et des structures - Fatigue of materials and structures, Hermes Paris (1997) (in French)
8. S. Suresh, Fatigue of Materials, Cambridge Solid State Science Series (1998)
9. O.H. Basquin, The exponential law of endurance tests, proceedings of the ASTM (1919), 30, 625
10. S. Bergamo, P. Schimmerling, F. Triboulet, P. Wilson, M.L. Facchinetti, M. Monin, F. Lefebvre, B. Weber, Préconisations pour les caractéristiques statistiques de résistance en fatigue - Applications aux aciers et autres matériaux utilisés dans la construction automobile – SIA (2017) (in French) - http://www.sia.fr/publications/494-preconisations-pour-caracteristiques-statistiques-resistance-fatigue?lng=fr
11. L. Navrátil, Fatigue under variable amplitude load, internship report, ENSTA Bretagne, Brest (2016)
12. C. Berger et al., Betriebsfestigkeit in Germany - an overview, International Journal of Fatigue (2002) 24, 603–625
13. P. Heuler, T. Bruder, H. Klätschke, Standardised load-time histories - A contribution to durability issues under spectrum loading, Materialwissenschaft und Werkstofftechnik. (2005), 36, 669 - 677
14. P. Heuler, H. Klätschke, Generation and use of standardised load spectra and load-time histories, International Journal of Fatigue (2005) 27, 974-990
15. C.M. Sonsino, Fatigue testing under variable amplitude loading, International Journal of Fatigue (2007), 29, 1080-1089
16. M. Martins de Oliveira, Dimensionnement à la fatigue des éléments d'une Liaison Au Sol, internship report, ENSTA Bretagne, Brest (2018)
17. I. Raoult, B. Delattre, private communication

Fatigue Life Extension of AA2024 Specimens and Integral Structures by Laser Shock Peening

Nikolai Kashaev[1*], *Sergey* Chupakhin[1], *Volker* Ventzke[1], *Manfred* Horstmann[1], *Stefan* Riekehr[1], *Alessandro* Barbini[1], *Jorge. F.* dos Santos[1], *Sören* Keller[1], *Benjamin* Klusemann[1, 2] and *Norbert* Huber[1]

[1]Institute of Materials Research, Materials Mechanics, Helmholtz-Zentrum Geesthacht, 21502 Geesthacht, Germany
[2]Institute of Product and Process Innovation, Leuphana University of Lüneburg, 21339 Lüneburg, Germany

Abstract. The goal of the present study is to understand the effects of laser shock peening (LSP)-induced residual stresses on the fatigue crack propagation (FCP) behaviour of the commonly used aircraft aluminium alloy AA2024 in T3 heat treatment condition. LSP treatment was performed using a pulsed Nd:YAG laser on compact tensile C(T)50-specimens with a thickness of 2.0 mm. LSP-treated specimens reveal a significant retardation of the fatigue crack propagation. The fatigue crack retardation effect can be correlated with the compressive residual stresses introduced by LSP throughout the entire specimen thickness. A possible application of the LSP process on a component like panel with three welded stringers representing a part of a fuselage structure was performed as well. The skin-stringer AA2024-AA7050 T-joints were realised through stationary shoulder friction stir welding (SSFSW), a variant of the conventional friction stir welding process. In this relatively new process, the shoulder does not rotate and therefore does not contribute to the heat generation. Consequently, a reduced and more homogeneous heat input leads to a less affected microstructure and better mechanical properties. The efficiency of the LSP process has been demonstrated resulting in an increase of 200 – 400% in fatigue lifetime.

1 Introduction

Laser shock peening (LSP) is a very promising technique for introducing deep compressive residual stresses in metallic materials [1-2]. The LSP process is essentially based on high-energy pulsed laser beams, which cause the propagation of shock waves in materials or components, and thereby of near-surface compressive residual stresses and work hardening [3-4]. LSP-induced residual stresses can be used to improve the damage tolerance behaviour of lightweight metallic structures. Several researches investigated the effects of LSP on the initiation and propagation of fatigue cracks in aluminium alloy specimens with various notch geometries [1,5-9]. The results clearly demonstrated that LSP can be used as an effective surface treatment technique for reducing or suppressing fatigue crack growth in aluminium alloys. However, application of a not optimised LSP process can produce high surface roughness without introducing noticeable compressive residual stresses [10]. In this case, LSP can reduce the fatigue life of critical components and structures. Therefore, much of the LSP process and its effect remain to be figured out as how the LSP process can be applied in an appropriate way.

The aim of the present study is to understand the effects of LSP-induced residual stresses on the fatigue crack propagation (FCP) behaviour of the commonly used aircraft aluminium alloy AA2024 in T3 heat treatment condition. Furthermore, a possible application of the LSP process is demonstrated on a component like panel with three welded stringers representing a part of a fuselage structure showing the application of LSP to large aircraft structures.

2 Experimental methods

2.1 Material, LSP treatment and SSFSW process

A 2.0 mm thick sheet of the aluminium alloy AA2024-T3 was used in this study. LSP treatment was performed using a Q-switched Nd:YAG laser operating at 10 Hz with a wave length of 1064 nm, pulse duration of 10 ns in case of C(T)-specimens and 20 ns in case of M(T)-panels. Two different diffractive optics were used to deliver the laser energy in a square spot of 1 mm x 1 mm and 3 mm x 3 mm on the specimen surface. In case of the diffractive optic of 3 mm x 3 mm, the specimen surface was covered with an aluminium foil. The LSP treatment was applied on the both surfaces of C(T)50-specimens (width 50 mm). The initial crack length, a_0 was 15 mm for the LSP-treated specimens using the diffractive optic 1 mm x 1 mm and 10 mm for the LSP-treated specimens using the diffractive optic 3 mm x 3 mm. The treated specimen area was 5 mm x 5 mm and 15 mm x 15 mm in case of the diffractive optics 1 mm x 1 mm and 3 mm x 3 mm, respectively. The sequence of LSP spots is depicted in Figure 1. The LSP treatment

* Corresponding author: nikolai.kashaev@hzg.de

was applied up to 5 times (5x overlapping) at the same location. The distance between the area of the LSP patterns and the initial crack tip was -1 mm and -2 mm for specimens treated using 1 mm x 1 mm and 3 mm x 3 mm diffractive optics, respectively (crack tip was in the area of the LSP patterns). The LSP-treatments for the C(T)50-specimens investigated in this study are summarized in Table 1. The LSP-treatment sequences were also investigated in the previous studies of the authors [15-16].

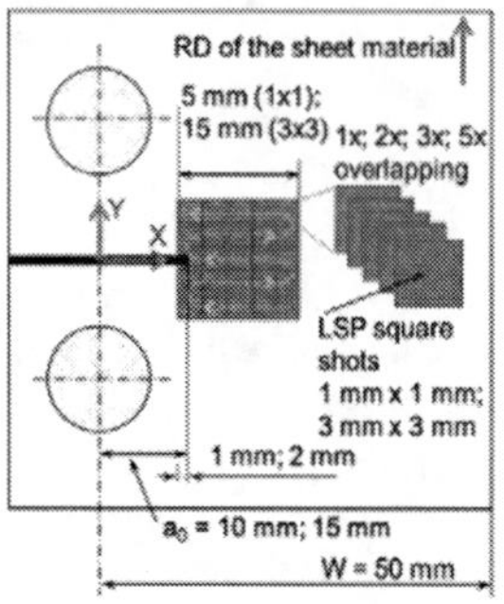

Fig. 1. Geometry of the C(T)50-specimen with positioning of the LSP patterns according to previous studies of the authors [15-16].

A possible application of the LSP process was also demonstrated on a component like panel. The skin-stringers T-joints were realised through stationary shoulder friction stir welding (SSFSW), a variant of the conventional friction stir welding (FSW) process. In this variant of the standard FSW process, the shoulder does not rotate, reducing significantly the heat generated and the area affected by the process. This lead to a considerable improvement of the mechanical properties in both static and fatigue tests when compared with FSW. Additionally, a smooth surface is created at the weld crown that does not require any post processing. To produce the T-joint, a design solution as shown in Figure 2(a) was used. The photo and macrograph of the stationary shoulder friction stir welded AA2024-AA7050 T-joint are shown in Figure 2(b-c). For details regarding the joining process, the reader is referred to [11].

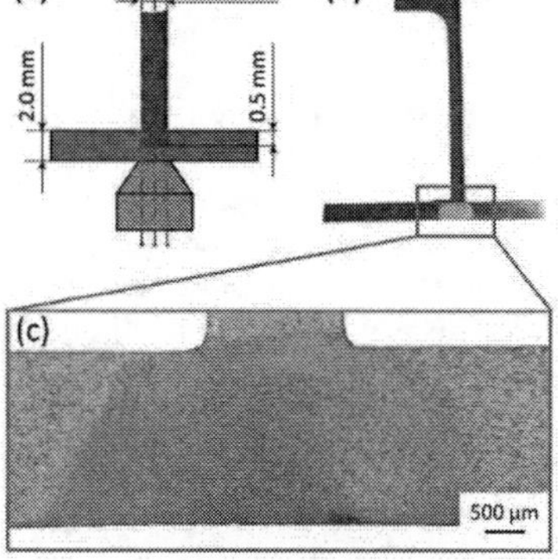

Fig. 2. (a) Sketch of the T-joint in two parts with partial penetration, (b) photo of the T-joint and (c) macrograph of the stationary shoulder friction stir welded skin-stringer T-joint configuration consisting of one skin sheet AA2024-T3 and one stringer AA7050-T7651 welded in one operation.

The LSP treatments on the stationary shoulder friction stir welded three-stringer panels were applied only on one side (stringer-side) using two different strategies (Figure 3 and Table 1).

Table 1. Specimen documentation.

Specimen identification	Number of LSP overlap	Distance of LSP patterns in front of the crack tip [mm]
BM	-	-
LSP 1.3J 1x1 1x	1	-1 (crack tip in LSP area, a_0 = 15 mm)
LSP 1.3J 1x1 2x	2	-1 (crack tip in LSP area, a_0 = 15 mm)
LSP 3J 1x1 1x	1	-1 (crack tip in LSP area, a_0 = 15 mm)
LSP 3J 3x3 3x	3	-2 (crack tip in LSP area, a_0 = 10 mm)
LSP 3J 3x3 5x	5	-2 (crack tip in LSP area, a_0 = 10 mm)
SSFSW	-	-
SSFSW-LSP1 (5J, 1x1 3x)	3	4 (a_0 = 3 mm)
SSFSW-LSP2 (5J, 3x3 3x)	3	4 (a_0 = 3 mm)

3 Results and discussion

3.1 Residual stress analysis

For the determination of depth resolved residual stresses, the hole drilling equipment "Prism" was used. "Prism" is equipped with an optical electronic speckle pattern interferometer (ESPI) system that provides high-quality full-field data for accurate residual stress determination. The technique is described in details in [12].

Figure 4 shows the residual stress measurement results obtained using the hole drilling technique. The measurements were performed at three positions in the middle of the LSP-area. Using the tool diameter of 2.0 mm, it was possible to obtain residual stress profiles up to half of the specimen depth (1 mm). BM specimen shows a non-uniform depth-resolved residual stress profile, where the residual stresses σ_{yy} are slightly higher than the residual stresses in x-direction (direction of crack propagation), σ_{xx}. Surface regions show tensile stresses that are compensated through compressive residual stresses in the core region of the specimen. This residual stress state can be caused through the production (rolling) process of the initial sheet material.

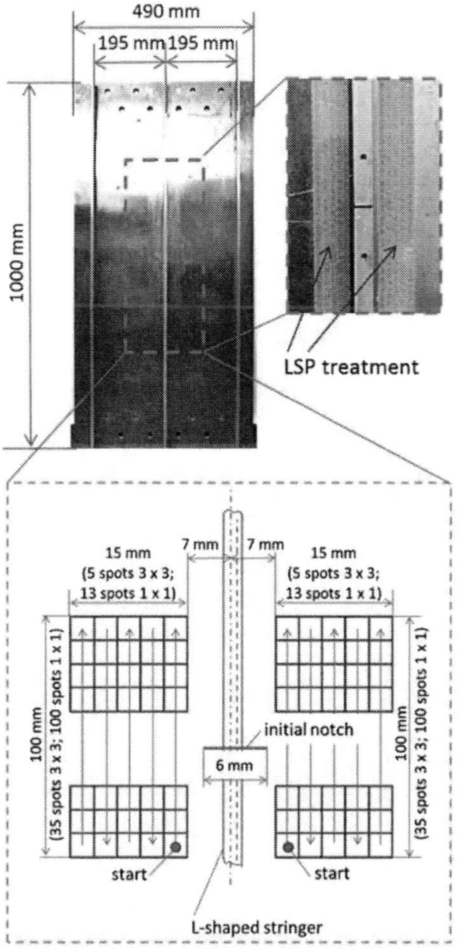

Fig. 3. Photo of the three-stringer-SSFSW-panel with two LSP-areas and sketch with the details of LSP-treatments.

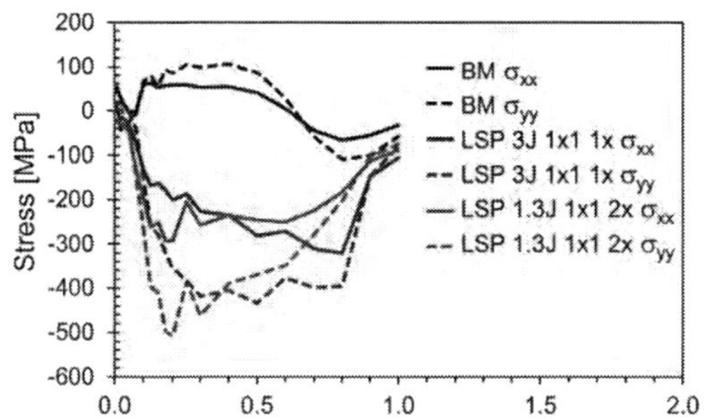

Fig. 4. Residual stress profiles measured with hole drilling method for C(T)50 BM specimen and two LSP-treated specimens. Average curves were calculated from three measurements. In case of LSP-treated specimens, the measurements were performed at three points in the middle of LSP-area.

The LSP-treated specimens show compressive residual stresses also in the core region of the specimens. The two LSP-treated specimens possess a comparable integral stress area due to the compressive residual stresses. The compressive residual stresses σ_{yy} are significant higher than the one in x-direction σ_{xx}. It should be mentioned, that the obtained maximum values of compressive residual stresses of approx. 500 MPa are significantly higher than the yield strength of the material investigated (370 MPa). As the used hole drilling technique overestimates residual stress values close or above the yield strength of the material, as investigated in [13], the real residual stress values should be lower. In case of equibiaxial residual stress profiles a correction methodology based on artificial neural network can be applied [14]. Nevertheless, the used hole drilling technique provides an useable qualitative result. It can be observed, that even in the centre of the specimens, compressive residual stresses are present. The results show, that by the use of the LSP treatment the generation of compressive residual stresses trough-the-thickness of thin sheet material is possible.

3.2. FCP test on C(T)-specimens

The specimens used for the FCP test are compact tension C(T)-specimens according to the standard ASTM E647-11 with a width W = 50 mm. All C(T)50-specimens were machined with the loading axis parallel to the rolling direction of the sheet material. Initial pre-cracks at a_0 = 10 mm and a_0 = 15 mm were introduced by electro discharge machining.

FCP tests were carried out using a servo-hydraulic machine with a capacity of 25 kN. Each specimen set was investigated under the same test conditions, i.e. the same cycle stress, applied load ratio $R_F = 0.1$, frequency of 10 Hz and room temperature. All specimens were polished in the expected area where the crack was supposed to grow to be able to observe the crack growth using an optical microscope. The crack growth was recorded automatically using a clip in the crack mouth and a second measurement consisting of periodic tracking of the locations of the crack tip using an optical microscope to magnify the crack. For both methods, the crack growth was recorded by the crack length data as a function of the number of cycles. An incremental second-order polynomial fitting method was used to filter the data in order to reduce the margin of error. A crack opening displacement (COD) clip was mounted on the front side of each C(T)50-specimen before the FCP tests. During the FCP tests, the applied load vs. COD curves were obtained.

In Figure 5 the FCP test results are presented. The LSP-treated specimens show a significant higher number of cycles required for propagating the fatigue crack up to the a/W-value of 0.6 (crack length, a divided by the C(T)50-specimen width from the pin holes, W = 50 mm). Figure 5(a) depicts the results of the FCP tests performed for LSP-treated specimens using 1 mm x 1 mm optic. The fatigue life of the LSP-treated specimens LSP 1.3J 1x1 1x increased by a factor of nearly two compared to the BM specimens. Further increase in the fatigue life by a factor of at least nine compared to the BM specimens is achieved for the LSP-treated specimens LSP 1.3J 1x1 2x and LSP 3J 1x1 1x. A comparable fatigue life increase can be achieved either by increasing the laser energy from 1.3 J to 3 J or by increasing the numbers of overlapping LSP treatments from 1x to 2x keeping the laser energy value constant at 1.3 J. The comparable fatigue life increase for the two LSP treatments can be explained in comparable integral stress areas of residual stress profiles for the two LSP treatments (Figure 4). The scatter in the fatigue life is higher for the specimens LSP 3J 1x1 1x which might be related to the scatter in the residual stresses induced through the LSP-treatment. The residual stress scatter-fatigue life scatter relationship is investigated in details in [15].

Figure 5(b) depicts the FCP test results performed for the LSP-treated specimens using 3 mm x 3 mm optic. The fatigue life for the LSP-treated specimen with 3x overlapping increased by a factor of nearly three compared to the BM specimen. Further fatigue life increase up to a factor of nearly four was obtained for the LSP-treated specimen with 5x overlapping.

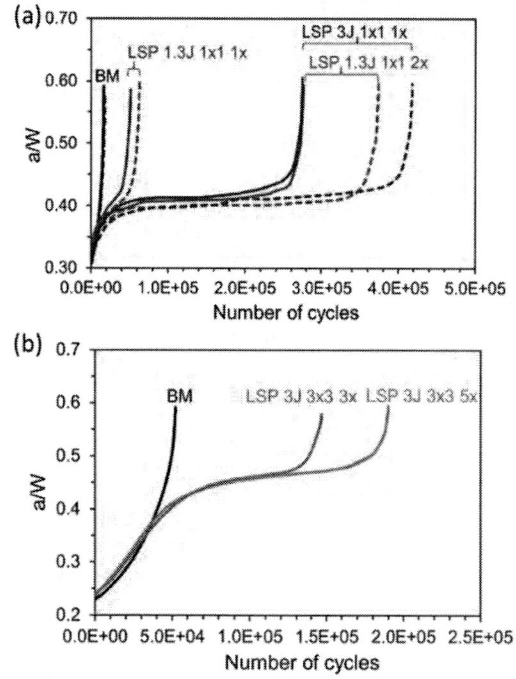

Fig. 5. FCP-test results for (a) 1 mm x 1 mm optic and (b) 3 mm x 3 mm.

The results of the FCP tests confirmed that significant retardation of fatigue crack propagation can be achieved for the LSP treatments performed using both diffractive optics. The positive effect of LSP can be increased by either increasing the laser energy or increasing the numbers of overlapping LSP treatments.

3.3. Consideration of LSP-induced residual stresses in the evaluation of FCP behaviour

To understand the retardation of the FCP in the LSP-treated specimens the applied load vs. crack opening displacement (COD) curves obtained at different crack lengths (a/W-values) were analysed. Throughout the analysed a/W-region from the beginning of the FCP test (a/W = 0.2) to the end of the FCP test (a/W = 0.6), the curves in case of the BM are linear. The curves do not show any noticeable inflection points at the beginning of the test at a/W = 0.2 mm and close to the end of the test at a/W = 0.56 (Figure 6(a)). In contrast to the BM specimen, the LSP-treated specimens show inflection points in the applied load vs. COD curves (Figure 6(b)). The curves obtained at different crack lengths show different curvatures for the first parts of the curves up to the inflection points. At the beginning of the FCP test, the applied vs. COD curves of the LSP-treated specimen shows a negative curvature (syncline) up to the inflection points. The gradients of the second parts of the curves are comparable to the gradients of the BM curves obtained at the same a/W-values. The lower gradients of the first parts of the curves indicate that tensile residual stresses were present. Tensile residual stresses increase the effective maximum load.

The curves for LSP-treated specimen obtained at a/W-values in the range between 0.47 and 0.59 show the positive curvature (anticline) in their first parts. The higher gradient of the first part of the curves indicate, that compressive residual stress were present and caused the crack closure that also increased the level of opening load, F_{op}. Taking into account the effective maximum load at the beginning of the FCP test (region with tensile residual stresses) and the increased values of the opening load in the region with compressive residual stresses, the effective load ratios, $R_{F\,eff}$ can be calculated. The proposed methodology for incorporation of tensile and compressive residual stresses in the calculating of effective load ratios is described in details in [16].

The applied (for BM specimen) and effective load ratios for two LSP-treated specimens are presented in Figure 7(a). The crack closure from the compressive residual stress for the LSP-treated specimen increased the level of opening load and therefore reduced the effective load amplitude. In the tensile residual stress regions (at the beginning of the FCP test), the effective load ratio was insignificantly reduced. The R_F-ratio for the BM specimen remained nearly constant during the FCP test. A similar retardation mechanism has been described also for FCP in laser heating induced residual stresses in [17], with the difference, that there tensile residual stresses are induced in the treated area, surrounded by compressive residual stresses.

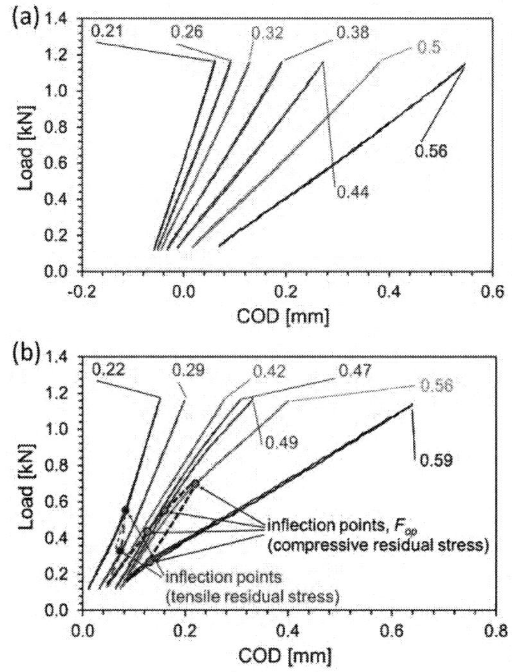

Fig. 6. Load vs. COD curves obtained at different a/W-values for (a) BM specimen (b) LSP-treated specimen (LSP 3J 3x3 5x).

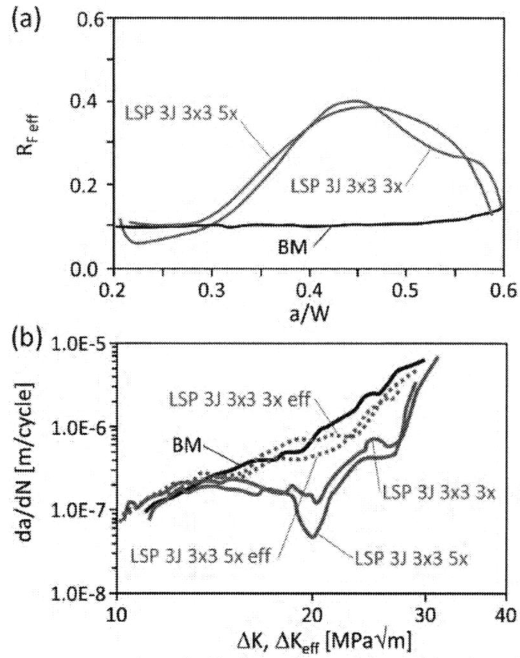

Fig. 7. Correction of effective values under consideration of the LSP-induced residual stresses. (a) $R_{F\ eff}$ vs. a/W and (b) da/dN vs. ΔK and ΔK_{eff}.

The FCP rates vs. applied stress intensity factor range, ΔK, are depicted in Figure 7(b). The BM specimen shows a typical linear behaviour in the logarithmic coordinates. The calculated ΔK values for the LSP treated specimens, considering the applied load levels, show minimum da/dN values at a ΔK value of approximately 20 MPa√m. By considering the effect of tensile and compressive residual stresses on the effective load range for the calculation of the stress intensity factor range, i.e., the so-called effective stress intensity factor range, ΔK_{eff} curves for the LSP-treated specimens are close to the curve of the BM specimen. It is evident that the applied corrections for calculating ΔK_{eff} are reasonable and consider the effects of the tensile/compressive residual stresses induced by the LSP treatment.

3.4 FCP test on M(T)-specimens

Constant amplitude FCP tests were carried out at load ratio, R_F of 0.1, using a servo-hydraulic machine at a frequency f ≤ 5 Hz in compliance with the standard ASTM E 647. It should be mentioned that the ASTM standard assumes homogeneous material properties; a condition, which may not necessarily be fulfilled for friction stir welded structures.

The stringer specimens were notched using an electro-discharge technique. A notch of $2a_0$ = 6 mm was placed in the centre of the panel (the stringer in the middle of the panel was cut, Figure 3). The crack length was measured optically using a travelling microscope.

The results of the FCP test for three-stringer panels are shown in Figure 8. A considerable effect of LSP-treatment in the FCP behaviour was observed. In case of the stationary shoulder friction stir welded specimen with two LSP-treated areas using the LSP process parameter set 2 (5J 3x3 3x, Table 1) the fatigue life is increased by a factor of more than two compared to the stationary shoulder friction stir welded specimen without LSP treatment. The LSP process parameter set 1 (5J 1x1 3x, Table 1) led to the fatigue life increase by a factor of more than four. The obtained results indicate that LSP is an effective process for extension of the fatigue life of aircraft structural components. It can be successfully implemented on thin-walled airframe structures to increase their resistance against fatigue crack propagation.

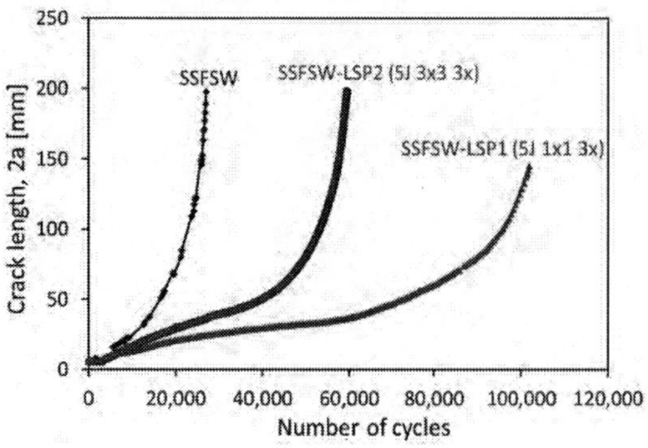

Fig. 8. Fatigue crack propagation test results of stationary shoulder friction stir welded 3-stringer panels with LSP treatments.

4 Conclusions

Based on the obtained results the following conclusions can be drawn:

1. LSP is an effective process for introducing high and deep compressive residual stresses in thin AA2024 sheets.
2. A significant retardation of the FCP through the presence of compressive residual stresses induced by the LSP treatment was observed.
3. A crack closure effect in the FCP tests was observed in specimens with the LSP treatment, which is caused by the presence of high compressive residual stresses.
4. The efficiency of the LSP process has been demonstrated on sub-component level resulting in an increase of 200 – 400% in fatigue lifetime.

Overall, LSP can be considered as a potential tool for the surface treatment of Al-structures in which a non-destructive inspection can detect cracks. The effects of the LSP treatment are the arrest of cracks and deceleration of the FCP.

This work was carried out within the frame of the LISA project, which was supported by the Light-weight materials Assessment, Computing and Engineering Centre (ACE) of the division Materials Mechanics of the Helmholtz-Zentrum Geeesthacht. The authors also thank the following project members for their valuable work and support: H. Tek and K. Erdmann (mechanical testing), L. Moura (using the hole drilling system) and F. Dorn (specimen preparation).

References

1. O. Hatamleh, M. Hill, S. Forth, D. Garcia, Mater. Sci. Eng. A **519**, 61 (2009)
2. S. Keller, S. Chupakhin, P. Staron, E. Maawad, N. Kashaev, B. Klusemann, J. Mater. Proc. Tech. **255**, 294 (2018)
3. L. Berthe, R. Fabbro, P. Peyre, E. Bartnicki, Eur. Phys. J. Appl. Phys. **3**, 215 (1998)
4. L. Berthe, R. Fabbro, P. Peyre, E. Bartnicki, J. Appl. Phys. **85**, 7552 (1999)
5. C. Correa, L. Ruiz de Lara, M. Diaz, J. A. Porro, A. Garcia-Beltran, J. L. Ocaña, Int. J. Fatigue **70**, 196 (2015)
6. Y. Tan, G. Wu, J. M. Yang, T. Pan, Fatigue Fract. Eng. Mater. Struct. **27(8)**, 649 (2004)
7. O. Hatamleh, J. Lyons, R. Forman, Int. J. Fatigue **29(3)**, 421 (2007)
8. O. Hatamleh, Int. J. Fatigue **31(5)**, 974 (2009)
9. E. Hombergsmeier, V. Holzinger, U. C. Heckenberger, Adv. Mater. Res. **891-892**, 986 (2014)
10. Z. Bergant, U. Trdan, J. Grum, Int. J. Fatigue **87**, 444 (2016)
11. A. Barbini, J. Carstensen, J. F. dos Santos, J. Mater. Sci. Tech. **34(1)**, 119 (2018)
12. M. Steinzig, E. Ponslet, Exp. Tech. **27(3)**, 43 (2003)
13. S. Chupakhin, N. Kashaev, N. Huber, J. Strain Anal. Eng. Des. **51(8)**, 572 (2016)
14. S. Chupakhin, N. Kashaev, B. Klusemann, N. Huber, J. Strain Anal. Eng. Des. **52(3)**, 137 (2017)
15. S. Chupakhin, B. Klusemann, N. Huber, N. Kashaev Int. J. Adv. Manuf. Tech., submitted
16. N. Kashaev, V. Ventzke, M. Horstmann, S. Chupakhin, S. Riekehr, R. Falck, E. Maawad, N. Schell, N. Huber, Int. J. Fatigue **98**, 223 (2017)
17. D. Schnubel, N. Huber N, Eng. Fract. Mech. **84**, 15 (2012)

Effect of Sandblasting on Low and High-Cycle Fatigue Behaviour after Mechanical Cutting of a Twinning-Induced Plasticity Steel

Antoni Lara[1,*], *Mercè* Roca[1], *Sergi* Parareda[1], *Núria* Cuadrado[1], *Jessica* Calvo[2] and *Daniel* Casellas[1,3]

[1]Fundació CTM Centre Tecnològic, Metallic and Ceramic Materials Department, 08243 Manresa, Spain
[2]Universitat Politècnica de Catalunya, Materials Science and Metallurgy Department, 08019 Barcelona, Spain
[3] Luleå University of Technology, Division of Mechanics of Solid Materials, 971 87 Luleå, Sweden

Abstract. In the last years, car bodies are increasingly made with new advanced high-strength steels, for both lightweighting and safety purposes. Among these new steels, high-manganese or TWIP steels exhibit a promising combination of strength and toughness, arising from the austenitic structure, strengthened by C, and from the twinning induced plasticity effect. Mechanical cutting such as punching or shearing is widely used for the manufacturing of car body components. This method is known to bring about a very clear plastic deformation and therefore causes a significant increase of mechanical stress and micro-hardness in the zone adjacent to the cut edge. To improve the cut edge quality, surface treatments, such as sandblasting, are often used. This surface treatment generates a compressive residual stress layer in the subsurface region. The monotonic tensile properties and deformation mechanisms of these steels have been extensively studied, as well as the effect of grain size and distribution and chemical composition on fatigue behaviour; however, there is not so much documentation about the fatigue performance of these steels cut using different strategies. Thus, the aim of this work is to analyse the fatigue behaviour of a TWIP steel after mechanical cutting with and without sandblasting in Low and High-Cycle Fatigue regimes. The fatigue behaviour has been determined at room temperature with tensile samples tested with a load ratio of 0.1 and load amplitude control to analyse High-Cycle Fatigue behaviour; and a load ratio of -1 and strain amplitude control to determine the Low-Cycle Fatigue behaviour. Samples were cut by shearing with a clearance value of 5%. Afterwards, a part of the cut specimens were manually blasted using glass microspheres of 40 to 95 microns of diameter as abrasive media. The results show a beneficial effect of the sandblasting process in fatigue behaviour in both regimes, load amplitude control (HCF) and strain amplitude control (LCF) tests, when these magnitudes are low, while no significant differences are observed with higher amplitudes.

Abstract. low-cycle fatigue, high-cycle fatigue, mechanical cutting, sandblasting, high manganese steel, TWIP steel

1 Introduction

Nowadays, lightweight construction is one of the major concerns in automotive industry to meet the increasing demands towards environment and safety related issues. A large part of body-in-white (BiW) components is currently manufactured using ultra high-strength steels (UHSS), with fracture strengths from 1000MPa upwards.

The cold forming of pieces of complex geometry is not an easy task with this type of steels, thus, there has been a growing interest for High Mn Steels (HMnS). The main characteristic of these steels is high Ultimate Tensile Strength (UTS) values, around 1000MPa, combined with high ductility, above 50%, and high strain hardening coefficient. This combination of properties is the result of the activation of specific deformation mechanisms such as TRIP and/or TWIP effects.

Currently, the industrialization of the production of HMnS and the reliability of their in-service properties have led to the first applications of HMnS for automotive components. Despite this, there are still aspects related to the performance of HMnS produced parts that need further evaluation in order to extend the implementation of these steels for industrial applications. In particular, there are still some concerns regarding the fatigue performance of components' cutting zones and the methodologies to improve it.

Regarding fatigue, failures are mainly triggered by pre-existing defects at the surface or inside the material. Internal defects are usually associated to non-metallic inclusions generated during steel manufacturing. Surface defects can be produced during cold rolling or part processing. Among the defects produced during forming, trimming produces surface irregularities at the cut edge, which may act as preferential sites for fatigue crack initiation and propagation. Several authors have assessed

* Corresponding author: toni.lara@ctm.com.es

the effect of cutting technologies on fatigue properties for different sheet steel grades: carbon and micro-alloyed steels [1]; UHSS with UTS up to 1400MPa [2] and deep drawing, high strength dual phase, austenitic and duplex stainless steels [3]. In all cases, it is concluded that the cut edge makes the fatigue strength of the metal decrease because of the roughness that the cutting process generates. It is also observed that the magnitude of the clearance was of minor importance, being more critical the tool wear.

If the effect of the different cutting methods is analyzed, the specimens with milled edges showed higher fatigue limit than mechanically cut edges. These authors also concluded that there were no significant differences between milled and water jet cutting, while the lowest fatigue behaviour was observed in laser cut specimens. Thus, the main conclusion of this and other works [4, 5] is that lower strength steels were less sensitive to the cutting method than the higher strength steels and that in most cases, shearing produced lower fatigue strengths than other cutting methods due to the presence of larger surface defects at the edges.

Regarding TWIP steels, there is relatively little work reported on their fatigue behaviour. High cycle fatigue [6, 7], as well as low cycle and extremely low cycle fatigue life [8], have been evaluated in some works. In general, the fatigue limit of TWIP steels correlates to their high ultimate tensile strength, with a ratio of fatigue limit to tensile strength around 0.4, similar to the one of stainless steels, but far from that obtained with other UHSS steels such as Dual or Complex Phase steels, around 0.5 or even higher [4, 9].

There is not much information on the effect of the cut edge in the fatigue life of these steels. Only Mateo et al. [10] evaluated the effect of the laser cutting on fatigue behaviour of two metastable steels, one of which was a TWIP steel, applying the staircase method. According to these results, TWIP steel is less sensitive than stainless steel to defects at the cut edge. However, no studies have been found on how the mechanical cutting affects the fatigue behaviour of these type of steels.

On the other hand, there are some works that report the negative effect of mechanical cut on forming and post-forming operations, such as hole expansion, due to the high amount of deformation introduced to the hole edge [11]. Thus, it is expected that this high degree of deformation introduced at the cutting edge could negatively affect fatigue behaviour.

In this sense, it has been demonstrated that some surface treatments can improve fatigue behavior through the retardation of fatigue crack initiation or propagation. These treatments, such as sandblasting or shot peening, are effective methods to enhance the fatigue properties, by inducing a compressive residual stress in the subsurface of the material [12, 13]. In sandblasting, the sample surface is blasted repeatedly by high-speed sand particles, leading to reduce the surface roughness, removal of surface scale and generation of local plastic deformation in the surface layer. In addition, a compressive residual stress layer is formed in the subsurface region. Microstructural changes due to sandblasting have also been reported in austenitic steels [14].

The aim of the present work is to increase the knowledge about the effect of the mechanical cutting on the fatigue performance of High Mn Steels (HMnS) and, propose industrial treatments to improve their fatigue behaviour. To achieve these objectives, after mechanical cutting, a part of the cut specimens have been manually blasted using glass microspheres of 40 to 95 microns of diameter as abrasive media. Materials with polished cut edges have been used as a reference in this study.

2 Materials

Two grades of TWIP steel were analysed: TWIP1.5 and TWIP3.0. The chemical compositions of these steels are shown in table 1 and table 2.

Table 1. Chemical composition of TWIP1.5 steel.

Element	C	Mn	Al	Si
Weight %	0,35	19,9	1,0	0,2
	P	N	S	Fe
	0,015	0,0126	0,0003	Balance

Table 2. Chemical composition of TWIP3.0 steel.

Element	C	Mn	Al	Si
Weight %	0,38	18,3	1,2	0,29
	P	N	S	Fe
	0,018	0,0104	0,0005	Balance

Then, samples of these steels were polished and etched following standard methods. Their microstructures can be seen in figure 1 and figure 2.

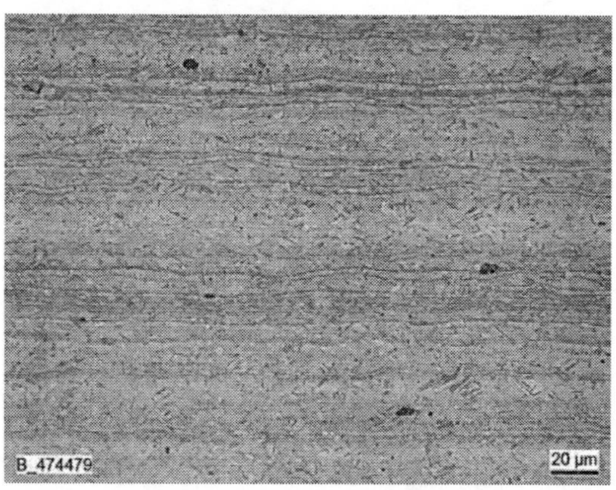

Fig. 1. Microstructure of TWIP1.5 at the centre (Nital etched).

Fig. 2. Microstructure of TWIP3 at the centre (Wallner etched).

As seen in the previous figures the two TWIP steels shows a homogeneous austenite matrix. The grain size is the only notable difference.

Conventional axial tensile tests were performed according to EN-ISO6892-1 with the specimens oriented transverse to the rolling direction. Table 3 shows the results.

Table 3. Mechanical properties of analysed TWIP steels.

Steel	YS [MPa]	UTS [MPa]	Elongation [%]
TWIP1.5	564	947	48.3
TWIP3.0	369	802	56.5

Table 3 shows, as would be expected, a higher Ultimate Tension Strength (UTS) and Yield Strength (YS) with a decreasing of grain size; whereas the uniform elongation and total elongation become smaller.

3 Experimental procedure

In order to evaluate the influence of the cutting process on the fatigue behaviour of these steels, the calibrated zone of specimens was cut using different strategies:

- Spark erosion (EDM) and polished, with a first grinding and a polishing with two types of abrasive pastes to get clean and shiny surfaces, identified as [REF] condition.
- Shearing using a cutting clearance of 5% of the sheet thickness and a punch radius of 30 microns, identified as [CUT] condition
- Shearing as previously commented and manually blasted using glass microspheres, identified as [SAN] condition.

Figure 3 shows the tool employed for cutting a part of the specimens tested.

The geometry of the edges was evaluated by cutting a specimen of each strategy in cross section and analysing them by optical microscopy, figure 4 and figure 5.

Vickers micro-hardness profiles were also performed to assess the effect of each cutting methodology and are shown in figure 6.

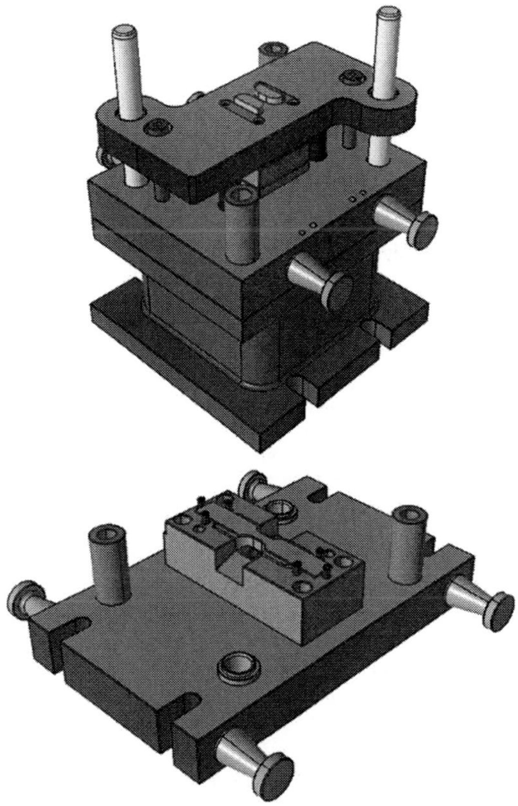

Fig. 3. Tool employed for cutting fatigue specimens.

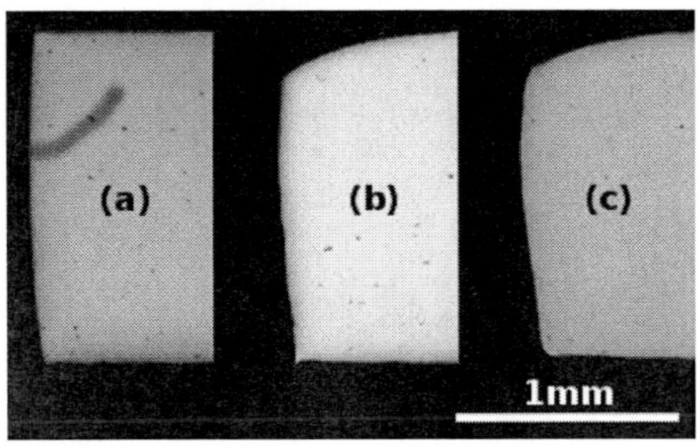

Fig. 4. TWIP1.5 cut edges: (a) REF, (b) CUT and (c) SAN.

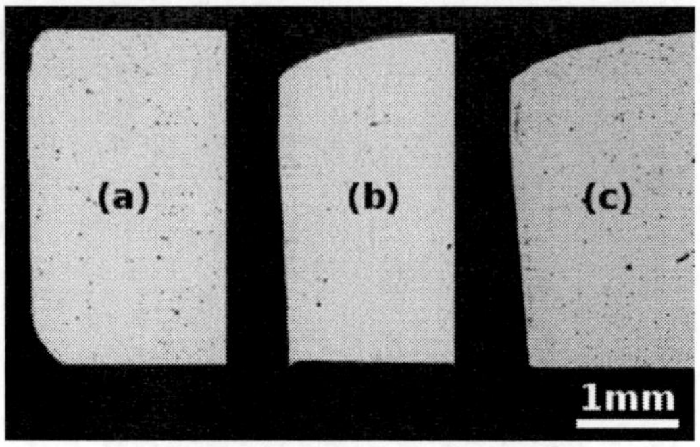

Fig. 5. TWIP3.0 cut edges: (a) REF, (b) CUT and (c) SAN.

Residual stresses were also evaluated on the surface of the samples by means of instrumented indentation using the methodology developed by Wang and Bao [15, 16]. In order to perform indentation tests, a Berkovich indenter was used with an applied penetration depth of 10 µm. Results of residual stresses, shown in figure 7, correspond to an average value from at least 20 indentations.

Regarding fatigue behaviour, tests were conducted in dual column servo-hydraulic testing machine. Low Cycle Fatigue (LCF) tests, to obtain εN curves, were performed according to ISO12106 standard [17] using an axial contact extensometer. Total strain versus number of cycles curves were then used to determine the Basquin and Coffin-Manson relationships. High Cycle Fatigue (HCF) tests, to obtain SN curves up to $2x10^6$ cycles, were obtained according to ISO1099 standard [18], using a load ratio (R) of 0.1. All tests were performed at room temperature. After these tests, some fracture surfaces were inspected by means of field emission scanning electron microscopy (FE-SEM).

4 Results

Sheet cut edges present different geometry depending on the cutting methodology employed. As shown in figure 4 and 5, polished samples present rounded edges without asperities, as expected (figure 4(a) and 5(a)). Sheared samples show the typical features of a sheared edge with rollover, smooth, fracture zones and burr in two materials (figure 4(b) and 5(b)). Once samples are submitted to the sandblasting operation, the roughness of the cutting edge is reduced and the burr removed.

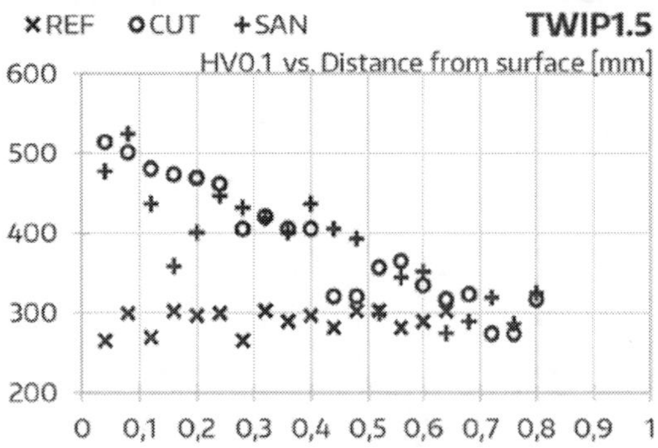

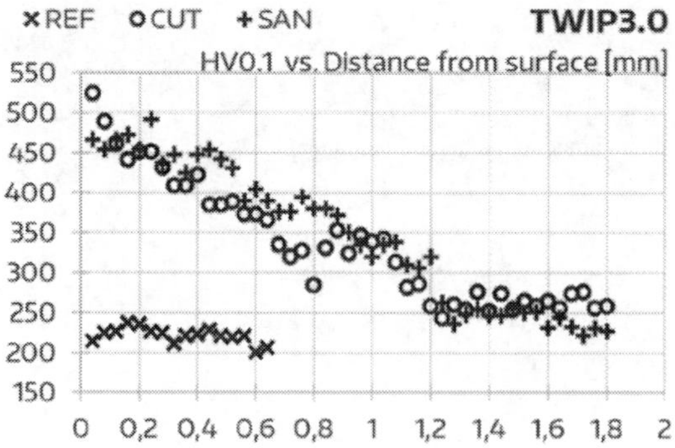

Fig. 6. Micro hardness profiles at the near cut edge.

Micro-hardness profiles shown at figure 6 show that the shearing process increases hardness near the cut edge, an increase that is not observed in polished specimens, for which the cut were done by EDM. The length of the cut-hardened area is deeper in the thicker steel, while no clear differences of hardness are observed between the sheared samples with and without sandblasting treatment.

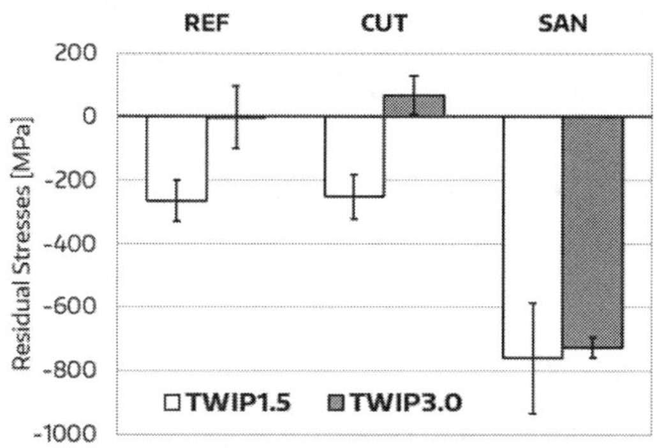

Fig. 7. Residual stresses obtained on the TWIP steels surface.

As can be seen in figure 7, sandblasting process induces high compressive stresses measurable even at penetration depth of 10 microns with instrumented indentation. The magnitude of these compressive stresses does not differ much from those reported by other authors that performed shot penning process with different TWIP steels, although the depth of the compressive layer was higher [19], 100 to 300 microns for a shot peening versus 25 to 75 microns for a sandblasting [20].

The main strain and stress–life fatigue results are summarized in figure 8 and 9, defined here as the strain and stress amplitude leading to 50% survival probability (experimental values).

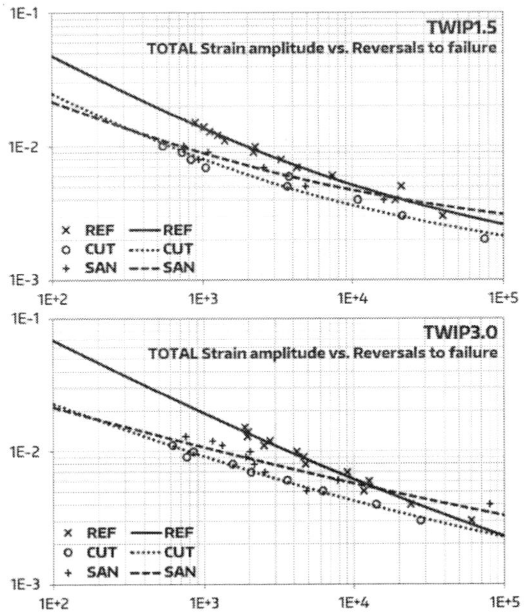

Fig. 8. εN curves obtained with TWIP steels

As shown in figure 8 and 9, fatigue performance depends on the cutting strategy employed. At low and high number of cycles (LCF and HCF tests), both strain and stress that can be applied for a given number of cycles, is higher for polished specimens compared with the sheared ones.

Concerning sandblasting specimens, the effect of this surface treatment on the fatigue performance depends on the applied stress level. Thus, the sandblasting process significantly increase the strain and stress-life behaviour when the stress level is low (low strain in LCF tests and low stress in HCF tests). On the contrary, at higher stress levels (high strain in LCF tests and high stress in HCF tests) sandblasting treatment does not introduce any noticeable improvement in the fatigue behaviour of this type of steels, since the fatigue results of sheared specimens are similar with and without sandblasting, at high stress levels.

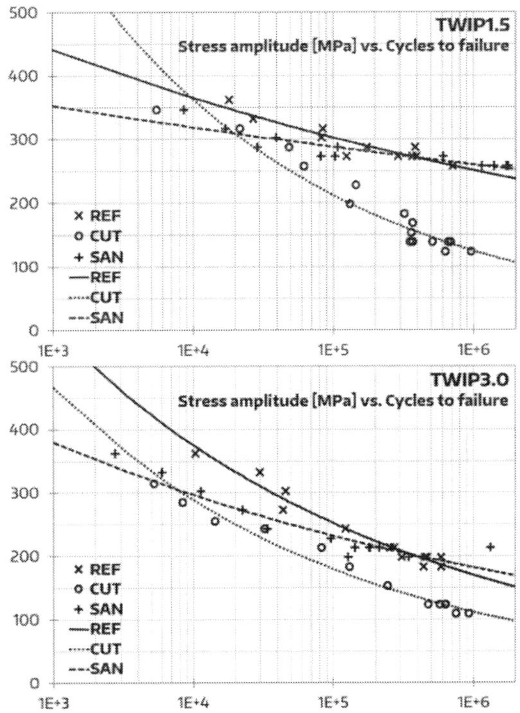

Fig. 9. SN curves obtained with TWIP steels.

These results may be explained as follows: sandblasting process to sheared specimens generates a compressive residual stress layer which is expected to influence the fatigue life. At low cyclic stresses, the favourable effects of the compressive residual stress layer results in life enhancement. As the cyclic stress increases, the residual stresses will relax quickly and the extent of life enhancement will decrease, as some authors support [21].

5 Discussion

Fatigue properties are extremely sensitive to the presence of pre-existent surface defects in the material, such as non-metallic inclusions, irregularities, asperities or even cracks originated during cutting or forming processes, because the crack nucleation step is then shortened, or even removed.

It is well accepted that materials with high ductility, or toughness, as TWIP steels, have a high crack propagation resistance and the fatigue life is mainly spent in the crack growth regime. Thus, tough materials are usually referred to as high defect tolerant since the crack nucleation step does not control the fatigue life. On the contrary, high strength materials have a low crack propagation resistance because of their limited toughness. In that case, crack nucleation may take an important part of the total fatigue life, because the fatigue crack growth rate is relatively high. Thus, high strength materials are more sensitive to pre-existent defects, and they are usually referred to as low defect tolerant materials.

Thus, to rationalize the origin of failure in each case, a detailed SEM investigation of one specimen of each cutting strategy was done. These SEM fractographies are shown in figures from 10 to 12 for TWIP1.5 and 13 to 15 for TWIP3.0.

Fig. 10. Fractography of failed TWIP1.5 polished sample, showing the fatigue initiation site.

The fatigue failure surfaces in most cases show a failure origin independent of the cutting strategy, with no clear differences between the polished, sheared and sandblasted specimens. In most of the surface specimens analysed by SEM, the failure origin occurs in non-metallic inclusions located near the edge of the specimens.

Thus, as discussed above, the important differences in fatigue performance observed are not conditioned by the nucleation stage, but by the propagation stage.

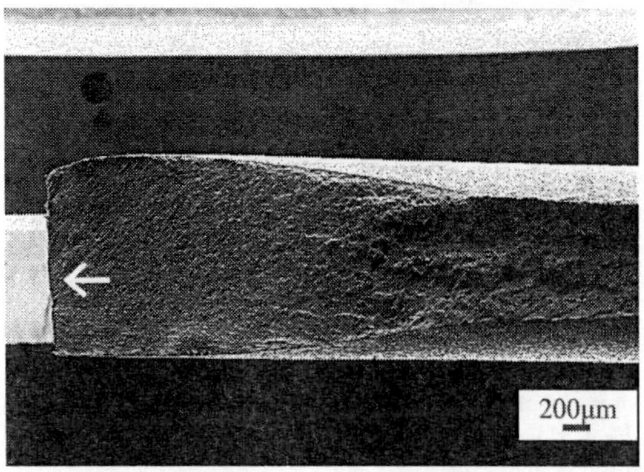

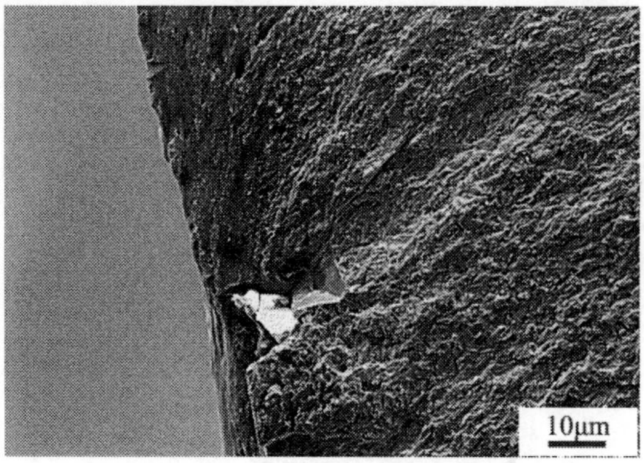

Fig. 11. Fractography of failed TWIP1.5 sheared sample, showing the fatigue initiation site.

Fig. 12. Fractography of failed TWIP1.5 sheared and sandblasted sample, showing the fatigue initiation site.

Fig. 13. Fractography of failed TWIP3.0 polished sample, showing the fatigue initiation site.

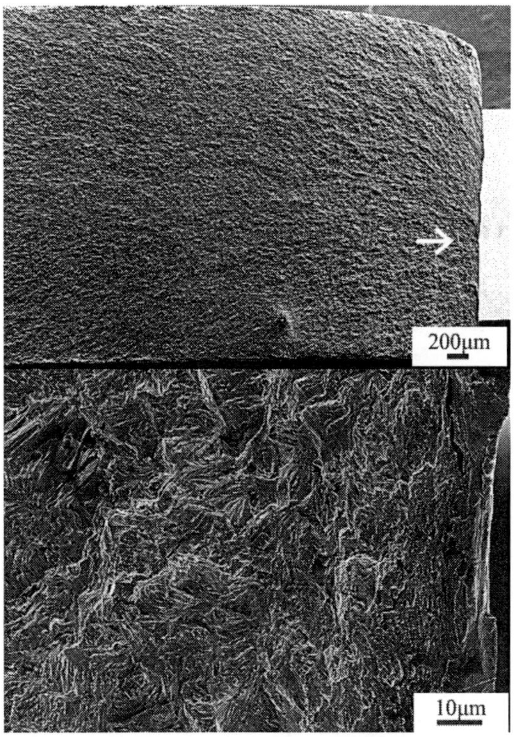

Fig. 14. Fractography of failed TWIP3.0 sheared sample, showing the fatigue initiation site.

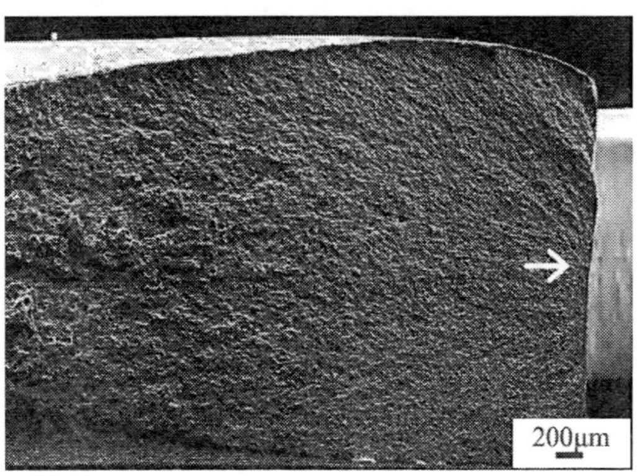

Fig. 15. Fractography of failed TWIP3.0 sheared and sandblasted sample, showing the fatigue initiation site.

These non-metallic inclusions, responsible for the crack nucleation, are probably the ones that cause that the fatigue performance of these two steels is not as high as expected. To know the nature of these non-metallic precipitates, an EDX analysis was performed, showing that most of them were aluminium nitrides.

Regarding fatigue performance between the different cutting strategies, and considering, as already advanced, that these are motivated by different crack propagation rates in each strategy, it seems clear that concerning the specimens sheared and with a sandblasting process after shearing, this different crack propagation ratio is due to the compressive stresses induced by sandblasting, as is shown in figure 7. These surface compressive stresses require greater tensile stresses to propagate the crack, which shifts the strain and stress-life curves to the right. As said before, this improvement is only observed at low strain or stress levels, since if cyclic stress increases, the residual stresses will relax quickly and the fatigue behaviour enhancement will not take place.

Differences between polished and sheared samples, and also with sandblasting specimens submitted at high strains or stresses, cannot be explained by differences in residual stresses, since, as observed with nano-indentation in figure 7, they are practically the same. In this case, this change in crack propagation rate has to be motivated by the microstructural change generated by the mechanical cut in the tool. Thus, in the specimens cut by spark erosion and then polished [REF], no deformation was introduced during cutting process, therefore the original microstructure, tougher, more tolerant to the defects and with a high crack propagation resistance is acting. In contrast, there are many works which support that during specimens shearing a high amount of deformation is introduced on the cut edge [11]. Thus, the microstructure in this region does not tolerate much more deformation and behaves like a material with a limited toughness and with a low crack propagation resistance.

6 Conclusions

Based on the experimental fatigue tests, surface analysis and fractographic observations, the following conclusions can be drawn:

- The fatigue behaviour of the TWIP steels analysed in this work is mainly governed by aluminium nitrides non-metallic inclusions located near the surface. Probably this phenomenon is the reason why fatigue performance of these two steels is not as high as would be expected.

- The presence of these defects near the surface makes the crack propagation step determinant in fatigue behaviour of these materials. Thus, to enhance fatigue performance is necessary to reduce the crack propagation rate.

- As seen above, a strategy to reduce the crack propagation rate is to use a cutting methodology which does not introduce a high amount of deformation on the microstructure, making the steel less defect tolerant because of a toughness reduction.

- Another strategy to reduce the crack propagation rate is the introduction of compressive residual stresses, which requires an increase of the applied tensile stresses to propagate the crack. Sandblasting process has proven to be efficient for this.

Acknowledgments

Part of the research leading to these results has received funding from the European Commission, Research Fund for Coal and Steel, under the grant agreement RFSR-CT-2015-00018 (iCUT project). Authors want to thank all consortium members for their collaboration and expertise.

References

[1] E. Maronne, A. Galtier, J.I. Robert, T. Ishikawa, Trans. Eng. Sci. 40 (2003)

[2] A. Björkblad, O. Skrinjar, Swerea KIMAB LättUHS Report 13 (2009)

[3] F. Meurling, A. Melander, J. Linder, M. Larsson, Scand. J. Metall. 30 (2001)

[4] A. Lara, I. Picas, D. Casellas, J. Mater. Process. Technol. 213 (2013)

[5] K. Mäntyjärvi, A. Väisänen, J.A. Karjalainen, Int. J. Mater. Form. 2 (2009)

[6] A.S. Hamada, A. Järvenpää, M. Honkanen, M. Jaskari, D.A. Porter, L.P. Karjalainen, J. Mater. Sci. Eng. A 517 (2009)

[7] J.J. Roa, G. Fargas, J. Calvo, E. Jiménez-Piqué, A. Mateo, Mater. Sci. Eng. A 628 (2015)

[8] C.W. Shao, P. Zhang, R. Liu, Z.J. Zhang, J.C. Pang, Z.F. Zhang, Acta Mater. 103 (2016)

[9] P. Matteis, G. Scavino, F. D'Aiuto, D. Firrao, Steel Research Int. 83 (2012)

[10] A. Mateo, G. Fargas, J. Calvo, J.J. Roa, Mater. Testing 57 (2015) 1-5.

[11] A. Lara, D. Frómeta, S. Molas, J. Rehrl, C. Suppan, D. Casellas, IDDRG 2016 Int. Conf. (2016)

[12] X.P. Jiang, X.Y. Wang, J.X. Li, D.Y. Li, C.-S. Man, M.J. Shepard, T. Zhai, Mat. Sci. Eng. A 429 (2006)

[13] X. You, Z. Wang, Q. Wang, M. Cao, M. Shen, W. Huang, Metals 7 (2017)

[14] X.Y. Wang, D.Y. Li, Wear 255 (2003)

[15] Q. Wang, K. Ozaki, H. Ishikawa, S. Nakano, H. Ogiso, Nucl. Instrum. Methods Phys. Res. B 242 (2006)

[16] Y. Bao, L. Liu, Y. Zhaou, Acta Mater. 53 (2005)

[17] ISO 12106:2017, Metallic materials -- Fatigue testing -- Axial-strain-controlled method

[18] ISO 1099:2017, Metallic materials -- Fatigue testing -- Axial force-controlled method

[19] C. Teichmann, L. Wagner, Shot Peener Mag. 28 (2014)

[20] A. Fatemi, 28th Forging Ind. Tech. Conf. (2011)

[21] N.K.R. Naidu, S.G.S. Raman, Int. J. Fatigue 27 (2005)

Cyclical fatigue of annealed and of thermally tempered soda-lime-silica glass

Jens Schneider[1] and *Jonas* Hilcken[1]

[1] ISM+D Institute of Structural Mechanics and Design, Glass Competence Center, TU Darmstadt, Germany

Abstract. We present experimental and theoretical investigations on the cyclic fatigue of annealed and of thermally tempered soda-lime-silica glass. Static fatigue due to subcritical crack growth at micro cracks significantly decreases the macroscopic strength of soda-lime-silica glass and causes a time-dependent strength reduction. A subsequent thermal tempering process is typically used to induce residual surface compression stresses, which inhibit the crack growth of surface cracks, and corresponding bulk tension stresses. From the experimental results we show that the existing models for static fatigue used in linear elastic fracture mechanics can be used for the lifetime prediction of cyclically loaded annealed glass and thermally tempered glass, although the (static) crack growth exponent slightly decreases in cyclic loading. The equivalent duration of tensile stress at the crack tip of a micro crack governs the crack growths and not the number of cycles. The threshold for subcritical crack growth determined from the cyclic experiments was found to be in good agreement with data from literature. But unlike in strength tests with singular and quasi-static re-loading, it could be found that periodic loading with load free intervals does not lead to a strength increase by crack healing effects. Based on the results, an engineering design concept for cyclically loaded glass is presented.

1 Introduction

1.1 Motivation

Flat glass, whose dimensions can reach up to 3.25 m x 20 m today, is increasingly taking over load-bearing functions in structures. For their design and dimensioning, verification methods are currently being developed to develop an *Eurocode Structural Glass* taking into account the brittle characteristics of glass [1]. Although periodic loading, e.g. from wind or live loads, can be decisive for glass design, very few scientific investigations are currently available on the cyclic fatigue of glass [2-5]. One reason is the complex experimental determination of the relevant parameters in cyclic fatigue tests. We recently determined the parameters for the most important glass in the building sector and the automotive industry (soda-lime-silicate glass) in extensive test series and, based on linear-elastic fracture mechanics, converted the findings into a design concept [6]. This article summarizes the most important findings of the experimental and theoretical investigations and the design concept, which can be integrated in the existing DIN 18008 [7] and the future *Eurocode* (now CEN-TS [1]).

1.2 Strength of Glass

The strength of glass preliminary depends on the presence of micro-cracks (in practice typically 4 µm to 100 µm) on the glass surface, the formation of which is unavoidable during production, transport, assembly and use. In the case of tensile stresses on the glass surface, the linear fracture mechanics and its *K*-concept provide a suitable model for determining the critical crack depths and geometries, which lead to (macroscopically) sudden failure due to supercritical, dynamic crack growth. The following failure condition must be fulfilled in mode I, which is the mode applicable for most glass cracks and loading in practice:

$$K_I = \sigma \cdot Y \cdot \sqrt{\pi a} < K_{Ic} \qquad (1)$$

where

K_I stress intensity factor
σ tensile stress perpendicular to crack direction
Y geometry factor
a crack depth
K_{Ic} critical stress intensity factor, K_{Ic}= 0,75 $MPa\ m^{1/2}$ for soda-lime-silica glass

However, this model of linear elastic fracture mechanics is only valid for conditions in which no water molecules can reach the crack tip. If this is not the case, as is the

* Corresponding author: schneider@ismd.tu-darmstadt.de

case for almost all applications in the building and automotive industry, crack growth can also occur below K_{Ic}. This depends on the stress intensity and the duration of the tensile stress at the crack tip of a micro-crack, since it depends on diffusion processes of the damaging medium (here: water) to the crack tip. However, it does not yet lead to instantaneous failure and is therefore called subcritical crack growth. The limit of stress intensity below which no subcritical crack growth takes place is called K_{I0} (fatigue threshold). For soda-lime silicate glass it ranges from K_{I0}= 0.2 to 0.4 $MPa\ m^{1/2}$ [8-11].

2 Experimental investigations

2.1. Defined pre-damage of the samples

The randomly distributed micro-cracks on the glass surface and their randomly distributed geometry and depth cause a very strong scattering and a size effect of the glass strength, thus larger areas under tensile stress having a larger probability of failure than smaller areas. This is very unfavourable for cyclic fatigue tests, as very short lifetimes (seconds) or very long lifetimes (years) can occur with real geometrical and material parameters. Therefore, a method for a defined pre-damage was developed in which the glass sample is indented and scratched with a conical diamond (120°, tip radius of r_S= 5 μm) at a defined vertical indentation force of 500 mN in an Universal Surface Tester (UST), see [12, 13]). The sample is scratched for a length of 2 mm (Fig. 1). The resulting initial vertical median cracks below the scratches thus are well reproducible and result in crack depths of a= 40-50 μm. This leads to a macroscopic tensile strength of about 45 MPa of annealed float glass (ANG) and of about 120 MPa for thermally fully tempered glass (FTG) in a coaxial double ring bending test applying a stress rate of 2 MPa/s. This corresponds to the 5%-fractile of the bending strength of ANG and FTG defined in product and design standards for glasses from soda-lime silicate glass [7, 14]. The coefficient of variation of the tensile strength applying this pre-damage method is less than 4 %.

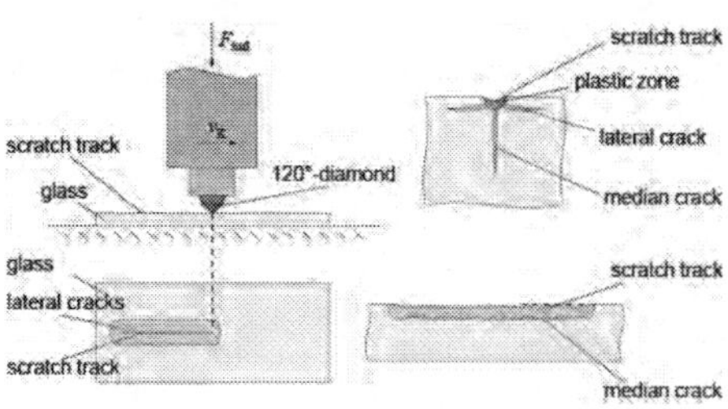

Fig. 1. Defined pre-damage by indentation and scratching using an Universal Surface Tester (UST)

2.2 Cyclic fatigue tests

The experimental investigations were carried out in a coaxial double ring bending test for annealed glass (ANG) and thermally tempered glass (FTG). FTG was also tested in a three-point bending test. By testing in the double ring bending test (Fig. 2), an influence of the orientation of the cracks and an edge influence from pre-damage of the glass edges can be excluded, because for small deformation of the sample, an equibiaxial tensile stress field is created within the load ring on the lower side of the specimen. With our test setup, however, only tension stresses (stress ratio $R>0$) and frequencies up to 1 Hz could be applied. Therefore, the additional three-point bending tests with frequencies up to 15 Hz and with alternating loads $(R<0)$ were carried out for FTG. The glasses tested were commercially available, annealed and thermally tempered soda-lime-silicate glasses with a nominal thickness of d=6 mm, which were produced in a float process. The so-called gas side was set under tension in order to exclude influences from surface damages from on the tin bath side of the float process, where the transportation rollers in the cooling section can induce significant micro-cracks.
Preliminary investigations had shown that - based on the chosen method of pre-damaging- neither the glass thickness nor the respective surface (roller side, the side facing to the rollers for thermally tempered glass) had any influence. The residual stresses were identified from stress optical measurements using a scattered light polariscope (SCALP-04, [17]).
The investigations were carried out at standard ambient conditions in a specially manufactured test chamber under defined humidity in order to specifically influence the subcritical crack growth. In addition, for each specimen subjected to cyclic fatigue, a reference specimen was tested until failure under quasi-static ramp loading at a stress rate of 2 MPa/s. Table 1 shows relevant investigation parameters and the most important observed effects.

Table 1: Experimental parameters and observations

Type	Parameter	Observation
Residual stresses	Residual surface compression stress FTG: up to -110 MPa	Increasing surface compression stresses increase the fatigue lifetime
Stress ratio R	-1 /0 /0,5 /0,73 /0,8	Increasing R reduces the fatigue lifetime
Form of load function	Sinus, Triangular, Trapezoidal function	Increasing fullness of the load function reduces the fatigue lifetime
Frequency	0,01 Hz to 15 Hz	No significant influence
Environmental conditions	0% - 100% *rel. humidity*	Increasing humidity reduces the fatigue lifetime
Load free intervals between load cycles	1 hour, 1 day, 1 week	No crack healing effects compared to continuous load cycles

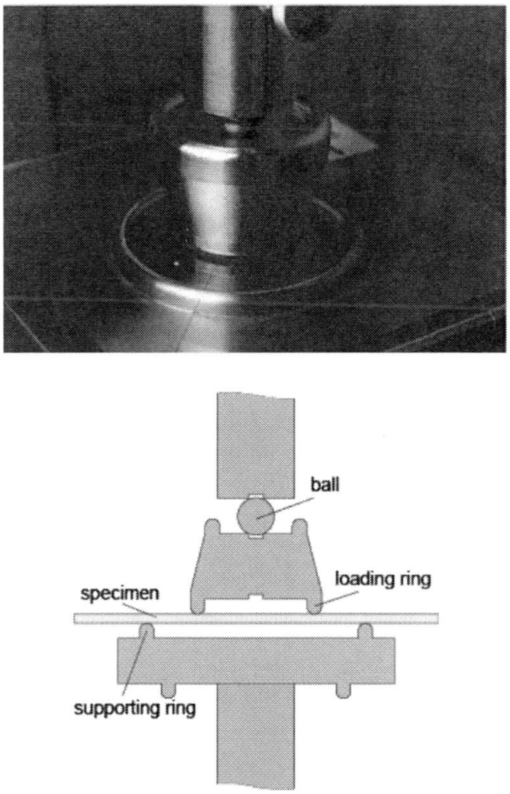

Fig. 2: Coaxial double ring bending test, $r_{loading\ ring}$ = 80 *mm*, $r_{supporting\ ring}$ *160 = mm* laboratory setup, section (below)

The investigations have clearly shown that it is not the number of load cycles that is decisive for failure, but the equivalent duration of tensile stress at the crack tip of a micro-crack. Therefore, the term *cyclic service lifetime* t_{fc} is used in the following. The results with regard to the form of the ambient conditions (Fig. 3) and of the load function (Fig. 4) as well as the influence of residual stresses (Fig. 5, 6) meet expectations, as they correspond to those of dynamic fatigue tests at a constant stress rate. In the diagrams, the cyclic failure stress σ_{max} is normalized to the quasi-static tensile strength σ_{fqs}.

In contrast to simple reloading under quasi-static loading, however, no strength increase due to crack healing effects could observed applying stress-free intervals by loading interruption during cyclic loading (Fig. 7, 8), which are frequently described in literature [14, 15, 18, 19].

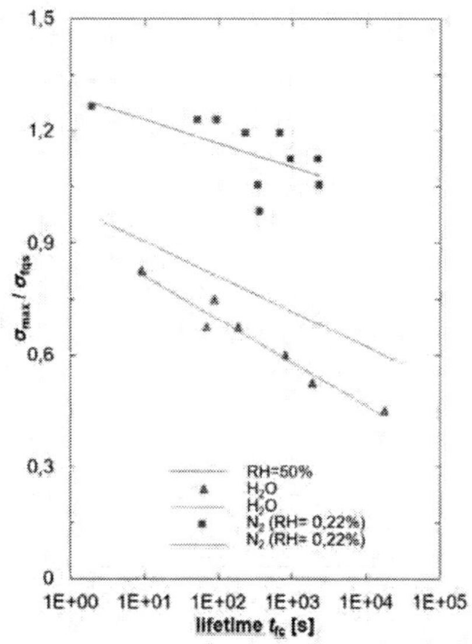

Fig. 3: Influence of the type of the ambient conditions on the lifetime of cyclically loaded annealed glass; *RH* rel. hum.; N_2 tested in nitrogen atmosphere; H_2O tested with liquid demineralized water on the sample

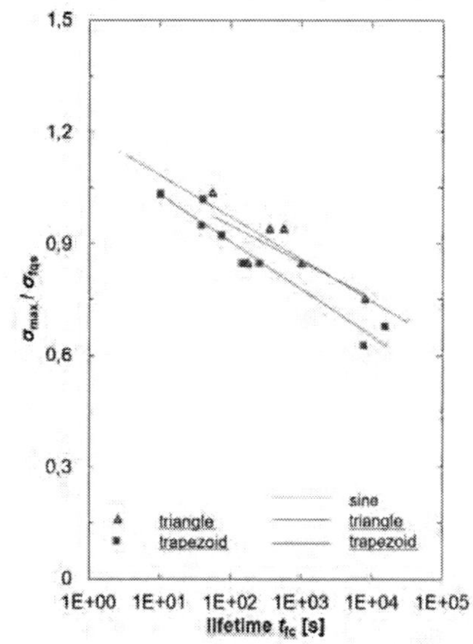

Fig. 4: Influence of the load function on the lifetime of cyclically loaded annealed glass

https://doi.org/10.1051/matecconf/201816518003

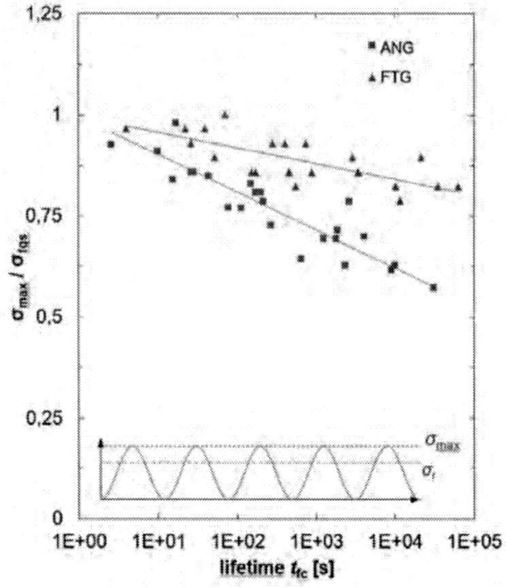

Fig. 5: Influence of thermally induced residual surface compressive stresses (ANG: squares, FTG: triangles)

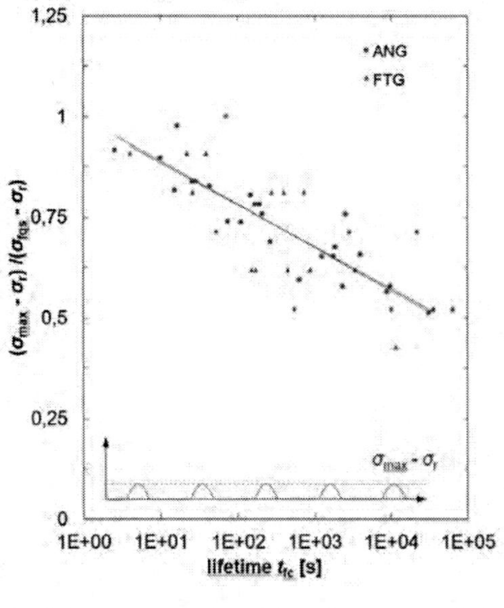

Fig. 6: Representation of Fig. 5 for "effective stresses" after deduction of the thermally induced residual surface compressive stresses σ_{res} measured with a scattered light polariscope [17]

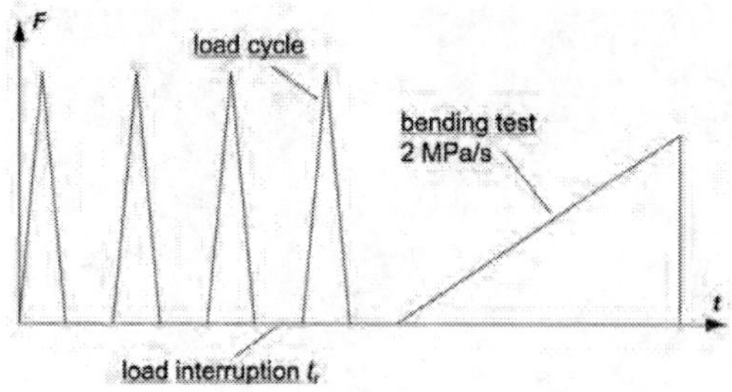

Fig. 7: Scheme for testing samples with stress free intervals by load interruption between load cycles and final ramp loading, load interruption 1 *h*, 1 *d*, 1 *week*

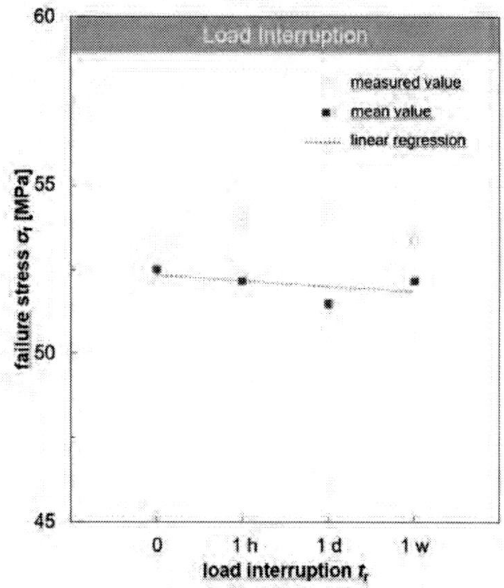

Fig. 8: Influence of stress free intervals on the tensile strength of annealed glass in ambient conditions (23°C, 50% RH)

3 Engineering model for cyclic fatigue lifetime prediction

To determine the service lifetime of the glass under cyclic fatigue, an empirical power law can be used that predicts the subcritical crack growth and has been very discussed in science for almost fifty years for dynamic fatigue at constant stress rates [8]. The decisive parameter is the crack propagation coefficient or crack growth exponent n, which represents the slope of the straight line for the region I in Fig. 10. The value of n is typically between n=10 and n=18 for glasses in dynamic fatigue. It can be converted into an equivalent constant load, which leads to failure by integrating an arbitrary stress function in analytical models (Fig. 9). Alternatively, the crack growth can be calculated in discrete time steps for the load function (Fig. 10). Fig. 11 shows that the lifetime prediction with the analytical model gives a very good prediction; Fig. 12 shows that the crack growth exponent n_c decreases for cyclic fatigue by about 10% compared to dynamic fatigue with a ramp loading.

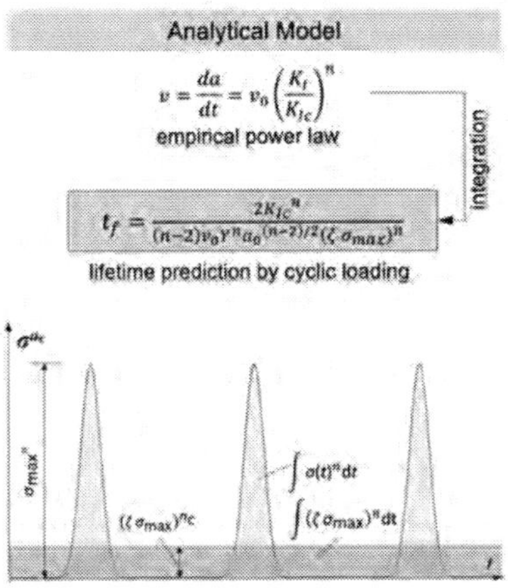

Fig. 9: Analytical for lifetime prediction based on crack growth for region I

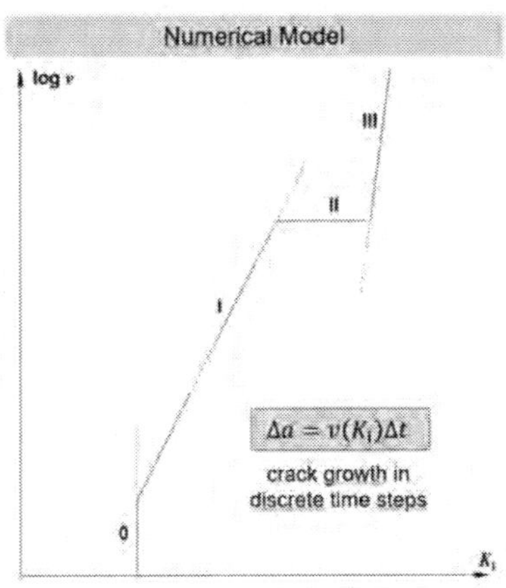

Fig. 10: Numerical model for lifetime prediction based on crack growth for region I

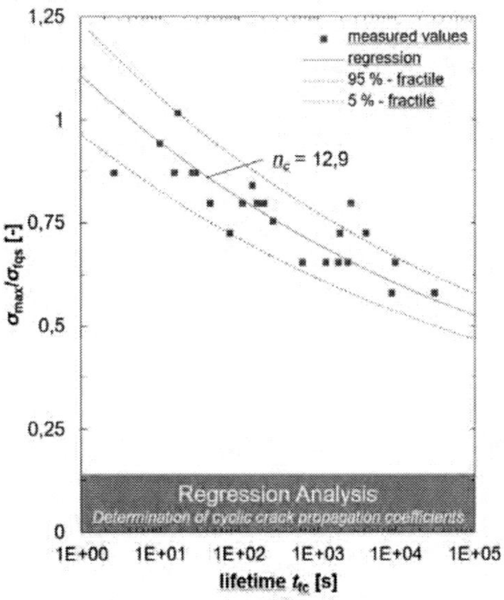

Fig. 11: Comparison between experimental data and analytical lifetime prediction with confidence intervals for n_c

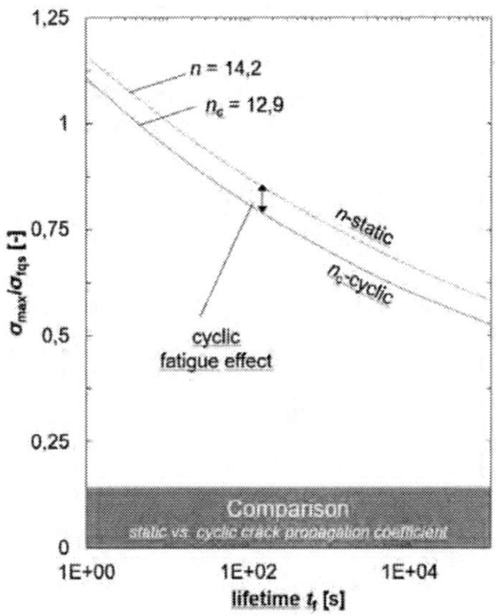

Fig. 12: Crack growth exponents for dynamic fatigue n and for cyclic fatigue n_c

4 Fatigue limit and design concept

Based on the numerical model, the fatigue limit below which crack growth stops can also be validated in comparison with the experimental data. A threshold of K_{I0}=0.26 $MPa\ m^{1/2}$ was derived from the experimental data using the numerical model (Fig. 13). This value is in very good agreement with literature values [8-11].

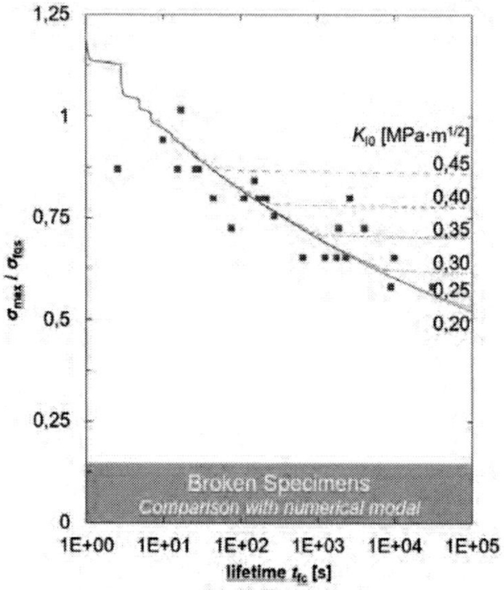

Fig. 13: Determination of the fatigue threshold K_{I0} with the numerical model

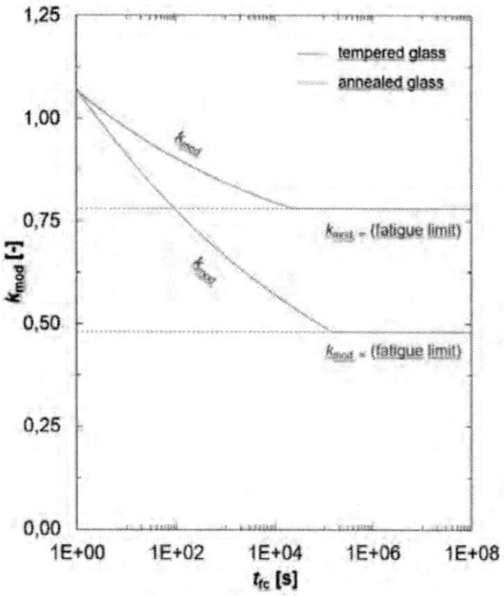

Fig. 14: Function of k_{mod} for the design concept according to DIN 18008 (eq. 3)

The design concept according to DIN 18008 [7] can be used for the transfer of the results into construction practice. According to DIN 18008-1, section 8.3.7, the design resistance of glass, expressed in stress values, reads:

$$R_d = \frac{k_{mod} \cdot k_c \cdot f_k}{\gamma_M} \tag{2}$$

where:

R_d design resistance (stress value)
k_{mod} modification factor for stress duration, see Fig. 14
k_c construction factor related to the failure consequence
f_k characteristic material strength from standardized tests (5%-fractile)
γ_M partial safety factor for the material strength

To take into account the influence of the type and duration of the cyclic action, the modification factor for stress duration k_{mod} can be determined as follows:

$$k_{mod} = \frac{c_1}{(t_{fc}\zeta^{n_c})^{1/n_c}} + c_2 \tag{3}$$

$$t_{fc} = T \cdot N \tag{4}$$

where:

c_1, c_2 constants (see table 2)
n_c crack growth exponent for glass in cyclic fatigue (see table 2 for soda-lime-silica glass)
ζ loading coefficient for the type of load function, on the safe side $\zeta = 1$
t_{fc} design lifetime [s]
T duration of a load cycle [s]
N number of load cycles

The stress coefficients ζ can be determined by integrating the effective stress of a representative load cycle. Evaluations of typical load functions can be found in [6]. To simplify $\zeta = 1$ is a safe assumption.

Table 2: Coefficients for the calculation of k_{mod}

Glass type	c_1	c_2	n_c
Annealed Glass (ANG)	0,90	0,09	12,9
Thermally Tempered Glass (FTG)	0,33	0,60	10,5

5 Conclusions

Based on the experimental work and theoretical models it could be shown that a lifetime prediction for soda-lime-silica glass subjected to cyclic fatigue is possible using crack growths models from dynamic loading, where a ramp load is typically applied. This finding was integrated in a simplified engineering design concept based on stress design. Still, the crack growths exponent for soda-lime-silica glass derived in the experimental campaigns with cyclic loading is about 10% smaller compared to dynamic loading. Moreover, it could be shown that unlike in strength tests with singular, quasi-static re-loading, a periodic loading with load free intervals does not lead to a strength increase by crack healing effects. Future works are planned to investigate the crack growths exponent in cyclic loading for other glasses like borosilicate glass and alumino-silicate glass and to study crack healing effects more in detail, also in combination with thermal and chemical tempering at elevated temperatures up to the transformation temperature of the glass.

References

1. CEN-TS *Structural Glass* Parts 1, 2, 3, Comité Européen de Normalisation (2018).
2. C. Gurney, S. Pearson, Proc. Roy. Soc. of London. Series A. Math. and Phys. Sciences **192** (1948) 1031, 537–544
3. B.-T. Lü, Theor. and Appl. Frac. Mech. **27** (1997) 2, 107–114
4. V.M. Sglavo, M.T. Gadott, T. Michelet, Fatigue & Frac. of Eng. Mat. &Struct **20** (1997) 8, 1225–1234
5. Ö. Bucak, H. Ehard, M. Feldmann, B. Hoffmeister, K. Langosch, F. Kemper, I. Mangerig, P. Ampunant, *Schallschutzelemente aus Glas an Eisenbahn-strecken.* Brückenbau - Construction & Engineering (2012), 16–21
6. J. Hilcken, *Zyklische Ermüdung von thermisch entspanntem und thermisch vorgespanntem Kalk-Natron-Silikatglas,* Dissertation, Technische Universität Darmstadt (2015)
7. DIN 18008-1, *Glas im Bauwesen – Bemessungs- und Konstruktionsregeln,* **Teil 1,** (2010)
8. S.M. Wiederhorn, L.H. Bolz, J. Am. Ceram. Soc. **53** (1970) 10, 543–548
9. E. Gehrke, C. Ullner, M. Hähnert, J. Mat. Science **26** (1991) 20, 5445–5455
10. K.-T. Wan, S. Lathabai, B.R. Lawn, J. Europ. Ceram. Soc **6** (1990) 4, 259–268
11. C. Kocer, R. Collins, J. Am. Ceram. Soc **84** (2001) 11, 2585–2593
12. J. Schneider, S. Schula, W.-P. Weinhold, Thin Solid Films. **520** (2012) 12, 4190–4198
13. S. Schula, *Charakterisierung der Kratzanfälligkeit von Gläsern im Bauwesen,* Dissertation, Technische Universität Darmstadt (2015)
14. EN 572-1, *Glass in building, Basic soda-lime silicate glass products,* **Part 1**, (2012)
15. C. Ullner, L. Höhne *Untersuchungen zum Festigkeitsverhalten und zur Rissalterung von Glas unter dem Einfluss korrosiver Umgebungsbedingungen,* Bericht Nr. 43 D (1993)
16. W.-T Han, M. Tomozawa, J. Am. Ceram. Soc. **72** (1989) 10, 1837–1843
17. GlassStress Ltd., *Scattered Light Polariscope SCALP-04* Instruction Manual Vers. 4.5.1
18. K.C. Datsiou, M. Overend, Glass Struct Eng **2** (2017) 105
19. A. Fink, *Ein Beitrag zum Einsatz von Floatglas als dauerhaft tragender Konstruktionswerkstoff im Bauwesen,* Dissertation, Technische Universität Darmstadt (2000)

https://doi.org/10.1051/matecconf/201816518004

Influence of surface residual stresses on the fatigue life and crack propagation behavior of turned Inconel 718 super-alloy

Hamid Javadi[1,*], *Walid* Jomaa[1], *Elvi* Dalgaard[2], *Myriam* Brochu[3], and *Philippe* Bocher[1]

[1]Ecole de Technologie Supérieure, Mechanical Engineering Department, 1100 Rue Notre-Dame Ouest, Montréal, H3C 1K3, Québec, Canada
[2]Pratt & Whitney Canada, 1000 Rue Marie-Victorin, Longueuil, J4G 1A1, Québec, Canada
[3]Ecole Polytechnique de Montréal, Mechanical Engineering Department, 2900 Boulevard Edouard-Montpetit, Montréal, H3T 1J4, Québec, Canada

Abstract. The effects of surface roughness parameters and residual stresses induced by finish turning on fatigue life and crack propagation behavior of Inconel 718 super-alloy are investigated. A criterion serving as a good indicator of surface quality is also discovered. This is carried out for some specific turning conditions in order to control surface characteristics including surface roughness and the bi-axial state of surface residual stresses and to investigate the effects of these characteristics on fatigue behavior. Rotating bending fatigue (RBF) tests are performed on the machined specimens at room temperature. Characterization techniques including X-ray diffraction (XRD), laser confocal microscopy, and opto-digital microscopy are used to investigate surface residual stresses in axial (ARS) and hoop (HRS) directions, surface roughness, and secondary fatigue cracks, respectively. The huge dispersion in fatigue life of smooth turned Inconel 718 with very low arithmetic average roughness (Ra) values indicates that surface finishing is not the only reliable criterion for quality control of machined components. Multivariate statistical analyses show that axial surface residual stress is the most important surface characteristic affecting the fatigue life of machined Inconel 718 for the conditions tested. Changing ARS from compressive to tensile increases the number of secondary cracks and also decreases the fatigue life of turned specimens.

1 INTRODUCTION

In today's aerospace industry, the main concerns include achieving an optimum, reliable, and reproducible manufacturing process with high predictability of component service life at the lowest production cost. Owing to their excellent mechanical properties at elevated temperatures, nickel-based super-alloys are commonly used to manufacture the hottest parts in aircraft engines by a share of over half by weight [1]. Some critical parts in engine hot compartments such as gas turbine disks are made of Inconel 718 super-alloy, which is arguably the most popular precipitation-hardened alloy in this class [2]. The manufacturing of these components involves various machining processes, namely milling, turning, and drilling, and each may have multiple steps, including roughing, semi-finishing, and finishing. It is known that finish machining operations significantly affect the surface integrity characteristics of Inconel 718 [2, 3]. Some of the main aspects of these alterations are topographical (surface finish), mechanical (residual stresses, plastic deformation), and thermal (material softening) [4-9].
Several studies have focused on the effects of machining processes on surface integrity of super-alloys [10-12]. Others have investigated the effects of surface integrity on fatigue lives of machined components [13-15]. Surface finish, in particular the arithmetic average roughness (Ra), is considered the main quality control criterion for manufacturing components. In the aerospace industry, Ra should be below 1 μm after the final machining steps for some critical parts [16]. Other surface integrity aspects such as residual stresses were found to have a significant effect on the fatigue lives of machined components [13, 17-19]. In particular, the low thermal conductivity and high strain rate sensitivity of Inconel 718 favor the generation of high-magnitude tensile residual stresses at the surface and deteriorate fatigue life considerably [13, 20].
It appears that, although it is the main machining process used in the manufacturing of rotating parts such as turbine disks, the effects of turning-induced surface integrity on the fatigue life of Inconel 718 have not been investigated and understood thoroughly. In the current paper, an analysis of rotating bending fatigue upon turned Inconel 718 superalloy is performed. To do so, fatigue specimens were carefully machined in order to produce various surface integrity characteristics while keeping arithmetic average roughness (Ra) below 1 μm. The main purpose of the present study is to find out which surface characteristics concerning surface roughness and residual stresses are the most influential

* Corresponding author: hamid.javadi.1@ens.etsmtl.ca

to fatigue life of turned Inconel 718 components with emphasis on fatigue crack behavior.

2 MATERIALS & METHODS

Fatigue specimens were made from 12 mm diameter Inconel 718 bars. The generic chemical composition of the Inconel 718 super-alloy is as follows: Ni-19Cr-18Fe-5Nb-3Mo-0.9Ti-0.5Al (wt. %). Before machining, the bars were solution treated and aged to reach a hardness of 41 ± 1 HRC and an average grain size of 25 µm. Rotating bending fatigue (RBF) specimens were machined by turning to produce the part in Figure 1. With this geometry, the entire length of the gage section (approximately 25 mm long) was stressed uniformly during fatigue tests. This latter is particularly useful for studying secondary fatigue cracks on the gage section, because the odds of crack nucleation are uniform for the whole length. Rotating bending fatigue (RBF) tests with a stress ratio R = -1 were performed at room temperature using a R.R. MOORE machine from INSTRON (SATP1962) under a stress amplitude of 448 MPa and a frequency of 30 Hz. Some selected fatigue specimens were mechanically fine-polished in order to remove machining-induced surface integrity and thereby evaluate the fatigue behavior of the bulk material (reference condition). Three finishing turning conditions were selected (Table 1) to produce multiple residual stress values in both the axial and hoop directions while keeping an arithmetic average roughness (Ra) below 1 µm. Four replications were machined for each turning condition in order to repeat fatigue tests and perform surface integrity characterizations.

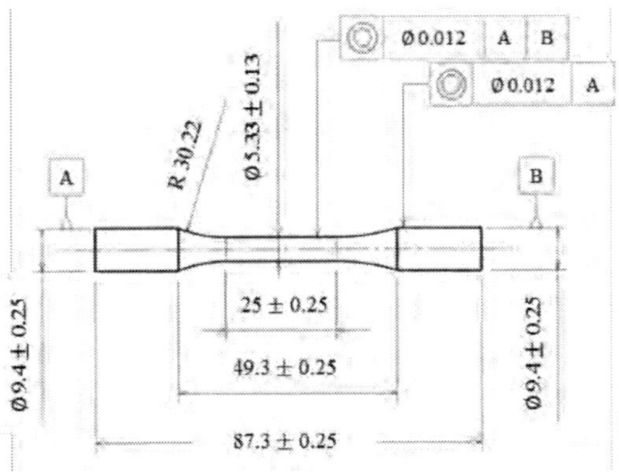

Fig. 1. Geometry of the straight-shank rotating bending fatigue (RBF) specimens (all in mm).

A laser confocal microscope LEXT OLS4100 from OLYMPUS was used to measure surface roughness parameters including maximum height of the roughness profile (Rt) and arithmetic average roughness (Ra) parameters. The residual stresses were measured using the X–ray diffraction technique according to $\sin^2\psi$ method with Proto iXRD1 system. In these measurements, lattice plane [311] was selected along with a Mn tube. Secondary cracks were observed on broken specimens using opto-digital microscope DSX500 from OLYMPUS.

3 RESULTS & DISCUSSION

Surface integrity characteristics and fatigue lives are displayed in Table 2. The industrial surface finishing criterion was met for all the conditions, as the arithmetic average roughness (Ra) was in all cases below 1 µm. A noticeable difference was observed in axial residual stress (ARS) for these low-roughness machined surfaces. Although arithmetic average roughness is kept below 1 µm, significant variation in Ra is also observed (from 0.02 to 0.74 µm). Hence, one can suspect that fatigue life can be influenced by the combined effects of surface roughness and residual stresses. Polishing the specimens further reduced the roughness (Ra = 0.02 µm). Polishing also successfully decreased the level of residual stresses and produced a free-stress surface resulting in a target fatigue life of 10^6 cycles.

In order to verify which of the two investigated surface integrity characteristics dominates fatigue life, a multivariate analysis was performed on the fatigue lives. A partial correlation matrix (Figure 2) was produced in which the correlation coefficient between two features is determined without considering the influences of the other characteristics [21]. The rule of thumb is the closer coefficient of correlation values to 1 or -1, the higher correlations exist between two characteristics, while values around zero means there is no linear correlation between the characteristics.

Figure 2 shows that among the surface integrity characteristics investigated, the axial residual stress had the greatest correlation with fatigue life (-0.61) while the hoop residual stress (HRS) and surface roughness parameters (Ra & Rt) had the lowest correlation coefficients (smaller than 0.5). Thus, under the experimental conditions investigated in the present study, the roughness parameters have less effect on the fatigue life of turned Inconel 718 than axial residual

Table 1. The surface conditions studied.

Run#	Cutting parameters			No. of specimens	
	Cutting speed S (m/min)	*Feed rate* f (mm/rev)	*Depth of cut* d (mm)	*Fatigue tests*	*Surface characterization*
Machined run #1	67	0.020	0.300	3	4
Machined run #2	89	0.098	0.230	3	4
Machined run #3	49	0.051	0.127	3	4
Polished	-	-	-	3	4

https://doi.org/10.1051/matecconf/201816518004

Table 2. Average values (four replications) of surface roughness, residual stresses, and average fatigue lives (three RBF tests) for the selected turning conditions and the polished specimens.

Run#	Surface roughness		Surface residual stresses		Fatigue life
	Ra (μm)	*Rt* (μm)	*Axial* (MPa)	*Hoop* (MPa)	*Nf* (Cycles)
Machined run #1	0.47 ± 0.11	4.51 ± 0.73	-275 ± 25	492 ± 28	1.85×10^6
Machined run #2	0.74 ± 0.13	4.84 ± 1.38	853 ± 32	996 ± 39	5.02×10^5
Machined run #3	0.28 ± 0.05	2.8 ± 0.32	146 ± 28	515 ± 22	8.33×10^5
Polished	0.020 ± 0.002	0.27 ± 0.05	-9 ± 30	-3 ± 43	9.85×10^5

stresses. Likewise, hoop residual stress is also less significant to fatigue life. It would appear that fatigue life is mainly affected by the axial surface residual stress for the conditions studied.

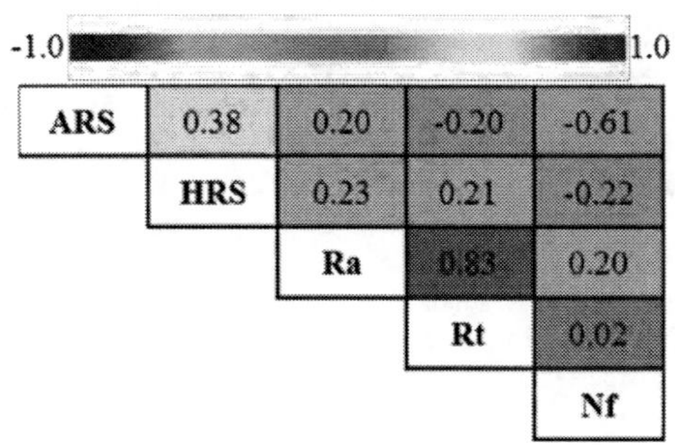

Fig. 2. Partial correlation matrix showing the degree of dependence between various surface characteristics (axial residual stress (ARS), hoop residual stress (HRS), arithmetic average roughness (Ra), and maximum height of the roughness profile (Rt)) and fatigue lives of machined specimens (Nf).

This result can be clearly seen in Figure 3, which depicts fatigue life versus ARS and HRS for all the specimens. Obviously, the greater ARS values diminish the fatigue lives of machined specimens; e.g., when ARS decreases from high tensile (849 MPa) to compressive (-412 MPa), the fatigue life is improved from 4.64×10^5 to 2.43×10^6 cycles. To illustrate the importance of ARS, comparing Run 1 and Run 3 with almost the same HRS values confirmed the two-fold increase in fatigue life was due to the compressive ARS for Run 1. On the other hand, the polished specimens had ARS values comparable with Run 3 and their HRS magnitudes were around 500 MPa different. However, their fatigue lives were very comparable, confirming that HRS has less effect on fatigue behavior than does ARS. This confirms the results seen in the partial correlation matrix. It appears that fatigue crack nucleation and propagation are significantly influenced by ARS.

In addition to the main fatigue crack causing the specimens' failures, secondary cracks were observed on the specimens after failure. Figure 4 shows typical secondary fatigue cracks for polished and machined specimens. The secondary crack on a polished surface, shown in Figure 4.a, nucleated on a point with the

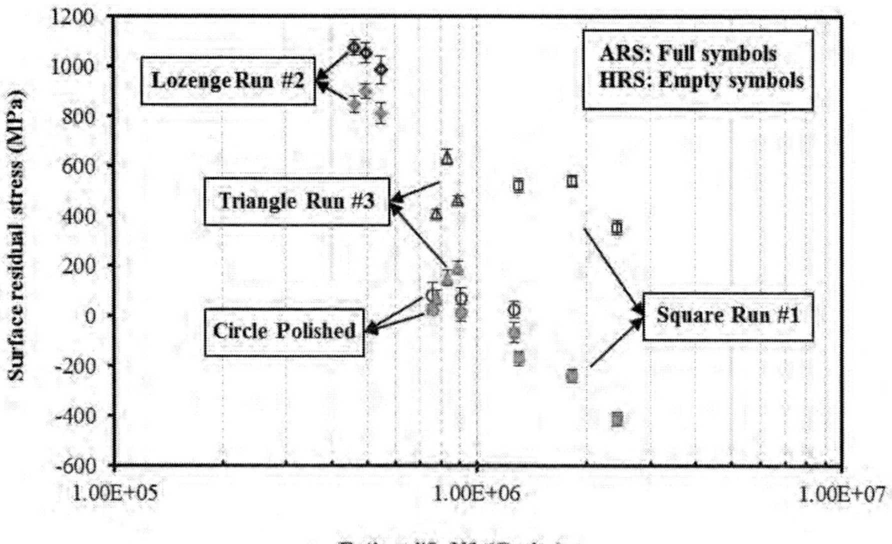

Fig. 2. Influence of machining-induced residual stresses in axial (full symbols) and hoop (empty symbols) directions on the fatigue lives of machined and polished specimens.

greatest local stress concentration during the RBF test and propagated in both sides inclined to the applied stress direction. The crack path is hypothetically determined by the crystallographic orientation of the surface grains and their activated slip systems. After passing a few grains (200 to 250 µm projected crack length), the crack propagation transited from the short crack to the long crack regime as growth direction became perpendicular to the applied stress. Since the machining-affected layer had been removed, the crack nucleation and propagation were not influenced by surface characteristics for polished specimens; the early stage of crack propagation is considered uniquely microstructure-dependent.

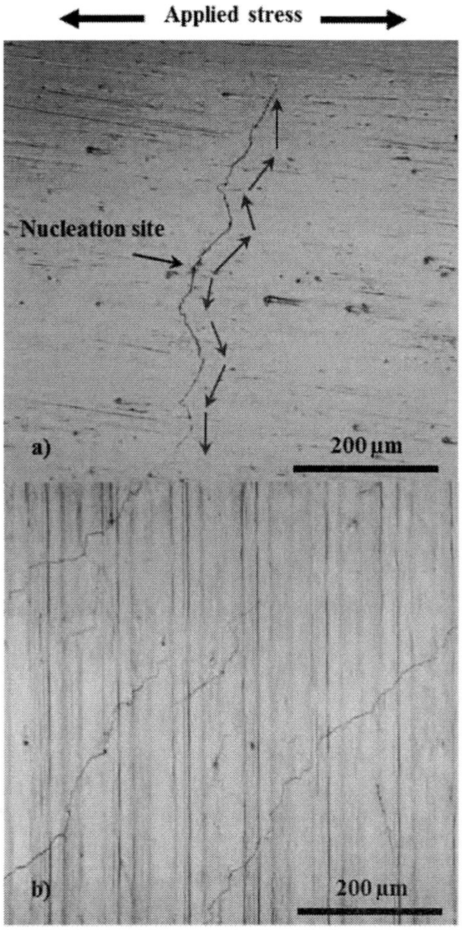

Fig. 4. Typical secondary cracks on fatigue specimens for a) polished and b) machined surfaces.

In contrast, Figure 4.b shows the secondary cracks for a machined surface with high tensile residual stresses (Run 2). It was difficult to find the nucleation sites for these cracks since machined surfaces are mainly formed by successive feed marks, scratches, and defects which could be potentially crack nucleation sites. High surface roughness and residual stresses may also act as stress raisers to promote fatigue crack nucleation.

The above cracks always propagated inclined to the loading direction while crossing the parallel feed marks. This may be because of the combined effects from axial and hoop residual stresses, which governs the stress state ahead of the secondary cracks and does not permit their paths to propagate perpendicularly to the loading direction. Moreover, most of these secondary cracks are long enough to be considered as the coalescence of few smaller cracks. It seems that for high tensile residual stresses, the secondary cracks grow faster while skipping the short fatigue crack propagation regime, progressing directly to the long crack regime.

The density of these secondary fatigue cracks was evaluated based on the microscopic observation of the entire gage section surface from a broken specimen (2 to 3 specimens per turning condition). The density of secondary fatigue cracks affected the fatigue behavior of the machined specimens; i.e., the greater the number of short cracks, the greater the probability of developing a fatal crack. To illustrate this, for three turning conditions (Runs 1 to 3), the average secondary crack densities were 12, 483, and 62 cm^{-2} and their average fatigue lives were 1.85×10^6, 5.02×10^5, and 8.33×10^5 cycles, respectively. Secondary crack density was minuscule for polished specimens (2 cm^{-2}) due to the low axial residual stress and defect-free surface.

Figure 5 illustrates the influence of machining-induced axial surface residual stresses on the density of secondary fatigue cracks for three selected turning conditions. The compressive stresses of the Run 1 specimen led to a lower secondary crack density and smaller crack size. Conversely, raising the residual stresses to highly tensile values (Run 2) caused a dramatic increase in secondary crack density and also resulted in longer cracks. This indicates that compressive residual stresses delay crack nucleation and short crack propagation.

4 CONCLUSIONS

The current study focuses on the effects of residual stresses and surface roughness on fatigue life and crack propagation behavior of turned Inconel 718 super-alloy. The specimens tested were prepared using several finish turning conditions in order to produce various surface residual stresses (from compressive to tensile stresses) while keeping arithmetic average roughness values below 1 µm (respecting industrial requirements for critical applications). High cycle fatigue lives (10^6 cycles) were targeted using a rotating bending fatigue (RBF) machine at room temperature with a fixed applied stress (about 40% yield strength). The results show that:

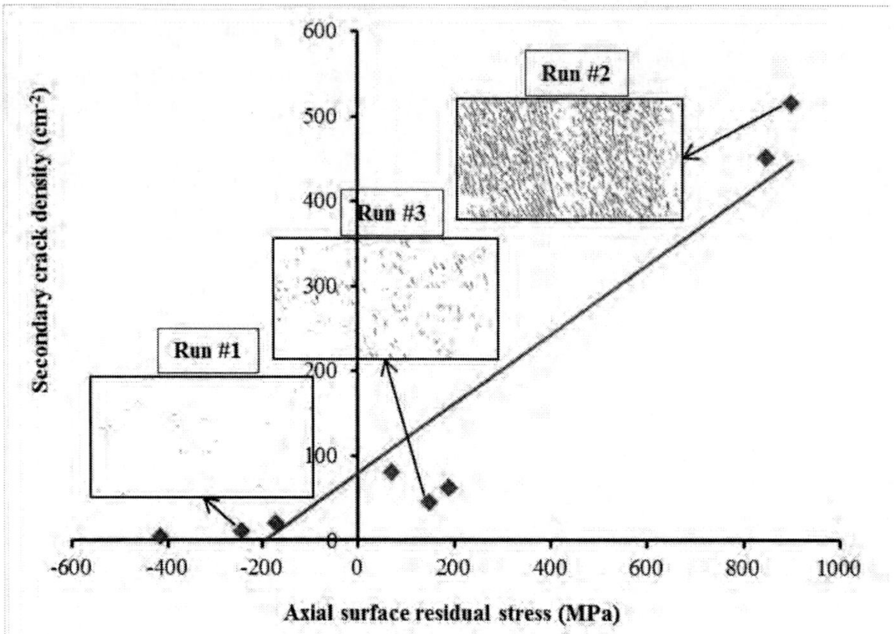

Fig. 5. Dependence of secondary crack densities on the axial surface residual stresses for three highlighted turning conditions by their secondary cracks map representation.

1- Elevated tensile axial surface residual stresses (849 MPa) lead to a significant (5-6 fold) reduction in the fatigue lives of the specimens (from 2.43×10^6 to 4.64×10^5 cycles) in comparison with compressive axial surface residual stresses (-412 MPa).

2- It is also found that the secondary crack densities rose with increased surface residual stresses, which supports the fact that compressive residual stresses delayed the crack initiation process, leading to longer fatigue life.

3- Multivariate statistical analysis showed that among the surface integrity characteristics studied, the axial residual stress has the greatest correlation with fatigue life and with crack density, while the arithmetic average roughness (Ra) has the lowest correlation coefficient if kept below a certain threshold.

4- Therefore, surface quality optimizations should be redefined in order to exploit the effect of surface residual stresses rather than surface roughness.

This research has been financially supported by the Natural Sciences and Engineering Research Council of Canada (NSERC) and the Consortium for Research and Innovation in Aerospace in Quebec (CRIAQ) under the project named MANU-510. The authors wish to express their sincere thanks to Pratt & Whitney Canada and Héroux-Devtek as industrial partners in this project, providing all in-kind materials and their technical and financial support.

REFERENCES

1. D. Ulutan, T. Ozel, Int. J. Mach. Tools & Manuf., **51**, 250 (2011)
2. R. M'Saoubi, J. Outeiro, H. Chandrasekaran, O. Dillon Jr, Int. J. Sustain. Manuf., **1**, 203 (2008)
3. J.P. Davim, *Surface Integrity in Machining* (Springer, 2010)
4. C. Schlauer, R.L. Peng, M. Odén, *Materials Science Forum*, **404**, 173 (2002)
5. R.M. Arunachalam, M.A. Mannan, A.C. Spowage, Int. J. Mach. Tools & Manuf., **44**, 879 (2004)
6. A.R.C. Sharman, J.I. Hughes, K. Ridgway, J. Mater. Process. Technol., **173**, 359 (2006)
7. A. Thakur, S. Gangopadhyay, Int. J. Mach. Tools & Manuf., **100**, 25 (2016)
8. D. Thakur, B. Ramamoorthy, L. Vijayaraghavan, Int. J. Adv. Manuf. Tech., **59**, 483 (2012)
9. A.B. Sadat, M.Y. Reddy, B.P. Wang, Int. J. Mech. Sci., **33**, 829 (1991)
10. E.O. Ezugwu, Z.M. Wang, C.I. Okeke, Tribol. Trans., **42**, 353 (1999)
11. R. Pawade, S.S. Joshi, P. Brahmankar, Int. J. Mach. Tools & Manuf., **48**, 15 (2008)
12. A.R.C. Sharman, J.I. Hughes, K. Ridgway, Mach. Sci. Technol., **8**, 399 (2004)

13. H. Sasahara, Int. J. Mach. Tools & Manuf., **45**, 131 (2005)
14. A. Sharman, D. Aspinwall, R. Dewes, D. Clifton, P. Bowen, Int. J. Mach. Tools & Manuf., **41**, 1681 (2001)
15. S. Bentley, A. Mantle, D. Aspinwall, Intermetallics, **7**, 967 (1999)
16. I.O.F. Standardization, ISO 1302: *Geometrical Product Specifications (GPS) - Indication of Surface Texture in Technical Product Documentation* (ISO, 2002)
17. G. Webster, A. Ezeilo, Int. J. Fatigue., **23**, 375 (2001)
18. D. Taylor, O. Clancy, Fatigue Fract. Eng. Mater. Struct., **14**, 329 (1991)
19. A. Pramanik, A. Dixit, S. Chattopadhyaya, M. Uddin, Y. Dong, A. Basak, G. Littlefair, Adv. Manuf., **5**, 59 (2017)
20. R. M'Saoubi, J. Outeiro, B. Changeux, J. Lebrun, A.M. Dias, J. Mater. Process. Technol., **96**, 225 (1999)
21. A.C. Rencher, *Methods of Multivariate Analysis* (John Wiley & Sons, 2003)

Experimental Life Improvement Quantification of Shot Peening and Fastener Modifications

Guillaume Renaud[1,*], *Pascal* Martin[2], and *Min* Liao[1]

[1]National Research Council Canada, 1200 Montreal Road, Ottawa (Ontario), K1A 0R6, Canada
[2]L-3 MAS, 10000 Helen Bristol, Mirabel (Québec), J7N 1H3, Canada

Abstract. Two coupon test programs were recently carried out at the National Research Council of Canada to certify shot peening and fastener modifications that aim at extending the lives of military aircraft components that are short of their revised usage goals. The life improvement associated with each modification was experimentally quantified by comparing the lives resulting from variants of each modification with that of pre-modification baseline configurations. This approach allowed determining the effects of several parameters, such as the applied stress level, geometric configuration, and modification specifications, on the life extension.

1 Introduction

Several military aircraft platforms currently undergo life extension retrofit programs to remove fatigue damage accumulation and to delay crack nucleation in aging airframes. These programs are tailored to each aircraft platform to ensure that both airworthiness and logistic risks are maintained at an acceptable level.

The benefit on life resulting from these modifications must typically be determined from tests that are either conservative, or representative of the most critical airframe locations to be reworked. This paper presents two coupon test programs that were recently carried out at the National Research Council of Canada (NRC) to certify modifications aimed at extending the lives of aircraft components short of their revised usage goals. Specifically, the life improvement resulting from robotic shot peening and from fastener replacements was experimentally quantified by comparing the life associated with variants of each modification with that of baseline configurations. This approach allowed evaluating the effect of modification parameters on the life extension, and determining the life benefits that can be included in life predictions for fleet management.

2 Test Programs

The efficiency of the tested life enhancement modification was quantified by comparing the "crack initiation" (CI) lives of coupons with and without the modification. The ratio between these two lives is herein defined as the "life improvement factor" (LIF). The CI life was defined as the time necessary for a fatigue crack to nucleate and grow to 0.01 inch in depth.

2.1 Robotic Shot Peening

First, a robotic shot peening modification was evaluated for the service life extension of major military airframe components. The advantage of using a robotic system for shot peening is to improve quality, repeatability and accessibility on the airframe where a technician has no or very limited access. This cold-working process produces a layer of beneficial compressive residual stress over the treated surface. Up to recently, these shot peening retrofits were carried out prior to 70% of the blueprint fatigue life of the considered aircraft, and assuming a LIF of 1.5.

Recently, new lifing requirements from the fleet operator made necessary to demonstrate a LIF of up to 3.0 in critical areas, along with a retrofit induction as late as 80% of the blueprint life. New tests were required because the available data was insufficient to certify that such LIF was achievable via shot peening for all the considered locations. These locations differed in terms of stress level, required incorporation time, geometry, and material grain orientation. Moreover, a thin layer of material had to be removed prior to shot peening to remove fatigue damage accumulated prior to the rework induction time. The parameters considered in the test plan are summarized in Table 1.

Default values were assumed for each of the physical characteristics to define "main certification tests" that comprised 30 baseline coupons and 30 peened coupons. In addition, "Conditional Tests" were defined in case the shot peening modification could not be certified for all critical locations. These tests, also comprising 30 baseline coupons and 30 peened coupons per series, aimed at identifying parameter values, other than the default, that would allow certification (e.g. earlier retrofit induction).

* Corresponding author: guillaume.renaud@nrc-cnrc.gc.ca

Table 1. Parameters considered in shot peening tests.

Characteristics	Test values considered
Material	Aluminum plate, 6 in thick
Stress level (ksi)	5 levels between 50 and 75
Flaw type	Surface in middle of radius and edge corner cracks
Fillet radius (in)	0.16, 0.25, and 0.38
Stress state	Pure uniaxial and some level of bending/shear
Grain direction	LT, TL and ST
Marker bands	Added or not added in spectrum
Spectrum	Historical and current spectra; Fleet 50th% to 90th% severity
Spectrum truncation	None or 30%
Manufacturing process	Early 1990's and today
Incorporation time	48% to 80% of fatigue life
Blend depth (in)	0.003-0.006 deep to 0.013-0.016 deep

"Sensitivity tests" were also defined to quantify the sensitivity of the shot peening repair effectiveness with respect to selected parameters, particularly when the effectiveness is close to the target value. These tests included between 7 and 15 coupons per series. Finally, "optional tests" were defined to cover other geometries and loading scenarios, if deemed necessary from the results obtained from the other tests.

In total, the number of coupons included in the test matrix ranged between 200 and 579 coupons, depending on how many conditional, sensitivity, and optional tests became required from the main certification tests. The two main coupon geometries considered in the test program are presented in Figure 1.

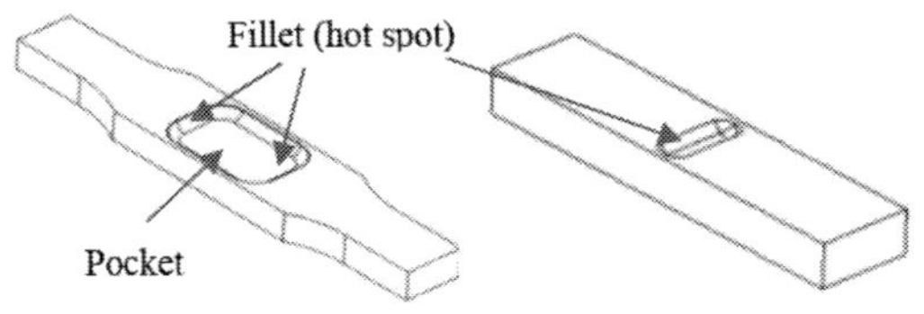

Fig. 1. Main geometries of shot peening test coupons.

2.2 Fastener Life Improvement Techniques

2.2.1 Blind Interference Fit Fasteners

Second, the fatigue performance resulting from the use of blind interference fit ground shank fasteners (I/F GS) were compared with that of legacy clearance fit blind fasteners with ring pad coining (RPC), that needed to be replaced, and with interference fit Hi-Lok fasteners (I/F HL) that can be used for locations that are accessible from both sides. These comparisons were required to assess retrofit modifications for which several fasteners had to be replaced and/or fastener holes had to be reworked, either for life improvement or for access purposes. However, the LIF of the I/F GS fasteners to be installed was limited by the fleet operator to 2.0. This suggested that the analytical CI life benefit that could be claimed from the use of these fasteners was lower than that of the RPC or I/F HL, which had an approved LIF of 3.0.

In cases where the local peak stress at the hole is more than 150% of the yield strength (1.5×Fty), the three types of fasteners were attributed a LIF of 1.0. However, this sudden shift from a LIF of 3.0 (or 2.0) to 1.0 when crossing the stress limit limitation was seen as arbitrary and not representative of the actual performance of these fastener systems. Further, assuming the same LIF before and after the modification appeared potentially un-conservative if the I/F GS fastener holes truly have a shorter life than that of the I/F HL or the RPC fasteners. For these reasons, a coupon test program was developed to evaluate the relative difference in LIF between the three fastener systems. The characteristics considered in this program are summarized in Table 2.

Table 2. Parameters considered in fastener tests.

Characteristics	Test values considered
Material	Aluminum plate, 6 in thick
Stress level	1.5×Fty and 2.0×Fty
Hole diameter (in)	0.25
Stress state	Through and bearing loads: no- and high- load transfer
Grain direction	LT
Marker bands	Added in spectrum
Spectrum	Current spectrum
Spectrum truncation	30%

Twelve mandatory and two conditional series of seven coupons were defined to cover baseline cases (open holes) and fastener configurations with and without load transfer. Moreover, each coupon allowed life determination from two identical holes. The geometry of the fastener tests coupons is illustrated in Figure 2.

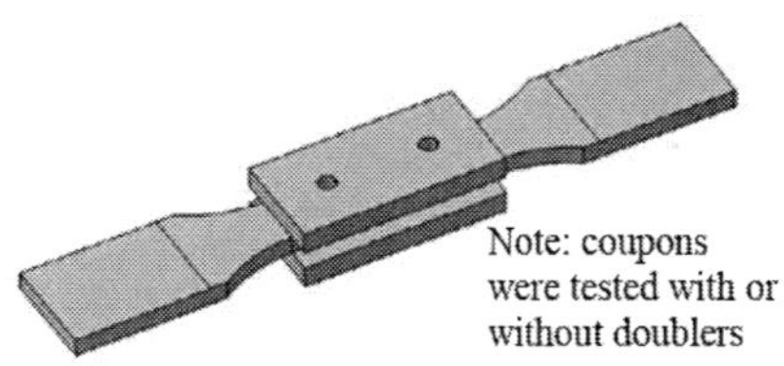

Fig. 2. Geometry of fastener test coupons (with doublers).

The three types of fasteners are presented in Figure 3 (the two blind fastener types look identical; only the coining and engineering fit differ).

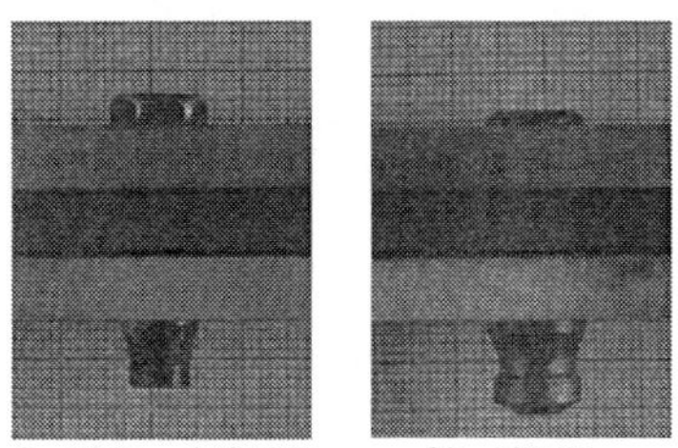

a) RPC and I/F GS b) I/F HL

Fig. 3. Fastener types (with doublers).

2.2.2 Bushing Repair Evaluation

One fastener location stood out amongst those that needed to be reworked as being the most critical from the in-service and full-scale test data reported to date. The complexity of the repair associated with that hole made the current analysis approach unreliable. The repair involves a significant oversizing of the hole, which reduces the benefit from RPC, and bushing installation, which shifts the edge of the hole significantly closer to the edge of the part (lower edge margin). For these reasons, a specific coupon test was developed to certify this repair configuration. In order to match the complex stress field, a 3D coupon was designed, as shown in Figure 4.

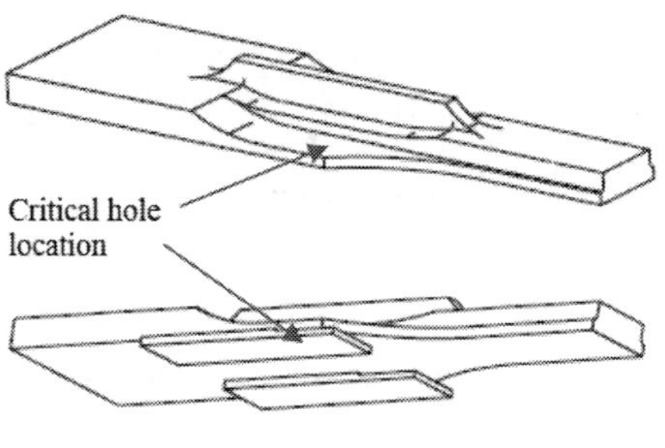

Fig. 4. 3D fastener test coupons.

The objective of these tests was to compare the performance of I/F bushing to replace I/F HL fasteners with and without RPC for the repair proposed at that location. First, the CI life for this case was to be determined for the two fastener systems. Then, coupons that would have been previously pre-cycled to a CI state, to simulate a late incorporation time, would be repaired by oversizing the hole and installing a bushing.

Testing of four series of seven coupons was still being performed at the time of writing this paper. As such, no results are reported herein.

2.2.3 Damage Removal Evaluation

Optional tests, not covered in this paper, were also defined to quantify the effect of fatigue damage removal for a nominal to next size fastener typical of the considered modification locations. The objective was to compare experimental fatigue damage removal life results with the current analysis approach. This approach removes fatigue damage by considering the increase in fatigue life from an increase in diameter using Miner's rule, no matter how much material is removed by a fastener rework. This assumption is potentially overly conservative, especially when enough material is removed such that fatigue life is virtually reset. Four conditional series of seven coupons were defined to cover that requirement.

In total, the number of coupons included in the test matrix for the fastener life improvement techniques ranged between 112 and 154 coupons, depending on how many conditional tests became required from the test results.

2.3 CI Life Improvement Factor Calculation

Fatigue life prediction for the airframe modifications considered in this work must consider uncertainty due to the variability observed in the tests. This variability is related to factors that include material properties, surface condition, rework process control, and geometry. The measure of this variability is given by the standard deviation σ of the log-lives observed in the various test series, which is used to calculate a scatter factor sf for a given probability of failure. This scatter factor is derived using the following equation [1]

$$sf = 10^{Z\sigma\sqrt{\frac{n+1}{n}}} \tag{1}$$

where Z is the normal variate for a given cumulative probability of failure, σ is the standard deviation of the observed log-CI lives, and n is the number of test results available. This equation is valid only if the standard deviation is known. In this work it is assumed to be calculated from the test series samples. The equation also assumes that the population from which the test samples are taken is log-normally distributed.

A cumulative probability of failure (CPOF), where failure in this context is "crack initiation" (0.01 inch), of 0.001 (therefore $Z = 3.09$) was assumed in the current work to derive the scatter factor.

The "safe life" calculated for a test series corresponds to the median life, calculated from a lognormal distribution fit of the test data, divided by the scatter factor. A "factorized" LIF can then be calculated as the ratio between the safe life of a life-enhanced configuration and that of the corresponding baseline (pre-mod) configuration. This approach is illustrated in Figure 5 for a typical shot peening LIF calculation.

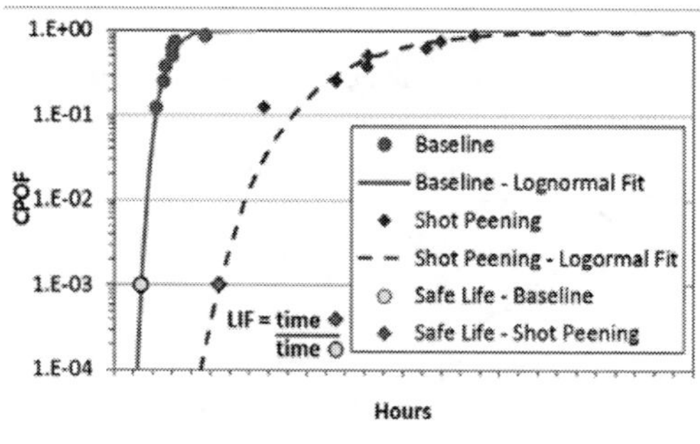

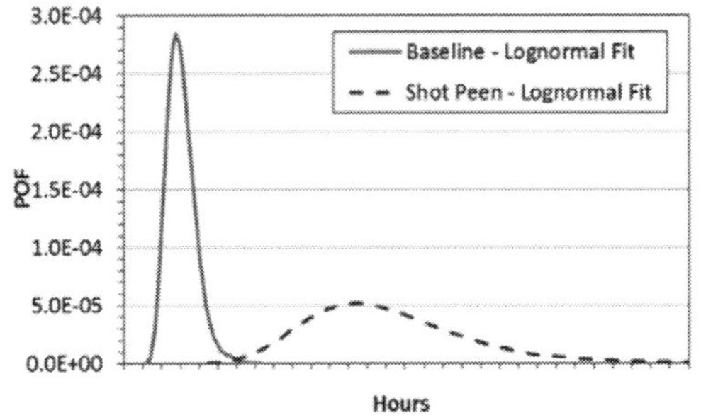

Fig. 5. Typical CI LIF calculation for shot peening coupons.

An unfactored LIF can also be calculated as the ratio of median coupon CI lives. For the fastener test coupons, life improvement factors were also calculated on crack growth from initiation to coupon failure. This was done to partly isolate the scatter arising from the fastener itself, which would mostly affect short cracks, from the other sources of scatter.

3 Test Requirements and Specifications

3.1 Coupon Etching

To represent the actual surface condition present on the aircraft, etching was performed on the coupons that assumed that the ion-vapor deposition (IVD) coating applied by the original equipment manufacturer was not removed. The IVD coating itself does not have any impact on the fatigue life but the pre-IVD etching is known to have an effect on crack nucleation.

The objective was to obtain pit depths in the range between 0.006 and 0.010 mm. The etching procedure was certified by measuring pits at various locations along the coupon critical locations. An example of such measurements is presented in Figure 6.

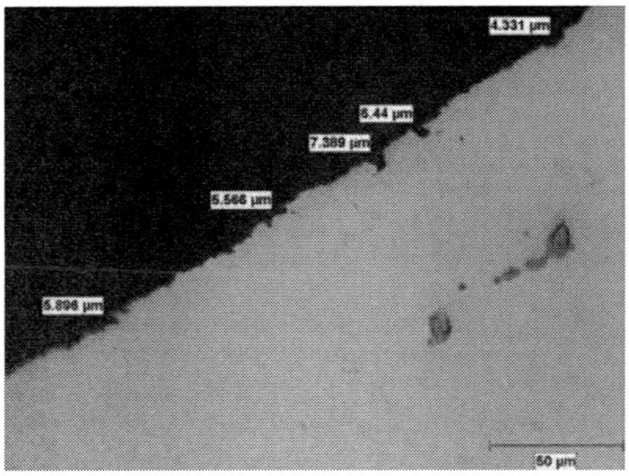

Fig. 6. Typical etch pits.

The shot peening test coupons were etched by complete immersion. This approach was later found to increase the risk of crack nucleation away from the critical fillet (hot spot) where cracks were expected to develop. In order to reduce that risk for the fastener test coupons, only a limited area around the fastener holes was exposed to the etching solution. An example of local etching is shown in Figure 7. This approach, combined with shot peening of the non-critical coupon surfaces, for instance the radii creating the transition between the coupon grip and gauge test sections, was successful in controlling crack nucleation within the hole area.

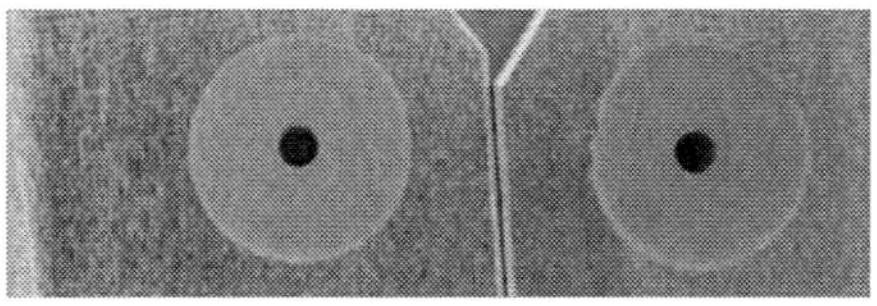

Fig. 7. Local etching around pilot holes (3D fastener coupons).

3.2 Initial Setup

The fatigue tests were performed at NRC on MTS test frames equipped with Arbitrary End-Level Compensation, calibrated as per ASTM E4-13 and following the ISO 9001:2008 Standard. Load frame alignment was performed using 12-gauge alignment transducers as per ASTM E1012 Class 5 criteria.

Load spectra were applied using a constant load rate approach, capped at a maximum equivalent frequency of 10 Hz for individual cycles. The selected load rate aimed at minimizing testing time while meeting absolute and relative accuracy of loading, demonstrated for a combination of load frames and coupon geometry, by sequential and statistical analysis of the error for each end point of the first passes of the spectrum. On average, the fatigue loads were applied at an equivalent frequency of approximately 8.25 Hz.

3.3 Periodic Inspections

The life improvement factors were based on the crack initiation life, defined as the time required for a 0.01 inch deep crack to be formed. Whenever possible, crack depth was to be estimated in-situ from periodic non-destructive inspections (NDI).

Initially, inspection of the shot peening coupons started at 80% of the predicted CI life, and then at every subsequent 5% until CI indications could be observed. These inspections were performed using a combination of surface inspection using eddy current testing (ECT) and enhanced liquid penetrant inspection (ELPI) [2]. Quantitative fractography was performed on several coupons of the same geometry and spectrum to calibrate crack depth measurement with these NDI techniques. Examples of such calibrations are presented in Figure 8.

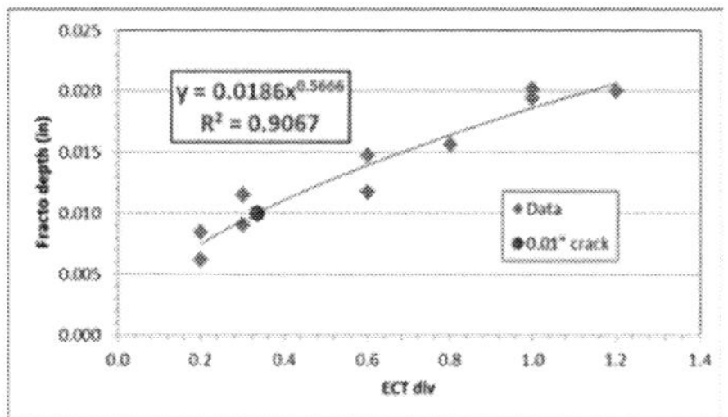

a) Crack depth vs. ECT divisions for baseline coupons

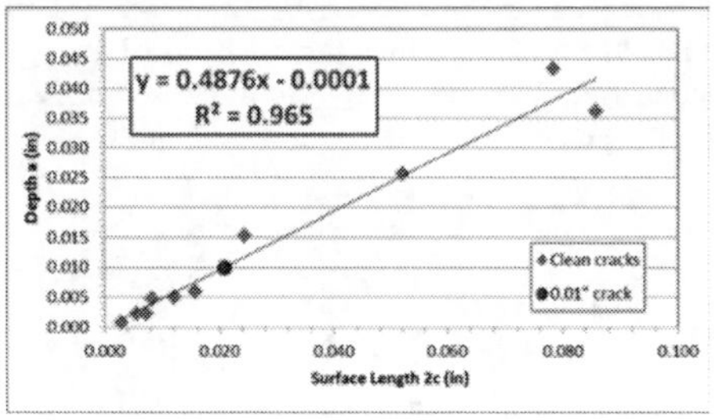

b) Crack depth vs. length (ELPI) for shot-peened coupons

Fig. 8. Calibration of non-destructive inspection techniques

In general, it was observed that ECI provided better crack depth estimations for baseline (non-peened) coupons, whereas ELPI was more efficient on shot-peened surfaces. This method for rapid crack detection has been shown to be able to detect cracks less than 0.004 inch in length in shot-peened surfaces, and able to reliably measure 0.01 inch cracks [2]. For un-etched baseline coupons however, where machining marks ran parallel to the cracks, small crack were difficult to measure with this technique. Examples of cracks observed using ELPI are presented in Figure 9.

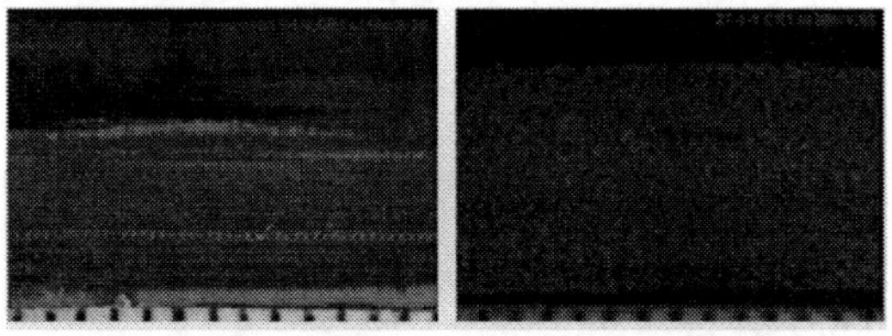

a) Baseline surface b) Shot-peened surface

Fig. 9. Examples of enhanced liquid penetrant inspection.

The two NDI techniques were fast but could provide only estimates of the crack depth. Moreover, crack detection was found to be difficult in some cases, resulting in potentially "missed" initiation. It was found that one to three additional measurements after CI could increase the confidence in the approximate initiation life. However, in some cases it was necessary to open the crack and determine the CI life from quantitative fractography. This technique is much slower but able to measure crack depth more accurately. Moreover, being able to measure the CI life after complete coupon failure, it makes periodic inspection unnecessary, which means that the test frames can be used more efficiently, without interruption.

3.4 Quantitative Fractography

Marker bands were added to the test spectra to allow quantitative fractography analysis of the coupons. The marker band type selected for the shot peening coupons was a series of two cycles with a peak and valley of 85% and -10% of the maximum spectrum load, respectively, followed by 200 cycles with a peak of 75% and valley of 50%. This pattern, which was repeated twice at the end of the spectrum block, was clearly visible for the historical spectrum but difficult to see for the current target spectrum, which self-marked similar bands.

A "bar-coding" type of marker band was used for the fastener tests [3]. In this case, eight different bar code patterns, combining cycles that ranged from 85% to 60% and cycles that ranged from 85% to 10%, were inserted over two passes of the spectrum. The use of different patterns, repeated only every two blocks, significantly increased the confidence in the time-stamping of the marker bands. The different bar code patterns are illustrated in Figure 10.

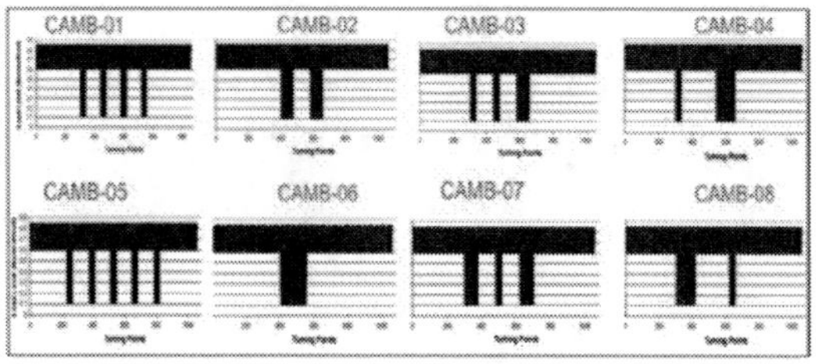

Fig. 10. Bar code marker band scheme (taken from [3]).

Because most of the fastener test coupons were assemblies with the test section being the fastener holes, it was not possible to detect CI using NDI. Indeed, for these coupons the nucleation points were either hidden under fastener collars or heads, or under doublers. Therefore, CI for the fastener test program was determined solely via quantitative fractography performed after complete coupon failure.

4 Results

4.1 Robotic Shot Peening

In total, 166 shot peening coupons were tested. The high life extension efficiency of the modification made possible to waive several test series and test less than the planned minimum of 200 coupons.

4.1.1 Correlation with Previous Test Program

First, a group of 42 coupons were tested to verify if a reasonably simple coupon shape (Figure 1, right) would provide results similar to a previous test program which was using a more complex 3D shape (Figure 11). These tests were carried out at three local stress levels (72.5 ksi, 62.9 ksi, 55.2 ksi) to quantify the sensitivity of the LIF with respect to the applied load.

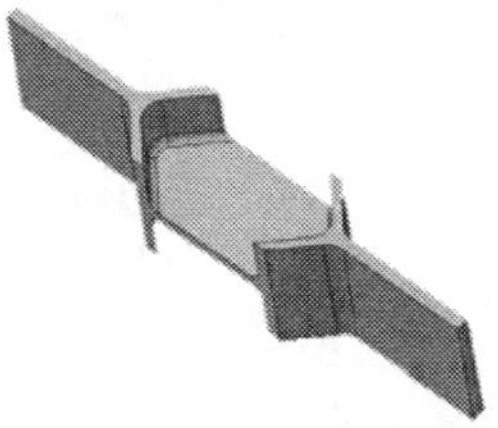

Fig. 11. Three-dimensional coupon from previous program.

Overall, the scatter observed on the new coupon geometry was higher than that on the previous 3D shape. This result was explained by the fact that the hot spot on the previous shape was a point, whereas it is a line on the new coupons. The calculated factored LIF was observed to increase with decreasing local stress, being at 3.89 for the highest load, and 8.17, and 12.18 for the moderate and lowest load, respectively. Furthermore, the scatter in baseline lives was also found to significantly increase when reducing the applied loads. The peened surface, on the other hand, made the life scatter much more consistent

Table 3. Correlation with previous test results.

Local Stress (ksi)	Scatter Factor (baseline / peened)	Unfactored LIF	Factored LIF
72.5	2.26 / 2.66	4.57	3.89
62.9	3.89 / 1.89	3.98	8.17
55.2	5.63 / 2.18	>4.71*	>12.18*

*Three of the peened coupons failed at the grips

These results show that under a local stress of 72.5 ksi the simple geometry was slightly conservative with respect to the previous 3D geometry, for which a factored LIF value of 4.74 had been determined. They confirmed that this stress level was applicable for the main certification tests.

4.1.2 Main Certification Tests

A group of 30 baseline and 30 peened coupons were tested for the main certification tests. Since some of the baseline coupons were rejected because they were manufactured out of tolerance, a second test point was extracted from the coupons, when available. In total, 49 test points were obtained for the baseline coupons, with a scatter factor of 2.08 in life. The manufacturing tolerance was tighten for the subsequent test groups.

The certification methodology was to pre-cycle the coupons to simulate a late rework introduction, at 80% of the blueprint aircraft life. The exact pre-cycling time was adjusted by taking into account the difference between the life scatter observed in the tests and fleet. The repair consisted in a 0.003 to 0.006 inch deep blend followed by robotic shot peening.

It was quickly realized that the rework performed so well at the hot spot that the coupons, not designed to sustain such high lives, started to crack in the grip area, due to fretting, and at the sharp edges of the pocket, which had been distorted at the microscopic scale by peening. These problems were partly resolved by developing a new protective layer to be installed between the grips and the coupons, which delayed and eventually eliminated fretting failure under the grips, and by adding a chamfer on the coupon pocket edge.

Overall, only five of the 30 coupons cracked in the hot spot, with an average factored LIF of 8.4. The other 25 coupons did not reach CI in the hot spot; ten failed from the pocket edge, and 15 failed in the grip area. Out of the five coupons that cracked at the hot spot, two showed clear NDI crack indications before rework and then were cycled up to a LIF of approximately 5.0 and 6.0. The other three cracks had much higher LIF values, between 10.0 and 14.8. Conservatively mixing the two CI lives for which cracks were present before the rework, the three other CI lives determined at the hot spot, and the lives of the 25 coupons that failed before CI was reached at the hot spot (conservative), an average life ratio of 8.4 was obtained, along with a standard deviation twice as large as that of the baselines series. Using these numbers, a conservative factored LIF of 5.0 was determined.

Therefore, although the main certification tests did not allow to perform an accurate statistical evaluation on the variability of extended life due to shot peening, they clearly showed that the considered rework performed as late as 80% of the blueprint life provides a LIF significantly above the required value of 3.0. With these results, several conditional, sensitivity, and optional tests were waived (deeper blend, earlier peen and lower stress, bending stress field, effect of marker bands, effect of spectrum truncation).

4.1.3 Sensitivity Tests

Several sensitivity tests were performed to quantify the effects of various parameters on the calculated LIF. First, the shot peening LIF for cracks on edges, as opposed to pocket radii, was evaluated using the coupons illustrated in Figure 12. All coupons in the baseline series were cycled to rupture, then analysed by quantitative fractography. The lives of the peened coupons, however, were so long that cycling was stopped before CI was detected. A minimum LIF of 16.0 was demonstrated for the edge crack sensitivity tests.

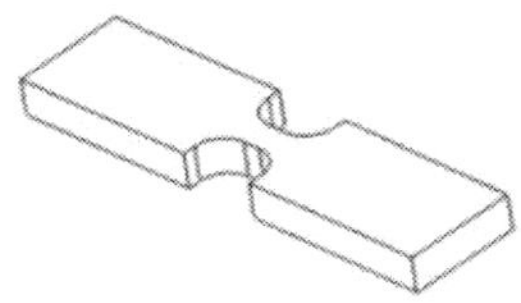

Fig. 12. Shot peened coupon for edge crack sensitivity test.

Another test series was performed to evaluate if the shot peening rework would be less efficient when the loading was in the short transverse (ST) grain direction. A minimum LIF of 9.0 was demonstrated.

Tests were performed to evaluate the effects of applying a 50th percentile usage spectrum severity vs. the 90th severity used for all the other tests. A minimum LIF of 8.9 was demonstrated.

Finally, tests were performed to simulate the scenario where 0.015 inch deep cracks were introduced before the rework. First, the coupons were pre-cycled until NDI suggested that the target crack depth was achieved. Then, the hot spot was modified by a 0.003 to 0.006 inch blend flowed by shot peening over the leftover cracks. The objective of these tests was to estimate the scatter and growth rate of such cracks when not completely removed before peening. As shown in Figure 13, kinks were observed in the crack growth curve between 0.010

inch and 0.012 inch, indicating the depth of the beneficial compressive stress layer. Once passed the layer, the crack growth rate was fairly uniform among peened coupons, and similar to that of the baseline coupons. Therefore, the balancing residual tensile stress did not significantly accelerate crack growth.

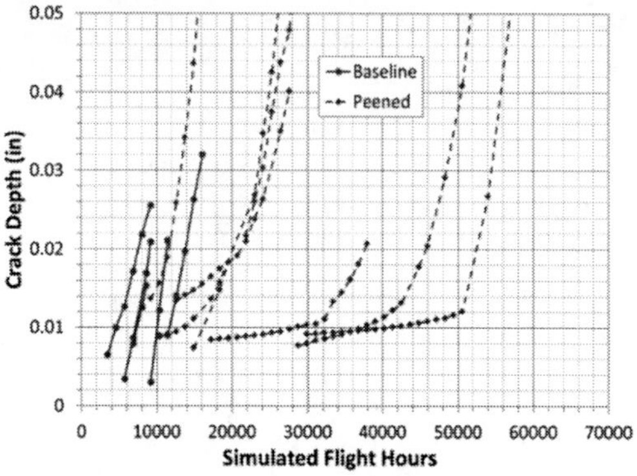

Fig. 13. Crack growth for retrofit on a 0.015 inch crack.

4.2 Fastener Life Improvement Techniques

At the time of writing this paper, testing of the 84 mandatory coupons forming the main fastener test group was completed, and testing of the more complex 3D coupons (bushing repair) was underway. For the fastener tests, all LIF values were calculated using CI lives determined from fractography. A typical example of the crack growth curves obtained is presented in Figure 14.

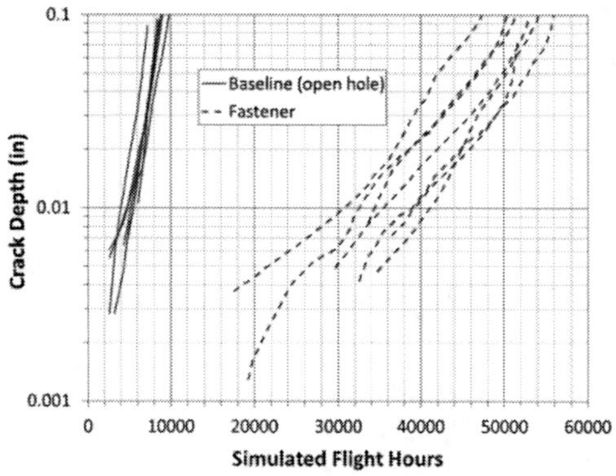

Fig. 14. Typical quantitative fractography on fastener coupons.

Factored and unfactored LIF values resulting from the three fastener types were calculated using baseline open hole coupons subject to the same peak local stress. A summary of the factored LIF values for crack initiation CI is presented in Table 4.

As assumed by the fleet operator, the CI LIF associated with the blind I/F GS fasteners is less than those obtained from the other two fastener systems. The only exception observed was for the of high stress case with load transfer, for which the I/F GS fasteners performed exceptionally well in the tests.

However, the assumption that the LIF is reduced beyond a stress level of 1.5×Fty up to 2.0×Fty cannot be confirmed from the test results. If fact, the test results suggest that the LIF increases when the local stress increases. This result is partly due the fact that the scatter factor is higher for low stress, which reduces the factored LIF.

Table 4. Fastener LIF comparison.

Fastener Type	Load Transfer	Local Stress (×Fty)	Factored CI LIF	Scatter Factor
RPC	Yes	1.5	In process	
		2.0	5.46	1.79
	No	1.5	7.94	1.38
		2.0	8.82	1.74
I/F HL	Yes	1.5	Conditional Tests	
		2.0		
	No	1.5	5.36	3.95
		2.0	8.26	3.19
I/F GS	Yes	1.5	3.27	4.42
		2.0	12.77	1.80
	No	1.5	2.06	4.99
		2.0	5.32	2.74

5 Conclusion

The objective of a shot peening coupon test program carried out recently at NRC was to demonstrate if a minimum LIF of 3.0 could be obtained from a robotic shot peening modification to be implemented at specific airframe critical locations. All test series demonstrated a LIF value above 3.0. Even for the main certification tests that simulated a modification as late as 80% of the blueprint life of the aircraft, the resulting LIF was significantly above the requirement. It was therefore recommended that a LIF of 3.0 could "safely" be assumed for the considered shot peening modifications.

The objective of another life extension certification program, still being carried out at NRC, was to determine if a life extension penalty should be considered when blind interference fit ground shank fasteners (I/F GS) were to be used instead of clearance fit blind fasteners with ring pad coining (RPC), or interference fit Hi-Lok (I/F HL) fasteners. Further, the life extension performance of these three fastener types were to be evaluated for highly loaded situations. Results obtained to date suggest that the I/F GS fasteners do not provide as much life extension as the other two fastener types. Furthermore, contrary to current assumption from the fleet operator, the factored LIF is not reduced when the stress goes from 1.5×Fty up to 2.0×Fty. More tests could be performed with the intent to identify the stress limit where the LIF starts to decrease.

This work was funded by the United States Navy Naval Air System Commands (NAVAIR). The etching process and marker band schemes were provided by the Defence Science and Technology (DST) Group of Australia.

References

1. N.I. Bullen, "A Note on Test Factors", A.R.C. Technical Report, R and M 3166 (1961)
2. M. Yanishevsky, "An Enhanced Fluorescent Liquid Penetrant Inspection Technique for Measurement of Surface Cracks", First Pan-American Conference for Nondestructive Testing, Toronto, Canada (1998)
3. M. Burchill, S. Barter, and M. McDonald, "The Development of Bar-Coded Marker Band Load Sequences to Enhance Fatigue Test Outcomes", 16th Australian International Aerospace Conference, (2015)

Effect of surface compressive residual stress introduced by surface treatment on fatigue properties of metallic material

Jinta Arakawa[1] , *Tatsuya* Hanaki[1], *Yoshiichirou* Hayashi[2], *Hiroyuki* Akebono[1*], and *Atsushi* Sugeta[1]

[1]Hiroshima University, Department of Mechanical Science and Engineering, 1-4-1 Kagamiyama Higashi-hiroshima Hiroshima, Japan
[2] Electric Power Development Co.,Ltd, Chigasaki Kanagawa, Japan

Abstract. This study considers shakedown in evaluating the fatigue limit of metals with compressive residual stress at the surface. We begin by applying tension-compression fatigue tests to ASTM CA6NM under conditions of controlled load and displacement to obtain fatigue limit diagram in compressive mean stress. The results imply that shakedown occurs under the condition of controlled displacement, therefore, shakedown should be considered when evaluating the fatigue limit of metals with compressive residual stress at the surface.

1 Introduction

To date, there have been many attempts to improve the fatigue properties of the materials used to construct industrial equipment, since these materials are subjected to compression and tension, bending, twisting and other forms of stress and strain. Quenching and heat treatment have been found to effectively improve the resistance of metals to compression and tension because the treatment is applied to the entire body of the metal [1-3]. In addition, carburizing, nitriding and shot peening are applied when bending and twisting are anticipated, because these processes increase surface fatigue resistance [4-10]. It is well known that shot peening leads to an elevated surface hardness in conjunction with a high degree of compressive residual stress on the surface of the material, both of which improve fatigue strength [11-14]. There are many different peening techniques, including shot, micro shot, ultrasonic shot, laser, water jet, oil jet and hammer peening. In addition, there have been many reports regarding the fatigue strength of metals subjected to peening, which have demonstrated that peening improves fatigue strength. As noted, peening imparts compressive residual stress to the material surface layer and hence improves fatigue strength. The present work examined the effectiveness of ultrasonic shot peening (USP) at improving the fatigue strength of ASTM CA6NM intended for use in hydraulic turbine runners [20]. According to this work, it is revealed that the shot peened material exhibits greater fatigue strength compared to the untreated metal, with a 60% increase in the fatigue limit. Furthermore, the reason of improved fatigue limit was studied by using modified Goodman line considering shakedown and taking into account for elevated hardness and compressive residual stress. As a result, the equivalent fatigue limit for the USP specimen was calculated to be 508 MPa, which is in good agreement with the experimental result of 540 MPa. Therefore, it is clear that the main factors responsible for improving the fatigue limit were the high hardness and elevated compressive residual stress for the peened metal. According to previous study, the fatigue limit of USP specimen was evaluated by using modified Goodman diagram considering shakedown in the compressive mean stress regime. However, to date, there are a lot of reports of experimental data which is relationship between the fatigue limit and the tensile mean stress, but, in contrast of its case, there have been a very few of reports of experimental data giving the relationship between the fatigue limit and the compressive mean stress. Thus, in order to precisely determine the fatigue strengths of metals based on their compressive residual stress, it is essential to obtain these data, and one goal of the work reported herein was to obtain these values for ASTM CA6NM. In the present study, a tension-compression cyclic test was carried out by using ASTM CA6NM for base material of hydraulic turbine runner to investigate the shakedown behavior.

2 Specimen and experimental procedure

The material employed during this work was cast stainless steel, ASTM CA6NM, and the mechanical properties and chemical composition of this metal are provided in Tables 1 and 2, respectively. Figure 1 presents the shape of the test specimens, which were made by mechanical machining. During the experimental trials, tension-compression fatigue tests were carried out for a mean stress value, σ_m, of -300, -200, -100 or 0 MPa using a hydraulic servo pulsar. The frequency, f, ranged from 10 to 15 Hz and the interrupted number of cycles, N_f, was 10^7.

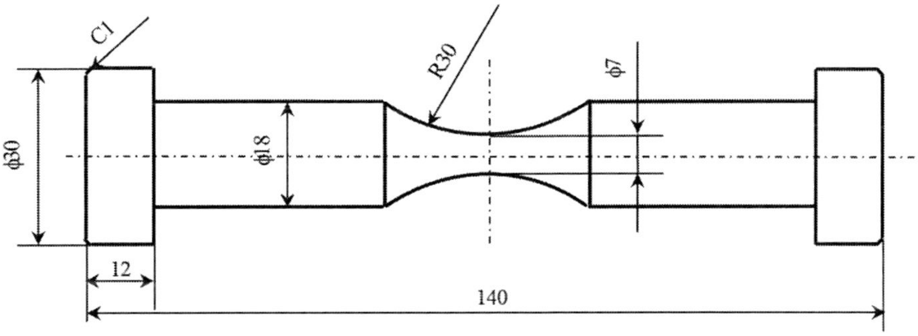

Fig. 1 Schematic diagram showing dimensions of test specimen used for fatigue tests.

Table 1. Mechanical properties of ASTM CA6NM.

HV	Yeild stress	Tensile strength
280	594 MPa	829 MPa

Table 2. Chemical composition of ASTM CA6NM (mass%).

C	Si	Mn	P	S	Ni	Cr	N	Al	Ca	O
0.049	0.5	0.83	0.04	0.004	3.62	12.82	0.027	0.01	<0.0005	0.00063

3 Experimental results

3.1. Cyclic hardening behaviour of ASTM CA6NM

In our study, in order to know the elastic-plasticity properties subjected to experimental cyclic, the multistage amplitude method was applied to ASTM CA6NM. The multistage amplitude method is that strain amplitudes are changed at the appropriate intervals. Figure 2 shows the schematic illustration of the multistage amplitude method. Furthermore, Fig. 3 indicates the results obtained by that method, and Fig. 3 collects the loop tips of hysteresis loops. According to Fig. 3, 0.2 % proof stress was 810 MPa, when the applied stress values of loop tips was saturated. The 0.2 % proof stress of 810 MPa is higher than the value of 594 MPa indicated in Table 1. Therefore, it was revealed that ASTM CA6NM exhibited cyclic hardening behaviour, by applying multistage amplitude method.

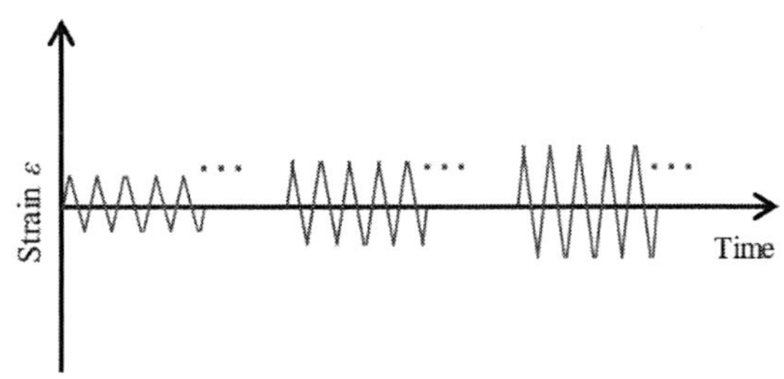

Fig. 2 Schematic illustration of multistage amplitude method.

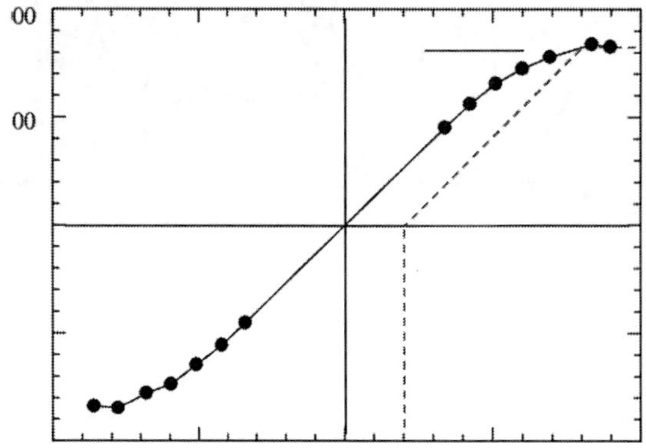

Fig. 3 Loop chips of the last cycle of each strain stage.

3.2. Fatigue limit diagram for ASTM CA6NM under controlled load condition

Stress amplitude versus number of cycle (*S-N*) curves (Fig. 4) were generated by applying a controlled load and σ_m values of -300, -200, -100 and 0 MPa. It is evident from these data that the fatigue limit increased as the mean stress decreased. And also, the fatigue limit was defined the value between a minimum stress amplitude in finite life regime and a maximum fatigue limit in experimental data, based on JSMS standard* ''Standard Evaluation Method of Fatigue Reliability for Metallic Materials: Standard Regression Method of *S–N* Curves''. The associated fatigue limit diagram is shown in Fig. 5, which demonstrates that fatigue limits in the negative mean stress values agree more with a modified Goodman line un-considering shakedown than with that line considering it. Figure 6 plots the relationship between piston displacement and applied stress up to 30 cycles at a σ_m value of -300 MPa and a σ_a value of 430 MPa. These data demonstrate that shakedown did not occur and that the sample was subjected to a fixed stress. Therefore, under conditions involving a controlled load, shakedown does not take place, and the fatigue limit diagram agrees with a modified Goodman line un-considering shakedown.

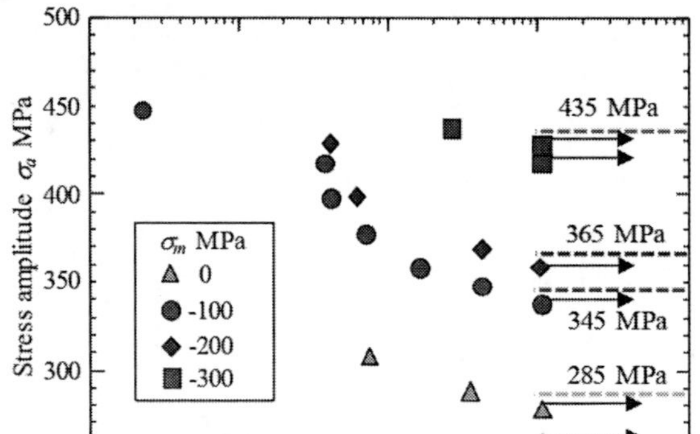

Fig. 4 *S-N* curves for ASTM CA6NM specimen.

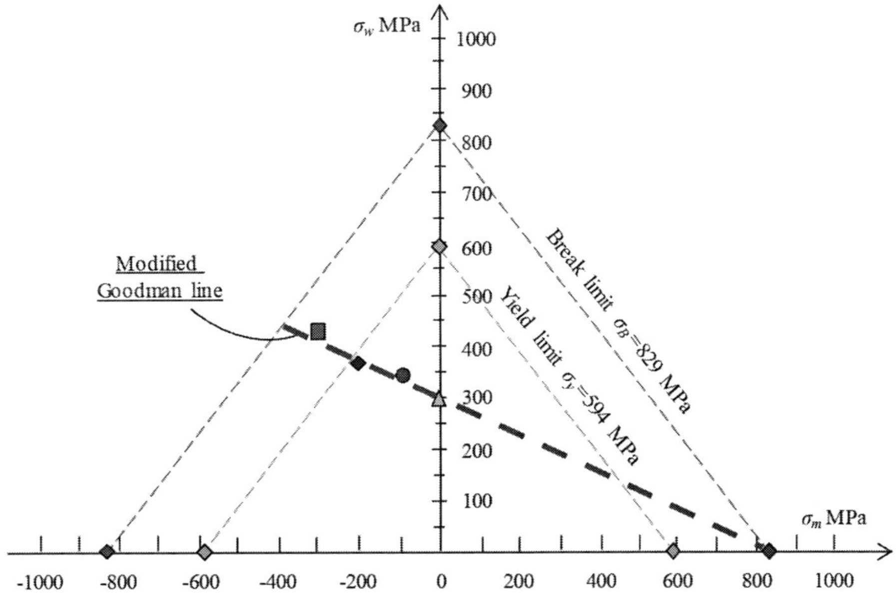

Fig. 5 Fatigue limit diagram for ASTM CA6NM specimen.

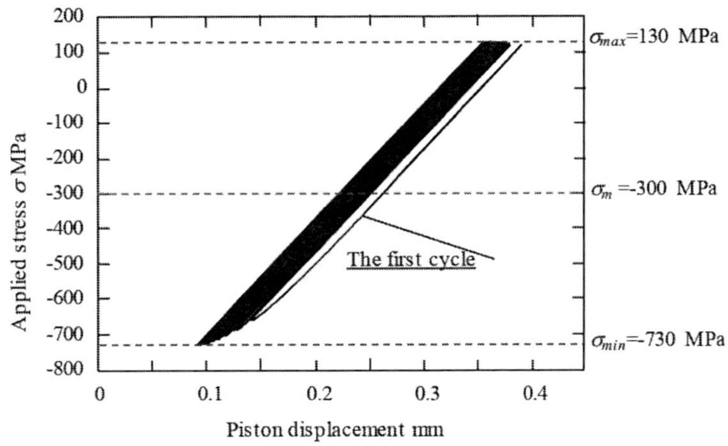

Fig. 6 The relationship between applied stress and piston displacement for ASTM CA6NM specimen.

3.2. Hysteresis loop for ASTM CA6NM under controlled displacement condition

In Section 3.1, the fatigue limit was estimated by tension-compression fatigue tests under controlled load conditions. In the trials reported in this section, hysteresis loops were obtained while applying a controlled displacement, in order to investigate the effect of shakedown. The displacement values were determined based on Fig. 6 by assuming complete elasticity. Figures 7(a), 7(b) show the relationships between applied stress and displacement for maximum piston displacements of -1.57×10^{-2} and -3.14×10^{-2} mm, minimum piston displacements of -2.59×10^{-1}, -2.75×10^{-1} mm and initial mean stress values of -425 and -475 MPa, respectively. Each experiment consisted of 30 cycles. It is evident from these data that the initial loop moved upward during subsequent cycles in whole loop. Therefore, it appears that shakedown occurred under these test conditions (that is, with controlled displacement). The mean stress values in Figs. 7(a) and 7(b) moved from -425 to -340 MPa, and from -475 to -350 MPa, respectively. Additional fatigue tests were conducted applying a σ_m value of -300 MPa and a σ_a value of 400 MPa in conjunction with controlled displacement. These trials resulted in an N_f of 7.2×10^5 cycles. These experimental conditions were equivalent to the fatigue limit under controlled load conditions. However, the material broke at 7.2×10^5 cycles under controlled displacement conditions. This early breakage is attributed to shakedown, meaning that the mean stress was transitioned to upper. Based on these data, it is apparent that estimations of the fatigue limit of a metal having a compressive residual stress at its surface layer, for the purposes of evaluating safety, should be conducted based on the associated modified Goodman line considering shakedown due to the appearance of it.

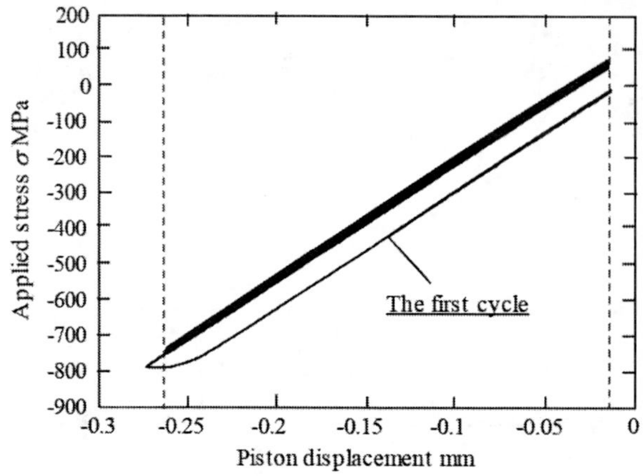

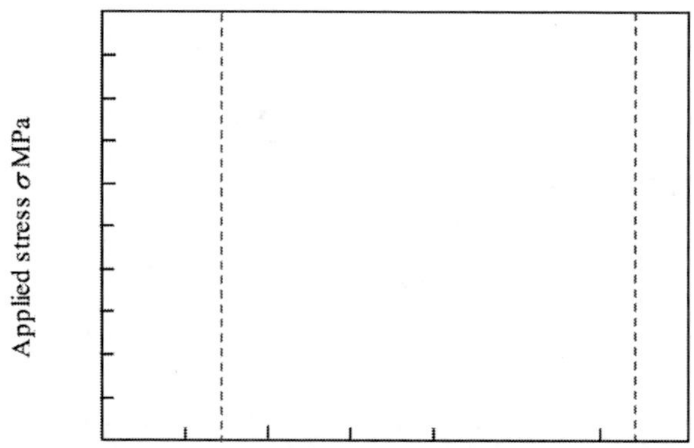

Piston displacement mm

(b) Maximum: -3.14×10^{-2} mm, Minimum: -2.75×10^{-1} mm

4 Conclusions

This study examined the effects of compressive residual stress on metals by conducting tension-compression fatigue tests by using ASTM CA6NM material under either controlled load or controlled displacement scenarios. The conclusions of this work are as below.

1. It is clear that the relationship between the fatigue limit and the mean stress agrees with modified Goodman line un-considering shakedown, when applied stress condition is controlled load. Therefore, in this condition, it is found that the sample is subjected to fixed applied stress and shakedown doesn't occur.

2. It is revealed that the shakedown occurs in the controlled displacement condition. Therefore, estimation of the fatigue limit of the material with compressive residual stress should be conducted by using modified Goodman line considering shakedown, in order to evaluate the fatigue limit safety.

References

1. Navid D, Atefe R, Seyyed M A, Mater. Sci. Eng. A, **683**, 1 (2017)
2. Bo W, Jianxun Z, Linjie Z, Young-Shik P, Ri-ichi M, Appl. Surf. Sci, **321**, 318 (2014)
3. Tatsuro M, Satoshi T, Susumu N, Mater. Sci. Eng. A, **669**, 127 (2016)
4. Farokhzadeh K, Edrisy A, Mater. Sci. Eng. A, **620**, 435 (2015)
5. Winck L B, Ferreira J L A, Araujo J A, Surf. Coat. Technol, **232**, 844 (2013)
6. Jörn B, Natalie S, Int. J. Fatigue, **82**, 35 (2016)
7. Fernández-Pariente I, Bagherifard S, Guagliano M, Ghelichi R, Eng. Fract. Mech, **103**, 2 (2013)
8. Edoardo C, Carlo G, Piermaria D, Marco B, Eng. Fail. Anal, **78**, 41 (2017)
9. Achintha M, Nowell D, Fufari D, Sackett E E, Bache M R, Int. J. Fatigue, **62**, 171 (2014)
10. Yoshinobu S, Koichiro N, Hitoshi I, Keiichiro T, Tomoyuki F, Tooru Y, Masamichi H, Int. J. Fatigue, **60**, 57 (2014)
11. He B Y, Soady K A, Mellor B G, Gary H, Reed P A S, Int. J. Fatigue, **82**, 280 (2016)
12. Zhou J Z, Huang S, Zuo L D, Meng X K, Sheng J, Tian Q, Han Y H, Zhu W L, Opt. Laser. Eng, **52**, 189 (2014)
13. Lihua Z, Yanjin G, Yanjie W, Zhendong X, Jun L, Jiqiang Z, Surf. Coat. Technol, **317**, 38 (2017)
14. Dalaei K, Karlsson B, Int. J. Fatigue, **38**, 75 (2012)
15. Jin G, Di S, Zhou W, Peng L, Weiguo W, J. Const. Steel. Research, **126**, 74 (2016)
16. Pant B K, Pavan A H V, Prakash R V, Kamaraj M, Int. J. Fatigue, **93**, 38 (2016)
17. You L, Liqun L, Lina S, Int. J. Fatigue, **98**, 121 (2017)
18. Ramos R, Ferreira N, Ferreira J A M, Capela C, Batista A C, Int. J. Fatigue, **92**, 87 (2016)
19. Azmir A, Christian S, Claudia G, Jens G, Appl. Surf. Sci, **388**, 468 (2016)
20. Jinta A, Motoki K, Yoshiichirou H, Ryota T, Hiroyuki A, Masahiko K, Atsushi S, Surf. Eng, **30**, 662 (2014)

Predictive Modeling of Arbitrary Loaded TMF Crack Growth in Superalloys

W. Steven Johnson [1], *Richard W.* Neu [1], *Andrew T.* Radzicki [2], *Benjamin* S. Adair [3], and *Vincent M.* Barker [4].

[1]*Georgia Tech - Atlanta, GA (United States of America)*
[2]*US Air Force - Colorado Springs, CO (United States of America)*
[3]*Self employed - Oskaloosa, IA (United States of America)*
[4]*Lockheed Martin - Marietta, GA (United States of America)*

Abstract. Hot sections in turbine engines are subjected to large variations in temperatures and mechanical/thermal loadings. As such, accurate predictions of fatigue crack growth must account for many physical phenomena: temperature dependent crack growth behavior, load interaction history effects, time at temperature effects, temperature history effects, and the effects of stress/temperature/time on the materials. Through extensive experimental work on superalloys, a very definite "temperature history effect" on the resulting crack growth behavior has been identified and modeled. This work also identified a Temperature Affected Zone (TAZ) that occurs ahead of the crack tip and affects subsequent crack growth rates. The size of the TAZ is dependent on temperature, hold time, and stress state. Measurements of the TAZ were made under various conditions. The changes that occur in this TAZ are a combination of oxidation and material microstructure evolution. Various simplified "hot section" engine spectra (changing temperatures and stress levels) were tested to determine resulting crack growth behavior. Correlation between the experiments and model predictions were good and generally conservative.

1 Introduction

The world is dependent on jet turbine aircraft for travel and commerce. Although the technology is relatively mature, there are still unexpected failures as a recent high publicity engine failure highlighted: an American Airlines flight in Chicago, Illinois in 2017 [1]. Final reporting blamed an undetected manufacturing flaw in one of the alloy 718 turbine disks that grew by low cycle fatigue until catastrophic failure [1]. As shown in Figure 1, the typical spectra turbine disks are exposed to are complex combinations of thermal and mechanical loading.

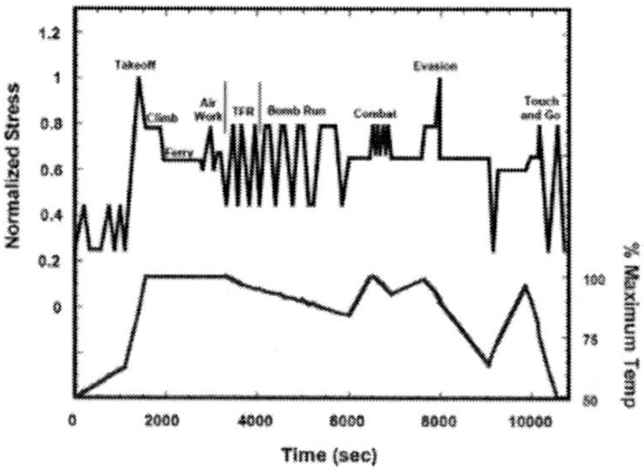

Fig. 1. Typical temperature and loading profile for military fighter aircraft high temperature rotor components [2]

Using a damage tolerant approach, where an initial flaw is assumed, predicting fatigue crack growth rate (FCGR) in these situations is especially difficult. The authors sought to develop a robust model for predicting crack growth in high temperature turbine engine alloys, specifically in nickel-base superalloys.

1.1 Background

Various studies have been undertaken to characterize the impact on FCGR of slow frequency loading at high temperature on Ni-base superalloys [3, 4, 5, 6]. In these situations, the combination of time and high temperature allows for the material ahead of the crack tip to become weakened and leads to intergranular and faster subsequent crack growth [6].

Though not the only active mechanism contributing to accelerated FCGR, oxidation ahead of the crack tip plays a significant role. The high stress at the crack tip exacerbates oxidation penetration, but not necessarily in a uniform manner. Pfaendtner and McMahan showed that in an oxidizing environment, certain grain boundaries are more susceptible to embrittlement [7]. The crack will tend to propagate along these embrittled grain boundaries faster than the rest of the crack front. This was corroborated by Viskari et al. [8]. Additionally, Viskari et al. provided a schematic of how uneven oxidation at the crack tip will lead to an uneven crack front, shown in Figure 2.

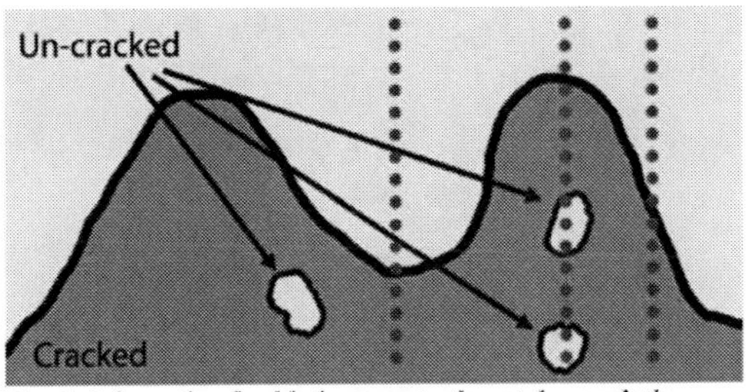

Fig. 2. Schematic of oxidation penetration at the crack tip leading to an uneven crack front and areas of uncracked material [8]

Work done on the depth of oxidation penetration at the crack tip by Miller et al. characterized how far ahead of the crack tip oxidation penetrated. They conducted a time-dependent crack growth test at 700°C and used X-ray photoelectron spectroscopy to analyze how far ahead of the crack tip oxidation had penetrated and showed an oxidation affected region extending 150μm ahead of the crack tip [9].

The concept of an area of material ahead of a crack tip, termed "damage zone" or "thermally affected zone," weakened by exposure to time spent at elevated temperature in a FCG situation for IN 718 has been explored by several authors [10, 11, 12, 13, 14]. In their work, the specific contributions of the different crack growth mechanisms (oxidation, creep, materials evolution, etc.) are not assigned, only the net result on the FCGR is taken into consideration.

It is important to note that though several of these authors use the term "damage zone," to avoid confusion with the concept of fatigue damage or plastically induced damage, the terminology of "temperature affected zone (TAZ)" was utilized for this effort. The concept states that while a crack is propagating through this TAZ, an increased FCGR will occur. Once the crack grows through the TAZ, the FCGR will return to the expected FCGR had no exposure to time at elevated temperature occurred. The authors use the term TAZ to indicate that the effect of time at temperature on crack growth is separated from the "normal" crack growth damage that occurs due to cyclic loading.

Work done by Storgards et al. showed that a crack propagating through this TAZ will have an intergranular appearance [14]. This intergranular appearance is similar to that for a low frequency, high temperature, and air environment situation [6]. This indicates that weakening of the grain boundaries plays a significant role in the effect of the thermally affected zone on FCGR. Along with areas of intergranular failure, areas of transgranular crack propagation are also present indicating that the area ahead of the crack tip is not fully dominated by one mechanism. Areas ahead of the crack tip not significantly weakened by exposure to time at high temperature (transgranular appearance) are left to hold back the crack tip when cycling resumes as the weakened material (intergranular appearance) puts up less resistance. Of course, whether the intergranular cracking happens depends upon the temperature, hold time, and material. Lundstrom et al. showed that for a specimen subject to fully time-dependent crack growth at 650°C, and then subsequently cycled, an increase in FCGR on the order of three-hundred times as fast is experienced [13].

2 Experimental Work

Robust experimental studies were undertaken to characterize the impact of spectrum loading on Ni-base superalloys. The effect of previous mechanical over- and under-loading was accomplished by Barker et. al. and Adair et. al. [10, 15]. Additionally, the depth of oxidation penetration at the crack tip for compressive and tensile holds of varying temperature, hold duration, and stress intensity at the crack tip was explored [16]. This paper focus on the latter experimental study.

2.1 Material Description

Studies were conducted on the powder metallurgy Ni-base superalloy Inconel 718, a polycrystalline alloy commonly used for turbine engine applications up to 650°C. Inconel 718 is primarily strengthened through precipitate hardening. As with other Ni-base superalloys, it is primarily strengthened by γ" and γ' phases in a γ matrix [17]. The material showed no preferred grain orientation and had an average grain diameter of 4μm. Additionally, the IN 718 exhibited a primarily intergranular distribution of globular δ precipitates. The nominal chemical composition of powder metallurgy IN 718 is shown in Table 1.

Table 1. Nominal chemical composition (wt. %) of powder metallurgy IN 718 [18]

	Ni	Cr	Fe	Nb	Mo	V	Ti	C
IN 718	Bal.	18.6	18.5	5.0	3.1	0.9	0.4	0.04

The material for this study came in the form of a forged turbine disks heat treated in a typical manner for a turbine disk application. The turbine disks were provided by the Pratt and Whitney- United Technologies Corporation. Single Edge Notched Tension (SENT) specimens were machined from the forged turbine disks such that the loading direction on the specimen was in the radial direction on the disk as shown in Figure 3.

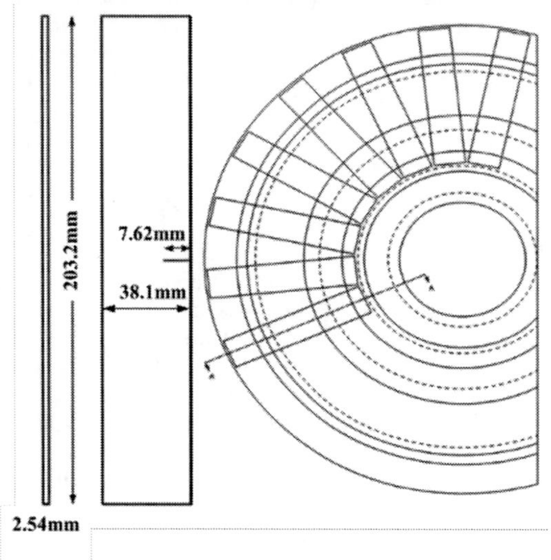

Fig. 3. Specimen geometry and the orientation they were machined from the turbine disk

Load controlled fatigue crack growth tests were conducted in accordance with ASTM Standard E647 utilizing an 89kN servo-hydraulic load frame utilizing a TestStar IIs controller and water cooled wedge grips [19]. The specimens were heated with a 3.5kW induction heater.

Cracks were measured both optically using a Questar QM-100 long distance microscope for all tests and using the direct current electric potential difference (EPD) method described in ASTM Standard E647 [19] for several test points. Power for the EPD crack measurements was proved by a Sorensen SRL10-25 power supply at a constant 5 amps. The signal was amplified with an Omni-Amp III set to 1000x and recorded with the TestStar IIs. The induction heater induced significant noise in the EPD signal, so readings for crack length were only taken utilizing the EPD system with the induction heater turned completely off (i.e. at room temperature). The EPD voltage signal was converted to crack length using HH Johnson's equation [20].

2.2 Thermomechanical Spectra

Several different spectra were designed to isolate the effect of time spent at elevated temperature. First, simple isothermal loading spectra were designed based on the work of Liu et al. to isolate the effects of hold duration, hold temperature, and hold stress intensity on the development of a TAZ and its effect on FCGR [12]. Similar creep-fatigue spectra, such as that used by Gustafsson et al. do not specifically isolate the effect of a single tensile hold at high temperature [11]. To isolate the impact of a single high temperature hold on the development of a TAZ, a test series including tensile holds at three temperatures: 550°C, 600°C, and 650°C, five hold durations: 30 seconds, 90 seconds, 300 seconds, 3600 seconds, and >15000 seconds, and three hold stress intensities: 25 MPa √m, 35 MPa √m, and 45 MPa √m was designed. These tests involved 1) cyclically growing the crack at least 0.25mm (R = 0.05) at low stress intensities (on the order of 9 MPA √m) to minimize effects of loading history on the tensile hold (the temperature for cycling was either 260°C (optical measurement) or room temperature (EPD crack measurement)), 2) executing the tensile hold at high temperature, and 3) cyclically growing the crack until it grew through the TAZ (R = 0.5) as shown in Figure 4. Thirty six tests were completed as shown in Table 2 below.

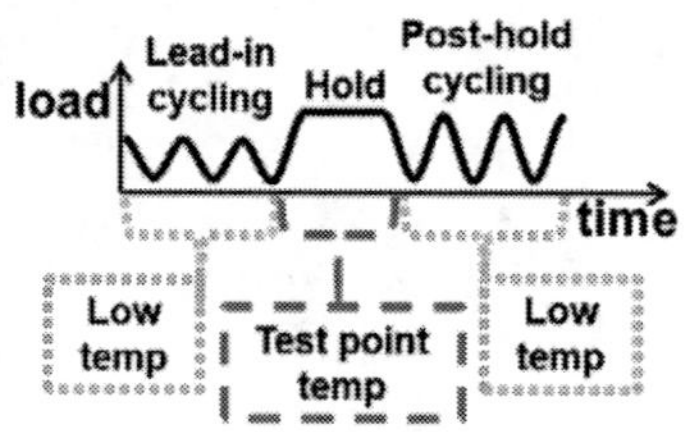

Fig. 4. Single TAZ experimental spectrum

Table 2. Isolated TAZ test matrix

Stress Intensity of Hold (MPa √m)	Hold Temp (°C)	Hold Duration (sec)
25	550	90, 300, 3600, 18200
	600	30, 90, 300, 3600
	650	30, 90, 300, 3600
35	550	90, 300, 3600, 18200
	600	30, 90, 300, 3600
	650	30, 90, 300, 3600
45	550	90, 300, 3600, 15675
	600	30, 90, 300, 3600
	650	30, 90, 300, 3600

Additionally, to better capture spectrum loading history effects, several spectra were developed. First, isothermal test points at 650°C were executed with holds at 0%, 25%, 50%, 75%, and 100% of the maximum cyclic load (R = 0.05). Second, out-of-phase (OP) thermomechnical fatigue (TMF) spectra with tensile holds at 50% of the cyclic maximum load were designed to highlight the effect of increasing or decreasing load leading into the tensile hold. Figure 5 shows both spectra.

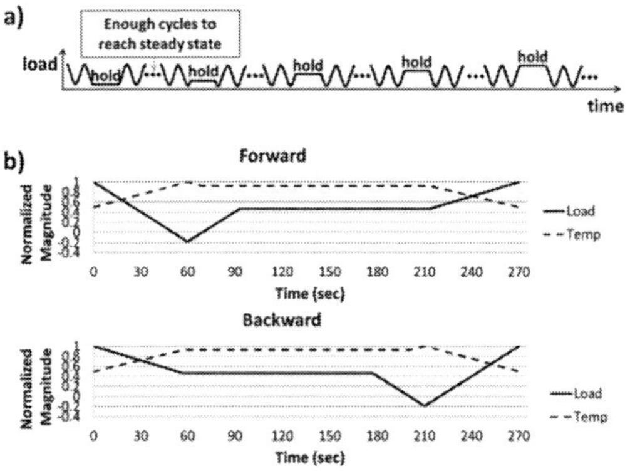

Fig. 5. Spectra to isolate spectra effects on TAZ: a) isothermal with intermittently imposed tensile holds b) OP TMF spectra

In addition to more fundamental TMF spectra used to characterize TAZ development and sequence effects, several different representative TMF aircraft spectra were developed with maximum temperatures of 650°C. Tested representative fighter and transport spectra are shown in Figure 6. Additionally, a spectrum developed by an aircraft engine consortium was tested and is shown in Figure 7 and included a compressive high temperature hold.

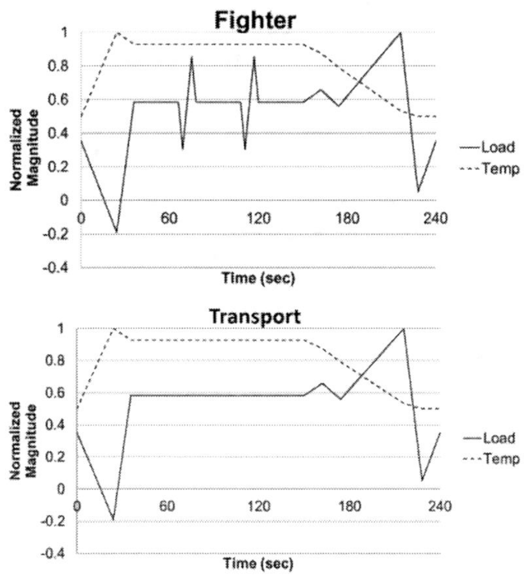

Fig. 6. Aircraft TMF spectra

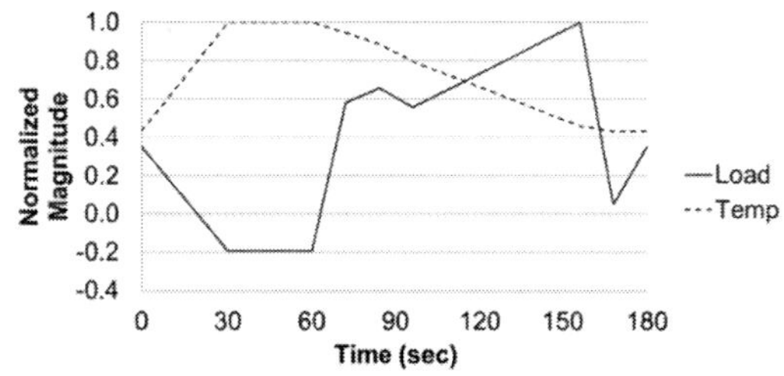

Fig. 7. Engine consortium developed TMF spectrum

3 Experimental Results

3.1 TAZ Fractography

Single tensile holds at the maximum cyclic load resulted in clearly visible change to the fracture surface. Fractographic investigation of the failure surface identified the characteristics of the TAZ as shown in Figure 8. It shows primarily transgranular crack propagation while the specimen was cycled at low temperature, as expected. The area ahead of the crack tip immediately following the hold however shows areas of intergranular as well as transgranular crack propagation. This agrees with the work of Storgärds et al. [21]. Additionally, secondary cracking was observed extending parallel to the crack growth direction. Another significant characteristic of the TAZ is that it extends further ahead of the crack tip in the center of the specimen (plane strain dominated) compared to the edges of the specimen (plane stress dominated). This indicates that it is more conducive for the TAZ to develop in an area of higher hydrostatic stress (plane strain). This higher level of stress strains the grain boundaries allowing oxygen to penetrate further than on the free surface.

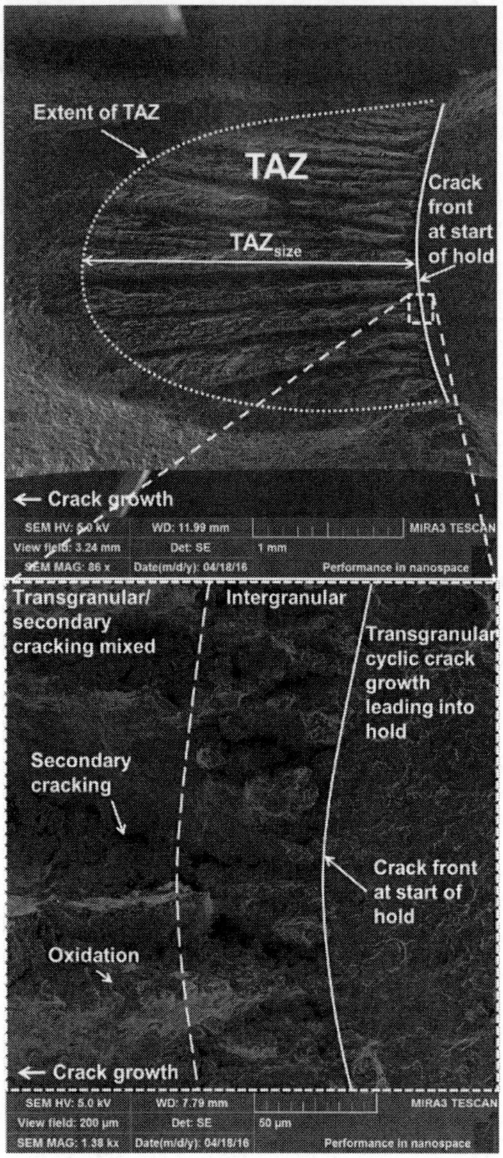

Fig. 8. TAZ appearance for K_{hold} = 45 MPa √m, Temp = 650°C and hold duration = 3600 sec

To some extent, these characteristics were present on the failure surface for all test points identified in Table 2. A qualitative comparison of the crack surface morphology for the test points is shown in Figure 9. As the stress intensity of hold, temperature of hold, and/or the duration of hold increases, the crack morphology transitions from one showing limited effects of weakening of the material ahead of the crack tip to one showing increased intergranular cracking, secondary cracking, and oxidation.

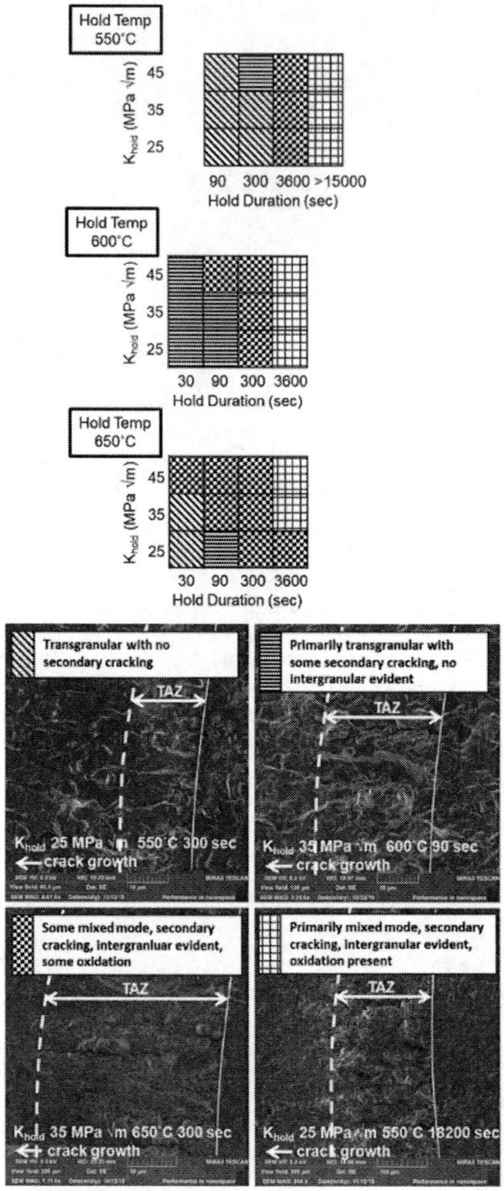

Fig. 9. Qualitative characteristics of the TAZ

3.2 TAZ Size Empirical Fit

To quantitatively characterize TAZ size, the depth of damaged material was measured at the center of the specimen as shown in Figure 8. Across the test points shown in Table 2, the TAZ size showed a power-law relationship as the tensile hold stress intensity, duration, and temperature were changed. Additionally, as diffusion of oxygen is a significant contributor to TAZ development, an Arrhenius type equation was used to fit TAZ development. The equation used to model TAZ size is given by:

$$TAZ_{size} = Ct^{n_1} K_{hold}^{\ \ n_2} e^{n_3 \frac{-Q}{RT}} \quad (1)$$

where TAZ_{size} is in mm, t is the hold duration in seconds, K_{hold} is the stress intensity of the hold in MPA √m, T is the temperature of hold in Kelvin, and R is the universal gas constant. Work by Chang showed the activation energy for high temperature oxidation of Ni-base superalloys as 25500 J/mol and was used for this model [22]. Regression analysis was used to determine the parameters C, n_1, n_2, and n_3. Table 3 shows the parameters used in this model.

Table 3. TAZ_{size} calculation parameters

C	0.291	$(sec)^{-n_1}(MPa\sqrt{m})^{-n_2}$
n_1	0.48	
n_2	2.6	
n_3	0.38	
Q	255000	J/mol
R	8.31446	J/mol*K

3.3 Impact of TAZ on FCGR

Initially, crack length measurements were collected only optically, however, as is evident in Figure 8, the TAZ does not reach the free surface of the specimen, and as this is where the optical method measures the crack length, no impact of FCGR was observed. Due to limited specimens available, only three test points were re-accomplished using both the optical method of crack length measurement as well as the EPD method. The FCGR data for these test points indicated that the crack growth rate does accelerate while it is growing through the TAZ as shown in Figure 10.

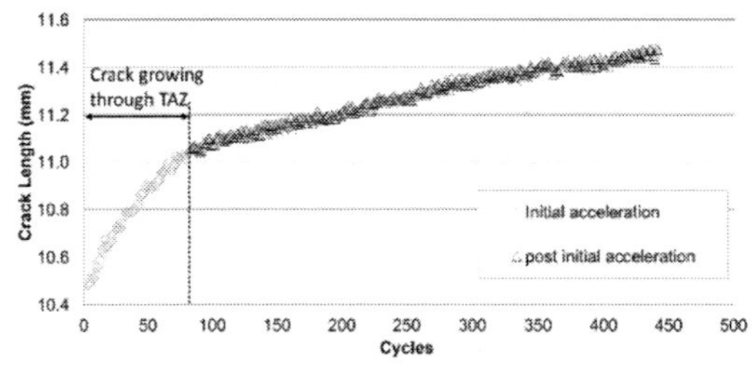

Fig. 10. Crack growth acceleration immediately following tensile hold at K_{hold} 45 MPA √m, hold duration = 3600 sec, hold temperature = 650°C

This increase in FCGR due to a TAZ was on the order of three to five times faster, not the three hundred times increase observed by Lundstrom et al. [13]. This acceleration more closely corresponds to the work of Barker et al. who observed four times increase in FCGR [10]. A TAZ acceleration factor (TAF) was developed to characterize the accelerating effect of the TAZ. The TAF is a multiplication factor on the crack growth rate and takes the form:

$$TAF = 1 + D\left(\frac{TAZsize_{current}}{r_p}\right)^{p} \quad (2)$$

where $TAZ_{sizecurrent}$ is the current TAZ size in mm, r_p is the plastic zone size in mm (based on plane stress/strain calculation in the model), and D and p are fitting parameters where D = 6.0 and p = 0.5. The measured FCGR accelerations due to TAZ showed a good correlation as shown in Figure 11.

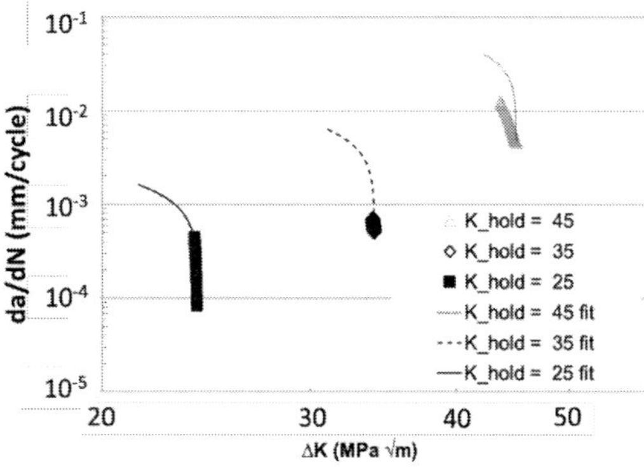

Fig. 11. TAZ impact on crack growth rate using equation (2)

3.4 Spectrum Effects on TAZ Development

Utilizing isothermal and TMF spectra as shown in Figure 5 and Figure 7, the impact of spectrum history on TAZ development was also characterized. Using the consortium spectrum, shown in Figure 7, compressive holds at elevated temperature were shown to have no accelerating effect on FCGR [23]. Additionally, using the isothermal spectrum, shown in Figure 5, tensile holds at 75% of the prior cyclic maximum load or less did not show development of a TAZ. This effect corresponds to the work of Ponnelle et al. [24].

The OP TMF spectra, shown in Figure 5, revealed the progression of loading (increasing or decreasing) leading into a tensile hold will determine if a high temperature tensile hold will develop a TAZ. If the load is increasing leading into a hold, the "forward" spectrum, a TAZ was observed to develop. Correspondingly, if the load is decreasing leading into a hold, the "backward" spectrum, a TAZ was not observed to develop.

4 FCG Modeling

4.1 Assumptions and Algorithm

Modeling of the TAZ's effect on FCGR was incorporated into a linear elastic fracture mechanics based crack growth model originally developed by Johnson, the Multi-Parameter Yield Zone model [25]. His isothermal model effectively accounted for the impact of previous over- and under-loading on subsequent fatigue crack growth. To expand this model to the thermomechanical spectra of rotor alloys, Baker et al. and Adair et al. completed an experimental study to characterize how mechanical over- and under-loads affect FCG when temperature is varying and incorporated their results to create the MPYZ-Thermomechanical Fatigue (MPYZ-TMF) model [10, 15].

The MPYZ-TMF model uses the stress intensity solutions for several common test specimen geometries. The user provides temperature dependent material properties, including: yield strength, fracture toughness, and crack growth rate parameters for either a Paris or Forman crack growth law with properties at intermediate temperatures being linearly interpolated. Additionally, load interaction and TAZ parameters are also provided by the user.

Experimental data suggests that TAZ develops when the crack is open at elevated temperatures [16]. Additionally, a general downward trend in loading will mitigate TAZ development. To account for these characteristics in the model, TAZ was only allowed to develop when:

1. Segment temperature is above 500°C (selected based on work of James et al. who showed negative effect of time at temp down to 538°C and Sadananda and Shahinian who showed no effect at 425°C and below [26, 27].
2. Stress intensity is greater than or equal to zero
3. Stress intensity is greater than or equal to 75% of any of the three previous segments.

Additionally, it is feasible that segments of a realistic spectrum may traverse these limits. To ensure TAZ is calculated to develop during portions of such segments, each segment was partitioned into 5 sections. With experimental data for the TAZ developed for high temperature hold durations from 30 to 18200 seconds, partitioning into too many sections could take the TAZ development calculation outside the range of the empirical TAZ data [16]. The use of five sections was a good balance between fidelity and computational efficiency. Figure 12 shows the sample fighter aircraft spectra from Figure 6 highlighting the portions where the model predicts TAZ will develop.

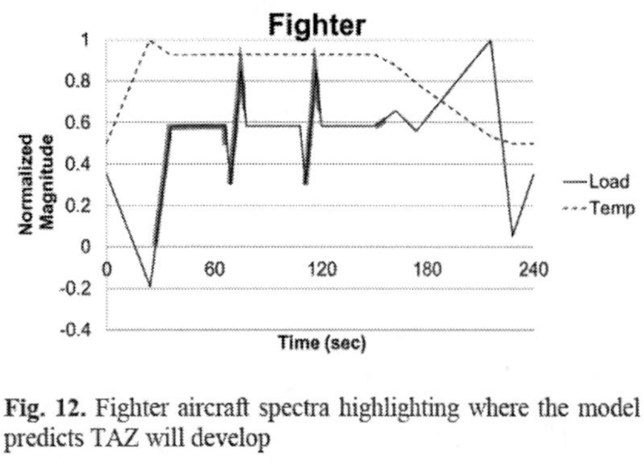

Fig. 12. Fighter aircraft spectra highlighting where the model predicts TAZ will develop

Since oxidation is a primary contributor to TAZ development, an equation similar to the parabolic oxygen diffusion law used by Ghonem et al. was used in the model to limit the addition of a new section's TAZ to existing TAZ [6]. TAZ portions are summed using:

$$TAZ_{size_{newtotal}} = TAZ_{size_{existing}} + \frac{TAZ_{size_{newsegment}}}{\left(TAZ_{size_{existing}}+1\right)^{TAC}} \quad (3)$$

where TAZ_{size} is in mm and the TAC is the TAZ accumulation coefficient (TAC) and was chosen as 2.0. This forces the contribution of new TAZ to diminish when a significant TAZ is already present and the trend is shown in Figure 13.

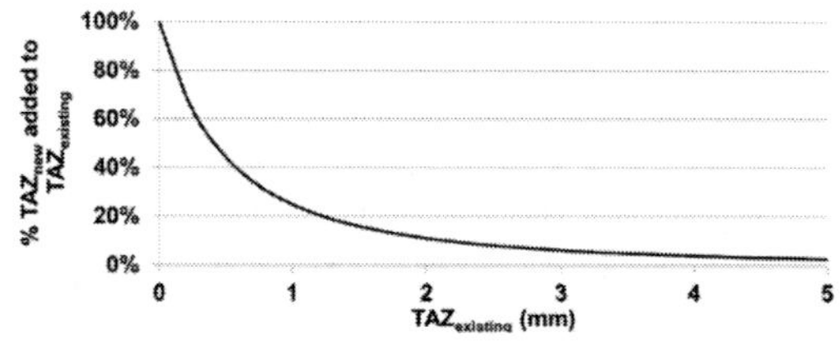

Fig. 13. Percentage of new TAZ that contributes to existing TAZ

One significant limitation of the MPYZ and the first iteration of the MPYZ-TMF model was its inability to account for complex TMF spectra such as that shown in Figure 1 without first simplifying by some method such as rainflow counting. To be able to account for such spectra, the current version of the model calculates crack growth load segment by load segment where crack propagation is only allowed to occur on segments with increasing stress intensity. The average temperature for the segment used for calculating crack growth. This also allows for the easy incorporation of TAZ development described earlier.

4.2 Predictions

For isothermal loading, the model predicts conservative results as shown in Figure 14. The model predicted more conservative results for fast frequency loading, where TAZ effects are not expected to impact FCGR, compared to slower loading which more closely matches experimental data, though is still conservative.

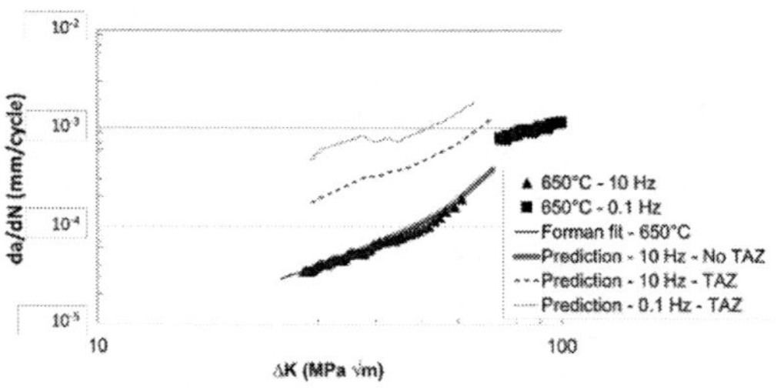

Fig. 14. MPYZ-TMF predictions for 650°C isothermal cyclic loading with and without TAZ effects compared to experimental data and Forman crack growth equation fit

The model predicted good results for the representative aircraft spectra as shown in Figure 15. In this case, the model also accurately predicts large TAZ development for the transport spectrum but a longer overall life because of the lack of significant subcycles to grow the crack as in the fighter spectrum.

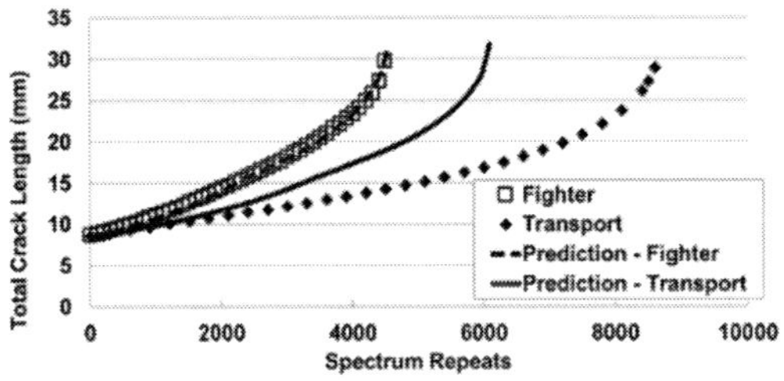

Fig. 15. MPYZ-TMF prediction – fighter and transport spectra

The model was also used to predict crack growth for a cylindrical specimen geometry as shown in Figure 16. Two strain-controlled tests were conducted using the consortium spectrum shown in Figure 7 with a hollow cylindrical IN 718 specimen with starter hole. The model predicted goods results and highlights the ability of the model to accommodate various specimen geometries.

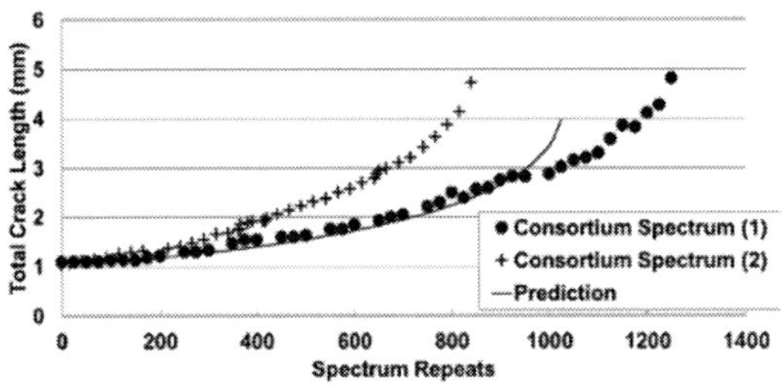

Fig. 16. Engine consortium experimental data and MPYZ-TMF prediction for hollow cylindrical specimen

5 Conclusions

The accelerating effect of time spent at elevated temperature in aircraft turbine disk spectra significantly affects the life of the component. The model developed provides a capability to account for the impact of previous over- and under-loads and the detrimental effect of time spent at elevated temperature without simplification of the spectra. Further, the unique approach of segmenting the TMF spectra to allow for the effects of time spent at elevated temperature to be incorporated matched well to experimentally observed trends. Aircraft operators depend on accurate modeling of crack growth and this model will serve to save costs in the form of better defined inspection intervals and removal from service decisions and improved safety.

References

1. National Transporation Safety Board, **NTSB/AAR-18/01** (2018)
2. J. M. Larsen, A. H. Rosenberger, G. A. Hartman, S. M. Russ, R. John, **Tech. Rep. ADP014134**, *AFRL* (2003)
3. T. Nicholas, T. Weerasooriya, N. E. Ashbaugh, *Fract. Mech. 16th symp.*, **ASTM STP 868**, 167-180 (1985)
4. T. Weerasooriya, S. Venkataraman, *Effects of load and thermal histories on mechanical behavior of materials* (1987)
5. A. Pineau, *Subcritical Crack Growth Due Fatigue, Corrosion and Creep,* 483-530 (1981)
6. H. Ghonem, D. Zheng, *Mat Sci and Engr: A,* **150**,151-160 (1992)
7. J. A. Pfaendtner, C. J. McMahon Jr, *Acta Mater.,* **49**, 3369-3377 (2001)
8. L. Viskari, S. Johansson, K. Stiller, *Materials at High Temperatures,* **28**, 336-341 (2011)
9. C. F. Miller, G. W. Simmons, R. P. Wei, *Scripta Mater.,* **44**, 2405-2410 (2001)
10. V. Barker, W.S. Johnson, B. Adair, S. Antolovich, A. Staroselsky, *Int. J. Fatigue,* **52**, 95-105 (2013)
11. D. Gustafsson, J. Moverare, K. Simonsson, S. Johansson, M. Hörnqvist, T. Månsson, S. Sjöström, *Procedia Eng.,* **10**, 2821-2826 (2011)
12. X. B. Liu, L. Z. Ma, K. M. Chang, E. Barbero, *Acta Metall. Sin.,* **18**, 55-64 (2005)
13. E. Lundström, K. Simonsson, D. Gustafsson, T. Månsson, *Eng. Fract. Mech.,* **118**, 17-30 (2014)
14. E. Storgärds, K. Simonsson, S. Sjöström, J. Moverare, *J. Eng. Gas Turb. Power,* **138**, 012503 (2016)
15. B. Adair, W. Johnson, S. Antolovich, A. Staroselsky, *Fatigue Fract. Eng. Mater. Struct.,* **36**, 217-227 (2013)
16. A. Radzicki, PhD Dissert., GA Inst Tech, (2016)
17. J. Radavich, *Superalloy 718--Metallurgy and Applications,* (1989)
18. G. Gessinger, *Powder Metallurgy of Superalloys: Butterworths Monographs Matrls*, Elsevier, 2013.
19. ASTM Standard, **E647-13AE1**, (2013)
20. H. Johnson, *Materials Research and Standards,* **5**, 442-445 (1965)
21. E. Storgärds, K. Simonsson, *Experimental Mechanics,* **55**, 559-568 (2015)
22. K. Chang, **GE-CRD 89CRD116** (1989)
23. A. Radzicki, W. Johnson, R. Neu, B. Annigeri, B. Ziegler, *Fatigue Fract. Eng. Mater. Struct.,* **40**, 1406-1415 (2017)
24. S. Ponnelle, B. Brethes, A. Pineau, *European Structural Integrity Society,* **29**, 257-266 (2002)
25. W. S. Johnson, **ASTM STP 748**, 85-102 (1981)
26. L. James, *Eng. Fract. Mech.,* **25**, 305-314 (1986)
27. K. Sadananda, P. Shahinian, *Creep-Fatigue-Environment Interactions,* 86-111 (1979)
28. E. Andrieu, R. Molins, H. Ghonem, A. Pineau,

Mat. Sci. Eng. A, **154**, 21-28 (1992)

Thermo-mechanical fatigue damage behavior for Ni-based superalloy under multiaxial loading

Dao-Hang Li and *De-Guang* Shang*

College of Mechanical Engineering and Applied Electronics Technology, Beijing University of Technology, 100124 Beijing, China

Abstract. The fatigue damage behavior was experimentally investigated in different axial-torsional thermo-mechanical loading conditions for Ni-based superalloy GH4169. The strain controlled tests were carried out with the same von Mises equivalent mechanical strain amplitude of 0.8% in the temperature range from 360°C to 650°C. The results show that the fatigue life is drastically reduced when the axial mechanical strain and the temperature are in-phase, which can be due to that the creep damage is induced by the tensile stress at high temperature. Moreover, the fatigue life is further decreased when the axial mechanical strain and the shear strain are out-of-phase, which can be attributed to that the non-proportional hardening can increase the creep and the oxidation damages. Furthermore, the tensile stress is crucial to the nucleation of creep cavities at high temperature compared with the shear stress. The tensile and shear stresses all can increase the creep damage under fatigue loading at high temperature. In addition, the oxidation damage can be induced during cyclic loading at high temperature, and it can be increased by the tensile mean stress caused in non-isothermal loading.

1 Introduction

In recent years the fatigue behavior of material under thermo-mechanical loading has become the focus of investigation, in which various damage mechanisms are existed simultaneously. For improving the accuracy of life prediction, the involved damage mechanisms during thermo-mechanical fatigue (TMF) loading need to be identified.

Many investigations [1-12] have assessed the fatigue damage behavior of material under uniaxial TMF loading. The results show that the thermal phase angle (between the axial mechanical strain waveform and the temperature waveform) has an important influence on fatigue life. For some superalloys [1-5], the fatigue life under in-phase (IP) TMF loading is obviously lower than that under out-of-phase (OP) TMF loading. During the IP TMF loading, the creep damage is caused besides the fatigue damage, which leads to the intergranular fracture of material, and this fracture mode can be the main reason of drastic decrease in fatigue life. For the OP TMF loading, the fatigue damage is dominant [1], in which the transgranular fracture is the primary failure mode. Moreover, for the above superalloys, the oxidation damage is also caused [6, 8], but it cannot play a dominant role. For some other superalloys [9-12], the OP TMF loading induces more serious damage compared with the IP TMF loading. The oxidation damage is dominant under OP TMF loading for these superalloys, and the high tensile stress corresponding low temperature can increase the oxidation damage. In common, the above investigations show that the damage mechanism is quite complex during TMF loading, and it is varied with the different loading conditions and materials. In addition, it is noteworthy that in the vast majority of the above investigations the tested thermal phase angle is only at 0° and 180°, thus the fatigue damage behavior under TMF loading in other thermal phase angle case needs to be further assessed.

Furthermore, the fatigue damage behavior under multiaxial TMF loading has been assessed in some investigations [13-18]. Axial-torsional thermo-mechanical fatigue (AT-TMF) tests can be used to understand some special fatigue behavior in non-isothermal multiaxial loading condition, in which the mechanical phase angle (between the axial mechanical strain waveform and the shear strain waveform) also needs to be controlled. So the effect of mechanical phase angle on fatigue damage behavior under AT-TMF loading has become a new investigation focus.

In addition, since the Ni-based superalloy has fine mechanical properties, corrosion resistance, good weld ability and long-term thermal stability [19], it is widely used in the structural components of modern gas turbine engines subjected to multiaxial thermo-mechanical fatigue loading. Thus the Ni-based superalloy is chosen to understand the fatigue damage behavior under AT-TMF loading.

In this paper, the main aim is to analyze the existed various damage mechanisms under AT-TMF loading. Moreover, the effects of thermal and mechanical phase angles on fatigue damage behavior and the influences of involved fatigue, creep and oxidation damages on fatigue life under AT-TMF loading will be discussed in detail.

* Corresponding author: shangdg@bjut.edu.cn

 https://doi.org/10.1051/matecconf/201816519002

2 Experimental

2.1 Material

The chemical composition of the Ni-based superalloy GH4169 used in this investigation is listed in Table 1. The material was subjected to a heat-treatment condition from reference [19], including solid solution at 970°C for 1h, air cooling to 720°C, first-stage aging at 720°C for 8 h, furnace cooling at cooling rate of 50°C/h to 620°C, second-stage aging at 620 °C for 8 h, and air cooling to room temperature.

Table 1. Chemical composition of superalloy GH4169 (wt %).

C	0.07
Cr	20.00
Ni	53.00
Co	0.70
Mo	3.00
Al	0.50
Ti	1.00
Nb	5.10
B	0.01
Mg	0.01
Mn	0.30
Si	0.32
P	0.01
S	0.01
Cu	0.28
Ca	0.01
Pb	0.00
Fe	Balance

2.2 Axial-torsional thermo-mechanical fatigue testing

Thin-walled tubular specimen, with the gauge section thickness of 1mm, was adopted in all AT-TMF tests, as shown in Fig. 1. All tests were carried out on a tension-torsion closed-loop servo-hydraulic testing machine, which utilizes radio-frequency induction heating unit and forced air cooling pipe allowing rapid heating and cooling, as shown in Fig. 2. A high temperature tension-torsion extensometer with the gauge length of 25mm was used to measure the strains. A chromel-alumel (type K) thermocouple was used to measure the temperature on the middle outer surface of the gauge section.

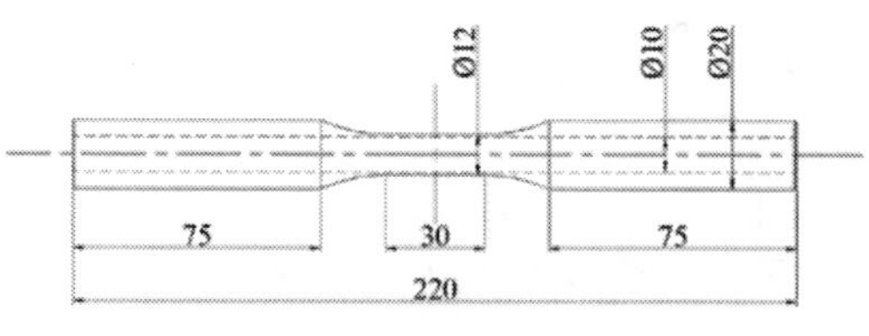

Fig. 1. Specimen geometry adopted in all AT-TMF tests.

Table 2. Loading conditions of AT-TMF tests.

Loading path	Loading waveform	Strain path
MIPTIP: Mechanical In-Phase, Thermal In-Phase	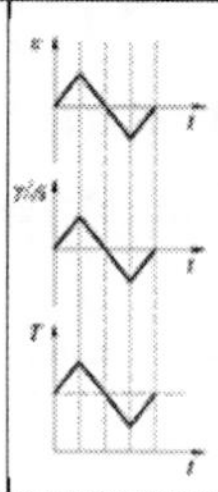	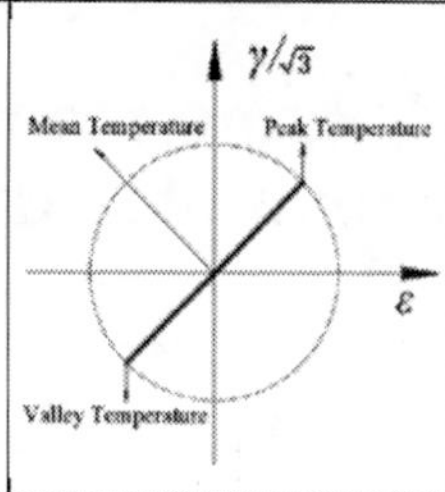
$MIPTOP^{90}$: Mechanical In-Phase, Thermal 90° Out-of-Phase	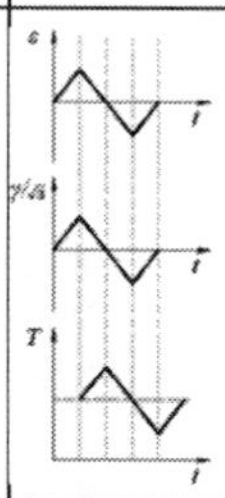	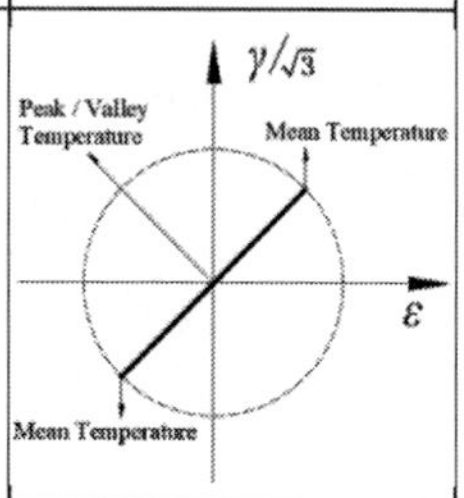
$MIPTOP^{180}$: Mechanical In-Phase, Thermal 180° Out-of-Phase	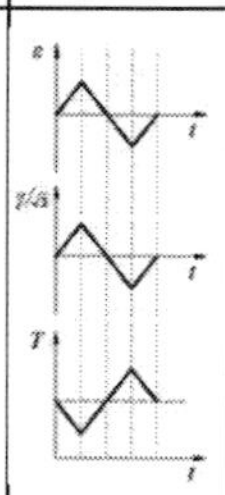	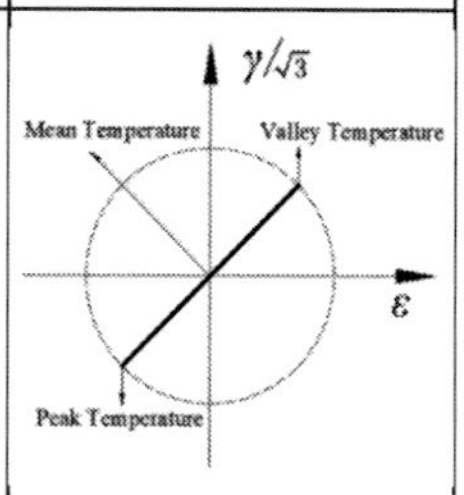
MOPTIP: Mechanical 90° Out-of-Phase, Thermal In-Phase	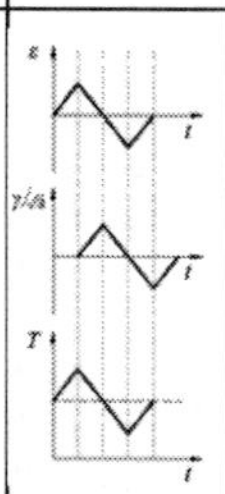	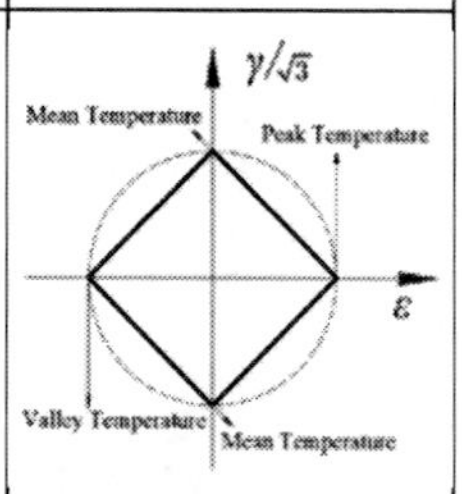
$MOPTOP^{90}$: Mechanical 90° Out-of-Phase, Thermal 90° Out-of-Phase	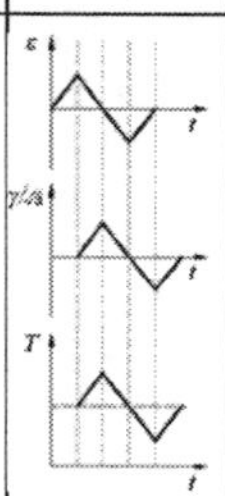	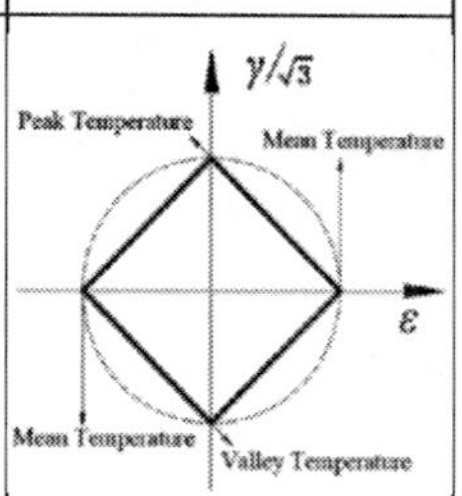
$MOPTOP^{180}$: Mechanical 90° Out-of-Phase, Thermal 180° Out-of-Phase	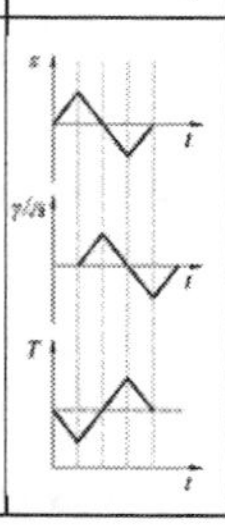	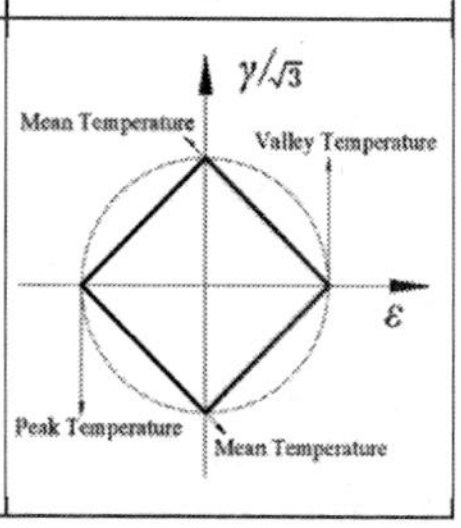

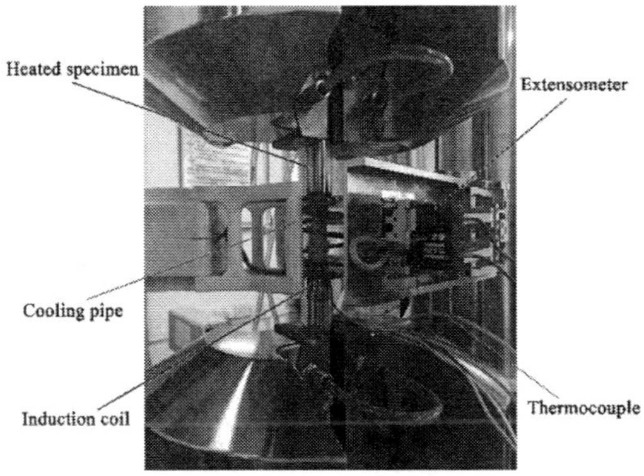

Fig. 2. Heated specimen, induction coil, cooling pipe, thermocouple and extensometer on the tension-torsion testing machine.

In AT-TMF tests, the axial thermal strain compensation was used [14], in which the axial total strain was calculated by adding the thermal strain to the mechanical strain ($\varepsilon_t = \varepsilon_m + \varepsilon_{th}$). For the torsional case, the shear strain is not influenced by the change of loading temperature [20], thus the torsional thermal strain compensation is not needed. Moreover, the equivalent mechanical strain (ε_{eq}) and the equivalent stress (σ_{eq}) at time t were calculated using the von Mises criterion [21]:

$$\varepsilon_{eq}(t) = \sqrt{[\varepsilon_m(t)]^2 + [\gamma(t)]^2/3} \quad (1)$$

where ε_m and γ are the axial mechanical strain and the shear strain, respectively.

$$\sigma_{eq}(t) = \sqrt{[\sigma(t)]^2 + 3[\tau(t)]^2} \quad (2)$$

where σ and τ are the axial stress and the shear stress, respectively. In addition, the thermal and mechanical phase angles both needed to be controlled for AT-TMF tests, and six different loading conditions were used in this investigation, as listed in Table 2.

The experimental parameters of AT-TMF tests are listed in Table 3. On the basis of the previous investigation [22], the non-proportional hardening was found to be obvious at the equivalent mechanical strain amplitude of 0.8%. For the purpose of effectively investigating the influence of non-proportional hardening on fatigue damage behavior under AT-TMF loading, the equivalent mechanical strain amplitude of all tests were selected at 0.8%. The ratios of mechanical shear and axial strains were 1.73 ($\sqrt{3}$) and 3.46 ($2\sqrt{3}$), which can be used to investigate the influence of strain ration on the fatigue damage behavior. Moreover, the tests were carried out in the temperature range from 360°C to 650°C. The cyclic period was 120s, and the loading frequency was about 0.0083Hz. The heating/cooling rate was calculated at about 4.83°C/s, which can maintain temperature uniformity in the gauge section of the specimen.

In addition, two thermal cycles with the mechanical loading of zero were implemented before the AT-TMF tests. The cyclic number corresponded to the decrease of 30% for stable peak stress was adopted as the fatigue failure life (N_f). After completing all tests, the scanning electron microscopy (SEM) was used to observe the morphology of fracture, and the energy dispersive x-ray analysis (EDAX) was employed to analyze the composition of fracture.

3 Experimental results and discussion

3.1 Varied fatigue lives under AT-TMF loading

Fig. 3 clearly shows that the fatigue lives are varied under AT-TMF loading even with the same equivalent mechanical strain amplitude of 0.8%. Moreover, thermal and mechanical phase angles have important influence on fatigue lives. The fatigue life is decreased when thermal phase angle is in-phase, and it is further reduced when mechanical phase angle is out-of-phase. As a result, the fatigue life under MOPTIP loading is the lowest. Furthermore, the fatigue life is relatively high under thermal 90° out-of-phase loading when the mechanical phase angle is the same, that is, MIPTOP^{90} > MIPTOP^{180} > MIPTIP, and MOPTOP^{90} > MOPTOP^{180} > MOPTIP, which means that the damage mechanisms are varied in different thermal phase cases. In addition, the strain ration also has important effect on fatigue lives. The fatigue life under MIPTIP loading with the strain ratio of 3.46 is higher than that with the strain ration of 1.73, which means that the fatigue damage behavior can be changed in different strain ratio cases. Therefore, it can be concluded that the varied damage mechanisms are the main reason of the obvious difference in fatigue lives under AT-TMF loading.

Table 3. Experimental parameters of AT-TMF tests.

Spec. no.	$\Delta\varepsilon_{eq}/2$ (%)	$\Delta\varepsilon_x/2$ (%)	$\Delta\gamma_{xy}/2$ (%)	Loading path	λ_ε	ΔT (°C)	t (s)
N1	0.800	0.566	0.980	MIPTIP	1.73	360-650	120
N2	0.800	0.566	0.980	MIPTOP^{90}	1.73	360-650	120
N3	0.800	0.566	0.980	MIPTOP^{180}	1.73	360-650	120
N4	0.800	0.800	1.386	MOPTIP	1.73	360-650	120
N5	0.800	0.800	1.386	MOPTOP^{90}	1.73	360-650	120
N6	0.800	0.800	1.386	MOPTOP^{180}	1.73	360-650	120
N7	0.800	0.358	1.239	MIPTIP	3.46	360-650	120

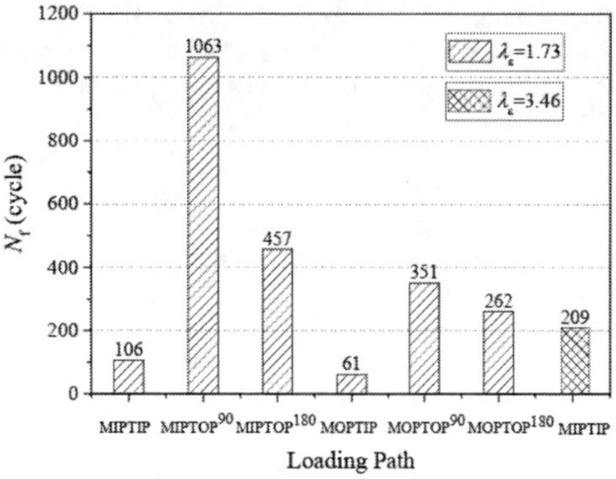

Fig. 3. Varied fatigue lives under AT-TMF loading ($\Delta\varepsilon_{eq}/2$=0.8%, ΔT=360°C-650°C, f=0.0083Hz).

3.2 Non-proportional hardening and mean stress under AT-TMF loading

Fig. 4 shows that the equivalent peak stress under MOPTOP^{90} loading is obviously higher than that under MIPTOP^{90} loading except the initial few cycles, which means that the non-proportional hardening can be induced in mechanical out-of-phase case. The non-proportional hardening can cause the enlargement of responded stress, which can increase the damage during fatigue loading process. It can be the main reason of that the fatigue lives under mechanical out-of-phase loading are lower than that under in-phase loading in the same thermal phase angle cases, that is, MOPTIP < MIPTIP, MOPTOP^{90} < MIPTOP^{90}, and MOPTOP^{180} < MIPTOP^{180}, as shown in Fig.3.

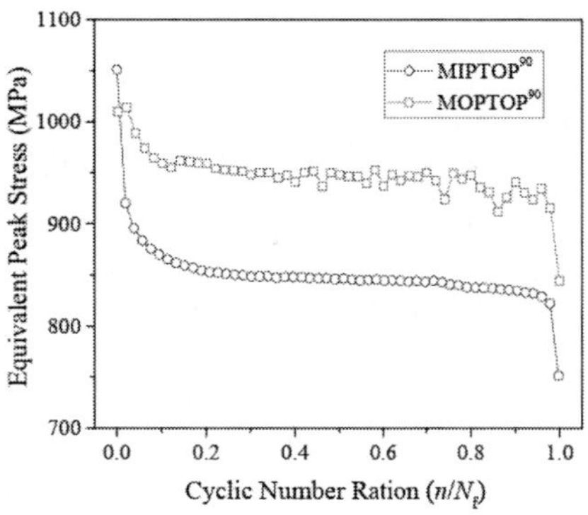

Fig. 4. Non-proportional hardening under AT-TMF loading ($\Delta\varepsilon_{eq}/2$=0.8%, λ_ε=1.73, ΔT=360°C-650°C, f=0.0083Hz).

Fig. 5 shows the special hysteresis loops under non-isothermal loading. The mean stress is induced when the peak and valley strain loadings correspond the different temperature points, that is, the negative mean axial stress under MOPTIP loading, and the positive mean axial stress under MOPTOP^{180} loading, as shown in Fig. 5 (a). And the mean stress is almost equal to zero when the peak and valley strain loadings correspond the same temperature point, as shown in Fig. 5 (b). It can be attributed to the dependence of responded stress on temperature. Moreover, the special deformation behavior under non-isothermal loading can induce the mean stress, which can influence the fatigue damage behavior.

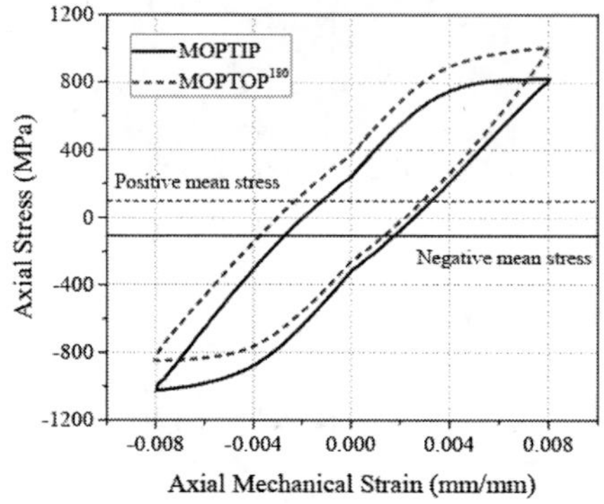

(a)

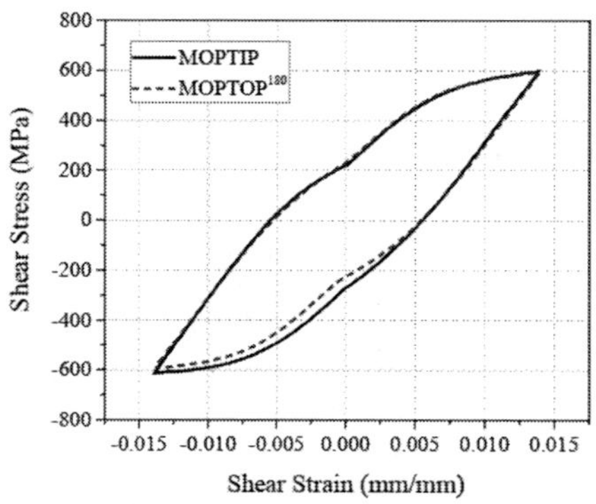

(b)

Fig. 5. Comparison of hysteresis loops under MOPTIP and MOPTOP^{180} loadings ($\Delta\varepsilon_{eq}/2$=0.8%, λ_ε=1.73, ΔT=360°C-650°C, f=0.0083Hz). (a) Axial hysteresis loop; (b)Torsional hysteresis loop.

3.3 Creep damage under AT-TMF loading

Fig. 6 shows that the dominant fracture mechanism is varied for different thermal phase angles even with the same mechanical phase angle. Fig 6 (a) and (c) show the visible morphology of grain boundaries, which means that the intergranular fracture is dominant under MIPTIP and MOPTIP loadings. The intergranular fracture implies that the creep damage is induced besides the fatigue damage, and it can be the main reason of the drastic decrease in fatigue life under thermal in-phase loading. However, the fatigue striation is evident in Fig.

6 (b) and (d), which means that the transgranular fracture is dominant under MIPTOP^{180} and MOPTOP^{180} loadings.

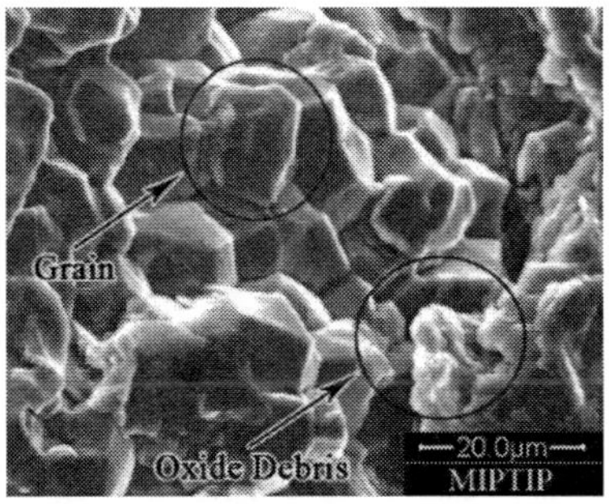

(a)

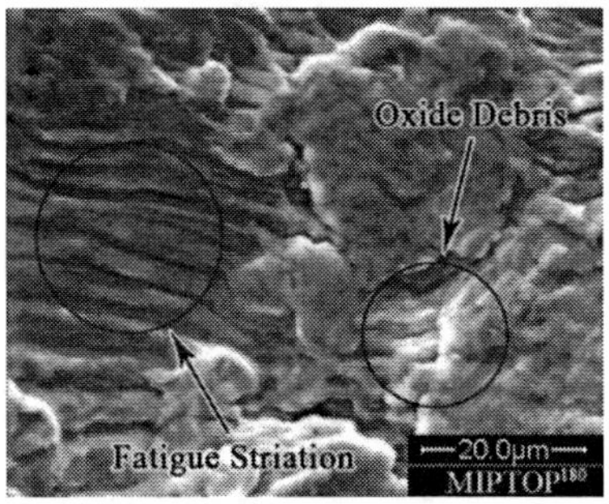

(b)

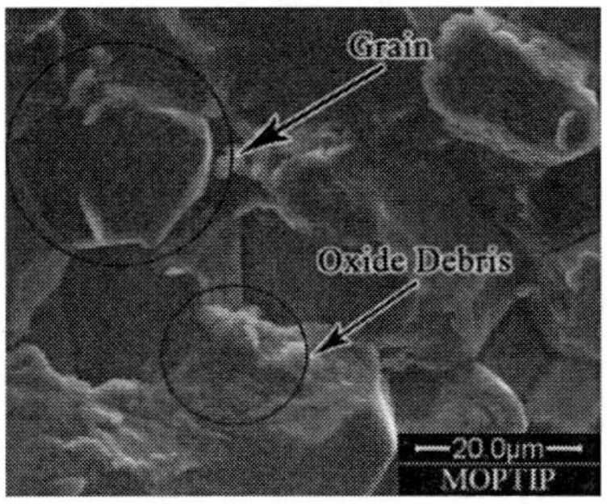

(c)

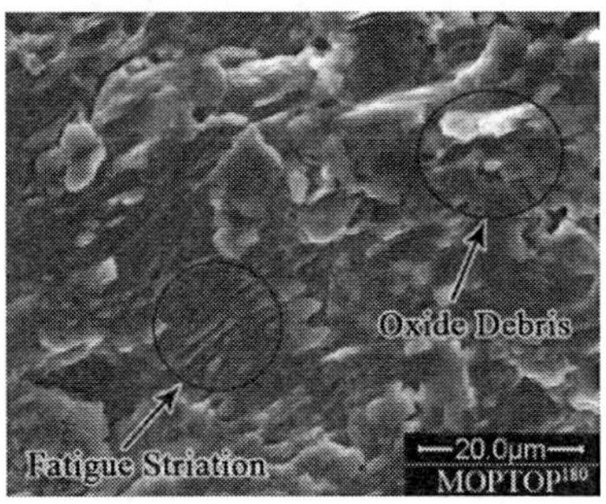

(d)

Fig. 6. Morphology of fracture under AT-TMF loading ($\Delta\varepsilon_{eq}/2$=0.8%, λ_ε=1.73, ΔT=360°C-650°C, f=0.0083Hz).
(a) Intergranular fracture under MIPTIP loading;
(b) Transgranular fracture under MIPTOP^{180} loading;
(c) Intergranular fracture under MOPTIP loading;
(d) Transgranular fracture under MOPTOP^{180} loading.

Thus it can be deduced that the creep damage is ignored, and the fatigue damage plays an important role in thermal out-of-phase loading. In conclusion, the creep damage is induced during thermal in-phase loading, which can cause the intergranular fracture, and the intergranular fracture can drastically decrease the fatigue life.

Fig. 7 (a) shows that the maximum tensile stress appears accompanied by the maximum temperature under MIPTIP loading. The creep damage can be induced by the tensile stress at high temperature, which is verified in the investigations concerning uniaxial TMF damage behavior [1-5]. Moreover, the creep damage can be further increased by the shear stress, since the shear stress can tear the creep cavities during cyclic loading. However, the maximum compressive stress appears when the shear stress and temperature reach their minimum values. The creep cavities cannot be sintered by the compressive stress due to the lower temperature [2]. Thus the creep cavities continuously grow and join during MIPTIP loading, and the intergranular fracture can be caused, which can be proved by the observed fracture mode in Fig. 6 (a).

Fig. 7 (b) shows that the tensile and shear stresses reach their maximum values when the temperature drops to its minimum value, and the maximum value of the temperature appears accompanied by the minimum values of compressive and shear stresses under MIPTOP^{180} loading. The creep damage cannot be induced in this condition, and the fatigue damage is dominant, which can be proved by the fracture feature in Fig. 6 (b).

Therefore, it can be concluded that the creep damage is induced by the tensile stress at high temperature, and it is increased by the shear stress during fatigue loading. It can be the main reason of that the fatigue life is decreased when thermal phase angle is in-phase, as shown in Fig. 3.

Moreover, the peak tensile stress under MIPTIP loading with the strain ration of 3.46 in Fig. 7 (c) is lower than that with the strain ratio of 1.73 in Fig. 7 (a). And the fatigue life under MIPTIP loading with strain ration of 3.46 is obviously higher than that with the strain ratio of 1.73, as shown in Fig. 3. Thus it can be deduced that the creep damage is lower at the strain ration of 3.46, and the decrease of tensile stress delays the nucleation of creep cavities at high temperature. However, the peak shear stress under MIPTIP loading with the strain ration of 3.46 in Fig. 7 (c) is higher than that with the strain ratio of 1.73 in Fig. 7 (a). Thus it can be deduced that the shear stress has a little influence on the nucleation of creep cavities at high temperature. In conclusion, the tensile stress has a significant influence on the nucleation of creep cavities at high temperature compared with the shear stress.

In addition, the extent of the creep damage depends on the magnitude of the stress and the temperature condition, and the non-proportional hardening can enlarge the responded stress, as show in Fig. 4. Thus it can be deduced that the non-proportional hardening can increase the creep damage, which can be a main reason of that the fatigue life under MOPTIP loading is obviously lower than that under MIPTIP loading, as shown in Fig. 3.

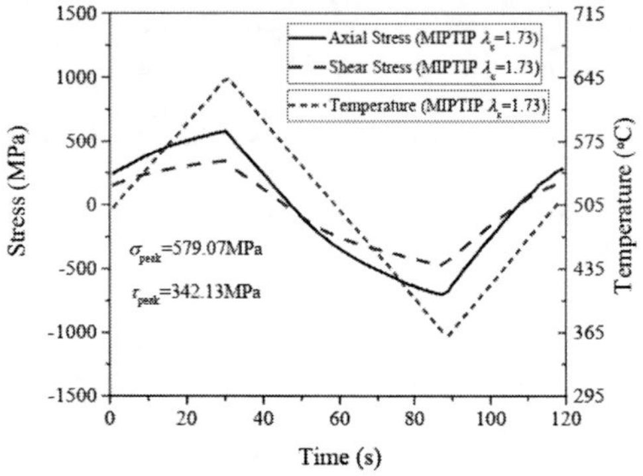

(a)

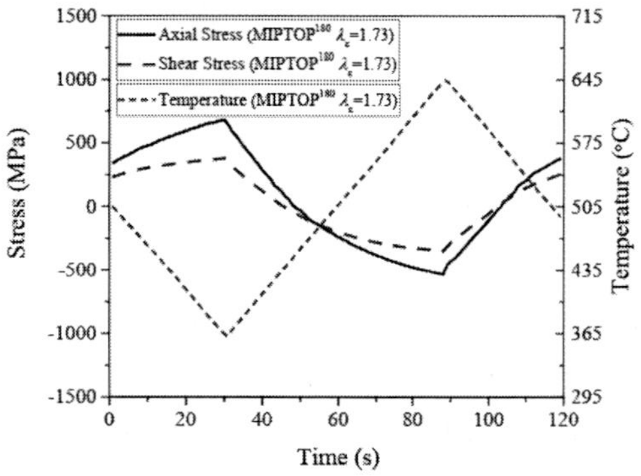

(b)

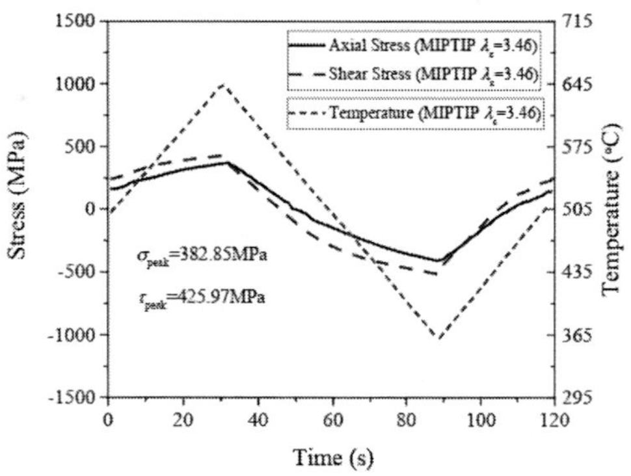

(c)

Fig. 7. Responded axial and shear stresses with temperature under AT-TMF loading ($\Delta\varepsilon_{eq}/2$=0.8%, ΔT=360°C-650°C, f=0.0083Hz). (a) MIPTIP, λ_ε=1.73; (b) MIPTOP^{180}, λ_ε=1.73; (c) MIPTIP, λ_ε=3.46.

3.4 Oxidation damage under AT-TMF loading

Fig. 6 also shows the oxide debris in all loading conditions, and the existence of the oxide debris can be proved by the appeared oxygen atom in the chemical composition of the fracture, as shown in Fig. 8. The observed oxide debris implies that the oxidation damage can be induced during fatigue loading at elevated temperature. Furthermore, the oxygen content is changed, which can express the varied degree of oxidation damage in different loading conditions. Compared with the mechanical in-phase loading, the oxidation is more serious under mechanical out-of-phase loading, that is, for oxygen content MOPTIP > MIPTIP, and MOPTOP^{180} > MIPTOP^{180}. And the oxidation damage is more serious under thermal 180° out-of-phase loading than that under thermal in-phase loading, that is, for oxygen content MIPTOP^{180} > MIPTIP, and MOPTOP^{180} > MOPTIP. Therefore, it can be concluded that the oxidation damage is varied with the different thermal and mechanical phase angles under AT-TMF loading.

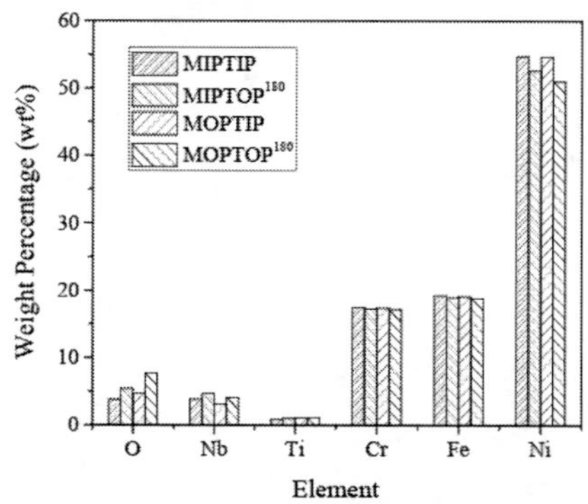

Fig. 8. Chemical composition of fracture under AT-TMF loading ($\Delta\varepsilon_{eq}/2$=0.8%, λ_ε=1.73, ΔT=360°C-650°C, f=0.0083Hz).

The extent of the oxidation damage also depends on the magnitude of stress and the temperature condition [23]. It can be deduced that the non-proportional hardening can increase the oxidation damage. As a result, the oxygen content under mechanical out-of-phase loading is higher than that under in-phase loading, as shown in Fig. 8. Furthermore, the increase of oxidation damage from non-proportional hardening can be a important reason of that the fatigue life is reduced when mechanical phase angle is out-of-phase, as shown in Fig. 3.

Moreover, the oxidation damage can be increased by the tensile mean stress induced during non-isothermal loading, which is verified in the investigations [9-12]. It can be the main reason of that the oxygen content under thermal 180° out-of-phase loading is higher than that under in-phase loading, as shown in Fig. 8.

In addition, the temperature corresponding the peak/vallley strain loading under thermal 90° out-of-phase loading (505°C) is lower than the temperature corresponding the valley strain loading under thermal 180° out-of-phase loading (650°C) in mechanical in-phase loading case. It can be deduced that the oxidation damage is lower under thermal 90° out-of-phase loading. As a result, for the fatigue life $MIPTOP^{90}$ > $MIPTOP^{180}$, and $MOPTOP^{90}$ > $MOPTOP^{180}$, as shown in Fig. 3.

Therefore, it can be concluded that the oxidation damage is induced during fatigue loading at high temperature. Moreover, the oxidation damage can be increased by the non-proportional hardening from mechanical out-of-phase loading and the tensile mean stress induced during non-isothermal loading.

4 Conclusions

The fatigue damage behavior was experimentally investigated for Ni-based superalloy under different AT-TMF loading conditions with the same equivalent mechanical strain amplitude. Some key conclusions are obtained as follows:

(1) The fatigue life is drastically decreased when the axial mechanical strain and the temperature are in-phase, which can be attributed to that the creep damage is induced by the tensile stress at high temperature.

(2) The fatigue life is further reduced when the axial mechanical strain and the shear strain are out-of-phase, since the non-proportional hardening can increase the creep and the oxidation damages.

(3) The tensile stress has an important effect on the nucleation of creep cavities at high temperature compared with the shear stress. However, the tensile and shear stresses both can increase the creep damage under cyclic loading at high temperature.

(4) The oxidation damage can be induced during fatigue loading at high temperature, and it can be increased by the tensile mean stress caused in non-isothermal loading.

The authors gratefully acknowledge the financial support of the National Natural Science Foundation of China (No. 51535001, 11572008).

References

1. A.G.D.L. Yedra, J.L. Pedrejón, A. Martín-Meizoso, R. Rodríguez, Exp. Techniques **40**, 777-787 (2014)
2. S. Guth, S. Doll, K.H. Lang, Mater. Sci. Eng. A **642**, 42-48 (2015)
3. G.V.P. Reddy, A. Nagesha, R. Sandhya, S. Sankaran, M.D. Mathew, K.B.S. Rao, Metall. Mater. Trans. A **46**, 695-707 (2015)
4. D. Lee, I. Shin, Y. Kim, J.M. Koo, C.S. Seok, Int. J. Fatigue **62**, 62-66 (2014)
5. Z.W. Huang, Z.G. Wang, S.J. Zhu, F.H. Yuan, F.G. Wang, Mater. Sci. Eng. A **432**, 308-316 (2006)
6. R. Petráš, V. Škorík, J. Polák, Mater. Sci. Eng. A **650**, 52-62 (2016)
7. S. Guth, R. Petráš, V. Škorík, T. Kruml, J. Man, K.H. Lang, J. Polák, Int. J. Fatigue **80**, 426-433 (2015)
8. X. Hu, D. Shi, X. Yang, Mater. Sci. Eng. A **674**, 451-458 (2016)
9. K. Prasad, C. Parlikar, D.K. Das, Int. J. Fatigue **92**, 107-115 (2016)
10. K. Prasad, R. Sarkar, K.B.S. Rao, M. Sundararaman, Metall. Mater. Trans. A **47A**, 1-18 (2016)
11. P. Schallow, H.J. Christ, Int. J. Fatigue **53**, 15-25 (2013)
12. A. Nagesha, R. Kannan, V.S. Srinivasan, R. Sandhya, B.K. Choudhary, K. Laha, Metall. Mater. Trans. A **47**, 1-18 (2016)
13. P.J. Bonacuse, S. Kalluri, AGARD CP **569**, 15 (1996)
14. S. Kalluri, P.J. Bonacuse, ASTM STP **1280**, 184-207 (1997)
15. J. Meersmann, F. Frenz, J. Ziebs, H.J. Kuhn, S. Forest, AGARD CP **569**, 19 (1996)
16. S.P. Brookes, H.J. Kühn, B. Skrotzki, H. Klingelhöffer, R. Sievert, J. Pfetzing, D. Peter, G. Eggeler, Mater. Sci. Eng. A **527**, 3829-3839 (2010)
17. A. Gosar, M. Nagode, Int. J. Fatigue **48**, 223-230 (2013)
18. P. Wang, L. Cui, A. Scholz, S. Linn, M. Oechsner, Int. J. Fatigue **67**, 220-227 (2014)
19. Editorial Committee of China Aeronautical Materials Handbook, China Aeronautical Materials Handbook **2nd ed.**, (2002)
20. C.E. Bakis, M.G. Castelli, J.R. Ellis, ASTM STP **1191**, 223-243 (1993)
21. X.W. Wang, D.G. Shang, Int. J. Fatigue **90**, 36-46 (2016)
22. D.G. Shang, G.Q. Sun, J.H. Chen, N. Cai, C.L. Yan, Mater. Sci. Eng. A **432**, 231-238 (2006)
23. F. Vöse, M. Becker, A. Fischersworring-Bunk, H.P. Hackenberg, Int. J. Fatigue **53**, 49-57 (2013)

Creep crack growth assessment in terms of continuum damage mechanics

Valery Shlyannikov[1,*] and *Andrey* Tumanov[1]

[1]Kazan Scientific Center, Russian Academy of Sciences, 420111 Kazan, Russia

Abstract. The stress, strain rate, and process zone with respect to the creep-crack growth in compact-tension C(T) specimen is analyzed by employing damage-evolution equations. The damage model for the fracture of the process zone is represented using a stress based formulation. Both damage free and defective creeping solids have been studied. The variations in the creep stress/strain and crack-tip governing parameter in terms of the creep stress intensity factors with respect to time and the evolution of creep damage are analyzed using an FE model for C(T) specimen. The creep-fatigue crack growth rate tests were performed on special designed program test-cycle. The interpretation of the experimental creep crack growth rate data was given in terms of introduced creep stress intensity factors based on undamaged and damaged stress/strain fields.

1 Introduction

In the past, the continuum damage mechanics (CDM) was used as a complementary approach in high-temperature fracture mechanics to analyze the damage and fracture process at different scales with respect to the changing material structure from the initiation of micro-cracks to the final fracture resulting from the macro-cracks [1, 2]. The CDM covered several microscopic mechanisms in the area near the macroscopic creep crack including void nucleation and growth on the grain-boundary facets, cavities coalescing to form a grain-size micro-crack, coalescing of micro-cracks, which finally lead to the propagation of creep macro-cracks.

In the present study the continuum damage mechanics is applied to assess the creep damage behavior. Special attention has been addressed to the influence of the creep damage accumulation on the on creep-fatigue crack growth rate. The interpretation of the experimental creep crack growth rate data at elevated temperature on C(T) specimens was given in terms of introduced creep stress intensity factors based on undamaged and damaged stress/strain rate fields. Comparison between the creep-fatigue crack growth rate characteristics indicates that they are a function of the continuum damage mechanics parameters.

2 Stress based model

Let ω denote in Eq.(1) the measure of damage with ω=0 denoting the undamaged state and ω = 1 the fully damaged state. The creep strain rate accumulation constitutive equation is generalized by the authors [1, 2] to multi-axial state of stress using J_2 - flow theory as follows

$$\frac{d\varepsilon_{eqv}}{dt}=\frac{3}{2}B\sigma_{eqv}^{n-1}S_{ij}\left(\frac{1}{1-\omega}\right)^{n} \quad (1)$$

where B and n are constants of the Norton power law equation. The simplest creep damage rate model, which is introduced by Kachanov [1], is a function of applied nominal stress and current accumulated damage has the following form

$$\frac{d\omega}{dt}=C\left[\sigma_f^{(\pm)}/(1-\omega)\right]^{m} \quad (2)$$

where C and m are material constants. Similarly Eq.(2), Shlyannikov and Tumanov [3] introduced a model for the rate of accumulation of stress-based creep damage as a function of multi-axial stress function $\sigma_f^{(\pm)}$ described by Eqs.(3,4) in the generalized form

$$\sigma_f^{(\pm)}=(1-\chi)\sigma_{kk/mp}^{(\pm)}+\chi\sigma_e,\ \sigma_e=\left[\frac{3}{2}S_{ij}S_{ij}\right]^{1/2} \quad (3)$$

$$\sigma_{kk}=\sigma_1+\sigma_2+\sigma_3=3\sigma_m,\ S_{ij}=\sigma_{ij}-\sigma_m\delta_{ij} \quad (4)$$

where χ is the experimental material constant that is determined as the ratio of uniaxial tensile to compression strength $\chi = \sigma_t/\sigma_c$. For brittle fracture χ = 0, while for ductile fracture χ = 1. Going back to Eq.(3) it should be noted that in Eq.(2) $d\omega/dt$ is assumed to be implicit function of the hydrostatic stress σ_m and the von Mises equivalent stress σ_e. The governing parameter in the form of the material constant χ helps in describing the

* Corresponding author: shlyannikov@mail.ru

effect of the multi-axial state of the stress-strain behavior of the material.

3 Governing parameters

To interpret the experimental data for the creep-fatigue crack growth rate in the C(T) specimen, the creep stress intensity factor (SIF) values are used based on the experimentally measured load-line deflection rate [4,5]

$$\bar{K}_{cr} = \left[\frac{P}{\sigma_0 B I_n^{FEM}(\theta,t,n,\omega)} \frac{\dot{V}_c}{bWL} \left(\frac{f'}{f} \right) \right]^{\frac{1}{n+1}} \quad (5)$$

where $\dot{V}_c$ - is the force-line displacement rate, b - is specimen thickness and f - is geometry dependent correction factor, W is specimen width, L is characteristic size. Equation (5) for the creep stress intensity factor includes a governing parameter in the form of I_n-integral, generally, for three-dimensional creeping solids. Shlyannikov et al. [4] modified the singular asymptotic fields in terms of the creep SIF under the secondary creep conditions to obtain the stress, strain, and displacement rates for elastic-nonlinear-viscous materials. Moreover, a numerical method was introduced [4] to determine the governing parameters of the creep-crack tip fields in terms of the I_n-integral for power-law creeping materials. This method was extended to analyze the creep-damaged material-fracture resistance characteristics. According to this approach, the I_n-integral value can be obtained directly from the FEA distributions of the dimensionless angular stress $\tilde{\sigma}_{ij}^{FEM}$ and displacement rate $\dot{\tilde{u}}_i$ functions

$$I_n^{FEM}(\theta,t,n,\omega) = \int_{-\pi}^{\pi} \Omega^{FEM}(\theta,t,n,\omega)\, d\theta \quad (6)$$

$$\Omega^{FEM}(\theta,t,n,\omega) = \frac{n}{n+1}\left(\tilde{\sigma}_e^{FEM}\right)^{n+1}\cos\theta -$$
$$\left[\tilde{\sigma}_{rr}^{FEM}\left(\dot{\tilde{u}}_\theta^{FEM} - \frac{d\dot{\tilde{u}}_r^{FEM}}{d\theta}\right) - \tilde{\sigma}_{r\theta}^{FEM}\left(\dot{\tilde{u}}_r^{FEM} + \frac{d\dot{\tilde{u}}_\theta^{FEM}}{d\theta}\right)\right]\sin\theta$$
$$-\frac{1}{n+1}\left(\tilde{\sigma}_{rr}^{FEM}\dot{\tilde{u}}_r^{FEM} + \tilde{\sigma}_{r\theta}^{FEM}\dot{\tilde{u}}_\theta^{FEM}\right)\cos\theta$$

where t is creep time, $\tilde{\sigma}_e$ is the von Mises equivalent stress and $\bar{\sigma}_e^{FEM} = \sigma_e^{FEM}/\sigma_0$. The numerical parameter I_n-integral and θ-variation angular functions of the suitably normalized functions $\tilde{\sigma}_{ij}$ and $\dot{\tilde{u}}_i$ depend on the creep exponent n and damage function ω. Unlike the conventional models used to predict the creep-crack growth wherein the I_n-integral is a function of only the creep exponent, in the present study, it is demonstrated that the governing parameter depends on the creep-damage function and creeping crack-tip stress–strain rate fields.

The particular case of the C(T) specimen FE-steady state analysis carried out using constitutive equation (1) with ω=0 ignores in the definition creep damage accumulation and growth. Since the stress and displacement rate ahead of the crack tip are a function only the creep time. Therefore post-processing governing parameters in the form of creep I_n-integral and creep stress intensity factor K_{cr} were calculated by substituting numerical values of stress tensor components $\sigma_{ij}^{FEM}(\omega=0)$ and displacement rates $\dot{u}_i^{FEM}(\omega=0)$ into Eqs.(5,6), respectively. This type of numerical results will denote as undamage fields. Accordingly, other types of basic calculations are performed for the elastic-nonlinear-viscous material with damages (Eq.1) that depend on the multi-axial of stress of state. In these cases the numerical values of stress tensor components $\sigma_{ij}^{FEM}(\omega,\chi)$ and displacement rates $\dot{u}_i^{FEM}(\omega,\chi)$ again were substituted into the left part of into Eqs.(5,6), respectively. As a result, the values of the post-processing governing parameters in the form of creep I_n-integral and creep stress intensity factor K_{cr} were determined. This type of numerical results will denoted as damage fields.

4 Results and discussion

Full-field FEA are performed using ANSYS finite element (FE) code to study the damage effect on the crack-front stress fields for C(T) specimen (Fig.1). Along the thickness direction, an identical planar mesh is repeated from the symmetry plane to the free surface. To catch the drastic change of the stress field near the free surface, the thickness of successive FE layers is exponentially reduced from the mid-plane toward the free surface. In the circumferential direction, 40 equally sized finite elements are defined in the angular region from 0 to π. The size of each ring gradually increases with the radial distance from the crack tip. The radial sizes of the finite elements are varied according to the geometric progression. Typical FE-mesh of the C(T) specimen contains 71640 elements. According to the symmetrical properties, one quarter of the actual structure was selected to establish the three-dimensional finite element model. The twenty-node quadrilateral brick isoparametric three-dimensional solid elements were used to model for the 3D C(T) specimen configuration.

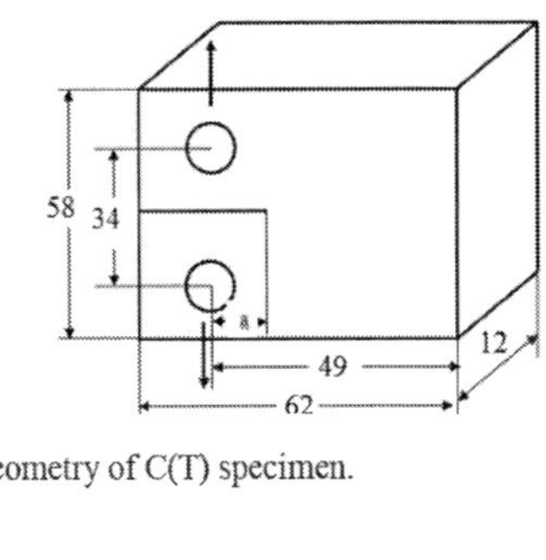

Fig. 1. Geometry of C(T) specimen.

By employing the constitutive Eq. (1) directly into the FEM rate-dependent formulation and using an explicit time integration procedure, we obtain a standard Runge–Kutta integration scheme wherein the finite-element stiffness matrix is derived from the elastic moduli. Thus, to determine the numerical stress/strain-rate fields, we need to first determine the damage-rate function by substituting the stress obtained in the initial iteration into Eq. (2). After determining the damage function ω using numerical integration and by substituting the creep strain rate $\dot{\varepsilon}_{ij}$ of Eq. (1) obtained from the resulting damage function ω into the ANSYS, the damage stress/strain rate fields are found. Finally, after obtaining the solution to the nonlinear problem, the ANSYS output file is used as an input data for the special code developed to determine the dimensionless stress-strain angular distributions, damage contour, and creep-stress intensity factor.

The numerical calculations and tests material is 12Cr1MoV steel which is used for main power plant components such as steam piping and reheat tube. The creep properties of analyzed material at the elevated temperature of 550°C are summarized in Table 1.

Table 1. The creep properties of material.

E [GPa]	σ_0 [MPa]	B	n	C	m
200	100	$1\text{x}10^{-14}$	5.0	$1\text{x}10^{-10}$	3.0

4.1 Creep-crack front stress fields

The important factor in this part of the present study is the influence of the realistic three-dimensional stress-strain state of the finite size cracked body on the creep-damage process. Hence, 3D full-field FE-analysis was performed on the standard compact type, C(T) specimen containing a crack subjected to pure mode I fracture. The specimen dimensions chosen are as recommended by the ASTM test standards with width w = 49 mm and thickness b =12.5 mm. To estimate the stress fields and damage variations, the couple deformation and damage constitutive equations of the form $\sigma(\dot{\varepsilon},\omega,\chi)$ (Eq. (1)), which include the proposed stress-based failure criterion (Eqs. (3) and (4)) are employed. The governing parameter χ, which is the ratio of the uniaxial tension strength to the compression strength, is equal to 0.8. These creep-damage constitutive equations are then incorporated into a finite-element calculation to study the influence of the damage on the crack-tip fields. For the creep in undamaged and damaged materials, we consider the stages of creep time varying from t = 1 h up to t = 10^3 h at the elevated temperature of 550°C. In the following figures, the distribution of stresses and damages will be presented in the thickness direction along the crack front of the C(T) specimen (Fig.1) from the free surface z/b = 0.0 to the mid-plane z/b = 0.5.

To assess the effects of damage formulation, comparisons under extensive-creep conditions t/t_T= 45.7 are made in Fig. 2 between the undamaged creep field and defective fields at the crack tip in the C(T) specimen. Fig. 2 shows the numerical dimensionless equivalent von Mises stress $\tilde{\sigma}_e$ angular distributions in the equivalent stress $\tilde{\sigma}_e$ at a crack tip distance of $\bar{r} = 1.384\cdot 10^{-3}$ (which indicate as R1) as a function of the crack front position in the thickness direction. Near the crack tip, the influence of the damage degree is evident and a sufficient difference is observed between the undamage and the creep damage fields. Furthermore, it is no longer true that θ-variations in the dimensionless stresses are independent of the creep-damage parameter ω. Moreover, the data presented in Fig. 2 demonstrate the 3D-effect of the finite thickness on the creep-damage field behavior at the crack tip. It follows from the comparison that in the undamaged material, the maximum of equivalent stress $\tilde{\sigma}_e$ is always located ahead of the crack θ =0, whereas in the damaged material the maximum stress depends on the position along the crack front z/b.

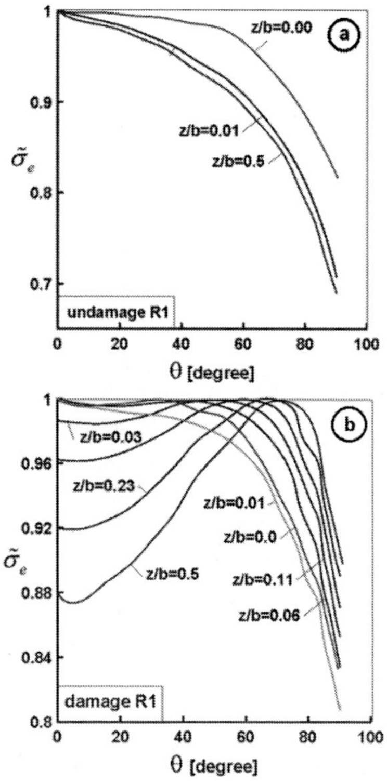

Fig. 2. Comparison dimensionless equivalent stress distributions for (a) undamaged and (b) damaged state.

4.2 Creep-damage zone contour and size

Figs. 3 and 4 show the contours of the creep damage zones around the crack tip and the maximum size of the creep damage zone behavior along of the crack front in the C(T) specimen, respectively, at creep time t/t_T = 45.7 for the different value of crack tip distance. Fig. 3 shows that the 3D stress-strain state affects the shape and size of the creep-damage contours.

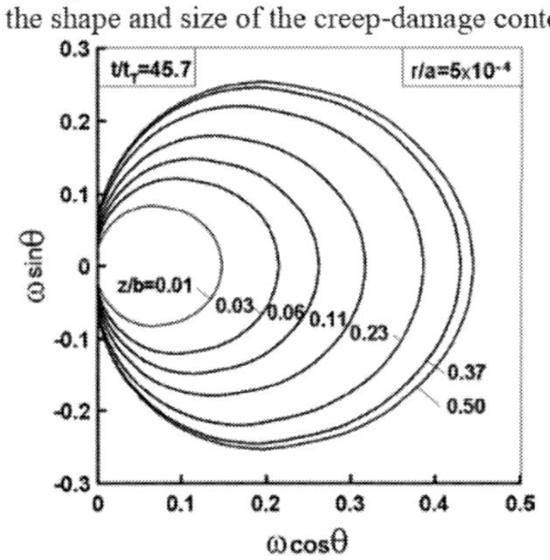

Fig. 3. Creep damage contour distributions along crack front.

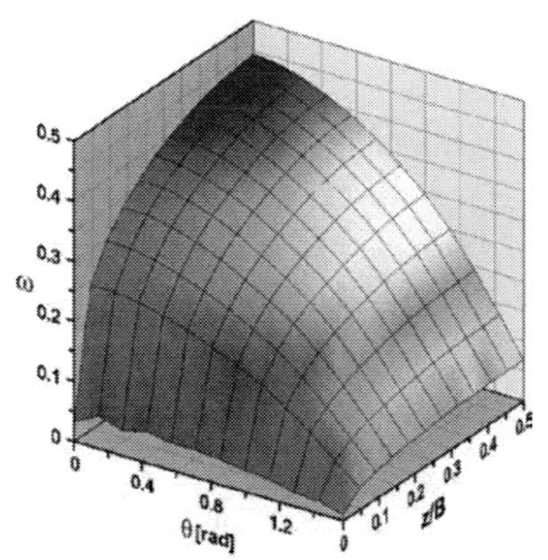

Fig. 4. Creep damage zone size distributions along crack front.

Moreover, the calculated contours strongly depend on the C(T) specimen thickness described by the dimensionless distance z/b. The shapes of the damages zones at the crack front distance r/a =5×10^{-4} in Fig.4 show that the maximum damage contour boundary from the crack front is located along the crack line. However, under the distance r/a =1.5×10^{-3} the maximum size of the creep-damage zone is located approximately $\theta = 60°$ from the crack line ahead of the crack tip (0° direction).

4.3 I_n-integral and creep SIF distributions

As it follows from numerical results, the main effects of the damage parameter ω and material properties χ on the dimensionless stress components $\tilde{\sigma}_{ij}$ and displacement rate components $\dot{\tilde{u}}_i$ arise through the normalizing factor in the form of the creep I_n -integral (Eqs. (6).

Figure 5 shows comparison creep I_n -integral distributions along the crack front for the stationary crack with dimensionless size a/w=0.5. It should be noted that the values for creep I_n -integral do not coincide with those of the plane strain undamaged and damaged fields at the same creep time. Furthermore, in the frame of the creeping solids formulation, the I_n -integral distributions as a function of the creep time for undamaged (Fig.5,a) and defective materials (Fig.5,b) are different and these distinctions increase with the increase in the holding time. In the definition, the plane strain solution is not sensitive to the finite thickness effect. In contrast, the proposed model (Eqs. (6)) can be used to better describe the full 3D conditions. It is clear that in the creeping solids, the behavior of the governing parameter of the crack-tip field I_n -integral is very sensitive to the variation in the material constitutive equation formulation.

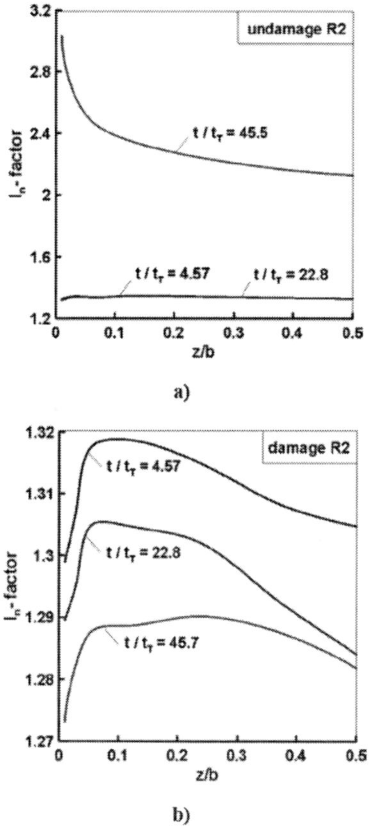

Fig. 5. I_n-factor distributions for (a) undamaged and (b) damage state of stationary crack.

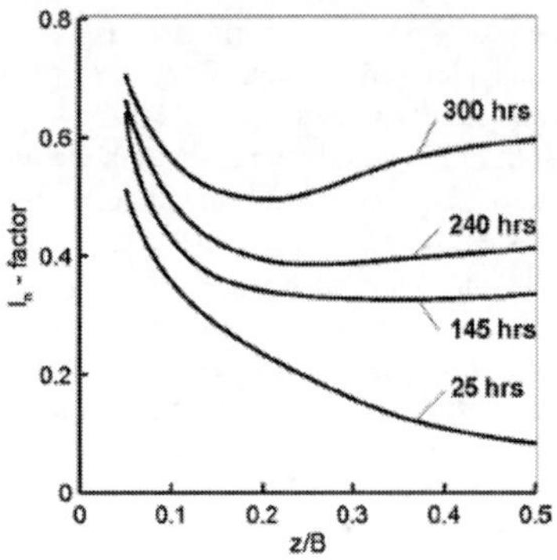

Fig. 6. I_n-factor distributions along growing crack front.

The behavior of the governing parameter of the power-law creep crack-tip fields I_n for the C(T) specimen is plotted in Fig.6 along the growing crack front towards the thickness direction for damage state. The results are presented in a series of plots at three creep times. The first is $t = 25$ hrs close to the transition time $t/t_T = 1$, the second is $t = 145$ hrs, the third $t = 240$ hrs and the fours $t = 300$ hrs, which are related to the extensive creep conditions. Fig.6 represent the variation in the I_n -integral distributions through the creep crack growth stage for the crack front distance $\overline{r} = 1.5 \cdot 10^{-3}$ (which indicate as R2). Based on Eq.(6), these distributions of the I_n -integral were used to calculate the creep-stress intensity factor in the C(T) specimen.

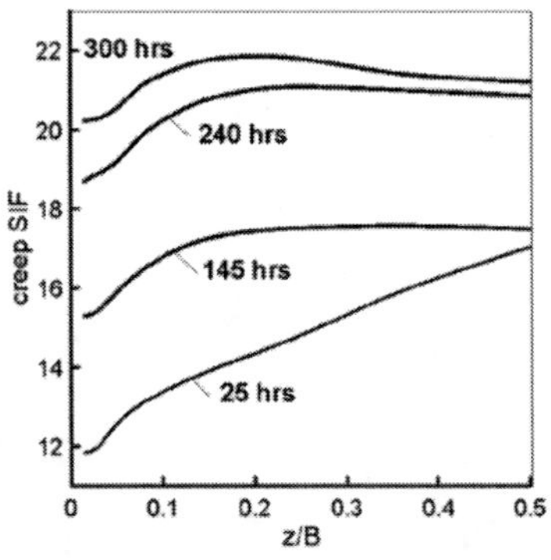

Fig. 7. Creep SIF distributions along crack front.

The distributions of the creep SIF along the crack front in the thickness direction is plotted in Fig.7 for the C(T) specimen. This parameter is plotted against the normalized specimen thickness z/b. The data presented in Fig.6 indicate that the level of the holding time has a significant impact on the distribution of the I_n-factor, and hence on the creep stress intensity factor. Recall that the position of each crack front corresponds to a certain number of loading cycles and the time creep. Analyzing the behavior of the creep SIF on Fig. 7 shows the influence of the crack front form through the I_n-factor as well as the value for the geometry dependent correction factor (f'/f) used in Eq. (5). It is possible to note that these distributions reproduce the experimentally observed crack front positions, i.e., the maximum value of the creep SIF is realized in the deepest point of the crack front whereas, the minimum of the creep SIF occurs in the surface point of the crack tip.

Fig. 8 shows the corresponding numerical results for the creep stress intensity factor behavior along the crack front in C(T) specimen as a function of the creep time based on Eq. (5). In these figures, the finite element creep SIF variations correspond to that normalized using the crack tip distance used is $\overline{r} = 1.384 \cdot 10^{-3}$. Fig.8 shows the comparisons of the creep SIF distributions for undamaged elastic-nonlinear-viscous material with damage parameter in constitutive Eq. (1) $\omega = 0$ and for damaged elastic-nonlinear-viscous material with damage parameter ω and constitutive law obtained using Eq. (2).

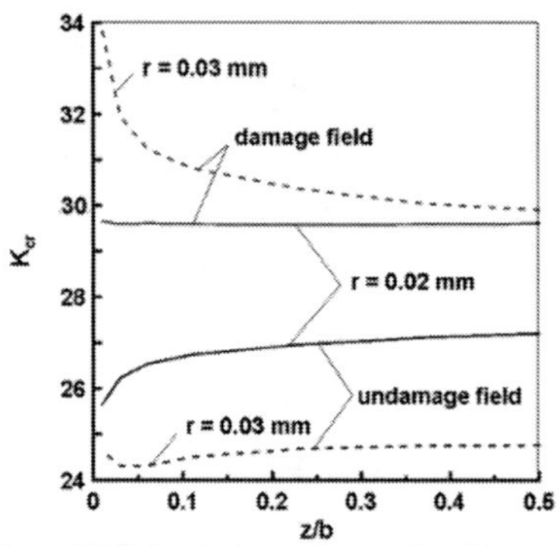

Fig. 8. Creep SIF behavior for undamaged and damaged state.

As observed in Fig.8, the creep SIF distributions as a function of the creep time for undamaged and defective materials are different and these distinctions depend on the crack front distance. Based on this comparison, we can conclude that the stress-based creep damage function (Eq. 2) provides a more accurate description of the actual near-tip fields for the elastic-nonlinear-viscous material.

4.4 Creep-fatigue crack growth rate

The description of the crack growth behavior under creep-fatigue conditions can be described in the terms of the elastic stress intensity factor K_I or the magnitude of the creep stress intensity factor K_{cr}. The elastic SIF is strictly valid only for linear elastic behavior, but it can be used as an approximation if the creep zone size near the crack tip is limited. Further, the linear fracture mechanics approach for characterizing creep-fatigue crack growth behavior relied on K_I for characterizing the crack growth rate per cycle, da/dN, while maintaining the loading frequency and the loading waveform constant. To use the elastic and creep stress intensity factors to characterize the creep crack growth of the tested specimens, it is necessary to measure crack length increment as a function of number of loading cycles.

The tests were conducted in trapezoidal waveform for constant triangular loading/unloading portions with hold times superimposed at maximum load. The potential drop (PD) and the unloading compliance methods were used to monitor crack length during the creep-fatigue tests. The waveforms for the loading and unloading portions were trapezoidal, and the loading/unloading times were held constant (5 s rise and decay times). A hold time of predetermined duration, 60 s, was superimposed on the trapezoidal waveforms at maximum load.

The tests were carried out at 550°C on special high-temperature test equipment with a load ratio, R, of 0.1and maximum load values, P_{max}, of 7 and 8 kN. The C(T) specimens were pre-cracked to an initial crack length to width ratio a/W of approximately 0.4 under cyclic loading at room temperature. Two times during the total creep-fatigue life was the test was stopped, and the C(T) specimen cooled to room temperature; then the specimen was again subjected to creep-fatigue loading at 550°C. This practice helped to determine the intermediate crack front positions during the creep-fatigue loading between the initial (pre-crack) and the final crack fronts. After total failure of the specimen, measurements of the crack sizes were taken for four positions of the crack front by means of an optical microscope. For each front, the crack size at five equally spaced points centered on the specimen mid-thickness line was measured along the crack front. The measured values of crack length on the free surface a_{fs} and at the mid-plane section a_{mp} of the specimen are related to the corresponding number of loading cycles and creep times.

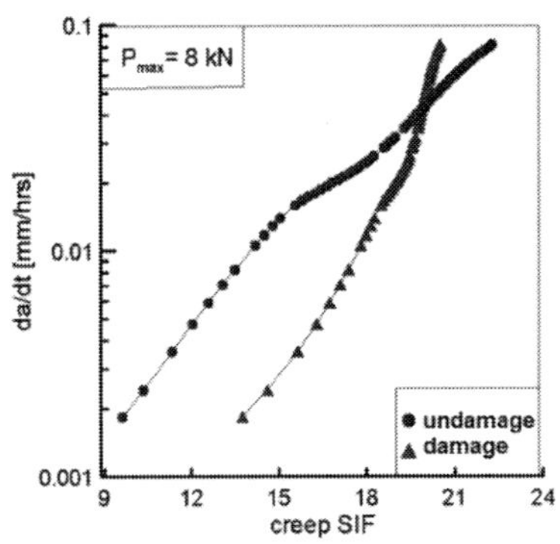

Fig. 9. Crack growth rate for different damage state.

Figure 9 represents the crack growth rate behavior of the 12Cr1MoV steel at 550°C as a function of the dimensionless creep stress intensity factor in the form of Eq.(5) for the hold time 60s. It is seen that the experimental data fall within a relatively narrow scatter band. Again the dependence of the crack growth rate, *da/dt*, on K_{cr} follows a nonlinear trend on a log-log scale at the final stage of creep-fatigue failure. To calculate the values of the creep SIF, the values of I_n-factor were used according to Eq. (6) and the geometry dependent correction factor *(f'/f)* corresponding to the intersection point of the crack front with a free surface of the sample.

In general, as expected, the crack growth rate increases with an increase in the elastic stress intensity factor, C(t)-integral and creep stress intensity factor range. An overall reasonable correlation between the crack growth rate and ΔK is an indication that dominantly quasi-elastic conditions were maintained throughout the testing. However, there is crack growth retardation at the final stage of failure due to blunting of the crack tip and high plasticity of the 12Cr1MoV steel at elevated temperature. These are limitations regarding the use of ΔK at an elevated temperature. It should be noted that the interpretation of the crack growth rate in terms of creep SIF permits better describe the impact of the material structure and properties on the crack growth resistance characteristics.

On Figure 9 one of lines is related to the undamage stress/strain rate state of the compact tension specimen, and the other belongs to the damage stress/strain rate fields along the crack front in the thickness direction. Unlike the definition for the elastic SIF and *C(t)* parameter, the creep stress intensity factor approach gives the possibility to obtain the crack growth rate as a function of the creep constitutive equation formulation for the C(T) specimen. The comparison shows that the crack growth rate along the crack front is different.

Comparison between the creep-fatigue crack growth rate characteristics indicates that they are a function of the continuum damage mechanics parameters. The application of the introduced parameters through experimental study of the creep-fatigue crack growth rate confirms effectiveness of the proposed approach based on stress damage model.

The authors gratefully acknowledge the financial support of the Russian Science Foundation under the Project 17-19-01614.

References

1. L.M.Kachanov, *Introduction to Continuum Damage Mechanics* (Martinus-Nijhoff, Dordrecht, 1986)
2. J.Lemaitre, *A Course on Damage Mechanics*, (Springer-Verlag, Berlin, 1996)
3. V.N.Shlyannikov, A.V.Tumanov, Fatigue Fract. Eng. Mater. Struct., DOI: 10.1111/ffe.12766 (2018)
4. V.N.Shlyannikov, A.V.Tumanov, N.V.Boychenko, Engng. Fract. Mech. **142**, 201–219 (2015)
5. V.N.Shlyannikov, A.V.Tumanov, N.V.Boychenko, A.M.Tartygasheva, Int. J. Press.Vess.Piping **139-140**, 86–95 (2016)

Investigations On Crack Propagation Under Cyclical Isothermal And Thermo-mechanical Loadings For A Type 304-L Stainless Steel Used For Pressurized Water Reactor

Cédric Gourdin[1,*], Gregory Perez[1], Patrick Le Delliou[2], and *Gaëlle Leopold*[2]

[1]DEN-Services d'Etudes Mécaniques et Thermiques (SEMT), CEA, University of Paris-Saclay, F-91191 Gif-sur-Yvette, France
[2]EDF, R&D, Site des Renardières, F-77818 Moret sur Loing Cedex, France

Abstract. The integrity of structures exhibiting flaws in Pressurized Water Reactor (PWR) has to be assessed to meet safety criteria. This paper deals with crack-propagation under cyclic thermo-mechanical loadings, as encountered in class I austenitic pipes of PWR's. To have a conservative and reliable assessment of the crack propagation due to the in-service loading, various codes and standards use simplified method. For example, the RSE-M Code introduces a plastic correction depending on the proportion of the mechanical loading. An improvement of the current method requires additional investigations. Moreover, components loaded with transient or thermal fluctuations are not really in load-controlled conditions. To this end, a device called PROFATH was designed. The specimen is a pre-cracked thick-walled tube undergoing a set of thermal cycles and loaded with a static mechanical force. During the first part of the thermal cycle, a high frequency induction coil heats the external wall of the tube. Then, the heating system stops and the specimen is cooled down by running water inside the tube. Finite element calculations show that only a region half-way along the tube should be heated to ensure adequate structural effect. In the heated zone, the machining of a sharp circumferential groove ensures the propagation of a unique crack. An electro-mechanical jack controls the level of the mechanical static load. Tests have been carried out, and these tests allow having an evaluation of the pertinence of the correction proposed by the RSE-M Code for a significant plasticity.

1 Introduction

In operating conditions, components such as train wheels, engine pistons, moulds, brazing of electrical devices or turbine blades are submitted to cyclic loadings which have thermal origins [1]. In nuclear power plants, thermal fatigue has firstly been observed in fast breeder reactor (FBR) and then, in some components of pressurized water reactors (PWR) [1 to 3]. In particular, cyclic thermal loadings may lead to multi-cracking, such as the thermal fatigue crazing [4]. In that frame, investigations on thermal loading have been undertaken using specific devices [5 to 8].

This present study focuses on the crack-propagation under cyclical thermo-mechanical loadings, as may be encountered in Pressurized Water Reactor pipes: thermal loadings may correspond to an incomplete mixing between cold and warm fluids, or thermal stratification... Mechanical loadings may result from water pressure itself, or may be bending moments induced by the boundary conditions and piping configuration. In order to estimate an eventual crack-propagation, the French RSE-M Code proposes to apply a plastic correction for ductile materials [9]; so, this correction concerns particularly the austenitic stainless steels (as AISI 304-L steel) [10]. In that frame, stress intensity factor range (ΔK) is deduced first from a linear elastic fracture mechanics analysis (LEFM), and is corrected after as:

$$\Delta K_{CP} = \alpha \Delta K \sqrt{(a + r_y)/a} \qquad (1)$$

where:

$$r_y = 1/6\pi(\Delta K \;\; /2S_y)^2 \qquad (2)$$

S_y is the conventional yield stress YS0.2 (0.2% yield stress).

In a case of pure thermal fatigue, α is simply taken equal to 1, whereas a α correction is proposed when the mechanical part of loading becomes significant ($K_{max}^{meca} > 0.2K_{max}$):

$$\alpha = 1 \; when \; r_y \le 0.05(W - a) \qquad (3)$$

$$\alpha = 1 + 0.15[r_y - 0.05 \cdot (W - a)/(0.035.(W - a))]^2$$

$$When \; 0.05(W - a) < r_y \le 0.085(W - a) \qquad (4)$$

$$\alpha = 1.15 \; when \; r_y > 0.085(W - a). \qquad (5)$$

* Corresponding author: cedric.gourdin@cea.fr

However, such methodology must be improved, particularly in the cases of cyclic thermo-mechanical loadings. For nuclear power plant components, thermal loading corresponds to restraint to thermal expansion, and thus mainly to strain-controlled loading. In that frame, crack propagation leads to some mechanical unloading in the un-cracked zone. In such way, methodologies based only on load-controlled tests cannot be very efficient.
In order to improve the estimation of crack propagation under cyclic thermal loading, a new device called PROFATH has been built in our laboratory. In that one, specimens are submitted in the same time to cyclic thermal loading and to constant tensile loading. This paper details the preparation of the tests and the first thermal maps obtained on this new device. However, present study begins with a crack propagation campaign performed for isothermal conditions on CT specimens.

2 Presentation of the thermo-mechanical device: PROFATH

- *Specimen geometry*: Specimen is a thick pipe (Fig 1). Despite cracks appear preferentially at the inner wall for real components (such as thermal fatigue crazing), crack is placed at the external wall. Such configuration allows better crack-opening measurements and facilitates the crack observations during the intermediate test stops. Calculation of mechanical fields with such geometry is relatively simple, since it corresponds to axisymmetric conditions. All the thermal and mechanical calculations have been made using finite element meshing; used software is the CEA code Cast3M [13]. Dimensions of specimen are taken as 60 mm outer diameter, 10 mm thickness and 300 mm length.

- *Thermal loading*: All the preliminary test calculations were made with a 30°C-300°C thermal cycle. In order to reduce the heating phase time, an external high-frequency eddy-current heating is placed around the crack plane. The maximum temperature of the outer wall is reached in about 30 s (Fig 1). Then, the temperature of the external wall is maintained until the temperature radial gradient goes down close to 10°C. After, heating is stopped and water-cooling of the inner wall is started. This second phase leads to a reduction of the temperature of the outer wall to 30°C. After a very brief air injection (0.5 s) to empty the inner wall of the specimen, a new cycle can be started.

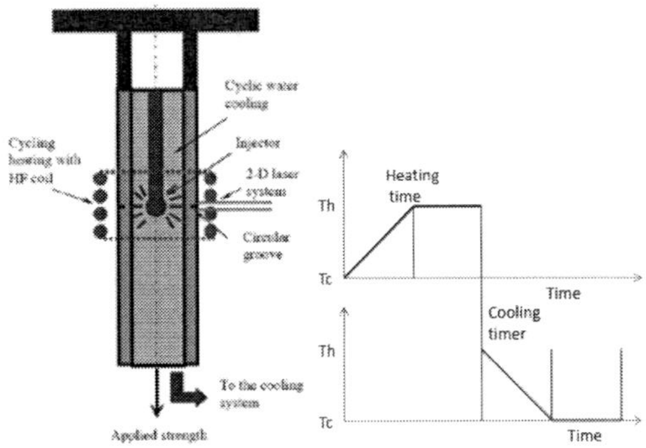

Fig 1. Test principle schematic and thermal cycle.

Stress calculations (Fig 2) clearly emphasize that a local heating is very well-adapted in the present case, since the external wall is mainly submitted to tensile stresses. In fact, such behavior results from component constraining effect: the un-heated part leads to severe constraining conditions, and induces bending stresses. Calculations show that maximum stresses and constraining effect are obtained for a 30 mm local length heating (15 mm for the half-specimen) as it is presented in Fig 2.

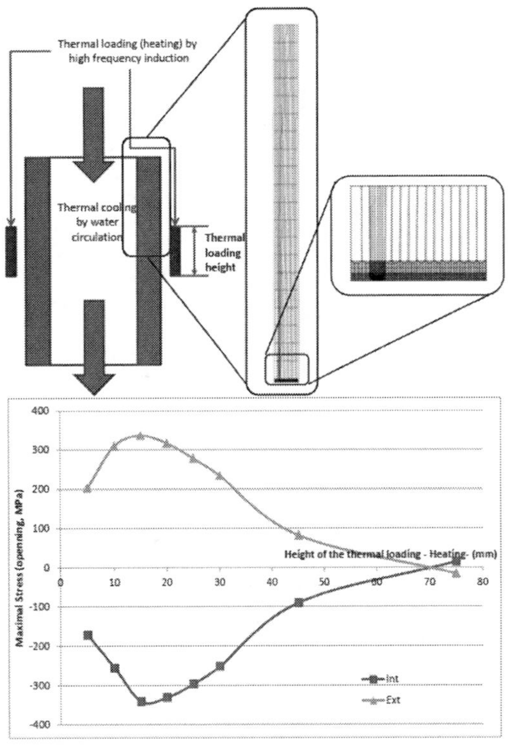

Fig 2: Finite-element meshing of the local heating design (CEA software Cast3M) and evolution of maximum stress as a function of heated half length

- Additional mechanical loading: Uniform tensile stress on specimen is simply obtained with the electric actuator of a DELTALAB tensile machine. PROFATH system is integrated between its columns, below the 200 kN capacity actuator, as shown in Fig 3.

- *Instrumentation*: Data acquisition and test control are provided by the joint use of a NATIONAL INSTRUMENT CompacDAQ device and CEA specific software written with LABVIEW.
Heating is devoted to a FIVE CELES high frequency generator. The 6 kW power is controlled by software through the 4/20 mA input of the generator. The water and air injection valves are also software controlled.
Accurate thermal mapping as a function of time are built from 0.5 mm diameter K-thermocouple measurements. Thermocouples can be placed at several depths (attached on outer wall, and 5 and 9 mm deep from outer wall), axial (7 values) and angular (0, 120, 240°) positions (Fig. 6).

- Mechanical centering: Strain gauges are positioned at the lower part of the specimen to allow an initial mechanical centering of the device.
- Crack length estimation: The use of an INSTRON extensometer placed above the crack plane makes a compliance measurement possible. Thanks to FE calculations, the equation linking compliance and crack length has been established.

Fig 3: PROFATH device.

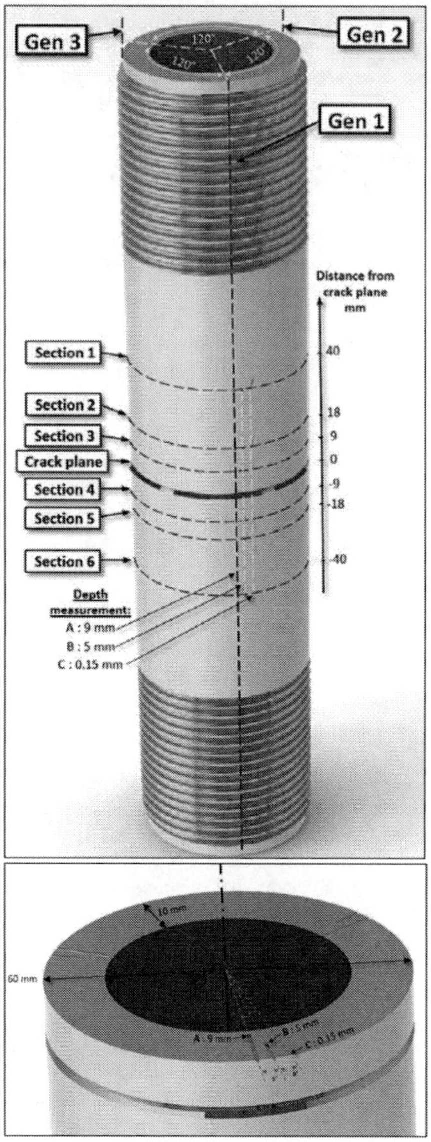

Fig 4: Location of the thermocouples.

3 Experimental device optimisation

- *Cooling*: Many preliminary tests were performed with two types of water cooling injection devices mentioned as "spherical injector" and "turning injector" and for different axial positions from the circular groove. Finally, chosen conditions are: a water flow rate of 12 l/min, a cooling temperature of 17°C or less, a "spherical injector" placed at the circular groove plane. Such conditions give a cooling time of about 50 seconds.
A Plexiglas specimen was used to identify the water behavior during injection with the two types of injector (Fig. 5). Pictures are obtained using a high speed camera (1000 pictures per second). Optical observations lead us to highlight that spherical injector brought less air in the

flow of water than the rotating one. This explains its better cooling efficiency.
- *Heater selection*: The purpose of the heating phase is to heat the entire thickness of the tube quickly before the thermal conduction spreads longitudinally. Initial tests have been performed with FIVE CELES to choose the generator and induction coil characteristics. Tests performed with specific specimen observed with a FLIR Infrared camera allowed us to choose a frequency of 100 kHz and a power of 6 kW.

Fig 5: Comparison of rotating and spherical injector's behaviour

- *Thermal map*: Establishment of experimental thermal maps is required in order to qualify accurately the real thermal loads. A specific specimen instrumented with as many as 20 thermocouples is used. The thermocouples are placed at outer surface and along wall thickness in some positions, and at three circumferential positions.
Test specimens are instrumented with 10 thermocouples. Only A (depth 9 mm) and C (surface) types are required in this case as explained in experimental part.

4 Experimental tests

Seven specimens have been tested. A test (n°5) was devoted to the study of the test specimen necking during testing.
The first test was devoted to develop and to validate the experimental process.
The experimental protocol has evolved through each test feedback. It can finally be synthesized through the following steps:
- Instrumentation of the specimen (extra thermocouples, strain gauges ...)
- Mechanical alignment of the specimen on the PROFATH device
- Thermal cycling and static mechanical load
- Mechanical compliance every 250/500 cycles to monitor the crack propagation: adjustment of test conditions for each step
- Mechanical fatigue post cracking is achieved by using a hydraulic device with a capacity of 500 kN. A cyclic displacement is imposed to the specimen with a ratio of 0.4,
- Optical analysis of the fracture surface.
An overview of the loading evolutions and the final fracture surfaces are presented for test 3 (Fig. 6).

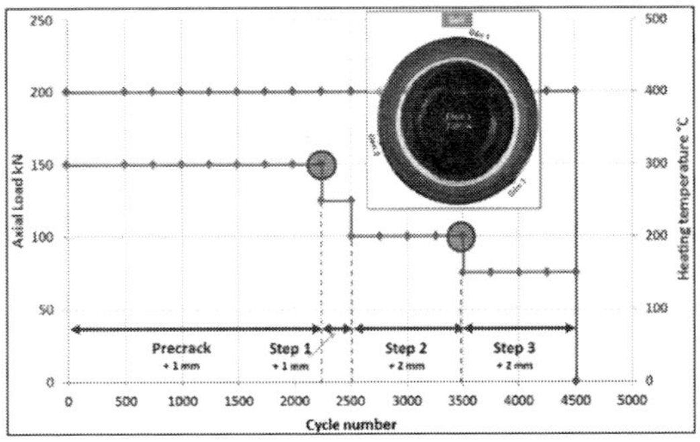

Fig 6: Overview of test n°3 with specimen 738C-A.

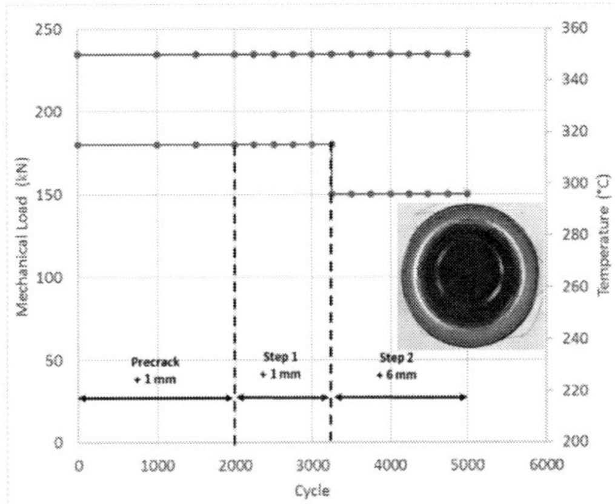

Fig 7: Overview of test n°6 with specimen 738D-B.

Application of dye penetrant product is performed for different intermediate steps during test. This one allows insuring optical measurements of the crack length after rupture. Let us note that such post-mortem values are in relative good agreement with estimates deduced from compliances as it is shown in Fig 7.

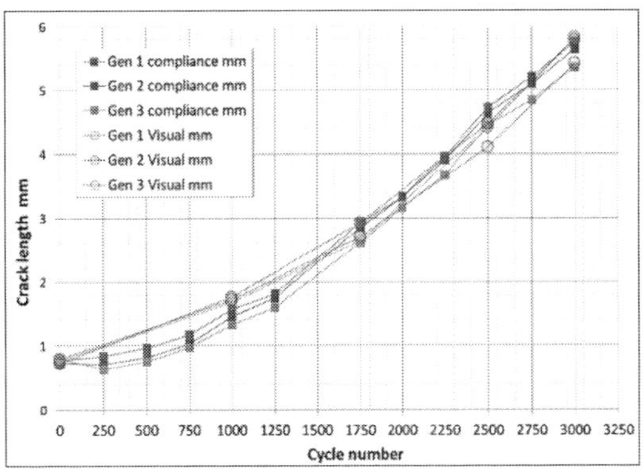

Fig 8: Crack propagation, comparison between estimates from compliance performed during the test, and optical measurements performed after post cracking (test n°2).

5 Analysis

- *Crack-growth rate tests in isothermal conditions*: In isothermal conditions, the crack-growth rate tests have been carried out at 20, 200 and 475°C under load controlled conditions with a load ratio R of 0.1 ($R = K_{min}/K_{max}$).

Specimens are compact tension specimens (CT25). During the test, crack-length is continuously evaluated from compliance measurements performed with a clip-gage. Let us note that such estimations deduced from compliance are in good agreement with measurements performed on final crack and on intermediate marks; 2 or 3 marks per test have been obtained by modifying temporary the load ratio (1000 cycles) and applying a product used for dye penetrant inspection. Fig 8 presents the crack-growth rate da/dN as a function of the effective stress intensity factor range (ΔK_{EFF}). This last value is directly deduced from the stress intensity factor range by applying a multiplying factor q including both crack-closure and mean stress effects [10]; the relation is:

$$\Delta KEFF = q*\Delta K \text{ with: } q = 1/(1-0.5*R) \text{ when } R \geq 0 \quad (6)$$

A Paris' law can be fitted for each investigated temperature; a significant influence of the temperature is clearly emphasized (Fig 8). Let us note that the Paris' law fitting remains acceptable, even when the un-cracked length becomes too small to verify the condition of validity proposed by ASTM E-647 (open symbols and NV).

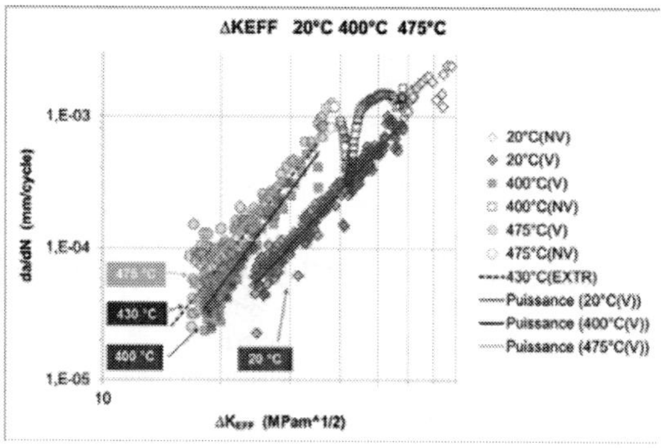

Fig 9: Crack-growth rate as a function of the effective stress intensity factor (ΔKEFF) at different temperatures; the open symbols correspond to non-valid results; the dotted line corresponds to an interpolation at 430 °C.

- *Analysis*: The thermo-mechanical calculations have been achieved using the CEA finite-element software Cast3M [13]. Only a thickness portion of the PROFATH specimen has been meshed, since asymmetrical condition can be considered. The meshing is detailed in Fig 1; elements are quadratics with 8 nodes (QUA8). At the crack tip, a very refined meshing is used, since the element size is reduced to only 30 microns.

A first step deals with thermal calculations; the aim is to reproduce the experimental information obtained with thermocouples and thermal maps obtained with FLIR Infrared camera (Fig. 10). Calculations are performed for the maximal temperatures of 400, 450 and 500°C chosen for the PROFATH program. Let us note that temperatures are measured with thermocouples placed at ± 9 mm from the crack plane. So, the maximal temperatures at the crack are systematically 30 °C higher (Fig 11); this difference is in good agreement with preliminary experimental measurements performed on a specific calibration specimen. Furthermore, an acceptable agreement can be observed between experimental measurements and F-E calculations at ± 9 mm (Fig 12).

Fig 10: Thermal map obtained with FLIR Infrared camera

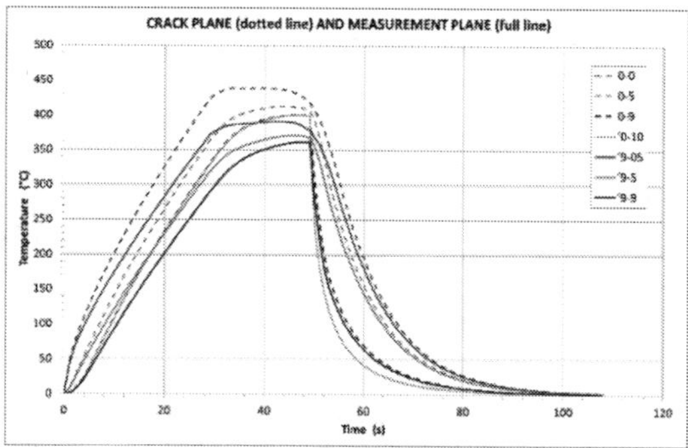

Fig 11: Cast3M calculations at 400 °C, estimated temperatures in the crack-plane (dotted line) and at ±9 mm from the crack plane (solid line), for outer wall, 5 and 9 mm deep.

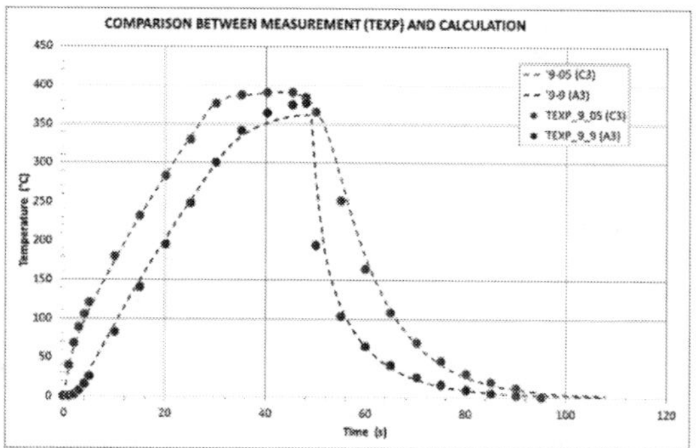

Fig 12: Cast3M calculations at 400 °C, comparison between estimated (solid line) and measured temperatures (dotted line) at ±9 mm from the crack plane and 9 mm deep.

In a second step, mechanical and fracture mechanics calculations are performed; crack-lengths are based on compliance estimations. As generally used for analysis on components and proposed by the RSE-M approach, calculations have been made with an elastic behavior. Initial geometry is considered in the present calculations; such analysis will be probably revised in the future when a best knowledge of the diameter contraction will be available. Application of a Paris' law requires both a crack deep enough (referred as "mechanically long crack") and intermediate SIF values (≥15 MPa√m). In this frame, the analysis begins when a crack-growth of 1 mm is reached from the initial machined groove.

The stress intensity factor (SIF) is deduced from the strain energy release rate using the G_THETA procedure of Cast3M; with a plane strain hypothesis. Fig 13 represents evolution of the SIF as a function of time. The maximum value is reached when the heating phase is ended; after, the SIF drops suddenly to a null value as soon as the water-injection starts; and it becomes equal to the mechanical component (K_{meca}) solely when the cycle is ended.

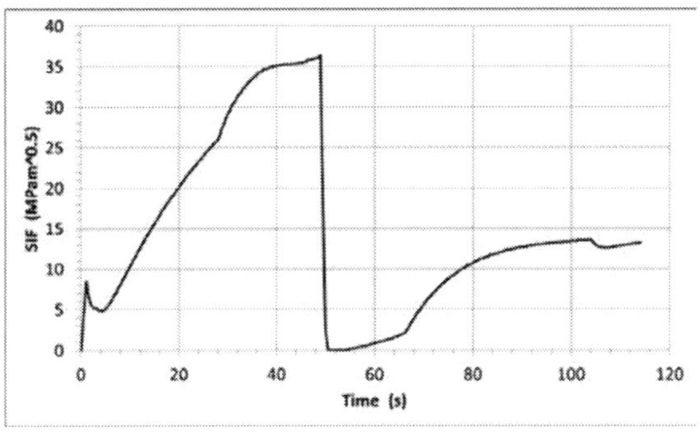

Fig 13: Cast3M calculations at 400 °C. Evolution of the SIF as a function of time; in this particular case the crack-length is 5.33 mm and the duration of cycle is 120 s.

Fig 14 confirms that an elastic-plastic correction is required, since applying a simple elastic approach conducts to completely unacceptable underestimates of the fatigue-life. Besides, it emphasizes a good agreement between experimental values and RSE-M predictions.

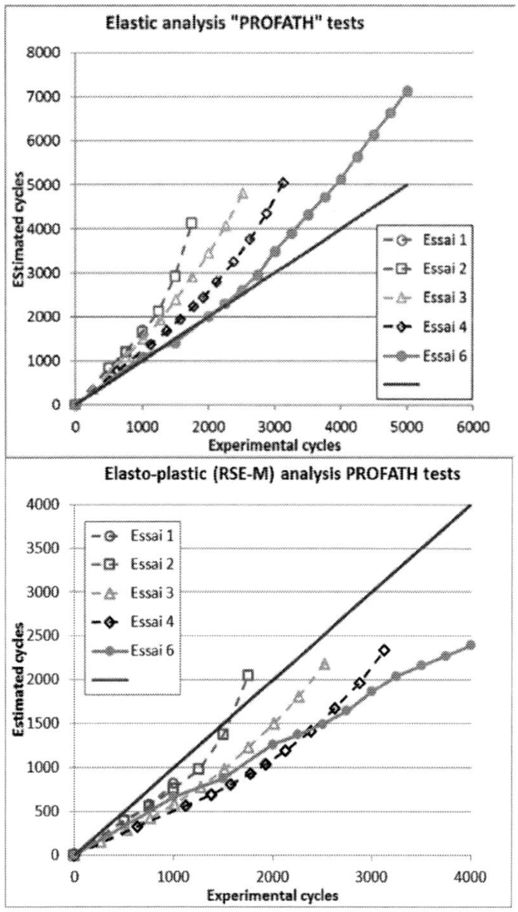

Fig 14: Number of cycles, comparison between predictions and experimental values for tests

- *Conclusions about prediction methodology*: A pure elastic fracture mechanics analysis conducts to severe underestimates of the fatigue-life. Consequently, such a methodology is completely not acceptable for thermo-mechanical loading, when the plasticity at the crack-tip is not negligible: when $0.05\,(W\text{-}a) < r_{yA} \leq 0.085\,(W - a)$ where W-a is the un-cracked length, r_{yA} the cyclic plastic zone calculated using an Irwin's relation.

An elastic-plastic analysis proposed by the RSE-M code conducts to good predictions; for this one a first step is based on elastic fracture mechanics calculation, a second step deals with application of plastic amplifying factor (α) and effective length ($a_{eff} = a + r_{yA}$).

6 Conclusions

In a first step, an isothermal test campaign was performed on CT-specimens with several loading conditions ($0.10 \leq R \leq 0.80$). It shows that application of a plastic correction (ry and α coefficient) is not indicated.
A new device called PROFATH was built. Specimen is a thick-walled tube, and cyclical loadings are applied: in a

first part of the cycle, the external wall is heated by a high frequency induction coil, and in a second part of the cycle, heating is stopped and the internal wall is cooled by running water. Finite element calculations show that only a median portion of the tube must be submitted to cyclical thermal loadings to guarantee an important component effect. To be sure that only one crack will propagated, a sharp circumferential groove is machined in the heated portion. A tensile mechanical loading may be also applied on specimen with an electro-mechanical jack.

Thermal mapping is deduced using a specific specimen instrumented with as many as 20 thermocouples: they are placed at outer surface and along wall thickness in some positions.

After an experimental optimization phase seven tests have been performed on the PROFATH device and they have been exploited:

- The elastic estimation seriously underestimates the crack propagation and is non-conservative,
- RSE-M estimation provides relative close and conservative results.

Mr Tournié Ivan and Godin Marvin are very greatly acknowledged for participation to the mounting device.

References

1. Fissolo, A., Fissuration en fatigue thermique des aciers inoxydables austénitiques, CEA Report R-5982, Mémoire d'Habilitation à Diriger les Recherches; 2001. Université des Sciences et Technologie de Lille.
2. Nulife project, Pilot project on thermal fatigue – item 3, Document D(08) 25383, December 2008.
3. Cipière, M.F., Le Duff, J.A., Thermal fatigue experience in French piping. International Institute Of Welding, Document n°XIII-1981-01, Lubjana, 2001.
4. Fissolo, A., Marini, B., Multiple Cracking in Thermal Fatigue, European Congress on Fracture 12th, Proceeding p. 31, Sheffield, UK, 1998.
5. Ancelet, O. and al, « Development of a test for the analysis of the harmfulness of a 3D thermal loading in tubes », International Journal of Fatigue, Vol. 29, 549-564, 2007.
6. Fissolo, A., Marini, B., Nais, G., Wident, P. "Thermal fatigue behaviour for a 316 L type steel" Journal of Nuclear Materials, 233 – 237 (1996), 156 – 161.
7. Fissolo, A., Gourdin, C., Vincent, L. "Crack-initiation under Thermal Fatigue: An Overview of CEA experience", 35th MPA Seminar, 9 October 2009, Stuttgart (Germany).
8. Bouin, P., Fissolo, A., Gourdin, C. "An experimental and numerical methodology to investigate on crack growth in a 304-L austenitic stainless steel pipe under thermal fatigue". Proceedings of the ASME 2010 Pressure Vessels and Piping Division/ K-PVP Conference, presented July 18 – 22, 2010, Bellevue, Washington, USA.
9. Code RSE-M – Modificatif 2005, Annexes 5.3 et 5.6.
10. RCC-MRx, « Règles de Conception et de Construction des Matériels Mécaniques des Installations Nucléaires applicables aux structures à haute température et à l'enceinte à vide ITER », AFCEN Code, Association Française pour les Règles de Conception et de Construction des chaudières Électronucléaires. www.afcen.com, 2012.
11. Produits métalliques – Pratique des essais de fissuration par fatigue. Norme AFNOR A03 – 404, 1991.
12. Standard Test Method for Measurement of Crack Growth Rates. American Society for Testing and Materials, ASTM 647, 1993.
13. Cast3M documentation. http://www-cast3m.cea.fr/.

Cyclic viscoplastic deformation modeling of a nickel-based single crystal superalloy with [001] orientation

Rongqiao Wang[1,2,3], *Bin* Zhang[1], *Dianyin* Hu[1,2,3,*], *Kanghe* Jiang[4], *Jianxing* Mao[1], *Fulei* Jing[5]

[1]School of Energy and Power Engineering, Beihang University, Beijing 100191, China
[2]Beijing Key Laboratory of Aero-Engine Structure and Strength, Beijing 100191, China
[3]Collaborative Innovation Center of Advanced Aero-Engine, Beijing 100191, China
[4]AECC Hunan Aviation Powerplant Research Institute, Zhuzhou 412000, China
[5]Aero Engine Corporation of China, Beijing 101304, China

Abstract. Accurate simulations of cyclic viscoplastic deformation behaviors of single crystal superalloys which are widely used for the manufacture of gas turbine blades are important for the effective design and safety assessment in practice. In this context, based on the in-phase thermomechanical fatigue (IP TMF) and out-of-phase thermomechanical fatigue (OP TMF) experiments of the nickel-based single crystal superalloy with [001] orientation, a modified constitutive model has been developed to describe the deformation behavior under thermomechanical loadings. The TMF experiment results indicate that stable hysteresis loops with remarkable ratcheting appear in both IP TMF and OP TMF. And it's worth noticing that the ratcheting growth direction of IP TMF and OP TMF are opposite. By introducing a Schmid stress rate related term to the back stress evolution equation, the slip-based Walker's constitutive model is modified in this study. And the simulation results of the deformation behavior reveal good agreement with the experiments under different IP TMF and OP TMF conditions.

1 Introduction

Due to excellent performance at elevated temperature, nickel-based single crystal superalloys have been widely used in turbine blades and vanes in gas turbine engines [1,2]. Generally, the preferred orientation of the single crystal turbine blade to subject the centrifugal loading is [001] orientation [1,3]. For gas turbine blades, TMF caused by the simultaneously centrifugal loadings and thermal loadings during the repeated engine operations is a major life-limiting factor [4,5]. And the first step in the calculation of fatigue life for the structural components, is the determination of stress-strain cycles [6]. Therefore, it is of great significance to investigate the TMF behavior and its accurate simulation method of the nickel-based single crystal superalloy with <001> orientation.

The unified viscoplastic constitutive models are widely applied to describe the viscoplastic deformation behaviors [7]. Previous researchers such as Walker, Chaboche, Bodner and Miller have proposed numerous viscoplastic models [8-11]. Since the nickel-based single crystal superalloys are orthotropic, it is necessary to consider the anisotropy in the viscoplastic constitutive model. A tensor-based mathematic approach is introducing a four order tensor in the constitutive model to characterize the anisotropy of the material [12-14]. Another method is a slip-based method which establishes flow rules and evolution equations of internal variables in the slip system based on the assumption that the inelastic deformation of the material is entirely caused by the slip [15-17]. Compared with the tensor-based method, the slip-based method could reflect the micro-deformation mechanism of the material and describe the phenomenon of slip in the deformation process of the nickel-based single crystal superalloy.

A series of fatigue and creep-fatigue tests clarify that the tensile ratcheting decreases the fatigue life [18]. Therefore, determination of the ratcheting is an important issue in the design of components in automotive, aerospace and aeronautics industries [19]. However, the accurate simulation of ratcheting in unified viscoplastic models is still a challenge [20-23]. In order to improve the accuracy of ratcheting simulation, a variety of modified approaches about kinematic hardening and dynamic recovery have been proposed, and achieved great results at specific conditions. [24-26]. However, there are few reports about the quantitative simulation of the IP TMF and OP TMF ratcheting behaviors which is observed under stress controlled loading with zero mean stress of the nickel-based single crystal superalloy.

In this regard, the objective of this paper is to develop a constitutive model to simulate the cyclic viscoplastic deformation of the nickel-based single crystal superalloy accurately. Firstly, stress-controlled TMF experiments were conducted to investigate the deformation behavior of the nickel-based single crystal superalloy with [001] orientation. Then, a modified slip-

* Corresponding author: hdy@buaa.edu.cn

based viscoplastic constitutive model was established through introducing a Schmid stress rate related term to the back stress evolution equation and the material constants were identified by fitting the experimental data. Finally, deformation behaviors of the nickel-based single crystal superalloy under LCF (low cycle fatigue) and TMF loadings were simulated, and the simulation results showed a good agreement with the experiment results.

2 Material and deformation behavior

2.1 Material

The material used in present work is a nickel-based single crystal superalloy casted along [001] orientation. The chemical composition and basic mechanical properties at different temperatures of the alloy are given in Table 1 and Table 2, respectively. According to the crystal orientation analysis, the crystal orientation deviations of the material less than 10° from [001] direction. And the materials have experienced the solution heat treatments at 1290 °C for 1h, then at 1300 °C for 2h, then at 1315 °C for 4h and then air-cooled. Subsequently, a two-stage aging treatment consisting of heat treatment at 1120 °C for 4h and at 870 °C for 32h is performed to achieve the designed γ/γ' microstructure (Fig 1). For nickel-based single crystal superalloy used in this study, the volume fraction of γ' precipitates dispersed in γ-matrix is typically about 70% at room temperature and the size of cuboidal γ'-particles is about 0.5 μm measured along the cube edge. In general, the main source of hardening in nickel-based single crystals is provided by the γ' precipitates.

Fig 1. γ/γ' microstructure of the nickel-based single crystal superalloy.

Table 1. Chemical composition of the material (wt. %).

Ni	Al	Ta	W	Co	Re	Hf	Cr	Mo
Balance	5.2~6.2	6.0~8.5	7.0~9.0	8.5~9.5	1.6~2.4	0.05~0.15	3.8~4.8	1.5~2.5

Table 2. Mechanical properties of the nickel-based single crystal superalloy in <001>direction.

Temperature (°C)	25	700	760	850	980	1070
Elastic modulus (GPa)	131.5	107.0	105.5	98.0	80.5	69.5
Shear modulus (GPa)	137.0	100.2	105.0	60.6	80.4	74.2
Poisson ratio	0.344	0.374	0.377	0.383	0.390	0.399

2.2 TMF deformation behavior

The hollow tubular uniaxial specimens used in the experiments were machined from the cast bars of the nickel-based single crystal superalloy, with axes parallel to [001] orientation. For the specimens, the gage section length was 25 mm, the outside diameter was 12.0 mm and the inside diameter was 9.0 mm. All of the TMF experiments were conducted under stress control with stress ratio R=-1 and frequency f=0.01.

The deformation behaviors under IP-TMF and OP-TMF conditions are shown in Fig 2. As can be seen, stable hysteresis loops with obvious ratcheting behaviors are found in both IP-TMF and OP-TMF, while the ratcheting direction is opposite. In addition, the hysteresis loop of the initial cycle is significantly asymmetric. Specifically, under IP TMF conditions, the tensile deformation is greater than the compressive deformation, which is opposite under OP TMF conditions. One reason is that the temperature at the tensile half-cycle of IP TMF and the compressive half-cycle of OP TMF is higher and the deformation behaviors of the materials at high temperature are greater than the deformation behaviors at low temperature. Of course, the nickel-based single crystal superalloy investigated in this study has the property of tension-compression asymmetry under isothermal cyclic loading [27], which would also have a certain influence on the deformation behavior. In addition, with the increase of stress amplitude, the ratcheting of both IP TMF and OP TMF increases obviously.

Based on the experimental results, it could be seen that the TMF deformation behavior simulation of the nickel-based single crystal superalloy mainly includes the accurate simulation of asymmetry of hysteresis loop and the ratcheting with opposite evolution direction in IP TMF and OP TMF. Therefore, the constitutive model established should accurately reflect these two aspects.

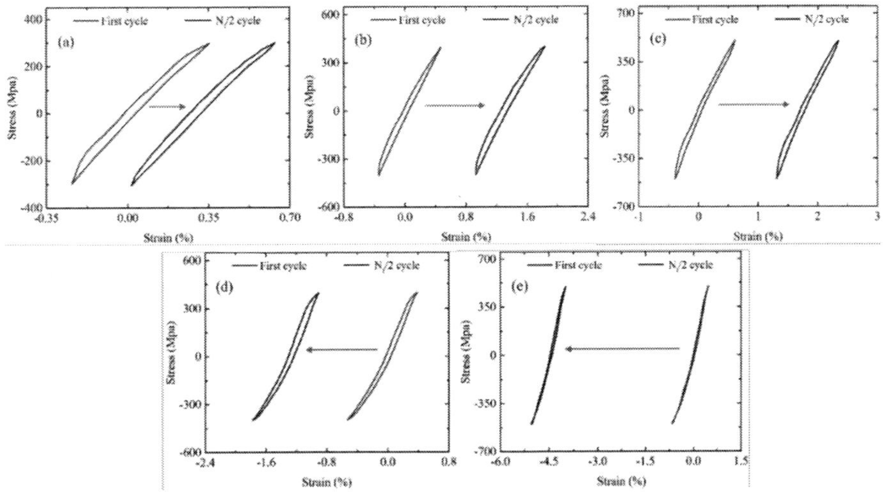

Fig 2. The TMF stress-strain hysteresis loops under 500°C-980°C: hysteresis loops of IP TMF with stress amplitude (a) 300 MPa, (b) 400 MPa, (c) 500 MPa; hysteresis loops of OP TMF with stress amplitude (d) 400 MPa; (e) 500 MPa. The green arrow indicates the direction of ratcheting.

3 Slip-based viscoplastic constitutive model

Slip-based Walker's constitutive model has been used to simulate the deformation behavior of the nickel-based single crystal superalloy such as PWA1480 and Hastelloy-X, and the simulation results show that this model could reflect the properties of anisotropy, cyclic hardening/softening and ratcheting accurately [15,28]. Thus we employ the slip-based Walker's constitutive model to describe the viscoplastic behavior in this paper.

In slip-based Walker's constitutive model, the total strain ε_m is composed of elastic strain ε_e and inelastic strain ε_{in} as given by

$$\varepsilon_m = \varepsilon_e + \varepsilon_{in} \tag{1}$$

On the basis of the Hooke's law, the elastic strain ε_e is governed as

$$\sigma = \boldsymbol{C} : \varepsilon_e \tag{2}$$

where σ is the macroscopic stress tensor and $\boldsymbol{C}$ is the stiffness tensor. For nickel-based single crystal superalloy with face centered cubic (FCC) structure, the stiffness tensor $\boldsymbol{C}$ can be given as follows:

$$\boldsymbol{C} = \begin{bmatrix} c_{11} & c_{12} & c_{12} & 0 & 0 & 0 \\ c_{12} & c_{11} & c_{12} & 0 & 0 & 0 \\ c_{12} & c_{12} & c_{11} & 0 & 0 & 0 \\ 0 & 0 & 0 & c_{44} & 0 & 0 \\ 0 & 0 & 0 & 0 & c_{44} & 0 \\ 0 & 0 & 0 & 0 & 0 & c_{44} \end{bmatrix} \tag{3}$$

$$c_{11} = \frac{(1-\mu)}{(1-2\mu)(1+\mu)} E \tag{4}$$

$$c_{12} = \frac{\mu}{(1-2\mu)(1+\mu)} E \tag{5}$$

$$c_{44} = G \tag{6}$$

where E, G and μ are elastic modulus, shear modulus and Poisson ratio, respectively. The values of them at different temperatures are shown in Table 2 [29].

According to the crystal plasticity theory, the deformation of the crystal is caused by the slip along a specific slip system. For the nickel-based single crystal superalloy, 12 octahedral slip systems {111} <011> and 6 cube slip systems {100} <011> are generally considered in the crystallographic model. The relationship between the inelastic strain rate $\dot{\varepsilon}_{in}$ and the slip rate $\dot{\gamma}^r$ in the rth slip system could be expressed as follows:

$$\dot{\varepsilon}_{in} = \sum_{r=1}^{18} \dot{\gamma}^r (\boldsymbol{m}^r \otimes \boldsymbol{n}^r) \tag{7}$$

where $\boldsymbol{m}^r$ and $\boldsymbol{n}^r$ are the vector of slip direction and normal vector of the slip plane, respectively. Meanwhile, the Schmid stress π^r in the rth slip system is given by

$$\pi^r = \boldsymbol{\sigma} : (\boldsymbol{m}^r \otimes \boldsymbol{n}^r) \tag{8}$$

K.P. Walker [15] assumed that the shear stress and the inelastic shear strain rate (slip rate) on the slip system followed the similar form of the isotropic Walker's viscoplastic constitutive mode [8], and established a viscoplastic constitutive model based on slip system. The simulation results of PWA1480 and Hastelloy-X show that slip-based Walker's constitutive model could simulate the deformation behavior of nickel-based single crystal superalloy accurately [15,24,28,30]. In general, for nickel-based single crystal superalloys, only octahedral slip systems are active when the uniaxial load is applied along [001] orientation. In slip-based Walker's constitutive model, the slip rate in the octahedral slip system is defined:

$$\dot{\gamma}^o = \left|\frac{\pi^o - \omega^o}{K^o}\right|^{n-1}\left(\frac{\pi^o - \omega^o}{K^o}\right) \tag{9}$$

where π^o, ω^o and K^o are the Schmid stress, back stress and drag stress in the octahedral slip system, respectively. In slip-based Walker's constitutive model, ω^o and K^o describe the kinematic hardening and isotropic hardening respectively. And they are defined as follows:

$$\dot{\omega}^o = \rho_1^o\dot{\gamma}^o - \rho_2^o\left|\dot{\gamma}^o\right|\omega^o - \rho_3^o\left|\omega^o\right|^{p-1}\omega^o \tag{10}$$

$$K^o = K_0^o + \rho_4^o\pi_{nz}^o + \rho_5^o\left|\Psi^o\right| \tag{11}$$

where ρ_1^o, ρ_2^o, ρ_3^o, ρ_4^o, ρ_5^o, n, p and K_0^o are material constants in octahedral slip system. In addition, π_{nz}^o and Ψ^o are non-Schmid stress and Takeuchi-Kuramoto stress, respectively. The calculation methods of them could be seen in reference [24]. Previous research shows that the slip-based Walker's constitutive model has poor accuracy of ratcheting simulation under TMF condition [24]. In order to improve the simulation accuracy, using the similar modification form in reference [24], the back stress evolution equation in this study is modified to

$$\dot{\omega}^o = \rho_1^o\dot{\gamma}^o + \rho_6^o\xi\left(\dot{\pi}^o\right)^3 - \rho_2^o\left|\dot{\gamma}^o\right|\omega^o - \rho_3^o\left|\omega^o\right|^{p-1}\omega^o \tag{12}$$

where $\dot{\pi}^o$ is the rate of Schmid stress, ρ_6^o is a material constant related to loading conditions and materials in the octahedral slip system, and ξ is a scaling factor which is defined:

$$\xi = \frac{1}{2}\left|sign(\dot{\gamma}^o) - \frac{\omega^o}{\rho_1^o/\rho_2^o}\right| \tag{13}$$

4 Simulation of cyclic viscoplastic behavior

4.1 Simulations under LCF loadings

The material constants of Walker's constitutive model listed in Table 1 are identified by fitting the uniaxial standard specimen experimental data which are from reference [29,31,32]. Using Walker's constitutive model with the material constants in Table 1, LCF deformation behaviors at 760°C and 980°C of the nickel-based single crystal superalloy are simulated. And the simulation results of hysteresis loops, cyclic responses of stress range and cyclic stress-strain curve are shown in Fig. 3~5 respectively. The results indicate that Walker's constitutive model could accurately describe the LCF deformation of the nickel-based single crystal superalloy in [001] orientation.

Table 3. Material constants for the constitutive model in the octahedral {111} <001> slip systems.

Material constants	ρ_1^o	ρ_2^o	ρ_3^o	ρ_4^o	ρ_5^o	n	p	K_0^o
Values at 760°C	1.0345 $\times10^6$	8.4504 $\times10^3$	5.3×10^{-7}	0	0.1	4.9	3.2	1.5862 $\times10^3$
Values at 980°C	1.9150 $\times10^6$	7.6600 $\times10^3$	1.0×10^{-8}	0	-0.5	2.7	1.5	1.8620 $\times10^3$

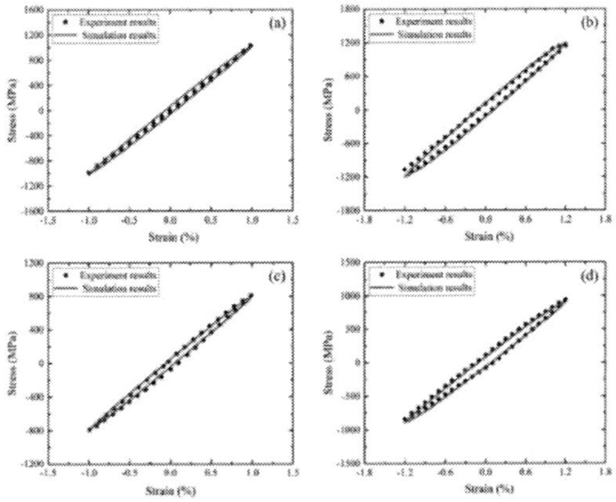

Fig 3. The experimental and simulated LCF stress-strain hysteresis loops under strain rates $5\times10^{-3}/s^{-1}$: (a) temperature: 760°C/ strain range: ±1.0%; (b) temperature: 760°C/ strain range: ±1.2%; (c) temperature: 980°C/ strain range: ±1.0%; (d) temperature: 980°C/ strain range: ±1.2%.

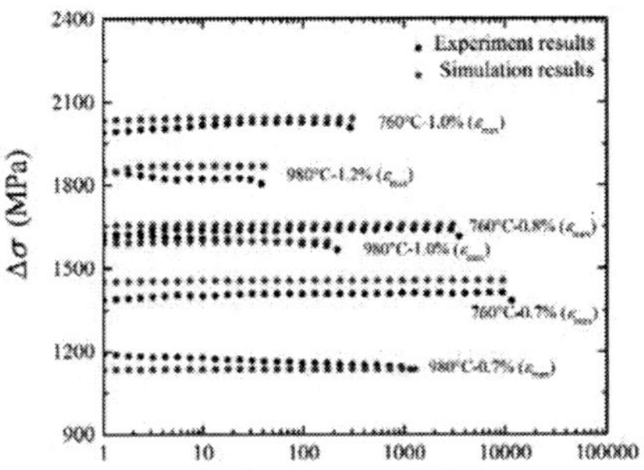

Fig 4. The experimental and simulated LCF cyclic responses of stress range at 760°C and 980°C with R=-1.

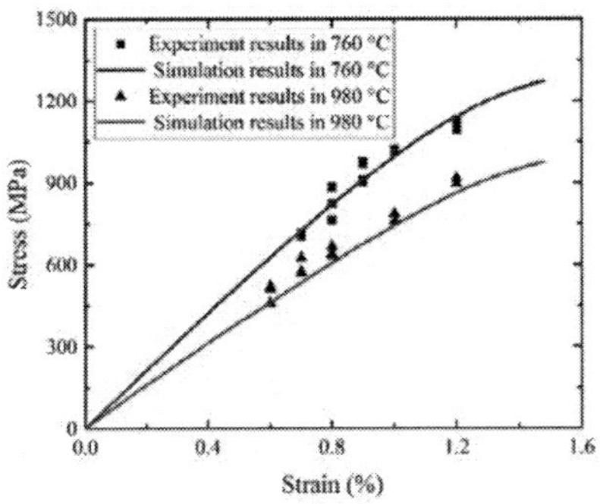

Fig 5. The experimental and simulated LCF cyclic stress-strain curve at 760°C and 980°C with R=-1.

4.2 Simulations under TMF loadings

Based on the TMF experimental results of the nickel-based single crystal superalloy in [001] orientation, it is found that the material constant ρ_6^o in octahedral {111} <001> slip systems could be expressed as follows:

$$\rho_6^o = \alpha^o \Delta\pi^o + \beta^o \tag{14}$$

where $\Delta\pi^o$ is the amplitude of Schmid stress. In addition, α^o and β^o are the material constants in octahedral slip systems. For IP TMF and OP TMF experiments at 500°C-980°C, the values of α^o and β^o are shown in Table 4. The experimental and simulated TMF deformation behaviors of the nickel-based single crystal superalloy is shown in Fig 6. As can be seen, the simulation results and experiment results under different experimental conditions show a good agreement.

Table 4. Material constants in modified constitutive model.

Material constants	α^o	β^o
Values in IP TMF	3.478×10^{-4}	-7.067×10^{-2}
Values in OP TMF	5.344×10^{-4}	-1.062×10^{-1}

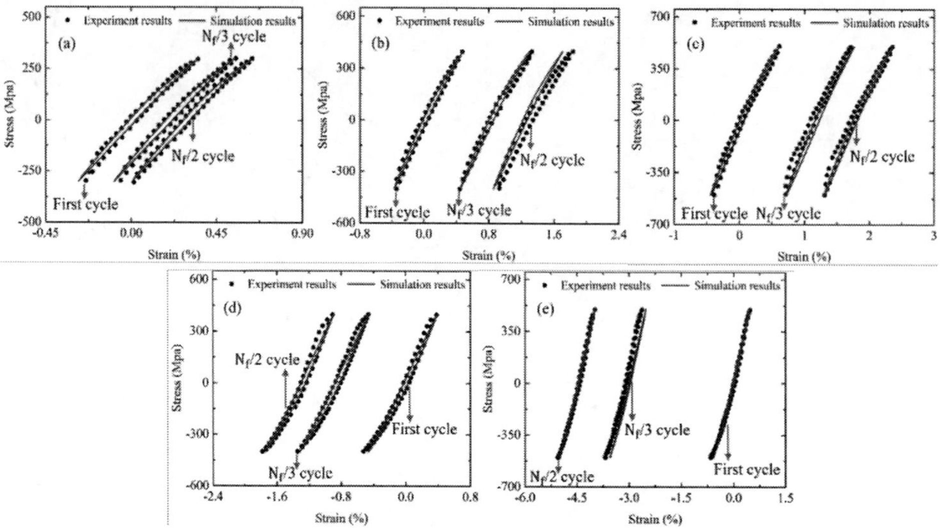

Fig 6. The experimental and simulated TMF stress-strain hysteresis loops under 500°C-980°C: hysteresis loops of IP TMF with stress amplitude (a) 300 MPa, (b) 400 MPa, (c) 500 MPa; hysteresis loops of OP TMF with stress amplitude (d) 400 MPa OP; (e) 500 MPa. N_f indicates the lifetime of the specimen.

4.3 Comparison of of classical and modified models

In order to verify the modification effect, the simulations of the TMF deformation behaviors are conducted through Walker's constitutive model and modified constitutive model respectively. Both the classical and modified constitutive model could simulate the TMF hysteresis loops of the nickel-based single crystal superalloy. Meanwhile, the ratcheting simulation results are in good agreement with the experimental results when the stress amplitude reaches 500 MPa. However, for Walker's constitutive model, when the stress amplitude is small (such as 300MPa and 400MPa), the simulated results in both IP TMF and OP TMF show excessive ratcheting, and the overestimation degree is reduced with the increase of the stress amplitude (see Fig. 7). After modification, the simulation capability of the constitutive model is much better in ratcheting simulation for the condition of 300 MPa and 400 MPa. It indicated that the modified method in this study has a good effect on improving the ratcheting simulation capability of Walker's constitutive model at lower stress amplitude.

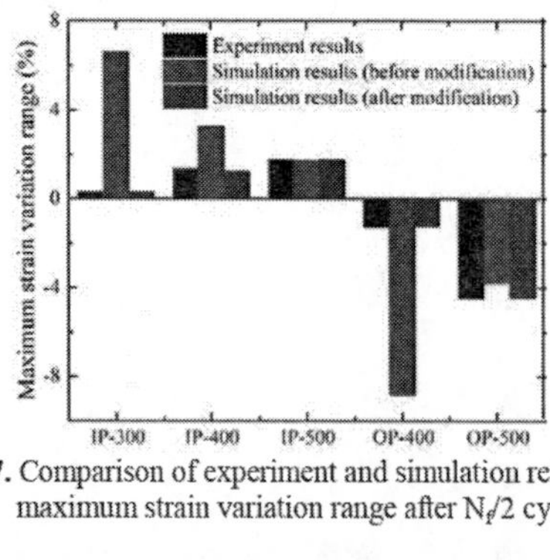

Fig 7. Comparison of experiment and simulation results of the maximum strain variation range after $N_f/2$ cycles.

5 Conclusions

Based on the TMF experiments conducted in this study, a modified constitutive model has been developed to describe the deformation behaviors of the nickel-based single crystal superalloys under thermomechanical loadings in this study. The detailed conclusions are summarized as follows:

(1) Stress-controlled IP TMF and OP TMF experiments of the nickel-based single crystal superalloy with [001] orientation was conducted. The hysteresis loops are stable in at least half-life period. And the remarkable ratcheting increases along the opposite direction in IP TMF and OP TMF, respectively.

(2) Based on the existing experimental data, the material constants of Walker's constitutive model were identified. And the simulation results of the deformation behaviors under LCF loadings are in good agreement with the experiment data, which verifies the strong simulation capability of the Walker's constitutive model in LCF deformation simulations.

(3) Combined with the deformation characteristics of the nickel-based single crystal superalloy under TMF loading, a Schmid stress rate related term is introduced to the back stress evolution equation of slip-based Walker's constitutive model for changing the evolution rate of the back stress. Through this method, a modified constitutive model has been developed, and the simulation results under different IP TMF and OP TMF loadings show a good agreement with the experiment results.

(4) Compared with the unmodified constitutive model, the modified constitutive model could inhibit the ratcheting under low stress amplitude conditions effectively. On the basis of the existing TMF experimental data, the accuracy of ratcheting simulation

improves significantly after modifying the constitutive model.

Acknowledgements

We acknowledge financial support from National Natural Science Foundation of China (NSFC) (Grants No. 51375031 and 51675024).

References

1. R.C. Reed, *The superalloys: fundamentals and applications* (Cambridge University Press, Cambridge, 2006)
2. A. Staroselsky, B.N. Cassenti, Int. J. Solids. Struct **48**, 13 2060-2075 (2011)
3. Z. Wen, D. Zhang, S. Li, et al, J. Alloy. Compd **692**, 301-312 (2016)
4. R.L. Amaro, S.D. Antolovich, R.W. Neu, A. Staroselsky, Procedia. Eng **2**, 1 815-824 (2010)
5. J.J. Moverare, S. Johansson, R.C. Reed, Acta. Mater **57**, 7 2266-2276 (2009)
6. O.U. Colak, Mater. Design **29**, 8 1575-1581 (2008)
7. A.V.J. Luis, C.M. Stewart, *Proceedings of the ASME 2015 International Mechanical Engineering Congress and Exposition*, (IMECE2015-50390, 2015)
8. K.P. Walker, *Research and development program for nonlinear structural modeling with advanced time-dependent constitutive relationships* (NASA-CR-165533, 1981)
9. J.L. Chaboche, Acad. Polon. Sci, Ser. Sci. Tech **25**, 1 33-39 (1977)
10. S.R. Bodner, Y. Partom, J. Appl. Mech **42**, 2 385-389 (1975)
11. A. Miller, J. Eng. Mater-T. ASME **98**, 2 97-105 (1976)
12. A. Baltov, A. Sawcauk, Acta. Mech 1, 2 81-92 (1965)
13. G.A. Swanson, I. Linask, D.M. Nissley, et al, *Life prediction and constitutive models for engine hot section anisotropic materials program: annual status report* (NASA-CR-174952, 1986)
14. G.A. Swanson, I. Linask, D.M. Nissley, et al, *Life prediction and constitutive models for engine hot section anisotropic materials program: second annual status report* (NASA-CR-174954, 1987)
15. K.P. Walker, E.H. Jordan, *Biaxial constitutive modelling and testing of a single crystal superalloy at elevated temperatures* (Mechanical Engineering Publications, London, 145-170, 1989)
16. D. Nouailhas, A.D. Freed, J. Eng. Mater. Technol **114**, 1 97-104 (1992)
17. L.T. Dame, D.C. Stouffer, J. Appl. Mech **55**, 2 325-331 (1988)
18. S. Date, H. Ishikawa, T. Otani, et al, Nucl. Eng. Des **238**, 2 336-346 (2008)
19. A. Navarro, J.M. Giráldez, C. Vallellano, Int. J. Fatigue **27**, 8 838-846 (2005)
20. D.L. Mcdowell, A.B. Lamar, *Advances in Plasticity*, (247-251, 1989)
21. J.L. Chaboche, Int. J. Plasticity **7**, 7 661-678 (1991)
22. J.L. Chaboche, *Advances in Plasticity*, (211-214, 1989).
23. E. Krempl, M.B. Ruggles, J. Mech. Phys. Solids **38**, 4 587-597 (1990)
24. E.H. Jordan, K.P. Walker, J. Eng. Mater-T. ASME **114**, 1 19-26 (1992)
25. J.L. Chaboche, Eur. J. Mech. A-Solid **13**, 501-518 (1994)
26. X. Chen, R. Jiao, Int. J. Plasticity **20**, 4 871-898 (2004)
27. Y. Li, B. Su, J. Mater. Eng, 5 7-10 (2002)
28. E.H. Jordan, S.X. Shi, K.P. Walker, Int. J. Plasticity **9**, 1 119-139 (1993)
29. H. Yu, X. Wu, et al, *Materials handbook for aeroengine design (fourth)* (Aviation Industry Press, Beijing, 2010)
30. D.M. Nissley, T.G. Meyer, K.P. Walker, *Life prediction and constitutive models for engine hot section anisotropic materials program* (NASA-CR-189223, 1992)
31. Z. Shi, X. Wang, S. Liu, et al, Prog. Nat. Sci **25**, 1 78-83 (2015)
32. B.Z. Wang, D.S. Liu, Z.X. Wen, et al, Mat. Sci. Eng. A-Struct **593**, 3 31-37 (2014)

Towards a temperature dependent and probabilistic lifetime concept for nodular ductile cast iron materials undergoing isothermal and thermo-mechanical fatigue

Elena Garcia Trelles[1,*], *Christoph* Schweizer[1], and *Stefan* Eckmann[1]

[1]Fraunhofer Institute for mechanics of materials IWM, Wöhlerstraße 11, 79108 Freiburg im Breisgau, Germany

Abstract. In this investigation, the fatigue behaviour of a ductile cast iron with high content of silicon and molybdenum, was experimentally characterized by performing isothermal low cycle fatigue (LCF) tests as well as out-of-phase thermomechanical fatigue (OPTMF) tests within the temperature range RT – 500 °C. The studied material shows an embrittlement at temperatures nearby 400 °C. A possible explanation for the observed lifetime reduction is intergranular embrittlement (IE). A mechanism based lifetime model is proposed for assessing the lifetime. The model is based on the assumption that the crack advance per cycle is correlated with the cyclic crack tip opening displacement (ΔCTOD) attributed to the crack tip blunting caused by accumulation of plastic and creep deformations ahead of the crack tip. Intergranular embrittlement is accounted for by introducing a temperature and strain rate dependent prefactor in the crack growth law, which only acts in a certain temperature range. The model is calibrated for a GJS material and successfully applied to predict the lifetime of this material when undergoing isothermal and non-isothermal mechanical loadings. A probabilistic interpretation of the scatter of the investigated material is presented in conjunction with the random nature of the initial defect size distribution.

1 Introduction

Nodular ductile cast iron materials (DCI) provide a good performance at high temperatures, especially DCI materials with high content of silicon and molybdenum. Due to their good mechanical properties, castability and cost efficient production processes, the use of these materials for fabricating components undergoing thermomechanical fatigue (i.e. cylinder heads, turbochargers, motor housings) is widely spread. In order to guarantee the safety of these components the damage mechanisms involved in the crack initiation and propagation regimes must be carefully investigated and integrated into lifetime concepts.

For non-isothermal conditions, the time and temperature dependent damage parameter D_{TMF} [1] was successfully applied for describing the fatigue damage of numerous high temperature resistant alloys. This parameter accounts for the temperature dependent material properties, the cyclic stresses and plastic strains, fatigue crack closure as well as for creep accelerated fatigue crack growth. The lifetime assessment is performed by assuming that the crack growth per cycle is proportional to the crack tip opening displacement.

However, this model formulation does not account for IE damage. The presence of IE damage is related to a shorter fatigue lifetime and the modelling of DCI materials suffering from IE damage demands a new formulation.

The IE damage mechanism was studied by several authors [2-6]. Nevertheless there is no clear explanation for this phenomenon. It is known to be dependent on microstructural features, such as the graphite size and the eutectic cell wall size, where greater sizes should promote intergranular fracture [2]. Based on tensile testing, it was also studied in [3] how increasing the percentage of silicon, changes the fracture mechanism from a mix of dimpled and intergranular fracture to cleavage and intergranular fracture. This change of fracture mechanism is thought to be related with the presence of MgO particles in the eutectic cell wall region serving as crack initiation sites [4]. Another investigation [5] concludes that MgO inclusions serve as crack initiation sites and the amount of phosphorus segregation on the grain boundaries dictates the fracture mode within the IE temperature of ferritic DCIs. In the same study the amount of segregated phosphor is connected to the solidification cooling rates where slower cooling rates should enhance phosphorus segregation. The kinematics of the process seem to be unclear, however all authors observed a drop in the ductility of the material at the IE temperature for the conducted tensile tests. [6] observed that the ductility drop derived from the IE damage is also dependent on the strain rate of the performed tensile tests. In this work the IE damage was modelled based on tensile tests performed at different temperatures and strain rates including the effect of strain rate on the IE damage process. [6] and [7] investigated the strain rate

* Corresponding author: elena.garcia.trelles@iwm.fraunhofer.de

dependency of the IE effect, concluding that increasing strain rates shift the IE temperature (corresponding to the ductility minimum) to higher temperature values

The first aim of this study is to propose a lifetime model including the damaging effect of IE to enable reliable lifetime assessment of DCIs. This is achieved with the introduction of the IE damage factor from [6] in the mechanism based damage model from [1] by modifying the original model formulation based on the D_{TMF} parameter in order to account for IE damage, which is expected to lead to accelerated crack growth and reproduces the temperature dependent lifetime reduction. The second aim is to validate the model through the LCF and OPTMF performed tests with the ferritic DCI GJS.

A probabilistic concept is introduced in order to provide an intuitive interpretation of the material scatter, attributing this inherent scatter to the fact that the initial flaw size causing crack initiation has not a fixed size. A more realistic approach to real conditions is to work with an initial flaw size probability distribution function.

2 Experimental

The ductile cast iron material concerning this study is referred as GJS due to confidentiality reasons. The microstructure of this spheroidal cast iron with ferritic matrix is illustrated in Fig. 1.

For characterizing the mechanical behaviour of the cast iron GJS, tensile tests, strain controlled LCF tests and OPTMF tests were carried out for representative strain amplitudes within the low cycle fatigue regime.

The tensile tests were performed at RT, 100 °C, 200 °C, 300 °C, 350 °C, 400 °C, 450 °C and 500 °C with a strain rate of $\dot{\varepsilon} = 10^{-4}\ s^{-1}$. Three further tensile tests were performed at 450 °C and one at 500 °C with $\dot{\varepsilon} = 10^{-3}\ s^{-1}$ in order to study the strain rate dependency of the mechanical properties.

The tensile tests were conducted with an electromechanical tensile testing machine Kappa SS from Zwick Roell. The specimens were heated using a three-zone radiation furnace from MTS. Type K thermocouples were used.

LCF tests were carried out at RT, 300 °C, 350 °C, 400 °C, 450 °C and 500 °C with $\dot{\varepsilon} = 10^{-3}\ s^{-1}$ to characterize the fatigue behaviour within the temperature range RT-500 °C.

OPTMF tests were performed to study the thermomechanical behaviour of GJS. For the thermal cycle a minimum temperature equal to 50 °C was fixed for all the specimens and the maximum temperature of the thermal cycles was set to 350 °C, 400 °C, 450 °C and 500 °C. The tests were conducted with a fixed heating rate of 5°C/s and air cooling was applied during cooling of the specimens. To study if the material exhibits time dependent behaviour, two OPTMF tests were conducted with a dwell time of 300 s applied at the minimum strain (dwell time in compression).

The LCF and OPTMF tests were performed on an Instron 8862 electromechanical testing machine. The specimens were heated using a high frequency induction heating system. Temperature control was achieved using type K thermocouples.

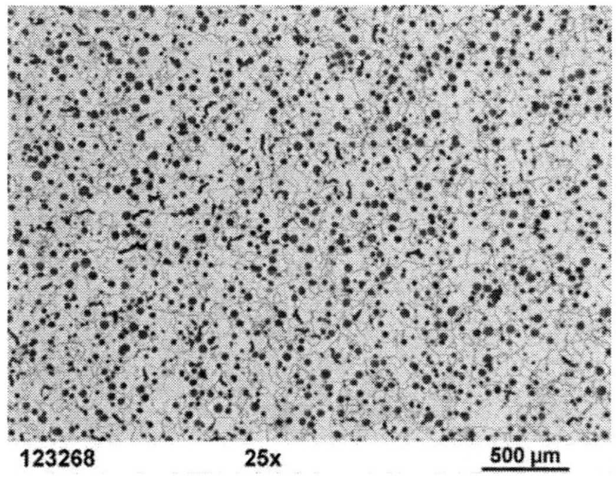

Fig. 1. Optical micrograph of the material GJS etched for 6 s with 3 % Nital.

3 Results and discussion

3.1 Monotonic material behaviour

In this section, the monotonic material properties evaluated from the conducted tests with $\dot{\varepsilon} = 10^{-4}\ s^{-1}$ and $\dot{\varepsilon} = 10^{-3}\ s^{-1}$ are presented in Fig. 2. It is observed that there is no significant strain rate effect on the tensile strength and monotonic yield stress within the studied temperature range. However, Fig. 3 illustrates that the temperature corresponding to the ductility minimum is dependent on the strain rate applied of the tensile test. It is observed how increasing the strain rate of the experiment, increases the temperature at which the ductility drop takes place. This agrees with the afore mentioned investigations from [6, 7].

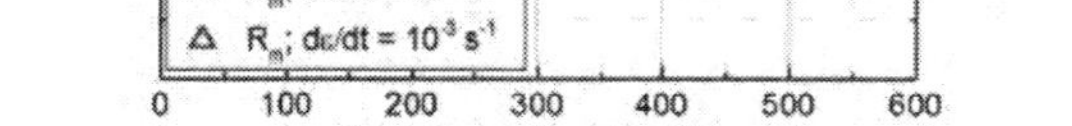

Fig. 2. Tensile strength and monotonic yield stress over temperature.

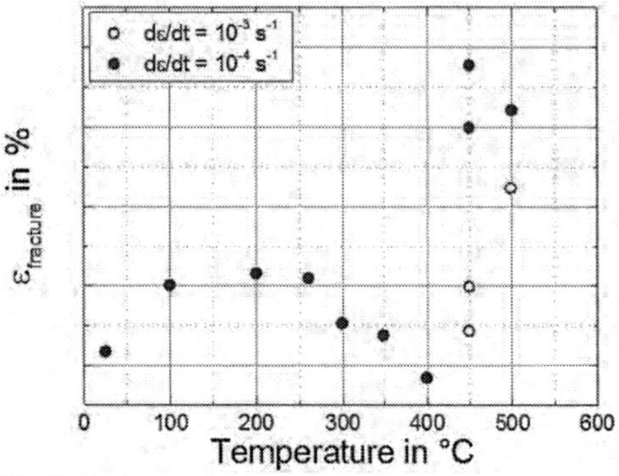

Fig. 3. Fracture strain over temperature.

3.2. Cyclic material behaviour

The cyclic material behaviour is investigated based on the LCF and OPTMF tests.

Based on the midlife cycle of each LCF test, an averaged Young´s modulus, cyclic yield stress and hardening exponent were obtained for each investigated temperature. The Young´s modulus decreases approximately linearly over temperature. The values of the cyclic yield stress over temperature stay relatively constant until 300 °C, dropping approximately linearly from this temperature until 500°C. The Ramberg-Osgood hardening exponent is almost three times higher at RT with approximately constant values between 300 °C and 500 °C.

The Wöhler curves of GJS illustrated in Fig. 4, point to lifetimes ten times shorter for the temperature range 400-450 °C in comparison to the lifetimes of tests conducted with similar strain amplitudes at RT and 500 °C. This temperature range coincides with the temperature range in Fig. 3, where the ductility drop is observed for the tensile tests conducted with $\dot{\varepsilon} = 10^{-3}\ s^{-1}$. For $\varepsilon_{a,mech}$ = 0.5 % an LCF test at 450 °C and two OPTMF tests conducted with a maximum temperature equal to 500 °C. It is observed that the OPTMF tests show significantly longer lifetimes than the corresponding LCF test. By comparing the two OPTMF tests, one performed with an applied dwell time of 300 s and the other without, it can be stated that no time dependent behaviour of the GJS material in the temperature range RT-500°C is observed.

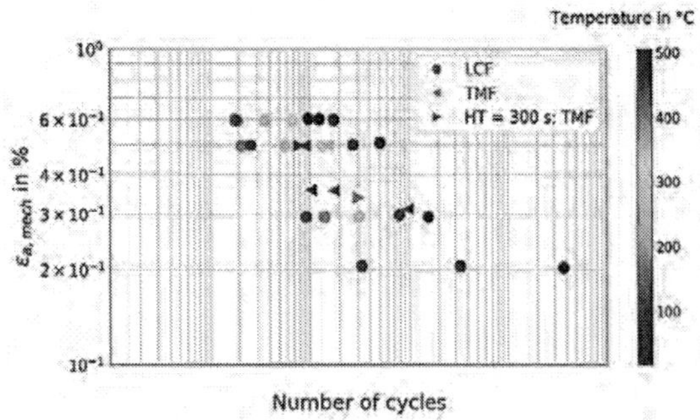

Fig. 4. Fatigue lives of LCF and OPTMF.

3.3. Lifetime model

3.3.1 Uniaxial lifetime model formulation and calibration

In this section a lifetime model including the main damage mechanisms causing fatigue failure is formulated.

For non-isothermal conditions, the time and temperature dependent damage parameter D_{TMF} [1] has been successfully applied for describing the fatigue damage of numerous high temperature resistant alloys. This parameter accounts for the temperature dependent material properties, the cyclic stresses and plastic strains, fatigue crack closure as well as for creep accelerated fatigue crack growth. The lifetime assessment is performed by assuming that the crack growth per cycle is proportional to the cyclic crack tip opening displacement as described in Eq.(1):

$$\frac{da}{dN_{fat}} = \beta \Delta CTOD^{B}. \tag{1}$$

The cyclic crack tip opening displacement, ΔCTOD, is calculated by means of fracture mechanics. For an elastic plastic material and a semicircular surface crack under monotonic loading, the J-Integral is calculated by [8] as:

$$J = \left(1.45\frac{\sigma^2}{E} + 2.4\frac{\sigma\varepsilon^{pl}}{\sqrt{1+3N'}}\right)a. \tag{2}$$

The first term of Eq. (2) accounts for the elastic contribution and the second term for the plastic contribution. σ is the stress, ε^{pl} is the plastic deformation and the parameter N' is the Ramberg-Osgood hardening exponent.

The crack tip opening displacement and the J-Integral are related by [9]:

$$CTOD = d_{N'}\frac{J}{\sigma_y}. \tag{3}$$

Where σ_y denotes the yield stress and $d_{N'}$ depends on the hardening behavior of the material. For power–law hardening materials, the crack-tip fields are the Hutchinson, Rice and Rosengren (HRR) singular fields [10]. For cyclic loading, the parameters in Eqs. (2) and (3) are replaced by their cyclic counterparts and ΔCTOD is computed as:

$$\Delta CTOD = d_{N'}\frac{\Delta J_{eff}}{\sigma_{cy}}. \tag{4}$$

σ_{cy} is the cyclic yield stress of the material and it is evaluated for a stabilized cycle. For a small semicircular surface crack, the cyclic J-integral in Eq. (4) can be estimated from the damage parameter according to Heitmann [11]:

$$Z_D = \left(1.45\frac{\Delta\sigma_{eff}^{\ 2}}{E} + 2.4\frac{\Delta\sigma\Delta\varepsilon^{pl}}{\sqrt{1+3N'}}\right). \tag{5}$$

Thus, Eq. (4) can be rewritten as:

$$\Delta CTOD = d_{N'}\frac{Z_D}{\sigma_{cy}}a. \tag{6}$$

This is equivalent to the expression:

$$\Delta CTOD = \frac{d_{N'}}{\sigma_{cy}}\left(1.45\frac{\Delta\sigma_{eff}^{\ 2}}{E} + 2.4\frac{\Delta\sigma\Delta\varepsilon^{pl}}{\sqrt{1+3N'}}\right)a, \tag{7}$$

where N' is the hardening exponent, obtained by fitting the Ramberg Osgood equation (Eq. 8) to a stabilized cycle.

$$\Delta\varepsilon = \frac{\Delta\sigma}{E} + 0.002\left(\frac{\Delta\sigma}{\sigma_{cy}}\right)^{\frac{1}{N'}} \tag{8}$$

The index eff implies that crack closure effects are included. In Eqs. (5) and (7), $\Delta\sigma$ and $\Delta\varepsilon^{pl}$ denote the stress range and the plastic strain range and can be taken from a stabilized hysteresis loop, e.g. the mid-life hysteresis loop.

Crack closure is accounted by the use of the effective stress range:

$$\Delta\sigma_{eff} = \sigma_{max} - \sigma_{op}. \tag{9}$$

σ_{max} is the maximum stress and can be determined from the stress-strain hysteresis loops as well. σ_{op} is the crack opening stress. The crack opening stress can be determined from empirical formulas or numerical calculations. In order to calculate the effective stress range, Schijve [9] and Schmitt [12] proposed the equation:

$$\Delta\sigma_{eff} = \Delta\sigma U(R_\sigma). \tag{10}$$

The function $U(R_\sigma)$ considers the load ratio dependency. The Newman crack closure model was chosen for modelling the investigated material. The main reason for this selection is the fact that in contrast with other empirical expressions available for modelling crack closure effects Newman proposed a dependency on the maximum stress, which was derived from systematic investigations with the strip yield model [13]:

$$U(R_\sigma, \sigma_{max}) = \frac{1 - \frac{\sigma_{op}}{\sigma_{max}}}{1 - R_\sigma}. \tag{11}$$

In Eq. (11), the term $\frac{\sigma_{op}}{\sigma_{max}}$ is calculated with the help of the polynomial functions given in Eqs. (12) and (13):

$$\frac{\sigma_{op}}{\sigma_{max}} = A_0 + A_1R_\sigma + A_2{R_\sigma}^2 + A_3{R_\sigma}^3 \; for \; R_\sigma \geq 0, \tag{12}$$

$$\frac{\sigma_{op}}{\sigma_{max}} = A_0 + A_1R_\sigma \; for \text{ -}1 \leq R_\sigma < 0, \tag{13}$$

$$A_0 = (0.825 - 0.34\alpha + 0.05\alpha^2)[cos(\pi\sigma_{max}/(2\sigma_y))]^{\frac{1}{\alpha}}, \tag{14}$$

$$A_1 = (0.415 - 0.071\alpha)\sigma_{max}/\sigma_y, \tag{15}$$

$$A_2 = 1 - A_0 - A_1 - A_3, \tag{16}$$

$$A_3 = 2A_0 + A_1 - 1, \tag{17}$$

where $\alpha = 1$ for plain stress and $\alpha = 3$ for plain strain. Newman's formula tends to predict decreasing opening stresses with increasing maximum stresses, which is in accordance with experimental observations.

A generalization of Eq. (6) for elastic-viscoplastic material behaviour under cyclic loading is achieved by the D_{TMF} damage parameter [12] and leads to the equation below:

$$\Delta CTOD = d_{N'}D_{TMF}a. \tag{18}$$

D_{TMF} is defined by the following equation:

$$D_{TMF} = \frac{Z_D}{\sigma_{cy}}F(t,\sigma,T). \tag{19}$$

The function $F(t,\sigma,T)$ describes the time dependent effects, which are a function of the loading history. F is defined as:

$$F(t,\sigma,T) = \left[1 + \alpha' * exp\left(\frac{Q_{cr}}{RT_0}\right)\int_{t_{start}}^{t_{end}} \sigma_{cy}^{n-2} * \delta\sigma * exp\left(-\frac{Q_{cr}}{RT}\right)dt\right]^{\frac{1}{n}}. \tag{20}$$

Where n is the Norton creep exponent, Q_{cr} is the activation energy, R is the universal gas constant, T_0 a reference temperature (typically $T_0 = 227$ °C). The parameter α' is an adjustable parameter.

The symbol "δ" denotes the stress increment in Eq. (20):

$$\delta\sigma = \sigma(t) - \sigma(t_{start}). \tag{21}$$

Here, $t_{start} \leq t \leq t_{end}$ holds. t_{start} and t_{end} denote the start and the end of the analyzed branch of the loading cycle.

For TMF loading, all temperature dependent material properties such as σ_{cy}, E and N' are taken as the average of their minimum and maximum values occurring within a branch (loading or unloading).

In order to receive an expression for the lifetime, it is assumed that the (short) crack growth phase dominates the lifetime. Inserting Eq. (18) into Eq. (1) one obtains:

$$\frac{da}{dN_{fat}} = \beta(d_{N'}D_{TMF}a)^B, \tag{22}$$

where β and B are material dependent parameters.

The crack tip opening displacement is calculated based on the D_{TMF} parameter by integrating the damage over the crack length for an initial crack size a_0 and a technical final crack size a_f. The number of cycles consumed by fatigue damage takes the form:

$$N_{fat} = \frac{{a_f}^{1-B} - {a_0}^{1-B}}{\beta * (1-B)} * (d_{N'}D_{TMF})^{-B}. \tag{23}$$

However, the D_{TMF} parameter does not always lead to satisfying lifetime correlations for DCI materials under low cycle and thermomechanical fatigue loading as depicted in Fig. 5, because it does not account for an embrittlement effect at intermediate temperatures. Hence, no temperature independent picture is obtained in Fig. 5 due to an underestimated damage contribution at temperatures close to the intergranular embrittlement temperature. With this motivation and the work from [6] in mind, the crack growth model based on D_{TMF} is modified to account for the embrittlement effect by integrating a temperature dependency into the parameter β from the crack growth law.

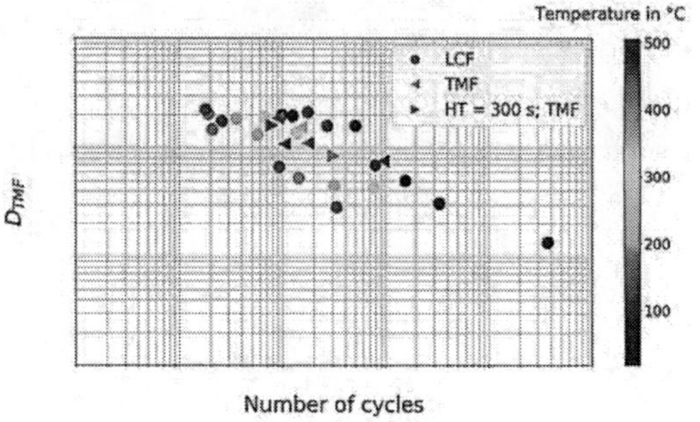

Fig. 5. D_{TMF} versus experimental lifetimes.

Kobayashi et al. [7] studied the ductility drop at the IE (intergranular embrittlement) temperature, which is thought to be related to the lifetime reduction of fatigue tests performed at temperatures near to the IE temperature [14]. He observed that the IE temperature (with the corresponding ductility drop) shifted to higher temperatures for tensile tests conducted with higher strain rates following an Arrhenius relationship, such that:

$$\dot{\varepsilon} = A_{IE} e^{-\frac{Q_{IE}}{RT_m}}. \tag{24}$$

T_m is the IE temperature corresponding to the strain rate applied for conducting the tensile test and R the universal gas constant. For modelling the IE damage of GJS, the activation energy was set to the value proposed in [6], $Q_{IE} = 290000$ KJ/mol, and A_{IE} was indirectly adjusted in order to match the experimentally observed IE temperature in Fig. 3, leading to $A_{IE} = 2 * 10^{19}\ s^{-1}$.

In [6], Wu et al. introduced a damage factor for considering the damage caused by intergranular embrittlement, with the following form:

$$D_{IE} = 1 + \frac{l}{\lambda}\varphi(T), \tag{25}$$

where Φ(T) is the temperature dependent IE damage function and Φ(T) ϵ [0,1]. The IE damage function can be modelled by means of a Gaussian function, whose parameters can be calibrated by fitting the Gauss function to the normalized rupture strain variation over temperature from tensile tests performed at different temperatures. This function relates the reduction of ductility over temperature with the lifetime reduction. The parameter l represents the eutectic cellular size in DCIs and λ the grain boundary Mo-rich eutectic cell thickness (from [6]: l = 100 μm and λ varies between 2 and 7 μm). $\frac{l}{\lambda}$ weights the effect of the IE damage. If φ(T) = 0, there is no IE damage acting, whereas for φ(T) = 1 Eq. (25) reaches its maximum value.

The advantage of describing the IE damage with a Gaussian function is that the maximum IE damage takes place at the IE temperature (T_m), therefore $\varphi(T_m) = 1$. On the other hand the center of symmetry of the Gauss function coincides with the IE temperature, which can be recalculated for different strain rates using Eq. (24) and thus, shift the φ(T) function consequently in order to recalculate the IE damage for the specific strain rate and temperature.

$$\varphi(T) = e^{-\frac{(T-T_m)^2}{w^2}}. \tag{26}$$

The only parameter left to calibrate is w, which gives a measure of the width of the adjusted Gaussian function, a broad bell shape would imply that the IE damage is present over a wider range of temperatures surrounding T_m, a narrow bell would suggest that the temperature interval where IE damage is acting, is smaller. Fig. 6 shows the calibrated IE damage function for GJS. For higher strain rates, T_m shifts to higher values.

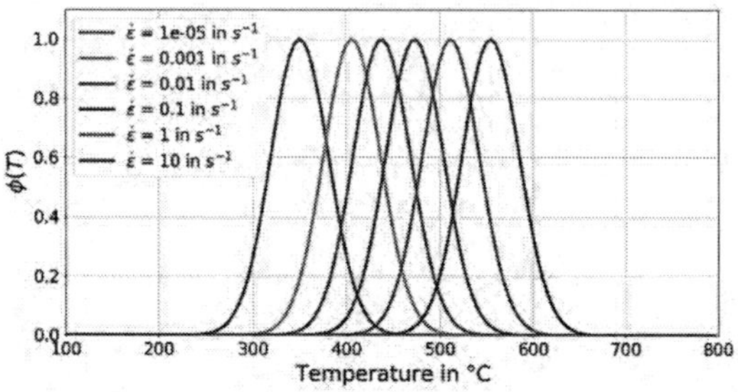

Fig. 6 Calibrated IE damage function for different strain rates.

Introducing the IE damage factor from Eq. (25) in Eq. (22) one obtains the desired expression:

$$\frac{da}{dN_{tot}} = \left(1 + \frac{l}{\lambda}\varphi(T)\right) * \beta(d_{N'}D_{TMF}a)^B. \tag{27}$$

In Eq. (27) the IE damage is only active if φ(T) > 0, and it leads to an acceleration of crack propagation within the temperatures affected by intergranular embrittlement.

By integrating the damage over the crack length for an initial crack size a_0 (i.e. grain size) and a technical crack size a_f, the number of cycles consumed by fatigue damage including intergranular damage results in:

$$N_{tot} = \frac{1}{\left(1+\frac{l}{\lambda}\varphi(T)\right)} * \frac{a_f^{1-B} - a_0^{1-B}}{\beta*(1-B)} * (d_{N'}D_{TMF})^{-B}. \tag{28}$$

Rearranging the terms in Eq. (27) the crack propagation per cycle could be rewritten as:

$$\boldsymbol{\frac{da}{dN_{tot}} = \beta_{IE}(T) * (d_{N'}D_{TMF}a)^B}, \tag{29}$$

where:

$$\boldsymbol{\beta_{IE}(T) = \beta * \left(1 + \frac{l}{\lambda}\varphi(T)\right)}. \tag{30}$$

This leads to a more compact expression for Eq. (28):

$$\boldsymbol{N_{tot} = \frac{a_f^{1-B} - a_0^{1-B}}{\beta_{IE}(T)*(1-B)} * (d_{N'}D_{TMF})^{-B}}. \tag{31}$$

The above generalized formulation of the lifetime model enables lifetime assessment of a material presenting time dependent behaviour and IE damage, however, it can also be applied for less complex material behaviours by deactivating the time dependency in case of an elasto-plastic material behaviour or by deactivating the IE damage, for materials were there is no ductility drop at the IE temperature.

https://doi.org/10.1051/matecconf/201816519006

Table 1 gives an overview of the model configuration for different material behaviours:

Table 1. Overview of the generalized lifetime model

Material behaviour	Generalized lifetime model configuration	Individualised lifetime formulation
Elasto-visco-plastic IE damage	$\beta_{IE}(T) = \beta_{IE}(T)$; $\varphi(T) > 0$ $D_{TMF} = \frac{Z_D}{\sigma_{cy}} F(t,\sigma,T)$	$N_{tot} = \frac{a_f^{1-B} - a_0^{1-B}}{\beta_{IE}(T) * (1-B)} * (d_{N'} D_{TMF})^{-B}$,
Elasto-visco-plastic No IE damage	$\beta_{IE}(T) = \beta$; $\varphi(T) = 0$ $D_{TMF} = \frac{Z_D}{\sigma_{cy}} F(t,\sigma,T)$	$N_{tot} = \frac{a_f^{1-B} - a_0^{1-B}}{\beta * (1-B)} * (d_{N'} D_{TMF})^{-B}$,
Elasto-plastic IE damage	$\beta_{IE}(T) = \beta_{IE}(T)$; $\varphi(T) > 0$ $D_{TMF} = \frac{Z_D}{\sigma_{cy}}$; $F(t,\sigma,T) = 1$	$N_{tot} = \frac{a_f^{1-B} - a_0^{1-B}}{\beta_{IE}(T) * (1-B)} * (d_{N'} \frac{Z_D}{\sigma_{cy}})^{-B}$,
Elasto-plastic No IE damage	$\beta_{IE}(T) = \beta$; $\varphi(T) = 0$ $D_{TMF} = \frac{Z_D}{\sigma_{cy}}$; $F(t,\sigma,T) = 1$	$N_{tot} = \frac{a_f^{1-B} - a_0^{1-B}}{\beta * (1-B)} * (d_{N'} \frac{Z_D}{\sigma_{cy}})^{-B}$,

Comparing Eq.(31) with Eq.(23) one observes that the effect of IE damage can be included in the lifetime model by just introducing a temperature dependency in the prefactor β.

In the selected material neither time dependent behaviour, nor oxidation effects are observed within the studied temperature range (RT-500°C). For this reason, the function $F(t,\sigma,T) = 1$ and the D_{TMF} parameter reduces itself basically to the Heitmann parameter. Therefore, the lifetime model for the investigated GJS can be written as:

$$N_{tot} = \frac{a_f^{1-B} - a_0^{1-B}}{\beta_{IE}(T) * (1-B)} * \left(\frac{Z_D}{\sigma_{cy}}\right)^{-B}. \qquad (32)$$

Where β and B are material parameters to be fitted in absence of IE, i.e. at RT. To this end, a more comfortable way of writing Eq. (32) for the parameters calibration at RT would be the following:

$$N_{tot} = A_{RT} * \left(\frac{Z_D}{\sigma_{cy}}\right)^{-B}. \qquad (33)$$

Rearranging Eq. (33) and linearizing the terms of the expression results in Eq. (34):

$$log\left(\frac{Z_D}{\sigma_{cy}}\right) = log\left(\frac{1}{A_{RT}}\right)^B + B * log(N_{tot}). \qquad (34)$$

This small reformulation of the Eq. (33) enables to obtain the material parameters B and A_{RT} directly from a Least Squares linear fit to fatigue data at RT. The fit of the slope leads to the value of parameter B and from the intercept value of the fit, A_{RT} can be obtained. Then, β can be calculated as:

$$\beta = \frac{a_f^{1-B} - a_0^{1-B}}{A_{RT} * (1-B)}, \qquad (35)$$

with a technical final crack length $a_f = 1000$ µm and an initial defect size $a_0 = 100$ µm.

By substitution of the adjusted material parameters in Eq. (32), it is now possible to calculate a predicted lifetime for each one of the available fatigue tests.

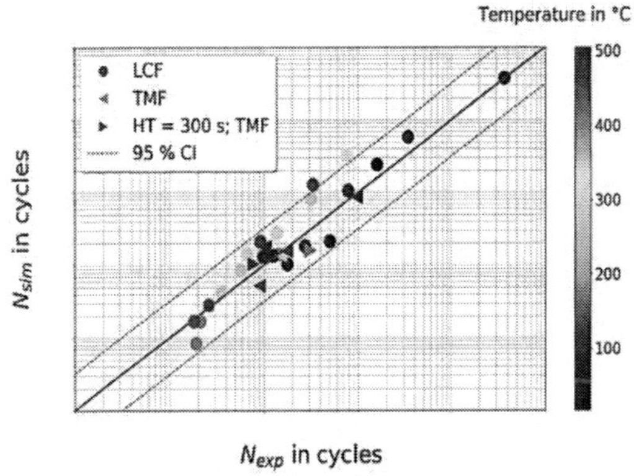

Fig. 7. Predicted lifetime vs experimental lifetime.

The quality of the model is illustrated by Fig. 7, where the predicted lifetime by the calibrated model is plotted against the corresponding experimental lifetime for every performed isothermal LCF test and OPTMF test.

3.3.2 Lifetime model validation and probabilistic interpretation of lifetime scatter

Assuming the proposed model is able to reproduce the damage mechanisms responsible for reducing the lifetime of GJS at intermediate temperatures, one can now interpret the scatter in Fig. 7 in terms of a defect distribution.

In Eq. (32) a fixed initial defect size was selected in order to integrate the lifetime. However the size of the initial defect is not equal for every test specimen. Material defects are usually represented by a probability distribution function (normal, Weibull, logarithmic…). The best way of determining distributions of defects serving as crack starters (i.e. graphite nodules, pores, shrinkage cavities) is by image processing of metallographic samples. Unfortunately this method is still very time and cost extensive, since casting defects are difficult to detect automatically within a fracture surface. In this section an indirect way of inferring the representative defect distribution probability density function is introduced.

The indirect inferred defect size was calculated for every test specimen by drawing a line of slope $-B$ passing through every single experiment when plotting $\frac{Z_D}{\sigma_{cy}}$ versus the experimental lifetime. Thereby, an intercept value is obtained for every test. From the intercept value, it is possible to calculate this fictive initial defect size by simply solving for the value of a_0 in Eq. (32):

$$a_{0,sim,i} = a_f - [\beta_{IE}(T) * A_i * (1-B)]^{\frac{1}{1-B}}, \qquad (36)$$

where the index "i" represents one tested specimen. A_i is derived from its respective intercept value for every test in a similar manner A_{RT} was determined in the previous section. Note, that for this calculation $\beta_{IE}(T)$ and not β must be used otherwise the initial defect size would be calculated without accounting for IE damage and would

https://doi.org/10.1051/matecconf/201816519006

lead to overestimated initial defect sizes for the temperatures were this mechanism is active.

The inferred initial defect sizes are shown in the histogram from Fig. 8. A 3-parameter Weibull distribution function (PDF) was fitted in order to reproduce the distribution of the initial size defects. In Fig. 9 the cumulative distribution function (CDF) based on the fitted 3-parameter distribution function is plotted against the empirical cumulative probability distribution of the initial defect size.

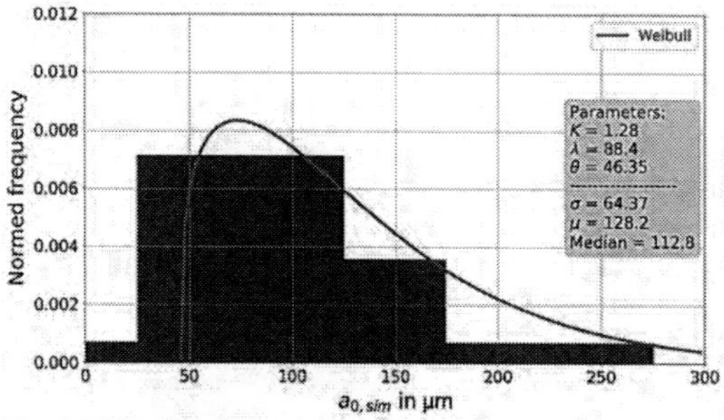

Fig. 8. PDF of the fictive initial defect size.

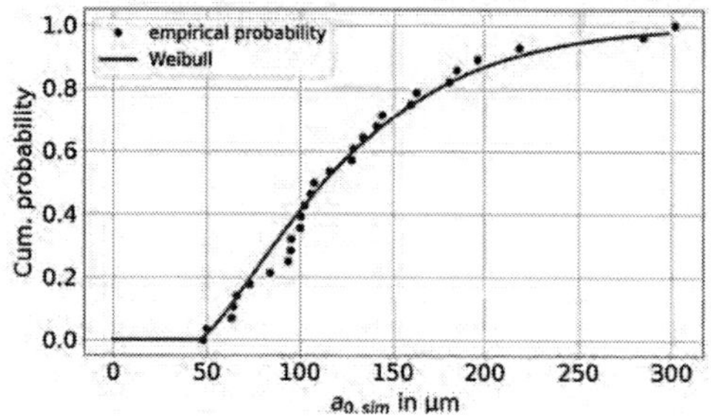

Fig. 9. CDF of the fictive defect size.

Figs. 10-16 illustrate the simulated Wöhler curves for different initial flaw size probabilities. For each selected probability: P = 0.95, P = 0.5 and P = 0.05, its corresponding initial defect size was extracted from the Weibull cumulative distribution function in Fig. 9. For each obtained initial defect the lifetime was newly integrated leading to a new point in the Wöhler diagram. This process was repeated for every test specimen and for each studied temperature a Wöhler diagram was generated.

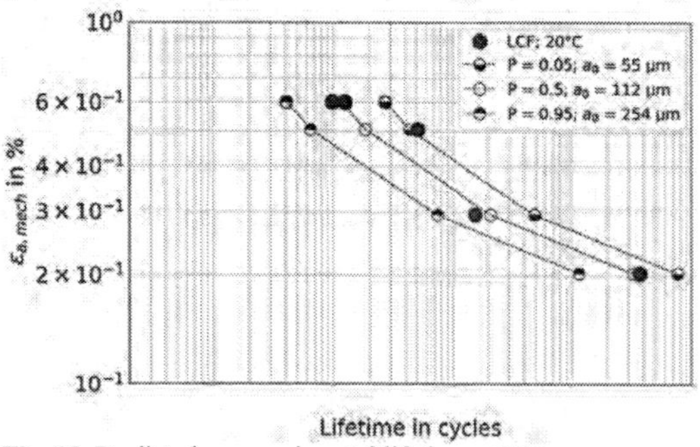

Fig. 10. Predicted vs experimental lifetimes at RT.

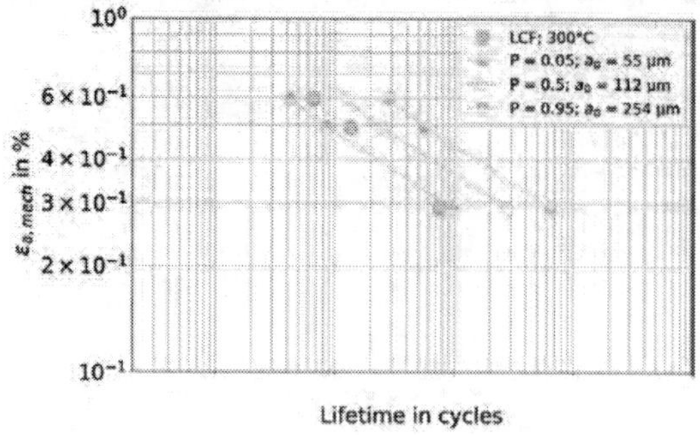

Fig. 11. Predicted vs experimental lifetimes at 300 °C.

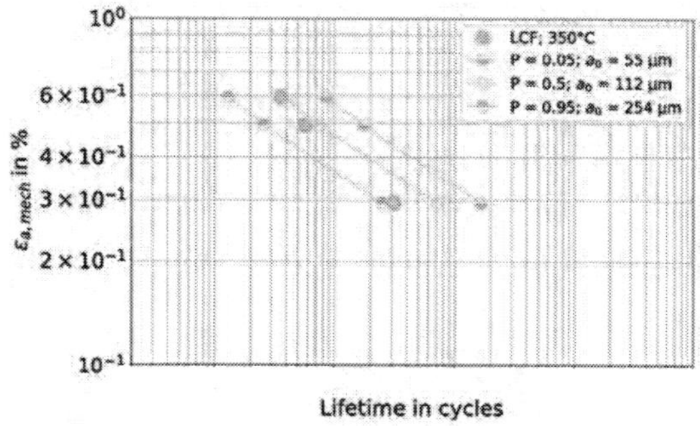

Fig. 12. Predicted vs experimental lifetimes at 350 °C.

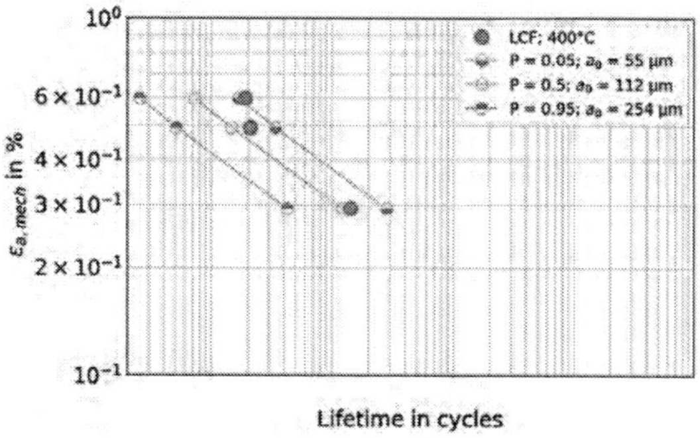

Fig. 13. Predicted vs experimental lifetimes at 400 °C.

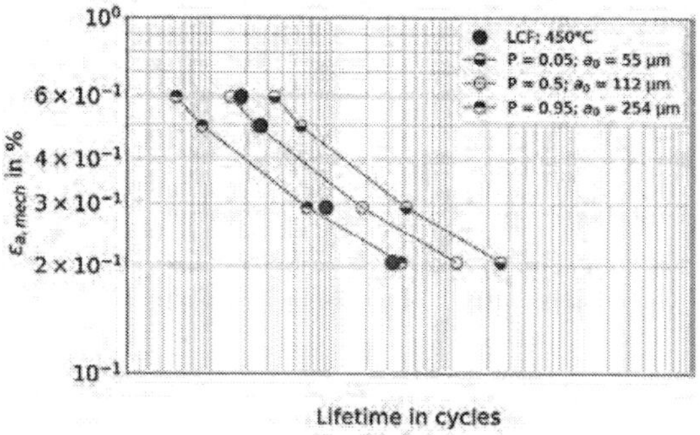

Fig. 14. Predicted vs experimental lifetimes at 450 °C.

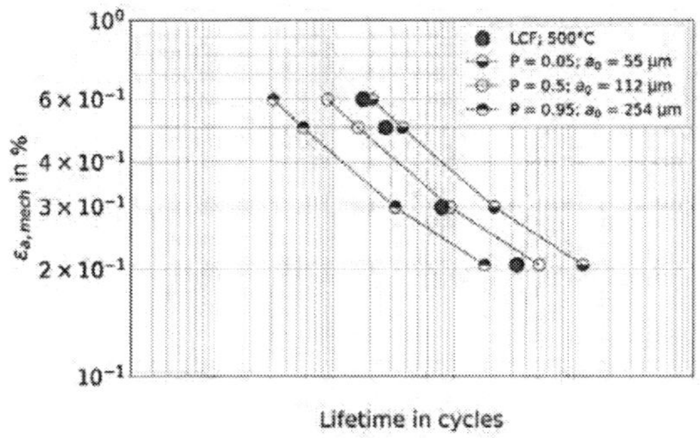

Fig. 15. Predicted vs experimental lifetimes at 500 °C.

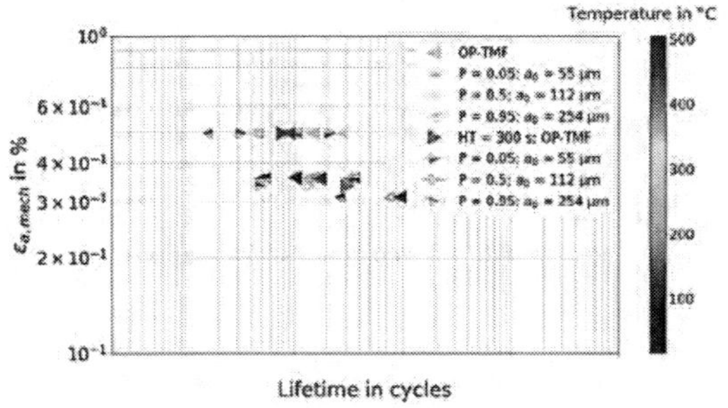

Fig. 16. Predicted vs experimental lifetimes for OPTMFs.

4 Summary

In this work it was assumed, that intergranular embrittlement is responsible for the observed lifetime reduction of LCF tests conducted at 400 °C and 450 °C and the ductility drop of tensile tests nearby 400 °C. A lifetime model based on the D_{TMF} damage parameter including intergranular embrittlement damage was formulated and calibrated. The predicted lifetimes are in very good agreement with the corresponding experimental lifetimes.

Further investigations are required in order to gain understanding about the process leading to IE damage at temperatures nearby 400°C and the strain rate dependency of the prefactor β from the crack growth law. To this end, fatigue crack growth tests should be carried out within the studied temperature range.

Due to the typical variability of defect size in DCIs materials (shrinkage cavities, pores or graphite size) a probabilistic assessment of the initial defect size was included, allowing a deterministic-probabilistic interpretation of the scatter from the experimental lifetimes.

ACKNOWLEDGEMENTS: This project has received funding from the European Union's Horizon 2020 research and innovation programme under grant agreement No 634135.

References

[1] T. Seifert, H. Riedel, "Fatigue Life Prediction of High Temperature Components in Combustion Engines and Exhaust Systems," EASC (2009)

[2] F. T. Shiao, T. S. Lui, L. H. Chen, and S. F. Chen, "Eutectic cell wall morphology and tensile embrittlement in ferritic spheroidal graphite cast iron," *Metall. Mater. Trans. A*, vol. 30, no. July, pp. 1775–1784 (1999).

[3] H.-M. Lin, T.-S. Lui, and L.-H. Chen, "Effect of Silicon Content on Intergranular Embrittlement of Ferritic Spheroidal Graphite Cast Iron Suffered from Cyclic Heating.," *Mater. Trans.*, vol. 44, no. 1, pp. 173–180 (2003)

[4] H.-M. Lin, T.-S. Lui, and L.-H. Chen, "Effect of Silicon Content on Intergranular Embrittlement of Ferritic Spheroidal Graphite Cast Iron Suffered from Cyclic Heating.," *Mater. Trans.*, vol. 44, no. 1, pp. 173–180 (2003)

[5] S. F. Chen, T. S. Lui, and L. H. Chen, "The effect of phosphorus segregation on the intermediate-temperature embrittlement of ferritic, spheroidal graphite cast iron," Metall. Mater. Trans. A, vol. 25, no. 3, pp. 557–561 (1994)

[6] X. Wu, G. Quan, R. MacNeil, Z. Zhang, and C. Sloss, "Failure Mechanisms and Damage Model of Ductile Cast Iron Under Low-Cycle Fatigue Conditions," Metall. Mater. Trans. A, vol. 45, no. 11, pp. 5085–5097 (2014)

[7] T. Kobayashi and K. Nishino, "Embrittlement of Ferritic Spheroidal Graphite Cast Iron by Magnesium," J. Japan Foundry Eng. Soc., vol. 70, p. .273-278 (1998)

[8] P. H. Riedel, "Fracture at high temperatures", Materials Research and Engineering. Springer, Berlin, Heidelberg (1987)

[9] V. Schivje, J, "Some formulas for the crack opening stress level", Eng. Fracture Mechanics., 14, 461-465 (1981)

[10] C.F. Shih, Tables of Hutchinson-Rice-Rosengren singular field quantities, Tech. rep. Brown University Report MRL E-147 (1983)

[11] H. H. Heitmann, H. Vehoff, and P. Neumann, "Life prediction for random load fatigue based on the growth behavior of microcracks," in Advances in fracture research 84 – proceedings of ICF6, pp. 3599–3606 (1984)

[12] W. Schmitt, R. Mohrmann, H. Riedel, A. Dietsche, A. Fischersworring-Bunk, Modelling the fatigue life of automobile components, In: Fatigue 2002 - Proceedings of the Eight International Fatigue Congress held 3-7 June 2002, Stockholm, Sweden (Ed. A.F. Blom), 781-788 (2002)

[13] J. C. Newman, A crack opening stress equation for fatigue crack growth, Int. J. Fract., Vol. 24, 131-135 (1984)

[14] T. S. Lui and C. G. Chao, "High-temperature properties of ferritic spheroidal graphite cast iron," J. Mater. Sci., vol. 24, no. 7, pp. 2503–2507 (1989)

Thermomechanical fatigue crack initiation in disc alloys using a damage approach

Daniel Leidermark[1,*], *Robert* Eriksson[1], *James P.* Rouse[2], *Christopher J.* Hyde[2], and *Svjetlana* Stekovic[3]

[1]Division of Solid Mechanics, Linköping University, 58183 Linköping, Sweden
[2]Gas Turbine and Transmission Research Centre, University of Nottingham, NG7 2RD, United Kingdom
[3]Division of Engineering Materials, Linköping University, 58183 Linköping, Sweden

Abstract. A fatigue crack initiation model based on damage accumulation via a fatigue memory surface in conjunction with a plastic strain energy parameter was evaluated for thermomechanical fatigue loading in a gas turbine disc alloy. The accumulated damage in each hysteresis loop was summed up, and it was assumed that the damage at the stable state is repeated until failure occurs. Crack initiation occurs when enough fatigue damage has been obtained, and the number of cycles can thus be directly determined. The fatigue damage is highly coupled to the constitutive behaviour of the material, where the constitutive behaviour was modelled using a non-linear hardening description. Based on this, a stable state was achieved and the obtained damage could be extracted. A user-defined material subroutine was implemented, incorporating both the constitutive description and the fatigue damage accumulation. The framework was adopted in a finite element context to evaluate the thermomechanical fatigue crack initiation life of the disc alloy RR1000. From the evaluation it could be seen that a good prediction of the thermomechanical fatigue life was achieved compared to performed experiments.

1 Introduction

In the field of energy production and aircraft propulsion, with the use of gas turbines, thermomechanical fatigue (TMF) is one of the most damaging conditions that can possibly arise in a component. Within this area, focusing on high-pressure turbine disc alloys, the fatigue crack initiation life is of high interest, as well as the subsequent crack propagation. It has been observed that fatigue crack initiation is related to early surface damage, where extrusions and intrusions form sharp surface crack-like defects [1]. Thus, indicating that fatigue crack initiation is attributed to the surface and its evolution.

A traditional way of evaluating crack initiation life of a component is to perform a post-process operation of the results obtained in a finite element (FE) analysis. By extracting a set of internal variables at a stable state, number of cycles to crack initiation according to some defined life function that has been calibrated to experiments can be predicted. This life function can be some kind of Coffin-Manson [2, 3] or Morrow [4] type of expression. One commonly used model to predict the fatigue crack initiation life is the Smith-Watson-Topper model [5]. It is an energy-based model that accounts for mean stress influence and assumes that damage is brought on by the maximum tensile stress. Another energy-density model that was used for low-cycle fatigue with dwell time at elevated temperatures is the model introduced by Ostergren [6]. In this model, damage is assumed to be caused by plastic deformation at tensile loading. Hence, the maximum tensile stress and the plastic strain range are used. It was shown by Kulawinski *et al.* [7] that both the Smith-Watson-Topper model and the Ostergren model were able to predict the TMF life with satisfactory results for Waspaloy, with a slightly better prediction using the Ostergren model. The model was further enhanced by combining the plastic stress-strain hysteresis with the initiation model by Zamrik and Renauld [8] and an Arrhenius term to yield even better correlation to the experiments.

An alternative approach, different to the post-processing method, is to incorporate a fatigue damage parameter in the FE-analysis, and accumulate damage until crack initiation occurs. Hence, closely linked to the constitutive model or directly implemented in a user-defined material model. One damage approach, proposed for high-cycle fatigue loading is the one proposed by Ottosen *et al.* [9], which is based on the load history by defining an endurance surface. The endurance surface moves with the loading using a "back-stress" term. A continuum damage approach based on the evolution of the associated endurance function defines the acquired damage, which eventually gives failure. Another fatigue damage model was proposed by Jiang [10]. It uses a plastic strain energy damage parameter coupled to a critical-plane [11] with a material memory surface to

* Corresponding author: daniel.leidermark@liu.se

 https://doi.org/10.1051/matecconf/201816519007

model fatigue crack initiation behaviour under multiaxial loading.

The aim of this paper is to define and evaluate a crack initiation model based on a damage accumulation approach and to predict the TMF crack initiation life from performed experiments of the turbine disc alloy RR1000. An engine relevant out-of-phase TMF loading cycle was considered.

2 Experiments

The studied material in question is the coarse grain powder processed γ' strengthened polycrystalline nickel-base turbine disc alloy RR1000. The chemical composition of the material, with decreasing weight percentage, is Ni-18.5Co-15Cr-5Mo-3.6Ti-3Al-2Ta-0.5Hf-0.06Zr-0.027C-0.015B. The manufacturing process steps and succeeding heat treatments of the material are given in Li *et al.* [12].

Five round, as-machined, specimens were tested at engine relevant conditions. The test specimens were machined from the pancake forging to a diameter of approximately 7.6mm. The applied engine relevant load cycle is a type of out-of-phase strain-controlled TMF cycle containing a hold-time of 30s. The temperature was cycled between 300–675°C and different mechanical strain ranges ($\Delta\varepsilon_{mech}$) were applied. The tests were performed in a hydraulic testing machine using induction heating and the strain ranges were controlled by a high-temperature extensometer with a gauge length of 12.75 mm. The number of cycles to TMF crack initiation (N_i) was determined by a 10% load-drop of the maximum tensile stress obtained in each cycle. The received crack initiation lives have been normalised with respect to the longest life obtained in the experiments due to confidentiality issues, see Table 1 for the experimental data.

3 Modelling

3.1 Constitutive model

For the purpose of describing the constitutive behaviour of the disc alloy in question, a rate dependent Perzyna viscoplastic material model [13] was implemented as a user-defined material subroutine in an FE-context. The constitutive model is able to describe the cyclic evolution of the material by two non-linear Armstrong-Frederick back-stress terms [14] and a linear as well as a saturated isotropic hardening term. In what follows, all tensors are presented in index notation, where second-order tensors are quantified by upper-case Roman or Greek-letters and scalar valued parameters are defined by lower-case Roman and Greek-letters. The following yield function was employed

Table 1. Experimental TMF data of the round specimens.

Specimen	Diameter [mm]	$\Delta\varepsilon_{mech}$ [%]	Normalised N_i
S1	7.61	0.8512	0.695
S2	7.60	0.8976	0.526
S3	7.62	0.8521	1.000
S4	7.60	0.8984	0.593
S5	7.61	0.8803	0.604

$$f = \sigma_{eq}^{vM}(\hat{\sigma}_{ij} - B_{ij}) - r - \sigma_Y \tag{1}$$

where $\hat{\sigma}_{ij}$ is the deviatoric stress tensor defined as $\hat{\sigma}_{ij} = \sigma_{ij} - \frac{1}{3}\sigma_{kk}\delta_{ij}$, δ_{ij} is the Kronecker's delta and B_{ij} is the total back-stress tensor. Isotropic hardening is described by r, σ_Y is the initial yield limit and σ_{eq}^{vM} is the von Mises equivalent stress, defined as

$$\sigma_{eq}^{vM} = \sqrt{\frac{3}{2}(\hat{\sigma}_{ij} - B_{ij})(\hat{\sigma}_{ij} - B_{ij})} \tag{2}$$

The total back-stress term is based on additively decomposition [15] of the two individual back-stress terms, $B_{ij} = \sum_k B_{ij}^k$, where the evolution law for each term is defined by

$$\dot{B}_{ij}^k = c_k\left(\frac{2}{3}a_k\dot{\varepsilon}_{ij}^{vp} - B_{ij}^k\dot{\lambda}\right), k = 1, 2 \tag{3}$$

with the individual material parameters c_k and a_k. To correlate the different stages of the cyclic evolution of the stress state, two isotropic hardening terms have been adopted; one linear and one saturated. These are summed up to the total isotropic hardening term ($r = r_1 + r_2$), in which the respective evolution law is given by

$$\dot{r}_1 = q_1\dot{\lambda} \tag{4}$$

$$\dot{r}_2 = h(q_2 - r_2)\dot{\lambda} \tag{5}$$

where q_1 describes the long term cyclic softening and the rapid initial softening is accounted for by h and q_2, which describes the rate of cyclic softening and the saturation.

The evolution law of the viscoplastic strain tensor is defined by the Perzyna flow rule based on the adopted yield function as

$$\dot{\varepsilon}_{ij}^{vp} = \begin{cases} 0 & , f \leq 0 \\ \dot{\lambda}\dfrac{\partial f}{\partial \sigma_{ij}} & , f > 0 \end{cases} \tag{6}$$

in which the viscoplastic multiplier is defined by the following power-law relation

$$\dot{\lambda} = \left(\frac{f}{\eta}\right)^m \tag{7}$$

https://doi.org/10.1051/matecconf/201816519007

where m and η are material parameters accounting for the viscous effects.

Material parameters for the above presented constitutive model and material were acquired from an optimization procedure based on stress partitioning and cycle jumping. An in-depth description of this procedure can be found in Rouse *et al.* [16], as well as generated material parameters. The parameters are temperature dependent and are defined continuous over the evaluated temperature range in the engine relevant TMF cycle.

3.2 Fatigue damage model

As crack initiation is mainly a surface dominant phenomena, the surface damage has to be evaluated. Hence, a fatigue damage model based on the memory surface concept [10] was used to evaluate the crack initiation life. The model consist of a fatigue damage term that incorporates the material memory and plastic strain energy in the point of interest, hence eliminating the need of a cycle-counting method.

In analogy with the yield surface in plasticity, a fatigue memory surface can be defined as

$$g = \sqrt{\hat{\sigma}_{ij}\hat{\sigma}_{ij}} - \sqrt{\frac{2}{3}}\sigma_{mem} \leq 0 \tag{8}$$

where σ_{mem} is the memory stress, which initially is set to the fatigue endurance limit of the material, σ_{end}. The evolution law of the memory stress is defined according to

$$\dot{\sigma}_{mem} = \sqrt{\frac{3}{2}}H(g)\langle\frac{\hat{\sigma}_{ij}\dot{\hat{\sigma}}_{ij}}{\sqrt{\hat{\sigma}_{ij}\hat{\sigma}_{ij}}}\rangle \\ -\beta(1-H(g))(\sigma_{mem}-\sigma_{end})\dot{\lambda} \tag{9}$$

The Heaviside function, $H(g)$, makes sure that the memory surface contracts when the stress state is lower than the memory stress ($g < 0$) and expands when it is on the memory surface ($g = 0$). The parameter β defines how much the surface contracts, and $\langle\cdot\rangle$ is the Macaulay bracket.

The damage accumulation is then directly coupled to the memory surface, using the memory stress to account for load-sequence effects. For proportional loading conditions the evolution of the damage is thus described by

$$\dot{\omega} = \langle\sigma_{mem} - \sigma_{end}\rangle^{\alpha}\,\dot{\psi}^{p} \tag{10}$$

where α is a material parameter and ψ^p is the plastic strain energy.

A simple multiaxial approach has been adopted for the plastic strain energy evolution law with inspiration of the work by Ostergren [6]

$$\dot{\psi}^{p} = H(\sigma_{kk})\sigma_{eq}^{vM}\dot{\lambda} \tag{11}$$

where $H(\sigma_{kk})$ accounts for micro-crack closure by making sure that damage is only accumulated during the tensile part from the loading cycle.

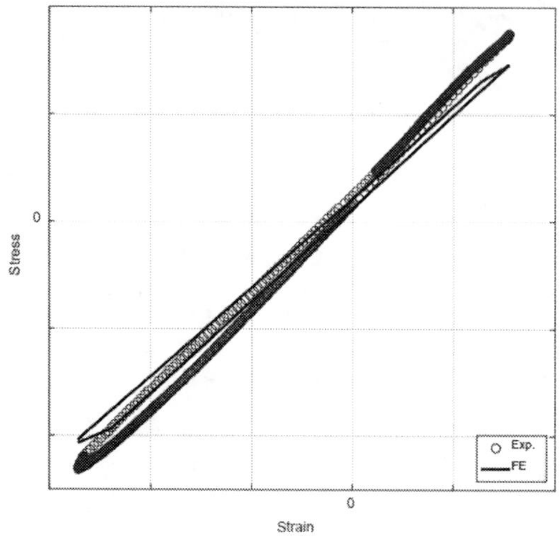

Figure 1. The stable experimental and simulated hysteresis loop of *S1*.

It have to be pointed out that the fatigue damage model for crack initiation is not coupled to the stress update in the constitutive model, as traditionally done for continuum damage models. The fatigue damage model is a stand-alone model with inputs directly from the constitutive model, as it is implemented as a subroutine within the above described user-defined material subroutine. Hence, the stress and strain state in each increment is directly feed into the fatigue damage subroutine. Thus, the history of the damage is accounted for and evolves in the subsequent iterations until the critical damage is reached, initiating a macroscopic crack. Consequently, the damage accumulation during each load cycle is summed up to the critical value ω_c, representing crack initiation. This is quantified as

$$\sum_{i=1}^{N_i} \Delta\omega_i = \omega_c \tag{12}$$

where $\Delta\omega_i$ is the damage accumulated during a load cycle.

4 Evaluation and calibration of TMF crack initiation

A quarter of the specimen was modelled and analysed, using only the straight part over the extensometer gauge length. This was due to that the material in question is isotropic and aspects in applying the boundary conditions to achieve the correct $\Delta\varepsilon_{mech}$ in the TMF cycle. The FE-model consists of 429 fully integrated 8-noded brick elements. Symmetry boundary conditions as well as locking in the axial direction at the bottom and loading at the top, according to the engine relevant TMF cycle, were applied. FE-simulations were performed using the FE-software LS-DYNA [17], version R7, with the user-defined material-fatigue model. Figure 1 shows an example of an obtained stable hysteresis loop from the

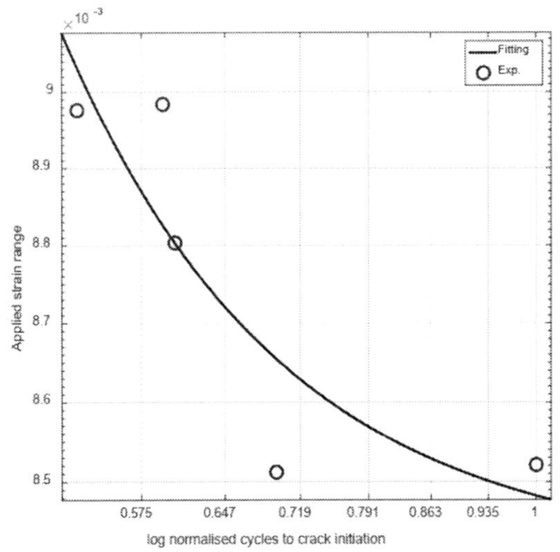

Figure 2. Applied strain range versus number of cycles to crack initiation.

FE-simulation compared to the experimental one, displaying the response of the *S1* specimen.

The fatigue endurance limit is present in the fatigue damage model, and this value can be taken from the Wöhler curve at infinite life (typically 10^6 cycles to failure) for the material in question. However, due to limiting resources of data for the present material, only rough estimates of the endurance limit at certain temperatures were available. Hence, instead of using these uncertain values, an evaluation of the endurance limit for the employed TMF cycle was determined based on the procedure suggested by Jiang [10]. Ranges were used rather than amplitudes due to the applied unsymmetrical TMF load cycle. The applied strain ranges were plotted against the number of cycles to crack initiation, see Figure 2, and a curve fit was performed using the following equation

$$(\Delta\varepsilon_{mech} - \Delta\varepsilon_{end})^d N_i = e \tag{13}$$

where $\Delta\varepsilon_{mech}$ and N_i are taken from Table 1 and $\Delta\varepsilon_{end}$, d and e are fitting parameters. One can interpret $\Delta\varepsilon_{end}$ as the endurance limit in strain range. Then, the generated $\Delta\varepsilon_{end}$ value was compared to the cyclic stress-strain curve to obtain the corresponding endurance limit of the material for this specific TMF cycle. This was done by evaluating the experimental Ramberg-Osgood relationship [18], hence the following expression was used

$$\frac{\Delta\varepsilon}{2} = \frac{\Delta\sigma}{2E} + \left(\frac{\Delta\sigma}{2K}\right)^{1/n} \tag{14}$$

where K and n are material parameters based on the experiments. These two parameters were evaluated by fitting Equation (14) to the stress and strain ranges of the

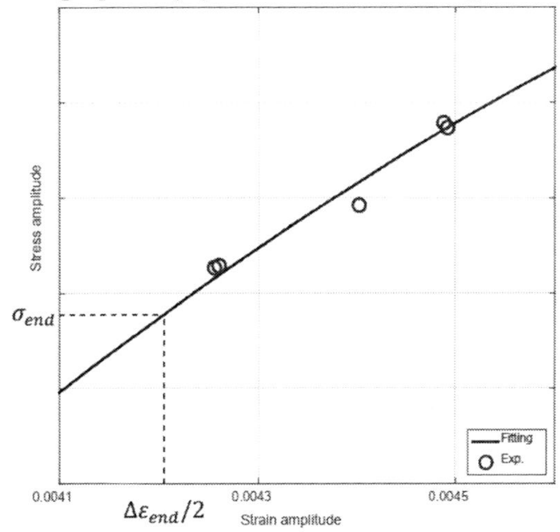

Figure 3. The Ramberg-Osgood curve and evaluation of the endurance limit.

performed experiments, and from this relationship the endurance limit σ_{end} could be determined, see Figure 3. The used Young's modulus E in the expression was taken as the experimental mean modulus, with regard to the temperature, during the on-loading in the TMF cycle.

Based on that only constant amplitude loading is present in the applied TMF load cycle, the parameter β was set equal to zero based on the work of Jiang [10]. This parameter is used for contracting the memory surface when going from a large amplitude to a lower in the succeeding cycle, hence changing the memory stress, and as no such loading is present in the load cycle no contribution will be obtained.

The critical damage was determined by post-processing the performed FE-simulations, in which the accumulated fatigue damage during each cycle was extracted in an element far away from the applied boundary conditions at the surface of the specimen. The extraction is only necessary up to the cycle in which a stable state is obtained, hence a simplification of the expression in Equation (12) was performed according to

$$\omega_c = \Delta\omega_1 + \cdots + \Delta\omega_x + \Delta\omega_{stable}(N_i - x) \tag{15}$$

where a stable state is reached after x number of cycles. In this manner, not all cycles up to the "real" crack initiation cycle need to be simulated in the FE-context, hence saving computational time. A difference of less than 5% between the accumulated fatigue damage from the previous cycle was quantified as a stable state. From this, the determination of the critical damage was done under the assumption that the value of α is known, which was not the case initially. Thus, a calibration involving α and the critical fatigue damage was conducted, by use of a best praxis fitting procedure based on the experimental results and FE-simulations. As the critical fatigue damage is to

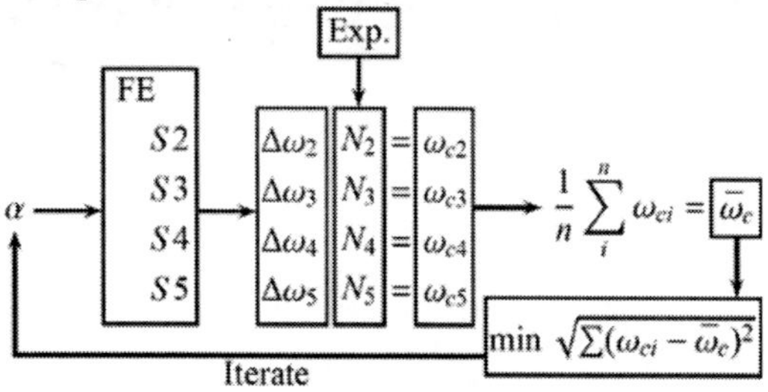

Figure 4. The best praxis fitting of the fatigue damage parameters.

be equal for any applied strain range in the material, a minimisation of the error with respect to the mean value of the critical fatigue damage for a given α value from each of the specimens was used, utilising Equation (15). Hence, both α and ω_c were obtained by coupling the experimental crack initiations lives with the fatigue damage model. A schematic display of the procedure is depict in Figure 4, where responses from *S2*, *S3*, *S4* and *S5* were used in the calibration and *S1* was used to verify the crack initiation model. The iterative procedure was stopped when an optimal solution of the comparable difference in the critical fatigue damage was reached, and a final mean value of these was used as the critical fatigue damage.

Based on the above described fitting procedure, α and ω_c could be determined. An FE-analysis of *S1* with the obtained values was performed to verify the adopted fatigue damage model. This can be seen in Figure 5, where the normalised predicted TMF crack initiation lives in the FE-context are compared to the ones obtained in the experiments. To access the quality of the model, a scatter band based on the standard deviation in the predicted lives from the specimens used in the calibration with respect to the experiments was added in the figure. One can see that approximately all calibrated specimens lie within the standard deviation and that the verifying specimen *S1* are captured within a factor of five standard deviations, corresponding to a factor of 1.3707 to the experimental TMF crack initiation life. The normalised predicted TMF crack initiation lives of the specimens and the relative factor between the predicted and experimental lives are summarised in Table 2.

5 Discussion

From the analyses it could be observed that the adopted fatigue damage model correlate well with the observed experimental behaviour, see Figure 5. As can be seen in Table 1, there is a noticeable different for the experimental lives between *S1* and *S3*. This gives a small misguidance in the calibration procedure of the fatigue parameters, as only *S3* is included. Hence, focus is given to the longer life at the comparable strain range level, which in hand yields a larger relative factor in the predicted TMF crack initiation life of *S1*, *cf.* Table 2. Thus, more performed tests would level out these kind of deviating results. Furthermore, from the results displayed

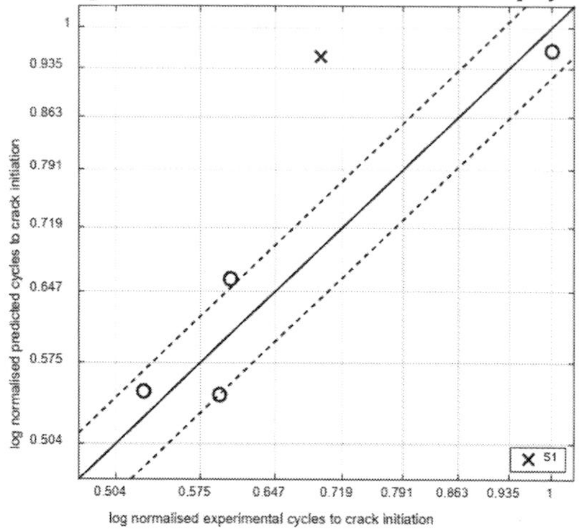

Figure 5. The normalised predicted TMF crack initiation lives versus the experimental ones.

in Figure 5 it can be seen that all the predicted crack initiation lives from the FE-analyses lie within a factor of 1.3707 from the experiments, corresponding to approximately five standard deviations from the calibrated set of specimens, where the verifying analysis of *S1* is the one setting this factor. This gives a sound interpretation that the proposed fatigue damage model enables to capture the correct behaviour of the investigated material under the applied engine relevant load conditions.

It is to be noted that a difference in Young's modulus is present in Figure 1 between the experimental curve and the one obtained from the FE-simulation. A reason is due to different material batches, as the identified material parameters in Rouse *et al.* [16] comes from a specific batch which is different from the one used in the evaluated TMF experiments. It is common that scatter in material properties are present between different batches of the same material. This is due to that different sizes of pancake forgings have been used, which were affected by the heat treatment, and as a consequence difference in grain size, grain size distribution, γ' size and distribution were obtained. However, the plastic strain range of the two hysteresis loops are approximately the same, which is important from a fatigue perspective as the fatigue damage model is based on the plastic strain energy. Furthermore, the difference in Young's modulus generates a little lower maximum and minimum stresses compared to the experiment.

Table 2. The predicted TMF crack initiation lives and the relative factor with respect to the experimental lives.

Specimen	Normalised predicted N_i	Factor
S1	0.9528	1.3707
S2	0.5490	1.0440
S3	0.9622	0.9622
S4	0.5459	0.9200
S5	0.6610	1.0952

The proposed method to evaluate the TMF crack initiation life is based on that the fatigue damage is extracted until a stable state is reach (5% difference) and then using the last value to predict the life with the critical damage, *cf.* Equation (15). This is a very straightforward, simple and computationally easy approach. On the other hand, this approach might not yield an entirely correct behaviour of the fatigue damage process, as the obtained damage will likely be further reduced with successive cycles and eventually reach a steady-state condition with no difference in damage. To capture this, a cycle-by-cycle FE-simulation has to be performed up to the cycle in which the critical damage is obtained by using Equation (12). This is a rather computational heavy procedure as many cycles need to be performed. Hence, to accelerate the simulation effort a cycle jumping procedure can be adopted, see *e.g.* Kontermann *et al.* [19], where the internal variables (including the memory stress) are extrapolated a number cycles forward. Though, one need to keep in mind that a cycle jumping procedure is in it self a complex element, and adoption of which might end up in a more complicated approach than the herein proposed simple method.

The performed experiments were conducted on smooth specimens, where a uniaxial state is present. This simplifies matters, as a uniaxial response is obtained in the FE-analyses and a uniaxial lifing criterion could have been adopted. As real component geometry contain stress-raising features, such as notches and holes, a multiaxial lifing criterion have to be used, which is the case in this work by utilising the von Mises equivalent stress in Equation (11). Hence, from a multiaxial perspective the implemented fatigue damage model can directly be utilised. The choice of the plastic strain energy parameter will likely change the level of fatigue damage, where *e.g.* a critical-plane assumption [11] can be adopted instead of the above simple multiaxial approach. Further development of the fatigue damage model can be to include a multiaxial critical-plane assumption, where the stress and strain are evaluated on a plane at *e.g.* a notched feature. This plane can also be used for the subsequent crack growth.

The herein defined and adopted endurance limit is not temperature dependent, as it is based on the applied engine relevant TMF load cycle. This means that the defined endurance limit is viewed as an approximate endurance limit with respect to the applied temperature field in this specific load cycle. One can of course reason that a temperature dependent endurance limit based on Wöhler curves is to be preferred, but due to lack of data this could not be done. By using the defined endurance limit, obtained from the Ramberg-Osgood relation, a slight difference compared to a temperature dependent limit will be generated within the fatigue damage accumulation, as the temperature difference of 375°C is present during the loading cycle. Another way, could have been to apply a reduction factor based on the temperature. Additionally, one could have used the temperature dependent ultimate tensile strength and scale it with a suitable factor to define the endurance limit.
Uncertainties are always present in modelling work, as each simplification adds up and are displayed in the final response. Based on the performed work, only five test specimens were used in evaluation of the fatigue damage model. This is not enough tests to generate a statistical good response, as one or two specimens might generate a TMF crack initiation life which are not entirely correct due to some uncontrollable factors occurring during the experiments, *cf. S1* and *S3*. Another uncertainty lies in the used material parameters, both for the constitutive model as well as the fatigue model. These are crucial for the final response, as a small variation will generate large repercussions in the results. As mentioned above, regarding the uncertainties in using different material batches, a difference in the simulated hysteresis loop will of course contribute to a discrepancy compared to the experimental one, and as a consequence a discrepancy will be obtained in number of cycles to crack initiation. Based on this, a large statistical foundation is needed to correctly evaluate the fatigue lives.

Disregarding the material parameters and different responses due to material batches, the constitutive model is what drives the fatigue damage model. The better conformance of the constitutive behaviour a more accurate TMF crack initiation life prediction will be generated. This, as the fatigue damage model is based on the responses obtained locally in the FE-model from the constitutive model.

6 Conclusions

The main conclusions that can be drawn from the presented work are:

- A satisfactory response of the constitutive behaviour compared to the experiments was obtained. The small difference is due to that a different material batch was used to calibrate the constitutive model.
- The adopted fatigue damage model predicts the TMF crack initiation lives within a factor of 1.3707 compared to experiments.
- The fatigue damage model facilitates a simple approach for accounting for load-sequence effects and eliminates cycle-counting methods.
- Further experiments need to be performed to reduce statistical uncertainties that can be drawn from the present work, as only four specimens were used in the calibration of the fatigue damage model. Hence,

minimizing the discrepancy in TMF crack initiation life for the verifying specimen.

The study has received funding from the Clean Sky 2 Joint Undertaking under the European Union's Horizon 2020 research and innovation programme under grant agreement No 686600.

References

[1] Polák J., Man J., Procedia Engineering **101**, 386-394 (2015)
[2] Tavernelli J.F., Coffin L.F., Journal of Basic Engineering **84**, 533-537 (1962)
[3] Manson S.S., Journal of Basic Engineering **84**, 537-541 (1962)
[4] Socie D.F., Morrow J.D., Risk and Failure Analysis for Improved Performance and Reliability, 141-194 (1980)
[5] Smith R.N., Watson P., Topper T.H., Journal of Materials **5**, 767-778 (1970)
[6] Ostergren W.J., Journal of Testing and Evaluation **4**, 327-339 (1976)
[7] Kulawinski D., Weidner A., Henkel S., Biermann H., International Journal of Fatigue **81**, 21-36 (2015)
[8] Zamrik S.Y., Renauld M.L., Thermo-mechanical Fatigue Behavior of Materials STP 1371 ASTM, 119-137 (2000)
[9] Ottosen N.S., Stenström R., Ristinmaa M., International Journal of Fatigue **30**, 996-1006 (2008)
[10] Jiang Y., Fatigue & Fracture of Engineering Materials & Structures **23**, 19-32 (2000)
[11] Karolczuk A., Macha E., International Journal of Fracture **134**, 267-304 (2005)
[12] Li H.Y., Sun J.F., Hardy M.C., Evans H.E., Williams S.J, Doel T.J.A., Bowen P., Acta Materialia **90**, 355–369 (2015)
[13] Perzyna P., Advances in Applied Mechanics **11**, 313-354 (1971)
[14] Frederick C.O., Armstrong P.J., Materials at High Temperatures **24**, 1-26 (2007)
[15] Chacboche J.L., International Journal of Plasticity **5**, 247–302 (1989)
[16] Rouse J.P., Hyde C.J., Leidermark D., Stekovic S., Jones J., Whittaker M., Lancaster R., Pattison S., Grant B., Fatigue 2018, MATEC Web of Conferences (2018)
[17] Hallquist J.O., *LS-DYNA Theory Manual* (Livermore Software Technology Corporation, Livermore, 2006) 680
[18] Ramberg W., Osgood W.R., Technical Notes National Advisory Committee for Aeronautics **902**, 1-29 (1943)
[19] Kontermann C., Scholz A., Oechsner M., Materials at High Temperatures **31**, 334–342 (2014)

Characterization and modelling of both high-cycle high-temperature fatigue behaviour of Stainless Steel Grades

Pierre-Olivier Santacreu[1,*] *Cloé* Prudhomme[1], *Benoit* Proult[1] and *Isabelle* Evenepoel[1]

[1]APERAM, Centre de Recherche, 62330 Isbergues, France

Abstract. In the same context of thermo-mechanical fatigue and high temperature applications of stainless steel, high-frequency vibration fatigue at high temperatures should be considered, in particular for automotive exhaust gas applications. In fact one of the most frequent incidents that can happen on exhaust components is an accumulation of low-cycle thermal fatigue and high-cycle fatigue. The prediction of the lifetime of a structure under such complex thermal and mechanical loading is therefore a constant challenge at high temperature due to the coupling of metallurgical, oxidation or creep effects. In order to better understand in a first approach, the high cycle fatigue of stainless steels at high temperatures, traction-compression tests were performed on flat specimens at 25Hz, under air and in isothermal conditions from ambient temperature to 850°C. Two different stress ratios, R=-1 and 0.1, are characterized with the objective to assess a multiaxial model for high temperature. Different criteria are used to predict the ruin of a structure under high-cycle fatigue but in general for ambient-around temperatures. In particular, multiaxial and stress-based DangVan criterion is today widely used to evaluate the risk of fatigue cracks initiation and it has been implemented recently in our fatigue life processor Xhaust_Life®. Therefore the Dang Van criterion was identified from the isothermal high cycle fatigue using the 2 stress ratio and its validity is discussed especially for temperatures higher than 500°C where strain rate and creep effects have increasing influence. Results are presented for two ferritic stainless steel grades used in high temperature exhaust applications: K41X (AISI 441, EN 1.4509) and K44X (AISI 444Nb, EN 1.4521).

1 Introduction

Endurance fatigue is one of the most current damage modes of component subjected to cyclic loadings, especially in transport (automotive, railway rolling-stock, aviation). For most of these applications, life prediction is a mandatory requirement and implies the determination of endurance limit and S-N curves. Because a structure is subjected to multiaxial loading, fatigue tests need to be carried out at minimum two stress ratios, generally at R=-1 and R=0.1, in order to take into account of this effect. In general room temperature endurance data are available to most standardized grades but less for high temperature (HT) conditions and different stress ratio conditions. High temperature fatigue is also often studied at low frequencies by Low-cycle fatigue using strain controlled test. On the other hand in many high temperature applications, the study of the thermomechanical or thermal fatigue makes more sense when a thermal cycle is applied on a constrained structure or when it is added to the cyclic loading. Finally in many real situations, accumulation of different types of high temperature damage processes has to be considered. Example of the exhaust manifold parts illustrates that situation where a coupling between thermo-mechanical fatigue, vibration fatigue and creep is often observed.

Several references are already available on the TMF of exhaust manifold [1-3]. In the present paper we are looking for both high temperature and high cycle fatigue in range from the ambient temperature to 850°C on two stabilized ferritic stainless steel grades for HT applications K41X (AISI 441, EN 1.4509) and K44X (AISI 444Nb, EN 1.4521) already presented in [4,5].
Multiaxial modelling of HCF is suitable for virtual design approach. In automotive, Dang Van criterion is very popular and has demonstrated its scientific relevance especially in automotive design [6]. However it is not so applied for HT, even if a tentative extension for non isothermal was proposed in [7]. On our side we have evaluated opportunities to apply multiaxial criterion including the Dang Van one.

2 Materials and Experimental Procedure

2.1. Ferritic stainless steel for high temperature applications

Ferritic stainless steels are chromium steel (AISI 400 series) containing a minimum of 11%Cr to ensure a

* Corresponding author: pierre-olivier.santacreu@aperam.com

proper corrosion resistance. Among ferritic stainless steel, grades used for high temperature applications are for most of them Ti or Nb stabilized grades and exhibit at least 15%Cr. Such chemical composition guaranties a stable microstructure for a large range of temperature and an improved oxidation resistance. Ferritic grades have also low coefficient of thermal expansion around 12.10^{-6}/°C compared to 19.10^{-6}/°C for refractory austenitic stainless steel. For all these reasons ferritic stainless steel grades are good candidates for high temperature applications like automotive exhaust systems, burner or solid oxide fuel cell interconnector.
The present study is focusing especially the AISI441 1.4509 (Aperam K41X) and an improved AISI 444Nb 1.4521 with 19%Cr and Nb-stabilized (Aperam K44X). Their compositions are given in Table 1. All tests presented hereafter have been carried out on flat samples in a standard 2B or 2D metallurgical state (cold rolled and annealed), with thickness between 1.5 and 2mm.

Table 1. Chemical Compositions in w% of the two ferritic grades studied in the present paper.

Grade	C	Cr	Si	Mn	Mo	N	Other
K41X-1.4509 **441**	0.02	17.8	0.6	0.3	-	0.015	Ti+Nb=0.6
K44X -1.4521HT **444Nb**	0.02	19	0.6	0.3	1.9	0.015	Nb=0.6

2.2 Tensile tests

High temperature tensile tests were carried out according to EN ISO 6892-2 standard on a 100KN INSTRON tensile machine with a 1000°C-resistance furnace (Fig.1). Specimens are cut in rolling direction. In general two or three samples are tested per condition. After a 2h heating procedure, tensile tests were performed at a constant displacement rate of 0.54mm min^{-1}.

2.3 High-Cycle fatigue tests

Endurance fatigue tests are carried out at high frequency (25Hz) under stress control. About 20 samples need to be tested to draw a fatigue S-N curve (or Wohler curve) and to determine the endurance limit at 2,000,000 cycles. In order to access a fatigue analysis of structural under multi-axial loadings, two stress ratios are selected: tensile-compression at R=-1 and alternating tension at R=0.1 where R is defined by:

$$R = \frac{\Sigma_{\min}}{\Sigma_{max}} \quad (1)$$

Specimen preparation is a major step and must be carefully done. Specimen geometry is taken in rolling direction and machined by milling with polishing of the edges. Because of the large quantity of results required for each condition, fatigue tests are subcontracted to METCUT SA that carried out these tests according to ASTM E-606-97 standard on a 100kN MTS fatigue machine. Temperature during the test is achieved by induction heating (Fig.2). Statistical data treatment performed at Aperam Research Center consists in determination of the endurance limit by the stair-case method and in fitting of the fatigue curve by different models (Wohler, Bastenaire, etc) with the help of ESOPE software.

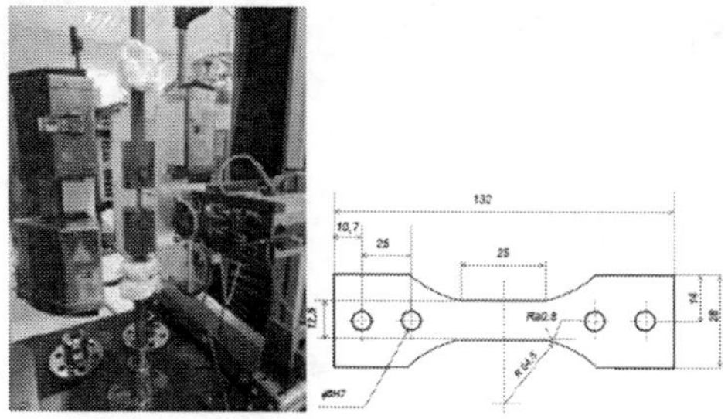

Figure 1 tensile test at high temperature: set-up on left, specimen geometry on right

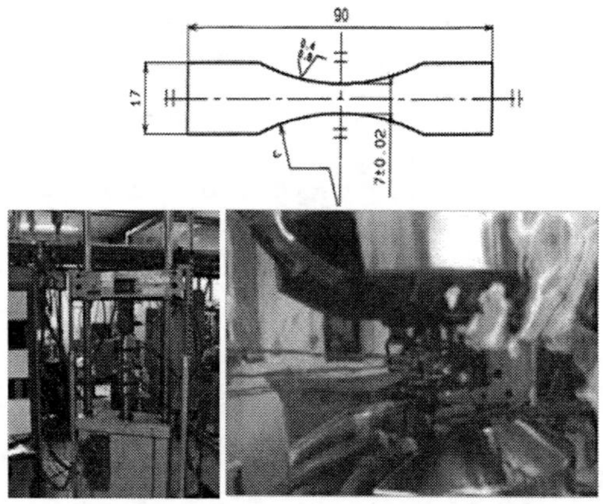

Figure 2 high-cycle fatigue at high temperature: specimen geometry on top and fatigue machine with induction heating (courtesy of METCUT)

2.4 Creep tests

Isothermal uniaxial creep tests are performed with an INSTRON 8802 tensile machine. Specimens are submitted to constant load while their deformation is monitored thanks to a high temperature extensometer up to 1% of creep strain. Tests are performed on flat specimens, cut along the rolling direction, between 750 and 950°C and up to 100h.

3 Experimental Results

3.1 Strengths and Endurance limits

Tensile properties evolution of ferritic stainless steel in function of the temperature is well illustrated on Fig.3 on the K44X 444Nb. Tensile strength and yield stress are dropping sharply after 600°C. At the reverse, elongation is increasing rapidly due to a striction of the specimen.

Fatigue limit are expressed by maximal fatigue strength Σ_{max}^{D} at 2,000,000 cycles but can be related to the stress amplitude Σ_a. Results obtained for temperature in range from room temperature to 850°C are given in Table 2 and are shown on Figure 4. It could be noticed that endurance limits for R=-1 and R=0.1 are crossing each other at around 650°C and so endurance limits in tensile-compression are becoming higher than endurance limit in alternating tension.

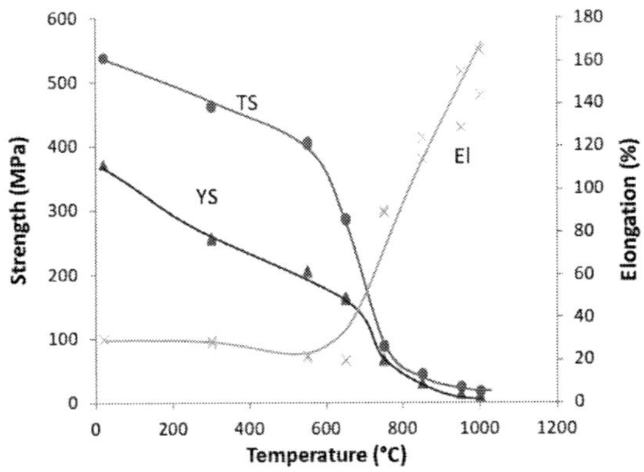

Figure 3. Tensile properties (TS, YS and El) in function of temperature for K44X-1.4521HT-444Nb

Table 2. Endurance limits Σ_{max}^{D} in MPa at 2,000,000 cycles in function of temperature.

Grades	R	23°C	300°C	650°C	750°C	850°C
444Nb	0,1	485 (6.9)	452.5 (0.2)	98.3 (4.1)		43.7 (1.8)
	-1	312.5 (11.6)	265 (0.5)	100 (4.5)		66.2 (1.7)
441	0,1	433 (9.5)	406.5 (11.3)		47.5 (12.4)	17.5 (0.2)
	-1	282.5 (3.5)	244.5 (7.4)		459 (1.4)	40.5 (0.1)

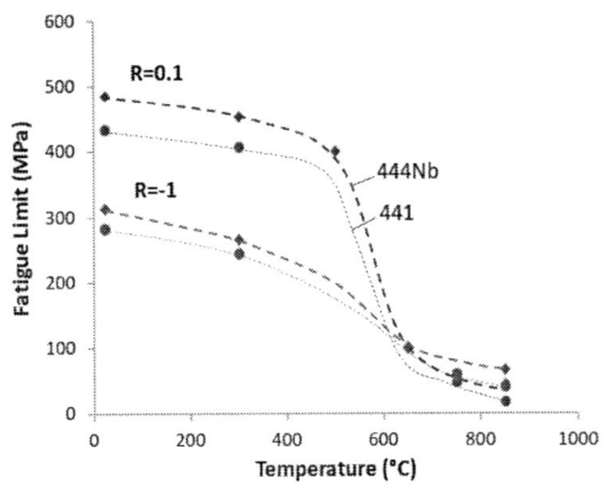

Figure 4 Endurance limit at 2,000,000 cycles, for R=-1 and R=0.1 in function of temperature for the 2 ferritic grades

3.2 Effect of temperature on fatigue to tensile strength ratio

The ratio between endurance limit and tensile strength is often calculated because it gives a constant value only dependent on steel family. This is true at room temperature but what happen when temperature is increasing?

For temperature range from room temperature to ~500°C fatigue strength ratio remains stable: 0.9-1 for R=0.1 and 0.5-0.6 for R=-1. These values are well correlated with usual values obtained on steels with a CC structure. For temperature high than 750°C, fatigue limit is affected by viscous effect, evidenced by a fatigue ratio greater than 1, i.e. a fatigue limit higher than tensile strength as shown for R=-1 (Fig.5). That could be surprising but tensile tests are carried out at low strain rate compared to HCF tests. Another interesting point already mentioned is the crossing of fatigue strength for the two stress ratio. The phenomenon can be attributed to a mean stress effect. Especially at R=0.1, fatigue fracture is linked with a severe necking of the sample area. For example, recorded measure of elongation on 441 sample gives 7mm after 15h of HCF test at conditions 25Hz, 850°C and 20MPa. In between at temperature around 650°C situations is more confusing due both viscous and both strain and precipitation hardening effect. In fact in that range of temperatures stabilized ferritic grades exhibit a precipitation of Fe-Nb intermetallics in aim to improved creep resistance.

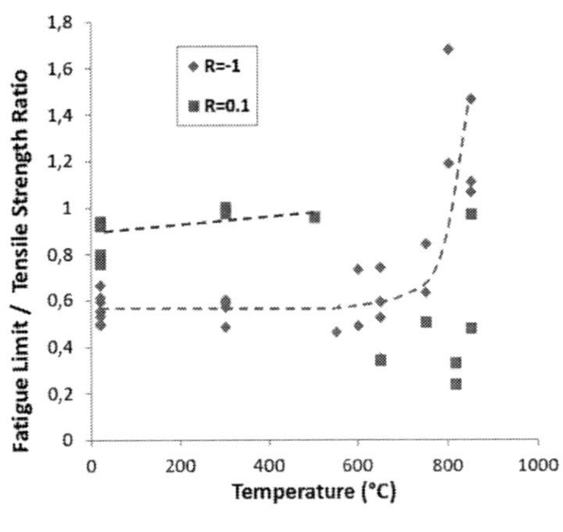

Figure 5 Ratio of fatigue limit on tensile strength for different stainless steel grades including 441 and 444Nb

4 Multi-axial fatigue modelling

4.1 From Haigh-Goodman to Dang Van Criterion

In first empiric approach, Haigh-Goodman diagram [8] can be drawn for the different temperature using stress amplitude Σ_a^D versus mean stress Σ_{mean}^D calculated from:

https://doi.org/10.1051/matecconf/201816519008

$$\Sigma_a^D = \frac{1}{2}\Sigma_{max}^D(1 - R) \quad (2)$$

$$\Sigma_{mean}^D = \frac{1}{2}\Sigma_{max}^D(1 + R) \quad (3)$$

The higher the temperature, the quicker the line rotation due to the more pronounced mean stress effect. Value of stress to creep fracture Σ_c for approximatively 20h (duration of the HCF test) are reported on mean stress axis for the 441 at 750°C and 850°C. They are quite relevant with fatigue limit at R=0.1 and confirm that creep mean stress effect is dominant at high temperature.

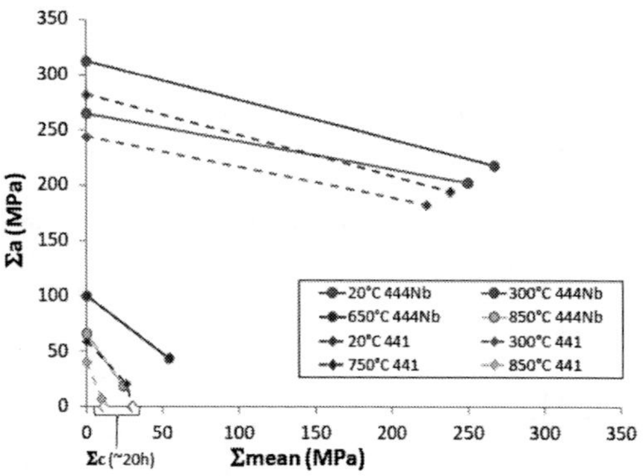

Figure 6. Haigh-Goodman diagram in range of temperatures from ambient to 850°C

An extension of macro-meso Dang Van approach can be established in case of isothermal or non-isothermal cyclic loadings. But it stipulates a local shakedown condition and so both mesoscopic residual stress ρ^* and plastic strain fields are independent of time. Therefore the fatigue criterion is expressed as a linear combination of local shear stress and hydrostatic stress at mesoscopic scale defined by:

$$\tau(t) = \frac{1}{2}.\Sigma_a.\left|\sin(2\pi.\frac{t}{\tau})\right| \quad (4)$$

$$p(t) = \frac{1}{3}\left(\Sigma_{mean} + \Sigma_a.\sin(2\pi.\frac{t}{\tau})\right) \quad (5)$$

At endurance limit the two quantities reach their maximal value, and the famous Dang Van line can be drawn from the two stress ratios fatigue results:

$$\tau_{max} = \frac{1}{4}.\Sigma_{max}^D.(1 - R) \quad (6)$$

$$p_{max} = \frac{1}{3}.\Sigma_{max}^D \quad (7)$$

The Dang Van lines for the different temperatures are shown on Fig.7. If the approach appears to be correct for 20°C and 300°C, the lines are rotating for the highest temperatures 650°C, 750°C and 850°C and the slopes are becoming positive. The hypothesis of local shakedown is definitively wrong for high temperatures.
Our proposal is to applied Dang Van criterion as long as the viscous/creep effect are weak and the ratio of fatigue limit on tensile strength is flat, i.e. up to about 500°C (about the third of the melting point). For the highest temperature, a new multiaxial approach appears to be necessary and based on accumulation of creep damage and HCF.

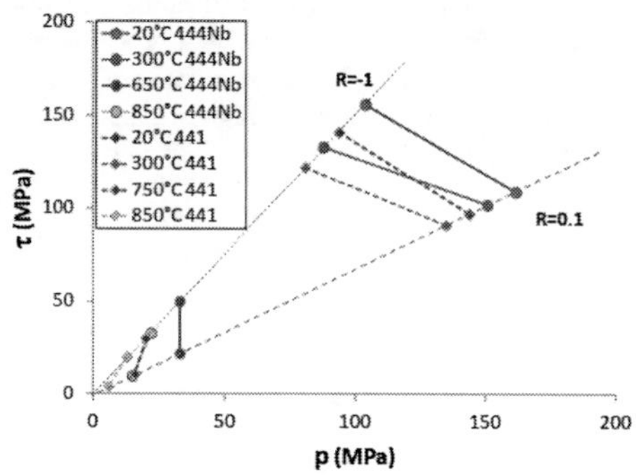

Figure 7. Dang Van diagram in range of temperatures from ambient to 850°C

4.2 Dang Van criterion in Xhaust_Life®

The Dang Van criterion was implemented into Aperam post processor of Abaqus, Xhaust_Life®, and then tested at low and intermediate temperatures in isothermal conditions. The mesoscopic stress field is related to the macroscopic stress field by

$$\sigma(t) = \Sigma(t) + \rho^* \quad (8)$$

Where the mesoscopic residual stress is determined according the algorithm proposed in [9]. Finally local shear $\tau(t)$and hydrostatic stresses $p(t)$ are easily computed from $\sigma(t)$. The distance between Dang Van line at 2,000,000 cycles and local stress cycles is calculated to determine the fatigue damage. The program, which highlights the area where there is a risk of fatigue cracks initiation, has been validated on uniaxial loading (Fig.8) and tested on biaxial loading. On the other hand validation tests in both multiaxial and intermediate temperature are not yet available.

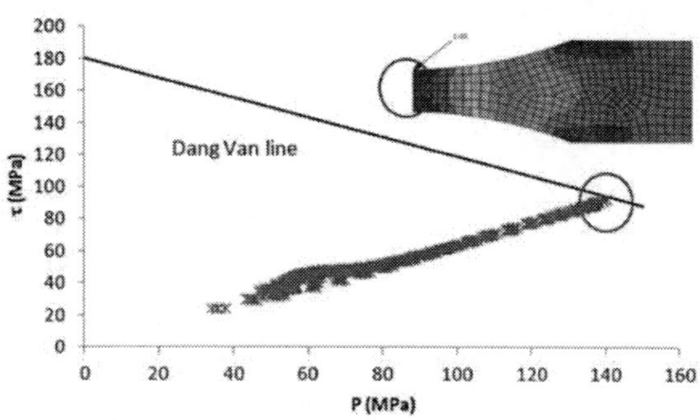

Figure 8. Example of Dang Van analysis on K41X fatigue specimen at 300°C

5 Discussion on a coupling with a creep damage model

A Larson-Miller (LM) creep model is also implemented and available in Xhaust_Life® post processor for creep damage risk evaluation. A multiaxial formulation is proposed in [11] and the model relates a creep stress Σ_c to a time to reach a strain of 1% given by:

$$P = T.(\log t_{1\%} + C) \quad (9)$$

$$\log \sigma_{eq} = B + m.P \quad (10)$$

Where P is the LM coefficient and C, B, m are material parameters identified from creep tests. σ_{eq} is the multiaxial equivalent stress.

A comparison between creep stress and HCF mean stress for an equivalent duration at 850°C evidences quite close values and therefore confirms a major influence of the creep in HCF damage at high temperature for R=0.1 (Tab.3).

Thus an idea could be to couple a multiaxial HCF model like Dang Van and the Larson-Miller creep model even if the accumulation rule for both HCF and creep damage has to be proposed.

Table 3. Comparison of mean stress at R=0.1 and creep stress (in MPa) to 1% elongation for same duration ~20h at 850°C.

Grade	Σ_{max}^D	Σ_{mean}	Σ_c
444Nb	43.7	24	22
441	17.5	9,6	15

5 Conclusion

Characterisations of HCF properties at high temperature were carried out at 2 stress ratios for multaxial modelling purpose. Due to viscous-creep effects, the analysis of the results for the highest temperatures, i.e. higher than 650°C, is tricky:

- A fatigue ratio greater than 1 is observed at high temperature due to strain rate effect (20Hz HCF test versus tensile test at 2% min^{-1} strain rate).
- A crossing of fatigue strength for the two stress ratios. The phenomenon can be attributed to a mean stress effect due to the creep damage.

Modelings using Goodman and especially Dang Van fatigue damage model were tested. If the Dang Van and Haigh-Goodman approach appear to be correct for 20°C and 300°C, the lines are rotating for the highest temperatures 650°C, 750°C and 850°C and the slopes are becoming positive in case of Dang Van. The hypothesis of local shakedown is definitively wrong for high temperature.

Once more, the mean stress effect due to creep damage is becoming dominant at R=0.1. So at high temperature the multiaxial HCF model must be reformulated.

Nomenclature

R stress ratio of high cycle fatigue test
TS tensile strength
YS yield stress
El elongation
Σ macroscopic stress field
σ,τ,p mesoscopic stress, shear and hydrostatic field
ρ* mesocopic residual stress
D fatigue / endurance limit supscript
c creep subscript
T temperature
t time
τ period of the cycle (reverse of frequency)
P, B,C,m Larson-Miller parameter

References

1. Watanabe et al., Thermal Fatigue Life prediction for stainless steel Exhaust Manifold, SAE Tech paper series 980841, 1998.
2. Santacreu PO et al. Design of Stainless steel Exhaust Manifold, La revue de metallurgie-CIT, pp615-620, 2004.
3. Thomas J.J et al. Thermomechanical design in Automotive industry, Fatigue Fracture Eng. Mater Structure 27, pp887-895, 2004.
4. P.O.Santacreu, C.Simon and A.Colman, Thermomechanical fatigue behavior of Stainless steel grades for Automotive Exhaust manifold applications, 4th vol. p.227-239 ASTM STP 1428, 2002
5. L.Faivre, P.O.Santacreu and A.Acher,A new ferritic stainless steel with improved
6. thermo-mechanical fatigue resistance for exhaust parts,Materials at High Temperature, Vol.30 (1) p.36-42, 2013.
7. K. Dang Van Comportement à la fatigue des métaux, Sciences et Techniques de l'Armement, Mémorial de l'Artillerie Francaise, p.113-116, 1973
8. H. Maitournam, T. M.L Nguyen-Tajan, Approche multi-échelle en fatigue polycyclique anisotherme C.R.Acad. Sci. Paris, t328, Série II b, p.373-379,2000.
9. Goodman J. Mechanics applied to engineering. 1st ed. London Longmans, Green and Co., 1899.
10. Dang Van K, Griveau B., and Message O, On a new multiaxial fatigue limit criterion: theory and applications, Biaxial and multiaxial fatigue, EGF 3, Mech. Eng. Publications, London, pp479-496, 1989.
11. P.O.Santacreu, L. Faivre and A.Acher, Life prediction approach for Stainless Steel Exhaust Manifold, SAE Int. J. Passeng. Cars-Mech. Syst. 5(2),2012 .

High cycle thermal fatigue of austenitic stainless steel

Yanjun Wang[1,2], *Ali* Charbal[1,2,+], *François* Hild[2] and *Ludovic* Vincent[1,*]

[1]DEN-Service de Recherches Métallurgiques Appliquées, CEA, Université Paris-Saclay, 91191, Gif sur Yvette, France
[2]LMT, ENS Paris-Saclay, CNRS, Université Paris-Saclay, 61 av. du Président Wilson, 94235 Cachan cedex, France
[+]now at Lehigh University, Packard Lab, 19 Memorial Drive West, Bethlehem, PA 18015, USA

Abstract. A new experimental setup called FLASH (THErmal Fatigue by LASer or Helium pulses) has been developed to perform thermal fatigue tests on materials used in structures of Sodium-cooled Fast Reactors (SFRs). A set of thermal loadings ranging from 150°C to 200°C prescribed by a high energy laser has been applied to A316L(N) stainless steel samples. Thermal fatigue tests are performed until a macrocrack is formed. The 3D displacement fields of the surface impacted by the laser beam are measured by a hybrid multiview system. A thermomechanical model is used to compute displacement fields that are compared with correlation measurements. These results are compared in terms of strain levels to the fatigue curve determined from standardized isothermal uniaxial mechanical fatigue tests.

1 Introduction

In a sodium cooled fast nuclear reactor, the primary pump circulates cold sodium (400°C) between fuel assemblies to remove the heat resulting from fission reactions. Several heat exchangers then transfer this energy from hot sodium (550°C) to steam generators in order to drive the turbines that eventually produce electricity [1]. There are different conditions where repeated thermal shocks may occur in the component walls of the reactor made of AISI 316L(N) austenitic stainless steel. When these loadings are repeated over a large number of cycles, initiation and propagation of a crack network may be induced at the surface of the components, which is referred to as thermal striping [2]. Design codes in the nuclear industry deal with thermal fatigue considering an equivalent strength (i.e. mechanical strain amplitude) as in standardized isothermal strain controlled push-pull mechanical fatigue tests. From previous thermal fatigue results obtained at CEA [3, 4], it appeared that this approach could be non-conservative, namely, the number of cycles to crack initiation in thermal fatigue experiments could be lower than in push-pull isothermal fatigue tests for the same levels of equivalent strain amplitude.

The present work focuses on the determination of crack initiation under thermal fatigue using the FLASH facility, which allows high cycle (about several 10^6) thermal fatigue tests to be performed with excellent stability of the thermal loading [5]. A pulsed laser is used to prescribe cyclic heat fluxes onto the surface of austenitic stainless steel samples. Thanks to the separation of the wavelengths of the laser (1.064µm) on the one hand and of an infrared camera [3.97-4.01µm] and two large-band pyrometers [1.55-2.2µm] on the other hand, measurements of the temperature fields are performed on the surface area of the specimens impacted by the laser where temperature variations are the largest and where crack initiation eventually occurs. This setup greatly improves the signal/noise ratio generally obtained in thermal fatigue experiments and then produces better estimations of the boundary conditions of the thermomechanical simulations required to interpret such tests. With the predictions of finite element analyses, it is deduced that the largest strain variations during cyclic loading are in the out-of-plane direction resulting in several micrometer out-of-plane displacements [6]. The 3D displacement fields are then measured by a hybrid multiview system composed of two visible light cameras and the infrared camera that also gives access to the Lagrangian temperature fields [7, 8]. Last, the equivalent Tresca strain variation obtained from finite element analyses is extracted in the region of crack initiation. This estimation is compared with results obtained from standardized isothermal uniaxial mechanical fatigue tests.

2 Experimental setup

FLASH is designed with the purpose of generating localized cyclic thermal shocks by pulsed laser or helium jets that are able to reproduce crack initiation in laboratory conditions (Fig. 1). The specimens are plates made of 316L(N) austenitic stainless steel (270 × 40 × 7 mm^3), which are continuously heated by an electrical current (Joule effect) until their center reaches 400°C, a temperature (controlled by a K-type thermocouple) representative of the cold primary sodium temperature. Static 4-point bend loading can be added so that the thermally shocked region will undergo a non-zero mean uniaxial stress.

* Corresponding author: ludovic.vincent@cea.fr

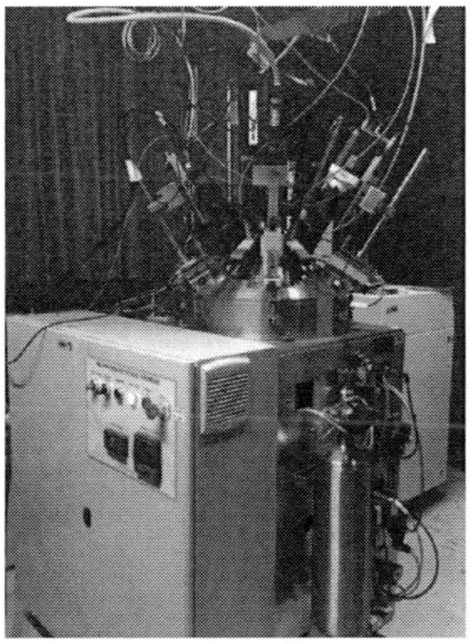

Fig. 1. Overview of FLASH facility

The thermal loadings are performed by a pulsed laser (TruPulse 156, Trumpf, $\lambda = 1064$nm, see Fig. 2-(1)) on the center of specimen at a frequency of 1Hz, the pulse duration is 50ms and the incident pulsed power is adapted to obtain the desired temperature variation on the surface of the specimen. A focusing optics allows a top-hat power density distribution to be obtained over a 5-mm in diameter disk. Due to the relatively low absorptivity of the polished and pre-oxidized surface, the laser beam is tilted to be reflected toward a power detector that gives access to the mean power reflected by the specimen surface.

An infrared camera (x6540sc FLIR, definition: 640×512 pixels, $\lambda \in [3, 5\mu\text{m}]$ reduced to $\lambda \in [3.97, 4.01\mu\text{m}]$ with an internal filter for high temperature measurements, Fig. 2.-(2)) is used to measure the 2D temperature field and the 3D displacement fields with a high magnification lens allowing for a pixel resolution of 15µm. Two fast pyrometers (KGA740-LO, $\lambda \in [3, 5\mu m]$, Fig. 2.-(3)) are used to measure and monitor within the impacted zone the temperature variation on a 2.5mm central area. The first visible light camera (MIRO M320S, Vision Research, definition: 1920×1080 pixels, Fig. 2.-(4)) with a pixel resolution of 10µm corresponds to the second device of the multiview system. Another visible light camera was recently introduced as the third device (pco.edge, definition: 2560×2160 pixels, Fig. 2.-(5)) with a higher pixel resolution of 6.5µm. All three cameras are synchronized on the same stroboscopic acquisition signal to compensate for the relatively lower acquisition frequency of the pco.edge camera, and at the same time to facilitate the comparison between simulated and experimental thermomechanical fields. The specimen is put inside an airtight chamber filled with helium atmosphere to eliminate possible excessive oxidation that would alter the surface emissivity, namely, the speckles to enable for DIC analyses, the heat exchange between the specimen and the laser beam (absorptivity) or helium atmosphere (convection).

Fig. 2. Experimental configuration with principal components (see main text for a detailed description of the components labelled in the picture)

3 Thermomechanical analyses

Three-dimensional finite element simulations are carried out with the assumption that the thermal and mechanical loadings are uncoupled. Such numerical simulations are necessary to interpret the results of the thermal fatigue test through equivalent strain variations (scalar) built from complete strain tensors. These simulations also give access to the 3D displacement field on the specimen surface impacted by the laser beam, which can be compared with the multiview correlation measurements. First, a thermal analysis is performed with the purpose of reproducing the temperature fields measured during cyclic loadings. Once the cyclic history of temperature at the shock center is stabilized, then the temperature field of the last cycle is used as an input of the thermo-mechanical calculation. Due to geometrical symmetry, only one quarter of the FLASH specimen (which corresponds to a simulation volume of $105 \times 20 \times 7$ mm^3) is modeled to minimize the computation time and numerical storage.

3.1 Heat transfer analyses

The purpose of this section is to present the finite element model of heat transfer to evaluate the temperature fields. The spatiotemporal temperature field is determined considering conductive and convective hear transfer processes

$$\rho c_p \frac{\partial}{\partial t} T(\mathbf{x}, t) = k \Delta T(\mathbf{x}, t) + s(\mathbf{x}) \qquad (1)$$

where T is the temperature distribution in the specimen, Δ the Laplacian, ρ the mass density ($kg \cdot m^{-3}$), c_p the specific heat capacity ($J \cdot kg^{-1} \cdot K^{-1}$), k the thermal conductivity ($W \cdot m^{-1} \cdot K^{-1}$), and s the volumetric heat source accounting for the heat flux induced by Joule effect. Other thermophysical parameters of the material are chosen from data reported in the French design code RCC-MRx [9] . The natural convection flux across the boundary layer on the surface of the specimen reads

$$q = -h(T - T_{He}) \qquad (2)$$

where q is the convection heat flux (W/m^2), h the heat convection coefficient ($W/m^2/°C$) [10] , T_{He} the temperature of He atmosphere (known to be equal to 30°C).

The specimen is initially heated to a central temperature at mid-thickness of 400°C using a low voltage, high current electrical power supply. This process is modeled by introducing only the Joule effect term, which is calibrated by adjusting the value of the volumetric source s so that the temperature at the central impacted zone stabilizes at 400°C after several minutes of transient phase. It is found that $s = 3.36 \times 10^7 W\,m^{-3}$, which is consistent with the experimental settings with 3kW electrical power. A temperature of 100°C is prescribed at the end of the simulated volume, which corresponds to the level measured at mid-thickness under the copper clamps during the test.

The resulting temperature field is introduced as the initial conditions. Then the cyclic thermal loading on the central zone of the surface is modeled by a super Gaussian profile [11]

$$I(y,z,p) = \alpha I_0 \frac{p4^{\frac{1}{p}}}{2\pi\Gamma(\frac{2}{p})} \exp(-2r(y,z)^p) \quad (3)$$

where α is the surface absorptivity, I_0 the chosen peak power of laser pulse known to be equal to 310W, Γ the Gamma function, p the shape parameter and r the reduced cross-sectional distance

$$r^2(y,z) = \left(\frac{y}{R_y}\right)^2 + \left(\frac{z}{R_z}\right)^2 \quad (4)$$

The parameters of the spatial power density distribution of the laser beam are determined by comparing experimental (as shown in Fig. 3) and numerical temperature fields from a thermal FE model updating procedure [12].

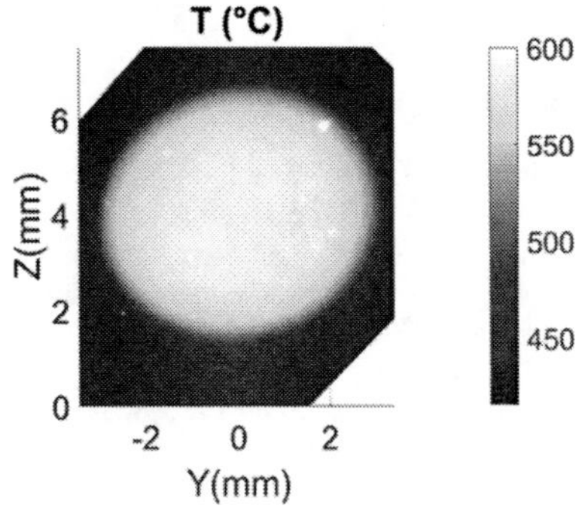

Fig. 3. Surface temperature field measured by IR camera at the end of a laser pulse

The values of identified parameters are summarized in Tab. 1.

Table 1. Identified parameters of spatial power density distribution of laser pulse

α	p	$R_y(mm)$	$R_z(mm)$
0.51	12	3.08	2.96

Based on the information described above, the temperature field at the end of a laser pulse calculated by the 3D transient model is shown in Fig. 4. The distribution of temperature resulting from the heating up phase is illustrated outside the zone impacted by the laser pulse. It shows that the temperature gradient is particularly steep along the thickness direction (labelled as X for the FE model), and lower along the other two directions in the plane of specimen surface (labelled as Z along the longitudinal direction, and Y along the transverse direction of the FE model), see Fig. 5.

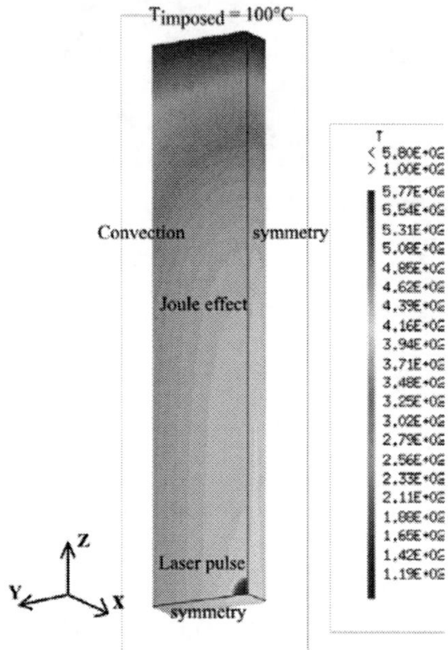

Fig. 4. 3D Temperature field at the end of a laser pulse with a prescribed temperature amplitude of 170°C in the central zone

This feature is directly linked to the relatively short laser pulse duration (50ms), the surface temperature change has occurred long before the shock effects have penetrated significantly through the thickness. A disk of almost 2mm in radius can be considered as being shocked with the same temperature variation and from which fatigue cracks initiate and propagate, which is confirmed from the optical microscopy observations reported in Section 4.

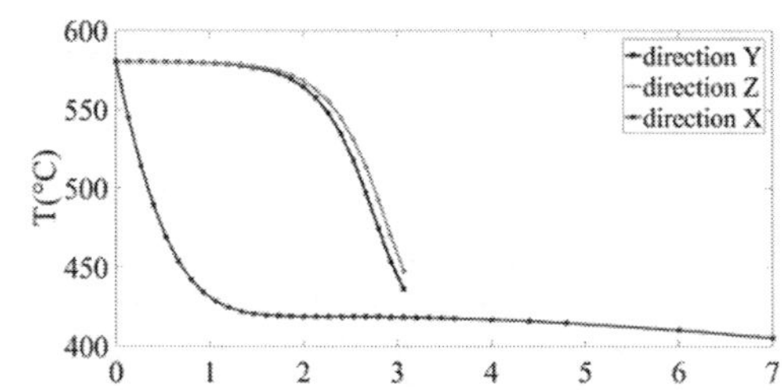

Fig. 5. Temperature gradient along three directions in the central beam impacted zone, at the end of a laser pulse

3.2 Mechanical analyses

The change of temperature induces significant thermal strain variations in the central part of the impacted zone, while the bulk remains in the initial state; such local expansions are then hindered by the bulk that is under tension during this laser pulse activation stage.

Thus a thermomechanical analysis is run by applying the stabilized cyclic thermal loading with the selected boundary conditions. First, there is no externally applied stress onto the specimen; and with symmetry considerations, no displacement on axial and transverse surfaces is expected.

With the assumption that the generated strain amplitudes are moderate, the total strain tensor is decomposed as follows

$$\underline{\underline{\varepsilon}}^{total} = \underline{\underline{\varepsilon}}^{meca} + \underline{\underline{\varepsilon}}^{ther} \quad (5)$$

where the thermal strain tensor is defined as

$$\underline{\underline{\varepsilon}}^{ther} = \alpha_T \cdot (T - T_{ini}) \underline{\underline{I}} \quad (6)$$

and α_T is the thermal expansion coefficient, $(T - T_{ini})$ the variation of temperature, and $\underline{\underline{I}}$ the identity tensor. The change of α_T as a function of temperature is extracted from the French design code RCC-MRx [9].

Within the small strain framework, the mechanical strain tensor is decomposed into elastic and plastic contributions

$$\underline{\underline{\varepsilon}}^{meca} = \underline{\underline{\varepsilon}}^{elas} + \underline{\underline{\varepsilon}}^{plas} \quad (7)$$

The elastic domain is defined by

$$f = J_2(\underline{\underline{\sigma}} - \underline{\underline{X}}) - R(T) \leq 0 \quad (8)$$

in which J_2 is the second invariant of the deviatoric stress tensor, R the temperature dependent initial yield stress, and $\underline{\underline{X}}$ the back-stress tensor.

As proposed in Ref. [3], a stabilized stress-strain behavior can be obtained after a short transient period; the cyclic stress change is described using a series of hardening laws. The thermomechanical behavior of the material is then described by a nonlinear kinematic hardening law [13]

$$\underline{\underline{X}} = \sum_{i=1}^{N=2} \underline{\underline{X_i}} \quad (9)$$

where each kinematic variable $\underline{\underline{X_i}}$ follows the growth law

$$\underline{\underline{\dot{X}_i}} = C_i(\tfrac{2}{3} A_i \underline{\underline{\dot{\varepsilon}^p}} - \underline{\underline{X_i}} \dot{p}) \quad (10)$$

The thermomechanical parameters R, C_i and A_i are calibrated on push-pull cyclic tests at 400°C and 500°C [12]. A consistent set of parameters is then obtained in which only R is a (linear) function of the temperature as

$$R(T) = 73.49 - 0.0735T \quad (11)$$

The values of identified parameters are listed in Tab. 2.

Table 2. Identified parameters for non-linear kinematic hardening laws

$A_1(MPa)$	C_1	$A_2(MPa)$	C_2
105.8	684.5	235.6	15.3

Mechanical simulations present nearly perfect equibiaxial stress distributions on the surface center as shown in Fig. 6.

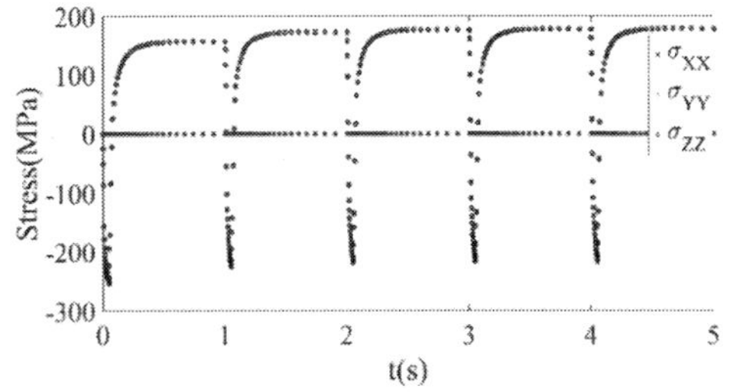

Fig. 6. Simulated stress history in the central zone of the specimen surface impacted by the laser beam

The histories of eigen strains (simulated and experimentally measured) are compared in Fig. 7. The levels of total strains measured by multiview correlation are well reproduced by the simulations along the two eigen directions in the plane of specimen surface.

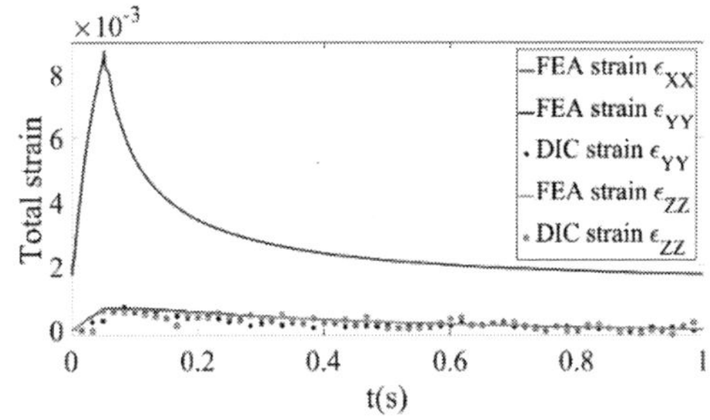

Fig. 7. Strain history in the central zone of the surface impacted by the laser beam. Solid lines stand for 3D simulation results, and solid symbols for experimental measurements

Moreover, the changes of out of plane displacement induced by the largest strain variation ϵ_{XX} during cyclic loading are consistent with the experimental measurements (Fig. 8).

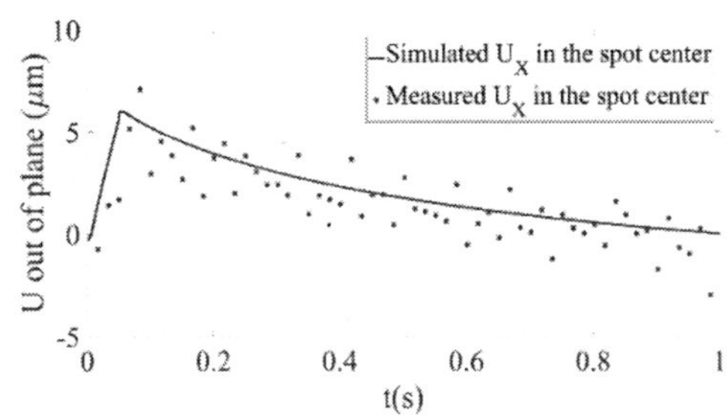

Fig. 8. Out of plane displacement in the central zone of the surface impacted by the laser spot. The solid line stands for 3D simulation results, and solid symbols for experimental measurements

4 Fatigue analysis and comparison with results from uniaxial isothermal fatigue tests

Since the out-of-plane strain variation is not experimentally accessible during thermal fatigue tests, the equivalent Tresca strain variation extracted from mechanical simulations is needed in order to interpret the thermal fatigue experiments and compare them with isothermal mechanical fatigue tests.

The Tresca equivalent total strain range is calculated by adding the contributions of the Tresca equivalent elastic and plastic strain ranges where the plastic strain range can be obtained with a methodology based on the correction of the Poisson effect in the elastic analysis of low cycle fatigue

$$\Delta\varepsilon^{tot}_{Tresca} = \Delta\varepsilon^{elastic}_{Tresca} + \Delta\varepsilon^{plastic}_{Tresca} \tag{12}$$

with

$$\Delta\varepsilon^{e,p}_{Tresca} = \frac{1}{1+\nu}\max\left(\left|\Delta\varepsilon_1^{e,p} - \Delta\varepsilon_2^{e,p}\right|, \left|\Delta\varepsilon_1^{e,p} - \Delta\varepsilon_3^{e,p}\right|, \left|\Delta\varepsilon_2^{e,p} - \Delta\varepsilon_3^{e,p}\right|\right) \tag{13}$$

where ν is the Poisson's ratio equal to 0.3 and 0.5 for elastic and plastic calculations, respectively. For this set of simulation, the Tresca equivalent total strain range is 0.45% in the central laser beam zone for a temperature variation of 170°C.

The number of cycles to initiate cracks in the fatigue test is deduced from singularities in the gray level residual maps of the correlation technique. The main fatigue crack initiates after 80,000 cycles (Fig 9). This sample has been observed periodically by optical microscopy during the thermal fatigue test in order to investigate the formation and growth of fatigue cracks at a scale hardly detectable by correlation techniques. After only 6,000 cycles, very intense slip bands are observed on the surface within the shocked region, and several microcracks with length of 50µm begin to initiate in the intersection of slip bands with grain boundaries.

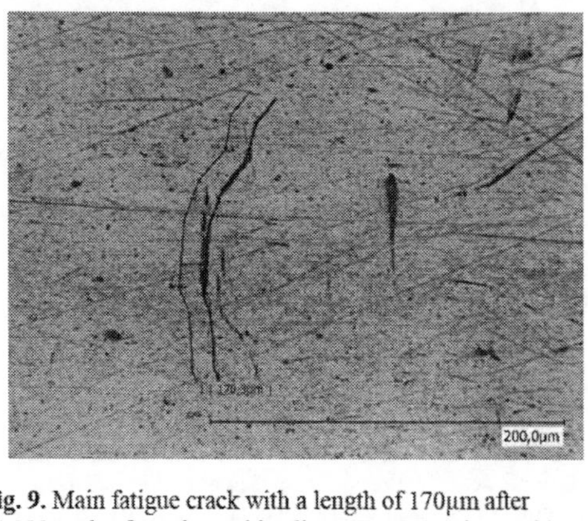

Fig. 9. Main fatigue crack with a length of 170µm after 80,000 cycles for a thermal loading ΔT=170°C observed by optical microscopy

The main fatigue crack is located just in the central zone of the surface impacted by the laser beam and initiated during the first 26,000 cycles. Then it propagated as shown in Fig. 10 until a 170µm crack length is clearly observed after 80,000 cycles. A second crack with another orientation (perpendicular to the first one) is observed after 42,000 cycles with an initial length of 40µm, and then it has grown up to 90 µm after 80,000 cycles. These observations confirm the biaxial state induced by thermal loading and the initiation of cracks on permanent slip bands induced by the plastic strains on the sample surface.

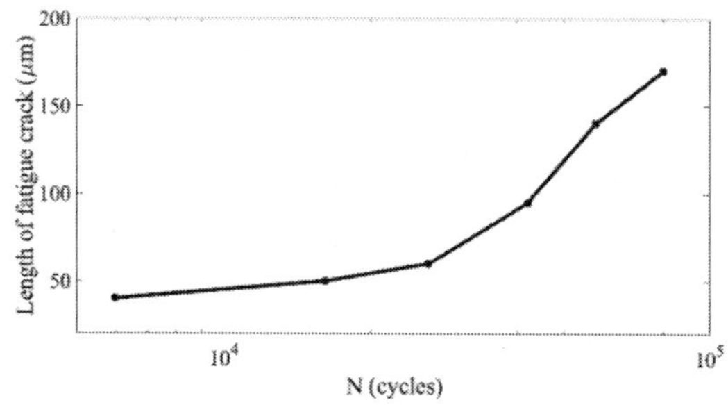

Fig. 10. Propagation of the main fatigue crack with a thermal loading ΔT=170°C observed by optical microscopy

In Fig. 11, the thermal fatigue results are compared with uniaxial isothermal fatigue data obtained at 400°C and 500°C for the same material. The ordinate of the graph is the Tresca strain variation, while the abscissa is the number of cycles to "initiate" 200µm cracks in thermal fatigue tests or the number of cycles to get a 25% decrease of the maximum stress in push-pull fatigue tests performed on 8mm cylindrical specimen.

With the FLASH data it is concluded that under the same equivalent strain ranges, that is to say under the same mechanical loading, the environmental effect has a minor influence on the fatigue life. This observation validates the explanations [14, 15] that the environmental effect is negligible compared to strain aging and dislocation motion at high temperatures (i.e., in ranges from 300°C to 800°C) for 316L austenitic stainless steel, especially for the frequency of pulsed laser (1Hz in FLASH tests) that leads to environmental interactions barely able to develop. Besides, FLASH results show higher number of cycles to initiation compared with uniaxial isothermal fatigue tests under the same strain range (0.5%). This difference may result from the mean bi-compressive stress states induced by hot thermal shocks, which is favorable for the fatigue resistance [16]. Yet it should be noted that this comparison is incomplete since push-pull fatigue test data under higher strain ranges are not available. Overall, the present thermal fatigue data are in good agreement with classical fatigue test results and are located well above the design curve of the RCC-MRx design code.

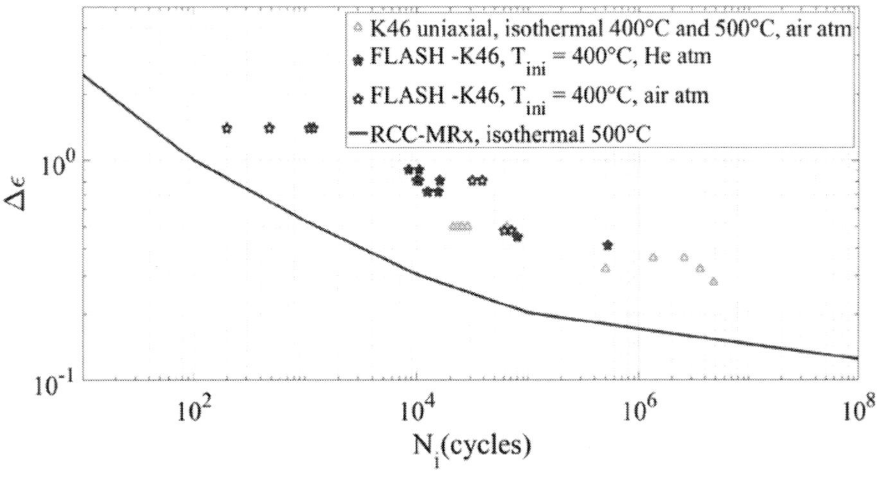

Fig. 11. Equivalent Tresca strain range as a function of number of cycles required to initiate 200μm cracks with FLASH (red symbols). Comparaison with number of cycles for isothermal fatigue tests (orange symbols) and with the fatigue design curve of RCC-MRx design code (solid line)

5 Conclusion

The FLASH facility allows thermal fatigue tests to be performed on specimen made of 316L(N) austenitic stainless steel with highly stable thermal amplitudes ranging from 150°C to 200°C prescribed by pulsed laser shocks. Full-field temperature and displacement measurements on the surface of the specimen are carried out by using infrared and visible light cameras. The equivalent Tresca strain range obtained from numerical simulations combined with crack initiation detection via correlation enable the fatigue strength of the studied material to be estimated.

The comparison with the French design code RCC-MRx shows a good conservative agreement. The next tests will be conducted with different thermal loading histories to build up an extensive and well-documented database. Furthermore, tests with helium pulses onto hot plates or with static 4-point bending should be performed to investigate the mean bi-tensile stress effect on the fatigue response.

References

1. P. Le Coz, J. Sauvage, J. P. Serpantié, *Proceedings of ICAPP* 2851. (France, 2011)
2. N. Robert, J. Economou, F. Cornuel, O. Volte, J.-M. Stephan, *International symposium on the contribution of materials investigations to improve the safety and performance of LWRs* (Fontevraud, France, 2006)
3. A. Fissolo, S. Amiable, O. Ancelet, F. Mermaz, J. M. Stelmaszyk, A. Constantinescu, C. Robertson, L. Vincent, F. Bouchet, Int J Fatigue, **31(3)**, 587-600. (2009)
4. A. Fissolo, S. Amiable, O. Ancelet, F. Mermaz, J. M. Stelmaszyk, A. Constantinescu, C. Robertson, L. Vincent, F. Bouchet, Int J Fatigue, **31(7)**, 1196-1210. (2009)
5. L. Vincent, M. Poncelet, S. Roux, F. Hild, D. Farcage, Procedia Eng, **66,** 669-675. (2013)
6. A. Charbal, L. Vincent, F. Hild, M. Poncelet, J. E. Dufour, S. Roux, D. Farcage, QIRT J, **13(1)**, 1-18. (2016)
7. A. Charbal, J. E. Dufour, F. Hild, M. Poncelet, L. Vincent, S. Roux, Exp. Mech, **56(5)**, 845-860. (2016)
8. Y. Wang, A. Charbal, J. E. Dufour, F. Hild, S. Roux, L. Vincent, *Photomechanics*, (France, 2018)
9. AFCEN, *RCC-MRx* (2012).
10. F. W. Giacobbe, Appl. Therm. Eng, **25(2)**, 205-225. (2005)
11. D. L. Shealy, J. A. Hoffnagle, Appl. Opt, **45(21)**, 5118-5131. (2006)
12. A. Charbal, *PhD thesis*, (University of Paris-Saclay, 2017)
13. J.-L. Chaboche, Int J Plasticity, **24(10)**, 1642-1693. (2008)
14. R. Alain, P. Violan, J. Mendez, Mater. Sci. Eng. A, **229(1-2)**, 87-94. (1997)
15. K. Sadananda, P. Shahinian, Metall. Mater. Trans. A, **11(2)**, 267-276. (1980)
16. L. Vincent, *HDR*, (Ecole Normale Supérieure de Cachan, 2012)

* Corresponding author: ludovic.vincent@cea.fr

Acceleration techniques for the numerical simulation of the cyclic plasticity behaviour of mechanical components under thermal loads

Jelena Srnec Novak[1,*], *Francesco* De Bona[1], *Denis* Benasciutti[2] and *Luciano* Moro[1]

[1]University of Udine, Politechnic Department of Engineering and Architecture (DPIA), via delle Scienze 208, 33100 Udine, Italy
[2]University of Ferrara, Department of Engineering, via Saragat 1, 44122 Ferrara, Italy

Abstract. Numerical simulations of components subjected to cyclic thermo-mechanical loads require an accurate modelling of their cyclic plasticity behaviour. Combined models permit to capture monotonic hardening as well as cyclic hardening/softening phenomena, that occur in reality. In principle the durability assessment of a component under thermal loads can be performed only if the cyclic behaviour is simulated until complete material stabilization. As materials stabilize approximately at half the number of cycles to failure, it follows that in case of small plastic strains a huge number of cycles must be considered and an unfeasible simulation time would be required. Accelerated models have thus been proposed in literature. The aim of this work is that of comparing the different acceleration techniques in the case a round mould for continuous casting loaded thermo-mechanically. It can be observed that the usual approach of using the stabilized stress-strain curve already from the first cycle could lead to relevant errors. An alternative method is that of increasing the value of the parameter that controls the speed of stabilization in the combined model. This approach permits the number of cycles to reach stabilization to be drastically reduced, without affecting the overall mechanical behaviour. Based on this approach, a simple design rule, that can be adopted, particularly when relatively small plastic strains occur, is finally proposed.

1 Numerical simulation of mechanical components under thermal loads

The choice of a suitable cyclic plasticity model to be used in numerical simulation of components subjected to cyclic thermo-mechanical loads is a crucial aspect in design. Some of typical examples are discussed in [1, 2, 3]. In fact, to perform a reliable thermal fatigue analysis a model able to capture accurately the behaviour of the material observed during experimental testing is required. On the other hand, durability assessment of a component under thermal loads can be done only if complete material stabilization is reached [4]. As materials stabilize approximately at half number of cycles to failure, it follows that in case of small plastic strain a huge number of cycles must be considered and an unfeasible computational time would be required.

In a recent work [5] dealing with a thermo-mechanical analysis of a squared mould showed that a huge number of cycles (around 60567) must be performed to reach the stabilized condition. Since the geometry of the squared mould requires a 3D finite element (FE) simulation, even taking into account symmetries and optimizing the mesh, ≈15 min is needed to compute 1 cycle or ≈630 day to obtain final results, according to the value (b≈4) of the speed of stabilization of the mould alloy, see Fig. 1. Therefore, to overcome this problem, some alternative models (accelerated, stabilized), proposed in [6], have been adopted and compared. It has been shown that accelerated models gave the most conservative results, as the lowest value of number of cycle to failure was obtained. On the other hand, in that work it was not possible to make a comparison with the combined (complete) model because a too long computational time was necessary to reach stabilization.

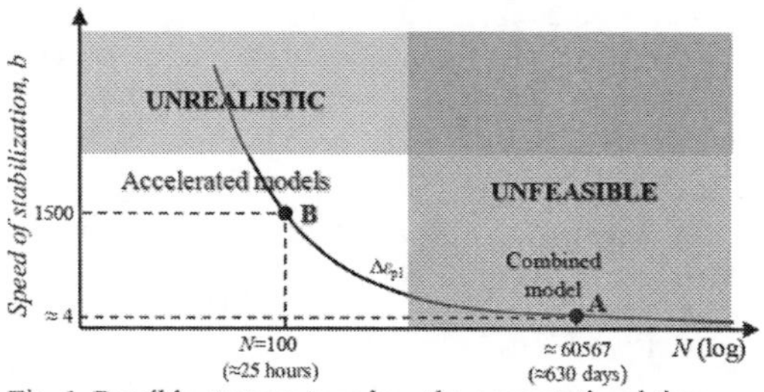

Fig. 1. Possible strategy to reduce the computational time.

Fig. 1 presents schematically the desirable diagram required to perform a correct design of the component. In principle the basic idea of the accelerated techniques is that of moving from point A to point B (i.e. from the purple "unfeasible" zone, characterized by too long computation time, to the white area, where this time is acceptable). This goal can be obtained by increasing a parameter of the material model, which governs the speed of stabilization. On the other hand, also too high speed of stabilization must be avoided, as it could lead to convergence problems and "unrealistic" results.

In order to perform a complete validation, a similar component with a simpler geometry (i.e. round mould) is considered in this work. Due to axi-symmetry, a plane model can be used thus permitting a faster simulation to be performed. The component behaviour can be followed till stabilization adopting the combined model. This case can be then taken as a reference to compare different accelerated models proposed in literature. The aim of this work is to validate the "design diagram" shown in Fig. 1, which permits a suitable speed of stabilization to be set-up, in order to reach stabilized conditions in a feasible computational time without affecting the consequent durability assessment.

2 Accelerated cyclic plasticity models

A combined (kinematic and isotropic) model captures simultaneously monotonic hardening (elasto-plastic) and cyclic hardening/softening behaviour of a material. The yield surface can be represented as [7]:

$$\sqrt{\frac{3}{2}(\boldsymbol{S}-\boldsymbol{X}):(\boldsymbol{S}-\boldsymbol{X})}-R-\sigma_0=0 \tag{1}$$

where $\boldsymbol{S}$ is the deviatoric stress tensor, $\boldsymbol{X}$ is the back stress tensor, R is the drag stress and σ_0 is the initial yield stress. In this case, the yield surface both translates (controlled by $\boldsymbol{X}$) and expands (controlled by R) with the plastic strain ε_p.

Different kinematic models have been proposed in literature to relate $\boldsymbol{X}$ to ε_{pl} [7]. Chaboche model (nonlinear kinematic) assumes that the increment of the back stress $d\boldsymbol{X}$ is expressed as a function of the plastic strain increment $d\varepsilon_{pl}$ and the accumulated plastic strain $d\varepsilon_{pl,acc}$ [7]:

$$\boldsymbol{X}=\sum_{i=1}^{n}\boldsymbol{X}_i \quad ; \quad d\boldsymbol{X}_i=\frac{2}{3}C_i d\varepsilon_{pl}-\gamma_i\boldsymbol{X}_i d\varepsilon_{pl,acc} \tag{2}$$

where C is the hardening modulus and γ is the non-linear recovery parameter that controls the decrease rate of the hardening modulus as plastic strain accumulates. The model with one pair of (C_1, γ_1) is known as the Armstrong and Frederick model. In addition, considering γ=0 the Prager model (i.e. linear kinematic) is obtained and relation (2) can be expressed as [7]:

$$d\boldsymbol{X}=\frac{2}{3}C_{lin}d\varepsilon_{pl} \tag{3}$$

The nonlinear isotropic model controls the homothetic expansion of the yield surface [7]:

$$dR=b(R_\infty-R)d\varepsilon_{pl,acc} \tag{4}$$

where R_∞ is the saturated drag stress, b is the parameter that controls the speed of hardening (R_∞>0) or softening (R_∞<0). Integration of (4) gives expression that links R_∞ to $\varepsilon_{pl,acc}$ [7]:

$$R=R_\infty\left[1-\exp\left(-b\varepsilon_{pl,acc}\right)\right] \tag{5}$$

Stabilized condition is obtained when R reaches R_∞. Cyclic hardening/softening behaviour of a material is mainly governed by the speed of stabilization b and the accumulated plastic strain $\varepsilon_{pl,acc}$. When the accumulated plastic strain is relatively small, a huge number of cycles is required to reach the stabilized condition as already shown in Fig. 1.

Some accelerated methods have thus been proposed in literature. In particular, in presence of creep and thermal fatigue, authors such in [8, 9, 10] suggest only a limited number of cycles to be simulated; even if this procedure seems not well defined, it has to be considered that the presence of visco-elasticity generally strongly reduces the stabilization time. On the other hand, some authors [11, 12] use the kinematic model with stabilized material condition from the beginning of simulation, at the same time neglecting the initial transient cyclic deformation. If creep constitutes the damage criteria, a more rigorous approach was proposed in [13], where an extrapolation technique is developed to speed up the simulation. An alternative approach was suggested in [6]. Since the accumulated plastic strain (see Eq. 5) depends on loading condition and cannot be changed, it was proposed to accelerate material stabilization by increasing (10-200 times) the speed of stabilization b.

3 Case study: round mould under variable thermal flux

A mould is a part of the continuous casting process where solidification of steel starts. During solidification, a huge thermal flux, q, is transferred from the molten steel, that is in contact to the inner surface, to the water-cooled outer surface of the mould, see Fig. 2. The thermal flux distribution varies between two conditions: q=0 when the plant is turned-off due to maintenance and $q=q_{max}$ when the plant operates [5].

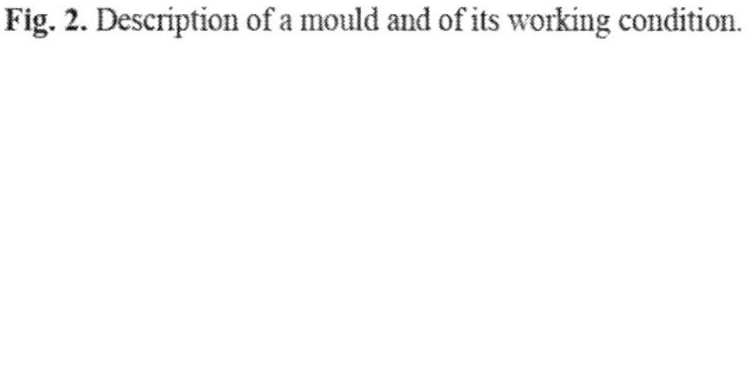

Fig. 2. Description of a mould and of its working condition.

 https://doi.org/10.1051/matecconf/201816519010

Generally, moulds are constituted by a hollow tube of copper alloys while according to the geometry of the final product, different types of cross sections (square, rectangular or rounded) can be adopted. Copper alloys are used because they provide the best combination of high thermal conductivity and adequate mechanical properties.

Table 1. Material properties for CuAg0.1 [15].

Temp. (°C)	E (MPa)	E_s (MPa)	σ_0 (MPa)	σ_{0*} (MPa)	C_1 (MPa)	γ_1	C_2 (MPa)	γ_2	C_3 (MPa)	γ_3	C_{lin} (MPa)	R_∞ (MPa)	b
20	119988	114763	113	86	25880	1627	24460	1624	15620	315.4	37439	-68	2.352
250	106080	94758	110	57	31310	1708	10240	343.6	5256	1748	18039	-75	3.894
300 °C	103800	94793	108	50	13170	1092	10700	398.2	10650	1155	18466	-77	5.293

In this work, a round mould 1000 mm long, 16 mm thick and with 200 mm inner diameter was studied, similarly to the component described in [14]. As material, it was assumed a CuAg0.1 alloy, whose properties, listed in Tab. 1, were experimentally obtained in [15, 16, 17].

3.1. Numerical simulation – FEM model

A two dimensional (2D) model was adopted due to axi-symmetry, thus strongly decreasing the computational time. The finite element model (FEM) had 760 elements and 2487 nodes. As can be seen in Fig. 3, the mesh was refined in the meniscus area close to the liquid free surface where the maximum thermal gradients occur

3.1.1 Thermal analysis and results

A thermal analysis was performed considering the thermal flux acting on the inner surface and imposing convection on the outer surface of the mould to simulate a water cooling. In the FEM model 8-node elements were used. The thermal flux proposed in [14] was 50% increased to reach a maximum temperature close to 300 °C for which material parameters are available in [15]. The convection coefficient is 48000 W/m^{2}K and temperature of the cooling water is 40 °C. Variation of the thermal flux schematically presented in Fig. 2 was simulated by a sequence of steady state analyses. A nonlinear solution was carried out to simulate the temperature dependence of thermal properties.

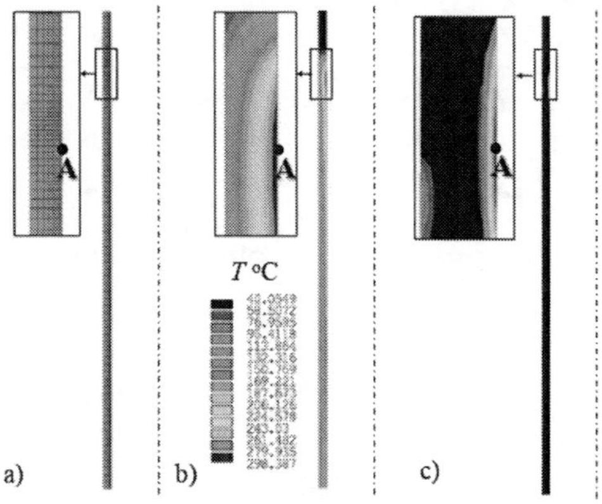

Fig. 3. FEM model a), temperature distribution b), and equivalent plastic strain c).

The highest temperatures and thermal gradients occur in the region close to meniscus. The maximum temperature 298 °C occurs in the position labelled with "A" (see Fig. 3b) that is also a representative point for the calculation presented in the following section.

3.1.2 Mechanical analysis and results

Temperature distribution calculated previously in the thermal analysis was the input data for the mechanical analysis. Since the component is free to expand, no mechanical constraints were imposed on the numerical model. The nonlinear mechanical analysis was performed assuming temperature dependence of material parameters and considering combined (nonlinear kinematic + nonlinear isotropic) model with parameters taken from [15].

The component is cyclically loaded until stabilization by using different material models. The comparison is performed considering the equivalent strain range $\Delta\varepsilon_{eq}$. Before proceeding with simulations, a criterion that defines when material stabilization actually occurs needs to be established.

Normalized maxima of the stress evolution as a function of the accumulated plastic strain are considered [7, 18]:

$$\frac{\sigma_{\max,i}-\sigma_{\max,1}}{\sigma_{\max,s}-\sigma_{\max,1}} \approx \frac{R}{R_\infty} = 1-\exp\left(-b\varepsilon_{\text{pl,acc}}\right) \quad (6)$$

where $\sigma_{\max,i}$, $\sigma_{\max,1}$ $\sigma_{\max,s}$ are the maximum stress for the Nth number of cycle, first and stabilized, respectively. Assuming that the plastic strain range $\Delta\varepsilon_{\text{pl}}$ is constant over cycles (see [7]):

$$\varepsilon_{\text{pl,acc}} \approx 2\Delta\varepsilon_{\text{pl}} N \quad (7)$$

the following relation proposed in [18] could be introduced:

$$2b\Delta\varepsilon_{\text{pl}} N \approx 5 \quad (8)$$

Eq. (8) assumes that a value close to 5 is good saturation criterion of the exponent in Eq. (6). The criterion adopted to determine the stabilized condition of a material is:

$$\frac{\sigma_{\max,i}-\sigma_{\max,1}}{\sigma_{\max,s}-\sigma_{\max,1}} \approx 0.993 \quad (9)$$

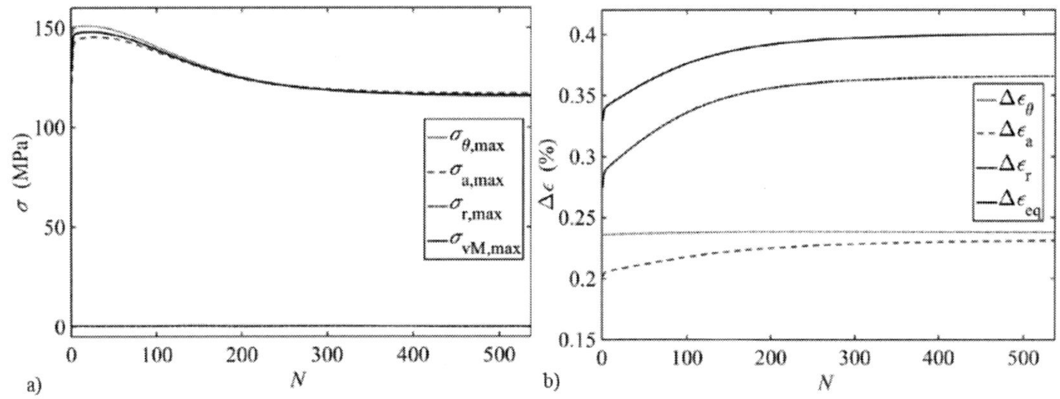

Fig. 4. Combined model: maximum stress a) and strain range b) versus number of cycles.

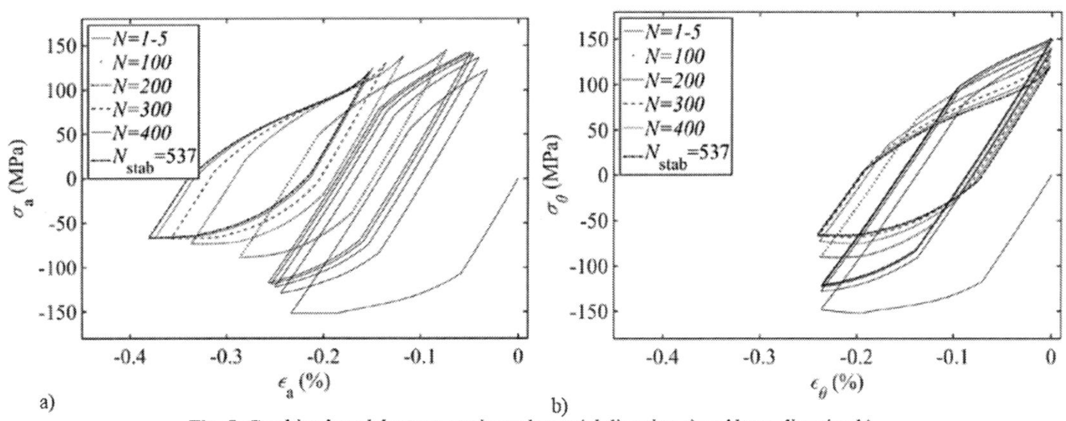

Fig. 5. Combined model: stress-strain cycles: axial direction a) and hoop direction b).

Results obtained with the combined model were compared with those achieved considering accelerated models with 8 different values of parameter b_a: 10b, 20b, 30b, 100b, 200b, 300b, 421b and 0.01b, covering a wider range with respect to that proposed in [6, 18]. Prager and nonlinear kinematic models with initial (E, σ_0) and stabilized (E_s, σ_{0*}) condition (where yield stress and Young's modulus were determined from 1st and stabilized cycle, respectively) were also implemented for comparison.

Firstly, the combined model was analysed. Adopting the stabilization criterion given by Eq. (9), the material stabilizes within 537 cycles. This results can be argued also from Fig. 4 and 6. Fig. 4a presents hoop, axial, radial and von Mises maximum stresses (computed at the critical point A) versus the number of cycles. It can be observed that a biaxial state of stress occurs. In Fig. 4b, the three components of strain range and the equivalent strain range, determined according to [4], versus the number of cycles are presented. It can be noticed that the hoop strain range is almost constant, whereas the other two components increase until stabilization. This behaviour can be clearly observed also in Fig. 5, where hoop and axial stress-strain evolution is presented. For the sake of clarity, only the first five cycles, the 100th, the 200th, the 300th, the 400th and the final stabilized cycles are presented.

Results obtained with several accelerated models are under investigation hereinafter. According to [6], stabilization of material can be reached in a shorter simulation time by increasing the original value of speed of stabilization b to b_a. On the other hand, in the same work [6] it was observed that in some cases, particularly when softening occurs, the accelerated solution can also be completely incorrect.

In this work a wide span of cases with different values of b_a were thus considered (b_a=10b, 20b, 30b, 100b, 200b, 300b). The lower bound case b_a=0.01b was also investigated. In this case, stabilization is reached in 8 cycles. This is due to the criterion defined by Eq. (9). Theoretically, models with speed of stabilization that tends towards zero requires a number of cycles to stabilization that tends towards infinite. Nevertheless, as in this case the contribution of the nonlinear isotropic model is negligible, cyclic decrease of stress from cycle to cycle is smaller than tolerance and therefore a fast stabilization seems to occur. The upper bound case (b_a=421b) was also investigated.

Fig. 6 shows a comparison between these cases in term of the maximum von Mises stress in each cycle. The number of cycles of each curve are normalized with respect to the corresponding number of cycles to stabilization N_{stab} in order to emphasize differences among the models occurring in first few cycles. Despite the different behaviour at the beginning, all the accelerated models reach almost the same value of von Mises stress. Obviously in the case b_a=0.01b a higher value of Von Mises stress is reached. In fact, as already explained, softening does not occur.

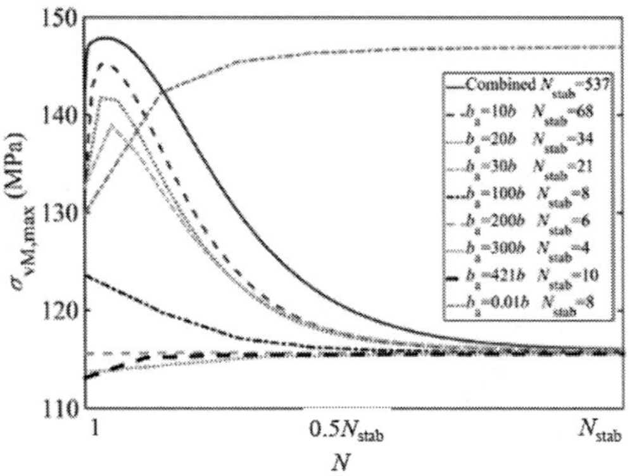

Fig. 6. Maximum von Mises stress evolution up to stabilization.

The evolution of the equivalent strain range from first to stabilized cycle is presented in Fig. 7. The equivalent strain range is calculated for all accelerated models and compared with results calculated considering the combined model (see Tab. 2). Results obtained with the combined model show a well know "*s-shaped*" curve which corresponds to the adopted nonlinear isotropic model. Quite similar "*s-shaped*" curves are also obtained with the accelerated models with a moderately increased speed of stabilization (b_a=10b÷30b), while this shape is lost for higher values of b_a (b_a=100b÷300b). Similar behaviour is observed also for the bound case b_a=0.01b, for which, as already shown, the model reaches stabilization in few cycles.

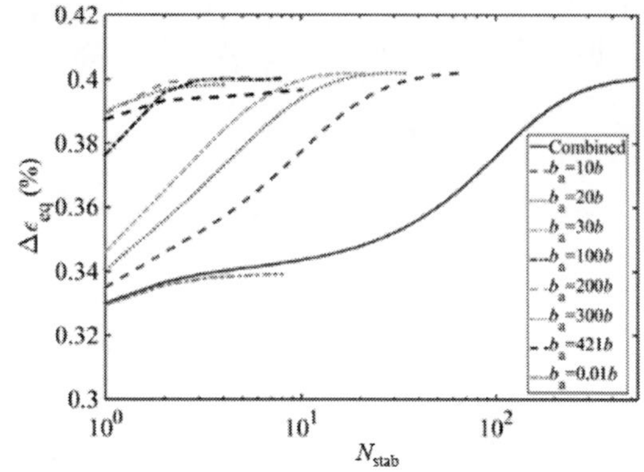

Fig. 7. Equivalent strain range versus number of cycles to stabilization.

In Fig. 7 also the case b_a=421b is reported, representing the upper bound. In fact, for higher values of b_a numerical problems do not permit the convergence to be reached. Due to the high speed of stabilization, the model wants to reach stabilization almost immediately somehow neglecting the nonlinear kinematic model that controls the monotonic hardening.

Fig. 8 shows the correlation between speed of stabilization and number of cycles to stabilization. As can be observed, plotting the data on a log-log scale gives almost a linear relation between b and N_{stab}. Accelerated models with b_a=0.01b and b_a=421b obviously deviate from this trend.

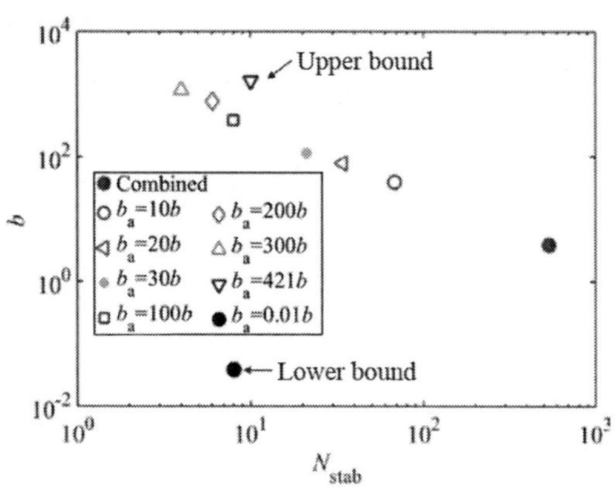

Fig. 8. Correlation between the speed of stabilization and number of cycles to stabilization.

Fig. 9 presents a "design diagram" calculated with Eq. (8) which correlates b, N and $\Delta\varepsilon_{pl}$, which is computed in the first cycles. All the three components of plastic strain range ($\Delta\varepsilon_{pl,a}$=0.0554% axial direction, $\Delta\varepsilon_{pl,\theta}$=0.0734% hoop direction and $\Delta\varepsilon_{pl,r}$=0.1289% radial direction) calculated by FE after the 5th cycle are considered and therefore three curves are presented.

Table 2. Number of cycles to stabilization and $\Delta\varepsilon_{eq}$ for point A (T=300 °C).

	Combined model	Accelerated models								Non. Kin initial	Non. Kin. stabilized	Prager model
		10b	20b	30b	100b	200b	300b	421b	0.01b			
N_{stab}	537	68	34	21	8	6	4	10	8	8	10	5
$\Delta\varepsilon_{eq}$ (%)	0.4003	0.4020	0.4019	0.4017	0.4002	0.4005	0.3984	0.3967	0.3391	0.3392	0.3658	0.3574
Δe (%)		0.41	0.40	0.36	-0.02	0.5	-0.48	-0.91	-15.30	-15.26	-8.60	-10.71

As can be noticed results obtained with all accelerated models except the two bound cases (b_a=0.01b and b_a=421b) show a good agreement with the adopted relation.

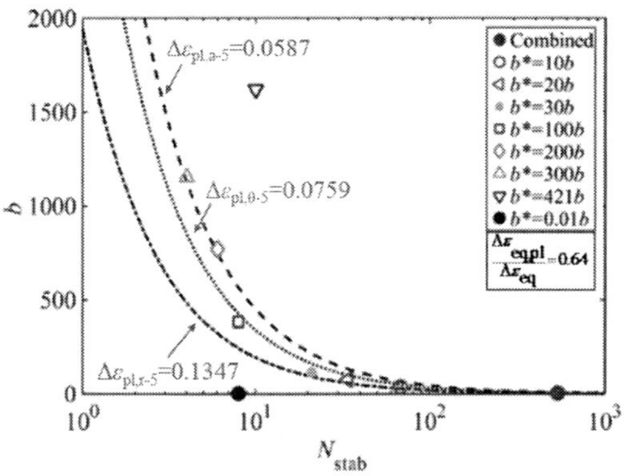

Fig. 9. Design diagram for T=300 °C.

As pointed out in Section 2, very often in literature it is suggested to perform the simulation only for few cycles. As shown in Fig. 7, this approach in this case would introduce a significant error. For instance, the value of the equivalent strain range obtained after 10 cycles is about 15% lower than the value obtained in stabilized condition. It follows that, entering in the Manson-Coffin curve [16], the number of cycles to failure would be overestimated. Also the usual approach of adopting the kinematic model in the stabilized condition could rise significant errors. Fig. 10 shows the equivalent strain range for all the adopted models. The case of the Prager and nonlinear kinematic (initial and stabilized) models also reported. The Prager model is often proposed in literature because parameter C_{lin} can be estimated by simply using monotonic uniaxial test. In the initial and stabilized nonlinear kinematic models the yield stress and the Young's modulus are estimated at the beginning of loading and at stabilization, respectively. Therefore, model parameters E and σ_0 are replaced by E_s and σ_{0*} estimated from stabilized cycles, while kinematic variables (C_i, γ_i) remain unaffected. All these approaches are based on a kinematic model and therefore capture only the monotonic hardening behaviour.

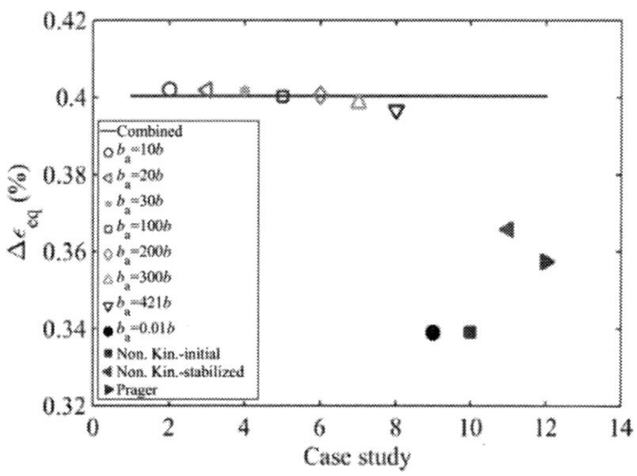

Fig. 10. Comparison between all adopted cyclic plasticity models in terms of equivalent strain range.

Tab. 2 summarizes the results shown in Fig.10. As can be noticed, neglecting the bound cases (b_a=0.01b and b_a=421b), $\Delta\varepsilon_{eq}$ does not show a dependence with the speed of stabilization and its value substantially coincides with that obtained with the combined model. A relative error Δe=($\Delta\varepsilon_{eq,a}$-$\Delta\varepsilon_{eq,c}$)/$\Delta\varepsilon_{eq,c}$ is calculated where $\Delta\varepsilon_{eq,a}$ and $\Delta\varepsilon_{eq,c}$ are the equivalent strain ranges for the considered accelerated and combined model, respectively. In the case of b_a=10b÷300b the error remains always in the range of -0.48÷0.5%. Result obtained with the nonlinear kinematic model that considers initial condition of a material (Non. Kin. initial) and neglects stabilized condition gives a high relative error (-15.26%) with respect to all adopted models. A quite similar error (-15.3%) is calculated with the accelerated model when b_a=0.01b, since the model stabilized after 8 cycles. Due to the strong simplification, Prager and nonlinear kinematic-stabilized (Non. Kin. stabilized) models provide, instead, a smaller but still significant relative error (-8.6% and -10.71%). This result is in good agreement with the conclusions in [6], where it was observed that the direct use of the stabilized model could lead to heavy mistakes. Based on the obtained results, it is possible to conclude that neglecting the initial as well as stabilized condition of material could give misleading results that can heavily affect durability assessment, in particular over-estimating the life of the component.

As the computation time is almost proportional to the number of cycles to stabilization, it is possible to conclude that the accelerated model permits a strong reduction of the computational effort, keeping the same accuracy as the combined model.

4 Design rule for practical use in FEM modelling

As Eq. (8) shows, for decreasing values of plastic strain range $\Delta\varepsilon_{pl}$, the number of cycles to reach stabilization N_{stab} increases proportionally. This case obviously falls in the "unfeasible" purple area depicted in Fig.1. It is of practical interest to assess a design rule to deal with these cases. For this purpose, the thermal flux proposed in [14] was only 30% increased to reach a maximum temperature close to 250 °C for which material properties are available.

Firstly, the numerical simulation with the combined model was performed up to the complete stabilization (1033 cycles) in order to have the reference case, see Tab. 3. The ratio between the equivalent plastic strain range $\Delta\varepsilon_{eq,pl}$ and the equivalent strain range $\Delta\varepsilon_{eq}$ was 0.57, while in the previous case this ratio was 0.64 (as expected, in the latter case the component is less plastically strained). The three components (axial, hoop and radial) of the plastic strain range computed after 5 cycles were used to evaluate the corresponding curves, according to Eq. (8), see Fig. 11.

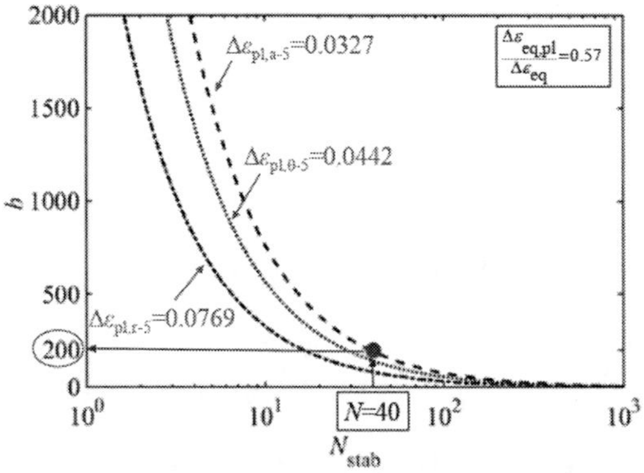

Fig. 11. Design diagram for T=250 °C.

Entering in the design diagram with the number of cycles to stabilization considered feasible to perform the numerical simulation, the speed of stabilization to be used in the accelerated model could be obtained. For example, assuming N_{stab}=40 gives b_a=200 (the curve relative to the axial strain component was used), see Fig 11. Running the FE analysis with the accelerated model with b_a=200 and performing a simulation of 40 cycles, a value of equivalent strain range $\Delta\varepsilon_{eq}$=0.3144% was obtained. With respect to the combined model, which requires to simulate 1033 cycles, in this case an error of only 2.27% was obtained with a huge reduction of the computational effort.

Table 3. Number of cycles to stabilization and $\Delta\varepsilon_{eq}$ for point A (T=250 °C).

	Combined model	Accelerated model
	b≈4	b_a=200
N_{stab}	1033	40
$\Delta\varepsilon_{eq}$ (%)	0.3075	0.3144
Δe (%)	–	2.27

5 Conclusions

The choice of the adopted cyclic plasticity model in numerical simulations is an important design aspect, particularly when dealing with components under cyclic thermo-mechanical loading. In principle, the material model should follow quite accurately the cyclic evolution of the material behaviour and the numerical simulation has to be performed up to stabilization. Quite often this goal is unfeasible and it is thus necessary to refer to simplified models or to acceleration techniques.

In this work, several models (combined, accelerated, Prager, nonlinear kinematic initial and stabilized models) have been considered and compared in terms of the equivalent strain range. Based on the obtained results, it is possible to conclude that the use of too simplified models (Prager, nonlinear kinematic initial and stabilized) neglecting initial or stabilized conditions may be inadequate, as they could lead to overestimate the fatigue life of the component. On the other hand, accelerated models give results that are always close to the fully-stabilized combined model, assumed as reference.

A design rule is finally proposed. According to this approach, particularly useful when the plastic strain range induced by the thermo-mechanical loads is relatively small, few cycles of simulation have to be firstly performed. According to the obtained value of plastic strain range and to the computational time, a curve can be calculated, permitting a suitable value of increased speed of stabilization to be introduced in the model to perform the final FEM analysis.

References

1. D. Benasciutti, F. De Bona, M. Gh. Munteanu, J. Strain Anal. Eng. Des. **51** (7), (2016)
2. L. Moro, D. Benasciutti, F. De Bona, M.Gh. Munteanu, Ironmak. Steelmak. DOI:10.1080/03019233.2017.1339482, (2017)
3. L. Moro, J. Srnec Novak, D. Benasciutti, F. De Bona, Ironmak. Steelmak. DOI: 10.1080/03019233.2017. 1371425, (2017)
4. S.S. Manson, *Thermal stress and low-cycle fatigue* (McGraw-Hill book company, 1966)
5. J. Srnec Novak, D. Benasciutti, F. De Bona, A. Stanojevic, P. Huter, Procedia Eng. **133** (2015)

6. J.L. Chaboche, G. Cailletaud. Comput. Struct. **23** (1986)
7. J. Lemaitre, J.L. Chaboche, *Mechanics of solid materials* (Cambridge University Press, Cambridge, 1990)
8. V.K. Arya, M.E. Melis, G.R. Halford, NASA Technical memorandum 102342, (1990)
9. P. Dufrenoy, D. Weichert, J. Therm. Stresses **26**, (2003)
10. S. Amiable, S. Chapuliot, A. Constantinescu, A. Fissolo, Fatigue Fract. Eng. Mater. Struct. **29**, (2006)
11. B. Li, L. Reis, M. de Freitas, Int. J. Fatigue **28**, (2008)
12. A. Campagnolo, F. Berto, C. Marangon, Theor. Appl. Fract. Mec. **81**, (2016)
13. C. Kontermann, A. Scholz, M. Oechsner, Mater. High Temp. **31**, (2014)
14. P. Galdiz, J. Palacios, B.G. Thomas, ECCC, (2014).
15. D. Benasciutti, J. Srnec Novak, L. Moro, F. De Bona, A. Stanojevic, Fatigue Fract. Eng. Mater. Struct. (published online) DOI: 10.1111/ffe.12783, (2018).
16. D. Benasciutti, J. Srnec Novak, L. Moro, F. De Bona, Fatigue Fracture Eng. Mater. Struct (published online) DOI: 10.1111/ffe.12781, (2018)
17. J. Srnec Novak, D. Benasciutti, F. De Bona, A. Stanojevic, A. De Luca, Y. Raffaglio, IOP Conf. Ser: Mater. Sci. Eng. **119**, (2016)
18. J.L. Chaboche, Int. J. Plasticity **24,** (2008)

The influence of tension mean stress to the crack initiation process at the very high cycle fatigue regime for a high strength steel

Thomas Rohm[1,2], *Karl-Heinz* Lang[1,*], and *Marcus* Korn[1]

[1]Karlsruhe Institute of Technology (KIT) Institute for Applied Materials (IAM), Engelbert-Arnold-Str. 4, 76131 Karlsruhe, Germany
[2]Karlsruhe University of Applied Sciences, Moltkestr. 30, 76133 Karlsruhe, Germany

Abstract. Strain controlled tests were performed on a high strength steel to investigate the influence of tension mean stress to the crack initiation process at the very high cycle fatigue regime. It is supposed that the variation of stress the state in the vicinity of a non-metallic inclusion results in a change to the energy transferred to the surrounding metallic matrix. To prove this, incremental step tests on cylindrical specimen were realized in which the tension mean stress was stepwise increased. It could be verified that the hysteresis area, which is a criterion of the work transferred to the specimen, decreases with increasing tension mean stress. The test results provide a reasonable physical model to estimate the influence of tension mean stress to the crack initiation process in the very high cycle fatigue regime.

1 INTRODUCTION

Some investigators like [1, 2, 3] have shown, that compressive residual stresses at the surface of specimens due to local plastic deformation or local plastic deformation and reduction of surface roughness, increases the lifetime also in the very high cycle regime. From these results it is known, that the compressive mean stress inhibits the crack initiation at or close to the surface, until the influence is compensated by crack initiation in the volume of the specimen. **A specific influence resulted from the rate of tension residual stress at the specimen volume was not mentioned.**

Considering the residual stress using the formula of Murakami [4] for the stress intensity factor (SIF) K_{max}, a particular effect can be assigned. To calculate K_{max} the following equitation (1) is used, with $\Delta\sigma/2$ = nominal stress and σ^{RS} = residual stress in MPa, S_{xz}= inclusion area in m^2 and $f = 0.5$ for subsurface defects or $f = 0.65$ for surface defects.

$$K_{max} = f \cdot (\Delta\sigma/2 + \sigma^{RS}) \cdot \sqrt{\pi\sqrt{S_{xz}}} \qquad (1)$$

Tension residual stresses increase the stress intensity factor at an inclusion. Therefore, the size of the optically dark area (ODA) [4] which is necessary to achieve the long crack growth threshold value is smaller as without residual stresses or even larger if compressive residual stresses are present as illustrated in fig. 1. But a different size of the ODA is related to a different lifetime of a specimen failing in the VHCF-regime [5, 6]. Accordingly, a modification of the residual stresses in the volume of a specimen will also cause a change in the lifetime. Higher tension residual stresses in the volume of a specimen due to higher or deeper reaching compressive residual stresses at the surface should result in a shorter lifetime at the very high cycle fatigue regime because in this lifetime regime crack initiation typically takes place in the volume of the specimen where tensile residual stresses are effective.

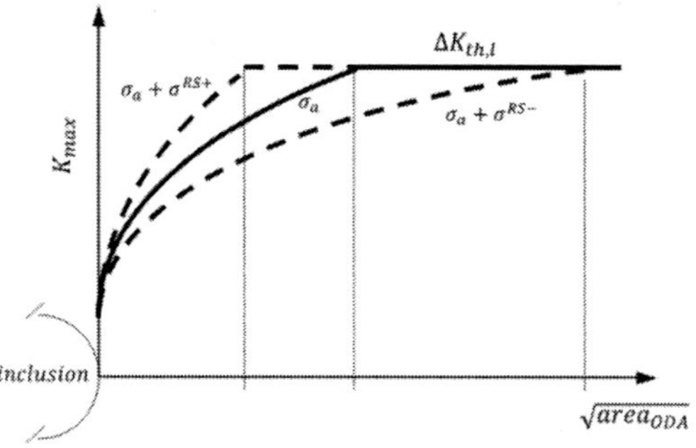

Fig. 1. Influence of residual stresses to the dimension of ODA until the long crack growth threshold is reached.

In [7] is reported, that a lager compressive residual stress fields resulting from shot peening lead to longer lifetime in the range between $N = 10^7$ to $N = 10^9$ cycles compared to slight compressive residual stresses produced by mechanical polishing. Similar results are reported in [8] where macro shot peened specimens exhibited longer lifetimes compared to micro peened specimens. Fig. 2 shows fatigue test results in the VHCF-regime [8], that if crack initiation at an inclusion in the volume with a formation of an ODA is present, higher residual tensile stresses in the volume as a result of higher compressive residual stress near the specimen surface dose not reduce the lifetime. These results are in contrast to the usual considerations of the impact of residual stresses to the lifetime behavior. This is due to the fact that residual stresses have different effects in the different phases of fatigue. Tensile residual stresses accelerate the growth of macroscopic cracks and thus reduce the lifetime. But in the VHCF-regime with

* Corresponding author: karl-heinz.lang@kit.edu

formation of an ODA at an inclusion the proportion of long crack growth of the total fatigue life is marginal [9]. Therefore, the influence of residual stresses on the crack initiation process (including the formation of ODAs) and on short fatigue crack growth must be considered. The impact of residual stresses in these phases can differ greatly from the effect on the long crack growth phase.

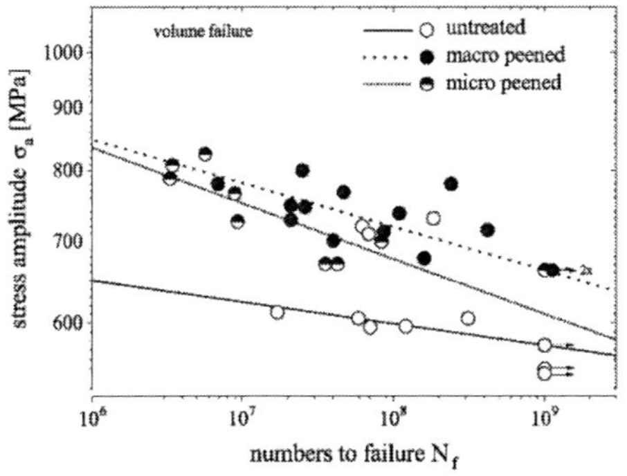

Fig. 2. Endurance test with the steel AISI 4140, R_m = 2145 MPa under tension compression load at $R = -1$. Surface condition: untreated, macro peened (0.29 mmA) and micro peened (glas beeds) [8].

The purpose of the present study is a selective evaluation of the impact of tension residual stresses on the crack initiation process. It is supposed that tension residual stresses which are effective in the surrounding of a non- metallic inclusion, where cracks are initiated in the VHCF-regime for high strength steels [10], leads to a reduction of the energy transferred to the specimen by the cyclic loading. Most of this work will dissipate by heat, but a small amount will be stored in the material, cause microstructural changes and thereby, induce fatigue damage in the material [11]. Because the crack initiation process in the VHCF-regime may take up to 99 % of the lifetime [9], the influence of residual stresses to this process dominates the lifetime behavior. Therefore, the working hypothesis is that tension residual stresses may extend the lifetime in the VHCF-regime by reducing the energy transferred to the matrix surrounding a non-metallic inclusion.

For the following tests specific assumptions are necessary: For the used steel in this study AlCaO oxides which have a globular shape are typically responsible for crack initiation [12]. It is assumed, that the shape of these inclusions is ideal spherical and that they are detached from the metallic matrix [13]. Therefore, the loading condition in the surrounding matrix is comparable to a notch when tensile stresses are acting. In the case of compressive stresses the existing inclusion can transfer loading and only minor effects of stress concentration appear as long as the different elastic properties of the matrix and the inclusion are neglected. For further information see [14]. Thus the external load ration at the specimen of $R = -1$ results in a higher load ration of about $R = -0.5$ at the inclusion if an elastic stress concentration of K_t = 2.045 for spherical cavity and with a Poisson's ratio $\nu = 0.3$ are assumed [15]. The influence of different residual stresses will be emulated by the variation of the mean load. In the case of over-elastic deformation the loading conditions in the notch root is determined by the elastic behavior of the non-plasticizing regions. Therefore, total strain control is required to emulate the loading conditions in the vicinity of a non-metallic inclusion on a smooth specimen.

2 EXPERIMENTAL PROCEDURES

The test material used in this study is the low alloyed steel 42CrMo4 (AISI 4140) in quenched and tempered condition. The chemical composition (mass percentage) of this steel is: 0.422 C, 1.062 Cr, 0.851 Mn, 0.162 Mo, 0.299 Si, 0.021 S, 0.016 P and balance Fe. The specimens were machined from a 15 mm rolled round bar to cylindrical specimens with a gauge section length of 10 mm and a diameter of 4 mm. The surface roughness was R_z = 2.79 µm ± 0.172 µm (average and standard deviation). After machining the specimens were austenitised at 850 °C in vacuum for 20 min, oil quenched and tempered at 180 °C for 180 min. The hardness was 611 ±11 HV0.1 (average and standard deviation), the ultimate tensile strength was 2145 MPa and the yield strength was 1730 MPa. To get an effective determination of the cyclic stress-strain-curve of the material in a low amount of testing time, the incremental step test is a common used investigation [16]. Because the influence of tension stress is focused, which is considered to be equivalent to mean stress, the test procedure is modified. Constant strain amplitude with variable mean strain will be executed and sequentially repeated to stabilize the stress-strain behaviour of the material.

The constant strain amplitude was $\varepsilon_a = 0.6$ %, the maximum total mean strain was $\varepsilon_{m,t} = 0.5$ % and for the test a frequency of $f = 0.5$ Hz was applied. To avoid an early failure of the specimen, the maximum total strain was chosen at a level, were a sufficient number of cycles are possible [17]. Theses investigations showed, that the material tends to cyclic softening until the final fracture, however it is not very distinct. The main softening process is the resolution of coherent ε-carbide precipitates due to cyclic loading, therefore an increased movement of slip of dislocations is possible. But this softening effect is limited, as result of [17] show. It is expected that the stabilization of the cyclic softening will develop after 10 or more sequence cycles. While the strain amplitude was constant, the total mean strain was stepwise exceeded. This variation represents the mean strain, which exists because of additional stress, likewise residual stress. At each mean strain level fife loops were performed to get closed hysteresis loops for the data evaluation. The area of the stress-strain hysteresis loop is equivalent to the transferred energy. Fig. 3 explores one sequence schematically and the test set up at the specimen for stain measurement.

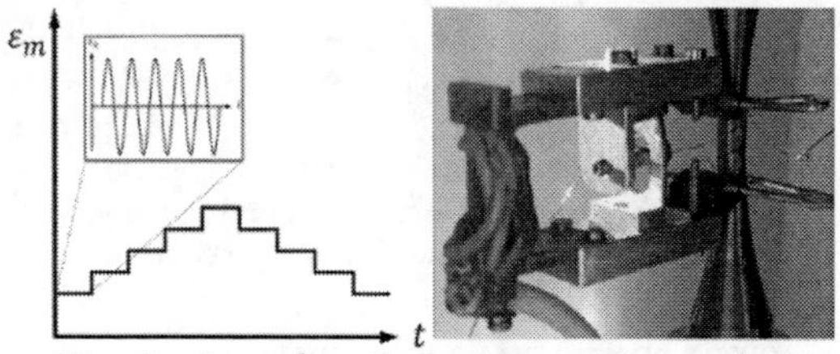

Fig. 3. One schematically test sequence and stain measurement setup at the specimen.

The stabilization was monitored comparing the maximum values of stress-strain-curves at the sequences. While the softening at the beginning was significant, after the 11th repetition the following sequences showed minor softening, so these sequences were evaluated. The stress and strain data were stored on a computer and automatically evaluated.

3 EXPERIMENTAL RESULTS AND DISCUSSION

Evaluating the data of the 13th sequence the plastic strain amplitude of the hysteresis loops is linear increasing with higher total mean strain, as fig. 4 illustrates. At the same time stress amplitude is decreasing, but with higher mean strain the reduction is clearly larger.

At higher mean stress the maximum stress exceeds the cyclic yield strength and hardening mechanisms occur. With increasing maximum total strain, a higher number of slip dislocations at different levels are activated at the polycrystalline material. The interaction of the dislocations and the pile up at micro structural barriers, like coherent ε-carbide precipitations, create a less increase of hardening. An extension of external stress enables dislocations to circumvent the carbides and cross slip enables their extinction.

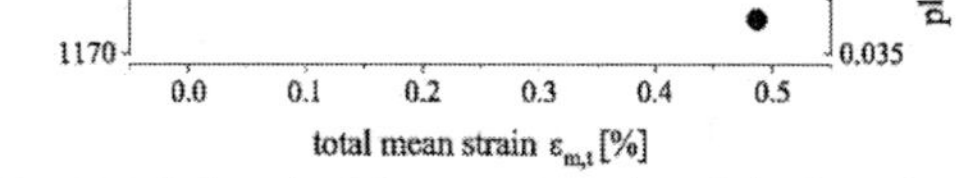

Fig. 4. Variation of maximum stress $\Delta\sigma/2$ and plastic strain amplitude $\Delta\varepsilon/2$ at different total mean strain.

As a result of cyclic stress-strain behavior of the material, the hysteresis loop area is increasing as the plastic strain amplitude increases, as fig. 5 illustrates. There the hysteresis area represents the specific work W which is transferred at the closed loop. At a total mean strain of $\varepsilon_{m,t} = 0.3$ % and a strain ratio of $R = -0.32$ the hysteresis area is unexpectedly decreasing. Macroscopic voids at the material surface could not be found and the analysis of the specimen stiffness provides not a reasonable explanation for this phenomenon.

It is supposed that a specific material softening is necessary to raise this effect. While the plastic strain is increasing constantly, an over proportional decrease of the stress amplitude with higher mean strain, results in a smaller hysteresis loop area.

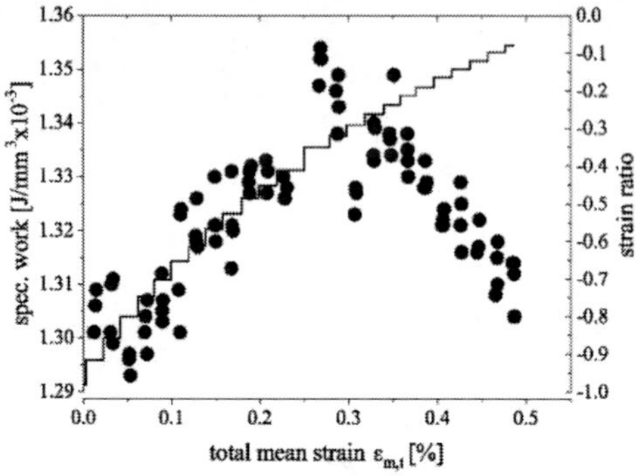

Fig. 5. Specific work W transferred of the hysteresis loop and strain ratio R depending on increasing total mean strain $\varepsilon_{m,t}$.

In contrary to the influence of tensile mean stress, where the growth of macroscopic cracks is accelerated and live time is decreased [18], this findings are evident suggests that the reduced hysteresis area could result in an increase of life time in the VHCF-regime. So less energy is transferred in the metallic matrix at the vicinity of the non-metallic inclusion per load cycle and a higher number of cycles is necessary to achieve the same amount of energy to initiate the crack growth.

A quantitative relationship between the minor specific work at the hysteresis loop, caused by increased mean strain and extended lifetime, derived from this results, cannot be provided at the moment. Further tests are necessary to investigated how much energy is dissipated within heat and how much is stored at the material for the crack initiation process in the VHCF-regime. With this information a more detailed correlation between residual tension stress and lifetime for at non-metallic inclusions initiated cracks with ODA-formation, could be realized.

4 SUMMARY

Within the scope of this research, it was shown that there is a sufficient influence of the tension residual stress which could result in a reduced transferred energy in the surrounding material of a spherical non-metallic inclusion. A significant incident could be identified that with increasing mean strain and constant strain amplitude less energy is transferred in the material. While the crack initiation process with ODA-formation is dominant in the VHCF-regime, the minor specific work per cycle can prolong the life time. Fig. 6 illustrates the influence of compressive residual stress near the surface and tension residual stress in the volume of the specimen, to the fatigue behavior in the VHCF-regime. The compressive stresses inhibit the crack initiation at or close to the surface (S) so a failure from the volume (VO) of the specimen can occur. But with higher compression residual stresses and therefor higher tension stresses at the volume, there will be an additional prolongation of lifetime, when ODA-formation in the VHCF-regime is present.

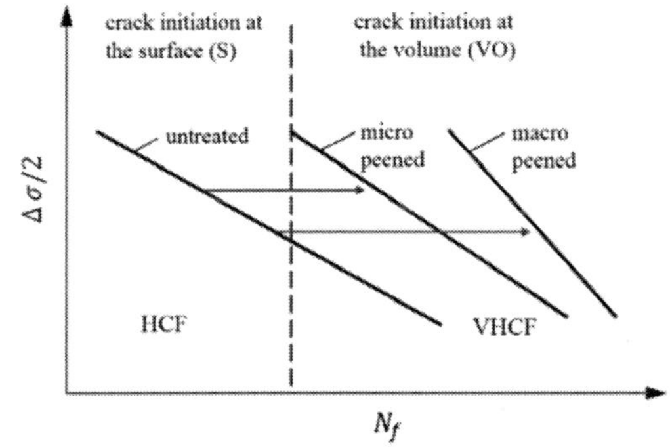

Fig. 6. Influence to the lifetime in the VHCF-regime due to low amount of residual stress by a procedure of micro peening and high amount of residual stress by macro peening.

This study provides a physically proven basis to assess the influence of mean stress in a new way for the VHCF-regime. The material behaviour is not fully investigated. The formation of the dislocations would be a helpful investigation to understand this phenomenon in detail. But this material configuration has already a very high dislocation density so any change is complex to detect and the evaluation would have a high uncertainty. Further investigations with a model material, like copper or pure iron are necessary. In an ideal material behaviour the softening would be constant with increasing mean strain so this influence could be eliminated. Also tests with minor changes of the stress ratio would be interesting, because results for higher stress ratio already exists [19].

ACKNOWLEDGEMENTS

The authors would like to thank the German Research Foundation (DFG) for the financial support at the SPP1466.

REFERENCES

1. C. R. Sohar, A. Betzwar-Kotas, C. Gierl, B. Weiss, H. Danninger, Materialwissenschaft und Werkstofftechnik **39** 248–257 (2008)
2. I.-S. Cho, A. Amanov, D.-W. Kim, H.-J. Park, B.-H. Kim, C.-S. Lee, Y.-S. Pyoun, I.-G. Park, VHCF5 341–346 (2011)
3. Y. Shimatani, K. Shiozawa, T. Nakada, T. Yoshimoto, Procedia Engineering **2** 873 – 882 (2010)
4. Y. Murakami, Metal Fatigue, (2002)
5. Y. Murakami, T. Nomoto, T. Ueda, Y. Murakami, Fatigue & Fracture of Engineering Materials & Structures **23** 893–902 (2000)
6. K. Shiozawa, L. Lu, Fatigue & Fracture of Engineering Materials & Structures **25** 813–822 (2002)
7. N. Shiraki, S. Abe, Y. Sugiyama, C. Braham, VHCF3, 601–608 (2004)
8. M. Korn, T. Rohm, K.-H. Lang, Springer Spektrum, (to be published)
9. Q. Wang, C. Bathias, N. Kawagoishi, Q. Chen, IJF **24** 1269–1274 (2002)
10. Y. Murakami, S. Kodama, S. Konuma, IJF **11** 291–298 (1989)
11. M. Bever, D. Holt, A. Titchener, Progress in Materials Science 5–177 **17** (1973)
12. K.-H. Lang, M. Korn, T. Rohm, Procedia Structural Integrity **2** 1133–1142 (2016)
13. Y. Murakami, N. N. Yokoyama, J. Nagata, Fatigue & Fracture of Engineering Materials & Structures **25** 735–746 (2002)
14. A. Stienon, A. Fazekas, J.-Y. Buffiere, Materials Science and Engineering: A **513–514** 376–383 (2009)
15. S. Timoshenko, Theory of elasticity, Engineering societies monographs, (1951)
16. H.-J. Christ, Werkstoff-Forschung und Technik **9** (1991)
17. P. N. Thielen, M. E. Fine, R. A. Fournelle, Acta Metallurgica **24** 1–10 (1976)
18. R. Stephens, A. Fatemi, R. Stephens, H. Fuchs, Metal Fatigue in Engineering, (2000)
19. Y. Furuya, T. Abe, Materials & Design **32** 1101–1107 (2011)

The influential factors on very high cycle fatigue testing results

*PENG Wen-jie** *XUE Huan*** *GE rui* *PENG zhou*
Wuhan Branch of Baosteel Central Research Institute
(R&D Center of Wuhan Iron & Steel Co.,Ltd.), Wuhan 430080, China

Abstract. When the fatigue cycle is extended from high cycle (10^5~10^7) to very high cycle (10^7~10^9), the fatigue testing results will be more sensitive to the influential factors such as specimen size, specimen surface roughness and the inclusion size. The influential factors on the very high cycle fatigue testing results are investigated in the present paper. Firstly, the design and control method for ultrasonic fatigue test were introduced for several specimen shapes. The effect of the shape, size and the surface roughness of specimen on the ultrasonic fatigue test results are investigated. Meanwhile, the effect of test frequency and the size of the inclusion on the fatigue test results are also investigated. It is concluded that: 1. the test results of specimen with different shape and size differ with each other, due to the risk volume is different. 2. There is a critical roughness for the specimen, depending on the hardness of tested metallic material. A larger roughness than the critical one will lead to a premature fracture. 3. The frequency effect is obvious for the low strength steel, however, is prone to vanish for the very high strength steel. 4. The very high cycle fatigue will be more sensitive to the inclusion size as the strength increases and the *S/N* curve character is strongly related to the size of the inclusion.

KEY WORDS: Very high cycle fatigue; size effect; frequency effect; inclusion size; surface roughness

Introduction

In industrial applications, many components are subjected to high-frequency, low-amplitude, very high cycle loading (10^8,10^9 or 10^{10}) in their working service [1-4]. Traditional 10^7 fatigue testing can't meet actual requirements. In order to investigate the fatigue properties of the components failure in very high cycle regime, ultrasonic fatigue test system had been developed and it offers an alternative and effective method for very high cycle fatigue testing [5-6].

For low cycle (LCF) and high cycle fatigue (HCF), the applied stress is relatively high and the failure is prone to initiate from surface. However, for very high cycle fatigue (VHCF), the cracks tend to start at internal defects or inclusions [7-8]. Consequently, the test results are more sensitive to the specimen size and the surface roughness. Meanwhile, when the fatigue cycle extends from high cycle to very high cycle, the fatigue property will be mainly related to the defect size of the specimen and the size of the defect will affect the *S-N* curve characteristics. Furthermore, when the fatigue frequency increased from usual frequencies to ultrasonic frequency, applying the ultrasonic test results directly may lead to unsafe results.

Therefore, in the present paper, a lot of fatigue experiments are conducted to investigate the effect of the specimen shape, size, surface roughness and loading

Corresponding author, *E-mail: pengwenjie@baosteel.com, ** E-mail: xue_huan@baosteel.com frequency on the test results. Besides, the relationship of the size of defect and the *S-N* curve characteristics are validated through the fatigue test results and the fracture surface scanning via SEM.

1. The design of the ultrasonic fatigue specimen

For ultrasonic fatigue machine, the vibration of the specimen is induced by a piezo-ceramic converter, which generates acoustical waves in the specimen through a power concentrator (horn) to obtain desired displacement and an amplification of the stress. The ultrasonic fatigue specimen must meet the condition of resonance with the experimental system.

Assuming the specimen axis is x axis, $U(x,t)$ is the longitudinal displacement of the section of the specimen at the t time, the resonant length of an ultrasonic fatigue specimen can be derived.

One-dimensional longitudinal wave equation is:

$$\frac{\partial^2 U(x,t)}{\partial t^2}=\frac{E}{\rho}\left(\frac{\partial^2 U(x)}{\partial x^2}+\frac{S'(x)}{S(x)}\frac{\partial U(x,t)}{\partial x}\right) \quad (1)$$

Assuming the specimen meeting the resonance condition, $U(x, t)$ can be separated as $U(x, t) = U(x)e^{iwt}$, Eq.1 can be rewritten as:

$$\frac{\partial^2 U(x)}{\partial x^2}+P(x)\frac{\partial U(x)}{\partial x}+k^2U(x)=0 \quad (2)$$

Where, $U(x)$ is the displacement formation of the specimen, $P(x)=\frac{S'(x)}{S(x)}$; $k=\frac{\omega}{c}$, c is the speed of the ultrasonic wave, $c=\sqrt{\frac{E}{\rho}}$, E is the tensile modulus; $\omega=2\pi f$, f is the frequency of the testing system.

Four different shape specimens are shown in Fig.1. The analytical solution of resonant length can be obtained according to the boundary conditions, continuity condition and the formation of the specimen cross section. The stress distribution and the maximum stress can be obtained via the derivative of the displacement function [9-10].

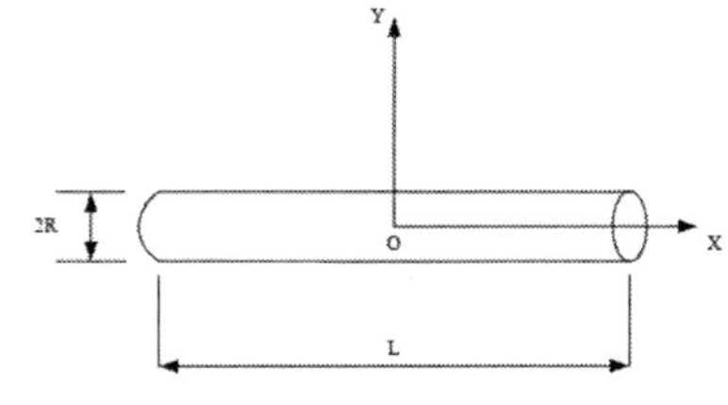

Fig.1a. Constant cross-section specimen

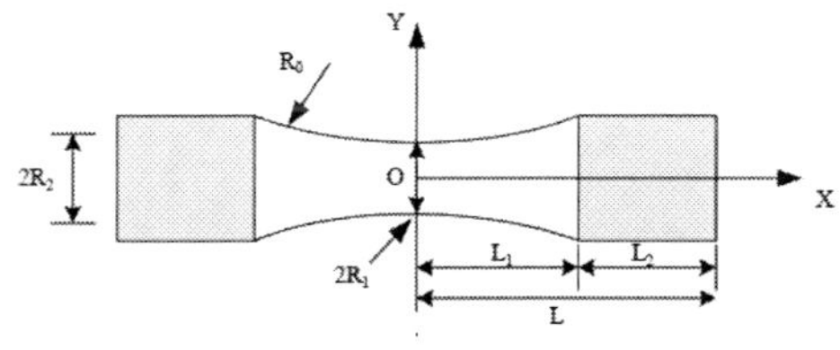

Fig.1b. Sandglass specimen

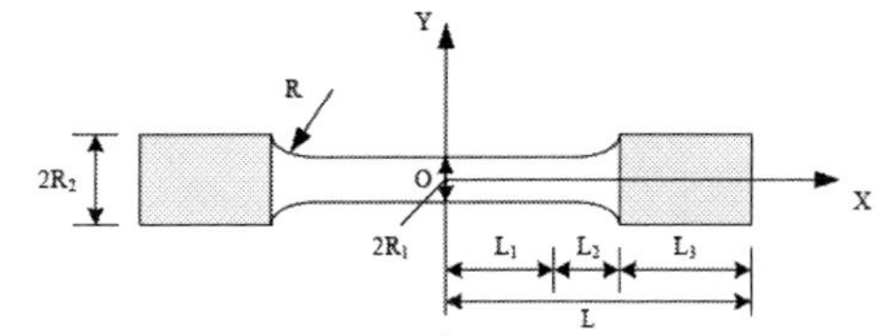

Fig.1c. Dog-bone specimen

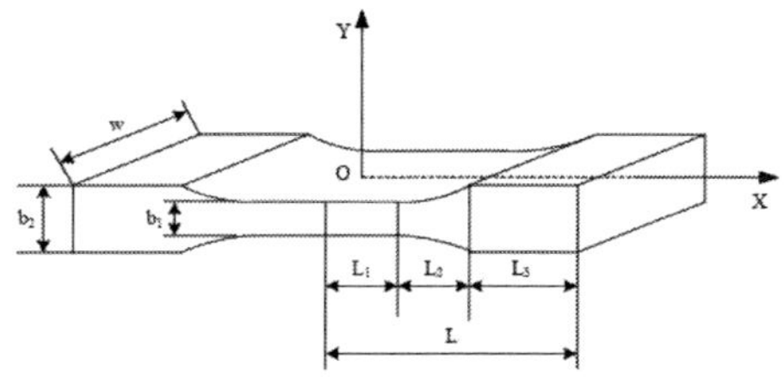

Fig.1d. Thin-plate specimen

1.1. The analytical solution of resonant length L of constant cross-section specimen in Fig.1a is as follows:

$$L=\frac{1}{2f}\sqrt{\frac{E}{\rho}} \quad (3)$$

The stress distribution formation of constant cross-section specimen is: $\sigma=E\frac{\partial U(x)}{\partial x}$ and the maximum stress is $\sigma_{max}=EkU_0$.

1.2. The analytical solution of resonant length L_2 of sandglass specimen in Fig.1b is as follows:

$$L_2=\frac{1}{k}\arctan\left\{\frac{1}{k}\left[\frac{\beta}{\tanh(\beta L_1)}-\alpha\tanh(\alpha L_1)\right]\right\} \quad (4)$$

Where, $\alpha=\frac{1}{L_1}\operatorname{arccos}h(\frac{R_2}{R_1})$, $\beta=\sqrt{\alpha^2-k^2}$.

The maximum stress is

$$\sigma_{max} = E\frac{\partial U(x)}{\partial x}\Big|_{x=0} = \beta E U_0 \varphi(L_1, L_2), \quad (5)$$

$$\varphi(L_1, L_2) = \frac{\cos(kL_2)\cosh(\alpha L_1)}{\sinh(\beta L_1)}$$

1.3. The analytical solution of resonant length L_3 of dog-bone in Fig.1c is as follows:

$$L_3 = \frac{1}{k}\arctan\left[\frac{1+\frac{\beta}{k}\tan(kL_1)\tanh(\beta L_2)}{\tan(kL_1)+\frac{k}{\beta}\tanh(\beta L_2)} - \frac{\alpha}{k}\tanh(\alpha L_2)\right] \quad (6)$$

Where, $k=\frac{\omega}{c}$, $c=\sqrt{\frac{E}{\rho}}$, $\omega = 2\pi f$,

$\alpha = \frac{1}{L_2}\operatorname{arccosh}(\frac{R_2}{R_1})$, $\beta = \sqrt{\alpha^2 - k^2}$

The maximum stress is: $\sigma_{max} = E\beta k\varphi(L_1, L_2, L_3)U_0$ (7)

Where,

$$\varphi(L_1, L_2, L_3) = \frac{\cos(kL_3)\cosh(\alpha L_2)}{\beta\sin(kL_1)\cosh(\beta L_2) + k\cos(kL_1)\sinh(\beta L_2)}$$

1.4. The analytical solution of resonant length L_3 of thin-plate specimen in Fig.1d is as follows:

$$\begin{cases} \alpha_1 = \frac{1}{2L_2}\ln\left(\frac{b_2}{b_1}\right), \ \beta_1 = \sqrt{\alpha_1^2 - k^2} \\ \varphi(L_1, L_2, L_3) = \frac{\cos(kL_3)\exp(\alpha_1 L_2)}{[\alpha_1\sin(kL_1) + k\cos(kL_1)]\sinh(\beta_1 L_2) + \beta_1\sin(kL_1)\cosh(\beta_1 L_2)} \end{cases} \quad (8)$$

Where,

$$L_3 = \frac{1}{k}\arctan\left\{\frac{\cos(kL_1)[\beta_1\cosh(\beta_1 L_2) - \alpha_1\sinh(\beta_1 L_2)] - k\sin(kL_1)\sinh(\beta_1 L_2)}{[\alpha_1\sin(kL_1) + k\cos(kL_1)]\sinh(\beta_1 L_2) + \beta_1\sin(kL_1)\cosh(\beta_1 L_2)}\right\} \quad (9)$$

The maximum stress is

$$\sigma_{max} = E\beta_1\varphi(L_1, L_2, L_3)kU_0 \quad (10)$$

The control software of the commercial ultrasonic fatigue testing machine USF-2000 contains only sandglass specimen. Nevertheless, the stress amplitude of other shape specimen can be converted to the stress amplitude of the sandglass specimen according to the principle of the ultrasonic fatigue testing machine [10]. For example, the stress amplitude of thin-plate specimen in Fig.1d can be converted via Eq.(11):

$$\sigma^0_{max} = \beta\varphi_0(L^0_1, L^0_2)\frac{\sigma_{max}}{\beta_1\varphi(L_1, L_2, L_3)k} \quad (11)$$

Where, σ_{max} is stress amplitude of thin-plate specimen, σ^0_{max} is stress amplitude of sandglass specimen under the same displacement amplitude condition.

2. The effect of the specimen shape and size

2.1 The effect of specimen shape

In order to investigate the effect of specimen shape on the ultrasonic fatigue test results, the same low carbon steel are processed into thin-plate specimen and sandglass specimen respectively, shown in Fig.2. The minimum thickness of the plate is 3mm and the minimum diameter of the sandglass specimen is also 3mm. The comparing test results are shown in Fig.3. Fig.3 shows that the fatigue lifetime of the sandglass specimen is much longer than thin-plate specimen under the same stress amplitude condition.

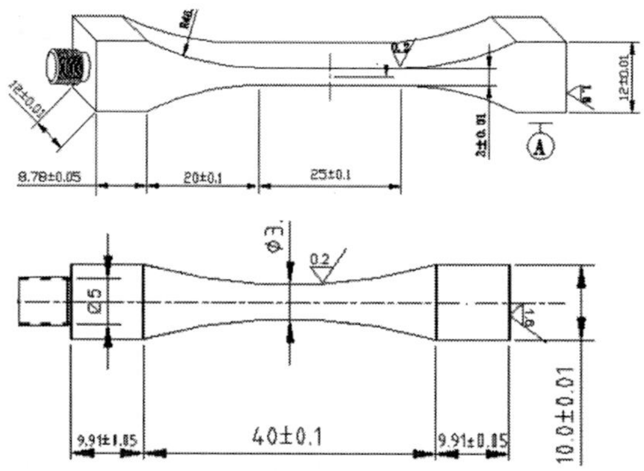

Fig.2. Thin-plate and sandglass specimen for comparing test

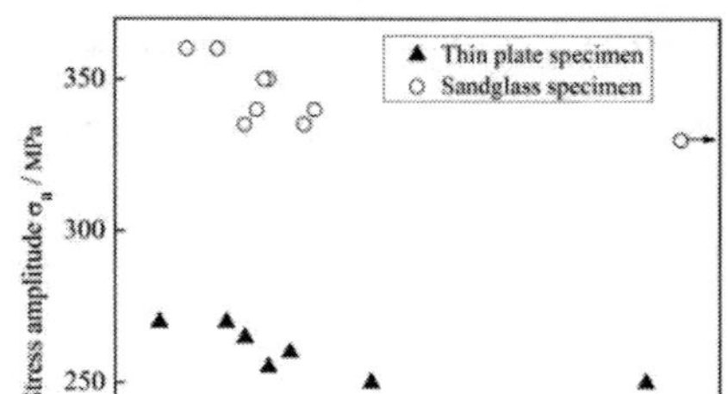

Fig.3.Test result of thin-plate and sandglass specimen

Dog-bone specimen and sandglass specimen of the same material shown in Fig.4 are also compared. The minimum diameter of the dog-bone specimen and the sandglass specimen are both 3mm.The comparing test results are shown in Fig.5. Fig.5 shows that the fatigue lifetime of the sandglass specimen is much longer than dog-bone specimen under the same stress amplitude condition.

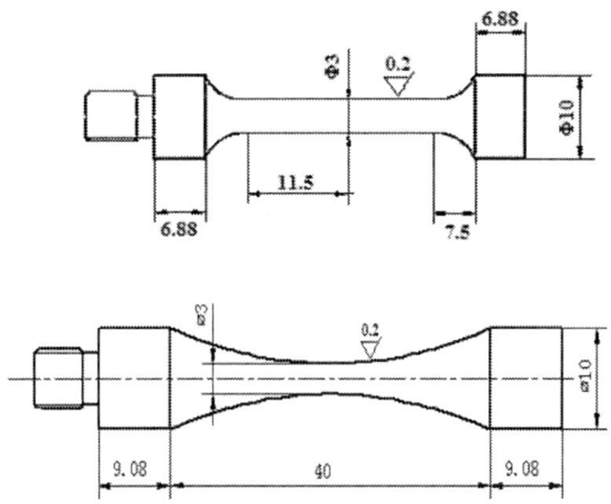

Fig.4. Dog-bone and sandglass specimen for comparing test

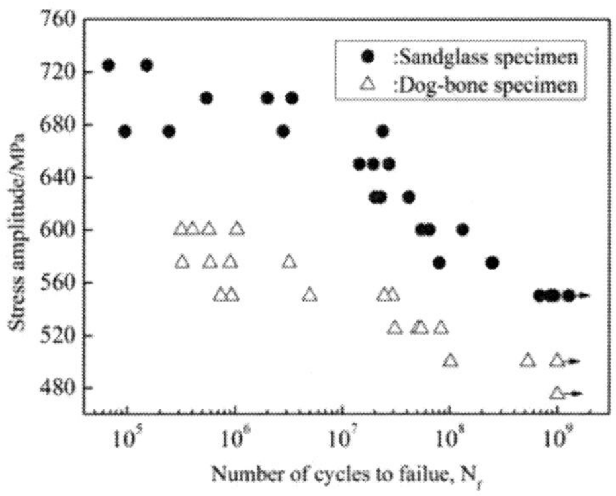

Fig.5. Test result of dog-bone and sandglass specimen

2.2 The effect of specimen size

In order to investigate the size effect on the ultrasonic fatigue property, ultrasonic fatigue tests are conducted on two sandglass specimen as shown in Fig.6 and the test results are compared. A specimen and B specimen have the same minimum cross-sectional area. However, the transition arc length of A specimen is longer than B specimen.

The comparing test results are shown in Fig.7. Fig.7 shows that for the low cycle region, the difference is not obvious. However for high cycle and very high cycle region, the lifetime of A specimen is shorter than B specimen under the same stress amplitude condition. Besides, the fatigue limit of the A specimen is lower than B specimen.

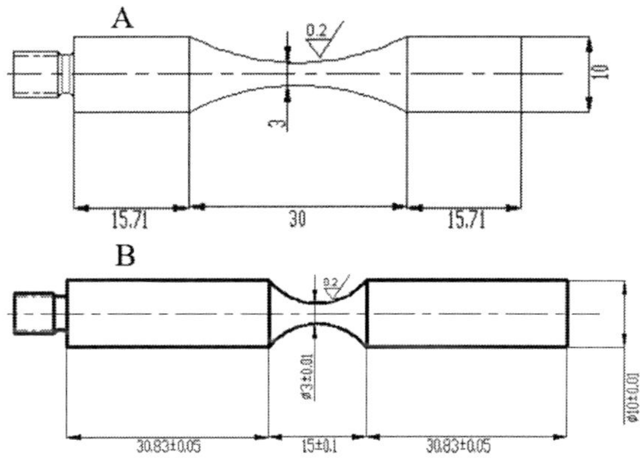

Fig.6. Sandglass specimen of different sizes for comparing

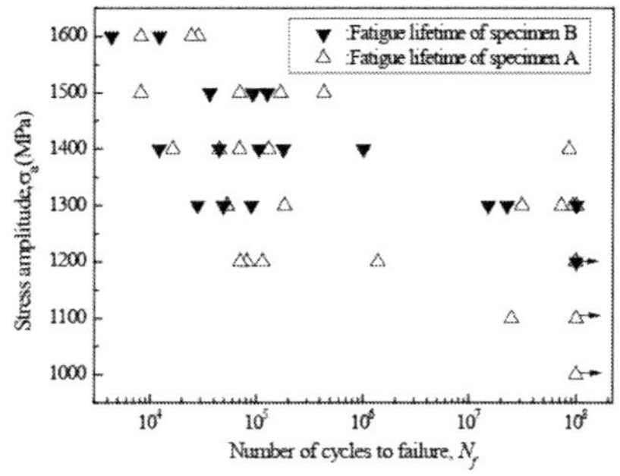

Fig.7. Test result of A and B type specimen

Another couple of sandglass specimens shown in Fig.8 are compared additionally. The minimum diameter of C specimen is 3mm and the surface roughness is 0.5μm. The minimum diameter of D specimen is 5mm and the surface roughness is 0.2μm. The comparing test results are shown in Fig.9. Fig.9 shows that although the surface roughness of C specimen is larger than D, the fatigue lifetime of C specimen is much longer than D specimen under the same stress amplitude condition as C specimen has less minimum diameter.

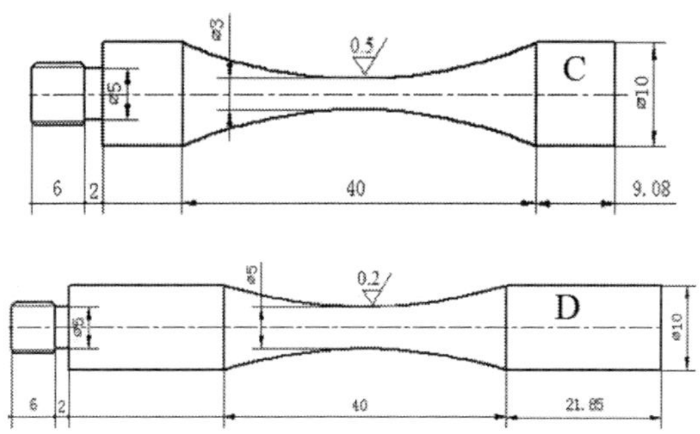

Fig.8. Sandglass specimen of different sizes for comparing

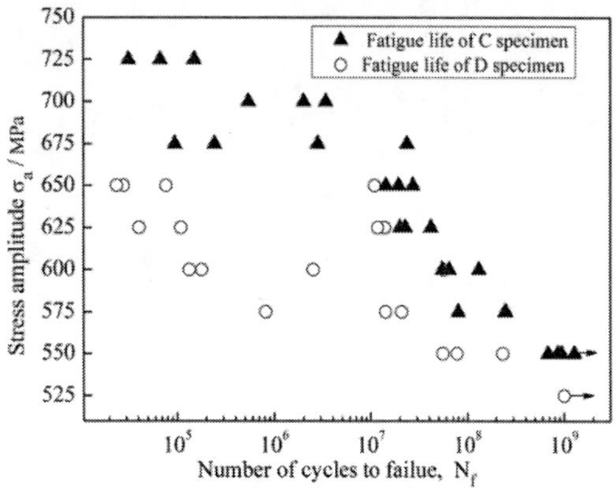

Fig.9. Comparison of the test result of C and D type specimen

It can be seen from Fig.3、Fig.5、Fig.7 and Fig.9 that the specimen shape and size affect the VHCF test results obviously, especially from Fig.3, Fig.5 and Fig.7. This is due to different risk volume which is estimated as the part of the specimen subjected to above 90% of the maximal stress [11-12]. LCF and HCF cracks mostly initiated from surface as the relatively high loading stress, hence, LCF and HCF are processes of finding defects from the surface. However, VHCF is a process of finding large defects inside the specimen volume [13]. The probability of large defects being present is likely to increase in response to the increase of the risk volume. As a result, VHCF is more sensitive to specimen size than LCF and HCF.

3. The effect of the specimen surface roughness

The specimen surface roughness is not only an influential factor on the HCF testing results, but also an influential factor on the VHCF [14]. If the roughness is too large, the fatigue strength will be dominated by surface roughness. However, if the roughness is reduced to a certain critical value, the surface can be considered to be "smooth". The formula of the critical value is expressed as [15]:

$$a_s \approx k(1+\frac{120}{\mathrm{HV}})^6,\quad k=0.171 \qquad (12)$$

Eq.(12) express the theoretical critical value of the surface roughness. The coefficient k are modified according to test results in the present paper:

$$a_s \approx k(1+\frac{120}{\mathrm{HV}})^6,\ k=0.08 \qquad (13)$$

It can be seen from Eq.(12) and Eq.(13), as the specimen material hardness or strength increases, the threshold becomes smaller. In other words, higher strength material requires lower roughness for VHCF testing.

In order to validate Eq.(13), a high strength steel (HV=380) is processed into sandglass specimens with the same size and different surface roughness. The critical roughness value is 0.42μm according to Eq.(13). The surface roughnesses of E group specimen are around 0.3μm and the surface roughnesses of F group specimen are around 0.74μm.

The VHCF results are shown in Fig.10. It can be seen that as the roughness of F group 0.74μm is much larger than the critical value 0.42μm, the fatigue lifetime of F group specimen is lower than E group at the same stress amplitude condition. Besides, only 2 specimens of F group fracture at very high cycle region, i.e. beyond 10^7 cycles, and only 3 specimens fracture from interior. The test results of F group are dominated by surface roughness and the larger roughness lead to premature fracture from surface. However, for E group, 11 specimens of fracture at very high cycle region and 13 specimens fracture from interior. The test results of F group are dominated by internal defects.

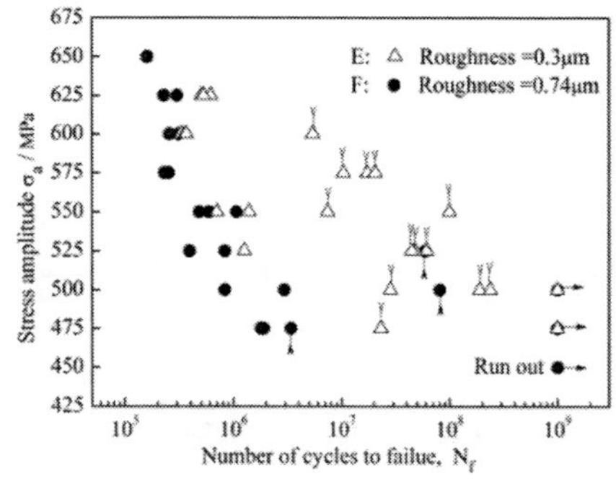

Fig.10. Test result of E and F group specimen (The vertical arrow denotes interior-initiated fracture)

4. The effect of the inclusion size

It can be know from above that VHCF usually fractured from interior and is dominated by internal defects. At the fatigue process, the surface defects (such as PSB or machining marks) and internal defects (such as inclusion) are a competition relation. When the applied loading is relatively high and the inclusion size is relatively small, the specimen tends to LCF from surface.

When the applied loading is relatively low, the specimen tends to VHCF from interior. When the applied loading is relatively high and the inclusion size is relatively large, the specimen either tends to VHCF from interior or tends to LCF from surface. Based on this, the relationship of *S/N* curve and inclusion size can be analyzed as below.

Case 1: The inclusion size ϕ is less than a certain critical value $\phi_{\text{inc}}^{\min}$. As the inclusion size is small enough, it does not affect the fatigue. At relatively low stress, the specimen will not fracture at very high cycle region. Therefore, the specimen either tends to fracture from surface at low cycle or high cycle region under high stress, or not fracture under low stress. The *S/N* curve character is indicated as the 1 type *S/N* curve in Fig.11.

Case 2: The inclusion size ϕ is large than the certain critical value $\phi_{\text{inc}}^{\min}$, however, less than a certain critical value $\phi_{\text{inc}}^{\max}$. The specimen fractured from surface at high stress. However, when the stress descends to a certain level, the surface crack of the specimen will be in balance with the internal crack competition. Therefore, the specimen either tends to LCF fracture from surface or VHCF fracture from interior inclusion at the balance stress. Thus, the *S/N* curve exhibits a horizontal line and the length of the horizontal line increases as the inclusion size descends. Then the *S/N* curve continuously descends at lower stresses as shown as the 2 type *S/N* in Fig.11.

Case 3: The inclusion size ϕ is large than a certain critical value $\phi_{\text{inc}}^{\max}$. As the inclusion size is large enough, the specimen will still fracture from the large inclusions at low stresses. As a result, the *S/N* curve exhibits a continuously descending shape and the horizontal asymptote is not obvious, as indicated as the 3 type *S/N* curve in Fig.11.

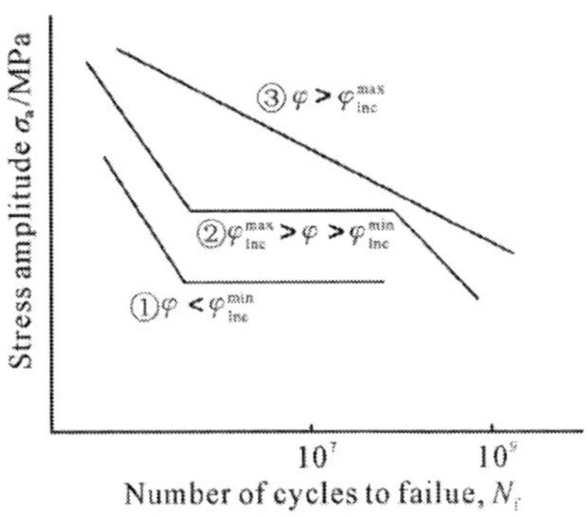

Fig.11. The relation of the *S/N* character and the inclusion size

For high strength steel, $\phi_{\text{inc}}^{\max}$ is 20μm[15].Assuming the inclusion shape is spherical, the critical value $\phi_{\text{inc}}^{\min}$ is calculated by using the following equation [15]

$$\phi_{\text{inc}}^{\min} = C_{\text{in}}(1+\frac{120}{\text{HV}})^6 \quad (14)$$

In which, C_{in} is a coefficient related with the location of the inclusion. C_{in} is 0.813,0.528,0.969 respectively, when the inclusion is located at the surface, sub-surface and interior as shown in Fig.12. In Fig.12, when the inclusion is located at the sub-surface, the competition between surface defects and internal defects turns into a cooperative relationship to jointly promote the production of fatigue crack, as a result, C_{in} is the least at this case.

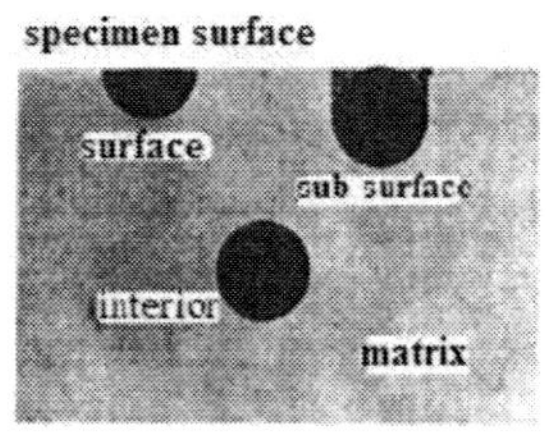

Fig.12. The locations of inclusion or defect in specimen

In order to validate the relationship of *S/N* curve and inclusion size analyzed above, VHCF *S/N* curve of a high strength steel is tested in this paper. The fracture surfaces of the fatigued sandglass specimen are scanned via SEM as shown in Fig.13a and the inclusion sizes are measured. The average of all the measured inclusion size ϕ is about 14.2μm. The critical value $\phi_{\text{inc}}^{\min}$ of the high strength steel is about 4.5μm according to Eq.(14). The average inclusion size is larger than the critical value $\phi_{\text{inc}}^{\min}$ and less than the critical value $\phi_{\text{inc}}^{\max}$, thus the *S/N* curve can be predicted to be the 2 type S/N in Fig.11. The prediction is validated through the tested *S/N* curve as shown in Fig.13b.

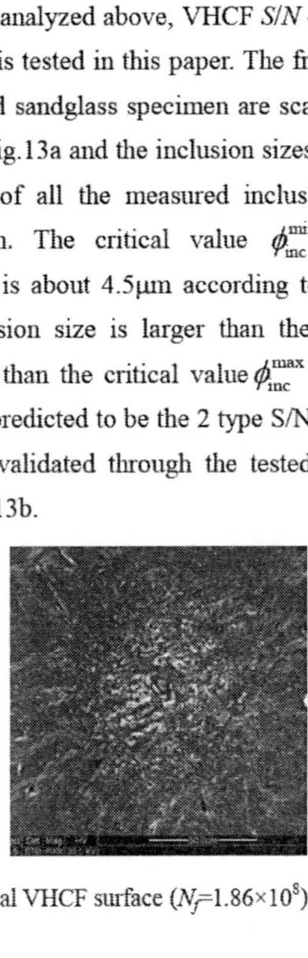

Fig.13a. A typical VHCF surface (N_f=1.86×10^8)

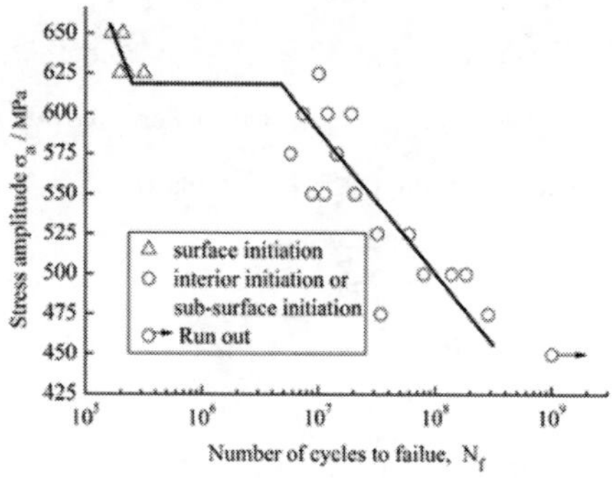

Fig.13b. The tested VHCF S/N curve of a high strength steel

5. Loading frequency effect

The very high frequency 20kHz would cause a drastic change of the fatigue behavior comparing with the usual frequency [17-18]. The change of loading frequency affects the test results through changing the strain rate. For specimens with the same size, the fatigue life tends to increase with an increase of the fatigue loading frequency as shown in Fig.14.

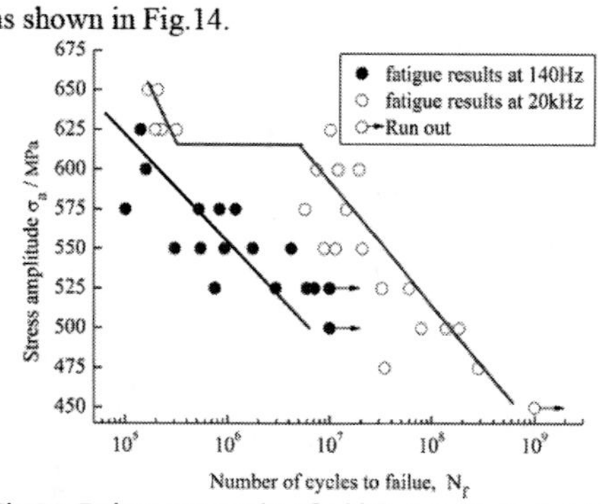

Fig.14. Fatigue test results of a high strength steel at 140Hz and 20kHz

Assuming the dislocations move freely, according to Johnston-Gilman [19], the distance a travels in dislocation a quarter of cycle is:

$$L=\int_0^{cycle/4} v\mathrm{d}t=\frac{A(n\sigma_0)^m}{2\pi f}\exp(-\frac{Q}{kT})\int_0^{\pi/2}(\sin\chi)^m\mathrm{d}\chi\propto\frac{1}{f} \qquad (15)$$

Q is the activation energy for dislocation, k is Boltzmann constant, A and m material constant, T is absolute temperature.

It is seen from Eq.(15) that the distance the dislocation travels decreases as the loading frequency increases. As a result, the fatigue life tends to increase with an increase of the fatigue loading frequency. The calculation is taken under the assumption that dislocations move freely. Practically for very high strength steel or alloy steel, there are a lot of obstacles such as fixed dislocations and grain boundary in materials and they will shorten the moving distance markedly. Hence the effect of load frequency may be very small or vanish in very high strength steels or alloy steel. This fact has been supported by author [20] and many other experiments [21-26].

Thus, the ultrasonic test results of very high strength steel or alloy steel can be used for practical. However, for low carbon steels, sensitive to the loading frequency, applying the ultrasonic test results directly will lead to unsafe results. The ultrasonic test results should be modified before applying. Wang H[27] used a modifying coefficient of load frequency to describe the effect of load frequency on fatigue behavior in ultrasonic fatigue testing. The modifying coefficient equals the ratio of fatigue strength coefficient of lower frequency load and high frequency load. Recently, Benjamin&Akira[28-29] point that S/N properties under usual frequencies of 0.2~140 Hz can be successfully normalized by the lower yield stress at the individual frequency. However, the S/N curve cannot be normalized as the same way for 20kHz as shown in Fig.15 as the crack initiation mode has been changed at 20kHz. So modifying the test results at 20kHz is a not easy thing as the mechanism of frequency effect should considered thoroughly and deeply.

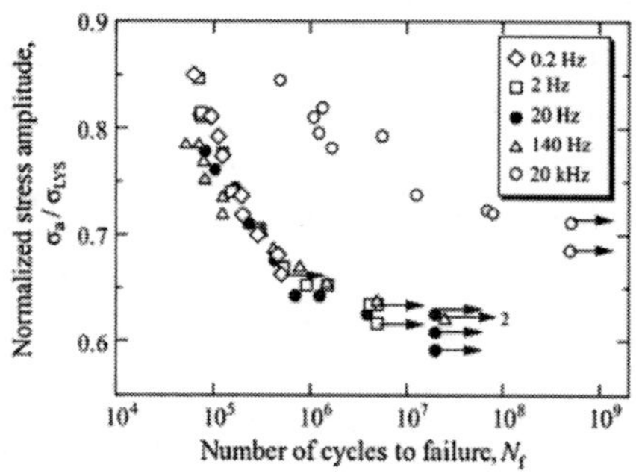

Fig.15.Normalized S/N curve by yield stress [28-29]

Conclusions

Fatigue tests of hundreds of specimen are conducted to investigate the effect of specimen shape, size, surface roughness, inclusion size and load frequency on the VHCF testing results. The following conclusions are dawn:

1. The fatigue test results differ for different specimen shape and size, as the risk volume differs. Furthermore, VHCF is more sensitive to specimen size than LCF and

HCF.

2. There is a critical roughness for the specimen, depending on the hardness of tested metallic material. A larger roughness than the critical one will lead to a premature fracture due to the surface defects and test result can not reveal the internal defects character.
3. The VHCF region is dominated by the internal defects for high strength steel and the VHCF *S/N* curve character is strongly related to the size of the inclusion. Thus, the inclusion size can be predicted through the VHCF *S/N* curve.
4. The frequency effect can be ignored for very high strength steel or alloy steel. However for low carbon steel, the fatigue life tends to increase as the loading frequency increases. Hence, the test results should be modified before used for practical. The modifying should be based on mechanism of frequency effect and it needs further research.

References

1. Marines I, Bin X, Bathias C. An understanding of very high cycle fatigue of metals. Int.J.Fatigue, 25(**9-11**),1101-7(2013)
2. Miller KJ, O'Donnell WJ. The fatigue limit and its elimination. Fatigue Fract Eng Mater Struct,**7**, 545-57(1999)
3. Bathias C, Drouillac L, Le Francois P. How and why the fatigue S-N curve does not approach a horizontal asymptote. Int.J.Fatigue, **23**,143-51(2001)
4. Murakami Y, Nomoto T, Ueda T. On the mechanism of fatigue failure in the super long life regime ($N>10^7$ cycles). Part I: influence of hydrogen trapped by inclusions. Fatigue Fract Eng Mater Struct, **11**,893-902(2000)
5. Mason W P. *Piezoelectric crystals and their application in Ultrasonic*. (New York: Van Nostrand, 1950)
6. Bathias C. Piezoelectric fatigue testing machines and devices. Int.J.Fatigue, **28**,1438-1445(2006)
7. Stanzl-Tchegg SE. Fracture mechanisms and fracture mechanics at ultrasonic frequencies. Fatigue Fract Eng Mater Struct,22(**7**),567-79(1999)
8. Wang QY, Berard JY, Dubarre A, Baudry G, Rathery S, Bathias C. Gigacycle fatigue of ferrousalloys.Fatigue Fract Eng Mater Struct,**8**,667-72(1999)
9. Xue Hongqian, et.al. The design of specimen for fatigue test at ultrasonic frequency. Acta Aeronautics et Astronautics sinica,25(**4**),425-428(2004)
10. Peng WJ, et.al. Ultrasonic fatigue test method of several kinds specimen. Journal of Wuhan Engineering Institute, 28(**2**),1-4(2016)
11. Y. Furuya. Specimen size effects on gigacycle fatigue properties of high-strength steel under ultrasonic fatigue testing. Script Materialia,**58**, 1014-1017(2008)
12. Y. Furuya, et al. 10^{10}-cycle fatigue properties of 1800MPa-class JIS-SUP7 spring steel. Fatigue Fract Engng Mater Struc,**26**,641-645(2003)
13. B. Pyttel, D. Schwerdt, C. Berger. Very high cycle fatigue - Is there a fatigue limit? Int.J.Fatigue, **33**,49-58(2011)
14. H. Itoga, K. Tokaji, M. Nakajima, H.-N. Ko. Effect of surface roughness on step-wise S–N characteristics in high strength steel. Int.J.Fatigue, **25**,379-385(2003)
15. Yang Zhenguo, et.al. Estimation of the critical size of inclusion in high strength steel under high cycle fatigue condition. Acta metallurgica sinica, 41(**11**),1136-1142(2005)
16. Yang Zhenguo, et.al. On the critical inclusion size of high strength steel under ultra-high cycle fatigue. Mater Sci Eng A , 2006, **427A**,167-174(2006)
17. Peng WJ, et.al. Ultrasonic fatigue tests on a high strength steel for welded structure. Advanced Materials Research, **503-504** :714-717 (2012)
18. Isamu Nonaka, Sota Setowaki, Yuji Ichikawa. Effect of load frequency on high cycle fatigue strength of bullet train axle steel. Int.J.Fatigue, 60(**1**),43-47(2014)
19. Zhang Yangyang, Duan Zheng, Shi HuiJi. Comparison of the very high cycle fatigue behaviors of INCONEL 718 with different loading frequencies. Science China Physics, Mechanics and Astronomy,56(**3**),617-623 (2013)
20. Peng WJ, et.al. An investigation of the fatigue property of ultra-high strength mould steel at 130 Hz and 20 kHz.Applied Mechanics and Materials, **239-240**, 96-99 (2013)
21. Aiguo Zhao, et.al. Effects of strength level and

loading frequency on very-high-cycle fatigue behavior for a bearing steel. Int.J.Fatigue, **38**,46-56(2012)

22. I.Marines, et.al. Ultrasonic fatigue tests on bearing steel AISI-SAE 52100 at frequency of 20 and 30 kHz. Int.J.Fatigue,**25**,1037-1046(2003)
23. Y. Furuya, S. Matsuoka, T. Abe, K. Yamaguchi. Gigacycle fatigue properties for high-strength low-alloy steel at 100 Hz, 600 Hz, and 20 kHz. Scripta Materialia,**46**,157-162(2002)
24. H. Mayer, M. Papakyriacou, R. Pippan, S. Stanzl-Tschegg. Influence of loading frequency on the high cycle fatigue propertiesof AlZnMgCu1.5 aluminium alloy.Materials Science and Engineering A ,**314**,48-54(2001)
25. Stanzl-Tschegg SE, Mayer H. Fatigue and fatigue crack growth of aluminium alloys at very high numbers of cycles. Int.J.Fatigue,**23**,231-237(2001)
26. Chen Q, et. al. Small crack behavior and fracture of nickel-based super alloy under ultrasonic fatigue. Int.J.Fatigue, **10**,1227-32(2005)
27. Wang Hong, Gao Qing. Effect of load frequency on fatigue behavior of material in ultrasonic fatigue testing. PTCA (Part: A PHYS.TEST), 41(**9**),433-435(2005)
28. Benjamin Guennec, et.al. Effect of the loading frequency on fatigue properties of JIS S15C low carbon steel and some discussions based on micro-plasticity behavior. Int.J.Fatigue,2014,**66**, 29-38(2014)
29. Benjamin Guennec, et.al. Dislocation-based interpretation on the effect of the loading frequency on the fatigue properties of JIS S15C low carbon steel. Int.J.Fatigue,**70**, 328-341(2015)

Fatigue crack initiation and propagation of 100Cr6 steel under torsional loading in very high cycle regime

Hongqian Xue[1,*], *Tao* Gao[1], *Zhidan* Sun[2], and *Xianjie* Zhang[1]

[1] Key Laboratory of Contemporary Design & Integrated Manufacturing Technology of Ministry of Education, Northwestern Polytechnical University, 710072, Xi'an, China
[2] ICD, P2MN, LASMIS, University of Technology of Troyes, UMR 6281, CNRS, Troyes, France

Abstract. Cyclic torsional fatigue properties of a high strength steel 100Cr6 are investigated using an ultrasonic torsional fatigue testing machine, and the results are compared with those obtained with fatigue tests under axial loading. Fatigue crack initiation and growth under torsion loading are observed in the very high cycle regime. Results show that fatigue cracks initiated from specimen surface as well as subsurface inclusions under torsion loading. However, subsurface MnS inclusions play a dominant role in crack initiation under torsion loading in the very high cycle regime. The initiation and early propagation of fatigue cracks are mostly controlled by the direction of the maximum shear stress. For surface crack initiation, cracks initiated in parallel to the longitudinal direction of the specimens. Once the shear crack propagated to a crack length of about 20-30 μm, crack branched to the angle close to the direction perpendicular to the remote maximum principal stresses. As for the subsurface fatigue crack initiation, the cracks parallel to the longitudinal direction of the specimens could not be observed, and crack propagation followed a spiral shape on a plane with an orientation of 45° with respect to the loading direction, which corresponds to the maximum principal stress plane.

1. Introduction

Structural components subjected to high frequency vibrations, such as those used in rotary parts in engines, are usually required to be designed using a lifetime failure-free criterion or an endurance limit for a very large number of cycles. Fatigue data obtained under uniaxial loading are not sufficient to assess very high cycle fatigue limit of mechanical components which are subjected to torsion loading over a wide range of cyclic frequency. In the rotary components such as transmission shafts, bearing structural components, fatigue failure occurs under the effect of combined bending and torsional loadings often at very high frequency. It is thus necessary to investigate the torsion fatigue performance of materials in very high cycle regime in order to appropriately design mechanical parts with respect to the life in service.

100Cr6 steel is a high strength steel usually used to build engine components in automotive industry. In service, the components are subjected to low amplitude, high frequency cyclic loading due to for example vehicle vibrations, and the number of cycles to failure is in the very high cycle regime. In the literature, several investigations of this steel under very high cycle fatigue have been reported. For example, Spriestersbach [1] studied the crack initiation mechanisms and threshold values of very high cycle fatigue failure for 100Cr6 steel. In reference [2], gigacycle fatigue tests were performed for 100Cr6 steel and prediction of fatigue failure was realized using a model based on the geometrical properties of inclusions. Mayer et al. [3] studied the very high cycle fatigue properties of a bainitic 100Cr6 steel, and in this work a special attention was paid to the crack initiation sites. However, most the ultrasonic fatigue tests conducted in these work are of uniaxial nature, and cyclic shear fatigue behavior and fatigue data of high strength steels in torsion are still limited in the literature. Yoshinobu [4] performed cyclic torsion fatigue test of a carburized alloy steel in very high cycle regime. They reported that crack deflection from mode II to mode I occurs under torsional loading. Akiniwa [5] also examined crack initiation and growth of several steels in the very high cycle regime. Similar to crack initiation and growth mechanisms observed for these steels, cracks initiation either perpendicular or parallel to the longitudinal direction of the specimen surface was reported under cyclic torsion loading. Recently, Mayer [6] performed ultra-high cycle torsion fatigue tests of a VDSiCr spring steel processed by shot peening at different load ratios. In this work, the interesting subsurface crack initiation has been observed under

* Corresponding author: xuedang@nwpu.edu.cn

torsional loading in spite of the maximum shear stress acting at the specimen. They pointed out that subsurface crack initiation occurs due to the effect of surface residual compression stress of specimen processed by shot peening, as there is no inclusions which could be observed at the subsurface crack initiation sites. Bathias and Xue [7] reported different findings from cyclic torsion fatigue test of 100Cr6 steel, and they indicated that fatigue crack initiates from subsurface inclusions of components in very high cycle regime. However, the role of inclusions on crack initiation under torsional loading in very high cycle regime has not been fully explained.

In order to evaluate the fatigue properties and resistance of the 100Cr6 steel in very high cycle regime, representative of service condition, torsion fatigue tests were conducted in this work. For the purpose of determining the fatigue strength in very high cycle regime, an ultrasonic torsion fatigue testing machine was used. To do this, four stress levels are chosen and 3–6 specimens are tested at each stress level. In addition, fracture surfaces were observed using Scanning Electronic Microscopy (SEM) in order to study the origin of fatigue crack initiation and the propagation mechanism, in relation to microstructural characteristics of the material.

2. Experimental aspects

2.1 Material

Fatigue tests were conducted with cylindrical dog-bone shaped specimens in 100Cr6 steel. The material was thermally treated with the following conditions: heating up to 850°C and kept constant during 30 minutes, then quenching at 220°C, tempering at 170° during 60 minutes. The chemical composition and the basic physical and mechanical properties of the studied 100Cr6 steel are shown in Tab. 1 and Tab. 2, respectively.

Table 1. Chemical composition of the 100Cr6 steel (wt.%)

C	Si	Mn	P	S	Mo	Ni	Cr
1.03	0.24	0.339	0.012	0.008	0.032	0.147	1.46

Table 2. Basic physical and mechanical properties of the 100Cr6 steel

E/GPa	ρ/kg . m^{-3}	$R_{e0.2\%}$/ MPa	σ_b/MPa	HV30
210	7870	1158	2300	780

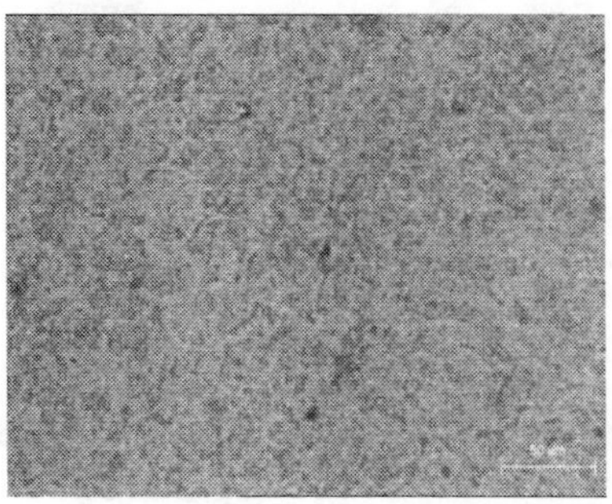

Figure 1. Optical microscopic observation of the microstructure for the 100Cr6 steel

The microstructure of the 100Cr6 steel was observed using optical microscopy, and it is shown in Fig.1. It can be seen that the ferrite and perlite phases are well distributed, and there is the presence of many large particles distributed in the material (Fig.1b).

2.2 Experimental procedure

In this study, fatigue tests were conducted with an ultrasonic fatigue testing machine under torsional loading, as described in [7]. Fatigue tests were carried out under constant amplitude, fully reversed loading conditions (stress ratio R=-1) at room temperature in air. In these tests, specimens are stimulated to resonance torsion vibrations, which causes fatigue loading of the specimens. The specimens were cooled with compressive air so as to reduce the temperature rise due to self-heating effect during fatigue test.

Before each test, the strain in the gage section of the specimen was calibrated using a strain gage bonded on the gage section. Under nominal elastic conditions for fatigue loading in very high cycle regime, there is a linear relationship between input displacement and the generated strain in the gage section. An example of system calibrating result is shown in Fig. 2. During fatigue tests, the test control software continuously records the displacement and controls the output of the power supplier and, therefore, indirectly controls the

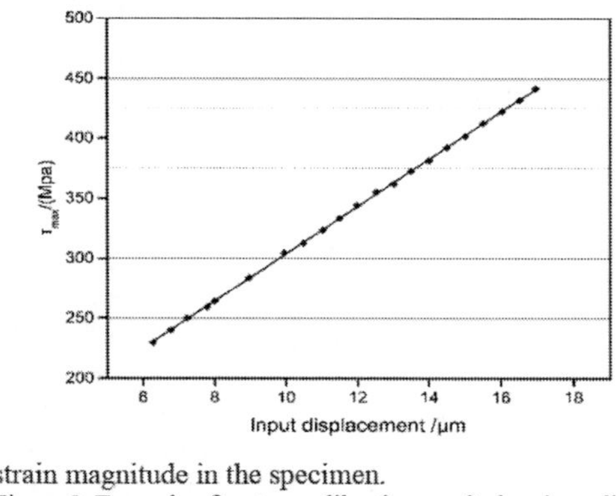

strain magnitude in the specimen.

Figure 2. Example of system calibrating result showing a linear relationship between the input displacement and the generated strain in the specimen

To detect fatigue crack initiation, the vibration frequency of the system was monitored. This is because the resonance frequency of the system could be dropped with the crack initiation and propagation. Fatigue cracks could be automatically detected when the natural frequency drops a lot in a short time. The fatigue tests were stopped either when the endured number of stress cycles exceeded 10^{10} cycles or when the specimen was broken. After fatigue tests, the scanning electron microscope (SEM) was used to observe the fatigue cracking features on both the specimen surfaces and the fracture surfaces.

https://doi.org/10.1051/matecconf/201816520003

2.3 Fatigue specimen

Specimen dimension for torsion fatigue test was determined using analytical or numerical method. The geometry of the torsion fatigue specimen of 100Cr6 steel is shown in Fig. 3. The diameter in the centre of the specimens is 5 mm for torsional fatigue tests. All the specimens were polished up to grade 2000 paper in order to remove machining marks and to observe crack initiation and propagation path.

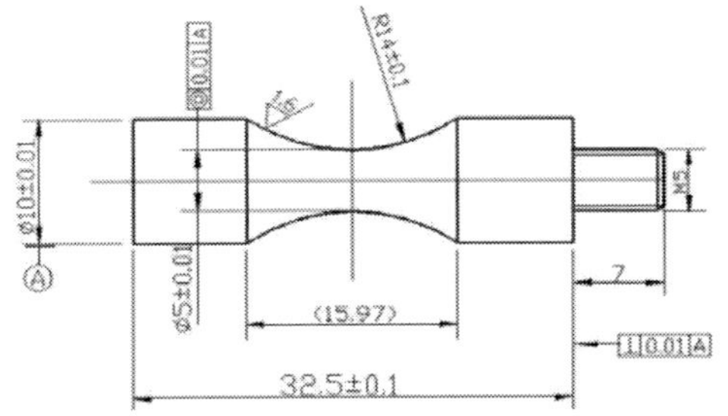

Figure 3. Dimension of 100Cr6 steel specimen tested using ultrasonic torsion fatigue testing machine

3. Experimental results

3.1 Fatigue strength

The results of the torsion fatigue experiments are presented in Fig. 4 as a plot of the torsion shear amplitude, τ_a, versus the number of cycles to failure, N_f. Hollow circle and solid circle symbols are used for fatigue crack initiation at surface or subsurface respectively. Triangles with arrows indicate run-outs. It is important to notice that the fatigue failure occurred even beyond 10^9 cycles for torsion fatigue tests, and there is thus no fatigue limit in very high cycle regime. According to Fig. 4, we can observe that the experimental data have relatively large dispersion at every stress level. At the stress amplitude of 620 MPa, the fatigue life ranges from $1.09\text{x}10^7$ cycles to 10^{10} cycles.

For the specimens failed below 10^7 cycles, there is only one specimen in which the crack initiation occurred at subsurface. However, among the 11 specimens that failed above 10^7 cycles, 9 specimens failed due to subsurface crack initiation and 2 due to surface crack initiation, which means that ultra-high cycle fatigue life shows a clear dependency on subsurface inclusions.

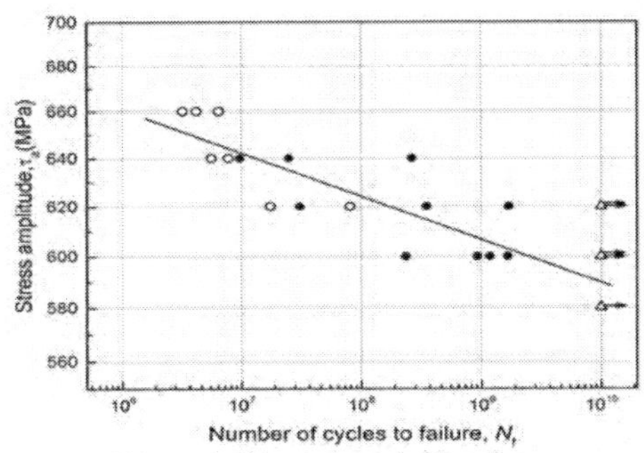

Fig.4 The S–N curves of torsion fatigue. Hollow circles indicate crack initiation at surface, while solid ones for subsurface.

In Fig. 5, the fatigue data of the material tested under axial loading presented in [7] are plotted to compare with the data tested under torsional loading. Specimens failed after 10^7 cycles under both torsional and axial fatigue loading, and it showed larger scatter for torsional fatigue tests of the steel than axial fatigue tests. The fatigue strength at 10^9 cycles of 100Cr6 is about 606 MPa, and the ratio between torsional and axial cyclic strength is 0.71.

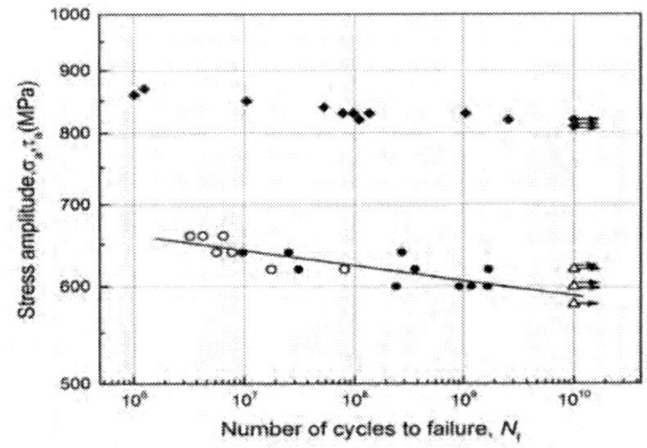

Fig.5 Comparison between the S-N curves for torsional loading (hollow and solid circles) and axial loading (solid squares)

3.2 Fatigue cracking mechanisms

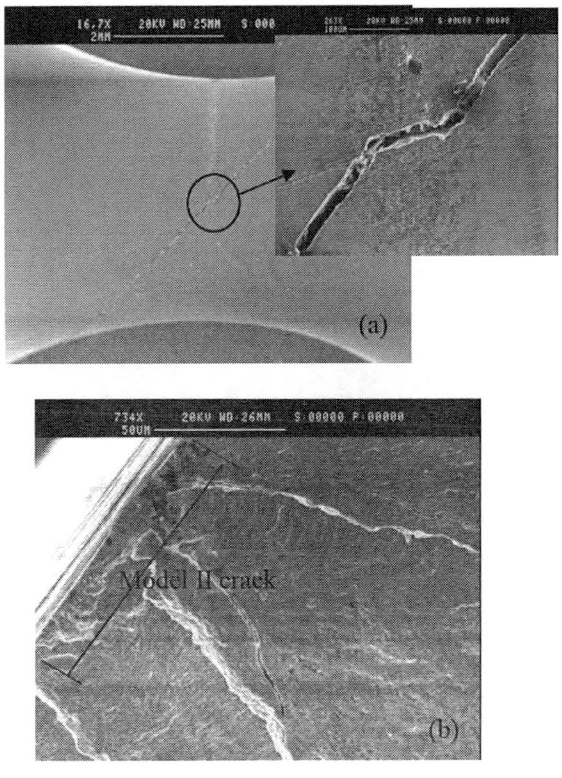

Fig. 6 Specimen surface (a) and fracture surface (b) of specimen broken after Nf = 1.75×10^7 under torsional loading at τ_a = 640 MPa

Both surface and subsurface crack initiation can be observed under cyclic torsional loading. Fig. 6 illustrates typical fracture surfaces for specimens with crack initiation occurring at surface. Fig. 6a shows the surface of a specimen which broke after N_f = 1.75×10^7 cycles under the torsion stress amplitude τ_a = 640 MPa. The micro-crack formed during the crack initiation stage was parallel to the longitudinal direction. After the model II crack propagated about 120 μm, it branched and propagated in a mode I manner. Fig. 6b shows the fracture surface of this specimen.

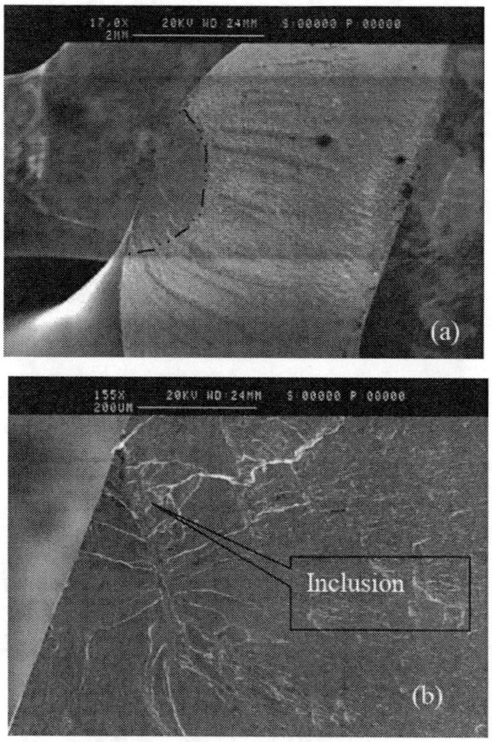

Fig.7 Fracture surface of a specimen: (a) global fatigue fracture surface with fisheye area, and (b) surface inclusion

If the fatigue fracture origin was a surface inclusion, the Mode I cracks propagated directly from the inclusion without significant Mode II growth. The macroscopic crack propagates perpendicularly to the direction of the maximum principal stress. Fig.7 shows the fracture surface of a specimen that failed after N_f = 7.85×10^6 cycles at the torsion stress amplitude τ_a = 660 MPa. The surface region indicates the process of early fatigue crack growth. The first stage of crack nucleation occurs in the maximum shear plane at the tip of the inclusion, which is the maximum loaded zone under both tension and torsion loadings (Fig.7a). The surface fisheye is observed around the elongated inclusion, which presents an elliptic area with lower roughness (the area indicated by red curve in Fig.7a). A change in color from inside to outside the fisheye area can be observed. The transition from dark to light in color is related to the change of crack propagation model. Inside the fisheye, the crack propagates in hybrid model with model I and model II crack, while the Mode I crack propagation is dominant outside the fisheye area.

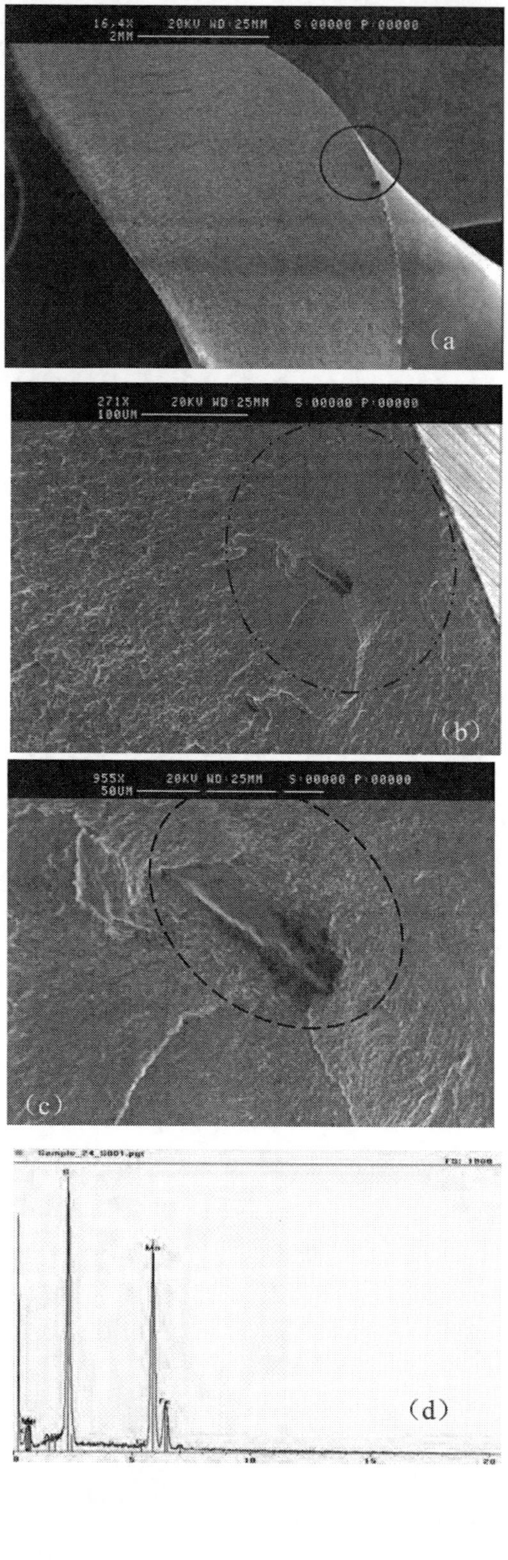

Fig.8 Fracture surface of specimen with crack initiation at subsurface: (a) global fatigue fracture surface, (b) fisheye area, and (c) FGA around inclusion, (d) chemical compositions of the inclusion analysed by EDX

Fig. 8 shows the surface and the fracture surface of a specimen that failed in the VHCF regime with a crack initiation at subsurface. A fisheye fracture surface with FGA around inclusion is visible, which was rarely reported in VHCF investigations. Fig. 8a shows stable crack propagation zone which almost covers the whole fracture surface. The fish-eye area is indicated by a red ellipse, as shown in Fig. 8b. The elliptic fish-eye area with length and width of approximately 260 μm and 210 μm respectively appears, although the boundary of the fish-eye is not very clear (see Fig. 8b). Magnification of the zone around inclusion shows that the FGA around the inclusion can be observed. The FGA area is also indicated by an ellipse in Fig. 8c, and the size of the slender inclusion inside the FGA is 54μm. The chemical composition of the inclusion on fracture surface was analyzed by Energy Dispersive Analysis of X-ray (EDX) equipped within the SEM. According to the results, this inclusion corresponds to a MnS inclusion.

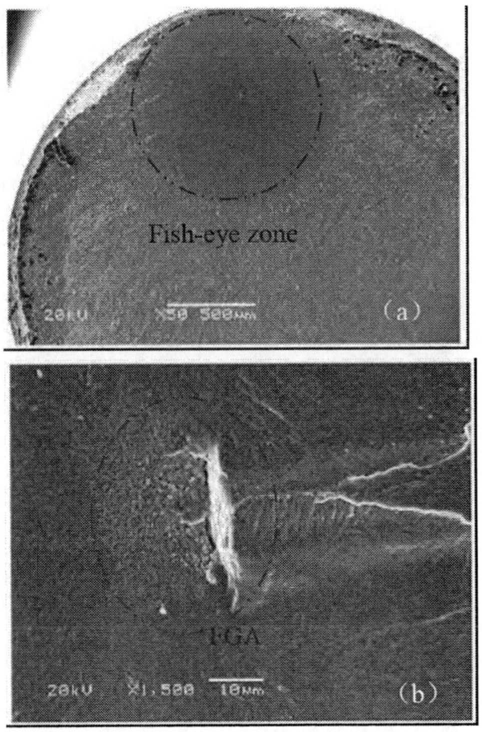

Fig. 9 Fracture surface of specimen with crack initiation at subsurface after Nf = 5.24×10^7 under axial loading with R = -1 at σ_a=820 MPa: (a) global fatigue fracture surface with fisheye area, and (b) FGA around inclusion

In the case of cyclic axial loading, stress distribution in the same cross section of specimen is uniform, and VHCF crack often initiates from interior inclusion for the high strength steel. Fig. 9 shows that fatigue crack initiation from a larger inclusion in VHCF regime under ultrasonic fatigue test with load ratio R = -1. It is noticed that a larger fish-eye area with clear boundary could be observed, and the inclusion is much far from the specimen surface than that in torsion fracture surface. On the other hand, unlike the spiral surface of torsion fracture surface, the crack growth under axial loading is macroscopically on the same plane.

4. Discussion

As presented in the literature, for high strength steels under uniaxial very high cycle fatigue loading, crack initiation often occurs at the subsurface due to the presence of metallurgical defects such as inclusions. As a matter of fact, in very high cycle regime the imposed stress amplitude is low. In this case, the macroscopic behavior of the material is in elastic regime and there is inelastic deformation only around defects such as flaws and inclusions. Since the presence probability of a defect is greater within a specimen than in its surface area, the typical crack initiation under very high cycle fatigue will be most often inside the specimen, instead of at the surface for a specimen under high cycle fatigue [7].

For a specimen subjected to torsional loading, the applied stress state is schematically presented in Fig. 10. Both the tensile and the compressive principal stresses are oriented 45° with respect to the specimen axis and they are mutually perpendicular. One shear stress component is parallel to the specimen axis, and the other is perpendicular to the specimen axis.

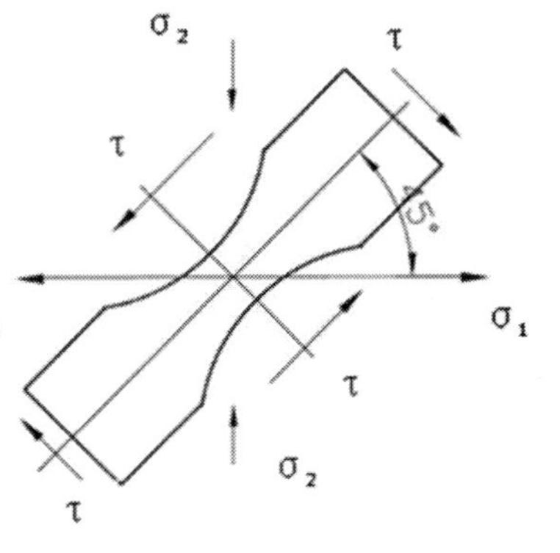

Figure 10. Schematic presentation of the stress state for a specimen under torsional loading

For all the fatigue tests conducted with a cyclic shear stress level just higher than the fatigue limit, a cracking behavior in the failed specimens has been observed. In this case, a small crack is nucleated on one of the maximum shear stress planes, and under further cyclic loading it propagated up to a length of several hundred microns. Then crack branching occurred with the increase of the number of cycles, which could generate multi-cracks. However, only one of these micro-cracks propagated following one of the planes of maximum normal stress, and lead to final fracture of specimen. Typical surface crack initiation in a specimen that failed in HCF regime (<107 cycles) is shown in Fig. 11. Crack grows in shear mode for first tens of microns, forming a

 https://doi.org/10.1051/matecconf/201816520003

small area with steps on the fracture surface perpendicular to the specimen's length direction. The second cracks are visible on the fracture surface. After the early propagation on the planes of the maximum shear stress, the fatigue cracks could change their paths to planes perpendicular to the maximum normal stress. It means that there is always a transformation trend from model II to model I crack due to the effect of the normal stress, as a result, forming a spiral surface with an orientation of 45° with respect to the loading direction, as shown in Fig.12.

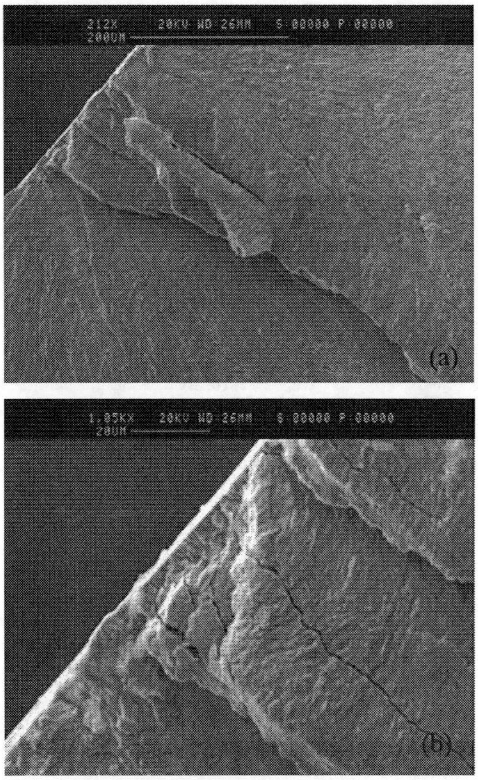

Fig. 11 Fracture surface of specimen with surface crack initiation (a), and its enlarged view (b)

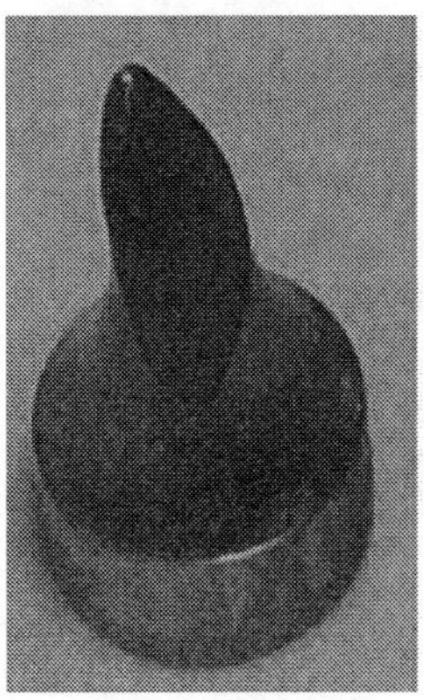

Fig. 12 Spiral fracture surface of specimen under cyclic torsion loading

In VHCF regime, fatigue crack initiation started likely at subsurface of specimen. In the literature, lots of work has been done concerning the transition of fatigue crack initiation from surface to subsurface inclusion when the imposed stress amplitude is decreased [8-10]. Mechanisms for the crack initiation at subsurface inclusions have been proposed [11-13]. In general, in the case of low stress amplitude, fatigue crack initiation of most specimens is linked to the subsurface inclusion. Although macroscopically the most loaded place is on the periphery surface of specimen, the presence of an inclusion in subsurface region changes the local stress state. There is thus a competition between the stress state modification caused by inclusions and the stress on periphery surface. The stress modification at an inclusion depends on its distance from the specimen surface, and on its size and shape. It is important to note that the nature of the inclusion determines directly the crack initiation behavior. For example, a brittle inclusion may be easy to break, which leads to crack initiation. For the high strength steel investigated in the study, the maximum slender inclusion size in the critical volume of a specimen is approximately 50 μm, which is much larger than the maximum defect size of 7.9 μm characterized in [5]. Hence, for the high strength investigated in this study, most fatigue fracture occurred from subsurface defects, namely, fatigue strength was determined by the MnS inclusion in very high cycle regime.

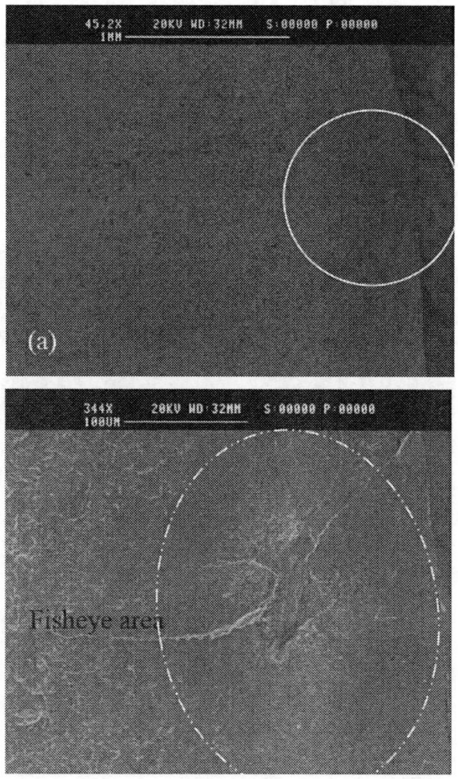

Fig.13. Illustration of an observed fracture surface showing fatigue crack initiation from the subsurface inclusion: (a) the smooth fish-eye surface, (b) enlarged view of the fish-eye area

Once a fatigue crack initiated from a subsurface inclusion, the local stress increased and complex stress states exist around the fatigue crack initiation region. Under further cyclic loading micro-crack growth took place surrounding the inclusion and progressively the cracking surface became a facet area. The surface or interior fisheye region indicates the process of early fatigue crack growth. Elliptical fish-eye area with smooth surface could be formed (Fig. 13), which is different from the stepped surface observed in [14]. Therefore, for titanium alloy in [14], crack initiation and early micro crack growth are in mode II dominated by shear loading. However, the cracks grow in mode I dominated by tensile stress in the rest of the fish eye area. For the high strength steel investigated in this study, the fish-eye fracture surface is inclined at small angle (less than 45°) perpendicular to the specimen's length axis with a successive surface. It is confirmed that the crack path inside the fish-eye is determined by both tensile and shear stress simultaneously. The difference of fisheye surface for these two kinds of materials is related to the large inclusion inside the steel. If the fatigue fracture origin was a subsurface inclusion, Mode I cracks propagated directly from the inclusion without significant Mode II growth [11]. Therefore, the facet perpendicular or parallel to the longitudinal direction due to crack initiation and early growth, could not be observed for this high strength steel in gigacycle regime.

Tensile stress on the 45° plane will exceed the tensile strength of the material before shear stress reaches the shear strength. In the very first stage of crack propagation, the fracture surface seems rather smooth. As in this stage the crack is short and of microscopic scale, the stress intensity factor is low, and the fatigue crack propagation is slow. It is important to indicate that the stage of slow crack propagation represents a large lifetime fraction of a whole fatigue test in the range of high cycle fatigue and very high cycle fatigue. When the micro-cracks reached the specimen surface, the fatigue crack propagated as surface cracks. The crack propagated along the plane loaded under cyclic tension stress, which is oriented 45° with respect to the axial direction. This orientation corresponds to the maximum principal stress plane. Although the fatigue cracking may appear to be on the maximum principal stress plane, the fatigue damage mechanism is in fact shear stress driven. As a result, the crack initiated from subsurface inclusion reveals a fracture surface oriented 45°. Thus, fatigue crack propagation occurred along the orientation normal to the 45° tensile plane, producing a spiral fracture surface. This investigation can be confirmed by Fig. 9 in which it is shown that the observation of broken specimen seems indicate that the crack propagated in 45° of axial direction.

5. Conclusion

In this work, torsion fatigue behavior of 100Cr6 steel in very high cycle regime was studied using an ultrasonic torsional fatigue machine. The fatigue test results indicate high scatter of obtained fatigue data due to the presence of a large number of inclusions in MnS.

Fatigue crack initiation started at the periphery surface at higher stress amplitudes. However, there is a transition of crack initiation site from surface to subsurface when the stress level was decreased and just slightly higher than very high cycle fatigue limit. Fatigue crack initiation of most specimens is linked to subsurface inclusion failure. As for the fatigue crack propagation, the fracture surface is rather smooth during the very first slow crack propagation stage. The overall crack orientation appears to follow a spiral shape on a plane with an orientation of 45° with respect to the loading direction, which corresponds to the maximum principal stress plane.

REFERENCES

1. D. Spriestersbach, P. Grad, E. Kerscher, Procedia Engineering **74**, 84-91 (2014)
2. G.M.D. Almaraz, Mech. Mater. **40**, (2008) 636-640
3. Mayer H, J Mater Sci **44**, 4919–4929 (2009)
4. Y. Shimamura, K. Narita, H. Ishii and et al, Int. J Fatigue **60**, 57-62 (2014)
5. Y. Akiniwa, S. Stanzl-Tschegg, H. Mayer, M. Wakita, K. Tanaka, Int. J Fatigue **30**, 2057–2063 (2008)
6. H. Mayer, R. Schuller, U. Karr and et al, Int. J Fatigue **70**, 322–327 (2015)
7. H.Q. Xue, C. Bathias. Eng Fract Mech **77**, 1866–1873 (2010)
8. B. Kaiser, B. Pyttel, C. Berger. Int. J Fatigue **33**, 23–32 (2011)
9. C. Bathias, Int. J Fatigue, **23**, 143-151 (2001)
10. Nasr, Y. Nadot, Ch. Bouraoui, R. Fathallah, M. Jouiad, Int. J Fatigue 32, 780–787 (2010)
11. S. Fujita, S. Matsuoka, Y. Murakami, G. Marquis, Int. J Fatigue **32**, 943–951 (2010)
12. H. Mayer, W. Haydn, R. Schuller, S. Issler, B. Furtner, M. Bacher-Hochst, Int. J Fatigue **31**, 242-249 (2009)
13. B. Pyttel, I. Brunner, B. Kaiser, C. Berger, M. Mahendran. Int. J Fatigue **60**, 101–109 (2014)
14. A. Nikitina, T. Palin-Lucc, A. Shanyavskiy, Procedia Structural Integrity **2**, 1125–1132 (2016)

Fatigue life assessment regarding different influences on the HCF/VHCF behavior of a martensitic steel

Igor Milošević[1, *], *Benjamin* Seisenbacher[1], *Gerhard* Winter[1], *Florian* Grün[1], *Martin* Kober[2]

[1]Chair of Mechanical Engineering, Montanuniversität Leoben, Franz-Josef-Strasse 18, Leoben 8700, Austria
[2]LEC GmbH, Inffeldgasse 19/II, Graz 8010, Austria

Abstract: Modern applications require a special treatment when the conventional specimen size is much larger than the component size. Additional to that, high sophisticated materials are used for highly loaded components. Often the conventional fatigue limit is exceeded and loads are applied in the VHCF regime. Focus was put on the lifetime calculation and the implementation of investigated fatigue data of a X5CrNiCuNb-16-4 type steel. Two specimen geometries with diameters $D_{7.5}$=7.5 mm and $D_{2.5}$=2.5 mm were tested at R=-1, at room temperature and up to 10^9 cycles to failure. The application of different software tools (FEMFAT, fe-safe) showed several issues based on the current results. Results showed a change of crack initiation mechanism to subsurface crack initiation at approx. $2x10^6$ cycles to failure. The gradient based correction of the reference fatigue data showed a good applicability up to $2x10^6$ cylces. The application of fe-safe allows the flexible modification of S/N parameters over the whole cycle range. The usage of the actual material configuration introduced several important questions regarding the fatigue data and the implementation into lifetime calculation tools.

1 Introduction

Nowadays modern applications require a number of superior characteristics (strength, ductility, etc.) served by high sophisticated structural materials. Precipitation hardened (PH) steels (Maraging) are well known as structural materials for chemical and power plants or as components in the aerospace and automotive industry.

Since the conventional fatigue limit has been questioned by numerous scientists, Bathias [1,2], Murakami [3,4] Mughrabi [5,6] and Miller [7] amongst others, so-called fish eye fractures introduced a new level of complexitiy to a long life fatigue behaviour. The change to an inclusion driven failure mechanism is accompanied by an increase of scatter and no infinite life.

Especially PH steels show a clear defect based life behaviour at longer lifetimes. [8] Several contributions were made from Schnitzer [9–11], Wang [12,13,13,14], Wu [14,15] and Schönbauer [16] concerning the microstructural and mechanical properties, study on temperature effects and the fatigue limit prediction by defect based models. It can be summarised that certain materials (type II, definition after [17]) with brittle inclusions (oxides) show a pronounced change of fatigue behaviour by inclusion based crack initiation.

In the process of fatigue life assessments fatigue data is used and implemented into common software tools (FEMFAT, fe-safe). These tools were not intended to cover changing failure mechanisms and different failure effects. Therefore a strategy has to be developed how to apply these tools up to the component based life assessment.

In most applications (lifetime estimation) the specimen geometry is usually smaller than the component size. In an actual case, the component size is much smaller than the specimen geometry.

In this work, fatigue results are shown regarding different influences and the application for a lifetime estimation process. Two software tools are compared presenting the pros and cons. Possibilities to cover certain influences are pointed out, and an outlook is given. The most important insights are concluded at the end.

2 Experimental procedures

2.1 Material

Fatigue experiments are performed with precipitation hardened stainless steel (X5CrNiCuNb16-4). The material specification is given in Table 1 below. The same material configuration was used in previous fatigue tests [8,18].

Table 1. Chemical composition (max. value) of selected chemical elements of a X5CrNiCuNb-16-4 steel (wt%, DIN EN 10088-3)

	C	Cr	Mo	Ni
X5CrNiCuNb16	0.07	17.0	0.6	5.0

* Corresponding author: igor.milosevic@unileoben.ac.at

The heat treatment of the material is austenitizing at approx. 1050 °C followed by a specific heat treatment, which induced intermetallic precipitations throughout the material matrix. Depending on the purpose (maritime applications, aerospace applications, etc.) a UTS of 1400 MPa can be reached besides an excellent corrosion behaviour.

The microstructure is characterised by a fine lath martensitic structure (<1 µm) without any significant (detectable) reverted austenite. The precipitation hardening increased strength and hardness.

2.2 Fatigue testing method

Fatigue tests in this work were conducted with two different testing machines. Common servo-hydraulic test rig (30 Hz testing frequency) as well as an own developed high frequency vibration technique (HVFT [19], 30 Hz, 928 ±8 Hz testing frequency) were used.

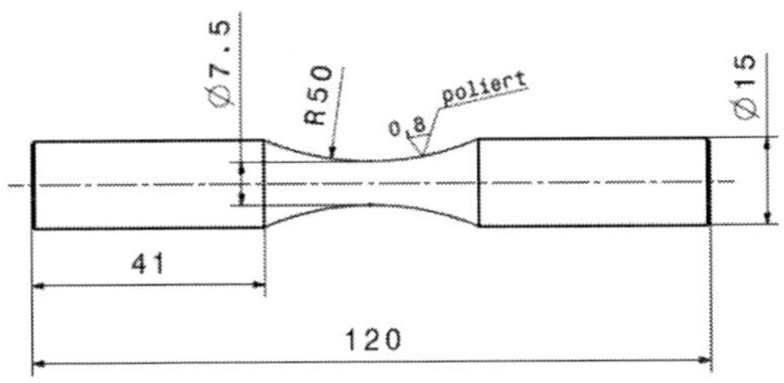

Fig. 1. Shape of standard hourglass ($D_{7.5}$=7.5 mm) specimens. The critical cross section was diamond polished to its final condition.

In Fig. 1 and Fig. 2 the specimen shapes are pointed out, which were used in presented investigations. HFVT specimens were also used and validated within a recent work [19]. All specimens were diamond polished in the critical cross section.

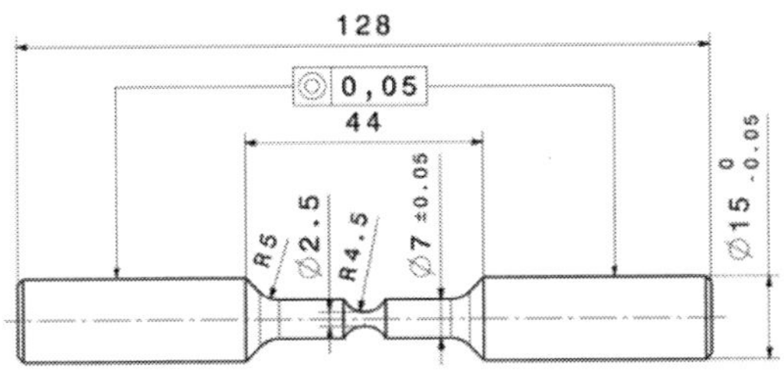

Fig. 2. Shape of a HFVT specimen (D=2.5 mm). The critical cross section was diamond polished to its final condition. [19]

Standard 30 Hz tests were carried out on the servo-hydraulic test rig Sinus 100.25 (Instron GmbH, ±100 kN) and high frequency tests (928 ±8 Hz, R=-1, T=25 °C) on the HFVT, which can be seen in Fig. 3.

Fig. 3. Different fatigue test systems a) HFVT [19] and b) Sinus 100.25 (Instron GmbH, ±100 kN)

2.3 Fractography

Representative fracture surfaces were analysed by SEM (EVO MA15, Carl Zeiss SMT). Due to the presence of a transition area (TA, [8]), where internal crack initiation was detected, EDS (Energy Dispersive X-Ray Spectroscopy) was used to determine the type of inclusion. No further defect-based investigations were carried out in this work.

3 Stress and life computation

Modern software tools are implemented in product development cycles, which mostly combine a finite element analysis (FEA, Abaqus/CAE) with certain post processing tools, such as FEMFAT or fe-safe. Due to a wide availability the presented study focuses on uniaxial damage models. This preliminary work is seen as an initial step for further component based implementations.

The FEA was carried out by a detailed stress analysis of presented geometries (Fig. 1, Fig. 2). Therefore twenty-node brick elements, C3D20, were used (quadratic order). Reduced integration was applied (2x2x2 integration points) due to several preliminary analyses published in [20,21].

Both specimens were loaded with a uniaxial nominal stress of 1 MPa. Material parameters (Table 2.) were given in the table below. No plastic deformation was applied.

Table 2. Material parameters E and ν used for stress calculations

	E in GPa	ν
Steel	210	0.3

The life computation was carried out by application of commonly known software tools FEMFAT and fe-safe.

The evaluation of local stress conditions, like the application of the relative stress gradient χ', lead to a modification of a referenced fatigue behaviour respectively S/N curve. Equation 1 briefly describes the calculation of n_χ where the tension/ compression fatigue limit σ_t, the rotating bending fatigue limit σ_{rb}, the relative stress gradient χ' and a material parameter K_D were implemented. A detailed description of this application is given by Eichlseder [22] and in recent studies by the authors [20,21].

$$n_\chi = 1 + \left(\frac{\sigma_{rb}}{\sigma_t} - 1\right) * \left(\frac{\chi'}{2/b}\right)^{K_D} \quad (1)$$

The local fatigue limit σ_f is calculated by Equation 2. Within the adjustment of the fatigue limit also the slope k and the cycles to the turn off point N_D are modified. This procedure was implemented by FEMFAT and therefore no additional description will be given.

$$\sigma_f = \sigma_t * n_\chi \quad (2)$$

Several procedures are suggested for damage calculations at stress amplitudes below the fatigue limit. The three most important approaches are Miner Modified, Miner Original and Miner Elementary. Miner Original assumes that a pronounced fatigue limit is present whereas Miner Elementary treats all stress amplitudes equally. Miner Modified proposes a second slope according to $k_2=2*k_1-1$.

The software tool fe-safe (developed by Dassault Systemes) is capable of lifetime calculations by different approaches. A major difference is obvious by implementation of S/N curves. Whereas FEMFAT uses parameters k, N_D and σ_t, fe-safe directly implements S/N datapoints. These datasets are connected and where no data is found an extrapolation is carried out. An according algorithm was chosen for lifetime calculations. Among others v. Mises, Normal Stress, etc. criteria are available with additional possibility of mean stress corrections.

A stress concentration based approach was not implemented. A major advantage is the variable implementation of S/N datapoints. Therefore a double kinked S/N curve, for example, allowed a more sophisticated application in the VHCF area.

4 Results

4.1 Fatigue and fractrography

In this section results are discussed with respect to specimen diameters $D_{7.5}$ and $D_{2.5}$. Fatigue tests were stopped at 10^7 cycles, unless otherwise stated, and specimens were categorised as run out tests.

S/N (Wöhler) curves displayed were generated by statistical evaluation of fatigue tests according to ASTM E739 [23] and arcsin $\sqrt{P}$ [24] representing a survival propability of P_S=50 %.

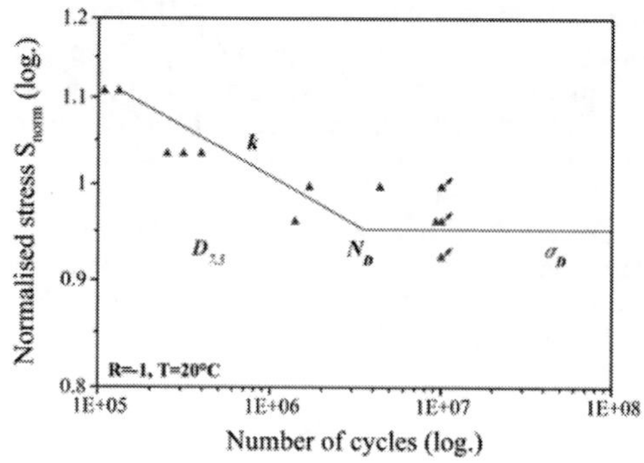

Fig. 4. Fatigue data from $D_{7.5}$ specimen geometries. $D_{7.5}$ specimens were tested at 30 Hz, R=-1 and at roomtemperature.

$D_{7.5}$ fatigue results were considered as reference fatigue tests. As pointed out in Fig. 4 tests were conducted at T=20°C and at a stress ratio R=-1. Parameters k=21.5, N_D=3,287,500 at a S_{norm}=0.95 were derived from statistical evaluations and implemented in FEMFAT.

Additional tests $D_{2.5}$ were conducted at 928 ±8 Hz, which were stopped after 10^9 cycles without failure. In the area up to $2x10^6$ cycles to failure the slope increased due to the geometrical specimen properties. As expected a higher overall fatigue strength was observed in comparison with 30 Hz results. Up to $2x10^6$ cycles to failure crack initiation mainly took place on the specimen surface.

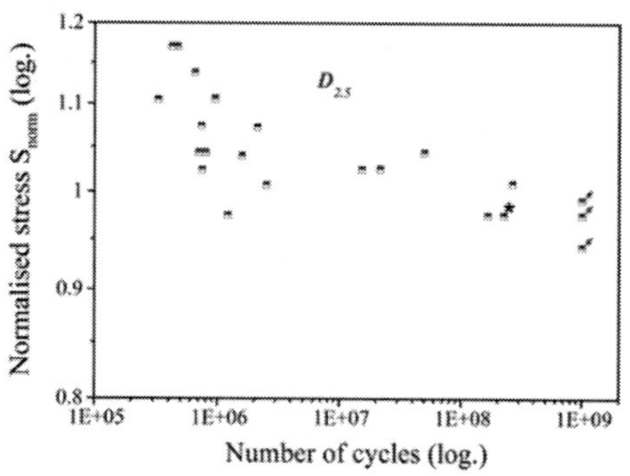

Fig. 5. Fatigue data from $D_{2.5}$ specimen geometries. Tests were conducted at 928 ±8 Hz, R=-1. [19]

After $2x10^6$ cycles to failure, a change of crack initiation mechanism could be detected. Furthermore, a increase of scatter could be detected over several orders of magnitude. In Fig. 5 several specimens were investigated by SEM to identify the type of defect.

In Fig. 6 (marked in Fig. 5 with *) a representative fracture surface is displayed. Almost centred an inclusion could be found, which showed subsurface crack initiation. Mainly internal inclusions showed cracks emanating from the subsurface region (Fig. 6), primary Al_2O_3 oxides (and their complex chemical combinations including MgO, etc.).

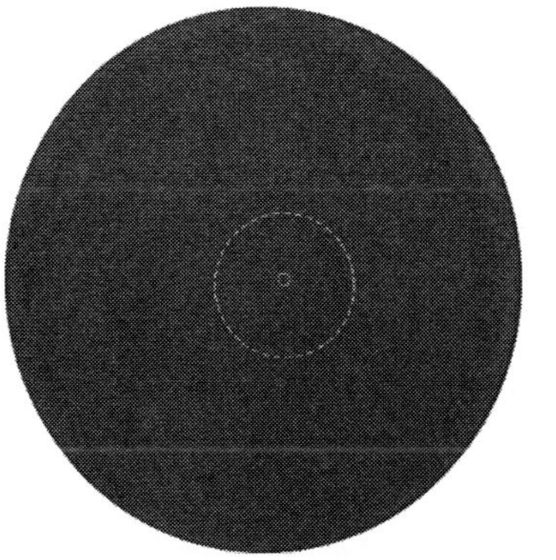

Fig. 6. SEM examinations revealed fish-eye fractures in $D_{2.5}$ (picture) and $D_{7.5}$ specimens, which showed Al_2O_3 inclusions without exception. Fracture surface presented at $2.25x10^8$ cycles to failure.

Selected $D_{7.5}$ specimens (30 Hz) showed failures just after 10^7 cycles, which were not taken into account for presented fatigue data up to 10^7 cycles to failure. Therefore further investigations at 928 ±8 Hz were carried out.

Fracture surfaces, above 10^8 cycles to failure, revealed FGA formations at a higher magnification (investigations in [19]). No FGA was found at lower lifetimes. A fatigue limit was examined to be nearly within the theoretical (extrapolated) fatigue limit of $D_{7.5}$ results.

4.2 Lifetime prediction

Data presented in Fig. 4 was implemented ($D_{7.5}$) into FEMFAT to ensure specimen lifetime calculations for $D_{2.5}$ specimens.

An identical dataset was implemented into fe-safe for comparing simulations. As mentioned fe-safe is not applying gradient based approaches. Analyses were carried out under consideration of the v.Mises criterion.

In Fig. 7 the FEMFAT modification is displayed regarding the modified slope $k_{\chi'}$, cycles to the turn off point $N_{D,\chi'}$ and the fatigue limit $\sigma_{D,\chi'}$. Due to the a relative stress gradient χ'=0.1804 at the surface of the specimen the modification, especially regarding the fatigue limit, was low.

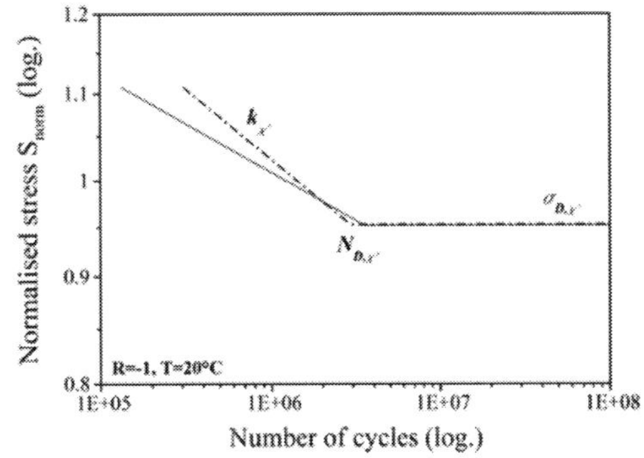

Fig. 7. Modification of S/N parameters according to FEMFAT 5.3. The relative stress gradient χ'=0.1804 was applied.

In Fig. 8 results of the lifetime calculation are displayed. The solid line represents the transition from a conservative (test life > simulation life) to a non-conservative (test life < simulation life) regime. The dashed line describes a scatter of 3 compared to the referenced data ($D_{7.5}$).

Due to the gradient approach FEMFAT results show a better result, almost within a scatter of 3. Fe-safe calculations showed a more conservative result as there is no modification of the initial fatigue data.

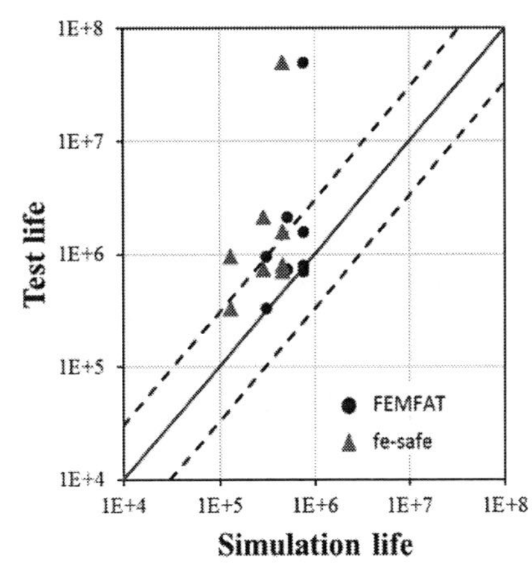

Fig. 8. Lifetime calculation results for three different load levels S_{norm}=1.1, 1.07 and 1.04.

Simulations were carried out at different load levels S_{norm}=1.1, 1.07 and 1.04. From a load level of S_{norm}=1.04 the scatter increased due to internal crack initiation. Lower load levels showed a very large deviation of results and therefore no calculations were presented.

5 Discussion

Here the results are discussed regarding different aspects of presented investigations. It was shown how data was applied regarding smaller specimen sizes.

Usually specimen geometries are smaller compared to actual component geometries. Regarding presented results a larger volume shows a decrease of fatigue strength compared to specimen fatigue tests. Presented investigations represent a preliminary study concerning the applicability of standard fatigue data to smaller volumes.

$D_{7.5}$ specimen fatigue data was used for many applications, always extrapolating the fatigue strength to larger component volumes. Therefore reduction factors were defined such as stress gradient and surface roughness.

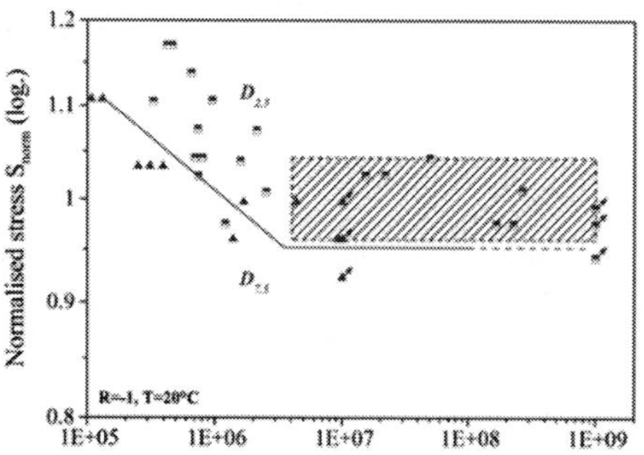

Fig. 9. Fatigue results of $D_{7.5}$ and $D_{2.5}$ tests and comparison of fatigue data. The critical area is displayed as shaded rectangle.

In Fig. 9 a summary is shown comparing two fatigue test results. Conventionally $D_{7.5}$ results (statisctical evaluation, fatigue limit) are extrapolated up to higher cycles. This approach is going to fail applied on $D_{2.5}$ results, especially taking into account fatigue tests failed at $> 10^8$ cycles. Some Materials exhibit a traditional fatigue limit evaluated at 10^7 cycles to failure. The presented material configuration, and of course various other materials, do not show a constant amplitude limit. This issue is still not considered within software tools, whereas FEMFAT offers damage calculation according to Miner's cumulative damage models.

The VHCF regime, which is defined to be from 10^7 to 10^9 cycles to failure, was analysed by additional tests. The shaded area in Fig. 9 represents the most critical area in terms of lifetime calculation. Up to now no ready-to-use approach was presented regarding characteristic values for a further implementation. Well-known defect based fatigue strength models were presented by Murakami [25,26] amongst others. Miner's rule lacks of statistical information, which is another issue in the actual investigation.

fe-safe provides the possibility of a more detailed application of S/N data. Every S/N relationship can be established and especially the application in the VHCF regime could be interesting as a multi kinked S/N curve could be realised. Therefore more detailed data is needed in the VHCF regime to evaluate the material's fatigue behavior beyond 10^7 cycles.

In the future additional tests will be carried out to improve fatigue data up to 10^9 cycles to failure. Future applications are going to reach higher lifetimes as a result of lightweight design and a higher power density. Further studies will concentrate on the different possibilities of fatigue data implementation and how this could be improved.

6 Conclusion

In this work, fatigue tests up tp 10^9 cycles to failure were presented under the consideration of decreasing specimen geometries. Additional to that the material behavior and subsurface crack initiation were presented. The fatigue data was implemented into common software tools and calculation results were summarised and discussed. Concluding remarks can be recorded as follows:

- Fatigue tests were presented taking into account the material properties and the change of the crack initiation mechanism.
- Tests showed the difference between classic fatigue data at 10^7 cycles to failure compared to data up to 10^9 cycles to failure.
- SEM analyses showed Al_2O_3 inclusions to be the main type of inclusion found in this material.
- Damage approaches were presented comparing two different software tools.
- Lifetime calculations showed the issues of conventional approaches and that the increase of scatter is not covered at all.

7 Acknowledgements

The authors would like to acknowledge the financial support of the "COMET - Competence Centres for Excellent Technologies Programme" of the Austrian Federal Ministry for Transport, Innovation and Technology (BMVIT), the Austrian Federal Ministry of Science, Research and Economy (BMWFW) and the Provinces of Styria, Tyrol and Vienna for the K1-Centre LEC EvoLET. The COMET Programme is managed by the Austrian Research Promotion Agency (FFG).

8 References

[1] Bathias, Fatigue & Fracture of Engineering Materials & Structures 7 ,559–565 (1999).
[2] C. Bathias ,163–171 (2000).
[3] Y. Murakami, International Journal of Fatigue 9 ,661–667 (1998).
[4] Y. Murakami, T. Nomoto, T. Ueda, Fatigue & Fracture of Engineering Materials & Structures 7 ,581–590 (1999).
[5] H. Mughrabi, Fat Frac Eng Mat Struct 8-9 ,755–764 (2002).
[6] H. Mughrabi, International Journal of Fatigue ,2–8 (2013).
[7] Miller, O'donnell, Fatigue Fract Engng Mater Struct 7 ,545–557 (1999).

[8] I. Milošević, C. Garb, G. Winter, F. Grün, M. Kober, Procedia Structural Integrity ,327–334 (2017).
[9] R. Schnitzer, Proceedings of the 8th International Tooling Conference Vol. I ,491–503 (2009).
[10] R. Schnitzer, R. Radis, M. Nöhrer, M. Schober, R. Hochfellner, S. Zinner, E. Povoden-Karadeniz, E. Kozeschnik, H. Leitner, Materials Chemistry and Physics 1 ,138–145 (2010).
[11] R. Schnitzer, S. Zinner, H. Leitner, Scripta Materialia 5 ,286–289 (2010).
[12] J. Wang, H. Zou, C. Li, S.-y. Qiu, B.-l. Shen, Materials Characterization 4-5 ,274–280 (2006).
[13] J. Wang, H. Zou, X.-y. Wu, S.-y. Qiu, B.-l. Shen, Nuclear Power Engineering Vol. 26. No.3 ,254–258 (2005).
[14] J.-H. Wu, C.-K. Lin, Materials Science and Engineering: A 1-2 ,291–298 (2005).
[15] J.-H. Wu, C.-K. Lin, Metall and Mat Trans A 6 ,1715–1724 (2002).
[16] B.M. Schönbauer, K. Yanase, M. Endo, International Journal of Fatigue ,205–216 (2016).
[17] H. Mughrabi, R. Wang, K. Differt, U. Essmann, ASTM International ,5-41 (1983).
[18] I. Milošević, P. Renhart, G. Winter, F. Grün, M. Kober, International Journal of Fatigue ,150–157 (2017).
[19] I. Milošević, P. Renhart, G. Winter, F. Grün, M. Kober (2018).
[20] I. Milošević, G. Winter, F. Grün, M. Kober, Frattura e Integrita Strutturale 42 (2017).
[21] I. Milošević, G. Winter, F. Grün, M. Kober, Procedia Engineering ,61–68 (2016).
[22] W. Eichlseder, *Rechnerische Lebensdaueranalyse von Nutzfahrzeugkomponenten mit der Finite Elemente Methode*, 1989.
[23] ASTM International, *Standard Practice for Statistical Analysis of Linear or Linearized Stress-Life (S-N) and Strain-Life (ε-N) Fatigue Data*, ASTM International, West Conshohocken, PA, 1998 (accessed on July 10, 2017).
[24] D. Dengel, Mat.-wiss. u. Werkstofftech. 8 ,253–261 (1975).
[25] Y. Murakami, T. Nomoto, T. Ueda, Fat Frac Eng Mat Struct (Fatigue Fracture of Engineering Materials and Structures) 11 ,893–902 (2000).
[26] Y. Murakami, Fatigue 90 ,377–382 (1990).

https://doi.org/10.1051/matecconf/201816520005

Influence of microstructural discontinuities on the behaviour of long cracks in the VHCF regime for the aluminium alloys EN AW 6082 and EN AW 5083

Tina Kirsten[1,*], *Fatih* Bülbül[2], *Marcel* Wicke[3], *Hans-Jürgen* Christ[2], *Angelika* Brückner-Foit[3], and *Martina* Zimmermann[1]

[1] Technical University of Dresden, Institute for Materials Science, 01069 Dresden, Germany
[2] University of Siegen, Institute for Materials Engineering, 57068 Siegen, Germany
[3] University of Kassel, Institute for Materials Engineering, 34125 Kassel, Germany

Abstract. In the present study two different aluminium alloys, the precipitation hardening alloy EN AW 6082 (peak-aged and overaged) and the work-hardening alloy EN AW 5083 (soft annealed) were examined. Fatigue cracks were initiated by means of a focused ion beam notch and a long-distance microscope was used for in-situ observation of the crack growth behaviour. The crack growth was investigated at constant stress intensity factors near the threshold regime. During the in-situ investigation a change in crack growth velocity was detected. It could be observed that the barrier function of grain boundaries and primary precipitations are the major reason for crack growth retardation despite the fact that the crack is in the long crack growth range. The microstructural influence becomes more important with decreasing ΔK values, meanwhile the average crack growth rate decreases simultaneously. Experimental results have shown that the Fe-based precipitates are influencing the crack growth rate for both aluminium alloys. Meanwhile, grain boundaries are causing a deceleration of the crack growth rate primarily in case of the work hardened aluminium alloy. This is assumed to be the reason for the smaller average crack growth rate in EN AW 5083 compared to that observed for the precipitation hardening alloy while applying comparable ΔK values.

1 Introduction

In previous studies the fatigue behaviour of metallic materials at very high number of load cycles ($N > 10^7$, VHCF) was thoroughly investigated (for example [1]). It has been shown that fatigue cracks can propagate even at amplitudes well below the classical fatigue limit (for example [2]). As for the life determining characteristics, these studies mainly focus on the crack initiation phase [3]. The crack growth phase at VHCF relevant low stress amplitudes is not as intensively investigated as of yet. The otherwise time and cost intensive testing in the VHCF regime can be addressed by using high frequency testing systems [4], such as the ultrasonic fatigue testing system used in this research.

The aim of the present work is to analyse the fatigue crack growth behaviour of two commercially available aluminium alloys (EN AW 6082 and EN AW 5083) in the near-threshold regime at constant stress intensity factors (ΔK) using an ultrasonic fatigue testing system in order to gain insight into the barrier function of microstructural inhomogeneities such as large precipitates or grain boundaries. The threshold ΔK_{th} is determined experimentally using ASTM E 647 [5]. This implies using the load-shedding method, in which the threshold is reached by continuously decreasing the applied load amplitude until the crack growth rate is less than 10^{-11} m/cycle. Previous studies could show that crack initiation for aluminium alloys may occur at subsurface regions in the VHCF regime [6]. In order to analyse the early crack growth behaviour it is necessary to add an artificial crack initiation site to the sample surface rather than rely on inherent crack initiation. Hence, a micronotch was inserted by means of focused ion beam technology (FIB).

2 Material and Testing

In this study the aluminium alloys EN AW 6082 and EN AW 5083 were investigated. Table 1 shows the chemical composition for both alloys.

EN AW 6082 is a precipitation-hardening alloy, whose microstructure is defined by a rolling texture with elongated grains of a length of ~500 µm and a width of ~50 µm (see Figure 1a). The material is characterized by primary (Mg_2Si, Al(Si,Mn,Fe)) and secondary (Mg_2Si) precipitates (see Figure 2). The alloy was examined in peak-aged (pa) and overaged (oa) condition. The heat treatment is given in Table 2.

EN AW 5083 is a work hardening alloy, which was investigated in the soft-annealed condition with nearly globular grains (around 40-50 µm, see Figure 1b). Besides the grain morphology the microstructure is

* Corresponding author: tina.kirsten@tu-dresden.de

defined by the primary precipitates of type Mg_2Si and $Al_6(Mn,Fe)$. Due to soft-annealing any influence of initial work hardening can be ruled out.

Table 1. Chemical composition (in weight %) of the materials investigated according to Bikar-Aluminium (EN AW 6082) and Hydro (EN AW 5083).

	EN AW 6082	**EN AW 5083**
Mg	0.6 - 1.2	4.0 - 4.49
Mn	0.4 - 1.0	0.4 - 1.0
Si	0.7 - 1.3	0.4
Fe	0.5	0.4
Zn	0.2	0.25
Cu	0.1	0.1
Cr	0.25	0.05 - 0.25
Ti	0.1	0.15
Al	Bal.	Bal.

Table 2. Precipitation hardening heat treatment.

	pa	**oa**
Homogenization	540 °C, 1 h	540 °C, 1 h
Quenching	Water	Water
Annealing	200 °C, 4 h	200 °C, 24 h

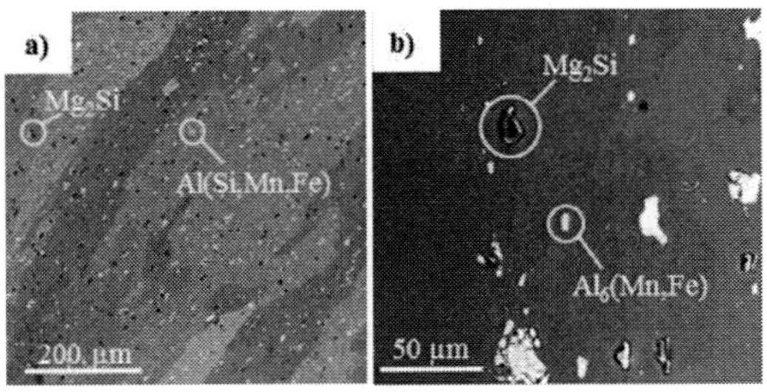

Figure 1. Microstructure with primary precipitates of the investigated material **a)** EN AW 6082 **b)** EN AW 5083.

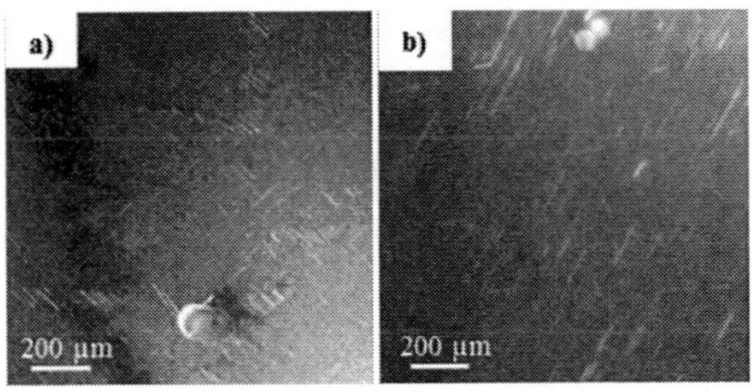

Figure 2. Transmission electron micrographs of the secondary precipitates of alloy EN AW 6082 in **a)** peak-aged and **b)** overaged condition.

The fatigue crack growth tests were performed using an ultrasonic fatigue testing system operating at a resonance frequency of about 20 kHz and a load ratio of R = -1. In front of the ultrasonic fatigue testing system, a long-distance microscope is installed, which has a maximum magnification of 750x and was provided by the workgroup of Prof. Brückner-Foit from the University of Kassel. The testing area is shown in Figure 3a. The samples are hour-glass shaped with a shallow notch in the centre area. Furthermore, there is a micro notch inserted on the ground of the shallow notch by means of focused ion beam technology. It acts as a definite crack initiation site for optical crack growth investigations (see Figure 3b).

The crack growth tests were carried out in pulse-pause mode to prevent the samples from heating. At first the thresholds for long crack growth were experimentally defined by means of the load shedding method. The threshold was assumed, when crack propagation was no longer measurable, which is defined as a drop of the crack growth rate below 10^{-11} m/cycle. Subsequent tests were performed at a quasi-constant stress intensity factor ΔK. This was achieved by repeatedly adapting the load amplitude $\Delta\sigma/2$ according to the optical measurements of the crack growth throughout the entire test.

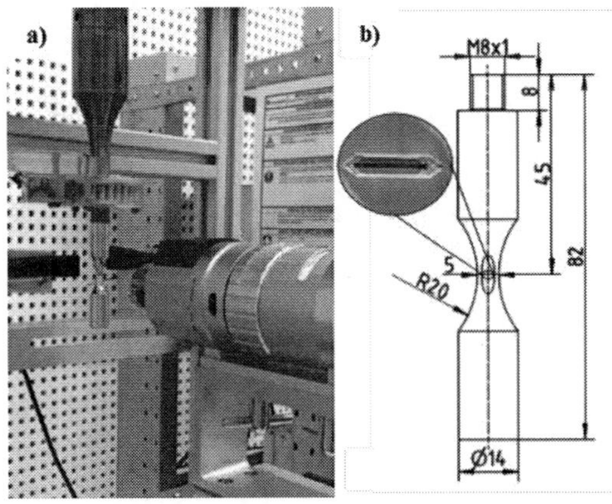

Figure 3. a) Ultrasonic fatigue testing system (BOKU Vienna) with long-distance microscope in front (Hirox) **b)** sample geometry with shallow notch and SEM image of FIB notch.

3 Experimental Results and Discussion

3.1 Threshold for long fatigue crack growth

First experiments were performed to determine the threshold for long fatigue crack growth. The results are shown in Table 3. It can be seen that the peak-aged material has a superior resistance against crack growth compared to the overaged material, which results in a higher threshold value ΔK_{th}. This is in accordance to the literature [7]. Furthermore Wicke et al. observed a similar tendency while measuring the threshold values at R = 0.1 for the same material [8]. The highest threshold however was measured for the work hardening alloy.

Applying the load-shedding method to define the long crack threshold can result in threshold values that are slightly overestimated since likely crack closure effects may become effective. However, because of the sample geometry restrictions due to the ultrasonic fatigue testing other methods, such as the method proposed by Tabernig and Pippan [9], were not applicable. Comparing the results presented in this manuscript with those of Stein et al., who applied the aforementioned method on the same material, it can be seen that the threshold values are in a similar range [10].

Table 3. Measured threshold values ΔK_{th} for the materials and heat treatment conditions investigated.

	EN AW 6082 pa	EN AW 6082 oa	EN AW 5083
$\Delta K_{th,1}$ [MPa√m]	1.39	1.31	1.45
$\Delta K_{th,2}$ [MPa√m]	1.44	1.35	1.56
$\Delta K_{th,3}$ [MPa√m]	1.56	1.38	1.69
Ø [MPa√m]	**1.46**	**1.35**	**1.57**

3.2 Fatigue crack growth at constant ΔK

In order to investigate the microstructural influence on the crack propagation in the VHCF regime, tests were carried out under constant ΔK values. The influence of the microstructure can be detected by changes of the crack growth rate (da/dN) over the crack length (a). Because of the used FIB notch, there are always two crack fronts, which are measured (left and right from the notch). If not otherwise specified, the following curves are considering always the overall fatigue crack growth on both sides of the FIB notch. Figure 4 shows the da/dN vs. a curves for all materials/heat treatment conditions investigated at ΔK values, which are less than 10 % above the determined average threshold values.

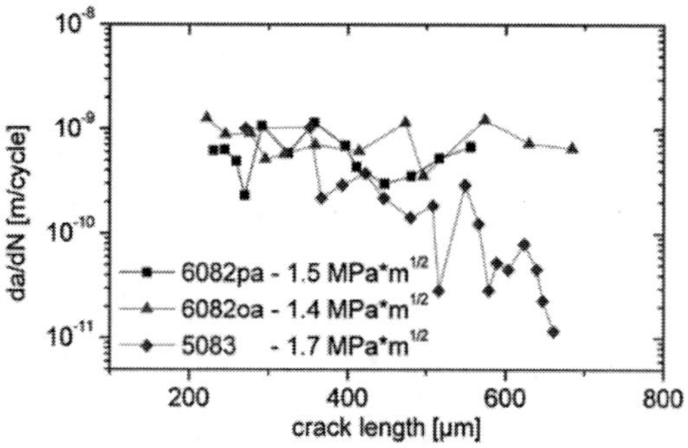

Figure 4. Crack growth rate vs. crack length for the materials/ heat treatment conditions investigated at constant ΔK values.

It can be seen that there is no great difference between the fatigue crack growth behaviour of the two heat treatment conditions of alloy EN AW 6082. Both curves are showing variations of the crack growth rate. It decreases at some points, which can be attributed to the microstructure (for further details see section 3.3). Compared to the precipitation-hardening aluminium alloy, the average crack propagation rate in EN AW 5083 is lower. Furthermore, a decrease of the crack growth rate can be observed throughout the entire test finally resulting in a rate below 10^{-11} m/cycle. According to Zerbst et al. the minimum crack growth rate is associated with the lattice arrangement of the atoms and the minimum displacement provided by the Burgers vector, which for aluminium is $2.86*10^{-10}$ m [11]. For this reason, the crack growth must have been stopped in between two measurements of the crack length, which is supposed to be caused by some microstructural barriers. The true crack growth rate should then be above the determined lower limit. A continuous crack growth seems not possible close to the threshold.

Similar tendencies can be detected for higher ΔK values (see Figure 5). The average crack propagation rate for both heat treatment conditions does not differ significantly, but in comparison the average crack growth rate for alloy EN AW 5083 is again well below that of EN AW 6082. Since the stress intensity factors for the curves given in Fig. 5 are comparable and the primary precipitates present in both alloys seem to have a stronger influence on the course of the crack growth curve (as will be shown later) than the overall growth rate, there must be another reason for the pronounced difference in growth rate. Hence, it is assumed that the smaller average grain size of alloy EN AW 5083 is associated with the lower crack propagation rates. In addition, a likely influence of a different size of the plastic zone formed at the crack tip has to be taken into consideration as well.

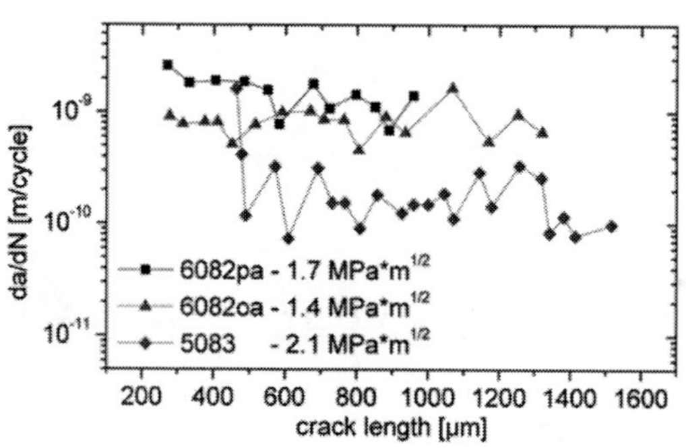

Figure 5. Crack growth rates vs. crack length at constant ΔK values for the alloys EN AW 5083 and EN AW 6082.

Figure 6 represents the da/dN vs. a curves for various ΔK values for the overaged condition of alloy EN AW 6082. There are two main characteristics visible. On the one hand the average crack growth rate increases with increasing ΔK value, while on the other hand the variations of the crack growth rate diminish at the same time. As a consequence, it can be assumed that the influence of the microstructure becomes smaller with increasing ΔK and is nearly neglectable at 2.2 MPa√m, which is 63 % above the threshold but still a very low crack growth rate.

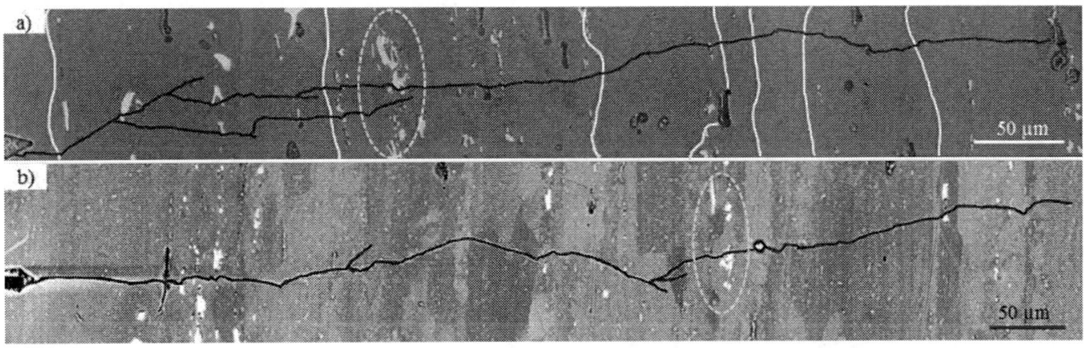

Figure 7. a) Crack path of alloy EN AW 6082 oa at ΔK = 1.4 µm (grain boundaries in white) **b)** Crack path of alloy EN AW 5083 at ΔK = 2.1 MPa√m.

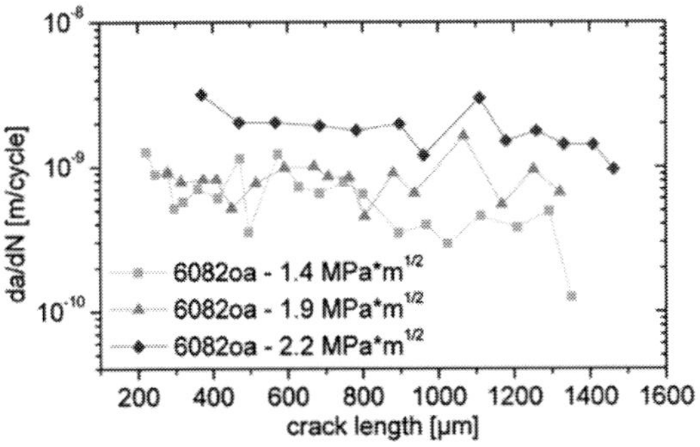

Figure 6. Crack growth rates vs. crack length at constant ΔK values for alloy EN AW 6082 in the overaged condition.

3.3 Correlation between microstructure and fatigue crack growth rate

By investigating the crack path on the surface of the samples, it becomes apparent that the decelerations of the crack growth rate are mainly caused by microstructural features. Especially the primary precipitates in alloy EN AW 6082 have a great influence on the long crack propagation in the VHCF regime.

Figure 7a shows the crack path for alloy EN AW 6082 oa at 1.4 MPa√m, corresponding to the da/dN vs. a curve shown in Figure 4. In this curve there are some decelerations of the crack growth rate visible, for example the decrease at a crack length of about 500 µm. This decrease is mainly caused by the right crack tip, which reaches at this point a cluster auf Fe-based primary precipitates (Figure 7a, yellow ellipse). At this point the precipitates act as a microstructural barrier. The crack growth rate decreases also slightly at the end of the test. This is also caused by primary precipitates (as can be seen in Figure 7a), but in this case by a large Mg_2Si precipitate (as was determined by EDS analysis). The right crack tip stops entirely at this point. The further slight decrease of the crack growth rate is caused by crack growth measurable on the left side of the FIB notch. As is was described earlier the crack growth curves are considering the overall crack propagation. All in all, it can be determined that both types of precipitates act as microstructural barriers in EN AW 6082.

Similar barrier effects can be seen for the peak-aged condition at 1.5 MPa√m (see Figure 8 and corresponding curve in Figure 4). The crack growth rate decreases after 400 µm, caused by a large cluster of Fe-based primary precipitates. After passing these precipitates the crack growth rate increases again.

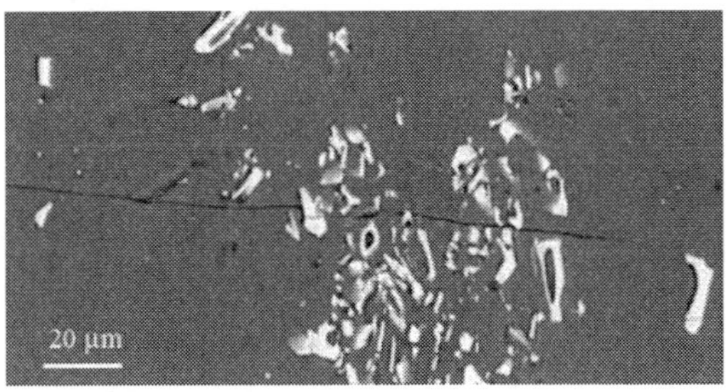

Figure 8. Crack path detail of alloy EN AW 6082 pa at ΔK = 1.5 MPa√m.

In EN AW 6082 grain boundaries seem not to have such a great influence on the crack growth as the primary precipitates. An example for their minor influence can be found in the overaged material (see Figure 9). After a crack growth of about 140 µm the crack growth rate decelerates. At this point, there is no obvious primary precipitate in the course of the crack, which could cause such a decrease, while a grain boundary can be detected at this point instead. However, the other grain boundaries do not have such a pronounced effect. Which role the orientation of the grain boundary relative to the crack path and the orientation of adjacent grains may possibly play, regarding the function as a microstructural barrier is still under investigation. In addition, the barrier function of the grain boundaries may also be affected by precipitates, because grain boundaries are known as preferred sites for precipitates [12].

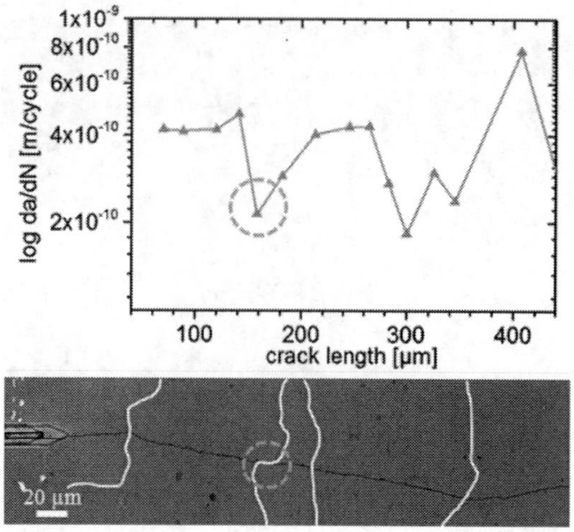

Figure 9. Crack growth rate (right to the FIB notch) for the overaged condition at $\Delta K = 1.9$ MPa√m with corresponding SEM image (grain boundaries in white).

Figure 7b shows the crack path of alloy EN AW 5083 at 2.1 MPa√m (corresponding curve in Figure 5). Obvious local minima were not visible in the da/dN vs. a curve, which can be correlated to only few primary precipitates in front of the crack tip in this test. While regarding only the right crack tip concerning the crack growth rate, there was a drop of the crack growth rate below 10^{-10} m/cycle measurable at a crack length of about 430 µm. This can be attributed to the Fe-based precipitate (Figure 7b, yellow ellipse). Furthermore, there are some kinks visible in the course of the crack, which can be associated to passing grain boundaries. When the crack path changes its growth direction this leads also to a deceleration in crack growth rate. Because of the smaller average grain size, this deceleration is not clearly visible in the curve but leads to a smaller average crack growth rate. Figure 10 shows an electron backscatter diffraction (EBSD) analysis of the microstructure surrounding the crack path at 1.7 MPa√m for alloy EN AW 5083. The end of the crack path, where more of the smaller grains are visible, corresponds to the decreasing crack growth rate (see Figure 4). Hence, the barrier effect of the grain boundaries became more effective at the end of the crack path observed in Figure 10. Determining the slip systems with the highest Schmid factor for this particular microstructural detail could show that the crack propagated sometimes in a shear-controlled manner (indicated by white arrows in Figure 10) rather than in mode-I, as is normally typical for long fatigue cracks. Some regions of shear-controlled crack propagation could be found for alloy EN AW 6082 as well, but rarely.

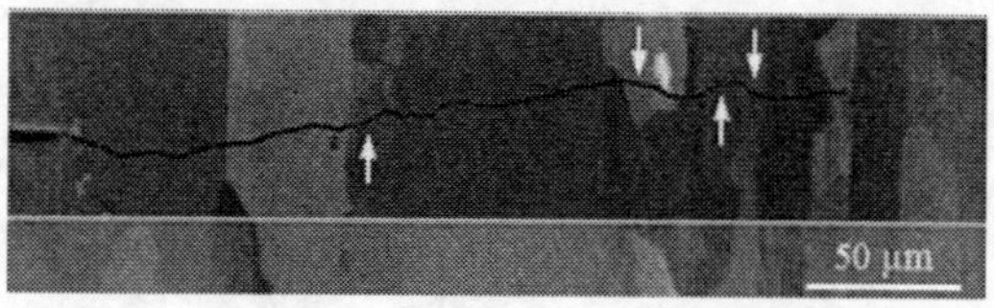

Figure 10. Crack path of alloy EN AW 5083 at 1.7 MPa√m with overlaying EBSD analysis.

3.4 Fracture surfaces

Figure 11 shows the fracture surfaces for the peak-aged and the overaged material. The semi-elliptical shape of the fracture surface, which is characteristic for surface cracks, is clearly visible. Only the ratio of crack length to depth (a/c) differs between the two samples. Alloy EN AW 6082 in the pa condition has a ratio of 0.67, whereas the overaged condition has a ratio of 0.78. Reason for this is supposed to be the higher ΔK value compared to the peak-aged condition. It can therefore be reasoned that the crack propagation at the surface is representative for the overall crack growth. The crack propagation on the surface, as well as the propagation into the material are adapting qualitatively to the different ΔK values.

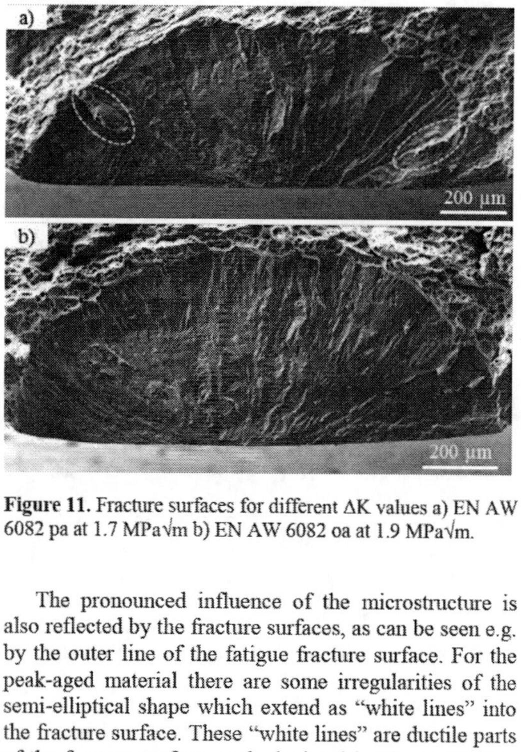

Figure 11. Fracture surfaces for different ΔK values a) EN AW 6082 pa at 1.7 MPa√m b) EN AW 6082 oa at 1.9 MPa√m.

The pronounced influence of the microstructure is also reflected by the fracture surfaces, as can be seen e.g. by the outer line of the fatigue fracture surface. For the peak-aged material there are some irregularities of the semi-elliptical shape which extend as "white lines" into the fracture surface. These "white lines" are ductile parts of the fracture surface, as is depicted in Figure 12. It is assumed that the crack front has been pinned at these sites. This feature can be found on the overaged fracture surface as well but is not as pronounced because of the higher ΔK value. By comparing the fracture surfaces at similar ΔK values the pinning is more pronounced in the overaged material. This has been also investigated by Stein et al. using flat samples [8]. The pinning effect is caused by primary precipitates, which however do not differ in alloy EN AW 6082 in both conditions. The reason for the higher pinning effect might be based on the coarsening of the secondary precipitates, which requires still further investigations.

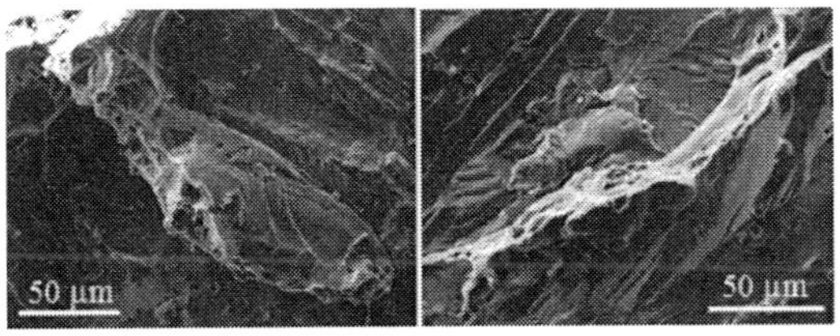

Figure 12. Detailed view of the fracture surface of Figure 11a (yellow ellipses).

Figure 13 shows the fracture surface of alloy EN AW 5083 at 2.1 MPa√m. The fracture surface itself has a much more ductile character than the ones of alloy EN AW 6082. It is assumed that the higher plastic fraction is influenced by a different size of the plastic zone in front of the crack tip, which might have caused already the lower crack propagation rate (see section 3.2). It is also apparent that the crack depth compared to the crack front is quite large; in detail the sample has a depth to length ratio of about 0.93.

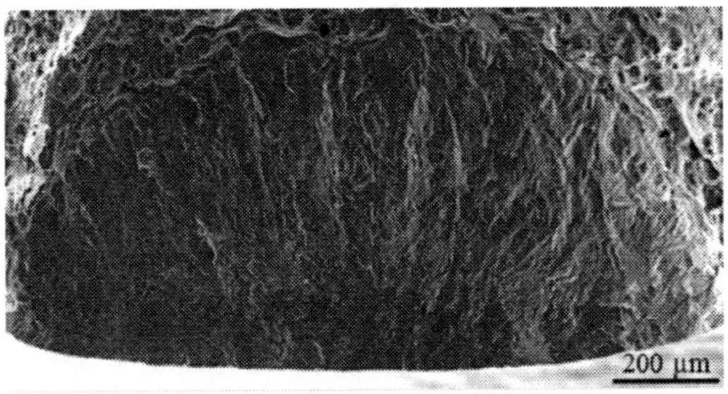

Figure 13. Fracture surface of alloy EN AW 5083 at $\Delta K = 2.1$ MPa√m.

4 Conclusion

The present research had the aim to experimentally investigate the long crack growth behaviour for two different aluminium alloys near the threshold regime and therefore at very small load amplitudes. The experiments have shown that the microstructure has a barrier function on the fatigue crack propagation for all three investigated material conditions in this region. For the fatigue crack growth behaviour of the two heat treatment conditions of alloy EN AW 6082 no significant differences could be observed. The primary precipitates, especially the Fe-based ones, have the strongest influence on the crack growth. Grain boundaries can decelerate the crack growth only in very rare occasion, while the role of the secondary precipitates is still under investigation. While analysing the fracture surfaces it seems that the primary precipitates have a greater pinning effect in the overaged condition. For alloy EN AW 5083 a smaller average crack growth rate can be attributed to the smaller grain size and the increasing barrier function of grain boundaries. Primary precipitates are mainly responsible for local changes in the crack growth rate, as was observed for EN AW 6082.

Acknowledgement

The authors would like to thank the German Research Foundation (DFG) for financial support.

References

[1] H.-J. Christ, M. Zimmermann. Proceedings of the 7th International Conference on Very High Cycle Fatigue. Siegener Werkstoffkundliche Berichte (2017)

[2] C. Berger, B. Pyttel, D. Schwerdt. Mat.-wiss. u. Werkstofftech. **39** (2008) 769-776

[3] H. Mughrabi. Fat. Fract. Engng. Mater. Struct. **22** (1999) 633-641

[4] S. Stanzl-Tschegg. Int. J. of Fatigue **60** (2014) 2-17

[5] ASTM E 647-11. Standard test method for measurement of fatigue crack growth rates. Philadelphia: American Society for Testing and Materials (ASTM); 2011.

[6] J. Bach, H.W. Höppel, M. Prell, M. Göken. Int. J. of Fatigue **60** (2014) 23-27

[7] E. Zaiken, R. O. Ritchie. Mat. Sci. & Eng. **70** (1985) 151-160

[8] M. Wicke, A. Brückner-Foit, T. Kirsten, M. Zimmermann, F. Bülbül, H.-J. Christ. Proc. Struct. Int. **7** (2017) 235-241

[9] B. Tabernig, R. Pippan. Eng. Fract. Mech. **69** (2002) 899-907

[10] T. Stein, M. Wicke, A. Brückner-Foit, T. Kirsten, M. Zimmermann, F. Bülbül, H.-J. Christ. J. Mat. Res. **23** (2017) 1-8

[11] U. Zerbst, M. Vormwald, R. Pippan, H.-P. Gänser, C. Sarrazin-Baudoux, M. Madia. Eng. Fract. Mech. **153** (2016) 190-243

[12] J. W. Martin. Precipitation Hardening – Second Edition. Butterworth-Heinemann (1998)

The Potential of Self-Tempered Martensite and Bainite in Improving the Fatigue Strength of Thermomechanically Processed Steels

Ulrich Krupp[1,*], *Mikhail* Solovev[1], *Felix* Honecker[1], *Bernhard* Adams[2], and *Jan-Christoph* Florian[3]

[1]Institute of Materials Design and Structural Integrity, University of Applied Sciences Osnabrück, 49009 Osnabrück, Germany

[2]Institute of Production Technology, University of Applied Sciences Osnabrück, 49009 Osnabrück, Germany

[3]Georgsmarienhütte GmbH, Neue Hüttenstraße 1, 49124 Goergsmarienhütte, Germany

Abstract. In contrast to a two-stage hardening and tempering process, the definition of optimized cooling routes after hot working of low-alloy Cr steel allows the adjustments of high-strength microstructures with a sufficient degree of ductility at the same time without any additional heat-treatment. While compressed air cooling after hot forging of micro-alloyed steel grades leads to the formation of lower bainite with fine-dispersed cementite platelets, quenching by water spray down to the martensite start temperature results in the formation of martensite, that is self-tempered during the subsequent slow-cooling in air. The precipitation of nano-sized cementite precipitates result in superior mechanical properties with respect to impact and tensile testing. Cyclic deformation and crack propagation tests being carried out using resonance testing (100Hz) and ultrasonic fatigue testing (20kHz) systems revealed a pronounced increase in fatigue strength by about 150MPa of the self-tempered martensite condition as compared to the bainitic modification. For the latter one, a steady decrease of the fatigue strength is observed rather than the existence of a real fatigue limit.

1 Introduction

Tempered martensitic steels are used in virtually all kind of engineering applications, which imply cyclic deformation at high load levels, e.g., power train in transportation, gear boxes or crack shafts in power generation. The excellent combination of strength and ductility can be attributed to (i) an optimum in size and distribution of carbides, (ii) a small prior austenite grain size and (iii) a sufficient ductility of the martensitic bulk material. These parameters are determined by the way of heat treatment. Typically, the steels are quenched after austenitization followed by a one- or two-stage tempering treatment. More recently, bainitic microstructures have been classified to exhibit a very similar microstructure than tempered martensitic steels (cf. [1]). During rapid quenching, the austenite remains supersaturated in carbon, eventually resulting in small ε-carbide platelets within the ferritic matrix, when quenching is passed to isothermal phase transformation (cf. Figs. 1 and 4).

Since the establishment of bainitic microstructures is much more cost-efficient than the processing of quenched and tempered steel, research has been focussed to optimize the fatigue properties of bainitic steels (e.g. [2]). With respect to the behaviour in the very-high-cycle fatigue regime (VHCF), the failure mechanisms are not fully understood yet. Even the question about the existence or non-existence of a real fatigue limit cannot be answered today. A characteristic feature of VHCF failures of high strength steels is the occurrence of crack initiation at internal microstructural defects, such as non-metallic inclusions. The defects are surrounded by a so-called fine-granular or optically dark area (FGA or ODA). The origin of this area is attributed to the accumulation of cyclic plasticity leading to local grain refining (polygonization, [3]) and/or hydrogen concentration [4]. Since the threshold stress intensity range for crack initiation decreases in a refined grain microstructure, the steel may fail at a very low stress amplitude, which is far below the macroscopic threshold for technical fatigue cracks (cf. [5,6]).

According to Murakami [7], the fatigue limit of high-strength steels can be correlated with the material's strength (Vickers hardness HV) and the size of the non-metallic inclusions (projected area), only, according to

$$\sigma_{FL} = \frac{C(HV+120)}{\left(\sqrt{area}\right)^{1/6}}\left(\frac{1-R}{2}\right)^{\alpha}, \qquad (2)$$

with R being the stress ratio, and C and α being material constants. From the correlation of different steel heat treatments with the respective fatigue properties it is

* Corresponding author: u.krupp@hs-osnabrueck.de

known that beside the material's strength and inclusion size, the bulk ductility, the prior austenite grain size as well as the size and distribution of carbides are significant in determining the VHCF behavior of heat-treated steel. Hence, it is the aim of the present study to optimize the microstructure of a micro-alloyed forging steel 16CrMnV 7 7 (German designation) in such a way that it can be used immediately after the forging process without an additional quenching and tempering treatment.

2 Experimental

The material studied in the present work is a micro-alloyed bainitic steel 16 CrMnV 7 7 (German designation), applied for forging applications in automotive industry, e.g., cam shafts. Its chemical composition is given in Table 1.

Table 1. Chemical composition of the micro-alloyed 16 CrMnV 7 7 grade steel (in wt.%).

Fe	C	Mn	Cr	Mn
Bal.	0.16	1.7	1.7	0.7

P	S	Ni	V+Nb+Al+N
0.013	0.017	0.16	0.17

The steel was delivered as round bars of 50mm diameter. Sections of these bars were heated up to 1250°C and 950°C, respectively, by means of an 70kW induction coil and hot-forged using a 10000N screw press (Weingarten). Immediately after deformation, the bars were quenched according to defined cooling conditions: (i) 1250°C => compressed air to 260°C => keeping isothermally in furnace at 260°C => air-cooling to RT (cooling rate $t_{8/5}$=103s), and (ii) water quenching to martensite start temperature (M_s) => air-cooling to RT (cooling rate $t_{8/5}$=56s). The two different cooling conditions are represented in the schematic time-temperature-transformation (TTT) diagram in Fig. 1, where (i) aims to isothermal bainite formation, and (ii) to martensite formation followed by self-tempering during decelerated cooling below M_s.

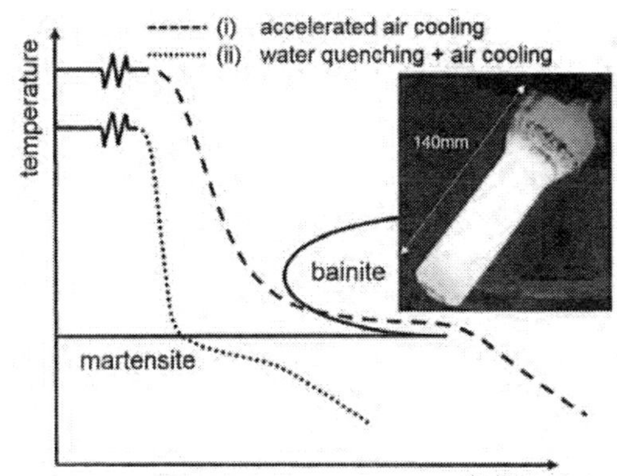

Fig. 1. Thermomechanical processing by means of two different cooling routes in a schematic TTT diagram for the steel 16CrMnV 7 7 applied to hot-forged bars (photograph).

From the thermomechanically processed bars (true deformation strain φ≅1), cylindrical specimens for ultrasonic fatigue testing at 19000Hz (UFTE system type BOKU Vienna) and for resonance fatigue testing at 95Hz (Rumul Testronic 100kN, Fig. 2) were machined. The latter ones were given a shallow notch to allow in-situ observation of surface damage by means of digital light microscopy (Hirox, Fig. 2). All tests were carried out under stress control at fully-reversed cycling (R=-1) and terminated when the resonance frequency dropped by 2%, which was attributed to macroscopic fatigue damage (end of life).

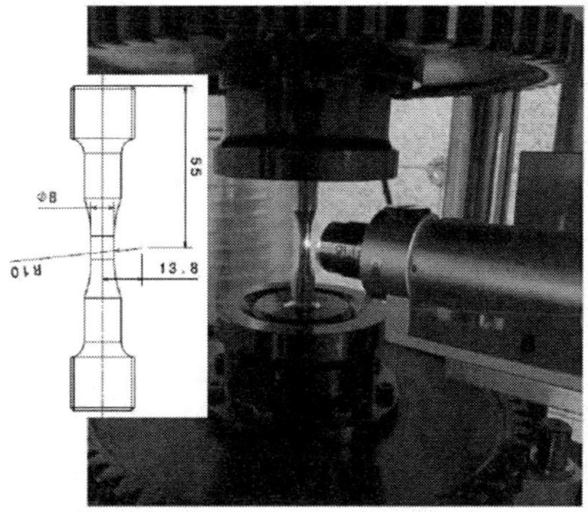

Fig. 2. Resonance fatigue testing (Rumul Testronic) with digital microscope for damage tracking using a shallow-notched specimen geometry.

After cycling, the fracture surfaces of the specimens were analyzed by scanning electron microscopy (SEM) in combination with energy-dispersive X-ray spectroscopy (EDS). In the case of internal fatigue crack initiation, the characteristic features, e.g., fine granular areas (FGA), were cut by focused ion beam milling (FIB SEM Zeiss Auriga FEG) to obtain cross-section information about local microstructure changes.

3 Results and Discussion

3.1 Steel Microstructure

After thermomechanical processing, the forged bars were cut and investigated with respect to (i) grain size, (ii) martensite/bainite lancets, and (iii) size and distribution of cementite precipitates. Fig. 3 shows the light micrographs of the bainitic microstructure (accelerated air cooling, Fig. 3a) and the martensitic microstructure (water-quenching + air cooling, Fig. 3b). Since the bainitic samples were forged at 1250°C, the prior austenite grain size is much larger (53μm, Fig. 3a)

 https://doi.org/10.1051/matecconf/201816520006

than in the case of the martensitic microstructure (8μm, Fig. 3b).

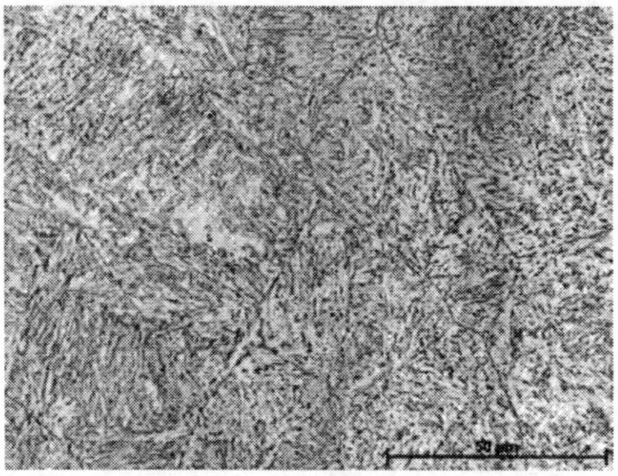

a

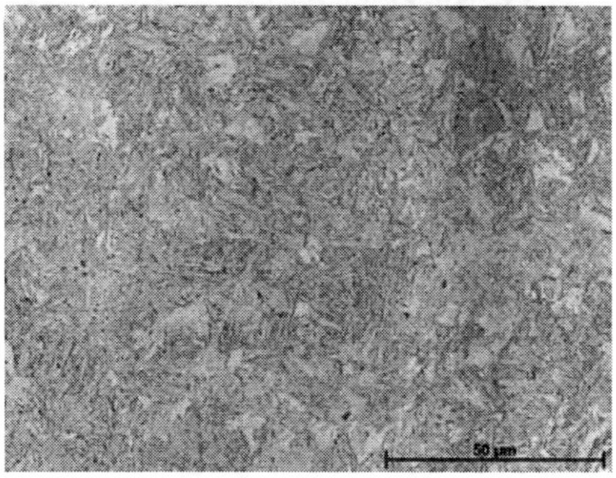

b

Fig. 3. Light micrographs of thermomechanically processed steel 16 CrMnV 7 7: (a) bainitic microstructure after accelerated air cooling, and (b) martensitic microstructure after water quenching and air cooling.

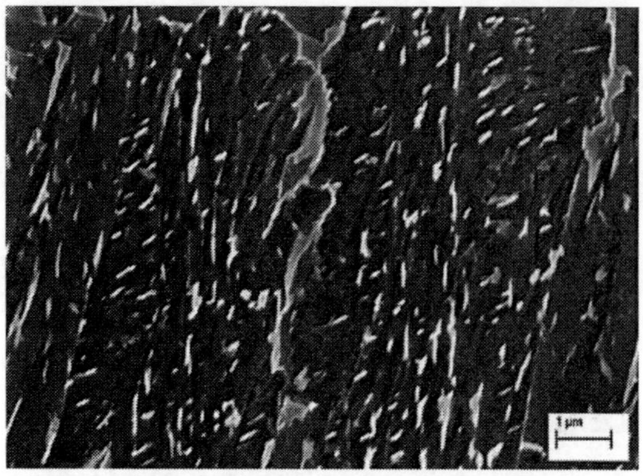

a

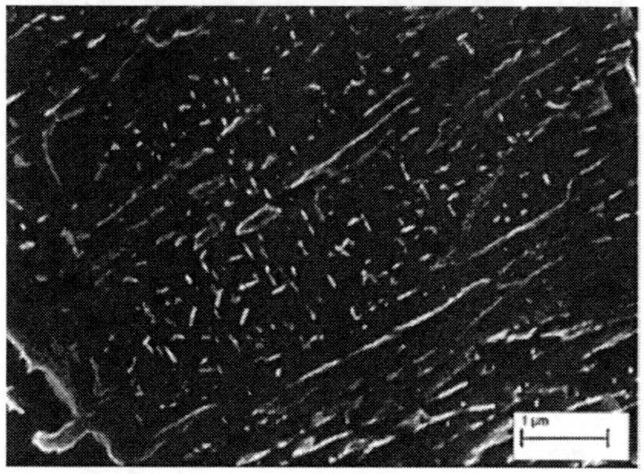

b

Fig. 4. SEM micrographs of thermomechanically processed steel 16 CrMnV 7 7: (a) large cementite precipitates along the bainite lancets and small cementite precipitates tilted by 60° towards the lancet axes, (b) martensitic microstructure showing nano-sized cementite precipitates as a result of self-tempering.

High-resolution SEM revealed a homogeneous distribution of small cementite precipitates. In the case of the bainitic microstructure, the ferrite lancets are separated by thin cementite films, while plate-shaped cementite precipitates of 100-200nm length are formed 60° tilted towards the lancet axes. The corresponding hardness was measured as 372HV30 and the yield strength as YS=870MPa. In the case of the martensite microstructure, 1μm sized cementite plates are formed along the martensite block boundaries, while tiny (<100nm) homogeneously distributed cementite precipitates were identified within the martensite blocks. Obviously, the transition from rapid quenching in water to slow-cooling in air immediately below the martensite start temperature M_s gave rise to self-tempering of the martensite. The hardness and the yield strength of the self-tempered martensite is substantially higher; 461HV30 and YS=1000MPam, respectively. Surprisingly, the ductility of the self-tempered material, measured as Charpy impact work, A_v=104J, is much higher than in the case of the bainitic material, Av=17J.

3.2 Fatigue Behavior

Fatigue testing of the thermomechanically processed specimens (according to section 3.1) supported the results of the monotonic tensile tests, i.e., the fatigue strength of the self-tempered martensitic material is substantially higher than that of the bainitic material. This is shown in the S/N Wöhler diagrams in Fig. 5 summarizing the results of the fatigue tests. For the bainitic material (Fig. 5a), no failure up to 1 billion cycles was observed at a stress amplitude of σ_a=375MPa (hence, this value can be understood as technical fatigue limit). In the case of the self-tempered martensite (Fig. 5b), the respective fatigue limit was identified at a stress amplitude of σ_a=600MPa. Since the analysis was focused to the fatigue behavior in the HCF and the VHCF regime, in most cases internal crack initiation was observed; except the four bainitic specimens loaded at stress amplitudes larger than 600MPa (Fig. 5a).

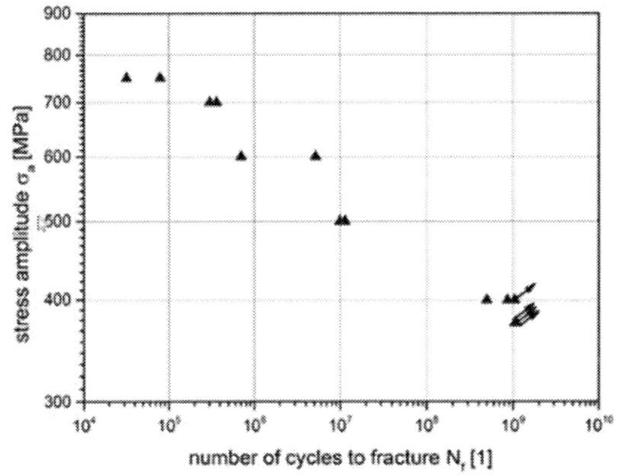

a

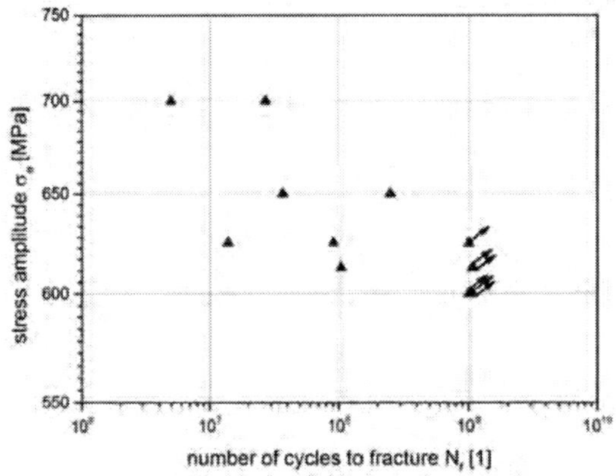

b

Fig. 5. S/N Wöhler diagram of specimens made of thermomechanically processed 16 MnCrV 7 7: (a) lower bainite microstructure (after accelerated air cooling), and (b) martensitic microstructure after water quenching and air cooling (UFTE testing was used for VHCF testing at N>10^7).

a

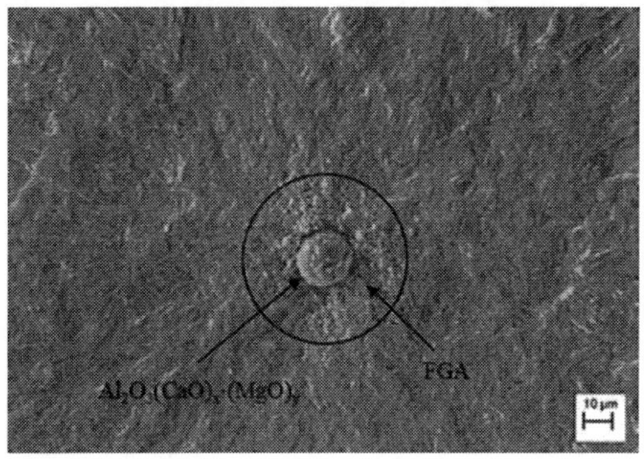

b

Fig. 6. FGA of a self-tempered martensitic specimen failed after 91 million cycles at a stress amplitude of σ_a=625MPa (cf. Fig. 5b): (a) fracture surface showing FGA at non-metallic inclusion and (b) corresponding FIB section through the FGA.

In the case of the self-tempered martensite, all VHCF-loaded specimens showed internal crack initiation (cf. Fig. 6a). By means of fracture surface analyses, non-metallic inclusions, typically of sizes between 10µm and 30µm, were identified as fatigue-crack-initiation sites. As it was reported by various authors before, the inclusions are surrounded by fine granular areas (FGA), the size of which increases with decreasing load amplitude (cf. [6]). Fig. 6b shows a typical example of a mixed Al-Ca-Mg oxide (analyzed by EDS) as VHCF-crack-initiating defect. Accounting to the Murakami equation (eq. 1), the inclusion shown in Fig. 6b ($area^{0.5}$=17.6$\mu m^{0.5}$) corresponds to an estimated fatigue limit of σ_{FL}=562MPa. Actually, the specimen was broken after almost 100 million cycles at a stress amplitude of σ_a=625 MPa. Hence, it seems that the Murakami equation provides a rather conservative estimation of the fatigue limit. The majority of the identified crack initiating defects are somewhat larger, i.e. $area^{0.5}$ between 17$\mu m^{0.5}$ and 56$\mu m^{0.5}$.

Within any of the fracture surfaces of the VHCF-loaded bainitic specimens, internal crack initiation and the formation of a FGA was observed as well. However, in no case a non-metallic inclusion was identified as crack-initiating defect (cf. Fig. 7a). This surprising result - the bainitic specimens were produced from the same steel batch than the martensitic ones – is attributed to both (i) the larger prior austenite grain size, probably leading to local stress concentrations that exceed the stress concentrations at the non-metallic inclusions, and (ii) the lower matrix strength, giving rise to a more pronounced cyclic plasticity around the crack-initiating defect. The nature of the non-inclusion FGA itself was proven as to be the same as in the case of the well-known inclusion FGA. FIB milling of the FGA shown in Fig. 7a revealed an approx. 2µm thick nano-grained layer (cf. Fig. 7b). It is worth mentioning that by serial FIB milling several micro debondings were identified, as it is shown in Fig. 7b. The observation of micro decohesion between carbides and matrix along prior austenite grain boundaries as origin of FGA formation was reported by Shiozawa et al. in [8]. Hence, it can be concluded that the observed micro-debonding effects are a consequence of microcracking, and are eventually the origin of the formation of a rough FGA layer. Therefore, micro debondings seem to play an important role during VHCF fatigue crack initiation and early growth process. Further analysis of their nature is subject of ongoing work.

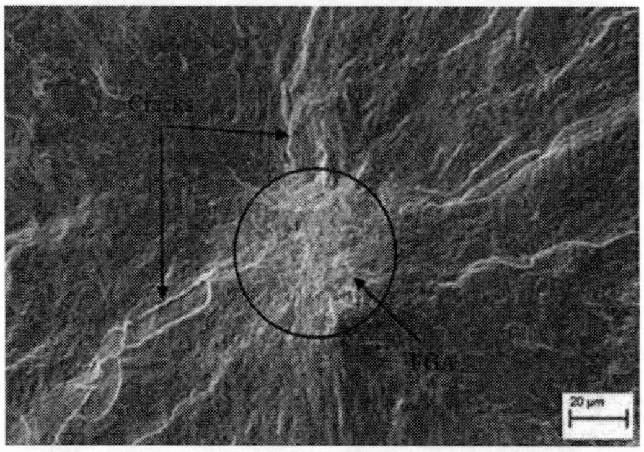

a

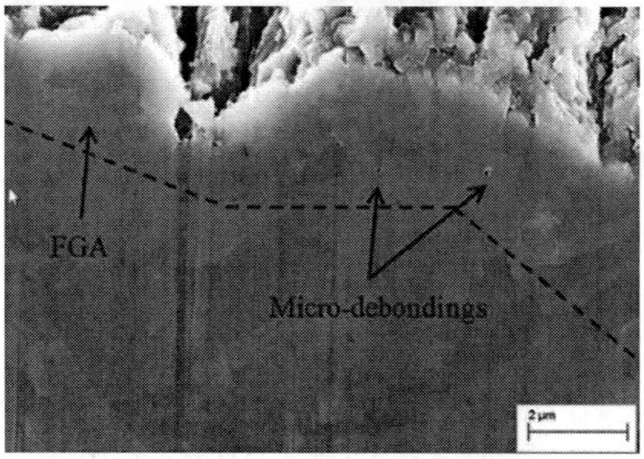

b

Fig. 7. FGA of a bainitic specimen failed after 870 million cycles at a stress amplitude of σ_a=400MPa (cf. Fig. 5a): (a) fracture surface showing FGA without any non-metallic inclusion and (b) corresponding FIB section through the FGA.

4 Conclusions

Thermomechanical processing of martensitic bainitic steel, i.e., hot forging followed by a defined cooling procedure, allows the establishment of tailored microstructures exhibiting a superior combination of strength and ductility. In the case of a micro-alloyed 0.16C-1.7Mn-1.7Cr forging steel, partial quenching in water followed by slow-cooling in air immediately below the martensite start temperature led to a self-tempering effect. The resulting homogeneously distributed nano-cementite platelets give the material not only a high yield strength and impact toughness, but also a high fatigue limit of at least 600MPa. In agreement to earlier work, VHCF failure is observed at numbers of cycles exceeding 100 million, typically initiating at internal defects surrounded by a fine granular area (FGA). Surprisingly, only in the case of the self-tempered martensitic structure the FGA is formed around non-metallic inclusions. In the case of the bainitic microstructure, only non-inclusion-type FGAs were identified; obviously in this case, the prior austenite grain boundaries are the weakest link leading to accumulated VHCF strain localization followed by polygonization and crack propagation into the FGA.

5 Acknowledgement

The financial support by the Volkswagen foundation (Lower Saxony Germany) in the framework of the research priority program OPTIHEAT and the supply of steel specimen material by Georgsmarienhütte GmbH are gratefully acknowledged.

References

1. H.K.D.H. Bhadeshia, *Bainite in Steels* (The Institute of Metals, London, 2001)
2. E. Kerscher, K.-H. Lang, Proc. Engng **2** 1731 (2010)
3. T. Sakai, J. of Solid Mech. Mater. Engng, **3** 425 (2009)
4. Y. Murakami, T. Nomoto, T. Ueda, Fatigue Fract. Engng. Mater. Struct. **23** 893 (2000)
5. P. Grad, B. Reuscher, A. Brodyanski, M. Kopnarski, E. Kerscher, Scripta Mater. **67** 838 (2012)
6. U. Krupp, A. Giertler, K. Koschella, Fatigue Fract. Engng. Mater. Struct. **40** (2017) 1731
7. Y. Murakami, Metal Fatigue: *Effects of Small Defects and Nonmetallic Inclusions* (Elsevier, Oxford 2002)
8. K. Shiozawa, Y. Morii, S. Nishino L. Lu, Intern. J. Fatigue **28** 1521 (2006)

Crack growth behaviour of aluminium wrought alloys in the Very High Cycle Fatigue regime

Fatih Bülbül[1,*], *Tina* Kirsten[2], *Marcel* Wicke[3], *Martina* Zimmermann[2], *Angelika* Brückner-Foit[3], and *Hans-Jürgen* Christ[1]

[1]Institut für Werkstofftechnik, Universität Siegen, 57068 Siegen, Germany
[2]Institut für Werkstoffwissenschaft, Technische Universität Dresden, 01069 Dresden, Germany
[3]Institut für Werkstofftechnik, Universität Kassel, 34125 Kassel, Germany

Abstract. Investigations have shown that in the regime of Very High Cycle Fatigue (VHCF) "natural" crack initiation often takes place underneath the material surface leading to crack propagation without contact to atmospheric components. In order to elucidate the environmental damage contribution and its effect on the VHCF long crack propagation, fatigue experiments with alternating environment (vacuum and laboratory air) were performed. An ultrasonic fatigue testing system (USFT) equipped with a small vacuum chamber was applied that enables the in-situ examination of the long fatigue crack propagation at a resonance frequency of about 20 kHz by using a long distance microscope. By means of the Focused-Ion-Beam technique, micro-notches were prepared in the USFT specimens. The tests were carried out on the aluminium alloys EN-AW 6082 and 5083 in different conditions. It has been found that the atmosphere has a significant influence on the VHCF long crack propagation which manifests itself in the crack path as well as in the crack growth rates. Because of pronounced single sliding in vacuum, shear-stress-controlled crack propagation was detected whereas in laboratory air normal-stress-controlled crack propagation occurred. Furthermore, it has been proven that the secondary precipitation state of the aluminium alloy significantly influences the VHCF long crack propagation in vacuum.

1 Introduction

In the regime of Very High Cycle Fatigue (VHCF), that means at very low stress amplitudes, "natural" cracks often initiate internally underneath the material surface [1, 2]. Internal cracks propagate without contact to atmospheric components, thus, crack propagation in vacuum can be assumed. However, up to now it is unknown how the fatigue crack propagation in the VHCF regime exactly takes place and it is questionable, if all the crack growth phases of the conventional crack growth curve occur depending on this condition.

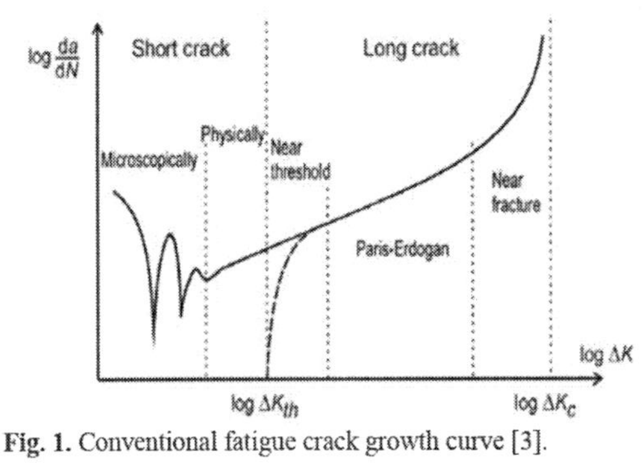

Fig. 1. Conventional fatigue crack growth curve [3].

Fig. 1 shows the conventional fatigue crack growth curve. In the short crack regime, the fatigue crack propagation is highly influenced by the microstructure, therefore, the crack propagation rates reveal a strong fluctuation. In the long crack regime, however, the fatigue cracks propagate continuously and can be described by the Paris-Erdogan law.

It is a well-known fact that the humidity of ambient air deteriorates the fatigue properties of aluminium alloys compared to an inert environment [4-6]. In this study, the atmospheric influence on the VHCF long crack propagation in the aluminium wrought alloys EN-AW 6082 and 5083 in different material conditions is investigated. Fatigue experiments with changing environment between laboratory air and vacuum were performed by means of the ultrasonic fatigue testing technique equipped with a small vacuum chamber. The detrimental effect of water vapour in ambient air as well as the effect of alloy microstructure on the long crack propagation will be discussed in this article.

2 Materials and methods

In this study, the aluminium wrought alloys EN-AW 6082 and EN-AW 5083 were investigated whereby the precipitation hardenable alloy was examined in the peak-aged (pa) as well as overaged (oa) condition. The heat treatment was performed at the Technische Universität Dresden. The solid solution strengthened aluminium

* Corresponding author: fatih.buelbuel@uni-siegen.de

 https://doi.org/10.1051/matecconf/201816520007

alloy EN-AW 5083 H111 has been strain-hardened and represents the precipitate-free condition.

Table 1. Ultimate tensile strength (R_m), yield stress ($R_{p0,2}$) and Vickers Hardness for the investigated aluminium alloys.

	R_m (MPa)	$R_{p0,2}$ (MPa)	Vickers Hardness
EN-AW 6082 (pa)	366	344	145 HV0.2
EN-AW 6082 (oa)	294	262	105 HV0.2
EN-AW 5083 H111	270	110	106 HV0.2

Figure 2 was generated at the Technische Universität Dresden by means of the transmission electron microscopy (TEM) showing secondary precipitates of the aluminium alloy EN-AW 6082 in peak-aged (Fig. 2a) and over-aged (Fig. 2b) condition. Due to the longer heat treatment, the secondary precipitates in the over-aged condition are coarsened.

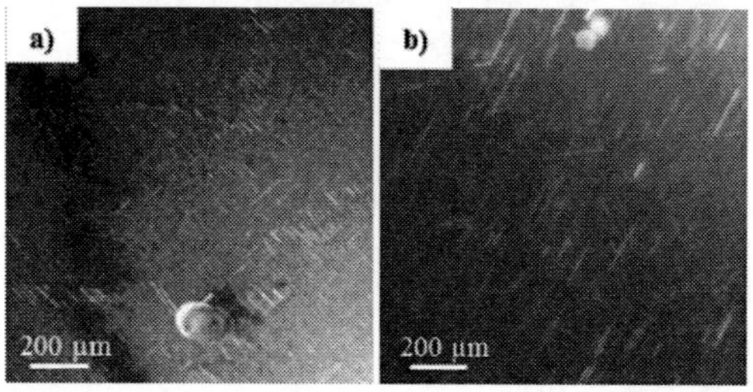

Fig. 2: Secondary precipitates of the aluminium alloy EN-AW 6082: a) peak-aged and b) over-aged condition (generated at the Technische Universität Dresden by means of TEM).

The fatigue crack growth experiments were carried out using an ultrasonic fatigue testing system (USFT) equipped with small vacuum chamber (Fig. 3).

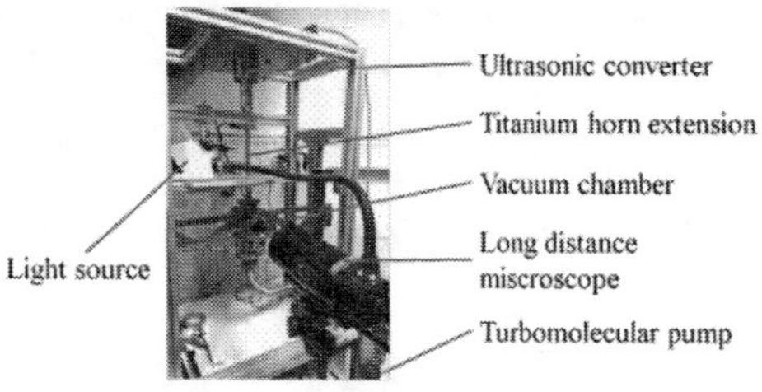

Fig. 3. In-situ setup of the USFT equipped with a small vacuum chamber.

The fatigue experiments were performed at resonance frequency of about 19.2 kHz under fully reversed tension-compression loading (R = -1). The pulse and pause sequences were adjusted for the fatigue in vacuum in order to prevent that the material temperature exceeds 30 °C. Fatigue experiments with changing environment (vacuum and laboratory air) were performed by means of interspersed testing sections in air. The ultrasonic specimens include a shallow-notch in order to monitor the fatigue crack propagation on the material surface by using a long distance microscope. Prior to the fatigue experiments a micro-notch was milled into the mechanically and elctrolytically polished ultrasonic specimens by means of the Focused-Ion-Beam (FIB) technique. In principle, the shape of the FIB micro-notch was chosen according to the research results of the Universität Kassel (Fig. 4a) [7].

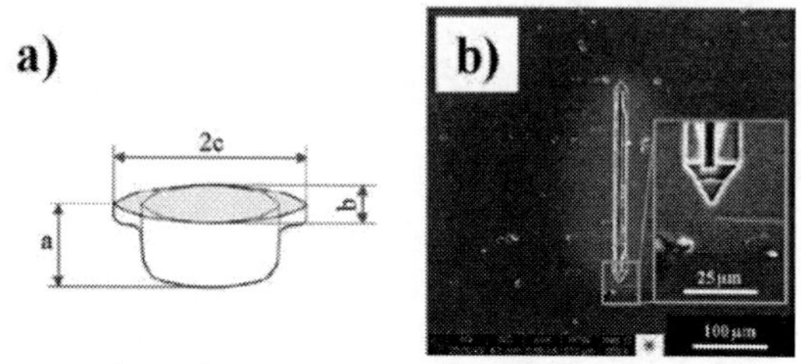

Fig. 4. Shape of micro-notch: a) schematic illustration of a micro-notch used at the Universität Kassel [7] and b) FIB micro-notch prepared at the Universität Siegen.

Figure 4b shows a detail view of the depth difference at the end of the FIB micro-notch leading to high local stress concentration during cyclic loading. Hence, crack initiation can be successfully realized even at very low stress amplitudes at both ends of the FIB micro-notch.

3 Results

Fatigue experiments with changing environment were performed for all the aluminium wrought alloys. Therefore, four characteristic areas result in the crack path named as Air 1, Vacuum 1, Air 2 and Vacuum 2.

3.1 Crack propagation behaviour of EN-AW 6082 (pa)

A FIB micro-notch with a length of 120 µm was prepared in an ultrasonic specimen. In Fig. 5 the crack path, the EBSD analysis and the corresponding crack growth rates are shown. Crack initiation took place under constant stress amplitude of 125 MPa in laboratory air. In the beginning part in section Air 1, the right crack tip started to propagate at an angle of about 45° to the loading axis and shows a correspondence to the calculated glide trace with the highest Schmidfactor. But after reaching a certain crack length, crack propagation perpendicular to the loading axis was detected. After that, the environmental condition was changed to Vacuum 1. In Fig. 5a, slip bands (blue ellipse) are clearly formed on the material surface corresponding on the right side to the glide trace with highest Schmidfactor. Furthermore, the crack path on both sides deviates significantly from the normal direction. On the left as well as right side, it can be seen that the crack followed the glide trace with the second highest Schmidfactor. Moreover, a comparable behaviour is shown in a sub-grain (marked as red rectangle in Fig. 5b) at the left crack path. A correlation between the slip band formation on the material surface and the calculated glide trace with the highest Schmidfactor is reconfirmed.

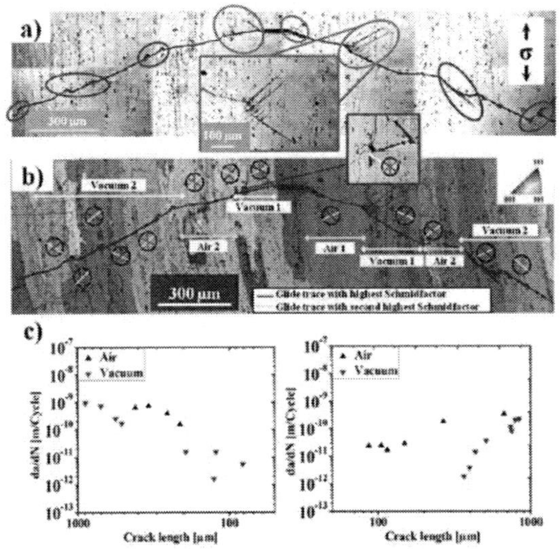

Fig. 5. Crack path in EN-AW 6082 (pa) at $\Delta\sigma/2$ = 125 MPa and changing environmental conditions: a) slip band formation on the material surface, b) EBSD-analysis with calculated glide traces and c) crack growth rates of the left and right crack tip.

In section Air 2, the crack propagated perpendicularly to the loading axis and no correlation between the crack path and the calculated glide traces is observed. In section Vacuum 2, the crack path again shows significant differences to the crack propagation in air. Within the grain on the right side of the crack tip, strong crack deflection and slip band extension of the crack path is detected (Fig. 5a and b). Moreover, a clear correlation between the crack path (left and right) and the calculated glide traces exists. The crack propagation rates are shown in Figure 5c. In vacuum the crack growth rates are significant smaller than those in air. The differences between air and vacuum can be also recognised in the fracture surface morphology which is shown in Fig. 6.

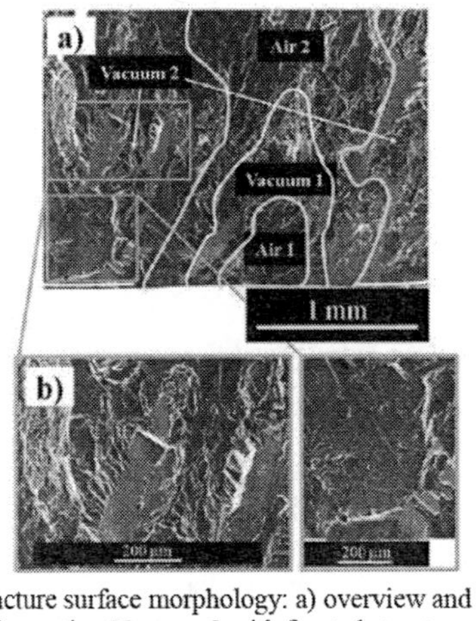

Fig. 6. Fracture surface morphology: a) overview and b) detail views of the section Vacuum 2 with faceted structures.

In Fig. 6a, the sections Air 1 and Air 2 show fatigue fracture surfaces which are homogeneous and differ significantly from those observed in the sections Vacuum 1 and Vacuum 2. In the detailed views in Fig. 6b in the section Vacuum 2, areas where the fracture surface morphology shows faceted structures are apparently visible.

3.2 Crack propagation behaviour of EN-AW 6082 (oa)

Prior to the fatigue experiment, a micro-notch with a length of 260 µm was introduced to an ultrasonic specimen of the aluminium wrought alloy EN-AW 6082 in the overaged condition. In Fig. 7, the crack path, the EBSD-analysis with the calculated glide traces for each grain and the crack growth rates for both crack tips are shown. A crack was initiated at constant stress amplitude of 120 MPa in laboratory air. In section Air 1, it can be seen that the crack path is pronounced perpendicular to the loading axis and does not show any correlation with the calculated glide traces. Due to the change of the environmental condition (Vacuum 1), slip bands (Fig. 7a blue ellipses) formed on the material surface corresponding clearly to the calculated glide traces of the individual grains. Moreover, on the left side of the crack path strong crack deflection was detected in some sections (Fig. 7a red arrows) whereby a correspondence between the crack path and the calculated glide trace with the highest Schmidfactor can be observed. On the right side, however, the crack propagated perpendicularly through the complete section.

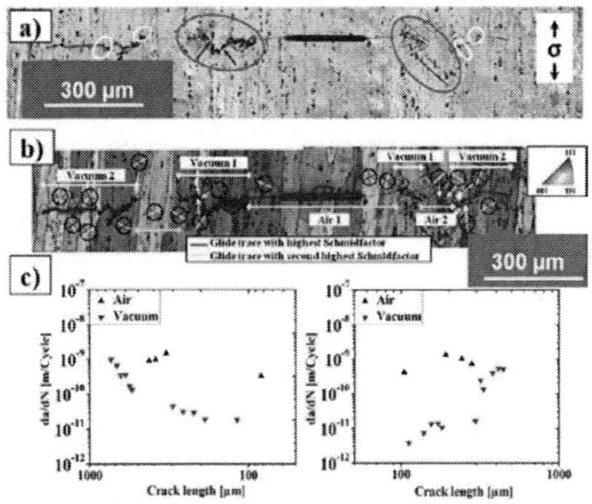

Fig. 7. Crack path in EN-AW 6082 (oa) at $\Delta\sigma/2$ = 120 MPa and changing environmental conditions: a) slip band formation on the material surface, b) EBSD-analysis with calculated glide traces and c) crack growth rates of the left and right crack tip.

In section Air 2, crack propagation perpendicular to the loading axis was detected again whereby the crack path does not show any correspondence with the calculated glide traces. In the final section Vacuum 2, the crack predominantly grew in the normal direction. However, slip band formation on the material surface (Fig. 7a white ellipses) took place only very occasionally and localized at the crack tip in specific sections. Here, a correlation with the calculated glide traces is observed. In Fig. 7c, the corresponding crack propagation rates are shown. Also here it can be noted that the crack growth rates in air are significant higher than those in vacuum. In Fig. 8 the fracture surface morphology of the fatigue experiment in Fig. 7 is shown.

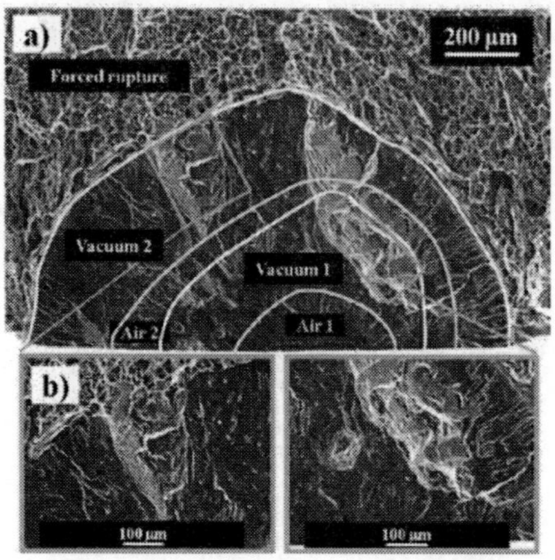

Fig. 8. Fracture surface morphology: a) overview and b) faceted structures in particular sections.

The fracture surface morphology in Air 1 as well as Air 2 is continuously smooth and homogeneous. Similar surface structures were also predominantly found in the area Vacuum 1 and Vacuum 2. However, there are sections where the fracture surface morphology deviates significantly and shows faceted structures (Fig. 8b).

3.3 Crack propagation behaviour of EN-AW 5083 H111

As for the aluminium alloy described above, a FIB micro-notch with length of 260 µm was introduced in the ultrasonic specimen of the aluminium wrought alloy EN-AW 5083 H111. The damage behaviour as a consequence of the material fatigue is shown in Fig. 9.

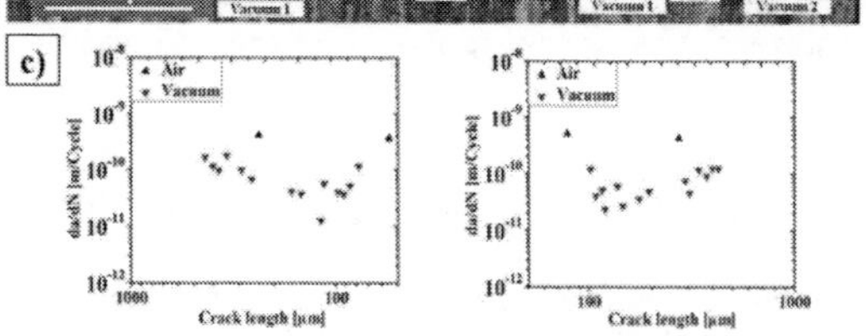

Fig. 9. Crack path in EN-AW 5083 H111 at $\Delta\sigma/2 = 120$ MPa and changing environmental conditions: a) slip band formation on the material surface with calculated glide traces, b) EBSD-analysis and c) crack growth rates of the left and right crack tip.

At the initial stage after crack initiation in laboratory air, the fatigue crack continuously propagated in normal direction through the whole crack path from the macroscopic perspective independent on the atmospheric condition (Fig. 9b). However, slip band formation took place in specific areas which clearly shows a correlation with the calculated glide traces (see Fig. 9a, section Vacuum1 and Vacuum 2). For instance, at the end of the left crack tip in Vacuum 2, slip band formation on the material surface corresponds clearly to the glide trace with the highest Schmidfactor. Furthermore, significant differences can be seen by comparing the crack propagation rates in air and vacuum. Obviously, the crack growth rates are higher in laboratory air than in vacuum by a factor of ten (Fig. 9c). Figure 10 shows the fracture surface of the crack path shown in Fig. 9.

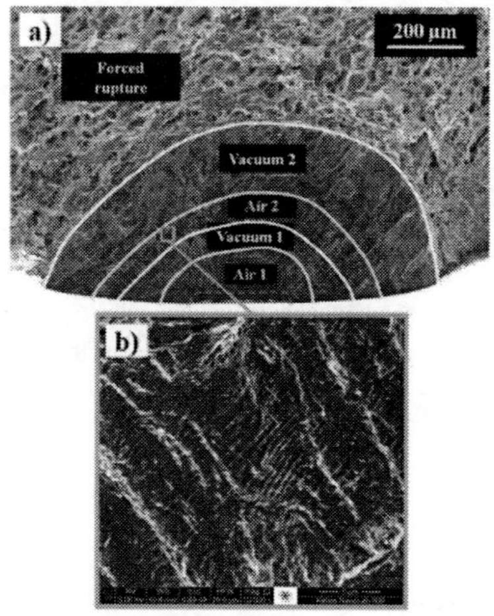

Fig. 10. Fracture surface morphology: a) overview of areas depending on the environmental condition and b) specific section in Air 2 showing striations.

At first glance, the fracture surface does not show significant differences between air and vacuum. However, there are particular areas in air where striations were locally found (see Fig. 10b). In general, the fracture surface is continuous smooth, homogenous and pronouncedly perpendicular to the applied normal stress.

4 Discussion

4.1 Atmospheric influence

The results of the fatigue experiments with changing environment clearly demonstrate an atmospheric effect on the VHCF long crack propagation. Especially in the aluminium alloy EN-AW 6082 in the peak aged condition, the differences can be detected in the crack path, the fracture surface and the crack growth rates. Due to the change of the environment between air and vacuum, it becomes obviously that not only the crack propagation rates clearly decrease in vacuum, but also the crack propagation mechanism changes significantly. Thus, after reaching a certain crack length in air the

VHCF long crack propagation takes place consistently normal-stress-controlled. The incoherence between the calculated glide traces and the crack path in Fig. 5b confirms this behaviour. Furthermore, the preferred Stage II crack propagation in air also manifests itself in the fatigue fracture surface in Fig. 6 showing very smooth and homogeneous structures pronouncedly perpendicular to the loading axis. The answer to the question whether the VHCF long cracks can propagate consistently in laboratory air, depends significantly on the spatial distribution of the primary precipitates and the stress intensity in front of the crack tip [8]. In contrast, very pronounced single sliding occurs in vacuum due to the absence of air humidity. Based on the results from the fatigue experiment shown in Fig. 5a, it can be concluded that slip band formation on the material surface and shear-stress-controlled VHCF long crack propagation along an activated slip band take place even if the crack is very long. Further, a clear correlation between the calculated glide traces, the slip band formation on the material surface and the crack path is confirmed (Fig. 5b). Moreover, in contrast to the fatigue crack propagation in air, the preferred Stage I long crack growth in vacuum is observed on the fatigue fracture surface which shows crystallographic facets corresponding to the {111} planes of aluminium. Therefore, it can be concluded that the damage mechanism during material fatigue in vacuum is significantly different from that in air. It is assumed that water vapour as a constituent of ambient air can rapidly reach the crack tip and adsorb on newly formed surfaces that form as a consequence of the crack progression resulting in a local chemical reaction. The formation of an aluminium oxide layer leads to water reduction at the interface between the oxide and the metal leading to the deposition of chemisorbed atomic hydrogen [5, 11]. A certain amount of the atomic hydrogen can diffuse through the plastically deformed zone in front of the crack tip during cyclic loading, if the crack propagation rate (below 10^{-9} m/cycle) is smaller than the mean diffusion distance of hydrogen at room temperature (3×10^{-9}m) [9, 10]. In this case, the hydrogen diffusion causes to an embrittlement effect [10, 11] at the crack tip which means that the atomic bonding can be locally reduced. The cyclic slip irreversibility will consequently be increased during cyclic loading so that multiple sliding is favoured in the plastically deformed zone. Thus, the VHCF long crack propagation takes place normal-stress-controlled without the slip band formation on the material surface. The fatigue properties of the investigated aluminium alloys deteriorate significantly due to the aforementioned mechanism and the resulting irreversible plastic deformation manifests itself in the crack growth rates. The values of the crack growth rates in air (above 10^{-10} m/cycle) correspond to the inter-atomic distance of aluminium (2.86×10^{-10} m) and hence, 1 Burgers Vector [8, 9]. In vacuum crack propagation rates between 10^{-12} and 10^{-11} m/cycle were detected. In other words, the reversibility of the plastic deformation in vacuum is much higher that explains the lower crack propagation rates below the inter-atomic distance of aluminium. As mentioned above, very pronounced single sliding is promoted in vacuum because of the absence of water vapour. This assumption is confirmed due to the experimental observations showing slip band formation on the material surface near the crack tip during material fatigue in the VHCF regime. However, the results of the fatigue experiments clearly demonstrate that the damage behaviour in vacuum is strongly affected by the secondary precipitation state of the hardenable aluminium wrought alloy EN-AW 6082.

4.2 Influence of the precipitation state on the VHCF long crack propagation in vacuum

The different heat treatments of the aluminium wrought alloy EN-AW 6082 lead to precipitation of a secondary phase. The particles of the secondary phase differ essentially in their shape and size. In particular, the microstructure in the peak-aged condition exhibits semi-coherent fine cuttable secondary precipitates with a preferred orientation, while non-coherent larger particles are presented in the over-aged condition. It can be assumed that the principal glide mechanism of dislocations during material fatigue in vacuum is significantly influenced by the secondary precipitated particles which, in turn, determine the VHCF long crack propagation. In the case of the peak aged condition, the cutting mechanism of the semi-coherent particles promotes a to-and-fro glide of dislocations on specifically favoured glide planes whereby the activation of the glide planes depends on the individual grain orientation to the loading axis [12]. Therefore, very intensive local plastic deformation takes place within individual slip bands along the crack front due to very pronounced single sliding causing shear-stress-controlled VHCF long crack propagation in EN-AW 6082 (pa). The coarsening of the secondary precipitated particles in the over-aged condition caused by the Ostwald ripening leads to a significant change of the interaction between the dislocations and particles. It is assumed that in this configuration dislocations preferentially bypass the non-coherent particles due to energetic reasons so that cross slipping is expected to be the main glide mechanism. Thus, several slip systems are activated leading to a uniformly deformation at the crack tip and, hence, normal-stress-controlled VHCF long crack propagation in vacuum. However, few areas with shear-stress-controlled crack propagation were also found in the crack path as well as in the fatigue fracture surface (Fig. 7a and 8b) where the cutting mechanism of the secondary particles was most probable preferred.

Similar findings were observed by Richard et al. [5] who summarized the results of different studies on the fatigue long crack propagation in metallic alloys. They concluded that two intrinsic regimes (named as intrinsic stage I-like and intrinsic stage II regime) exist in high vacuum which differ significantly in respect to the crack growth mechanism. Thus, crack propagation in the intrinsic stage I-like regime preferentially takes place in microstructures which contain fine cuttable precipitates leading to heterogeneously localized deformation within slip bands in the near of the crack tip. As a result, faceted fracture surfaces and strong crack deflection in the crack path occur. The intrinsic stage II crack propagation preferably takes place in microstructures containing semi-coherent or non-coherent precipitates. In this microstructure configuration, several slip systems are activated during cyclic loading resulting in a homogeneous deformation in front of the crack tip and a smooth crack path.

The microstructure of the aluminium alloy EN-AW 5083 H111 as the secondary precipitation free condition exhibits an increased dislocation density due to strain-hardening. As a consequence of the solid solution hardening, in general the aluminium alloy is expected to prevent wavy slip due to the decrease of the stacking-fault energy [12]. However, it is assumed that the high dislocation density in the material has a dominant effect on the plastic deformation in front of the crack tip leading to the activation of several slip systems during cyclic loading. Consequently, only normal-stress-controlled crack propagation takes place leading to a smooth crack path and a fatigue fracture surface pronouncedly perpendicular to the loading axis (Fig. 9b and 10).

5 Conclusion

Fatigue experiments with a changing environment between air and vacuum were performed on the aluminium wrought alloys EN-AW 6082 in the peak-aged and over-aged condition as well as EN-AW 5083 H111 by means of the ultrasonic fatigue testing technique. The main findings are summarized as follows:

- The fatigue experiments with a changing environment clearly show that the atmospheric conditions have a strong effect on the material damage. In vacuum, very pronounced single sliding takes place leading to shear-stress-controlled VHCF long crack propagation in the aluminium alloy EN-AW 6082 in the peak-aged condition. The VHCF long crack propagation in vacuum in the peak-aged condition differs, therefore, significantly from the conventional long crack growth in the Low Cycle or High Cycle Fatigue regime.

- The stage I crack propagation in EN-AW 6082 (pa) in vacuum is independent on anisotropic effects as rolling texture and takes place due to the absence of the water vapour in ambient air.

- It is assumed that water vapour in ambient air can rapidly reach newly formed crack surfaces during material fatigue leading to an embrittlement effect in the plastically deformed area, if the crack propagation rate is smaller than the mean diffusion distance of hydrogen. As a result, normal-stress-controlled VHCF long crack propagation (Stage II) takes place after reaching a certain crack length in laboratory air.

- The larger irreversibility of the plastic deformation in laboratory air compared to that in vacuum manifests itself in the crack propagation rates. Thus, crack propagation rates in the range of 10^{-10} m/cycle were detected in air corresponding to the inter-atomic distance of aluminium. The value of the crack growth rates in vacuum (10^{-12} - 10^{-11} m/cycle) are related to the larger reversibility of the plastic deformation in front of the crack tip.

- The results clearly demonstrate that the VHCF long crack propagation in vacuum is strongly influenced by the secondary precipitations in EN-AW 6082. Thus, very intensive and localized plastic deformation within slip bands takes place in the peak aged condition leading to stage I crack propagation. In this case, the cutting mechanism of the semi-coherent particles enables a to-and-fro glide of dislocations on specifically favoured glide planes. Due to the coarsening of the secondary particles in the over-aged condition, dislocations preferentially bypass the non-coherent precipitates. This leads to a homogeneous plastic deformation along the crack front and, hence, normal-stress-controlled crack propagation. However, shear-stress-controlled VHCF long crack propagation in the over-aged condition can also occur in specific areas, if the secondary particles are locally cuttable.

- In the solid solution and strain-hardened aluminium alloy EN-AW 5083 H111 basically multiple slipping takes place in the plastically deformed area in front of the crack tip supported by the increased dislocation density in the material. As a result, only stage II crack propagation was detected in this material condition.

Acknowledgements

The financial support by the Deutsche Forschungsgemeinschaft (DFG) is grateful acknowledged.

References

1. D. Schwerdt, B. Pyttel, C. Berger, Int. J. Fatigue **33**, 33-41, (2011)
2. H.W. Höppel, M. Prell, L. May, M. Göken, Proc. Engng. **2**, 1025-1034, (2010)
3. J. Pokluda, P. Šandera, *Micromechanisms of Fracture and Fatigue (Springer-Verlag)*, (2010)
4. R.P. Wei, Engng. Fract. Mech. **1**, 633-651, (1970)
5. S. Richard, C. Gasquères, C. Sarrazin-Baudoux, J. Petit, Engng. Fract. Mech. **77**, 1941-1952, (2010)
6. J.T. Burns, J.J. Jones, A.D. Thompson, J.S. Locke, Int. J. Fatigue **106**, 196-207, (2018)
7. W. Wessel, *Dissertation (Universität Kassel)*, (2011)
8. M. Wicke, A. Brueckner-Foit, T. Kirsten, M. Zimmermann, F. Buelbuel, H.-J. Christ, Proc. Struct. Integrity 7, 235-241, (2017)

9. S.E. Stanzl, E.K. Tschegg, in. J.M. Wells, O. Buck, L. D. Roth, I.K. Tien (Eds.), Ultrasonic Fatigue, The Met. Soc. of AIME, 553-561, (1982)
10. H. Mayer, M. Papakyriacou, R. Pippan, S. Stanzl-Tschegg, Mat. Sci. Engng. **A314**, 48-54, (2001)
11. J.R. Scully, G.A. Young Jr, S.W. Smith, *Gaseous hydrogen embrittlement of material in energy technologies **1***, 707-768, (2012)
12. F. Ostermann, *Anwendungstechnologie Aluminium (Springer-Verlag)*, (2006)

Influence of surface morphology on the very high cycle fatigue behavior of metastable and stable austenitic Cr-Ni steels

Annika Boemke[1,*], *Marek* Smaga[1], and *Tilmann* Beck[1]

[1]Institute of Materials Science and Engineering, TUK, 67663 Kaiserslautern, Germany

Abstract. The present study investigates conventional and cryogenically turned specimens of metastable austenitic steel AISI 347 and stable austenitic steel AISI 904L in the VHCF regime. The cryogenic turning process includes cooling by CO_2 snow and generates a surface layer on the specimens of metastable austenitic steel, which is characterized by a phase transformation from paramagnetic fcc - austenite to ferromagnetic bcc - martensite and grain refinement. The stable austenitic steel retains its purely austenitic structure after cryogenic turning, but also shows grain refinement in the surface layer. The specimens with different surface morphology were cyclically loaded at ambient temperature using an ultrasonic fatigue testing system developed and built at the authors' institute. The testing machine operates at frequencies of approx. 20 kHz to achieve high numbers of load cycles within a reasonable time. To avoid self heating of the specimen, the tests were performed in pulse-pause mode. All specimens were tested with a load ratio of R = -1. During cyclic loading, the metastable austenitic steel partially transformed from paramagnetic fcc - austenite to ferromagnetic bcc - martensite, while no phase transformation could be detected in the stable austenitic steel.

1 Introduction

Investigations in the very high cycle fatigue (VHCF) regime on steels are necessary to characterize fatigue behavior above 10^7 load cycles [1]. Metallic materials can be classified into two different types in the VHCF regime: i) type I materials, which are ductile single-phase materials without inner defects and ii) type II materials, which have inner defects such as inclusions [2]. Type I materials show predominant crack initiation from the surface whereas crack initiation in type II materials changes from surface to subsurface cracks in the VHCF regime, often characterized by a typical fish-eye structure.

In type I materials, which have a mostly fcc structure, an increasing surface roughness occurs during fatigue loading due to cyclic slip irreversibilities. In a further state, persistent slip bands (PSB) can be formed at the surface, if the PSB threshold is exceeded, and can lead to a crack initiation at the surface [3]. This effect shows that the influence of the surface plays a significant role in fatigue behavior even in the VHCF regime.

Until now the influence of surface modification has been investigated up to the high cycle fatigue regime [4–8]. Especially in metastable austenitic steels the effect of transformation from fcc - austenite to bcc - martensite is used to generate martensitic surface layers, e.g. with shot peening or deep rolling [9, 10]. A positive influence of cryogenic turning on the lifetime of metastable austenitic steel is determined as well [5]. Subsequent polishing of the cryogenically turned specimen to eliminate the influence of roughness induced by the cryogenic turning process further increases fatigue life of the specimen [4]. So far, the influence of cryogenic turning has not been investigated in the VHCF regime. Many investigations focus on the VHCF behavior of austenitic stainless steels [11–13]. Stable austenitic steels also show failure above 10^8 load cycles [14]. Metastable austenitic steels show a true fatigue limit in the VHCF regime, if bcc - martensite is developed during fatigue testing [15]. Some investigations focus on predeformed metastable austenites containing bcc - martensite before testing. Predeformed specimens with bcc - martensite content of 27 vol % show a true fatigue limit as well, whereas predeformed specimens with bcc - martensite content of 54 vol % show a second decrease in the Wöhler-curve in the VHCF regime [16].

To investigate the influence of surface layers on the VHCF behavior of austenitic steels, this study focused on VHCF tests of stable and metastable austenitic steels with an ultrasonic testing system up to $N = 1x10^9$. The surface morphology is generated by the cryogenic turning process, which induces bcc - martensite in the surface layer.

2 Materials and experimental setup

2.1 Materials

The investigated materials were the metastable austenitic steel AISI 347 (X6CrNiNb1810, solution-annealed T = 1050 °C, t = 35 min, quenched in He) and the stable

* Corresponding author: boemke@mv.uni-kl.de

austenitic steel AISI 904L (X1NiCrMoCu25205, solution-annealed T = 1100 °C, t = 30 min, quenched in H_2O). The chemical compositions are given in table 1.

Table 1. Chemical composition of AISI 347 and AISI 904L in wt %

	C	N	Cr	Ni	Mo	Cu	Mn
904L	0.03	0.06	19.92	24.34	4.22	1.42	0.95
347	0.02	0.02	17.19	9.44	0.23	0.11	1.55

The metastability of both materials is characterized by M_{d30} temperature according to Angel [17] as well as martensite start temperature M_S according to Eichelmann [18]. The characteristic temperatures as well as mechanical properties of both materials are given in table 2 and confirm the metastability of AISI 347 and the high stability of AISI 904L. Both materials consist of a pure austenitic microstructure at initial state. The average grain size equivalent diameter is 17 µm for AISI 347 and 38 µm for AISI 904L including twins.

Table 2. Characteristic temperatures and mechanical properties of both materials

	$M_{d30,Angel}$ in °C	$M_{S,Eichelmann}$ in °C	Young's modulus in MPa	$R_{p0.2}$ in MPa	UTS in MPa
904L	-220	-1156	187	307	631
347	46	-87	179	225	603

Grain orientation maps determined by electron backscatter diffraction (EBSD) at initial state of both materials are shown in Fig. 1. Both materials do not have a preferred orientation.

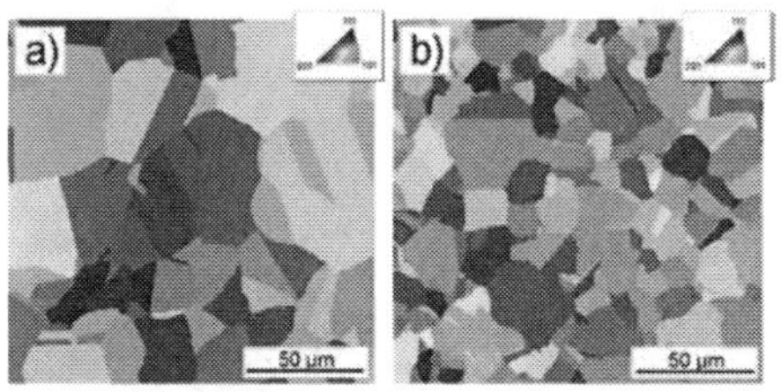

Fig. 1. Grain orientation of AISI 904L (a) and AISI 347 (b) at solution-annealed state

2.2 Experimental setup

The fatigue tests were performed at an ultrasonic VHCF testing system built at the authors' institute (Fig. 2). The testing system consists of an ultrasonic generator, which generates frequencies in the range of f = 19.5 to 20.5 kHz. Electrical oscillation is transformed to a mechanical oscillation by a converter using piezoceramic elements with an attached booster to increase the mechanical vibration amplitude. The specimen is fixed at the booster on one side. The testing setup is controlled by a Windows based computer and a LabVIEW based real time system. The systems are also used for data collection during fatigue testing.

The VHCF testing system features a Compact Laser Vibrometer to measure the displacement amplitude ($displ._a$) of the lower end of specimen. Using a multifunction I/O device with a sample rate of 500 kS/s it is possible to resolve the sinusoidal oscillation of 20 kHz. Software developed at the authors' institute, based on LabView, counts the real number of cycles. The software also acquires the frequency of the oscillation of the specimen. An infrared sensor measures the temperature of the gauge length of the specimen. The difference between the maximum temperature and the start temperature before pulsing is calculated as change in temperature ΔT (see Fig. 6). To avoid overheating the specimens, all tests were performed in pulse-pause mode, depending on the stress level. Three compressed air coolers were also applied around the specimen.

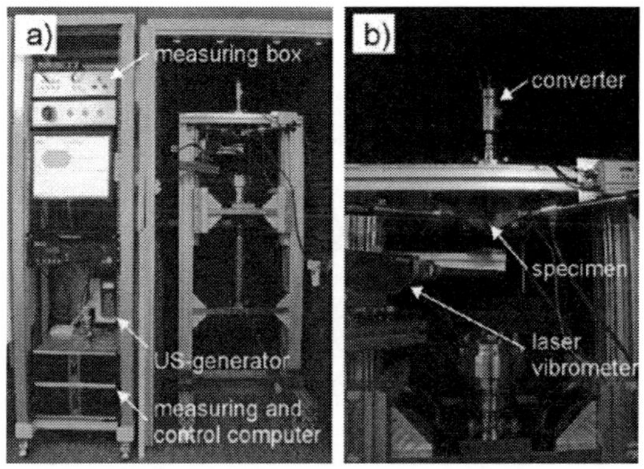

Fig. 2. VHCF testing system including US generator, measuring computers and measuring box (a) and close up view (b) of installed specimen with measuring equipment

The geometry of the specimen was designed with the finite element method (FEM) program ABAQUS using the mechanical properties of each material to ensure the eigenfrequency of the specimen was in the range of the working frequency of the ultrasonic generator. Thus the specimen oscillated symmetrically with the maximum of displacement at both ends and the stress maximum in the gauge length. The stress amplitude was calculated with FEM based on the elastic properties of the material and the displacement amplitude measured by laser vibrometer. Due to the high passive forces of the cryogenic turning process [19] the diameter of the gauge length may not be less than 4 mm. The geometry of the specimen is given in figure 3.

https://doi.org/10.1051/matecconf/201816520008

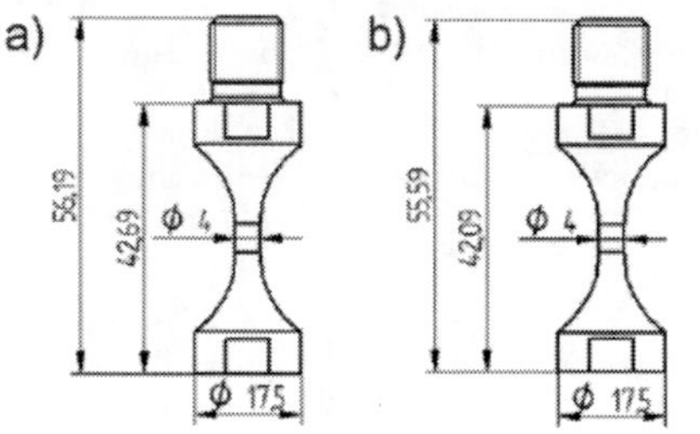

Fig. 3. Specimen geometry of AISI 904L (a) and AISI 347 (b)

To measure bcc - martensite in the material a Feritscope™ sensor was used. This sensor detects ferromagnetic parts in the specimen. The values are given in Fe % and can be converted in vol % of α′-martensite by a calibration factor [20]. In this study the values in Fe % were used.

3 Results

3.1. Influence of cryogenic turning on surface morphology

To investigate the influence of surface morphology on VHCF behavior, four surface morphologies were produced. For reference, conventionally turned and subsequently mechanically polished specimens were tested. The specimens of metastable austenite AISI 347 were additionally electrolytically polished to eliminate the remnant bcc - martensitic content of 0.2 Fe % induced on the surface in the production process. After electrolytic polishing, no ferromagnetic phases were detected. Hence, AISI 904L specimen consisted of a stable austenitic surface layer after polishing (SSL_p) while AISI 347 specimen consisted of an austenitic surface layer after polishing (ASL_p). To produce further morphologies, a cryogenic turning process was used. Two CO_2 nozzles were attached to the CNC lathe to cool the specimens' surface during turning. The following turning parameters were used: cutting depth $a_p = 0.2$ mm, feed $f = 0.35$ mm/rev, cutting speed $v_c = 30$ m/min [21]. The cryogenically turned specimens were labelled as follows: SSL_{t035} for the stable surface layer after cryogenic turning of AISI 904L, and MSL_{t035} for the martensitic surface layer after turning of AISI 347.

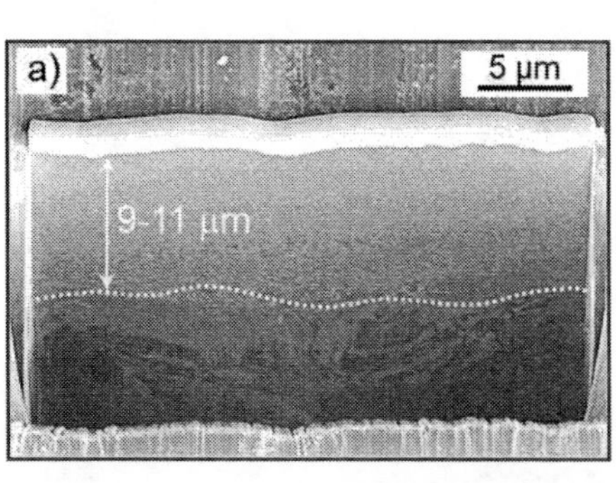

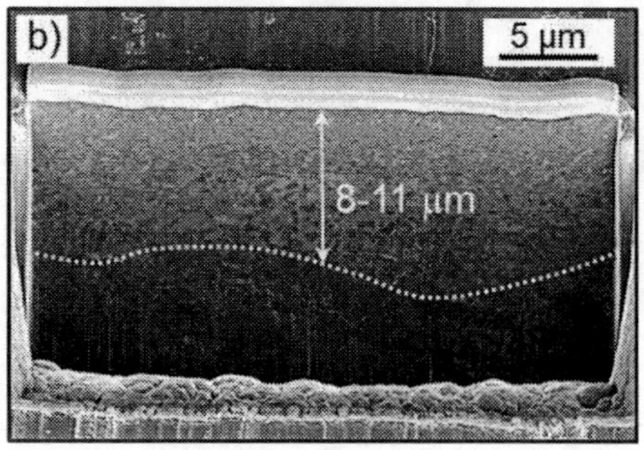

Fig. 4. Focused Ion Beam cuts of cryogenically turned surfaces of AISI 904L (a) and AISI 347 (b)

The surface layers of the cryogenically turned specimens were characterized using focused ion beam (FIB) microscopy. Figure 4 shows micrographs of FIB cuts up to a depth of about 20 µm for both materials with near surface microstructure acquired by detecting secondary electrons resolved by the ion beam. AISI 904L as well as AISI 347 show a nanocrystalline layer directly beneath the surface. The depth of the layer for both materials is up to 11 µm.

The metastable austenitic steel AISI 347 partially transformed from paramagnetic fcc γ-austenite to deformation induced ferromagnetic bcc α'-martensite. Measurements by a Feritscope™ sensor, which detects ferromagnetic phases in the specimen, confirmed the presence of bcc - martensite. The stable austenite AISI 904L maintained its austenitic structure, which was also confirmed by Feritscope™ measurements.

The topography of the specimens with different morphologies was characterized by confocal microscopy (Fig. 5) using the software µSoft control. The analysis was performed with the software MountainsMap Premium. The average roughness R_z was for the polished specimens (SSL_p, ASL_p) very low ($R_z < 1$ µm) but higher for the cryogenically turned specimens (MSL_{t035}, SSL_{t035}), approximately $R_z = 11$ µm.

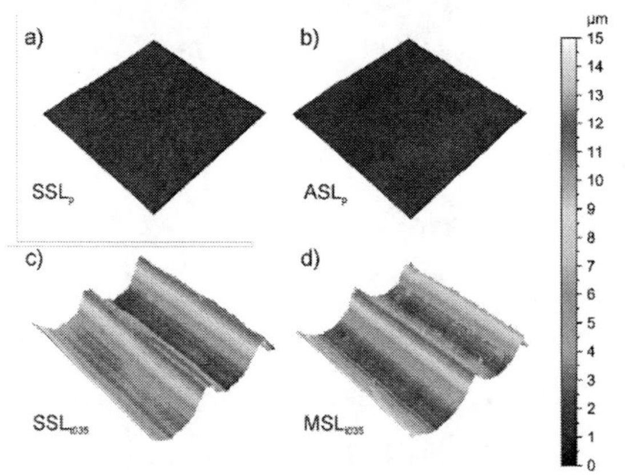

Fig. 5. Surface topography of AISI 904L in SSL_p (a) and SSL_{t035} state (c) and AISI 347 in ASL_p (b) and MSL_{t035} state (d) by confocal microscopy

3.2 VHCF results of the conventionally turned specimens

To investigate the influence of surface morphology on the very high cycle fatigue behavior of metastable and stable austenitic Cr-Ni steels, the polished specimens from both materials were firstly tested at the ultrasonic testing system. The Feritscope™ measurements indicated 0.00 Fe % in both specimens before testing.

Both materials showed strong self heating during the load pulses. Even after finishing the pulse, further surface heating could be detected with the infrared pyrometer (Fig. 6). The change of temperature ΔT during a pulse was calculated as the difference between the maximum temperature and the temperature at the start of each pulse. The higher the displacement amplitude, the higher the self heating of the gauge length. Therefore, the pulse-pause ratio had to be adjusted for each displacement level. All cycles above 90% of the target amplitude were counted and considered in the evaluation of fatigue life.

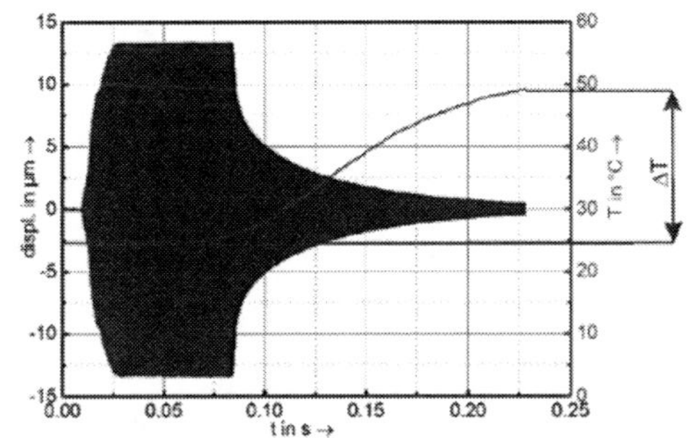

Fig. 6. Displacement and temperature of AISI 904L during pulse

The oscillation behavior of AISI 904L differed from that of AISI 347 due to the high austenite stability of AISI 904L, i.e. the displacement amplitude of AISI 904L increased only slightly during testing as shown in figure 7 for a single step test compared to AISI 347 (Fig. 8).

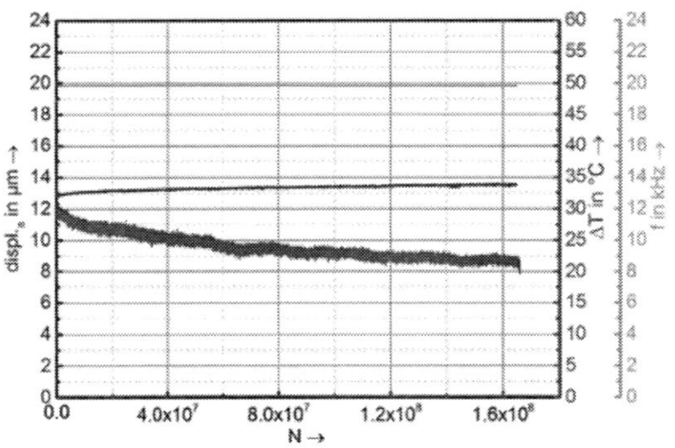

Fig. 7. Displacement amplitude, temperature difference and frequency of AISI 904L during testing at $\sigma_a = 315$ MPa

Due to cyclic hardening, the temperature difference decreased. The decrease flattened during testing and temperature reached a nearly constant level up from $1x10^8$ cycles until specimen failure. The changes in temperature and the increase of displacement amplitude were less pronounced than for AISI 347. For the calculation of the stress amplitude in the gauge length the average value of displacement amplitude (13.4 µm) at $1x10^7$ cycles was used. The frequency of the oscillation measured with the laser vibrometer was constant during the whole test.

The AISI 347 specimen also showed strong self heating during the load pulses. Even after finishing the pulse further surface heating could be detected, similar to AISI 904L (comp. Fig 6). Accordingly, the pulse time had to be reduced to 40 ms and the pause time extended up to 10 s. During the test, the oscillation of the specimen changed to higher amplitudes despite constant control voltage for generating the oscillation. The first pulses were characterized by an oscillation plateau with a constant amplitude level during each pulse as shown in figure 8a). The target amplitude is marked with a red colored dashed line. During further cycling, this plateau shape changed to pulses with an increasing displacement level above the target amplitude due to the transient behavior of the material (Fig. 8b). Readjusting of the control parameters, e.g. PID-Parameter, was necessary to modify the oscillation to a form with nearly constant displacement level (Fig. 8c). This readjusting led to a rapid increase of the displacement curve as shown in figure 8 at $8.5x10^6$ cycles. The continuous increase of the amplitude correlated with a decrease of ΔT as a characteristic result of cyclic hardening. The frequency measured by laser vibrometer was constant during the whole test.

The continuous increase of the displacement amplitude prohibited determination of an average value and consequently the calculation of stress amplitude in the gauge length of the AISI 347 specimen was not possible.

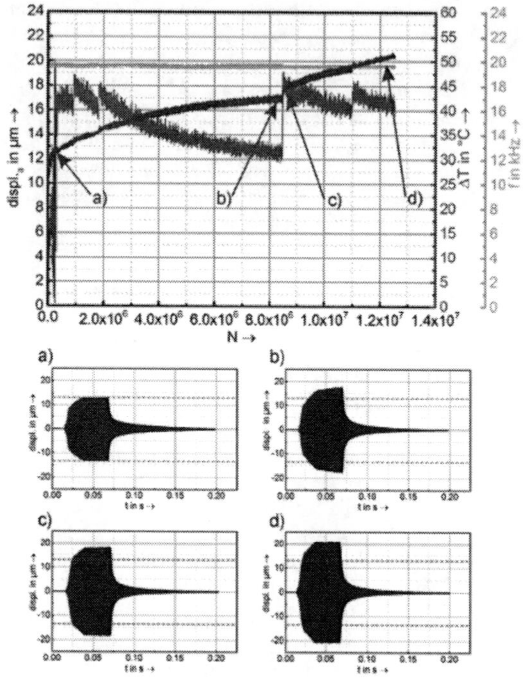

Fig. 8. Displacement amplitude, temperature difference and frequency of AISI 347 during ultrasonic testing without controlling the displacement amplitude

Next the displacement amplitude of the specimen of AISI 347 was controlled from the beginning by readjusting the control parameters to prevent the increasing displacement level during pulses as shown in the test before (comp. Fig. 8b). The control voltage was constant during the fatigue test. By readjusting the control parameters, the increase of the displacement amplitude was kept within acceptable margins from the beginning of the test as shown in figure 9. At the same time, the temperature increase dropped sharply as described for the previous test (see Fig. 8). The fatigue test was interrupted at certain intervals so that the ferromagnetic phase contents could be measured with a Feritscope™ sensor in the gauge length as an indication of bcc - martensite formation. The bcc - martensite fraction increased as temperature increase during each pulse decreased. The martensitic phase has a bcc crystallographic structure and higher strength than fcc austenite [22], which led to less damping of the oscillation and resulted in higher displacement amplitude. At $4x10^5$ cycles, the target displacement amplitude was achieved and stayed constant until the ultimate cycle number of 10^9 was reached.

Further cyclic loading at the target displacement amplitude led to increasing cyclic hardening of the material. The temperature decreased and the bcc - martensite content increased. Simultaneously the displacement amplitude rose moderately, so readjusting of the parameters was necessary again at $1.5x10^6$ cycles. At this point the pulse-pause ratio was increased to 48 ms pulse and 8 s pause in order to reach higher effective frequencies on condition that the maximum temperature of the specimens gauge length would not exceed 50 °C. Another readjusting of the parameters led to an abrupt rise in the ΔT course. Further cyclic hardening caused higher α'-martensite contents which led to further readjusting actions during the test associated with abrupt steps in ΔT. With increasing bcc - martensite transformation the decrease of ΔT became flatter and reached a nearly stable level. Further readjusting of the control parameters could now be used to optimize pulse-pause relations to achieve an effective frequency up to more than f = 1 kHz.

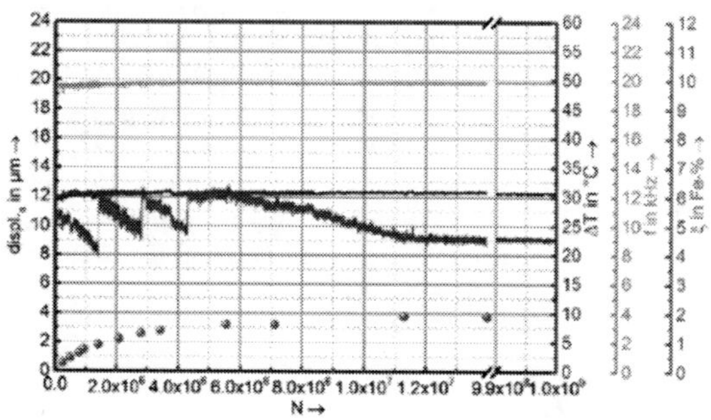

Fig. 9. Displacement amplitude, temperature difference, frequency and α'-martensite development (ξ) of AISI 347 during ultrasonic testing with controlling displacement amplitude

3.3 VHCF results of cryogenically turned specimens

In the next step, the cryogenically turned specimens were fatigued at the ultrasonic testing system. The gauge length of the specimens was turned with the parameters as described in chapter 3.1.

Before testing, Feritscope™ measurements of the stable austenite AISI 904L indicated 0.0 Fe %. The stress amplitude, calculated according to constant displacement amplitude was 315 MPa for the SSL_p and SSL_{t035} specimens. Figure 10a) shows the displacement amplitudes of both specimens as well as ΔT. Whereas the bar representing the conventionally turned and polished specimen with a stable austenitic surface layer (SSL_p) is grey, the lifetime of the specimen with a stable surface layer after cryogenic turning (SSL_{t035}) is shown in red color. The cryogenically turned specimen showed lower self heating than the conventionally turned specimen, although the pulse time was the same because the cyclic hardening characterized by decreasing ΔT values was less pronounced in the cryogenically turned specimen, due to the nanocrystalline structure produced by the turning process as shown in Fig 4. However, fatigue life of the cryogenically turned specimen was distinctly shorter (Fig. 10b) due to the higher roughness of the surface (comp. Fig. 5a and c).

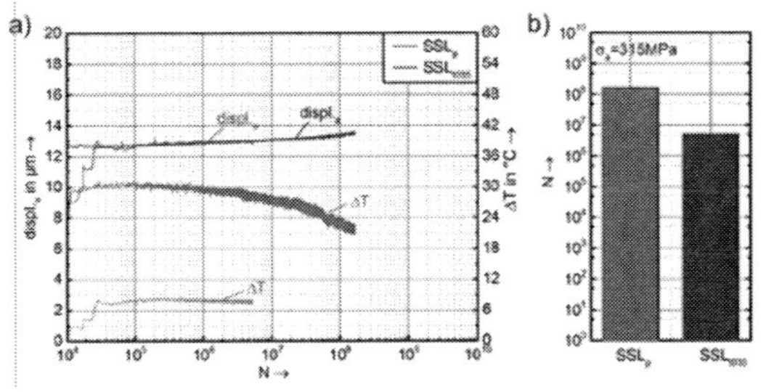

Fig. 10. Displacement amplitude and temperature difference of AISI 904L in SSL_p and SSL_{t035} state (a) and number of cycles to failure of both states (b)

Scanning electron microscopy fractographs of both specimens are shown in Fig. 11. The mechanism of failure changed from subsurface crack initiation in the SSL_p specimen to surface crack initiation in the specimen in SSL_{t035} state. The fracture surface of SSL_p had a fine structured area around the crack initiation site, as shown in the middle of figure 11 a), whereas SSL_{t035} had the crack initiation from the surface, as shown in figure 11 b).

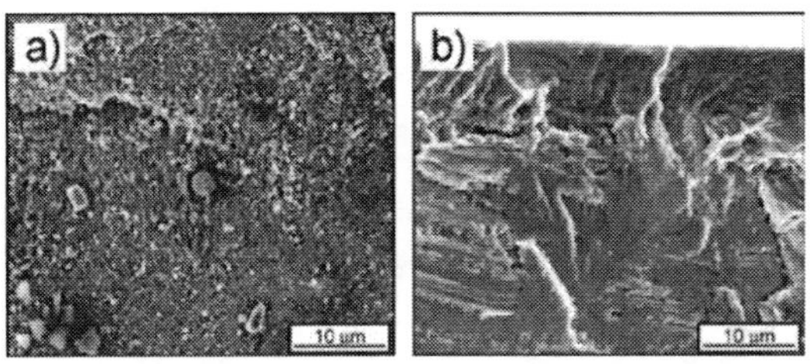

Fig. 11. Scanning electron microscope images of AISI 904L in SSL_p state (a) and SSL_{t035} state (b)

The metastable austenite AISI 347 was also characterized by Feritscope™ measurements before fatigue testing. The conventionally turned specimen with austenitic surface layer after polishing (ASL_p) did not contain ferromagnetic phases, i.e. 0.0 Fe % was measured on the surface. The cryogenically turned specimen with a martensitic surface layer after turning (MSL_{t035}) showed values of 2.8 Fe % indicated by the Feritscope™ sensor before testing. The specimen in the ASL_p state took $1.7x10^4$ cycles to achieve the target amplitude as described in chapter 3.2 due to its transient behavior. Subsequently, the amplitude was nearly constant. The steps in temperature were caused by adjusting parameters and optimizing pulse-pause ratio. At $5.7x10^5$ load cycles, specimen failure occurred associated with abrupt decreasing ΔT and displacement amplitude values due to crack initiation at the surface. The cryogenically turned specimen reached the target displacement amplitude twice as fast as the ASL_p specimen. Up from this point the amplitude was nearly stable for the rest of the fatigue test. The ΔT values were lower at the MSL_{t035} state, because of the nanocrystalline surface layer and the martensitic phase induced by the cryogenic turning process. Because ΔT and displacement amplitude were stable until $7x10^6$ load cycles and further changes in the test were not expected, the pulse-pause ratio was optimized to raise effective frequency. The resulting development of ΔT and displacement amplitude of both specimens are given in figure 12.

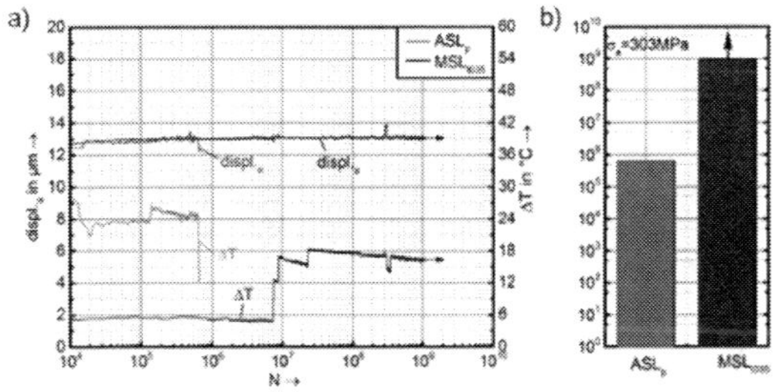

Fig. 12. Displacement amplitude and temperature difference of AISI 347 in ASL_p and MSL_{t035} state (a) and number of cycles to failure of both states (b)

The cryogenically turned specimen from the metastable AISI 347 reached the ultimate number of cycles without failure contrary to the stable AISI 904L, despite the same roughness on the surface (comp. Fig. 5 c and d). This suggests that the α'-martensite phase in the surface layer overcompensates the detrimental effect of high roughness and leads to longer specimen lifetime. The α'-martensitic content of the cryogenically turned specimen was stable during the whole test at 2.8 Fe %.

Scanning electron microscope investigations of the fracture surface of the ASL_p specimen are shown in Figure 13. Crack initiation is starting at the surface of the specimen, which is typical for a failure at cycle numbers up to 10^6.

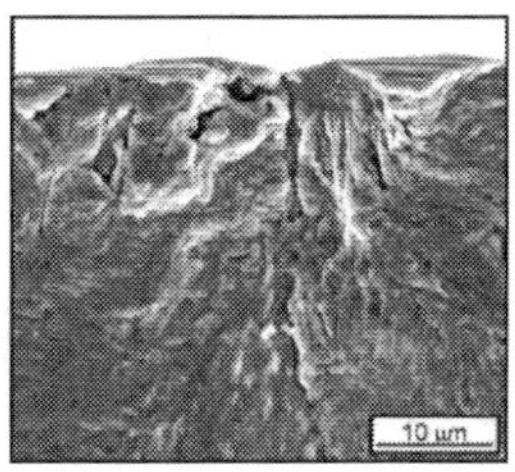

Fig. 13. Scanning electron microscope image of AISI 347 in ASL_p state

4 Conclusion

The present work investigates the very high cycle fatigue (VHCF) behavior of a stable and a metastable austenitic Cr-Ni steel on ultrasonic testing systems with a frequency of f = 20 kHz as well as the influence of surface morphology on fatigue behavior produced by a cryogenic turning process compared to conventionally turned and subsequently polished specimens. Both materials consisted of pure austenitic structure at the as received state confirmed by SEM/EBSD and Feritscope™ measurements. The stable AISI 904L showed low cyclic hardening during fatigue testing and thus a slight increase of displacement amplitude. At $1.66x10^8$ cycles, specimen failure occurred from subsurface crack initiation. Even after failure no bcc - martensite could be detected.

In contrast to that, the metastable AISI 347 specimens did show a pronounced increase of displacement amplitude during testing without adjusting parameters due to their transient behaviour which also causes a remarkable decrease of the specimen self-heating during each load pulse ΔT. The cyclic hardening indicated by decreasing ΔT correlated in metastable austenites with the formation of deformation induced α'-martensite. The ferromagnetic bcc - martensite was confirmed with Feritscope™ measurements. Without intervention, i.e. regular readjusting control parameters, no single step test at AISI 347 was possible with the ultrasonic testing system. To realize fatigue tests at constant displacement amplitudes, the parameters were readjusted until a nearly stable state of the material was achieved.

The influence of cryogenic turning on VHCF behavior of stable AISI 904L and metastable AISI 347 steel was investigated. The cryogenically turned specimens were fatigued at the same displacement amplitude as their polished counterparts with the same pulse pause ratios to compare the results. FIB cuts show the machining induced changes in microstructure under the surface, with a nanocrystalline layer formed in the cryogenic turning process. The cryogenically turned specimen of AISI 904L as well as that of AISI 347 showed lower self heating during the load pulses compared to their polished counterparts. Furthermore, the AISI 904L specimen achieved a lower number of cycles to failure in the cryogenically turned state due to higher surface roughness resulting from the turning process. In contrast to that, the cryogenically turned specimen of AISI 347 achieved the ultimate number of cycles of 10^9 in spite of higher roughness than their polished counterpart which failed at $6.4\text{x}10^6$ cycles under the same loading conditions. This leads to the conclusion that the influence of bcc α'-martensite induced by cryogenic turning overcompensates the detrimental effects of roughness and leads to a higher lifetime of the metastable austenitic specimen.

The authors thank and are grateful for the support of the German Research Foundation (DFG) within the CRC926 *Microscale Morphology of Component Surface*, and the Nanostructuring Center (NSC) of the University of Kaiserslautern for performing the FIB-investigations. The cryogenically turned specimens were turned at the Institute for Manufacturing Technology and Production Systems. The authors thank Prof. J.C. Aurich as well as H. Hotz for their support. We also thank Prof. J. Seewig, that we could perform the confocal microscopy investigations at the Institute for Measurement and Sensor-Technology.

References

1. C. Bathias, Fatigue Fract. Eng. Mater. Struct. **22**, 559–565 (1999)
2. H. Mughrabi, Fatigue Fract. Eng. Mater. Struct. **25**, 755–764 (2002)
3. H. Mughrabi, Int. J. Fatigue **28**, 1501–1508 (2006)
4. R. Skorupski, M. Smaga, D. Eifler, Adv. Mat. Res. **891-892**, 464–469 (2014)
5. M. Smaga, R. Skorupski, P. Mayer, B. Kirsch, J. C. Aurich, I. Raid, J. Seewig, J. Man, D. Eifler, T. Beck, *Procedia Structural Integrity* **5**, 989–996 (ScienceDirect, Funchal, 2017)
6. A.M. Eleiche, M.M. Megahed, N.M. Abd-Allah, J. Mater. Process. Technol. **113**, 502–508 (2001)
7. R. K. Nalla, I. Altenberger, U. Noster, G. Y. Liu, B. Scholtes, R. O. Ritchie, Mater. Sci. Eng. A **355**, 216–230 (2003)
8. T. Klotz, D. Delbergue, P. Bocher, M. Lévesque, M. Brochu, Int. J. Fatigue **110**, 10–21 (2018)
9. I. Altenberger, B. Scholtes, U. Martin, H. Oettel, Mater. Sci. Eng. A **264**, 1–16 (1999)
10. C. Ye, S. Suslov, D. Lin, G. J. Cheng, Philos. Mag. **92**, 1369–1389 (2012)
11. M. Smaga, A. Sorich, D. Eifler, T. Beck, *VHCF7 Seventh International Conference on Very High Cycle Fatigue*, Siegener werkstoffkundliche Berichte **14**, 249–254 (Dresden, 2017)
12. K. Takahashi, T. Ogawa, Journal of Solid Mechanics and Materials Engineering **2**, 366–373 (2008)
13. T. Naoe, Z. Xiong, M. Futakawa, J. Nucl. Mater. **468,** 331–338 (2016)
14. J. V. Carstensen, H. Mayer, P. Brønsted, Fatigue Fract. Eng. Mater. Struct. **25**, 837–844 (2002)
15. A. C. Grigorescu, P. M. Hilgendorff, M. Zimmermann, C. P. Fritzen, H. J. Christ, Int. J. Fatigue **93**, Part 2, 250–260 (2016)
16. C. Müller-Bollenhagen, M. Zimmermann, H. J. Christ, Procedia Eng. **2**, 1663–1672 (2010)
17. T. Angel, Journal of the Iron and Steel Institute, **177**, 165–174 (1954)
18. G. H. Eichelmann, T. C. Hull, Trans. Am. Soc. Met. **45**, 77–104 (1953)
19. P. Mayer, R. Skorupski, M. Smaga, D. Eifler, J. C. Aurich, Proc. CIRP **14**, 101–106 (2014)
20. J. Talonen, P. Aspegren, H. Hanninen, Mater. Sci. Technol. **20**, 1506–1512 (2004)
21. J. C. Aurich, P. Mayer, B. Kirsch, D. Eifler, M. Smaga, R. Skorupski, CIRP Ann. - Manuf. Techn. **63**, 65–68 (2014)
22. M. Smaga, R. Skorupski, D. Eifler, T. Beck, J. Mater. Res. **32**, 4452–4460 (2017)

Rotary bending as a mean for improving micro-cleanliness of stainless steels for high demanding applications

Thomas Sourisseau[1,*], *Dominique* Ferrière[1], *Frédéric* Moser[1], *Maximilien* Libert[1], *Pierre* Chemelle[1], and *Christian* Deville-Cavellin[1]

[1]Research Center, UGITECH, avenue Paul Girod, 73403 UGINE Cedex, France

Abstract. Stainless steels are used for automotive or medical applications which require a high fatigue resistance correlated to a high level of micro-cleanliness. A methodology based on rotary bending tests carried out on wires or bars has been defined to determine the material's endurance limit (after 100 millions cycles) and identify the largest subcutaneous inclusions or precipitates where failures initiate. This methodology has been applied to EN 1.4568 spring wires. Failures were found to initiate both at oxide inclusions and *AlN* precipitates. For the same size, *AlN* precipitates were observed to be more critical towards crack initiation than oxide inclusions, due to their angular shape and lower thermal expansion at high temperatures. However, oxide inclusions larger than the *AlN* maximum size strongly impact the material's fatigue limit, and their density and size should be reduced.

1 Introduction

Austenitic stainless steel wires and bars are used in a wide range of applications which require both corrosion resistance and mechanical properties. For some specific applications like medical implants (mainly EN 1.4441 or EN 1.4472 stainless steel grades) or springs dedicated to automotive injectors (where EN 1.4568 stainless steel grade is used), they also have to sustain very high cycle (100 millions or more) fatigue phenomena in moderately aggressive environments. Hence, the endurance limit of those materials (which may be defined as the maximum stress below which the material may sustain an infinite number of cycles) should be optimized to support high pressures.

This characteristic is closely related to bar or wire surface finish (*i.e.* roughness and presence of geometrical surface irregularities on one hand, residual stresses due to surface work-hardening on the other hand) as well as its mechanical properties [1]. Several authors [1-2] have proposed relationships between endurance limit (σ_D in MPa), tensile strength (R_m in MPa), yield strength ($Rp_{0.2}$ in MPa) and elongation at break (A in %) of steels. Equation (1) can be used in the case of a large range of steels and stainless steels to roughly estimate the value of σ_D [1]:

$$\sigma_D = 0.41 R_m + 2A \qquad (1)$$

High tensile strengths (up to 1700 MPa and more) can be achieved on targeted austenitic stainless steels like EN 1.4568 by a combination of:

- high carbon content,
- metastable austenite, converted into hard martensite by work-hardening during wire-drawing,
- precipitation hardening (by a heat treatment at 480°C) which makes *NiAl* precipitate within the martensite.

Beside mechanical properties and surface finish, the high cycle fatigue resistance of stainless steels is strongly correlated to their micro-cleanliness, and mostly to inclusion morphologies and sizes, inclusion locations and compositions, elastic and plastic properties of inclusions relatively to those of the surrounding matrix, and finally adhesion between inclusions and matrix[1][3-8]. Those effects are particularly enlightened in the case of automotive springs whose surfaces have been either machined or shaved to remove surface defects and lower their roughness.

Inclusions mainly lower fatigue resistance by increasing stresses in their surrounding area which is then work-hardened. Micro-cracks may initiate either by tearing of inclusion/matrix interfaces, or at the periphery of those work-hardened areas [1]. Inclusion size is of prime importance [1,3,8], but its effect strongly depends on the level of applied stresses. In oligocyclic fatigue, micro-cracks may initiate at numerous small inclusions, due to high applied stresses, which makes failure follow preferentially a way passing near those small but close one to another inclusions, in the maximum stress area. Hence, high dispersions of small inclusions must be avoided if a high oligocyclic fatigue resistance is targeted. On the opposite, for high cycle fatigue with low applied stresses, the biggest inclusions are most harmful, since they are preferential sites for crack initiation, those cracks propagating until final fracture occurs [1].

For high cycle fatigue, Murakami[3,8] has observed that the endurance limit of a material is not necessarily

* Corresponding author: thomas.sourisseau@ugitech.com

correlated to the results of standard micro-cleanliness quotations (according to ASTM E45 or ISO 4967 for example) and has found relationships between the fatigue limit σ_D (MPa) of the material, its hardness Hv (kgf/mm^2) and the parameter $\sqrt{A_{max}}$ (µm), where A_{max} is equal to the area of the largest inclusion projected on the plan perpendicular to maximum main stress. In the case where the largest inclusion is subcutaneous or internal, the following equation has been proposed:

$$\sigma_D = \frac{1.56(H_v + 120)}{\left(\sqrt{A}\right)^{1/6}} \tag{2}$$

MURAKAMI has also defined a methodology [3,8] based on Extreme Value Method and rotary bending tests, which determines the value of A_{max} in a given volume of material.

Based on this approach, the present work aims at setting up rotary bending equipment adapted to wires or bars, and defines protocols for sample preparation and test parameter calculation. Finally, this methodology was applied to determine the material's fatigue limit and also identify the largest subcutaneous precipitates formed in EN 1.4568 spring wires used for demanding automotive applications, for which a high fatigue resistance is requested. The analysis of the largest inclusions should help understanding the mechanism of their formation and hence proposing solutions to improve the endurance limit of the material by reducing the density of those inclusions.

2 The rotary bending equipment

Rotary bending machines may belong to two main types, depending on the evolution of the applied moment along the sample:

- those with a linear moment in relation to a two-point bending system; in this case, applied stress is maximum at the middle of the sample;
- those with a constant moment along a part of the sample, thanks to a three- or four-point bending system.

The equipment used in this study is a four-point bending system which aims at applying a constant moment at the surface of a stainless steel wire over a length of 200 mm (which is close to the length of current injector springs). The volume of tested material also corresponds to the first superficial millimeter of the wire over a length of 200 mm (see figure 1). Testing 20 samples along a coil or a batch of bars gives a tested volume which can be representative of the complete coil or batch [3,8].

Standard rotary bending machines are not able to test such long samples. Moreover, the other originality of this equipment comes from its ability to test a sample of stainless steel wire or bar with its industrial surface state, unlike standard benches whose samples are machined.

Figure 2 shows a general view of the rotary bending equipment which has been supplied by ZARIAN GmbH. Wire or bar samples of diameters between 2 and 7 mm can be tested, which is not the case with standard equipment. The sample is mounted in a jaw and turns at a constant frequency thanks to a 2,5 kW brushless engine. This latter is set on a platform which can freely rotate to avoid any unexpected additional bending nor twisting of the sample.

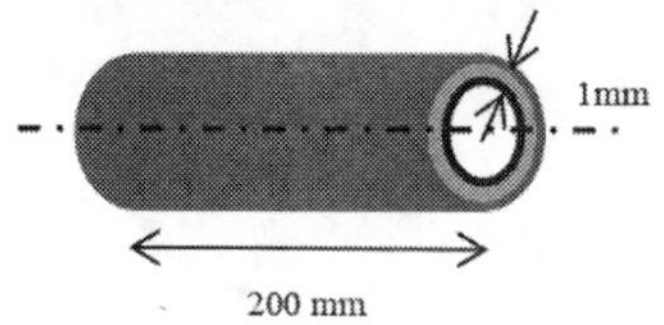

Fig. 1. Sample tested volume

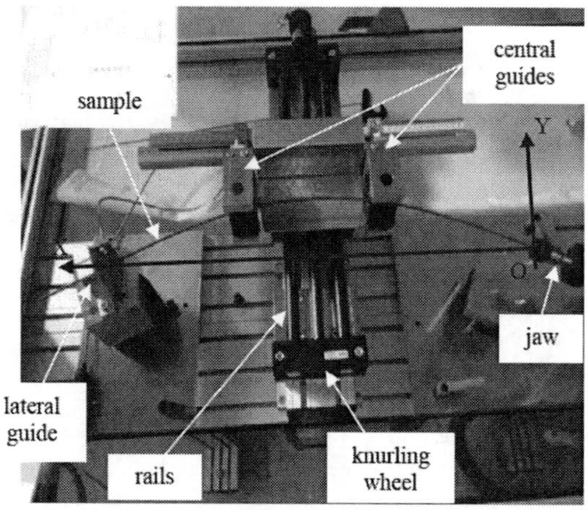

Fig. 2. Rotary bending machine for stainless steel wires

Two central guides may translate along both X and Y axis (on rails), and one lateral guide (at the sample extremity) may translate along X axis (see figure 2). Their positions, which can be precisely measured by verniers, give the targeted curvature to the sample. Positions of central guides along Y axis are set by a knurling wheel. Initial alignment of both the three guides and the jaw (when no curvature is applied to the sample) has been checked to avoid any bias in the test results.

Furthermore, a gauge set on the engine counts the number of sample rotations and both engine and counting stop when the sample breaks (*via* a laser detector placed near the lateral guide which detects when sample end stops moving) or when the targeted number of cycles (*i.e.* 100 millions) is reached.

Among the most critical points are the conditions of contact between sample and the three guides. These latters should keep friction as low as possible in order to avoid sample twisting and heating. This is achieved by using seals of specific material and internal diameter. Finally, seal length has been optimized to minimize additional bending of the sample in those contact zones.

3 Preparation and characterization

Trials have been carried out with 3.60 mm diameter drawn wire of grade EN 1.4568, produced by UGITECH from wire-rod coils.

3.1 Chemical analysis and wire-rod metallurgy

Chemical analysis of the tested material is given in table 1. Blooms have been hot-rolled into wire-rod coils which have been then shaved and drawn to final diameter 3.60 mm.

Table 1. Chemical analysis of EN 1.4568 samples

	C	Si	Mn	Ni	Cr	Mo	S	P	Cu	Al
EN 10088-3	≤ 0,090	≤ 0,7	≤ 1,0	6,5 - 7,8	16,0 - 18,0	-	≤ 0,015	≤ 0,040	-	0,7 - 1,5

Grade EN 1.4568 solidifies as primary ferrite, and then transforms into austenite during hot-rolling, according to figure 5. Residual ferrite in wire-rod is less than 1%. A typical austenitic microstructure of wire-rod is shown in figure 3.

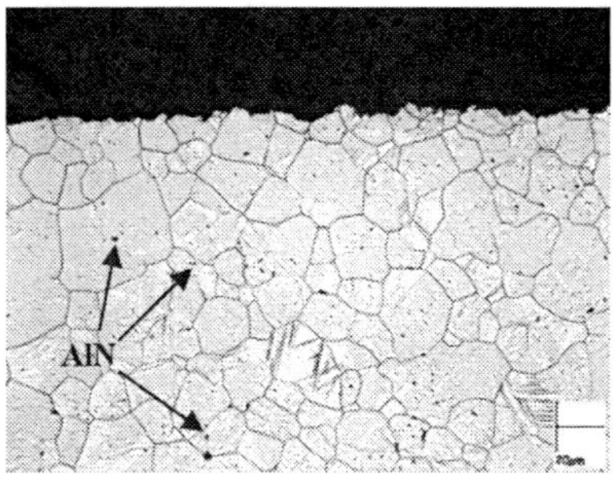

Fig. 3. Austenitic microstructure of EN 1.4568 wire-rod with dispersed *AlN* precipitates

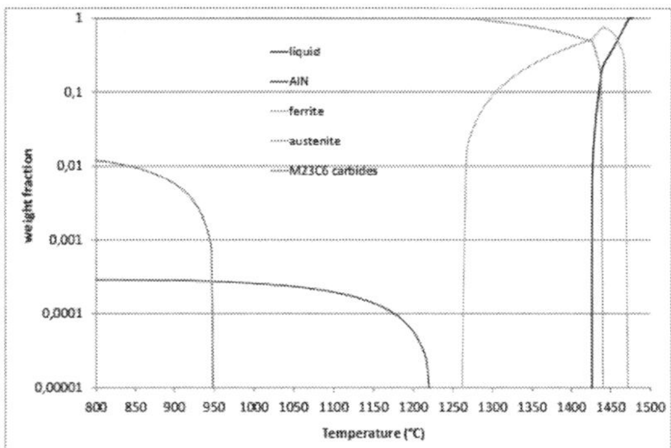

Fig. 4. Thermodynamic phase diagram of the studied EN 1.4568 material (according to Thermocalc® calculations)

Since EN 1.4568 grade is characterized by a high Aluminium content, melting process has been carried out under conditions which should minimize Oxygen and Nitrogen contents. Indeed, Aluminium is very reactive towards these elements, leading to the formation of both oxide inclusions (before and during solidification) as well as *AlN* precipitates which appear below 1230°C, once material has been completely solidified, according to figure 4. Even if Oxygen and Nitrogen contents are very low, oxide inclusions and *AlN* precipitates have been observed on micrographs, as illustrated by figure 3. Furthermore, Titanium content is also very low which prevents angulous Titanium nitrides (favoring crack initiation during fatigue) from appearing in the studied material.

3.1 Preparation of drawn wire samples

Surface preparation is a key point to get successful results with rotary bending testing. Since the objective of the present study is to analyze stainless steel micro-cleanliness and its influence on high cycle fatigue resistance, failures should initiate at subcutaneous inclusions (in the first superficial millimeter) and not at surface defects. The wire-rod coils were hence shaved to remove all the surface defects which might have been induced by hot-rolling. Since the shaved wire had a very smooth surface, it then underwent a second operation aiming at restoring a sufficient surface roughness for the following wire-drawing step. Indeed, this roughness helped carrying lubricant into the drawing die when the wire section was reduced, ensuring a good lubrication during wire-drawing and preventing new surface defects (like scratches) from appearing.

Wire-drawing gives high mechanical properties to the wire by work-hardening, which contributes to its high fatigue resistance. Regarding grade EN 1.4568, wire-rod is almost fully austenitic, but austenite turns into hard martensite during wire-drawing. Martensite contents up to 60% can be achieved on wires drawn with reduction rates close to 70%, which makes tensile strength as high as 1500 MPa. This level of tensile strength is required to ensure sufficient spring stiffness.

600mm long samples have been cut in one coil of 3.60 mm diameter drawn wire of grade EN 1.4568. Those samples have been straightened with a block of straightener rolls in order to get a satisfying straightness. This latter was required to avoid inaccurate rotary bending results.

Furthermore, straightened samples have been heat treated for 1 h at 480°C in order to relax high internal stresses present in the martensitic phase, and to harden the material by precipitation of *NiAl* particles (whose size is of a few tens of nanometers) in the martensite. This hardening phenomenon gave additional 200 to 300 MPa to the tensile strength of the drawn wire. This heat-treatment has been carried out under vacuum in order to avoid sample oxidation.

Finally, samples needed to be shot-peened to get compressive stresses at their surface in order to avoid failure initiation at the surface, since applied stresses are maximal at the wire surface during bending. A shot-peening apparatus, equipped with four identical nozzles (set at the four cardinal points) and with a sample driving system, has been specially designed by S.F.G.P. (Société Française de Grenaillage de Précontrainte) to shot-peen wire samples as homogeneously as possible on the whole wire periphery and over the whole sample length. Different shot-peening conditions (type of shot, nozzle pressure and Almen intensity, driving speed) have been tested and final conditions have been selected which optimized both compressive residual stresses and surface roughness.

Mechanical properties of the final samples have been measured by tensile testing at room temperature: the corresponding values have been compiled in table 2. The YOUNG modulus value of table 2 agrees with the decreasing of this parameter reported in the literature [9] when the product is work-hardened. Martensite content was measured by saturation magnetization and found equal to 58%.

Table 2. Tensile properties of studied EN 1.4568 samples

Tensile strength R_m (MPa)	Yield strength $Rp_{0.2}$ (MPa)	Elongation at break A (%)	YOUNG modulus E (GPa)
1656	1482	1	175

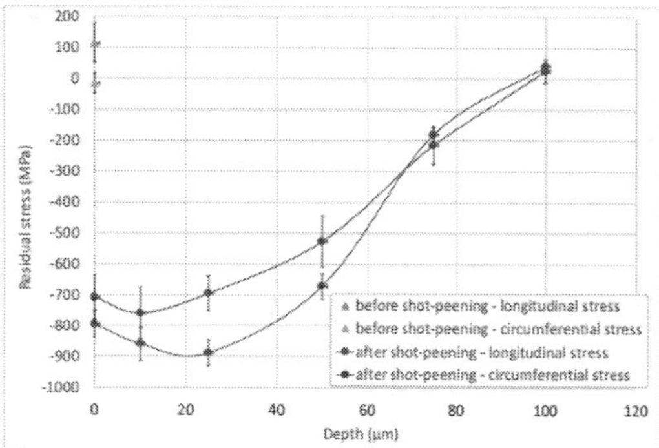

Fig. 5. Longitudinal and circumferential residual stresses measured by X-ray diffraction in the studied EN 1.4568 samples

Furthermore, sample roughness has been measured with a profilometer and a cut-off value equal to 17.5. Values of R_a and R_z (along the wire axis) have been found equal to respectively 1,8 µm and 11 µm.

Finally, longitudinal and circumferential residual stresses have been measured on final samples with a X-ray SEIFERT XRD3003 diffractometer and by $\sin^2(\psi)$ method[10]. For stress profiles, successive layers of material (of defined depths) have been removed by chemical dissolution of the material immersed in a *HNO₃/HCl/HF* electrolyte for variable times. Since measurements in austenite phase are not exploitable due to effects of preferential orientations, only measured values found in the martensite phase have been compiled [10]. Figure 5 shows that the pre-stressed depth is close to 80 µm, with maximum compressive stresses equal to respectively –750 MPa and –900 MPa for circumferential and longitudinal stresses, at a depth close to 20 µm. It is to be noticed that surface residual stresses of the drawn wire after heat treatment at 480°C and before final shot-peening are close to zero, meaning that complete relaxation occurred during heat treatment.

4 Conditions of rotary bending tests

Figure 6 describes a stressed sample (with diameter ∅) during a rotary bending test, with the theoretical profile of applied bending moment. Positions A and C represent contact points between sample and central guides. Sample deflection is maximal at point B (sample half length).

Since targeted samples are round with rather low diameters, measurement of applied stresses at sample surface with a strain gauge is difficult, so that applied stresses need to be calculated from the sample curvature.

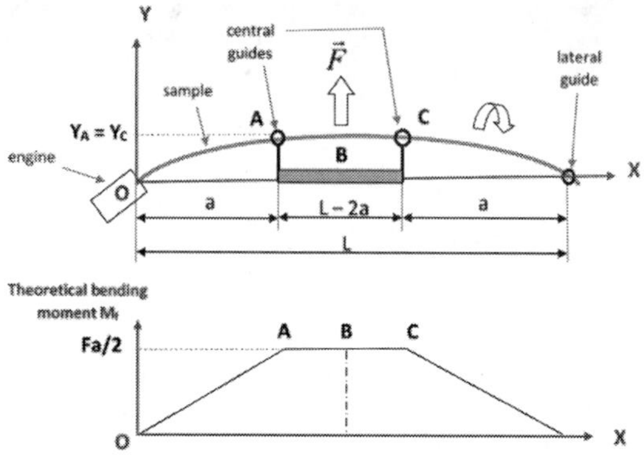

Fig. 6. Applied bending load F, and theoretical bending moment M_f during a rotary bending test

In this study, rotary bending tests have been carried out with a 3.60 mm diameter sample of EN 1.4568 grade with a length L equal to 600 mm, bent with different y_A values (positions of both central guides along Y axis) ranging from 55 to 95 mm, and with a constant value of a = 200 mm (see figure 6). Rotating speed was constant and equal to 4500 rpm.

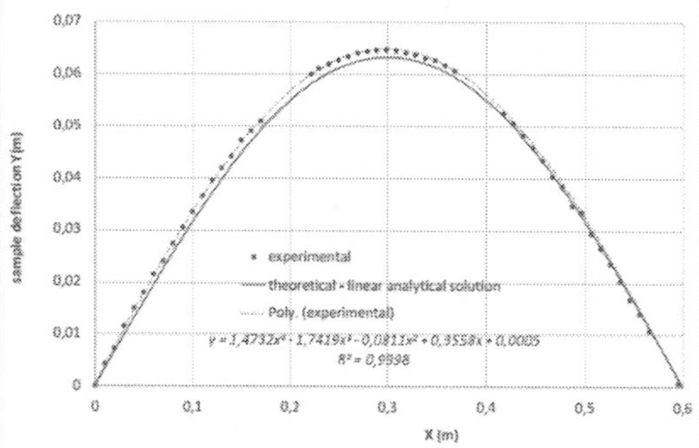

Fig. 7. Comparison between experimental and theoretical profiles y(x) of 3.60 mm diameter stainless steel samples bent with y_A = 55 mm

For a given y_A (or y_C) value, calculation of the testing parameters (bending moment and load, maximum and minimum applied stresses) requires the following steps, according to M. AYADA and al[11]:

1/ measurement of the sample deflection y(x) along its length; example of measurements obtained for y_A = 55 mm have been reported on figure 7;

2/ calculation of the resultant bending load F by minimization of the potential energy (sum of the work of load F and sample deformation energy) which gives the following relationship between load F (N) and maximum deflection y_B (m) at point B [11]:

$$F = EI\alpha y_B\left[1 + \beta y_B^2 + \lambda y_B^4\right] \quad (3)$$

where:

E (GPa) is the sample YOUNG modulus;

I (m^{-4}) is the sample moment of inertia of area;

α, β and λ are L- and a- dependent parameters.

3/ calculation of reaction forces N_1 and N_2 on the one hand, friction forces μN_1 and μN_2 on the other hand (μ being the friction coefficient) at contact points O (between sample and jaw) and A (between sample and central guide), according to figure 8. Those forces can be deduced from the force equilibrium equation projected on X and Y axis, which gives the following relationships:

$$N_1 = \frac{F}{2(\cos\theta_1 + \mu\sin\theta_1)} \quad (4)$$

$$N_2 = \frac{F}{2(\cos\theta_2 - \mu\sin\theta_2)} \quad (5)$$

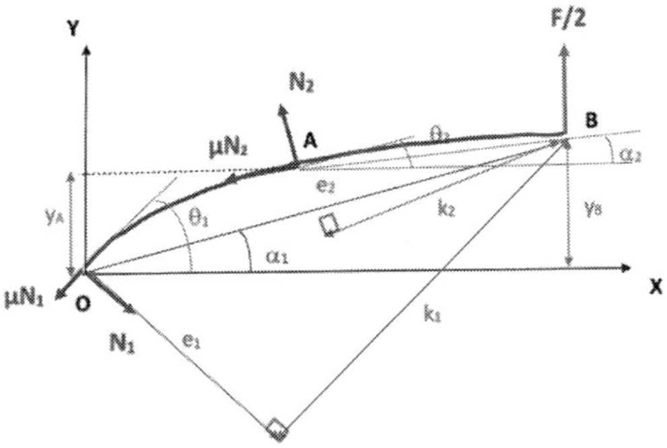

Fig. 8. Description of the loading system during rotary bending test (first half of sample length)

4/ calculation of the global bending moment M_B applied at point B, equal to the sum of the bending moments of the contact forces N_1, N_2, μN_1 and μN_2 :

$$M_B = N_1 k_1 + \mu N_1 e_1 - N_2 k_2 + \mu N_2 e_2 \quad (6)$$

where k_1, k_2, e_1 and e_2 are geometrical parameters deduced from the measured sample profile according to figure 8;

5/ finally, determination of maximum applied stress S_{max} (at sample extrado) and minimum applied stress S_{min} (at sample intrado), both calculated at sample half-length (point B), thanks to the following relationships:

$$S_{max} = \frac{M_B\phi}{2I} + \frac{4H}{\pi\phi^2} \quad (7)$$

$$S_{min} = \frac{-M_B\phi}{2I} + \frac{4H}{\pi\phi^2} \quad (8)$$

where H is the global force resulting from the different contact and friction forces, equal to :

$$H = \mu N_1\cos\theta_1 - N_1\sin\theta_1 + \mu N_2\cos\theta_2 + N_2\sin\theta_2 \quad (9)$$

It has to be pointed out that no sample heating has been observed during rotary bending test, even near the contact points with the central guides, showing that friction coefficient μ is very low. A value μ = 0,02 corresponding to a very well lubricated contact has also been chosen for all the calculations reported in this article.

Furthermore, sample profile can be calculated by the linear analytical method (under small deflections), by double integrating the theoretical bending moment (represented in figure 6), and by considering that this moment is constant between the central guides. Under those assumptions, sample profile y(x) follows polynomial functions:

$$y(x) = \frac{6y_A}{a^2(3L-4a)}\left[\frac{x^3}{6} - a\left(\frac{L}{2} - \frac{a}{2}\right)x\right] \quad \text{if } x\in[0;a] \quad (10)$$

and:

$$y(x) = \frac{6y_A}{a(3L-4a)}\left[\frac{x^2}{2} - \frac{L}{2}x + \frac{a^2}{6}\right] \text{if } x\in[a;(\frac{L}{2} - a)] \quad (11)$$

However, experimental measurements of sample profiles are best fitted with a 4-order polynomial function (see figure 7) and some differences between experimental and theoretical profiles may become significant for high deflections. Moreover, the linear analytical approach may engender discrepancies in the calculation of maximum and minimum applied stresses in comparison to those based on experimental measurement of sample profile (discrepancies which become significant for high deflections), as shown in table 3. For this reason, values of S_{max} reported in the following paragraphs have been calculated based on measured sample profiles.

Finally, despite those differences, table 3 shows that S_{max} values are almost constant between points A and B, which means that this parameter is almost constant in the sample part located between both central guides.

Table 3. Values of rotary bending parameters calculated by the linear analytical approach or by experimental measurement of sample profile

y_A (mm)	$y_B - y_A$ (mm)		S_{max} (MPa) at sample half-length (point B)		S_{min} (MPa) at sample half-length (point B)		S_{max} (MPa) at the contact point between sample and central guide (point A)	
	by linear analytical method	with measured sample profile	by linear analytical method	with measured sample profile	by linear analytical method	with measured sample profile	by linear analytical method	with measured sample profile
55	8,3	9,8	489	503	-490	-504	487	500
65	9,8	11,3	572	583	-572	-583	568	578
85	12,8	14,0	755	772	-756	-773	746	761
95	13,8	15,5	876	942	-877	-944	862	921

5 Results of rotary bending tests and inclusion analyzes

The results of rotary bending tests carried out on 3.60mm diameter samples of EN 1.4568 grade are shown in WÖHLER's curve of figure 9.

Over the 25 sample failures observed during the trials, only three initiated at sample surface, showing that the material preparation is adapted to the testing conditions. The other failures have initiated at subcutaneous inclusions or precipitates, located in the first 0.5 mm of sample depth. Particles have been analyzed with a JEOL 7001F SEM, equipped with a SDD-EDX analyzer, run at 15kV. Two populations of particles have been identified as initiating sites (see figure 10):

- Aluminium nitrides whose sizes lay between 5 and 9 µm and which represent the majority of the cases;
- oxide inclusions whose sizes can reach up to 23 µm and which represent about 20% of the observed particles.

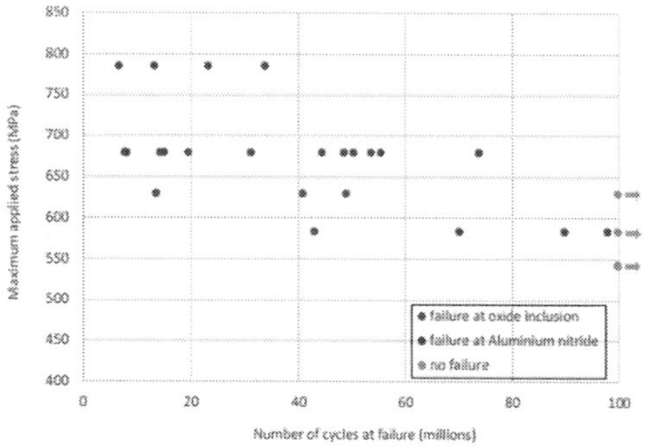

Fig. 9. Results of rotary bending tests carried out on 3,60mm diameter drawn wires of EN 1.4568 grade

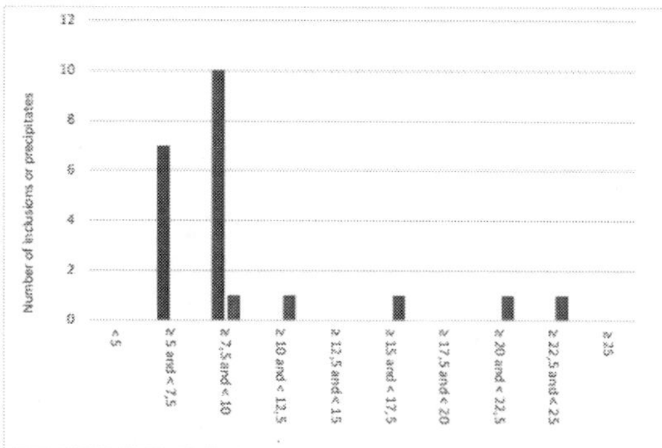

Fig. 10. Size distribution of particles observed after rotary bending tests carried out on EN 1.4568 drawn wire

Oxide inclusions have rather globular shapes whereas *AlN* are more flat and angulous, according to figures 11 and 12. SEM observations combined with X-Ray analyzes and mapping have shown that oxide inclusions are complex and multiphase oxides, with high global contents of both Al_2O_3 and CaO, but also containing MgO in variable amounts (from 0.5% up to 75%) depending on the analyzed area of the same inclusion (see figure 12). Some inner or outer parts are locally enriched in MgO. Global contents of those inclusions are close to that of $C_{12}A_7$ compound.

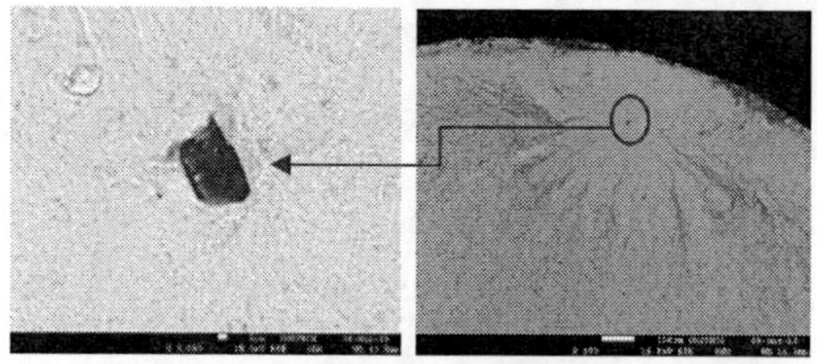

Fig. 11. Typical Aluminium nitride precipitate identified as failure initiation site during rotary bending test carried out on EN 1.4568 drawn wire (SEM observation)

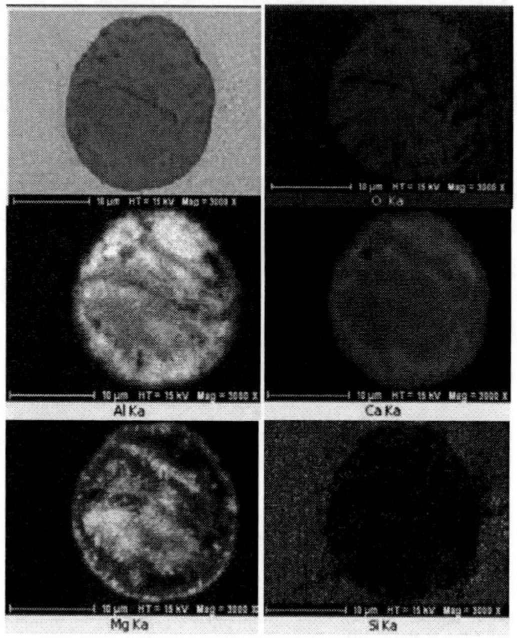

Fig. 12. Typical oxide inclusion as failure initiation during rotary bending test carried out on EN 1.4568 drawn wire; backscattered electron SEM image and X-Ray maps of the following chemical elements: Oxygen (red), Aluminium (white), Calcium (magenta), Magnesium (yellow) and Silicon (orange)

Beside their size, shape and location, the criticity of inclusions towards crack initiation is also linked to their relative plasticity at high temperatures and their relative thermal expansion to that of the stainless steel matrix. Indeed, both factors may promote local tensile residual stresses and micro-crack initiation at the inclusion/matrix interface [1,6]. Values of such factors have been collected in table 4 for oxide inclusions and precipitates which have been observed in this study [6,12-14]. In this table, the plasticity index refers to the parameter defined by R. Kiessling[6]: the lower this index, the more critical the inclusion towards crack initiation. Comparing *AlN* and *Al-Ca* oxides for the same size and location, the figures suggest that *AlN* precipitates are more likely to promote cracks than *Al-Ca* oxides because of their angular shape and their lower thermal expansion coefficient at high temperatures. No value has been reported regarding the plasticity index of *AlN* at high and room temperatures, but this index may be very low, close to zero, as it is the case for *Al-Ca* oxides. Those observations agree with figure 10 which shows that failure initiations during rotary bending tests have occurred at *AlN* in the large majority of the cases, and that oxide inclusions have been involved in failure initiations only when their size was greater than the maximum size of *AlN* precipitates, *i.e.* oxides inclusions of diameter higher than 9 µm. Also, those results agree with the fact that oxide inclusions of diameter higher than 9 µm are preferential sites for failure initiation in the sample. Furthermore, it can be noticed that according to WÖHLER's curve shown in figure 9, oxide inclusions larger than 9 µm are correlated with lower number of cycles at failure than Aluminium nitrides for a given value of maximum applied stress. When the subcutaneous area of the material is free of such inclusions and only oxides of lower diameter are present, failures initiate on largest Aluminium nitride precipitates.

The effect of inclusion or precipitate size on the fatigue limit can be substantiated by MURAKAMI's approach. Values of calculated endurance limit, via equation (2) in the case of *AlN* or oxide inclusions of different sizes, have been collected in table 5. Firstly, calculated value of σ_D with equation (2) in the case of oxide inclusion maximum size (equal to 23 µm) is close to the experimental value found to be around 580 MPa. On the other hand, the presence of oxide inclusions of diameter 9 mm (the minimum diameter of oxide inclusions observed as failure initiation sites) leads to σ_D value very close to that obtained in the presence of maximum size *AlN* precipitates. Finally, a σ_D increase of 100 MPa could be expected by suppressing oxide inclusions with diameters higher than 9 µm.

Table 4. Values of mechanical and physical properties relative to inclusions or precipitates [6,12-14]

	$C_{12}A_7$	MgO	AlN	EN 1.4568 grade
T_{fusion} (°C)	1455	2800	2200	1475
Vickers Hardness (HV) at room T°	≈ 900	≈ 700	550 to 1150	210 to 510
Plasticity index[6] when T° ∈ [800;1400°C]	≈ 0	0 to 0,5		1
Thermal expansion between 20 and 1000°C ($x10^{-6}$/°C)	8	11,5	5	15

Table 5. Values of fatigue limit σ_D calculated with different inclusion/precipitate sizes according to Murakami's work[3] (see equation (2))

	AlN	**oxyde**	
inclusion/precipitate shape	rectangular	round	
inclusion/precipitate size (µm)	9 x 7	Ø9	Ø23
inclusion/precipitate cross-section area (µm²)	63	63,5	415
Vickers hardness (Hv)	510	510	510
fatigue limit σ_D (MPa)	**696**	**695**	**595**

An improvement of the fatigue limit of EN 1.4568 drawn wires also requires a reduction of the number of oxide inclusions larger than 9 µm. Figure 12 shows that those oxides are complex compounds with local areas enriched in *MgO*. Some investigations are under progress to explain the mechanism of their formation and hence propose modifications of the melting process to reduce their density and size.

Furthermore, improving the endurance limit of EN 1.4568 drawn wires requires also to lower the maximum size of Aluminium nitride precipitates, which is closely related to the Nitrogen content of the steel. In order to lower Nitrogen content, inerting during both melting and remelting processes should be improved.

6 Conclusions and perspectives

A methodology has been defined to study the influence of micro-cleanliness on the high cycle fatigue resistance of stainless steel wires or bars. It is based on MURAKAMI's work and rotary bending tests carried out up to 100 millions cycles. It has been applied to EN 1.4568 drawn wires of diameter equal to 3,60 mm. The first trials have validated the setting of rotary bending equipment, the protocol for sample preparation and the calculation of test parameters.

Furthermore, they have shown that failures occurring during fatigue tests initiated preferentially at subcutaneous oxide inclusions of diameters higher than 9µm. Those inclusions are complex and multiphase oxides, with high global contents of both Al_2O_3 and *CaO*, but also containing *MgO* in locally variable amounts (from 0.5% up to 75%) inside the same inclusion. Some investigations are under progress to explain the mechanism of their formation and hence propose modifications of the melting process to reduce their density and size. According to MURAKAMI's work, their suppression could lead to a 100 MPa increase of the material's fatigue limit.

When the sample subcutaneous area is free of oxide inclusions whose diameters are higher than 9 µm, failures are seen to initiate at *AlN* precipitates, whose sizes are lower than 9 µm, due to their angular shapes, and their relative low plasticity and thermal expansion at high temperatures in comparison to the stainless steel matrix. The size reduction of those *AlN* precipitates is mainly correlated to the steel Nitrogen content, also to the inerting conditions during both melting and remelting operations.

Authors' acknowledgements go to UGITECH's project team (S. THÉVENET, R. SCHNEIDER, P. TROLLIET, JC. CHRISTIN, A. LANGLET, K. BENCHALLAL), as well as persons involved in the development of remelted products (C. PHILIBERT, F. BONIFASSI, R. REYNAUD, A. CHÈZE, G. PELLISSIER, F. CAYER-BARRIOZ, N. SOCQUET-JUGLARD, N. CRÉPEL-PROVOST). Other people from UGITECH R&D are associated to this study (C. BOURGIN, ML. MANTOUX, V. MARIE, D. JULIANO, D. CHOULET, G. DUNAND-MARTIN), as well as T. QUIVY from Imphy drawing plant. Authors are also grateful to the supplier ZARIAN GmbH, represented by S. BLAETTERLEIN and B. LAUTERBACH. The authors want to thank Y. FOLLEZOU, JM. DUBOIS and F. PROVENT from S.F.G.P. for their contribution.

https://doi.org/10.1051/matecconf/201816520009

References

1. H.P. Lieurade, *Propriétés d'emploi des aciers – la rupture par fatigue des aciers*, IRSID OTUA, Paris (1987)
2. T. Zouhair, *Essais de fatigue par la méthode des "blocs programmes"*, PhD, INSA Lyon (1979)
3. Y. Murakami, *Metal fatigue : effects of small defects and nonmetallic inclusions,* Elsevier (1993)
4. D. Rousseau, L. Séraphin, R. Tricot, *Internal Report n° 55793*, Ugitech Research Center (1974)
5. H. Nordberg, *Swedish symposium on non-metallic inclusions in steel,* Swedish Institute for Metal Research, Uddeholm (1981)
6. R. Kiessling, *Non-metallic inclusions in steel – Part III,* The Iron and Steel Institute, (1968)
7. C. Bertrand and al, *Improving fatigue life of special steels by modifying their inclusion engineering*, Final Report of European Com. Tech. Steel Research (2002)
8. Y. Murakami, *Journal of Research of the National Institute of Standards and Technology*, **99**, 4 (1994)
9. T. Manninen, *Influence of cold-work on the elastic properties of austenitic stainless steels,* Proceedings of the 7th European Stainless Steel Conference ESSC 2011, Como, Italy, September 2011
10. M. Libert, *Internal Analysis Report n° 16_159,* Ugitech Research Center, September 2016
11. M. Ayada, N. Takamura, *Transactions of Japan Society of Spring Engineers,* **1992**, 37, p59-64 (1992)
12. Website of CTTC association (Centre de Transfert de Technologies Céramiques) : http://www.cttc.fr
13. P. Klocek, *Handbook of Infrared Optical Materials*, Library of Congress Cataloging in-publication data, NY, USA, 1991
14. C.E. Cicutti and al, *Ironmaking and Steelmaking*, **24**, 2, p155-159 (1997)

Fatigue behavior of 590MPa Tensile Strength galvanneal coated sheet steel laser welded blanks

Shrikant P. Bhat[1,*], *Ronald* Soldaat[1], and *Gagan* Tandon[2]

[1]ArcelorMittal Global R&D, 3001 East Columbus Drive, East Chicago, IN 46312, USA
[2]ArcelorMittal Tailored Blanks, 8650 Mt Elliot, Detroit, MI 48211, USA

Abstract. The use of Laser Welded Blanks (LWBs) with different grade/gauge combinations in automotive body structures is well established; however, the acceptance of LWBs in fatigue critical chassis and underbody components has been slower because of lack of reliable models for durability assessment of laser welded joints (LWJ). Most prior fatigue studies of LWBs are carried out in tension - tension loading mode, making it difficult to relate it to the cyclic deformation and fatigue behaviors of the substrate steel grade. In contrast, in this study, LWJ is conceptualized as a "notch" to estimate the local stresses from the strain - life data of the parent grade and the fatigue notch concentration factor (K_f) is estimated from the nominal stress values of LWJ. The method is illustrated with strain controlled fatigue data for 1.4 mm galvanneal coated 590 MPa steel and fully reversed, fatigue data for homogeneous and heterogeneous LWB combinations. The results indicate that for both homogeneous and heterogeneous LWJ configurations, K_f increases with fatigue life, but tends to saturate at life levels greater than about million cycles. Considering 10^5 cycles to failure as an example, K_f is estimated as 1.07 for the homogeneous and 1.25 for the heterogeneous combinations.

1 Introduction

1.1 Industrial Relevance

Over the last several decades, increasingly stringent safety requirements, demands for higher fuel economy and the need to reduce greenhouse gas emissions for cleaner environment all at a competitive price (raw material and conversion cost included) have spurred the development of new high strength steel grades and more importantly newer processes to better utilize steel grades in automotive applications. One such technology introduced in the early 1980's for better steel utilization is Tailor Welded Blanks (TWB). While there are many welding methods to produce TWBs, this report will focus on tailored blanks produced by laser welding (LWB). The advantages of laser welding to produce TWBs include ease of welding different steels (grade, thickness, and coating), narrow weld and heat affected zone (HAZ), relatively low geometric distortion, relatively high welding speed and high penetration [1]. Over the last several years, LWB has gained wide acceptance in body structure applications (for example, door rings, door inner panels, B-pillars, sills, floors, etc.). However, the use of LWBs in chassis, suspension and other underbody components (for example, subframes) has been rather minimal. The paucity stems from the fact that while thinner (or lower strength) material can be placed in areas with less critical stiffness, strength and crashworthiness requirements, it may be difficult to find "less critical" areas of a component designed for fatigue/durability resistance [2]. In addition, many of the underbody components tend to be heavier gauge hot rolled or hot-rolled & coated products for which laser-welding (to produce LWB) may not be as straight forward.

This research was conducted to promote high strength steels in front or rear subframes of a passenger vehicle via LWB to be weight competitive with alternate materials. Road-load durability assessment is a critical performance evaluation step for subframes. Following the preliminary part design and manufacturing feasibility evaluation, it was necessary to address the durability concerns, both with respect to component testing and a method of estimating it via finite element analysis (FEA) codes for the LWB joint. This report summarizes the fatigue test results and an estimation of notch effects due to LWB in support of this overall effort on one steel grade of interest to the sub-frame.

1.2 Technical Challenges – Conceptualizing LWB as a Notch

Fatigue failures almost always initiate at geometric discontinuities (shape change, thickness change or joints) that act as stress or strain concentrations. In general, under cyclic loading, the geometric discontinuities are less effective than predicted by K_t (static theoretical stress concentration factor) in raising the local stress or strain and therefore, the effects of notches in fatigue is usually described by K_f, Fatigue Notch Factor or Fatigue Strength Reduction Factor. In this investigation, LWB zone is conceptualized as a "notch" and the focus is to determine K_f for LWB.

* Corresponding author: shrikant.bhat@arcelormittal.com

The fatigue notch factor K_f is the ratio of un-notched fatigue strength to notched fatigue strength at a given life level [3]:

$$K_f = (\sigma_{UNNOTCHED}) / (\sigma_{NOTCHED})$$

Although this concept of K_f is relatively simple, experimental determination is difficult because it depends on notch geometry (LWB zone), life level, loading mode, load ratio, etc making the applicability of data from one experimental protocol to another questionable and often difficult.

In the automotive industry with emphasis on finite-life design, the use of low cycle fatigue (LCF) concepts for component durability estimation is rather common. The focus is on crack initiation and the crack propagation life is ignored or considered to provide a safety factor against catastrophic failures. Therefore ideally, the estimation of K_f should take into account the LCF behavior of the steel grade.

There have been many studies elucidating the effects of laser welding on fatigue behavior of steels; differences in laser welding technology applied to part assembly and that used in tailored blank production should be kept in mind in reviewing these results. For example, Wang and Ewing [4] showed that the tensile fatigue strengths of laser welds are superior to those of spot welds in lap shear joint configuration for DQSK steel. Daneshpour et al [5] showed that laser spot welds exhibited better fatigue life compared to spot welds, but only at "high load" levels (or short life) for dual phase 780 but not in DQSK type of steel. Asim et al [6] found that in laser welded lap shear joints of HSLA steel, crack initiation was near the boundary of the base metal and heat affected zone (HAZ) under static conditions, but initiated at pre-existing crack tips under cyclic loading (both low cycle and long-life regimes). Anand et al [2] reported that zinc-coated steels with differential thickness TWB joints exhibited a lower fatigue limit compared to similar TWB joints of uncoated materials. Most of these earlier studies were carried out in tension-tension fatigue mode (often with load ratio R = 0.1) and most often these studies do not report the LCF behavior of the parent grade of steel, making any generalization extremely difficult. In the current program, LCF behavior of the parent steel grade and an estimation of K_f for two different LWB combinations will be emphasized.

2 Experimental Details

2.1 Materials and Experimental Procedure

The steel grade used in this study is GA 590R, indicating a minimum tensile strength of 590 MPa with a relatively high yield ratio (Yield strength/Tensile strength) and galvannealed (GA) coating produced via hot dip process. The coating and the substrate metallurgy details of producing such products can be found elsewhere [7,8]. Two commercially produced coils at 1.4 mm and 1.0 mm thicknesses were secured for the present study. The chemistry and the coating details of the product are listed in Table 1 while the tensile properties (determined following the JIS protocols) are listed in Table 2. No measurable differences were found for the two thicknesses.

Table 1. Chemistry and coating characteristics of GA 590R

SUBSTRATE CHEMISTRY, (wt %)							GA COATING (Top/Bottom)	
C	Mn	Si	P	S	Al	N	Wt, g/m²	% Fe
0.104	2	0.13	0.014	0.001	0.02	0.005	43/38	9/9

Table 2. Tensile properties of GA 590R

TENSILE TEST RESULTS (JIS No. 5 Transverse)		
YIELD STRENGTH, MPa	TENSILE STRENGTH, MPa	TOTAL ELONGATION, %
494	656	25

Rectangular LWBs were produced in the homogeneous (1.4 /1.4 mm) and heterogenous (1.4/ 1.0 mm) combinations using a commercial CO_2 laser. A cross-section of one of the laser welded joints is shown in Figure 1 and demonstrates good laser weld quality.

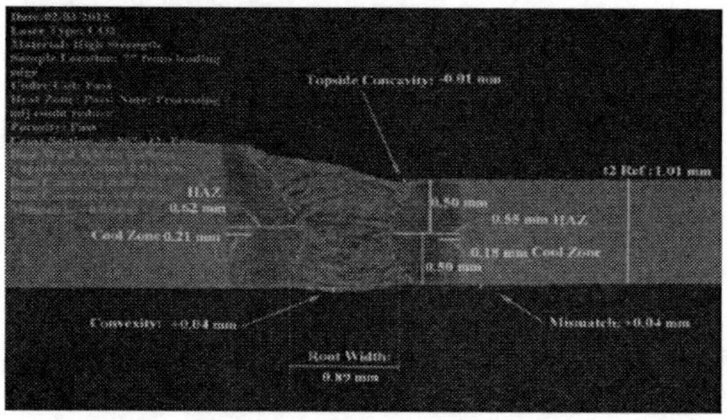

Fig. 1. Heterogeneous (1.4 / 1.0 mm) LWJ of GA 590R

The three primary areas of investigation included: a) Tensile and fatigue testing of monolithic GA 590R (1.4 mm); b) Tensile and fatigue testing of homogeneous LWB coupons (1.4/1.4 mm). and c) Tensile and fatigue testing of heterogenous LWB coupons. Tensile tests were carried out on an Instron machine and the axial, tension-compression, fully reversed (R = -1) fatigue tests were carried out on an MTS servo-hydraulic test system following the ASTM E 606 procedure adapted for thin sheet steels.

3 Results and Discussion

3.1. Tensile test results

Tensile test results are summarized in Table 3 for monolithic, as well as homogeneous, and heterogeneous LWB combinations (weld line perpendicular to loading direction). The data indicate that the laser welding process has very little influence on the yield and tensile strengths of the steel but it greatly reduces the "apparent" ductility parameters - total elongation (TE) and uniform elongation (UE) values. This is illustrated in Figure 2 via the engineering stress – strain curves for the three conditions investigated in this work.

Table 3. Tensile test results for LWB of GA 590R

DESCRIPTION	GAUGE, mm	YS, MPa	TS, MPa	TE, %	UE, %
Monolithic	1.4	494	656	24.7	12.9
Homogeneous	1.4/1.4	497	652	18.6	9.4
Heterogeneous	1.4/1.0	492	658	13.7	6.8

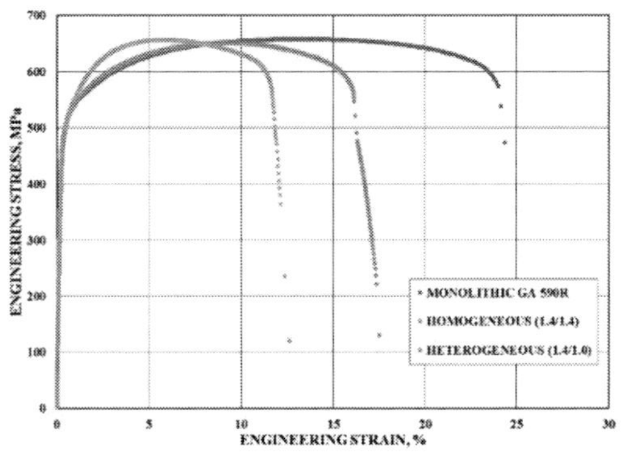

Fig.2. Engineering stress -strain curves for monolithic and two LWB GA 590R

The apparent loss of ductility is attributable to the higher hardness of the weld region in the center portion of the gauge length of the tensile coupon. The Vickers hardness profiles from the center of the weld for the three conditions are shown in Figure 3. The weld hardness (380 – 400 VHN) is approximately twice that of the base metal (190 – 210 VHN). The weld hardness values are reasonable with the fully martensitic structure with a steel chemistry of 0.1C, and 2 Mn (this hardness corresponds to a tensile strength of about 1220 MPa).

It should be noted that the weld and the HAZ in LWBs are comparatively narrow at about 2 mm (weld zone width about 1 mm and the HAZ width about 1 mm, see Figure 1). This high hardness zone acts as a barrier to deformation during the tensile deformation. In other words, the martensite in the weld region is expected to be in the elastic regime, even when other portions of the specimen are well into the plastic deformation regime.

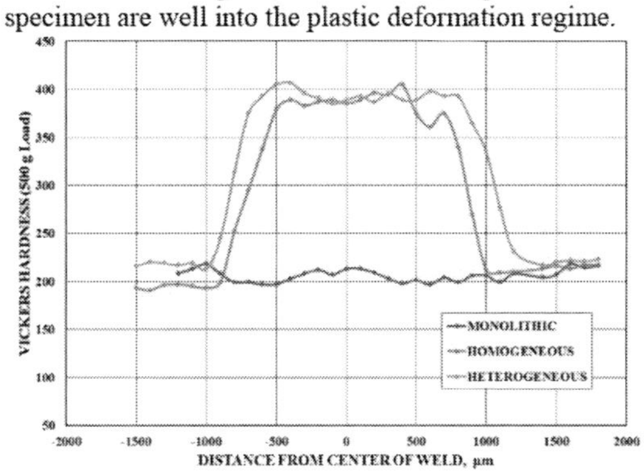

Fig 3. Microhardness profile of base steel and LWB coupons

Thus, the weld region may be considered to act as an artificial "grip" region, but the machine records the total extension and the original gauge length is used to calculate the engineering total elongation and uniform elongation values. Further, Figure 4 shows the fracture profiles of two LWB coupons after tensile testing with final failure occurring in the base metal, significantly far away from the weld zone, in both homogeneous and heterogeneous LWBs.

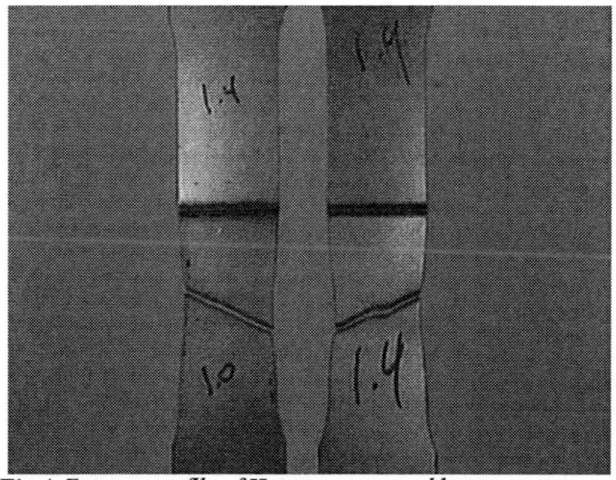

Fig 4. Fracture profile of Heterogeneous and homogeneous LWB tensile coupons

Therefore, it is emphasized that the ductility parameters determined using the normal tensile specimen protocols are "apparent" values and do not reflect the true values. The weld zone is narrow in LWB and the hardness remains unaffected in other areas. This "high" hardness of the weld zone concentrates the deformation zone away from the weld area and the final fracture occurs due to strain localization in the matrix. In this regard, it is important to highlight that in laser welding of GA 590R, the high hardness in the weld zone decreases gradually in the HAZ reaching that of the base metal without any observable softening (no valley in the hardness profile) in the HAZ. In contrast, many earlier research reports indicate that the tensile failures in LWBs with DP steels originate in the HAZ due to local softening or hardness valleys [9 – 11]. Such was not the case and no HAZ softening was noted in GA 590R after LWB production.

3.2. Fatigue test results

In this investigation, fatigue behavior of GA 590R is characterized under four different experimental protocols: a) Strain controlled fatigue characteristics of the base steel (monolithic); b) Load controlled fatigue behavior for base steel (monolithic); c) Load controlled fatigue behavior of homogeneous laser welded coupons (1.4 /1.4 mm combination) and d) Load controlled fatigue behavior of heterogeneous laser welded coupons (1.4/1.0 mm combination). These results are presented first and then the data are used to estimate the fatigue notch factors of LWB joints in this steel grade.

3.2.1 Strain Controlled Low Cycle Fatigue (LCF) of GA 590R

LCF characterization of the steel was carried out through axial, fully reversed, constant amplitude, total strain controlled tests; at least two specimens were tested at each strain level in the life range of ~ 2000 to 5,000,000 cycles to failure. Both cyclic deformation and life aspects as determined through strain controlled tests are highlighted below.

Cyclic Response Curves: Macroscopic stability of a steel under fatigue loading can be represented as a plot of the stress needed to impose the strain limits as a function of the number of cycles elapsed throughout the life. Cyclic response curves (CRC) for GA 590R at three representative strain amplitudes are shown in Figure 5, in which the resistance to strain cycling is represented as the averages of the tensile and compressive maximum flow stress values (without regard to sign). Cycling continuously modifies the stress response throughout fatigue life at most strain amplitudes. At very low strain amplitude (0.15%), the stress response is very stable through most of life. However, at higher strain amplitudes (for example, 0.6 %), specimens exhibit gradual softening throughout life and only a weak indication of saturation can be detected late in life (approximately at half-life).

Cyclic Stress – Strain (CSS) Curve: An appropriate stress – strain relationship is needed to account for the inelastic behavior of steel in a structure, for example near a notch. The results presented in Figure 5 indicate that cyclic loading can alter the stress – strain relationship significantly from its virgin form. This change in behavior is depicted in Figure 6 by comparing the monotonic and cyclic stress – strain behavior for low strain levels (up to 1% total strain). The CSS curve deviates from the monotonic curve significantly and shows a large cyclic softening in the yield point region, due to the removal of the yield point i.e. generation of fresh unpinned dislocations [12]. From an engineering perspective, the use of tensile stress – strain curve for fatigue analysis may lead to significant errors.

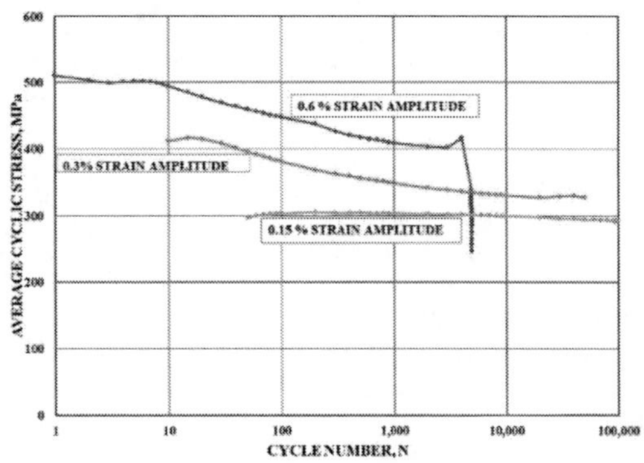

Fig. 5. Cyclic response curves at three representative strain amplitudes for GA 590R.

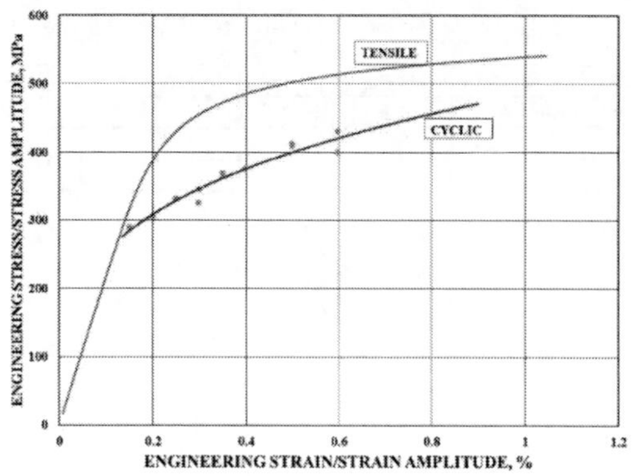

Fig. 6. Comparison of cyclic and tensile stress – strain curves for GA 590R

Strain – Life Curves: Strain – life data from constant amplitude tests for GA 590R are shown in Figure 7; this figure includes total strain amplitude (test control variable) as well as the component strain amplitudes (Plastic and Elastic) determined from approximately half-life hysteresis loops. The two component strains when plotted against fatigue life on a log – log scale lead to straight lines and of course the total strain – life plot is not. Following the procedures detailed in ASTM E739, the data were analyzed with the fatigue life as the dependent variable and the plastic and elastic strain components as the independent variables to determine Fatigue Parameters, using the equations given below:

$$(\Delta\varepsilon_{pl}/2) = \varepsilon_f^{'} (2N_f)^c$$

Plastic Strain – Life (Coffin-Manson Equation)

$$(\Delta\varepsilon_{el}/2) = (\sigma_f^{'}/E)\,(2N_f)^b$$

Elastic Strain – Life (Basquin Equation)

$$(\Delta\sigma/2) = k'(\Delta\varepsilon_{pl}/2)^{n'}$$

Cyclic Stress – Strain Equation

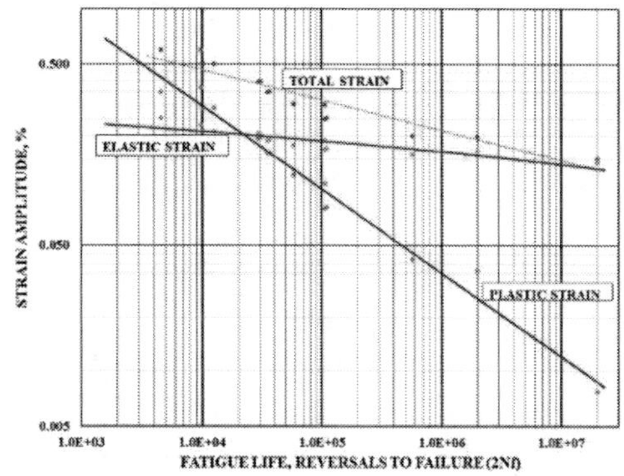

Fig. 7. Strain -Life fatigue plots for GA 590R

The fatigue parameters determined from linear regression analyses of the data (on transformed logarithmic variables) are listed in Table 4. Correlation coefficients are also listed for each of the equations to indicate robustness of the data. It must be kept in mind that all the data come from one coil and one location and thus they do not reflect the variability within the grade. Comparison of data in Table 3 to those in Table 4 show that the cyclic yield strength of 376 MPa is substantially lower than that of the tensile yield strength of 494 MPa for this steel. Thus, the use of the tensile YS for calculations under fatigue loading situations may be too optimistic.

Table 4. Summary Strain – Life fatigue parameters for GA 590R

1. CYCLIC STRESS -STRAIN EQUATION	
Cyclic Strength Coefficient, k', MPa	**681**
Cyclic Strain Hardening Exponent, n'	**0.096**
Correlation Coefficient, R^2	**0.85**
Cyclic Yield Strength (CYS), MPa	**376**
2. PLASTIC STRAIN - LIFE EQUATION	
Fatigue Ductility Coefficient, ε'_f	**0.282**
Fatigue Ductility Exponent, c	**-0.491**
Correlation Coefficient, R^2	**0.98**
3. ELASTIC STRAIN - LIFE EQUATION	
Fatigue Strength Coefficient, σ'_f, MPa	**849**
Fatigue Strength Exponent, b	**-0.069**
Correlation Coefficient, R^2	**0.86**

3.2.2 High Cycle Fatigue (HCF) for monolithic GA 590R

Long life fatigue data generated under stress control are shown in Figure 8 as the Stress – Life (S-N) plot with 10^7 cycles taken as the run-outs (tests were terminated if the specimen survived this life level). The data refer to axial loading and fully reversed (Load Ratio, R = -1) condition and the tests were run at a constant frequency of 20 or 30Hz. Using these data, fatigue limit is estimated to be 310 MPa and a fatigue ratio (Fatigue Limit / Tensile Strength) of 0.47.

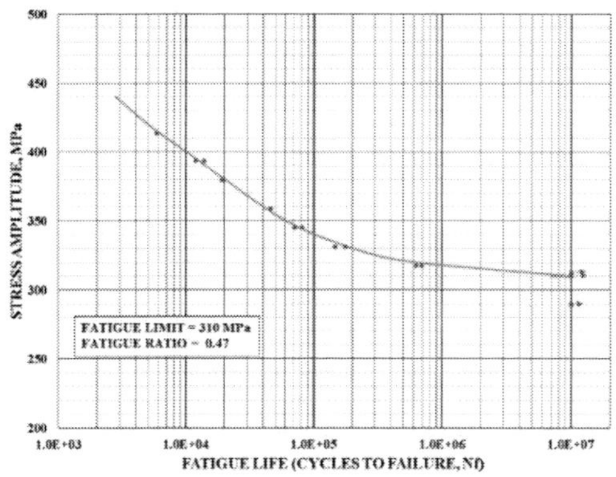

Fig. 8. S-N curve for monolithic GA 590R (1.4 mm)

3.2.3 S-N Curve for Homogeneous LWB of GA 590R

Attempts to carry out strain controlled fatigue tests under axial, fully reversed mode were not successful, because of excessive buckling due to stress concentration at the weld line. Therefore, only the load controlled fatigue tests were carried out for both homogeneous and heterogeneous laser welded coupons. For all the fatigue tests, the loading axis was perpendicular to the weld line. Square cross section of the specimens and careful alignment (of load train) were critical to achieve fully reversed loading condition. A critical distinction of the current fatigue work on laser welded coupons is important to emphasize. Most prior fatigue work after laser welding [2, 4-6, 9-11] is carried out in tension – tension mode (typical load ratio, R = + 0.1) rather than in fully reversed condition (R = -1) to minimize local buckling issues. However, it is well known that only the R = -1 loading condition does not introduce significant mean stress at notch roots [13] during fatigue tests and makes it possible to relate these data to those of the smooth specimen (monolithic) strain -life data. Because of mean stress effects at the notch root, it is difficult to relate the data generated under other load ratios (for example R= 0.1) to smooth specimen LCF data (strain control tests are not usually carried out with mean strain). Thus, to the best of our knowledge, the current investigation is the first one to record the fatigue data for laser butt-welded joints under fully reversed, axial loading condition.

S–N data for homogeneous (1.4/1.4 mm) laser welded GA 590R coupons are presented in Figure 9. Estimated fatigue limit of 300 MPa and the Fatigue Ratio of 0.46 are only slightly below that of the monolithic GA 590R (~ 4 % drop) and it may be concluded that the influence of laser butt welding on the axial, fully reversed fatigue is to decrease the fatigue limit by 10 MPa or ~ 4% for this steel.

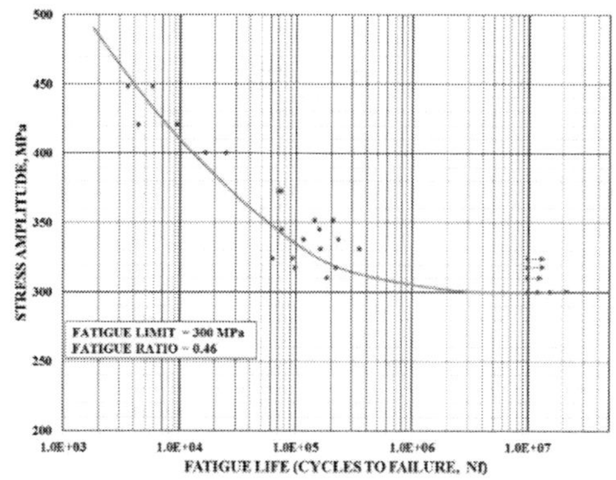

Fig.9. S-N curve for homogeneous LWB of GA 590R (1.4 / 1.4 mm); Axial, fully reversed.

3.2.4 S-N Curve for Heterogeneous Laser Welded GA 590R

Stress controlled fatigue tests were repeated for heterogeneous (1.4 mm/1.0 mm) laser welded GA 590R with the same test protocols. To maintain axiality and to avoid fractures in the specimen shoulders, it was necessary to add thin metal shims on the thinner side of the specimen. All the stress calculations are based on the thinner gauge section. The results are presented in Figure 10. Fatigue limit data for the three test conditions

https://doi.org/10.1051/matecconf/201816521001

are summarized in Table 5. Estimated fatigue limit of 240 MPa for the heterogeneous LWB and a fatigue ratio of 0.36 are substantially lower than those of the base material and the homogeneous LWB. This decrease can be attributed to the local stress concentration introduced by gauge differences at the weld. Keeping in mind that even for heterogeneous combination, tensile strengths are the same as that for the base material, this demonstrates the influence of local "notches" on the fatigue behavior of joints. This will be illustrated further in the following section.

Table 5. S-N data summary for GA590R

FATIGUE LIMIT DATA FOR GA 590R LWB (AXIAL, FULLY REVERSED, R = -1)			
SPECIMEN CONDITION	FATIGUE LIMIT (10^7 CYCLES)	TS, MPa	FATIGUE RATIO (Fatigue Limit /TS)
MONOLITHIC	310	656	0.47
HOMOGENEOUS (1.4/1.4)	300	652	0.46
HETEROGENEOUS (1.4/1.0)	240	658	0.36

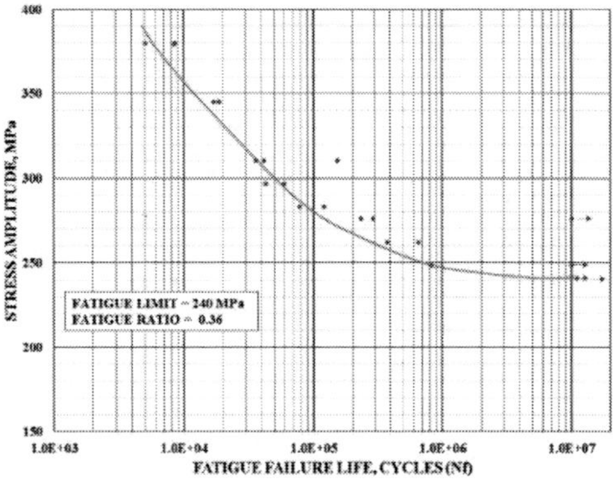

Fig 10. S-N curve for heterogeneous LWB (1.4/1.0 mm) of GA590R

3.3. Estimation of Fatigue Notch Factor for LWB

Ever since Wholer found almost 150 years ago that adding material to a railway axle might in fact make it weaker in fatigue, notch effects have been a thorny problem both for researchers and design engineers [14, 15]. The primary focus in notch fatigue research has been to estimate the local (notch root) stress (σ) and the local strain (ε) from nominal values of stress (S) and strain (e). It is particularly more challenging when the nominal stress exceeds the local cyclic yield strength of the material (the use of static yield strength is not appropriate for steels exhibiting cyclic softening as in the case of GA 590R). Over the years, several approximate analytical methods have been advanced, none too satisfactory [16, 17]. Several years ago, one of the present authors advanced the concept of estimating the local stresses from smooth specimen strain controlled fatigue data for estimating the fatigue strength reduction factor (K_f) for mechanical notches [18]. In this study, these concepts for estimating the local stresses are extended to both homogeneous and heterogeneous LWB combinations.

The basic premise of the methodology is that the fatigue life of a specimen with a laser weld joint (LWJ) under nominal stress control and that of a smooth specimen undergoing the simulated local strain reversals are the same. Then, the local cyclic stress for the smooth specimen giving the same life as that of a specimen with LWJ can be estimated from the strain controlled fatigue parameters of the base material. The equations used to estimate these are:

$$\sigma_n = \sigma_f^{'} (2N_f)^b$$

$$\sigma_n = k^{'}(\varepsilon_f^{'})^{n'} (2N_f)^{cn'}$$

The two equations are equivalent, the first equation uses the elastic strain – life parameters and the second makes use of the plastic strain –life and cyclic stress – strain parameters. Obviously, the errors in estimating these parameters from experimental data (see table 4 for R^2 values) will result in slightly different notch root stress values (σ_n). The ratio of the two stresses (σ_n/S) yields the fatigue notch factor, K_f.

In analyzing the data for LWJ under nominal stress control, the run-out data were excluded for this portion of the analysis. Regression analyses were carried out on finite life data (on transformed variables) to obtain Basquin type coefficients. Figure 11 represents such data for the heterogeneous LWJ condition. The calculated K_f values at several life levels are summarized in Table 6 for homogeneous and heterogeneous LWJ conditions using the elastic life parameters of the base steel to estimate the local stresses. The use of the plastic – life parameters for estimating the local stresses will give slightly different K_f numbers, but the general trend remains the same.

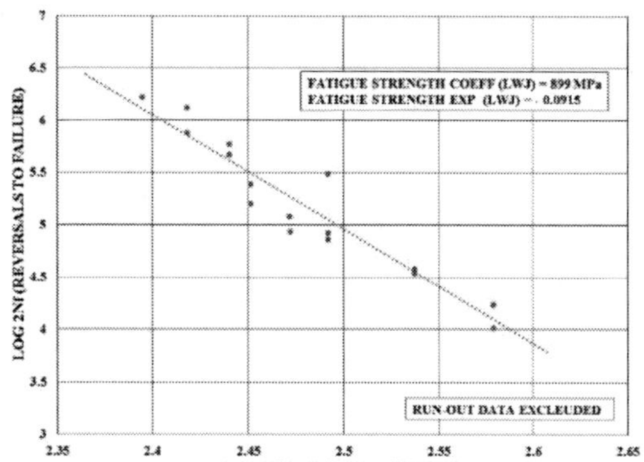

Fig.11. Linearized S-N plot for Heterogeneous LWJ of GA 590R

Table 6. Fatigue notch factor for laser welded GA 590R

	FATIGUE PARAMETERS		STRESS AMPLITUDE (MPa) FOR FATIGUE LIFE (REVERSALS TO FAILURE)				
STEEL CONDITION	$\sigma_f^{'}$	b	20,000	100,000	200,000	1,000,000	2,000,000
GA 590 MONOLITHIC	849	-0.0686	430	385	367	329	314
HOMOGENEOUS LWB	1058	-0.0922	425	366	343	296	278
HETEROGENEOUS LWB	899	-0.0915	363	314	294	254	238
FATIGUE NOTCH FACTOR							
K_f^{HOMO}			1.01	1.05	1.07	1.11	1.13
K_f^{HETERO}			1.18	1.23	1.25	1.30	1.32

Figure 12 summarizes the variation of K_f with fatigue life for both homogeneous and heterogeneous LWJ conditions. It is worth noting that K_f is not a constant, but rather that it increases with increasing fatigue life and tends to "saturate" beyond about 10^6 cycles to failure.

This trend is similar to that reported for several high strength sheet steels with geometric notches [18]; however, the variation of K_f as a function of life appears to be both material and notch geometry (K_t) dependent and as yet generalizations are difficult [13]; determination of K_t to represent the laser welding process is even more difficult.

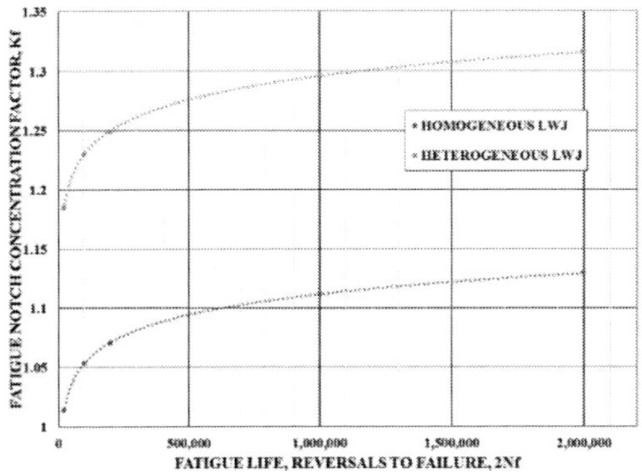

Fig.12. K_f vs fatigue life for LWB of GA 590R

While K_f depends on life, it is instructive to compare the values at one life for the two LWJ combinations. For example, at 10^5 cycles to failure (or 2x 10^5 reversals), K_f can be considered as 1 for the monolithic GA 590R (base steel) and it increases to 1.07 for the homogeneous (1.4/1.4 mm) and to 1.25 for the heterogeneous (1.4/1.0 mm) laser welded joint configurations. In other words, while LWJ by itself reduces the fatigue strength by 7%, when this technology is used to weld two gauge combinations (differential of 0.4 mm), the fatigue strength decreases by 25% at the same life level. In this investigation, the same laser welding process protocols were used for both homogeneous and heterogeneous welds, and further optimization of the process (especially for heterogeneous combination) may provide different results. In addition, the present data suggest that decreasing the gauge differential, (e.g.1.4/1.2 mm combination) may also decrease the K_f value for the joint, but the weight reduction potential will go down as well. Of course, another approach might be to look at different steel grades (higher strength or better ductility) to improve the base fatigue properties as well as "decrease" the LWJ notch sensitivity. Further research efforts are required to provide specific guidelines to these questions.

Working with a component supplier, strain controlled fatigue parameters and K_f values for the heterogeneous geometry were utilized in a preliminary durability analysis using the commercially available CAE codes. Reasonable agreements with component testing and CAE predictions were obtained. Thus, the LWB concept for automotive chassis applications and durability assessment protocols with LWJ appear to have potential.

4 Summary and Conclusions

The aim of this research work was to examine the potential of LWB concepts for chassis applications, where durability concerns have been a major hurdle. The experimental program consisted of laboratory fatigue testing of the base steel and the homogeneous and heterogeneous laser welded blank combinations of two different gauges and modeling the Laser Welded Joint (LWJ) as a "notch" to determine the fatigue notch factors. The salient findings include:

1) Yield and tensile strengths of both homogeneous (same gauge) and heterogeneous LWJs are comparable to that of the base metal GA 590R (for the LWB process used here). However, the measured TE and UE are lower in both LWB conditions compared to the base metal and this apparent decrease is attributed to the hard (martensitic) weld zone in the LWJ acting as artificial "grip" region inside the gauge section of the tensile coupons.

2) For GA 590R steel, cyclic loading significantly alters the stress – strain relationship from its virgin form. The steel exhibits a large cyclic softening, for example, the cyclic yield strength of 376 MPa is about 118 MPa lower than the monotonic yield strength. Thus, the use of the tensile stress – strain data under fatigue loading conditions may be too optimistic.

3) Axial, fully reversed Strain – Life and Stress – Life data for monolithic GA 590R (1.4 mm), and axial, fully reversed Stress – Life data for the homogeneous (1.4 /1.4 mm) and heterogeneous (1.4 /1.0 mm) combinations of LWJs were carried out in this investigation. The test program (R=-1) is unique for not introducing residual stresses during cyclic loading for characterizing the LWB concept for fatigue resistance.

4) Fatigue limit of GA 590R of 310 MPa decreases to 300 MPa for homogeneous and to 240 MPa for heterogeneous LWJ configurations.

5) Central essence of this program has been to conceptualize LWJ as a "notch" which allows the estimation of local stresses at LWJ from strain – life data of the base steel and subsequently relate them to the nominal stress – life data of LWB coupons. From these, fatigue notch factors (K_f) for both homogeneous and heterogeneous LWJs have been determined. Such estimation of K_f is unique for fatigue work with laser welding.

6) For both homogeneous and heterogeneous LWJ configurations, K_f increases with fatigue life, but tends to saturate at life levels greater than about one million cycles; thus, the use of a single constant in durability estimation regardless of life levels may lead to significant errors.

Acknowledgements

Authors acknowledge the support received from F.tech R&D North America, during this investigation. It is also a pleasure to acknowledge the diligent work of George Girman throughout the experimental program. Permission of ArcelorMittal is highly appreciated.

References

1. E. Bayraktar, D. Kaplan, and B. S. Yilbas, J. Mater Process Technol, **204**, 440 (2008).
2. D. Anand, D. L. Chen, S. D. Bhole, P. Andreychuk and G. Boudreau, Mater Sci and Engg, **A 420**, 199 (2006)
3. M. R. Mitchell, *Fatigue and Microstructure*, (ASM, Metals Park, Ohio 1979)
4. P. C. Wang and K. M. Ewing, Welding Journal **70, (No. 10)**, 43 (1991)
5. S. Daneshpour, M. Kocak, S. Riekehr, and C. Gerristen, Welding in the World, **53 (no. 9/10),** R221 (2009)
6. K. Asim, K. Sripichai and J. Pan, Int J Fatigue, **61,** 283 (2014)
7. D. Bhattacharya, N. Fonstein, O. Girina, I. Gupta, O. Yakubovsky, *45th MWSP Conf Proceedings* (ISS, Warrendale, Pa, 2003)
8. A. R. Mardera, Prog Mater Sci. **45,** 191 (2000)
9. M. Ono, A. Yoshitake and M. Ohmura: NLL Tech Review, **86,** 13 (July 2002)
10. H. Gong, S. Wang, P. Knysh and Y. Korkolis, Mater and Design; **90**, 1115 (2016)
11. D. Dong, Y. Liu, Y. Yang, J. Li, M. Ma, and T. Jian, Mater Sci and Engg. **A 594,** 17 (2014)
12. S. P. Bhat, *HSLA Steels Technology & Applications* (ASM, Metals park, OH, 1983)
13. B. M. Wundt: *Effect of Notches on Low-cycle Fatigue: A Literature Survey* (ASTM, Philadelphia, PA, 1972)
14. N. E. Frost, K. J. Marsh and L. P. Pook, *Metal Fatigue* (Clarendon Press, 1974)
15. H. O. Fuchs and R. I. Stephens, *Metal Fatigue in Engineering* (John Wiley & Sons, 1980)
16. T. H. Topper, R. M. Wetzel, and Jo Dean Morrow J Materials, **4**, 200 (1969)
17. K. Molski and G. Glinka, Mater Sci Engg, **50,** 93 (1981)
18. S. P. Bhat, *High Strength Steel for Automotive Use* (SAE, Warrendale, PA 1983)

Changing in Fatigue Life of 300 M Bainitic Steel After Laser Carburizing and Plasma Nitriding

Antonio J. Abdalla[1,2,*], *Douglas* Santos[1], *Getúlio* Vasconcelos[2], *Vladimir H.* Baggio-Scheid[2], and *Deivid F.* Silva[1]

[1]*Faculdade de Engenharia de Guaratinguetá - FEG/UNESP, Guaratinguetá SP, Brazil*
[2]*Instituto de Estudos Avançados - IEAv, Div. Fotônica, São José dos Campos SP, Brazil*

Abstract. In this work 300M steel samples is used. This high-strength steel is used in aeronautic and aerospace industry and other structural applications. Initially the 300 M steel sample was submitted to a heat treatment to obtain a bainític structure. It was heated at 850 °C for 30 minutes and after that, cooled at 300 °C for 60 minutes. Afterwards two types of surface treatments have been employed: (a) using low-power laser CO_2 (125 W) for introducing carbon into the surface and (b) plasma nitriding at a temperature of 500° C for 3 hours. After surface treatment, the metallographic preparation was carried out and the observations with optical and electronic microscopy have been made. The analysis of the coating showed an increase in the hardness of layer formed on the surface, mainly, among the nitriding layers. The mechanical properties were analyzed using tensile and fatigue tests. The results showed that the mechanical properties in tensile tests were strongly affected by the bainitic microstructure. The steel that received the nitriding surface by plasma treatment showed better fatigue behavior. The results are very promising because the layer formed on steel surface, in addition to improving the fatigue life, still improves protection against corrosion and wear.

1 Introduction

The 300 M is an important steel grade developed by 4340 steel to improve the tensile strength and toughness of the steel. This proposal was reached with the increased of percentage of silicon and introduction of vanadium[1]. This steel is used in aircraft landing gear, rocket motor envelopes, junctions of some parts of satellite launch vehicles and other structural purposes in the automotive and equipment industries [2].

Among the possible structures of 300 M steel, the formation of bainite is one of the most indicated ways to improve the toughness of steel. In this work, a heat treatment was carried out at 300 °C leading to formation of a bainitic structure. This microstructure improved the yield and tensile strength [3,4].

The good properties of the bainitic structure result from the fine or ultra fine grains of the bainitic ferrite with fine uniformly distributed secondary phase structures. This fact has led to a growing industrial demand for the use of bainitic steels in recent decades [5,6].

When working with steels it is always important to prevent corrosion and wear. In this way, surface thermochemical treatments contribute to the formation of a hard layer and a corrosion resistant sample. In this research two types of treatments were used: plasma nitriding and laser carburizing. These processes can still improve fatigue life because it hinders the crack initiation and delays its propagation, due to residual compression stresses introduced into the surface [7-10].

In the last two decades there has been a lot of progress in the technology used in surface treatment processes. Among these developments are the plasma nitriding method in vacuum oven and ion implantation. Another important process is associated to the use of laser, such as laser carburizing [11-13].

In this work, the changes in the mechanical properties of tensile, fatigue and hardness of a 300 M steel were evaluated. These properties were evaluated before and after the steel passed through heat treatment to form a bainitic structure and after receiving the surface treatments of plasma nitriding and laser carburizing.

2 Test methods

The 300 M steel for test is taken from the shapes with 3mm; the chemical composition is shown inTab.1.

Table 1. Chemical composition of 300 M steel in wt%

Elements	C	Si	Mn	P	S	Ni
% Weight	0,39	1,83	0,82	0,008	0,006	1,68
Elements	**Cr**	**Mo**	**Cu**	**Al**	**Ti**	**V**
% Weight	0,77	0,39	0,14	0,036	0,0037	0,08

* Corresponding author: abdalla@gmail.com.br

2.1 Specimens and heat treatment

Tensile specimens were made according to ASTM E8M and fatigue specimens in according to ASTM E466. These were initially subjected to the heat treatment to obtain the bainitic structure by an isothermal treatment. This was done from the austenitization of the specimens at 850 °C for 30 minutes and isothermal cooling in a salt bath at 300 °C for 60 minutes.

The specimens of the 300M steel were divided into four sets: (a) annealed steel - A, (b) steel with bainitic structure - B, (c) bainitic structure with plasma nitriding surface treatment - BN and (d) bainitic structure with surface treatment of laser carburing - BC.

The samples in each condition were metallographically prepared and etched with Nital 2% to characterization by optical or scanning electron microscopy.

2.2 Surface treatments

2.2.1 Laser Carburizing

After the isothermal treatment the samples were ground and sanded to 1000 mesh, washed and dried. The coating with carbon black was applied by spraying and irradiated via low power CO_2 laser. A carbon black solution used was prepared with 10 g of carbon black and 0.1 g of carboxymethylcellulose (CMC) in 100 ml of ethanol.

The laser used was a CO_2 laser with wavelength of 10.6 μm, output power equal to 125W, diameter of the laser beam 300μm (brand: Synrad and model: J48-5W), and nitrogen flow to act as a protector in the irradiated layer, preventing oxidation [14-15].

The application of the CO_2 laser has as parameters: resolution 500 dpi (dots per inch), 125W power and velocity of 600 mm/s. The image of the steel after coating with carbon black and application of CO_2 laser is showing in Fig.1. This treatment was applied in the useful area of the specimens for fatigue tests.

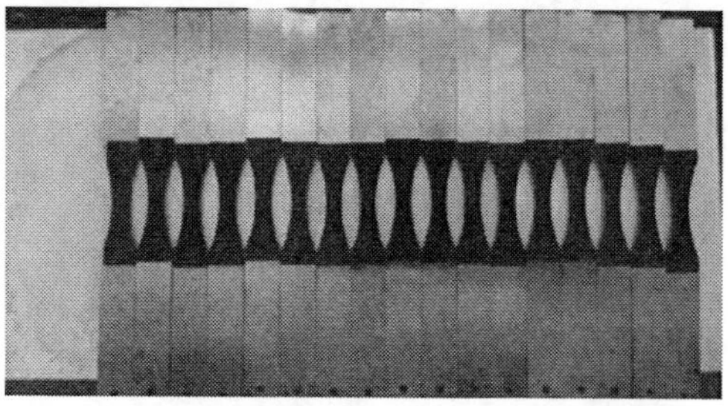

Fig. 1. Fatigue specimens with laser carburizing (black).

2.2.2 Plasma nitriding

The plasma nitriding treatment took place at a temperature of 500 ° C for 3 hours. The reactor employed for sample used a dc- pulsed source of approximately 370 V. The samples were treated in a gas environment containing 75% N_2 and 25% H_2, the pressure was kept constant at 3 mbar.

2.3 X-ray diffraction

X-ray diffraction analysis was performed on the surfaces of the samples studied. These analyzes were performed on an equipment using copper tube with graphite monochromator. A voltage of 40kV and a current of 30mA were used. The measurements were made by sweeping the samples from $2\theta = 20°$ to 95°, with angular pitch of 0.05° and counting time of 1s per step.

2.4 Mechanical Properties

For tensile and fatigue tests an MTS 810.23M machine was used, with a load cell with a capacity of 250 kN. Three specimens per condition were used in the tensile tests. The tensile tests have been performed at a speed of 1mm/min. For the fatigue tests, the load ratio was 0.1 and the frequency of 25 Hz was used.

Vickers hardness measurements were performed by microindentation with the objective of obtaining the microhardness profile of the samples from the surface towards the substrate, the load used was 50 gf for a time of 10 s.

3 Results and Discussion

The micrographs of steel 300M without the application of heat treatment (Condition A) are shown in Fig.2. The microstructure shows the typical formation of perlite with ferrite (gray) and cementite in lamellar form (white).

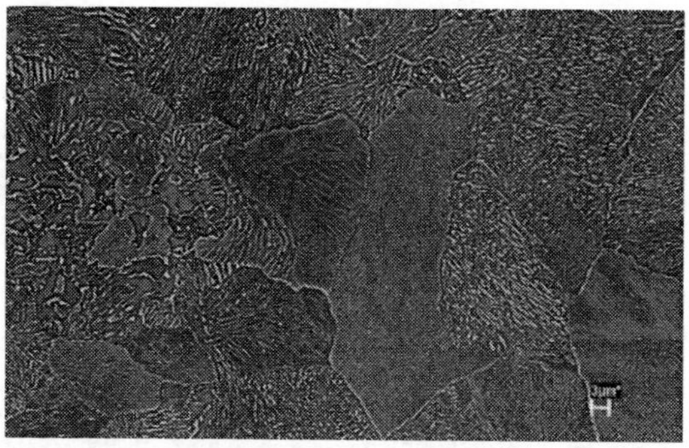

Fig. 2. SEM show 300M steel without heat treatment (annealed steel). Etching: nital 2%. Condition A.

The micrographs of steel after the heat treatment to form a bainitic structure (Condition B) are shown in Fig.3. Besides the bainitic structure, small portions of other phases such as martensite (M) or ferrite (F) are present.

The applied heat treatment produced the formation of a predominant bainitic structure, leading to an increase in the ultimate strength (σ_u) and yield strength (σ_y) and a reduction in ductility as measured by elongation.

https://doi.org/10.1051/matecconf/201816521002

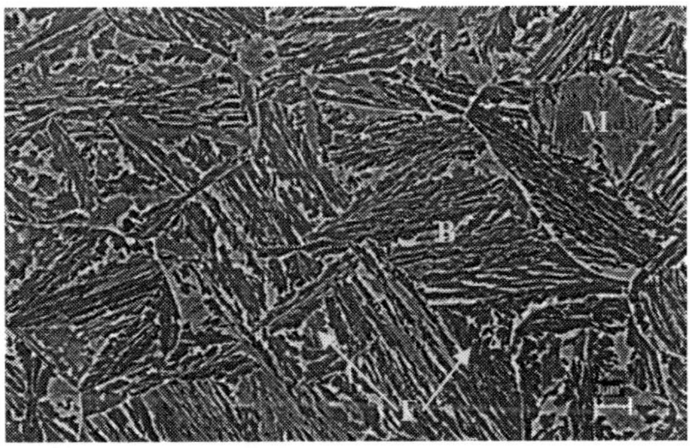

Fig. 3. SEM - shows 300M steel with heat treatment (bainitic structure). Etching: nital 2%. Condition B. B=bainite; M=martensite, F= Ferrite.

In table 2 the results of the tensile tests for both conditions: bainitic (B) – formed by isothermal treatment and annealed condition (A) – as received material are shown.

Table 2. Mechanical properties of 300M steel.

Heat Treatment	Structure	σ_y (MPa)	σ_u (MPa)	Elongation (%)
Annealed Cond. A	Perlite	951±12	1496±25	16,2±1,8
Isothermal Cond. B	bainite	1067±11	1615±25	10,0±0,9

After de laser carburizing a thin white layer about 20 μm thick was formed. Just below it another layer was formed where heating occurred due to the application of the laser (heat affected zone - HAZ) of about 60 μm. In Fig.4 a white layer with a high concentration of carbon and iron on the surface is shown. Below is the heat affected zone (HAZ), on which the occurrence of localized partial quenching is observed. Below these regions lies the base metal composed mainly of bainite.

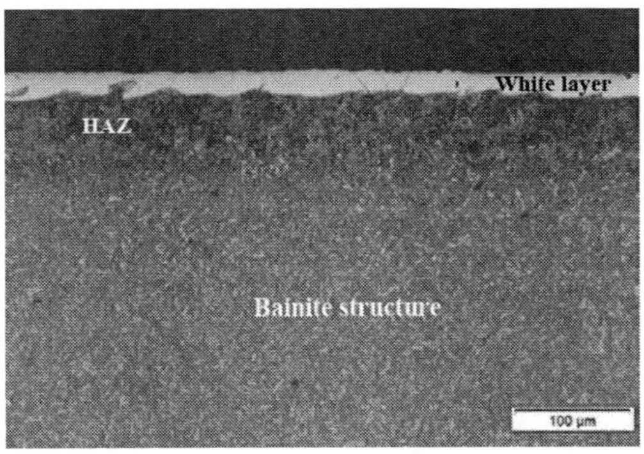

Fig. 4 - Optical Microscopy: Typical zones formed in surface after laser treatment.

It was observed that the hardness of the white layer increased to about 596 ± 21 HV and the HAZ remained around 464 ± 39 HV, similar to that found in the bainitic structure 476 ± 28 HV.

In Fig.5, a micrograph of 300 M steel with bainitic structure that received surface treatment of plasma nitriding is shown. The formation of a thin layer of nitrides on the steel surface, about 3 mm thick, can be noticed. This layer has a high hardness, with 815 ± 25 HV. This increase in hardness occurred due to the formation of hard nitrides of two types, $Fe_{2\text{-}3}N$ and Fe_4N.

The hardness measurements found for the various regions of the steel subjected to surface treatments can be seen in Table 2. Below the white layer, the diffusion of the nitrogen atoms occurs, narrow visible in the image, not withstanding detected by the increase of hardness. This diffuse bed of nitrogen reaches about 80 μm. In this region a gradient decreases in hardness from the nitride layer (white) until the substrate hardness is found at about 80 μm from the surface. In this region the average hardness value was 728 ± 74 HV.

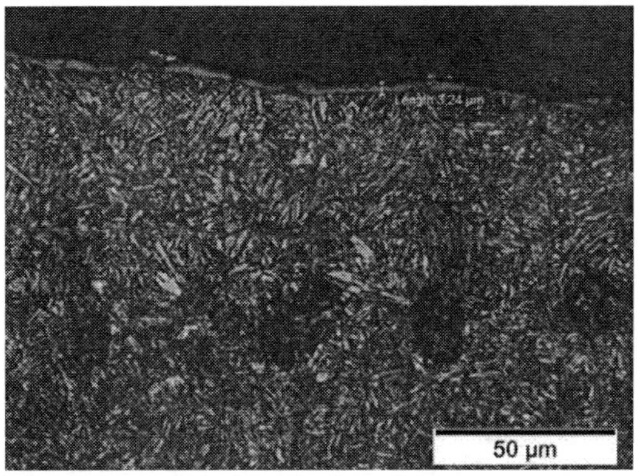

Fig. 5 - Optical Microscopy: Typical zones formed on surface after plasma nitriding treatment.

In Table 2, it is shown that the base material had little influence on hardness due to surface treatment, the significant change occurred only due to the heat treatment for the formation of the bainite structure. The major change in surface hardness occurred due to the treatment with plasma nitriding.

Table 2. Hardness values after surface treatments.

	With surface treatment		Without surface treatment	
	Laser Carburizing (HV)	Plasma nitriding (HV)	As received (HV)	Bainitic Structure (HV)
White layer	596±21	815 ±25	287±9	473 ±12
HAZ	464±39	728±74		
Base material	476±28	489 ±18		

In Fig.6 the X-ray diffraction for the three conditions studied in this work is shown: 300 M steel with bainitic structure without surface treatment and steel with surface treatment with plasma nitriding and laser carburizing.

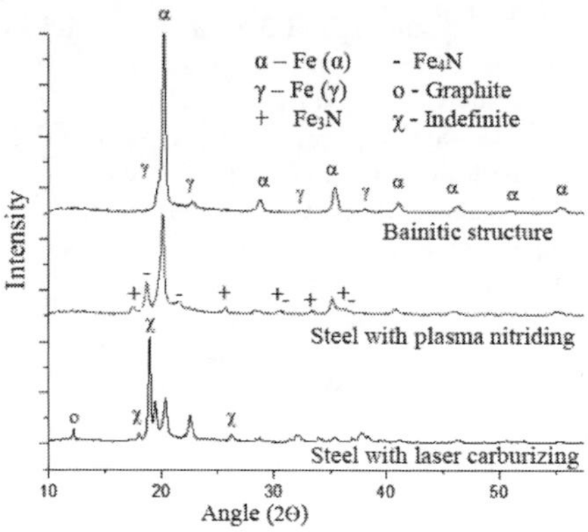

Fig.6 - X-ray diffraction on the surface of the steel for the conditions: bainitic structure, plasma nitriding and laser carburizing.

The mechanical properties observed in tensile are shown in table 3. After the heat treatment in the material as received, the microstructure was transformed from ferritic-pearlite to bainitic, increasing the value of the yield and ultimate strength. Surface treatments did not produce significant changes on the mechanical properties of the steel, however, a reduction in the ductility due to the treatment with plasma nitriding was observed.

Table 3 - Mechanical Properties

Heat treatment	Yield Strength (MPa)	Ultimate Strength (MPa)	Elongation (%)
As received	843 ±15	973 ±0,2	16 ±1,8
Bainitic	913 ±13	1364 ±16	13 ±0,6
Laser carburizing	907 ±23	1353 ±42	12 ±0,9
Plasma nitriding	951 ±23	1356 ±31	09 ±0,3

In Fig. 7, the three S-N fatigue curves for the studied steels are shown: steel with bainitic structure, with laser carburizing and plasma nitriding treatment. It is noticed that there is a fatigue life increase for the steel that was nitrided and a reduction in the fatigue performance for the carburizing steel. One of the factors that influenced the reduction of the life of the carbonated steel was the increase on the surface roughness. In the bainitic and carburizing conditions the roughness values were similar: Ra (0.16 ± 0.04) and Rz (1.09 ± 0.28). For the carburizing condition, due to laser irradiation, these values increased to Ra (0.36 ± 0.06) and Rz (3.41 ± 0.69). The surface of the carburizing steel was also more irregular, with the formation of several phases due to heating and a region of low surface hardness due to the formation of graphite. On the other hand, on the surface of the plasma nitrided steel there was a hard layer and a deeper atomic diffusion zone that contributed to delay the start of fatigue cracking. The tests were interrupted by reaching the mark of 1,000,000 cycles (run out).

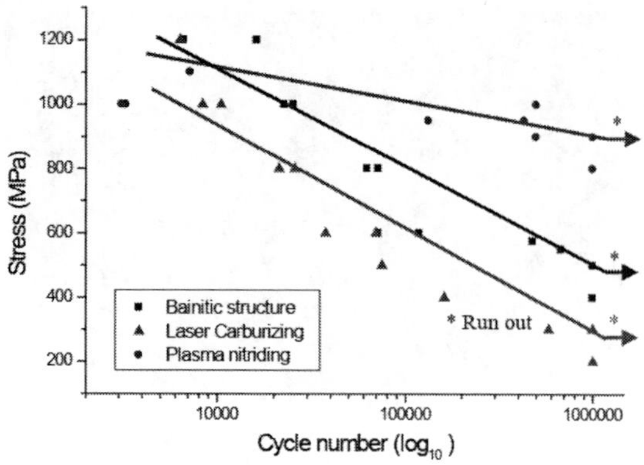

Fig.7 - Fatigue curves (S-N) for the conditions: 300 M Steel with bainitic structure and steel with surface treatment of plasma nitriding and laser carburizing.

4 Conclusions

The plasma nitriding of the 300 M steel with bainitic structure provided the formation of a homogeneous layer with high hardness, contributing to improve fatigue performance for a high number of cycles.

The results of the tensile tests were similar for all the studied conditions; there was no significant change in the yield or ultimate strength. There was a reduction in ductility as measured by elongation to the nitrided plasma condition.

The fatigue performance for the steel subjected to the laser carburizing treatment was inferior to the steel with bainitic structure. The hardness of the carburizing region was similar to that observed in the bainitic structure and did not contribute to delay the appearance of surface microcracks. There was also an increase in surface roughness, a factor that contributed to worsen performance in fatigue.

References

[1] Santos, D.; Vasconcelos, G.; Abdalla, A. J.; Lima, M. S.F.; Souza Neto, F.; *Surface Characterization in a 300 M Bainitic Steel Laser Carburizing*. ScienceDirect.*Proc. Engin*. **114** [2015] 322–329.

[2] Cardoso, A.S.M.; Abdalla, A.J.; Baptista, A.C.R.; Lima, M.S.F. *Comparison of High Cycle Fatigue in 4340 and 300M Steels Welded with Fiber Laser*. Advanced Mat. Research Vols. **891-892** [2014] pp 1507-1512.

[3] Yoozbashi, M.N.;Yazdani, S.; Wang, T.S. *Design of a new nanostructured, high-Si bainitic steel with lower cost production*. Mat. and Design, v.**32** [2011] p.3248-3253.

[4] Caballero, A.; Vuorinen, E.; Sundin, K.G.; Prakash, B.; Sourmail, T.; Smanio, V.; Caballero, F.G.; Garcia-Mateo, C.; Elvira, R. *Wear of nanostructured carbide-free bainitic steels under dry rolling–sliding conditions*. Wear, v.**298–299** [2013] p.42–47.

[5] Yoozbashi, M. N.; Yazdani, S.; Wang, T. S. *Design of a new nanostructured, high-Si bainitic steel with*

lower cost production. Materials and Design, Tabriz, v.**32** [2011] p.3248-3253.

[6] Zhu, K.; Oberbilling, C.; Musik, C.; Loison, D.; Iung, T. *Effect of B and B + Nb on the bainitic transformation in low carbon steels*. Materials Science and Engineering A, v.**528** [2011] p.4222-4231.

[7] Sirin, S.Y., Sirin, K., Kaluc, E., Materials Characterization **59** [2008] p. 351-358.

[8] Lima, M.S.F. *Lasers in Material Processing*. In: Lackner, *Lasers in Chemistry: Influencing Matter* [2008] p.1195-1209.

[9] Scheid V. H. B., Abdalla A. J., Vasconcelo G., *Effect of heating post-treatment on nitrided stainless steel.* Surface and Coatings Technology , v.**201** [2006] p. 4058-4061.

[10] Reis,A.G.; Reis, D.A.P.; Abdalla, A.J.; Couto, A.A.; Otubo, J. *Short-term Creep Properties and Fracture Surface of 18 Ni (300) Maraging Steel Plasma Nitrided.* Materials Research, v.**01** [2017] p.01-08.

[11] Ueda M.; Silva M.M.; Lepienski C.M.; Soares Jr. P. C.; Gonçalves J.A.N.; Reuther H. *High temperature plasma immersion ion implantation of Ti6Al4V*, Surface and Coatings Technology, v. **201** [2007] p. 4953-4956.

[12] Ueda, M.; Silva, A.R.; Pillaca, E.J.D.M.; Mariano, S.F.M.; Rossi, J.O.; Oliveira, R.M.; Pichon, L.; Reuther, H. *New possibilities of plasma immersion ion implantation (PIII) and deposition (PIII&D) in industrial components using metal tube fixtures*. Surface & Coatings Technology, v. **312** [2017] p. 37-46.

[13] Cunha, M.C.; Lima, M.S.F. *The influence of laser surface treatment on the fatigue crack growth of AA 2024-T3 aluminum alloy alclad sheet.* Surface & Coatings Technology, v. **329** [2017] p. 244-249.

[14] Chagas, D. C.; Dias. A.N.; Antunes, E.F.; G. Vasconcelos. *Aplicação de lubrificantes sólidos (negro de fumo) e tratamento térmico superficial de têmpera em matrizes de aço AISI 4340 via laser* CO_2, IV SICI -IEAv, vol.**1** [2010] S.J.Campos, São Paulo, Brasil.

[15] Ganev, R.A. "*Low-power laser hardening of steels*", Journal of Mat.Proc.Tech., v.**121** [2002], p. 414-419.

MATEC Web of Conferences **165**, 21003 (2018) https://doi.org/10.1051/matecconf/201816521003
FATIGUE 2018

Notch effect of welded joint

Vladimír Chmelko[1*], *Matúš* Margetin[1], and *Michal* Harakal'[1]

[1]Faculty of Mechanical Engineering, Department of Applied Mechanics and Mechatronics, Námestie slobody 17, 812 31 Bratislava, Slovakia

Abstract. In general, the welded joints are multifactorial notches. This study is focused to separate the geometrical influence (when the weld can be seen on the surface) and technological factors (internal defects of the weld, structural changes of the material and residual stress) on its fatigue life. Proposed methodology of this separation comes out from the cyclic test of three specimens – group with existing weld, group without weld joint (base material) and another group where was removed outer geometry of the weld.

1 Page layout

The weld joint has often the same strength as the strength of the base material. This statement does not apply at all to fatigue strength (Fig. 1). Euro code 3 „Design of steel structures“ states allowable stress amplitudes for different welding joints [1]. These values are significantly lower than the fatigue strength values at the harmonic loading of low-carbon structural steels. The reason is in the very essence of a welding joint - a joint that changes the structure of the material in the locality of the joint and its geometry. Notch effects of welds are a very wide problem [2-7]. In more detail, we will analyse the problem of two components of the notch effect of the welding joint.

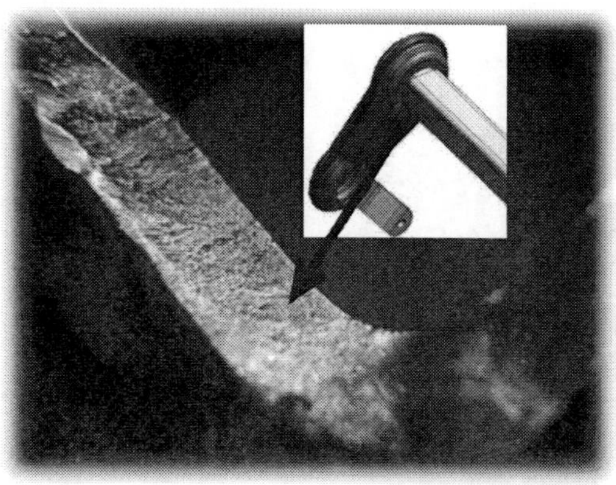

Fig. 1. Fatigue fracture caused by welding of the wheel arm with axle at high stress levels

From the fatigue strength assessment point, we can consider the weld joint as a stress and deformation concentrator, which contains two fatigue strength reducing sources and the lifetime of the structure [2]:

- As technological notch, we describe a subset of subsurface discontinuities (cracks, interstices, porosities, bubbles) and non-homogeneities of material like (structural changes, residual stresses ...) that negatively affect the fatigue behaviour of the weld joint – Fig. 2;
- As geometrical notch, we mean the sum of all surface geometrical irregularities (welding beads, surface welding errors) that create stress concentration and deformation on the free surface of the component, thereby reducing its fatigue strength.

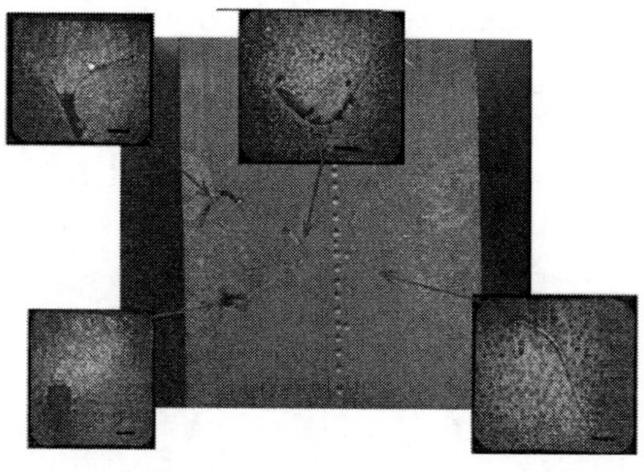

Fig. 2. Set of errors in the welding microstructure: top left - recast error characterized by imperfect joining of two layers of weld bead; top right - imperfectly welded root of the weld (this error occurred due to too large a drop of weld metal at the beginning of the welding); bottom left - bubble in welding metal; Bottom right - inclusion between the weld metal and the base material

2 The notch coefficient of the weld joint

* Corresponding author: vladimir.chmelko@stuba.sk

https://doi.org/10.1051/matecconf/201816521003

The total notch coefficient of the weld joint can be defined by comparing the fatigue curves at the Fig. 3.-5. in the following shape

$$\beta_{\sigma(2Nf)} = \frac{\sigma^{H}_{a(2Nf)}}{\sigma^{Zv}_{a(2Nf)}} , \qquad (1)$$

where $\sigma_{a(2Nf)}$ is a stress amplitude corresponding to chosen number of cycle to the fracture of the base material specimen,

$\sigma^{W}_{a(2Nf)}$ is a stress amplitude corresponding to a certain number of cycles to the fracture of specimen with the weld joint.

For three different welds, cyclic test of the samples were performed in two separated groups (Fig. 3) :

- specimens with the weld joint in the middle
- specimens of base material

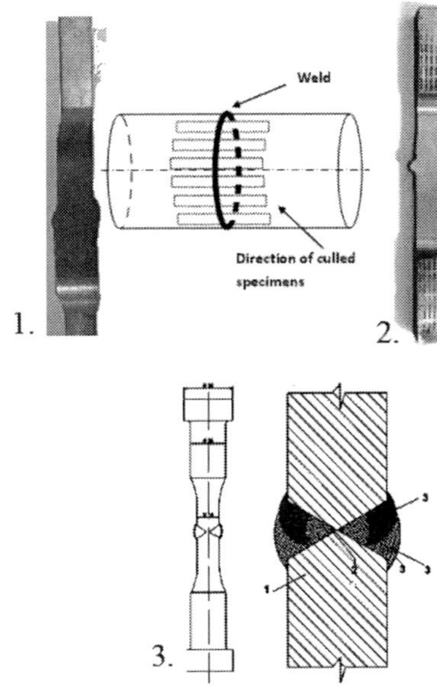

Fig. 3 The expertise of three type of welds

The first set was the specimens from end V-weld culled directly from courtyard of pipe DN700 of the transit gas pipeline. The material represented the rolled pipeline steel the Mannesmann steel, equivalent with the steel St52. By gradual milling of the side faces, the test specimens were narrowed in the working part, with the original surface of the pipe retained on the front surfaces, with the welding on both sides. The obtained Basquin dependence for specimens with the weld joint and for base material of the pipeline are at the Fig.4.

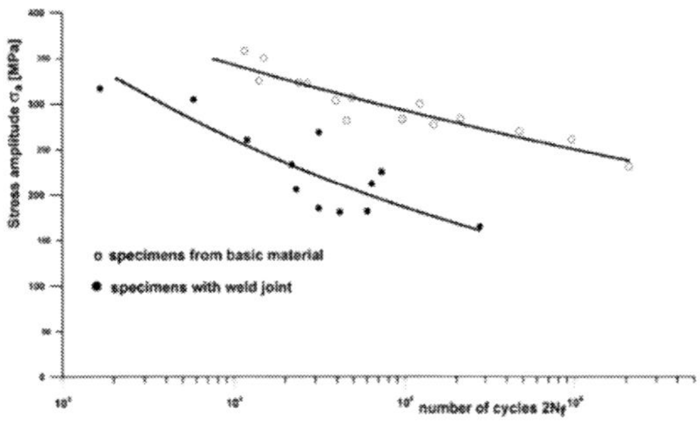

Fig. 4. Basquin dependency σ_a = f(2Nf) [8] for smooth specimens of pipeline material (St 52) and specimens of pipeline with weld joint

From the regression lines in Fig. 4, was calculated the notch coefficient defined by Eq. 1 as the number of cycles to the fracture in the shape

$$\beta^{1}_{\sigma} = 0.64219(2N_f)^{0.077767} \qquad (2)$$

From the graphical process at Fig. 5, it can be seen that the values of the notch effect in the region of 10^4 cycles are about $\beta \approx 1{,}3$ and in the regions over the 10^6 cycles reach and then gradually exceed the value $\beta \approx 2$.

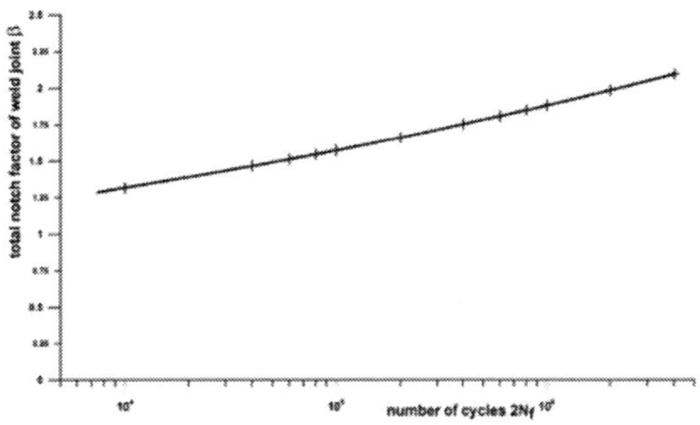

Fig. 5 Dependency of the notched effect of the pipe welding joint (St52) on the number of cycles to the fracture

The second set was a rounded V-weld made as the front weld of two cylindrical parts. The Basquin curves of the sample material and the samples with the weld in the middle of the working part are at the Fig.6.

Weld joint was made in two steps: first, the welding root was formed in its axis in a special preparation ensuring the consistency of the two cylindrical parts of the sample; in the second step, during continual rotation was made up the weld up to the outer cylindrical parts of the specimen, while the specimen was still in a special preparation maintaining the co-existence of the resulting sample with the weld.

 https://doi.org/10.1051/matecconf/201816521003

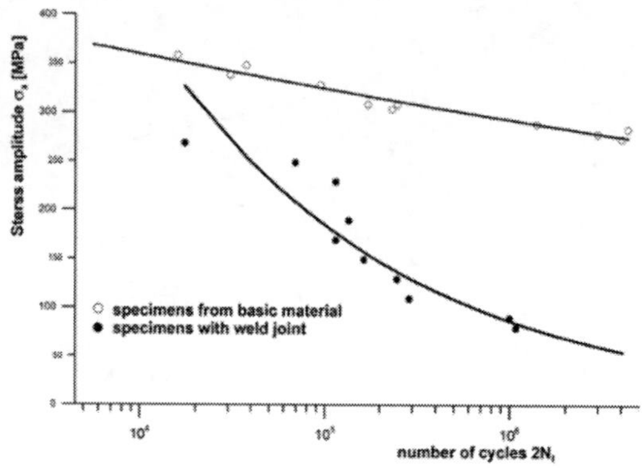

Fig. 6 The Basquin dependency $\sigma_a = f(2N_f)$ for smooth specimens of steel and specimens with weld joint

From the regression lines in Fig. 6, was calculated the notch coefficient defined by Eq. 1, as the number of cycles to the fracture in the following shape

$$\beta_\sigma^2 = 0.06768(2N_f)^{0.283} \qquad (3)$$

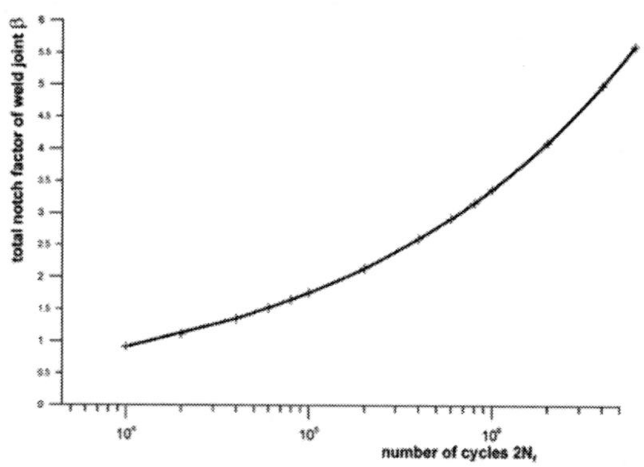

Fig.7 The dependency of notch effect of the V-weld weld to the number of cycles to fracture

From its graphical diagram Fig.7, it can be seen that the notch effect in region of 10^4 cycles are about $\beta \approx 1$ (even smaller) and in regions over the 10^6 cycles reach and then gradually exceed the value $\beta \approx 4$.

The third experimental set consisted of samples with a frontal V-weld, as well as Group 1. They were selected from the weld of two short sections of the pipeline DN500 in the longitudinal direction. The material represented the rolled pipeline steel that is equivalent with the steel S355J2. By gradual milling of the side faces, the test specimens were narrowed in the working part, with the original surface of the pipe retained on the front surfaces, with the welding on both sides. The obtained Basquin dependence for specimens with the weld joint and for sample of the base pipeline material are at the Fig.8.

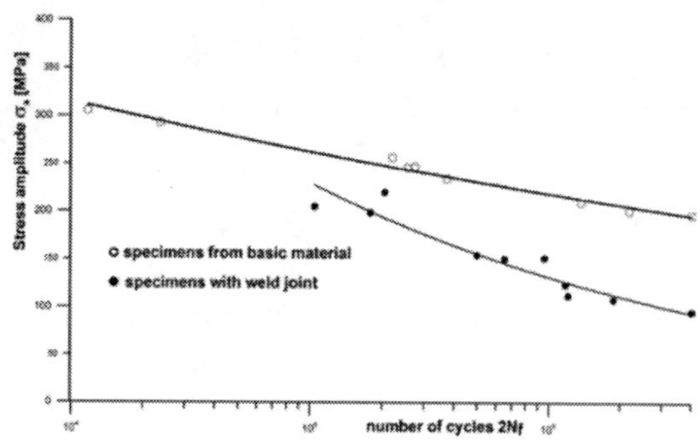

Fig. 8 The Basquin dependency $\sigma_a = f(2N_f)$ for smooth specimens of steel with the weld join

From the regression line in Fig. 8, was calculated notch coefficient defined by Eq. 1 as the number of cycles to the fracture in the following form

$$\beta_\sigma^3 = 0.17841(2N_f)^{0.16137} \qquad (4)$$

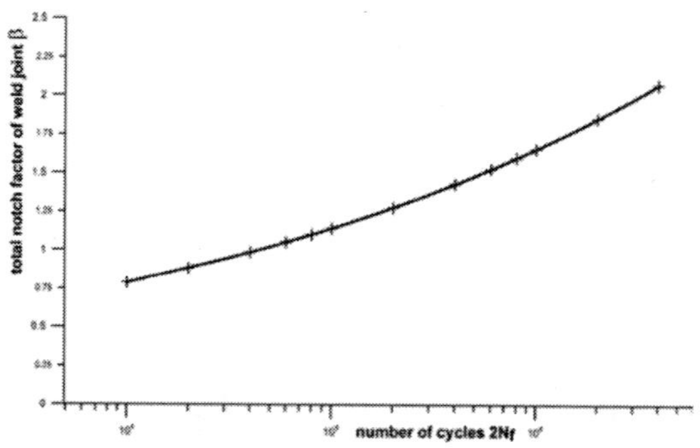

Fig.9 Dependency of notch effect of the V-weld weld joint to the number of cycles to the fracture

From its graphical process in Figure 9, it can be seen that the notch effect values in the region of 10^4 cycles are about $\beta \approx 1$ (even smaller) and in regions over the 10^6 cycles reach and then gradually exceed the value $\beta \approx 2$.

Comparison of the notch coefficients obtained by the above described processes for three different types of welds, depending on the number of cycles to the fracture, is shown in Figure 10.

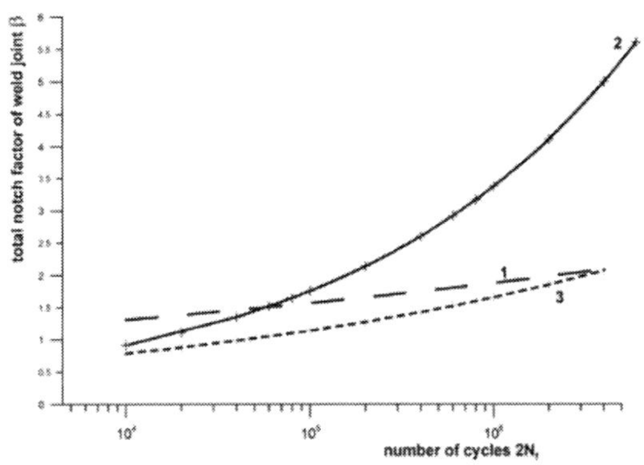

Fig. 10 Comparison of notch coefficients for three experimental sets: 1 - end V-weld culled directly from transit gas pipeline DN700; 2 - rounded V-weld; 3 - end V-weld culled from pipeline DN500 made special for this test

3 Methodology of separation of the technological and geometric part of the welding joint

The total notch coefficient of the weld joint can be expressed as the sum of two of its parts, where each of these parts has different physical cause, in the form

$$\beta_{\sigma} = \beta_{\sigma,T} + \beta_{\sigma,G} \quad , \qquad (5)$$

where β_{σ} – total weld notch coefficient

$\beta_{\sigma,T}$ – technological weld notch coefficient

$\beta_{\sigma,G}$ – geometrical weld notch coefficient

The technological part of the notch is mainly caused by internal defects and the structure of the heat-affected zone as a result of the welding joint manufacturing technology. The geometric part is caused by the classical concentration of the force flow at the point of the weld metal elevation relative to the base material.

The technological part of the notch effect can be influenced by the overall choice and parameters of the technological process. After its application on the welding process, the technological part of the friction effect on the fatigue life is already given. The geometrical part of the notch can be affected even after the welding joint has been created by additional surface treatment. Therefore, it may be useful in practice to know its size.

From the point of view of the methodology, it is necessary to create three groups of samples (fig. 11):
- specimens of base material (1)
- specimens with weld joint (2)
- specimens with removed geometry of weld joint (3)

Fig. 11 The groups of specimens for separation of geometric and technological part of the notch.

The cyclic tests of such prepared specimen groups allow quantifying the individual notch effects. Fatigue curves of life in a mode with controlled force in the form of Basquin dependence $\sigma_a = f(2Nf)$ are displayed in Fig. 12 for the type of weld 2 (fig 3). Expect the longest fatigue life for the selected amplitude stress level σ_a achieves smooth samples of the base material, the shortest life of the sample with the full notch effect of the weld. However, the effect of removing the geometric welding notch (dashed line) is positive - it is positive in regions of the high life cycles - above the intersection P (in this case approximately 1.6 x 10^5 cycles) or so called high-cycle fatigue. On the other hand, in the short life span, the effect of the removal of welding geometry is even negative and leads to shorter lifetimes than to untreated weld joints.

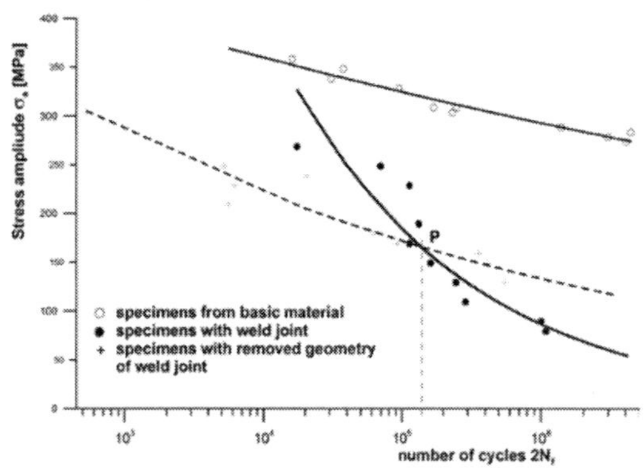

Fig. 12 Basquin curves σ_a = $f(2Nf)$ for welded specimens, specimens without weld and specimens with removed geometry of the weld

As shown in [9], it is possible to quantify and display the individual components of the overall notch effect of the weld joint, as shown in Fig. 13.

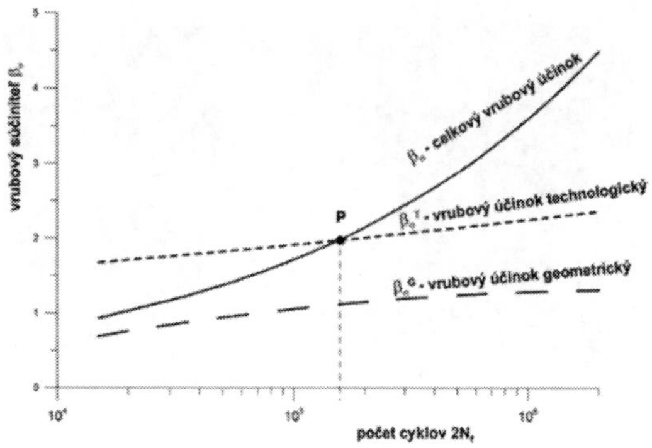

Fig. 13 Separation of individual notch effects of the weld joint 2 displayed in Fig.3

4 Evaluation of results

The study is devoted to three types of welding joints (Fig 3). For each weld joint, were obtained the Basquin dependency by experiment and compared with the same dependence obtained on the specimens of the correspondent base materials. In all three cases, the shortening of the fatigue life due to the welding was demonstrated only at a certain value of the amplitude load, not in its entire range. Likewise, the notched coefficients, expressed as a function of the number of cycles to the fracture, achieved more significant values, up to the number of orders 10^3 i.e. 10^4 cycles to the fracture. The notch effects of weld joints 1 and 3 exceeded the value 2 up to the regions of long fatigue lifetimes of the order 10^6 cycles.

The presented methodology of cyclic test of three groups of specimens allowed the separation of the overall notch effect on its two main components - technological and geometric. Results were obtained for weld type 2 (fig.3).

The notch coefficient including the welding process factors is virtually constant throughout the lifetime ($\beta_{\sigma T} \cong 2$). The notch coefficient including the influence of the welding geometry increases the overall notched impact effect of the weld only in the area of life above point P (so-called high-cycle fatigue). Under P, that means. in the short lifetime, its values are less than 1, that means reduces the overall notch effect of the weld. This surprising phenomenon may be due to the presence of local plastic deformations at the root of the welding geometry (high stress amplitudes) that lead to a decrease in the stress concentrator and the formation of pressure residual stresses in the welding heel. These factors positively influence the fatigue life and thus counteract the effect of technological factors [10].

This work was supported by the Research & Development Operational Programme funded by the ERDF ITMS: 26240220084 Science city Bratislava.

References

1. Eurocode 3. *Design of Steel Structures*. Part 1–9: Fatigue. UNI ENV 1993-1-9. Brussels, CEN 2015
2. Radaj, D., Sonsino, C. M., Fricke, W.: *Fatigue Assessment of Welded Joints by Local Approaches*. 2nd edt. , Woodhead Publishing Ltd. (2006)
3. Haagensen, P. J.: Fatigue Design ofWelded Structures – Norsok and Eurocode 3. Trondheim, NTNU 2011.
4 Hobbacher, A. (Ed.).: Recommendations for Fatigue Design of Welded Joints and Components. Document IIW-1823-07, ex XIII-2151r4-07/XV-1254r4-07. Paris, International Institute of Welding 2008.
5. Hobbacher, A. F.: Int. J. Fatigue, 31, 2009, p. 50.
6.Margetin, M.: Contribution to Fatigue Life Time Evaluation of Welded Joints. [Ph.D. thesis]. Bratislava, Slovak University of Technology in Bratislava 2012
7. M. Kepka, M. Kepka, Jr., Calculations of fatigue life of a welded joint in the construction of the trolleybus rear axle. Proceedia Structural Integrity, **5**, 103(2017), p. 1409-1416.
8. Basquin, O. H.: Proceedings ASTM, 10, 1910, p. 625.
9. V. Kliman, V. Chmelko, M. Margetin, Analysis of the notch effect of welded joint and of grinding effect. In Kovové materiály. Metallic materials. **53**, 6 (2015), p. 429-441
10.Kliman, V. - Bílý, M. - Prohácka, J.: Int. J. Fatigue **15**, 2 (1993), p. 93-107.

Residual stress redistribution during elastic shake down in welded plates

Jazeel R. Chukkan[1,2,*], *Guiyi* Wu[3], *Michael E.* Fitzpatrick[1] , *Elvin* Eren[3], *Xiang* Zhang[1], and *Joe* Kelleher[4]

[1]Coventry University, Faculty of Engineering, Environment and Computing, Priory Street, Coventry, UK
[2]NSIRC, TWI Ltd, Granta Park, Great Abington, Cambridge, CB21 6AL, UK
[3]TWI Ltd, Granta Park, Great Abington, Cambridge, CB21 6AL, UK
[4]ISIS Pulsed Neutron & Muon Source, Harwell Campus, Didcot, Oxfordshire, UK

Abstract. Residual stresses are a consequence of welding in various structures such as ships and offshore structures. Residual stresses can be relaxed or redistributed according to the load levels during operation. The elastic shakedown phenomenon can be considered as one of the reasons for this change. This paper studies the relaxation/redistribution of weld residual stress during different levels of shakedown in a butt-welded plate chosen according to ship design and welding procedures. Welding was performed on DH36, a ship structural steel. Neutron diffraction was used to measure residual stresses in these plates in the as-welded state and after different levels of shakedown. A mixed hardening model in line with the Chaboche model is determined for both weld and base material. A numerical model is developed to estimate the shakedown limit on butt-welded plate. Further, the redistribution of residual stress in a numerical weld model according to the different levels of shakedown limit is studied. Based on the shakedown limit of the butt-welded plate, a shakedown region is determined, where the structure will undergo elastic shakedown in the presence of an existing residual stress field if the maximum stress on the load section after a few initial cycles is in the shakedown region.

1. Introduction

Welding is the primary joining technique used in many offshore structures which results induces in complex residual stresses in many components. Residual stresses in welded structures which are exposed to complex cyclic loads are sometimes observed to be different from the as-welded state [1]. One of the main reasons for this change is elastic shakedown. Elastic shakedown is defined as a plastic deformation causing a change in plasticity induced residual stress during the first few load cycles, followed by an elastic response which is associated with a limit called the shakedown limit [2]. On a structure, a primary load above the shakedown limit will result in reverse plasticity or ratchetting [3]. Generally, for a rigid–perfectly-plastic solid the load limit can be determined by using Melan's lower bound and Koiter's upper bound theorems [4,5]. These theorems in their original form assume linear kinematics and elastic-perfectly plastic material. However, linear or nonlinear isotropic and linear kinematic classics are two ideal models that are not representative for realistic structural materials subjected to cyclic loadings [6]. It is recommended to adopt analysis with mixed hardening models such as Chaboche [6, 7].

There has been much research on the shakedown analysis of components under thermal loads which are associated with a residual stress field following initial plastic deformation [3, 8]. However, little research is available on the shakedown analysis of components with a pre-existing weld residual stress field. These problems are generally considered as a stress relaxation or stress redistribution problem [9]. One of the reasons could be that in the presence of a residual stress field, elastic shakedown can result in relaxation or redistribution of the existing residual stress. However, shakedown analysis on these components could be used for understanding the effect of relaxation or redistribution of residual stresses on the state of elastic shakedown. It is worth noting that shakedown investigation has been used for predicting thresholds of a cracked body [10]. Also, the shakedown analysis of a ring-stiffened cylinder model was used to study variation in the ultimate strength of the component [11]. The shakedown limit on a welded plate is determined by considering absorbed plastic deformation energy as shakedown criterion [12].

Plastic deformation redistributes residual stresses thereby changing the pattern to achieve internal equilibrium. A welded structure can be subjected to a variety of loading during operation such as tensile, compression or alternating loading. In this study, the redistribution of residual stresses following different numbers of tensile load cycles were investigated.

2. Governing equations

2.1 Constitutive model

The numerical modelling is established on an elastic-plastic model initially developed by Lemaitre and Chaboche [7]. The model is implemented with a combination of one isotropic hardening R and three kinematic hardenings α_1, α_2 and α_3. Total strains ε_{ij}^{t} are decomposed into elastic strains ε_{ij}^{e} and plastic strains ε_{ij}^{p}.

* Corresponding author: chukkanj@uni.coventry.ac.uk

Table 1:Constitutive model parameters for DH36 steel in the FE model.

DH36	E/GPa	ν	σ_y/MPa	Q	b	C_1/MPa	γ_2	C_1/MPa	γ_2	C_1/MPa	γ_2
BM	200	0.3	350	-48.6	87.5	4360	16.4	38524	116	8000	40
WM	200	0.3	400	-102	14	8912	29.65	102300	400	8000	40

The elastic domain is described by a typical von Mises yield criterion f=0, where f is defined as;

$$f = J(\sigma - \alpha_1 - \alpha_2 - \alpha_3) - R - \sigma_y \quad (1)$$

where σ_y is the yield strength of the material, σ is the Cauchy stress tensor and J is the von Mises equivalent stress defined as;

$$J(\sigma) = \sqrt{\frac{3}{2}\sigma_d : \sigma_d} \quad (2)$$

where σ_d is the deviatoric stress tensor. The isotropic hardening R and the kinematic hardening α_i are defined as below;

$$R = b(Q - R)\dot{p} \quad (3)$$

$$\dot{\alpha}_1 = \frac{2}{3}C_1\dot{\varepsilon}_p - \gamma_1\dot{\alpha}_1\dot{p} \quad (4)$$

$$\dot{\alpha}_2 = \frac{2}{3}C_2\dot{\varepsilon}_p - \gamma_2\dot{\alpha}_2\dot{p} \quad (5)$$

$$\dot{\alpha}_3 = \frac{2}{3}C_3\dot{\varepsilon}_p - \gamma_3\dot{\alpha}_3\dot{p} \quad (6)$$

where $\dot{p}$ is the rate of accumulated plasticity, $b, Q, C_1, C_2, C_3, \gamma_1, \gamma_2$ and γ_3 are material parameters. Strain-controlled fatigue tests were carried out to calibrate material parameters. Numerical minimisation technique between the Finite Element (FE) model response and the stable hysteresis obtained from experiments is used to calibrate the required parameters. The material parameters obtained for DH36 base material and weld material are shown in Table 1.

2.2 Shakedown analysis

The applied load on the component should be below the shakedown limit, to achieve elastic-shakedown state. Above this limit, a continuous plastic flow develops or constantly repeats until the structure collapses due to low-cycle fatigue.

On a component under cyclic loading, the plastic strain increment at the end of each load cycle over a time interval nT to $(n+1)T$ can be expressed as:

$$\Delta\varepsilon_{ij}^p = \int_{nT}^{(n+1)T} \dot{\varepsilon}_{ij}^p \mathrm{d}t \quad (7)$$

After each load cycle, the total work done on a body with volume V and generalised displacement ΔD can be determined as:

$$\int_{nT}^{(n+1)T} P\,\dot{D}dt = W^e + W^p \quad (8)$$

$$W^e = \int_{nT}^{(n+1)T} \mathrm{d}t \int_{nT}^{(n+1)T} \sigma_{ij}\dot{\varepsilon}_{ij}^e \mathrm{d}V \quad (9)$$

$$W^p = \int_{nT}^{(n+1)T} \mathrm{d}t \int_{nT}^{(n+1)T} \sigma_{ij}\dot{\varepsilon}_{ij}^p \mathrm{d}V \quad (10)$$

where P is the generalised load, W^e and W^p are the elastic and plastic work done respectively.

After a specific number of cycles, if $\dot{\varepsilon}_{ij}^p = 0$, W^p will be zero and upon unloading, the right-hand side of equation (8) vanishes and there will exist a time-independent residual stress field ρ_{ij} depending on the initial plastic deformation. This suggests that the structural response is elastic and hence the lower bound theorem is satisfied. Hence it can be concluded that W^p defined in equation (11) can be used as a shakedown criterion.

$$W^p = \sum_{i=1}^{N} \int_{V^e} \sigma_{ij}\Delta\varepsilon_{ij}^p \mathrm{d}V^e \quad (11)$$

where V^e is the volume of an element and N is the number of elements in the area of region.

3. Methodology

3.1 Numerical methods

3.1.1 Weld model

Thermo-Mechanical analysis of multi-pass Gas Metal Arc welding of DH36 steel was performed in a sequentially coupled analysis. This study uses an equivalent static heat source in conjunction with the ‘chunking’ method explained in section III.15.4.5.3 of [13]. The model was initially developed on a rectangular plate with dimensions of 400 × 140 × 12.7 mm^3. The rectangular numerical weld model was then cut to the dimension of a mechanical load model as shown in Fig.1 The cutting was simulated by FEA using the element removal technique.

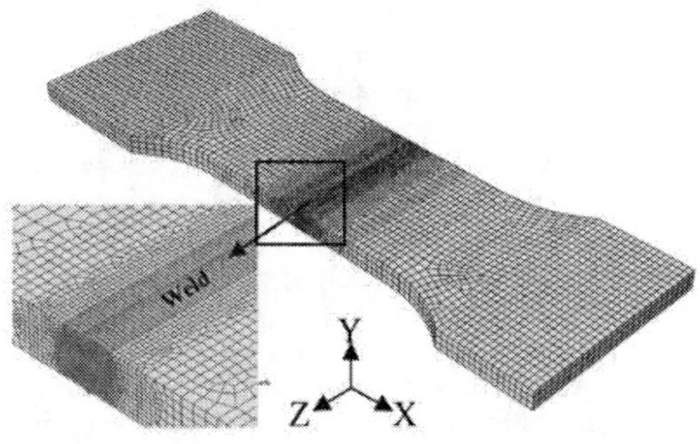

Fig. 1: Weld numerical model following element removal

3.1.2 Shakedown limit

A numerical model was developed to estimate the shakedown limit of the butt weld at hand by considering plastic work done as the shakedown criterion. Consider a load domain P defined as a region in the variable space which consists of load-time history. For N independent generalised loads, P_1, P_2, ... P_N, the domain can be represented as

$$P = [P_1, P_2, \dots. P_N]^T = \lambda\{\psi_N p_N\} \quad (12)$$

where $\psi_N$is the amplitude factor of $p_N$which has an upper and lower limit and λ is the generalised load factor. The objective of the limit analysis is to determine λ.

Initially, the shakedown limit of the butt-welded plate without weld residual stress was determined using a step-by-step iterative procedure illustrated in Fig.2. A Python script was used with the FE package Abaqus to implement the procedure.

Let $\lambda_i(i = 1, \dots\dots. n)$ be load factors where n is the number of increments. The largest value of λ_i will be the load at which the structure can achieve shakedown with imposed boundary conditions. The procedure was implemented for uni-axial loading condition. The shakedown limit analysis was performed on a half model of the FE design used for the weld simulation.

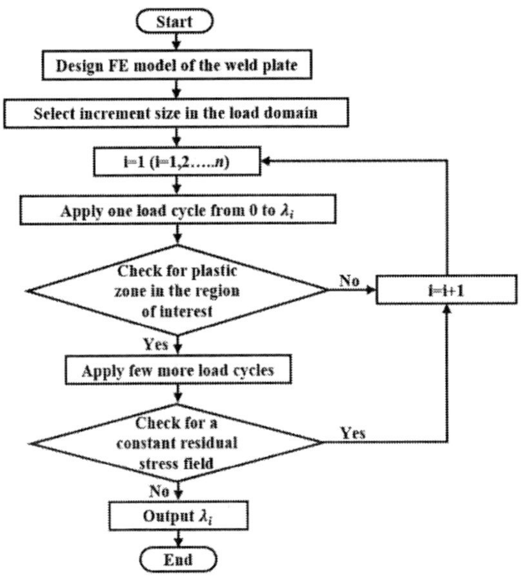

Fig. 2: Flow chart showing modelling procedure for shakedown limit analysis

An appropriate level of increment was selected according to the problem and the computational efficiency. The load variation used for the limit analysis is the same as explained in the next section. Initially, the structure is loaded from 0 to λ_i and then unloaded, which constitutes a load cycle. The structure is investigated for any plastic zones developed following a single load cycle. If there exists a plastic zone, the structure is subjected to a few more load cycles until a stable residual stress field is obtained. If there is no plastic zone, then the load level is increased by an increment.

Stable residual stress is assessed by calculating plastic work done at each load step at a given λ_i. If the total plastic work of successive load cycles is equal, then change in plastic strain is zero and hence change in residual stress is zero.

3.2 Experimental details

3.2.1 Specimen manufacturing and cyclic loading

Gas Metal Arc Welding was employed to manufacture a weld plate of dimensions $400 \times 140 \times 12.7$ mm^3 with four passes with a procedure specification qualified in accordance with the Lloyd's Register classification. The plate was in full restraint during the welding process to represent welding in ship and offshore structures. Welding was performed along the width of the plate as shown in Fig. 3a. Thermocouples and high-temperature strain gauges were used to monitor transient heat and strain variations respectively. Thermocouples and strain gauge data were later used to validate heat transfer model in the sequentially coupled welding simulation.

Following welding, the weld plate was cut using EDM to prepare mechanical test specimens. After cutting, the plate was subjected to a few load cycles to study redistribution of weld residual stresses. Load cycles applied were tensile (R-ratio=0) half-sine cycles with 0.25Hz frequency. The maximum applied stress used on each plate was equivalent to achieving 75% of the yield strength of the parent material (350 MPa) to ensure that there is some residual stress redistribution. The load was applied across the weld and along the transverse residual stress component as shown in Fig.4.

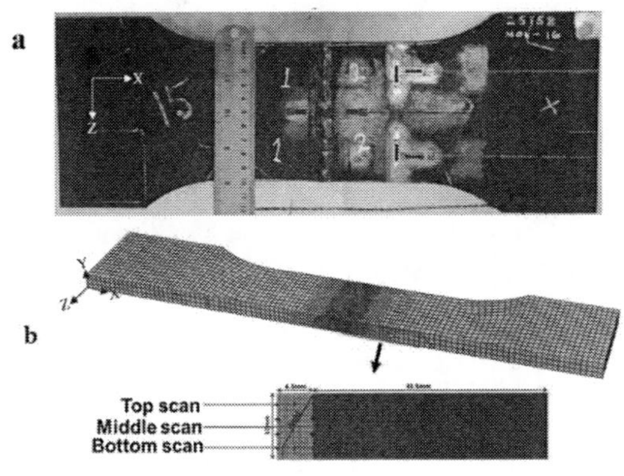

Fig. 3: a) Weld plate top view b) Front view of the measurement location

3.2.2 Neutron Diffraction

Neutron diffraction was used to measure residual stresses in the plates in the as-welded state and after different levels of shakedown. Neutron diffraction is a non-destructive technique which can determine strains through the thickness of a component. The ENGIN-X instrument at the UK's ISIS pulsed neutron source was used to measure the lattice spacing at each location using time-of-flight neutron diffraction technique.

The plate was set up in two different orientations to obtain all three principal stress components. With two detector banks at ±90° to the incident beam, it was possible to measure strains in two directions simultaneously. A gauge volume of 3 × 3 × 3 mm^3 was used to obtain measurements from 30 points at mid-thickness across the weld and through the thickness at selected locations on the plate as shown in Fig. 3b. The measurements were performed in the as-welded state, after one load cycle and after three load cycles.

Fig. 4: The loading set-up

4. Results and discussions

4.1 Shakedown limit

A half of the weld plate is modelled for the limit analysis for computational efficiency. The FE model is subjected to increasing loads starting from 50% of yield strength of the parent material. The loading is gradually increased until a constant residual stress ceases to exist in the plate. After each load cycle, the equivalent plasticity at the end of a cycle is used to check the formation of plasticity. After about 40 many iterations, the shakedown limit λ estimated by the FE model is 1.17. Now this implies that in the absence of residual stress field any cyclic load within 1.17 times the yield strength applied in this set-up will achieve elastic-shakedown state.

4.2 Residual stress redistribution

4.2.1 As welded residual stress

Fig. 5 and 6 compare the numerically predicted longitudinal and transverse stress components and the neutron diffraction measurements in the as-welded state. The comparison is made across the weld taken from the top scan and bottom scan. Longitudinal stress components on and around the weld are as high as the yield strength of the corresponding material. The bottom scan of transverse component consisting of the root pass and near locations are slightly compressive. Less filler metal deposition and the low rate of cooling can be the reason for this. This distribution is consistent with plates welded with high restraint.

The numerical predictions have a similar trend with the measured values, and a good agreement was seen between the measured and numerically predicted values. However, the deviation in values may be due to the post machining performed on the weld plate before stress measurements along with the experimental errors and assumptions in the numerical simulations.

4.2.2 Residual stress redistribution

Fig. 7 and 8 compare the redistribution of longitudinal stress components from neutron diffraction measurements and numerical model after the application of three load cycles in the top scan and bottom scan respectively. As mentioned earlier, the applied load is along the transverse residual stress component, hence little relaxation on the longitudinal stress component is observed.

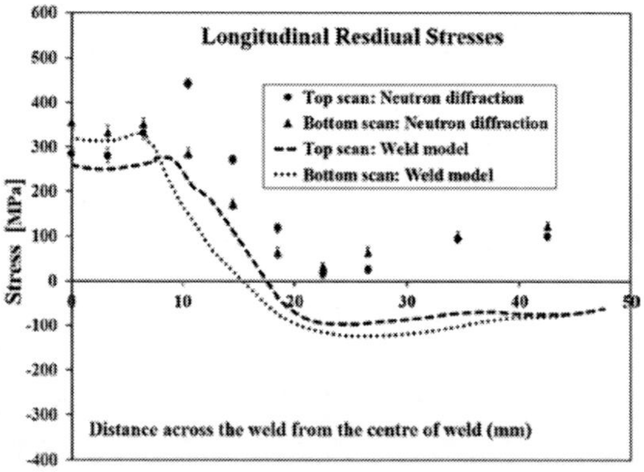

Fig. 5: As-welded longitudinal residual stress component comparison between experimental measurement and FE model

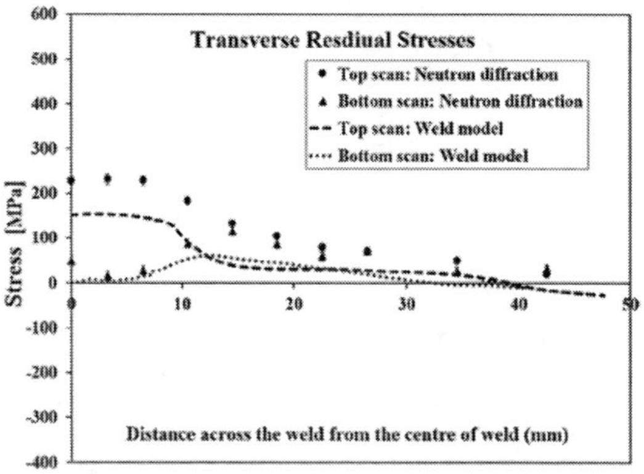

Fig. 6: As-welded transverse residual stress component comparison between experimental measurement and FE model

0.7 times limit load was applied across the weld for the investigation of stress redistribution experimentally. The experimental results for all neutron diffraction measurements show that the redistribution in the longitudinal component is minimal for a load applied across the weld as shown in fig. 7 and 8. A slight relaxation of about 100 MPa is noticed in the top scan near the weld toe moving away from the weld centre. The thickness and weld specification of the plate is selected based on the flat bar stiffener member in bottom deck of a ship structure. Since the capacity of the in-situ loading rig at ENGIN-X is not enough to introduce residual stress shakedown, loading was performed at TWI Ltd, between neutron measurements. Even though at most care was taken during the set-up, small errors are expected in consequence.

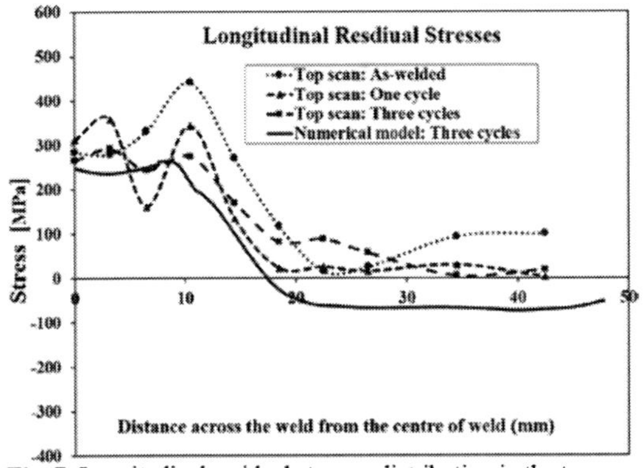

Fig. 7: Longitudinal residual stress redistribution in the top scan

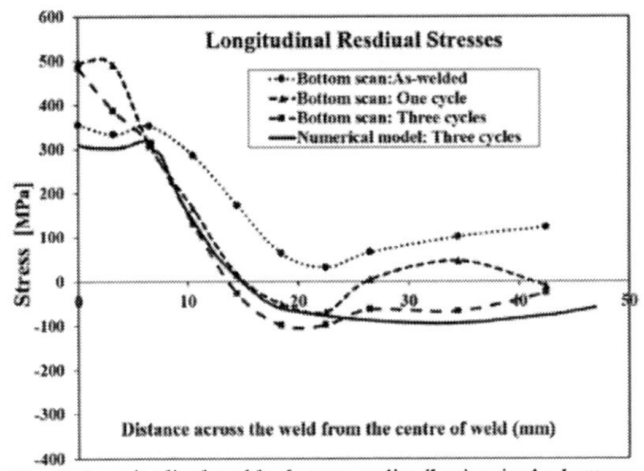

Fig. 8: Longitudinal residual stress redistribution in the bottom scan

Fig. 9 and 10 compare the redistribution of transverse stress components from experimental measurement and numerical model after the application of three load cycles in the top scan and bottom scan respectively.

Although the transverse residual stress following three load cycles shows redistribution, there are variations with which redistribution evolve in top scans and bottom scans. In the case of top scan, after three load cycles, experimental measurements show about 90% relaxation, and numerical results show about 70% relaxation in the weld toe region. This can be attributed to the underprediction of initial weld residual stresses which is an influencing factor in the amount of relaxation. Percentage of relaxation after *N* load cycles is calculated using equation (13). Additionally, the amount of redistribution is maximum after the first load cycle as shown in fig.9. This is in line with previous work in the literature.

$$S(\%) = \frac{Initial\ RS - \ RS\ at\ Nth\ Cycle}{Initial\ weld\ stress} \times 100 \quad (13)$$

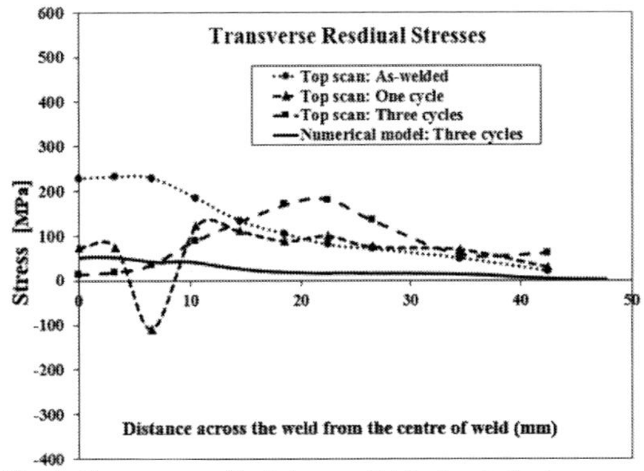

Fig. 9: Transverse residual stress redistribution in the top scan

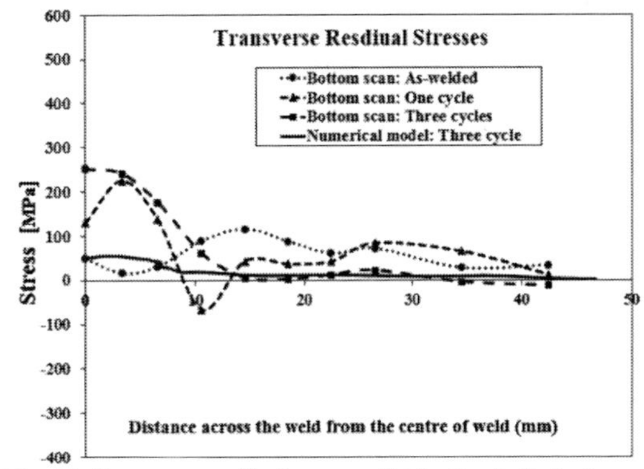

Fig. 10: Transverse residual stress redistribution in the bottom scan

On the other hand, the bottom scan of transverse component has moved more into the tensile zone than the as-welded state after the application of three load cycles. This small increase could be the variation of transverse component through the thickness to maintain internal equilibrium following relaxation in the top scan. The increase seen in the tensile stress in the neutron diffraction measurement is not seen in the bottom scan of the numerical model. However, there exists a tensile region with maximum stress of 90 MPa at the mid-thickness following relaxation.

It is noted that preloading weld plates for relieving tensile stresses can be useful on the top surface or weld toe of the plate (which is considered as detrimental), but it can also introduce tensile stresses through the thickness.

4.2.3 Effect of shakedown

A parametric study was performed in FE where the load cycles with magnitudes of 0.5, 0.7 and 0.8 times limit load were applied on the same plate with the same weld residual field. Fig. 11 shows the state of residual stress in the shakedown regime after the first three cycles. The shakedown region is defined by normalised applied stress and residual stress, both normalised by yield strength of the parent material.

With applied load levels at 0.5, 0.7 and 0.8 λ, there is about 35, 70 and 100% relaxation in transverse stress respectively in the weld toe region field after three load cycles. If the residual stress after first few cycles and the applied load is within the shakedown region it can be said that the structure will enter elastic-shakedown state and failure will be dominated by high cycle fatigue. Since the shakedown limit is calculated based on the same material model and the load variation, the evolution of accumulated plasticity will fall below the limit, given that there is adequate amount of relaxation in the first few cycles. It is defined as adequate relaxation because one should be careful in assuming shakedown state by just considering the relaxation in the initial residual stress. According to R5 procedure, for a structure to undergo shakedown state, it should have the maximum plastic zone developed in the section less than 20% of the section thickness [14]. If the applied load does not relax the stresses below the shakedown region in the first few cycles, it is likely to undergo reverse plasticity or ratcheting because of the accumulated plasticity over each load cycle.

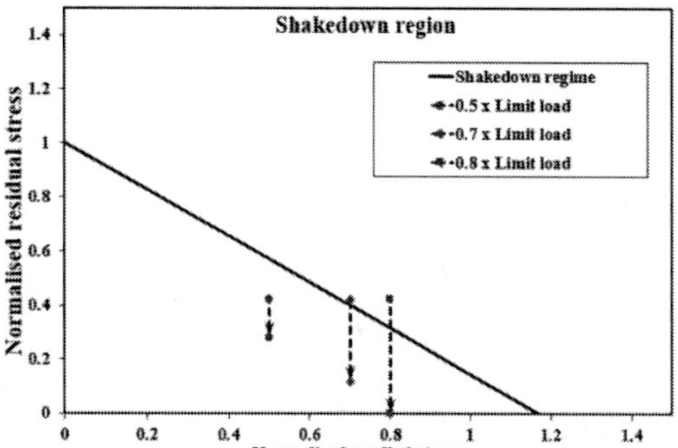

Fig. 11: Shakedown region after the first three load cycles

The authors are currently investigating the redistribution of residual stress on this plate after ten load cycles by experimental and numerical methods. Future study would be to investigate the effect of cyclic tensile loading on the longitudinal residual stresses in butt welds and on different weld joint configuration such as a T-joint configuration.

5. Conclusions

In this work, the effects of residual stress redistribution on elastic shakedown and redistribution of residual stress at different applied load levels are discussed. The conclusions drawn from this study are:

- Shakedown limit of any component can be estimated using a simplified method where the change in plastic work done during each load cycle can be used as a shakedown criterion.
- Experimental measurement of residual stress redistribution after three load cycles of 70% limit load showed about 90% relaxation in transverse stress component at about 2 mm below the weld toe. However, 2.5mm above the bottom surface shows an increase in tensile stresses near weld toe.
- In the presence of residual stresses, any component subjected to a cyclic load will achieve shakedown state if the combined applied stress and maximum residual stress after a few load cycles is below the shakedown region as shown in fig.11.

Acknowledgements: - This publication was made possible by the sponsorship and support of Coventry University and Lloyd's Register Foundation, a charitable foundation helping protect life and property by supporting engineering-related education, public engagement, and the application of research. The work was enabled through, and undertaken at, the National Structural Integrity Research Centre (NSIRC), a postgraduate engineering facility for industry-led research into structural integrity established and managed by TWI through a network of both national and international Universities.

References

1. J. Eckerlid, A. Ulfvarson, Mar. Struct. 8: 385-406 (1995).
2. M. Abdel-Karim, Int. J. Pressure Vessels Piping 82: 427- 58 (2005).
3. J. Bree, The J. of Strain Analysis for Eng. Design 2: (1967).
4. E. Melan, Arch Appl Mech 9: 116-26 (1938).
5. W. T. Koiter, Proc. Koninkl. Ned. Akad. Wetenschap B59: (1956).
6. R5, 3rd ed. (2014).
7. J. Lemaitre, and J. Chaboche, Mechanics of Solid Materials, (1994).
8. T. Li, H. Chen, W. Chen, J. Ure, Int. J. Pressure Vessels Piping 88, 301- 10 (2011).
9. K. Yuan, Y. Sumi, Analysis and Design of Mar Struct, (2013).
10. Y. Huang, E. Stein, Eng. Fract. Mech. 54: 107-12 (1996).
11. B. C. Cerik, S. Cho, J. Mar. Sci. Technol. 18: 524-34 (2013).
12. Y. Sun, S. Shen, X. Xia, and Z. Xu, Mater Des 47: 106-14 (2013).
13. British Energy, R6 Revision 4: Assessment of the Integrity of Structures Containing Defects, (2014).
14. D. Knowles, Procedia Engineering 86, 315-26 (2014).

A numerical analysis of multiaxial fatigue in a butt weld specimen considering residual stresses

Iñigo Llavori[1*], *Unai* Etxebarria[1], *Arkaitz* López-Jauregi[1], *Ibai* Ulacia[1], *Done* Ugarte[1], *Jon Ander* Esnaola[1], and *Miren* Larrañaga[1*]

[1]Escuela Politécnica Superior – Mondragon Unibertsitatea, Loramendi 4, 20500 Arrasate-Mondragon (Gipuzkoa). Spain.

Abstract. Residual Stress (RS) pattern changes considerably depending on the width of the plates and the welding parameters, having effect on the fatigue strength. Most of the standards do not consider them and in some works, yield stress is taken as residual stress value. It results in a very conservative estimation of fatigue life. Authors developed recently a numerical model to predict more properly the value of RS pattern depending on the plate thickness. In a welded joint, considering the RS and alternating axial loads, the evolution of the stresses is multiaxial, becoming necessary its study. Therefore, the aim of this work is to analyse different fatigue indicator parameters (Smith-Watson-Topper, Fatemi-Socie, and Critical Plane implementation of the Basquin equation) in order to predict the fatigue behaviour of butt-weld components. For that purpose, the numerical model to predict the RS pattern in welded joint developed by this research group is used.

1 Introduction

Welding process is one of the most used joining methods in the metal manufacturing industry [1]. Gas metal arc welding (GMAW) is widely used, due to its high productivity. Specially, spray transfer mode multipass welding is suitable in the case of high thickness structures [2, 3] because of the high rate of metal transfer, good arc stability, absence of weld spatter and uniform and regular metal transfer to the workpiece.

However, in welding joints, there are different variables that have a negative effect on the fatigue life [4], such as residual stresses, the inhomogeneous geometry of the welding cord or the generation of areas with different mechanical properties. Therefore, the main failure mode of welded joints is the fatigue fracture [5, 6].

Residual stresses (RS) are generated due to the high thermal gradients of the welding process, where there are non-uniform heating and cooling cycles [4]. The thermal expansion of the material is constrained by the adjacent areas that are not at the same temperature; in addition, these areas at higher temperature have lower strength, and consequently, can suffer compressive plastic deformation. During the cooling down process, the yielded areas limit the elastic springback of not yielded areas, generating internal stresses that remain on the welded component. These RS can be tensile or compressive, depending on the limitations imposed by the surrounding areas. The final RS pattern relies on different factors, such as, mechanical restraints, welding sequence, preparation of the weld groove, the number of weld passes or structure dimensions [7, 8].

On the one hand, the estimation of RS pattern due to welding process is complex due to the multi-physics phenomena as heat, electricity or mechanical work that take part in the process. On the other hand, the experimental measurement of RS nowadays presents some limitations as the methods available are not fully reliable and imply huge time and economic cost [8, 9]. Therefore, most of the fatigue life prediction approaches do not take into account RS or consider the yield stress value as RS magnitude [10, 11, 1]. This results in very conservative life estimation, and thus, in oversized designs.

In recent works, authors have presented a numerical methodology for RS pattern estimation in spray transfer multipass welding [12, 13]. This methodology enables an accurate estimation of the RS generated, based on the heat source and the welding speed. This is the methodology employed in this work to consider the RS of a butt-welded joint.

Because of the presence of the RS, when we apply a remote uniaxial load to a welded component, a multiaxial evolution of the loading-unloading cycle is generated. In other words, there is a multiaxial stress state independently of the applied load.

In this work, we have used different multiaxial parameters to estimate the fatigue life of a butt-welded component under uniaxial loading conditions, considering the RS. The objective is to evaluate and to analyse the suitability of each multiaxial parameter for the studied case.

* Corresponding author: mlarranaga@modragon.edu , illavori@mondragon.edu

2 Numerical methodology for residual stress estimation

In order to solve the thermal and mechanical fields along the welding process, we have used an uncoupled thermos-mechanical model implemented in the simulation software ABAQUS FEA (Fig. 1). This approach is accepted as dimensional changes in the welding process can be neglected and mechanical deformation energy is insignificant compared to the thermal energy from the welding arc [14]. Both equation systems, thermal and mechanical, are solved by using the implicit direct integration method.

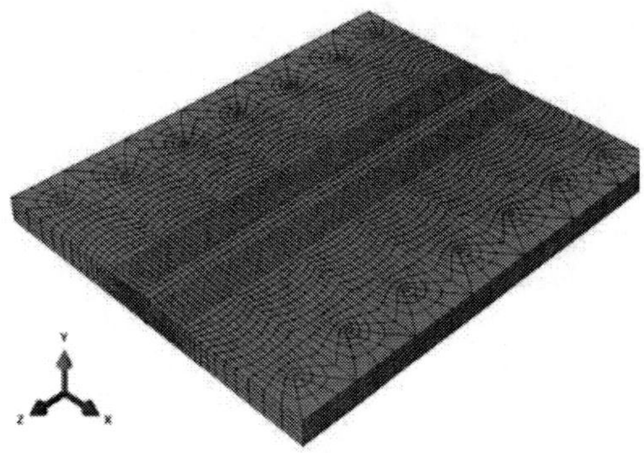

Fig. 1. Modelling of the simulation process [12]

For both models (thermal and mechanical) full integration continuum hexahedral elements and the addition of filler material through all passes is modelled by using the kill/rebirth method [2, 3]. In this method, all the weld bead elements are initially inactive and, consequently, eliminated (killed) from the equation system. Then, according to the welding speed and material addition discretization size, elements are reactivated (rebirthed) simulating the welding torch pass (Fig. 2) [15].

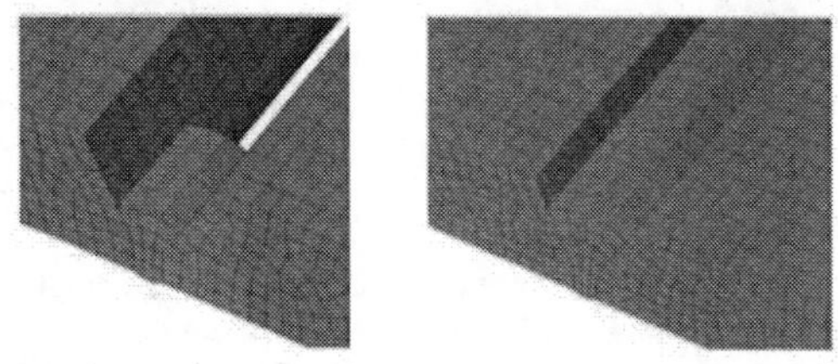

Fig. 2. Simulation of the welding process

The main input parameters needed for the numerical simulation are the welding speed and the heat source power. These parameters have been defined according to the analytical procedure proposed by the authors in [12].

The results obtained by this numerical method were compared to experimental results, obtaining a very good agreement. Therefore, we can say that this numerical prediction method is widely validated [12, 13, 16, 17].

3 Multiaxial fatigue

Material fatigue refers to a progressive degradation of a material caused by loading and unloading cycles. The stress fluctuations suffered over time weakens or breaks the material even at stresses lower than the yielding value. Accordingly, lot of effort has been directed at developing fatigue life prediction models. Fatigue is characterized with a high scatter of the lifetime. Probabilistic approaches are recently arising in the literature to address this problem. However, the majority of the models currently used analyses fatigue in a deterministic way, *i.e.* a structure fails if a given parameter reaches a critical value.

Nowadays, a variety of different approaches for fatigue life prediction exists, such as approaches based on multiaxial fatigue criteria, damage mechanics or micromechanics, which are extensively reviewed in literature. Multiaxial fatigue criteria reduces the multiaxial stresses (usually computed by FEM analysis) to an equivalent uniaxial stress state. This way, the results can be compared to an experimental fitting curve obtained from uniaxial fatigue data. A crucial step when selecting a multiaxial criterion is to check whether the simplification from multiaxial stress state to an equivalent uniaxial stress state are acceptable or valid. This task is not simple and requires the detailed study of the evolution of stresses and strains along the loading cycle.

The book published by Socie and Marquis [18] presents a wide and detailed study about the principal multiaxial parameters, also known as Fatigue Indicator Parameters (FIPs). Those parameters can be broadly classified into three groups: strain-based, stress-based and energy-based FIPs. Strain-based FIPs are generally related with Low-Cycle Fatigue (LCF) where plastic deformation may be predominant. Stress-based FIPs are associated with High-Cycle Fatigue (HCF) where the stresses usually remains in the elastic domain. Finally, energy-based models relates the product of stresses and strains to quantify fatigue life, which generally are applicable to both LCF and HCF regime.

For these complex stresses or loading states, other approaches such as the critical plane method are more suitable. The critical plane method has been developed from the experimental observation of nucleation and crack growth under multiaxial loading. The critical plane models include the dominant parameters that govern the type of crack initiation and propagation. An adequate model must be one that estimates correctly both fatigue life and the dominant failure plane. However, several failure modes exists, and there is not a unique parameter that suits all. Nonetheless, the most popular parameters are the energetic criteria known as Fatemi-Socie (*FS*) [19] and Smith-Watson-Topper (*SWT*) [20].

The *SWT* parameter is applied in those materials where the crack growth occurs in mode I. The critical plane is defined as the one where the product of maximum normal stress ($\sigma_{n,max}$) and normal strain amplitude ($\varepsilon_{n,a}$) is maximum.

$$SWT = \left(\sigma_{n,max}\varepsilon_{n,a}\right)_{max} = \frac{\sigma_f'^2}{E}(2N_f)^{2b} \quad (1)$$

where σ_f' is the fatigue strength coefficient, b is the fatigue exponent, and N_f is the number of cycles to failure.

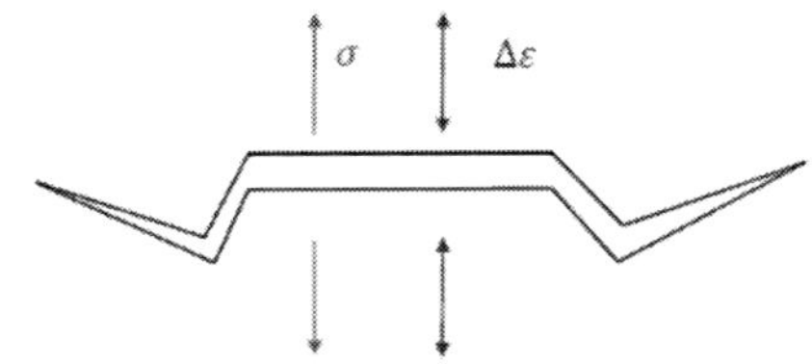

Fig 2. Physical base of the *SWT* parameter.

Under shear loading condition, crack lip surfaces generate frictional forces that reduce stresses at the crack tip, thus increasing fatigue life. However, tensile stresses and strains will separate the crack surfaces, reducing the friction forces. The energetic FIP *FS* can be understood as the cyclic shear strain to include the crack closure effect multiplied by normal stress to take into account the opening of the crack.

$$FS = \frac{\Delta\gamma_{max}}{2}\left(1 + k_{FS}\frac{\sigma_{n,max}}{\sigma_y}\right) = (1+\nu)\frac{\sigma_f'}{E}(2N_f)^b + \frac{k_{FS}}{2}(1+\nu)\frac{\sigma_f'^2}{E\sigma_y}(2N_f)^{2b} \quad (2)$$

where $\Delta\gamma_{max}$ is the maximum range of shear strain on any plane, $\sigma_{n,max}$ is the maximum normal stress in that particular plane, σ_y is the material yield stress, k_{FS} is a material dependent factor.

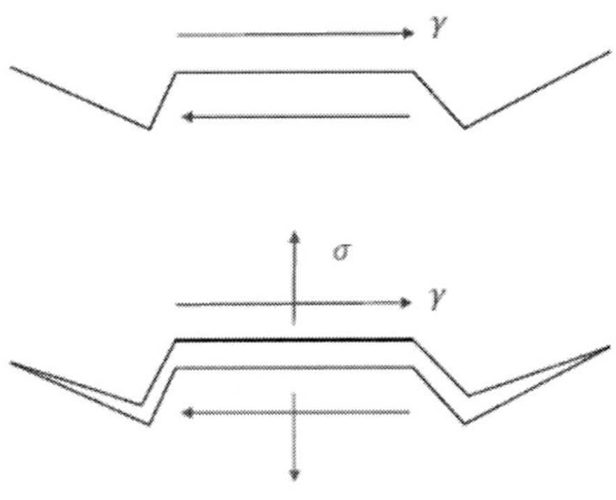

Fig 3. Physical base of the FS parameter.

The last parameter used in this work is the procedure used previously by the authors. In this case, it was assumed that the fatigue life estimation of welded structures can be conducted considering only the crack propagation [1]. For this reason, maximum normal stress criterion [21] based on Papadopoulos *et al.* [22] where influence of the shear stress is neglected, was used. This method is based on the critical plane implementation of the Basquin equation. The linear interaction between the stress amplitude and mean stress defined by Goodman were used to take into account the influence of RS on the fatigue strength.

$$\frac{\sigma_{a\text{-}res}}{\sigma_e} + \frac{(\sigma_{mean} + \sigma_{res})}{\sigma_u} = 1 \quad (3)$$

where σ_e is the endurance limit at 10^7 cycle for the base material, and σ_u is the ultimate tensile stress (525 MPa for the studied case), $\sigma_{a\text{-}res}$ is the stress amplitude considering RS, σ_{mean} is the mean stress and σ_{res} is the numerically estimated welding RS value in the critical plane of the studied zone. For fully reversed case (R = -1), the previously defined equation is written as follows.

$$\frac{\sigma_{a\text{-}res}}{\sigma_e} + \frac{\sigma_{res}}{\sigma_u} = 1 \quad (4)$$

4 Case study

The component analysed in this work is formed by two butt-welded S275JR 10 mm-thick plates. Filler material 1.2 mm diameter PRAXAIR M-86 filler wire, according to the AWS/ASNE SFA 5.18 ER70S-6 standard, and quasi-constant weld pass section were considered.

We have obtained the numerical estimation of the RS pattern by the methodology described in section 2 (Fig. 3).

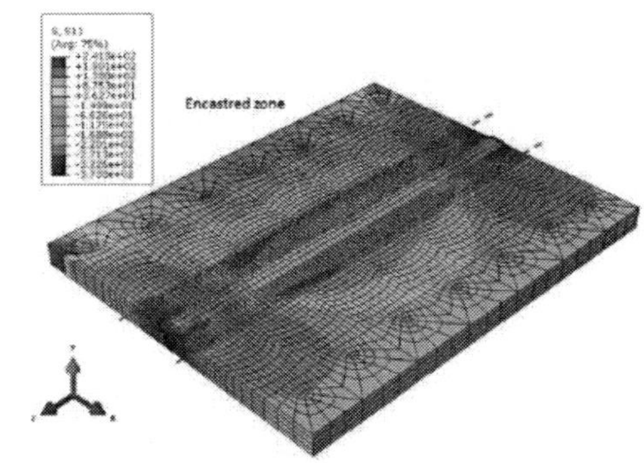

Fig. 3. Residual stress pattern numerically estimated [12].

We have analysed the RS stress pattern along the whole welding cord. As can be observed in Fig. 3, the maximum stresses appear at the middle of the cord. We have chosen the most stressed element in order to do this analysis.

After obtaining the RS pattern due to welding process, we applied a uniaxial remote load perpendicular to the welding cord. We recorded the evolution of stresses and strains of a complete loading-unloading cycle for the most stressed element in the welding cord.

5 Results

The experimental results for numerical correlation have been taken from previous work done by the authors [13], and they are summarized in Table 1. Additionally, HCF calibrated parameters for the base material S275JR is summarized in Table 2.

Table 1. Experimental results of butt-weld specimens at R = -1, performed by force control in a conventional fatigue machine MTS 810.

Test	σ_a [MPa]	N_f [cycles]
1	125	133880
2	125	139649
3	125	233312
4	125	292178
5	125	965456
6	115	355750
7	115	371078
8	110	538638
9	110	1159103
10	110	1243530
11	110	1352999
12	110	1461791
13	110	1579331
14	90	1080407
15	90	1431281
16	90	1977803
17	90	2244631
18	80	9895092

Table 2. Calibrated Basquin parameters for S275JR base material.

σ_f' [MPa]	b [-]
415.48	-0.047

Fig. 3 shows the prediction of the employed FIPs comparing with experimental results. As previously commented, the fatigue life estimation results are based on the stresses and strains of a specific element at the middle of the welding cord.

On the one hand, it can be observed that the *SWT* parameter shows the best correlation because it takes into account the effect of mean stress (through the maximum stress) and strain amplitude.

On the other hand, the worst correlation is the one presented by the *FS* parameter. This may be because in these tests, the applied load is uniaxial and the effect of the residual stresses on the shear stress is negligible. Finally, the prediction of critical plane implementation of the Basquin model is the most conservative due to the use of the Goodman equation that is described in the literature as a conservative [23].

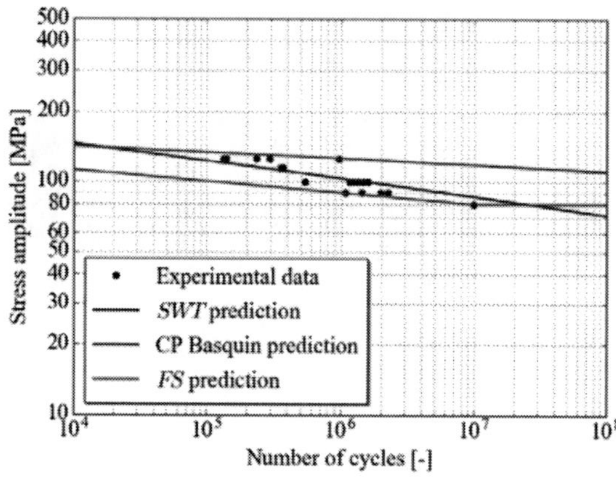

Fig. 4. Fatigue Indicator Parameters prediction vs.experimental tests.

Fig. 4, Fig. 5 and Fig. 6 show the correlation between the predicted stress amplitude (σ_{predic}) and the experimental ones (σ_{exp}) for the 3 parameters analysed. The mean absolute percentage error (*MAPE*) has been calculated as:

$$MAPE = \frac{100}{m}\sum_{j=1}^{m}\frac{\left|\sigma_{exp,j} - \sigma_{predic,j}\right|}{\sigma_{exp,j}} \quad (5)$$

The computed error therefore for each parameter is:

- *SWT* = 6.19%
- *FS* = 24.21%
- *CP* Basquin = 10.45%.

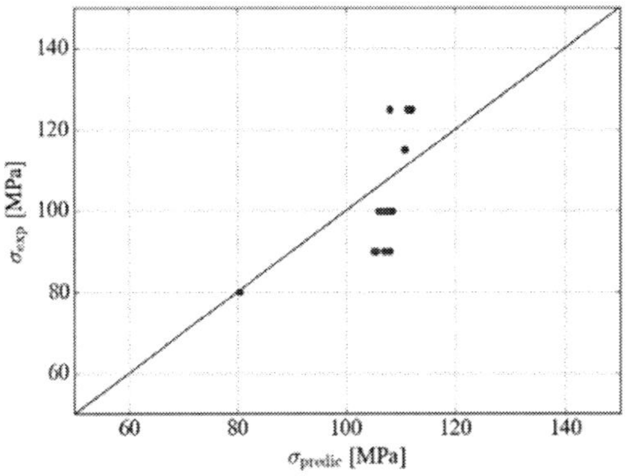

Fig. 5. Experimental stress amplitude, σ_{exp}, vs predicted stress amplitude, σ_{predic}, by *CP* Basquin parameter.

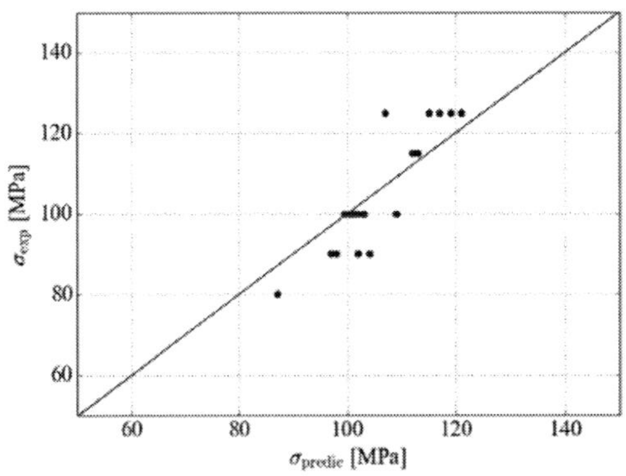

Fig. 6. Experimental stress amplitude, σ_{exp}, vs predicted stress amplitude, σ_{predic}, by *SWT* parameter.

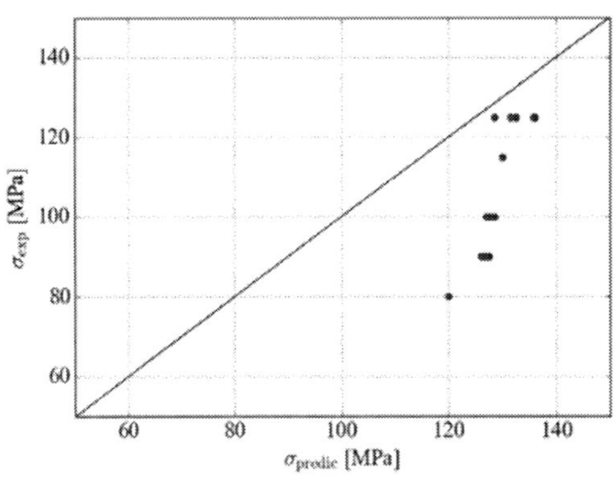

Fig. 7. Experimental stress amplitude, σ_{exp}, vs predicted stress amplitude, σ_{predic}, by *FS* parameter.

These results suggest that the *SWT* parameter can be a good indicator to predict the fatigue lifetime of butt-weld specimens.

However, the results obtained are based on the stress and strain state of only one element. In order to obtain more representative results, the next step of this research will be to evaluate the FIPs at all the elements of the welding cord.

In addition, we have analysed one loading case (uniaxial loading). As we have observed in the results, each FIP may be more suitable for a different loading condition. Thus, a wider analysis is recommended, taking into account different loading cases.

6 Conclusions

In this study different fatigue indicator parameters (*SWT*, *FS*, and *CP* Basquin) have been used in order to predict the fatigue life of butt-weld specimens. This study has been analysed with a numerical procedure proposed by the authors in order to predict the effect of RS on fatigue life.

The numerical results suggest that the *SWT* parameter is a suited FIP to predict the fatigue life of butt-weld specimens.

References

1. T. Lassen, N. Recho, *Fatigue life analyses of welded structures*. Iste London (2006).
2. M. Mousavi, M. Haeri, ISA T **50** (3), 409-418 (2011).
3. P. E. Murray, Weld. J. **81** (7), 125-131 (2002).
4. S. J. Maddox, *Fatigue strength of welded structures*, Woodhead publishing (1991).
5. A. Carpinteri, A. Spagnoli, S. Vantadori, Int J Fatigue **31**, 188-196 (2009).
6. P. Livieri, P. Lazzarin, Int. J. Fracture **133** (3): 247-276 (2005).
7. B. Brickstad, B. L. Josefson, Int. J. Pres. Ves. Pip. **75** (1):11-25 (1998).
8. M. Asadi, J. Goldak, J. Nielsen, J. Zhou, S. Tchernov, D. Downey, Int. J. Mech. Mater. Des. **5**, 353-364 (2009).
9. A. Paradowska, J. W. Price, R. Ibrahim, T. Finlayson, J. Mater. Process. Technol. **164**, 1099-1105 (2005).
10. C. M. Sonsino, Int. J. Fat. **31** (1), 173-187 (2009).
11. D. Radaj, C. M. Sonsino, W. Fricke, *Fatigue assessment of welded joints by local approaches*, Woodhead publishing (2006).
12. A. Lopez-Jauregi, I. Ulacia, J.A. Esnaola, D. Ugarte, I. Torca, J. Larrañaga, Int. J. Adv. Man. Tech. **76**, pp. 2117-2129 (2014).
13. A. Lopez-Jauregi, J. A. Esnaola, I. Ulacia, I. Urrutibeaskoa, A. Madariaga, Int. J. Fat. **79**, pp. 75-85 (2015).
14. Safety data sheet for Stargon 82, http:/www.cialgas.com/wpcontent/uploads/2012/09/STARGON-82.pdf. Accessed 11 June 2014.
15. J.A.Esnaola, I. Ulacia, A. Lopez-Jauregi, D. Ugarte. Residual Stress Pattern Prediction in Spray Transfer Multipass Welding by Means of Numerical Simulation. Book chapter. ISBN 978-953-51-5765-6. Accepted.

16. J. A. Esnaola, I. Ulacia, D. Ugarte, A. Lopez-Jauregi, I. Torca, J. Larrañaga, *Proceedings of the ASME IMECE* (2016).
17. J.A. Esnaola, A. Lopez, I. Ulacia, D. Ugarte, I. Torca, J. Larrañaga, *Materials today*, Interjoin (2016).
18. D. Socie, G. Marquis, *Multiaxial Fatigue*, SAE (2000).
19. A. Fatemi, D.F. Socie, Fatigue Fract Eng M. 187(1): 149-165 (1988).
20. R.N. Smith, P. Watson, T.H. Topper. J. Mater. **5**, No. 4, pp. 767-778 (1970).
21. A. Karolczuk, E. Macha, Int. J. Fract. **134**, 267-304 (2005).
22. I. Papadopoulos, Fatigue Fract. Eng. Mater. Struct. **21**, 269-85 (1998).
23. N. Meyer, Phd Thesis, University of Toledo, USA (2015).

A numerical analysis to investigate the spot weld local influence

Florian Mainnemare[1, 2, *], *Benoit* Delattre[1], *Ida* Raoult[1], *Olivier* Villars[1], *Pierre-Alain* Boucard[2], *Pierre-Alain* Guidault[2]

[1]PSA groupe, Scientific and Future Technologies Department, 78943 Vélizy-Villacoublay, France
[2] LMT - ENS Cachan - CNRS - Université Paris-Saclay, 61 avenue du Président Wilson, 94235 Cachan, France

Abstract. Relations between existing spot-weld finite element models and the mechanical behaviour of spot-weld are investigated for fatigue design application. Specifically, the different zones that constitute the numerical model are interrogated with respect to physical observation. In order to identify the Mechanical Influence Area of the Spot-Weld here designated as "MIASW" we propose a methodology to determine numerically its size. Hence, we develop a finite element coupon representing a spot-weld under a basis of four unit load cases. The coupon provides mechanical quantities of interest that allows us to characterize the spot-weld response that shell element are unable to represent. This response is crucial to be captured for further fatigue analysis.

1 Introduction

1.1. Background

Existing spot-weld models in literature have been developed for three main applications in the automobile industry. The first one is NVH (Noise, Vibration and Harshness), in a first approximation, it requires accuracy in the structural stiffness and kinematic [8]. The second application is fatigue design; it focuses on different quantities of interest such as the local stiffness and the stress distribution accuracy. The third application is crash analysis and requires nonlinear behavior for large deformation. Its complexity makes it very hard to unify a model with the first two applications, without developing a costly numerical model. However, there are "polyvalent" spot-weld models for both NVH and fatigue applications but those are usually a compromise between the two domains. The challenge is to represent both local and global stiffnesses, within an industrial constraint of minimizing the numerical cost of the model. In our case, although our main objective is to develop such a polyvalent model, we focus in this paper on spot weld models for static purpose and fatigue prediction. In this domain, the existing industrial models commonly represent sheets by shell elements because of their low numerical cost. Hence, we consider as spot-weld model all the numerical entities at the spot-weld location that are not part of the sheet model (it can be 1D or 3D element, kinematic coupling, different mechanical behavior).

This study aims to investigate the spot-weld local influence on structural sheet. For that purpose, we develop volumetric finite element models subject to classical in-service mechanical load. We introduce kinematic and static quantities in order to discuss the choices made in the literature. A final objective statement about the relevance in modelling different zones within the model is formulated.

1.2. Definitions

Geometrically, the spot-weld nugget is defined as the area delimited by the notch root. However, it has been observed physically that the spot-weld has a metallurgical influence zone on the nearby sheet larger than the nugget, through the Heat Affected Zone (HAZ) [7]. Thus, the spot-weld might affect the local behavior of the structure within a mechanical influence zone as well, through side effects for example. This influence zone could be larger than the nugget. At this point, we define such a zone as the Mechanical Influence Area of the Spot-Weld so-called "MIASW" (see Fig. 1.)

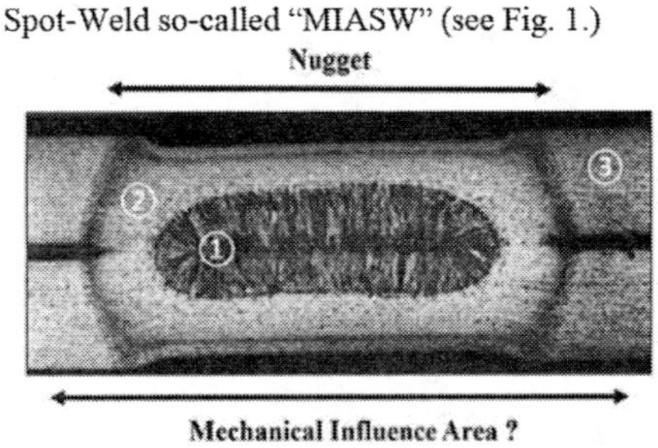

Fig. 1. Sectional view of a spot-weld for automobile industry purpose. 1/ Coarse grained Heat Affected Zone (HAZ) 2/ Sub-Critical HAZ 3/ Base metal.[7] The nugget is defined in between the notch roots. The Mechanical Influence Area of the Spot-Weld ("MIASW") cannot be defined geometrically but requires further mechanical analysis.

This raises the necessity to develop an accurate numerical footprint of the spot-weld model that characterizes the mechanical response unable to be

* Corresponding author: florian.mainnemare@mpsa.com

represented by shell element. We define the Footprint of the Spot-Weld Model ("FSWM" from now on) as the projection area of all numerical entities that constitute the spot-weld model. (see Fig 2.)

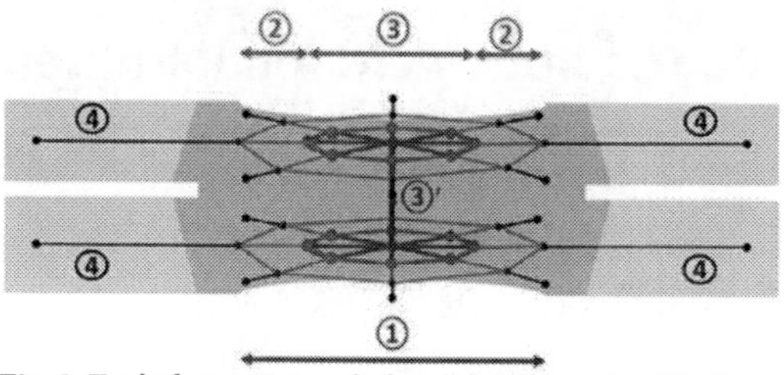

Fig. 2. Typical recent numerical model of the spot-weld where is represented: 1/ Footprint of the Spot-Weld Model ("FPSWM"). 2/ Peripheral zone. 3 & 3'/ Both compose the rigid core but could be defined with different numerical parameters. 4/ Shell elements representing the nearby sheet.

We observed so far in literature that the footprint of the existing models was a very scattered data. Indeed, it is mainly used as a calibration parameter to reach the reference stiffness for the joint. And it is mostly evaluated through a volumetric finite element analysis or experimental test. Therefore, the FSWM for fatigue analysis was shown to depend more on the mesh (size and orientation) than on the physical spot-weld influence area. To qualify this specificity we investigate in part 2.1 the choices made around the FSWM in literature. In part 3 we conduct a numerical study in order to estimate the MIASW both statically and cinematically. And thus, what should be the minimum dimension of the FSWM to be able to represent the MIASW. At this point, we could affirm that we need FSWM greater or equal to MIASW in order to represent accurately the mechanical response of the spot-weld that cannot be captured by the shell finite element model. However, we do not investigate here methods that could propose to grant a part of the mechanical accuracy encapsulated inside the spot-weld numerical model. But that would enrich by extrapolation for example, through an off-line additional step, the measured mechanical fields.

Another key point concerns the Rigid Core within the Spot-Weld Model ("RCSWM" from now on). introduced by multiple studies to represent numerically the geometrical stiffness induced by the spot-weld. The dimension of this rigid core is usually justified only through the spot-weld geometry aspect, without additional mechanical investigation to characterize such a behavior. In this study, after recalling in part 2.2 the different divisions between the rigid core and the peripheral zone within the FSWM. We propose in part 3 a method to justify the rigid core dimensions using mechanical quantities of interest.

1.3. Boundaries

Our approach is defined in a quasi-static environment with the hypothesis of linear displacement, and linear homogeneous and isotropic material. Justified by the assumption made in fatigue analysis.

Furthermore, we consider an academic problem with the assumption of no geometrical perturbation in certain vicinity of the spot-weld.

2 State of the art around spot-weld models

2.1. Existing dimensions of FSWM

We decided to distinguish the existing FSWM with respect to their dimension relatively to the nugget dimension, as it is a clear measurable entity. We distinguished three categories: no footprint, a footprint equal to the nugget and larger than the nugget.

2.1.1 No FSWM

The first group gathers the so-called Point to Point (P2P) models, joining the sheets through a 1D beam or rigid element. It results in a much localized influence area of the link. The beam parameter as the diameter, the second moment of inertia, the Young modulus are usually calibrated in order to match the global stiffness of volumetric [4] or experimental reference loading. These models involve an erroneous local stress distribution, overcome by analytical stress function or volumetric refined model associated to the loading case of the joint [9]. Moreover, a major drawback of this method concerns the numerical processing for the mesh congruent requirement.

2.1.2 FSWM equal to the nugget

The second group includes the models defined with a FSWM taken equal to the spot-weld nugget diameter. This connection area is usually defined with no more justification than geometry even if the axial symmetry of the nugget is not respected [10].

2.1.3 FSWM larger than the nugget

The third group collects the models with a FSWM larger than the spot-weld nugget diameter. For most of them, the extra area allows the spot weld model to transmit bending rigidity to the sheet. They use Rigid Body Element (RBE), kinematic coupling or shell elements to define the contact area [1]. The stiffness calibration is commonly based on the mechanical response of a volumetric structure or an experimental test. [13], [2]. The model developed by [6] could be considered to affect a larger zone as well as it uses eXtended Finite-Element to improve the spot-weld kinematic through analytical displacement functions spread below the nugget.

2.2 Rigid core inclusion

For geometrical reasons, the structural stiffness of the spot-weld nugget is higher than the rest of the structure. Hence, in the literature a large diversity exists in the modelling of this phenomenon. It appears that different material behaviors are sometimes attributed to the different regions of the spot-weld model. In regards to this distinction, we could classify the existing models into three categories.

2.2.1 Rigid core and elastic peripheral zone

The first category distinguish an infinitely rigid core on the one side and a peripheral elastic zone on the other side. It is for example the case for the Spider model where a central part is composed of a rigid element to link the facing sheets and 16 beam elements connected to the surrounding shell elements. We found a few comparable models in the literature but less widely used for industrial applications [12]. Similarly, the "umbrella" or "ACM1" model is made of a rigid central element and beam elements connect the core to the sheets [11]. In the same way, the Bolt Spot model developed by Renault is composed of a stiff beam and a shell core with adjustable properties (thickness and/or Young modulus) possibly different from the regular shell elements describing the sheets [3].

2.2.2 Elastic core and peripheral zone

We could introduce a second category, which distinguishes a spot-weld core and a peripheral zone, but eventually characterized by different behaviors. For example the "ACM2" model developed by (Heiserer et Al) for the convenience of numerical implementation is made of a solid elastic hexahedral element to characterize the spot-weld core and is connected to the rest of the sheet through rigid body elements. The MSC Cweld model developed by NASTRAN software enters this category as well. It is an elastic beam connected to the two sheets through patches with local kinematical constraint.

2.2.3 No peripheral zone

Finally, a last category can be introduced for the models with no distinction between a spot-weld core and a zone of influence. The early developed and low cost "P2P" models are part of this category The "single brick" model does not make any area distinction neither. It is defined as a hexahedral element connecting two facing elements of both sheets. It requires a congruent mesh as well[10]. The model developed by[13] is characterized by a complete rigid zone linked to the surrounding shell elements of the structure. Although it is enhanced with analytical stress distribution function, no differentiation of area is proposed. Similarly, the model developed by [5] extends the linking area with analytical displacement function but does not explicitly distinguish a central part from a crown in the spot-weld model.

3 Method

In our study we first investigate whether the Footprint of the Spot-Weld Model ("FSWM") could be related to the Mechanical Influence Area of the Spot-Weld ("MIASW"). For that purpose, we use a fully volumetric finite element mesh of a circular coupon. This model is made of two sheets linked at their center by a spot-weld. We apply four unit loading cases to the sheet edges. Then, we extract both kinematic and static quantities in order to conclude on the size of the MIASW. As a second step, we question how to determine the size of the RCSWM. We proposed a method of determination of this zone through kinematic and stress distribution analyses.

3.1. Coupon characteristics

3.1.1 Determination of the MIASW

For the coupon dimensions, we elected a representative case of what could be found in an industrial context. The coupon has a total diameter of 34 mm for a 6 mm nugget diameter at the centre. This is based on the typical minimal distance from a spot-weld to another. Upper and lower sheet are respectively 1.5mm and 1.3mm thick, with a 0.2 mm inter-sheet length. The nugget diameter is 6 mm. Material parameters are taken equal to the one used for the sheet metal structure. The finite element model is constituted of 60096 solid quadratic hexahedral elements of type C3D20 in Abaqus/Standard (Version 6.16, Simulia Corp., RI, USA). It results in a little more than 1.5 million of degrees of freedom. The radial mesh size is 1 mm until the spot-weld radius at which it is refined to 0.2 mm. The mesh is axisymmetric above a radius of 0.1 mm. Finally, each sheet is made of 8 elements within the thickness.

3.1.2 RCSWM

The same numerical model is used for this part except that a kinematic coupling is defined between all the nodes within a cylindrical central volume. This cylindrical central volume shares the same axis as the coupon, with a variable diameter and a height from the top surface to the bottom surface of the coupon.

3.2. Boundary conditions

Boundary conditions are defined identically for the two analyses.

3.2.1 Loading of the coupon

The following procedure is defined in order to avoid any residual load in the loading zone area.
For both the upper and lower sheet individually (see Fig 3.a), b))., we define a kinematic coupling constraint between a central geometrical point and all the nodes at the radius: $R = R_{Coupon}$.. The two geometrical points lie at the same position: the intersection of the coupon axis and the middle plane of the inter-sheet. But they undergo different couplings. Structural loading is applied through the point coupled to the upper sheet. While the dual degree of freedom of the point coupled to the lower sheet is set equal to zero. The constrained degree of freedom through the kinematic coupling is defined as it is opposed to the loading.

Rigid body movements are locked up through isolated nodes (N_x and N_y) with respect to a minimum additional boundary conditions (see Fig 3.) The resulting loads within these isolated nodes are verified to be close to zeros in order to confirm that they lock only rigid body movement without inducing interfering loads.

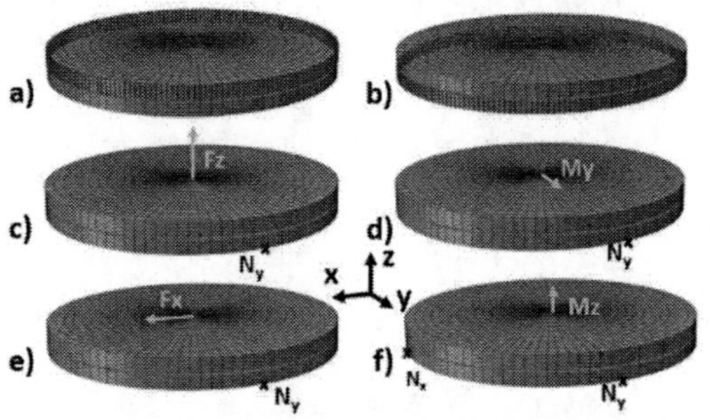

Fig. 3. Coupon boundary conditions and loading cases of the coupon. a) Upper crown connected to geometrical center point. b) Lower crown connected to geometrical center point. c) Tensile loading case in z direction (Fz). d) Bending loading case in y direction (My). e) Shear loading case in x direction (Fx). f) Torsion loading case in z direction (Mz)

3.3. Quantities of interest

In this part, we introduce mechanical quantities of interest in order to be able to quantify the MIASW. The aim is to investigate in which area the necessary conditions for the shell theory to apply are not satisfied. Thus, we decide to focus on two major necessary conditions of the shell theory application. The first one is based on the kinematic hypothesis for shell theory for both Kirchhoff-Love and Reissner-Mindlin cases. It stipulates that every fiber normal to the mean surface of the shell remains straight after deformation. The second hypothesis states that stress distribution is planar. More precisely, regarding the Kirchhoff-Love shell theory definition, out-of-plane components of the stress tensor are negligible compared to the in-plane stress components. Thus, the stress tensor becomes:

$$\underline{\underline{\sigma_{KL}}} = \begin{bmatrix} \sigma_{xx} & \sigma_{xy} & 0 \\ & \sigma_{yy} & 0 \\ sym & & 0 \end{bmatrix} \quad (1)$$

Differently, for the Reissner-Mindlin theory, the out-of-plane shear stress is taken into account. It leads us to:

$$\underline{\underline{\sigma_{RM}}} = \begin{bmatrix} \sigma_{xx} & \sigma_{xy} & \sigma_{xz} \\ & \sigma_{yy} & \sigma_{yz} \\ sym & & 0 \end{bmatrix} \quad (2)$$

In order to evaluate the validity of these conditions, we introduce the following quantities of interest (4) & (5).

3.3.1 Kinematic hypothesis validity

The coupon is meshed to obtain aligned nodes along the thickness of the sheets. Therefore on the deformed model, we are able to compute the linear regression of the fibres normal to the mean plane for each sheet. We designate them as "numerical thickness fibres" as they are the consequence of the numerical finite element modelling. The linear fitting regression is carried out through the Principal Components Analysis method. Then we are able to extract the equation of the best linear fit and indicators about the quality of the linear regression (disparity, standard deviation, Pearson's correlation coefficient). We decide to elect a first order statistical indicator to represent with clarity the deviation between the thickness fibre and its linear fit. Hence, we introduce the distance "d" between the numerical thickness fibre and its linear regression under the following definition: it is the distance between a point within the thickness fibre and its orthogonal projection on the linear (see figure 4.). In the following, we will thus quantify the amplitude of the buckling of the normal fibre through the distance quantity δ:

$$\forall j \in [1, \dots, N_{Fibres}],\ \delta_j = \max\{d_i, i = [1, \dots, N^j_{nodes}]\} \quad (3)$$

Where N_{Fibres} is the number of fibres of the coupon and N^j_{nodes} is the number of nodes of fibre j.

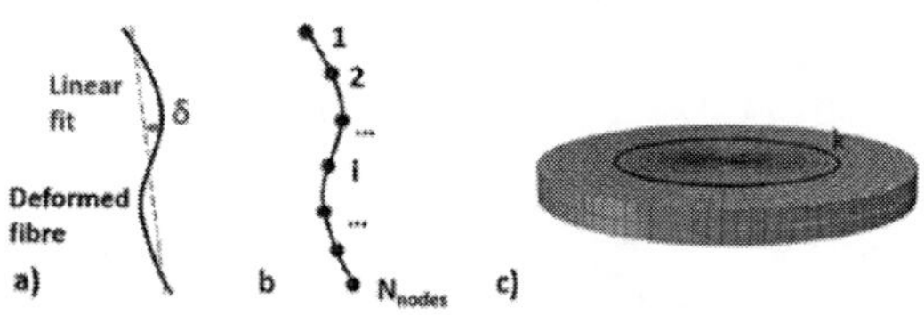

Fig. 4. a) Measure of the distance between a deformed thickness fibre and its linear fit. b) Nodes distribution along a thickness fibre j included in a fixed radius k. c) fixed radius k of the upper sheet

3.3.2 Static hypothesis validity

In order to evaluate the validity of the planar stress hypothesis, we need to introduce a quantity of interest that indicates the amplitude ratio between in-plane and out-of-plane stresses. Thus for every thickness fibre at a fixed radius, we evaluate the value of the highest ratio between the maximum out-of-plane stress to the maximum in-plane stress of each node of the thickness fibre. In other terms, the stress ratio S_{OP} is defined as:

$$S_{OP} = \max_j(\max_i[\frac{\sigma_{zz}^{ij}}{\sigma_{IP\,maxi}^{ij}}]) \quad (4)$$

Where σ_{zz}^{ij} and $\sigma_{IP\,maxi}^{ij}$ are respectively the σ_{zz} component of the stress tensor and the maximum In-Plane stress components. For the node i belonging to the thickness fibre j (see Fig 4.b).

Similarly, we introduce:

$$S_{OPS} = \max_j(\max_i[\frac{\sigma_{OPS\,maxi}^{ij}}{\sigma_{IP\,maxi}^{ij}}]) \quad (5)$$

Where $\sigma_{OPS\,maxi}^{ij}$ is the maximum Out-of-Plane stress component of the node i belonging to the thickness fibre j (see Fig 4.b). Including the Out-of-Plane Shear (OPS) stress components. Hence, we could evaluate S_{OP} and S_{OPS} for each radius along the coupon.

4 Results

4.1. Mechanical quantities of interest for MIASW estimation

4.1.1 Kinematic hypothesis validation

In this part we aim at evaluating the area around the spot weld where the kinematic hypothesis of a thickness fibre remaining straight during deformation is not respected. First, we plot the evolution of the distance quantity δ along the coupon, for every loading case.

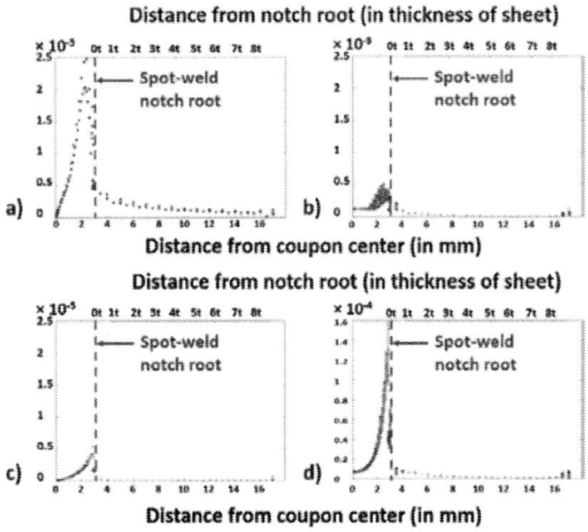

Fig. 5. Amplitude of the quantity of interest δ, according to the radial distance to the middle of the coupon. Radial distance is both expressed in millimetres and in thickness of the sheet unity. The response is given for the four load cases: a) tensile load; b) bending load; c) torsion load; d) shearing load.

As a first observation, the phenomenon seems to have a relatively small order of magnitude. Indeed, we observe that the shear loading case induce the largest thickness fibre buckling. For that case the quantity δ reaches a maximum of 1.6 10^{-4} mm of deformation for a 1.5 mm thick sheet. The thickness fibre buckling is mostly localized within the nugget area and drops quickly beyond the spot-weld notch root. We decide to analyse the buckling relatively to the global deformation of the coupon. Thus, we plot (Fig 6.) the evolution of the ratio between the quantity of interest δ and the maximum displacement of the structure for each loading case.

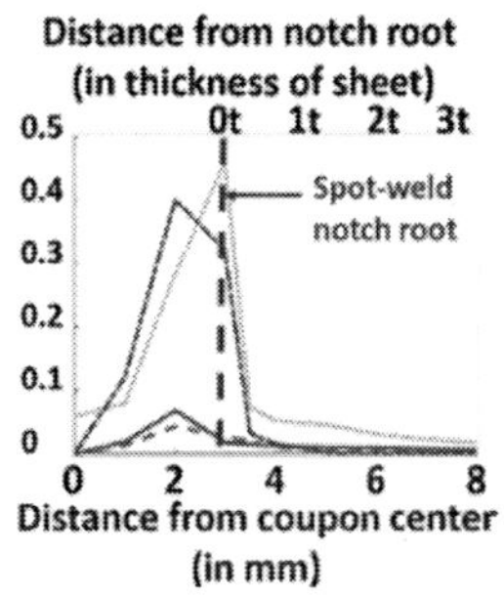

Fig. 6. Ratio (in %) between the quantity of interest δ to the maximum displacement of the structure each loading case. Amplitude of the quantity of interest δ, according to the radial distance to the middle of the coupon. Radial distance is both expressed in millimetres and in thickness of the sheet unity. The response is given for the four loading cases: ——: Tension; — —: Bending ; ······· : Shear load ; ‒ ‒ ‒ : Torsion

The ratio is found to reach at maximum 0.46% for the shear load but drops to 0.05% at a distance of one sheet thickness from the spot-weld notch root. For both tensile, bending and torsion loads, the ratio beyond one time the sheet thickness from the notch root is lower than 0.01%.

4.1.2 Static hypothesis validation

Similarly to the kinematic approach, we aim to evaluate in which area around the spot-weld the hypothesis of planar stress distribution is not respected. For that purpose, we plot in Fig 7. the evolution of the stress ratios S_{OP} (Fig 7.a) & b)) and S_{OPS} (Fig 7.c) & d)) along the coupon, for every loading case.

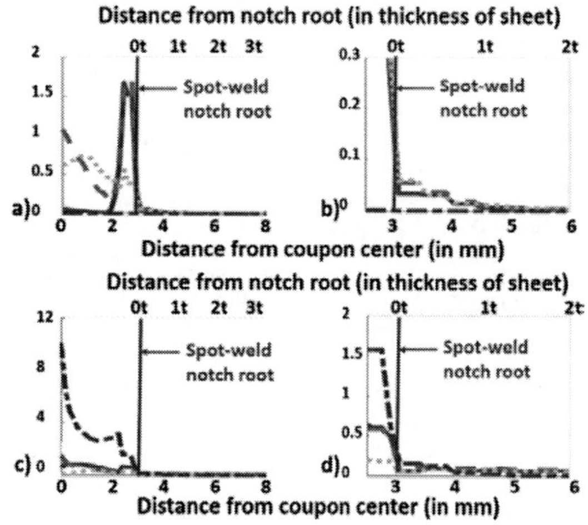

Fig. 7. a) & b): Amplitude of the stress ratios S_{OP} for each loading case according to the radial distance to the middle of the coupon. b) is simply a zoom of a)
c) & d): Amplitude of the stress ratios S_{OPS} for each loading case according to the radial distance to the middle of the coupon. d) is simply a zoom of c)
▬▬ : Tensile load; ▬ ▬: Bending load ; ······ : Shearing load ; ▬▬▬ : Torsion load
Radial distance is both expressed in millimetres and in thickness of the sheet unity.

We need to analyse separately the two stress quantities of interest.

The stress ratio amplitude of S_{OP} drops drastically beyond the spot-weld notch root. At one time the sheet thickness it is equal to 0.01±0.004 for the tensile, bending and shearing loading cases. For the torsion load this ratio is equal to 1.7 10^{-5}. Beyond one time the sheet thickness, S_{OP} decrease until reaching zero. For example it is equal to 7.2 $10^{-7} \pm 0.1\ 10^{-7}$ at three times the sheet thickness beyond the notch root.

The stress ratio amplitude of S_{OPS} drops drastically beyond the spot-weld notch root as well. However, at one time the sheet thickness beyond the notch root the stress ratio amplitude is found to be equal to 0.081±0.016 for the tensile, bending and shearing loading cases. For the torsion loading case, this ratio is equal to 0.02. Furthermore, when moving away to three times the sheet thickness beyond the notch root, the stress ratio S_{OPS} is found to be equal to 0.052 ± 0.014 for the tensile, bending and shearing loading cases. It remains constant until the end of the coupon which is six times the sheet thickness beyond the notch root. For torsion, the stress ratio S_{OPS} is found to remain close to zero from two times the sheet thickness beyond the notch root where it is equal to 1.3 10^{-5}.

4.1. Mechanical quantities of interest for RCSWM estimation

The aim of this part is to evaluate the equivalent size of the RCSWM that represents the mechanical response of the spot-weld correctly. For that purpose, we consider the coupon defined in part 3.1.1 as a numerical reference of the coupon defined in part 3.1.2. Then, the loading cases described in part 3.2.1 are applied. The aim is to find the rigid central cylinder that leads to a similar kinematic and load distribution than the reference coupon. Once the rigid core area is modelled with rigid elements, the numerically elected MIASW has to remain unchanged. Results for this study are expected soon to conclude on this topic.

5 Conclusion

In this study, we first raise a summary of the existing spot-weld models for fatigue design, in the automotive industry.

Then we describe a methodology to evaluate the dimension of the MIASW. We evaluate the area in which hypothesis of the shell theory are not respected. It allows us to give a numerical justification on the dimension of the MIASW. We are able to conclude that, with respect to the Reissner-Mindlin, thick shell theory, the MIASW lies approximately until one sheet thickness beyond the notch root of the spot-weld. Beyond this area, the shell theory hypotheses are validated. However, the results we obtained lead to a different conclusion if Kirchhoff-Love shell theory is used for shell modelling. In the far vicinity of the spot-weld (up to six times the sheet thickness beyond the notch root) the planar stress distribution hypothesis appears to be less clearly respected. The maximum Out-of-Plane stress amplitude represents 5% of the In-Plane maximum stress amplitude.

In literature, RCSWM is found to lead to different modelling within different areas of application. We introduce a methodology that could characterize its dimension with respects to mechanical quantities of interest. However, results concerning the investigation on the RCSWM dimension remain to be expected soon. It will make it possible to model this area with rigid elements for example, reducing the numerical cost of the model.

Finally, with this study we were able to propose a MIASW area of the spot-weld. However, the rest of the MIASW needs to be properly modelled in order to characterize the mechanical fields necessary for fatigue analysis. Indeed we know that this region of interest is crucial for fatigue analysis. Thus, the topic will be largely covered in further studies.

6 Bibliography

[1] *MSC.Nastran 2003: Linear Static Analysis Users Guide, MSC.Software Corporation, Santa Ana, CA 92707 USA.*
[2] J. Andersson and J. Deleskog. Fatigue life and stiffness of the spider spot weld model. Master's thesis, Chalmers University of Technology, 2014.
[3] Y. Baik. spot weld modeling for vehicule durability performance with ansa. In *4th ANSA & META International Conference*, 2011.
[4] J-F. Beniguel. *Modélisation du comportement mécanique des assemblages de structures soudées*. PhD thesis, Cachan, Ecole normale supérieure, 2002.
[5] P. Burry. *Quelques techniques d'analyse mécanique pour les structures automobiles*. PhD thesis, Ecole Polytechnique, 2007.
[6] P. Burry, M. Habibou, and G. Billotey. Eléments-finis étendus pour la modélisation des structures soudées par points. *Mechanics & Industry*, 9(2):103–108, 2008.
[7] S. Dancette, D. Fabregue, V. Massardier, J. Merlin, T. Dupuy, and M. Bouzekri. Experimental and modeling investigation of the failure resistance of advanced high strength steels spot welds. *Engineering Fracture Mechanics*, 78(10):2259–2272, 2011.
[8] RO. De Alba Alvarez, NS. Ferguson, and BR. Mace. A robust spot weld model for structural vibration analysis. *Finite Elements in Analysis and Design*, 89:1–7, 2014.
[9] F. Jean-Luc. *Dimensionnement à la fatigue polycyclique de structures soudées*. PhD thesis, Palaiseau, Ecole polytechnique, 1996.
[10] K. Pal and D.L. Cronin. Static and dynamic characteristics of spot welded sheet metal beems. *Transactions of the ASME*, 117:316–322, 1995.
[11] M. Palmonella, M. I. Friswell, J. E. Mottershead, and Lees A. W. Finite element models of spot welds in structural dynamics: review and updating. *Computers and Structures*, 83 (8-9):648–661, 2005.
[12] M. M. Rahman, R. A. Bakar, M. M. Noor, M.R.M Rejab, and MSM. Sani. Fatigue life prediction of spot-welded structures: a finite element analysis approach. *European Journal of Scientific Research*, 22(3):444–456, 2008.
[13] P. Salvini, E. Scardecchia, and F. Vivio. Fatigue life prediction on complex spot welded joints. *SEA 1997 Trans J Mater Manufact*, 106:967–75, 1997.

Seam weld shell element model for thin walled structure FE fatigue design

D. Turlier[1,*], *M.L. Facchinetti*[2], *S. Wolf*[1], *I. Raoult*[3], *B. Delattre*[3], *A. Magnin*[1], *N. Grimonprez*[2]

[1]LOHR Industrie, Duppigheim, France
[2]PSA Groupe, Voujeaucourt, France
[3]PSA Groupe, Vélizy Villacoublay, France

Abstract. In automotive industry, the FE fatigue analysis of mechanical structures made of steel thin walled parts and seam welded assemblies uses a dedicated technique based on shell element modelling for components and on 1D rigid elements for welds. This method has been validated with several intensive fatigue test campaigns using gas metal arc welded samples with different assemblies, for both bending and torsional loads. The fatigue results are relative to crack initiation at the weld toe with bending normal stress or with longitudinal shear stress for several load ratios. The purpose of the current work is the transposition of the initial method to another FE welded model. In a recent IIW guideline for the assessment of weld root fatigue, a shell element weld model has been proposed for seam weld fatigue assessment in case of weld root crack initiation. The idea is to analyse the possible extension of this FEA weld element model for weld toe fatigue analysis and several comparisons of stress results are detailed and discussed. Different fatigue criteria are used to verify the correlation with the fatigue test results. First, the approach is based on maximum shear stress, then structural stress is calculated and results are compared to IIW S/N curves.

1 Introduction

In automotive industry, the FE fatigue analysis of mechanical structures made of steel thin walled parts and seam welded assemblies uses a dedicated technique based on shell element modelling for components and on 1D rigid elements for welds [1, 2]. The stress tensor components are calculated at the weld toe location and the fatigue assessment is performed using a Dang Van maximum shear criterion. This method has been validated with several intensive fatigue test campaigns using gas metal arc welded samples with different assemblies, for both tensile, bending and torsional loads. The fatigue results are relative to crack initiation at the weld toe with bending normal stress or with longitudinal shear stress for several load ratios. The reports of those test campaigns contain all accurate details useful for a numerical correlation. The purpose of the current work is the transposition of the initial method to another FE welded model. In a recent IIW guideline for the assessment of weld root fatigue [3], a shell element weld model has been proposed for seam weld fatigue assessment in case of weld root crack initiation. The weld element model has been validated mainly for asymmetric fillet T joint weld assemblies [4, 5]. This kind of assemblies are widely used in industry. An inclined shell element row represents the seam weld with element nodes located at the mid-height of the weld legs. Weld elements are connected to the metal parts with a series of 1D rigid and MPC elements combination. The weld leg imprints on the metal sheet allow to define the leg contour. The idea is to analyse the possible extension of this FEA weld element model for weld toe fatigue analysis.

2 Weld FE Models

The models for the weld assembly use either Fayard's model which involves 1D rigid connections [2, 3], called "model 1" in the sequel, or the Seamsim method [4] for which the fillet weld is modelled using shell elements over the weld throat mid-height ribbon surface, named "model 2. Thin sheet structures are modelled using shell elements over their mid-surfaces.

2.1 model 1

PSA Group currently uses this method for weld seam representation. It involves a specific meshing rule and rigid links between welded steel sheets, as showed in the following figure (Fig.1).

* Corresponding author: virtual.testing@lohr.fr

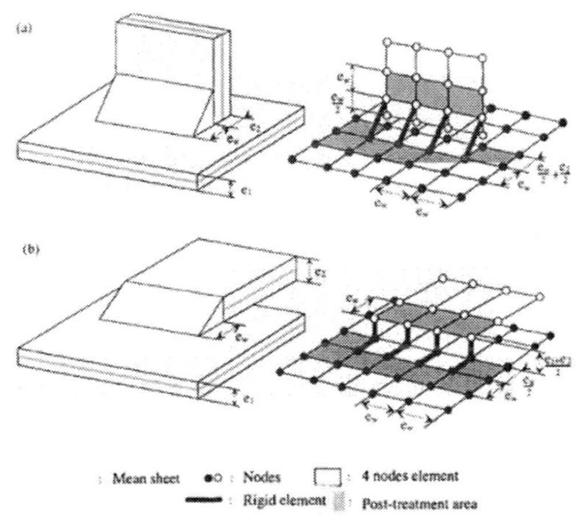

Fig. 1. Fayard model. (a) T-joint, (b) overlap joint.

For fatigue assessment, the stress tensor components are extracted from the centroid of the elements adjacent to the seam weld. The amplitude of the mesoscopic resolved shear stress τ_a is considered in combination with the maximum hydrostatic stress P,

$$\tau_{eq} = \tau_a + \alpha P \tag{1}$$

where α is a material parameter equal to 0.33, and the resulting equivalent stress is compared to the fatigue limit.

2.2 Model 2

Volvo car company did propose a different approach to model the weld seam connection by using a direct shell element connection at the assembly plates weld toe locations [6, 7].

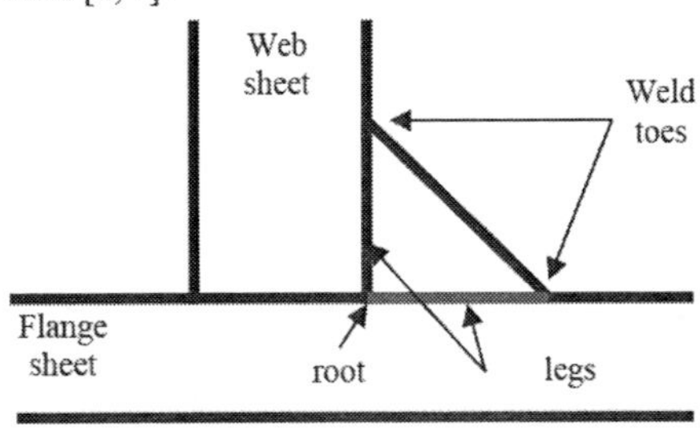

Fig. 2. Weld T joint.

The structural stress calculation is performed based on shell element nodal forces extraction. This technique reduces the dependency to the mesh size and the element quality. The technique has been enhanced by using a shell element row with connections at the weld leg mid-height location in order to extend the structural stress analysis for the weld root fatigue assessment [4, 5]. In order to explain the model, T joint assembly example is used (Fig. 2).

The assembly is composed of two steel sheets called web and flange, which are linked by a weld seam. First, metal sheet neutral fibres are extracted within the steel sheets and also the weld seam. The weld leg contours are projected over the steel sheet neutral fibre surfaces in order to delimit the weld toe contour (Fig.3).

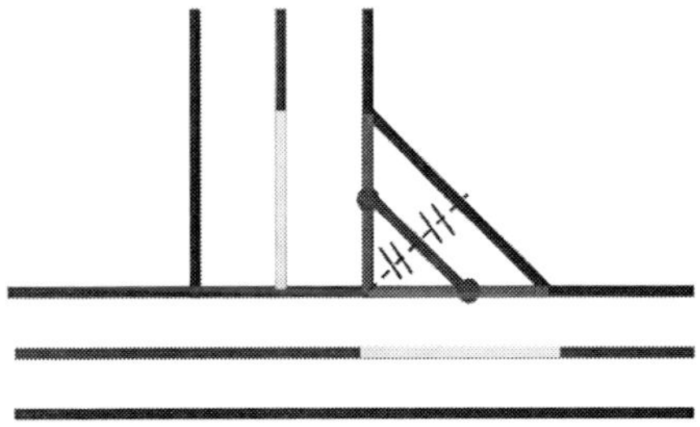

Fig. 3. Model preparation.

Neutral fibre surfaces are meshed using shell elements and the weld connection is performed by using both 1D rigid elements in red colour in Fig.4 and MPC – Multi Point Constraint connections in cyan colour.

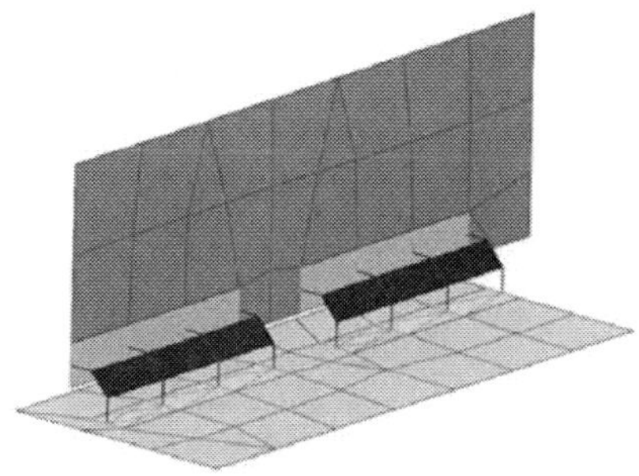

Fig. 4. Assembly model.

For fatigue assessment, at the weld toe, the structural stress approach is used as defined in [8] with shell elements. The structural stress is derived from the through thickness linear gradient stress value at the weld toe. It takes into account the assembly effect but it excludes the weld toe local singularity. As mentioned in [8], stress is calculated based on nodal force and moment extraction.

The fatigue structural stress amplitude is compared to a fatigue limit value.

3 Fatigue specimens and stress analysis

3.1 Specimens description

According to [1], the specimens have been tested with different loading conditions. They are structures made of mild steel sheets (σ_y = 170 MPa) and they are presented in Fig. 5. Residual stresses due to the process have been relieved by a thermal treatment. Specimen D only is made of high strength steel (σ_y = 450 MPa) and in this case two configurations have been tested, with the residual stresses either relieved or not.

- Specimen A is a double fillet weld T joint of a simple square plate 5 mm thick welded onto a 2 mm thin steel plate.
- Specimen B is a single side fillet weld joint of a U shape folded 3 mm thick steel sheet welded onto a 2 mm thick steel plate.
- Specimen C is a traversing "Tube" 3.5 mm thick and 39 mm extern diameter welded onto a 2 mm thick steel plate with a single side fillet welded joint.
- Specimen D is a U shape folded sheet 5 mm thick welded onto a 2.5 mm thick and 50 mm extern diameter tube.

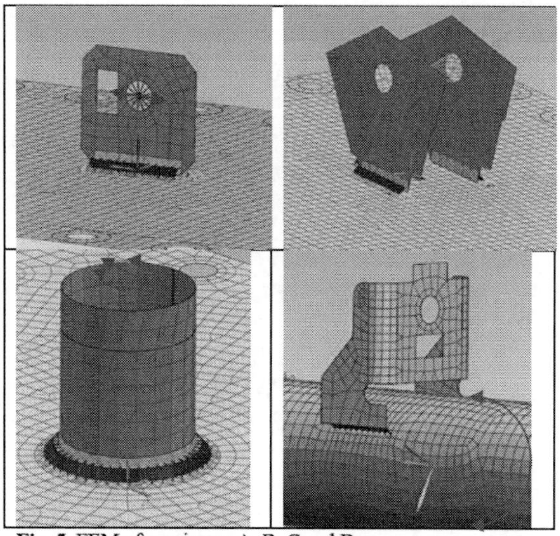

Fig. 5. FEM of specimens A, B, C and D.

Under a constant amplitude cyclic load, the failure criterion is expressed by the cycle number N corresponding to a significant increase in the crack growth rate, moreover correlated to a 30% decrease of the signal of a strain gauge located at a distance of 3 mm from the hot-spot.

3.2 Load and boundary conditions

In each case (except for specimen D), the support steel sheet is clamped on its edges by a series of screws.

Fig. 6. Specimen B constrain conditions.

In the case of specimen D, a fixed constraint has been applied on a side of the support tube (left side of the picture D, Fig.5.) and bending loads are applied on the tube's other side so that the applied force is orthogonal to the tube and in the direction of the drilled hole folded plate face.

3.3 Loading conditions

In specimen A, the load is applied through rigid connections at the centre node of the drill holes. The specimen has been tested in bending only and for load ratios $R=Fmin/Fmax$ equal to -1 and to 0, respectively. The specimen B has rigid connections between nodes on drill edges and a central node located on the drilling axis between the two parallel sides. Loads are applied on this central node. This specimen has been tested in bending and in tensile conditions for R equal to -1 and -0.5. Specimen C loads are applied on a defined cylindrical area beginning at a 44 mm distance from the tube basis to the end of the tube. This area is a cylinder on which bending forces are applied according to a geometrical repartition. Torques are also applied on this superior tube area. This specimen has been tested in bending and in torsion for load ratio equal to -1. The specimen D has been tested in bending only and for load ratios R equal to -0.2; -1 ; and -2 with stress relieved samples and for a load ratio equal to -1 for as welded samples.

3.4 Hot spot location

Cracks appear at weld seam toes in the thinner steel sheet for each specimen type.

In order to compare the stress states obtained with the two models within the cracking zone, the stress amplitude is evaluated over the curvilinear abscissa along the weld seam toe contour.

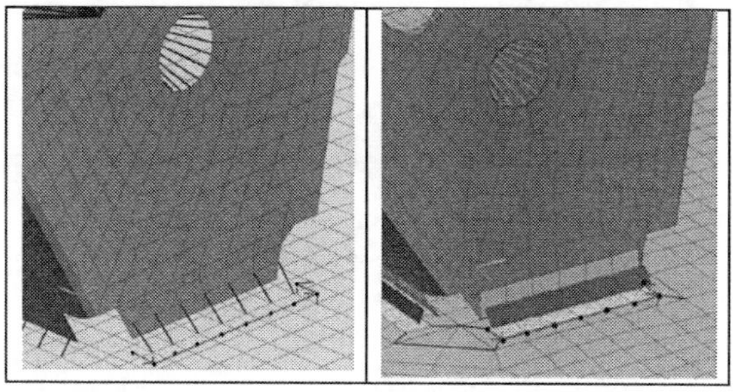

Fig. 7. Specimen B weld toe contour for both methods.

Based on model 1, the stress tensor is extracted at the weld toe contour at the corresponding element centroid. The maximum principal stress is used in case of tensile mode or bending mode tests. From maximum principal stress, the maximum stress amplitude can be evaluated considering minimum principal (out of plane) stress as null, so that:

$$\tau_a = \sigma_I /2 \quad (2)$$

For the torsional load specimen, the in plane shear stress is used as τ_a

$$\tau_a = (\sigma_I - \sigma_{II})/2 \quad (3)$$

Using the weld shell element in model 2, the structural stress is calculated thanks to the bending moment and the normal force perpendicular to the weld line. From nodal forces and moments at the node at the weld toe contour, the load densities are calculated by inverse matrixes of the element shape functions. This results in the line bending moment density m_x and a line normal force f_y (Fig.8.). The membrane normal stress σ_m is calculated accordingly (4); and the bending moment leads to the bending normal stress σ_b (5), where t is the steel sheet thickness. The structural stress is the combination of the two normal stress components.

$$\sigma_m = f_y/t \quad (4)$$

$$\sigma_b = (6 * m_x)/t^2 \quad (5)$$

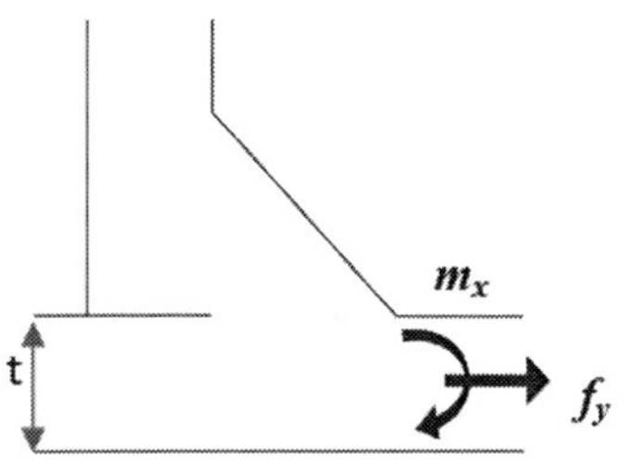

Fig. 8. Plate loads at the weed toe.

For the two methods, the corresponding design stress is plotted along the weld toe curvilinear abscissae, and both indicate the same distribution with the extremum value at the crack initiation location as shown in Fig.9., for specimen B, where hot spot is located at the seam weld termination.

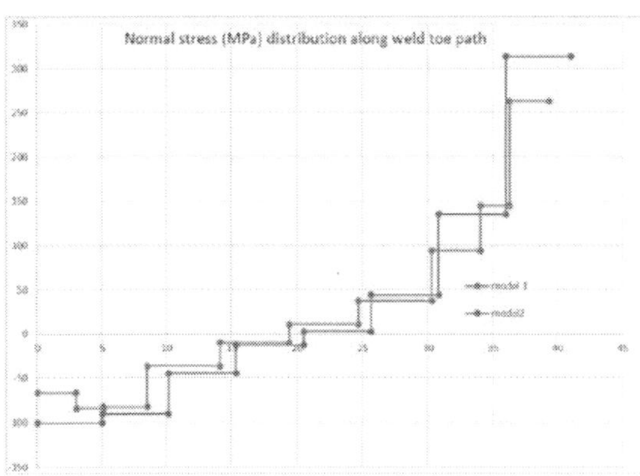

Fig. 9. Specimen B weld toe stress distribution.

The two methods give similar design stress value for specimens A and B. In the case of the torsional specimen C, the shear stress at the weld side outer face of the shell element, a different stress value was observed. In model 1, the calculated stress is much higher than in model 2. For a 1000 Nm torque application the shear stress distributions over the weld toe contour are illustrated in Fig.10. In fact, due to cantilever rod connection, the torsion torque transmission generates nodal moments as illustrated in Fig.11. The longitudinal shear load transmission with the induced nodal moment generates a circumferential waving perturbation and in-plane opposite bending nodal moments at the shell element edge at the 1D connection side.

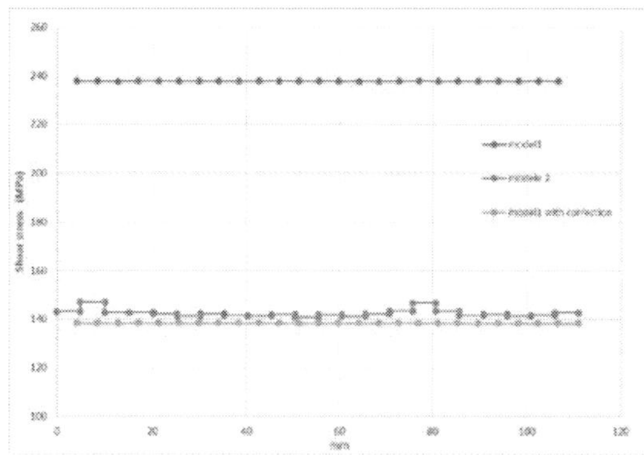

Fig. 10. 1000 Nm torsion shear stress distribution specimen D.

This numerical effect may be vanished by extracting stress tensor over the shell element mid-surface. Comparing mid-surface shear results between the two models, the shear stress distributions are now similar as illustrated in stress value with correction in Fig.10.

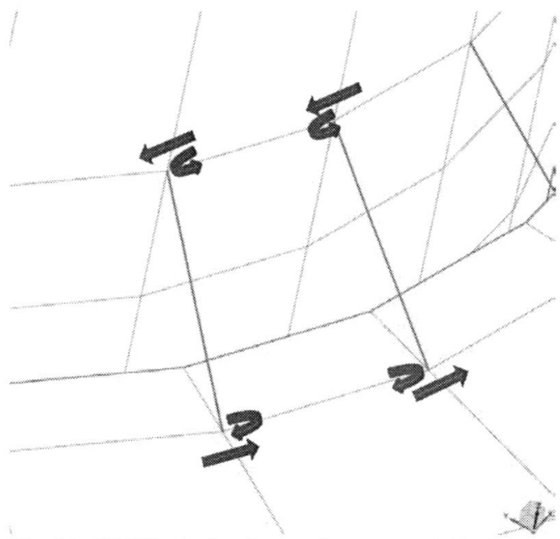

Fig. 11. 1000 Nm torsion, force and moment nodal loads.

This effect is reduced when using the model 2 where the 1D element node projection is only half of the plate thickness.

In order to compare the two models, the principal stress values are extracted at each specimen hot spot location. For bending and tensile load specimen, values are extracted on the top face of the element at the weld joint side and for torsional load, stress is extracted at the mid height shell surface. In order to compare the two models, the principal stress tensors at the hot spot location are evaluated and the values are given in Tab.1.

Tab.1. Principal stress tensors at hot spot location.

		load 100 daN		
		model 1	model 2	%
A bending	σI	310.0	290.56	-6.27
	σII	180.0	152.1	-15.5
B bending	σI	330.0	318.6	-3.4
	σII	210.0	216.7	3.2
B tensile	σI	273.0	230.0	-15.7
	σII	210.0	180.8	-13.9
C bending	σI	299.0	302.6	1.2
	σII	110.0	131.0	19.1
D bending	σI	98.0	112.0	14.3
	σII	23.0	20.0	-13.0

The first four load cases of table 1 show a consistent correlation between model 1 and model 2 results. Both models are able to represent the same multiaxial stress state at the weld toe within an overall scatter above 19%. The historical proposal of Fayard is definitely proven effective for fatigue computation despite of the use of alternative seam weld FE model and solver, respectively. Even if it is the calibration of the model versus experimental results which most makes a design criterion effective, here knowing that the model is robust is a good point to note.

3.4 τa-N Graphs

The τ_a-N graphs can be drawn, where τ_a is the maximal shear stress amplitude in the critical zone from the specimen and where N is the cycle number after which one gauge located in those same critical zones during test phase has lost 30% of the initial signal.

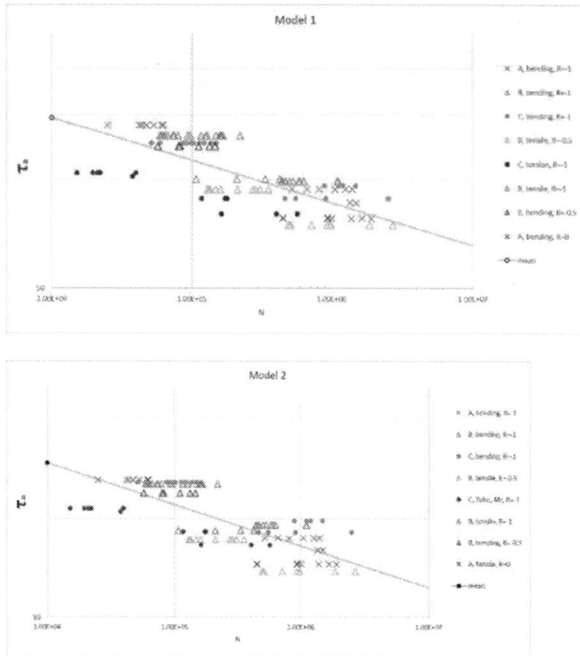

Fig. 12. τ_a-N for both models.

We notice that the range level between the two methods are quite similar. Average slopes also are similar: 8.55 for model 1 versus 7.94 for model 2. The two modelling methods lead to very similar results.

3.5 S-N Graphs

The aim of this section is to analyse the test results in terms of structural stress; only normal stress at the weld toe hot spot is taken into account or mid surface shear stress for the torsional specimen. For each specimen type, each solicitation type and each load ratio, an S/N curve is plotted. Comparisons between specimens but also between different load ratios and solicitation type can be provided.

3.5.1 Mean stress effect

In order to study the mean stress influence, the data from specimen D are analysed because it offers the most diversified load ratio results; R=-0.2, -1, -2 with relieved residual stress and R=-1 without relieved residual stress.

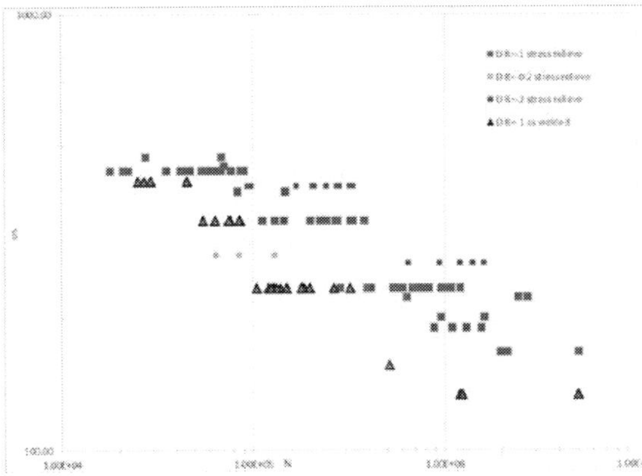

Fig. 13. σ_s-N specimen D.

The S/N curve slope for no stress relieved specimen is m=4, for stress relieved specimen m=5. For each load ratio category an S/N curve is established and a 2 000 000 cycles stress amplitude is evaluated and given in Tab.2.

Tab.2. 2 million cycles stress amplitude by load ratio.

	load ratio	m	σ_s
stress relieved	R=-1	5	201.64
stress relieved	R=-2	5	249.74
stress relieved	R=-0,2	5	151.58
as welded	R=-1	4	134.39

The mean stress influence factors from the test results are compared to IIW recommendations [8]. For stress relieved specimen, the mean stress correction factor is -0.4 R+1.2 when R varies from 0.5 to -1. The test result for R=-1 is taken as reference, the correction is applied and it fits the test result when R equals to -0.2, see Fig.14.

IIW correction limits mean stress effect factor to 1.6 when R reaches-1, but tests show that fatigue mean stress influence is still present for higher compressive mean stress when R=-2.

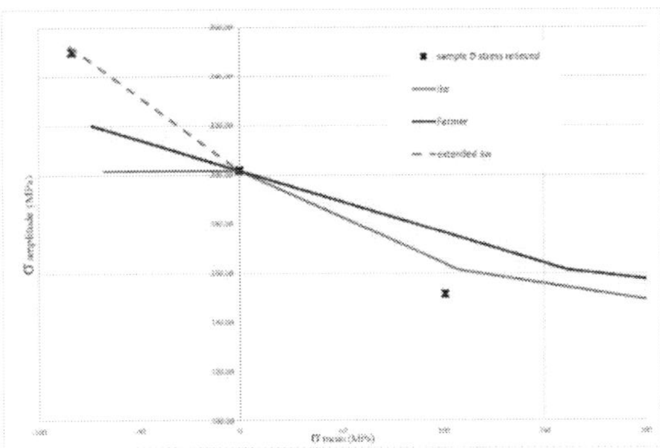

Fig. 14. Mean stress effect corrections and specimen D results.

In [6], Fermer established a correction method for load ratios for two R domains; R being negative or R being positive, with no specific R value limits. This mean stress influence is illustrated in Fig.12 and it does not have enough effect to fit the test results.

A proposal is to extend IIW recommendation R correction formula down to R=-2. The resulting correction does fit with results, and the factor related to R=0.2 to R=-2 is 1.56 for a test result factor equals to 1.65.

For R=-1, the stress relieved effect is quantified by a factor 1.3 for thin metal sheet assembly and 1.6 for 3D thick component in [8], the test leads to a factor equals to 1.56, which is consistent.

When considering IIW mean stress correction the estimated structural stress amplitude for 2 million cycles would be for R=0.5, 126 MPa.

3.5.2 IIW structural stress comparison

For structural stress, the IIW guideline, [9], proposes FAT90 as design S/N curve for load carrying fillet weld and FAT100 for no load carrying fillet welds. Those FAT values correspond to a stress range limit (max-min) for 2 million cycles, a load ratio R=-0.5 and a survival probability of 97.5% and a slope $m = 3$.

To compare with the test results, mean stress effect correction for R=0.5 is performed and corrected structural stress results are displayed in Fig.15.

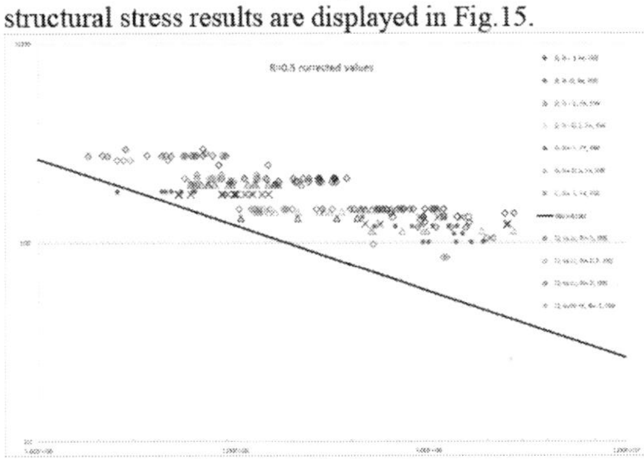

Fig. 15. Structural stress amplitude vs N.

Test results show much higher stress limits than IIW design values. For low cycle number, FAT90 is not conservative. In case of low strength steel, as mentioned in IIW recommendations, the design curves are appropriate on the condition that the stress range should be within the limits of the elastic properties of the material. The range of the design values of the stress range shall not exceed 1.5 times the yield strength.

High value for high cycle number is dedicated to the automotive thin sheet assembly, and the associated weld quality and may not be transposed to other applications. At the hot spot locations, the metal sheet thickness is either 2 or 2.5 mm, the thin walled structure has a thickness improve effect which might be investigated.

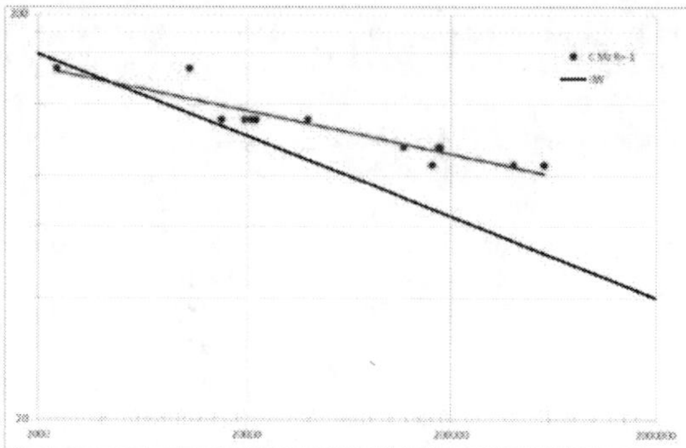

Fig 16. Shear stress amplitude τa-N.

For shear stress, IIW S/N curve is FAT80 with $m = 5$ and it is given for nominal stress for fillet weld in [8]. Torsion test shear stress mid-plane values are given in Fig. 16. The fitted S-N curve slope is nearly 9. IIW design S-N curve is also conservative as long as the number of cycles is higher than 100 000 cycles.

4 Conclusions

For fillet weld fatigue assessment, Fayard model (model 1) has been validated on the basis of an intensive testing campaign using four welded specimen configurations.

A recent fillet weld shell element model (model 2) has been proposed and validated for weld root fatigue assessment using structural stress approach. The aim of the work has been to verify its use for automotive weld assembly fatigue assessment.

Based on the four specimens, the comparison is performed by generating the stress density distribution at the weld toe and the hot spot location is correlated.

In the case of tensile and bending test configurations, the principal stress components calculation show a consistent correlation between model 1 and model 2 results. Both models are able to represent the same multiaxial stress state at the weld toe within an overall scatter of 19%.

The calculated fatigue shear stress Woehler curves give very similar distribution for the two modelling techniques. The historical proposal of Fayard is definitely proven effective for fatigue computation despite of the use of alternative seam weld FE model and solver, respectively. Even if it is the calibration of the model versus experimental results which most makes a design criterion effective, here knowing that the model is robust is a good point to note. Conversely, the last configuration relative to torsion load specimen highlights a major difference indeed. This is due to two different shortcomings already detailed in §3.3, i.e. the shear stress distribution within the tubular member thickness (Fig.11) and a circumferential waving perturbation at the two sides of the weld toe connection, respectively. Thus, in this very particular load case, the historical proposal of Fayard suffers from a relevant over-estimation of the stress state, mainly driven by a low-level numerical instability of the FE model. Actually, at present it would be an easy and hasty judgment to simply announce that the Fayard framework is non-conservative. Further impact analysis is expected before stating a definite conclusion. In this case, the correlation of the two modelling approaches is obtained by using a local-nominal shear stress value at the connected plate from either the element mid-plane stress tensor or by the shear nodal force transformation.

The analysis of the tests result using the structural stress method leads to several issues. The mean stress influence factor proposed by IIW recommendation may be extended for high compressive mean stress with a continuously benefit effect. Test results are compared to IIW design S-N curves which are very conservative and not appropriate for automotive applications. This is due to the use of specific steel grades in thin steel sheet assembly with high quality process.

1. J.L. Fayard, "Dimensionnement à la fatigue polycyclique de structures soudées ",Thèse Ecole Polytechnique (1996)
2. K Dang Van, A Bignonnet, J.L. Fayard, "Assessment of welded structures by a structural multiaxial fatigue approach", Biaxial/Multiaxial Fatigue and Fracture, (Elsevier Science Ltd. And ESIS. Publication 31, 2003)
3. W. Fricke, "IIW guideline for the assessment of weld root fatigue". Weld World 57(6):753–791, 2013 doi:10.1007/s40194-013-0066-y
4. D. Turlier,P. Klein, F Bérard, "Seam Sim" method for seam weld structural assessment within a structure FEA". Proc. Int. Conf. AWST 2010, (GEDIK Education foundation, Istanbul, pp 651–658 2010)
5. D. Turlier, P. Klein, F. Bérard, "FEA shell element model for enhanced structural stress analysis of seam welds", Weld World (2014) 58: 511. doi:10.1007/s40194-014-0134-y
6. M. Fermér, M. Andreasson, B. Frodin, "Fatigue life prediction of MAG-welded thin-sheet structures", (SAE Technical Paper 982311, 1998).
7. M. Fermér, H . Svensson, "Industrial experiences of FE-based fatigue life predictions of welded automotive structures", (Blackwell Science Ltd. Fatigue Fract Engng Mater Struct 24, 489–500, 2001)
8. A. Hobbacher, "Recommendations for Fatigue Design of Welded Joints and Components", 2nd edition, Springer International Publishing, 2016. doi: 10.1007/978-3-319-23757-2
9. E. Niemi, W. Fricke, S.J. Maddox, Structural Hot-Spot Stress Approach to Fatigue Analysis of Welded Components - Designer's Guide, Second Edition, Springer Nature Singapore Pte Ltd. 2018. doi:10.1007/978-981-10-5568-3

Numerical Simulation of the Fatigue Behaviour of a Friction Stirred Channel Aluminium Alloy

Catarina Vidal[1], *Ricardo* Baptista[2], and *Virgínia* Infante[3,*]

[1]Instituto de Plasmas e Fusão Nuclear, Instituto Superior Técnico, Universidade de Lisboa, Av. Rovisco Pais, 1049-001 Lisboa, Portugal
[2]ESTSetúbal, Instituto Politécnico de Setúbal, Campus do IPS, Estefanilha, 2914-508 Setúbal, Portugal
[3]LAETA, IDMEC, Instituto Superior Técnico, Universidade de Lisboa, Av. Rovisco Pais, 1049-001 Lisboa, Portugal

Abstract. Friction stir channelling (FSC) is one of the most promising innovations based on the friction stir concepts. FSC uses a non-consumable tool, similar to that used to perform the friction stir welding (FSW) process, to produce continuous and integral channels, with any path, in monolithic aluminium alloys components in a single step. This paper presents the mechanical behaviour of friction stirred channel components in particular the fatigue resistance of the channels analysed. Finite element analyses were performed taking into account the maximum stress values applied in experimental fatigue tests of the aluminium components. The energy release rate for different crack lengths was obtained using the software ABAQUS. The fatigue crack growth curve was established according to the Paris Law, and the crack propagation direction was calculated using the maximum tangential stress (MTS) criterion. During the analysis it was observed that the critical areas are located in the vicinity of the channels corners, as it was observed in the fracture surfaces of the experimental fatigue test results.

1 Introduction

The disruptive nature of the friction stir welding (FSW) process has led to many developments from the research and technological development institutions and industry. The FSW technology has been subjected to the most demanding quality standard requirements and used in challenging industrial applications over a wide range of structural and non-structural components, mainly in aluminium alloys. The disruptive character of the FSW process is emphasized by the numerous technic-scientific papers and patents published and the several friction stir based technologies that are being invented and developed since the original patent has been issued [1-4].
The friction stir channelling (FSC) technology, proposed for the first time in 2005 by Mishra as a method of manufacturing heat exchanging devices [5], is one of the most promising innovations based on the friction stir concepts. FSC uses a non-consumable tool, similar to that used to perform the FSW process, to produce continuous and integral channels, with any path, in monolithic aluminium alloys components in a single step.
The FSC technology involves complex interactions among a multiplicity of simultaneous thermo-mechanical processes. These interactions affect the heating and cooling rates, the plastic deformation and flow, the dynamic recrystallization phenomena and the mechanical integrity of the channel.
A typical FS channel cross-section macrograph is presented in Figure 1.

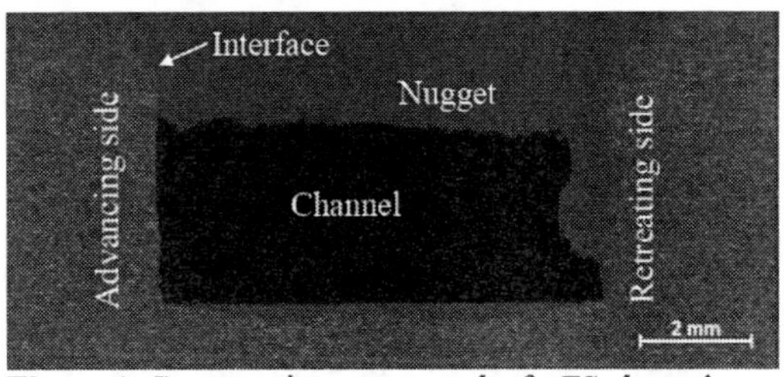

Figure 1. Cross-section macrograph of a FS channel.

The fatigue process can in general be divided into two main steps [6]: initiation of micro-cracks due to local accumulation of dislocations, high stresses at local points, plastic deformation around inhomogeneous inclusions or other imperfections; crack propagation, which causes permanent damage. The position and mode of fatigue crack initiation depends on the microstructure of the material, the type of the applied stress and geometry of the specimen [7]. The crack initiation periods can be very different and cracks can be initiated either on or under surface. Fatigue cracks initiation can be observed in stress concentration points where the

* Corresponding author: virginia.infante@tecnico.ulisboa.pt

resistance area is smaller. The number of cycles, N_f required for fracture can be determined from the number of stress cycles, N_i, required for the appearance of the initial crack in the material and the number of stress cycles, N_p required for a crack to propagate from the initial to the critical crack length, when the final failure can be expected to occur ($N_f = N_i + N_p$).

The interface between two different materials, is considered to show a weak behaviour, when compared with the base material or the added material. Whether one is considering a crack between a base material and a coating [8], a crack between to dissimilar material joined by friction stir welding [9] or a crack between two non-metallic materials [10], crack propagation rate on the interface will always be higher. This is attributed by Shah et al. [11] to the mismatch in the Young's modulus between the different materials. Therefore, studying the crack propagation along material interfaces is a complex mater and requires the use of different fracture parameters, as the energy release-rate, when considering crack propagation between to different elastic layers [11] or friction stir welds [9].

When a crack propagates along an interface between two different materials, the crack always behaves in mixed mode. Even in a pure mode I problem, the mismatch between the two materials Young's modulus, disrupts the symmetry of the problem. The crack tip is then always subjected to normal and tangential stresses. Therefore, crack propagation will occur in Mode I and Mode II, but these modes cannot be uncoupled. In other words, crack propagation will always be depended on the crack length and the materials properties mismatch. Unlike linear homogeneous conditions, the stress field near the crack tip does not show an inverse square root singularity. The stresses near the interface crack tip show an oscillatory behaviour (eq. 1), characterized by an oscillatory index (eq. 2), which is a function of the two materials properties.

$$\sigma \sim r^{-\frac{1}{2}+\epsilon} \tag{1}$$

$$\epsilon = \frac{1}{2\pi} ln\left(\frac{1-\beta}{1+\beta}\right) \tag{2}$$

Where parameters β and α are interface parameters defined by Dundurs [12]. The elastic mismatch parameter β, is a function of the Poisson's ratio and the material shear modulus (eq. 3). While parameter α measures the relative materials stiffness (eq. 4).

$$\beta = \frac{\mu_1(1-2\vartheta_2)-\mu_2(1-2\vartheta_1)}{2(\mu_1(1-\vartheta_2)+\mu_2(1-\vartheta_1))} \tag{3}$$

$$\alpha = \frac{E_1-E_2}{E_1+E_2} \tag{4}$$

The behaviour of the stress field along the crack interface, was defined by Rice [13]. The oscillatory behaviour of the stress singularity, can be define by eq. 5, where the stress intensity factors for mode I and mode II, cannot be decoupled.

$$\sigma_{yy} + i\sigma_{yy} = \frac{K_I+iK_{II}}{\sqrt{2\pi r}} r^{i\epsilon} \tag{5}$$

Crack propagation on materials interface is therefore more complex to analyse, as it requires the know the solution for the previous coupled stress intensity factors. Kakiuchi et al. [9] and Xuan et al. [14] have successfully used the energy release rate, G, to study interface crack propagation. This fracture parameter can be easily calculated by finite element method (FEM) and is more suitable for this type of problem. Considering the energy release rate for crack propagation, one can use the Paris Law in the following form:

$$\frac{da}{dN} = A\Delta G^n \tag{6}$$

Where A and n are material parameters.

In this work finite element analyses were performed to determine the energy release rate for a given crack length. Considering the alternative version of the Paris Law in eq. 6, the elapsed number of fatigue cycles for a given crack increment can be calculate. This iterative process was used to predict fatigue crack propagation behaviour on the materials interface.

2 Material and experimental details

2.1. Material

The strain hardened aluminium alloy AA5083 in the H111 condition was used in this investigation. Its chemical composition and main mechanical properties obtained at room temperature are presented, respectively, in Tables 1 and 2.

Table 1. Chemical composition of AA5083-H111 [15].

Element	Al	Mg	Mn	Fe	Cr
Weight (%)	93.38	5.26	1.02	0.19	0.15

Table 2. Mechanical properties of AA5083-H111 at room temperature [15].

Vickers hardness	92HV0.5
Yield strength	210 MPa
Ultimate tensile strength	375 MPa
Elongation	18.25 %
Young's modulus	68.9 GPa

2.2. Friction stirred channel production

The friction stirred channel was produced on 15mm-thick rolled plate along the rolling direction using an ESAB LEGIO™ 3UL numeric control equipment. Plunge and dwell periods (zero tool travel speed) were performed under vertical position control and processing period (non-zero tool travel speed) was carried out under vertical force control.

In order to obtain a plate with an internal channel, a selected set of FSC parameters was implemented as shown in Table 3. A modular H13 steel tool that enables internal forced refrigeration was used. This tool is based on three main components: body, shoulder and probe. All tool's arrangement used had a cylindrical probe with 8mm diameter and a 19mm diameter plane shoulder with one spiral. The cylindrical probe used had left handed threads along its length, with a trapezoidal profile, a thread pitch of 3mm and a depth of cut of 0.7mm. The geometric features of the shoulder used are presented in Table 4. The tool was rotated in the counter clockwise direction. The tool tilt angle was 0° for all the runs.

Table 3. FSC parameters.

Probe length (mm)	Tool rotation speed (rpm)	Tool travel speed (mm/min.)	Vertical force (kN)
8	1100	50	1.5

Table 4. Geometric features of the tool shoulder.

Spiral pitch	Spiral height (mm)
1	1

In order to model the real geometry of the channel for the numerical simulations, its geometric parameters were measured (Figure 3).

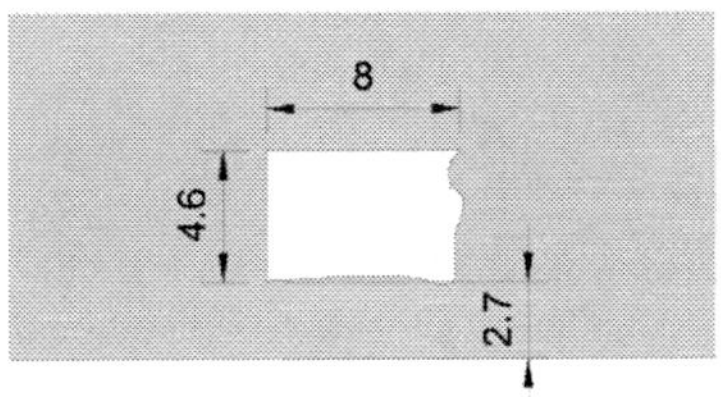

Figure 3. Geometry and dimensions of the channel produced by FSC.

2.3. Fatigue testing

Specimens for fatigue testing were produced according to the standard E 290-97a requirements as shown in Figure 4.

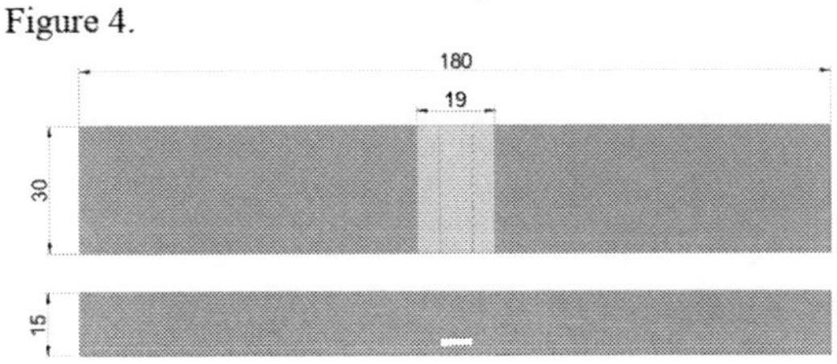

Figure 4. Geometry and dimensions of the fatigue specimens.

Fatigue tests were performed at room temperature on a servo-hydraulic Instron® 8502 testing machine with a load cell of 100kN. The run-out limit was set to $3x10^6$ cycles. The loading regime was a load controlled, sinusoidal constant amplitude waveform with a min/max stress ratio of R= 0.1, a frequency range of [5, 10] Hz and a four-point bending set up.
The applied load, P, was calculated by eq. 7:

$$P = \frac{\sigma b h^2}{3w} \quad (7)$$

where σ is the maximum stress, b is the specimen's width, h is the specimen's thickness and w is the smallest span distance (30mm).
The specimens were tested with the processed surface under tensile stress and loads were applied out of the processed zone. Tests were carried out with the maximum stress levels ranging from 30-150MPa, in order to determine the mean S-N curves.
After fatigue testing, fracture surfaces of failed specimens were observed in a scanning electron microscope (SEM).

3 Numerical simulation

3.1. Finite element model

Crack propagation on the nugget/thermo-mechanically affected zone (TMAZ) interface on the advancing side (AS) was analysed using the commercial FEM program ABAQUS. The specimen was modelled using a two-dimensional solid model, with a constant 30mm depth. The channel and the nugget geometries were interpolated from a sample image, using the spline function to reconstruct the channel geometry (Figure 5). The 180x30mm section was subjected to four-point bending loads and boundary conditions, considering a 90mm span.

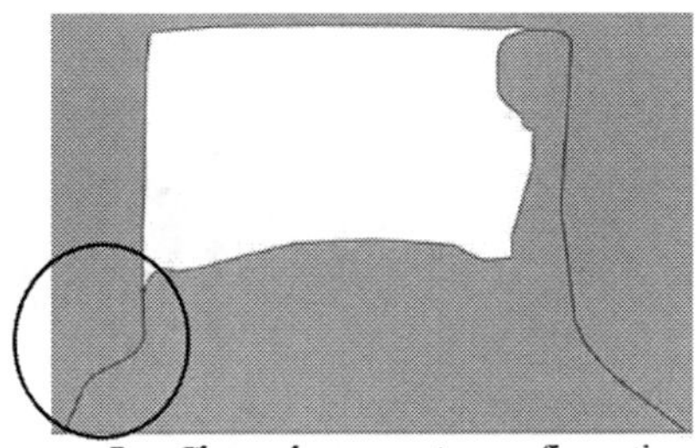

Figure 5. Channel nugget configuration. Crack propagation location along the interface.

3.2. Fatigue failure simulation

Different load amplitudes were applied for fatigue crack propagation simulation. The boundary conditions and loads were applied in four nodes, after verification that this simplification does not influence the results, when compared with the more complex simulation of the supports-specimen contact problem. By removing the supports and applying the boundary conditions in four nodes, the FEM analysis is linear and can be solved faster.
Both the base material and the channel nugget material where considered to be linear elastic, with properties shown in Table 2 and Table 5. Quadratic plane strain elements were used to model the geometry. A special meshing technique was used to define the crack tip and

model stress singularity around the crack tip. Special wedge elements with collapse nodes, were used to determine the first contour integral, as show in Figure 6. A maximum of 50000 nodes were used. The energy release rate value, used for crack propagation calculation, was determined by the contour integral method. Crack propagation was allowed along the materials interface, using eq. 8. A constant crack increment (Δa) was considered, and therefore the number of elapsed cycles between increments is calculated according with:

$$N = \frac{\Delta a}{A \Delta G^n} \quad (8)$$

The parameters for the Paris Law were estimated from the results obtained by Ilman et al. [17]. These authors. studied the crack propagation along a similar base material and friction stir welded material (Table 6), but not along the materials interface. Crack propagation was simulated with these two models and compared with the experimental results.

Table 5. Linear elastic properties for the base material and FSW material, used in FEM analysis [15].

	Young's modulus (GPa)	**Poisson rate**
Nugget	69.8	0.3

Table 6. Fatigue crack propagation parameters for the base material and FSW material, adapted from [17].

	A (m/cycle)	**n**
Base material	1.365×10^{-1}	2.222
FSW	1.116×10^{-4}	1.078

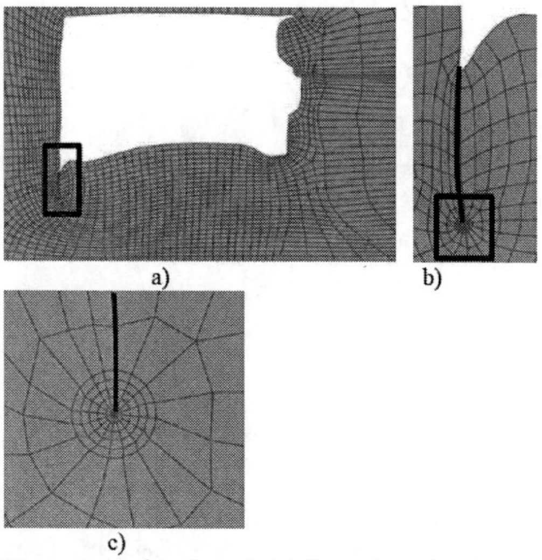

Figure 6. Mesh and crack details. a) channel and nugget, b) crack and mesh along the crack, c) spider-web mesh for crack tip contour integral calculation.

4 Results

4.1. Fatigue testing

Table 7 presents the fatigue test results for the FSC specimens analysed.

Table 7. Fatigue test results.

Fatigue Specimen ID	σ_{max} **(MPa)**	**Freq. (Hz)**	N_f, **Cycles to failure**	**Obs.**
FSC_1	150	3	15259	
FSC_2	125	4	17094	
FSC_3	100	4	46165	
FSC_4	100	4	67293	
FSC_5	80	5	165572	
FSC_6	80	5	120796	
FSC_7	65	8	773309	
FSC_8	65	8	808165	
FSC_9	50	8	3000000	Run-out

The S-N curves are presented in *log-log* scale. Each point represents an experimental test and the curves were obtained by fitting a linear regression eq. 8, assuming N_f as the dependent variable:

$$\sigma_{max}^{m} N_f = K_0 \quad (8)$$

where σ_{max} is the maximum stress, N_f is the number of cycles to failure and m and K_0 are empirical constants, namely the exponent and the coefficient of the S-N curve, respectively. The correlation lines fitted by eq. 8 represent the average regression curve with 50% of probability of failure. The regression lines are expressed by:

$$\log(N_f) = \log(K_0) - m \log(\sigma_{max}) \quad (9)$$

In order to compare the fatigue test results, the S-N curves were described in terms of the fatigue strength at 2×10^6 cycles and the slope m, which is a recommended procedure to assess the fatigue strength of welded aluminium joints when the S-N curves derive directly from experimental data [18]. In addition, K_0 values and the standard error of the estimate (*SEE*) were also calculated for the regression line estimated. Results are presented in Table 8.

Table 8. Parameters of the mean curves.

Condition ID	m	K_0	***SEE* (in log)**	**Fatigue strength at 2×10^6 cycles (MPa)**
BMa	3.867	1.10×10^{14}	0.0651	100
FSC	4.453	4.95×10^{13}	0.1717	46

4.3. Numerical simulations

[a] Specimens without channel.

In order to study the fatigue crack propagation, a crack on the materials interface was allowed to propagate. The crack was simulated between the closest corner of the channel and the free surface. The initial crack size was considered to be 0.1 mm and the end condition occurred when the crack reached the free surface (until the channel loss of tightness). Figure 7 shows the evolution of the calculated energy release rate along vs the crack length. These values were used to calculate the number of elapsed fatigue cycles between a given crack increment. FSC specimen allow for a 3.3 mm crack length, as the crack changes direction along the interface between the base material and the nugget (Figure 5). As the crack opening mode changes from mode I to mode II, the energy release rate decreases slightly, slowing the crack propagation. Finally, the crack propagation rate will increase, as the crack length reaches the critical value of 21 J/mm^2.

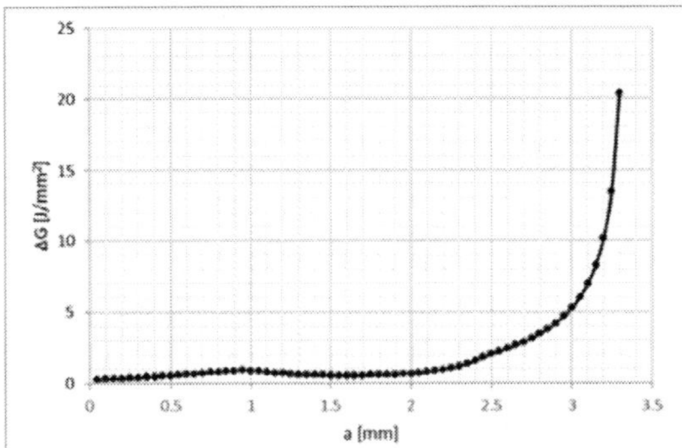

Figure 7. Energy release rate vs crack length, for a fatigue load with a maximum normal stress of 50 MPa.

Simulations of fatigue crack propagation were performed for different fatigue loads, and considering crack propagation properties of both, the base material and the FSW material. Table 9 summarizes the results. As concluded by Ilman et al. [17], the base material has a higher fatigue propagation resistance. Therefore, the obtained fatigue lives are always higher than the those obtained for the FSW material.

Table 9. Numerical fatigue simulation results.

σ_{max} (MPa)	FSW - N (Cycles)	BM - N (Cycles)
150	25728	40676
125	38115	91454
100	61662	246520
80	99756	664510
65	156080	1672000
50	274780	5365000

5 Discussion

From the global analysis of the fatigue test results is possible to conclude that FSC specimens exhibit a very low fatigue strength comparing to the unprocessed base material, even though they have no defects.
As presented in Table 8 the scattering of the results is quite low which demonstrates that, on one hand it is possible to predict the fatigue behaviour of FSC components and, on the other hand, the technological process is stable and the channel is reproducible.
A thicker closing layer contributes to longer fatigue crack propagation periods, although it has been observed that for FSC specimens, the crack propagation period is residual when compared with the number of cycles required for crack initiation.
The fracture of all specimens took place in the nugget/TMAZ interface on the advancing side, as represented in Figure 8, where the dashed line symbolizes the crack propagation path. This interface is a region of stress intensity between very fine and coarse grains which is amplified by the channel corner geometry, as well as, presents the smallest resistant area.

Figure 8. Crack path identification.

When considering both material models presented by Ilman et al. [17], one can see (Figure 9) that the fatigue lives predicted by the model presented herein and the base material properties, are higher than the experimental ones. This shows that this material model is not suited for the fatigue crack propagation of cracks on the interface between the base material and the FSW material. As mentioned this interface is considered to show a weak behavior. When considering the FSW material model, as seen in Figure 10, the predicted fatigue life represents 9% to 187% of the experimental life, as the fatigue load increases. Therefore the FSW material model is suitable for simulating fatigue crack propagation when the applied fatigue nominal stress has a maximum value of 100 MPa.
Within this loading range the model is in good agreement with the experimental results. The fatigue crack propagation life represents a maximum of 70% of the experimental life, for an 80 MPa applied fatigue load. Nevertheless, this value is very high, as the experiments show that the crack initiation period represented the majority of the fatigue life.

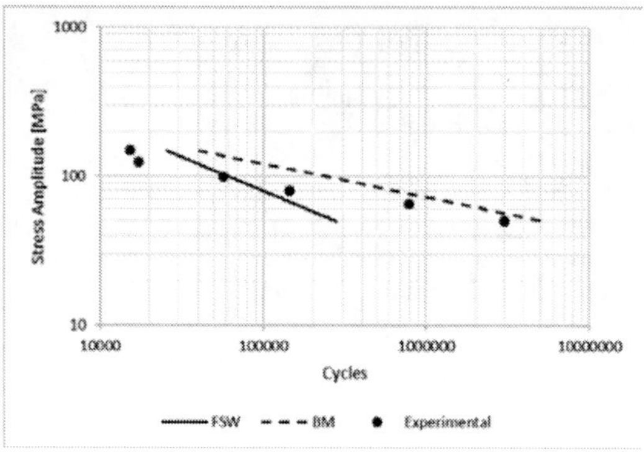

Figure 9. Fatigue life simulations for the FSC specimens.

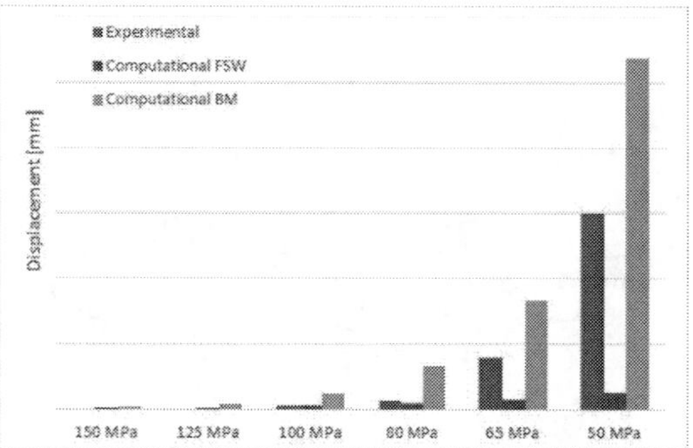

Figure 10. Comparison of experimental and computational values.

6 Conclusions

The global analysis of the fatigue test results have shown that FSC specimens exhibit a very low fatigue strength comparing to the unprocessed base material, even though they have no defects. A thicker closing layer contributes to longer fatigue crack propagation periods, although it has been observed that for FSC specimens, the crack propagation period is residual when compared with the number of cycles required for crack initiation. The fracture of all specimens took place in the nugget/TMAZ interface on the advancing side where a smallest resistant area is observed.
Based on the computational results one can conclude that the material model used is not suited for the fatigue crack propagation of cracks on the interface between the base material and the FSW material, when the applied fatigue loading is higher than 100 MPa. The fatigue crack propagation life obtained in the computational simulations represents a maximum of 70% of the experimental life, for an applied fatigue loads of 80 MPa. Nevertheless, this value is very high, as the experiments show that the crack initiation period represented the majority of the fatigue life. Future work should be done to quantify the fatigue crack initiation and propagation periods, in order to better choose the correct material model for the fatigue crack propagation simulations.

Acknowledgements

The authors would like to acknowledge the Portuguese Foundation for the Science and Technology (FCT) for its financial support through the PhD scholarship FCT SFRH/BD/62963/2009.

References

1. Thomas, W.M., Friction stir welding and related friction process characteristics, 7th International Conference Joints in Aluminium, Inalco, TWI, Cambridge, 1998.
2. Thomas, W.M., Sylva, G., Developments of friction stir welding, ASM Materials Solutions 2003, Conference and Exposition, Pittsburgh, Pennsylvania, USA, 2003.
3. Thomas, W.M., Nicholas, D.E., Staines, D.G., Tubby, P.J., Gittos, M.F., FSW Process Variants and Mechanical Properties, Doc IIW-1664-04, Welding in the World 49 (3/4), pp 4-11, 2005.
4. Thomas, W.M., Norris, I.M., Staines, D.G., Clarke, P.J., Horrex, N.L., Friction stir welding - variants and process techniques, The First International Conference Joining of Aluminium Structures, Moscow, Russian, 2007.
5. Mishra, R.S., Integral channels in metal components and fabrication thereof, Patent no. US 6923362 B2, August 2, 2005.
6. Fajdiga, G., Sraml, M., Fatigue crack initiation and propagation under cyclic contact loading, Engineering Fracture Mechanics, 76 (20069), pp. 1320-1335.
7. Cheng, W., Cheng, H.S., Mura, T., Keer, L.M:, Micromechanics modelling of crack initiation under contact fatigue, ASME Journal Tribology, 1994, 116, pp. 2-8.
8. Zhang, H., Yang, J., Hu, J., Li, X., Li, M., & Wang, C. (2018). An experimental and simulation study of interface crack on zinc coating/304 stainless steel. Construction and Building Materials, 161, 112–123. https://doi.org/10.1016/j.conbuildmat.2017.11.022
9. Kakiuchi, T., Uematsu, Y., & Suzuki, K. (2016). Evaluation of fatigue crack propagation in dissimilar Al/steel friction stir welds. Procedia Structural Integrity, 2, 1007–1014. https://doi.org/10.1016/j.prostr.2016.06.129
10. Shah, S. G., Ray, S., & Chandra Kishen, J. M. (2014). Fatigue crack propagation at concrete-concrete bi-material interfaces. International Journal of Fatigue, 63, 118–126. https://doi.org/10.1016/j.ijfatigue.2014.01.015
11. Suo, Z. G., & Hutchinson, J. W. (1990). Interface Crack Between 2 Elastic Layers. International Journal of Fracture, 43(5), 1–18. https://doi.org/Doi 10.1007/Bf00018123

12. Dundurs J. Edge bonded dissimilar orthogonal elastic wedges under normal and shear loading. J Appl Mech 1969;36:650.
13. Rice JR, Sih GC. Plane Problems of Cracks in Dissimilar Media. ASME. J. Appl. Mech. 1965;32(2):418-423. doi:10.1115/1.3625816.
14. Xuan, F. Z., Tu, S. T., & Wang, Z. (2005). A modification of ASTM E 1457 C* estimation equation for compact tension specimen with a mismatched cross-weld. Engineering Fracture Mechanics, 72(17), 2602–2614. https://doi.org/10.1016/j.engfracmech.2005.05.002
15. Vidal, C., Infante, V., Vilaça, P., Fatigue behaviour at elevated temperature of friction stir channelling solid plates of AA5083-H111 aluminium alloy, International Journal of Fatigue, Volume 62, May 2014, pp. 85-92, https://doi.org/10.1016/j.ijfatigue.2013.10.012
16. Vidal, C., Infante, V., Vilaça, P., Assessment of Performance Parameters for Friction Stir Channelling, Proceedings of the IIW 2011 International Conference on Global Trends in Joining, Cutting and Surfacing Technology, Chennai, India, 21-22 July, 2011, ISBN 978-81-8487-152-4, paper IC_99.
17. M.N.Ilman, W. (2013). Fatigue crack growth behaviour of shot peened 5083 aluminium alloy friction stir welds. International Conference on Joining Materials, (May 2013).
18. Maddox, S.J., Review of fatigue assessment procedures for welded aluminium structures, International Journal of Fatigue 25, pp 1359 1378, 2003.

Undercut tolerances in industry from a fracture mechanic perspective

Ceferino Steimbreger[1,*], and *Mirco* Chapetti[1]

[1]Laboratory of Experimental Mechanics (LABMEX), INTEMA (Institute for Material Science and Technology), CONICET - University of Mar del Plata, J.B. Justo 4302, (B7608FDQ), Mar del Plata, Argentina.

Abstract. Fatigue is an important damage mechanism that particularly affects welded components, since they are likely to present residual stresses, inhomogeneities and stress raisers. Assessment of cyclic load effects on welds has concerned both industries and scientist for decades; unexpected failure must be prevented and at the same time, structures must withstand design loads with minimum requirements of material. All these facts together with economic issues have lead to the creation of normative that rule designing and construction of welded components. Particularly, toe undercuts are generally found in large structures, and large scatter and disagreement exists towards their significance and effects. Documents usually limit only their depth without considering radius, width or length, and there is currently no explanation to that fact. Understanding the damaging process will also help to set less conservative tolerances, with consequent cost reduction due to less demanding inspection. The present paper deals with a fracture mechanic approach that uses the Resistance Curve concept to predict fatigue limit of welded components with undercuts. Results revealed that depth is the most influencing variable, and it can be used as the limiting parameter in design regulations. Moreover, good correlation was obtained with FAT values normally assigned to this kind of defect.

1 Introduction

Tolerances for weld defects have been established in many codes of construction, in order to keep failure risk to a minimum. Initially, some of these documents were intended as means of communication between the welder and the engineer and they were based mainly on good workmanship and experience. Nowadays, their scope is much more important and their usage is widespread all over the world. Furthermore, some standards have recently included weld defect evaluation under cyclic loading, evincing the increasing interest of parties in achieving safe structures.

In spite of all these developments in the field of welded constructions, lack of agreement between codes and standards with respect to tolerances is evident. Moreover, limits are sometimes over-conservative and hard to measure for the unaided eye, which considerably delays inspection process. Therefore, an evaluation of tolerances with a scientific basis is needed.

In general, acceptance levels for weld defects are determined by considering quality control levels or weld quality systems, such as those in ISO 5817 [1]. They allow the existence of flaws in a component, provided that their size is less severe than certain limits. In cases where fatigue is an important damaging mechanism, welds are classified into pre-defined details associated to a specific S-N curve that defines its safe cyclic behaviour. Although this proved to work well in a variety of situations, no consistent relation can be established between tolerances for defects and the actual damaging phenomenon, which constitutes the main drawback of weld quality systems.

If a flaw in a weld is found to be out of the limits provided by the quality system, immediate rejection can be delayed until a deeper analysis confirms the risk of maintaining that defect in service. In this regard, fitness-for-purpose/service (FFP/FFS) analyses are normally employed [2-4]. These methods determine in a rational manner, whether a structure or component containing flaws is able to continue in service safely.

In parallel with development of fitness-for-purpose guidelines, fracture mechanic proved to be a powerful tool to determine severity of a variety of discontinuities, and remaining fatigue life of structures. Pertinent documents have included guidelines on fracture mechanics, although some parameters, like threshold stress intensity factor, are usually disregarded or adopt very conservative values [2-6].

Previous works from the authors employed a fracture mechanic methodology based on the resistance curve concept, to determine fatigue strength of welded joints containing undercuts [7]. Results are summarised in Figure 1, in the form of Frost diagram. It was suggested that this methodology may serve as a tool to predict safe acceptance limits to undercuts and other defects.

* Corresponding author: cefesteim@gmail.com

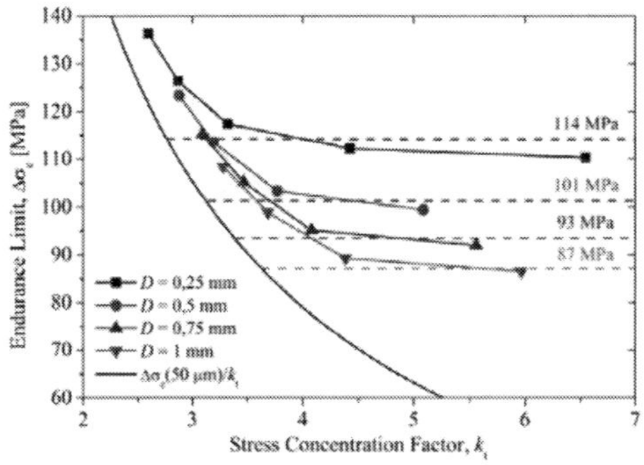

Fig. 1. Frost diagram for a reference plain fatigue limit of $\Delta\sigma_e$ = 315 MPa (a_i = 50 μm) [7].

In order to compare these outcomes with current regulation in the subject, a brief description of relevant normative is carried out in the following paragraphs.

1.1 AWS D1.1 [8]

Acceptance criteria for WPS Qualification, and also for Welder and Welding Operator Qualification, limit undercut depth to 1/32" (1 mm) when performing visual inspection or macroetch tests. Additionally, all welds must meet specified visual acceptance criteria that, in the case of non-tubular connections, depend on whether the structure is statically or cyclically loaded. Under the last mode, undercut depth is limited to 0.25 mm in primary members under any design loading condition, when the weld is transverse to tensile stress.

1.2 ASME BPVC [9-11]

As a result of static strength considerations, undercuts whose depth is beyond a minimum section thickness are forbidden. In the case of finished longitudinal and circumferential joints in boilers [9], depth must not exceed 0.8 mm, or 10% of the wall thickness, whichever is less. Moreover, a smooth transition between the surfaces being joined is required. Defects found to be rejectable must be removed, rewelded and re-examined.

An important fact in the ASME BPVC is that tolerances for undercut depth are similar in all sections. However, some differences can be found in relation to thickness dependence and undercut length. For example, Subsection NF in Section III [10] presents less stringent acceptance limits for shorter undercuts (smaller length) or thicker members. Both variations rely on static strength considerations and on reduction of required section thickness. This situation is held throughout the code and although some sections do include fatigue considerations, no clear relation between undercut tolerances and cyclic loading can be established. It is worth mentioning, that measurement of reduction in thickness due to undercutting, is not required if there is no disagreement between the involved parties about its acceptability [11].

If fracture mechanic analyses or refined fatigue assessment are necessary, API 579-1/ASME FFS-1 [3] is referred throughout the code.

1.3 SS-EN ISO 5817 [1]

Quality levels are designated by letters B, C and D, with the former corresponding to the highest requirements on the final weld. These groups are based on production and good workmanship.

Limits proposed in this document are directly applicable to visual testing of weldments. In the case of welds with thickness over 3 mm, quality level B limits undercut depth to $D \leq 0.05\ t$ but max. 0.5 mm; level C to $D \leq 0.1\ t$ but max. 0.5 mm; and level D accepts undercuts with $D \leq 0.2\ t$ but max. 1 mm.

If loading mode is cyclic, then quality levels must additionally meet fatigue requirements in terms of fatigue classes (FAT). FAT value refers to the stress range at 2.10^6 cycles for a two-sided survival probability of 95 %, based on mean curves proposed by IIW Recommendations [5]. The basis of these additional requirements is that limits for imperfections in quality level C and B must be adjusted in order to compliment fatigue class FAT 63 and FAT 90, respectively. Supplementary fatigue level FAT 125 can be assigned to weld quality B, for some imperfections, although this is not generally achieved in the as-welded condition. Static limits for undercuts satisfy fatigue requirements for levels C63 and B90. However, B 125 demands removal of this type of imperfection.

It can also be highlighted, that a smooth transition is required, in order to achieve good-workmanship standards. This restricts the application of the document to solely blunted undercuts (i.e. not too sharp).

1.4 BS 7608 [12]

Quality control can sometimes be assessed by considering the same acceptance criteria in production as those adopted for procedure or weld qualification [8, 13] or in BS EN ISO 5817 [1]. Such provisions rely on arbitrary quality criteria and they may not be suitable for higher fatigue class needs. Moreover, their requirements are likely to incur additional cost for lower fatigue classes. Therefore, a fitness-for-purpose assessment is generally preferred to specify quality requirements.

Tolerance for undercuts at the toe of welds subjected to cyclic loading can also be determined based on fatigue test results, particularly those obtained by Petershagen [14]. Undercuts will therefore have an associated S-N curve, depending on its depth/thickness ratio. Acceptance or rejection of the flaw is decided by comparing this actual quality category with the required quality category. The latter can be determined with the required stress range and design fatigue life, or alternatively with the weld quality category of the reference detail (e.g. butt joint). Requirements for undercuts are shown in Table 1. Correspondence with BS 7910 quality categories [2] is also displayed.

Table 1. Fatigue based tolerances for undercuts in transversely stressed butt-welds [2, 12].

Required class (FAT)	Equivalent BS 7910 quality category	Undercut depth, *D*
		Butt welds
C	-	Not permitted
D	Q1	$0.025\ t, \leq 1$ mm
E	Q2	$0.05\ t, \leq 1$ mm
F	Q3	$0.075\ t, \leq 1$ mm
F2	Q4	$0.1\ t, \leq 1$ mm
G	Q5	$0.1\ t, \leq 1$ mm

Some validity boundaries apply to Table 1. First of all, it is restricted to shallow undercuts at the toe of perfectly aligned joints. In the case of sharp defects, where depth cannot be measured accurately, probability of crack nucleation is higher, and they must therefore be rejected according to this standard. Additionally, plate thickness must be in the range of 10 to 40 mm. Undercuts formed in materials out of this limits should be assessed as planar flaws following BS 7910 procedures [2]. Validity limitations are necessary due to the restricted database available [14]. In cases where stresses act parallel to weld direction, there is no limiting size for undercut depth, since they do not affect fatigue behaviour under this condition.

As it was mentioned earlier, limits shown in Table 1 were deduced from large experimental fatigue data in transversely-stressed butt and fillet steel joints, in which undercuts were either naturally generated or artificially machined at the toe. These results were analysed statistically assuming a log-normal distribution of fatigue life to obtain the lower 95% confidence limit, corresponding to 97.7% probability of survival. Quality category was determined by comparing these data with that of flawless welds. Particularly, reduction in fatigue strength due to undercut depth was quantified in terms of steps in the grid of quality category S-N curves. Parameter D/t was found to best reduce scatter in experimental data [14, 15].

1.5 API 1104 [16]

Undercuts must fulfil specific requirements when visual inspecting: their depth at the toe of the final bead on the outside of the pipe shall not exceed 0.8 mm or 12.5% of the pipe wall thickness, whichever is smaller. Moreover, there must not be more than 50 mm of undercutting in any continuous 300 mm length of weld.

In addition to these requirements, depth of undercuts found in the cover or root bead shall not be larger than values given in Table 2, when inspection is carried out visually or by mechanical means. When employing other non-destructive techniques like radiography, magnetic particles or ultrasonic method, undercuts at the cover pass or root bead must only satisfy length restrictions.

Previous tolerances for undercuts are based on empirical criteria for good workmanship. Although these limits have proven to be reliable in pipeline systems throughout the years, if a more stringent analysis is desired and it is approved by the company, fitness-for-purpose assessment can be used [2].

Table 2. Maximum depth of undercuts, for visual and mechanical inspection [16].

Depth	Length
$D > 0.8$ mm or $D > 0.125\ t$, whichever is smaller	Not acceptable
0.4 mm $< D \leq 0.8$ mm or $0.06\ t < D \leq 0.125\ t$, whichever is smaller	50 mm in a continuous 300 mm weld length or one-sixth the weld length, whichever is smaller
$D \leq 0.4$ mm or $D \leq 0.06\ t$, whichever is smaller	Acceptable, regardless of length

In summary, high cycle fatigue considerations are lightly accounted for in this standard, and it is not clear weather tolerances are defined for static or cyclic loading conditions. In spite of this, limit values adopted throughout the document are useful for comparing with other standards and codes.

1.6 IIW Recommendations [5, 17]

Undercuts are regarded in the guideline as additive imperfection, which means that they are adding their effect on fatigue behaviour (e.g., undercut and toe radius). For rapid assessment of these flaws, Table 3 is given, which is based on experimental tests and results from literature [14, 18]. However, IIW recommends the effective notch stress approach or fracture mechanics for deeper analyses of undercut´s effect.

Table 3. Acceptance levels for Weld toe undercuts in steel, according to IIW recommendations [5, 17].

Fatigue class (FAT)	Allowable undercut, *D/t*	
	Butt welds	Fillet welds
100	0.025	Not applicable
90	0.05	Not applicable
80	0.075	0.05
71	0.1	0.075
63 and lower	0.1	0.1

1.7 VOLVO STD 181-0004 [19]

VOLVO made the first attempts to include fatigue consideration in old weld quality systems, which were originally based on good workmanship in fabrication [20]. Although, normal workmanship conditions are related to a certain extent to adequate weld performance, a proper analysis based on fitness-for purpose was not established. Particularly, structures subjected to fatigue loading needed special attention. As a result, this standard was developed in 2008, by considering IIW Recommendations [5]. Later, IIW published its own guideline to weld quality systems [17], based on VOLVO standard. In parallel, ISO added fatigue consideration to its latest version in 2014 [1].

Most requirements in STD 181-004 are based on notch-stress analyses. Jonsson et al. [21] proposed acceptance levels for undercuts based on the effective notch stress approach [22] that assigns a radius of 1 mm to the undercut root. Tolerances were obtained by considering that the worst acceptable defect is given by a stress value two standard deviations above the stress level of a normal weld, free from undercuts [21]. This limit, or safe value of the stress, is later compared to the stress at the root of an undercut for different depths. Intersection of the two variables determines the limit for undercut depth.

By applying this procedure to butt and fillet welds, limiting tolerances in terms of D/t were obtained. An average of the two was considered for the weld quality system. Same methodology can be extended to high quality welds. STD 181-004 adopted less stringent requirements for undercut depth in qualities VD and VC. These classes need to satisfy additional demands in terms of the "outer transition radius", which should not be lower than 0.3 mm and 1 mm, respectively. This condition differs from ISO 5817, that solely request a smooth transition between the weld and the base material.

It is interesting to note, that results are in accordance with IIW recommendations [5, 17], whose acceptance limits are based on Petershagen's work [14]. Both researches are based on stress analysis and no crack consideration is involved.

1.8 DNVGL RP-C203 [6]

It is recommended to construct a proper WPS to avoid deep undercuts in production. Common workmanship standards are considered in design S-N curves, and default tolerance for this type of defect is normally limited to 0.6 mm [23]. However, specific requirements can be assigned depending on the desired fatigue strength. Tolerances for undercuts are then limited to null for S-N curves better than D, 0.5 mm for classes F to D, and 1 mm for class F1 and lower. In the case of butt welds with desired fatigue behaviour better than class D, ISO level B125 applies (ISO 5817).

2 Assumptions

In order to determine stress intensity factors associated to a crack growing from the undercut root, a finite element software [24] was employed. Two-dimensional linear elastic model was considered and symmetry was assumed. Weld geometry is displayed in Figure 2, together with symmetric model. Thickness is 19 mm and weld with a reinforcement angle is 147°. Load configuration and boundary conditions are also shown. Four point bending was considered. Minor and major span are represented by s and L, respectively, t is the plate thickness and P is the applied load. All these variables define the maximum nominal stress on the surface that is used in the calculations.

Analysis carried out in welds with undercuts and its usage for safe tolerance determination in standards, is limited to a 19 mm stress-relieved weld, made of structural steel (A36) and loaded in four point bending scheme [7]. In order to extend these results to other butt-welds, it should be verified that these outcomes are conservative, or at least similar to those for other materials, thicknesses and loading conditions.

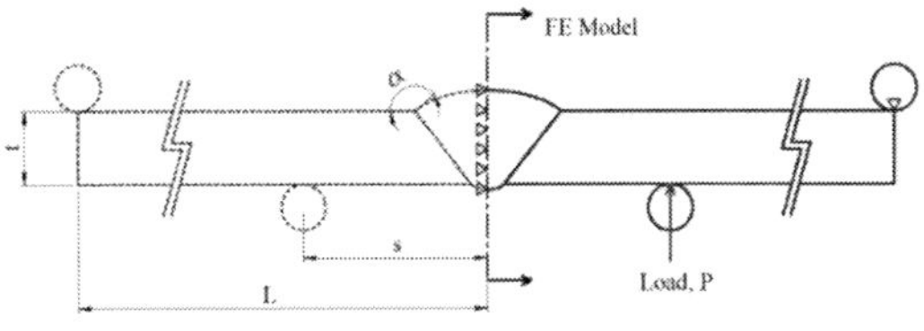

Fig. 2. Geometry, load configuration and boundary conditions of the finite element model. t is the plate thickness, α is the reinforcement angle, s is half the minor span, L is half the major span and P is the applied load.

3 Results and discussions

In order to compare acceptance limits for undercut at the toe of welded connections in current regulations, it is important to mention, that the preferred parameter to do this is undercut depth, D. It can be found in literature, that this measure provided the best fitting of experimental results. The work of Petershagen [14] revealed that D/t reduced even more the scatter in large amount of data. Some documents demand additional requirements in terms of undercut length [8, 16] and weld toe radius [19], and others distinguish between different types of joints and loading directions. Herein, the most dangerous situation for a butt connection is considered, i.e. an undercut along the whole length of the weld bead, loaded transversely.

Figure 3 summarises design fatigue curves, which are relevant for assessing cyclic behaviour in a butt-welded joint. FAT 80, 90 and 100 corresponding to IIW recommendations, present a change in slope to m = 22 for cycles above 1.10^7. BS 7608 assigns a curve similar to FAT 90, but the knee point in the former case determines the fatigue strength for infinite life and they do not present a slope to account for ultra-high cycle fatigue. Likewise, AWS follow a FAT 90 trend until 4.10^6, where it becomes horizontal.

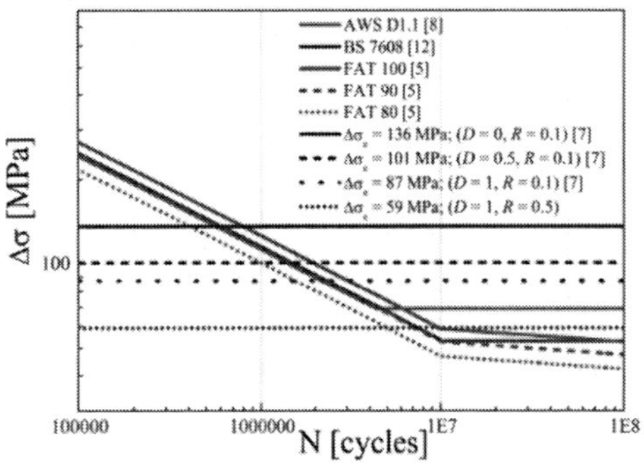

Figure 3: S-N curves employed in different codes and standards, and fatigue limit results from prediction.

In the same plot depicted in Figure 3, 4 horizontal curves are also provided. They correspond to fatigue limits from predictions for a stress-relieved 19 mm butt weld without undercut (straight line), a stress-relieved 19 mm butt weld with an undercut 0.5 mm deep (slashed line), a stress-relieved 19 mm butt weld with an undercut 1 mm deep (dotted line) and a butt weld with an undercut 1 mm deep but including residual stresses (short dotted line). The latter was obtained as explained below. Note that the value of fatigue strength at 2.10^6 cycles is similar to predictions for welds with undercuts. On the other hand, fatigue limit in theoretical curves defined as the knee point, are somewhat closer to predictions for a weld with 1 mm undercut, and considering residual stresses ($R = 0.5$).

Majority of codes and standards that are relevant to fatigue design make use of a reference stress in the design S-N curve for fatigue assessment. Definition of this limit depends on each document, but it is generally related to the FAT value and in some cases to the stress where a change in slope in the S-N curve is defined.

Table 4 summarises acceptance limits found in normative, when structures and components withstand cyclic loads. Some documents assign different tolerances to different desired fatigue strength, but others do not consider alternative levels of conservatism [8-11, 16]. As it is expected, codes of construction like AWS and ASME adopt the most conservative values. In the former case, it limited undercut depth to 0.25 mm for more than 30 years without any strength consideration.

Fatigue class corresponding to each weld designation or quality system, is displayed in column two of Table 4. Additionally, FAT values (strength in MPa at 2.10^6 cycles) are provided, based on design S-N curves from each document. In this regard, it must be mentioned that although IIW reference S-N curves prevail in the majority of documents studied in the present work, not all regulations adopt those profiles, as indicated in Figure 3. Therefore, fatigue limits, considered as the "knee-point" in IIW curves, or the stress value at which infinite fatigue life is obtained in others, are listed in column 4.

Analysis carried out by the authors presented results of fatigue limit for butt-welded joints with different undercut configurations (see Figure 1). This means that infinite fatigue life is obtained in those welds if loaded below that stress. Although comparison of results with fatigue limits (see column 4 of Table 4) is reasonable from a safe-life design point of view, this methodology can be over-conservative in cases where fatigue is not the primary damage mechanism, like in nuclear reactors or oil piping systems. Moreover, even in situations where severe fatigue damage is expected, design is generally based on FAT values, with satisfactory results. Therefore, the use of FAT values is justified and it is valuable to compare tolerances in relevant normative with results from previous work from the authors [7].

Figure 4 illustrates results from Table 4, and shows maximum tolerance for undercut depth, for several desired fatigue strength, in terms of FAT values. It must be highlighted, that connection between points was made by a straight line rather than a stepped profile, as can be found in relevant literature [14]. This was preferred since in real structures a continuous decrease in fatigue strength for increasing undercut depths is expected.

Table 4: Undercut tolerances in codes and standards, for a butt-welded joint, 19mm thick.

Document	Fatigue class	FAT	Fatigue limit		Tolerance [mm]
			S [MPa]	N ($x10^6$)	
AWS [8]	C	90	69	4	0.25
BS 7608 [12]	C	125	78	10	0
	D	90	53	10	0,475
	E	80	47	10	0,95
	F	71	40	10	1
	F2	63	35	10	1
	G	50	29	10	1
SS-EN ISO 5817 [1]	B125	125	73,1	10	0
	B90	90	52,7	10	0,5
	C63	63	36,9	10	0,5
API [16]	-	-	-	-	0,4
ASME [9-11]	C	90	70	>2	0,8
DNVGL [6]	C2	100	58,48	10	0
	D	90	52,63	10	0,5
	E	80	46,78	10	0,5
	F	71	41,52	10	0,5
	F1	63	36,84	10	1
	G	50	29,24	10	1
IIW [5, 17]	FAT 125	125	73.1	10	0
	FAT 100	100	58,5	10	0,475
	FAT 90	90	52,7	10	0,95
	FAT 80	80	46,8	10	1
	FAT 71	71	41,5	10	1
	FAT 63	63	36,9	10	1
	FAT 56	56	32,8	10	1
VOLVO [19]	VB	125	73,1	10	0
	VC	100	58,5	10	0,475
	VD	80	46,8	10	0,95
	VE	63	36,9	10	1
Fracture Mechanic [7]	D=0	136	136	-	0
	D=0,25	114	114	-	0,25
	D=0,5	101	101	-	0,5
	D=0,75	93	93	-	0,75
	D=1	87	87	-	1

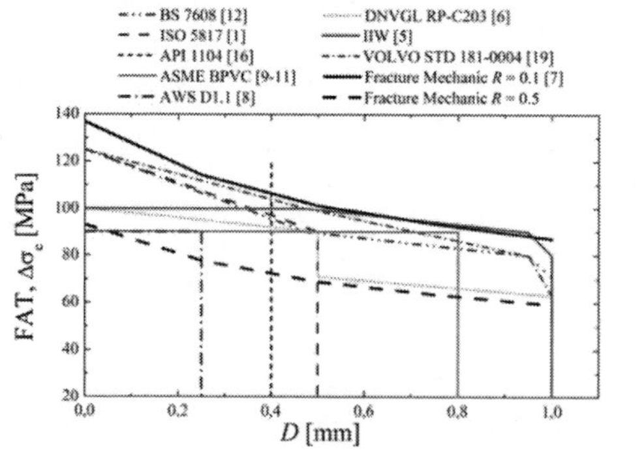

Figure 4: Results from predictions and comparison with regulations. Butt joint, t = 19 mm, FAT values.

It can be seen in Figure 4 that AWS D1.1 contains the most conservative tolerances, since no undercuts deeper than 0.25 mm are permitted. BS 7608, VOLVO STD 181-004 and IIW recommendations suggest similar tendencies, limiting maximum tolerable undercut to 1 mm. ISO 5817 follows a similar trend as VOLVO and BS, up to 0.5 mm, which is the maximum admissible undercut depth in a 19 mm thick butt-joint. On the contrary, DNVGL RP-C203 shows more conservative strength beyond D = 0.5 mm, but also accepts a maximum permissible depth of 1 mm.

In the same plot, results from author´s predictions are displayed with a black straight line. This curve gives the fatigue limit of the weld (136 MPa), when D = 0, and decreases continuously with undercut depth. Note that it slightly differs from regulations, and it accurately describes the trend followed by most regulations. However, it must be mentioned that predictions developed in this work were calculated for a stress ratio R = 0.1 and they do not account for residual stresses. Normally, design S-N curves are based on extensive experimental data in real welds, and hence they include residual stresses in their definition, as it was pointed out in the introduction.

3.1 Residual stress correction

In order to unify criteria in Figure 4 and consider welds containing residual stresses, the change in the resistance curve for different values of stress ratio, R was analysed. As an example, a semi-circular undercut with D = 0.5 mm was used. Resistance curve and critical condition in A36 steel is shown with black lines in Figure 5.

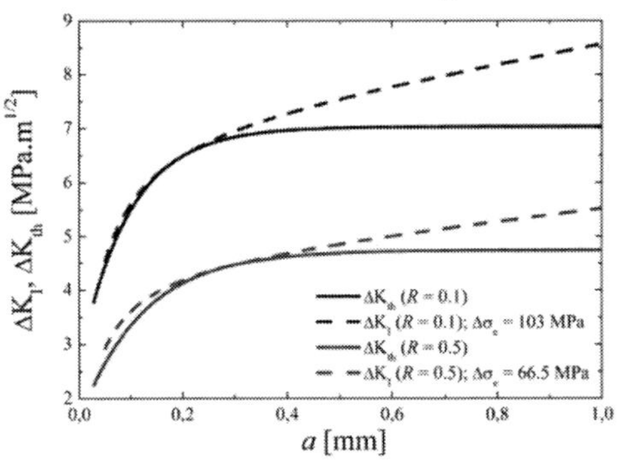

Figure 5: Variation of resistance curve and critical applied stress intensity factor in A36 weld with undercuts 0.5 mm deep, at different values of R.

In order to assess the effect of residual stresses, dependence of long crack propagation threshold with stress ratio must be known. It was found [25] that the following relation applies to A36 steel:

$$\Delta K_{thR} = 7.6\text{-}5.7\, R \qquad (1)$$

A quick examination of Eq. (1) and Figure 5 reveals that plateau given by long crack propagation threshold, is shifted to lower values of ΔK, when R is increased. Therefore, a reduction in fatigue strength is likewise expected. However, a deeper analysis must consider the change of the plane fatigue limit, $\Delta\sigma_{eR}$, with R, which in consequence produces a variation of the microstructural fatigue threshold, ΔK_{dR}, according to Eq. (2).

$$\Delta K_{dR} = Y\, \Delta\sigma_{eR}\, (\pi d)^{1/2} \qquad (2)$$

where d is value of the strongest microstructural barrier, which is usually determined by ferrite grain size and bainite or martensite lath length [26].

A conservative estimation of the fatigue limit for different values of R can be obtained by means of Goodman´s relation. It is easy to demonstrate that Eq. (3) relates stress amplitude for R = -1, $\sigma_{e\text{-}1}$, with stress amplitude at a desired asymmetric stress ratio, R, σ_{AeR}.

$$\sigma_{AeR} = 1/\{(1/\sigma_{e\text{-}1}) + (1+R)/[\sigma_{UTS}\,(1\text{-}R)]\} \qquad (3)$$

where σ_{UTS} is the ultimate tensile strength of the base material (472 MPa for A36).

Since the value of $\Delta\sigma_{eR}$ = 360 MPa used in predictions corresponds to R = 0.1, then the inverse procedure should be followed to find $\sigma_{e\text{-}1}$. By doing so, it can be obtained that $\sigma_{e\text{-}1}$ = 337 MPa. After this, stress amplitude for other stress ratios can be deducted directly from Eq. (3). Particularly, for R = 0.5, σ_{AeR} = 107 MPa, and therefore $\Delta\sigma_{AeR}$ = 214 MPa. Finally, by recalling Eq. (2), ΔK_{dR} = 2.25 MPa√m is obtained. Resistance curves for R = 0.1 and R = 0.5 are depicted in Figure 5. It is also indicated in this plot, the curve for the applied stress intensity factor range in the limiting condition (i.e. where it touches the threshold curve in a single point) for R = 0.1 (black dotted line) and an increased stress ratio R = 0.5 (red dotted line). Fatigue strengths resulting in both cases are 103 and 66.5 MPa, respectively. This shift is equivalent to a decrease around 35 %, which means that a limit load (or equivalently, a limit stress or fatigue limit) 35 % smaller than that for R = 0.1 could be allowed. These results are displayed with a black broken line in Figure 4, and a clear lower bound to tolerances in industry can be seen.

At the same time, relationship in Eq. (1), gives a value of 7.03 MPa√m when R = 0.1 and 4.75 MPa√m when R = 0.5. Then, a reduction in ΔK_{thR} around 32% is deduced. It must be highlighted that both quantities (that obtained by the resistance curve method and the one calculated directly with Eq. (1)) are similar. Therefore, it can be said that the stress corresponding to the contact point between applied driving force and threshold, is reduced in approximately the same amount as long crack propagation threshold, for two different values of R. Therefore, control of this property is relevant when assessing the effect of residual stresses.

3.2 Thickness correction

Since the 70´s, it has been known that fatigue strength increases with decreasing thickness. A potential relation was found to best fit experimental results, with the following equation [27, 28]:

$$\sigma_t / \sigma^{tR} = (t_R / t)^n \qquad (4)$$

Where σ_{tR} is the fatigue limit for a weld on a plate with a reference thickness t_R, and σ_t is the fatigue limit at another thickness t. n is the exponent that was found to be close to 0.25.

Right term in Eq. (4) is defined as the thickness correction/reduction factor in codes and standards. It should be multiplied by the FAT value in order to consider thickness effect in design [5, 6, 12]. This is only done for $t > t_R$, since the benefit of an enhanced fatigue limit for smaller thicknesses is usually disregarded in standards and recommendations.

IIW Recommendations [5] specifies a reference thickness of $t_R = 25$ mm and a value of n that depends on the type of joint to be assessed. In the case of transverse as-welded butt joints, $n = 0.2$ corresponds. Value of the reference thickness to be used in Eq. (4) is still at open debate. But in general, $t_R = 25$ mm and $n = 0.2$ is employed in design documents [5, 6, 12].

On the other hand, resistance curve concept has proved to properly describe changes in fatigue limit due to thickness variations [25]. Additionally, due to the fact that the model employed to determine stress intensity factor is linear elastic, the effect of thickness can be assessed using the same configuration used for $t = 19$ mm. By similarity principles, geometry and load scheme can be scaled down in order to have geometrically similar stress distribution and the same nominal stress, but for a different thickness. Many authors [29-31] made use of this method to derive fracture mechanic results from a reference configuration to another.

In the presence of undercuts, scaling down geometry in Figure 2 can be accomplished by considering a constant ratio D/t, or keeping undercut depth and shape constant with varying t. Results from both situations are shown in Figure 6. Also depicted in this chart is the theoretical relation in Eq. (4), considering $t_R = 25$ mm and $n = 0.2$. It must be highlighted that results corresponding to semi-circular undercuts correspond well with this theoretical curve.

On the other hand, when analysing data for $D/t = 0.5/19$ = constant it can be seen a sharp change in slop around 22 mm, associated to a shift in the non-propagating crack length. The latter presents a sharp step at $t = 22$ mm, above which $a_{np} = a_i = 0.05$ mm, in correspondence with Figure 6.

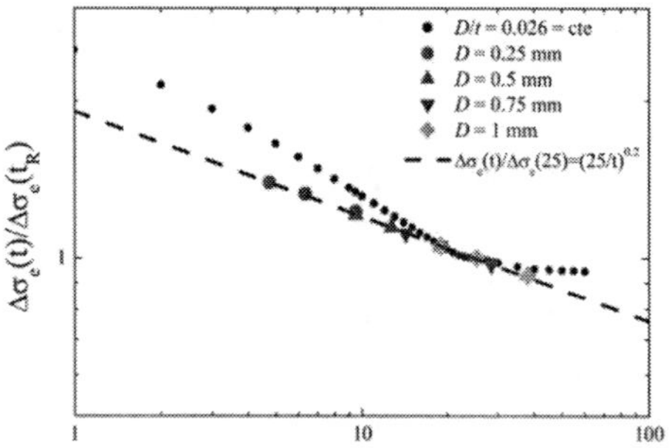

Figure 6: Normalised fatigue limit as a function of thickness, for D/t = cte and D = 0.25, 0.5, 0.75 and 1 mm. $t_R = 25$ mm.

The aforementioned results verify that fatigue limit prediction in a 19 mm butt-weld with undercuts is a conservative approach for tolerance determination in industry, when thicknesses are below 19 mm. For thicker plates, it is reasonable to apply a correction factor according to Eq. (4). An exponent $n = 0.2$ and $t_R = 25$ is found to best fit behaviour of defective welds. In contrast, as-welded undercut-free joints may be better described by an exponent $n = 0.25$ [25]. However, it must be highlighted that flawless welds are seldom found in real components, and it is therefore reasonable to adopt 0.2 in codes and regulations.

3.3 Loading correction

In order to further verify extension of previous results to a wider range of situation, loading mode was changed from four points bending to traction, and the resistance curve method was applied. This was done for a single undercut depth, $D = 0.5$ mm, giving predictions displayed in the Frost diagram in Figure 7. Three different initial crack lengths were analysed, showing flattering of the curve with increasing crack size. According to obtained outcomes, fatigue limits of welded joints containing undercuts under traction are very similar to those resulting from bending.

The fact that the loading mode does not generate significant variation in the fatigue limit of defective butt welds supports previous discussion about tolerances in industry.

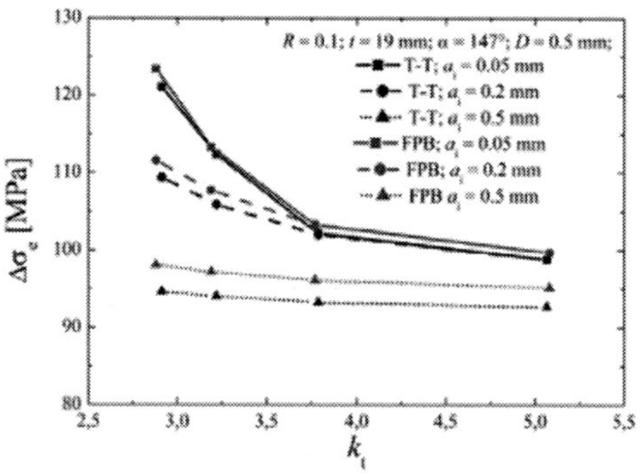

Figure 7: Frost diagram comparing fatigue limit of butt-weld with undercuts under traction (T-T) and bending (FPB).

4 Conclusions

A fracture mechanic approach that uses the resistance curve method was employed to predict fatigue limit of welded components containing undercuts. Methodology proved to be a useful tool to quantify the effect of different variables, like defect dimensions, overall weld geometry, load scheme and residual stresses, among others. Consequently, parametric studies can be developed to compare the importance of involved variables. Less meaningful variables can be simplified, focusing the attention to significant parameters.

Results were compared with tolerances adopted in current regulation for undercuts. It was demonstrated that undercut depth is the most important variable

describing weld behaviour, and not undercut radius, width or length, as it is assumed in many studies.
It was shown that good correlation exists between predictions and FAT values considered in documents. A decrease in tolerances is seen when expected fatigue strength is higher. Trend is properly described by outcomes obtained, and the majority of standard tolerances lay between predictions for stress-relieved welds (R = 0.1) and those considering residual stresses. Real welds may experience stress relief during fatigue damage, and therefore, associated curve might be located at an intermediate level. Correlation of fatigue limit obtained from theoretical S-N curves and prediction was best satisfied by an undercut 1 mm deep, at R = 0.5. However, S-N curves are known to be very conservative for long fatigue lives.
Methodology was also used to assess the effect of thickness, showing that predictions are safe for $t \leq 19$ mm. However, for thicker welds, a correction/reduction factor should be accounted for, as mentioned in normative. Exponent n = 0.2 best describes the trend for welds with undercuts, in accordance with literature.
Finally, the change of the loading mode from bending to traction does not seem to generate significant changes to predicted fatigue strength.

Authors express their gratitude to the funding provided by CONICET (Consejo Nacional de Investigaciones Científicas y Técnicas), and by Agencia Nacional de Promoción Científica Tecnológica, Argentina (PICT 2013 Nro.0110).

References

1. SS-EN ISO 5817. *Welding. Fusion-welded joints in steel, nickel, titanium and their alloys. Quality levels for imperfections*, ISO/TC 44/SC 10 (2014)
2. British Standard, *BS 7910:2013+A1:2015 Guide to methods for assessing the acceptability of flaws in metallic structures*, BSI Standards Limited (2013)
3. API Standard, *579-1/ASME FFS-1 2007, Fitness-For-Service*, New York: ASME (2007)
4. British Energy Generation Limited, *Assessment of the integrity of structures containing defects, R6 Revision 4, Amendment 7* (2009)
5. A.F. Hobbacher, *Recommendations for Fatigue Design of Welded Joints and Components*, 2nd ed., Springer International Publishing (2016)
6. Det Norske Veritas, *DNVGL-RP-C203. Fatigue design of offshore steel structures* (2016)
7. C. Steimbreger, M.D. Chapetti, Int. J. Fatigue, **105**, 296-304 (2017)
8. AWS, *AWS D1.1/D1.1M:2015. Structural Welding code-Steel*, American Welding Society (2015)
9. ASME, *ASME BPVC - Section I, Rules for Construction of Power Boilers*, American Society of Mechanical Engineers (2015).
10. ASME, *ASME BPVC - Section III, Division 1, Subsection NF: Supports*, American Society of Mechanical Engineers (2010).
11. ASME, *ASTM BPVC – Section VIII, Division I, Rules for Construction of Pressure Vessels*, American Society of Mechanical Engineers (2015).
12. British Standard, *BS 7608-2014: Guide to fatigue design and assessment of steel products*, BSI Standards Limited (2014)
13. British Standard, *BS-EN ISO 15609-1:2004: Specification and qualification of welding procedures for metallic materials.* BSI Standards Limited (2004)
14. H. Petershagen, Weld. World, **28**(7/8), 114-125 (1990)
15. K. Iida, M. Sato, M. Nagai, Weld. World, **39**(5): 262-271 (1997)
16. API, *API 1104: Welding of Pipelines and Related Facilities* , 21st ed., API Publishing Services, Washington, DC (2013)
17. B. Jonsson, G. Dobmann, A.F. Hobbacher, M. Kassner, G. Marquis, *IIW Guidelines on weld quality in relationship to fatigue strength*, IIW Collection, Springer International Publishing, Switzerland (2016)
18. J.R. Spadea, K.H. Frank, *Fatigue strength of fillet-welded transverse stiffeners with undercuts,* Research Report 0-4178-1: Fillet weld detailing for stiffener, Texas Department of Transportation (2004)
19. Volvo Group. *STD 181-0004. Weld Quality Standard* (2016)
20. T. Björk, J. Samuelsson, G.B. Marquis, *Weld. World*, **52**(1/2): 34-46 (2008)
21. B. Jonsson, J. Samuelsson, G.B. Marquis, *Weld. World*, **55**(11/12): 79-88 (2011)
22. W. Fricke, *IIW Recommendations for the fatigue assessment of welded structures by notch stress analysis IIW-2006-09*, Woodhead Publishing Limited, Cambridge, England (2012)
23. Det Norske Veritas, *DNVGL-CG-0129. Class Guideline: Fatigue Assessment of Ship Structures* (2015)
24. Dassault Systèmes, Simulia. *Abaqus 6.13 User manual*, Providence, RI, USA (2013)
25. M.D. Chapetti, J. Belmonte, T. Tagawa, T. Miyata, Sci. Tech. Weld. Join. **9**(5), 430-438 (2004)
26. M.D. Chapetti, Int. J. Fatigue, **25**(12), 1319–26 (2003)
27. T.R. Gurney, *Fatigue of welded structures*, 2nd ed., London: Cambridge UP (1979)
28. S.J. Maddox, *Fatigue strength of welded structures*, 2nd ed., Abington, Cambridge (1991)
29. S. Berge, Eng. Frac. Mechs., **21**(2), 423-435 (1985)
30. A.M. Al-Mukhtar, J. Fail. Anal. Preven., **13**, 63-71 (2013)
31. I. Lotsberg, Int. J. Fatigue, **66**, 86-100 (2014)

Through-process modelling of welding and service of 9Cr steel power plant components under load-following conditions

Padraig Mac Ardghail[1,2*], *Noel Harrison*[1,2], and *Sean B Leen*[1,2]

[1]Mechanical Engineering, College of Engineering and Informatics, NUI Galway, University Road, Galway, Ireland, H91 HX31.

[2]Ryan Institute for Environmental, Marine and Energy Research, NUI Galway, University Road, Galway, Ireland, H91 HX31.

Abstract. This paper is concerned with the development of a through-process model to predict the in-service performance of high-temperature, 9Cr steel, power plant components. A multi-pass welding simulation is conducted using the finite element software Abaqus. A user-material subroutine, including microstructure evolution and a physically-based constitutive model, is employed to predict the mechanical response of the material during welding and to predict welding residual stresses. Points are sampled from the FE geometry and their microstructure parameters and residual stress values are used in a uniaxial code to predict the relative in-service lives of the different weld regions under load-following power plant operating conditions. It is shown that post-weld heat treatment significantly improves predicted life and that there is a strong correlation between predicted microstructure before service and the predicted in-service life.

1 Introduction

Thermal power plant has traditionally operated under base-load conditions, where operating steam pressures and temperatures were highly consistent throughout the life of the power plant. However, energy demand is not consistent. More energy is used during the day than at night. Businesses require more energy during weekdays than at weekends. Advances in technology allow for real-time monitoring of energy demand and a more responsive energy network. Next-generation thermal power plant can adjust output to avoid oversupply of energy with associated wasted fuel and added costs. This responsive operating condition is known as load-following operation.

Load-following operation increases plant efficiency but it can lead to early failure of current-generation plant components. Increases and decreases in operating temperatures and pressures lead to fluctuating thermal, mechanical and thermomechanical loadings. This fluctuation exposes plant components to high-temperature fatigue, creep-fatigue and thermomechanical fatigue.

Fatigue loading causes early failure in 9Cr steels (e.g. P91, P92) due to strain-induced microstructure degradation, as shown by Sauzay et al. [1]. Welded components are particularly vulnerable to microstructure degradation due to microstructure changes which occur during the welding process.

9Cr steel parent metal (PM) has good high-temperature strength due to its hierarchical microstructure and precipitate nano-strengthening. Carefully-controlled heat-treatment processes create a PM with optimally-sized prior-austenite grains (PAG), precipitates of optimal size and with optimal spacing and with martensitic laths of optimal widths. Welding subjects this PM microstructure to high temperatures and heating-rates, which compromise the microstructure and thus the high-temperature strength.

The present study considers five main regions in a welded joint, namely the weld metal (WM), the coarse-grained, fine-grained and intercritical heat-affected zones (CGHAZ, FGHAZ, ICHAZ), and the PM. The ICHAZ and FGHAZ are of particular importance as they are traditionally the regions in which failure occurs in welded 9Cr steel components. This failure is known as Type IV failure [2]. Most work on Type IV cracking has focused on creep-dominated failure but high-temperature fatigue also produces ICHAZ failure, according to Farragher et al. [3] and Shankar et al. [4].

This work presents a through-process modelling methodology for simulation of welding and in-service operation of 9Cr steel components. The through-process methodology is implemented in four stages:

- A thermal model simulates the laying of separate weld beads and the heating and cooling experienced by the component.

* Corresponding author: p.macardghail1@nuigalway.ie

- A microstructure evolution model uses the temperature history from the thermal model to predict the evolution of solid-state phases, PAGs, precipitates, laths and hardness.
- A physically-based, temperature-dependent constitutive model uses the temperature history from the thermal model to apply thermal, metallurgical and transformation strains and the microstructure evolution to define the constitutive modelling parameters during welding.
- The 9Cr steel component is subjected to mechanical loading (load-following operation in the current work) and the physically-based constitutive model with its weld-predicted modelling parameters is used to predict mechanical performance.

2 Method

The first part of the through-process model is a purely thermal analysis. The FE geometry is described in Fig. 1. First, the whole FE model is heated to 350°C to represent inter-pass temperatures. All of the weld passes except for the first one are deactivated using Abaqus's 'model change' interaction. The first weld bead is heated using a temperature boundary condition which heats it from 0°C to 1505°C in 1s and holds at 1505°C for 0.5s. The boundary condition is then deactivated and the weld bead is allowed to cool for 45s. The next bead is activated over a period of 0.01s and then its temperature boundary condition is applied and so on until the final weld bead. The final weld bead has a longer cooling time, followed by a cooling boundary condition which reduces the temperature linearly to room temperature. This final cooling is done to expedite the analysis. Heat transfer is controlled by assigning the thermal properties from Yaghi et al. [5].

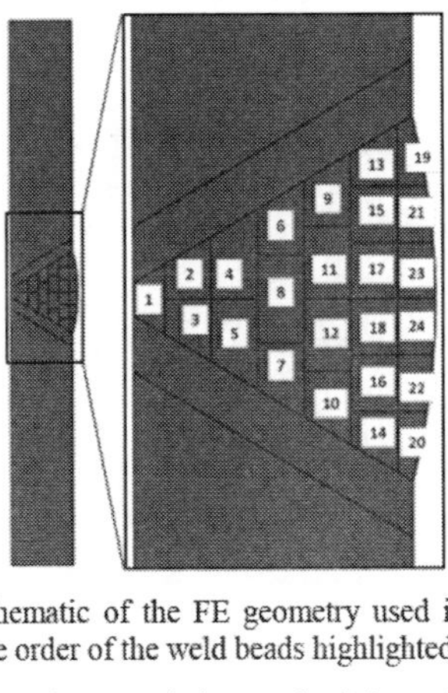

Fig. 1. A schematic of the FE geometry used in the present study with the order of the weld beads highlighted.

The second part of the methodology involves the microstructural model. There are five key microstructure parameters modelled in the present work. The first part involves phase evolution, namely of austenite, delta-ferrite, and untempered martensite. The PM is assumed to exist in a tempered state before welding.

Austenite transformation, a rapid process in this model, is described using a simple lever rule relating austenite volume fraction, V_γ, to temperature:

$$V_\gamma = 1 - \frac{A_{c3} - T}{A_{c3} - A_{c1}} \quad (1)$$

T is the current temperature between A_{c1} (830°C) and A_{c3} (924°C) [6]. The formation of delta ferrite follows a similar equation but with different transformation temperatures A_{c4} (1250°C) and A_{c5} (1450°C), based on the phase-transformation diagram for 9Cr steels [7]. If the grain size is predicted to be small (less than 1 μm) then the delta ferrite will resist transformation back to austenite upon cooling and become retained delta-ferrite in the weld. Austenite can similarly be retained as it cools, otherwise it transforms into martensite, which is considered 'untempered':

$$V_{\alpha'} = \lambda(1 - \exp[-0.011(M_s - T)]) \quad (2)$$

T is the current temperature below M_s (400°C) and λ is a constant used to ensure the volume fraction reaches a value of 1.

When the material transforms to austenite, the grain size evolves. Recrystallisation, represented by the recrystallized volume fraction, S, occurs according to:

$$\Delta S = \lambda_2 \Delta T \quad (3)$$

Due to the rapid nature of recrystallisation during welding, the authors considered a linear relationship using the constant λ_2 to be sufficient for the model. The volume fraction, S, controls grain size during recrystallization:

$$d = (d_0 - d_{min})(1 - S) + d_{min} \quad (4)$$

In Eqn. 4, d is the grain size, d_0 is the original grain size, d_{min} is a minimum grain size (to prevent grains of zero or negative sizes). After the recrystallization is complete, the grains grow as follows:

$$\dot{d} = (\lambda_3 - \lambda_4)(d_{lim} - d) + \lambda_5 \quad (5)$$

In Eqn. 5, d_{lim} is an asymptotic value of d towards which d rapidly converges. The rate term is divided into and isothermal term, λ_3, and a thermally-transient term, λ_4. During welding, λ_4 dominates and during heat treatment (HT), λ_3 dominates. After d reaches the value of d_{lim}, further growth continues according to λ_5. Eqns. 3, 4 and 5 were calibrated against published heat treatment and simulated welding data from the literature [8, 9].

Material hardness evolves according to:

$$\dot{H} = \lambda_6 (H_{lim} - H) \quad (6)$$

Eqn. 6 is what will be described in the rest of the paper as an 'asymptotic' equation, including a rate-term, here λ_6, and a limit term, H_{lim}. H_{lim} can be a large number at high temperature, e.g. during normalising, which will increase the hardness, H, to produce a hardened, untempered martensite state. At lower, tempering temperatures, H_{lim} may be smaller than H, leading to softening. Eqn. 6 was calibrated against HT data for P91 [10].

The evolution of lath-width is described as follows:

$$\dot{L}_w = \lambda_7 \left(L_{w\lim} - L_w \right) - \lambda_8 \dot{H} \tag{7}$$

The first term in Eqn. 7 is an asymptotic term to account for rapid lath-coarsening during PWHT or tempering. This part of Eqn. 7 was calibrated against data from [9]. The second term accounts for longer-term lath-coarsening during heat-treatment. This term was calibrated against data from [11].

The precipitate size evolution is described as follows:

$$\dot{m} = \lambda_9 \left(m_{\lim-1} - m \right) + \lambda_{10} \left(m_{\lim-2} - m \right) \tag{8}$$

The first asymptotic term in Eqn. 8 is for HT and the second asymptotic term is for ageing and creep. The first asymptotic term was calibrated against data from [9] and the second asymptotic term was calibrated against data from Zielinska et al. [12] and Guguloth et al. [13].

The third part of the through-process methodology, the mechanical model, has three aspects – monotonic stress-strain response, cyclic softening and long-term, creep-type damage. The three aspects are present in the main visco-plastic equation, shown here in uniaxial form:

$$\dot{\varepsilon}_p = \alpha \sinh\left(\beta f \right) \tag{9}$$

where f is the yield function, described as:

$$f = \frac{|\sigma - \chi|}{1 - D - D_c} + R - \sigma_y \tag{10}$$

where D is plasticity-damage, D_c is creep-type-damage, R is isotropic softening and χ is kinematic hardening. The kinematic hardening is based on a Taylor-type, dislocation-based formulation:

$$\dot{\chi} = \frac{\varsigma M G b \dot{\rho}}{\sqrt{\rho}} \tag{11}$$

where M is the Taylor factor, G is the material's shear modulus, b is the magnitude of the Burgers vector and ζ is a multiplicative constant. Dislocation-density evolution occurs according to:

$$\dot{\rho} = \frac{\rho_0 \dot{\bar{\rho}}}{\left(1 - \bar{\rho} \right)^2} \tag{12}$$

ρ_0 is the initial dislocation density and $\bar{\rho}$ is a normalised dislocation density, which has a value between 0 and 1, according to:

$$\bar{\rho} = 1 - \frac{\rho_0}{\rho} \tag{13}$$

This definition of normalised dislocation-density was used by Li et al. [14].

Normalised dislocation-density evolves according to:

$$\dot{\bar{\rho}} = c_1 \left(\bar{\rho}_{\lim} - \bar{\rho} \right) \left| \dot{\varepsilon}_p \right| \tag{14}$$

$\bar{\rho}_{lim}$ is a temperature-dependent value between 0 and 1, calibrated against 9Cr steel test data and c_1 is a temperature-dependent rate-term.

As similar set of equations is adopted for the cyclic evolution of R, as follows:

$$\Delta R = \frac{\varsigma M G b \Delta \rho_{cyc}}{\sqrt{\rho_{cyc}}} \tag{15}$$

$$\Delta \rho_{cyc} = \frac{\rho_0 \Delta \bar{\rho}_{cyc}}{\left(1 - \bar{\rho}_{cyc} \right)^2} \tag{16}$$

$$\bar{\rho}_{cyc} = 1 - \frac{\rho_0}{\rho_{cyc}} \tag{17}$$

ρ_{cyc} is taken as a cyclic dislocation-density, which evolves upon load-reversal. The dislocation-density accumulated during plastic loading (Eqn. 10) is used to update the 'cyclic' dislocation density:

$$\Delta \bar{\rho}_{cyc} = c_{1-cyc} \left(\bar{\rho}_{\lim-cyc} - \bar{\rho}_{cyc} \right) \bar{\rho} \tag{18}$$

This distinction between dislocation density within a cycle and between cycles is described schematically in Fig. 2, showing the change in $\bar{\rho}$ and the definition of $\Delta \bar{\rho}_{cyc}$.

Damage evolution is controlled via a linear relationship with plastic strain:

$$\dot{\omega}_1 = a_1 \left| \dot{\varepsilon}_p \right| \qquad \varepsilon_p > a_2 \tag{19}$$

This damage term was calibrated against monotonic data. Equation 19 describes post-necking damage, hence the threshold plastic strain, a_2. A second damage term of a similar form describes final catastrophic damage, as follows:

$$\dot{\omega}_2 = a_3 \left| \dot{\varepsilon}_p \right| \qquad \varepsilon_p > a_4 \tag{20}$$

The justification for Eqns. 19 and 20, as well as parameters a_2 and a_4, is displayed schematically in Fig. 3. Combined, the damage term is as follows:

$$\dot{D} = \dot{\omega}_1 + \dot{\omega}_1 \tag{21}$$

Equation 21 corresponds to plasticity damage primarily associated with the manufacturing process, as opposed to in-service operation. For in-service conditions, creep damage is important. Creep test data from [13], was analysed to approximate creep-type damage, as follows:

$$\dot{D}_c = \frac{a_5 \dot{m}}{L_w} \qquad e_c^{Ideal} > 0.045 \tag{22}$$

This damage term is microstructure-driven. In this case, precipitate-coarsening $\dot{m}$ is the main driving force, although narrow, untempered laths accelerate the damage evolution. e_c^{Ideal} is a threshold value above which long-term, creep-type damage occurs. This threshold value is derived from creep strain data recorded in [13]. The tertiary creep, assumed to correspond to the onset of damage, appeared to occur at an approximate creep-strain of 0.045 under three different conditions. The present work does not attempt to simulate creep strain, which would require its own constitutive model, e.g. a power-law model, such as in [15], or a hyperbolic sine model, such as in [16]. Instead, the creep-strain data was correlated with microstructure and microstructure-evolution data and is presented here as an idealised creep-strain, as follows:

$$\dot{e}_c^{Ideal} = \frac{c_3 \dot{m}}{L_w} \tag{23}$$

The identification of Eqns. 22 and 23 is illustrated in Fig. 4, with Eqn. 23 corresponding to Fig. 4b and Eqn. 22 corresponds to Fig. 4c.

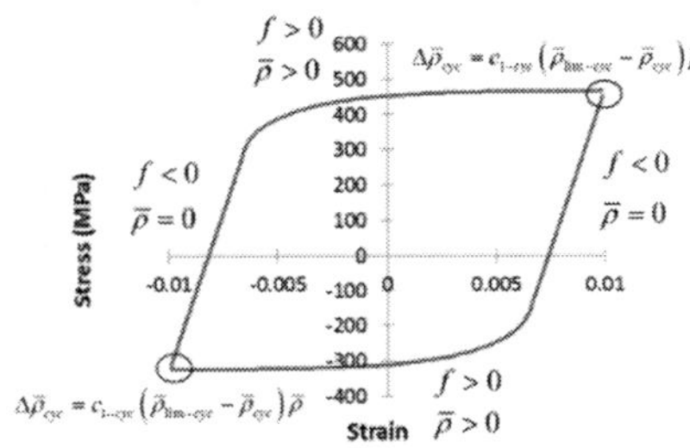

Fig. 2. A schematic representation of the dislocation-density evolution within and between fatigue cycles.

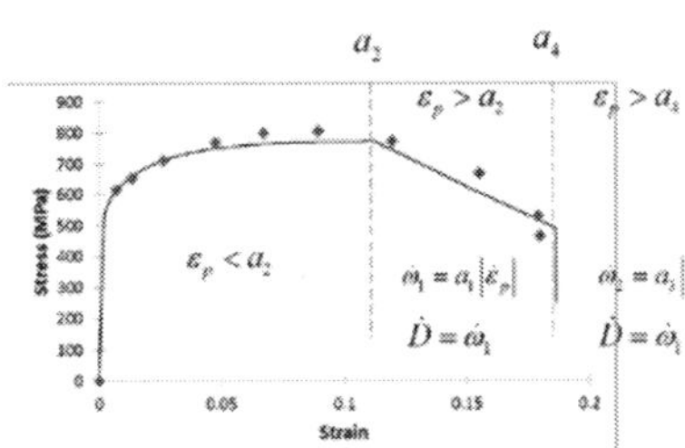

Fig. 3. A schematic representation of the separate plasticity-damage parameters acting on a sample monotonic stress-strain curve (Data points from [6]).

After the FE model is welded, PWHT was applied to the whole geometry (750°C for 70 minutes, as per ISO/TR 14745). The constitutive model was able to predict residual stresses. The residual stresses relaxed during PWHT if the material hardness was reducing due to PWHT (Eqn. 6). In this case, the stress reduced according to a similar asymptotic equation, using the same rate-term as in Eqn. 6 but with an asymptotic value of 0 MPa.

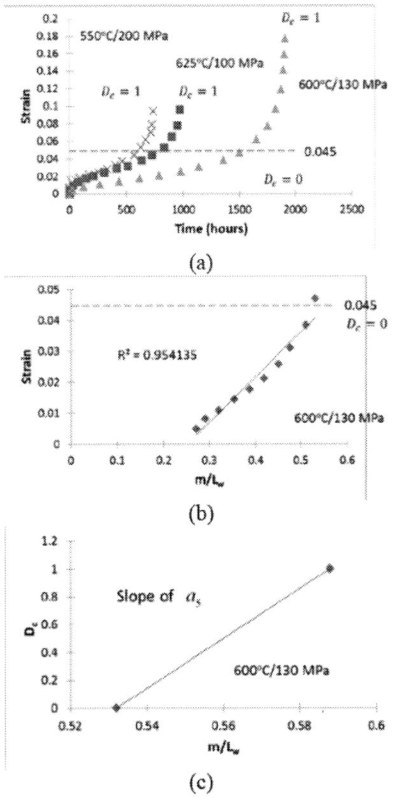

Fig. 4. (a) Creep-test data, from [13], showing the assumptions of threshold creep strain, and initial and final creep-damage values, (b) the correlation between creep-strain, the precipitate diameter, m, and the lath-width, L_w, up to the threshold value for one (sample) set of creep-test data, and (c) the correlation between creep-damage, the precipitate diameter, m, and the lath-width, L_w for one (sample) set of creep-test data.

After PWHT was simulated, an internal pressure load and a temperature boundary condition were applied, based on plant data. Based on the data, the material was cycled between 616°C and 570°C and between 8.5 MPa and 12.5 MPa (see Fig. 5a and 5b), with drops to 80°C and 0 MPa to simulate shutdowns. Each operating cycle, from start-up to shutdown, was modelled as lasting 57 hours.

Several FE elements were sampled along the inside surface of the FE pipe to gather data for microstructure and residual stress in the PM, HAZ and at the weld centreline (WCL). Due to the duration of in-service loading, the authors decided to expedite the simulation by using the sampled data in a uniaxial code.

https://doi.org/10.1051/matecconf/201816521010

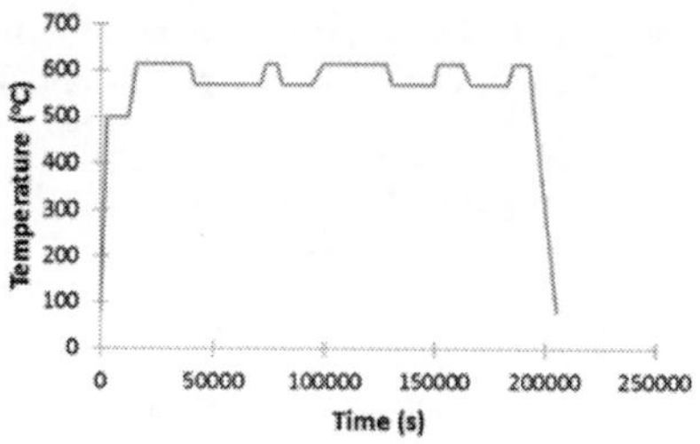

(a)

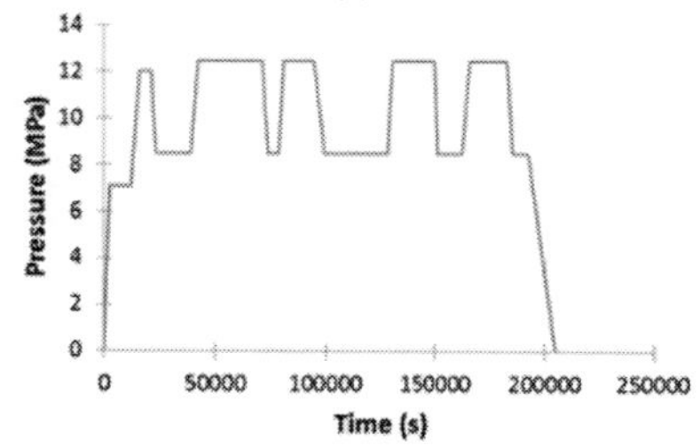

(b)

Fig. 5. The proposed cyclic operating conditions for the simulated plant component, (a) temperature and (b) pressure

The internal pressure (Fig. 5b), was converted into an equivalent stress using Lamé's equations for a thick-walled pipe. This equivalent stress was converted to an equivalent strain for the purposes of the material model, which is a strain-controlled. Microstructure equations were retained to enable in-service microstructure changes. The as-welded and PWHT cases for the pipe were both simulated.

Each load cycle is approximately 57 hours long (Fig. 5). It was decided to simulated one year's worth of service (8766 hours, approx 154 cycles) and to compare the results for the different sampled points.

3 Results

3.1. Microstructure evolution - Welding

The results of the microstructure model are as follows. Figure 4 shows the as-welded case. In Fig. 6a, the PM PAG size (38 μm) is unaffected; a thin, fine-grained (9 to 13 μm) region results from the welding, consistent with the FGHAZ, and then the PAG sizes grow until they reach their largest size in the WM (55 μm). In Fig. 6b, the WM is predicted to be a harder region than the PM (447 kgf/mm^2, 226 kgf/mm^2, respectively), although it is clear that the later weld passes softened the earlier passes via auto-tempering, consistent with results found in [6][6]. In Fig. 6c, the predicted precipitate sizes are displayed showing clear evidence of auto-tempering, with precipitates in the earlier weld beads being larger than those in the final weld beads. The as-welded laths in the WM and HAZ are narrower than the unaffected PM, as expected (Fig. 6d).

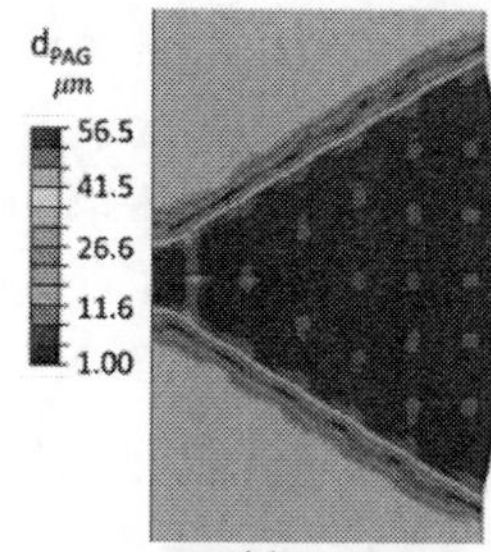

(a)

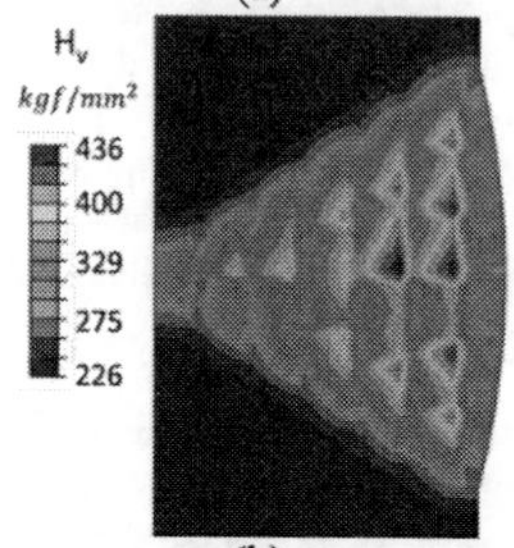

(b)

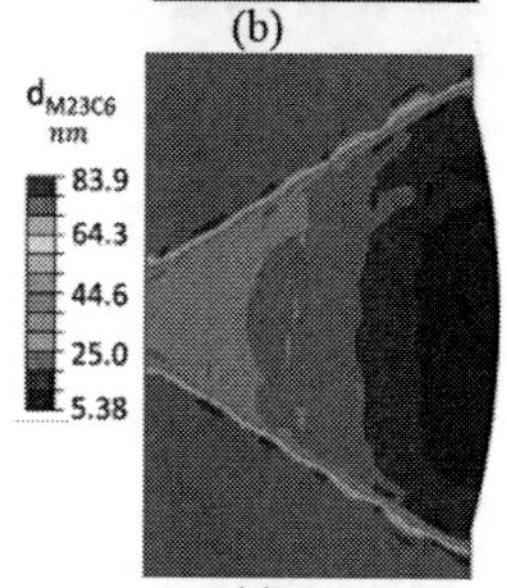

(c)

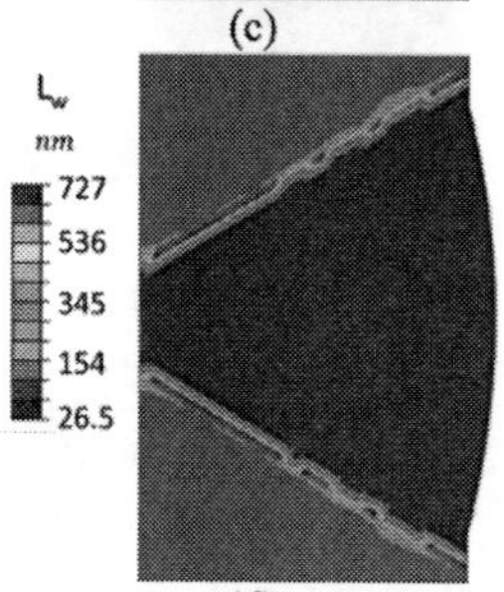

(d)

Fig. 6. As-welded microstructure predictions for (a) prior-austenite grains, (b) Vickers hardness, (c) precipitates and (d) lath width.

Following PWHT (Fig. 7), the hardness values equalise across the HAZ, with values ranging from 251 kgf/mm^2 to 208 kgf/mm^2. Laths in the WM grow from 30 nm to 500 nm, while the PM laths grow from 603 mm to 643 mm, though the FGHAZ still sees some relatively narrow laths (380 nm). The precipitates grow from between 8 and 28 nm in the WM to 69 nm, compared to the PM's 76 nm. Thus PWHT is predicted to homogenise the HAZ.

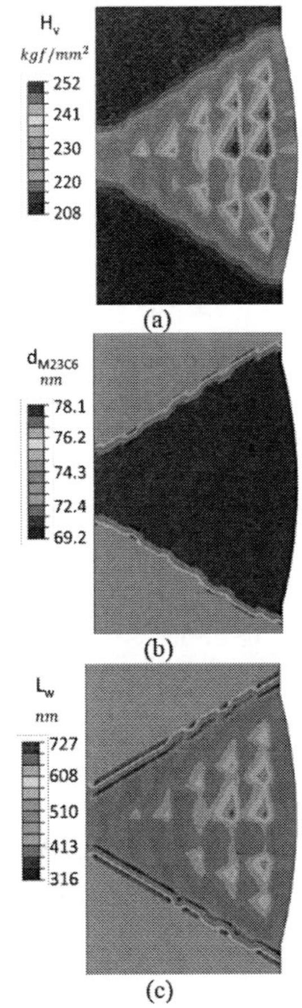

Fig. 7. Microstructure predictions for (a) hardness, (b) precipitates and (c) lath width after PWHT.

3.2. In-service modelling

The sampled points (Fig. 8) for the in-service analysis are described in Tables 1 and 2, for the as-welded and PWHT cases, respectively. Each value is the average value of the four integration points in the sampled element. The stress values correspond to equivalent residual stress.

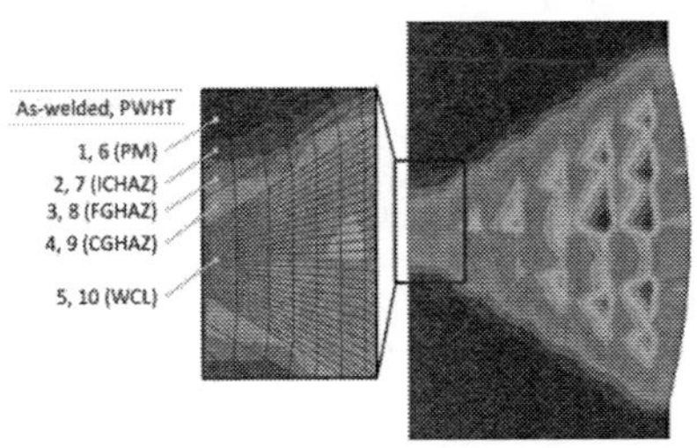

Fig. 8. An illustration of the sampled points (1 to 5 for the as-welded case, 6 to 10 for the PWHT case) extracted from the FE simulation for use in in-service modelling.

The PM was identified by the fact that its microstructure was unaffected by the welding process. The ICHAZ was identified because its slightly reduced PAG size compared to the PM indicated partial recrystallization, and therefore only partial austenite transformation.

Table 1. A description of the sampled points along the pipe wall interior in the as-welded condition

Point	PAG (μm)	Position	Stress (MPa)	Hardness (kgf/mm^2)	Lath width (nm)	Precip. Dia. (nm)
1	38	PM	237	226	603	76
2	27.5	ICHAZ	401	239	575	74
3	9.65	FGHAZ	404	257	189	31.6
4	42.1	CGHAZ	436	273	41	8.46
5	55.2	WCL	473	293	29	8.48

Table 2. A description of the sampled points along the pipe wall interior in the PWHT condition

Point	PAG (μm)	Position	Stress (MPa)	Hardness (kgf/mm^2)	Lath width (nm)	Precip. Dia. (nm)
6	38	PM	43.6	208	643	76.3
7	27.5	ICHAZ	76.7	210.5	638	75
8	9.65	FGHAZ	75.2	214	449	70.3
9	42.1	CGHAZ	79.1	217	443	69.2
10	55.2	WCL	83.1	220	452	69.2

The FGHAZ and CGHAZ both fully transformed to austenite during welding but they could be distinguished by their PAG sizes. The location of the WCL was known from the geometry of the weld rather than by its microstructure.

The predicted time to failure, t_f, in hours, for each of the sample points are displayed in Fig. 9 and Table 3. Table 3 includes the relative increase in life from the as-welded condition, t_f^{AW}, to the PWHT condition, t_f^{PWHT}.

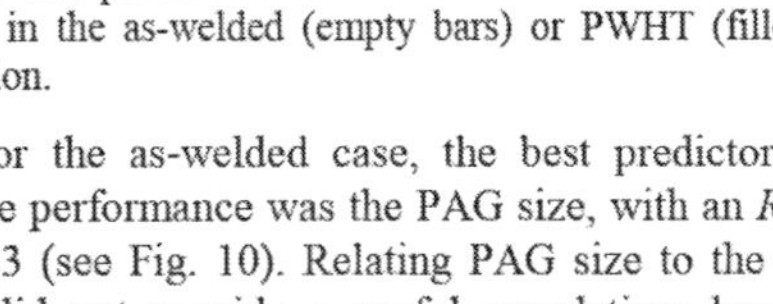

Fig. 9. The predicted time to failure for each of the sample points in the as-welded (empty bars) or PWHT (filled bars) condition.

For the as-welded case, the best predictor of in-service performance was the PAG size, with an R^2 value of 0.83 (see Fig. 10). Relating PAG size to the PWHT case did not provide a useful correlation; however a better correlation was found by using hardness (R^2 =

0.80), the lath-widths (R^2 = 0.87) or the precipitate size (R^2 = 0.93, see Fig. 11).

Table 3. The predicted times to failure for each of the sample points, and the relative increase in life caused by PWHT.

Point	t_f^{AW}	Point	t_f^{PWHT}	$\frac{t_f^{PWHT}}{t_f^{AW}}$
1	1026	6	3500	3.4
2	456	7	2590	5.7
3	218	8	1614	7.4
4	1015	9	1553	1.5
5	1018	10	1564	1.5
3	**218**	9	**1553**	**7.1**

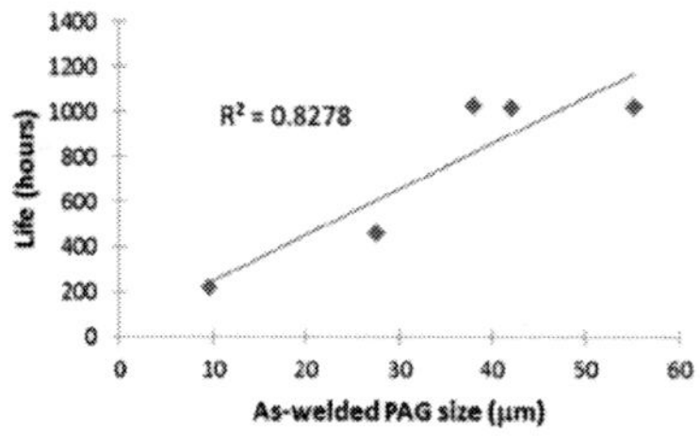

Fig. 10. The correlation between the predicted time to failure for each of the sample points in the as-welded condition and the predicted PAG sizes.

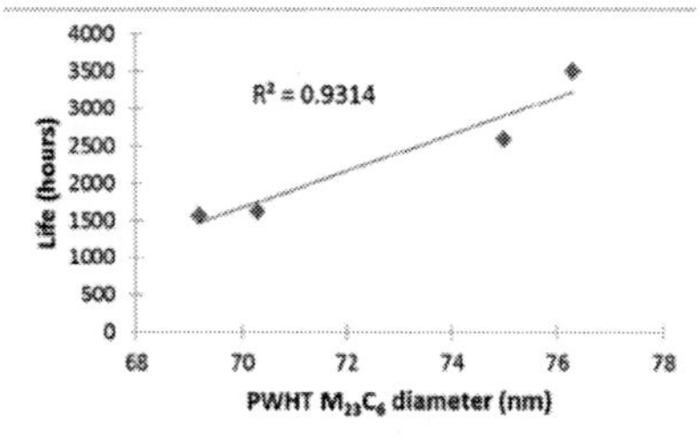

Fig. 11. The correlation between the predicted time to failure for each of the sample points in the PWHT condition and the predicted carbide diameters.

4 Discussion

The thermal model follows a well-established methodology of using block-dumped weld-bead elements to heat the surrounding PM. Heat-transfer properties from previously-published work [5] were adopted. The sequentially-coupled thermal and mechanical analysis is a common way to model welding processes [17, 18], and the use of an axi-symmetric model is a common way of modelling pipe components, which has been shown to give reasonable results compared to 3D models [19, 20].

The microstructure evolution model is based on test data from the literature. This gives the authors confidence that it can accurately represent microstructure evolution; however the model has a large number of modelling parameters which could be reduced via the use of even more physically-based models, e.g. the lath-width evolution model in [21].

The constitutive model is based on a range of conventional modelling approaches, e.g. Taylor hardening, Kocks-Mecking dislocation-density evolution, hyperbolic sine functions, etc. Relating constitutive modelling parameters to microstructure features is typical of physically-based modelling.

The in-service aspect of the model was applied uniaxially to sample points rather than to the entire FE geometry (Fig. 6). Ideally, the load-following conditions would be applied to the FE geometry but the authors found this to be too computationally-expensive and time-consuming for the current work. A full FE in-service simulation will be attempted in future work.

The as-welded pipe is predicted to have significantly lower service life than the PWHT pipe. This is an expected and encouraging result. There were some interesting predictions made as to the relative performance of different weld regions. The location with the lowest life in the PWHT pipe is predicted to be the CGHAZ, contrary to expectations regarding Type IV cracking, although there is little difference between predicted IC-FGHAZ, CGHAZ and WM lives. This indicates that the PWHT successfully removed any regions which were obviously weak compared to other regions.

Failure was dominated by the long-term damage term (Eqn. 22) compared to plasticity damage (Eqn. 21). For the most part, yield did not occur in the material under the prescribed loading conditions and when it did, the plastic strain was insufficient to activate plasticity damage.

The weakest point in the as-welded pipe was the FGHAZ, as expected [2, 22]. PWHT extended the life of this region by a factor of 7.4. The as-welded CGHAZ and WM both had similar predicted lives, as expected from their similar microstructures (Table 1).

The as-welded CGHAZ and WM might seem as though they should predict the shortest lives but their high hardness allows for in-service tempering and relaxation of residual stresses. The in-service tempering allowed the lath-width to increase, thus slowing the onset of long-term damage (Eqns. 22, 23). The FGHAZ is not sufficiently hard for tempering to occur and its detrimental microstructure takes its toll on the predicted in-service life.

PWHT appears to have been most beneficial for the PM and ICHAZ regions by conferring them with the widest laths. This indicates that damage and failure are less likely to occur in these regions. This is somewhat contrary to expectations concerning the ICHAZ; however, ICHAZ failure occurs in fine-grained ICHAZ regions [22], whereas the ICHAZ region predicted by this work has coarser grains.

The predicted in-service lives correlated well with PAG size in the as-welded case, mainly because it was possible for the other microstructure features to evolve during service. For the PWHT samples, however, the microstructures were more-or-less in their final form and less likely to be affected by in-service loading, so all microstructure features except PAG sizes could serve as indicators of performance.

Quantitatively, the model predicts very short lives for all sampled regions. The longest predicted life was just 3500 hours, or 146 days, or 0.4 years. This is most likely due to the idealisations and assumptions introduced in Eqns. 22 and 23. Future work will require a more detailed and rigorous creep-type damage model but, for the present study, Eqns. 22 and 23 provide the through-process methodology with some capability to predict in-service performance. The qualitative trends predicted by this work are promising. Overall, this work presents a step towards a through-process model for power plant steels and a step towards a weld design tool for industry.

5 Conclusions

- Multi-pass welding and PWHT was simulated for a thick-walled, 9Cr steel, pipe component.
- Microstructure evolution, including grain size evolution, distribution of tempered and untempered regions, hardness, martensitic laths and precipitate size, was modelled.
- Auto-tempering effects were captured in the model and PWHT was predicted to equalise the HAZ microstructure.
- The welding simulation predicted a fine-grained region a short distance from the fusion zone, consistent with the IC-FGHAZ. PWHT does not fully recover the microstructure of this region, leaving relatively narrow laths.
- Realistic load-following plant operating conditions were simulated for discrete points sampled from the pipe interior to represent the PM, ICHAZ, FGHAZ, CGHAZ and WM in both the as-welded and PWHT conditions.
- The as-welded pipe was predicted to have a significantly shorter service life than the PWHT pipe, with the fine-grained region being the first to fail. The PWHT pipe predicted a similar life for the FGHAZ, CGHAZ and WM and that failure initiates between the FGHAZ and the WM.

6 Acknowledgements

This research is funded by Science Foundation Ireland grant number SFI/14/IA/2604. The authors would like to acknowledge the contributors to this research; NUI Galway, the Ryan Institute, General Electric (UK), ESB International, University of Limerick, Imperial College, London and Fraunhofer IWM, Freiburg. The authors would also like to thank Mr. S. Scully for supplying load-following plant temperature and pressure data for use in this work.

7 References

1. M. Sauzay, H. Brillet, I. Monnet, M. Mottot, F. Barcelo, B. Fournier, A. Pineau, Mat. Sci. Eng. A **400-401** (2005)
2. J. A. Francis, W. Mazur, H. K. D. H. Bhadeshia, Mat. Sci. Tech. **22**, 12 (2006)
3. T. P. Farragher, S. Scully, N. P. O'Dowd, C. J. Hyde, S. B. Leen, J. P. V. T. **136** (2014)
4. V. Shankar, R. Sandhya, M. D. Mathew, Mat. Sci. Eng. A, **528** (2011)
5. A. H. Yaghi, T. H. Hyde, A. A. Becker, J. A. Williams, W. Sun, J. Mat. Proc. Tech. **167** (2005)
6. A. H. Yaghi, T. H. Hyde, A. A. Becker, W. Sun, J. P. V. P. **111-112** (2013)
7. H. Cerjak, Proc. International Conference Safety and Reliability of Welded Components (2008)
8. L. Milovic, T. Vuherer, I. Blacic, M. Vrhovac, M. Stankovic, Materials and Design **46** (2013)
9. P. J. Ennis, A. Czyrska-Filemonowicz, O. M.M. I. **1** (2002)
10. G. Potirniche, I. Charit, K. Rink, F. Barlow, NEUP 2009 Project, **09-835** (2013)
11. D. R. Barbadikar, G. S. Deshmukh, L. Maddi, K. Laha, P. Parameswaran, A. R. Ballal, D. R. Peshwe, R. K. Paretkar, M. Nandagopal, M. D. Mathew, I. J. P. V. P. **132-133** (2015)
12. A. Zielinski, M. Miczka, B. Boryczko, M. Sroka, Archives of Civil and Mechanical Engineering, **16** (2016)
13. K. Guguloth, J. Swaminathan, N. Roy, R. N. Ghosh, Mat. Sci. Eng. A, **684** (2017)
14. H. Li, J. Lin, T. A. Dean, S. W. Wen, A. C. Bannister,. I. J. Plasticity, **25** (2009)
15. S. Khayatzadeh, D. W. J. Tanner, C. E. Truman, P. E. J. Flewitt, Eng. Frac. Mech., **175** (2017)
16. C. Ó Murchú, S. B. Leen, P. O'Donoghue, R. A. Barrett, Mat. Sci. Eng. A, **682** (2017)
17. K. Abburi Venkata, C. E. Truman, R. C. Wimpory, T. Pirling, I. J. P. V. P. (2017) (In press)
18. L. E. Lindgren, Computer methods in applied mechanics and engineering, **195** (2006)
19. P. Dong, J. P. V. T. **123** (2001)
20. D. Deng, H. Murakawa, Comp. Mat. Sci., **37** (2006)
21. E. I. Galindo-Nava, P. E. J. Rivera-Diáz-del-Catillo, Acta Materialia, **98** (2015)
22. J. S. Lee, K. Maruyama, Metallurgical Materials International, **21** (2015)

https://doi.org/10.1051/matecconf/201816521011

Failure Assessment Diagram (FAD) analysis of fatigue test results for X65 welded joints

Hsin Jen Hoh*, *John Hock Lye* Pang, and *Kin Shun* Tsang

School of Mechanical & Aerospace Engineering, Nanyang Technological University, 50 Nanyang Avenue, Singapore 639798, Singapore

Abstract. Offshore pipelines transfer oil and gas from seabed to production facility on the surface. The long pipelines are formed by welding of pipe segments, where these welded joints are a source of stress concentration and defects from which fatigue cracks can grow. This work aims to study the behaviour of deep fatigue cracks. In this current work, finite-element based on a parametric study of four-point bending is used to assess the stress intensity factors (SIFs) of deep surface cracks in X65 specimens, while considering local limit load the remaining load bearing ligament. These deep cracks take on a non-regular shape and have widths that exceed that of the specimen. They will be compared to empirical expressions from derived standards such as British Standards BS7910, which may be more conservative. The existing large flaw is also assessed via the failure assessment diagram (FAD). The effects of limit load solutions and reference stresses used to determine the FAD diagram will be discussed.

1 Introduction

The degradation of transmission pipelines systems due to fatigue by environmental loads or corrosion leads to formation of cracks that initiate from regions of high stresses such as weld joints. These joints connect pipe segments to form a long pipeline, but are sources of stress concentration and defects from which fatigue cracks can grow.

1.1 Background

In prior testing [1], butt-welded X65 offshore pipes (outer diameter 32" and 1" thickness) were cut into fatigue coupons (t= 25.4 mm, b= 100 mm). A four-point bending test (L= 240 mm, l= 120 mm) was setup (Fig. 1) such that tensile stresses occurred at the pipe surface with weld cap. The four-point bending fatigue test was conducted at various loads (load ratio 0.1) until the coupons could no longer support fatigue cyclic loading. The load and their corresponding fracture surface were recorded in Table 1.

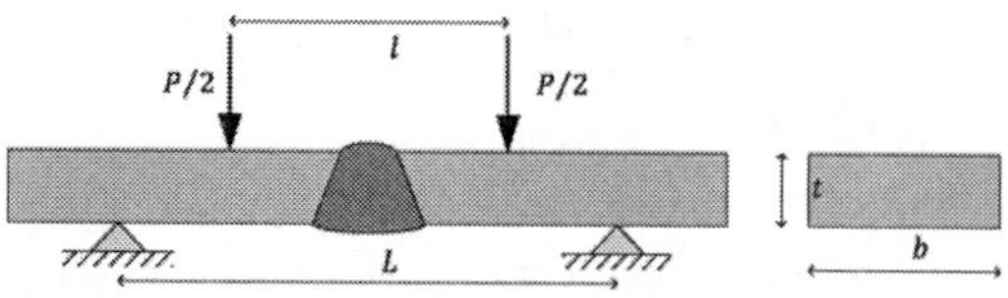

Fig. 1. Four-point fatigue bending setup of X65 steel coupons [1].

While linear elastic fracture mechanics approaches such as the solutions for semi-elliptical cracks in plates or pipes [2-3] apply for smaller sized defects, they do not apply to these deep or larger sized cracks, when plastic collapse becomes significant.

This work will employ the Failure Assessment Diagram (FAD) to determine if an existing deep flaw in X65 steel pipe is acceptable by: investigating SIFs of a varying depth edge crack that has progressed beyond specimen width; and understanding the effect of plasticity in such large flaws.

Table 1. Bending fatigue fracture surfaces [1]

No	Stress range MPa	Fracture surfaces
1	292.5	
2	320	
3	337.5	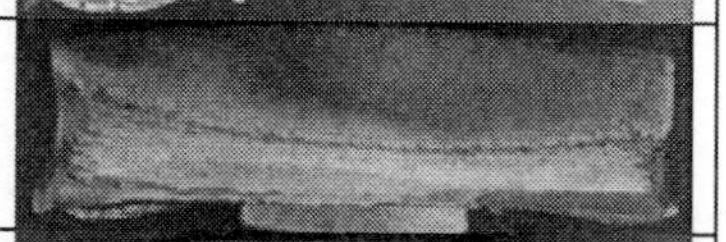
4	382.5	
5	450	

* Corresponding author: hjhoh@ntu.edu.sg

https://doi.org/10.1051/matecconf/201816521011

1.2 Failure Assessment Diagram (FAD)

The BS7910:2013 [4] outlines fracture assessment by means of a Failure Assessment Diagram (FAD). The FAD establishes the relationship between brittle fracture (by fracture parameter K_r, equation 1) with local plasticity (plastic collapse parameter L_r, equation 2). Plastic collapse could be in the form of global or local collapse [5].

$$K_r = \frac{K_I}{K_{IC}} \quad (1)$$

$$L_r = \frac{\sigma_{app}}{\sigma_{limit}} = \frac{\sigma_{ref}}{\sigma_Y} \quad (2)$$

The fracture parameter is an indicator of fracture under linear elastic conditions, while the plastic collapse parameter could be defined as ratio of applied load to limit load (equation 2), where limit load refers to the remote load that will cause spread of plasticity. Equation 2 also show the relation of the limit load to the parameter reference stress σ_{ref} parameter, also commonly used to determine the plastic collapse parameter.

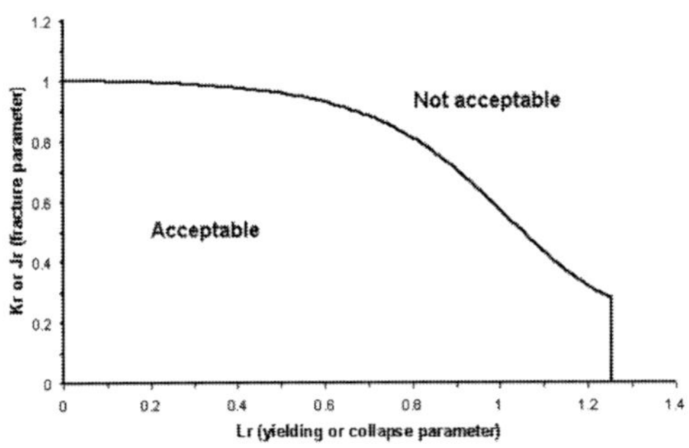

Fig. 2. A typical failure assessment diagram (FAD) [6].

The FAD typically comprises an assessment envelope given by the equation of a curve $K_r = f(L_r)$, and a cut-off value of L_r, $L_{r,max}$ (Fig. 2). If the assessment point plotted lies within the area bounded by the axes, assessment line and $L_{r,max}$ line, the flaw is acceptable. The simplest (option 1) FAD in the BS7910:2013 does not require stress-strain data. The governing equations of the assessment envelope for this case are

$$f(L_r) = (1 + L_r^2/2)^{-\frac{1}{2}}\left[0.3 + 0.7e^{-\mu L_r^6}\right] \text{ for } L_r \le 1 \quad (3)$$

$$f(L_r) = L_r^{(N-1)/(2N)} \quad \text{for } 1 < L_r < L_{r,max} \quad (4)$$

$$f(L_r) = 0 \quad \text{for } L_r \ge L_{r,max} \quad (5)$$

where

$$\mu = \min\left(0.001\frac{E}{\sigma_Y}, 0.6\right) \quad (6)$$

$$N = 0.3\left(1 - \frac{\sigma_Y}{\sigma_u}\right) \quad (7)$$

$$L_{r,max} = \frac{\sigma_Y + \sigma_u}{2\sigma_Y} \quad (8)$$

The more complicated options outlined (options 2 and 3) require more input in terms of material and stress analysis with higher accuracy. For example, the option 2 approach requires detailed stress-strain data, especially at strains below 1%. The FAD envelope needs to be determined at least for certain values of L_r. Option 3 requires elastic and elastic-plastic analysis to determine J-integral values.

The analysis requires material properties such as yield strength σ_Y, ultimate tensile strength σ_u, elastic modulus E and fracture toughness K_{IC}. The yield strength and ultimate tensile strength of the X65 steel were obtained via relation to hardness [7] (σ_Y = 550 and σ_u = 633 MPa), while the fracture toughness of 280 MPa $\sqrt{m}$ was used [8].

To determine the location of the evaluation point on the FAD, requires K_I as well as limit load or reference stress. Some solutions for reference stress are available from the BS7910:2013. For the current study of large flaws, the closest crack geometries would be that of a semi-elliptical surface crack (Fig. 3) and an extended surface flaw (Fig. 4).

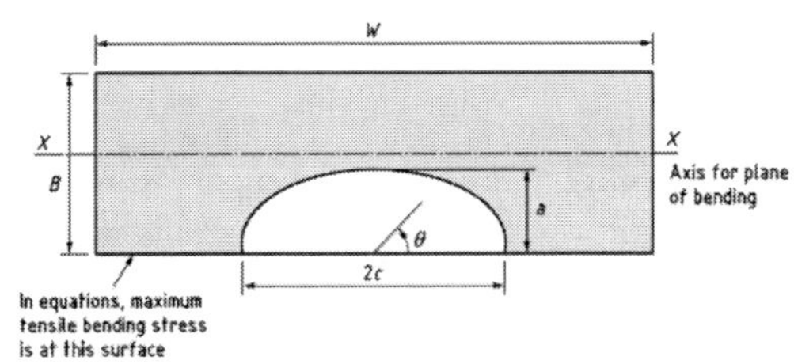

Fig. 3. Geometry of semi-elliptical surface crack in plate [4].

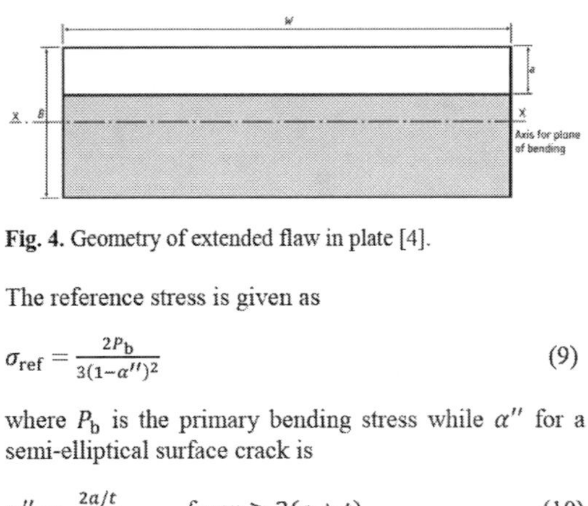

Fig. 4. Geometry of extended flaw in plate [4].

The reference stress is given as

$$\sigma_{ref} = \frac{2P_b}{3(1-\alpha'')^2} \quad (9)$$

where P_b is the primary bending stress while α'' for a semi-elliptical surface crack is

$$\alpha'' = \frac{2a/t}{1+\left(\frac{t}{c}\right)} \quad \text{for } w \ge 2(c+t) \quad (10)$$

$$\alpha'' = 2\left(\frac{a}{t}\right)\left(\frac{c}{w}\right) \quad \text{for } w < 2(c+t) \quad (11)$$

while α'' for an extended surface flaw is

$$\alpha'' = \frac{a}{B} \quad (12)$$

However, these current empirical expressions from the BS7910:2013 in determining this reference stress are overly conservative.

In the current work, the K_I is determined from linear elastic fracture mechanics finite element analysis, while the limit load is determined from an iterative finite element analysis.

2 Fracture parameter

The X65 fatigue bending specimens considered have developed flaws that are large (beyond plate width, and usually more than half of plate thickness). The irregular-shaped flaws will be studied via finite element approach.

2.1 Stress intensity factor solutions of interest

In determining the fracture parameter, several available stress intensity factor (SIF) solutions were of interest. This includes BS7910:2013 SIF solutions for the surface semi-elliptical crack in a plate and extended surface flaw [4] and another solution of extended surface flaw under bending [9]. However, these solutions are limited to smaller sized flaws.

2.2 Finite element model of plate with deep flaws

The extended finite element method (XFEM) was used to study the deep flaws in the specimen. The approach simplifies conventional finite element crack modelling by allowing for an approximation of cracks without the need for intricate meshing of the crack front.

Measurements were taken at the side edges of the crack front, one-quarter length from the sides, as well as the final crack depth. Based on the available information, the crack profile is built to approximate the actual crack profile (Fig. 5).

The test specimen (cut from a pipe) was modelled as a plate, ignoring effects of pipe curvature as it had a large radius relative to thickness. The length of the model corresponds to the constant bending moment section in four-point bending. One end of the plate model was fixed, while the surface on the other end was constrained to a single point (Fig. 6a). The bending moment was applied to this point.

The mesh was composed of 8-noded reduced integration cubic elements (C3D8R). A finer mesh was used around the crack front (Fig. 6b).

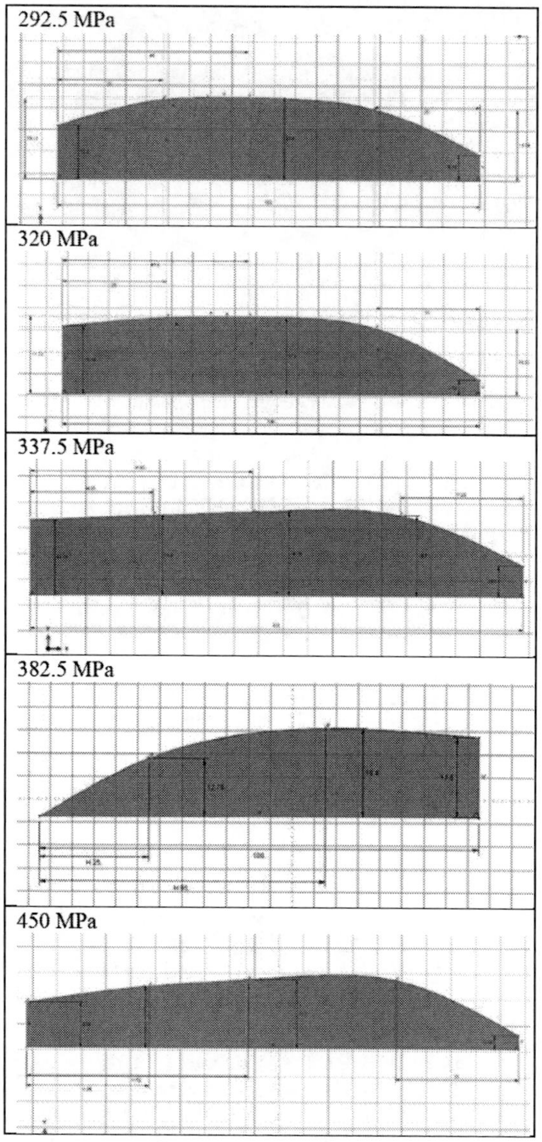

Fig. 5. Measured crack profiles used in finite element model.

(a) Solid model and boundary condition setup.

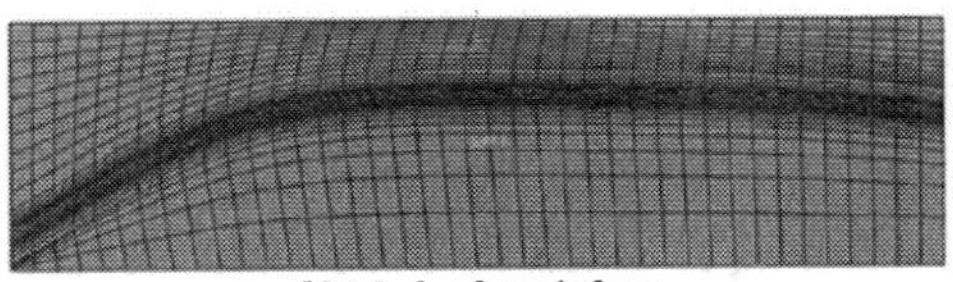

(b) Mesh of crack front

Fig. 6. Model of Specimen 2, of stress range 320 MPa.

A sample von Mises stress distribution at the crack plane is shown in Fig. 7. The resulting Y-factors are shown in Table 2, with the SIFs determined from

$$K_I = Y\,\sigma\sqrt{\pi a_i} \tag{9}$$

where a_i is the depth of the crack front location from the free surface (Fig. 8).

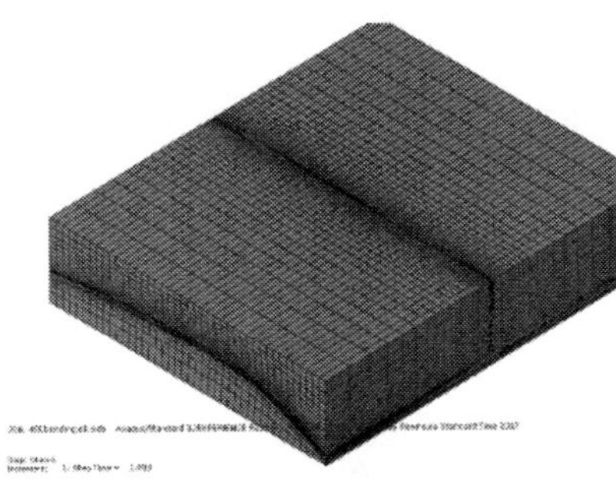

(a) Finite element model

(b) Stress distribution at crack plane

Fig. 7. Model of Specimen 3, of stress range 337.5 MPa.

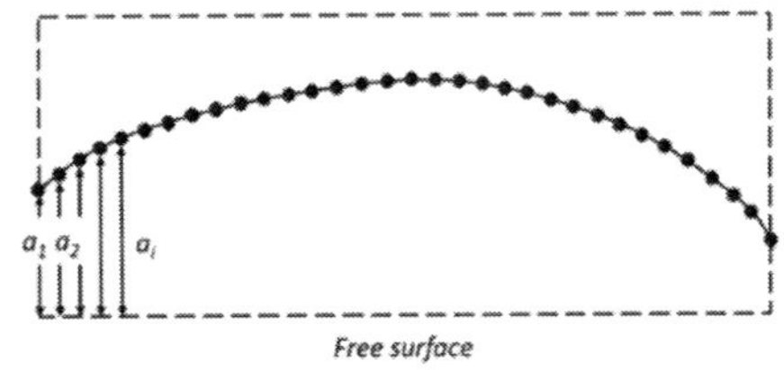

Fig. 8. Depth of crack front location from free surface.

Table 2. Stress range and stress intensity factors (bending)

No	Stress range MPa	Shape Factor Y
1	292.5	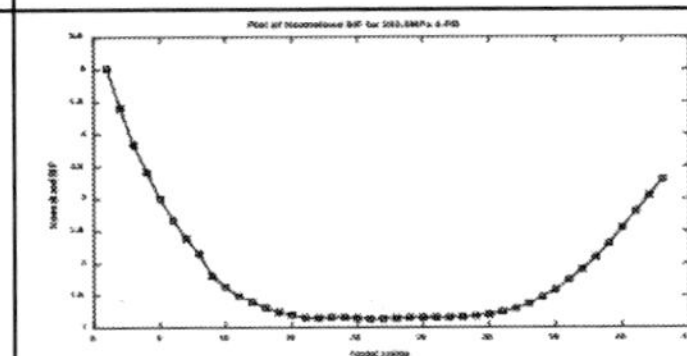
2	320	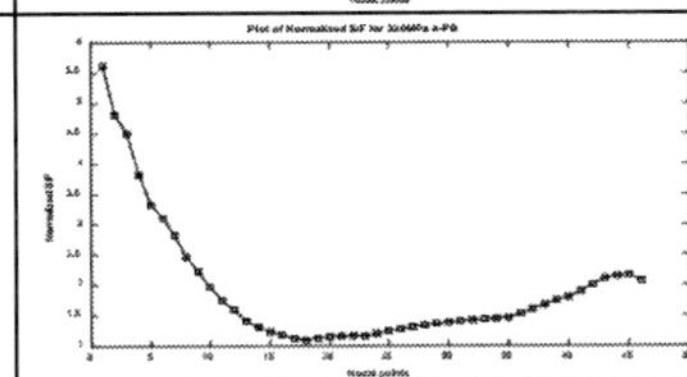
3	337.5	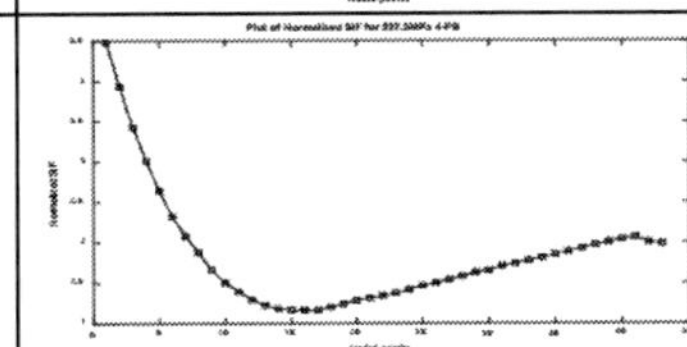
4	382.5	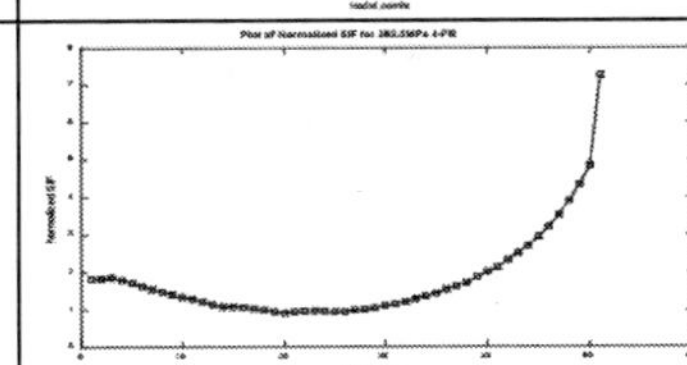
5	450	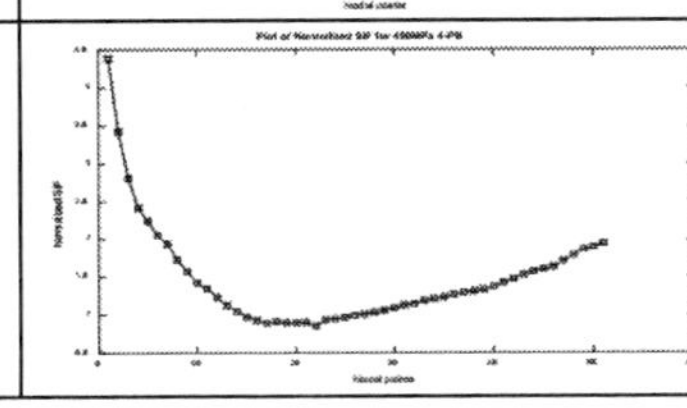

3 Plastic collapse parameter

The global limit load refers to the load that would allow unbounded displacements throughout the structure, and is equivalent to the load capacity of the structure. The local limit load refers to the load required to cause complete plasticity in the remaining ligament within the crack plane.

https://doi.org/10.1051/matecconf/201816521011

A possible approach to determine the limit load uses load increment and iteration up to yield [10]. The approach is accurate but computing-intensive.

For each model, incremental load steps were run iteratively until the entire ligament undergoes complete yielding. An example of the final stresses at full ligament yield are shown in Fig. 7. The final stresses at crack profile and sides are shown in Fig. 5. The limit loads determined are tabulated in Table 3.

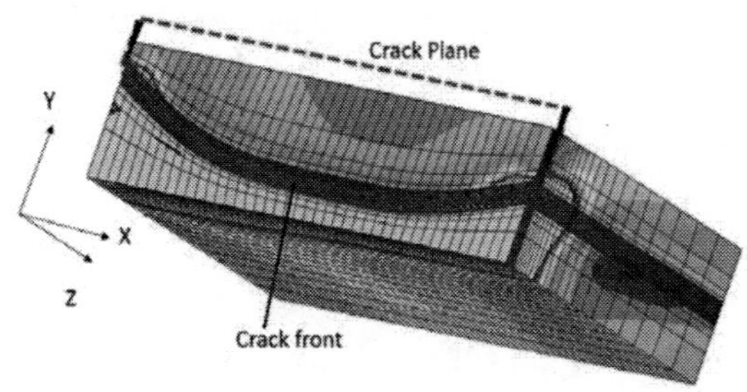

Fig. 7. Stress distribution at full ligament yield (Specimen 1 292.5 MPa)

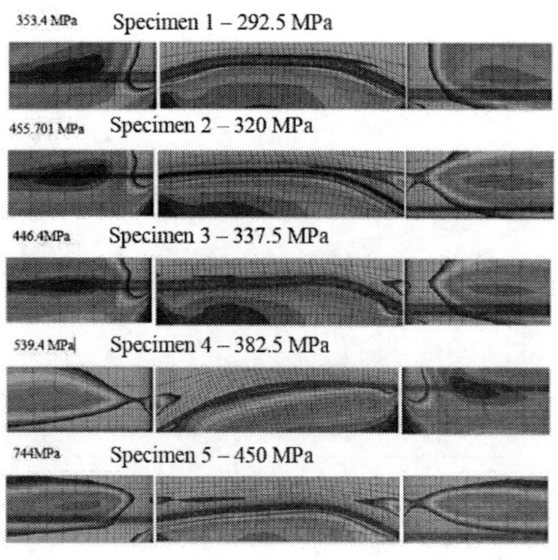

Fig. 8. Crack plane section view to determine local limit load (bending)

Table 3. Limit loads determined by iterative approach

No	Stress range MPa	Limit load (bending) /MPa
1	292.5	353.4
2	320	455.7
3	337.5	446.4
4	382.5	539.4
5	450	744

4 Failure assessment of X65

The FAD can be constructed from equations (3) to (7), while the assessment points can be determined from results in Tables 2 and 3, giving Table 4. The stresses were determined for the deepest points and at the free surfaces with highest SIFs at the different loads.

Table 4. Assessment points of X65 steel under bending

No	Stress range MPa	K_r (free surface)	K_r (deepest point)	L_r
1	292.5	0.779	0.327	0.919
2	320	0.655	0.357	0.78
3	337.5	0.702	0.352	0.84
4	382.5	0.603	0.354	0.787
5	450	0.520	0.310	0.672

The assessment points on the FAD (Fig. 9) show that under bending, the cracks at the surface are close to failure or unacceptable, while the points at the crack depth are deemed acceptable.

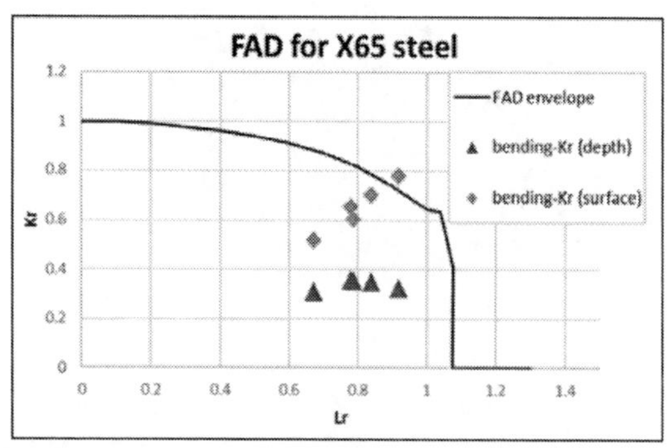

Fig. 9. Failure assessment diagram for X65 with large flaws.

5 Conclusion

The BS7910:2013 solutions are sufficiently accurate for flaws that are small, but become very conservative when large cracks are considered. The current assessment of previously tested X65 specimens with large cracks show local through-thickness yielding is likely to occur before global ligament yielding. The current work demonstrates a more accurate approach to assess large flaws by considering the SIFs and limit loads by finite element analysis.

References

1. J.H.L. Pang, H.J. Hoh, K.S. Tsang, J. Low, S.C. Kong, W.G. Yuan, Int. J. Fatigue **94** 158-165 (2017)
2. J.C. Newman Jr, I.S. Raju, Eng. Fract. Mech. **15** 185-192 (1981)
3. H.J. Hoh, J.H.L. Pang, K.S. Tsang, Int. J. Fatigue **87** 279-287 (2016)
4. British Standards Institution, BS7910:2013 (2013)
5. S.J. Garwood, R.H. Leggatt, T. Ingham, Fatigue Fract. Eng. Mater. Struct. **17** (1994)
6. M.J. Cheaitani, J. Pipe Eng. **8** 149-166 (2009)
7. S.H. Hashemi, Mater. Sci. Eng. A **528** 1648-1655 (2011)
8. J. Capelle, J. Furtado, Z. Azari, S. Jallais, G. Pluvinage, Eng. Fract. Mech. **110** 270-280 (2013)
9. B. Gross, J.E. Srawley, NASA TN D-2603 (1965)
10. I. Sattari-Far, Int. J. Pres. Ves. & Piping **57** 237-243 (1994)

Failure mechanisms and fatigue strength reduction factor of a Cr-Ni-Mo-V steel welded joint up to ultra-long life regime

Ming-Liang Zhu*, *Fu-Zhen* Xuan

Key Laboratory of Pressure Systems and Safety, Ministry of Education; School of Mechanical and Power Engineering, East China University of Science and Technology, Shanghai 200237, China

Abstract. It is known that welded joint is much "weaker" than base metal due to discontinuities of geometry, materials and residual stresses. It seems current international design rules do not adopt a uniform approach to weld efficiency, which is often defined as the ratio of the strength of a welded joint to the strength of base metal, in their guidance for creep and fatigue design of welds. This appears to be a great barrier for the application of nuclear welded structures which has a prolonged design lifetime of 60 years. In this work, fatigue strength reduction factor of a Cr-Ni-Mo-V steel welded joint, machined from welded steam turbine rotors for nuclear power plant, was investigated by performing axially push-pull cyclic loads tests with both cross-weld and pure base metal specimens up to very high cycle fatigue regime under ultrasonic frequency at ambient temperature. The effects of residual stress, strain localization, and micro-defects in mismatched steels on failure mechanisms of welds were discussed thoroughly. Results show that fatigue strength reduction factor is varied in the range of 0.95-0.975, and is found to be dependent on fatigue lifetime for the first time. It is indicated that variation of fatigue strength reduction factor are associated with transition of crack initiation from specimen surface in high cycle fatigue regime to interior micro-defects in very high cycle fatigue regime. Comparing existing codes and standards for fatigue design of welds with experimental data indicates the over-conservativeness of present code-based design method. This implies a micro-defect based fatigue design approach is required for long life safe and reliability of weldments.

1 Introduction

The weld efficiency or joint efficiency is often defined as the ratio of the strength of a welded joint to the strength of base metal, and is thus inclusive of tensile, fatigue and creep strength. The weld efficiency can be related to strength reduction factor of welded joint, as welded joint is much "weaker" than base metal due to discontinuities of geometry, materials and residual stresses. It seems that current international design rules do not adopt a uniform approach to weld efficiency in their guidance for creep and fatigue design of welds. Creep design of engineering structures at elevated temperatures based on ASME Section III [1] defines creep strength reduction factor (CSRF) based on weld metal only, while the fast reactor design code RCC-MR [2] defines weld efficiency based on welded joint. In case of the fatigue strength reduction factor (FSRF), the RCC-MR code recommends that the fatigue curve of weld joint has a scaling factor of 1.25 lower than the fatigue curve of base metal, indicating a weld efficiency of 80%, whereas the ASME Subsection NH suggests a constant FSRF factor of 2.0 on the fatigue strength of welded joint (weld efficiency of 50%). A precise and unified evaluation of weld efficiency with sound mechanistic representation and associated proper recognition of its role in joint design are of great significance in balancing design safety and conservatism of modern engineering components and structures under extreme service conditions.

It is highly recommended by many code and standards that the creep strength reduction factor needs to be determined by experiments. Otherwise, in case of difficulty in creep experiments or availability of materials, single empirical values are often employed in design analysis, i.e., the value of 0.8 (80%) is proposed by EN13445-3, DIN EN13480-3, and ASME Section III. Define CSRF as a simple value and treatment of strength reduction factor as constant have long been challenged in creep damage assessment of welds. Tu and Sandström [3, 4] evaluated the influence of stress multiaxility and the corresponding stress redistribution process of welded components on the determination of CSRF by both experimental and numerical simulations. The CSRF was later found to be a function of testing temperature and time, especially in some heat resistance steels where the time dependent damage of microstructures are indicative of CSRF variation [5]. Tabuchi and Takahashi [6] reported the reduction factor of 100,000 hours creep rupture strength of welded joint to base metal was 0.75 at 600°C and 0.7 at 650°C for a Modified 9Cr-1Mo Steel. A similar presentation of CSRF as a function of

* Corresponding author: mlzhu@ecust.edu.cn

temperature was also reported by them in welded joints of Grade 122 steel [7].

The design against fatigue of welds is currently enabled by various codes and recommendations with well-developed design and assessment approaches, such as safe life design and damage tolerant design [8]. The geometrical and structural factors of welded joint has developed into a notch-based fatigue design approach [9], also termed local stress or strain approach [10, 11], which, generally leads to costly over-design, has witnessed great effectiveness in both low cycle fatigue (LCF) and high cycle fatigue (HCF) regimes. It is generally accepted that there exists a transition of crack initiation modes from specimen surface in LCF and HCF regimes to interior micro-defects in very high cycle fatigue (VHCF) regime which is characteristic of ultra-long fatigue life (fatigue life higher than 10^7 cycles) under lower stress levels often below the traditional fatigue limit [12]. The change of the shape of S-N curves has attracted great interests in fatigue design [13, 14]. Sonsino [15] gave material and manufacturing dependent recommendations for design of components submitted to loadings below the knee point of the conventional S-N curve. Murakami [16] thought of material defects as the basis of fatigue design. The weld efficiency in LCF and HCF regimes is thus inferred different from the one in VHCF regime, a point that is lack of research and verification. Moreover, the distribution and maximum permissible size of weld imperfections, unfortunately, are not explicitly taken into account in current fatigue design standards.

Therefore, in the present work, axially loaded cyclic tests using both welded joint and base metal specimens of Cr-Ni-Mo-V steel welds were carried out up to the VHCF regime to evaluate the weld efficiency and its underlying mechanisms. The fatigue design of welds up to the VHCF regime was then discussed by comparing various existing codes and standards. Finally, potential improvements by incorporating the micro-defects into the code are discussed based on the present results.

2 Materials and experiments

2.1 Materials and specimens

The material investigated was a welded joint with the 25Cr2Ni2MoV steel being the base metal (BM). The weldment was prepared by the submerged arc welding (SAW) technique using a weld metal (WM) rich of Ni [17]. The forged BM was quenched and tempered with microstructures dominated by lathy tempered martensites. After the SAW process and a post weld heat treatment (PWHT) at 580°C for 10 hours (furnace cooling), the yield strength (YS) and ultimate tensile strength (UTS) of the welded joint at room temperature was 726 MPa and 778 MPa, respectively. The microstructures of the WM mainly contain tempered bainites with long and lathy shape. The heat-affected zone (HAZ) shows a combined tempered martensite and bainites. A detailed observation of microstructures, and the tensile and strain hardening behavior have been reported in our previous paper [17].

Cross-weld specimens for VHCF testing were machined from the welded joint, and BM samples were also prepared for comparison purposes. Fig. 1 shows the shape and dimensions of the specimens for fatigue testing. For the cross-weld samples, it is noted that the WM, HAZ, and some parts of the BM constitute the parallel section to ensure the same stress during cyclic test. The same dimensions were employed for pure BM specimens, as indicated in Fig. 1b.

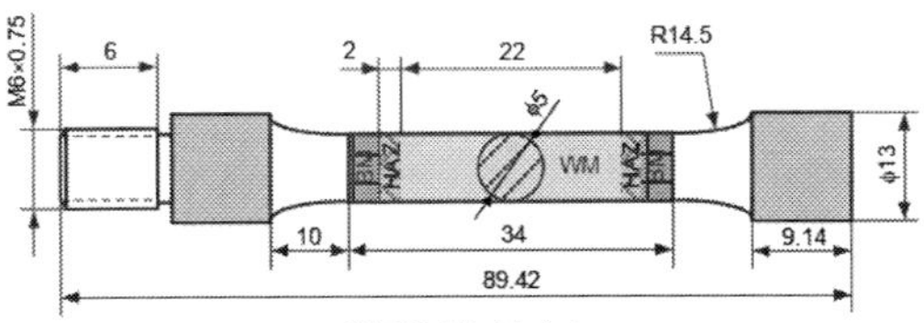

(a) Welded joint

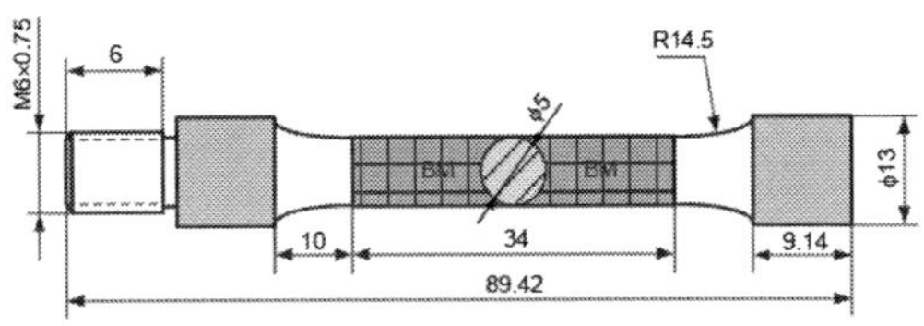

(b) Base metal

Fig. 1 Shape and dimensions of fatigue specimens for testing at 20 kHz: (a) cross-weld specimen, (b) pure base metal (dimensions in mm)

2.2 Fatigue testing methods

Prior to testing, all the specimens were mechanically polished. The final surface roughness of the specimen, R_a, is lower than 0.2 μm. Fatigue tests were then conducted on an ultrasonic system (USF-2000, Shimadzu, Japan) at ambient temperature with a frequency around 20 kHz. A pre-calibration process for the ultrasonic machine is required to establish a linear relationship between the horn side amplitude (in μm) and amplified output amplitude (in %), on which the determination of cyclic stress level is based. The load ratio of the test is -1. Compressive air was used to cool the middle part of the specimen to prevent the specimen from over-heating. In addition, an intermittent loading condition, i.e., 500 ms of pulse followed by 1000 ms of pause, was employed to minimize the thermal effect. As reported in [18], the thermal effect could be noticably reduced by employing cooling measures in a lower strength steel. After testing, all fracture surfaces of failed specimens were observed by scanning electron microscopy (SEM) in order to identify the crack initiation modes.

3 Results

3.1 *S-N* curve

Fig. 2 shows the *S-N* curves of the fatigue tests. It is noted that the *S-N* curves for both the two kinds of tests show a continuously decreasing mode. The traditional fatigue limit is not apparently observed. With the increasing of fatigue life, there exists a transition of crack initiation from specimen surface to interior micro-defects for cross-weld samples. While for the pure BM specimens, surface related fatigue cracking seems to be dominant as failure can occur at surface defect even at N_f of 10^8 cycles. Data from cross-weld samples have a more severe scatter than those of pure BM specimens. This is because, each part of the cross-weld samples has the potential to fatigue cracking, while the BM specimens can only fail at the BM. Apart from the *S-N* curve, another important difference between the two types of testing is the fatigue strength. It is obvious that the pure BM material has larger fatigue strength than that of cross-weld samples. This indicates the influence of welding process, i.e., welded joints always contain micro-defects, residual stress, and in particular, the mismatched structures. The strained welded joint will have more severe strain localization at weaker zone. A detailed discussion on the weld efficiency will be presented later. For comparison of fatigue strength, the results for pure BM and cross-weld specimens are then fitted as equations (1) and (2), respectively, as shown below.

$$\lg(\sigma_a) = 2.92367 - 0.04723\lg(N_f) + 0.00275\lg(N_f)^2 \quad (1)$$

$$\sigma_a = 602.3(N_f)^{-0.0089} \quad (2)$$

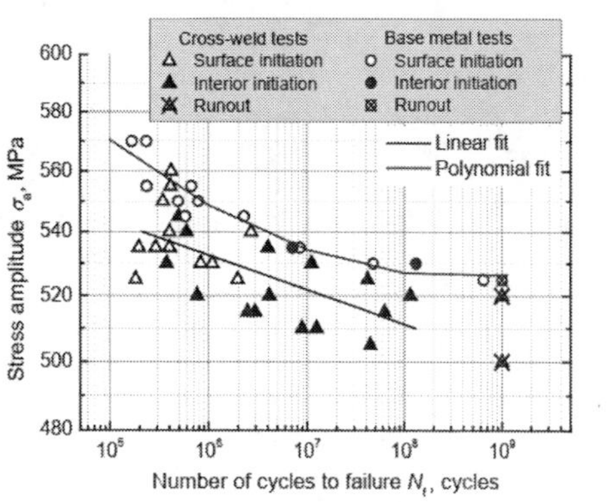

Fig. 2 *S–N* curves of the welded joint and pure base metal

3.2 Crack initiation behavior

Fig. 3 shows the morphologies of fatigue crack initiation of cross-weld specimens. A fatigue crack initiated from surface inclusion is observed at BM with the N_f of 4×10^5 cycles, as shown in fig. 3a; under similar N_f, the fatigue crack nucleated from interior pore with its boundary clearly shown in fig. 3b is observed at WM. Fig. 3c presents an interior fatigue cracking from non-metallic inclusion at WM in the VHCF regime where fatigue lifetime is longer. It is believed that the interior micro-defects being the crack initiation site is responsible for the disappearance of traditional fatigue limit. An enlarged observation around the inclusion illustrates the occurrence of fine granular area (FGA), according to Sakai et al. [19], with the characteristic of rough but fine surface. Here, the width of FGA at the radical direction is measured to be about 35 μm. The formation mechanisms of this kind of FGA in lower strength steels have been discussed in our previous paper with the help of microscopic observation based on FIB technique [20]. It was concluded that the FGA was actually polycrystalline grains formed due to continuously developed plastic energy around micro-defects [20].

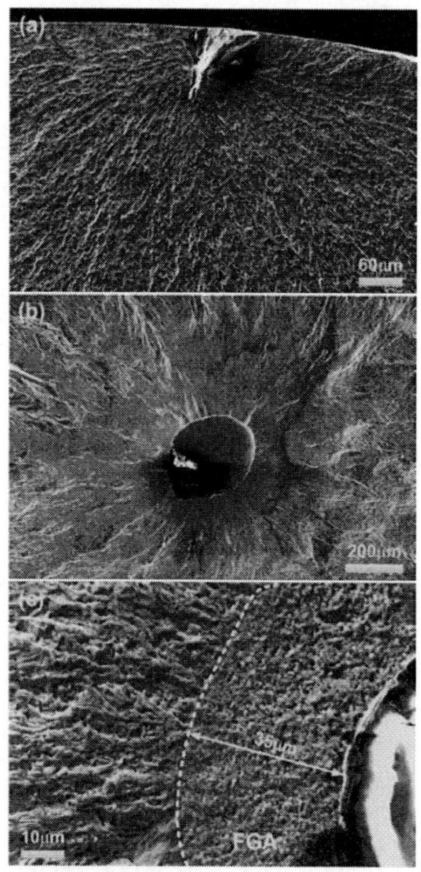

Fig. 3 Morphologies of fatigue crack initiation for cross-weld specimens: (a) crack initiation at the surface inclusion at BM (σ_a=540 MPa, N_f=4×10^5 cycles), (b) crack initiation at the interior pore at WM (σ_a=540 MPa, N_f=6×10^5 cycles) and (c) crack initiation at the non-metallic inclusion at WM (σ_a=525 MPa, N_f=4.15×10^7 cycles)

Fig. 4 shows the crack initiation behaviour of pure BM samples. As mentioned before, surface related fatigue cracking can even occur at extremely longer fatigue lifetimes, such as the one shown in fig. 4a, where a crack is formed from surface micro-defect with a N_f of

https://doi.org/10.1051/matecconf/201816521012

6.4×10^8 cycles. This may due to there are no larger interior defects. Fig. 4b presents crack initiation from a sub-surface inclusion which is different from the commonly observed fish-eye pattern. Another subsurface inclusion induced fatigue failure is shown in fig. 4c where the sample has the same stress level as the one in fig. 4a. This indicates that under such load amplitude, the fatigue crack can either initiate from specimen surface or subsurface micro-defect, which is the factor that dominates the variability of fatigue lifetime, and has the implication for the occurrence of fatigue limit for this kind of steel. A magnified view of the crack growth region on the fracture surface clearly indicates the striations, as illustrated in fig. 4d.

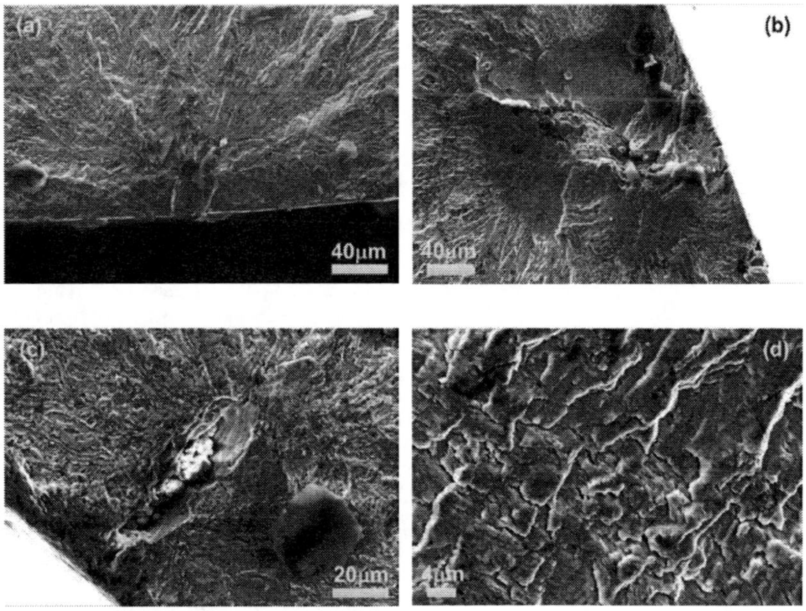

Fig. 4 Morphologies of fatigue crack initiation of pure BM samples: (a) surface inclusion (σ_a=525 MPa, N_f=6.4×10^8 cycles), (b) subsurface inclusion (σ_a=535 MPa, N_f=7.12×10^6 cycles), (c) subsurface inclusion (σ_a=525 MPa, N_f=4.15×10^7 cycles) and (d) striations formed at higher crack growth rate

3.3 Residual stress distribution along the welds

The X-ray diffraction method was applied to characterize the residual stress distribution along the welds. A XRD X350A residual stress analyser was used with a Cr Kα radiation tube with 2θ in the range of 163-149°. The residual stress was measured at selected points along the parallel section of the welds at the exposed surface. Moreover, a particular amount of surface layer was removed with the help of electrolytic polishing to measure residual stress at different depths, i.e., 10, 20 and 40 μm. Both axial and circumferential stresses were measured, and corresponding results are presented in fig. 5a. It is observed that the surface residual stress is mainly compressive in both the two directions. Axial stress varies from -190 MPa to -140 MPa, with higher values at the WM, whereas the circumferential stress is more compressive with the highest value (-252 MPa) also at the WM. Fig. 5b presents the residual stress distribution versus the depth of measurement at selected micro-zones along the welded joint. It is obvious that the compressive residual stress changes to positive at a distance of approximately 40 μm to the specimen surface, regardless of the points measured. This depth value has perfect consistence to the location of fatal inclusions shown in figs. 4b and 4c, which at least partly justifies the occurrence of sub-surface crack initiation behaviour. This also helps rationalize the transition of fatigue cracking from specimen surface to interior micro-defects [21]. Note the results reported here is a mixed case of residual stress originated from both welding and sample machining processes. The role of machining is to alter the initial residual stress state after welding, and after the stress relief heat treatment. The residual stress data presented here are original ones that do not consider the relaxation during cyclic loadings.

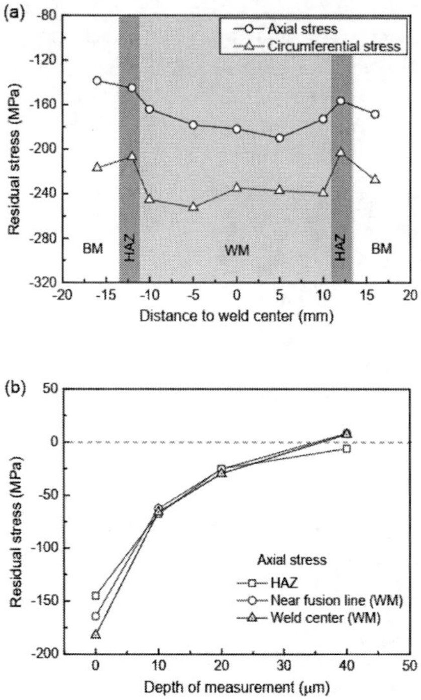

Fig. 5 Distribution of residual stress at specimen surface along the welded joint (a), and variation of axial residual stress with depth of measurement (b)

4 Discussion

4.1 Weld efficiency

As shown in fig. 2, the welded joint samples have lower fatigue strength than the pure BM ones. Fig. 6 depicts the FSRF, F, versus the fatigue lifetime. Here, F is calculated to be the ratio of fatigue strength value of welded joint samples to pure BM specimens. The value of F thus denotes the weld efficiency in terms of fatigue strength. It is apparent that the weld efficiency is not a constant value, but is varied with the fatigue lifetime. This is similar to the CSRF which is varied with temperature and time. Here, F is in the range of 0.95 to 0.976 for N_f in the range of 10^5 cycles to 10^9 cycles. It is observed that the F values in LCF and HCF regimes are different from those at ultra-long life regime. Most importantly, the F is found to be a function of fatigue lifetime. This has the implication of improving the current rules that are simply assuming constant FSRF. As shown in Fig. 6, with the increasing of N_f, F shows an increasing trend to the highest value of 0.976 at N_f of 9×10^6 cycles, after which the F is decreased. This has great significance in fatigue design of engineering structures and components in the VHCF regime where the conservativeness or allowance can be improved.

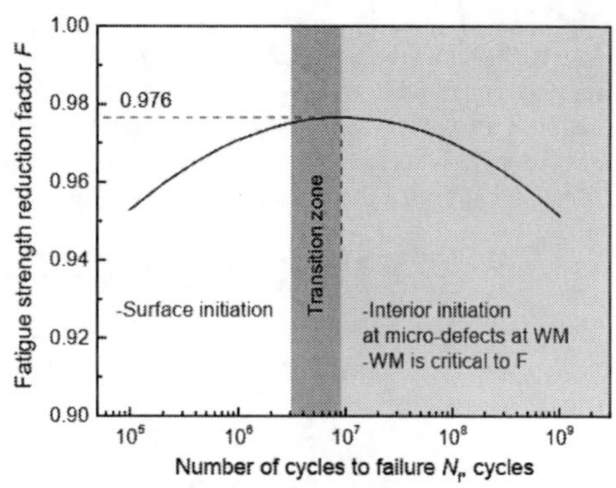

Fig. 6 Variation of fatigue strength reduction factor, F, with fatigue lifetime.

In this work, the underlying mechanisms for variation of weld efficiency with fatigue lifetime can be related with fatigue crack initiation modes of welded joint. It is interesting to find that the F is increased gradually in surface crack initiation dominated region, and continues increasing in the surface-to-interior transition region, and is finally decreased slowly for the dominant interior crack initiation at micro-defects at the WM. This means that the WM is the most critical to the FSRF in the VHCF regime, which implies that the design of components with ultra-long life requirement should consider the role of micro-defects in fatigue failure. *This is the first time to propose the role of micro-defect in fatigue design of a welded joint in the VHCF regime.* The questions of whether it is safe, whether it is conservative or non-conservative, when applying current fatigue design methods by extending the design life to VHCF regime, is yet unknown, unless with a full comparison.

4.2 Fatigue design of welds in current codes and standards

Current fatigue design codes for welded structures mainly include ASME Boiler and Pressure Vessel Code (i.e., Section III and Section VIII) [1], French code RCC-MR [2], BS 7608 [22], IIW recommendation [8], EN 1993-1-9: 2005 [23] and DNV-RP-C203 [24]. In this work, existing codes and standards on fatigue design of welded joint are compared based on the present test results.

4.2.1 ASME Boiler and Pressure Vessel Code

In the ASME Boiler and Pressure Vessel Code [1], fatigue design curves are based on *S-N* curves generated from smooth specimens of BM. In Section III, the fatigue design curves are obtained by applying a factor of 20 to fatigue cycles and 2 to stresses, based solely on engineering judgments. Moreover, weld efficiency is regarded as a stress intensification factor which accounts for the effects of local structural discontinuity (stress concentration) on the fatigue strength. While actually, for welded joint, apart from the stress concentration due to geometry factor, fatigue strength reduction should also take account of weld imperfections, and weld residual stress. It is noted that, the practice of ASME Code is different from the case in other popular codes and standards where fatigue design of a particular welded joint is based on choosing one from a series of parallel *S-N* curves, in which process the FSRF is not fully represented. Dong et al. [25] tried to propose a master *S-N* curve for fatigue design of welded structures based on the discussion of fatigue strength reduction factor in ASME Code.

4.2.2 BS 7608

In the guide to fatigue design and assessment of steel products, BS 7608 [22], a general equation for *S-N* curve is listed as follows,

$$S_r^m N = C_0 \quad \text{(BS7608)} \tag{3}$$

where S_r is stress amplitude, N is the number of cycles, C_0 is a parameter defining the mean line S_r-N relationship, and m is the negative inverse slope of *S-N* curve. Here, the parameters for class C welds are selected and listed in Table 1. In case of fatigue design up to the ultra-long life regime, apart from the m and C_0 values, the standard provides a S_r of 102.27 MPa at N_f of 10^7 cycles, and keeps constant for longer fatigue lifetime. In the standard, the role of weld imperfections on the shape of *S-N* curves is not considered.

4.2.3 IIW recommendation

It is known that the fatigue resistance of a welded joint is limited by the fatigue resistance of the BM. As pointed out in IIW recommendation [8], when the fatigue assessment of classified structural details and welded joints is based on the nominal stress range, a fatigue class (FAT) of 112 is chosen here for transverse loaded butt welds. In case of very high cycle applications, m=3 for $1\times10^4 \leq N \leq 1\times10^7$ cycles and m=22 for $N>1\times10^7$ cycles is defined for the *S-N* curves [8]. By considering the residual stress effect, a fatigue enhancement factor $f(R)$ of 1.3 is applied here in case of R=-1 with medium residual stress, therefore, FAT is 145.6 MPa. The IIW recommendation also provides the guidelines for assessing the acceptance levels for porosity and inclusions in steel welds. The parameter for assessing inclusions is the maximum length. For a stress relieved welded joint, the maximum length of inclusion is 7.5 mm in case of the FAT is 100 MPa. In this work, the size of critical inclusions that are fatigue crack initiation sites is at the micro-meter level, as presented in Figs. 3 and 4. It is therefore inferred that the fatigue class should be higher than the highest value listed in the recommendation, that is, FAT >100 MPa.

In assessing the fatigue welds when considering the weld imperfections, for inner inclusions, the IIW recommendation offers a FAT of 71 for weld length less than 10 mm. This seems very conservative for fatigue assessment of welds as not all the inner micro-defects will be the failure site in the VHCF regime. The IIW recommendation does not provide the information of weld efficiency in case of fatigue assessment based on *S-N* curves of BM. It can be concluded that the current application of IIW recommendation to fatigue assessment of steel welds in the VHCF regime is either conservative or the role of micro-defects/weld efficiency is not adequately considered.

4.2.4 EN 1993-1-9:2005

The standard of EN 1993-1-9:2005 [23] also provides fatigue design curves up to the VHCF regime, where several parameters are defined. As listed in Table 1, the m is 3 for $N \leq 5\times10^6$ cycles, and increased to 5 for $5\times10^6 \leq N \leq 1\times10^8$ cycles. For N higher than 1×10^8 cycles, the fatigue resistance is determined by the cut-off limit $\Delta\sigma_L$, which is 0.549 times of conventional fatigue limit at constant amplitude load, $\Delta\sigma_D$. According to the code, a detail category of 112, similar to the FAT 112 by IIW recommendation, is selected for transverse butt welds checked by NDT. Similarly, the EN code keeps the conservative style of fatigue assessment but keeps the role of micro-defect in the VHCF regime and weld efficiency unconsidered.

4.2.5 DNV-RP-C203

For the recommended practice of DNV-RP-C203 [24], the basic *S-N* curve is given as

$$\log N = \log \bar{a} - m \log \Delta\sigma \tag{4}$$

where $\log\bar{a}$ is intercept of log$N$-axis by *S-N* curve, which takes the standard deviation of log$N$ into account. As listed in Table 1, m is 3 for $N \leq 10^7$ cycles while m is increased to 5 for $N \geq 10^7$ cycles. For transverse butt welds without backing strip, the weld detail category of C1 is selected here for *S-N* curves at air environment. The practice applies the traditional fatigue limit, and set the value at 65.5 MPa at N of 10^7 cycles. Note that choosing detail category of C1 is based on the understanding that overfill at weld toe is dressed flush without stress concentration, and failure is mainly related with weld defects. This means the *S-N* curve of the DNV code, though conservative for fatigue design of welds, considered the failure mode from micro-defects. No information of weld efficiency values are given in the recommended practice.

4.3 Fatigue design of welds for ultra-long life requirement

Fig. 7 shows a comparison of fatigue design curves of welds in current code and standards up to the VHCF regime. Note the fatigue data is specimen-based while in actual cases structural, environmental factors are involved. It is observed that all the design curves based on existing codes and standards are below the fatigue data of Cr-Ni-Mo-V steel welded joint. Severe conservativeness exists in BS7608, IIW recommendation, EN 1993-1-9: 2005 and DNV-RP-

C203, whose design curves are lying below the ASME and RCC-MR codes where fatigue design of welds is based on experimental data of BM. This means the design practice in ASME and RCC-MR codes are more reasonable than the simple procedure of selecting one among categorised *S-N* curves in other standards. This clearly indicates that design by current code and standards are very conservative, which will lead to costly over-design of welded structures and components. In other words, the current design practice is overloaded with safety factors by sacrificing engineering economics. This would partly explain the reasons why existing engineering components is still in use in a safe condition even the original design life has passed for several years.

The current code-based design needs improving due to over-conservatism. According to the tests results, design according to several stages of *S-N* curves with varied m values, i.e., the treatment in IIW recommendation, are not necessarily suitable for application into welded joints because the actual *S-N* curves may present a continuously decreasing shape. Whether it is a general rule or it is specimen-dependent has to be answered with in-depth further research. It seems there lacks a sound link between experimental results and modelling methods in codes. In this regard, how to incorporate the weld imperfections into modelling design curves is still an open issue. The criterion for defect assessment needs to be improved with an aim to differentiate the non-damaging defects from damaging ones under service loading conditions. The material defect based fatigue design for ultra-long life requirement, as proposed by Murakami [13, 16], is likely to reduce the conservativeness for fatigue design, if it could be extended into structure and component scales.

Table 1. Parameters in various codes and standards for fatigue design of steel welded joints

Codes and standards	**Parameters**			
ASME Section III	Weld efficiency of 50% for fatigue design of welded joint			
RCC-MR	Weld efficiency of 80% for fatigue design of welded joint			
BS7608, Class C	m	C_0	S_r, N_f=10^7 cycles	S_r, N_f=5×10^7 cycles
	3.5	1.082×10^{14}	102.27	102.27
IIW, steel welded joint	m=3	10^4<N<1×10^7 cycles		
	m=22	N>1×10^7 cycles		
	FAT 112 for transverse loaded butt weld, $f(R)$=1.3, and weld imperfections are considered			
EN 1993-1-9: 2005, welded joint	m=3	$N\leq5\times10^6$ cycles	$\Delta\sigma_D$ is constant amplitude fatigue limit at N=5×10^6 cycles, and $\Delta\sigma_L$ is cut-off limit at N=1×10^8 cycles.	
	m=5	$5\times10^6\leq N\leq1\times10^8$ cycles		
	$\Delta\sigma_L$=0.549$\Delta\sigma_D$	$N\geq1\times10^8$ cycles		
	Detail category: 112 for transverse butt welds, and checked by NDT			
DNV-RP-C203, welded joint	m=3	$N\leq10^7$ cycles	Fatigue limit at N=10^7 cycles is 65.5 MPa;	
	m=5	$N\geq10^7$ cycles	$\log\overline{a_1}$ =12.449, $\log\overline{a_2}$ =16.081	
	Detail category: C1 for transverse butt welds without backing strip, and the overfill is dressed flush without stress concentration, failure is mainly related with weld defects			

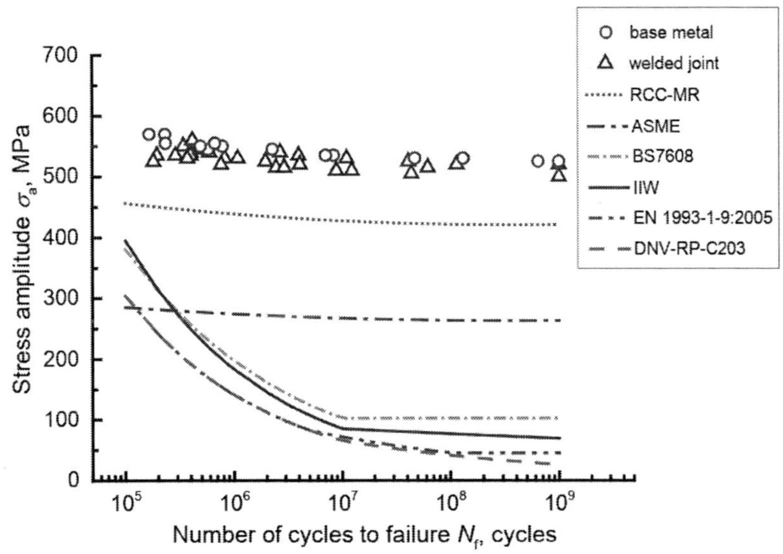

Fig. 7 A comparison of fatigue design curves in existing codes and standards up to the VHCF regime

5 Conclusions

In this paper, axially push-pull cyclic tests on a Cr-Ni-Mo-V steel welded joint were performed up to the VHCF regime under ultrasonic frequency at ambient temperature. Weld efficiency and fatigue design of the welded joint was discussed by comparing fatigue data with current codes and standards. The following conclusions can be reached.

(1) The weld efficiency in terms of fatigue strength reduction factor is found to be varied with fatigue lifetime for the first time. It increases from 0.95 in surface crack initiation dominated regime to the maximum value of 0.976 followed by gradually decreasing in interior micro-defects induced crack initiation in the VHCF regime.

(2) Current code-based fatigue design of welded joint is so conservative that would lead to costly over-design of engineering components, when comparing with experimental fatigue data. Any aims for improving the accuracy of current codes and standards should take account of weld imperfections for designing against fatigue of welds up to the ultra-long life regime.

Acknowledgements

The authors are grateful for the supports provided by the National Natural Science Foundation of China (51575182).

References

[1] ASME Boiler and Pressure Vessel Code, Section VIII, The American Society of Mechanical Engineers, 2004
[2] RCC MR Design Code, Section 1, French Society for Design and Construction Rules for Nuclear Island Components, 2002
[3] R. SandströM, S.T. Tu, J. Pres. Ves. Technol., 116, 76-80 (1994)
[4] S.-T. Tu, R. Sandström, Int. J. Pres. Ves. Piping, 57, 335-344 (1994)
[5] M.D. Mathew, S. Latha, K.B.S. Rao, Mater. Sci. Eng. A, 456, 28-34 (2007)
[6] M. Tabuchi, Y. Takahashi, Evaluation of Creep Strength Reduction Factors for Welded Joints of Modified 9Cr-1Mo Steel (P91), in: ASME 2006 Pressure Vessels and Piping/ICPVT-11 Conference, Vancouver, BC, Canada, 2006, pp. 529-534.
[7] Y. Takahashi, M. Tabuchi, J. Pres. Ves. Technol., 133, 021401-021401-021405 (2011)
[8] A. Hobbacher, *Recommendations for fatigue design of welded joints and components* (Springer, 2016)
[9] C.M. Sonsino, W. Fricke, F. De Bruyne, A. Hoppe, A. Ahmadi, G. Zhang, Int. J. Fatigue, 34, 2-16 (2012)
[10] D. Radaj, C.M. Sonsino, W. Fricke, Int. J. Fatigue, 31, 2-11 (2009)
[11] D. Radaj, C.M. Sonsino, W. Fricke, *Fatigue assessment of welded joints by local approaches* (Woodhead publishing, 2006)
[12] B. Pyttel, D. Schwerdt, C. Berger, Int. J. Fatigue, 33, 49-58 (2011)
[13] A. Roiko, Y. Murakami, Int. J. Fatigue, 41, 140-149 (2012)
[14] P.R. Fry, Eng. Fail. Anal., 46, 179-187 (2014)
[15] C.M. Sonsino, Int. J. Fatigue, 29, 2246-2258 (2007)
[16] Y. Murakami, Int. J. Fatigue, 41, 2-10 (2012)

[17] M.-L. Zhu, F.-Z. Xuan, Mater. Des., 65, 707-715 (2015)
[18] H.Q. Xue, E. Bayraktar, C. Bathias, J. Mater. Proces. Technol., 202, 216-223 (2008)
[19] T. Sakai, N. Oguma, A. Morikawa, Fatigue Fract. Eng. Mater. Struct., 38, 1305-1314 (2015)
[20] M.-L. Zhu, F.-Z. Xuan, Mech. Mater., 100, 198-208 (2016)
[21] M.-L. Zhu, F.-Z. Xuan, Y.-N. Du, S.-T. Tu, Int. J. Fatigue, 40, 74-83 (2012)
[22] BS 7608:2014+A1:2015, Guide to fatigue design and assessment of steel products, The British Standards Institution, 2015
[23] BS EN 1993-1-9:2005, Eurocode 3: Design of steel structures----Part 1-9: Fatigue, The British Standards Institution, 2005
[24] Recommended Practice DNV-RP-C203, Fatigue Design of Offshore Steel Structures, DET NORSKE VERITAS, 2008
[25] P. Dong, J.K. Hong, D. Osage, M. Prager, Welding in the World, 47, 31-43 (2003)

Fatigue life and crack growth behavior of post welded Aluminum 5183 alloy

Vidit Gaur[1,*], *Manabu* Enoki[1], *Toshiya* Okada[2] and *Syohei* Yomogida[2]

[1]Department of Materials Engineering, The University of Tokyo, 7-3-1 Hongo, Bunkyo-ku, Tokyo, 113-8656, Japan
[2]Research & Development Division, UACJ Corporation, 3-1-12 Chitose, Minato-ku, Nagoya-city, Aichi Pref. 455-8670, Japan

Abstract. In this study we investigated the fatigue life and crack growth behavior of Al-5183 alloy. Microscopic analysis revealed nearly equi-axed grains and no texture in longitudinal or cross-sectional plane of the welded plates. Gas porosities with an average size of 45 μm, comparable to grain size (55 μm), were present and often initiate fatigue failures. Load-controlled cyclic tests at different stress-ratios (R = -1, -0.5, 0.1, 0.5. 0.7 and 0.8) revealed decrease in fatigue lives with increase in R-ratio. At R > 0.7, no fatigue failure could be observed, suggesting a probable mean-stress saturation effect on fatigue. Mean stress also tends to control the crack initiation sites: surface initiated failures at low mean stresses while sub-surface pores induced failures at higher mean stresses. Fatigue-crack growth tests on CT specimens at different R-ratios (0.1, 0.5 and 0.8) revealed reduction in crack growth rates (and in threshold values) with increasing R-ratio. The ΔK applied for pores responsible for fatigue failures were often lower than or near to the threshold values and also, the size of such pores was of order of magnitude of grain size, thus crack initiated from pores are short cracks and further tests are progress.

1 Introduction

Welded structures are generally weaker under fatigue than the base material because of the stress-concentration factor arising from weld shape or non-welded root gaps, weld defects and change in properties of weld area. Many welding techniques exist such as Friction stir welding (FSW) [1], fiber laser-Metal Inert Gas (MIG) hybrid welding [2], MIG welding, Tungsten inert gas (TIG) welding. Among them, FSW (a solid-state welding technique), is the most effective one. TIG and MIG welding often results in reduction of fatigue strength [3-4] due to welding defects such as porosity, hot cracking, tensile residual stresses, distortion, etc. Porosities are most commonly encountered defects and act as stress raisers thus lowering down the crack nucleation period significantly depending upon their size, shape, position and reduce the fatigue lives [5]. Most of the experimental studies on fatigue behavior of metal alloys containing defects revealed the crack initiation from pores, specifically located at or near to the surface [6-7]. Nearly all of these studies were done at constant R-ratio and less data is available on influence of mean stress on fatigue behavior and damage mechanisms for fatigue cracks initiating from natural defects [8-9].

In this study we investigated the fatigue life and crack growth behavior of Al-5183 alloy, post welding of Al-5083 (base metal). Al-5xxx alloys are commonly used, high-strength and non-heat treatable aluminum-magnesium alloys. MIG welding is still preferable choice for such type of alloys [10]. 50 mm thick Al-Mg alloy (Al-5083) plates with 90° weld groove angle of double-V type butt joint were welded using single pass MIG welding and using Al-5183 as filler-wire material.

Load-controlled fatigue tests were conducted on the cylindrical specimens taken out from the weld bead (fusion part), at different stress ratios (R-ratios = $\sigma_{min}/\sigma_{max}$), followed by fracture surface examination. A modified Walker's model [11] provided the best prediction of fatigue lives and endurance limit than any of the existing models. For crack growth tests, standard CT specimens were extracted from the welded part such that the notch of the CT specimen lies along the weld bead and the crack propagation plane is same as in cylindrical specimens. Both ΔK-decreasing and ΔK-increasing tests were done in the range of 10^{-10} m/cycle to 10^{-6} m/cycle.

The overall aim of this study is to investigate the composition of crack initiation and crack growth in total fatigue life of the material containing pores (or defects) and finally be able to predict the total fatigue life

* Corresponding author: gaur@rme.mm.t.u-tokyo.ac.jp

including the mean stress correction and to develop an extensible computational framework by integrating the advanced computational techniques such as crystal plasticity modelling for crack initiation prediction, extended finite element method for crack growth etc.

2 Material

The material investigated in this study is Al-5183 alloy (filler metal) post-welding of Al-5083 (base metal). The chemical composition of base metal (Al-5083) and filler material (Al-5183) is shown in Table 1 and the respective tensile properties are reported in Table 2. The micro-structure was analyzed along the weld center line after electro-polishing the weld cross-section using Barker's reagent (1.8% Fluoroboric acid in water) in as fabricated state. The material exhibit a dendritic type micro-structure with nearly spherical magnesium (Mg) precipitates of an average size of 30 μm. The average grain size was nearly 55 μm. After analyzing the EBSD data, it was found that no significant texture was present in both longitudinal and cross-sectional direction. The respective pole figures are shown in Fig. 1.

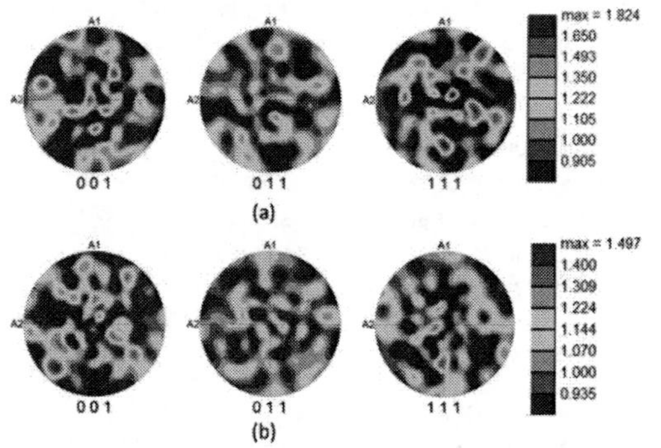

Figure 1: Pole figures of (a) cross-section and (b) longitudinal section of weld.

As expected, pores were present in the fusion part of the weld. For characterization of porosity, several high resolution images were taken and then using the open source matlab script [12], these images were converted into binary images and Watershed algorithm was used for identifying pores. A pore size data set was generated from the input of given set of images and an appropriate threshold of 30 μm was applied to distinguish between the data corresponding to voids left by Mg precipitates while polishing and the pores resulted from welding. The statistical analysis showed that the average size of pores is 45 μm, which is comparable to average grain size (55 μm). These pores were often responsible for fatigue crack initiation as illustrated and discussed later.

Table 1: Chemical composition (wt. %)

Material	Cu	Si	Fe	Mn	Mg	Zn	Cr	Ti
Al-5083	0.03	0.14	0.25	0.68	4.32	0.01	0.11	0.02
Al-5183	0.00	0.07	0.17	0.66	5.03	0.00	0.07	0.07

Table 2: Mechanical properties of Al-5183.

Proportional Limit (MPa)	0.2% Yield stress (MPa)	Tensile stress (MPa)	Hardness, H_V
98	140	285	77

3 Experimental Procedures

Al-5083 plates of 50 mm thickness were cut in double V-shape with 90° groove angle (as illustrated in Fig. 2a). Then, single layer MIG welding was performed using Al-5183 as filler material at relative humidity of 40%, welding current of 700-800 A, welding voltage of 33-34 V, welding speed of 20-30 cm/min and heat input speed of 6 kJ/min. After welding, cylindrical specimens of 5 mm diameter and 10 mm gage length (Fig. 2b) were machined from the weld bead along the longitudinal direction and then polished until the surface roughness (R_a) of 0.2 μm is achieved.

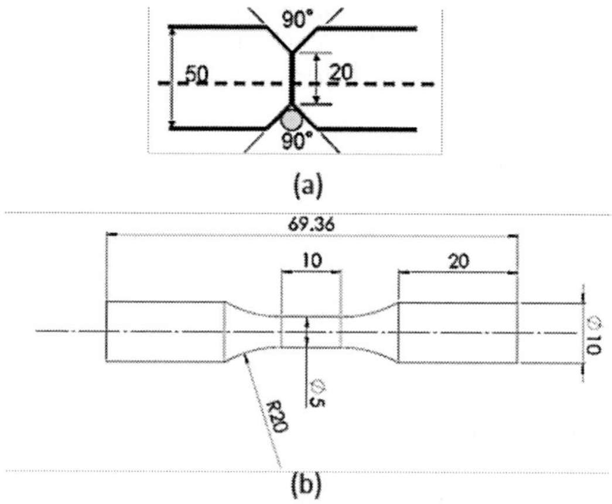

Figure 2: (a) Specifications of weld shape. Yellow circle denotes the location from where the specimens for fatigue tests were extracted (b) Geometry of cylindrical specimens used for fatigue tests.

A 50 kN servo-hydraulic Shimadzu fatigue testing machine with a Shimadzu 4830 controller and basic windows software was used for fatigue tests. All tests were done in ambient air, at a temperature of 22 °C and at relative humidity of around 40%.

To study and model the hysteresis behavior of the material, few strain-controlled fully reversed low-cycle fatigue tests were done using triangular waveform. A constant strain rate of 10^{-3} /sec was maintained during the tests. Three different strain amplitudes (ε_a) were applied using an extensometer of 8 mm gage length: 0.35%, 0.50% and 0.80%.

To study the high-cycle fatigue behavior and characterize the mean stress effect on fatigue damage, several load-controlled push-pull tests were done at different R-ratios:

$$R = \frac{\sigma_{min}}{\sigma_{max}} = -1,\ -0.5,\ 0.1,\ 0.5\ 0.7,\ 0.8 \qquad (1)$$

 https://doi.org/10.1051/matecconf/201816521013

Sinusoidal waveform was used to apply cyclic loading. All tests were done in air and at a frequency of 30 Hz. The fatigue tests were continued until the complete failure of specimen into two halves or until 10^7 cycles, whichever comes first. Usually, endurance limit is defined as the stress amplitude at which 50% of the tested specimens survived a given number of cycles but due to limited number of specimens available, repetition of fatigue tests at same stress amplitude was not possible. Thus, the stress amplitude at which specimen survived 10^7 cycles was considered as the endurance limit for that R-ratio.

For long crack growth tests, standard CT specimens of width (W) 39 mm and 7 mm thick (Fig. 3) were extracted from the welded part such that the notch of the CT specimen lies along the weld bead i.e. the notch face is parallel to the weld bead and the crack propagation plane is same as in cylindrical specimens. The notch shape was slightly modified so as to allow the mounting of clip gage. A pre-crack of about 3 mm was made by load shredding procedure such that the average crack growth rate was lower than 10^{-8} m/cycle.

Both ΔK-decreasing and ΔK-increasing tests were done in the range of 10^{-10} m/cycle to 10^{-6} m/cycle. ΔK-decreasing tests were usually followed by ΔK-increasing tests as described in ASTM E-647-00 standard. For ΔK-decreasing tests, stepwise load shredding procedure was followed and the crack was allowed to crack by 0.5 mm before decreasing the load for next step.

Tests were done at three different R-ratios (R = 0.1, 0.5, 0.8) and at a constant frequency of 50 Hz. The crack length was measured by the travelling optical microscope (KEYENCE VHX-6000). The threshold stress intensity factor range (ΔK_{th}) for a particular R-ratio was defined when the crack growth rate of 10^{-10} m/cycle was achieved.

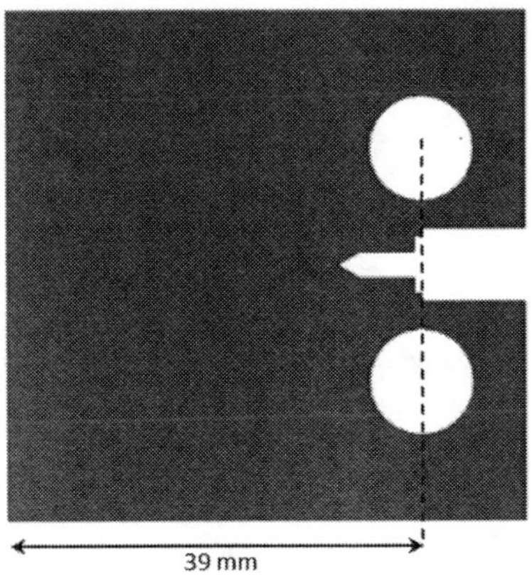

Figure 3: Geometry of CT specimen (7 mm thick) used for long crack growth tests.

JEOL JCM-6000 Scanning Electron Microscope (SEM) was used to examine the fracture surfaces of broken specimens. A Keyence digital optical microscope and a laser microscope Olympus OLS-4000 was used for topographic measurements and porosity measurements.

4 Results

4.1. Load-controlled high-cycle fatigue

The fatigue life data obtained from load-controlled high-cycle fatigue tests has been plotted in terms of stress amplitude or stress range against the total number of cycles to failure (Δσ-N plot, also known as Wöhler curve) and is shown in Fig. 4a. Some scatter in fatigue life data was observed but not significant enough and a single power law could be easily fitted except for R = 0.1, for which bilinear plot was required to fit the experiment data. The fitting to experimental data points was done using regression analysis.

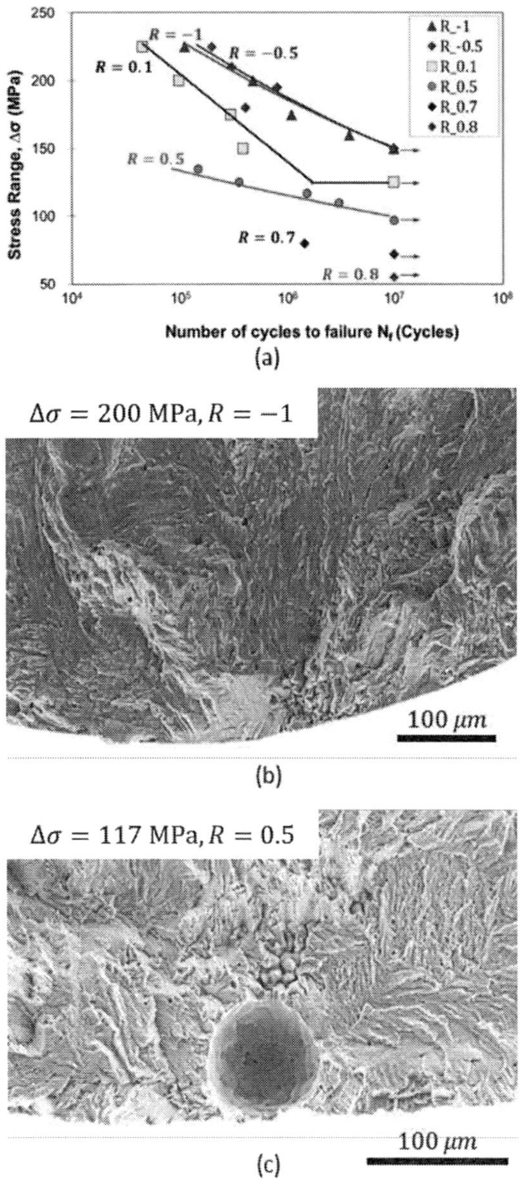

Figure 4: Whöler curve (a) and types of crack initiation: surface-induced (b) and defect-induced (sub-surface pore).

Fatigue lives and the endurance limits, both, were decreasing with increase in R-ratio, implying the detrimental effect of mean stress on fatigue lives. However, the detrimental effect for R = -0.5 was negligible as compared to that for R = -1. A non-monotonic trend in slopes of S-N curves was observed: first decreased when the R-ratio was increased from R = -1 to R = 0.1, then started to increase after R = 0.1. Beyond a certain R-ratio (R > 0.7), no fatigue failure could be observed within the investigated range of fatigue lives (10^7 cycles).

Majority of fatigue failures at lower R-ratios (< 0.1) initiated from surface without any defects (Fig. 4b) while at higher R-ratios (≥ 0.1), most of the fatigue failures initiated from welding defects (pores) (Fig. 4c). For the cases where the fatigue failure occurred from defects, it was observed that the defects were nearly spherical in shape and were located on/near to the surface, with a typical range for pore size between 50 μm to 140 μm.

4.2. Crack growth

The crack growth points obtained for ΔK-decreasing and ΔK-increasing followed nearly the same trend as shown in Fig. 5. However, it is recommended in ASTM-E-647 that in such a case, the point of ΔK-increasing should only be considered for analysis and duplicate points obtained from ΔK-decreasing should be deleted. Two different crack growth tests were conducted for each R-ratio so as to see the accuracy and preciseness of the obtained experimental results and it was found that the crack growth results obtained are precise and reliable.

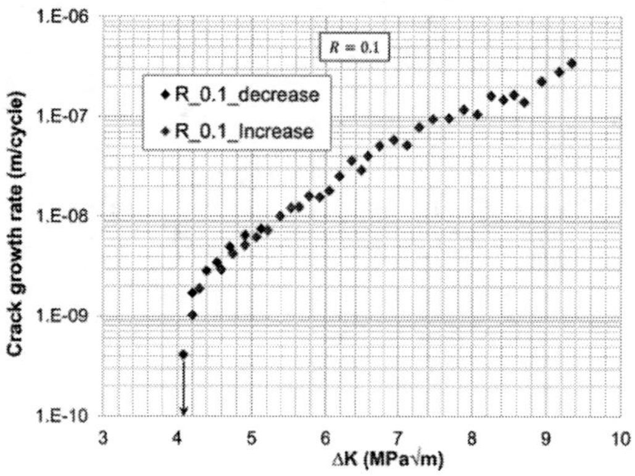

Figure 5: Crack growth data for ΔK-decreasing and ΔK-increasing for R = 0.1.

Fig. 6 compares the crack growth rate at different R-ratios and as expected, decrease in R-ratio decreases the fatigue crack growth rate and the threshold ΔK. However, the drop in crack growth rates was less from R = 0.5 to R = 0.8 as compared to that from R = 0.1 to R = 0.5. For R = 0.1 and 0.5, a definite Paris regime could be observed spanning along the significant range of ΔK while for R = 0.8, Paris regime was limited to a smaller ΔK range and a stage-III type of crack propagation was observed.

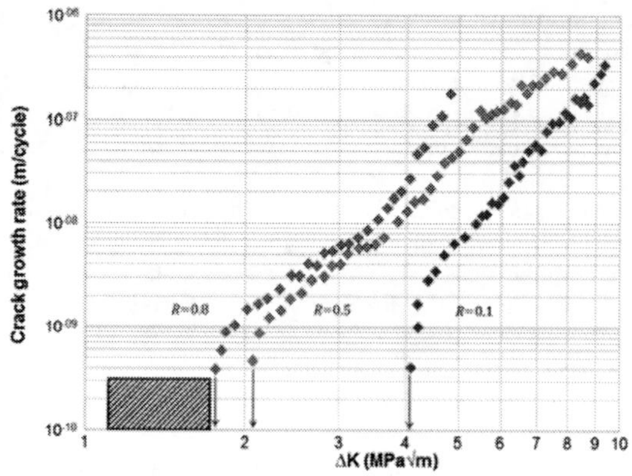

Figure 6: Long crack growth data at different R-ratios.

5 Discussion

5.1 Fatigue lives

Increase in mean stress decreases both the endurance limits and the fatigue lives with some scatter. However, plotting the true stresses instead of nominal stresses in Wöhler curve (S-N curve) did not reduce the scatter in fatigue data. The experimental fatigue data was also plotted in Goodman diagram (Fig. 7a). Data are plotted as broken (red dots) and unbroken (green dots) specimens only. Some popular empirical models available in literature, like Gerber's parabola, Smith-Watson-Topper's model, Walker's model and the modified Walker's model proposed in [11] are also plotted in same plot and compared with each other. It appears that the modified Walker's model when combined with mean ductile fracture line predicts better the evolution of endurance limit.

At very high stress ratios (R > 0.7), no fatigue failure could be observed within the investigated range of fatigue lives thus suggesting a probable saturation effect of mean stress above this R-ratio. To precise this critical R-ratio, two tests were done at R = 0.7 (two different stress amplitudes: one broken test and one non-broken). Fig. 7b compares the evolution of mean strain during a fatigue test at R = 0.8 (no fatigue failure) and R = 0.7 (no fatigue failure). Mean strain was calculated from strain amplitude which was recorded during the fatigue tests using an extensometer. Although, the specimen did not fail until 10^7 cycles and the tests were stopped but mean strain was continuously increasing with time for R = 0.8 while for R = 0.7, a steady state was reached. It implies the continuous accumulation of plastic strain (probably due to ratchetting) at stress ratio above R = 0.7. It is thus reasonable to say that above this R-ratio (R = 0.7), mean stress effect on fatigue is indeed saturated but it does not necessarily imply non-failure of specimens. Above this R-ratio (transition point), material is expected to fail due to ratchetting if the tests were continued beyond 10^7 cycles.

Similar mean stress saturation effects have already been reported in some studies for e.g. in Ti alloys [13] and in steels [14]. Kovacs [14] also reported the failure of martensitic steels (X10CrNiMoV12-2-2) in fatigue at high mean stresses with a typical ratchetting failure at a very high stress ratio (R = 0.7) and thus consistent with interpretation and observation in this study.

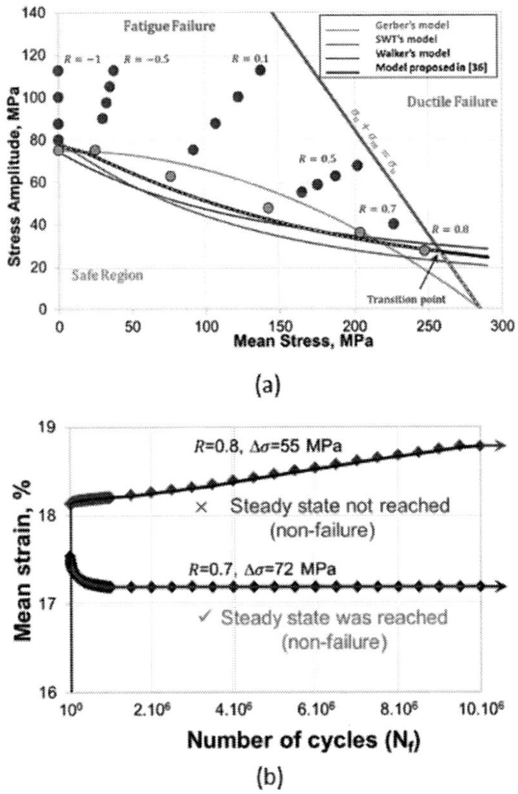

Figure 7: (a) Goodman diagram for combined experimental fatigue data. (b) Evolution of mean strain at high R-ratios.

5.2 Crack initiation

SEM observations of fracture surfaces revealed two types of crack initiation mechanisms:

- Surface-initiated crack initiation, dominant at lower R ratios (< 0.1)
- Defect-initiated crack initiation, dominant at high R-ratios (> 0.1).

All defects responsible for crack initiation appeared to be gas porosities resulting from welding, owing to their nearly spherical shapes. Such porosities could have resulted from ingression of various gases into the weld pool and most susceptible one is the hydrogen gas considering its sources like moisture, hydrocarbons from contaminants on parent material and many others. But the exact mechanism is not yet fully understood. The fraction of specimens with defect-initiated failures was calculated for each R ratio by dividing the total number of specimens initiated from pores to the total number of fatigue tests done at that R-ratio and a dependence of crack initiation site on R-ratio was found: % of defect-initiated failures increased with increase in R-ratio (Fig. 8).

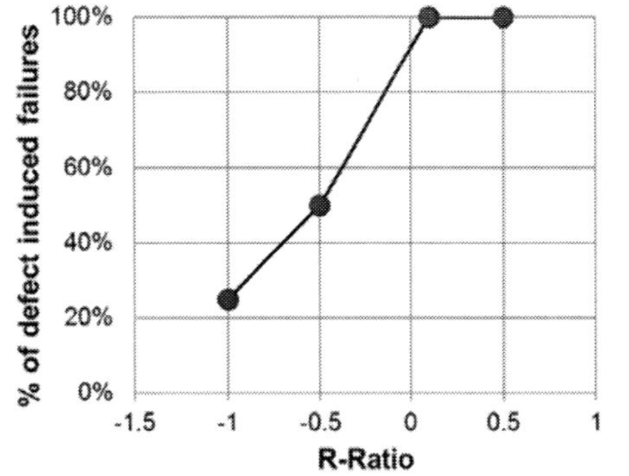

Figure 8: Evolution of fraction of defect-induced crack initiations at different R-ratios.

Of all the defects present in the material, only the ones located on or near to the surface were mainly responsible for crack initiation, implying the significant role of surface or sub-surface pores on fatigue damage. The severity of sub-surface pores has also been studied numerically by many authors and it was concluded that the pores touching the surface are always more prone to initiate damage as compared to other pores [11, 15]. It is already been shown that the stress concentration factor (K_t) varies with the position of pore with respect to surface. The K_t value is highest when the pore is located near to the surface as compared to the one located far away. Such a variation of K_t with pore position does not change with pore size. Higher K_t induces in-homogeneous stress-strain field around the pores and higher cyclic plasticity at the edge of pore, thereby making it more prone to fatigue damage initiation. Also, high R-ratio, which means higher peak stress, promotes local plastic flow around the pore. These arguments can probably explain preferential crack initiation from defects located on or near to the surface in cases of defect induced failures and their dominance at high R-ratios.

Usually it is expected that increase in mean-stress should decrease crack initiation time because of the plastic flow that occurred during the first ramp load of the cycle. Also, the presence of pores could have reduced the crack nucleation time and thus the combined effect should reduce the crack initiation time significantly. It is thus expected that at R = 0.8, a crack should have initiated but no failure was observed, probably because of the insufficient crack driving force (ΔK) so that the crack, if initiated, could not have propagated and the same could be verified using crack growth characteristics as discussed next.

5.3 Crack growth

Mean stress certainly seems to affect crack growth rates and reasons for this has already been known i.e. crack closure effects: plasticity induced crack closure, roughness-induced crack closure and oxide-induced crack closure.

SEM observations of fracture surfaces at different stress ratios revealed effect of mean stress on fracture surfaces. For low stress ratios (R < 0.1), smooth and clear fatigue striations could be observed (Fig. 9a) while for higher stress ratios (R > 0.1), such smooth striations were difficult to locate (Fig. 9b). The probable reason could be the lower ΔK at high R-ratios and could imply the role of plasticity-induced crack closure [16].

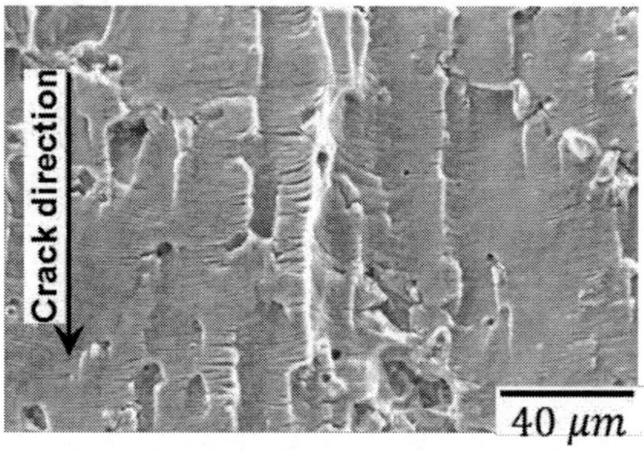

(a)

(b)

Figure 9: Striations as observed on fracture surfaces at different R-ratios: (a) R = -0.5 and (b) R = 0.7.

The fracture surface roughness was also compared at different stress ratios. The fracture surface at higher R-ratio appeared to be rougher as compared to that at lower R-ratio and thus suggest the role of asperity-induced or roughness induced crack closure [17]. An attempt was made to quantify this change in surface roughness using the Olympus LASER microscope. The ductile failure region of the broken specimens was carefully removed so as to avoid any interference with the lens of the microscope. Standard procedures were adopted for measurements using the recommended cut-off value (2.5 mm) to filter longer wavelength of the roughness (ISO 4288-1996). Five measurements were done at each stress ratio (R-ratio) and then an average roughness value (R_a) was calculated and is plotted in Fig. 10 for different R-ratios. It is evident that the roughness value (R_a) tends to increase with increase in R-ratio with saturating effect at very high R ratios.

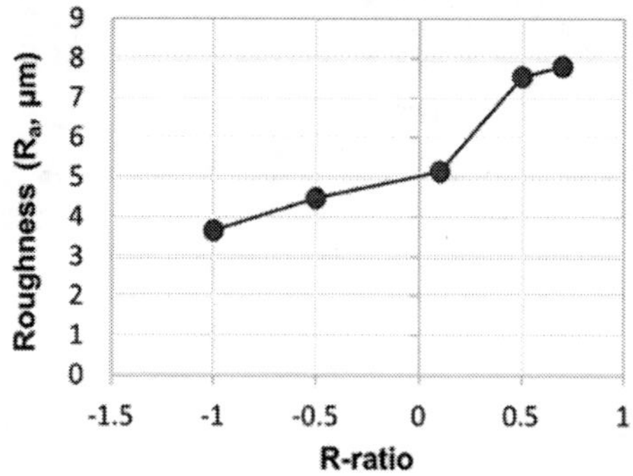

Figure 10: Estimated trend in mean value of roughness (Ra, μm) of fracture surfaces at different R-ratios.

To understand better the crack growth from defects, stress intensity factor was calculated for each defect responsible for fatigue crack initiation using the Murakami's "√area" model [8]:

$$\Delta K = \beta\, \Delta\sigma\, \sqrt{\pi\sqrt{area}} \quad (2)$$

Where β = 0.5 for internal defect and β = 0.65 for surface touching defect. The stress range (Δσ) for negative R-ratios was calculated considering only the tensile part of the load cycle, since compressive load cycle do not contribute to crack growth. The initial stress intensity factor for the defects typically lies within the range of 1.1-1.7 MPa√m for different R-ratios. The same has also been marked by hatched black box in Fig. 6. It is quite clear that the ΔK for these defects lie below the respective ΔK_{th} at each R-ratio.

Considering the largest observed defect responsible for fatigue failures (i.e. 100 μm) and the largest possible stress range that can be applied at R = 0.8 (i.e. 56 MPa), the stress intensity factor calculated using Eq. 2 came out to be 0.6 MPa√m. This initial stress intensity factor range is way lower than the ΔK_{th} for R = 0.8 and probably is also lower than the intrinsic threshold value for crack propagation for this material. Thus at this R-ratio, even though it is expected that a crack should have initiated but it may not propagate due to very low crack driving force.

But given the fact that the size of defects are comparable to that of average grains size, the crack initiating from these defects cannot analyzed using long crack growth data. The crack initiating from defects are the microstructurally short crack and short crack growth data may be required to explain initiation from these defects. Many studies [18-19] have shown that the short crack can grow at a faster rate than the long cracks, even below ΔK_{th}, which might also be the case here. Therefore, to understand better the contribution of crack initiation and crack propagation in total fatigue live for defect-induced failures, it is required to have some data on short crack growth. Such short crack growth tests are still in progress.

6 Conclusions and further work

Based on the work done till now, following conclusions could be drawn:

- Cyclic loading with mean-stress decrease the fatigue lives, due to combine effects of the crack closure effects and also probably due to the reduction in crack-nucleation time.
- A saturation effect of mean-stress on fatigue was observed at R > 0.7, with no fatigue failure above this R-ratio.
- Mean stress also tends to control crack initiation sites in metals containing defects: defect-induced failures were dominant at high R-ratios, with nearly all of the defects located on/near the surface.
- The probable reason for higher fraction of defect-induced failures at higher R-ratio could be enhanced local cyclic plasticity due to high stress concentration factor associated with them.

Long crack growth data alone could not possibly explain the crack growth behavior of cracks initiation from naturally occurring defects inside the material. It is so because the size of defects was comparable to the average grain size and thus should be treated as microstructurally short cracks. Short crack growth tests are in progress.

The authors are thankful to the Japanese research funding agency (JST) for funding this project through its Cross-ministerial Strategic Innovation Promotion (SIP), "Structural Materials for Innovation" program.

References

1. R. S. Mishra, Z.Y. Ma, Mater. Sci. Eng. R Rep. **50**, 1-78 (2005)
2. I. Bunaziv, O.M. Akselsen, A. Salminen, A. Unt, J. Mater. Process. Technol. **233**, 107-114 (2016)
3. H. Jakubczak, G. Glinka, Int. J. Fatigue **8**, 51-57 (1986)
4. U.Brandt, F. Lawrence, C.M. Sonsino, Fatigue Fract. Eng. Mater. Struct. **24**, 117-126 (2001)
5. H.R. Ammar, A.M. Samuel, F.H. Samuel, Mater. Sci. Eng., A **473**, 65-75 (2008)
6. M. Couper, A. Neeson, J. Griffiths, Fatigue Fract. Eng. Mater. Struct. **13**, 213-227 (1990)
7. Q.G. Wang, D. Apelian, D.A. Lados, J. Light Met. **1**, 73-81 (2001)
8. Y. Murakami, *Metal defects: Effect of small defects and nonmetallic inclusions* (2002)
9. Z.Lei, Y. Hong, J. Xie, C. Sun, A. Zhao, Mater. Sci. Eng., A **558**, 234-241 (2012)
10. P. Threadgill, A. Leonard, H. Shercliff, P. Withers, Int. Mater. Rev. **54**, 49-93 (2009)
11. V. Gaur, M. Enoki, T. Okada, S. Yomogida, Int. J. Fatigue **107**, 119-129 (2018)
12. A. Rabbani, S. Jamshidi, S. Salehi, J. Pet. Sci. Eng. **123**, 164-171 (2014)
13. R.J. Morrissey, D.L. McDowell, T. Nicholas, Mech. Time-Depend. Mater. **2**, 195-210 (1998)
14. S. Kovacs, T. Beck, L. Singheiser, Int. J. Fatigue **49**, 90-99 (2013)
15. Z. Xu, W. Wen, T. Zhai, Metall. Mater. Trans. A **43**, 2763-2770 (2012)
16. W. Elber, Eng. Fract. Mech. **2**, 37-45 (1970)
17. S. Suresh, R. Ritchie, Metall. Mater. Trans. A **13**, 1627-1631 (1982)
18. K.J. Miller, Fatigue Fract. Eng. Mater. Struct. **10**, 75-91 (1987)
19. A. McEvily, Trans. on Eng. Sci. **13**, 93-107 (1996)

High Temperature Fatigue Crack Growth Rate Studies in Stainless Steel 316L(N) Welds Processed by A-TIG and MP-TIG Welding.

Manuel Thomas[1], *Raghu* V. Prakash[1*], *Ganesh Sundara Raman* S[3] and *Vasudevan*, M.[4]

[1]Department of Mechanical Engineering, IIT Madras, 600036 Chennai, India
[2]Department of Mechanical Engineering, IIT Madras, 600036 Chennai, India
[3] Department of Metallurgical and Materials Engineering, IIT Madras, 600036 Chennai, India
[4] Indira Gandhi Centre for Atomic Research, 603102, Kalpakkam, India

Abstract. Welded stainless steel components used in power plants and chemical industries are subjected to mechanical load cycles at elevated temperatures which result in early fatigue failures. The presence of weld makes the component to be liable to failure in view of residual stresses at the weld region or in the neighboring heat affected zone apart from weld defects. Austenitic stainless steels are often welded using Tungsten Inert Gas (TIG) process. In case of single pass welding, there is a reduced weld penetration which results in a low depth-to-width ratio of weld bead). If the number of passes is increased (Multi-Pass TIG welding), it results in weld distortion and subsequent residual stress generation. The activated flux TIG welding, a variant of TIG welding developed by E.O. Paton Institute, is found to reduce the limitation of conventional TIG welding, resulting in a higher depth of penetration using a single pass, reduced weld distortion and higher welding speeds. This paper presents the fatigue crack growth rate characteristics at 823 K temperature in type 316LN stainless steel plates joined by conventional multi-pass TIG (MP-TIG) and Activated TIG (A-TIG) welding process. Fatigue tests were conducted to characterize the crack growth rates of base metal, HAZ and Weld Metal for A-TIG and MP-TIG configurations. Micro structural evaluation of 316LN base metal suggests a primary austenite phase, whereas, A-TIG weld joints show an equiaxed grain distribution along the weld center and complete penetration during welding (Fig. 1). MP-TIG microstructure shows a highly inhomogeneous microstructure, with grain orientation changing along the interface of each pass. This results in tortuous crack growth in case of MP-TIG welded specimens. Scanning electron microscopy studies have helped to better understand the fatigue crack propagation modes during high temperature testing.

1 Introduction

The Austenitic Stainless steels are known to possess adequate high-temperature physical properties and excellent corrosion resistance. By virtue of this, Austenitic Stainless steels are routinely used as structural members in chemical industries and power plants. The low carbon grade of austenitic stainless steels (316L, 304L), normally designated as L grades is commonly used in applications where high resistance to intergranular corrosion is desirable [1]. The addition of nitrogen in a small amount (up to 0.08wt%) is known to improve the mechanical strength and impart improved creep resistance [2-5]. The improved creep resistance in the temperature range of 500-600°C makes the type 316L(N), as the most widely used material for the power plant components.

Welding is often being used in the fabrication of power plant structural components. The presence of weld can compromise the structural integrity, as the weld often acts as a site for the defect. When exposed to the cyclic loads, it can act as a fatigue crack initiation site, which leads to further propagation by fatigue and fast fracture, thus, making it necessary to estimate the crack growth rates in the welds for life prediction.

The available literature presents the details of the effect of different welding process on the fatigue behavior for variety of austenitic stainless-steel weldments [6]. These studies indicate a reduction in crack growth rates in the weldments compared to the base metal. Mills [1] has conducted a detailed review of the fracture toughness of austenitic stainless steels (SS304 and SS316) and it welds, and has observed the significant variation in the mechanical properties and fracture toughness of the material, depending on the welding process adopted. Shahinian and James[7-10] observed significant disparity in the crack growth rates of SS316 welds prepared using different flux and filler materials, thus, indicating the effect of welding technique on the fatigue crack growth rate. As many of the power plant components work at a high temperature for a considerable period of time, it is important to understand the fatigue response of the SS304L(N) and SS316L(N) material and its welds to better estimate the life.

The present work presents the high-temperature fatigue crack growth in type 316L(N) stainless steel welds, joined using two processes: a) conventional multi-pass TIG (MP-TIG) and b) Activated Flux TIG (A-TIG) welding process. It may be noted that the welding of metals using conventional TIG process is

* Corresponding author: raghuprakash@iitm.ac.in

limited by the thickness of material that can be joined as one needs to avoid lack of penetration. The use of conventional TIG for the joining the stainless steel is limited to plates having a maximum thickness of 3 mm, which is a serious limitation [12] for many power plant component building. Further, it is noted from the literature that variable penetration rates are exhibited by the stainless steels of different heats (or batches) during TIG welding. The use of multiple passes for achieving the increased penetration may result in low productivity, weld distortion, shrinkage and residual tensile stress.

Activated flux TIG (A-TIG) welding is a variant of TIG welding, developed by E.O. Paton Electric Welding Research Institute [13]. Higher depth of weld penetration is possible in A-TIG welding, making it suitable for joining thicker plates in a single pass, thereby increasing the productivity [13]. The A-TIG welding process involves the use of an electronegative flux, applied to the surface to be welded which tends to constrict the arc, thereby increasing the current density. This results in the higher weld penetration and faster welding.

For the present study, an indigenously developed multi-component flux is used for the A-TIG welding. This study is intended to evaluate the fatigue crack performance of welds produces using indigenously developed A-TIG welding and compare it with the conventional MP-TIG weld.

2 Experimental Methods

The chemical composition of the base metal and weld is presented in Table l. The welded plates of 5 mm thickness 316L(N) were prepared using MP-TIG, A-TIG welding technique. A layer of flux having patented composition has been applied on the surface of the plates to be welded. It may be noted that the A-TIG welding process was completed in a single pass with out the use of any filler wires. The filler wires of 316L were used in the preparation of 316L(N) MP-TIG joints. The detailed procedure of the process and parameters used is reported in Ref. 13.

Fatigue crack growth experiments were conducted in accordance to ASTM E647, using a MTS 810 servo-controlled hydraulic actuator system. Compact tension-type specimens extracted from the weld plates, from the location of Heat affected zone (HAZ) and weld metal bead as indicated in the schematic (Fig. 1). Figure 2 presents the schematic of the C(T) specimen used for fatigue crack growth rate experiments. The orientation of the machined notch was parallel to the direction of the welding to ensure that the fatigue crack grew along the weld bead (or) in the HAZ that is parallel to the weld bead.

The fatigue crack growth rate testing was conducted at a temperature of 823K (550°C). A three-zone split furnace was used to heat the specimen and maintain the temperature. All the tests were conducted at a constant positive stress ratio (Ratio of minimum to maximum load) of R=0.1. Crack growth rate was estimated based on compliance technique using an Epsilon make High Temperature crack opening displacement gage/The knife edges for mounting the High temperature COD gage was made integral to the C(T) specimen geometry by wire-EDM process. The load-COD response during fatigue cycling was recorded at periodic intervals of (~ 750 cycles) at a low frequency of 0.5 Hz, over 5 cycles in an otherwise constant amplitude load cycling of the specimen at a typical test frequency of 10 Hz.,

The crack length was calculated using unloading compliance method as described by ASTM E-647 test standards. The failed samples were examined using a Quanta-400 FE SE Scanning Electron Microscope to understand the fracture mode details.

3 Results and Discussion

3.1. Microstructure

The weld macrographs provided in figures 3 and 4 indicates the extent of penetration and the cross-section of the weld bead. A relatively narrow weld bead can be observed for the A-TIG joints, while the MP-TIG joints exhibit wider weld geometry. The optical micrograph for the two types of welds indicates that primary austenitic solidification mode was observed irrespective of the type of welding in both the types of welds considered.

Table 1. Chemical composition (wt%) of type 316L(N) and welds

Material	C	Mn	Si	Cr	Ni	Mo	N	S	Fe
SS316L(N)	0.03	1.65	0.44	17.5	12.2	2.3	0.09	0.01	Balance
MP-TIG weld	0.03	1.50	0.53	19.0	11.0	2.3	0.09	0.01	Balance
A-TIG weld	0.03	1.45	0.50	17.5	12.0	2.4	0.09	0.01	Balance

The microstructure of the resultant weld joint was analyzed with the help of a Nikon optical microscope. The samples were polished and electrolytically etched to reveal the microstructure. The microhardness profile was obtained over the weld, using Vickers microhardness apparatus. A load of 1.75 N was applied, with a dwell of 15s. Indentations were made at base metal, heat affected zone and weldment at regular spacing.

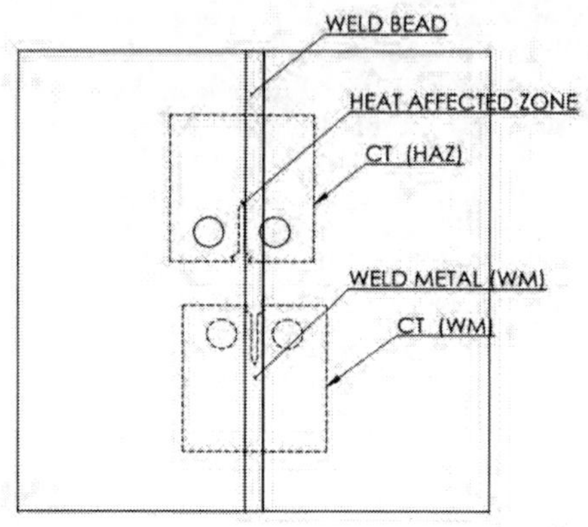

Fig. 1. Compact tension specimen extraction lay out from weld plates.

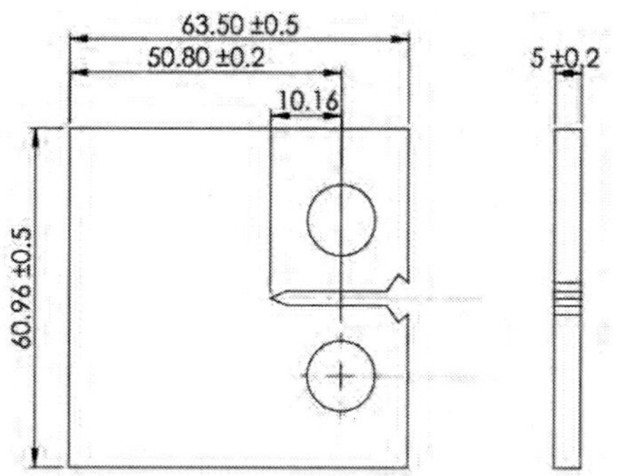

Fig. 2. Schematic of C(T) specimen. All dimensions are in mm.

In case of A-TIG weld metal, fusion zone contained equiaxed growth of grains running parallel to the direction of weld obtained in each weld. Columnar grains are located near the base metal zone, oriented in the transverse direction of welding. Further, it may be noted that in case of A-TIG welding, during solidification, maximum thermal gradient is observed in the transverse direction and the same is indicated from the orientation of columnar grains. The enhanced depth of penetration accounts for the presence of equiaxed grains running through the entire depth of weld. The high heat input in single pass during the welding has resulted in coarse grains in the weld.

In case of MP-TIG joints, the microstructure exhibited random discontinuous distribution of columnar and equiaxed grains. The repeated heating cycles during each pass in welding process has resulted in the highly incoherent distribution of grains (as shown in Figure 3). Thus the effect of thermal cycling due to multi-pass welding is predominant near the region of each weld pass, and the interface of each pass shows a clear change in microstructure. The preferential alignment of grains can be seen normal to each weld pass interface. The grains are observed to be fine in the MP-TIG as the heat input in each pass is lower.

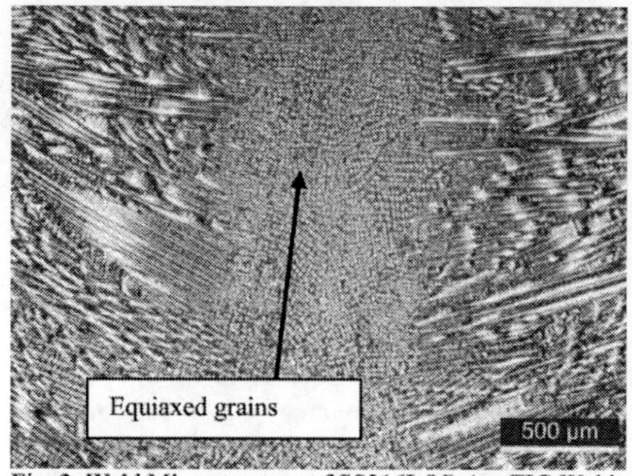

Fig. 3. Weld Microstructure of SS316L(N) A—TIG Weld.

Fig. 4. Microstructure of SS316L(N) MP-TIG Weld.

3.2 Micro-hardness

The variation in microhardness over the weld profile is shown in Figures 5 and 6. One can observe a direct correlation between the micro-hardness values and the microstructure. The fusion zone in the A-TIG joints show a reduced hardness value compared to the base metal region. Maximum variation in hardness was observed near the fusion boundary, which corresponds to the region of heat affected zone. Equiaxed grains present at the center of the welded joint showed a maximum hardness, while the adjacent columnar grains showed the minimum value. The slower rate of cooling in A-TIG welds has resulted lower hardness value than the corresponding MP-TIG welds.

The MP-TIG welded joints had higher hardness than its base metal. Higher degree of variation in the hardness can be observed across the weld profile, which can be due to the discontinuous distribution of the grains in MP-TIG welding. The zone adjacent to base metal exhibited an abrupt change in hardness, confirming the presence of heat affected zone. The hardness value obtained for MP-TIG welds are appreciably higher than its A-TIG counterpart. Repeated thermal cycle during each pass and faster cooling rates has promoted formation of finer grains in MP-TIG welds, thereby increasing hardness value.

The MP-TIG welded joints had higher hardness than its base metal. Higher degree of variation in the hardness can be observed across the weld profile, which can be due to the discontinuous distribution of the grains in MP-TIG welding. The zone adjacent to base metal exhibited

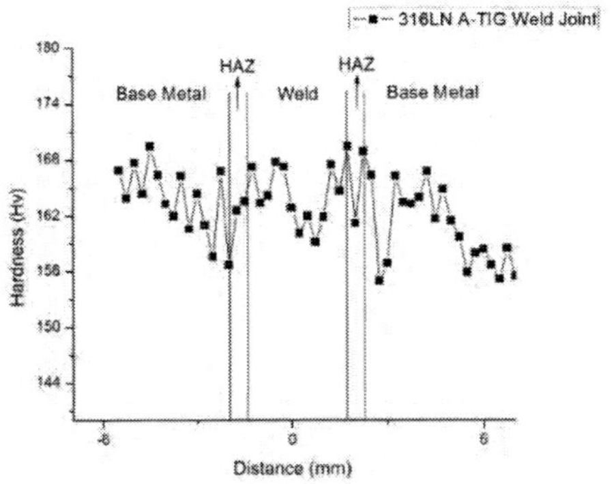

Fig. 5. Microhardness profile of SS316L(N) A-TIG Weld.

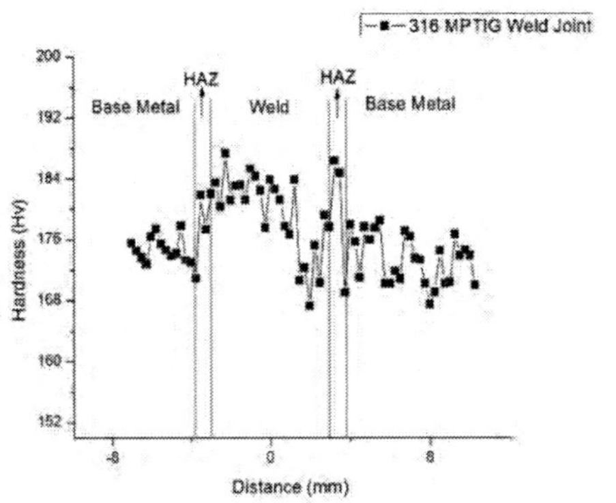

Fig. 6. Microhardness profile of SS316L(N) MP-TIG Weld.

an abrupt change in hardness, confirming the presence of heat affected zone. The hardness value obtained for MP-TIG welds are appreciably higher than its A-TIG counterpart. Repeated thermal cycle during each pass and faster cooling rates has promoted formation of finer grains in MP-TIG welds, thereby increasing hardness value.

2.3 Fatigue crack growth at high temperature

Fatigue crack growth rate (da/dN) estimated from crack length versus cumulative number of cycles i.e. a-N curves from experiments conducted at a stress ratio (R) of 0.1, is plotted as a function of applied stress intensity range. Figure 7 presents the crack growth rate obtained for the base metal, MP-TIG, A-TIG welds at 823K. The crack growth characteristic for the material in the Heat affected zone region is depicted in Figure 8. It was observed that at the specific high temperature, the crack growth rates exhibited by the weld prepared using MP-TIG and A-TIG technique was lower than the parent base metal. Between the MP-TIG and A-TIG weld process, the resistance to the crack growth rate was higher in case of specimens prepared using MP-TIG technique. This could be due to the presence of finer grains in the MP-TIG welded samples. A similar slow down in the crack growth rate was exhibited by the heat affected zone of MP-TIG weld specimens. It is interesting to note that that the fatigue rack growth rates in HAZ of A-TIG weld samples are almost the same as that of the base metal. Thus, the fatigue crack growth rate of weld metal region in A-TIG samples is lower than that of HAZ of A-TIG specimens.

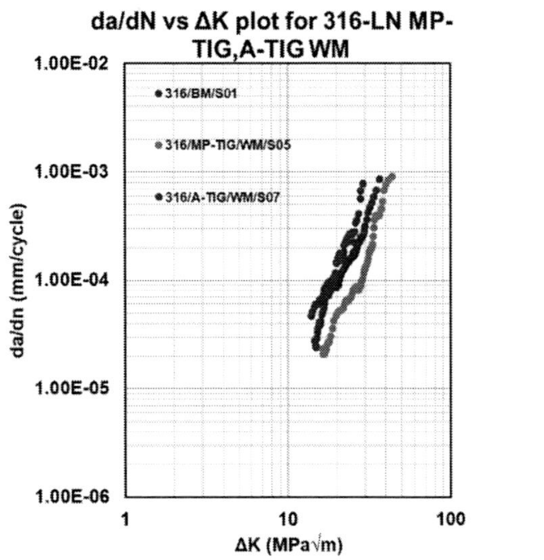

Fig. 7. Fatigue crack growth rate plot for SS316L(N) base metal, SS316L(N) MP-TIG weld metal and SS316L(N) A-TIG weld metal.

The increased resistance to the crack growth in the MP-TIG welds can be due to the fine-grained microstructure. It could also be due to the repeated thermal cycling during the multi-pass welding that has resulted in highly refined randomly oriented grains, which imparts resistance to the crack growth. The slowdown in fatigue crack growth rate observed at the heat affected zone can be due to the high hardness exhibited by the MP-TIG heat affected zone.

Examination of the fracture surface using scanning electron microscope reveals trans-granular mode of failure in both the welds. Figures 9 and 10 represent the observed fracture surface for the welds joined using MP-TIG and A-TIG technique respectively. At regions of low crack growth rate or near the chevron notch initiation zone, the fracture surface is predominantly marked by the presence of river lines and cleavage facets.

https://doi.org/10.1051/matecconf/201816521014

Figures 11 and 12 present the SEM image where the fatigue striations can be seen all along the fracture surface. The fracture surface was found to be rougher for the MP-TIG welds, with herringbone pattern of tear ridges. The MP-TIG welds near the initiation zone exhibit typical quasi-cleavage fracture which is a combination of cleavage facets and tear ridges. The quasi cleavage is typically observed in fine grained structure that is fractured at high temperature. It can further be observed from these SEM images that there is tear ridges between the planes containing striations that run parallel to the direction of crack. The striations are more predominant at intermediate crack growth rates.

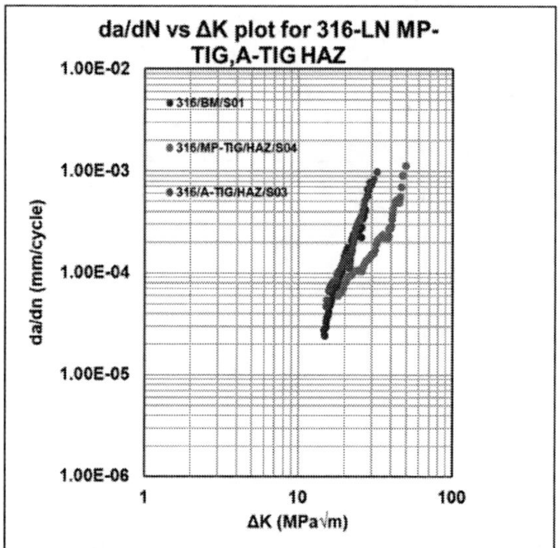

Fig. 8. Fatigue crack growth rate plot for SS316L(N), SS316L(N) MP-TIG weld and SS316L(N) A-TIG in the Heat Affected Zone.

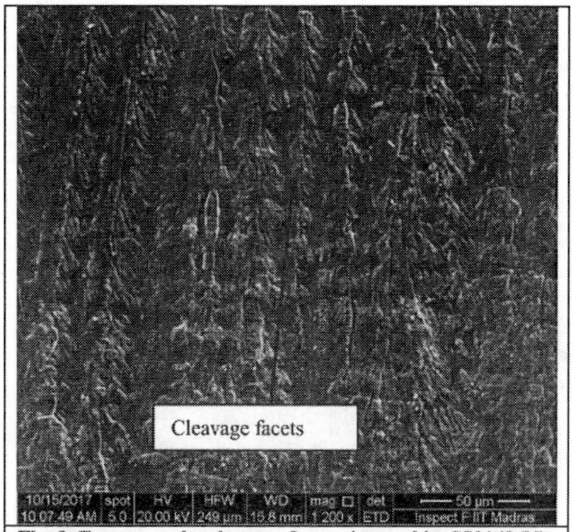

Fig. 9. Trans-granular cleavage facets observed in SS316L(N) MP-TIG welds, at a distance of ~2 mm from initiation (ΔK ~ 15 MPa√m)

Figure 13 indicates the fracture surface region near final failure. The final failure in all the specimens was due to ductile mode, marked by fibrous appearance of the fracture surface. With an increase in crack growth rates near the fast fracture region, increased presence of dimples was noticed; void coalescence begins to appear along with the striations. The fast fracture region had considerable presence of dimple structure. The voids that arise from the inclusion or coarser precipitates, get enlarged and during further yielding, can get sheared at the location, giving a fibrous appearance.

Fig. 10. River line pattern of fracture surface for SS316 A-TIG joint, located at a distance of ~2 mm from the Chevron notch initiation zone.

Fig. 11. Observed striation pattern for SS316L(N) A-TIG joint, located at ~10mm from initiation zone.

3 Conclusions

The high-temperature fatigue crack growth in SS316L(N) metal welds fabricated using MP-TIG and A-TIG technique was investigated experimentally. Ancillary metallographic and fractographic evaluation of the surface was performed to identify the failure mechanism. In all the test cases, the mode of failure observed was trans-granular in nature. The rate of crack growth in parent metal (base metal) was found to be faster than that in weld metal, with MP-TIG weld showing better resistance to the crack growth. The weld metal of A-TIG showed reduced crack growth rates for a given stress intensity factor range compared to HAZ region as well as base metal. In the case of MP-TIG weld, the HAZ crack growth rates were considerably slower than base metal. From these perspectives, it would appear reasonable to have MP-TIG has the process for improving the fatigue crack growth rate resistance of material.

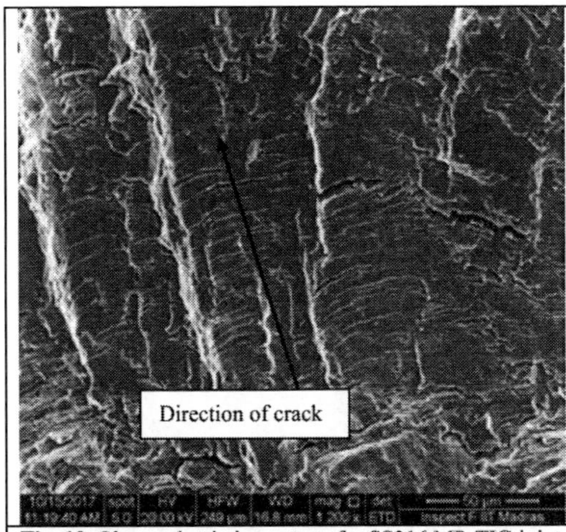

Fig. 12. Observed striation pattern for SS316 MP-TIG joint, located 12mm from initiation zone.

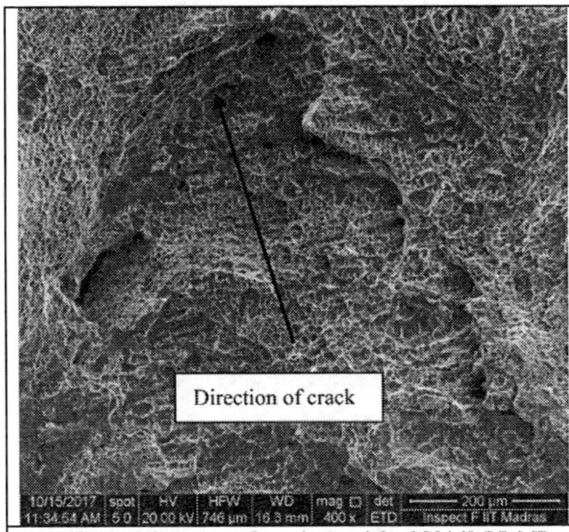

Fig. 13. Final fracture surface observed for SS316L(N) MP-TIG

References

1. W. J. Mills, International Materials Reviews **42.2** (1997)
2. V. Ganesan, M. D. Mathew, P. Parameswaran, K. Laha, *Procedia Engineering* **55** (2013)
3. V. Ganesan, M. D. Mathew, P. Parameswaran, K. Bhanu Sankara Rao. *Transactions of the Indian Institute of metals* **63, No. 2-3** (2010)
4. M. D. Mathew, K. Laha,V. Ganesan,*Materials Science and Engineering*: **A 535** (2012)
5. M. D. Mathew, G. Sasikala, K. Bhanu Sankara Rao, S. L. Mannan, *Materials Science and Engineering*: **A 148, No. 2** (1991)
6. A.C. Pickard, R. O. Ritchie J. F. Knott.*Metals Technology* **2.1** (1975)
7. P. Shahinian, *Welding Journal* **57, No. 3** (1978)
8. P. Shahinian, H. H. Smith, H. E. Watson, Fatigue at Elevated Temperatures, *ASTM International*, 1973
9. L.A James, *Welding Journal* **56, No. 12** (1977)
10. L.A. James, W. J. Mills, *Welding Journal* **66, No. 8** (1987)
11. K.C. Mills, B. J. Keene, *International Materials Reviews* 35, no. 1 (1990)
12. M.T. Tanaka,T. Shimizu, T. Terasaki, M. Ushio, F. Koshiishi, Yang C.-L, *Science and technology of welding and joining* **5, No. 6** (2000)
13. M. Vasudevan, *Journal of Materials Engineering and Performance* **26, No. 3** (2017).

Low-cycle fatigue behavior of extruded Al-7Zn-2Mg-1.5Cu-0.2Sc-0.1Zr alloy at room and low temperatures

Zhu Qingyan[1], *Chen* Lijia[1], *Xu* Chengji[2], *Che* Xin[1], and *Li* Feng[1]

[1]School of Materials Science and Engineering, Shenyang University of Technology, 110870 Shenyang, China
[2]Liaoning Equipment Manufacturing Vocational and Technical College, 110161 Shenyang, China

Abstract: Due to the low density and high specific strength, aluminum alloys have been considered for automotive and aerospace applications. The aluminum components usually service in the conditions of low temperature and dynamic loading. Therefore, the research on the low temperature fatigue performances of Al alloys has great significance. The low-cycle fatigue tests for the extruded Al-7Zn-2Mg-1.5Cu-0.2Sc-0.1Zr alloy subjected to solution plus aging treatment have been conducted at 25°C and -40°C, respectively. The strain ratio and cyclic frequency applied in the low-cycle fatigue test were -1 and 0.5Hz, respectively. The experimental results show that at 25°C, the alloy exhibits the cyclic hardening at the total strain amplitudes of 1.0% and 1.2%, and the cyclic stabilization at the total strain amplitudes of 0.4%, 0.6% and 0.8%. At -40°C, however, the cyclic stability is observed during whole fatigue deformation at the total strain amplitudes of 0.4%, 0.5%, 0.6%, 0.7% and 0.8%. The relationship between the elastic strain amplitude, plastic strain amplitude and reversals to failure can be described by Basquin and Coffin-Manson equations, respectively. In addition, the observation results of fatigue fracture surfaces reveal that the cracks initiate at the free surface of fatigue specimen and propagate in a transgranular mode.

1 Intrduction

The Al-Zn-Mg-Cu series aluminum alloys, as a structural material widely used in the field of aerospace, rail transit, automobile and ship, have good application prospect due to the high strength associated with low density[1-3]. The optimum properties of aluminum alloys are generally attained by various kinds of alloying additions and heat treatments[4, 5]. To further improve the mechanical properties of Al-Zn-Mg-Cu alloys, many attempts have been made by various methods. At present, the research of aluminum alloy with Sc and Zr is closely watched by many scientists and engineers[6-8]. Liu et al[9] investigated the effect of trace amount of Sc and Zr on microstructure and mechanical properties of novel Al-Zn-Mg-Cu-Sc-Zr alloys. The results indicateed that abundant nanometer-sized $Al_3(Sc, Zr)$ particles effectively pined dislocations and subgrain boundaries, exhibiting excellent antirecrystallization behavior and precipitates strengthening effect. Besides, with the additions of Sc and Zr increased, a higher density of precipitates was detected obviously with excellent refinement effect. The grain refinement strengthening and precipitation strengthening play a dominant role in the additional strength of the Al-Zn-Mg-Cu-Sc-Zr alloys. In addition, heat processing deformation is also an effective method to improve the properties of alloys, such as rolling, extrusion and forging[10, 11].

In fact, the structural parts usually suffer from different working conditions and fatigue loads during normal service. The complex kinds of service condition usually include different temperature and corrosive environments, which often influence the lifetime of the components. Therefore, the fatigue properties of aluminum alloy were studied by scientific researchers. In order to approximate the actual service conditions, fatigue tests were designed to be different strain ratios, corrosion environment conditions and so on[12-15]. Sreenivasan et al[16] have investigated the role of accumulation of ratcheting strain on low cycle fatigue behavior of aluminum 7075-T6 alloy, the results indicated that ratcheting strain was detrimental to fatigue life of aluminum 7075-T6 alloy by comparing the low cycle fatigue behavior of the investigated alloy with and without ratcheting indicated. And fatigue life of the alloy decreased with the increase of previously imposed ratcheting strain. However, only few researchers studied the influence of low temperature on low cycle fatigue life of the material due to the winter temperatures are generally low in some region. Hence, it needs an study on low temperature-fatigue interaction of aluminum alloys.

In the present paper, the strain-controlled low cycle fatigue(LCF) tests were carried out on specimens of Al-7Zn-2Mg-1.5Cu-0.2Sc-0.1Zr alloy at room and low temperature. The effect of low temperature on the low cycle fatigue behavior of the alloy was delineated, and the damage mechanism related to the low temperature was also characterized in this study.

2 Material and experiments

The Al-7Zn-2Mg-1.5Cu-0.2Sc-0.1Zr alloy ingots after homogenization treatment, were subjected to hot extrusion deformation at 450°C with an extrusion ratio

* Corresponding author: chenlj-sut@163.com

of 40. The as-extruded alloy bars was were solution treated at 475°C for 1h, and then artificial aging at 120°C for 44h. The monotonic tensile yield strength and ultimate tensile strength of the alloy at finally state were 601.9MPa and 700.3MPa, and the elongation is 8.5%.

Total strain-controlled low cycle fatigue test was performed on a MTS Landmark 370.10 computer-controlled servo-hydraulic fatigue testing machine at room temperature and -40°C in an air environment. The cooling method was liquid nitrogen refrigeration. The low-cycle fatigue experiment was loaded with sinusoidal waveform, the strain-ratio R was -1, imposed cyclic frequency was 0.5Hz, and the specified strain amplitudes were from 0.4% to 1.2%. The samples size for room and low temperature fatigue tests are shown in Fig. 1. Specimens were tested up to failure, and the corresponding cycling number was defined as the low-cycle fatigue life. The fracture morphology of fatigue specimens was analyzed using S 3400 scanning electron microscope(SEM).

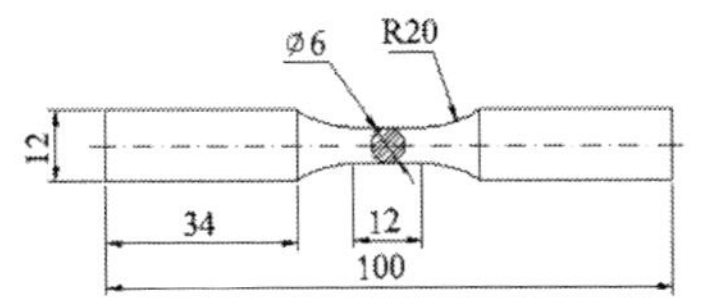

(a) fatigue sample size for room temperature

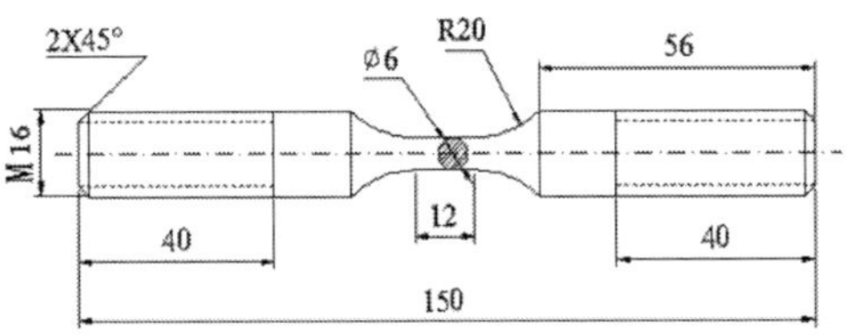

(b) fatigue sample size for low temperature

Fig. 1. Specimen designs for fatigue tests (mm)

3 Results and discussions

3.1 Cyclic stress response behavior

Fig. 2 shows the variety of cyclic stress amplitude with different strain amplitude of the alloy under room and low temperature. Obviously, as the total strain amplitude increase, the stress amplitude also increase. Besides, it can be seen from Fig. 2 that the cyclic stress response behavior of the alloy is related to the total strain amplitude and the experimental temperature. The cyclic stress response behavior of the alloy at room temperature is shown in Fig. 2a, the alloys exhibit obviously cyclic strain hardening characteristic during the fatigue deformation period at high total strain amplitudes of 1.0% and 1.2%. But at low total strain amplitudes of 0.4%, 0.6% and 0.8%, the stress response behavior of the alloy present cyclic stability, where the cyclic stress amplitude remain essentially constant as cyclic deformation progress. However, the stress response behavior of the alloy have distinctly different at low temperature in Fig. 2b, the alloy display continuously cyclic stabilization under all the total strain amplitudes of 0.4%-0.8% until the cyclic stress fall fast due to the initiation and propagation of fatigue cracks.

The cyclic hardening of the alloy can be understood as: dislocation source can appear in the alloy internal during low cycle fatigue deformation, dislocation source can generate a large number of dislocations to make dislocation density increase. At the same time, the movement of dislocation can produce interaction to form dislocation tangles and also can pile up in ahead of grain boundary or sub-boundary, which hinder the movement of dislocation and improve the cyclic deformation resistance of the alloy[17, 18]. In addition, Al_3Sc, Al_3Zr and $Al_3(Sc, Zr)$ particles can be formed in the alloy due to the addition of Zr and Sc elements, and these particles have a blocking effect on the dislocation motion resulting in cyclic hardening. But the main strengthening phase having enhanced effect is nano-scale η'($MgZn_2$) phase in Al-Zn-Mg-Cu alloy. These small precipitation can be cutted damage by moving dislocation during the process of cyclic loading, leading to the strengthening phase decrease in alloy and the strength of the alloy reduce, finally cause the cyclic softening. Besides, the dislocation in the process of movement and interaction, can rearrange to form regular array of dislocations and dislocation network. The annihilation of dislocations take place when opposite sign dislocations meet each other, which reduce dislocation density and decrease the

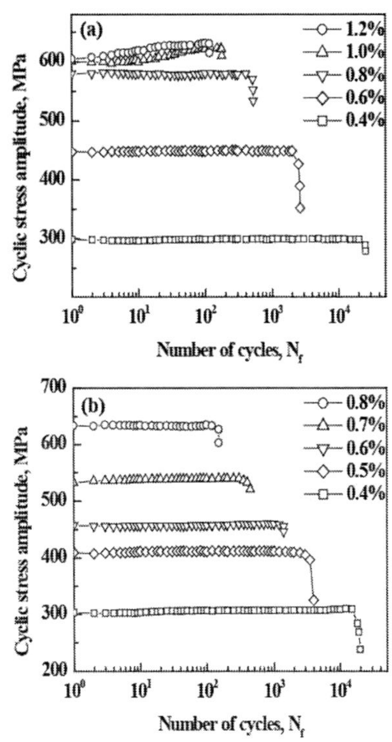

Fig. 2. Cyclic stress response curves of Al-Zn-Mg-Cu-Sc-Zr alloy
(a) room temperature; (b) -40°C temperature

slip resistance of dislocations resulting in the cyclic softening[19, 20]. In fact, cyclic hardening and cyclic softening are two competing relationships. When the hardening effect is offset by the softening effect, the stable cyclic response can be attained.

Fig. 3 shows comparison diagram of cyclic stress response curves for the alloy at the same total strain amplitude under the room and -40°C temperature. It can be seen that at the same strain amplitude, the stress amplitudes of the alloy at low temperature are higher than those at room temperature. And the gap of the stress amplitude between room and -40°C temperature gradually increase, as the increase of strain amplitude. In addition, the stress amplitude at strain amplitudes of 0.8% in low temperature close to the stress amplitude at strain amplitudes of 1.2% in room temperature. It illustrates that the reduction of temperature improve the

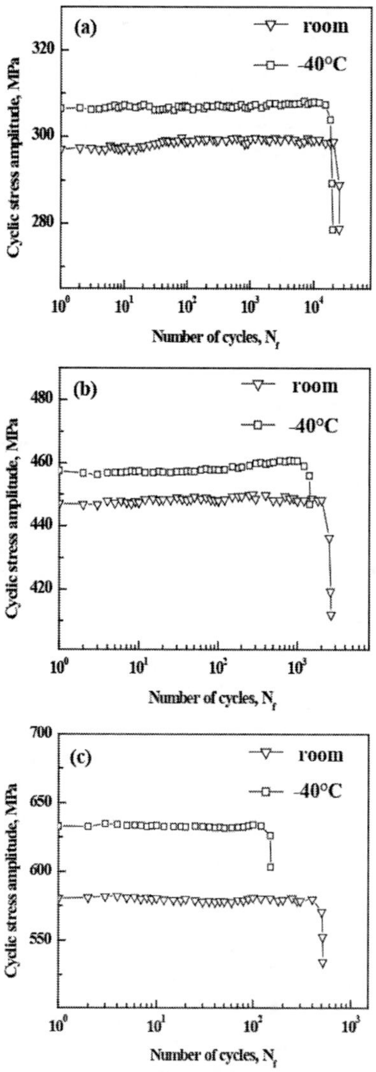

Fig. 3. The comparison diagram of cyclic stress response curves at room and -40°C temperature
(a) $\Delta\varepsilon_t/2$=0.4%; (b) $\Delta\varepsilon_t/2$=0.6%; (c) $\Delta\varepsilon_t/2$=0.8%

deformation resistance of alloy.

3.2 Low-cycle fatigue life behavior

Fig. 4 shows the total strain amplitude ($\Delta\varepsilon_t/2$) versus fatigue life (N_f) curves for the alloy at room and -40°C temperature. It can be seen that the fatigue life of the alloy at room temperature is higher than that at low temperature. And the gap of fatigue life between room and -40°C temperature gradually increase, as the increase of strain amplitude. This phenomenon is the same as the cyclic stress amplitude changes. It indicates that the enhanced deformation resistance influence on the fatigue life significantly as the decrease of the temperature.

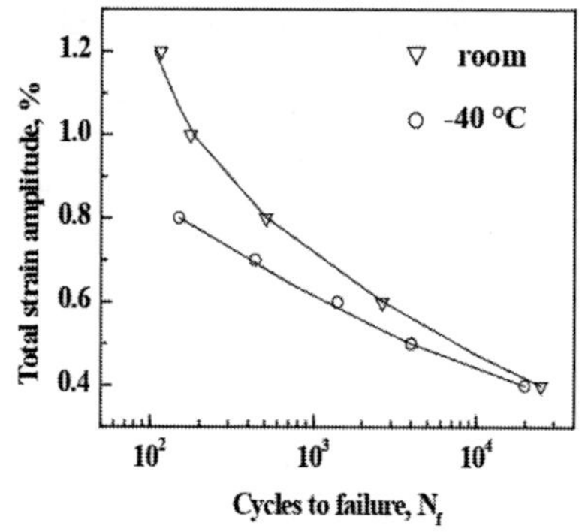

Fig. 4. Total stain amplitude vs fatigue live curves for the alloy at room and -40°C temperature

In the LCF regime, materials usually undergo cyclic loading with a relatively high strain and stress, the total strain amplitude normally consists of elastic strain amplitude ($\Delta\varepsilon_e/2$) and plastic strain amplitude ($\Delta\varepsilon_p/2$). The relationship between elastic strain amplitude and fatigue life (N_f) can be described by the following Basquin equation, additionally, plastic strain amplitude and fatigue life can be described by Coffin-Manson equation. Which are espressed respectively as

$$\Delta\varepsilon_e/2=\frac{\sigma'_f}{E}(2N_f)^{-b} \quad (1)$$

$$\Delta\varepsilon_p/2=\varepsilon'_f(2N_f)^{-c} \quad (2)$$

where ε'_f is the fatigue ductility coefficient and c is the fatigue ductility exponent, E is the elastic modulus, σ'_f is the fatigue strength coefficient, b is fatigue strength exponent.

Fig. 5 shows the curve of relation between strain amplitud and fatigue life in reversals to failure ($2N_f$). It can be seen from this figure that elastic the strain amplitude and plastic strain amplitude as a linear function of fatigue life in reversals to failure for the alloy, where the elastic strain amplitude and plastic strain amplitude are taken from the hysteresis loops at mid-life. And thus, the elastic and plastic strain amplitudes for the fatigue life satisfies the Basquin and Coffin-Manson formula, respectively. Furthermore, the strain fatigue parameters ε'_f, c, σ'_f and b for the alloy can be obtained comparing with the room temperature, the reduction of

Table 1. Strain fatigue parameters at room and -40°C temperature

Temperature	ε'_f	c	σ'_f(MPa)	b	K'(MPa)	n'
room	0.62	1.069	1473.7	0.144	1306.5	0.113
-40°C	1.24	1.290	1177.7	0.120	1371.6	0.109

through the linear regression analysis, and their values are given in Table 1. It can be seen from Table 1 that temperature can improve ε'_f and c values as well as decrease σ'_f and b values.

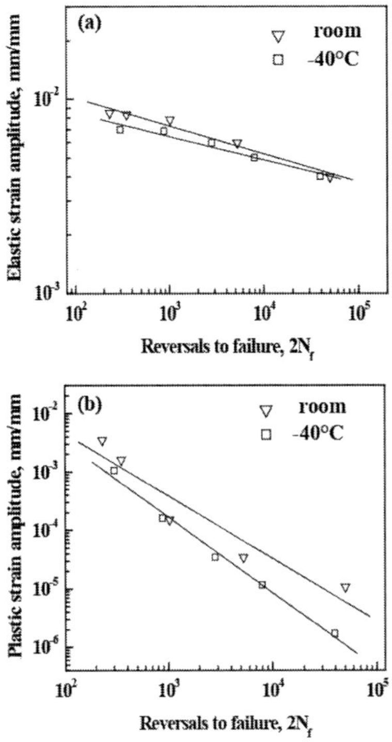

Fig. 5. Strain amplitude vs fatigue life curves at room and -40 °C temperature
(a) elastic strain amplitude; (b) plastic strain amplitude

3.3 Cyclic stress-strain behavior

The cyclic stress-strain response depicts the relation between the cyclic stress amplitude and plastic strain amplitude, and is useful for understanding the strain-controlled cycic deformation behavior of material. Usually, the cyclic stress-strain parameters are characterized using power law fits by the following relation

$$\Delta\sigma/2 = K'(\Delta\varepsilon_p/2)^{n'} \tag{3}$$

where $\Delta\sigma/2$ is the stress range, K' is the cyclic strength coefficient and n' is the cyclic strain hardening exponent. This equation provides a measure of the response of the material to cyclic straining. The cyclic stress-strain curve is plotted by joining the vertices of the hysteresis loops at half life and illustrated in Fig. 6. The K' and n' values

of the both alloys can be determined through the linear regression analysis, and are also listed in Table 1. It can be seen from Table 1 that the reduction of temperature can increase the K′ and decrease n′ values of the alloy.

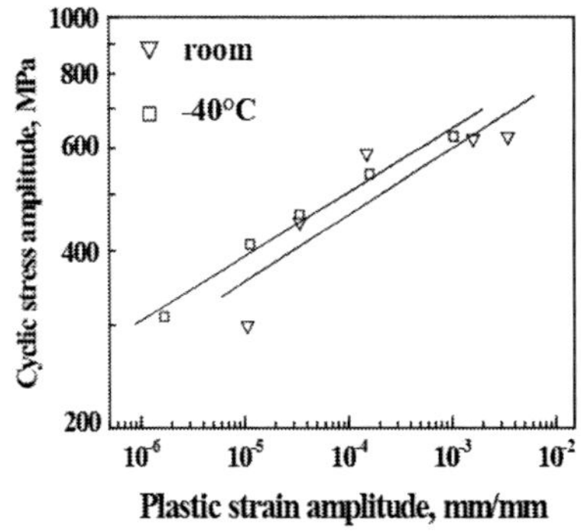

Fig. 6. Cyclic stress vs strain curve at room and -40°C temperature

3.4 Fatigue fractograph

In order to analyze the fatigue fracture mechanism of the alloy, the SEM observation was used to investigate the features of fracture morphologies of the low cycle fatigue samples. Two representative samples which have been tested at total strain amplitude of 0.6% under room and -40°C temperature , are chosen to examine their fractography. Fig. 7 shows typical fractographs in the region of the fatigue crack initiation on the alloy at room and -40°C temperature. It can be seen that fatigue crack source located at the free surface of specimens by a transgranular manner on the top right of images. Furthermore, the fracture surface exists a large number of tearing ridges at room temperature. But at low temperature, the fracture surface shows more cleavage characteristics. It indicates that the alloy has better toughness at room temperature.

Fig. 8 shows the characteristics of the fatigue crack propagation on the alloy at the temperature of room and -40°C. It can be clearly observed that a mass of fatigue strip exist in the crack propagation zones, and also exist some subtle differences. The fatigue crack propagation area is rough and have a large number of tear ridges on the fracture surface at room temperature. And yet it is relatively smooth at low temperature. In addition, the secondary cracks can be observed on the fracture surfaces. It indicates that the mode of fatigue crack

https://doi.org/10.1051/matecconf/201816522001

propagation is a transgranular way for the alloy both room and low temperature. It indicates that the fatigue fracture converts from ductile to brittle gradually with the temperature decreasing.

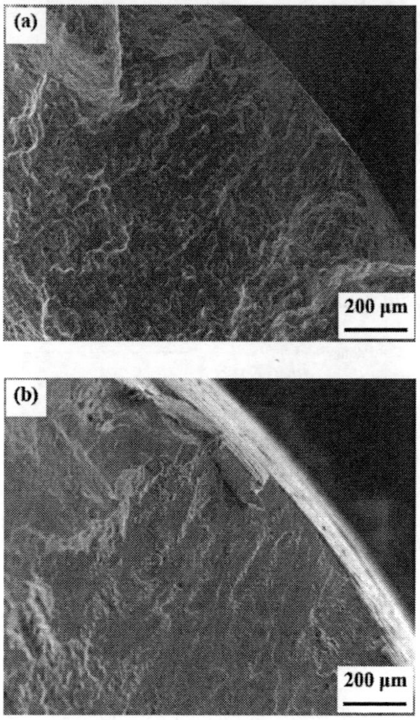

Fig. 7. Fatigue crack initiation sites of Al-Zn-Mg-Cu-Sc-Zr alloy
(a) room temperature; (b) -40°C temperature

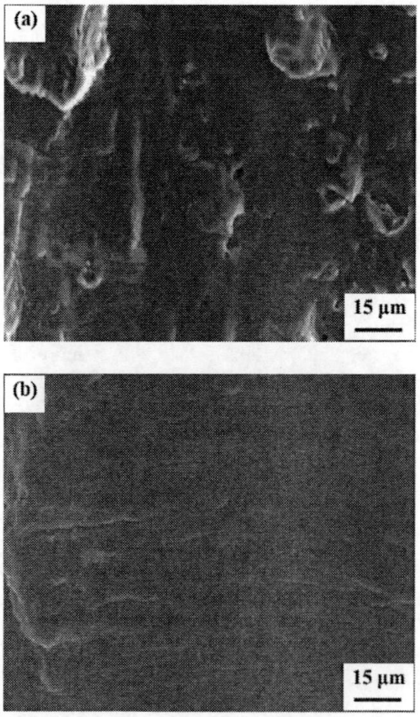

Fig. 8. Fatigue crack propagation region of Al-Zn-Mg-Cu-Sc-Zr alloy
(a) room temperature; (b) -40°C temperature

4 Conclusions

(1)During the low-cycle fatigue deformation, the Al-Zn-Mg-Cu-Sc-Zr alloy exhibit the cyclic hardening and cyclic stability, mainly depending on the imposed total strain amplitude and the test temperature. The cyclic stress amplitudes of the Al-Zn-Mg-Cu-Sc-Zr alloy at low temperature are higher significantly under all total strain amplitudes.

(2)The fatigue life of Al-Zn-Mg-Cu-Sc-Zr alloy is higher at room temperature than that at low temperature. The relatation between elastic strain amplitude and plastic strain amplitude with fatigue life in reversals to failure for the alloy can be described by the Basquin and Coffin-Manson equations, respectively.

(3)The fatigue cracks of Al-Zn-Mg-Cu-Sc-Zr alloy initiate from the specimen free surface. The mode of fatigue crack propagation is a transgranular way with some secondary cracks and tear ridges for the alloy under the condition of low cycle fatigue loading at room and -40°C temperature.

References

1. Z. Huda, P. Edi, Mater. Des. **46(4)**, 552 (2013).
2. R.G. Guan, T. Di, Acta Metall. Sin. (Engl. Lett.) **30(5)**, 409(2017).
3. G.F. Xu, X.Y. Peng, Y.L. Duan, Y. Deng, Z.M. Yin, Trans. Nonferr. Metal. Soc. **26(8)**, 1577(2016). (in Chinese)
4. X.M. Zhang, Y.L. Deng, Y. Zhang, Acta Metall. Sin. **51(3)**, 257(2015). (in Chinese)
5. P.A. Rometsch, Y. Zhang, S. Knight, Trans. Nonferr. Metal. Soc. **24(7)**, 2003(2014).
6. R. Guan, Y. Shen, Z. Zhao, X. Wang, J. Mater. Sci. Technol. **33(3)**, 215(2017).
7. Y. Buranovaa, V. Kulitskiyb, M. Peterlechnera, A. Moguchevab, R.K aibyshevb, S.V. Divinskia, G. Wildea, Acta Mater. **124(2)**, 210(2017).
8. Z. Chen, W. Yang, S. Wang, J. Shu, Rare Met. Mater. Eng. **39(8)**, 1499(2010). (in Chinese)
9. J. Liu, P. Yao, N.Q. Zhao, C.S. Shi, H.J. Li, X. Li, D.S. Xi, S. Yang, J. Alloys Compd. **657(1)**, 717 (2016).
10. K.F. Adam, Z. Long, D.P. Field, Metall. Mater. Trans. A **48(4)**, 1(2017).
11. W.T. Huo, J.T. Shi, L.G. Hou, J.S. Zhang, J. Mater. Process. Tech. **239(1)**, 303(2017).
12. H. Yamada, M. Tsurudome, N. Miura, K Horikawa, N. Ogasawara, Mater. Sci. Eng. A **642(8)**, 194(2015).

13. V. Sabelkin, S. Mall, H. Misak, J. Mater. Eng. Perform. **26(6)**, 1(2017).
14. Z.Y. Ye, D.X. Liu, X.H. Zhang, X.M. Zhang, M.X. Lei, Z. Yang, Acta Metall. Sin. (Engl. Lett.) **28(8)**, 1047(2015).
15. J.T. Burns, J.J. Jones, A.D. Thompson, J.S. Locke, Int. J. Fatigue, **106(1)**, 196(2018).
16. S. Sreenivasan, S.K. Mishra, K. Dutta, Mater. Sci. Eng. A **698(6)**, 46(2017).
17. K.L. Fan, X.S. Liu, G.Q. He, H. Chen, Mater. Sci. Eng. A **632(4)**, 127(2015).
18. A.H. Feng, D.L. Chen, Z.Y. Ma, Metall. Mater. Trans. A **41(10)**, 2626(2010).
19. X.Q. Ding, G.Q. He, C.S. Chen. J. Mater. Sci. **45(15)**, 4046(2010).
20. X. Che, X. Liang, L. Chen, F. Li, Acta Metall. Sin. **(50)9**, 1046(2014). (in Chinese)

Hydrogen Embrittlement Mechanism in Fatigue Behaviour of Austenitic and Martensitic Stainless Steels

Sven Brück[1], Volker Schippl[1], Hans-Jürgen Christ[1] and Claus-Peter Fritzen[2]
[1]*Institut für Werkstofftechnik, Universität Siegen, D-57068 Siegen, Germany*
[2]*Institut für Mechanik und Regelungstechnik - Mechatronik, Universität Siegen, D-57068 Siegen, Germany*

Abstract. In the present study, the influence of hydrogen on the fatigue behaviour of the high strength martensitic stainless steel X3CrNiMo13-4 and the metastable austenitic stainless steels X2Crni19-11 with various nickel contents was examined in the low and high cycle fatigue regime. The focus of the investigations was the changes in the mechanisms of short crack propagation. The aim of the ongoing investigation is to determine and quantitatively describe the predominant processes of hydrogen embrittlement and their influence on the short fatigue crack morphology and crack growth rate. In addition, simulations were carried out on the short fatigue crack growth, in order to develop a detailed insight into the hydrogen embrittlement mechanisms relevant for cyclic loading conditions.

1 Introduction

The demand for more efficient and cleaner technologies leads to the impulse to establish hydrogen as an energy carrier, for example in the automotive sector. This field of applications is already in the centre of research for a long time, where an important focus is put on a reliable and safe fatigue life prediction for weight-optimized and cyclically loaded components.

Mechanically loaded components in automotive such as pressure vessels, pipes, valves and compressors, are often exposed to hydrogen environments. The combination of mechanical stress and hydrogen environment can lead to more rapid material failure resulting from hydrogen embrittlement effects.

Various theories are known from the literature that try to explain and describe these hydrogen effects and their basic mechanisms, such as HELP, HEDE, AIDE or HESIV (see for example the overview article by Lynch [1]). These mechanisms define idealized behaviour for special conditions and their relevance is strongly affected by stress and medium. They can occur separately or in combination, so that different mechanisms can take the dominant role during crack initiation and crack growth.

Most of the research on hydrogen embrittlement of stainless steels deals with the characterization of long fatigue crack growth behaviour [2, 3] or the effect of hydrogen on monotonic properties obtained in tensile tests [4, 5]. There is only little information available about the mechanisms of crack initiation and early crack growth.

Hence, the aim of this study is the identification of the changing microstructural mechanisms of fatigue crack initiation and crack propagation of microstructurally small fatigue cracks resulting from the presence of hydrogen in one martensitic and in two metastable austenitic stainless steels.

The findings are intended to serve as a basis for a physically based simulation model which describes quantitatively the growth of microstructurally small fatigue cracks [6]. Short crack growth is determined by characteristic microstructural processes such as the growth of cracks along grain boundaries or slip bands, which is decisively influenced by the diffusion of hydrogen in the microstructure as well as by the phase transformation of the fcc γ-austenite to bcc α'-martensite at the crack tip. The interaction of these effects must be integrated in the model in a realistic manner.

Metastable austenitic stainless steels

Metastable austenitic stainless steels show a deformation-induced martensite formation during fatigue. This phase transformation of γ-austenite (fcc) into α'-martensite (bcc) occurs spontaneously, without any diffusion and a lattice-distorting effect, in which the respective atoms remaining neighbours in the metal lattice. For the phase transformation, the martensite starting temperature M_S (equation (1)) must be underran.

A threshold value for the transformation, which is determined by the difference of the free energies of both phases ΔG, is reached. A temperature-induced

* Corresponding author: Sven.Brueck@uni-siegen.de

transformation can only proceed below M_S, above M_S a mechanical deformation energy ΔG_{mech} must additionally be applied. Above a high temperature T_0, no phase transformation is possible. The martensite starting temperature M_S can be estimated using the following empirical equation [7]:

Ms=1350-1665 (%C+%N)-28 %Si-33 %Mn-42 %Cr-61 %Ni (1)

In addition to the martensite starting temperature, the equations for the Schaeffler diagram can also be used to characterize the material with regard to its tendency to martensitic phase transformation. It represents the phases present at room temperature in CrNi steels, for which nickel and chromium equivalents are calculated and plotted in the diagram. According to equation (2) and (3), the effect of the main alloying elements on the phase transformation is taken into account [7].

Ni-equivalent=%Ni+30 %C+0.5 %Mn (2)

CR-equivalent=%Cr+%Mo+1.5 %Si+0.5 %Nb+2 Ti (3)

2. Experimental setup and materials

The experiments were carried out on one martensitic X3CrNiMo13-4 (X3) and two metastable austenitic stainless steels. The basic composition of the austenitic stainless steels correspond to the X2CrNi19-11 (304L) and were alloyed (Table 1) with two different nickel contents (9.07% (X2-9) and 12.36% (X2-12)). The grain sizes of the X2-9 and X2-12 was on average 205 µm. The nickel equivalent is 10.35 for the X2-9 and 13.715 for the X2-12. The martensite starting temperature M_S is calculated to -276.86°C for the X2-12 and -82.59°C for the X2-9. Both the calculated values of the starting temperature and the nickel equivalents show that the X2-9 has a stronger tendency for a martensitic transformation than the X2-12. The composition of the X3CCrNiMo13-4 is also shown in Table 1

Table 1: Chemical composition of the two metastable austenitic (X2CrNi19-X) and one martensitic (X3CrNiMo13-4) stainless steels studied (weight %)

alloy	C	Si	Mn	P	S	Cr	Ni	others
X3CrNiMo13-4 (X3)	0.023	0.42	0.61	0.018	0.001	12.38	4.41	Mo 0.368
								Nb -
								Ti 0.004
								N -
X2CrNi19 (X2-12)	0.016	0.48	1.75	0.012	0.008	17.78	12.36	Mo 0.02
								Nb 0.01
								Ti 0.005
								N 0.017
X2CrNi19 (X2-9)	0.014	0.53	1.71	0.010	0.009	18.05	9.07	Mo 0.02
								Nb 0.01
								Ti0.005
								N 0.016

Fatigue tests (R = -1) were performed with X2-9 and X3 specimens at f=1 Hz using a servohydraulic testing machine and with X2-12 specimens in a miniature test machine (Figure 1, described in detail in [8]) in laboratory air at room temperature. These tests were performed with precharged and as reference with non-charged specimens. The fatigue tests were carried out at different testing machines and stress amplitudes because the crack initiation period should be monitored in the LCF / HCF regime despite their different fatigue strengths.

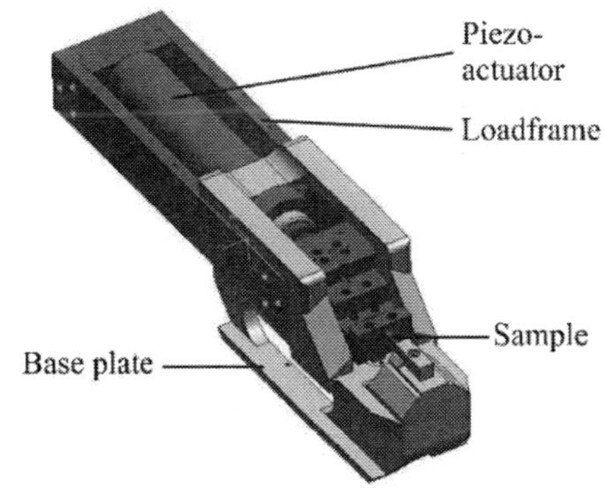

Fig 1: Functional representation a), installation of this system into the confocal laser-microscope b) [8]

The crack propagation was detected *in-situ* by means of a far-field microscope (X2-9, X3) and a confocal laser microscope (X2-12). In fixed intervals of 5,000 cycles, microstructure and crack length were recorded to analyse the side of crack initiation and crack growth rates. The crack initiation is limited to a smaller observation area by placing shallow notches on the front and back of the specimens (Figure 2).

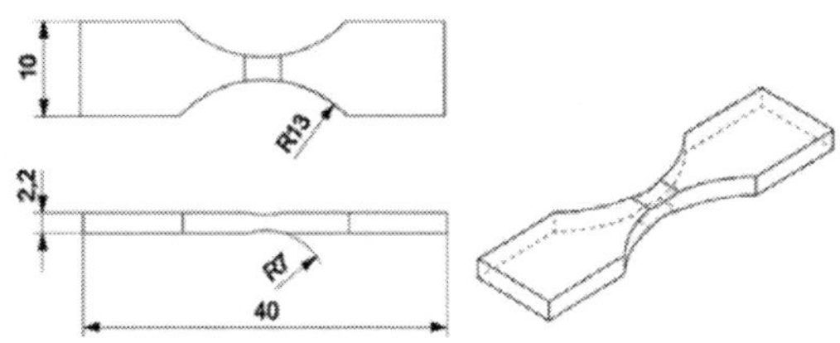

Fig 2: Geometry of the in-situ specimen

The observation of the microstructure was performed using a high-resolution SEM / FIB Helios Nanolab 600 with backscattered electron diffraction (EBSD).

In order to characterize the influence of internal hydrogen on crack propagation, specimens were precharged with hydrogen from the gas phase (approx. 180 °C, 180 bar H_2, 28 days) in the autoclave at the Materialprüfungsanstalt Stuttgart (MPA). The reached hydrogen contents were measured in Siegen using a hydrogen analyser LECO by means of hot gas extraction. Furthermore, specimens without hydrogen were tested as reference under identical conditions.

3. Results, discussion and modeling concept

3.1 Results: austenitic stainless steels

The fatigue tests were carried out at different stress amplitudes because the crack initiation period should be maintained in the LCF/HCF transition range, despite deviating fatigue strengths. In order to compare the crack growth of X2-9 and X2-12 in all loading conditions, the results of the fatigue tests were summarized in the da/dN-ΔK diagrams in figure 3. Both stainless steels show similar da/dN values in the considered ΔK range; however, the curves allow different statements about the crack growth mechanisms. The oscillating curves correspond to the expectation of short crack growth curves according to literature [6], due to a strong interaction of the propagation of microstructural short cracks with the microstructural features. In the case of X2-9, it can be observed that the curves of the precharged specimens have higher da/dN values both at low and higher ΔK values; thus short crack growth is significantly accelerated by hydrogen.

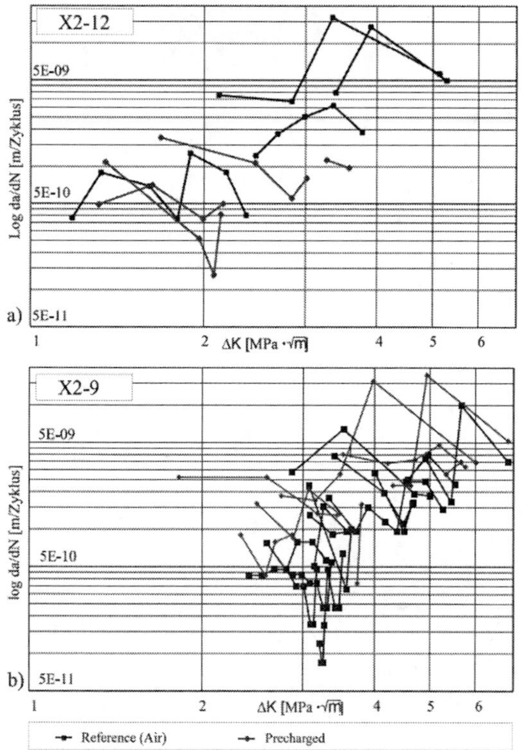

Fig 3: da/dN-ΔK representation a) for X2-12 and b) X2-9 obtained on hydrogen precharged specimens (red) and reference specimens (black)

Furthermore, the curves of the X2-9 reference specimens show a stronger dependence on microstructural barriers, which manifests itself in the form of more variable crack growth rates. In contrast, the precharged X2-9 curves show a reduced barrier effect, which means that the crack growth rate increases more quickly and shows less oscillation. The reduced barrier effect can be explained clearly by the HELP mechanism, because it facilitates the movement of the dislocations locally.

The X2-12 samples show a significantly higher scatter compared to the X2-9 and thus no clear tendency of a hydrogen effect on the short crack propagation. However, it can be assumed that the crack grows even more slowly. This can be explained by the changing of the mechanism of the crack growth from single to alternating sliding, which occurs earlier under the influence of internal hydrogen. Furthermore, it is known from the literature [9] that austenitic stainless steels with increasing nickel content show a reduction of the hydrogen susceptibility.

Due to the *in-situ* experiments, the characteristic crack initiation sites of the precharged and the reference X2-12 specimens could be identified (Figure 4). An analysis of the slip band morphology and the crack initiation shows significant differences between the reference and precharged specimens. The average distance between the slip bands is about 4.7 µm and 1.92 µm in the precharged specimens and in the reference specimens, respectively. While the minimum height is identical, the highest slip bands in the precharged specimens are higher compared with those in the reference specimens (0.14 µm to 0.08 µm, respectively). Also, Aubert et al. have found this effect in stainless steel 316L specimens [10]. For AISI 310s stainless steel specimens, Robertson shows an increase of the slip band height because of hydrogen [11]. These observations show that the present hydrogen leads to a localization of plastic deformation and formation of the higher slip bands, likely resulting from the HELP mechanism. Hydrogen reduces the flow shear stress and localizes the dislocation motion, which is responsible for the displacement in the shear planes.
In the reference specimens, the cracks initiated mostly at grain boundaries, often in slip bands inside a grain. In contrast, for hydrogen-precharged specimens it was found that cracks mostly form at grain boundary triple points. This observation suggests a predominant hydrogen mechanism. Because grain boundaries [12] represent deep traps, a very large fraction of the hydrogen will be stored there after the precharging. This holds true in particular for major defects such as grain boundary triple points. Following the HEDE mechanism, in such places a local decohesion occurs, and therefore cracks can preferably initiate at these locations.

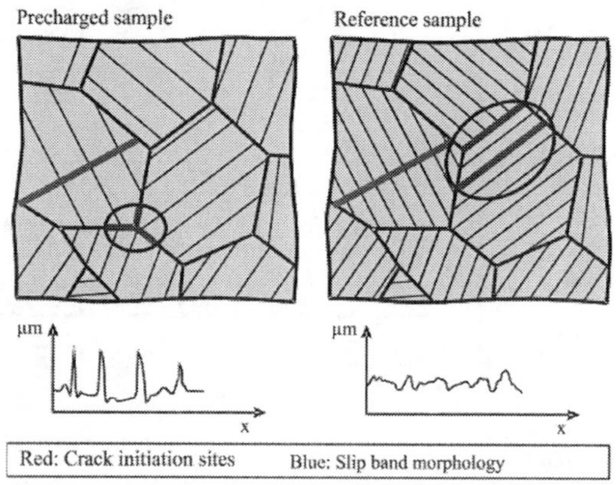

Fig. 4: Crack initiation and slip band morphology in precharged specimens (left) and reference specimens (right) of X2-12

3.2 Results: martensitic stainless steel

To characterize the influence of hydrogen on the fatigue behaviour of the X3, fatigue tests were carried out on the servohydraulic testing machine. For this, uncharged specimens were tested both in a laboratory air atmosphere and in vacuum (3.4×10^{-5} mbar) and were observed with a far-field microscope. Thus the influence of the atmosphere could be taken into account. The tested reference samples showed a basic hydrogen content of about 2 wppm after preparation. The hydrogen content of the precharged specimens was about 4-5 wppm. Figure 6 shows the influence of the atmosphere and the internal hydrogen on the crack growth rate of the X3. The cycle of the crack initiation could not be detected exactly during every fatigue test, but the crack lengths at the first observation show a clear image of the harmful influence of the internal hydrogen. Up to approx. 10.000 cycles, nearly all reference samples and the precharged specimens have already formed cracks at air. First cracks initiate in the vacuum specimens approx. 10-15.000 cycles later. The detected cracks in the precharged specimens are already significantly longer than those at laboratory air.
The respective crack morphologies show a clear influence of the atmosphere. Cracks in vacuum are less branched in the specimens than in the reference and precharged specimens. The crack propagation in the precharged samples is more transgranular than in the reference samples (more intergranular) (see Figure 5).

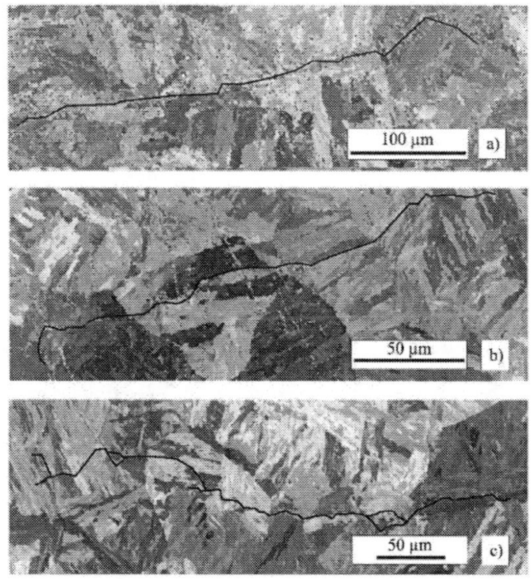

Fig. 5: EBSD scan of the X3 specimen: a) in vacuum (75.000 cycles), b) at laboratary air (50.000 cycles) and c) precharged (5 wppm)

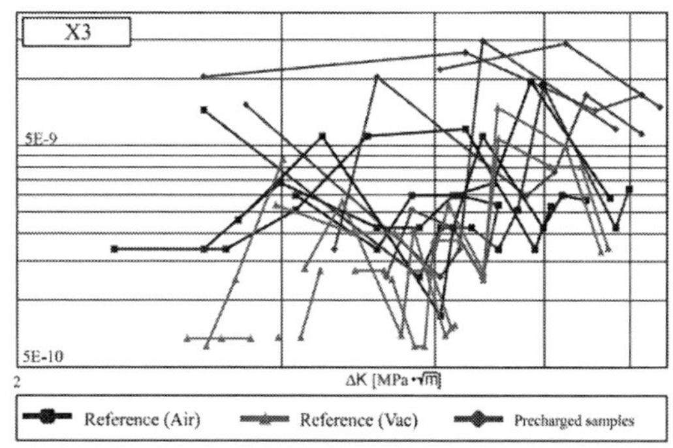

Fig. 6: Crack growth rate of precharged specimens (5 wppm) and reference specimens (laboratory air, vacuum) of the X3, R=-1, $\Delta\sigma/2$=640 MPa (Vac), $\Delta\sigma/2$=665 MPa (Air, precharged)

3.3 Modeling strategy

The short fatigue crack growth model considers the plastic deformation in shear bands and the deformation-induced martensitic transformation at the crack tip and is based on several previous works [13, 14]. To consider the influence of hydrogen on the shear deformation along slip bands, the local hydrogen concentration in the microstructure is calculated after every crack growth step.

Short crack growth & shear band model

The short crack growth model (summarised in Figure 7) is based on the stage-I crack growth model by Navarro and de los Rios [15] and extended to simulate short fatigue crack propagation in a real two-dimensional polycrystalline microstructure by [13, 14]. The model describes the deformation of the crack and the deformation along shear bands at the crack tip. The crack can freely open and slide, but crack closure effects are also considered in the simulation. Plastic deformation as sliding along shear bands occurs if the friction stress $\tau^{F,0}$ exceeds the maximum resolved shear stress in a slip system at the crack tip. For the numerical solution, a boundary element method is used and the crack and the plastic zone (shear bands) are discretized with boundary elements. According to Navarro and de los Rios [15], the crack growth per loading cycle is calculated by the determined range of the cyclic crack tip slide displacement $\Delta CTSD$ by the boundary element method (eq. 4):

$$\frac{da}{dN}=C_{CTSD}\cdot\Delta CTSD \tag{4}$$

with the material dependent constant C_{CTSD}, representing the irreversible part of plastic deformation at the crack tip. As the sliding along the shear bands is decreased as the crack tip approaches a grain boundary, the crack growth is decreased until a shear band in the next grain is generated. This slip transmission through a grain boundary occurs if a critical stress intensity or critical shear stress on a slip system in the neighboring grain, calculated at the sensor elements located at a distance r from the grain boundary, is exceeded ($\tau > \tau^{c,0}$).
The grain boundary resistance for slip transmission can be determined according to [15] by assuming that at the endurance limit σ^D, a growing crack stops at the grain boundary in the first grain:

$$\tau^{c,0}=\frac{1}{2}\sqrt{\frac{D_m}{r}}\,\sigma^D\cdot S_{max}+\tau^{F,0} \tag{5}$$

with the mean grain diameter D_m, the distance from the sensor element to the grain boundary r and the maximum Schmid factor S_{max}.
The sliding along shear bands can be characterized by the friction stress $\tau^{F,0}$ and can be determined with the cyclic yield stress $\sigma^{F,zykl}$ with equation (6):

$$\tau^{F,0}=(\sigma^{F,zykl}-\sigma^D)\cdot S_{max}. \tag{6}$$

The crack growth model is capable of reproducing the decreasing crack growth rate near a grain boundary, as seen during stage I crack growth. When the crack tip approaches a grain boundary, the crack growth decreases until a shear band is formed in the neighboring grain. The crack tip slide displacement as well as the crack growth rate increases due to the increasing plastic zone size. With this modelling approach the well-known oscillating crack growth rate in the stage-I crack propagation can be characterized. This stage where only one shear band is activated at the crack tip is called stage Ia [16]. The transition stage from a crack growth controlled by shear stresses (stage Ia) to a crack growth controlled by the maximum normal stress (stage II) is called stage Ib. In the stage Ib, additional shear bands are activated at the crack tip and the crack path is determined by vectorially adding the sliding along the two activated shear bands [16].
To simulate intercrystalline crack propagation along grain and phase boundaries a critical normal stress σ^c at the crack tip has to be exceeded. The crack growth along grain boundaries is assumed to be driven by the normal stresses resp. the range of the crack tip opening displacement $\Delta CTOD$ (eq. 7):

$$\frac{da}{dN}=C_{CTOD}\cdot\Delta CTOD. \tag{7}$$

Again, the material dependent constant C_{CTOD} represents the irreversible part of plastic deformation at the crack tip. The plastic deformations at the crack tip – occurring in the neighboring grains – are modelled by projecting these deformations to a yield strip along the grain boundary if the von-Mises stress σ^{VM} exceeds the cyclic yield stress σ^F.

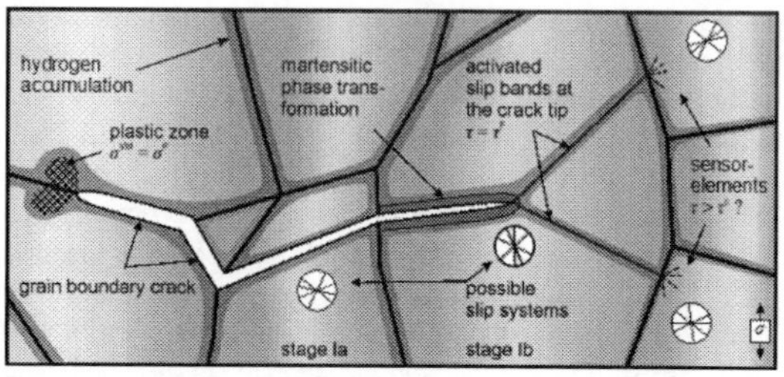

Fig. 7: Schematic representation of the short fatigue crack growth model

Martensitic transformation model

To simulate the crack growth in a metastable austenitic stainless steel, the model considers the deformation-induced martensitic transformation at the crack tip [14, 17]. According to Bogers & Burgers [18] and Olson & Cohen [19], the martensitic transformation (from γ to α') occurs by shearing on two separate shear bands compatible to the Bogers & Burgers model. This criterion is evaluated after each crack growth cycle and the newly formed martensitic area at the crack tip extends in the direction of the calculated crack growth. The amount of shear deformation along each shear band determines the size of the generated martensite domain. The martensitic transformation is associated with a transformation-induced volume expansion within the domain and contains a simple shear and a uniaxial dilation (invariant-plane strain) to describe the true shape deformation.

Modelling the influence of hydrogen on shear bands

The experimental observations showed that hydrogen influences the shear band morphology, thus the sliding along shear bands as well as the slip transmission through grain boundaries (see Figure 4). To model this effect, the friction stress along the slip bands (τ^F) as well as the activation of new shear bands (τ^c) are assumed to be dependent on the local hydrogen concentration $c_H(x)$ [cf. 22]:

$$\tau^c(x)=\tau^{c,0}\cdot(1+\varsigma_c\cdot c_H(x)), \tag{8}$$

$$\tau^F(x)=\tau^{F,0}\cdot(1+\varsigma_F\cdot c_H(x)). \tag{9}$$

The parameters ς_c and ς_F have to be adjusted to appropriately predict the experimentally observed fatigue crack growth rates. Due to this approach, the local hydrogen concentration is determined and calculated based on the following diffusion model.

https://doi.org/10.1051/matecconf/201816522002

Diffusion model

In the simulation, it is assumed that the hydrogen dependent material parameters (τ^c and τ^F) are constant during the loading cycle. After the loading cycle, the slip along shear bands is further evaluated to calculate the crack growth rate, see eq. (4). After this step, the local hydrogen concentration is calculated based on the diffusion model and the parameters controlling the slip band activation and sliding (τ^c and τ^F) are updated. The time increment Δt_c for the diffusion processes are determined by the previously calculated incremental loading cycles ΔN for the crack growth and the loading frequency f:

$$\Delta t_c = \Delta N / f \tag{10}$$

The diffusion model is based on the influence of the hydrostatic stresses on the chemical potential in the microstructure [20]. By applying Fick's first and second law the following differential equation is used to describe the transient redistribution of the hydrogen concentration in the microstructure:

$$\frac{\partial \boldsymbol{c}_\mathbf{H}}{\partial t} = \boldsymbol{D} \cdot \nabla^2 \boldsymbol{c}_\mathbf{H} + \frac{V_\mathrm{H} \boldsymbol{D}}{RT} \nabla \boldsymbol{\sigma}_\mathbf{h} \nabla \boldsymbol{c}_\mathbf{H} + \frac{V_\mathrm{H} \boldsymbol{D}}{RT} \nabla^2 \boldsymbol{\sigma}_\mathbf{h} \boldsymbol{c}_\mathbf{H}. \tag{11}$$

The transient hydrogen concentration $\boldsymbol{c}_\mathbf{H}$ depends on the transient evolution of hydrostatic stresses $\boldsymbol{\sigma}_\mathbf{h}$ as well as the diffusion coefficient of hydrogen in the material $\boldsymbol{D}$, the partial, molar volume of hydrogen V_H, the gas constant R, the temperature T and the time t.
The transient hydrostatic stress $\boldsymbol{\sigma}_\mathbf{h}(t)$ is evaluated in the short crack growth model and is calculated using the Einstein summation convention (i=1, 2, 3) by:

$$\boldsymbol{\sigma}_\mathbf{h}(t) = -\sigma_{ii}(t)/3. \tag{12}$$

As the model evaluates the stresses at discrete time steps Δt, it is assumed that the stresses are constant in this interval, which is schematically represented in figure 8.

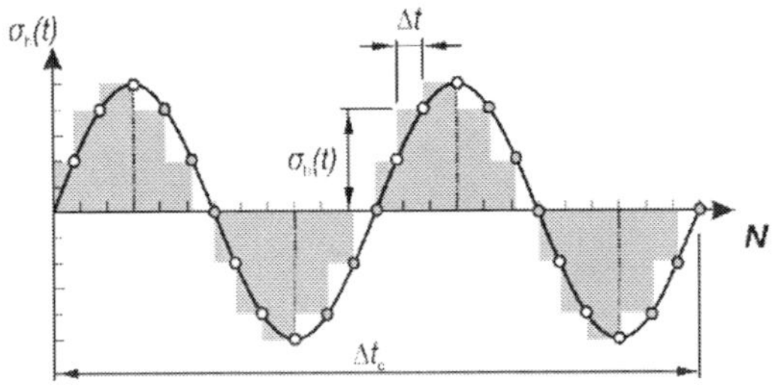

Fig. 8: Schematic of the constant hydrostatic stresses $\sigma_\mathrm{h}(t)$ in the time interval Δt in the diffusion model

To solve the partial differential equation in (11), the streamline upwind Petrov/Galerkin finite element method is used to stabilize convection-dominated influences of the hydrostatic stress on the diffusion.

Simulation results and discussion

To verify the presented model, the simulation results are compared with experimentally observed cracks and crack growth rates.

Simulation results of uncharged specimens in X2-12

As seen in figure a)-c) the experimentally observed crack path and crack growth is compared with the simulation result. The crack growth is only calculated by equation (4) as no crack growth along a grain boundary is modelled. The irreversibility constant C_{CTSD} is determined with 0.2 µm/cycle by comparing the calculated with the experimentally observed crack growth in Figure 9 c). The parameter for the slip band mucleation is evalutated from equation (5) with the endurance limit σ^D=250 MPa and the mean grain diameter D_m=205 µm and the friction stress chosen with τ^F=80 MPa according to [21].
It can be seen that the crack growth rate of the left crack tip increases with increasing crack length. But this effect is less pronounced in the pffigredicted crack path compared to the experimental data. This can be explained by the insufficient prediction of the right crack path.

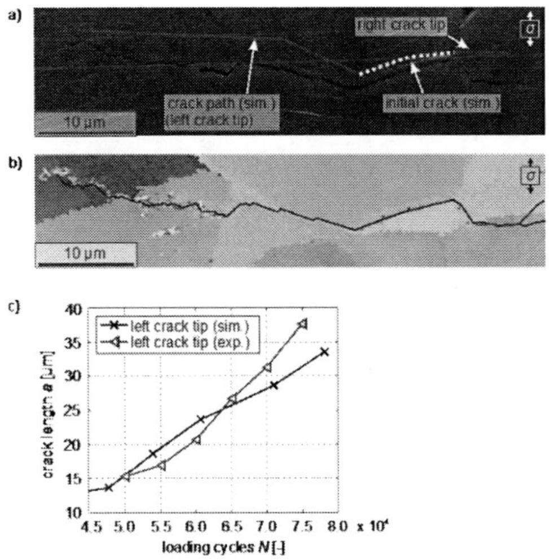

Fig. 9: SEM-image with the calculated crack path a), IPF-image b) and comparison of the calculated crack length (left crack tip) c) during cyclic loading with experimental data of the uncharged X2-12, $\Delta\sigma/2$=365 MPa, f=1 Hz

Simulation results of precharged specimens in X2-12

Following the simulation of the crack path in an uncharged specimen, the crack path and crack growth as well as the martensitic transformation in a precharged specimen are calculated (Figure 10). The calculated right crack path has a good correlation whereas the left crack tip stopped at the grain boundary. The simulated martensitic phase at the crack tip is also shown in figure 10 a. The transformation only occurs if two compatible shear bands at the crack tip are activated. As this condition is not met during each crack growth, the simulated martensitic phase is not continuously

generated along the simulated crack path in contrast to figure 10 b.
The simulation predicts that the crack path does not follow one slip system, but grows by alternate sliding along two activated shear bands at the crack tip which is facilitated in the presence of hydrogen compared to the crack growth behavior in uncharged specimens. This effect supports the assumption that hydrogen facilitates the activation of shear bands at the crack tip.

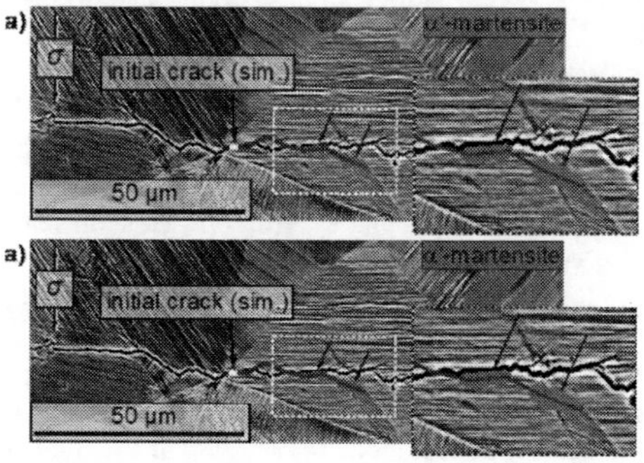

Fig. 10: SEM-image with the calculated crack path and the calculated martensite phase (green) a), phase-image of the precharged X2-12, $\Delta\sigma/2$=365 MPa, f=5 Hz, c_H=40 wppm b) [cf. 19]

Conclusions

The results of *in-situ* fatigue tests were presented, which were performed on two metastable austenitic stainless steels and one martensitic stainless steel at a test frequency of 1 Hz under varying tensile-compressive load in the LCF / HCF regime. The focus of the investigation was on the characterization of the hydrogen mechanisms which influence the crack initiation and the short crack propagation.

The analysis of crack initiation and short crack growth behaviour of the metastable austenite samples showed a significant hydrogen effect, the X2-12 showing less susceptibility than the X2-9 due to its high nickel content.

An influence on the slip band morphology was also discovered. Hydrogen changes the crack initiation sites from slip bands (transgranular crack initiation) to grain boundary triple points (intergranular crack initiation). This change can be explained by the HEDE mechanism. Microstructural small cracks grow with a significantly higher crack growth rate in hydrogen-precharged specimen and show a significantly lower tendency to oscillating curves in the da/dN-ΔK diagram. The change in the crack propagation can be explained by a reduced barrier effect of microstructural barriers, caused by a facilitated displacement movement, according to the HELP mechanism.

The material X3 shows an even greater dependence of the fatigue properties on the test atmosphere and the hydrogen content. Cracks propagate faster in the precharged samples due to the hydrogen content. The crack morphology changes due to the hydrogen content and hydrogen effects (HELP) from more transgranulary to a more intergranulary crack propagation in the precharged specimen compared to the reference specimen. In Summary, the larger the martensitic phase, the higher the inceasing of crack growth rate caused by hydrogen.

The simulation results show that the presented model is capable of predicting the experimentally observed crack path and the crack growth behavior. Based on the experimental observation of the changing slip band morphology under the influence of hydrogen, it is assumed that the activation and sliding of shear bands is facilitated depending on the local hydrogen concentration. In the simulation, the hydrogen concentration in the microstructure is calculated based on a diffusion model taking into account the influence of the hydrostatic stress on the hydrogen distribution. The influence of the hydrogen on the slip band formation affects the crack path as seen in the precharged X2-12 leading to facilitated activation of a second shear band. The crack growth in the martensitic stainless steel is simulated in the uncharged condition where the crack growth is dominated by cracking along the martensite lath boundaries. A good correlation with the simulated crack path is achieved. The crack path could only be predicted if the stress for the activation of new shear bands is increased compared to X2-12. Further, the increasing grain boundary density supports facilitated crack growth along grain boundaries. Further simulations on the crack growth for the metastable austenitic stainless steel as well as the for the martensitic stainless steel for precharged specimens and specimens in compressed hydrogen could give a more profound understanding on the influence of hydrogen on the crack propagation.

Acknowledgements

The authors wish to thank Deutsche Forschungsgemeinschaft (DFG) and Forschungsvereinigung Verbrennungskraftmaschinen e.V. (FVV) for the financial support of this research project titled *Fatigue life reduction of steels in pressurized hydrogen as a consequence of changes in small crack propagation mechanisms* (Reference numbers CH92/48-1 (DFG) and 1185 (FVV)).

References

1. Lynch S., Cor. Rev. **30**, 3-4 (2012)
2. Murakami, Y., Kanezaki, T. and Mine, Y., Metall. Mater. Trans.A **41**, 10 (2010)
3. Skipper, C., Leisk, G., Saigal, A. and San Marchi, C, Proc. Int. Hyd. Conf., Somerday, B., Sofronis, P. and Jones, R. (editors) (2009)
4. San Marchi, C., Michler, T., Nibur, K. A. and Somerday, B. P., Int. J. Hydrogen Energy **35** (2010)
5. Yagodzinski, Y., Saukkonen, T., Andronova, E., Rissanen, L. and Hänninen, H., Proc. Int. Hydr. Confer., Somerday, B., Sofronis, P. and Jones, R. (editors) (2009)

6. Schippl, V., Brück, S., Schwarz, M., Proc. Int. Hydr. Confer., Somerday, B., Sofronis, P. and Jones, R. (editors) (2016)
7. Roth, I., Dissertation, H.-J. Christ (Hrsg.), Siegener werkstoffkundliche Berichte, 5 (2012)
8. Brück, S., Schippl, V., Schwarz, M., Tagungsband zur Werkstoffprüfung, Borsutzki, M. and Moninger, G. (editors) (2015)
9. San Marchi, C., Sandia Report (2012)
10. Aubert, I., Saintier, N., Olive, J.M., Scripta Mat. **66** 9 (2012)
11. Robertson, I. M., Lillig, D., Ferreira, P.J., Proc. Int. Hydr. Confer., Somerday, B., Sofronis, P. and Jones, R. (editors) (2009)
12. Ratte, E., Dissertation, Shaker Verlag **8** (2007)
13. Schick, A., Dissertation, Siegen, VDI-Verl. (Fortschritt-Berichte / VDI Reihe **18**, 292) (2004)
14. Kübbeler, M., Dissertation, Universität Siegen (2017)
15. Navarro, A.; de los Rios, E. R., Phil.Mag. A 57 (1) (1988)
16. Künkler, B., Dissertation, VDI-Verl. (Fortschritt-Berichte VDI Reihe **18**, 312) (2007)
17. Hilgendorff, P.-M.; Grigorescu, A.; Zimmermann, M.; Fritzen, C.-P.; Christ, H.-J., Proc. Mater. Sci. **3** (2014)
18. Bogers, A. J., Burgers, W. G., Acta Metall. **12** (2) (1964)
19. Olson G. B.; Cohen M., (1975), Metall. Trans. A **6** (4) (1975)
20. Li J. C. M.; Oriani R. A.; Darken L. S., Zeitschrift für Physikalische Chemie Neue Folge, **49** (1966)
21. Müller-Bollenhagen, C., Dissertation, Siegener Werkstoffkundliche Berichte **3** (2011)

Fatigue investigation of complex weldments by the means of the local strain energy density approach

Luigi Mario Viespoli[1], *Antonio* Alvaro[2], *Bård* Nyhus[2], and *Filippo* Berto[1]

[1]Department of Mechanical and Industrial Engineering, Norwegian University of Science and Technology (NTNU), Norway

[2]Department of Materials and Nanotechnology, SINTEF Industry, Norway

Abstract. The paper investigates the use of an energetic approach based on the strain energy density failure criteria to predict the fatigue life of welded joints in aluminum alloy. The cited criterion has already been proved valid to assess the failure of components in presence of sharp and blunt notches, and several results are present in the literature for different materials. The geometry tested in the present work is a double V-grooved full penetration butt weld, subsequently heat treated, and loaded orthogonally to the welding direction with load ratio R=0. This configuration makes the weld toe a notch of great opening angle. The aim of the paper is to verify the soundness of the energetic criteria for this class of welded joints, comparing the tests results with the numerical predictions. In the computational part, the energy in a given volume can be computed directly from the nodal displacements, thus not needing the stress values. Differently from the stress intensity factor approach, this property allows to a fast computation of the strain energy density by the means of a coarse mesh. The results of different configurations of geometry and meshing are compared to find the simplest modeling scheme capable of providing an accurate estimate of the fatigue life of the joint.

1 Introduction

The fatigue life assessment of welded components can be executed with different techniques [1] of which, due to their great simplicity, commonly used are the Nominal Stress and the Hot Spot Stress approaches [2]. The drawback of this simplicity, which makes them ideal in complex structures with a great number of welds, the detail they provide regarding the stress field in the stress intensification area is poor, so they are strongly related to the determination of the right fatigue class for every welded detail. In the presence of a sharp notch, that is a weld toe or root, the static and fatigue resistance can be evaluated using the Notch Stress Intensity Factors [3]. Such procedure presents two drawbacks: the necessity of an accurate evaluation of the stress field [4], so a very refined discretization and a non-constant critical value, but function of the opening angle through the William's eigenvalues [5]. The use of special elements [6], capable of describing the asymptotic behavior in proximity of the stress singularity point, can be helpful in the reduction of the number of elements, but often this is not implemented in commercial codes. The adoption of the mean Strain Energy Density for the fatigue life prediction [7] allows to a rapid and accurate assessment and to the computation *a-posteriori* of the Notch Stress Intensity Factors [8] with a closed-form analytical correlation. Since the SED is computed directly from the stiffness matrix and nodal displacement, without the need of the stress solution, its value is accurate also for very coarse meshes [9], providing an evident computational advantage. The local energetic criterion states that the failure occurs when the mean SED, W, averaged over a control volume of precise radius R, equals a critical value W_c, not dependent on the geometrical characteristics of the notch, but only on the material properties, as it is the radius R. Dealing with a sharp V-notch and considering a bi-dimensional problem, the energy is computed over a circular area Ω of radius R which center is the notch tip and symmetric with respects to the notch bisector. In the hypothesis of validity of the William's solution, the mean SED over the critical area is:

$$\bar{W} = \frac{e_1}{E}\left[\frac{K_1}{R^{1-\lambda_1}}\right]^2 + \frac{e_2}{E}\left[\frac{K_2}{R^{1-\lambda_2}}\right]^2 \quad (1)$$

Being λ_1 and λ_2 the William's eigenvalues, E the Young's modulus and e_1 and e_2 functions of the notch's opening angle and of the stress state. The Notch Stress

Intensity Factors K_1 and K_2 can be computed from the stress field as:

$$K_1 = \sqrt{2\pi}\lim_{r\to 0}(\sigma_\theta)_{\theta=0} r^{1-\lambda_1} \quad (2)$$

$$K_2 = \sqrt{2\pi}\lim_{r\to 0}(\tau_{r\theta})_{\theta=0} r^{1-\lambda_2} \quad (3)$$

The fatigue life assessment by the means of the SED has been validated for welded joints and both sharp and blunt notches [10,11]. If the opening angle is greater than 103°, the Mode II stress field is not singular, so its contribution to the strain energy around the notch tip is negligible compared to the contribution of the Mode I, allowing to express the mean SED as:

$$\bar{W} = \frac{e_1}{E}\left[\frac{K_1}{R^{1-\lambda_1}}\right]^2 \quad (4)$$

An estimate of the radius R of the control area Ω can be estimated in the hypothesis of validity of the Beltrami failure criterion, $W_c=\Delta\sigma/2E$, and applying it to tests on butt welded joints. The critical SED so found is then compared to the mean SED in the case of a welded joint with opening angle greater than 103°, obtaining:

$$R = \left(\frac{\sqrt{2e_1}\,\Delta K_1}{\Delta\sigma}\right)^{\frac{1}{1-\lambda_1}} \quad (5)$$

It is important to point out the dependence of the critical radius to the failure criterion applied. In the case of aluminum welded joints, the radius found with a number of experimental tests is $R=0.12\ mm$ [12].

2 Joint fatigue testing

2.1 Specimens

The main goal of the fatigue testing of the welded samples is to evaluate the fatigue performance of double V-groove butt welds of heat-treated aluminum alloy at room and low temperature. The specimens have been manufactured joining two AA6082-T6 plates with a MIG welding process and then cutting and machining the single specimens to a width of 30 mm. The shielding gas used is argon and two series of samples have been tested: 5 and 20 mm of thickness. The size of the plates was originally 180x20x2000 mm and 150x5x2000 mm. No post welding treatment has been performed.

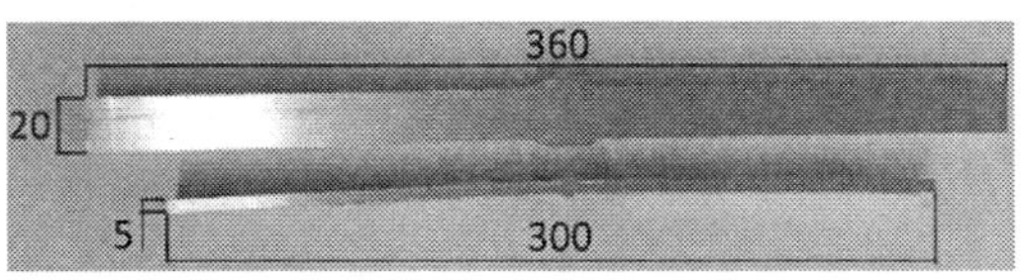

Figure 1. Two of the tested specimens, all cut to a width of 30 mm. It is evident the pronounced bending induced by the welding process, especially for the 5 mm thick sample.

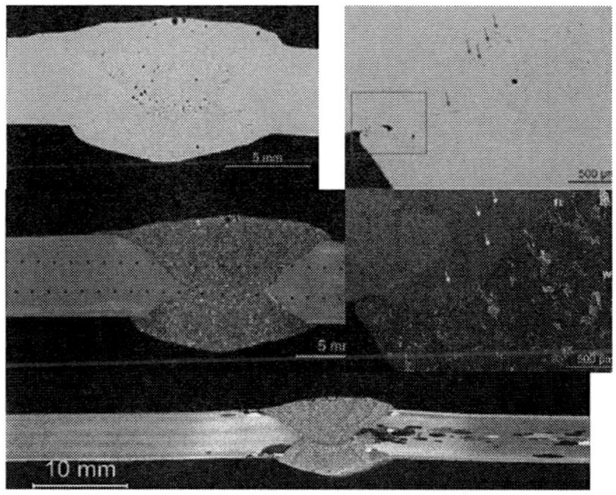

Figure 2. A metallographic investigation of the junction, in the pictures for the 5 mm thick joints, clearly shows the presence of defects such as porosity in the penetration zone, cracks in the HAZ (arrows) and a root defect (frame). The base material can be observed to have partly recrystallized.

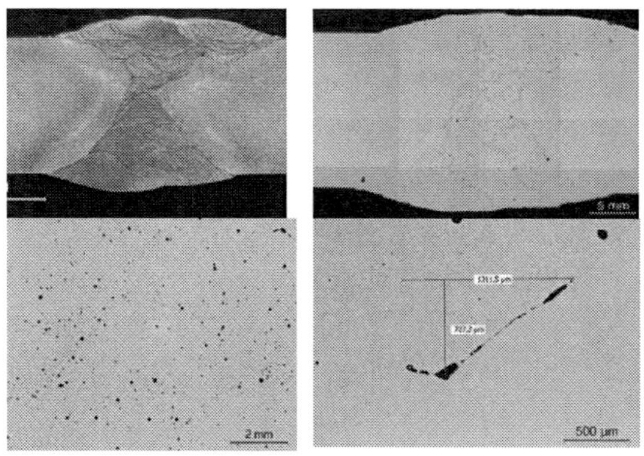

Figure 3. As in figure 2, also the 20 mm joints present porosity and cracks.

After production, the samples present a noticeable angular misalignment, particularly pronounced in the case of the 5 mm joints. This geometrical characteristic, due to the rigid clamping system, heavily influences the effective load acting on the specimen. This effect is taken into account in the computation according to the recommendations of IIW, as explained in the following chapter.

2.2 Testing and fatigue curves

The fatigue texting has been executed at a load ratio R=0 in air with a sinusoidal load cycle, at room temperature and at -60 °C. The equipment used is a MTS servo-hydraulic machine with a load capacity of 100 kN. The nominal stress increment due to the angular misalignment between flat plates, in the case of fixed ends, is obtained multiplying the nominal longitudinal stress imposed by the machine times a coefficient k_m,

 https://doi.org/10.1051/matecconf/201816522003

function of the geometrical and elastic characteristics of the joint.

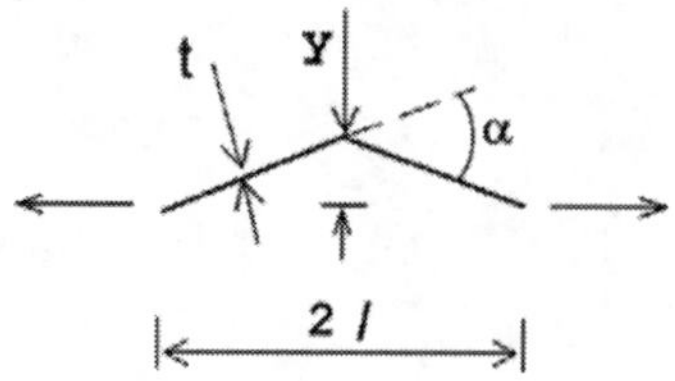

$$k_m = 1 + \frac{3y}{t} \cdot \frac{\tan(\beta/2)}{\beta/2} \quad (6)$$

With:

$$\beta = \frac{2l}{t}\sqrt{\frac{3\sigma_m}{E}} \quad (7)$$

Figure 4. Corrective factor for angular misalignment according to IIW.

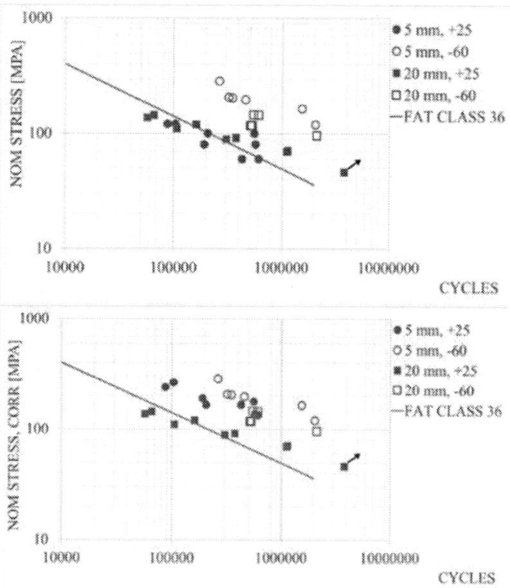

Figure 5. Confront between the fatigue curve of the FAT CLASS 36, the resistance proposed by IIW for V-groove non-grounded butt welds in aluminum alloy. The results are reported both for the non-corrected (top) and the corrected (bottom) nominal traction stress. The run out is marked by an arrow.

The results obtained in terms of fatigue life are in accordance with the minimum duration obtained by the fatigue prediction with nominal stress and the correct fatigue class stress. It is evident, by the comparison of the non-corrected and the corrected results, the importance of taking into account the distortion of the joints due to the welding process. From the corrected plot, Figure 5 left, two effects are evident: an increase in fatigue life for the tests at low temperature and a better behavior of the joints realized with thinner plates. This thickness effect, explained in paragraph 4, is due to the increase of SED in the critical volume associated to an increase in thickness for two geometrically similar joints.

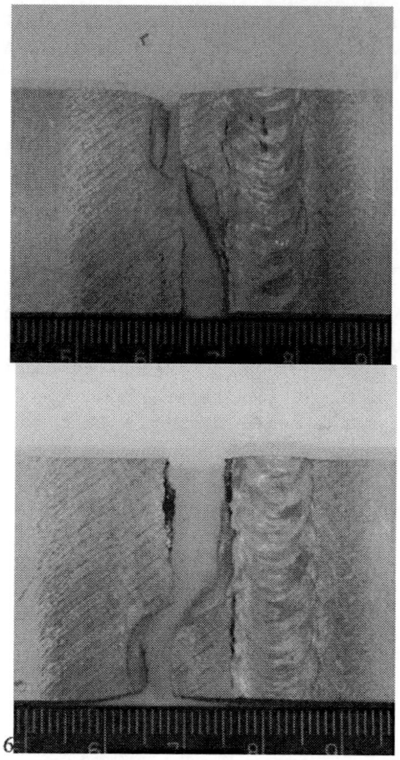

Figure 6. Crack initiation points (hot spots) for the 5 mm thick plate at room (top) and low temperature (bottom). The fatigue crack initiates at the stress intensification point and follows approximately the Heat Affected Zone, with final plastic fracture, in the case of both testing temperatures.

3 SED analysis

3.1 FEM modeling

The local mean strain energy density has been proved to be a valid and reliable tool of fatigue prediction in the presence of stress intensification due to the geometry of the component [4]. The main objective of this paper is to verify whether this tool can be applied to the case of a butt-weld joint, in which the notch opening angle and the weld size have a high degree of variability. The property of the SED of being independent of the notch opening angle plays a key role for the fitness of the method to be applied to this problem. For the finite element analysis the joint is modeled as an ideal linear elastic continuum, thus neglecting the material's defects and residual stresses after the welding process. In fact, in a similar way to what is done for the fatigue assessment with

https://doi.org/10.1051/matecconf/201816522003

nominal stress and fatigue class, these effects are already considered in the fatigue band of the material. Regarding modelling the geometry, the real dimensions of the joint, particularly its deflection and width and height of the weld have been measured and averaged. The weld has been modeled then as an ideal arc passing through the two weld toes and the center point of the weld. Assuming symmetrical geometry, only a quarter of the joint has been modeled and it has been subject to a unit traction load. The distortion/misalignment is not taken into account in the model, which therefore simulate a straight joint, but is used to compute the corrected stress from the nominal stress according to IIW. The mean SED at the weld toe for a generic nominal load is then easily computed by multiplying the mean SED of the unit load times the square of the nominal load applied. For this type of joint, the geometrical weld toe stress concentrator induce crack initiation. To predict the crack initiation life, approximated to the total life of the specimen, the critical volume of radius R=0.12 mm is centered in the weld toe. The very low sensibility of the mean SED to the mesh refinement allows to the use of a coarse discretization, while the stress state has been set to plain strain. The load application edge in the model is set 50 mm distant from the edge of the specimen, according to the clamping distance.

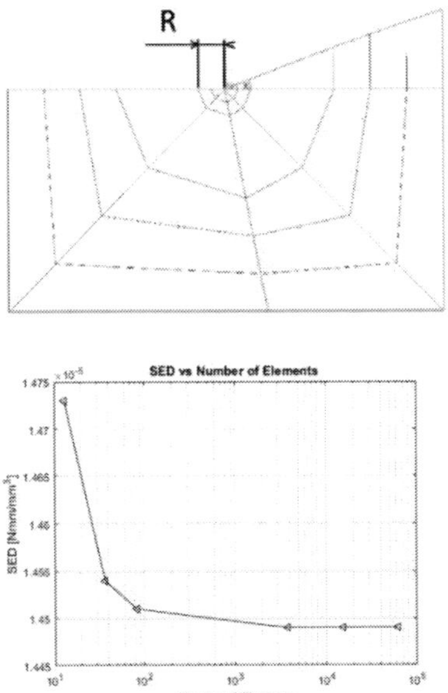

Figure 7. Coarse mesh around the weld toe and critical volume of radius R=0.12 mm (top). Mean SED for unit traction stress as a function of the number of elements: it is evident how a fine mesh is not necessary to the local energetic approach (bottom). Both figures refer to the 5 mm thick plates.

3.2 SED fatigue band

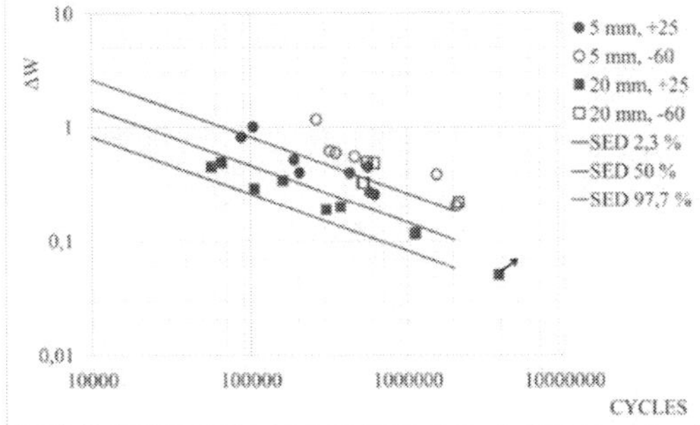

Figure 8. Summary of the results obtained in terms of mean strain energy density computed from the corrected stress.

Comparing the summary of the results obtained in terms of mean SED computed from the corrected stress to the results in terms of nominal stress the scatter is decreased. The results at room temperature are concentrated in the fatigue band available in the literature for welded aluminum joints, with a higher resistance for the 5 mm thick plates. A strong increase in fatigue life is observed for the specimens tested at low temperature, due to the slower crack initiation and propagation.

4 Thickness effect

From the analysis of the results, it is clear an influence of the thickness of the joint on the total fatigue life, with a better performance for thinner plates. Even though this is not reported for the detail tested in the IIW document, the effect has important impact on the joint's behavior and can be easily explained recurring to an analytical description based on the theory of the Strain Energy Density [3]. Being "a" and "b" two geometrically similar joints, of characteristic dimensions t_a and t_b, being "i" 1 or 2 according to the loading mode and being λ_i the corresponding William's eigenvalue, the N-SIFs of the two joints are correlated by:

$$K_{i,b} = K_{i,a}\left(\frac{t_b}{t_a}\right)^{1-\lambda_i} \quad (8)$$

The mean SED can be re-written in terms of N-SIFs and geometrical parameters:

$$\Delta\overline{W} = \frac{1}{E}e_1K_1^2R^{2(\lambda_1-1)} + \frac{1}{E}e_2K_2^2R^{2(\lambda_2-1)} \quad (9)$$

If the notch opening angle is higher than 102°, only the Mode I loading is singular and the Mode II is negligible. The mean SED for the two joints is expressed as:

$$\begin{cases} \Delta\bar{W}_a = \frac{1}{E} e_1 K_{1,a}^2 R^{2(\lambda_1 - 1)} \\ \Delta\bar{W}_b = \frac{1}{E} e_1 K_{1,b}^2 R^{2(\lambda_1 - 1)} \end{cases}$$

Considering the thickness of the samples tested, the correlation between the mean SED for the two types is:

$$\Delta\bar{W}_b = \Delta\bar{W}_a \left(\frac{t_b}{t_a}\right)^{2(1-\lambda_1)} = \Delta\bar{W}_a \left(\frac{20}{10}\right)^{2(1-0.674)} = \Delta\bar{W}_a \cdot 1.57$$

Considering the result of the FEM analysis, the mean SED for a unit traction stress at the weld toe for the 20 mm and the 5 mm joints, are 2.38 E-5 and 1.45 E-5 respectively, in a ratio of 1.64, close to the value found by the analytical approximation. This explains how two similar joints of different size can have a significantly different fatigue life when subjected to the same nominal stress.

5 Conclusions

In the present paper, the results of a series of longitudinal traction fatigue tests performed on two different thicknesses, 5 and 20 mm, butt weld aluminum joints are presented. The tests have been executed both at room and at low temperature (-60 °C), in as welded condition. The main goals of the study are two: understanding the effect of size, dimensional deviations and temperature on the fatigue life and verifying the fitness of the application of the local energetic method of the mean SED to the rapid fatigue assessment for butt-welded joints. Three main conclusions are drawn from the tests executed. The mean SED approach is suitable to predict the fatigue behavior for V-groove butt welds, with a simple 2D model based on the average size and distortion of the joints, providing more accurate results than the Nominal Stress approach. For the alloy and the welding technology object of the study, the thicker plates (20 mm) show a reduced fatigue life compared to the thinner ones (5 mm), how explained in paragraph 4. Moreover, the testing at low temperature shows an important increase in fatigue life compared to the results at room temperature. The average increase in fatigue life at low temperature is of 3 and 6 times for the 5 mm and 20 mm thick joints respectively, but the number of samples tested so far doesn't allow to a safe and certain numerical quantification of the effect.

The authors wish to thank the Research Council of Norway for funding through the Petromaks 2 Programme, Contract No.228513/E30. The financial support from ENI, Statoil, Lundin, Total, Scana Steel Stavanger, JFE Steel Corporation, Posco, Kobe Steel, SSAB, Bredero Shaw, Borealis, Trelleborg, Nexans, Aker Solutions, Marine Aluminum, FMC Kongsberg Subsea, Hydro and Sapa are also acknowledged.

References

[1] Eurocode 3: Design of steel structures - Part 1-9: Fatigue, EUROPEAN COMMITTEE FOR STANDARDIZATION Ref, No, EN 1993-1-92005: E

[2] Hobbacher A., RECOMMENDATIONS FOR FATIGUE DESIGN OF WELDED JOINTS AND COMPONENTS IIW document IIW-1823-07

[3] Lazzarin P, Tovo R. A notch intensity factor approach to the stress analysis of welds. Fatigue & Fracture of Engineering Materials & Structures 1998; 21: 1089-1103.

[4] Lazzarin P, Zambardi R. A finite-volume-energy based approach to predict the static and fatigue behavior of components with sharp V-shaped notches. International Journal of Fracture 112: 275–298, 2001.

[5] Williams ML. Stress singularities resulting from various boundary conditions in angular corners of plates in tension. J Appl Mech 1952;19:526–8.

[6] Henshell RD, Shaw KG. Crack tip finite elements are unnecessary. Int J Numer Methods Eng 1975; 9:495–507

[7] Filippo Berto, Paolo Lazzarin, The volume-based Strain Energy Density approach applied to static and fatigue strength assessments of notched and welded structures, Procedia Engineering 01 (2009) 155–158

[8] Lazzarin P, Berto F, Gomez F J, Zappalorto M. Some advantages derived from the use of the strain energy densityover a control volume in fatigue strength assessments of welded joints. International Journal of Fatigue 30 (2008) 1345–1357.

[9] Lazzarin P, Berto F, Zappalorto M. Rapid calculations of notch stress intensity factors based on averaged strain energy density from coarse meshes: Theoretical bases and applications. International Journal of Fatigue 32 (2010) 1559–1567.

[10] Lazzarin, P., Berto, F., 2005. Some expressions for the strain energy in a finite volume surrounding the root of blunt V-notches. International Journal of Fracture, 135(1-4), 161–185.

[11] Lazzarin, P., Berto, F., Elices, M., Gómez, J., 2009. Brittle failures from U- and V-notches in mode I and mixed, I + II, mode: A synthesis based on the strain energy density averaged on finite-size volumes. Fatigue & Fracture of Engineering Materials & Structures, 32(8), 671–684.

[12] Livieri, P. & Lazzarin, P. Int J Fract (2005) 133: 247. https://doi.org/10.1007/s10704-005-4043-3

Fatigue behaviour of an industrial synthetic rubber

Thomas Balutch[1,3], *Bertrand* Huneau[1,*], *Yann* Marco[2], *Pierre* Charrier[3], and *Clément* Champy[3]

[1]Institut de Recherche en Génie Civil et Mécanique (GeM), UMR CNRS 6183, Ecole Centrale de Nantes, 44300 Nantes, France
[2]Institut de Recherche Dupuy de Lôme (IRDL), FRE CNRS 3744, ENSTA Bretagne, 29200 Brest, France
[3]Vibracoustic Nantes, CAE Durability Department, 1 Rue du Tertre, 44470, Carquefou, France

Abstract. For some automotive anti-vibration applications, for instance exhaust hangers, center bearing bushes or torsional vibration dampers, temperature constraints make the use of synthetic rubbers, such as EPDM, necessary because of their better heat aging resistance compared to natural rubber. The aim of this paper is to understand the features of the fatigue behaviour of an industrial EPDM compared to the well-known natural rubber. To do so, fatigue tests are conducted on hourglass-shaped specimens, and fracture surfaces are analysed using optical microscopy and scanning electron microscopy (SEM). It appears that every samples exhibit only one root cause of failure. Thus, two types of precursors are identified as responsible of the final fracture of samples: material's inclusions and mold flaws. Interrupted fatigue tests are then performed and fatigued samples are observed with SEM. The built procedure allows us to follow fatigue cracks initiation and propagation along cycles, and to propose local damage mechanisms for each type of precursors. A global damage scenario is finally considered and compared to the one of natural rubber described in the literature.

1 Introduction

Anti-vibration solutions in automotive are crucial to ensure comfort, safety and durability in vehicle systems. Most of the parts that are designed in Vibracoustic to provide these functions are made of rubber, and particularly of natural rubber (NR), known for its good fatigue resistance that is often attributed to its faculty to crystallize under strain [1]. Because of the significant heat experienced by some parts near car's hot-spots, natural rubber can no longer be used (long periods with T>90°C). As a consequence, synthetic rubbers such as EPDM are preferred because of better heat aging resistance.

However, fatigue prediction for this material is difficult due to an important scattering of the lifetime results that can reach more than 4 decades (see Fig. 1).

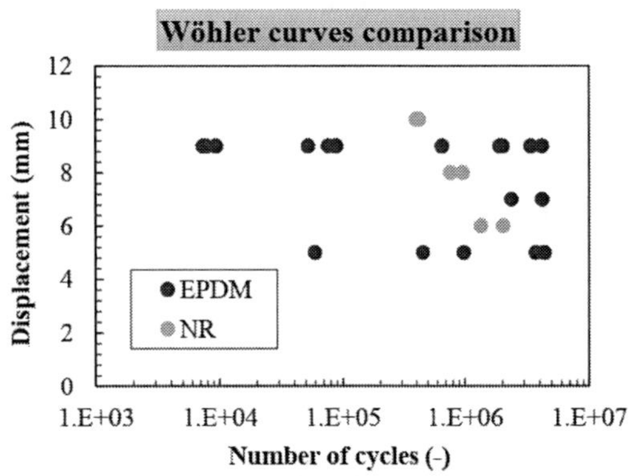

Fig. 1. Fatigue results scattering of EPDM compared to natural rubber (data from Flamm et al. [2]).

Moreover, this rubber matrix presents little or no strain-induced crystallization which leads to a lower resistance to fatigue crack growth when compared to natural rubber [3-5]. Finally, it has to be underlined that compared to natural rubber, only few scientific papers are dealing with the fatigue behaviour of EPDM [2-10].

Thus, the aim of this paper is to highlight the specificities of the fatigue behaviour of an EPDM, and to compare them with the ones of the more known natural rubber.

2 Experiments

* Corresponding author: bertrand.huneau@ec-nantes.fr

 https://doi.org/10.1051/matecconf/201816522004

2.1. Material and samples

The material of this study is a fully formulated industrial EPDM, filled with carbon black and crosslinked with sulfur. As it is used for real mass-produced parts, the quantitative results will not be provided.

The samples used to perform the mechanical tests are known as AE2 samples [11]. It is an hourglass-shaped specimen with a notch radius of 2mm. The important strain gradient induced by this geometry focuses fatigue crack initiations close to the skin of the median area. On top of that, to be representative of real parts' process, injection molding is used to manufacture this specimen. As a consequence, all AE2 present a specific mold flaw on the zone that corresponds to the junction of the two parts of the mold (see Fig. 2).

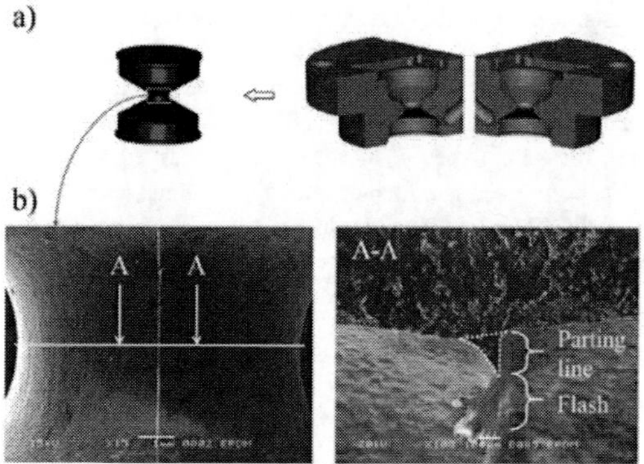

Fig. 2. a) Scheme of the injection molding of an AE2 sample and b) SEM observations of its related mold flaws.

2.2 Observations tools

In order to analyse fracture surfaces and fatigue cracks on the surface of AE2 samples, two different observations techniques are used.

The first one is stereo microscopy. The stereo microscope Carl Zeiss STEMI 2000C available in Vibracoustic can take numerical pictures with magnification from x6.5 to x50.

The second one is scanning electron microscopy (SEM). Two microscopes are used in this study. A JEOL 6060-LA with an EDS (Energy Dispersive Spectroscopy) probe to perform quantitative chemical analysis. It is located at "Ecole Centrale de Nantes" and used for fracture surfaces analysis. The other one is a JEOL JCM 6000plus used for interrupted fatigue tests analysis at Vibracoustic. As the scale of interest for this study goes from some micrometres to several millimetres, the highest magnification reached here is approximately x2000.

Thanks to carbon black, the rubber is sufficiently conductive to use secondary electrons in high vacuum without using any metal deposit. Depending on the test, SEM parameters have to be set accordingly. For fracture surfaces analysis, the acceleration voltage is set to 20kV with a medium spot size in order to have a good quantitative EDS analysis. For interrupted fatigue tests, the energy of the electron beam must be reduced in order not to damage the rubber, which could influence the fatigue life and the related damage scenario. To overcome this difficulty, the acceleration voltage is set to 10kV, the spot size is small, the magnification for pictures remains low, and a special attention is given to the exposure time in the SEM. The samples are stretched of 3mm when put in the SEM, in order to open the cracks.

2.3 Fatigue tests

The fatigue tests are conducted on a hydraulic bench in which 5 samples can be set up at once. The mechanical load is displacement controlled, and the frequencies are adapted for each load level in order to limit the self-heating of the rubber however keeping a correct amount of experimental time. Only relaxing conditions are tested in this study (R=0), at 25°C. Higher testing temperature is for sure a perspective of the study.

First, a Wöhler curve is built with five load levels, and five samples per load levels. After that, an additional test internally called "radar test" is conducted. This test consists in breaking 25 samples at the same load level to better estimate the scattering of the lifetime results.

The loss of effective stiffness is used as the end of life criterion (see Ostoja-Kuczynski et al. [11] for more details). The end of life is defined as the detected initiation (N_i number of cycles, see Fig. 3), and corresponds approximately to the presence of a 2mm long crack on AE2 samples. This criterion has the advantages of being easy to use with automated post-treatment, and more representative of the failure of real parts, that will lose their mechanical properties before total fracture. As a fatigue parameter, the maximum principal strain in the notched area is used to plot all the fatigue life results.

2.4 Interrupted fatigue tests

The purpose of these tests is to analyse the initiated cracks on samples that have undergone some fatigue cycles, before total failure. In the literature, several authors built protocols for different rubbers [12-15]. Because of the important scattering of the fatigue results, and the fast fatigue crack propagation in EPDM, it is not possible to use the same protocol as Le Cam et al. [12] and to wait for a crack of a certain size. Indeed, the step between no visible cracks and total failure is too short and easy to miss. Moreover, the use of a certain percentage of the mean duration life for different specimens cannot be considered as in Huneau et al. [15]. Samples could be completely broken or not damaged at all after fatigue testing and this would make the analysis irrelevant. Therefore, the same samples need to be monitored along fatigue cycles in order to be able to link the damage of each sample to its own duration life.

In consequence, 10 samples are chosen to be representative of the scattering of the results. SEM pictures of the full median side surface are taken and specimens are then subjected to X number of cycles. These 2 steps are repeated until complete fracture of all

https://doi.org/10.1051/matecconf/201816522004

the samples. The load level is the one of the "radar test", because it is where the material's fatigue behaviour is the best described. As mentioned in section 2.2, the electron beam can have an influence on the fatigue behaviour, so a special care is given to the acquisition of SEM pictures. This is also why the use of repeated tomography measurements has been rejected, considering the needed long exposure time and high energetic x-rays beam for one complete scan.

At last, the number of cycles X needs to be correctly chosen, to be sure not to miss the propagation of a main crack. In Fig. 3 is plotted the maximum force versus the number of cycles curve, which we use to defined X thanks to equations (1) and (2).

$$N_p = N_f - N_i \quad (1)$$

$$X = N_p / 2 \quad (2)$$

With N_f the number of cycles until failure, N_i the number of cycles until the detection of initiation, and N_p the number of cycles of macroscopic propagation. This protocol is justified by the fact that N_p is almost not scattered compared to N_i as it is shown later on Fig. 7.

The worst possible case would be that no cracks initiations are visible on the surface before N_i number of cycles, which is possible because of the important crack growth rate of EPDM. By relating X to half of the macroscopic propagation, at least one observation of the main crack before total failure is possible.

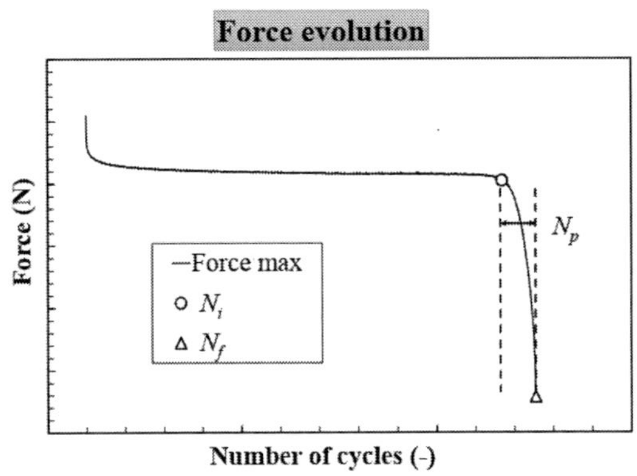

Fig. 3. Maximum force evolution of an AE2 sample for "radar test" load level versus the number of cycles.

3 Results and discussion

3.1 Fractographic analysis

Once a sample is totally broken, it is possible to find the root causes of failure by analysing its fracture surface, which is very different depending on the material. The fracture surfaces of the studied EPDM and a natural rubber from another internal work are presented in Fig. 4.

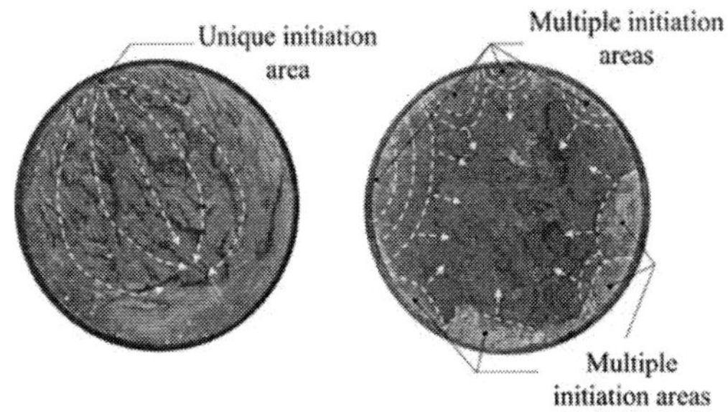

Fig. 4. Fracture surfaces analysis of an EPDM (left) and a natural rubber (right) with stereo microscopy.

For natural rubber, the surface appears very rough, and semi-circular arcs, which can be linked to initiation sites, are highlighted all around the skin. The final failure occurs at the centre of the fracture surface. Thus, the fatigue life seems to be governed by multiple cracks initiation and fatigue cracks coalescence.

For EPDM, the fracture surface appears different. The smooth part in the bottom right corresponds to the final "brittle" fracture. The rest of the surface is associated with fatigue "rivers" that converge to one spot only, that is the single root cause of failure. The specific roughness decreases when approaching the root cause. At the end, every EPDM's fracture surfaces exhibit only one root cause of failure.

When a focus is made on the initiation spot, it is possible to split precursors in two categories: inclusions, and parting line region. Inclusions are agglomerates from ingredients of the rubber's formulation and parting line region is the mold flaw presented in Fig. 2, including the parting line and the flash. This split is made because the physical origin of these two flaws are very different: one is intrinsic to the material, and the other is purely geometrical, and only process dependent.

The initiation spot is easy to find with stereo microscopy, but SEM is needed to check the presence of inclusion and to analyse it with EDS. Moreover, SEM enables us to confirm that parting line region is really responsible of initiation and not an inclusion close to this region. Fig. 5 shows an example of an inclusion initiating a fatigue crack. Thanks to all SEM pictures, it is also possible to measure the dimensions of the precursors. The geometrical parameter used to represent the global precursor's size is $\sqrt{Area}$. It has been initially proposed by Murakami and Endo [16] for fatigue of metals, and is still used nowadays, for example in references [17, 18].

However, an inclusion is not systematically found for initiation outside the parting line region. One explanation for this can be that the inclusion responsible of the failure fell from the fracture surface. Another explanation can be that an inclusion is below the surface, covered with rubber, that makes it non visible with surface observations like SEM.

https://doi.org/10.1051/matecconf/201816522004

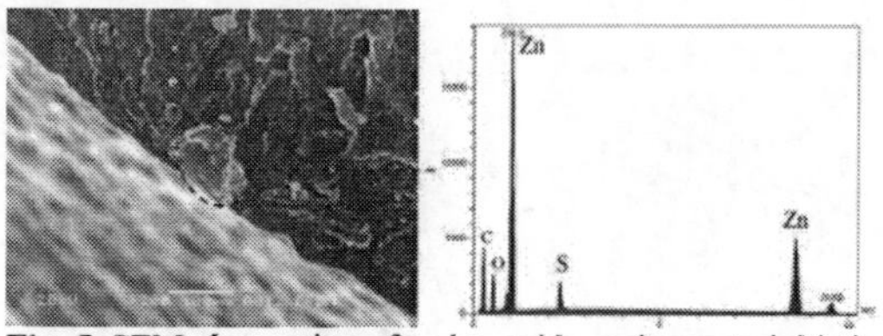

Fig. 5. SEM observation of a zinc oxide agglomerate initiating a crack (left) and its related EDS analysis (right).

3.2 Lifetime results

Fatigue tests are conducted according to section 2.3 and all the lifetime results are plotted on Fig. 6. To keep reasonable test durations, some tests have been interrupted before total failure, and are represented with an arrow on the right. As expected, there is an important scattering (more than 2 decades), yet the mean duration lives of every load level are aligned well enough to fit a correct power law.

N_p is measured for every sample, and Fig. 7 shows the evolution of this number against the load level. It appears that there is very little scattering on this number compared to N_i on Fig. 6. This suggests that the scattering of the fatigue results comes rather from the initiation step than the propagation one.

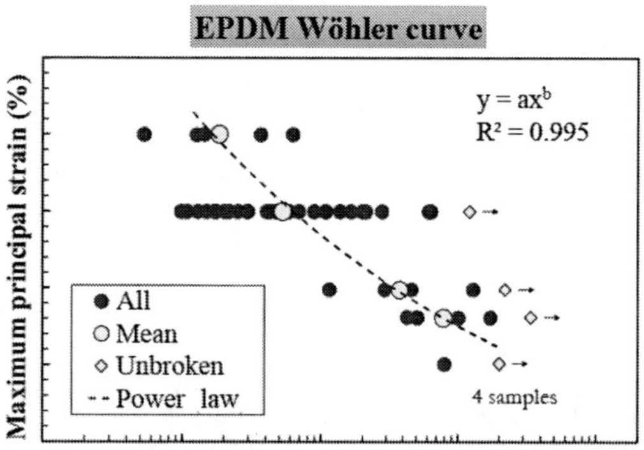

Fig. 6. Wöhler curve of the studied EPDM including "fatigue radar" data.

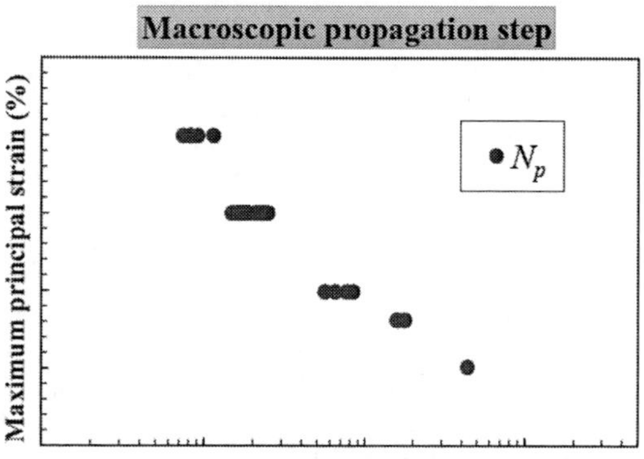

Fig. 7. Number of cycles for macroscopic propagation.

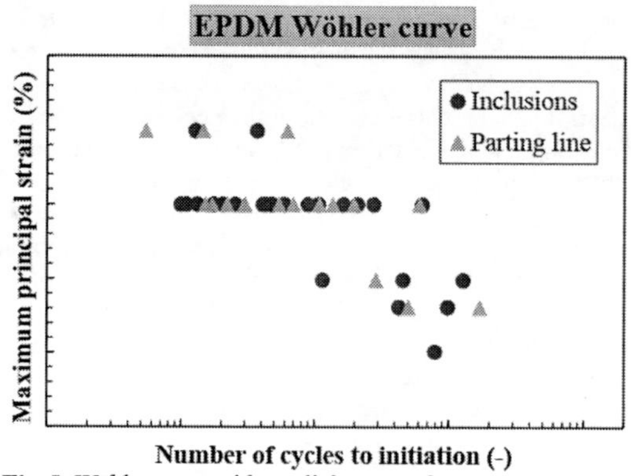

Fig. 8. Wöhler curve with a split by type of precursors.

Thanks to the fractographic analysis described in section 3.1, the fatigue results can be split by type of precursors on Fig. 8: inclusions or parting line region. It seems that there is no real tendency on the fatigue results between initiation on purely geometrical flaws that are process dependent, and on natural intrinsic flaws. Globally, 60% of failures occur on inclusions, and 40% on parting line. This means that the geometry of the parting line region is almost as critical as the inclusions present in the material. This results is only valid for the mold geometry considered here, and could be different for another sample geometry, or even for AE2 manufactured with a different mold.

On Fig. 9, the influence of the size of inclusions versus the number of cycles is plotted, considering only the "radar test" load level. Fig. 9 reveals that there is no direct correlation between the size of an inclusion and the duration life. The scattering of the fatigue results cannot be explained only by a scattering in the size of inclusions, and other key parameters must be considered, as the chemical composition, the exact location of the inclusion in the sample, or its 3D shape.

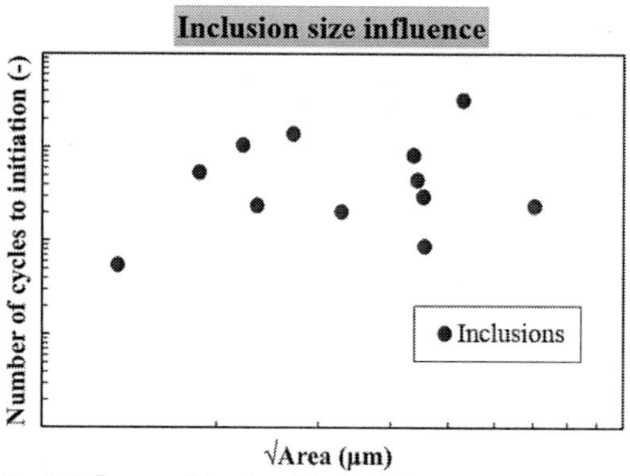

Fig. 9. Influence of the size of the inclusions.

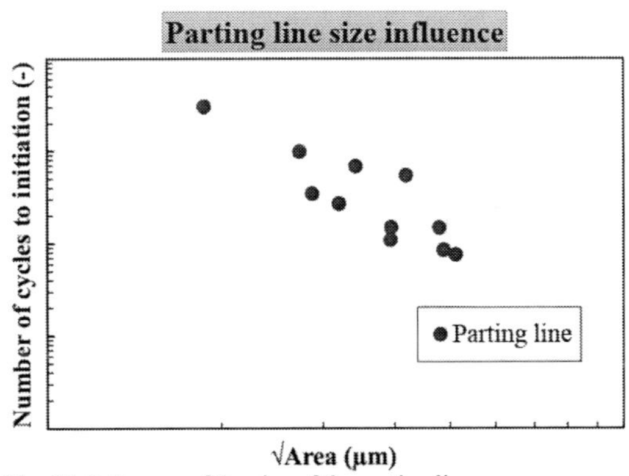

Fig. 10. Influence of the size of the parting line.

Conversely, the same thing has been done for parting line by taking the area defined by the dashed line in Fig. 2. The results in Fig. 10 show a good correlation between the size of the parting line and the duration life. It appears that the duration life decreases with an increasing size of the parting line. As a consequence, the scattering of the fatigue results for initiation on parting line region can be explained by the one of the parting line geometry. This result proves that small fluctuations of the precursor's geometry can have a huge influence on the lifetime results. In fact, the parting line geometry is assumed to be highly dependent on all the process parameters, and could be very difficult to control.

Finally, it has to be noted that a split according to the chemical nature of the inclusions could not have been made, except for zinc oxide agglomerates. For the rest of the inclusions, EDS spectrum were very similar and presented not enough differences to build categories.

3.3 Damage scenario

3.3.1 Crack tip

Thanks to interrupted fatigue tests, the crack tip is pictured along fatigue cycles. Fig. 11 shows that the crack propagation mechanisms are associated with ligaments. Similar crack tip has been observed for natural rubber, silicone rubber and EPDM in references [19-21]. These ligaments break under tension and cause the roughness of the fracture surface. It also appears on Fig. 11 that the dimensions of the ligaments increase when the crack grow, which is in accordance to an increasing roughness on the fracture surface (see section 3.1). Le Cam et al. [22] also noticed this evolution of the roughness on fracture surfaces of natural rubber.

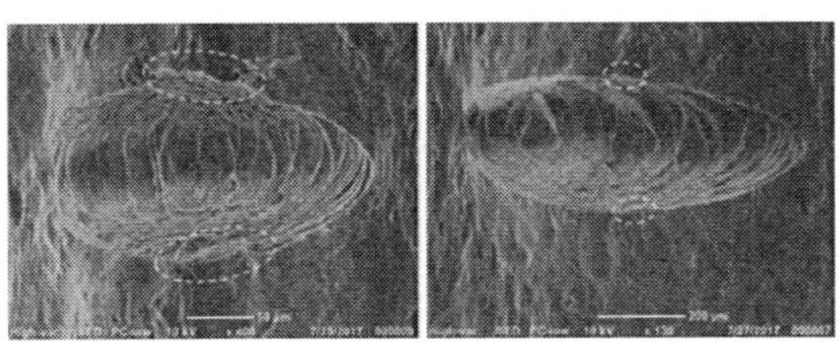

Fig. 11. Crack tip shape evolution before (left, x400) and after (right, x130) some fatigue cycles.

3.3.2 Crack initiation on inclusions

For crack initiation on inclusions, two different mechanisms have been observed. They will be referred as mechanism A and B for the rest of the article.

Mechanism A is related to inclusions that have high internal cohesion compared to the bonding with the matrix and can be described as follow (see Fig. 12):

- Stage 1: debonding at one pole;
- Stage 2: opening on the sides;
- Stage 3: propagation into the bulk.

Mechanism B is related to inclusions exhibiting low internal cohesion and is defined by (see Fig. 13):

- Stage 1: inclusion's fracture;
- Stage 2: opening on the sides;
- Stage 3: propagation into the bulk.

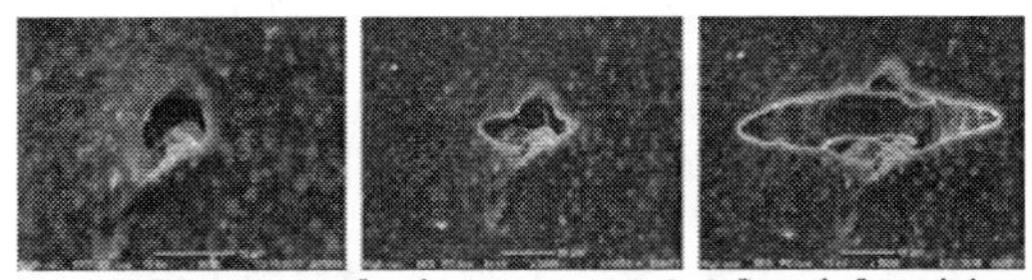

Fig. 12. Damage mechanism A, stage 1, 2, 3 from left to right.

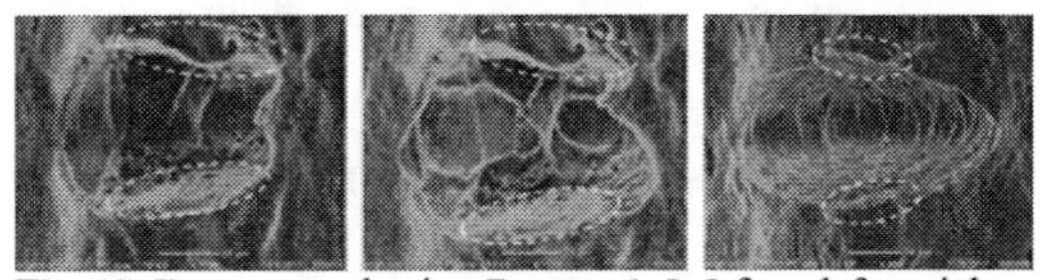

Fig. 13. Damage mechanism B, stage 1, 2, 3 from left to right.

The internal cohesion and interface with the matrix is assumed to depend on the type and size of the inclusion, as well as on the rubber matrix. In this study, cracks initiate from zinc oxide agglomerates following mechanism A, and following mechanism B for all other inclusions. These two mechanisms are schematized on Fig. 14. The main difference between these two mechanisms is the stage 1. Stage 3 corresponds to a "long" crack that is no longer influenced by its precursor.

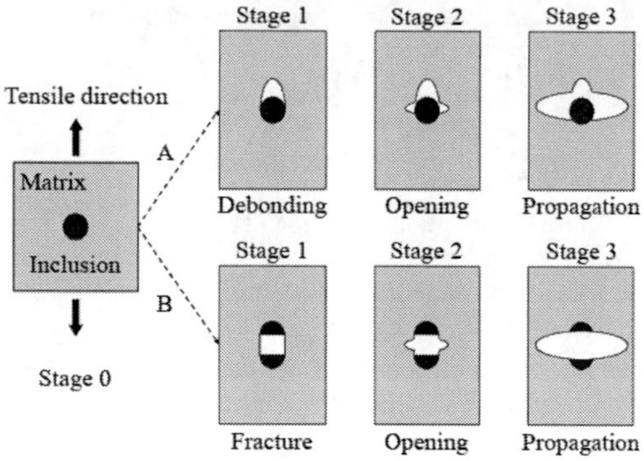

Fig. 14. Scheme of the initiation mechanisms on inclusions.

For natural rubber, Huneau et al. [15] have also proposed these two initiations mechanisms. Mechanism A was observed for crack initiation on rigid inclusions such as carbon black agglomerates (from 9μm to 17μm), and mechanism B for small zinc oxide agglomerates (average size of 7μm), but most of the time, the authors only witnessed the stage 1 of mechanism B. Perhaps they would have seen the stages 2 and 3 more often, if their zinc oxide agglomerates were bigger than 10μm, as it is the case for the inclusions following mechanism B in this study. Furthermore, the fact that 2 different damage mechanisms (A for EPDM, and B for natural rubber) are observed for inclusions of the same type (zinc oxide agglomerates) in different rubber matrix, proves that the balance between internal cohesion and inclusion/matrix interface properties could be very complex to evaluate.

To sum-up, it appears that the considered local damage mechanisms are similar for EPDM and natural rubber, but that the correspondence between these mechanisms and the inclusions initiating cracks is not the same.

3.3.3 Crack initiation on the parting line

Fatigue crack initiation related to the parting line region has been followed for every sample with a special care, because it is known from the fractographic study that it can be critical for the fatigue behaviour of this EPDM. The proposed damage scenario is presented on Fig. 15 and can be described as:

- Stage 1: tearing of the flash;
- Stage 2: propagation into the parting line;
- Stage 3: propagation into the bulk.

Thanks to all observations, it is assume that stage 1 is mostly dependent on the flash's geometry. The step between stage 2 and stage 3 appears to be more dependent on the parting line geometry. This last step could be the critical one, because once a crack is in the bulk, it propagates until the final failure. When a crack initiates on the flash, the parting line could act like the last barrier preventing the crack propagation into the bulk. Thus, the fatigue failure would be more dependent on the parting line features than on the flash ones, and these observations could confirm the good correlation between the size of the parting and the fatigue results in Fig. 10.

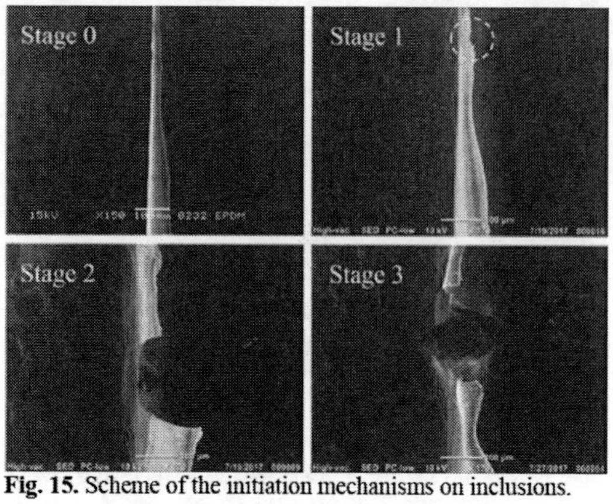

Fig. 15. Scheme of the initiation mechanisms on inclusions.

Finally, a two steps initiation mechanism can also be considered. As the parting line region is created by a flow of rubber between the two parts of the mold, an inclusion can be stuck in the parting line region by a funnel effect. This creates a high stress/strain concentration that eventually initiates a crack in the parting line region, but due to an inclusion.

These mechanisms have been observed for natural rubber [23], even if not described in details. The initiation mechanisms on the parting line therefore seem very similar for EPDM and natural rubber.

3.3.4 Competition between fatigue cracks

Fig. 16 illustrates that over a whole AE2 sample, several stage 3 cracks can be observed at the same time. These long cracks, initiated from inclusions and in the parting line region, are quiet distant so that they do not influence each other. For every tested sample, one of these initiated cracks finally takes the lead and propagates until the complete fracture of the specimen, without any cracks coalescence. This confirms the fractographic analysis in section 3.1 where only one root cause of failure is found. As shown on Fig. 16, stage 3 cracks were seen only belatedly compared to the duration life of the considered sample.

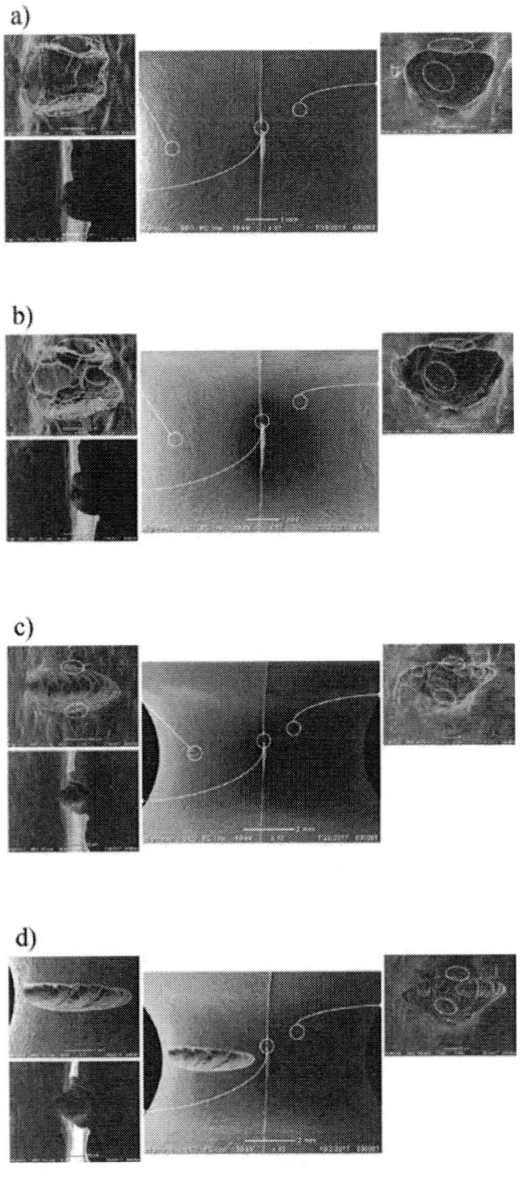

Fig. 16. Fatigue cracks evolution in EPDM at different duration lives, and different magnifications: a) 30% of the total duration life, b) 40%, c) 85%, d) 100%.

Overall, the considered damage scenario is the following:

- Few stage 1 initiations both on inclusions and in the parting line region;
- Some stage 2 and stage 3 cracks, but far from each other and without fatigue cracks coalescence. This step occurs belatedly during the fatigue life of the sample;
- One of the cracks eventually propagates until final fracture of the specimen.

Thus, the fatigue failure of EPDM samples seems to be driven only by the initiation of a crack on a “critical” flaw, considering the mechanical load. The scattering of the fatigue results could be explained by the fluctuation of this flaw from one sample to the other. In fact, for material’s inclusions, not only the type and shape but also the precise location in the sample could have an important influence. For instance, a big inclusion far from the skin will be harmless compared to a smaller one stuck into the parting line. Moreover, one can imagine that the geometry of the parting line region is highly dependent not only on every parameter of the process, but also on how the sample is removed from the mold, or manipulated after removal. All of these small variations could create an important scattering in the fatigue results.

Considering the literature of the fatigue behaviour of natural rubber [12-15, 21-25], the global damage scenario is very different. Indeed, for a natural rubber, cracks are far more numerous (see Fig. 17). Furthermore, it appears that they can initiate during the first fatigue cycles, and this step is followed by a slow crack propagation. Interrupted fatigue tests, as well as fracture surfaces analysis show that cracks coalescence finally causes the final fracture of natural rubber specimens (see Fig. 4). Thus, even if a harmful inclusion is present close to the skin of the median area, it will not be the only root cause of failure.

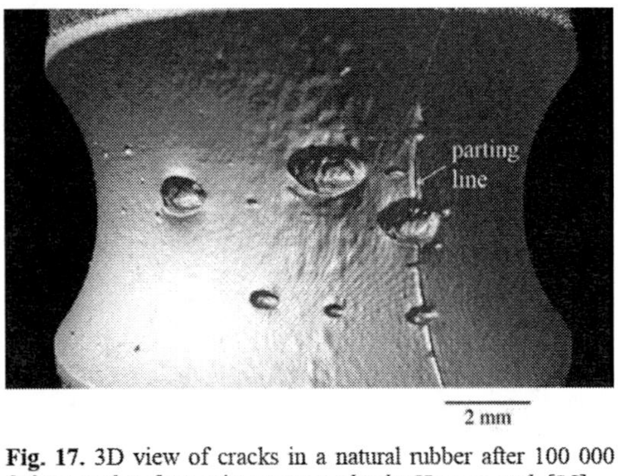

Fig. 17. 3D view of cracks in a natural rubber after 100 000 fatigue cycles, from microtomography, by Huneau et al. [25].

As a results, fatigue failure of natural rubber is driven by an early multi-initiation, a multi-propagation, and cracks coalescence, whereas for EPDM, only one critical flaw which exhibits a late initiation is responsible of the final fracture. Accordingly, the propagation step, which is known to be little scattered, represent an important part of the total lifetime of natural rubber. Conversely, as EPDM presents an elevated fatigue crack growth rate [3-5], the initiation step seems to be the key to understand its fatigue behaviour. These main differences between the two considered rubbers could explain why there is so little scattering in the fatigue results of natural rubber compared to EPDM.

At last, it has to be noted that the fatigue lifetime results are highly influenced by the chosen sample's geometry.

4 Conclusion

The important scattering of the EPDM fatigue results is confirmed by the build of a Wöhler curve with AE2 samples. It appears that the macroscopic propagation is almost not scattered, and this suggests that the scattering of the lifetime results comes from initiation step rather than propagation one. The related fracture surface exhibit only one root cause of failure that can be either inclusions of the rubber's formulation, or purely geometrical flaws created by the process. Approximately 60 % of the failure occurs on inclusions, and 40 % on parting line region but without any clear influence on the lifetime results. It has to be noted that this result is highly dependent on the mold, and sample geometry. A good correlation appears between the $\sqrt{Area}$ of the parting line and the fatigue life results, so the scattering in the fatigue failure could be explained by a fluctuation in the parting line geometry. However, no tendencies are noticed on the fatigue results for the $\sqrt{Area}$ of inclusions. This could be explained by the important influence of other key parameters as the nature of the inclusion, its precise 3D shape, or its precise location, for instance the distance to the median zone, and the distance to the skin.

Thanks to interrupted fatigue tests with regular SEM observations, local damage mechanisms are proposed for each type of precursors. It appears that they are very similar to what has been proposed for natural rubber [15, 23]. Nevertheless, the main differences come from the global damage scenario. For natural rubber, it has been shown [12-15, 21-25] that an important number of cracks initiate during the first fatigue cycles. This multi-initiation followed by a slow propagation and cracks coalescence could explain the low scattering of the fatigue results of natural rubber. On the opposite, for the studied EPDM, only a few number of cracks initiate far from each other, and one of them eventually propagates until complete fracture of the specimen without any cracks coalescence. Thus the fatigue failure of a given sample seems to be driven only by its most "critical" flaw regarding fatigue, and this could explain the important scattering in the fatigue results compared to natural rubber. Furthermore, the fact that stage 3 cracks, whatever the origin of the precursor, occurs only belatedly in the duration life of the sample, proves that the total lifetime cannot be reduced to crack propagation of initial flaw, and this needs further investigations.

References

1. W. Mars, A. Fatemi, Rubber Chem. Technol. **77**, 391-412 (2004)
2. M. Flamm, T. Steinweger, J. Spreckels, T. Brüger, ECCMR V, 233-238 (2008)
3. F. Abraham, G. Clauß, T. Alshuth, J. Kroll, Kaut. Gummi Kunstst. **58**, 595-599 (2005)
4. W. W. Chung, Y. W. Chang, Korea Polym. J. **9**, 319-326 (2001)
5. J. Royo, Polym. Test. **11**, 325-344 (1992)
6. F. Abraham, T. Alshuth, S. Jerrams, Mater. Design **26**, 239-245 (2005)
7. H. W. Chou, J. S. Huang, S. T. Lin, J. Appl. Polym. Sci. **103**, 1244-1251 (2007)
8. C. Gögelein, H. J. H. Beelen, M. Van Duin, Soft Matter **13**, 4241-4251 (2017)
9. H. Ismail, N. S. Che Mat, N. Othman, J. Vinyl Addit. Techn. **31**, 5-12 (2017)
10. S. Jerrams, J. Hanley, N. Murphy, H. Ali, Rubber Chem. Technol. **81**, 638-649 (2008)
11. E. Ostoja-Kuczynski, P. Charrier, E. Verron, G. Marckmann, L. Gornet, G. Chagnon, ECCMR III, 41-47 (2003)
12. J. B. Le Cam, B. Huneau, E. Verron, L. Gornet, Macromolecules **37**, 5011-5017 (2004)
13. V. Le Saux, Y. Marco, S. Calloch, P. Charrier, Polym. Eng. Sci. **51**, 1253-1263 (2011)
14. K. Legorju Jago, Rubber Chem. Technol. **85**, 387-407 (2012)
15. B. Huneau, I. Masquelier, Y. Marco, V. Le Saux, S. Noizet, C. Schiel, P. Charrier, Rubber Chem. Technol. **89**, 126-141 (2016)
16. Y. Murakami, M. Endo, Eng. Fract. Mech. **17**, 1-15 (1983)
17. M. I. Houria, Y. Nadot, R. Fathallah, M. Roy, D. M. Maijer, Int. J. Fatigue **80**, 90-102 (2015)
18. V.D. Le, F. Morel, D. Bellett, N. Saintier, P. Osmond, Mat. Sci. Eng. A-Struct. **649**, 426-440 (2016)
19. S. Beurrot, B. Huneau, E. Verron, J. Appl. Polym. Sci. **117**, 1260-1269 (2010)
20. S. V. Hainsworth, Polym. Test. **26**, 60-70 (2007)
21. M. Flamm, J. Spreckels, T. Steinweger, U. Weltin, Int. J. Fatigue **33**, 1189-1198 (2011)
22. J. B. Le Cam, B. Huneau, E. Verron, Int. J. Fatigue **52**, 82-94 (2013)
23. I. Masquelier, PhD thesis, UBO, 2014
24. N. Saintier, G. Cailletaud, R. Piques, Int. J. Fatigue **28**, 61-72 (2006)
25. B. Huneau, I. Masquelier, Y. Marco, O. Brzokewicz, P. Charrier, ECCMR VIII, 393-398 (2013)

 https://doi.org/10.1051/matecconf/201816522005

Modeling of hydrogen effects on short crack propagation in a metastable austenitic stainless steel (X2CrNi19-11)

Volker Schippl[1,*], *Sven* Brück[1], *Hans-Jürgen* Christ, *Claus-Peter* Fritzen[2], *Martina* Schwarz[3], and *Stefan* Weihe[3]

[1]Institut für Werkstofftechnik, Universität Siegen, 57068 Siegen, Germany
[2] Institut für Mechanik und Regelungstechnik - Mechatronik, Universität Siegen, 57068 Siegen, Germany
[3]Materialprüfungsanstalt, Universität Stuttgart, 70174 Stuttgart, Germany

Abstract. The experimentally observed short fatigue crack growth rate of an uncharged specimen tested in air is compared with results obtained from specimens tested in 10 MPa hydrogen atmosphere and specimens previously charged with hydrogen. To further discuss the hydrogen related short propagation mechanisms, a simulation approach for predicting short fatigue crack growth is presented. The boundary element method is used for calculating stresses and displacements in an anisotropic elastic solid. The hydrogen concentration is assumed to be homogeneously distributed in the microstructure. Based on this modelling approach, it could be concluded that hydrogen leads to an increasing short fatigue crack growth rate due to increasing irreversible deformation processes at the crack tip and also promotes grain boundary cracking in specimens tested in 10 MPa hydrogen atmosphere.

1 Introduction

Due to the global effort to reduce the emission of CO_2, the need for renewable and sustainable energy resources arises. Hydrogen is a promising candidate in the automobile industry as a fuel with zero-emissions during the electrochemical reaction in the fuel cell.

During operation, the components responsible for transporting the hydrogen gas to the fuel cell experience cyclic loading due to service loading conditions. Thus, to safely design structural components with long product lives, the effect of hydrogen on the material behaviour during cyclic loading has to be considered. Commonly, materials are used which show little or no effect on hydrogen. This includes materials such as austenitic stainless steels containing a high amount of nickel, e. g. AISI 316. These austenitic stainless steels have a high resistance against corrosion and are very ductile, but are relatively expensive due to the high nickel content. In order to be able to safely use austenitic stainless steels with a lower nickel content, the effects of hydrogen on the fatigue damage has to be clarified. A lot of the published research highlights the influence of hydrogen on the fatigue damage in the long crack growth regime [1, 2]. Fukuyama et al. investigated the fatigue properties of AISI 304 tested in hydrogen pressure up to 4.0 MPa [1]. They observed an increasing fatigue crack growth rate with increasing hydrogen pressure. Murakami et al. observed fatigue cracks up to 1000 μm in specimens previously charged with hydrogen [2]. They have also shown that the presence of hydrogen leads to an increasing fatigue crack growth rate in these specimens.

Thus, it is well known that hydrogen can lead to a significant reduction of fatigue life by changing the crack propagation mechanisms in the long crack growth regime. But during cyclic loading in high cycle fatigue, the initiation and growth of microstructurally short cracks may be crucial to determine the fatigue life. It is well established that under these conditions, short crack growth can take up to 90% of the total fatigue life [3].

The crack propagation of short cracks is highly influenced by microstructural characteristics. For austenitic stainless steels under low amplitude loading, it could be observed that short cracks propagate along shear bands which formed at the crack tip due to locally high stresses. Wilkinson et al. proposed a theoretical model for describing the crack propagation of these microstructurally short cracks [4]. In 1996, they explained that the cyclic crack advance is caused by irreversible deformation processes at the crack tip. Dislocations and vacancy dipoles nucleating in the direction of the activated slip system in front of the crack tip are absorbed by the crack during cyclic loading leading to crack growth. During this crack propagation, grain and phase boundaries may act as barriers for dislocations and thus hinder the further propagation of slip bands/shear bands and short cracks to the next grain. This leads to a decreasing crack growth rate near grain boundaries and is characteristic for the stage I crack growth regime.

The question now arises how hydrogen changes the deformation and crack growth mechanisms. To elucidate

* Corresponding author: volker.schippl@uni-siegen.de

the mechanisms responsible for the observed increasing fatigue crack growth rate in the presence of hydrogen, a metastable austenitic stainless steel is tested as received in air as well as in 10 MPa hydrogen atmosphere and in a condition precharged with hydrogen. With fatigue tests in these conditions, the short crack propagation mechanisms related to hydrogen can be further investigated and distinguished from those observed in laboratory air. The interpretation of these experimental observations will be supported by simulation studies on the short crack propagation. The simulation results will be validated by comparing the observed crack growth rates and crack paths in real microstructures. Furthermore, it is possible to relate these results to the proposed theories on the effects of hydrogen on the deformation and crack growth mechanisms.

A well accepted theory states a hydrogen-enhanced localized plasticity (HELP) mechanism [5]. Birnbaum et al. proposed that hydrogen "shields" the interaction between dislocations and short range stress fields and may also decrease the stability of dislocation dipoles. They further stated that this effect is pronounced where dislocations are closely packed e. g. at dislocation pile-ups. This is the case near grain boundaries or at the vicinity of a crack. Another established theory is the hydrogen-enhanced decohesion (HEDE) mechanism [6, 7, 8]. It is proposed that hydrogen e. g. residing at grain boundaries weakens the strength and facilitates cracking. It is further stated that this mechanism holds for hydrogen occupying interstitial normal lattice site where it decreases the interatomic potential. In 2004, Nagumo proposed the hydrogen-enhanced stress-induced vacancy (HESIV) mechanism [9]. The observed decreasing crack growth resistance in presence of hydrogen during static tensile tests in austenitic stainless steels AISI 304 and 316L is explained by the increasing density of voids at the crack tip [10]. For a more detailed description on different theories on the effect of hydrogen on deformation processes, the reader is directed to an overview by Lynch [11].

The proposed theories described above can be seen as the fundamental mechanisms explaining the influence of hydrogen on deformation processes in the atomic scale which causes hydrogen related deformation mechanisms at the mesoscopic scale e. g. when short crack growth processes are observed. An example of this theoretical approach is the explanation of increased fatigue crack growth rate in the long crack growth regime in carbon steel JIS-SM490B by the hydrogen-enhanced successive fatigue crack growth (HESFCG) model [12]. It is proposed that hydrogen accumulates at the crack tip causing a localization of slip deformation due to the HELP mechanism. This reduces the capability of the crack to blunt which further leads to successive crack growth.

The investigation of the short fatigue crack propagation in the presence of hydrogen builds upon previous work [13-17]. This study emphasises on relating the observed fatigue crack growth rates with respect to the testing configuration. With this approach, it is possible to reveal new insights into hydrogen-enhanced fatigue crack growth mechanisms at the mesoscopic scale and allows conclusions on the dominant fundamental hydrogen mechanisms acting in austenitic stainless steels.

In the following section a detailed description of the investigated material as well as on the experimental results is given. These observations lead to the presented short fatigue crack growth model used in this study which is solved by the boundary element method (BEM). The simulation results will be presented and discussed highlighting the acting hydrogen-enhanced short crack propagation mechanisms.

2 Experimental results

The investigated metastable austenitic stainless steel is based on the chemical composition of X2CrNi19-11 (AISI 304L) and is alloyed according to Table 1 where the yield strength R_{p02}, the calculated stacking fault energy (SFE) [18] as well as the M_{d30}-temperature [19] are given.

Three different fatigue tests were conducted: (i) on uncharged specimens in laboratory air, (ii) on precharged specimens in laboratory air and (iii) on uncharged specimens in 10 MPa hydrogen atmosphere. The fatigue tests on uncharged and precharged specimens have been carried out using a miniature testing machine, further described in [20], at a test frequency of $f = 1$ Hz, a stress ratio of $R = -1$ and with a stress amplitude $\sigma_a = 365$ MPa. The geometry of the specimen used is shown in Fig. 1a. The fatigue tests in 10 MPa hydrogen atmosphere were conducted on a specimen as shown in Fig. 1b applying a stress amplitude of $\sigma_a = 270$ MPa, a test frequency of $f = 1$ Hz and a stress ratio of $R = -1$.

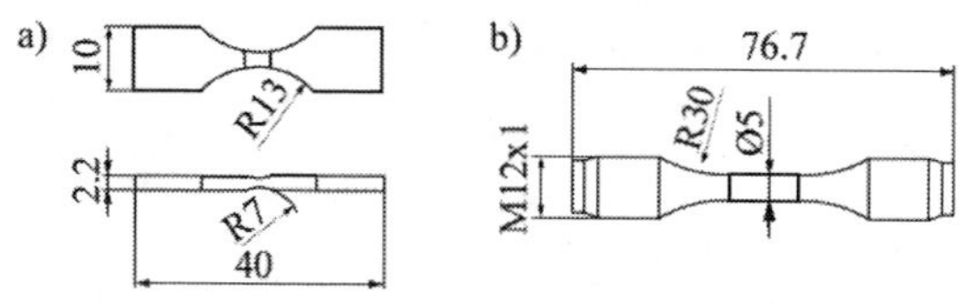

Fig. 1. (a) Geometry of the in situ specimen used for the uncharged and precharged fatigue tests and (b) specimen for fatigue tests in 10 MPa hydrogen atmosphere

The precharged specimens were exposed to 10 MPa hydrogen atmosphere at 180°C for 28 days. The hydrogen concentration was measured by the hot gas extraction method. A steady state hydrogen concentration of $c_H = 40$ wppm was reached resulting in a diffusion coefficient $D = 1.8 \cdot 10^{-8}$ cm²/s which is in a good agreement with obtained results on AISI 304 [21]. Fig. 2 shows the experimentally observed crack path in

Table 1. Chemical composition (wt. %), yield strength (R_{p02}), stacking fault energy (SFE) and M_{d30}-temperatures of the material

C	Si	Mn	P	S	Cr	Ni	Mo	Nb	Ti	N	R_{p02} [MPa]	SFE [mJ/m²]	M_{d30} [°C]
0.016	0.48	1.75	0.012	0.008	17.78	12.36	0.02	0.01	0.005	0.017	199	31.33	-85.91

the uncharged specimen (Fig. 2a-b), the precharged specimen (Fig. 2c-d) and the uncharged specimen in 10 MPa hydrogen atmosphere (Fig. 2e-f).

Cracks predominantly initated at grain boundaries in the uncharged specimen and in specimen in 10 MPa hydrogen atmosphere as often observed in uncharged austenitic stainless steels [22, 23]. In precharged specimens, the crack initation site shifted towards triple junctions [24]. As earlier mention, hydrogen residing at grain boundaries could weaken the interfacial strength leading to premature fracture. This effect is more pronounced at triple junctions, thus leading to facilitated crack initiation at these sites.

The crack further propagates through grains in uncharged and precharged specimens in the absence of any or with partially very confined martensite formation at the crack tip. The crack paths in specimens in 10 MPa hydrogen atmosphere cleary deviated from the crack paths observed in uncharged and in precharged specimens. The crack propgation is often restricted to grain boundaries. In uncharged and precharged specimens, the transgranular crack propagates by alternate sliding on two alternating slip systems as can be seen in Fig. 2a and c. These observations can be explained by the different distribution of hydrogen in the precharged specimens and in the specimens in 10 MPa hydrogen atmosphere. In the precharged condition, a homogeneous steady-state hydrogen concentration at the surface was established due to the charging conditions. Thus, the difference of the hydrogen occupying normal interstitial lattice sites compared to the hydrogen residing at the grain boundaries is less significant. It follows that neither trans- nor intergranular crack growth is favored and the crack path is comparable to uncharged specimens such that the crack initiation from triple junctions is followed by transgranular crack propagation.

For the specimens in 10 MPa hydrogen atmosphere facilitated hydrogen entry at grain boundaries may play a key role in explaining the different crack propagation paths. This results in a locally increasing hydrogen content at grain boundaries during cyclic loading. Following the HEDE mechanism, this promotes the crack propagation along grain boundaries.

It can also be seen that during cyclic loading a localization of plastic strain occurs as indicated by the emerging slip markings on the specimen surface as often observed for steels with low stacking fault energies. To determine the influence of hydrogen on the slip morphology, the average slip band heigth and average distance between slip bands were determined in the vicinity of the crack [24]. It could be observed that the maximum slip band heigth in precharged specimens (0.14 µm) was significantly increased as compared to measurements in uncharged specimens (0.08 µm). The average distance between slip bands increased from 1.92 µm in uncharged specimens to 4.7 µm in precharged specimens. Similar results could be obtained by Aubert et al. in AISI 316L [25] and by Abraham et al. in AISI 310s [26]. Both showed that in presence of hydrogen, an increase in slip band heigths and spacing of slip bands can be investigated. These observations further support the acting HELP mechanism in these metastable austenitic stainless steels. As hydrogen "shields" the interaction between dislocations and short range stress fields, the spacing between dislocation is reduced and the mobility of dislocations increases. This results in the observed increase in the maximum slip band heigth.

In the following section, the theory on modelling short fatigue crack propagation in the investigated austentic stainless steel is presented based on the mentionend experimental observations. The simulation emphasises on the deformation process in the vicinity of the crack. The influence of hydrogen on the crack propagation mechanisms will be revealed based on the simulation results of the predicted crack paths and crack growth rates of the experimentally observed short cracks depicted in Fig. 2.

3 Simulation model

In the simulation model, real microstructures are represented in the mesoscopic scale as a two-dimensional anisotropic elastic solid including microstructural deformation processes. The investigated microstructure consists of several grains with distinctive crystallographic orientations and crystallographic structure of the corresponding phase which further determine the orientation of the different slip systems.

The propagation of microstructurally short cracks is significantly affected by microstructural characteristics such as grain or phase boundaries, the predominant tendency of dislocation structure arrangements such as evolving persistent slip bands and phase transformations processes e. g. the deformation induced martensite formation. In the simulation, the crack path is predicted starting from a pre-existing initial crack located at the nucleation site determined by scanning electron microscope (SEM) images. To predict the crack growth

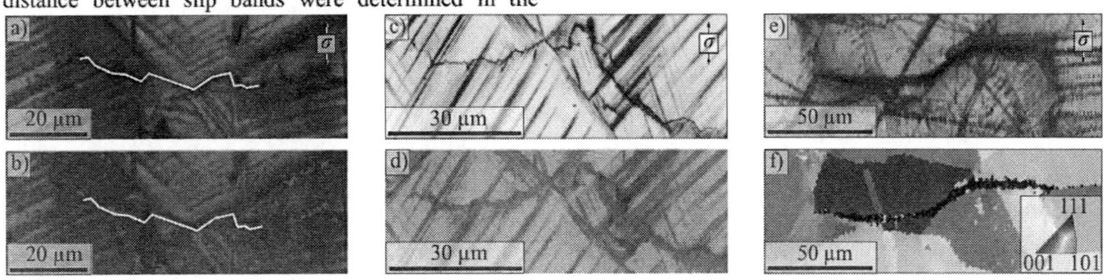

Fig. 2. (a) Crack morphology and (b) EBSD scan of the uncharged specimen tested in laboratory air (crack depicted as a white line), (c), (d) of the precharged specimen (c_H = 40wppm) and (e), (f) of the uncharged specimen in 10 MPa hydrogen atmosphere

in the simulation, two different crack propagation mechanisms have to be distinguished: (i) transgranular crack growth and (ii) intergranular crack growth.

The model for determining transgranular crack growth is based on the short crack growth model by Wilkinson et al. [4]. The crack growth rates of these microstructurally short cracks are calculated by relating the maximum and the minimum crack tip slide displacement during cyclic loading ($\Delta CTSD = CTSD^{max} - CTSD^{min}$) to the crack growth rate da/dN [27]

$$\frac{da}{dN} = C_{CTSD} \cdot \Delta CTSD \quad (1)$$

where the parameter C_{CTSD} is a measure for the irreversible deformation processes at the crack tip. The crack tip slide displacement (*CTSD*) is mainly driven by the shear deformation along slip bands nucleating at the crack tip. Thus, only these shear bands are modelled in the simulation including only short range effects on the crack deformation processes. To model the shear band evolution during cyclic loading, the nucleation of these shear bands as well as the sliding along these shear bands is explicitly modelled. For nucleating shear bands at the crack tip, it is assumed that a critical shear stress (τ^c) has to be exceeded. The plastic shear deformation along these shear bands is realised by applying an ideally elastic-plastic material law with a characteristic critical friction stress (τ^F). It has further to be mentioned that only up to two slip systems are allowed to nucleate at the crack tip in a grain as commonly experimentally observed, e. g. see Fig. 2c. A slide deformation along these shear bands significantly alters the crack tip slide displacement *CTSD* and the crack growth rate da/dN. With this approach, it is possible to simulate the experimentally observed decreasing crack growth rate when the crack tip approaches grain boundaries. The crack tip slide displacements as well as the crack growth rate decrease until the critical shear stress for slip transmission through grain boundaries (τ^c) is exceeded resulting in new shear band nucleation in the neighbouring grain and in a sudden additional plastic deformation along these newly created shear bands that further increases the crack tip slide displacement. Besides to the determination of the crack growth rate, the direction of the crack advance has to be modelled. It is assumed that the crack propagates in the direction of the activated shear bands by vectorially adding the corresponding deformation components along the active slip systems.

If the crack tip reaches the grain boundary, it can further propagate either along this grain boundary or into the next grain by activating new shear bands. This problem is solved by introducing the critical normal stress for crack growth along a grain boundary (σ^c) [15]. If the critical stress for nucleation of shear bands as wells as for crack growth along a grain boundary is exceeded, the condition $\xi = \tau^c/\sigma^c \cdot |\sigma^{GB}/\tau^{MRSS}|$ determines the new crack path with σ^{GB} as the calculated normal stress along the specific grain boundary and τ^{MRSS} represents the maximum resolved shear stress in the next grain. Thereby, for $\xi > 1$, the crack follows the grain boundary.

To model the crack growth along grain boundaries, a similar approach to eq. (1) is used. The crack growth rate in this case is calculated by the relative crack tip opening displacement ($\Delta CTOD$) [15]

$$\frac{da}{dN} = C_{CTOD} \cdot \Delta CTOD \quad (2)$$

where the parameter C_{CTOD} is again a measure for the irreversible deformation processes at the crack tip. The crucial parameter for crack growth along grain boundaries is the crack tip opening displacement. This approach can be justified by Fig. 2e where the crack propagates in the normal direction to the applied exterior load. This results in increasing normal stresses along the grain boundary such that the tangential component of the crack displacement becomes insufficient for describing the crack growth processes.

The plastic deformations at the vicinity of the grain boundary crack is modelled according to the Dugdale model [28]. The plastic zone is projected to yield strips along the active grain boundary [15]. The condition for yielding is met if the von-Mises stress (σ^{vM}) exceeds the cyclic flow stress (σ^F) of the material.

The crack growth mechanisms described above are assumed to be valid in the presence of hydrogen as can be observed in Fig. 2a-e. The simulation of short cracks based on the presented mechanisms will further reveal the acting hydrogen related mechanisms during short crack propagation.

To solve the model, an elastostatic simulation is carried out and solved numerically with the boundary element method (BEM). In this simulation, the fatigue loading is sequentially applied. During each loading step, the microstructure is iteratively solved and adjusted e. g. in the case of new shear bands nucleated at the crack tip or the iterative determination of the plastic deformation along shear bands or grain boundaries. The boundary element method used for this simulation is presented in the following section.

4 Numerical method

Using the boundary element method compared to the finite element method (FEM) has several advantages in the case for solving the short crack propagation. One obvious reason is the effort for discretization which is significantly less compared to the FEM. Especially when taking a closer look at the modelled deformation processes in the microstructure which are restricted to the representative shear bands such that only the solution at these sites are relevant. The elastic solution inside a grain is obtained by only discretizing the grain boundary (Γ^b). Furthermore, with the used BEM the relative displacements and stresses in the crack elements as well as along shear bands can directly be calculated. In the presented BEM, only the upper flank (Γ^{s+}) of the crack or a shear band is modelled. Thus, to calculate displacements and stresses, the following displacement and stress boundary integral equations have to be solved where greek subscripts take the values 1 or 2 and latin subscripts are in the range from 1 to 3.

$$c_{ij}u_i = \int_{\Gamma^b} \left[u_{ij}^* t_i - t_{ij}^* u_i\right] \mathrm{d}\Gamma^b + \int_{\Gamma^{s+}} t_{ij}^* \Delta u_i \mathrm{d}\Gamma^{s+} \qquad (3)$$

$$\sigma_{j\gamma} = \int_{\Gamma^b} \left[d_{ij\gamma}^* t_i - s_{ij\gamma}^* u_i\right] \mathrm{d}\Gamma^b + \int_{\Gamma^b} s_{ij\gamma}^* \Delta u_i \mathrm{d}\Gamma^{s+} \qquad (4)$$

By defining appropriate boundary conditions, eq. (3) is used to calculate the displacements u_i and the boundary stresses t_i along the discretized grain boundaries as well as the relative displacements Δu_i along the crack and shear band elements. In a post-processing stage, the stresses $\sigma_{j\gamma}$ inside the grains are determined. In eq. (3), u_{ij}^* and t_{ij}^* are the displacement and traction fundamental solutions both with a weak singularity and in eq. (4) $d_{ij\gamma}^*$, and $s_{ij\gamma}^*$ are the stress and the higher-order stress fundamental solutions. These are given by Wang [29] who derived these fundamental solutions for a general anisotropic elastic solid.

With the general plane stress assumption, the shear deformations and stresses in the slip bands can be calculated in the true direction of the activated slip system.

In the present study, the collocation formulation of the boundary element method is used. In contrast to the Galerkin formulation, the test function is defined by the Dirac function. The evaluation of the integrals in eq. (3) and (4) are obtained analytically by discretizing the grain boundary as well as the crack or shear bands by elements with quadratic shape functions [17].

In the case of modelling a polycrystalline microstructure the individual homogeneous grains have to be coupled at the neighbouring boundaries of each grain. In the BEM, this is achieved by enforcing the continuity conditions along these boundaries [30]. A similar approach is used for modelling cracks along grain boundaries such that relative displacements can directly be calculated [16].

By using the presented BEM, it is finally possible to represent and solve the crack growth propagation in a two-dimensional microstructure consisting of individual grains with different anisotropic elastic properties and the changing short crack growth mechanisms in the presence of hydrogen can be observed.

5 Simulation of HCF short crack growth in the presence of hydrogen

In the following, three different simulations on the short crack propagation are presented and compared with the experimentally observed crack path and crack growth rates, (i) in an uncharged specimen, (ii) in a specimen precharged with hydrogen and (iii) in a specimen tested in 10 MPa hydrogen atmosphere.

To reveal the influence of hydrogen on the short crack propagation, the following modelling strategy is used. At first, the presented crack growth model is validated in the uncharged specimen. Subsequently, simulations in the specimen precharged with hydrogen and in the specimen in 10 MPa hydrogen atmosphere is carried out with the same validated modelling and simulation parameters as in the uncharged case. By adjusting appropriate parameters such that the crack paths as well as the crack growth rates in the presence of hydrogen are correctly predicted, the changing short crack growth mechanisms related to hydrogen will be revealed.

The model parameters chosen to calculate the crack propagation in the uncharged condition are given in Table 2. These parameters can be calculated according to [13-15] with the endurance limit of the material defined as $\sigma^D = 250$ MPa and the experimentally obtained mean grain size of $D_{mean} = 205$ µm.

Table 2.
Model parameters of microstructural deformation processes

Parameter	Uncharged condition
τ^c	108 MPa
τ^F	80 MPa
σ^c	216 MPa
σ^F	410 MPa

Fig. 3 shows the simulation results obtained in the uncharged condition. In Fig. 3a, it can be seen that the predicted crack path in the simulation correlates well with the experimentally observed crack path. The parameters C_{CTSD} and C_{CTOD} describing the irreversible crack growth processes are determined by comparing the predicted crack growth rate with experimentally obtained

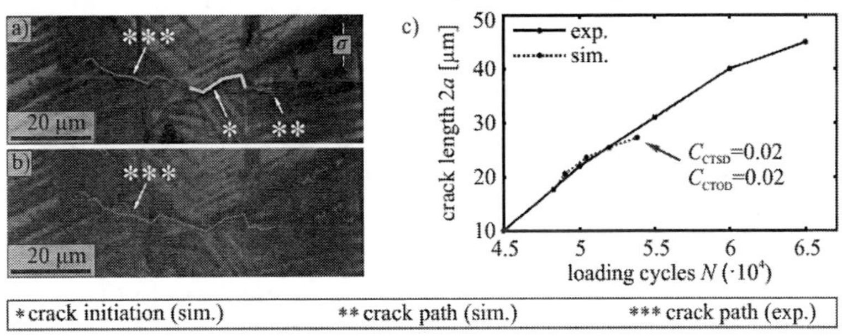

Fig. 3. (a) Crack morphology and (b) EBSD scan of the crack path in an uncharged specimen tested in laboratory air compared with the predicted crack path in the simulation starting from a predefined inital crack and (c) comparison of the calculated crack length $2a$ related the loading cycles N with experimentally obtained data

results. Similar values for the irreversibility constants are determined by Kübbeler [16] for short crack growth simulations in the metastable austenitic stainless steel X2CrNi18-9 ($C_{CTSD} = 0.018$). With these results, the assumptions in the crack growth model can be justified for this material such as the modelling of up to two representative shear bands at each crack tip. Based upon the simulation of the short crack propagation in the uncharged specimen, the changing short crack growth mechanisms in the presence of hydrogen can be discussed.

In Fig. 4, the simulation results of the short crack propagation in the presence of hydrogen are demonstrated. The model parameters given in Table 2 are not adjusted to predict the crack path in the precharged specimen (Fig. 4a). During the early phase of crack growth, the predicted crack path can reasonably be described, but later deviates from the experimentally observed crack path. This can be explained by taking a closer look at the activated shear bands and their deformation during the simulation as further emphasized in Fig. 4a. First of all, it can be seen that the simulation provides a good result of the activated shear bands in comparison with the observed slip markings. The amount of shear deformation along these simulated shear bands is depicted by the size of the straight lines. As stated before, the direction of the crack growth in the simulation is governed by the shear deformation and is obtained by vectorially adding these deformations. The deviating crack path during the later stage of crack growth for example at the left crack tip is a consequence of the underestimation of the slip deformation on the second activated shear band. Nevertheless, during early crack propagation a good agreement can also be obtained for the crack growth rate prediction (Fig. 4b). It is important to note that the irreversible processes at the crack tip are significantly increased in the presence of hydrogen as indicated by the increasing parameters C_{CTSD} as well as C_{CTOD}. Based on the short crack model by Wilkinson et al. [4] stating that dislocations and vacancy dipoles absorbed by the crack tip are the dominant mechanism for short crack growth, the presented results supports the theory of Birnbaum et al. [5] that hydrogen enhances these local processes at the crack tip. The prevailing HELP mechanism in the crack growth process in precharged austenitic stainless steels is further supported by [2, 31, 32].

Fig. 4c demonstrates the simulation result for the predicted crack path in 10 MPa hydrogen atmosphere. In Fig. 4d, the EBSD map of the microstructure is added to better distinguish the trans- and intergranular crack path. The calculated crack path could only correctly be described by significantly lowering the critical normal stress for crack growth along a grain boundary σ^c (from 216 MPa to 116 MPa). Similar to the results obtained in the precharged specimen, the irreversibility parameters have to be increased to achieve a reasonable correlation between the predicted and the experimentally observed crack growth (Fig. 4e). It has to be mentioned that the number of cycles leading to the initial crack size defined in the simulation could not accurately be detected. The most conservative assumption results in the definition that the crack initiated at $N = 0$. As a consequence, the effect of hydrogen on the crack growth is underestimated. Even though, the effect of hydrogen on increasing the crack growth rate is obvious (see Fig. 4e). Based on these results, it can be concluded that the short crack propagation in 10 MPa hydrogen atmosphere is

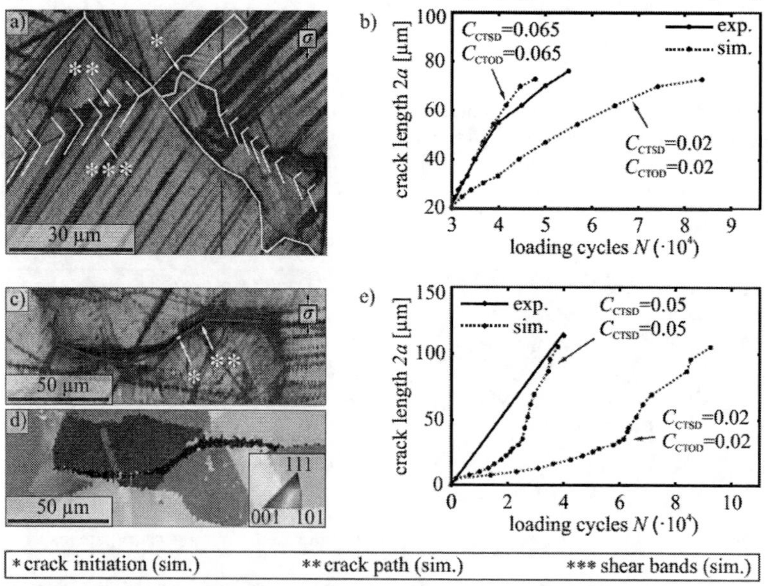

Fig. 4. (a) Crack morphology of the crack path in a precharged specimen ($c_H = 40$wppm) compared with the predicted crack path in the simulation starting from a predefined inital crack, (b) comparison of the calculated crack length $2a$ related the loading cycles N with experimentally obtained data and (c) crack morphology and (d) EBSD scan of the crack path in a specimen tested in 10 MPa hydrogen atmosphere compared with the predicted crack path in the simulation starting from a predefined inital crack and (e) comparison of the calculated crack length $2a$ related the loading cycles N with experimentally obtained data

increased by two mechanisms. At first, the transgranular crack growth is facilitated by the aforementioned hydrogen-enhanced irreversible crack growth processes at the crack tip. And on the other hand, the cracking along the grain boundaries is promoted by hydrogen. This can be explained by the HEDE model. The entry of hydrogen at the grain boundaries is facilitated. As grain boundaries can act as strong barriers, hydrogen atoms tend to reside at these sites leading to an increased local hydrogen concentration. This results in weakening the strength of grain boundaries and promoting hydrogen related cracking, as observed in [1]. Thus, the cause for the increasing irreversibility C_{CTOD} is the effect of weakened grain boundary strength rather than facilitated irreversible plastic deformation processes at the crack tip as in the case for transgranular crack growth along shear bands.

7 Conclusions

In this study, the cyclic short crack propagation of a metastable austenitic stainless steel was investigated in an uncharged specimen, in a specimen precharged with hydrogen and in a specimen in 10 MPa hydrogen atmosphere. The experimental observations were supported by simulations on the short crack growth in these different conditions. From these results, it could be found that the prevailing hydrogen mechanism for short crack growth is dependent on the presence of hydrogen.

1. In precharged specimen, the short crack propagation is predominantly transgranular. The irreversible plastic deformations at the crack tip are locally enhanced by hydrogen according to the hydrogen-enhanced localized plasticity (HELP) model. Birnbaum et al. [5] proposed that hydrogen leads to decreased interactions between dislocations and short range stress fields. Based on the simulation results, it could be found that hydrogen enhances the short crack growth mechanisms as stated by Wilkinson et al. [4] where the crack propagates by absorbing dislocations or vacany dipoles at the crack tip and further results in an increased fatigue crack growth rate.
2. In specimen tested in 10 MPa hydrogen atmosphere, the short crack propagation is predominantly intergranular. The simulation supports the prevailing hydrogen-enhanced decohesion (HEDE) model such that hydrogen promotes grain boundary cracking in the investigated metastable austenitic stainless steel. Furthermore, hydrogen also increases the irreversible crack growth process in the case for transgranular crack propagation in this condition.

Acknowledgement

The authors gratefully acknowledge financial support of this study by Deutsche Forschungsgemeinschaft (DFG) and Forschungsvereinigung Verbrennungskraft-maschinen e.V. (FVV). Reference numbers: CH92/48-1 (DFG) and 1185 (FVV).

References

1. S. Fukuyama, K. Yokogawa, K. Kudo, M. Araki, Trans. Jpn. Inst. Met., **26**, 5, 325-331 (1985)
2. Y. Murakami, T. Kanezaki, Y. Mine, S. Matsuoka, Metall. Mater. Trans. A, **39A**, 1327-1339 (2008)
3. H. H. Heitmann, H. Vehoff, P. Neumann, Pergamon Press Ltd. **5**, 599-606 (1984)
4. A. J. Wilkinson, S. G. Roberts, Scripta Mater. **35**, 1365-1371 (1996)
5. H. K. Birnbaum, P. Sofronis, Mater. Sci. Engng. **A176**, 191-202 (1994)
6. A. R. Troiano, Trans. ASM **52**, 54-80 (1960)
7. R. A. Oriani, P. H. Josephic, Acta Metall. **25**, 979-988 (1977)
8. R. A. Oriani, Corrosion **43**, 390-397 (1987)
9. M. Nagumo, Mater. Sci. Tech. **20**, 940-950 (2004).
10. M. Hatano, M. Fujinami, K. Arai, H. Fujii, M. Nagumo, Acta Mater. **67**, 342-353 (2014)
11. S. Lynch, Corrosion Reviews **30** (3-4), 105-123 (2012)
12. S. Matsuoka, N. Tsutsumi, Y. Murakami, Trans. Jpn. Soc. Mech. Eng. A **74** (748), 44-53 (2008)
13. A. Schick, Dissertation, Universität Siegen, 2004
14. B. Künkler, Dissertation, Universität Siegen, 2007
15. P. Köster, Dissertation, Universität Siegen, 2015
16. M. Kübbeler, Dissertation, Universität Siegen, 2017
17. P.-M. Hilgendorff, Dissertation, Universität Siegen, 2017
18. F. B. Pickering, Proc. Conf. Stainl. Steels **84**, 2-28 (1984)
19. K. Nohara, Y. Ono, N. Ohashi, J. Iron Steel Inst. Jpn. **63**, 212-222 (1977)
20. I. Roth, Dissertation, Universität Siegen, 2011
21. T.-P. Perng, C. J. Altstetter, Acta metall. **34** (9), 1771-1781 (1986)
22. A. Heinz, P. Neumann, Acta Met. Mat. **38**, 1993-1940 (1990)
23. I. Roth, M. Kübbeler, U. Krupp, H.-J. Christ, C.-P. Fritzen, Proc. Engn. **2** (1), 941-948 (2010)
24. S. Brück, V. Schippl, C.-P. Fritzen, H.-J. Christ, M. Schwarz, S. Weihe, ASME, 297-306 (2017)
25. I. Aubert, N. Saintier, J. M. Olive, Scripta Mater. **66** (9), 698-701 (2012)
26. D. P. Abraham, C. J. Altstetter, Metall. Mater. Trans. **26A**, 2859-2871 (1995)
27. A. Navarro, E. R. de los Rios, Phil. Mag. A **57** (1), 15-36 (1988)
28. D. S. Dugdale, J. Mech. Phys. Solids **8**, 100-104 (1960)
29. C.-Y.Wang, Int. J. Solids Struct. **31**, 2591-2597 (1994)
30. M. Kübbeler, I. Roth, U. Krupp, C.-P. Fritzen, H.-J. Christ, Eng. Fract. Mech. **78**, 462-468 (2011)
31. K. Nygren, A. Nagao, P. Sofronis, I. M. Robertson, ASME, 270-277 (2017)
32. M. L. Martin, P. Sofronis, I. M. Robertson, T. Awane, Y. Murakami, Int. J. Fat. **57**, 28-36 (2013)

Effect of heat treatment on fatigue crack initiation of laser powder bed fusion stainless steel 316L

Meng Zhang[1], *Chen-Nan* Sun[2], *Xiang* Zhang[3], *Phoi Chin* Goh[4], *Jun* Wei[2], *David* Hardacre[4], and *Hua* Li[1,*]

[1]Singapore Centre for 3D Printing, School of Mechanical and Aerospace Engineering, Nanyang Technological University, 50 Nanyang Avenue, Singapore 639798
[2]Singapore Institute of Manufacturing Technology, ASTAR, 73 Nanyang Drive, Singapore 637662
[3]Faculty of Engineering, Environment & Computing, Coventry University, Coventry CV1 5FB, UK
[4]Lloyd's Register Global Technology Centre, 1 Fusionopolis Place, #09-11 Galaxis, Singapore 13852

Abstract. Laser powder bed fusion (L-PBF) is an emerging additive manufacturing technique for building structural components. L-PBF processing defects, such as lack of fusion pores, promote fatigue crack initiation and shorten the fatigue life. With well-controlled processing, critical pores can be avoided such that the microstructure-driven intergranular crack initiation mode becomes operative. In this work, the fatigue crack initiation behaviours of as-built and solution annealed L-PBF stainless steel 316L were studied. Crack initiation of the as-built samples is driven by de-bonding of the dendritic grain boundaries. High temperature annealing results in the formation of thermally-induced defects, possibly via the reheat cracking mechanism and the nucleation of pre-existing gas pores. As heat treating could have led to recrystallization and annihilation of the original grain boundary defects, the thermally-induced defects became the new sites for crack initiation. In addition, heat treatment incurred significant reduction in yield strength, such that the interaction of fatigue and ratcheting strain accumulation dominated the deformation behaviour of the material. The resulting fatigue strength in the finite life regime was reduced by about 13% but the fatigue endurance limit was not affected.

1 Introduction

In laser powder bed fusion (L-PBF) processing, parts are built layer-by-layer via the fusion of metal powder by high power laser. Laser movements are controlled based on input from computer-aided design models such that near-net-shape parts could be produced. Such design and manufacturing flexibility makes L-PBF the ideal platform for various creative applications, including high performance lattice structures [1], sophisticated components such as injection moulding tools [2] and novel heat transfer devices [3], as well as customised medical products [4, 5].

The implementation of this process for actual part production is met with difficulties rooted in the complex processing-structure-property relations. The L-PBF technique is affected by a large number of variabilities, including laser settings, power characteristics, build orientations etc., all of which could affect the evolution of microstructure, trigger the formation of processing-defects and impact the mechanical performance [6-8]. Considering the current lack of field experience with additive manufactured parts and appropriate non-destructive inspection method for detecting micron-size defects [9], questions remain in the integrity and reliability of L-PBF components, especially if they were to be used for safety-critical applications.

For conventionally-manufactured parts, such as welded [10] and cast [11] structures, heat treatment is often employed as the remediation measure for promoting uniformity of composition and microstructure, and to prevent unexpected part failure. The applicability of this approach to L-PBF needs to be examined in relation to the nature of the process-induced microstructure and defects.

For instance, as a result of rapid cooling, high residual stress, exceeding the material's yield strength, had been observed in as-built L-PBF materials [12]. Local plastic deformation could lead to geometrical distortion and cracking [13]. Besides preheating the build platform for reducing the thermal gradient during processing [14], post-processing stress-relieving heat treatment [15] could also be performed to ameliorate the residual stress. Such procedures had been shown to be effective in improving the fracture toughness of Ti6Al4V [16] and the tensile properties of a nickel-base superalloy [17].

Another key processing issue is porosity. Deviations from the optimum build settings and poor process maintenance, e.g. laser miscalibration or the use of damaged recoater blade, could affect the energy available for consolidating the metal powder and promote the formation of lack of fusion defects. As these defects are formed unintentionally and thereby difficult to be controlled, they remain in the as-built parts and increase the risk of component failure. Results from experimental fatigue test had shown that un-melted regions are the primary cause for fatigue crack initiation

* Corresponding author: lihua@ntu.edu.sg

 https://doi.org/10.1051/matecconf/201816522006

and degraded fatigue strength [18-20]. In response, hot isostatic pressing (HIP), which involves the simultaneous application of heat and pressure, could be employed for removing residual porosity. The fatigue resistance of materials such as TiAl6V4 [21, 22] was found to improve after HIP because of the elimination of failure-critical lack of fusion defects. However, materials with complex microstructures, e.g. Inconel 718 [23] and maraging steel [24], could be prone to failure from second phase particles, and thereby require tailored post-processing heat treatments.

Stainless steel 316L is one of the highly investigated alloys that showed good manufacturability via the L-PBF process. Its fatigue properties are of general interest because of the material's suitability for various fatigue-related applications, e.g. biomedical devices [25]. Prior works on this material showed that its fatigue resistance is not significantly altered by low temperature stress-relieving heat treatment [26, 27]. To further understand the influence and applicability of heat treatment, higher temperature annealing was performed in this work for examining the effect of more pronounced changes in microstructure on the fatigue failure behaviour of this alloy.

2 Experimental setup

Samples were built on a L-PBF machine (EOS M290) equipped with a 400 W Yb-fiber laser. Commercial stainless steel 316L powder with a predominantly spherical shape was used as the raw material for processing. L-PBF build parameters were optimised for a layer thickness of 20 μm and built on a preheated platform.

As shown in Figure 1a, samples were fabricated as rectangular blocks, with the long edge being oriented at a small angle from the recoating direction to prevent uneven powder deposition due to interaction between the build part and the recoater blade [28]. After being removed from the baseplate, the blocks were machined into test specimen geometry by electrical discharge machining (EDM) wire cutting. Eight specimens were sliced from each block along the height such that the loading directions of the specimens are perpendicular to the build direction. Flat fatigue test specimens were designed according to requirements of the ASTM E466 and the dimensions are as shown in Figure 1b.

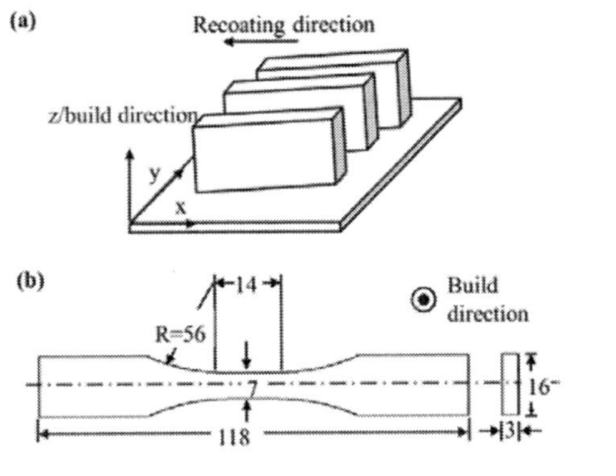

Fig. 1. (a) Orientation of rectangular blocks on the L-PBF build platform. (b) Geometry of the fatigue test specimen. Eight test specimens were machined from each rectangular block by wire cutting.

Some as-built specimens were subjected to annealing heat treatment in accordance with AMS2750 at a holding temperature of 1093 °C for 25 mins in a vacuum furnace and subsequently gas quenched. Final grinding step was performed prior to mechanical testing to remove any surface contaminants that could have been generated from the heat treatment.

Load-controlled fatigue tests were performed by applying sinusoidal loading at frequency of 5 Hz and load ratio of 0.1 on an MTS 810 hydraulic testing machine under ambient condition. Specimens that did not fail after 10^6 cycles were considered as runouts. For microstructural analysis, specimen cross sections parallel to the build direction were polished and observed using optical microscope. Scanning electron microscope (SEM) was employed for examination of the fracture surfaces.

3 Results and discussion

3.1 Crack-initiating defects

SEM fracture images of the as-built and heat-treated samples are displayed in Figure 2, where the crack initiation zones are marked by dotted lines. The as-built sample is characterised by intergranular crack initiation, where fine solidification dendrites, or sub-grains, are clearly shown in Figure 2b. Cracking of this nature is indicative of intergranular brittleness which, for L-PBF stainless steel 316L, could be attributed to the presence of high dislocation density and elemental segregation at grain boundaries. Specifically, studies done by Zhong et al. [29] had shown that the cellular boundaries are enriched with molybdenum atoms and nano-sized inclusions, which could create local stress concentration and promote cracking. Details of the fatigue fracture mechanisms for as-built L-PBF stainless steel 316L could be found in the authors' prior work [30].

A different type of defect was observed at the crack origin of the heat-treated sample (Figure 2c). The defect adopts a flat and 'compressed' appearance; the distinct line marking the boundary between the defect and the crack propagation region indicates that the defect could be pre-existing in the material after heat treatment.

To further analyse the origins of the defects, optical micrographs of the samples were taken, as shown in Figure 3. For the as-built condition (Figure 3a), small isolated pores with equivalent circular diameters on the order of 5 μm are present. Pores of this nature are unavoidable for the L-PBF process due to the entrapment of gas bubbles, which could have originated from the evaporation of low melting point elements or oxygen gas pre-existing in the powder feedstock [31]. Relative to the grain boundary defects in Figure 2b, such process-

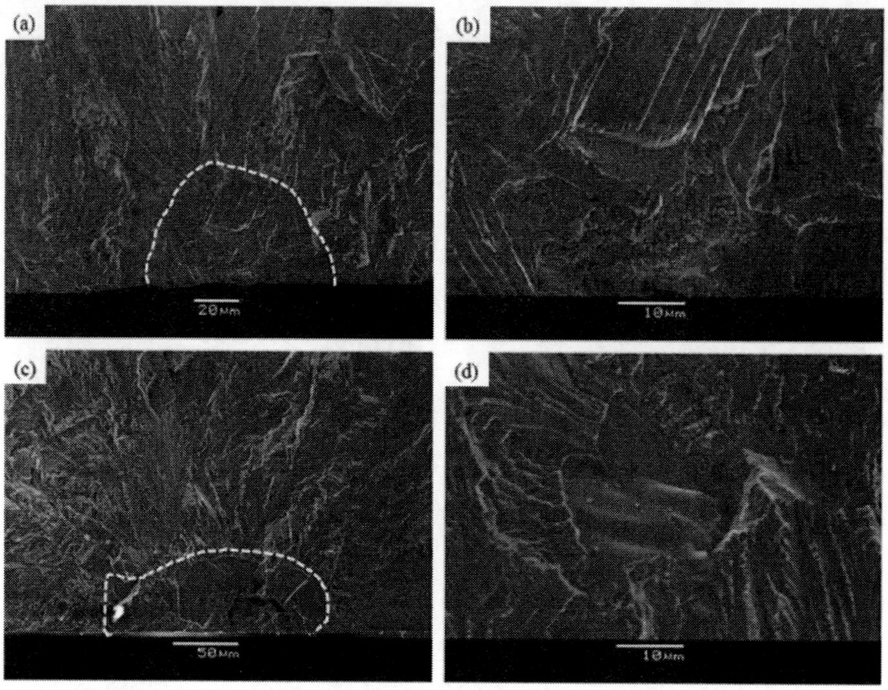

Fig. 2. SEM fracture images of samples subjected to cyclic loads. (a) Crack initiation region (dotted line) of as-built sample; (b) enlarged view of (a) showing solidification dendrites revealed by intergranular cracking; (c) crack initiation region of a heat-treated sample; (d) similar but smaller defect in the interior of a heat-treated sample.

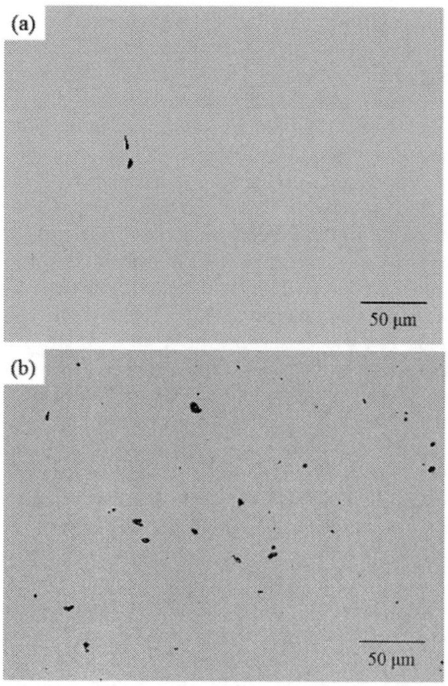

Fig. 3. Optical micrographs of the (a) as-built and (b) heat-treated samples. Black pixels indicate voids.

induced pores are smaller in size and could have incurred weaker stress concentration under the cyclic loading, such that they were not critical in causing crack initiation for samples in the as-built condition.

Defect condition worsened after heat treatment. As shown in Figure 3b, irregularly-shaped voids are prevalent in the heat-treated sample. While the increase in size of the defects is marginal (from about 5 µm to 10 µm), there is a significant increase in number. These defects are smaller in comparison with that found at the fatigue fracture origin in Figure 2c. To verify if they are the same type of defects, the fractographs were examined more closely. Similar defects of smaller size were found in the interior of the sample, as shown in Figure 2d. Since fatigue failure seeks the largest-occurring defect, it is possible that the use of optical microscope for inspection had missed detecting the larger defects. Moreover, quantification of three dimensional defect by the two dimensional microscopic method is affected by stereological error. As the defects are flat and thin, polishing on sections perpendicular to the thickness direction could have easily caused them to be removed. Hence, it is possible that the crack-initiating defects are the same type of defect as those shown on the optical micrograph, and are therefore generated by the heat treatment.

This result indicates that the currently tested heat treatment condition produces defects that are detrimental to fatigue failure. Specifically, as the high temperature heat treatment could have caused the annihilation of dislocations at the grain boundaries [32] and recrystallization, any remnant intergranular defects were no longer detrimental to cracking. As a result, the thermally-induced defects became operative and initiated fatigue cracks. The possible mechanisms for the formation of the thermally-induced defect is explained in the next section.

3.2 Mechanisms of defect formation

For welded austenitic stainless steel structures containing heat affected zones (HAZ), reheating is known to produce intergranular cracking [33, 34]. The relaxation of residual stress is opposed by the bulk of the grains, leading to the accumulation of intense plastic strain at the grain. When deformation occurs via the sliding of grain boundaries, grain boundary cracks would be formed.

The defects observed in the heat-treated samples could potentially be attributed to such reheat cracking mechanism. Characteristics of the L-PBF stainless steel 316L microstructure render the material susceptible to reheat cracking, as discussed in the following paragraphs.

Metallurgically, the walls of the cellular grains in L-PBF stainless steel 316L are packed with a high concentration of dislocation network [35]. While the high dislocation density is desirable in terms of creating a strengthening effect by resisting dislocation motion, locally, it could lead to reduced ductility and embrittled cellular grain boundaries, as reflected in the intergranular crack initiation behaviour of the as-built samples. With rising temperature during annealing, residual stress could not be easily released by intragranular slip as it is impeded by the pre-existing dislocation network. The release of residual stress thus occurs intergranularly across the cellular grain boundaries, promoting the formation of cracks.

Mechanically, L-PBF involves rapid solidification which induces high residual stress levels. For materials with solidification microstructures, local variations in residual stress could be present due to difference in material phases and thermal history [36]. As the consolidation process in L-PBF is characterised by competitive grain growth at the solid-liquid interface [37], local differences in stress state are inevitable. Moreover, with successive laser passes, regions of the material experience repeated heating and cooling such that residual stress of different extent is generated along the build direction [38]. Since the relaxation of residual stress during heat treatment is accompanied by elastic recovery of various regions in the bulk material, it promotes the opening of high stress concentration areas [33] such as pre-existing defects, in this case grain boundaries or gas pores. As temperature in the furnace ramped up to the holding temperature during annealing, stress relaxation could have caused intense plastic strain at the tip of the defects and provoked cracking.

Alternatively, the fusion of pre-existing porosity could also have resulted in defects enlargement after heat treatment. However, as the coalescence of pores under such circumstance takes place via the mechanism of surface tension and thermal equilibrium [39], the resulting porosity tends to be spherical in shape, rather than as the flat and irregular defects observed in this work. Nevertheless, it is possible that neighbouring pores could have diffused into the reheat cracks, expanding the space between the cracked surfaces. Adjacent intercellular spaces could also have diffused out and coalesced, contributing volume to the reheat cracks.

3.3 Influence on high cycle fatigue properties

Figure 4a shows the S-N curves for the as-built and heat-treated samples adapted from the authors' prior work [40]. Heat treatment caused a near parallel downward shift of the S-N curve, where the fatigue lives and fatigue strength of the heat-treated samples are 60% and 13% lower than the as-built samples respectively. The fatigue limit, as represented by the stress levels of the runout samples, however are similar regardless of heat treatment.

As besides impacting the crack initiation mechanism via the generation of failure-critical defects, heat treatment also alters the material microstructure which influences the fatigue deformation behaviour, the effects of both factors thus need to be considered for assessing the observed high cycle fatigue properties. For stainless steel 316L, annealing causes the breaking down of the pre-existing intercellular dislocation network which is instrumental in providing the high strength of the material [35]. For the samples tested in this work, the yield strength of the as-built samples is about 590 MPa and the elongation to failure is 43%; after heat treatment, yield strength dropped by nearly 30%, to 420 MPa, while ductility increased by approximately the same extent to 56% [40].

Since the samples were tested over the same stress range, plastic deformation and the accumulation of plastic strain could have taken place via the ratcheting mechanism in some of the heat-treated samples. To analyse this behaviour, the ratcheting strain ϵ_r was calculated using the relation:

$$\epsilon_r = 0.5(\epsilon_{max} + \epsilon_{min})$$

where ϵ_{max} and ϵ_{min} are the maximum and minimum strain values of a particular cycle.

Figure 4b-c shows the plot of ϵ_r against the cycles to failure values of samples tested at maximum applied cyclic stress of 438 MPa and 365 MPa respectively. It can be seen that under the higher stress condition, the as-built and heat-treated samples exhibit distinctly different cyclic strain responses. For the as-built sample, strain stabilised after a relatively short life and remained constant throughout before catastrophic failure took place. The heat-treated sample, however, shows continuously increasing strain with the progression of asymmetric cycling. For the lower stress condition, ratcheting strain stabilised for both as-built and heat-treated samples, and failure did not occur when the tests were terminated at 10^6 cycles.

The behaviour of the heat-treated sample under the high stress loading is associated with ratcheting strain accumulation, where under stress-controlled cyclic loading with positive mean stress, inelastic strain builds up with each cycle. Besides fatigue damage, which arose from the alternating stress, permanent deformation in the direction of loading also took place in the bulk material, increasing crack density and promoting necking of the sample. Locally, at the crack-initiating defect, ratcheting can trigger microcrack nucleation and shorten the fatigue initiation life [41]. As heat treatment caused significant strength degradation, the applied cyclic stress exceeded the yielding stress, giving rise to the static deformation mode.

It is interesting to note that in comparison with the as-built condition, heat treatment incurred much lower reduction in fatigue strength (13%) than yield strength (30%), and most of the heat-treated samples survived long lives, i.e. > 10^5 cycles, despite the ratcheting mechanism. This could be attributed to the higher ductility after heat treatment, which partially offset the effect of strength reduction on fracture toughness and allowed the material to sustain greater damage prior to fracture.

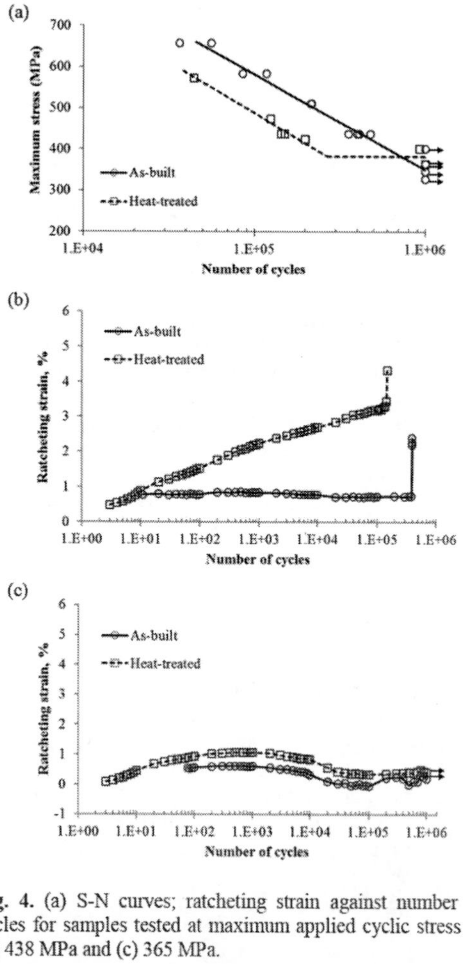

Fig. 4. (a) S-N curves; ratcheting strain against number of cycles for samples tested at maximum applied cyclic stress of (b) 438 MPa and (c) 365 MPa.

When the cyclic stress is lower than the yield strength, strain accumulation did not take place for both the as-built and heat-treated samples as global deformation is mostly in the elastic domain. This explains the similar fatigue limits in spite of heat treatment. In fact, crack growth experiments done by Riemer et al. [26] showed that L-PBF stainless steel 316L subjected to high temperature thermal treatment in the form of HIP has better near threshold performance because of coarsened grains, which act as obstacles to crack growth. Hence, it is expected that the long life fatigue properties of the heat-treated material are to be comparable, if not superior to the as-built form.

In summary, high temperature heat treatment of L-PBF stainless steel 316L generates critical fatigue failure defects and microstructure transformation, both of which affect the high cycle fatigue properties. In the finite life regime, fatigue deformation is complemented by ratcheting strain accumulation as a result of the reduced strength after heat treatment; the effect of crack initiating defects could play a secondary role in altering the fatigue stress-life relation since failure-critical defects are present in both the as-built and the heat-treated forms. In the low stress high cycle regime, where deformation is dominated by fatigue activities, the fatigue properties of both types of samples converge, leading to similar long life performance. Further experiments could be performed to evaluate the effect of thermally-induced defect and microstructure on the fatigue crack initiation and failure behaviours in the very high cycle fatigue regime.

4 Conclusions

The high cycle fatigue properties and crack initiation behaviours of L-PBF stainless steel 316L in the as-built and solution-annealed conditions were examined in this work. Based on the experimental results, it can be concluded that:

1. High temperature annealing leads to the formation of thermally-induced defects which become operative under high cycle fatigue loading and cause fatigue crack initiation.
2. The thermally-induced defects could have been formed due to reheat cracking during residual stress relaxation. The diffusion and coalescence of neighbouring porosities and intercellular spaces could also have led to the enlargement of the reheat cracks. The critical annealing conditions, in terms of temperature and duration, leading to the formation of the defects remain to be further investigated.
3. The annealing condition tested in this work produces simultaneous effects on crack initiation and the bulk deformation behaviour. Fatigue-ratcheting interactions were responsible for the fatigue properties of the heat-treated samples under the tested stress condition and the reduced fatigue resistance. Better fatigue properties are expected for the heat-treated samples in the long life regime due to reduced intergranular brittleness and coarsened grains which impede crack growth.

Acknowledgement

This work was supported by the Singapore Economic Development Board (EDB) Industrial Postgraduate Programme (IPP).

References

1. C. Yan, L. Hao, A. Hussein, and D. Raymont, Int. J. Mach. Tools Manuf., **62**, 32-38 (2012)
2. P. Razvan and P. Ancuta, *New Trends in 3D Printing*, (InTech, 2016).
3. M. Wong, I. Owen, C.J. Sutcliffe, and A. Puri, International Journal of Heat and Mass Transfer, **52**(1), 281-288 (2009)
4. B. Vandenbroucke and J.-P. Kruth, Rapid Prototyping Journal, **13**(4), 196-203 (2007)
5. M. Javaid and A. Haleem, Alexandria Journal of Medicine, (2017)
6. P. Hanzl, M. Zetek, T. Bakša, and T. Kroupa, Procedia Engineering, **100**, 1405-1413 (2015)
7. K. Guan, Z. Wang, M. Gao, X. Li, and X. Zeng, Mater. Des., **50**, 581-586 (2013)
8. G. Yu, D. Gu, D. Dai, M. Xia, C. Ma, and K. Chang, Appl. Phys. A, **122**(10), 891 (2016)
9. M. Gorelik, Int. J. Fatigue, **94**, 168-177 (2017)
10. X. Huang, M.C. Chaturvedi, and N.L. Richards, Metall. Mater. Trans. A, **27**(3), 785-790 (1996)
11. M.H. Bina, *Metallurgy - Advances in Materials and Processes*, (InTech, Rijeka. 2012).
12. P. Mercelis and J.-P. Kruth, Rapid Prototyping Journal, **12**(5), 254-265 (2006)
13. L. Parry, I.A. Ashcroft, and R.D. Wildman, Additive Manufacturing, **12**, 1-15 (2016)
14. K. Kempen, B. Vrancken, L. Thijs, S. Buls, J. Van Humbeeck, and J.-P. Kruth. in *Solid Freeform Fabrication Symposium Proceedings*. 2013)
15. J.J. Lewandowski and M. Seifi, Annu. Rev. Mater. Res, **46**, 14.1-14.36 (2016)
16. V. Cain, L. Thijs, J. Van Humbeeck, B. Van Hooreweder, and R. Knutsen, Additive Manufacturing, **5**, 68-76 (2015)
17. T. Vilaro, C. Colin, J.D. Bartout, L. Nazé, and M. Sennour, Mater. Sci. Eng. A, **534**, 446-451 (2012)
18. E. Wycisk, A. Solbach, S. Siddique, D. Herzog, F. Walther, and C. Emmelmann, Physics Procedia, **56**, 371-378 (2014)
19. T.M. Mower and M.J. Long, Mater. Sci. Eng. A, **651**, 198-213 (2016)
20. A. Yadollahi, N. Shamsaei, S.M. Thompson, A. Elwany, and L. Bian, Int. J. Fatigue, (2016)
21. D. Greitemeier, F. Palm, F. Syassen, and T. Melz, Int. J. Fatigue, **94**, 211-217 (2017)
22. G. Kasperovich and J. Hausmann, J. Mater. Process. Technol., **220**, 202-214 (2015)
23. V. Popovich, E. Borisov, A. Popovich, V.S. Sufiiarov, D. Masaylo, and L. Alzina, Mater. Des., **131**, 12-22 (2017)
24. Y. Bai, Y. Yang, D. Wang, and M. Zhang, Mater. Sci. Eng. A, (2017)
25. M.M. Dewidar, K.A. Khalil, and J. Lim, Transactions of Nonferrous Metals Society of China, **17**(3), 468-473 (2007)
26. A. Riemer, S. Leuders, M. Thöne, H. Richard, T. Tröster, and T. Niendorf, Eng. Fract. Mech., **120**, 15-25 (2014)
27. S. Leuders, T. Lieneke, S. Lammers, T. Tröster, and T. Niendorf, J. Mater. Res., **29**(17), 1911-1919 (2014)
28. P.J. da Silva Bartolo, A.C.S. de Lemos, A.M.H. Pereira, A.J.D.S. Mateus, C. Ramos, C. Dos Santos, D. Oliveira, E. Pinto, F. Craveiro, and H.M.C. da Rocha Terreiro. in *Proceedings of the 6th International Conference on Advanced Research in Virtual and Rapid Prototyping*. (Portugal, 2013)
29. Y. Zhong, L. Liu, S. Wikman, D. Cui, and Z. Shen, J. Nucl. Mater., **470**, 170-178 (2016)
30. M. Zhang, C.-N. Sun, X. Zhang, P.C. Goh, J. Wei, D. Hardacre, and H. Li, Mater. Sci. Eng. A, **703**, 251-261 (2017)
31. B. Zhang, Y. Li, and Q. Bai, Chinese Journal of Mechanical Engineering, **30**(3), 515-527 (2017)
32. K. Saeidi, X. Gao, Y. Zhong, and Z.J. Shen, Mater. Sci. Eng. A, **625**, 221-229 (2015)
33. A. Vinckier and A. Dhooge, J. Heat. Treat., **1**(1), 72-80 (1979)
34. R. Skelton, I. Goodall, G. Webster, and M. Spindler, Int. J. Press. Vessels Pip., **80**(7), 441-451 (2003)
35. L. Liu, Q. Ding, Y. Zhong, J. Zou, J. Wu, Y.-L. Chiu, J. Li, Z. Zhang, Q. Yu, and Z. Shen, Mater. Today, (2017)
36. P.J. Withers and H.K.D.H. Bhadeshia, Mater. Sci. Technol., **17**(4), 366-375 (2001)
37. A.R.A. Dezfoli, W.-S. Hwang, W.-C. Huang, and T.-W. Tsai, Scientific Reports, **7**, 41527 (2017)
38. Y. Liu, Y. Yang, and D. Wang, Int. J. Adv. Manuf. Technol., **87**(1-4), 647-656 (2016)
39. H. Zhao and T. DebRoy, Welding Journal, **80**(8), 204-210 (2001)
40. M. Zhang, C.-N. Sun, X. Zhang, P.C. Goh, J. Wei, H. Li, and D. Hardacre, JOM, **70**(3), 390–395 (2018)
41. P. Lukáš and L. Kunz, Int. J. Fatigue, **11**(1), 55-58 (1989)

https://doi.org/10.1051/matecconf/201816522007

Identification of fatigue damage evolution in 316L stainless steel using acoustic emission and digital image correlation

Farhan Tanvir[1,2*], *Tariq* Sattar[1], *David* Mba[3], *Graham* Edwards[4], *Elvin* Eren[4], and *Yoann* Lage[5]

[1]London South Bank University, 103 Borough Rd, London SE1 0AA, UK
[2]NSIRC, TWI Ltd, Granta Park, Great Abington, Cambridge, CB21 6AL, UK
[3]De Montfort University, Gateway House, Leicester LE1 9BH, UK
[4] TWI Ltd, Granta Park, Great Abington, Cambridge, CB21 6AL, UK
[5]Mistras Group Ltd, Norman Way, Over, Cambridge CB24 5QE, UK

Abstract. One of the main objectives of Acoustic Emission (AE) monitoring is to identify approaching critical stage of damage in the structure before it fails. State-of-the-art AE analysis is done on the features in both the time and frequency domains. Many features such as centroid frequency, duration, rise-time, count and energy are dependent on acquisition settings; threshold and timing parameters. Incorrect acquisition settings may result in inaccurate classification of the AE source. This work proposes a new feature in the time domain signal based on 2nd order Renyi's entropy, which proves to be efficient in identifying different stages of damage. Renyi's entropy is a measure of uncertainty or randomness of the signals and is directly derived from the distribution of signal amplitude. Therefore, it is independent of threshold and timing parameters. The validity of the proposed parameter is investigated by performing AE monitoring during fatigue endurance test of 316L stainless steel. Digital Image Correlation (DIC) and global strain monitoring was carried out to relate material damage with AE activity. The result shows Renyi's entropy to be an effective measure to identify critical stages of damage in the material.

1. Introduction

Fatigue damage has been a matter of interest for years. It is expected that almost half of mechanical failure is due to fatigue. In Liquid Natural Gas (LNG) containment, sloshing is the dominant fatigue damage mechanism that causes structural failure. Therefore, for safety and reliability, it is important to monitor fatigue damage in these structures.

Acoustic Emission (AE) [1], is capable of identifying different stages of damage in the material and has shown potential as a Structural Health Monitoring (SHM) method [2] [3]. In comparison with other Non-Destructive Test (NDT) methods, AE can provide location and characterization of the damage in real time. AE monitoring utilizes sensors and data acquisition equipment to detect the elastic waves generated in the material as a result of damage. AE signals can be measured using several parameters including rise-time, duration, energy, count, peak amplitude, etc. Figure 1 below highlights the definitions of these parameters.

AE has been shown to be sensitive to different stages in damage growth. There are several sources of damage that contribute to AE, including plastic yielding, micro facture, crack initiation, crack propagation and crack rubbing [4]. The crack initiation has been shown to be accompanied by a significant increase in the peak amplitude, cumulative count and energy of the AE [5]. Han et al. [6] [7], suggested AE features such as count rate, peak amplitude and rise-time are sensitive to three different stages of fatigue; stage 1 fatigue crack initiation, stage 2 slow fatigue crack propagation and stage 3 rapid crack propagation. Han et al. [7] [8] distinguished other cracking mechanisms based on peak amplitude and rise time observation such as twinning, crack extension, cleavage facture, plastic zone activity, ligament shearing between micro-voids and micro-cracks. The change in facture mode from tensile to shear has been shown to be accompanied by a sharp increase in the duration and rise time of the signal [9], Robert et al. [10], demonstrated the prediction of crack length based on the count rate of the signal.

Most of the studies conducted above are based on parametric analysis of AE signals. Except for peak amplitude, AE parameters (explained in fig 1) are, dependent on acquisition settings such as threshold and timing. Different settings will result in different AE calculated parameters.

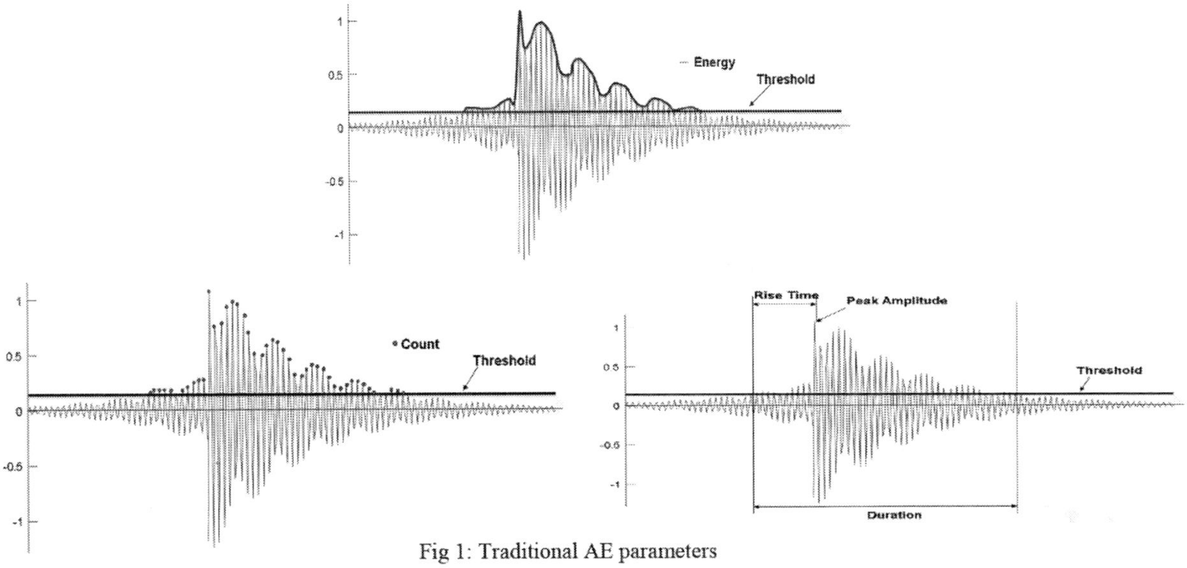

Fig 1: Traditional AE parameters

Recently, entropy of a signal in the time domain has been shown to be independent of acquisition settings. This is because it is derived from the probability mass in the amplitude distribution of signals and has shown to be effective in AE monitoring during fatigue [11] [12]. However, the work performed in [11] [12] is on the basis of Shannon's entropy, which is based on a linear averaging operator. The entropy computation in this research is based on second order Renyi's entropy. Renyi's entropy is based on the second functional class called the non-linear averaging operator, which has been proved to be more flexible and accurate than Shannon's entropy. Renyi's entropy computed in this work is compared with traditional AE parameters and has proved to be equally effective in discriminating damage in the material. Entropy of a signal is a measure of disorder and randomness in the signal. It is shown in this work that signals from material damage are associated with a higher value of entropy (higher randomness and disordered) than that of noise.

2. Methodology

2.1. Experimental Procedure

Dog-bone sample of 316L stainless steel, designed according to the standard E466-15 was used for the test (figure 2). Tables 1 and 2 show the chemical composition and mechanical properties of the specimen respectively. Fatigue endurance tests on the specimen was carried out with an Instron servo-hydraulic testing machine at ambient temperature. The specimen was tested under sinusoidal cyclic loading, with constant frequency of 5Hz, peak load of 24KN and loading ratio of 0.1.

AE signals from the fatigue test were recorded and analysed by an AMSY-6 data acquisition system controlled by Vallen software. Two AE sensors with a peak sensitivity frequencies between 100 kHz to 450 kHz and preamplifiers with a gain of 34dB were used. The distance between the sensors for the test was 90mm. Hit Definition Time (HDT) of 400μs, Re-Arm-Time of 1ms and threshold of 40dB were used as the data acquisition settings. Duration based Transient Recording (TR) was used for the test with a maximum TR page length of 26,214μs. A 1D location processor for 1D linear source localization was used with a wave velocity calibration of 5000m/s. The localisation facilitates the elimination of the noise generated from metal to metal contact at the grip, by recording signals generated only from the gauge section of the specimen.

In order to relate material response with AE activity, the specimen was monitored simultaneously with other NDT techniques. Two tests were conducted in total. The first test was conducted with simultaneous AE and global strain measurement. The second test was conducted with simultaneous AE, global strain and Digital Image Correlation (DIC) measurement. In order to perform DIC measurement, blocks of pixels were introduced in the gauge section of the specimen on the side of the specimen opposite the AE sensors. An ARAMIS-5m DIC system with a frame rate of 15Hz up to 29Hz and camera resolution of 2448 x 2050 px was used to capture the 2D displacement field of the pixel blocks. The frequency of the images taken by the DIC system was 0.005Hz.

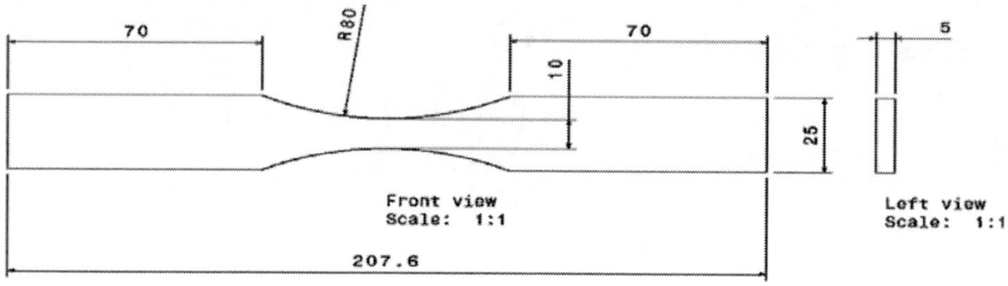

Fig 2: Dog-bone fatigue endurance specimen

Table1: Chemical Composition

C	Mn	P	S	Si	Cu	Ni	Cr	Mo	Nb
0.025	1.32	0.03	0.002	0.48	0.36	10.03	16.58	2.02	0.041

Table2: Mechanical Properties

Young's Modulus	193GPA
Yield Strength (Longitudinal) **Rp** 1%	347MPA
Tensile Strength (Longitudinal)	613MPA

2.2. Renyi's Entropy

In the 1950s, Alfred Renyi introduced a parametric family of information theory called Renyi's entropy that generalizes Shannon's entropy [13]. Renyi pointed out that Shannon's entropy estimation is based on the first functional class which is the linear averaging operator($g(x) = cx$). He proposed an entropy formulation based on the second functional class called the non-linear averaging operator ($g(x) = c2^{(1-a)x}$) which proved to be more flexible than that of Shannon's formulation [14].

Entropy is based on the measure of uncertainty in a probability distribution function and is widely used in mathematics and physics [15]. Given a random sequence$\{x_1 x_2\, x_3, \ldots, x_{n,}\}$, Renyi's entropy is calculated according to equation (1), as:

$$H_a(x) = \frac{1}{1-a}\log\left(\sum_{k=1}^{n}(P(x_k))^a\right) \quad (1)$$

Where:-

$H_a(x)$ is Renyi's entropy

$P(x_k)$ is the probability mass associated with the value x_1.

a is the order of the equation, which is selected to be 2 in this paper.

Log to the base 2 is used in this case in order to measure the entropy in bits. Log to the base 10 or $\log_e$ can also be used in order to measure entropy in bins and bans respectively. For simplicity, Renyi's entropy will be referred to as AE entropy in this paper. The fundamental reason for using AE entropy is its independency from acquisition settings such as threshold and timing, unlike traditional AE parameters. The calculation procedure of AE entropy begins with recording AE waveform generated from the experiment using the AE monitoring system. The waveform captured from the experiment is converted into a txt file containing millivolt and microsecond values. The discrete probability distribution of the waveform is then calculated. Several methods have been proposed to calculate the probability of mass in order to calculate the entropy of a digital waveform [16] [17]. The probability of mass calculation in this work is based on taking into account the original spectrum of the signal, explained in [16]. Probability of mass for entropy calculation, by taking into account the original spectrum of the signal has been widely used in signal processing [18] [19]. After probability of mass calculation, AE entropy is calculated using eqn 1. Figure 3 below shows the AE entropy calculation steps.

https://doi.org/10.1051/matecconf/201816522007

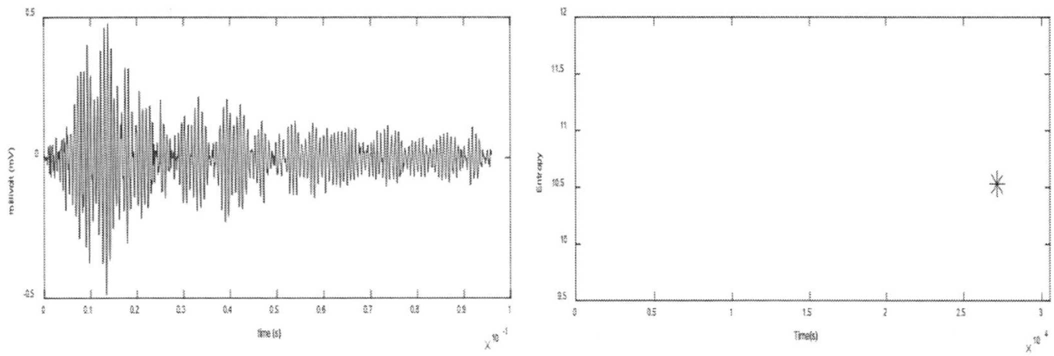

x_1	0.009134
x_2	0.003045
x_3	0.003045
....	....
x_{4799}	0.015223
x_{4800}	0.012178
x_{4801}	0.009134

X=x		P(X=x)
0	0.009134	0.0016
1	0.003045	0.0002
2	0.003045	0.0002
...	....	....
4799	0.015223	0.0045
4800	0.012178	0.0029
4801	0.009134	0.0016

Fig 3: Entropy calculation steps

3. Results and Discussions

3.1. AE data collection with %elongation of the specimen

A plot of peak amplitude of the AE waveform and % elongation of the specimen versus time was recorded during the cyclic loading of the test machine (figure 4).

The initial part of the test is accompanied by a significant increase in the peak amplitude (up to 92dB) of the signal and % elongation of the specimen. At this stage of the test, signals in the lower amplitude range are associated with elastic to plastic transformation of the material [20] [21], whereas signals in the higher amplitude range (above 60dB) could originate from initial metal to metal contact noise [22]. Subsequently, signals with lower amplitude (below 55dB) start to appear up to about 14,000s. The time from the onset of loading until 14,000s can be regarded as the damage accumulation or crack nucleation stage.

It can be observed in figure 5, a plot of count rate and %elongation versus time that up to 14,000s, except for the onset of loading, there is a low count rate, suggesting that at this stage there is damage accumulation or crack nucleation [23] [24]. The initial increase in count rate at the onset of loading is due to the enhanced micro-structural activity due to elastic to plastic transformation and noise due to loading train. At the end of this stage the damage becomes localized, micro cracks start to appear and join to form the short cracks [24]. Figure 5 also shows a sudden increase in count rate of the signals after 14,000s. The increase in count rate here is associated with micro damage of the material such as growth and coalescence of micro crack [24]. By comparing with figure 4, an increase in count rate is also accompanied by high amplitude signals at 14,000s and 18,000s (above 55dB) as a result of the micro damage in the material.

 https://doi.org/10.1051/matecconf/201816522007

After 25,000s the damage becomes visibly localized and both count rate and peak amplitude increase significantly, due to macro fatigue crack initiation, growth and ultimate failure. Peak amplitude of the signal is a useful parameter for identifying damage mechanisms in a material. However, it is difficult in this experiment to accurately identify damage mechanism such as growth and coalescence of micro crack based on peak amplitude observation alone, as it is possible to confuse the initial jump in peak amplitude, at the onset of loading, to be from the material damage.

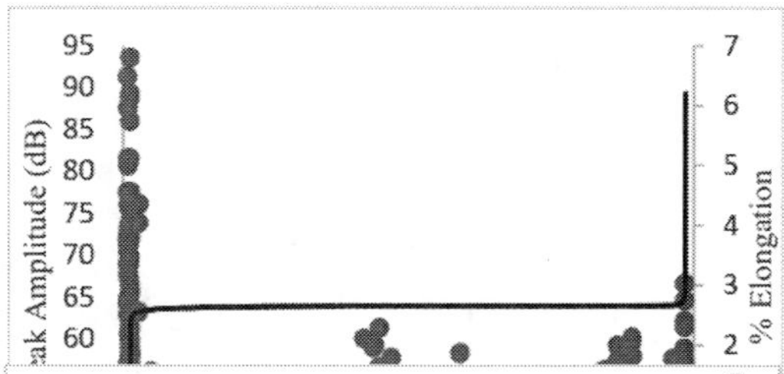

Fig 4: AE peak amplitude and % elongation versus time

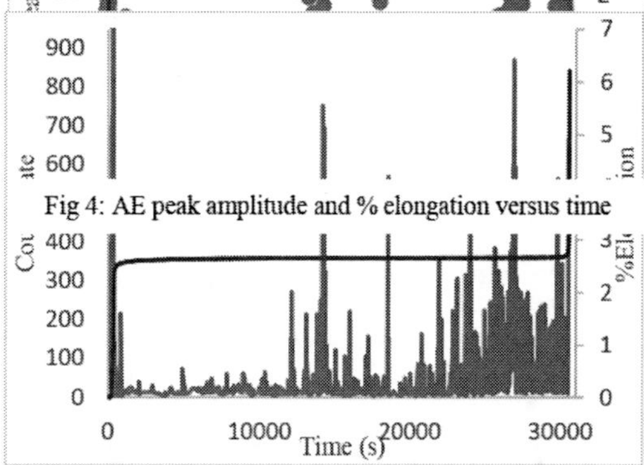

Fig 5: AE Count Rate rate and % elongation versus time

As an alternative to peak amplitude parameter, a threshold independent parameter called AE entropy was calculated from the recorded waveforms, as explained in section 1. The plot of AE entropy and % elongation versus time is shown in figure 6.

It is evident from figure 6 that waveform with higher peak amplitude in the beginning of the test is associated with a lower value of AE entropy (below 11). During the damage accumulation stage, up to 14,000s, entropy variation turned out to be stable. The waveform captured at 14,000s showed a sudden rise in AE entropy, higher than that at the beginning, which was caused by elastic to plastic transformation and mechanical noise. The sudden rise in AE entropy after 14,000s is also accompanied by a sudden rise in count rate of the signal, observed in figure 5, suggesting these signals to be originating from enhanced micro-structural activity, such as growth and coalescence of micro crack [23] [24].

Waveforms captured at 18,000s also showed an increase in AE entropy accompanied by a sharp increase in the count rate. However, the increase in AE entropy at 18,000s is not very significant as compared to the AE entropy at 14,000s. This could be due to a small magnitude of micro damage as compared to the one at 14,000s.

AE Entropy was stable until 23,000s. At approximately 25,000s damage in the specimen was visibly localized and was accompanied by a significant increase in both the count rate and AE entropy of the signal leading to the macro fatigue crack initiation, growth and ultimate failure of the specimen. A significant number of signals after 25,000s, were associated with AE entropy that was higher than the entropy at the onset of cyclic loading.

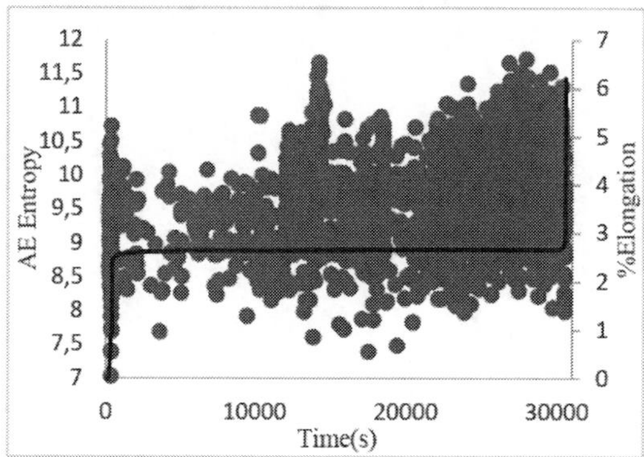

Fig 6: AE entropy and % elongation versus time

Figure 7 illustrates a comparison of cumulative count, cumulative energy and cumulative entropy for the entire length of the test. Cumulative energy and count have been widely used in fatigue damage evaluation during AE monitoring [6] [20] [25]. Therefore, these parameters have been chosen to compare the cumulative trend of AE entropy. It is evident in figure 7 that all three curves show the same trend. The initial rise in the cumulative parameters in region A, is due to the elastic to plastic transformation and mechanical noise. Region B is the damage accumulation stage and is accompanied by a very small increase in the cumulative parameters. A sudden rise in the cumulative parameters in regions C and E, is due to the growth and coalescence of micro cracks. Regions D and F are associated with a stable rise in the cumulative parameters. The stable rise in the cumulative parameters is associated with stable accumulation of micro damage such as micro crack growth. Region G is accompanied by a significant rise in the cumulative parameters. The sharp increase in the cumulative parameters in region G form 25,000 to 27,000s is due to macro fatigue crack initiation as suggested in [6] [7] [25] and the stable increase in the cumulative parameter from 27,000 to 29,500s is associated with stable macro fatigue crack growth. The sharp increase in the cumulative parameters after 29,500s until the end is due to the result of rapid crack propagation, suggested in [6] [7] [25].

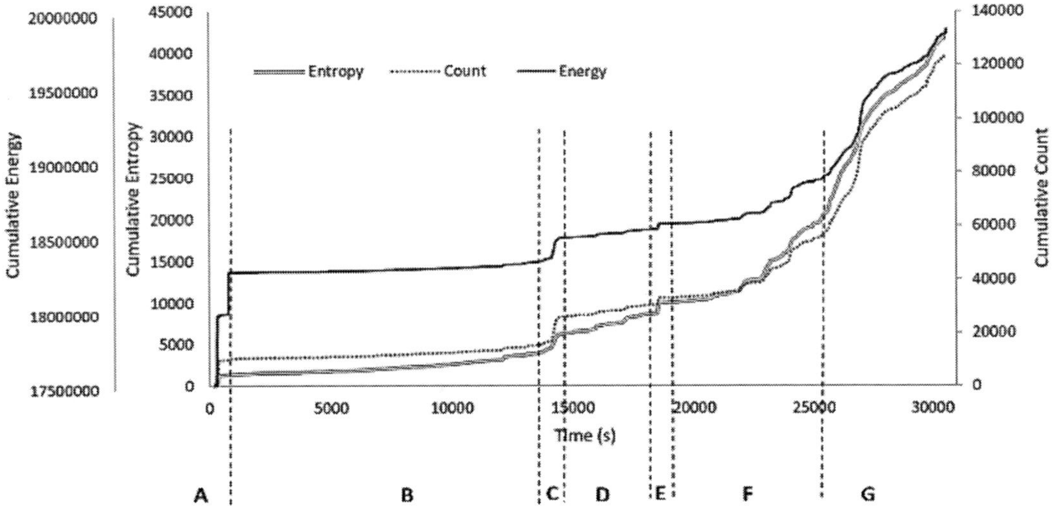

Fig 7: Cumulative entropy, cumulative energy and cumulative count versus time

The similar increasing trend in all three parameters showed that cumulative entropy can also be used as fatigue damage evaluation parameter instead of cumulative count and cumulative energy.

The above analysis indicates that AE entropy is sensitive to the fatigue damage evaluation in 316L stainless steel. The high peak amplitude distribution of the signal in the beginning of the test was associated with a lower value of AE entropy. Material damage, such as growth and coalescence of micro crack, macro fatigue crack initiation, propagation and ultimate failure was accompanied by a significant increase in AE entropy. Since AE entropy is a measure of uncertainty or randomness in a waveform, this suggest that waveform from material damage is associated with more uncertainty or randomness than that of noise. Moreover, cumulative entropy had a similar trend as the cumulative count and cumulative energy suggesting that the cumulative entropy can also be used as a fatigue damage evaluation parameter.

3.2. AE data collection with DIC

The secondary monitoring technique used in the first test was the %elongation of the material, which did not show any change in trend of fatigue damage evaluation except for the onset of loading and ultimate failure. To improve the performance of the secondary monitoring technique, the second test was conducted with Digital Image Correlation (DIC) measurement alongside AE monitoring.

DIC provided a full surface displacement and strain image over the entire gauge length of the specimen. There were several images taken during the test.

Figure 8 shows a few of the images captured by DIC during cyclic loading of the specimen. Image (a) in figure 8 is the reference image. This image was taken prior to loading of the specimen. Image (b) was taken at 11,188s after the start of the test. Image (c) was taken at 18,664s. It can be seen in image (c) that the macrostrain is localized in a small volume, marked by the black arrow. Localized macrostrain occurs long after the accumulation of micro crack initiation, coalescence and growth in the persistence slip band. In image (d) and (e) at 18,665s and 19,004s macro fatigue crack propagation can be identified from the localized macrostrain seen in (c). Image (f) at 19,400s shows the ultimate failure of the specimen captured by DIC.

DIC provides crucial information regarding fatigue damage evaluation. Firstly, information is obtained regarding macrostrain localization, because of micro fatigue crack initiation, coalescence and growth. Secondly, information is obtained regarding macro fatigue crack propagation leading to the ultimate failure of the specimen.

No	a	b	c	d	e	f
Images	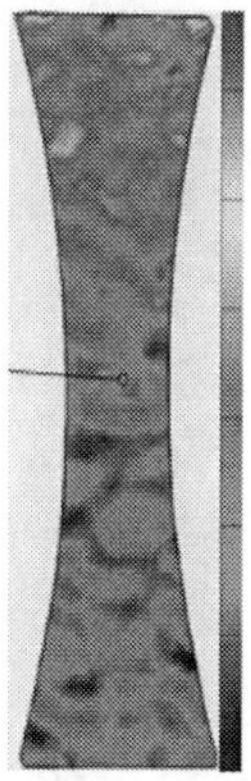	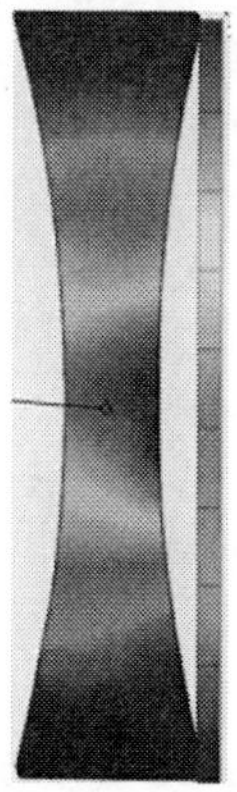	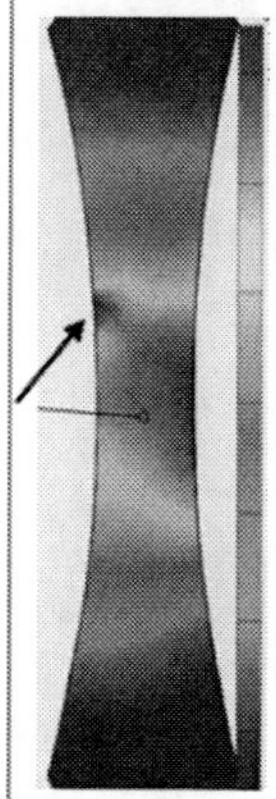	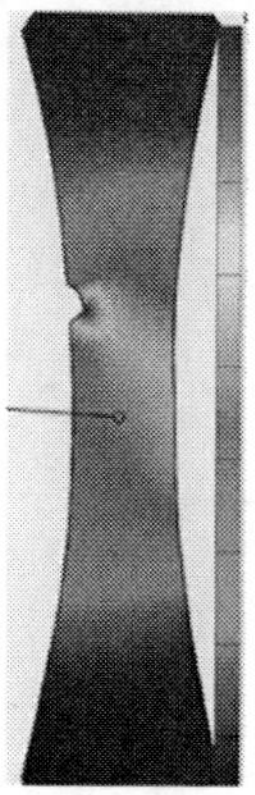	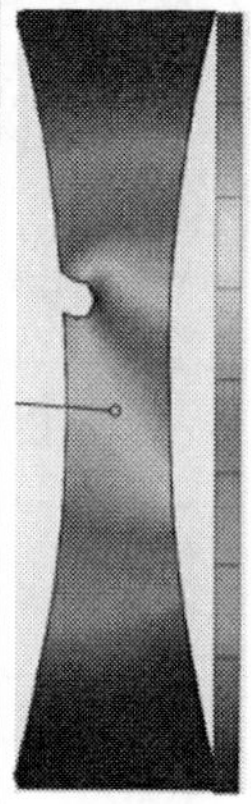	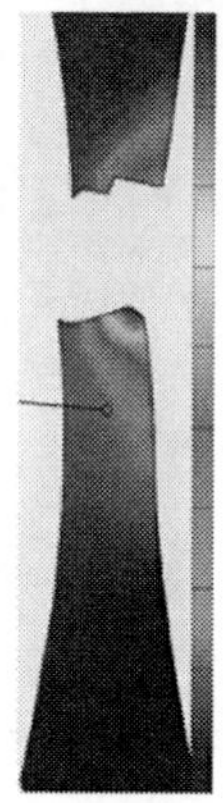
Time	Reference (0s)	11188s	18664s	18665s	19004s	19400s

Fig 8: Digital Image Correlation images

Figure 9 shows a plot of cumulative count, cumulative energy and cumulative entropy for the entire length of the second test. Like the first test, it is clear from the graph that all curves have the same trend. A significant increase in the cumulative parameters in region A, can be attributed to the elastic to plastic transformation and mechanical noise originating from the metal to metal contact at the specimen grips. The damage accumulation stage in region B is accompanied by a minimum increase in the cumulative parameters. A sharp increase in the cumulative parameters in region C and E can be associated with micro crack coalescence and growth, whereas a stable increase in the cumulative parameters in region D and F is due to a stable accumulation of micro crack growth in the material. After the onset of macrostrain localization, macro fatigue crack initiates, which is accompanied by a sharp increase in the cumulative parameters. This trend is also consistent with other research [6] [7]. After macro fatigue crack initiation, the cumulative parameters show a stable variation between 18,700s to 19,300s. This region is associated with stable macro crack propagation. The stable increase in cumulative parameters due to stable macro fatigue crack propagation is also consistent with other researchers [6] [7] [21]. The sharp increase in the cumulative parameter at the end, after 19,300s, is due to an unstable crack propagation, suggested by [6] [7] [21].

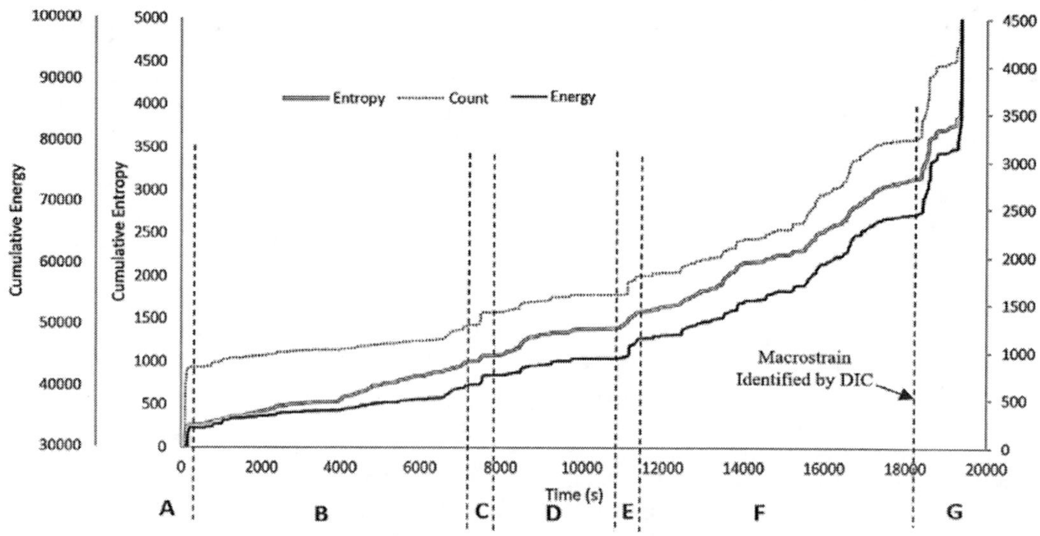

Fig 9: Cumulative entropy, cumulative energy and cumulative count versus time

4. Conclusion

In this paper, a new qualitative parameter called AE entropy, based on Renyi's entropy has been proposed. The new parameter proved effective in discriminating different stage of damage during a fatigue endurance test, leading to ultimate failure of 316L stainless steel. AE entropy is a measure of uncertainty of the amplitude distribution of the waveform. Therefore it is independent of time driven parameters and threshold. The following can be concluded form the results obtained during these experiments:

- A comparison between AE entropy and traditional AE peak amplitude measurements concluded that the initial increase in the peak amplitude of the signal as a result of elastic to plastic transformation and mechanical noise, is associated with a lower value of AE entropy. Growth and coalescence of micro crack, macro crack initiation, macro crack propagation and ultimate failure is associated with a higher entropy value. This could provide an effective method for characterizing damage in the material.

- Cumulative AE entropy, cumulative count and cumulative energy followed the same trend during the fatigue endurance test of 316L stainless steel. The traditional AE cumulative analysis during fatigue based on count and energy, can be replaced with AE entropy, because of its independent nature in terms of time driven parameter and threshold.

Acknowledgements: - This publication was made possible by the sponsorship and support of London South Bank University and Lloyd's Register Foundation, a charitable foundation helping protect life and property by supporting engineering-related education, public engagement, and the application of research. The work was enabled through, and undertaken at, the National Structural Integrity Research Centre (NSIRC), a postgraduate engineering facility for industry-led research into structural integrity established and managed by TWI through a network of both national and international Universities.

References

1. C.J.Hailer, *Handbook of Nondestructive Evaluation-Chapter 10*,(The McGraw-Hill Companies, 2001).
2. M. David, R.B.K.N. Rao, Shock and Vibration Digest, **38**, 3-16, (2006).
3. C. Boller, *Sixth European Workshop on Structural Health Monitoring*, (Dresden 2012).
4. A.C.E. Sinclair, D.C. Connors,C.L. Formby, Material Science and Engineering, **28**, 263-273, (1977).
5. R. Roy, N. Parida, S.Sivaprasad, S. Tarafder. K.K.Ray, Meterial Science and Engineering A, **486**, 562-571, (2008).
6. Z. Han, H. Lou, J. Cao and H. Wang, Materials Science and Engineering A, **528**, 7751– 7756, (2011).
7. Z. Han, H. Luo, Y. Zhang and J. Cao, Materials Science & Engineering A, **559**, 534-542, (2012).
8. H. Zhiyuan, L. Hongyun, S. Chuankai, L. Junrong and P. Mayorkinos, Material Science & Engineering A, **597**, 270-278, (2014).
9. D.G.Aggelis, E.Z.Kordatos and T.E.Matikas, Machenics Research Communication, **38**, 106-110, (2011).
10. T.M.Roberts, M.Talebzadeh , *J*ournal of construction steel reserach, **59**, 679-694, (2003).
11. C. Mengyu, Z. Zhang, D. Quan and Y. Song, International Journal of Fatigue, **109**, 145-156, (2018).
12. C. Mengyu, Z. Zaoxiao and D. Quan, Mechanical sytems and signal processing, **100**, 617-629, (2018).
13. A. Renyi, Proc. of the 4th Berkeley Symp. Math. Statist. Prob, **1**, 457-561, (1960).
14. D. Xu and E. Deniz, *Information Theoretic Learning* (47-102, New York, NY: Springer, 2010).
15. Gray. M, Robert. *Entropy and Information Theory*, (Boston, MA: Springer, 2011).
16. K. Ekˇstein, T. Pavelka, In *Proceedings of PhD Workshop Systems & Control*, University of West Bohemia, (2004).
17. A.Vahaplar,C.C.Celikoglu,M.Ozgoren,Mathematical and Computational Applications, **16**,43-52, (2011).
18. V. Vijean, M. Hariharan, S. Yaacob, M. Nazri, B. Sulaiman and A. Adom, Computers and Electrical Engineering, **39**,1549–1560, (2013).
19. V.Vijean, H.M, S, Yaacob, M.N.B.Sulaiman, Biocybernetics and Biomedical Engineering, **34**,169-177, 2014.
20. O. A. Amer, A.-L. Gloanec, S. Courtin, C. Touze, *Procedia Engineering*,**66**,651-660, (2013).
21. K.S.Han,K.H.Oh, Key Engineering Materials,**306-308**, 271-278, (2006).
22. M. Chai, Z. Zhang, Q. Duan and Y. Song, International Journal of Fatigue, **109**, 145-156, (2018).
23. D. Barktova, F. Vlasic, P. Mazal and O. Dvoracek, *In Metal 23rd International Conference on Metallurgy and Materials*, 1-6 (2014).
24. P. Mazal, F. Vlasic and V. Koula, *Procedia Engineering*, **133**, 379-388,(2015).
25. M.Chai,J.Zhang, Z.Zhang, Q.Duan, G.Cheng. Applied Acoustics,**126**,101-113, (2017).

Effect of Temperature on Strawbale Wall Rendered with Laterite and Clay Nanoparticle

Adeola A. Adedeji[3*], *Aminu* Ibiyeye[2], *Abiola M* Adedeji[3]
Department of Civil Engineering, Faculty of Engineering and Technology, University of Ilorin, Nigeria

Abstract. In this study, the influence of the plastering materials on the thermal performance of three (3) straw bale building models were evaluated. The plastering materials considered are laterite mortal, laterite mortal with nanoparticle clay screeding and cement-sand mortal. The results were based on the experimentally measured thermal conductance, thermal resistance and thermal transmittance of the building model according to ASTM C1045-90 standard specification. The analysis of the measured data revealed that the laterite plaster significantly improve the thermal resistivity of the building model compared to cement-sand mortal. However, clay nano particle screeding on the laterite plaster does not significantly affect the thermal performance of the straw bale building model.

1. INTRODUCTION

With the ever increasing population of people in the world today and the global economic recession plaguing many countries of the world today, of which Nigeria (which is the most populous country in Africa and that with the highest annual growth rate in Africa) is not left out. Thus giving rise to the periodic increase in the demands for housing and the escalating cost of conventional building materials. Hence, the need to construct buildings with viable low cost materials has become a necessity in our fast growing society in other to overcome the problem in the housing sector [1]. With this problem at hand, there is need to develop alternatives, which have to be cost effective and create an ecological friendly room for the comfortability of the dwellers, due to the fact that some building materials may impair the ecological friendly condition of our rooms i.e. more economical, ecological friendly and readily available thus making this research important. One material that meets with these requirements is straw bale. It is an agricultural waste that is strong, cheap and readily available. Typical straw bale is shown in Fig. 1. The oldest surviving straw bale building is 107 years old, as of 2010 (US DOE,[2]). This structure, along with dozens of other straw bale buildings, was built by European settlers in the Sand Hills region of western Nebraska, USA at the turn of the 20th century.

Fig. 1 Guinea corn straw bale

However, the environmental, physical and chemical properties of straws forming the bale strongly affect the durability of straw bale building. Part of the environmental conditions include temperature, relative humidity, Oxygen and pH. If these conditions are conducive for micro-organism to thrive inside the bale, it leads to straw deterioration and negatively affect the durability of straw building [3]. Therefore, selection of suitable plastering material is very important for straw bale constructions. Hence, the thrust of this work was to study the thermal conductivity through the wall of a straw bale building plastered with laterite and clay nanoparticle materials.

Clay is one of the oldest building materials on Earth, among other ancient, naturally-occurring geologic materials such as stone and organic materials like wood. Between one-half and two-thirds of the world's population, in both traditional societies as well as developed countries, still live or work in buildings made with clay, often baked into brick, as an essential part of its load-bearing structure. Also a primary ingredient in many natural building techniques, clay is used to create adobe, cob, cordwoodand rammed earth structures and building elements such as wattle and daub, clay plaster, clay render case, clay floors and clay paints. It has also been used as a mortar in brick chimney sand stone walls where protected from water. Clays exhibit plasticity when mixed with water in certain proportions. However, when dry, it becomes firm and when fired in a kiln, permanent physical and chemical changes occur. These changes convert the clay into a ceramic material. Because of these properties, it is used for making pottery, both utilitarian and decorative, and construction products such as bricks, wall and floor tiles. Different types of clay, when used with different minerals and firing conditions, are used to produce earthenware, stoneware, and porcelain. Prehistoric humans discovered the useful properties of clay.

Currently, 45% of all the energy consumed in the world is used in the manufacture and transportation of such building materials (Earth Garden [4]). In addition, most standing structures are not in the least bit energy efficient. In the United States alone, as much as 70% of the electricity consumed and 40% of carbon dioxide emitted by residential homes is wasted due to the fact that many structures are poorly insulated and need

* Corresponding author: aaadeji@unilorin.edu.ng

excessive amounts of electricity and fuel to balance their energy inefficacies [5]. Energy and resource conservation have become a popular priority in today's construction industry. Whether it is to save energy costs or to genuinely act more sustainably, both commercial and non-for-profit interest groups have been growing more concerned on making buildings more "green". Green characteristics include efficient energy usage, water efficacy, decreased carbon dioxide emissions, increasing the utilization of natural light, and improvements in indoor air qualities. Current sustainable design and construction efforts can make energy efficient homes up to 90% less energy intensive than standard structures that are built to the same building code [5]. One method of building energy efficient structures is through straw bale construction. Building with straw bales has remarkable advantages that building with conventional materials such as wood, steel and concrete lack in terms of cost, abundance, and sustainability. However, the most distinct advantage of straw bale is that it is a highly efficient thermal insulator.

As this is true for individual grains of straw, baled straw is actually a very effective construction material because of its high density. To clarify the distinction between straw and straw bales, it can be compared to sheets of paper and a heavy bounded book. As sheets of paper are very weak and unstable individually, they work in the contrary when bounded together

Fig. 3: Typical straw bale wall [4]

Another misconception concerning straw bale is that it is hazardous in terms of combustion. However, because straw bales are densely compacted, they are actually very fire resistant as there is limited oxygen within the bale to permit combustion. Fig. 3 shows a typical straw bale wall.

2. Materials and Methods

2.1 Description of Straw bale building Model

In order to evaluate the influence of plastering material on the thermal conductivity in a straw bale building, three building models (A, B and C) of the same scaled dimension of 400 mm x 300 mm in plan and 400 mm in height on scale 1:100 representing a standard room of 4 m x 3 m x 4 m, were constructed of straw bale wall. The thickness of the wall is 110 mm.

2.2 Materials

The materials used for the study are: dry guinea corn straw with maximum diameter of 20 mm, binding wire, laterite, earth clay, ordinary Portland cement of grade 42.5 conforming to BS 12 [6], naturally occurring river sand, portable water conforming to [7], wood for the window and door frame, asbestos for the ceiling and aluminium roofing sheets.

2.3 Soil Characterisation

The particle size distribution for the laterite material used for the rendering was determined by standard sieve analysis test while clay materials passing Nano sieve was used for the plastering. The moisture content of the soil samples was determined in accordance with BS 1377 [8].

The straws whose average diameter is 20 mm were cut into equal lengths of 400 mm to achieve the height of the scaled building model except at the window and door locations where the lengths were adjusted to fit the model geometry. The baling was achieved by vertically stacking the straws in a wooden mould until it is no longer possible to add more straws. The packed straws were tied together with binding wire at near the top quarter along the straw length. The baled straws were then carefully pushed out of the mould with the wooden base plate specially designed to achieve this, thereafter another binding wire were tied in turn at the middle and bottom quarter along the straw length to ensure adequate rigidity and firmness of the baled straws. Fig. 4 shows some of the baled straws

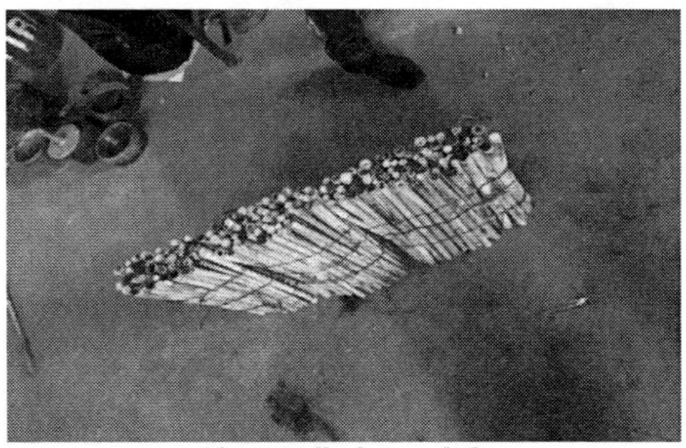

Fig. 4 Stalked straw bale

The straw bale walls in building model A was plastered with laterite mortal (laterite and water). The optimum moisture content for the laterite, determine in accordance with BS 1377 [8], was used to form the laterite mortal. Model B was similar to model A except that the laterite plaster was rubbed with nano particle clay mortal to achieve smooth surface. The control model, C was plastered with cement-sand mortal of 1:5 mix proportion. The thickness of the plastering for all the models was 15 mm. The plastered straw bale wall was cured for 7 days before joining with mortal to form the building model. Fig.s 5 and 6 show the straw bale building model under construction.

3.6 Thermal Conductivity Test

The thermal conductivity through the walls of the building models were measured in accordance with [9] standard specifications. The heating was generated through hot plate while the temperatures were recorded using a data logger with sensors positioned at surfaces of the wall (inside and outside), centre of the building model as well outside the building model. The test was carried out for both isolated (window and door closed) and open room conditions. The temperatures were recorded at intervals of 10 minutes over a period of one hour. Figs 5 and 6 show the experimental setup for the building model A and model C.

Fig. 5 Temperature test on model plastered with laterite

Fig. 6 Temperature test on model b plastered with laterite and rubbed with clay

The thermal conductivity (C), thermal resistivity (R) and thermal transmittance otherwise known as overall coefficient of heat transfer (U) for the models were obtained using the following models as given is ASTM C 1045[9]:

$$C=QA(Th-Tc) \quad (1)$$

$$R=C-1 \quad (2)$$

$$U=QA(T1-T2) \quad (3)$$

Where: Q is the time rate of one-dimensional heat flow through the metering area of the test apparatus, W = 1200 W, T_h is the area-weighted temperature of specimen hot surface, K, Tc is the area-weighted temperature of the specimen cold surface, K, T_1 is the area-weighted air temperature 75 mm or more from the hot side surface, K, T2 is the area-weighted air temperature 75 mm or more from the cold side surface, K, A is the wall area, m^2

4.0 Results and Discussions

4.1 Properties of the lateritic soil

4.1.1 Moisture content and Dry density

Fig. 7 illustrates the graph of dry density against the moisture content of the laterite material used for the plastering. From the Fig., the optimum moisture content for the laterite is 8.7 % while the maximum dry density is 17.77 kg/m^3.

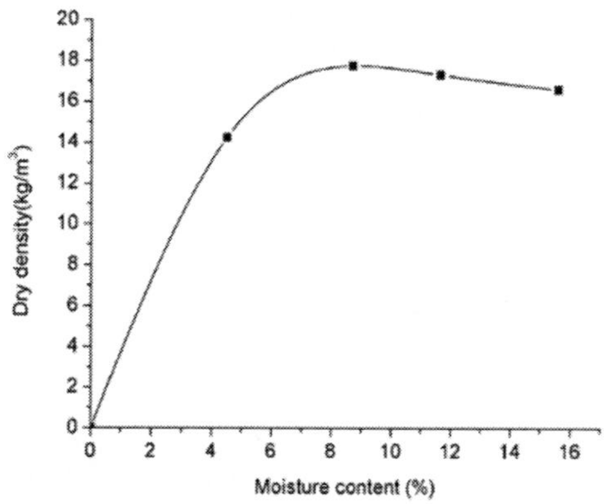

Fig. 7: Dry density versus moisture content for the Laterite material

4.1.2 Particle size distribution

The result of particle size distribution is necessary for soil classification. The particle size distribution for the laterite material is shown Fig. 4. From the graph, the following parameters were obtained:

Effective size D10 = 0.26 mm
D30 = 0.43mm
D60 = 0.74 mm

Coefficient of uniformity, CU = D60 / D10 = 2.85

Coefficient of curvature, CC = $(D30)^2$ / (D60 * D10)= 0.96

Fineness modulus, FM = (summation of % cumulative weight retained between 8mm to 0.15mm sieve size) / 100, FM = 2.82

Therefore, the soil is classified as "Gravelly lateritic soil" and "A-2-6 Laterite according to the AASHTO classification system.

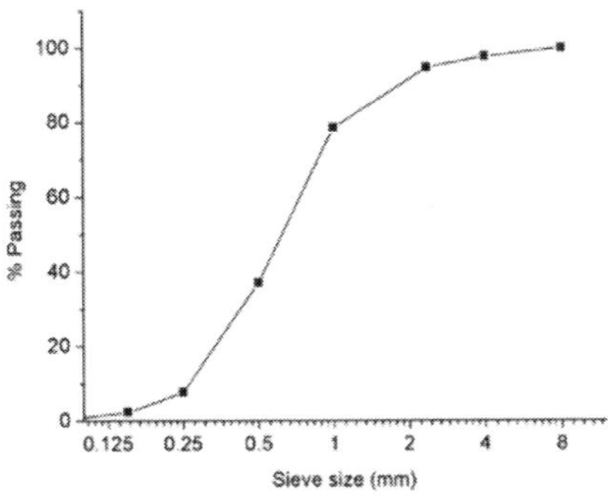

Fig. 8: Laterite particle size distribution curve

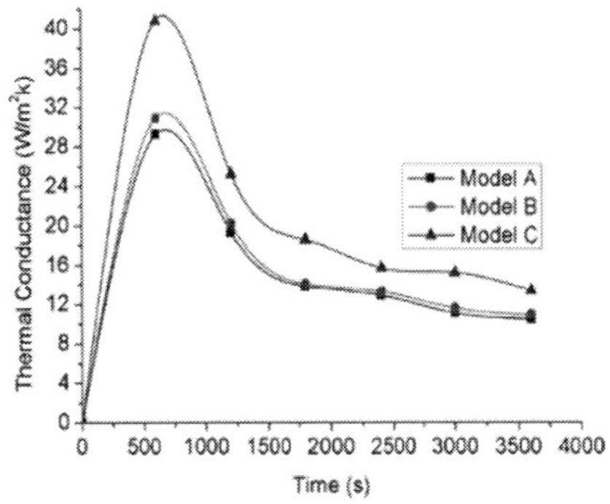

Fig. 9 Thermal conductivity versus time for closed/isolated strawbale building

4.2 Thermal Conductivity

Fig.s 10 and 11 compared the results of thermal conductivity tests on the straw baled building models for closed and open room conditions. From the graph, the straw baled building plastered with cement-sand mortal (model C) has the highest thermal conductivity. The implication is that the thermal resistance of the cement - sand mortal is lower compared to that of clay materials (model A and model B).

Therefore, the laterite mortal is more thermal effective than the cement-sand mortal. Furthermore, rubbing laterite render with clay nanoparticles does not significantly influence the thermal conductivity of the straw bale building models. Therefore, laterite plastered building (model A) and laterite plastered model rubbed with nano particle clay (model B) showed similar thermal performance.

4.2 Thermal Transmittance

Fig.s 8 and 9 also compared the thermal tranmittance values for the straw baled building models under closed and open room conditions. The result is similar to that of thermal conductivity. The straw baled building plastered with cement-sand mortal (model C) showed the highest thermal transmittance.

This implies that the laterite material provides beter thermal insulation for both the straws inside the walls as well as the surrounding environment. However, the thermal performane of laterite plastered building (model A) and laterite plastered model rubbed with nano particle clay (model B) are not significantly different. This shows that the thermal insulating properties of both the clay and the latrite are very similar.

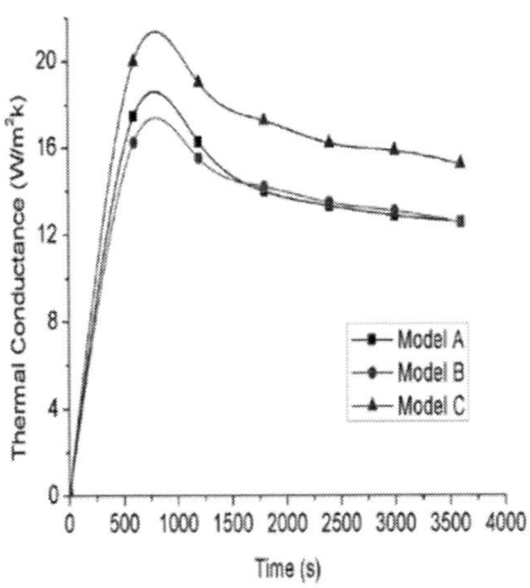

Fig. 10 Thermal conductivity versus time for opened strawbale building

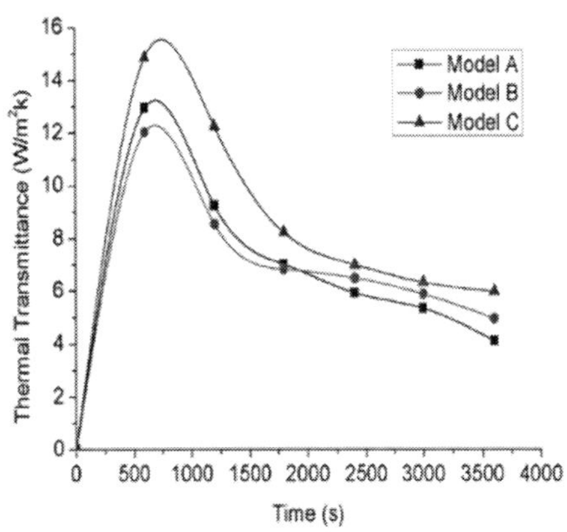

Fig. 10 Thermal conductivity versus time for opened strawbale model

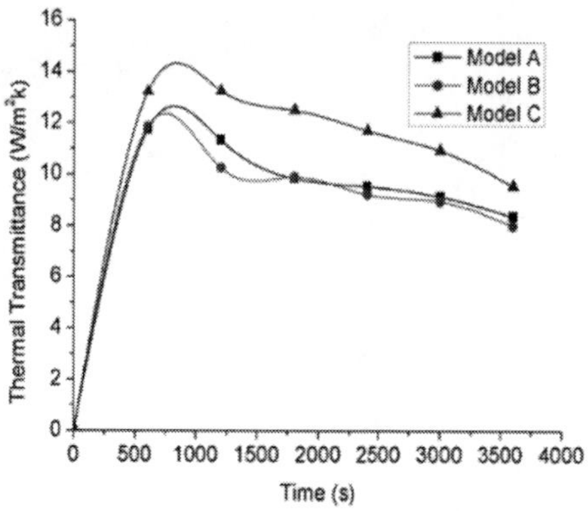

Fig. 10 Thermal conductivity versus time for closed/isolated strawbale building model

5. Conclusion

In the present study, the thermal performance of straw bale building plastered with cement-sand mortal and laterite mortal were evaluated. From the study, it is concluded that: The thermal performance of a straw bale building depends on the plastering material. The thermal resistivity of the laterite plastered wall was found to be greater than that of cement-sand mortal. The thermal transmittance in cement-sand plastered straw bale building is higher compared to straw bale building plastered with laterite mortal. Clay screeding does not significantly influence the thermal performance of laterite plastered straw bale wall.

References

1. A. A. Adedeji. Introduction and design of straw bale masonry, Olad publishers Ent., Ilorin. (2007).
2. U.S Department of energy, straw bale construction and building code, Assessed onlin: http.//www.thelaststraw.com.
3. Ashour, T., Georg, H., & Wu, W. (2011). Performance of straw bale wall: A case of study. *Energy and Buildings, 43*(8), 1960-1967, (2011)
4. Earth Garden. Straw Bale Home Building. Maryborough, Australia: McPherson's Printing Group.greenhomebuilding.com. (n.d.). Retrieved September 7, 2010, Assessed online: http://www.greenhomebuilding.com/strawbale.httm, (2014).
5. 5. Jafferiji H., Karina R. and Yao W., Straw bale Construction: The Application in Massachusetts. Thesis submitted to the faculty of Worcester Polytechnic Institute (WPI), Worcester. (2011).
6. FWS, The sandhills: building partnerships for an Ecosystem. U.S. Fish and Wildfish Service. Retrieved1 June 2010 from http://www.fws.gove/mountain-prairie/pfw/ne/ne4.httm. (2010).
7. 7. ASTM C1045 Standard Practice for Calculating Thermal Transmission Properties Under SteadyState Conditions, American Society for Testing and Materials ASTM, West Conshohocken, PA. (1997).
8. 8. A. A. Adedeji, Evaluating Thermal Resistivity of a Cement Plastered Straw Bale Wall: A Review" Department of Civil Engineering, University of Ilorin , Ilorin, Nigeria.(2011).
9. Mehmet F. A, Ahmet E.T and Omer F.C. Strawbale Houses: A Quick Solution to Accomodation Problem.'3rd International Conference on Sustainable Construction Materials and Technologies, Assessed from http://www.classie.info/Proceedings.htm on 4th February, 2017.

Hysteresis Analysis of Prestressed Brick Frame with Strawbale Masonry Infill Subjected to Seismic Loads

Adeola A. Adedeji[1,*], *Abiola* Adedeji[2], *Aminu* Ibiyeye3
[1,2,3]Department of Civil Engineering, University of Ilorin,, Ilorin, Nigeria
*aaadeji@unilorin.edu.ng

Abstract : This study looks at designing and evaluating strength of typha strawbale wall cross section rather than the assumption of edge column acting as axially loaded members that should resist all vertical members from the loads acting on the wall. Based on the aforementioned claim, the objective of this work is to provide the average design thickness for the cement-plastered typha strawbale that stiffens the prestressed brick frame. Data on strength and deformation of the structure are the input for the analytical models. Pseudo-dynamic earthquake response tests, of a one quarter (¼) scale model low rise storeys prestressed brick frames stiffened with cement plastered strawbale masonry was conducted. The structure is idealized as plane frames. The analysis utilised the hysteresis models for members' models as time-independent. The force-displacement relationship of the members' models has been evaluated by the approximate method on the basis of the material properties and structural geometry. A model was investigated with straw bale infill panel to determine the hysteretic parameters, stiffness deterioration and strength degradation due to seismic forces.
Keywords: seismic; masonry-infill; typha strawbale, finite element; degradation, brick

1. Introduction

A number of past studies ([1],[2],[3]) focused on evaluating the experimental behavior of masonry infilled frames to obtain formulations of limit strength and equivalent stiffness. A more rigorous analysis of structures with masonry infilled frames requires an analytical model of the force – deformation response of masonry infill. While a number of finite element models have been developed to predict the response of infilled frames ([4]. [5], [6]), such micro modeling is time-consuming for analysis of large structures. Alternatively, a macro model allows treatment of the entire infill panel as a single unit.

Saneinejad and Hobbs [7] developed a method based on the equivalent diagonal strut approach for the analysis and design of frames with masonry infill walls subjected to in-plane forces. The UBC required that the shear walls acting independently of the ductile moment-resisting portions of the frame shall resist total required seismic forces. It is also required that design force distributions along elevations of the structure for both axial- flexural and shear design shall be the same. It has been observed from the research carried out by [8] the UBC later claim could be realized by shear design based on calculation that demands that consider actual flexural capacity of the walls. The actual contribution of the edge members (columns and beams) panel RC/ prestressed concrete should be considered in evaluating available shear strength. Therefore the study has recommended that in designing and evaluating of strength of wall, cross section should be considered rather than the assumption of edge column acting as axially loaded members that should resist all vertical members from the loads acting on the wall.

Based on the aforementioned claim, the objective of this paper was to provide the average design thickness for the cement-plastered strawbale stiffens the prestressed burnt-brick frame. Data on strength and deformation of the structure are the input for the analytical models. The plastered strawbale wall was connected to the prestressed brick frame with mortar (1:6 cement-sand mix) to constitute a simple support.

2. Modelling of Structural Members

2.1 Frame: Beam and Column Members

A one-dimensional model was used for beam and column in this paper. The beam or column member was idealized as a perfectly idealized elastic massless line element with two nonlinear rotational springs at the two ends. The model could have two rigid zones outside the rotational spring.(Fig. 1).

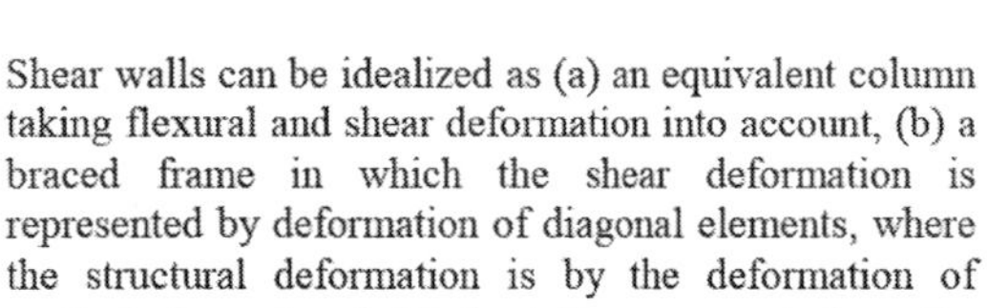

Fig. 1. One-dimensional model for beam and column

2.2 Analytical Model of strawbale Infill

Shear walls can be idealized as (a) an equivalent column taking flexural and shear deformation into account, (b) a braced frame in which the shear deformation is represented by deformation of diagonal elements, where the structural deformation is by the deformation of vertical element and (c) short line represent along the height with each short segment with hysteretic characteristics. These models have advantages and disadvantages. In most cases the horizontal boundary beams are assumed to be rigid.

*aaadeji@unilorin.edu.ng

The proposed analytical model assumes that the contribution of the straw bale wall infill panel to the response of the infilled frame can be modelled by replacing the panel by a system of two diagonal straw bale wall compression struts. Since the tensile strength of masonry is negligible, the individual straw bale strut is considered to be in effective in tension. However, the combination of both diagonal struts provides a lateral load resisting mechanism for the opposite lateral directions of loading. The lateral force-deformation relationship for the structural straw bale infill panel is assumed to be a smooth curve bounded by a bilinear strength envelope until the yield force and then on a post yield degraded stiffness until the maximum force is reached.

2. 2.1 Strawbale Wall as Infill Panel

The use of strawbale infill for the construction of a building in place of conventional materials is its cheapness, availability and flexibility in terms of workability and strength.

Though the individual straw bale strut is considered to be in effective in tension, yet strawbale masonry has a high tensile strength. The load resisting mechanism of infill frames is idealized as a combination of moment resisting frame system formed by the frame and a pin-jointed system formed by the strawbale. However, because of the absence of a realistic, yet simple analytical model, the plastered straw bales as infill panels might be neglected in the non-linear analysis of building structures. Such an assumption may lead to substantial inaccuracy in predicting the lateral stiffness, strength, and ductility of the structure.

Saneinejad and Hobbs [7] developed a method based on the equivalent diagonal strut approach for the analysis and design of steel or concrete frames with masonry infill walls subjected to in-plane forces. The method takes into account the elastoplastic behaviour of infilled frames considering the limited ductility of infill materials. The formulation provides only extreme or boundary values for design purposes. In the case of straw bale panel, the aspect ratio, shear stresses at the interface between the infill and frame, together with the frame strength were accounted for and the formulation expresses the boundary values for design purposes.

3. Equivalent Strut Model

The description, in brief, of the formulations for predicting the parameters of the strawbale infill-panel is presented in this section. Considering the straw-bale infilled frame as shown in Fig. 2, the maximum lateral force and corresponding displacement in the infill straw bale panel are considered.at lateral dimension of the wall panel for the masonry strength at the inclination of the diagonal strut to the horizontal at the shear strength (V= average of 0.347 N/mm^2), of straw bale wall. The area and length of the equivalent diagonal struts and are denoted as A_d and L_d, respectively.

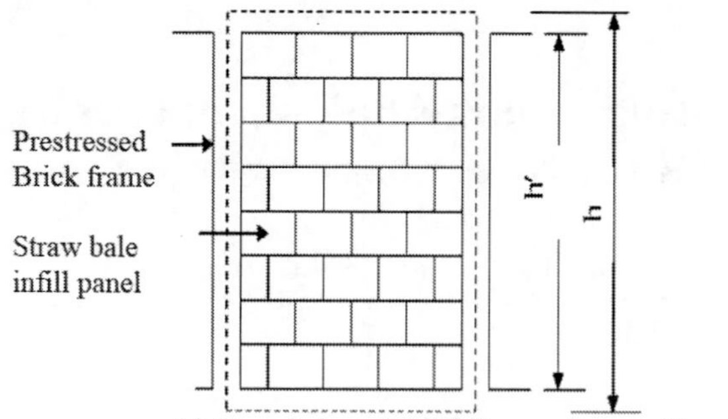

(a) Straw bale wall infill frame sub assemblage

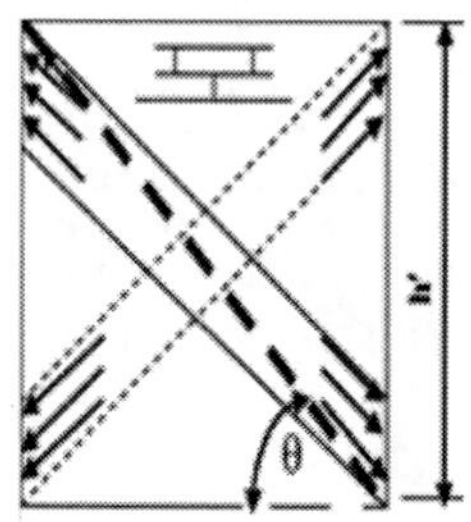

(b) Straw bale wall infill panel

Fig. 2. Wall equivalent diagonal strut

4.1 Time-independent smooth hysteresis model

A smooth hysteretic model proposed by [9] is used for the structural straw bale masonry infill panel. The model, which was developed based on the Bouc-Wen model for hysteresis behaviour furnishes a smooth hysteresis force-displacement relationship between force V and displacement U is,

$$V_i = V_y\,[am + (1 - a)\,Z_i] \tag{1}$$

Where ductility is calculated as Ui/Uy, subscript i = instantaneous values, subscript y = yield values, and Z = hysteretic component determined by solving the following differential equation by [10] as;

$$dZi = [a_{eff} - 1\ Z_i\ \text{In}\ [p\ \text{sgn}(d_m\ Z) + y]]\ di \tag{2}$$

where signum function sgn(x) = 1 for x > 0 and = -1 for x < 0; a_{eff} = 3 , and y = constants that control the shape of the generated hysteretic loops (Assumed values a_{eff} = 1, and p = y = 0.5. At x = 0, the effect of cyclic loading is neglected) and n controls the rate of transition from the elastic to yield state.

4.2. Stiffness Decay and Strength Degradation

The yielding system in general is the loss of stiffness due to deformation beyond yield point. The stiffness decay is incorporated directly in the hysteretic model by including the control parameter r) in equation (1) for hysteretic parameter Z in which 77 is obtained by pivotal deterioration method [11]. Degrading systems such as straw bale wall infill panels will also exhibit loss of strength in the inelastic range. The strength deterioration is modelled by reducing the yield force V_y from the original value V_{yo} at each step k,

$$V_{yk} = V_{yk} (1 - DI) \tag{4}$$

in which DI = cumulative damage parameter dependent on the maximum attained ductility, $I_{,max}$, and the cumulative energy dissipated [11, 12, 13].

4.4. Cracking slip model
Opening and closing of masonry cracks resulting in the pinching of hysteresis loops is a commonly observed phenomenon in masonry structural systems subjected to cyclic loading. The concept of slip lock element proposed by [9] was adopted in this study to formulate a hysteretic model. A slip-lock element is incorporated in series so that displacement ductility $j_u = j_{UJ} + j_{u2}$, with the smooth degrading element.

5. Experimental Tests

5.1 Tested Structure and properties of materials

The tested structure represented a three-storey one-bay (span prestressed concrete ductile moment resisting frame filled with straw bale masonry), built at a scale of ¼. The overall geometry of the test structure is depicted in Fig. 3. Analytical example was carried out for a full scale version of the test structure and all pertinent quantities were scaled using the length factor of 0.25. Each tested structure consists of strawbale wall (made of 127x78x61 mm^3 unit size; elephant grass bale plastered with 1: 6 cement-sand mix ratio; average density of 2200 kg/m3; 12% absorption and moisture content of 3.61. Its average compressive and flexural strengths are 2.83 and 1.56 N/mm^2) respectively) whose principal plane was parallel with the direction of the input motion. Design gravity loads comprised of 70 kN (self weight and live load). Seismic design effects were determined using modal spectral analysis. The design spectral ordinates were set so that the first-mode shear base was equal to the design base shear required by the UBC 1982 for a ductile moment-resisting space frame. Also, the response spectrum was selected to reflect soil condition that superseded by a layer of sandy soil. Earthquake design actions were determined with a three dimensional elastic analysis model based on the cross section of columns and beams.

The three dimensional structure was idealized as a pseudo-three-dimensional model in which only the planar modes of the main structural systems (walls and frames) were considered. The considerations of stiffness and strength of frame elements perpendicular to the plane of the transverse walls and frames were not considered. Perfect base fixity of the frames was assumed. Horizontal floor level responses have been recoded in lateral and main direction. At 3.2 m/s^2, the structure did not increase its response. The maximum base shear of 129 kN is obtained at the roof displacement of 12.3 mm.

. The input signals to the shaking table modelled accelerated history for uniaxial tests are as indicated in Table 1 with a total of four earthquake records. These values were used as input ground motions in the dynamic analysis. These values are notations used for individual ground motion, the components of earthquake motion record, peak ground acceleration (a_{eff}) in terms acceleration due to gravity (g), peak ground velocity (v_g), significant duration of the accelerogram (t_{SD}) and the normalized characteristics intensity (In).

Table 1 Characteristics of selected input earthquake ground motion

Ground motion	Notation	Peak ground acceleration (a_{eff})	Peak ground velocity (v_g) m/s	Significant duration (t_{SD}), s	normalized characteristics intensity (I_n)
El Centro	ELCENT	0.35g	0.34	24	2.66
Haruka-oki	HARU	0.42g	0.46	6	2.12
Silmar	SYLM	0.84g	0.61	47	1.71
Hacinohe	HACH	0.23g	0.34	28	1.52

6. Typical Analysis Example

6.1. Sections properties

Beam and column sectional area are shown in Fig. 6 . Second moment of area (I_b) = 2.278 x 10-3 m4(Ic) = 6.75 x 10 -4 mm4 respectively. The straw bale wall is measured acting as diagonal members of the frame in this analysis. It acts as a stiffening element and the area and length of the equivalent diagonals are as calculated below, Ld equivalent diagonal strut (Fig. 3)., Ad = 14235.0 mm2.6681.6mm and area of

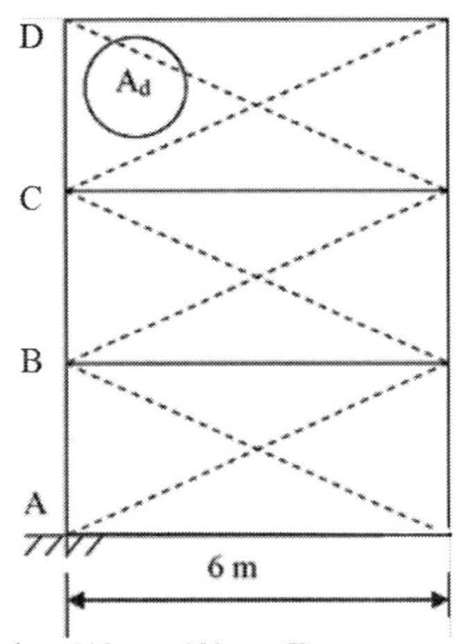

Beam section: 300 mm 450 mm T
Column section: 300 mm 450 mm
Fig..3. A three-storey building

6.2 Earthquake Analysis and Loading

For the absorption and dissipation of the energy through its motion, the frame of the building must be sufficiently ductile. For calculating ductility and other parameters it is assumed that the tests were entered to the ductile mode. It is assumed that during a strong ground motion the inertial of a building results in a horizontal shear force at the base and proportional to the weight of the building and the imposed ground motion, so that:

$$VD = C_s M g \quad (1)$$

where VD = total dynamic base shear, M = total mass of the building and its contents (kg), g = acceleration due to gravity, C_s = seismic coefficient, a_{eff} = dependent effective peak acceleration, S = seismic response factor, R = factor related to ductility of structure or force reduction factor, and In= normalized characteristics intensity. A good example of the seismic record characteristics is given by Moroni et al (1996) from 1985 Chilean earthquake. Assumed for the prestressed brick frame is the ductility factor R = 5 (braced frame, with effective peak acceleration aeff = 0.1 – 0.2g, seismic response factor S = 2 and the intensity factor = 1, so that the seismic coefficient Cs = 0.04. The total design base shear is 143.3kN.. The lateral forces on the building are distributed over its height accordingly. The sum of the lateral forces is equal to the total base shear. The lateral forces on the building frame are shown in Table 2.

Table 2 Seismic forces on Building frame

Storey level	Equivalent lateral force (kN)	Equivalent lateral shear
Roof	41.45	41.45
2	71.65	113.10
1	143.3	256.40
Total dynamic		256.4

The maximum lateral force and corresponding displacement in the infill straw bale panel are as follows: At 32° at V_m+ (V_m) = 14486N, while the corresponding displacement U_m+ (U_m) = 23.6mm. The initial stiffness Ko of the infill panel is estimated K_o =1227.63N/mm, the lateral yield force and displacement of the infill panel is 14193 .4 N and U_y +(U_{y-}) = 11. 7mm. Using different straw bale thickness, the result of the analysis for the stiffness deterioration and shear decay are shown in CASES II to IV.

CASE II, t = 275mm, Ad = 13048.8mm², V_m = 13279.2N, U_m = 23.6mm, K_O=1125. 4N/mm, V_y = 13145.1 N, U_y = 11.7mm.The result of the analysis is shown in Table.2 At the instantaneous value 6, other cases (CASES III and IV) with varying wall thicknesses (t) are:

CASE III at t = 250mm,
A_d = 11862.53mm², V_m = 12071.9N, U_m = 23.6mm, K_o=1023. 05N/mm, and V_y = 11949.9 N, Uy = 11.7mm .
CASE IV, t = 200mm,
Ad = 10961.62mm², V_m = 11155.2N, U_m = 23.6mm, K_o=945. 36N/mm and V_y= 11042. 5N, Uy = 11.7mm.

6.5 Deflection check

Maximum deflection of masonry wall is 24 mm. This value is greater than 23.7mm that was obtained in the analysis. Thus, the maximum deflection of the straw bale infill is less than the maximum allowed for masonry walls with the same dimensions (6.000 x 3.500 m).

Table 3. Observed and measured response maxima for base isolated motion (Table acceleration =1.07m2/t)

Response	Peak acceleration	Base shear kN	Total weight of the structure kN	Base moment kNm	Roof displacement Mm
Observed (prototype)	0.58	129	210	95.8	12.3
Calculated	0.1	143	365.62	158.6	23.6

6.6. Dynamic response results

At calculated Z =1, the stiffness decay is incorporated in the hysteretic model by including the control parameter T] for the hysteretic parameter (r)i > 1.0). The default value of Sk = 5 is recommended. The results in Table 4 shows a typical example of iteration of dynamic response characteristics from the analysis are as for 200 and 250 only.

Table 4 Dynamic response characteristics of straw bale infill thickness (t) = 200mm

Displacement (U_i)	Ductility (Hi)	Control values (r_{j_i})	Hysteretic parts (dZ_i)	Lateral force (V_i), N	Shear decay (V_m−V_i)
11.7	1.0	1	1	11042.5	112.70
11.8205	1.0103	0.9983	1.0017	11.062.22	92.98
11.9210	1.0189	0.9969	1.0031	11078.66	76.54
12.0045	1.0260	0.9957	1.0043	11092.32	62.88
12.0737	1.0319	0.9948	1.0053	11103.64	51.56
12.1311	1.0368	0.9940	1.0061	11113	42.20
12.1782	1.0409	0.9933	1.0067	11120.72	34.48
12.2171	1.0442	0.9928	1.0073	11127.09	28.11
12.2491	1.0469	0.9923	1.0077	11132.33	22.87
12.2754	1.0492	0.9920	1.0081	11136.64	18.56
12.971	1.0510	0.9917	1.0084	11140.13	15.03
12.3148	1.0526	0.9914	1.0087	11143.08	12.12

Other dynamic response results for t (thickness) = 275 and 300 are not shown here, but are represented in the graphical forms. The strength deterioration is modelled by reducing the yield force V_y from V_{yo} at each step k.

7 Results and Discussion

It was observed, from the result of the analysis, that the maximum lateral force the straw bale infill can be subjected to increases with the thickness of the infill panel. For thickness of 200mm, V_m = 11155.2N; for t = 250mm, V_m = 12071.9N; for t = 275mm, V_m =13279.2N and for t = 300mm, V_m = 14486N. Also, the stiffness and strength (hysteretic properties) of the straw bale wall infill decreases after the yield force. The yield force (V_y) being 14193.4N for t=300mm; 13145.1N for t = 275mm; 11949.9N for t = 250mm and 11042.5N for t = 200mm. The maximum deflection of the infill was obtained to be 23.7mm, which was less than the maximum 24mm specified for masonry infill walls. Also, it was found out that the hysteretic parameters deteriorate at a rate proportional to the thickness of the infill pane

Deflection (U,mm)

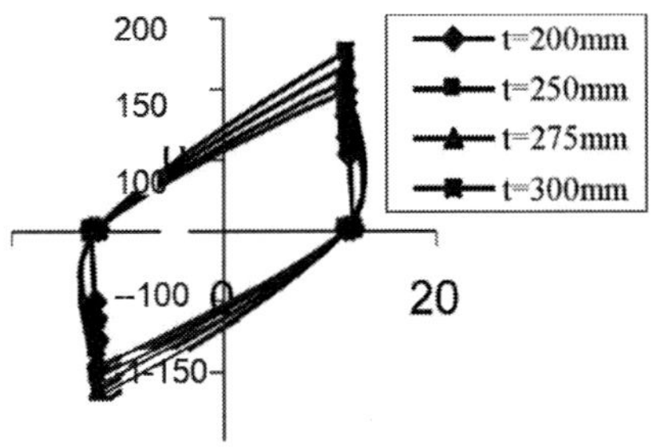

Fig. 4 Force-displacement relationship for strawbale infill (at t = 200 to 300mm)

Deflection (U,mm)

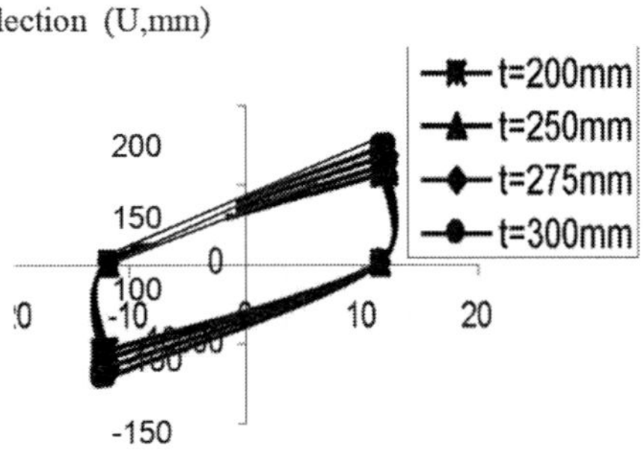

Fig. 5 Hysteresis curve with stiffness decay

Deflection (U,mm)

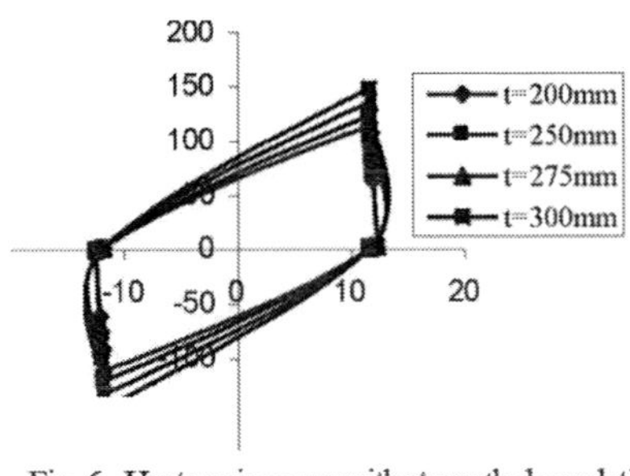

Fig. 6 Hysteresis curve with strength degradation

8. Concluding Remarks

The use of straw bale wall as an infill panel for the proposed hysteretic model can compete effectively with other conventional materials, such as earthwall (Adedeji, 2002), since its maximum deflection Um, is less than that specified for masonry walls. Though a perfect base fixity of the frames was assumed, the structure did not increase its response at 3.2 m/s^2. The maximum base shear of 129 kN is obtained at the roof displacement of 12.3 mm. these are far less values than the analysis results. The computed force-deformation response can be used to assess the overall structural damage and its distribution to a sufficient degree of accuracy.The structural materials characteristics were nominal values and may be different from the prescribed values. Such characteristics of materials for further analysis should be based on the confined and unconfined prestressed concrete.

References

1. R. E. Klinger and Bertero, V. V. Earthquake resistance of infilled frames, Journal of Structural Engineering, ASCE.104 (6), (1978),
2. V. V. Bertero and S. Broken, Infills in Seismic resistant building. Journal of Structural Engineering., ASCE. 109 (6): 1337-1361. (1983).
3. J. B. Mander and B. Na . Seismic resistance of brick-infilled steel frames with and without retrofit, The masonry J. 12 (2): 24-37. (1994).
4. M. Dhanasekar and A. W. Page, The influence of brick masonry infill properties on behaviour of infilled frames, Proc, Instin. of Civ. Engrs., London England, Part 2. 593-605 (1986).
5. Morom Mosalam, K.M. (1996). Modelling of the non-linear Seismic behaviour of gravity load designed infilled frames, 1995 EERI student Paper, Los Angeles, Calif. 1788. (1996).
6. P. K. Malnotra, Cyclic demand spectrum. Earthquake Engineering and StructuralDynamics, (31):1441-1457. (2002).
7. N. Buffalo, N. Y. Saneinejad, A., and Hobbs, B. Inelastic design of infilled frames. Journal of Structural Engineering. ASCE, 121 (4): 634-650.(1995).
8. E. Ahmet., Aktan, M., Vitelmo, V and Bertero, E., Seismic response of R/C frame-wall structures, Journal of Structural Engineering, 111(8), 1803 – 1821. (1984).
9. T. T. Berber and Noori, M. N.,. Random Vibration of degrading pinching system, Journal of Engineering Mech, ASCE 11(8): 1011-1026. (1985).
10. A. Madan, Reinhorn, A.M, Mander, J.B, Modeling of masonry infill panels for structural analysis, Journal of Structural Engineering, ASCE 123(10), (1997). (1997).
11. A. M. Rein, A. M., Madan, A.,Valles, R. E., Reichman, Y and Mander, J. B., Modelling of masonry infill panels for structural analysis, NCEER-95-0018, Nat.Centre of Earthquake Engineering, New York, (1995).
12. A. A. Adedeji, Hysteresis analysis of framed structures stiffened with earth wall infill subjected to seismic Impact, Conference Proc. SUSI–2002, UK. (2001)

https://doi.org/10.1051/matecconf/201816522009

13. A. Adedeji, Analytical approach to framed structures stiffened with earth wall infill subjected to seismic loads, LAUJET Journal, 1(1).(2003).

Effects of Elevated Temperatures on Mechanical's concrete specimen behaviour

Nesrine Khodja[1,*], and *Hadda* Hadjab[2]

[1]PhD student, University of science and Technology Houari Boumediene16111 Bab Ezzouar Alger, Algiers
[2]Professor, University of Science and Technology Houari Boumediene16111 Bab Ezzouar Alger, Algiers

Abstract. This paper presents an experimental study on the performance of concrete, subjected to high temperatures. Investigation was carried out by using two mixtures: an Ordinary Concrete (OC) and High Performance Concrete (HPC) with 10% of Silica Fume (SF) replaced of cement weight and 1,5% of super plasticizer, the water-binder (w/b) ratio used was 0,5 and 0,32 respectively. The produced concrete specimens are heated at a rate of 7 °C/min up to different temperatures (150, 300, 450, 600 and 900°C). In order to ensure a uniform temperature through the specimen, the temperature is held constant at the temperature stage for one hour before cooling. Mechanical properties at ambient temperature and residual mechanical properties after heating have already been determined. We examined also the effect of high temperature on the aggregate-cement paste interface and aggregate microstructure of the concrete specimens by scanning electron microscopy (SEM) method to reveal changes occurred after the cycle of heating-cooling. The results revealed that relative strength of the concrete specimens decreased as the exposure temperature increased and reaches about a quarter of its initial strength at 900°C. Moreover SEM examinations showed that, increasing temperature caused weakening of the adherence of aggregate and cement matrix by the appearance of cracks and micro cracks.

1 Introduction

At high temperatures or under thermal situations such as fire in tunnels (Channel, Mont Blanc, Great Belt Link, Tauern) and nuclear reactors, prediction of concrete performances is of great importance for safety evaluation of concrete structures. HPC is considered as a revolutionary material but its behavior is not yet mastered when exposed to high temperatures. Under these conditions, physical and chemical changes in the microstructure of concrete affect widely its mechanical behavior. Furthermore knowledge of the microscopic behavior of high temperature concretes is interesting to demonstrate the physico-chemical transformation and observe the influence of the different constituents of the concrete. Dehydration of the matrix components from 110°C to about 550°C causes shrinkage and thus cracking of the concrete [1, 2, 3, 4]. Similarly aggregates undergo changes following a high temperature rise, such as the decarbonation of calcareous aggregates at about 800°C-900°C [4, 5, 6]. The aim of this work was to study the effect of high temperatures (ranging from 20°C to 900°C) on the compressive strength of a high performance concrete comparing with a reference concrete, in addition to microstructural observations who allowed showing on a very good scale the appearance of cracks in the cement paste and around the aggregates.

2 Materials and Testing

The cement used in the present study was Portland cement (CEM II/B 42.5) with specific weight 3.15 and the specific surface area of PC measured with Blaine method was 4235 cm^2/g. Two types of sand are used, (S1) fine sand (0/1), with a modulus of fineness Mf = 0.97 with specific gravity 2.58, made it possible to carry out the correction of the granulometry of (S2) the coarse sand (0/4) of Mf = 3.1 with specific gravity 2.66, Making it possible to obtain a sand mixture of Mf = 2.5 and having a granulometric curve lying inside the normalized zone. One type of coarse aggregate was employed to carry out this research, Silicate-limestone aggregate (G) with specific gravity 2.71, the considered silica fume (SF) used contains a silica of 95% with a specific weigh 2.2. The superplasticizer used was incorporated in the concrete mixture to improve workability. The specific weight of super plasticizer was 1.07 g/cm^3.

The distribution of coarse aggregate (gravel) in comparison with the sand aggregate are shown in (Fig.1), Tap water at room temperature was used in all mixes.

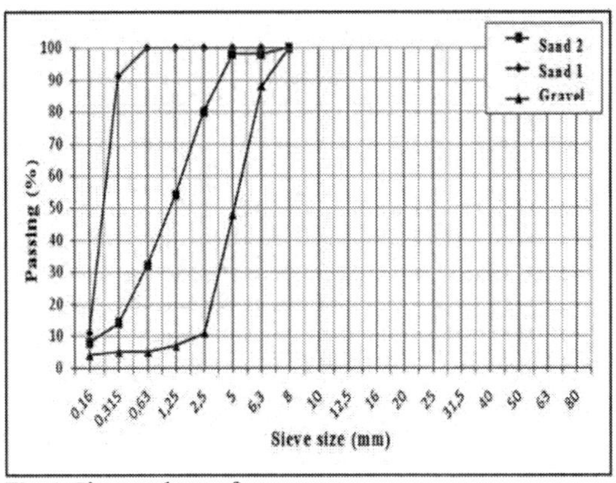

Fig. 1. Sieve analyses of aggregates.

*Corresponding author: khodja.nesrine@gmail.com

2.1 Mix proportion

In this study two groups of concrete mixtures were used. In group 1, Ordinary concrete (OC) was formulated according to Dreux-Gorisse method, the water–cement (w/c) ratio used in mixture was 0.5. On the other hand, group 2, mixture design was made with according to the formulation method of the University of Sherbrook, which makes it possible to formulate (HPC) without air entrained according to the standard ACI 211-1[7]. The water–binder (w/b) ratio used in mixture was 0.32. SF was used at 10% replacement by weight of cement. The proportions of concrete mixes are given in Table 1, Table 2 respectively.

Table 1. Composition of Ordinary concrete (kg/m3).

component	(%)	Density	Volume (l)	weight (Kg)
Sand S1	10%	2,58	63,6	164
Sand S2	25%	2,66	163,54	435
Gravel	65%	2,72	421,83	1147
cement	-	3,1	129,03	400
Water	-	1	208	208

Table 2. Composition of High Performance concrete (kg/m3).

component		(%)	Density	Volume (l)	weight (Kg)
Sand S1		11%	2,58	88,9	229
Sand S2		29%	2,66	228,6	608
Gravel		60%	2,72	476,25	1295
équivalent binder	C	90%	3,1	145,16	450
b = C+2SF	SF	10%	2,2	11,36	50
Water		5.8%	1	161,28	161
Superplasticizer (1.5% of b)		1.5%	1,07	2,26	7,5

2.2 Test procedure

Parallelepipedic specimens (4x4x16 cm) were prepared then cured in lime saturated water at 20 ± 2 °C until the time of the testing. After 28 days the specimens of concrete were placed in an electric oven for performing the fire tests (maximum temperature of 1200 °C).The used furnace (PROETI HD-230) could achieve a high heating rate speed that is somewhat similar to actual fire conditions. The dimensions of the oven hole are of 200 mm · 200 mm · 200 mm, provided with insulator material and the outer body is stiff steel, to safeguard the user in the case of concrete specimen explosion. Specimens are heated to desired temperature of 150, 300, 450, 600 and 900°C (Figure. 2). The heating rate was set at 7°C/min for all cases, which is consistent with RILEM 129-MHT standard recommendation [8, 9] and In order to ensure a uniform temperature through the specimen, this last is held constant at the temperature stage for one hour before cooling to room temperature at the air (Figure. 3).

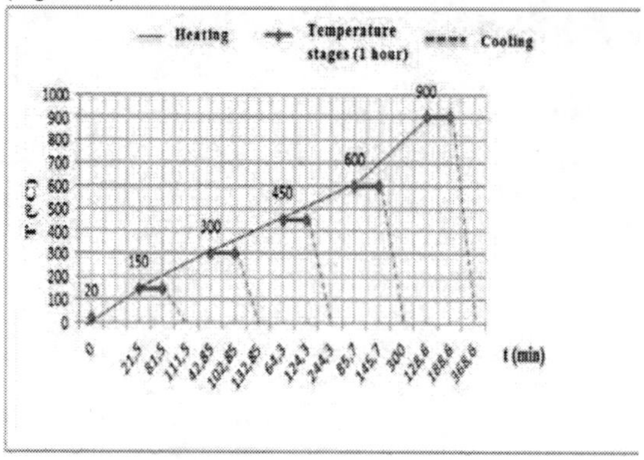

Fig. 2. Heating and cooling cycle.

Fig. 3. Cooling the test specimens in air.

The compressive strength of concrete specimens was measured owing to uniaxial compression instrument (Figure.4) with a capacity of 3000 kN and loading rate was 0.5 MPa/s. The compressive strength tests were carried out on the specimens after exposed to elevated temperatures. Three specimens were used for each temperature and each cooling regimes to perform compressive strength tests. And the average of three measurements of specimens was presented and discussed in the study.

Fig. 4. Compressing test on prismatic specimen.

After strength testing, the samples were collected and marked in air tight containers to undergo Scanning Electron Microscopy (SEM) for changes to their microstructure. In this work, microstructural examinations and analyzes were carried out using the Scanning Electron Microscope (SEM - QUANTA 600). The device is equipped with an EDS microanalyzer and can work under conditions of environmental pressure. Figure.5 gives an overview of this installation.

The microstructural analysis was carried out at well-defined temperatures, namely: 20°C, 300°C and 600°C. The samples heated to 900° C. were severely damaged by the heat shock resulting from air cooling so that these samples were not subjected to microstructural analysis therefore it was impossible to characterize the cracking.

Fig. 5. Scanning Electron Microscope QUANTA 600.

3 Results and discussion

3.1 Residual compressive strength

The concrete specimens produced in this study were exposed to the elevated temperatures at 150, 300, 450, 600 and 900°C, then their residual compressive strength values were determined. The residual compressive strengths of concrete specimens are presented in (Fig.6), for ordinary and high performance concrete successively. The residual compressive strength ratios at elevated temperatures as compared to the original strength before heating (20°C) are given in (Fig.7).

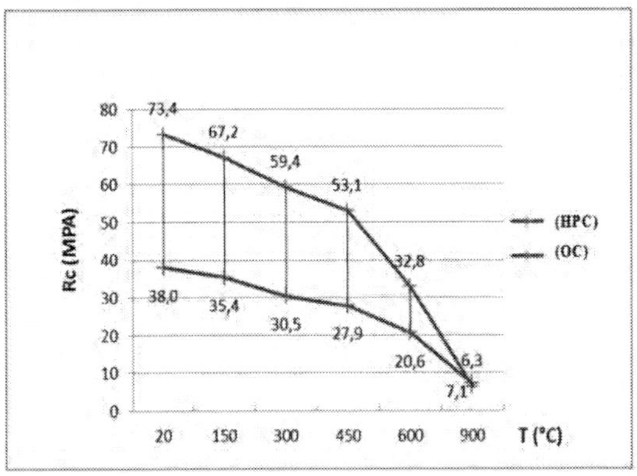

Fig. 6. Residual compressive strengths with increasing temperature

There are two concrete behavior zones. In the first zone, which varies from room temperature to 450 ° C, there is a decrease in resistance but in all cases of small magnitude. In the second zone starting from the intermediate limit to 900 ° C, there is a continual decrease in the compressive strength and this regardless of the type of concrete.

It was seen from (Fig.6) that the compressive strength value of high performance concrete decreases significantly with increasing temperature passing from 73.44 MPA at 20°C until reaching an ultimate value of 6.3 MPA at 900°C. It exhibited approximately 91% of the strength value which was measured at room temperature. In comparison with normal concrete who loses about 81% of its initial resistance. As reported in the literature [10, 11]. This decrease in compressive strength may be the occurrence of micro and macro-cracks in concrete because of thermal stress formations resulting from the temperature differences between surface and core high temperature [12].

At 150 °C, (HPC) and (OC) mixture was not affected significantly (Fig.7). It exhibited approximately 91.5 % and 93% respectively of the strength value which was measured at room temperature. Referring to [14] the decrease in resistance is due to an increase in the chemical energy of the water, which, by increasing the number of layers adsorbed on the surface of the solids, increases the disjunction pressures between the Different CSH layers.

When the temperature reached to 450 °C, the compressive strength of specimens decreased between 20% and 27.65%. This temperature is critical because calcium hydroxide dehydration takes place about 550 °C [14]. In addition, this result causes the shrinkage and strength loses of concrete [15].

The strength losses (41.81%, 55.3% respectively) which were occurred in the ordinary and high concrete specimens respectively increased significantly when the temperatures were reached to 600 and 900°C due to the decomposition of C–S–H gel [15,16]. The concrete specimens were completely disintegrated at 900 °C.

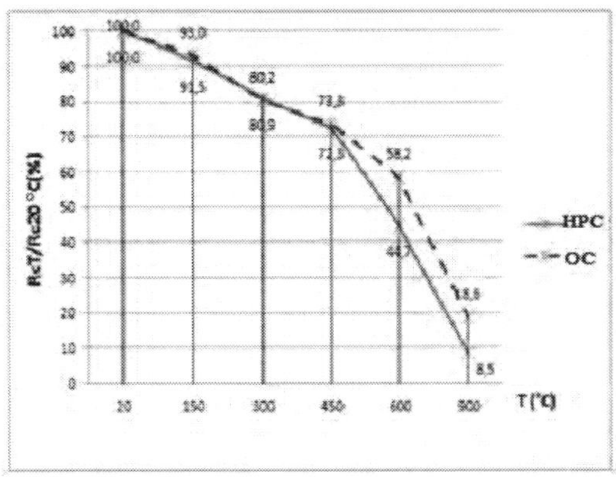

Fig. 7. Relative compressive strengths with increasing temperature

One possible explanation of this reduction in strength may be the incompatibilities between the different phases of concrete (matrix and aggregates). During cooling the stresses caused by the thermal incompatibilities of these materials induce cracking resulting in lower strengths [13].

3.2 Scanning Electron Microscopy (SEM)

3.2.1 At 20°C

Scanning Electron Microscopy (SEM) was used to obtain micrographs of (HPC) and (OC) samples before heating (at 20°C) and after exposure to high temperatures (300°C and 600°C) .

Fig. 8. SEM micrographs of OC sample at 20°C, 100x.

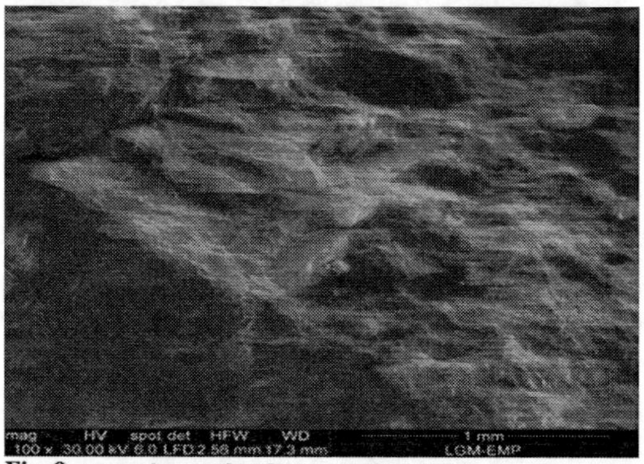

Fig. 9. SEM micrographs of HPC sample at 20°C, 100x.

The surface of the ordinary concrete sample that has not been subjected to high temperatures (Fig.8) does not have any facies of alteration and no longer a particular crack. The dense and amorphous structure of (HPC) testifies that this material has not been thermally degraded.

Moreover, we note that the (OC) has a large number of pores or voids that can initially be air bubbles, characterized by black cavities as shown in the micrograph (Fig.8) thus generating a less dense area and more permeable. Unlike that of (HPC) which has an amorphous structure (Fig.9), and much less porous than conventional concrete.

For (HPC) at 20 ° C with an enlargement of 400x. It is noted that there is a very good adhesion to the interface cement -granulate (Fig.11), it is almost perfect compared to that of ordinary concrete, as we can see the stripping (Fig.10) of materials due to the polishing operation, or evaporation of water, or air in the sample (OC). The exposed concrete in (Fig.11) containing silica fume has a stronger cohesion and is more compact compared to the BO, the fineness of the silica fume particles allows them to fit between the grains of cement and at the paste aggregate interface by creating additional CSH due to the pozzolanic reaction with portlandite. As a result, a denser paste, a finer porosity and better adhesion with coarse aggregates result.

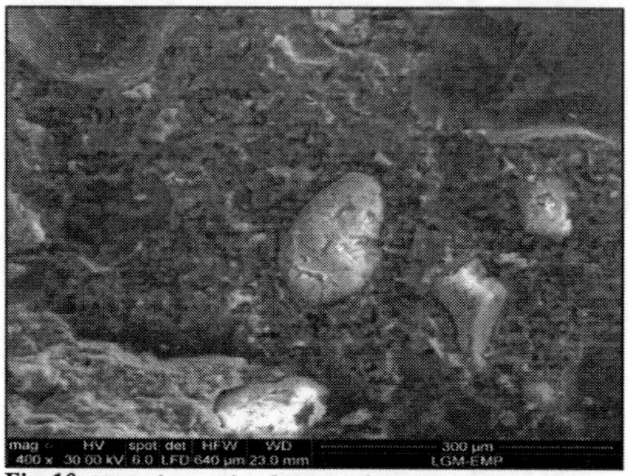

Fig. 10. SEM micrographs of OC sample at 20°C, 400x.

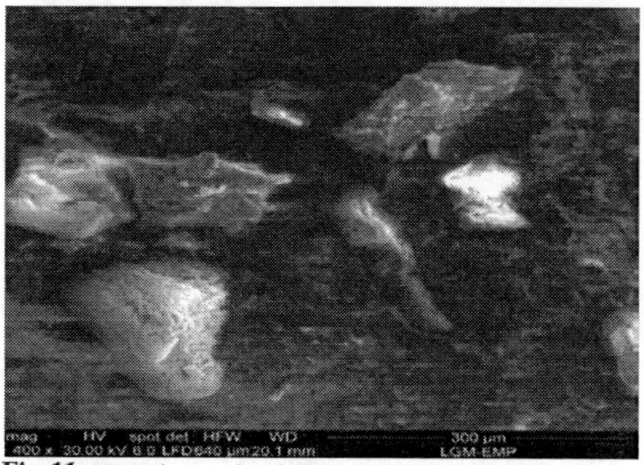

Fig. 11. SEM micrographs of HPC sample at 20°C, 400x.

For (HPC), there is certain homogeneity between the grains of sand and the cement. In addition, the mineral Silica Fume (SF) leads to a densification of the microstructure, which has the effect of reducing the porosity of the transition halo. All these modifications improve the transfer properties and therefore the durability of (HPC).

3.2.2 At 300°C

Aggregate surface texture is one of the most important factors affecting bond strength, for example rough surfaces usually have a higher bond than sawn surfaces. It is known that the transition zones (interfaces) are the weakest link of the concrete and crack growth usually starts at the aggregate–cement matrix interfaces [17].

In addition the adherence of aggregate and cement matrix of (OC) was observed as quite well (Fig.10). However, the adherence was not the same of specimen exposed to the temperature of 300°C, capillary fissure which did not wraparound the aggregate completely was observed at the aggregate–cement matrix inter-faces of some aggregates (Fig.12).

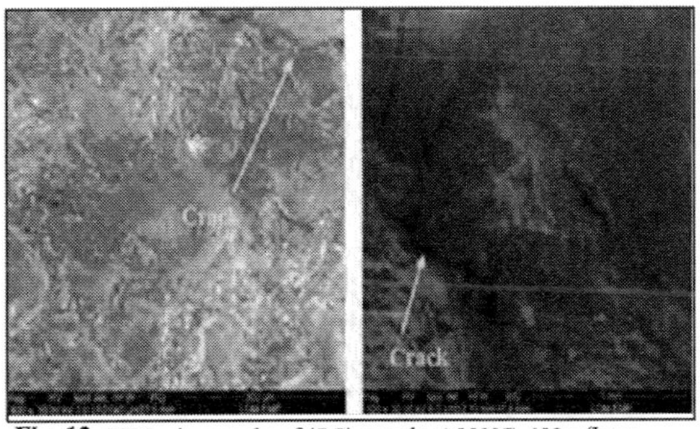

Fig. 12. SEM micrographs of (OC) sample at 300°C, 400x *(Inter-granular cracks)*

In high performance concrete cracks of different types are present. Trans-granular fissures passing through the aggregates (Fig.13, Fig 14) and inter-granular cracks (Fig.15), generally those of quartz nature. It is possible that these cracks are due to the phenomenon of cleavage which characterizes these aggregates at a temperature of 300° C. In addition to the Intergranular cracks that protrude at the paste-aggregate interface (Fig.14).

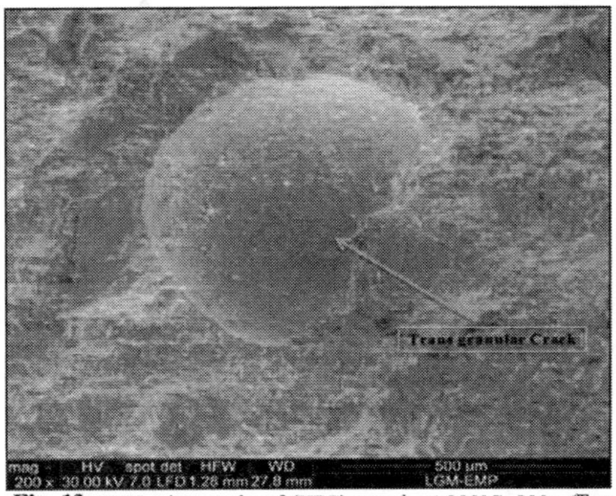

Fig. 13. SEM micrographs of (HPC) sample at 300°C, 200x *(Trans-granular cracks)*

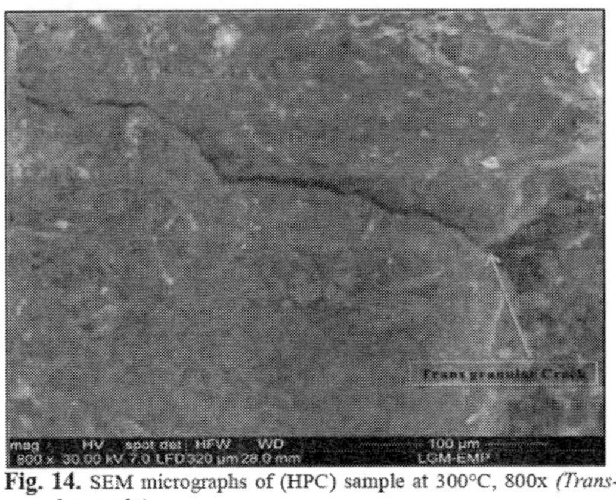

Fig. 14. SEM micrographs of (HPC) sample at 300°C, 800x *(Trans-granular cracks)*

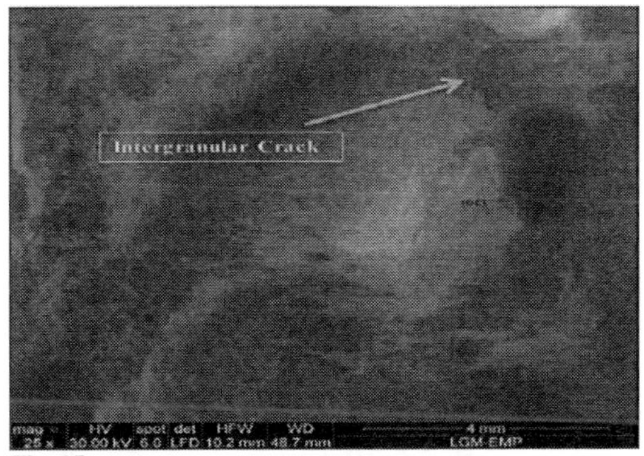

Fig. 15. SEM micrographs of (HPC) sample at 300°C, 25x.

According to (Fig.16), it is clear that the hydration products formed up to 300 ° C, CSH; have a massive structure. And the porosity of the HPC remains relatively low as the microstructure of the latter is not so damaged.

Fig. 16. SEM micrographs of (HPC) sample at 300°C, 400x.

The microstructure of high performance concrete is denser and more compact (Fig.16). Homogeneity of the matrices is however significantly different.

Fig. 17. SEM micrographs of (OC) sample at 300°C, 400x.

3.2.3 At 600°C

Cracking is much more pronounced for heated samples (Fig.18) and (Fig.19), especially at 600 ° C. At these temperatures, cracks are observed at the dough aggregate interface as at 300 ° C but also within the dough and in the aggregates.

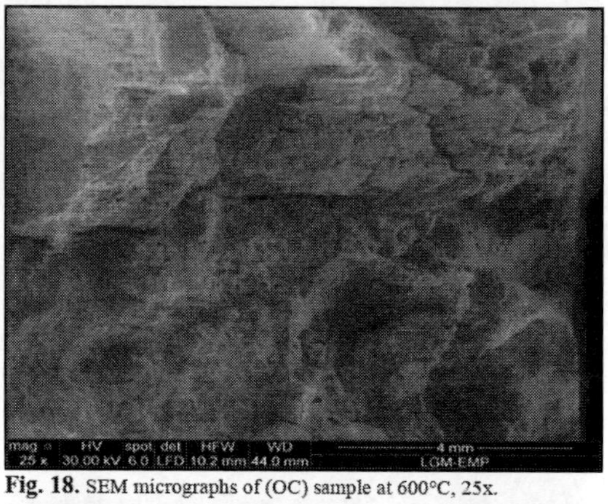

Fig. 18. SEM micrographs of (OC) sample at 600°C, 25x.

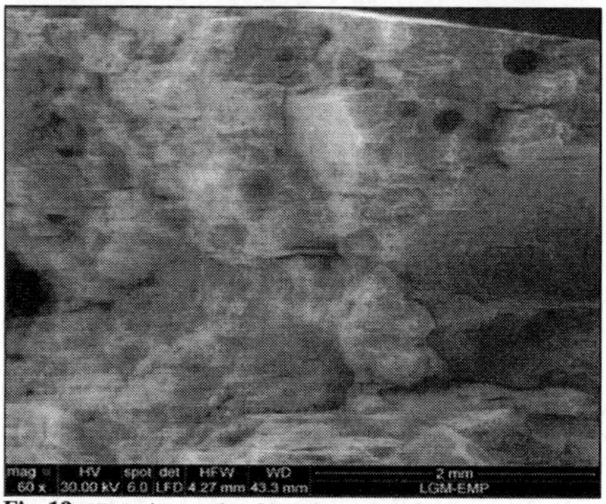

Fig. 19. SEM micrographs of (HPC) sample at 600°C, 60x.

From the two previous figures (Fig.18) and (Fig.19), it can be seen that the crack density is much greater in the two types of concrete than that resulting from the previous temperatures.

For (HPC) These cracks are fine (of the order of 1 micron) (Fig.19), (Fig.21) and form a kind of cracking while they are wide (of the order of 10 microns), cross the cement matrix (Fig.18), (Fig.20) and bypass the aggregates in ordinary concrete (BO). The origin of this behavior is not fully understood. Nevertheless, the authors, Noumowé et al [18] believe that the physico-chemical transformations that take place during the heat treatment and after cooling and because of the porosity generated are at the origin of the formation of cracks.

The deterioration of the cement paste after heating to 600 ° C is evident. We can observe the cracks that leave perpendicularly from the surface of a siliceous granulate (Fig.22). Observing an increase in the number of pores and micro cracks from 300 °C on. At 600 °C the author [19] identified the presence of sintered areas, cracks, and small ettringite crystals, probably formed after the samples were cooled.

Fig. 20. SEM micrographs of (OC) sample at 600°C, 100x.

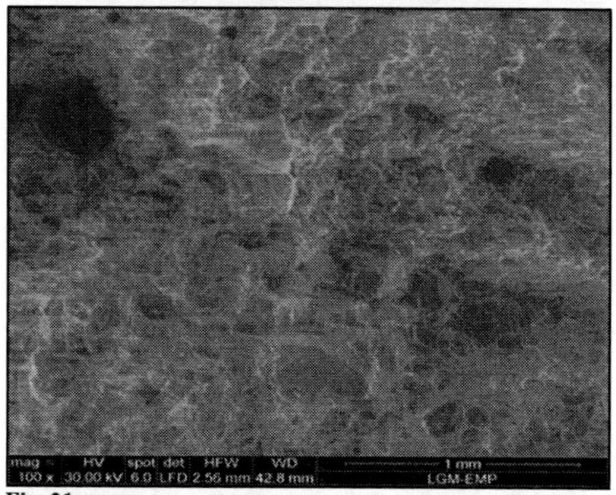

Fig. 21. SEM micrographs of (HPC) sample at 600°C, 100x.

Fig. 22. SEM micrographs of (HPC) sample at 600°C, 800x.

In addition, it is noted that the fracture surfaces of the samples, are intergranular in ordinary concretes (Fig.20), and are trans-granular in (HPC) (Fig.22). This can be explained by the fact that the cementitious matrix of the (OC) is less resistant because of weak bonds with the granulate which has a more permeable halo containing a large amount of Ca (OH)2 portlandit, unlike that of the (HPC) containing mainly CSH which gives to concrete a very strong bond making the dough more rigid than the aggregate, and gives it a low capillary porosity which leads to good resistance to chemical attack or mechanical action.

3 Conclusions

Based on mechanical's tests and SEM examinations made on concrete specimens subjected to elevated temperature, the following conclusions can be drawn:

1. The residual compressive strength values of concrete mixture decreased significantly depending on increasing temperatures. The strength losses which were occurred in the concrete specimens increased when the temperatures reached 600°C and 900°C due to the decomposition of C–S–H gel.

2. The study led to a deeper understanding of the phenomena occurring in concrete when it is heated to a high temperature and to reveal differences in behavior between ordinary and high performance concrete. The fact of cooling the concrete in the open air, the latter undergoes a thermal shock seeing its mechanical properties drop in a very fast way which is of the order of 81% for the (OC) against 91% for the (HPC) compared to initial resistance.

3. Inclusion of 10% SF into concrete (as a replacement of cement weight) significantly improved the compressive strength at 20°C and reduced it markedly with increasing temperature. The strength reduction starts from elevating temperature 450°C to 900 °C with sharp drop which may be due to internal cracks formation as a result of volume expansion of concrete matrix.

4. The significant drop in mechanical and microstructural properties may be the consequence of phenomena occurring during the fast cooling method.

5. Temperature causes chemical changes in the hydration products, leading to the appearance of cracks, voids and intense paste shrinkage. These effects, added to the differential movement between the paste and aggregates, promote intense degradation of concrete properties.

6. According to results of SEM examination, the adherence of aggregate and cement paste decreased depending on increasing temperature. Therefore, the most of the cracks develop at the weak interface around the aggregate.

7. Based on our previous observations, it is concluded that high performance concrete may have cracks of different types. First, the trans-granular cracks that passes through aggregates, usually those of a quartz nature. It is possible that these cracks are due to the cleavage phenomenon that characterizes these aggregates. Second, the cracks surrounding the aggregates. These are mainly related to the gradients of dimensional variations between the cement matrix and the aggregates. In this temperature range, the aggregates expand as the matrix shrinks by dehydration.

8. The mode of cracking is bound the interface area paste-aggregate which is considered the weak point of the concrete. This zone of high porosity plays an important role and its composition depends strongly on the nature of granulate and the properties of the cement paste.

9. The effect of heating is more pronounced for ordinary concrete, namely an increase in porosity due mainly to a transformation of small pores in large pores, presumably by rupture of the partitions of the capillaries under the effect of the pressure of water vapor generated by increasing the temperature.

10. High performance concrete is more vulnerable than ordinary concrete at high temperature.

References

[1] M. Abramowicz, R. Kowalski, *The influence of short time water cooling on the mechanical properties of concrete heated up to high temperature*, Journal of civil engineering and management. **Vol XI**, **N°2**, pp. 85–90, (2005).

[2] CEN/TC, 250/SC2 N 0466 prEN 1992-1-2 (Draft for stage 49) Eurocode 2, *Design of concrete structures*- **Part 1.2**: General rules – Structural fire design, (2002).

[3] I. Hager, *High temperature behavior of high performance concretes- evolution of the main mechanical properties*, Doctorat thesis of the national school of bridges and highways, France, (2004).

[4] P. Pimienta, I. Hager, *Fire behavior of high performance concrete with fibers*, report HPC **2000**, (2003).

[5] SK. Hando, S. Agarwal, SK. Agarwal *Physicochemical, mineralogical, and morphological characteristics of concrete exposed to elevated temperatures*, Cem Concr Res , **32:1009**-18, (2002).

[6] BM. Luccioni, MI. Figueroa, RF. Danesi, *Thermo-mechanical model for concrete exposed to elevated temperatures*, Eng Struct; **25:729**-42.

[7] CP. Aitcin, *High -performance concrete*, modern concrete technology **5**, University of Sherbrook, Canada, pp 216-220, (2001).

[8] RILEM TC129-MHT, Rilem Technical Committees 129-MHT - *test methods for mechanical properties of concrete at high temperatures*, part 1 : Introduction, part 2 : *Stress-strain relation*, part 3 : *Compressive strength for service and accident conditions*. Materials and Structures, **Vol. 28**, , no 181 :410-414 (1995).

[9] RILEM TC129-MHT, *Test methodes for mechanical properties of concrete at high temperatures*, Materials and structures, **Vol.33**, pp 219-223 (May 2000).

[10] H.Yang ,Y. Lin, C.Hsiao, J-Y. Liu, *Evaluating residual compressive strength of concrete at elevated temperatures using ultrasonic pulse velocity*, Fire Safety J, **44**:121–30, (2009).

[11] Ö.Arıöz, *Effects of elevated temperatures on properties of concrete*, Fire Safety J, **42**:516–22, (2007).

[12] H. Tanyıldızı, A. Cosßkun, *Performance of light weight concrete with silica fume after high*

temperature, Constr Build Mater, **22**:2124–9, (2008).

[13] M Talha Junaid, A. Khennane1, O. Kayali1, *Investigation into the Effect of the Duration of Exposure on the Behaviour of GPC at Elevated Temperatures*, MATEC Web of Conferences 11, University of New South Wales Australia, 5p, (2014).

[14] Georgali B, Tsakiridis PE. *Microstructure of fire-damaged concrete.* A casestudy. Cem Concr Compos,**27**:255–9, (2005).

[15] Ö. Arıöz, *Effects of elevated temperatures on properties of concrete*. Fire Saf;42:516–22, (2007).

[16] N.Yüzer, Aköz F, Öztürk LD. *Compressive strength-color change relation inmortars at high temperature*. Cem Concr Res, **34**:1803–7, (2004).

[17] G.Giaccio, R.Zerbino, *Failure mechanism of concrete combined effects of coarse aggregates and strength level*, Adv Cem Based Mater, 7:41–8, (1998).

[18] A.Noumowé, *Effet de hautes températures (20-600°C) sur le béton. Cas particulier du béton a hautes performances* , Lyon, 233p Institut National des Sciences Appliquées de Lyon, (1995).

[19] M.Sousa. *Estudo experimental do comportamento mecânico e da microestrutura de um concreto convencional após simulação das condições de incêndio*. Belo Horizonte. Thesis (Master) - Universidade Federal de Minas Gerais, 126 p, (2009).

 https://doi.org/10.1051/matecconf/201816522011

The effects of shot peening on subsurface layer properties and fatigue performance of case-hardened 18CrNiMo7-6 steel

Hsin Shen Ho, *Junfeng* Xie, *Lingli* Sun, *Penghui* Niu, and *Erliang* Zhang*

School of Mechanical Engineering and Henan Key Engineering Laboratory of Anti-Fatigue Manufacturing Technology, Zhengzhou University, Science Road 100, 450001 Zhengzhou, China.

Abstract. The present study is conducted with a dual-aim; firstly, examine the effect of several single shot peening conditions on the subsurface layer properties and fatigue performance of the case-hardened 18CrNiMo7-6 steel. By carrying out the subsurface integrity analysis and fatigue testing, the underlying relationships among the peening process, subsurface layer property and fatigue performance are investigated, the way peening conditions affect the fatigue life and its associated scatter for the case-hardened 18CrNiMo7-6 steel is quantitatively assessed.

1 Introduction

High strength steels are massively used in the aeronautical and automotive applications for their high-loading capacity. With the increasing demand for high reliability products, the dual properties of high strength and high ductility are expected for such steel materials to give a longer service lifetime of engineering machinery. This motivates the design of a new class of material, referred to as gradient nanostructured (GNS) high strength steel [1]. As fatigue failures generally occur on the material surface and propagate into the interior, a structural architecture comprising a GNS subsurface region and a coarse-grained interior is considered optimal for enhanced fatigue behavior. The GNS subsurface layer can be formed through surface strengthening treatments by means of mechanical method such as shot peening (SP), which entails impacting a surface with small, round and hard particles with force sufficiently high to create localized plastic deformation for the formation of a deep work-hardened and compressed subsurface layer in metals and alloys [2]. Despite the development of many new types of peening techniques such as laser shot peening and ultrasonic shot peening [3-4], conventional shot peening method remains one of the most important means in the industrial production due to its high flexibility and productivity, low cost and environmental friendliness. For these reasons, shot peening process is of particular interest in the present work.

Mechanical methods can be combined with other surface strengthening treatments such as thermo-chemical methods for further improvement in fatigue strength and fatigue life of steel components [5-7]. However, the hybrid process cannot always be advantageous to the application of steels. For instance, it is shown in Ref. [6] that when both carbo-nitrided gear and carbo-nitrided plus peened gear made of AISI 4130 steel are subjected to contact fatigue testing, fatigue damages, which are ought to be in different modes, are found to occur after about the same testing time for both types of methods. This could most probably be a consequence of the use of inappropriate shot peening condition, the extreme shot bombardment on the surface can cause roughness increment and flaw occurrence, and therefore outweigh the beneficial effect brought by the hybrid process.

Thus, understanding of how the mechanical process parameters affect the heat-treated surface and subsurface layer properties, which in turn influence the fatigue performance is the ultimate aim of the present work. To tackle this issue, high strength 18CrNiMo7-6 steel, widely employed in high-speed and heavy-duty gear applications, is selected as the target material. The carburizing and shot peening are respectively selected as thermo-chemical and mechanical methods since they are the versatile surface strengthening treatments commonly used in the automotive industry. In the present work, a few single shot peening conditions with different Almen intensities and shot diameters are carried out on 18CrNiMo7-6 steel posterior to the carburizing process. By carrying out the surface and subsurface layer property analysis and fatigue testing, this work is aimed to establish qualitative inter-relationships between peening process conditions, surface and subsurface layer properties and fatigue performance.

2 Material and methods

2.1. Gear material

* Corresponding author: erliang.zhang@zzu.edu.cn

In the present study, the target material is 18CrNiMo7-6 steel with its chemical composition summarized in Table 1. All the specimens are carburized in a controllable gas carburizing furnace, then subsequently quenched in oil to room temperature, and finally followed by tempering and cooling in air. The thermo-chemical conditions are utilized in agreement to gear heat treatment standard ANSI/AGMA B89-2004 [8].

Table 1. Chemical composition of 18CrNiMo7-6.

Element	Composition (wt %)
C	0.15-0.21
S	≤ 0.035
Si	≤ 0.40
Mn	0.50-0.90
P	≤ 0.025
Cr	1.50-1.80
Ni	1.40-1.70
Mo	0.25-0.35
Fe	balance

2.2 Shot peening treatments

The shot peening treatment is performed with the air blast shot peening machine. The specimens are treated by shot peening with different intensities and shot diameters. The intensities of peening are measured by the arc height of Almen specimens, varying from 0.25 mmA – 0.45 mmA . Cast steel balls with diameters of 0.3 mm and 0.6 mm are selected as shot blasting media in accordance to ISO 26910-1 [9]. The coverage of shot peening is 200 % for all peening processes, which is obtained by doubling the exposure time needed to achieve full coverage (100%). The different shot peening configurations of immediate interest are presented in Table 2.

Table 2. Shot peening parameters for case-hardened 18CrNiMo7-6 steel.

Variant	Shot diameter (mm)	Coverage (%)	Almen intensity (mmA)
SP0	-	-	-
SP1	0.3	200	0.25
SP2	0.6	200	0.25
SP3	0.6	200	0.45

2.3 Measurement methods

The cross-section of the specimen is first mechanically polished using standard metallographic techniques, then chemically-etched by 8% Nital. The cross-section microstructure of the specimen is subsequently investigated by scanning electron microscopy (SEM) using Zeiss Auriga FIB-SEM at room temperature.

The microhardness is measured along the depth of the cross-section of the specimen using a Vickers microhardness tester with a load of 200g and a holding time of 10s.

The residual stress along the depth of the cross-section of the specimen is measured by the X-ray stress analyzer (LXRD, Proto, Canada) using Cr Kα radiation and martensitic {211} crystalline plane. The voltage and current are respectively 30 kV and 25 mA. The measured interference peaks are evaluated by the sin 2ψ method with the diffraction angle (2θ) varying from -45° and 45°.

The surface state of the specimen is examined by Veeco Wyko NPFLEX optical profiler. The measured surface topography is characterized by amplitude surface roughness parameters in agreement to ISO 25178-2012 [10].

2.4 Fatigue testing

Following the set-up and procedure for conducting the SAE single tooth bending fatigue test using a servo-hydraulic fatigue testing machine [11], the applied cyclic fatigue load is 0.82 of the yield stress of the bulk material, the stress ratio $R = 0.01$ and the nominal cyclic frequency is 89 Hz at room temperature. Each test series includes six gear tooth specimens. All the specimens are tested until occurrence of fracture, else the tests are terminated at 3×10^6 cycles.

3 Results and discussions

3.1. Microstructure

In Figure 1, the cross-sectional surface by SEM is presented for the carburized specimen with different microstructures.

Fig. 1. Cross-sectional SEM of SP0; the microstructure of (a) top surface region and (b) bulk material.

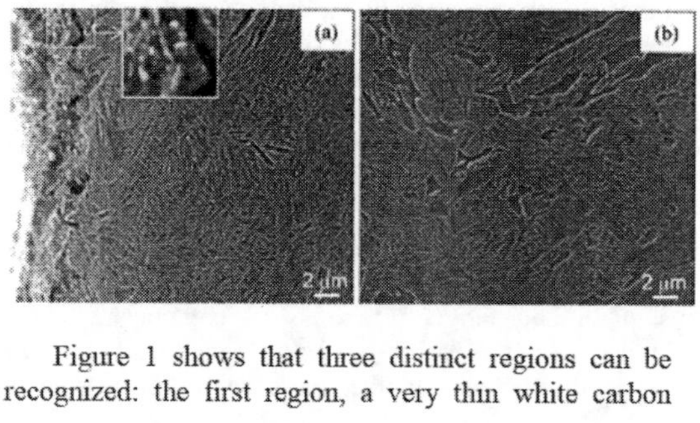

Figure 1 shows that three distinct regions can be recognized: the first region, a very thin white carbon oxide layer which is hard and brittle with the thickness lying between a few microns with the presence of the pores, carbides and micro-cracks within this oxidized layer; the second region, the case-hardened zone which consists mainly of martensite microstructure with a very small quantity of retained austenite and finely dispersed undissolved spherical carbide particles; and the third region, the bulk material, i.e. the substrate without any evidence of microstructural change.

Figure 2, illustrating SEM micrographs of the cross-section of 18CrNiMo7-6 steel subjected to the combined method, shows that shot peening performed after carburizing can not only suppress the porous structure at the outer oxidized layer but also change the microstructure of the near surface region of 18CrNiMo7-6 steel.

Fig. 2. Cross-sectional SEM of SP3 specimen; the microstructure of (a) top surface region and (b) bulk material.

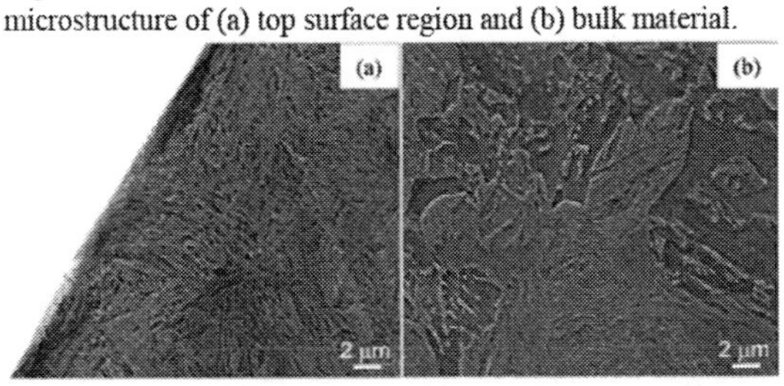

In fact, shot peening is a process which consists of projecting high-velocity shot particles by the action of an external force. So, when the external force reaches the critical stress required for dislocation movement, dislocation slip is generated and well-defined slip band structure is therefore formed. On the same slip plane, when the amount of dislocations impeded at the grain boundaries increases, the pile-ups of dislocations are formed [12]. Consequently, the phenomenon of stress concentration can easily occur in the vicinity of grain boundaries due to the heavy accumulation of the dislocations pile-ups against the boundaries. When a critical stress value is reached, the grain boundaries are destroyed and a high-density nano-scale deformation twins is formed. The intensive interactions between twin boundaries and dislocations may then lead to subsequent transformation of two-dimensional twin lamellae structures into three-dimensional nanograins with their boundary characteristics lying in the form of low-angle and high-angle. Herein, there are a considerable fraction of low-angle grain boundaries and a high density of dislocations resided in the non-equilibrium high-angle grain boundaries [13]. In addition to plastic deformation-induced grain refinement, strain-induced phase transformation is another mechanism by which nano-surface forms during shot peening. The strain-induced nucleation of martensite transformation can take place more dominantly at higher-energy grain boundaries due to their atomic structures and energies. Besides, the strain localization bands and deformation twins that accumulate high strain energy storage could also promote martensitic transformation. All these are elaborated here to show that the principal microstructural changes induced by shot peening consist mainly of plastic deformation-induced grain refinement and strain-induced phase transformation.

3.2 Hardness

Figure 3 presents the distributions of in-depth microhardness of all specimens. For SP0 carburized specimen, the maximum microhardness appears on the surface, then gradually decreases with increasing depth and finally reaches a stabilized value, which is well-accepted to be a consequence of the declination of the carbon content from the surface to the center. Meanwhile, when shot peening is performed on case-hardened steel, regardless of shot peening condition, a similar trend of the microhardness distribution is also observed, which is generally found to be attributed to the shot peening-induced reduction of plastic deformation from the outer surface to the center.

Fig. 3. Microhardness distribution curves of all surface treated specimens.

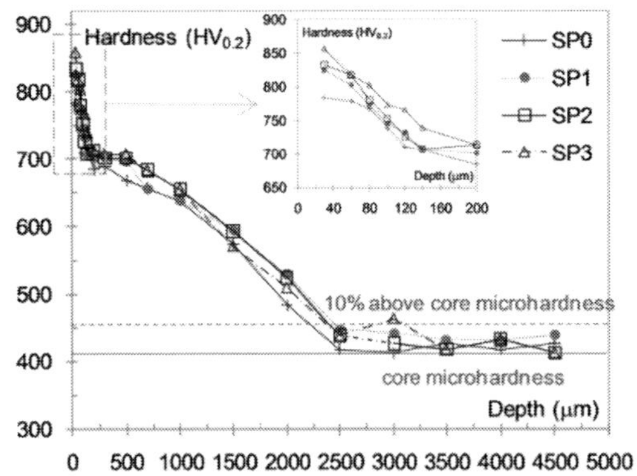

Case depth after thermo-chemical treatments is very often considered as a matter of convention. When fatigue characteristics are regarded, a microhardness value of 10% above the core microhardness is generally used to characterize the case depth [7]. The case depth is approximated to be ~2250 μm, as exhibited in Figure 3.

The hybrid process of carburizing and shot peening has caused an increment in surface microhardness from the top surface up to a limited depth approximated to be ~140 μm. Compared to the maximum surface microhardness of carburized SP0 specimen (784 $HV_{0.2}$), that of carburized and peened are estimated to increase by 9.3% for SP3 specimen (857 $HV_{0.2}$), by 6.3% for SP2 specimen (833 $HV_{0.2}$), and by 5.2% for SP1 specimen (825 $HV_{0.2}$), which can be attributed to the presence of shot peening-induced structure consisting mainly of the martensitic state. The increase in the maximum surface microhardness caused by the increase of shot diameter is quite comparable to that resulted by the increase of Almen intensity. Indeed, the values of the maximum surface microhardness which are found to increase by ~2.9% with the increment of the Almen intensity and by ~1.0% with the increasing of the shot diameter shows that the maximum surface microhardness-increment is restricted to some extent, which infers that the plastic deformation of 18CrNiMo7-6 steel subjected to the combined processing of shot peening and carburizing has reached a critical condition.

The results hereby obtained show that the thermo-chemical surface treatment is kind of advantageous in terms of microhardness distribution compared to mechanical surface treatment, in accordance with the results obtained in Ref. [14].

3.3 Residual stress

In Figure 4, the residual stress in-depth distributions of all the surface treated specimens are depicted. It is shown that in the range of testing, all the specimens have similar trend of obvious compressive residual stress distribution; the top surface compressive residual stress first increases to a maximum (denoted as maximum compressive residual stress), then decreases gradually along the depth direction of the hardened layer and remains constant thereafter.

Fig. 4. Residual stress distribution curves of all surface treated specimens

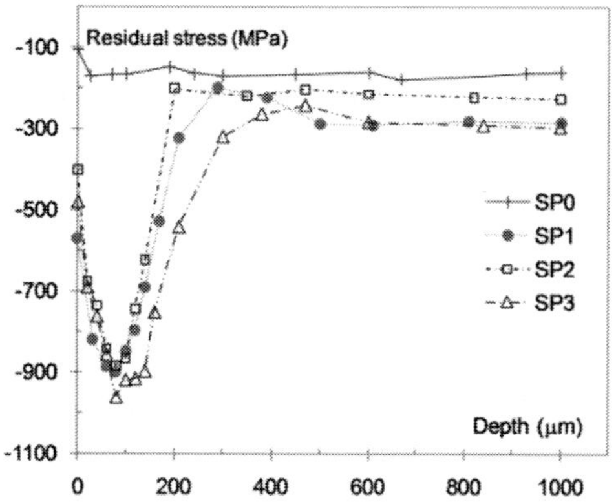

As can be seen from the Figure 4, the increment of the process-induced top surface compressive residual stress and maximum compressive residual stress are much more important for the combined processing of carburizing and shot peening than the individual carburizing process only. The observation shows that shot peening process is highly beneficial in generating compressive residual stress as compared with the carburizing process.

Regardless of the shot peening process parameter, the values of the compressive residual stress are exhibited to be lower at the top surface than at the peak position and the maximum compressive residual stresses are revealed to be located in the subsurface region at the depth of ~90 μm. The former is found to be resulted from the counteracting effects of not only the tensile residual stress but also the high surface roughness produced by prior machining. And the latter suggests that the steel material which undergoes plastic deformation has most probably reached a critical plastic deformation limit, and thus, the continuous impingement of the shot particles would not further deepen the compressed subsurface region.

In Figure 4, the values of the compressive residual stress observed at the top surface and peak position are increasing with the increase of Almen intensity. This is because the increase in Almen intensity, which involves the increase in shot velocity, produces more kinetic energy and thereby, induces a greater amount of plastic deformation on the hardened surface region, and as a consequence the compressive residual stress is elevated. However, with an increase of the shot diameter, the top surface and maximum compressive residual stresses are on the decrease. This is because when larger diameter shots are projected on the surface of the specimen with a small-scale finite kinetic energy, a larger impact area is created, resulting thereby in a reduction of equivalent plastic strain and consequently, the compressive residual stress is lowered.

It can be seen at the top surface region that the effect of shot diameter is more significant than that of Almen intensity on the compressive residual stress; the decrease of shot diameter induces an increase of compressive residual stress by 42.0% while the increase of Almen intensity causes an increase of compressive residual stress by only 19.7%. It can also be observed from Figure 4 that the values of the compressive residual stress at the peak position (~ 90 μm) appear to be less severely affected by shot diameter rather than Almen intensity; the increase of Almen intensity leads to an increase in compressive residual stress by 8.5% but the compressive residual stress grows by only 1.8% with the decrease of shot diameter.

3.4 Roughness

In what follows, the roughness characterization results are presented. The arithmetic mean height parameter (S_a), which is a commonly used roughness parameter for characterization of measured surface texture, is used to discuss the surface states of treated specimens. With the S_a value of ~2.51 μm, the surface of the carburized specimen can be considered to be rather rough. In this instance, the roughness can be ascribed to the roughness left by machining (*c.f.* Figure 5(a)) but also to the formation of the oxide layer produced by carburizing (*c.f.* Figure 1(a)).

Table 3. S_a values of all surface treated specimens.

Variant	S_a (μm)
SP0	2.51
SP1	1.97
SP2	2.69
SP3	1.89

It can be seen from Table 3 that the magnitude of roughness due to shot peening induced at the caburized specimen strongly depends on the applied shot peening conditions. When the Almen intensity is low, the impact of shot peening using smaller shot blasting particles is equivalent to the effect of surface polishing and as a consequence, surface roughness is reduced; the surface roughness of SP1 specimen (with $S_a \approx 1.97$ µm, *c.f.* Figure 5(b)) is exhibited to decrease by 21.5% compared to that of SP0 specimen (with $S_a \approx 2.51$ µm, *c.f.* Figure 5(a)). Though, the use of smaller shot media is unable to provide sufficient plastic deformation that could modify the surface texture left by the heat treatments and machining operations (*c.f.* Figure 5(b)). If shot peening is performed using larger shot blasting particles, the surface roughness is increased; the roughness of SP2 specimen (with $S_a \approx 2.69$ µm, *c.f.* Figure 5(c)) is found to be 7.2% higher than that of SP0 specimen. Though, the use of larger shot media producing a higher level of plastic deformation would allow obtaining a more evenly distributed texture (*c.f.* Figure 5(c)).

Fig. 5. Surface topographies at the root surface of (a) SP0, (b) SP1, (c) SP2 and (d) SP3.

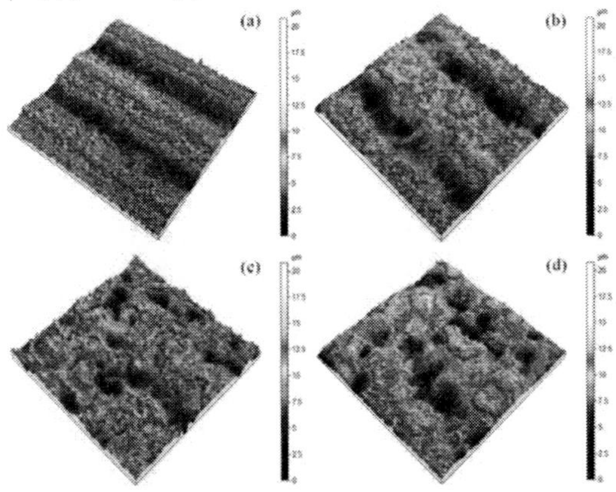

With increasing Almen intensity from 0.25 mmA to 0.45 mmA, the surface state of the specimen is improved and thus, the surface roughness is reduced; the S_a value of SP3 specimen ($\approx$ 1.89 µm, *c.f.* Figure 5(d)) is approximated to be 24.7% less than that of SP0 specimen. This can be explained by the fact that the very hard surface layer has been subjected to the "intense" shot peening process. It is worth mentioning that using a high Almen intensity at 200% coverage can be interpreted as if every portion of the hard surface has been treated with very high impact energy shot blasting particles for at least two times. Thereby, less plastic deformation is generated as the surface roughness is hardly altered and consequently, the surface roughness is considerably being lowered. This observation is found to be in accordance with the results exhibited in the literature [10].

3.5 Fatigue test data

The fatigue tests are carried out to examine the fatigue performance of all the surface treated specimens. In Figure 6, the fatigue lives (N_f) of these surface treated specimens are depicted. N_f for case-hardened specimens is determined in the range of 14.8 - 57.9 × 10^3 cycles. It is observed in the figure that the shot peening process has considerably improved the fatigue lives of case-hardened specimens, which is attributed to the principal effects induced by shot peening, e.g. microstructure refinement, retained austenite reduction, microhardness amelioration, compressive residual stress generation and surface roughness alteration. It is worth noting that the trend of fatigue life improvement is highly dependent on the surface and subsurface layer properties of 18CrNiMo7-6 steel, which in return, are directly influenced by the shot peening conditions.

Fig. 6. Bending fatigue lives of surface treated specimens.

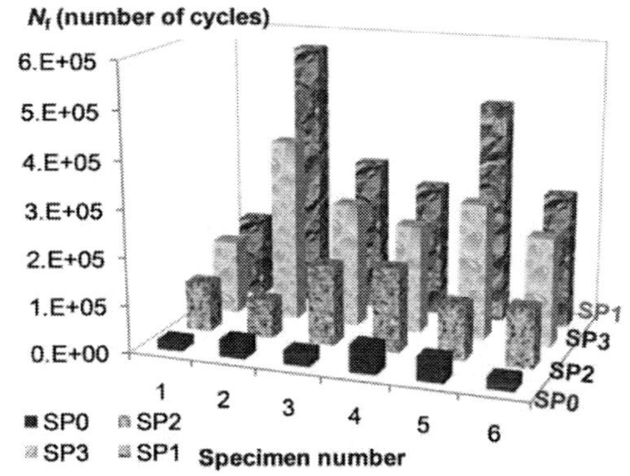

It is clearly shown in Figure 6 that N_f for SP2 specimens is the lowest and it is improved with an increment in the Almen intensity, as is exhibited for SP3 specimens. This considerable improvement is solely a consequence of the presence of deeper compressed work-hardened subsurface layer with ameliorated surface roughness in SP3 specimens. Meanwhile, SP1 specimens are found to have the highest N_f in spite of having moderate work hardening and compressive residual stress. The marked increase in N_f for SP1 specimens can be related to the fact that the magnitude of the surface roughness is the lowest among all peened specimens (*c.f.* Table 3). Nevertheless, the dispersion degree of the fatigue life of the SP1 specimen is found to be rather high, i.e. the repeatability rate of the SP1 specimen test sequences is rather low. This might be a consequence of the presence of the scatter in fatigue strength endowed by the surface and subsurface layer properties, which is unable to completely cause a shift in fatigue crack initiation site from a surface to subsurface location, and therefore cause the simultaneous occurrence of both surface and subsurface fatigue crack initiation in a fatigue testing series. This certainly intensifies the dispersion of the fatigue life.

4 Conclusions

In this paper, the effect of different shot peening conditions on the surface and subsurface layer properties, and the fatigue performance of case-hardened 18CrNiMO7-6 steel has been investigated. The analysis of the surface and subsurface integrity reveal that the subsurface region of the case-hardened 18CrNiMO7-6 steel can no further accumulate enormous localized plastic deformation during shot peening process as a consequence of the attainment of a critical limit of the plastic deformation for 18CrNiMo7-6 steel due to the presence of a very high dislocation density and carbon content induced by the prior carburizing process. Based on the assessment of fatigue performance, the inter-relationships between peening process – subsurface layer property - fatigue performance are qualitatively established; shot peening application with lower intensity and smaller shots causing moderate subsurface layer properties results in an increase in both fatigue life and its associated scatter, while shot peening application with higher intensity and larger shots producing superior subsurface layer properties exhibits comparable estimate of fatigue life with decreased scatter in fatigue data.

Acknowledgement. The authors are grateful to National Natural Science Foundation of China - International Cooperation and Exchange Program (Grant No. 51650110502) and Henan Provincial Research Bureau (Grant No. 17A460025) for their financial support.

References

1. K. Lu, Science **345**, 1455 (2014)
2. Y.K. Gao, X.R. Wu, Acta Mater. **59**, 3737 (2011)
3. S. Huang, Y. Zhu, W. Gui, P. Peng, H.C. Qiao, X.G. Diao, P.K. Chu, J Mater. Eng. Perform. **25**, 813 (2017)
4. V. Singh, M. Marya, J Mater. Eng. Perform. **25**, 338 (2016)
5. Y. Lv, L.Q. Lei, L.N. Sun, Mater. Sc. Eng. A **658**, 77 (2016)
6. A.C. Batista, A.M. Dias, J.L. Lebrun, J.C. Le Flour, G. Inglebert, Fatigue Fract. Eng. Mater. Struct. **23**, 217 (2000)
7. F. Ashrafizadeh, Surf. Coat. Tech. **174-175**, 1196 (2003)
8. ANSI/AGMA B89. Gear material and heat treatment manual (2004)
9. ISO 26910-1. Springs -- Shot peening – Part 1: general procedures (2009)
10. ISO 25178-2. Geometrical product specifications (GPS) -- Surface texture: areal -- Part 2: terms, definitions and surface texture parameters (2012)
11. R.W. Buenneke, M.B. Slane, C.R. Dunham, M.P. Semenek, M.M. Shea, J.E. Tripp, Gear single tooth bending fatigue test, SAE Technical Paper 821042 (1982)
12. S.M. Hassani-Gangaraj, K.S. Cho, H.J.L. Voight, M. Gugliano, C.A. Schuh, Acta Mater. **97**, 105 (2015)
13. X.H. An, Q.Y. Lin, S.D. Wu and Z.F. Zhang, Mater. Res. Lett. **3**, 135 (2015)
14. S.M. Hassani-Gangaraj, A. Moridi, M. Guagliano, A. Ghidini and M. Boniardi, Int. J. Fatigue **62**, 67 (2014)

Investigation of fatigue crack growth under loading sequence effects using in-situ SEM testing

Shan Jiang[1], *Wei* Zhang[1,*], *Liang* Cai, and *Zili* Wang[1]

[1]Science and Technology on Reliability and Environmental Engineering Laboratory, School of Reliability and Systems Engineering, Beihang University, Beijing 100191, China

Abstract. Fatigue damage is one of the most important failure mechanisms in engineering components. The excited structures are usually subjected to spike loads in the fatigue weakness area during their service lives. The nonlinear loading sequence effects due to overloads are significant in the crack propagation process. In this paper, an in-situ scanning electron microscope (SEM) testing is performed to analyse the mechanisms of nonlinear fatigue crack growth affected by the load sequence. The crack tip behaviours under constant amplitude loading cycles superimposed by tensile overload were observed. The SEM experiment results reveal that the overload effects include the transient weakened area (shear bands and micro-cracks) and the relatively long-term retardation. Additionally, the observation loading sequence influence region is larger than the theoretical value. According to these SEM testing analyses, the Willenborg is modified considering the nonlinear loading sequence effects. In this approach, the damage zone concept is introduced to account for the instantaneous acceleration. Moreover, the loading sequence effect area is defined as the whole plastic zone due to overload rather than part of it. The proposed algorithm is validated by experiment data of 350WT steel and Al 2024-T351 specimens under constant loading with overloads. Good agreements are observed.

1 Introduction

The damage-tolerance design is widely used in many engineering projects. The integrity of damage-tolerance structures subjected to variable amplitude loading during service depends on understanding the fatigue crack growth behavior under such condition. The constant amplitude (CA) loading with overload is a typical scenario in variable loading cases. When the loading spectrum includes multiple overload cycles, the high nonlinearity loading sequence effects have a much stronger influence on fatigue crack growth behavior than under CA loading. Therefore, this investigation focuses on fatigue crack tip behavior and life prediction under constant loading superimposed by overloads.

Many recent monographs present the state-of-the-art in knowledge of fatigue crack growth behavior under overload effects in metallic materials by using experimental investigation and theoretical analysis [1-4]. J. Lankford. JR. and D. L. Davidson examined the fracture surfaces by using the scanning electron microscope (SEM), and further analyzed the plastic zones associated with single overloads [5]. James Lankford and David L. Davidson also performed the SEM overload fatigue tests, in which a special in-situ servo-controlled hydraulic loading stage is used [6]. In their study, the retardation phenomenon was observed directly and the crack tip opening displacements were measured. D. Damri and J. F. K nott observed the fatigue surfaces under the SEM [7]. They aimed to identify the retardation mechanism associated with single overload. These SEM testing observations provide an adequate description of the transient fatigue crack growth behavior under loading sequence effects. Once an overload is applied, the crack growth rate initially increases and then retards in a longer time.

Furthermore, attention is focused on the quantitative prediction of the fatigue life. Wheeler proposed a series of empirical models to account for the overload effects, in which the plastic zone size is a crucial influencing factor of fatigue crack propagation [8]. However, the Wheeler method has several adjustable parameters, which are used to tune the model prediction to fit the actual testing data. And then Willenborg modified these formulations based on physical mechanisms [9]. In this method, the effective stress intensity factor is defined to depict the retardation phenomenon. Many modifications of these two approaches have been proposed in the literatures [10, 11]. Although most of the existing approaches can quantify the fatigue crack growth retardation after overload, they still offer the most potential improvements. These formulations fall short of calculating the aforementioned initial crack growth acceleration. Therefore, a SEM fatigue testing under CA loading with overload is conducted in this paper. Once an overload is applied, a "damage zone" in the vicinity of the crack tip appears where the potential damage can be seen. After the overload cycle, the crack transient

* Corresponding author: zhangwei.dse@buaa.edu.cn

acceleration and long-time retardation phenomenon are observed. In the Willenborg model, the overload effects are assumed to be vanished when the current plastic zone reaches the boundary of the previous large plastic zone. Nevertheless, the SEM fatigue tests found that the realistic loading sequence influence area is greater than the theoretical solution. Therefore, based on these SEM experiment results, a modified Willenborg model is proposed. This algorithm can not only explain the retardation phenomenon, but also evaluate the initial acceleration. In this approach, the subsequent loading cycles are under the influence of overload until the crack tip reaches the boundary of the large plastic zone. The testing data of 350WT and aluminum alloy 2024-T351 are employed to validate the proposed approach.

The paper is organized as follows. First, the SEM fatigue testing under CA loading with overload is illustrated. Next, the modified Willenborg model is developed. And then, the prediction results are compared with the fatigue testing data of 350WT and aluminium alloy 2024-T351. Comparison between the proposed approach and the Willenborg model is also performed. Finally, some conclusions are given based on the current investigation.

2 Experiments

2.1. Experiment set-up and procedure

Crack growth testing were conducted using the edge-crack tension specimens make of Al 2024-T3 in longitudinal (LT) orientation. The configurations of the specimens are: length= 15 mm, width= 10 mm, thickness= 1 mm. The initial crack length is around 1.3 mm. The material components of Al 2024-T3 are listed in Table 1. The mechanical properties are as following: σ_y =345 MPa, σ_{UTS} =483 MPa, Elongation at break =18%.

Table 1. material component of Al 2024-T3 (Wt.%)

Al	Cr	Cu	Fe	Mg	Mn	Si	Ti	Zn
90.7-94.7	Max 0.1	3.8-4.9	Max 0.5	1.2-1.8	0.3-0.9	Max 0.5	Max 0.15	Max 0.25

A symmetrical loading stage is installed in the Quanta FEG 650 SEM chamber for the high-resolution in-situ observation, as shown in Fig 1. The load capacity of the tensile stage is 2 kN. During the testing, the vacuum level is remained to be less than 3×10^{-8} MPa.

Fig. 1. In-situ SEM fatigue testing setup

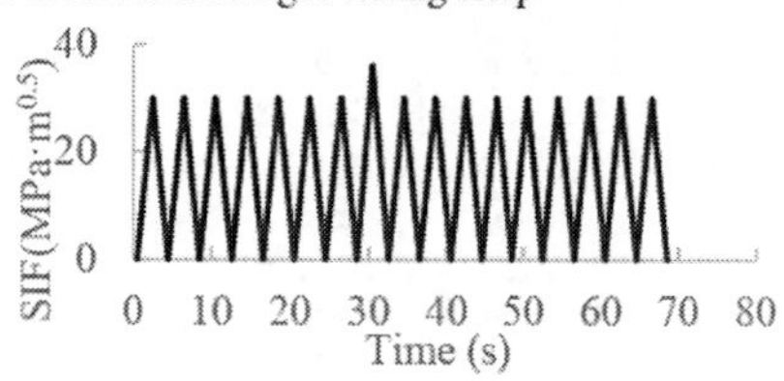

Fig. 2. The loading condition in the SEM fatigue testing

The loading condition is shown in Fig 2. The stress intensity factor (SIF) solution can be expressed as:

$$K=Y\cdot\sigma\cdot\sqrt{\pi a}$$

$$Y=1.122-0.231\cdot\left(\frac{a}{b}\right)+10.55\cdot\left(\frac{a}{b}\right)^2-21.71\cdot\left(\frac{a}{b}\right)^3+30.382\cdot\left(\frac{a}{b}\right)^4 \quad (1)$$

where σ is the applied stress level, *a* is the crack length, *b* is the width of the specimen. Firstly, the specimen is subjected to a tension-tension cyclic CA loading sequence with P_{max}=1 kN, R=0. And then an overload is applied. The overload ratio R_{OL} is 1.2. Next, continue applying the loads which are same as the previous CA stage. The crack tip behaviour is recorded by the observation system which consists of a metallurgical microscope and a high-resolution imaging acquisition system.

2.2 Experimental results and discussion

The SEM fatigue testing results for the specimen subjected to CA loading with overload are shown in this section. In the initial CA loading condition, Fig 3 shows the fatigue surface under peak loading, which demonstrates a smooth and stable crack growth process. The crack plane is approximately flat and perpendicular to the direction of the external applied stress. The average crack growth rate during CA loading is 0.5 μm/cycle, which is a baseline for the following analysis.

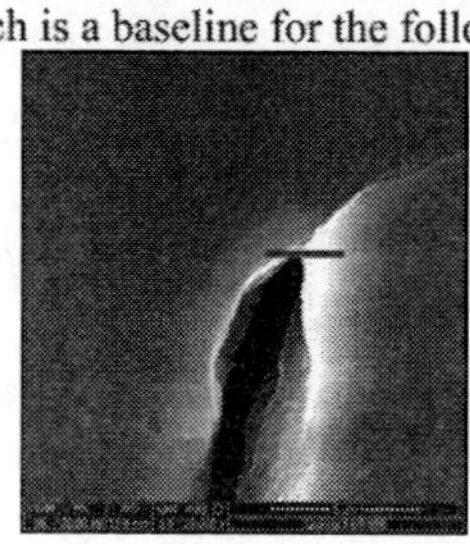

Fig. 3. The crack tip behaviour under peak loading

Fig 4 shows the crack tip behaviour in the overload cycle. A distinct from fatigue surfaces under CA loading, it is apparent that the crack growth zone consisting of potential damage and shear bands are greater. The weaken area in the vicinity of crack is characterized as the damage zone term. This observation implies that instantaneous large load variation enhances the sharply tensile effect to lead to crack growth instability.

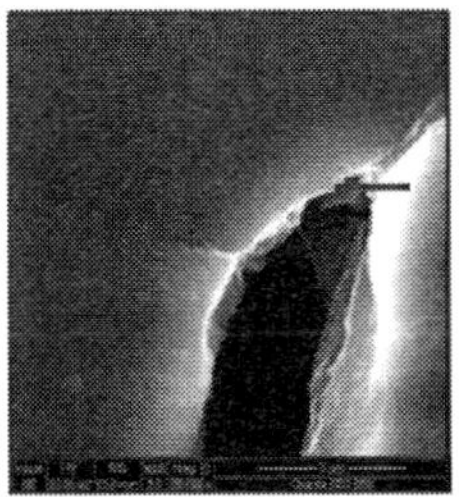

Fig. 4. The crack tip behaviour in the overload cycle

Figs 5.(a)-(c) show the crack tip behaviour during four load cycles after the overload. Under the first peak loading after overload, the crack tip is at point 'A' in Fig 5.(a). A micro-crack on the left side of the main crack can be seen, which keeps growing in the next loading cycle (Fig 5.(b)). Ultimately, this micro-crack links up with the principal crack in the fourth cycle, and the crack tip becomes at point 'C' in Fig 5.(c), which extremely speed up the crack propagation. The average crack growth rate during these four cycles (3.04 μm/cycle) is much faster than before (0.5 μm/cycle). The appearance of the micro-crack and the linking up process manifest the overload gives rise to the transient crack growth acceleration.

(a) the first peak loading after the overload

(b) the second peak loading after the overload

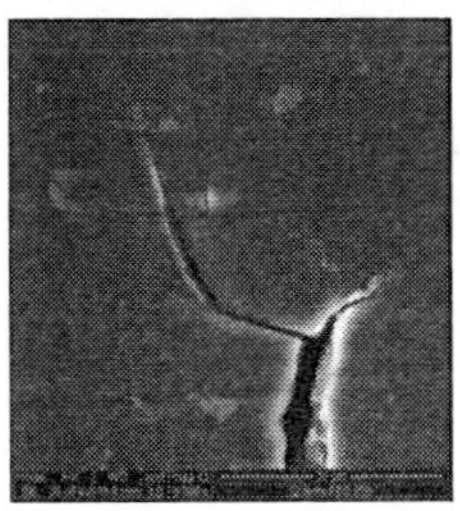

(c) the fourth peak loading after the overload

Fig. 5. The crack tip behaviour right after the overload

Fig 6 shows the crack tip behaviour in the several cycles after the spike load. After 40 cycles, the crack tip reaches point 'D' from point 'C' in Fig 6. This implies that crack penetrates the aforementioned damage zone and grows slowly.

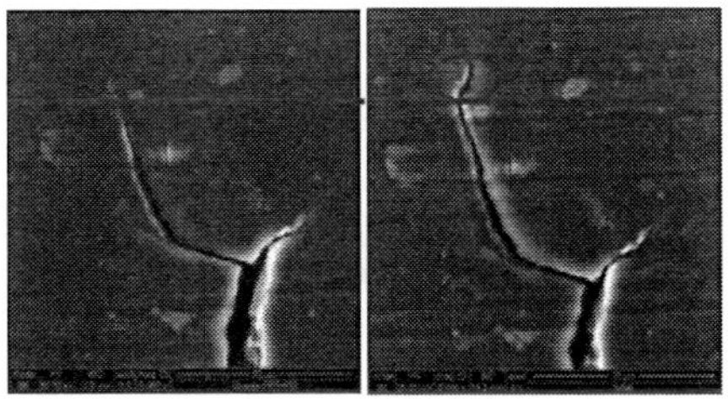

Fig. 6. The crack tip behaviour in several cycles after the overload

The fatigue surface during the crack growth retardation process is magnified, as shown in Fig 7. The magnifying power increases to 20000 from 5000. The crack tip behaviour in 10 cycles are investigated. Although the applied stress levels are the same as the previous CA loading condition, the average crack growth rate during these 10 cycles (0.27 μm/cycle) is awfully smaller than before (0.5 μm/cycle). As a consequence, this SEM fatigue testing reveals that overload will lead to the appearance of damage zone, which elevates the crack growth rate within a transient time. After the acceleration, the crack growth will retard in a relatively longer time.

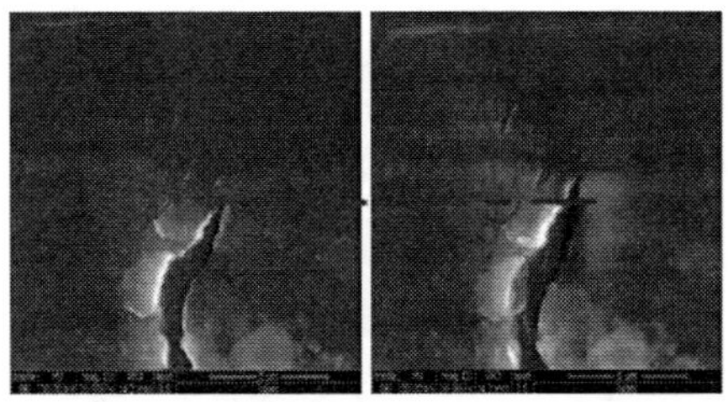

Fig. 7. The crack tip behaviour in 10 cycles during the retardation process

Keeping loading, the crack growth rate eventually returns back the original trend under CA loading condition. As shown in Fig 8, the da/dN-ΔK data are collected and the Willenborg model is used to calculate the crack growth under CA loading. Δa is the crack increment under the effects of the historical overload, which is around 300 μm. The corresponding calibration parameters are: C=0.0064e-6 and m=3.334. K_c is calculated to be 83 MPa·$m^{0.5}$. In the Willenborg model, the overload plastic zone is calculated by using the relationship:

$$Z_{OL,W} = \frac{1}{\eta\pi}\cdot\left[\frac{(K_{\max})_{OL}}{\sigma_y}\right]^2 \tag{1}$$

where the value of η is established to be equal to 3 through the analysis [16]. The predicted crack increments under overload effects are 100 μm, which is much smaller than the experimental value.

 https://doi.org/10.1051/matecconf/201816522012

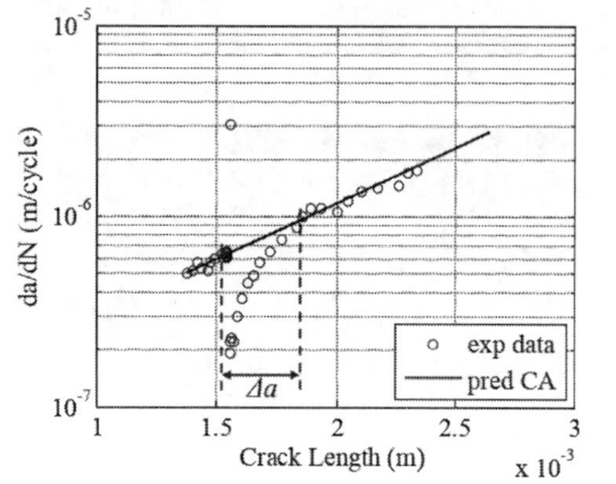

Fig. 8. *a-da/dN* curve collected from the SEM fatigue testing

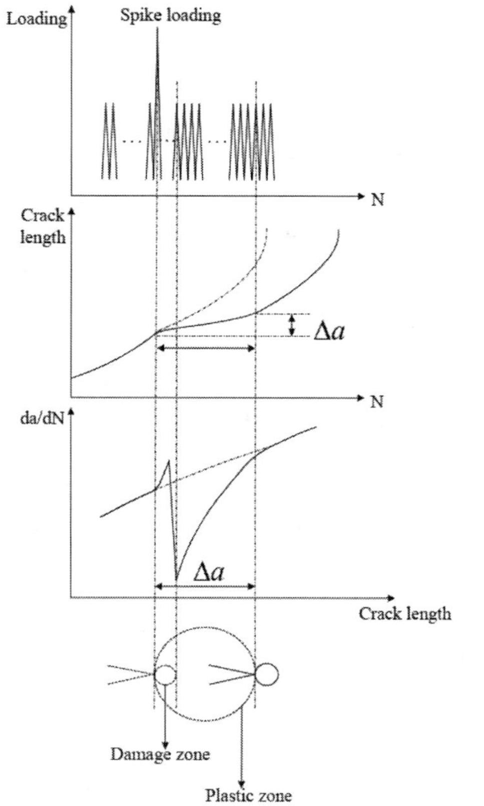

Fig. 9. Schematic illustration of interaction effect

3 Model development

According to the SEM fatigue testing observations, the loading sequence effects under CA loading with overloads are investigated, as shown in Fig 9. *Δa* represents the crack increment under the loading sequence effects. Right after the overload, a damage zone appears ahead of the crack tip, in which the crack growth rate instantaneously jumps to a high level. Meanwhile, a large plastic deformation generates, which makes the fatigue crack growth retard in the following cycles. When the crack penetrates the damage zone, the crack growth rate reaches to the minimum value. And then the crack grows through the large plastic zone, the retardation effects gradually relieves and eventually vanishes entirely when the crack tip reaches the boundary of the large plastic zone. The loading sequence effect region is modified to be the whole large plastic zone caused by overload. Afterwards, the crack growth will return back to the original trend.

Willenborg developed a series models based on plasticity physical mechanisms [9]. Current plastic zones are created inside the large plastic zone due to an overload. The retardation phenomenon is described quantitatively by a reduction in stress intensity factor. In this paper, the damage zone concept is used to further modified the Willenborg model, which aims to evaluate the initial acceleration crack propagation process and predict the loading sequence effects more precisely. The effective stress intensity factor can be expressed as:

$$K_{\max.eff} = K_{\max} - K_{red} \tag{2}$$

$$K_{\min.eff} = \begin{cases} K_{\min} - K_{red} & K_{\min} > K_{red} > 0 \\ 0 & K_{\min} \le K_{red} \le 0 \ or \ K_{red} \le 0 < K_{\min} \end{cases} \tag{3}$$

$$\Delta K_{eff} = K_{\max.eff} - K_{\min.eff} \tag{4}$$

where $K_{max.eff}$ and K_{max} are the maximum effective and apparent SIF under CA loading, respectively. $K_{min.eff}$ and K_{min} are the corresponding minimum values; ΔK_{eff} is the effective stress intensity factor range; K_{red} is the modified stress intensity factor, which characteraizes the loading sequence effects. When K_{min} is larger than K_{red} and both of these two values are positive, ΔK_{eff} is equal to ΔK (the SIF range). Otherwise, ΔK_{eff} is equal to $K_{max.eff}$. The algorithm of K_{red} is modified as:

$$K_{red} = \begin{cases} [(K_{\max})_{OL} - K_{\max}] \cdot (1 - \frac{D_{OL}}{d + \Delta a})^{0.1} \cdot (1 - \frac{\Delta a}{Z_{OL}})^{0.5} & d + \Delta a < D_{OL} \\ (K_{\max})_{OL} \cdot (1 - \frac{\Delta a}{Z_{OL}})^{0.5} - K_{\max} & d + \Delta a \ge D_{OL} \text{ and } \Delta a \le Z_{OL} \\ 0 & \Delta a \le Z_{OL} \end{cases} \tag{5}$$

where $(K_{max})_{OL}$ is the SIF of the overload cycle; Δa is the amount crack increment since the overload cycle; Z_{OL} and D_{OL} are the plastic zone and damage zone sizes created by overload, respectively; d is the current damage zone. The formulation proposed by Johnson is employed to calculate the plastic zone size, which can be written as [12]:

$$Z_{OL} = \alpha \left[\frac{(K_{\max})_{OL}}{\sigma_y} \right]^2 \tag{6}$$

where σ_y is the tensile yield strength, and α is the plastic zone size factor. Voorwald et al. presented a parametric function for α, which is validated by an elastic plastic finite element analysis [13, 14]. This equation is given below:

$$\alpha = \begin{cases} \dfrac{1}{6\pi} & t \geq 2.5\left(K_{max}/\sigma_y\right)^2 \\ \dfrac{1}{\pi} & t \leq \dfrac{1}{\pi}\left(K_{max}/\sigma_y\right)^2 \\ \dfrac{1}{6\pi} + \dfrac{5}{6\pi}\left(\dfrac{2.5 - t\left(K_{max}/\sigma_y\right)^{-2}}{2.5 - \pi^{-1}}\right) & \dfrac{1}{\pi}\left(K_{max}/\sigma_y\right)^2 \leq t \leq 2.5\left(K_{max}/\sigma_y\right)^2 \end{cases} \tag{7}$$

where t is the specimen thickness. The damage zone size is assumed to be proportion to the plastic area size, which can be calculated by the following equation:

$$D_{OL} = \beta \cdot Z_{OL} = \beta \cdot \alpha \left[\frac{\left(K_{max}\right)_{OL}}{\sigma_y} \right]^2 \tag{8}$$

where β is the proportionality coefficient. Many experiment results and simulations indicate β=0.05 is a reasonable evaluation. The Forman equation in conjunction with the Willenborg model can be expressed as [15]:

$$\frac{da}{dN} = C \cdot \frac{(\Delta K_{eff})^m}{(1 - R_{eff})K_C - \Delta K_{eff}} \tag{9}$$

$$R_{eff} = \frac{K_{min.eff}}{K_{max.eff}} \tag{10}$$

where C and m are the material parameters; R_{eff} is the effective stress ratio; K_C is the critical SIF, which is defined as [15]:

$$K_C = Y \cdot \sigma \cdot \sqrt{\pi a_c} \tag{11}$$

where Y is the geometry factor of SIF; a_c is the critical crack length.

4 Model validation

4.1 Crack growth prediction in 350WT under constant loading with overloads

An experimental fatigue program was conducted by Taheri et al [16]. The crack growth data on 350WT steel under CA loading with overloads are collected, which are used to validation the proposed model. The rectangular specimens of 300 length × 100 width ×5 thickness mm is used. The initial centre crack length is 2a=20 mm. the mechanical properties of 350WT in longitudinal (LT) orientation are as follows: σ_y =350 MPa, σ_{UTS} =524 MPa, Modulus of elasticity=205 GPa, Poisson's ration=0.30. K_c is equal to 110 MPa·m$^{0.5}$.

Before any prediction can be given, there are still two unknown parameters in Eq.(9). The *da/dN-ΔK* data with R=0.3 in Ref.[17] are used to calibrate these parameters, as shown in Fig 10. The fitting coefficients are C=3.793E-9 and m=2.365.

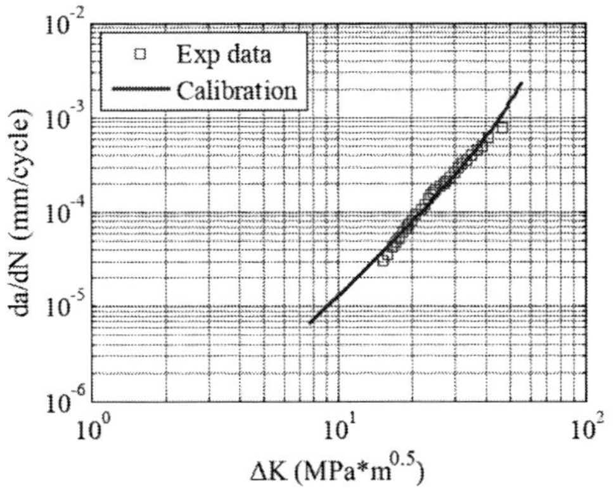

Fig. 10. *da/dN-ΔK* calibration of 350WT

The prediction results are shown in Fig 11. The circles are the testing data. The single overloads are applied at the crack length of 15 and 25mm. The pink line represents the prediction under CA loading as a baseline. The dashed line is the prediction of Willenborg model, while the full line is the calculation result of modified Willenborg approach. For CA loading condition, there is no different between these two methods. It is clear that the modified approach can give the better predictions under CA loading with overloads.

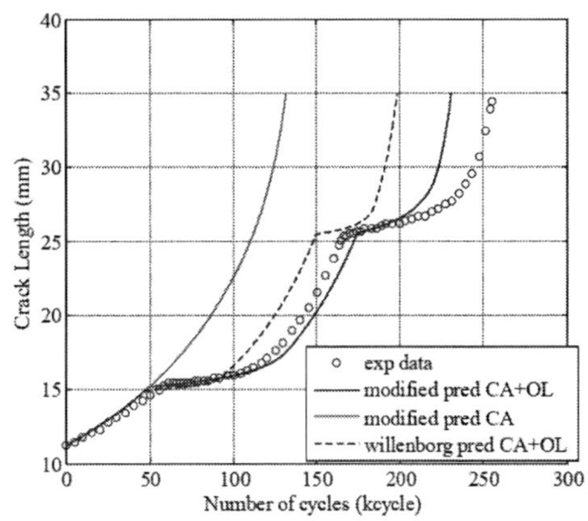

Fig. 11. a-N curves under constant loading with spike load

More detailed information can be obtained in the da/dN-ΔK curve, as shown in Fig 12. The Willenborg model falls short of accounting for the initial crack growth acceleration. Additionally, the retardation effects region calculated by Willenborg model is less than testing data. Compared with the Willenborg method, the proposed approach can distinctly match the experimental data better. It is noted that the modified approach is applicable to calculate the fatigue crack growth under the CA loading with overloads.

 https://doi.org/10.1051/matecconf/201816522012

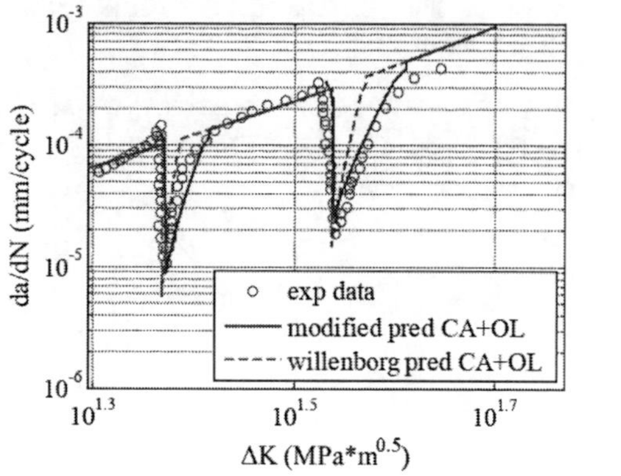

Fig. 12. ΔK -da/dN curve for constant loading with spike load

4.2 crack growth prediction in Al 2024-T351 under constant loading with multiple overloads

The experiment data of 2024-T351 aluminium alloy are also used to validate the presented method [18]. The specimen dimensions are as follows: length=250 mm, width=100 mm and thickness=6 mm. Initial centre crack length is 2a=20 mm. In addition, the mechanical properties of Al 2024-T351 in longitudinal (LT) orientation are as follows: σ_y =365 MPa, σ_{UTS} =466 MPa. The value of K_c is 62.2 MPa·m$^{0.5}$.

Following the same procedure, the parameters are calibrated by the da/dN-ΔK data under CA loading with R=0.3, as shown in Fig 13 [17]. The calibration results are C=1.75E-9 and m=3.6504.

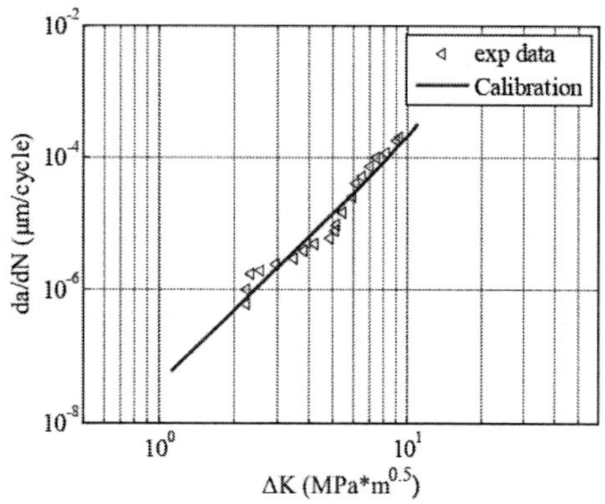

Fig. 13. da/dN-ΔK calibration of Al 2024-T351

The comparison between the predicted a-N curves and testing data in Al 2024-T351 is shown in Fig 14. The overloads are applied at the crack length of 20 mm, 25 mm and 32 mm. There are three obvious retarded regions after the overload. These predictions are slightly conservative and follows the same trend as the experiment data.

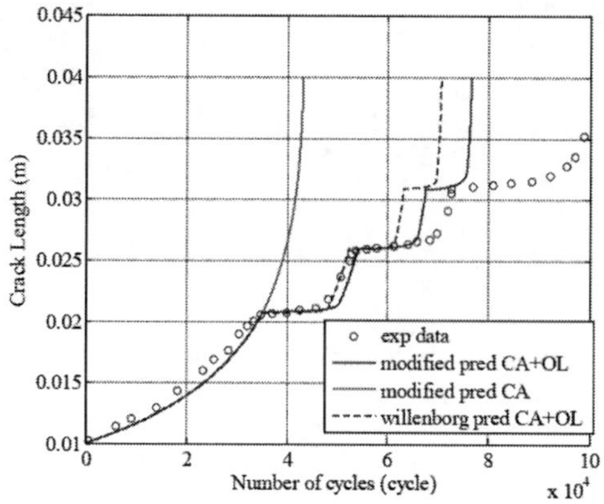

Fig. 14. a-N curves under constant loading with multiple overloads

Similarly, a-da/dN curves are used for further model validation [19], as shown in Fig 15. It is clear that the prediction of the Willenborg model (dashed line) returns back much earlier than the testing data. Therefore, the modified Willenborg method based on the SEM fatigue testing observations can capture the major trends of fatigue life curves under CA loading with overloads.

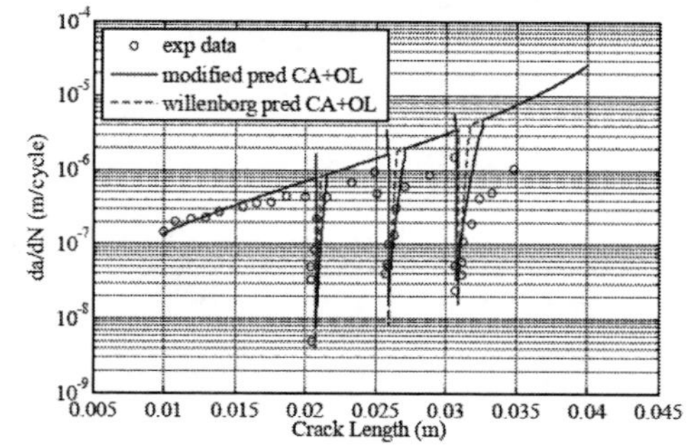

Fig. 15. a-da/dN curve for constant loading with multiple overloads

5 Conclusions

An in-situ SEM fatigue testing is conducted to investigate the loading sequence effects caused by overload. Based on these experiment observations, a modified Willenborg model is derived to calculate the fatigue crack growth under CA loading with overloads. This approach is validated by the extensive experiment data of steel and aluminium alloy. The following conclusions can be drawn:

The in-situ SEM fatigue testing under CA loading with overload reveals that a spike load could lead to the transient crack growth acceleration and long-time retardation. Moreover, the observation overload effect area is obviously larger than the theoretical calculation result.

The modified Willenborg approach can not only describe the retardation phenomenon due to overload, but also predict the initial crack growth acceleration. In addition, it manifests that the predicted overload effect area matches the testing data better than the Willenborg model by the comparisons. The loading sequence effects could be quantificationally described well by the proposed method.

The research is financially supported by the National Natural Science Foundation of China (Grant No. 51405009).

References

1. C. Q. Wang, J. J. Xiong, R. A. Shenoi, et al. Int J Fatigue. **83**, 280-287 (2016)
2. G. R. Chanani, Effect of thickness on retardation behavior of 7075 and 2024 aluminum alloys, *Flaw Growth and Fracture*, 365-387 (ASTM International, 1977)
3. T. Zhao, J. Zhang, Y. Jiang, Int J Fatigue. **30**(7), 1169-1180 (2008)
4. S. Y. Lee, P. K. Liaw, H. Choo, et al. Acta Mater. **59**(2), 485-494 (2011)
5. J. Lankford, D. L. Davidson, J Eng Mater-T ASME. **98**(1), 17-23 (1976)
6. J. Lankford, D. L. Davidson, The effect of overloads upon fatigue crack tip opening displacement and crack tip opening closing loads in aluminium alloys. ICF5, Cannes (France) (1981)
7. D. Damri, J. F. Knott, Fatigue Fract Eng M. **14**(7), 709-719 (1991)
8. O. E. Wheeler, J Basic Eng-T ASME. **94**(1), 181-186 (1972)
9. J. Willenborg, R. M. Engle, H. A. Wood. A crack growth retardation model using an effective stress concept. No. AFFDL-TM-71-1-FBR, Air Force Dynamics Lab Wright-Patterson Afb On, (1971)
10. B. K. C. Yuen, F. Taheri. Int J Fatigue. **28**(12), 1803-1819 (2006)
11. J. P. Gallagher, A generalized development of yield zone models. No. AFFDL-TM-FBR-74-28, AIR FORCE FLIGHT DYNAMICS LAB WRIGHT-PATTERSON AFB OH, (1974)
12. W. S. Johnson, Multi-parameter yield zone model for predicting spectrum crack growth. Methods and Models for Predicting Fatigue Crack Growth Under Random Loading. ASTM STP 748, J. B. Chang and C. M. Hudson, Eds., American Society for Testing and Materials, 85-102 (ASTM International, 1981)
13. H. J. C. Voorwald, M. A. S. Torres, C. C. E. P. Júnior. Int J Fatigue. **13**(5), 423-427 (1991)
14. X. Huang, M. Torgeir, W. Cui. Int J Fatigue. **30**(1), 2-10 (2008)
15. R. G. Forman, V. E. Kearney, R. M. Engle. J Basic Eng-T ASME. **89**(3), 459-463 (1967)
16. F. Taheri, D. Trask, N. Pegg. Mar Struct. **16**(1), 69-91 (2003)
17. X. Huang, M. Torgeir, W. Cui. Int J Fatigue. **30**(1), 2-10 (2008)
18. F.J. McMaster, D. J. Smith. Int J Fatigue. **23**, 93-101 (2001)
19. J. T. Liu, P. A. Du. Zhang Z Y. Sci China Technol Sc. **55**(3), 673-683 (2012)

Evaluation of the horizontal load-carrying capacity of a thin steel bridge pier by means of the Damage Subloading Surface model

Fincato Riccardo[1,*], *Tsutsumi* Seiichiro[1], and *Momii* Hideto[2]

[1]JWRI, Osaka University, 5670047 11-1 Mihogaoka, Ibaraki, Japan
[2]Nikken Engineering Corporation, 822-0003 4727-1 Kamitonno, Nogata, Fukuoka, Japan

Abstract. The evaluation of the load-carrying capacity of a structure is an important aspect of the structural design. In particular, large seismic events can significantly compromise the performances of structures, or parts of them, due to the formation and accumulation of plastic deformation. In the present paper, we analyse the seismic performance of a thin steel bridge column by means of an unconventional elastoplastic and damage model. The idea is to evaluate the horizontal load-carrying capacity of the pier for different seismic events, analysing the conditions that can compromise the stability of the structure.

1 Introduction

The evaluation of the capacity of the structure to undergo seismic events, without compromising the stability, is a crucial point for the design. According to international standards, the first step is the identification of the seismic phenomenon in terms of the characterization of the loading wave hitting the structure. One complication in the definition of the load is due to the fact that seismic excitations are quite complex. Their intensity and direction are quite random and it is difficult to define a standard loading wave. Moreover, in most of the cases, the structures are designed to bear the maximum load which reduces the evaluation on one single loading event. One example of experimental work was carried out by Nishikawa *et al.* [1] after the Kobe earthquake of January 1995, where around 1% of the steel piers of national roads were significantly damaged. Gao *et al.* [2] try to model the Nishikawa's experiment using the two-surface model, developed by the Nagoya University, with a good approximation of the inelastic cyclic behaviour of the steel columns. Two loads are applied to the steel pier in both works of Nishikawa *et al.* and of Gao *et al.*. The first one is a compressing axial load, which simulates the weight of the superstructure. The second one is cyclic lateral displacement with increasing amplitude through the cycles.

Recently, the authors published a work [3] reproducing the same experimental results carried out by Nishikawa and extending the investigations on non-proportional cyclic loading along two horizontal directions. In [3] the authors adopted an unconventional plasticity model, named Extended Subloading Surface model (i.e. S-S model) [4], with the inclusion of the so-called 'tangential plasticity' [5, 6]. The numerical results were in good agreement with the real behaviour of the thin wall pier. Moreover, the extension to a bi-directional load offered additional information about the load-carrying capacity of the structure.

In the present paper, a different constitutive model is adopted, describing the material behaviour within the framework of the *Continuum Damage Mechanics* (CDM).

An additional internal scalar variable D, describing the ductile damage, is coupled with the constitutive equations of the Extended Subloading Surface according to the approach described in [7]. The idea of the present study is to evaluate the structural response under different seismic waves. According to the standard design, the seismic performance in large earthquakes is often evaluated with the maximum loading wave that can hit the structure. However, in most cases, more than one seismic event was registered in the days after the first earthquake, defining the so-called seismic swarm. The application of a series of cyclic loadings, with a smaller amplitude than the one prescribed for the design of the structures, can also jeopardize the stability, due to the plastic strain and damage accumulation.

2 The Damage S-S model

2.1. Constitutive equations

The Damage Subloading Surface model (i.e. Damage S-S) was developed coupling the ductile damage variable with the constitutive equations of the S-S model following the approach proposed by Lemaitre [8]. The constitutive equations are not dealt in this paper, for details the reader is referred to [7], and to [9, 10] for some numerical examples.

The choice of an unconventional plasticity model is due to the ability of the S-S model to catch a realistic accumulation of plastic deformation during cyclic

* Corresponding author: fincato@jwri.osaka-u.ac.jp

loading, as shown in [11, 12], or in fatigue life investigations [13, 14, 17]. Briefly, a new surface, i.e. the subloading surface, is obtained from the conventional plastic surface, here named normal-yield surface (see Fig. 1), by means of a similarity transformation. The subloading surface always passes through the current stress state and it contracts or expands, following the loading or unloading of the material. Moreover, the similarity centre is not fixed but it can move in the stress space following the plastic strain rate. This feature allows obtaining close hysteresis loops in the unloading-reloading process [12] with a realistic description of the mechanical ratcheting.

The present work assumes a finite elastoplasticity framework with a hypoelastic based plasticity. The strain rate tensor is decomposed additively into an elastic part $\mathbf{D}^e$ and a plastic part $\mathbf{D}^p$

$$\mathbf{D} = \mathbf{D}^e + \mathbf{D}^p \tag{1}$$

The coupling between the elasticity and damage is assumed as in [8], through the factor (1 - D), that leads to the definition of the corotational rate of the Cauchy stress $\overset{\circ}{\boldsymbol{\sigma}}$ as:

$$\overset{\circ}{\boldsymbol{\sigma}} = (1-D)\mathbf{E} : \mathbf{D}^e = (1-D)\mathbf{E} : (\mathbf{D} - \mathbf{D}^p) \tag{2}$$

where $\mathbf{E}$ is the tensor of the elastic constant and D is the isotropic scalar damage variable. The coupling between the plastic strain and the damage is obtained through the consistency condition. The expressions of the normal-yield and subloading surface in the Damage S-S model can be written as:

$$\begin{gathered} f(\hat{\boldsymbol{\sigma}}) = (1-D)F(H) \\ f(\overline{\boldsymbol{\sigma}}) = (1-D)RF(H) \\ \hat{\boldsymbol{\sigma}} = \boldsymbol{\sigma} - \boldsymbol{\alpha} \end{gathered} \tag{3}$$

where f is a generic stress function, assumed as the von Mises criterion in this study, $\boldsymbol{\sigma}$ is the Cauchy stress, $\boldsymbol{\alpha}$ is the back stress, F is the isotropic hardening function, H is the isotropic hardening variable (i.e. cumulative plastic strain variable) and R is the similarity ratio. $\overline{\boldsymbol{\sigma}}$ in the second of Eq. (3) represents the Cauchy stress observed by the centre of the subloading surface.

Finally, the expression of the plastic strain rate can be obtained by the material time derivative of the second of Eq. (3) as:

$$\mathbf{D}^p = \frac{\lambda}{(1-D)} \frac{\partial f(\overline{\boldsymbol{\sigma}})}{\partial \boldsymbol{\sigma}} \tag{4}$$

where λ is the plastic multiplier defined as in [7]. Both isotropic and kinematic hardening laws are considered for the material description. The isotropic hardening law is a modified version of the law presented in [15], where an additional linear hardening term is added through the material constant K:

$$F(H) = F_0 + K(H - H_d) + F_0 h_1 \left[1 - e^{h_2 (H - H_d)}\right] \tag{5}$$

where F_0 is the initial size of the normal-yield surface, h_1 and h_2 are two material parameters defined by the user and H_d is a threshold introduced to simulate the stress plateau. The kinematic hardening law is defined according to the approach suggested by Chaboche [16] with the sum of N non-linear Armstrong–Frederick kinematic laws:

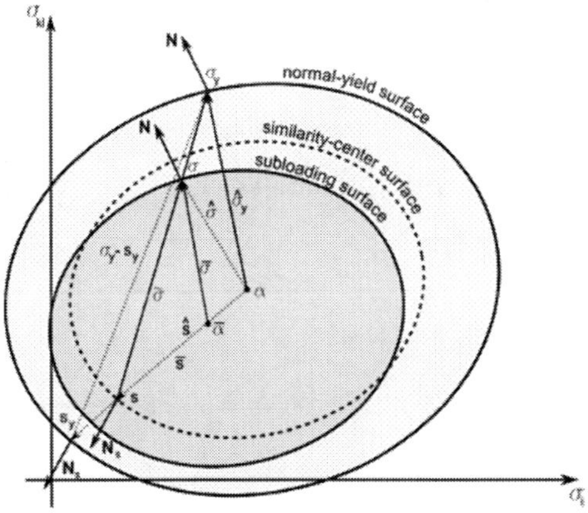

Fig. 1. Sketch of the normal-yield and subloading surface.

$$\begin{gathered} \overset{\circ}{\boldsymbol{\alpha}}_i = \left(C_i \mathbf{D}^p - \lambda B_i \boldsymbol{\alpha}_i\right) \\ \overset{\circ}{\boldsymbol{\alpha}} = \sum_{i=1}^{n} \overset{\circ}{\boldsymbol{\alpha}}_i, \quad n = 1,\ldots N \end{gathered} \tag{6}$$

where $\overset{\circ}{\boldsymbol{\alpha}}$ is the corotational rate of the back stress and C_i, B_i are material constants. The definition of the material parameters is discussed in the following section 3.2.

2.2 Ductile damage criteria

It is well established that the mechanism that leads to the progressive failure of the metallic materials is due to the presence of impurities in the media. The application of external solicitations causes the formation of micro voids around the impurities, their growth and coalescence until the formation of a crack. Nowadays, in the panorama of the CDM, several theories try to reproduce this phenomenon with more or less complicated models. One common aspect is represented by the individuation of the key factors that affect the damage evolution: the stress triaxiality and the Lode angle.

Previous works of the authors [7, 9] implemented the original Lemaitre's damage evolution law, which can consider the exponential decay of the ductility with the increase of the stress triaxiality but cannot describe the dependency of the damage evolution to the different loading conditions. An attempt to include the Lode angle effect on the Lemaitre's law was carried out in [10], where the authors reproduced the failure behaviour of Q460 steel round notched bars and flat grooved plates [18].

Recent works of Bai and Wierzbicki [19] and Algarni *et al.* [20, 21] implements a modified Mohr-Coulomb (M-C) criterion for the ductile fracture. This failure criterion has been extensively used for the description of porous media, such as rocks and soils, however, its ability to consider the stress triaxiality η and Lode angle parameter $\overline{\theta}$ effects makes it suitable for the description of metallic materials too. The definition of the equivalent strain to fracture can be, in fact, expressed as a function of these two parameters only, which are analytically defined as:

$$\sigma_m = -\frac{1}{3}(\sigma_1+\sigma_2+\sigma_3)$$
$$\eta = \sigma_m / \sigma_{Mises} \quad (7)$$
$$\bar{\theta} = 1-\frac{6\theta}{\pi}$$

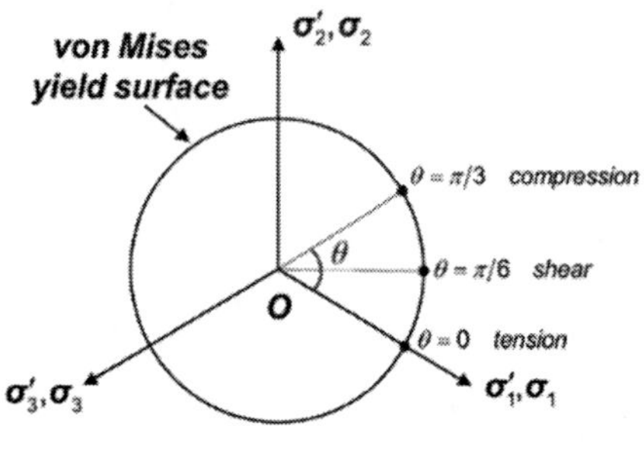

Fig. 2. Schematic representation of a generic stress state in the principal stress space.

where $\sigma_1, \sigma_2, \sigma_3$ are the principal stresses $(\sigma_1 \geq \sigma_2 \geq \sigma_3)$, σ_m is the mean stress, σ_{Mises} is the von Mises stress and θ is the Lode angle as reported in Fig. 2.

The set of admissible values for the stress triaxiality and the Lode angle parameter and the equivalent strain to fracture defines the 'fracture envelope' $\bar{\varepsilon}_f$. Bai and Wierzbicki [9] defined $\bar{\varepsilon}_f$ analytically as:

$$\bar{\varepsilon}_f = \left\{ \frac{A}{c_2}\left[\sqrt{\frac{1+c_1^2}{3}}\cos\left(\frac{\bar{\theta}\pi}{6}\right) + c_1\left(\eta + \frac{1}{3}\sin\left(\frac{\bar{\theta}\pi}{6}\right)\right)\right]\right\}^{-\frac{1}{N}} \quad (8)$$

where A, c_1, c_2, N are four material constants that have to be defined through an experimental campaign. In detail, the formula reported in Eq. (8) is valid uniquely if a von Mises potential is used, otherwise, the general expression requires the calibration of other four material parameters.

This simplification gives a realistic approximation of the ductile behaviour whenever the Lode angle effect on plasticity is limited and the effect of the pressure dependency on the yield surface can be neglected [19]. An application of the M-C fracture criterion with a von Mises plastic potential can be found in [22]. Based on Eq. (8) the ductile damage increment can be written as:

$$\dot{D} = \frac{\sqrt{2/3}\,|\mathbf{D}^p|}{\bar{\varepsilon}_f(\eta,\bar{\theta})} \mathrm{H}(H-d_1)$$
$$\mathrm{H}(H-d_1) = \begin{cases} 0, & H-d_1 < 0 \\ 1, & H-d_1 \geq 0 \end{cases} \quad (9)$$

Where d_1 is an additional material parameter and H represent the Heaviside step function, which was introduced by the authors to allow the damage to evolve whenever the cumulative plastic strain exceeds the parameter d_1.

3 Numerical analyses

The constitutive equations discussed in the previous section 2 were implemented via user subroutine for the commercial code *Abaqus* (ver. 6.14-5). The calibration of the material parameters for the Damage S-S model was carried out reproducing the experimental results in [1], analysing the horizontal load-carrying capacity of a steel bridge pier subjected to a constant axial compressing load and a cyclic horizontal solicitation. Nishikawa *et al.* investigated the seismic response of nine steel columns, characterized by different materials and geometry. The analyses reported in this work refer to the sample number 8 in [1], see Fig. 3 and Table 1.

The parameters H_y and δ_y in Table 1 represent the horizontal load and horizontal displacement when the specimen yields close to the base (see Fig. 4) and P_y is the squash load. The values of the geometrical and structural parameters were taken directly from [1].

Table 1 Structural parameters of the specimen.

h [mm]	3,403
t [mm]	8.70
D [mm]	891.3
P/P_y	0.124
H_y [kN]	414.9
δ_y [mm]	10.5

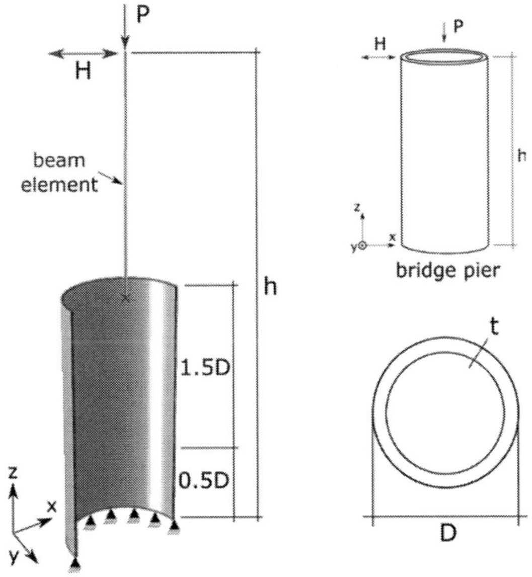

Fig. 3. Model and geometry of the bridge pier.

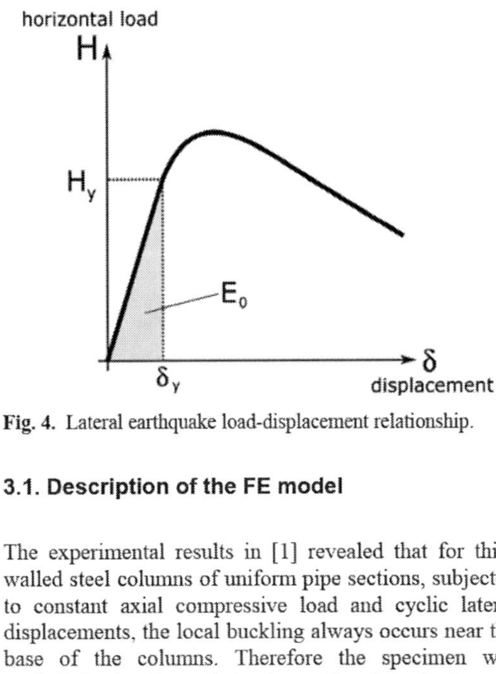

Fig. 4. Lateral earthquake load-displacement relationship.

3.1. Description of the FE model

The experimental results in [1] revealed that for thin-walled steel columns of uniform pipe sections, subjected to constant axial compressive load and cyclic lateral displacements, the local buckling always occurs near the base of the columns. Therefore the specimen was modelled using the real circular section from the base of the column to a height equal to two times the external diameter and with a rigid beam for the remaining part, see Fig.3. A similar strategy was also adopted in [2]. Moreover, in order to reduce the computation effort, only half of the column was considered, applying symmetric boundary conditions. The lower part of the pier was meshed with eight-node hexahedral elements (i.e. Abaqus C3D8 elements) and the nodes of the top cross section were connected to the beam with rigid links. A mesh refinement was considered close to the base of the pier, where the maximum accumulation of plastic deformation is expected. The minimum element size is 9.8 mm (circumference) × 2.25 mm (thickness) × 15 mm (axial), comparable with the minimum element size used by Van Do *et al.* [24], who conducted a mesh size sensitivity analysis on the same study case. The total number of elements amount to 51841.

The steel pier is compressed with a constant axial load P, simulating the weight of the superstructure. The compressive load is applied at the beginning of the analysis and kept constant during the whole simulation. Four different kinds of horizontal cyclic loads were applied. The first one reproduced the experimental cyclic loading in [1] (i.e. *L* 0 in Fig. 5), the other three simulate the effect of more than one seismic event in sequence. The details about these last three cyclic loading conditions are discussed in section number 3.3.

3.2. Calibration of the Damage S-S model

The Damage S-S model requires the calibration of 13 material constants for the elastoplastic behaviour and 4 parameters for the ductile damage behaviour.

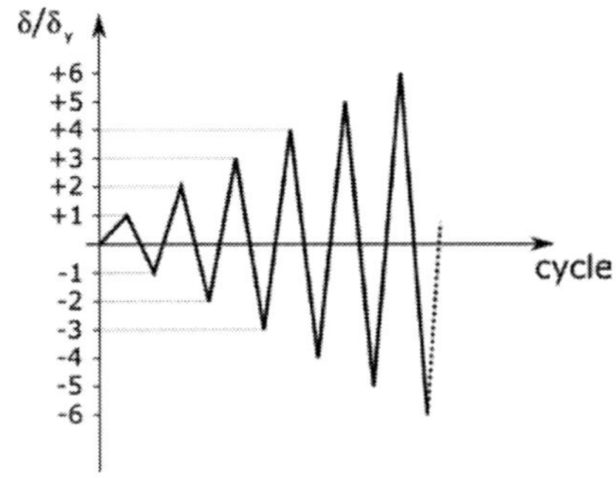

Fig. 5. Loading sequence for the model calibration (*L 0*).

The Young's modulus, the Poisson's ratio and the yield stress were assumed directly from the Nishikawa *et al.*'s paper, considering a mild SS400 structural steel. The isotropic and kinematic hardening constants of Eqs. (5) and (6) and the 4 material constants for the ductile damage evolution were obtained minimizing the differences between the experimental results in [1] (sample n.8) and the numerical curve obtained with the present model. The values of the elastoplastic constants are reported in the following Table 2, where R_e, u, c and χ are four constants proper of the subloading surface model, the details are available in [4, 25]. Fig. 6 and Fig. 7 report graphically the calibration of the Damage S-S model for the steel column. As it can be seen the numerical simulation is in good agreement with the experimental results, catching well the normalized horizontal load-normalized horizontal displacement behaviour.

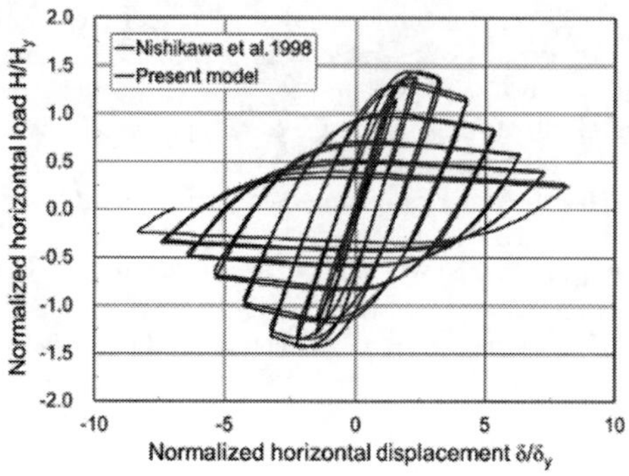

Fig. 6. Hysteresis loops for the pier in Fig. 3 (present model).

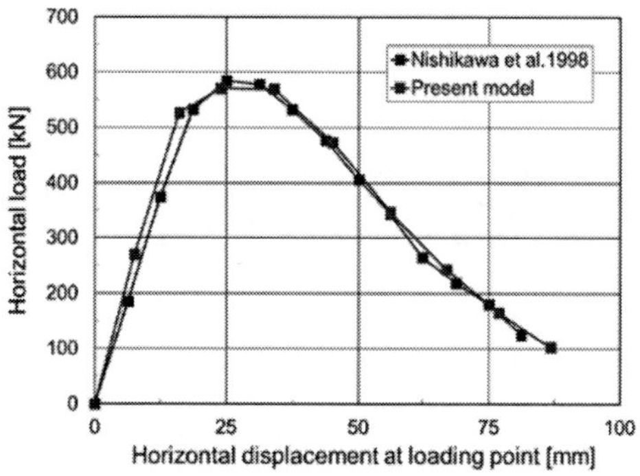

Fig. 7. Envelope of the positive side.

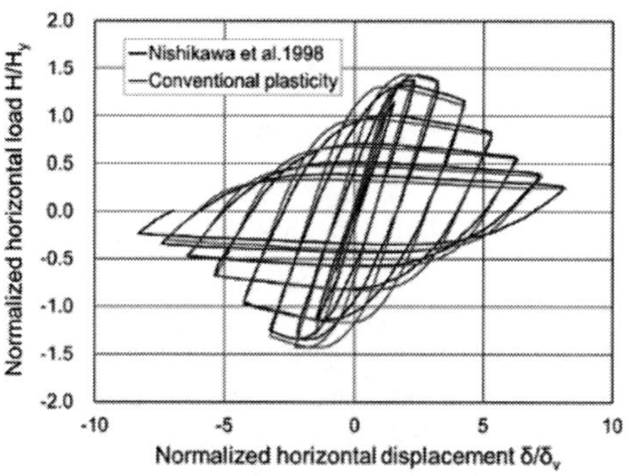

Fig. 8. Hysteresis loops for the pier in Fig. 3 (conventional plasticity).

Moreover, Fig. 7 displays the envelope (positive side) of the horizontal load – horizontal displacement at loading. The load peak value and its horizontal displacement are caught quite well by the model. There is a slight overestimation of the material stiffness in the first part of the curve, probably due to the simplifications and assumptions of the FE model and uncertainties on the parameters. Overall the degree of approximation is satisfactory. Fig. 8 shows the response obtained using a conventional elastoplastic and damage model. In this case, the response of the model seems to be excessively stiff, overestimating the real material performance.

As an additional check, the SS400 steel material parameters in Table 2 were adopted for a uniaxial monotonic extension on a cubic element (C3D8R element). The true stress-cumulative plastic strain is shown in Fig. 9. The value of the H_d, for the stress plateau, was assumed accordingly to [23]. The present model seems to overestimate the material performances compared with the uniaxial stress-cumulative plastic strain reported in Van Do *et al.* [24], but they seem to be in good agreement with the curves reported in [23].

The absorbed energy per loop by the steel pier can be represented by the area contained inside the hysteresis loop as schematically depicted in Fig. 10. Fig. 11 reports the comparison of the absorbed energy per loop in [1] against the absorbed energy obtained in the FE simulation, normalized against the elastic energy E_0. The value of the elastic energy is simply given by the work of the horizontal load at yield H_y on the displacement δ_y (see Fig. 4):

$$E_0 = \left(H_y \delta_y\right)/2 \tag{10}$$

The response obtained with a conventional elastoplastic model overestimate the absorbed energy in the first six loops. This is due to the fact that the material response during cycles is considered as purely elastic until the stress state reached the plastic potential, generating an overestimation of the hysteresis loops area visible in the previous Fig. 8. Moreover, the normalized absorbed energy of the loop five is almost the same as the one of the loop six anticipating the peak of the absorption for the pier.

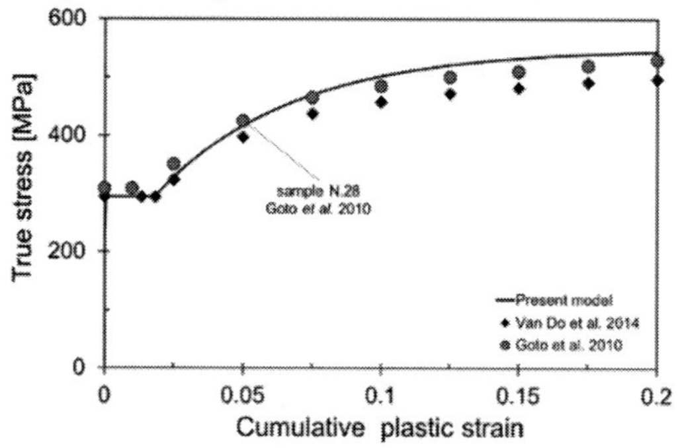

Fig. 9. Uniaxial behaviour of the SS400 steel.

Table 2 Damage S-S parameters.

Young's modulus [MPa]	206000
Poisson's ratio	0.3
F_0 [MPa]	294
u, c, χ	500, 200, 0.9
h_1, h_2	0.36; 15.0
K [MPa]	75
C_1 [MPa], B_1	3000; 23.01
C_2 [MPa], B_2	588.6; 22.84
H_d	0.183

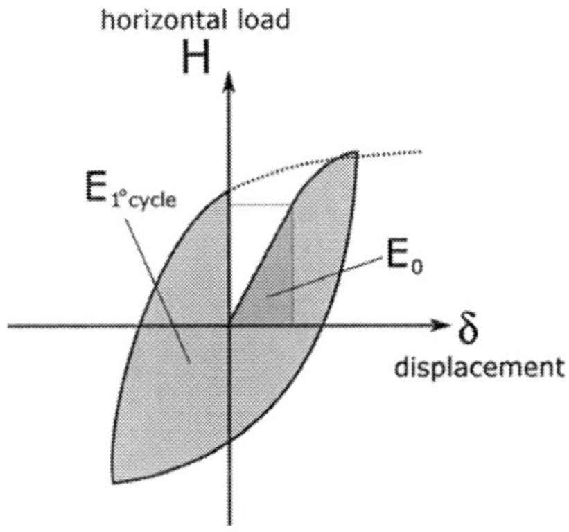

Fig. 10. Example of the absorbed energy during the first cycle.

The overlap between the curves in Fig. 11, for the Damage S-S model, is quite good, with a slight underestimation of the absorbed energy after the sixth cycle. The energy absorption keeps increasing after the peak load is reached (around δ=25 mm, see Fig. 7) and reaches it maximum during the sixth cycle. This point can be considered as the limit state from the standpoint of energy absorption capacity for the structure and it can be used to evaluate the seismic performance of the bridge columns.

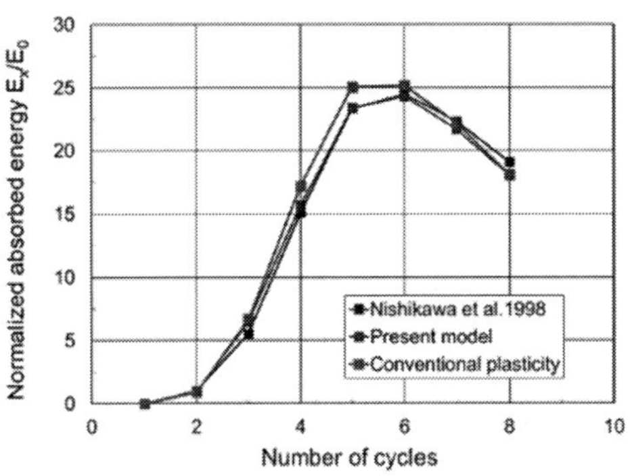

Fig. 11. Energy absorption for the loading sequence of Fig. 5.

The characterization of the M-C criterion for the ductile behaviour of SS400 steel is still missing, therefore in this study the four constants for the definition of the equivalent strain to fracture in Eq. (8) were obtained by fitting the experimental values of the hysteresis loops in Fig. 6. The damage parameters are reported in the following Table 3. In order to make the simulation more realistic, the constants were calibrated starting from the reference values obtained in a different work of the authors (currently under submission [26]) for SM490Y steel. The value of d_1 was set to 2.5% in order to simulate the real behaviour of steels which don't show any mechanical degradation under low plastic deformation. Moreover, a critical value for the damage D_c was set equal to 0.15, considering that value as the starting point for the crack opening. The critical damage parameter is included in the present formulation in order to avoid pathological mesh dependency which is the result of the so-called *local continuum* assumption [27], typical of many models within the CDM.

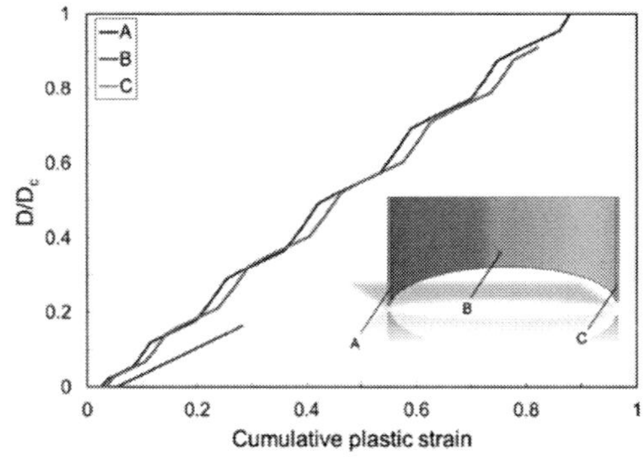

Fig. 12. Damage evolution (element centroid) for three points at the base of the steel pier.

Table 3 M-C parameters.

A [MPa]	650
N	0.25
c_1	0.15
c_2 [MPa]	540

Fig. 12 reports the damage evolution against the cumulative plastic strain at the centroid of three elements located around the base of the steel column, where the local buckling takes place. As expected, the elements A and C show a higher damage accumulation due to the fact that they are aligned along the loading direction *x*. The damage evolution shows a clear difference in accumulation during the compressive and tensile loading of the pier, due to the asymmetric shape of the fracture envelope (see Fig. 13).

The damage concentrate around the base of the column, where the plastic deformation accumulates. Moreover, the damage contour field shows that the accumulation is higher on the internal wall of the pier rather than on the external wall. The following Fig. 14 and Fig. 15 show the evolution of the stress triaxiality and of the Lode angle parameter for the element A. The dashed blue lines are representative of the average values that are computed using the integral form reported by Bai and Wierzbicki in [9].

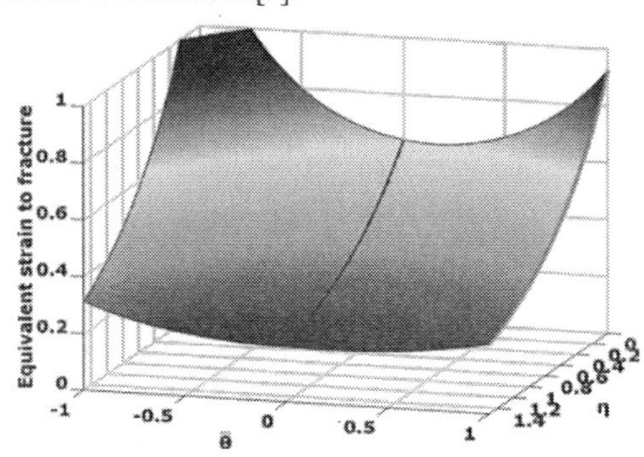

https://doi.org/10.1051/matecconf/201816522013

Fig. 13. M-C failure envelope.

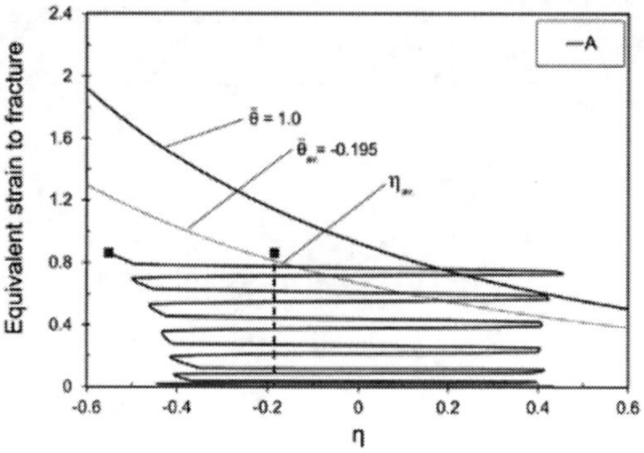

Fig. 14. Stress triaxiality evolution (and its average value) for the element A of Fig. 12 (centroid).

3.3. Evaluation of the horizontal load-carrying capacity

The graph of Fig. 11 gave a criterion that allows judging the capability of the structure to absorb energy. The work done by the horizontal solicitation increases until it reaches its peak in the sixth cycle. After that point, the structure is not able to absorb more energy and the area inside the hysteresis loops becomes gradually smaller.

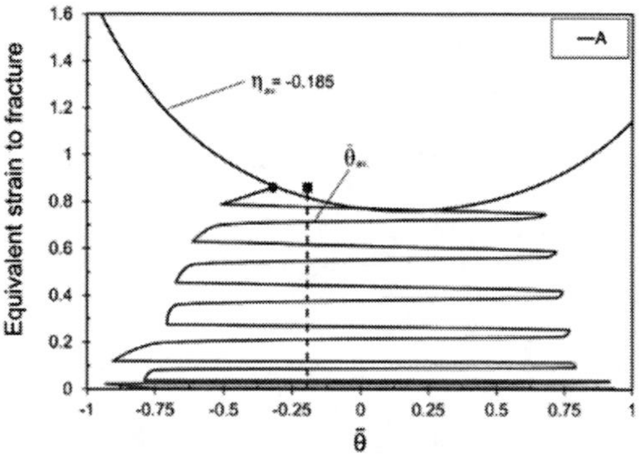

Fig. 15. Lode angle parameter evolution (and its average value) for the element A of Fig. 12 (centroid).

The idea, therefore, is to consider the structure compromised after the sixth cycle.

In reality, earthquakes are characterized by a series of wave loads that take place in a window of time of few days after the first seismic event. In general, the first shock wave is the most intense but often the aftershock waves have an almost similar magnitude (i.e. the earthquake in Tōhoku in 2011 or the central Italy earthquake in 2016). The following Fig. 16 displays the three different loading conditions used to evaluate the horizontal load-carrying capacity of the steel column subjected to a series of shock waves with smaller intensity compared to the one in Fig. 5. Instead of using a monotonically increasing lateral displacement (i.e. *L 0* loading condition), the three new loading conditions are represented by a sort of harmonic loading, composed by a series of shock waves with different maximum lateral displacements (i.e. $5\delta_y$, $4\delta_y$, $3\delta_y$ respectively).

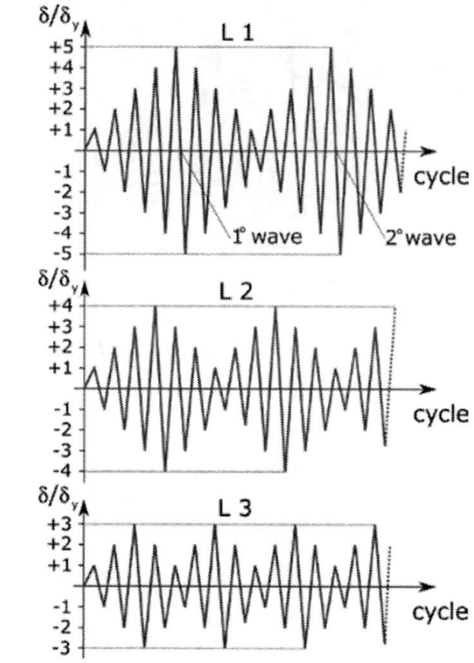

Fig. 16. The three loading sequence for the evaluation of the seismic response of the pier.

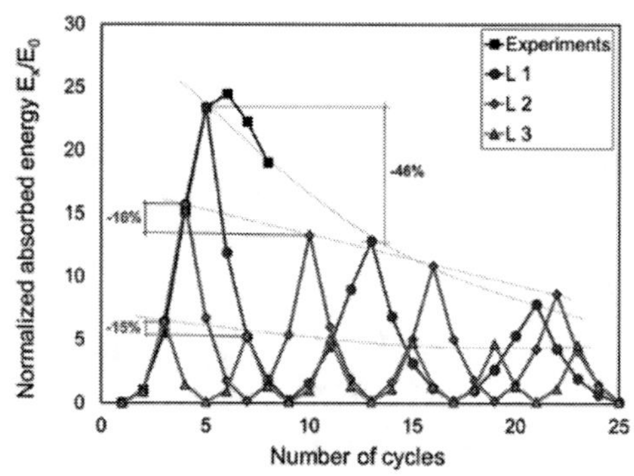

Fig. 17. Absorbed energy per loop for the loading conditions *L 1*, *L 2* and *L 3*.

Fig. 17 reports the absorbed energy per loop for the three loading conditions of Fig. 16. A maximum number of 25 cycles is displayed to make the graph clearer. The loading condition *L 1* shows a significant decrease (i.e. -46%) in the absorbed energy at the maximum lateral displacement between the first and the second wave. Smaller decreases are instead obtained for the condition *L 1* and *L 2*: -16% and 15% respectively. Moreover, the trend for the decrease of the peak seems to be exponential in *L 1* and almost linear for the other two solicitations.

The definition for the criteria of the load-carrying capacity for the pier is more complicated in case of these new loading conditions since it is not easy to establish a point where the absorbed energy decrease as in Fig. 11.

However, in a first approximation, it is still possible to establish an energy criteria for the evaluation of the structural performance in terms of cumulative absorbed energy. It is here assumed that the cumulative absorbed energy, in a monotonically increasing loading condition and up to the point when the absorbed energy per loop is maximized, can represent the criteria to evaluate different loading conditions. In other words, the total absorbed energy for the *L 1*, *L 2* and *L 3* loading cases are compared with the total absorbed energy up to the sixth cycle of the monotonically increasing load $E_{crit.}$.

$$E_{crit.} = \sum_{i=1}^{6} E_{i-loop}^{L0} \tag{11}$$

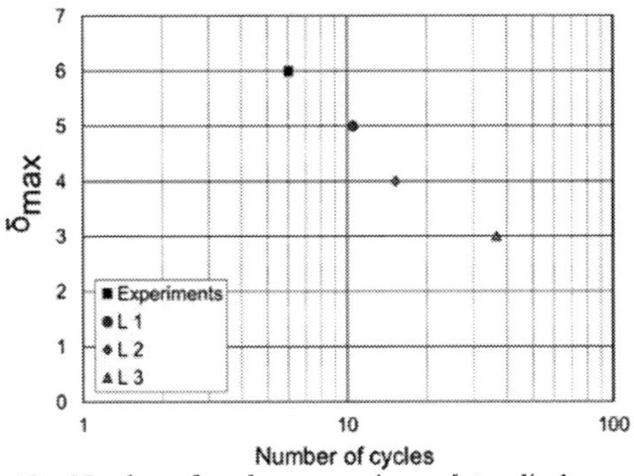

Fig. 18. Number of cycles vs maximum later displacement when the critical value of the total absorbed energy is reached for the *L 1*, *L 2* and *L 3* loading conditions.

Fig. 18 displays the number of cycles when the critical absorbed energy, defined in Eq. (11), is reached for the *L 1*, *L 2* and *L 3* loading conditions. As expected, the number of cycles increases with the decrease of the maximum imposed displacement δ_{max}. The black, the red and green points are almost aligned, showing a linear increase in the number of cycles. When the loading amplitude becomes smaller the number of cycles sensibly increase, due to the fact that the material tends to behave elastically. It is worth mentioning, at this point, that the energy criteria defined in Eq. (11) is subjected to some uncertainties. First, the ductile damage law defined in Eq. (9) is characterized by a linear damage evolution which overestimates the damage accumulation during cycles. This leads to a smaller absorbed energy per cycle due to the material softening and therefore to an overestimation of the number of cycles to reach the critical energy. Moreover, the ductile damage accumulation law should be also provided with a term that takes into account the non-proportionality of the load, since it has been shown that this factor affects the ductility of metals [20, 28]. On the other hand, the criteria can provide, in first approximation, a valid tool to judge the seismic performance of steel bridge columns under unidirectional seismic loading.

4 Conclusions

In this paper, the performance of a steel bridge columns in terms of horizontal load-carrying capacity was evaluated. The pier behaviour under a constant compressive load and horizontal cycling loading were investigated by means of an unconventional elastoplastic and damage model named Damage S-S model. The results can be summarized in the following points:

- The Damage S-S model can catch a realistic structural response of the pier. The graphs reporting the horizontal load vs the horizontal displacement (Fig. 6), the positive load envelope vs lateral displacements (Fig. 7) and the absorbed energy per loop (Fig. 11) showed good agreement with the experimental data reported in [1]. On the other side, the use of a conventional plasticity model overestimate the material performances.
- The damage is localized at the base of the column and it can provide information on the location of the crack formation.
- A criterion for the load-carrying capacity of the structure is provided in terms of cumulative absorbed energy and it is used to judge the seismic performance of the structure subjected to different harmonic loadings. The results showed an increasing number of cycles for a decrease of the maximum imposed displacement.

Future works will focus on the definition of a new ductile damage accumulation law, more suitable for the description of the ductility behaviour under non-proportional cyclic loading.

References

1. K. Nishikawa, S. Yamamoto, T. Natori, K. Terao, H. Yasunami and M. Terada, Eng. Struct., **20**, 540, (1998)
2. S. Gao, T. Usami, H. Ge, J. Eng. Mech., **124**, 260 (1998).
3. H. Momii, S. Tsutsumi, R. Fincato, Transaction of JWRI, **44**, 41 (2015)
4. K. Hashiguchi, *Elastoplasticity Theory* (Springer, Berlin, 2009)
5. K. Hashiguchi, S. Tsutsumi, Int. J. PLasticity **117**, 117 (2001)
6. S. Tsutsumi, K. Hashiguchi, Int. J. Plasticity, **21**, 1941 (2005)
7. R. Fincato, S. Tsutsumi, Int. J. Num. Meth. Eng. **113**, 1729 (2017)
8. J. Lemaitre, J. Engng. Mat. Tech **107**, 83 (1985)
9. R. Fincato, S. Tsutsumi, Fracture and Structural Integrity **38**, 231 (2016)
10. R. Fincato, S. Tsutsumi, Q. J. Jpn. Weld. Soc. **35**, 185s (2017)

11. K. Hashiguchi, M. Ueno, T. Ozaki, Acta Mech. **223**, 985 (2012)
12. R. Fincato, S. Tsutsumi, Marine Struct. **55**, 104 (2017)
13. S. Tsutsumi, K. Murakami, K. Gotoh, M. Toyosada, JASNAOE **7**, 243 (2008)
14. S. Tsutsumi, M. Toyosada, F. Dunne, Procedia Engineering **2**, 139 2010.
15. K. Hashiguchi, T. Yoshimaru, Int. J. Plast. **11**, 347 (1995)
16. J.L. Chaboche, Int. J. Plast. **2**, 149 (1986)
17. S. Tsutsumi, K. Morita, R. Fincato, H. Momii, Fracture and Structural Integrity **38**, 240 (2016)
18. W. Li, F. Liao, T. Zhou, H. Askes, J. of Constr. Steel Res. **123**, 1 (2016)
19. Y. Bai, T, Wierzbicki, Int. J. Fract **161**, 1 (2010)
20. M. Algarni, Y. Choi, Y. Bai, Int. J Fatigue **96**, 162 (2017)
21. M. Algarni, Y. Bai, Y. Choi, Eng. Fract. Mech. **147**, 140 (2015)
22. S. Tsutsumi, T. Kitamura, R. Fincato, Proceedings of TEAM2017, Osaka, Japan
23. Y. Goto, G.P. Kumar, N. Kawanishi, J. Struct. Eng. **36**, 1413 (2010)
24. V.N. Van Do, C.H. Lee, K.H. Chang, Comp. Mech. **53**, 1209 (2014)
25. S. Tsutsumi, M. Toyosada, K. Hashiguchi, J. of. App. Mech. JSCE, **9**, 455 (2006)
26. S. Tsutsumi, T. Kitamura, R. Fincato, Eng. Fract. Mech. (to be published) (2018).
27. J. Besson, Int. J. Dam. Mech. **19**, 3 (2010)
28. J. Papasidero, V. Doquet, D. Mohr, Int. J. Solids Struct **69-70**, 459 (2015)

Design Optimization and Fatigue Analysis for Seal Joint of Control Rod Drive Mechanism

*Shao Xuejiao**, *Dujuan*
Nuclear Power Institute of China ,Chengdu,610213,China

Abstract. Two kinds of seal joint thickness of control rod drive mechanism seal house were analyzed on stress and fatigue by using finite element method, and it provides analytical basis for equipment optimization, design manufacturing and problem handing. In the fatigue computing, all the transients were grouped into several groups, so as to separate the transients not at the same stage obviously. Meanwhile, the transients can be respectively combined in every group to reduce the redundant conservatism of computation. The elastoplastic strain correction factor was modified by analyzing thermal and mechanical load separately referring the rules of RCC-M 2002.By comparing the stress of two thickness joints, the results are found that the fatigue results of the thinner thickness are safer, but this is contrary to the rule of other results. The fatigue usage factor can be reduced by using modified elastoplastic strain correction factor when the amplify of the primary and secondary stress is large produced by both thermal load and mechanical load.

1. INTRODUCTION

The margin of safety will decrease if the aging of react pressure vessel is not be controlled, this will increase the risk of health and safety in public. The effects of LWR coolant environments on fatigue life in pressure vessel cannot be ignore.

The LWR coolant environments contain the temperature; dissolved–oxygen (DO) level in water; strain rate; strain amplitude which are the key parameters that influence fatigue life. In the stage of designing engineering, the conservatism which was considered by fatigue strain data is not enough.

Two approaches have been proposed for incorporating the environmental effects：(a) develop new fatigue design curves for LWR applications (b) use an environment correction factor to account for the effects of the coolant environment,

The second approach, proposed by M.Higuchi and K.Iida[1,2], considers the effects of reactor coolant environments on fatigue life in terms of an environmental correction factor Fen, which is the ratio of fatigue life in air at room temperature to that in water under reactor operating conditions. Specific expressions for Fen, based on statistical models and on the correlations developed by the Environmental Fatigue Data Committee of the Thermal and Nuclear Power Engineering Society of Japan, have been proposed.

The fatigue of key parts on control rod drive mechanisms was analyzed considering the influence of environmental fatigue cased by the coolant in the react pressure vessel. The elastoplastic strain correction factor was modified by analyzing thermal and mechanical loads separately referring the rules of RCC-M 2002[3].

2.MODEL AND CALCULATION APPROACH

2.1 Finite element model

The axi-symmetric model was created for the analysis which comprises the rop section of the adapter, the lower section of the latch housing seal house, the assembly threading and the thin-lip seal. A detailed view of the mesh density including the cut locations is shown in Fig. 1.The mechanical loads considered in this calculation are structural temperatures, internal pressure transient, operating basis earthquake(OBE). Only a vertical constraint is necessary to restrain the bottom of the model.

In the fatigue calculation, the transient thermal and pressures boundary condition were applied to the finite element model, the node stress caused by temperature grads and pressure will combine with the earthquake loads, the time of the peak value was determined as the event of fatigue analysis. If the primary plus secondary stress range S_n exceeds the $3S_m$ limit, the peak value was modified by the elastoplastic strain correction factor. Based on the fatigue design curves for structural and the number of transients, the usage factor can be calculated by the software of ANSYS.

* *Shao Xuejiao*: shaoxuejiao@163.com

https://doi.org/10.1051/matecconf/201816522014

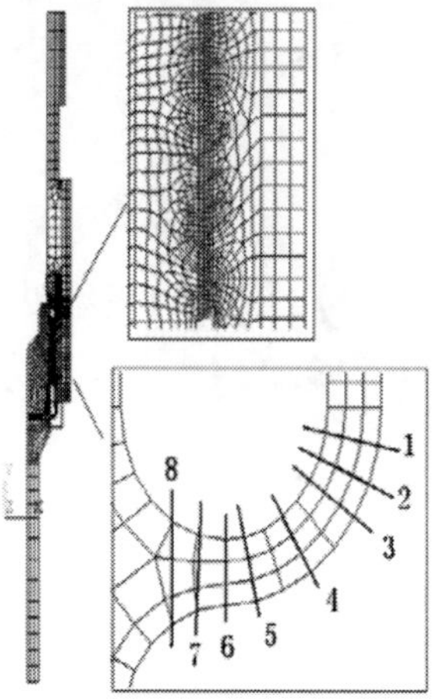

Figure 1 The finite element model and stress cut location

2.2 Fatigue analysis approach and Ke modifiy

In the ANSYS, the elastoplastic strain correction factor (K_e) is determined from the S_n value as follows:

$$S_n \leqslant 3S_m,\ K_e=1$$
$$3S_m \leqslant S_n \leqslant 3S_m,\ K_e=1+(1-n)(S_n/S_m-1)/n(m-1) \quad (1)$$
$$S_n \geqslant 3S_m,\ K_e=1/n$$

The values for m and n are 1.7 and 3 for austenitic stainless steel. The K_e factor is determined from the S_n range and the total stress range S_p, yielding to the Salt stress range used for fatigue analyses:

$$S_{alt}=K_eS_p/2 \quad (2)$$

According to the edition of RCCM 2002, The value of S_{alt} takes the following value[4-6]:

$$S_{alt}=(K_{emech}S_{pmech}+K_{etherm}S_{ptherm})/2 \quad (3)$$

Where $S_{p(mech)}$ is the range of the mechanical part of the stresses, $S_{p(therm)}$ is the range of the thermal part of the stresses. Elastoplastic stress correction factor for the thermal part, calculated in the case of austenitic stainless steels, by the following formula:

$$K_{etherm}=\max\{1.86[1-1/(1.66+S_n/S_m)],1\} \quad (4)$$

To sum up, the computing method of ANSYS is over conservation ,the new fatigue results were get using the expression (3) by user programming.

2.3 Environmental fatigue analysis

The first approach, the temporary fatigue design curve for LWR application supplied by NUREG/CR-5999(1993) [7] is shown in Fig.2 used to analyze the key parts on control rod drive mechanisms.

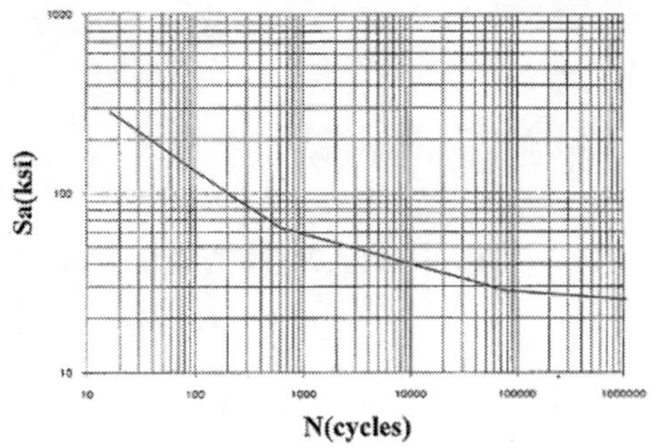

Figure 2 The temporary fatigue data for austenitic stainless steel

The second approach, proposed by Higuchi and Iida, considers the effects of reactor coolant environments on fatigue life in terms of an environmental correction factor Fen, which is the ratio of fatigue life in air at room temperature to that in water under reactor operating conditions. The Fen of austenitic stainless steels is represented by

$$Fen=\exp(0.935-T'\varepsilon'O') \quad (5)$$

where T′, O′, and $\dot{\varepsilon}'$ are transformed temperature, DO, and strain rate, respectively, defined as follows:

$$T'=0 \quad (T<150℃)$$
$$T'=(T-150)/175 \quad (150\le T<325℃) \quad (6)$$
$$T'=1 \quad (T\ge 325℃)$$

$$\varepsilon'=0 \quad (\dot{\varepsilon}>0.4\%/s)$$
$$\varepsilon'=\ln(\dot{\varepsilon}/0.4) \quad (0.0004\le\dot{\varepsilon}\le 0.4\%/s) \quad (7)$$
$$\varepsilon'=\ln(0.0004/0.4) \quad (\dot{\varepsilon}<0.0004\%/s)$$

$$O'=0.26 \quad (8)$$

The critical parameters that influence fatigue life are summarized :(1)Strain Rate: fatigue life decreases with decreasing strain rate.(2)Dissolved Oxygen in Water: the fatigue lives of austenitic are decreased significantly in low–DO water. The decrease in life is greater at low strain rates and high temperatures.(3) *Temperature*: the data suggest a lower threshold temperature of 150°C.

3. SIMULATIONS AND DISCUSSIONS

3.1 Effect of Ke

The fatigue results of the cut locations were listed In Table 1 which show the results exceed 1 only using the ANSYS, the result can reduce if adopt the user programming according the rules of RCCM 2002. The results show that the precision of fatigue usage factor can be elevated by using modified K_e when the amplify of the primary and secondary stress is large to some extent.

Table 1. The effect of Ke on fatigue usage

Section	$u(K_e)$		u(modified K_e)	
	in	out	in	out
1	1.152	0.000	0.357	0.001
2	1.805	0.003	0.646	0.008
3	1.481	0.001	0.663	0.003
4	0.012	0.000	0.022	0.000
5	0.000	0.008	0.000	0.011
6	0.482	2.236	0.292	0.513
7	1.127	2.059	0.330	0.496
8	0.041	0.109	0.041	0.077

3.2 Effect of environmental fatigue

In the section, two approaches were adopted in evaluating the third section with one transient. The usage factor is 0.144 using the design fatigue curve, but the usage factor is 0.6173 if using the temporary fatigue curve considering the effects of the coolant environment. This approach need compute the every key part of equipment once again ,the computing cost is high to some extent.

For the second approach, the second, the forth and the fifth column of the table 2 show the strain rate influence the Fen obviously when in the high temperature. If the temperature is higher than 150℃, the effect of temperature on Fen is sensitive. This approach can make use of existing result effectively.

Table 2. The effect of Fen on fatigue usage

/	1	2	3	4	5
temperature/ ℃	154.7	154.7	20	290.8	290.8
Strain rate/s^{-1}	5.98×10^{-6}	2.0×10^{-6}	5.98×10^{-6}	2.0×10^{-6}	4.0×10^{-3}
T'	0.027	0.027	0	0.805	0.805
O'	0.26	0.26	0.26	0.26	0.26
ε '	-6.506	-6.908	-6.506	-6.908	0
Fen	2.66	2.67	2.55	10.81	2.55
Initial usage	0.144	0.144	0.144	0.144	0.144
Modified usage	0.383	0.384	0.367	1.56	0.367

4. CONCLUSIONS

(1) The precision of fatigue usage factor can be elevated by using modified K_e when the amplify of the primary and secondary stress is large to some extent.

(2) The difference of results for the austenitic stainless steels is obvious if considering the coolant environmental fatigue comparing with the results in air.

References

1. M.Higuchi and K.Iida, "Fatigue Strength Correction Factors for Carbon and Low-Alloy Steels in Oxygen-Containing High-Temperature Water", Nucl. Eng. Des. 129, 293-306(1991).
2. K.Iida, T.Bannai, M.Higuchi, K.Tsusumi, and K.Sakaguchi, "Comparison of Japanese MITI Guideline and Other Methods for Evaluation of Environmental Fatigue Life Reduction", in Pressure Vessel and Piping Codes and Standards, PVP Vol.419, M.D. Rana,ed., American Society of Mechanical Engineers, New York,pp.73-82(2001).
3. Afcen. RCC-M: Editon 2000 and Addendum June 2002, Section:Subsection B , 2002.
4. Prost J P , Arnaudeau F. Elastic plastic strain correction factor in stainless steel pipes and Nozzles Subjected to Thermal Shocks . 4th ICPVT ,1980: 80-90.
5. Grandemange J M., Héliot J., Vagner J. Improvements on Fatigue Analysis Methods for the Design of Nuclear Components Subjected to the French RCC-M Code.WRC Bulletin , 1991(2):361.
6. FAMATOME ANP, APPENDIX A -Pastic strain correction factors.
7. S.Majumdar, O.K.Chopra, and W.J.Shack, " Interim Fatigue Design Curves for Carbon, Low-Alloy, and Austenitic Stainless Steels in LWR Environments", NUREG/CR-5999,1993.

Increasing fatigue performance in AHSS thick sheet by surface treatments

Sergi Parareda[1,*], *Antoni* Lara[1], *Henrik* Sieurin[2], *Héber* D'Armas[1], *Daniel* Casellas[1,3]

[1]Fundació CTM Centre Tecnològic, Metallic and Ceramic Materials Department, 08243 Manresa, Spain
[2]SCANIA AB, Materials Technology Department SE-151 87 Södertälje, Sweden.
[3]Luleå University of Technology, Division of Mechanics of Solid Materials, 971 87 Luleå, Sweden

Abstract. Advanced High Strength Steels (AHSS) have been widely applied in the automotive industry as an affordable solution for car lightweighting, mainly in parts subjected to crash requirements. Heavy duty vehicle (HDV) can also benefit from the expertise learned in cars, but parts must be designed considering fatigue resistance, especially on trimmed areas, and stiffness. Mechanical surface treatments, as blasting or shot peening, help increasing fatigue life of AHSS in trimmed areas and will allow weight reduction in HDV through gauge downsizing. The expected decrease in stiffness through thickness reduction can be improved by design changes. However, scarce information about the effect of mechanical surface treatments on AHSS are available. Thus, the aim of this work is to evaluate the increment in fatigue life of two different steel grades (350 MPa, and 500MPa of yield strength) in thick sheet by means of mechanical surface treatment – sandblasting. High Cycle Fatigue [HCF] tests were conducted at alternating load [R=-1]. Residual stresses were measured by an X-ray tensometry prior fatigue tests. Also the surface roughness [Rz] and form is measured using an optical non-contact 3D microscope. On the other hand, the fracture surfaces of the test specimens were observed via scanning electron microscope (SEM) in order to determine the crack initiation points. The evaluation of fatigue life in terms of SN curves is also discussed, analysing how the sandblasting process modifies the surface roughness and introduce compressive residual stresses on the external layer of the material. Both phenomena enhance the fatigue strength of the evaluated steel grades.

Keywords. High-cycle fatigue, thick sheet, sandblasting, HSS, AHSS, HSLA steel, residual stresses, X-ray analysis.

1 Introduction

In the past years high strength steels and lightweight design have gained wide popularity to improve crashworthiness and reduce weight on passenger cars. Mainly efforts have been concentrated in the lightweight of the body in white (BiW) to achieve the GHG reduction targets. Passenger cars accounted for about 60% of global emissions in road transport in 2012 and it is expected to decrease to 45% by 2030. On the other hand, heavy duty vehicles (HDV) accounted for about 30% of the overall GHG emissions at 2012 and it is expected to increase beyond 40% at 2030 [1]. In this context, CO_2 standards, green taxation and the increasing alternatives in materials and forming processes for lightweighting are helping to reduce the global emissions for cars but emissions for trucks and buses are far to be decreased. Since the expected increase of HDV emissions is important enough to not be ignored, weight reduction in the truck and bus sector is the indicated way to make it cleaner. Although, the application of AHSS is less utilized in trucks and buses, there is a great weight reduction potential in exchanging traditional steel grades with AHSS also in HDV. Multiple key aspects such as lower energy (fuel) consumption, payload increase and material resource savings are directly related with a more efficient use of resources. A simple example is given by a supposed case in which if a 10% higher stress can be allowed by a material upgrading, the sheet thickness could be reduced by 5%. Gauge downsizing is straight related with payload increase for weight dimensioned transport, thus in this case payload is increased by a 5% which means an improvement in freight transport and reduction of GHG emissions.

Owing to multiple loading situations during service, HDV parts have to satisfy high mechanical demands. The elastic behaviour and crashworthiness of the truck are satisfied with a high yield strength (YS) and high ultimate strength (UTS) respectively. Truck stability solicitation is accomplished by an appropriate design with a utilization of complex shapes to get the desired stiffness. On the other side, fatigue strength is a crucial property to satisfy. Since many parts are exposed to high cycle fatigue Al alloys are not a material choice due to its limited fatigue resistance. Accordingly, advanced high strength steels (AHSS) are the suitable material family candidate to achieve high weight reduction in

* Corresponding author: sergi.parareda@ctm.com.es

HDV while improving part durability at affordable cost and environmental impact.

Fatigue behaviour is influenced by factors as steel strength, surface conditions, presence of non-metallic inclusions, welding defects, mean stress, residual stress, plate thickness, temperature, corrosion, number of cycles or applied stresses. Fatigue failures are often triggered by surface defects that can easily be generated during forming process. Several authors have studied the effect of cutting effect on fatigue features for carbon and microalloyed steels [2, 3].

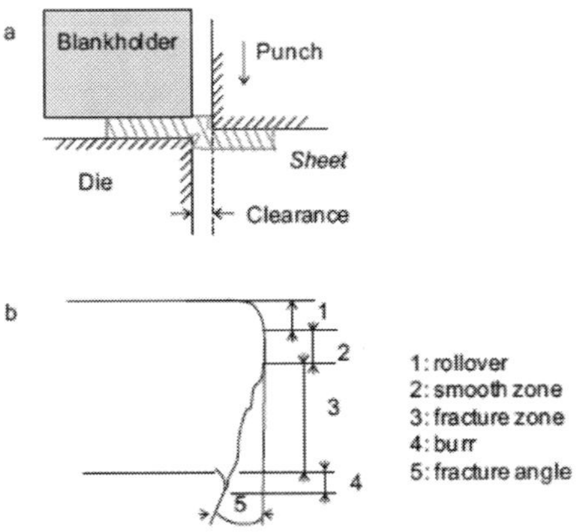

Fig. 1. (a) Illustration of trimming process and definition of cutting clearance. (b) Characteristics regions generated at the cut edge in mechanical cutting.

In this sense, the shearing process used in the automotive industry produce surface irregularities at the cut edge strongly affected by material type, material thickness, condition of cutting edges (sharp/worn), blankholder force, geometry of the cut and clearance (i.e. distance between punch and die). As a result of trimming process a typical cut edge shape can be observed on both the parent material and the cut-off. Five characteristics regions can be identified on the cut edge: rollover, smooth zone, fracture zone, burr and fracture angle which act as preferential sites to crack nucleation [4].

Infinite life design criterion based on SN or Wöhler curves is usually used to study the effect of cut edge on fatigue strength [5]. The associated fatigue life phenomenon is the crack nucleation, thus a way to enhance the fatigue strength is to delay the crack initiation. There are several factors that influence the crack initiation life: Surface roughness (Rz), surface defects, surface treatments, residual stresses at the surface and notch effect (K_t). Subsequent operations such as shot blasting or shot penning are suitable to modify and improve several of these factors [6, 7].

Shearing process is used in HDV industry due to the higher production rates that can be achieved and a fast cost amortization of the mechanical tools. Moreover, sandblasting operation is used as a pretreatment of trimmed parts to clean and smooth the surface before painting. Thus, keeping in mind the necessity to reduce the GHG emissions of HDV sector, the mechanical solicitations on chassis parts and the forming processes used, the present work aims at evaluating the effect of surface treatment as sandblasting on fatigue performance in AHSS. To achieve this objective, two different hot rolled HSLA steel grades with and without sandblasting process have been analyzed by constant stress amplitude alternate bending test. Materials with cut edge have been as reference in this study. Compressive residual stresses measurements have been carried out by X-ray tensometry [8] prior fatigue test. Besides, edge features have been analyzed by topographical 3D analysis.

2 Experimental procedure

Two hot rolled HSLA steel sheets of a thickness of 3mm studied had a nominal YS of 350MPa and 500MPa respectively. Chemical composition and sheet thickness of the studied steels are listed in Table 1.

Table 1. Chemical composition (in wt%) and sheet thickness of studied steels.

Steel grade	350 - 3mm	500 - 3mm
%C	0.067	0.0738
%Si	0.0167	0.0225
%Mn	0.587	0.798
%P	0.0061	0.0069
%S	<0.0010	<0.0010
%Cr	0.0266	0.0361
%Mo	<0.0050	<0.0050
%Ni	0.0306	0.0339
%Al	0.0430	0.0429
%Nb	0.0243	0.0336

The delivered 350 steel grade shows a homogeneous distribution of pearlite in a fine ferritic matrix (Fig. 2). 500 steel grade also shows a homogeneous distribution of pearlite in a ferritic matrix (Fig. 3), in this case much finer than the 350 grade. Fig 4 shows an oxide scale on the surface of both materials probably from steelmaking process.

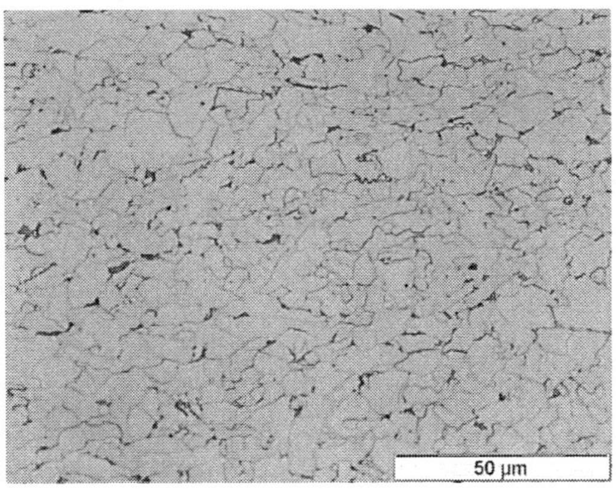

Fig. 2. Microstructure of 350 steel grade, ferrite-pearlite microstructure.

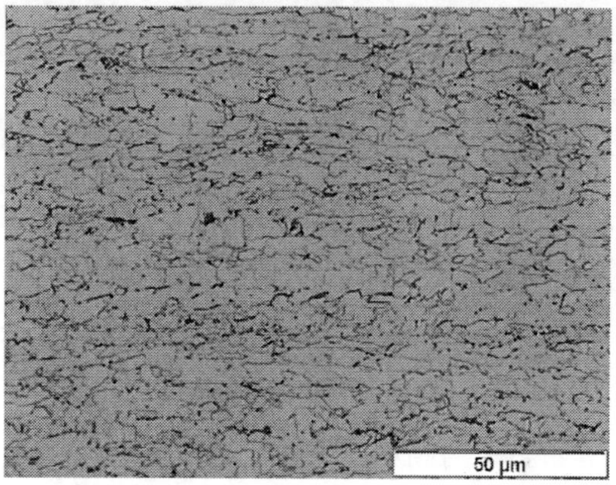

Fig. 3. Microstructure of 500 steel grade, finer ferrite-pearlite microstructure.

In order to evaluate the influence of the trimming process on the mechanical properties of the steel sheet, the calibrated zone of fatigue specimens (Fig. 6) was cut using a cutting clearance of 10% of the sheet thickness and a punch with sharp edges (radius close to 0 mm). Shearing process conditions (i.e. clearance and punch radius) are defined according to the cutting conditions employed in the automotive industry.

Specimens were manually blasted after trimming using glass microspheres of 40 to 95 µm of diameter as abrasive media and compressed air pressure of 6 bar. The distance between the workpiece and the nozzle was set at 20mm. The blasting direction was always perpendicular to the surface.

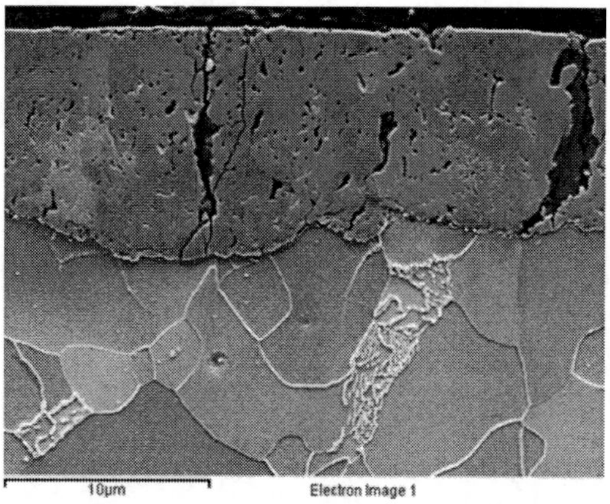

Fig. 4. Oxide scale in the 350 steel grade. It should be noted that the scale cracks do not penetrate the substrate.

Thus, two edge conditions are defined:

- Group 1 (G1): Trimmed edge
- Group 2 (G2): Trimmed edge and subsequent sandblasting operation

The edge quality resulting from each condition was evaluated by different techniques: (a) cross section inspection by optical microscopy (Fig. 5) (b) edge roughness (Rz) by, inspecting optical non-contact 3D microscope. Specimens cut in cross-section were etched with Nital at 2% to analyse the shear zone close to the edge. Compressive residual stresses were measured by the X-ray diffraction (XRD) technique at the surface of the calibrated zone after trimming and sandblasting processes. The area for the measurements of residual stress was 2 mm in diameter.

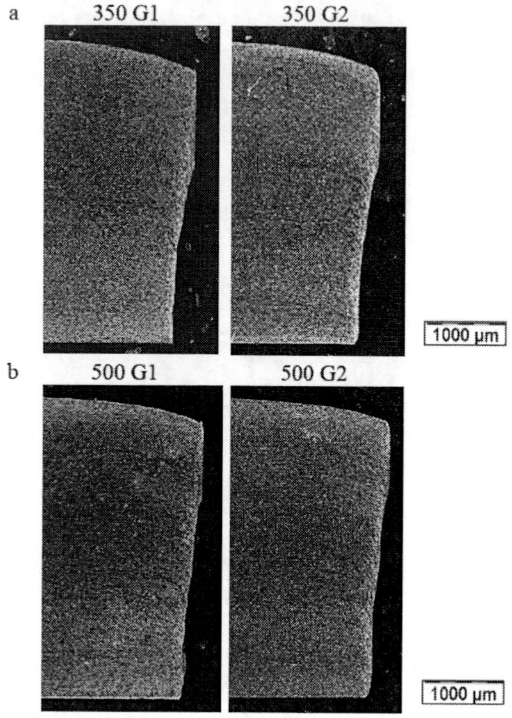

Fig. 5. Cross-section of trimmed edges observed by optical microscopy. (a) 350 steel grade, and (b) 500 steel grade. Deformed microstructure can be observed in the cut edge.

Mechanical properties of the studied materials were evaluated by means of tensile tests using specimens extracted perpendicularly to the rolling direction. The UTS, YS and the total elongation (TE) of each material were evaluated using tensile test under monotonic loading conditions in a universal testing machine (not standardized specimen). The applied load rate was 2.5mm/min.

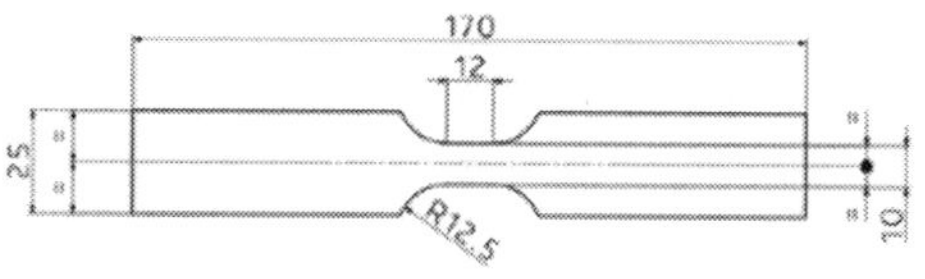

Fig. 6. Trimmed specimen geometry and dimensions.

Fatigue tests were conducted in a servo-hydraulic testing machine at four-point alternative bending (R=-1) at room temperature. This test configuration is selected because bending is the most common stress state in HDV parts. The fatigue limit (σ^{FAT} – stress amplitude) was evaluated at $1x10^5$, as it is the range used in the design process. At least 15 specimens for each material and condition were tested at a frequency of 10Hz. The maximum stress on the specimen was calculated using the four-point bending equation (1). Fracture surfaces were inspected by means of field emission scanning electron microscopy (FE-SEM).

$$\sigma_f = \frac{3}{2} \cdot F_a \cdot \frac{(l_1 - l_2)}{b \cdot h^2} \quad (1)$$

Table 2 summarizes tensile test under monotonic loading and bending test under cyclic loading performed with the 350 and 500 steel specimens and different edge quality.

Table 2. Mechanical tests (monotonic (M) and cyclic (C)) for the two sheet steels and edge quality (G1 and G2).

Edge	G1		G2	
Test	**M**	**C**	**M**	**C**
350	√	√	-	√
500	√	√	-	√

3 Results

Monotonic and cyclic test results are summarized in Table 3 for all the tested materials and edge conditions employed.

Table 3. Values of yield stress (YS), ultimate tensile strength (UTS), total elongation at 10mm (TE) and fatigue limit (σ^{FAT} – stress amplitude) for the studied steels.

Material	YS (MPa)	UTS (MPa)	TE_{10mm} (%)	σ^{FAT} (MPa)
350-G1	464	477	58.4	362±17
350-G2	-	-	-	365±18
500-G1	641	651	43.3	421±21
500-G2	-	-	-	465±29

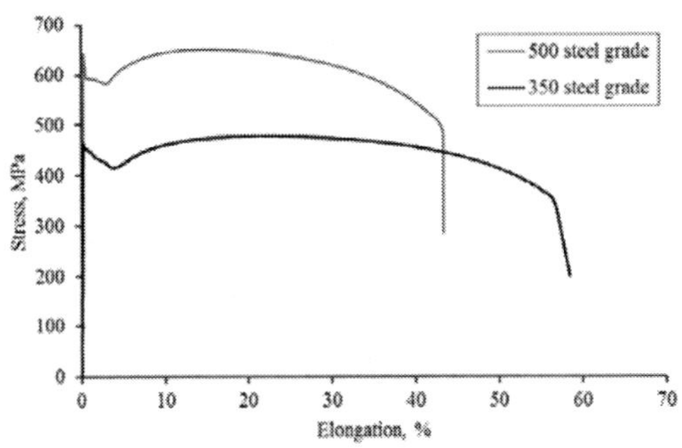

Fig. 7. Tensile test curves for the studied steels.

As observed in Fig.7, if the UTS increases the elongation decreases due to the differences observed in the microstructure. A fine microstructure can reach high strength but is not possible to maintain the elongation of the material.

Sandblasting process introduces compressive stresses on the surface. The residual stresses results from X-ray tensometry measurement of each steel grade are shown in Fig. 8.

Table 4. Compressive residual stresses measured in the surface for the studied steels.

Material	σ_c (MPa)
350-G2	-339,7±27.5
500-G2	-348.4±15.6

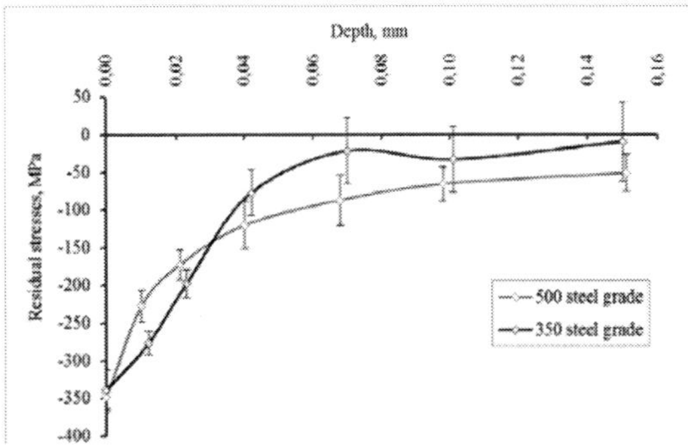

Fig. 8. Residual compressive stresses in the blasted surface for the studied steels.

As can be seen in Fig. 9(a)(b) fatigue strength is highly improved at high number of cycles for both materials because the crack nucleation is delayed by the compressive residual stresses of the sandblasting process.

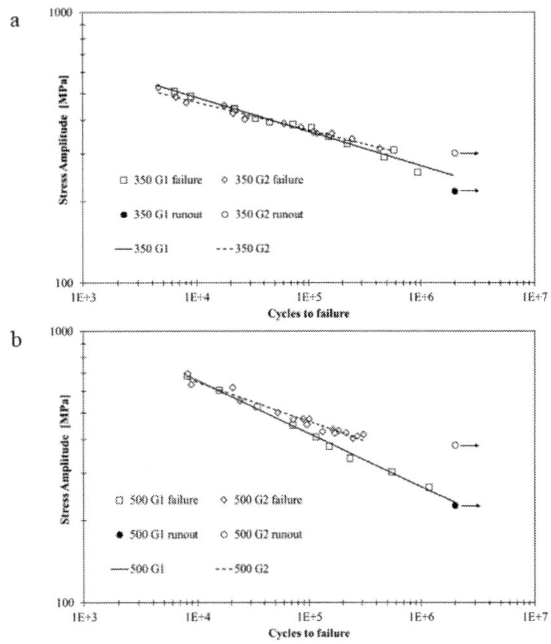

Fig. 9. Fatigue test curves for the studied steels. (a) 350 steel grade (b) 500 steel grade.

Surface roughness is measured close to the upper and lower surface for each edge condition and material by an optical non-contact 3D microscope. Table 5 summarizes the Rz values in the smooth zone and in the fracture zone for the studied conditions.

Table 5. Surface roughness values (Rz) in the smooth zone and fracture zone for the studied materials.

Rz (μm)	350-G1	350-G2	500-G1	500-G2
Smooth	6±1	5±1	7±1	5±1
Fracture	40±1	7±1	40±1	14±2

4 Discussion

Although all the edges present the characteristics regions generated at the cut edge by mechanical cutting, different characteristics were observed depending on the strategy used. As can be seen in Fig. 5 G1 specimens have an irregular surface, sharp corners and also small burr. On contrary, in G2 specimens the surface is smooth without sharp areas, the corners are rounded and the burr is folded. The rollover region is similar for the two materials. The smooth zone is shorter as the strength of the material increase, due to a decrease in elongation of the material (Fig. 7). The fracture angle is bigger as the strength of the material decreases, because of the material elongation capacity increases.

Tensile residual stresses can be detrimental since it is equivalent to apply a tensile mean stress on the component that reduce the necessary applied stress for the crack initiation/crack growth. On contrary, when residual compressive stresses are present, the applied stress necessary to crack initiation/growth must be higher to negate the residual compressive stresses. Thus, residual compressive stresses on the surface are beneficial to increase the fatigue life. Although, focusing on 350 steel grade, there is no difference between fatigue strength of each edge condition at $1x10^5$ cycles. In this case the surface smoothening and the compressive residual stresses are not beneficial to increase fatigue strength at the analysed number of cycles. In contrast, 500 steel grade shows a noticeable increment in fatigue strength on G2 edge condition at $1x10^5$ cycles. Compressive residual stresses and changes on edge shape geometry are beneficial to increase the fatigue performance of this steel grade. The difference observed between G1 conditions can explained by grain size, i.e. the finer the microstructure is (more grain boundaries) the higher energy for crack propagation is needed (cycles required). This means that the higher steel grade has more resistance to crack propagation, i.e. higher damage resistance capacity [9]. On the other hand, sandblasting process has no effect on fatigue strength at low cycle fatigue regime because the residual stresses are easily cancelled by the high stresses in this regime.

Fatigue origins are related to surface irregularities and load ratio (R). Fig. 10 shows a typical fracture observed after the fatigue test. Multiple fatigue initiation sites (arrows in Fig. 10) are identified by FE-SEM. Alternative bending localizes the maximum stresses at the upper and lower surface of the specimen. It should be noted that in the centre of the specimen the applied stress is null. For this reason, the multiple crack origins are localized close to the surfaces, where stress is higher and not on the centre of the edge as might occur in tensile tests. It is in agreement with results from previous works [4]. FE-SEM images show multiple crack initiation sites. The cracks with small striations length are related to low stresses, which means that were the first cracks to nucleate and grew at low stresses until the stresses on the specimen were high enough to nucleate and propagate secondary cracks. On the other side, the cracks with wide striations are related to high stress, which means that the crack underwent a fast growth.

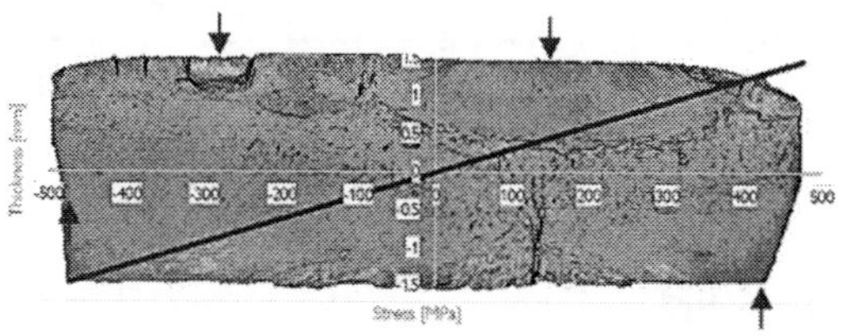

Fig. 10. Fracture and bending stress distribution of failed specimen.

Sandblasting introduces compressive residual stresses in the surface and also plastically deforms the edge reducing the surface roughness, folding the burr and smoothening the surface which decreases the notch effect (K_t). Moreover, blasting operation removes the oxide scale observed in the studied specimens.

The fracture of each specimen was studied to determine the initial crack zone, see Fig 11-14. In G1 specimens crack initiation zone was mainly located in the fracture zone, because of the roughness and also the burr (stress concentrator) in this zone is highly stressed. Nevertheless, in G2 specimens the crack initiation zone is often observed in both zones, in the smooth zone and also in the fracture zone respectively.

Fig. 11. Fractography of failed 350 steel grade specimen. Crack initiation in the burr of G1 specimen.

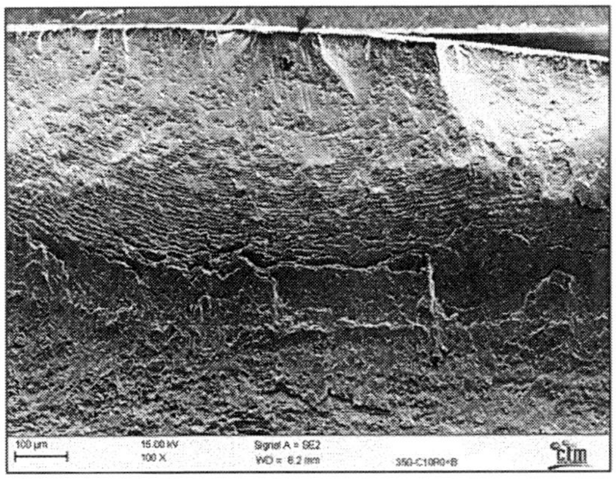

Fig. 12. Fractography of failed 350 steel grade specimen. Crack initiation in the surface of G2 specimen

Fig. 13. Fractography of failed 500 steel grade specimen. Crack initiation in the burr of G1 specimen

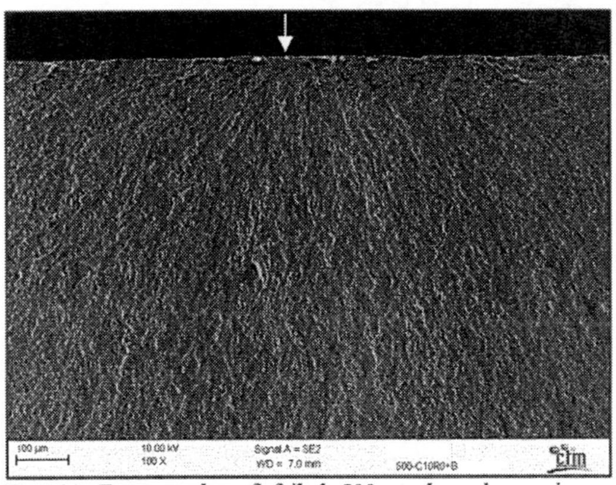

Fig. 14. Fractography of failed 500 steel grade specimen. Crack initiation in the surface of G2 specimen

5 Conclusions

The conclusions of this study based on the experimental tests, fractography analysis and shearing theory mentioned above are given below:

- Steel microstructure affects the characteristics of the sheared edge. A high strength steel shows small rollover area and smooth zone due to a low elongation of the material.

- Sandblasting process rounds the trimmed edges, folds the burr on the material and reduces the edge roughness. It also introduces compressive residual stresses that increase the fatigue performance.

- At low cycles the sandblasting has no effect on fatigue strength. On contrary, at high cycles (low stresses) the sandblasting increases the fatigue life.

- Bending tests generate the maximum stresses in the top and bottom surfaces of the specimen, whereas stress is null at specimen centre. For that reason, the crack nucleation tends to be localized in higher stresses areas (upper or lower surface). It implies that defects at surface (rollover and burr) dominates fatigue behaviour, whereas defects located at the edge fracture surface have a minimum impact on fatigue life.

6 References

[1] Study by CE Delft, "Marginal Abatement Cost Curves for HDVs", 2012 - http://eur-lex.europa.eu/legal-content/EN/TXT/PDF/?uri=CELEX:52014DC0285&from=EN

[2] E. Maronne, A. Galtier, J.I. Robert, T. Ishikawa, Trans. Eng. Sci. 40 (2003)

[3] M. Kinefuchi, T. Kitamura; J. Soc. Mater. Sci., Japan, Vol.62, No.12, pp.764-769, (2013)

[4] A. Lara, I. Picas, D. Casellas, J. Mater. Process. Technol. 213 (2013)

[5] T. Stenberg, E. Lindgren, Z. Barsoum, I. Barrnicho; Materialwissenschaft und Werkstofftechnik 48(6):556-569 (2017)

[6] D. Arola, A. E. Alade, W. Weber; Macch. Sci. tech. 10:197:218 (2006)

[7] H. Fujimoto, H. Ueda, R. Ueji, H. Fuji; ISIJ Inter. Vol 56, No. 7, pp. 1276-1284 (2016)

[8] K. Kolaøík, N. Ganev, Z. Pala, J. Drahokoupil; Mat. Struct., Vol. 22, No. 2 (2015)

[9] H. SuzukiA. J. Mcevily ; Met. Trans. Volume 10, Issue 4, pp 475-481 (1979)

Thermo-mechanical fatigue prediction of a steam turbine shaft

Martin Nesládek[1,*], *Michal* Bartošák[1], *Josef* Jurenka[1], *Jan* Papuga[1], *Milan* Růžička[1], *Petr* Měšťánek[2] and *Jan* Džugan[3]

[1]Czech Technical University in Prague, Faculty of Mechanical Engineering, Technická 4, 166 07 Prague 6, Czech Republic
[2]Doosan Škoda Power s.r.o., Tylova 1/57, 301 28 Plzeň, Czech Republic
[3]COMTES FHT a.s., Průmyslová 995, 334 41 Dobřany, Czech Republic

Abstract. The increasing demands on the flexibility of steam turbines due to the use of renewable energy sources substantially alters the fatigue strength requirements of components of these devices. This paper presents Thermo-Mechanical Fatigue (TMF) design calculations for the steam turbine shaft. The steam turbine shaft is exposed to complex thermo-mechanical loading conditions during the operating cycle of the turbine. An elastic-plastic structural Finite Element Analysis (FEA) of the turbine shaft is performed for the turbine operating cycle on the basis of calculated temperature fields obtained in a previous transient thermal FEA. The temperature dependent material parameters, which are used in the elastic-plastic FEA, are obtained from the uniaxial tests. Consequently, the TMF is predicted for the steam turbine shaft. Several fatigue criteria are used for the identifications of the critical domain and for the TMF damage assessment of the turbine shaft.

1 Introduction

The energy industry has been experiencing significant changes recently. The overall share of renewable sources in the Czech Republic is around 13 percent. In Germany it is even more, approximately 30 percent and the trend is that this share will continue to grow significantly in the future. However, instability of the power output due to these sources may cause problems in energy supplies if it is not compensated by some back-up sources. Currently, there are no reliable ways to save energy for power shortages in the network in such a large scale. This role must be taken over by traditional thermal power sources such as gas-fired power plants. The problem is that these plants already have to or will have to cover shortages in the daily regime. Increased number of start-ups and shutdowns as long as increasing demands on speed-up of the start-ups set new challenges in the gas and steam turbine design.

Components of steam turbines are often exposed to thermo-mechanical loading due to variable loading conditions, including start-up, full load, load change and stop phases. There have been increasing demands for reliable Low-Cycle Fatigue (LCF) and TMF fatigue predictions [1-3] of components, in order to provide safety, guaranteed function, durability and in order to avoid over-dimensioning of the components. The main damage mechanisms of TMF are often referred to as oxidation, fatigue and creep [4, 5]. Constitutive material model derived from the material testing, finite element analysis and suitable fatigue criterion is required for the reliable damage assessment of components subjected to cyclic thermo-mechanical loading conditions [3].

The Flexturbine project aims to revise the design of key turbine components and propose new design methodologies so that the turbines can cope with the changing situation in the transmission grids. The paper illustrates the work done on the prediction methodology for thermo-mechanical fatigue design of steam turbine shafts. The fatigue damage is assessed for the turbine shaft on the basis of several fatigue criteria. The turbine shaft is exposed to complex thermo-mechanical loading due to variable service conditions.

The paper presents experimental results of the uniaxial material tests. The temperature dependent constitutive material model is calibrated on the basis of the obtained experimental data. The material model is elastic-plastic and is used in the structural FEA of the steam turbine shaft, which is performed on the basis of the simulated temperature fields that are obtained from the previous transient thermal FEA. The finite element calculations for the steam turbine shaft are performed in the commercial finite element software ANSYS. Consequently, TMF is predicted for the turbine shaft on the basis of the calculated stress, strain and temperature histories.

The first fatigue criterion used for the TMF life prediction of the turbine shaft is the Damage Operator Approach (DOA) [6, 7]. The DOA is based on the classical strain-life approach. Cycle equivalent temperature or separated rain-flow counting are no longer required with the DOA. The oxidation effect is taken into account indirectly, because experiments were carried out under ambient conditions. The mean-stress effect is taken into account in the DOA as well, and is represented by the Smith, Watson and Topper damage parameter. The other fatigue criteria are the uniaxial

* Corresponding author: martin.nesladek@fs.cvut.cz

SWT [8] and Nihei [9] methods. The classical SWT fatigue damage model is often used in the energy industry. However, the original SWT and the Nihei criteria are not formulated to handle the loading cycles with variable temperatures. An appropriate cycle reference temperature has to be selected if these methods are to be applied. All the applied methods are based on the standard isothermal low-cycle fatigue data, i.e. strain-life curves obtained under isothermal loading conditions. From a practical point of view, the prediction models are implemented as the in-house codes (C++ and APDL scripts). The TMF results obtained for the steam turbine shaft with the DOA, the SWT and the Nihei models are compared.

2 The material

2.1. Experimental data

The steam turbine shaft usually operates under a wide temperature range. If accurate material response calculation and fatigue prediction are to be performed, the temperature dependence of the respective material model parameters must be known. The experiments should be carried out in such a way that the exhibited temperature range of the investigated component is fully covered. The investigated material of the steam turbine shaft is chrome-molybdenum creep resistant steel. The following uniaxial tests were conducted in order to identify temperature dependent material behaviour:

- Static tensile tests,
- Low-cycle fatigue tests.

The uniaxial tests were performed under ambient isothermal loading conditions in the temperature range of 20 to 600 °C. The low-cycle fatigue tests were performed on the servo-hydraulic MTS 810 machine. The split furnace Mellen was used for the heating of the specimens. The force was measured by the load cell. The LCF tests were performed as fully reversed at high strain rates and as strain controlled. High temperature extensometer with ceramic rods was used for the strain control. The tensile and the LCF tests were performed for the cylindrical specimens.

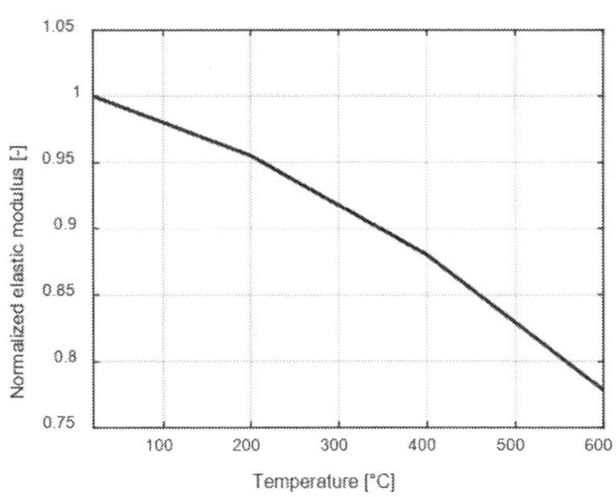

Fig. 1. The temperature dependent elastic modulus.

The temperature dependent elastic modulus E, the Cyclic Stress-Strain Curves (CSSC), and the Manson-Coffin and the Basquin fatigue curves were derived from the low-cycle fatigue tests. The CSSC constitutes the material stress-strain response under cyclic loading and has the following mathematical form, known as the Ramberg-Osgood relation:

$$\varepsilon_a = \frac{\sigma_a}{E} + \left(\frac{\sigma_a}{K'}\right)^{1/n'}, \tag{1}$$

where ε_a and σ_a are the strain and stress amplitudes, respectively. K' is the cyclic hardening coefficient and n' is the cyclic strain-hardening exponent. The CSSS were obtained from the stabilized hysteresis loops at mid-life.

The Manson, Coffin and Basquin curves are often necessary for the damage assessment for the LCF and the TMF loading conditions. The Manson, Coffin and Basquin curves are written in the common mathematical form as follows:

$$\varepsilon_a = \frac{\sigma_f'}{E}\left(2N_f\right)^b + \varepsilon_f'\left(2N_f\right)^c, \tag{2}$$

where N_f is the number of cycles to failure, σ_f' and b are the fatigue strength coefficient and exponent, respectively. ε_f' is the fatigue ductility coefficient and c is the fatigue ductility exponent. The number of cycles to failure was determined in a classical way, when 10% drop of the tensile stress occurred in comparison with the stabilized state. The parameters of the Manson, Coffin and Basquin curves are temperature dependent.

Fig. 2. The temperature dependent cyclically stable cyclic

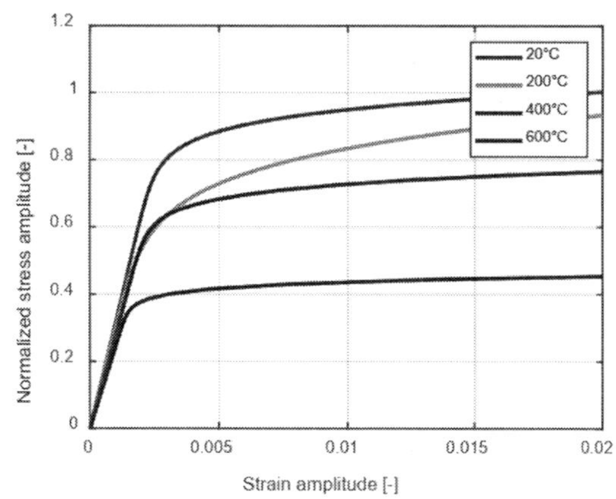

stress-strain curves.

The temperature dependent elastic-modulus for the investigated material is presented in Fig.1. The elastic modulus was determined from the stabilized hysteresis loops at mid-life. The CSSC for the selected temperatures are presented in Fig.2. The Manson, Coffin and Basquin curves are presented in Fig. 3. The specimens showed monotonically increasing durability as well as the ductility for the high strain ranges with the increasing temperature.

In our future work, the experimental data will be extended with low-cycle fatigue tests with hold time in a tension. The obtained experimental results are consequently used for the selected constitutive material model calibration.

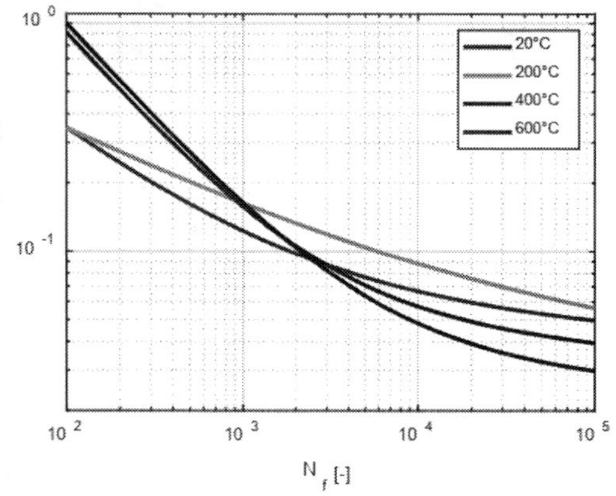

Fig. 3. The temperature dependent Manson, Coffin and Basquin curves.

2.2 Constitutive material model

A constitutive material model must be able to describe the cyclic material behaviour for the temperature range of the steam turbine shaft that is exhibited. The temperature-dependent Chaboche non-linear kinematic hardening model [10, 11] is selected for the elastic-plastic FEA.

The stress tensor $\boldsymbol{\sigma}$ is obtained on the basis of the generalized Hooke law, as follows:

$$\boldsymbol{\sigma} = \mathbf{C}: (\boldsymbol{\varepsilon} - \boldsymbol{\varepsilon}_p - \boldsymbol{\varepsilon}_{th}), \quad (3)$$

where $\boldsymbol{\varepsilon}$ is the total strain tensor, $\boldsymbol{\varepsilon}_{th}$ is the thermal strain tensor, and $\boldsymbol{\varepsilon}_p$ is the plastic strain rate tensor. $\mathbf{C}$ denotes the fourth order isotropic elastic tensor. The elastic tensor components are defined by the temperature-dependent elastic modulus E, and by the Poisson ratio v.

The model assumes associated plastic flow:

$$\dot{\boldsymbol{\varepsilon}}_p = \dot{\bar{\varepsilon}}_p \frac{\partial f(\boldsymbol{\sigma}-\boldsymbol{\alpha})}{\partial \boldsymbol{\sigma}}, \quad (4)$$

where $\dot{\boldsymbol{\varepsilon}}_p$ represents the plastic strain rate tensor, and $\dot{\bar{\varepsilon}}_p$ is the equivalent plastic strain rate, defined as:

$$\dot{\bar{\varepsilon}}_p = \sqrt{\frac{2}{3}\dot{\boldsymbol{\varepsilon}}_p : \dot{\boldsymbol{\varepsilon}}_p}. \quad (5)$$

The pressure-independent von Mises yield function, allowing for kinematic hardening, is defined as:

$$f(\boldsymbol{\sigma} - \boldsymbol{\alpha}) = k, \quad (6)$$

where k is the material dependent and the temperature dependent yield stress, and $f(\boldsymbol{\sigma} - \boldsymbol{\alpha})$ denotes the equivalent Mises stress with respect to the backstress $\boldsymbol{\alpha}$, defined as:

$$f(\boldsymbol{\sigma} - \boldsymbol{\alpha}) = \sqrt{\frac{2}{3}(\mathbf{S} - \boldsymbol{\alpha}'): (\mathbf{S} - \boldsymbol{\alpha}')}, \quad (7)$$

where $\mathbf{S}$ denotes the deviatoric stress tensor, and $\boldsymbol{\alpha}'$ is the deviatoric part of the backstress. Assuming the non-linear kinematic hardening model, the overall backstress is composed of multiple backstress components as follows:

$$\boldsymbol{\alpha} = \sum_{k}^{3} \boldsymbol{\alpha}_k. \quad (8)$$

The non-isothermal evolution law of the backstress component is defined as:

$$\dot{\boldsymbol{\alpha}}_k = \frac{2}{3} C_k \dot{\boldsymbol{\varepsilon}}_p - \gamma_k \boldsymbol{\alpha}_k \dot{\bar{\varepsilon}}_p + \frac{1}{C_k}\frac{\partial C_k}{\partial T}\boldsymbol{\alpha}_k \dot{T}, \quad (9)$$

where C_k and γ_k are temperature dependent material parameters. The model with three backstresses is selected. γ_3 is considered as a constant, and is equal to zero. Ratcheting is not studied in this paper.

In this study, the temperature dependent parameters k, C_k and γ_k were obtained by the standard least-square method from the LCF test data for each test temperature indepedently, and by interpolation for the temperatures between the test temperatures.

3 Fatigue damage assessment

3.1 Fatigue damage calculation concept

Damage calculation forms a part of post-processing, and is based on the results of previous mechanical FEA. The following fatigue criteria are applied in order to predict the lifetime for the steam turbine shaft:

- The Damage Operator Approach (DOA also known as the Nagode method)
- The SWT method
- The Nihei method

From a practical point of view, the fatigue criteria were implemented as post-processing code. The APDL script intended especially for finding the critical domain in the turbine shaft was coded. A user can choose from a wide range of possible approaches – the maximum or range of the stress or strain tensor components, the stress or strain tensor invariants or the Manson-McKnight method. The selected criterion is evaluated in the corner nodes of the element set, which needs to be first predefined in the turbine shaft FE model in ANSYS.

The script creates several types of the outputs. First, the text file with the node numbers that are sorted according to the value of the criterion that is written. The criterion value and time when maximum and minimum occur are also recorded. A separate APDL script may then be applied to upload the obtained results into the original **.rst* file. Second, for the individual nodes from the selected element set, separate data files containing time histories of the stress and the strain components and the temperatures are written. Then, these files are used for the fatigue analysis using the Nagode, SWT or the

Nihei method, which are implemented in *C++* code. The results expressed as the local fatigue damage and the lifetime are stored in the separate text file, which is then used for the data upload back into the source file **.rst*. As well as the data obtained by the APDL script, these results are mapped on the FE model of the turbine shaft and could be visualised in ANSYS, for instance, as contour plots. The above mentioned conception of the program codes for the fatigue prediction of the turbine shaft is represented by the flowchart in Fig. 4.

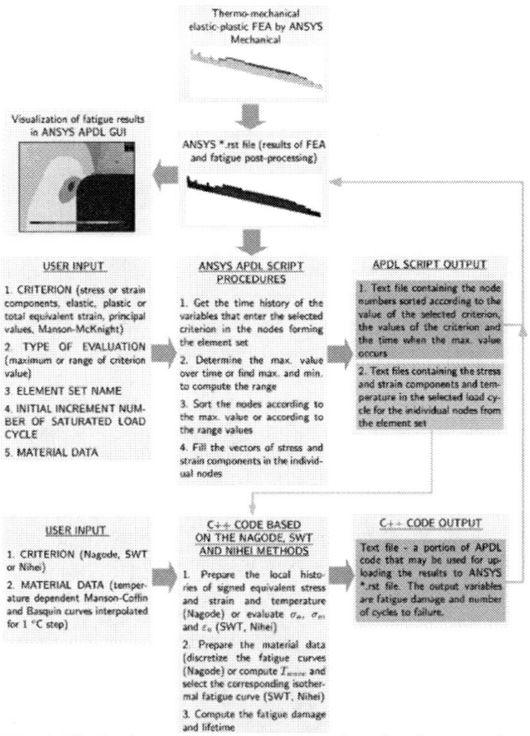

Fig. 4. The basic conception of the developed software codes for the fatigue prediction of the steam turbine shafts.

3.2 The SWT and the Nihei methods

To take the influence of variable stress and strain into account in fatigue prediction the following parameter is widely used:

$$P_{SWT} = \sqrt{E\varepsilon_a(\sigma_a + \sigma_m)}, \tag{10}$$

which is the well-known SWT parameter [8]. E is the Young's modulus, ε_a is the total mechanical strain amplitude and σ_a and σ_m is the stress amplitude and mean, respectively.

By considering the Walker mean stress correction, Nihei et al. 9 have introduced a modification of the original SWT formula in this form:

$$P_{Nih} = \sqrt{\sigma_a^w(\sigma_a + \sigma_m)^{1-w}\varepsilon_a E}. \tag{11}$$

Recommended value for the w exponent is 0.5. Evaluation of the stress and strain amplitudes may be done according to the Manson-McKnight method [12] by evaluating means and amplitudes of the individual stress or strain tensor components separately. For the stress tensor components we may write

$$\sigma_{ij,a} = \frac{max_t(\sigma_{ij}) - min_t(\sigma_{ij})}{2}, \tag{12}$$

$$\sigma_{ij,m} = \frac{max_t(\sigma_{ij}) + min_t(\sigma_{ij})}{2}. \tag{13}$$

The equivalent stress amplitude and mean is then computed by considering the von Mises norm:

$$\sigma_a = \frac{\sqrt{2}}{2}\sqrt{\begin{gathered}\left(\sigma_{xx,a} - \sigma_{yy,a}\right)^2 + \left(\sigma_{yy,a} - \sigma_{zz,a}\right)^2 + \\ +\left(\sigma_{xx,a} - \sigma_{zz,a}\right)^2 + \\ +6\left(\sigma_{xy,a}^2 + \sigma_{xz,a}^2 + \sigma_{yz,a}^2\right)\end{gathered}} \tag{14}$$

and

$$\sigma_m = \frac{\sqrt{2}}{2}\sqrt{\begin{gathered}\left(\sigma_{xx,m} - \sigma_{yy,m}\right)^2 + \left(\sigma_{yy,m} - \sigma_{zz,m}\right)^2 + \\ +\left(\sigma_{xx,m} - \sigma_{zz,m}\right)^2 + \\ +6\left(\sigma_{xy,m}^2 + \sigma_{xz,m}^2 + \sigma_{yz,m}^2\right)\end{gathered}} \tag{15}$$

The equivalent mean stress value σ_m is signed according to the sign of the maximum principal stress over the entire analysed time history.

The resulting value of fatigue lifetime obtained by these methods depends on the local load history represented by the parameter P on the one hand and on the material fatigue properties involved by the Manson-Coffin and Basquin parameters of regression $\sigma_f', b, \varepsilon_f', c$ on the other hand. However, these parameters are temperature dependent, which in case of thermo-mechanical loading sets additional requirements on the computational procedure - a decision process that determines the local reference temperature over the entire cycle has to be implemented. For this purpose, selection of the local reference temperature is based on the time-weighted average over the assumed cycle.

3.3 The Damage Operator Approach

The DOA is used for continuous fatigue damage calculations of the steam turbine shaft. The DOA and is an extension of the strain-life approach to loading cycles with variable amplitude and temperature. There is no longer a need for cycle equivalent temperature and separate rain-flow cycle counting [6].

The stress tensors from structural FEA are transferred into the equivalent stress histories by applying different multiaxial criteria. The signed von Mises equivalent stress, or the critical plane approach are frequently used with the DOA. In this study, the signed von Mises equivalent stress is selected for the steam turbine shaft.

In this paper the temperature histories, $T_i(t_j)$, the signed von Mises equivalent stress histories, $\sigma_i(t_j)$, and as well the equivalent mechanical strain histories, $\varepsilon_i(t_j)$, are obtained from the previous structural FEA of the steam turbine shaft.

The constitutive material model used in the structural FEA of the steam turbine shaft is elastic-plastic, therefore the equivalent mechanical strain, $\varepsilon_i(t_j)$, is the elastic-plastic strain, $\varepsilon^{ep}(t_i)$, that contributes to the fatigue damage, as follows:

$$\varepsilon^{ep}(t_i) = \varepsilon_i(t_i). \quad (16)$$

Next, the equivalent stress, $\sigma_i(t_i)$, and the elastic-plastic strain, $\varepsilon^{ep}(t_i)$, are transferred into the selected damage parameter $P(t_i)$. If no mean-stress correction is used, then:

$$P(t_i) = \varepsilon^{ep}(t_i). \quad (17)$$

In this paper, the mean stress effect is included in the Smith-Watson-Topper damage parameter, which is defined as:

$$P_{SWT}(t_i) = \sqrt{(\sigma_m(t_i) + \sigma_a(t_i))E(T_i)\varepsilon_a^{ep}(t_i)}, \quad (18)$$

where $E(T_i)$ is the temperature dependent instantaneous elastic modulus. The SWT damage parameter used in the DOA requires online calculation of the mean stress, $\sigma_m(t_i)$, and the stress amplitude, $\sigma_a(t_i)$, and also the strain amplitude $\varepsilon_a^{ep}(t_i)$ [11].

The fatigue damage $D_f(t_i)$ can be expressed as the total variation, as follows:

$$D_f(t_i) = D_f(t_{i-1}) + |\mathcal{D}(t_i) - \mathcal{D}(t_{i-1})|. \quad (19)$$

$\mathcal{D}(t_i)$ is the damage operator. The damage operator introduces the cyclic damage evolution and follows the Masing and memory rules. The damage operator is expressed as:

$$\mathcal{D}(t_i) = \sum_{j=1}^{n_p} \mathcal{D}_j(t_i) = \sum_{j=1}^{n_p} \gamma_j(T_i)\, P_{\gamma j}(t_i) \quad (20)$$

for $0 \le t_1 \le t_2 \le \ldots \le t_i \le \ldots$, where $P_{\gamma j}(t_i)$ is the play operator with a general initial value given as:

$$P_{\gamma j}(t_i) = max\left\{P(t_i) - p_j\, , min\left\{P(t_i) + \right.\right.$$
$$\left.\left. +p_j\, , \frac{\gamma_j(T_{i-1})}{\gamma_j(T_i)}\, P_{\gamma j}(t_{i-1})\right\}\right\}. \quad (21)$$

$P_{\gamma j}(t_i)$ represents the backstress and follows kinematic hardening. p_j are the yield stresses of the segment sliders. $\gamma_j(T_k)$ are the temperature-dependent Prandtl densities, and can be obtained explicitly as follows:

$$\gamma_j(T_k) = \frac{1}{p_{j+1}-p_j}\left(\frac{d_{fj+1}(T_k)}{4} - -\sum_{i=1}^{j-1}\gamma_i(T_k)(p_{j+1}-p_j)\right). \quad (22)$$

$\gamma_j(T_k)$ are obtained from the temperature dependent strain-life curves, Fig.3, or from a damage parameter - life curves, which need to have been transformed previously from the P-N_f curves to the corresponding P-d_f curves. d_f is a damage of one cycle, obtained as follows:

$$d_f = \frac{1}{N_f}. \quad (23)$$

The fatigue damage is modelled with the intensive use of the hysteresis operators. The DOA is a continuous rain-flow method for cycles with variable temperature [13].

Assuming a doubled equivalent stress, strain and temperature history, the predicted number of cycles to failure for a selected damage parameter is calculated as follows [7]:

$$N_f = \frac{1-D_1+D_2}{D_2}, \quad (24)$$

where D_1 is the fatigue damage of the first run, and D_2 is the fatigue damage of the subsequent run. The limit damage value is assumed to be equal to 1. If $D_1 \ge 1$ then $N_f = 1$.

It should be remarked that the stresses obtained from the elastic-plastic FEA should relax. The real material behaviour is viscoplastic. In our future work, the viscoplastic approximation will be used [7]. The fatigue damage model will be extended with a creep damage model. With the DOA, creep and fatigue damage calculations are separated and the oxidation effect is taken into account indirectly [6]. The DOA is suitable for the engineering TMF calculations and for the complex stress, strain and temperature histories.

4 Finite element analysis of the steam turbine shaft

The overall objective of the finite element analyses (FEAs) was to suggest a reliable and effective way how to perform simulations of the elastic-plastic material behaviour in the critical domains of turbine shafts with respect to the complex operating modes. A model case of the turbine shaft was selected and all calculations as part of the development of the methodology were done on this component.

Heat resulting from dissipative processes in material is negligible in comparison with convective heat. Therefore sequential heat transfer-stress analysis was used to calculate mechanical response. The temperature fields obtained from the transient thermal analysis of the steam turbine shaft are subsequently used in the elastic-plastic FEA of the shaft.

The assumption of axisymmetry of the shaft geometry and loads is adopted for the finite element (FE) model. The following types of loads are considered as dominant and therefore applied in the model:

- Thermal loads caused by the constrained thermal expansion
- Inertia loads due to rotating mass of the shaft and blades
- Loads due to the steam flow impact

In this study, the time distribution of the loads representing the turbine cold-start process is assumed. The steam turbine cold-start procedure consists of several phases. During the pre-warming phase the turbine is rotated by a turning gear while being heated by warming steam. The acceleration phase consists of controlled increase of the speed of rotation to its nominal value in vacuum conditions. Loading phase is then a sequence of several controlled steps, during which the steam mass flow, temperature and pressure reach their nominal characteristics. After this initial set of load regimes, the turbine operates in approximately steady inlet and outlet conditions and the rotor temperature field reaches its steady state. The shutdown procedure is divided into three phases: unloading to 50% and run-out, during which the speed of rotation goes from maximum to zero, and cooling in the final stage. Due to the casing insulation, it may take several weeks to cool down the turbine to the initial state.

The thermal response of the shaft material is simulated by the sequence of ten separate transient thermal analyses defined in the ANSYS Mechanical FE code. In general, convective heat transfer conditions are defined as couples of the heat transfer coefficient values and steam temperatures. Moreover, the blade roots are also considered in the model for accurate determination of the temperature field in the vicinity of the blade grooves (Fig. 5).

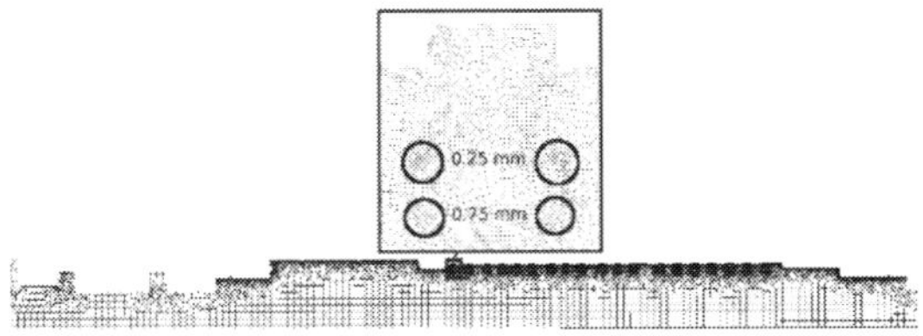

Fig. 5. The finite element model mesh. The element size in the blade grooves is illustrated.

Mechanical response of the material is analysed by taking the calculated temperature fields within the ten individual load regimes into account. Frictional contacts are set between the blade roots and grooves and the inertial forces due to rotating blade aerofoils are replaced by concentrated forces in the aerofoil centroids. The steam flow effect is replaced by the horizontal force component acting in the centroids as well.

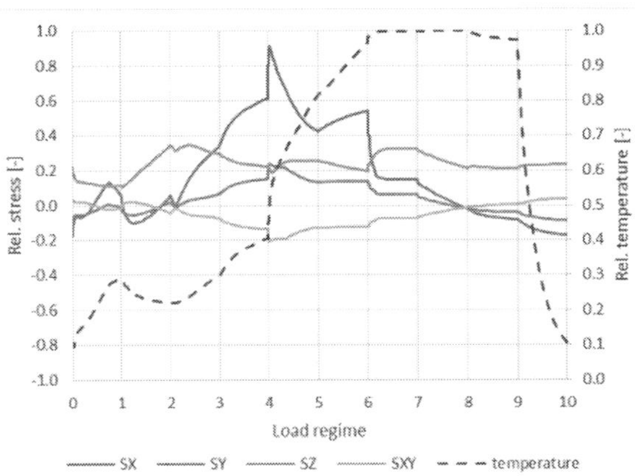

Fig. 6. Sample local stress and temperature history.

To obtain the saturated stress-strain response of the steam turbine shaft, three cold-start cycles were analysed consecutively. The dominant stress component in the potentially critical domains is the normal stress in the radial direction, and is presented in Fig. 6. It occurs in a particular time instant during the start-up when the maximum speed of rotation is reached while the material is not yet fully warmed up.

5 Fatigue damage assessment for the steam turbine shaft

For the purpose of this study, the fatigue prediction is focused on the critical site detection and on the assessment of the thermo-mechanically induced fatigue damage evaluated by the three methods. The FE model loaded by the set of loads as mentioned in the previous section was analysed. The fatigue analysis was done by using the SWT and the Nihei methods, as well as the DOA by Nagode et al. The work and the results published here extend and refine those presented in [14].

Based on the results of the FE simulations and the overall expertise, potentially critical domains were defined (Fig. 7). These domains are primarily the regions of the blade grooves and the main internal seals. The element set spanning these domains was defined for the purpose of performing fatigue analysis using the in-house codes. Fig. 8 to Fig. 10 show the fatigue damage contours evaluated by the mentioned criteria in the domain, which is framed in red in Fig. 7.

Fig. 7. The potential critical domains for the steam turbine

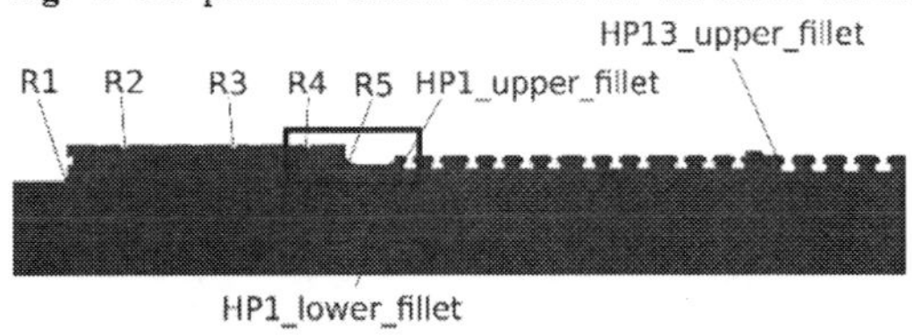

shaft.

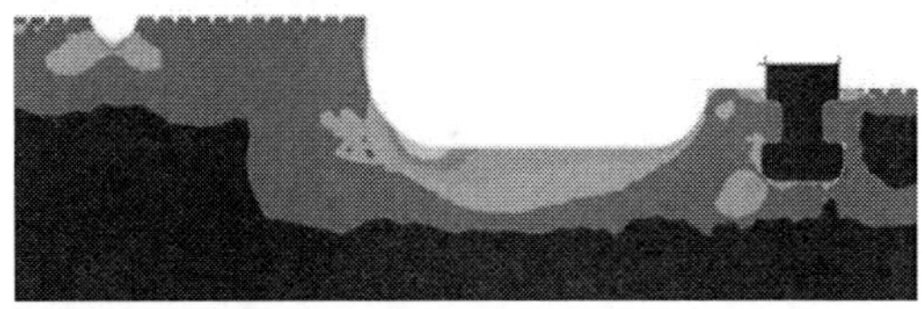

Fig. 8. Distribution of the fatigue damage computed by the DOA.

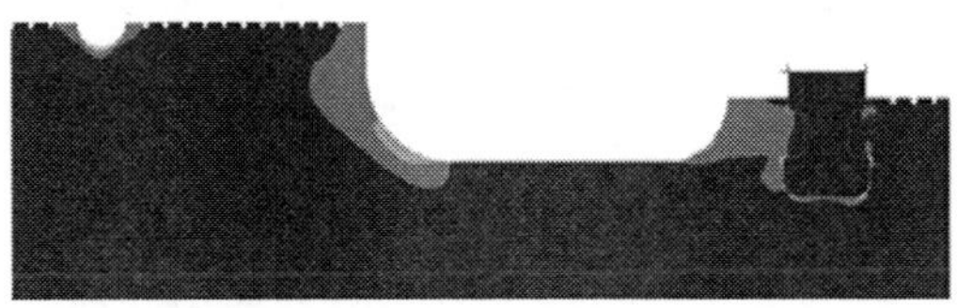

Fig. 9. Distribution of the fatigue damage computed by the SWT method.

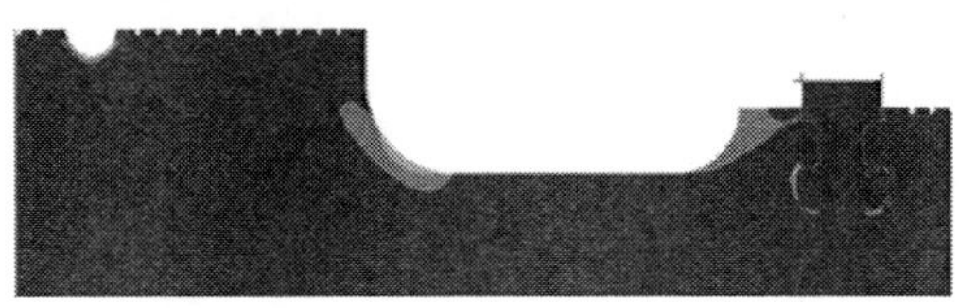

Fig. 10. Distribution of the fatigue damage computed by the Nihei method.

	R1	R2	R3	R4	R5	HP1_upper_fillet	HP1_lower_fillet	HP13_upper_fillet
N by Nagode [-]	5.3E+04	2.1E+04	2.6E+03	5.7E+03	6.2E+03	4.3E+04	6.9E+03	7.3E+04
N by SWT [-]	1.3E+06	5.9E+07	5.8E+06	4.4E+07	5.7E+08	8.6E+11	3.0E+06	1.8E+08
N by Nihei [-]	1.7E+06	1.1E+08	1.1E+07	8.6E+07	1.2E+09	3.4E+12	4.3E+06	1.5E+09

Table 1. Results of fatigue analysis in the selected domains of the turbine shaft.

Table 1 provides the numerical results in terms of the fatigue life values in the selected domains. The most critical place according to the DOA is the "R3" domain in the main internal sealing. Predicted life of this place is 2600 cold-start cycles. It is also the place of the maximum equivalent mechanical strain range. The Nihei and the SWT methods predict critical site in the "R1" domain, however the predicted life is two orders of magnitude higher than in the case of the DOA results. This observation reveals the expected weakness of the two isothermal methods, which is the nonrigorous approach to selecting the cycle reference temperature.

It should be noted that the presented results are preliminary and will be further refined based on the progress of experimental programme conducted to determine the material behaviour especially with focus on the viscoplastic response.

6 Conclusions

The work done within the Flexturbine project was introduced in this paper, and some results concerning the fatigue prediction methodology applicable to steam turbine rotors were presented. The strategy for TMF prediction of this structure is based on 2D axisymmetric FE model for simulating the time variable thermo-mechanical stress-strain response. The in-house codes for fatigue analysis that are capable of processing the ANSYS **.rst* files are based on a number of criteria, including the Manson-McKnight, the DOA and other methods. It is possible to determine the critical domain including the extent of the damage or the number of cycles to initiation. The obtained results can be mapped to the original FE model to obtain an overview of the distribution of the resulting values.

The used elastic-plastic material with kinematic hardening is capable of describing temperature dependent cyclic plasticity, but cannot capture time dependent material behaviour. However, the advantage of the elastic-plastic material model is computational speed in comparison with viscoplastic material models.

The advantage of the DOA is that the approach uses only isothermal data that are easy to obtain. The continuous fatigue damage calculation is convenient for the complex thermo-mechanical loading cycles as in the case of the steam turbine shaft.

In our future work, the viscoplastic approximation will be used for the elastic-plastic FEA results of the steam turbine shaft. For that reason, the LCF tests will be extended with LCF tests with a hold time. The DOA results for the signed von Mises equivalent stress will be compared with results obtained with the critical plane approach. A creep damage model will be added to the fatigue damage.

Acknowledgement: This project has received funding from the European Union's Horizon 2020 research and innovation programme under the Grant Agreement No. 653941.

References

1. E. Charkaluk A. Bignonnet, A. Constantinescu, K. Dang Van. Fatigue design of structures under thermomechanical loadings. Fatigue Fract Eng M **25** (2002): 1199-1206.
2. H. Sehitoglu. Thermal and thermomechanical fatigue of structural alloys. ASM International, Member/Customer Service Center, Materials Park, OH 44073-0002, USA (1996): 527-556.
3. A. Constantinescu, E. Charkaluk, G. Lederer, L. Verger. A computational approach to thermomechanical fatigue. Int J Fatigue, **26**, 8 (2004): 805-818.
4. R.W. Neu, H. Sehitoglu. Thermomechanical Fatigue, Oxidation and Creep: Part I. Damage Mechanisms. Metall Trans A, **20**, 9A (1989): 1755-1767.
5. R.W. Neu, H. Sehitoglu. Thermomechanical Fatigue, Oxidation and Creep: Part II. Life Prediction. Metall Trans A, **20**, 9 (1989): 1769-1783.
6. M. Nagode, M. Hack, M. Fajdiga. Low cycle thermo-mechanical fatigue: Damage operator approach. Fatigue Fract Eng M **33** (2009): 149-160.
7. M. Nagode, F. Längler, M. Hack. A time-dependent damage operator approach to thermo-mechanical fatigue of Ni-resist D-5S. Int J Fatigue, **33**, 5 (2011): 692-699.
8. K.N. Smith, P. Watson, T.H. Topper. A stress-strain function for the fatigue of metals. J Mater, JMLSA, **5**, 4 (1970): 767-778.
9. M. Nihei, P. Heuler, Ch. Boller, T. Seeger. Evaluation of mean stress effect on fatigue life by use of damage parameters. Int J Fatigue, **8**, 3 (1986): 119-126.
10. J.-L. Chaboche, Time-independent constitutive theories for cyclic plasticity, Int J Plasticity **2**, 2 (1986): 149-188.
11. J.-L. Chaboche, Constitutive equations for cyclic plasticity and cyclic viscoplasticity, Int J Plasticity **5**, 3 (1989): 247-302.
12. J. Papuga, M. Vargas, M. Hronek. Evaluation of uniaxial fatigue criteria applied to multiaxially

loaded unnotched samples. Engineer Mechan **19**, 2/3 (2012): 99-111.

13. M. Nagode, Continuous damage parameter calculation under thermomechanical random loading, MethodsX **1** (2014): 81-89.
14. M. Nesládek, J. Jurenka, M. Lutovinov, M. Růžička, P. Měšťánek, J. Džugan. An assessment of thermo-mechanically induced fatigue damage of a steam turbine shaft. Procedia Struct Integrity, **7** (2017): 190-197.

Case study of multiaxial criteria for rolling contact fatigue of bearing steels

Miroslav Vaculka[1,2*], *Libor* Nohál[1,2], and *Petr* Vosynek[1]

[1] Faculty of Mechanical Engineering, Brno University of Technology, Technická 2896/2, 616 69 Brno, Czech Republic
[2] ZKL – Výzkum a vývoj, a.s., Jedovnická 2346/8, Líšeň, 628 00 Brno, Czech Republic

Abstract. A rolling contact fatigue (RCF) differs from the classic fatigue in a stress-state. Nowadays, a prediction of the RCF is still not on the sufficient level. A lot of researchers tried to apply different multiaxial fatigue criteria (MFC) to the RCF, respectively they modified some or even proposed new one. Our paper focuses on assessment of bearing life estimation based on mentioned methods with experimental validation in laboratory. Comparison and summarization of different methods used in MFC life estimation is presented, with inclusion of fatigue material properties, hardness and probability. Mainly bearing steels are used for evaluation and region of high-cycle and giga-cycle fatigue.

Introduction

The rolling bearing is exposed to many types of failure during operation. These types are well described in the publication of Harris [1] who presents analytic methods for predict these failures. This study exclusively deals with the failure named Rolling Contact Fatigue (RCF) which leads to pitting (Fig. 1). It is not yet too well-known to what extent the classic fatigue and the RCF are related. The difference is that the crack initiates on the surface of part within classic fatigue, whereas the crack initiates in subsurface layer in the RCF (macropitting) (Fig. 3). It is caused by cyclic contact loading which leads to a specific distribution of stress state and maximal shear stress beneath the surface (Fig. 2). Publications [2-5] describe an analogy between the classic and the rolling contact fatigue, on the basis of which Multiaxial Fatigue Criteria (MFC) can be used for the RCF, even though they were originally derived for the classic fatigue.

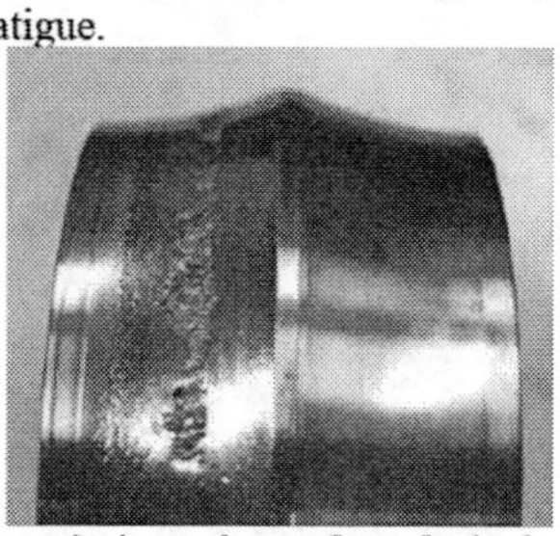

Fig. 1. Pitting on the inner ring surface of spherical roller bearing

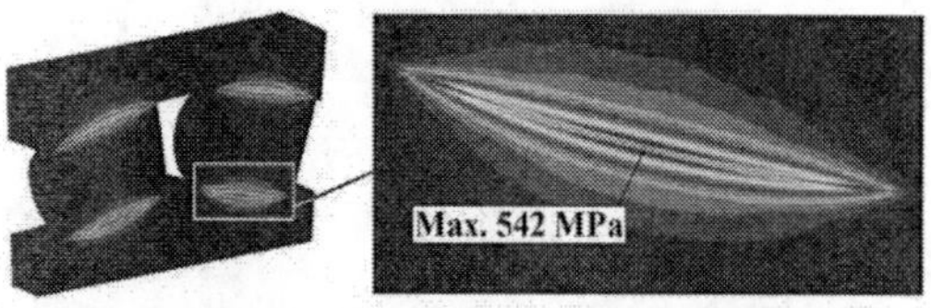

Fig. 2. Stress state in the contact area of roller and ring; Max Shear Stress in [MPa] is below the surface of raceway

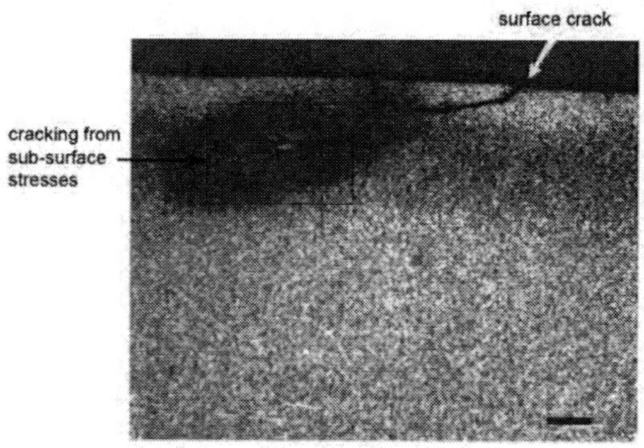

Fig. 3. Crack due to rolling contact fatigue [6]

Nowadays, it is well-known that due to stress raisers on the surface of bearing raceways, such as scratches, surfaces inclusions or other inhomogeneity, RCF cracks can initiate [7-9]. In this case, the failure is called micropitting. Micropitting is typical for modern high strength steels or bearing steels.

RCF can occur, for example, at the contact of a train wheel with rails or at gears. RCF at rolling bearings is specific because of the fact that the chrome bearing steel shows such resistance to fatigue that cracks occur in the gigacycle (ultra-high) region of fatigue.

Commonly used methodology for the assessment of service-life of rolling bearings is based on ISO 281 or more recently on the ISO 16 281 standard. They have certain limitations which proposed methodology eliminates. The main benefit is the opportunity of using the Finite Element Method (FEM), which is how the stress state in rolling bearing is analyzed. Using MFC and this stress state is possible to evaluate the service-life. There are some studies of appropriateness of MFC for this problem [10-12] but it has not been established yet which is the best. Therefore, a verification of MFC was proposed.

* Corresponding author: miroslav.vaculka@vutbr.cz

1 Multiaxial fatigue criteria

Multiaxial fatigue criteria can be divided into several groups. This study primarily deals with the criteria for high-cycle fatigue. They divide into stress-based (or stress invariant based), based on critical plane approach and energy-based (or integral approach). According to the research of appropriateness, some of them are described below.

1.1 Dang Van

In literature, for simulations of life ended by RCF, we can frequently encounter the Dang Van criterion [16], proposed in form (1) [10-11, 13-15]. $\tau_{\max}(t)$ is the instantaneous maximal shear stress, $\sigma_h(t)$ is the instantaneous hydrostatic stress, τ_W is the fatigue limit in reversal torque and σ_W is the fatigue limit of fully reversed tension-compression or bending of rotation. Even though this criterion is used very often, it is inappropriate (too benevolent) [14, 17, 18], because it does not consider the significant influence of negative hydrostatic stress.

$$\tau_{\max}(t) + 3\left(\frac{\tau_W}{\sigma_W} - \frac{1}{2}\right)\sigma_h(t) \leq \tau_W \tag{1}$$

1.2 Dang Van with locus by Desimone et. al

As mentioned above, the Dang Van criterion does not provide appropriate results for life, so Desimone et. al [14] proposed an approach for its refinement. A failure locus for negative σ_h in (1) modifies τ_W into a constant value (2).

$$\tau_W = \frac{1}{2}\sigma_W \tag{2}$$

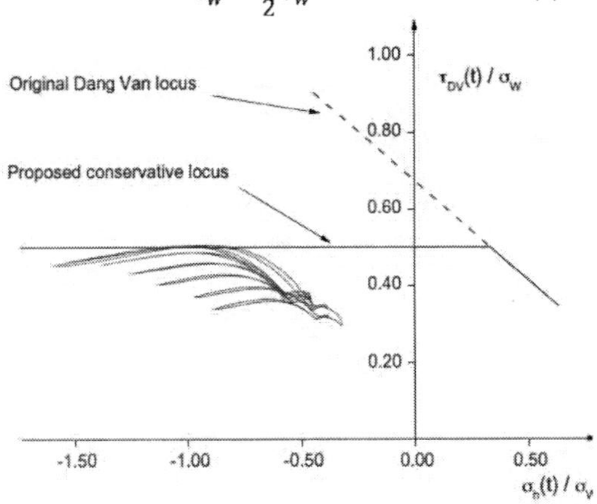

Fig. 4. Conservative locus for the Dang Van criterion applied to rolling/sliding contact [11]

This modification creates the failure locus illustrated in Fig. 4. Recent papers confirmed this work with experiments even for bearing steels [11, 19].

1.3 Liu and Mahadevan

The Liu and Mahadevan criterion [20, 21] provides a good match of life results with an experiment even for bearing steel. [11] provides a proof. The Liu and Mahadevan criterion is based on two characteristic planes – the critical plane, which is a plane of maximal amplitude of shear stress, and the plane of fatigue failure, where the macrocrack occurs. An angle difference between these two planes α can be calculated using (3) and (4).

$$s = \frac{\tau_W}{\sigma_W} \tag{3}$$

$$\cos(2\alpha) = \frac{-2 + \sqrt{4 - 4\left(\frac{1}{s^2} - 3\right)\left(5 - \frac{1}{s^2} - 4s^2\right)}}{2\left(5 - \frac{1}{s^2} - 4s^2\right)} \tag{4}$$

Using (5) and (6), the necessary parameters can be calculated and finally, using (7), the criterion is set up. σ_{eq} is the amplitude of equivalent stress, $\sigma_{a,c}$ is the amplitude of normal stress, $\sigma_{m,c}$ is the mean value of normal stress, $\tau_{a,c}$ is the amplitude of maximal shear stress, and $\sigma_{a,c}^H$ is the amplitude of hydrostatic stress.

$$s \geq 1 \Rightarrow \begin{cases} \beta = s \\ k = 9(s^2 - 1) \\ \eta = 1 \end{cases} \tag{5}$$

$$s < 1 \Rightarrow \begin{cases} \beta = \sqrt{(\cos 2\alpha)^2 s^2 + (\sin 2\alpha)^2} \\ k = 0 \\ \eta = \frac{3}{4} + \frac{1}{4}\left(\frac{\sqrt{3} - \frac{1}{s}}{\sqrt{3} - 1}\right) \end{cases} \tag{6}$$

$$\sigma_{eq} = \frac{1}{\beta}\sqrt{\left[\sigma_{a,c}\left(1 + \eta\frac{\sigma_{m,c}}{\sigma_W}\right)\right]^2 + \left(\frac{\sigma_W}{\tau_W}\right)^2(\tau_{a,c})^2 + k\left(\sigma_{a,c}^H\right)^2} \tag{7}$$

For the evaluation of a safety fatigue factor or even for the evaluation of the number of cycles up to a failure, a comparison with fatigue limit or with the S-N curve can be used (8).

$$\sigma_{eqv} \leq \sigma_W \tag{8}$$

1.4 Crossland

Crossland [22] and Sines are stated as simple criteria [23], but Crossland is the more successful one [24, 25]. For the Crossland criterion, s_m (9), which is the centre of the smallest hypersphere circumscribing the load path in the deviatoric space (more in [26]), and stress deviator at the generic instant $s(t)$, must be evaluated. Now, it is possible to calculate (10) the amplitude of the square root of the second invariant of the deviatoric component of the stress tensor $\sqrt{J_{2,a}}$. Then, the final form can be stated as (11).

$$s_m = \min_{s'} \max_t \sqrt{\frac{1}{2}(s(t) - s')(s(t) - s')} \tag{9}$$

$$\sqrt{J_{2,a}} = \max_t \sqrt{\frac{1}{2}(s(t) - s_m)(s(t) - s_m)} \tag{10}$$

$$\sqrt{J_{2,a}} + \left(\frac{3\tau_W}{\sigma_W} - \sqrt{3}\right)\sigma_{h,max} \leq \tau_W \tag{11}$$

1.5 Papadopoulos

This criterion [27], based on mesoscopic scale, is proposed as the best agreement with out-of-phase bending and torsion experimental results for hard metals, where in (3) equation (12) applies for s. The criterion is estimated as (13).

$$0.557 \leq s \leq 0.8 \tag{12}$$

$$\sqrt{\langle T_a^2 \rangle} + \left(\frac{3\tau_W}{\sigma_W} - \sqrt{3}\right)\sigma_{h,max} \leq \tau_W \tag{13}$$

In a simplified form, it is possible to say that $\sqrt{\langle T_a^2 \rangle}$ is *"the volumetric root mean square of the amplitude of the resolved shear stress on all possible planes"* [10].

1.6 Roessle–Fatemi

One of the aims is the inclusion of the surface hardness to the calculation of rolling bearing durability. Poeppelman [28] used the Roessle–Fatemi criterion [29] for the fatigue evaluation of high hardness steels with appropriate results. In this study, this criterion is verified even for bearing steel.

The criterion is in fundamentals different from the others mentioned. It is based on the Coffin-Manson curve (strain-life approach). This means that it presumes plastic deformations. The criterion uses the Brinell hardness HB to calculate the fatigue strength coefficient (14) and the fatigue ductility coefficient (15).

$$\sigma_f' = 4.25(HB) + 225 \tag{14}$$

$$\varepsilon_f' = \frac{0.32(HB)^2 - 487(HB) + 191000}{E} \tag{15}$$

σ_f' is in MPa and E is the modulus of elasticity, also in MPa. Fatigue strength exponent b was determined to be -0.09 and the fatigue ductility exponent c to be 0.56. The final criterion involves only the material hardness and the modulus of elasticity (16).

$$\varepsilon_a \leq \frac{\sigma_f'}{E}\left(2N_f\right)^{-0.09} + \varepsilon_f'\left(2N_f\right)^{-0.56} \tag{16}$$

[30] describes an inclusion of multiaxiality in this criterion. They use von-Mises, Tresca and maximum principal stress criteria.

2 Proposed methodology of numerical solution

The first part is the evaluation of stress state in the subsurface layer of raceway. This can be achieved using the Hertz theory at simple tasks. At more complex tasks, using ISO 16 281 standard or using finite element method (FEM) is needed. The application of MFC follows. Criteria intended for high-cycle fatigue are proposed in form given by equation (17). σ_{eqv} is the reduced stress characteristic for each criterion, and λ is the limit value, e.g. fatigue limit. However, in this form they provide only the information of the fatigue safety factor. This study suggests using methodology proposed by Liu and Mahadevan [20] to extend the inequation (17) by the dependence on the number of cycles up to a failure (18), which allows a prediction of service-life.

$$\sigma_{eqv} \leq \lambda \tag{17}$$

$$\sigma_{eqv}(N_f) \leq \lambda(N_f) \tag{18}$$

For example, the Dang Van criterion (mentioned in (1)) can be converted into being the dependent on number of cycles up to a failure using (19) and (20) as follows.

$$\sigma_W \rightarrow \sigma_W(N_f) = f_{-1} \tag{19}$$

$$\tau_W \rightarrow \tau_W(N_f) = t_{-1} \tag{20}$$

t_{-1} replaces τ_W. t_{-1} is a designation of shear stress in which the failure occurs for given number of cycles according to the fatigue test. Therefore, it is important to know such a part of S-N curve that describes the region of fatigue in which the failure of rolling bearings is expected.

A generalized form of the algorithm of numerical calculation of life may look as follows:

```
F = 1
ε … chose deviation
Nf … chose fatigue rate limit
While F > 0
    F = σeqv(Nf) − λ(Nf) … from eq. (18)
    Nf = Nf − ε
Loop
  ⇨ Nf
```

It means that at the beginning we define a variable $F > 0$, then we choose ε as a step that we use for finding N_f as a trial and error method. If the inequation (18) does not apply, the algorithm is ended and the final value N_f is found. The use of logarithm coordinates is preferable.

It is important to affect the difference between the number of revolutions, which is stated by basic durability L_{10}, and the number of cycles to which the bearing is exposed during this durability. A description of the calculation of the number of cycles is in [31]. It depends on the size and number of rolling elements. After several mathematical adjustments, it is possible to calculate it as (21). The parameters are described in Fig. 5, Z is the number of rolling elements (or balls).

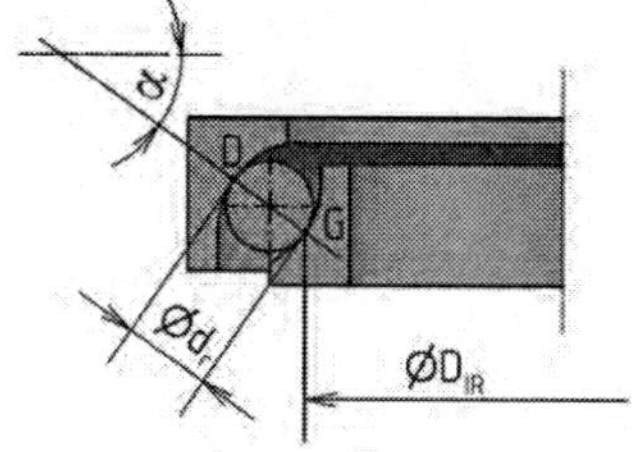

Fig. 5. The schema of bearing with important parameters for calculation. D, G are contact points

$$N_f = L_{10}\frac{Z}{1 - \frac{D_{IR}}{2(D_{IR} - d_r\cos(\alpha))}} \tag{21}$$

2.1 Application of the multiaxial fatigue criteria using material characteristics based on hardness

This methodology is purely empirical, but it provides good results (see the verification below). S-N curve is determined using equations (14) and (15). The necessary coefficients describing the S-N curve can be calculated as (22). The final form of the S-N curve is presented in (23). Variable N is number of cycles and it is in logarithm form.

$$A_\sigma = \sigma_f' \cdot 2^{-0.09} \quad (22)$$

$$\sigma_a = A_\sigma \cdot (10^N)^{B_\sigma} \quad (23)$$

Using these material characteristics, all criteria described in this study are applied with the same methodology.

3 Verification

Based on the described principle, some criteria where tested. For the verification, the experiment AXMAT II was chosen. FEM model of integral parts of the AXMAT test rig was created. The experiment and the numerical analysis are described in this section.

3.1 Experiment

The experimental RCF apparatus employed in this study is a flat washer-type RCF test-rig with acoustic emission and vibration monitoring systems. This special test-rig, shown in Fig. 6, was designed for life tests of thrust bearings (smaller size) and an evaluation of the rolling contact fatigue resistance of the material [32].It consists of a mechanical loading lever, an electrical motor, a specimen holder, a catch driver, a supporting frame and a monitoring system.

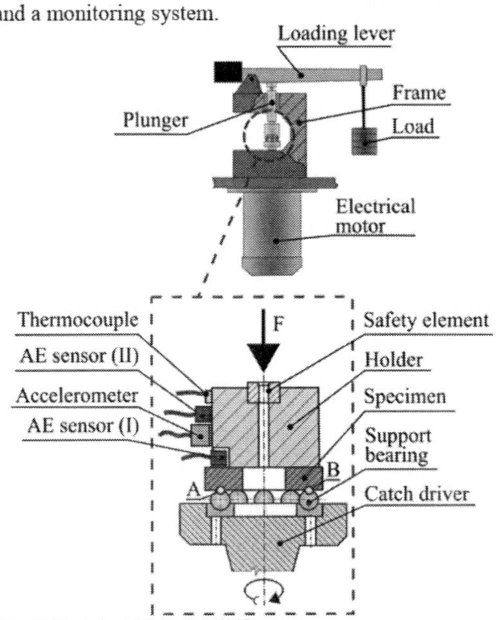

Fig. 6. Test rig AXMAT II [32]

The speed of the electrical motor can be adjusted by a frequency converter to the required level. This allows standard RCF tests to be performed, including tests at low speeds. The upper ring of the test bearing is clamped in the holder and the lower ring is fastened in the catch driver. In the case of testing material specimens, the specimen was fastened in the holder in place of the upper bearing ring (the geometry of the specimen is shown in Fig. 7). The holder is stationary, and the catch driver is driven by a shaft attached to an electrical motor. For a standard RCF test, the speed was set at 2150 min^{-1}. The holder is equipped with a polyamide safety element to offload in case of specimen overheating. The Hertzian contact stress for rolling contact tests can be set using the combination of weights in the range from 2000 to 6000 MPa.

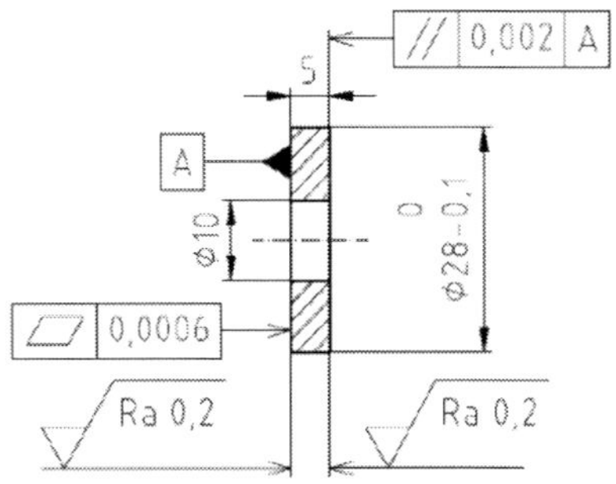

Fig. 7. Geometry of the specimen [32]

Table 1. Material characteristics (means) of applied material of hardened bearing steel 100Cr6MnSi6-4

Yield Stress	$R_{p0.2}$ [MPa]		2 021
Ultimate Strength	R_m [MPa]		2 040
Elongation	A [%]		0.3
Fatigue Limit	σ_f [MPa]		945
Brinell Hardness	HB	μ	647
		σ	10

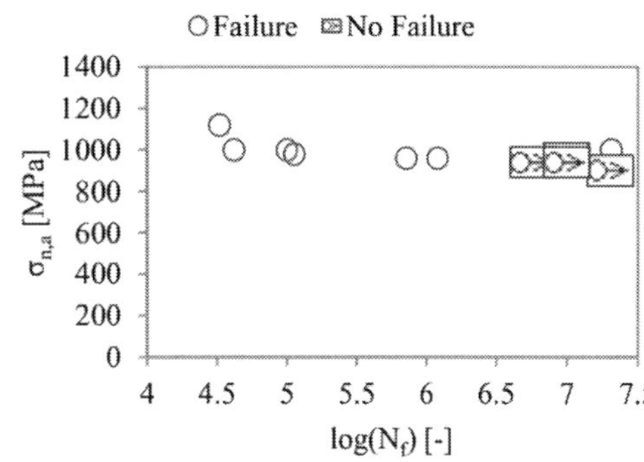

Fig. 8. Fatigue test of hardened bearing steel 100Cr6MnSi6-4. S-N curve is not determined due to brittleness of material

In this study, the specimen was loaded by force which should cause the contact stress of 5000 MPa according to Hertzian calculation. This load causes plastic deformations, but the number of cycles up to failure is still in the high-cycle region. The reason of higher value of the load is that the experiment was very time consuming (100 – 140 h per one specimen). Material characteristics of the tested material are in Table 1. This bearing steel was hardened and it is brittle (for brittle materials is in Eq. (3) $s \geq 1$ [11].). Therefore, the performance of fatigue tests was problematic, as shown in Fig. 8. Fatigue limit was determined (also in Table 1).

https://doi.org/10.1051/matecconf/201816522017

3.2 Numerical analysis

The finite-element model of integral parts of the AXMAT test rig is created. These parts were housing washer, balls and the specimen. Thrust loading of this assembly was cyclic symmetric, therefore, only this symmetrical part of assembly was modelled (see Fig. 9). The material of the specimen was ideal elastic-plastic with bilinear behaviour. The elastic modulus $E = 210\,GPa$ and the Poisson's ratio $\nu = 0.3$ were used. For contact solution, augmented Lagrangian formulation was used.

Due to the plasticity, deformation shows a dent on the raceway surface of the specimen (Fig. 10). Therefore, contact pressure did not reach the stated value $5000\,MPa$, but $4179\,MPa$ (Fig. 11).

In the next part of the durability solution, the stress state of the specimen in the form of principal stresses was exported from the strain-stress analysis to the fatigue analysis. In z-axis only data under the contact point were used (Fig. 12). The stress results are shown in Fig. 13. Due to the plastic strain, peak-to-peak amplitude of maximal shear stress was not variable from a certain point beneath the surface.

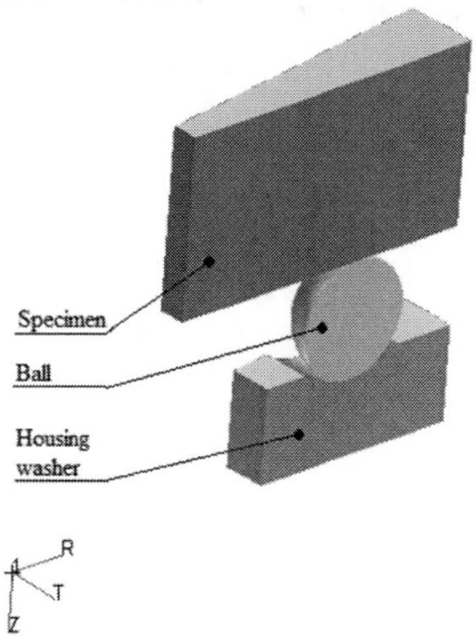

Fig. 9. Model of AXMAT specimen and support bearing used for FEM analysis

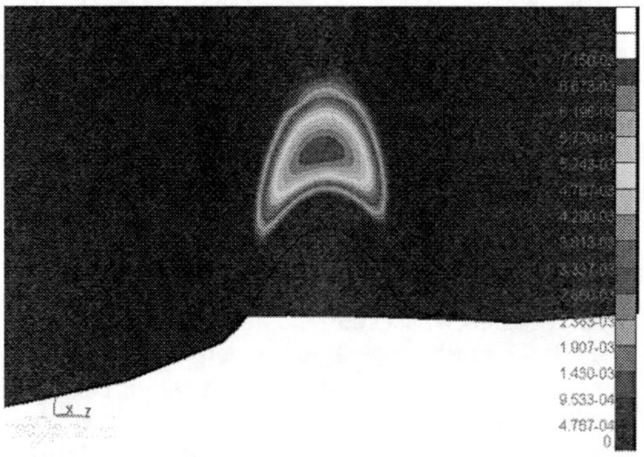

Fig. 10. Deforation of the area near the contact (20x magnification), variable: plastic strain [-]

Fig. 11. Contact stress on specimen in $[MPa]$. Due to occuring plastic strains, contact stress $\mathbf{5000\ MPa}$ is not achieved

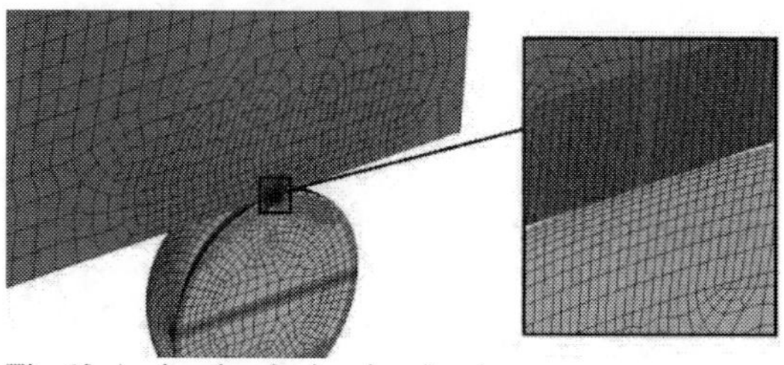

Fig. 12. Analysed nodes in subsurface layer

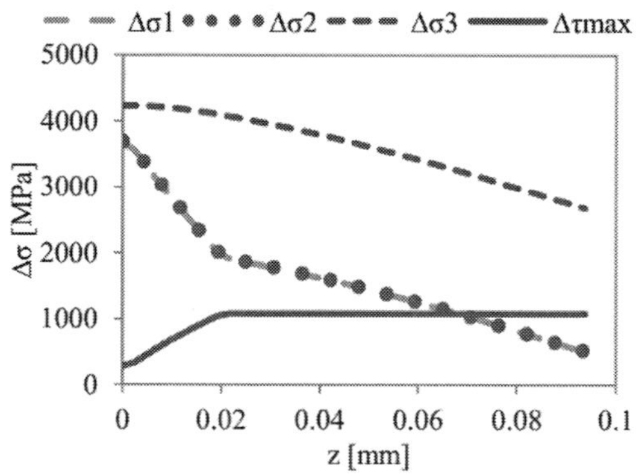

Fig. 13. Results of stress in subsurface layer due to elastic-plastic material

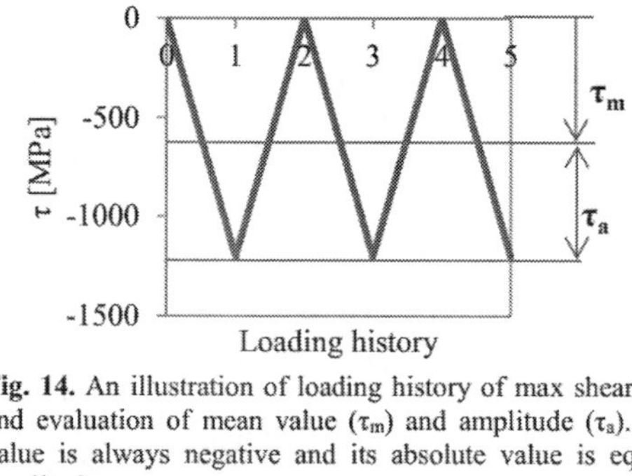

Fig. 14. An illustration of loading history of max shear stress and evaluation of mean value (τ_m) and amplitude (τ_a). Mean value is always negative and its absolute value is equal to amplitude

It was presumed that the thrust loading of test rig causes uniform distribution of load on all balls. Thus, the amplitude and mean stress are always the same and the fatigue estimation is not influenced by variable loading history. After each load of the analysed point, complete unload follows. Therefore, load history appears as shown in Fig. 14 and the amplitude and the mean stresses can be easily calculated. These data provided input for the criteria. The final results of the number of cycles up to failure was recalculated to a number of rotations using equation (21). Here, angle α was equal to $\pi/2$, thus, the number of rotations were always equal to number of cycles divided by half of number of balls. Then, number of rotations is recalculated on time in hours.

From fatigue limit of the material of the specimen was established:

- No failure – criterion Dang Van, Crossland, Papadopoulos and Dang Van with modified locus
- Failure – criterion Liu Mahadevan

However, this study proposed an approach to life estimation in combination with setting the S-N curve from hardness and following calculation using MFC. The Gaussian distribution for hardness input is included. Therefore, results are given with their probability of failure. It is chosen probability 0.5, 0.9, 0.95 and 0.99. Results from this methodology are shown in Fig. 15.

3.3 Discussion

Measuring the S-N fatigue curve material characteristic of the 100Cr6MnSi6-4 bearing steel is problematic. Therefore, MFC cannot be used for life estimation, only for the failure prediction. Only the Liu and Mahadevan criterion predicts the failure in this case. MFC in combination with hardness properties can be effectively used for life estimation. From these results it is obvious that the Liu Mahadevan criterion is the most appropriate from all tested criteria.

Yet, there are several influences which were not included in the analysis. The most critical is that torsional fatigue tests were not performed. Also, the AXMAT experiment was not properly set up for time reasons (100 – 140 h per specimen). Contact pressure exceeded the appropriate limit and calculations could be influenced by plastic strains. A better way is set the loading with no following plastic deformation, but time can be significantly increased. Then, influences of macro- and micro-slippages and influences of roughness of surfaces still are not included in the analysis.

Due to the suggested verification of this study, more levels of loading were not tested. Also, more different types of material and different hardness of specimen surfaces should be tested in the future.

4 Conclusion

The methodology of the evaluation and the verification of the durability of rolling bearing has been proposed. Durability was evaluated as quantification of service-life, which finishes together with the pitting occurrence. For the verification, the use of test-rig named AXMAT II was proposed. The 100Cr6MnSi6-4 hardened bearing steel was tested. Weibull curve was drawn from these tests.

The procedure of durability calculation was described. Stress state or strain state respectively in area beneath the surface of raceways was input for calculation. This state together with material characteristics gained from fatigue tests input to multiaxial fatigue criteria. They were used in calculation and results were compared with the Weibull curve. The Liu and Mahadevan criterion seems to be the most appropriate for this application at this moment. But with the increasing number of tests, other criteria can be more appropriate for the rolling contact fatigue of rolling bearing.

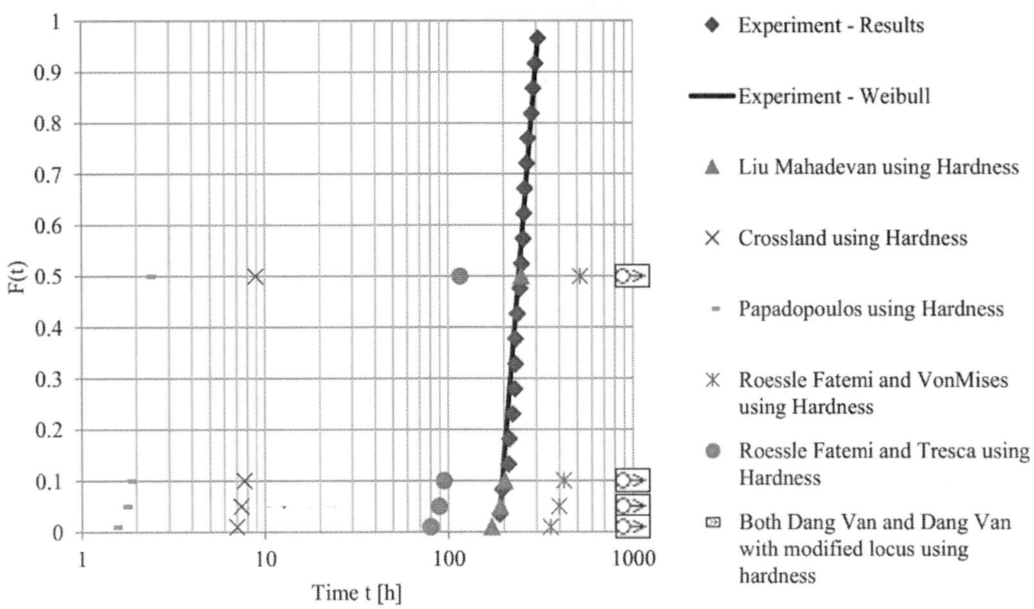

Fig. 15. Comparison of experimental results of 100Cr6MnSi6-4 bearing steel and MFC (using hardness as input) results

This work is an output of project NETME CENTRE PLUS (LO1202) created with financial support from the Ministry of Education, Youth and Sports under the "National Sustainability Programme I."

References

1. T. Harris, M. Kotzalas, *Rolling bearing analysis*, 5th ed., CRC/Taylor & Francis, Boca Raton, FL (2007).
2. S. Way, *Pitting due to rolling contact*, J Appl Mech. (1935) A49-A58.
3. P. Rycerz, A. Olver, A. Kadiric, *Propagation of surface initiated rolling contact fatigue cracks in bearing steel*, International Journal of Fatigue. vol. **97** (2017) 29-38.
4. A. Olver, *The Mechanism of Rolling Contact Fatigue: An Update*, Proceedings of the Institution of Mechanical Engineers, Part J: Journal of Engineering Tribology. vol. **219** (2006) 313-330.
5. C. Chue, H. Chung, *Pitting formation under rolling contact*, Theoretical and Applied Fracture Mechanics. vol. **34** (2000) 1-9.
6. G. Dambaugh, *Fatigue Considerations of High Strength Rolling Bearing Steels*, Schaeffler Group USA, Inc. (2006) 33.
7. N. Soda, T. Yamamoto, *Effect of Tangential Traction and Roughness on Crack Initiation/Propagation During Rolling Contact*, A S L E Transactions. vol. **25** (2008) 198-206.
8. S. Borgese, *An Electron Fractographic Study of Spalls Formed in Rolling Contact*, Journal of Basic Engineering. vol. **89** (1967) 943-.
9. J. Martin, A. Eberhardt, *Identification of Potential Failure Nuclei in Rolling Contact Fatigue*, Journal of Basic Engineering. vol. **89** (1967) 932-.
10. M. Ciavarella, F. Monno, *A comparison of multiaxial fatigue criteria as applied to rolling contact fatigue*, Tribology International. vol. **43** (2010) 2139-2144.
11. S. Foletti, S. Beretta, M. Tarantino, *Multiaxial fatigue criteria versus experiments for small crack under rolling contact fatigue*, International Journal of Fatigue. vol. **58** (2014) 181-192.
12. E. Conrado, S. Foletti, C. Gorla, I. Papadopoulos, *Use of multiaxial fatigue criteria and shakedown theorems in thermo-elastic rolling–sliding contact problems*, Wear. vol. **270** (2011) 344-354.
13. B. Alfredsson, *Applying multiaxial fatigue criteria to standing contact fatigue, International Journal of Fatigue*. vol. **23** (2001) 533-548.
14. H. Desimone, A. Bernasconi, S. Beretta, *On the application of Dang Van criterion to rolling contact fatigue*, Wear. vol. **260** (2006) 567-572.
15. M. Ciavarella, F. Monno, G. Demelio, *On the Dang Van fatigue limit in rolling contact fatigue*, International Journal of Fatigue. vol. **28** (2006)
16. V. Dang, G. Cailletaud, J. Flavenot, A. Douaron, H. Liurade, *Criterion for high cycle fatigue failure under multiaxial loading*, ICBMFF2. (1986) 459-78.
17. A. Bernasconi, P. Davoli, M. Filippini, S. Foletti, *An integrated approach to rolling contact sub-surface fatigue assessment of railway wheels*, Wear. vol. **258** (2005) 973-980.
18. A. Ekberg, E. Kabo, H. Andersson, *An engineering model for prediction of rolling contact fatigue of railway wheels*, Fatigue html_ent glyph="@amp;" ascii="&"/ Fracture of Engineering Materials and Structures. vol. **25** (2002) 899-909.
19. L. Houpert, F. Chevalier, *Rolling Bearing Stress Based Life—Part I: Calculation Model*, Journal of Tribology. vol. **134** (2012) 021103-.
20. Y. Liu, S. Mahadevan, *Multiaxial high-cycle fatigue criterion and life prediction for metals*, International Journal of Fatigue. vol. **27** (2005) 790-800.
21. Y. Liu, B. Stratman, S. Mahadevan, *Fatigue crack initiation life prediction of railroad wheels*, International Journal of Fatigue. vol. **28** (2006) 747-756.
22. B. Crossland, *Effect of large hydrostatic pressure on the torsional fatigue strength of an alloy steel*, Proc. Int. Conf. on Fatigue of Metals, Institution of Mechanical Engineers. (1956) 138-149.
23. G. Sines, *Behavior of metals under complex static and alternating stresses*, Metal Fatigue. (1959) 145-469.
24. A. Banvillet, *A volumetric energy based high cycle multiaxial fatigue citerion*, International Journal of Fatigue. vol. **25** (2003) 755-769.
25. A. Carpinteri, A. Spagnoli, *Multiaxial high-cycle fatigue criterion for hard metals*, International Journal of Fatigue. vol. **23** (2001) 135-145.
26. A. Bernasconi, *Efficient algorithms for calculation of shear stress amplitude and amplitude of the second invariant of the stress deviator in fatigue criteria applications*, International Journal of Fatigue. vol. **24** (2002) 649-657.
27. I. Papadopoulos, *A comparative study of multiaxial high-cycle fatigue criteria for metals*, International Journal of Fatigue. vol. **19** (1997) 219-235.
28. C. Poeppelman, *Axial and Torsion Fatigue of High Hardness Steels*, A Thesis (2011).
29. M. Roessle, *Strain-controlled fatigue properties of steels and some simple approximations*, International Journal of Fatigue. vol. **22** (n.d.) 495-511.
30. D. McClaflin, *Torsional deformation and fatigue of hardened steel including mean stress and stress gradient effects*, International Journal of Fatigue. vol. **26** (2004) 773-784.
31. M. Vaculka, *Návrh metodiky výpočtu životnosti valivého ložiska*, Diploma Thesis (2006).
32. L. Nohál, F. Hort, J. Dvořáček, P. Mazal, *An experimental investigation of rolling contact fatigue of steels using acoustic emission method*. Insight - Non-Destructive Testing and Condition Monitoring vol. **55** (2013)

An ad-hoc fretting wear tribotester design for thin steel wires

Iñigo Llavori[1,*], *Alaitz* Zabala[1], *Mikel Aingeru* Urchegui[2], Wilson Tato[1], Andrea Aginagalde[1], Iñaki Garate[1], and Xabier Gómez[1]

[1]Surface Technologies, Mondragon University, Loramendi 4, 20500 Arrasate-Mondragón, Spain
[2]Orona EIC, Orona Ideo, Jauregi bidea s/n, 20120, Hernani, Gipuzkoa, Spain

Abstract. Steel wire ropes experience fretting wear damage when the rope runs over a sheave promoting an oscillatory motion between the wires. Consequently, wear scars appear between the contacting wires leading to an increase of the stress field and the following rupture of the wires due to fatigue. That is why the understanding and prediction of the fretting wear phenomena of thin wires is fundamental in order to improve the performance of steel wire ropes. The present research deals with the design of an ad-hoc fretting wear test machine for thin wires. The test apparatus is designed for testing thin wires with a maximum diameter of 1.0 mm, at slip amplitudes ranging from 5 to 300 μm, crossing angle between 0-90°, and contacting force ranging from 0,5 to 5 N. The working principle of displacement amplitude and contacting force as well as the crossing angle between the wires are described. Preliminary studies for understanding the fretting wear characteristics are presented, analysing 0.45 mm diameter cold-drawn eutectoid carbon steel (0.8% C) wires (tensile strength higher than 3000 MPa).

1 Introduction

Many mechanical components such as, bearing housings, flexible couplings and spines or wire ropes are simultaneously subjected to a fretting wear and fatigue damage. Fretting phenomenon arises when two bodies are in contact subjected to relative movement of small amplitude, producing damage to the contact surface [1]. Depending on the magnitude of stresses, fretting phenomenon can cause catastrophic failure of such mechanical components.

Steel wire ropes exhibit high axial strength and bending flexibility. These properties are largely dependent on their construction and the manufacturing process of the wires. Despite their outstanding mechanical properties, steel wire ropes are prone to fretting wear damage due to the oscillatory motion between wires. The wear scars appearing at the contact interface increase the stress fields, leading to the rupture of the wires due to fatigue, and eventually the failure of the rope. The correct selection of wire material and rope configuration is therefore crucial in order to increase the rope life, which demands the characterization of the steel wire behaviour.

Steel wire's fretting wear behaviour characterization requires the use of a fretting tribometer. A tribometer is a test machine capable of measuring tribological parameters between surfaces in contact, such as friction force, coefficient of friction or wear volume. This work presents the design of the developed ad-hoc tribotester, designed for the characterization of thin steel wires under highly stable displacement strokes and contact forces.

2 Development of a fretting wear tribotester

2.1 Design requirements

The main challenge on the design of a fretting test lies on the precise control of the displacement amplitude and the contact force. On the one hand, fretting stroke is established between 0 and 300 μm [1]. On the other hand, previous works identified that premature rupture of the wires occurred for a contact force higher than 3 N [2,3]. Furthermore, it should be considered that the construction of a rope presents multiple wire contacts at different angles. Therefore, the test rig should be able to reproduce those crossing angles. Accordingly, the design requirements are defined as follows:

- Displacement amplitude ranging from 5 to 300 μm.
- Crossing angle ranging from 0 to 90 degrees.
- Contact force ranging from 0 to 5 N

2.2 Displacement amplitude module design

Servo-hydraulic [4], piezo-electric [5] or electro-mechanical actuators [6,7] can be used to generate a short cyclic displacement. The main characteristics of each actuator can be summarized as follows:

- Servo-hydraulic actuators are generally used in high demanding machines, since they provide a very stiff driving system with a precise control of the stroke.

* Corresponding author: illavori@mondragon.edu

- Piezo-electric actuators are ideal for a low mass, high frequency and short stroke applications. Fatigue life of the piezo-electric actuator is dependent on the displacement amplitude.
- Electro-mechanical systems converts the rotational motion into linear motion. They provide a simple and cost effective solution.

Fig. 1 shows the camshaft mechanism designed for the tribotester displacement system. A variable eccentric connected to an electro-mechanical motor converts the rotational motion into a reciprocating motion. Aditionally, two oscillating rods connected in series allow to reduce the displacement stroke into a highly stable short movement.

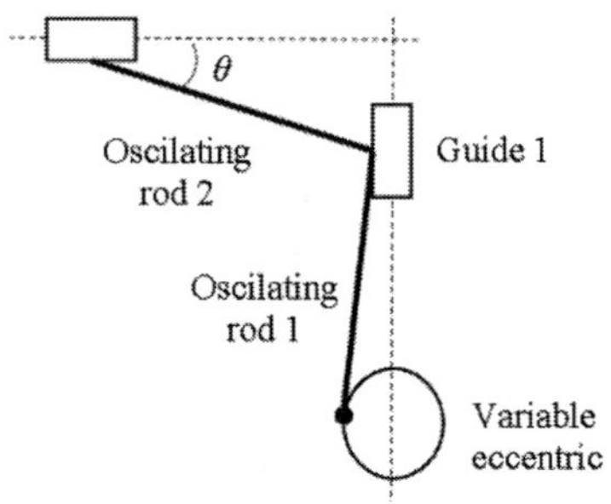

Fig. 1. The camshaft mechanism designed for the tribotester displacement.

Note that the displacement reduction is dependent on the angle θ: lower values of cyclic amplitude can be obtained with lower angles. The actual amplitude of cyclic displacement can be calculated from:

$$\delta = l_{or2}\left(\cos\theta_1 - \cos\theta_2\right) \tag{1}$$

where δ is the displacement amplitude, l_{or2} is the oscillating rod 2 length and θ_1 and θ_2 are the maximum and minimum angles obtained with the variable eccentric.

2.3 Contact force and wire fixation module design

Contact force can be applied by dead weight system [6,7,8], linear actuators [9] or a mechanical screw type actuator with an adjustable spring [10]. It should be noted that low mass loading systems such as spring or pneumatic actuators produce less severe conditions than dead weight loading systems, since the latter gives rise to shock loading.

The adopted solution in the tribotester is therefore the screw type actuator with an adjustable spring. This mechanical system in combination with a load cell allows to measure in real-time the contact force, avoiding the necessity of doing the hypothesis of a constant normal force.

Low contact forces between thin steel wires generate high stresses in the contact interface. Therefore, the need for reliable low force measurement is critical in the present application. Generally, contact and friction force measurements are decoupled, i.e. multiple uniaxial sensors are used to monitor each force [6,7,8,11]. This concept is simple in design, nonetheless it presents the following disadvantages:

- Large (space limitation).
- Misalignment and angle errors due to installation problems.
- Lack of stiffness which leads to linearity errors.
- Difficult compensation for the crosstalk effect.

Fig. 2 shows the multiaxial load cell capable of monitoring forces in three axis adopted in the designed tribotester and the schematic of the configuration. The benefits of the multiaxial sensors are listed below:

- Compact design.
- Same reference point for all force components.
- Easy compensation for the crosstalk effect.

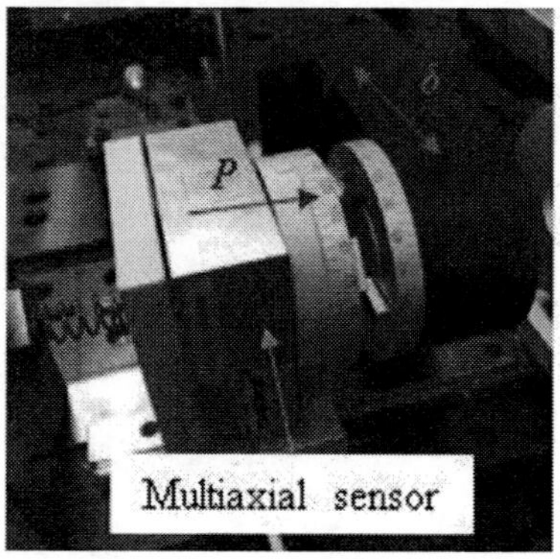

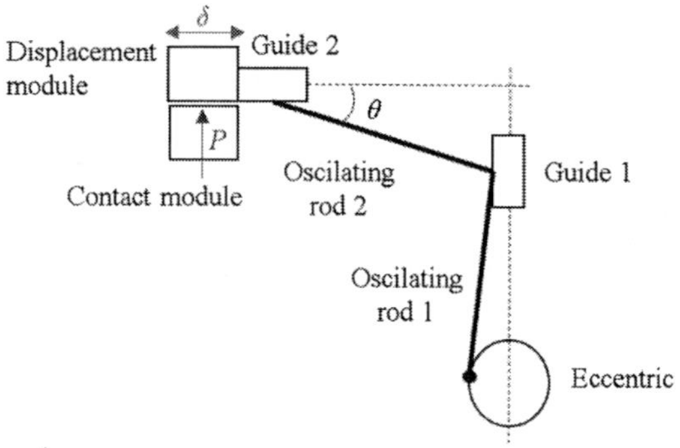

Fig. 2. Upper figure, Multiaxial load cell capable of monitoring forces in three axis adopted in the designed tribotester. (Lower figure, Schematic of the tribotester configuration.

Fig. 3 shows the developed wire fixation system, which avoids non desired movements. As can be noted in the figure, the wire presents a toroid shape, being the radius of the support considerably superior comparing to the wire diameter. Therefore, the contact achieved between the two wires can be considered a cross cylinder configuration. It should be noted that the fixing system allows to locate the wire at different angles, thus being possible to simulate different contacts present in ropes.

Fig. 3. Thin wires fixation system.

3 Verification tests

Fretting loops show the relationship between slip displacement and friction force along a fretting cycle. The shape of the fretting loop gives information on the fretting regime: where an elliptic shape is related with partial slip regime (Figure 4 (a)), and shifted rectangle is associated with gross slip regime (Figure 4 (b)). Furthermore, the area of the fretting loop can be used to determine the the coefficient of friction and wear.

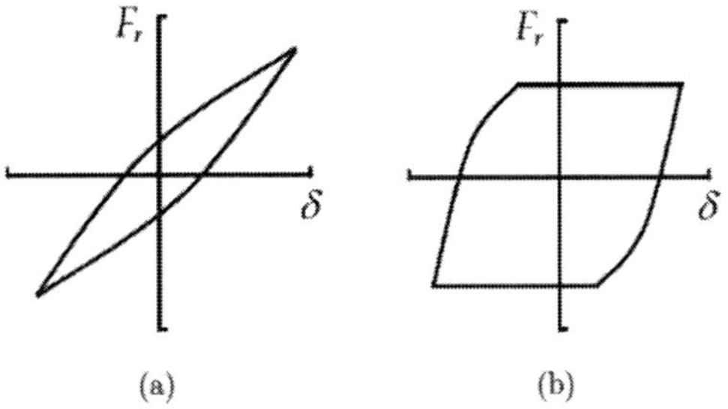

Fig. 4: Fretting loop shape. (a) partial slip regime; (b) gross slip regime [12].

Preliminary studies for understanding the fretting wear characteristics were conducted on 0.45 mm diameter cold-drawn eutectoid carbon steel (0.8% C) with a tensile strength higher than 3000 MPa. Fig. 5 and Fig. 6 show the unfiltered measurements of the contact (P) and friction (Q) forces and the monitored fretting loop respectively. The obtained results show a highly stable curves with very low signal to noise ratio, thus verifying the correct behaviour of the developed tribotester. It is expected that further analysis using the test rig will allow a better understanding of wear behaviour in different contact configurations.

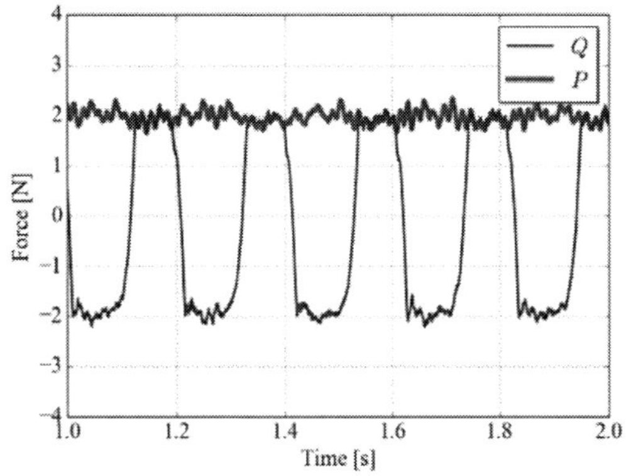

Fig. 5. Measured data. Contact (P) and friction (Q) forces.

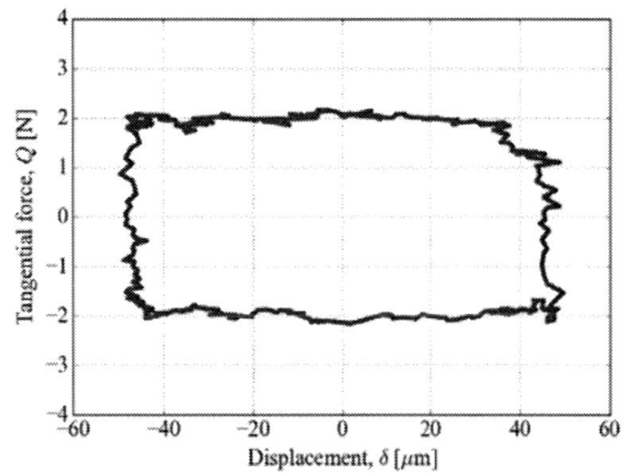

Fig. 6. Measured data. Fretting loop in gross slip regime.

4 Conclusions

A fretting wear tribotester for thin steel wires was developed. Highly stable data were measured validating the designed test rig. It is expected that further analysis using the test rig will allow a better understanding of fretting-wear behaviour of thin wires in different contact configurations, which may contribute to foster advancements in the design of better performing ropes.

Acknowledgment

The authors gratefully acknowledge the financial support given by the Basque Government under the "MEDECA ELKARTEK"

References

1. R.B. Waterhouse, *Wear*, **100**: pp. 107-118 (1984).
2. A. Cruzado, M. Hartelt, R. Wäsche, M.A. Urchegui, X. Gómez, *Wear*, **268**: pp. 1409-1416 (2010).
3. A. Cruzado, M. Hartelt, R. Wäsche, M.A. Urchegui, X. Gómez, *Wear*, **273**: pp. 60-69 (2011).

4. S. Fouvry, Ph. Kapsa, L. Vincent, *Wear*, **247**: pp. 41-54 (2001).
5. E. Marui, H. Endo, N. Hasegawa, H. Mizuno, *Wear*, **214**: pp. 221-230 (1998).
6. D. Klaffke, *Wear*, **104**; pp. 337-343 (1985).
7. R. Ramesh, R. Gnanamoorthy, *Materials & Design*, **27**: pp. 141-146 (2006).
8. D. Wang, D. Zhang, S. Ge, *Tribology International*, **52**: pp. 178-189 (2012).
9. H. Mohrbacher, J.P. Celis, J.R. Roos, *Tribology International*, **28**: PP. 269-278C (1995).
10. B.U. Wittkowsky, P.R. Birch, J. Dominguez, S. Suresh. *FFEMS*, **22**: pp. 307-320 (1999).
11. J. Vázquez, Phd Thesis, Universidad de Sevilla. (2009).
12. L. Vincent, Y. Berthier, M. Godet, ASTM, pp. 33-48 (1992).

Crack nucleation in solid materials under external load - simulations with the Discrete Element Method

Piotr Klejment[1,*], *Wojciech* Dębski[1]

[1]Institute of Geophysics Polish Academy of Sciences, Księcia Janusza 64, 01-452 Warsaw, Poland

Abstract. Numerical analysis of cracking processes require an appropriate numerical technique. Classical engineering approach to the problem has its roots in the continuum mechanics and is based mainly on the Finite Element Method. This technique allows simulations of both elastic and large deformation processes, so it is very popular in the engineering applications. However, a final effect of cracking - fragmentation of an object at hand can hardly be described by this approach in a numerically efficient way since it requires a solution of a problem of nontrivial evolving in time boundary conditions. We focused our attention on the Discrete Element Method (DEM), which by definition implies "molecular" construction of the matter. The basic idea behind DEM is to represent an investigated body as an assemblage of discrete particles interacting with each other. Breaking interaction bonds between particles induced by external forces imeditelly implies creation/evolution of boundary conditions. In this study we used the DEM approach to simulate cracking process in the three dimensional solid material under external tension. The used numerical model, although higly simplified, can be used to describe behaviour of such materials like thin films, biological tissues, metal coatings, to name a few.

1 Introduction

Cracking of materials is an extremely complicated process that includes processes in scales from atomic (breaking intermolecular bonds) up to a scale of thousands of kilometres in the event of catastrophic earthquakes (in the energy scale from individual eV to 10 24 J). [1] Such a large span of the scale raises a lot of questions, in particular about scalability of cracking processes, existence of factors determining the final size of the fracture area (on a macroscopic scale), course of the preceding and occurring processes during material destruction, etc [2].

The aim of this research was to try to look at the cracking processes on a scale typical for engineering and seismology (millimetres to meters) using micro-physics methods. The proposed research methodology was based on large-scale simulations using the Discrete Element technique. We mainly focused on cracking hypothetical three- dimensional materials subjected to uniaxial stretching with constant velocity and sample deformation. The above assumptions underlying the simulations may seem quite unrealistic. In fact, however, they quite well allow to describe the behaviour of cracking structures such as thin films (eg biological structures), metal coverings (eg aircraft fuselages), tailoring materials, etc.

2 Crack propagation

The well-known fact is that cracking solid bodies are determined by the structure of a given material, its atomic and micro-structural structure, but also by the way of applying external forces leading eventually to its destruction and fragmentation. [3] Usually, the rupture source is described by a single crack or dislocation, following the pioneering vision of Griffith [4]. The dynamic crack propagation causes a relaxation of stresses and energy release, leading in the consequence to material failure. It was shown experimentally that the microdestruction leads to macrodestruction.

Correct analysis of complexity of the fracture process or/and interactions of microcracks at big concentrations typical for prefracture state is possible only in terms of statistical models. The kinetic model of evolution of crack population was introduced by Czechowski [5] and developed in [6,7,8,9,10]. It lies at a level intermediate between the purely statistical approach and the fully microscopic treatment. The elementary objects are microcracks which can nucleate, propagate and coalesce. The problem of crack interaction and fusion is faced in its simplest aspects (binary interaction) but avoids its most delicate features by introducing extramechanical probabilistic assumptions. The kinetic approach operates on crack size distribution function that evolution is governed by the modified coagulation equation (mesoscopic level). Relations with the macroscopic picture, concerning the stress field evolution and the relationship between the time to fracture and the applied stress, were derived.

3 Discrete Element Method

Numerical analysis of cracking processes require an appropriate numerical technique. Classical engineering approach to the problem has its roots in the continuum mechanics and is based mainly on the Finite Element Method. [11] This technique allows simulations of both elastic and large deformation processes, so it is very popular in the engineering applications. In the process of cracking, the object is fragmented - disintegration into many independent bodies, and this effect is that no

* Corresponding author: pklejment@igf.edu.pl

numerical technique based directly on the mechanics of continuous media is clearly unable to adequately describe. Fragmentation of an object can hardly be described by this approach in a numerically efficient way since it requires a solution of a problem of nontrivial evolving in time boundary conditions.

For this reason to describe solid material fragmentations we focused our attention on the Discrete Element Method (DEM), which by definition implies "molecular" construction of the matter. [12,13] The basic idea behind DEM is to represent an investigated body as an assemblage of discrete particles interacting with each other. Breaking interaction bonds between particles induced by external forces imeditelly implies creation/evolution of boundary conditions – the most important element of the fragmentation proces. In this study we used the DEM approach to simulate cracking process in the three dimensional solid material under external tension. The used numerical model, although highly simplified can be used to describe behaviour of such materials like thin films, biological tissues, metal coatings, to name a few.

DEM is a computer simulation technique introduced by Cundall and Strack [14] and widely used in chemistry and engineering. In geotechnical application it was mostly applied to modeling granular media (soil, etc). The method is particularly convenient for rupture process study because it accommodates, in a natural way, the medium fragmentation effects. The method consists on representation of the medium by an ensemble of finite size particles (real atoms in the case of molecular dynamic simulation), which interact according to a specified interaction forces. [15,16]

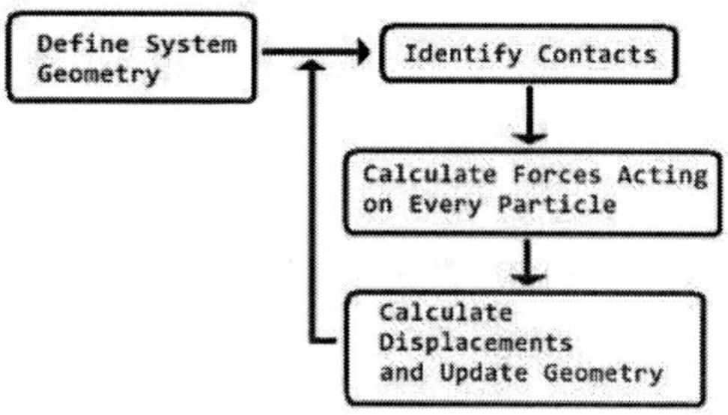

Fig. 1. The simplified scheme of DEM simulation.

The evolution of the system is achieved by sequential evaluating (Figure 1), for each time step, the positions, momenta and circular momenta of all particles, using Newton dynamical equation with prescribed interaction forces and assumed initial and boundary condition. [17,18] Although the approach is computationally time consuming and requires advanced HPC computational techniques, it provides the most direct insight into fully 3D dynamics of rupturing process. In this research we used the ESyS-Particle implementation of the DEM method developed at Queensland University, Australia [19] and released for the research application as the open source software.

4 Creation of the tensile fracture due to tensile forces in heterogeneous material

4.1 Simulation settings

We conducted simulations assuming that the tested materials can be described as a collection of "tight" packed in a given area of spherical particles with different radii (statistical inhomogeneity of materials) interacting with each other harmonic forces that can be represented as springs connecting the means of "molecules". The simulations performed concern a relatively simple mechanism of cracking solids under the influence of uniaxial stretching.

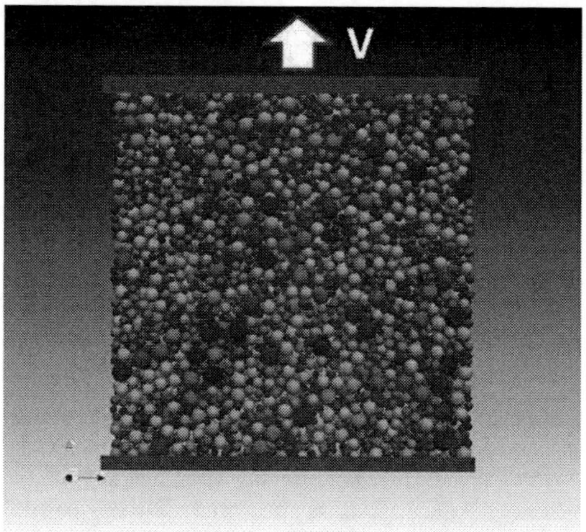

Fig. 2. The sample used in the simulations which consists of 3587 randomly distributed particles with radii 0.1 mm – 0.3 mm.

However, they showed complete compliance with experimental observations and "wider life experience". In spite of their simplicity and simplicity of the used model of interactions, they have already provided valuable information on the characteristics of the area around the tip of the developing cohesive zone.

We studied development of the kinetic model for populations of interacting different type cracks and computer simulations of models dynamics, as well as fracturing process with different regimes of generating internal stresses.

In simulations, we inserted the block of particles into the simulation object and bonded the particles together and we specified the type of interactions between bonded particles and initialise the interaction parameters. We created non-rotational elastic bonds between bonded particles as specified in the NRotBondPrms [19] parameter set:

```
bondGrp = sim.createInteractionGroup(
NRotBondPrms(
name = "sphereBonds",
normalK = 10000.0,
breakDistance = 0.5,
tag = 1,
scaling = True
))
```

NRotBondPrms contains five parameters: a bond tag specifying which bonded particles will undergo this interaction, a unique name for the interaction group, the elastic stiffness (normalK) of the bonds, a boolean (scaling) to specify whether to scale the stiffness with particle size, and a breakDistance specifying the separation distance that must be exceeded in order to break a bond between two particles. When a particle-pair exceeds the breakDistance the bond is removed and those two particles thereafter interact according to the interactions specified for unbonded particle pairs. [19]

4.2 Maximum uniaxial stress

As it is shown in the Table 1 we prepared various samples with different elastic stiffness of the bonds between particles. Our sample (Figure 2) had 3576 particles with radii from 0.1 mm to 0.3 mm and dimensions 10mm x 10 mm x 1mm. Firstly, we conducted series of measurements with uniaxial stretching velocity equal to 1mm/s. The aim was to calculate macroscopic Young's Modulus of the samples. This is so called calibration of the DEM model. In this way we can recognize what kind of real elastic material our samples correspond to. The results are visible in the Table 1.

Table 1. Micro- and macro-parameters of the simulations. Maximum uniaxial stress and type of cracking are shown for different uniaxial stretching velocities and elastic stiffness of the bonds.

Velocity [mm/s] / Elastic stiffness [N/mm]	5	10	50
500 (Young's Modulus 1.57 GPa)	732 MPa, Type 1	1465 MPa, Type 1	7325 MPa, Type 1
1000 (Young's Modulus 2.56 GPa)	733 MPa, Type 1	1467 MPa, Type 1	7335 MPa, Type 1
5000 (Young's Modulus 10.81 GPa)	742 MPa, Type 2	1483 MPa, Type 1	7414 MPa, Type 1
10000 (Young's Modulus 20.41 GPa)	814 MPa, Type 3	1503 MPa, Type 3	7510 MPa, Type 1
20000 (Young's Modulus 39.54 GPa)	1578 MPa, Type 2	1634 MPa, Type 2	7696 MPa, Type 3
50000 (Young's Modulus 95.01 GPa)	3890 MPa, Type 2	4034 MPa, Type 2	8209 MPa, Type 3
200000 (Young's Modulus 363.89 GPa)	14852 MPa, Type 2	15176 MPa, Type 2	16302 MPa, Type 2

Then, after calibration, we carried out series of measurements with three uniaxial stretching velocities: 5mm/s, 10mm/s and 50mm/s. For each one of them we calculated the value of maximum stress when main crack occurs. The results are also in Table 1. What is interesting, maximum stress weakly depends on elastic stiffness for smaller values of this parameter (500 – 20000 for velocities 10 and 50, 500 – 10000 for velocity 5). In this area character of cracking seems to be similar. However, for higher values of elastic stiffness there is significant increase in the maximum value of stress. At the same time, for the same values of elastic stiffness different uniaxial stretching velocities weakly influence on maximum stress.

4.3 Three types of cracking

On the basis of stress-strain curves for our results we were able to distinguish three different types of cracking: 1) type 1 when cracking is dispersed in time and space, it is impossible to extract main event or specific time of it, 2) type 2 when there is significant one big crack and some smaller 'aftercracks' and 3) type 3 with only one large crack that divides the material into two parts. These types were marked in the Table 1 and we are able to detect a certain regularity. Type 1 is always the first one, in the middle type 3 occurs and final stage is type 2. The higher the velocity the longer type 1 appears (type 3 and 2 require stronger elastic stiffness of bonds).

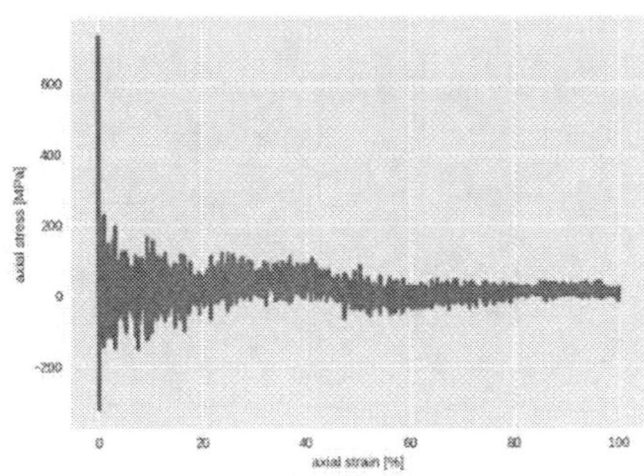

Fig. 3. First type of stress-strain curve for uniaxial stretching.

https://doi.org/10.1051/matecconf/201816522019

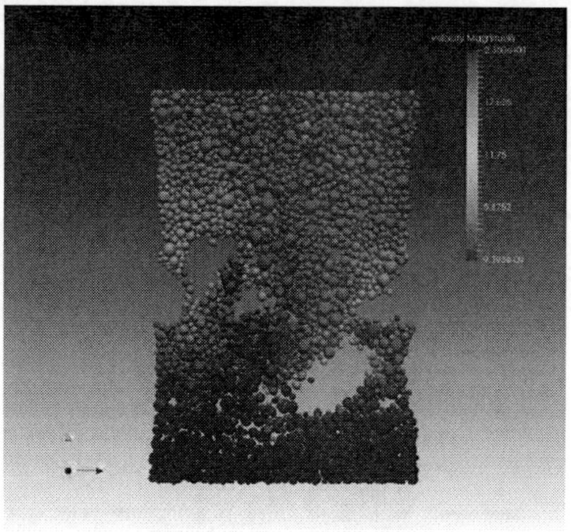

Fig. 4. Cracks propagation for stress-strain curve from Fig.3.

In the Figure 3 is shown stress-strain curve typical for type 1 of cracking. It is impossible to extract single events, this is rather continuous process. Visualization is presented in the Figure 4. During uniaxial stretching a lot of cracks nucleation foci appeared in the whole area of the sample. They become bigger over time and join with each other.

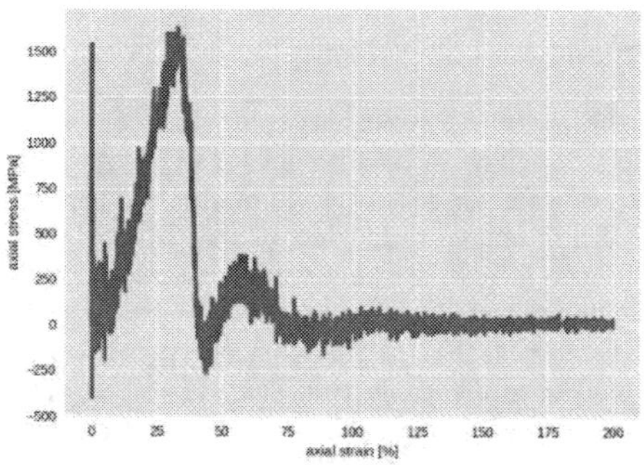

Fig. 5. Second type of stress-strain curve for uniaxial stretching.

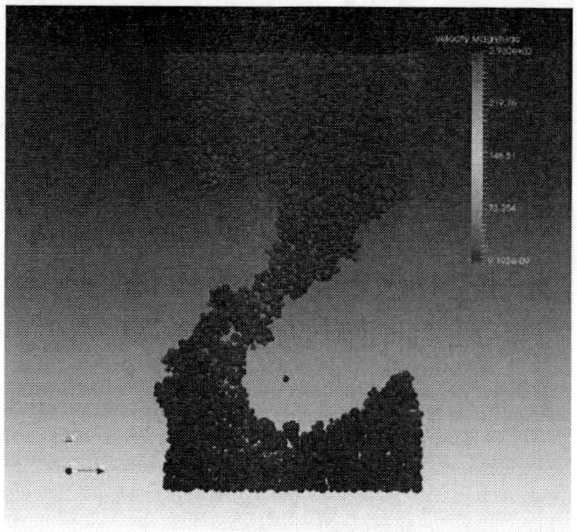

Fig. 6. Cracks propagation for stress-strain curve from Fig.5.

Second type of cracking is presented on the basis of stress-strain curve in the Figure 5. One main crack can be distinguished and some 'aftercracks' after it. In simulation (Figure 6) it is seen as one big crack which propagate quickly along sample and some smaller nucletion sources in other parts of the material which join together with main crack after some time.

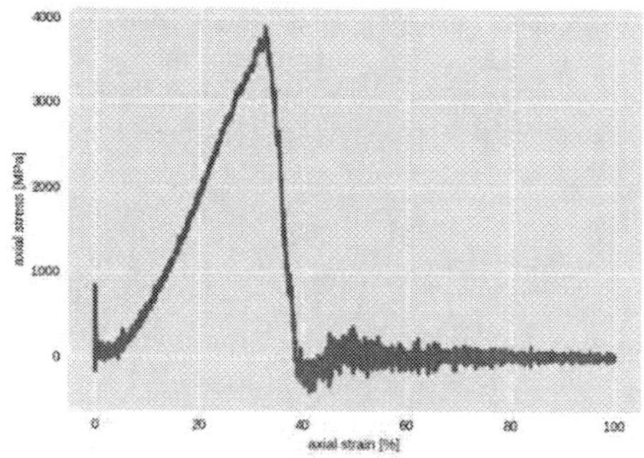

Fig. 7. Third type of stress-strain curve for uniaxial stretching.

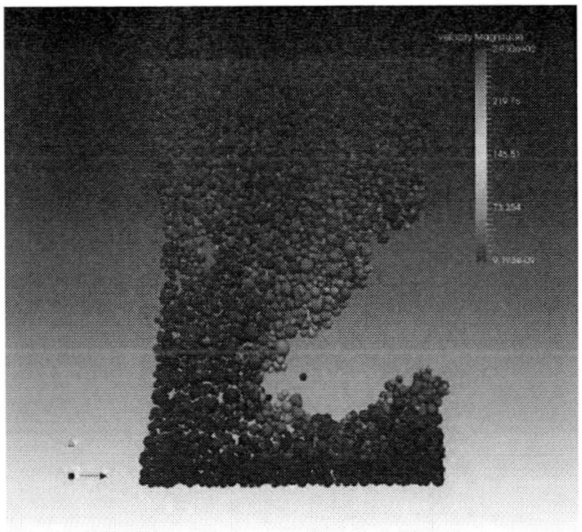

Fig. 8. Cracks propagation for stress-strain curve from Fig.7.

In the Figure 7 third type of cracking is presented on the basis of stress-strain curve. In this type of stretching one single crack appears, easily seen in the chart in the form of huge peak. In the Figure 8 one can observe that this type of crack starts from one side of the sample and propagate very fast along it. After this event the stress drops quickly to zero and sample is divided into two separated parts.

4.4 Kinetic energy

Kinetic energy provides information about total kinetic energy of all particles in every time step. This energy is connected only with linear movement of the particles because rotational ones are not considered in our simulations. On the x axis is presented strain scaled to the maximum value during simulation.

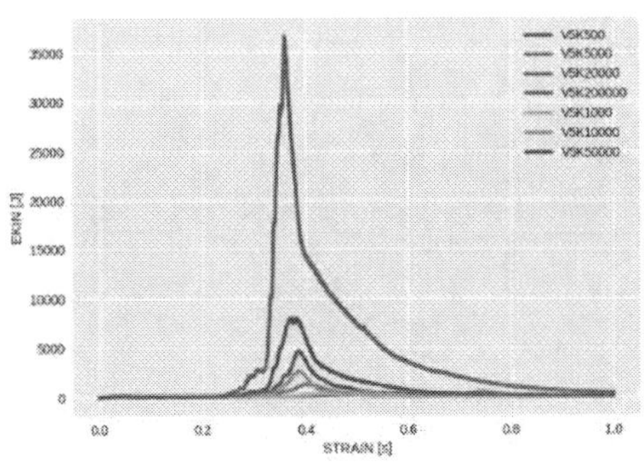

Fig. 9. Kinetic energy of all particles as a function of scaled strain for uniaxial stretching with 5 mm/s velocity.

The higher elastic stiffness of the bonds the higher kinetic energy during uniaxial stretching (Figure 9). What is interesting, for the same velocity 5mm/s all curves have almost the same shape and occur for the same strain.

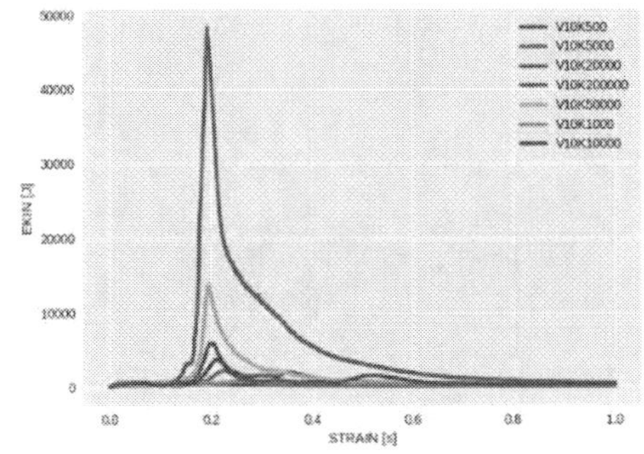

Fig. 10. Kinetic energy of all particles as a function of scaled strain for uniaxial stretching with 10 mm/s velocity.

In comparison to the Figure 9, in the Figure 10 is shown dependence for 10mm/s velocity. The energy of the cracking process is bigger, however, not two times as one would expect on the basis of velocity change from 5mm/s to 10mm/s.

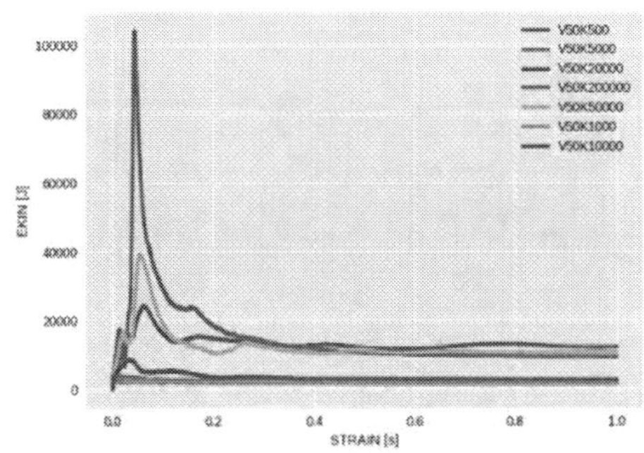

Fig. 11. Kinetic energy of all particles as a function of scaled strain for uniaxial stretching with 50 mm/s velocity.

What is interesting and can be seen on the basis of Figures 9,10 and 11 is that strain, when peak of the kinetic energy occurs, seems not to depend on the stretching velocity and elastic stiffness of bonds. In mentioned three figures strain is scaled to maximum value, however, in real numbers, the strain is almost the same for all simulations.

4.5 Potential energy

Potential energy describes all kind of interactions connected with bonds between particles. As in case of kinetic energy this is total value in every time step. Results are shown in the Figures 12,13,14. Two interesting issues can be observed. Firstly, as in the case of kinetic energy the strain for which peak of energy occurs seems not to depend on stretching velocity and elastic stiffness of bonds. Additionally, the maximum value of potential energy strongly depends on elastic stiffness, however, does not depend very much on velocity. For instance, for elastic stiffness 200000 N/mm total maximum potential energy is equal to about 80000 J

https://doi.org/10.1051/matecconf/201816522019

and only slightly changes when velocity increases from 5 mm/s to 50 mm/s.

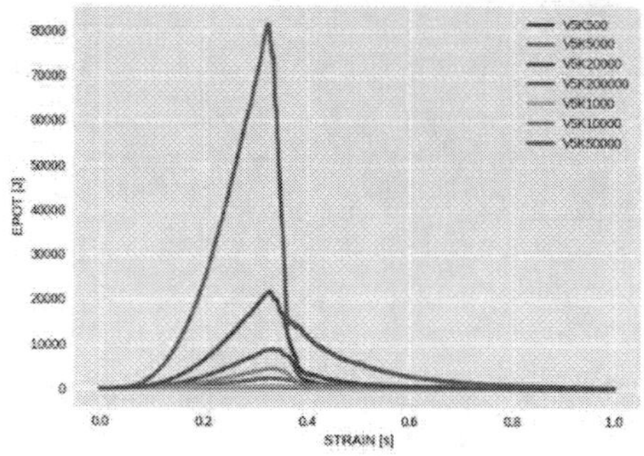

Fig. 12. Potential energy of all bonds between particles as a function of scaled strain for uniaxial stretching with 5 mm/s velocity.

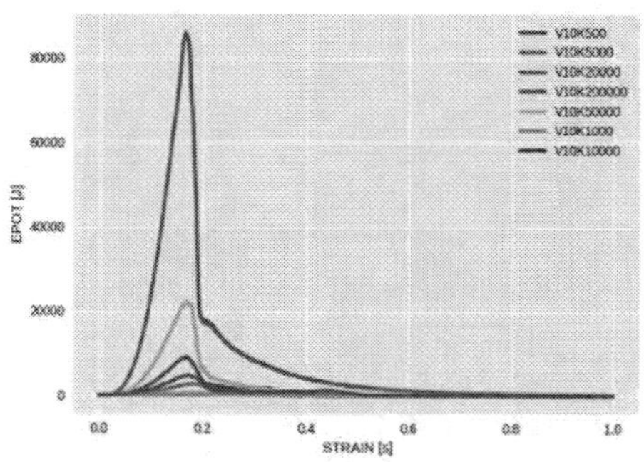

Fig. 13. Potential energy of all bonds between particles as a function of scaled strain for uniaxial stretching with 10 mm/s velocity.

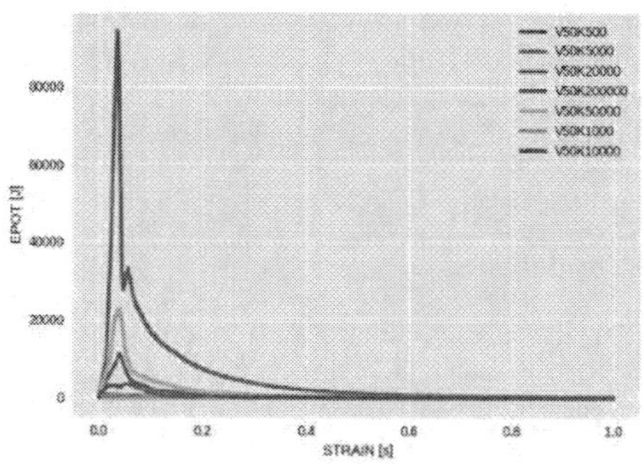

Fig. 14. Potential energy of all bonds between particles as a function of scaled strain for uniaxial stretching with 50 mm/s velocity.

4.3 Number of bonds

The initial number of bonds between particles inside the material is about 9700. The dependence between number of bonds as well as elastic stiffness and velocity is shown in the Figures 15,16,17. For velocity 5 mm/s, paradoxically, the biggest loss of bonds occurred for elastic stiffness 1000. It clearly shows that there is no clear dependence between this parameters. It is very similar for velocity 10 mm/s in the Figure 16. Quite interesting is Figure 17 where velocity 50 mm/s is shown. For this type of stretching the change in number of bonds is the most rapid, after the cracking quickly become constant.

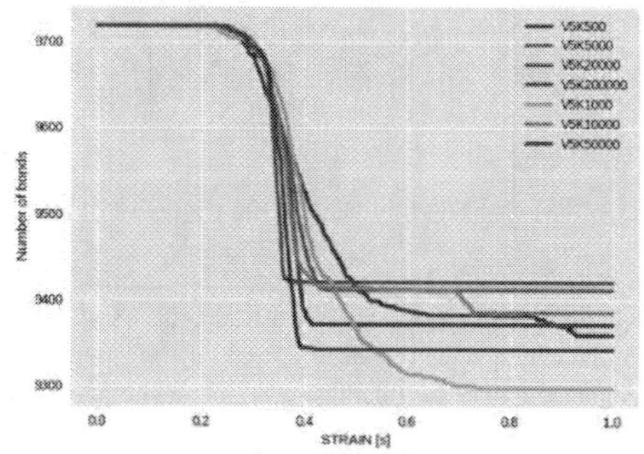

Fig. 15. Number of bonds between particles as a function of scaled strain for uniaxial stretching with 5 mm/s velocity.

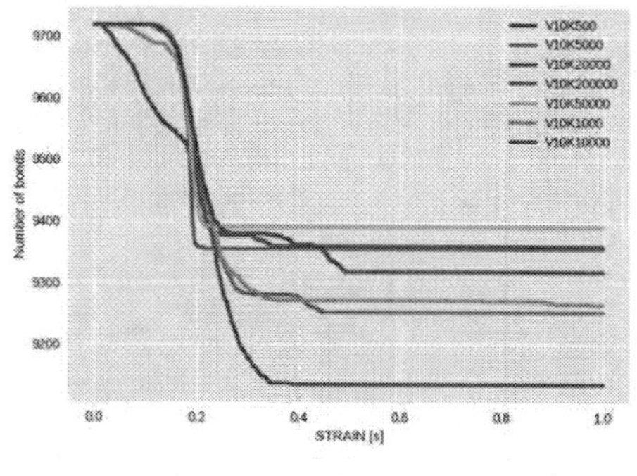

Fig. 16. Number of bonds between particles as a function of scaled strain for uniaxial stretching with 10 mm/s velocity.

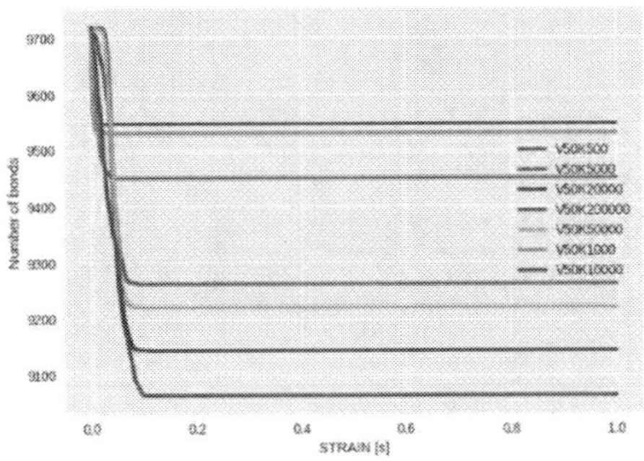

Fig. 17. Number of bonds between particles as a function of scaled strain for uniaxial stretching with 50 mm/s velocity.

5 Summary

In our research we investigated numerous parameters connected with uniaxial stretching of the sample for different elastic stiffness of the bonds between particles and velocities of stretching. We were able to distinguish three different types of cracking for materials and we calculated macroscopic Young's Modulus of the whole sample as well as maximum stress when main crack occurs. Additionally, we checked the dependence between kinetic energy of particles linear movement, potential energy of bonds between particles and number of bonds between particles as a function of stretching velocity and elastic stiffness of the bonds.

Breaking and fragmentation of solid materials is an extremely complex process, involving scales ranging from an atomic one up to thousands of kilometers in case of catastrophic earthquakes. Such a large scale span of breaking processes opens many questions concerning, for example, scaling of breaking processes, existence of factors controlling the final size of broken area, existence of precursors, dynamics of fragmentation, to name a few.

We trust that such simulations can help, soon, to understand better different processes in geophysics. There is a growing seismological evidence based on modern, high quality seismic data providing that, in spite of the large scale of failures observed in the earthquake dynamics, the breaking of rocks building the Earth's crust is controlled by processes at the micro scale. [20, 21] This suggests that during rupturing there are ongoing interactions and synchronizations of processes on significantly different spatial and energy scales leading to the final post-failure state. [22] Thus, the exhaustive description of the breaking process should take these effects into account. Unfortunately, such a unified description is currently still impossible.

Acknowledgements

- Young Scientists Grant Institute of Geophysics PAS no (5f/IGF PAN/2016mł)
- Young Scientists Grant Institute of Geophysics PAS no (6d/IGF PAN/2017mł)
- Computational Reseach Grant of Interdisciplinary Centre for Mathematical and Computational Modelling, Warsaw University – Project GES 17
- Computational Reseach Grant of Interdisciplinary Centre for Mathematical and Computational Modelling, Warsaw University – Project GC70-6
- Computational Reseach Grant of Interdisciplinary Centre for Mathematical and Computational Modelling, Warsaw University – Project GB72-7
- Interdisciplinary Polar Studies of Centre for Polar Studies „Polar-KNOW"
- Grant Opus of National Science Centre no 2015/17/B/ST10/01946

References

[1] Teisseyre R. (ed.) Theory of Earthquake Premonitory and Fracture Processes, PWN, Warszawa, 1995.

[2] Christensen, R.M., The Theory of Materials Failure, Oxford Univeristy Press, 2013.

[3] Jaeger, J., N. G. Cook, and R. Zimmerman (2007), Fundamentals of Rock Mechanics, pp. 488, Wiley-Blackwell, Malden, MA.

[4] Griffith A.A., The Phenomena of Rupture and Flow in Solids Philosophical Transactions of the Royal Society of London. Series A, Containing Papers of a Mathematical or Physical Character, Vol. 221,163-198, 1921.

[5] Czechowski Z., A kinetic model of crack fusion, Geophys. J. Int., 104, 419-422, 1991.

[6] Czechowski Z., A kinetic model of nucleation, propagation and fusion of cracks, J. Phys. Earth, 41, 127-137, 1993.

[7] Czechowski Z., Dynamics of fracturing and cracks [w] „Theory of Earthquake Premonitory and Fracture Processes", R. Teisseyre [ed], PWN, Warszawa, 447-469, 1995.

[8] Czechowski Z., A kinetic model of macrocrack propagation, Acta Geophys. Pol. 45, no 3, 193-204, 1997.

[9] Czechowski Z., A generalized model of three component mixture of crack, Acta Geophys. Pol, 46, no 4, 375-388, 1998.

[10] Czechowski Z., A stochastic approach to the cracktip propagation model, Acta Geophys. Pol. 48, no.1, 57-68, 2000.

[11] Onate E., Rojek J., Combination of discrete element and Finite element methods for dynamic analysis of geomechanics problems (2004). Computer methods in applied mechanics andengineering, 193, 3087-3128

[12] Klejment P., Dębski W., „Brazilian Test – A Microscopic Point of View on The Tensile Fracture Generation", conference proceedings, V International

Conference on Particle-based Methods – Fundamentals and Application
[13] Klejment P., Kosmala A., Foltyn N., Dębski W., „Symulacje komputerowe parametrów mikroskopowych i makroskopowych materiałów przy pomocy Metody Elementów Dyskretnych", „Innowacje w polskiej nauce w obszarze matematyki i informatyki", p.5-19
[14] Cundall, P. A., and O. D. L. Strack (1979), A discrete numerical model for granular assemblies, Geotechnique, 29(1), 47–65, doi:10.1680/geot.1979.29.1.47
[15] Egholm, D. L. (2007), A new strategy for discrete element numerical models: 1. Theory, J. Geophys. Res., 112, B05203, doi:10.1029/2006JB004557
[16] Fakhimi, A., and E. Alavi Gharahbagh (2011), Discrete element analysis of the effect of pore size and pore distribution on the mechanical behavior of rock, Int. J. Rock Mech. Min. Sci., 48(1),77–85, doi:10.1016/j.ijrmms.2010.08.007
[17] O'Sullivan C., Particulate Discrete Element Modeling: A Geomechanics Perspective, S.1.: Spon Press/Taylor & Francis, 2011
[18] Potyondy, D., and P. Cundall (2004), A bonded-particle model for rock, Int. J. Rock Mech. and Min. Sci., 41 (8), 1329–1364, doi:10.1016/j.ijrmms.2004.09.011
[19] Abe S., Boros V., Hancock W., Weatherley D., ESyS – Particle Tutorial and User's Guide Version 2.3.1, 2014 https://launchpad.net/esys-particle
[20] Li Y. G., Imaging, Modeling and Assimilation in Seismology, De Gruyter, Berlin 2012
[21] Udias U., R. Madariaga, E. Buforn, Source Mechanisms of Earthquakes: Theory and Practice, Cambridge Univ. Press, 2014.
[22] Newman W., Gabrielov A., Durand T., Phoenix S.L. and D. Turcotte, An exact renormalization model for earthquakes and material failure: statics and dynamics, Physica D 77, 200-216, 1994. Peirce F.T., J. Text. Ind. 17, 355, 1926.

20 kHz 3-point bending fatigue of automotive steels

Mohamed Sadek[1,*], *Jens* Bergström[1], *Nils* Hallbäck[1] and *Christer* Burman[1]

[1]Karlstad University, Department of Engineering and Physics, SE-658 88 Karlstad, Sweden.

Abstract. The 20 kHz load frequency enables fatigue tests for very high cycle fatigue life, 10^9-10^{13} cycles, within conveniently short time. In automotive applications, many components are subjected to flexural loading and hence bending fatigue is an important test mode. Ultrasound fatigue test instruments have been used successfully in several assessments of fatigue strength and more commonly in uniaxial loading. Here, a 3-point bending fatigue test rig operating in resonance at 20 kHz load frequency has been designed to test plane specimens at R=0.1 loading. The test rig design and stress calculations are presented. Testing for fatigue strength was conducted using the staircase method with 15 specimens of each steel grade, specimens reaching 10^8 cycles were considered run-outs giving fatigue strength at 10^8 cycles. Additional 15 specimens of each grade were tested for S-N curves with the upper limit above 10^9 cycles. Two different common automotive steels, 38MnSiV5, a micro-alloyed ferritic-pearlitic steel, and 16MnCr5, a carburizing martensitic steel, were tested. The fatigue strengths achieved from the staircase testing are 340 and 419 MPa stress amplitudes for the 38MnSiV5 and 16MnCr5 steels, respectively. The S-N curves of the steels appear to be quite flat in the tested life range 10^7 – 10^9.

Keywords: Three-point bending, VHCF, automotive steels, ferritic-pearlitic steel, carburized steel

1 Introduction

Many components in automotive mechanical systems are subjected to cyclic flexural loading like bending or torsion. Hence the importance of investigating fatigue properties of automotive materials in these specific loading modes. During the last couple of decades, the development of the new fast fatigue testing system, the *Ultrasonic Fatigue Testing System*, have been taking place. The 20 kHz load frequency of the system enables fatigue tests for very high cycle fatigue life, 10^9-10^{13} cycles, within a conveniently short time, or significantly reduced test time and costs at shorter life. Ultrasound fatigue test instruments have been used successfully in several assessments of fatigue strength and more commonly in uniaxial loading. However, the development of special test rigs for other than uniaxial load modes in ultrasonic fatigue testing system, e.g. bending and torsion, has been taking place in various laboratories [1-4].

Components exposed to surface stress gradients are also often conveniently bestowed enhanced performance by surface strengthening using mechanical, thermal or thermochemical processes, and carburizing is one such method often used [5].

The aim using carburized steels is to obtain high-carbon plate type martensitic case with excellent wear and fatigue resistance and tough low-carbon core. Jo et al. [6] have studied the fatigue behaviour of carburized Cr-Mo gear steels subjected to multiaxial stress states. Axial and rotating bending have been performed at fully reversed loading (R=-1) and the results have been compared. The S-N curves they have achieved showed that the fatigue strength (at 10^6 cycles) in axial loading was ca 70% of the fatigue strength of rotating bending with surface initiation in the early lives (<10^5 cycles) and sub-surface initiations at longer lives (>10^5 cycles).

The fatigue strength of carburized steel depends on several factors, e.g. increased surface strength, residual stresses, modified microstructure, stress gradient, material hardness, surface roughness and surface oxidation.

Bomas et al. [7] have used Weibull's weakest link concept to calculate a predicted fatigue limit for case-hardened 16MnCr5-steel grade. This model calculates

* Corresponding author: mohamed.sadek@kau.se

the 50% survival probability as the product of survival probability of the surface and the survival probability of the volume/core, Eq. 1.

$$P_s = P_s(A) * P_s(V) \quad (1)$$

Since the weakest link concept is based on the assumption that material strength is related to material imperfections, the calculated survival probability is depending not only on the load, rather also on the stressed volume.

In another study, Bomas et al. [8] have analysed the VHCF behaviour of 20MnCr5-steel grade with two different carburizing heat treatments yielding one superclean (nearly free of oxides) and one with normal content of non-metallic inclusions. The carburized layer depth (case-hardened depth, CHD) is 0,7 mm. The tests were conducted using an electro-magnetic resonant system (180-190 Hz) at two different loading ratios, R=-1 and R=0. Two different notched specimen geometries have been used, K_t=1,14 and K_t=1,21. Finite life and reduced fatigue strength was observed up to 2x10^8 cycles. The initiation sites (regarding failure in VHCF regime) have been depending on several factors. Initiations were obtained exclusively in the interior at both non-defect sites and at non-metallic inclusions for the K_t=1,14. But for K_t=1,21, initiations in the carburized surface have also been observed. The weakest-link concept has again been used to predict the fatigue limit combined with the properties of the core and the carburized layer.

In the present work, a three-point bending fatigue test rig is designed and constructed to fit into the ultrasonic fatigue testing equipment to test plane specimens at R=0.1 loading. Two different automotive steels are investigated here; 38MnSiV5, a ferritic-pearlitic high strength micro-alloyed steel, and 16MnCr5, a carburizing martensitic steel grade. The complications with the three-point bend testing are most importantly holding the specimen in a fixed position to achieve the correct stress distribution in the specimen with maximal stress level at the specimen midsection.

2 Experimental

Two different steel grades are tested in the newly developed ultrasonic three-point bending fatigue testing system. The testing was aided by analytical and FEM computation of the deformation, strain and stress distribution in the specimen.

2.1. Materials

Two different common automotive steels were tested; 38MnSiV5 - a ferritic-pearlitic high strength micro-alloyed steel (Fig.1), in which normal contents of sulphide and oxide inclusions have been observed and 16MnCr5 - a carburizing martensitic steel grade (Fig.2) containing relatively high content of small sized sulphides and oxides to improve machinability. The carburizing, hardening and tempering cycle used was 900 °C/3.5 h-860 °C/2.5 h-oil quench 80 °C-160 °C/2 h, and this heat treat cycle exempt of carburizing was used on tensile test specimens to obtain the carburized core properties. The chemical composition of the two steel grades is presented in Table 1. Mechanical properties are presented in Table 2 (regarding the carburizing steel the given values are of the core).

Table 1. Chemical composition of the two steel grades.

Grade	C	Mn	Si	P	S	Cr
38MnSiV5	0,37	1,46	0,68	0,011	0,050	0,12
	Ni	Mo	V	Cu	Al	Sn
	0,10	0,03	0,10	0,12	0,014	0,009
	C	Mn	Si	P	S	Cr
16MnCr5	0,16	1,10	0,27	0,011	0,021	1,02
	Ni	Mo	V	Cu	Al	Sn
	0,11	0,03	0,01	0,14	0,021	0,009

Figure 1. Microstructure of the ferritic-pearlitic 38MnSiV5-steel grade.

https://doi.org/10.1051/matecconf/201816522020

Figure 2. Microstrucure of the carburizing martensitic 6MnCr5-steel grade.

Table 2. Mechanical properties of the two steel grades. Core properties for the carburized 16MnCr5-steel grade.

Grade	Tensile Strength (MPa)	Yield Strength (MPa)	Hardness (HV30)
38MnSiV5	867	589	267
16MnCr5*	1175	876	438

As mentioned, the 16MnCr5-steel grade is a carburized martensitic steel. The carburizing treatment results in a higher carbon content at the surface elevating the surface hardness. The Vickers hardness profile is displayed in Fig. 3. The hardness profile indicates a carburized layer depth of ca 0.6 mm, corresponding to the depth of half of the hardness drop.

The surfaces of the specimen were ground and finally polished with 6 μm and then 3 μm polishing paste.

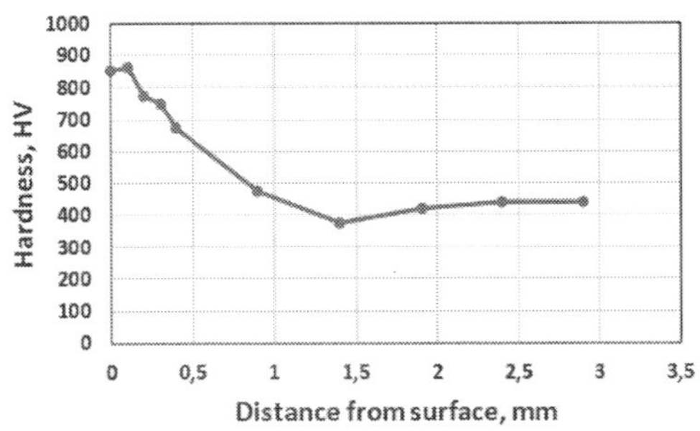

Figure 3. Vickers hardness profile for the carburized 16MnCr5-grade. Measured with a 25 gram load.

2.2. Instruments

A newly designed and build specimen holder is mounted on a self-aligning sliding bearing support (Fig. 4). The function of the holder is to keep the specimen in a fixed position regarding all directions.

The self-aligning support with the specimen holder are mounted into the ultrasonic fatigue testing system (Fig. 5). The low friction sliding bearing provides the alignment of the specimen holder in a plane perpendicular to the applied force direction.

Figure 4. Three-point bending specimen holder mounted on the Self-aligning supporter.

Figure 5. Specimen holder and Self-aligning supporter inserted into the ultrasonic fatigue testing system.

A specially designed metal tip is attached to the horn and transfers the load down to the specimen (Fig. 6). It is designed so that the load is transferred evenly across the specimen width [3] and meets the resonance requirements of the system.

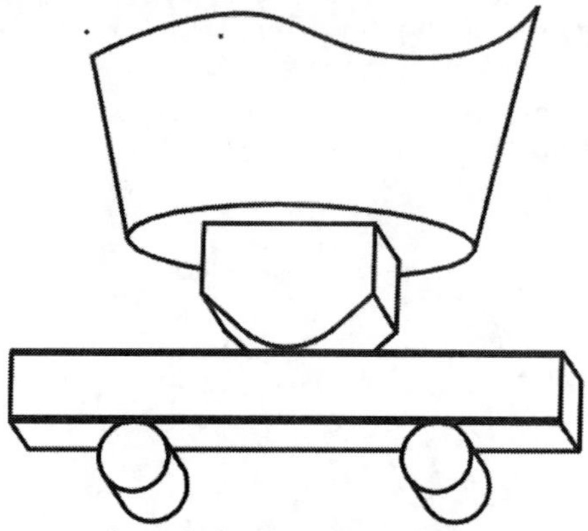

Figure 6. Load-transferring tip mounted into the horn.

For varying the load ratio *(R)* a mean static load is required. This static load is superimposed by mounting the whole load train together with the specimen holder and the self-aligning supporter into a mechanical screw test machine and loaded in compression. The two supporters have a diameter of 4 mm and are placed in tracks of approximately the same diameter allowing them to rotate freely.

Pressurized undercooled air is used to cool down the specimen and the surrounding parts during the testing.

2.2.1 Three-point bending specimen

The specimen used in this work is a rectangular bar specimen with 4x8 mm^2 cross-section and 31,8 mm length. To eliminate stress concentrations at the edges of the specimen, chamfers (0,2 mm deep at 45° angle) are milled at the edges.

2.3. Modelling and computation

The specific dimensions of the specimen are carefully chosen and tested with FEM for the steel specimen to have a flexural mode resonance frequency of 20 kHz. Analytical analysis of the specimen displacement (when under a cyclic load) has been derived elsewhere [4] and is presented in Eq. 2.

$$U(x) = A_0\left(\frac{\cosh(kL)}{\cos(kL)+\cosh(kL)}\right) * \left(\cos(kx) + \frac{\cos(kL)}{\cosh(kL)}\cosh(kx)\right) \quad (2)$$

where A_0 is the displacement at the middle of the specimen, L is the length of the specimen and k is the frequency dependent parameter given in Eq. 3.

$$k = \left(\frac{12\omega^2\rho}{Eh^2}\right)^{1/4} \quad (3)$$

where $\omega=2\pi f$ is the angular frequency, ρ is the material density, E is the elastic modulus and h is the height of the specimen. From the same analytical procedure, the specimen's resonance length *(L)* and the required distance between the supporters (L_0) are defined as in Eq. 4-5.

$$2L = 0{,}506925\left(\frac{Eh^2}{\rho f^2}\right)^{1/4} \quad (4)$$

$$2L_0 = 0{,}27966\left(\frac{Eh^2}{\rho f^2}\right)^{1/4} \quad (5)$$

These expressions were used as guidelines to select the proper test and specimen geometries in the present study. Final calculations of the test set-up were done using FEM and the Abaqus software. As well, FEM was used to compute the stress amplitudes and distribution in the specimen (Fig. 7). For varying loading ratio, both static and dynamic stress components are required, i.e. a mean load (σ_m) and a sinusoidal stress amplitude (σ_a), Eq. 6.

$$\sigma(t) = \sigma_m + \sigma_a \sin(\omega t) \quad (6)$$

Both the static mean load and the dynamic stress amplitude are related to the maximum displacement of the specimen (at the mid-section of the specimen directly under the load-transferring tip), calculated by static and dynamic FEM simulations respectively. Fig. 8 shows the stress distribution and the maximum stresses for a displacement of 20 µm. For the cyclic stress amplitude (σ_a) the so-called m-factor is needed to control testing. The m-factor correlates the maximum stress amplitude to the displacement, Eq. 7, since the control parameter during testing is the displacement at the bottom of the horn (this displacement is calibrated against the ultrasound generator voltage input to the oscillator directly controlled by the operator).

$$m_{amp} = \frac{\sigma_{a.max}}{U} = \frac{226\ MPa}{20\ \mu m} = 11{,}3\ MPa/\mu m \quad (7)$$

The mean stress (σ_m) is controlled by the force input in the tensile testing machine. Hence the m-factor for the mean load is a correlation between the statically determined maximum stress and the force acting on the specimen, Eq. 8.

$$m_{mean} = \frac{\sigma_{m.max}}{F} = \frac{220{,}5\ MPa}{1104\ N} = 0{,}1997\ MPa/N \quad (8)$$

Here, a load ratio of $R=0,1$ was used in all cases.

https://doi.org/10.1051/matecconf/201816522020

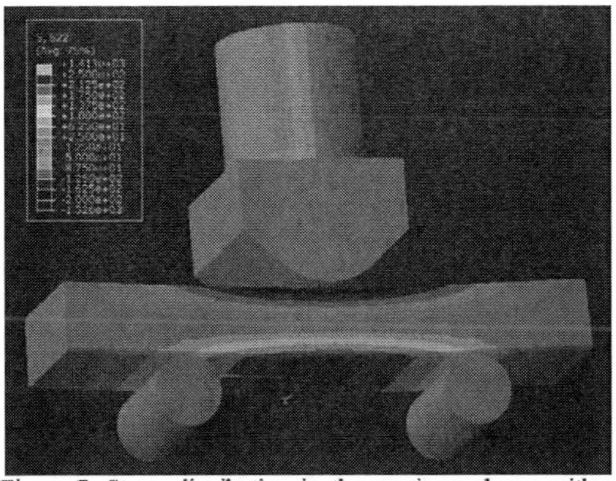

Figure 7. Stress distribution in the specimen shown with a FEM model.

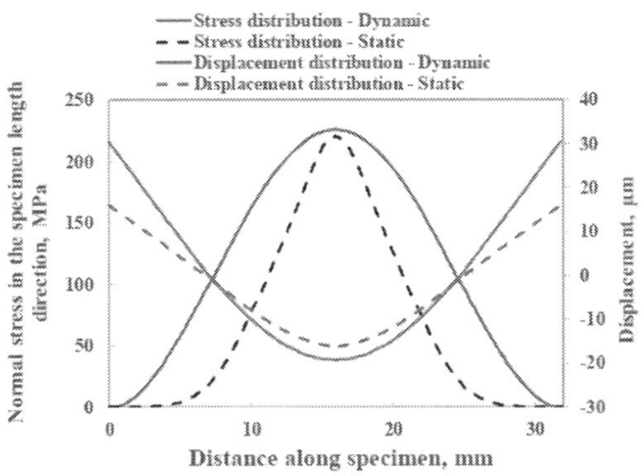

Figure 8. Stress and displacement distribution along specimen's length in dynamic and static simulations.

2.4. Testing

Testing for fatigue strength was conducted using the staircase method with 15 specimens of each steel grade. The tests were stopped at 10^8 cycles and the specimens surviving were considered run-outs, which defines the obtained fatigue strength at 10^8 cycles [9]. If a specimen fails prior to the run-out life, the next specimen is tested at a lower stress level, but if the specimen survives until run-out the next specimen is tested at a higher stress level. The increments are usually around 5% of the mean fatigue limit. For the data reduction technique, here the Dixon-Mood [10] technique was used, yielding both the fatigue strength and the standard deviation.

Additional 15 specimens of each steel grade were tested for the S-N curves with the upper limit above 10^9 cycles. The evaluation of the S-N curves was done according to the SS-ISO12107 standard [11] Linear Fatigue Response, as in Eq. 9,

$$Log_{10}(N) = b_0 + b_1\, Log_{10}(S) \qquad (9)$$

from where also the regression parameters b_0 and b_1 were estimated.

3 Results

The new constructed rig for three-point bending fatigue testing at 20 kHz worked appropriately. The specimen was held in a fixed position and the loading was performed correctly.

The fatigue strength *(σ_{FL})* at 10^8 cycles and the standard deviation yielded by the staircase testing are presented in Table 3.

Table 1. Fatigue strength at 10^8 load cycles and standard deviation for the two steel grades.

Steel grade	σ_{FL}	Standard deviation
38MnSiV5	340 MPa	±10 MPa
16MnCr5	419 MPa	±14 MPa

The S-N curves of the two materials are displayed in Fig. 9. The evaluation according to the SS-ISO12107 standard were performed on the results of both materials. The red line in Fig. 9 represent the predicted fatigue strength for the 38MnSiV5-steel grade according to Eq. 9 with life as depending variable. For the 16MnCr5-steel grade the predicted fatigue strength showed a significant deviation from the results, hence it is not plotted here. Instead, a regression line using life as independent variable is inserted. The regression parameters according to the standard are presented in Table 4.

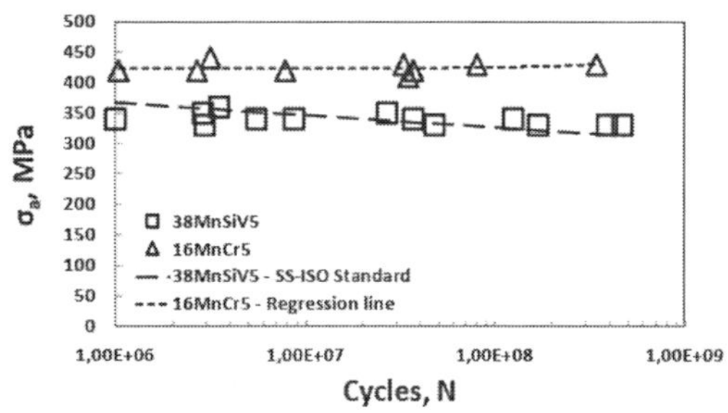

Figure 9. S-N curves of the two steel grades and the predicted S-N curve for the 38MnSiV5-grade according to Eq. 9.

Table 2. Regression parameters according to the SS-ISO12107 standard.

Grade	b_0	b_1	R^2
38MnSiV5	105,08	-38,63	0,28
16MnCr5	-16,41	8,99	0,01

As observed in Fig. 9, the S-N curves of both steel grades has a flat appearance in the 10^6 – 10^9 range, thus

https://doi.org/10.1051/matecconf/201816522020

the fatigue life range is quite large at that level of fatigue strength.
A study of the fracture surfaces was performed using a high resolution FEG-SEM LEO scanning electron microscope. The initiation sites were found to be at the flat surface, in the corners and even in the interior (close to the surface) for the 38MnSiV5-steel grade, but exclusively in the corners for the 16MnCr5-steel grade. Figures 10-12 shows some examples of typical fracture surfaces.

Figure 10. Crack initiation at the corner, 38MnSiV5-steel grade. σ_a=340 MPa and N_f=10^6 cycles.

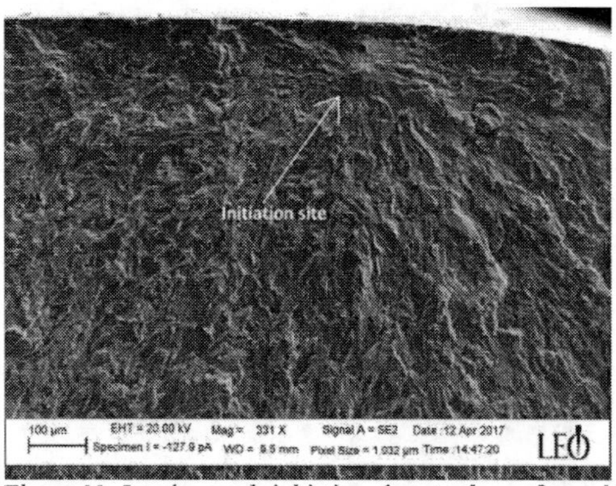

Figure 11. Interior crack initiation close to the surface with a FGA formation, 38MnSiV5-steel grade. σ_a=330 MPa and N_f=3,9*10^8 cycles.

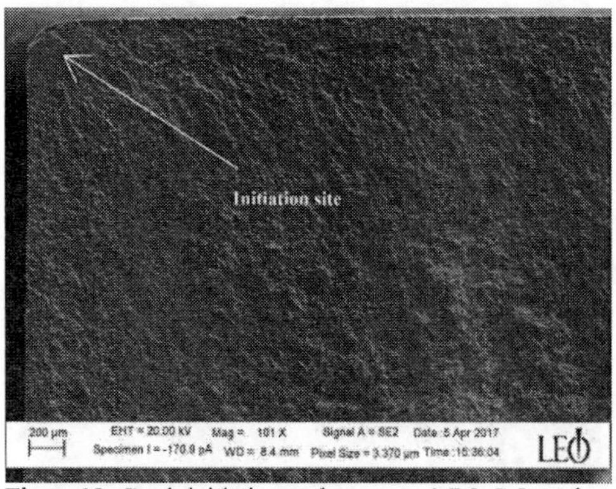

Figure 12. Crack initiation at the corner, 16MnCr5-steel grade. σ_a=420 MPa and N_f=7,8*10^6 cycles.

4 Discussion

During the development of the new three-point bending rig, several complications were encountered. One problem was the alignment of the specimen perpendicular to the load direction. This was elegantly solved by mounting the specimen holder onto a self-aligning support at the bottom of the load train (Fig. 4-5) with the eucentric rotation at the point of load tip contacting the specimen surface. With this solution, the specimen is aligned during the application of the mean load and keeps the position during the rest of the test. Furthermore, the friction between the specimen and the supports and the surrounding position constraining parts caused wear and elevated temperatures. This problem was handled by delicate positioning of the surrounding constraining parts.

The precision of the sinusoidal loading *(σ_a)* is depending on the step size of the voltage input, which is 0,1 V in the current ultrasonic fatigue testing system. With the calculated m-factor (Eq. 7) and the displacement calibration, the precision of stress amplitude *(σ_a)* for the used setup with the specific horn dimensions is calculated to ±5 MPa (corresponding to 0,5 µm displacement). The maximum possible stress amplitude *(σ_a)* with the current setup is ca 950 MPa (corresponding to 84 µm displacement) due to the maximum voltage input of ca 9 V. The precision percentage is hence 0,5/84=0,6%.

Evidently, the S-N curves of the steels appears to be quite flat in the tested life range $10^6 - 10^9$. These shapes have been seen in the results from 20 kHz uniaxial fatigue testing of the same materials at the same loading ratio *(R=0,1)* [12]. It is speculated whether fatigue crack initiation mechanisms controlling life are different at short and long lives. Further comparison to the uniaxial loading results, shows that the S-N curve of the 38MnSiV5-steel grade is quite higher (ca 60 MPa) for the three-point bending tests. For the 16MnCr5-steel grade however, the S-N curves of three-point bending and uniaxial testing showed good agreement. Commonly, one expects higher fatigue strength in bending than in tension fatigue, as [6] have found where the results showed that the fatigue strength of the uniaxial loading tests were about 70% of the fatigue strength of the rotating bending testing. However, surface conditions may play an important role. Thus, the less defect sensitive 38MnSiV5 will gain in bending, while the opposite holds for the carburized 16MnCr5 steel. As a consequence, the 16MnCr5 fatigue strength in tension is 1.6 times larger than that that of the 38MnSiV5 steel, but in bending only 1.2 times larger.

The fractography study showed that the initiation sites for the 38MnSiV5-steel grade were distributed between the bottom surface, bottom corners and even some interior initiations close to the surface. While for the 16MnCr5-steel grade, the carburized layer elevates the material strength and possibly creates some residual stresses at the surface. The effect of this was restriction of the initiation sites to the bottom corners of the specimen. The fact that, for carburized specimens, the initiation sites are rather found in the interior is confirmed by [8, 12] where surface initiations were observed only on specimens with a relatively high stress concentration factor, K_t.

5 Conclusion

The development and exploration of a new bending fatigue test rig operating at 20 kHz load frequency has been presented. Careful design of operation proved the test rig to run efficiently and to produce reliable results. Finite fatigue lives between 10^6 and 10^9 load cycles were obtained of the ferritic-pearlitic and carburized martensitic steels, and similar to tests in tension fatigue flat SN-curves were obtained. The mechanisms responsible of this behaviour are still left to be determined. The increase in fatigue strength using the carburized grade rather than the ferritic-pearlitic steel was much smaller in bending than in tension fatigue, contrary to what expected.

6 Acknowledgements

The research leading to these results has received funding from the EuropeanUnion's Research Fund for Coal and Steel (RFCS) research programme under grant agreement no. [RFS-CT-2013-00015 (FREQTIGUE)].

References

1. W. Hoffelner, Journal of Physics, E: Scientific Instruments, **13(6):** p. 617 (1980)
2. W. Hoffelner, P. Gudmundson, Engineering fracture mechanics, **16(3)**: p. 365-371 (1982)
3. H.Q. Xue, et al., International Journal of Fatigue, **29(9-11):** p. 2085-2093 (2007)
4. C. Bathias, P.C. Paris, *Giga cycle fatigue in mechanical practice*, Marcel Dekker (2005)
5. Bergström, J., et al, Final report RFS-CT-2013-00015 (FREQTIGUE). 2018.
6. B. Jo, S. Sharifimehr, Y. Shim, A. Fatemi, International Journal of Fatigue, **100**: p. 454-465 (2017)
7. H. Bomas, P. Mayr, M. Schleicher, Materials Science and Engineering **A234-236:** p. 393-396 (1997)
8. H. Bomas, K. Burkart, H.W. Zoch, International Journal of Fatigue **60:** p.63-73 (2014)
9. Y.-L. Lee, *Fatigue testing and analysis: theory and practice*, Elsevier (2005)
10. W.J. Dixon, A.M. Mood, Journal of the American Statistical Association, **43(241):** p. 109-126 (1948)
11. *SS-ISO 12107, Metallic materials – Fatigue testing – Statistical planning and analysis of data* (2012)
12. M. Sadek, J. Bergström, N. Hallbäck, C. Burman, R. Elvira, B. Escauriaza, *Seventh International conference on Very High Cycle Fatigue (VHCF7) Proceedings*, **7:** p. 193-198 (2017)

Sensitivity analysis for effects of displacement amplitude and loading frequency on low-cycle fatigue lifetime in carbon/epoxy laminated composites

Mohammad Azadi[1,*], *Mohsen* Alizadeh[2], and *Hassan* Sayar[1]

[1]Faculty of Mechanical Engineering, Semnan University, Semnan, Iran
[2]Faculty of Aerospace Engineering, Semnan University, Semnan, Iran

Abstract. In the present article, the low-cycle fatigue lifetime of carbon/epoxy laminated composites has been investigated. The sensitivity analysis has been also performed to study effects of the displacement amplitude and the loading frequency on the low-cycle fatigue lifetimes in composites. For such objective, displacement-control low-cycle fatigue testing has been done on the open-hole standard specimen. Fatigue tests included 4 different displacement amplitudes at 200 mm/min of the loading frequency and 4 different loading frequencies under 7 mm of the displacement amplitude. The sensitivity analysis was carried out by the MINITAB software, considering a linear function for fitting experimental data by the predicting model. Experimental results showed that by increasing the displacement amplitude, the low-cycle fatigue lifetime decreased, as expected. In addition, when the loading frequency enhanced, the low-cycle fatigue lifetime of composites decreased. Besides, the maximum stress had a reverse behavior, compared to the fatigue lifetime. The sensitivity analysis depicted that the displacement amplitude was sensitive on both the fatigue lifetime and the maximum stress. The loading frequency was sensitive on the maximum stress and was not sensitive on the fatigue lifetime.

1 Introduction

Nowadays, composite structures have been widely in the aerospace industry. Such components are exposed to cyclic loadings, which lead to the fatigue fracture. Therefore, to know the fatigue behavior of composites is essential for engineers to design an appropriate structure, based on boundary conditions.

In this field of study, including the fatigue behavior of laminated polymer matrix composites (PMC), different researches have been presented until now, which details can be found in the literature. De Baere et al. [1] studied on the tension-tension fatigue behavior of a carbon reinforced thermoplastic to evaluate a dumbbell-shaped specimen. They depicted that there was only a small influence of the frequency on the fatigue lifetime for higher stress levels, whereas for stress levels lower than 675 MPa, a lower frequency yielded a larger number of cycles to failure. Broughton et al. [2] investigated the open-hole tension-tension fatigue behavior of a glass fiber-reinforced polymer (GFRP) laminated composite. Constant amplitude tests were conducted at two different R-ratios and the block amplitude tests involved two or more stress levels. They noted that decreasing the test frequency to 1 Hz reduced the surface temperature of the specimen, considerably. This reduction in temperature was accompanied by a significant reduction in the fatigue lifetime.

Yudhanto et al. [3] evaluated open-hole fatigue characteristics and the damage growth of stitched plain weave carbon/epoxy laminates. They found that the round stitch reduced the fatigue limit of carbon/epoxy plain weave laminates under tension-tension fatigue loading. Besides, the parallel stitch had no effect on the S-N curve of carbon/epoxy laminates. De Baere et al. [4] have a research on the interlaminar behavior of a carbon fabric reinforced thermoplastic lap shear specimen under quasi-static and tension-tension fatigue loading. They showed that no crack growth was visible during quasi-static experiments, except near the lifetime end, whereas significant crack growth was present during fatigue experiments. Besides, they considered two loading frequencies and observed no significant influence on the behavior and the fatigue lifetime.

Brunbauer et al. [5] checked the dependency of the fiber volume content on mechanical properties, the fatigue damage and the microstructure of carbon/epoxy laminates. They depicted that the damage behavior of off-axis specimens under tension-tension fatigue testing changed with increasing the fiber volume content and the applied cyclic load. While matrix cracking and interfacial debonding were dominating damage mechanisms in specimens with low fiber content, fiber bridging and pull-out were monitored with increasing fiber content. Dai et al. [6] characterized the open-hole quasi-static and fatigue behavior of 3D woven carbon/epoxy composites. The

* Corresponding author: m_azadi@semnan.ac.ir

surface crack initiation and progression during tension-tension fatigue loading was identified using thermo-elastic stress analysis, which revealed that the orthogonal weave had larger surface damage area than the angle-interlock weave.

Ma et al. [7] presented the off-axis tensile fatigue assessment based on residual strength for the carbon fiber-reinforced composite, at room temperature. They conducted tension-tension cyclic fatigue tests under load controlling at a sinusoidal frequency of 10 Hz to obtain the stress-fracture cycles (S-N) relationship. They showed that the more flexible matrix and the weaken fiber-matrix interface after the fatigue phenomenon were responsible for the modulus decrease and the strain increase. Meanwhile, no acute interfacial/matrix damage in resins during fatigue was responsible for the sustaining strength. Feng et al. [8] studied on tension-tension fatigue performances and the reliability in the fatigue lifetime of T700/MTM46 laminated composites. They indicated that failure modes of fatigue specimens were quite different with that of static specimens. The delamination phenomenon tended to be the main failure mode for fatigue specimens, while the brittle facture was dominated for static specimens.

Fujimoto et al. [9] investigated the low-cycle fatigue behaviour of carbon fiber reinforced plastic (CFRP) laminated composites under repeated out-of-plane loading. They found that the fatigue fracture occurred in a brittle manner with almost no prior deformation. Based on the fracture surface observation, the fracture at the carbon fiber/epoxy layer was predominant. Horrmann et al. [10] studied the effect of the fiber waviness on the fatigue lifetime of carbon fiber-reinforced polymer (CFRP) materials, under axial loadings. They showed that there had a severe influence on the static strength and on the fatigue lifetime of specimens.

Korkiakoski et al. [11] checked the influence of the specimen type and the reinforcement on the tension-tension fatigue lifetime of unidirectional glass fiber-reinforced plastic (GFRP) laminates. Their results depicted that the test specimen type had a significant effect on the failure mode and the fatigue lifetime. Besides, a significantly higher fatigue lifetime was measured for the composite with the powder-bound fabric reinforcement, when compared to the composite with the stitched reinforcement. Cole et al. [12] found the interphase mechanical behavior of carbon fiber-reinforced polymer under cyclic loading, by presenting a micro-/nano-scale technique to characterize the damage formation. They utilized a finite element model to determine a material compliance effect in fibers, which provided a basis for linking local interphase mechanical behavior to the global fatigue behavior of composites. Shen et al. [13] had an experimental investigation on the damage evaluation in cross-ply glass fiber-reinforced plastic (GFRP) laminates, subjected to quasi-static and fatigue loadings. They illustrated that fatigue loading with lower stress level accelerated the damage process and generated damages more thoroughly at failures.

Although some researches have been presented based on the frequency effect on the fatigue lifetime of composites, but still requires to investigate. Therefore, as a novel research, effects of the displacement amplitude and the loading frequency have been represented on the low-cycle fatigue lifetime in carbon/epoxy laminated polymer matrix composites. For such objective, the sensitivity analysis on experimental data was carried out by the MINITAB software. Then, obtained results have been reported in figures and tables.

2 Materials and Experiments

2.1 Laminated Composite

Standard Specimens were made of unidirectional carbon fibers (Sika-Wrap-230C) and the CR-80 epoxy (including the CH-80-2 hardener, with the weight percent of 30%). The thickness of each unidirectional fabric was 0.13 mm and 14 (or 7 symmetric) layers have been utilized for each composite plate (250 mm × 350 mm). The total thickness of each laminate was 3.2 mm, which was manufactured by the hand lay-up method. Composite layers were considered as $[0_3/90_2/0_2]_s$, including 14 layers, as mentioned.

Then, based on the ASTM-D5766 standard [14], open-hole specimens have been cut by water-jetting from the initial composite plate. The dimension of each standard specimen was 36 mm × 250 mm × 3.2 mm, with the hole diameter of 6 mm. The standard specimen geometry can be seen in Fig. 1.

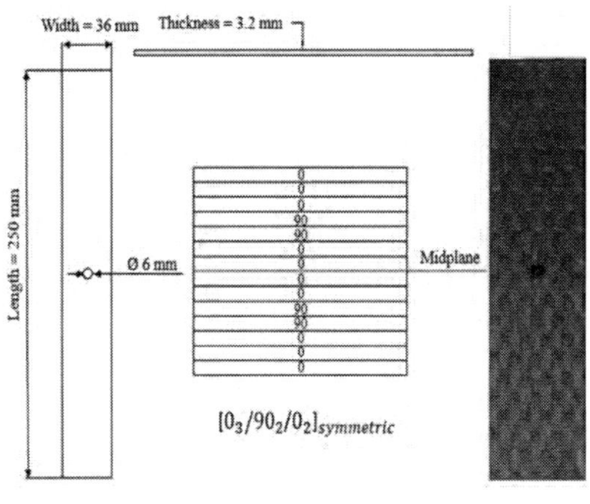

Fig. 1. The geometry of the standard specimen.

2.2 Fatigue Testing

For fatigue testing, a universal tension-compression device, STM-150 Santam Company, was utilized with the loading capacity of 15 tons. Fig. 2 shows the equipment of fatigue testing. Cyclic loadings were applied on standard specimens under a tension-tension condition. Fatigue tests were displacement-controlled and the displacement ratio ($R_d = d_{min}/d_{max}$) was zero, means that the minimum displacement was zero. Two parameters were considered to change for low-cycle fatigue test conditions, including the displacement amplitude and the loading frequency. When the loading frequency was 200 mm/min, the displacement amplitude was 5.5, 6.0, 6.5 and 7.0 mm. When the displacement amplitude was 7.0 mm, the loading frequency was 25, 50, 100 and 200 mm/min.

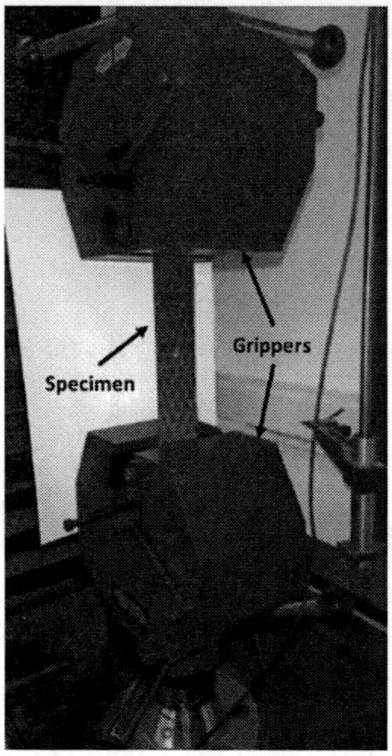

Fig. 2. The equipment of fatigue testing.

The scanning electron microscopy, XL30 Philips Company, was also used to investigate the fracture surface of specimens. Before, each specimen was coated by silver under a vacuum environment, since the composite was not conductive.

2.3 Regression Analysis

Generally in the statistical analysis, the F-value and the P-value can be used for finding the effectiveness of parameters. The F-value is the ratio of residual variances in a model, without the predictor versus with the predictor. Values closed to unity show that the predictor does not affect the residual variance. Large values indicates a reduction in the residual variance that can be attributed to the predictor. However, the P-value shows that the predictor is uncorrelated to the response [15].

Before using a model (or a formulation) for the objective function based on factors, the sensitivity analysis should be carried out based on statistical data. For this objective, the effectiveness of each parameter (or factor or predictor) could be calculated by comparing F_0 to $F_{\alpha,a-1,N-a}$. It should be noted that $a-1$ and $N-a$ are the degree of freedom of the treatment and the error, respectively [15]. Besides, α is the risk value, which was considered as 0.05, based on 95% of the confidence level.

The value of F_0 can be calculated based on mean squared values (MS) of treatments (including factor levels) and the error [15-16].

$$F_0 = MS_{treatment} / MS_{error} \quad (1)$$

Then also, $F_{\alpha,a-1,N-a}$ can be found from the analysis of variance (ANOVA tables) [15]. The condition of the effectiveness of factors can be written as follows,

$$F_0 > F_{\alpha,a-1,N-a} \quad (2)$$

As mentioned, the F-value is based on the ratio of mean squares. The F-value is used to test the overall significance for a regression model, to compare fits of different models, to test specific regression terms, and to test the equality of means [15].

Another parameter, which is the P-Value, could be considered for finding the effectiveness of factors. The P-value can be utilized to determine the statistical significance, in a hypothesis test. One factor would be effective, when the P-Value is lower than 0.05 [15-16]. All these calculations was performed in the MINITAB software. The fitted model for the objective function, N_f (the low-cycle fatigue lifetime) and S_{max} (the maximum stress) were considered as follows,

$$N_f = C_0 + C_1 \times (d \text{ or } f) \quad (3)$$

$$S_{max} = D_0 + D_1 \times (d \text{ or } f) \quad (4)$$

Where d and f were the displacement amplitude and the loading frequency, respectively. Besides, C_0, C_1, D_0 and D_1 were statistical constants, which could be calculated from the regression analysis. It should be noted that the regression analysis is to find a mathematical relationship between variables. Ideally in the regression analysis, one tries to keep changes in factors, independent of each other [15].

To check the accuracy of predicted models, besides relative errors between calculated and experimental data, the coefficient of determination (R^2) could be used. It should be noted that the adjusted R^2 value is to determine how well the model fits data, when the number of predictors in the model is adjusted. The adjusted R^2 value incorporates the number of predictors in the model to help to choose the correct model. In addition, the predicted R^2 value is to determine how well the model predicts responses for new observations. Larger values of the predicted R^2 value indicate models in a greater predictive ability. Since the predicted R^2 value is calculated by systematically removing each observation from the data set, estimating the regression equation, and determining how well the model predicts the removed observation [15].

3 Results and Discussions

Obtained results including the stress history versus the low-cycle fatigue lifetime can be seen in Figs. 3 to 6. In these figures, the stress history included the maximum stress, the minimum stress, the stress amplitude and the mean stress, all versus the low-cycle fatigue lifetime. As a result, by increasing the loading frequency, the stress increased and the fatigue lifetime decreased, under a constant displacement amplitude. In addition, when the displacement amplitude decreased, the stress also decreased and therefore, the fatigue lifetime increased, under a constant loading frequency, as expected.

Effects of the displacement amplitude and the loading frequency on the low-cycle fatigue lifetime and the maximum stress of carbon/epoxy composites can be seen in Figs. 7 and 8.

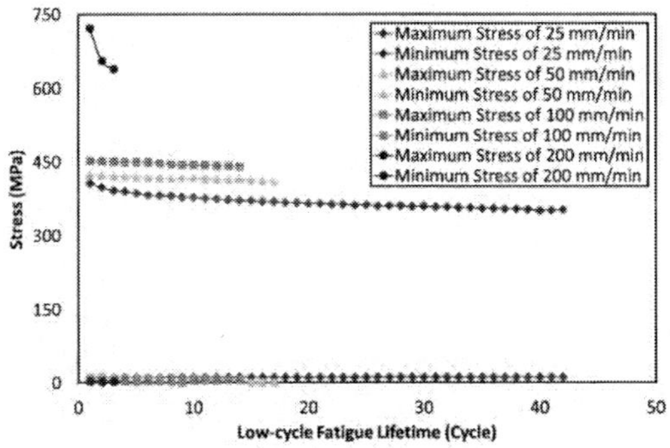

Fig. 3. The stress history (the maximum stress and the minimum stress) versus the low-cycle fatigue lifetime, under the displacement amplitude of 7 mm.

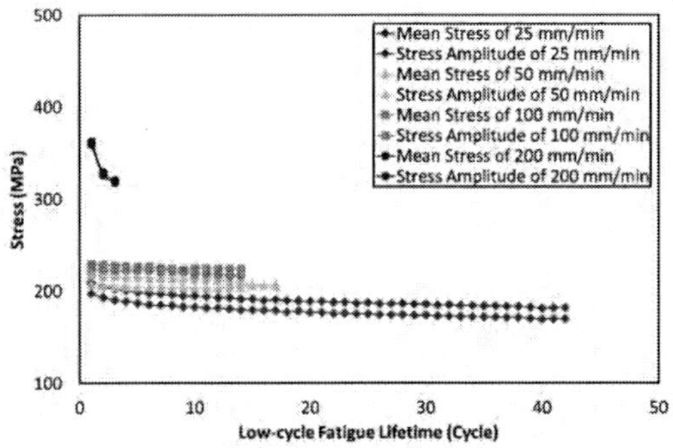

Fig. 4. The stress history (the mean stress and the stress amplitude) versus the low-cycle fatigue lifetime, under the displacement amplitude of 7 mm.

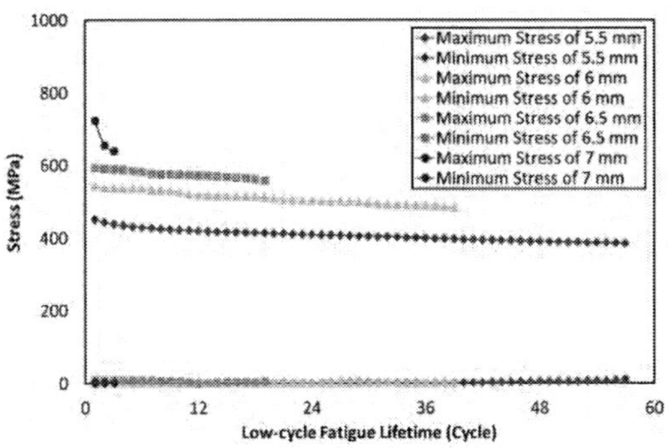

Fig. 5. The stress history (the maximum stress and the minimum stress) versus the low-cycle fatigue lifetime, under the loading frequency of 200 mm/min.

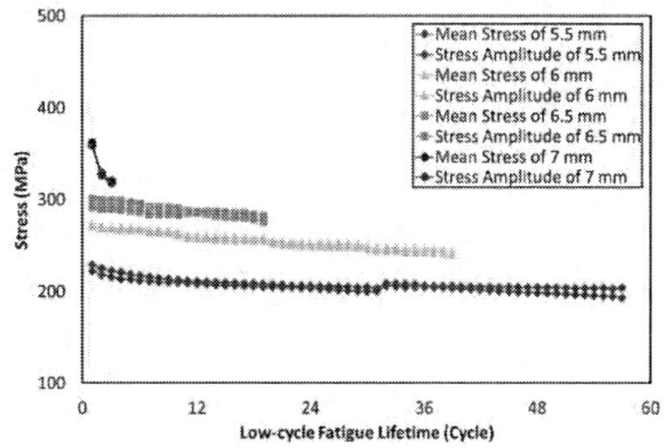

Fig. 6. The stress history (the mean stress and the stress amplitude) versus the low-cycle fatigue lifetime, under the loading frequency of 200 mm/min.

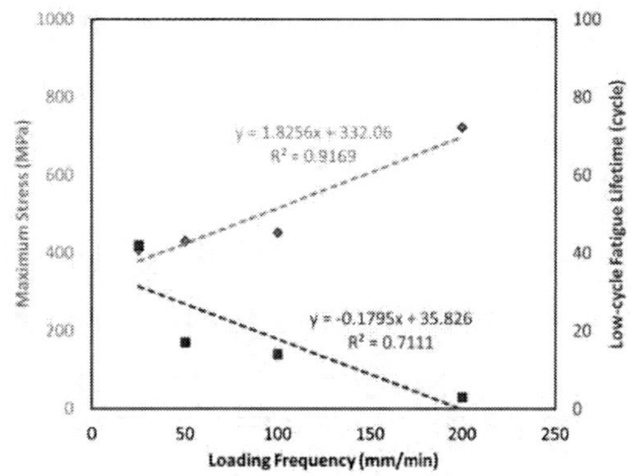

Fig. 7. The effect of the loading frequency on the maximum stress and the low-cycle fatigue lifetime, under the displacement amplitude of 7 mm.

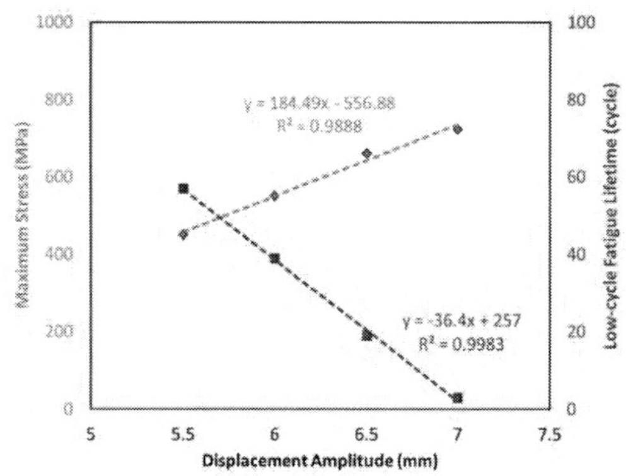

Fig. 8. The effect of the displacement amplitude on the maximum stress and the low-cycle fatigue lifetime, under the loading frequency of 200 mm/min.

According to Figs. 7 and 8, R^2 values for linear curve fitting were more than 92%, unless one condition, including the fatigue lifetime versus the loading frequency, which was about 71%, as a proper value.

Based on obtained results from the regression analysis, Tables 1 to 4 present F-value, P-value and coefficients of determination (R^2), for effects of the displacement amplitude and the loading frequency on the fatigue lifetime and the maximum stress. As mentioned in the previous part, the P-value should be less than 0.05 [15-16] to consider a factor as an effective or sensitive parameter. Based on the objective, the fatigue lifetime and the maximum stress, the displacement amplitude was an effective parameter. Since the P-value was obtained as 0.001 and 0.006, respectively. However, based on the fatigue lifetime as an objective, the P-value was not proper and more than 0.05. This means that the loading frequency was not effective. When the objective was the maximum stress, the loading frequency was effective, since the P-value was 0.042, less than 0.05. It should be noted that R^2 values in Tables 1 to 4 were the same as R^2 values in Figs. 7 and 8. As the last result, adjusted and predicted R^2 values were proper for the regression analysis of the relation between the fatigue lifetime or the maximum stress and the displacement amplitude. However, these results were not proper for the regression analysis of the relation between the fatigue lifetime or the maximum stress and the loading frequency.

Table 1. The regression analysis for the displacement amplitude effect on the fatigue lifetime.

	DF	Adj. *SS*	Adj. *MS*	F-Value	P-Value
Regression	1	1656.2	1656.2	1183.0	0.001
Displacement	1	1656.2	1656.2	1183.0	0.001
Error	2	2.8	1.4	-	-
Total	3	1659.0	S=1.18	R^2 = 99.83%	
$R^2_{adjusted}$ = 99.75%			$R^2_{predicted}$ = 99.28%		

*DF: Degree of freedom, Adj.: Adjusted.

Table 2. The regression analysis for the displacement amplitude effect on the maximum stress.

	DF	Adj. *SS*	Adj. *MS*	F-Value	P-Value
Regression	1	42544.4	42544.4	176.7	0.006
Displacement	1	42544.4	42544.4	176.7	0.006
Error	2	481.7	240.8	-	-
Total	3	43026.1	S=15.52	R^2 = 98.88%	
$R^2_{adjusted}$ = 98.32%			$R^2_{predicted}$ = 94.00%		

*DF: Degree of freedom, Adj.: Adjusted.

Table 3. The regression analysis for the loading frequency effect on the fatigue lifetime.

	DF	Adj. *SS*	Adj. *MS*	F-Value	P-Value
Regression	1	578.8	578.8	4.92	0.157
Frequency	1	578.8	578.8	4.92	0.157
Error	2	235.2	117.6	-	-
Total	3	814.0	S=10.85	R^2 = 71.11%	
$R^2_{adjusted}$ = 56.66%			$R^2_{predicted}$ = 0.00%		

*DF: Degree of freedom, Adj.: Adjusted.

Table 4. The regression analysis for the loading frequency effect on the maximum stress.

	DF	Adj. *SS*	Adj. *MS*	F-Value	P-Value
Regression	1	59884	59884	22.08	0.042
Frequency	1	59884	59884	22.08	0.042
Error	2	5425	2712	-	-
Total	3	65309	S=52.08	R^2 = 91.69%	
$R^2_{adjusted}$ = 87.54%			$R^2_{predicted}$ = 16.19%		

*DF: Degree of freedom, Adj.: Adjusted.

The formulation for the fatigue lifetime (N_f) versus the displacement amplitude (d) can be written as follows,

$$N_f\,(\text{cycle}) = 257.00 - 36.4\ d\ (\text{mm}) \quad (5)$$

The formulation for the maximum stress (S_{max}) versus the displacement amplitude (d) can be written as follows,

$$S_{max}\,(\text{MPa}) = -556.9 + 184.5\ d\ (\text{mm}) \quad (6)$$

The formulation for the fatigue lifetime (N_f) versus the loading frequency (f) can be written as follows,

$$N_f\,(\text{cycle}) = 35.83 - 0.1785\,f\,(\text{mm/min}) \quad (7)$$

The formulation for the maximum stress (S_{max}) versus the loading frequency (f) can be written as follows,

$$S_{max}\,(\text{MPa}) = 332.1 + 1.826\,f\,(\text{mm/min}) \quad (8)$$

As a discussion, Staehler et al. [17] observed the dependency of the loading frequency on the high-cycle fatigue lifetime between 40 and 375 Hz, in carbon fiber-reinforced silicon carbide composites. They showed that lower loading frequency (4 to 40 Hz) had no effects on the fatigue lifetime. It means that there was no difference in cycles to failure, between 4 or 40 Hz.

In general, they found mechanisms, which were responsible for the reduction in the fatigue lifetime of composites, under highest loading frequency, including the macroscopic behavior (such as the stress-strain response during cycling), microscopic characteristics (failure and damage mechanisms), and also temperature effects. They indicated that one reason can be the specimen surface temperature, which increased due to the internal heat generation from the sliding friction between constituents of the composite, under cyclic loading. This enhancement was directly related to the loading frequency and/or the applied cyclic stress level. However, they showed that there was no clear indication that the loading frequency greatly impacted either the stress-strain response or the overall appearance of fracture surfaces. However, a closer examination of specimens, cycled at the highest frequency (375 Hz), illustrated an evidence of the localized oxidation at fiber surfaces that might have attributed to the reduction in the fatigue lifetime at this frequency [17].

Mall et al. [18] repeated such study, but for the elevated temperature of 550°C on the same material, carbon fiber-reinforced silicon carbide composites, under the tension-tension high-cycle fatigue loading at three frequencies of 0.1, 10 and 375 Hz.

They indicated that there was an increase in cycles to failure, as the loading frequency changed from 0.1 to 375 Hz. This trend was different at room temperature. Their analysis of damage mechanisms illustrated that the oxidation of carbon fibers was the major difference between room and elevated temperatures. Such phenomenon caused to a reduction in cycles to failure, under lower loading frequencies at the elevated temperature, in comparison to that at room temperature. In other words, by increasing the loading frequency, because of the internal friction between fibers and the matrix, caused to an increase in the internal temperature of the composite. Matrix cracks from the mismatch between coefficients of the thermal expansion in carbon fibers and the silicon carbide matrix initiated at higher temperatures, due to higher loading frequencies [18].

4 Conclusions

In this article, the sensitivity analysis for effects of the displacement amplitude and the loading frequency on the low-cycle fatigue lifetime in carbon/epoxy laminated composites was conducted. Experimental obtained results can be mentioned as follows,

- By increasing the displacement amplitude, the low-cycle fatigue lifetime decreased, as expected.
- When the loading frequency enhanced, the low-cycle fatigue lifetime of composites decreased.
- The maximum stress had a reverse behavior, compared to the fatigue lifetime.
- The displacement amplitude was effective or sensitive on both the fatigue lifetime and the maximum stress.
- The loading frequency was effective or sensitive only on the maximum stress and was not effective or sensitive on the fatigue lifetime.

As an acknowledgement, authors should thank Irankhodro Powertrain Company (IPCo.), in Tehran, Iran, for the financial support on this project.

References

1. I. De Baere, W. Van Paepegem, C. Hochard, J. Degrieck, Polym. Testing **30**, 663 (2011)
2. W.R. Broughton, M.R.L. Gower, M.J. Lodeiro, G.D. Pilkington, R.M. Shaw, Compos. Part A **42**, 1310 (2011)
3. A. Yudhanto, Y. Iwahori, N. Watanabe, H. Hoshi, Int. J. Fatigue **43**, 12 (2012)
4. I. De Baere, W. Van Peapegem, I. Degrieck, Polym. Testing **32**, 1273 (2013)
5. J. Brunbauer, H. Stadler, G. Pinter, Int. J. Fatigue **70**, 85 (2015)
6. S. Dai, P.R. Cunningham, S. Marshall, C. Silva, Compos. Struct. **131**, 765 (2015)
7. Y. Ma, Y. Zhang, T. Sugahara, S. Jin, Y. Yang, H. Hamada, Compos. Part A **90**, 711 (2016)
8. Y. Feng, C. Gao, Y. He, T. An, C. Fan, H. Zhang, Compos. Struct. **136**, 64 (2016)
9. K. Fujimoto, M. Hojo, A. Fujita, Proc. Struct. Integ. **2**, 182 (2016)
10. S. Horrmann, A. Adumitroaie, C. Viechtbauer, M. Schagerl, Int. J. Fatigue **90**, 139 (2016)
11. S. Korkiakoski, P. Brondsted, E. Sarlin, O. Saarela, Int. J. Fatigue **85**, 114 (2016)
12. D.P. Cole, T.C. Henry, F. Gardea, R.A. Haynes, Compos. Sci. Technol. **151**, 202 (2017)
13. H. Shen, W. Yao, W. Qi, J. Zong, Compos. Part B **120**, 10 (2017)
14. ASTM D5766, *Standard test method for open-hole tensile strength of polymer matrix composite laminates* (ASTM, 2011)
15. D.C. Montgoery, *Design and analysis of experiments* (John Wiley and Sons, 2012)
16. M. Azadi, M. Iziy, A. Marbout, M. Azadi, A. Hajiali Mohammadi, J. Engine Res. **43**, 63 (2016)
17. J.M. Staehler, S. Mall, L.P. Zawada, Compos. Sci. Technol. **63**, 2121 (2003)
18. S. Mall, J.M. Engesser, Compos. Sci. Technol. **66**, 863 (2006)

 https://doi.org/10.1051/matecconf/201816522022

Rapid determination of the high cycle fatigue properties of high temperature aeronautical alloys by self-heating measurements

Vincent Roué[1,2,*], *Cédric* Doudard[1], *Sylvain* Calloch[1], *Frédéric* Montel[1], *Quentin* Pujol D'Andrebo[2] and *Fabien* Corpace[2]

[1]Institut de Recherche Dupuy de Lôme (IRDL) UMR 6027, ENSTA Bretagne, 29200 Brest, France
[2]Safran Aircraft Engines, Villaroche, 77550 Moissy-Cramayel, France

Abstract. The determination of high cycle fatigue (HCF) properties of a material with standard method requires a lot of specimens, and could be really time consuming. The self-heating method has been developed in order to predict S–N–P curves (*i.e.*, amplitude stress – number of cycles to failure – probability of failure) with only a few specimens. So the time-saving advantage of this method has been demonstrated on several materials, at room temperature. In order to reduce the cost and time of fatigue characterization at high temperature, the self-heating method is adapted to characterize HCF properties of a titanium alloy, the Ti-6Al-4V (TA6V), at different temperatures. So the self-heating procedure is adjusted to conduct tests with a furnace. Two dissipative phenomena can be observed on self-heating curves. Because of this, a two-scale probabilistic model with two dissipative mechanisms is used to describe them. The first one is observed for low amplitudes of cyclic loading, under the fatigue limit, and the second one for higher amplitudes where the mechanisms of fatigue damage are activated and are dissipating more energy. This model was developed on steel at room temperature. Even so, it is used to describe the self-heating curves of the TA6V at several temperatures.

1 Introduction

Performance of turbojet can be improved by increasing the operation temperature or by reducing the mass of the components. Consequently, jet engine parts are subjected to increasing stress load in environments in which the temperature increases. In the context of aircraft parts, HCF (high cycle fatigue) design method must be as accurate as possible and so a lot of fatigue tests are needed to study the influence of some parameters (temperature, mean stress,..) on the HCF properties of materials.

The material tested in this study is a titanium alloy the TA6V (Ti-6Al-4), widely used in aircraft industry. This material is often used for the construction of jet engine compressors. The pressure and also the temperature rise in the direction of the flow passing through the compressor, so multiple conditions of loads and temperatures must be studied to correctly design in fatigue the different parts of the compressor. Indeed, stator vanes and discs are subjected to vibratory loadings, so to a load ratio of around R=-1. Rotor blades are subjected to a high mean stress due to centrifugal forces in addition to the vibratory loadings, meaning the load ratio increases to around R=0.5 or R=0.7 depending on the case. Thus, the fatigue properties at several load ratios (the ratio between the minimal and the maximal cyclic stress applied on the specimen) and at different temperatures are needed for the TA6V.

However the determination method of the fatigue limit, arbitrarily set at the stress amplitude for a lifetime of 10^7cycles, with the Staircase method requires a lot of specimens and time. Even more specimens are needed to correctly evaluate the dispersion of the fatigue properties. So this process is quite long and expensive, just to determine the HCF properties (limit and dispersion) for a given load ratio and temperature. So the purpose of this study is to develop a fast prediction method of HCF properties at high temperature (up to 450°C for the TA6V) based on self-heating measurements under cyclic loadings.

This method is based on the monitoring of the stabilized temperature of the studied specimen during cyclic loadings. Several consecutive blocks of a few thousand cycles with increasing stress amplitude are applied on the specimen. For each block (i.e. for each stress amplitude), the stabilized temperature of the specimen is determined. An empirical interpretation of this evolution allows the obtaining of an estimation of the endurance limit [1,2]. Indeed, the fatigue mechanisms are also dissipative mechanisms. So when the stress amplitude applied on the specimen is higher than the endurance limit, theses mechanisms are dissipating energy and the temperature of the specimen rises, which allows us to estimate the endurance limit of the material.

The self-heating method, or a similar method, has been used on different materials : a wide range of steel [3,4], composite materials [5] and elastomeric materials

* Corresponding author: vincent.roue@safrangroup.com

[6]. It has also been used for different load ratios [7] and extended to multiaxial loadings [8,9]. In all of these cases, the link between self-heating measurements and fatigue properties (at least the mean endurance limit) is established. Self-heating methods could also be used to quickly quantify the influence of various parameters (such as mean stress effect [7] or plastic prestrain [3]) on fatigue properties.

In this paper, we propose an extension of the use of self-heating measurement to the prediction of HCF properties of the TA6V at different temperatures (up to 450°C). Moreover, different load ratios are studied to demonstrate the capability of the HCF properties predictions with the self-heating method. The purpose is to quickly obtain a Goodman-Haigh diagram for endurance limits at the different temperature used during the fatigue design of the components. This study seeks to reduce the time and cost of the characterization of fatigue properties characterization of materials, in order to fulfil industrial needs.

Moreover, this method allows the study of fatigue mechanisms without creep influence. Indeed, during classical fatigue tests at high temperature, specimen rupture could be due to creep damage because of low frequencies used during fatigue testing (around 100Hz), while actual parts are subjected to high frequencies (around 20kHz). So the parts are less subjected to creep damage than the specimens used to determine fatigue properties. However, during self-heating measurements, only a few thousand cycles are applied on the specimen, which is enough to reach the stabilization of the measured temperature. So, the endurance limit could be determined with this method without creep influence.

This paper is divided into three parts. The first part describes the development and performance of self-heating tests at high temperature. It presents the adjustments made to the procedure to conduct tests with a furnace. The self-heating results are also presented for several conditions of temperature and load ratio. The second part presents the two-scale probabilistic model with the two dissipative phenomena developed by Munier et al. [3], and used here as a first approach to analyze the self-heating curves. This section describes the principle of the model and presents the calculation of the dissipation. Thus, the heat conduction equation can be solved and the experimental self-heating measurements can be described. The link with the HCF properties is also presented. Finally, in the third section, the model is applied on the self-heating results to determine the HCF properties of the material. The empirical determination of the mean endurance limit (corresponding to a probability of failure of 50% at 10^7cycles) is presented. Finally, the results predicted by the model are compared to classical experimental fatigue results.

2 Self-heating test procedure at high temperature

The self-heating test procedure is well established to perform tests at room temperature on metal alloys [2,3]. The procedure is adapted here to conduct tests with a furnace and to be able to measure a low rise of the specimen temperature (the self-heating temperature) at high temperature (up to 450°C).

In this section, the specimen and the experimental device are presented. So adjustments to conduct self-heating measurement with the furnace are briefly explained. Then the procedure of the self-heating measurements under cyclic loadings is presented and applied to one example of temperature and load ratio (T=250°C and R=-1).

2.1 Experimental device

In order to be able to measure a low rise of the specimen temperature at high temperature, the procedure and the experimental device are adapted to improve and optimize the temperature signals. Some precautions are needed to reduce signals noise. Moreover the heating system and its regulation must be chosen to have the best environmental temperature, with small temperature variations and a good temperature stabilization. The regulation of the system must not be linked directly to the specimen temperature. Indeed, if it is the case, the self-heating of the specimen could disrupt the regulation of the environmental temperature.

The aim of this subsection is to propose an experimental protocol to study the effect of the temperature on the HCF properties. All the specimens are tested under the same conditions, with the same procedure and device.

2.1.1 Specimen and experimental set-up

The experimental set-up is composed of the hydraulic machine, the furnace split in three different zones allowing a better regulation of the temperature and high temperature reverse-stress pullrods with water-cooled collet grips (Figure 1).

The geometry of the specimen in the zone of interest is a cylindrical shape (Figure 2). The ends of the specimen are threaded and interface parts had been developed to enable cyclic loadings at high temperature. The specimen geometry had been developed by Safran and it is the geometry used during classical fatigue tests. All specimens were provided by Safran, and all underwent the same manufacturing process.

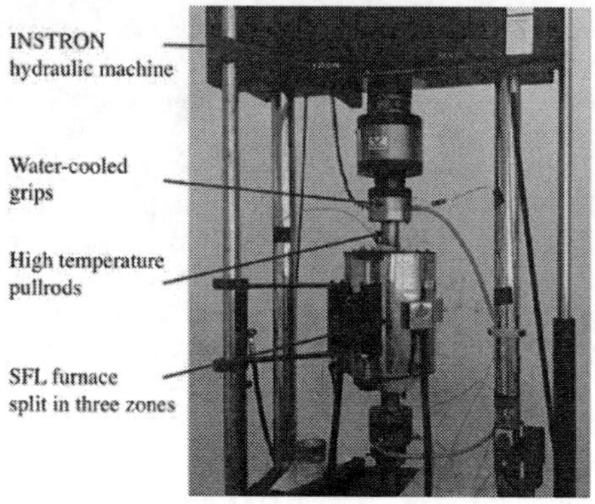

Fig. 1. Experimental device for the self-heating test at high temperature.

This heating technology and so this furnace were chosen because it enables the environment to maintain a stable temperature, with a certain inertia. The regulation of the furnace has a long characteristic time, which allows us to consider the environmental temperature during the self-heating procedure as constant. Indeed, the duration of one block of cyclic loading is shorter than the furnace characteristic time.

The pullrods must be cooled to protect the cell and the hydraulic system of the machine. So one of the main sources of thermal losses is through conduction in the pullrods. Other thermal losses, such as convective losses, are reduced as much as possible by controlling the thermal boundary conditions. So the thermal environment is as stable as possible to reduce the noise on thermal signals.

2.1.2 Temperature acquisition

A 0D approach is adopted in this paper (*i.e.* only the mean temperature elevation of the specimen is considered). Therefore, a thermocouple is sufficient to measure the mean temperature of the specimen in the region of interest. In the same way as the self-heating procedure at room temperature, three thermocouples are required, one is fixed on the center of the specimen and the other two are fixed on the high temperature grips (Figure 2). The thermocouples of the grips are close enough to the specimen to measure the variation of temperature in its immediate environment, without measuring the rise of temperature during its self-heating.

To reduce the thermal noise as much as possible and reduce the variations of the environmental temperature, we must control the thermal boundary conditions. The most important issue is to avoid stack effect, which greatly deteriorates the thermal signal, hence the precautions taken to block the air flow (rock wool, ceramic fiber cord, etc.).

The relative accuracy of the measuring system, from the type K thermocouples used and through the electronic conditioner, is lower than 0.01 K. Its response time is of the order of one second, so the acquisition frequency is set at 1 Hz. Thus, the temperature elevation of the specimen θ is given by

$$\theta = T_S - \frac{T_{UG} + T_{LG}}{2}, \tag{1}$$

with T_S the temperature in the region of interest of the specimen, T_{UG} the temperature of the uppergrip and T_{LG} the temperature of the lowergrip. Unlike the procedure at room temperature, an offset is often needed to correct the elevation of temperature of the specimen θ. Indeed, the temperatures of the grips are lower than the temperature in the center of the specimen as a result of the water-cooled system and the conduction losses through the pullrods (Figure 4). The θ temperature is rapidly stabilized and reaches a constant temperature. It allows us to predict the temperature drift and perform self-heating measurements even before the complete stabilization of the temperature in the furnace. The self-heating temperature is the temperature θ more or less the offset of the stabilized value of θ.

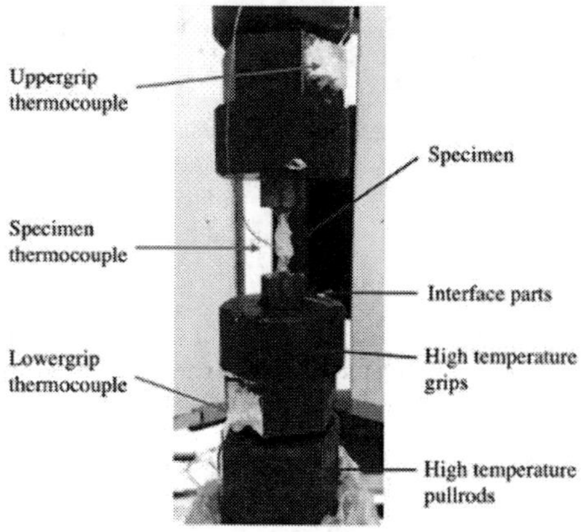

Fig. 2. Geometry of the specimen and thermocouples fixing.

Moreover, the specimen is quite small in comparison with the pullrods and the furnace, and so the self-heating of the specimen does not disrupt the environmental temperature. It allows us to measure the self-heating of the specimen ranging from small rises in the specimen temperature (around 0.04 K), with the optimization of the temperature signals due to the precautions taken for temperature acquisition and noise reduction of the environmental temperature, to quite important rises of the specimen temperature (around 10 K), without any interference in the furnace regulation.

So after stabilization of the θ temperature, the relative resolution is still lower than 0.01 K. The thermal noise of the measuring system is around 0.03 K. The stabilized temperature is still oscillating due to the regulation of the furnace, external perturbations and losses. The amplitude of the oscillations is around 0.1 K, but its frequency is sufficiently low to be predictable and perform a block of cycling loadings during a self-heating procedure.

2.2 Self-heating procedure

Given that the measuring system is set up and optimized to seek small elevation of temperature at high

https://doi.org/10.1051/matecconf/201816522022

temperature, self-heating measurements of the specimen can now be done at high temperature.

So in this subsection, the self-heating method under cyclic loadings is presented. The procedure is applied to one condition of temperature and load ratio (T=250°C and R=-1) as an example.

2.2.1 Loading conditions

A complete self-heating test consists of the application of a successive series of cyclic loadings, with increasing stress amplitude Σ_0 (Figure 3). All the blocks of cyclic loading are performed at the same frequency, 30 Hz, and with the same number of cycles, 6000 cycles. Moreover, they are performed at the same load ratio R (depending of the chosen conditions for each specimen), given by

$$R = \frac{\sigma_{min}}{\sigma_{max}}, \qquad (2)$$

with σ_{min} the minimum stress applied during the block of cycling loading and σ_{max} the maximum stress. For the example, the load ratio is fixed at R=-1, so $\sigma_{min} = -\Sigma_{0_i}$ and $\sigma_{max} = \Sigma_{0_i}$.

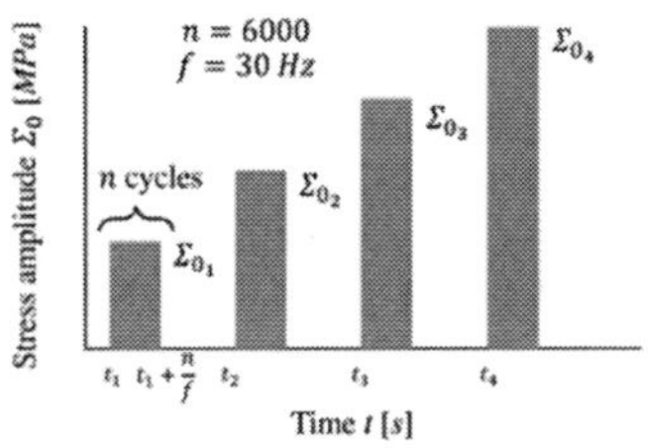

Fig. 3. Successive series of cyclic loadings with increasing stress amplitudes.

2.2.2 Self-heating temperature

During each cyclic loading block, the temperature elevation θ is registered. Few examples of thermal responses are presented on Figure 4. The stress $\Sigma_{\max}$ is used to normalize all the stresses represented in this paper. The value was set arbitrarily.

Figure 4 shows the evolution of temperature elevation for few loading steps. During each of them, a "steady-state" temperature $\bar{\theta}$ is determined. In fact, the temperature elevation does not reach a stabilized value for every loading step. More cycles should have been done to possibly reach a steady-state temperature. Actually, some longer cyclic loadings had been done, and the elevation of temperature does not seem to stabilize. Anyway, the "steady-state" temperature $\bar{\theta}$ is evaluated during the same period of time at the end of each loading step, given that the thermal conditions are the same. Indeed the characteristic time to reach thermal balance (evaluated during the cooling of the specimen) is the same for every loading step. Few errors could be made with this analysis, but we can still study the evolution of this steady-state temperature $\bar{\theta}$ as a function of the stress amplitude Σ_0 applied during the cyclic loading. Indeed, any errors would be constant in each loading step, and we are only interested by the evolution of this steady-state temperature $\bar{\theta}$. Thus, Figure 4 shows that the steady-state temperature $\bar{\theta}$ increases with the stress amplitude Σ_0.

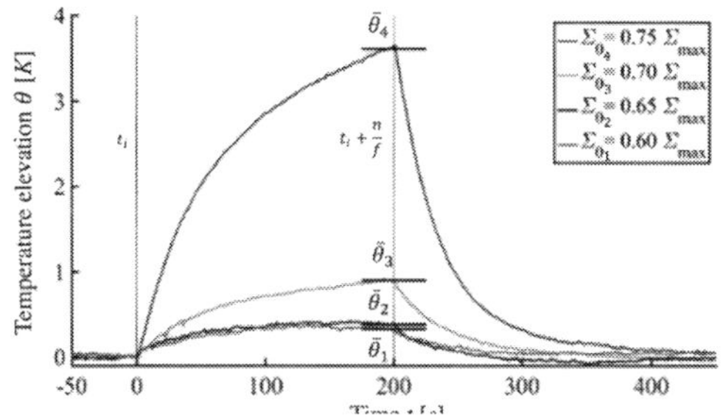

Fig. 4. Evolution of the mean temperature elevation during several blocks of cyclic loadings.

2.2.3 Self-heating curve

At the end of the self-heating test, a self-heating curve is obtained. It represents the steady-state temperature $\bar{\theta}$ relating to the stress amplitude Σ_0 (Figure 5). The self-heating curve is also plotted in a logarithmic scale to clarify the presence of the two self-heating regimes : the first for low amplitudes is called the primary regime and the second one for highest amplitudes [3,4].

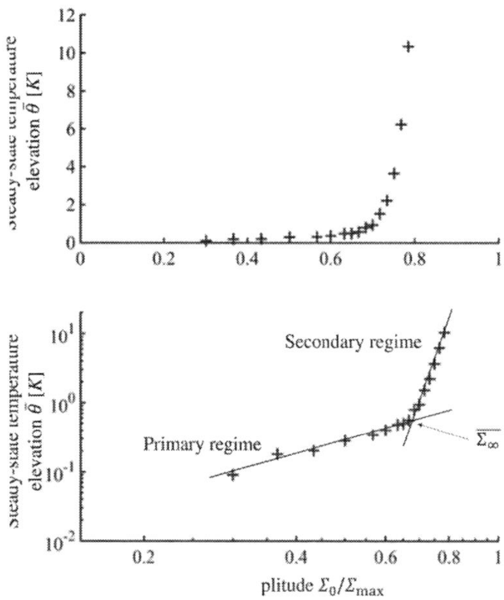

Fig. 5. Self-heating curve for the condition T=250°C and R=-1.

The primary regime has a slope of two in the logarithmic diagram, in accordance with the observations of Munier on steels at room temperature [3,4]. The secondary regime has a higher slope in the logarithmic diagram. This regime is associated with the progressive appearance of microplasticity [2]. Indeed, this regime is associated with the appearance of fatigue damage. In the model, it is used to determined fatigue properties of the material.

The empirical determination of the mean endurance limit proposed by Munier concerned an endurance limit set at the order of 10^6 cycles. For aeronautical applications, the endurance limit is defined for a number of cycles of 10^7. Thus, the empirical determination of the mean endurance limit exposed in this paper is slightly different. This limit $\overline{\Sigma_\infty}$ is determined as the amplitude for which the secondary regime appears (Figure 5).

2.3 Self-heating results at several temperatures

In this subsection, the self-heating curves obtained for several conditions of temperatures and loading ratios are presented. The curves are briefly described, and the main observations made during the tests are noted.

2.3.1 Load ratio effect on self-heating curves

Figure 6 shows the effect of the load ratio on self-heating curves. Three load ratios are presented, R=-1, R=0 and R=0.5 for the temperature T=250°C.

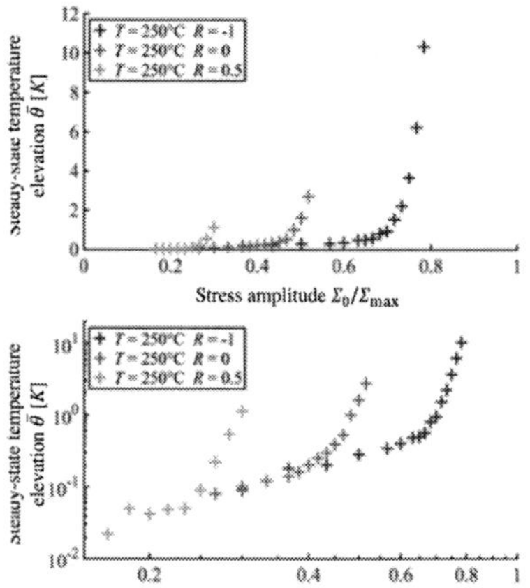

Fig. 6. Influence of the loading ratio R on the self-heating curves for the temperature T=250°C.

The self-heating curves obtained at different load ratios seem to have the same primary regime. Moreover, the secondary regime is activated at lower stress amplitude when the load ratio increases. This regime is correlates with the fatigue damage, meaning the fatigue properties of the TA6V decrease when the load ratio increases. Indeed, when the load ratio is R=0 or R=0.5, a mean stress is applied and the maximum stress is higher ; so the endurance limit is lower.

2.3.2 Temperature effect on self-heating curves

Figure 7 shows the evolution of self-heating curves with the temperature. Two temperatures (250°C and 450°C) were tested at a load ratio of R=0.

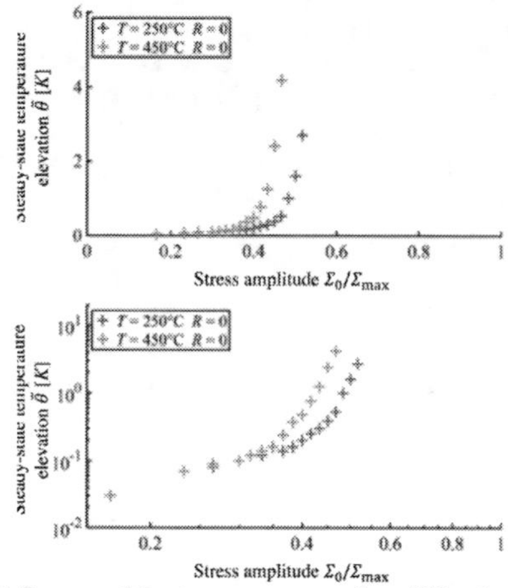

Fig. 7. Influence of the temperature on the self-heating curves for the loading ratio R=0.

The secondary regime is activated at lower stress amplitude when the temperature increases, so the fatigue properties decrease when the temperature is higher.

All the self-heating curves presented in this paper seem to have the same primary regime. It also has a slope value of two in the logarithmic diagram, in accordance with Munier's observations on steel at room temperature [4].

However, the beginning and the intensity of the secondary regime depend on the conditions of testing. The self-heating curves show that the secondary regime is activated at lower stress amplitude when the load ratio or the temperature increase. Indeed, the fatigue properties are affected with the presence of a higher mean stress or a higher environmental temperature.

3 The two-scale probabilistic model with dissipative mechanisms

The aim of this section is to present and summarize the two-scale probabilistic model developed by Munier et al. [3] on a wide range of steels at room temperature. Indeed, the presence of two regimes on the self-heating curves of the TA6V and other experimental observations (influence of the frequency, etc.) impel us to use this model as a first approach, although the dissipation mechanisms for the TA6V are not yet clearly determined.

Firstly, the principle of the model is displayed. Then the dissipated energy calculation is presented. It enables us to solve the heat conduction equation and describe the self-heating curves. Finally, the link with fatigue properties is presented.

3.1. Principle of the model and Representative Elementary Volume

The description of the two regimes requires the presence of two dissipative mechanisms. So the Representative Elementary Volume (REV) is defined as an elasto-plastic matrix containing a set of inclusions [3,4]. The set of inclusions represent microplasticity sites, with an additional plastic hardening. So under low amplitude cyclic loadings, the inclusions have the same behavior as the matrix, and so its dissipation is the same. Above a given stress amplitude threshold (different for each inclusion), the inclusion is activated and another dissipation appears. This stress amplitude threshold is defined as a probabilistic variable, and it represents the population of inclusions in the REV. So the dissipation associated with the matrix corresponds to the primary regime. The secondary regime is then associated to the dissipation of the inclusions. The REV is represented in Figure 8.

The sites activation represents the process of fatigue damage initiation. Thus, the analysis of the secondary regime allows us to determine fatigue properties of the material.

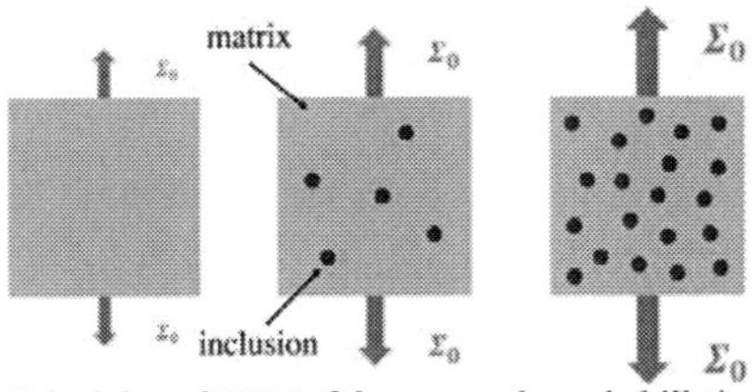

Fig. 8. Principle and REV of the two-scale probabilistic model.

3.2. Thermodynamic framework and dissipated energy calculation

In this subsection, the matrix behavior and the behavior of one inclusion are presented. Then the activation scenario of the sites, and so the behavior of the set of inclusions, is presented.

3.2.1 Dissipation in the matrix

The dissipation of the matrix is associated with the primary regime, which has a slope of two in a logarithmic diagram. So Munier et al. [3,4] proposed a framework for which the mean steady-state temperature $\bar{\theta}$ is a quadratic function of stress amplitude Σ_0. This result is obtained by defining a non-linear kinematic hardening for the matrix and by introducing a specific pseudo-potential.

So the dissipation can be calculated for one cycle of loading. Then the cyclic dissipative energy density (integration of the intrinsic dissipation over one cycle) can be approximated by a quadratic function of the stress amplitude

$$E_d^{mat} \approx K\,\Sigma_0^2, \tag{3}$$

with K depending on several model parameters.

3.2.2 Dissipation in one inclusion under cyclic loading

One inclusion is defined by its stress amplitude threshold σ_y^μ. If the stress applied within the inclusion (which is not the same as the macro-stress amplitude Σ_0) is lower than this threshold, the behavior of the inclusion is the same as the matrix. In this part, let us assume that the stress applied within the inclusion is greater than that of the threshold. An additional hardening is activated, and the dissipation of the inclusion is no longer the same as the matrix.

A localization law is needed to describe the relation between the stress applied on the inclusion and the macro-stresses applied on the REV. This relation is deduced from Eshelby analysis [10]. The activated hardening of the inclusion is defined by a linear kinematic hardening [2,3]. Then the cyclic dissipative energy density can be calculated. Indeed, during a tension-compression cyclic loading and if $\Sigma_0 > \sigma_y^\mu$, it can be written as

$$E_d^{inc}(\sigma_y^\mu) \approx \frac{4\,\sigma_y^\mu}{h^\mu}\langle \Sigma_0 - \sigma_y^\mu \rangle, \tag{4}$$

with h^μ depending on the hardening parameters and the localization law parameters.

3.2.3 Activation of the population of inclusions and dissipation in the REV

The activation threshold σ_y^μ is a probabilistic variable. It is assumed to be described by a Point Poisson Process, so no spatial correlations are considered. Thus the number of active sites in a domain Ω of volume V is given by

$$N(\Omega) = \lambda V, \tag{5}$$

with λ the mean density of active sites in the volume. It is also assumed that λ follows a power law of the stress amplitude, given by [2]

$$\lambda = \frac{1}{V_0}\left(\frac{\Sigma_0}{S_0}\right)^m, \tag{6}$$

with m and $V_0 S_0^m$ two material parameters.

Experimental observations of persistent slip bands on some steel have shown that a power-law dependence could represent this appearance of sites [3]. Moreover, when this choice is made in addition to the weakest link assumption, the probability of failure is the same as the Weibull model [2].

So the cyclic dissipative energy density can be now calculated in the REV, it is given by

$$E_d = K\,\Sigma_0^2 + \int_0^{\Sigma_0} E_d^{inc}(\Sigma)\frac{V_s}{V}\,dN_a(\Sigma)\,d\Sigma, \tag{7}$$

with V_s the volume of one site and dN_a the number of sites activated between a stress amplitude of Σ and $\Sigma + \mathrm{d}\Sigma$. This number is obtained with the mean density of active sites λ,

$$dN_a(\Sigma) = V\left(\lambda(\Sigma + \mathrm{d}\Sigma) - \lambda(\Sigma)\right) = \mathrm{V}\frac{d\lambda}{d\Sigma}d\Sigma. \tag{8}$$

Thus the cyclic dissipative energy density is finally given by

$$E_d = K\,\Sigma_0^2 + \frac{4}{h^\mu}\frac{V_s}{V_0 S_0^m}\frac{\Sigma_0^{m+2}}{(m+1)(m+2)}. \tag{9}$$

3.3. Description of self-heating curves

https://doi.org/10.1051/matecconf/201816522022

In order to describe self-heating curves, the heat conduction equation must be solved. With the 0D approach, this equation is given by [4]

$$\dot{\theta} + \frac{\theta}{\tau_{eq}} = \frac{f}{\rho c} E_d, \tag{10}$$

with τ_{eq} the characteristic time of thermal equilibrium (depending on the testing conditions, specimen geometry, etc.), f the loading frequency, ρ the material density and c the specific heat. Therefore, the mean steady-state temperature elevation $\bar{\theta}$ can be written as a function of the stress amplitude Σ_0 and is given by

$$\bar{\theta} = \alpha \left(\frac{\Sigma_0}{\Sigma_{\max}}\right)^2 + \beta \left(\frac{\Sigma_0}{\Sigma_{\max}}\right)^{m+2}, \tag{11}$$

with α and β two constants defined from the previous parameters and results. This equation allows us to describe a self-heating curve, in accordance with the experimental observations, with three parameters :

- the intensity of the primary regime α, related to the cyclic dissipation of the matrix;
- the intensity of the secondary regime β, associated with the cyclic dissipation of the population of inclusions;
- the slope of the secondary regime m, related to the gradual appearance of active sites (from the definition of λ).

3.4. Prediction of fatigue properties

According to the model, it is assumed that the activation of the second kinematic hardening within the inclusion governs the process of fatigue damage initiation. So the evaluation of fatigue properties is done with the secondary regime of the self-heating curve. Thus the probability of finding k active sites in a domain Ω of volume V is given by [11]

$$P_k(V) = \frac{(\lambda k)^k}{k!} e^{-\lambda V}. \tag{12}$$

Moreover, the weakest link theory is assumed in the model. So the probability of failure P_F is equal to the probability of finding at least one active site,

$$P_F = P_{k\geq 1} = 1 - P_{k=0}. \tag{13}$$

By using the definition of λ, the probability of failure P_F is given by

$$P_F = 1 - \exp\left(-\frac{V}{V_0}\left(\frac{\Sigma_0}{S_0}\right)^m\right). \tag{14}$$

This equation corresponds to the Weibull model [12], with m representing the Weibull modulus. It is well known that this model takes into account the effect of volume on HCF properties. In order to take the stress heterogeneities effect into account, the concept of effective volume is introduced [13]. From the definition of the probability of failure, the mean endurance limit and the corresponding standard deviation are given by

$$\overline{\Sigma_\infty} = S_0 \left(\frac{V_0}{V_{eff}}\right)^{\frac{1}{m}} \Gamma\left(1 + \frac{1}{m}\right) \tag{15}$$

and

$$\overline{\overline{\Sigma_\infty}} = S_0 \left(\frac{V_0}{V_{eff}}\right)^{\frac{1}{m}} \sqrt{\Gamma\left(1 + \frac{2}{m}\right) - \Gamma^2\left(1 + \frac{1}{m}\right)}, \tag{16}$$

where Γ is the Euler gamma function and V_{eff} is the effective volume of the specimen. These last two equations show that the scatter of fatigue properties is only dependent of the Weibull modulus m, which can be directly identified on a self-heating curve. Indeed, it represents the slope of the secondary regime [2,3]. The energetic criterion used to describe S-N-P curves is not discussed in this paper.

4 Self-heating results and validation

In this last section, the two-scale probabilistic model is applied on the self-heating curves. Then the predictions are compared to some classical fatigue tests.

4.1. Identification of the model parameters

The primary regime is the same for all self-heating curves, so the intensity of the primary regime α is identify on all the curves and has an identical value for all of them.

The self-heating curves obtained with an applied mean stress (*i.e.* R=0 and R=0.5) could be altered by other dissipative mechanisms for high stress amplitude. The maximum stress for the last loading steps is often really close to the macroscopic elasticity limit, so the slope of the secondary regime is only identified for the self-heating curve obtained at R=-1. This allows us to identify the slope of the secondary regime m, which is also fixed for all the other curves. Indeed, it represents the set of inclusions in the specimens, which are all the same (Figure 9).

The only parameter left is β, the intensity of the secondary regime which depends on the testing conditions (*T* and *R*). This parameter is identified for all the curves to fit with the last points of the self-heating curves.

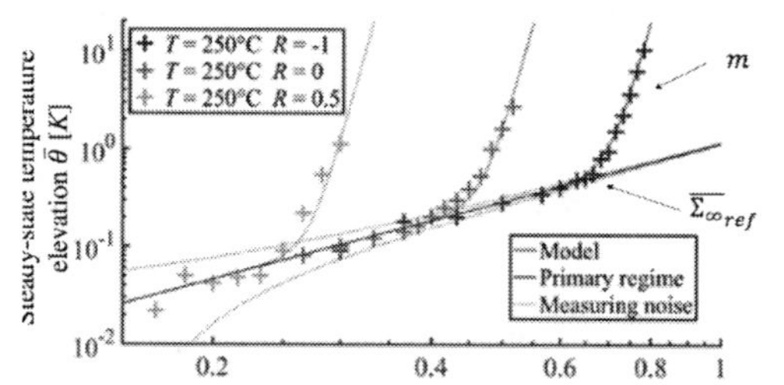

Fig. 9. Analysis of the self-heating curves and identification of the model parameters.

4.2. Empirical determination of the mean endurance limit

Indeed, it is possible to determine the mean fatigue limit with the parameter β with the equation

$$\overline{\Sigma_\infty} = \overline{\Sigma_\infty}_{ref} \left(\frac{\beta_{ref}}{\beta}\right)^{\frac{1}{m}}. \quad (17)$$

Thus, a couple of values $(\beta_{ref}, \overline{\Sigma_\infty}_{ref})$ are needed to determine the fatigue properties for all conditions of temperature and load ratio.

These two values are determined by the identification of the parameter β_{ref} on a self-heating curve and with an empirical determination of its mean endurance limit $\overline{\Sigma_\infty}_{ref}$. The chosen reference self-heating curve is the curve for the condition R=-1 (for the same reasons as previously) and T=250°C. Figure 9 shows the empirical determination of $\overline{\Sigma_\infty}_{ref}$ as the stress amplitude for which the secondary regime begun.

This method allows us to determine the mean endurance limit for each presented condition in this paper. It is possible to represent these results in a Haig diagram (Figure 10). Each point represents one specimen and was obtained in only one day.

4.3. Comparison with classical fatigue tests

The comparison with classical fatigue tests must still be done to validate the method. Figure 11 shows the comparison with some fatigue tests from Safran database. These fatigue tests were performed on specimens with identical geometry, but obtained from a different process.

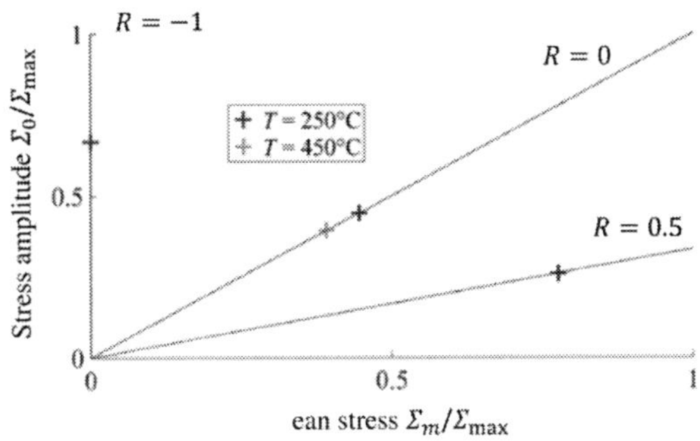

Fig. 10. Haig diagram obtained by self-heating measurements.

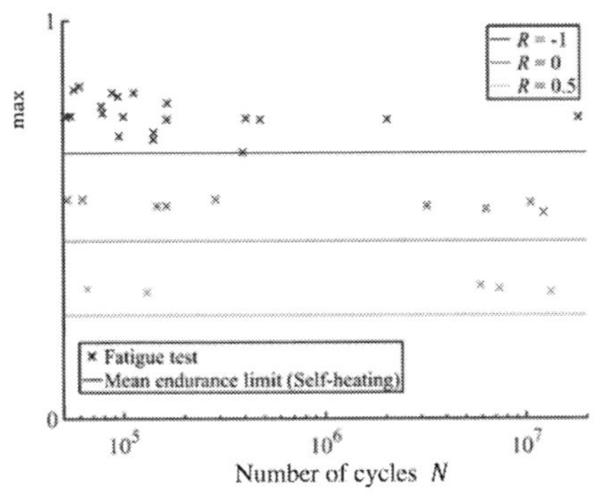

Fig. 11. Comparison of mean endurance limits obtained by the self-heating method with classical fatigue tests at 250°C.

Complementary fatigue tests must be performed to properly compare these self-heating results. However, there is a clear trend and the self-heating method at high temperature looks promising in reducing the time and cost of fatigue characterization at high temperature.

The scatter of the fatigue limit must also be analysed through classical fatigue tests to compare with the slope of the secondary regime and the parameter m.

5 Conclusion and perspectives

The present paper was devoted to the extension of the use of self-heating method under cyclic loadings for the prediction of HCF at different temperatures (up to 450°C). The purpose is to quickly obtain a Goodman-Haigh diagram for endurance limits at different temperatures. This study seeks to reduce the time and cost of the characterization of fatigue properties of materials, to fulfill industrial requirements.

The self-heating procedure and temperature acquisition were adapted to measure small increases in temperature corresponding to the self-heating of the specimen at high temperature. This paper has presented some examples of self-heating curves for different combinations of temperature and load ratio. These curves demonstrates the influence of the loading ratio and the temperature; the HCF properties identified with the self-heating curves can be seen to decrease when the temperature increases and in the presence of a mean stress. Indeed, we are able to measure the self-heating of specimens at high temperature. Dissipative mechanisms are well activated. For the TA6V, complementary studies and observations must be done to clearly identify these mechanisms. However, the presence of two regimes on the self-heating curves impels us to use a two-scale probabilistic model, developed on steel at room temperature, as a first approach.

A new empirical method is proposed to determine the mean endurance limit, set at the stress amplitude of a 10^7cycles lifetime. Complementary fatigue tests must be performed to validate the self-heating results, but thus far, the results have been promising. Self-heating measurements under other conditions of temperatures and loading ratios will be tested on the TA6V to quickly obtain a Haigh diagram.

Moreover, some other aspects of the two-scale probabilistic model must be studied. For example, the link between the fatigue dispersion and the slope of the secondary regime. The application of an energetic criterion could also be used to describe S-N-P curves.

The self-heating method at high temperature is a promising possibility for reducing the time and cost of the characterization of HCF properties. The use of this method could also be interesting at greater temperature, for the fatigue design of turbine parts.

References

1. C. E. Stromeyer. Proc Roy Soc London, A **90**, 411-425 (1914)

2. C. Doudard, S. Calloch, F. Hild, P. Cugy, A. Galtier. Fatigue Fract Eng Mat struct, **3**, 279-288 (2005)
3. R. Munier, C. Doudard, S. Calloch, B. Weber. Int J Fatigue, **63**, 46-61 (2014)
4. R. Munier. PhD Thesis (2012)
5. L. Jegou, Y. Marco, V. Le Saux, S. Calloch. Int J Fatigue, **47**, 259-267 (2013)
6. V. Le Saux, Y. Marco, S. Calloch, C. Doudard, P. Charrier. Int J Fatigue, **32**, 1582-1590 (2010)
7. G. La Rosa, A. Risitano. Int J Fatigue, **22**, 65-73 (2000)
8. C. Doudard, M. Poncelet, S. Calloch, C. Boué, F. Hild, A. Galtier. Int J Fatigue, **29**, 748-757 (2007)
9. M. Poncelet, C. Doudard, S. Calloch, B. Weber,A. Galtier. J Mech Phys Solids, **58**, 578-593 (2010)
10. E. Kröner. Acta Met, **9**, 155-161 (1984)
11. R. Gulino, S. L. Phoenix. J Mater Sci, **11**, 3107-3118 (1991)
12. W. Weibull. J Appl Mech, **18**, 293-297 (1951)
13. F. Hild, R. Billardon, D. Marquis. C R Acad Sci Paris, II **315**, 1293-1298 (1992)

Fatigue Tests – Important Part of Development of New Vehicles

Miloslav Kepka[1,*], *Miloslav* Kepka Jr.[1], *Pavel* Zlabek[1], *Petr* Heller[1], *Jan* Chvojan[1], *Vaclav* Mentl[1] et al.

[1]University of West Bohemia, Faculty of Mechanical Engineering, Univerzitni 8, 30614 Pilsen, Czech Republic

Abstract. In city of Pilsen (Czech Republic) modern transport engineering is developed. The Skoda Transportation (production company) has successfully been producing rail and road vehicles for many years (electric locomotives, trams, metro cars, trolleybuses, battery buses). This producer cooperates in developing these vehicles with the Research and Testing Institute (commercial research institute) and with the University of West Bohemia (public university). Fatigue tests are carried out by the Dynamic Testing Laboratory at the Research and Testing Institute and by the Regional Technological Institute, the research center of the Faculty of Mechanical Engineering at the university. The paper describes various fatigue tests and presents their practical realization in the mentioned laboratories.

1 Introduction

Fatigue tests can be classified with respect to various criteria. For instance, the following sorting is a common one:

- by the size of the structure tested (test specimen of a material, structural node, vehicle substructure or component, and the entire vehicle);
- by the controlling quantity (force, displacement, strain, acceleration, temperature);
- by the number of controlling quantities (a single controlling quantity, multiple controlling quantities);
- by the number of load directions (uniaxial loading vs. multi-axial loading);
- by the load-time profile (harmonic profile, a range of harmonic cycle blocks, a random process).

Fatigue tests may, however, be classified in other ways as well, depending on their role in research and development:

- tests intended to provide new fundamental scientific findings in the field of fatigue of materials and structures;
- tests intended to provide input data for expert reports and fatigue life calculations;
- tests mapping the impact of metallurgical production or subsequent processing upon final fatigue strength and life;
- tests conducted as part of a product development process (designers typically use these to validate design assumptions, results of computations, and computational models);
- prescribed fatigue tests (unless these are completed successfully, the product must not be introduced to the market or put into service);
- tests aimed to identify and remove causes of fatigue failures in service.

A special case of fatigue testing are virtual fatigue tests. Various computational programs for multi-body simulation, finite element method and dynamics are used to implement these virtual tests. A suitable fatigue postprocessor is necessary for this type of activity.

The most of the listed types of fatigue tests apply to the development of rail and road vehicles. The paper (and the poster) presents interesting examples of fatigue tests, which have been performed in collaboration of the Skoda Transportation (production company) with laboratories of the Research and Testing Institute and the Regional Technological Institute.

The authors of this paper have worked in the Skoda Transportation or the Research and Testing Institute in recent years, and today they continue to collaborate with these companies as academic or university researchers.

2 Rail vehicles

2.1 Activities of Research and Testing Institute

In city of Pilsen there is a long-standing tradition of testing the undercarriage frames of rolling stock. The Research and Testing Institute has a dominant position in this field and provides these services to domestic and foreign customers. Some examples are in Fig. 1.

A majority of bogie frames are made by welding or fabricated as a combination of cast parts and weldments. They are complex and intricate-shaped structures with a number of geometric notches and weakened locations. In general, boogie frames are heavy-duty parts of railway vehicles whose operation must be trouble-free.

Bogies operate under severe static, quasi-static and dynamic loads and their service failure would threaten many lives and lead to material losses.

* Corresponding author: kepkam@rti.zcu.cz

https://doi.org/10.1051/matecconf/201816522023

Hence, the development, manufacturing and testing of bogies demands maximum attention from the outset, through strength calculations and optimization of the structural design, to the final experimental verification.

Prior to the fatigue test, stresses in the bogie frame under static loads are experimentally analysed. Exceptional loads, service loads and partial service loads are the load categories for which the stress and deformation responses are measured and evaluated.

Fatigue tests of bogie frames of railway vehicles fall into the group of prescribed tests. The objective of the prescribed procedures is to demonstrate that the structure does not sustain permanent deformation under exceptional or service loads and that no cracks form during fatigue testing. The fatigue test simulates all major forces, which are caused in service by the vehicle's mass, track irregularities, the vehicle's running curves, acceleration, braking, etc.

Fig. 1. Fatigue tests of bogies and undercarriage frames.

With the development of the testing technology, able to combine individual control electronic channels of hydraulic load actuators to multichannel units, began to fulfill dreams of some engineers about an experimental thorough assessment of all the operational and exceptional loading impacts on the bogie frames design. Similarly, as on their computational model.

With the development of the test technique, the dreams of some engineers have begun to perform a thorough laboratory simulation of all operational and exceptional loads that affect the bogie frame in operation. Similarly, as on their computational model.

All accelerations of bogie weights like motors, gear boxes, brake units etc. should by physically simulated in all orthogonal directions. The number of loading actuators rises up (for example up to twenty-two). The authors have already discussed within the professional public increasing customer demands for the number of load forces.

The complications associated with this phenomenon cannot be overlooked. It is relatively easier to compute a detailed part of the bogie frame, but practically much more complicated. Each actuator has to have minimum space around itself.

Fig. 2. Test of locomotive frame with twenty-two actuators.

The complicated fatigue test consists many procedures and partial tasks like test stand design, construction of appropriate fixture parts and their manufacturing, assembly of the test stand, technical preparation – installation of strain gauges and other sensors, performing static tests, measurement of all required variables, their analysis and evaluation, cycling etc. Fixture parts and their assembling for the connection for a "special" force introducing can be difficult, complicated and challenging, the impact of this force on the fatigue strength of the bogie frame negligible. It is the key and experienced decision of designers if they want to use - simpler test stand or more complicated. Both can be good and both can be wrong. Examples of tests with a radically different number of load forces are in Fig. 2 and Fig. 3.

2.2 Activities of Regional Technological Institute

The university research center focuses more on research and development, but strength and fatigue tests are also an integral part of these activities.

The development of tramway bogie for narrow track gauge is one of the interesting project. A simplified fatigue test under the relevant UIC directive was also part of the development process. The tram bogie was subjected to four decisive forces in this test. A photo of the test is in Fig. 3.

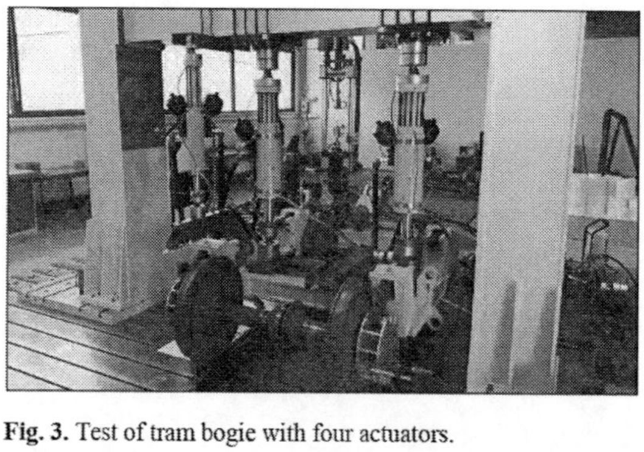

Fig. 3. Test of tram bogie with four actuators.

The next interesting projects of the Regional Technology Institute was the development of the hybrid body of a metro car. The "political" idea of the project was a successful demonstration that even technical university can transform own expertise to a functional product and can manage this type of project.

As a demonstrator, the real sample of the hybrid body was designed, manufactured and tested. This body was made of a higher strength steel and special composite panels. The design was supported by FEM calculations, fundamental research of mechanical properties of all used materials. The fatigue tests of important structural details were organised in collaboration with the Research and Testing Institution. Fig. 4 shows testing the node connecting composite sides and bottom part of body. A photo of the manufacture of the body is in Fig. 5. Static strength tests of the whole body according to the relevant standards were realised in testing hall of the Skoda Transportation. The photo of the test is in Fig. 6.

The potential for weight saving of 800 kg was confirmed. The body has lower weight and manufacturing labour more intensive than the existing one. Thus, it can save production and operating costs. Some technologies was patented during the solution of the project.

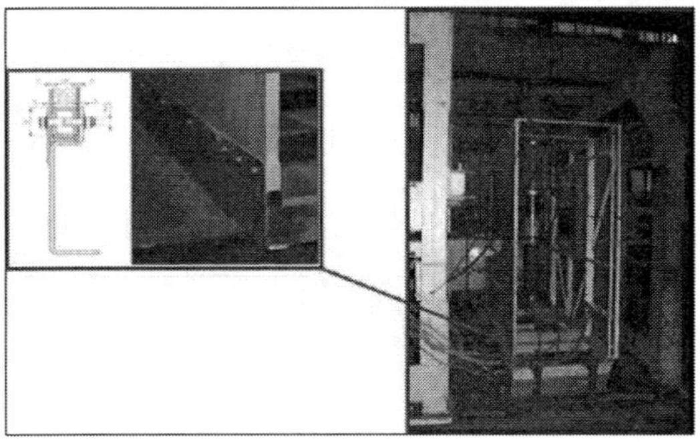

Fig. 4. Fatigue test of the node connecting composite sides and bottom part of body.

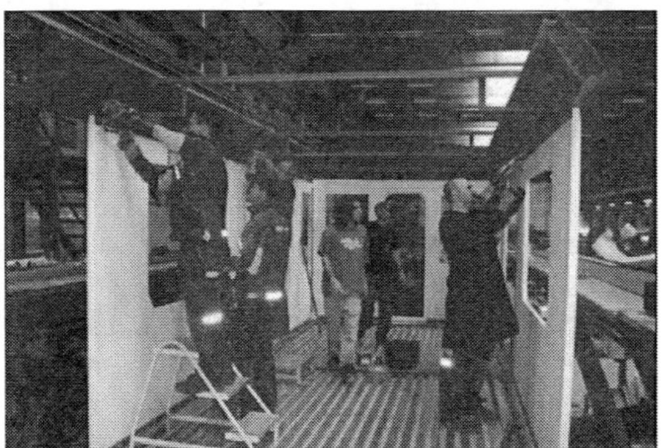

Fig. 5. Manufacture of hybrid body.

Fig. 6. Static strength test of hybrid body.

3 Road vehicles

3.1 Activities of Research and Testing Institute

City of Pilsen also has a long-standing tradition of testing bodyworks and undercarriage parts of road vehicles for urban public transport. The Research and Testing Institute provides for various domestic manufacturers not only fatigue tests, but also dynamic calculations, strength calculations by finite element method, measurement of operational stresses on model and real (city) routs.

The original methodology was applied at development of series of Skoda buses (mainly trolleybuses), Fig. 7.

Fig. 7. Skoda trolleybuses supported by broad fatigue testing.

The methodology, which has been successfully applied for many years in the development of Skoda trolleybuses and buses, has been presented in the literature in the past. This methodology is now being developed in connection with the production of battery buses.

The relationship between testing vehicles on test tracks and on a test stand is schematically illustrated in Fig. 8.

At the Research and Testing Institute, it is possible to test whole vehicles. Stand tests are an important part of the process of optimizing the mechanical structure of the vehicle. Testing complete bodywork on a test stand is an efficient method for choosing the best design with regard to the desired operational strength and fatigue life.

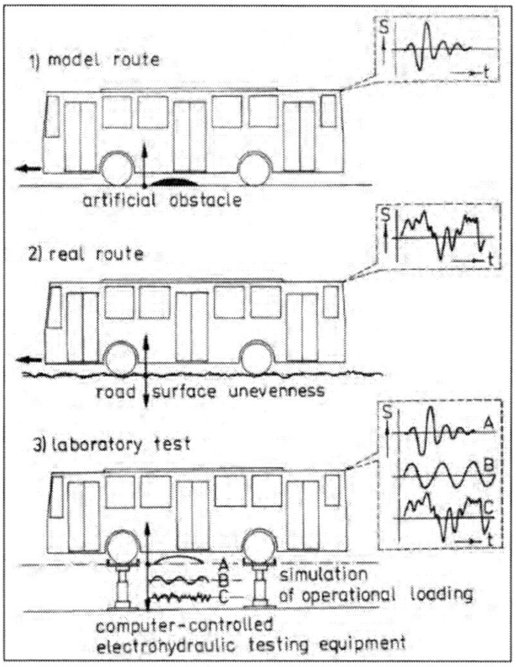

Fig. 8. Testing vehicles on test tracks and on test stand.

The design variants to be compared are usually prepared with a support of computational methods. The experimental programme on the test stand is also a test on the production technology and quality of labour. The test stand at the Research and Testing Institute is a computer-controlled electro-hydraulic device, consisting of several independent excitation cylinders where the test object is mounted for test. This sophisticated equipment ensures great flexibility in terms of test programmes and, at the same time, absolute reproducibility of the tests.

Experiments and tests on the stand are aimed at investigating the stress response in tested body related to excitation forces or deformations applied to the wheels. Three typical test modes are available: static, dynamic, dynamic-strength.

Static tests are used to determinate the stress fields and changes in the body resulting from torsion deformation from the up and down movement of individual wheels.

Dynamic tests use harmonic excitation whereby both acceleration and mechanical stresses are monitored at selected points on the body, response frequencies are determinated and parameters of the damping elements (such as bellow, dampers and tyres) are tested.

Dynamic-strength test are used to determinate the stress response to shock experienced in situations where the vehicle rides over a bump. The test stand makes it possible to simulate a driving on a real road characterized by random distribution of obstacles of a given statistical nature.

Fig. 9 is an example of possibilities of operational load simulations on test stands: simulation of vertical road unevenness and simulation of force acting in connection between the front and rear parts of the articulated vehicle.

Fig. 9. Possible simulations of operational loading on test stand.

Fig. 10 shows the most complex test performed on test site of the Research and Testing Institute. The construction of the body and undercarriage frame of the first fully low-floor articulated Skoda trolleybus was very complex tested. The operation of this trolleybus showed absolute reliability in terms of strength and fatigue life.

Fig. 10. Structure of articulated trolleybus on test stand.

In order to get a comprehensive view on the future operational reliability of the vehicle, it is necessary to deal with the detailed investigation of the fatigue strength and lifetime of the important construction nodes and vehicle components.

Laboratory tests on important parts and sub-assemblies aim at choosing the most advantageous design and technology with regard to the desired operational strength and lifespan of the whole body. Tests and experiments are typically performed on parts and their configurations that appear frequently in the design, e.g. joints between supporting sections in the body, or on parts exposed to extreme stress during operation (front spring holders, rear axle spring holders, articulated coupling between twin coaches, etc.)

During the tests, the decisive operational stresses are simulated by means of electro-hydraulic tests equipment. Parts to be tested are geared with strain gauges and mounted on the test stand. Measured data are subsequently used to analyse stress during laboratory tests and establish limit value states decisive for the fatigue strength and lifespan of the parts. By exposing different design variants to the same test conditions, it is possible to choose the best solution for the given operating environment.

Careful planning of the test programme, including in interface conditions for parts under test, is essential for establishing straightforward relationships between the test and real operating conditions.

The laboratory tests can be divided into two main groups: static tests and tests of fatigue strength and fatigue life.

Static tests consist of detailed analysis of the stress and deformation fields relating to the given test conditions.

Fatigue tests are used to establish zones and points of critical stress in the construction, i.e. points where fatigue may cause early failures.

The measured data are used for deriving various reliability parameters of the design by calculation. Various fatigue test methods are available (some possibilities are schematically illustrated in Fig. 11.) with different requirements as to the information on the real operational stresses, testing times, number of samples and overall costs.

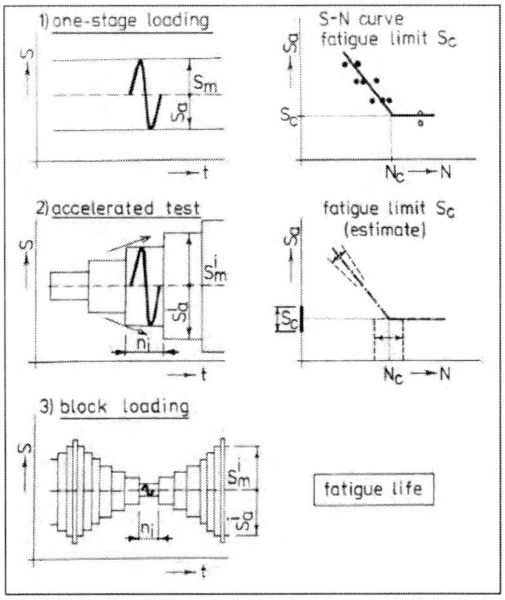

Fig. 11. Industry-friendly test load layout for fatigue testing of vehicle components.

The classic fatigue test with harmonic loading is used to derive the S-N curve and the fatigue limit of the part under test.

The accelerated fatigue test, where the complex stress succession in increased in steps, may be used in cases where the results, e.g. the fatigue limits for given critical point, are needed urgently and where poorer accuracy of information is acceptable. Compared to the classic fatigue tests, the accelerated methods of testing are faster, less costly and require fewer test samples.

The test on operational fatigue life uses a test programme based on a stress profile determinated in the real operating conditions of the vehicle. Thus, it is possible to derive very realistic estimates of the lifespan for a chosen critical part of the body. Disadvantages of this method are: high cost, long testing time and lesser general, applicability of the results.

The laboratory tests of critical vehicle parts yield the following information:

- fatigue limit for given critical point in a part/sub-assembly,
- response characteristics in the above points measured during test runs of vehicles on the both model and real test tracks.

In evaluating, the following criteria can be observed:

- maximum stress amplitudes are within the range of infinite fatigue life: results satisfactory, no changes in design necessary,
- maximum stress amplitudes exceed significantly the fatigue limit: results unsatisfactory, the design must be improved and the changes verified experimentally,
- maximum stress amplitude are in a region of the fatigue limit: further analysis is needed to assess the effect of the real service conditions on the given critical points, with regard to the lifespan.

Many of the laboratory tests were precisely planned and implemented during the development of Skoda trolleybuses. Examples are in Fig.12 to Fig. 15.

Fig. 12. Fatigue test of a rear axle.

Fig. 13. Fatigue test of a rear suspension beam.

Fig. 14. Fatigue test of a front suspension holder.

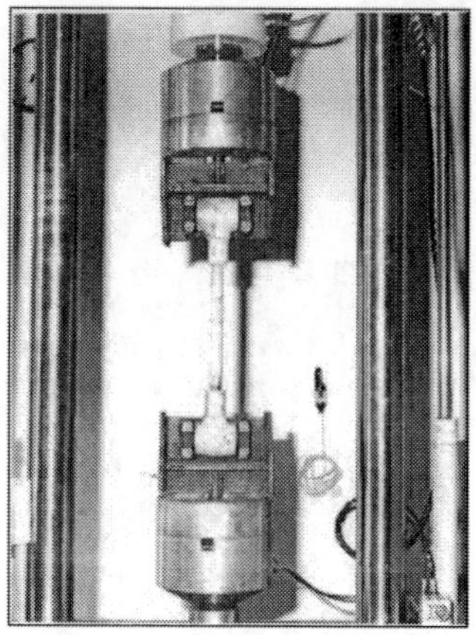

Fig. 15. Fatigue test of a stabilizing guide bar.

3.2 Activities of Regional Technological Institute

The Regional Technological Institute is more involved in fatigue life calculations than in the fatigue testing of road vehicles and its parts and components. For these calculations, nCode software is used. The classic calculations of fatigue damage, as well as vibration fatigue, are solved.

The effort is focused on probability dimensioning, taking into account an existing scatter of fatigue properties of materials and random loads of assessed components. In Fig.16, predicted fatigue life distribution functions of a vehicle component are compared to failures that occurred during the vehicle operation.

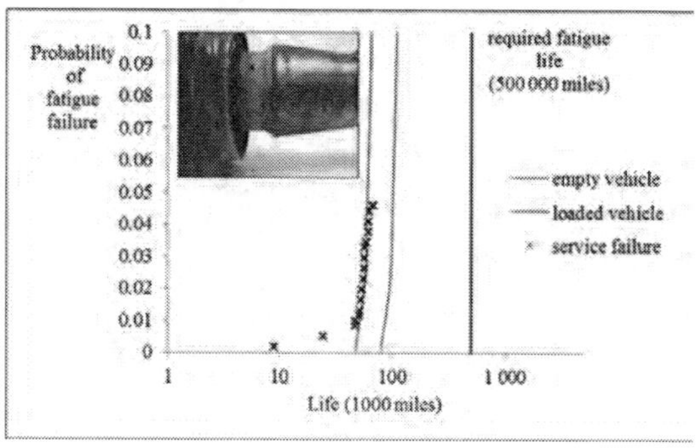

Fig. 16. Predicted fatigue life distribution functions and service failures (radius rod pin).

Figure 17 shows the design of a test rig to be used for multiple testing of welded body profile joints.

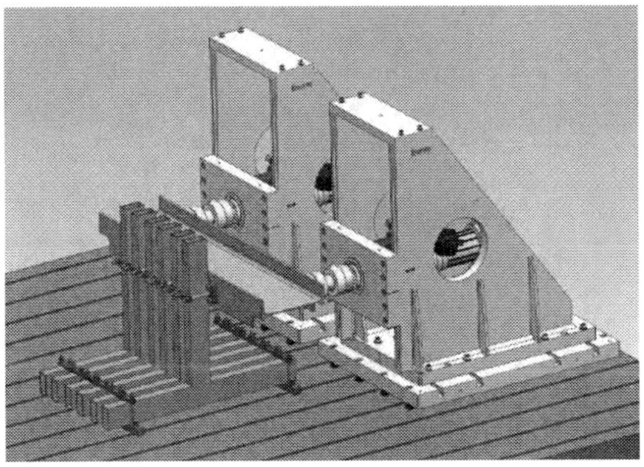

Fig. 17. Predicted fatigue life distribution functions and service failures (radius rod pin).

CONCLUSIONS

The fatigue tests play a major role in the development of mechanical parts of road and rail vehicles. The contribution demonstrates their variants. The presented examples were prepared and realised in collaboration of producer Skoda Transportation with dynamic testing laboratories at the Research and Testing Institute and the Regional Technological Institute in Pilsen, Czech Republic. The results of these fatigue tests and fatigue life investigations are described in non-public testing reports.

The contribution has been prepared under project LO1502 'Development of the Regional Technological Institute' under the auspices of the National Sustainability Programme I of the Ministry of Education of the Czech Republic aimed to support research, experimental development and innovation.

A simple expression to estimate the fatigue endurance of welded joints

Mirco Daniel Chapetti* and *Ceferino* Steimbreger

Laboratory of Experimental Mechanics (LABMEX), INTEMA (Research Institute for Material Science and Technology), CONICET – National University of Mar del Plata, J.B. Justo 4302, (B7608FDQ), Mar del Plata, Argentina.

Abstract. A simple expression is proposed to estimate the fatigue endurance of welded joints that can be used to understand and analyze in a simple way the influence of the main geometrical, mechanical and material effects (weld geometry, local geometry, material properties, residual stresses and size of defects). The proposed expression was derived from the results of the analysis of the fatigue strength of welds studied by means of a fracture mechanics approach that takes into account the fatigue behavior of short cracks by using the resistance curve method. For that purpose numerical simulation of transversely stressed butt, T and cruciform joints were performed.

1 Introduction

There exist several calculation methods for fatigue life analysis of welded joints, including the nominal stress method, hot spot method, notch stress method, local strain method, and fracture mechanics methods. Most of these methods are currently already either officially standardized [1, 2] or at least in internationally accepted recommendations [3]. However, there still exists a dominance of the traditional nominal stress method, which has been prevailing for over 45 years in the fatigue design of welded structures. To ensure that the full effect of the three key features dominating the fatigue life of welded joints (geometric stress concentrations, welding flaws, and residual stresses) are allowed for in design, most fatigue design rules consist of series of $\Delta\sigma$–N curves based on data obtained from constant amplitude fatigue tests on actual weldments [3-5], and employ classification methods of specifying design curves in terms of the fatigue strength at a given number of cycles (e.g. $2x10^6$ or 10^7). Since the stress concentration effect of the welded joint geometry is included, $\Delta\sigma$ refers to the nominal stress adjacent to the weld. Furthermore, to allow for the influence of residual stresses, the full stress range is used.

An important consequence of the geometric stress concentrations associated with most welded joints, the severity of which is usually compounded by the presence of welding flaws, is that fatigue cracks readily initiate and the life is dominated by fatigue crack growth [3-4]. This accounts for the drastic reduction in fatigue life resulting from the presence of a weld. It also explains why fracture mechanics is so relevant to the fatigue assessment of weldments. However, a simple form of the Paris-Erdogan power law is usually applied, given linear $\Delta\sigma$-N curves on a log-log basis with a slope m equal to that of the Paris law, compatible with the fatigue crack low for the material. Since m is approximately 3 for most steels, $\Delta\sigma$-N curves with slope 3 are widely adopted [1-3,6]. However, this procedure does not allow analyzing the influence of the different parameters defining the fatigue behavior.

To avoid some limitations on the analysis, a fracture mechanics methodology based on the resistance curve concept was proposed by Chapetti *et al* to analyze welded joint, accounting for the fatigue behavior of short cracks and considering the fatigue crack propagation threshold as a function of crack length [7,8]. Details of the method and some applications can be found on reference [7,8], [9] and [10]. In this work a simple expression to estimate the fatigue endurance of welded joints is proposed on the bases of the results of previous analysis and results [7-10].

2 The fracture mechanics approach

The method estimates the threshold for fatigue crack propagation as a function of crack length, ΔK_{th}, and the fatigue crack propagation rate as a function of the difference between the applied driving force ΔK and ΔK_{th}, as follows [11]:

$$\frac{da}{dN} = C\left(\Delta K - \Delta K_{th}\right)^m \qquad (1)$$

The threshold for fatigue crack propagation as a function of crack length a is estimated by using the model proposed by Chapetti [12], as follows:

$$\Delta K_{th} = \Delta K_{dR} + \left(\Delta K_{thR} - \Delta K_{dR}\right)\left[1 - e^{-k(a-d)}\right] \qquad a \geq d \quad (2)$$

where d is the average microstructural dimension (e.g. grain size, lath size, etc) and ΔK_{dR} and k are given by [12]:

* Corresponding author: mchapetti@fi.mdp.edu.ar

$$\Delta K_{dR} = Y\,\Delta\sigma_{eR}\,\sqrt{\pi\, d} \tag{3}$$

$$k = \frac{1}{4\,d}\,\frac{\Delta K_{dR}}{\left(\Delta K_{thR} - \Delta K_{dR}\right)} \tag{4}$$

where $\Delta\sigma_{eR}$ is the plain fatigue limit of the material. Figure 1 shows schematically the threshold curve given by equation (2) as a function of the square root of the crack length a.

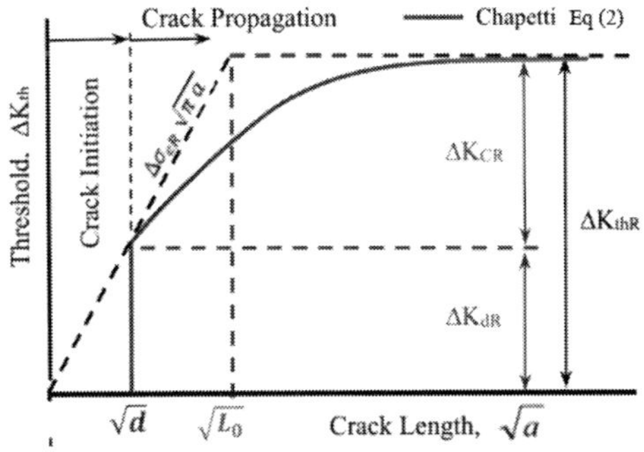

Fig. 1. ΔK_{th} as a function of the square root of a.

3 Previous analysis and results

Previous works have been done by using the proposed method in order to analyze the influence of different weld geometry parameters on the fatigue resistance of welded joints, like plate thickness, reinforcement angle, weld toe geometry, defect sizes and undercuts [7-10]. Results showed that the method allows understanding the influence of the different mechanical, geometrical and microstructural parameters on the definition of the fatigue resistance of the joint. In this way, it is possible to define which the main parameters for a given configuration are and to found the worst cases or limit values for them, according to the welding procedures and qualities. For instance, in a recent publication [10] we have shown that depth is the principal variable affecting fatigue behavior of butt-welded components with undercuts. Even though analysis based on continuum mechanics generally consider stress concentration factor as the controlling parameter, they fail to explain the importance of undercut depth in current regulations. In contrast, tolerable values of the undercut depth D can be obtained with the method proposed for a given desired fatigue strength (endurance limit) of the welded joint. This is done by considering the tendency in Frost diagram for sharp notches, as can be observed in Fig 2 that shows the fatigue limit prediction as a function of the stress intensity factor k_t of the configuration for four different undercut depths D. Estimations were done considering the existence of defects of about 200 μm.

Curves plotted in Fig. 2 correspond to different undercut depths: 0.25, 0.5, 0.75 and 1 mm. For each depth, by varying notch radius the stress concentration factor k_t is modified. Likewise, different undercut geometry affect the shape of the applied driving force ΔK and a new value of the fatigue limit corresponds. It is easy to understand that the higher is the stress concentration factor the lower is the fatigue limit. However, it can be seen in Fig. 2 that the fatigue limit tends asymptotically to a minimum value for each undercut depth, being lower for deeper undercuts (size effect). If the notch radius is supposed to tend to zero, the undercut can be considered a crack of length equal to the undercut depth. Therefore, it can be thought that the minimum value is related to the fatigue limit of the welded joint without undercut, with an initial crack of length $a_i = D$. A deeper analysis must consider the size of the non-propagating crack associated to the configuration ΔK-ΔK_{th} at the fatigue limit, which can be as long as 200-300 μm.

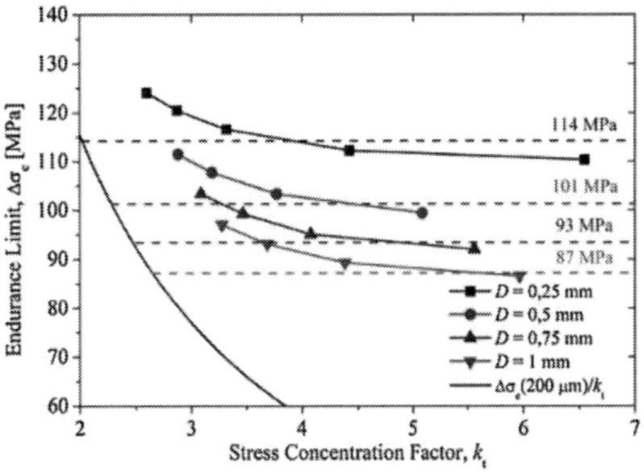

Fig. 2. Estimated fatigue limit or endurance ($\Delta\sigma_e$) of the joint as a function of k_t for butt welds with undercuts.

It is necessary to emphasize that predictions for high k_t (asymptotic trend in Fig 2) correspond to the worst scenario for a given undercut depth. Very high k_t values are seldom found in real undercuts but they allow to safely predicting tolerances.

Results exposed in reference [10] highlight the importance of not only undercut depth, but also threshold for fatigue crack propagation and non-propagating crack length, when designing against fatigue damage. In search of a conservative simplification for the estimation of the lowest fatigue limit of butt-welds with undercuts, a simple relation was then established between fatigue limit, crack propagation threshold, notch depth and maximum non-propagating crack length [10]:

$$\Delta\sigma_e = \frac{\Delta K_{th}}{F\left[\pi\left(D + a_{np-\max}\right)\right]^{f}} \tag{5}$$

where $a_{np\text{-}max}$ is the maximum non-propagating crack length for a given notch depth, as indicated before, and ΔK_{th} is the threshold value when $a = D + a_{np\text{-}max}$. F and f are constants that depend on the weld detail without undercut, i.e. reinforcement angle and thickness. Note that if reinforcement is ground (180°), then $F = 1.12$ and $f = 0.5$, as in a "through-thickness" crack. It is also important to note that in the analysis ΔK_{th} was roughly constant for $a > 0.5$ mm and equalize the fatigue threshold for long cracks (ΔK_{thR}).

Finally, it is important to say that although fixed global geometry and loading condition were used in the analysis of the influence of undercuts, the proposed method can scientifically explain the effect and relevance of most of the mechanical and geometrical parameters, as was shown in references [7], [8], [9] and [10].

4 The simplifying hypotheses

The analysis carried out in references [7-10] includes all details related with the configuration and use only a few hypotheses related with the applied models. In the present work the idea is to use the different results obtained in previous analysis and look for a simplification to be use in standards in a simple way. The detailed analysis should help us to further improve these simplifications for future applications. The following hypotheses are applied to develop the simplification:

- The existence of crack-like defects in welded joints can be normally assumed, which eliminate the crack initiation stage of fatigue life. A review by Grover [13] suggested that even high-quality welds contain flaws up to a depth of about 100 μm. Other works observed initial crack-like defect depths of about 10-120 μm [14], 20-150 μm [15] or 10-400 μm [16], according to the welding conditions. In IIW guidelines [17] initial crack length of 50-150 mm is recommended for fracture mechanics applications. Radaj and Sonsino [18] have recommended an initial crack size a_i = 100-250 μm for life prediction in welded structures. Such defect depths clearly fall within the short crack regime. In this work, initial crack length a_i = 200 μm is considered for the analysis, but other values can be assumed.

- The weld toe is considered as a sharp corner. In this case the theoretical elastic stresses near to the surface would tend to infinity (singularity), but that is acceptable as crack-like initial defects exist and therefore the stresses at the surface are not required. This hypotheses was confirmed in references [7,8], that showed that the stress concentration at a given depth x (k_{tx}) does not depend on the notch root radius (ρ) when $x > 0.2$ ρ. For instance, if the initial crack length is 100 mm the influence of notch root radius seems to have no important effect on the applied stress distribution for notch root radii less than 0.5 mm. A greater toe radius would give lower stress concentrations, so that in such cases the assumption would be conservative.

- Through thickness cracks are conservatively considered. The cracks, initiated at weld toes and guided by them along the surface, quickly develop a small aspect ratio, so that it is possible to conservatively consider the geometric factor Y equal to 1.12.

- In order to estimate fatigue limit (or endurance for high cycle fatigue), the threshold for fatigue crack propagation as a function of crack length should be necessary considered. The threshold should include the short crack regime, where it depends on the crack length. This threshold curve depends mainly on the threshold for long crack and the minimum microstructural threshold associated to the fatigue limit of the material where the crack nucleates. This curve can be simplified according to the configuration associated with the fatigue limit, which sometimes includes non-propagating cracks. This should be considered in the analysis in order to look for further simplifications.

5 The analysis

Numerical simulations of transversely stressed butt, T and cruciform joints were carried out for the analysis. Tensile and bending loading were considered for all three joints, getting six different configurations for the analysis. The fracture mechanics approach introduced in section 2 and detailed in references [7-10] was applied. A 19 mm thick plate was considered in the analysis for all cases. Reinforcement angle was 33° for the butt-welded joint and 45° for the T and cruciform welds. Gusset in T and cruciform models was 50 mm long and 19 mm thick, whereas fillet leg was 10 mm. A hole with a radius of 1 mm was adopted at the root. Simulation was purely geometric and residual stresses resulting from the welding process were not accounted for.

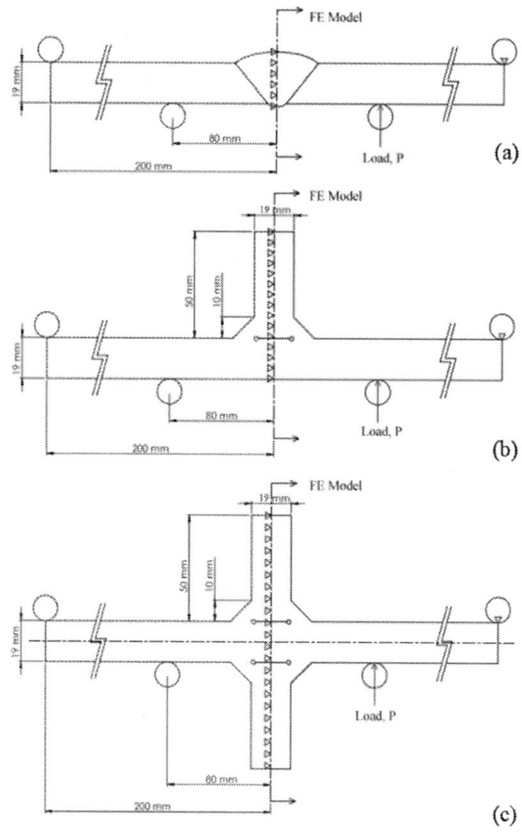

Fig. 3. FE model for (a) butt joint, (b) T-joint and (c) Cruciform joint.

Fig. 3 illustrates symmetry and schematically displays load configurations, boundary conditions and overall geometry. Four point bending and pure traction were applied to the samples. Two dimensional linear elastic analyses were used and symmetry was assumed in all the models.

Stress intensity factor calculations were achieved following software procedure for fracture mechanic simulations [ABAQUS]. Then, cracks were introduced as "seam cracks" at the weld toe, where the crack is more likely to nucleate. Mesh consisted in 6-node quadratic plane strain triangles at the crack front, which use a modified second-order interpolation and 8-node bi-quadratic plane strain quadrilateral elements in the rest of the model. Since the model is linear elastic, it is recommended to include a square root singularity at the crack tip, which improves accuracy of the stress intensity factor calculation.

Table 1 indicates the six analyzed configurations. The parameter $M_{k\text{-}200}$ represents the stress concentration generated by the configuration (geometry and loading) at a depth equal to 200 μm, which is taken as the initial crack length a_i in the present analysis. It means that a maximum crack-like defect of 200 μm is considered.

Table 1. Analyzed configurations.

Joint	Loading	Configuration	$M_{k\text{-}200}$
Butt	Tensile	B-T	1,48
	Bending	B-B	1,51
T	Tensile	T-T	1,58
	Bending	T-B	1,73
Cruciform X	Tensile	X-T	1.9
	Bending	X-B	1,52

6 Results and proposal

Table 2 shows the main results we need for the proposal: the estimated fatigue limit (endurance) for each configuration ($\Delta\sigma_e$), the recommendation given by IIW for Butt, T and cruciform joints, and the parameter M_{k200}.

Table 2. Results.

Joint	$\Delta K_{ai=200}$ [MPa m$^{1/2}$]	$\Delta\sigma_e$ [MPa]	$\Delta\sigma_e$ IIW [MPa]	$M_{k\text{-}200}$
B-T	Tensile	95	90	1,48
B-B	Bending	93		1,51
T-T	Tensile	89	85	1,58
T-B	Bending	81		1,73
X-T	Tensile	92	71	1,52
X-B	Bending	74		1,9

The analysis was carried out considering residual stresses and an effective load ratio R equal to 0.5, with a threshold for fatigue crack propagation for a crack length equal to 200 μm of about 4 MPa m$^{1/2}$. Details of the numerical analysis and results will be published soon. The fatigue endurance is estimated by making the applied ΔK equal to the threshold ΔK_{th} for the initial crack length a_i, using the following expression:

$$\Delta K = Y\, M_{k-a_i}\, \Delta\sigma_e \sqrt{\pi\, a_i} = \Delta K_{th} \tag{6}$$

From which we get the following general expression for fatigue endurance estimation for welded joints:

$$\Delta\sigma_e = \frac{\Delta K_{th}}{Y\, M_{k-a_i} \sqrt{\pi\, a_i}} \tag{7}$$

With a_i = 200 μm and Y = 1,12 (through thickness cracks) we get the following simplification:

$$\Delta\sigma_e = \frac{35\ \Delta K_{th}}{M_{k-200}} \tag{8}$$

Fig 4 shows the fatigue endurance proposed by the IIW document for the three different joints, together with the estimated ones for the six analyzed configurations given by expression (8) for ΔK_{th} = 4 MPa m$^{1/2}$, all as a function of the parameter $M_{k\text{-}200}$.

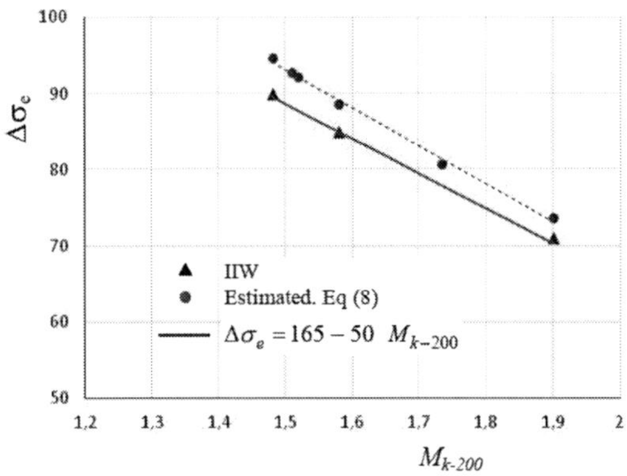

Fig. 4. Fatigue endurance as a function of the macroscopic stress concentration $M_{k\text{-}200}$.

Fig 4 shows a clear linear relationship between the endurance limit and the parameter $M_{k\text{-}200}$. For the fatigue endurances recommended by IIW the following simple expression can be proposed to explain the trend:

$$\Delta\sigma_e = 165 - 50\ M_{k-200} \tag{9}$$

where the fatigue endurance $\Delta\sigma_e$ is in MPa.

7 Concluding remarks

After different studies and analysis done by using fracture mechanics methodologies to understand the influence of the different mechanical, geometrical and material parameters on the fatigue behavior of welded joints, important conclusions and hypotheses can be related in order to look for simple and useful expressions and procedures to estimate their fatigue endurances. Results show that the fatigue endurance of welded joints can be explain in the simple way by accounting for the stress concentration at a given depth from the fatigue crack initiation point, given by the macro-geometry of the joint ($M_{k\text{-}ai}$). These results open an important door to look for new phenomenological and theoretical way to develop a simple procedure to estimate the fatigue endurance of welded joints for engineering applications. We think that much effort should be done in this way.

References

1. *Fatigue design and assessment of steel structures code of practice*, British Standard BS 7608 (1993).
2. *Guide on methods for assessing the acceptability of flaws in metallic structures*, British Standard BS 7910 (1999).
3. A. Hobbacher. IIW XIII-2151r1-07 XV-1254r1/07 (2007).
4. T. A. Gurney. *Fatigue of welded structures*, Cambridge University Press (1978).
5. *AWS structural welding code*, 9th edn, AWS, Miami, FL, USA (1985).
6. Guideline for fatigue design of steel constructions, JSSC, Gihodo Ltd, Tokyo, Japan, 1993.
7. M.D. Chapetti, J. Belmonte, T. Tagawa and T. Miyata. IIW X-1545-03 XIII-1971-03 (2003)
8. M.D. Chapetti, J. Belmonte, T. Tagawa and T. Miyata. J Sci Tech Weld J, **9,** 430 (2004).
9. M.D. Chapetti and L. Jaureguizahar. Int J Fatigue **43**, 43 (2012).
10. C. Steimbreger and M.D. Chapetti. Int. J. Fatigue **105**, 296 (2017).
11. C. Molina and M.D. Chapetti. Int J Fatigue **108**, 47 (2018).
12. M.D. Chapetti. Int J Fatigue **25**, 1319 (2003).
13. J.L. Grover. Int *Conference on Fatigue of Welded Construction*, Brighton, 275 (1987).
14. I.F. Smith and R.A. Smith. Fatigue Fract Eng Mater Struct **5**, 151 (1982).
15. E.G. Signes, R.G. Baker, H.D. Harrison and F.M. Burdekin. Weld J **14**, 108 (1967).
16. F. Watkinson, P.H. Bodger and J.D. Harrison. *Int Conference on Fatigue Welded Structures*. The WI, 97 (1970).
17. B. Jonsson. *IIW Guidelines on weld quality in relationship to fatigue strength* (2016).
18. D. Radaj and M. Sonsino. *Fatigue assessment of welded joints by local approaches*. Abington Publishing (1998).

Influence of Weld Toe Radii on Fatigue Life Prediction

Kin Shun Tsang[1,*], *John H. L.* Pang[1], *Hsin Jen* Hoh[1]

[1]School of Mechanical & Aerospace Engineering, Nanyang Technological University, 50 Nanyang Avenue, Singapore 639798, Singapore

Abstract. A study was carried out to investigate the influence of the weld toe radii on the fatigue strength of butt welded joints loaded in bending and tension. Fatigue analysis starting from weld toe cracks in marine and offshore welded pipe specimens were conducted using cut-out four-point bend fatigue test specimens. Fatigue life can be enhanced by reducing the local stress concentration generated by weld toe radius effects. This study investigated the effect of different weld toe radii on the stress intensity factor at the region of the weld toe through Finite Element Analysis (FEA). FEA was used to model a butt welded steel plate extracted from a pipe subject to tension and four-point bend loading. Semi-elliptical surface (SESC) cracks were modeled at the weld toe region with different SESC crack depth and surface crack length. Four weld toe radii and two modes of loading were investigated. The stress intensity factors, weald toe magnification factors, or M_{kc} and M_{ka} are reported and used for fatigue life prediction.

1 Introduction

With the complex nature of dealing with welds in marine and offshore structures, it is essential to understand the fatigue life prediction research of welded joints better. The fatigue life of welded joints is dependent on the weld toe radius effects on local stress concentration on fatigue life of welded joints in addition to the global structural stress concentration effect of the welded joint geometry.

The fatigue life of welded structures with initial weld toe cracks can be predicted by the linear elastic fracture mechanics approach covered in BS7910 [1]. Fatigue crack growth prediction can be made using the stress intensity factor range (ΔK) and crack propagation rate (da/dN) [2] through a Paris law [3]. The Stress Intensity Factor (SIF) solutions can be determined from finite element analysis [4]. The ratio of the SIF of a plate with welded attachment as compared to the SIF of a plain plate is defined as the weld magnification (M_k) factor [5]. M_k factors are influenced by the size of the welded attachment, as well as the size and depth of the weld toe crack.

Niu and Glinka [6] employed weight function methods that included weld toe radius as a parameter in assessing the stress intensity factor of welded joints. However, this study only allows for the assessment of stress intensity factors for the deepest point of the semi-elliptical crack, with no mention of the effect of the weld toe radius on the crack ends. Improving on Niu and Glinka's weight function, Brennan et al. [7] conducted a study to generate parametric equations for T-butt weld toe stress intensity factor, which took into account the weld toe radius effect, among other factors. However, the methodology employed the usage of two-dimensional finite element analysis to evaluate the stress distribution caused by the opening of the crack, before using that as an input into a weight function solution for determining stress intensity factors. While these equations are validated against a wide spectrum of values, the usage of two-dimensional finite element analysis may not be as accurate. Studies by Zhang et al. [8] also concluded that the crack growth rate will be overestimated if two-dimensional M_k solutions were used instead of three-dimensional solutions.

In BS7910 [1], in the section that deals with surface cracks, it is found that M_k is a function of the crack depth, a, overall width of the attachment measured from left weld toe to right weld toe, and the thickness of the plate, t. Figure 1 shows a schematic of our four-point bend test specimen extracted from an offshore pipe. In this test configuration, multiple weld toe cracks will be initiated at the butt weld-to-parent metal weld toe region at the bottom weld toe sites.

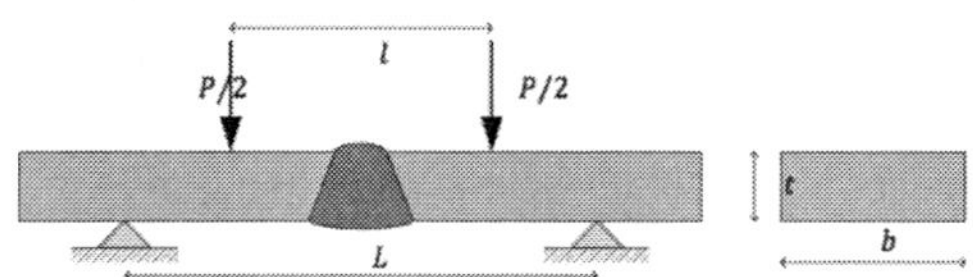

Figure 1. Butt weld, used in four point bend loading.

In BS7910, the Mk factors for weld toe cracks is reported for a specific weld toe radii citing the reference by Bowness and Lee [9]. The effect of the weld toe radii on the stress intensity factor was not the focus of the study. Hence, there is a need to investigate the effect of weld toe radii on the magnification factor using three-dimensional finite element analysis.

* Corresponding author: kstsang@ntu.edu.sg

This work will investigate the influence of the weld toe radii on the M_{kc} and M_{ka} factors for butt welded joints loaded in tension and bending. The derived M_k factors are then used to predict the fatigue life of a four-point bend fatigue test specimen result taking into account a sharp weld toe radius and a weld toe radius of 1.0mm in the fatigue life prediction example. ABAQUS was used to model semi-elliptical surface crack (SESC) in a steel plate and the model is then modified to develop the butt weld geometry over it to determine the M_{kc} and M_{ka} factors. Different semi-elliptical surface crack geometries (a, c) were modeled at the weld toe. Four weld toe radii were investigated for sharp toe with zero radius, and 3 radii sizes of 0.1, 0.5 and 1.0mm respectively. The FEA work looks at two modes of loading for tension and four-point bending.

2 Weld toe crack model

This section describes the methodology and modelling procedure of the weld toe crack model.

2.1 Methodology

XFEM was used in this study to produce a finite element model of the crack, to investigate the effect of different weld toe radii on the M_k factors. A parametric study using plane butt joint geometry was carried out using the ABAQUS. Figure 2 shows the plain plate and plate with attachment that was modelled.

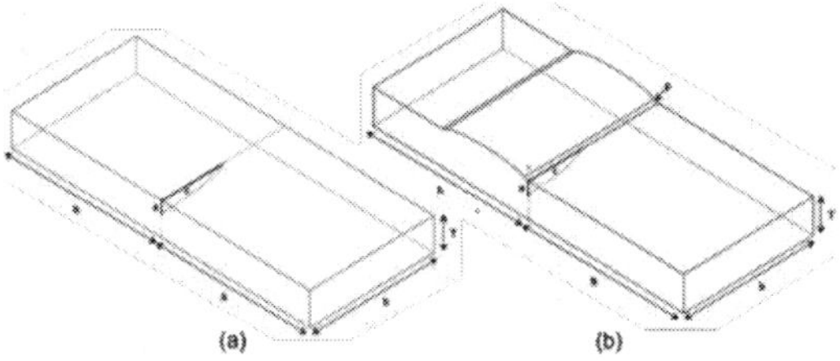

Figure 2. a) Plain plate b) Plate with butt weld geometry.

The analysis matrix covered is shown in Table 1.

Table 1. Matrix of model parametered investigated.

a/T	a/c	ρ (mm)
0.05	0.1, 0.5, 1.0	0, 0.1, 0.5, 1.0
0.1	0.1, 0.5, 1.0	0, 0.1, 0.5, 1.0
0.5	0.1, 0.5, 1.0	0, 0.1, 0.5, 1.0

Table 1 lists the different parameters that will be used to compare the effect of the weld toe radii. The base value of ρ=0 is used to compare the effect of a sharp weld toe compared to a weld toe with different weld toe radii, ρ values of 0.1, 0.5 and 1.0mm. Two analyses was carried out for applied tension (membrane) loading, and for four-point bend loading.

2.2 Modelling Procedure

A 3D half model was used, such that the crack was symmetrical about the x axis, due to the symmetry. At the area nearest to the crack, the model was partitioned according to the shape of the crack, with a finer mesh. The mesh then became coarser progressively away from the semi-elliptical surface crack. This allowed for a faster computational time and accuracy was calibrated with SIF solutions reported by Newman and Raju [12]. .

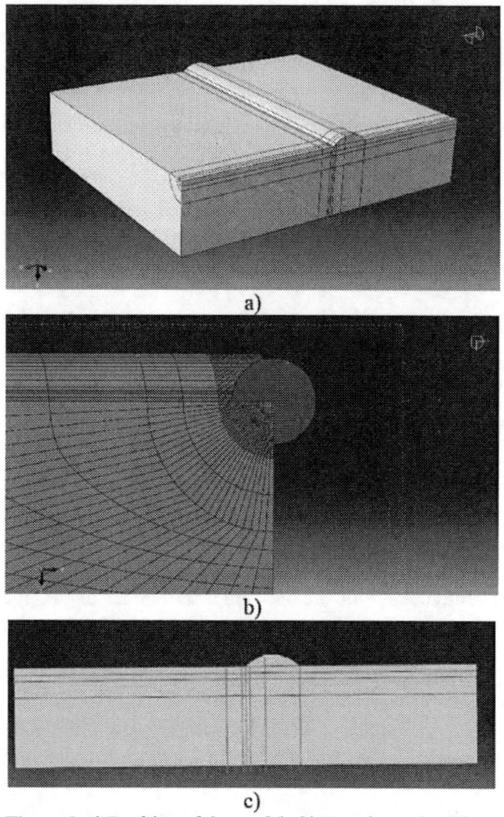

Figure 3. a) Partition of the model ; b) Sample mesh of the model near the crack represented by the circle; c) Profile of the plate with butt weld feature with a sharp weld toe, ρ= 0.

The Elastic Modulus of 210 GPa and a Poisson's ratio of 0.3 for the steel material was used. The model is for a four-point bend fatigue specimens reported by Pang et al. [10]. The dimensions of the steel plate are shown in Table 2.

Table 2. Dimensions of the steel plate.

Dimension	Length
Length of steel plate ($2h$)	240 mm
Width of steel plate ($2b$)	200 mm
Height of steel plate (T)	25 mm
Height of weld attachment (h_{att})	3 mm
Length of weld attachment (L)	14 mm

2.2.1 Tensile Load

https://doi.org/10.1051/matecconf/201816522025

For tensile loading, a load was applied on one end of the model. A fixed displacement boundary condition was applied on the opposite face from which the load was applied.

Figure 4. Sample load on a model undergoing tensile loading.

As a half-crack model was used, a boundary condition to account for symmetrical conditions was employed.

2.2.2 Bending Load

The bending load is based on the four-point bend loading conditions (Figure 1).

The bending stress applied and bending moment of relationship is given in equation 1.

$$\Delta\sigma = -\frac{6M_z}{bT^2} \quad (1)$$

Figure 5. Four point bend loading illustration.

To obtain the SIF value along the crack front, 8 contour integrals were used. Only the 3rd to 7th integral are used in evaluating the average SIF value at each location for reliable results compared to Newman's reported results [11].

3 Effect of Weld Toe Radius on SIF

This section describes the results of the simulation.

3.1 Tensile Loading

To observe the effect of the weld toe radii on the stress intensity factor on tensile loading, the results will be separated into that at the crack ends, as well as that at the deepest point of the crack, which will be discussed in the next section.

3.1.1 Crack Ends

The results are sorted according to its crack aspect ratio (a/c). The result from each simulation is given in the table below:

Table 3. $M_{k,c}$ values for crack ends under tensile load.

a/c	*a/T*	$M_{k,c}$			
		$\rho = 0$	$\rho = 0.1$	$\rho = 0.5$	$\rho = 1.0$
1.0	0.05	1.6802499	1.6037761	1.4578918	1.4054604
	0.1	1.4218269	1.4056439	1.3105523	1.2595960
	0.5	1.1658519	1.1744366	1.1532133	1.1280553
0.5	0.05	1.6862815	1.6341801	1.5023521	1.4529703
	0.1	1.4416781	1.4229711	1.3383768	1.2887207
	0.5	1.1404797	1.1442619	1.1319429	1.132051
0.25	0.05	1.8243668	1.7622910	1.5815262	1.4937924
	0.1	1.5572065	1.5350833	1.4345385	1.3698200
	0.5	1.2010747	1.2045454	1.1945812	1.1628974

The M_k results are plotted into graphical form according to the crack depth ratio (*a/T*), crack aspect ratio (*a/c*) and crack radii (*ρ*).

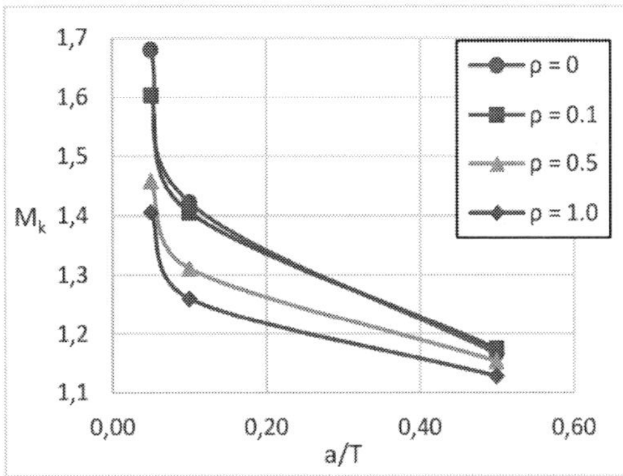

a)

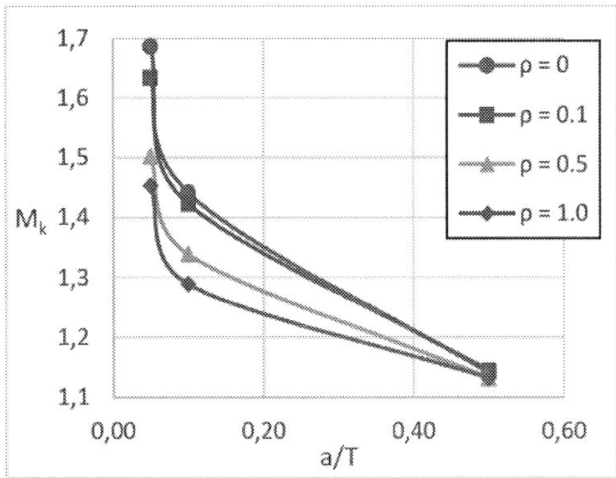

b)

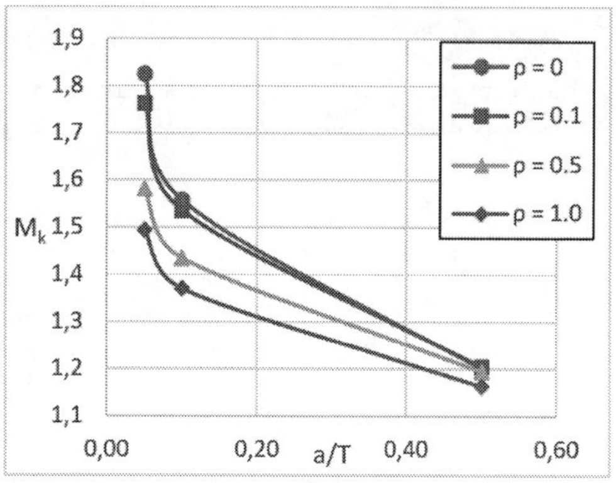

c)

Figure 6. Plot of $M_{k,c}$ values for crack ends under tensile loading: a) a/c = 1.0; b) a/c = 0.5; c) a/c = 0.25.

From the graphs (in Figure 6), it can be seen that with an increasing weld toe radius, the $M_{k,c}$ value at the crack ends decreases. This is especially true for smaller cracks. The decrease in $M_{k,c}$ factor is as much as 16.3% in the case of $a/c = 1$, $a/T = 0.05$. From this set of results, it can be inferred that the weld toe radius does help in reducing the stress intensity factor at the weld toe for small cracks.

3.1.2 Deepest Point of Crack

Similarly, the data will be sorted according to its crack aspect ratio (*a/c*) to allow for meaningful discussion of the results. The results from each simulation are given in the table and Figure 7 below:

Table 4. $M_{k,a}$ values for deepest point of crack under tensile load.

a/c	*a/T*	$M_{k,a}$			
		ρ = 0	**ρ = 0.1**	**ρ = 0.5**	**ρ = 1.0**
1.0	0.05	1.0367309	1.0163758	1.0264768	1.0506581
	0.1	0.9504995	0.9386921	0.9430820	0.9557977
	0.5	0.9617063	0.9614247	0.9622694	0.9655075
0.5	0.05	1.0481380	1.0337193	1.0388283	1.0617620
	0.1	0.9666330	0.9656253	0.9677527	0.9705520
	0.5	1.0107441	1.0107441	1.0115259	1.0058580
0.25	0.05	1.0485092	1.0479340	1.0511935	1.0528233
	0.1	1.0047072	1.0065705	1.0075512	1.0096106
	0.5	0.9809211	0.9844022	0.9848038	0.9825947

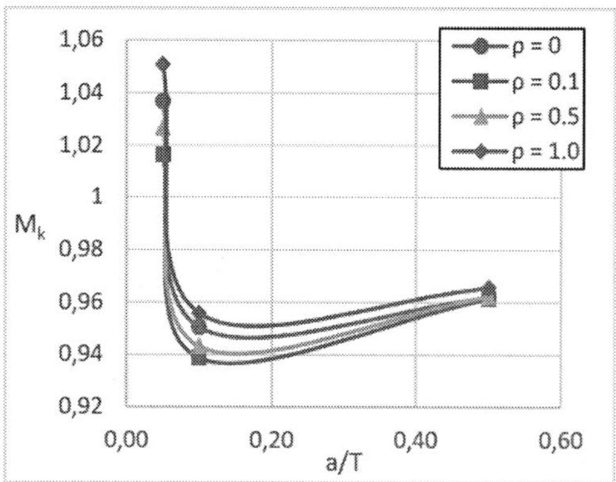

a)

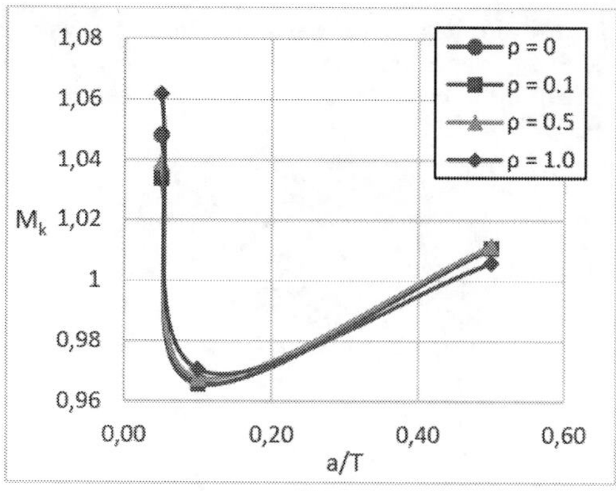

b)

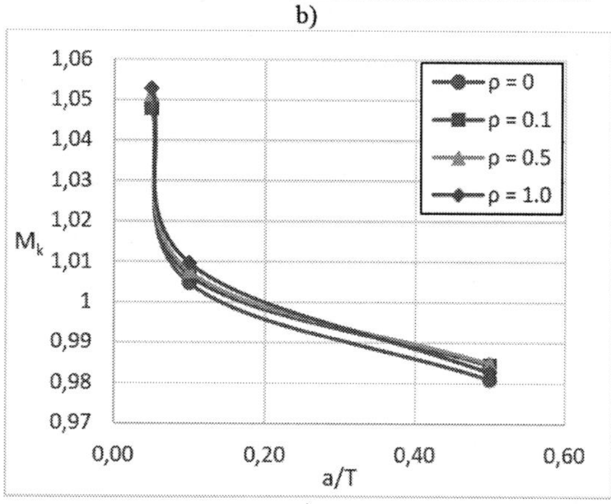

c)

Figure 7. Plot of $M_{k,a}$ values for deepest point of crack under tensile loading: a) a/c = 1.0; b) a/c = 0.5; c) a/c = 0.25.

For the deepest point of the crack under tensile loading, there is little effect of the weld toe radius on the $M_{k,a}$ values. As can be seen from Figure 7, for each *a/T* parameter, the $M_{k,a}$ values are clustered closely together with little variation. The deepest point of the crack is far from the weld toe, and the weld toe radius has little to no effect on the $M_{k,a}$ values.

3.2 Four Point Bend Loading

This section will discuss the effect of weld toe radii on the Mk factor subject to four-point bend loading.

3.2.1 Crack Ends

The results are given in Table 5 and Figure 8.

Table 5. $M_{k,c}$ values for crack ends under four point bend load.

a/c	*a/T*	M_k			
		ρ = 0	**ρ = 0.1**	**ρ = 0.5**	**ρ = 1.0**
1.0	0.05	1.7187543	1.7020807	1.5376877	1.4474300
	0.1	1.4603804	1.4189133	1.3266730	1.2809634
	0.5	1.1777423	1.1971851	1.1606210	1.1385664
0.5	0.05	1.6860383	1.6343649	1.4999299	1.4390141
	0.1	1.4847930	1.4604336	1.3690509	1.3114615
	0.5	1.1545724	1.1575397	1.1447261	1.1273266
0.25	0.05	1.7963111	1.7310344	1.5587810	1.4564554
	0.1	1.6251058	1.5852956	1.4758597	1.4070811
	0.5	1.2156430	1.2462131	1.1914073	1.2232167

The results were plotted in graphical form according to the crack aspect ratios (a/c).

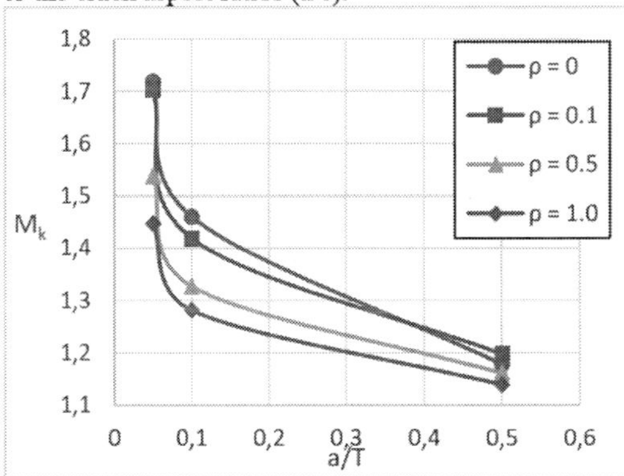

a)

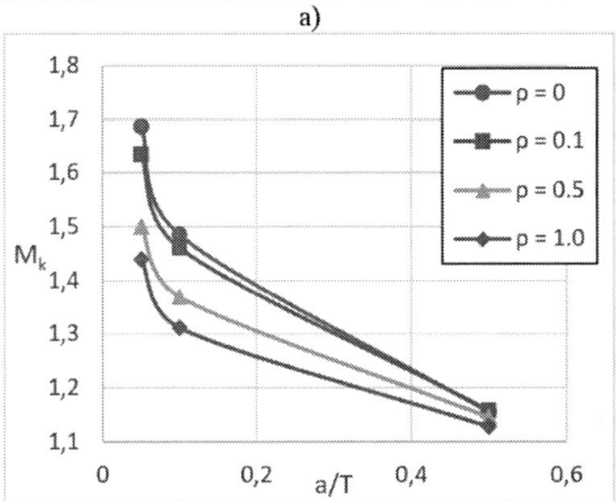

b)

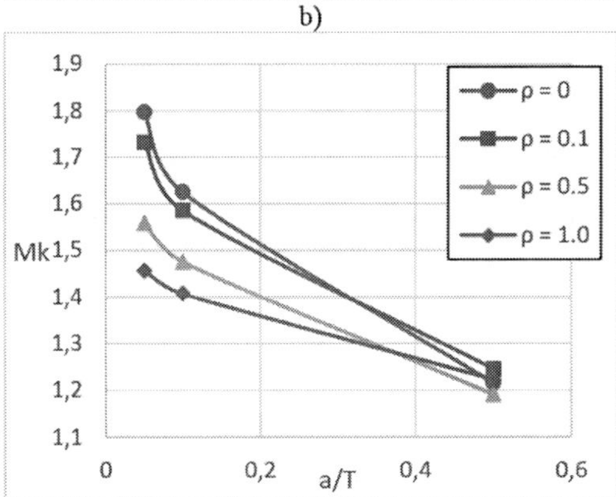

c)

Figure 8. Plot of $M_{k,c}$ values for crack under four point bend loading: a) a/c = 1.0; b) a/c = 0.5; c) a/c = 0.25.

The $M_{k,c}$ values of the crack ends subjected to four-point bend loading is similar to that of the crack ends subjected to tensile loading. There is a decreasing trend of $M_{k,c}$ values with increasing weld toe radius, for smaller cracks. For bending, the decrease in $M_{k,c}$ factor is as much as 18.9% in the case of $a/c = 0.25$, $a/T = 0.05$.

For larger cracks ($a/T = 0.5$), the weld toe radius plays a limited role in reducing the stress intensity factor.

3.2.2 Deepest Point of Crack

The results are given in Table 6 and Figure 9.

Table 6. $M_{k,a}$ values for deepest point of crack for bend load.

a/c	*a/T*	$M_{k,a}$			
		ρ = 0	ρ = 0.1	ρ = 0.5	ρ = 1.0
1.0	0.05	1.2029432	1.2465108	1.2604977	1.2350739
	0.1	1.0101311	1.0053868	1.0129676	1.0289566
	0.5	0.8951208	0.9038336	0.8870623	0.8911142
0.5	0.05	1.1131830	1.1022148	1.1065656	1.1247377
	0.1	0.9807743	0.9811518	0.9839140	0.9869814
	0.5	0.9504534	0.9502521	0.9516799	0.9535723
0.25	0.05	1.0407916	1.0628861	1.0648652	1.0664868
	0.1	1.0246193	1.0474590	1.0484325	1.0498063
	0.5	0.9751157	0.9923764	0.9631603	0.9938617

The results are shown in graphical form according to the crack aspect ratios (*a/c*).

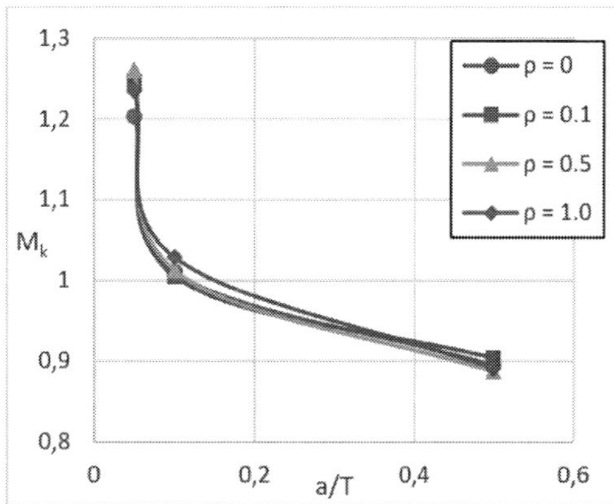

a)

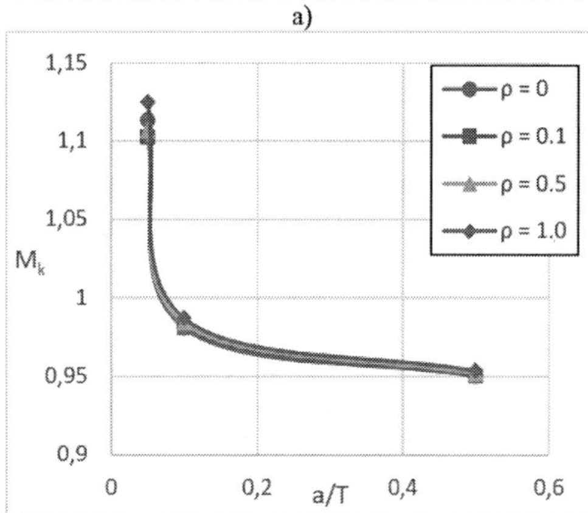

b)

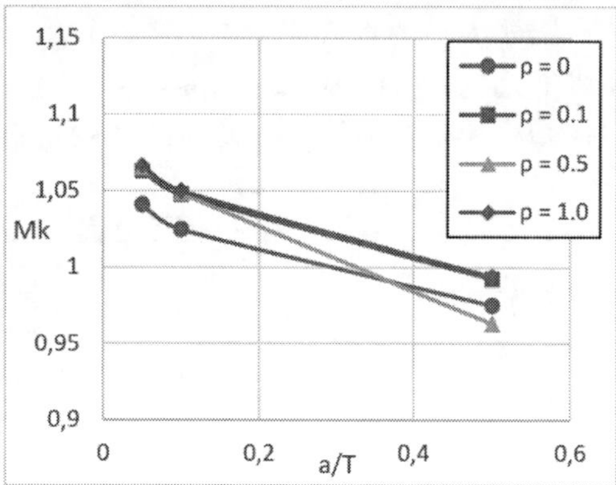

c)

Figure 9. $M_{k,a}$ values for deepest point of crack under four point bend loading: a) a/c = 1.0; b) a/c = 0.5; c) a/c = 0.25.

The weld toe radius has a negligible effect on the $M_{k,a}$ values, and therefore the stress intensity factor at the deepest point of the crack.

4 Fatigue Life prediction

4.1 M_k factor solutions

With the results obtained in the above section, the M_k factors obtained can be curve fitted into solutions, as a function of *a/T* ratios, expressed in the form of a power law (Eqn. 2). For this current fatigue life prediction, a crack aspect ratio, *a/c* = 0.5, will be used, with only bending load considered.

The M_k factor solutions follow the format of the following equation:

$$M_k = x\left(\frac{a}{T}\right)^y + z \quad (2)$$

where the coefficients *x*, *y* and *z* can be found in Table 7.

Table 7. M_k coefficients *x*, *y* and *z* for *a/c* = 0.5

Loading	Radius, ρ	$M_{k,c}$		
		x	y	z
$M_{k,c}$	0.0	0.4141	-0.3088	0.6416
	1.0	0.1388	-0.4266	0.9408
$M_{k,a}$	0.0	0.0001235	-2.399	0.9498
	1.0	0.0001602	-2.329	0.9528

4.2 Fatigue life prediction

For the fatigue life prediction, the material was modeled following X65 steel, similar to a previous experiment [10]. The yield strength and ultimate tensile strength of the base material was obtained via relation to hardness (σ_Y = 550 and UTS = 633 MPa). The following fatigue law constants were used: design values of $C = 5.21 \times 10^{-13}$ and $m = 3$ as given in the BS 7910 standard [1]. Small semi-elliptical surface cracks of initial crack depth (*a*= 0.1 mm) and aspect ratio (*a/c*= 0.5) were used. The SIFs of plain plates obtained by Newman and Raju [12] are used in the analysis. The locations of each crack initiation sites were reported in Pang et al [10]. The loading type applied was bending, at a load of 320MPa.

The fatigue life prediction is conducted through a C++ program developed called **Fatigue Crack Growth Analysis (FCGA)** software. This was developed to evaluate the fatigue life of welded structures with many weld toe semi-elliptical surface cracks (any number of sites and different crack sizes). The FCGA software can accommodate fatigue threshold constants, crack closure based Paris Laws. $M_{k,c}$ and $M_{k,a}$ factors can be curve-fitted for FCGA entry format. An illustration of the FCGA software results is shown in Figure 10. The multiple weld toe cracks coalesce and emerge as a single fatigue crack leading to final crack size as the four-point bending fatigue test was stopped when the crack depth exceeds 70% of the plate thickness.

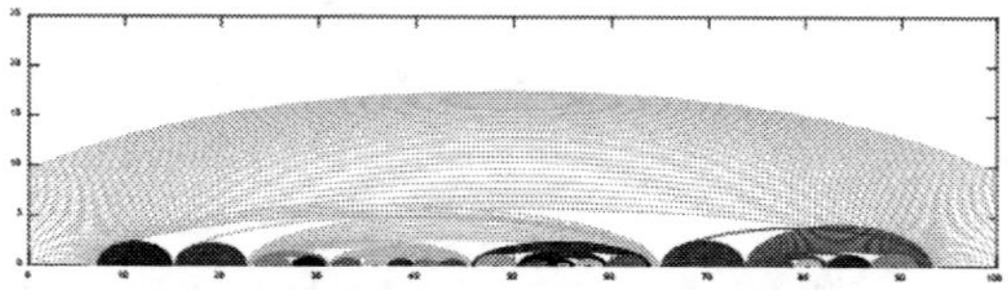

Figure 10. Many crack initiation sites from weld toe

The result of the fatigue life prediction is shown in Table 8, for the sharp weld toe versus the weld toe with a radius of 1.0 mm. The percentage difference in predicted fatigue life is 67 % longer with respect to a sharp weld toe condition compared to the case with a 1.0mm weld toe radius.

Table 8. Fatigue life prediction results

Radius, *ρ* (mm)	Fatigue Life, (cycles)
0.0	123, 500
1.0	206, 000
% diff	67%

The results show that the weld toe radius has a significant effect on the predicted fatigue life of the test specimen. Life prediction using M_k factors from sharp weld toe models gives conservative predict fatigue life. On-going studies are in progress for weld-toe grinding radius effect on Mk factors and predicted fatigue lives.

The actual four-point bend fatigue test life was 377,921 cycles. This indicates that the overall total fatigue life from the fatigue test result is longer. Hence, crack initiation or early crack growth effects may be needed to model the total fatigue life, crack initiation and propagation life of the test specimen.

5 Conclusion

The effect of weld toe radius on the fatigue behavior of a cut-out butt weld specimen was investigated. Four weld toe radii and two modes of loading were reported.

From this study, the weld toe radius of 1.0 mm can reduce the stress intensity factor at the weld toe for small cracks (*a/T* = 0.05). The decrease in $M_{k,c}$ factor was 16.3% in the case of *a/c* = 1, *a/T* = 0.05 for tensile load, and 18.9% in the case of *a/c* = 0.25, *a/T* = 0.05 for bending.

The results of the fatigue life prediction show that the weld toe radius has a significant effect on the predicted fatigue life of the butt weld joint. The fatigue life with the radius of 1.0mm increases the fatigue life by 67% compared to the case with a sharp toe with zero toe radius. The fatigue life prediction employs the FCGA software which is able to model the multiple weld toe cracks found in the test result. The weld toe radius dependent $M_{k,c}$ factor reported in this work provide useful data and insights on how to incorporate weld toe radius effects in fatigue life prediction applications for welded joints.

References

1. British Standards Institution, *BS 7910:2013.* (2013)
2. S. Tanaka, T. Kawahara, and H. Okada, Mar. Struct., **39**, 315–334, (2014)
3. P. C. Paris and F. Erdogan, J. Fluids Eng., **85**, 528, (1963)
4. W. He, J. Liu, and D. Xie, Mar. Struct., **35**, 45–69, (2014)
5. S. J. Maddox, Int. J. Fract., **11**, 2, 221–243, (1975)
6. X. Niu and G. Glinka, Int. J. Fract., **40**, 255, (1989)
7. F. P. Brennan, W. D. Dover, R. F. Karé, and A. K. Hellier, Int. J. Fatigue, **21**, 1051–1062, (1999)
8. Y. H. Zhang, S. J. Maddox, and G. R. Razmjoo, 36134, 525–533, (2002)
9. D. Bowness and M. M. K. Lee, Int. J. Fatigue, **22**, 369–387, (2000)
10. J. H. L. Pang, H. J. Hoh, K. S. Tsang, J. Low, S. C. Kong, and W. G. Yuan, Int. J. Fatigue, **94**, 158–165, (2017)
11. J. H. L. Pang, K. S. Tsang, and H. J. Hoh, Mar. Struct., **48**, 1–14, (2016)
12. J. C. Newman and I. S. Raju, *Eng. Fract. Mech.*, **15**, 185–192, (1981)

Complex fibers orientation distribution evaluation in short glass fiber-reinforced thermoplastic (PA66 GF50)

Prashanth Santharam[1, 2], *Thomas* Parenteau[1, 2], *Pierre* Charrier[1], *Denis* Taveau[1], *Vincent* Le Saux[2], and *Yann* Marco[2]

[1]Vibracoustic, CAE and Durability Prediction Department, 44474 Carquefou, France
[2]Institut de Recherche Dupuy de Lôme (IRDL), UMR CNRS 6027, ENSTA Bretagne, France

Abstract. The use of composites made of polyamide 6.6 matrix and short-glass-fibers in automobile industry is progressive due to its low density and cost. The injection molded short glass fiber-reinforced (SGFR) thermoplastics structural parts such as intake, manifold and engine mount housing, induce complex fiber orientation distributions (FOD). This microstructure governs the macroscopic properties such as the mechanical stiffness and fatigue resistance. To estimate the FOD on such industrial parts at complex angles and ribs, we rely on simulation results and micro tomography analysis. The interest of this paper is to develop a semi-automated, quick and efficient orientation tensor identification approach from 2D microscopic images, which is capable of observing relatively larger surface compared to micro tomography. We finally conclude by comparing it with micro tomography and simulation results. Furthermore, we investigate its relevance with fatigue service ability.

1 Introduction

The urge to replace heavy metallic parts had made the car manufacturers look for light weight and environment friendly alternatives. Wherein, short glass fiber-reinforced (SGFR) thermoplastics fit the exact requirements. They are cost efficient solutions, which combine sufficient stiffness for many structural components and large freedom of shapes provided by injection molding. During the process of injection, the SGFR exhibits specific microstructure. It is characterized by a heterogeneous orientation of the fibers (that leads to an anisotropic mechanical behavior) along the structural part, and also in thickness due to the well-known skin-core effect [1-3]. Due to the macroscopic behavior impact, mechanical response and fatigue resistance of the part are affected to a considerable extent [1]. Currently, the FOD predicted by commercial simulation software are used by the companies in their design loop to improve their parts. The difference between the reality seen on the parts and these simulation results are questionable, thereby creating a demand to clarify the results with experimental analysis. In some cases, even micro-tomography analysis is seen as a replacement. The evaluation of these tensors is clearly a key point for the design loop, representative with the geometric details met on industrial parts. In order to provide a better or faster description, this paper presents a semi-automated approach, which could be used to analyze large areas of the sample. This 2D microscopic images stores less space than that of tomography analysis. Similarly, the time taken to process the images is also significantly lower. First, the experimental protocols and the image computing procedures are detailed and then a comparison to micro-tomography result is achieved on a simple case. Then, the analysis is extended to describe the orientation tensors on a wider area, and for more complex orientation on structural samples defined previously [1]. Moreover, the identification of additional features like cracks, porosities and clusters of fibers are also presented. Finally, the orientation tensors identified are compared on one hand, to the ones predicted by Moldflow and on the other hand, to the thermal fields and crack locations observed during fatigue tests.

2 Experiments

2.1. Material and samples

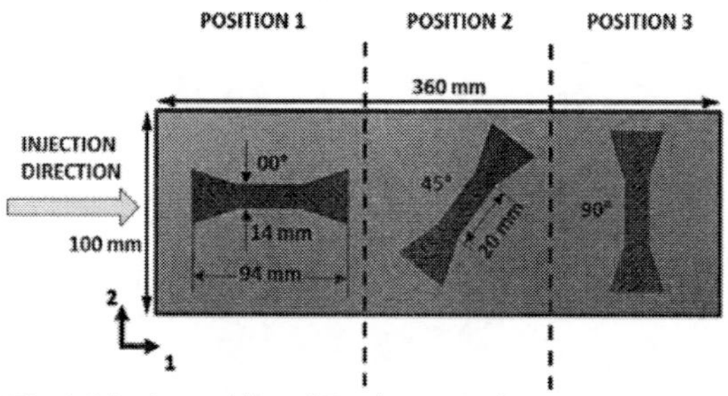

Fig. 1. Injection molding of Dog bone samples.

The material investigated here is a thermoplastic, reinforced with 50% in mass glass fibers (PA66 GF50)

prashanth.santharam@vibracoustic.com

 https://doi.org/10.1051/matecconf/201816522026

provided by Solvay company. The samples were tested at an ambient temperature in a conditioned state (RH50).Two types of samples were used in this study. The first one is called the Dog-bone specimen and is quite standard even if here, we use the Solvay geometry. The Dog-bone specimens are cut from the ~3mm thick injection plates as shown in Fig.1. These plates have homogenous and distinct layers though the thickness. For this reason, it's a suitable sample to perform one dimensional analysis of the orientation tensor. It's also necessary to keep in mind that the location *(position 1, 2 or 3 shown in Fig. 1)* of the sample within the plate has an influence on the orientation tensor evolution at the skin [1]. The orientation of the cut allows for varying the tensile direction versus the main injection direction. In the following, only 0 degree/position 2 samples will be presented.

The second kind of sample is much more specific and is defined to be representative of some geometric details (and therefore injection flows) met on industrial parts. The geometry of the samples called T-Bone is presented in Fig. 2. For this same geometry, two injection locations are possible, providing a very interesting way to modify the obtained microstructure, especially at the flow bifurcation. The first gate location is on the side and is presented in Fig. 2, referenced as T-Bone 1B. The second gate location is from the tip of the central rib and is referenced as T-Bone 1A.

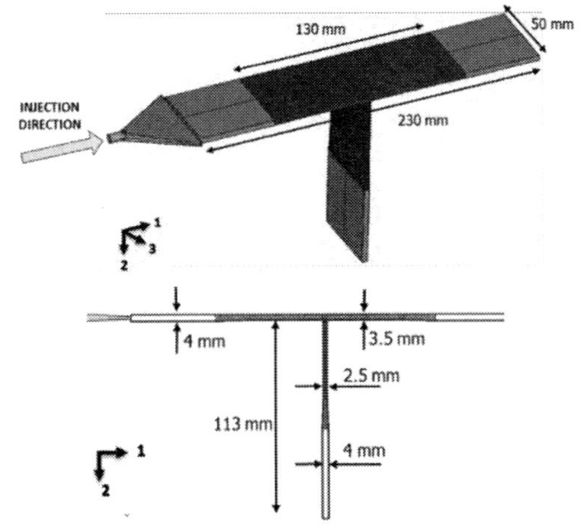

Fig.2 Injection molding of T bone 1B samples.

2.2 Microstructural observations

2.2.1 Optical observation

The Keyence VHX5000 microscope was used to perform 2D micrograph analysis. The samples are cut to the required location and angle, after which they could be grouped together in a resin (to be time efficient) for the polishing protocol. The resin samples undergo the polish protocol, starting from larger grain size to smaller ones; with polish time ranging from 4 min to 20 min. The images taken after every polish step are shown in Fig. 3. The edges of the fibers should be preserved with care, as they are necessary to identify their full length. The polish protocol should be enough to distinguish the fibers and matrix as shown in Fig. 3 after the final 3μm polish.

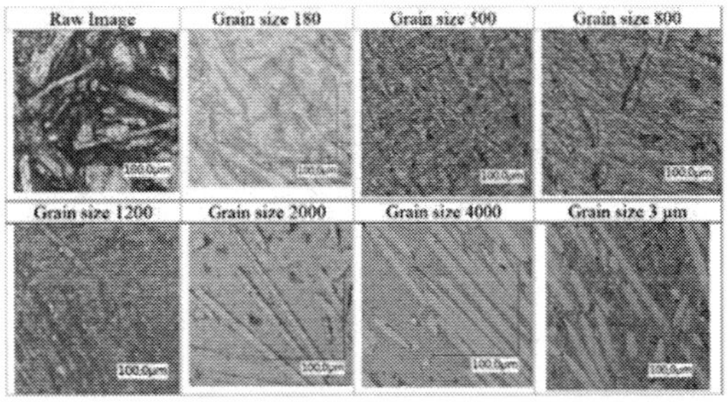

Fig. 3. Specimen surfaces after the various steps of the polishing protocol.

After the polish step, the Keyence microscope is used to take the microscopic 2D images of the sample. Fig. 4 shows the setup of the Keyence microscope observing the samples fixed in the resin. In addition, Fig. 4 indicates the location on the corresponding specimen. Since it is not possible to observe a representative surface with one image, the Keyence image stitching option is used to get a larger view of the specimen. External tools can also be used to serve the purpose [6].

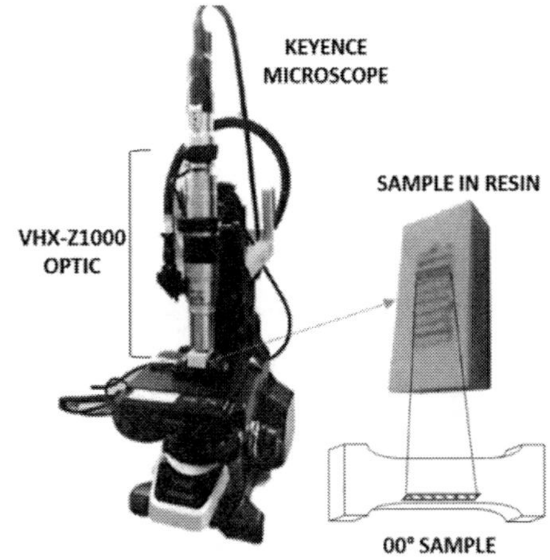

Fig. 4. Polished sample with Keyence microscope.

2.2.2 μ-Tomography Observation

The CT scan is used to perform micro tomography analysis on the same material, 0 degree position 2, to obtain orientation tensor evolution along the thickness of the sample. Target is to illustrate and quantify the skin-core effect of such samples. The calculation of the orientation tensor and its components are well described in [1-3], [7-10]. For simplicity purpose, only the two components of the second order orientation tensor (a11 and a22) are plotted in Fig. 2 [1], wherein direction 1 represents the injection direction and 2 is the one perpendicular to the injection direction as shown in Fig.1. The motivation of presenting this data is to get reliable data that can be compared to the outputs of our image analysis.

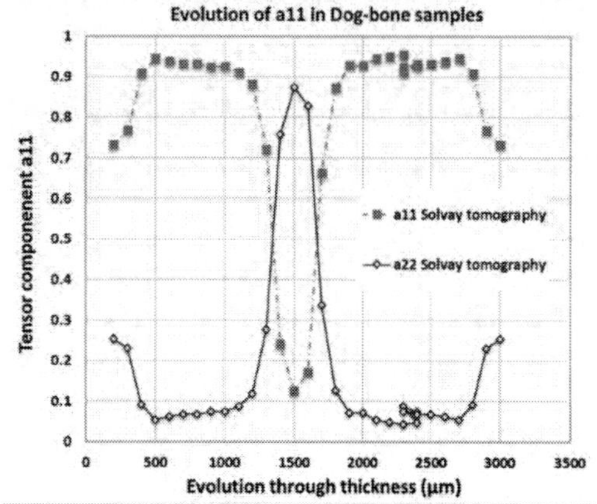

Fig. 5. µ tomography Dog-Bone orientation tensor

2.3 Mechanical Test on T-Bone 1B

The T-Bone 1B sample has been tested under cyclic flexural conditions. The test was displacement controlled, with load ratio and the test frequency as R=0 and 1 Hz respectively. The details of these tests can be found in [1, 4]. From experience gained from previous experimental observations, the dissipation fields of the T-bone 1B sample indicated that the primary location of failure originates on the right arm (opposite to the injection direction) of the sample almost at the mid-point along the specimen as shown in Fig. 6 [1]. Fig. 6 shows the initial thermal field and the dissipated energy in the region of interest (dashed red box in Fig. 6). During these tests, the thermal fields were recorded by an infrared camera (FLIR SC7600-BB) using a 50mm lens. The thermal resolution is improved with a pixel-wise calibration, taking into account the housing temperature [4]. The calculation of dissipation fields are discussed more clearly in [1, 4].

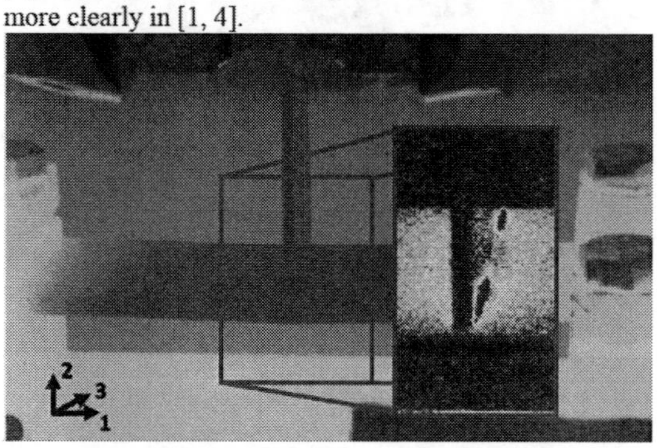

Fig. 6. Thermal field and dissipated energy of T-bone 1B

In order to focus on the location of interest, the T-bone 1B sample was cut into two halves as shown in Fig. 7 (left). During the test, the central rib is fixed in place while the left and right arms are moved simultaneously with the help of an experimental set up shown in Fig. 7.

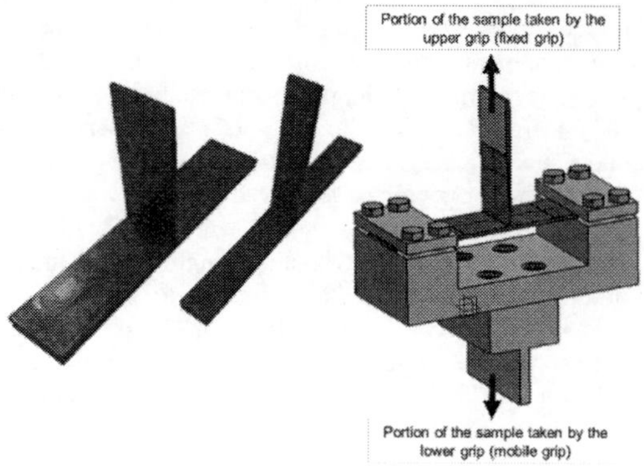

Fig. 7. Half and full T-Bone 1B sample with testing apparatus

3 Tools for microscopic analysis

3.1. Measurement of Orientation Tensor on Dog-bone samples

The orientation tensor is a second order tensor used to describe the orientation of fibers in composites. Based on the requirements, element based average values of these tensors are used for analysis. Each fiber can be represented by a unit vector $\underline{p}$ carried by its principal axis. It is thus possible to define the orientation of a fiber by the projection of this vector in a global coordinate system (Fig. 9). The components of this vector can be evaluated thanks to the angles (θ, φ) [1-3], [6, 9].

$$\underline{p} = \begin{cases} p_1 = sin\theta cos\phi \\ p_2 = sin\theta sin\phi \\ p_3 = cos\theta \end{cases} \quad (1)$$

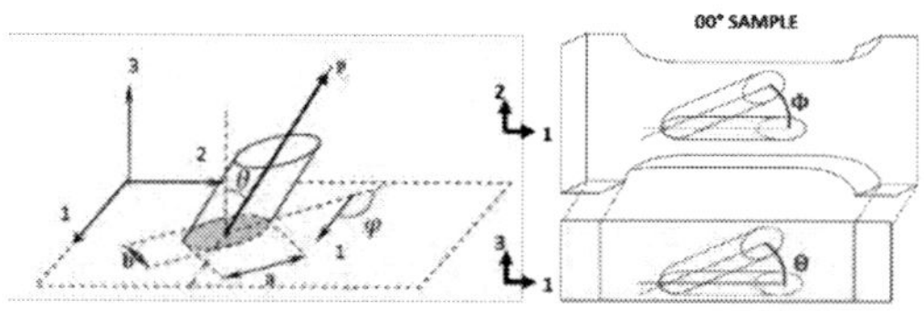

Fig. 8. Orientation tensor calculation [2]

The second order tensor orientation of a fiber k is determined as the double tensor product of $\underline{p}$ as given by equation (2) [1-3], [9].

$$\underline{\underline{a}} = \begin{pmatrix} \sin^2\theta^k\cos^2\phi^k & \sin^2\theta^k sin\phi^k cos\phi^k & sin\,\theta^k cos\theta^k cos\phi^k \\ \sin^2\theta^k sin\phi^k cos\phi^k & \sin^2\theta^k sin^2\phi^k & sin\,\theta^k cos\theta^k sin\phi^k \\ sin\,\theta^k cos\theta^k cos\phi^k & sin\,\theta^k cos\theta^k sin\phi^k & \cos^2\theta^k \end{pmatrix}$$

(2)

In order to define this tensor by optical analysis (i.e. by observing the surface), initially, we detect every fibers in the image by identifying the ellipse formed by the cutting plane and the fiber as shown in Fig. 8. Algorithms such as diffusion and water-shed were used in the process. Each ellipse is characterized by its two axes, 'a'- the major and 'b'- the minor as shown in Fig. 8 [4, 5]. The angle θ is defined by equation (3) [2, 9, 10].

$$\theta = \cos^{-1}\left(\frac{b}{a}\right) \quad (3)$$

The angle between the axis 1 and the major axis of the ellipse shown in Fig. 8 [2], corresponds to φ angle and it depends on the second-moments of the ellipse. The second order orientation tensor can thus be determined for a fiber. To reconstruct the global tensor orientation, it's also necessary to take into account that the probability of cutting a fiber perpendicular to the surface which depends of the length-'l' of the considered fiber, its diameter-'d' and its orientation. This function F^k is defined in [2], [7-10].

3.1. Measurement of Orientation Tensor on Dog Bone samples

Now that the orientation tensor is calculated for every fiber in the given image, the next step would be to group the fibers. Based on the aspect ratio of the given image, it is subdivided into bins [6]. For example if the given image has equal number of pixels in X and Y directions, then the number of bins can also be equal in X and Y directions (columns and rows). Fig. 9 is an illustration for the image with bins, it also explains that the fibers are classified based on the centroid location of its bounding box. For the example given in Fig. 9, the fiber is related to grid (2, 2) since the centroid of the fiber is in it. The tensor value of every grid is calculated by a simple average of all the fibers (its centroids) located in the grid.

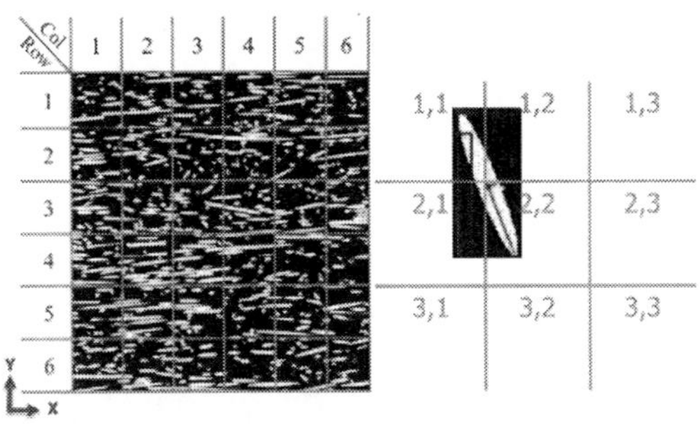

Fig. 9. Classification of fibers using grids

The Fig. 10 is an example of an image (equal pixels in both directions) from 0 degree skin sample position 2 polished from in direction 3 (refer Fig. 1) until the depth of 500μm (approximately in the skin of the sample). Since it's a square image, it was divided into 50 bins in both directions, which are latter plotted as classical histogram. As mentioned earlier in section 1.2, the layers are almost homogeneous, which can be clarified by column wise average tensor value plot shown in Fig. 10. Wherein, the total average value of tensor component $a_{11} = 0.89$.

Similarly, on viewing the sample from direction 2, a scattered view of the tensor through the thickness of the

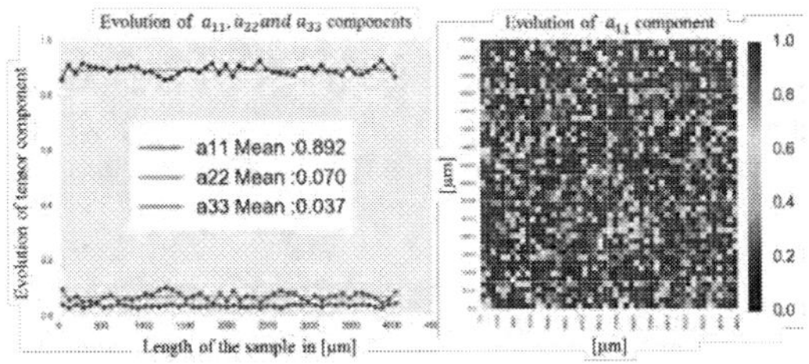

Fig. 10. Tensor Map for top view sample

sample is observed as shown in Fig. 11.The values at the skin are not clearly predicted because the side view (direction 2 view) did not have the good estimation of the tensor component a_{12}, which corresponds to the phi (ϕ) angle in Fig. 8. For this reason, the phi (ϕ) angles are estimated by the top view (direction 3 view as explained in Fig. 10) to rectify the tensor component obtained from the side view (direction 2 view). Which is basically reconstructing the tensor components, considering the phi (ϕ) values obtained from the plane 1, 2 and theta (θ) obtained from plane 2, 3 (refer Fig. 11). The a_{11} *and* a_{22} tensor components obtained by this calculation is compared with the orientation tensor components obtained by Solvay [1-3] in Fig. 11. The trend of the evolution is quite comparable except for the values at the skin for both a_{11} *and* a_{22} components.

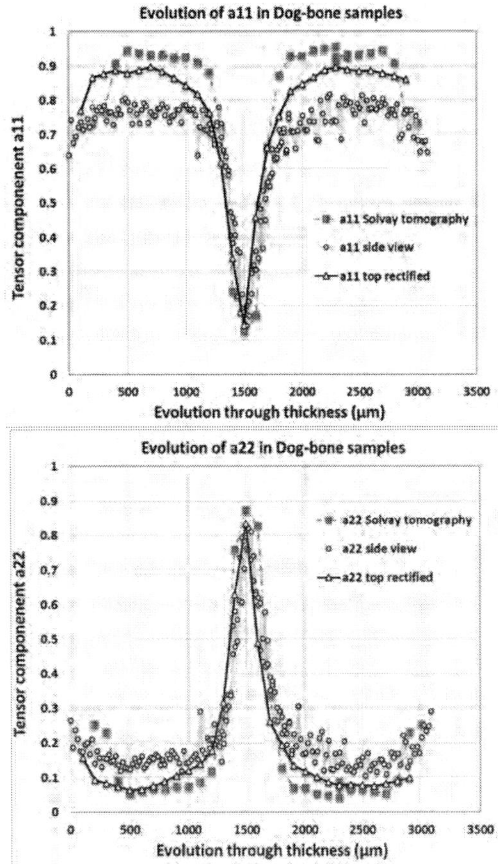

Fig. 11. Evolutions of a_{11} *and* a_{22} for 0 degree position 2 sample

 https://doi.org/10.1051/matecconf/201816522026

3.2. Measurement of Orientation Tensor on T bone

In order to extend the orientation calculation approach on T-bone 1B sample at the region of interest (section 2.3), four microscopic images are obtained from the Keyence, which are combined during the processing step by recognizing the patters in the images [6]. This led to 300*200 bin histogram data, and the a_{11} component of T-bone 1B is the plot is shown in Fig. 12.

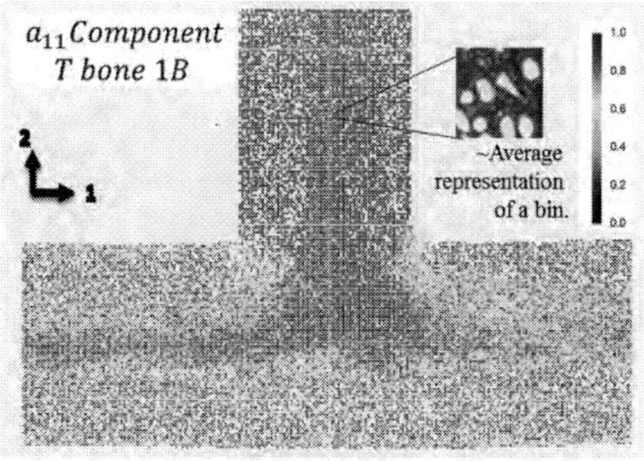

Fig. 12. Orientation tensor map for T-bone 1B sample

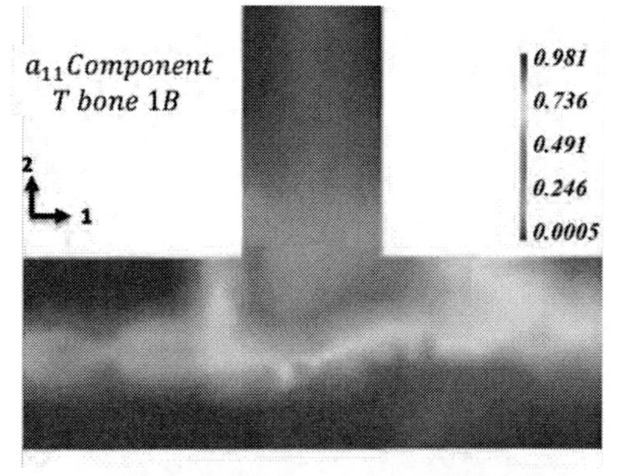

Fig. 13. Orientation tensor for T-bone 1B - Moldflow

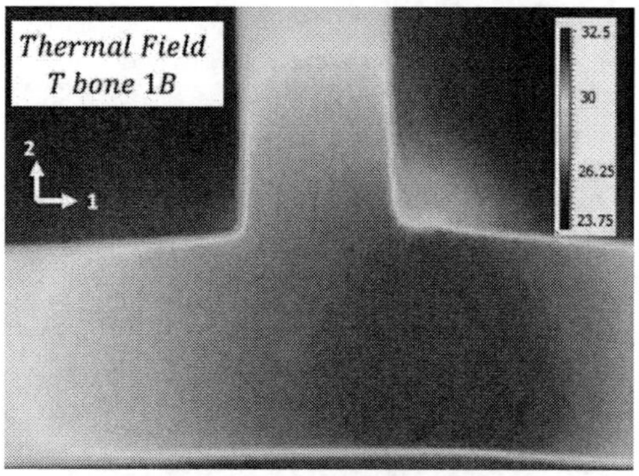

Fig. 14. Thermal field for T-Bone 1B sample

In the figure, the average representation of one bin in the histogram is highlighted. It is possible to notice that there are many empty bins, which means that no fiber were detected or highlights the presence of fiber cluster or void or crack which have replaced the fibers in that location. Detection of data other than fibers are discussed in section 3.3.

3.3. Measurement of additional features

When attempts were made to analyze the microscopic images even further, it was possible to distinguish the voids and cracks from the rest of the matrix and fibers based on grey scale as shown in Fig. 15 (grey scale of Keyence Image (A) in Fig. 16) [5, 6]. The dark pixels in the image were identified either as cracks or voids, therefore the peak at the darker region were identified.

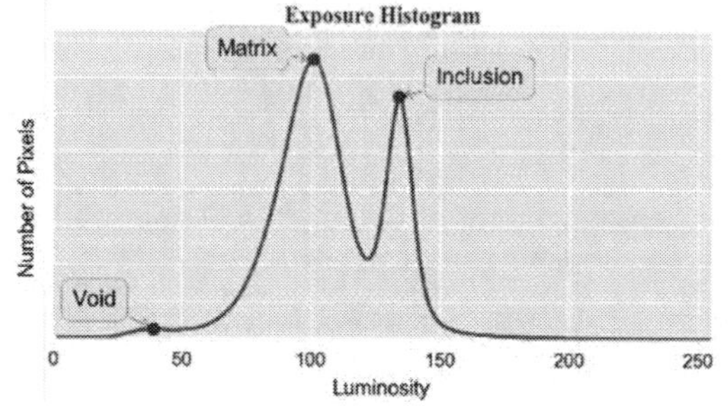

Fig. 15. Grey scale of the total image used for analysis

In Fig. 16, the Keyence Image (A) show a closer view of the T-Bone 1B sample, the Processed Image (A) shows the processed data which has identified the voids as pixels are presented (cracks can also be located). With this information, it is possible to have a good estimation of distribution and size of the voids and cracks. The T-bone 1B sample, many such voids shown in Fig. 16 were clearly located at the flow bifurcation region. For the same sample, when viewed from a different direction as shown in Fig. 17, many clusters and voids are noticed. The clusters are normally identified if the total volume of the detected contour is larger than the maximum possible area of the fiber [5, 6]. Similar to voids, it is possible to quantify the location, distribution, size in addition the orientation of the clusters.

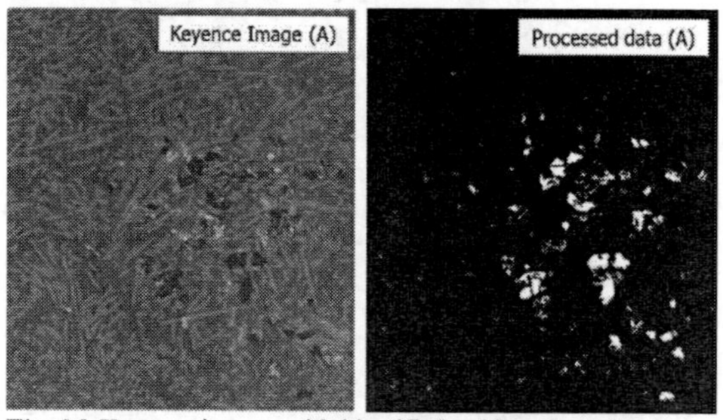

Fig. 16. Keyence images with identified voids and cracks

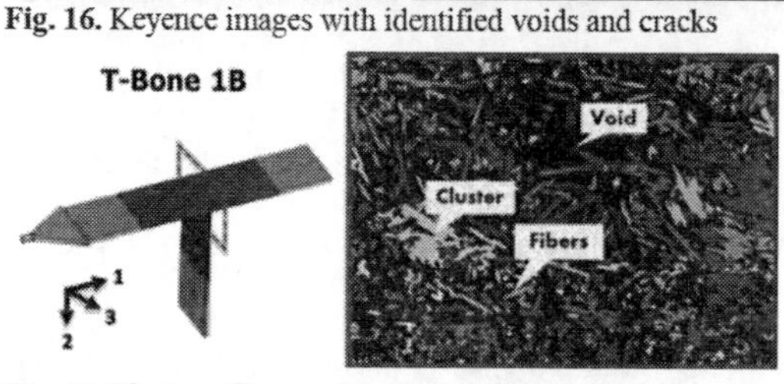

Fig. 17. Clusters, fibers and voids in T-Bone 1B sample

4 Comparison to numerical predictions and to fatigue observations

4.1. Comparison to the orientation tensor predicted by injection simulation

The Moldflow simulation results for the same T-bone 1B sample is presented in Fig. 13 [1], wherein the injection is along direction 1. On the first view, the trend of the orientation tensor is quite similar to Fig. 12 obtained by microscope observations. The first remark is that, in both cases, the orientations in the left and right arms are completely different, at least on the regions that are close to the bifurcation. This is of course induced by the unbalanced filling history, as the gate is located on the left arm along direction 1. The second remark is related to the mesh size and quality, Fig. 13 shows a Moldflow mesh with 12 element layer (through thickness) with 1.7 million elements in full model. This quality of mesh was enough to describe the overall orientation gradient. If we now come specifically to the comparison between the simulations and the microscopic observations, the following points are observed. The skin of the left arm appears to be well oriented in both the cases. Even the aftermath of the bifurcation is clearly visible by change of directions in the orientation tensor. On closer examination, the skin-core effect is more pronounced in the experiment than in the simulation. The skin of the central rib is well described with lower values in the a_{11} component and higher values in a_{22} component. Even the values at the core are comparable. It therefore seems that the injection simulations give a reasonable evaluation of the global shape of the orientation tensor, but fails to identify the local gradients.

4.2. Comparison to the fatigue observations

The thermal fields were observed on regular intervals around the region of interest during the fatigue test. The fatigue test followed the test discipline mentioned in section 2.3. Further details of the self-heating tests are mentioned in [1]. For fatigue tests on different loading conditions, the T-bone 1B sample always failed on the right arm (opposite to the injection direction). This was repetitive and even the location of failure was consistent [1]. The orientation tensor could explain this observation, as the combination of the mechanical fields

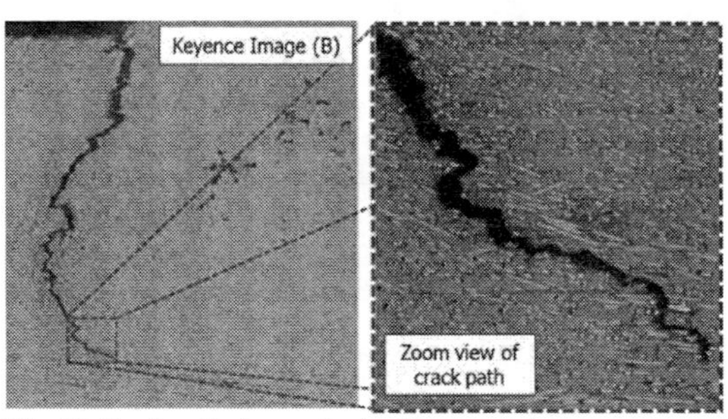

Fig. 18. Crack path observed using Keyence microscope

induced by the stress concentration in the vicinity of the radius will lead to different failure consequences as the local orientations are clearly different. This point would require a deeper investigation using Finite Elements simulations. The last point to highlight is that the crack propagation path appears to follow the direction of orientation as seen in the Keyence Image (B) at Fig. 18 for T-bone 1A sample.

5 Conclusions

This paper presented a semi-automated approach to determine the SGFR orientation tensors, based on 2D microscopic images and on principles already developed elsewhere. This technique exhibited a faster analysis protocol than the ones using micro-tomography. Moreover, a larger area can be efficiently analyzed thanks to the combination of several observation zones. Beyond the orientation tensor, the developed tools are also giving access to other interesting microstructural features (cracks, porosities and clusters of fibers). Finally, the identified orientation tensors have been compared to the ones predicted by Moldflow and to the thermal fields and crack locations observed during the fatigue tests. It appears that these comparisons provide very precious understanding and validation cases. The perspectives are wide on this topic, from measuring the thermal fields at the microstructure's scale, to finer investigation of the mechanical failure dependency on the microstructure characteristics.

References

1. L. Serrano, *Caractérisation thermomécanique du comportement en fatigue des thermoplastiques renforcés de fibres de verre courtes*, PhD UBO, Brest (2015)
2. A. Megally, *Étude et modélisation de l'orientation de fibres dans des thermoplastiques renforcés*, PhD ENSAM, Paris (2005)
3. D. Dray, *Prediction of the thermoplastic properties of an injection molded composite,* PhD ENSAM, Paris (2008)
4. V. Le Saux, C. Doudard, *Infrared Physics and Technology* **80**, 83-92 (2017)
5. L. Grady, *IEEE Trans. Pattern Analysis and Machine Intelligence* **28,** 1768-1783 (2016)
6. J. Erik Solem, *Programming Computer Vision with Python*, (2012)
7. R.S. Bay and C.L. Tucker III, *Polymer Engineering & Science*, **32**(4), 240-253, 317, 332 (1992)
8. C. Eberhardt and A. Clarke, *Composites Science and Technology*, **61**(10), 1389-1400 (2001)
9. S.G. Advani and C.L. Tucker III, *Journal of Rheology*, 31(8), 751-784 (1987)
10. M. Gupta and K.K Wang, *Polymer Composite* **33**, 410 (1993)

Research of crack initiation in tension test specimen made of nickel alloy EI961sh

Kamaliddin Karimbaev[1,*], *Ivan* Pleshcheev[1], and *Elena* Bredihina[1]

[1] Central Institute of Aviation Motors, Strength and dynamics Department, 111116, Russia

Abstract. In this paper method of numerical computations using explicit scheme, implemented in LS-Dyna (Ansys) software complex, is introduced and verified. Obtained solution explains experimental fact mentioned by P. Ludwik, that cracking in cylindrical test specimen made of plastic material starts in the middle of the smallest cross-section of the specimen. Introduced method allows verifying law, obtained by N.N. Davidenkov during unique experimental research in which he studied logarithmic strain in specimen's neck by pickling of cross-sections in this zone. Additionally it is possible to estimate amount of heat, generated during rupture of specimen, using obtained solution and law of conservation of energy. For this purpose tensile tests, in which thermal camera was used for temperature measuring, were conducted. It was shown, that all deformation energy apart from elastic and shape-forming energy in volume element, calculated in the moment before rupture is transformed to heat. Also tensile tests with various rate of loading were conducted for more detailed research of rupture process.

1.Introduction

Most strength properties of materials are obtained from standard tension test. Ductile materials usually crack after initiation of neck, when severe plastic deformation occurs. Strength limit is chosen as the largest stress, occurred during experiments. However, in various constructions (for example in disks) cracking occurs without necking. Therefore,other methods of calculating ultimate strengthare proposed. For example, in [1] true stress is proposed in form:

$$S_k=\sigma_{fr}/(1-\psi) \quad (1)$$

where, $S_k = P/A$ - ultimate fracture stress, $\sigma_{fr}= P/A_0$ – engineering fracture stress, ψ – plasticity parameter, P – axial load, A, A_0 – actual and initial cross-section area respectively.

P. Ludwik [2]and other researchers showed by means of X-ray equipment, that cracking in cylindrical test specimen made of plastic material starts in the middle of the smallest cross-section of the specimen. After necking the stress state at the shrinking cross-section of the neck is not uniform and uniaxial anymore. That is why true stress obtained by (1) is not really true ultimate stress limit, but it is average axial stress. Experimental and numerical research of stress distribution in neck were introduced in this paper. Maximum axial stress in the middle of the smallest cross-section before fracture was computed. This value is proposed to be true ultimate stress limit.

In the secondsection method of numerical computations using explicit scheme, implemented in LS-Dyna (Ansys) software complex, is introduced and verified. Explicit scheme allows to adequately simulate severe plastic deformation, because it uses equation of motion instead of equilibrium equations. Due to the fact that in explicit scheme small integration time step is needed for achieving numerical stability, implicit scheme was used for calculations in elastic zone, while explicit scheme was used only after plastic yielding appeared. Computed stress distribution was compared with distribution, obtained by N.N. Davidenkov and Spiridonova [3] during experimental research in which they studied logarithmic strain in specimen's neck by pickling of cross-sections in this zone.

In the third section scale factor using specimens with different diameters of cross-section was investigated.

In the fourthsection formula based on the law of conversation of energy, for estimating amount of heat generated during rapture was proposed. To verify this formula tensile tests, in which thermal camera was used for temperature measuring, were conducted.

In the fifthsection dependence between true stress and loading rate was investigated

In the sixth section Poisson's ratio in deep plasticity was investigated using optical deformation measurements method (Vic-3D).

In the seventh section critique of modern methods of fracture mechanics is introduced. Usually crack is introduced in finite-element model after elastic solution is conducted, which might be inaccurate, because in plasticity stress distribution may change.

2. Method of identification of stress state in the smallest cross-section after necking.

Method of calculating true ultimate stress proposed in this paper is based on the calculation of stress state using finite-element analysis of tension test of specimen. Finite element model of cylindrical test specimen is introduced in the figure 1.

Fig. 1. Axisymmetric finite element model of specimen

Axisymmetric formulation of finite elements is used in this model. To archive necking small narrowing is introduced in the model (about 1% reduction of radius of cylinder).As was mentioned before cracking of specimen occurs during severe plastic deformation, that is why it is necessary to use equation of motion. To use reduce computational time solution was conducted in two stages: implicit scheme was used for elastic deformation and explicit scheme for plastic yielding.In the explicit analysis in LS-Dyna static solution is used as preloading. Special numerical computation was conducted to assure that both explicit and implicit schemes provide the same results for elastic deformation.

Main difference of explicit analysis from implicit is that in explicit schemeuncoupled equation of motion is used and step by step solution is conducted. Equations used in Ls-Dyna explicit scheme are:

$$\mathbf{U}_{t+\Delta t}= \mathbf{U}_t + \mathbf{V}_{t+\Delta t/2}\Delta t_{t+\Delta t/2} \quad (2)$$

$$\mathbf{V}_{t+\Delta t/2}= \mathbf{V}_{t-\Delta t/2} + \mathbf{a}_t\Delta t_t \quad (3)$$

$$\mathbf{a}_t = [\mathbf{M}]^{-1} (\mathbf{F}_t^{ext} - \mathbf{F}_t^{int}) \quad (4)$$

$$\mathbf{x}_{t+\Delta t} = \mathbf{x}_0 + \mathbf{U}_{t+\Delta t} \quad (5)$$

$$\Delta t_{cr}= 2/p_{max} \quad (6)$$

Where $\mathbf{U}_t$, – displacement vector for time t, $\mathbf{V}_{t+\Delta t/2}$ – velocity vector for time $t+\Delta t/2$, $\mathbf{a}_t$– acceleration vector. $[\mathbf{M}]^{-1}$– inverse mass matrix, $\mathbf{F}_t^{ext}$,$\mathbf{F}_t^{int}$– external and internal force vector, Δt_{cr}– critical time step, p_{max}– maximum frequency of model.

The base specimen is cylindrical specimen, made of EI961sh alloy with radius r=4 mm. Stress-strain diagram obtained from experiment is shown in figure 2. Loading rate is 0.65 mm/s

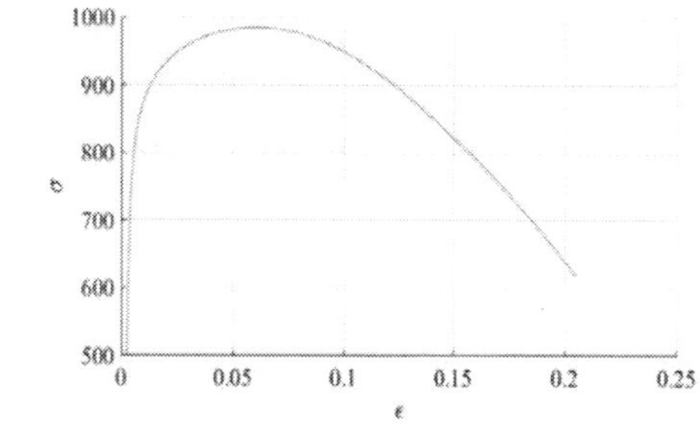

Fig.2. Experimental stress-strain diagram.

Finite element model (see figure 1) contains 3200 elements and 3381 nodes. Standard piecewise material model was used. True stress-strain diagram was obtained from experimental stress-strain diagram

However fracture stress and strain were adjusted so that the average axial stress before fracture and reduction at fracture were equal to those obtained from experiment. Average axial stress before fracture is calculated according (1) and is equal to 1630 MPa

Distribution of axial stress across the smallest cross-section before fracture is showed on the figure3. Ratio of maximum axial stress to average axial stress is 1.23. It is suggested that maximum axial stress in the smallest cross-section is the actual true cracking stress and independent from size and construction of specimen.

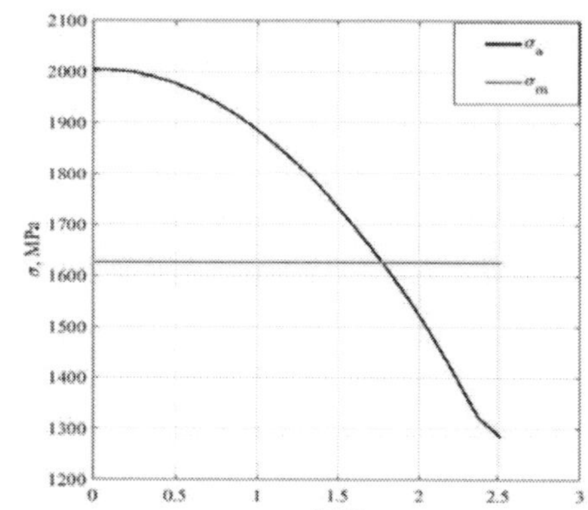

Fig.3. Axial stress distribution and average stress in the smallest cross-section

Davidenkov and Spiridonova, derived following formulas for stress distribution in the smallest cross-section:

$$\sigma_a = \sigma_i(a/2R) \cdot (1 - r^2/a^2 + 2R/a) \quad (7)$$

$$\sigma_r = \sigma_t = \sigma_i(a/2R) \cdot (1 - r^2/a^2) \quad (8)$$

 https://doi.org/10.1051/matecconf/201816522027

Stress distribution from finite-element analysis was compared with distribution, derived by Davidenkov (see figure 4, 5). Neck curative radius R for Davidenkov's formula of distribution was calculated from FE model. Davidenkov's stress distribution is very closeto stress distribution derived by finite element analysis at the time t=4.94 seconds which is Δt = 3.06 seconds before fracture. However for later times Davidenkov's stress distribution is much different from stress distribution obtained from finite element analysis.

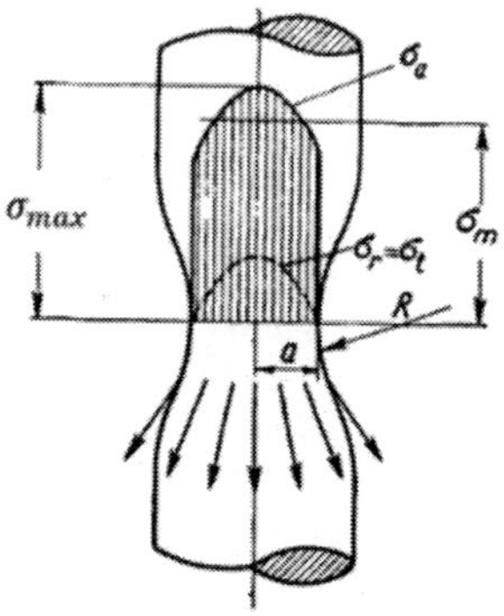

Fig.4. Stress distribution according Davidenkov's formula.

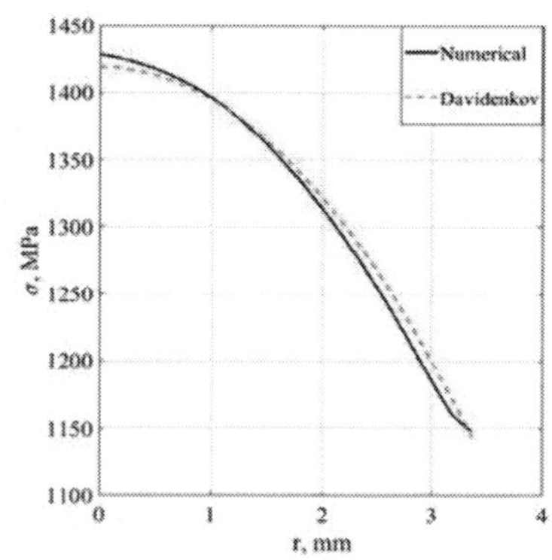

Fig.5. Axial stress distribution according numerical analysis and Davidenkov in the smallest cross-section

Since plastic deformation is dynamic process, stress is a function of time. That is why ratio of maximum axial stress in the specimen to average axial stress is not constant value and is changing during deformation.

3. Scale factor

It is widely known that in various materials true stress, calculated from (1) is decreasing as specimen radius increasing. Usually this fact is explained by statistics, the larger specimen, larger probability of having defects. However in this article it is shown that even in ideal material decreasing of medium axial stress with increasing radius will occur. It could be explained bythe fact that fracture occurs when the true stress limit is reached in the center of cross section. While true stress limit calculated by (1) is averaged axial stress, which means that it will be decreased when the area by which it is evaluated increases.

Figure shows axial stress distribution for different radiuses of specimen, obtained by numerical solution.

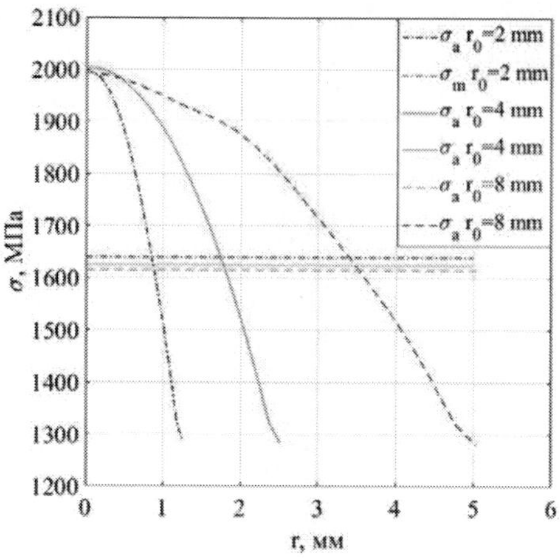

Fig.6. Axial stress distribution before rupture for various radiuses of specimen.

4. Estimation of heat generated during rupture of specimen

To estimate heat generation per volume unit during rupture following expression is proposed:

$$Q = \sigma_{pr}(\varepsilon_{fr} - \varepsilon_{pr}) \qquad (9)$$

where$Q$– heat generation per volume unit, σ_{pr} – proportionality limit, ε_{fr} – rupture deformation, ε_{pr} – elastic deformation

This expression could be obtained by subtracting elastic energy and plastic distortion energy from full deformation energy:

$$U_\Sigma = \int \sigma d\varepsilon \qquad (10)$$

$$U_{el} = \sigma_{pr}\varepsilon_{pr} \qquad (11)$$

$$U_{dis} = \int(\sigma - \sigma_{pr})d\varepsilon \qquad (12)$$

where U_Σ, U_{el}, U_{dis} – full, elastic and distortion deformation energy respectively.

Graphical representation of the result is shown on figure 7.

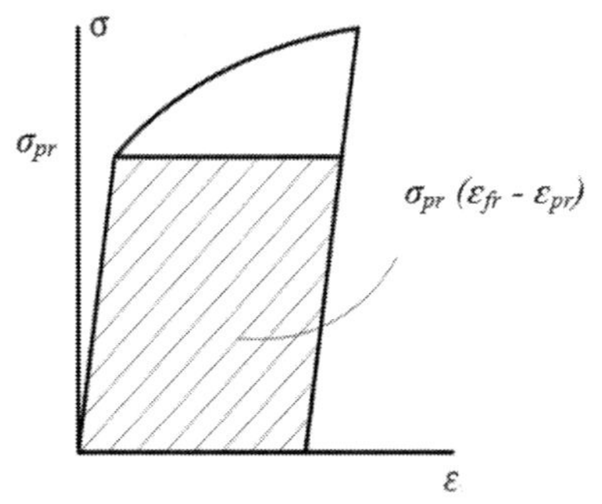

Fig.7. Illustration to formula (9)

To verify this formula tensile test, in which thermal camera was used for temperature measuring, was conducted.Frame frequency of thermal camera was equal to p_{cam}=9 Hz

Figure 8 shows temperatures measured by thermal camera during tests. Fracture occurred during time $0<t_{frac}<1/p_{cam}$. To estimate possible error of temperature measurements extrapolation was made (see Figure 8). Value of maximum measured temperature could be estimated within boundaries $54°C<T_{max}<59°C$. Value of temperature calculated by formula (9) is T_{cal}=55.4°C.

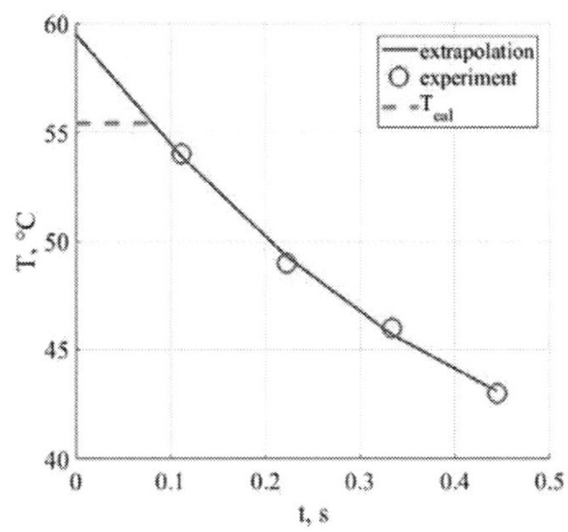

Fig.8. Temperature extrapolation.

5. Effect of loading rate on true ultimate stress limit

Loading rate could effect on true ultimate stress limit. Figure 9 shows 2 stress-strain curve obtained from experiment with various loading rates (0.3 mm/min and 0.65 mm/s). Radius of specimen's cross-section is r=4 mm. Average axial stress before fracture was calculated (1) and equals 1620 MPa and 1630 MPa respectively.

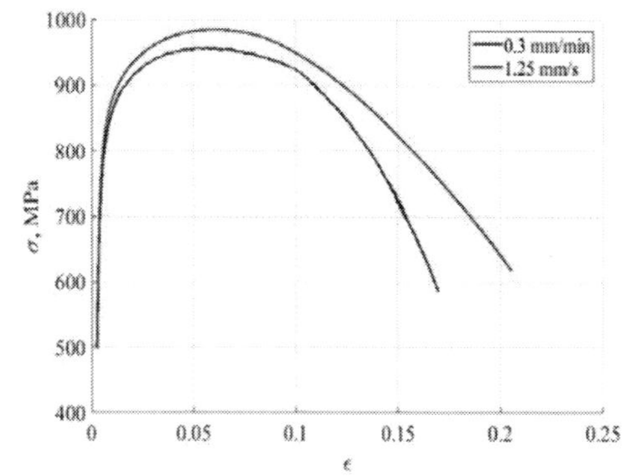

Fig.9. Experimental stress-strain diagram for various load rates.

Numerical experiments were conducted to investigate effect of loading rate upon true ultimate stress limit. Result is shown on the figure 10.

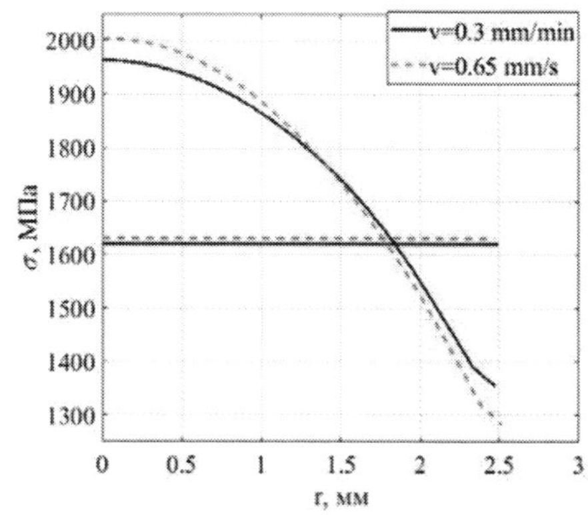

Fig.10. Axial stress distribution and average stress in the smallest cross-section for two different loading rates.

According to numerical simulations and experiments loading rate does not effect on true ultimate stress limit significantly.

6. Poisson's ratio in deep plasticity

Poisson's ratio in plasticity theory is usually suggested to be equal to ν=0,5. In other words material in plastic state behaves like incompressible liquid. However this assumption is still to be verified.

To verify this assumption local axial and traverse strains should be measured. Standard tools of measurement like extensometers gauges are inapplicable in this case. Extensometers cannot measure local deformation. Strain gauges measure local deformation in a given point in the measuring surface area, however, it is very difficult to predict the point of yielding and the necking area. That is why optical measurement system was used (see figure 11). The system can measure local axial and traverse strains during tensile test (see figure 12).

Fig.11. Vic-3D measurement system.

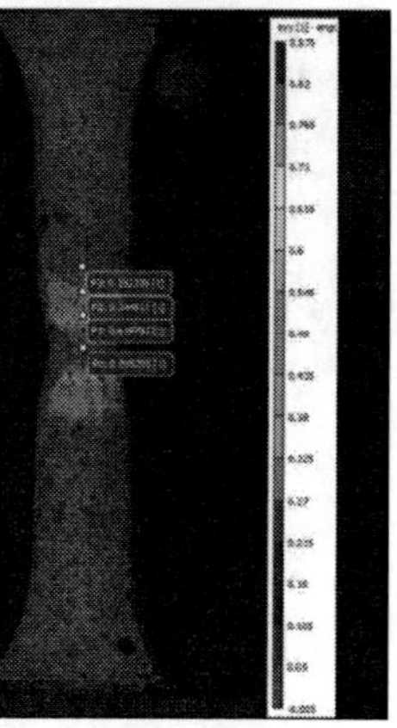

Fig.12. Picture of the specimen taken by optical system

Ratio of traverse strain to axial strain, obtained by optical measurement system for four surface points versus axial strain is shown in the figure 13.

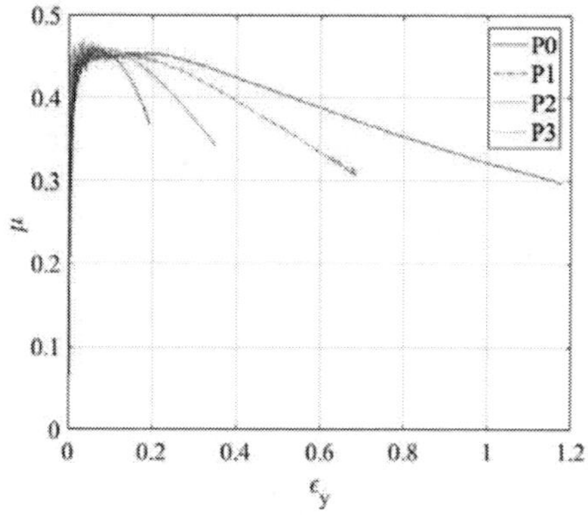

Fig.13. Poisson's ratio versus axial stress.

As can be seen Poisson's ratio changes during loading. In the end of loading Poisson's ratio is decreasing and is very much different than suggested ν=0.5. That is why strain state calculated on this assumption might be inaccurate for EI916sh alloy.

7. Conclusion (about using plasticity analysis in fracture mechanics)

To evaluate life expectancy of construction with possible defects fracture mechanics is used.Modern fracture mechanics is based on elasticity theory to investigate stress state near the crack tip. In elastic approach crack tip becomes singularity point, in which infinite value of stress occurs.

To conduct solution correctly and consistently often computation of construction stress state without cracks is evaluated in elastic zone. After that cracks are introduced into zones, with largest stress values.It is known that stress distribution in elasticity theory is function, which is subject to the Cauchy conditions. Such function reaches an extremum at the boundaries of body. So after solving elastic problem for construction maximum stress occurs only at the boundariesof the body.

However, as one can see in the tension test specimen fracture starts in the center of specimen not at the boundaries. It means that if plasticity is reached during loading, stress redistribution occurs. Therefore stress distribution may be much different from distribution, obtained from elastic solution and maximum stress may occur in different location.

That is why for proper crack analysis plasticity should be taken into account. Though it is worthy of note that in real constructions presence of torque and bending load in elastic-plastic analysis leads to stress distribution with maximum stress at the body boundaries.

Acknowledgments

We wish to greatly acknowledge to Denis Palchikov for conducting experiment using optical measurement system and to Mikhail Volkov for thermal camera measurements of heat generation during rupture.

References

1. I.A. Birger, B.F. Shorr, (eds) (1993), Raschet na prochnost detaley mashin [Strength analysis of machine parts], Mashinostroenie, Moscow, Russia
2. P. Ludwik, Z. Ver. Deut. Ing., **71** (1927), 1532.
3. N.N. Davidenkov, N.I. Spiridonova, Mechanical methods of testing. Analysis of the state of stress in neck of a tension test specimen, American Society for Testing and Materials, **46**, pp. 1147-1158, 1947

A new method for evaluating the influences of surface topography on fatigue propriety of the random machined surfaces

Lin Ma[1,2], *GuangTao* Xu[1,2*], *Gang* Wang[1,2] ,*and MingHao* Zhao[1,2,3]

[1]School of Mechanical Engineering, Zhengzhou University, Zhengzhou, 450001, China
[2]Henan Key Engineering Laboratory for Anti-fatigue Manufacturing Technology, Zhengzhou, 450001, China
[3] School of Mechanics & Engineering Science, Zhengzhou University, Zhengzhou, Henan, 450001, China

Abstract. The Stress concentration factor (SCF) induced by the machined surface is more complex than that resulting from macro-geometry discontinuities and has great effect on fatigue life of structure. However, another important parameter, stress gradient (SG), was always ignored. The notch roots or valleys of the wave surface constitute fatigue hot points, where cracks occur, so it is essential to study the SCF and SG at valleys rather than just the root-mean-square SCF variable. In this work, a new method for evaluating the influences of surface topography on fatigue propriety of the random machined surfaces was given. An analytical method using Fourier transformation to simulate machined surface topography is presented. Analytical formulae for SCF and SG for random machined surfaces are derived subjected to a general loading and validate these formulae via finite element method (FEM). Joint probability-distribution function for SCF and SG at the valleys of the random machined-surface topography of the machined sample was obtained after different cycles fatigue test. This method gave us how the surface topography effect the fatigue properties of machined components. Fatigue test of machined sample for a single crystal nickel based alloy were established for validated this method. The obtained results should be useful in studying and evaluating fatigue properties of machined components.

1 Introduction

The stress concentration factor (SCF) is an important parameter in the application of mechanical engineering and knowledge of this parameter and the stress distribution around notches is important for engineers engaged in the design or fatigue-life prediction of engineering components [1-4].

However, the stress concentration induced by the machined surface is more complex than that resulting from macro-geometry discontinuitiesand is characterized by the occurrence of (among other features) multiple hotspots. Neuber's rule characterized stress concentrations introduced by machined surface topography [3]. Castagnetti and Dragoni [5] proposed a method for modifying this rule, by changing the expression of the depth-reduction factor, to increase the range of applicability and improve the corresponding accuracy. Gao [6] analyzed the stress concentration for slightly undulating surfaces by employing elastic Green's functions. The first-order solution of the SCF was obtained by treating the surface as a perturbation from a reference plane. Recently, Medina [7] derived a stress concentration formula for an arbitrary shallow surface using the first-order perturbation approach, which can be applied in various cases (including undulating surfaces, parabolic notches, and Gaussian notches) according to Rice [8], for the analysis of the stress intensity resulting from a varying crack front. Medina and Hinderliter [9] obtained an analytical solution for the SCF of a slightly roughened random surface by combining Gao's method, the Hilbert transform and the energy conservation principle.

Stress concentrations induced by the surface topography have a significant effect on the fatigue life of structures [10, 11]. The Stress Gradient (SG) near the stress concentration may also influence the fatigue life of components [4, 12-14]. The relative SG ahead of notches, as calculated (with a high degree of accuracy) by Filippini [15], exhibited a strong dependence on only the notch radius. The volumetric method can be used to determine the fatigue life from the relative SG [16]. Moreover, the influence of the size effect and the SG on the high-cycle fatigue strength of a DIN EN 10088-3 martensitic stainless steel has been experimentally determined and simulated via the finite element method [17]. The corresponding results confirmed that SG has a significant effect on the fatigue strength.

There are many roughness parameter used to evaluate the fatigue life[18], such as, the average roughness (Ra), peak to valley height roughness (Ry) and10-points roughness (Rz) to characterize roughness of a machined surface. And surface quality is positive with increasing fatigue cycles. However, the reduction in fatigue life with increasing Ra did not occur in a consistent manner [19,20]. The fatigue stress concentration factor, related to SG and SCF, can describe the relationship between surface topography and fatigue life. Hovever, there was few work for using the SC and SG to evaluate the

*Corresponding author: xgtzzu@zzu.edu.cn, xgtzzu@163.com

https://doi.org/10.1051/matecconf/201816522028

influences of surface topography on fatigue propriety of the random machined surfaces.

Therefore, SG and SCF, which both are dependent on the surface topography of the machined surface, must be simultaneously considered during fatigue-property evaluation. As such, in the present work, we propose a method to evaluate the influences of surface topography on fatigue propriety of the random machined surfaces. This paper is organized as follows: an approach for simulations of the machined surface from the measured data was given in the section 2, the analytical solutions of SCF and SG, obtained via Fourier transformation in section 3, in the section 4, validation of these solutions under uniform loading via finite element simulations was presented, a 2D joint probability-distribution function of SCF and SG associated with the machined-surface topography was given in section 5, different fatigue test cycles for the proposed method to evaluate the influences of surface topography on fatigue propriety of the random machined surfaces in the section 6. Finally, the conclusions of the study are presented.

2 Analytical simulation of random machined surfaces

The random machined-surface topography can be determined by measuring a series of discrete data points corresponding to location and height, using a BRUKER NPFLEX 3D Optical Microscope (vertical resolution: ~3 nm.) The topography is expressed by superposing a series of cosine components on the surface as follows [21]:

$$y(x) = \sum_{i=1}^{n} a_i \cos\left(\frac{2\pi x}{\lambda_i} + \theta_i\right), \tag{1}$$

where $y(x)$ is the measured surface topography, n is the number of series components, and a_i , λ_i and θ_i are the amplitude, wave-length, and phase angle , respectively, of the cosine series. Using a FFT algorithm code in MATLAB, we can transform the measured topography into a cosine series with known parameters a_i , λ_i and θ_i .

Determining the number of cosine components as shown in Eq.(1) at each point is essential for accurate characterization of the topography. We denote the measured machined-surface topography and the simulated-surface topography as $R(x)$ and $W(x)$, respectively. The difference between these topographies decreases with increasing number of terms in the series, and extremely high frequency components may be neglected, owing to the presence of the noise signal. We defined the following criterion:

$$\left|\frac{R(x)-W(x)}{R(x)}\right| < \delta, \tag{2}$$

where δ is the preset relative error between the original measured machined surface and the simulated surface topography. In this paper, we assume that δ =1%.

As Fig. 1 shows, the simulated surface topography corresponds closely to the measured original machined-surface topography.

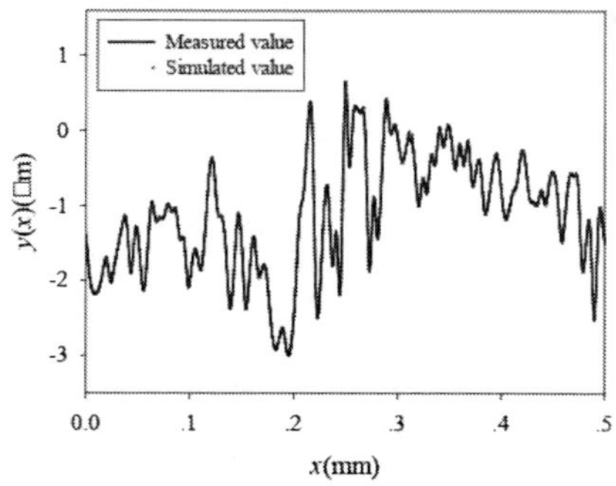

Fig.1. Measured surface topography and simulated surface topography.

3 Stress concentration factor and stress gradient of surface topography

Gao [6] conducted a first-order perturbation analysis of a slightly undulating surface of a half plane subjected to uniform tension T and obtained the following expression for the stress:

$$\sigma_{xx}(x,y) = T\left\{1 + \frac{2}{\pi} \int_{-\infty}^{\infty} Re\left[\frac{x'-x}{\left(x'-x-i[y-a(x)]\right)^3}\right][a(x')-a(x)]dx'\right\}, \tag{3}$$

where $a(x)$: random surface and $|y-a(x)|$: vertical distance from the observation point (x,y) to the surface $y=a(x)$.

Using Fourier transformation, the random surface can be described by Eq. (1) and substituting this expression into Eq. (3), yields get the stress component $\sigma_{xx}(x,d)$ acting at a depth d from the surface:

$$\sigma_{xx}(x,d) = T\left[1 + \sum_{i=1}^{n} \frac{4\pi a_i}{\lambda_i}\left(1 - \frac{\pi d}{\lambda_i}\right) e^{-\frac{2d\pi}{\lambda}} \cos\left(\frac{2\pi x}{\lambda_i} + \theta_i\right)\right]. \tag{4}$$

By inserting $d=0$ into Eq. (4), we obtain the analytical expressions for SCF and SG at any point on the random surface. These expressions are given as follows:

$$K_t(x)=\frac{\sigma_{xx}}{T}=1-4\pi\sum_{i=1}^{n}\frac{a_i}{\lambda_i}\cos\left(\frac{2\pi x}{\lambda_i}+\theta_i\right), \qquad (5a)$$

$$\chi(x)=\frac{\partial\sigma_{xx}}{\partial d}/T=-12\pi^2\sum_{i=1}^{n}\frac{a_i}{\lambda_i^2}\cos\left(\frac{2\pi x}{\lambda_i}+\theta_i\right). \qquad (5b)$$

The SCF obtained for a single cosine wave with $a/\lambda<0.1$, differed a maximum of 8% from that determined via the finite element method [22]. Eqs. (5a) and (5b) are based on and, hence, has the same precision as Eq. (3).

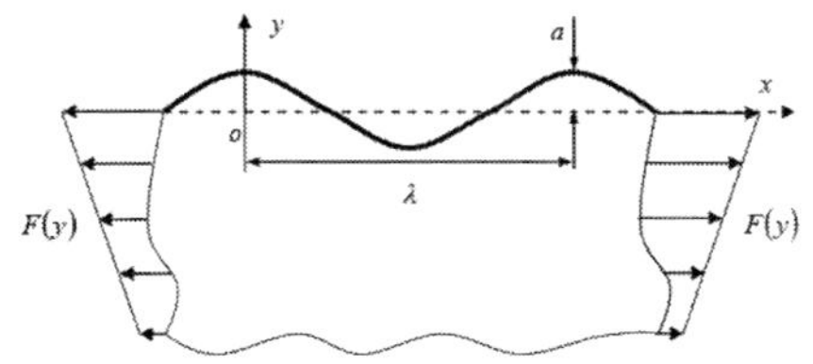

Fig.2. Single cosine wave surface schematic model under non-uniform loading.

4 Validation of the analytical solutions via finite element simulations

4.1. SCF and SG under non-uniform loading

As a numerical example for the section 3, we consider a half plane with a machined surface, subjected to a linear load, $F(y)$, as shown in Fig. 2, to validate the analytical solutions (5a) and (5b). Non-uniform loading (such as a bending load) is frequently encountered in engineering activities. Therefore, the effect of linear loading on the SCF and SG of machined surface must be determined.

The surface amplitude is negligible compared with the characteristic size of the components. Therefore, the SCF is assumed to be the same as that associated with uniform stress, and is described by the first-order solution given in Eq. (5a), where

$$T=F(0), \qquad (6a)$$

whereas the SG is calculated from:

$$\chi_B=\chi_T+\chi_\sigma, \qquad (6b)$$

where χ_T is the SG associated with a uniform stress load, and χ_σ is the gradient of the applied linear load

$$\chi_\sigma=\frac{1}{T}\left(\left.\frac{dF(y)}{dy}\right|_{y=0}\right). \qquad (7)$$

For the single wave, the SCF (Fig. 3) and SG (Fig. 4) are determined via the analytical formulae and the finite element method. Furthermore, maximum relative error values of 8% and 3% for the SCF and SG, respectively, are realized when the gradient of the applied linear load is 2 mm^{-1}.

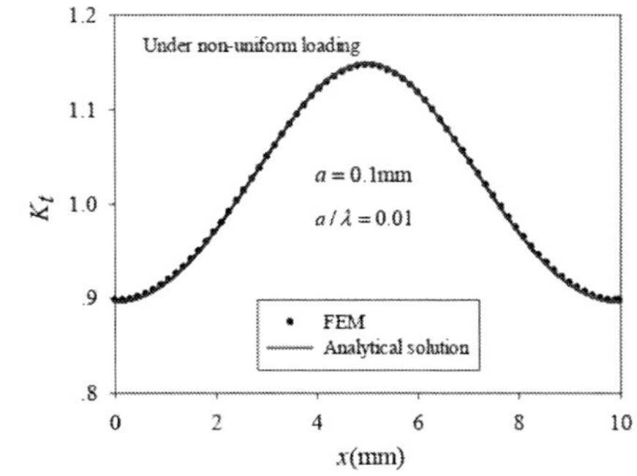

Fig.3. Stress concentration factor under non-uniform loading.

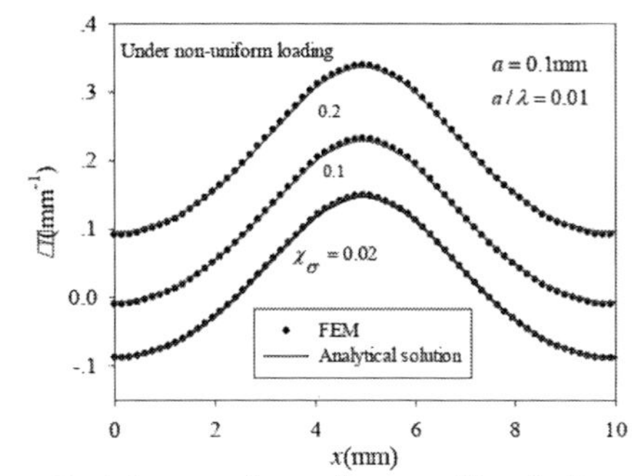

Fig.4. Stress gradient under non-uniform loading.

4.2. Random machined surfaces

The stress of a semi-infinite plate with the measured machined surface was determined following the procedure described in section 3. The SCF (Fig. 5) and SG (Fig. 6) were determined by the analytical solution and via finite element model with a triangular-element mesh. The close correspondence obtained verifies that the analytical formula, see Eq. (5), can be used to calculate the SCF and SG induced by random machined-surface topography.

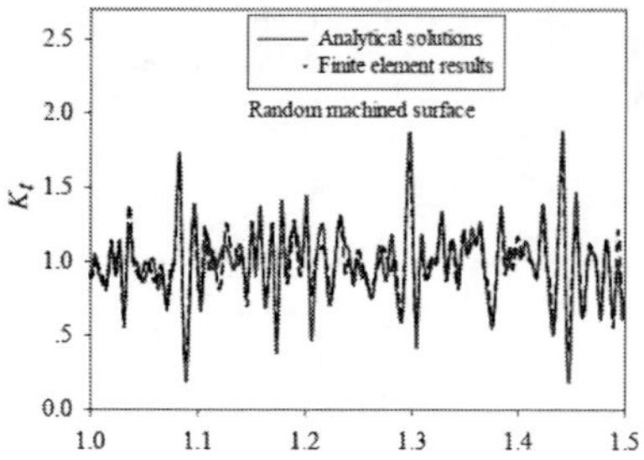

Fig.5. Stress concentration factor of the machined original surface.

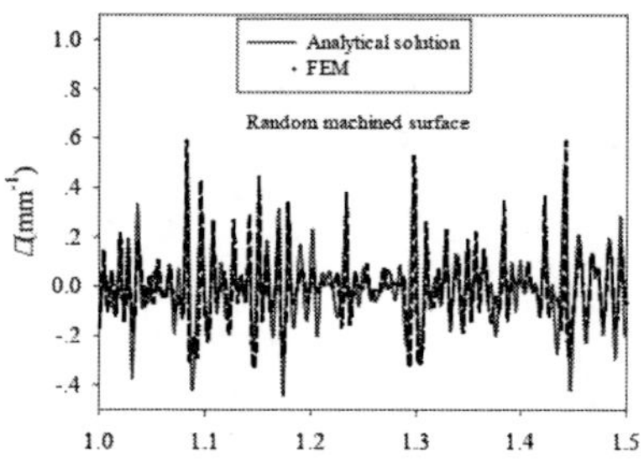

Fig.6. Stress gradient of the machined original surface.

5 Two-dimensional joint probability distribution function for SCF and SG of machined-surface topography

As reported in a previous study, the notch root or valley of the wave surface constitutes a fatigue hot point, where cracks occur [23, 24]. Therefore, we simultaneously consider the stress concentration at the valleys along the random surface, and the SG rather than just the single-random variable SCF.

We measured eight groups of the 2D machined surface (4 mm length with $Ra = 0.2\,\mu m$), which are perpendicular to the direction of the machining grooves. To prevent the effects associated with inhomogeneity, we chose an evaluation length of 4 mm, in accordance with the ISO 4288-1996 standard. The SCF and SG of the machined surface topography were obtained using the respective analytical formulae (see Eq. (5)).

After assuming a normal distribution, for both random variables, we then determined the relativity by calculating the covariance matrix. The covariance of both random variables is always >0. Using mathematical statistics [25], we determined the 2D normal density function by calculating the mean, variance, standard deviation, and relative parameter of the samples. The 2D normal density function is given as follows:

$$f(K_t,\chi)=\frac{1}{2\pi\sigma_1\sigma_2\sqrt{1-\rho^2}}exp\left\{-\frac{1}{2(1-\rho^2)}\right.$$

$$\left.\left[\left(\frac{K_t-\mu_1}{\sigma_1}\right)^2-2\rho\frac{K_t-\mu_1}{\sigma_1}\cdot\frac{\chi-\mu_2}{\sigma_2}+\left(\frac{\chi-\mu_2}{\sigma_2}\right)^2\right]\right\}, \quad (8)$$

in this example, the variance $\sigma_1^{\ 2} = 0.0786$, $\sigma_2^{\ 2} = 0.0511$, mean $\mu_1 = 1.34$, $\mu_2 = 0.281$, and relative coefficient $\rho = 0.906$ (see Fig. 7).

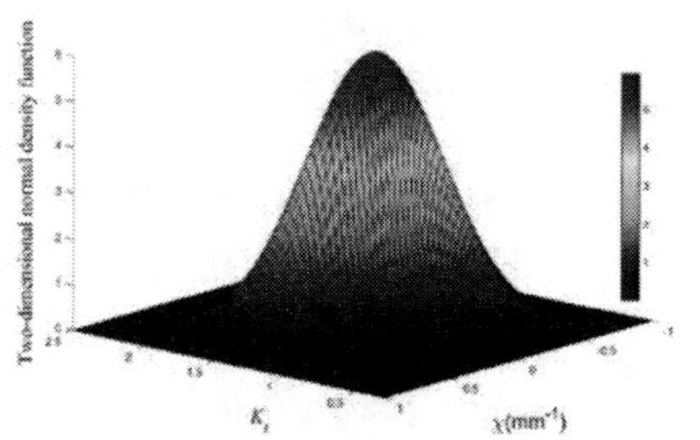

Fig. 2 Two-dimensional normal density function.

6. Evolution the influences of surface topography on fatigue propriety of the random machined surfaces

In order to further investigate how the surface topography effect on fatigue life and simultaneously introduce SCF and SG these two-dimensional random variable, we did SC nickel-based super alloy fatigue test in the same environment with different surface topography, decreasing grain boundary and material microstructure influence factors as far as possible.

6.1 Specimen preparation

Cylindrical test specimens, as shown in Fig. 8, have been machined by grinding with lubricant. Test specimens were prepared in such a way that maximum load is applied perpendicular to the rolling direction. To characterize roughness of a machined surface, average roughness (Ra) is generally used. The machined surfaces (Ra = $0.4\,\mu m$) specimen, were produced by varying the feed rate. Measurement surface topography of every specimen sign region, which is 4mm in length in the middle part of the sample, as shown in Fig. 8 red box, still used a BRUKER NPFLEX 3D Optical Microscope as initial surface topography.

6.2 Experimental procedures

The tensile strength properties of the SC, as shown in Table 1, were obtained from specimens prepared according to ASTM E-8 Standard [26]. The tensile tests were performed at a strain rate of 0.001 /s in a 15kN servo-hydraulic MTS machine.

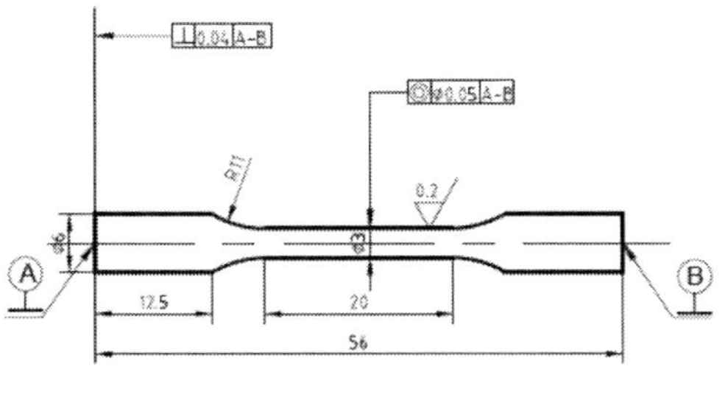

Fig.3.Specimen geometry (dimensions in mm).

Table 1 Tensile mechanical properties of SC nickel-based superalloy.

σ_{YS} (MPa)	σ_{UTS} (MPa)
963	996

(σ_{UTS} —ultimate ensile stress ; σ_{YS} —yield stress)

Axial fatigue tests have been performed at frequency of 6Hz in ambient conditions utilising a loading ratio of R = 0.1 under sinusoidal waveform at room temperature according to ASTM E-466 [27]. The specimen were under load controlled condition using above MTS machine, respectively 0.5 σ_{YS} and 0.4 σ_{YS}, which was expected to result in a fatigue life of between 10^4 and 10^6 cycles. During the cyclic testing, the specimen with 2e+5 cycles, 3e+5 cycles, 4e+5 cycles, 5e+5 cycles. When fracture occured in the effective location and tests were stopped if the specimen did not fail at 10^7 cycles, the number of cycles to failure was recorded.

6.3 Evolution of the surface topography

By observing the same region of the specimen, Fig.9 shows surface topography was changing with cycles increasing, where x is positions and z is heights of the surface. Fig.10 and Fig.11 show the evolutions of SCF and SG, and linear regression curves increase with the cycles as a function of the number of fatigue cycles. SCF and SG was positive with increasing fatigue cycles, which might causally be involved peaks and valleys as the same reason in [28].

Form Fig.12, Fig.13 and Fig.14 we observed parameters the variance, the mean, and the relative coefficient in Eq.(8) evolution. The average of SCF and SG were increasing with fatigue cycles, however, SCF increased more quickly than SG. The relative coefficient always kept the value of 0.9, which showed SCF and SG has sustained high relativity.

7 Conclusions

An efficient approach, was developed for simulating the surface topography of random machined rough surfaces. The SCF and SG induced by the original machined surface are essential for characterizing the surface

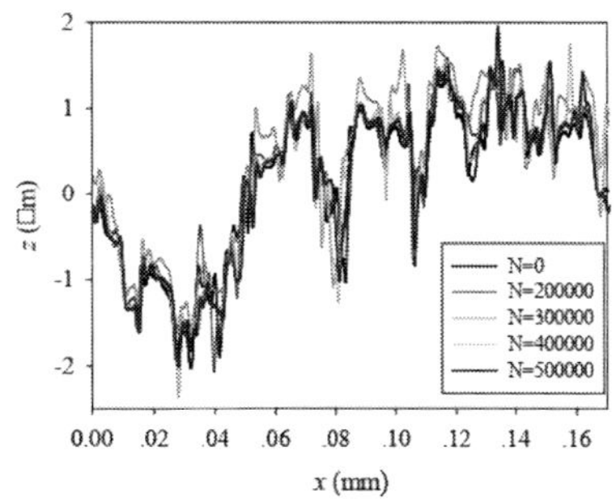

Fig.4. Surface topography for the different fatigue test cycles of the specimen.

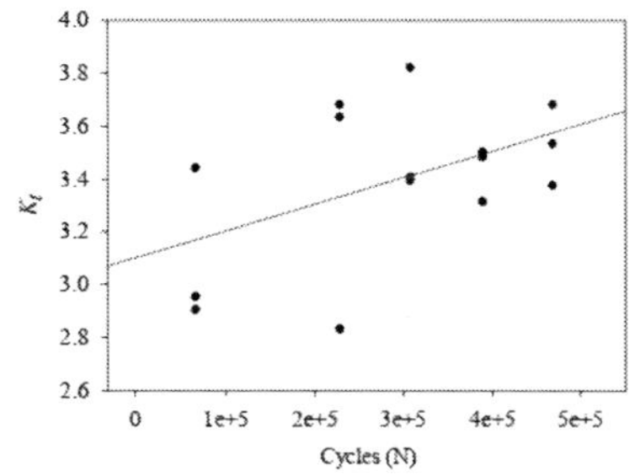

Fig.5. The SCF for the different fatigue test cycles.

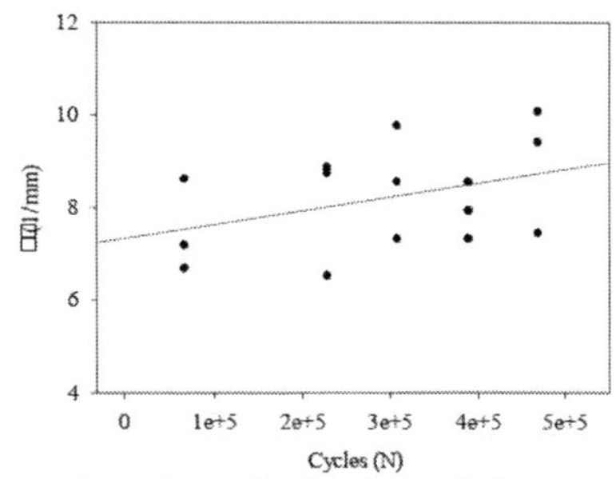

Fig.6. The SG for the different fatigue test cycles.

topography and, hence, analytical formulae for these parameters were presented. The accuracy of these formulae was validated via finite element analyses for the case of non-uniform loading, such as when linear loads are applied. We also determined the two-dimensional joint probability-distribution functions for SCF and SG at the valleys along the random surface of the machined-surface topography. Finally, we eliminate the variance, the mean, and the relative coefficient as far as possible, in fatigue testing to study the evolution of surface topography on fatigue life. The obtained results should be useful in studying and evaluating fatigue properties of machined components.

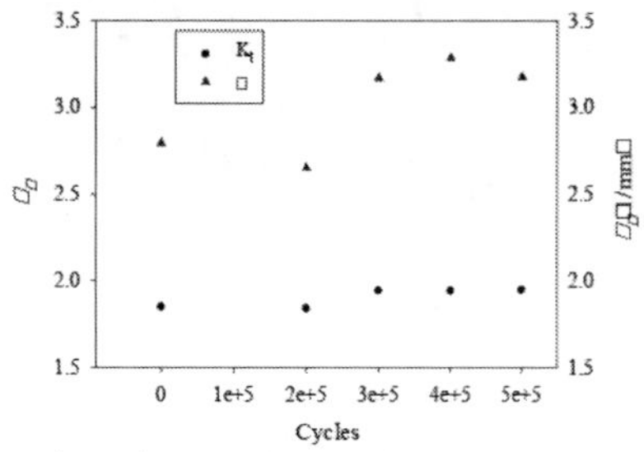

Fig. 7 The mean for the different fatigue test cycles.

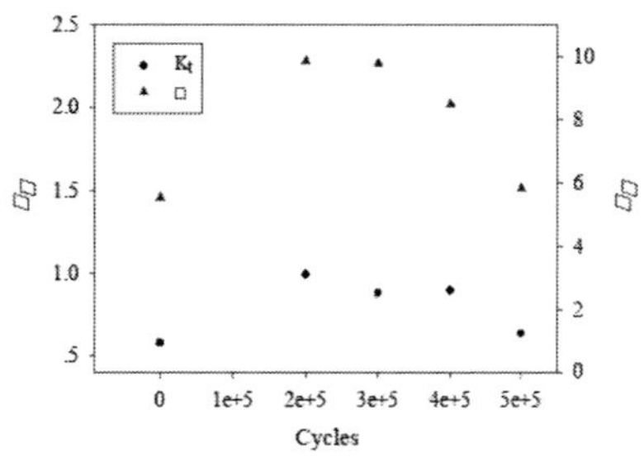

Fig. 8 The variance for the different fatigue test cycles .

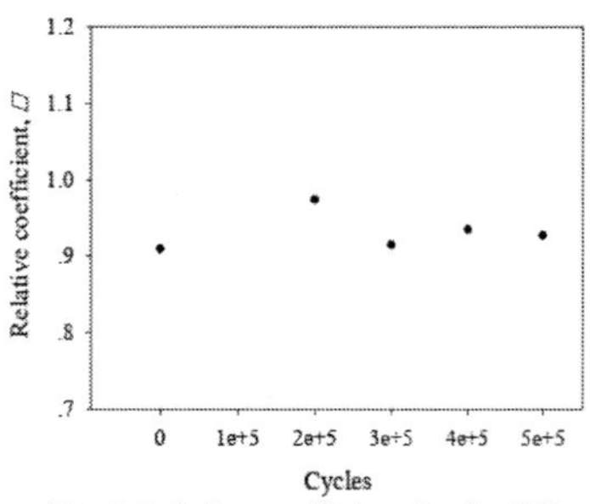

Fig. 9 Relative coefficient for the different fatigue test cycles

Acknowledgements

This work was supported by the Key Research Project of Higher Education Institution of Henan Province (No 16A460024).

References

1. Medina HE, Pidaparti R, Hinderliter B. Appl. Mechs. Rev. **67**: 010802(2014)
2. Pilkey W, Pilkey D. *Peterson's Stress Concentration Factors.* 3rd ed., New York :Wiley.(2008)
3. Neuber H. *Kerbspannungslehre.*2nd ed., Springer-Verlag, Berlin: Germany.(1958)
4. Peterson RE, *Stress concentration factors.* New York :Wiley(1974)
5. Castagnetti D, Dragoni E. Materialwissenschaft Und Werkstofftechnik.44: 364-71.(2013)
6. Gao H. J Mech. Phys. Solids.**39**: 443-58 (1991)
7. Medina H. Int. J. Solids. Struct. 69-70:86-93(2015)
8. Rice JR. J Appl. Mech. **52**: 571, (1985)
9. Medina H, B Hinderliter. Int. J. Solids. Struct. **51** (10) :2012-2018,(2014)
10. Taylor D, Clancy OM. Fatigue. Fract. Eng. M.**14**: 329-36.(2007)
11. Ås SK, Skallerud B, Tveiten BW. Int. J. Fatigue . **30**: 2200-2209(2008)
12. Schijve J. Fatigue. Engng. Mater. Struct. **3**: 325–338(1980)
13. Gates N, Fatemi A. Theor. Appl. Fract. Mec.**84**:3-25(2016)
14. Schijve J. Fatigue Engng Mater Struct 1982;5: 77-90.
15. Filippini M. Int. J. Fatigue. **22**:397-409.(2000)
16. Adib-Ramezani H, Jeong J. Comp. Mater. Sci. **39**: 649-63(2007)
17. Milošević I, Winter G, Grün F. IProcedia Eng. **160**:61-68(2016)
18. ZHANG G, LIU Jun, Liu YS, Yue ZF.Journal of mechanical strength,**32**(1):110-115.(2010)
19. Yao CF, WU DX, Jin QC , et al. Trans. Nonferrous Met. Soc. China.**23**:650–660(2013)
20. Ardi DT, Li YG, Chan KHK, Blunt L, Bache MR. *Procedia Cirp.***13:**19-24(2014)
21. Aoto Y, Noguchi H. Int. J. Fract. **131**:59-78(2005)
22. Gao H. Int. J. Solids. Struct. 28:703-25(1991)
23. Rusk DT, W Hoppe, W Braisted, N Powar. Int. J. Fatigue. **31**:1454-63(2009)
24. Smith RA, Miller KJ. Int. J. Mech. Sci. **20**: 201-06(1978)
25. Rice J. *Mathematical Statistics and Data Analysis.* Duxbury Press.(1994)
26. ASTM E-8M-98. (1995)
27. ASTM E-466. (2004)
28. Lacerda JCD, Martins GD, Signoretti VT , Teixeira RLP. Int. J. Fatigue. **102**:143-148(2017)
29. Shahzad M, Chaussumier M, Chieragatti R, Mabru C , et al. J. Mat. Processing. Tech. **210**:1821-1826(2010)

Optimization of Contact Edge Profile for Minimizing Contact Pressure in a Press-fitted Shaft

Dong-Hyung Lee[1,*], *Ha-Young* Choi[2], *Seok-Jin* Kwon[1], *and Jeong-Won* Seo[1]

[1]Fatigue and Fracture Research Team, Korea Railroad Research Institute, Korea
[2]Department of Mechanical Engineering, Dongyang Mirae University, Korea

Abstract. In the shrink or press-fitted shafts such as railway axles, the rotor of a steam turbine or coupling, a high-stress concentration takes place in the close of contact edge due to relative slip between shaft and boss in a press-fitted shaft and this is a major cause of fatigue failure of the shaft. The object of this paper is to build a finite element analysis model for analysing press-fitted and bending load condition in a press-fitted assembly and is to propose a hub shape optimization method to minimize a contact pressure in the close of shaft contact edge. Numerical asymmetric-axisymmetric finite element model was developed to predict the contact stress state of the press-fitted shaft. Global optimization method, genetic algorithm, and local optimization method, sequential quadratic programming, was applied to the press-fitted assembly to optimize the hub contact edge geometry. The results showed that the maximum contact pressure with optimized hub shape decreased more than 60% compared to conventional hub shape, the maximum contact stress affecting fatigue life reduced about 47%. In addition, hub shape optimization design could be a useful tool, able to increase the load capabilities of press fits concerning wear and fatigue behaviour.

1 Introduction

The wheelset of the railway vehicle is a structure in which an axle and a wheel are press-fitted, and fretting damage occurs at the end of a fitted region between the axle and the wheel. Fatigue strength of press-fitted shaft is reduced by 50% compared to normal fatigue strength due to fretting occurring at the axle [1]. The study on the improving fatigue strength of press-fitted shaft can be distinguished mainly into three categories such as a study on the design effects of the press-fitted shaft edge, a study on the effect of the residual stress of shaft, and a study on the damage reduction of fretting [1]. The study on the effect of residual stress is to improve fatigue strength by generating compressive residual stress on the shaft through carburizing or induction heat treatment. The research on the reduction of fretting damage is about preventing or reducing fretting corrosion through coating contact surface or changing the material properties of the contact part.

There are some studies on the design effects of the press-fitted shaft edge. Peterson and Wahl [2] represented creating a groove to relieve stress concentration at the end of the hub is useful for fatigue strength improvement through the photoelastic experiment. Nishioka and Komatsu [3] tested different shape for the pressed-fitted shaft structures and showed that fatigue limit with overhanging hub structure is increased by 50%. Kubota et al. [4] reported that proper grooving of the shaft is effective for improving the fretting fatigue limit. Lee [5] applied a taper to the end of hub contact edge and reported that the appropriate taper is effective in improving the fatigue life. Biron [6] proposed an optimization method for an overhang type press-fitted shaft, which minimizes the weight and maximizes the fatigue life of the contact and fillet parts. The above series of studies to improve fatigue strength in the press-fitted shaft contributed significantly to increase the fatigue life on the axle. However, making grooves on the shaft may cause the crack in the groove area, and the fatigue life is decreased. In addition, in case of the axle, a study is required to improve fatigue strength without significantly changing the shape of the shaft or hub during the maintenance phase.

Therefore, in this study, the contact edge shape of the hub is optimized to minimize contact pressure, which directly affects the fatigue life, through the finite element analysis and the numerical analysis for the press-fitted shaft. FEA model of the press-fitted shaft subjected to fatigue bending loads has been built to analyse contact pressure and contact stress. Using the FEA results and the approximate optimization method, the optimal hub design that minimizes contact pressure of the contact edge of a press-fitted shaft is derived. The effectiveness of the optimization procedure is verified by comparing contact pressures and stresses of the press-fitted shaft with the optimized hub shape and traditional hub shape.

2 Analysis methods

2.1. Finite element analysis

* Corresponding author: dhlee@krri.re.kr

https://doi.org/10.1051/matecconf/201816522029

The analysis model is for the press-fitted test specimen subjected to the fretting fatigue test, and the structure and shape of the specimen are shown in Fig. 1. In order to calculate the stress distribution of the press fit and the loaded condition, we used a commercial FE code ABAQUS [7], which is widely used in non-linear numerical analyses.

The press-fitted specimen is modelled using an asymmetric-axisymmetric element (CAXA4R4), and for the details of the modelling method, we refer the reader to the authors' previous report [5]. As the stress distribution at the end of contact area changes quickly due to micro-slip, the minimum element at the contact end is divided by 0.015 mm as shown in Fig.2. Shrinkage allowance on the contact surface ensures that the Lamé's nominal contact pressure [8] is 90 MPa. The nominal bending stress of 135 MPa, which is fatigue limit of the press-fitted shaft specimen, is applied, and for the friction interpretation, Lagrange multiplier friction formulation and 0.6 for the coefficient of friction are used.

2.2 Shape optimum design method

When the load is applied to the press-fitted shaft, the contact pressure is calculated by FEA, and the approximate model for optimization is constructed. Approximate optimization has the advantage of reduction of time consumption compared to the optimization method through the numerical analysis or experiment directly. The approximate model is created

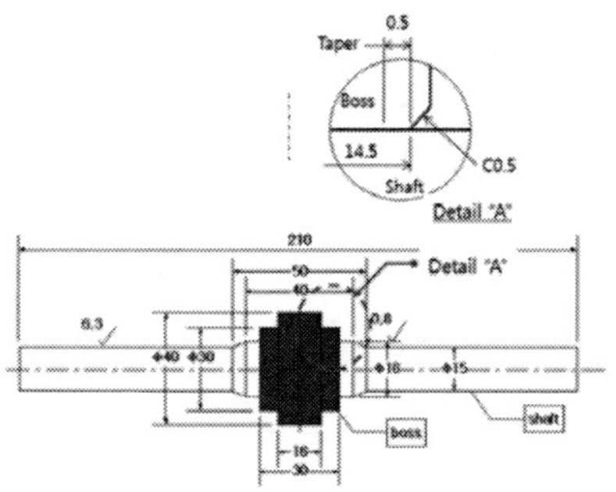

Fig. 1 Geometry of the press-fitted fatigue specimen.

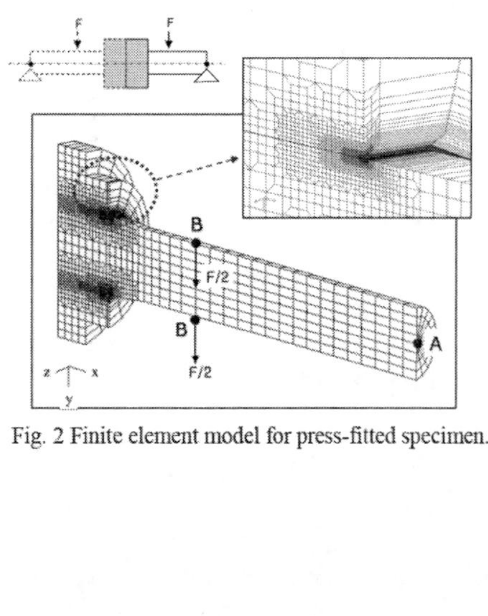

Fig. 2 Finite element model for press-fitted specimen.

by the Response Surface Method (RSM) [9], and as the optimization method, Genetic Algorithm (GA) [10] that is global optimization method and Sequential Quadratic Programming (SQP) [11] that is local optimization method, is applied.

Fig. 3 represents the optimization procedure of hub shape design. The analytical samples are extracted by Latin Hypercube Sampling (LHS), and the values of contact pressure along the contact surface are obtained from the finite element analysis according to the change of contact edge shape. After generating the approximate model using the response surface method, the optimum design is performed through the generated approximate model and optimization algorithms. The objective function of the optimization is defined as the contact pressure along the contact surface of the press-fitted shaft, and the design variables are formulated as shown in Fig. 4. In Fig. 4, the location of p_2 is the end of the hub, and the location of p_0 is the inner region of the hub at a distance 0.5 mm from the end of the hub. Because fatigue damage occurs at 0.2~0.3 mm inner region from the end of contact edge, the variable, p_0, is set to 0.5 mm.

According to the result of the previous study [5], the magnitude and location of the maximum contact pressure vary by the amount of taper at the end of

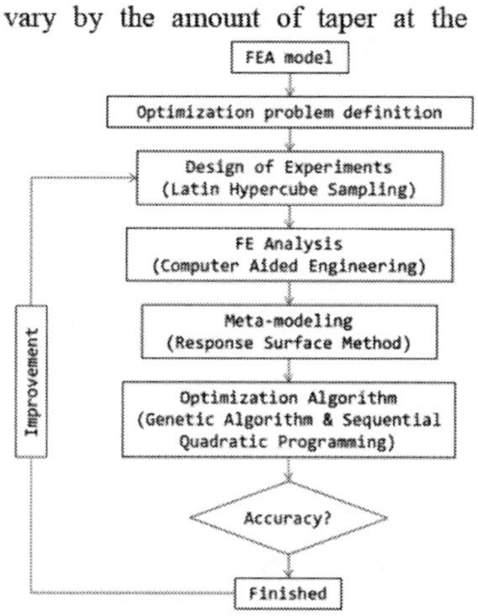

Fig. 3 Workflow of optimization process.

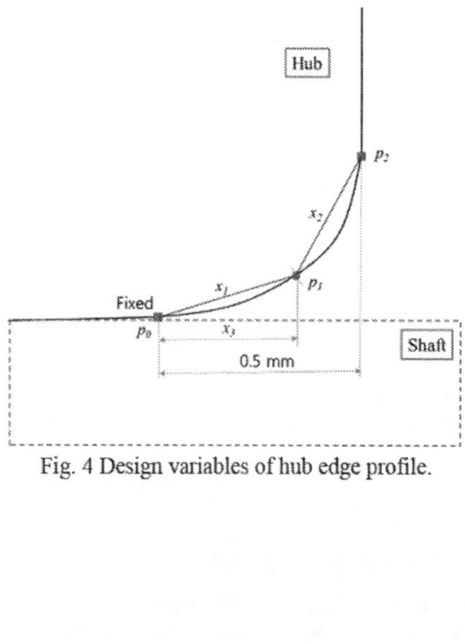

Fig. 4 Design variables of hub edge profile.

contact edge, so the inclination of hub contact edge is set as a variable. The two gradients, x_1 and x_2, and the horizontal coordinates of the location where gradient changes, x_3, are defined as design variables. The inner contact profile of the hub connecting the three points p_1, p_2 and p_0 is created with curve by using Piecewise Cubic Hermite Interpolating Polynomial (PCHIP). The location p_1 is located between p_0 and p_2, and the gradient of the two points, p_1 and p_0, is x_1, and the gradient of the two points, p_1 and p_2, is x_2. The final formulation for optimization is defined as follows:

Minimize $F_{contact\ pressure}$
Design variables

$$0.000 \le x_1 \le 0.010$$
$$0.010 \le x_2 \le 0.015$$
$$0.000 \le x_3 \le 0.450$$

The approximate model generation for the optimization problem of this study uses the experimental planning method, Latin Hypercube Sampling, which is widely used in the field of engineering. The approximate model equation is established using the Response Surface Method with Latin Hypercube Sampling and Finite Element Analysis, and its accuracy is shown in Fig. 5.

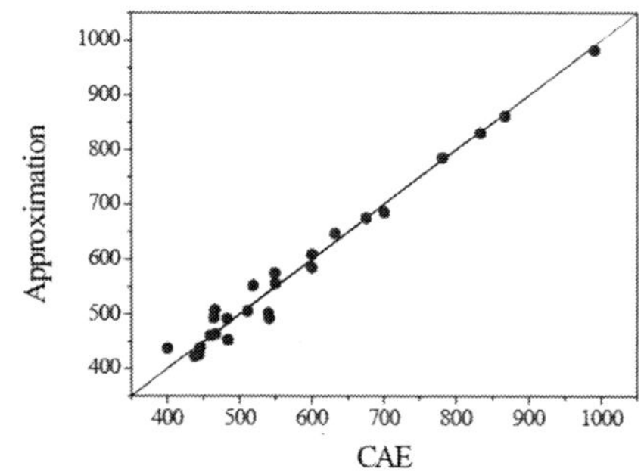

Fig. 5 Accuracy of real CAE and RSM results.

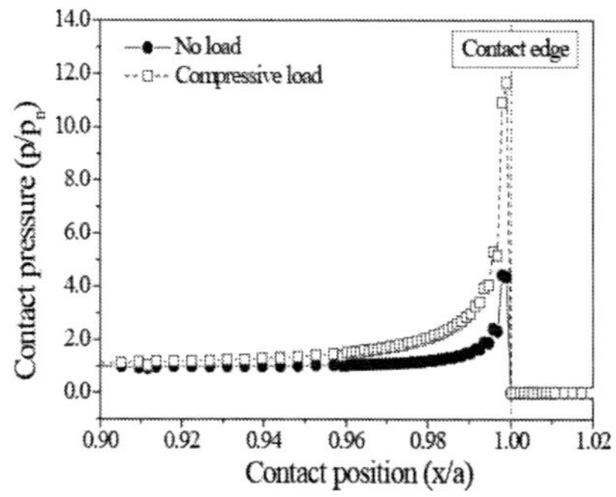

Fig. 6 Contact pressure distributions of press-fitted shaft.

3 Results and reviews

3.1 Finite element analysis results

Fig. 6 shows the analysis results of the distribution of contact pressure at the end of the shaft contact edge when the compressive load and no load are applied to the press-fitted shaft. In this figure, the x- and y-axes show the normalized values through the y-axis is divided by the nominal contact pressure (90 MPa) and the x-axis is divided by the contact length 'a' (14.5 mm). As shown in the figure, even when no load is applied to the press-fitted shaft, the contact pressure at the end of contact edge is approximately four times higher than the nominal contact pressure because of press fitting. When the bending load is applied to the press-fitted shaft, the maximum contact pressure is higher than that of press-fitted state (no load) and is approximately 11 times of the nominal contact pressure. Therefore, the significant contact pressure is generated at the end of the contact edge and it can be seen that the stress concentration in this region causes the fretting damage.

3.2 Optimization results

The results of optimizing the hub design of the press-fitted shaft using genetic algorithms and sequential quadratic programming are shown in Table 1.

With the approximate model that is created by response surface method, the optimization with the genetic algorithm is performed by setting parameters as follow; the number of generations to 100, number of population of 100, mutation rate of 30 %, and crossover rate of 10 %. Since the SQP algorithm is influenced by the position of the initial design variables, the initial design variables are set to three points.

The optimal and lowest maximum contact pressure is obtained from the analysis case of SQP-3 and the

Table 1. Optimization results

		Original	Optimization			
			GA	SQP-1	SQP-2	SQP-3
x_1 (x 10^{-3}mm)		-	3.41	3.43	3.39	3.38
x_2 (x 10^{-3}mm)		-	12.59	12.55	12.45	12.53
x_3 (x 10^{-3}mm)		-	269.85	273.39	270.73	267.14
Contact pressure (MPa)	CAE	1,050	393.5	398.6	399.6	393.3
	Approx.	-	371.8	372.0	371.9	371.9
Initial Values (x 10^{-3}mm)		-	-	x_1: 2.50 x_2: 11.25 x_3: 112.50	x_1: 5.00 x_2: 12.50 x_3: 225.00	x_1: 7.50 x_2: 13.85 x_3: 337.50

maximum contact pressure is reduced by 63% from that of original design in compressive loading condition. The approximate value of the maximum contact pressure derived from the optimized design is 371.9 MPa and the value of finite element analysis is 393.3 MPa, and there is 6 % in error of contact pressure.

The results of the SQP with three initial values have almost same values with those of the optimization with GA. The program running time with SQP is about 60 times shorter than that with the genetic algorithm, and for this optimal design, it is clear that the performing the optimization with SQP is more efficient than that with GA.

Fig. 7 and Fig. 8 shows the distribution of contact stress (von Mises) on the shaft contact edge when the compressive load is applied to the normal and optimal press-fitted assembly, respectively. For normal press-fitted assembly, high stress up to about 1,000 MPa is produced at the end of contact edge of the shaft. For the optimized press-fitted assembly, due to the curved inner geometry of hub, a maximum stress of 530 MPa is occurred at 0.25mm position inside the shaft-hub contact end, and the maximum stress is reduced by about 47 % compared with the conventional one. From these results, it can be concluded that the fatigue life of optimal designed press-fitted shafts is greatly increased, which is effective in reducing the fretting damage.

4 Conclusions

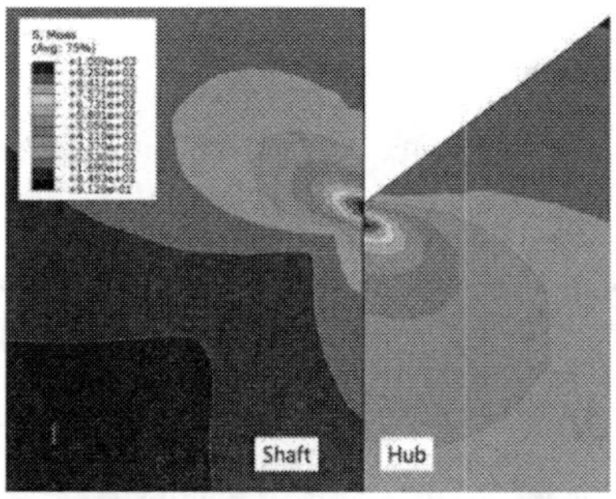

Fig. 7 Contact stress (von Mises) distribution of conventional press-fitted assembly.

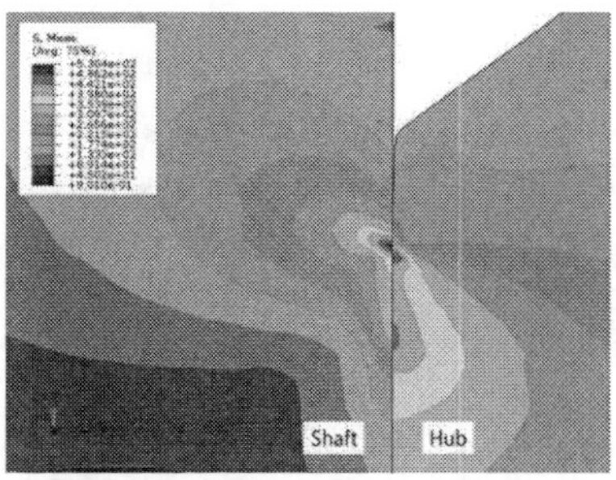

Fig. 8 Contact stress (von Mises) distribution of optimized press-fitted assembly.

A finite element analysis model is constructed for a press-fitted shaft, which is widely used in general industrial machinery, and the optimal design method of hub shape is proposed by using the approximation optimization method.

As a result of applying the genetic algorithm and the sequential quadratic programming to optimize the shape of the press-fitted shaft, there is no difference in the optimization results, and the analysis time with SQP algorithm was much shorter than that with the genetic algorithm.

The contact pressure of the optimized press-fitted shaft decreased by about 60% compared to the normal one, and the contact stress that affects the fatigue life is reduced by about 47%. From these results, it can be concluded that the proposed hub shape optimization method is effective for improving fatigue strength in a press-fitted assembly.

This research was supported by a grant from R&D Program of the Korea Railroad Research Institute, Republic of Korea.

References

1. Hirakawa, K., Toyama, K. and Kubota, M., "The Analysis and Prevention of Failure in Railway Axles," Int. J. Fatigue, Vol. 20, No. 2, pp. 135~144, 1998.
2. Peterson, R.E., and Wahl, A.M., "Fatigue of shafts at fitted members with a related photoelastic analysis," J Appl Mech, Vol. 2, No. 1, pp.A1–A11, 1935.
3. Nishioka, K. and Komatsu, H., "Researches on increasing the fatigue strength of press-fitted shaft assembly," Bull JSME, Vol. 10, pp.880–889, 1967.
4. Kubota, M., Kataoka, S. and Kondo, Y., "Effect of stress relief groove on fretting fatigue strength and index for the selection of optimal groove shape," Int. J. of Fatigue, V. 31, Issue 3, pp. 439-446, 2009.
5. Lee, D.H., Kwon, S.J., Seo, J.W., Kwon, S.T. and You, W.H., "Evaluation of Fatigue Crack Initiation Life according to the Hub Contact Shape in a Press-fitted Shaft," KSPE Spring conference, 2010.
6. Biron, G., Vadean, A. and Tudose, L., "Optimal design of interference fit assemblies subjected to fatigue loads," Struct Multidisc Optim, 47, pp. 441-451, 2013.
7. Abaqus® Version 6.10 Analysis User's Manual, Dassault Systèmes, 2010.
8. Ugural, A.C. and Fenster, S.K., "Advanced strength and applied elasticity," Prentice-Hall, Inc. 1995.
9. Hong, K. J., Jeon, K. K., Cho. Y. S., Choi, D. H., Lee, S. J., "A Study on the Construction of Response Surface for Design Optimization," Trans. of the KSME(A), 24(6), 1408-1418, 2000.
10. Holland, J.H., "Adaptation in natural and artificial systems," Ann Arbor, University of Michigan Press, 1975.

11. Fletcher, R., "Practical methods of optimization," John Wiley and Sons, 1987.
12. Simpson, T. W., Lin, D. K. J., Chen, W., "Sampling Strategies for Computer Experiments: Design and Analysis," International Journal of Reliability and Applications, 2(3), 209–240, 2001.
13. Fritsch, F. N. and Carlson, R. E., "Monotone Piecewise Cubic Interpolation," SIAM J. Numerical Analysis, Vol. 17, No. 2, pp. 238-246, 1980.

Interface damage model for fatigue-driven fracture

Feiyu Geng*, *Akke S. J.* Suiker

Department of the Built Environment, Eindhoven University of Technology, P.O. box 513, 5600 MB Eindhoven, The Netherlands

Abstract. A fatigue interface damage model is presented that is based on combining a fatigue evolution law with a static interface damage model. The fatigue model elegantly enables the simulation of crack initiation and propagation in a computationally efficient and accurate way, accounting for mixed-mode loading conditions and fatigue loading of variable amplitude. The main features of the fatigue model are explained and demonstrated by means of an illustrative numerical example.

1 Introduction

Cohesive zone models have been widely used to numerically predict discrete cracking phenomena in brittle and ductile materials subjected to static loading conditions, see for example, [1-4]. At the crack tip of the cohesive zone the traction across the crack surfaces is related to the relative displacement of the crack faces (= crack face separation) by means of a traction-separation law. Accordingly, the ultimate traction determines the critical stress level at which the crack nucleates, and the area under the traction-separation law, which equals the fracture toughness, characterizes the propagation characteristics of the crack.

In the literature various suggestions have been made to extend static cohesive zone models towards the case of fatigue. When based on Paris law, this requires the calculation of the crack driving force at the crack tip, which corresponds to the energy released per unit advance of crack length, see [5-8]. The crack driving force can be straightforwardly established for a single crack with a well-defined crack tip, but cannot be uniquely computed in case of complex fracture patterns characterized by (various) crack coalescence and crack bifurcation events. In order to circumvent this problem, various investigators have defined the fatigue growth rate as a function of the local deformation in a material point of the crack [9-12]. In the approach of Khoramishad et al. [9,12], the resulting fatigue damage parameter degrades the static traction-separation diagram that is used for computing the traction and the effective damage under the actual cyclic loading conditions applied. This local approach for the computation of the strength degradation during fatigue has the advantage that complex fracture patterns composed of multiple cracks can be analyzed in a natural and robust fashion.

Inspired by the approach of Khoramishad et al. [9,12], this paper presents a fatigue interface damage model that is based on combining the fatigue evolution law presented in [12] with the static interface damage model presented in [1]. The fatigue model elegantly enables the simulation of crack initiation and propagation in a computationally efficient and accurate way, accounting for mixed-mode loading conditions and fatigue loading of variable amplitude.

2 Fatigue interface damage model

The interface damage model used for simulating fracture under fatigue loading is constructed by combining the fatigue damage law suggested in [12] with the static interface damage model proposed in [1]. The combination of the two models is described below, followed by listing the main features of the fatigue interface damage model.

2.1. Review of the static interface damage model

In the static interface damage model proposed in [1] the cohesive tractions $\boldsymbol{t}$ in an interface material point are related to the effective relative displacements $\boldsymbol{v}$ as

$$t_i = (1 - D)K v_i - DK\delta_{1i}\langle -v_1 \rangle \text{ where } i \in \{1,2\} \quad (1)$$

with the indices '1' and '2' denoting the components in normal and shear directions, respectively. Further, K is the elastic stiffness (with dimension of force×length^{-3}), and δ_{ij} is the Kronecker delta symbol. Note that the Macauley brackets $\langle x \rangle = (x + |x|)$ are used to avoid interpenetration of the crack faces when v_1 is negative. The damage parameter $D = 0$ refers to an undamaged state and $D = 1$ represents the fully damaged state of a material point. A typical bi-linear traction-separation law, schematized in Fig. 1, is adopted to specify the softening behavior experienced during damage. Here, κ is a history parameter, whereby the onset of damage relates to $\kappa = v^0$, and completion of damage is reflected by $\kappa = v^u$.

* Corresponding author: f.geng@tue.nl

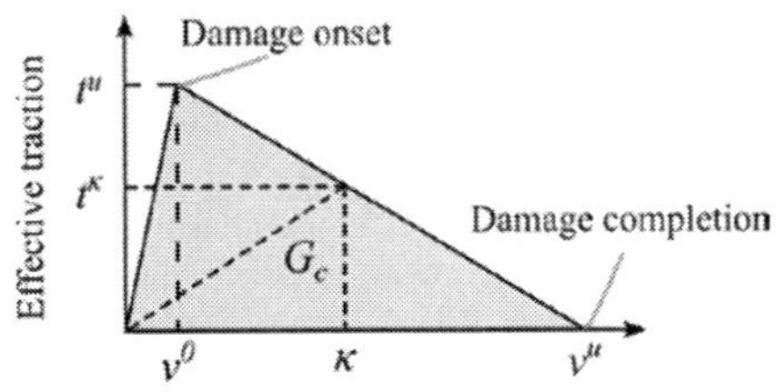

Fig. 1. A bi-linear traction-separation law under static loading (taken from [1]).

The evolution of the damage parameter $D = \widehat{D}(\kappa)$ occurs in accordance with the rate-dependent kinetic law

$$\dot{D} = \begin{cases} \frac{\hat{F}(\lambda,\kappa)}{\eta} & \text{for } \lambda \geq \kappa \text{ and } v^0 \leq \kappa < v^u \\ 0 & \text{for } 0 \leq \lambda < \kappa \text{ or } \kappa = v^u \end{cases} , \quad (2)$$

where λ is the effective relative (crack face) displacement $\lambda = \|\mathbf{v}\| = \sqrt{v_1^2 + v_2^2}$, and η is a relaxation parameter characterizing the rate dependency of the damage process. Further, the damage loading function has the form

$$\hat{F}(\lambda,\kappa) = \hat{f}(\lambda) - \widehat{D}(\kappa) = \frac{v^u(\lambda - v^0)}{\lambda(v^u - v^0)} - \frac{v^u(\kappa - v^0)}{\kappa(v^u - v^0)} . \quad (3)$$

The upper expression in Eq.(2) reflects the damage rate when the effective relative displacement exceeds the damage history variable κ, with $v^0 \leq \kappa < v^u$; the lower expression sets the damage rate to be zero when $0 \leq \lambda < \kappa$, which corresponds to: (i) the threshold value for damage nucleation has not (yet) been reached, or (ii) the material point is experiencing unloading or (iii) the material point has been fully damaged.

To simulate fracture processes under arbitrary loading conditions, a mode-mixity parameter is defined as

$$\beta = \frac{|v_2|}{|v_2| + \langle v_1 \rangle} , \quad (4)$$

with $\beta = 0$ for pure mode I loading conditions ($v_2 = 0$) and $\beta = 1$ for pure mode II loading conditions ($v_1 = 0$). By starting from a common energy-based, mixed-mode failure criterion [13]

$$\frac{G_I}{G_{I,c}} + \frac{G_{II}}{G_{II,c}} = \frac{G}{G_c} , \quad (5)$$

where G_I, G_{II} and G are the mode I, mode II and total energy release rates and $G_{I,c}$, $G_{II,c}$ and G_c are the corresponding critical values, i.e., the fracture toughnesses, the effective displacement at damage onset and damage completion can be, respectively, derived as [1]:

$$v^0 = \hat{v}^0(\beta) = v_1^0 v_2^0 \sqrt{\frac{1 + 2\beta^2 - 2\beta}{\left(\beta v_1^0\right)^2 + \left((1-\beta) v_2^0\right)^2}} , \quad (6)$$

and

$$v^u = \hat{v}^u(\beta) = \frac{2(1 + 2\beta^2 - 2\beta)}{K v^0} \left[\left(\frac{(1-\beta)^2}{G_{I,c}} \right) + \left(\frac{\beta^2}{G_{II,c}} \right) \right]^{-1} . \quad (7)$$

Here, $v_1^0 = t_1^u / K$ and $v_2^0 = t_2^u / K$ are the effective displacements at which damage nucleates under pure mode I and pure mode II loading conditions, respectively, with t_1^u and t_2^u the ultimate tractions in mode I and mode II, respectively.

2.2 Fatigue formulation

2.2.1 Fatigue evolution law

In the present section the static interface damage model above is extended towards a fatigue formulation. For his purpose, a fatigue damage parameter D_f is introduced, which incrementally evolves by ΔD_f under a load cycle increment ΔN as [12]:

$$\frac{\Delta D_f}{\Delta N} = \begin{cases} \alpha[(\lambda_{max} - \lambda_{th})\gamma^n]^p & \text{for } \lambda_{max} > \lambda_{th} \\ 0 & \text{for } \lambda_{max} \leq \lambda_{th} \end{cases} , \quad (8)$$

with the load factor γ given by

$$\gamma = \frac{(1-R)/2}{1 - \left[\frac{P_{max}}{2P_S}(1+R)\right]^m} . \quad (9)$$

Here, the load ratio is given by $R = P_{min}/P_{max}$, with P_{min} the minimum amplitude of the fatigue loading, P_{max} is the maximum amplitude of the fatigue loading, and P_s the ultimate static strength obtained from a preliminary static analysis. Further, α, m, n and p are calibration parameters, and λ_{max} and λ_{th} are the maximum effective displacement and the corresponding threshold value above which fatigue is activated, respectively, with λ_{max} given by

$$\lambda_{max} = \frac{v_1}{2} + \sqrt{\left(\frac{v_1}{2}\right)^2 + \left(\frac{v_2}{2}\right)^2} , \quad (10)$$

where v_1 and v_2 are the relative crack face displacements in the normal and shear directions, respectively.

Similar to the damage parameter D introduced in Section 2.1, the fatigue damage parameter D_f is bounded as $0 \leq D_f \leq 1$. As shown by Eqs.(8) and (10), the evolution of D_f is set by the local, relative crack face displacements in a material point of the crack. The upper expression in Eq.(8) represents the incremental change of the fatigue damage parameter when the effective deformation surpasses the threshold value; otherwise, the incremental fatigue damage is set to zero, as indicated by the lower expression in Eq.(8).

2.2.2 Degrading traction-separation law

Fig. 2 illustrates the scheme used for combining the static interface damage model reviewed in Section 2.1 with the fatigue damage evolution law, Eq.(8). The procedure essentially consists of two steps:

(a) In the first step the maximum amplitude of the fatigue loading, P_{max}, is quasi-statically imposed on the actual structure in a load-controlled fashion. The possible cracking generated under this loading is computed with the static interface damage model outlined in Section 2.1.

(b) In the second step cracking can (further) develop, but now in a load-cycle driven fashion. Accordingly, a number of load cycles ΔN is applied incrementally, which, via Eq.(8), results in incremental fatigue damage ΔD_f if the effective relative separation λ_{max} computed in the first, quasi-static loading step is larger than the threshold value λ_{th}. The subsequent incremental update D_f of the fatigue damage parameter is used to reduce the strength and toughness parameters defining the traction-separation law of the static interface damage model, in accordance with

$$\zeta_i = (1 - D_f)\zeta_i^0 \quad \text{where } i \in \{1,2\} \ , \qquad (11)$$

with ζ_i representing the normal (index "1") and shear (index "2") components of the strength and toughness parameters, given by t_1^u, $t_2^u, G_{I,c}$ and $G_{II,c}$. The initial values ζ_i^0 of these parameters are represented by $t_1^{u0}, t_2^{u0}, G_{I,c}^0$ and $G_{II,c}^0$. In this way, the fatigue damage parameter D_f effectively *degrades* the static traction-separation law. From the degraded traction-separation law the traction is subsequently computed using the static interface damage model outlined in Section 2.1. This update procedure is iteratively repeated until the mechanical system has converged towards equilibrium, after which a new load cycle increment ΔN is imposed and step 2 is repeated.

The ultimate separation of a material point in a crack is reached either when the effective damage parameter D reaches unity, $D = 1$, or when the fatigue damage parameter reaches unity $D_f = 1$. Note that $D = 1$ can be reached when (i) $D_f = 1$ or (ii) $0 < D_f < 1$.

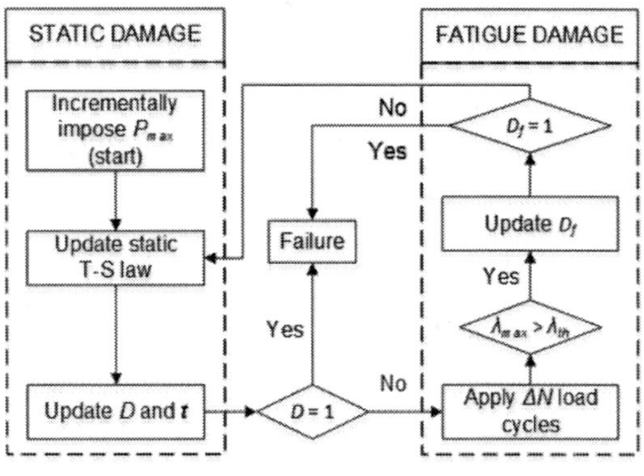

Fig. 2. The computational scheme for a fatigue simulation, whereby the static interface damage model outlined in Section 2.1 is combined with the fatigue damage evolution law, Eq.(8). T-S denotes the traction-separation law.

It is instructive to consider the sequence of fatigue damage initiation and propagation in an interfacial crack generated in a double-cantilever beam subjected to pure model I (tensile) loading, as shown in Fig. 3. Specific material points in the interface of the double-cantilever beam specimen are indicated by circle, square and cross symbols. The material point indicated by a circle is considered to be the reference material point, which successively experiences three different types of response, as explained below.

(a) Initially the amplitude of the fatigue loading is applied in a quasi-static fashion, see Fig. 3a, whereby the response in the interface of the double-cantilever beam specimen follows the static interface damage model presented in Section 2.1. Observe that the reference material point indicated by a circle is in the elastic regime. Conversely, the material point marked with a square already experiences damage development, whereby its response lies on the softening curve. The material point marked with a cross has completely damaged, and the crack faces at this material point have fully separated.

(b) Once the fatigue amplitude has been imposed, the process becomes driven by load cycles, see Fig. 3b. The number of load cycles incrementally applied equals $\Delta N = N1$, which will induce fatigue cracking in the interfacial material point if the corresponding value λ_{max} exceeds the threshold λ_{th}. The fatigue damage reduces the strength and toughness in accordance with Eq.(11), leading to the degraded traction-separation diagram indicated in the middle figure of Fig. 3b. The reference material point indicated by the circle now experiences damage, whereby both the static damage parameter and the fatigue damage parameter lie between zero and one, i.e. $0 < D < 1$ and $0 < D_f < 1$. The material point marked by the square symbol has fully damaged, with $D = 1$ and $0 < D_f < 1$.

(c) The load cycle process is continued by applying a new load cycle increment, $\Delta N = N2 - N1$, as shown in Fig. 3c. Due to this new load cycle increment the strength and toughness have decreased in accordance with Eq.(8), as a result of which the traction-separation law has further degraded. Accordingly, the reference material point approaches the fully damaged state, whereby the relative separation of the crack faces is close to the ultimate separation v^u.

2.2.3 Specific features of fatigue model

Some specific features of the proposed fatigue interface damage model are summarized below:

- Arbitrary cracking paths can be simulated by placing interface elements in between the continuum elements modeling the bulk response of the structure, in accordance with the approach originally advocated in [14]. This naturally enables the simulation of fatigue fracture characterized by relatively complex fracture scenarios, including crack bifurcation and crack coalescence events.
- The fatigue interface damage model is able to describe discrete fracture under arbitrary loading conditions, due to the fact that the static interface damage model rigorously accounts for mixed-mode cracking effects.
- The fatigue interface damage model can be applied under variable amplitude loading conditions, whereby load cycle blocks of different amplitude are subsequently imposed in an incremental fashion.

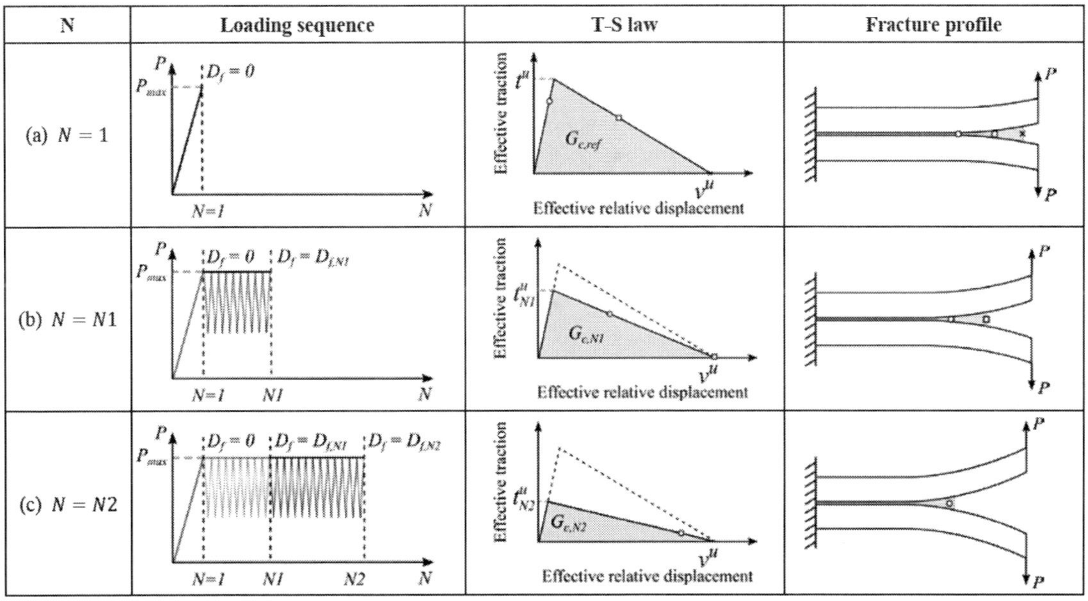

Fig. 3. Evolution of traction-separation (T-S) law and fracture profile during fatigue loading process. Specific material points in the interface of the double-cantilever beam specimen are indicated by the circle, square and cross symbols.

- The load-cycle driven approach used in the fatigue model is computationally efficient. Instead of describing the precise crack growth in each load cycle, it considers the average crack growth during a number of load cycles. Hence, one load cycle increment may consist of a large number of load cycles, thereby allowing for the economical computation of high-cycle fatigue processes.

3 Case study

The discretized equations of the fatigue interface damage model were implemented in a User-defined MATerial (UMAT) model within the commercial finite element program Abaqus. The basic characteristics of the fatigue model are illustrated below by performing a numerical analysis on a relatively simple mode I specimen, see Fig. 4.

3.1. Problem description

The specimen shown in Fig. 4 is composed of two continuum elements sandwiching a single interface element. The width and height of the specimen are equal to 10 mm, the thickness of the interface element is set as 0.05 mm, two supports are placed at the bottom nodes of the interface element and a tensile (mode I) fatigue load P is applied at the top of the specimen, which is characterized by a maximum amplitude of $P_{max} = 4.56 \times 10^3$ N and a load ratio $R = P_{min}/P_{max} = 0.1$.

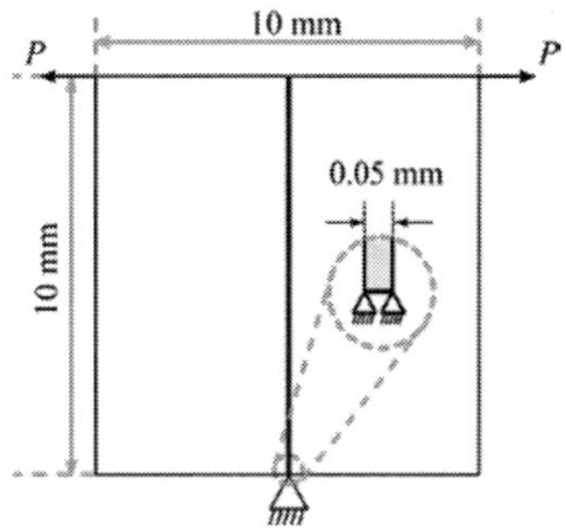

(a) Geometry and boundary conditions.

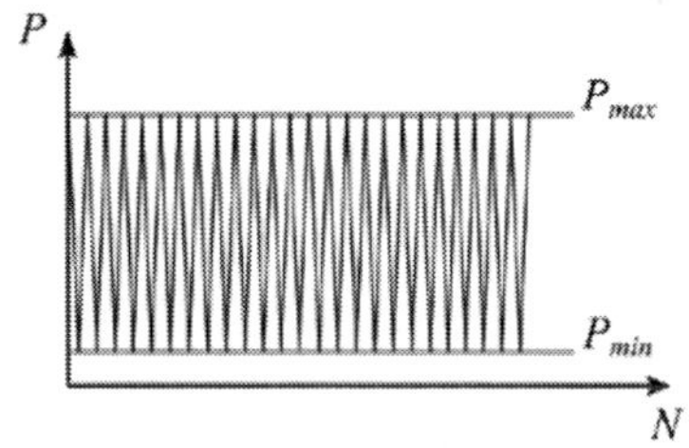

(b) Cyclic loading characteristics.

Fig. 4. Geometry and boundary conditions for a simple mode I specimen load subjected to a tensile fatigue loading *P* at the top.

The material used for the specimen is a T300/977-2 carbon-fiber reinforced epoxy laminate. The elastic properties are taken from [6] and are listed in Table 1, whereby the "1", "2" and "3" material directions correspond to the vertical, horizontal and out-of-plane directions of the specimen.

Table 1. Elasticity parameters of the specimen.

parameter	value	parameter	value
E_{11}	150.0 GPa	G_{12}	6.0 GPa
E_{22}	11.0 GPa	G_{13}	6.0 GPa
E_{33}	11.0 GPa	G_{11}	3.7 GPa
ν_{12}	0.25	ν_{13}	0.5
ν_{23}	0.45		

The parameters of the interface damage model are listed in Table 2. With these parameters, the *static* failure load was computed from a preliminary quasi-static analysis as $P_s = 9.11 \times 10^3$ N. Note that the maximum amplitude of the fatigue loading is half of the static failure load, $P_{max}/P_s = 0.5$.

Table 2. Parameters of the interface damage model.

Parameter	Symbol	Value
Elastic stiffness	K	1×10^6 N/mm^3
Normal strength	t_1^u	45 MPa
Shear strength	t_2^u	45 MPa
Mode I fracture toughness	$G_{I,c}$	0.268 kJ/m^2
Mode II fracture toughness	$G_{II,c}$	0.632 kJ/m^2
Relaxation parameter	η	2×10^{-6} s
Calibration parameter	a	90
Calibration parameter	p	1.2
Deformation threshold	λ_{th}	2×10^{-5} mm
Calibration parameter	m	2
Calibration parameter	n	2

3.2. Results

After the load amplitude is imposed on the specimen in a quasi-static fashion, load cycles increments are applied up to the stage at which the specimen fails, which occurs at $N = 8990$ load cycles. The deformation profile at fatigue failure is illustrated in Fig. 5, indicating that the fatigue crack has fully developed across the specimen height. The evolutions of the effective damage parameter D, the fatigue damage parameter D_f, the ultimate traction t^u and the effective relative displacement λ_{max} in a specific material point of the interface element are illustrated in Fig. 6. It can be observed that the interface element remains elastic during the initial, quasi-static application of the maximum amplitude of the fatigue load, since the effective damage parameter D remains zero during this phase. After this phase has finished ($N = 1$), the specimen is subjected to load cycle increments, during which the fatigue damage parameter D_f increases linearly with N. As a result of the linear increase, the ultimate traction t^u of the traction-separation curve decreases linearly with N, in accordance with Eq.(11). The effective damage D of the interface element starts to develop at $N = 7089$. During the subsequent load cycles, the integrity of this element diminishes rapidly, eventually leading to a fully developed crack at $N = 8990$, at which $D = 1$ and $D_f = 0.26$. The degradation behavior of the traction-separation law is illustrated in Fig. 7 by showing this diagram at three different loading stages labeled "1" ($N = 1$), "2" ($N = 3589$) and "3" ($N = 7089$). Stage "1" thus refers to the initial, static traction-separation diagram and stage "3" to the traction-separation diagram at fatigue failure, whereby it can be observed that the ultimate mode I traction t_1^u indeed eventually has been reduced by $D_f \times 100\% = 26\%$, in accordance with Eq.(11).

Fig. 5. Deformation profile (not on scale) of mode I specimen at fatigue failure ($N = 8990$).

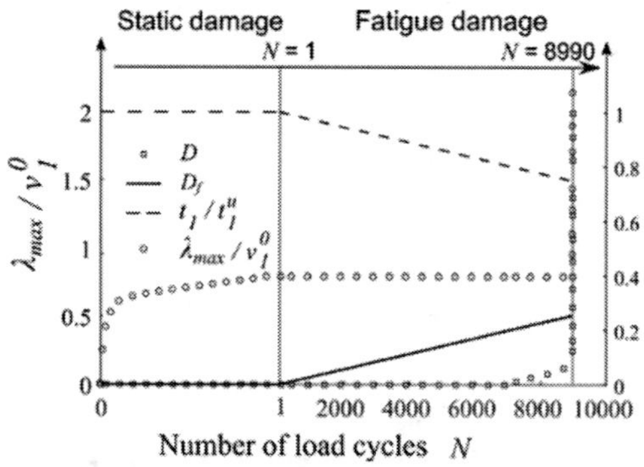

Fig. 6. Effective damage parameter D, fatigue damage parameter D_f, normalized ultimate traction t^u/t_1^u and normalized maximum effective deformation λ_{max}/v_1^0 as a function of the number of load cycles N.

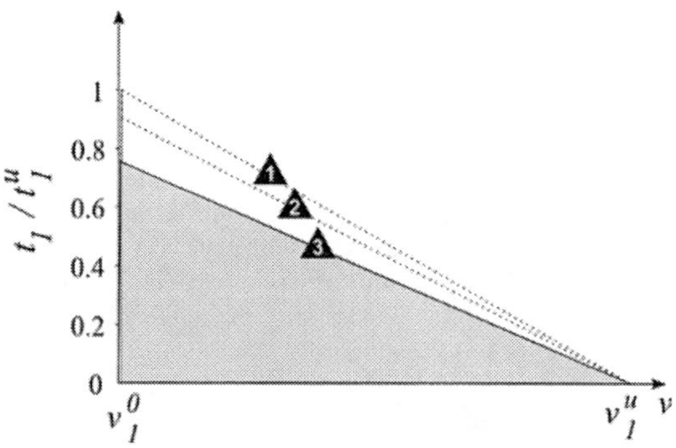

Fig. 7. Traction-separation diagrams at three different loading stages "1" ($N = 1$), "2" ($N = 3589$) and "3" ($N = 7089$).

4 Conclusions

In this paper a static interface damage model is combined with a fatigue evolution law to construct a fatigue interface damage model. The fatigue interface damage model can be applied under variable amplitude loading conditions, whereby load cycle blocks of different amplitude are subsequently imposed in an incremental fashion. The model is able to describe discrete fracture under arbitrary loading conditions, due to the fact that the static interface damage model rigorously accounts for mixed-mode cracking effects. Future research will focus on the application of the model for the determination of the fatigue resistance of the blades of wind turbines, whereby complex fracture scenarios, including crack bifurcation and crack coalescence, will be accounted for.

The financial support from the Chinese Scholarship Council (CSC) to F.G is gratefully acknowledged.

References

1. M.V. Cid Alfaro, A.S.J. Suiker, R. de Borst, J.J.C. Remmers, Eng. Fract. Mech. **76**, 761-780 (2009)
2. G. Alfano, M. Crisfield, Int. J. Numer. Meth. Eng. **50**, 1707-1736 (2001)
3. P.P. Camanho, C.G. Davila, M.F. De Moura, J. Compos. Mater. **37**, 1415-1438 (2003)
4. H. Li, N. Chandra, Int. J. Plast. **19**, 849-882 (2003)
5. B.L.V. Bak, C. Sarrado, A. Turon, J. Costa, Appl. Mech. Rev. **66**, 060803 (2014)
6. A. Turon, J. Costa, P.P. Camanho, C.G. Dvila, Composites Part A **38**, 2270-2282 (2007)
7. P.W. Harper, S.R. Hallett, Int. J. Fatigue **32**, 1774-1787 (2010)
8. A. Pirondi, F. Moroni, J. Adhes. **86**, 501-521 (2010)
9. H. Khoramishad, A.D. Crocombe, K.B. Katnam, I. A. Ashcroft, Int. J. Fatigue **32**, 1146-1158 (2010)
10. M.H.J.W. Paas, P.J.G. Schreurs, W.A.M. Brekelmans, Int. J. Solids. Struct. **30**, 579-599 (1993)
11. R.H.J. Peerlings, W.A.M. Brekelmans, R. de Borst, M.G.D. Geers, Int. J. Numer. Methods Eng. **49**, 1547-1569 (2000)
12. H. Khoramishad, A.D. Crocombe, K.B. Katnam, I.A. Ashcroft, Int. J. Adhes. **30**, 513-521 (2010)
13. J. Hutchinson, Z. Suo, Adv. Appl. Mech. **29**, 63-189 (1992)
14. X.P. Xu, A. Needleman, J. Mech. Phys. Solids. **42**, 1397-1434 (1994)

https://doi.org/10.1051/matecconf/201816522031

The study of the relationship between life limiting factor and stress level for FGH96

Didi. Yang[1], *Yi. Shi*[1], *Guolei. Miao*[2], *Xiaoguang. Yang*[1] and *Duoqi.Shi*[1]

[1]Beihang University - Beijing (China)
[2]Chengdu Holy Industry & Commerce Corp. Ltd. (Group) - Chengdu (China)

Abstract. FGH96 is a Chinese made powder metallurgy nickel based superalloy. This reserach aims to investigate the effect of stress on the microstructure of the worst fatigue life at 600℃. The specimens were first polished by abrasive paper and then electropolished to elimate the effect of surface residual stress. The fatigue tests were conducted at 600℃, and the loading ratio was 0.05 at the frequency of 5Hz. Tests were conducted at 900, 1000, 1100, 1200MPa respectively. The fatigue life under the same condition were compared and the fractography of specimens were examined under SEM to further identify the life-limiting factors of the material. The initiation mode transferred from facet initiated to inclusion initiated from lower stress level to higher stress level. Meanwhile, the initiation position transferred from internal to surface with the increase of stress level. This means at higher stress level, fatigue life is mainly limited by inclusion at surface while at lower stress level the limit is controlled by internal facet.

1 Introduction

FGH96 is one of the active-disc materials. Many stu-dies have been conducted on the fatigue life of FGH96 materials. Large to full-size and sub-dimensions[1], down to sample level[2], there are many researches on factors affecting fatigue life. In order to deeply study the micro-structural factors that affect fatigue life, this paper cond-ucts SEM studies on fatigue fractures, by comparing the impact factors of the worst fatigue life under the same load, and the difference in the worst fatigue life factors under different loads.

Fatigue life is determined by crack initiation life and crack propagation life. The aim of this study is to explore the effect of fatigue crack initiation modes and position and the crack propagation path on fatigue life. The main modes of crack initiation investigated this time are facet and inclusion initiation. Under the normal stress conditions, facet initiation may occur. In the manu-facturing process, nonmetallic and heterogenous metal inclusions will be introduced in FGH96. Stress concen-tration caused by inclusions make cracks appear. Crack initiation positions can be divided into surface, subsurface, and internal. The crack propagation path is mainly investi-gated by macroscopic fractures. The crack propagation path is influenced by environmental factors and stress factors.

2 Testing material and experiment procedure

2.1 materials

The material tested is a high-temperature nickel-base superalloy named FGH96. FGH96 is China's second-generation polycrystalline powder metallurgy superalloy, mainly used for aero-engine turbine discs. The chemical compositions is given in Table 1 FGH96 was manufac-tured by vacuum smelting alloy, atomizing alloy followed by isostatically pressing, extruding and isothermally forging[3].

EBSD measurements were used to map the grain size and characterize the grain orientation. The obtained microstructure is shown in Fig. 1. Macroscopically, the material is isotropic and contains twins. The average grain size is 23mm. There is no defect information such as in-clusions and pores been observed, for the limited scanning area of EBSD. Detailed information has been introduced in previous work[4].

Table 1. Chemical composition of experimental steels (wt%).

Cr	Co	W	Mo
15.5～16.5	12.5～13.5	3.8～4.2	3.8～4.2
Al	Nb	Ti	Ni
2.0～2.4	0.6～1.0	3.5～3.9	Bal.

* Corresponding author: yangdidi4521@163.com

https://doi.org/10.1051/matecconf/201816522031

Fig. 1. Microstructure of the FGH96 (grain orientation and size)

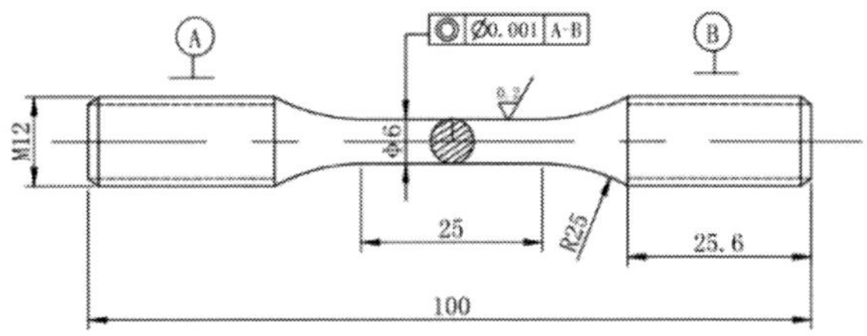

Fig. 2. The shape and dimensions of the specimen

2.2 experiment procedure

All fatigue specimens were machined from the same disk. The geometry of the specimen is shown in Fig. 2. To eliminate the impact of processing, e.g. surface residual stress, surface scratches and roughness, specimens were roughly polished using increasingly fine abrasive paper up to 2000 grit followed by electropolishing. Electro-polishing was conducted using an electrolyte of 55% ethanol with 35% butyl cellusolve and 10% perchloric acid at 25V and 20 ℃. The measurement results show that the surface of the specimen participates in stress below -100 MPa, and the effect of residual stress can be ignored.

The fatigue tests were conducted at 600 °C using a SHIMADZU servo-hydraulic test system according to the test standard. The tests were performed in stress control at the frequency of 5 Hz, the ratio (R) of 0.05. Nowadays the maximum stress of the aero-engine turbine disk is approximately 1100 Mpa[5] (with the rotation speed 10000r/min). And considering the existing test data[6,7], the stress levels were set at 1100MPa and 1200MPa for the low cycle fatigue and 900MPa and 1000MPa for the high cycle fatigue respectively.

* Corresponding author: yangdidi4521@163.com

3 Experiment results

3.1 stress and life

With the increase of the stress, the corresponding life reduced rapidly. And the variability of the fatigue life showed a trend of decreasing first and then increasing. The reason was given in previous work[4].

3.2 fatigue fracture analysis

To study the factors that limit the fatigue life, the fatigue fracture surfaces of all the tests were investigated by SEM analysis, as shown in Fig. 3, Fig. 4, Fig. 5 and Fig. 6. The main object of the study is the stage of crack initiation. At the same time, the state of the crack propagation stage under different stresses has also been studied.

Through the analysis of the fracture map, there are two initiation modes: (a) facet initiated; (b) inclusion initiated, and three initiation positions: (a) internal initiated (b) sub-surface initiated and surface initiated. With the increase of the stress level, the crack initiation modes of the worst-life changes from facet initiation to inclusion initiation. In addition, the crack initiation position starts from the internal to the surface. Table 2 shows the initiation modes of the worst life test specimen at each stress level.

At lower stress level, the facet initiation position mostly appear in the internal. While, at the higher stress level, they mostly appear on the surface, consistent with the reaction of inclusions to stress. It has been mentioned in previous work that when the crack initiation position approaches the surface, the crack growth will be influenced by environmental factors (oxygen, etc.), so the life is shorter than the internal crack. But at 900MPa, there was a completely opposite result. As the crack approached the surface, the fatigue life increased (as shown in Fig. 8) and the lifetime of the crack near the subsurface was 18 times that of the internal.

In addition, at higher stress level, the shape of the crack no longer maintains an arc-shaped outward extension at the later stage of the extension. The shorter the fatigue lives are, the narrower the crack surface is, showing a "bamboo shoot" shape, as shown in Fig. 7. The “thinner” the bamboo shoots are, the shorter the fatigue life is.

4 Discussion

The relationship between stress and life is plotted in semi-logarithmic coordinates, as shown in Fig. 9 and Fig. 10. The S-N curve obtained by exponential calculation is a straight line. Using the 3-sigma theory, the worst life curve with a confidence of 99.87% can be calculated. To study the factors limiting the fatigue life, focusing on the part between the S-N curve and the worst life curve. The microscopic factors that limit fatigue life will be explored separately from the crack initiation modes and the initi-ation position.

4.1 Initiation modes

As shown in the Fig. 9, the initiation modes transfer from facets to inclusion when the stress increasing. At 900MPa and 1000MPa, the initiation modes of cracks are facet, as shown in Fig. 3 and Fig. 4. And at 1100MPa and 1200MPa, the initiation modes of cracks are inclusion, as shown in Fig. 5 and Fig. 6. Crack initiation mechanism of the inclusion initiation and facet initiation are very different. The facet crack initiation is that under the normal stress, the crystal plane with the weakest atomic bonding cracks. Meanwhile, there are usually three cond-

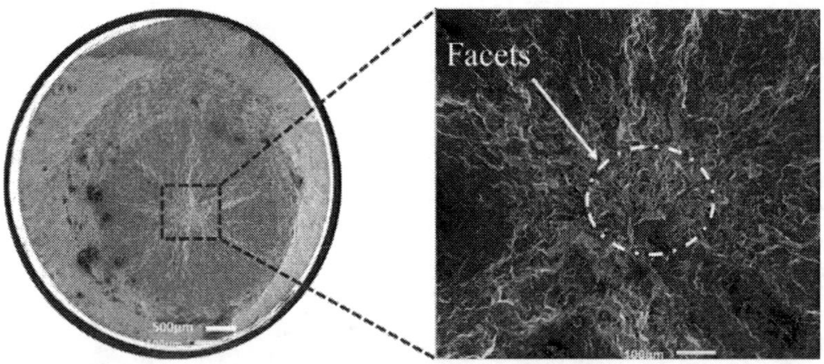

Fig. 3. SEM image of FGH96 fatigue fracture at 900 Mpa (left: low magnification. right: high magnification)

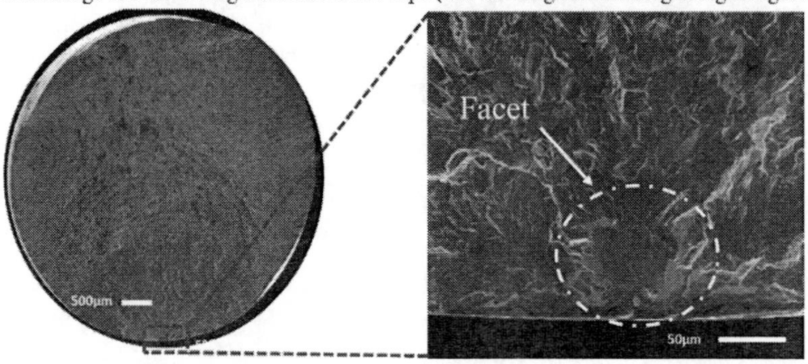

Fig. 4. SEM image of FGH96 fatigue fracture at 1000 Mpa (left: low magnification. right: high magnification)

https://doi.org/10.1051/matecconf/201816522031

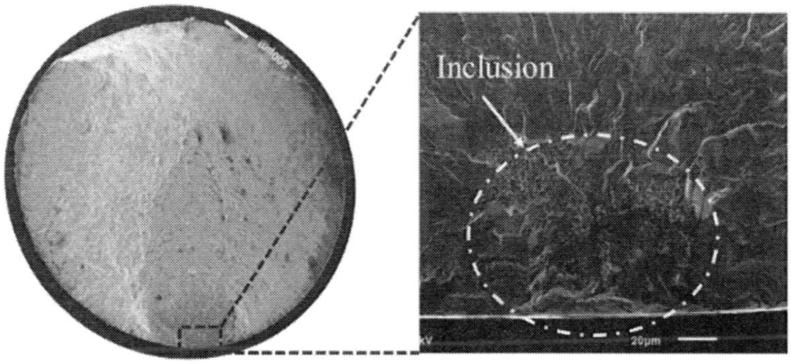

Fig. 5. SEM image of FGH96 fatigue fracture at 1100 Mpa (left: low magnification. right: high magnification)

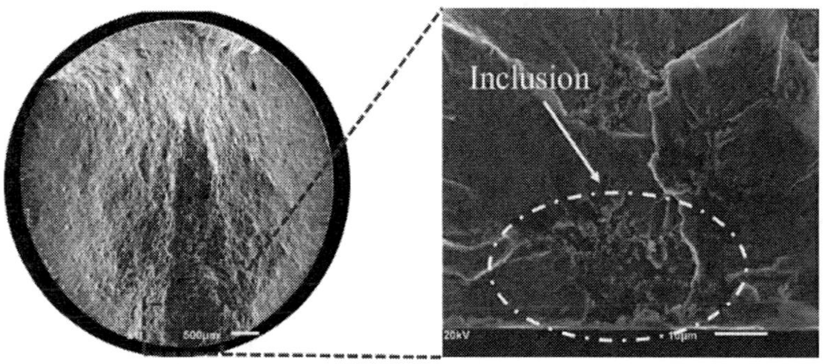

Fig. 6. SEM image of FGH96 fatigue fracture at 1200 Mpa (left: low magnification. right: high magnification)

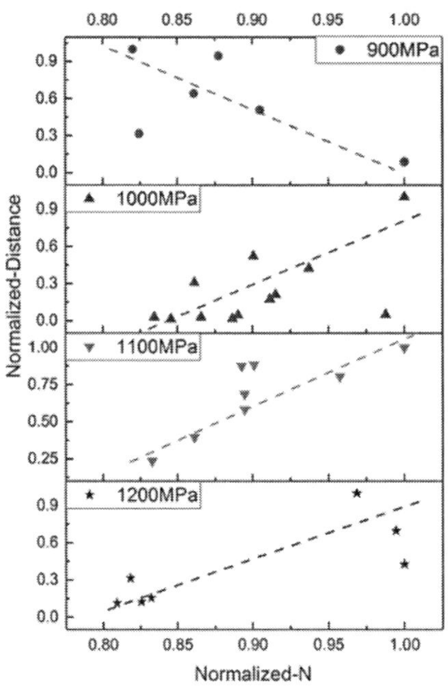

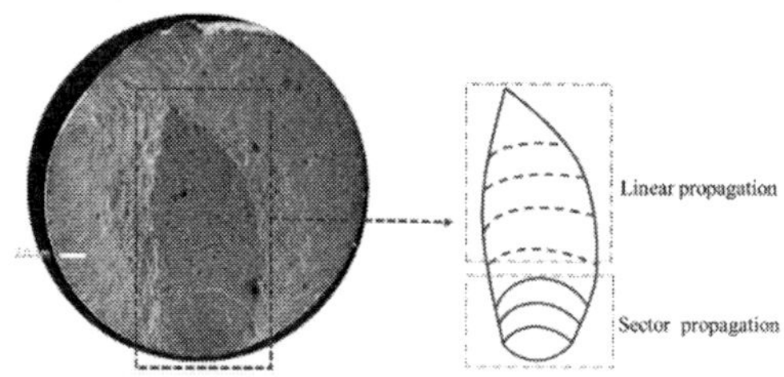

Fig. 7. Schematic diagram of the "bamboo shoot" shape at 1200MPa

Table 2. Crack initiation information at four stresses

Stress Level (Mpa)	Minium Fatigue Life (cycles)	Initiati on Modes	Initiation Position
900	557563	Facet	Internal
1000	100587	Facet	Surface
1100	27772	Inclusion	Surface
1200	8581	Inclusion	Surface

Fig. 8. Normalized distance (crack initiation position to surface) and fatigue life

https://doi.org/10.1051/matecconf/201816522031

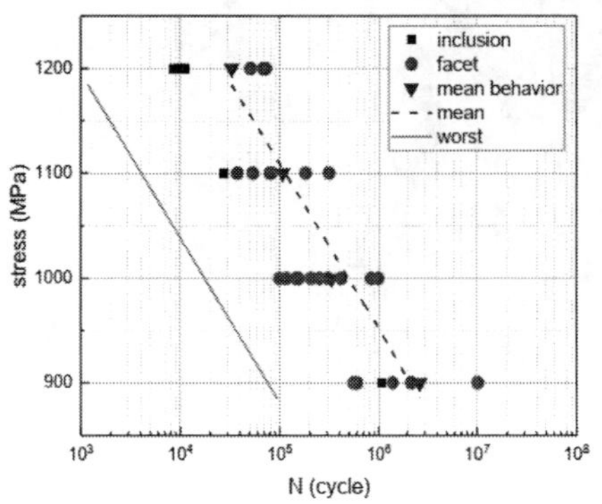

Fig. 9. Comparing the initiation modes at four stresses

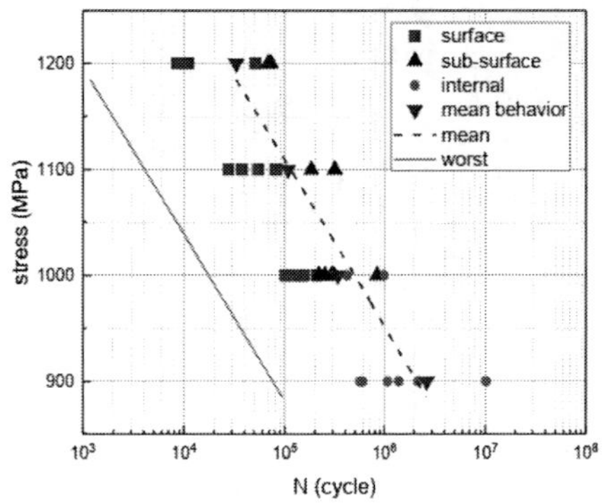

Fig. 10. Comparing the initiation modes at four stresses

itions that Nickel-based powder superalloy initiating cracks at inclusions: (a) cracking of the inclusion itself, (b) cracking of the inclusion/substrate interface, (c) cracking of the substrate adjacent to the inclusion[8].

Forming a crack surface requires absorbing enough energy, according to Griffith theory. Energy absorption is related to load and time. Stress concentration occurs around the inclusions. But, when the external load is small, the stress concentration caused by the inclusion does not have a great influence. Meanwhile, the energy required to break the atomic bonding force forming a facet is smaller than the energy needed to initiate crack near an inclusion. So the facet crack initiation occurs first at lower stress level. As the load increases, the stress around the inclu-sion caused by stress concentration increases rapidly, and the difference with the average stress increased. Then the material close to the inclusion first obtains enough energy to initiate cracks. This can explain the initiation of cracks around inclusions at higher stress level.

4.2 Initiation position

The initiation position transfers from internal to surface when the stress increases. At higher stress levels, the crack initiation positions are mainly surface and subsurface, as shown in Fig. 4, Fig. 5, Fig. 6. As analyzed in 4.1 section, with the increase of stress, inclusions become the main micro-structural factor that limits the fatigue life. And the stress concentration factor caused by inclusions becomes more evident with the inclusion approaching the surface[9], thus the surface and subsurface of the material are more prone to cracks. When the stress level is not too high (at 900MPa), the area (volume) of the inclusion can offset the influence of the stress concentration factor caused by the position. Therefore, the position of the crack initiation is random and may even appear inside, as shown in Fig. 3.

When at lower stress level is, such as 900MPa, the main mode of initiation is facet. The facet occurs mainly at large grains and grain boundaries, and these features are common within the specimen. So cracks usually initiated in the internal of the material.

4.3 crack growth

The SEM result of the worst life specimen in 1200MPa was shown in Fig. 11.The narrower the crack surface is, the faster the critical length can be reached at the same energy input, making the remaining distance of the specimen (i.e. the length of the line in the rectangular frame in Fig.11) less than the safety distance. Then the break occurred. On the other hand, sharp crack surfaces increase stress concentration, making crack growth faster and fatigue life worse.

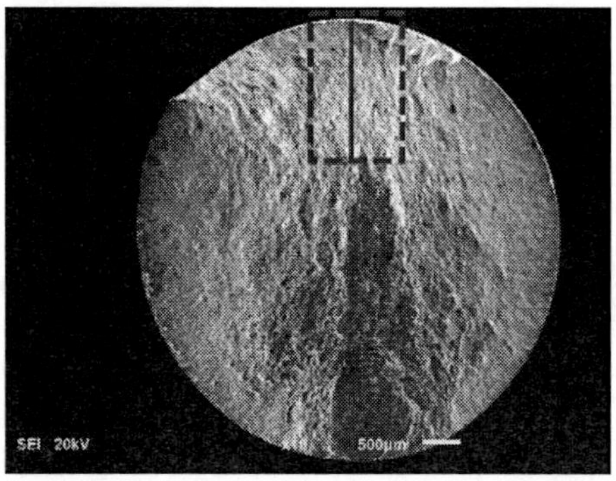

Fig. 11. Fracture diagram of worst life at 1200MPa

Conclusion

By comparing FGH96 fatigue life and fracture micro-structure, the conclusions are as follows:

1) The main limiting factors of fatigue life are inclusion defects and crack initiation caused by facets. At higher stress levels, inclusions are the main microscopic factors limiting fatigue life; at lower stress levels, facet is the microscopic factor limiting fatigue life;

2) At higher stress levels, fatigue cracks are mostly initiated on the surface, while at lower stress levels, cracks are initiated more internally.

3) The main factor limiting the fatigue life is the relation-ship between inclusions and stress. At lower stress levels, inclusions do not substantially affect the worst life, and the inherent properties of the material (atomic bond weakest resultant force) cause crack initiation, limit-ing fatigue life. At higher stress levels, inclusion limits fatigue life.

4) At the higher stress level, the closer the initiation position is to the surface, the shorter the life is; while at the lower stress level, the closer the initiation position is to the surface, the longer the lifetime is.

The authors would like to thank the sponsorship by National Program on Key Basic Research Project (973 Program) (No. 2015CB057401).

References

1. Y. Feng, C. Wu, W. Gao, W. Chen, J. Aerosp. Power. **27**:628-634(2012)
2. D. Wei, X. Yang, Y. Wang, et al, J. Aerosp. Power, 22: 425-430(2007)
3. China Aviation Material Handbook Committee. *China Aviation Material Handbook*. **5**, 29 (2002)
4. G. Miao, X. Yang, D. Shi, Mater. Sci. Eng., A, **668**:66-72(2016)
5. T. Liu, Q. Deng, Y. Liu, J. Mech. Res. Appl, **28**:94-96. (2015)
6. Cashman G T, Int. J. Fatigue, **32**:492-496 (2010)
7. S.K. Jha, M.J. Caton, J.M. Larsen. *AFRL-RX-WP-TM-2010-4084*, 57-72 (2008)
8. Y. Zeng, M. Zhang, J. Dong, et al, J. Mater. Eng,**353-358**:1185-1190 (2005)
9. S. Yuan, Beijing: BUAA (2014)

Ability of multiaxial fatigue criteria accounting for stress gradient effect for surface defective material

Wichian Niamchaona[1,*], *Fabienne* Pennec[1], *Kévin* Tihay[2], *Michel* Duchet[2], *Bastien* Weber[2], and *Jean-Louis* Robert[1]

[1]Institut Pascal, IUT d'Allier, 03100 Montluçon, France
[2]ArcelorMittal, Maizières Research, 57283 Maizières-Les-Metz Cedex, France

Abstract. New high strength steels are widely used nowadays in many industrial areas as in automotive industry. These steels are more resistant and provide higher fatigue limits than latter ones but they are also more sensible to small defects. Natural defects that outcome from metallurgy (as shrinkage, inclusion, void) are not considered in this study. We focus on small manufacturing defects such as cutting edge defects generated by punching or other surface defects due to stamping. These defects are harmful on the material fatigue behaviour due to high stress concentration at defects root. They also generate stress gradient that is beneficial from the fatigue strength point of view. This study focusses on the stress gradient (it does not account for the size effect) from cylindrical defect on specimen edge. Practically a normal stress gradient is added in multiaxial fatigue criteria formulation. Both critical plane approach and integral approach are involved in the present study. This gradient is calculated from stress states at defects root by using FEM. Criteria fatigue function at *N* cycles is used to assess the material fatigue strength. Obviously multiaxial fatigue criteria accounting for stress gradient give more precise fatigue functions than criteria that do not consider the gradient influence.

1 Introduction

High strength steels are used in automotive structures such as flask of seat, door bar or suspension arm to reduce their own weight and their energy consumption and to improve their durability. These components are generally made by punching from steel sheets and then by forming through stamping or folding. These processes may generate small defects (figure 1) that can occur on cut edge or external surfaces.

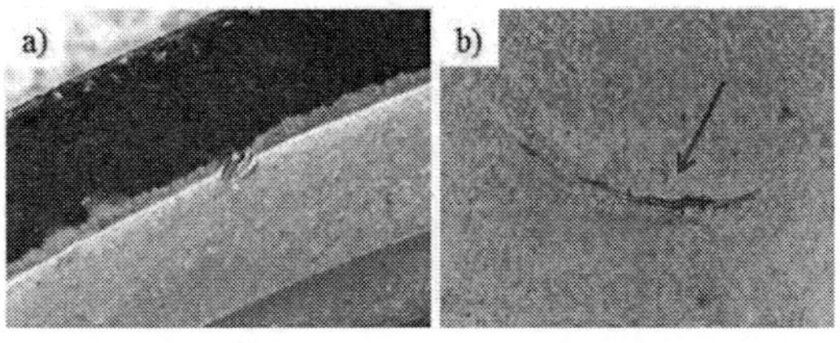

Fig. 1. Small defects occurrence a) on cut edge and b) on external surface of suspension arms.

The small defect may be a cause of fatigue failure in case of variable or constant amplitude loading. Practically it reduces the structure lifetime. It generates some stress concentration that is detrimental from the fatigue point of view. As an evidence, the defect reduces the local fatigue strength but it also generates one stress gradient in the vicinity of the defect. The stress gradient becomes a beneficial phenomenon for the fatigue behaviour of smooth specimens subjected to different fatigue loading types [1]. In the case of tension compression loading no stress gradient occurs as there is one uniform stress field all over the cross section. In the case of rotative or cantilever bending the stress distribution generates some gradient (figure 2) that makes increase the specimen fatigue limit up to 5% and 10% respectively.

This study focusses only on the stress gradient induced by the defect without accounting for the size effect of the sample. The stress gradient is integrated in multiaxial fatigue criteria [2] both for critical plane approach [3, 4] and integral approach [5, 6]. The normal stress gradient is considered as it is much more predominant on fatigue behaviour than shear stress gradient [7, 8, 9].

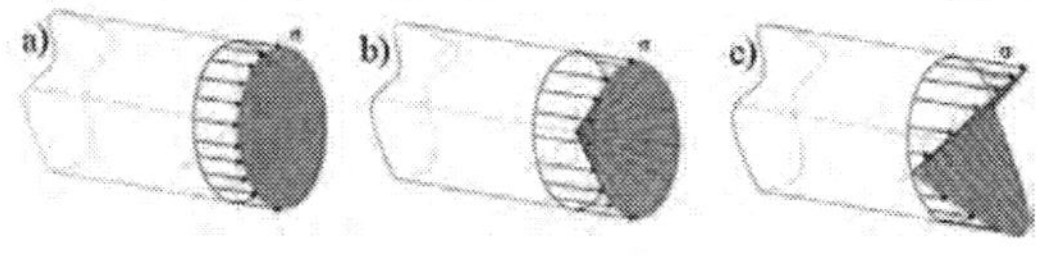

Fig. 2. Stress distribution over the beam cross section for a loading of a) tension-compression b) rotative bending and c) cantilever bending [1].

The stress gradient may be obtained by parabolic interpolation of the stress states over three points (i.e. nodes) at the defect root. The assumption is that fatigue crack initiates at the defect root where some high stress

* Corresponding author: niamchaona.w@gmail.com

concentration occurs. Some researchers [10, 11, 12] consider the stress gradient to be assessed from one zone which length is equal to $\sqrt{area}$. The parameter $\sqrt{area}$ is in fact the square root of the area obtained by the projection of the defect volume over one plane perpendicular to the loading direction [13]. The stress states are obtained for this study by the Femap Nastran Finite Element code.

The outcome of one multiaxial fatigue criterion is a fatigue function E that allows to compare the analysed multiaxial stress cycle to the material fatigue strength at N cycles. In this study the fatigue strength is generally assessed at 2.10^6 cycles, as it is commonly referenced in the automotive industry. The purpose of this paper is mainly to evaluate how accounting for stress gradient gives more precise prediction ability to multiaxial criteria of the actual fatigue strength in the case of surface defect occurrence.

2 Experimentation

This section aims to exhibit the experimental fatigue strength reduction that is observed in the case where one surface defect is present. Different defect depths are studied. The experimental fatigue test conditions are described below.

2.1 Specimen and fatigue tests

The samples are obtained by cutting from one CP800 high strength steel sheet 2 mm thick. Sample edges were polished before fatigue testing. Then the cylindrical defect is machined on the edge of the specimen minimal cross section by electro-erosion with diameter wire equal to 200 μm. The geometry of the defective fatigue specimen is shown in figure 3.

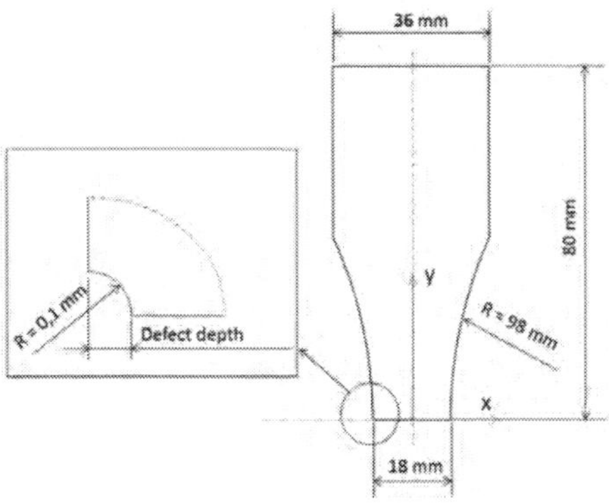

Fig. 3. Geometry of the fatigue specimen with one defect on its edge.

The defect depths vary from 30 μm to 300 μm. The defect radius remains identical for any depth (R = 0.1 mm). The geometrical sample has a stress concentration factor equal to 1.07. The 810MTS fatigue machine is used. Its capacity is a ±100 kN tensile-compressive load.

The test frequency is 20 Hz and fatigue solicitation is tension with loading ratio equal to 0.1. The fatigue lifetime threshold is 2.10^6 cycles when no fatigue failure occurs. The defects are observed by a VHX-900F Keyence optical microscope. It allows to measure the real shape and size of the defect after machining. It allows also to observe the fractured section of the sample to identify the crack initiation site and the crack propagation area up to ductile fracture. Other defects shapes and locations have been studied (figure 4). However we focused in this paper on cylindrical defects located on the specimen edge.

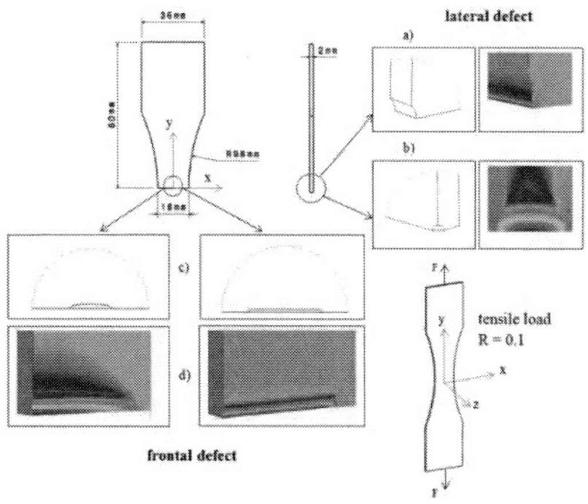

Fig. 4. Geometry and location of studied defects a) on the edge b) in the corner c) and d) on the front surface.

2.2 Defects fatigue influence

The results from experiments are shown by S-N curves plotted in figure 5.

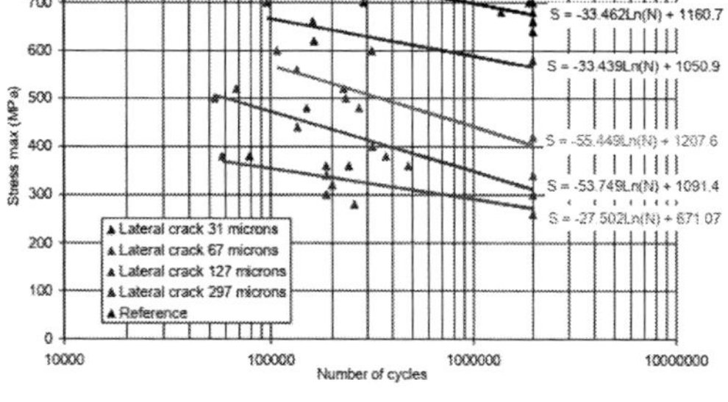

Fig. 5. S-N curves of CP800 steel versus defect depth under tensile load (R = 0.1).

The fatigue strength is reduced when the defect depth increases. The deeper the defect is the most detrimental its influence is from the fatigue resistance point of view. The table 1 gives the experimental fatigue strength up to 2.10^6 cycles for each defect depth.

One has to notice that the defect induced fatigue strengths at 2.10^6 cycles are used as the corresponding

 https://doi.org/10.1051/matecconf/201816522032

maximum nominal stress applied to the defected specimen for Finite Element Method calculations.

Table 1. Fatigue strengths at 2.10^6 cycles.

Defect depth (µm)	Fatigue strength (MPa)
no defect	675
31	566
67	403
127	312
297	272

3 Numerical simulations

In this section are presented the results of FEM calculation obtained by using Femap Nastran software. Stress fields and stress gradients are determined around defects for fatigue analysis purpose.

3.1 Stress field near defect root

The first stress field calculation is realised under a purely elastic stress-strain response assumption. The sample model is meshed with hexahedral elements (QUAD). The mesh size is about 3 µm at defect root and then increases afar from defect root in order to reduce calculation time consumption. Symmetry planes are also used so that a quarter of the sample is finally calculated. The tensile load is applied to the specimen along y direction. The σ_{yy} stress field around the defect root is presented in figure 6.a.

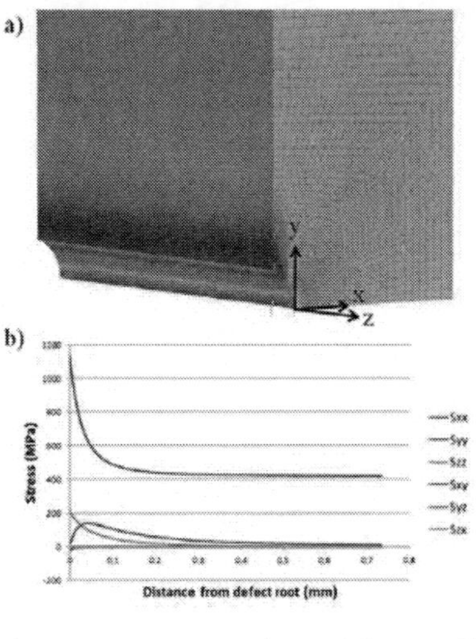

Fig. 6. a) σ_{yy} stress concentration at defect root b) elastic response stress components along x-direction at defect root for the sample with a 67 µm deep defect.

Stress states are thus multiaxial in the vicinity of the defect (figure 6.b). The stress concentration factor of σ_{yy} component is related to the defect depth as summarised in table 2.

Table 2. Maximum σ_{yy} stress value and stress concentration factors.

Defect depth (µm)	Nominal stress (MPa)	Maximum elastic stress (MPa)	K_t
without defect	675	720	1.07
31	566	1267	2.24
67	403	1141	2.83
127	312	1104	3.54
297	272	1328	4.88

Stress states at root defect exhibit triaxiality effect. The highest stress components are σ_{yy}, σ_{xx} and σ_{zz} respectively. Other stress components are close to zero. The stress concentration factor increases with the depth of the defect. The maximum elastic stress at the defect root is beyond the CP800 yield stress R_e that is equal to 830 MPa. The FEM analysis with elastoplastic stress-strain behaviour is thus the next step. The tensile monotonic elastoplastic stress-strain response for the next numerical calculations is given by figure 7.

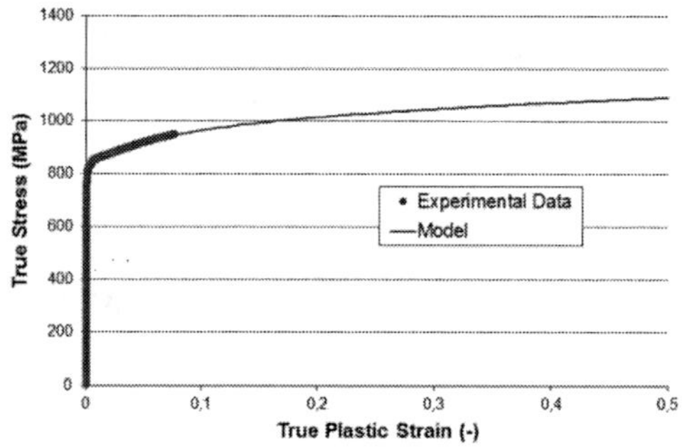

Fig. 7. Tensile monotonic elastoplastic stress-strain response of the CP800 steel.

3.2 Stress gradient at defect root

The stress gradient is calculated in the most fatigue-prone area of the defective material i.e. at the defect root. The stress gradient is calculated for all the stress components along directions x, y and z. The x direction is normal to the defect root (transverse direction) whereas z corresponds to the sample thickness direction. One 3-points parabolic interpolation is used to determine all the stress gradients (figure 8).

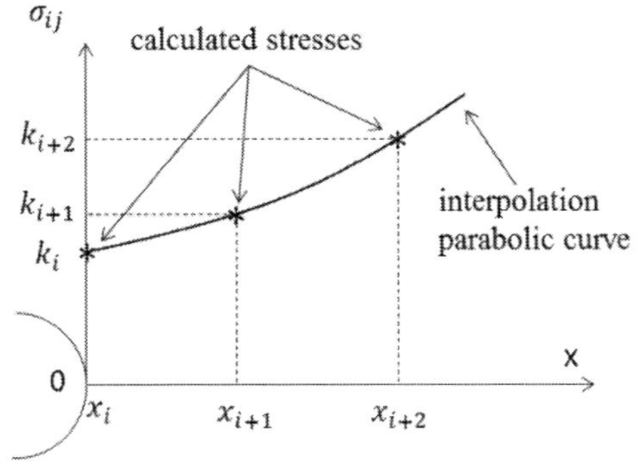

Fig. 8. Parabolic interpolation method used for assessing stress gradients along x direction at defect root.

The stress component distribution along x direction is assumed to be modelled by parabolic model, i.e. :

$$k = ax^2+bx+c \quad (1)$$

Practically the first three nodes at the defect root are retained for the local stress gradient assessment. The actual stress distribution is assumed to check the stress distribution i.e. :

$$k_i = a(x_i)^2+b\, x_i +c \quad (2)$$

$$k_{i+1} = a(x_{i+1})^2+b\, x_{i+1}+c \quad (3)$$

$$k_{i+2} = a(x_{i+2})^2+b\, x_{i+2}+c \quad (4)$$

From equation (2), (3) and (4), one obtains :

$$a = \frac{(k_{i+1}-k_i)(x_{i+2}-x_i)-(k_{i+2}-k_i)(x_{i+1}-x_i)}{(x_{i+1}-x_i)(x_{i+2}-x_i)(x_{i+1}-x_{i+2})} \quad (5)$$

$$b = \frac{(k_{i+1}-k_i)-a(x_{i+1}^2-x_i^2)}{x_{i+1}-x_i} \quad (6)$$

The gradient of one given stress component along direction x is determined by :

$$\left(\frac{\partial\sigma}{\partial x}\right)_{x=x_i} = \left(\frac{\partial k}{\partial x}\right)_{x=x_i} = 2a(x_i) + b \quad (7)$$

The calculation procedure is carried out for all the stress components of the stress states in the vicinity of the defect root. By this way the following gradient matrix is obtained from the stress gradients of stress components in directions x, y and z.

$$[G]_\sigma = \begin{bmatrix} \frac{\partial\sigma_{xx}}{\partial x} & \frac{\partial\sigma_{yy}}{\partial x} & \frac{\partial\sigma_{zz}}{\partial x} & \frac{\partial\sigma_{xy}}{\partial x} & \frac{\partial\sigma_{yz}}{\partial x} & \frac{\partial\sigma_{zx}}{\partial x} \\ \frac{\partial\sigma_{xx}}{\partial y} & \frac{\partial\sigma_{yy}}{\partial y} & \frac{\partial\sigma_{zz}}{\partial y} & \frac{\partial\sigma_{xy}}{\partial y} & \frac{\partial\sigma_{yz}}{\partial y} & \frac{\partial\sigma_{zx}}{\partial y} \\ \frac{\partial\sigma_{xx}}{\partial z} & \frac{\partial\sigma_{yy}}{\partial z} & \frac{\partial\sigma_{zz}}{\partial z} & \frac{\partial\sigma_{xy}}{\partial z} & \frac{\partial\sigma_{yz}}{\partial z} & \frac{\partial\sigma_{zx}}{\partial z} \end{bmatrix}_{x,y,z} \quad (8)$$

The fatigue influence of the stress gradient is implemented in most multiaxial fatigue criteria through the gradient normal stress σ_{hh} that is acting on the material plane which unit normal vector is denoted as h. One has :

$$\sigma_{hh} = [h]^t \cdot [\sigma] \cdot [h] \quad (9)$$

The gradients of this normal stress along axes x, y and z are expressed by equations (10), (11) and (12) respectively.

$$\frac{\partial\sigma_{hh}}{\partial x} = h_x^2\left(\frac{\partial\sigma_{xx}}{\partial x}\right) + h_y^2\left(\frac{\partial\sigma_{yy}}{\partial x}\right) + h_z^2\left(\frac{\partial\sigma_{zz}}{\partial x}\right) + 2h_xh_y\left(\frac{\partial\sigma_{xy}}{\partial x}\right) + 2h_yh_z\left(\frac{\partial\sigma_{yz}}{\partial x}\right) + 2h_zh_x\left(\frac{\partial\sigma_{zx}}{\partial x}\right) \quad (10)$$

$$\frac{\partial\sigma_{hh}}{\partial y} = h_x^2\left(\frac{\partial\sigma_{xx}}{\partial y}\right) + h_y^2\left(\frac{\partial\sigma_{yy}}{\partial y}\right) + h_z^2\left(\frac{\partial\sigma_{zz}}{\partial y}\right) + 2h_xh_y\left(\frac{\partial\sigma_{xy}}{\partial y}\right) + 2h_yh_z\left(\frac{\partial\sigma_{yz}}{\partial y}\right) + 2h_zh_x\left(\frac{\partial\sigma_{zx}}{\partial y}\right) \quad (11)$$

$$\frac{\partial\sigma_{hh}}{\partial z} = h_x^2\left(\frac{\partial\sigma_{xx}}{\partial z}\right) + h_y^2\left(\frac{\partial\sigma_{yy}}{\partial z}\right) + h_z^2\left(\frac{\partial\sigma_{zz}}{\partial z}\right) + 2h_xh_y\left(\frac{\partial\sigma_{xy}}{\partial z}\right) + 2h_yh_z\left(\frac{\partial\sigma_{yz}}{\partial z}\right) + 2h_zh_x\left(\frac{\partial\sigma_{zx}}{\partial z}\right) \quad (12)$$

Finally the gradient of the normal stress along the direction perpendicular to the considered plane is given by equation (13) :

$$G_{hh} = \frac{\partial\sigma_{hh}}{\partial h} = h_x\left(\frac{\partial\sigma_{hh}}{\partial X}\right) + h_y\left(\frac{\partial\sigma_{hh}}{\partial Y}\right) + h_z\left(\frac{\partial\sigma_{hh}}{\partial Z}\right) \quad (13)$$

As a matter of fact multiaxial fatigue criteria take into account this latter gradient from its absolute value $|G_{hh}|$.

4 Multiaxial fatigue criteria

Stress states at defect root are interpreted as the actual multiaxial fatigue strength of CP800 steel up to N cycles by the use of stress gradient dependent multiaxial fatigue criteria. The two most common approaches in multiaxial fatigue, i.e. the critical plane and integral approaches are involved in this study as they are the two main ways for fatigue assessment in multiaxial fatigue. The concept of one multiaxial fatigue criterion is to calculate its own fatigue function that is equal to unity when the analysed multiaxial stress cycle reaches the material fatigue strength up to N cycles. This property may be formulated as :

$$E\{[\sigma_{ij}(t)], \sigma_{-1}(N), \tau_{-1}(N), f_{-1}(N), \sigma_0(N), \dots\} = 1 \quad (14)$$

where $[\sigma_{ij}(t)]$ is the analysed multiaxial stress cycle. $\sigma_{-1}(N), \tau_{-1}(N), f_{-1}(N), \sigma_0(N), \dots$ are particular fatigue strengths up to N cycles (for fully reversed tension-compression load, fully reversed torsion solicitation, fully reversed bending load,…). When the multiaxial stress cycle corresponds to the material fatigue strength up to N cycles, the calculated fatigue function of the criterion can be interpreted as a validity checking of its prediction ability. The E value may be interpreted according to the following manner :

$E < 1$: the criterion is non conservative as it assesses the multiaxial stress cycle under the fatigue strength of the material,

$E > 1$: the criterion is conservative as it states that the multiaxial stress cycle is greater than the material fatigue strength,

$E = 1$: the criterion is fair as its prediction properly corresponds to the experiment : it agrees the fact that the analysed cycle correspond to material fatigue strength up to N cycles.

4.1 Conventional fatigue criteria

This section describes the two employed fatigue criteria. The first one is Dang Van criterion that is based on the so call critical plane approach. This criterion was proposed in 1973 and its formulation uses two stress parameters which are the alternate shear stress $\tau_{ha}(t)$ and hydrostatic pressure $P_H(t)$. The damage indicator of the criterion for the material plane which unit normal vector is h is given by :

$$E_h = \max_t \left\{\frac{\tau_{ha}(t)+\alpha P_H(t)}{\beta}\right\} \quad (15)$$

The critical plane approach concept can be stated as follows : the so-called critical plane is the most solicited one, with respect to the criterion formulation and drives the material fatigue behaviour. The criterion fatigue function is consequently equal to the criterion damage indicator related to this critical plane. This is written as :

$$E_{DV} = \max_h(E_h) \quad (16)$$

Two material parameters α and β have to be determined from two fatigue limits under different loading types (i.e. tension-compression and torsion) or under two different loading ratio. The determination of the parameters α and β is one step that is called the criterion calibration. When the fatigue limits (σ_{-1} and τ_{-1}) under fully reversed tension-compression and torsion tests (R = -1) are used, one has :

$$\alpha = 3\left(\frac{\tau_{-1}}{\sigma_{-1}} - \frac{1}{2}\right) \quad \text{and} \quad \beta = \tau_{-1} \quad (17)$$

When the fatigue strengths ($\sigma_{-1}(N)$ and $\tau_{-1}(N)$) under fully reversed tension-compression and torsion tests up to N cycles are used, one states :

$$\alpha = 3\left(\frac{\tau_{-1}(N)}{\sigma_{-1}(N)} - \frac{1}{2}\right) \quad \text{and} \quad \beta = \tau_{-1}(N) \quad (18)$$

The second fatigue criterion is based on the so-called integral approach. Zenner criterion is used for that purpose. The criterion was proposed in 1993. It defines a fatigue function that is the quadratic mean value of the damage indicator related to all the possible material planes. By this way all these material planes contribute to the fatigue function assessed by the criterion. This concept is clearly distinct from the critical plane approach where only one plane brings its contribution to the fatigue damage assessment. The criterion use four stress parameters : the shear stress amplitude τ_{ha}, the mean shear stress τ_{hm}, the normal stress amplitude σ_{hha} and the mean normal stress σ_{hhm}. The damage indicator E_h related to the h normal vector plane is expressed as :

$$E_h = a\tau_{ha}^2(1 + m\tau_{hm}^2) + b\sigma_{hha}^2(1 + n\sigma_{hhm}) \quad (19)$$

The Zenner criterion fatigue function is obtained from :

$$E_{ZN} = \frac{1}{\sigma_{-1}}\sqrt{\left(\frac{15}{2}\right)\frac{1}{S}\int_S E_h dS} \quad (20)$$

The integral concerns all the differently oriented possible material planes by considering a unit radius sphere which area S is equal to 4π.

Four material parameters a, b, am and bn have to be determined. The criterion calibration is generally proposed from four fatigue limits : σ_{-1} (under fully reversed tension-compression load), τ_{-1} (under fully reversed torsion load), σ_0 (under repeated tension load) and τ_0 (under repeated torsion load) [6]. The expression of the four criterion parameters are given by equation (21) to (24).

$$a = \frac{1}{5}\left[3\left(\frac{\sigma_{-1}}{\tau_{-1}}\right)^2 - 4\right] \quad (21)$$

$$b = \frac{1}{5}\left[6 - 2\left(\frac{\sigma_{-1}}{\tau_{-1}}\right)^2\right] \quad (22)$$

$$am = \frac{\sigma_{-1}^2 - \left(\frac{\sigma_{-1}}{\tau_{-1}}\right)^2\left(\frac{\tau_0}{2}\right)^2}{\frac{12}{7}\left(\frac{\tau_0}{2}\right)^4} \quad (23)$$

$$bn = \frac{\sigma_{-1}^2 - \left(\frac{\sigma_0}{2}\right)^2 - \frac{4}{21}am\left(\frac{\sigma_0}{2}\right)^4}{\frac{15}{14}\left(\frac{\sigma_0}{2}\right)^3} \quad (24)$$

The Dang Van and Zenner criteria formulations presented in this section are stress gradient free. They correspond to the classical or conventional formulation of the criteria as initially proposed by their author. In the next section the stress gradient dependency is introduced in the criteria formulation.

4.2 Fatigue criteria formulation accounting for stress gradient

The stress gradient is taken into account by adding a term related to the normal stress gradient to the conventional criteria formulation. This method is inspired from Weber's work in 1999 [1]. This normal stress gradient term introduces a supplementary material parameter γ to be determined when the criterion is calibrated. The stress gradient dependent criterion derived from Dang Van formulation is proposed as follows :

$$E_h = \max_t \left\{\frac{\tau_{ha}(t)+\alpha P_H(t)+\gamma|G_{hh}|}{\beta}\right\} \quad (24)$$

The fatigue function of the criterion is still obtained on the critical plane by stating :

$$E_{DVG} = \max_h(E_h) \quad (25)$$

The material parameters α and β are identical to those of the conventional criterion. The parameter γ is obtained by the criterion calibration from the fatigue limit obtained for one defective specimen. One has to note that the defect size is a reference value. Generally we choose this reference size as the average value of all the fatigue tested defect sizes. The criterion calibration is realized by iterative method as shown in figure 9.

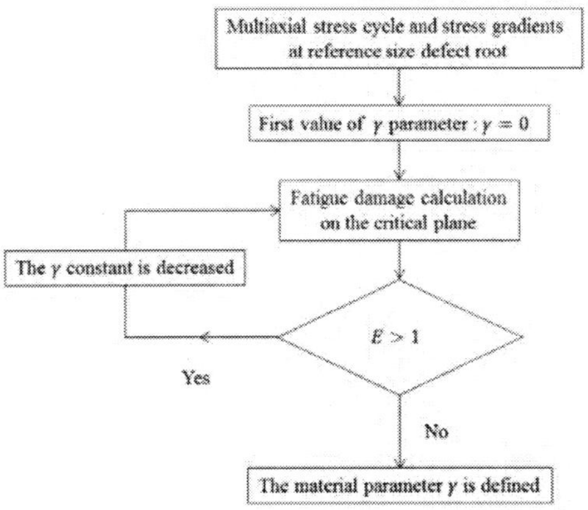

Fig. 9. Iterative method to calibrate the criterion from a defect dependent fatigue test.

For the Zenner criterion the same process is run to integrate within its formulation the stress gradient dependency. Its gradient dependent damage indicator is given by :

$$E_h = a\tau_{ha}^2\left(1 + m\tau_{hm}^2\right) + b\sigma_{hha}^2\left(1 + n\sigma_{hhm}\right) + \gamma|G_{hh}| \quad (26)$$

The Zenner fatigue function remains analogous. It is expressed by :

$$E_{ZNG} = \frac{1}{\sigma_{-1}}\sqrt{\left(\frac{15}{2}\right)\frac{1}{S}\int_S E_h dS} \quad (27)$$

The material parameters a, b, am and bn are still the same as for conventional formulation of the criterion. The material parameter γ is determined by the calibration of the criterion on one fatigue strength obtained for specimen with an average size defect. The same iterative method than for Dang Van criterion is employed to determine γ.

4.3 Validation of the gradient dependent criteria

The fatigue functions of the Dang Van and Zenner modified criteria are calculated for fatigue strengths at 2.10^6 cycles. They are compared with those obtained with the conventional multiaxial criteria. The fatigue strengths were obtained from R=0.1 uniaxial tension fatigue tests. The Haigh's diagram according to Gerber parabolic model is then used to assess the fatigue strength of CP800 steel under fully reversed (R=-1) and repeated (R=0) tension. The two torsion fatigue strengths under fully reversed load and repeated load are obtained from :

$$\tau_{-1} = \frac{\sigma_{-1}}{\sqrt{3}} \quad (28)$$

$$\tau_0 = \frac{4\sigma_{-1}}{\frac{2\sigma_{-1}}{\sigma_0}+1} \quad (29)$$

The fatigue strength for defective material that is used for the criteria calibration has a 127 μm deep defect (that is the average depth of fatigue tested defective specimen). The fatigue functions are given in table 3.

Table 3. Fatigue functions assessment from conventional and gradient dependent criteria.

Defect depth (μm)	Fatigue function at 2.10^6 cycles			
	Dang Van	DVG	Zenner	ZNG
31	1.49	0.97	1.41	0.98
67	1.47	0.96	1.40	0.95
127	1.47	1.00	1.40	1.00
297	1.54	1.05	1.45	1.08

The fatigue functions assessment obtained for the different experimentally tested defect depths are plotted on figure 10. Criteria integrating the stress gradient effect give a strongly better accuracy of the CP800 steel fatigue strength when the specimens contain defects whatever the size. Conventional fatigue criteria strongly underestimate the real fatigue strength of the material with cylindrical defects located on the specimen edge.

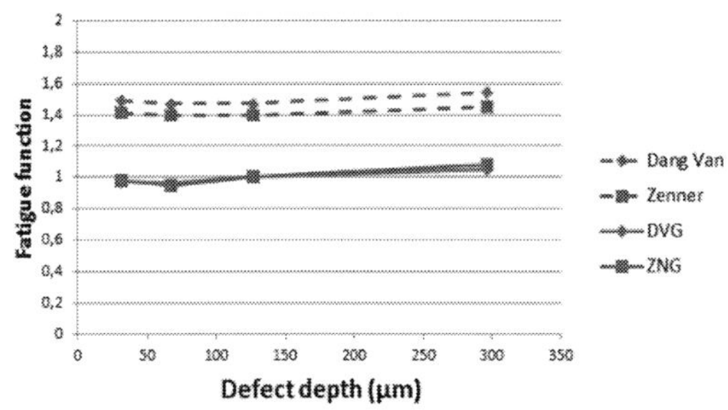

Fig. 10. Comparison of fatigue predictions between conventional and stress gradient dependent fatigue criteria against fatigue test results of defective specimen.

5 Conclusion

The CP800 high strength steel that is mainly devoted to automotive application has been studied against fatigue with the presence of surface defect. The defect reduces the fatigue strength of the steel because of the local stress concentration factor that exists in the vicinity of the defect root. Moreover this fatigue strength decrease cannot be properly predicted without accounting for the stress gradient induced in the highly stressed material volume.

The purpose of the presented work is to assess how multiaxial fatigue criteria are able to predict the steel fatigue behaviour is such cases of high stress gradient. Two multiaxial fatigue approaches, the critical plane and the integral concepts, are investigated to account for stress gradient and to predict the fatigue strength of the defective steel. Dang Van and Zenner criteria are the most commonly used multiaxial stress criteria and have been modified in order to integrate the stress gradient effect in their formulation. The ability of these two criteria is scanned over the whole range of experimental fatigue test results, with defect depth from 30 μm up to 300 μm. Prediction obtained for gradient dependent criteria are compared to those realized with the conventional criteria.

A quite good precision is obtained for the gradient dependent criteria formulation whereas conventional criteria underestimate the material fatigue strength with an amount about 40%. An other peculiarity concerns the stress gradients that are observed around the defect root. They are really higher than stress gradients that can be observed on smooth specimen for particular loading as bending or torsion solicitation. The ability of multiaxial criteria to correctly depict the stress gradient influence on the material fatigue behaviour around defect requires to make the criterion calibration from one fatigue test realized on a defective specimen.

References

1. B. Weber, *Fatigue multiaxiale des structures industrielles sous chargement quelconque*, Thesis INSA Lyon, (1999)
2. B. Weber, B. Kenmeugne, J.C. Clement, J.L. Robert, Comp. Mat. Sci., **15** (1999)
3. K. Dang Van, A. Le Douaron, H.P. Lieurade, *Advances in Fracture Research*, (New Dehli, 1984)
4. K. Dang Van, B. Griveau, O. Message, Mec. E. Pub., (1989)
5. J. Liu, H. Zenner, Mat.-wiss. u. Werksofftech, **24** (1993)
6. H. Zenner, A. Simbürger, J. Liu, Int. J. Fat., **22** (2000)
7. I.V. Papadopolos, V.P Panoskaltsis, E. Frac. Mec., **55,** 4 (1996)
8. D.H. Luu, M.H. Maitournam, Q.S Nguyen, Int. J. Fat., **61** (2014)
9. M. Zepeng, P. Le Tallec, H. Maitournam, Proc. E., **133** (2015)
10. T. Billaudeau, *Fatigue multiaxiale des matériaux à défauts : mécanismes et critère d'endurance*, Thesis ENSMA Poitiers, (2002)
11. Y. Nadot, T. Billaudeau, E. Frac.Mec., **73** (2006)
12. A. Nadot, P. Mu, C. Nadot-Martin, A. Steinwandel, *The Tenth International Conference on Multiaxial Fatigue & Fracture,* (ICMFF10, Kyoto, 2013)
13. Y. Murakami, *Metal fatigue : Effect of small defects and nonmetallic inclusions*, (2002)

Safety margin against the fatigue fracture under random loading

Martin Garan[1] , *Vladimír* Chmelko[2*], and *Ervin* Schafer[1]

[1]Slovak University of Technology, Faculty of Mechanical Engineering, Department of Applied Mechanics and Mechatronics, Námestie Slobody 17, 812 31 Bratislava, Slovakia
[2] Faculty of Mechanical Engineering, Regional Technological Institute, University of West Bohemia, Univerzitní 22, 306 14 Plzeň, Czech republic

Abstract. The fatigue lifetime of the structure in operation is express in the number of working cycles, kilometers or time depending units. By harmonic process of loading is possible to express the safety margin as reserve against the required lifetime in the form of number cycles or amplitude of loading. By non-harmonic character of loading process in operation is difficult to express the level of safety margin as the reserve of loading (safety margin in loading). On the case of towing trailer will be discussed more ways of expression the reserve (safety margin) of structure against the fatigue fracture for required fatigue lifetime. For experimental acquired loading process in the most loaded cross-section will be discussed concept of determination of the level safety as reserve in the form of loading parameters. In the case, when the peaks of loading process exceeding the material yield stress (in the most loaded point of the structure) will be discussed concept of expression of loading in the form of time-varying process of strain. By using the cyclic material properties obtained in mode with controlled total strain is possible the safety margin to express also as the multiple of process loading parameters.

1 Concept of fatigue lifetime assessment

We will use the "safe life" approach to assess the fatigue life of a real construction. This approach is schematically illustrated in Fig. 1

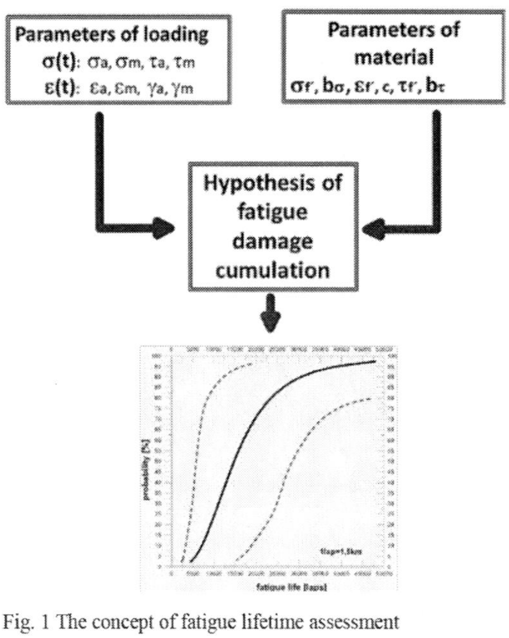

Fig. 1 The concept of fatigue lifetime assessment

Load input parameters are the time deformation curves respectively. tension in the form of individual components. The most credible way to get them is to measure them directly during real-time construction. Before performing the measurement, it is necessary to identify the critical cross-sections i.e. critical points of the structure. They are the places with the highest variable stress component, which determines the minimum fatigue life of the structure as a whole. At the critical point of the construction (one or more), it is then necessary to obtain the desired time-varying processes of the stress of the representative length and content by direct measurement. The second set of inputs required are the cyclic properties of the material that are required by the chosen criterion of fatigue damage cumulation. Due to the scattering of the cyclic properties of the material as well as the variation of the individual realizations of the loading process, it is necessary to use the statistic approach for estimating the fatigue lifetime developed in [1] and practically used e.g. in [2]. As a result of such an approach for fatigue lifetime evaluation (including scattering of individual input parameters), is then the fatigue lifetime distribution function, which may take account of the randomness of the occurrence of the material parameters or of the load parameters or of the two groups of parameters. In the Fig. 2 is an example of a fatigue lifetime distribution function, where for required fatigue lifetime probability of survival (e.g., 95%) is marked the guaranteed value of the fatigue lifetime of the critical place of assessed structure.

* Corresponding author: vladimir.chmelko@stuba.sk

 https://doi.org/10.1051/matecconf/201816522033

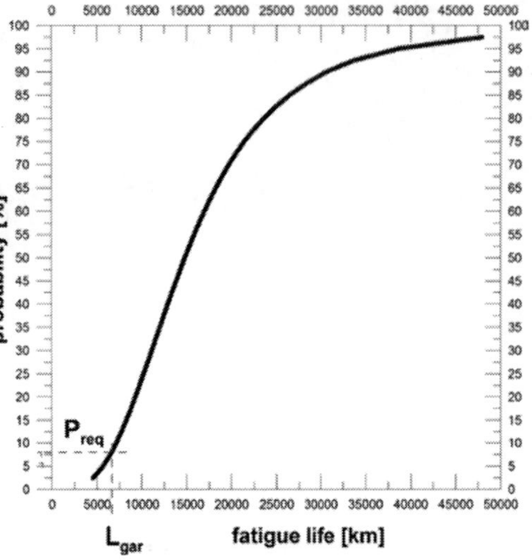

Fig. 2 The guaranteed fatigue lifetime for the required probability of survival

2 The measurement of loading parameters

The subject of assessment is the axle that serves as the chassis of a caravan. It consists of a main beam of cross section 82x4 mm from the steel St52, at the ends of which are the arms with the wheel units. On both sides at a distance of 180 mm from the wheel's axis, a trapezoidal bracket is attached to the beam to fix the spring bars. The borders on both sides of the console serve to correctly attach the axle to the chassis and drive the caravan. To fix the mutual position of the axle with the caravan undercarriage, there are 4 screw connections on each of the brackets. As the axles used in caravans have to be braked, the center of the beam is a rack and cantilever beam for rear brake. (Fig .3).

Fig. 3. The assessed axle

The axle loading diagram is shown in Fig. 4. The most stressed cross-sections are the area between the points of attachment of the spring bars, where a combination of vertical bending and caravan weight and torque from the spring bars occurs. Less stress is the strain from side forces (tilting), vertical forces (tens) and bending from pulling force. The strain gauges were distributed around the center of the axle at the marked cross-section. The layout and strain gauge connection was chosen as shown in (Fig. 5) to determine the individual stresses in the cross-section and to find the point with the highest stress values [3, 4].

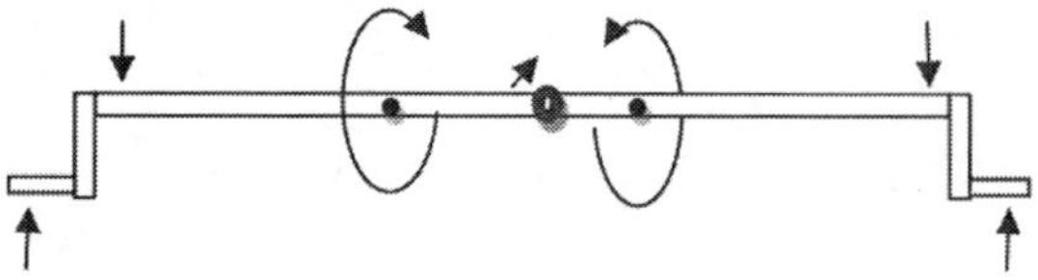

Fig. 4 Axle loading diagram. Black points indicate the location of the anchor bars and the red circle indicates the position of the sensor location

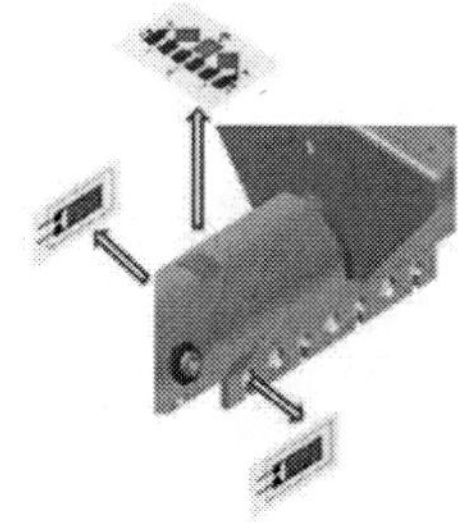

Fig. 5 Strain gauges arrangement around cross section

The stress measurement was performed during the test at the testing polygon. The typical stresses in the most stressed cross-sectional area of the measured cross-section for each type of road and two total load weights are shown in Fig. 6. Diagrams of the dominant stresses (vertical bending and torsion) are time-depending naturally and acting proportionally in phase.

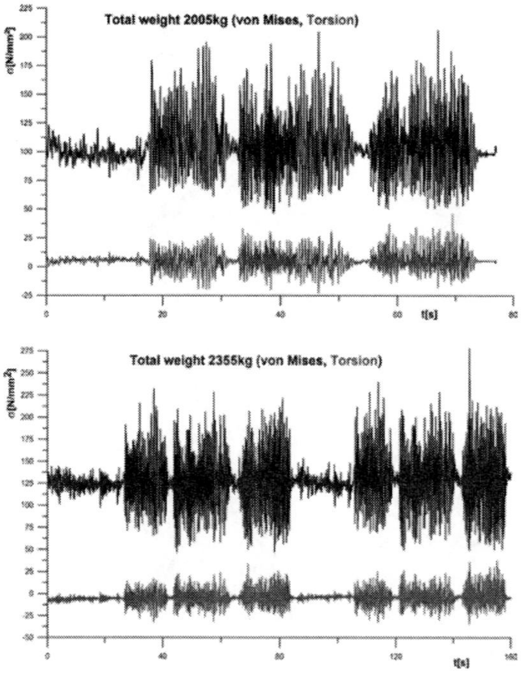

Fig. 6 Typical stresses at the most stressed cross-section point during driving on different road reliefs

3 The assessment of the fatigue lifetime safety by multiple of loading

A sufficiently long loading record allowed a relevant assessment of fatigue lifetime at the most stressed point of axle. The calculation procedure considering the variance of the cyclic properties of material (St52) of the axle is obvious from the following scheme displayed in Fig.7.

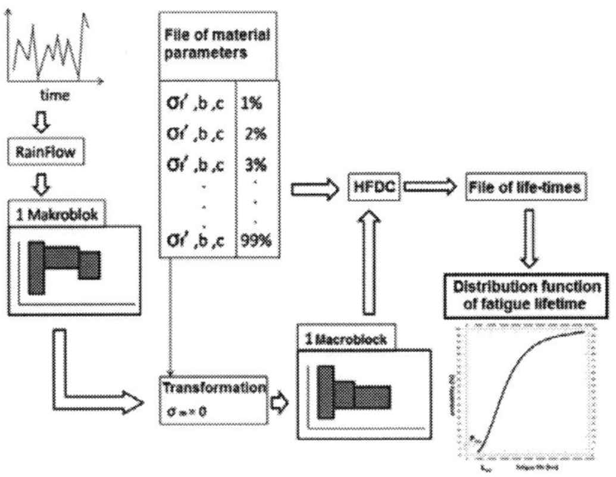

Fig. 7 The scheme of fatigue lifetime assessment with considering the variance of the material parameters

As criterion of the fatigue damage cumulation was used the energetic criterion by Kliman [4] in the form

$$D = \frac{1}{N_{f\,Min}} \sum n_i \left(\frac{\sigma_{ai}}{\sigma_{a\,Max}}\right)^{\frac{b+c}{b}}, \qquad (1)$$

where σ_{ai} is an individual stress amplitude

n_i is frequency of its occurrence

σ_{amax} is the highest stress amplitude in the loading block

N_{fmin} is the number of cycles to the fracture corresponding to the highest stress amplitude

Fatigue lifetimes in kilometres of traffic operation for individual road reliefs and for selected probabilities of occurrence of material parameters (confidence limits, prediction limits) are displayed in Tab.1.

Tab. 1 Fatigue lifetimes for individual road reliefs (axle of a caravan)

Asphalt road		Belgium paving	
2005kg [10^6km]		**2005kg [10^6km]**	
CI 95%	4123072274	CI 95%	114.5
CI 5%	125199630	CI 5%	3.17
PI 95%	8961754744.5	PI 95%	313.8
PI 5%	76277760	PI 5%	1.67
Reg.line	933444075.6	Reg.line	27
2350kg [10^6km]		**2350kg [10^6km]**	
CI 95%	4941080.6	CI 95%	1.7
CI 5%	91402.1	CI 5%	0.0275
PI 95%	12391753.8	PI 95%	5.2
PI 5%	53486.2	PI 5%	0.014
Reg.čiara	970959.1	Reg.čiara	0.34
Patch asphalt		**Potholes**	
2005kg [10^6km]		**2005kg [10^4km]**	
CI 95%	207.9	CI 95%	6885.3
CI 5%	6.6	CI 5%	242.2
PI 95%	565.4	PI 95%	18638
PI 5%	3.4	PI 5%	1265
Reg.line	51.9	Reg.line	1787.6
2350kg [10^6km]		**2350kg [10^4km]**	
CI 95%	4.4	CI 95%	11.4
CI 5%	0.089	CI 5%	0.17
PI 95%	13.17	PI 95%	37.4
PI 5%	0.046	PI 5%	0.0843
Reg.line	0.95	Reg.line	2.27

The fatigue lifetime values are calculated for two axle loadings - 2000kg and 2350kg. From the measured stresses, it can be seen the increase of the peaks during increasing the loading and also shortening the calculated fatigue lifetime. A small (rarely) disputation is to express the degree of safety against fatigue fracture for operating loadings with time-depending stress diagrams. The degree of construction safety for this type of operation can be expressed in several ways:

- as a multiple of the required number of kilometers (i.e. time of operation) against the guaranteed one for the required risk of premature fracture (Fig.8)

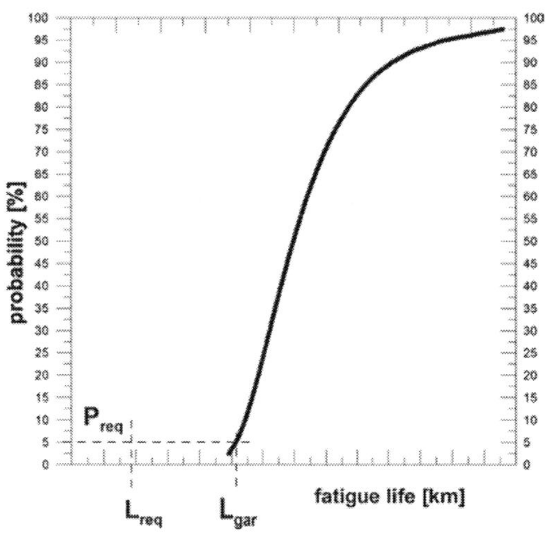

Fig. 8 The expression of the safety level $L_{guaranteed}/L_{required}$ at the time of operation for the permissible risk of premature fracture

- as multiple the loading to which the structure was designed
- ...

There is no direct cause between the peak points at the critical point of the structure and the weight of the load (e.g. in the case of the assessed axle). Increased loading weight increases the mean value of the cycle and can also increases the cycle amplitudes as dynamic components of the loading process. The multiple of these increases is likely different. A slight increase in loading parameters could be, for example, multiplicity of the standard deviation of the loading process as the parameter of the dynamic loading component. Determining the value of such a multiple (loading safety) is not, however, possible directly. However, it is possible to create additional processes from the original loading process by multiplying its peaks by selected multiples. For each such obtained loading process, to calculate the fatigue lifetime according to the scheme in Fig. 7 and progress of the fatigue lifetime as the multiple of standard deviation display graphically. From obtained dependence with suitable regression curve, it is then possible to identify the multiple of the standard deviation of the original loading process for the required fatigue life value (in the case of the required number of kilometres – Fig.9)

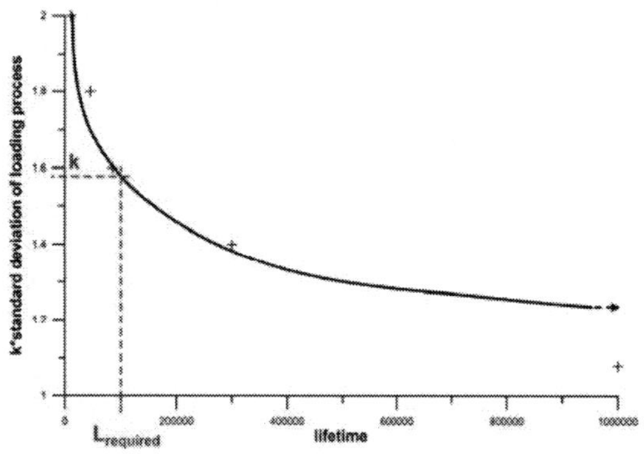

Fig. 9 Finding a multiple *k* of the standard deviation of the loading process for required fatigue lifetime

Gradual increasing of the peak multiplicity in the loading process, that means its standard deviation, however, leads to the need to correct the peaks obtained on the elasto-plastic state of the material at least from the moment when the slope value is reached. This correction can be performed using a Neuber's hyperbola resulting from an equivalent surface in σ-ε diagram (Fig. 10).

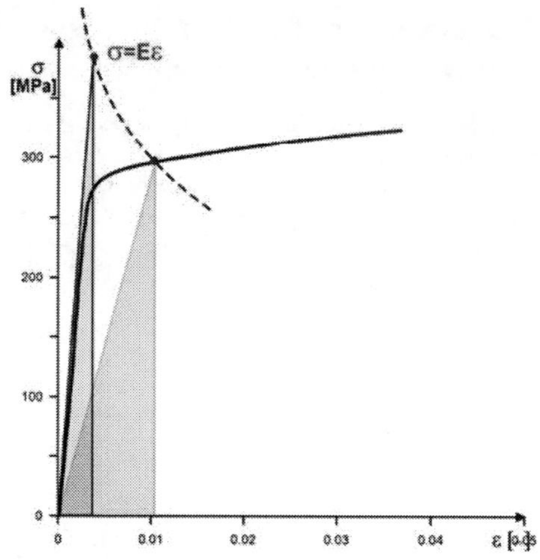

Fig. 10 Neuber method of correcting "linear stresses" on the elastic-plastic state

As material characteristics for elastic-plastic zone can be used cyclic deflection curve, or the static tensile curve. However, after exceeding the "critical" value of the multiplicity of peaks in the loading process, the fatigue lifetime value after the previous reduction begins to increase (Fig. 11).

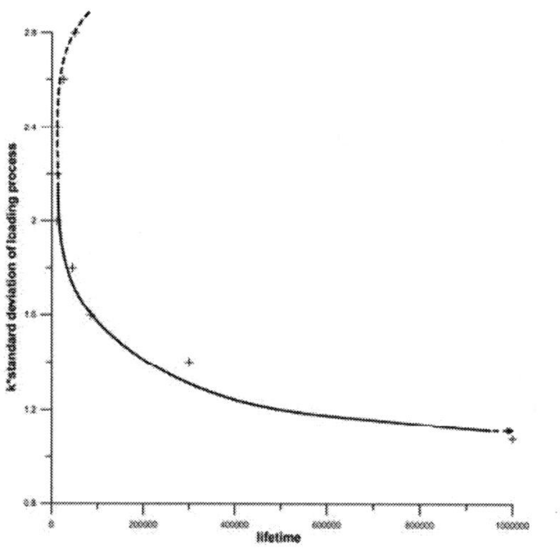

Fig. 11 Development of the standard deviation multiple *k* of the stress loading process on the fatigue lifetime

This incorrect phenomenon occurs due to the slower increase in peak stresses of peak values in the cycles (corrected stresses in the elastic-plastic state) and the faster increase in lower stresses in cycles (linear multiple). In fact, it starts decreasing the amplitude values in the largest cycles occurs in the overall increase of the mean cycle value. Solving of this disproportion can be the expression of the loading process in the form of amplitudes of strain which, by multiplying their peaks, need not to be corrected on the elastic-plastic state.

Using the overall strain process with respect of time for calculation of the fatigue damage requires using of corresponding material properties [4] which can be acquired from the life time curve in the following form

$$\varepsilon_{at} = f(2N_f) \quad (2)$$

By measuring on specimens created from material of the axle St52 (in the mode of the controlled amplitude of the total strain) is possible to acquire relationship displayed in (Fig. 12).

For the purpose simplification the calculation of the fatigue damage is in this case approximated the life time curve by the following expression

$$\varepsilon_{at} = \varepsilon_d'(2N_f)^d \quad (3)$$

This entitles us to a high coefficient of correlation in the regression based on the relationship (Eq. 3) (see Fig. 12).

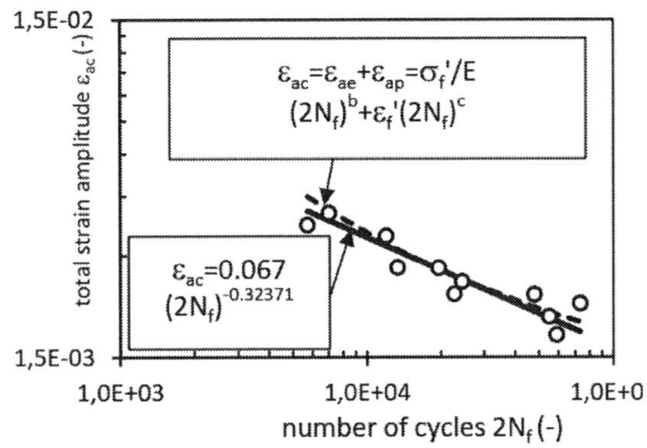

Fig.12. Comparison of the fatigue lifetime curve with experimental results. (o - experimental points)

4 Conclusions

The paper presented the analysis of fatigue lifetime and safety against fatigue fracture in the case of caravan axle. For loading processes at the most loaded cross-section point obtained by direct measurement in the form of stresses, the calculated fatigue lifetime values for corresponding pre-fracture risk was calculated with considering the variance of the material parameters. To determine the degree of stress fatigue safety, a multiple of the standard deviation of the process (implemented as a multiple of the peak of the process) was used. For created disproportion of increasing the fatigue lifetime by necessary correction of stress peak values on elastic-plastic state, a new approach of assessment has been proposed. This lies in expressing the loading process in the form of the time-depending process of the overall strain amplitude. Then the curve of the dependence of the overall strain amplitude on the number of cycles to the fracture is used for the calculation of the fatigue lifetime. The desired loading multiple by respected requirement for required fatigue lifetime in the form of multiples of the overall strain process can be shown analogous to Fig. 9. In the case of the zero-mean value of the individual loading amplitudes, this multiple of the standard deviation of the process peaks leads to determining the safety of the operation as a direct multiple of the loading amplitudes.

The presented contribution has been prepared under project LO1502 "Development of the Regional Technological Institute" under the auspices of the National Sustainability Program of the Ministry of Education of the Czech Republic aimed to support research, experimental development and.

References

1. V. Kliman,: Prediction of the random load fatigue life distribution. Fatigue Design, ESIS 16, Edited by J. Solin et al, 1993, Mechanical Engineering Publications, London, p.241-224
2. M. Kepka, M. Kepka, jr., Parametric calculations of fatigue life of critical part of trolleybus rear axle. Procedia Engineering. 213, (2018), s. 227-238.
3. V. Chmelko, V. Kliman, M. Garan, In-time monitoring of fatigue damage. In Procedia Engineering. 101, (2015), p. 93-100
4. V. Chmelko, M. Garan, Long-term monitoring of strains in a real operation of structures. proceedings of the 14th IMEKO TC10 Workshop, Milano (2016), p. 333-336
5. V. Kliman, Fatigue Life Estimation under Random Loading Using the Energy Criterion. Int. J. Fatigue, 7, no. 1 (1985), p.39-44

https://doi.org/10.1051/matecconf/201816522034

The Effect of Overload on Crack Growth Behaviour in Biaxial Fatigue

Mehdi Mokhtarishirazabad*, *Pablo* Lopez-Crespo, *Belen* Moreno

Department of Civil and Materials Engineering, University of Malaga, C/Dr Ortiz Ramos s/n, 29071, Malaga, Spain

Abstract. This paper focus on the study of biaxial fatigue cracks based on a methodology that enables evaluation of the overload effect in fatigue. The methodology is based on the experimental evaluation via full-field technique of digital image correlation of the effective stress intensity factor and the crack opening displacement (COD). The tests were performed on cylindrical specimens made of low carbon steel subjected to tension-tension and a combination of tension and torsion loads. The experimental data was used to compare the crack growth behaviour with an overload cycle under uniaxial and biaxial loading. A hybrid method was also employed to evaluate the value of the Stress Intensity Factor (SIF) before and after applying the overload. The results showed a reliable estimation of closure level and SIF for both uniaxial and biaxial loading condition. A new procedure previously developed for uniaxial loads was adapted to the study of biaxial loads. This new procedure was used to identify important differences produced by the overload event.

* Corresponding author: mokhtari@uma.es

1 Introduction

Stress intensity factor is known as one of the most important fracture parameters for characterising the crack behaviour in the engineering components. Since 1957 when Irwin [1] formulized the stress intensity factor, a considerable works have been done to improve our understanding on its importance in crack growth behaviour under static and cyclic loading [2,3]. Finite element method had been employed widely during last decades for detailed analysis of the crack tip fields (stress, strain, displacement) to evaluate the SIF and phenomenon such as crack closure [4–6]. In addition, some mathematical models have been introduced describing the crack tip field such as Westergaard's [7], Williams' [8], and recently CJP [9] models. Different techniques have been developed to obtain an full-field measurement of the crack tip field, experimentally, such as Moire interferometry [6], photo-elasticity[10], thermos-elasticity[11], and digital image correlation (DIC) [12].

Among these techniques, DIC is widely using due to a number of advantageous such as being easy to implement, easy sample preparation, and being a scale-free method. The latter one means that it can measure on the scale ranging from a few meters [13] to micro-meters [14,15]. Measured displacement data by DIC can be fitted to one of abovementioned models using Multipoint over-deterministic method [16] to estimate the most important fracture parameters of the material such as SIF[12,17–19].

Many structures in real applications are often subjected to a mixed mode loading rather pure mode I of loading [20,21]. Therefore, considering the mixed-mode loading in the evaluation of SIF will be more representative of the material fracture behaviour under the actual working condition. Yoneyama et al [22], evaluated the mixed mode SIFs of polymethyl-methacrylate by post-processing the displacement field derived by DIC method. A method based on the Lagrangian conservation law for mixed-mode SIFs estimations used by Réthoré et al [23] to estimate the SIFs using the concept of virtual crack extension and auxiliary field. Lopez-Crespo et al. [24] fitted Muskhilishvili's model to the displacement data derived by DIC to evaluate the SIFs in a sample made of Al 7010 alloy under mode I and mixed mode (I+II) loading. They reported a good agreement between nominal and experimental evaluation of SIFs.

In addition, many engineering components are subjected to variable amplitude loads. To estimate the fatigue life of these components, it is essential to quantify the crack growth rate which is, in turn, affected by the plastic deformation at the crack tip. To consider the effect of external load variation on crack growth rate, usually the effects of a single overload are examined [25,26]. Nowell et al. [6] employed the moire interferometry and DIC to determine the presence of crack closure in propagating fatigue crack by examining the measured in-plane displacement around the crack tip. They used a similar method used previously by de Matos [27] to monitor the displacement of some pairs of points close to the crack tip for determining the closure level.

DIC also used by Yusof et al. [28] to determine the closure load using crack opening displacement (COD) and evaluated SIF. They used Muskhilishvili's model to evaluate the experimental SIF. Others have conducted experimental and numerical investigation on the effect of an overload cycle on fatigue crack growth under mixed mode loading [29–31]. Nevertheless, there are very little knowledge, if any, about the application of optical methods to study the effect of overload cycle on the crack growth rate under tension-tension (uniaxial) and tension-torsion (biaxial) cyclic loading is available. Recently, the author has conducted an experiment to study the influence a single overload cycle on the closure level and fatigue crack growth rate in hollow cylinder samples under in-phase tension-torsion loading [32]. In the present work, DIC is employed to evaluate the crack closure and mixed mode SIFs for samples under tension-tension and tension-torsion loading.

2 Materials and Methods

Experiments are conducted on a low carbon steel (St 52-3N). Due to its good weldability and high ductility of this alloy, this alloy is oftenused in offshore platforms and marine structures. Table 1 shows the chemical composition of the material. The mechanical behaviour of the alloy is showed in Table 2. This material combines very good fatigue properties with low environmental impact for non-transport applications [33].

Microstructural of the alloy consists of ferrite and pearlite bands as vertical white and black bands, respectively (Fig.1). Fig. 2 illustrates a schematic of the geometry of the sample.

Table 1. Chemical composition of St 52-3N alloy

C	Si	Mn	P	S	Cr	Ni	Mo
0.17	0.22	1.23	0.01	>0.0001	0.07	0.06	0.16

Table 2. Monotonic properties of St-52-3N steel

Yield stress, σ_y	386 MPa
Ultimate tensile stress, σ_u	639 MPa
Young's modulus, E	206 GPa
Shear Modulus	78 GPa

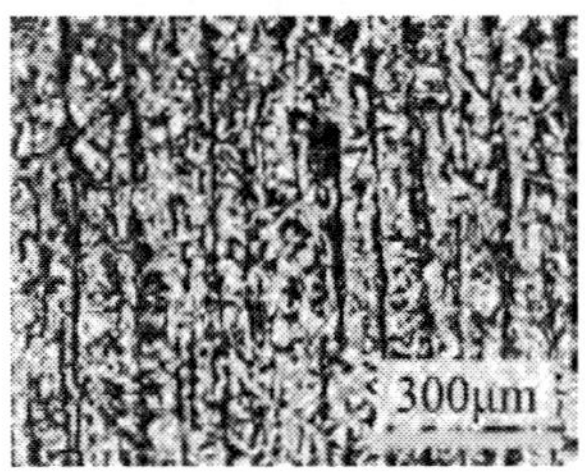

Fig. 1. The microstructure of St52-3N steel. Black and white vertical bands are showing the pearlite and ferrite bands, respectively [34].

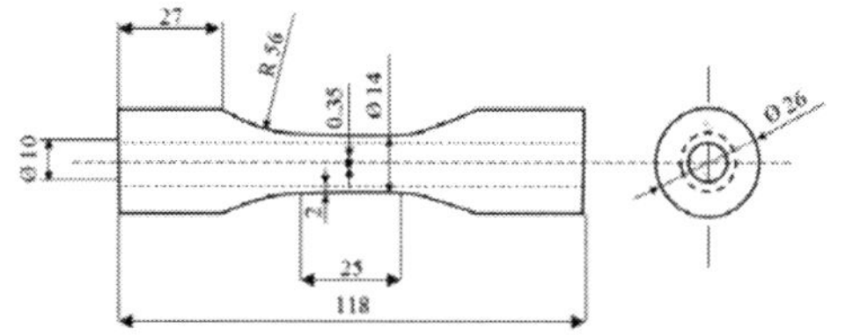

Fig. 2. The geometry of the hollow cylinder specimen with a central hole. All dimensions are in mm.

Axial and biaxial loads under stress control mode were conducted by a MTS 809 servo-hydraulic loading rig coupled by a biaxial extensometer Epsilon 3550. Uniaxial tension-tension cyclic loading test was conducted on Sample1 with sinus signal and axial load ratio of 0.1 (R_a = 0.1), while an in-phase torsional load with load ratio of -1 (R_t=-1) superimposed to the axial loadings on Sample2 for the biaxial loading case. The range of axial load was 0.56 of the yielding strength of the alloy. Fig. 3 shows the loading path for Sample2. To enforce the crack to nucleate inside the field of view, a through hole with a diameter of 0.378 and 0.432 mm was drilled in the outer surface of the Sample1 and Smaple2, respectively. A zero-offset in the number of cycles was applied when the crack length was 0.638 and 0.430 mm for Sample1 and Sample2, respectively. It should mention that crack length determined by adding the radius of hole to the length of the crack from the crack initiation point at the circumference of the hole to the crack tip in one side of the hole. Visual inspection, showed an equal crack length at both sides of the hole.

In order to study the effect of the overload on the crack propagation behaviour, a 40% overload was applied on both samples when the crack length was 0.99 mm in Sample1 and 0.796 mm in Sample2. That is, the load range in the overload cycle was 1.4 times larger than load range during the rest of the test. The cyclic loading then continued until the crack length reached 1.5 mm. Table 3 shows the nominal value of applied loads for both samples. The crack length was evaluated from the crack-tip position with optical microscopy. The related average error in measuring the crack growth rate was 0.002 μm/cycle. The incremental polynominal technique as recommended in ASTM E647-15 [35] was employed to examine the cyclic crack growth rate. Table 3 shows the loading condition for both samples.

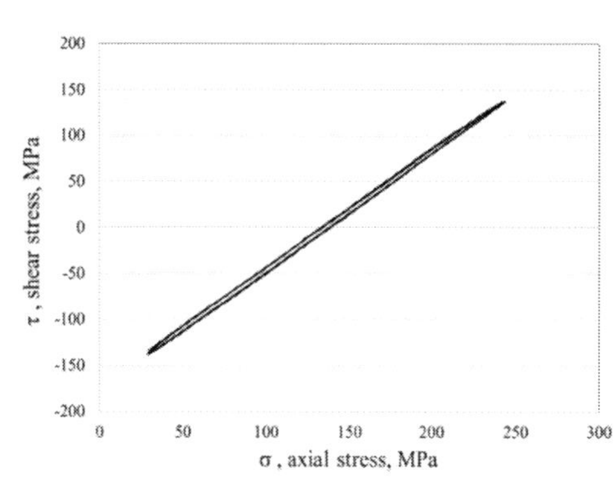

Fig. 3. Loading path for Sample2.

Table 3. Axial and shear stress values for specimens with and without overload cycle.

Specimen	Crack length at OL, μm	$\Delta\sigma$ MPa	$\Delta\tau$ MPa	$\Delta\sigma_{OL}$ MPa	$\Delta\tau_{OL}$ MPa
Sample1	0.99	216	277	-	-
Sample2	0.79	216	277	302.4	388

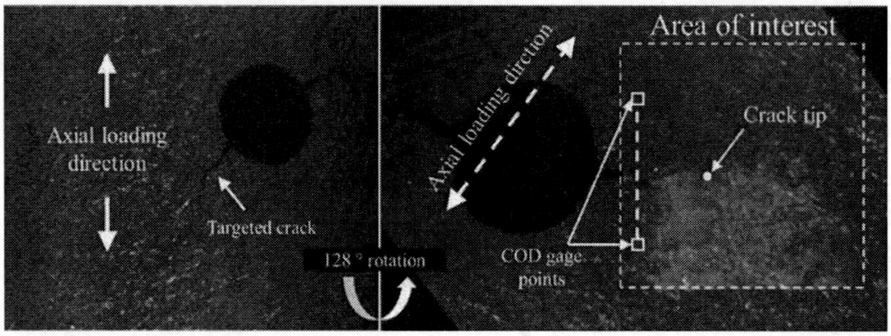

Fig. 4. The field of view of the Sample2 and the vertical displacement contour around the crack tip derived by DIC. The image is rotated 128 degree counter clockwise to obtain a horizontal crack line. The position of COD gage points is illustrated.

Crack tip displacement fields were captured during cyclic loading with a 5 MP CCD camera coupled with a macro Navitar lens and adaptor tube which provided a field of view (FOV) was 3.55 × 2.97 mm^2. The contrast required for DIC technique was introduced by finely abrading the surface so that the bare crack tip region is imaged directly [36] This allowed faster and easier positioning of the crack tip [37]. In this way, a resolution of 1.45 micron per pixel was obtained. To take sufficient number of images, the loading frequency was reduced to 0.3 Hz to take 75 images during a complete cycle. Obtained images processed by VIC 2D V6 software [38].Recommendations given in [39] were followed to improve the quality of the SIF estimations. Accordingly, a high order interpolation scheme of optimized 8-tap spline was used for DIC analysis to achieve sub-pixel accuracy. The subset size was adjusted to 39×39 pixels. In addition, just 25% of the crack line was considered in the area of interest (the area where the displacement data were measured by DIC). The relative error in the estimated displacement (and consequently in COD) was 2.95%. This relative error was estimated from the quality of the correlation of the images taken at different loads. By taking to account the size of the area of interest (0.8 ×0.8 mm^2), two terms in the Williams' solution were used as suggested in [39]. Since the crack propagation direction for Sample2 was 50 ± 3 °, to extract the vertical and the horizontal displacements with respect the axial loading axis, captured images have been rotated so that the crack appears horizontal in all images. Fig. 4 illustrates the vertical displacement contour for an area of interest around a crack with a length of 0.55 mm (Sample2).

To monitor the behaviour of the left-side crack, the images were rotated 128° counter clockwise so that the crack line was horizontal. Displacement data points inside an area of 0.8×0.8 mm^2 were extracted and fitted to Williams' solution in order to calculate SIFs.

In order to determine the presence of crack closure, crack opening displacement (COD) was measured by DIC at 0.3 mm behind the crack tip as follows:

$$COD = \sqrt{u_{x\,bot}^2 - u_{y\,bot}^2} - \sqrt{u_{x\,top}^2 - u_{y\,top}^2} \quad (1)$$

where u_x and u_y are the horizontal and vertical displacement respectively. The subscripts "top" and "bot" refer to the position of the virtual extensometer points relative to the crack line. These points are shown in Fig. 4. In order to improve the quality of presented COD data, the Fourier series with 1 and 3 terms was fitted to the loading signal and COD data, respectively.

To estimate the SIFs, derived displacement data by DIC were fitted into Williams' series using the multipoint over-deterministic method. For the sake of brevity, the related equations and details of the method are not presented here. It has been explained in details in our previous work [32]. SIF in mode I and mode II then reported for various crack lengths before and after overload for both uniaxial and biaxial loading condition.

3 Results and Discussion

3.1 Fatigue Crack Growth Rate

Fig. 5 shows the evolution of crack growth before and after applying a 40% overload on Sample1 and Sample2 under uniaxial and biaxial loading, respectively. There are two main difference between the crack behaviour under uniaxial and biaxial loading regime. First, the post-overload crack growth of Sample1 (Fig.5.a) consists a delayed retardation stage, while it is not appreciable for Sample2 (Fig. 5.c) under biaxial loading. Such delay in retardation has been observed in our previous study for biaxial loading only when the range of baseline axial and shear stress was 25% and 17% lower than thereof for Sample2 [32]. Secondly, crack growth rate is considerably higher (by a factor of about 20) for

Sample2 in which an in-phase torsional load was superimposed on the axial cyclic loads. The duration of retardation, the distance that the crack advances during the transient behaviour after overload, was 0.385 and 0.275 mm which is corresponds to 41000 and 2550 cycles for Sample1 and Sample2 respectively (Fig. 5.b and d).

The noticeable difference between the crack growth rate for Sample1 and Sample2 can be attributed to two main difference between the applied loads. First, by calculating the principal stresses corresponds to the applied axial and torsional loads, it was observed that the principal stress value normal to the crack line (σ1) and consequently the equivalent von Mises stress (σe) at the maximum axial load were 303 and 338 MPa which is 26% and 40% higher than applied axial load in Sample1.

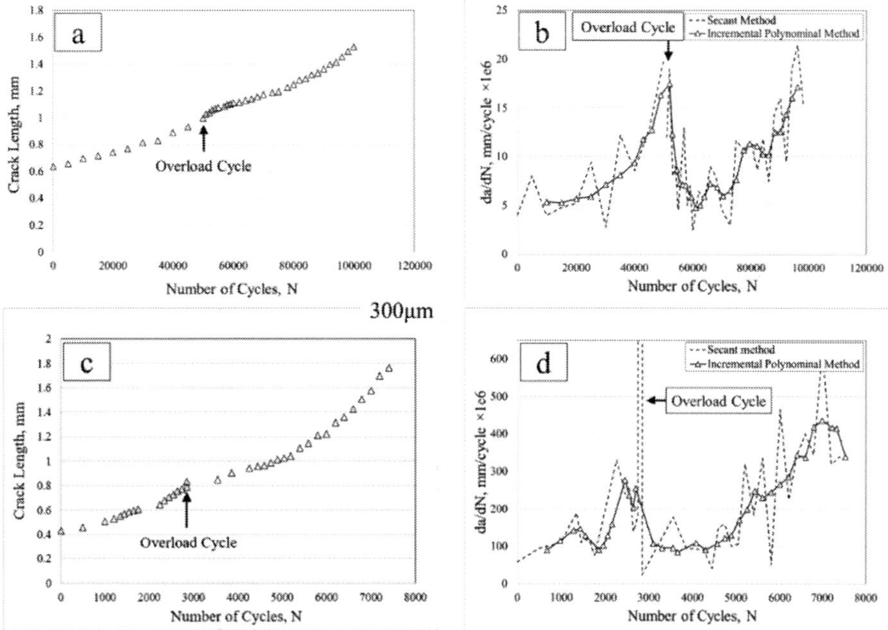

Fig. 5. Evolution of crack length and crack growth rate as a function of number of cycles for Sample 1 (a and b) and Smaple2 (c and d).

Although the stress amplitude (σ_a) was almost at the same scale as pure axial loading condition, the mean stress has been increased form 136 MPa in Sample1 to 231 MPa in Sample2. In addition, the orientation of the principal axis with respect to the loading axis (θ) is not fixed during a complete cycle. It is a witness of non-proportional loading. It is believed that non-proportional loading is generally more damaging than proportional loading [40]. Fig. 6 shows the variation of applied loads during two cycles for Sample2.

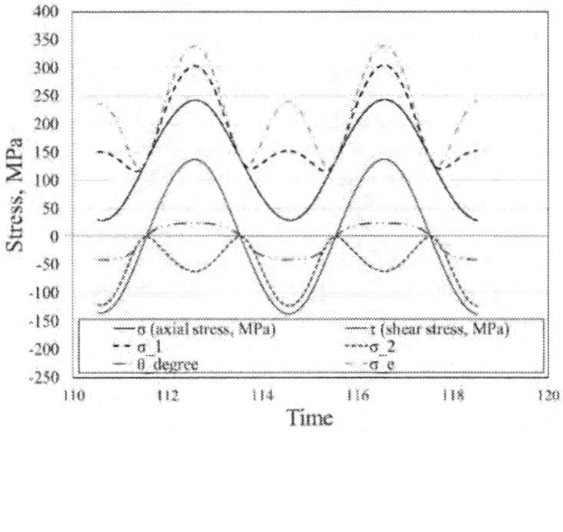

Fig. 6. Variation of nominal, principal and von Mises equivalent stresses during 2 complete cycles in Smaple2.

3.2 COD Examination

To determine the presence of crack closure, the displacement between two points at the top and bottom pf the crack line was monitored in different stages of the test (e.g. before and after overload cycle). All COD measurements were derived from the external surface of the specimen. Since the specimen thickness in the crack region was small (2mm), no large differences are expected along the thickness [41]. Fig. 7 shows how the COD changes during a complete cycle for both samples. It can be seen that the loading and unloading path in Sample2 is considerably deviated from linear behaviour. In addition, right after overload cycle, the maximum value of COD for Sample2 has been reduced by 0.95 μm (18%), while it has been increase by a negligible value of 0.1 μm for Sample1. On the other hand, from fig. 7.c and d, it can be observed that the slope of COD curve has been changed at lower portion of the graph by increasing the crack length for Sample1 which was under pure axial loading. This change in the slope often referred to as "knee" [42].

A compliance based method [43] was followed to quantify the opening load. Fig. 8 shows how opening load has been determined for Sample1. In Fig 8.b , COD-offset refers to the distance between the fitted least squares line at loads higher than 8 kN (50% of loading range) and the COD curve at each point. The estimated opening load for Sample1 at crack length of 1.03, 1.195 and 1.527 mm was 3. 47, 4.35 and 4.62 kN, respectively. No closure was observed before and after overload cycle in Sample2.

3.3 Experimental Evaluation of SIF by DIC

DIC employed to extract the displacement field around the crack tip. Following the method described in [17,32], the stress intensity factor in mode I and mode II was determined experimentally.

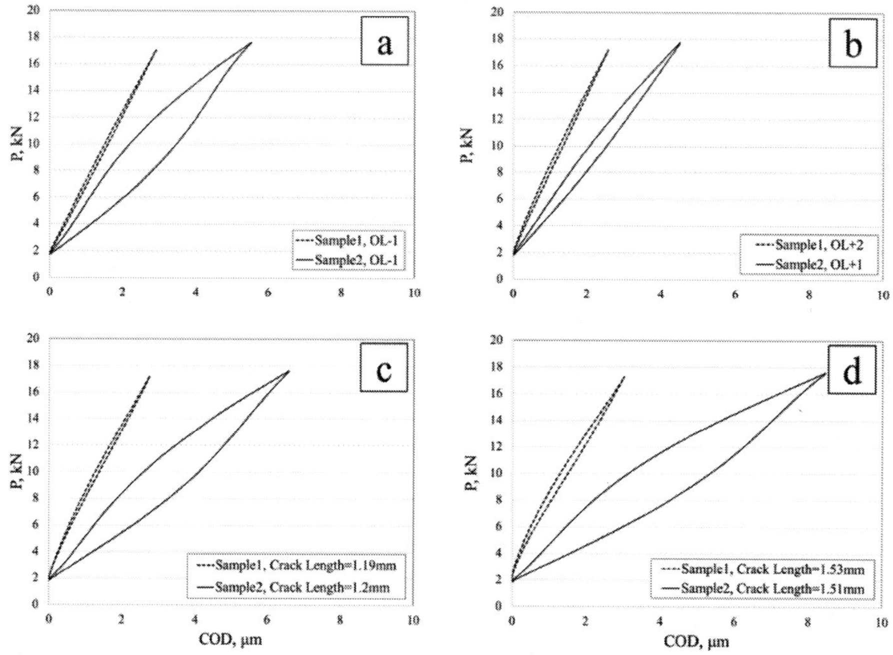

Figure 7. COD behaviour for Sample1 and Sample2 at different stages of loading (different crack lengths)

The method includes fitting the experimental displacement data to Williams' model using a multipoint over-deterministic method. Fig. 4 shows an example of the displacement field extracted from the surface of the Sample1. The experimental estimations of SIFs for Sample1 under uniaxial loading, were compared with the following theoretical solutions of SIF for a circular through crack in a cylinder as described in the AFGROW DTD handbook [44]:

$$\Delta K_{theo} = F_0 \sigma_a \sqrt{\pi a} \qquad (2)$$

where F_0 is the geometry factor, σ_a is the nominal stress amplitude and a is the crack length which is the sum of hole diameter and measured crack length from the crack initiation point at the circumference of the hole to the crack tip in both side of the hole. Effective stress intensity factor (ΔK_{eff}) defined by considering the effective applied load ($\Delta P_{eff}=P_{max}-P_{op}$) in the SIF calculations. The results are summarized in Table 3.

A good agreement between ΔK_{exp} and ΔK_{eff} can be seen for crack length (a) larger than 2 mm. The poor experimental evaluation of SIF for short cracks can be attributed to the fact that for short cracks validation of Linear Elastic Fracture Mechanic (LEFM) is under question, while Williams' model is based on LEFM. It can also be seen that after applying the overload, ΔK_{eff} has decreased from 17.6 to 15.4 MPa√m (about 12%). Similar behaviour was also observed under uniaxial load [45]. It is interesting that both of ΔK_{exp} and ΔK_{eff} are slightly smaller after 20,000 cycles after overload cycle. It should be mentioned that the difference between reported values is very little which is inside the error bands of the evaluated ΔK_{exp} (±0.5 MPa√m). As it is expected from uniaxial loading, the estimated value for ΔK_{II} was negligible.

With respect to Sample2, only the experimental evaluation of ΔK_I and ΔK_{II} are reported (Table 4) due to the lack of proper analytical method for evaluation of the SIF for the geometry and loading condition of the sample. It can be seen that right after overload cycle ΔK_I and ΔK_{II} decreased by 7.8 and 6.6 MPa√m. Moreover, $\Delta K_{I\,exp}$ for Sample2 was more than two times higher than thereof for Sample1. This can justify higher crack growth rate under biaxial loaded sample rather pure axial loaded one.

Table 3. Comparison between evaluated ΔKI by theoretical and experimental methods for Sample1.

	, *a*, mm	ΔP, N	Δσ, MPa	ΔK_{theo}	ΔK_{eff}	ΔK_{exp}
OL-50kCycle	1.28	15718	221	17.6	17.6	11.3
OL-1Cycle	1.99	15261	220	17.6	17.6	15.0
OL+2Cycle	2.06	15254	220	17.7	15.4	15.7
OL+25kCycle	2.39	15347	223	18.0	15.2	15.5
OL+50kCycle	3.05	15425	228	18.7	15.3	16.9

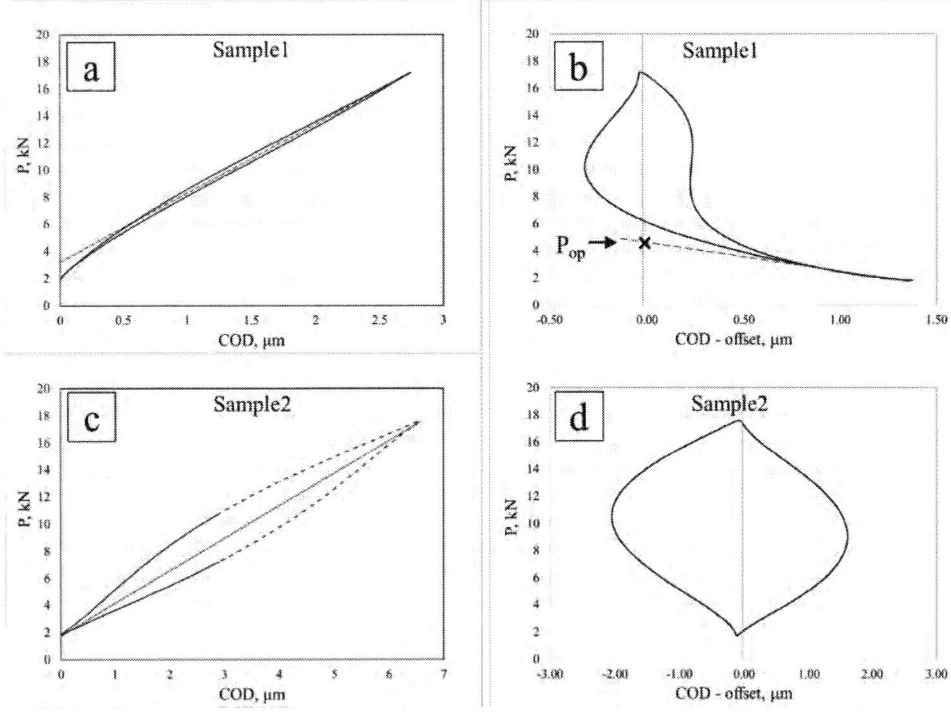

Fig. 8. Evaluation of opening load for Sample1 and Sample2 after overload at crack length of 1.195 and 1.209 mm, respectively.

4 Conclusion

An investigation on the effect of overload cycle of 40% on the crack growth rate for uniaxial and biaxial loading was conducted by employing an optical - analytical method. It was observed that the crack growth rate increased by a factor of about 20 by superimposing an in-phase torsional load in Sample2. COD examinations also showed higher values for Sample2 under biaxial load. Crack closure was observed immediate after overload cycle for pure axial loaded sample, while no closure was detected for sample under biaxial loads. A hybrid method including extracting full-field displacement data by DIC and fitting it to the analytical solution of Williams' were used to evaluate the experimental SIFs. A good agreement between experimental and theoretical solution for SIF was achieved.

Acknowledgements. Financial support of Ministerio de Economia y Competitividad through grant reference MAT2016-76951-C2-2-P is greatly acknowledged.; and the University of Malaga through Campus de Excelencia Internacional del Mar (CEIMAR) through Líneas Emergentes program and for providing PhD scholarship is greatly acknowledged.

References

[1] Irwin GR. Analysis of stresses and strains near the end of a crack traversing a plate. J Appl Mech ASME 1957;**E24**:351–69.

[2] Anderson TL. Fracture Mechanics Fundamentals and Applications. 2nd Ed. CRC Press; (1995).

[3] Paris PP, Erdogan FF. A Critical Analysis of Crack Propagation Laws. ASME J Basic Eng (1963);**85**:528–33.

[4] Sutton MA, Orteu J-J, Schreier HW. Image correlation for shape, motion and deformation measurements. New York: Springer; (2009).

[5] de Matos PFP, Nowell D. Numerical simulation of plasticity-induced fatigue crack closure with emphasis on the crack growth scheme: 2D and 3D analyses. Eng Fract Mech (2008);**75**:2087–114.

[6] Nowell D, Paynter RJH, De Matos PFP. Optical methods for measurement of fatigue crack closure: moiré

interferometry and digital image correlation. Fatigue Fract Eng Mater Struct (2010);**33**:778–90.
[7] Westergaard HM. Bearing pressures and cracks. J Appl Mech (1939);**61**:A49–53.
[8] Williams ML. On the stress distribution at the base of a stationary crack. J Appl Mech (1957);**24**:109–14.
[9] Christopher CJ, James MN, Patterson EA, Tee KF. Towards a new model of crack tip stress fields. Int J Fract (2007);**148**:361–71.
[10] Ramesh K, Gupta S, Kelkar AA. Evaluation of stress field parameters in fracture mechanics by photoelasticity - revisited. Eng Fract Mech (1997);**56**:25-41-45.
[11] Díaz FA, Yates JR, Patterson EA. Some improvements in the analysis of fatigue cracks using thermoelasticity. Int J Fatigue (2004);**26**:365–76.
[12] McNeill SR, Peters WH, Sutton MA. Estimation of stress intensity factor by digital image correlation. Eng Fract Mech (1987);**28**:101–12.

[13] Yoneyama S, Kitagawa A, Iwata S, Tani K, Kikuta H. Bridge deflection measurement using digital image correlation. Exp Tech (2007);**31**:34–40.
[14] Sutton MA, Li N, Joy DC, Reynolds AP, Li X. Scanning electron microscopy for quantitative small and large deformation measurements Part I: SEM imaging at magnifications from 200 to 10,000. Exp Mech (2007);**47**:775–87.
[15] Carroll JD, Abuzaid W, Lambros J, Sehitoglu H. High resolution digital image correlation measurements of strain accumulation in fatigue crack growth. Int J Fatigue (2013);**57**:140–50.
[16] Sanford RJ, Dally JW. A general method for determining mixed-mode stress intensity factors from isochromatic fringe patterns. Eng Fract Mech (1979);**11**:621–33.
[17] Yates JR, Zanganeh M, Tai YH. Quantifying crack tip displacement fields with DIC. Eng Fract Mech (2010);**77**:2063–76.
[18] Hamam R, Hild F, Roux S. Stress intensity factor gauging by digital image correlation: Application in cyclic fatigue. Strain (2007);**43**:181–92.
[19] Roux S, Hild F. Stress intensity factor measurements from digital image correlation: post-processing and integrated approaches. Int J Fract (2006);**140**:141–57.
[20] Qian J, Fatemi A. Mixed mode fatigue crack growth: A literature survey. Eng Fract Mech (1996);**55**:969–90.
[21] Socie D, Marquis, Gary. Multiaxial Fatigue. SAE International; (1999).
[22] Yoneyama S, Ogawa T, Kobayashi Y. Evaluating mixed-mode stress intensity factors from full-field displacement fields obtained by optical methods. Eng Fract Mech (2007);**74**:1399–412.
[23] Réthoré J, Gravouil a., Morestin F, Combescure a. Estimation of mixed-mode stress intensity factors using digital image correlation and an interaction integral. Int J Fract (2005);**132**:65–79.
[24] López-Crespo P, Burguete RL, Patterson EA, Shterenlikht A, Withers PJ, Yates JR. Study of a crack at a fastener hole by digital image correlation. Exp Mech (2009);**49**:551–9.
[25] Carlson RL, Kardomateas GA, Bates PR. The effects of overloads in fatigue crack growth. Int J Fatigue (1991);**13**:453–60.
[26] Salvati E, O'Connor S, Sui T, Nowell D, Korsunsky AM. A study of overload effect on fatigue crack propagation using EBSD, FIB-DIC and FEM methods. Eng Fract Mech (2015);**167**:1–14.
[27] de Matos PFP, Nowell D. Experimental and numerical investigation of thickness effects in plasticity-induced fatigue crack closure. Int J Fatigue (2009);**31**:1795–804.
[28] Yusof F, Lopez-Crespo P, Withers PJ. Effect of overload on crack closure in thick and thin specimens via digital image correlation. Int J Fatigue (2013);**56**:17–24.
[29] Sander M, Richard H. Experimental and numerical investigations on the influence of the loading direction on the fatigue crack growth. Int J Fatigue (2006);**28**:583–91.
[30] Sander M, Richard HA. Finite element analysis of fatigue crack growth with interspersed mode I and mixed mode overloads. Int J Fatigue (2005);**27**:905–13.
[31] Srinivas V, Vasudevan P. Study of the influence of mixed mode overload on mode I fatigue crack propagation. Int J Press Vessel Pip (1993);**56**:409–17.
[32] Mokhtarishirazabad M, Lopez-Crespo P, Moreno B, Lopez-Moreno A, Zanganeh M. Optical and analytical investigation of overloads in biaxial fatigue cracks. Int J Fatigue (2017);**100**:583–90.
[33] Chaves V. Ecological criteria for the selection of materials in fatigue. Fatigue Fract Eng Mater Struct (2014);**37**:1034–42.
[34] Lopez-Crespo P, Moreno B, Lopez-Moreno A, Zapatero J. Study of crack orientation and fatigue life prediction in biaxial fatigue with critical plane models. Eng Fract Mech (2015);**136**:115–30.
[35] ASTM E647-15, Standard Test Method for Measurement of Fatigue Crack Growth Rates. West Conshohocken, PA: (2015).
[36] Lopez-Crespo P, Moreno B, Lopez-Moreno A, Zapatero J. Characterisation of crack-tip fields in biaxial fatigue based on high-magnification image correlation and electro-spray technique. Int J Fatigue (2015);**71**:17–25.
[37] Zanganeh M, Lopez-Crespo P, Tai YH, Yates JR. Locating the crack tip using displacement field data: a comparative study. Strain (2013);**49**:102–15.
[38] Vic-2D V6 Reference Manual, Correlated Solutions Incorporated (C.S.Inc).
[39] Mokhtarishirazabad M, Lopez-Crespo P, Moreno B, Lopez-Moreno A, Zanganeh M. Evaluation of crack-tip fields from DIC data: A parametric study. Int J Fatigue (2016);**89**:11–9.
[40] S. Suresh. Fatigue of Materials. 2nd ed. New York: Cambridge University Press; (1998).
[41] Withers PJ, Lopez-Crespo P, Kyrieleis a., Hung Y-C. Evolution of crack-bridging and crack-tip driving force during the growth of a fatigue crack in a Ti/SiC composite. Proc R Soc A Math Phys Eng Sci (2012);**468**:2722–43.

[42] Joyce JA, Sutton GE. An automated method of computered-controlled low-cycle fatigue crack growth testing using the eslastic-plastic parameter cyclic J. In: Cullen WH, Landgraf RW, Kaizand LR, Underwood JH, editors. Autom. test methods Fract. fatigue crack growth, American Society for Testing and Materials; (1985), 225–47.
[43] Skorupa M, Beretta S, Carboni M, Machniewicz T. An algorithm for evaluating crack closure from local compliance measurements. Fatigue Fract Eng Mater Struct (2002);**25**:261–73.
[44] DTD AFGROW Handbook. n.d.
[45] Lopez-Crespo P, Mostafavi M, Steuwer A, Kelleher JF, Buslaps T, Withers PJ. Characterisation of overloads in fatigue by 2D strain mapping at the surface and in the bulk. Fatigue Fract Eng Mater Struct (2016);**39**:1040–8.

https://doi.org/10.1051/matecconf/201816522035

A new local approach model for cleavage fracture in ferritic steels and its validation

Guian Qian[1,*], *Wei-Sheng* Lei[2,*], *Markus* Niffenegger[1]

[1] Paul Scherrer Institute, Nuclear Energy and Safety Department, Laboratory for Nuclear Materials, OHSA/06, 5232 Villigen PSI, Switzerland
[2]Applied Materials, Inc., 974 East Arques Avenue, Sunnyvale, CA 94085, USA

Abstract. The paper discusses and applies a statistical approach to correlate the fracture behavior of a notched and fracture mechanics specimen. The method can be used for fatigue analysis. The random nature of cleavage fracture process determines that both the microscopic fracture stress and the macroscopic properties including fracture load, fracture toughness and the ductile to brittle transition temperature are all stochastic parameters. This understanding leads to the proposal of statistical assessment of cleavage induced notch toughness of ferritic steels according to a new local approach to cleavage fracture. The temperature independence of the two Weibull parameters in the new model induces a master curve to correlate the fracture load at different temperatures. A normalized stress combining the two Weibull parameters and the yield stress is proposed as the deterministic index to measure notch toughness. This proposed index is applied to compare the notch toughness of a ferritic steel with two different microstructures.

1 Introduction

Ferritic steels are commonly used to fabricate reactor pressure vessels (RPVs) in the commercial light water reactors. Ferritic steels have body centered cubic crystal structures that possess the ductile-to-brittle transition temperature (DBTT) characteristic **[1, 2]**. This leads to the inherent susceptibility of RPVs to stochastic transgranular cleavage induced brittle fracture in the actual service temperature range due to geometrical discontinuities at notches and cracks, and in the occurrence of pressurized thermal shock loadings. Both deterministic and statistical approaches have been developed to evaluate structural integrity of RPVs against brittle fracture. As reviewed in **[3]**, the conventional transition temperature method, fracture mechanics, and fracture physics approaches are deterministic; the Master Curve (MC) method **[4]** and most of the Local Approach (LA) **[5]** are statistical. The MC method empirically describes cleavage fracture toughness distribution using a three-parameter Weibull distribution with fixed values of the modulus (m = 4) and the threshold toughness ($K_{min} = 20MPa\sqrt{m}$), with only the scale parameter (K_0) to be determined by fitting fracture toughness (K_{Ic}) data **[4]**. LA aims to develop a predictive, rather than descriptive, statistical model inspired by a pertinent physical understanding of cleavage process. The Beremin model **[5]** was the pioneering work in LA for cleavage fracture. The Beremin model is essentially a two-parameter Weibull distribution as below:

$$P = 1 - exp[-(\sigma_W/\sigma_0)^m] \quad (1)$$

$$\text{with } \sigma_W = \left(\int_{V_p} \sigma_1^m \cdot dV/V_0\right)^{1/m} \quad (2)$$

where P is the cumulative probability of failure, V_p denotes the volume of the plastic deformation zone as the cleavage fracture process zone, m and σ_0 are the two model parameters known as Weibull modulus and the scale parameter, respectively, σ_1 is the maximum tensile principal stress, V_0 is an elementary volume representing the mean volume occupied by each micro-crack in a solid, dV is the differential volume, σ_W denotes the Weibull stress.

The Beremin model has suffered from the ambiguity in model parameter calibration, i.e., the variation of the model parameters with temperature and geometrical constraint **[6-10]**.

Recent studies **[3, 13, 14]** have identified the fundamental defects of the Beremin model, which are summarized in **Table 1**. Note that some of the highlighted fundamental defects are inherited by those modifications to the Beremin model, such as the incompliance with the physical assumption of plastic yielding as a prerequisite to cleavage fracture due to the adoption of a ***fixed-value*** threshold stress σ_{th} (including the case of $\sigma_{th} = 0$) **[6-10]**, and the violation to the normality axiom of probability owing to the adoption of Eq. (6a) as the basic formulation of cumulative probability P **[10]**. An example was elaborated in detail in the commentary **[15]** to the work in **[10]**. In **Table 1**, the necessary corrections to the Beremin model are also

* Corresponding author: guian.qian@psi.ch, Leiws2008@gmail.com

provided to ensure the mathematical rigorousness and the physical compliance with the five assumptions below:

(1).The uniform spatial distribution of microcracks
(2).The weakest-link postulate of brittle fracture
(3).Plastic yielding as a prerequisite for cleavage fracture
(4). The maximum tensile principal stress criterion for cleavage fracture (Eq.(8))
(5).The power-law distribution of microcrack size (Eq.(9))

Basically, equations are defined in **Table 1** to translate the five assumptions into mathematical formulations and check the validity of each step for formulating the Beremin model. As a result of these necessary corrections, the Beremin model in Eq. (1) or Eq.(8a) is reformulated to Eq.(8b) in strict conformity to the five assumptions.

Accordingly, by adopting a three-parameter Weibull distribution density function of microscopic cleavage fracture stress S in place of the power law distribution of microcrack size (a) (Eq.(9) in **Table 1**) as assumption (5),

$$g(S) = m \cdot [(S-\sigma_{th})^{m-1}/\sigma_0^m] \cdot \exp[-(S-\sigma_{th})^m/\sigma_0^m] \quad (11)$$

Table 1. Comparison of Beremin model and its corrections.

Beremin model and corrections			**Equation / inequation Number**
	Beremin Model	**Corrections to Beremin Model**	
Model assumption #1	Uniform spatial distribution of microcracks		
Model assumption #2	Weakest-link model		
Basic formulation of cumulative probability P(V)	$P(V) = 1 - \exp\left[-\int_V p(\sigma,V_0)\frac{\delta V}{V_0}\right]$	$P(V) = 1 - \exp\left\{\int_V \ln[1-p(\sigma,V_0)]\frac{\delta V}{V_0}\right\}$	(3a,b)
Self-consistency check	$V = V_0$: Eq. (1a)$\Rightarrow P(V_0) = 1 - \exp[-p(\sigma,V_0)]$ *False!*	$V = V_0$: Eq. (1b) $\Rightarrow P(V_0) = p(\sigma,V_0)$ *True!*	(4a,b)
Model assumption #3	Plastic yielding as a prerequisite for cleavage fracture $\Rightarrow$ Initial yielding as the threshold stress state for cleavage fracture.		
Model assumption #4	Microscopic fracture criterion : $\sigma = \sigma_1 \geq S$		(5)
Model assumption #5	Power-law distribution of microcrack size: $f(a) = ka^{-\upsilon}(k > 0, \upsilon > 1)$, $m = 2\upsilon - 2$		(6)
Fracture probability of a volume element V_0, $p(\sigma,V_0)$	$p(\sigma,V_0) = \int_{a(\sigma_1)}^{\infty} f(a)da = \left(\frac{\sigma_1}{\sigma_0}\right)^m$	$p(\sigma,V_0) = \int_{a(\sigma_1)}^{a_{max}(\sigma_{th}=\sigma_{1,0})} f(a)da = \frac{\sigma_1^m - \sigma_{1,0}^m}{\sigma_0^m - \sigma_{1,0}^m}$	(7a,b)
Cumulative probability	$P(V) = 1 - \exp\left[-\int_V \left(\frac{\sigma_1}{\sigma_0}\right)^m \frac{\delta V}{V_0}\right]$	$P(V) = 1 - \exp\left[\int_V Ln\left(\frac{\sigma_0^m - \sigma_1^m}{\sigma_0^m - \sigma_{1,0}^m}\right)\frac{\delta V}{V_0}\right]$	(8a,b)
Model applicable range	$0 \neq \left(\frac{\sigma_{ys}}{\sigma_0}\right)^m \leq p(\sigma,V_0) \leq 1 \Rightarrow \sigma_{ys} \leq \sigma_1 \leq \sigma_0$ *Violate normality axiom of probability.*	$0 \leq p(\sigma,V_0) \leq 1 \Rightarrow \sigma_{1,0} \leq \sigma_1 \leq \sigma_0$	(9a,b)
Assessment of cumulative probability	$0 \leq P(V) \leq 1 - \exp(-V/V_0) \neq 1$ *Violate normality axiom of probability*	$0 \leq P(V) \leq 1$	(10a,b)
Key references	[5,10]	[3,13,14]	
Notes	a – microcrack size; S – strength of volume V_0 with an embedded microcrack of size a; V- volume of plastic deformation zone as fracture process zone; V_0- average volume occupied by a microcrack; σ- effective stress to induce cleavage fracture; σ_1- maximum tensile principal stress; σ_{th} – threshold strength ($S \geq \sigma_{th}$); $\sigma_{1,0}$- maximum tensile principal stress of a differential volume element (δV) at initial yielding; σ_{ys}-yield stress; k, υ, m, σ_0 –model parameters as material constants.		

a new local approach to cleavage fracture was proposed **[3, 13]**:

$$p(\sigma, V_0) = \int_{\sigma_{th}=\sigma_{1,0}}^{\infty} g(S)dS = 1 - \exp\left(\frac{\sigma_1-\sigma_{1,0}}{\sigma_0}\right)^m \quad (12)$$

$$P = 1 - \exp\left\{\int_{V_p} \ln[1 - p(\sigma, V_0)]\, dV/V_0\right\} = 1 - exp[-(\sigma_W/\sigma_0)^m] \quad (13)$$

$$\sigma_W = \left[\int_{V_p} (\sigma_1 - \sigma_{1,0})^m \cdot dV/V_0\right]^{1/m} \quad (14)$$

where σ_W is the Weibull stress. Note that the essential difference between Eq.(14) and Eq. (3) lies in that $\sigma_{1,0}$ in Eq.(14) varies with temperature, strain rate and stress state, while σ_{th} in Eq.(3) is a fixed-value independent of temperature, strain rate and stress state.

The present work tries to explore this aspect by applying the new local approach model in Eq. (12) to assess the published datasets for notch toughness that was already analyzed by deterministic approaches. The focus is to statistically assess and compare the notch toughness. Further, both the deterministic and the statistical approaches will be used to analyze the notch toughness.

2. Experimental data and method of calibration

2.1 Experimental data

The experimental data of a rolled C-Mn pressure vessel steel 16MnR reported by Wang et al. in **[18]** are used. Cleavage initiation in notched specimens with carbides and inclusions was investigated at 77K and 143K. The same steel went through different heat treatments to obtain the same ferrite grains but with fine carbide (FC) and coarse carbide (CC) particles, respectively. The mechanical tests, scanning electron microscopy (SEM) analysis and measurements, and finite element analysis (FEA) were employed to compare the notch toughness of the specimens with FC and CC particles.

The static four point bending (4PB) tests were conducted at 77K and 143K on 45°-angle V-notched prismatic beams in **Fig.1a**. The general yield load P_{gy} of the 4PB specimens was calculated by

$P_{gy} = 0.7045\sigma_{ys}B(W-a)^2/L$, where B is specimen thickness, W the height, a = 4.25 mm the notch depth, and L the bending span, with L = B = W = 12.7 mm. **Fig.2** summarizes the measured fracture loads P_f.

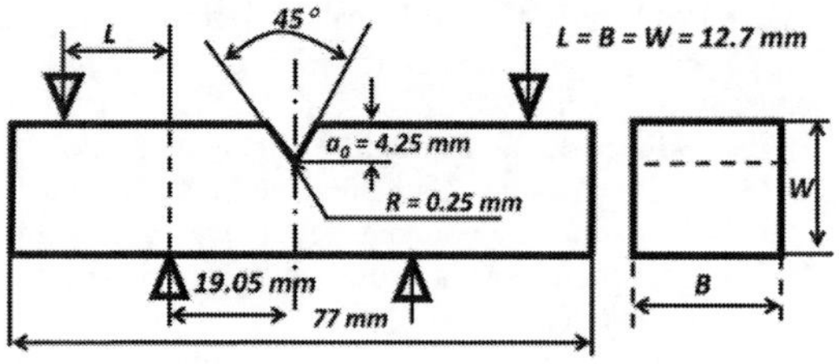

Fig. 1a. Four-point bending specimen: geometry [18].

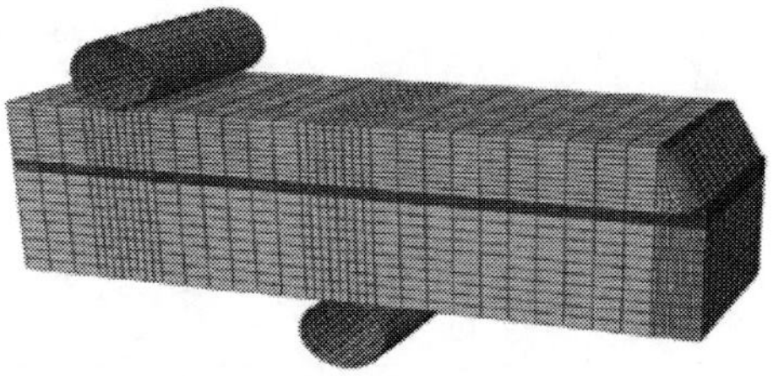

Fig. 1b. Four-point bending specimen: finite element model.

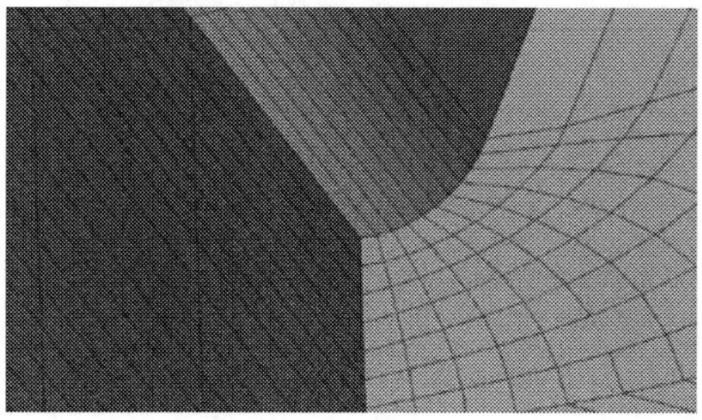

Fig. 1c. Four-point bending specimen: layout of finite elements at notch tip.

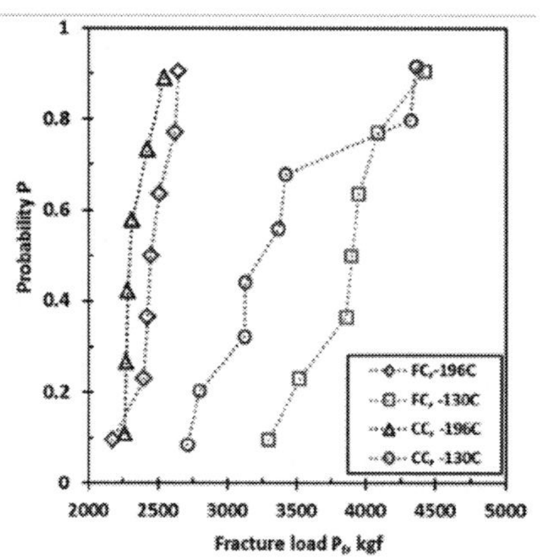

Fig. 2. Experimental results of fracture loads P_f **[18]**.

2.2 Method of calibration

In the present study, finite element analysis (FEA) is conducted using ABAQUS software version 6.14 for stress distribution in the 4PB specimen. 20-node brick elements are used. Only a quarter of the specimen was modeled due to symmetry with a total of 15940 elements as shown in **Fig.1b** and **Fig.1c**.
Eq. (13) is rewritten as

$$LnLn[1/(1-P)] = mLn(\sigma_W) - mLn(\sigma_0) \qquad (15)$$

The calibration method for Eq. (15) was developed in **[19]**. **Fig.3** shows the flow chart of calibration. Here we just briefly highlight a few key points for the calibration. In addition to a subroutine being written to extract the stress and strain distributions for different elements to calculate the newly defined Weibull stress σ_W in Eq. (14), the key challenge is to properly calculate the threshold stress $\sigma_{1,0}$ for the initial yielding of a finite volume element ΔV to enable the calculation of σ_W. As elaborated in the ***Introduction*** section, due to the non-uniform stress distributions in a notched specimen, the moment for each volume element to start initial plastic yielding and the amplitude of the stress $\sigma_{1,0}$ at the initial yielding therein are different. In theory, we can determine the stress $\sigma_{1,0}$ in each finite volume element ΔV by FEA with the experimentally measured true elastic-plastic hardening stress-strain curve of the steel. Since during the loading process, the stress state in each volume element keeps changing. At each loading step, while some elastic volume elements reach the initial plastic yielding state, the other already plastically yielded volume elements in the preceding loading step will go through strain hardening phase. This demands a sufficiently small loading step in order to precisely capture the moment of initial plastic yielding state of each volume element. This would translate into a significant amount of numerical calculations to determine the values of $\sigma_{1,0}$ in all relevant volume elements. Therefore, this path is not practically desirable. An alternative and cost-effective approach to determine the values of $\sigma_{1,0}$ in different volume elements has been thus developed in **[19]**. In essence, it adopts an elastic-perfectly plastic material model for FEA to determine the value of $\sigma_{1,0}$ in each volume element. The resultant value of stress $\sigma_1$in a plastically deformed volume element is stored as $\sigma_{1,0}$ for that particular volume element. Then the experimentally measured true elastic-plastic hardening stress-strain curve of the steel in **Table 1** is used for FEA to calculate the stress $\sigma_1$in each volume element at a given load P_f. Next, substituting the values of the instant stress σ_1 and the threshold stress $\sigma_{1,0}$ for exactly the same volume element one by one into Eq. (14) to calculate σ_W at each load as input to Eq. (15). Then the same conventional calibration procedure for Eq. (15) in **[19]** is applied to obtain the values of m and σ_0. The proposed approach to determining the values of $\sigma_{1,0}$ in different volume elements according to an elastic-perfectly plastic material model significantly reduces the amount of numerical calculation. The stress redistribution between a perfectly plastic volume element and an elastic volume can be different from that between a plastic hardening volume element and an elastic volume. However, since we adopt the same values of elastic modulus, Poisson's ratio and yield stress for the real material in the elastic-perfectly plastic material model, we expect the difference in the values of $\sigma_{1,0}$ determined using an elastic-perfectly plastic material model and using a true elastic-plastic hardening material model is insignificant. The reader is reminded that in contrast to the conventional fixed-value threshold, the proposed approach to the determination of $\sigma_{1,0}$ pertinently captures the most critical characteristics of threshold stress, that is, the temperature and stress state dependence of threshold. In our future work, we expect to conduct a direct comparison of the values of $\sigma_{1,0}$ determined using the afore cited two different material models.

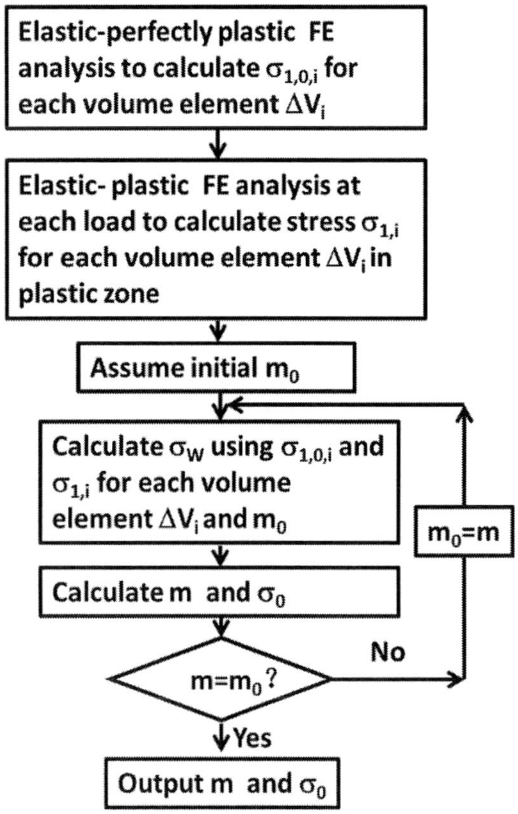

Fig. 3. Flow chart for the calibration of Eq.(15).

3. Results

Using the experimental data in **Fig.2** as input, Eq.(15) was calibrated at 77K for FC and CC specimens, respectively. **Fig.4** shows the calibration results at 77K. For FC specimens, $m = 25.6, \sigma_0 = 228.3MPa$; For CC specimens, $m = 11, \sigma_0 = 161.1MPa$. With the calibrated parameters, Weibull stress at different loadings are calculated for the temperature of 77K and 143K, as shown in **Fig. 5a** and **5b**.

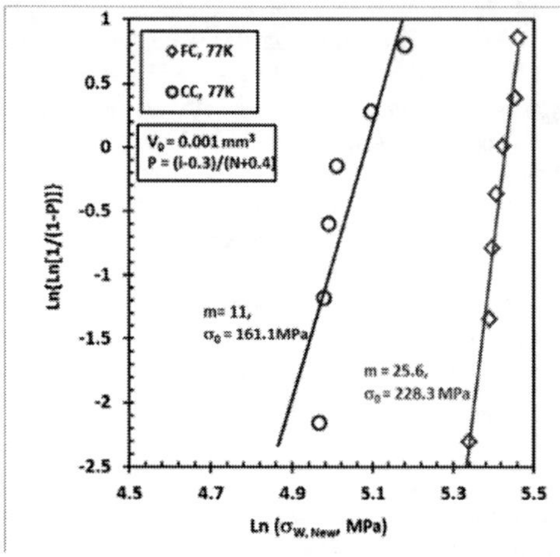

Fig. 4. Summary of calibration results of Eq.(18) at 77K.

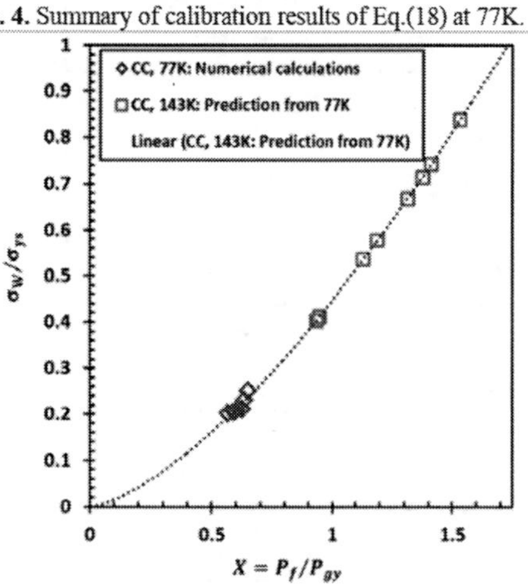

Fig. 5a. Weibull stress at 77K and 143K for the CC material.

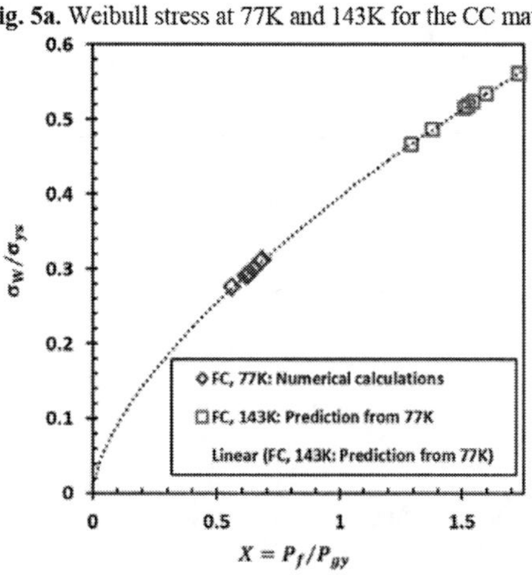

Fig. 5b. Weibull stress at 77K and 143K for the FC material.

The new model Eq.(13) assumes temperature independence of model parameters, so we did not calibrate Eq. (15) at 143K, but will justify this in the following. As discussed in detail in **[19]**, according to the definition of the new Weibull stress σ_W in Eq.(14), if the two model parameters (m and σ_0) are indeed intrinsic material properties independent of temperature, for a given material and a fixed specimen geometry, σ_W is a function of the fracture load P_f at a fixed temperature (T):

$$\sigma_W(T) = F(P_f(T)) \tag{16}$$

Alternatively, Eq. (16) can be converted to the following format that allows considering temperature effect:

$$\frac{\sigma_W(T)}{\sigma_{ys}(T)} = F(\frac{P_f(T)}{P_{gy}(T)}) \tag{17a}$$

$$\text{or} \quad \sigma_W(T) = \sigma_{ys}(T) \cdot F(\frac{P_f(T)}{P_{gy}(T)}) \tag{17b}$$

where the general yield load P_{gy} is defined in Section ***2.1 Experimental data***.

In order to validate the temperature independence of Weibull modulus m, we now substitute Eq. (17b) in Eq. (13) to get

$$P = 1 - exp\left[-\left(\frac{\sigma_{ys}}{\sigma_0}\right)^m F^m(\frac{P_f(T)}{P_{gy}(T)})\right] \tag{18}$$

Eq. (18) is further rewritten as,

$$Y = \frac{1}{\sigma_{ys}} \cdot \left\{ln\left[\frac{1}{(1-P)}\right]\right\}^{\frac{1}{m}} = \frac{1}{\sigma_0} \cdot F(\frac{P_f(T)}{P_{gy}(T)}) \tag{19}$$

A master curve behavior between the compound parameters $Y = \frac{1}{\sigma_{ys}} \cdot \left\{ln\left[\frac{1}{(1-P)}\right]\right\}^{\frac{1}{m}}$ and $X = \frac{P_f(T)}{P_{gy}(T)}$ is thus expected regardless of the test temperature, only if the temperature independence of two model parameters (m and σ_0) holds. According to Eq. (19), the experimental data in **Fig.2** at two different temperatures are transferred into **Fig.6a** and **b** for the expected correlation with the calibrated values of $m = 25.4$ for FC and $m = 11$ for CC in sequence.

As shown in **Fig.6a** and **b**, the experimental results at both temperatures fall onto a single "master curve" that correlates the compound parameters $Y = \frac{1}{\sigma_{ys}} \cdot \left\{ln\left[\frac{1}{(1-P)}\right]\right\}^{\frac{1}{m}}$ and $X = \frac{P_f(T)}{P_{gy}(T)}$ for each material. What underpins the "master curve" behavior is the temperature independence of Weibull parameters m and σ_0. Therefore, **Fig.6 a** and **b** validate the assumption of Eq. (13) on the temperature independence of model parameters m and σ_0.

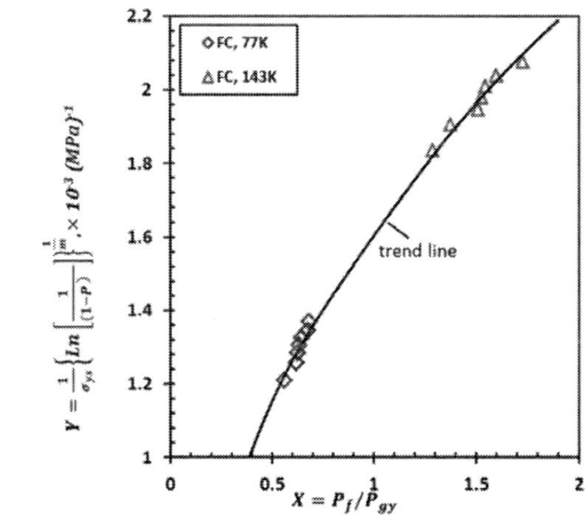

Fig. 6a. Correlation between the compound parameters $Y = \frac{1}{\sigma_{ys}} \cdot \left\{ln\left[\frac{1}{(1-P)}\right]\right\}^{\frac{1}{m}}$ and $X = \frac{P_f(T)}{P_{gy}(T)}$ at two different temperatures for specimens FC.

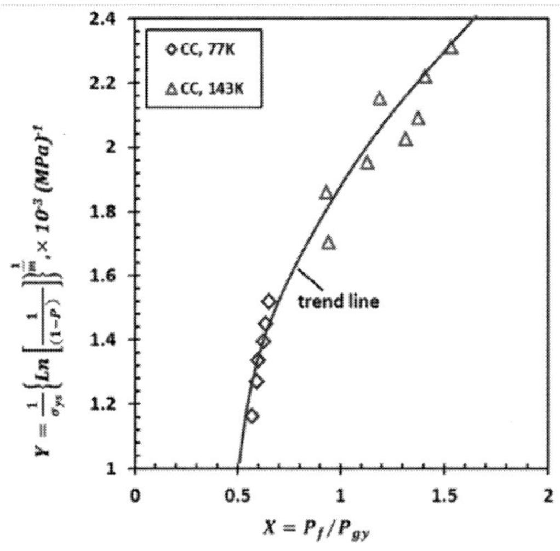

Fig. 6b. Correlation between the compound parameters $Y = \frac{1}{\sigma_{ys}} \cdot \left\{ln\left[\frac{1}{(1-P)}\right]\right\}^{\frac{1}{m}}$ and $X = \frac{P_f(T)}{P_{gy}(T)}$ at two different temperatures for specimens CC.

5. Conclusions

As an effort to adopt a fracture physics based statistical approach for correlating fracture properties of a simple and less expensive notched specimen and a costly fracture mechanics specimen, a new statistical model of cleavage fracture is discussed and applied to assess notch toughness of ferritic steels against cleavage fracture by analyzing published cleavage fracture datasets of 16Mn steel with fine and coarse carbides (FC and CC) and at two test temperatures in **[18]**. The following conclusions can be drawn:

A new local approach to fracture and the calibration method are discussed. The proposed calibration method for a new local approach to cleavage fracture is proved feasible using experimental data.

The proposed calibration method for a new local approach to cleavage fracture is proved feasible using experimental data.

There exists a master curve to correlate the compound parameters $Y = \frac{1}{\sigma_{ys}} \cdot \left\{ln\left[\frac{1}{(1-P)}\right]\right\}^{\frac{1}{m}}$ and $X = \frac{P_f(T)}{P_{gy}(T)}$ at two different temperatures for each microstructure (FC and CC). This validates the temperature independence of two model parameters (m and σ_0).

It should be noted that the weakest link based approach can be used for the fatigue analysis, especially for the very-high-cycle fatigue analysis where crack initiation from the inhomogeneity inside the material occurs.

References

1. Zhang, Z., and Qian, X. (2017) Fatigue Fract. Eng. Mater. Struct., 40, 1128-1142.
2. Chao, Y.J., Yang, S., and Sutton, M.A. (1994) J. Mech. Phys. Solids, 42, 629-647.
3. Lei, W.-S. (2016) Philos. Mag., 96, 3586-3631.
4. Wallin, K. (2002) Eng. Fract. Mech., 69, 451-481.
5. Pineau, A. (2006) Int. J. Fract., 138, 139-166.
6. Bakker, A., and Koers, R.W.J. (1991) ESIS/EGF9, J.G. Blauel and K.H. Schwalbe, eds., Mechanical Engineering Publications, London, pp. 613-632.
7. Hausild, P., Berdin, C., and Bompard, P. (2005) Mater. Sci. Eng. A, 391, 188-197.
8. Moattari, M., Sattari-Far, I., Persechino, I., and Bonora, N. (2016) Mater. Sci. Eng. A, 657, 161–172.
9. Petti, J.P. and Dodds, R.H. (2005) Eng. Fract. Mech., 72, 91-120.
10. Ruggieri, C., Savioli, R.G., Dodds, R.H. (2015) Eng. Fract. Mech., 146, 185–209.
11. Heerens, J., Pfuff, M., Hellmann, D., and Zerbst, U. (2002) Eng. Fract. Mech., 69 , 483-495.
12. Lei, W.-S. (2016) Mech. Mater., 101, 81–92.
13. Lei, W.-S. (2016) Mech. Mater., 93, 184–198.
14. Lei, W.-S. (2016) Philos. Mag. Lett., 96, 101-111.
15. Lei, W.-S. (2017) Eng. Fract. Mech., 178, 527-534.
16. Lei, W.S., Li, D.M., Yao, M. (1996) Mater. Chem.Phys., 46, 1–5.
17. Kotrechko, S., Gryschchenko, V., Kozak, V., Dlouhy, I. (2016) Solid State Phenomena, 258, 281-285
18. Wang, G. Z., Liu, Y. G., Chen, J. H. (2004) Mater. Sci. Eng. A, 369, 181-191.
19. Qian, G., Lei, W.-S. , Niffenegger, M. (2017) Mater. Sci. Eng. A, 694, 10-12.

AUTHOR INDEX

AUTHOR INDEX

AUTHOR INDEX

AUTHOR INDEX

AUTHOR INDEX

AUTHOR INDEX

AUTHOR INDEX

EDP Sciences – Web of Conferences
17, Avenue du Hoggar
Parc d'Activité de Courtabouf
BP 112
F-91944 Les Ulis Cedex A
France

ISBN 978-1-5108-6386-6

9 781510 863866

CURRAN ASSOCIATES INC.
proceedings.com

American College of Veterinary Internal Medicine (ACVIM) Forum 2017

National Harbor, Maryland, USA
8 - 10 June 2017

Volume 1 of 2

ACVIM
American College of Veterinary Internal Medicine